Knobil and Neill's

Physiology of Reproduction

THIRD EDITION

Knobil and Neill's Physiology of Reproduction

THIRD EDITION

Volume 1

Editor-in-Chief

Jimmy D. Neill, Ph.D.
Distinguished Professor
Department of Physiology and Biophysics
The University of Alabama at Birmingham
Birmingham, Alabama

Section Editors

Tony M. Plant, Ph.D.
University of Pittsburgh School of Medicine
Department of Cell Biology and Physiology
Pittsburgh, Pennsylvania

Donald W. Pfaff, Ph.D.
Department of Neurobiology and Behavior
The Rockefeller University
New York, New York

John R. G. Challis, D.Sc., F.R.S.C.
Departments of Physiology, Ob/Gyn, and Medicine
University of Toronto
Toronto, Ontario
Canada

David M. de Kretser, M.D., A.O.
Monash Institute of Medical Research
Monash University
Melbourne, Australia

JoAnne S. Richards, Ph.D.
Department of Molecular and Cellular Biology
Baylor College of Medicine
Houston, Texas

Paul M. Wassarman, Ph.D.
Brookdale Department of Molecular, Cell and Developmental Biology
Mount Sinai School of Medicine
New York, New York

AMSTERDAM • BOSTON • HEIDELBERG • LONDON
NEW YORK • OXFORD • PARIS • SAN DIEGO
SAN FRANCISCO • SINGAPORE • SYDNEY • TOKYO

Academic Press is an imprint of Elsevier

ELSEVIER
ACADEMIC
PRESS

11830 Westline Industrial Drive, St. Louis, MO 63146, USA
525 B Street, Suite 1900, San Diego, CA 92101-4495, USA
84 Theobald's Road, London WC1X 8RR, U.K.

KNOBIL AND NEILL'S PHYSIOLOGY OF REPRODUCTION ISBN-13: 978-0-12-515401-7
ISBN-10: 0-12-515401-1

ISBN-13: 978-0-12-515401-7
ISBN-10: 0-12-515401-1

Acquisitions Editor: *Tari Paschall*
Developmental Editor: *Kirsten Funk*
Publishing Services Manager: *Julie Eddy*
Project Manager: *Kelly E. M. Steinmann*
Marketing Manager: *Patricia Howard*
Book Designer: *Bill Drone*

For all information on all Elsevier Academic Press Publications
visit our Web site at www.books.elsevier.com

Printed in the United States of America.

Last digit is the print number: 9 8 7 6 5 4 3 2 1

Contents

Contributors ix
Foreword xv
In Memoriam: Ernst Knobil (1926–2000) xxi
Preface to the First Edition xxv
Preface to the Second Edition xxvii
Preface xxix

VOLUME 1

Gametes, Fertilization, and Embryogenesis
Paul M. Wassarman, Editor

1. The Spermatozoon 3
E. M. Eddy

2. Fertilization in Mammals 55
Harvey M. Florman and Tom Ducibella

3. Gamete and Zygote Transport 113
Susan S. Suarez

4. Implantation 147
Sudhansu K. Dey and Hyunjung Lim

5. Anatomy and Genesis of the Placenta 189
Graham J. Burton, Peter Kaufmann, and Berthold Huppertz

6. Sex Determination and Differentiation 245
Amanda Swain

7. Preimplantation Embryogenesis 261
Kathleen H. Burns and Martin M. Matzuk

Female Reproductive System
JoAnne S. Richards, Editor

8. Embryology and Genetics of the Mammalian Gonads and Ducts 313
Keith L. Parker and Bernard P. Schimmer

9. Oviduct and Endometrium: Cyclic Changes in the Primate Oviduct and Endometrium 337
Alexandra P. Hess, Nihar R. Nayak, and Linda C. Giudice

10. Follicular Development: Mouse, Sheep, and Human Models 383
Aleksandar Rajkovic, Stephanie A. Pangas, and Martin M. Matzuk

11. Ovulation 425
Lawrence L. Espey and JoAnne S. Richards

12. Structure, Function, and Regulation of the Corpus Luteum 475
Richard L. Stouffer

13. Growth Hormone, Insulin-Like Growth Factors, and the Ovary 527
Carolyn A. Bondy, Jian Zhou, and Jose A. Arraztoa

14. Gonadotropin Signaling in the Ovary 547
Mary Hunzicker-Dunn and Kelly Mayo

15. Steroid Receptors in the Ovary and Uterus 593
John F. Couse, Sylvia C. Hewitt, and Kenneth S. Korach

16. Physiology and Molecular Biology of the Relaxin Peptide Family 679
Ross A. D. Bathgate, Aaron J. W. Hsueh, and O. David Sherwood

Male Reproductive System
David M. de Kretser, Editor

17. Anatomy, Vasculature, and Innervation of the Male Reproductive Tract 771
B. P. Setchell and W. G. Breed

18. Cytology of the Testis and Intrinsic Control Mechanisms 827
J. B. Kerr, K. L. Loveland, M. K. O'Bryan, and D. M. de Kretser

19. The Sertoli Cell 949
Michael D. Griswold and Derek McLean

20. Physiology of Testicular Steroidogenesis 977
Douglas M. Stocco and Michael J. McPhaul

21. Endocrine Regulation of Spermatogenesis 1017
Liza O'Donnell, Sarah J. Meachem, Peter G. Stanton, and Robert I. McLachlan

22. The Epididymis 1071
Bernard Robaire, Barry T. Hinton, and Marie-Claire Orgebin-Crist

23. Physiology of the Male Accessory Sex Structures: The Prostate Gland, Seminal Vesicles, and Bulbourethral Glands 1149
Gail P. Risbridger and Renea A. Taylor

24. Male Sexual Function 1173
Shalender Bhasin and George S. Benson

25. Immunophysiology of the Male Reproductive Tract 1195
M. P. Hedger and D. B. Hales

Pituitary and Hypothalamus
Jimmy D. Neill, Editor

26. Pituitary and Hypothalamus: Perspectives and Overview 1289
*John W. Everett**

27. Anatomy of the Hypothalamo–Hypophysial Complex 1309
Robert B. Page

28. Physiology of the Gonadotropin-Releasing Hormone Neuronal Network 1415
Allan E. Herbison

29. Gonadotropes and Lactotropes 1483
Gwen V. Childs

30. Gonadotropins: Chemistry and Biosynthesis 1581
George R. Bousfield, Li Jia, and Darrell N. Ward

---

*Deceased.

31. Gonadotropin-Releasing Hormone Regulation of Gonadotropin Biosynthesis and Secretion . 1635
Kyeong-Hoon Jeong and Ursula B. Kaiser

32. Prolactin: Structure, Function, and Regulation of Secretion 1703
Karen A. Gregerson

VOLUME 2

Reproductive Behavior and Its Control
Donald W. Pfaff, Editor

33. Neurobiology of Male Sexual Behavior 1729
Elaine M. Hull, Ruth I. Wood, and Kevin E. McKenna

34. Hormonal, Neural, and Genomic Mechanisms for Female Reproductive Behaviors, Motivation, and Arousal 1825
D. W. Pfaff, Y. Sakuma, L.-M. Kow, A. W. L. Lee, and A. Easton

35. Maternal Behavior 1921
M. Numan, A. S. Fleming, and F. Levy

36. Communicative Behaviors, Hormone–Behavior Interactions, and Reproduction in Vertebrates 1995
John C. Wingfield

37. Pheromones and Mammalian Reproduction 2041
John G. Vandenbergh

Reproductive Processes and Their Control
Tony M. Plant, Editor

38. Puberty in the Rat 2061
Sergio R. Ojeda and Michael K. Skinner

39. Puberty in the Sheep 2127
Douglas L. Foster and Leslie M. Jackson

40. Puberty in Nonhuman Primates and Humans 2177
Tony M. Plant and Selma F. Witchel

41. Seasonal Regulation of Reproduction in Mammals 2231
Benoît Malpaux

42. Neuroendocrine Control of Mating-Induced Ovulation 2283
Alexander S. Kauffman and Emilie F. Rissman

43. Neuroendocrine Control of the Ovarian Cycle of the Rat 2327
Marc E. Freeman

44. Neuroendocrine Control of the Ovarian Cycle of the Sheep 2389
Robert L. Goodman and E. Keith Inskeep

45. Control of Follicular Development, Corpus Luteum Function, the Maternal Recognition of Pregnancy, and the Neuroendocrine Regulation of the Menstrual Cycle in Higher Primates 2449
Anthony J. Zeleznik and Clifford R. Pohl

46. Suckling and the Control of Gonadotropin Secretion 2511
Alan S. McNeilly

47. Physiological Mechanisms Integrating Metabolism and Reproduction 2553
Stephanie M. Krasnow and Robert A. Steiner

48. Stress and the Reproductive System 2627
Michel Ferin

49. Aging in the Hypothalamic–Pituitary–Testicular Axis 2697
David J. Handelsman

50. Aging in the Hypothalamic–Pituitary–Ovarian Axis 2729
Genevieve Neal-Perry and Nanette F. Santoro

Pregnancy
John R. G. Challis, Editor

51. Immunobiology of Human Pregnancy 2759
Joan S. Hunt, Ramsey H. McIntire, and Margaret G. Petroff

52. Placental Transfer 2787
D. E. Atkinson, R. D. H. Boyd, and C. P. Sibley

53. Placental Endocrine Function 2847
Felice Petraglia, Pasquale Florio, and Michela Torricelli

54. Maternal Adaptation to Pregnancy 2899
Kent L. Thornburg, Susan P. Bagby, and George D. Giraud

55. Parturition 2925
William Gibb, Stephen J. Lye, and John R. G. Challis

56. Developmental Origins of Health and Disease: Theoretical Considerations, Mechanisms, and Implications 2975
Peter D. Gluckman and Mark A. Hanson

Lactation
Jimmy D. Neill, Editor

57. Lactation and Its Hormonal Control 2993
Margaret C. Neville

58. Oxytocin: Synthesis, Secretion, and Reproductive Functions 3055
J. Peter H. Burbach, Larry J. Young, and John A. Russell

59. Milk Ejection and Its Control 3129
J. B. Wakerley

Index 3191

Contributors

Jose A. Arraztoa, M.D.
Developmental Endocrinology Branch, National Institute of Child Health and Human Development, Bethesda, Maryland

D. E. Atkinson, Ph.D.
School of Biological Science, University of Manchester, Manchester, U.K.

Susan P. Bagby, M.D.
Division of Nephrology & Hypertension and Heart Research Center, Department of Medicine, Oregon Health & Science University, Portland, Oregon

Ross A. D. Bathgate, Ph.D.
Howard Florey Institute, University of Melbourne, Victoria, Australia

George S. Benson, M.D.
6414 Fannin Street, Suite g150, Houston, Texas

Shalender Bhasin, M.D.
Division of Endocrinology, University of the City of Los Angeles, Los Angeles, California

Carolyn A. Bondy, M.D.
Developmental Endocrinology Branch, National Institute of Child Health and Human Development, Bethesda, Maryland

George R. Bousfield, Ph.D.
Department of Biological Sciences, Wichita State University, Wichita, Kansas

R. D. H. Boyd, M.B.
School of Biological Science, University of Manchester, Manchester, U.K.

W. G. Breed, D.Phil.
Department of Anatomical Sciences, University of Adelaide, Adelaide, Australia

J. Peter H. Burbach, Ph.D.
Department of Pharmacology and Anatomy, Rudolf Magnus Institute of Neurosciences, Utrecht, The Netherlands

Kathleen H. Burns, M.D., Ph.D.
Department of Pathology, The Johns Hopkins Hospital, Baltimore, Maryland

Graham J. Burton, M.D., D.Sc.
Department of Anatomy, University of Cambridge, Cambridge, U.K.

John R. G. Challis, D.Sc., F.R.S.C.
Research and Associate Provost Professor of Physiology, Ob/Gyn and Medicine, University of Toronto, Toronto, Ontario, Canada

Gwen V. Childs, Ph.D.
Department of Neurobiology and Developmental Sciences, College of Medicine, University of Arkansas for Medical Sciences, Little Rock, Arkansas

John F. Couse, Ph.D.
Laboratory of Reproductive and Developmental Toxicology, National Institute of Environmental Health Sciences, National Institutes of Health, Research Triangle Park, North Carolina

D. M. de Kretser, M.D., A.O.
Monash Institute of Reproduction and Development, Monash Medical Centre, Melbourne, Australia

Sudhansu K. Dey, Ph.D.
Department of Pediatrics, Cell and Developmental Biology and Pharmacology, Director, Division of Reproductive and Developmental Biology, Vanderbilt University Medical Center, Nashville, Tennessee

Tom Ducibella, Ph.D.
Department of Obstetrics and Gynecology, Tufts University, New England Medical Center, Boston, Massachusetts

A. Easton, Ph.D.
Department of Neurobiology and Behavior, The Rockefeller University, New York, New York

E. M. Eddy, Ph.D.
Gamete Biology Section, Laboratory of Reproductive and Developmental Toxicology, National Institute of Environmental Health Sciences, National Institutes of Health, Research Triangle Park, North Carolina

Lawrence L. Espey, Ph.D.
Department of Biology, Trinity University, San Antonio, Texas

John W. Everett*
Department of Neurobiology, Duke University School of Medicine, Durham, North Carolina

Michel Ferin, M.D.
Department of Obstetrics/Gynecology, Columbia University, College of Physicians and Surgeons, New York, New York

A. S. Fleming, Ph.D.
Department of Psychology, University of Toronto at Mississauga, Mississauga, Ontario, Canada

Pasquale Florio, Ph.D.
Department of Pediatrics, Obstetric and Reproductive Medicine, University Siena, Siena, Italy

Harvey M. Florman, Ph.D.
Department of Cell Biology, University of Massachusetts Medical School, Worcester, Massachusetts

Douglas L. Foster, Ph.D.
Reproductive Sciences Program, Department of Obstetrics and Gynecology, Department of Ecology and Evolutionary Biology, University of Michigan, Ann Arbor, Michigan

Marc E. Freeman, Ph.D.
Department of Biological Science, Florida State University, Biomedical Research Facility, Tallahassee, Florida

William Gibb, Ph.D.
Department of Obstetrics and Gynecology, and Department of Cellular and Molecular Medicine, University of Ottawa, Ottawa, Ontario, Canada

George D. Giraud, M.D., Ph.D.
Department of Medicine, Heart Research Center, Oregon Health & Science University, Portland, Oregon

Linda C. Giudice, M.D., Ph.D.
Center for Research on Reproduction and Women's Health and Genomic Medicine, Department of Obstetrics and Gynecology, Stanford University School of Medicine, Stanford, California

Peter D. Gluckman, F.R.S.
Liggins Institute, University of Auckland, Grafton, Auckland, New Zealand

Robert L. Goodman, Ph.D.
Department of Physiology and Pharmacology, West Virginia University, Morgantown, West Virginia

Karen A. Gregerson, Ph.D.
University of Cincinnati, College of Pharmacy, Cincinnati, Ohio

Michael D. Griswold, Ph.D.
School of Molecular Biosciences, Washington State University, Pullman, Washington

D. B. Hales, Ph.D.
Department of Physiology and Biophysics, University of Illinois at Chicago, Chicago, Illinois

David J. Handelsman, M.B., B.S., Ph.D.
ANZAC Research Institute, Sydney, Australia

Mark A. Hanson, D.Phil.
Centre for Developmental Origins of Health and Disease, University of Southampton, Princess Anne Hospital Level F (887), Southampton, U.K.

M. P. Hedger, Ph.D.
Monash Institute of Medical Research, Monash University, Melbourne, Australia

Allan E. Herbison, M.B.Ch.B., Ph.D.
Center for Neuroendocrinology and Department of Physiology, Otago School of Medical Sciences, Dunedin, New Zealand

Alexandra P. Hess, M.D.
Center for Research on Reproduction and Women's Health and Genomic Medicine, Department of Obstetrics and Gynecology, Stanford University School of Medicine, Stanford, California

*Deceased.

Sylvia C. Hewitt, M.S.
National Institute of Environmental Health Sciences, National Institutes of Health, Research Triangle Park, North Carolina

Barry T. Hinton, Ph.D.
Department of Cell Biology, University of Virginia Health System, Charlottesville, Virginia

Aaron J. W. Hsueh, Ph.D.
Division of Reproductive Biology, Department of Obstetrics and Gynecology, Stanford University School of Medicine, Stanford, California

Elaine M. Hull, Ph.D.
Psychology Department, Florida State University, Tallahassee, Florida

Joan S. Hunt, Ph.D.
Department of Anatomy and Cell Biology, University of Kansas Medical Center, Kansas City, Kansas

Mary Hunzicker-Dunn, Ph.D.
Department of Cell and Molecular Biology, Center for Reproductive Science, Northwestern University Medical School, Chicago, Illinois

Berthold Huppertz, Ph.D.
Department of Anatomy II, University Hospital RWTH Aachen, Aachen, Germany

E. Keith Inskeep, Ph.D.
Division of Animal and Veterinary Sciences, West Virginia University, Morgantown, West Virginia

Leslie M. Jackson, Ph.D.
Reproductive Sciences Program, Department of Obstetrics and Gynecology, Department of Ecology and Evolutionary Biology, University of Michigan, Ann Arbor, Michigan

Kyeong-Hoon Jeong, Ph.D.
Division of Endocrinology, Diabetes, and Hypertension, Brigham and Women's Hospital and Harvard Medical School, Boston, Massachusetts

Li Jia, Ph.D.
Department of Biological Sciences, Wichita State University, Wichita, Kansas

Ursula B. Kaiser, M.D.
Division of Endocrinology, Diabetes, and Hypertension, Brigham and Women's Hospital and Harvard Medical School, Boston, Massachusetts

Alexander S. Kauffman, Ph.D.
Department of Biochemistry and Molecular Genetics, University of Virginia School of Medicine, Center for Research in Reproduction, Charlottesville, Virginia

Peter Kaufmann, M.D.
Department of Anatomy II, University Hospital RWTH Aachen, Aachen, Germany

J. B. Kerr, Ph.D.
Department of Anatomy and Cell Biology, Faculty of Medicine, Nursing and Health Sciences, Monash University, Victoria, Australia

Kenneth S. Korach, Ph.D.
Program Director, Environ Disease Medicine Program Chief, National Institute of Environmental Health Sciences, Research Triangle Park, North Carolina

L.-M. Kow, Ph.D.
Department of Neurobiology and Behavior, The Rockefeller University, New York, New York

Stephanie M. Krasnow, Ph.D.
Departments of Physiology and Biophysics, Obstetrics and Gynecology, and Biology, University of Washington, Seattle, Washington

A. W. L. Lee, Ph.D.
Department of Neurobiology and Behavior, The Rockefeller University, New York, New York

F. Levy, Ph.D.
Responsable de l équipe de Comportement, Physiologie de la Reproduction et des Comportements, INRA-CNRS-Université de Tours, Centre INRA de Tours, Nouzilly, France

Hyunjung Lim, Ph.D.
Departments of Obstetrics and Gynecology and Cell Biology and Physiology, Washington State University School of Medicine, St. Louis, Missouri

K. L. Loveland, Ph.D.
Monash Institute of Reproduction and Development, Monash Medical Centre, Monash University, Victoria, Australia

Stephen J. Lye, Ph.D.
Department of Physiology, Department of Obstetrics and Gynecology, University of Toronto; Division of Child Health and Human Development, Samuel Lunenfeld Research Institute, Mount Sinai Hospital, Toronto, Ontario, Canada

Benoît Malpaux, Ph.D.
UMR INRA-CNRS-Université de Tours, Physiologie de la Reproduction et des Comportements, Centre INRA de Tours, Nouzilly, France

Martin M. Matzuk, M.D., Ph.D.
Department of Pathology, Baylor College of Medicine, Houston, Texas

Kelly Mayo, Ph.D.
Department of Biochemistry, Cellular and Molecular Biology, Center for Reproductive Science, Northwestern University, Evanston, Illinois

Ramsey H. McIntire, Ph.D.
Department of Anatomy and Cell Biology, University of Kansas Medical Center, Kansas City, Kansas

Kevin E. McKenna, Ph.D.
Northwestern University Medical School, Chicago, Illinois

Robert I. McLachlan, Ph.D.
Prince Henry's Institute of Medical Research, Monash Medical Centre, Victoria, Australia

Derek McLean, Ph.D.
Department of Animal Sciences, Washington State University, Pullman, Washington

Alan S. McNeilly, D.Sc., F.R.S.E.
Centre for Reproductive Biology, The University of Edinburgh, Edinburgh, U.K.

Michael J. McPhaul, M.D.
Department of Internal Medicine, University of Texas Southwestern Medical Center, Dallas, Texas

Sarah J. Meachem, Ph.D.
Prince Henry's Institute of Medical Research, Monash Medical Centre, Victoria, Australia

Nihar R. Nayak, Ph.D.
Center for Research on Reproduction and Women's Health and Genomic Medicine, Department of Obstetrics and Gynecology, Stanford University School of Medicine, Stanford, California

Genevieve Neal-Perry, M.D., Ph.D.
Department of Obstetrics and Gynecology, Albert Einstein College of Medicine, The Bronx, New York

Margaret C. Neville, Ph.D.
University of Colorado Health Sciences Center at Fitzsimmons, Aurora, Colorado

M. Numan, Ph.D.
Department of Psychology, Boston College, Chestnut Hill, Massachusetts

M. K. O'Bryan, Ph.D.
Monash Institute of Reproduction and Development, Monash University, Monash Medical Centre, Victoria, Australia

Liza O'Donnell, Ph.D.
Prince Henry's Institute of Medical Research, Monash Medical Centre, Victoria, Australia

Sergio R. Ojeda, D.V.M.
Division of Neuroscience, Oregon Regional Primate Research Center, Beaverton, Oregon

Marie-Claire Orgebin-Crist, Ph.D.
Center for Reproductive Biology Research, Vanderbilt School of Medicine, Nashville, Tennessee

Robert B. Page, M.D.
Emeritus Professor of Neurosurgery/Department of Neurosurgery, Milton S. Hershey Medical Center, Pennsylvania State University, Hershey, Pennsylvania

Stephanie A. Pangas, Ph.D.
Departments of Pathology and Molecular and Cellular Biology, Baylor College of Medicine, Houston, Texas

Keith L. Parker, M.D., Ph.D.
Department of Internal Medicine, Division of Endocrinology and Metabolism, University of Texas Southwestern Medical Center, Dallas, Texas

Felice Petraglia, M.D.
Department of Pediatrics, Obstetrics and Reproductive Medicine, University Siena, Siena, Italy

Margaret G. Petroff, Ph.D.
Department of Anatomy and Cell Biology, University of Kansas Medical Center, Kansas City, Kansas

D. W. Pfaff, Ph.D.
Department of Neurobiology and Behavior, The Rockefeller University, New York, New York

Tony M. Plant, Ph.D.
Departments of Cell Biology and Physiology and Obstetrics, Gynecology, and Reproductive Sciences, University of Pittsburgh School of Medicine, Pittsburgh, Pennsylvania

Clifford R. Pohl, Ph.D.
School of Health Sciences, Duquesne University, Pittsburgh, Pennsylvania

Aleksandar Rajkovic, M.D., Ph.D.
Departments of Obstetrics and Gynecology, Baylor College of Medicine, Houston, Texas

JoAnne S. Richards, Ph.D.
Department of Molecular and Cellular Biology, Baylor College of Medicine, Houston, Texas

Gail P. Risbridger, Ph.D.
Centre for Urology Research, Monash Institute of Reproduction and Development, Monash University, Victoria, Australia

Emilie F. Rissman, Ph.D.
Department of Biochemistry and Molecular Genetics, University of Virginia School of Medicine, Charlottesville, Virginia

Bernard Robaire, Ph.D.
Department of Pharmacology and Therapeutics, Department of Obstetrics and Gynecology, McGill University, Montréal, Québec, Canada

John A. Russell, M.B.Ch.B., Ph.D.
Division of Biomedical Sciences, University of Edinburgh, Edinburgh, U.K.

Y. Sakuma, M.D., Ph.D.
Department of Physiology, Nippon University School of Medicine, Tokyo, Japan

Nanette F. Santoro, M.D.
Department of Obstetrics and Gynecology and Women's Health, Albert Einstein College of Medicine, The Bronx, New York

Bernard P. Schimmer, Ph.D.
Banting and Best Department of Medical Research and Department of Pharmacology, University of Toronto, Toronto, Ontario, Canada

B. P. Setchell, Ph.D., Sc.D.
Department of Anatomical Sciences, University of Adelaide, Adelaide, Australia

O. David Sherwood, Ph.D.
Department of Molecular and Integrative Physiology and College of Medicine, University of Illinois at Urbana-Champaign, Urbana, Illinois

C. P. Sibley, Ph.D.
School of Biological Science, University of Manchester, Manchester, U.K.

Michael K. Skinner, Ph.D.
Center for Reproductive Biology, School of Molecular Biosciences, Washington State University, Pullman, Washington

Peter G. Stanton, Ph.D.
Prince Henry's Institute of Medical Research, Monash Medical Centre, Victoria, Australia

Robert A. Steiner, Ph.D.
Department of Physiology and Biophysics, School of Medicine, University of Washington, Seattle, Washington

Douglas M. Stocco, Ph.D.
Department of Cell Biology and Biochemistry, Texas Tech University Health Sciences Center, Lubbock, Texas

Richard L. Stouffer, Ph.D.
Senior Scientist and Head, Division of Reproductive Sciences, Oregon National Primate Research Center/Oregon Health & Science University, Beaverton, Ohio

Susan S. Suarez, Ph.D.
Department of Biomedical Sciences, Cornell University, College of Veterinary Medicine, Ithaca, New York

Amanda Swain, Ph.D.
Institute of Cancer Research, University of London, London, U.K.

Renea A. Taylor, Ph.D.
Centre for Urology Research, Monash Institute of Reproduction and Development, Monash University, Victoria, Australia

Kent L. Thornburg, Ph.D.
Department of Medicine, Heart Research Center, Oregon Health & Science University, Portland, Oregon

Michela Torricelli, M.D.
Department of Pediatrics, Obstetrics and Reproductive Medicine, University Siena, Siena, Italy

John G. Vandenbergh, Ph.D.
Department of Zoology, North Carolina State University, Raleigh, North Carolina

J. B. Wakerley, Ph.D.
Department of Anatomy, University of Bristol, School of Medical Sciences, Bristol, U.K.

Darrell N. Ward, Ph.D.
Department of Biological Sciences, Wichita State University, Wichita, Kansas

John C. Wingfield, Ph.D.
Department of Zoology, University of Washington, Seattle, Washington

Selma F. Witchel, M.D.
Department of Pediatrics, University of Pittsburgh School of Medicine, Pittsburgh, Pennsylvania

Ruth I. Wood, Ph.D.
Department of Cell and Neurobiology, USC School of Medicine, Los Angeles, California

Larry J. Young, Ph.D.
Yerkes Primate Center, Emory University, Atlanta, Georgia

Anthony J. Zeleznik, Ph.D.
Department of Cell Biology and Physiology, University of Pittsburgh School of Medicine, Pittsburgh, Pennsylvania

Jian Zhou, M.D., Ph.D.
Developmental Endocrinology Branch, National Institute of Child Health and Human Development, National Institutes of Health, Bethesda, Maryland

Foreword*

I am pleased and honored to have been asked to prepare the Foreword to this volume of work depicting the progress in research on the physiology of reproduction as well as the resulting gains in understanding made over the past few years. The expertise that is represented by the numerous contributors to this work is so impressive that I am humbled even to contemplate adding anything of note. It is only by virtue of having personally witnessed a very large segment of twentieth century research on reproduction that I am emboldened to reflect on the byways and the trail blazings that have brought this field to its present proud state of enlightenment with regard to the long sought-after means of controlling the procreative process in humankind. Clearly, there are many important and knotty problems yet to be resolved, but the pace of progress over the past several years has quickened to the extent that one is left in expectant wonderment as to where and when the next revolutionizing development will occur.

The experimental method of studying reproduction was initiated in 1849 with Berthold's discovery of a blood-borne activity that came from the testis and stimulated growth of distant organs such as the comb and wattles. In so doing he utilized one of the most fundamental means of demonstrating the function of an endocrine organ, namely, surgical removal to determine what deficiencies follow, coupled with implantation or transplantation to ascertain whether the deficiencies were repaired. At that time it was not possible to take the next step, namely, preparation of an active extract of the testes, because nothing was known about the nature of the bioactivity. Forty years later, Brown-Séquard claimed to have prepared an active extract of dog testes; however, as is well known, his enthusiastic claims for restoration of his own sexual activity at an advanced age were not substantiated. Actually, these simple means of studying reproductive physiology persisted well into the twentieth century, including the studies of such pioneering stalwarts as Marshall, Heape, Prenant, Bouin, Ancel, Loeb, Cushing, and Aschner. Observations otherwise were limited to cyclic and seasonal changes in sexual behavior among common laboratory and small domestic animals. This type of eyeball research remained in vogue through the early 1920s and overlapped the extension of visualization to the microscopic level. The latter revealed, for the first time, the precise timing of events in the ovarian cycle through microscopically observable cellular changes in the vaginal fluid. My point in mentioning these early studies is to emphasize that although the tools and techniques were inordinately primitive by present standards, the results established a firm base of knowledge on which to build.

The study of cyclic changes in the vaginal smear in rats and the findings of estrogenic activity in follicular fluid during the early 1920s led to an explosion of interest in the study of reproduction. The field was fortunate in attracting to its ranks a small band of exceedingly able biologists and biochemists who, in 1932, were to become authors of the classic first edition compendium, *Sex and Internal Secretions*, a volume overwhelmingly devoted to reproductive endocrinology. It was this landmark of progress that finally gave propriety to the study of reproduction and put it on a par with the study of other major bodily systems. Incredible as it may seem, it was only a decade earlier that a distinguished panel of the National Research Council had declared that sex research was not a fitting topic for scientific study.

Lest our pride in today's spectacular pace of progress unduly bedazzle the mind, it should not be overlooked that the developments recorded in the ten-year span from 1926 to 1936 may never be equaled. Among those monumental achievements, all of the native sex steroid hormones were

*This Foreword by Roy O. Greep is reprinted from the 1988 (first) and 1994 (second) editions of this work (E. Knobil and J. D. Neill, *The Physiology of Reproduction*. Raven Press, New York. Copyright Elsevier). This now-deceased author wrote a remarkably prescient account of the future of research in the field that remains as relevant today as it did in the first two editions.

brought to light, their structures were determined, their functions were defined, and they were made available in pure form for research and therapy. Similarly, all of the pituitary, placental, and urinary tropic hormones were identified, and their functions were defined. Like today's competition for priority rights, publicity, and potential financial gain, these earlier periods also were times for intense rivalries, but rarely with prospects for financial rewards. It would be difficult to overstate the boost that was given to basic and clinical research in reproduction as a result of the availability of estradiol-17β, testosterone, and progesterone in pure form and of known potency. The replacement of homemade extracts and such elastic entities as rat units, mouse units, capon units, and so forth, with micrograms of pure hormone was revolutionizing and allowed the study of reproduction on a quantitative basis.

Prior to World War II the thrust of research on reproduction dealt predominantly with the steroid hormones. This was the heyday of steroid biochemistry. After World War II the emphasis shifted to the protein and peptide hormones, where it still remains strong. This prolonged and difficult effort yielded many biochemical triumphs. Most notable among these were the isolation of the pituitary, placental, and urinary gonadotropins, as well as the determination of their primary structure as glycoproteins comprised of two dissimilar and covalently bonded subunits, the isolation and synthesis of the gonadotropin-releasing hormone (GnRH) of hypothalamic origin, and the isolation and structural characterization of relaxin.

The availability of pure protein and polypeptide hormones made possible the production of hormone-specific antibodies as well as the application of immunological techniques to the study of reproduction. An outcome of great consequence was the development of radioimmunoassay as the new means of measuring all of the hormones relating to reproduction. The sensitivity of this new technique was so great that it made possible, for the first time, the measurement of all these hormones in the body fluids. It had the further distinct advantage of requiring such a small amount of fluid that the monitoring of blood levels of the hormones of reproduction could be done throughout an estrous or menstrual cycle by close serial sampling. This revealed still another and most unexpected finding, the pulsatile pattern of secretion.

Identifying the homeostatic mechanism(s) responsible for maintaining a steady state in various physiologic systems of the body has been fraught with many challenging problems, but these pale in comparison with the difficulties encountered in trying to elucidate the mechanisms maintaining a constantly changing system, a characteristic of the reproductive system of female mammals. The earliest piece of evidence suggested the existence of a "push-pull" mechanism that later came to be known as *negative feedback*. It was based on the demonstration that an estrogenic extract administered to immature rats would maintain the ovaries in an infantile state. This was quickly followed by conclusive evidence that estrogen acted to inhibit pituitary follicle-stimulating hormone (FSH) stimulation of follicular growth and maturation; however, the effect on luteinizing hormone (LH), ovulation, and luteinization remained unsettled. Gaps continued to exist in all proposed explanations of reproductive cycles. None of these explanations took into account the influence of photoperiodicity on seasonal breeders, nor did they account for the role of the stimulus of mating in nonspontaneous ovulators. Following the discovery of the hypothalamic control of pituitary function, estrogen was shown to exert its action on both the pituitary and the hypothalamus; however, the problem of accounting for cyclicity remained. Adding to the complexity, radioimmunoassay revealed an unexpectedly high level of blood estrogen just prior to ovulation, an event not in keeping with the negative feedback concept.

Finally, after many years of searching for a way out of this frustrating situation, a glimmer of light appeared at the end of this long dark tunnel—light that soon turned to brilliance. In 1969, Goding and associates found that the administration of large doses of estrogen to ewes at the time of estrus did not block, but instead entrained, ovulation. Shortly thereafter, in more elaborate examination of the relationship of blood estrogen levels and ovulation in rhesus monkeys in Knobil's laboratory, it was revealed that elevated estrogen levels preceded and appeared to trigger ovulation. On further examination, Knobil and colleagues found that when blood estrogen reached a critical level the feedback mechanism switched from a negative to a positive, or stimulative, action. This utterly new finding greatly advanced our understanding of the endocrine mechanism governing reproductive cycles. There still remain, however, some uncertainties: Why does the switch in feedback action occur; to what extent and at what stage of the cycle does estrogen act at the level of the pituitary or the hypothalamus, or both; and lastly, what role, if any, do the ovarian peptides, especially inhibin, play in controlling reproductive cycles?

The progress of research on reproduction has been chronicled in numerous review articles by individual authors. Many have appeared in *Recent Progress in Hormone Research, Volumes 1 to 42*. Other major sources include the multiple editions of such titles as: *Marshall's Physiology of Reproduction, Fourth Edition* (1990); *Sex and Internal Secretions*, whose third and last edition was issued in 1961; two volumes on the *Female Reproductive System* (1973), and one on the *Male Reproductive System* (1975) in *Section 7 of the Handbook of Physiology*, published by the American Physiological Society; and four serial volumes on reproductive physiology in the *International Review of Physiology*, the last one being issued in 1983. The present volume will provide comprehensive coverage and meet the current needs of the field of reproductive physiology, a field that is rapidly gathering momentum from the application of new and highly sophisticated tools and techniques.

In viewing the vast literature dealing with research on the male and female reproductive systems and considering the rate at which it is accumulating, one might ask whether this staggering proliferation of books and articles is essential to progress; the answer is an emphatic "Yes!" The yardstick by which progress is measured in this or any other field is not in the number of articles published or the amount of financial support but in improved understanding. Such gains are generally marked by sharp peaks at indeterminate intervals, separated by avalanches of incremental gains, as recorded in an ever-growing list of journals. The point to remember is that without this persistent chipping away at a major problem there would be no solutions and no quantum leaps forward. In research very little comes from out of the blue. Part of the driving force in research is its adventuresome nature and ever-present possibility that one's efforts will pay off in an important manner. It may not be entirely fair, but in research (as in most human activities), the spoils go to the victor in the form of kudos, prizes, awards, public attention, and, increasingly in the present technological age, monetary gains—sometimes of great magnitude. What effect this latter may have, if any, on the long-cherished sanctity of science has not been determined, but it has become a matter of concern.

This volume bears the title *The Physiology of Reproduction*. Physiology, by traditional consensus, is that branch of science which studies the functions of a living organism or any of its parts and includes the basic underlying processes. It will be understood that most of the studies reviewed here will be based more on holistic research than on research at the submicroscopic or molecular level. It is unfortunate that the excitement generated by recent fantastic advances in molecular biology and development has tended to downgrade the value of whole-animal research, and physiology in particular is sometimes looked upon as passé. Actually, the two categories of research are complementary, and both are essential for maximum advancement of knowledge. Whole-animal research cannot become outdated because it is the quintessence of biological relevance and the means by which molecular findings must ultimately be evaluated.

In the same vein, no one immersed in reproductive endocrinology can be unaware of the current tendency to regard research at the molecular level as representative of exceptional scientific talent. This is a common consequence of the opening of a new arena of investigation. I recall an incident that happened at a scientific meeting back in the 1930s. The first three papers in a session chaired by an eminent embryologist were on endocrine topics—mine was the third. That being ended, the chairman took pains to assure the audience that the meeting could now turn to considerations of more fundamental nature. One of the other three papers was given by Herbert M. Evans, who bristled noticeably but held his fire. There was also an earlier period when one either worked on steroid biochemistry or something of lesser appeal like biology. Anyone who remembers the 1950s will recall a flash-in-the-pan ignited by cybernetics, a study of automatic control systems both neural and physical. The gurus of cybernetics captured the attention of the press and of audiences throughout the land, but eventually this obsession suffered the fate of other passing preoccupations. My own observation is that the closer one approaches the molecular level of research, the more one becomes dependent on highly sophisticated instrumentation to make the observations and to read out results that are often quite free of extraneous variables. Toward the obverse situation, one's dependence on an extensive background of experience and physiological increases as does the unavoidable complex of *in vivo* variables that must be taken into account. In either case we have today the availability of far more diverse approaches to a given problem in any field of biomedical research than has ever existed before. In Berthold's day there was only one experimental method available; today's number is untold but is probably in the hundreds, perhaps thousands. This is an exceedingly promising situation and one to which investigators of all persuasions must adjust. Open minds will experience exhilaration over substantive achievements at any point on this observational spectrum.

One of the major factors influencing research on reproduction has been the availability of funds or lack thereof. Prior to the institution of federal funding (i.e., prior to the middle of the twentieth century), reproductive research was sparsely supported by university departmental funds, industry, small grants from the Committee for Research in Problems of Sex within the National Research Council, and some aid from the Rockefeller Foundation. The National Institutes of Health were slow in providing significant support of research on reproduction because of restrictions on the support of work related in any way to birth control. This occurred despite the simultaneous postwar baby boom. What kept research afloat during this critical period was major support by the Ford Foundation plus lesser contributions by other major foundations. It was not until the establishment in 1968 of the Center for Population Research in the NICHD that major governmental funding in this area became available, but the boost was short-lived. As a result of the imposition of fiscal restraints in the early 1970s, federal support dwindled and has remained at a minimal level ever since. Support from all sources is woefully incommensurate with the distressing expansion of the human population and the need for safe, effective, economical, and readily available means of limiting human fertility.

The physiology of reproduction is predominantly under hormonal control. The first essential step in studying reproduction was identification of the hormones involved and the functions they serve. This having been accomplished, efforts turned to a detailed analysis as to how hormones act within the body. During the 1980s there has been a rising tide of interest in the binding of steroid, protein, or peptide hormones to receptors on specific target cells. Much effort is currently being directed toward the isolation and chemical characterization of these receptors. They are known to be composed of a protein or proteins, and some information has already been gained as to their partial or provisional structure. This, however, is only a preliminary step in the complex process whereby hormone action results in an end response such as growth, secretion of a target cell hormone, or altered behavior. The curtain has already been raised on the climatic and final chapter of the story on how hormones act. This involves linkage of the hormone-receptor complex with the nuclear genetic apparatus leading through a now well-defined series of processes to the manifestation of a physiological response in the living organism. Genes that bring about the expression of certain hormonal signals are being isolated, modified, transferred between species, and also inserted into bacteria where they direct the biosynthesis of specific hormones in large quantity. Thus genes are being manipulated in ways that raise the potential of altering the reproductive process. It is largely as a result of developments in endocrinology at the molecular level that bewildering possibilities loom on the horizons of reproductive research—they are within reach; they are science, not fiction; and they stagger the imagination.

It being granted that nothing succeeds like success, then the new edition of this highly successful two-volume compendium on *The Physiology of Reproduction* is destined for an illustrious fate. This second edition will maintain the same high standards of the first and again fulfill an existing need in a field that is experiencing rapid growth and exhilarating progress. Like the first edition, this one will provide a critical assessment of the state-of-the-art in every aspect of research on the physiology of reproduction by eminent authorities.

In the years intervening between this edition and the last, notable changes have taken place in the study of reproduction. These stem largely from major advances in technology. Remarkable new instruments, techniques, and methods have enabled investigators to probe ever deeper into the interaction between hormones and genes, thereby eliciting *in vivo* responses. New parameters are being added to the target tissues of the classical reproductive hormones as revealed by the presence of receptor sites in tissues, the physiologic significance of which often remains tantalizingly obscure. Similarly, newly identified substances of endocrine or paracrine nature are being added to this domain of research with persisting frequency. Some of these substances—the endothelins, interleukins, activins, inhibins and prorenin, to name a few—also exhibit a puzzling array of effects on extraneous tissues. Their study is being aided by the fact that their structure is known and, though rare, they are available.

Great strides are also being made in many other aspects of research on reproduction. Much work is being done on the structure of receptors and the loci of binding sites on segments of the folded gonadotropic molecules. A full-scale effort is underway seeking an elucidation of the neural mechanism underlying pulsatile secretion. Neuroendocrinologists are closing in on an elusive pulse generator located in the central nervous system. This looms as another landmark discovery in reproductive biology.

Research on reproduction is flourishing and the future appears bright. The taboos are gone. All aspects of the reproductive process are an open book. One area that has taken a quantum leap forward is the clinical application of an important body of relevant new knowledge gained in both basic and clinical spheres. Expanded opportunities have been opened by greatly improved diagnostic procedures, more effective treatment of disorders, and new methods of controlling fertility. Contributing greatly to this explosive development is the dissemination of information on reproductive matters to the lay public by the mass media. Concerned individuals have been made aware of the existing new means of manipulating the male and female reproductive systems for enhancement or inhibition of fertility. The joys and comforts that accrue respectively to these opposing modes of fertility control have enriched the lives of a grateful public. To that end I may note that it was by virtue of these frontier reproductive measures that my own progeny includes a new grandson and namesake.

Roy O. Greep

In Memoriam: Ernst Knobil (1926–2000)

Ernst Knobil died on April 13, 2000, in his 73rd year of life, thus ending a remarkable career of outstanding scientific accomplishments, of leadership positions in physiology and endocrinology, and of mentorship to numerous students and fellows. He was noted for his clarity of thought, his endless pursuit of excellence, and his abiding interest in integrative biology.

Dr. Knobil was a visionary leader and a pioneer in many areas of endocrinology, including growth and reproduction. Dr. Knobil's now classic contributions include the species-specific effects of GH, a model for positive and negative estrogen feedback control of the menstrual cycle, and elucidation of the hypothalamic GnRH pulse generator. His discovery that pulsatile GnRH stimulates LH, whereas continuous GnRH desensitizes pituitary LH secretion, has forever altered the field of reproductive endocrinology. This remarkable experimental observation unmasked a pivotal role for pulsatile secretion as a mechanism of hormonal control.

The elder son of an Austrian father (Jakob Knobil) and a German mother (Regina Seidmann), Ernst was born in Berlin, Germany, on September 20, 1926. When Ernst was about six years old, the Knobil family moved to Paris due to the deteriorating political conditions in Germany. Then, in 1940, when the Germans invaded Paris and Ernst was 13, the family emigrated to New York City.

Ernst entered the New York State College of Agriculture at Cornell in 1942, at the age of 15. He chose Animal Science as his major due to interests developed from time spent on farms in France during the summers, and from attending the Kinderhook Farm Camp, after moving to New York. Upon graduating from Cornell in 1948 (including a two-year interruption for service in the U.S. Army), he entered graduate school in zoology upon recommendation of the late Sidney Asdell, author of the well-known reference work, *Patterns of Mammalian Reproduction*. He worked in the laboratory of Professor Samuel L. Leonard, who, as a member of Hisaw's laboratory at the University of Wisconsin, had participated in the initial identification of LH. After completing his Ph.D., Ernst accepted a postdoctoral position with Roy O. Greep at the Harvard School of Dental Medicine from 1951 to 1953. In Greep's laboratory, he was introduced to the

Endocrine Reviews 22(6):721–723

rhesus monkey as an experimental animal, and he performed studies on adrenal cortical function and on fetal–maternal interrelationships. While a fellow, he assumed Greep's teaching duties in endocrinology and rapidly gained recognition as a gifted and scholarly teacher. Because of this, in 1953, he was appointed Instructor in the Physiology Department of the Harvard Medical School. In 1957, he was promoted to Assistant Professor after having been selected by Harvard Medical School for the prestigious Markle Scholar in Academic Medicine for the years 1956–1961.

Ernst's research first gained international recognition for his discovery of the "species specificity" of GH during his research in 1954–1959. His initial studies in hypophysectomized monkeys confirmed that bovine and porcine GH preparations were inert in the monkey, as had been reported for humans. However, he found that GH extracted from monkey pituitary glands was fully effective in stimulating various parameters of growth (1). These findings were quickly confirmed in humans and gave birth to the program of the National Pituitary Agency to collect human pituitary glands from which GH was purified for treatment of GH deficiency. This program was later discontinued when human GH became the first protein to be produced for clinical use using recombinant DNA technology.

In 1961, Ernst was appointed Chairman of the Department of Physiology at the University of Pittsburgh School of Medicine, a position he held for 20 years. There he initiated studies on regulation of reproductive cycles in rhesus monkeys because primates had been little studied in this regard, and because new methodologies for measuring hormones in blood were becoming available. First, he established the Center for Research in Primate Reproduction and built the Pittsburgh Primate Center to supply animals and to train postdoctoral fellows. Next, he developed RIAs for progesterone, LH, FSH, and E2 "after long and arduous methodological studies" (2) to permit description of the time courses of circulating hormones during the menstrual cycle of the rhesus monkey. These and subsequent studies revealed that E2, rather than progesterone, was the ovarian hormone that stimulated the mid-cycle LH surge. Moreover, he observed that low tonic levels of E2 were inhibitory to LH secretion, whereas prolonged elevations of E2 were stimulatory to LH secretion. Progesterone was shown to be primarily inhibitory, synergizing with low levels of E2 and at higher levels, antagonizing the stimulatory effect of estradiol (2). These studies were monumentally important because they established for the first time the exact roles of E2 and progesterone in regulating gonadotropin secretion and, hence, the reproductive cycle.

Another crucial discovery was the episodic nature of the interaction between the hypothalamus and pituitary. Pulses of LH secretion were observed at about hourly intervals and, hence, were denominated "circhoral oscillations." The LH pulses were correctly assumed to be the consequence of pulsatile GnRH secretion by the hypothalamus into the pituitary portal circulation which, in turn, gave rise to the notion of an oscillator or signal generator in the central nervous system ("GnRH pulse generator") (2,3). A decade or more of study revealed that pulsatile LH secretion was a crucial regulator of the menstrual cycle. For prolonged LH secretion, GnRH had to be administered in a pulsatile fashion; continuous administration resulted in desensitization of LH secretion (3), an observation that forms the basis of a currently popular treatment of prostate cancer in humans. Knobil further observed that the hourly pulsatile infusions into monkeys deprived of endogenous GnRH by lesioning of the arcuate nucleus reinitiated menstrual cycles. These cycles were characterized by normal LH and FSH levels and were accompanied by normal follicular growth, ovulation, and corpus luteum function as signified by normal patterns of E2 and progesterone secretion. From these studies, he argued that the ovary rather than the hypothalamus was the "zeitgeber" of the menstrual cycle; an extension of this notion was that, unlike the rat, the monkey did not require a midcycle surge of GnRH secretion to stimulate LH secretion, the heightened responsiveness of the pituitary to GnRH occasioned by E2 being sufficient (3,4).

Extension of these studies to normal prepubertal monkeys led to the initiation of menstrual cycles characterized by normal gonadotropin and E2 and progesterone levels. From these studies he concluded "that neither adenohypophysial nor ovarian competence is limiting in the initiation of puberty in the rhesus monkey." Rather, "puberty is normally initiated by the activation of hypothalamic mechanisms that control the circhoral pulsatile release of GnRH into the pituitary portal circulation" (3).

A final breakthrough in the understanding of the primate menstrual cycle was the multiunit recording in the medial basal hypothalamus of electrical activity associated with each pulse of LH secretion. Initially, these studies were confined to ovariectomized animals because restraint in the apparatus necessary for electrical recordings from the hypothalamus suppressed menstrual cycles in intact monkeys. This problem was circumvented by adopting telemetry that permitted electrical recordings of hypothalamic activity in unrestrained monkeys exhibiting normal menstrual cycles (4,5).

This very brief description of his studies hardly conveys the transforming nature of the findings. To understand their impact, it is only sufficient to note that we understand the regulation of reproductive cycles in rhesus monkeys better than for any other species. This monumental result can be attributed to several personal characteristics: 1) the logical force that is evident throughout his research career; 2) a level of clear and incisive thought that makes Occam's razor seem dull by comparison; 3) insistence on the development of methods appropriate to the experimental question being asked, no matter how arduous or time consuming; and 4) a high level of scholarship in all matters, at all times, and in all places.

In 1981, Ernst accepted the Deanship of the University of Texas Medical School at Houston. There he continued to explore regulation of the reproductive cycles in macaques; indeed, most of the studies described above on electrical recordings of the hypothalamus were conducted in Houston. After his tenure as Dean ended in 1984, he continued as Director of the Laboratory of Neuroendocrinology until 1997 when he closed his laboratory. More than 80 fellows and students studied in his laboratories in Boston, Pittsburgh, and Houston, and contributed to the studies described earlier.

Ernst received many awards, including the highest ones awarded by the Society for the Study of Reproduction (Carl G. Hartman Award, 1983), The Endocrine Society (Fred Conrad Koch Award, 1982), and the American Physiological Society (Walter B. Cannon Memorial Lecture, 1997). He was elected to numerous positions of leadership including the Presidencies of The Endocrine Society (1976), the American Physiological Society (1979), and the International Society of Endocrinology (1984–1988). A noteworthy achievement of his Endocrine Society presidency was the establishment of *Endocrine Reviews*. He was a member of many other U.S. and foreign scientific societies' review boards, NIH study sections, and the editorial boards of numerous scientific journals.

Dr. Knobil was a member of the U.S. National Academy of Science (1986), the American Academy of Arts and Sciences, a foreign associate of the French Academy of Science, the Italian National Academy of Science and the Belgian Royal Academy of Medicine. He received several honorary degrees, including the University of Bordeaux (1980), the Medical College of Wisconsin (1983), the University of Liege (1994), and the University of Milan (2000). In addition to being the author of 217 scientific papers, he was the editor of several reference books in endocrinology and reproduction, including *The Handbook of Physiology* (1974), *The Physiology of Reproduction* (1988, 1994), and *The Encyclopedia of Reproduction* (1998). In addition to his numerous scientific contributions, Ernst was a visionary leader in the field of endocrinology and a wonderful mentor and friend to the many people who were privileged to work closely with him.

He is survived by his wife, Julane Hotchkiss, four children, and three grandchildren.

ACKNOWLEDGMENTS

The editorial assistance of Dr. Adolph Friedman, Staff Consultant for the History Project of The Endocrine Society, is gratefully acknowledged.

Jimmy D. Neill
Distinguished Professor
Department of Physiology and Biophysics
University of Alabama School of Medicine
Birmingham, Alabama 35294

Address all correspondence and requests for reprints to: Jimmy D. Neill, Department of Physiology and Biophysics, University of Alabama School of Medicine, Birmingham, Alabama 35294.

REFERENCES

1. **Knobil E, Greep RO** 1959 The physiology of growth hormone with particular reference to its action in the rhesus monkey and the "species specificity" problem. Recent Prog Horm Res 15:1–69
2. **Knobil E** 1974 On the control of gonadotropin secretion in the rhesus monkey. Recent Prog Horm Res 30:1–46
3. **Knobil E** 1980 The neuroendocrine control of the menstrual cycle. Recent Prog Horm Res 36:53–88
4. **Hotchkiss J, Knobil E** 1994 The menstrual cycle and its neuroendocrine control. In: Knobil E, Neill JD, eds. The Physiology of Reproduction. Vol. 2, ed. 2 New York: Raven Press; 711–749
5. **Knobil E** 1997 The wisdom of the body revisited (1997 Walter B. Cannon Memorial Lecture). News Physiol Sci 14:1–11

Preface to the First Edition

This work was undertaken, after much deliberation, in an attempt to fill a need for a comprehensive, scholarly treatise on the physiology of mammalian reproduction. A major inspiration for this effort has been the volume *Sex and Internal Secretions,* a factual and conceptual beacon which guided generations of reproductive biologists from the time of its first publication in 1932 to the appearance of its last edition over a quarter of a century ago.

The book is divided into five major sections, and these, in turn, are loosely arrayed in two domains. The first covers the components of the reproductive system, and the second discusses reproductive processes and their physiological control. In the latter, we have included reproductive behavior in the conviction that this fundamental aspect of reproduction clearly belongs in the physiological realm and will remain a demanding challenge long after all the other mysteries in the field have been resolved.

In our discussion of reproductive systems, we have been aware of the profound differences among mammals in the way some fundamental processes, such as the ovarian cycle, are controlled. We have addressed this issue, in part, by providing separate coverage of major mammalian groups where this seemed appropriate. It has been left to the reader to ascertain the similarities and differences among them. In any case, we must ever be mindful in considering reproductive processes, from the control of ovulation to the initiation of parturition, not to extrapolate from one species to another without due reflection.

It is hoped that this book will be useful to all serious students of reproductive physiology be they scientists, teachers, or physicians.

THE EDITORS

Preface to the Second Edition

The second edition of this work represents the labor of some 61 groups of authors and is, therefore, marked by inevitable redundancies and lacunae. The six years that have elapsed between the first and second editions have seen dramatic and often unanticipated developments in some aspects of reproductive biology, with only little new understanding in others. But, as expected, the quantity and difficulty of the questions raised has increased manifold. We remain markedly ill-informed of the complex control systems that govern reproductive processes and surprised by the striking species differences in the accomplishment of common, fundamental reproductive tasks. The control of ovulation, the advent of puberty, and the initiation of parturition are but three cases in point.

We are deeply saddened by the untimely loss of Larry L. Ewing who contributed so much to the construction of the first edition with his wisdom and good humor, to say nothing of his hard work and devotion to the enterprise.

As these volumes were in the final stages of completion, the field also lost one of its great pioneers and most sagacious contributors, John W. Everett. We are privileged to count one of his last publications among the chapters of this work.

We hope that our expectations for the first edition have also been achieved in the second: scholarly, comprehensive examinations of the principal issues in reproductive physiology that will remain useful for a decade, at least, to all serious students of reproductive physiology.

THE EDITORS

Preface

In the decade that has passed since the last appearance of this work, the study of reproductive physiology has undergone monumental changes. Chief among these advances are organismal cloning with the attendant discovery of a diverse set of stem cells, and the cloning and sequencing of genomes with the resulting identification of proteomes. Therefore, the demanding challenge to our authors was incorporating the effect of these new findings into our understanding of the physiology of reproduction. Indeed, the 59 groups of authors of the current work have met this challenge in splendid fashion, providing a synthesis of the new information at molecular, cellular, and organismal levels of organization. The bibliographies of the chapters are extensive, and taken together with earlier editions of the work, provide coverage of all of the relevant literature of the topics being reviewed.

Given the explosion of new information, how can we claim the current work is comprehensive, when earlier editions had similar lengths (60 and 61 chapters) to the present one (59 chapters)? As noted by others (B. Alberts, et al. [1983]. *Molecular Biology of the Cell*. Garland Publishing, New York and London), this is because "there is a paradox in the growth of scientific knowledge. As information accumulates in ever more intimidating quantities, disconnected facts and impenetrable mysteries give way to rational explanations, and simplicity emerges from chaos." Whether this dictum can survive studies of proteomes with their large numbers of interacting proteins in regulatory networks remains to be seen.

The major changes in our understanding of reproduction and the passage of time have necessitated changes in the list of authors and editors. Of the 59 groups of authors, 44 are new to this edition. Only two editors remain from the first edition (Pfaff and Neill) so that five of seven section editors are new; separate sections and editors were also added to cover "Pregnancy" and "Lactation." Perhaps the most substantial change in this edition is the absence of Ernst Knobil as the editor-in-chief, due to his untimely death in the year 2000. This edition is dedicated to his memory, and is so signified by including a memorial article about him, and renaming the work *Knobil and Neill's Physiology of Reproduction*. His wise counsel has been missed, but we have endeavored to achieve his expectations expressed in the second edition that the work be a "scholarly, comprehensive examination of the principal issues in reproductive physiology" that will be useful "to all serious students of reproductive physiology."

A feature of the new edition that has not changed is the Foreword by Roy O. Greep. This pioneer in the discovery and characterization of pituitary hormones wrote a remarkably prescient account of the future of research in the field in 1988 that remains relevant even now eighteen years since it was written.

We welcome a new publisher, Academic Press, an imprint of Elsevier, to this third edition of the work. Particular thanks go to Dr. Jasna Markovac, Senior Vice-President of the Science and Technology Division at Elsevier, whose interest in this work dates to preparation of the second edition as its executive editor. The current edition owes its existence to her.

Finally, we thank the Senior Development Editor of this edition, Kirsten Funk, whose day-to-day work on these volumes has brought them to fruition. Her remarkable persistence and constant cheerfulness have sustained the prodigious effort required to produce such a voluminous work.

Jimmy D. Neill
Editor-in-Chief

Gametes, Fertilization, and Embryogenesis

CHAPTER 1

The Spermatozoon

E. M. Eddy

Introduction, 3
The Sperm Plasma Membrane, 4
The Mosaic Sperm Surface, 4
Identification of Domains, 6
Localized Sperm Membrane Components, 7
Formation and Maintenance of Domains, 8
Sperm Plasma Membrane Lipids, 9
The Sperm Head, 11
The Sperm Nucleus, 11
The Sperm Head Cytoskeleton, 13
The Acrosome, 14
The Flagellum, 18
Connecting Piece, 19
Axoneme, 20
Mitochondrial Sheath, 21
Outer Dense Fibers, 22
Fibrous Sheath, 26
Abnormal Flagella, 30
Flagellar Motion, 32
Summary, 37
References, 38

ABSTRACT

This chapter examines the novel structural features of the spermatozoon, their composition and assembly, and the mechanisms regulating their function. The spermatozoon is a highly specialized cell with an array of novel structural features and functional characteristics which provide it with the singular capability of delivering the male genome to the egg. The head of a spermatozoon contains a highly condensed nucleus and the acrosome, an enzyme-filled vesicle whose contents are released as the sperm approaches or reaches the egg surface to facilitate fertilization. The flagellum is a highly novel feature that provides the spermatozoon with the ability to propel itself through liquid medium. Substantial advances have occurred within the last decade in identifying the genes responsible for many of these novel structural features and functional characteristics. Many of the genes are expressed only in spermatids and encode novel proteins that are unique to spermatozoa. They include proteins that package the DNA, form the sperm head cytoskeleton, constitute the enzymes and matrix components of the acrosome, assemble into the novel structural components of the flagellum, become ion channels involved in regulating motility, and are the adenylyl cyclase that produces cyclic adenosine monophosphate (cAMP) to trigger the signaling cascades that regulate spermatozoon functions. Because such proteins are essential to spermatozoa, mutations that disrupt their synthesis or function can result in male infertility. The gene knockout approach has identified many proteins required for normal sperm functions in mice that are likely to be required in humans as well. However, many critical questions remain to be answered about the composition, organization, and function of spermatozoa.

INTRODUCTION

The spermatozoon is the end product of the process of spermatogenesis proceeding through successive mitotic, meiotic and postmeiotic phases within the seminiferous tubules of the testis. During the mitotic phase, the progeny of germ line stem cells undergo a series of divisions to expand the spermatogonial population.

Gamete Biology Section, Laboratory of Reproductive and Developmental Toxicology, National Institute of Environmental Health Sciences, National Institutes of Health, Research Triangle Park, North Carolina

The meiotic phase begins with the last cell cycle S phase and culminates in two meiotic divisions that rapidly occur without DNA replication to produce haploid spermatids. While these two phases are crucial for the development of the male gamete, it is during the post-meiotic phase that spermatozoa form. This phase is characterized by extensive remodeling of spermatids into sperm by formation of the acrosome, nuclear condensation, flagellar development, and loss of most of the cytoplasm. These events result in a cell highly differentiated in structure and function, but developmentally totipotent and able to combine with an egg to begin the process that gives rise to the next generation.

The two main components of a sperm are the *head* and *flagellum*, joined by the *connecting piece* (Fig. 1). The head contains the nucleus, acrosome, cytoskeletal structures, and a small amount of cytoplasm. The *nucleus* contains highly condensed chromatin and is capped anteriorly by the *acrosome*, a membrane-enclosed cytoplasmic vesicle containing hydrolytic enzymes. From the connecting piece, the flagellum is divided successively into the midpiece, principal piece and end piece regions. It contains a central complex of microtubules forming the *axoneme*, surrounded in turn by *outer dense fibers* extending from the neck into the principal piece. The midpiece contains the *mitochondrial sheath*, a tightly wrapped helix of mitochondria

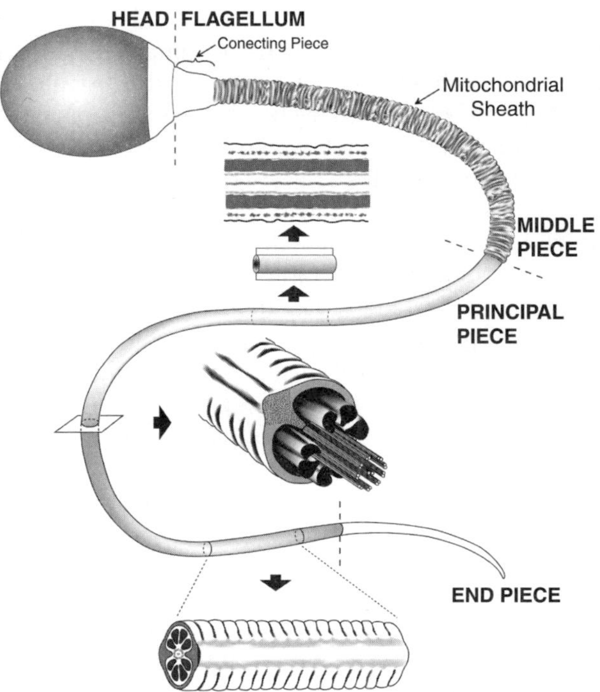

FIG. 1. General features of the mammalian spermatozoon. The head of the sperm is attached to the flagellum by the connecting piece. The regions of the flagellum are the middle piece, the principal piece, and the end piece. The middle piece contains the mitochondrial sheath, while the principal piece contains the fibrous sheath. Longitudinal and cross-sectional views of the principal piece and a segment of fibrous sheath are indicated by arrows; the internal components of the flagellum are identified in Fig. 6.

surrounding the outer dense fibers and axoneme. Most of the length of the flagellum is made up of the principal piece, defined by the presence of a *fibrous sheath* surrounding the axoneme and outer dense fibers. The outer dense fibers and the fibrous sheath are cytoskeletal structures novel to the sperm flagellum in higher vertebrates and may have evolved with the development of internal fertilization (1). The flagellum, like the head, is tightly enclosed by the plasma membrane and contains a sparse amount of cytoplasm. While most mammalian spermatozoa have these general characteristics, there are substantial species-specific differences in the size and shape of the head, and in the length and relative amount of the different components of the flagellum. Sperm from invertebrates usually have a head with an acrosome and a flagellum containing an axoneme and mitochondria, but lacking accessory cytoskeletal structures (2).

The specialized structural features of mammalian spermatozoa reflect the unique functions of this cell type. The acrosome contains enzymes essential for penetrating the investments of the egg to achieve fertilization, while the flagellum contains the energy sources and machinery to generate the motility necessary for the sperm to reach the egg. These functions are essential for delivery of the genetic material contained in the sperm nucleus to the cytoplasm of the egg, where combination of the haploid male and female pronuclei occurs to produce the zygote and initiate development. In most mammals, the sex chromosome carried in the haploid sperm nucleus determines the sex of the resulting animal (3). Both a maternal and a paternal genome are required for normal development to proceed to term, usually due to differential imprinting of genes during gametogenesis in males and females (4,5).

This chapter focuses on the characteristics of mammalian spermatozoa (6), with particular attention to the molecules currently known to contribute to their structure and function. The major topics considered are: the organization and composition of the sperm plasma membrane, the structural components of the head of the sperm, and the organization and composition of structural features of the flagellum. Other chapters in this volume are concerned with spermatogenesis, capacitation and fertilization, and early embryonic development. They should be consulted for additional information necessary for understanding the role of the structural and functional features of sperm in reproductive processes.

THE SPERM PLASMA MEMBRANE

The Mosaic Sperm Surface

A characteristic feature of spermatozoa is the subdivision of the plasma membrane into regional domains that differ in composition and function. A variety of approaches were used to identify the heterogeneous nature of the sperm plasma membrane, including detection of surface charge distribution, labeling with lectins, application of freeze fracture methods, use of membrane intercalating agents, and labeling with antibodies. Knowledge that the organization and composition of the plasma membrane varies between different regions of the sperm surface led to the concept that the sperm plasma membrane is a mosaic of restricted domains that reflect the specialized functions of surface and cytoplasmic components of spermatozoa (7–10). These domains are dynamic features that undergo changes in organization and composition during the life of the cell (11,12).

The major regions of the plasma membrane on the sperm head of most mammals are the *acrosomal region* (anterior head) and the *postacrosomal region* (posterior head) (Fig. 2). The plasma membrane of the acrosomal region can usually be subdivided into the: (a) *marginal segment domain* (apical segment domain, anterior band domain, peripheral rim domain) over the anterior margin of the acrosome, (b) *principal segment domain* (anterior acrosomal domain) over the major portion of the acrosome, and (c) *equatorial segment domain* (posterior acrosomal domain) over the posterior part of the acrosome. The size and shape of these domains vary between species. The marginal and anterior acrosomal domains together are sometimes referred to as the *acrosomal cap*. The central crescent domain separates the marginal and principal segment domains in the guinea pig and possibly other species (11).

The *postacrosomal region domain* (posterior head, postacrosomal segment) includes the plasma membrane between the posterior margin of the acrosome and the connecting piece. The margin between the acrosomal and postacrosomal regions is delimited in some species by the *serrated band* (subacrosomal ring), which girdles the sperm head at the posterior margin of the equatorial segment.

The *posterior ring* (nuclear ring, striated ring) is located at the junction between head and connecting piece and believed to form a tight seal between the cytoplasmic compartments of the two main portions of the spermatozoon. The plasma membrane of the flagellum is divided into the *middle piece domain*, *principal piece domain*, and *distal piece domain*. The middle and principal piece domains are separated by the *annulus*, a fibrous ring that is a component of the flagellar cytoskeleton and closely applied to the inner surface of the plasma membrane.

Most sperm-surface domains probably are established during spermiogenesis as spermatids are remodeled into spermatozoa. However, spermatozoa undergo additional shape and surface changes during

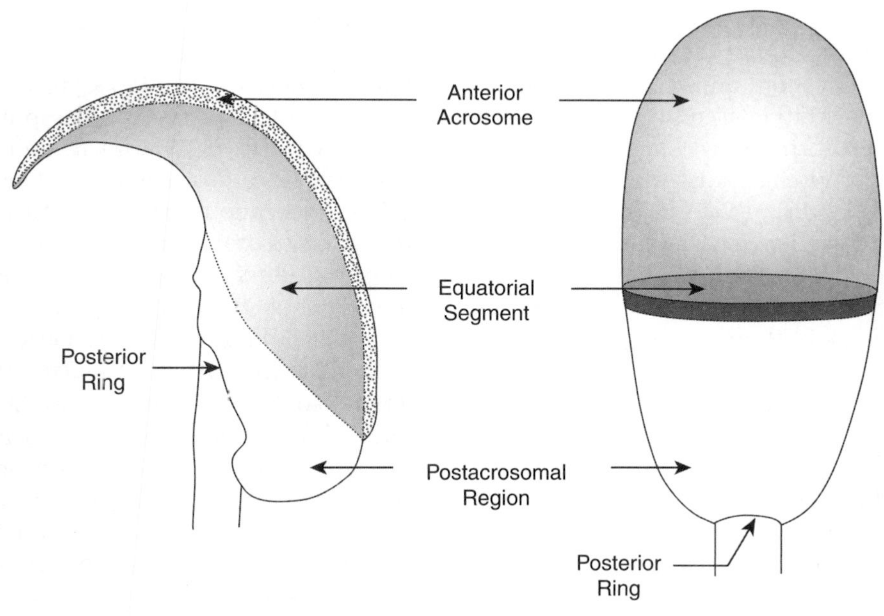

FIG. 2. General features of the head of mouse and rabbit spermatozoon. The major regions of the sperm head are the acrosomal region and the postacrosomal region. The acrosomal region of the falciform-shaped mouse sperm head has a relatively small anterior acrosome compared to the spatulate-shaped rabbit sperm head. However, the equatorial segment of the mouse sperm is more expansive than the same region on the rabbit sperm. The postacrosomal region is that portion of the sperm head not covered by the acrosome. The posterior ring defines the boundary between the postacrosomal domain of the head and the connecting piece. The differences in the size and distribution of these regions are compared in Fig. 4.

epididymal maturation and some domains acquire their final form and composition after spermiogenesis. The middle piece domain is modified as sperm transit the epididymis by the cytoplasmic droplet migrating from the anterior to the posterior end and being shed. In addition, the shape of the sperm head can change (12,13), the acrosome can undergo a reduction in size (13,14), and proteins on the surface of the sperm head can migrate to other domains (15).

A correlation occurs between the time of appearance of some surface components during spermatogenesis and their final localization on guinea pig sperm (16). It was hypothesized that the temporal regulation of surface expression of surface components directs them to their correct surface domains, with newly synthesized proteins being inserted into anterior-most domains. They might not diffuse into more posterior domains because they either are anchored in the anterior domain, or barriers exist to prevent diffusion between domains. While this model suggests that localization of surface components to specific domains might not require a specific sorting mechanism, signals might exist that identify them for removal from inappropriate domains (16).

Identification of Domains

Surface Charge

The heterogeneity of the sperm plasma membrane first was suggested by studies of surface charge. Spermatozoa from ram, rabbit, and bull suspended in medium between oppositely charged electrodes were drawn tail first towards the anode, suggesting that sperm have a net negative charge associated predominantly with the tail (17–19). This was supported by studies of rabbit and stallion sperm using electron microscopy to demonstrate that greater binding of positively charged colloidal iron hydroxide occurred on the surface of the flagellum than on the head (20–22).

Lectins

Domains also were observed using lectins that bind saccharide molecules with relatively high specificity (23,24). Lectins labeled with radioactive iodine, fluorescent tags, or ultrastructural markers demonstrated that different saccharides are heterogeneously distributed on the sperm surface and that lectin binding sites are generally present in higher density on the head than on the tail [e.g., (25,26)]. Lectin agglutination assays also found that regional differences occur in the location and amount of specific saccharides on the sperm surface (27,28). Electron microscopy of rabbit and hamster sperm treated with ferritin- or hemocyanin-labeled lectins also found differences in binding on the head and tail (25,29). However, these studies often indicated that the distribution of some lectin-binding sites was not confined to specific domains.

Freeze-Fracture

Freeze-fracture, freeze-etch, and surface replica studies detected regional domains similar to those seen in labeling studies. Although species-specific differences in the size and distribution of intramembranous particles occurred, characteristic patterns in the head often were seen associated with the marginal and anterior acrosomal domains, the posterior margin of the posterior acrosomal domain, postacrosomal domain, and the posterior ring. In the flagellum of some species, particles were found in the plasma membrane in diagonal arrays over the mitochondrial helix in the middle piece, in circumferential arrays over the annulus, and in a longitudinal zipper-like array in the principal piece, overlying outer dense fiber number one (30).

Membrane-intercalating Agents

Studies using membrane-intercalating agents followed by freeze-fracture or agents tagged with fluorescent labels also suggested that the composition of the plasma membrane differs between domains. Filipin is a polyene that complexes β-hydroxysterols with sterol. Guinea pig sperm treated with filipin developed filipin-sterol complexes between plaques of intramembranous particles in the anterior and posterior acrosomal domains (31–33), implying that these domains are sterol-rich. The plasma membrane of the postacrosomal region had less than one-fourth as many filipin-sterol complexes as the acrosomal region and the pattern produced suggested that sterols are present mainly in the outer half of the bilayer (8,33). The plasma membrane in the acrosomal region of stallion and golden hamster sperm also appeared to have significantly higher amounts of sterols than the postacrosomal region (34,35). The area between the anterior and posterior acrosomal domains of guinea pig sperm often contained circular patches of membrane cleared of sterols and intramembranous particles (36,37).

Another agent used to identify domains was the antibiotic polymyxin B, which binds anionic phospholipids and produces crenellation of anionic phospholipid-rich membranes (38,39). Changes in guinea pig sperm treated with polymyxin B suggested that anionic phospholipid concentration is high in the plasma membrane over the marginal segment of the acrosomal region (38) and lower in the principal and equatorial segments (8,38,40). In addition, domains were observed using the fluorescent lipid analog, 1,1′-dihexadecyl-3,3′-tetramethyl-indocarbocyanine perchlorate (C_{16}diI), which intercalates into the outer leaflet of the plasma membrane. The acrosomal region of ram sperm labeled more intensely with this agent than the postacrosomal region (41). The differences in affinity of the probe for these regions apparently result from its interactions with lipids and proteins heterogeneously distributed within the plane of the membrane.

Antibodies

Antibodies proved to be particularly useful for identifying distinct sperm surface domains and for localizing and isolating specific surface molecules. They can be conjugated directly with various labels or detected indirectly with secondary antibodies that carry labels visible by light or electron microscopy. Some antisera prepared against sperm or spermatogenic cells bound to specific regions, such as the principal segment domain (42), the postacrosomal region (43), or the principal piece and end piece domains of the flagellum (44). Others bound to multiple regions of the sperm surface, including the equatorial and postacrosomal domains (45), the acrosomal and middle piece regions (44), the head and midpiece regions (46), the entire flagellum (46), or the entire sperm surface (47). However, the multiple and variable specificities of polyclonal antisera to whole cells or mixtures of antigens often limited their usefulness for dissecting the distribution of specific sperm-surface components, for defining the biochemical characteristics and functional roles of such components, and for identifying the molecules recognized.

Some of these problems were circumvented using monoclonal antibodies to identify molecules localized to subdomains of the sperm surface. Their use confirmed that the surface of sperm from many different species consists of well-defined domains and demonstrated that different proteins and glycoproteins may be either confined to individual domains or shared by multiple domains. However, attempts to use immunoprecipitation with monoclonal antibodies to identify specific proteins can also result in the isolation of multiple proteins (9) which might either be subunits of a molecular complex (48,49) or functionally unrelated molecules that share a common epitope (50).

Localized Sperm Membrane Components

The presence of a mosaic sperm surface is reflected in the localized distribution of different proteins. Calcium channel proteins are localized to specific regions of the sperm plasma membrane and involved in regulating calcium homeostasis required for various sperm functions. CatSper1 is a six trans-membrane-spanning, voltage-gated Ca^{2+}-channel protein involved in regulating calcium entry, and is present in the plasma membrane of the principal piece region of the mouse sperm flagellum (51). Disruption of the gene for CatSper1 resulted in male infertility due to failure of the sperm to achieve hyperactivated motility (51). A related calcium channel protein, CatSper2, is located in the plasma membrane of the flagellum and disruption of the gene for CatSper2 likewise resulted in infertility and failure to achieve hyperactivated motility (52). These results suggest that a direct or indirect interaction between the two proteins might be required for hyperactivation to occur. Another group of voltage-gated Ca^{2+}-channel proteins is present in the plasma membrane of the principal piece ($Ca_v1.2$, $Ca_v2.2$, and $Ca_v2.3$), the acrosome ($Ca_v1.2$ and $Ca_v2.2$) and the postacrosomal region ($Ca_v2.3$) of mouse sperm (51). Disruption of the gene for a subunit (alpha 1E) of $Ca_v2.3$ resulted in increased straight-line velocity and linearity parameters of sperm motility as determined by computer-assisted sperm analysis (53).

Plasma membrane Ca^{2+}/calmodulin-dependent ATPase 4 (PMCA4) is widely expressed, but is abundant in the testis and localized primarily to the principal piece and to a lesser extent to the acrosome of mouse sperm. PCMA4 is a member of a family of proteins that move calcium from the cytosol across the plasma membrane. Disruption of the gene for PMCA4 resulted in a substantial increase in sperm intracellular calcium concentrations and in male infertility due to impaired sperm motility (54,55).

Other proteins present in restricted regions of the sperm surface were identified by immunostaining and cDNA cloning approaches. Glucose transporter 8 (GLUT8, *Slc2a8*) is present in other cell types, but expressed at high levels in the testis and predominantly associated with the acrosome region of mouse and human sperm (56). Sperad (PM52, G11) accumulates over the head of late acrosomal phase spermatids and the middle piece of the flagellum of guinea pig sperm, but is localized to the periacrosomal region by the time they reach the cauda epididymis (57). In addition, gene-targeting studies have shown that basegin (EMMPRIN, CE9, HT7) is involved in cell–cell interactions in the testis essential for the completion of spermatogenesis in mice (58,59).

However, it also is associated primarily with the principal piece of sperm from the caput epididymis and the middle piece of sperm from the cauda epididymis (60).

Formation and Maintenance of Domains

The separation of different membrane proteins into specific domains presumably allows the compartmentalization of responses by spermatozoa to external conditions and signals. In some cases, barriers to diffusion of plasma membrane proteins develop during spermiogenesis at the same time as membrane domains (61). However, the specific mechanisms responsible for forming and maintaining sperm-surface domains are not well defined. Most surface domains overlie distinct cytoplasmic organelles or features (Table 1), suggesting that morphogenetic processes that establish the shape and organization of the spermatozoon have a major role in determining the location of surface domains. Furthermore, since transmembrane proteins generally are stabilized by linkage through membrane skeleton proteins to cytoskeletal structures (62,63), such associations may be important for defining the boundaries and contents of different sperm surface domains.

Actin and Associated Proteins

Actin and a number of actin-associated proteins possibly involved in producing or maintaining domains have been identified in spermatozoa of several mammals. By using antibodies, actin filaments were found in the subacrosomal space in rat sperm (64), postacrosomal region of rat, rabbit, bull, and boar sperm (65–76) and in the connecting piece of hamster, rabbit, bull, and human sperm (73). They also were detected in the posterior region of the head and in the connecting piece, middle piece, and principal piece of human sperm (66,73), and along the concave margin of the hamster sperm head (69,74). Other proteins frequently found in association with actin are myosin and spectrin. Myosin was identified in the acrosomal region (66) and in the neck region of human sperm (67). Spectrin was associated with the acrosome of spermatids in the mouse, but not detected in epididymal sperm (77,78). However, human spermatozoa contained spectrin in the anterior acrosome and principal piece regions (78) and rabbit sperm contained spectrin in the anterior acrosome region and postacrosomal segment of the head and the principal piece of the tail (72).

The differences reported for the immunolocalization of actin in sperm made it difficult to determine

TABLE 1. *Sperm head domains and associated cytoskeletal proteins*

Domain	Protein	Gene	Features	References
Whole head	T-actin 2	*Actl7b*	Actin-like protein, postmeiotic, testis specific	(79)
Acrosomal	ARC	*Arc*	Activity regulated cytoskeleton-associated protein, postmeiotic, highly expressed in testis, also in neck and principal piece	(80)
	CASK	*Cask*	Calcium/calmodulin-dependent serine protein kinase, postmeiotic, widely expressed	(204)
	Testis fascin	*Fscn3*	Fascin homolog 3, postmeiotic, testis specific	(205)
Perinuclear theca *Subacrosomal region*	PERF 15	*Fabp9*	Fatty acid binding protein 9, meiotic and postmeiotic, testis specific	(206)
	SubH2Bv		H2B variant PT15, postmeiotic, testis-specific	(207)
Perinuclear theca *Postacrosomal region*	Cylicin I	*Cylc1*	Basic protein of sperm head cytoskeleton 1, postmeiotic, testis specific	(208)
	Cylicin II	*Cylc2*	Basic protein of sperm head cytoskeleton 2, postmeiotic, testis specific	(209,210)
	Calicin	*Ccin*	Actin-binding protein, postmeiotic, testis specific	(84,208,211)
	CPα3	*Capza3*	Actin-capping protein alpha subunit 3, postmeiotic, testis specific	(84,212–214)
	CPβ3	*Cabzb*	Actin-capping protein beta subunit, postmeiotic, tissue specific isoform (splice variant), high in testis and lower in brain	(85,215)
	Arp-T1	*Actrt1*	Actin-related protein T1, postmeiotic, testis specific	(82,216)
	Arp-T2	*Arpm2*	Actin-related protein M2, postmeiotic, testis specific	(82,216)
	CYPT1	*Cypt1*	Cysteine-rich basic protein, postmeiotic, testis specific	(217)
	STAT4	*Stat4*	Signal transducer and activator of transcription, postmeiotic, expressed in testis and a few other tissues	(218)

if it has a significant role in sperm function. Some of these differences might have been due to the varying specificities of antibodies and others to variations between species. However, more recent studies identified genes for several actin-like and actin-associated proteins present in spermatozoa (Table 1). Most of these proteins were testis-specific and expressed during the postmeiotic phase of spermatogenesis. Some were found associated with the acrosomal region of the head, including the testis-specific actin-like proteins T-actin 2 (79) and ARC (80), and a testis-specific fascin paralog (81). Others were found in the perinuclear theca, including actin-related proteins Arp-T1 and Arp-T2 (82), the actin-binding protein calicin (83), and the CPα3 and CPβ3 subunits of actin-capping protein (84,85). While these findings support a role for actin in sperm, it remains to be determined if the proteins participate in establishing and maintaining surface domains. It is likely that many of these proteins are involved in the rearrangement and shaping of nuclear and cytoplasmic components in elongating spermatids (86) and might participate in the acrosome reaction (78,81,84,87).

Maintenance of Surface Domains

The maintenance of sperm surface domains and the segregation of specific proteins into domains probably involves one or more processes, including: (a) restriction of mobility of surface molecules in their final domain, (b) existence of barriers to movement of surface components at the domain boundary, and (c) thermodynamic partitioning of molecules into specific regions (88). There is evidence that restriction of mobility of transmembrane components can occur by molecular interactions outside the cell. For example, fertilin is evenly distributed over the entire head of testicular sperm, but becomes restricted to the posterior head as sperm transit the epididymis in the guinea pig (89). A brief treatment of testicular sperm with trypsin also results in the localization to the posterior head, suggesting that proteolytic processing of the β fertilin subunit in the epididymis releases a constraint and allows fertilin redistribution (89).

Associations between the plasma membrane and internal structures are believed to form barriers between domains and serve to restrict the movement of some proteins. The plasma membrane is closely associated with the annulus at the junction of the middle piece and principal piece, and with the posterior ring at the junction between the head and connecting piece. The PT-1 antigen exhibited free diffusion within the plasma membrane of the distal tail domain, but was prevented from migrating into the midpiece, possibly by the annulus (90). Two antigens (2B1, 2D6) present over the whole surface of the flagellum of rat sperm showed antibody-induced patching, but did not migrate onto the head (91), perhaps due to the posterior ring. However, other sperm surface components apparently were not restricted from redistribution by the posterior ring and annulus (92).

Other structures that might serve to constrain or anchor plasma membrane components are the groups of intramembranous particles seen in freeze-fracture studies. These include the zipper-like array of particles associated with the ribs of the fibrous sheath (30) and oblique strands of intramembranous particles associated with the underlying mitochondria (30). In addition, the particle-bare and filipin-complex sparse plasma membrane between the anterior acrosome and the equatorial segment domains might help to define a barrier (93,94).

Sperm Plasma Membrane Lipids

Lipid Composition

There are variations in lipid composition of the sperm plasma membrane between species, but common features distinguish sperm plasma membranes from those of other cell types. While the cholesterol or glycolipid content of sperm plasma membranes is not unusual, they contain relatively high amounts of plasmalogens (20%–40%), other ether-linked phospholipids (95–98), and lipids with long, polyunsaturated aliphatic chains (99). Phospholipids make up about 70% of the total plasma membrane lipid in boar sperm (100), and choline phospholipids account for almost two-thirds of phospholipids in the plasma membrane of the anterior head of ram sperm (96). Sterols are the next most abundant lipid, resulting in a cholesterol/phospholipid molar ratio of about 0.12 (100). It was suggested that the high cholesterol to phospholipid ratio accounts for the lack of thermotropic phase transitions in the plasma membrane of live human sperm detected by Laurdan fluorescence spectroscopy (101). Freeze-fracture studies with filipin suggested that the amount of sterol in the anterior acrosome is about four times that present in the postacrosomal region in guinea pig and bull spermatozoa (8,36,102). The postacrosomal region contains few sterols or anionic lipids in guinea pig spermatozoa (40), but the cytoplasmic droplet of the middle piece is probably rich in both (8,33). Also, cholesterol sulfate makes up only a small fraction of the total sterol content, but is a major component of the plasma membrane over the acrosome of human sperm (103).

Free fatty acids comprise a relatively small amount of the lipid in boar spermatozoa, whereas diacylglycerols are present in about the same amounts as glycolipids (100). The phospholipid/protein ratio is approximately 0.68 on a weight basis in plasma membranes isolated from boar spermatozoa (100), suggesting that the amounts of total lipid and protein are about the same. However, this is for the total plasma membrane and the amounts and types of lipids and the lipid/protein ratios probably vary between different domains.

Although glycolipids are not an abundant component of the sperm plasma membrane, those present might have a significant role in lateral domain organization (100,104–108). The major glycolipid in sperm is sulfogalactosylglycerolipid (SGG), an unusual form also found in low amounts in the brain (107,109). SGG is present in both the head and tail fractions of spermatozoa (107).

Diffusion of Lipids

Although lipid is free to diffuse in the plasma membrane of most cells (110), fluorescence recovery after photobleaching (FRAP) measurements indicate that a large fraction of the plasma membrane lipids of mammalian sperm is not free to diffuse laterally (111,112). The nondiffusing lipid pool develops during spermiogenesis (113), but can increase to more than 50% of the lipid during epididymal maturation (111). The diffusion rate was not significantly affected when ram sperm were treated with pronase, suggesting that interactions with surface proteins generally do not restrict plasma membrane lipids (108). FRAP studies showed that the lipid diffusion rates were significantly higher over the acrosome and postacrosomal region than on the midpiece and principal piece in live bull, boar, ram, and mouse spermatozoa (114). FRAP studies also observed a free exchange of a lipid analog by lateral diffusion in the plasma membrane between the head and midpiece regions of ram and boar sperm (115) and between the midpiece and distal tail of ram sperm (111). However, studies using the method of fluorescence loss in photobleaching (FLIP) detected a diffusion barrier to ~200 nm particles between the equatorial segment and postacrosomal domains, but not between the principal and equatorial segments (116).

When sperm were subjected to hypo-osmotic shock to produce surface blebs, the diffusion rate for lipids were comparable on blebs and on untreated sperm, suggesting that interactions of plasma membrane lipids with underlying cytoskeletal elements do not affect diffusion (108). In addition, when artificial bilayers were formed from sperm plasma membrane lipid extracts, the fraction of nondiffusing lipid was about the same as that on intact sperm, suggesting that lipid–lipid interactions are responsible for restricting lateral lipid movement (108). Differential scanning calorimetry studies indicated that significant amounts of gel-phase lipids are present in the anterior region of the head of ram sperm which allow sperm plasma membrane lipids to segregate by lateral phase separation into coexisting fluid and gel domains at physiological temperatures (95). The gel domains may contain the nondiffusing lipid fraction present in the sperm plasma membrane.

Lipid Rafts

Some lipids and proteins associate in the plasma membrane to form microdomains, known as lipid rafts, which are insoluble in low concentrations of nonionic detergents and rich in cholesterol, glycosphingolipids, and GPI-anchored proteins (117). A marker of lipid rafts, caveolin-1 (CAV1), localized to the anterior acrosome and principal piece in mouse and guinea pig sperm (118). In addition, the transient receptor-1 homolog (TRPC1), a putative capacitative Ca^{2+}-channel protein, co-immunolocalized with CAV1 on mouse sperm (119). Other lipid raft markers, the GPI-anchored CD59 protein, GM1 ganglioside, and integral membrane protein flotillin-2 (FLOT2) were found on human sperm, and FLOT2 was localized mainly on the posterior head and middle piece (120). A serine protease, PRSS21, another GPI-anchored lipid raft component, was localized in the head, cytoplasmic droplet, and midpiece of mouse sperm (121). These studies suggest that lipid rafts are present throughout the surface of mammalian sperm and not restricted to particular domains.

Changes in Lipids with Maturation and Capacitation

Changes in the lipid content and distribution in the spermatozoon plasma membrane occur during epididymal maturation and capacitation, and these may have substantial effects on the composition and function of the membrane in different domains. The total lipid content of sperm decreases during epididymal maturation in boar, bull, ram, and rat (122–129), and the cholesterol content decreases in ram, rat, and hamster sperm (130–132). In addition, the cholesterol/phospholipid ratio and concentration of phosphatidylserine, phosphatidylethanolamine, cardiolipin and ethanolamine plasmalogen decreases with maturation in ram sperm (123,130). However, increases occur in the amount of sulfo-conjugated sterols in hamster and human sperm (131,132) and in unsaturated fatty acids in whole ram sperm (130). Studies using plasma membranes isolated from boar spermatozoa confirmed earlier results that the total lipid content decreases in sperm during epididymal maturation (99). There was a decrease in the level of fatty acids and an increase in diacylglycerol, but no change in the degree of saturation of fatty acids. Plasma membrane from the anterior head region of ram sperm is particularly rich in ethanolamine and choline phosphoglycerides and the amount of dermosterol and ethanolamine in this region of the plasma membrane decreases, while the cholesterol to phospholipid ratio increases, during epididymal maturation (96).

Changes in the amount and composition of lipids in the plasma membrane of sperm during maturation are thought to explain why ejaculated sperm are more sensitive to cold shock than are testicular sperm (134,135). These changes may also account for the maturation-dependent decrease in charge density at the phospholipids/water interface of ram spermatozoa, detected by electron spin resonance (133), and the decrease in membrane fluidity of bull spermatozoa, seen by fluorescence polarization spectroscopy (135). Analysis of testicular and ejaculated ram spermatozoa by FRAP indicated that there are regional differences in the decrease of plasma membrane fluidity (111). During maturation, the diffusion rate of a fluorescent lipid analog increased in all regions of the sperm except the midpiece.

There is a general efflux of cholesterol from the sperm plasma membrane during capacitation, which in turn alters the fluidity of the membrane and the lateral mobility of associated proteins (136). A sulphogalactolipid analogue incorporated into the marginal segment of ram sperm migrated to the equatorial segment with capacitation (137). Capacitation also resulted in the redistribution of the DE glycoprotein on rat sperm (138) and fertilin on guinea pig sperm (9,139). In addition, freeze-fracture studies of filipin-treated guinea pig sperm observed a loss of cholesterol during capacitation from the plasma membrane overlying the acrosome (11,40). Similar studies with human sperm observed development of cholesterol-free patches in the plasma membrane over the acrosome with capacitation, presumably facilitating fusion between the plasma membrane and acrosomal membrane (140).

THE SPERM HEAD

The mammalian sperm head contains the nucleus and acrosome surrounded by moderate amounts of cytoskeletal components and cytoplasm (Fig. 3). The acrosome caps the anterior end of the nucleus

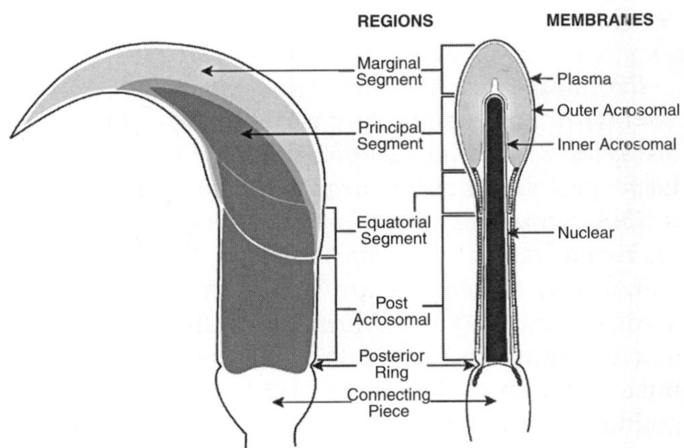

FIG. 3. The distribution of regions and membranes on a falciform-shaped sperm head. The regions include the marginal segment and principal segment of the anterior acrosome, the equatorial segment of the posterior acrosome, and the postacrosomal region between the posterior margin of the acrosome and the posterior ring. The outermost plasma membrane encloses the entire spermatozoon, while the outer acrosomal and inner acrosomal membranes form a continuous enclosure of the acrosomal contents. The nuclear envelope is a double membrane enclosing the nucleus. The subacrosomal cytoskeleton (perforatorium) is located between the inner acrosomal membrane and the nuclear envelope. The postacrosomal cytoskeleton is located between the plasma membrane and the nuclear envelope. Together these are referred to as the perinuclear theca. [Modified from (12).]

and cytoskeletal components lie in the narrow space between the acrosome and the nucleus and between the acrosome and the plasma membrane. Sperm of most mammalian species have a spatulate head (Fig. 2), with the nucleus and acrosome flattened in the plane of the anterior-posterior axis of the sperm and the acrosome and nucleus are symmetrical structures. However, in the guinea pig and some other animals, the distal part of the principal piece bends out of the flattened plane of the sperm head. In contrast, the sperm of some rodents have a falciform-shaped head (Fig. 2), with the acrosome overlying the convex margin of the nucleus. Although sperm are uniform in size and shape within most species, a third or more of human sperm are abnormal in size and shape.

The Sperm Nucleus

The chromatin of the sperm nucleus is highly condensed and its volume is considerably less than the nuclear volume in a somatic cell (~5%). The organization and amount of DNA and the arrangement and composition of the nucleoproteins are unique features of the sperm nucleus. The two meiotic divisions that occur during spermatogenesis result in the sperm containing only one copy of each chromosome.

Nuclear Proteins

Protamines are the major nuclear proteins associated with mammalian sperm DNA (141–144). These are relatively small (27–65 amino acids) and highly basic proteins rich in arginine and cysteine. The mRNAs encoding mouse protamines are transcribed from the haploid genome in round spermatids and translation is delayed until spermatids are undergoing elongation (144). Most mammals have only one active protamine gene, but a few species, including mice and humans have two (142,143–147). Gene targeting in mice was used to determine if the second protamine provides redundancy to the DNA condensation process or if both are necessary (148). It was found that both protamines are essential for fertility and that nuclear formation and sperm DNA stability was disrupted by a decrease in the amount of either protamine (148,149). It remains to be determined if a mutation in one copy of the gene for either protamine 1 or protamine 2 is detrimental to human fertility.

There are two general models for the association of protamines with DNA. The more widely accepted model suggests that protamines are present in an extended configuration and lie in the major or minor groove of the DNA helix, causing the DNA to become packaged into toroidal loops (150,151). The other model suggests that protamine is packaged into α-helical cylinders (152). These cylinders are thought to lie in the major or minor DNA groove and to facilitate orderly DNA condensation. They may subsequently cross-link with neighboring cylinders to effect stabilization. Both models indicate that the chromatin is stabilized by the formation of covalent disulfide linkages between protamines on adjacent DNA strands.

Relatively little is known about other proteins associated with sperm chromatin. It is reported that as much as 15% of the DNA in human sperm remains complexed with histones (153). These include H2A, H2B, and H3 variants (153,154), and it was suggested that the histones in human sperm might be part of a sequence-specific component of the genome which is programmed for expression in early development (154). In addition, the heterochromatin protein M31 was detected in mouse total sperm proteins on Western blots and presumably was associated with centromeres, but histone H1 was not detected (155). Furthermore, hamster sperm nuclei appear to contain a DNase (156), and enzymes involved in de novo pyrimidine biosynthesis were detected in the nucleus of human sperm (157).

While results of studies using freeze-fracture (30,93,158), birefringence (159), and physical methods (160) suggested that the chromatin in sperm of some mammals is stacked in lamellar plates, other studies suggested that the chromatin of mouse and human spermatozoa have a random fibrogranular organization (161,162). There also is evidence that chromosomes and centromeres have a preferred, nonrandom, spatial localization within the nucleus of human sperm (163).

Nuclear Envelope

The sperm nucleus is enclosed by an unusual nuclear envelope. Over most of the nucleus, nuclear pores are absent and the inner and outer membranes of the nuclear envelope are 7–10 nm apart, while in most other cells nuclear pores are abundant and the two nuclear membranes are 40–60 nm apart (6,30,93,164). Caudal to the posterior ring, in the "redundant nuclear envelope," sperm nuclear pores are abundant and arranged in a hexagonal pattern. The membranes of the anterior part of the nuclear envelope contain a rich array of randomly distributed intramembranous particles (165), whereas the closely apposed membranes of the nuclear envelope in the implantation fossa contain large, closely packed particles surrounding particle-free areas (30).

Nuclear Lamina

The nuclear lamina is a protein meshwork lining the inner surface of the nuclear envelope and forms part of the nuclear skeletal network (karyoskeleton) anchoring the chromatin (166–170). It consists primarily of lamins, members of the intermediate filament protein family. In general, expression of A-type lamins is developmentally regulated and B-type lamins are constitutively expressed (171). There is an isoform of lamin A, the C2 splice variant, that is present during meiosis and spermiogenesis in mice (172) and rats (173), but is not detected in sperm (171,173). A knockout of the gene (*Lmna*) for lamin A resulted in a high level of apoptosis in pachytene spermatocytes and the disruption of spermatogenesis (174).

The B1 and B2 lamins are products of separate genes (*Lmnb1, Lmnb2*) and two B-type lamins are present during spermatogenesis. Lamin B1 is present in nuclei throughout spermatogenesis in mice (175) and rats (176,177). While it was not detected in rat sperm (178), it was localized to a narrow domain in the nucleus of mouse sperm (175) and to the postacrosomal region of the nucleus of human, bull and rabbit sperm (171). A variant of lamin B2 generated by differential splicing and alternative polyadenylation, termed lamin B3 (179), was detected in pachytene spermatocytes, but not in sperm from mice (179) or rats (173,176).

Other proteins reported to be associated with the nuclear lamina during spermiogenesis include lamina-associated polypeptide-2 (LAP2) in the rat (178) and lamin B receptor (LBR) in the mouse (180). LAP2 is associated with the inner nuclear membrane of cells and binds chromatin and lamin B1 (181). The TPβ variant of LAP2 was detected in rat sperm (178). In elongating spermatids, protamine 1 docks through LBR to the nuclear envelope (180). LBR presumably binds to lamin B1, since it is the only lamin detected in elongating spermatids. However, it remains to be determined if LBR, like lamin B1, is retained after spermiogenesis and present in mouse sperm.

The Sperm Head Cytoskeleton

Cytoskeletal structures are located in three regions of the head of mammalian sperm. The *subacrosomal cytoskeleton* is located between the acrosome and nucleus, the *postacrosomal cytoskeleton* lies between the nucleus and the plasma membrane posterior to the acrosome, and the *para-acrosomal cytoskeleton* is present between the anterior tip and convex surface of the acrosome and the plasma membrane of falciform sperm (Fig. 3).

Structure of the Sperm Head Cytoskeleton

The subacrosomal and postacrosomal cytoskeletons can be isolated together (161) and are referred to as the *perinuclear theca* (182,183). This complex covers the external surface of the nuclear envelope and consists of multiple proteins (161). The isolated perinuclear theca retains the shape of the sperm nucleus in the absence of DNA and the acrosome. It was suggested that the perinuclear theca may be an extrinsic determinant of nuclear shape (182).

The subacrosomal cytoskeleton (perforatorium, subacrosomal layer, perinuclear material) is composed of amorphous, electron-dense material and occupies the narrow space between the inner acrosomal membrane and the outer membrane of the nuclear envelope (184). It is a prominent feature of the falciform-shaped heads of rodent sperm, including rat, mouse, and hamster. In these species, the subacrosomal cytoskeleton is referred to as the *perforatorium* (185) because of its similarity in appearance to a structure by this name in toads (186) and birds (187). It is thought to have a mechanical role in egg penetration (188,189). In rat sperm, the perforatorium is a curved triangular rod anterior to the apex of the head. It splits into one dorsal and two ventral interconnected prongs that taper posteriorly to become continuous with the postacrosomal dense lamina (183). It becomes more resistant to being solubilized during epididymal transit, apparently due to extensive disulfide bond formation (184,190,191). The resistance to solubilization was used to develop a procedure for isolating the perforatorium from mature rat sperm (184,191). However, the subacrosomal cytoskeleton is a minor component of the spatula-shaped heads present in sperm of most mammalian species (192,193).

The postacrosomal cytoskeleton is referred to as the postacrosomal sheath (postnuclear sheath, postacrosomal dense lamina, postnuclear cap, postnuclear body) (6,93,164,185,194–196). It consists of the portion of the perinuclear theca that lies between the nuclear envelope and the plasma membrane of the postacrosomal segment of the sperm head. It is continuous with the posterior end of the subacrosomal cytoskeleton at the posterior margin of the acrosome and extends caudally to the posterior ring. It is a 10- to 15-nm thick layer containing an ordered array of filaments (93,164,197–200). The basal region is a zone of close adhesion between the plasma membrane, the postacrosomal cytoskeleton, and the nuclear envelope (201).

The para-acrosomal cytoskeleton (filamentous cytoskeletal complex) is present in hamster sperm, lying between the acrosome and the plasma membrane (76). It is a tripartite structure, consisting of a cone at the anterior tip and a bifurcated sheet on the convex surface of the head. It is formed of filaments similar in size to intermediate filaments. Filaments have also been seen along the convex side of the acrosome of vole sperm (199), and striations were noted along the ventral surface of the acrosome of rat sperm (198), suggesting that the para-acrosomal cytoskeleton may be present in other species.

Composition of the Sperm Head Cytoskeleton

The resistance of the perinuclear theca to solubilization in nondenaturing detergents and high salt buffers (184,191) has been quite useful in the biochemical characterization of some of the components of the sperm head cytoskeleton and the identification of their genes (Table 1). In addition, it is likely that some components are lost during the isolation procedure. Some studies identified proteins described as having an anterior sperm head or acrosomal distribution, which probably corresponds to the subacrosomal cytoskeleton. These include an actin-like protein (T-actin 2) (202), an activity-regulated cytoskeleton-associated protein previously found to be induced in neurons in response to stimuli (ARC) (203). There is also a member of a family of membrane-associated proteins with multiple domains associated with kinase activity and protein-protein interactions (CASK) (204), and a member of a family of actin-bundling proteins (testis fascin) (205). Proteins reported to be localized to the subacrosomal region of the perinuclear theca include a fatty acid binding protein (PERF15) (206), and a variant of the H2B histone (subH2Bv) (207).

Components of the postacrosomal region of the perinuclear theca include two basic proteins, cyclin 1 and 2 (208–210); an actin-binding protein, calicin (211); two components of an actin capping protein complex, CPα3 and CPβ3 (212–215); two actin-related proteins, Arp-T1 and Arp-T2 (216); a cysteine-rich basic protein, CYPT1 (217); and a signal transducer and activator of transcription, STAT4 (218). Eleven of these 15 proteins are reported to be testis-specific, and all but one (PERF15) are expressed only during the postmeiotic phase of spermatogenesis (Table 1). This underscores the novelty of the postacrosomal theca to spermatogenic cells. During epididymal maturation, a variety of sperm proteins are disulfide cross-linked, probably including proteins that are located in the perinuclear theca (190).

Other proteins associated with the perinuclear theca include actin, spectrin, and calmodulin (219,220). In addition, antibodies have localized several proteins to the perinuclear theca, but they remain to be identified or characterized using molecular methods. These include the thecins (221), the Dp71f isoform of dystrophin, β-dystrobrevins (222), and antigen AJ-p90 (223).

Roles of the Sperm Head Cytoskeleton

Sperm-head cytoskeletal components have a structural role in defining the shape of the sperm head (188,189), and a functional role in aiding sperm penetration of the egg and its investments at fertilization (201,224). However, the individual parts of the perinuclear theca may have different roles. The perforatorium develops as a distinct structure at the very end of spermiogenesis (192,225). Because sperm have largely acquired their final form by that time, the perforatorium probably does not produce the falciform shape, but might provide structural reinforcement to stabilize that shape.

The postacrosomal sheath is slightly more prominent in spatulate sperm than in falciform sperm. It too develops after elongation and flattening of the spermatid head. It might help to maintain the asymmetric shape, but the perinuclear theca can be solubilized without greatly affecting the shape of the sperm nucleus (226). Another possibility is that the perinuclear theca provides stiffening and support for the posterior part of the sperm head to prevent its flexure during sperm motion. An additional role of the perinuclear theca might be to link together and stabilizes the association of the acrosome, nucleus, and postacrosomal plasma membrane (182,192,194,226). Other potential roles of the postacrosomal sheath are to generate or maintain the distinctive properties of the overlying plasma membrane (227) or bind the plasma membrane to the sperm (228). These are attractive possibilities, but they remain to be tested experimentally.

The Acrosome

The acrosome is a unique sperm organelle that originates from the Golgi complex and contains enzymes necessary for the sperm to penetrate through the investments of the egg to achieve fertilization. It is a membrane-enclosed vesicle sitting as a cap over the nucleus in the anterior part of the sperm head. The acrosome is highly conserved throughout evolution and along with the axoneme is a hallmark of spermatozoa of animals in many phyla. Most of the acrosomal components are produced during spermiogenesis, but a few begin to be synthesized in late pachytene spermatocytes.

Structure of Acrosome

The inner acrosomal membrane overlies the anterior part of the outer membrane of the nucleus (Fig. 3). It is continuous with the outer acrosomal membrane, which lies close to the inner surface of the plasma membrane of the anterior sperm head. The acrosome consists of two segments, the acrosomal cap (anterior acrosome) and the equatorial segment (posterior acrosome), which correspond in distribution to the plasma membrane domains with the same names. During the acrosome reaction, the outer acrosomal membrane and the plasma membrane fuse and vesiculate, and most of the acrosomal contents are discharged. The inner acrosomal membrane and the equatorial segment persist until sperm-egg fusion in most species. Acrosome shape and size varies widely between species (Figs. 2 and 4), and the distribution and relative prominence of these two segments differ accordingly (185).

The equatorial segment forms a band that approximately overlies the equator of the head of spatulate spermatozoa. In sperm possessing a falciform head, the equatorial segment may cover much of the lateral surfaces of the head. However, in species such as the wooly opossum (229), which have a discoid sperm head flattened in the plane perpendicular to the axis of the tail ("carpet tack-shaped head"), an equatorial segment may not be identifiable. The portion of the acrosomal cap that extends beyond the

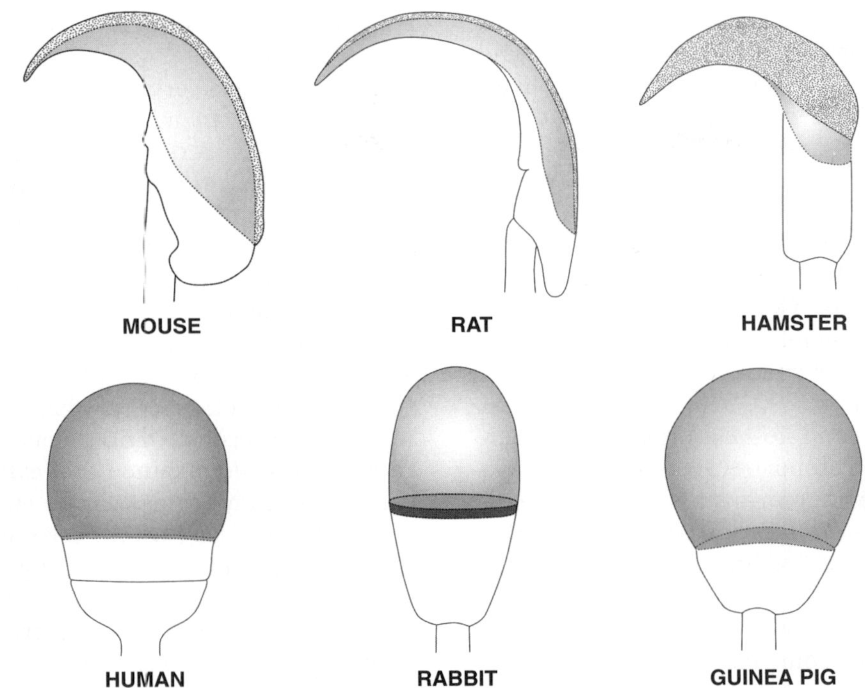

FIG. 4. Lateral views of the falciform-shaped heads of mouse, rat, and hamster spermatozoa; ventral views of spatulate-shaped human, rabbit and human spermatozoa. The size and shape of the spermatozoon head is specific to each species, but varies considerably between species.

anterior margin of the nucleus is referred to as the marginal segment (apical segment, anterior band, peripheral rim), and the portion overlying the nucleus is referred to as the principal segment (acrosomal segment). In the human, monkey, bull, boar, rabbit, and bat the acrosome is relatively small, with no appreciable extension beyond the nucleus, whereas in the guinea pig, chinchilla, and ground squirrel the acrosome has a large apical segment (13,94,185,230).

Electron microscopy reveals that the acrosome, particularly the marginal segment, often has a more complex shape than is obvious by examining sperm smeared onto a slide. The shape of the acrosome is characteristic of the species (185,230). The final shape of the acrosome may be influenced by extrinsic forces generated by cytoskeletal elements in the spermatid and/or Sertoli-cell cytoplasm (76,230) or by forces intrinsic to the nucleus (179,232). However, in some species it also appears that forces intrinsic to the acrosome are involved. Acrosomes of guinea pig, chinchilla, and hamster sperm continue to undergo morphological differentiation after spermatogenesis, and the definitive shape is not achieved until sperm reaches the distal portion of the epididymis (13,231). As might be expected from species differences, genetic factors also influence acrosome formation and shape. For example, structurally defective acrosomes form in pink-eyed sterile mutant mice (233,234). Acrosomes also fail to form in blind sterile mutant mice, even though proacrosomal granules, the manchette, and flagellar structures form and some nuclear elongation and chromatin condensation occurs (195,235).

The acrosome sometimes shows an internal lamellar or crystalline structure (30), an ordered substructure (236), or a cobblestone-like pattern (200,237). A 4.2-nm periodicity is present in the cortical region of the acrosome of rat sperm, lying just deep to the outer acrosomal membrane on the convex surface (30,238). A similar pattern is also present in the acrosome of human sperm (239). Other evidence that the acrosome has a substructure comes from studies on hamster spermatozoa disrupted by nitrogen cavitation. This treatment results in loss of much of the acrosomal matrix, but components immediately underlying the outer acrosomal membrane remain intact (76). These components are present in two areas: one is a larger and looser layer of fibrous material over the dorsal and lateral surfaces of the acrosome, whereas the other is a more compact fibrous layer adjacent to the anterior margin of the acrosome.

The membrane of the acrosome, particularly the equatorial segment, contains particles that form a crystalline array, giving the membrane a highly regular, granular appearance. These may be an indication of ordered structure in the underlying acrosomal components. Such features have been reported in rabbit (93), bull (200,240), human (162), rat (30,241), mouse (165), guinea pig (30,240,241), degu (240), and rhesus monkey (240). This pattern was seen by freeze-etch, freeze-fracture, and on replicas of air-dried and critical-point-dried spermatozoa. Although the outer acrosomal membrane appears to be fragile and easily displaced or disrupted at the time of the acrosome reaction, it also has a thickened appearance because of an electron-dense coating on the inner surface (242–246). This inner surface coat of the outer acrosomal membrane has been isolated from bull sperm and shown to be composed mainly of three large glycoproteins (290, 280, and 260 kDa) as well as 115-, 81-, 58-, and 46-kDa proteins. In addition, there is a set of proteins between 34 and 12 kDa (246). Lectins bind to the inner surface of the membrane and WGA is observed to bind to the 46-kDa component. It was suggested that glycosylated molecules at this site might help to stabilize the membrane or play a functional role in the membrane fusion events of the acrosome reaction (246). The 200 and 58 kDa components are phosphorylated in a cAMP-independent manner, whereas the proteins between 34 and 12 kDa appear to include calmodulin-binding proteins (246).

Inner acrosomal membrane development begins in early spermatids when the membrane of the proacrosomal granule abuts and then flattens against the nuclear envelope. The granule spreads over the apical end of the nucleus during acrosomal development and nuclear elongation (186,238). The inner acrosomal membrane is exposed and becomes continuous with the plasma membrane when the acrosome reaction occurs. The inner acrosomal membrane in mouse and rabbit spermatozoa is quite resistant to chemical and physical disruption, including treatment with nonionic detergents and sonication (247,248). However, boar-sperm inner acrosomal membrane is sensitive to proteinase treatment (249), and lectins are found to bind to the inner acrosomal membrane of sperm from hamster (29,250) and guinea pig (251), indicating that glycoproteins are present. Bridges 7 nm wide and with 7 nm center-to-center spacing were reported to be present between the inner and outer acrosomal membranes in boar sperm, apparently holding these structures together (249). However, the inner acrosomal membrane appears to be fluid because antigens recognized by monoclonal antibody PH-20 migrate from the plasma membrane of the postacrosomal region of the guinea pig sperm head to the inner acrosomal membrane following the acrosome reaction (252–254). It was suggested that in the equatorial segment the inner acrosomal membrane is associated with an extensive scaffolding network possibly transmembrane in nature (190).

Contents of Acrosome

The acrosome serves a critical role in the process of fertilization in mammals. Multiple enzymes are present in the acrosome (Table 2), including acid hydrolases commonly found in lysosomes, and other enzymes specific to spermatogenic cells. Although it has been described as a specialized lysosome (255–257), the acrosome also has the characteristics of a regulated secretory vesicle. During the acrosome reaction, acrosomal contents are released by calcium-mediated exocytosis in response to specific signals. However, the precise molecular mechanisms responsible for initiating and completing the acrosome reaction still remain to be determined. Following the release and activation of acrosomal enzymes, spermatozoa penetrate the zona pellucida surrounding the oocyte, a process that can be blocked by protease inhibitors (258,259).

The best-characterized constituent of the acrosome is acrosin (preproacrosin, proacrosin), a member of the serine protease superfamily that is present only in spermatogenic cells (Table 2). This trypsin-like protease differs from similar enzymes in other tissues in molecular weight, substrate specificity and inhibitor specificity (260–264). The nucleotide sequences have been determined for proacrosin cDNAs from several mammalian species, including boar (265,266), human (267,268), mouse (269,270), and rat (271). Although proacrosin synthesis occurs predominantly in round spermatids, proacrosin mRNA or protein have been detected in pachytene spermatocytes of mouse (272), guinea pig (273), and human (274). Immunolocalization studies suggest that proacrosin is present mainly in the anterior segment of the acrosome in human, boar, bull, and rabbit spermatozoa (275), but it also has been localized to the inner acrosomal membrane (276–278). Because serine protease inhibitors block fertilization in vitro, acrosin was though to be responsible for limited proteolysis of the zona pellucida to allow sperm to reach the egg. However, mice with a targeted mutation in the gene encoding acrosin (*Acr*) are fertile (279), indicating that other serine proteases participate in this process. Acrosin does appear to contribute to zona penetration because sperm from mice lacking acrosin were moderately slower to fertilize than sperm from wild type mice (280).

Acrosin is rapidly released following the acrosome reaction and the bulk of the proacrosin is in the soluble acrosomal fraction (281). The acrosomal protein sp32 (Table 2) binds preproacrosin and proacrosin, but not mature acrosin and was suggested to be involved in packaging the precursor forms of acrosin in the acrosome (282). In addition, acrosin inhibitors are present both within the acrosome (283) and in

TABLE 2. *Acrosome components*

Protein	Gene	Expression	Features	References
N-acetylglucosaminidase	N.D.	Ubiquitous	Soluble compartment; putative testis-specific isoform	(297)
Acrogranin, granulin	*Grn*	Ubiquitous	Soluble compartment; cysteine-rich	(322,323)
Acrosin, proacrosin, preproacrosin	*Acr*	Testis-specific, mRNA and protein low in spermatocytes, high in spermatids	Soluble compartment; serine protease	(260–287)
AM50 (p50, apexin)	N.D.	Testis-specific	Soluble compartment; similar to neuronal pentraxin II, calcium-dependent binding	(327)
CRISP-2 (TPX1)	*CRISP2, Crisp2*	Testis-specific, protein in spermatocytes	Soluble compartment; cysteine-rich protein	(328)
β-galactosidase	N.D.	Ubiquitous	Soluble compartment; putative testis-specific isoform, differs in size and enzymatic properties from lysosomal form	(295–297)
Hyaluronidase	*Hyal*	Ubiquitous	Soluble compartment and matrix associated; located mainly in principal segment, some bound to inner acrosomal membrane	(288–294)
MC41	N.D.	Testis-specific epitope	Matrix of cortical region of marginal segment; putative ZP2-binding protein	(334)
SAMP14	*SPACA4*	Testis-specific	Matrix and inner and outer membranes; similar to Ly6/urokinase plasminogen activator receptor superfamily protein	(329)
SAMP32	*SPACA1*	Testis-specific	Inner membrane in principal and equatorial segments; novel protein	(340)
SLLP1	*SPACA3, Spaca3*	Abundant in testis, low in pancreas and Raji cells	Matrix, principal and equatorial segments; C lysozyme-like protein	(339)
sp32	*Acrbp*	Testis-specific	Matrix; proacrosin-binding protein	(338)
sp56 (AM67)	*Zp3r*	Testis-specific, protein in late pachytene spermatocytes	Matrix; marginal segment; putative zona binding protein	(335,336)
Sp-10	*ACRV1, Acrv1*	Testis-specific	Matrix, principal segment and posterior equatorial segment; putative zona binding protein	(337)
Synaptotagmin IV	*SYT6*	Ubiquitous	Outer acrosomal membrane; calcium-binding protein	(330)
TESP1	*Tesp1*	Testis-specific	Matrix, principal segment; serine protease, putative role in zona penetration	(316)
TESP2	*Tesp2*	Testis-specific	Matrix, principal segment; serine protease, putative role in zona penetration	(316)
TESP4	*Prss2*	Ubiquitous	Matrix, principal segment; pancreatic trypsin-like serine protease, putative role in zona penetration	(317)
Zonadhesin	*ZAN*	Testis-specific	Inner and outer membranes in spermatids, matrix in sperm; zona binding protein	(331,332)

N.D., not determined.

seminal plasma (284,285). Sperm-associated inhibitors in the boar resembling Kazal-type inhibitors are present in seminal plasma (286). Another serine protease inhibitor closely related to the plasma protein C inhibitor has been identified in the acrosome of human spermatozoa (287).

Other enzymes in the acrosome include a spermatogenic cell-specific hyaluronidase that is different from the common lysosomal form (288–290). It is located predominantly in the principal segment in bovine and ram sperm (276,291–293). However, ram sperm denuded of plasma membrane and outer acrosomal membrane retain half the amount of hyaluronidase present in intact sperm, and some of the enzyme may be bound to the inner acrosomal membrane (294).

Two forms of β-galactosidase have been isolated from rat testis and spermatozoa, a lysosomal form and a smaller acrosomal form with distinct physicochemical and enzymatic properties (295,296). A 67-kDa β-galactosidase has been identified in rabbit sperm acrosomes, and additional larger species are detected in testis homogenates (297). Although acrosomal β-galactosidase was originally identified as a distinct isozyme (295,296), further studies are needed to determine if the sperm enzyme is derived from the larger form present in the testis. Similarly, the acrosomal hydrolase, β-N-acetylglucosaminidase, isolated from sperm is smaller than isoforms isolated from the testis (296,297).

A variety of other hydrolytic enzymes have been reported to be present in the acrosome, most of which were identified using enzymatic assays or immunostaining. Proteinases reported include a collagenase-like peptidase (298), a cathepsin D-like protease (299), cathepsin H (300), trypsin-like proteases distinct from acrosin (301), and dipeptidyl peptidase II (302,303). In addition, calpain II (304), neuraminidase (305), nonspecific esterases (306), arylamidase (307), aspartylamidase (308), arylsulfatase A (298,309), acid phosphatase (310), β-N-acetylglucosaminidase (311), a phospholipase C (312,313), and phospholipase A_2 (314,315) have been reported to be present in the acrosome. It is generally believed that these enzymes are soluble components that are released when the acrosome reaction occurs. However, the genes encoding most of these enzymes have not been identified and they remain to be characterized at the molecular level. Nearly all are members of enzyme families, and it is unknown which family member is present in the acrosome, or if these enzymes are products of testis-specific or ubiquitously expressed genes. The exceptions are TESP1 and TESP2, testis-specific serine proteases (316), and TESP4, a ubiquitously expressed homologue of pancreatic trypsin (317). The proteases might participate along with acrosin to facilitate penetration of the zona pellucida by spermatozoa.

Other proteins identified as acrosomal constituents might function during the acrosome reaction. Inhibitory guanine nucleotide-binding regulatory proteins (G_i proteins), identified by immunocytochemistry, are associated with the developing acrosome (318) and are retained in the acrosomal region of mature spermatozoa (319,320). Immunoreactivity is lost following induction of the acrosome reaction (320), consistent with the hypothesis that G_i proteins function as signal transducing elements mediating this exocytotic event (321). Acrogranin (Table 2), a 67-kDa glycoprotein, was originally identified as an acrosomal constituent in guinea pig spermatogenic cells and sperm (322). Amino acid sequences deduced from cDNA sequences indicated that mouse and guinea pig acrogranins have extensive homology with granulins and epithelins, growth modulating peptides previously identified in somatic tissues (323). Other peptides that have been localized in the acrosome by immunocytochemical methods include proenkephalin in rats (324), gastrin in humans (325) and the related cholecystokinin peptide in other mammals (326). The functional roles of these peptides during spermatogenesis or fertilization have not been determined.

Other nonenzymatic acrosomal components have been identified which often are testis-specific proteins whose roles remain to be determined (Table 2). Some are soluble components that are released when the acrosome reaction occurs. They include AM50 (p50, apexin) found in the guinea pig and similar to neuronal pentraxin II (327), and the cystein-rich protein CRISP-2 (328). Others are associated with the acrosomal membranes, including SAMP32 (329), synaptotagmin IV (330), and zonahesin (331,332). The majority of the nonenzymatic acrosomal proteins are components of the acrosomal matrix, the portion of the acrosome that is relatively insoluble in nonionic detergents and largely remains associated with the sperm after the acrosome reaction (333). In addition, some components might be involved in zona binding, including MC41 (334), sp56 (335,336), and Sp-10 (337). Other matrix proteins that have been identified are sp32, a proacrosin-binding protein (338,339), and SAMP14, with regions of sequence similarities to the plasminogen activators (340).

THE FLAGELLUM

The flagellum of the mammalian spermatozoon consists of four distinct segments: the connecting piece (neck), the middle piece, the principal piece, and the end piece (Fig. 1). The main structural components within the flagellum of the mammalian sperm are the axoneme, the mitochondrial sheath, the outer dense fibers, and the fibrous sheath (Fig. 6). The axoneme is composed of a "9 + 2" complex of microtubules that extends the full length of the flagellum. The outer dense fibers are adjacent to the axoneme and extend from the connecting piece to the posterior portion of the principal piece. The characteristic feature of the middle-piece segment is the mitochondrial sheath, while the fibrous sheath defines the extent of the principal-piece. The mitochondrial sheath immediately surrounds the outer dense fibers in the middle piece, whereas the fibrous sheath surrounds the outer dense fibers in the principal piece. The base of the flagellum abuts the nucleus at the junction between connecting piece and head (341,342).

The flagellum provides the motile force necessary for the sperm to reach the egg surface and achieve fertilization. The different elements of the flagellum generate and shape the waves of bending that produce this force and propagate the waves from the base to the tip. The human sperm is about 60 μm long, and the flagellum is 55 μm of this length (343). However, sperm vary considerably in length between species. Rabbit sperm are 46 μm long, mouse sperm are 120 μm long, rat sperm are 190 μm long, and Chinese hamster sperm are 250 μm long (344). The flagellum of a human sperm is greater than 1 μm in diameter in the connecting-piece segment and tapers towards the posterior tip (343).

Connecting Piece

The main components of the connecting piece (Fig. 5) are the capitulum (the dense fibrous plate-like structure that conforms to the shape of the implantation fossa) and the segmented columns (12). The implantation fossa develops with the apposition of the nuclear envelope and the basal plate, a dense plaque of material which is adherent to the outer nuclear membrane of the nuclear envelope. The interspace between the two membranes of the nuclear envelope in this region contains a regular array of periodic densities 6 nm wide and 6 nm apart (6). Freeze-fracture studies indicate that the membrane of the nuclear envelope lining the implantation fossa contains a dense population of large and regularly spaced intramembranous particles surrounding a central particle-free region (30). Fine filaments traversing the narrow region between capitulum and basal plate presumably are responsible for attaching the capitulum of the flagellum to the basal plate of the head (164,165). Trypsin treatment appears to cleave heads from tails at the plane between the capitulum and basal plate (345,346), but decapitation of sperm with primary amines or sodium dodecyl sulfate usually results in cleavage between the inner nuclear membrane and the outer nuclear membrane (347). Heads and tails can also be separated by sonication, but the cleavage site is not predictable (348,349).

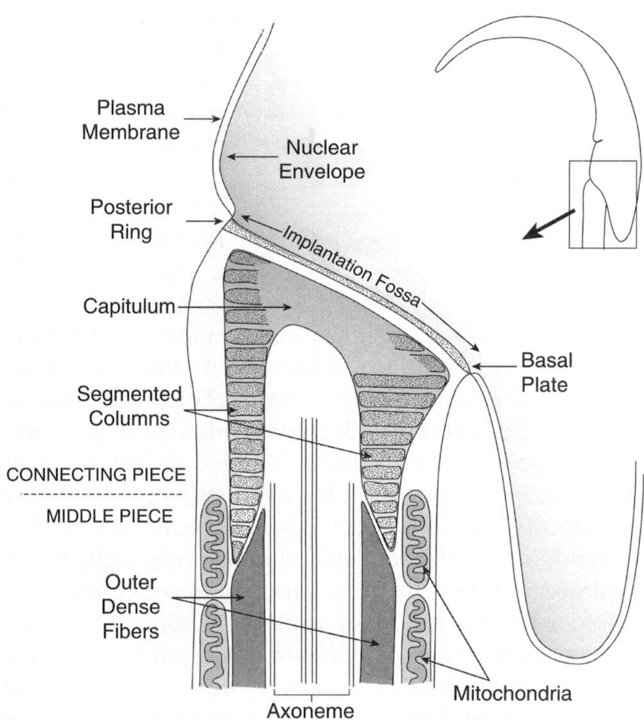

FIG. 5. Connecting piece region of a spermatozoon. The basal plate is adherent to the nuclear envelope, defining the implantation fossa and forming the site of attachment of the flagellum to the sperm head. The connecting piece is topped by the capitulum, with the segmented columns extending from it caudally to fuse with the outer dense fibers. [Modified from (353).]

The basal plate and capitulum are composed of proteins that are soluble in ionic detergent containing a disulfide-bond reducing agent (349,350). Although the composition of these structures is not known, they may be related to ciliary rootlets (6), which contain ankyrin, a 250- and 230-kDa protein dimer (343). A genetic defect in a bull caused the flagellum of most mature sperm to be detached from the heads (256). Detachment began during late spermatogenesis and continued in the epididymis; the resulting detached flagella are motile, metabolically active, and able to penetrate cervical mucus (351).

Extending posteriorly from the capitulum usually are two major and five minor segmented columns 1–2 μm in length (Fig. 5). The two major columns split into two columns each and along with the other five columns, fuse to the nine outer dense fibers extending throughout most of the remaining length of the flagellum. However, the segmented columns and outer dense fibers have different origins, and the continuity between them develops late in spermiogenesis (230,341). The segmented columns of the connecting piece are cross-striated, with a typical periodicity of 6.65 nm between segments. Each segment, in turn, has nine or 10 horizontal bands (230). During development of the flagellum, a transversely or obliquely oriented proximal centriole lies between the longitudinally oriented distal centriole and a depression in the capitulum (230). In round spermatids, the distal centriole is continuous with the axoneme and other accessory structures form around it as the connecting piece develops. In many species, the distal centriole disintegrates late in spermiogenesis, while the proximal centriole disintegrates soon thereafter or as the sperm transits the epididymis (94,164,195,341,352). It appears that the centrioles are involved in development of the connecting piece and the axoneme, but they clearly are not required for sperm motility (6,230,353).

Axoneme

Microtubules

The organization of the axoneme (axial filament complex) of the mammalian sperm tail (Fig. 6) is the same as that of cilia and flagella of most plants and animals. It consists of two central microtubules surrounded by nine microtubule doublets (354). Each doublet consists of a complete A microtubule, onto which is attached a "C-shaped" B microtubule. Two arms extend from the A microtubule toward the B microtubule of the adjacent doublet. When axonemes are viewed from base to tip, the arms project clockwise (355). In the rat spermatozoon, the central pair of microtubules extends into the connecting piece to the capitulum, whereas the other microtubules appear to end on the remnants of the base of the distal centriole (353,356). Radial spokes project helically towards the central pair of microtubules from the seven outer doublets of microtubules that surround the central pair (357). The doublets are numbered one through nine, with number one being the doublet situated on a plane perpendicular to that bisecting the microtubules of the central pair. Doublet number two is adjacent to the arms of doublet number one, and so on in a clockwise direction, as viewed from the distal towards the proximal end of the flagellum (Fig. 6).

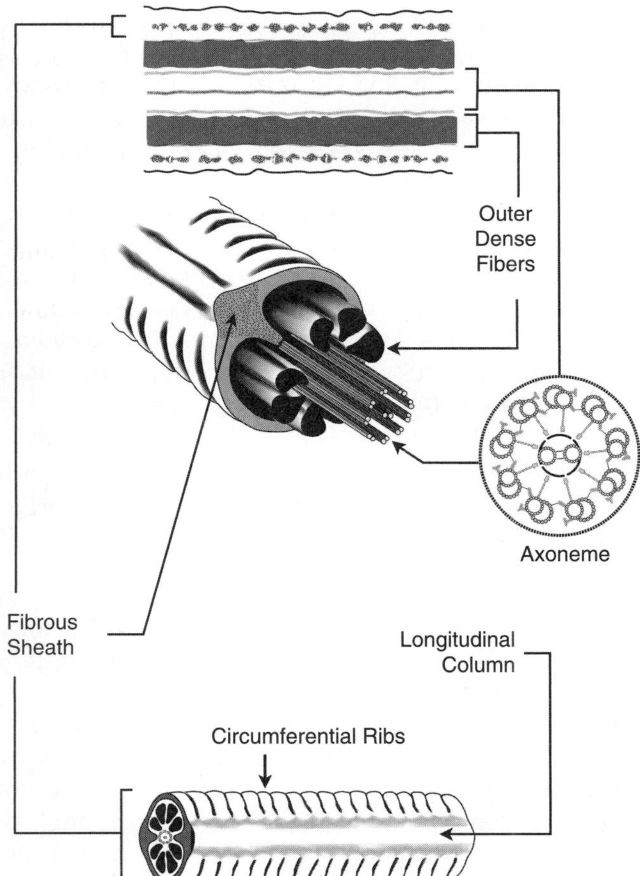

FIG. 6. The cytoskeletal components of the sperm flagellum. The axoneme extends from the connecting piece to the distal tip of the flagellum (Fig. 1). It consists of nine outer doublets of microtubules surrounding a central pair of microtubules. The outer doublets consist of microtubules A and B. Inner and outer dynein arms extend from the A microtubule towards the B microtubule of the adjacent doublet. The outer dense fibers extend from the connecting piece into the posterior portion of the principal piece (Fig. 1). They lie between the mitochondrial sheath and the axoneme in the middle piece and between the fibrous sheath and axoneme in the principal piece. The fibrous sheath is composed of two longitudinal columns connected by circumferential ribs and surrounds the axoneme and outer dense fibers in the principal piece region of the flagellum (Fig. 1).

The microtubules are composed of α-tubulin and β-tubulin, closely related proteins of approximately 56 and 54 kDa, respectively. Multiple α-tubulin and β-tubulin genes are expressed during spermiogenesis (358), including testis-specific isoforms, and the diversity of these proteins are further expanded by posttranslational modifications (359–361). One of the first postmeiotically expressed genes identified was for a spermatid α-tubulin with a unique sequence (360,361). This led to the prescient suggestion that the gene encoded a novel α-tubulin involved in determining special properties of the sperm axoneme (360).

Dyneins

The dyneins are a large family of motor proteins associated with microtubules, including the outer doublet microtubules in axonemes of cilia and flagella. Their ATPase activity (362) generates force unidirectionally towards the minus (–) end of microtubules to generate sliding forces between adjacent doublets of microtubules during flagellar movement (363). Dyneins consist of heavy, intermediate, light intermediate, and light chains. The motor activity is associated with heavy chains and axonemal dyneins are present as heterotrimers, heterodimers and monomers of heavy chains (364). Most of what is known about the function of axonemal dyneins comes from studies on ciliated unicellular organisms (365) and sea urchin sperm (362,363) and is presumed to apply to mammalian sperm. Indeed, the targeted disruption of the gene for dynein heavy chain 7 (*Mdhc7*) in mice resulted in drastically altered sperm motility and male infertility (366). In addition, mRNA sequences for axonemal dyneins cloned from human, mouse and other mammals are present in GenBank.

It is thought that the central pair of microtubules (central apparatus) and radial spokes have a key role in regulating dynein activity and flagellar motility (367,368). Eight proteins have been identified that are associated with the central apparatus and seven are known to be associated with the radial spokes (367). They include microtubule-binding proteins, calcium binding proteins, and signaling pathway components, several of which have been identified in mammalian sperm (367).

Tektins

The tektins were identified when it was observed that dissociation of the sea urchin sperm flagellum released ribbons of two to four protofilaments (369). These could be further fractionated into filaments 2–3 nm in diameter (370) with physical and chemical properties similar to intermediate filaments (371). Three tektin proteins were identified in the filaments and found to be located throughout the entire length of the axoneme (371). These studies indicated that tektins are structural components of the microtubule doublets in the sea urchin sperm flagellum and possibly involved in the assembly or function of axonemal microtubules (371). Antibodies to tektins in sea urchin sperm cross-reacted with proteins in the axonemes of cilia and in the basal bodies in male germ cells of marine invertebrates (372,373). It also appeared that tektins are present in centrioles in mammalian tissue culture cells (372).

The three tektin genes are highly conserved and orthologues are expressed in mammals, principally in male germ cells and in ciliated cells. The first mammalian tektin gene cloned was mouse *Tekt1* (374). The gene is expressed highly during spermatogenesis and an antibody to TEK1 labeled the centrosome in spermatids and the caudal end of the sperm head in elongating spermatids, but no signal was detected in epididymal sperm (375). The absence of staining in sperm was thought to be due to either loss of the protein or masking of the epitope (375). Expression of the mouse *Tekt2* gene was observed only in testis during the postmeiotic period. The protein was detected by immunohistochemistry in elongating spermatids and was localized to the flagellum of spermatozoa (376). The human *TEK2* gene was found to be expressed at the highest level in testis and at lower levels in trachea, lung, and brain. The *TEKT2* protein was detected by immunofluorescence in the region of the basal body and in the principal piece of the flagellum of ejaculated sperm (377). Targeted disruption of the *Tekt2* gene resulted in male infertility, abnormal sperm morphology and motility, altered formation or attachment of the dynein inner arm in the sperm axoneme, and altered morphology of the dynein arms in tracheal cilia (377). In addition, the mouse *Tekt3* gene is expressed at high levels in adult testis and at low levels in brain (378).

Mitochondrial Sheath

The mitochondria are helically wrapped around the outer dense fibers in the middle piece of the sperm tail (Fig. 1). They are generally arranged end-to-end, but the number of parallel helices, the number of gyres, and the length of the middle piece vary between species. In the mouse, the mitochondria are usually arranged in two parallel helices, with an average of 87 windings around the flagellum (379). The morphogenesis of this arrangement in the spermatid occurs by: (a) the migration of the annulus from the connecting piece down the flagellum to the beginning of the fibrous sheath, (b) the formation of a dextral helix of elongated mitochondria around the flagellum, (c) the division of these mitochondria into spherical mitochondria (d), and finally, the elongation and end-to-end apposition of these mitochondria into two tight sinistral helices (379). In the mouse, the mitochondria are usually of variable length and abut end-to-end at random along the helix. Genetic selection has been used to produce sub-strains of mice with midpieces longer and shorter than the average length of 21 μm (380). In most species there is a high degree of uniformity in the organization of the mitochondrial sheath. Mitochondrial sheaths composed of a precise number of mitochondria of identical size and shape are present in the little brown bat (381), wooly opossum (382), and Chinese hamster (240). In the common brown bat there are two mitochondria of identical size in each turn of the gyre, and their ends always meet on a plane passing through the central pair of microtubules in the axoneme (381). In the rhesus monkey there is a tendency for the demarcation between mitochondria to occur in longitudinal register in alternate gyres (240).

The length of the middle piece and the disposition of mitochondria in spermatozoa have been determined for several mammalian species using surface replica methods (240). The middle pieces differ considerably in length and the organization of mitochondria is quite variable between species. The midpiece length is related to the length of the sperm and in most mammals is between 8 and 12 μm (8.5 μm in rabbit, 9 μm in guinea pig, 10 μm in rhesus monkey, and 12 μm in bull), but is 64 μm in rat and 100 μm in Chinese hamster. The number of mitochondrial gyres in the midpiece is related to its length as well, but there is not an obvious correlation between the disposition of mitochondria and length. The mitochondria are organized into about 41 gyres arranged in a quadruple or quintuple helix in rabbit, form in an irregular concentric pattern in guinea pig, lie in about 40 gyres arranged in a single or double helix in rhesus monkey, and there are about 64 gyres usually organized in three helices in bull sperm. There are around 362 mitochondrial gyres in rat sperm and the gently spiraling mitochondria fit together in intricate patterns around the large outer dense fibers. Finally, the mitochondria in Chinese hamster sperm appear to form a double helix with each mitochondrion wrapping one-third the circumference of the middle piece (240).

The mitochondrial sheath adheres to an underlying complex of filaments named the sub-mitochondrial reticulum (SMR) (76,383). The SMR complex is a network of ribbons of filamentous material that are laterally interconnected and fuse with the annulus at the junction between the midpiece and the principal piece. Fractions isolated from bull sperm that are enriched for the SMR complex are resistant to solubilization and are enriched for 56- and 54-kDa proteins (383) that remain to be characterized. The SMR may function in assembling and maintaining the ordered array of mitochondria in the midpiece (383).

Another component of the midpiece is a 17- to 15-kDa fibrous protein that forms an insoluble capsule around sperm mitochondria. It originally was called mitochondrial capsule protein (MCP) (384), and contains an unusually high percentage of proline and cysteine as well as a low percentage of hydrophobic amino acids (385). The MCP protein is synthesized in elongating spermatids in the rat (386) and the mRNA encoding this protein is transcribed in late pachytene spermatocytes and at higher levels in spermatids in the mouse (387). The sperm midpiece binds selenium and selenium deficiency causes abnormal development of the sperm midpiece leading to disorganization of the mitochondrial sheath (388,389). MCP was believed to be the selenoprotein affected by selenium deficiency (388). However, more recent studies showed that the gene encoding MCP does not contain potential selenocysteine codons (390,391), and the protein was renamed sperm mitochondria-associated cysteine-rich protein (SMCP). Mice with a targeted disruption of the *Smcp* gene produced sperm with normal head, mitochondrial, and flagellar morphology as determined by electron microscopy. When the mutation was on a 129/Sv genetic background, the male were infertile due to defects in sperm motility (392). However, male C56BL/6 mice with the disrupted gene were fertile, implying that the *Smcp* gene interacts with an unknown modifier gene (392).

An additional sperm mitochondria-associated protein is spergen-1, a small protein with a mitochondria-targeting signal that is translated and transcribed in spermatids (393). It localizes to the midpiece of rat sperm and results in mitochondrial aggregation when transfected into COS-7 cells (393). The localization of spergen-1 in sperm suggests that it is a component of, or associated with the mitochondrial sheath, and it was hypothesized that spergen-1 might function as an adhesive molecule involved in assembling the mitochondrial sheath during spermiogenesis.

Outer Dense Fibers

General Features

The nine outer dense fibers surround the axoneme, forming a "9 + (9 + 2)" complex throughout the length of the middle piece and most of the principal piece of the flagellum of mammalian sperm (Figs. 1 and 6). The outer dense fibers are numbered corresponding to the adjacent microtubule doublet. Although similar structures are present in sperm of nonmammalian species, it is unknown at the present time if they are encoded by orthologues of genes for outer dense fiber proteins in mammalian sperm. The basic structural plan of the sperm flagellum shows little variation in mammals, but the outer dense fibers differ considerably in size and shape between species. The fibers are often teardrop shaped in cross section, with the peripheral margin of the fiber being rounded and the centrally projecting portion tapering toward a doublet of the axoneme (6). The outer dense fibers also differ among themselves in shape and size, with fibers 1, 5, and 6 (and sometimes 9) being larger than the others. The fibers are thickest in the proximal part of the middle piece and gradually taper in size toward the distal tip. They terminate in the proximal half of the principal piece in human, macaque, and bat sperm, but extend to near the end of the principal piece in rat, hamster, guinea pig, and ground squirrel sperm (6). In human (394), bull, and rat sperm (395,396), the outer dense fibers occupy 60% of the length of the principal piece. In most species, fibers 3 and 8 terminate in the first part of the principal piece, and their place is taken by inward extensions of the longitudinal columns of the fibrous sheath. The larger fibers (1, 5, and 6) are usually the last to terminate (396).

Formation

The formation of outer dense fibers begins in spermatids, with the appearance of a small dense fiber immediately adjacent to the outer aspect of each microtubule doublet (397–399). In the rat, outer dense fibers first appear in the most proximal part of the future middle piece in step 8 of spermatid development and extend into the principal piece by step 14 (400). During step 15 the outer dense fibers increase in diameter and change shape, with fibers 1, 5, and 6 developing ahead of the other fibers. During step 16 the fibers enlarge rapidly along their entire length and take on their nearly final form, with slight growth continuing during steps 17 to 19 (400). Immunocytochemical studies have confirmed the proximal to distal direction of assembly of the outer dense fibers (401,402). Studies using radioautography to localize metabolically incorporated proline and cysteine suggest that rapid protein synthesis accompanies the growth of the outer dense fibers during step 16 (400).

Structure

The outer dense fibers usually appear homogeneous with a slightly less dense cortical layer when examined by electron microscopy. The cortical layer stains intensely with tannic acid (6) and phosphotungstic acid (403). Surface replicas of sperm from rat or mouse outer dense fibers show striations in the outer contour of the cortex, with a periodicity of 40 nm (404). When fixed in the presence of ruthenium red, the fibers appear to be composed of a single lamina of 6- to 7-nm globular particles (405). These striations have the same 70 to 80 degree "sinistral obliquity" as the mitochondrial helix in the middle piece (404). The striations appear to be composed of a double linear array of 6- to 8-nm diameter globular subunits, and a central depression in each striation probably accounts for an apparent periodicity of 20 nm seen on negatively stained specimens (405,406). The periodic substructure is confined to the cortex in rat sperm (405); however, this substructure is present in both cortical and medullary elements in human and bull sperm (406).

Composition

The flagellum contains cysteine-rich proteins that are cross-linked by disulfide bonds during epididymal maturation (190,350). The resulting low solubility of outer dense fibers allowed their isolation in detergent solutions lacking disulfide bond-reducing agents (407) and the characterization of their major proteins. Outer dense fibers isolated from rat sperm were found to contain 4 to 6 major proteins between 90 and 11 kDa (405,409–411), from bull sperm to contain three major proteins between 85 and 11 kDa (406,408,412), and from human sperm to contain two major proteins of 67 and 55 kDa (413). The amino acid compositions of outer dense fibers from bull and rat sperm were similar (412), but differences were found between individual proteins in amino acid composition, N-terminus amino acid, peptide maps, net charge, and phosphorylation (405,406,409,412). Infrared spectrum analysis suggested that the higher molecular weight outer dense fiber components of various species have a α-helical configuration (406). In addition, it was determined that zinc is required for spermatogenesis (414), incorporated into the flagellum in late spermatids (415), and localized in the outer dense fibers (406). The smaller components were found to be capable of binding zinc to their sulfhydryl groups (416).

A better understanding of the composition of outer dense fibers has resulted from the cloning of genes for several constituent proteins (Table 3). The first gene identified was *Odf1* (outer dense fiber of sperm tails 1). The gene is transcribed in spermatids and encodes a cysteine-proline-rich 27-kDa protein that is the major protein in rat outer dense fibers (417–419). It contains a repetitive motif (cysteine-glycine-proline) present in *Drosophila* sperm proteins (420) and also an α-crystallin domain characteristic of the small heat-shock protein family (421). The orthologous gene has been identified in humans (422), mice (423), pigs, and cattle (424). The human *ODF1* gene appears to be expressed in embryos and the mouse *Odf1* gene in adult heart and mid-gestation embryos, based on the presence of expressed sequence tags (EST) in these tissues in public databases.

Three other genes have been identified for proteins that are constituents of "outer dense fiber of sperm tails" (ODF2, ODF3, and ODF4). They are not homologous to each other or to ODF1 and their names relate only to their original source. The rat *Odf2* gene was cloned using ODF1 as bait in a yeast two-hybrid screen, and the two proteins interact through leucine zipper motifs (425). The genes for ODF2 are expressed during the postmeiotic phase and encode a major 84-kDa outer dense fiber protein in rat (426), mouse (427), and human (428). While ODF1 was detected only in the medulla, ODF2 was detected in both the cortex and the medulla of the outer dense fibers (426). Although ODF2 is abundant in spermatogenic cells, it is also a cell-cycle dependent scaffold component of centrosomes in somatic cells (429,430). However, the mouse *Odf3* gene is testis-specific and expressed only during the postmeiotic phase of spermatogenesis (431). The deduced amino acid sequence encodes a highly conserved 28-kDa protein. The deduced amino acid sequence of human ODF3 is 93% identical and contains the same

TABLE 3. *Outer dense fiber-associated proteins*

Protein	Gene	Features	References
ODF1 (RT7, Odf27, rts5/1)	*Odf1*	Outer dense fiber of sperm tails 1, structural protein, present only in medulla of ODF, postmeiotic, high in testis and low in a few other tissues	(416–424)
ODF2 (Odf84)	*Odf2*	Outer dense fiber of sperm tails 2, structural protein, also present in connecting piece of sperm and centrosomes of other tissues, interacts with ODF1, meiotic and postmeiotic, highly expressed in testis	(425–430)
ODF3 (SHIPPO1)	*Odf3*	Outer dense fiber of sperm tails 3, structural protein, postmeiotic, high expression in testis, low in epididymis and brain	(431)
ODF4 (OPPO 1, Tisp-62)	*Odf4*	Oppo (from Japanese word for tail) 1, outer dense fiber of sperm tails 4, structural protein, meiotic and postmeiotic, testis specific	(432–433)
Spag4	*Spag4*	Sperm associated antigen 4, structural protein, interacts with ODF1, present in manchette and axoneme, not in ODF, postmeiotic, testis specific	(434–435)
Spag5	*Spag5*	Sperm associated antigen 5, structural protein, interacts with ODF1, meiotic and postmeiotic, testis specific	(436–437)
OIP1	*Rnf38*	ODF1 interacting protein, ring finger protein 38, structural protein, interacts with ODF1, localizes to sperm tail, meiotic and postmeiotic, high expression in testis, present in several other tissues	(438)
Tpx-1	*Crisp2*	Testis specific protein 1, cysteine-rich secretory protein 2, also present in acrosome and connecting piece and longitudinal columns of FS, meiotic and postmeiotic, testis specific	(439–441)
Sak57	*Krt5*	Sperm associated keratin of mass 57kDa, keratin 5, keratin complex 2 gene 5, structural protein, also found in manchette and acroplaxome, meiotic and postmeiotic, expressed in many tissues	(442)
VDAC2	*Vdac2*	Voltage dependent anion channel 2, ubiquitously expressed	(443)
VDAC3	*Vdac3*	Voltage dependent anion channel 3, ubiquitously expressed	(443,444)
KLC3	*Klc3*	Kinase light chain 3, motor protein, interacts with ODF1, postmeiotic, expressed in multiple tissues	(446–447)
Cdk5·p35	*Cdk5r1*	Cyclin-dependent kinase 5 regulatory subunit (p35) 1, regulator of proline directed serine/threonine kinase, associated with developing ODF, ubiquitously expressed	(448)

number of amino acids as the mouse protein (431). The mouse *Odf4* gene also is expressed only in the testis and it encodes a novel 33-kDa protein that is relatively resistant to extraction from outer dense fibers. Transcription begins during the meiotic phase, but the protein is first detected during the postmeiotic phase (432). The human *ODF4* gene encodes a 30-kDa protein and differs from the mouse gene in genomic structure and in exon/intron usage (433). The human gene also appears to be expressed only in the testis (433).

Three other ODF1-interacting proteins have been identified in rat sperm outer dense fibers (SPAG4, SPAG5, and OIP1) using ODF1 as bait in yeast two-hybrid screens. The gene for SPAG4 (sperm associated antigen 4), like ODF2, has a leucine zipper that interacts with a leucine zipper in ODF1 (434). The *Spag4* gene is expressed specifically in round spermatids and leads to the synthesis of a 49-kDa protein in elongating spermatids. However, immunoelectron microscopy revealed that SPAG4 interacts with two microtubule-containing structures in spermatids, the manchette and the axoneme, but not outer dense fibers in epididymal sperm. It was suggested that SPAG4 is involved in outer dense fiber assembly by forming a link between ODF1 and axonemal microtubules (434). The human *SPAG4* gene has been cloned as well (435).

SPAG5 (sperm associated antigen 5) also interacts with ODF1 via leucine zippers (436). However, ODF2 does not interact with either SPAG4 or SPAG5, indicating that their interactions with ODF1 involve more than leucine zipper binding. The *Spag5* gene is expressed only in the testis and encodes a 200-kDa protein that is synthesized in pachytene spermatocytes and spermatids. SPAG5 has a 73% similarity with Deepest, a putative mitotic spindle protein of unknown function (436). However, targeted disruption of the *Spag5* gene in mice did not affect spermatogenesis or fertility (437) and the role of this protein also remains unknown.

OIP1 (ODF1 interacting protein 1) interacts with Cys-Gly-Pro repeats in the C-terminus of ODF1 (438). OIP1 is a RING-finger family protein and its gene has been named *Rnf38*. The *Rnf38* gene appears to have biphasic expression during spermatogenesis, with one peak in spermatogonia and another in round spermatids. The approximately 70-kDa OIP1 protein is present in the outer dense fibers of rat sperm. While the *Rnf38* gene is expressed at a high level in testis, it is also expressed at lower levels in epididymis, heart, and smooth muscle (438).

Another group of proteins found in outer dense fibers are also either associated with other sperm components (TPX1) or are present in numerous other tissues (SAK57, VDAC2, VDAC3, KLC3, and the CDK5/p35 complex). The TPX1 protein is a member of the cysteine-rich secretory protein family and is encoded by the *Crisp2* gene. This gene is expressed only in testis and its transcripts are abundant in spermatocytes and spermatids (439,440). TPX1 is a component of outer dense fibers (440,441), but also is present in the developing acrosome in spermatids, and in the perinuclear theca, connecting piece, and longitudinal columns of the fibrous sheath in rat sperm (441). TPX1 is presumed to be a structural protein of the outer dense fibers, but it role has not yet been experimentally determined.

SAK57 is a filamentous cytoskeletal protein encoded by *Krt5* (keratin complex 2 gene) and was identified by peptide sequencing of a 57-kDa protein isolated from rat sperm outer dense fibers (442). It also was found in the axoplaxome, a cytoskeletal plate anchoring the developing acrosome to the nuclear envelope in spermatids. The human *KRT5* gene is expressed in numerous tissues including testis, but little is known about the expression of its orthologues in other species.

VDAC2 and VDAC3 (voltage-dependent anion channels 2 and 3, porin 2 and 3) also were identified by peptide sequencing, starting with a 30- to 32-kDa protein isolated from bovine sperm outer dense fibers (443). Antisera to human recombinant VDAC2 and VDAC3 were used in immunofluorescence and immunoelectron microscopy assays to verify that the two proteins are present in the outer dense fibers of bovine sperm (443). Other studies have demonstrated that VDAC3 has an important role in sperm. Targeted disruption of the *Vdac3* gene in mice often resulted in loss of an outer microtubule doublet (usually doublet 7) in the axoneme and loss of the associated outer dense fiber, markedly reduced sperm motility, and male infertility (444). VDAC2 and VDAC3 are ubiquitously expressed, but the only other effects reported in these mice were unusually shaped mitochondria and reduced respiratory chain activity in skeletal muscle (444).

KLC3 (kinesin light chain 3) was serendipitously identified in a yeast two-hybrid assay using kinase-dead c-*mos* as bait (446). The *Klc3* is expressed at low levels in spermatocytes and high levels in spermatids, as well as at lower levels in several other tissues. Although characterized initially as a microtubule binding protein (446), subsequent studies determined that KLC3 binds to the surface of outer dense fibers in spermatids, isolated outer dense fibers, and to ODF1 in a microtubule-independent fashion (447). KLC3 was lost from the outer dense fibers during isolation, suggesting that it is associated with the outer dense fibers, but not an integral component. It also was found that the other two KLC isoforms, KLC1 and KlC2, are not detectable in spermatids, suggesting that KLC3 is an essential component of the kinesin motor during flagellar formation (447).

The *Cdk5* gene is ubiquitously expressed and encodes a proline-directed serine/threonine kinase that pairs with an activator, either p35 or p39. The substrates for Cdk5 in the brain include cytoskeletal proteins, actin proteins, and microtubule-associated proteins. In mice and rats, *Cdk5* and the gene (*Cdk5r1*) for p35 are expressed in the testis at the highest levels in elongating spermatids (448). Immunolabeling of p35 progressed from proximal to distal in the flagellum as spermiogenesis advanced and Cdk5 staining increased in the flagellum during this same period (448). Immunoelectron microscopy demonstrated that both were present in the outer dense fibers and in the basal plate, capitulum, and striated collar of the connecting piece as well (335).

Satellite Fibers

Satellite fibers are present in the flagellar matrix between the outer dense fibers (6,185,450). These fibers form in step 19 of spermiogenesis in the rat, just before spermatozoa are released from the seminiferous epithelium (400). They are present in limited numbers in most species, but are highly developed in the ground squirrel (185) and bandicoot (451), which have unusually thick sperm tails (185). Their role is unclear, but they may be accessory tensile elements of the motor apparatus (185). The satellite fibers appear to arise by separating from the edge of the outer dense fibers (10), suggesting that they may be composed of outer dense fiber proteins.

Function

The outer dense fibers were speculated to be contractile because of their close association with the axoneme, their coincident appearance during phylogeny with internal fertilization, and a concomitant increase in the size of the mitochondrial sheath (185).

High-speed cinematography of mouse, human, rabbit, and opossum sperm indicated that they are relative flexible, forming arcs with a small radius of curvature as they beat. However, rat and Chinese hamster sperm appear very stiff while beating and have a large radius of curvature (344). There is a correlation between the radius of curvature and the size of the dense fibers in these species, suggesting that the dense fibers might influence the form of the beat by determining the elastic properties of the sperm tail (344). Measurement of the relative tensile strengths of sperm of seven mammalian species indicates that tensile strength correlates with the size of the outer dense fibers (449). It is suggested that the outer dense fibers provide added strength to protect sperm from damage by shear forces encountered during epididymal transit or ejaculation (449).

It was reported that outer dense fibers are antigenically similar to actin (403) and that an ATPase is associated with the outer dense fibers (450,452). However, biochemical studies have not found a similarity in composition of outer dense fibers to actin, myosin, or tubulin (404). Other investigators have been unable to detect ATPase activity associated with outer dense fibers (406). This suggests that the outer dense fibers might not play an active role in flagellar motion. However, stabilization of outer dense fiber proteins by abundant disulfide cross-linking may give them significant passive elastic properties that serve to stiffen or provide elastic recoil for the sperm tail (6). In addition, the presence of voltage-dependent anion channels (444), and a serine/threonine kinase and its activator (448) suggests that the outer dense fibers may serve as scaffolds for proteins involved in regulating sperm function.

Fibrous Sheath

Structure

The fibrous sheath defines the extent of the principal piece (Figs. 1 and 6), by far the longest segment of the flagellum (185). It is a cytoskeletal structure that is probably unique to the flagellum of sperm in mammals and some birds. It closely underlies the plasma membrane and is a tapering cylinder formed by two *longitudinal columns* connected by *circumferential ribs* (Fig. 6). The columns are formed by longitudinally oriented, loosely packed filamentous structures 15–20 nm in diameter. The longitudinal columns run peripheral to microtubule doublets 3 and 8 of the axoneme and are attached to outer dense fibers 3 and 8 in the proximal part of the principal piece (397). Distal to the termination of these two dense fibers, the tapered central edges of the longitudinal columns attach to ridges projecting from microtubule doublets 3 and 8 (185). The columns lie approximately in a plane sagittal to both of the inner doublet microtubules of the axoneme and have been referred to as *dorsal* (nearest outer doublet 3) and *ventral* (nearest outer doublet 8) *columns* in sperm containing a falciform shaped head (185). The size and shape of the longitudinal columns vary considerably between species. They are narrow and inconspicuous in the common brown bat and guinea pig, but are large in the Chinese hamster and opossum and result in prominent ridges that give the principal piece an elliptical profile in cross section (185).

The ribs of the fibrous sheath are composed of closely packed, circumferentially oriented filaments (185). The ribs broaden toward their ends, where they merge with the longitudinal columns and with each other. They are closely spaced and may bifurcate and anastomose with adjacent ribs. This occurs frequently in the mouse, resulting in broad bands instead of slender separate ribs. The ribs also vary in shape and size between species, being broad and flat in the bat and greatly expanded at their ends in the opossum. The thickness of the ribs diminishes toward the distal end of the principal piece, and the ribs and outer dense columns end abruptly at the posterior margin of the principal piece (6,185). In human spermatozoa, the ribs are 10–20 nm apart and about 50 nm wide (343).

Formation

Formation of the fibrous sheath takes place throughout much of spermiogenesis in the rat and proceeds from distal to proximal, opposite to the direction of outer dense fiber development (401,402,453). The longitudinal columns first appear as thin rods joined to microtubule doublets 3 and 8 near the distal end of the flagellum in step 2 of spermiogenesis. The columns increase in length during steps 2 to 10. In step 11, evenly spaced and circumferentially arranged pairs of spines form against the plasma membrane in the distal part of the flagellum. During steps 12 to 14, these structures associate with the columns in a distal to proximal direction. In early step 15, additional material is deposited between the pairs of spines to form the definitive ribs, and the longitudinal columns because thickened. During steps 15 to 16, additional ribs are formed from posterior to anterior, along the remaining proximal segment of the principal piece (453). Fibrous sheath proteins continue to be synthesized during the entire 15-day period of fibrous sheath formation (453). Filamentous material present in the cytoplasm of earlier stage spermatids appears to be the rib precursor in human (368,454), marmoset (455), and bandicoot sperm (399).

Composition

The fibrous sheath is highly resistant to acid solubilization (456) and disulfide bonds cross-link the proteins and stabilize the fibrous sheath (192). These characteristics were utilized to develop a procedure for isolating the fibrous sheath from rat sperm (457,458). Isolated rat fibrous sheaths contain 3 to 7 major proteins from 115- to 11-kDa (410,458–460). An 80-kDa protein is phosphorylated and peptide map analysis indicates that 80- and 11.5-kDa proteins are unrelated to proteins of similar size in the outer dense fibers (459). The major insoluble tail component synthesized during spermiogenesis in the mouse is a 74-kDa protein (461) that may be a fibrous sheath component. The isolated mouse sperm fibrous sheath contains five major proteins of 112- to 24-kDa (460). The major proteins in the isolated human sperm fibrous sheath are 68-kDa and 54- to 51-kDa (462).

The genes for several fibrous sheath proteins now have been cloned and the developmental expression, localization, and functional properties of the proteins characterized (Table 4). Three of the fibrous sheath proteins are cAMP-dependent protein kinase (PKA) anchoring proteins (AKAP) that probably are found only in spermatogenic cells (463). One is AKAP4, a ~80-kDa protein that is the most abundant fibrous sheath protein and a major structural component (464,465). Transcription of *Akap4* mRNA in mouse begins in early round spermatid development (steps 2–6), but translation is delayed until the development of elongating spermatids (step 14) (466). Two alternative transcripts encode ~110-kDa precursor forms which are processed in the flagellum to identical ~80-kDa mature forms (467). Immunoelectron microscopy determined that AKAP4 is present in both the longitudinal columns and the ribs of the fibrous sheath of mouse sperm (467). AKAP4 has two regions of homology with PKA-anchoring domains characteristic of AKAP proteins, one domain with dual specificity for binding either the RIα or RIIα subunit of PKA and one domain that binds only RIα (468). Targeted disruption of the *Akap4* gene did not affect sperm numbers, but the sperm failed to develop progressive motility and male mice were infertile. The framework of the fibrous sheath formed, but the

TABLE 4. *Fibrous sheath-associated proteins*

Protein	Gene	Features	References
AKAP4 (p82, FSC1, AKAP82, FS 75)	*Akap4*	"PKA anchoring protein 4," major structural protein, two PKA anchoring sites, postmeiotic, testis specific	(464–469)
AKAP3 (AKAP110, FSP95)	*Akap3*	"PKA anchoring protein 3," structural protein, PKA anchoring site, postmeiotic, testis specific	(470–471)
TAKAP-80	N.D.	"Rat testis-specific, developmentally regulated, RII-binding protein," postmeiotic, testis and other tissues	(472)
FSIP1	*Fsip1*	Fibrous sheath interacting protein 1, binds to AKAP4, postmeiotic, testis specific	(466)
FSIP2	*Fsip2*	Fibrous sheath interacting protein 2, binds to AKAP4, meiotic and postmeiotic, testis specific	(466)
ASP	*Ropnl1*	"AKAP-associated sperm protein," ropperin 1-like protein, binds to PKA anchoring site on AKAP3, testis specific	(473)
Ropporin (RPPN, ODF5)	*Ropn1*	Rhophilin-associated protein 1, AKAP-binding sperm protein, postmeiotic, testis specific	(473)
rhophilin	*Rhpn1*	Rho GTPase binding protein 1, meiotic and postmeiotic, high expression in testis and low in some other tissues	(474–476)
Sp17	*Spa17*	Sperm protein 17, meiotic and postmeiotic, mainly in principal piece, detectable in middle piece and dorsal acrosome, testis specific	(477–478)
CABYR (fibroshethin 2)	*Cabyr*	Calcium-binding tyrosine phosphorylation regulated protein, mainly expressed in testis	(479)
FS39 (DDC8)	*Ddc8*	39 kDa fibrous sheath protein, postmeiotic, testis specific	(480)
Tsga10		Testis specific 10, meiotic and postmeiotic, testis and other tissues	(481)
HK1-S	*Hk1*	Hexokinase 1-S, spermatogenic cell HK-1 splice variants, meiotic and postmeiotic, mainly in principal piece, testis specific	(482–484)
GAPD-S (GAPD-2)	*Gapds*	Glyceraldehyde 3-phosphate dehydrogenase-S, sperm-specific glycolytic enzyme, postmeiotic, testis specific	(484–489)
GSTM5	*Gstm5*	Glutathione S-transferase mu 5, meiotic and postmeiotic, high expression in testis and low in other tissues	(490)
Sptrx-1	*Txndc2*	Sperm thioredoxin-1, associated with assembling longitudinal columns of fibrous sheath in spermatids, not detected in sperm, postmeiotic, testis specific	(491)
Sptrx-2	*Txndc3*	Sperm thioredoxin-2, present in longitudinal columns and ribs in fibrous sheath in sperm, postmeiotic, testis specific	(492)
36K-A	*Pdhb*	Pyruvate dehydrogenase E1 component β subunit, possible testis specific splice variant, meiotic and postmeiotic, multiple tissues	(493)

fibrous sheath remained incompletely developed, the flagellum was shortened, and proteins usually associated with the fibrous sheath were absent or substantially reduced in amount (469).

The *Akap3* gene also is expressed during the postmeiotic phase of spermatogenesis and encodes another of the major proteins of the fibrous sheath (470,471). The ~110-kDa AKAP3 protein has a single PKA RIIα binding site. In the mouse, *Akap3* mRNA and AKAP3 protein were both detected in early round spermatids (step 2), but AKAP3 protein was not detected in the flagellum until the beginning of spermatid elongation (step 8) (466). In human sperm, the AKAP3 protein was detected by immunoelectron microscopy only in the circumferential ribs (471). These results suggest that AKAP3 is involved in formation of the framework of the fibrous sheath, and that AKAP4 is deposited on this framework late in spermiogenesis (469).

The third AKAP present in the fibrous sheath is TAKAP-80, also expressed during spermiogenesis (472). It was identified in the rat, but probably is present in the mouse because the rat cDNA hybridized with mRNA from mouse testis similar in size to the rat transcript. The deduced protein sequence of TAKAP-80 indicated that it is a 55-kDa protein with a single RIIα binding site (472).

Yeast two-hybrid screens were used to identify additional fibrous sheath proteins that interact with AKAP4 or AKAP3. Yeast two-hybrid screens using AKAP4 as bait determined that it binds AKAP3 as well as and two other spermatogenic cell-specific proteins, FSIP1 and FSIP2 (fibrous sheath interacting proteins 1 and 2) (466). Yeast two-hybrid assays indicated that AKAP3 binds to the C-terminal end of AKAP4 and AKAP4 binds to the middle region of AKAP3 (466). FSIP1 is a novel ~50-kDa protein of unknown function. Expression of the *Fsip1* gene is restricted to testis and begins in round spermatids at about the same time as *Akap4* expression. FSIP1 binds to the C-terminal portion of AKAP4, and AKAP4 binds to a region at the C-terminal end of FSIP1 (466). FSIP2 is also a testis-specific protein and *Fsip2* gene expression begins during the latter part of pachytene spermatocyte development. The *Fsip2* transcript encodes a ~760-kDa protein of unknown function. It remains to be determined if FSIP1 and FSIP2 are involved in fibrous sheath formation, are bound to AKAP4 in the intact fibrous sheath, or link AKAP4 to other fibrous sheath components (466). However, the FSIP2 binding site on AKAP4 encompasses the dual RIα/RIIα-binding domain, suggesting that FSIP2 might have a role in regulating PKA anchoring to AKAP4.

Two proteins (ropporin and ASP) were found to interact with the PKA-binding domain of AKAP3 (473). Both proteins have N-terminal regions similar to the AKAP-binding domain of the PKA RII subunit. Ropporin was identified previously as a rhophilin-binding protein that is testis-specific and localized to the principal piece of mouse sperm (474). It was found that the gene encoding ropporin, *Ropn1*, is first transcribed in round spermatids, but the protein is not detected until late in the postmeiotic phase. By immunoelectron microscopy, ropporin appeared to be most abundant on the inner surface of the fibrous sheath (474). Rhophilin contains a binding motif for Rho, a small GTPase that triggers reorganization of the cytoskeleton and regulates motility and cell-substrate adhesion (475). Rhophilin is expressed at high levels in the testis, is first transcribed during spermatogenesis near the end of the meiotic phase, but is not testis specific 476). It is tightly bound to the fibrous sheath and was localized to the outer surface of the fibrous sheath by immunoelectron microscopy (474). ASP (AKAP-associated sperm protein) is 39% identical to ropporin and thought to be found only in testis (473). However, the EST expression profile for the mouse gene (*Ropnl1*) suggests that it is more widely expressed.

Database searches also determined that two other proteins contain regions similar to the N-terminus of the AKAP binding region of the RII PKA subunit (473). One was SP17, a protein found mainly in the principal piece, but also present in low amounts in the middle piece and throughout the equatorial region of the acrosome of mouse sperm (477). It was shown subsequently that SP17 co-localizes with AKAP3 in the fibrous sheath of mouse sperm and binds to AKAP3 in vitro (478). The other protein with an RII-like domain was fibrousheathin II. It is expressed mainly in the testis and is known currently as CABYR (calcium-binding tyrosine phosphorylation regulated protein) (479). Multiple variants of this 86-kDa protein were identified in human testis, not all of which contained the RII-like domain or the calcium-binding EF-hand domain. Some of the variants gained calcium-binding capacity when phosphorylated during capacitation (479). It is unknown how binding of these proteins to AKAP3 might affect sperm function. However, it was suggested that they might be involved in regulating sperm motility (473), presumably by modulating RII binding to AKAPs.

Two additional structural proteins (FS39, Tsga10) were reported to be associated with the fibrous sheath (Table 4). FS39 is testis-specific in the mouse and has similarities to myosin and other filamentous proteins (480). While transcription began early in the postmeiotic phase, the protein was first detected in elongating spermatids (step 13) (480). Tsga10 contains a myosin-like domain similar to the Ezrin/radixin/moesin family domain. The gene is expressed at high levels in mouse testis, but the transcript is also present in some fetal tissues and tumor cells (481). Transcription begins in pachytene spermatocytes and the 67-kDa form of the protein was first detected in spermatids in the rat. In addition to the 67-kDa form, a 27-kDa processed form of the protein was present in purified rat fibrous sheath (481).

Several testis-specific isozymes or proteins associated with enzymatic activity also are known to be associated with the fibrous sheath (Table 4), including glycolytic enzymes (HK1-S, GAPD-S), a glutathione-S transferase (GSTM5), two thioredoxins (SPTRX-1, SPTRX-2), and a pyruvate dehydrogenase E1 subunit (PDHB).

Alternate transcripts of HK1 (hexokinase type 1) are present in spermatogenic cells and the N-terminal porin-binding domain usually present in HK1 is absent and replaced with a spermatogenic cell-specific domain in the testis (482). There are at least three alternatively spliced transcripts, but they encode the same protein. The HK1-S protein is found in the principal piece of the flagellum (483), but hexokinase activity is solubilized from mouse sperm with 0.5% Triton X-100 (484), indicating that it is not tightly bound to the fibrous sheath.

A protein present in the principal piece and in the fibrous sheath isolated from mouse sperm, initially referred to as ATC (485), was found to be a testis-specific orthologue of GAPD (glyceraldehyde 3-phosphate dehydrogenase) encoded by the *Gapds* gene (486). Transcription begins in round spermatids (487) and the 47-kDa GAPDS protein is first seen in elongating spermatids (488). The motility of sperm from mice with a targeted disruption of the *Gapds* gene was reduced to a weak, twitching motion and males are infertile (489). Although oxygen consumption by sperm from the mutant mice was comparable to that of sperm from wild-type mice, ATP levels were greatly reduced. These results strongly suggest that the energy required for sperm motility is derived primarily from ATP produced by glycolysis in the principal piece and not by oxidative phosphorylation in mitochondria (489). While two enzymes in the glycolytic pathway, GAPDS and HK1-S, have been identified in the principal piece, it will be important to demonstrate that the other enzymes that function during glycolysis are present there as well.

A member of the mu-class glutathione S-transferase family is a major 26-kDa protein in the fibrous sheath of mouse sperm (490). The *Gstm5* gene is expressed at highest levels in testis and GSTM5 protein were found in pachytene spermatocytes and spermatids, but the EST expression profile indicates that the gene is expressed at lower levels in other tissues as well. The GST proteins are scavengers of cytotoxic substances, such as electrophilic compounds and peroxides and GSTM5 might serve to protect sperm from damage by such agents.

Two members of the thioredoxin family of proteins are expressed only in spermatogenic cells and during the postmeiotic phase. Thioredoxins are able to reduce disulfide bonds in the presence of thioredoxin reductase. The 90-kDa SPTRX1 protein contains a 15-amino-acid sequence repeated 23 times in the N-terminal region and a thioredoxin motif in the C-terminal region (491). It is seen by immunoelectron microscopy to be associated primarily with the longitudinal columns of the fibrous sheath during flagellar formation in rat spermatids. However, little SPTRX1 was detected in rat epididymal sperm by this method or by Western blotting and it was suggested to be a transient component whose role is to participate in the formation of the longitudinal columns of the fibrous sheath (491). The SPTRX2 protein contains a N-terminal thioredoxin domain followed by three NDP domains. While the *Sptrx2* gene has a pattern of expression similar to *Sptrx1*, the SPTRX2 protein is incorporated into both the longitudinal columns and ribs of the fibrous sheath late in spermiogenesis and remains an integral component of the fibrous sheath in epididymal sperm (492).

The 36K-A and 36K-B proteins isolated from hamster sperm flagella are phosphorylated at serine residues in a cAMP-dependent manner in association with activation of motility. The 36K-A protein was localized to the principal piece and present in extracts of the fibrous sheath, but not in extracts of outer dense fibers (493). Using a proteomics approach, eight peptide fragments were found to match regions covering 32% of the sequence of pyruvate dehydrogenase E1 component β (PDHB) (493). The PDH complex usually is considered to be a component of the mitochondrial matrix involved in aerobic energy production and it is not obvious what role PDHB would have in the fibrous sheath. However, the cDNA needs to be cloned and sequenced to provide definitive evidence that 36K-A is pyruvate dehydrogenase E1 component β and not a homologous protein.

Function

The classical view has been that the function of the fibrous sheath is to modulate the plane of the flagellar beat by imposing a restraint to sliding of axonemal doublets and to flagellar bending (463). The attachment of axonemal doublets 3 and 8 to the longitudinal columns of the fibrous sheath would seem to restrict their participation in microtubule sliding and axoneme binding during flagellar motion. The longitudinal columns themselves might also limit bending of the flagellum in this same plane, whereas these features would not restrict flagellar bending perpendicular to this plane (6). Although sperm appear to swim with a planar effective stroke and recovery stroke somewhat like cilia, they also rotate (344), and the propagated waves have a three-dimensional component in the distal portion of the tail when viewed in a microscope (6). However, it is not known how the fibrous sheath might influence flagellar movement under the significantly different conditions present in the female reproductive tract.

It has become apparent that the fibrous sheath also serves as a scaffold for components involved in signal transduction, energy production, and other functions (463). The fibrous sheath serves as a docking site for signal transduction components that are candidates for regulating key process of sperm function leading up to fertilization. The cAMP-signaling pathway participates in the regulation of processes associated with sperm maturation, motility, capacitation, hyperactivation and the acrosome reaction (494–496). PKA is the predominant target for cAMP, synthesized from ATP by adenylate cyclase in response to external stimuli, and PKA in turn phosphorylates proteins on serine and threonine residues. The downstream targets for PKA in this cascade are tyrosine kinases, yet to be identified, that modify proteins in the flagellum associated with these processes (496). The presence of AKAP4, AKAP3, and TAKAP-80 in the fibrous sheath strongly suggests that one of its main roles is to anchor the PKA involved in such processes.

Components of the Rho-GTPase signaling pathway also are associated with the fibrous sheath. Ropporin is a spermatogenic cell-specific protein that binds to rhophilin, both of which are associated with the fibrous sheath (474–476). Rhophilin binds to Rho, a small GTPase that acts through effector molecules to modulate reorganization of cytoskeletal components (475). Ropporin binds to the RII-anchoring site on AKAP3 and contains an N-terminal domain similar to the AKAP-binding region of the RII subunit of PKA (473). Other proteins associated with the fibrous sheath that also have similar domains are ASP, P17, and CABYR (473). This led to the suggestion that AKAPs serve as scaffolding proteins for the Rho-GTPase pathway and that Rho activates kinases and phosphatases involved in regulating sperm motility (473). Although Rho GTPases regulate important cellular processes associated with cell movement and adhesions, their roles in spermatogenesis and sperm function remain largely unexplored (497).

An additional role for the fibrous sheath is to anchor enzymes of the glycolytic pathway. Although glycolysis is a highly conserved process, several of the glycolytic enzymes in sperm have biochemical or enzymatic properties that differ from those of glycolytic enzymes in somatic cells (498). Only a few of the glycolytic enzymes in sperm have been examined closely to determine the basis for these differences. These include two glycolytic enzymes in sperm, PGK2 (499) and GAPDS (486), encoded by novel genes, and HK-1S, a novel isoform encoded by an alternate transcript (483). Sperm require glucose for in vitro fertilization and are not able to use pyruvate or lactate as energy substrates (500,501). In addition, blocking mitochondrial oxidative phosphorylation with oligomycin does not inhibit fertilization (500). Such results suggest that sperm do not depend on mitochondria for their energy. This possibility was strongly supported by studies showing that targeted disruption of the *Gapds* gene drastically reduced ATP levels and sperm motility, leading to male infertility (489).

Abnormal Flagella

Spermatozoa with abnormal flagella are present in low numbers in many mammals, but are relatively common in humans. For example, approximately 14% of sperm from a group of 22 fertile men had abnormal tails (502). In addition, asthenozoospermia, or a preponderance of sperm with low motility, is found frequently in infertile men. A study of 1,085 sperm samples from infertile men found that 81% had defects in motility, while 19% had asthenozoospermia without other defects in sperm number or morphology (503). Another study found that alterations in flagellar structure are responsible for the motility defects in 70% of infertile men (504).

Two categories of flagellar defects have been defined (505). One category includes nonspecific or nonsystematic defects, seen as heterogeneous combinations of alterations affecting variable percentages of sperm among different individuals. These are usually acquired secondary to pathological conditions that affect the testis or duct system. The second category includes systematic defects, often seen as a particular sperm phenotype that predominates in a sample from an individual and resembles the phenotype of sperm from other individuals. These are likely to have a genetic origin and tend to show family clustering (505).

Heritable Effects on Human Sperm Structure

One hereditary condition causing sperm flagellar defects is referred to as immotile-cilia syndrome (ICS) (506), Kartagener's syndrome, or primary ciliary dyskinesia (PCD) (507). It often results in male sterility, situs inversus, and chronic respiratory problems (506). These conditions are directly or indirectly a consequence of an autosomal recessive trait that makes cilia or flagella unable to perform normal and coordinated movements (508). Sperm from men with the syndrome have defective axonemes with such alterations as a partial or complete lack of dynein arms, a lack of outer dynein arms, a lack of inner arms, or abnormally short spokes and a lack of the central sheath (508). Because of the different types of defects seen in sperm and cilia, mutations in multiple genes might contribute to the syndrome (506). Over 200 axonemal proteins are present in human cilia (509) and many of these probably are present in the sperm flagellum. However, the relatively low incidence of ICS suggests that a smaller number of genes are involved (510). Two strategies being applied to identify such genes include screening for mutation in candidate genes selected on the basis of their expected function in axonemes, and genetic linkage analysis in families with ICS to map chromosomal regions associated with the disease (510). Such studies have led to the identification of two genes for components of the outer dynein arms that are associated with ICS and the elimination of other candidate genes (510).

A different type of flagellar defect probably of genetic origin is known as dysplasia of the fibrous sheath (DFS) (511). This syndrome is characterized by severe asthenozoospermia and by the sperm having hyperplastic and disorganized fibrous sheaths, and short, rigid, thick, and immotile tails (505,512). Other associated changes seen include distortion or severe disruption of the axoneme, extension of outer dense fibers 3 and 8 into the principal piece, and failure of the annulus to migrate caudally. The occurrence of this syndrome in brothers suggests that DFS can be caused by recessive mutations in autosomal genes involved directly or indirectly in the assembly and organization of the fibrous sheath (512). If DFS is the only symptom present, the mutations might occur in spermatogenic cell-specific genes expressed in spermatids. However, about 20% of DFS patients have respiratory symptoms from early childhood, suggesting either that their sperm defects might be due to ISC (504), that mutations in some genes might cause both ISC and DFS, or that these individuals have mutations in multiple genes associated with these syndromes.

An additional type of flagellar defect that occurs in sperm from infertile men is called flagellar dyskinesia (394). It has been observed in brothers, suggesting that it can have a genetic cause (513). The sperm have normal axonemes, but altered placement of periaxonemal structures. These include abnormal extension of individual dense fibers along the axoneme, altered order of termination of these structures, and a modified number and location of longitudinal columns of the fibrous sheath (394). The outer dense fibers have abnormal positions with respect to each other, being placed symmetrically with respect to a plane that passes through microtubule doublet 1 and between doublets 5 and 6. The most common defect of the longitudinal columns of the fibrous sheath is the presence of only one column adjacent to doublets 3 or 8 (394). One suggestion was this might be due to a defect in the A microtubules of the outer doublets that affects the adjacent formation or assembly of the outer dense fibers and longitudinal columns of the fibrous sheath (503).

Heritable Effects on Mouse Sperm Structure

Several spontaneous mutations affecting formation of the flagellum have been identified in mice. The mutations usually are pleiotropic, causing defects in other tissues in addition to the seminiferous epithelium (234,514). This is because they were identified by other effects of the mutation and subsequently found to cause defects in sperm. Homozygous mice with the Wobbler mutation apparently produce normal numbers of sperm, but few sperm are motile (515). The sperm appear to have average length tails and normal heads, but 70% of the sperm in the vas deferens have ultrastructural defects in the flagellum. These defects include the absence of one to four outer doublets and the corresponding dense fibers, most commonly those from positions 4 through 7. Other defects seen less frequently are supernumerary microtubules or the absence of central pair tubules in the axoneme (514). Defective sperm are common in T/t mice (516), with the most frequent flagellar defects seen in sterile mice being a lack of four outer dense fibers and doublets (517). The male-sterile mutation, hydrocephalic-polydactyl (*hpy*), results in the absence of a flagellum or in partially assembled axonemal structures and/or poorly organized aggregates of other tail components (518). The axoneme usually is absent or abnormally formed and the outer dense fibers and fibrous sheath are morphologically atypical, when present. The axoneme dysgenesis might be due to failure to form stabile structures because of defective subunits (518).

A defect often seen in an abnormal or degenerating flagellum is the absence of microtubule doublets 4 through 7 and the associated dense fibers (515,517,519). Adding ATP to detergent-extracted rat sperm tails produces the same alterations. Doublets 8 through 3 and the associated dense fibers appear to remain firmly attached to the fibrous sheath, while doublets 4 through 7 are extruded from the sperm tail (356). This suggests that sperm seen with this defect may be degenerating. In some situations, the flagellar axoneme appears to form normally and then to become progressively disorganized, as occurs during spermiogenesis in quaking (520) and in *hpy/hpy* mice (518). However, in Wobbler mice, the percentage of abnormal sperm increases as they transit the male tract. Only about 5% of sperm tails are abnormal in the testis, while nearly 70% of sperm from the vas deferens have abnormal flagella (515).

Gene Knockouts Affecting Mouse Sperm Structure and Function

The use of targeted gene disruption to produce gene knockout mice has led to the identification of genes expressed only in the testis that encode proteins required for sperm motility and fertility. The effects of these knockouts on the assembly, structural characteristics, or functional properties of the flagellum vary considerably, probably due to the complex structural and functional features of the flagellum. While some gene knockouts have identified the roles of specific molecules in the structure and function of sperm, others have been less informative.

Gene knockouts can cause obvious structural changes in the fibrous sheath or axoneme that disrupt sperm motility. Examples include the *Akap4* and *Tekt2* genes. AKAP4 is a major structural protein of the fibrous sheath and sperm from *Akap4* knockout mice had a shortened flagellum, an underdeveloped fibrous sheath, and a lack of forward motility (469). Tektin-t is a component of the axoneme and a knockout of the *Tekt2* gene resulted in disruption of the microtubule doublet dynein inner arms (378).

However, some knockouts of genes encoding proteins in the fibrous sheath, axoneme or middle piece have disrupted sperm motility without causing noticeable changes in the structure of the flagellum. Examples include the *Gapds*, *Dnahc1*, and *Smcp* genes. GAPDS is located in the fibrous sheath and disruption of the *Gapds* gene caused a dramatic reduction in sperm motility without causing obvious changes in the structure of the fibrous sheath (489). Dynein heavy chain 7 is a component of the axonemal doublet inner arms. A striking reduction in motility occurred in *Dnahc1* gene knockout mice without producing noticeable changes in the ultrastructure of the dynein arms (521). The mitochondrion-associated cysteine-rich protein is located in the middle piece. Sperm motility was reduced in *Smcp* gene knockout mice on a 129/Sv genetic background. No structural changes were apparent, and there was no effect of sperm motility in mice on a mixed genetic background (C57BL/6 × 129/Sv) (392).

In addition, knockouts can produce altered sperm motility, but also result in other sperm defects and developmental anomalies. Examples include the *Ube2* and *Spage5* genes. The knockout of the *Ube2b* gene encoding a component of the ubiquitin pathway caused the longitudinal columns of the fibrous sheath to be associated with the wrong axonemal doublets and some outer dense fibers to be abnormally long, but it also caused head shape anomalies (522). The knockout of the gene for sperm-associated antigen 6 (*Spag6*), an orthologue of a gene in *Chlamydomonas* for an axonemal protein, resulted in marked motility defects, but also produced hydrocephalus (523).

Finally, knockouts of genes for flagellar components are not always informative. An example was the knockout of the *Spag5* gene. SPAG5 binds to ODF1, a major component of outer dense fibers, but male *Spag5* gene knockout mice were fertile and it was suggested that another protein might compensate for the absence of SPAG5 (437).

These results emphasize that a substantial number and variety of genes encode components essential for spermatogenesis (524) or the integrity of the structure and function of the flagellum. While some are expressed only in the testis, others are also expressed at additional sites in the body. This suggests that a mutation in any one of a number of genes might lead to defects in the flagellum that result in asthenozoospermia and other sperm defects.

Flagellar Motion

The Flagellar Wave

The flagellar wave is propagated in a plane perpendicular to the central pair of microtubules of the axoneme. It passes through doublet 1 and between doublets 5 and 6, with the active stroke being toward doublet 1 (525). The flagellum twists during bend propagation (526,527), with the plane of bending moving toward doublet 2 (526). The part of the axoneme containing doublets 1 through 5 is thought to be active during this phase of flagellar motion (526). The outer dense fibers and fibrous sheath appear to play a passive role in flagellar motion (528–531). The tapering of the dense fibers is believed to decrease local resistance to bending (394,532) and might explain the progressive increase in the amplitude of curvature observed during flagellar wave propagation (530,532, 533). The small flagellar amplitude observed in certain abnormal human sperm (534) may be due to constraints imposed by the abnormal arrangement of dense fibers in those sperm (394).

Sperm from the caput epididymidis in most mammals have a vibratory or slow and ineffective beat that often results in circular swimming patterns (131). In contrast, sperm from the cauda epididymidis usually move with a vigorous motion that results in rapid forward movement, referred to as progressive motility. In correlation with this change in swimming pattern, the flagellum appears to become more rigid and to beat with a reduced arc of curvature (535). This may be due to the increased disulfide bond formation that occurs in outer dense fibers and the fibrous sheath during epididymal maturation (190). However, cauda sperm from different species show different patterns of motility, flagellar beat, and flagellar rigidity (344).

Regulation of Sperm Motility

Sperm undergo two types of motility, progressive motility and hyperactivated motility. They acquire the capacity for progressive motility during epididymal maturation, but do not become motile until released from the epididymis. This type of motility is characterized by vigorous and relatively symmetrical flagellar motion and results in rapid forward movement. After a period of time in culture or in the female reproductive tract, sperm display hyperactivated motility, characterized by high amplitude whiplike beating of the flagellum, asymmetrical flagellar bends, and a circular or erratic swimming trajectory (281,536). Hyperactivation occurs during the process of capacitation, but it has not been determined definitively whether hyperactivation is a consequence of capacitation or whether hyperactivation and capacitation are separate processes that occur in parallel. While hyperactivation is a readily observable change, capacitation is a process during which sperm gain the ability to undergo the acrosome reaction and is observable after it has occurred. Several factors involved in triggering or maintaining sperm motility are also involved later in capacitation and the acrosome reaction. This suggests that different mechanisms regulate how and when the factors participate in specific processes.

Activation of Motility. The typical pathway for stimulus-induced activation of a cellular process occurs when an extrinsic stimulus triggers a receptor to initiate a signal transduction cascade. It involves changes in the conformation, phosphorylation, and/or localization of proteins in the pathway and in the cellular process being activated. Key factors involved in the initiation of progressive motility and induction of hyperactivated motility in sperm are known. These include calcium ions (Ca^{2+}), bicarbonate (HCO_3^-), and cyclic adenosine monophosphate (cAMP). However, it is likely that other factors involved in triggering and regulating the coordinated flagellar activity responsible for sperm motility remain to be identified.

One possible mechanism that has been suggested for the initiation of motility is that it occurs as a result of release from the influence of an inhibitor of motility. Sperm motility quiescence factors were reported to be present in bovine cauda epididymal fluid (537) and in the rat epididymis (538). In addition, immobilin, a high-molecular weight glycoprotein present in epididymal fluid, was suggested to prevent motility in the epididymis of some species by producing viscoelastic drag (538,539). Another motility inhibitor was reported to be present in seminal plasma, being highest in fluid from seminal vesicles, but also present in prostatic fluid of bull, rat, and rabbit (540). This 15-kDa protein inhibited the reactivation of motility in detergent-extracted sperm and reduced the motility of previously reactivated sperm. An activator might in turn regulate this motility inhibitor (540). In addition, a forward motility factor was reported to be present in bovine epididymal fluid (541). However, these factors remain to be identified and characterized at the molecular level.

There is limited information about receptors on the sperm surface that might be involved in triggering progressive or hyperactivated motility. Possible candidates include olfactory receptors and a GABA receptor. One olfactory receptor (MORE23) that has been studied is expressed in the olfactory epithelium and testis (542). Exposure of mouse sperm to a cognate ligand for this receptor caused an increase in intracellular Ca^{2+} and changes in flagellar beating (542). Another olfactory receptor on human sperm (hOR17-4) has been implicated in chemotactic responses associated with fertilization. Exposure of human sperm to a cognate ligand for this receptor, the floral scent bourgeonal, evokes changes in intracellular Ca^{2+} levels and swimming behavior (543). In addition, a GABA receptor also has been implicated as a possible target for extrinsic signals. Most studies have focused on the role of this receptor in the acrosome reaction, but exposure of human sperm to GABA was reported to cause changes in sperm motility parameters and hyperactivation (544).

Calcium and Motility. Calcium has important roles in multiple aspects of sperm function, including motility. Recent studies with knockout mice have demonstrated that at least four components that participate in the regulation of intracellular Ca^{2+} levels (CatSper1, CatSper2, Ca_v2.3, and PMCA4) are involved in the initiation of sperm motility. CatSper1 is a testis-specific voltage-gated Ca^{2+} entry channel localized to the sperm principal piece. One laboratory reported that male *Catsper1* gene knockout mice were infertile and their sperm had reduced progressive motility (545). Another research group also reported that male mice lacking CatSper1 were infertile.

However, they found that sperm lacking CatSper1 demonstrated progressive motility with an increased flagellar beat frequency, but failed to develop hyperactivated motility (51). These differences in findings were suggested to be due to differences in the genetic backgrounds of the mice (51). CatSper2 is another voltage-gated ion channel, is closely related to CatSper1, and is present in the flagellum. Sperm from *Catsper2* gene knockout mice showed a slight decrease in flagellar amplitude during progressive motility, but like sperm lacking CatSper1 they failed to develop hyperactivated motility (52).

Disruption of the gene for the $a_1$2.3 subunit of the voltage-dependent Ca^{2+} channel (VDCC) resulted in significant increases in straight-line velocity and linearity in sperm from mutant mice compared to wild-type mice (53), indicating this calcium channel also has a role in regulating sperm motility. In addition, disruption of the gene for PMCA4, a Ca^{2+}/calmodulin-dependent ATPase involved in Ca^{2+} efflux, also resulted in male infertility. One research group reported there were no differences in motility of sperm with or without PMCA4 when incubated in a medium that did not induce capacitation, but that sperm without PMCA4 became immotile when incubated in capacitation medium (54). However, another group reported that the majority of sperm lacking PMCA4 were immotile (55). In addition, the latter group indicated that the motile sperm present had decreased average velocity, progressive velocity, and track speed, and intracellular calcium levels over two times higher than normal sperm (55).

Other regulators of ion movements across the plasma membrane also have been shown to have important roles in sperm motility. Disruption of the gene for a voltage-dependent anion channel, VDAC3, resulted in markedly reduced sperm motility, but also frequently caused alterations in the structure of the axoneme (445). In addition, a novel sperm-specific sodium-hydrogen exchanger (NHE) involved in regulating intracellular pH was found to be located in the principal piece. Disruption of its gene (*Slc9a10*) resulted in male infertility and markedly reduced sperm motility (546). Furthermore, a unique a4 subunit isoform of the Na, K-ATPase found only in testis and located in the midpiece, also is required for sperm motility. This ATPase is responsible for translocating Na^+ out and K^+ into the cell. Specific inhibition of the a4 isoform with the cardiac glycoside ouabain substantially reduced the percentage of motile sperm over time (547). There also is evidence that an inositol 1,4,5-triphosphase receptor-gated release of Ca^{2+} from an intracellular store has an important role in regulating hyperactivated motility (548).

cAMP and Motility. Cyclic adenosine monophosphate is a key second messenger in the regulation of sperm motility. An increase in cAMP levels occurs when adenylyl cyclase is stimulated to convert ATP to cAMP. A major result of this in sperm is the activation of cAMP-dependent kinase A (PKA), which phosphorylates serine and threonine residues on neighboring proteins to trigger a cascade of protein phosphorylation events. This occurs primarily in the flagellum and results predominantly in phosphorylation of proteins on tyrosine residues [e.g., (549,550)].

The adenylyl cyclases present in most cells are plasma membrane-bound (mAC) and regulated by G proteins in response to external stimuli. There is evidence for G-protein activated mACs in mouse sperm that are involved in the acrosome reaction (551) and in human sperm that are involved in chemotaxis and hyperactivation (543). However, sperm appear to depend primarily on a more recently discovered soluble adenylyl cyclase (sAC) that lacks a putative transmembrane domain and is not G-protein regulated (552). It was found to be most abundant in testis by Northern and in situ analysis (553), present in lower amounts in kidney and choroid plexus (554), and detectable in several other tissues by RT-PCR (553). The mRNA was first observed in the mouse testis in pachytene spermatocytes and was present in high amounts in round spermatids (553). The protein was detected in the rat testis by immunohistochemistry in late pachytene spermatocytes and spermatids (554), and in sperm by Western blotting (554,555). Two alternatively spliced transcripts are expressed in rat testis, which encode a 187-kDa and a C-terminal truncated 52-kDa-form (552,554–556). The 187-kDa-form is present predominantly in soluble fractions from rat testis, but tightly bound to a 4M urea-insoluble fraction in sperm, possibly cytoskeletal components of the flagellum (554).

It was known that HCO_3^- and Ca^{2+} are involved in the cAMP-regulated activation of sperm motility (559), but the mechanisms were not understood. However, recent studies using sAC isolated from rat sperm or recombinant rat and human sAC have shown that this occurs by HCO_3^- acting directly on sAC (555,557,559). HCO_3^- increased sAC activity in vitro in two ways, by increasing enzyme velocity and by relieving substrate inhibition that occurs at high ATP-Mg^{2+} concentrations (559). Since the HCO_3^-concentration in the epididymis is low, it was suggested that sAC activity would be limited by substrate inhibition in sperm stored in the epididymis (559). These studies also demonstrated that Ca^{2+} acts directly on sAC and does so synergistically with HCO_3^- (557,559). Although it was thought that calmodulin and cytoplasmic pH were involved in the regulation of cAMP production in sperm, these studies suggested that neither had a significant role in regulating the process (557,559).

The effect of HCO_3^- on sperm motility is quite rapid. Stop-motion and waveform analysis indicated that upon addition of HCO_3^- the flagellar beat frequency of mouse sperm began increasing within 5 seconds and was near maximal by 30 seconds (560). This treatment also facilitated opening of voltage-gated Ca^{2+} channels, but at a slower rate than the increase in beat frequency. The addition of PKA inhibitors prevented the increase in flagellar beat frequency and Ca^{2+} channel responses to HCO_3^-, strongly suggesting that HCO_3^- influences these processes by raising cAMP levels and stimulating PKA-mediated protein phosphorylation (560).

The most convincing evidence that sAC has a critical role in sperm motility came from the characterization of mice with a knockout of the gene encoding sAC. The mice showed no overt abnormalities and the testis appeared to be normal, but male mice were infertile and their sperm showed only a trembling movement in the midpiece and no forward motility (561). However, when these sperm were loaded with cAMP they underwent a time-dependent recovery of both overall motility and vigorous forward motility (561).

PKA and Motility. PKA phosphorylates proteins on serine and threonine residues which in turn activate a signaling cascade leading to tyrosine phosphorylation of flagellar proteins. The presence of proteins in the fibrous sheath with PKA anchoring sites (AKAP3, AKAP4, and TAKAP-80) strongly suggests that one of the major roles of this structure is to anchor PKA in the principal piece region of the flagellum. The defining characteristic of an AKAP is the presence of an amphipathic helix region that binds PKA (562), usually in a subcellular location that places it close to its target proteins (563). The PKA holoenzyme consists of two catalytic (C) subunits and two regulatory (R) subunits. The genes have been identified for four R subunits (RIα, RIβ, RIIα, and RIIβ) in human and mouse, and three C subunits (Cα, Cβ, and Cγ) in human and two C subunits (Cα, Cβ) in mouse. There are at least three isoforms for Cβ and two isoforms for Cα in mouse, one of which (Cα2) lacks the N-terminal myristoylation signal and is present only in the testis (564). The R subunits dimerize by their N-terminal domains and bind to AKAPs by this region. The R and C subunits have a cAMP-binding site and when cAMP binds to the R subunits, the C subunits are released and their catalytic activity is triggered by cAMP.

The distribution of the different R and C subunits in sperm is consistent with PKA being involved in the regulation of motility. RI subunits are usually in the detergent soluble fraction and RII subunits in the detergent resistant fraction of cells. The majority of RII was present in the detergent-resistant fraction of rat sperm and tightly attached to the flagellum (565). Using immunohistochemistry, RIIα was seen to be present mainly in the axonemal region of the flagellum of bovine sperm (566) and in the middle piece and distal portion of the principal piece of the flagellum of mouse sperm (657), while RI was seen in association with outer dense fibers and the fibrous sheath of boar sperm (568).

While treatment of bovine sperm with a membrane permeable peptide that inhibits PKA anchoring caused an arrest in motility, other inhibitors of PKA catalytic activity had minimal effect on motility. This was suggested to indicate that RII interaction with AKAPs, and not PKA catalytic activity is essential for motility (566). However, in RIIα knockout mice the majority of the PKA was not anchored in the flagellum and sperm motility and fertility were not changed, indicating that the peptide inhibitor might have additional nonspecific effects (567). While RI appears to be present in low amounts in the flagellum (567–569), it binds to a RIα-specific domain and a dual RIα/RIIα domain of AKAP4 in vitro (570), suggesting that it might be present in significant amounts in specific locations within the flagellum. However, the RIα knockout is an embryonic lethal (571) and the role of RIα in sperm motility has not been determined.

Immunohistochemistry showed that Cα2 is present in the mouse testis in pachytene spermatocytes and spermatids, and in the midpiece and axoneme of ovine sperm and mouse sperm, while Western blotting determined that it is in the cytosol of spermatids (572). A knockout of the gene for Cα lead to early postnatal death for the majority of mice, but a small number of runted mice made it to adulthood (573). Although spermatogenesis in these mice appeared to progress normally, PKA activity was less than 10% of normal in testis. RIα and RIIα were not detected in sperm, more than a quarter of the sperm showed morphological abnormalities, and only a small percentage showed progressive motility (573). However, it was difficult to rule out that these effects were secondary to other changes. This was clarified when a knockout of the spermatogenesis-specific Cα2 isoform was performed. Male mice lacking Cα2 produced normal numbers of sperm with progressive motility, but were infertile. Although Cα was not required for progressive motility, the sperm failed to hyperactivate or to accumulate tyrosine-phosphorylated proteins (574). The level of RIα in sperm was reduced considerably, while the level of RIIα was reduced somewhat. The RIIα remaining was most abundant in the principal piece of the flagellum, suggesting that it anchored to AKAPs in that region in the absence of C subunits. In spite of these defects, the sperm were able to fertilize zona-free eggs in vitro (574).

Phosphorylation and Motility. The steady-state phosphorylation status of a protein is determined by the relative activity of the protein kinases and phosphatases acting on that protein. Increases in sperm protein phosphorylation have been implicated in the initiation of progressive motility, hyperactivated motility, capacitation, the acrosome reaction, and fertilization. Most sperm proteins which become phosphorylated between release from the epididymis and fertilization are in the flagellum and potentially involved in motility. In an individual spermatozoon some of these functional changes probably occur sequentially and others occur concurrently. However, they occur asynchronously within a population of spermatozoa and most phosphorylation assays give an average status for a population, confounding efforts to relate a change in phosphorylation of a specific protein to a change in motility or other sperm functions.

PKA phosphorylates proteins on serine and threonine residues, which in turn activates a signaling cascade leading to robust tyrosine phosphorylation of flagellar proteins. Most of what is currently known about changes in phosphorylation in sperm comes from studies on tyrosine phosphorylation. Studies defining the correlation between capacitation and pattern of changes in tyrosine phosphorylation (575) were an important advance and provided a valuable approach for monitoring capacitation. This has greatly facilitated studies leading to a better understanding of the roles of BSA, Ca^{2+}, HCO_3^- and other factors involved in the regulation of protein phosphorylation during capacitation (576,577).

A number of changes need to occur in the sperm before the tyrosine phosphorylation begins and the time course is relatively slow (576). Newly phosphorylated proteins are first seen in mouse sperm after 30 to 45 minutes in capacitation medium and the maximum level of phosphorylation is reached at about 90 minutes (575). While some of the increases in phosphorylation might be involved in regulating hyperactivated motility, initiation of progressive motility occurs within a few seconds of exposure to medium containing HCO_3^- (560). This suggests that rapid PKA-mediated serine/threonine phosphorylation events are involved in initiation of progressive motility.

A few sperm proteins have been identified that are serine/threonine phosphorylated in response to HCO_3^-. These include 96-, 64-, and 59-kDa bands recognizable on Western blots of boar sperm extracts with an antibody against the consensus target sequence for PKA (578). Proteomics approaches indicated that the 96- and 59-kDa bands were ODF2. It was suggested that PKA phosphorylation of target sites within a leucine zipper domain might result in changes in protein–protein interactions to alter mechanical properties of outer dense fibers (578). In addition, AKAP4, AKAP3, and valosin-containing protein (VSP), the homologue of a SNARE-interacting protein, were found to be serine and tyrosine phosphorylated in human sperm (579); but the sperm were capacitated overnight and the relationship of these changes to motility is unknown. Another sperm protein that is serine phosphorylated is glycogen synthase kinase-3 (GSK-3). Stimulation of bovine sperm motility with a general phosphodiesterase inhibitor (IBMX) or a serine/threonine-specific phosphatase inhibitor (calyculin A) resulted in a dramatic increase in GSK-3 serine phosphorylation and decrease in its catalytic activity (580). In addition, the pyruvate dehydrogenase E1 component β (PDHB) associated with the fibrous sheath in hamster sperm is serine phosphorylated in a cAMP-dependent manner in association with the initiation of motility (493). However, it is not obvious how serine/threonine phosphorylation of these proteins affects motility.

Several of the proteins that undergo changes in tyrosine phosphorylation during capacitation have been identified. They include the fibrous sheath proteins AKAP3 (470,471,579,581,582), AKAP4 (464,471,579), and CABYR (479). Others reported are valosin-containing protein (VCP) (579,583) and a homologue of the SNARE-interacting protein NSF (579). Fifteen additional human sperm proteins that undergo tyrosine phosphorylation were identified by micro-sequencing (579). These included proteins that are axonemal (α-tubulin), and fibrous sheath or outer dense fiber components (ODF1, VDAC2, GSTM3, PDHB) (Tables 3 and 4). Some of these also were identified (previous paragraph) as serine and threonine phosphorylated proteins.

Kinases responsible for tyrosine phosphorylation during capacitation must be present in sperm that are either direct targets of PKA or in the downstream signaling cascade. A few tyrosine kinases have been identified in sperm. A 42-kDa tyrosine kinase (sp42) was isolated and purified from ejaculated boar sperm (584). Immunocytochemistry indicated that it is male germ cell-specific and localized by immunocytochemistry to the boar sperm midpiece, but it has not been further characterized. Another tyrosine kinase isolated from boar sperm (TK-32) was shown by an in-gel tyrosine kinase activity assay to require calcium and bicarbonate in the incubation medium (585). It will be interesting to learn more about the function and molecular identity of these tyrosine kinases. An additional tyrosine kinase identified is present in the connecting piece and principal piece of the boar sperm flagellum (586). This 72-kDa protein is not extracted by NP-40 and appears to be bound to cytoskeletal components of the sperm. It is recognized by an anti-phospho-Sky antibody and is likely to be Sky, a *Src* family protein tyrosine kinase. It was shown to respond to cAMP-PKA signaling by tyrosine phosphorylating proteins mainly in the connecting and middle pieces (587). In addition, c-*yes*, another member of the *Src* family has been identified in human sperm (588). However, it was shown by immunohistochemistry to be present in the sperm head and unlikely to be involved directly in regulation of motility.

While the primary signaling pathway for protein phosphorylation passes through PKA, other routes for the initiation of this process are present. Stimulation of sperm motility occurred when human sperm were treated with an inhibitor (LY294002) of phosphatidylinositol 3-kinase (PI3-kinase). This resulted in an increase in cAMP production and tyrosine phosphorylation of AKAP3, leading to the suggestion that PI3-kinase is a negative regulator of sperm motility (581). Treating sperm with the inhibitor also resulted in stimulation of binding of PKA to AKAP3 through RIIβ, apparently due to tyrosine phosphorylation of AKAP3. This was interpreted to indicate that PI3-kinase negatively regulates sperm motility by interfering with binding of PKA to AKAP3 (582). PI3-kinase also might be an upstream regulator of glycogen synthase kinase-3 (GSK-3), also a negative regulator of sperm motility. In somatic cells, PI3-kinase-generated phosphoinositol lipids activate PDK1, causing it to phosphorylate Akt, which in turn phosphorylates GSK [e.g., (589,590)]. Treating bovine sperm with protein phosphatase inhibitors (calyculin A, IBMX) resulted in phosphorylation of serine 21 of GSK-3α, reduced GSK-3 kinase activity, and initiation and stimulation bovine sperm motility (591). Thus, GSK-3 also acts as a negative regulator of motility. GSK-3 is located in the principal piece region (and postacrosomal region) of bovine sperm where it could participate in the regulation of motility.

In addition, the proline-directed serine/threonine kinase Cdk5 and its activator p35 are present in the outer dense fibers and in the basal plate, capitulum, and striated collar of the connecting piece in mouse sperm (335,448). The substrates for Cdk5 in the brain include actin, microtubule-associated proteins and other cytoskeletal proteins, suggesting that Cdk5 could have a significant role in modifying axonemal proteins in association with initiation of progressive or hyperactivated motility.

The inhibition of phosphatase activity results in the stimulation of motility, strongly suggesting that phosphatase have an important role in regulating this and other processes in sperm. While phosphatases are important components of signaling and regulatory pathways in other cell types, little is known about phosphatases in sperm. A sperm phosphatase isoform that has received some attention is PP1γ2. It is a proline-directed serine/threonine phosphatase and thus could serve to buffer the serine/threonine kinase activity of PKA. It is a testis-specific isoform and the only PP1 present in spermatozoa of several mammalian species (591). There is an inverse relationship between PP1γ2 activity and the motility of sperm from different regions of the epididymis, suggesting that PP1γ2 limits motility of sperm in the epididymis. PP1γ2 is regulated by a homologue of the yeast protein phosphatase binding protein sds22, and PP1γ2 complexed with sds22 is catalytically inactive (592). A phosphorylated form of PP1γ2 is present predominantly in the sperm head and may be involved in signaling events associated with fertilization (593). A knockout of the PP1γ gene was done, but the mice could not be used to gain useful information about the role of PP1 in sperm because spermatogenesis was disrupted severely and only a few motile sperm were present in the epididymis (594).

SUMMARY

A rich knowledge of the structural and functional features of spermatozoa was gained in the past in classical studies using light and electron microscopy. This information is the foundation on which current studies are building an understanding of the molecular composition and functional mechanisms responsible for the unique properties of spermatozoa. The use of newer molecular tools during the last decade has led to substantial advances in identifying the molecular components responsible for the novel structural and functional features of spermatozoa. Although the earlier studies provide information valuable for interpreting the results of current studies, they often are no longer readily accessible and are overlooked. This chapter included information gained both in earlier and in current studies, related the two bodies of information, and identified some of the major advances that have occurred and some of the questions that remain.

The role of the spermatozoon is of course to deliver the male haploid genome to the egg so that it can combine with the female haploid genome to initiate development of the next generation. The novel structural and functional features of spermatozoa are exquisitely well designed for this purpose. The major features of spermatozoa were divided here into three arbitrary categories: the plasma membrane, the head, and the flagellum. They can be considered analogous in some ways to a space ship: (a) the plasma membrane forms a barrier to a hostile environment, (b) the head contains the payload being delivered, (c) and the flagellum provides the propulsion required for delivery of the payload.

The most significant advances in knowledge about spermatozoa in the last decade have come from the use of the tools of molecular biology and proteomics to identify the genes and proteins responsible for its composition, and of gene targeting methods for determining the role of specific proteins in sperm function. Some of these advances are: (a) the identification of calcium channels that have an essential role in sperm motility; (b) the identification of a novel bicarbonate activated adenylyl cyclase; (c) the finding that capacitation is accompanied by the tyrosine phosphorylation of a characteristic subset of flagellar proteins; (d) the identification of genes encoding proteins that contribute to the structure and function of cytoskeletal elements in the head and flagellum; (e) the identification of heritable effects on the structure and function of the flagella of human spermatozoa and the genes responsible for similar effects on mouse spermatozoa; (f) and the demonstration that many of the proteins that comprise the features specific to spermatozoa are the products of novel genes or alternative transcripts that are expressed during the postmeiotic phase of spermatogenesis.

There also are large holes in our knowledge of how different molecular components are assembled into the complex structures of spermatozoa, the mechanisms responsible for sperm function, and how these mechanisms are regulated. Some of the questions remaining include: (a) How are sperm surface domains established and maintained? (b) What mechanisms are involved in delivery of surface components to specific domains? (c) What is the role of surface domain and lipid raft components in plasma membrane function? (d) What are the mechanisms responsible for producing and maintaining the species-specific sperm head shapes? (e) What are the roles of the sperm head cytoskeletal components? (f) Are the acrosomal enzymes identical to lysosomal enzymes, or are they products of novel genes or alternative transcripts? (g) How is assembly of the complex cytoskeletal structures of the flagellum regulated? (h) Are the components of the outer dense fibers and fibrous sheath found only in mammals or are there homologues in nonmammalian vertebrates? (i) What mechanisms regulate the initiation of progressive motility and hyperactivated motility? (j) What are the targets of PKA and what are the components of the downstream signal transduction pathways that lead to initiation of motility, hyperactivation, capacitation, and the acrosome reaction?

Although there have been considerable advances in knowledge, clearly there is a great deal more to be learned that is important to human male reproductive health. More knowledge is required to understand the long-term effects of assisted reproductive technologies, particularly intracytoplasmic sperm injection, on the integrity of the sperm genome and the risk of transmitting genes causing infertility in males to their male offspring. There is a need to identify the gene mutations detrimental to sperm function to allow identification of patients who will or will not benefit from further medical treatment. A better knowledge of the composition, organization, and function of spermatozoa is also needed to allow development of highly specific approaches to regulating sperm function in ways beneficial to family planning, either to improve sperm function and male fertility or to inhibit sperm function and disrupt male fertility. Furthermore, additional knowledge about mechanisms regulating function of spermatozoa is also essential for identifying the effects of environmental agents on male fertility and the genetic integrity of spermatozoa. There clearly are many important reasons for learning much more about the structure and function of the mammalian spermatozoon.

REFERENCES

1. Baccetti, B. (1986). Evolutionary trends in sperm structure. *Comp. Biochem. Physiol.* 85A, 29–23.
2. Roosen-Runge, E. (1977). *Developmental and Cell Biology Series, Vol. 10, The Process of Spermatogenesis in Mammals*, Cambridge University Press, Cambridge.
3. Segal, S. (1985). Sexual differentiation in vertebrates. In *MBL Lectures in Biology, Vol. 7, The Origin and Evolution of Sex* (H. O. Halvorson and A. Monroy, Eds.), pp. 263–270. Alan R. Liss, New York.
4. Surani, M. A., Allen, N. D., Barton, S. C., Fundele, R., Howlett, S. K., Norris M. L., and Reik, W. (1990). Developmental consequences of imprinting of parental chromosomes by DNA methylation. *Phil. Trans. Roy. Soc. (Lond.) B.* 326, 313–327.
5. Kelly, T. L. J., and Trasler, J. M. (2004). Reproductive epigenetics. *Clin. Genet.* 65, 247–260.
6. Fawcett, D. W. (1975). The mammalian spermatozoon. *Dev. Biol.* 44, 394–436.
7. Koehler, J. K. (1978). The mammalian sperm surface: studies with specific labeling techniques. In *International Review of Cytology* (G. H. Bourne and J. F. Danielli, Eds.), Vol. 54, pp. 73–108. Academic Press, New York.
8. Friend, D. S. (1982). Plasma-membrane diversity in a highly polarized cell. *J. Cell. Biol.* 93, 243–249.
9. Primakoff, P., and Myles, D. G. (1983). A map of the guinea pig sperm surface constructed with monoclonal antibodies. *Dev. Biol.* 98, 417–428.
10. Holt, W. V. (1984). Membrane heterogeneity in the mammalian spermatozoon. In *International Review of Cytology* (G. H. Bourne and J. F. Danielli, Eds.), Vol. 87, pp.159–194. Academic Press, New York.
11. Bearer, E. L., and Friend, D. S. (1960). Morphology of mammalian sperm membranes during differentiation, maturation, and capacitation. *J. Electron Microsc. Tech.* 16, 281–297.
12. Toshimori, K. (1998). Maturation of mammalian spermatozoa: modifications of the acrosome and plasma membrane leading to fertilization. *Cell. Tissue Res.* 293, 177–187.
13. Fawcett, D. W., and Hollenberg, R. D. (1963). Changes in the acrosomes of guinea pig spermatozoa during passage through the epididymis. *J. Reprod. Fertil. Suppl.* 6, 276–292.
14. Jones R. C. (1971). Studies of the structure of the head of boar spermatozoa from the epididymis. *J. Reprod. Fertil. Suppl.* 13, 51–64.

15. Hunnicutt, G. R., Koppel, D. E., and Myles, D. G. (1997). Analysis of the process of localization of fertilin to the sperm posterior head plasma membrane domain during sperm maturation in the epididymis. *Dev. Biol.* 191, 46–59.
16. Cowan, A. E., and Myles, D. G. (1993). Biogenesis of surface domains during spermiogenesis in the guinea pig. *Dev. Biol.* 155, 124–133.
17. Bangham, A. D. (1961). Electrophoretic characteristics of ram and rabbit spermatozoa. *Proc. R. Soc. Lond. (Biol.)* 155, 292–305.
18. Nevo, A. C., Michaeli, I., and Schindler, H. (1961). Electrophoretic properties of bull and of rabbit spermatozoa. *Exp. Cell. Res.* 23, 69–83.
19. Bedford, J. M. (1963). Changes in the electrophoretic properties of rabbit spermatozoa during passage through the epididymis. *Nature* 200, 1178–1180.
20. Cooper, G. W., and Bedford, J. M. (1971). Acquisition of surface charge by the plasma membrane of mammalian spermatozoa during epididymal maturation. *Anat. Rec.* 169, 300–301.
21. Yanagimachi, R., Noda, Y. D., Fujimoto, M., and Nicolson, G. (1972). The distribution of negative surface charges on mammalian spermatozoa. *Am. J. Anat.* 135, 497–520.
22. Lopez, M. L., de Souza, W., and Bustos-Obregon, E. (1987). Cytochemical analysis of the anionic sites on the membrane of the stallion spermatozoa during the epididymal transit. *Gamete Res.* 18, 319–332.
23. Sharon, N., and Lis, H. (1974). Use of lectins for the study of membranes. In *Methods in Membrane Biology* (E. D. Korn, Ed.), Vol. 3, pp. 147–199. Academic Press, New York.
24. Nicolson, G. (1974). The interactions of lectins with animal cell surfaces. In *International Review of Cytology* (G. H. Bourne and J. F. Danielli, Eds.), Vol. 39, pp. 89–190. Academic Press, New York.
25. Nicolson, G., and Yanagimachi, R. (1974). Mobility and restriction of mobility of plasma membrane lectin-binding components. *Science* 184, 1294–1296.
26. Millette, C. F. (1977). Distribution and mobility of lectin binding sites on mammalian spermatozoa. In *Immunobiology of Gametes* (M. Edidin and M. H. Johnson, Eds.), pp. 51–71. Cambridge University Press, Cambridge.
27. Kashiwahara, T., Tanaka, R., and Matsomoto, T. (1965). Tail to tail agglomeration of bull spermatozoa by phytoagglutinins present in soy beans. *Nature* 207, 831–832.
28. Nicolson, G., and Yanagimachi, R. (1972). Terminal saccharides on sperm plasma membranes: identification by specific agglutinins. *Science* 177, 276–279.
29. Kinsey, W. H., and Koehler, J. K. (1976). Fine structural localizations of Concanavalin A binding sites on hamster spermatozoa. *J. Supramol. Struct.* 5, 185–189.
30. Friend, D. S., and Fawcett, D. W. (1974). Membrane differentiations in freeze-fractured mammalian sperm. *J. Cell. Biol.* 63, 641–664.
31. Bradley, M. P., Ryans, D. G., and Forrester, I. T. (1980). Effects of filipin, digitonin, and polymyxin B on plasma membrane of ram spermatozoa—an EM study. *Arch. Androl.* 4, 195–204.
32. Elias, P. M., Friend, D. S., and Goerke, J. (1979). Membrane sterol heterogeneity: freeze-fracture detection with saponins and filipin. *J. Histochem. Cytochem.* 27, 1247–1260.
33. Friend, D. S. (1984). Membrane organization and differentiation in the guinea-pig spermatozoon. In *Ultrastructure of Reproduction* (J. Van Blerkom and P. M. Motta, Eds.), pp. 75–85. Martinus Nijhoff Publishers, The Hague.
34. Toshimori, K., Higashi, R., and Oura, C. (1987). Filipin-sterol complexes in golden hamster sperm membranes with special reference to epididymal maturation. *Cell. Tissue Res.* 250, 673–680.
35. Lopez, M. L., and de Souza, W. (1991). Distribution of filipin-sterol complexes in the plasma membrane of stallion spermatozoa during the epididymal maturation process. *Mol. Reprod. Dev.* 28, 158–168.
36. Friend, D. S. (1980). Freeze-fracture alterations in guinea-pig sperm membrane preceding gamete fusion. In *Membrane-membrane Interactions* (N. B. Gilula, Ed.), pp. 153–165. Raven Press, New York.
37. Friend, D. S., and Heuser, J. E. (1981). Orderly particle arrays on the mitochondrial outer membrane in rapidly-frozen sperm. *Anat. Rec.* 159, 198–199.
38. Bearer, E. L, and Friend, D. S. (1980). Anionic lipid domains: correlation and functional topography in a mammalian cell membrane. *Proc. Nat. Acad. Sci. U S A* 77, 6601–6605.
39. Friend, D. S., and Bearer, E. L. (1981). α-hydroxysterol distribution as determined by freeze-fracture cytochemistry. *Histochem. J.* 13, 535–546.
40. Bearer, E. L., and Friend, D. S. (1982). Modifications of anionic lipid domains preceding membrane fusion in guinea pig sperm. *J. Cell. Biol.* 92, 604–615.
41. Wolf, D. E., and Voglmayr, J. K. (1984). Diffusion and regionalization in membranes of maturing ram spermatozoa. *J. Cell. Biol.* 98, 1678–1684.
42. O'Rand, M. G., and Romrell, L. J. (1980). Appearance of regional surface autoantigens during spermatogenesis: comparison of anti-testis and anti-sperm antisera. *Dev. Biol.* 75, 431–441.
43. Koehler, J. K. (1974). Studies on the distribution of antigenic sites on the surface of rabbit spermatozoa. *J. Cell. Biol.* 67, 647–659.
44. Millette, C. F., and Bellvé, A. R. (1977). Temporal expression of membrane antigens during mouse spermatogenesis. *J. Cell. Biol.* 74, 86–97.
45. Koehler, J. K., and Perkins, W. D. (1974). Fine structure observations on the distribution of antigenic sites on guinea pig spermatozoa. *J. Cell. Biol.* 60, 789–795.
46. Tung, K. S. K., Han, L-B. P., and Evan, A. P. (1979). Differentiation autoantigen of testicular cells and spermatozoa in the guinea pig. *Dev. Biol.* 68, 224–238.
47. Tung, P. S., and Fritz, I. B. (1978). Specific surface antigens on rat pachytene spermatocytes and successive classes of germinal cells. *Dev. Biol.* 64, 297–315.
48. Primakoff, P., Cowan, A., Hyatt, H., Tredick-Kline, J., and Myles, D. G. (1988). Purification of the guinea pig sperm PH-20 antigen and detection of a site-specific endoproteolytic activity in sperm preparations that cleaves PH-20 into two disulfide-linked fragments. *Biol. Reprod.* 38, 921–934.
49. Blobel, C. P., Myles, D. G., Primakoff, P., and White, J. M. (1990). Proteolytic processing of a protein involved in sperm-egg fusion correlates with acquisition of fertilization competence. *J. Cell. Biol.* 111, 69–78.
50. Villarroya, S., and Scholler, R. (1986). Regional heterogeneity of human spermatozoa detected with monoclonal antibodies. *J. Reprod. Fertil.* 76, 435–447.
51. Carlson, A. E., Westenbroek, R. E., Quill, T., Ren, D., Clapham, D. E., Hille, B., Garbers, D. L., and Babcock, D. F. (2003). CatSper1 required for evoked Ca^{2+} entry and control of flagellar function in sperm. *Proc. Natl. Acad. Sci. U S A* 100, 14864–14868.
52. Quill, T. A., Sugden, S. A., Rossi, K. L., Doolittle, L. K., Hammer, R. E., and Garbers, D. L. (2003). Hyperactivated sperm motility driven by CatSper2 is required for fertilization. *Proc. Natl. Acad. Sci. U S A* 100, 14869–14874.
53. Sakata, Y., Saegusa, H., Zong, S., Osanai, M., Murakoshi, T., Shimizu, Y., Nota, T., Aso, T., and Tanabe, T. (2002). Ca(v)2.3 (alpha 1E) Ca^{2+} channel participates in the control of sperm function. *FEBS Lett.* 516, 229–233.
54. Okunade, G. W., Miller, M. L., Pyne, G. J., Sutliff, R. L., O'Connor, K. T., Neumann, J. C., Andringa, A., Miller, D. A.,

Prasad, V., Doetschman, T., Paul, R. J., and Shull, G. E. (2004). Targeted ablation of plasma membrane Ca^{2+}-ATPase (PMCA) 1 and 4 indicates a major housekeeping function of PMCA1 and a critical role in hyperactivated sperm motility and male fertility for PMCA4. *J. Biol. Chem.* 279, 33742–33750.
55. Schuh, K., Cartwright, E. J., Jankevics, E., Bundschu, K., Liebermann, J., Williams, J. C., Armesilla, A. L., Emerson, M., Oceandy, D., Knobeloch, K-P., and Neyses, L. (2004). Plasma membrane Ca^{2+} ATPase is required for sperm motility and male fertility. *J. Biol. Chem.* 279, 28220–28226.
56. Schürmann, A., Axer, H., Scheepers, A., Doege, H., and Joost, H-G. (2002). The glucose transport facilitator GLUT8 is predominantly associated with the acrosomal region of mature spermatozoa. *Cell. Tissue Res.* 307, 237–242.
57. Olson, G. E., Winfrey, V. P., Westbrook, V. A., and Melner, M. H. (1998). Targeting of the domain-specific integral membrane protein PM52 to the periacrosomal plasma membrane during guinea pig spermatogenesis. *Mol. Reprod. Dev.* 50, 103–112.
58. Igakura, T., Kadomatsu, K., Kaname, T., Muramatsu, H., Fan, Q. W., Miyauchi, T., Toyama, Y., Kuno, N., Yuasa, S., Takahashi, M., Senda, T., Taguchi, O., Yamamura, K., Arimura, K., and Muramatsu, T. (1998). A null mutation in basigin, an immunoglobulin superfamily member, indicates its important roles in peri-implantation development and spermatogenesis. *Dev. Biol.* 194, 152–165.
59. Kuno, N., Kadomatsu, K., Fan, Q-W., Hagihara, M., Senda, T., Mizutani, S., and Muramatsu, T. (1998). Female sterility in mice lacking the basigin gene, which encodes a transmembrane glycoproteins belonging to the immunoglobulin superfamily. *FEBS Lett.* 425, 191–194.
60. Saxena, D. K., Oh-oko, T., Kadomatsu, K., Muramatsu, T., and Toshimori, K. (2002). Behavior of a sperm surface transmembrane glycoproteins basigin during epididymal maturation and its role in fertilization in mice. *Reproduction* 123, 435–444.
61. Cowan, A. E., Nakhimovsky, L., Myles, D. G., and Koppel, D. E. (1997). Barriers to diffusion of plasma membrane proteins form during early guinea pig spermiogenesis. *Dev. Biol.* 73, 507–516.
62. Burridge, K., and Feramisco, J. R. (1982). a-actinin and vinculin from nonmuscle cells: calcium-sensitive interactions with actin. In *Cold Spring Harbor Symposia on Quantitative Biology*, Vol. XLVI, Part 2, pp. 587–597. Cold Spring Harbor Laboratories, New York.
63. Marchesi, V. T. (1985). Stabilizing infrastructure of cell membranes. *Annu. Rev. Cell. Biol.* 1, 531–561.
64. Vogl, A. W., Genereux, K., and Pfeiffer, D. C. (1993). Filamentous actin detected in rat spermatozoa. *Tissue Cell.* 25, 33–48.
65. Welch, J. E., and O'Rand, M. G. (1985). Identification and distribution of actin in spermatogenic cells and spermatozoa of the rabbit. *Dev. Biol.* 109, 411–417.
66. Campanella, C., Gabbiani, G., Baccetti, B., Burrini, A. G., and Pallini, V. (1979). Actin and myosin in the vertebrate acrosomal region. *J. Submicrosc. Cytol.* 11, 53–71.
67. Clarke, G. N., and Yanagimachi, R. (1978). Actin in mammalian sperm heads. *J. Exp. Zool.* 205, 125–132.
68. Greenberg, B. J., and Tamblyn, T. M. (1981). Actin from mature bovine spermatozoa. *J. Cell. Biol.* 91, 191a.
69. Flaherty, S. P., Winfrey, V. P., and Olson, G. E. (1986). Localization of actin in mammalian spermatozoa: a comparison of eight species. *Anat. Rec.* 216, 504–515.
70. Peterson, R. N., Russell, L. D., Bundman, D., and Freund, M. (1978). Presence of microfilaments and tubular structure in chemically induced acrosome reactions of boar spermatozoa. *Biol. Reprod.* 19, 459–465.
71. Tamblyn, T. M. (1980). Identification of actin in boar epididymal spermatozoa. *Biol. Reprod.* 22, 727–734.
72. Camatini, M., Colombo, A., and Bonfanti, P. (1991). Identification of spectrin and calmodulin in rabbit spermiogenesis and spermatozoa. *Mol. Reprod. Dev.* 28, 62–69.
73. Flaherty, S. P., Winfrey, V.P., and Olson, G. E. (1988). Localization of actin in human, bull, rabbit, and hamster sperm by immunoelectron microscopy. *Anat. Rec.* 221, 599–610.
74. Clarke, G. N., Clarke, F. M., and Wilson, S. (1982). Actin in human spermatozoa. *Biol. Reprod.* 26, 319–327.
75. Talbot, P., and Kleve, M. G. (1978). Hamster sperm cross-react with antiactin. *J. Exp. Zool.* 204, 131–136.
76. Olson, G. E., and Winfrey, V. P. (1985). Substructure of a cytoskeletal complex associated with the hamster sperm acrosome. *J. Ultrastruct. Res.* 92, 167–179.
77. Damjanov, I., Damjanov, A., Lehto, V-P., and Virtanen, I. (1986). Spectrin in mouse gametogenesis and embryogenesis. *Dev. Biol.* 114, 132–140.
78. De Cesaris, P., Filippini, A., Ziparo, E., Russo, M. A., and Stefanini, M. (1989). Distribution of analogues of spectrin, fodrin and protein 4.1 in rat spermatogenic cells. *Prog. Clin. Biol. Res.* 296, 149–152.
79. Tanaka, H., Iguchi, N., Egydio de Carvalho, C., Tadokoro, Y., Yomogida, K., and Nishimune Y. (2003). Novel actin-like proteins T-ACTIN 1 and T-ACTIN 2 are differentially expressed in the cytoplasm and the nucleus and mouse haploid germ cells. *Biol. Reprod.* 69, 475–482.
80. Maier, B., Medrano, S., Sleight, S. B., Visconti, P. E., and Scrable, H. (2003). Developmental association of the synaptic activity-regulated protein Arc with the mouse acrosomal organelle and the sperm tail. *Biol. Reprod.* 68, 67–76.
81. Tubb, B., Mulholland, D. J., Vogl, W., Lan, Z-J., Niederberger, C., Cooney, A., and Bryan, J. (2002). Testis fascin (FNCN3): a novel paralog of the actin-bundling protein fascin expressed specifically in the elongate spermatid head. *Exp. Cell. Res.* 275, 92–109.
82. Heid, H. W., Figge, U., Winter, S., Kuhn, C., Zimbelmann, R., and Franke, W. W. (2002). Novel actin-related proteins Arp-T1 and Arp-T2 as components of the cytoskeletal calyx of the mammalian sperm head. *Exp. Cell. Res.* 279, 177–187.
83. Lécuyer, C., Dacheux, J-L., Hermand, E., Mazeman, E., Rousseaux, J., and Rousseaux-Prévost, R. (2000). Actin-binding properties and colocalization with actin during spermiogenesis of mammalian sperm calicin. *Biol. Reprod.* 63, 1801–1810.
84. Hurst, S., Howes, E. A., Coadwell, J., and Jones, R. (1998). Expression of a testis-specific putative actin-capping protein associated with the developing acrosome during rat spermiogenesis. *Mol. Reprod. Dev.* 49, 81–91.
85. von Bülow, M., Rackwitz, H-R., Zimbelmann, R., and Franke, W. W. (1997). CP ß3, a novel isoform of an actin-binding protein, is a component of the cytoskeletal calyx of the mammalian sperm head. *Exp. Cell. Res.* 233, 216–224.
86. Kierszembaum, A. L., Rivkin, E., and Tres, L. L. (2003). The actin-based motor myosin Va is a component of the acroplaxome, an acrosome-nuclear envelope junctional plate, and of the manchette-associated vesicles. *Cytogenet. Genome Res.* 103, 337–344.
87. Brener, E., Rubinstein, S., Cohen, G., Shternall, K., Rivlin, J., and Breitbart, H. (2003). Remodeling of the actin cytoskeleton during mammalian sperm capacitation and acrosome reaction. *Biol. Reprod.* 68, 837–845.
88. Myles, D. G., and Primakoff, P. (1985). Sperm surface domains. In *Hybridoma Technology in the Biosciences and Medicine* (T. A. Springer, Ed.), pp. 239–250. Plenum Press, New York.
89. Phelps, B. M., Koppel, D. E., Pimakoff, P., and Myles, D. G. (1990). Evidence that proteolysis of the surface is an initial step in the mechanism of formation of sperm cell surface domains. *J. Cell. Biol.* 111, 1839–1847.

90. Myles, D. G., Primakoff, P., and Koppel, D. E. (1984). A localized surface protein of guinea pig sperm exhibits free diffusion in its domain. *J. Cell. Biol.* 98, 1905–1909.
91. Gaunt, S. J., Brown, C. R., and Jones, R. (1983). Identification of mobile and fixed antigens on the plasma membrane of rat spermatozoa using monoclonal antibodies. *Exp. Cell. Res.* 144, 275–284.
92. Toshimori, K., Araki, S., Tanii, I., and Oura, C. (1992). Masking the cryptodeterminant on the 54-kilodalton mouse sperm surface antigen. *Biol. Reprod.* 47, 1161–1167.
93. Koehler, J. K. (1970). A freeze-etch study of rabbit spermatozoa with particular reference to head structures. *J. Ultrastruct. Res.* 33, 598–614.
94. Fawcett, D. W. (1965). The anatomy of the mammalian spermatozoon with particular reference to the guinea pig. *Z. Zellforsch.* 67, 279–296.
95. Wolf, D. E., Maynard, V. M., McKinnon, C. A., and Melchior, D. L. (1990). Lipid domains in the ram sperm plasma membrane demonstrated by differential scanning calorimetry. *Proc. Natl. Acad. Sci. U S A* 87, 6893–6896.
96. Parks, J. E., and Hammerstedt, R. H. (1985). Developmental changes occurring in the lipids of ram epididymal spermatozoa plasma membrane. *Biol. Reprod.* 32, 653–668.
97. Agrawal, P., Magargee, S. F., and Hammerstedt, H. (1988). Isolation and characterization of the plasma membrane of rat cauda epididymal spermatozoa. *J. Androl.* 9, 178–189.
98. Poulos, A., Voglmayr, J. K., and White, I. G. (1973). Phospholipid changes in spermatozoa during passage through the genital tract of the bull. *Biochim. Biophys. Acta* 306, 194–202.
99. Evans, R. W., Weaver, D. E., and Clegg, E. D. (1980). Diacyl, alkenyl and alkyl ether phospholipids in ejaculated, in utero-, and in vitro-incubated porcine spermatozoa. *J. Lipid Res.* 21, 223–228.
100. Nikolopoulou, M., Soucek, D. A., and Vary, J. C. (1985). Changes in the lipid content of boar sperm plasma membranes during epididymal maturation. *Biochim. Biophys. Acta* 1815, 486–498.
101. Palleschi, S., and Silvestroni, L. (1996). Laurdan fluorescence spectroscopy reveals a single liquid-crystalline lipid phase and lack of thermotropic phase transitions in the plasma membrane of living human sperm. *Biochim. Biophys. Acta* 1279, 197–202.
102. Bradley, M. P., Ryans, D. G., and Forrester, I. T. (1980). Effects of filipin, digitonin, and polymyxin B on plasma membrane of ram spermatozoa—an EM study. *Arch. Androl.* 4, 195–204.
103. Langlais, J., Zollinger, M., Plante, L., Chapdelaine, A., Blea, G., and Roberts, K. D. (1981). Localization of cholesterol sulfate in human spermatozoa in support of a hypothesis for the mechanism of capacitation. *Proc. Nat. Acad. Sci. U S A* 78, 7266–7270.
104. Parks, J. E., Arion, J. W., and Foote, R. H. (1987). Lipids of plasma membrane and outer acrosomal membrane from bovine spermatozoa. *Biol. Reprod.* 37, 1249–1258.
105. Nikolopoulou, M., Soucek, D. A., and Vary, J. C. (1986). Lipid composition of the membrane released after an in vitro acrosome reaction of epididymal boar sperm. *Lipids* 21, 566–570.
106. Mack, S. R., Zaneveld, L. J., Peterson, R. N., Hunt, W., Russell, L. D. (1987). Characterization of human sperm plasma membrane: glycolipids and polypeptides. *J. Exp. Zool.* 243, 339–346.
107. Murry, R. K., Narasimhan, R., Levine, M., Shirley, M., Lingwood, C. A., Schachter, H. (1980). Galactoglycerolipids of mammalian testis, spermatozoa and nervous tissues. In *ACS Symposium Series no. 128, Cell Surface Glycolipids* (C. Sweeley, Ed.), pp. 105–125. American Chemical Society Press, Washington, DC.
108. Wolf, D. E., Lipscomb, A. C., and Maynard, V. M. (1988). Causes of nondiffusing lipid in the plasma membrane of mammalian spermatozoa. *Biochemistry* 27, 860–865.
109. Ishizuka, I., and Yamakaw, T. (1985). Glycoglycerolipids. In *Glycolipids* (H. Wiegandt, Ed.), pp. 101–197. Elsevier, New York.
110. Edidin, M. (1981). Molecular motions and membrane organization and function. In *Comprehensive Biochemistry* (J. B. Finian and R. H. Michell, Eds.), pp. 37–82. Elsevier/North Holland Biomedical Press, Amsterdam.
111. Wolf, D. E., and Voglmayr, J. K. (1984). Diffusion and regionalization in membranes of maturing ram spermatozoa. *J. Cell. Biol.* 98, 1678–1684.
112. Wolf, D. E., Hagopian, S. S., Lewis, R. G., Voglmayr, J. K., and Fairbanks, G. (1986). Lateral regionalization and diffusion of a maturation dependent antigen in the ram sperm plasma membrane. *J. Cell. Biol.* 102, 1826–1831.
113. Wolf, D. E., Scott, B. K., and Millette, C. F. (1986). The development of regionalized lipid diffusibility in the germ cell plasma membrane during spermatogenesis in the mouse. *J. Cell. Biol.* 103, 1745–1750.
114. Wolfe, C. A., James, P. S., Mackie, A. R., Ladha, S., and Jones, R. (1998). Regionalized lipid diffusion in the plasma membrane of mammalian spermatozoa. *Biol. Reprod.* 59, 1506–1514.
115. Mackie, A. R., James, O. S., Ladha, S., Jones, R. (2001). Diffusion barriers in ram and boar sperm plasma membranes: directionality of lipid diffusion across the posterior ring. *Biol. Reprod.* 64, 113–119.
116. James, P. S., Hennessy, C., Berge, T., and Jones, R. (2004). Compartmentalisation of the sperm plasma membrane: a FRAP, FLIP and SPFI analysis of putative diffusion barriers on the sperm head. *Cell. Sci.* 117, 6485–6495.
117. Edidin, M. (2003). The state of lipid rafts: from model membranes to cells. *Annu. Rev. Biophys. Biomol. Struct.* 32, 57–83.
118. Travis, A. J., Meridushev, T., Vargar, L. A., Jones, B. H., Purdon, M. A., Nipper, R. W., Galatioto, J., Moss, S. B., Hunnicutt, G. R., and Kopf, G. S. (2001). Expression and localization of caveiolin-1, and the presence of membrane rafts, in mouse and guinea pig spermatozoa. *Dev. Biol.* 240, 599–610.
119. Treviño, C. L., Serrano, C. J., Beltrán, C., Felix, R., and Darszon, A. (2001). Identification of mouse *trp* homologs and lipid rafts from spermatogenic cells and sperm. *FEBS Lett.* 509, 119–125.
120. Cross, N. L. (2004). Reorganization of lipid rafts during capacitation of human sperm. *Biol. Reprod.* 71, 1367–1373.
121. Hondo, A., Yamagata, K., Suriura, S., Watanabe, K., and Baba, T. (2002). A mouse serine protease TESP5 is selectively included into lipid rafts of sperm membrane presumably as a glycosylphosphatidylinositol-anchored protein. *J. Biol. Chem.* 277, 16976–16984.
122. Dawson. R. M. C., and Scott, T. W. (1964). Phospholipid composition of epididymal spermatozoa prepared by density gradient centrifugation. *Nature* 202, 292–293.
123. Quinn, P. J., White, I. G. (1967). Phospholipid and cholesterol content of epididymal and ejaculated ram spermatozoa and seminal plasma in relation to cold shock. *Atus. J. Biol. Sci.* 20, 1205–1215.
124. Grogan, D. E., Mayer, D. T., Sikes, J. D. (1966). Quantitative differences in phospholipids of ejaculated spermatozoa and spermatozoa from three different levels of the epididymis of the boar. *J. Reprod. Fertil.* 12, 431–436.
125. Poulos, A., Voglmayr, J. K., White, I. G. (1973). Phospholipid changes in spermatozoa during passage through the genital tract of the bull. *Biochim. Biophys. Acta* 306, 194–202.
126. Poulo, A., Brown-Woodman, P. D. C., White, I. G., and Cox, R. I. (1975). Changes in phospholipids of ram spermatozoa during migration through the epididymis and possible origin

of prostaglandins F_2 in testicular and epididymal fluid. *Biochim. Biophys. Acta* 388, 12–21.
127. Terner, C., MacLaughlin, J., Smith, B. R. (1975). Changes in lipase and phospholipase activities of rat spermatozoa in transit from the caput to the cauda epididymis. *J. Reprod. Fertil.* 45, 1–8.
128. Evans, R. W., and Setchell, B. P. (1979). Lipid changes in boar spermatozoa during epididymal maturation with some observations on the flow and composition of boar rete testis fluid. *J. Reprod. Fertil.* 57, 189–196.
129. Aveldano, M. I., Rotstein, N. P., and Vermouth, N. (1992). Lipid remodelling during epididymal maturation of rat spermatozoa. *Biochem. J.* 283, 35–241.
130. Scott, T. W., Voglmayr, J. K., and Stechell, B. P. (1967). Lipid composition and metabolism in testicular and ejaculated ram spermatozoa. *Biochem. J.* 102, 456–461.
131. Bleau, G., and VandenHeuvel, W. J. A. (1974). Desmosteryl sulfate and desmosterol in hamster epididymis. *Steroids* 24, 549–556.
132. Lalumiere, G., Bleau, G., Chapdelaine, A., and Roberts, K. D. (1976). Cholesterol sulfate and sterol sulphatase in the human reproductive tract. *Steroids* 27, 247–260.
133. Hammerstedt, R. H., Keith, A. D., Hay, S., Deluca, N., and Amann, R. P. (1979). Changes in ram sperm membranes during epididymal transit. *Arch. Biochem. Biophys.* 196, 7–12.
134. Voglmayr, J. K., Scott, T. W., Setchell, B. P., and Waite, G. M. H. (1967). Metabolism of testicular spermatozoa and characteristics of testicular fluid collected from conscious rams. *J. Reprod. Fertil.* 14, 87–99.
135. Vijayasarathy, S., and Balaram, P. (1982). Regional differentiation in bull sperm plasma membranes. *Biochem. Biophys. Res. Commun.* 108, 760–764.
136. Travis, A. J., and Kopf, G. S. (2002). The role of cholesterol efflux in regulating the fertilization potential of mammalian spermatozoa. *J. Clin. Invest.* 110, 731–736.
137. Gadella, B. M., Lopes-Cardozo, M., van Golde, L. M., Colenbrander, B., and Gadella, T. W., Jr. (1995). Glycolipid migration from the apical to the equatorial subdomains of the sperm head plasma membrane precedes the acrosome reaction. Evidence for a primary capacitation event in boar spermatozoa. *J. Cell. Sci.* 108, 935–946.
138. Rochwerger, L., and Cuasnicu, P. S. (1992). Redistribution of a rat sperm epididymal glycoprotein after in vitro and in vivo capacitation. *Mol. Reprod. Dev.* 31, 34–41.
139. Cowan, A. E., Koppel, D. E., Vargas, L. A., and Hunnicutt, G. R. (2001). Guinea pig fertilin exhibits restricted lateral mobility n epididymal sperm and becomes freely diffusing during capacitation. *Dev. Biol.* 236, 502–509.
140. Tesarik, J., and Flechon, J. E. (1986). Distribution of sterols and anionic lipids in human sperm plasma membrane: effects of in vitro capacitation. *J. Ultrastruct. Mol. Struct. Res.* 97, 227–237.
141. Grimes, S. R., Jr. (1986). Nuclear proteins in spermatogenesis. *Comp. Biochem. Physiol.* 83B, 495–500.
142. Hecht, N. B. (1989). Mammalian protamines and their expression. In *Histones and Other Basic Nuclear Proteins* (L. Hnilica, G. Stein, and J. Stein, Eds.), pp. 347–373. CRC Press, Boca Raton, FL.
143. Oliva, R., and Dixon, G. H. (1991). Vertebrate protamine genes and the histone-to-protamine replacement reaction. In *Progress in Nucleic Acid Research and Molecular Biology* (M. E. Cohn and K. Moldave, Eds.), Vol. 40, pp. 25–94. Academic Press, New York.
144. Hecht, N. B. (1999). Protamine gene expression: a model for post-transcriptional gene regulation in male germ cells. In *The Male Gamete: From Basic Knowledge to Clinical Applications* (C. Gagnon, Ed.), pp. 5–10. Cache River Press, Vienna, IL.
145. Bellvé, A. R., and Carraway, R. (1978). Characterization of two basic chromosomal proteins isolated from mouse spermatozoa. *J. Cell. Biol.* 79, 177a.
146. Mayer, J. F., Chang, T. S. K., and Zirkin, B. R. (1981). Spermatogenesis in the mouse. 2. Amino acid incorporation into basic nucleoproteins of mouse spermatids and spermatozoa. *Biol. Reprod.* 25, 1041–1051.
147. Balhorn, R., Weston, S., Thomas, C., and Wyrobek, A. J. (1984). DNA packaging in mouse spermatids. Synthesis of protamine variants and four transition proteins. *Exp. Cell. Res.* 150, 298–308.
148. Cho, C., Willis, W. D., Goulding, E. H., Jung-Ha, H., Choi, Y. C., Hecht, N. B., and Eddy, E. M. (2001). Haploinsufficiency of protamine-1 or -2 causes infertility in mice. *Nat. Genet.* 28, 82–86.
149. Cho, C., Jung-Ha, H., Willis, W. D., Goulding, E. H., Stein, P., Xu, Z., Schultz, R. M., Hecht, N. B., and Eddy, E. M. (2003). Protamine 2 deficiency leads to sperm DNA damage and embryo death in mice. *Biol. Reprod.* 69, 211–217.
150. Balhorn, R., Cosman. M., Thjornton, K., Krishman, V. V., Corzett, M., Bench, G., Kramer, C., Lee, IV J., Hud, N. V., Allen, M., Preito, M., Meyer-Ilse, W., Brown, J. T., Kirz, J., Zhang, X., Bradbury, E. M., Maki, G., Braun, R. E., and Breed, W. (1999). Protamine mediated condensation of DNA in mammalian sperm. In *The Male Gamete: From Basic Knowledge to Clinical Applications* (C. Gagnon, Ed.), pp. 55–70. Cache River Press, Vienna, IL.
151. Balhorn, R. (1982). A model for the structure of chromatin in mammalian sperm. *J. Cell. Biol.* 93, 298–305.
152. Warrant, R. W., and Kim, S-H. (1978). α-Helix-double helix interaction shown in the structure of a protamine-transfer RNA complex and a nucleoprotamine model. *Nature* 271, 130–135.
153. Tanphaichitr, N., Sobhon, P., Taluppeth, N., and Chalermisarachai, P. (1978). Basic nuclear proteins in testicular cells and ejaculated spermatozoa in man. *Exp. Cell. Res.* 117, 347–356.
154. Gatewood, J. M., Cook, G. R., Balhorn, R., Schmid, C. W., and Bradbury, E. M. (1990). Isolation of four core histones from human sperm chromatin representing a minor subset of somatic histones. *J. Biol. Chem.* 265, 20662–20666.
155. Hoyer-Fender, S., Singh, P. B., and Motzkus, D. (2000). The murine heterochromatin protein M31 is associated with the chromocenter in round spermatids and is a component of mature spermatozoa. *Exp. Cell. Res.* 254, 72–79.
156. Sotolongo, B., Line, E., and Ward, W. S. (2003). Ability of hamster spermatozoa to digest their own DNA. *Biol. Reprod.* 69, 2029–2035.
157. Carrey, E. A., Dietz, C., Glubb, D. M., Loffler, M., Lucocq, J. M., and Watson, P. F. (2002). Detection and location of the enzymes of de novo pyrimidine biosynthesis in mammalian spermatozoa. *Reproduction* 123, 757–768.
158. Koehler, J. K. (1966). Fine structure observations in frozen-etched bovine spermatozoa. *J. Ultrastruct. Res.* 16, 359–375.
159. Bendet, I. J., Bearden, J., Jr. (1972). Birefringence of bull sperm. II. Form birefringence of bull sperm. *J. Cell. Biol.* 55, 501–510.
160. Sipski, M. R., and Wagner, T.E. (1977). The total structure and organization of chromosomal fibers in eutherian sperm nuclei. *Biol. Reprod.* 16, 428–440.
161. Bellvé, A. R. (1982). Biogenesis of the mammalian spermatozoon. In *Prospects for Sexing Mammalian Sperm* (R. P. Amann and G. E. Seidel, Jr., Eds.), pp. 69–102. Associated University Press, Boulder, CO.
162. Koehler, J. K. (1972). Human sperm head ultrastructure: a freeze-etching study. *J. Ultrastruct. Res.* 39, 520–539.
163. Zalenskaya, A., and Zalensky, A. O. (2004). Non-random positioning of chromosomes in human sperm nuclei. *Chromosome Res.* 12, 163–73.

164. Pedersen, H. (1972). The postacrosomal region of man and *Macaca artoides*. *J. Ultrastruct. Res.* 40, 366–377.
165. Stackpole, C. W., and Devorkin, D. (1974). Membrane organization in mouse spermatozoa revealed by freeze-etching. *J. Ultrastruct. Res.* 49, 167–187.
166. Gerace, L., Comeau, C., and Benson, M. (1984).Organization and modulation of nuclear lamina structure. *J. Cell. Sci.* (Suppl.) 1, 137–160.
167. Krohne, G., and Benavente, R. (1986).The nuclear lamins. A multigene family of proteins in evolution and differentiation. *Exp. Cell. Res.* 162, 1–10.
168. Gerace, L., Blum, A., and Blobel, G. (1978). Immunocytochemical localization of the major polypeptides of the nuclear pore complex lamina fraction. Interphase and mitotic distribution. *J. Cell. Biol.* 79, 546–566.
169. Hancock, R., and Baulikis, T. (1982). Functional organisation of the nucleus. In *International Review of Cytology* (G. H. Bourne and J. F. Danielli, Eds.), Vol. 79, pp. 165–214. Academic Press, New York.
170. Lebkowski, Y. S., and Laemmli, U. K. (1982). Non-histone proteins and long-range organization of HeLa interphase DNA. *J. Mol. Biol.* 156, 21–141.
171. Goldman, R. D., Gruenbaum, Y., Moir, R. D., Shumaker, D. K., and Spann, T. P. (2002). Nuclear lamins: building blocks of nuclear architecture. *Genes Dev.* 16, 533–547.
172. Furakawa, K., Ikanagi, H., and Hotta, Y. (1994). Identification and cloning of an mRNA coding for a germ cell-specific A-type lamin in mice. *Exp. Cell. Res.* 212, 426–430.
173. Alsheimer, M., and Benavente, R. (1996). Change of karyoskeleton during mammalian spermatogenesis: expression pattern of nuclear lamin C2 and its regulation. *Exp. Cell. Res.* 228, 181–188.
174. Alsheimer, M., Liebe, B., Sewell, C.L., Scherthan, H., and Benavente, R. (2004). Disruption of spermatogenesis in mice lacking A-type lamins. *J. Cell. Sci.* 117, 1173–1178.
175. Moss, S. B., Burnham, B. L., and Bellvé, A. R. (1993). The differential expression of lamin epitopes during mouse spermatogenesis. *Mol. Reprod. Dev.* 34, 164–174.
176. Smith, A., and Benavente, R. (1992). Identification of a short nuclear lamin protein selectively expressed during meiotic stages of rat spermatogenesis. *Differentiation* 52, 55–60.
177. Vester, B., Smith, A., Krohne, G., and Benavente, R. (1993). Presence of a nuclear lamina in pachytene spermatocytes of the rat. *J. Cell. Sci.* 104, 557–563.
178. Alsheimer, M., Fecher, E., and Benavente, R. (1998). Nuclear envelope remodeling during rat spermiogenesis: distribution and expression pattern of LAP2/thymopoietins. *J. Cell. Sci.* 111, 2227–3224.
179. Furukawa, K., and Hotta, Y. (1993). cDNA cloning of a germ cell specific lamin B3 from mouse spermatocytes and analysis of its function by ectopic expression in somatic cells. *EMBO J.* 12, 97–106.
180. Mylonis, I., Drosou, V., Brancorsini, S., Sassone-Corsi, P., and Giannakouros, T. (2004). Temporal association of protamine 1 with the inner nuclear membrane protein lamin B receptor during spermiogenesis. *J. Biol. Chem.* 279, 11626–11631.
181. Cowen, P., and Burke, B. (1996). Cytoskeleton-membrane interactions. *Curr. Opin. Cell. Biol.* 8, 56–65.
182. Bellvé, A. R., and O'Brien, D. A. (1983). The mammalian spermatozoon: structure and temporal assembly. In *Mechanisms and Control of Animal Fertilization* (H. F. Hartmann, Ed.), pp. 55–137. Academic Press, Orlando, FL.
183. Bellvé, A. R., Chandrika, R., Martinova, Y. S., and Barth, A. H. (1992). The perinuclear natrix as a structural element of the mouse sperm nucleus. *Biol. Reprod.* 47, 451–465.
184. Olson, G. E., Hamilton, D. W., and Fawcett, D. W. (1976). Isolation and characterization of the perforatorium of rat spermatozoa. *J. Reprod. Fertil.* 47, 293–297.
185. Fawcett, D. W. (1970). A comparative view of sperm ultrastructure. *Biol. Reprod.* Suppl. 2, 90–127.
186. Burgos, M. H., and Fawcett, D. W. (1956). An electron microscopic study of spermatid differentiation in the toad. *Bufo arenarum* Hensel. *J. Biophys. Biochem. Cytol.* 2, 223–240.
187. Nagano, T. (1962). Observations on the fine structure of the developing spermatid in the domestic chicken. *J. Cell. Biol.* 14, 193–205.
188. Clermont, Y., Einberg, E., Leblond, C. P., and Wagner, S. (1955). The perforatorium—an extension of the nuclear membrane of the rat spermatozoon. *Anat. Rec.* 121, 1–12.
189. Yanagimachi, R., and Noda, Y. D. (1970). Ultrastructural changes in the hamster sperm head during fertilization. *J. Ultrastruct. Res.* 31, 465–485.
190. Calvin, H. I., and Bedford, J. M. (1971). Formation of disulfide bonds in the nucleus and accessory structures of mammalian spermatozoa during maturation in the epididymis. *J. Reprod. Fertil.* Suppl. 13, 65–75.
191. Austin, C. R., and Bishop, M. W. H. (1985). Some features of the acrosome and perforatorium in mammalian spermatozoa. *Proc. R. Soc. Lond. (Biol.)* 149, 234–240.
192. Lalli, M., and Clermont, Y. (1981). Structural changes in the head component of the rat spermatid during late spermatogenesis. *Am. J. Anat.* 160, 419–434.
193. Huang, T. T. F., and Yanagimachi, R. (1985). Inner acrosomal membrane of mammalian spermatozoa: its properties and possible functions in fertilization. *Am. J. Anat.* 174, 249–268.
194. Courtens, J. L., Courot, M., and Fléchon, J. E. (1976). The perinuclear substance of boar, bull, rams and rabbit spermatozoa. *J. Ultrastruct. Res.* 57, 54–64.
195. Fouquet, J-P., Valentin, A., and Kann, M-L. (1992). Perinuclear cytoskeleton of acrosome-less spermatids in the blind sterile mutant mouse. *Tissue & Cell* 24, 655–665.
196. Nicander, L., and Bane, A. (1966). Fine structure of the sperm head in some mammals with particular reference to the acrosome and subacrosomal substance. *Z. Zelforsch.* 72, 496–515.
197. Phillips, D. M. (1977). Surface of the equatorial segment of mammalian acrosome. *Biol. Reprod.* 16, 128–137.
198. Phillips, D. M. (1975). Cell surface structure of rodent sperm heads. *J. Exp. Zool.* 191, 1–8.
199. Koehler, J. K. (1978). Observations on the fine structure of vole spermatozoa with particular reference to cytoskeletal elements in the mature sperm head. *Gamete Res.* 1, 247–257.
200. Plattner, H. (1971). Bull spermatozoa: a re-investigation by freeze etching using widely different cryofixation procedures. *J. Submicrosc. Cytol.* 3, 19–32.
201. Olson, G. E., Noland, T. D., Winfrey, V. P., and Garbers, D. L. (1983). Substructure of the postacrosomal sheath of bovine spermatozoa. *J. Ultrastruct. Res.* 85, 204–218.
202. Tanaka, H., Iguchi, N., Egydio de Carvalhho, C., Tadokoro, Y., Yomogida, K., and Nishimune, Y. (2003). Novel actin-like proteins T-ACTIN 1 and T-ACTIN 2 are differentially expressed in the cytoplasm and nucleus of mouse haploid germ cells. *Biol. Reprod.* 69, 475–482.
203. Maier, B., Medrano, S., Sleight, S. B., Visconti, P. E., and Scrable, H. (2003). Developmental association of the synaptic activity-regulated protein Arc with the mouse acrosomal organelle and the sperm tail. *Biol. Reprod.* 68, 67–76.
204. Burkin, H. R., Zhao, L., and Miller, D. J. (2004). CASK is in the mammalian sperm head and is processed during epididymal maturation. *Mol. Reprod. Dev.* 68, 500–506.
205. Tubb, B., Mulholland, D. J., Vogl, W., Lan, Z-J., Niederberger, C., Cooney, A., and Bryan, J. (2002). Testis fascin (FSCN3): a novel paralog of the actin-bundling protein fascin expressed specifically in the elongate spermatid. *Exp. Cell. Res.* 275, 92–109.

206. Korley, R., Pouresmaeili, F., and Oko, R. (1997). Analysis of the protein composition of the mouse sperm perinuclear theca and characterization of its major protein constituent. *Biol. Reprod.* 57, 1426–1432.
207. Aul, R. B., and Oko, R. J. (2001). The major subacrosomal occupant of bull spermatozoa is a novel histone H2B variant associated with the forming acrosome during spermiogenesis. *Dev. Biol.* 239, 376–387.
208. Hess, H., Heid, H., and Franke, W. W. (1993). Molecular characterization of mammalian cylicin, a basic protein of the sperm head cytoskeleton. *J. Cell. Biol.* 122, 1043–1052.
209. Hess, H., Heid, H., Zimbelmann, R., and Franke, W. W. (1995). The protein complexity of the cytoskeleton of bovine and human sperm heads: the identification and characterization of cylicin II. *Exp. Cell. Res.* 218, 174–182.
210. Rousseaux-Prévost, R., Lécuyer, C., Drobecq, H., Sergheraert, C., Dacheax, J-L., and Rousseaux, J. (2003). Characterization of boar sperm cytoskeletal cylicin II as an actin-binding protein. *Biochem. Biophys. Res. Commun.* 303, 182–189.
211. Lécuyer, C., Dacheux, J-L., Hermand, E., Mazeman, E., Rousseaux, J., and Rousseaux-Prévost, R. (2000). Actin-binding properties and colocalization with actin during spermiogenesis of mammalian sperm calicin. *Biol. Reprod.* 63, 1801–1810.
212. Hurst, S., Howes, E. A., Coadwell, J., and Jones, R. (1998). Expression of a testis-specific putative actin-capping protein associated with the developing acrosome during rat spermiogenesis. *Mol. Reprod. Dev.* 49, 81–91.
213. Howes, E. A., Hurst, S. M., and Jones, R. (2001). Actin and actin-binding proteins in bovine spermatozoa: potential role in membrane remodeling and intracellular signaling during epididymal maturation and the acrosome reaction. *J. Androl.* 22, 62–72.
214. Miyagawa, Y., Tanaka, H., Iguchi, N., Kitamura, K., Nakamura, Y., Takahashi, T., Matsumiya, K., Okuyama, A., and Nishimune, Y. (2002). Molecular cloning and characterization of the human orthologue of male germ cell-specific actin capping protein α3 (*cpaα3*). *Mol. Human Reprod.* 8, 531–539.
215. von Bülow, M., Rackwitz, H-R., Zimbelmann, R., and Franke, W. W. (1997). CP ß3, a novel isoform of an actin-binding protein, is a component of the cytoskeletal calyx of the mammalian sperm head. *Exp. Cell. Res.* 233, 216–224.
216. Heid, H. W., Figge, U., Winter, S., Kuhn, C., Zimbelmann, R., and Franke, W. W. (2002). Novel actin-related proteins Arp-T1 and Arp-T2 as components of the cytoskeletal calyx of the mammalian sperm head. *Exp. Cell. Res.* 279, 177–187.
217. Kitamura, K., Iguchi, N., Kaneko, Y., Tanaka, H., and Nishimune, Y. (2004). Characterization of a novel postacrosomal perinuclear theca-specific protein CYPT1. *Biol. Reprod.* 71, 1927–1935.
218. Herrada, G., and Wolgemuth, D. J. (1997). The mouse transcription factor Stat4 is expressed in haploid male germ cells and is present in the perinuclear theca of spermatozoa. *J. Cell. Sci.* 110, 1543–1553.
219. Kann, M. L., Feinberg, J., Rainteau, D., Dadoune, J. P., Weinman, S., and Fouquet, J. O. (1991). Localization of calmodulin in perinuclear structures of spermatids and spermatozoa: a comparison of six mammalian species. *Anat. Rec.* 230, 481–488.
220. Juárez-Mosqueda, M. L., and Mújida, A. (1999). A perinuclear theca substructure is formed during epididymal guinea pig sperm maturation and disappears in acrosome reacted cells. *J. Struct. Biol.* 128, 225–236.
221. Bellvé, A. R., Chandrika, R., and Barth, A. (1990). Temporal expression, polar distribution and transition of an epitope domain in the perinuclear theca during mouse spermatogenesis. *J. Cell. Sci.* 96, 745–756.
222. Hernández-González, E. O., Martínez-Rojas, D., Mornet, D., Rendón, A., and Mújica, A. (2001). Comparative distribution of short dystrophin-related superfamily products in various guinea pig spermatozoa domains. *Eur. J. Cell. Biol.* 80, 792–798.
223. Jassim, A., Foxon, R., Purkis, P., Gray, A., and Al-Zuhdi, Y. (1993). AJ-p90: a novel protein of the perinuclear theca in human sperm subacrosome. *J. Reprod. Immunol.* 23, 169–188.
224. Czaker, R. (1985). Morphogenesis and cytochemistry of the postacrosomal dense lamina during mouse spermiogenesis. *J. Ultrastruct. Res.* 90, 26–39.
225. Oko, R., and Clermont, Y. (1988). Isolation, structure and protein composition of the perforatorium of rat spermatozoa. *Biol. Reprod.* 39, 673–687.
226. Longo, F. J., Krohne, G., and Franke, W. W. (1987). Basic proteins of the perinuclear theca of mammalian spermatozoa and spermatids: a novel class of cytoskeletal elements. *J. Cell. Biol.* 105, 1105–1120.
227. Olson, G. E., and Winfrey, V. P. (1988). Characterization of the postacrosomal sheath of bovine spermatozoa. *Gamete Res.* 20, 329–342.
228. Oko, R., and Clermont, Y. (1988). Isolation, structure and protein composition of the perforatorium of rat spermatozoa. *Biol. Reprod.* 39, 673–687.
229. Phillips, D. M. (1970). Development of spermatozoa in the wolly opossum with special reference to the shaping of the sperm head. *J. Ultrastruct. Res.* 33, 369–380.
230. Fawcett, D. W., and Phillips, D. M. (1969). Observations on the release of spermatozoa and on changes in the head during passage through the epididymis. *J. Reprod. Fertil.* Suppl. 6, 405–418.
231. Russell, L. D. (1984). Spermiation: the sperm release process: ultrastructural observations and unresolved problems. In *Ultrastructure of Reproduction* (J. Van Vlerkom and P. M. Motta, Eds.), pp. 46–66. Martinus Nijhoff Publishers, The Hague.
232. Fawcett, D. W., Anderson, W. A., and Phillips, D. M. (1971). Morphogenetic factors influencing the shape of the sperm head. *Dev. Biol.* 26, 220–251.
233. Hunt, D. M., and Johnson, D. R. (1971). Abnormal spermiogenesis in two pink-eyed sterile mutants in the mouse. *J. Embryol. Exp. Morphol.* 26, 111–121.
234. Bryan, J. H. D. (1977). Spermatogenesis revisited: III. The course of spermatogenesis in a male-sterile pink-eyed mutant type in the mouse. *Cell. Tissue Res.* 180, 173–186.
235. Sotomayor, R. E., and Handel, M. A. (1986). Failure of acrosome assembly in a male sterile mutant. *Biol. Reprod.* 34, 171–182.
236. Wooding, F. B. P. (1973). The effect of Triton X-100 on the ultrastructure of ejaculated bovine sperm. *J. Ultrastruct. Res.* 42, 502–516.
237. Fléchon, J. E. (1974). Freeze-fracturing of rabbit spermatozoa. *J. Submicrosc. Cytol.* 19, 59–64.
238. Hermo, L., Rambourg, L. A., and Clermont, Y. (1980). Three-dimensional architecture of the cortical region of the Golgi apparatus in rat spermatids. *Am. J. Anat.* 157, 357–373.
239. Pedersen, H. (1972). Further observations on the fine structure of the human spermatozoon. *Z. Zellforsch.* 123, 305–315.
240. Phillips, D. M. (1977). Mitochondrial disposition in mammalian spermatozoa. *J. Ultrastruct. Res.* 58, 144–154.
241. Koehler, J. K. (1975). Periodicities in the acrosome or acrosomal membrane: some observations on mammalian spermatozoa. *Biol. J. Linnean Soc.* Suppl. 1, 337–342.
242. Zahler, W. L., and Doak, G. A. (1975). Isolation of the outer acrosomal membrane from bull spermatozoa. *Biochim. Biophys. Acta* 406, 479–488.

243. Russell, L., Peterson, R., and Freund, M. (1979). Direct evidence for formation of hybrid vesicles by fusion of plasma and outer acrosomal membranes during the acrosome reaction in boar spermatozoa. *J. Exp. Zool.* 208, 41–56.
244. Noland, T. D., Olson, G. E., Garbers, D. L. (1983). Purification and partial characterization of plasma membranes from bovine spermatozoa. *Biol. Reprod.* 29, 987–998.
245. Topfer-Petersen, E., and Schill, W. B. (1981). A new separation method of subcellular fractions of boar spermatozoa. *Andrologia* 13, 174–176.
246. Olson, G. E., Winfrey, V. P., Garbers, D. L., and Noland, T. D. (1985). Isolation and characterization of a macromolecular complex associated with the outer acrosomal membrane of bovine spermatozoa. *Biol. Reprod.* 33, 761–779.
247. Thakkar, J. K., East, J., Seyler, D., and Fanson, R. C. (1983). Surface-active phospholipase A_2 in mouse spermatozoa. *Biochim. Biophys. Acta* 754, 44–50.
248. Rahi, H., Sheikhnejade, G., and Srivastava, P. N. (1983). Isolation of the inner acrosomal-nuclear membrane complex from rabbit spermatozoa. *Gamete Res.* 7, 215–225.
249. Cornwall, G. A., Vreeburg, J. T., Holland, M. K., and Orgebin-Crist, M-C. (1990). Interactions of labeled epididymal secretory proteins with spermatozoa after injection of ^{35}S-methionine in the mouse. *Biol. Reprod.* 43, 121–129.
250. Jimenez, C., Lefrancois, A. M., Ghyselinck, N. B., and Dufaure, J. P. (1992). Characterization and hormonal regulation of 24 kDa protein synthesis by the adult murine epididymis. *J. Endocrinol.* 133, 197–203.
251. Schwarz, M. A., and Koehler, J. K. (1976). Alterations in lectin binding to guinea pig spermatozoa accompanying in vitro capacitation and the acrosome reaction. *Biol. Reprod.* 21, 1295–1307.
252. Myles, D. G., and Primakoff, P. (1984). Localized surface antigens of guinea pig sperm migrate to new regions prior to fertilization. *J. Cell. Biol.* 99, 1634–1641.
253. Cowan, A. E., Myles, D. G., and Koppel, D. E. (1991). Migration of the guinea pig sperm membrane protein PH-20 from one localized surface domain to another does not occur by a simple diffusion-trapping mechanism. *Dev. Biol.* 144, 189–198.
254. Cowen, A. E., Primakoff, P., Myles, D. G. (1986). Sperm exocytosis increases the amount of PH-20 antigen on the surface of guinea pig sperm. *J. Cell. Biol.* 103, 1289–1297.
255. Fawcett, D. W. (1975). Morphogenesis of the mammalian sperm acrosome in new perspective. In *The Functional Anatomy of the Spermatozoon* (B. A. Afzelius, Ed.), pp. 199–210. Pergamon Press, Oxford.
256. Mann, T., and Lutwak-Man, C. (1981). *Male Reproductive Function and Semen*. Springer-Verlag, New York.
257. Allison, A. C., and Hartree, E. F. (1970). Lysosomal enzymes in the acrosome and their possible role in fertilization. *J. Reprod. Fertil.* 21, 501–515.
258. Stambaugh, R., Brackett, B. G., and Mastroianni, L. (1969). Inhibition of in vitro fertilization of rabbit ova by trypsin inhibitors. *Biol. Reprod.* 1, 223–227.
259. Beyler, S. A., and Zaneveld, L. J. D. (1982). Inhibition of in vitro fertilization of mouse gametes by proteinase inhibitors. *J. Reprod. Fertil.* 66, 425–431.
260. Polakoski, K. L., and Parrish, R. F. (1977). Boar proacrosin. Purification and preliminary activation studies of proacrosin isolated from ejaculated boar sperm. *J. Biol. Chem.* 252, 1888–1894.
261. Tobias, P. S., and Schumacher, G. F. B. (1977). Observation of two proacrosins in extracts of human spermatozoa. *Biochem. Biophys. Res. Commun.* 74, 434–439.
262. Brown, C. R., and Harrison, R. A. P. (1978). The activation of proacrosin in spermatozoa from ram, bull and boar. *Biochim. Biophys. Acta* 526, 202–217.
263. Mukerji, S. K., and Meizel, S. (1979). Rabbit testis proacrosin. Purification, molecular weight estimation, and amino acid and carbohydrate composition of the molecule. *J. Biol. Chem.* 254, 11721–11728.
264. Müller-Esterl, W., and Fritz, H. (1981). Sperm acrosin. In *Methods in Enzymology* (L. Lorand, Ed.), Vol. 80, pp. 621–632. Academic Press, New York.
265. Baba, T., Kashiwabara, S-I., Watanabe, K., Itoh, H., Michikawa, Y., Kimura, K., Takada, M., Fukamizu, A., and Arai, Y. (1989). Activation and maturation mechanisms of boar acrosin zymogen based on the deduced primary structure. *J. Biol. Chem.* 264, 11920–11927.
266. Adham, I. M., Klemm, U., Maier, W-M., Hoyer-Fender, S., Tsaousidou, S., and Engel, W. (1989). Molecular cloning of preproacrosin and analysis of its expression pattern in spermatogenesis. *Eur. J. Biochem.* 182, 563–568.
267. Baba, T., Watanabe, K., Kashiwabara, S-I., and Arai, Y. (1989). Primary structure of human proacrosin deduced from its cDNA sequence. *FEBS Lett.* 244, 296–300.
268. Adham, I. M., Klemm, U., Maier, W-M., and Engel, W. (1990). Molecular cloning of human preproacrosin cDNA. *Hum. Genet.* 84, 125–128.
269. Kashiwabara, S-I., Baba, T., Takada, M., Wantanabe, K., Yano, Y., and Arai, Y. (1990). Primary structure of mouse proacrosin deduced from the cDNA sequence and its gene expression during spermatogenesis. *J. Biochem.* 108, 785–791.
270. Klemm, U., Maier, W-M., Tsaousidou, S., Adham, I. M., Willison, K., and Engel, W. (1990). Mouse preproacrosin: cDNA sequence, primary structure and postmeiotic expression in spermatogenesis. *Differentiation* 42, 160–166.
271. Klemm, U., Flake, A., and Engel, W. (1991). Rat sperm acrosin: cDNA sequence derived primary structure and phylogenetic origin. *Biochim. Biophys. Acta* 1090, 270–272.
272. Kashiwabara, S-I., Arai, Y., Kodaira, K., and Baba, T. (1990). Acrosin biosynthesis in meiotic and postmeiotic spermatogenic cells. *Biochem. Biophys. Res. Comm.* 173, 240–245.
273. Anakwe, O. O., Sharma, S., Hardy, D.M., and Gerton, G. L. (1991). Guinea pig proacrosin is synthesized principally by round spermatids and contains O-linked as well as N-linked oligosaccharide side chains. *Mol. Reprod. Dev.* 29, 172–179.
274. Escalier, D., Gallo, J-M., Albert, M., Meduri, G., Bermudez, D., David, G., and Schrevel, J. (1991). Human acrosome biogenesis: immunodetection of proacrosin in primary spermatocytes and of its partitioning pattern during meiosis. *Development* 113, 779–788.
275. Garner, D. L., and Easton, M. P. (1977). Immunofluorescent localization of acrosin in mammalian spermatozoa. *J. Exp. Zool.* 200, 157–162.
276. Morton, D. B. (1975). Acrosomal enzymes: immunochemical localization of acrosin and hyaluronidase in ram spermatozoa. *J. Reprod. Fertil.* 45, 375–378.
277. Morton, D. B. (1977). Lysosomal enzymes in mammalian spermatozoa. In *Immunobiology of Gametes* (M. Edidin and M. H. Johnson, Eds.), pp. 115–155. Cambridge University Press, London.
278. Green, D. P. L., and Hockaday, A. R. (1978). The histochemical localization of acrosin in guinea-pig sperm after the acrosome reaction. *J. Cell. Sci.* 32, 177–184.
279. Baba, T., Azuma, S., Kashiwabara, S., and Toyoda, Y. (1994). Sperm from mice carrying a targeted mutation of the acrosin gene can penetrate the oocyte zona pellucida and effect fertilization. *J. Biol. Chem.* 269, 31845–31849.
280. Adham, I. M., Nayernia, K., and Engel, W. (1997). Spermatozoa lacking acrosin protein show delayed fertilization. *Mol. Reprod. Dev.* 46, 370–376.

281. Yanagimachi, R. (1981). Mechanisms of fertilization in mammals. In *Fertilization and Embryonic Development in Vitro* (L. Mastroianni and J. D. Biggers, Eds.), pp. 81–182. Plenum Press, New York.
282. Baba, T., Niida, Y., Michikawa, Y., Kashiwabara, S-i., Kodaira, K., Takenaka, M., Kohno, N., Gerton, G. L., and Arai, Y. (1994). An acrosomal protein, sp32, in mammalian sperm is a binding protein specific for two proacrosins and an acrosin intermediate. *J. Biol. Chem.* 269, 10133–10140.
283. Flörke-Gerloff, S., Tschesche, H., Müller-Esterl, W., and Engel, W. (1984). Intra-acrosomally located acrosin-inhibitors: evolution and developmental patterns in mammals. *Gamete Res.* 10, 327–337.
284. Möritz, A., Lilja, H., and Fink, E. (1991). Molecular cloning and sequence analysis of the cDNA encoding the human acrosin-trypsin inhibitor (HUSI-II). *FEBS Lett.* 278, 127–130.
285. Jonáková, V., Cechová, D., Töpfer-Petersen, E., Calvete, J. J., and Veselksý, L. (1991). Variability of acrosin inhibitors in boar reproductive tract. *Biomed. Biochim. Acta* 50, 691–695.
286. Jonáková, V., Calvete, J. J., Mann, K., Schäfer, W., Schmid, E. R., and Töpfer-Peterson, E. (1992). The complete primary structure of three isoforms of a boar sperm-associated acrosin inhibitor. *FEBS Lett.* 297, 147–150.
287. Moore, A., Penfold, L. M., Johnson, J. L., Latchman, D. S., and Moore, H. D. M. (1993). Human sperm-egg binding is inhibited by peptides corresponding to core region of an acrosomal serine protease inhibitor. *Mol. Reprod. Dev.* 34, 280–291.
288. Zaneveld, L. J. D., Polakoski, K. L., and Schumacher, G. F. B. (1973). Properties of acrosomal hyaluronidase from bull spermatozoa. Evidence for its similarity to testicular hyaluronidase. *J. Biol. Chem.* 248, 564–570.
289. Yang, C.-H., and Srivastava, P. N. (1975). Purification and properties of hyaluronidase from bull sperm. *J. Biol. Chem.* 250, 79–83.
290. Goldberg, E. (1977). Isozymes in testes and spermatozoa. In *Isozymes: Current Topics in Biological and Medical Research* (M. Ratazzi, J. Scandalios, and G. Whitt, Eds.), Vol. 1, pp. 79–124. Alan R. Liss, New York.
291. Mancini, R. E., Alonso, A., Barquet, J., and Nemirovski, B. (1964). Histo-immunological localization of hyaluronidase in bull testis. *J. Reprod. Fertil.* 8, 325–330.
292. Gould, S. F., and Bernstein, M. H. (1975). Localization of bovine sperm hyualuronidase. *Differentiation* 3, 123–132.
293. Brown, C. R. (1975). Distribution of hyaluronidase in the ram spermatozoa. *J. Reprod. Fertil.* 45, 537–539.
294. Hardy, D. M., Oda, M. N., Friend, D. S., and Huang, T. T. F., Jr. (1991). A mechanism for differential release of acrosomal enzymes during the acrosome reaction. *Biochem. J.* 275, 759–766.
295. Majumder, G. C., and Turkington, R. W. (1974). Acrosomal and lysosomal isoenzymes of β-galactosidase and *N*-acetyl-β-glucosaminidase in rat testis. *Biochemistry* 13, 2857–2864.
296. Majumder, G. C., Lessin, S., and Turkington, R. W. (1975). Hormonal regulation of isoenzymes of N-acetyl-β-glucosaminidase and β-galactosidase during spermatogenesis in the rat. *Endocrinology* 96, 890–897.
297. Nikolajczyk, B. S., and O'Rand, M. G. (1992). Characterization of rabbit testis β-galactosidase and arylsulfatase A: purification and localization in spermatozoa during the acrosome reaction. *Biol. Reprod.* 46, 366–378.
298. Koren, E., and Milkov, S. (1973). "Collagenase-like" peptidase in human, rat and bull spermatozoa. *J. Reprod. Fertil.* 32, 349–356.
299. Erickson, R. P., and Martin, S. R. (1974). The relationship of mouse spermatozoal to mouse testicular cathepsins. *Arch. Biochem. Biophys.* 165, 114–120.
300. Haraguchi, C. M., Ishido, K., Kominami, E., and Yokota, S. (2003). Expression of cathepsin H in differentiating rat spermatids: immunoelectron microscopic study. *Histochem. Cell. Biol.* 120, 63–71.
301. Arboleda, C. E., and Gerton, G. L. (1988). Studies of three major proteases associated with guinea pig sperm acrosomes. *J. Exp. Zool.* 244, 277–287.
302. Talbot, P., and DiCarlantonio, G. (1985). Cytochemical localization of dipeptidyl peptidase II (DPP II) in mature guinea pig sperm. *J. Histochem. Cytochem.* 33, 1169–1172.
303. DiCarlantonio, G., Talbot, P., and Dudenhausen, E. (1986). Partial purification and characterization of dipeptidyl peptidase II (DPP II) from guinea pig testes. *Gamete Res.* 15, 161–175.
304. Schollmeyer, J. E. (1986). Identification of calpain II in porcine sperm. *Biol. Reprod.* 34, 721–731.
305. Srivastava, P. N., and Abou-Issa, H. (1977). Purification and properties of rabbit spermatozoal acrosomal neuraminidase. *Biochem. J.* 161, 193–200.
306. Bryan, J. H. D., and Unithan, R. R. (1972). Non-specific esterase activity in bovine acrosomes. *Histochem. J.* 4, 413–419.
307. Meizel, S., and Cotham, J. (1972). Partial characterization of a new bull sperm arylaminidase. *J. Reprod. Fertil.* 28, 303–307.
308. Bhalla, V. K., Tillman, W. L., and Williams, W. L. (1973). Presence of β-aspartyl N-acetylglucosamine amido hydrolase in mammalian spermatozoa. *J. Reprod. Fertil.* 34, 137–139.
309. Dudkiewicz, A. B. (1984). Purification of boar acrosomal arylsulfatase A and possible role in the penetration of cumulus cells. *Biol. Reprod.* 30, 1005–1014.
310. Gonzales, L. W., and Meizel, S. (1973). Acid phosphatases of rabbit spermatozoa. II. Partial purification and biochemical characterization of the multiple forms of rabbit spermatozoan acid phosphatase. *Biochim. Biophys. Acta* 320, 180–194.
311. Stambaugh, R., and Buckley, J. (1970). Comparative studies of the acrosomal enzymes of rabbit, rhesus monkey and human spermatozoa. *Biol. Reprod.* 3, 275–282.
312. Ribbes, H., Plantavid, M., Bennet, P. J., Chap, H., and Douste-Blazy, L. (1987). Phospholipase C from human sperm specific for phosphoinositides. *Biochim. Biophys. Acta* 919, 245–254.
313. Hinkovska-Galchev, V., and Srivastava, P. N. (1992). Phosphatidylcholine and phosphatidylinositol-specific phospholipases C of bull and rabbit spermatozoa. *Mol. Reprod. Dev.* 33, 281–286.
314. Meizel, S. (1984). The importance of hydrolytic enzymes to an exocytotic event, the mammalian sperm acrosome reaction. *Biol. Rev.* 59, 125–157.
315. Rönkkö, S. (1992). Immunocytochemical localization of phospholipase A_2 in the bovine seminal vesicle and on the surface of the ejaculated spermatozoa. *Int. J. Biochem.* 24, 869–876.
316. Kohno, N., Yamagata, K., Yamada, S., Kashiwabara, S-i., Sakai, Y., and Baba, T. (1998). Two novel testicular serine proteases, TESP1 and TESP2, are present in the mouse sperm acrosome. *Biochem. Biophys. Res. Comm.* 245, 658–665.
317. Ohmura, K., Kohno, N., Kobayashi, Y., Yamagata, K., Sato, S., Kashiwabara, S-i., and Baba, T. (1999). A homologue of pancreatic trypsin is localized in the acrosome of mammalian sperm and is released during acrosome reaction. *J. Biol. Chem.* 274, 29426–29432.
318. Karnik, N. S., Newman, S., Kopf, G. S., and Gerton, G. L. (1992). Developmental expression of G protein α subunits in mouse spermatogenic cells: evidence that $G_{\alpha i1}$ is associated with the developing acrosome. *Dev. Biol.* 152, 393–402.

319. Garty, N. B., Galiani, D., Aharonheim, A., Ho, Y-K., Phillips, D. M., Dekel, N., and Salomon, Y. (1988). G-proteins in mammalian gametes: an immunocytochemical study. *J. Cell. Sci.* 91, 21–31.
320. Glassner, M., Jones, J., Kligman, I., Woolkalis, M. J., Gerton, G. L., and Kopf, G. S. (1991). Immunocytochemical and biochemical characterization of guanine nucleotide-binding regulatory proteins in mammalian spermatozoa. *Dev. Biol.* 146, 438–450.
321. Kopf, G. S., and Gerton, G. L. (1991). The mammalian sperm acrosome and the acrosome reaction. In *Elements of Fertilization* (P. M. Wassarman, Ed.), Vol. 1, pp. 153–203. CRC Press, Boca Raton, FL.
322. Anakwe, O. O., and Gerton, G. L. (1990). Acrosome biogenesis during meiosis: evidence from the synthesis and distribution of an acrosomal glycoprotein, acrogranin, during guinea pig spermatogenesis. *Biol. Reprod.* 42, 317–328.
323. Baba, T., Hoff, H. B. III, Nemoto, H., Lee, H., Orth, J., Arai, Y., and Gerton, G. L. (1993). Acrogranin, an acrosomal cysteine-rich glycoprotein, is the precursor of the growth-modulating peptides, granulins, and epithelins, and is expressed in somatic as well as male germ cells. *Mol. Reprod. Dev.* 34, 233–243.
324. Kew, D., Muffly, K. E., and Kilpatrick, D. L. (1990). Proenkephalin products are stored in the sperm acrosome and may function in fertilization. *Proc. Natl. Acad. Sci. U S A* 87, 9143–9147.
325. Schalling, M., Persson, H., Pelto-Huikko, M., Odum, L., Ekman, P., Gottlieb, C., Höfelt, T., and Rehfeld, J. F. (1990). Expression and localization of gastrin messenger RNA and peptide in spermatogenic cells. *J. Clin. Invest.* 86, 660–669.
326. Persson, H., Rehfeld, J. F., Ericsson, A., Schalling, M., Pelto-Huikko, M., and Höfelt, T. (1989). Transient expression of the cholecystokinin gene in male germ cells and accumulation of the peptide in the acrosomal granule: possible role of cholecystokinin in fertilization. *Proc. Natl. Acad. Sc. U S A* 86, 6166–6170.
327. Noland, T. D., Friday, B. B., Meulit, M. T., and Gerton, G. L. (1994). The sperm acrosomal matrix contains a novel member of the pentaxin family of calcium-dependent binding proteins. *J. Biol. Chem.* 269, 32607–32614.
328. Foster, J. A., and Gerton, G. L. (1996). Autoantigen 1 of the guinea pig sperm acrosome is the homologue of mouse Tpx-1 and human TPX1 and is a member of the cysteine-rich secretory protein (CRISP) family. *Mol. Reprod. Dev.* 44, 221–229.
329. Hao, Z., Wolkowicz, M. J., Shetty, J., Klotz, K., Bolling, L., Sen, B., Westbrook, V. A., Coonrod, S., Flickinger, C. J., and Herr, J. C. (2002). SAMP32, a testis-specific, isoantigenic sperm acrosomal membrane-associated protein. *Biol. Reprod.* 66, 735–744.
330. Michaut, M., De Blas, G., Tomes, C. N., Yunes, R., Fukuda, M., and Mayorga, L. S. (2001). Synaptotagmin VI participates in the acrosome reaction of human spermatozoa. *Dev. Biol.* 235, 521–529.
331. Hardy, D. M., and Garbers, D. L. (1994). Species-specific binding of sperm proteins to the extracellular matrix (zona pellucida) of the egg. *J. Biol. Chem.* 269, 19000–19004.
332. Olson, G. E., Winfrey, V. P., Bi, M., Hardy, D. M., and NagDas, S. K. (2004). Zonadhesin assembly into the hamster acrosomal matrix occurs by distinct targeting strategies during spermiogenesis and maturation in the epididymis. *Biol. Reprod.* 71, 1128–1134.
333. Olson, G. E., Winfrey, V. P., and NagDas, S. K. (2003). Structural modification of the hamster sperm acrosome during posttesticular development in the epididymis. *Micros. Res. Tech.* 61, 46–55.
334. Tanii, I., Oh-oko, T., Yshingaga, K., and Toshimori, K. (2001). A mouse acrosomal cortical matrix protein, MC41, has ZP2-binding activity and forms a complex with a 75-kDa serine protease. *Develop. Biol.* 238, 332–341.
335. Foster, J. A., Friday, B. B., Maulit, M. T., Blobel, C., Winfrey, V. P., Olson, G. E., Kim, K-S., and Gerton, G. L. (1997). AM67, a secretory component of the guinea pig sperm acrosomal matrix, is related to mouse sperm protein sp56 and the complement component 4-binding protein. *J. Biol. Chem.* 272, 12714–12722.
336. Kim, K-S., Cha, M. C., and Gerton, G. L. (2001). Mouse sperm protein sp56 is a component of the acrosomal matrix. *Biol. Reprod.* 64, 36–43.
337. Foster, J. A., Klotz, K. L., Flickinger, C. L., Thomas, T. X., Wright, R. M. Castillo, J. R., and Herr, J. C. (1994). Human SP-10: acrosomal distribution, processing and fate after the acrosomal reaction. *Biol. Reprod.* 51, 1222–1231.
338. Baba, T., Niida, Y., Michikawa, Y., Kashiwabara, S-i., Kodaira, K., Takenaka, M., Kohno, N., Gerton, G. L., and Arai, Y. (1994). An acrosomal protein, sp32, in mammalian sperm is a binding protein specific for two proacrosins and an acrosin intermediate. *J. Biol. Chem.* 269, 10133–10140.
339. Mandal, A., Klotz, K. L., Shetty, J., Jayes, F. L., Wolkowicz, M. J., Bolling, L. C., Coonrod, S. A., Black, M. B., Diekman, A. B., Haystead, T. A. J., Flickinger, C. J. and Herr, J. C. (2003). SLLP1, a unique, intra-acrosomal, non-bacteriolytic, c lysozyme-like protein of human spermatozoa. *Biol. Reprod.* 68, 1525–1537.
340. Shetty, J., Wolkowicz, M. J., Digilio, L. C., Klotz, K. L., Jayes, F. L., Diekman, A. B., Westbrook, V. A., Farris, E. M., Hao, Z., Coonrod, S. A., Flickinger, C. J., and Herr, J. C. (2003). SAMP14, a novel, acrosomal membrane-associated, glycosylphosphatidylinositol-anchored member of the Ly-6/urokinase-type plasminogen activator receptor superfamily with a role in sperm-egg interaction. *J. Biol. Chem.* 278, 30506–30515.
341. Zamboni, L., and Stefanini, M. (1971). The fine structure of the neck of mammalian spermatozoa. *Anat. Rec.* 169, 155–172.
342. Oura, C. (1971). The ultrastructure and development of the neck region of the golden hamster spermatozoon. *Monitore Zool. Ital.* 5, 253–264.
343. Baccetti, B. (1984). The human spermatozoon. In *Ultrastructure of Reproduction* (J. Van Blerkom and P. M. Motta, Eds.), pp. 110–126. Martinus Nijhoff, The Hague.
344. Phillips, D. M. (1972). Comparative analysis of mammalian sperm motility. *J. Cell. Biol.* 53, 561–573.
345. Edelman, G. M., and Millette, C. F. (1971). Molecular probes of spermatozoon structures. *Proc. Nat. Acad. Sci. U S A* 68, 2436–2440.
346. Millette, C. F., Spear, P. G., Gall, W. E., and Edelman, G. M. (1973). Chemical dissection of mammalian spermatozoa. *J. Cell. Biol.* 58, 662–675.
347. Young, R. J., and Cooper, G. W. (1979). Separation of the head and tail of mammalian spermatozoa by primary amines: evidence for their junction by Schiff bases. In *The Spermatozoon* (D. W. Fawcett and J. M. Bedford, Eds.), pp. 391–394. Urban & Schwarzenberg, Baltimore.
348. Calvin, H. I. (1976). Isolation and subfractionation of mammalian sperm heads and tails. In *Methods in Cell Biology* (D. M. Prescott, Ed.), Vol. 13, pp. 85–104. Academic Press, New York.
349. Bellvé, A. R., Anderson, E., and Hanley-Bowdoin, L. (1975). Synthesis and amino acid composition of basic proteins in mammalian sperm nuclei. *Dev. Biol.* 47, 349–365.
350. Bedford, J. M, and Calvin, H. I. (1974). Changes in the -S-S-linked structures of the sperm tail during epididymal maturation with comparative observations in sub-mammalian species. *J. Exp. Zool.* 187, 181–204.

351. Blom, E., and Birch-Anderson, A. (1970). The ultrastructure of the bull sperm. *Nord. Vet. Med.* 17,193–212.
352. Illison, L. (1966). Fine structure of the mature spermatozoan head and neck of the mouse. *J. Anat.* 100, 949–950.
353. Woolley, D. M., and Fawcett, D. W. (1973). The degeneration and disappearance of the centrioles during the development of the rat spermatozoon. *Anat. Rec.* 177, 289–302.
354. Fawcett, D. W., Porter, K. R. (1954). A study of the fine structure of ciliated epithelia. *J. Morph.* 94, 221–281.
355. Gibbons, I. R., and Grimstone, A. V. (1960). On flagellar structure in certain flagellates. *J. Biophys. Biochem. Cytol.* 7, 697–716.
356. Olson, G. E., and Linck, R. W. (1977). Observations of the structural components of flagellar axonemes and central pair microtubules from rat sperm. *J. Ultrastruct. Res.* 61, 21–43.
357. Bryan, J., and Wilson, L. (1971). Are cytoplasmic microtubules heteropolymers? *Proc. Nat. Acad. Sci. U S A* 8, 1762–1766.
358. Hecht, N. B., Kleene, K. C., Distel, R. J., and Silver, L. M. (1984). The differential expression of the actins and tubulins during spermatogenesis in the mouse. *Exp. Cell. Res.* 153, 275–280.
359. Kierszenbaum, A. L. (2002). Sperm axoneme: a tale of tubulin posttranslational diversity. *Mol. Reprod. Dev.* 62, 1–3.
360. Distel, R. J., Kleene, K. C, and Hecht, N. B. (1984). Haploid expression of a mouse testis α-tubulin gene. *Science* 224, 68–70.
361. Hecht, N. B., Bower, P. A., Waters, S. H., Yelick, P. C., and Distel, R. J. (1986). Evidence for haploid expression of mouse testicular genes. *Exp. Cell. Res.* 164, 183–190.
362. Gibbons, I. R., and Rowe, A. J. (1965). Dynein: a protein with ATPase activity from cilia. *Science* 149, 424–426.
363. Gibbons, I. R., and Fronk, E. (1972). Some properties of bound and soluble dynein from sea urchin flagella. *J. Cell. Biol.* 54, 365–381.
364. Burgess, S. A., and Knight, P. J. (2004). Is the dynein motor a winch? *Curr. Opin. Struct. Biol.* 14, 138–146.
365. Satir, P. (1999). The cilium as a biological nanomachine. *FASEB J.* 13, S235–S237.
366. Neesen, J., Kirschner, R., Ochs, M., Schmiedl, A., Habermann, B., Mueller, C., Holstein, A. F., Nuesslein, T., Adham, I., and Engel, W. (2001). Disruption of an inner arm dynein heavy chain gene results in asthenozoospermia and reduced ciliary beat frequency. *Hum. Mol. Genet.* 10, 1117–1128.
367. Smith, E. F., and Yang, P. (2004). The radial spokes and central apparatus: mechano-chemical transducers that regulate motility. *Cell. Motil. Cytoskeleton* 57, 8–17.
368. Wemmer, K. A., and Marshall, W. F. (2004). Flagellar motility: all pull together. *Curr. Biol.* 14, D992–R993.
369. Linck, R. W. (1976). Flagellar doublet microtubules: fractionation of minor components and α-tubulin from specific regions of the A-tubule. *J. Cell. Sci.* 20, 405–539.
370. Linck, R. W., and Langevin, G. L. (1982). Structure and chemical composition of insoluble filamentous components of sperm flagellar microtubules. *J. Cell. Biol.* 58, 1–22.
371. Linck, R. W., Amos, L. A., and Amos, W. B. (1985). Localization of tektin filaments in microtubules of sea urchin flagella by immunoelectron microscopy. *J. Cell. Biol.* 100, 126–135.
372. Steffen, W., and Linck, R. W. (1988). Evidence for tektins in centrioles and axonemal microtubules. *Proc. Natl. Acad. Sci. U S A* 85, 2643–2647.
373. Stephens, R. E., and Lemieux, N. A. (1998). Tektins as structural determinants in basal bodies. *Cell. Motil. Cytoskel.* 40, 379–392.
374. Norrander, J., Larsson, M., Ståhl, S., Höög, C., and Linck, R. W. (1998). Expression of ciliary tektins in brain and sensory development. *J. Neurosci.* 18, 8912–8918.
375. Larsson, M., Norrander, J., Gräslund, S., Brundell, E., Linck, R., Ståhl, and Höög, C. (2000). The spatial and temporal expression of Tekt1, a mouse tektin C homologue, during spermatogenesis suggests that it is involved in the development of the sperm tail basal body and axoneme. *Eur. J. Cell. Biol.* 79, 718–725.
376. Iguchi, N., Tanaka, H., Fujii, T., Tamura, K., Kanelo, Y., Nojima, H., and Nishimune, Y. (1999). Molecular cloning of haploid germ cell-specific tektin cDNA and analysis of the protein in mouse testis. *FEBS Lett.* 456, 315–321.
377. Wolkowicz, J. J., Naaby-Hansen, S., Gamble, A. R., Reddi, P. P., Flickinger, C. J., and Herr, J. C. (2002). Tektin B1 demonstrates flagellar localization in human sperm. *Biol. Reprod.* 66, 241–250.
378. Tanaka, H., Iguchi, N., Toyama, Y., Kitamura, K., Takahashi, T., Kaseda, K., Maekawa, M. and Nishimune, Y. (2004). Mice deficient in the axonemal protein tektin-t exhibit male infertility and immotile-cilium syndrome due to impaired inner arm dynein function. *Mol. Cell. Biol.* 24, 7958–7964.
379. Roy, A., Yan, W., Burns, K. H., and Matzuk, M. M. (2004). Tektin3 encodes an evolutionarily conserved putative testicular microtubules-related protein expressed preferentially in male germ cells. *Mol. Reprod. Dev.* 67, 295–302.
380. Woolley, D. M. (1970). Selection for the length of the spermatozoan midpiece in the mouse. *Genet. Res.* 16, 225–228.
381. Fawcett, D. W., and Ito, S. (1965). The fine structure of bat spermatozoa. *Am. J. Anat.* 116, 567–610.
382. Phillips, D. M. (1970). Ultrastructure of spermatozoa of the woolly opossum Caluromys philander. *J. Ultrastruct. Res.* 33, 381–397.
383. Olson, G. E., and Winfrey, V. P. (1990). Mitochondria–cytoskeletal interactions in the sperm midpiece. *J. Struct. Biol.* 103, 13–22.
384. Calvin, H. I., Cooper, G. W., and Wallace, E. W. (1981). Evidence that selenium in rat sperm is associated with a cysteine-rich structural protein of the mitochondrial capsule. *Gamete Res.* 4, 139–149.
385. Pallini, V., and Bacci, E. (1979). Bull sperm selenium is bound to a structural protein of mitochondria. *J. Submicr. Cytol.* 11, 165–170.
386. Calvin, H. I., Grosshans, K., Musicant-Shikora, S. R., and Turner, S. I. (1987). A developmental study of rat sperm and testis selenoproteins. *J. Reprod. Fertil.* 81, 1–11.
387. Kleene, K. C., Smith, J., Bozorgzadeh, A., Harris, M., Hahn, L., Karimpour, I., and Gerstel, J. (1990). Sequence and developmental expression of the mRNA encoding the selenoprotein of the sperm mitochondrial capsule in the mouse. *Dev. Biol.* 137, 395–402.
388. Calvin, H. I., Wallace, E. W., and Cooper, G. W. (1981). Role of selenium in the organization of the mitochondrial sheath in the organization of the mitochondrial sheath in rodent spermatozoa. In *Selenium in Biology and Medicine* (J. E. Spallholz, J. L. Martin, and H. E. Ganther, Eds.), pp. 319–324. AVI, Westport, CT.
389. Olson, G. E., Winfrey, V. P., Hill, K. E., and Burke, R. F. (2004). Sequential development of flagellar defects in spermatids and epididymal spermatozoa of selenium-deficient rats. *Reproduction* 127, 3335–3342.
390. Adham, I. M., Tessman, D., Soliman, K. A., Murphy, D., Kremling, H., Szpizer, C., and Engel, W. (1996). Cloning, expression, and chromosomal localization of rat mitochondrial capsule selenoprotein (MCS): the reading frame does not contain potential UGS selenocysteine codons. *DNA Cell. Biol.* 15, 159–196.
391. Cataldo, L., Baig, K., Oko, R., Mastrangelo, M. A., and Kleene, K. C. (1996). Developmental expression, intracellular localization, and selenium content of the cysteine-rich

protein associated with the mitochondrial capsules of mouse sperm. *Mol. Reprod. Dev.* 45, 320–331.
392. Nayernia, K., Adham, I. M., Burkhardt-Göttges, E., Neesen, J., Rieche, M., Wolf, S., Sancken, U., Kleene, K., and Engel, W. (2002). Asthenozoospermia in mice with targeted deletion of the sperm mitochondrion-associated cysteine-rich protein (*Smcp*) gene. *Mol. Cell. Biol.* 22, 3046–3052.
393. Doiguchi, M., Mori, T., Toshimori, K., Shibata, Y., and Iida, H. (2002). Spergen-1 might be an adhesive molecule associated with mitochondria in the middle piece of spermatozoa. *Dev. Biol.* 252, 127–137.
394. Serres, C., Feneux, D., and Jouannet, P. (1984). Abnormal distribution of the periaxonemal structures in a human sperm flagellar dyskinesia. *Cell. Motil. Cytoskel.* 6, 8–76.
395. Lindemann, C. B., Fentie, I., and Rikmenspoel, R. (1980). A selective effect of Ni^{2+} on wave initiation in bull sperm flagella. *J. Cell. Biol.* 87, 420–426.
396. Telkka, A., Fawcett, D. W., and Christensen, A. K. (1961). Further observations on the structure of the mammalian sperm tail. *Anat. Rec.* 141, 231–246.
397. Fawcett, D. W., and Phillips, D. W. (1969). The fine structure and development of the neck region of the mammalian spermatozoon. *Anat. Rec.* 165, 153–184.
398. de Kretser, D. M. (1969). Ultrastructural features of human spermiogenesis. *Z. Zellforsch.* 98, 229–236.
399. Sapsford, C. S., Rae, C. A., and Cleland, K. W. (1970). Ultrastructural studies on the development and form of the principal piece sheath of the Bandicoot spermatozoon. *Aust. J. Zool.* 8, 21–48.
400. Irons, M. J., and Clermont, Y. (1982). Formation of the outer dense fibers during spermiogenesis in the rat. *Anat. Rec.* 202, 463–471.
401. Oko, R., and Clermont, Y. (1989). Light microscopic immunocytochemical study of fibrous sheath and outer dense fiber formation in the rat spermatid. *Anat. Rec.* 225, 46–55.
402. Clermont, Y., Oko, R., and Hermo, L. (1990). Immunocytochemical localization of proteins utilized in the formation of outer dense fibers and fibrous sheath in rat spermatids: an electron microscope study. *Anat. Rec.* 227, 447–457.
403. Gordon, M., and Bensch, K. G. (1968). Cytochemical differentiation of the guinea pig sperm flagellum with phosphotungstic acid. *J. Ultrastruct. Res.* 24, 33–50.
404. Woolley, D. M. (1971). Striations in the peripheral fibers of rat and mouse spermatozoa. *J. Cell. Biol.* 49, 936–939.
405. Olson, G. E., and Sammons, D. W. (1980). Structural chemistry of outer dense fibers of rat sperm. *Biol. Reprod.* 22, 319–332.
406. Baccetti, B., Pallini, V., and Burrini, A. G. (1976). The accessory fibers of the sperm tail. III. High-sulfur and low-sulfur components in mammals and cephalopods. *J. Ultrastruct. Res.* 57, 289–308.
407. Pihlaja, D. J., and Roth, L. E. (1973). Bovine sperm fractionation. II. Morphology and chemical analysis of tail segments. *J. Ultrastruct. Res.* 44, 293–309.
408. Baccetti, B., Pallini, V., and Burrini, A. G. (1973). The accessory fibers of the sperm tail. I. Structure and chemical composition of the bull coarse fibers. *J. Submicrosc. Cytol.* 5, 237–256.
409. Vera, J. C., Brito, M., Zuvic, T., and Burzio, L. O. (1984). Polypeptide composition of rat sperm outer dense fibers. A simple procedure to isolate the fibrillar complex. *J. Biol. Chem.* 259, 5970–5977.
410. Price, J. M. (1973). Biochemical and morphological studies of outer dense fibers of rat spermatozoa. *J. Cell. Biol.* 59, 272a.
411. Oko, R. (1988). Comparative analysis of proteins from the fibrous sheath and outer dense fibers of rat spermatozoa. *Biol. Reprod.* 39, 168–182.
412. Brito, M., Figueroa, J., Vera, J. C., Cortes, P., Hott, R., and Burzio, L. O. (1986). Phosphoproteins are structural components of bull sperm outer dense fiber. *Gamete Res.* 15, 327–336.
413. Henkel, R., Stalf, T., and Miska, W. (1992). Isolation and partial characterization of the outer dense fiber proteins from human spermatozoa. *Biol. Chem. Hoppe-Seyler* 373, 685–689.
414. Gunn, S. A., and Gould, T. C. (1970). Cadmium and other mineral elements. In *The Testis* (A. D. Johnson, W. R. Gomes, and N. L. Vandemark, Eds.), Vol. III, pp. 377–481. Academic Press, New York.
415. Miller, M. J., Vincent, N. R., and Mawson, C. A. (1961). An autoradiographic study of the distribution of zinc-65 in rat tissues. *J. Histochem. Cytochem.* 9, 111–125.
416. Calvin, H. I. (1979). Electrophoretic evidence for the identity of the major zinc-binding polypetides in the rat sperm tail. *Biol. Reprod.* 21, 873–882.
417. van der Hoorn, F. A., Tarnasky, H. A., Nordeen, S. K. (1990). A new rat gene RT7 is specifically expressed during spermatogenesis. *Dev. Biol.* 142, 147–154.
418. Burfeind, P., and Hoyer-Fender, S. (1991). Sequence and developmental expression of a mRNA encoding a putative protein of rat sperm outer dense fibers. *Dev. Biol.* 148, 195–204.
419. Morales, C. R., Oko, R., and Clermont, Y. (1994). Molecular cloning and developmental expression of an mRNA encoding the 27 kDa outer dense fiber protein of rat spematozoa. *Mol. Reprod. Dev.* 37, 29–240.
420. Schäfer, M., Börsch, D., Hülster, A., and Schäfer, U. (1993). Expression of a gene duplication encoding conserved sperm tail proteins is translationally regulated in *Drosophila melanogaster*. *Mol. Cell. Biol.* 13, 1708–1718.
421. Kappe, G., Franck, E., Verschuure, P., Boelens, W. C., Leunissen, J. A., and de Jong, W. W. (2003). The human genome encodes 10 alpha-crystallin-related small heat shock proteins: HspB1-10. *Cell Stress Chaperones* 8, 53–61.
422. Gastmann, O., Burfeind, P., Günther, Hameister, H., Szpirer, C., and Hoyer-Fender, S. (1993). Sequence, expression and chromosomal assignment of a human sperm outer dense fiber gene. *Mol. Reprod. Dev.* 36, 407–418.
423. Hoyer-Fender, S., Burfeind, P., and Hameister, H. (1995). Sequence of mouse Odf1 and its chromosomal localization: extension of the linkage group between human chromosome 8 and mouse chromosome 15. *Cytogenet. Cell. Genet.* 70, 200–204.
424. Kim Y., Adham, I. M., Haack, T., Kremling, H., and Engel, W. (1995). Molecular cloning and characterization of the bovine and porcine outer dense fibers cDNA and organization of the bovine gene. (1995). *Biol. Chem. Hoppe-Seyler* 376, 431–435.
425. Shao, X., Tarnasky, H. A., Schalles, U., Oko, R., and van der Hoorn, F. A. (1997). Interactional cloning of the 84-kDa major outer dense fiber protein Odf84. *J. Biol. Chem.* 272, 6105–6113.
426. Schalles, U., Shao, X., van der Hoorn, and Oko, R. (1998). Developmental expression of the 84-kDa ODF sperm protein: localization to both the cortex and medulla of outer dense fibers and to the connecting piece. *Dev. Biol.* 199, 250–260.
427. Hoyer-Fender, S., Petersen, C., Brohmann, H., Wolgemuth, D. J. (1998). Mouse Odf2 cDNAs consist of evolutionary conserved as well as highly variable sequences and encode outer dense fiber proteins of the sperm tail. *Mol. Reprod. Dev.* 51, 167–175.
428. Petersen, C., Füzesi, L., and Hoyer-Fender, S. (1999). Outer dense fibre proteins from human sperm tail: molecular cloning and expression analyses of two cDNA transcripts encoding proteins of ~70 kDa. *Mol. Hum. Reprod.* 5, 627–635.
429. Nakagawa, Y., Yamane, Y., Okanoue, T., Tsukita, S., Tsukita, S. (2001). Outer dense fiber 2 is a widespread centrosome scaffold component preferentially associated with mother centrioles: its identification from isolated centrosomes. *Mol. Cell. Biol.* 12, 1687–1697.

430. Donkor, F. F., Mönnich, M., Czirr, E., Hollemann, T., and Hoyer-Fender, S. (2004). Outer dense fibre protein 2 (ODF2) is a self-interacting centrosomal protein with affinity for microtubules. *J. Cell. Sci.* 117, 4643–4651.
431. de Carvalho, C. E., Tanaka, H., Iguchi, N., Ventelä, S., Nojima, H., and Nishimune, Y. (2002). Molecular cloning and characterization of a complimentary DNA encoding sperm tail protein SHIPPO1. *Biol. Reprod.* 66, 785–795.
432. Nakamura, Y., Tanaka, H., Koga, M., Miyagawa, Y., Iguchi, N., de Carvelho, C. E., Yomogida, K., Nozaki, M., Nojima, H., Matsumiya, K., Okuyama, A., and Nishimune, Y. (2002). Molecular cloning and characterization of *oppo* 1: a haploid germ cell-specific complementary DNA encoding sperm tail protein. *Biol. Reprod.* 67, 1–7.
433. Kitamura, K., Miyagawa, Y., Iguchi, N., Nishimura, H., Tanaka, H., and Nishimune, Y. (2003). Molecular cloning and characterization of the human orthologue of the *oppo 1* gene encoding a sperm tail protein. *Mol. Hum. Reprod.* 9, 237–243.
434. Shao, X., Tarnasky, H. A., Lee, J. P., Oko, R., and van der Hoorn, F. A. (1999). Spag4, a novel sperm protein binds outer dense-fiber protein Odf1 and localizes to microtubules of manchette and axoneme. *Dev. Biol.* 211, 109–123.
435. Tarnasky, H., Gill, D., Murthy, S., Xhao, X., Demetrick, D. J., and van der Hoorn. (1998). A novel testis-specific gene, SPAG4, whose product interacts specifically with outer dense fiber protein ODF27, maps to human chromosome 20q11.2. *Cytogenet. Cell. Genet.* 81, 65–67.
436. Shao, X., Xue, J., and van der Hoorn, F. A. (2001). Testicular protein Spag5 has similarity to mitotic spindle protein Deepest and binds outer dense fiber protein Odf1. *Mol. Reprod. Dev.* 59, 410–416.
437. Xue, J., Tarnasky, H. A., Rancourt, D. E., and van der Hoorn, F A. (2002). Targeted disruption of the testicular SPAG5/ Deepest protein does not affect spermatogenesis or fertility. *Mol. Cell. Biol.* 22, 1993–1997.
438. Zarsky, H. A., Cheng, M., and van der Hoorn, F. A. (2003). Novel RING finger protein OIP1 binds to conserved amino acid repeats in sperm tail protein ODF1. *Biol. Reprod.* 68, 543–552.
439. Kasahara, M., Gutknecht, J., Brew, K., Spurr, N., and Goodfellow, P. M. (1989). Cloning and mapping of a testis-specific gene with sequence similarity to a sperm-coating glycoproteins gene. *Genomics* 5, 527–534.
440. O'Bryan, M. K., Loveland, K. L., Herszfeld, D., McFarlane, J. R., Hearn, M. T. W., and de Kretser, D. M. (1998). Identification of a rat testis-specific gene encoding a potential rat outer dense fibre protein. *Mol. Reprod. Dev.* 50, 313–322.
441. O'Bryan, M. K., Sebire, K., Meinhard, A., Edgar, K., Keah, H-H., Hearn, M. T. W., and de Kretser, D. M. (2001). Tpx-1 is a component of the outer dense fibers and acrosome of rat spermatozoa. *Mol. Reprod. Dev.* 56, 116–125.
442. Kierszenbaum, A. L., Rivkin, E., Fefer-Sadler, S., Mertz, J. R., and Tres, L. L. (1996). Purification, partial characterization, and localization of *Sak57*, an acidic intermediate filament keratin present in rat spermatocytes, spermatids, and sperm. *Mol. Reprod. Dev.* 44, 382–394.
443. Kierszenbaum, L. L., Rivkin, E., and Tres, L. L. (2003). Acroplaxome, an F-actin-keratin-containing plate, anchors the acrosome to the nucleus during shaping of the spermatid head. *Mol. Biol. Cell.* 14, 4628–4640.
444. Hinsch, K-D., De Pinto, V., Aires, V. A., Schneider, X, Messina, A., and Hinsh, E. (2004). Voltage-dependent anion-selective channels VDAC2 and VDAC3 are abundant proteins in bovine outer denser fibers, a cytoskeletal component of the sperm flagellum. *J. Biol. Chem.* 279, 15281–15288.
445. Sampson, M. J., Decker, W. K., Beaudet, A. L., Ruitenbeek, W., Armstrong, D., Hicks, M. J., and Craigen, W. J. (2001). Immotile sperm and infertility in mice lacking mitochondrial voltage-dependent anion channel type 3. *J. Biol. Chem.* 276, 39206–39212.
446. Junco, A., Bhullar, B., Tarnasky, H. A., and van der Hoorn, F. A. (2001). Kinesin light-chain KLC3 expression in testis is restricted to spermatids. *Biol. Reprod.* 64, 1320–1330.
447. Bhullar, B., Zhang, Y., Junco, A., Oko, R., and van der Hoorn, F. A. (2003). Association of kinesin light chain with outer dense fibers in a microtubule-independent fashion. *J. Biol. Chem.* 278, 16159–16168.
448. Rosales, J. L., Lee, B-C., Modarressi, M., Sarker, K. P., Lee, K-Y., Jeong, Y-G., Oko, R., and Lee, K-Y. (2004). Outer dense fibers serve as a functional target for Cdk·p35 in the developing sperm tail. *J. Biol. Chem.* 279, 1224–1232.
449. Baltz, J. M., Williams, P. O., and Cone, R. A. (1990). Dense fibers protect mammalian sperm against damage. *Biol. Reprod.* 43, 485–491.
450. Nagano, T. (1965). Localization of adenosine triphosphatase activity in the rat sperm tail as revealed by electron microscopy. *J. Cell. Biol.* 65, 25:101–112.
451. Cleland, K. W., and Lord Rothschild. (1959). The bandicoot spermatozoon: an electron microscopic study of the tail. *Proc. Roy. Soc. (Biol.)* 150, 24–42.
452. Nelson, L. (1958). Cytochemical studies with the electron microscope. I. Adenosine triphosphatease in the rat spermatozoa. *Biochem. Biophys. Acta* 27, 634–641.
453. Irons, M. J., and Clermont, Y. (1982). Formation of the outer dense fibers during spermiogenesis in the rat. *Anat. Rec.* 202, 463–471.
454. Wartenberg, H., and Holstein, A. F. (1975). Morphology of the "Spindle-shaped body" in the developing tail of human spermatids. *Cell. Tissue Res.* 159, 435–443.
455. Rattner, J. B., and Brinkley, B. R. (1970). Ultrastructure of mammalian spermiogenesis. I. A tubular complex in developing sperm of the cottontop marmoset *Sequinus oedipus*. *J. Ultrastruct. Res.* 32, 316–322.
456. Bradfield, J. R. G. (1955). Fibre patterns in animal flagella and cilia. *Symp. Soc. Exp. Biol.* 9, 306–334.
457. Olson, G. E. (1979). Isolation of the fibrous sheath and perforatorium of rat spermatozoa. In *The Spermatozoon* (D. W. Fawcett and J. M. Bedford, Eds.), pp. 395–400. Urban & Schwarzenberg, Baltimore.
458. Olson, G. E., Hamilton, D. W., and Fawcett, D. W. (1976). Isolation and characterization of the fibrous sheath of rat epididymal spermatozoa. *Biol. Reprod.* 14, 517–530.
459. Brito, M., Figueroa, J., Maldonado, E. U., Vera, J. C., and Burzio, L. O. (1989). The major component of the rat sperm fibrous sheath is a phosphoprotein. *Gamete Res.* 22, 205–217.
460. Eddy, E. M., O'Brien, D. A., Fenderson, B. A., and Welch, J. E. (1991). Intermediate filament-like proteins in the fibrous sheath of the mouse sperm flagellum. *Ann. N. Y. Acad. Sci.* 637, 224–239.
461. O'Brien, D. O., and Bellvé, A. R. (1980). Protein constituents of the mouse spermatozoon. II. Temporal synthesis during spermatogenesis. *Dev. Biol.* 75, 405–418.
462. Beecher, K. L., Homyk, M., Lee, C.-Y. G., and Herr, J.C. (1993). Evidence that 68-kilodalton and 54-51-kilodalton polypeptides are components of the human sperm fibrous sheath. *Biol. Reprod.* 48, 154–164.
463. Eddy, E. M., Toshimori, K., and O'Brien, D. A. (2004). Fibrous sheath of mammalian spermatozoa. *Micros. Res. Tech.* 61, 103–115.
464. Carrera, A., Gerton, G. L., and Moss, S. B. (1994). The major fibrous sheath polypeptide of mouse sperm: structural and

functional similarities to the A-kinase anchoring proteins. *Dev. Biol.* 165, 272–284.
465. Fulcher, K. D., Mori, C. Welch, J. E., O'Brien, D. A., Klapper, D. G., and Eddy, E. M. (1995). Characterization of Fsc1 cDNA for a mouse sperm fibrous sheath component. *Biol. Reprod.* 52, 41–49.
466. Brown, P. R., Miki, K., Harper, D. B., and Eddy, E. M. (2003). A-kinase anchoring protein 4 binding proteins in the fibrous sheath of the sperm flagellum. *Biol. Reprod.* 68, 2441–2448.
467. Johnson, L. R., Foster, J. A., Haig-Ladewig, L., VanScoy, H., Rubin, C., Moss, S. B., and Gerton, G. B. (1997). Assembly of AKAP82, a protein kinase A anchor protein, into the fibrous sheath of mouse sperm. *Develop. Biol.* 192, 340–350.
468. Miki, K., and Eddy, E. M. (1998). Identification of tethering domains for protein kinase A type Iα regulatory subunits on sperm fibrous sheath protein FSC1. *J. Biol. Chem.* 273, 34384–34390.
469. Miki, K., Willis, W. D., Brown, P, R., Goulding, E. H., Fulcher, K. D., and Eddy, E. M. (2002). Targeted disruption of the Akap4 gene causes defects in sperm flagellum and motility. *Dev. Biol.* 248, 331–342.
470. Vijayaraghavan, S., Liberty, G., Mohan, J., Winfrey, V., Olson, G., and Carr, D. (1999). Isolation and molecular characterization of AKAP110, a novel, sperm-specific protein kinase A–anchoring protein. *Mol. Endo.* 13, 705–717.
471. Mandal, A., Naaby-Hansen, S., Wolkowicz, M., Klotz, K., Shetty, J., Retief, J., Coonrod, S., Kinter, M., Sherman, N., Cesar, F., Flickinger, and Herr, J. C. (1999). FSP95, a testis-specific 95-kilodalton fibrous sheath antigen that undergoes tyrosine phosphorylation in capacitated human spermatozoa. *Biol. Reprod.* 61, 1184–1197.
472. Mei, X., Singh, I. S., Erlichman, J., and Orr, G. A. (1997). Cloning and characterization of a testis-specific, developmentally regulated A-kinase-anchoring protein (TAKAP-80) present on the fibrous sheath of rat sperm. *Eur. J. Biochem.* 246, 425–432.
473. Carr, D. W., Fujita, A., Stentz, C L., Liberty, G. A., Olson, G. E., and Narumiya, S. (2001). Identification of sperm-specific proteins that interact with A-kinase anchoring proteins in a manner similar to the type II regulatory subunit of PKA. *J. Biol. Chem.* 276, 17332–17338.
474. Fujita, A., Nakamura, K., Kato, T., Watanabe, N., Ishizaki, T., Kimura, K., Mizoguchi, A., and Narumiya, S. (2000). Ropporin, a sperm-specific binding protein of rhophilin that is localized in the fibrous sheath of sperm flagella. *J. Cell. Sci.* 113, 103–112.
475. Bishop, A. L., and Hall, A. (2000). Rho GTPases and their effector proteins. *Biochem. J.* 348, Pt. 2, 241–255.
476. Nakamura, K., Fujita, A., Murata, T., Watanabe, G., Mori, C., Fujita, J., Watanabe, N., Ishizaki, T., Yoshida, O., and Narumiya, S. (1999). Rhophilin, a small GTPase Rho-binding protein, is abundantly expressed in the mouse testis and localized in the principal piece of the sperm tail. *FEBS Lett.* 445, 9–13.
477. Kong, M., Richardson, R. T., Widgren, E. E., and O'Rand, M. G. (1995). Sequence and localization of the mouse sperm autoantigenic protein, SP17. *Biol. Reprod.* 53, 579–590.
478. Lea, I. A., Widgren, E. E., and O'Rand, M. G. (2004). Association of sperm protein 17 with A-kinase anchoring protein. *Reprod. Biol. Endo.* 2, 57.
479. Naaby-Hansen, S., Mandal, A., Wolkowicz, J. M., Sen, B., Westbrook, A., Shetty, J., Coonrod, S. A., Klotz, K. L., Kim, Y-H., Bush, L. A., Flickinger, C. J., and Herr, J. C. (2002). CABYR, a novel calcium-binding tyrosine phosphorylation-regulated fibrous sheath protein involved in capacitation. *Dev. Biol.* 242, 236–254.
480. Catalano, R. D., Hillhouse, E. W., and Vlad, M. (2001). Developmental expression and characterization of FS39, a testis complementary DNA encoding an intermediate filament-related protein of the sperm fibrous sheath. *Biol. Reprod.* 65, 277–287.
481. Modarressi, M. H., Behnam, B., Cheng, M., Taylor, K. E., Wolfe, J., and van der Hoorn, F. A. (2004). *Tsga*10 encodes a 65-kilodalton protein that is processed to the 27-kilodalton fibrous sheath protein. *Biol. Reprod.* 70, 608–615.
482. Mori, C., Welch, J. E., Fulcher, K. D., O'Brien, D. A., and Eddy, E. M. (1993). Unique hexokinase messenger ribonucleic acids lacking the porin-binding domain are developmentally expressed in mouse spermatogenic cells. *Biol. Reprod.* 49, 191–203.
483. Mori, C., Nakamura, N., Welch, J. E., Gotoh, H., Goulding, E. H., Fajioka, M., and Eddy, E. M. (1998). Mouse spermatogenic-cell specific type 1 hexokinase (*mHk1-s*) transcripts are expressed by alternative splicing from the *mHk1* gene and the HK1-S protein is localized mainly in the sperm tail. *Mol. Reprod. Dev.* 49, 374–385.
484. Travis, A. J., Foster, J. A., Rosenbaum, N. A., Visconti, P. E., Gerton, G. L., and Moss, S. B. (1998). Targeting of a germ cell-specific type 1 hexokinase lacking a porin-binding domain to the mitochondria as well as to the head and fibrous sheath of murine spermatozoa. *Mol. Biol. Cell.* 9, 263–276.
485. Fenderson, B. A., Toshimori, K., Muller, C. H., Lane, T. F., and Eddy, E. M. (1988). Identification of a protein in the fibrous sheath of the sperm flagellum. *Biol. Reprod.* 38, 345–357.
486. Welch, J. E., Schatte, E. C., O'Brien, D. A., and Eddy, E. M. (1992). Expression of a glyceraldehyde 3-phosphate dehydrogenase gene specific to mouse spermatogenic cells. *Biol. Reprod.* 46, 869–878.
487. Mori, C., Welch, J. E., Sakai, Y., and Eddy, E. M. (1992). In situ localization of spermatogenic cell-specific glyceraldehyde 3-phosphate dehydrogenase (*Gapd-s*) messenger ribonucleic acid in mice. *Biol. Reprod.* 46, 859–868.
488. Bunch, D. O., Welch, J. E., Magyar, P. L., Eddy, E. M., and O'Brien, D. A. (1998). Glyceraldehyde 3-phosphate dehydrogenase-S protein distribution during mouse spermatogenesis. *Biol. Reprod.* 58, 834–841.
489. Miki, K., Qu, W., Goulding, E. H, Willis, W. D., Bunch, D. O., Strader, L. F., Perreault S. D., Eddy, E. M., O'Brien D. A. (2004). Sperm motility and male fertility require GAPDS, a testis-specific glycolytic enzyme. *Proc. Natl. Acad. Sci. U S A* 101, 16501–16506.
490. Fulcher, K. D., Welch, J. E., Klapper, D. G., O'Brien, D. A., and Eddy, E. M. (1995). Identification of a unique μ-class glutathione S-transferase in mouse spermatogenic cells. *Mol. Reprod. Dev.* 4, 415–424.
491. Yu, Y., Oko, R., and Miranda-Vizuete, A. (2002). Developmental expression of spermatid-specific thioredoxin-1 protein: transient association to the longitudinal columns of the fibrous sheath during sperm tail formation. *Biol. Reprod.* 67, 1546–1554.
492. Miranda-Vizuete, A., Tsang, K., Yu, Y., Jiménez, A., Pelto-Huikko, M., Flickinger, C. J., Sutovsky, P., and Oko, R. (2003). Cloning and developmental analysis of murid spermatid-specific thioredoxin-2 (SPTRX-2), a novel sperm fibrous sheath protein and autoantigen. *J. Biol. Chem.* 278, 44874–44885.
493. Fujinoki, M., Kawamura, T., Toda, T., Ohtake, H., Ishimoda-Takagi, T., Shimizu, N., Yamaoka, S., and Okuno, M. (2004). Identification of 36 kDa phosphoprotein in fibrous sheath of hamster spermatozoa. *Comp. Biochem. Physiol.* Pt. B, 137, 509–520.
494. Hyne, R. V., and Garbers, D. L. (1979). Regulation of guinea pig adenylate cyclase by calcium. *Biol. Reprod.* 21, 1135–1142.

495. Yanagimachi, R. (1994). Fertilization. In *The Physiology of Reproduction* (E. Knobil and J. D. Neill JD, Eds.), Vol. 1, pp. 189–317. Raven Press, New York.
496. Kopf, G. S., Ning, X. P, Visconti, P. E., Purdon, M., Galantino-Homer, and Fornés, H. (1999). Signaling mechanisms controlling mammalian sperm fertilization competence and activation. In *The Male Gamete: From Basic Science to Clinical Applications* (C. Gagnon, Ed.), pp. 106–118. Cache River Press, Vienna, IL.
497. Lui, W-Y., Lee, W. M., and Cheng, C. Y. (2003). Rho GTPases and spermatogenesis. *Biochim. Biophys. Acta* 1593, 121–129.
498. Eddy, E. M., Welch, J. E., Mori, C., Fulcher, K. D., and O'Brien, D. A. (1994). Role and regulation of spermatogenic cell-specific gene expression: enzymes of glycolysis. In *Function of Somatic Cells in the Testis* (A. Bartke, Ed.), pp. 362–372. Springer-Verlag, New York.
499. McCarrey, J. R., and Thomas, K. (1987). Human testis-specific PGK gene lacks introns and possesses characteristics of a processed gene. *Nature* 326, 501–505.
500. Fraser, L. R., and Quinn, P. J. (1981). A glycolytic product is obligatory for initiation of the sperm acrosome reaction and whiplash motility required for fertilization in the mouse. *J. Reprod. Fertil.* 61, 25–35.
501. Mahadevan, M. M., Miller, M. M., and Moutos, D. M. (1997). Absence of glucose decreases human fertilization and sperm movement characteristics in vitro. *Hum. Reprod.* 12, 119–123.
502. Kubo-Irie, M., Matsumiya, K., Iwamoto, T., Kaneko, S., and Ishijima, S. (2004). Morphological abnormalities in the spermatozoa of fertile and infertile men. *Mol. Reprod. Dev.* 70, 70–81.
503. Curi, S. M., Ariagno, J. I., Chenlo, P. H., Mendeluk, G. R., Pugliese, M. N., Sardi Segovia, L. M., Repetto, H. E. H., and Blanco, A. M. (2003). Asthenozoospermia: analysis of a large population. *Archiv. Androl.* 49, 343–349.
504. Chemes, H. E. (1991). The significance of flagellar pathology in the evaluation of asthenozoospermia. In *Comparative Spermatology 20 Years Later* (B. Baccetti, Ed.), Serono Symposium Publications, Vol. 75, pp. 815–819. Raven Press, New York.
505. Chemes, H. E., and Rawe, V. Y. (2003). Sperm pathology: a step beyond descriptive morphology. Origin, characterization and fertility potential of abnormal sperm phenotypes in infertile men. *Hum. Reprod. Update* 9, 405–428.
506. Afzelius, B. A. (1981). Genetical and ultrastructural aspects of the immotile-cilia syndrome. *Am. J. Hum. Genet.* 33, 852–864.
507. Rossman, C. M., Forrest, J. B., Less, R. M., Newhouse, A. F., and Newhouse, M. T. (1981). The dyskinetic cilia syndrome: abnormal ciliary motility in association with abnormal ciliary ultrastructure. *Chest* 80, 860–865.
508. Afzelius, B. A., and Eliasson, R. (1970). Flagellar mutants in man: on the heterogeneity of the immotile-cilia syndrome. *J. Ultrastruct. Res.* 69, 43–52.
509. Ostrowski, L. E., Blackburn, K., Radde, K. M., Moyer, M. B., Schlatzer, D. M., Moseley, A., and Boucher, R. C. (2002). A proteomic analysis of human cilia. Identification of novel components. *Mol. Cell. Proteomics* 1, 451–465.
510. Geremek, M., and Witt, M. (2004). Primary ciliary dyskinesia: genes, candidate genes and chromosomal regions. *J. Appl. Genet.* 45, 347–361.
511. Chemes, H. E., Brugo, S., Zanchetti, F., Carrere, C., and Lavieri, J. C. (1987). Dysplasia of the fibrous sheath: an ultrastuctural defect of human spermatozoa associated with sperm immotility and primary sterility. *Fertil. Steril.* 48, 664–669.
512. Chemes, H. E. (2000). Phenotypes of sperm pathology: genetic and acquired forms in infertile men. *J. Androl.* 21, 799–808.
513. Escalier, D. (2003). New insights into the assembly of the periaxonemal structures in mammalian spermatozoa. *Biol. Reprod.* 69, 373–378.
514. Searle, A. G. (1982). The genetics of sterility in the mouse. In *Genetic Control of Gamete Production and Function* (P. G. Crosignai, B. L. Rubin, and M. Fraccaro, Eds.), pp. 93–114. Grune & Stratton, New York.
515. Leestma, J. E., and Sepsenwol, S. (1980). Sperm tail axoneme alterations in the Wobbler mouse. *J. Reprod. Fertil.* 58, 267–270.
516. Samant, S. A., Fossella, J., Silver, L. M., and Pilder, S. H. (1999). Mapping and cloning recombinant breakpoints demarcating the hybrid sterility 6-specific sperm tail assembly defect. *Mamm. Genome* 10, 88–94.
517. Olds, P. J. (1971). Effect of the T locus on sperm ultrastructure in the house mouse. *J. Anat.* 109, 31–37.
518. Bryan, J. H. D. (1977). Spermatogenesis revisited: IV. The course of spermiogenesis in mice homozygous for another male-sterility-inducing mutation, hpy (hydrocephalic-polydactyl). *Cell. Tissue Res.* 180, 187–201.
519. Cooper, T. G., and Hamilton, D. W. (1977). Observations on destruction of spermatozoa in the cauda epididymides and proximal vas deferens of non-seasonal male animals. *Am. J. Anat.* 149, 93–110.
520. Bennett, W. I., Gall, A. M., Southard, J. L., and Sidman, R. L. (1971). Abnormal spermiogenesis in quaking. A myelin-deficient mutant mouse. *Biol. Reprod.* 5, 30–58.
521. Neesen, J., Kirschner, R., Ochs, M., Schmiedl, A., Habermann, B., Mueller, C., Holstein, A. F., Nuesslein, T., Adham, I., and Engel, W. (2001). Disruption of the inner arm dynein heavy chain gene results in asthenozoospermia and reduced ciliary beat frequency. *Hum. Mol. Genet.* 10, 1117–1128.
522. Escalier, D., Bai, X-Y., Silvius, D., Xu, P-X, and Xu, X. (2003). Spermatid nuclear and sperm periaxonemal anomalies in the mouse *Ube2b* mutant. *Mol. Reprod. Dev.* 65, 298–308.
523. Sapiro, R., Kostetskii, I., Olds-Clarke, P., Gerton, G. L., Radice, G. L., and Strauss, J. F. III. (2002). Male infertility, impaired sperm motility, and hydrocephalus in mice deficient in sperm-associated antigen 6. *Mol. Cell. Biol.* 22, 6298–6305.
524. Matzuk, M. M., and Lamb, D. J. (2002). Genetic dissection of mammalian fertility pathways. *Nat. Med.* 8, S33–S40.
525. Woolley, D. M. (1977). Evidence for twisted plane undulation in golden hamster sperm tails. *J. Cell. Biol.* 67, 159–170.
526. Yeung, C. H., and Woolley, D. M. (1984). Three-dimensional bend propagation in hamster sperm models and the direction of roll in free-swimming cells. *Cell. Motil.* 4, 215–226.
527. Woolley, D.M., and Osborn, I. W. (1984). Three-dimensional geometry of motile hamster spermatozoa. *J. Cell. Sci.* 67, 159–170.
528. Phillips, D. M., and Olson, G.E. (1975). Mammalian sperm motility. Structure in relation to function. In *The Functional Anatomy of the Spermatozoon* (B. A. Afzelius, Ed.), pp. 117–126. Pergamon Press, New York.
529. Lindemann, C. B. (1980). Requirements for motility in mammalian sperm. In *Testicular Development, Structure, and Function* (A. Steinberger and E. Steinberger, Eds.), pp. 473–479. Raven Press, New York.
530. Yeung, C. H., and Woolley, D. M. (1983). A study of bend formation in locally reactivated hamster sperm flagella. *J. Muscle Res. Cell. Motil.* 4, 625–645.
531. Rikmenspoel, R. (1984). Movements and active moments of bull sperm flagella as a function of temperature and viscosity. *J. Exp. Biol.* 108, 205–230.
532. Serres, C., Feneux, D., Jouannet, P., and David, G. (1984). Influence of the flagellar wave development and propagation on the human sperm movement in seminal plasma. *Gamete Res.* 9, 183–195.
533. Gray, J. (1958). The movement of the spermatozoa of the bull. *J. Exp. Biol.* 35, 96–108.

534. Feneux, D., Serres, C., and Jouannet, P. (1985). Sliding spermatozoa: a dyskinesia responsible for human infertility? *Fertil. Steril.* 44, 508–511.
535. Bedford, J. M. (1975). Maturation, transport and fate of spermatozoa in the epididymis. In *Handbook of Physiology, Vol. 5, Endocrinology, Section 7, Male Reproductive System* (R. O. Greep, Ed.), pp. 303–317. American Physiological Society, Washington, DC.
536. Suarez, S. S., and Ho, H. C. (2003). Hyperactivated motility in sperm. *Reprod. Domest. Anim.* 38, 119–124.
537. Carr, D. W., and Acott, T. S. (1984). Inhibition of bovine spermatozoa by cauda epididymidal fluid: I. Studies of a sperm motility quiescence factor. *Biol. Reprod.* 30, 913–925.
538. Turner, T.T., and Giles, R. D. (1982). A sperm motility-inhibiting factor in the rat epididymis. *Am. J. Physiol.* 242, R199–R203.
539. Usselman, M. C., and Cone, R. A. (1983). Rat sperm are mechanically immobilized in the cauda epididymidis by "immobilin," a high molecular weight glycoprotein. *Biol. Reprod.* 29, 1241–1253.
540. de Lamirande, E., and Gagnon, C. (1984). Origin of a motility inhibitor within the male reproductive tract. *J. Androl.* 5, 269–276.
541. Brandt, H., Acott, T. S., Johnson, D. J., and Hoskins, D. D. (1978). Evidence for the epididymal origin of bovine sperm forward motility protein. *Biol. Reprod.* 19, 830–835.
542. Fukuda, N., Yomogida, K., Okabe, M., and Touhara, K. (2004). Functional characterization of a mouse testicular olfactory receptor and its role in chemosensing and in regulation of sperm motility. *J. Cell. Sci.* 117, 5835–5845.
543. Spehr, M., Schwane, K., Riffell, J. A., Barbour, J., Zimmer, R. K., Neuhaus, E. M., and Hatt, H. (2004). Particulate adenylate cyclase plays a key role in human sperm olfactory receptor-mediated chemotaxis. *J. Biol. Chem.* 279, 40194–40203.
544. Calogero, A. E., Hall, J., Fishel, S., Green, S., Hunter, A., and D'Agata, R. (1996). Effects of gamma-aminobutyric acid on human sperm motility and hyperactivation. *Mol. Hum. Reprod.* 2, 733–738.
545. Ren, D., Navarro, B., Perez, G., Jackson, A. C., Hsu, S., Shi, Q., Tilly, J. L., and Clapham, D. E. (2001). A sperm ion channel required for sperm motility and male fertility. *Nature* 413, 603–609.
546. Wang, D., King, S. M., Quill, T. A., Doolittle, L. K., and Garbers, D. L. (2003). A new sperm-specific Na^+/H^+ exchanger required for sperm motility and fertility. *Nature Cell. Biol.* 5, 1117–1122.
547. Woo, A. L., James, P. F., and Lingrel, J. B. (2000). Sperm motility is dependent on a unique isoform of the Na,K-ATPase. *J. Biol. Chem.* 275, 20693–20699.
548. Ho, H-C., and Suarez, S. S. (2001). An inositol 1,4,5-trisphosphate receptor-gated intracellular Ca^{2+} store is involved in regulating sperm hyperactivated motility. *Biol. Reprod.* 65, 1606–1615.
549. Leclerc, P., De Lamirande, E., and Gagnon, C. (1996). Cyclic adenosine 3′,5′ monophosphate-dependent regulation of protein tyrosine phosphorylation in relation to human sperm capacitation and motility. *Biol. Reprod.* 55, 684–692.
550. Si, Y., and Okuno, M. (1999). Role of tyrosine phosphorylation of flagellar proteins in hamster sperm hyperactivation. *Biol. Reprod.* 61, 240–246.
551. Leclerc, P., and Kopf, G. S. (1999). Evidence for the role of heterotrimeric guanine nucleotide-binding regulatory proteins in the regulation of the mouse sperm adenylyl cyclase by the egg's zona pellucida. *J. Androl.* 20, 126–134.
552. Buck, J., Sinclair, M. L., Schapal, L., Cann, M. J., and Levin, L. R. (1999). Cytosolic adenylyl cyclase defines a unique signaling molecule in mammals. *Proc. Natl. Acad. Sci. U S A* 96, 79–84.
553. Sinclair, M. L., Wang, X-Y., Mattia, M., Conti, M., Buck, J., Wolgemuth, D. J., and Levin, L. R. (2000). Specific expression of soluble adenylyl cyclase in male germ cells. *Mol. Reprod. Dev.* 56, 6–11.
554. Fang, X., and Conti, M. (2004). Expression of the soluble adenylyl cyclase during rat spermatogenesis: evidence for cytoplasmic sites of cAMP production in germ cells. *Dev. Biol.* 265, 196–206.
555. Chen, Y., Cann, M. J., Litvin, T. N., Iourgenko, V., Sinclair, M. L., Levin, L. R., and Buck, J. (2000). Soluble adenylyl cyclase as an evolutionarily conserved bicarbonate sensor. *Science* 289, 625–628.
556. Jaiswal, B. S., and Conti, M. (2001). Identification and functional analysis of splice variants of the germ cell soluble adenylyl cyclase. *J. Biol. Chem.* 276, 31698–31708.
557. Jaiswal, B. S., and Conti, M. (2003). Calcium regulation of the soluble adenylyl cyclase expressed in mammalian spermatozoa. *Proc. Natl. Acad. Sci. U S A* 100, 10676–10681.
558. Garbers, D. L., and Kopf, G. S. (1980). The regulation of spermatozoa by calcium and cyclic nucleotides. In *Advances in Cyclic Nucleotide Research* (P. Greengard and G. A. Robison, Eds.), pp. 251–306. Raven Press, New York.
559. Litvin, T. N., Kamenetsky, M., Zarifyan, A., Buck, J., and Levin L. R. (2003). Kinetic properties of "soluble" adenylyl cyclase. Synergism between calcium and bicarbonate. *J. Biol. Chem.* 278, 15922–15926.
560. Wennemuth, G., Carlson, A. E., Harper, A. J., and Babcock, D. F. (2003). Bicarbonate actions on flagellar and Ca^{2+} channel responses: initial events in sperm activation. *Development* 130, 1317–1326.
561. Esposito, G., Jaiswal, B. S., Xie, F., Krajnc-Franken, M. A., M., Robben, T. J. A. A., Strik, A. M., Kuil, C., Philipsen, R. L. A., van Duin, M., Conti, M., and Gossen, J. A. (2004). Mice deficient for soluble adenylyl cyclase are infertile because of a severe sperm-motility defect. *Proc. Natl. Acad. Sci. U S A* 101, 2993–2998.
562. Carr, D. W., Stofko-Hahn, R. E., Fraser, I. C., Bishop. S. M., Acott, T. S., Brennan, R. G., and Scott, J. D. (1991). Interaction of the regulatory subunit (RII) of cAMP-dependent protein kinase with RII-anchoring proteins occurs through an amphipathic helix binding motif. *J. Biol. Chem.* 266, 14188–14192.
563. Pawson, T., and Scott, J. D. (1997). Signaling through scaffold, anchoring, and adaptor proteins. *Science* 278, 2075–2080.
564. Desseyn, J-L., Burton, K. A., and McKnight, G. S. (2000). Expression of a nonmyristylated variant of the catalytic subunit of protein kinase A during male germ-cell development. *Proc. Natl. Acad. Sci. U S A* 97, 6433–6438.
565. Horowitz, J. M., Toeg, H., and Orr, G. A. (1984). Characterization and localization of cAMP-dependent protein kinases in rat caudal epididymal sperm. *J. Biol. Chem.* 259, 832–838.
566. Vijayaraghavan, S., Olson, G. E., NagDas, S., Winfrey, V. P., and Carr, D. W. (1997). Subcellular localization of the regulatory subunits of cyclic adenosine 3′ ,5′-monophosphate-dependent protein kinase in bovine spermatozoa. *Biol. Reprod.* 57, 1517–1523.
567. Burton, K. A., Treash-Osio, B., Muller, C. H., Denphy, E. L., and McKnight, G. S. (1999). Deletion of type II alpha regulatory subunit delocalizes protein kinase A in mouse sperm without affecting motility or fertilization. *J. Biol. Chem.* 272, 24131–24136.
568. Moos, J., Peknicová, J., Geussova, G., Philimonenko, V., and Hozák, P. (1998). Association of protein kinase A type I with detergent-resistant structures of mammalian sperm cells. *Mol. Reprod. Dev.* 50, 79–85.

569. Visconti, P. E., Johnson, L. R., Oyaski, M., Fornés, M., Moss, S. B., Gerton, G. L., and Kopf, G. S. (1997). Regulation, localization, and anchoring of protein kinase A subunits during mouse sperm capacitation. *Dev. Biol.* 192, 351–363.
570. Miki, K., and Eddy, E. M. (1998). Identification of tethering domains for protein kinase A type Iα regulatory subunits on sperm fibrous sheath protein FSC1. *J. Biol. Chem.* 273, 34384–34390.
571. Amieux, P. S., and McKnight, G. S. (2002). The essential role of RI alpha in the maintenance of regulated PKA activity. *Ann. N. Y. Acad. Sci.* 968, 75–95.
572. San Agustin, J. T., and Witman, G. B. (2001). Differential expression of the C_s and $C\alpha 1$ isoforms of the catalytic subunit of cyclic 3′,5′-adenosine monophosphate-dependent protein kinase in testicular cells. *Biol. Reprod.* 65, 151–164.
573. Skålhegg, B. S., Huang, Y., Su, T., Idzerda, R. L., McKnight, G. S., and Burton, K. A. (2002). Mutation of the Cα subunit of PKA leads to growth retardation and sperm dysfunction. *Mol. Endo.* 16, 630–639.
574. Nolan, M. A., Babcock, D. F., Wennemuth, G., Brown, W., Burton, K. A., and McKnight, G. S. (2004). Sperm-specific protein kinase A catalytic subunit C_{a2} orchestrates cAMP signaling for male fertility. *Proc. Natl. Acad. Sci. U S A* 101, 13483–13488.
575. Visconti, P. E., Bailey, J. L, Moore, G. D., Pan, E., Olds-Clarke, P., and Kopf, G. (1995). Capacitation of mouse spermatozoa. I. Correlation between the capacitation state and protein tyrosine phosphorylation. *Development* 121, 1129–1137.
576. Visconti, P. E., and Kopf, G. S. (1998). Regulation of protein phosphorylation during sperm capacitation. *Biol. Reprod.* 59, 1–6.
577. Urner, F., and Sakkas, D. (2003). Protein phosphorylation in mammalian sperm. *Reproduction* 125, 17–26.
578. Harrison, R. A. P. (2004). Rapid PKA-catalysed phosphorylation of boar sperm proteins induced by the capacitating agent bicarbonate. *Mol. Reprod. Dev.* 67, 337–352.
579. Ficarro, S., Chertihin, O., Westbrook, V. A., White, F., Jayes, F., Kalab, P., Marto, J. A., Shabanowitz, J., Herr, J. C., Hunt, D. F., and Visconti, P. E. (2003). Phosphoproteome analysis of capacitated human sperm. *J. Biol. Chem.* 278, 11579–11589.
580. Somanath, P. R., Jack, S. L., and Vijhayaraghavan, S. (2004). Changes in sperm glycogen synthase kinase-3 serine phosphorylation and activity accompany motility initiation and stimulation. *J. Androl.* 25, 605–617.
581. Luconi, M., Carloni, V., Marra, F., Ferruzzi, P., Forti, G., and Baldi, E. (2003). Increased phosphorylation of AKAP by inhibition of phosphatidylinositol 3-kinase enhances human sperm motility through tail recruitment of protein kinase A. *J. Cell. Sci.* 117, 1235–1246.
582. Luconi, M., Porazzi, I., Ferruzzi, P., Marchiani, S., Forti, G., and Baldi, E. (2005). Tyrosine phosphorylation of the A kinase anchoring protein 3 (AKAP3) and soluble adenylate cyclase are involved in the increase of human sperm motility. *Biol. Reprod.* 72, 22–32.
583. Geussova, G., Kalab, P., and Peknicova, J. (2002). Valosine containing protein is a substrate of cAMP-activated boar sperm tyrosine kinase. *Mol. Reprod. Dev.* 63, 366–375.
584. Berruti, G., and Borgonovo, B. (1996). sp42, the boar sperm tyrosine kinase, is a male germ cell-specific product with a highly conserved tissue expression extending to other mammalian species. *J. Cell. Sci.* 109, 851–858.
585. Tardif, S., Dubé, C., and Bailey, J. L. (2003). Porcine sperm capacitation and tyrosine kinase activity are dependent on bicarbonate and calcium but protein tyrosine phosphorylation is only associated with calcium. *Biol. Reprod.* 68, 207–213.
586. Harayama, H., Muroga, M., and Miyake M. (2004). A cyclic adenosine 3′,5′-monophosphate-induced tyrosine phosphorylation of Syk protein tyrosine kinase in the flagella of boar spermatozoa. *Mol. Reprod. Dev.* 69, 436–447.
587. Harayama, H., (2003). Viability and protein phosphorylation patterns of boar spermatozoa agglutinated by treatment with a cell-permeable cyclic adenosine 3′,5′-monophosphate analog. *J. Androl.* 24, 831–842.
588. Leclerc, P., and Goupil S. (2002). Regulation of the human sperm tyrosine kinase c-*yes*. Activation by cyclic adenosine 3',5'-monophosphate and inhibition by Ca^{2+}. *Biol. Reprod.* 67, 301–307.
589. Brazil, D. P., and Hemmings, D. A. (2001). Ten years of protein kinase B signaling: a hard Akt to follow. *Trends Biochem. Sci.* 26, 657–664.
590. Cantrell, C. A. (2001). Phosphoinositide 3-kinase signaling pathways. *J. Cell. Sci.* 114, 1439–1445.
591. Smith, G. D., Wolf, D. P., Trautman, K. C., and Vijayaraghavan, S. (1999). Motility potential of macaque epididymal sperm: the role of protein phosphatase and glycogen synthase kinase-3 activities. *J. Androl.* 20, 47–53.
592. Huang, Z., Khatra, B., Bollen, M., Carr, D. W., and Vijayaraghavan, S. (2002). Sperm PP1γ2 is regulated by a homologue of the yeast protein phosphatases binding protein sds22. *Biol. Reprod.* 67, 1936–1942.
593. Huang, Z., and Vijayaraghavan, S. (2004). Increased phosphorylation of a distinct subcellular pool of protein phosphatases, PP1γ2, during epididymal sperm maturation. *Biol. Reprod.* 70, 439–447.
594. Varmuza, S., Jurisicova, A., Okano, K., Hudson, J., Boekelheide, K., and Shipp, E. B. (1999). Spermiogenesis is impaired in mice bearing a targeted mutation in the protein phosphatase 1cγ. *Dev. Biol.* 205, 98–110.

Knobil and Neill's Physiology of Reproduction, Third Edition edited by Jimmy D. Neill, Elsevier © 2006

CHAPTER 2

Fertilization in Mammals

Harvey M. Florman[1] and Tom Ducibella[2]

Introduction, 55
Acquisition of Fertilization Competence by Gametes, 56
Pre-Ovulatory Acquisition of the Egg's Ability to Undergo Activation, 56
Sperm Capacitation, 56
The Acrosome Reaction, 62
Sperm–Egg Interaction, 65
Early Events Preceding Gamete Membrane Fusion, 65
Sperm–Egg Plasma Membrane Interactions, 79
Incorporation of the Fertilizing Sperm into the Egg and the Fate of Sperm Components, 82
Events Following Sperm–Egg Membrane Fusion, 83
A Brief Note on the Applications of Fertilization Biology, 93
Acknowledgments, 93
References, 94

INTRODUCTION

Fertilization is the process by which haploid gametes, sperm and egg, unite to produce a genetically distinct individual. Interest in this area has exploded in recent years. Here are some numbers: the PubMed database (http://www.ncbi.nlm.nih.gov/entrez/query.fcgi) lists 46,203 publications on *fertilization*, of which 254 appeared during the decade 1950–1959 and 874 were published in 1960–1969, while in 2004 alone there are 2,022. There is a long history of inquiry into the mechanisms that underlie reproduction (1–3), but several factors can be identified that contribute to this recent swelling of interest. It is recognized that fertilization is the consequence of a precisely ordered sequence of cellular interactions and that the unraveling of the underlying mechanisms can provide insight into basic cellular processes. There are also societal reasons for widening interest. Models of fertilization inform efforts to design new strategies to limit population growth. As the estimated world population rises beyond 6.4 billion (4), there is increasing need to develop new methods of contraception, and an accompanying drive to explore the molecular basis of fertilization (5). Conversely, there is a growing awareness that failures of sperm–egg interaction are the root lesion of certain types of infertility (6,7). Finally, the clinical use of assisted reproductive technology continues to expand while concerns about certain unintended consequences of those methods are being raised (8). This has redoubled efforts to understand the physiological process of fertilization.

Inspection of the publication database also provides a rationale for producing this review at this time. Approximately 40% of the fertilization citations collected in PubMed (18,449 of 46,203) were published since the second edition of this text appeared in 1994. It is therefore reasonable to reconsider this field in light of recent progress. In doing so, it is recognized that this subject, defined most broadly, begins with the acquisition of competence by sperm within the epididymis and by oocytes within follicle, and includes the events within the zygote that

[1]Department of Cell Biology, University of Massachusetts Medical School, Worcester;
[2]Department of Obstetrics and Gynecology, Tufts–New England Medical Center, Boston, Massachusetts.

culminate in the union of parental genetic material. Instead, the present work is more narrowly focused on the events of sperm–egg interaction. This scope is considerably reduced from the chapter on fertilization by Yanagimachi in previous editions (9,10). Readers should still consult those magisterial reviews for topics that are not covered here, such as epididymal maturation of sperm and the late events of zygote formation, as well as for a thorough treatment of the pre-1994 literature in other areas. In addition, the present review focuses almost exclusively on the events of gamete interaction in mammals. Readers interested in specific aspects of these topics can consult a number of other monographs and reviews (11–15).

ACQUISITION OF FERTILIZATION COMPETENCE BY GAMETES

Pre-Ovulatory Acquisition of the Egg's Ability to Undergo Activation

Oocytes acquire full ability to activate just prior to ovulation. (The female germ cell that fuses with sperm is, in the case of most mammals, an oocyte arrested at metaphase II. However, this is not the case in all mammals or in many nonmammalian species. We will refer to this ovulated germ cell as an egg in this review and reserve the term oocyte for events that occur prior to ovulation.) That is, gonadotropin-stimulated antral follicles have pre-ovulatory oocytes that are full grown, but unable to undergo normal activation if retrieved and injected with inositol-1,4,5-trisphosphate (IP_3) (16) or fertilized in vitro (17). Although the most obvious sign of immaturity is their stage of meiotic maturation (e.g., prophase I instead of metaphase II if the LH surge has not yet taken place), the stage of the oocyte chromatin does not actually explain the failure of normal activation. For example, eggs of different organisms are fertilized at different stages of meiosis, from prophase I to the haploid (female) pronuclear stage, and undergo normal development (18).

Instead, important changes in the mammalian oocyte cytoplasm take place during luteinizing hormone (LH)-induced meiotic maturation and are likely to be responsible for the development of activation competence. Activation competence refers to the ability of the mature egg to undergo cortical granule secretion, pronuclear formation, and completion of meiosis. Like uncapacitated sperm, pre-ovulatory oocytes do not undergo normal signaling and secretion. Two deficiencies in these fertilized oocytes are the failure to normally elevate intracellular Ca^{2+} (Ca^{2+}_i) (19) and to undergo cortical granule secretion, which is not restored by experimentally inducing fertilization-like Ca^{2+} oscillations (20). In addition, fertilized pre-ovulatory oocytes are not able to induce normal decondensation of the sperm chromatin (21), convert it into a functional state (22), and undergo pronuclear envelope formation (23,24).

The developmental acquisition of these abilities coincides with a meiotic maturation-associated increases in specific molecules involved in these processes, such as the IP_3 receptor (IP_3R), glutathione, and calmodulin-dependent protein kinase II (CaMKII), as well as several crucial cell cycle kinases (all of which are discussed later). Translationally mediated increases in expression of such proteins (25) are considered to be important for normal egg activation and onset of development, e.g., as demonstrated by RNAi down-regulation (26,27).

Sperm Capacitation

Definitions and General Comments

Sperm in the cauda epididymis have not yet completed the process of functional maturation and remain limited in their functional capacity. The requirement for additional maturation was recognized by M. C. Chang (28) and C. R. Austin (29,30), who observed in 1951 that freshly ejaculated mammalian sperm were unable to fertilize eggs in vivo until they had resided for a period of time within the female reproductive tract. Sperm that have completed this process are referred to as "capacitated" (30).

Capacitation was initially defined in terms of fertilizing potential and this presented two experimental difficulties. First, it encompassed all modifications of sperm within the female reproductive tract, irrespective of whether those modifications directly related to the acquisition of fertility. Second, fertilization assays in vivo and in vitro require sperm:egg ratios of >1,000:1, but only directly assess the capacitation status of the few sperm that penetrate eggs. Mammalian sperm are functionally heterogeneous (10,31–33) and so it was difficult to correlate the physiological status of a few fertilizing sperm with population-averaged biochemical or molecular properties.

Fertilization remains the definitive test of sperm capacitation, yet more narrow definitions have been proposed to facilitate experimentation. For example, Austin noted in his initial studies that sperm penetration of the egg zona pellucida (zona) in vivo was specifically regulated by capacitation (29). It was later recognized that sperm must undergo the acrosome reaction to enter the zona (34), and that this exocytotic event is triggered by the zona (35).

Accordingly, capacitation can be operationally defined as those maturational events by which sperm acquire the ability to acrosome react in response to stimulatory signals from the zona (36). Other definitions have also been proposed based on enhanced tyrosine phosphorylation of sperm proteins or other end points, although a satisfactory demonstration that these modifications are accurate markers for the completion of capacitation is still awaited (10,37). Such alternative definitions reflect biases on the part of investigators as to the important events in capacitation, but nevertheless can be useful as long as results are correlated with sperm fertilizing potential.

Capacitation was initially detected due to its protracted time course which, in some species, may take several hours (10). In contrast, sperm of many invertebrates and some nonmammalian vertebrates fertilize eggs quickly following release from the male reproductive tract. This is often interpreted to mean that capacitation is a uniquely mammalian phenomenon and that, at least with regard to these events, there are fundamental differences in the control of fertilization between mammalian and nonmammalian organisms. It may be misleading though to focus on relative time courses. Capacitation is generally understood to mean the regulation of sperm functionality by factors associated with the egg or with the female reproductive tract. Similar processes also occur in many nonmammalian animals and in some plants, albeit with much more rapid time courses. Factors associated with eggs activate sperm metabolism and motility, or otherwise modulate sperm function in algae, ciliaphores, echinoderms, and amphibians (38–41). Capacitation may then represent the adaptation in mammals of a general strategy in which fertilization is regulated by the female.

We consider the physiological process of capacitation and current thoughts regarding the underlying molecular mechanisms. Yanagimachi provides a more comprehensive discussion of these and other issues, including the related process of epididymal maturation (10). Other reviews provide either a general perspective or focus on specific aspects of this process, including questions of assay (42–47).

Summary of Changes in Sperm Associated with Capacitation

Capacitation is associated with widespread changes in the cellular physiology and biochemistry of sperm. These include alterations in: surface properties, such as peripheral membrane protein composition, antigen localization and surface charge; plasma membrane properties, such as membrane potential, lipid composition and transmembrane phospholipids asymmetry, and the lateral diffusion of lipids and proteins; metabolism; apparent intracellular pH as well as in the cytosolic activities of calcium and other ions; altered cyclic nucleotide metabolism; and changes in protein phosphorylation (see Table 1). This extensive physiological reprogramming takes place in both the head and flagellum.

Many of these alterations have been discussed extensively by Yanagimachi (10) and will only be described here briefly and in reference to thoughts regarding underlying mechanisms. Several general points should be noted first.

(i) As a consequence of capacitation, sperm express the ability to interact with eggs due to alterations occurring largely in the head region and also exhibit specific types of flagellar motility. Many of the individual biochemical and biophysical changes that accompany capacitation are likely to relate to one or both of these two general functional modifications.

(ii) Capacitation is not a unitary process but instead the sum of multiple individual reactions. These component reactions proceed with different time courses and have distinct environmental requirements in vivo and in vitro (10). For example, mouse sperm are unable to adhere to the zona immediately upon release from the cauda epididymis and acquire this ability during capacitation in vitro, but do so much more rapidly and with simpler media requirements than they develop full fertility (48,49).

(iii) The time course for the component reactions of capacitation, as well as for the completion of the process (as assessed by fertilizing ability), differs widely among mammalian species (10). It has been suggested that these differences may be related to the availability of eggs within the female reproductive tract (50).

(iv) Sperm populations do not capacitate in a highly synchronized manner. Large fractions of the population are not capacitated in vitro and may not do so in vivo. This may reflect an underlying regulatory process that maintains a low representation of capacitated sperm at any one time and continually replenishes this pool, as suggested by Eisenbach and co-workers (33). In addition, in vitro culture conditions may not yet be optimized. In either case, this complicates the use of population-based assays.

(v) Capacitation in vitro and in vivo may differ in important regards. Studies with some animal models use cauda epididymal sperm that have not been mixed previously with seminal fluids. However, such fluids contain negative regulators of capacitation (so-called "decapacitation factors") and may also contain positive regulators (10,51). Even when ejaculated sperm are used, in vitro culture conditions will represent a dramatic simplification of the environment within the female reproductive tract.

For example, Yanagimachi noted that the ionic composition of the female reproductive tract fluids differs from the standard culture conditions used for capacitation in vitro and this is expected to influence sperm physiology (10). In addition, female reproductive tract fluids contain a range of bioactive molecules that may affect sperm function. It is therefore essential to confirm that in vitro models also apply in vivo.

Mechanism(s) of Capacitation

Capacitation occurs in vivo within the female reproductive tract. The site at which this process begins is not fixed, with variations among mammalian species correlating with the site of initial sperm deposition; however, capacitation in vivo is completed in the oviduct (10,52). There also does not appear to be a strict requirement for factors unique to the female reproductive tract, as demonstrated by the capacitation of sperm in vivo in ectopic sites, and in vitro in media containing either no biological fluids or fluids drawn from ectopic sites (e.g., blood serum) (10). Yet, female reproductive tract factors can alter the kinetics of capacitation and so apparently provide signals that regulate this process even if those signals are not unique to this region. Several processes occur during the hours required for capacitation. These will be reviewed briefly and are summarized in Table 1.

Desorption Processes. Epdidymal and seminal fluids contain proteins and other macromolecules that adsorb to the sperm surface. These factors may either occlude key sites on sperm or otherwise suppress sperm functional activity. The desorption of these factors correlates temporally with the acquisition of fertilizing ability and conditions such as hypertonic incubation that accelerate desorption also accelerate capacitation. It has been suggested that the environment of the female reproductive tract may be particularly efficient in driving desorption (51,53–56).

Ionic Changes. HCO_3^- is required for capacitation in vitro in a number of species (57,58). This anion is present in female reproductive tract fluid (10). There is no direct evidence yet regarding a role in vivo although a recent study is consistent with secretion of HCO_3^- by an epithelial cystic fibrosis transmembrane regulator and a role in capacitation (59,60). HCO_3^- contributes to pH regulation, which is essential for capacitation in some species (61),

TABLE 1. *Sperm changes associated with capacitation or occurring during incubation under capacitating conditions*

Change	Effect	Ref.
Surface effects		
Proteins		
β-Galactosyltransferase	Unmasked	(49)
Fertilin (ADAM1, 2)	Becomes mobile	(655)
Various epitopes	Distribution or accessibility changes	(10)
Lipids		
Cholesterol, other sterols	Decrease	(45)
Phospholipids	Asymmetry decrease	(93–95)
	Increased diffusion	(10)
Glycolipids	Redistribution	(660)
Surface charge	Less negative	(10)
Membrane potential	Hyperpolarize	(65)
Ion concentration or content		
Ca^{2+}	Increase	(62, 661)
Zn^{2+}	Decrease	(662)
K^+	Decrease, no change	(10, 65)
Na^+	Increase	(10)
pH	Increase	(61)
Ion channels and transporters		
Na^+/K^+ ATPase	Increased activity	(10)
Ca^{2+} ATPase	Decreased activity	(664)
Ca^{2+} ATPase - PMCA4	Hyperactivation defect in null sperm	(660)
Na^+/Ca^{2+} exchanger	Increased activity	(10, 665)
CatSper2	Hyperactivation defect in null sperm	(131)

Change	Effect	Ref.
Enzyme activities (non-ATPase)		
Adenylyl cyclase	Increase	(10)
Protein kinase A	Increase	(654)
Protein kinase C	Increase	(10)
ERK kinase	Increase	(656)
Protein tyrosine phosphorylation		
AKAP3, 4	Increased phosphorylation	(657, 658)
CABYR	Increased phosphorylation	(659)
VDAC2	Increased phosphorylation	(112)
VCP/p97	Increased phosphorylation	(112)
ODF1	Increased phosphorylation	(112)
Other proteins	Increased phosphorylation	(112)
Other effects		
ATP levels	Decrease	(10)
Reactive oxygen species	Increase	(663)
Functional effects		
Motility	Hyperactivation	(10)
	Chemotaxis	(112)
Zona binding	Acquired	(10)
Acrosome reaction	Acquired	(10)

but its major role appears to be in the regulation of adenylyl cyclase activity (see below).

Other ionic changes include elevations of intracellular Ca^{2+} and hyperpolarization of membrane potential (Table 1) (10,32,62). Elevated Ca^{2+}_i may be due either to enhanced influx through a regulated pathway, such as a channel, or decreased efflux (63). In this regard, the Ca^{2+}/calmodulin-dependent Ca^{2+}-ATPase, PMCA4, is located in the sperm flagellum and has been linked to capacitation-associated changes in motility (64).

Membrane potential hyperpolarization was specifically associated with the subpopulation of mouse sperm that capacitate in vitro, when capacitation was defined as the ability to undergo a zona-induced acrosome reaction (32). This shift is associated with an apparent enhanced contribution of K^+ permeability to membrane potential (65), possibly due to conductance through an inwardly rectifying K^+ channel (66). A small increase in $(K^+)_i$, the estimated intracellular K^+ concentration, was noted during capacitation but this was not statistically significant (65). There is not necessarily inconsistent with a contribution of K^+ channels since, as Hille reminds us, even large changes in membrane potential, such as occur during a neuronal action potential, do not require significant changes in the bulk concentration of intracellular ions (67). In addition, three factors contribute to membrane potential: transmembrane bulk ion activity gradients, membrane dipole moment, and surface charge (67). In this regard, the study that correlated membrane potential and capacitation state in single sperm utilized a class of fluorescent probes that are more sensitive to dipole moment and possibly surface charge, and less responsive to bulk ion activity gradients (68).

The function of hyperpolarization is uncertain although a plausible model was forwarded (32) based on the proposed role of voltage-sensitive ion channels in fertilization (see below), and the role of membrane potential in controlling channel activation and inactivation (67). Specifically, channels such as the low voltage-activated, or T type, Ca^{2+} channels (Ca_v3 class) exhibit voltage-dependent inactivation. In these cases, a relatively depolarized membrane potential maintains the channel in the nonconducting inactivated state. In order to evoke a current, the channel must first be returned to the closed state by a priming hyperpolarization prior to opening in response to a subsequent depolarization (67,69). A similar mechanism has been proposed for the low voltage-activated channel of mouse sperm, in which the channel is inactivated by the relatively depolarized membrane potential in uncapacitated sperm and capacitation-dependent hyperpolarization then recycles the channel to the closed state from which it can respond to stimulatory agonists (32). As pointed out by Arnoult et al., these shifts in membrane potential may help to minimize spontaneous acrosome reactions until after capacitation is complete (32). However, this model has not yet been rigorously tested.

Adenylyl Cyclase and Protein Kinase A Pathways. cAMP levels increase during capacitation, while membrane permeant cAMP analogues produce or accelerate some aspects of capacitation. This has lead to the general conclusion that capacitation is dependent upon cAMP production (10,39,70). Two types of adenylyl cyclase activity are expressed during spermatogenesis and present in mammalian sperm. First, members of the transmembrane adenylyl cyclases are present. These enzymes share a domain architecture and are typically stimulated following receptor activation (71). All nine members of the transmembrane adenylyl cyclase family were detected in mammalian sperm, with class VIII prominent in the flagellar principal piece and class III in the head (72). Some evidence has linked transmembrane adenylyl cyclase family members either to events that follow capacitation, such as the initiation of acrosome reactions, or to the odorant receptor-mediated chemotaxis process, where the capacitation state of sperm is not determined (see below), but there is no compelling data supporting their role in capacitation (73,74,76,617).

In addition to the transmembrane enzymes, a "soluble" adenylyl cyclase (sAC) activity was partially purified from testis and sperm in 1975 (77) and more recently was cloned and characterized by Levin, Buck, and coworkers and by others (78–80). This enzyme differs from the transmembrane adenylyl cyclase in that it is insensitive to G protein or forskolin modulation, but instead is regulated by HCO_3^- and Ca^{2+} (78,81–83). sAC contributes an essential component, and possibly all, of the capacitation-associated increase in cAMP production. As noted previously, HCO_3^- is required for capacitation and is believed to have its effects, at least in part, from its modulation of sAC activity (84). sAC activity was purified from testis cytosol but in other tissues is found in the particulate fraction and associated with a number of organelles or compartments. Zippin and co-workers have noted that this would permit local domains of HCO_3^-/Ca^{2+}-cAMP signaling, which could have significant implications for sperm function (85).

The cAMP-dependent protein kinase (protein kinase A, or PKA) is a major target of elevated cAMP during capacitation. Regulatory and catalytic subunits of PKA have been identified in sperm, including a novel sperm-specific splicing isoform of the widely distributed C_α isoform which is designated C_s or $C_{\alpha 2}$ (86,87). The activation of PKA through sAC is rapid. For example, porcine sperm respond to the addition of HCO_3^- by generating a peak cAMP response within 60 seconds (88) and PKA-dependent protein phosphorylation peaks within 90 seconds of the addition of HCO_3^- (89). As is discussed below, this response may be specifically associated with the control of motility during capacitation (86,90). Other cAMP/PKA responses develop with a more protracted time course, including membrane lipid reorganization and protein tyrosine phosphorylation. These late events will also be considered below.

Plasma Membrane Changes. Phospholipids of somatic cells are distributed asymmetrically between the lipid leaflets of the plasma membrane. The outer leaflet is enriched in sphingomyelin and phosphatidylcholine, while most of the phosphatidylethanolamine and essentially all of the phosphatidylserine are on the inner leaflet (91). Asymmetry is maintained by three phospholipid transferase activities: a flippase activity that transfers the aminophospholipids, phosphatidylserine and phosphatidylethanolamine, from the outer- to the inner-leaflet; a floppase that nonspecifically transfers phospholipids to the outer leaflet; and a scramblase that transfers all phospholipids in both directions (92). Asymmetry can break down either during cell stimulation or necrotic conditions. A similar asymmetric distribution occurs in sperm and apparently collapses during capacitation (93–95). Addition of HCO_3^- at levels similar to those within the female reproductive tract resulted in a collapse of phospholipid asymmetry that can be detected within 2 minutes and peaked within the next 5–20 minutes. This response is mediated by PKA (95).

The collapse of phospholipid asymmetry facilitates a decrease in sperm membrane cholesterol content during capacitation (31,95–97). The relationship between extractable cholesterol and cholesterol sulfate content and capacitation was first noted by Davis (98), extended by Langlais and Roberts (99) and by Go and Wolf (100), and then examined thoroughly by Kopf and his co-workers (45). The evidence for a role of cholesterol efflux can be summarized as follows: there is a quantitative decrease in sperm membrane sterol after capacitation; addition of cholesterol to the medium slows or inhibits capacitation in vitro; and capacitation in vitro is accelerated or requires a sterol acceptor, such as albumin, high-density lipoprotein, or β-cyclodextrins (95,100–103).

The precise role of cholesterol abstraction in sperm capacitation is not yet known. It is now recognized that cholesterol is enriched in membrane microdomains, or lipid rafts, which also concentrate a variety of signal transduction proteins, and that the function of these rafts is modified by loss of cholesterol (104,105). Emerging data indicates that lipid rafts are also present in sperm membrane and the modulation of their function may account for aspects of capacitation (106,107). Other plausible roles of cholesterol efflux include effects on membrane thickness, and hence the function of membrane proteins (68), as well as an enhanced ability of membranes to fuse as a result of altered localization of exocytotic proteins or due to the control of membrane biophysical properties (108).

Protein Phosphorylation—Late Events. The rapid activation of PKA during sperm capacitation in vitro was noted previously. Visconti and co-workers reported that the tyrosine phosphorylation state is enhanced in an array of mouse sperm proteins during capacitation in vitro, and that this process is dependent upon PKA (109,110). This observation has been extended to a number of other mammalian species (111). Several capacitation-dependent phosphoproteins have been identified, including the PKA anchoring proteins AKAP3 and AKAP4; and the valosin-containing protein VCP/p97, which is related to the exocytotic protein NSF (Table 1) (112). Yet, there is little information regarding the relevant protein kinases and/or phosphatases that mediate this process. The appearance of these phosphoproteins is a late event in the capacitation process and correlates both temporally and in terms of media requirements in vitro with the onset of fertility (that is, complete capacitation) and with the development of the ability to undergo a zona-induced acrosome reaction (109,111).

Functional Changes Associated with Capacitation

As noted previously, the major functional consequences of capacitation include changes in flagellar motility and in the ability of sperm to interact with eggs.

Hyperactivated Motility. Following release from the cauda epididymis into either seminal fluids or physiological media in vitro, sperm exhibit flagellar bends that are relatively symmetrical and describe a low-amplitude, long-wavelength wave. The result is linear, forwardly progressive swimming. In contrast, sperm that were incubated under capacitating conditions in vitro, or were recovered from the oviduct at a time when they were expected to be capacitated in vivo, display a hyperactivated pattern of motility that consists of an increase in velocity; a decrease in forwardly progressive swimming with a corresponding increase in lateral displacement of the head; and deeper, more asymmetric flagellar bends (Fig. 1) (10,113–115). Hyperactivation can be

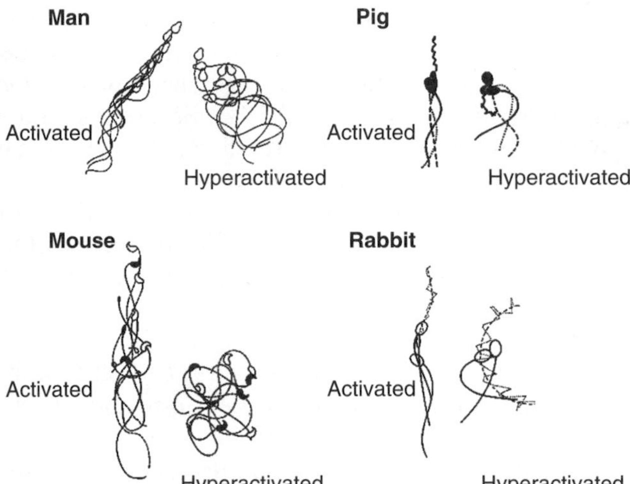

FIG. 1. Capacitation results in a hyperactivated pattern of sperm flagellar motility. Activated motility is characteristic of uncapacitated sperm while hyperactived motility is observed after capacitation in vitro or in sperm recovered from the oviduct. Note that hyperactivation includes a transition to a high-amplitude, short-wavelength flagellar wave and the resulting path of the sperm head is less linear and sweeps out a larger area than seen with activated motility. [Figure is adapted from (10) with permission of Lippincott Williams & Wilkins. Courtesy of Dr. Ryuzo Yanagimachi.]

dissociated from other aspects of capacitation that are required for fertility in vitro by selective media requirements (58,116) or by genetic approaches (117). Thus, hyperactivation represents a component reaction of capacitation that is manifest in the flagellum.

Several potential roles for hyperactivated motility mode have been suggested. First, hyperactived sperm enter viscoelastic substances more efficiently (118). In this regard, sperm must penetrate the viscous environment of oviduct mucus and, later, during interaction with eggs they must sequentially move through the hyaluronan-enriched extracellular matrix of the cumulus oophorus around the egg and then the zona. Second, hyperactivation may assist ascent up the oviduct to the site of fertilization by permitting sperm to dissociate from transient adhesion to the oviductal epithelium (10,118–121). Hyperactivation then acts as a filtering agent that permits access to the site of fertilization only to those sperm that have completed some portion of the capacitation process.

An understanding of the mechanisms of hyperactivation has been slowed by the absence of specific ligands that initiate this motility. There have been suggestions that the transition from activated to hyperactivated motility modes may be regulated by factors in the female reproductive tract at the time of fertilization, including follicular fluid or factors released by the cumulus oophorus complex that surrounds the ovulated egg, but no consensus has emerged regarding the identities of active factors or even the nature of the regulation (10,122–125). Some insight into the intracellular signaling processes is provided by the recognition that HCO_3^- is required for the onset of hyperactivation in vitro (58). Relevant targets of HCO_3^- are not known, but may include either direct or indirect modulation of Ca^{2+} entry mechanisms (90), or stimulation of the HCO_3^-/Ca^{2+}-responsive adenylyl cyclase activity, sAC, with the subsequent activation of either CNG channels or the cAMP-dependent protein kinase (81,86,90,126,127).

$Ca^{2+}{}_i$ controls flagellar bending asymmetry. Bending waves may be initiated by release of Ca^{2+} from internal stores at the top of the flagellum (128) and may be shaped locally during propagation by influx through Ca^{2+} channels along the flagellum (129,130). In particular, sperm of mice bearing a targeted disruption of the catsper2 gene do not exhibit hyperactivated motility and are infertile. The catsper family of genes encodes putative Ca^{2+}-conducting ion channels that are specific to sperm. The simplest, though not the only explanation for this phenotype is that Ca^{2+} influx is required for hyperactivation (131). In addition, modulation of Ca^{2+} clearance from the flagellar cytoplasm by the plasma membrane Ca^{2+}-ATPase, PMCA4, may contribute to the control of hyperactivation (64).

Chemotaxis. Chemotaxis is the reorientation of motility in response to a chemical gradient. Sperm chemotaxis toward factors derived from eggs or from the female reproductive tract has been observed in marine species, amphibians and possibly in mammals (11,40). The case in mammals, which is the particular focus here, is based on extensive physiological evidence, largely from Eisenbach (11). Sperm of several mammalian species exhibit chemotaxis toward factors from follicular fluid, eggs or cumulus complexes in well-controlled assay systems (11,132,133). As shown in Fig. 2, human sperm will reorient motility toward a medium conditioned by human eggs (and contained follicular fluid) (132). The proportion of chemotactically responsive sperm in a population was typically low (<25%), with individual sperm residing in a responsive state for a limited period of time and then apparently being replenished by recruitment from the nonresponsive pool. It has been suggested that chemotactic response depends upon the capacitation state of sperm (11). Interestingly, Bahat and colleagues recently found that sperm could also orient motility in the shallow thermal gradients that exist along the length of the mammalian oviduct (134). They suggested that sperm may be guided to the site of fertilization by the sequential steps of thermotaxis, which operates on centimeter-scale distances, followed by chemotaxis when sperm are within millimeters of the egg/cumulus complex (11).

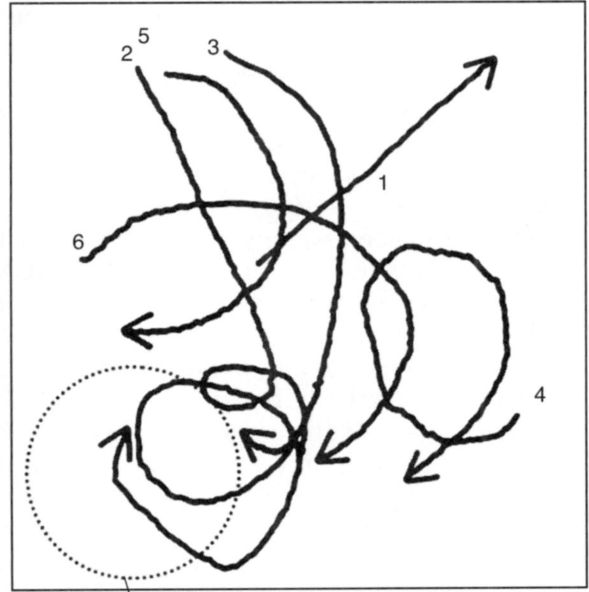

FIG. 2. Capacitation is associated with sperm chemotaxis in response to eggs and follicular fluid. Human oocytes and cumulus were recovered by aspiration of ovarian follicles and incubated in vitro. Culture medium conditioned by incubation with oocytes, and also containing ~5% follicular fluid (which was recovered with oocytes during aspiration from follicles), was added to a field of human sperm (site of addition is shown by *dashed line*). Human sperm motility was tracked using video microscopy. Paths of six sperm are shown, of which five reorient toward the conditioned medium. [This figure is taken from (132) with permission of The National Academies Press and is provided through the courtesy of Dr. Michael Eisenbach.]

The current status of this model is complicated by difficulties in the isolation of active factors.

Human and mouse sperm also exhibit chemotaxis towards chemically defined odorants (72,135,136), consistent with the expression of odorant receptors (137,138), of appropriate G proteins, and of the olfactory-type transmembrane adenylyl cyclase (139). However, the capacitation state of sperm preparations was not determined directly in these studies (72,135,136). Moreover, the fraction of human sperm that exhibit odorant responses (136) is much larger than that which display chemotactic responses toward female reproductive tract factors or are expected to have completed capacitation (11). It is therefore difficult at the moment to determine whether odorant responses underlie the chemotactic response to female reproductive tract factors or represent a different process.

Regulation of Sperm–Egg Interaction by Capacitation. The ability of sperm to interact with eggs is regulated at several points by capacitation. As a result of capacitation sperm acquire the following functional capabilities: to penetrate the cumulus matrix, while uncapacitated sperm adhere to the outer edge of the cumulus and fail to penetrate (140); to adhere to the zona, while uncapacitated sperm cannot bind (48,49); and to undergo a zona-evoked acrosome reaction (36,141). In some species there is also an increase in the rate of spontaneous (that is, agonist-independent) acrosome reactions as a result of capacitation. Moreoever, since the acrosome reaction is a requirement for the late events of fertilization then capacitation will control the overall process (10).

Capacitation: Summary and Conclusions

Capacitation regulates the ability of sperm reach the site of fertilization and to interact with eggs. As such, the underlying mechanisms are central to an understanding of mammalian fertilization. However, any model that, in 2005, attempts to link biochemical or biophysical changes in sperm with key functional consequences (such as acquisition of hyperactivated motility or the ability to undergo acrosome reactions) would be highly speculative.

However, there have been significant changes in our understanding of capacitation since the publication of the previous edition of this text (142). For example, it is now understood that protein tyrosine phosphorylation is an important process and efforts are underway to identify these phosphoproteins. It is anticipated that the relevant protein kinases and phosphatases will also soon be known. These and other biochemical processes (for example, sAC regulation) are increasingly being coupled with functional consequences of capacitation. It may be that a true molecular of capacitation will be available for the next edition of this text.

The Acrosome Reaction

Description of the Acrosome Reaction

The sperm of many animal species, including all mammals, contain a single large secretory vesicle, or acrosome, in the apical region of the anterior head overlying the nucleus (Fig. 3). Despite variations in vesicular size and shape among the mammalian species (Fig. 3), there are a number of general features of the acrosome. It is an acidic vesicle that contains a variety of proteins, including: proteolytic and other enzymatic activities, bioactive peptides and proteins, and other proteins of uncertain function (10,13). A catalog of these constituents is provided by Yanagimachi (10). These proteins and peptides are embedded in a reticular network. Regional subdomains can be differentiated within the acrosome based on histochemical or ultrastructural appearance as well

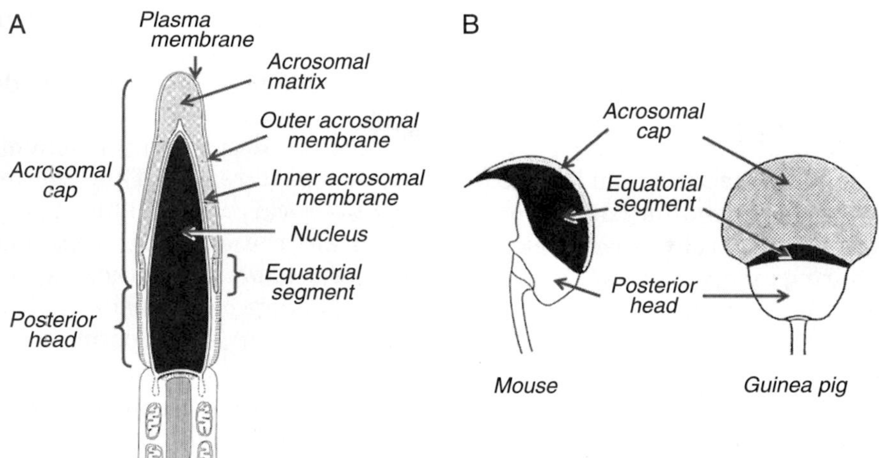

FIG. 3. The organization of the mammalian sperm head is illustrated. **A:** Internal organelles in the head consist of the nucleus and acrosome. Acrosomal membrane domains (inner- and outer-acrosomal membranes) and matrix compartment are shown. The cell surface is divided into zones that correspond to underlying membrane structures. The acrosomal cap overlies the outer acrosomal membrane; the equatorial segment overlies that region where the inner and outer acrosomal membranes merge; and the posterior head, which overlies nuclear membrane but not acrosomal membrane. **B:** All mammalian sperm share the general features outlined in panel **A,** but exhibit considerable variation in size and shape of the head, as well as in the shape of surface domains (acrosomal cap, equatorial segment, and posterior head). Stereotypic examples are shown of sperm with hook-shaped heads (mouse, but also morphologically similar to hamster and rat) and paddle-shaped heads (guinea pig, but also morphologically similar to man, bull, and ram). [Figure is adapted from (10) with permission of Lippincott Williams & Wilkins. Courtesy of Dr. Ryuzo Yanagimachi.]

as upon the distribution of specific proteins. These contents are delimited by a continuous vesicular membrane, which is further divided into three regions: an inner acrosomal membrane lying against the nuclear membrane, an outer acrosomal membrane subjacent to the plasma membrane, and an equatorial segment where these membrane domains join together (Fig. 3) (10,143,144).

The contents of this vesicle are released by an exocytotic event known as the acrosome reaction (Fig. 4). This consists of controlled membrane fusion–fission processes between the outer acrosomal

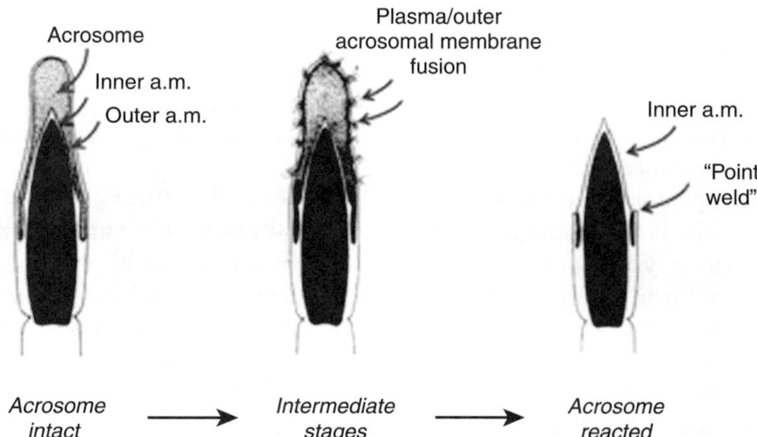

FIG. 4. Sequence of events during the mammalian sperm acrosome reactions. *Acrosome intact* panel: Acrosome intact sperm have a continuous acrosomal membrane (*a.m.*) and plasma membrane. *Intermediate stage* panel: The initial stages of exocytosis are characterized by fusion between the outer acrosomal membrane and plasma membrane at multiple sites, forming hybrid vesicles. *Acrosome reacted* panel: As exocytosis proceeds, the acrosomal matrix disperses and hybrid vesicles are sloughed. The inner acrosomal membrane forms the apical sperm surface in acrosome reacted sperm. A "point weld" between the acrosomal plasma membrane forms in the equatorial segment and maintains cytoplasmic integrity in the posterior head. [Figure is adapted from (10) with permission of Lippincott Williams & Wilkins. Courtesy of Dr. Ryuzo Yanagimachi.]

membrane and the plasma membrane. The result is the formation of vesicles consisting of both membrane domains and containing sperm cytosol trapped in the intravesicular space. These hybrid vesicles eventually are sloughed. The acrosomal matrix disperses, a process driven by acrosomal protease activity, and protein contents are released. The regional organization of the matrix is reflected in a temporally ordered release of vesicular contents during exocytosis (10,145–147).

The acrosome reaction has the characteristic features of regulated exocytosis, including the storage of secretory vesicles for prolonged periods of time in the cytoplasm prior to release (148). Regulated exocytotic systems are driven by a stimulatory agonist or signal and might be expected to exhibit little or no secretion until that stimulatory signal arrives. However, many regulated exocytotic systems in fact show a low rate of basal secretion that is enhanced by agonist. For example, miniature postsynaptic potentials represent a continuous basal release of neurosecretory vesicles that is increased by depolarization-induced Ca^{2+} entry (67,149). Similarly, sperm populations undergo a basal release process that is referred to as the spontaneous acrosome reaction and, as will be discussed later, agonists in the zona stimulate this rate of exocytosis.

This low rate of spontaneous exocytosis should be considered in light of two other issues regarding sperm. First, sperm have only one secretory vesicle. In other exocytotic cells, such as neurons and eggs, there are large numbers of secretory vesicles and so low levels of spontaneous release can be tolerated. In contrast, an individual sperm once acrosome reacted cannot undergo a second round of exocytosis. Second, sperm that have already completed the acrosome reaction cannot penetrate the cumulus, as discussed previously. Taken together, these observations suggest that sperm may require a stringent control of spontaneous exocytosis. Factors such as the coupling of membrane potential shifts and voltage-inactivated T channels may contribute to this control (32).

The fusion–fission process of the acrosome reaction is spatially controlled in that it terminates in the equatorial segment during the physiological acrosome reaction. This termination point is essential in order to prevent extensive cytosolic loss throughout the sperm ("point weld"; Fig. 4). In contrast, dying sperm undergo a generalized and uncontrolled breakdown of the plasma membrane that is accompanied by a dispersal of acrosomal contents. The problems presented by this "false acrosome reaction" are discussed in detail by Yanagimachi (10). Most assays detect component(s) of the vesicular matrix and so will report both those sperm that have completed a regulated, physiological acrosome reaction and also those necrotic sperm that have completed a false acrosome reaction.

Significance of the Acrosome Reaction

The cellular consequences of the acrosome reaction then include the externalization of acrosomal vesicle contents and the display of a new cell surface domain in the apical region of the sperm. There are several physiological consequences of exocytosis. Sperm must complete the acrosome reaction in order to penetrate the zona. In addition, the acrosome reaction is an obligatory prerequisite for sperm–egg membrane fusion (10,150,151). Sperm that have completed the acrosome reaction do not interact with the zona under physiological circumstances, but can do so experimentally following removal of the zona or by drilling holes in the zona with proteases or local application of acidic media. Such studies show that sperm with intact acrosomes may contact the egg plasma membrane but will not fuse until the acrosome reaction has occurred (10). The acrosome reaction then acts as a functional switch that converts sperm into a fusion-competent state.

How the acrosome reaction affects fusion competence is not yet completely resolved. One possibility is that key membrane domains are only exposed as a consequence of the acrosome reaction. Mammalian sperm apparently fuse with the egg by means of membranes in the posterior head and equatorial segment; that is, membranes that were already accessible to eggs prior to the acrosome reaction. However, as will be discussed in more detail later, an obligatory step in the gamete membrane adhesion process that serves as a docking step prior to fusion may involves the sperm inner acrosomal membrane, which is only exposed at the time of the acrosome reaction (10,152). Another possible mechanism involves the autocrine effects due to release of acrosomal hormones or bioactive peptides.

Site of the Acrosome Reaction

Sperm die during migration through the female reproductive tract and undergo the false acrosome reaction associated with dying cells. However, the great majority of living sperm within the oviduct, but not in contact with the cumulus or with eggs, have intact acrosomes (10,153–157). When sperm are instead observed during passage through the cumulus matrix the acrosomes were, in most cases, intact (10,158,159). Some sperm within the cumulus exhibited a swelling in the anterior head. Cummins and Yanagimachi suggest that this is a sign of an imminent acrosome reaction (140). Alternatively, this may reflect either the consequences of a metastable fusion pore state of kiss-and-run exocytosis, as occurs in somatic secretory systems (160,161) and may occur in sperm (162), or a final preliminary process that precedes the membrane fusion event. As will be discussed below, there are likely to be factors within the cumulus that can stimulate sperm and may act as final priming agents prior to the initiation of the acrosome reaction. In summarizing this area, Yanagimachi concluded that "the balance of evidence indicates that fertilizing spermatozoa do not initiate the true acrosome reaction in vivo until they come in contact with the zona pellucida" (10). A discussion of the signal transduction mechanisms of the acrosome reaction will be deferred then to the section on sperm interaction with the zona.

SPERM–EGG INTERACTION

Early Events Preceding Gamete Membrane Fusion

Fertilization typically occurs in the oviduct ampulla. Sperm at this site are thought to have completed capacitation. The egg is encased within a zona and surrounded by layers of cells of the cumulus oophorus (or cumulus). Sperm must penetrate these vestments to contact and fuse with the egg. In this section, we consider the structure of these vestments and their suggested roles in fertilization. A current model of the events of fertilization that precede gamete membrane fusion is presented in Fig. 5.

Sperm Interaction with the Cumulus Oophorus

The cumulus surrounds the egg during the later stages of folliculogenesis and is assigned functions both prior to ovulation and during capture of ovulated eggs by the oviduct (10,52,163,164). In eutherian mammals the cumulus is retained after sperm entry into the oviduct and is believed to participate in the early events of sperm–egg interaction.

Biochemical Composition of the Cumulus Oophorus. The cumulus is composed of several thousand ovarian granulosa cells embedded within a

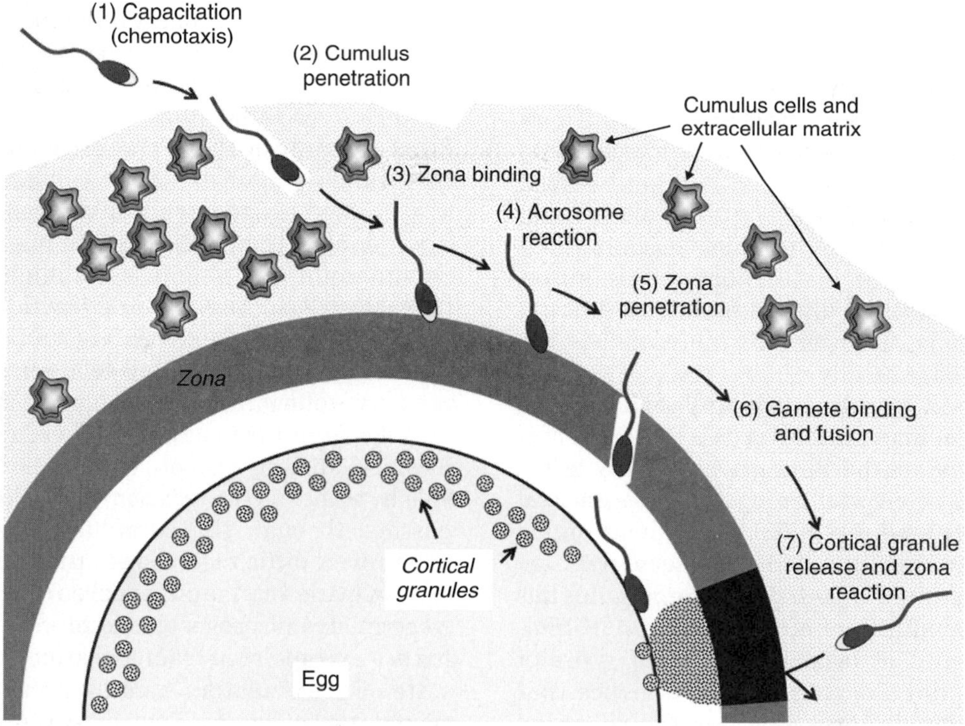

FIG. 5. Model of the early events of mammalian sperm–egg interaction. An *egg* is shown with *cortical granules* in the peripheral cytoplasm. Surrounding the egg are a zona pellucida (*zona*) and a cumulus oophorus complex of *cumulus cells* embedded in an extracellular *matrix*. According to present models, the process of sperm–egg interaction proceeds by the following stages. **1:** Sperm capacitation is completed in the oviduct. Part of this process may involve the development of a chemotactic response. **2:** Sperm penetrate the extracellular matrix of the cumulus complex. At this time the sperm is believed to retain an intact acrosome (yellow crescent in anterior sperm head). The cumulus environment may contain soluble and insoluble components that stimulate sperm. **3:** Acrosome-intact sperm reach the zona and bind to ZP3. **4:** ZP3 triggers the acrosome reaction. **5:** Sperm penetrate the zona. **6:** Sperm bind to and fuse with the egg plasma membrane. **7:** Cortical granule exocytosis is an early event of egg activation. The released cortical granule contents modify the zona (the "zona reaction") and form part of the block to polyspermic fertilization.

complex extracellular matrix. The major component of this matrix is hyaluronan, an unbranched (β1, 4-glucuronic acid: β1,3-*N*-acetylglucosamine)$_n$ polymer in which the number of disaccharide repeat units may be >2,000 and relative molecular mass >1 MDa (165,166). Proteoglycans containing chondroitin sulphate, heparan sulphate, and dermatin sulphate glycans covalently attached to core proteins are also present (10,167). Protein components include elements common to many extracellular matrices, such as collagen, laminin, fibronectin, tenascin-C (168), and a number of uncharacterized components (169). In addition, members of the inter-α-trypsin inhibitor family (IαI) are present (170). IαI is a multimeric protein complex that enters the cumulus matrix from serum. It consists of a light chain, the protease inhibitor bikunin, attached by means of a chondroitin-4-sulfate linkage to IαI heavy chains. IαI heavy chains are then enzymatically transferred to hyaluronan, thereby stabilizing the cumulus matrix (171–173). This extracellular protein/glycan network is anchored to cumulus cells by means of hyaluronan binding to the cell surface hyaluronan binding protein, CD44 (174,175) and, possibly, by fibronectin binding to cell surface integrins (176). In this regard, both CD44 and integrins can initiate signal transduction processes by binding and activating a wide range of intracellular regulatory or scaffold proteins (165,177,178). There may then be a dynamic interaction in which alterations in the extracellular matrix may be expected to regulate the physiology of cumulus cells. It can be speculated then that the gradual loosening and dispersal of the cumulus matrix that occurs within the oviduct may signal an altered functional state of those cumulus cells, however this question has not yet been examined critically.

Role(s) of the Cumulus during Fertilization. Marsupial eggs lack a cumulus at the time of fertilization. In addition, eutherian eggs are readily fertilized in vitro and in vivo following the experimental removal of this vestment (10,52). Thus, the cumulus does not appear to be essential for gamete interaction. Nonetheless, it has been reported that the cumulus has a subtle beneficial affect on fertilization (10,179,180). This may be due to at least two different types of processes. First, there is considerable evidence that the cumulus controls sperm access to the zona and to the egg. Sperm populations exhibit morphological, biochemical, and functional heterogeneity. It has been suggested that the efficiency of fertilization may depend on preventing sperm that have compromised functional ability from reaching the egg surface. Such compromised sperm include those with abnormal morphology, but also those that are incompletely capacitated or otherwise incapable of undergoing an acrosome reaction as well as those with sub-optimal motility (10,181). A number of studies show that the cumulus acts as a selective barrier, arresting both uncapacitated sperm as well as sperm that have completed the acrosome reaction, while permitting capacitated, acrosome-intact sperm to penetrate and engage in the later steps of fertilization (10,140, 182–184). The mechanism of sperm trapping that underlies filter selectivity is unknown.

Second, the cumulus may provide factors that regulate sperm function and so enhance fertilization. Exposure of sperm to either cumulus cells or to cumulus-conditioned medium increased the duration of motility, the forward velocity, and the force generated by flagellar beating (10,122,185–187). In addition, it has been suggested that cumulus cells may release a sperm chemoattractant (133). In no case has a soluble factor been isolated from cumulus preparations and identified.

In this regard, it is necessary to consider the case of progesterone. It is synthesized and released from the ovarian follicle prior to ovulation, present in follicular fluid, and then released from the follicle at ovulation (188,189). It is now understood that progesterone can act on sperm through a nongenomic receptor to activate phospholipase C (PLC), elevate $Ca^{2+}{}_i$, and regulate several aspects of cellular behavior (190–195). Specifically, progesterone: (a) can induce the hyperactivated flagellar motility (196) that, as discussed previously, is characteristic of sperm from the oviduct ampulla that are capacitated; (b) acts as a "priming" agent in that it can render sperm more responsive to the induction of the acrosome reaction by the zona (197); and (c) at high concentrations (10^{-6} M) drives the acrosome reaction in the absence of other stimulatory agonists (190). It has therefore been argued that cumulus cells retain steroidogenic capacity following ovulation, release progesterone into the cumulus matrix, and thereby regulate sperm function during gamete interaction. According to this hypothesis, the acrosome reaction occurs during passage through the cumulus matrix. However, it has proven difficult to assess the steroidogenic state of post-ovulatory cumulus cells from the oviduct and so estimates of progesterone produced by the cumulus are extrapolated from assessment of the synthetic state of pre-ovulatory cumuli (188,189). Yet, as discussed previously, such extrapolations must be viewed with caution in light of the dynamic state of the cumulus matrix following ovulation and its potential effects on cumulus cell function. It remains unresolved whether the effects of progesterone on sperm are physiologically significant, and whether this steroid is released from ovulated cumuli or with follicular fluid.

Similarly, cumulus complexes obtained from superovulated mice release various prostaglandins in vitro, including PGE1, PGE2, and PGF2-α, and a role for PGE1 and PGE2 was proposed (198). Schaefer and coworkers reported a PGE receptor on human sperm that bound prostaglandins with a novel selectivity. This receptor was functionally active as prostaglandin binding drove $Ca^{2+}{}_i$ elevations in human sperm through a pertussis toxin-sensitive pathway (199). However, the physiological role of this intriguing receptor has not been studied extensively.

The discussion to this point has focused on soluble components released by the cumulus. However, extracellular matrix components may also regulate sperm. Sperm are expected to interact with hyaluronan during penetration through the cumulus. It is then notable that this glycosaminoglycan activates sperm signal transduction pathways, driving an elevation in $Ca^{2+}{}_i$ levels (200) and in sperm protein phosphorylation state (201). As a consequence of hyaluronan treatment, sperm exhibit altered patterns of flagellar motility (201,202), an enhanced acrosome reaction response to stimulatory agonists from the zona pellucida (203), and possibly the direct initiation of the acrosome reaction (204). These effects appear to be due to the interaction of hyaluronan with the sperm surface hyaluronidase, PH-20 (see below) (200). These observations are consistent with a local regulation of sperm function by insoluble components of the matrix.

Sperm Penetration through the Cumulus Oophorus. It has been known for some time that sperm contained a cumulus-dispersing activity (205). When high concentrations of sperm are used, as occurs routinely during in vitro fertilization, the cumulus matrix loosens and breaks up. In contrast, low sperm concentrations are maintained in vivo during fertilization, where there may only be 10–100 sperm surrounding a clutch of eggs and where sperm:egg ratios are in the range of 1–10:1 (155,206). The cumulus is not disaggregated by these physiological sperm concentrations and so presents a barrier that must be traversed in order for the later events of fertilization to occur (181,207).

The first model that attempted to account for cumulus penetration drew on observations that the cumulus matrix-dispersing activity of sperm was due to hyaluronidase (208–211); that hyaluronidase was detected in the acrosome (212,213); and that acrosome-reacted sperm were detected within the cumulus matrix (153 214). As a result, it was proposed that the acrosome reaction necessarily occurred while sperm were entering the cumulus matrix, and that the subsequent exocytotic release of acrosomal hyaluronidase locally dispersed that matrix and facilitated sperm passage to the zona pellucida (10,215).

This model has been reevaluated in light of more recent observations. First, as discussed previously, sperm that have completed the acrosome reaction cannot penetrate the cumulus efficiently. Sperm that had completed exocytosis prior to reaching the cumulus become bound to the outer edge, while those that complete the acrosome reaction within the cumulus become trapped and cannot proceed (10,140). This is consistent with the detection of sperm with intact acrosomes within the cumulus during fertilization in vivo (157–159) and also suggests that acrosome reacted-sperm that had been observed within the cumulus in ultrastructural studies (153,214) may be trapped and not, as first believed, in the process of penetration through the matrix (207). Second, it is now understood that the acrosome reaction is not absolutely required for expression of hyaluronidase activity towards extracellular substrates. Some evidence had suggested the presence of a pool of hyaluronidase on the sperm surface in addition to that present in the acrosome (216,217). Subsequently, it was recognized that PH20, a glycosyl-phosphatidylinositol-linked protein in the sperm plasma membrane (218,219), has a hyaluronidase activity (220–222) that is essential for penetration of the cumulus matrix (223). Third, as will be discussed later, a number of lines of evidence support the notion that sperm with intact acrosomes preferentially associate with the zona pellucida and that an acrosome reaction-inducing agonist is present in the zona pellucida (10,150). Finally, sperm observed by light microscopy during transit through the cumulus matrix often exhibit a swelling in the anterior head that precedes the acrosome reaction (10,140). However, it has not been shown that such sperm have initiated exocytosis. Recently, it has been suggested that a kiss-and-run exocytotic mechanism (160,161) may be utilized during the acrosome reaction, permitting the transient exposure of vesicular contents (162). Taken together, these observations suggest that sperm with intact acrosomes are capable of traversing the cumulus matrix through a process that depends upon the hyaluronidase activity of PH20 (222).

Sperm Adhesion to the Zona Pellucida

During fertilization in vivo, sperm that have traversed the cumulus then encounter the zona. This matrix is porous with regard to small molecules, proteins such as antibodies, and even virus particles, but nevertheless presents a barrier to sperm. In order to reach the egg plasma membrane, sperm adhere to outer edge of the zona, undergo acrosomal exocytosis that is regulated by the zona, and then move into and through the zona matrix. Estimates for the time course of these events in vivo vary from only a few minutes to >15 minutes. Yanagimachi noted that sperm are infrequently observed in the process of zona penetration, suggesting that this is a rapid process (10). Following gamete fusion, modifications in the functional properties of the zona form an aspect of the block to polyspermy. In this section, we consider sperm interaction with the zona prior to gamete fusion, while modifications of the zona following gamete fusion will be deferred to a later section. In addressing the structure and function of zona proteins, the authors identify genes and their protein products according to nomenclature approved for the human genome by the HUGO Gene Nomenclature Committee (http://www.gene. ucl.ac.uk/ nomenclature/) and, where necessary, will also indicate species.

Structure and Biochemical Composition of the Zona Pellucida. The zona is a specialized extracellular matrix that is synthesized and secreted by the growing oocyte (150,224–226). The four proteins that are components of the zona are designated ZP1, ZP2, ZP3, and ZP4 (150,225,227,228). ZP1-3 were first identified in 1980 (225) and were subsequently characterized directly by biochemical analysis of native proteins and also by molecular cloning in several species (150,226,229–232). ZP4 was identified in genomic databases in 1999 (233) and the presence of a protein encoded by this gene in a native zona emerged only in 2004 (227). Analysis of protein sequences reveals that ZP4 is most closely related to ZP1 and that these genes likely arose from a common ancestor by a gene duplication process (233,234). All of the zona proteins are synthesized by oocytes and posttranslationally modified by glycosylation at serine/threonine (O-linked) and at asparagine (N-linked) residues (226,235–239).

These proteins share several sequence motifs and a common molecular architecture (Fig. 6). Each of the four zona proteins consists of an N-terminal signal sequence, followed in sequence moving towards the C-terminus by a ZP domain, a consensus furin cleavage site, and a single predicted transmembrane domain. Functional roles for several of these motifs have either been demonstrated experimentally or can be inferred with confidence from examination of similar sequences in somatic cell proteins. Hence, the signal sequence targets nascent zona proteins to the oocyte secretory pathway (91); the transmembrane domain is believed to anchor these polypeptides to secretory pathway membrane; and proteolysis at the consensus furin cleavage site permits the release of zona protein from the membrane (240–242) [however, also see (243)].

In contrast, the role of the ZP domain [accession numbers: SMART (http://smart.embl-heidelberg.de/), SM00241; pfam (http://www.sanger.ac.uk/software/pfam/), PF00100] is has only emerged recently. This is a large domain of ~260 amino acids, including eight conserved cysteine residues that most likely participate in the formation of disulfide bonds that

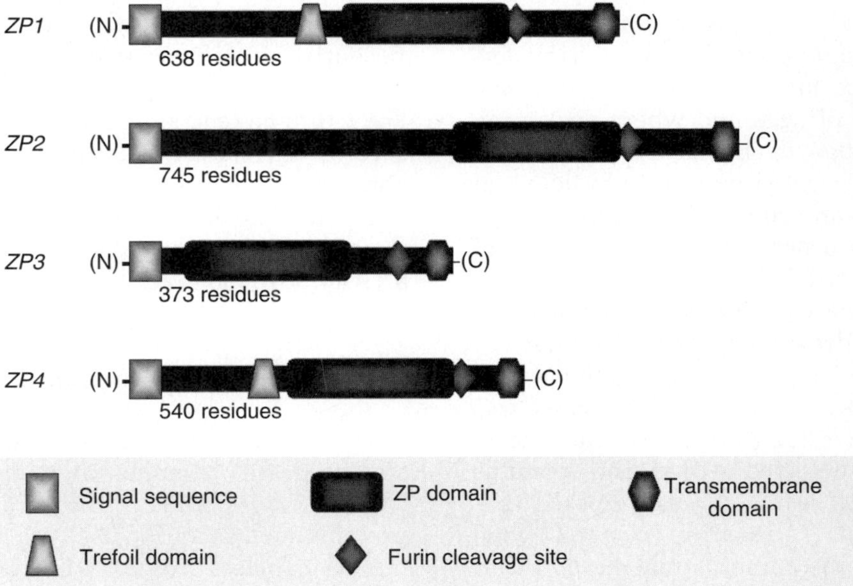

FIG. 6. Molecular architecture of zona pellucida proteins. Each protein has an N-terminal signal sequence, a zona pellucida domain (ZP), a consensus furin cleavage site to the C-terminal side of the ZP domain, and a predicted transmembrane domain. ZP1 and ZP4 also have a trefoil domain. The N- and C-termini are indicated (N- and -C, respectively). Lengths of human protein sequences are indicated as amino acid residues: ZP1, P60852; ZP2, Q05996; ZP3, NP_009086; ZP4, CAI22066.

are essential for tertiary structure (244,245). It is present in the mammalian zona proteins and in some egg coat proteins in birds, fish, and certain invertebrates, but is also widely distributed in metazoan glycoproteins of somatic cell origin. ZP domain proteins include extracellular matrix components such as the α- and β-tectorins of the inner ear, renal uromodulin (Tamm–Horsfall protein) and cuticulin; as well as membrane proteins such as the transforming growth factor (TGF)-β receptor type III and estrogen-regulated gene type I (ERG1) (150,244,246). Several lines of evidence point to a role of the ZP domain in protein–protein interactions, and particularly in the polymerization of extracellular matrix filaments, a role that would particularly relevant in the assembly of a zona (245,247).

These proteins form a reticular meshwork that surrounds the egg to a depth of several micrometers. In the mouse, where this matrix has been examined by biochemical and electron microscopic methods, the zona is composed of ZP2/ZP3 filaments that are held together by non-covalent interactions (Fig. 7). Filaments are crosslinked at intervals by ZP1. The recognition that mouse ZP1 is a homodimer, composed

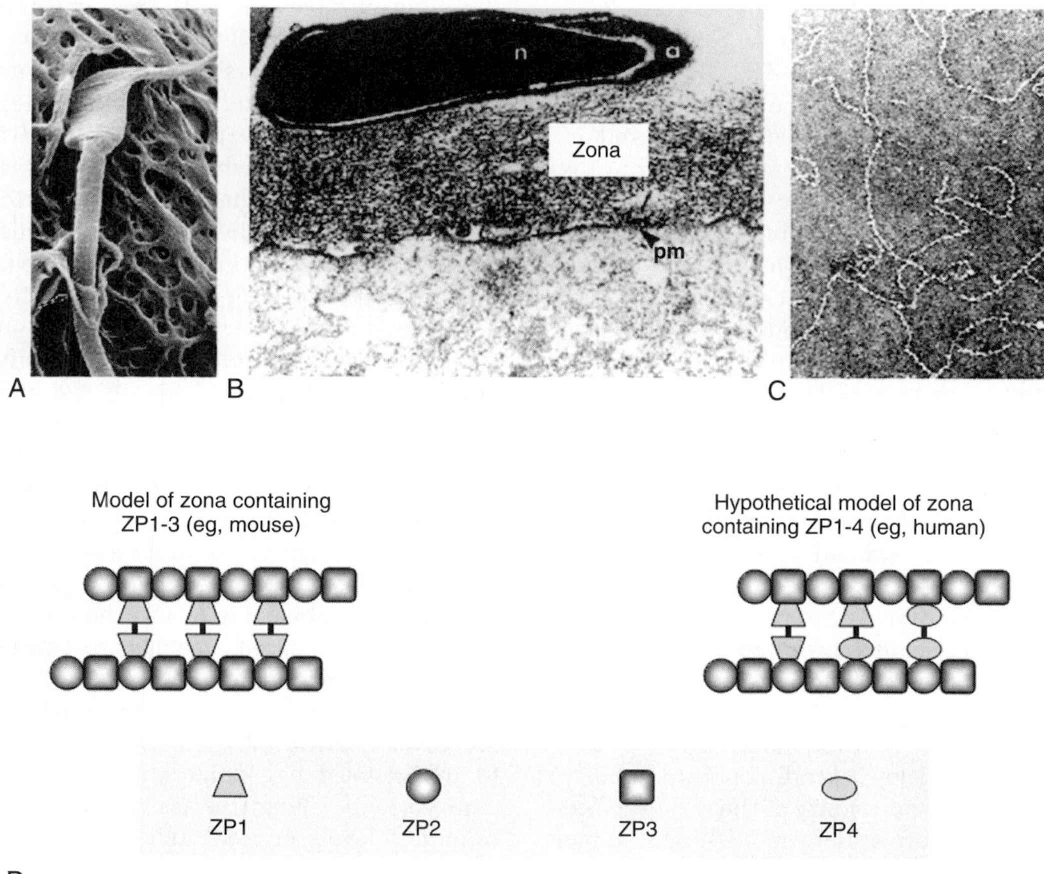

FIG. 7. Model of the structural organization of the zona pellucida. **A:** Scanning electron micrograph of hamster sperm on the surface of the zona pellucida. Note the filamentous organization of the zona pellucida matrix. **B:** Transmission electron micrograph of mouse zona pellucida (*zona*) laying outside the egg plasma membrane (*pm, arrowhead*). Note the apparent filamentous structure of the zona pellucida. An acrosome-intact sperm is bound to the outer surface of the zona pellucida (*n*, sperm nucleus; *a*, acrosome). **C:** Mouse zona pellucida is solubilized and filaments are visualized by electron microscopy. These filaments are composed of ZP2 and ZP3. **D:** Model of the molecular organization of the zona pellucida. *Left panel:* Model of the mouse zona, which is composed of ZP1–3. ZP2/ZP3 filaments are cross-linked by ZP1 homodimers. *Right panel:* A hypothetical model of a zona containing ZP1–4 is shown. This speculative model assumes that: (a) basic structural elements of the zona are conserved between species; and (b) that the relationship between ZP1 and ZP4 (see text) bespeaks a conserved function. It proposes that a zona containing ZP1–4 retains a ZP2/ZP3 filament structure similar that of the mouse, but may have different types of filament crosslinkers. Three crosslinkers are formally possible: ZP1 homodimers, ZP4 homodimers, and ZP1/ZP4 heterodimers. [Panel A adapted from (10) with permission of Lippincott Williams & Wilkins; courtesy of Dr. Ryuzo Yanagimachi. Panel B adapted from (650) with permission of Elsevier; courtesy of Dr. Paul Wassarman. Panel C adapted from (248). Panel D adapted from (650); courtesy of Dr. Paul Wassarman.]

of 120 kDa subunits that are stabilized by interchain disulfide bonds, is consistent with a proposed role as a crosslinker (225,226,248). The ZP4 protein is not present in the mouse zona as the locus is a pseudogene due to the accumulation of stop codons. In contrast, expression of all four zona genes has been demonstrated directly by analysis of the zona of human and rat (227,233,239). This indicates that the presence of ZP4 is not necessary for the assembly of a functional zona matrix in vivo, at least if ZP1 is present. It will be necessary to determine whether the model of zona structure in the mouse also applies to species in which four zona genes are expressed. Given the evolutionary relationship between ZP1 and ZP4 (233,234), it is tempting to suggest that the construction of filaments from ZP2/ZP3 repeats may be a common feature of the zona, but that these filaments can be crosslinked by ZP1 and/or ZP4 (Fig. 7).

Finally, minor components may associate loosely with the zona in vivo. These include glycans such as hyaluronan, and proteins such as oviductin, that may be adsorbed from either the cumulus extracellular matrix or from oviductal and follicular fluids (207,249–252). The functional significance of these minor components has been difficult to evaluate.

Assays of Sperm Binding to the Zona Pellucida. Sperm recognize and adhere to the outer margin of the zona before beginning the penetration process (Fig. 4). This initial interaction takes place in vivo in the restricted space beneath the cumulus. However, these events are examined in vitro after first removing the cumulus, thereby affording sperm direct contact with the zona. Under these conditions, sperm can adhere tenaciously to the zona of unfertilized eggs (10). (This form of interaction is referred to as "binding" and will be considered in greater detail below.) Binding occurs rapidly, with sperm residing in the vicinity of the zona surface for 0.05–0.1 seconds before either establishing a stable contact or moving away (253).

Before turning to the mechanism of sperm-zona adhesion, it is useful to consider the types of assays available. It should be recalled that cells form specific, interactions with other cells or with noncellular substrates, but also stick to those surfaces nonspecifically. The total binding, which is the experimentally determined parameter, is the sum of these specific and nonspecific interactions (254,255). In this regard, sperm–zona interaction is typically studied in vitro in the absence of a cumulus, where there are no barriers to sperm contact with or escape from the ZP. However, the cumulus may alter the conditions for adhesion by orienting sperm towards the ZP and by constraining the ability of sperm to move away from the zona surface. This leads to the question of whether tenacious, rapid adhesion is a feature of sperm–egg interaction or only a requirement imposed by in vitro assay conditions.

Two types of evidence argue strongly for the relevance of tenacious adhesion. First: sperm that are tightly bound to the zona surface can proceed to the later stages of gamete interaction. This was demonstrated clearly by Saling and co-workers, who allowed sperm to bind to the zona and then subjected them to shear forces sufficient to remove all but tightly bound sperm. Those sperm were able to undergo the acrosome reaction and penetrate the zona (256). Thus, tenaciously bound sperm are not trapped at the zona surface; instead, this is a physiologically relevant interaction. Second: the ability to form tenacious bonds correlates with the known behavior of sperm during fertilization. For example, capacitated sperm, which can fertilize eggs, bind to the zona tenaciously in vitro whereas uncapacitated, infertile sperm fail to bind tightly (257–259). Similarly, sperm bind tenaciously to the zona of unfertilized eggs but not to those of 2-cell embryos, consistent with other information regarding the zona-block to polyspermy (see below) (10).

Assays were developed to quantify tenacious binding after removal of loosely bound sperm from the zona. In some cases, arbitrary or defined shear forces were applied [e.g., (see 260)]. Recognition that sperm only establish loose, readily reversible contacts with the zona of 2-cell embryos permitted development of a biologically defined assay in which sperm are incubated with unfertilized eggs and with two-cell embryos. Shear forces are then applied until few (<5) sperm remained attached to two-cell embryos. As noted previously, total binding is the sum of specific (that is, tenacious) and nonspecific binding. Moreover, sperm can bind both tenaciously and loosely to unfertilized egg zona but only bind loosely to embryo zona. Using this assay, it is possible to determine the levels of total and nonspecific binding by counting sperm on unfertilized eggs and two-cell embryos, respectively, and then to calculate specific binding levels (261).

Several points should be made regarding this assay. First, it is always possible to determine the relative tenacity of adhesion. However, if binding strength is to be correlated with biological processes then the assay must be carried out using two-cell embryos controls of the homologous species, and then is only interpretable in species that exhibit a strong zona block to polyspermy. Second, the assay can be used effectively for analysis of the interaction of homologous gametes but is not necessarily valid when comparing heterologous interactions across species or taxa. Differences in sperm shape and size, as well as the density of adhesion molecules, are expected to influence the response to shear forces and complicate interpretation.

Sperm Adhesion to the Zona Pellucida. The following sequence of events has been established regarding the adhesion to the zona. Sperm first establish a weak contact with the zona of unfertilized eggs which is readily disrupted by gentle aspiration or by other shear forces. This contact is referred to as "attachment" (257,259,262). Uncapacitated sperm can attach to the zona, as can sperm of heterologous species. Despite being a low-affinity, low-specificity interaction, attachment may nevertheless be mediated by specific proteins on sperm and zona. For example, candidate low-affinity or low-selectivity adhesion sites are present in both sperm and zona, and may mediate attachment (263–266).

Attachment develops into a more tenacious form of contact referred to as "binding." The first direct evidence that binding was mediated by specific adhesion molecules was provided by competition binding experiments, in which soluble extracts of zona of unfertilized eggs inhibited sperm binding to intact ZP whereas similar extracts of embryo zona lacked this activity (Fig. 8) (267).

ZP3 has a number of the anticipated characteristics of a sperm adhesion molecule from the zona. First, purified ZP3 inhibits sperm–zona adhesion in the competition binding assay in both mouse and hamster systems, and completely accounted for the inhibitory activity found in soluble, unfractionated extracts of the zona. The inhibitory potency of soluble ZP3, as expressed as the concentration required to inhibit sperm adhesion by 50% (IC_{50}) is ~1–2 nM. In contrast, purified ZP1 and ZP2 do not retain similar activity following purification (261,268). Second, ZP3 purified from 2-cell embryos did not inhibit sperm-zona binding, as expected in a species such as the mouse where there is a strong zona block to polyspermy (261). Third, ZP3 selectively binds to the sperm head, which is the region that interacts with the zona (269,270).

Based on these results, it is widely accepted that ZP3 is the only component of the zona that retains the characteristics of a sperm adhesion molecule following purification. This conclusion is drawn largely, but not exclusively, from the mouse system. Some data may imply that other zona glycoproteins may also participate in sperm binding, either acting alone or in combination with ZP3 (271–275). However, zona glycoproteins from some species, including the bovine and porcine, can only be purified following a preliminary glycosidase treatment of unfractionated preparations (271,272,275). Since zona glycan moieties are believed to be important in sperm binding (see below) then this raises questions regarding the effects of glycosidase treatment. Alternatively, the

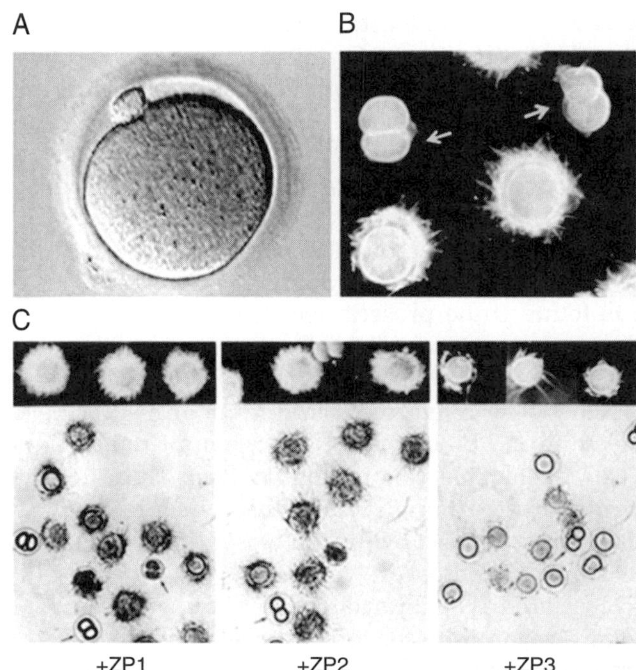

FIG. 8. ZP3 has the anticipated characteristics of a sperm adhesion protein. **A:** Mouse egg with polar body within a zona. **B:** Mouse sperm bind to zonae of eggs but not to those of 2-cell embryos (*arrows*), as visualized by darkfield optics. **C:** Zona binding assays were performed as in panel B except that sperm were added in the presence of soluble ZP1, ZP2, or ZP3. Soluble ZP3 inhibited sperm binding to intact zonae in a competitive manner, whereas ZP1 and ZP2 did not exhibit inhibitory activity. Panel shows darkfield and brightfield views. Note the presence of 2-cell embryo controls (*arrows*). [Panel B adapted from (650) with permission of Elsevier; and panel C from (261) with permission of Cell Press. Courtesy of Dr. Paul Wassarman.]

effects of targeted gene disruption and replacement experiments may indicate a role for other zona proteins in sperm binding (274). A more complete analysis of the phenotype of this mouse is required before the significance of these results can be appreciated (266). At present, the burden of evidence indicates that ZP3 is the sperm adhesion molecule of the zona.

What is the mechanism of tenacious adhesion of sperm to the zona? To address this question, zona glycoprotein domains were modified and assessed for a functional role in adhesion. These experiments were initially carried out in the mouse system. As discussed previously, soluble extracts of the zona inhibit sperm binding to the zona competitively (261,267). This adhesion activity was not destroyed when zonae were treated with pronase, a crude mixture of endo- and exo-proteases (276,277). The products of pronase digestion of the zona are small glycopeptides, suggesting that protein conformation or specific protein sequence motifs were unlikely to be essential determinants of the adhesion domain and pointing to a role of nonpeptide moieties such as glycans (277).

This was consistent with other observations that the adhesion receptor activity of the zona was resistant to protein denaturing conditions such as boiling, repetitive freeze/thaw cycles, or treatment with denaturing detergents [reviewed in (226)]. Proteolysis of purified zona glycoproteins showed that the pronase-resistant adhesion activity was associated with ZP3 of unfertilized eggs but not with either egg ZP1 or ZP2, or with 2-cell embryo ZP3 (277). This activity was found to be present specifically in an O-linked glycan fraction of ~3.9 kDa, as assessed by size-exclusion chromatography; that is, an oligosaccharide attached to the hydroxyl residues of either serine or theonine on ZP3 through N-acetylglucosamine (237). Two conclusions can be drawn from these experiments. First, zona glycoproteins, which are synthesized and processed by the egg biosynthetic machinery coordinately, must nevertheless be differentially glycosylated. The alternative model suggests that all zona glycoproteins have identical glycans but these are buried within the tertiary structure of ZP1 and ZP2 while exposed in ZP3. This would predict that isolated glycans from all these glycoproteins would have adhesion bioactivity, and so cannot account for the presence of such activity only in ZP3 glycans. Second, the loss of adhesion activity following fertilization, which is an essential element of the block to polyspermy (see below), cannot be due exclusively to conformational changes in ZP3 that result in the burial of active glycan domains.

There is wide agreement from studies in a number of species that tenacious adhesion is due to sperm interaction with a zona glycan (10,226,278). These studies have been extended in a number of directions: by glycosidase experiments to determine the sequence of the adhesion glycan (276,279); by use of model oligosaccharides to deduce glycan information (280–282); by mutational and biochemical analysis of ZP3 structure (226,283–286); and by the ongoing effort to characterize zona glycans by biochemical methods and, or recently, by mass spectrometry (287–292). At present, these approaches have not defined the nature of binding sites on ZP3 for sperm or of the molecular interactions that mediate adhesion.

What is the nature of the cognate adhesion molecule on sperm? The anticipated characteristics of an "ideal" adhesion molecule can be summarized as follows: it should interact directly with the zona or ZP3; this interaction should recapitulate the observed regulation of adhesion, including changes during capacitation, following fertilization, and species specificity; and it should be restricted to the sperm head, and likely to the plasma membrane (162), as this is the region which binds the zona.

Using ^{125}I-ZP3 binding methods it was determined that the receptor is on the sperm head (269) and that there are in the range of 10,000–50,000 zona binding sites/sperm (263,269). A high-affinity binding site, detected by equilibrium binding methods, had an apparent affinity of ZP3 of 0.72 nM, as expressed as a dissociation constant (K_d) (263). This is similar to the observed potency of ZP3 as an inhibitor of gamete binding, where the half-maximal inhibition of adhesion (IC_{50}) was 1–2 nM ZP3 (261,277), indicating that this high-affinity site may mediate tenacious sperm binding to the zona.

Candidate adhesion sites on sperm have been suggested based on direct interaction with zona or ZP3, as in the cases of sp56 (293,294), zonadhesin (295), p95 (296,297), and β-1,4-galactosyltransferase (298), or on the inhibition of adhesion with highly specific antibody probes, as in the case of PH-20 (299,300). In addition, a number of other candidate zona adhesion molecules have been proposed based on the effects of less precise probes (10,150). At present, none of these proteins have all of the anticipated characteristics of a sperm surface adhesion receptor, either because they lack the anticipated cellular localization or binding specificity or because targeted deletion of these genes fail to produce the anticipated infertile phenotype.

Induction of the Acrosome Reaction by the Zona Pellucida

According to our understanding, the acrosome of the fertilizing sperm remains within tens of nanometers of its target membrane but not secreted during the period of transport through the epididymis and through the female reproductive tract. Stimulatory signals delivered at contact with the zona initiate the process of exocytosis. These events are discussed here and summarized in Fig. 9.

ZP3 Receptor(s) Coupled to the Acrosome Reaction. Observations that acrosome-intact sperm bind to the zona tenaciously and undergo the acrosome reaction (256,301) led to the suggestion that the zona actively induced exocytosis. Agonist activity was subsequently demonstrated using solubilized zona (35) and the active factor identified as ZP3 (302). The maximal response to solubilized zonae differs among mammalian species: in mice, >50% of the sperm population may undergo acrosome reactions in response to zona stimulation (35,302), whereas in bovine and porcine sperm only 10%–25% of the population may respond (141,303). These differences are likely due to varying efficiencies of in vitro capacitation protocols since sperm must be extensively capacitated in order to respond to zona stimulation (36,141).

Small glycopeptides produced by proteolysis of ZP3 (237,277) or of unfractionated zonae (276) retain the

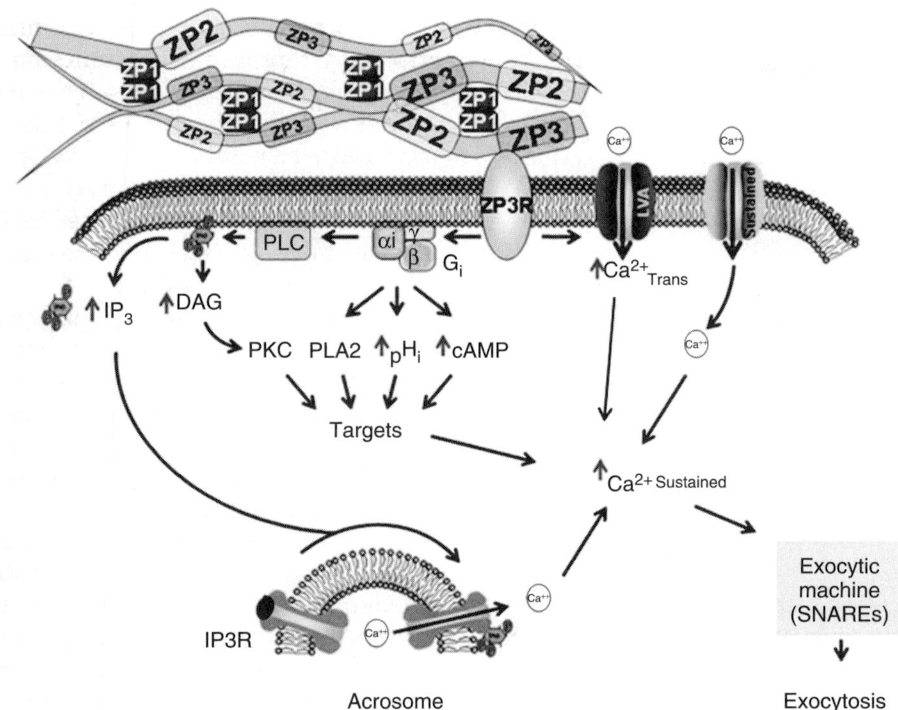

FIG. 9. Model of ZP3-evoked acrosome reaction. Initial stages of ZP3 signal transduction include the activation of a pertussis toxin-sensitive G_i protein and a transient Ca^{2+} entry through low voltage-activated (*LVA*) channels. G protein activation is believed to lead to a rise in pH_i, cAMP, and the stimulation of a phospholipase C (*PLC*) and phospholipase A2 (*PLA2*). Intermediate stages of signaling include the production of IP_3 and diacylglycerol (*DAG*), leading to Ca^{2+} release from internal stores through an IP_3 receptor (IP_3R) channel as well as Ca^{2+} entry through a sustained influx pathway (*sustained*). The resulting persistent elevations of $Ca^{2+}{}_i$ lead, possibly through several intermediate steps, to the activation of an exocytotic machinery. See text for details.

characteristics of a sperm adhesion domain, as discussed previously. However, these glycopeptides did not retain the ability to induce acrosome reactions that was present in native ZP3 (277). Two models could be proposed to account for these observations. In one case, ZP3 is conjugated with multiple copies of a sperm-binding glycan that crosslink sperm receptors, with the resulting patching of sperm surface sites triggering exocytosis (304,305). Alternatively, ZP3 may have two functional sperm interaction domains: a glycan adhesion domain and a second exocytotic agonist domain that is dependent upon peptide motifs (306). These models suggest that there may either be a single, unitary ZP3 receptor that mediates both adhesion and initiation of exocytosis, or separate receptors for each process. It is not yet possible to differentiate between these models or to unequivocally identify a receptor.

Early Events of ZP3 Signal Transduction Leading to a Sustained Ca^{2+} Response. Despite difficulties in identifying a ZP3 receptor, there is considerable information emerging regarding the downstream effectors that couple receptor activation to exocytosis. The discussion here will be simplified in two ways in order. First, it will be restricted almost exclusively to those signal transduction pathways where direct ZP3 or zona regulation has been demonstrated. Some elements of these pathways are listed in Table 2. The authors are aware of a larger literature regarding mechanisms that drive the agonist-independent acrosome reactions which occur spontaneously in sperm population, as well as those that are triggered either by pharmacological agents or by agents other than zonae. These pathways, which include a wide range of protein kinases, phospholipases, and other signaling intermediates, have been reviewed elsewhere (307,308). It is likely that elements of these pathways will also participate in ZP3 signaling, yet at present their role remains theoretical. Second, discussion of ZP3 signal transduction will be organized in large measure as consisting of three phases: upstream events leading to a sustained $Ca^{2+}{}_i$ elevation; characterization of the $Ca^{2+}{}_i$ response; and downstream effectors of elevated $Ca^{2+}{}_i$. It should be noted that signaling events are sorted into these sequential phases, in most instances, by pharmacological rather than kinetic criteria; events within a specific phase do not necessarily exhibit

TABLE 2. *Sperm responses to stimulation with ZP3 or with unfractionated zonae pellucidae*

Signal	Change	References
Ions and ion channels		
Membrane potential	Depolarization	(314)
$Ca^{2+}{}_i$	Elevation	(312)
Channels and ions		
T channels	Activation	(342)
TRPC2 channels	Activation	(370)
IP_3 receptor	Activation	(316)
pH_i	Alkalinization	(312)
HCO_3^-	Required	(57)
Enzymes		
Adenylyl cyclase	Activation	(75)
Phospholipase A2	Activation	(339)
Phospholipase C	Activation	(75)
Signal transducers		
G_{i1}, G_{i2}	Activation	(309,311)
Calmodulin	Activation	(315)

similar time courses. Other aspects of signaling that do not readily fit into this construction will be introduced at a later point.

Several initial events of ZP3 signaling have been identified. The heterotrimeric G proteins, G_{i1} and G_{i2}, are activated by ZP3 (309) and Bordetella pertussis exotoxin (PTx), the membrane-penetrating ADP-ribosyltransferase inhibitor of this class of G proteins, prevents the zona- or ZP3-driven acrosome reaction (310–313). Direct information regarding the molecular nature of the downstream targets of either of α_i or G_i-derived $\beta\gamma$ subunits is not yet available, although PTx studies indicate that G_i activation is necessary for the subsequent stimulation of PLC, alkalinization of pH_i, and that sustained $Ca^{2+}{}_i$ response (312,314). Some data suggests that, under specialized conditions, zona activates sperm adenylyl cyclase through a pertussis toxin-sensitive, $\beta\gamma$ subunit-mediated mechanism (75). The resulting production of cAMP may plausibly act through protein kinase A. There is strong evidence for a role of protein kinase A in capacitation, but as yet little compelling support for it's activation by ZP3 (315). Other mechanisms of cAMP action are available (127). In addition, other G proteins, including $G_{q/11}$ (316) and possibly G_s (317) are present in the acrosomal region of the sperm head, but have not yet been shown to be activated by ZP3.

As noted, stimulation with zonae or ZP3 produces a rise in sperm pH_i. Fluorescent probes report that bovine sperm pH_i increases by ~0.25 units following stimulation in vitro and that this is blocked by PTx, indicating the essential role of G protein signaling in this pathway (312,318). When G_i signal transduction was blocked and then sperm were treated with zona, it was observed that the additional step of alkalinizing pH_i with a permeant weak base was sufficient to bypass the PTx inhibition and to drive the acrosome reaction (314). This hints that the pH_i rise may represent an essential step in ZP3-evoked G protein signaling. The precise role of alkalinization is not yet apparent: it enhances the magnitude of the $Ca^{2+}{}_i$ rise following experimental depolarization (314,319) and may play a similar role during ZP3-evoked depolarization (see below). In this regard, sperm have several pH_i regulators, including a Na^+-H^+ antiporter, a HCO_3^-–Cl^- exchanger, and other pathways that can drive alkalinization (320–323), but it is not yet known which of these or other pathways is activated by ZP3.

A third early signal transducer is PLC. Roldan and co-workers had detected an enhanced production of the products of PLC catalysis in mouse sperm following treatment with zona, thereby demonstrating enzyme recruitment (324). Recent attention has focused on the PLC-δ4 isoform. There are several splice products of this gene that are expressed uniquely in the spermatogenic lineage, of which ALTII is the predominant form in sperm (325–327). The targeted deletion of this gene results in sperm that, when stimulated with zona, do not produce sustained Ca_i^{2+} elevations and do not acrosome react (326,327). The prevalence of several of the other PLC isoforms that are present in sperm is not noticeably altered in the sperm of PLC-δ4–/– mice. Thus, these other isoforms fail to complement the δ4 enzyme with regard to ZP3 signaling (326,327), suggesting that δ4 has either a subcellular localization, an activation mechanism, or catalytic features that permit a unique role in ZP3 signaling (328,329). Ca^{2+} is generally believed to regulate PLC-δ isoforms by binding to both the EF-hand domain and to the C2 domain, thereby resulting in translocation to membranes (330). This may be complicated in the specific case of the δ4 enzyme, where the C2 domain exhibits novel, Ca^{2+}-independent membrane binding (331). An alternative possibility may be that G_i, the sperm G protein activated by ZP3, regulates δ4. Murase and Roldan found that the zona-induced production of diacylglycerol in sperm was inhibited by PTx, indicating that major PLC that was activated was a downstream target of G_i (332). PLC-δ isoforms are not typically controlled by G_i, although some reports from somatic cells point to this possibility (333). Thus, there is strong evidence for a role of PLC, and specifically of the δ4 isoform, in the acrosome reaction, but the mechanism by which ZP3 activates this enzyme is not yet appreciated.

The products of phosphatidylinositol-specific PLC-catalyzed hydrolysis of phosphatidylinositol-4,5-bisphosphate (PI-4,5-P_2) are diacylglycerol and inositol-1,4,5-trisphosphate (IP_3). Both of these second messengers participate in ZP3 signal transduction.

Diacylglycerol binds to C1 domains in a large number of proteins to regulate cell function (334), although its most prominent targets are the classical and novel families of protein kinase C (PKC). Other agents, such as the phorbol esters, also bind to the C1 domain and activate classical and novel PKCs. Evidence for a role for classical PKCs in the ZP3-driven acrosome reaction include: the presence of these kinases in the sperm head (335,336); the initiation of PKC activation and of acrosome reactions by phorbol esters in the absence of zona in some (335) but not all species (337); and observations that classical PKC activators either stimulate (338) or produce a complex acceleration of the kinetics of the zona-induced acrosome reaction (337). Participation of classical PKC isoforms is also consistent with the observed early elevations of $Ca^{2+}{}_i$ following ZP3 stimulation (see below) (32). Other targets of diacylglycerol may also participate in exocytosis: for example, zona stimulation of guinea pig sperm leads via a PTx- and diacylglycerol-sensitive mechanism to the activation of phospholipase A2 and from that point to the potential regulation of arachidonate signaling pathways (339). Such alternative diacylglycerol mechanisms have not been extensively examined.

The second product of PLC action, IP_3, may play a role in the ZP3-driven acrosome reaction. The major action of IP_3 is to bind to and open IP_3R Ca^{2+} release channels, thereby promoting the release of Ca^{2+} from sequestered intracellular stores into the cytosol (340,341). IP_3R is present at high site density in the acrosomal membrane of the sperm of many mammalian species and may participate in ZP3 signal transduction, as will be discussed in the next section (316).

The final element in this discussion of the early events in ZP3 signal transduction is the generation of a transient $Ca^{2+}{}_i$ signal. A rapid Ca^{2+} spike was observed when mouse sperm were treated with ZP3. $Ca^{2+}{}_i$, as reported by intracellular ion-selective fluorescent probes, rose from 10% maximal response to 90% maximal response in <40 msec and then relaxed to basal levels during the next 200 msec. Transients were specifically triggered by ZP3 and only observed in sperm that undergo the acrosome reaction. Pharmacological agents that block the transient also block the ZP3-induced acrosome reaction (32,342). Both the activation kinetics and pharmacology suggest that this transient response was carried by T-type voltage-sensitive Ca^{2+} channels. In this regard, the three genes that encode the $\alpha 1$ subunit of T-type Ca^{2+} channels (or, $Ca_v3.1$–3.3) are all present on mature sperm, with $Ca_v3.1$ and 3.2 located in the head where ZP3 signal transduction occurs (Table 3) (343–350). Interestingly, several of the other ZP3-dependent transducers that were discussed in this section, including PLC-δ4 and the classical PKCs, are regulated by Ca^{2+}. One possible role of the $Ca^{2+}{}_i$ transient may be to modulate, either directly or indirectly, the activity of such ZP3 transducers.

How are T currents evoked by ZP3? Small depolarizations of membrane potential can, under appropriate conditions, activate T currents (67). ZP3 causes a 20–30 mV depolarization of mouse and bovine sperm membrane potential, due apparently to the activation of a poorly selective cation channel (314). The magnitude of this depolarization is sufficiently large to activate T channels if starting from an appropriately hyperpolarized resting potential (67).

The signaling components described here are classified as upstream effectors of ZP3 action since they either have been shown or are expected to occur prior to the ZP3-induced sustained Ca^{2+} response (see below). For example, inhibition of G protein activation with PTx (312) or of the $Ca^{2+}{}_i$ transient with T channel antagonists (32,314) prevents the sustained Ca^{2+} response. The current understanding of these early effectors may be summarized as follows: several threads of signaling pathways are evident, however a continuous pathway linking a primary transducer (such as a G protein) to either to exocytosis or to the intermediary event of sustained Ca^{2+} elevation is not available.

Mechanisms of the Sustained Ca^{2+} Response to ZP3. An essential role for $Ca^{2+}{}_i$ in stimulus-secretion coupling was first suggested in endocrine cells and in neurosecretion (351,352), and extended to many, if possibly not all (353), cell types (149,354). A specific role in the egg-induced acrosome reaction was first suggested in echinoderms (355) and later in mammals (356). However, the heterogeneity of sperm populations with regard to capacitation state and ZP3 response has complicated population-based assays, while the small cytosolic volumes and complex geometry has been an obstacle to electrophysiological approaches using either intracellular electrodes or patch clamp approaches. The development in the 1980s of Ca^{2+}-selective fluorescent probes that could be passively introduced into small cells, and of optical systems to acquire and analyze low light-level probe emission from single living sperm cells, permitted progress in this area.

Studies, first in bovine (312,357) and extended subsequently to other mammalian species, including mouse (314,327) and hamster (358), found that either crude zona extracts or purified ZP3 stimulated a sustained $Ca^{2+}{}_i$ elevation in sperm; that is, an elevation that persists for minutes in the continued presence of agonist, as distinct from the transient, T channel-based response that is complete in ~250 msec. (More precisely, "sustained" is used in the sense that the time-averaged $Ca^{2+}{}_i$ value is above basal levels for a period of minutes. This level may be constant or consist of high frequency $Ca^{2+}{}_i$ spikes.) The sustained $Ca^{2+}{}_i$ elevation is essential for the ZP3-evoked acrosome reaction. In this regard, $Ca^{2+}{}_i$ signals are encoded with amplitude and duration information and cells can discriminate between these different signals to produce different effects (359,360). It may be speculated that a requirement for a sustained $Ca^{2+}{}_i$ rise to drive the acrosome reaction permits sperm to use $Ca^{2+}{}_i$ transients to regulate other functions without producing acrosome reactions.

It is well established that the sustained Ca^{2+} response to ZP3 requires external Ca^{2+} and is inhibited by Ca^{2+} entry blockers, pointing to an essential role of influx across the plasma membrane (10,361,362). Ca^{2+} indicator dye studies show that the $Ca^{2+}{}_i$ rises begin in the sperm head, and more specifically, are first apparent in the equatorial segment (312,327,358). This does not necessarily mean that the equatorial segment is where ZP3 signaling begins or even where Ca^{2+} channels first open. The response of indicator dyes can be influenced by local factors such as protein concentration (363) or by factors that influence how rapidly ion concentrations rises locally following channel opening: such conditions include the concentration of Ca^{2+} binding proteins and other buffers; cytoplasmic volume, which may be very restricted in the equatorial segment (144); and local Ca^{2+} efflux mechanisms (359,364). If we instead focus on the sperm head as the site of sperm-zona interaction and the location of ZP3 binding sites on sperm (see above), then we find a large number of Ca^{2+}-conducting ion channels that are candidate mediators of ZP3 signaling: seven voltage-sensitive Ca^{2+} channels (four high voltage-activated and three low voltage-activated), at least four trpc Ca^{2+}- conducting cation channels, and an IP_3R on the acrosomal membrane (Table 3). Any of these channels could, in theory, contribute to the ZP3-evoked $Ca^{2+}{}_i$ response and drive the acrosome reaction.

It was initially suggested that the sustained $Ca^{2+}{}_i$ response was due to influx of extracellular Ca^{2+} through a Ca_v1 (L-type) channel. Evidence supporting this notion included the presence of Ca_v1 channels in sperm (129,365) and the observation that the many (but not all) agonists and antagonists of Ca_v1

TABLE 3. *Calcium conducting channels in mammalian sperm. This list includes channel proteins documented in mammalian sperm. Channels may be in the plasma membrane or in intracellular organelle membranes. Excluded are channel transcripts that are detected in spermatogenic cells or testis, as such channel proteins may not be retained in mature sperm*

Channel	Pore-forming subunit detected	Localization	Evidence of function	References
Plasma membrane channels				
Voltage-gated				
High-voltage				
activated	$Ca_v1.2$	ac, eq, pp	-	(129,666,667)
$Ca_v2.1$		eq, ph, mp, pp	-	(129)
$Ca_v2.2$		ac, ph, pp	A	(130)
$Ca_v2.3$		ac, ph, pp	A	(130,349,668)
Low-voltage				
activated	$Ca_v3.1$	ac	B	(349)
$Ca_v3.2$		ac	B	(349)
$Ca_v3.3$		ac, mp	B	(349)
TRPC	trpc1	ac, pp	-	(669,670)
trpc2		ac, eq	C	(370)
trpc3		ph, pp	-	(669–671)
trpc4		mp	-	(669,671)
trpc5		ac	-	(670)
trpc6		neck	-	(669)
CNG	CNGA3	mp, pp	D	(126,127)
CNGB1		pp	D	(126,127)
CatSper	CatSper1	pp	E	(672)
CatSper2		mp, pp	E	(673)
Organelle membrane channels				
IP_3 receptor	IP_3R type 1,3	ac, ph, mp, pp	F	(128,674)

ac, acrosomal cap of head; eq, equatorial segment of head; ph, posterior head; mp, flagellar midpiece; pp, flagellar principle piece.
(A) Depolarization produces Ca^{2+} dye signal that is partially blocked by selective (Cav2.2) or non-selective (Cav2.3) inhibitors (668). (B) ZP3 produces Ca^{2+} dye signal that is blocked by a several Ca_v3 inhibitors (32), but it is not known which one or several family members are involved. (C) ZP3 produces Ca^{2+} dye signal that is blocked by a gene-specific antibody (370). (D) Cyclic nucleotide-evoked current in bovine sperm, as measured by excised patch clamp methods (126). (E) Channel activity not demonstrated but catsper -/- mice are infertile (672, 673). (F) IP_3 releases $^{45}Ca^{2+}$ from digitonin-permeabilized sperm (316).

channels could modulate the sustained response (361). Sperm also have Ca_v3 (T-type) channels (342,344,366,367) and many aspects of the pharmacology of Ca_v1 and Ca_v3 channels overlap (342). The sustained influx pathway was subsequently isolated experimentally and found to exhibit a pharmacology that differed from that of both Ca_v1 and Ca_v3 channels (362). The results can be explained most readily by the sequential activation of a Ca_v3 channel followed by a second influx pathway that directly mediates the sustained Ca^{2+}_i response, with the opening of the sustained entry pathway in response to ZP3 depending on the upstream activation of a Ca_v3 channel (32,362). As a result, Ca_v3 channels, acting indirectly at an upstream site, can account for the observed pharmacology of the sustained response and the role of Ca_v1 channels is being reevaluated. While four members of the Ca_v1 channel family are present in the sperm head (Table 3 and references therein), there is no information available yet as to whether they are regulated by ZP3, or even whether ZP3 can produce the strong depolarizations required to open the $Ca_v1.2$, 2.1, and 2.2 (L-, P/Q-, and N-types) channels.

What is the molecular nature of the sustained Ca^{2+} entry pathway? With the recognition that PLC is a downstream transducer of the ZP3 signal (see above), attention shifted to the role of trpc channels. The stimulation of PLC results in a sustained entry of Ca^{2+} in many animal cells. The trpc proteins are candidate subunits of the channels that mediate this influx process (329,368). Several members of the trpc family (329,369) are expressed in sperm (Table 3 and references therein). Jungnickel and co-workers have assigned a role to trpc2 in ZP3 signal transduction in mouse sperm. A function-blocking antibody directed against an extracellular domain of this protein inhibited the sustained phase of Ca^{2+}_i signaling during ZP3 stimulation and also inhibited the acrosome reaction. The localization of this channel in the anterior head also supported the proposed function (370).

The mechanism(s) by which PLC activation leads to an ion current through trpc channels is not yet completely understood (329,368). However, two types of regulation have been proposed for trpc channels, including trpc2, that are consistent with our current understanding of ZP3 signaling. First, trpc2 in the mouse vomeronasal organ may be gated by diacylglycerol (371) and, as has been noted, diacylglycerol production is enhanced by zona stimulation (197). While plausible, this mechanism has not yet been verified in sperm.

Second, it has been suggested that trpc2, as well as some other but not all trpc channels, may be activated either directly or indirectly by the IP_3R channel (329,368). Direct activation may be accomplished by IP_3R binding to trpc2, possibly through a conformational coupling mechanism (372). Alternatively, some trpc channels, including trpc2, may be gated through a store-operated mechanism; that is, the depletion of intracellular Ca^{2+} stores provides a signal which opens the plasma membrane channel (329,368,373). Several types of data support a role for IP_3R in ZP3 signaling. As noted previously, IP_3R Ca^{2+} release channels are present in the acrosomal membrane (316), Ca^{2+} stores within the acrosome have been documented (374), and sperm are known to synthesize IP_3 in response to zona stimulation (324). This strongly argues for the activation of IP_3R and Ca^{2+} release from internal stores. Other observations, including studies in permeabilized sperm and using Ca^{2+} store-depleting agents (316,362, 374–377), are consistent with this model. Lacking is direct demonstration of the emptying of acrosomal Ca^{2+} pools during ZP3 signaling.

While ZP3 may promote release of sequestered Ca^{2+}, it is unlikely that this is sufficient to produce a sustained Ca^{2+}_i elevation or drive exocytosis. Pharmacological agents that are expected to cause extensive emptying of intracellular Ca^{2+} stores appear to drive (ZP3-independent) acrosome reactions in the absence of extracellular Ca^{2+} (374). Yet, it is well established that the ZP3-activated process requires external Ca^{2+} and is inhibited by Ca^{2+} entry blockers, pointing to an essential role of influx across the plasma membrane (10,361,362).

The most parsimonious model at the moment, given available data, is that the sustained Ca^{2+}_i phase of ZP3 signal transduction is due both to release of Ca^{2+} from internal stores and influx of extracellular ion. A role is assigned to trpc2 in mouse sperm. However, the situation is complicated by the fact that this channel cannot account for ZP3 responses in humans and certain other species, where the trpc2 locus has accumulated stop codons and become a pseudogene (373,378,379), and by the fertility of mice with targeted deletions of the trpc2 gene (380,381). In this regard, the trpc2 antibody blocked most but not all of the ZP3-evoked sustained Ca^{2+}_i response (370), and so other channels in the sperm head (Table 3), including other members of the trpc family, may also contribute to ion influx. The tools to analyze this situation in greater detail, such as highly selective channel antagonists or mice in which genes for several of these channels are simultaneously disrupted, are not presently available.

Downstream Events of ZP3 Signal Transduction. A sustained Ca^{2+}_i elevation leads to acrosomal exocytosis. The downstream events ultimately entail the recruitment and activation of an exocytotic machinery. Our present understanding of the composition and function of this exocytotic machinery began in the genetic studies of constitutive secretion in yeast and in the biochemical analysis of mammalian intra-Golgi vesicle transport (382,383), and progressed to describe a highly conserved system that is believed to account for membrane fusion (384). Components of this universal system include SNARE proteins on vesicle and target membranes, SNARE regulatory proteins, Rab proteins and effectors, Ca^{2+} binding proteins, and selected other proteins. These components form complexes that bridge between two membranes, providing the energy to overcome repulsive forces and driving membrane fusion (385–387).

Elements of this system were first identified in sea urchin sperm by Schulz and coworkers (388,389) and in mammalian sperm by Ramalho-Santos et al. (390). Subsequent studies have established that the key members of the exocytotic machinery are present in mammalian sperm: the SNARE proteins such as synaptobrevin, a number of syntaxin isoforms, and VAMP (390–394); the SNARE regulators SNAP and NSF (395–397); Rab3a (398); and a variety of isoforms of the Ca^{2+} sensor protein, synaptotagmin (390,391,394). Finally, functional studies indicate that rab3a can drive the acrosome reaction in permeabilized sperm (398). As noted earlier, in many cell types a rise in Ca^{2+}_i is the major trigger of exocytosis. The exocytotic machinery identified in sperm includes at least one Ca^{2+} sensor, the C2 domain protein synaptotagmin (399), and so may provide one downstream target for the sustained Ca^{2+}_i response. Given the widespread use of this protein assembly in membrane fusion processes, it is likely that it also mediates the acrosome reaction. The coupling steps, if any, between the elevation of Ca^{2+}_i and the activation of the exocytotic apparatus in sperm have not been described.

Final Events of the Acrosome Reaction. Fusion between the apical plasma membrane and the underlying outer acrosomal membrane occurs in discrete zones in some species while in other cases it may occur along the whole region. The internal structure of the acrosome consists of a reticular matrix and soluble proteins, organized into morphologically or biochemically defined domains (400). During the final stages of the acrosome reaction this matrix disperses in an organized fashion and is associated with the differential, time-dependent release of acrosomal proteins (146,147). Acrosin, one of the major serine proteases of sperm, has been assigned a role in the dispersal of the acrosomal matrix, and this process is greatly slowed in the sperm of mice with targeted disruptions of the acrosin gene (145).

Sperm Penetration of the Zona Pellucida. Sperm initially contact the zona by means of plasma membrane in the anterior head region. As the acrosome reaction progresses this membrane domain vesiculates and is sloughed. Sperm must remain tethered to the zona during this process. Adhesion sites on the sperm plasma membrane that mediated zona binding remain present as the plasma membrane vesiculation process begins and can maintain binding during the early stages of the acrosome reaction. In some species, these hybrid vesicles can be found affixed to the zona surface, attesting to the continued functionality of adhesion proteins (10). Proteins within the acrosome can the bind the zona with high affinity and may anchor sperm in place during the intermediate and late stages of exocytosis: candidate proteins include zonadhesin (401,402), sp56 (146), and proacrosin, the enzymatically inactive zymogen form of acrosin (10,403). After dispersal of acrosomal contents, sperm can begin penetration of the zona. In vitro evidence suggests that at this point adhesion may be mediated by binding sites on the sperm inner acrosomal membrane that interact with ZP2 (270,404–406). In this regard, zona filaments are believed to consist of repeating ZP2/ZP3 dimers and so a transition from ZP3-dependent primary contact to ZP2-dependent secondary interactions is plausible within those current models.

Sperm that have completed the acrosome reaction then proceed into the zona. Three hypotheses have been forwarded to account for sperm penetration through this matrix. The mechanical hypothesis, that sperm enter the zona due solely to thrusting force provided by flagellar motility, has been rejected on experimental and theoretical grounds (407,408). Two other models propose that zona penetration requires the participation of sperm factors as well as flagellar motility. In one case, the sperm factor acts in a nonenzymatic manner to produce a local disruption of zona filament structure, possibly through induced alterations in the conformation of zona proteins. This type of process occurs during fertilization in abalone (409,410) and but has not been studied extensively in mammals. Finally, sperm proteases are suggested to assist penetration by proteolytic cleavage of zona proteins. A variety of evidence suggested that acrosin might act as a mammalian zona-lysin (10), however results of the acrosin gene disruption experiment as well as other data have shown that this protease is not required for zona penetration. Baba and coworkers have pointed to the presence of a wide range of acrosomal and membrane proteases in sperm that may participate in zona penetration (145). This process is currently being reevaluated in light of these other candidates and also the increased understanding of egg coat penetration in noneutherian and nonmammalian organisms (409–411).

Sperm–Egg Plasma Membrane Interactions

Plasma Membrane Domains

Specialized regions of the plasma membrane participate in gamete binding and fusion. Mere contact anywhere at their cell surfaces is not sufficient to ensure fertilization. Regionalization is highly developed in the sperm membrane and specific domains come into very close association with the egg surface. This apparent initial adhesion is followed by fusion in neighboring regions (10). Additional evidence has come from antibody studies demonstrating the localization of proteins and inhibition of fertilization (412,413).

Indirect evidence for an initial binding step prior to fusion has come from studies in which bound, nonfused sperm on eggs can be identified and/or the two processes (binding and fusion) can be dissociated. After in vitro fertilization of zona-free eggs, fusion failure of residual surface-bound sperm is inferred from the absence of the transfer of a DNA dye from the egg cytoplasm to the sperm nucleus (414,415). Certain experimental conditions have been found in which gamete binding occurs, but fusion does not. For example, in the absence of glucose or the presence of inhibitors of glucose or its metabolites sperm–egg binding is observed without fusion, as discussed by Yanagimachi (10). Reduced binding occurs in the absence of extracellular Ca^{2+}, but those sperm bound do not undergo fusion. The restoration of Ca^{2+} or glucose allows some of the bound sperm to fuse. The remaining sperm that fail to fuse following the reestablishment of standard culture conditions may not have bound to the egg plasma membrane in a physiologically relevant manner (413,416). In this regard, physiologically relevant interactions require the participation of appropriate membrane domains on acrosome-reacted sperm and this determination is difficult in light microscopic studies.

Electron microscopic (EM) investigations of the acrosome reaction and fertilization made possible the more precise identification of the membrane interactions and structures involved in initial interaction of sperm and egg membranes (417–420). After binding to the zona, the fertilizing sperm undergoes the acrosome reaction, which exposes the inner acrosomal membrane (IAM) on the sperm head. Following passage through the zona, it is this region that becomes closely associated with the egg membrane prior to fusion (421). Subsequently, the sperm's equatorial segment and posterior head regions become closely associated with the egg surface and undergo fusion with the egg plasma membrane (10,422). Interestingly, it is these more posterior regions that undergo fusion and not the IAM, which appears to be engulfed by the egg (423,424). Regarding the egg, the rodent egg membrane is regionalized into a nonmicrovillous domain over the metaphase II spindle and a highly microvillous region which is reported to be the predominant or exclusive site of sperm–egg fusion (10,416,425). However, these morphological domains are not observed on all mammalian egg species. Although careful time-course studies have been undertaken with only a limited number of mammalian species (e.g., guinea pig and hamster), these studies provided crucial information indicating where to look for the molecules involved in sperm–egg binding and fusion.

Fruitful biochemical studies to characterize gamete binding proteins had interesting historical origins, which demonstrate how fundamental observations on gamete interactions eventually led to the identification of the genes for these proteins. In the 1970s it was discovered that different sperm membrane domains, even within the sperm head itself, had different membrane particle arrays after freeze fracture EM, suggesting the domain-dependent segregation of proteins. This biochemical morphology coupled with the knowledge (above) of the localization of membrane functional domains for binding and fusion led investigators to look for a means to generate domain-specific probes. Since research on functional domains coincided with the widespread use of monoclonal antibody technology to study cell surface proteins in the late 1970s and 1980s, investigators set out to determine candidate binding and fusion proteins (426). Their strategy was not only to look for antibodies localizing to the binding and fusion domains on sperm, but also for function-blocking activity in binding and fusion in vitro fertilization assays. These studies, in turn, led to the cloning and sequencing of diverse families of candidate binding and fusion proteins.

Candidate Sperm Proteins for Binding and Fusion

In 1987, the sperm protein, fertilin, was discovered as the antigen of a monoclonal antibody screen. Fertilin is a membrane-anchored heterodimer (α, β) that is localized to the posterior or equatorial membrane, depending on the species (413,427). While fertilin is found on the entire head of testicular sperm, it is concentrated on the posterior head during epididymal maturation, appropriately localized for participation in fertilization. The β subunit has a "disintegrin" domain, in which an RGD-like QDE/TDE sequence [for the single letter amino acid code, see (224)] is present, that is a candidate integrin binding site, and a variety of integrins are present on eggs as discussed below. RGD-like peptides, antibodies to them as well as native fertilin-β, and recombinant fertilin inhibit sperm–egg binding and fusion in zona-free in vitro fertilization assays (428–430). Binding was also reduced in in vitro fertilization assays of sperm from fertilin-β knockout mice (431); while male mice are sterile, the phenotype is complex. For example, fertilin-α is not detected and sperm migration to the oviduct and sperm-zona binding in vitro are also reduced. It should be noted that that although in vitro gamete binding assays are useful, the extent of physiological relevance is under considerable discussion (163). Regarding fertilin-α, less extensive evidence suggests that it also has a role in gamete interaction (428). Although it has a hydrophobic region similar to viral "fusion peptide" domains, it is not found in found in all mammals (412,413).

After being identified as gamete recognition proteins, fertilin-α and -β were found to be members of a wide-spread group of membrane-anchored proteins, called the ADAM family, containing ***a*** ***d***isintegrin ***a***nd ***m***etalloprotease domain (432). In the 1990s, cloning and sequencing led to the discovery of additional mouse sperm ADAM members, most notably cyritestin, which has testis-specific expression and protein localization on the sperm head. Although its overall structure resembles fertilin-β, cyritestin has a different disintegrin loop sequence and sequence specific peptides from cyritestin have greater inhibitory activity in in vitro fertilization assays than those from fertilin-β (430,433). Like the fertilin-β knockouts, cyritestin knockout mice are infertile and have reduced sperm-egg and -zona binding. However, the rate of gamete fusion in these animals is normal (434,435). It is important to note that either of these two knockouts also reduces the level of expression of the other protein and, in addition, fertilin-α is not detected (163,435). Moreover, sperm from null males undergo a reduced level of sperm–egg fusion in vitro, suggesting that other sperm fusion proteins are present.

Other gamete interaction proteins made in the epididymis are members of the CRISP family (***c***ysteine-***ri***ch ***s***ecretory ***p***roteins) [see (436) and citations therein]. One of the CRISP-1 proteins, called DE, initially associates with the dorsal region of the rat sperm head, and migrates to the equatorial segment upon the acrosome reaction. In other mammals, it is found on the posterior region of the sperm head. Unlike ADAM family members, it is not a transmembrane protein but is surface-associated. Although the majority of DE is lost during capacitation, the remaining DE is considered to be involved in sperm–egg fusion because of the inhibitory effects of purified native DE, recombinant DE, and antibodies to DE in in vitro fertilization assays. The primary effect appears to be on gamete fusion rather than adhesion (437). Male rats immunized with DE show reduced fertility and sperm from these males adhere to eggs, but fusion is inhibited (438). A human orthologue has been reported (439).

In summary, until recently the strongest sperm candidate binding proteins for the egg were those of the ADAM and CRISP families, while other many other lesser characterized candidates have been proposed based primarily on antibody studies (412, 413,426). However, the uncertainty of the roles of the ADAMs and their likely integrin partners (see below) leaves room for the discovery of new sperm adhesion and fusion proteins.

Candidate Egg Proteins for Binding and Fusion

Although the search for egg-binding and fusion proteins has been hindered by the small amount of material available, insights from the sperm ADAM family proteins provided a rationale for investigating integrin expression on the egg surface. Integrins are a large family of heterodimeric adhesion proteins, mediating cell-extracellular matrix and cell–cell interactions, some of which have RGD-like binding sites (91). Early studies demonstrated that RGD-coated beads bound to hamster and human eggs and these peptides had some inhibitory activity in animal in vitro fertilization assays (440,441). At least three integrins ($\alpha_5\beta_1$, $\alpha_6\beta_1$, and $\alpha_V\beta_3$) are detected in the plasma membranes of mouse, hamster, and human eggs (412,442–445). Of these, $\alpha_6\beta_1$ has been studied most extensively, beginning with the report that a function-blocking antibody to α_6 reduced sperm–egg binding (444).

Various approaches to inhibit or remove integrin adhesive activity on eggs supports their involvement in gamete binding. In contrast, the targeted disruption of integrin genes either singly or in combination failed to produce the anticipated results, such that the relative roles of different integrin isoforms and which interact with fertilin in vivo will require further investigation (for reviews see references 446–448). Anti-β_1 antibodies moderately inhibit sperm–egg interactions in the mouse, pig, and human (449–451). Although antibody and cross-linking studies are consistent with the binding of $\alpha_6\beta_1$ to fertilin, other studies make it possible that fertilin may bind to a different integrin. Although egg integrins are not required for fertilization (20,428, 452,453), the redundancy of gamete interaction proteins to promote fertilization success remains a distinct possibility. The search for other classes of egg proteins required for binding and fusion continues. For example selective removal from eggs of most GPI-anchored proteins results in infertility in mice (454).

A different type of egg membrane protein, CD9, is required for gamete fusion in mammals. In the tetraspanin family, CD9 was identified based on the significantly reduced fertility of females with this protein knocked out and its role in membrane fusion in myoblasts. In these animals, zona penetration and sperm binding to the egg plasma membrane are similar to those from control females (455–457). However, eggs from mice bearing targeted disruption of the CD9 gene rarely fused with wild-type sperm or were activated, even in zona-free in vitro fertilization assays. After intracytoplasmic sperm injection (ICSI), egg$^{(CD9-)}$ activation is apparently normal and embryos develop to term (457). Moreover, expression of normal CD9 in eggs$^{(CD9-)}$ restores the capacity for normal gamete fusion and calcium oscillations after zona-free in vitro fertilization (458). The extracellular loop of the protein is required for gamete fusion (459) and has a binding site for PSG17, a member of the immunoglobulin superfamily. Mutation of this site or addition of PSG17 to gamete fusion assays suggests an important role for a protein-like PSG17, pending its identification on one of the gametes (460). In addition, a sperm-specific immunoglobulin superfamily family member, Izumo, was recently identified. Sperm from mice with a targeted disruption of the Izumo gene can penetrate the zona and bind to the egg plasma membrane, but fail to fuse with the egg. Fertility is restored by expression of Izumo in these mice or by ICSI, which bypasses the gamete membrane fusion step (461). Finally, there is evidence that CD9 and $\alpha_6\beta_1$ associate in the mouse egg membrane (412,457) and that other tetraspanin genes are expressed by mammalian eggs (462–464).

Models for Binding and Fusion

In some lower organisms, elegant genetic studies have demonstrated sex-specific proteins required for gamete union, a process requiring adhesion, activation (signaling), and fusion in *Chlamydomonas* (465). In mammals, the roles and mechanisms of specific proteins are not as well established. Current evidence (discussed above) is consistent with a model in which both gametes have a variety of binding proteins that could interact during the adhesive phase of fertilization (Fig. 10). Some of these may function to stabilize the relatively small areas of sperm and egg membranes that initially interact while others may bring the membranes within molecular approximation required for fusion or to trigger the fusion process itself (Fig. 11). The crystal structure of a putative abalone sperm fusion protein coating the acrosomal process reveals both hydrophobic and electrostatic domains proposed to account for its function (466). In mammals, fertilization may not

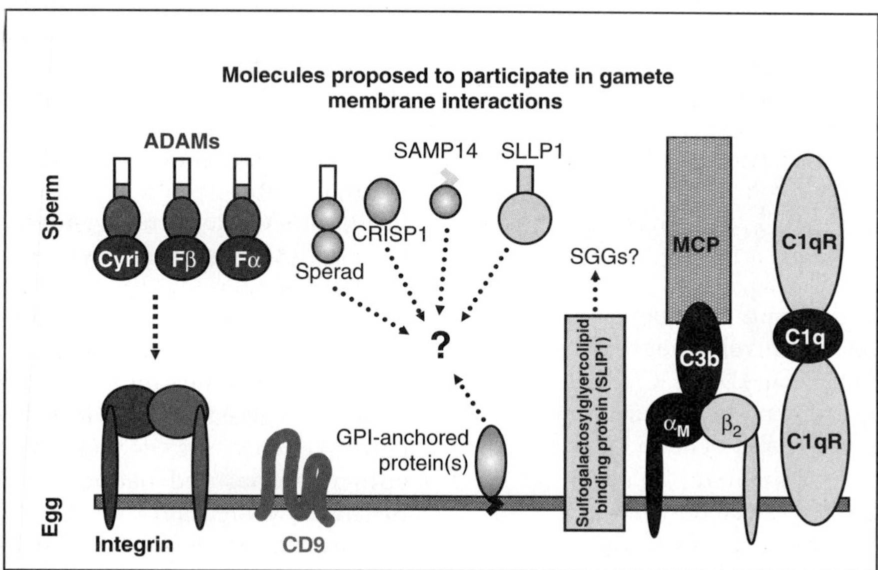

FIG. 10. A schematic compilation of molecules reported to be involved in the plasma membrane adhesion or fusion of the mammalian sperm and egg. Elucidation of the molecular interactions between these proteins requires further investigation. In general, those on the left side are more extensively characterized biochemically than those on the right. [For those not discussed in this chapter and further information see (412,413,417). Courtesy of Dr. Janice Evans.]

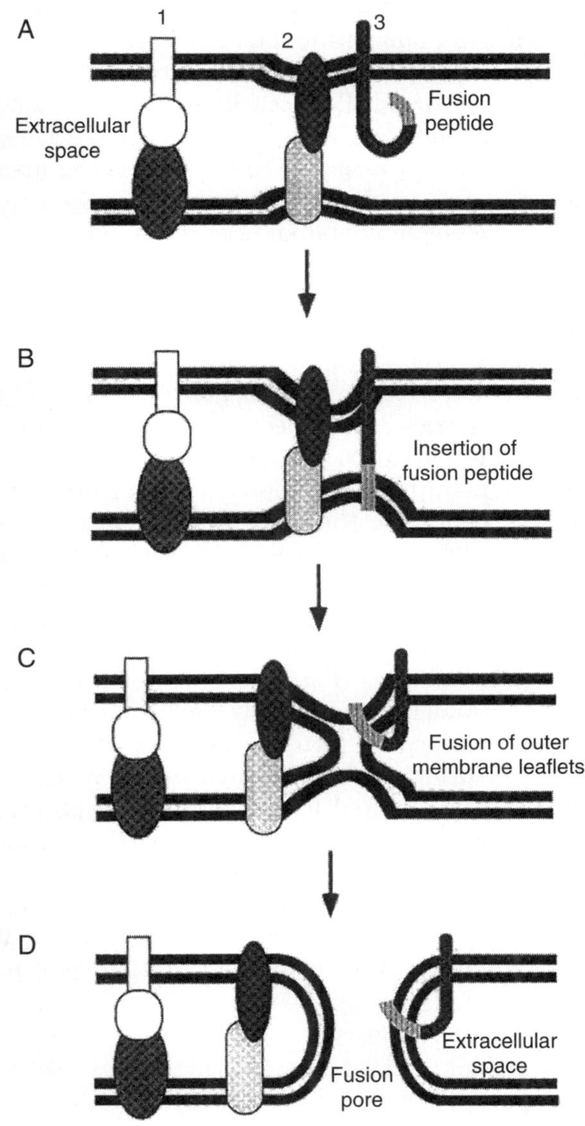

FIG. 11. Hypothetical model of the fundamental steps in cell fusion based on studies of viral entry into eukaryotic cells. The process involves adhesion, membrane approximation, and two fusion steps. Although this model is expected to apply to gamete fusion, relatively little direct evidence has been reported. [From (412), by permission of Oxford University Press. Courtesy of Dr. Janice Evans.]

require protein families and mechanisms unique to gametes, based on the involvement of "shared" families of proteins (e.g., CD, ADAM, CRISP), as well as the universal requirement of extracellular Ca^{2+}, for fusion of gametes (467) and of somatic cells.

Incorporation of the Fertilizing Sperm into the Egg and the Fate of Sperm Components

Very soon after sperm–egg fusion, perhaps within a matter of seconds, sperm tail motility ceases (467–469). Since initial fusion occurs on the sperm head, an important signal appears to be sent to the axonemal apparatus to down-regulate dynein-mediated microtubule sliding. In many species, the entire sperm tail is incorporated into the egg cytoplasm, e.g., in rabbit eggs within 20 minutes (470). Although lacking the volcano-shaped actin-filled "fertilization cone" of marine eggs, fertilized mammalian eggs have a network of cortical actin filaments (471). There is evidence both for (472,473) and against (474,475) a role for actin filaments in the incorporation of the sperm nucleus and organelles into the egg cytoplasm. In place of the fertilization cone, the sperm head appears to be engulfed by an extension of the egg plasma membrane, resulting in a vesicle derived from the inner acrosomal and egg plasmalemmal membranes (Fig. 12) (476).

An intriguing question has been whether most of the sperm components that are incorporated into the fertilized egg have a role in early development (477). With increasing importance attached to maternal and paternal genomic imprints, the chromatin from both gametes is required for normal development to term. For example, experimentally induced parthenogenetic diploid, gynogenetic development without sperm results in a low percentage of terminally arrested rabbit fetuses with beating hearts and somites (478). In addition, the sperm centrosome, which functions as a microtubule organizing center, and associated microtubules have a role in pronuclear migration (479) which brings both genomes in close proximity to facilitate chromosome congression on the first metaphase plate prior to the first mitotic cleavage. Centrosomes and specific gamete centrosomal proteins may be required for normal spindles and chromosome segregation during early cleavage in primates, but differences among mammalian species have been observed (477,479,480). Unlike the egg's microtubule organizing centers which lack centrioles, those of most mammalian sperm have one or two centrioles [although absent in some rodent eggs, as discussed by Shatten (479)]. Since gynogenetic diploid embyros undergo cleavage, blastocyst development and early fatal development without a fertilizing sperm, centrosomes appear to be more important than centrioles during very early development. Spindle poles, each with a pair of centrioles, arise during early development. The precise role at fertilization of mammalian paternal centrioles, as well as other components of both maternal and paternal centrosomes, requires further investigation.

However, evidence for the importance of many other sperm components has not been as well established. In most of the species investigated to date, the entire sperm tail with its microtubules, associated proteins, and dense fibers eventually disperses inside the egg. Most sperm mitochondria appear to

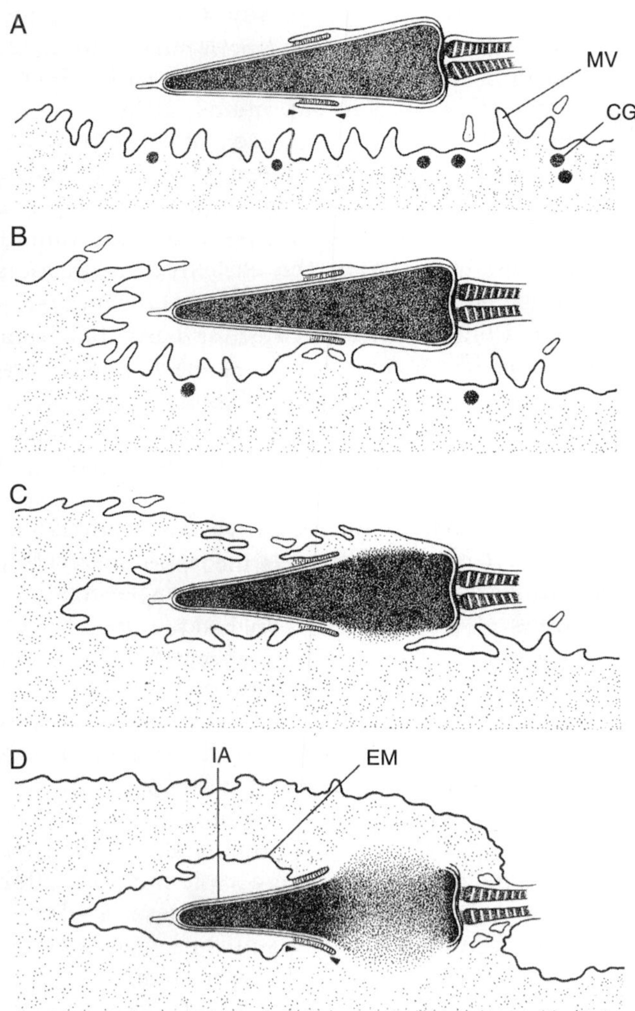

FIG. 12. Diagram of the early stages of mammalian sperm–egg interaction. Although the inner acrosomal membrane (IA) of the sperm may make initial contact with egg microvilli and possibly undergo adhesion, this region has not been observed to undergo fusion. The equatorial segment of the head of an acrosome-reacted sperm undergoes the initial fusion with the egg plasma membrane. The amount of continuity between the sperm and egg cytoplasm increases in a posterior direction on the sperm head. Anteriorly, the inner acrosomal membrane and egg plasma membrane (EM) form a vesicle inside the egg cytoplasm. CG, cortical granule; MV, microvillus. [Reprinted from (417) with permission from Elsevier. Courtesy of Dr. Michael Bedford.]

degenerate quickly or later from ubiquitination in the egg cytoplasm (481), although some may persist (10). After fertilization, components of the sperm plasma membrane are still detected on the surface of marine and mammalian eggs (482,483) and some appear to disperse into the egg membrane (484). Their function in development, if any, is unknown and appears doubtful in light of the success of human ICSI. It should be noted that paternal effect mutants have been reported in lower organisms and some may yet have relevance to mammalian fertilization and/or early development.

Events Following Sperm–Egg Membrane Fusion

The onset of normal development is dependent upon a series of so-called "events of egg activation" that are induced by fertilization and occur prior to first cleavage. The early events are the secretion of egg cortical granules, the blocks to polyspermy, and onset of paternal chromatin decondensation. In most mammals, these events are well underway within the first hour of fertilization. Later events include the completion of maternal meiosis, maternal chromatin decondensation, male and female pronuclear formation, and the recruitment of maternal mRNAs resulting in the expression of new proteins. Before first cleavage, the zygote also undergoes the onset of genomic reprogramming and limited zygotic genome activation. Many, if not most of the aspects of activation, are directly or indirectly dependent upon a Ca^{2+}-driven signaling pathway and subsequent changes in the activities of specific protein kinases.

Block(s) to Polyspermic Fertilization

The "block(s)" to polyspermic fertilization (485) are necessary to prevent the incorporation of two sperm nuclei into the zygote's genome, which in most animals results in abnormal development. In most mammals, dispermic fertilization is usually prevented by a combination of the fertilized egg's blocks to polyspermy and low sperm number in the oviduct. However, it is estimated that 1%–3% of clinically recognized natural human conceptions (486,487) result in triploid embryonic abortions, many from dispermic fertilization (488–490). Many reproductive technologies (e.g., intrauterine insemination and in vitro fertilization) that are used to treat infertility, produce domestic animals, or save endangered species utilize much higher sperm concentrations than those in the oviduct during natural conception. For example, in the case of human in vitro fertilization, sperm concentrations are 100–1,000 times higher, with 5%–10% polyspermy rates (491). To prevent polyspermy, fertilized mammalian eggs utilize blocks at the zona and/or plasma membrane. The relative contributions of these two blocks vary in different species (10) (Table 4).

In mammalian eggs, little is known about the biochemical basis of the membrane block (10) which does not appear to have the membrane depolarization component observed in marine eggs (485,492). The membrane block is detected as early as 1–6 minutes after sperm–egg fusion in eggs of mice (493) and hamsters (494), although other reports suggest a slower block (495). It appears to prevent sperm–egg fusion because in vitro fertilization of zona-free eggs

TABLE 4. *Classification of animals based on the mechanisms of polyspermy block*

	Group		
	I	II	III
Species	Golden hamster Dog Sheep Field vole Ferret Human	Rat Mouse Guinea pig Cat	Rabbit Pika Mole Bat
Presence of Sperm in PVS	Very Rare	Often 1 Sperm	Often ≥ a few
Primary Source of Polyspermy Block	ZR	ZR and PMB	PMB

PMB, plasma membrane block; PVS, perivitelline space; ZR, zona reaction.
Adapted from Table 13 (10).

results in large numbers of bound sperm, but >50% of eggs remain monospermic, even after many hours of culture. If the plasma membrane and zona blocks were to fail, triploidy in the embryo is sometimes prevented by the elimination of extra sperm chromatin in human eggs (490).

In theory, the zona block could involve specific modifications in key sperm-binding residues (150,496), changes in zona supramolecular structure (497–499), and/or the addition of new proteins after the cortical reaction (500). In fact, ultrastructural studies report fertilization-associated changes in zona filaments and structure in human eggs (501,502). The molecular changes resulting from differential effects on various layers (501,503) within the zona remain to be established. At the biochemical level, one model of the zona block involves modifications of specific zona proteins that could account for the prevention of additional sperm binding to the zona, undergoing the acrosome reaction, and penetrating the zona. The model is largely derived from elegant biochemical studies of isolated zona proteins (504), although the relative contributions of each of these proteins in the native zona of fertilized eggs are not known with certainty (505). After fertilization, ZP3 appears to undergo an important functional modification(s) which is likely to be in the carbohydrate portion (237) involved in sperm binding and acrosome reaction induction, because purified ZP3 isolated from embryos has lost its ability to bind sperm and induce the acrosome reaction (261,269,302). The precise molecular change in ZP3 has not been demonstrated, although evidence implying a role for glycosidase activity (see below) is consistent with the carbohydrate modification model. The model includes an additional mechanism to prevent the passage of acrosome-reacted, zona-bound sperm from completing zona penetration. Two lines of evidence are that acrosome-reacted sperm bind ZP2 from unfertilized eggs but not from activated eggs, and that ZP2 undergoes proteolytic cleavage after fertilization in the mouse (497) and probably the human (Fig. 13) (506). Transgenic mouse eggs in which mouse ZP2 is replaced by the human isoform bind mouse (not human) sperm (274). Interestingly, in these eggs, human ZP2 proteolysis is not detected by gel electrophoretic methods after fertilization and mouse sperm remain associated with the zona. In human in vitro fertilization, ZP2 cleavage is observed in some inseminated eggs that fail to fertilize (507), whereas eggs that fail to activate after ICSI often do not show this ZP2 alteration (508). It has been speculated that ZP2 proteolytic cleavage results in a functionally important change in the supramolecularstructure of the zona (497,498).

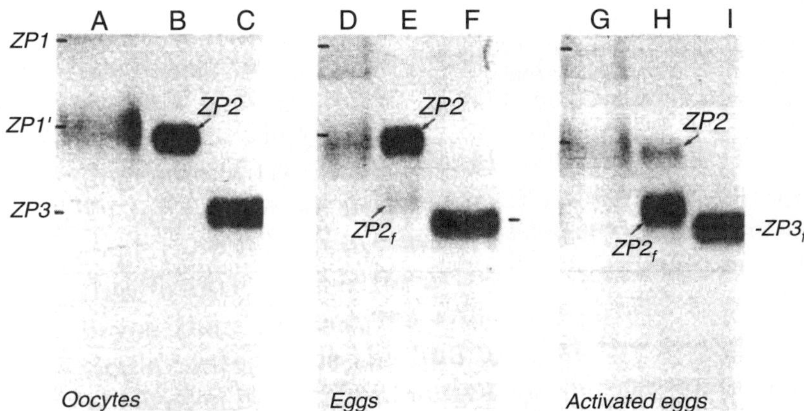

FIG. 13. Biochemical change in the mouse egg zona pellucida protein ZP2 after fertilization. Separation of purified ^{125}I-labeled proteins from germinal vesicle stage oocytes, metaphase II eggs, and fertilized eggs was performed by discontinuous SDS-polyacrylamide gel electrophoresis (under reducing conditions). Note the shift in the position of ZP2 whereas a change in ZP3 is not detected. [Reprinted from (497) with permission from Elsevier. Courtesy of Dr. Paul Wassarman.]

Several lines of evidence indicate that the zona block results from the action of proteins secreted by the egg upon fertilization. First, the role of secreted protein in the extracellular matrix-level block to polyspermy is well established in marine and amphibian eggs (509). A major source of secretion is the exocytosis of thousands of egg cortical granules. Within 15 minutes of mammalian fertilization, the egg initiates the exocytosis of cortical granules (the "cortical reaction"), and, perhaps, undergoes other cortical or surface changes that have yet to be identified. In mammalian eggs, thousands of 0.1–1 μm-diameter, membrane-delimited cortical granules are found in the peripheral cortex underlying the plasma membrane. Cortical granule contents are thought to be released by exocytosis following sperm–egg fusion and diffuse across the perivitelline space, where they remove or mask sperm receptors in the egg's extracellular investments, or otherwise modify those investments (500,510). It has also been proposed that some cortical granule-derived macromolecules remain in the perivitelline space forming a "cortical granule envelope" (500,511), whose function and molecular composition warrant further investigation.

Compared to molecules from sea urchin egg cortical granules, less is known in mammals about the identities of cortical granule proteins, their biochemical activities, and the resulting specific molecular modifications responsible for the zona block. Although early studies of factors released by activated eggs suggested that cortical granules contain a trypsin-like activity (512,513) and an ovoperoxidase activity (514), the identity of these factors remains to be established in mammalian eggs.

More recent studies of proteins secreted by artificially activated mouse eggs have identified approximately six major species of approximately 20, 28, 32, 34, 45, and 70 kDa (515–518). Regarding the zona block, of notable interest is an exudate fraction with a 21–34 kDa protein and proteinase activity that cleaves ZP2 (515), consistent with models of the zona block to polyspermy. Also consistent with the model is the evidence the secretion of N-acetylglycosaminidase activity (519) that has been proposed to modify ZP3. If the zona blocks are indeed due to the action of released factors, they should be found within cortical granules prior to egg activation. To date, only the 32 kDa protein, the 75 kDa protein, and the N-acetylglycosaminidase activity have been localized to intact cortical granules. Such studies have been handicapped by the small amount of cortical granule material available from mammalian eggs, unlike their marine counterparts. Genomic and proteomic studies are likely to play major roles in leading to the elucidation of the structure, biochemical function, and species diversity of mammalian cortical granule components, as well as to what extent they contribute to various blocks to polyspermy.

Cell Cycle Progression and Completion of Meiosis

Cell cycle progression represents the important transition from meiosis to mitosis and is triggered by fertilization. In temporal order, these include the completion of meiosis, formation of male and female pronuclei, DNA duplication of each of the pronuclear genomes, pronuclear envelope breakdown, and the first mitotic metaphase prior first cleavage. This review will focus more attention on the mechanisms responsible for early events in this series because they are more directly regulated by fertilization (Fig. 14).

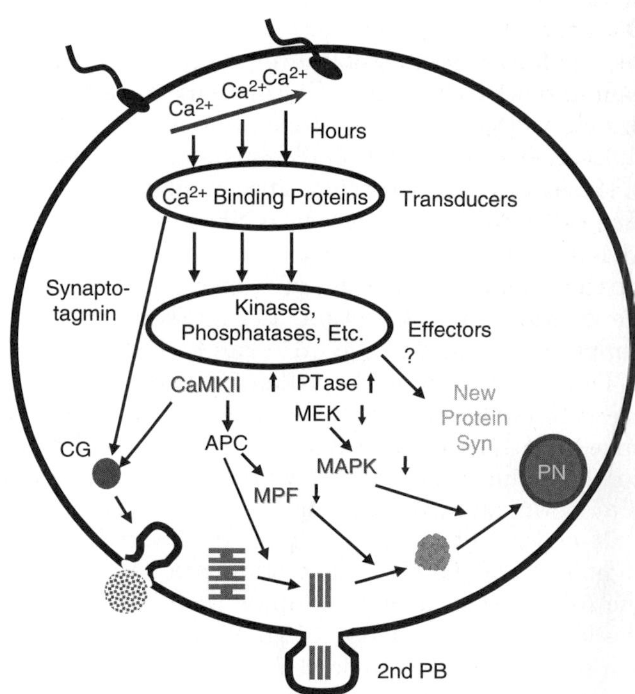

FIG. 14. Signaling diagram of egg activation. The model includes the pathway downstream of multiple elevations of Ca^{2+}_i and the relative temporal order of the biological events at the bottom of the egg from left (earlier) to right (later). Some Ca^{2+}-senstive tranducers (e.g., calmodulin, not shown) and effectors (CaMKII) have important roles (see text). Others, for example those promoting new protein expression which is Ca^{2+}-dependent, are yet to be identified. Small vertical arrows indicate whether the activity of an effector increases or decreases to promote the pathway with which it is associated. New protein synthesis (Syn) results from the recruitment of maternal mRNAs. APC, anaphase promoting complex; Ptase, phosphatase; MAPK, MAP kinase; MEK, MAPK kinase; 2nd PB, second polar body; PN, pronucleus; CG, cortical granule; CaMKII, calmodulin-dependent protein kinase II; MPF, maturation promoting factor. Synaptotagmin is a Ca^{2+}-sensitive protein that regulates secretion in other cells. [Modified from (552) with permission from Elsevier.]

After ovulation, fertilization of most (but not all) mammalian eggs occurs at metaphase II (MII) of meiosis. The failure to complete meiosis prior to sperm fusion is not unusual and the meiotic stage at which the egg is fertilized depends on the animal species (18). The cell cycle of ovulated mammalian eggs is arrested transiently at MII by elevated activity of maturation promoting factor (MPF, a dimer of $p34^{cdc2}$/cyclin B) (520). MPF activation is necessary for switching cells from interphase to metaphase (521). For example, in early mitotic cleavages after fertilization, MPF levels rise very transiently during metaphase and then decrease from cyclin degradation with each cell cycle. Before fertilization, MPF activation promotes the oocyte's transition from the GV stage to metaphase I. Between metaphase I and II, its activity does not undergo a mitotic-like decrease, but drops slightly, explaining the absence of Ca^{2+}_i interphase during this period. The persistent elevation of MPF activity at MII is maintained by the so-called "cytostatic factor" or CSF within the egg (522–525), as well as continued cyclin synthesis. Still under investigation, CSF activity in vertebrate eggs involves several cell cycle kinases that prevent the inactivation of the $p34^{cdc2}$ kinase and activation of the APC (anaphase-promoting complex/cyclosome) (Fig. 15). Knocking out one CSF component, the MOS protein, results in the absence of MII arrest and parthenogenetic activation of mouse eggs (526,527). Thus, prior to fertilization, the egg's cell cycle is temporarily locked in the MII state by CSF.

Fertilization-induced oscillatory increases of egg Ca^{2+}_i stimulate exit from MII, visualized as anaphase onset and subsequent extrusion of the second polar body. Calcium signals are transduced by calmodulin (CaM), which, in turn, associates with and activates CaM-dependent protein kinase II (CaMKII). Ca^{2+} signaling and CaMKII activation are required for the APC-mediated destruction of cyclin, resulting in the loss of MPF activity (528,529). A more detailed working model is provided in Fig. 11. The APC, which ubiquitinates proteins for destruction, also removes the inhibitor of the protease that cleaves the linkers between sister chromatids, allowing anaphase onset (530) (Fig. 16). This working model is derived from studies of other cells, including *Xenopus* eggs, which are also arrested in MII. More recent investigations of mammalian fertilization also demonstrate that CaMKII activity as well as the APC and its upstream activators are important for the exit from MII (531–535).

Given the cortical location of the second meiotic spindle and small constriction required for extrusion, the second polar body forms relatively rapidly after fertilization. The mechanisms for the cortical location and fertilization-induced rotation of the meiotic spindle in mammalian eggs remain unsolved. Although the polar body is discarded following fertilization, it can be used for the experimental induction of diploidy (by suppression of extrusion) in investigations of parthenogenetic development (478,536), for egg orientation during sperm injection in human ICSI [however, see Silva and co-workers (537)], and for chromosomal biopsy to infer the genetic [however, see Angell (538)] or the likely telomeric content of the egg.

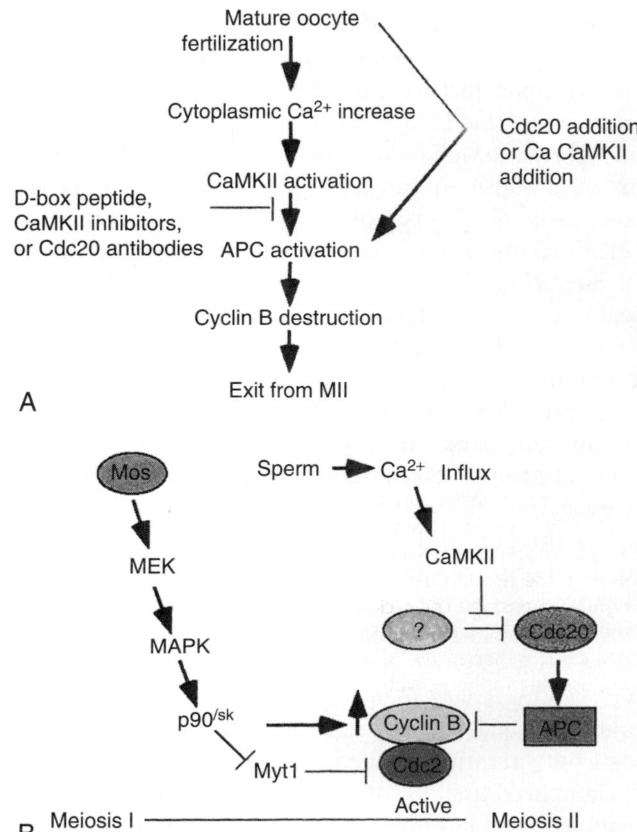

FIG. 15. Pathways involved in meiotic arrest and exit for vertebrate eggs. The evidence for these pathways came first from *Xenopus* eggs (525,528) and studies to date have confirmed many aspects of the model in mouse eggs (see text). **A:** Simplified pathway of metaphase II (MII) exit with experimental evidence. Cdc 20 and cyclin are cell cycle regulatory proteins. **B:** Model incorporating both the mechanisms of meiotic arrest at metaphase II of meiosis and the signals to exit this arrested state upon fertilization. The development of arrest requires the Mos protein and MAPK (MAP kinase) pathway resulting in an elevated and temporally maintained level of MPF (maturation or M-phase promoting factor) activity. This pathway becomes functional between metaphase I and II of meiosis. Exit from metaphase II requires the fertilization-induced elevation of Ca^{2+}_i, resulting in the CaMKII-mediated activation of the APC, the anaphase promoting complex. The APC ubiquitinates cyclin B, resulting in its destruction and the loss of MPF activity required for metaphase arrest. The protein(s) designated as "?" are under investigation. A cell division inhibitor, Emi2 (651) may be involved, while Emi1 is not necessary (652). For recent mammalian studies, see (534,535). [Modified from (653). Original courtesy of Dr. Peter Jackson.]

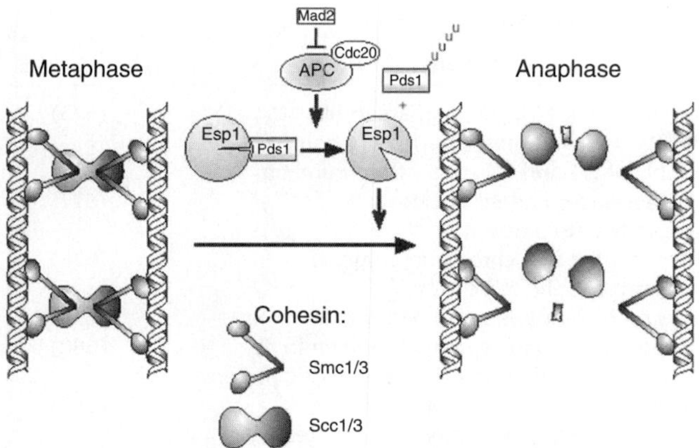

FIG. 16. Model of sister chromatid separation at anaphase in yeast, which is consistent with studies of activated *Xenopus* eggs undergoing the metaphase to anaphase transition at meiosis II. For example, note the early regulatory step involving the CDC20 protein and the APC. The APC is activated by CDC20, which in turn is dependent upon the rise in Ca^{2+} and CaMKII activity (see Fig. 15), and by progression past the cell cycle check point represented by the Mad2 protein, an APC inhibitor. During metaphase, the protein Esp1 (separin) is inhibited by Pds1. APC activation releases this inhibition, resulting in the activation of Esp1, which cleaves cohesin and allows anaphase onset. [Reprinted from (530) with permission from Nature Publishing Group. Courtesy of Dr. Kim Nasmyth.]

The Pronuclear Stage

In mammalian fertilization, the haploid genome of each gamete becomes surrounded by its own nuclear ("pronuclear") envelope, in which the first mitotic "S" phase takes place to generate two copies of each chromosome. Thus, there are two distinct pronuclei. Examination of this pronuclear stage has been crucial for human in vitro fertilization programs, because the large size of the pronucleus allows for the rapid identification of fertilized eggs and removal of abnormal polyspermic or triploid embryos (with >3 pronuclei; discussed previously). Following this transient pronuclear stage, the pronuclear envelopes disassemble and chromosome condensation reoccurs to set up the first mitotic cleavage division.

Pronucleus formation is dependent upon factors in the egg cytoplasm. For the fertilizing sperm, this process begins with the disassembly of the sperm's nuclear envelope. In sea urchin eggs, in vitro studies have demonstrated that the egg kinase, PKC, phosphorylates and solubilizes the sperm nuclear lamina (539). Although lamin protein phosphorylation is correlated with destabilizing the nuclear envelope in many cells, the nuclear envelope of mammalian sperm appears to lack an inner lamin protein layer (470,540) and PKC inhibitors do not block paternal chromatin decondensation (541). Although the precise cause for sperm nuclear envelope breakdown remains to be established, it is possible that transiently elevated levels of cell cycle kinases play a role, e.g., MAPK activity in mouse eggs prevents pronuclear envelope assembly (506).

Because the volume of the male pronucleus is 100–1,000 times greater than that of the original sperm nucleus, there is a dramatic remodeling of the condensed sperm chromatin (542). Mammalian sperm DNA is largely associated with sperm-specific basic proteins (protamines) in a crystalline-like complex (543,544), which is the most highly condensed chromatin in eukaryotes and arises during late spermatogenesis via the replacement of most histones by protamines. After fertilization, paternal chromatin remodeling is thought to occur by a multistep process of protamine removal, nucleosome assembly, and resetting the chromatin status so that it is appropriate for early development (542).

Protamine replacement by histones is thought to involve the reduction of protamine disulfide bonds by reducing agents in the egg cytoplasm, phosphorylation, and egg proteins that chaperone or exchange the sperm chromatin proteins for new ones (Table 5). After fertilization, glutathione, a disulfide bond reducing agent, is thought to reduce the disulfide linker bonds between the sperm protamines which are involved in sperm chromatin condensation (21,545). Inhibition of the meiotic maturation-associated increase in oocyte glutathione prevents sperm chromatin decondensation. In addition, another oocyte protein, nucleoplasmin, facilitates protamine removal, chromatin decondensation, and histone replacement in amphibian eggs (546,547). The mammalian homologue of one nucleoplasmin isoform (NPM2) accumulates in the oocyte nucleus, and, although fertilization appears to take place, knockout females have fertility defects in chromatin organization during failed preimplantation development (548). Protamine removal occurs by the time that the maternal chromatin has reached the end of anaphase II of meiosis (549–551). Studies in other organisms provide evidence for egg proteins that assemble specific histones into sperm chromatin nucleosomes (Table 5). However, many of these have not been identified in mammalian eggs. The transformation of sperm chromatin is shown in Fig. 17.

Although the formation of the female pronucleus also requires chromatin decondensation, maternal DNA is not under protamine constraint. At the time of fertilization, MII-stage chromatin is condensed from the high levels of CSF activity previously discussed and a pronuclear envelope is absent. Even after fertilization-induced MII exit, maternal chromatin remains condensed during the completion of meiosis and extrusion of the second polar body. Decondensation and entry into interphase require a sustained decrease in MPF activity driven by multiple oscillations of

TABLE 5. *Egg factors implicated in sperm chromatin remodeling*

Factor	Source	Activity	References
Nucleoplasmin	*Xenopus*	Removal of X and Y from, and assembly of H2A and H2B onto, sperm chromatin	(675)
N1	*Xenopus*	Assembly of H3 and H4 onto new chromatin	(676)
HIRA	*Xenopus*	Replication-independent nucleosome assembly onto plasmids	(677)
Template-activating factor Iβ	*Xenopus*	Dispersion of sperm chromatin, removal of sperm-specific basic proteins	(678)
Nucleosome-assembly protein-1	*Drosophila*	Dispersion of *Xenopus* sperm chromatin in vitro	(679)
P22	*Drosophila*	Dispersion of *Xenopus* sperm chromatin in fly oocyte extract; release of sperm-specific proteins	(680)
DF 31	*Drosophila*	Dispersion of *Xenopus* sperm chromatin in vitro; release of sperm-specific proteins	(681)
Sésame gene product	*Drosophila*	Formation of male pronucleus	(682)
–	*Mytilus* (surf clam)	Decondenses spermatozoa in egg extracts	(683)
–	*Spisula solidissima* (mussel)	Protein characteristics similar to nucleoplasmin	(684)

Table adapted from (542).

$Ca^{2+}{}_i$ (552). For example, a transient decrease in MPF activity from an insufficient number of Ca^{2+} oscillations allows exit from MII and polar body extrusion, but chromatin decondensation fails to take place.

Despite the different pathways involved in chromatin remodeling, the formation of the male and female pronuclear envelopes occurs at approximately the same time. Pronuclear envelope assembly begins at the periphery of decondensing chromatin (470), and in sea urchins, forms by ATP-dependent recruitment of vesicles to chromatin and GTP-dependent fusion (553). The latter studies have evidence that targeting vesicles to chromatin is mediated by a lamin B receptor-like integral membrane protein. Pronuclei appear to draw on a reserve of egg endoplasmic reticulum to generate the large surface area that each pronucleus requires, which is not surprising given the size of the prophase I nucleus (germinal vesicle) of the oocyte and the extensive endoplasmic reticulum that is present at that stage. Pronuclear envelope assembly also requires a decrease in MAPK activity (506) which follows the drop in MPF activity by several hours

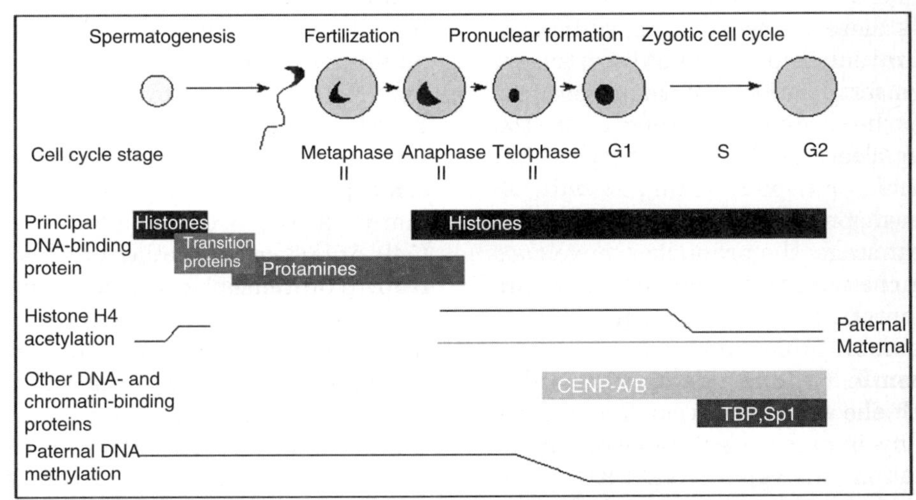

FIG. 17. Summary of sperm chromatin remodeling after fertilization in mammals. Protamines replace sperm histones during spermatogenesis. After fertilization, histones in the egg cytoplasm replace these protamines during the completion of meiosis. After this exchange, 5-methyl cytosine is no longer detected on the paternal DNA. During pronuclear formation, both paternal and maternal chromatin decondense in separate pronuclei (not shown), and deacetylation of histones takes place on paternal chromatin. Other egg proteins (centromeric CENP-A/B, and transcription factors TBP and Sp1) are imported into the pronuclei and modify the chromatin. DNA replication in individual pronulei is followed by pronuclear envelope breakdown and the first mitotic cleavage. [Reprinted from (542) with permission from Reproduction Online. Courtesy of Dr. Hugh Clarke.]

or more, depending on the species. The decrease in MAPK activity and formation of pronuclei are calcium dependent and require a greater number of Ca^{2+} oscillations than do the decrease in MPF activity and anaphase II onset (552).

The pronuclei are far from merely a passive staging area for the two parental genomes prior to assembling the zygotic genome. Both sets of chromatin undergo asynchronous DNA synthesis at pronuclear stage, generating the 4C amount of DNA required for the ensuing first mitosis. In addition, changes in DNA methylation take place, with the paternal genome undergoing a decrease unlike the maternal genome (554). This paternal decrease appears to be an active, rather than passive, demethylation since it occurs before "S" phase and when replication is blocked (555). The paternal pronucleus also has a higher level of histone acetylation (556) and a lower level of methylation of histone H3 at lysine-9 (557). These differences in methylation, acetylation, or, perhaps other factors, may be the basis for the larger burst of transcriptional activity from the male than female pronucleus in the mouse (556,558,559). The acquisition of transcriptional permissiveness is facilitated by protein synthesis (560) and recruitment of maternal mRNAs (561). The significance of this gene activity at the 1-cell stage, prior to the onset of major zygotic gene activation at the 2-cell stage in the mouse and at somewhat later cleavage stages in large mammals, remains to be established. Epigenetic changes in chromatin early in development may a play a role in regulating early zygotic transcription (562).

The pronuclear stage ends when MPF levels rise again, resulting in nuclear envelope breakdown and chromosome condensation. Since meiosis has now been completed, the duplicated chromosomes line up for the first mitotic cleavage. It is noteworthy that the spindle is centrally placed resulting, after first cleavage, in two blastomeres of similar size, in contrast to the previous meiotic divisions. It is likely that a post-fertilization change in the egg cortex and/or spindle apparatus proteins is involved in this crucial shift in spindle position.

Other Cytoplasmic Events. Early cytoplasmic events, such as Ca^{2+} elevation and second PB extrusion, are followed by important cytoskeletal-driven changes, the translation of maternal mRNAs, and the first cleavage division. After fertilization, changes in the organization of microfilaments/actin, intermediate filaments/cytokeratins, microtubules/tubulin have been observed. While the function of actin alterations has been debated (discussed previously), intermediate filament changes are thought to be mediated by PKC (563) and functionally important later during preimplantation development (564). Microtubule polymerization plays important roles in pronuclear migration (565) and formation of the first mitotic spindle. In eggs of other organisms, the cell cortex and cytoskeleton are involved in localizing maternal information used after fertilization (18).

Because the egg is such a large cell and the two pronuclei often form in opposite hemispheres of the egg, migration of the pronuclei is necessary to bring the maternal and paternal genomes together. This is accomplished, in part, by the sperm or egg centrosomes, depending on the species (477,479). As in the sea urchin, centrosome-associated microtubules are thought to play an important guidance role in the pronuclear migration. The molecular mechanism by which the cytoskeleton actually translocates these relatively large pronuclei, compared to other organelles, a distance of 50–100 μm from opposite directions remains to be established. However, it is likely to involve cytoplasmic, microtubule-associated motor proteins found in eggs (565,566).

Many other organisms undergo specification of the body axis and/or germ cell orientation soon after fertilization, such as amphibians, ascidians, and nematodes. However, the fertilization-induced mobilization of maternal information (asymmetrically arranged in the egg) to direct or mark cell lineages has not been as firmly established in mammals. Despite the fact that experimental blastomere deletion, addition and repositioning give rise to normal mice, recently indirect evidence in mouse eggs has been presented for positional information that is used in the organization of cells in the blastocyst (567–569) [however, see (570)]. However, the molecular and/or cytoskeletal determinants responsible have yet to be established.

Signaling Pathways Responsible for the Events of Egg Activation

Ca^{2+} Is the Prominent Messenger of Activation. The mechanism by which fertilization initiates the events of egg activation has become a fascinating story of cellular signaling, which is still in progress. In its broadest sense, these events include transformation of not only the egg, but also the sperm. In theory, sperm transformations could result from egg cytoplasmic factors (e.g., glutathione or various kinases) acting on sperm proteins, such as the protamines or proteins arranged in an intact nuclear envelope, both of which are absent in the unfertilized egg. Egg transformations could require the reciprocal process (sperm factors transforming egg structures) and/or the initiation of signals to activate dormant egg factors that would drive the events of egg activation, as well as relieve any inhibitory influences, such as CSF, present in the gametes prior to fertilization. In fact, while there is evidence for all of these mechanisms, a Ca^{2+} signaling system has been found to drive the majority of the events of egg activation (excluding the initial decondensation of sperm chromatin) (571). As we shall see, one prominent theory in mammals is that the union of the two gametes also brings together different components, reconstituting a functional Ca^{2+} signaling system.

From a historical perspective, almost 100 years ago frog eggs were parthenogenetically activated with a needle and the stimulus was later traced to the influx of Ca^{2+} from the extracellular medium (572–574). However, in the 1970s, although the calcium theory of egg activation was supported further with the advent of calcium ionophore A23187, it became apparent that internal calcium stores of hamster eggs were sufficient for ionophore-induced egg activation (575). Not long afterward, an increase in cytosolic Ca^{2+} from fertilization or ionophore treatment was demonstrated more directly by imaging eggs with the Ca^{2+} probe aequorin (576). With the advent of improved imaging technology and a variety of Ca^{2+} probes, a wave of Ca^{2+}_i release from the point of sperm–egg interaction was observed in many species (577), including mammals (578,579).

In contrast to some lower species whose eggs have a single rise in Ca^{2+}_i upon fertilization, mammalian eggs undergo a long series of transient Ca^{2+}_i rises (referred to as "oscillations") for many hours (579,580). The duration of individual rises are on the order of a few minutes with oscillation frequencies of 1 per 10–40 minutes reported in different species. Information is apparently encoded in the number of Ca^{2+} rises as different egg activation events require different numbers of oscillations for their initiation (552). Moreover, for most events, more oscillations are required for event initiation than for completion. Changing the Ca^{2+} oscillation parameters also alters the extent of parthenogenetic mammalian development (478,536). These periodic Ca^{2+} increases generate nearly synchronous oscillations of activity of the kinase, CaMKII (533). CaMKII inhibitors prevent the completion of meiosis in activated *Xenopus* and mouse eggs, as well as delay cortical granule secretion (528, 531–533).

The central role of the "explosive" initial rise in egg Ca^{2+}_i (574) and the subsequent Ca^{2+}_i oscillations in fertilization, and particularly in driving the completion of meiosis, cortical granule secretion, and the appearance of some new proteins, has produced intense interest in the underlying mechanisms. A variety of hypotheses have been put forth, including sperm–egg contact- or receptor-mediated activation and various fusion models, invoking the addition of sperm membrane ion channels, or the passage of Ca^{2+} from the sperm into the egg, or the release of a sperm signaling protein (Fig. 18).

IP_3 and the IP_3R. The role of IP_3-evoked release of sequestered Ca^{2+} through the activation of an IP_3R channel is key feature in signal transduction in somatic cells (581–583) and has featured in most of the models of egg activation (Fig. 18) (584,585). A brief recounting of this pathway begins with observations of a distinctive and dense network of endoplasmic reticulum in the egg cortex, which matures during hormone-induced meiotic maturation in many species (586,587). Since in other cells the endoplasmic reticulum is a well known intracellular storage site for calcium and Ca^{2+} release therefrom is regulated by ligand-gated ion channels in the endoplasmic reticulum membrane (e.g., the ryanodine- and IP_3-receptors) then the endoplasmic reticulum of the egg appeared to be a prime source of Ca^{2+} for egg activation events and optimally positioned close to the site of sperm–egg interaction. Thus, it was logical to test ryanodine and IP_3 for their ability to induce egg activation. Although both agents cause a rise in Ca^{2+}_i in mammalian eggs, IP_3 is generally accepted to be the major effector (579, 585,588). The microinjection of IP_3 into eggs causes a rise in Ca^{2+} and cortical granule exocytosis in marine, amphibian, and mammalian eggs. Interestingly, a single physiological dose of IP_3 did not cause all Ca^{2+}-dependent events of egg activation in mammals, such as cell cycle resumption and the recruitment of maternal mRNA (589), probably due to the absence of prolonged Ca^{2+} oscillations normally present after fertilization (552).

Metabolic studies in sea urchin and amphibian eggs provided important evidence for the production, turnover, and increase in mass of IP_3 at fertilization (590–593). In *Xenopus* eggs, IP_3 production is required for the fertilization-induced Ca^{2+} wave and most likely derived from PLC-induced PIP_2 cleavage (phosphatidylinositol-4,5-bisphosphate) since antibodies against PIP_2 inhibit the Ca^{2+} wave (594). In mammalian eggs, it has not been possible to measure IP_3 production in single eggs, but the release of caged IP_3 over time results in Ca^{2+} oscillations and supports the hypothesis that a continuous low level of IP_3 is made during fertilization (595). A relatively nonhydrolyzable analogue of IP_3 and potent agonist of the IP_3R, adenophostin, causes long-standing Ca^{2+} oscillations in mouse eggs (596).

The ability of IP_3 to induce the release of $(Ca^{2+})_i$ and its production at fertilization led to studies demonstrating that a functional IP_3R was present and necessary for Ca^{2+} release as well as egg activation. Although all three isoforms of the IP_3R are expressed in mammalian eggs (597,598), the primary isoform expressed is the IP_3R type-1 and it is preferentially

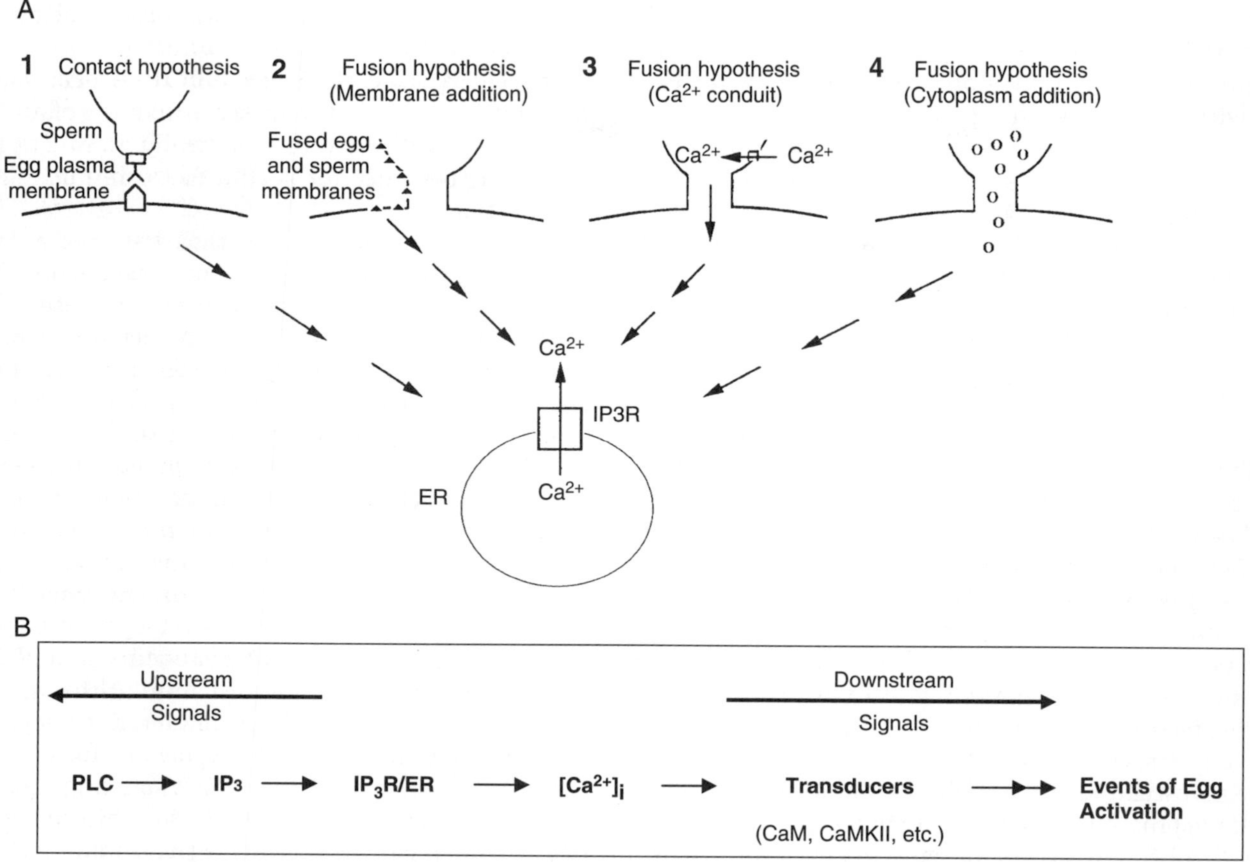

FIG. 18. Mechanisms regulating the elevation Ca^{2+} at fertilization. **A:** Diagram summarizing various hypotheses to explain the initiation of the increase in Ca^{2+}_i from the endoplasmic reticulum (ER) in the egg. (*1*) Contact hypothesis, e.g., via gamete surface receptors. (*2*) Fusion adds an activating factor to the egg membrane. (*3*) A catalytic amount of Ca^{2+} from the sperm is provided after fusion. (*4*) Cytoplasmic addition of a sperm factor (e.g., PLC-ζ). [Reprinted from (585) with permission from Elsevier. Courtesy of Dr. Linda Runft.] **B:** Simplified diagram of the major signaling steps immediately upstream and downstream of the elevation of Ca^{2+}_i. These have been well supported by experimental studies (see text). Note that the hypotheses in **A** must take into account the introduction and/or activation of PLC.

localized to the egg cortex. Importantly, inhibition of the IP_3R-1 by the injection of a function-blocking antibody or heparin prevents oscillations of Ca^{2+} after fertilization or sperm extract injection in mammalian eggs (599,600). IP_3R degradation increases before the pronuclear stage, likely in response to prolonged IP_3 production (601,602).

Egg Phospholipase C and Src Family Kinases. The involvement of IP_3, the IP_3R and diacylglycerol in egg activation led investigators to work further upstream in the signaling mechanism of fertilization. PLC was considered an attractive upstream effector because it generates the latter two second messengers and specific PLC isoforms are tailored to respond to different types of signals that activate cells (603). For example, binding of specific ligands to heterotrimeric G-protein–coupled surface receptors leads to PLC-β activation, whereas the binding of certain growth factors results in the activation of tyrosine kinase receptors or cytosolic kinases that stimulate PLC-γ (91). Although initial studies indicated the presence and potential involvement of PLC-β in egg activation under specialized conditions [see, for example, (604,605)], direct experimental approaches have not provided evidence that it is required for egg activation at fertilization [reviewed in (606–608)].

The potential involvement of PLC-β was supported by reports that mouse eggs expressing exogenous m1 muscarinic receptor underwent egg activation and cleavage after treatment with acetylcholine (604,609). This indicated the likely presence of a G-protein signaling machinery, although a candidate receptor was not identified. Furthermore, the injection of the G-protein activator, GTP-γS, stimulates Ca^{2+} oscillations in mammalian eggs, whereas GDP-βS, a nonhydrolyzable GDP analog that inhibits G-protein function, prevented sperm-induced Ca^{2+} oscillations (610,611). In this regard, the mouse egg expresses at least two PLC-β isoforms (605,612).

However, other studies did not support the hypothesis that heterotrimeric, G-protein–coupled receptors drove egg activation in mammals. Two ADP-ribosyltransferase exotoxins, the cholera toxin activator of G_S function and the pertussis toxin antagonist of G_i, failed to affect egg activation (609,613,614). Attention focused on other G-protein family members (G_q, G_{11}, G_{14}, and G_{15}) that were expressed in eggs (615), but function-blocking antibodies to G_q and G_{11} did not block fertilization-induced Ca^{2+} oscillations (616). Monomeric small GTP-binding proteins can also activate PLC-β, although by a mechanism different than that of heterotrimeric G-proteins. One such candidate, Rho GTPase, appears to be involved in sperm–egg fusion (473). In addition, the effects of PLC-β gene targeting on early development has been discussed (612).

PLC-γ deserves mention because it is found in mouse eggs (605,617) and its activity increases after fertilization in sea urchin and *Xenopus* eggs (618,619). Moreover, in elegant experiments, injected SH2 domain dominant negatives of PLC-γ inhibited or delayed fertilization-induced Ca^{2+} release in sea urchin and ascidian eggs (620–623). However, some of the same investigators did not observe inhibition using a similar approach in *Xenopus* and mouse eggs (617,624). As with PLC-β, it cannot yet be ruled out that PLC-γ is activated by a different mechanism at some point during the process of Ca^{2+} oscillations in mammalian eggs (612). In this regard, injection of PLC-γ1 into mouse eggs is followed by fertilization-like oscillations (625).

If Ca^{2+} oscillations at fertilization are mediated by egg PLC, elucidation of the upstream pathway responsible would lead to a better understanding of the how the sperm actually triggers egg activation. PLC-γ is activated by Src family kinases very soon after fertilization in several nonmammalian species (585), but not in mouse eggs (626). Src family kinases appear to be required for fertilization-induced elevation of egg Ca^{2+} in these species, but a role in mammalian fertilization has not been firmly established. Injection of recombinant c-Fyn (a member of the Src family) induces mouse egg activation (627). c-Fyn has been detected in mouse eggs (628) and it has been proposed that it associates with the truncated c-kit tyrosine kinase of sperm and activates egg PLC-γ (629). However, there is evidence that this pathway differs from the one during mammalian fertilization (585,612).

Sperm Phospholipase C. The failure to firmly identify a PLC in mammalian eggs required for fertilization-induced activation has lead to serious consideration of a factor from sperm that is itself a PLC or activates PLC, similar to truncated c-kit (630). This "sperm factor" hypothesis derived initially from experiments by Dale and co-workers showing that "injection of a soluble sperm extract into sea urchin eggs triggers the cortical reaction" (631). Although a role for a sperm factor in sea urchin egg fertilization remains to be established firmly, there has been increasing evidence for such a factor in mammalian fertilization.

This hypothesis requires that the sperm factor is transferred into the egg by gamete fusion. Thus, fusion should precede the massive increase in egg Ca^{2+}_i: this, in fact, has been reported for mouse eggs (632,633). In the mouse egg, high resolution Ca^{2+} imaging studies do not support the idea that there is a large pool of sperm cytosolic Ca^{2+} that is transferred inside the egg upon gamete fusion (632). Since the injection of physiological amounts of Ca^{2+} or of IP_3 into mammalian eggs does not result in fertilization-like Ca^{2+}_i oscillations [reviewed in (634)], they are also unlikely to account for the sperm factor activity.

The sperm factor activity should be capable of eliciting a long period of Ca^{2+} oscillations and egg activation, similar to those observed in fertilized eggs. In mammals, the first reports of such activity came from injection of sperm extracts into hamster, mouse and rabbit eggs (612,635). Many studies have demonstrated that this activity stimulates the phosphoinositide pathway, the events of egg activation and cleavage (636,637). Sperm factor activity is associated with demembranated sperm heads and is lost therefrom after residence in egg cytoplasm (638–640). Function blocking antibodies to the IP_3R inhibit sperm factor-mediated Ca^{2+} rises, indicating that it acts upstream of the IP_3R (641).

The activity is dependent upon a protein component(s) (636,637) and many candidate proteins have been proposed [reviewed in (607,608)]. Although various proteins have been proposed to have activity, some have been ruled out and the evidence for others is relatively limited (585,612). A PLC from sperm that was highly active in the egg, in contrast to the relatively inactive PLCs in the unfertilized egg, would fit the requirement of generating IP_3. Sperm factor fractions in many studies have PLC activity and a PLC inhibitor, U73122 (albeit at high levels), prevents the Ca^{2+} oscillations induced by this activity (585). Although the PLC-δ4 protein is primarily expressed in the testis and is involved in Ca^{2+} mobilization prior to the acrosome reaction, injection into mouse eggs failed to elicit Ca^{2+} oscillations (326,327,642).

However, an attractive candidate is another isoform of PLC, the sperm-specific PLC-ζ. This was found in the rodent, pig, and primate by genomic database mining, Northern analysis, and Western blot (643,644). PLC-ζ is localized in the part of the sperm head that enters the egg first following gamete membrane fusion (645). Other evidence for the role of PLC-ζ as a component of sperm factor include: the mass estimates of partially purified SF activity (30–70 kDa) are consistent with that of human and mouse PLC-ζ (70.4 and 74 kDa, respectively); expression of PLC-ζ mRNA in mouse eggs results in fertilization-like Ca^{2+}_i oscillations and parthenogenetic egg activation; and immunodepletion of PLC-ζ from SF fractions removed their ability to induce Ca^{2+} rises. Finally, sperm that had been depleted of PLC-ζ by transgenic RNAi methods can fuse with eggs and initiate Ca^{2+}_i oscillations, but the number of subsequent oscillations was reduced, as was the generation of pronuclei, which is consistent with the absence of transgenic offspring (646). Future work should address whether the amount and location of PLC-ζ (during early fertilization) are consistent with its proposed function.

In summary, the production of IP_3, binding of IP_3 to its receptor, and mobilization of intracellular Ca^{2+} appear to be a near universal mechanism for initiating Ca^{2+} release, which, in turn, drives many of the events of egg activation. Investigations to date suggest that the mechanism of PLC activation may vary in different animals. Despite the predicted importance of PLC-ζ in mammals, roles for egg PLCs and perhaps other gamete molecules in egg activation have not been completely ruled out, either downstream or in parallel with PLC-ζ (612). Interestingly, from a clinical point of view, the SF hypothesis helps to explain the success of human ICSI (intracytoplasmic sperm injection), in which surface receptor interaction of the two gametes appears to be bypassed.

A BRIEF NOTE ON THE APPLICATIONS OF FERTILIZATION BIOLOGY

Studies of fertilization in a wide variety of invertebrate and nonmammalian vertebrate animals, where complex culture conditions are not required, led to studies of mammalian fertilization in vitro (10). The ability to fertilize eggs of echinoderms in simple sea water (3,647) suggested that mammalian fertilization might occur in simple media. Subsequent crucial developments were the recognition that mammalian sperm must be capacitated in order to become fertile (28–30) and that only fertilized mature eggs, and not pre-ovulatory oocytes, are developmentally competent (discussed at the beginning of this chapter). A clear experimental line can be drawn from the realization in the early 1950s that sperm must be incubated in vivo in order to fertilize eggs, to the development of culture conditions that support fertilization and cleavage of eggs in vitro in the 1960s and 1970s, to the extension of these methods without major modification to human in vitro fertilization (648). The birth of Louise Brown in 1978, the first successful human in vitro fertilization case, initiated a series of new methods that would become widely used in many countries and have, to this point, resulted in the birth of over 1,000,000 babies.

Assisted reproductive technologies include in vitro fertilization, subzonal sperm injection, gamete intrafallopian transfer, and ICSI. Although the advent of successful ICSI (649) in 1992 seemed to defy the expected process of fertilization-associated signaling since it by-passed sperm–egg adhesion and fusion, the more recent discoveries of sperm factor activity and PLCζ are likely to provide an explanation for the mechanism of ICSI-mediated egg activation. While considerable success has been achieved in overcoming infertility for many patients, little is known about the extent of clinical impact of these artificial procedures over the entire lifespan of the offspring produced (8).

Other technologies that have been made possible due to our understanding of fertilization are animal cloning and the production of certain types of stem cell cells from mouse blastocysts (18). In both cases, the egg chromosomes are removed and replaced a somatic cell nucleus, followed by artificial activation of the egg and cleavage. In the case of animal cloning, embryo transfer to a host is performed, whereas for stem cell production, outgrowth of the blastocysts in vitro can generate stem cell lines with the genotype of the donor. Animal cloning has the potential to save endangered species, or, coupled with gene transfer, to produce quantities of important eukaryotic proteins. However, existing cloning methods have a very low success rate and have been widely condemned by scientists, clinicians, and society for use in humans. Regarding stem cell lines, if transferred back to the donor, they have the potential to replace cells that have undergone aging, cancer, disease, or destruction (the process of "therapeutic cloning"). Although therapeutic cloning of human stem cells has great medical potential, it is subject to ethical discussions about the use of human eggs and preimplantation embryo-like cell masses prior to cell line derivation in vitro. Thus, fertilization research has broad implications and rigorous evaluation of new infertility- and stem cell-related applications is necessary to evaluate their use in humans.

ACKNOWLEDGMENTS

The authors thank the following colleagues for their insightful suggestions or for reading parts of the manuscript: J. Michael Bedford, Hugh Clarke,

Michael Eisenbach, Janice Evans, Laurinda Jaffe, Stella Markoulaki, Diana Myles, Paul Primakoff, Richard Schultz, and Bayard Storey. We also appreciate Sara Matson's comments and assistance. This work was supported by the National Institute of Child Health and Human Development. Due to space constraints this work is not as expansive in scope or as comprehensive in literature citations as those in previous editions.

REFERENCES

1. Aristotle. (1970). *History of Animals, Books IV–VI*. Harvard University Press, Cambridge, MA.
2. Aristotle. (1991). *History of Animals, Books VII–X*. Harvard University Press, Cambridge, MA.
3. Lillie, F. R. (1919). *Problems of Fertilization*. University of Chicago Press, Chicago.
4. U. S. Census Bureau World Popclock. Available at http://www.census. gov/cgi-bin/ipc/popclockw.
5. Nass, S. J., and Strauss, J. F. III. (2004). *New Frontiers in Contraceptive Research. A Blueprint for Action*. National Academies Press, Washington, DC.
6. Liu, D. Y., and Baker, H. W. G. (2000). Defective sperm-zona pellucida interaction: a major cause of failure of fertilization in clinical in-vitro fertilization. *Hum. Reprod.* 15, 702–708.
7. Liu, D. Y., and Baker, H. W. (2003). Disordered zona pellucida-induced acrosome reaction and failure of in vitro fertilization in patients with unexplained infertility. *Fertil. Steril.* 79, 74–80.
8. Schultz, R. M., and Williams, C. J. (2002). The science of ART. *Science* 296, 2188–2190.
9. Yanagimachi, R. (1988). Mammalian fertilization. In *The Physiology of Reproduction* (E. Knobil and J. Neill, Eds.), pp. 135–185. Raven Press, New York.
10. Yanagimachi, R. (1994). Mammalian fertilization. In *The Physiology of Reproduction* (E. Knobil and J. Neill, Eds.), pp. 189–317. Raven Press, New York.
11. Eisenbach, M. (2004). *Chemotaxis*. Imperial College Press, London.
12. Yanagimachi, R. (2003). Fertilization and development initiation in orthodox and unorthodox ways: from normal fertilization to cloning. *Adv. Biophys.* 37, 49–89.
13. Hardy, D. M. (2002). *Fertilization*. Academic Press, San Diego.
14. Darszon, A., Beltran, C., Felix, R., Nishigaki, T., and Trevino, C. L. (2001). Ion transport in sperm signaling. *Dev. Biol.* 240, 1–14.
15. Trounson, A., and Gosden, R. G. (2003). *Biology and Pathology of the Oocyte: Its Role in Fertility and Reproductive Medicine*. Cambridge University Press, Cambridge.
16. Fujiwara, T., Nakada, K., Shirakawa, H., and Miyazaki, S. (1993). Development of inositol trisphosphate-induced calcium release mechanism during maturation of hamster oocytes. *Dev. Biol.* 156, 69–79.
17. Ducibella, T. (1996). The cortical reaction and development of activation competence in mammalian oocytes. *Hum. Reprod. Update* 2, 29–42.
18. Gilbert, S. (2003). *Developmental Biology*. Sinauer Associates, Inc., Sunderland, MA.
19. Mehlmann, L. M., and Kline, D. (1994). Regulation of intracellular calcium in the mouse egg: calcium release in response to sperm or inositol trisphosphate is enhanced after meiotic maturation. *Biol. Reprod.* 51, 1088–1098.
20. Abbott, A. L., Fissore, R. A., and Ducibella, T. (1999). Incompetence of preovulatory mouse oocytes to undergo cortical granule exocytosis following induced calcium oscillations. *Dev. Biol.* 207, 38–48.
21. Perreault, S. D., Barbee, R. R., and Slott, V. L. (1988). Importance of glutathione in the acquisition and maintenance of sperm nuclear decondensing activity in maturing hamster oocytes. *Dev. Biol.* 125, 181–186.
22. McLay, D. W., and Clarke, H. J. (1997). The ability to organize sperm DNA into functional chromatin is acquired during meiotic maturation in murine oocytes. *Dev. Biol.* 186, 73–84.
23. Iwamatsu, T., and Chang, M. C. (1972). Sperm penetration in vitro of mouse oocytes at various times during maturation. *J. Reprod. Fertil.* 31, 237–247.
24. Niwa, K., and Chang, M. C. (1975). Temporal relationship between insemination and sperm penetration in immature rats, with special reference to the effect of incubation of spermatozoa. *Fertil. Steril.* 26, 1266–1272.
25. Mendez, R., and Richter, J. D. (2001). Translational control by CPEB: a means to the end. *Nat. Rev. Mol. Cell Biol.* 2, 521–529.
26. Stein, P., Svoboda, P., and Schultz, R. M. (2003). Transgenic RNAi in mouse oocytes: a simple and fast approach to study gene function. *Dev. Biol.* 256, 188–194.
27. Xu, Z., Williams, C. J., Kopf, G. S., and Schultz, R. M. (2003). Maturation-associated increase in IP3 receptor type 1: role in conferring increased IP3 sensitivity and Ca2+ oscillatory behavior in mouse eggs. *Dev. Biol.* 254, 163–171.
28. Chang, M. C. (1951). Fertilizing capacity of spermatozoa deposited into fallopian tubes. *Nature* 168, 697–698.
29. Austin, C. R. (1951). Observations on the penetration of the sperm into the mammalian egg. *Aust. J. Scient. Res.* 4, 581–596.
30. Austin, C. R. (1952). The 'capacitation' of the mammalian sperm. *Nature* 170, 326.
31. Gadella, B. M., and Van Gestel, R. A. (2004). Bicarbonate and its role in mammalian sperm function. *Anim. Reprod. Sci.* 82–83, 307–319.
32. Arnoult, C., Kazam, I. G., Visconti, P. E., Kopf, G. S., Villaz, M., and Florman, H. M. (1999). Control of the low voltage-activated calcium channel of mouse sperm by egg ZP3 and by membrane hyperpolarization during capacitation. *Proc. Natl. Acad. Sci. U S A* 96, 6757–6762.
33. Cohen-Dayag, A., Tur-Kaspa, I., Dor, J., Mashiach, S., and Eisenbach, M. (1995). Sperm capacitation in humans is transient and correlates with chemotactic responsiveness to follicular factors. *Proc. Natl. Acad. Sci. U S A* 92, 11039–11043.
34. Austin, C. R., and Bishop, M. W. H. (1958). Role of the rodent acrosome and perforatorium in fertilization. *Proc. R. Soc. Lond. B.* 149, 241–248.
35. Florman, H. M., and Storey, B. T. (1981). Inhibition of in vitro fertilization of mouse eggs: 3-quinuclidinyl benzilate specifically blocks penetration of zonae pellucidae by mouse spermatozoa. *J. Exp. Zool.* 216, 159–167.
36. Ward, C. R., and Storey, B. T. (1984). Determination of the time course of capacitation in mouse spermatozoa using a chlortetracycline fluorescence assay. *Dev. Biol.* 104, 287–296.
37. Suarez, S. S., and Ho, H. C. (2003). Hyperactivation of mammalian sperm. *Cell Mol. Biol. (Noisy-le-grand)* 49, 351–356.
38. Hoshi, M., Nishigaki, T., Kawamura, M., Ikeda, M., Gunaratne, J., Ueno, S., Ogiso, M., Moriyama, H., and Matsumoto, M. (2000). Acrosome reaction in starfish: signal molecules in the jelly coat and their receptors. *Zygote* 8, Suppl. 1, S26–S27.
39. Garbers, D. L., and Kopf, G. S. (1980). The regulation of spermatozoa by calcium and cyclic nucleotides. In *Advances in Cyclic Nucleotide Research* (P. Greengard and G. A. Robison, Eds.), pp. 251–306. Raven Press, New York.
40. Vacquier, V. D. (1998). Evolution of gamete recognition proteins. *Science* 281, 1995–1998.
41. Quill, T. A., and Garbers, D. L. (2002). Sperm motility activation and chemoattraction. In *Fertilization* (D. M. Hardy, Ed.), pp. 29–55. Academic Press, San Diego.

42. Eisenbach, M. (1999). Mammalian sperm chemotaxis and its association with capacitation. *Dev. Genet.* 25, 87–94.
43. Eisenbach, M. (1999). Sperm chemotaxis. *Rev. Reprod.* 4, 56–66.
44. Tulsiani, D. R., and Abou-Haila, A. (2004). Is sperm capacitation analogous to early phases of Ca2+-triggered membrane fusion in somatic cells and viruses? *BioEssays* 26, 281–290.
45. Visconti, P. E., Galantino-Homer, H., Moore, G. D., Bailey, J. L., Ning, X., Fornes, M., and Kopf, G. S. (1998). The molecular basis of sperm capacitation. *J. Androl.* 19, 242–248.
46. Cohen-Dayag, A., and Eisenbach, M. (1994). Potential assays for sperm capacitation in mammals. *Am. J. Physiol.* 267, C1167–C1176.
47. Olds-Clarke, P. (2003). Unresolved issues in mammalian fertilization. *Int. Rev. Cytol.* 232, 129–184.
48. Saling, P. M., Storey, B. T., and Wolf, D. P. (1978). Calcium-dependent binding of mouse epididymal spermatozoa to the zona pellucida. *Dev. Biol.* 65, 515–525.
49. Shur, B. D., and Hall, N. G. (1982). Sperm surface galactosyltransferase activities during in vitro capacitation. *J. Cell Biol.* 95, 567–573.
50. Giojalas, L. C., Rovasio, R. A., Fabro, G., Gakamsky, A., and Eisenbach, M. (2004). Timing of sperm capacitation appears to be programmed according to egg availability in the female genital tract. *Fertil. Steril.* 82, 247–249.
51. Bedford, J. M. (1970). Sperm capacitation and fertilization in mammals. *Biol. Reprod. Suppl.* 2, 128–158.
52. Bedford, J. M. (2004). Enigmas of mammalian gamete form and function. *Biol. Rev. Camb. Philos. Soc.* 79, 429–460.
53. Chang, M. C., Austin, C. R., Bedford, J. M., Brackett, B. G., Hunter, R. H. F., and Yanagimachi, R. (1977). Capacitation of spermatozoa and fertilization in mammals. In *Frontiers in Reproduction and Fertility Control: A Review of Reproductive Sciences and Fertility Control* (R. O. Greep, and M. A. Koblinsky, Eds.), pp. 434–451. MIT Press, Cambridge, MA.
54. Oliphant, G., and Brackett, B. G. (1973). Capacitation of mouse spermatozoa in media with elevated ionic strength and reversible decapacitation with epididymal extracts. *Fertil. Steril.* 24, 948–955.
55. Oliphant, G., and Brackett, B. G. (1973). Immunological assessment of surface changes of rabbit sperm undergoing capacitation. *Biol. Reprod.* 9, 404–414.
56. Reyes, A., Oliphant, G., and Brackett, B. G. (1975). Partial purification and identification of a reversible decapacitation factor from rabbit seminal plasma. *Fertil. Steril.* 26, 148–157.
57. Lee, M. A., and Storey, B. T. (1986). Bicarbonate is essential for fertilization of mouse eggs: Mouse sperm require it to undergo the acrosome reaction. *Biol. Reprod.* 34, 349–356.
58. Boatman, D. E., and Robbins, R. S. (1991). Bicarbonate: carbon-dioxide regulation of sperm capacitation, hyperactivated motility, and acrosome reactions. *Biol. Reprod.* 44, 806–813.
59. Wang, X. F., Zhou, C. X., Shi, Q. X., Yuan, Y. Y., Yu, M. K., Ajonuma, L. C., Ho, L. S., Lo, P. S., Tsang, L. L., Liu, Y., Lam, S. Y., Chan, L. N., Zhao, W. C., Chung, Y. W., and Chan, H. C. (2003). Involvement of CFTR in uterine bicarbonate secretion and the fertilizing capacity of sperm. *Nat. Cell Biol.* 5, 902–906.
60. Sutton, K. A., Jungnickel, M. K., and Florman, H. M. (2003). Of fertility, cystic fibrosis and the bicarbonate ion. *Nat. Cell Biol.* 5, 857–859.
61. Parrish, J. J., Susko-Parrish, J. L., and First, N. L. (1989). Capacitation of bovine sperm by heparin: Inhibitory effect of glucose and role of intracellular pH. *Biol. Reprod.* 41, 683–699.
62. Baldi, E., Casano, R., Falsetti, C., Krausz, C., Maggi, M., and Forti, G. (1991). Intracellular calcium accumulation and responsiveness to progesterone in capacitating human spermatozoa. *J. Androl.* 12, 323–330.
63. Wennemuth, G., Babcock, D. F., and Hille, B. (2003). Calcium clearance mechanisms of mouse sperm. *J. Gen. Physiol.* 122, 115–128.
64. Okunade, G. W., Miller, M. L., Pyne, G. J., Sutliff, R. L., O'Connor, K. T., Neumann, J. C., Andringa, A., Miller, D. A., Prasad, V., Doetschman, T., Paul, R. J., and Shull, G. E. (2004). Targeted ablation of plasma membrane Ca2+-ATPase (PMCA) 1 and 4 indicates a major housekeeping function for PMCA1 and a critical role in hyperactivated sperm motility and male fertility for PMCA4. *J. Biol. Chem.* 279, 33742–33750.
65. Zeng, Y., Clark, E. N., and Florman, H. M. (1995). Sperm membrane potential: hyperpolarization during capacitation regulates zona pellucida-dependent acrosomal secretion. *Dev. Biol.* 171, 554–563.
66. Munoz-Garay, C., de la Vega-Beltrán, J. L., Delgado, R., Labarca, P., Felix, R., and Darszon, A. (2001). Inwardly rectifying K^+ channels in spermatogenic cells: functional expression and implication in sperm capacitation. *Dev. Biol.* 234, 261–274.
67. Hille, B. (2001). *Ionic Channels of Excitable Membranes*. Sinauer Associates Inc., Sunderland, MA.
68. Bedlack, R. S., Wei, M.-D., Fox, S. H., Gross, E., and Loew, L. M. (1994). Distinct electric potentials in soma and neurite membranes. *Neuron* 13, 1187–1193.
69. Tse, A., Tse, F. W., and Hille, B. (1994). Calcium homeostasis in identified rat gonadotrophs. *J. Physiol.* 477, 511–525.
70. Morton, B., and Albagli, L. (1973). Modification of hamster sperm adenyl cyclase by capacitation in vitro. *Biochem. Biophys. Res. Commun.* 50, 697–703.
71. Ludwig, M. G., and Seuwen, K. (2002). Characterization of the human adenylyl cyclase gene family: cDNA, gene structure, and tissue distribution of the nine isoforms. *J. Recept. Signal. Transduct. Res.* 22, 79–110.
72. Spehr, M., Schwane, K., Riffell, J. A., Barbour, J., Zimmer, R. K., Neuhaus, E. M., and Hatt, H. (2004). Particulate adenylate cyclase plays a key role in human sperm olfactory receptor-mediated chemotaxis. *J. Biol. Chem.* 279, 40194–40203.
73. Fraser, L. R., Adeoya-Osiguwa, S. A., and Baxendale, R. W. (2003). First messenger regulation of capacitation via G protein–coupled mechanisms: a tale of serendipity and discovery. *Mol. Hum. Reprod.* 9, 739–748.
74. Leclerc, P., de, L. E., and Gagnon, C. (1996). Cyclic adenosine 3′, 5′monophosphate-dependent regulation of protein tyrosine phosphorylation in relation to human sperm capacitation and motility. *Biol. Reprod.* 55, 684–692.
75. Leclerc, P., and Kopf, G. S. (1999). Evidence for the role of heterotrimeric guanine nucleotide-binding regulatory proteins in the regulation of the mouse sperm adenylyl cyclase by the egg's zona pellucida. *J. Androl.* 20, 126–134.
76. Liguori, L., Rambotti, M. G., Bellezza, I., and Minelli, A. (2004). Electron microscopic cytochemistry of adenylyl cyclase activity in mouse spermatozoa. *J. Histochem. Cytochem.* 52, 833–836.
77. Braun, T., and Dods, R. F. (1975). Development of a Mn-2+-sensitive, "soluble" adenylate cyclase in rat testis. *Proc. Natl. Acad. Sci. U S A* 72, 1097–1101.
78. Buck, J., Sinclair, M. L., Schapal, L., Cann, M. J., and Levin, L. R. (1999). Cytosolic adenylyl cyclase defines a unique signaling molecule in mammals. *Proc. Natl. Acad. Sci. U S A* 96, 79–84.
79. Sinclair, M. L., Wang, X.-Y., Mattai, M., Conti, M., Buck, J., Wolgemuth, D. J., and Levin, L. R. (2000). Specific expression of soluble adenylyl cyclase in male germ cells. *Mol. Reprod. Dev.* 56, 6–11.
80. Jaiswal, B. S., and Conti, M. (2001). Identification and functional analysis of splice variants of the germ cell soluble adenylyl cyclase. *J. Biol. Chem.* 276, 31698–31708.
81. Chen, Y., Cann, M. J., Litvin, T. N., Iourgenko, V., Sinclair, M. L., Levin, L. R., and Buck, J. (2000). Soluble adenylyl cyclase as an evolutionarily conserved bicarbonate sensor. *Science* 289, 625–628.
82. Litvin, T. N., Kamenetsky, M., Zarifyan, A., Buck, J., and Levin, L. R. (2003). Kinetic properties of "soluble" adenylyl cyclase.

Synergism between calcium and bicarbonate. *J. Biol. Chem.* 278, 15922–15926.

83. Jaiswal, B. S., and Conti, M. (2003). Calcium regulation of the soluble adenylyl cyclase expressed in mammalian spermatozoa. *Proc. Natl. Acad. Sci. U S A* 100, 10676–10681.
84. Harrison, R. A. (1996). Capacitation mechanisms, and the role of capacitation as seen in eutherian mammals. *Reprod. Fertil. Dev.* 8, 581–594.
85. Zippin, J. H., Chen, Y., Nahirney, P., Kamenetsky, M., Wuttke, M. S., Fischman, D. A., Levin, L. R., and Buck, J. (2003). Compartmentalization of bicarbonate-sensitive adenylyl cyclase in distinct signaling microdomains. *FASEB J.* 17, 82–84.
86. Nolan, M. A., Babcock, D. F., Wennemuth, G., Brown, W., Burton, K. A., and McKnight, G. S. (2004). Sperm-specific protein kinase A catalytic subunit Calpha2 orchestrates cAMP signaling for male fertility. *Proc. Natl. Acad. Sci. U S A* 101, 13483–13488.
87. San Agustin, J., Leszyk, J. D., Nuwaysir, L. M., and Witman, G. B. (1998). The catalytic subunit of cAMP-dependent protein kinase of ovine sperm flagella has a unique amino-terminal sequence. *J. Biol. Chem.* 273, 24874–24883.
88. Harrison, R. A., and Miller, N. G. (2000). cAMP-dependent protein kinase control of plasma membrane lipid architecture in boar sperm. *Mol. Reprod. Dev.* 55, 220–228.
89. Harrison, R. A. (2004). Rapid PKA-catalysed phosphorylation of boar sperm proteins induced by the capacitating agent bicarbonate. *Mol. Reprod. Dev.* 67, 337–352.
90. Wennemuth, G., Carlson, A. E., Harper, A. J., and Babcock, D. F. (2003). Bicarbonate actions on flagellar and Ca^{2+}-channel responses: initial events in sperm activation. *Development* 130, 1317–1326.
91. Alberts, B., Johnson, A., Lewis, J., Raff, M., Roberts, K., and Walter, P. (2002). *Molecular Biology of the Cell.* Garland Science, New York.
92. Kol, M. A., de, K. B., and de Kroon, A. I. (2002). Phospholipid flip-flop in biogenic membranes: what is needed to connect opposite sides. *Semin. Cell Dev. Biol.* 13, 163–170.
93. Nolan, J. P., Magargee, S. F., Posner, R. G., and Hammerstedt, R. H. (1995). Flow cytometric analysis of transmembrane phospholipid movement in bull sperm. *Biochemistry* 34, 3907–3915.
94. Harrison, R. A., Ashworth, P. J., and Miller, N. G. (1996). Bicarbonate/CO2, an effector of capacitation, induces a rapid and reversible change in the lipid architecture of boar sperm plasma membranes. *Mol. Reprod. Dev.* 45, 378–391.
95. Harrison, R. A., and Gadella, B. M. (2005). Bicarbonate-induced membrane processing in sperm capacitation. *Theriogen.* 63, 342–351.
96. Flesch, F. M., Brouwers, J. F., Nievelstein, P. F., Verkleij, A. J., van Golde, L. M., Colenbrander, B., and Gadella, B. M. (2001). Bicarbonate stimulated phospholipid scrambling induces cholesterol redistribution and enables cholesterol depletion in the sperm plasma membrane. *J. Cell Sci.* 114, 3543–3555.
97. Gadella, B. M., and Harrison, R. A. (2000). The capacitating agent bicarbonate induces protein kinase A-dependent changes in phospholipid transbilayer behavior in the sperm plasma membrane. *Development* 127, 2407–2420.
98. Davis, B. K. (1981). Timing of fertilization in mammals: sperm cholesterol/phospholipid ratio as a determinant of the capacitation interval. *Proc. Natl. Acad. Sci. U S A* 78, 7560–7564.
99. Langlais, J., and Roberts, K. D. (1985). A molecular membrane model of sperm capacitation and the acrosome reaction of mammalian spermatozoa. *Gamete Res.* 12, 183–224.
100. Go, K. J., and Wolf, D. P. (1985). Albumin-mediated changes in the sperm sterol contents during capacitation. *Biol. Reprod.* 32, 145–152.
101. Visconti, P. E., Ning, X., Fornes, M. W., Alvarez, J. G., Stein, P., Connors, S. A., and Kopf, G. S. (1999). Cholesterol efflux-mediated signal transduction in mammalian sperm: cholesterol release signals an increase in protein tyrosine phosphorylation during mouse sperm capacitation. *Dev. Biol.* 214, 429–443.
102. Visconti, P. E., Galantino-Homer, H., Ning, X., Moore, G. D., Valenzuela, J. P., Jorgez, C. J., Alvarez, J. G., and Kopf, G. S. (1999). Cholesterol efflux-mediated signal transduction in mammalian sperm. Beta-cyclodextrins initiate transmembrane signaling leading to an increase in protein tyrosine phosphorylation and capacitation. *J. Biol. Chem.* 274, 3235–3242.
103. Cross, N. L. (1996). Effect of cholesterol and other sterols on human sperm acrosomal responsiveness. *Mol. Reprod. Dev.* 45, 212–217.
104. Golub, T., Wacha, S., and Caroni, P. (2004). Spatial and temporal control of signalling through lipid rafts. *Curr. Opin. Neurobiol.* 14, 542–550.
105. Edinin, M. (2001). Membrane cholesterol, protein phosphorylation, and lipid rafts. *Sci. STKE* 2001, e1–e3.
106. Osheroff, J. E., Visconti, P. E., Valenzuela, J. P., Travis, A. J., Alvarez, J., and Kopf, G. S. (1999). Regulation of human sperm capacitation by a cholesterol efflux-stimulated signal transduction pathway leading to protein kinase A-mediated up-regulation of protein tyrosine phosphorylation. *Mol. Hum. Reprod.* 5, 1017–1026.
107. Travis, A. J., Merdiushev, T., Vargas, L. A., Jones, B. H., Purdon, M. A., Nipper, R. W., Galatioto, J., Moss, S. B., Hunnicutt, G. R., and Kopf, G. S. (2001). Expression and localization of caveolin-1, and the presence of membrane rafts, in mouse and Guinea pig spermatozoa. *Dev. Biol.* 240, 599–610.
108. Salaun, C., James, D. J., and Chamberlain, L. H. (2005). Lipid rafts and the regulation of exocytosis. *Traffic* 5, 255–264.
109. Visconti, P. E., Bailey, J. L., Moore, G. D., Pan, D., Olds-Clarke, P., and Kopf, G. S. (1995). Capacitation of mouse spermatozoa. I. Correlation between the capacitation state and protein tyrosine phosphorylation. *Development* 121, 1129–1137.
110. Visconti, P. E., Moore, G. D., Bailey, J. L., Leclerc, P., Connors, S. A., Pan, D., Olds-Clarke, P., and Kopf, G. S. (1995). Capacitation of mouse spermatozoa. II. Protein tyrosine phosphorylation and capacitation are regulated by a cAMP-dependent pathway. *Development* 121, 1139–1150.
111. Visconti, P. E., Westbrook, V. A., Chertihin, O., Demarco, I., Sleight, S., and Diekman, A. B. (2002). Novel signaling pathways involved in sperm acquisition of fertilizing capacity. *J. Reprod. Immunol.* 53, 133–150.
112. Ficarro, S., Chertihin, O., Westbrook, V. A., White, F., Jayes, F., Kalab, P., Marto, J. A., Shabanowitz, J., Herr, J. C., Hunt, D., and Visconti, P. E. (2003). Phosphoproteome analysis of capacitated human sperm. Evidence of tyrosine phosphorylation of A kinase-anchoring protein 3 and valosin containing protein/P97 during capacitation. *J. Biol. Chem.* 278, 11579–11589.
113. Yanagimachi, R. (1969). In vitro capacitation of hamster spermatozoa by follicular fluid. *J. Reprod. Fertil.* 18, 275–286.
114. Yanagimachi, R. (1970). The movement of golden hamster spermatozoa before and after capacitation. *J. Reprod. Fertil.* 23, 193–196.
115. Wong, M. M. Y., and Foskett, J. K. (1991). Oscillations of cytosolic sodium during calcium oscillations in exocrine acinar cells. *Science* 254, 1014–1016.
116. Neill, J. M., and Olds-Clarke, P. (1987). A computer-assisted assay for mouse sperm hyperactivation demonstrates that

bicarbonate but not bovine serum albumin is required. *Gamete Res.* 18, 121–140.

117. Olds-Clarke, P. (1989). Sperm from $t^{w32/+}$ mice: capacitation is normal, but hyperactivation is premature and nonhyperactivated sperm are slow. *Dev. Biol.* 131, 475–482.
118. Suarez, S. S. (1996). Hyperactivated motility in sperm. *J. Androl.* 17, 331–335.
119. Stauss, C. R., Votta, T. J., and Suarez, S. S. (1995). Sperm motility hyperactivation facilitates penetration of the hamster zona pellucida. *Biol. Reprod.* 53, 1280–1285.
120. DeMott, R. P., and Suarez, S. S. (1992). Hyperactivated sperm progress in the mouse oviduct. *Biol. Reprod.* 46, 779–785.
121. Suarez, S. S., Dai, X. B., DeMott, R. P., Redfern, K., and Mirando, M. A. (1992). Movement characteristics of boar sperm obtained from the oviduct or hyperactivated in vitro. *J. Androl.* 13, 75–80.
122. Fetterolf, P. M., Jurisicova, A., Tyson, J. E., and Casper, R. F. (1994). Conditioned medium from human cumulus oophorus cells stimulates human sperm velocity. *Biol. Reprod.* 51, 184–192.
123. Tesarik, J., Mendoz Oltras, C., and Testart, J. (1990). Effect of the human cumulus oophorus on movement characteristics of human capacitated sperm. *J. Reprod. Fertil.* 88, 665–675.
124. Oehninger, S., Sueldo, C., Lanzendorf, S., Mahony, M., Burkman, L. J., Alexander, N. J., and Hodgen, G. D. (1994). A sequential analysis of the effect of progesterone on specific sperm functions crucial to fertilization in vitro in infertile patients. *Hum. Reprod.* 9, 1322–1327.
125. Kay, V. J., Coutts, J. R., and Robertson, L. (1994). Effects of pentoxifylline and progesterone on human sperm capacitation and acrosome reaction. *Hum. Reprod.* 9, 2318–2323.
126. Weyand, I., Godde, M., Frings, S., Weiner, J., Muller, F., Altenhofer, W., Hatt, H., and Kaupp, U. B. (1994). Cloning and functional expression of a cyclic-nucleotide-gated channel from mammalian sperm. *Nature* 368, 859–863.
127. Wiesner, B., Weiner, J., Middendorff, R., Hagen, V., Kaupp, U. B., and Weyand, I. (1998). Cyclic nucleotide-gated channels on the flagellum control Ca^{2+} entry into sperm. *J. Cell Biol.* 142, 473–484.
128. Ho, H. C., and Suarez, S. S. (2003). Characterization of the intracellular calcium store at the base of the sperm flagellum that regulates hyperactivated motility. *Biol. Reprod.* 68, 1590–1596.
129. Westenbroek, R. E., and Babcock, D. F. (1999). Discrete regional distributions suggest diverse functional roles of calcium channel alpha1 subunits in sperm. *Dev. Biol.* 207, 457–469.
130. Wennemuth, G., Westenbroek, R. E., Xu, T., Hille, B., and Babcock, D. F. (2000). Ca_V2. 2 and Ca_V2. 3 (N- and R-type) Ca^{2+} Channels in Depolarization-evoked Entry of Ca^{2+} into Mouse Sperm. *J. Biol. Chem.* 275, 21210–21217.
131. Quill, T. A., Sugden, S. A., Rossi, K. L., Doolittle, L. K., Hammer, R. E., and Garbers, D. L. (2003). Hyperactivated sperm motility driven by CatSper2 is required for fertilization. *Proc. Natl. Acad. Sci. U S A* 100, 14869–14874.
132. Ralt, D., Goldenberg, M., Fetterolf, P., Thompson, D., Dor, J., Mashiach, S., Garbers, D. L., and Eisenbach, M. (1991). Sperm attraction to a follicular factor(s) correlates with human egg fertilizability. *Proc. Natl. Acad. Sci. U S A* 88, 2840–2844.
133. Sun, F., Bahat, A., Gakamsky, A., Girsh, E., Katz, N., Giojalas, L. C., Tur-Kaspa, I., and Eisenbach, M. (2005). Human sperm chemotaxis: both the oocyte and its surrounding cumulus cells secrete sperm chemoattractants. *Hum. Reprod.* 20, 761–767.
134. Bahat, A., Tur-Kaspa, I., Gakamsky, A., Giojalas, L. C., Breitbart, H., and Eisenbach, M. (2003). Thermotaxis of mammalian sperm cells: a potential navigation mechanism in the female genital tract. *Nat. Med.* 9, 149–150.
135. Fukuda, N., Yomogida, K., Okabe, M., and Touhara, K. (2004). Functional characterization of a mouse testicular olfactory receptor and its role in chemosensing and in regulation of sperm motility. *J. Cell Sci.* 117, 5835–5845.
136. Spehr, M., Gisselmann, G., Poplawski, A., Riffell, J. A., Wetzel, C. H., Zimmer, R. K., and Hatt, H. (2003). Identification of a testicular odorant receptor mediating human sperm chemotaxis. *Science* 299, 2054–2058.
137. Parmentier, M., Libert, F., Schurmans, S., Schiffmann, S., Lefort, A., Eggerickx, D., Ledent, C., Mollereau, C., Gerard, C., Perret, J., Grootegoed, A., and Vassart, G. (1992). Expression of members of the putative olfactory receptor gene family in mammalian germ cells. *Nature* 355, 453–455.
138. Vanderhaeghen, P., Schurmans, S., Vassart, G., and Parmentier, M. (1993). Olfactory receptors are displayed on dog mature sperm cells. *J. Cell Biol.* 123, 1441–1452.
139. Defer, N., Marinx, O., Poyard, M., Lienard, M. O., Jegou, B., and Hanoune, J. (1998). The olfactory adenylyl cyclase type 3 is expressed in male germ cells. *FEBS Lett.* 424, 216–220.
140. Cummins, J. M., and Yanagimachi, R. (1986). Development of ability to penetrate the cumulus oophorus by hamster spermatozoa capacitated in vitro, in relation to the timing of the acrosome reaction. *Gamete Res.* 15, 187–212.
141. Florman, H. M., and First, N. L. (1988). The regulation of acrosomal exocytosis. I. Sperm capacitation is required for the induction of acrosome reactions by the bovine zona pellucida in vitro. *Dev. Biol.* 128, 453–463.
142. Ganitkevich, V. Y., and Isenberg, G. (1993). Membrane potential modulates inositol 1, 4, 5-trisphosphate-mediated Ca^{2+} transients in guinea pig coronary myocytes. *J. Physiol.* 470, 35–44.
143. Meizel, S., and Deamer, D. W. (1978). The pH of the hamster sperm acrosome. *J. Histochem. Cytochem.* 26, 98–105.
144. Eddy, E. M., and O'Brien, D. A. (1994). The spermatozoon. In *The Physiology of Reproduction* (E. Knobil and J. Neill, Eds.), pp. 29–77. Raven Press, New York.
145. Honda, A., Siruntawineti, J., and Baba, T. (2002). Role of acrosomal matrix proteases in sperm-zona pellucida interactions. *Hum. Reprod. Update* 8, 405–412.
146. Foster, J. A., Friday, B. B., Maulit, M. T., Blobel, C., Winfrey, V. P., Olson, G. E., Kim, K. S., and Gerton, G. L. (1997). AM67, a secretory component of the guinea pig sperm acrosomal matrix, is related to mouse sperm protein sp56 and the complement component 4-binding proteins. *J. Biol. Chem.* 272, 12714–12722.
147. Kim, K. S., Foster, J. A., and Gerton, G. L. (2001). Differential release of guinea pig sperm acrosomal components during exocytosis. *Biol. Reprod.* 64, 148–156.
148. Burgess, T. L., and Kelly, R. B. (1987). Constitutive and regulated secretion of proteins. *Annu. Rev. Cell Biol.* 3, 243–293.
149. Stevens, C. F. (1993). Quantal release of neurotransmitter and long-term potentiation. *Cell* 72 Suppl, 55–63.
150. Wassarman, P. M., Jovine, L., and Litscher, E. S. (2001). A profile of fertilization in mammals. *Nat. Cell Biol.* 3, E59–E64.
151. Evans, J. P., and Florman, H. M. (2002). The state of the union: the cell biology of fertilization. *Nat. Cell Biol.* 4, Suppl, s57–s63.
152. Allen, C. A., and Green, D. P. L. (1997). The mammalian acrosome reaction: gateway to sperm fusion with the oocyte? *BioEssays* 19, 241–247.
153. Bedford, J. M. (1972). An electron microscopic study of sperm penetration into the rabbit egg after natural mating. *Am. J. Anat.* 133, 213–254.

154. Bryan, J. H. D. (1974). Capacitation in the mouse: response of murine acrosomes to the environment of the female reproductive tract. *Biol. Reprod.* 10, 414–421.
155. Cummins, J. M., and Yanagimachi, R. (1982). Sperm–egg ratios and the site of the acrosome reaction during in vivo fertilization in the hamster. *Gamete Res.* 5, 239–256.
156. Overstreet, J. W., and Cooper, G. W. (1979). The time and location of the acrosome reaction during sperm transport in the female rabbit. *J. Exp. Zool.* 209, 97–104.
157. Suarez, S. S., Katz, D. F., and Overstreet, J. W. (1983). Movement characteristics and acrosomal status of rabbit spermatozoa recovered at the site and time of fertilization. *Biol. Reprod.* 29, 1277–1287.
158. Crozet, N. (1984). Ultrastructural aspects of in vivo fertilization in the cow. *Gamete Res.* 10, 241–251.
159. Crozet, N., and Dumont, M. (1984). The site of the acrosome reaction during in vivo penetration of the sheep oocyte. *Gamete Res.* 10, 97–105.
160. Valtorta, F., Meldolesi, J., and Fesce, R. (2001). Synaptic vesicles: is kissing a matter of competence? *Trends Cell Biol.* 11, 324–328.
161. Wrightman, R. M., and Haynes, C. L. (2004). Synaptic vesicles really do kiss and run. *Nat. Neurosci.* 7, 321–322.
162. Kim, K. S., and Gerton, G. L. (2003). Differential release of soluble and matrix components: evidence for intermediate states of secretion during spontaneous acrosomal exocytosis in mouse sperm. *Dev. Biol.* 264, 141–152.
163. Talbot, P., Shur, B. D., and Myles, D. G. (2003). Cell adhesion and fertilization: steps in oocyte transport, sperm-zona pellucida interactions, and sperm–egg fusion. *Biol. Reprod.* 68, 1–9.
164. Mahi-Brown, C. A., and Yanagimachi, R. (1983). Parameters influencing ovum pickup by oviductal fimbria in the golden hamster. *Gamete Res.* 8, 1–10.
165. Toole, B. P. (2004). Hyaluronan: from extracellular glue to pericellular cue 1. *Nat. Rev. Cancer* 4, 528–539.
166. Weigel, P. H., Hascall, V. C., and Tammi, M. (1997). Hyaluronan synthases 39. *J. Biol. Chem.* 272, 13997–14000.
167. Rodgers, R. J., Irving-Rodgers, H. F., and Russell, D. L. (2003). Extracellular matrix of the developing ovarian follicle 4. *Reproduction* 126, 415–424.
168. Familiari, G., Verlengia, C., Nottola, S. A., Renda, T., Micara, G., Aragona, C., Zardi, L., and Motta, P. M. (1996). Heterogeneous distribution of fibronectin, tenascin-C, and laminin immunoreactive material in the cumulus-corona cells surrounding mature human oocytes from IVF-ET protocols—evidence that they are composed of different subpopulations: an immunohistochemical study using scanning confocal laser and fluorescence microscopy. *Mol. Reprod. Dev.* 43, 392–402.
169. Ball, G. D., Wieben, E. D., and Byers, A. P. (1985). DNA, RNA, and protein synthesis by porcine oocyte-cumulus complexes during expansion. *Biol. Reprod.* 33, 739–744.
170. Odum, L., Andersen, C. Y., and Jessen, T. E. (2002). Characterization of the coupling activity for the binding of inter-alpha-trypsin inhibitor to hyaluronan in human and bovine follicular fluid. *Reproduction* 124, 249–257.
171. Zhuo, L., Yoneda, M., Zhao, M., Yingsung, W., Yoshida, N., Kitagawa, Y., Kawamura, K., Suzuki, T., and Kimata, K. (2001). Defect in SHAP-hyaluronan complex causes severe female infertility. A study by inactivation of the bikunin gene in mice. *J. Biol. Chem.* 276, 7693–7696.
172. Zhuo, L., Salustri, A., and Kimata, K. (2002). A physiological function of serum proteoglycan bikunin: the chondroitin sulfate moiety plays a central role. *Glycoconj. J.* 19, 241–247.
173. Hess, K. A., Chen, L., and Larsen, W. J. (1999). Inter-alpha-inhibitor binding to hyaluronan in the cumulus extracellular matrix is required for optimal ovulation and development of mouse oocytes. *Biol. Reprod.* 61, 436–443.
174. Campbell, S., Swann, H. R., Aplin, J. D., Seif, M. W., Kimber, S. J., and Elstein, M. (1995). CD44 is expressed throughout pre-implantation human embryo development. *Hum. Reprod.* 10, 425–430.
175. Yokoo, M., Miyahayashi, Y., Naganuma, T., Kimura, N., Sasada, H., and Sato, E. (2002). Identification of hyaluronic acid-binding proteins and their expressions in porcine cumulus-oocyte complexes during in vitro maturation. *Biol. Reprod.* 67, 1165–1171.
176. Sutovsky, P., Flechon, J. E., and Pavlok, A. (1995). F-actin is involved in control of bovine cumulus expansion. *Mol. Reprod. Dev.* 41, 521–529.
177. Ponta, H., Sherman, L., and Herrlich, P. A. (2003). CD44: from adhesion molecules to signalling regulators. *Nat. Rev. Mol. Cell Biol.* 4, 33–45.
178. Boudreau, N. J., and Jones, P. L. (1999). Extracellular matrix and integrin signalling: the shape of things to come. *Biochem. J.* 339, Pt 3, 481–488.
179. Chen, L., Russell, P. T., and Larsen, W. J. (1993). Functional significance of cumulus expansion in the mouse: roles for the preovulatory synthesis of hyaluronic acid within the cumulus mass. *Mol. Reprod. Dev.* 34, 87–93.
180. Testart, J., Lassalle, B., Frydman, R., and Belaisch, J. C. (1983). A study of factors affecting the success of human fertilization in vitro. II. Influence of semen quality and oocyte maturity on fertilization and cleavage. *Biol. Reprod.* 28, 425–431.
181. Austin, C. R. (1960). Capacitation and the release of hyaluronidase from spermatozoa. *J. Reprod. Fertil.* 3, 310–311.
182. Suarez, S. S., Katz, D. F., and Meizel, S. (1984). Changes in motility that accompany the acrosome reaction in hyperactivated hamster spermatozoa. *Gamete Res.* 10, 253–265.
183. Cherr, G. N., Lambert, H., Meizel, S., and Katz, D. F. (1986). In vitro studies of the golden sperm acrosome reaction: Completion on the zona pellucida and induction by homologous soluble zonae pellucidae. *Dev. Biol.* 114, 119–131.
184. Corselli, J., and Talbot, P. (1987). *In vitro* penetration of hamster oocyte-cumulus complexes using physiological numbers of sperm. *Dev. Biol.* 122, 227–242.
185. Bradley, M. P., and Garbers, D. L. (1983). The stimulation of bovine caudal epididymal sperm forward motility by bovine cumulus-egg complexes *in vitro*. *Biochem. Biophys. Res. Commun.* 115, 777–787.
186. Fetterolf, P. M., Sutherland, C. S., Josephy, P. D., Casper, R. F., and Tyson, J. E. (1994). Preliminary characterization of a factor in human follicular fluid that stimulates human spermatozoa motion. *Hum. Reprod.* 9, 1505–1511.
187. Westphal, L. M., el, D., I, Shimizu, S., Tadir, Y., and Berns, M. W. (1993). Exposure of human spermatozoa to the cumulus oophorus results in increased relative force as measured by a 760 nm laser optical trap. *Hum. Reprod.* 8, 1083–1086.
188. Osman, R. A., Andria, M. L., Jones, A. D., and Meizel, S. (1989). Steroid induced exocytosis: the human sperm acrosome reaction. *Biochem. Biophys. Res. Commun.* 160, 828–833.
189. Yamashita, Y., Shimada, M., Okazaki, T., Maeda, T., and Terada, T. (2003). Production of progesterone from de novo-synthesized cholesterol in cumulus cells and its physiological role during meiotic resumption of porcine oocytes. *Biol. Reprod.* 68, 1193–1198.
190. Thomas, P., and Meizel, S. (1989). Phosphatidylinositol 4, 5-bisphosphate hydrolysis in human sperm stimulated with follicular fluid or progesterone is dependent upon Ca^{2+} influx. *Biochem. J.* 264, 539–546.

191. Blackmore, P. F., Neulen, J., Lattanzio, F., and Beebe, S. J. (1991). Cell surface-binding sites for progesterone mediate calcium uptake in human sperm. *J. Biol. Chem.* 266, 18655–18659.
192. Meizel, S., and Turner, K. O. (1991). Progesterone acts at the plasma membrane of human sperm. *Mol. Cell. Endocrinol.* 11, R1–R5.
193. Blackmore, P. F., Fisher, J. F., Spilman, C. H., Im, W. B., and Bleasdale, J. E. (1995). Effects of steroids on calcium fluxes in sperm. In *Human Acrosome Reaction* (P. Fenichel and J. Parinaud, Eds.), pp. 165–177. Colloque INSERM, John Libbey Eurotext, Ltd., Montrouge, France.
194. Blackmore, P. F. (1993). Rapid non-genomic actions of progesterone stimulate Ca2+ influx and the acrosome reaction in human sperm. *Cell Signal* 5, 531–538.
195. Baldi, E., Luconi, M., Bonaccorsi, L., Muratori, M., and Forti, G. (2000). Intracellular events and signaling pathways involved in sperm acquisition of fertilizing capacity and acrosome reaction. *Front. Biosci.* 5, E110–E123.
196. Jaiswal, B. S., Tur-Kaspa, I., Dor, J., Mashiach, S., and Eisenbach, M. (1999). Human sperm chemotaxis: is progesterone a chemoattractant? *Biol. Reprod.* 60, 1314–1319.
197. Roldan, E. R. S., Murase, T., and Shi, Q.-X. (1994). Exocytosis in spermatozoa in response to progesterone and zona pellucida. *Science* 266, 1578–1581.
198. Viggiano, J. M., Herrero, M. B., Cebral, E., Boquet, M. G., and de Gimeno, M. F. (1995). Prostaglandin synthesis by cumulus–oocyte complexes: effects on in vitro fertilization in mice. *Prostaglandins Leukot. Essent. Fatty Acids* 53, 261–265.
199. Schaefer, M., Hofmann, T., Schultz, G., and Gudermann, T. (1998). A new prostaglandin E receptor mediates calcium influx and acrosome reaction in human spermatozoa. *Proc. Natl. Acad. Sci. U S A* 95, 3008–3013.
200. Cherr, G. N., Yudin, A. I., Li, M. W., Vines, C. A., and Overstreet, J. W. (1999). Hyaluronic acid and the cumulus extracellular matrix induce increases in intracellular calcium in macaque sperm via the plasma membrane protein PH-20. *Zygote* 7, 211–222.
201. Ranganathan, S., Bharadwaj, A., and Datta, K. (1995). Hyaluronan mediates sperm motility by enhancing phosphorylation of proteins including hyaluronan binding protein. *Cell Mol. Biol. Res.* 41, 467–476.
202. Kornovski, B. S., McCoshen, J., Kredentser, J., and Turley, E. (1994). The regulation of sperm motility by a novel hyaluronan receptor. *Fertil. Steril.* 61, 935–940.
203. Vandevoort, C. A., Cherr, G. N., and Overstreet, J. W. (1997). Hyaluronic acid enhances the zona pellucida-induced acrosome reaction of macaque sperm. *J. Androl.* 18, 1–5.
204. Sabeur, K., Cherr, G. N., Yudin, A. I., and Overstreet, J. W. (1998). Hyaluronic acid enhances induction of the acrosome reaction of human sperm through interaction with the PH-20 protein. *Zygote* 6, 103–111.
205. Pincus, G., and Enzmann, E. (1936). The comparative behavior of mammalian eggs in vivo and in vitro. II. The activation of tubal eggs in the rabbit. *J. Exp. Zool.* 73, 195–206.
206. Russo, M. J., Bayley, H., and Toner, M. (1997). Reversible permeabilization of plasma membranes with an engineered switchable pore. *Nat. Biotechnol.* 15, 278–282.
207. Talbot, P. (1985). Sperm penetration through oocyte investments in mammals. *Am. J. Anat.* 174, 331–346.
208. Kurzrok, R., Leonard, S. L., and Conrad, H. (1946). Role of hyaluronidase in human infertility. *Amer. J. Med.* 1, 491–506.
209. McClean, D., and Rowlands, I. W. (1942). The role of hyaluronidase in fertilization. *Nature* 150, 627–628.
210. Werthessen, N., Bergman, S., Greenberg, B., and Gargill, S. (1945). A technique for the assay of hyaluronidase in human semen and its correlation with sperm concentration. *J. Urol.* 22, 540–565.
211. Austin, C. R., and Braden, A. W. H. (1952). Passage of the sperm and the penetration of the egg in mammals. *Nature* 170, 919–921.
212. Mancini, R. E., Alanso, A., Barquet, J., and Nerimovsky, N. R. (1964). Histo-immunological localization of hyaluronidase in bull testis. *J. Reprod. Fertil.* 8, 325–330.
213. Morton, D. B. (1975). Acrosomal enzymes: Immunochemical localization of acrosin and hyaluronidase in ram spermatozoa. *J. Reprod. Fertil.* 45, 375–378.
214. Yanagimachi, R., and Noda, Y. (1972). Ultrastructural changes in the hamster sperm head during fertilization. *J. Ultra. Res.* 31, 465–485.
215. Austin, C. R. (1965). *Fertilization*. Prentice Hall, Englewood Cliffs, NJ.
216. Metz, C. B., Seiguer, A. C., and Castro, A. E. (1972). Inhibition of the cumulus dispersing and hyaluronidase activities of sperm by heterologous and isologous antisperm antibodies. *Proc. Soc. Exp. Biol. Med.* 140, 776–781.
217. O'Rand, M. G., and Metz, C. B. (1974). Tests for rabbit sperm surface iron-binding protein and hyaluronidase using the "exchange agglutination" reaction. *Biol. Reprod.* 11, 326–334.
218. Myles, D. G., and Primakoff, P. (1984). Localized surface antigens of guinea pig sperm migrate to new regions prior to fertilization. *J. Cell Biol.* 99, 1634–1641.
219. Phelps, B. M., Primakoff, P., Koppel, D. E., Low, M. G., and Myles, D. G. (1988). Restricted lateral diffusion of PH-20, a PI-anchored sperm membrane protein. *Science* 240, 1780–1782.
220. Gmachl, M., and Kreil, G. (1993). Bee venom hyaluronidase is homologous to a membrane protein of mammalian sperm. *Proc. Natl. Acad. Sci. U S A* 90, 3569–3573.
221. Gmachl, M., Sagan, S., Ketter, S., and Kreil, G. (1993). The human sperm protein PH-20 has hyaluronidase activity. *FEBS Lett.* 336, 545–548.
222. Myles, D. G., and Primakoff, P. (1997). Why did the sperm cross the cumulus—to get to the oocyte: functions of the sperm surface proteins PH-20 and fertilin in arriving at and fusing with the egg. *Biol. Reprod.* 56, 320–327.
223. Lin, Y., Mahan, K., Lathrop, W. F., Myles, D. G., and Primakoff, P. (1994). A hyaluronidase activity of the sperm plasma membrane protein PH-20 enables sperm to penetrate the cumulus cell layer surrounding the egg. *J. Cell Biol.* 125, 1157–1163.
224. Bleil, J. D., and Wassarman, P. M. (1980). Synthesis of zona pellucida proteins by denuded and follicle-enclosed mouse oocytes during culture in vitro. *Proc. Natl. Acad. Sci. U S A* 77, 1029–1033.
225. Bleil, J. D., and Wassarman, P. M. (1980). Structure and function of the zona pellucida: identification and characterization of the proteins of the mouse oocyte's zona pellucida. *Dev. Biol.* 76, 185–202.
226. Wassarman, P. M. (1988). Zona pellucida glycoproteins. *Annu. Rev. Biochem.* 57, 415–442.
227. Lefievre, L., Conner, S. J., Salpekar, A., Olufowobi, O., Ashton, P., Pavlovic, B., Lenton, W., Afnan, M., Brewis, I. A., Monk, M., Hughes, D. C., and Barratt, C. L. (2004). Four zona pellucida glycoproteins are expressed in the human. *Hum. Reprod.* 19, 1580–1586.
228. Conner, S. J., Lefievre, L., Hughes, D. C., and Barratt, C. L. (2005). Cracking the egg: increased complexity in the zona pellucida. *Hum. Reprod.* 20, 1148–1152.
229. Wassarman, P. M. (1999). Mammalian fertilization: molecular aspects of gamete adhesion, exocytosis, and fusion. *Cell* 96, 175–183.
230. Rankin, T., Soyal, S., and Dean, J. (2000). The mouse zona pellucida: folliculogenesis, fertility and pre-implantation development. *Mol. Cell Endocrinol.* 163, 21–25.

231. Castle, P. E., and Dean, J. (1996). Molecular genetics of the zona pellucida: implications for immunocontraceptive strategies. *J. Reprod. Fertil. Suppl.* 50, 1–8.
232. Harris, J. D., Hibler, D. W., Fontenot, G. K., Hsu, K. T., Yurewicz, E. C., and Sacco, A. G. (1994). Cloning and characterization of zona pellucida genes and cDNAs from a variety of mammalian species: the ZPA, ZPB and ZPC gene families. *DNA Seq.* 4, 361–393.
233. Hughes, D. C., and Barratt, C. L. (1999). Identification of the true human orthologue of the mouse Zp1 gene: evidence for greater complexity in the mammalian zona pellucida? *Biochim. Biophys. Acta* 1447, 303–306.
234. Connor, S. J., and Hughes, D. C. (2003). Analysis of fish *ZP1/ZPB* homologous genes – evidence for both genome duplication and species-specific amplification models of evolution. *Reproduction* 126, 347–352.
235. Greve, J. M., Salzmann, G. S., Roller, R. J., and Wassarman, P. M. (1982). Biosynthesis of the major zona pellucida glycoprotein secreted by oocytes during mammalian oogenesis. *Cell* 31, 749–759.
236. Salzmann, G. S., Greve, J. M., Roller, R. J., and Wassarman, P. M. (1983). Biosynthesis of the sperm receptor during oogenesis in the mouse. *EMBO J.* 2, 1451–1456.
237. Florman, H. M., and Wassarman, P. M. (1985). O-linked oligosaccharides of mouse egg ZP3 account for its sperm receptor activity. *Cell* 41, 313–324.
238. Yurewicz, E. C., Pack, B. A., and Sacco, A. G. (1992). Porcine oocyte zona pellucida M_r 55, 000 glycoproteins: identification of O-glcosylated domains. *Mol. Reprod. Dev.* 33, 182–188.
239. Hoodbhoy, T., Joshi, S., Boja, E. S., Williams, S. A., Stanley, P., and Dean, J. (2005). Human sperm do not bind to rat zonae pellucidae despite the presence of four homologous glycoproteins. *J. Biol. Chem.* 280, 12721–12731.
240. Litscher, E. S., Qi, H., and Wassarman, P. M. (1999). Mouse zona pellucida glycoproteins mZP2 and mZP3 undergo carboxy-terminal proteolytic processing in growing oocytes. *Biochemistry* 38, 12280–12287.
241. Qi, H., Williams, Z., and Wassarman, P. M. (2002). Secretion and assembly of zona pellucida glycoproteins by growing mouse oocytes microinjected with epitope-tagged cDNAs for mZP2 and mZP3. *Mol. Biol. Cell* 13, 530–541.
242. Williams, Z., and Wassarman, P. M. (2001). Secretion of mouse ZP3, the sperm receptor, requires cleavage of its polypeptide at a consensus furin cleavage-site. *Biochemistry* 40, 929–937.
243. Dangott, L. J., and Garbers, D. L. (1984). Identification and partial characterization of the receptor for speract. *J. Biol. Chem.* 259, 13712–13716.
244. Bork, P., and Sander, C. (1992). A large domain common to sperm receptors (Zp2 and Zp3) and TGF-b type III receptor. *FEBS Lett.* 300, 237–240.
245. Jovine, L., Darie, C. C., Litscher, E. S., and Wassarman, P. M. (2005). Zona pellucida domain proteins. *Ann. Rev. Biochem.* 74, 83–114.
246. Sutton, K. A., Jungnickel, M. K., and Florman, H. M. (2002). If music be the food of love. *Nat. Cell Biol.* 4, E154–E155.
247. Jaffe, R. C., Stevens, D. M., and Verhage, H. G. (1985). The effects of estrogen and progesterone on glycogen and the enzymes involved in its metabolism in the cat uterus. *Steroids* 45, 453–462.
248. Greve, J. M., and Wassarman, P. M. (1985). Mouse egg extracellular coat is a matrix of interconnected filaments possessing a structural repeat. *J. Mol. Biol.* 181, 253–264.
249. Talbot, P. (1984). Hyaluronidase dissolves a component in the hamster zona pellucida. *J. Exp. Zool.* 229, 309–316.
250. Kan, F. W., St-Jacques, S., and Bleau, G. (1989). Immunocytochemical evidence for the transfer of an oviductal antigen to the zona pellucida of hamster ova after ovulation. *Biol. Reprod.* 40, 585–598.
251. Malette, B., Paquette, Y., Merlen, Y., and Bleau, G. (1995). Oviductins possess chitinase- and mucin-like domains: a lead in the search for the biological function of these oviduct-specific ZP-associating glycoproteins. *Mol. Reprod. Dev.* 41, 384–397.
252. Verhage, H. G., Fazleabas, A. T., Mavrogianis, P. A., O'Day-Bowman, M. B., Schmidt, A., Arias, E. B., and Jaffe, R. C. (1997). Characteristics of an oviductal glycoprotein and its potential role in fertility control. *J. Reprod. Fertil. Suppl.* 51, 217–226.
253. Baltz, J. M., and Cardullo, R. A. (1989). The number and rate of formation of sperm-zona bonds in the mouse. *Gamete Res.* 24, 1–8.
254. Cuatrecasas, P., and Hollenberg, M. D. (1976). Membrane receptors and hormone action. *Adv. Prot. Chem.* 30, 251–451.
255. Lauffenburger, D. A. (1991). Models for receptor-mediated cell phenomena: adhesion and migration. *Ann. Rev. Biophys. Biophys. Chem.* 20, 387–414.
256. Saling, P. M., Sowinski, J., and Storey, B. T. (1979). An ultrastructural study of epididymal mouse spermatozoa binding to zonae pellucidae in vitro: sequential relationship to the acrosome reaction. *J. Exp. Zool.* 209, 229–238.
257. Gwatkin, R. B. L. (1976). *Fertilization Mechanisms in Man and Mammals*. Plenum Press, New York.
258. Hartmann, J. F., and Gwatkin, R. B. L. (1971). Alteration of sites on the mammalian sperm surface during capacitation. *Nature* 234, 479–481.
259. Hartmann, J. F., and Hutchison, C. F. (1984). Nature of the prepenetration contact interactions between hamster gametes *in vitro*. *J. Reprod. Fertil.* 36, 49–57.
260. Ducibella, T., Kurasawa, S., Rangarajan, S., Kopf, G. S., and Schultz, R. M. (1990). Precocious loss of cortical granules during mouse oocyte meiotic maturation and correlation with an egg-induced modification of the zona pellucida. *Dev. Biol.* 137, 46–55.
261. Bleil, J. D., and Wassarman, P. M. (1980). Mammalian sperm–egg interaction: identification of a glycoprotein in mouse egg zonae pellucidae possessing receptor activity for sperm. *Cell* 20, 873–882.
262. Hartmann, J. F., Gwatkin, R. B. L., and Hutchison, C. F. (1972). Early contact interactions between mammalian gametes *in vitro*: evidence that the vitellus influences adhesion between sperm and zona pellucida. *Proc. Natl. Acad. Sci. U S A* 69, 2767–2769.
263. Thaler, C. D., and Cardullo, R. A. (1996). The initial molecular interaction between mouse sperm and the zona pellucida is a complex binding event. *J. Biol. Chem.* 271, 23289–23297.
264. Thaler, C. D., and Cardullo, R. A. (1996). Defining oligosaccharide specificity for initial sperm-zona pellucida adhesion in the mouse. *Mol. Reprod. Dev.* 45, 535–546.
265. Ensslin, M. A., and Shur, B. D. (2003). Identification of mouse sperm SED1, a bimotif EGF repeat and discoidin-domain protein involved in sperm–egg binding. *Cell* 114, 405–417.
266. Jungnickel, M. K., Sutton, K. A., and Florman, H. M. (2003). In the beginning: lessons from fertilization in mice and worms. *Cell* 114, 401–404.
267. Gwatkin, R. B. L., and Williams, D. T. (1976). Receptor activity of the solubilized hamster and mouse zona pellucida before and after the zona reaction. *J. Reprod. Fertil.* 49, 55–59.
268. Moller, C. C., Bleil, J. D., Kinloch, R. A., and Wassarman, P. M. (1990). Structural and functional relationships between mouse and hamster zona pellucida glycoproteins. *Dev. Biol.* 137, 276–286.

269. Bleil, J. D., and Wassarman, P. M. (1986). Autoradiographic visualization of the mouse egg's sperm receptor bound to sperm. *J. Cell Biol.* 102, 1363–1371.
270. Mortillo, S., and Wassarman, P. M. (1991). Differential binding of gold-labeled zona pellucida glycoproteins mZP2 and mZP3 to mouse sperm membrane compartments. *Development* 113, 141–149.
271. Yurewicz, E. C., Sacco, A. G., Gupta, S. K., Xu, N., and Gage, D. A. (1998). Hetero-oligomerization-dependent binding of pig oocyte zona pellucida glycoproteins ZPB and ZPC to boar sperm membrane vesicles. *J. Biol. Chem.* 273, 7488–7494.
272. Yonezawa, N., Fukui, N., Kuno, M., Shinoda, M., Goko, S., Mitsui, S., and Nakano, M. (2001). Molecular cloning of bovine zona pellucida glycoproteins ZPA and ZPB and analysis for sperm-binding component of the zona. *Eur. J. Biochem.* 268, 3587–3594.
273. Bagavant, H., Yurewicz, E. C., Sacco, A. G., Talwar, G. P., and Gupta, S. K. (1993). Block in porcine gamete interaction by polyclonal antibodies to a pig ZP3 beta fragment having partial sequence homology to human ZP3. *J. Reprod. Immunol.* 25, 277–283.
274. Rankin, T. L., Coleman, J. S., Epifano, O., Hoodbhoy, T., Turner, S. G., Castle, P. E., Lee, E., Gore-Langton, R., and Dean, J. (2003). Fertility and taxon-specific sperm binding persist after replacement of mouse sperm receptors with human homologs. *Dev. Cell* 5, 33–43.
275. Kudo, K., Yonezawa, N., Katsumata, T., Aoki, H., and Nakano, M. (1998). Localization of carbohydrate chains of pig sperm ligand in the glycoprotein ZPB of egg zona pellucida. *Eur. J. Biochem.* 252, 492–499.
276. Shur, B. D., and Hall, N. G. (1982). A role for mouse sperm surface galactosyltransferase in sperm binding to the egg zona pellucida. *J. Cell Biol.* 95, 574–579.
277. Florman, H. M., Bechtol, K. B., and Wassarman, P. M. (1984). Enzymatic dissection of the functions of the mouse egg's receptor for sperm. *Dev. Biol.* 106, 243–255.
278. Miller, D. J., Shi, X., and Burkin, H. (2002). Molecular basis of mammalian gamete binding. *Recent Prog. Horm. Res.* 57, 37–73.
279. Bleil, J. D., and Wassarman, P. M. (1988). Galactose at the nonreducing terminus of O-linked oligosaccharides of mouse egg zona pellucida glycoprotein ZP3 is essential for the glycoprotein's sperm receptor activity. *Proc. Natl. Acad. Sci. U S A* 85, 6778–6782.
280. Hanna, W. F., Kerr, C. L., Shaper, J. H., and Wright, W. W. (2004). Lewis x-containing neoglycoproteins mimic the intrinsic ability of zona pellucida glycoprotein ZP3 to induce the acrosome reaction in capacitated mouse sperm. *Biol. Reprod.* 71, 778–789.
281. Litscher, E. S., Juntunen, K., Seppo, A., Penttila, L., Niemela, R., Renkonen, O., and Wassarman, P. M. (1995). Oligosaccharide constructs with defined structures that inhibit binding of mouse sperm to unfertilized eggs in vitro. *Biochemistry* 34, 4662–4669.
282. Kerr, C. L., Hanna, W. F., Shaper, J. H., and Wright, W. W. (2004). Lewis x-containing glycans are specific and potent competitive inhibitors of the binding of ZP3 to complementary sites on capacitated, acrosome-intact mouse sperm. *Biol. Reprod.* 71, 770–777.
283. Chen, J., Litscher, E. S., and Wassarman, P. M. (1998). Inactivation of the mouse sperm receptor, mZP3, by site-directed mutagenesis of individual serine residues located at the combining site for sperm. *Proc. Natl. Acad. Sci. U S A* 95, 6193–6197.
284. Kinloch, R. A., Sakai, Y., and Wassarman, P. M. (1995). Mapping the mouse ZP3 combining site for sperm by exon swapping and site-directed mutagenesis. *Proc. Natl. Acad. Sci. U S A* 92, 263–267.
285. Liu, C., Litscher, E. S., and Wassarman, P. M. (1997). Zona pellucida glycoprotein mZP3 bioactivity is not dependent on the extent of glycosylation of its polypeptide or on sulfation and sialylation of its oligosaccharides. *J. Cell Sci.* 110, 745–752.
286. Boja, E. S., Hoodbhoy, T., Fales, H. M., and Dean, J. (2003). Structural characterization of native mouse zona pellucida proteins using mass spectrometry. *J. Biol. Chem.* 278, 34189–34202.
287. Dell, A., Morris, H. R., Easton, R. L., Patankar, M., and Clark, G. F. (1999). The glycobiology of gametes and fertilization. *Biochim. Biophys. Acta* 1473, 196–205.
288. Dell, A., Chalabi, S., Easton, R. L., Haslam, S. M., Sutton-Smith, M., Patankar, M. S., Lattanzio, F., Panico, M., Morris, H. R., and Clark, G. F. (2003). Murine and human zona pellucida 3 derived from mouse eggs express identical O-glycans. *Proc. Natl. Acad. Sci. U S A* 100, 15631–15636.
289. Easton, R. L., Patankar, M. S., Lattanzio, F. A., Leaven, T. H., Morris, H. R., Clark, G. F., and Dell, A. (2000). Structural analysis of murine zona pellucida glycans. Evidence for the expression of core 2-type O-glycans and the Sd(a) antigen. *J. Biol. Chem.* 275, 7731–7742.
290. Hirano, T., Takasaki, S., Hedrick, J. L., Wardrip, N. J., Amano, J., and Kobata, A. (1993). *O*-Linked neutral sugar chains of porcine *zona pellucida* glycoproteins. *Eur. J. Biochem.* 214, 763–769.
291. Mori, E., Takasaki, S., Hedrick, J. L., Wardrip, N. J., Mori, T., and Kobata, A. (1991). Neutral oligosaccharide structures linked to asparagines of porcine zona pellucida glycoproteins. *Biochemistry* 30, 2078–2087.
292. Takasaki, S., Mori, E., Hirano, T., Furukawa, K., Amano, J., Mori, T., Hedrick, J. L., Wardrip, N. J., and Kobata, A. (1993). Structures of sugar chains included in porcine zona pellucida glycoproteins. *J. Reprod. Dev.* 39, 39–40.
293. Bleil, J. D., and Wassarman, P. M. (1990). Identification of a ZP3-binding protein on acrosome-intact mouse sperm by photoaffinity crosslinking. *Proc. Natl. Acad. Sci. U S A* 87, 5563–5567.
294. Bookbinder, L. H., Cheng, A., and Bleil, J. D. (1995). Tissue- and species-specific expression of sp56, a mouse sperm fertilization protein. *Science* 269, 86–89.
295. Hardy, D. M., and Garbers, D. L. (1995). A sperm membrane protein that binds in a species-specific manner to the egg extracellular matrix is homologous to von Willebrand factor. *J. Biol. Chem.* 270, 26025–26028.
296. Leyton, L., and Saling, P. (1989). 95 kd sperm proteins bind ZP3 and serve as tyrosine kinase substrates in response to zona binding. *Cell* 57, 1123–1130.
297. Burks, D. J., Carballada, R., Moore, H. D., and Saling, P. M. (1995). Interaction of a tyrosine kinase from human sperm with the zona pellucida at fertilization. *Science* 269, 83–86.
298. Miller, D. J., Macek, M. B., and Shur, B. D. (1992). Complementarity between sperm surface b-1, 4-galactosyltransferase and egg-coat ZP3 mediates sperm–egg binding. *Nature* 357, 589–593.
299. Primakoff, P., Hyatt, H., and Myles, D. G. (1985). A role for the migrating sperm surface antigen PH-20 in guinea pig sperm binding to the egg zona pellucida. *J. Cell Biol.* 101, 2239–2244.
300. Hunnicutt, G. R., Primakoff, P., and Myles, D. G. (1996). Sperm surface protein PH-20 is bifunctional: one activity is a hyaluronidase and a second, distinct activity is required in secondary sperm-zona binding. *Biol. Reprod.* 55, 80–86.
301. Saling, P. M., and Storey, B. T. (1979). Mouse gamete interactions during fertilization in vitro. Chlortetracycline as a fluorescent probe for the mouse sperm acrosome reaction. *J. Cell Biol.* 83, 544–555.
302. Bleil, J. D., and Wassarman, P. M. (1983). Sperm–egg interactions in the mouse: sequence of events and induction of

the acrosome reaction by a zona pellucida glycoprotein. *Dev. Biol.* 95, 317–324.

303. Berger, T., Turner, K. O., Meizel, S., and Hedrick, J. L. (1989). Zona pellucida-induced acrosome reaction in boar sperm. *Biol. Reprod.* 40, 525–530.
304. Leyton, L., and Saling, P. (1989). Evidence that aggregation of mouse sperm receptors by ZP3 triggers the acrosome reaction. *J. Cell Biol.* 108, 2163–2168.
305. Macek, M. B., Lopez, L. C., and Shur, B. D. (1991). Aggregation of b-1, 4-galactosyltransferase on mouse sperm induces the acrosome reaction. *Dev. Biol.* 147, 440–444.
306. Endo, Y., Mattei, P., Kopf, G. S., and Schultz, R. M. (1987). Effects of a phorbol ester on mouse eggs: dissociation of sperm receptor activity from acrosome reaction-inducing activity of the mouse zona pellucida protein, ZP3. *Dev. Biol.* 123, 574–577.
307. Breitbart, H. (2003). Signaling pathways in sperm capacitation and acrosome reaction. *Cell Mol. Biol. (Noisy-le-grand)* 49, 321–327.
308. Roldan, E. R. S. (1999). Signaling for exocytosis: lipid second messengers, phosphorylation cascades, and cross-talk. In *The Male Gamete: From Basic Science to Clinical Applications* (C. Gagnon, Ed.), pp. 127–138. Cache River Press, Vienna, IL.
309. Ward, C. R., Storey, B. T., and Kopf, G. S. (1994). Selective activation of Gi1 and Gi2 in mouse sperm by the zona pellucida, the egg's extracellular matrix. *J. Biol. Chem.* 269, 13254–13258.
310. Endo, Y., Lee, M. A., and Kopf, G. S. (1988). Characterization of an islet-activating protein sensitive site in mouse sperm that is involved in the zona pellucida-induced acrosome reaction. *Dev. Biol.* 129, 12–24.
311. Endo, Y., Lee, M. A., and Kopf, G. S. (1987). Evidence for a role of a guanine nucleotide-binding regulatory protein in the zona pellucida-induced mouse sperm acrosome reaction. *Dev. Biol.* 119, 210–216.
312. Florman, H. M., Tombes, R. M., First, N. L., and Babcock, D. F. (1989). An adhesion-associated agonist from the zona pellucida activates G protein-promoted elevations of internal Ca and pH that mediate mammalian sperm acrosomal exocytosis. *Dev. Biol.* 135, 133–146.
313. Lee, M. A., Check, J. H., and Kopf, G. S. (1992). A guanine nucleotide-binding regulatory protein in human sperm mediates acrosomal exocytosis induced by the human zona pellucida. *Mol. Reprod. Dev.* 31, 78–86.
314. Arnoult, C., Zeng, Y., and Florman, H. M. (1996). ZP3-dependent activation of sperm cation channels regulates acrosomal secretion during mammalian fertilization. *J. Cell Biol.* 134, 637–645.
315. Lopez-Gonzalez, I., de la Vega-Beltrán, J. L., Santi, C. M., Florman, H. M., Felix, R., and Darszon, A. (2001). Calmodulin antagonists inhibit T-type Ca^{2+} currents in mouse spermatogenic cells and the zona pellucida-induced sperm acrosome reaction. *Dev. Biol.* 236, 210–219.
316. Walensky, L. D., and Snyder, S. H. (1995). Inositol 1, 4, 5-trisphosphate receptors selectively localized to the acrosomes of mammalian sperm. *J. Cell Biol.* 130, 857–869.
317. Baxendale, R. W., and Fraser, L. R. (2003). Immunolocalization of multiple Galpha subunits in mammalian spermatozoa and additional evidence for Galphas. *Mol. Reprod. Dev.* 65, 104–113.
318. Meldolesi, J. (1998). Calcium signaling. Oscillation, activation, expression. *Nature* 392, 863–866.
319. Babcock, D. F., and Pfeiffer, D. R. (1987). Independent elevation of cytosolic (Ca^{2+}) and pH of mammalian sperm by voltage-dependent and pH-sensitive mechanisms. *J. Biol. Chem.* 262, 15041–15047.
320. Zeng, Y., Oberdorf, J. A., and Florman, H. M. (1996). pH regulation in sperm. The role of separate Na^+, Cl^-, and HCO_3^--dependent and arylaminobenzoate-dependent mechanisms in the control of internal pH in mouse sperm. *Dev. Biol.* 173, 510–520.
321. Demarco, I. A., Espinosa, F., Edwards, J., Sosnik, J., De la Vega-Beltran, J. L., Hockensmith, J. W., Kopf, G. S., Darszon, A., and Visconti, P. E. (2003). Involvement of a Na^+/HCO_3^- cotransporter in mouse sperm capacitation. *J. Biol. Chem.* 278, 7001–7009.
322. Garcia, M. A., and Meizel, S. (1999). Regulation of intracellular pH in capacitated human spermatozoa by a Na+/H+ exchanger. *Mol. Reprod. Dev.* 52, 189–195.
323. Wang, D., King, S. M., Quill, T. A., Doolittle, L. K., and Garbers, D. L. (2003). A new sperm-specific Na(+)/H(+) Exchanger required for sperm motility and fertility. *Nat. Cell Biol.* 5, 1117–1122.
324. Roldan, E. R., Murase, T., and Shi, Q. X. (1994). Exocytosis in spermatozoa in response to progesterone and zona pellucida. *Science* 266, 1578–1581.
325. Lee, S. B., and Rhee, S. G. (1996). Molecular cloning, splice variants, expression, and purification of phospholipase C-delta 4. *J. Biol. Chem.* 271, 25–31.
326. Fukami, K., Nakao, K., Inoue, T., Kataoka, Y., Kurokawa, M., Fissore, R. A., Nakamura, K., Motoya, K., Mikoshiba, K., Yoshida, N., and Takenawa, T. (2001). Requirement of phospholipase Cd4 for the zona pellucida-induced acrosome reaction. *Science* 292, 920–923.
327. Fukami, K., Yoshida, M., Inoue, T., Kurokawa, M., Fissore, R. A., Yoshida, N., Mikoshiba, K., and Takenawa, T. (2003). Phospholipase Cd4 is required for Ca2+ mobilization essential for acrosome reaction in sperm. *J. Cell Biol.* 161, 79–88.
328. Irino, Y., Cho, H., Nakamura, Y., Nakahara, M., Furutani, M., Suh, P. G., Takenawa, T., and Fukami, K. (2004). Phospholipase C delta-type consists of three isozymes: bovine PLCdelta2 is a homologue of human/mouse PLCdelta4. *Biochem. Biophys. Res. Commun.* 320, 537–543.
329. Montell, C. (2005). The TRP superfamily of cation channels. *Sci. STKE.* 2005, re3.
330. Rhee, S. G. (2001). The regulation of phosphoinositide-specific phospholipase C. *Ann. Rev. Biochem.* 70, 281–312.
331. Ananthanarayanan, B., Das, S., Rhee, S. G., Murray, D., and Cho, W. (2002). Membrane targeting of C2 domains of phospholipase C-delta isoforms. *J. Biol. Chem.* 277, 3568–3575.
332. Murase, T., and Roldan, E. R. S. (1996). Progesterone and the zona pellucida activate different transducing pathways in the sequence of events leading to diacylglycerol generation during mouse sperm acrosomal exocytosis. *Biochem. J.* 320, 1017–1023.
333. Murthy, K. S., Zhou, H., Huang, J., and Pentyala, S. N. (2004). Activation of PLC-delta1 by Gi/o-coupled receptor agonists. *Am. J. Physiol. Cell Physiol.* 287, C1679–C1687.
334. Brose, N., Betz, A., and Wegmeyer, H. (2004). Divergent and convergent signaling by the diacylglycerol second messenger pathway in mammals. *Curr. Opin. Neurobiol.* 14, 328–340.
335. Lax, Y., Rubinstein, S., and Breitbart, H. (1997). Subcellular distribution of protein kinase C alpha and betaI in bovine spermatozoa, and their regulation by calcium and phorbol esters. *Biol. Reprod.* 56, 454–459.
336. Breitbart, H., Lax, J., Rotem, R., and Naor, Z. (1992). Role of protein kinase C in the acrosome reaction of mammalian spermatozoa. *Biochem. J.* 281, 473–476.
337. Lee, M. A., Kopf, G. S., and Storey, B. T. (1987). Effects of phorbol esters and a diacylglycerol on the mouse sperm acrosome reaction induced by the zona pellucida. *Biol. Reprod.* 36, 617–627.
338. Liu, D. Y., and Baker, H. W. (1997). Protein kinase C plays an important role in the human zona pellucida-induced acrosome reaction. *Mol. Hum. Reprod.* 3, 1037–1043.

339. Yuan, Y. Y., Chen, W. Y., Shi, Q. X., Mao, L. Z., Yu, S. Q., Fang, X., and Roldan, E. R. (2003). Zona pellucida induces activation of phospholipase A2 during acrosomal exocytosis in guinea pig spermatozoa. *Biol. Reprod.* 68, 904–913.
340. Berridge, M. J. (1993). Inositol trisphosphate and calcium signalling. *Nature* 361, 315–325.
341. Patterson, R. L., Boehning, D., and Snyder, S. H. (2004). Inositol 1, 4, 5-trisphosphate receptors as signal integrators. *Annu. Rev. Biochem.* 73, 437–465.
342. Arnoult, C., Cardullo, R. A., Lemos, J. R., and Florman, H. M. (1996). Activation of mouse sperm T-type Ca^{2+} channels by adhesion to the egg zona pellucida. *Proc. Natl. Acad. Sci.* 93, 13004–13009.
343. Jagannathan, S., Punt, E. L., Gu, Y., Arnoult, C., Sakkas, D., Barratt, C. L., and Publicover, S. J. (2002). Identification and localization of T-type voltage-operated calcium channel subunits in human male germ cells. *J. Biol. Chem.* 277, 8449–8456.
344. Lievano, A., Santi, C. M., Serrano, J., Trevino, C. L., Bellve, A. R., Hernandez-Cruz, A., and Darszon, A. (1996). T-type Ca^{2+} channels and a_{1E} expression in spermatogenic cells, and their possible relevance to the sperm acrosome reaction. *FEBS Lett.* 388, 150–154.
345. Santi, C. M., Darszon, A., and Hernandez-Cruz, A. (1996). A dihydropyridine-sensitive T-type Ca^{2+} current is the main Ca^{2+} current carrier in mouse primary spermatocytes. *Amer. J. Physiol.* 271, C1583–C1593.
346. Son, W.-Y., Lee, J.-H., Lee, J. H., and Han, C.-T. (2000). Acrosome reaction of human spermatozoa is mainly mediated by a1H T-type calcium channels. *Mol. Hum. Reprod.* 6, 893–897.
347. Stamboulian, S., Kim, D., Shin, H. S., Ronjat, M., De Waard, M., and Arnoult, C. (2004). Biophysical and pharmacological characterization of spermatogenic T-type calcium current in mice lacking the CaV3. 1 (alpha1G) calcium channel: CaV3. 2 (alpha1H) is the main functional calcium channel in wild-type spermatogenic cells. *J. Cell Physiol.* 200, 116–124.
348. Serrano, C. J., Trevino, C. L., Felix, R., and Darszon, A. (1999). Voltage-dependent Ca(2+) channel subunit expression and immunolocalization in mouse spermatogenic cells and sperm. *FEBS Lett.* 462, 171–176.
349. Trevino, C. L., Felix, R., Castellano, L. E., Gutierrez, C., Rodriguez, D., Pacheco, J., Lopez-Gonzalez, I., Gomora, J. C., Tsutsumi, V., Hernandez-Cruz, A., Fiordelisio, T., Scaling, A. L., and Darszon, A. (2004). Expression and differential cell distribution of low-threshold Ca(2+) channels in mammalian male germ cells and sperm. *FEBS Lett.* 563, 87–92.
350. Felix, R. (2005). Molecular physiology and pathology of Ca^{2+}-conducting channels in the plasma membrane of the mammalian sperm. *Reproduction* 129, 251–262.
351. Douglas, W. W., and Rubin, R. P. (1961). The role of calcium in the secretory response of the adrenal medulla to acetylcholine. *J. Physiol. (Paris)* 159, 40–57.
352. Katz, B., and Miledi, R. (1967). The timing of calcium action during neuromuscular transmission. *J. Physiol.* 189, 535–544.
353. Hille, B., Billiard, J., Babcock, D. F., Nguyen, T., and Koh, D.-S. (1999). Stimulation of exocytosis without a calcium signal. *J. Physiol.* 520, 23–31.
354. Neher, E., Fernandez, J. M., and Lindau, M. (1987). The calcium dependence of vesicle exocytosis. *Res. Publ. Assoc. Res. Nerv. Ment. Dis.* 65, 103–110.
355. Dan, J. C. (1954). Studies on the acrosome. III. Effect of calcium deficiency. *Biol. Bull.* 107, 335–349.
356. Yanagimachi, R., and Usui, N. (1974). Calcium dependence of the acrosome reaction and activation of guinea pig spermatozoa. *Exp. Cell Res.* 89, 161–174.
357. Florman, H. M. (1994). Sequential focal and global elevations of sperm intracellular Ca^{2+} are initiated by the zona pellucida during acrosomal exocytosis. *Dev. Biol.* 165, 152–164.
358. Shirakawa, H., and Miyazaki, S. (1999). Spatiotemporal characterization of intracellular Ca^{2+} rise during the acrosome reaction of mammalian spermatozoa induced by the zona pellucida. *Dev. Biol.* 208, 70–78.
359. Berridge, M. J., Bootman, M. D., and Lipp, P. (1998). Calcium: a life and death signal. *Nature* 395, 645–648.
360. Crabtree, G. R. (1999). Generic signals and specific outcomes: signaling through Ca^{2+}, calcineurin, and NF-AT. *Cell* 96, 611–614.
361. Florman, H. M., Corron, M. E., Kim, T. D. H., and Babcock, D. F. (1992). Activation of voltage-dependent calcium channels of mammalian sperm is required for zona pellucida-induced acrosomal exocytosis. *Dev. Biol.* 152, 304–314.
362. O'Toole, C. M. B., Arnoult, C., Darszon, A., Steinhardt, R. A., and Florman, H. M. (2000). Ca^{2+} entry through store-operated channels in mouse sperm is initiated by egg ZP3 and drives the acrosome reaction. *Mol. Biol. Cell* 11, 1571–1584.
363. Baylor, S. M., and Hollingworth, S. (2000). Measurement and interpretation of cytoplasmic (Ca^{2+}) signals from calcium-indicator dyes. *News Physiol. Sci.* 15, 19–26.
364. Neher, E. (1998). Vesicle pools and Ca^{2+} microdomains: new tools for understanding their roles in neurotransmitter release. *Neuron* 20, 389–399.
365. Goodwin, L. O., Leeds, N. B., Hurley, I., Mandel, F. S., Pergolizzi, R. G., and Benoff, S. (1997). Isolation and characterization of the primary structure of testis-specific L-type calcium channel: implications for contraception. *Mol. Hum. Reprod.* 3, 255–268.
366. Hagiwara, S., and Kawa, K. (1984). Calcium and potassium currents in spermatogenic cells dissociated from rat seminiferous tubules. *J. Physiol.* 356, 135–149.
367. Santi, C. M., Darszon, A., and Hernandez-Cruz, A. (1996). A dihydropyridine-sensitive T-type Ca2+ current is the main Ca2+ current carrier in mouse primary spermatocytes. *Am. J. Physiol.* 271, C1583–C1593.
368. Parekh, A. B., and Putney, J. W. Jr. (2005). Store-operated calcium channels. *Physiol. Rev.* 85, 757–810.
369. Minke, B., and Cook, B. (2002). TRP channel proteins and signal transduction. *Physiol. Rev.* 82, 429–472.
370. Jungnickel, M. K., Marrero, H., Birnbaumer, L., Lemos, J. R., and Florman, H. M. (2001). Trp2 regulates entry of Ca^{2+} into mouse sperm triggered by egg ZP3. *Nat. Cell Biol.* 3, 499–502.
371. Lucas, P., Ukhanov, K., Leinders-Zufall, T., and Zufall, F. (2003). A diacylglycerol-gated cation channel in vomeronasal neuron dendrites is impaired in TRPC2 mutant mice: mechanism of pheromone transduction. *Neuron* 40, 551–561.
372. Zhu, M. X., and Tang, J. (2004). TRPC channel interactions with calmodulin and IP3 receptors. *Novartis. Found. Symp.* 258, 44–58.
373. Vannier, B., Peyton, M., Boulay, G., Brown, D., Qin, N., Jiang, M., Zhu, X., and Birnbaumer, L. (1999). Mouse trp2, the homologue of the human trpc2 pseudogene, encodes mTrp2, a store depletion-activated capacitative Ca^{2+} channel. *Proc. Natl. Acad. Sci. U S A* 96, 2060–2064.
374. Herrick, S. B., Schweissinger, D. L., Kim, S. W., Bayan, K. R., Mann, S., and Cardullo, R. A. (2005). The acrosomal vesicle of mouse sperm is a calcium store. *J. Cell Physiol.* 202, 663–671.
375. Blackmore, P. F. (1993). Thapsigargin elevates and potentiates the ability of progesterone to increase intracellular free calcium in human sperm: possible role of perinuclear calcium. *Cell Calcium* 14, 53–60.
376. Meizel, S., and Turner, K. O. (1993). Initiation of the human sperm acrosome reaction by thapsigargin. *J. Exp. Zool.* 267, 350–355.
377. De Blas, G., Michaut, M., Trevino, C. L., Tomes, C. N., Yunes, R., Darszon, A., and Mayorga, L. S. (2002). The intraacrosomal

calcium pool plays a direct role in acrosomal exocytosis. *J. Biol. Chem.* 277, 49326–49331.
378. Liman, E. R., and Innan, H. (2003). Relaxed selective pressure on an essential component of pheromone transduction in primate evolution. *Proc. Natl. Acad. Sci. U S A* 100, 3328–3332.
379. Wes, P. D., Chevesich, J., Jeromin, A., Rosenberg, C., Stetten, G., and Montell, C. (1995). TRPC1, a human homolog of a *Drosophila* store-operated channel. *Proc. Natl. Acad. Sci. U S A* 92, 9652–9656.
380. Stowers, L., Holy, T. E., Meister, M., Dulac, C., and Koentges, G. (2002). Loss of sex discrimination and male–male aggression in mice deficient for TRP2. *Science* 295, 1493–1500.
381. Leypold, B. G., Yu, C. R., Leinders-Zufall, T., Kim, M. M., Zufall, F., and Axel, R. (2002). Altered sexual and social behaviors in trp2 mutant mice. *Proc. Natl. Acad. Sci. U S A* 99, 6376–6381.
382. Rothman, J. E. (1994). Mechanisms of intracellular protein transport. *Nature* 372, 55–63.
383. Pryer, N. K., Wuestehube, L. J., and Schekman, R. (1992). Vesicle-mediated protein sorting. *Annu. Rev. Biochem.* 61, 471–516.
384. Ferro-Novick, S., and Jahn, R. (1994). Vesicle fusion from yeast to man. *Nature* 370, 191–193.
385. Rettig, J., and Neher, E. (2002). Emerging roles of presynaptic proteins in Ca++-triggered exocytosis. *Science* 298, 781–785.
386. Ungar, D., and Hughson, F. M. (2003). SNARE protein structure and function. *Annu. Rev. Cell Dev. Biol.* 19, 493–517.
387. Sudhof, T. C. (2004). The synaptic vesicle cycle. *Annu. Rev. Neurosci.* 27, 509–547.
388. Schulz, J. R., Sasaki, J. D., and Vacquier, V. D. (1998). Increased association of synaptosome-associated protein of 25 kDa with syntaxin and vesicle-associated membrane protein following acrosomal exocytosis of sea urchin sperm. *J. Biol. Chem.* 273, 24355–24359.
389. Schulz, J. R., Wessel, G. M., and Vacquier, V. D. (1997). The exocytosis regulatory proteins syntaxin and VAMP are shed from sea urchin sperm during the acrosome reaction. *Dev. Biol.* 191, 80–87.
390. Ramalho-Santos, J., Moreno, R. D., Sutovsky, P., Chan, A. W., Hewitson, L., Wessel, G. M., Simerly, C. R., and Schatten, G. (2000). SNAREs in mammalian sperm: possible implications for fertilization. *Dev. Biol.* 223, 54–69.
391. Hutt, D. M., Cardullo, R. A., Baltz, J. M., and Kgsee, J. K. (2002). Synaptotagmin VIII Is localized to the mouse sperm head and may function in acrosomal exocytosis. *Biol. Reprod.* 66, 50–56.
392. Katafuchi, K., Mori, T., Toshimori, K., and Iida, H. (2000). Localization of a syntaxin isoform, syntaxin 2, to the acrosomal region of rodent spermatozoa. *Mol. Reprod. Dev.* 57, 375–383.
393. Ramalho-Santos, J., Terada, Y., and Schatten, G. (2002). VAMP/synaptobrevin as an acrosomal marker for human sperm. *Fertil. Steril.* 77, 159–161.
394. Hutt, D. M., Baltz, J. M., and Ngsee, J. K. (2005). Synaptotagmin VI and VIII and syntaxin 2 are essential for the mouse sperm acrosome reaction. *J. Biol. Chem.* 280, 20197–20203.
395. Ramalho-Santos, J., and Schatten, G. (2004). Presence of N-ethyl maleimide sensitive factor (NSF) on the acrosome of mammalian sperm. *Arch. Androl.* 50, 163–168.
396. Tomes, C. N., Michaut, M., Blas, G. D., Visconti, P., Matti, U., and Mayorga, L. S. (2002). SNARE complex assembly is required for human sperm acrosome reaction. *Dev. Biol.* 243, 326–338.
397. Tomes, C. N., De Blas, G. A., Michaut, M. A., Farre, E. V., Cherhitin, O., Visconti, P. E., and Mayorga, L. S. (2005). alpha-SNAP and NSF are required in a priming step during the human sperm acrosome reaction. *Mol. Hum. Reprod.* 11, 43–51.
398. Yunes, R., Michaut, M., Tomes, C., and Mayorga, L. S. (2000). Rab3A triggers the acrosome reaction in permeabilized human spermatozoa. *Biol. Reprod.* 62, 1084–1089.
399. Sudhof, T. C., and Rizo, J. (1996). Synaptotagmins: C2-domain proteins that regulate membrane traffic. *Neuron* 17, 379–388.
400. Olson, G. E., and Winfrey, V. P. (1994). Structure of acrosomal matrix domains of rabbit sperm. *J. Struct. Biol.* 112, 41–48.
401. Bi, M., Hickox, J. R., Winfrey, V. P., Olson, G. E., and Hardy, D. M. (2003). Processing, localization and binding activity of zonadhesin suggest a function in sperm adhesion to the zona pellucida during exocytosis of the acrosome. *Biochem. J.* 375, 477–488.
402. Olson, G. E., Winfrey, V. P., Bi, M., Hardy, D. M., and Nagdas, S. K. (2004). Zonadhesin assembly into the hamster sperm acrosomal matrix occurs by distinct targeting strategies during spermiogenesis and maturation in the epididymis. *Biol. Reprod.* 71, 1128–1134.
403. Jones, R. (1990). Identification and functions of mammalian sperm–egg recognition molecules during fertilization. *J. Reprod. Fertil. Suppl.* 42, 89–105.
404. Bleil, J. D., Greve, J. M., and Wassarman, P. M. (1988). Identification of a secondary sperm receptor in the mouse egg zona pellucida: role in maintenance of binding of acrosome-reacted sperm to eggs. *Dev. Biol.* 128, 376–385.
405. Tsubamoto, H., Hasegawa, A., Inoue, M., Yamasaki, N., and Koyama, K. (1996). Binding of recombinant pig zona pellucida protein 1 (ZP1) to acrosome-reacted spermatozoa. *J. Reprod. Fertil. Suppl.* 50, 63–67.
406. Tsubamoto, H., Hasegawa, A., Nakata, Y., Naito, S., Yamasaki, N., and Koyama, K. (1999). Expression of recombinant human zona pellucida protein 2 and its binding capacity to spermatozoa. *Biol. Reprod.* 61, 1649–1654.
407. Green, D. P. L., and Purves, R. D. (1984). Mechanical hypothesis of sperm penetration. *Biophys. J.* 45, 659–662.
408. Green, D. P. L. (1988). Sperm thrusts and the problem of penetration. *Biol. Rev.* 63, 79–105.
409. Kresge, N., Vacquier, V. D., and Stout, C. D. (2001). Abalone lysin: the dissolving and evolving sperm protein. *BioEssays* 23, 95–103.
410. Vacquier, V. D., Carner, K. R., and Stout, C. D. (1990). Species-specific sequences of abalone lysin, the sperm protein that creates a hole in the egg envelope. *Proc. Natl. Acad. Sci. U S A* 87, 5792–5796.
411. Bedford, J. M. (1998). Mammalian fertilization misread? Sperm penetration of the eutherian zona pellucida is unlikely to be a lytic event. *Biol. Reprod.* 59, 1275–1287.
412. Evans, J. P. (2002). The molecular basis of sperm–oocyte membrane interactions during mammalian fertilization. *Hum. Reprod. Update* 8, 297–311.
413. Primakoff, P., and Myles, D. (2002). Gamete fusion in mammals. In *Fertilization* (D. Hardy, Ed.), pp. 303–318. Academic Press, San Diego.
414. Hinkey, R. E., Wright, B. D., and Lynn, J. W. (1986). Rapid visual detection of sperm–egg fusion using the DNA-specific fluorochrome Hoechst 33342. *Dev. Biol.* 118, 148–154.
415. Luttmer, S., and Longo, F. (1986). Examination of living and fixed gametes and early embryos stained with supravital fluorochromes (Hoechst 33342 and 3,3′-dihexyloxacarbocyanine iodide). *Gamete Res.* 15, 267–283.
416. Stein, K. K., Primakoff, P., and Myles, D. (2004). Sperm–egg fusion: events at the plasma membrane. *J. Cell Sci.* 117, 6269–6274.
417. Bedford, J., and Cooper, G. (1978). Membrane fusion events in fertilization of vertebrate eggs. In *Membrane Fusion* (G. Poste and G. Nicolson, Eds.), pp. 65–125. North Holland, Amsterdam.
418. Phillips, D. M., and Shalgi, R. (1982). Sperm penetration into rat ova fertilized in vivo. *J. Exp. Zool.* 221, 373–378.

419. Shalgi, R., and Phillips, D. M. (1980). Mechanics of in vitro fertilization in the hamster. *Biol. Reprod.* 23, 433–444.
420. Shalgi, R., Phillips, D. M., and Jones, R. (1989). Status of the rat acrosome during sperm-zona pellucida interactions. *Gamete Res.* 22, 1–13.
421. Huang, T. T. F., and Yanagimachi, R. (1985). Inner acrosomal membrane of mammalian spermatozoa: its properties and possible functions in fertilization. *Amer. J. Anat.* 174, 249–268.
422. Myles, D., Cho, C., Yuan, R., and Primakoff, P. (1999). A current model for the role of ADAMs and integrins in sperm–egg membrane binding and fusion in mammals. In *The Male Gamete: From Basic Science to Clinical Applications* (C. Gagnon, Ed.), pp. 249–255. Cache River Press, Vienna, IL.
423. Shalgi, R., and Phillips, D. (1980). Mechanics of sperm entry in cycling hamsters. *J. Ultra. Res.* 71, 154–161.
424. Moore, H., and Bedford, J. (1983). The interaction of mammalian gametes in the famale. In *Mechanism and Control of Animal Fertilization* (J. Hartmann, Ed.), pp. 453–497. Academic Press, New York.
425. Yanagimachi, R. (1988). Sperm–egg fusion. In *Current Topics in Membranes and Transport* (N. Duzgunes and F. Bronner, Eds.), pp. 3–43. Academic Press, San Diego.
426. Myles, D. G. (1993). Molecular mechanisms of sperm–egg membrane binding and fusion in mammals. *Dev. Biol.* 158, 35–45.
427. Primakoff, P., Hyatt, H., and Tredick-Kline, J. (1987). Identification and purification of a sperm surface protein with a potential role in sperm–egg membrane fusion. *J. Cell Biol.* 104, 141–149.
428. Evans, J. P., Schultz, R. M., and Kopf, G. S. (1997). Characterization of the binding of recombinant mouse sperm fertilin alpha subunit to mouse eggs: evidence for function as a cell adhesion molecule in sperm–egg binding. *Dev. Biol.* 187, 94–106.
429. Myles, D. G., Primakoff, P., and Bellve, A. R. (1981). Surface domains of the guinea pig sperm defined with monoclonal antibodies. *Cell* 23, 433–439.
430. Yuan, R., Primakoff, P., and Myles, D. G. (1997). A role for the disintegrin domain of cyritestin, a sperm surface protein belonging to the ADAM family, in mouse sperm–egg plasma membrane adhesion and fusion. *J. Cell Biol.* 137, 105–112.
431. Cho, C., Bunch, D. O., Faure, J.-E., Goulding, E. H., Eddy, E. M., Primakoff, P., and Myles, D. G. (1998). Fertilization defects in sperm from mice lacking fertilin b. *Science* 281, 1857–1859.
432. Wolfsberg, T. G., and White, J. M. (1996). ADAMs in fertilization and development. *Dev. Biol.* 180, 389–401.
433. Linder, B., and Heinlein, U. A. (1997). Decreased in vitro fertilization efficiencies in the presence of specific cyritestin peptides. *Dev. Growth Differ.* 39, 243–247.
434. Shamsadin, R., Adham, I. M., Nayernia, K., Heinlein, U. A., Oberwinkler, H., and Engel, W. (1999). Male mice deficient for germ-cell cyritestin are infertile. *Biol. Reprod.* 61, 1445–1451.
435. Nishimura, H., Cho, C., Branciforte, D. R., Myles, D. G., and Primakoff, P. (2001). Analysis of loss of adhesive function in sperm lacking cyritestin or fertilin beta. *Dev. Biol.* 233, 204–213.
436. Ellerman, D. A., Da, R., V, Cohen, D. J., Busso, D., Morgenfeld, M. M., and Cuasnicu, P. S. (2002). Expression and structure-function analysis of DE, a sperm cysteine-rich secretory protein that mediates gamete fusion. *Biol. Reprod.* 67, 1225–1231.
437. Cohen, D. J., Ellerman, D. A., and Cuasnicu, P. S. (2000). Mammalian sperm–egg fusion: evidence that epididymal protein DE plays a role in mouse gamete fusion. *Biol. Reprod.* 63, 462–468.
438. Ellerman, D. A., Brantua, V. S., Martinez, S. P., Cohen, D. J., Conesa, D., and Cuasnicu, P. S. (1998). Potential contraceptive use of epididymal proteins: immunization of male rats with epididymal protein DE inhibits sperm fusion ability. *Biol. Reprod.* 59, 1029–1036.
439. Cohen, D. J., Ellerman, D. A., Busso, D., Morgenfeld, M. M., Piazza, A. D., Hayashi, M., Young, E. T., Kasahara, M., and Cuasnicu, P. S. (2001). Evidence that human epididymal protein ARP plays a role in gamete fusion through complementary sites on the surface of the human egg. *Biol. Reprod.* 65, 1000–1005.
440. Bronson, R. A., and Fusi, F. (1990). Evidence that an Arg-Gly-Asp adhesion sequence plays a role in mammalian fertilization. *Biol. Reprod.* 43, 1019–1025.
441. Fusi, F. M., Vignali, M., Busacca, M., and Bronson, R. A. (1992). Evidence for the presence of an integrin cell adhesion receptor on the oolemma of unfertilized human oocytes. *Mol. Reprod. Dev.* 31, 215–222.
442. Fusi, F. M., Vignali, M., Gailit, J., and Bronson, R. A. (1993). Mammalian oocytes exhibit specific recognition of the RGD (Arg-Gly-Asp) tripeptide and express oolemmal integrins. *Mol. Reprod. Dev.* 36, 212–219.
443. Tarone, G., Russo, M. A., Hirsch, E., Odorisio, T., Altruda, F., Silengo, L., and Siracusa, G. (1993). Expression of beta 1 integrin complexes on the surface of unfertilized mouse oocyte. *Development* 117, 1369–1375.
444. Almeida, E. A., Huovila, A. P., Sutherland, A. E., Stephens, L. E., Calarco, P. G., Shaw, L. M., Mercurio, A. M., Sonnenberg, A., Primakoff, P., and Myles, D. G. (1995). Mouse egg integrin alpha 6 beta 1 functions as a sperm receptor. *Cell* 81, 1095–1104.
445. Evans, J. P., Schultz, R. M., and Kopf, G. S. (1995). Identification and localization of integrin subunits in oocytes and eggs of the mouse. *Mol. Reprod. Dev.* 40, 211–220.
446. DeLisi, C. (1981). The magnitude of signal amplification by ligand-induced receptor clustering. *Nature* 289, 322–323.
447. Moore, E. D. W., Becker, P. L., Fogarty, K. E., Williams, D. A., and Fay, F. S. (1990). Ca^{2+} imaging in single living cells: theoretical and practical issues. *Cell Calcium* 11, 157–179.
448. Roe, M. W., Lemasters, J. J., and Herman, B. (1990). Assessment of Fura-2 measurements of cytosolic free calcium. *Cell Calcium* 11, 63–73.
449. Evans, J. P., Schultz, R. M., and Kopf, G. S. (1997). Characterization of the binding of recombinant mouse sperm fertilin alpha subunit to mouse eggs: evidence for function as a cell adhesion molecule in sperm–egg binding. *Dev. Biol.* 187, 94–106.
450. Ji, Y. Z., Wolf, J. P., Jouannet, P., and Bomsel, M. (1998). Human gamete fusion can bypass beta1 integrin requirement. *Hum. Reprod.* 13, 682–689.
451. Linfor, J., and Berger, T. (2000). Potential role of alphav and beta1 integrins as oocyte adhesion molecules during fertilization in pigs. *J. Reprod. Fertil.* 120, 65–72.
452. Miller, B. J., Georges-Labouesse, E., Primakoff, P., and Myles, D. G. (2000). Normal fertilization occurs with eggs lacking the integrin alpha6beta1 and is CD9-dependent. *J. Cell Biol.* 149, 1289–1296.
453. He, Z. Y., Brakebusch, C., Fassler, R., Kreidberg, J. A., Primakoff, P., and Myles, D. G. (2003). None of the integrins known to be present on the mouse egg or to be ADAM receptors are essential for sperm–egg binding and fusion. *Dev. Biol.* 254, 226–237.
454. Alfieri, J. A., Martin, A. D., Takeda, J., Kondoh, G., Myles, D. G., and Primakoff, P. (2003). Infertility in female mice with an oocyte-specific knockout of GPI-anchored proteins. *J. Cell Sci.* 116, 2149–2155.
455. Kaji, K., Oda, S., Shikano, T., Ohnuki, T., Uematsu, Y., Sakagami, J., Tada, N., Miyazaki, S., and Kudo, A. (2000).

The gamete fusion process is defective in eggs of CD9-defective mice. *Nature Gen.* 24, 279–282.
456. Le Naour, F., Rubinstein, E., Jasmin, C., Prenant, M., and Boucheix, C. (2000). Severely reduced female fertility in CD9-deficient mice. *Science* 287, 319–321.
457. Miyado, K., Yamada, G., Yamada, S., Hasuwa, H., Nakamura, Y., Ryu, F., Suzuki, K., Kosai, K., Inoue, K., Ogura, A., Okabe, M., and Mekada, E. (2000). Requirement of CD9 on the egg plasma membrane for fertilization. *Science* 287, 321–324.
458. Kaji, K., Oda, S., Miyazaki, S., and Kudo, A. (2002). Infertility of CD9-deficient mouse eggs is reversed by mouse CD9, human CD9, or mouse CD81; polyadenylated mRNA injection developed for molecular analysis of sperm–egg fusion. *Dev. Biol.* 247, 327–334.
459. Zhu, G. Z., Miller, B. J., Boucheix, C., Rubinstein, E., Liu, C. C., Hynes, R. O., Myles, D. G., and Primakoff, P. (2002). Residues SFQ (173–175) in the large extracellular loop of CD9 are required for gamete fusion. *Development* 129, 1995–2002.
460. Ellerman, D. A., Ha, C., Primakoff, P., Myles, D. G., and Dveksler, G. S. (2003). Direct binding of the ligand PSG17 to CD9 requires a CD9 site essential for sperm–egg fusion. *Mol. Biol. Cell* 14, 5098–5103.
461. Inoue, N., Ikawa, M., Isotani, A., and Okabe, M. (2005). The immunoglobulin superfamily protein Izumo is required for sperm to fuse with eggs. *Nature* 434, 234–238.
462. Neilson, L., Andalibi, A., Kang, D., Coutifaris, C., Strauss, J. F. III, Stanton, J. A., and Green, D. P. (2000). Molecular phenotype of the human oocyte by PCR-SAGE. *Genomics* 63, 13–24.
463. Stanton, J. L., and Green, D. P. (2001). A set of 840 mouse oocyte genes with well-matched human homologues. *Mol. Hum. Reprod.* 7, 521–543.
464. Takahashi, Y., Bigler, D., Ito, Y., and White, J. M. (2001). Sequence-specific interaction between the disintegrin domain of mouse ADAM 3 and murine eggs: role of beta1 integrin-associated proteins CD9, CD81, and CD98. *Mol. Biol. Cell* 12, 809–820.
465. Misamore, M. J., Gupta, S., and Snell, W. J. (2003). The Chlamydomonas Fus1 protein is present on the mating type plus fusion organelle and required for a critical membrane adhesion event during fusion with minus gametes. *Mol. Biol. Cell* 14, 2530–2542.
466. Kresge, N., Vacquier, V. D., and Stout, C. D. (2001). The crystal structure of a fusagenic sperm protein reveals extreme surface properties. *Biochemistry* 40, 5407–5413.
467. Yanagimachi, R. (1978). Sperm–egg association in animals. *Curr. Top. Dev. Biol.* 12, 83–105.
468. Wolf, D. P., and Armstrong, P. (1978). Penetration of the zona-free mouse egg by capacitated epididymal sperm: cinemicrographic observations. *Gamete Res.* 1, 39–46.
469. Sato, K., and Blandau, R. J. (1979). Time and process of sperm penetration into cumulus-free mouse eggs fertilized in vitro. *Gamete Res.* 1, 39–46.
470. Longo, F. J., Krohne, G., and Franke, W. W. (1987). Basic proteins of the perinuclear theca of mammalian spermatozoa and spermatids: a novel class of cytoskeletal elements. *J. Cell Biol.* 105, 1105–1120.
471. Webster, S. D., and McGaughey, R. W. (1990). The cortical cytoskeleton and its role in sperm penetration of the mammalian egg. *Dev. Biol.* 142, 61–74.
472. Gundersen, G. G., Gabel, C. A., and Shapiro, B. M. (1982). An intermediate state of fertilization involved in internalization of sperm components. *Dev. Biol.* 93, 59–72.
473. Kumakiri, J., Oda, S., Kinoshita, K., and Miyazaki, S. (2003). Involvement of Rho family G protein in the cell signaling for sperm incorporation during fertilization of mouse eggs: inhibition by Clostridium difficile toxin B. *Dev. Biol.* 260, 522–535.
474. Maro, B., Johnson, M. H., Pickering, S. J., and Flach, G. (1984). Changes in actin distribution during fertilization of the mouse egg. *J. Embryol. Exp. Morphol.* 81, 211–237.
475. Schatten, G., Schatten, H., Spector, I., Cline, C., Paweletz, N., Simerly, C., and Petzelt, C. (1986). Latrunculin inhibits the microfilament-mediated processes during fertilization, cleavage and early development in sea urchin and mice. *Exp. Cell Res.* 166, 191–208.
476. Yanagimachi, R., and Noda, Y. D. (1970). Ultrastructural changes in the hamster sperm head during fertilization. *J. Ultra. Res.* 31, 465–485.
477. Hewitson, L., Simerly, C. R., and Schatten, G. (2002). Fate of sperm components during assisted reproduction: implications for infertility. *Hum. Fertil. (Camb).* 5, 110–116.
478. Ozil, J. P. (1990). The parthenogenetic development of rabbit oocytes after repetitive pulsatile electrical stimulation. *Development* 109, 117–127.
479. Schatten, G. (1994). The centrosome and its mode of inheritance: the reduction of the centrosome during gametogenesis and its restoration during fertilization. *Dev. Biol.* 165, 299–335.
480. Simerly, C., Dominko, T., Navara, C., Payne, C., Capuano, S., Gosman, G., Chong, K. Y., Takahashi, D., Chace, C., Compton, D., Hewitson, L., and Schatten, G. (2003). Molecular correlates of primate nuclear transfer failures. *Science* 300, 297.
481. Sutovsky, P., Moreno, R. D., Ramalho-Santos, J., Dominko, T., Simerly, C., and Schatten, G. (2000). Ubiquitinated sperm mitochondria, selective proteolysis, and the regulation of mitochondrial inheritance in mammalian embryos. *Biol. Reprod.* 63, 582–590.
482. O'Rand, M. G. (1977). The presence of sperm-specific surface isoantigens on the egg following fertilization. *J. Exp. Zool.* 202, 267–273.
483. Gabel, C. A., Eddy, E. M., and Shapiro, B. M. (1979). After fertilization, sperm surface components remain as a patch in sea urchin and mouse embryos. *Cell* 18, 207–215.
484. Longo, F. J. (1989). Incorporation and dispersal of sperm surface antigens in plasma membranes of inseminated sea urchin (Arbacia punctulata) eggs and oocytes. *Dev. Biol.* 131, 37–43.
485. Jaffe, L. A., and Gould, M. (1985). Polyspermy-preventing mechanisms. In *Biology of Fertilization* (C. B. Metz and A. Monroy, Eds.), pp. 223–250. Academic Press, New York.
486. Boue, J., Boue, A., and Lazar, P. (1975). The epidemiology of human spontaneous abortions with chromosomal anomalies. In *Aging Gametes: Their Biology and Patholgy* (R. J. Blandau, Ed.), pp. 330–348. Karger Press, New York.
487. Golbus, M. (1981). Chromosome aberrations and mammalian reproduction. In *Fertilization and embryonic Development In Vitro* (J. D. Biggers and L. Mastroianni, Eds.), pp. 1–25. Plenum Press, New York.
488. Beatty, R. (1978). The origin of human triploidy: an integration of qualitative and quantitative evidence. *Ann. Hum. Genet.* 41, 299–314.
489. Jacobs, P. A., Angell, R. R., Buchanan, I. M., Hassold, T. J., Matsuyama, A. M., and Manuel, B. (1978). The origin of human triploids. *Ann. Hum. Genet.* 42, 49–57.
490. Kola, I., and Trounson, A. O. (1989). Dispermic human fertilization: violation of expected cell behavior. In *The Cell Biology of Fertilization* (H. Schatten and G. Schatten, Eds.), pp. 277–293. Academic Press, New York.
491. Hammitt, D. G., Syrop, C. H., Van Voorhis, B. J., Walker, D. L., Miller, T. M., and Barud, K. M. (1993). Maturational asynchrony between oocyte cumulus-coronal morphology and

nuclear maturity in gonadotropin-releasing hormone agonist stimulations. *Fertil. Steril.* 59, 375–381.

492. Jaffe, L. A., Sharp, A. P., and Wolf, D. P. (1983). Absence of an electrical polyspermy block in the mouse. *Dev. Biol.* 96, 317–323.
493. Sato, K. (1979). Polyspermy-preventing mechanisms in mouse eggs fertilized in vitro. *J. Exp. Zool.* 210, 353–359.
494. Stewart-Savage, J., and Bavister, B. D. (1991). Time course and pattern of cortical granule breakdown in hamster eggs after sperm fusion. *Mol. Reprod. Dev.* 30, 390–395.
495. McAvey, B. A., Wortzman, G. B., Williams, C. J., and Evans, J. P. (2002). Involvement of calcium signaling and the actin cytoskeleton in the membrane block to polyspermy in mouse eggs. *Biol. Reprod.* 67, 1342–1352.
496. Wassarman, P. M. (1992). Mouse gamete adhesion molecules. *Biol. Reprod.* 46, 186–191.
497. Bleil, J. D., Beall, C. F., and Wassarman, P. M. (1981). Mammalian sperm–egg interaction: fertilization of mouse eggs triggers modification of the major zona pellucida glycoprotein, ZP2. *Dev. Biol.* 86, 189–197.
498. Dean, J. (2004). Reassessing the molecular biology of sperm–egg recognition with mouse genetics. *BioEssays* 26, 29–38.
499. Hoodbhoy, T., and Dean, J. (2004). Insights into the molecular basis of sperm–egg recognition in mammals. *Reproduction* 127, 417–422.
500. Hoodbhoy, T., and Talbot, P. (1994). Mammalian cortical granules: contents, fate, and function. *Mol. Reprod. Dev.* 39, 439–448.
501. Familiari, G., Nottola, S. A., Macchiarelli, G., Micara, G., Aragona, C., and Motta, P. M. (1992). Human zona pellucida during in vitro fertilization: an ultrastructural study using saponin, ruthenium red, and osmium-thiocarbohydrazide. *Mol. Reprod. Dev.* 32, 51–61.
502. Nikas, G., Paraschos, T., Psychoyos, A., and Handyside, A. H. (1994). The zona reaction in human oocytes as seen with scanning electron microscopy. *Hum. Reprod.* 9, 2135–2138.
503. Keefe, D., Tran, P., Pellegrini, C., and Oldenbourg, R. (1997). Polarized light microscopy and digital image processing identify a multilaminar structure of the hamster zona pellucida. *Hum. Reprod.* 12, 1250–1252.
504. Wassarman, P. M. (1990). Regulation of mammalian fertilization by zona pellucida glycoproteins. *J. Reprod. Fertil. Suppl.* 42, 79–87.
505. Amleh, A., and Dean, J. (2002). Mouse genetics provides insight into folliculogenesis, fertilization and early embryonic development. *Hum. Reprod. Update* 8, 395–403.
506. Moos, J., Visconti, P. E., Moore, G. D., Schultz, R. M., and Kopf, G. S. (1995). Potential role of mitogen-activated protein kinase in pronuclear envelope assembly and disassembly following fertilization of mouse eggs. *Biol. Reprod.* 53, 692–699.
507. Ducibella, T., Dubey, A., Gross, V., Emmi, A., Penzias, A. S., Layman, L., and Reindollar, R. (1995). A zona biochemical change and spontaneous cortical granule loss in eggs that fail to fertilize in in vitro fertilization. *Fertil. Steril.* 64, 1154–1161.
508. Gross, V., Dubey, A., Penzias, A. S., Layman, L., Reindollar, R., and Ducibella, T. (1996). Biochemical study of individual zonae from human oocytes that failed to undergo fertilization in intracytoplasmic sperm injection. *Mol. Hum. Reprod.* 2, 959–965.
509. Wessel, G. M., Brooks, J. M., Green, E., Haley, S., Voronina, E., Wong, J., Zaydfudim, V., and Conner, S. (2001). The biology of cortical granules. *Int. Rev. Cytol.* 209, 117–206.
510. Ducibella, T. (1991). Mammalian egg cortical granules and the cortical reaction. In *Elements of Mammalian Fertilization* (P. M. Wassarman, Ed.), pp. 205–230. CRC Press, Boca Raton, FL.
511. Dandekar, P., and Talbot, P. (1992). Perivitelline space of mammalian oocytes: extracellular matrix of unfertilized oocytes and formation of a cortical granule envelope following fertilization. *Mol. Reprod. Dev.* 31, 135–143.
512. Gwatkin, R. B., Williams, D. T., Hartmann, J. F., and Kniazuk, M. (1973). The zona reaction of hamster and mouse eggs: production in vitro by a trypsin-like protease from cortical granules. *J. Reprod. Fertil.* 32, 259–265.
513. Wolf, D. P., and Hamada, M. (1977). Induction of zonal and egg plasma membrane blocks to sperm penetration in mouse eggs with cortical granule exudate. *Biol. Reprod.* 17, 350–354.
514. Gulyas, B. J., and Schmell, E. (1980). Ovoperoxidase activity in ionophore treated mouse eggs. I. Electron microscopic localization. *Gamete Res.* 3, 267–278.
515. Moller, C. C., and Wassarman, P. M. (1989). Characterization of a proteinase that cleaves zona pellucida glycoprotein ZP2 following activation of mouse eggs. *Dev. Biol.* 132, 103–112.
516. Pierce, K. E., Siebert, M. C., Kopf, G. S., Schultz, R. M., and Calarco, P. G. (1990). Characterization and localization of a mouse egg cortical granule antigen prior to and following fertilization or egg activation. *Dev. Biol.* 141, 381–392.
517. Pierce, K. E., Grunvald, E. L., Schultz, R. M., and Kopf, G. S. (1992). Temporal pattern of synthesis of the mouse cortical granule protein, p75, during oocyte growth and maturation. *Dev. Biol.* 152, 145–151.
518. Gross, V. S., Wessel, G., Florman, H. M., and Ducibella, T. (2000). A monoclonal antibody that recognizes mammalian cortical granules and a 32-kilodalton protein in mouse eggs. *Biol. Reprod.* 63, 575–581.
519. Miller, D. J., Gong, X., Decker, G., and Shur, B. D. (1993). Egg cortical granule N-acetylglucosaminidase is required for the mouse zona block to polyspermy. *J. Cell Biol.* 123, 1431–1440.
520. Jones, K. T. (2004). Turning it on and off: M-phase promoting factor during meiotic maturation and fertilization. *Mol. Hum. Reprod.* 10, 1–5.
521. Nebreda, A. R., and Ferby, I. (2000). Regulation of the meiotic cell cycle in oocytes. *Curr. Opin. Cell Biol.* 12, 666–675.
522. Masui, Y., and Markey, C. M. (1971). Cytoplasmic control of nuclear behavior during meiotic maturation of frog oocytes. *J. Exp. Zool.* 177, 129–145.
523. Sagata, N. (1996). Meiotic metaphase arrest in animal oocytes: its mechanisms and biological significance. *Trends Cell Biol.* 6, 22–28.
524. Abrieu, A., Doree, M., and Fisher, D. (2001). The interplay between cyclin-B-Cdc2 kinase (MPF) and MAP kinase during maturation of oocytes. *J. Cell Sci.* 114, 257–267.
525. Tunquist, B. J., and Maller, J. L. (2003). Under arrest: cytostatic factor (CSF)-mediated metaphase arrest in vertebrate eggs. *Genes Dev.* 17, 683–710.
526. Colledge, W. H., Carlton, M. B., Udy, G. B., and Evans, M. J. (1994). Disruption of c-mos causes parthenogenetic development of unfertilized mouse eggs. *Nature* 370, 65–68.
527. Hashimoto, N., Watanabe, N., Furuta, Y., Tamemoto, H., Sagata, N., Yokoyama, M., Okazaki, K., Nagayoshi, M., Takeda, N., Ikawa, Y., et al. (1994). Parthenogenetic activation of oocytes in c-mos–deficient mice. *Nature* 370, 68–71.
528. Lorca, T., Cruzalegui, F. H., Fesquet, D., Cavadore, J. C., Mery, J., Means, A., and Doree, M. (1993). Calmodulin-dependent protein kinase II mediates inactivation of MPF and CSF upon fertilization of Xenopus eggs. *Nature* 366, 270–273.
529. Nixon, V. L., Levasseur, M., McDougall, A., and Jones, K. T. (2002). Ca(2+) oscillations promote APC/C-dependent cyclin B1 degradation during metaphase arrest and completion of meiosis in fertilizing mouse eggs. *Curr. Biol.* 12, 746–750.
530. Uhlmann, F., Lottspeich, F., and Nasmyth, K. (1999). Sister-chromatid separation at anaphase onset is promoted by cleavage of the cohesin subunit Scc1. *Nature* 400, 37–42.

531. Johnson, J., Bierle, B. M., Gallicano, G. I., and Capco, D. G. (1998). Calcium/calmodulin-dependent protein kinase II and calmodulin: regulators of the meiotic spindle in mouse eggs. *Dev. Biol.* 204, 464–477.
532. Tatone, C., Delle, M. S., Iorio, R., Caserta, D., Di, C. M., and Colonna, R. (2002). Possible role for Ca(2+) calmodulin-dependent protein kinase II as an effector of the fertilization Ca(2+) signal in mouse oocyte activation. *Mol. Hum. Reprod.* 8, 750–757.
533. Markoulaki, S., Matson, S., and Ducibella, T. (2004). Fertilization stimulates long-lasting oscillations of CaMKII activity in mouse eggs. *Dev. Biol.* 272, 15–25.
534. Madgwick, S., Nixon, V. L., Chang, H. Y., Herbert, M., Levasseur, M., and Jones, K. T. (2004). Maintenance of sister chromatid attachment in mouse eggs through maturation-promoting factor activity. *Dev. Biol.* 275, 68–81.
535. Chang, H. Y., Levasseur, M., and Jones, K. T. (2004). Degradation of APCcdc20 and APCcdh1 substrates during the second meiotic division in mouse eggs. *J. Cell Sci.* 117, 6289–6296.
536. Ozil, J. P., and Huneau, D. (2001). Activation of rabbit oocytes: the impact of the Ca2+ signal regime on development. *Development* 128, 917–928.
537. Silva, C. P., Kommineni, K., Oldenbourg, R., and Keefe, D. L. (1999). The first polar body does not predict accurately the location of the metaphase II meiotic spindle in mammalian oocytes. *Fertil. Steril.* 71, 719–721.
538. Angell, R. R. (1994). Polar body analysis: possible pitfalls in preimplantation diagnosis of chromosomal disorders based on polar body analysis. *Hum. Reprod.* 9, 181–183.
539. Stephens, S., Beyer, B., Balthazar-Stablein, U., Duncan, R., Kostacos, M., Lukoma, M., Green, G. R., and Poccia, D. (2002). Two kinase activities are sufficient for sea urchin sperm chromatin decondensation in vitro. *Mol. Reprod. Dev.* 62, 496–503.
540. Schatten, G., Maul, G. G., Schatten, H., Chaly, N., Simerly, C., Balczon, R., and Brown, D. L. (1985). Nuclear lamins and peripheral nuclear antigens during fertilization and embryogenesis in mice and sea urchins. *Proc. Natl. Acad. Sci. U S A* 82, 4727–4731.
541. Ducibella, T., and Lefevre, L. (1997). Study of protein kinase C antagonists on cortical granule exocytosis and cell-cycle resumption in fertilized mouse eggs. *Mol. Reprod. Dev.* 46, 216–226.
542. McLay, D. W., and Clarke, H. J. (2003). Remodelling the paternal chromatin at fertilization in mammals. *Reproduction* 125, 625–633.
543. Ward, W. S., and Coffey, D. S. (1991). DNA packaging and organization in mammalian spermatozoa: comparison with somatic cells. *Biol. Reprod.* 44, 569–574.
544. Balhorn, R., Cosman, M., Thorton, K., Krishnam, V., Corzett, M., Bench, G., Kramer, C., Lee, J., Hud, N., Allen, M., Prieto, M., Meyer-Ilse, W., Brown, J., Kirz, J., Zhang, X., Bradbury, E., Maki, G., Braun, R. E., and Breed, W. G. (1999). Protamine mediated condensation of DNA in mammalian sperm. In *The Male Gamete: From Basic Science to Clinical Applications* (C. Gagnon, Ed.), pp. 55–70. Cache River Press, Vienna, IL.
545. Zirkin, B. R., Perreault, S. D., and Naish, S. (1989). Formation and function of the male pronucleus during mammalian fertilization. In *The Molecular Biology of Fertilization* (H. Schatten and G. Schatten, Eds.), pp. 91–114. Academic Press, New York.
546. Ohsumi, K., and Katagiri, C. (1991). Characterization of the ooplasmic factor inducing decondensation of and protamine removal from toad sperm nuclei: involvement of nucleoplasmin. *Dev. Biol.* 148, 295–305.
547. Arnan, C., Saperas, N., Prieto, C., Chiva, M., and Ausio, J. (2003). Interaction of nucleoplasmin with core histones. *J. Biol. Chem.* 278, 31319–31324.
548. Burns, K. H., Viveiros, M. M., Ren, Y., Wang, P., DeMayo, F. J., Frail, D. E., Eppig, J. J., and Matzuk, M. M. (2003). Roles of NPM2 in chromatin and nucleolar organization in oocytes and embryos. *Science* 300, 633–636.
549. Ecklund, P. S., and Levine, L. (1975). Mouse sperm basic nuclear protein. Electrophoretic characterization and fate after fertilization. *J. Cell Biol.* 66, 251–262.
550. Kopecny, V., and Pavlok, A. (1975). Autoradiographic study of mouse spermatozoan arginine-rich nuclear protein in fertilization. *J. Exp. Zool.* 191, 85–96.
551. Rodman, T. C., Pruslin, F. H., Hoffmann, H. P., and Allfrey, V. G. (1981). Turnover of basic chromosomal proteins in fertilized eggs: a cytoimmunochemical study of events in vivo. *J. Cell Biol.* 90, 351–361.
552. Ducibella, T., Huneau, D., Angelichio, E., Xu, Z., Schultz, R. M., Kopf, G. S., Fissore, R., Madoux, S., and Ozil, J. P. (2002). Egg-to-embryo transition is driven by differential responses to Ca(2+) oscillation number. *Dev. Biol.* 250, 280–291.
553. Collas, P., and Poccia, D. (1998). Methods for studying in vitro assembly of male pronuclei using oocyte extracts from marine invertebrates: sea urchins and surf clams. *Methods Cell Biol.* 53, 417–452.
554. Mann, M. R., and Bartolomei, M. S. (2002). Epigenetic reprogramming in the mammalian embryo: struggle of the clones. *Genome Biol.* 3, REVIEWS 1003.
555. Dean, W., Santos, F., Stojkovic, M., Zakhartchenko, V., Walter, J., Wolf, E., and Reik, W. (2001). Conservation of methylation reprogramming in mammalian development: aberrant reprogramming in cloned embryos. *Proc. Natl. Acad. Sci. U S A* 98, 13734–13738.
556. Adenot, P. G., Mercier, Y., Renard, J. P., and Thompson, E. M. (1997). Differential H4 acetylation of paternal and maternal chromatin precedes DNA replication and differential transcriptional activity in pronuclei of 1-cell mouse embryos. *Development* 124, 4615–4625.
557. Cowell, I. G., Aucott, R., Mahadevaiah, S. K., Burgoyne, P. S., Huskisson, N., Bongiorni, S., Prantera, G., Fanti, L., Pimpinelli, S., Wu, R., Gilbert, D. M., Shi, W., Fundele, R., Morrison, H., Jeppesen, P., and Singh, P. B. (2002). Heterochromatin, HP1 and methylation at lysine 9 of histone H3 in animals. *Chromosoma* 111, 22–36.
558. Ram, P. T., and Schultz, R. M. (1993). Reporter gene expression in G2 of the 1-cell mouse embryo. *Dev. Biol.* 156, 552–556.
559. Aoki, F., Worrad, D. M., and Schultz, R. M. (1997). Regulation of transcriptional activity during the first and second cell cycles in the preimplantation mouse embryo. *Dev. Biol.* 181, 296–307.
560. Wang, Q., Chung, Y. G., deVries, W. N., Struwe, M., and Latham, K. E. (2001). Role of protein synthesis in the development of a transcriptionally permissive state in one-cell stage mouse embryos. *Biol. Reprod.* 65, 748–754.
561. Aoki, F., Hara, K. T., and Schultz, R. M. (2003). Acquisition of transcriptional competence in the 1-cell mouse embryo: requirement for recruitment of maternal mRNAs. *Mol. Reprod. Dev.* 64, 270–274.
562. Latham, K. E., and Schultz, R. M. (2001). Embryonic genome activation. *Front. Biosci.* 6, D748–D759.
563. Gallicano, G. I., Yousef, M. C., and Capco, D. G. (1997). PKC: a pivotal regulator of early development. *BioEssays* 19, 29–36.
564. Gallicano, G. I., and Capco, D. G. (1995). Remodeling of the specialized intermediate filament network in mammalian eggs and embryos during development: regulation by protein kinase C and protein kinase M. *Curr. Top. Dev. Biol.* 31, 277–320.

565. Payne, C., Rawe, V., Ramalho-Santos, J., Simerly, C., and Schatten, G. (2003). Preferentially localized dynein and perinuclear dynactin associate with nuclear pore complex proteins to mediate genomic union during mammalian fertilization. *J. Cell Sci.* 116, 4727–4738.
566. Scholey, J. M., Porter, M., Lye, R., and McIntosh, J. (1989). Cytoplasmic microtubule-associated motors. In *The Cell Biology of Fertilization* (H. Schatten and G. Schatten, Eds.), pp. 139–163. Academic Press, New York.
567. Gardner, R. (1997). The early blastocyst is bilaterally symmetrical and its axis of symmetry is aligned with the animal-vegetal axis of the zygote in the mouse. *Development* 124, 289–301.
568. Zernicka-Goetz, M. (2002). Patterning of the embryo: the first spatial decisions in the life of a mouse. *Development* 129, 815–829.
569. Gardner, R., and Davies, T. (2003). The basis and significance of pre-patterning in mammals. *Philos. Trans. R. Soc. Lond. B. Biol. Sci.* 358, 1331–1339.
570. SeGall, G. K., and Lennarz, W. J. (1979). Chemical characterization of the component of the jelly coat from sea urchin eggs responsible for induction of the acrosome reaction. *Dev. Biol.* 71, 33–48.
571. Kline, D., and Kline, J. T. (1992). Repetitive calcium transients and the role of calcium in exocytosis and cell cycle activation in the mouse egg. *Dev. Biol.* 145, 80–89.
572. Jaffe, L. A. (1985). The role of calcium explosions, waves, and pulses in activating eggs. In *Biology of Fertilization* (C. B. Metz and A. Monroy, Eds.), pp. 127–165. Academic Press, New York.
573. Whitaker, M., and Steinhardt, R. (1985). Ionic signaling in the sea urchin egg at fertilization. In *Biology of Fertilization* (C. B. Metz and A. Monroy, Eds.), pp. 167–225. Academic Press, New York.
574. Gilkey, J. C., Jaffe, L. F., Ridgway, E. B., and Reynolds, G. T. (1978). A free calcium wave traverses the activating egg of the medaka, Oryzias latipes. *J. Cell Biol.* 76, 448–466.
575. Steinhardt, R. A., Epel, D., Carroll, E. J., Jr., and Yanagimachi, R. (1974). Is calcium ionophore a universal activator for unfertilised eggs? *Nature* 252, 41–43.
576. Cuthbertson, K. S., Whittingham, D. G., and Cobbold, P. H. (1981). Free Ca2+ increases in exponential phases during mouse oocyte activation. *Nature* 294, 754–757.
577. Stricker, S. A. (1999). Comparative biology of calcium signaling during fertilization and egg activation in animals. *Dev. Biol.* 211, 157–176.
578. Deguchi, R., Shirakawa, H., Oda, S., Mohri, T., and Miyazaki, S. (2000). Spatiotemporal analysis of Ca(2+) waves in relation to the sperm entry site and animal-vegetal axis during Ca(2+) oscillations in fertilized mouse eggs. *Dev. Biol.* 218, 299–313.
579. Miyazaki, S., Shirakawa, H., Nakada, K., and Honda, Y. (1993). Essential role of the inositol 1, 4, 5-trisphosphate receptor/Ca^{2+} release channel in Ca^{2+} waves and Ca^{2+} oscillators at fertilization of mammalian eggs. *Dev. Biol.* 158, 62–78.
580. Jones, K. T. (1998). Ca2+ oscillations in the activation of the egg and development of the embryo in mammals. *Int. J. Dev. Biol.* 42, 1–10.
581. Streb, H., Irvine, R. F., Berridge, M. J., and Schulz, I. (1983). Release of Ca2+ from a nonmitochondrial intracellular store in pancreatic acinar cells by inositol-1, 4, 5-trisphosphate. *Nature* 306, 67–69.
582. Berridge, M. J., and Irvine, R. F. (1984). Inositol trisphosphate, a novel second messenger in cellular signal transduction. *Nature* 312, 315–321.
583. Berridge, M. J., Lipp, P., and Bootman, M. D. (2000). The versatility and universality of calcium signalling. *Nat. Rev. Mol. Cell Biol.* 1, 11–21.
584. Williams, C. J. (2002). Signalling mechanisms of mammalian oocyte activation. *Hum. Reprod. Update* 8, 313–321.
585. Runft, L. L., Jaffe, L. A., and Mehlmann, L. M. (2002). Egg activation at fertilization: where it all begins. *Dev. Biol.* 245, 237–254.
586. Mehlmann, L. M., Terasaki, M., Jaffe, L. A., and Kline, D. (1995). Reorganization of the endoplasmic reticulum during meiotic maturation of the mouse oocyte. *Dev. Biol.* 170, 607–615.
587. Shiraishi, K., Okada, A., Shirakawa, H., Nakanishi, S., Mikoshiba, K., and Miyazaki, S. (1995). Developmental changes in the distribution of the endoplasmic reticulum and inositol 1, 4, 5-trisphosphate receptors and the spatial pattern of Ca2+ release during maturation of hamster oocytes. *Dev. Biol.* 170, 594–606.
588. Ayabe, T., Kopf, G. S., and Schultz, R. M. (1995). Regulation of mouse egg activation: presence of ryanodine receptors and effects of microinjected ryanodine and cyclic ADP ribose on uninseminated and inseminated eggs. *Development* 121, 2233–2244.
589. Kurasawa, S., Schultz, R. M., and Kopf, G. S. (1989). Egg-induced modifications of the zona pellucida of mouse eggs: effects of microinjected inositol 1, 4, 5-trisphosphate. *Dev. Biol.* 133, 295–304.
590. Ciapa, B., and Whitaker, M. (1986). Two phases of inositol polyphosphate and diacylglycerol production at fertilisation. *FEBS Lett.* 195, 347–351.
591. Ciapa, B., Borg, B., and Whitaker, M. (1992). Polyphosphoinositide metabolism during the fertilization wave in sea urchin eggs. *Development* 115, 187–195.
592. Stith, B. J., Goalstone, M., Silva, S., and Jaynes, C. (1993). Inositol 1, 4, 5-trisphosphate mass changes from fertilization through first cleavage in Xenopus laevis. *Mol. Biol. Cell* 4, 435–443.
593. Turner, P. R., Sheetz, M. P., and Jaffe, L. A. (1984). Fertilization increases the polyphosphoinositide content of sea urchin eggs. *Nature* 310, 414–415.
594. Nuccitelli, R., Yim, D. L., and Smart, T. (1993). The sperm-induced Ca2+ wave following fertilization of the Xenopus egg requires the production of Ins(1, 4, 5)P3. *Dev. Biol.* 158, 200–212.
595. Jones, K. T., and Nixon, V. L. (2000). Sperm-induced Ca(2+) oscillations in mouse oocytes and eggs can be mimicked by photolysis of caged inositol 1, 4, 5-trisphosphate: evidence to support a continuous low level production of inositol 1, 4, 5-trisphosphate during mammalian fertilization. *Dev. Biol.* 225, 1–12.
596. Sato, Y., Miyazaki, S., Shikano, T., Mitsuhashi, N., Takeuchi, H., Mikoshiba, K., and Kuwabara, Y. (1998). Adenophostin, a potent agonist of the inositol 1, 4, 5-trisphosphate receptor, is useful for fertilzation of mouse oocytes injected with round spermatids leading to normal offspring. *Biol. Reprod.* 58, 867–873.
597. Mehlmann, L. M., Mikoshiba, K., and Kline, D. (1996). Redistribution and increase in cortical inositol 1, 4, 5-trisphosphate receptors after meiotic maturation of the mouse oocyte. *Dev. Biol.* 180, 489–498.
598. He, C. L., Damiani, P., Ducibella, T., Takahashi, M., Tanzawa, K., Parys, J. B., and Fissore, R. A. (1999). Isoforms of the inositol 1, 4, 5-trisphosphate receptor are expressed in bovine oocytes and ovaries: the type-1 isoform is down-regulated by fertilization and by injection of adenophostin A. *Biol. Reprod.* 61, 935–943.
599. Miyazaki, S., Yuzaki, M., Nakeda, K., Shirakawa, H., Nakanishi, S., Nakade, S., and Mikoshiba, K. (1992). Block of Ca^{2+} wave and Ca^{2+} oscillation by antibody to the inositol 1, 4, 5-trisphosphate receptor in fertilized hamster eggs. *Science* 257, 251–255.

600. Wu, H., He, C. L., and Fissore, R. A. (1997). Injection of a porcine sperm factor triggers calcium oscillations in mouse oocytes and bovine eggs. *Mol. Reprod. Dev.* 46, 176–189.
601. Brind, S., Swann, K., and Carroll, J. (2000). Inositol 1, 4, 5-trisphosphate receptors are downregulated in mouse oocytes in response to sperm or adenophostin A but not to increases in intracellular Ca(2+) or egg activation. *Dev. Biol.* 223, 251–265.
602. Jellerette, T., He, C. L., Wu, H., Parys, J. B., and Fissore, R. A. (2000). Down-regulation of the inositol 1, 4, 5-trisphosphate receptor in mouse eggs following fertilization or parthenogenetic activation. *Dev. Biol.* 223, 238–250.
603. Turner, P., and Jaffe, L. A. (1989). G-proteins and the regulation of oocyte maturation and fertilization. In *The Cell Biology of Fertilization* (H. Schatten and G. Schatten, Eds.), pp. 297–318. Academic Press, New York.
604. Moore, G. D., Kopf, G. S., and Schultz, R. M. (1993). Complete mouse egg activation in the absence of sperm by stimulation of an exogenous G protein-coupled receptor. *Dev. Biol.* 159, 669–678.
605. Dupont, G., McGuinness, O. M., Johnson, M. H., Berridge, M. J., and Borgese, F. (1996). Phospholipase C in mouse oocytes: characterizatiuon of b and g isoforms and their possible involvement in sperm-induced Ca^{2+} spiking. *Biochem. J.* 316, 583–591.
606. Perez-Reyes, E., Wei, X., Castellano, A., and Birnbaumer, L. (1990). Molecular diversity of L-type calcium channels. Evidence for alternative splicing of the transcripts of three non-allelic genes. *J. Biol. Chem.* 265, 20430–20436.
607. Shai, Y., Bach, D., and Yanovsky, A. (1990). Channel forming properties of synthetic pardaxin and analogues. *J. Biol. Chem.* 265, 20202–20209.
608. Leong, D. A. (1989). Intracellular calcium levels in rat anterior pituitary cells: single-cell techniques. *Meth. Enzymol.* 168, 263–284.
609. Williams, C. J., Schultz, R. M., and Kopf, G. S. (1992). Role of G proteins in mouse egg activation: stimulatory effects of acetylcholine on the ZP2 to $ZP2_f$ conversion and pronuclear formation in eggs expressing a functional m1 muscarinic receptor. *Dev. Biol.* 151, 288–296.
610. Miyazaki, S. (1988). Fertilization potential and calcium transients in mammalian eggs. *Dev. Growth Differ.* 30, 603–610.
611. Fissore, R. A., and Robl, J. M. (1994). Mechanism of calcium oscillations in fertilized rabbit eggs. *Dev. Biol.* 166, 634–642.
612. Kurokawa, M., Sato, K., and Fissore, R. A. (2004). Mammalian fertilization: from sperm factor to phospholipase Czeta. *Biol. Cell* 96, 37–45.
613. Miyazaki, S. (1988). Inositol 1, 4, 5-trisphosphate-induced calcium release and guanine nucleotide-binding protein-mediated periodic calcium rises in golden hamster eggs. *J. Cell Biol.* 106, 3455–353.
614. Moore, G. D., Ayabe, T., Visconti, P. E., Schultz, R. M., and Kopf, G. S. (1994). Roles of heterotrimeric and monomeric G proteins in sperm-induced activation of mouse eggs. *Development* 120, 3313–3323.
615. Williams, C. J., Schultz, R. M., and Kopf, G. S. (1996). G protein gene expression during mouse oocyte growth and maturation, and preimplantation embryo development. *Mol. Reprod. Dev.* 44, 315–323.
616. Williams, C. J., Mehlmann, L. M., Jaffe, L. A., Kopf, G. S., and Schultz, R. M. (1998). Evidence that Gq family G proteins do not function in mouse egg activation at fertilization. *Dev. Biol.* 198, 116–127.
617. Mehlmann, L. M., Carpenter, G., Rhee, S. G., and Jaffe, L. A. (1998). SH2 domain-mediated activation of phospholipase Cgamma is not required to initiate Ca2+ release at fertilization of mouse eggs. *Dev. Biol.* 203, 221–232.
618. Rongish, B. J., Wu, W., and Kinsey, W. H. (1999). Fertilization-induced activation of phospholipase C in the sea urchin egg. *Dev. Biol.* 215, 147–154.
619. Sato, K., Tokmakov, A. A., Iwasaki, T., and Fukami, Y. (2000). Tyrosine kinase-dependent activation of phospholipase Cgamma is required for calcium transient in Xenopus egg fertilization. *Dev. Biol.* 224, 453–469.
620. Carroll, D. J., Ramarao, C. S., Mehlmann, L. M., Roche, S., Terasaki, M., and Jaffe, L. A. (1997). Calcium release at fertilization in starfish eggs is mediated by phospholipase Cgamma. *J. Cell Biol.* 138, 1303–1311.
621. Shearer, J., De, N. C., Emily-Fenouil, F., Gache, C., Whitaker, M., and Ciapa, B. (1999). Role of phospholipase Cgamma at fertilization and during mitosis in sea urchin eggs and embryos. *Development* 126, 2273–2284.
622. Runft, L. L., and Jaffe, L. A. (2000). Sperm extract injection into ascidian eggs signals Ca^{2+} release by the same pathway as fertilization. *Development* 127, 3227–3236.
623. Carroll, D. J., Albay, D. T., Terasaki, M., Jaffe, L. A., and Foltz, K. R. (1999). Identification of PLCgamma-dependent and -independent events during fertilization of sea urchin eggs. *Dev. Biol.* 206, 232–247.
624. Runft, L. L., Watras, J., and Jaffe, L. A. (1999). Calcium release at fertilization of Xenopus eggs requires type I IP3 receptors, but not SH2 domain-mediated activation of PLCgamma or G(q)-mediated activation of PLCb. *Dev. Biol.* 214, 399–411.
625. Mehlmann, L. M., Chattopadhyay, A., Carpenter, G., and Jaffe, L. A. (2001). Evidence that phospholipase C from the sperm is not responsible for initiating Ca(2+) release at fertilization in mouse eggs. *Dev. Biol.* 236, 492–501.
626. Mehlmann, L. M., and Jaffe, L. A. (2005). SH2 domain-mediated activation of an SRC family kinase is not required to initiate Ca2+ release at fertilization in mouse eggs. *Reproduction* 129, 557–564.
627. Sette, C., Bevilacqua, A., Geremia, R., and Rossi, P. (1998). Involvement of phospholipase Cgamma1 in mouse egg activation induced by a truncated form of the C-kit tyrosine kinase present in spermatozoa. *J. Cell Biol.* 142, 1063–1074.
628. Talmor, A., Kinsey, W. H., and Shalgi, R. (1998). Expression and immunolocalization of p59c-fyn tyrosine kinase in rat eggs. *Dev. Biol.* 194, 38–46.
629. Sette, C., Paronetto, M. P., Barchi, M., Bevilacqua, A., Geremia, R., and Rossi, P. (2002). Tr-kit–induced resumption of the cell cycle in mouse eggs requires activation of a Src-like kinase. *EMBO J.* 21, 5386–5395.
630. Whitaker, M., and Swann, K. (1993). Lighting the fuse at fertilization. *Development* 117, 1–12.
631. Dale, B., Defelice, L. J., and Ehrenstein, G. (1985). Injection of a soluble sperm extract into sea urchin eggs triggers the cortical reaction. *Experimentia* 41, 1068–1070.
632. Jones, K. T., Soeller, C., and Cannell, M. B. (1998). The passage of Ca2+ and fluorescent markers between the sperm and egg after fusion in the mouse. *Development* 125, 4627–4635.
633. Lawrence, Y., Whitaker, M., and Swann, K. (1997). Sperm–egg fusion is the prelude to the initial Ca2+ increase at fertilization in the mouse. *Development* 124, 233–241.
634. Smrcka, A. V., Hepler, J. R., Brown, K. O., and Sternweis, P. C. (1991). Regulation of polyphosphoinositide-specific phospholipase C activity by purified G_q. *Science* 251, 804–807.
635. Swann, K., and Lai, F. A. (1997). A novel signalling mechanism for generating Ca2+ oscillations at fertilization in mammals. *BioEssays* 19, 371–378.
636. Swann, K. (1990). A cytosolic sperm factor stimulates repetitive calcium increases and mimics fertilization in hamster. *Development* 110, 1295–1302.

637. Stice, S. L., and Robl, J. M. (1990). Activation of mammalian oocytes by a factor obtained from rabbit sperm. *Mol. Reprod. Dev.* 25, 272–280.
638. Kuretake, S., Kimura, Y., Hoshi, K., and Yanagimachi, R. (1996). Fertilization and development of mouse oocytes injected with isolated sperm heads. *Biol. Reprod.* 55, 789–795.
639. Kimura, Y., Yanagimachi, R., Kuretake, S., Bortkiewicz, H., Perry, A. C., and Yanagimachi, H. (1998). Analysis of mouse oocyte activation suggests the involvement of sperm perinuclear material. *Biol. Reprod.* 58, 1407–1415.
640. Knott, J. G., Kurokawa, M., and Fissore, R. A. (2003). Release of the Ca^{2+} oscillation-inducing sperm factor during mouse fertilization. *Dev. Biol.* 260, 536–547.
641. Oda, S., Deguchi, R., Mohri, T., Shikano, T., Nakanishi, S., and Miyazaki, S. (1999). Spatiotemporal dynamics of the (Ca2+)i rise induced by microinjection of sperm extract into mouse eggs: preferential induction of a Ca2+ wave from the cortex mediated by the inositol 1, 4, 5-trisphosphate receptor. *Dev. Biol.* 209, 172–185.
642. Fukami, K., Inoue, T., Kurokawa, M., Fissore, R. A., Nakao, K., Nagano, K., Nakamura, Y., Takenaka, K., Yoshida, N., Mikoshiba, K., and Takenawa, T. (2003). Phospholipase Cd4: from genome structure to physiological function. *Adv. Enzyme Regul.* 43, 87–106.
643. Cox, L. J., Larman, M. G., Saunders, C. M., Hashimoto, K., Swann, K., and Lai, F. A. (2002). Sperm phospholipase Czeta from humans and cynomolgus monkeys triggers Ca2+ oscillations, activation and development of mouse oocytes. *Reproduction* 124, 611–623.
644. Saunders, C. M., Larman, M. G., Parrington, J., Cox, L. J., Royse, J., Blayney, L. M., Swann, K., and Lai, F. A. (2002). PLC zeta: a sperm-specific trigger of Ca(2+) oscillations in eggs and embryo development. *Development* 129, 3533–3544.
645. Fujimoto, S., Yoshida, N., Fukui, T., Amanai, M., Isobe, T., Itagaki, C., Izumi, T., and Perry, A. C. (2004). Mammalian phospholipase Cz induces oocyte activation from the sperm perinuclear matrix. *Dev. Biol.* 274, 370–383.
646. Knott, J. G., Kurokawa, M., Fissore, R. A., Schultz, R. M., and Williams, C. J. (2005). Transgenic RNA interference reveals role for mouse sperm phospholipase Cz in triggering Ca^{2+} oscillations during fertilization. *Biol. Reprod.* 72, 992–996.
647. Lillie, F. R. (1913). The mechanism of fertilization. *Science* 38, 524528.
648. Bavister, B. D. (2002). How animal embryo research led to the first documented human IVF. *Reprod. Biomed. Online* 4, Suppl 1, 24–29.
649. Palermo, G., Joris, H., Devroey, P., and Van Steirteghem, A. C. (1992). Induction of acrosome reaction in human spermatozoa used for subzonal insemination. *Hum. Reprod.* 7, 248–254.
650. Wassarman, P. M., Jovine, L., Qi, H., Williams, Z., Darie, C., and Litscher, E. S. (2005). Recent aspects of mammalian fertilization research. *Mol. Cell Endocrinol.* 234, 95–103.
651. Tung, J. J., Hansen, D. V., Ban, K. H., Loktev, A. V., Summers, M. K., Adler, J. R. III, and Jackson, P. K. (2005). A role for the anaphase-promoting complex inhibitor Emi2/XErp1, a homolog of early mitotic inhibitor 1, in cytostatic factor arrest of Xenopus eggs. *Proc. Natl. Acad. Sci. U S A* 102, 4318–4323.
652. Ohsumi, K., Koyanagi, A., Yamamoto, T. M., Gotoh, T., and Kishimoto, T. (2004). Emil-mediated M-phase arrest in Xenopus eggs is distinct from cytostatic factor arrest. *Proc. Natl. Acad. Sci. U S A* 101, 12531–12536.
653. Reimann, J. D., and Jackson, P. K. (2002). Emil is required for cytostatic factor arrest in vertebrate eggs. *Nature* 416, 850–854.
654. Visconti, P. E., Johnson, L. R., Oyaski, M., Fornes, M., Moss, S. B., Gerton, G. L., and Kopf, G. S. (1997). Regulation, localization, and anchoring of protein kinase A subunits during mouse sperm capacitation. *Dev. Biol.* 192, 351–363.
655. Cowan, A. E., Koppel, D. E., Vargas, L. A., and Hunnicutt, G. R. (2001). Guinea pig fertilin exhibits restricted lateral mobility in epididymal sperm and becomes freely diffusing during capacitation. *Dev. Biol.* 236, 502–509.
656. Luconi, M., Barni, T., Vannelli, G. B., Krausz, C., Marra, F., Benedetti, P. A., Evangelista, V., Francavilla, S., Properzi, G., Forti, G., and Baldi, E. (1998). Extracellular signal-regulated kinases modulate capacitation of human spermatozoa. *Biol. Reprod.* 58, 1476–1489.
657. Carrera, A., Moss, J., Ning, X. P., Gerton, G. L., Tesarik, J., Kopf, G. S., and Moss, S. B. (1996). Regulation of protein tyrosine phosphorylation in human sperm by a calcium/calmodulin-dependent mechanism: identification of A kinase anchor proteins as major substrates for tyrosine phosphorylation. *Dev. Biol.* 180, 284–296.
658. Mandal, A., Naaby-Hansen, S., Wolkowicz, M. J., Klotz, K., Shetty, J., Retief, J. D., Coonrod, S. A., Kinter, M., Sherman, N., Cesar, F., Flickinger, C. J., and Herr, J. C. (1999). FSP95, a testis-specific 95-kilodalton fibrous sheath antigen that undergoes tyrosine phosphorylation in capacitated human spermatozoa. *Biol. Reprod.* 61, 1184–1197.
659. Naaby-Hansen, S., Mandal, A., Wolkowicz, M. J., Sen, B., Westbrook, V. A., Shetty, J., Coonrod, S. A., Klotz, K. L., Kim, Y. H., Bush, L. A., Flickinger, C. J., and Herr, J. C. (2002). CABYR, a novel calcium-binding tyrosine phosphorylation-regulated fibrous sheath protein involved in capacitation. *Dev. Biol.* 242, 236–254.
660. Gadella, B. M., Lopes-Cardozo, M., van Golde, L. M., Colenbrander, B., and Gadella, T. W., Jr. (1995). Glycolipid migration from the apical to the equatorial subdomains of the sperm head plasma membrane precedes the acrosome reaction. Evidence for a primary capacitation event in boar spermatozoa. *J. Cell Sci.* 108, 935–946.
661. Singh, J. P., Babcock, D. F., and Lardy, H. A. (1978). Increased calcium-ion influx is a component of capacitation of spermatozoa. *Biochem. J.* 172, 549–556.
662. Andrews, J. C., Nolan, J. P., Hammerstedt, R. H., and Bavister, B. D. (1994). Role of zinc during hamster sperm capacitation. *Biol. Reprod.* 51, 1238–1247.
663. Aitken, J., and Fisher, H. (1994). Reactive oxygen species generation and human spermatozoa: the balance of benefit and risk. *BioEssays* 16, 259–267.
664. Roldan, E. R., and Fleming, A. D. (1989). Is a Ca2+-ATPase involved in Ca2+ regulation during capacitation and the acrosome reaction of guinea-pig spermatozoa? *J. Reprod. Fertil.* 85, 297–308.
665. Rufo, G. A., Schoff, P. K., and Lardy, H. A. (1984). Regulation of calcium content in bovine spermatozoa. *J. Biol. Chem.* 259, 2547–2552.
666. Carlson, A. E., Westenbroek, R. E., Quill, T., Ren, D., Clapham, D. E., Hille, B., Garbers, D. L., and Babcock, D. F. (2003). CatSper1 required for evoked Ca2+ entry and control of flagellar function in sperm. *Proc. Natl. Acad. Sci. U S A* 100, 14864–14868.
667. Benoff, S. (1998). Voltage-dependent calcium channels in mammalian spermatozoa. *Front. Biosci.* 3, d1220–d1240.
668. Svoboda, K., and Mainen, Z. F. (1999). Synaptic (Ca^{2+}): intracellular stores spill their guts. *Neuron* 22, 427–430.
669. Trevino, C. L., Serrano, C. J., Beltran, C., Felix, R., and Darszon, A. (2001). Identification of mouse trp homologs and lipid rafts from spermatogenic cells and sperm. *FEBS Lett.* 509, 119–125.
670. Sutton, K. A., Jungnickel, M. K., Wang, Y., Cullen, K., Lambert, S., and Florman, H. M. (2004). Enkurin is a novel

calmodulin and TRPC channel binding protein in sperm. *Dev. Biol.* 274, 426–435.

671. Castellano, L. E., Trevino, C. L., Rodriguez, D., Serrano, C. J., Pacheco, J., Tsutsumi, V., Felix, R., and Darszon, A. (2003). Transient receptor potential (TRPC) channels in human sperm: expression, cellular localization and involvement in the regulation of flagellar motility. *FEBS Lett.* 541, 69–74.
672. Ren, D., Navarro, B., Perez, G., Jackson, A. C., Hsu, S., Shi, Q., Tully, J. L., and Clapham, D. E. (2001). A sperm ion channel required for sperm motility and male fertility. *Nature* 413, 603–609.
673. Quill, T. A., Ren, D., Clapham, D. E., and Garbers, D. L. (2001). A voltage-gated ion channel expressed specifically in spermatozoa. *Proc. Natl. Acad. Sci. U S A* 98, 12527–12531.
674. Kuroda, Y., Kaneko, S., Yoshimura, Y., Nozawa, S., and Mikoshiba, K. (1999). Are there inositol 1, 4, 5-triphosphate (IP3) receptors in human sperm? *Life Sci.* 65, 135–143.
675. Philpott, A., Krude, T., and Laskey, R. A. (2000). Nuclear chaperones. *Semin. Cell Dev. Biol.* 11, 7–14.
676. Dilworth, S. M., Black, S. J., and Laskey, R. A. (1987). Two complexes that contain histones are required for nucleosome assembly in vitro: role of nucleoplasmin and N1 in Xenopus egg extracts. *Cell* 51, 1009–1018.
677. Ray-Gallet, D., Quivy, J. P., Scamps, C., Martini, E. M., Lipinski, M., and Almouzni, G. (2002). HIRA is critical for a nucleosome assembly pathway independent of DNA synthesis. *Mol. Cell* 9, 1091–1100.
678. Matsumoto, K., Nagata, K., Miyaji-Yamaguchi, M., Kikuchi, A., and Tsujimoto, M. (1999). Sperm chromatin decondensation by template activating factor I through direct interaction with basic proteins. *Mol. Cell Biol.* 19, 6940–6952.
679. Ito, T., Bulger, M., Kobayashi, R., and Kadonaga, J. T. (1996). Drosophila NAP-1 is a core histone chaperone that functions in ATP-facilitated assembly of regularly spaced nucleosomal arrays. *Mol. Cell Biol.* 16, 3112–3124.
680. Kawasaki, K., Philpott, A., Avilion, A. A., Berrios, M., and Fisher, P. A. (1994). Chromatin decondensation in Drosophila embryo extracts. *J. Biol. Chem.* 269, 10169–10176.
681. Crevel, G., and Cotterill, S. (1995). DF 31, a sperm decondensation factor from Drosophila melanogaster: purification and characterization. *EMBO J.* 14, 1711–1717.
682. Loppin, B., Berger, F., and Couble, P. (2001). The Drosophila maternal gene sesame is required for sperm chromatin remodeling at fertilization. *Chromosoma* 110, 430–440.
683. Rice, P., Garduno, R., Itoh, T., Katagiri, C., and Ausio, J. (1995). Nucleoplasmin-mediated decondensation of Mytilus sperm chromatin. Identification and partial characterization of a nucleoplasmin-like protein with sperm-nuclei decondensing activity in Mytilus californianus. *Biochemistry* 34, 7563–7568.
684. Herlands, L., and Maul, G. G. (1994). Characterization of a major nucleoplasmin-like germinal vesicle protein which is rapidly phosphorylated before germinal vesicle breakdown in Spisula solidissima. *Dev. Biol.* 161, 530–537.

Knobil and Neill's Physiology of Reproduction, Third Edition
edited by Jimmy D. Neill,
Elsevier © 2006

CHAPTER 3

Gamete and Zygote Transport

Susan S. Suarez

Introduction, 113
Sperm Transport, 113
Sperm Competition and Cooperation, 113
Timing of Sperm Deposition with Respect to Ovulation, 114
Site of Semen Deposition, 114
Vaginal Defenses against Sperm, 117
Sperm Transport through the Cervix, 117
Sperm Transport through the Uterus, 120
Transport through the Uterotubal Junction, 122
Sperm Transport through the Oviduct, 125
Oocyte and Embryo Transport in the Oviduct, 132
Oocyte Pickup, 133
Oocyte Transport in the Oviduct, 135
Embryo Transport in the Oviduct, 135
Cigarette Smoke Interferes with Oocyte and Embryo Transport, 136
Summary, 136
Acknowledgments, 136
References, 136

INTRODUCTION

How do mammalian sperm and oocytes reach each other? It might seem a simple matter of chance, because millions of sperm are usually deposited in the female by the male. Nevertheless, there is much evidence for the existence of mechanisms that regulate the passage of sperm through the female reproductive tract. These mechanisms ensure not only that fertilization is successful but also that sperm with normal morphology and vigorous motility are the ones to succeed.

In mammals, oocytes are usually fertilized within hours of ovulation (1). Sperm, however, may linger within the female for days or even months, awaiting arrival of the oocyte in the oviduct. In mice, fertilization takes place within 7 hours of mating (2,3), but in some bats, mating takes place during winter and fertilization occurs in the spring (4–6). Because sperm are terminally differentiated cells deprived of an active nucleus and a synthetic apparatus, they must survive the wait without benefit of reparative mechanisms available to many other cells. Sperm are subjected to physical stresses during ejaculation and some phases of transport through the female tract, and they may sustain oxidative damage. Furthermore, because sperm are allogenic to the female, they must endure or avoid the defenses of the female immune system (7). Thus, to succeed, sperm must somehow overcome these challenges.

For the unfertilized oocyte, problems of survival may be minor, because fertilization usually occurs soon after release from the ovary. The major consideration for oocyte transport is to get the oocyte directly and quickly from the ovary into the oviduct, thereby avoiding ectopic fertilization and the dangerous consequences of ectopic implantation (8).

In the previous edition of this chapter, Michael J. K. Harper (9) discussed some of the earlier work on gamete transport in greater detail.

SPERM TRANSPORT

Sperm Competition and Cooperation

Survival of sperm in the female tract is not enough to ensure success. Sperm may be forced to compete with the sperm of other males to pass on their genetic

Department of Biomedical Sciences, Cornell University, Ithaca, New York.

complement (10–13). They may also compete with other sperm from the same ejaculate if mutations occur in germ cells that affect function and are not expressed in all sperm. Intraejaculate sperm competition has been proposed to be one factor driving the production of so many sperm by the male (14). The existence of competition implies a race to the oocyte.

On the other hand, in some species, sperm within an ejaculate cooperate during at least part of their passage through the female tract. In the wood mouse, *Apodemus sylvaticus*, sperm have small hooks on their heads that they use to latch onto each other and form trains of hundreds or thousands of cells. The trains swim faster than solitary sperm (15). There are other examples of sperm joining to each other by their heads. Sperm of new world marsupials, such as the opossum species *Didelphis virginiana* (16) and *Monodelphis domestica* (17), form pairs that break apart shortly before fertilization. Pairing enhances passage of marsupial sperm through viscous media and thus may provide an advantage for sperm swimming in thick oviductal fluid (18). Guinea pig (*Cavia porcellus*) sperm heads stack together in rouleaux in the epididymis and the rouleaux are maintained in the female tract, even into the oviduct (19–21). Sperm of flying squirrels (*Glaucomys volans*) (22) and armadillos (*Cabassous unicinctus*) (23) also form rouleaux. Hamster sperm (*Mesocricetus auratus* and *Cricetulus griceus*) retrieved from the caudal epididymis agglutinate by their heads when incubated under capacitating conditions in vitro and then break apart as they become hyperactivated (24–26); it is not known whether this behavior occurs in vivo. Whether or not these last examples represent sperm cooperation, it is evident that cooperation has evolved in sperm of diverse species.

Timing of Sperm Deposition with Respect to Ovulation

Sperm may be required to survive for quite a long time in species with long estrous periods or with long periods between estrus and ovulation. In such cases, mating early in estrus necessitates storage of sperm. Mares (*Equus caballus*) ovulate about 5 days after the onset of estrus and must therefore be able to store sperm for that period (27). Many species of hibernating vesperilionid and rhinolophid bats mate before they enter hibernation in the autumn and their sperm are stored until spring, when ovulation occurs shortly after arousal (12,28).

One solution to shorten the length of time that sperm must wait for the oocyte is coitus-induced ovulation. Female rabbits, for example, remain in estrus for long periods of time until they mate, which triggers ovulation about 12 hours later (29). In the Sumatran rhinoceros (*Dicerorhinus sumatrensis*), mating activity induces a luteinizing hormone surge within 1–2 hours, followed by ovulation (30). Other induced ovulators include domestic cats and other feline species (31,32), camels (33), the American black bear (*Ursus americanus*) (34), the grasscutter (*Thryonomys swinderianus*) (35), and several species of shrews (31,36,37) and voles (31). Thus, this strategy has been adopted by a broad spectrum of mammals.

Site of Semen Deposition

Vaginal Semen Deposition

The site of semen deposition is not easy to establish, because it must be determined by examining the female immediately after coitus and by considering the anatomy of the penis, vagina, and cervix during coitus. This has been accomplished for humans, in which semen has been observed pooled in the anterior vagina near the cervical os shortly after coitus (Fig. 1). Within minutes of vaginal deposition, however, human sperm begin to leave the seminal pool and enter the cervix (38). Semen of rabbits (*Oryctolagus cuniculus*) is also deposited in the anterior vagina, but a few million sperm have been recovered from the cervix 1 minute after coitus (39).

Rats (*Rattus norvegicus*) and hamsters (*Mesocricetus auratus*) have been classified by some as uterine depositors of semen, because their sperm can be found in the uterine horns minutes after coitus. Nevertheless, results of a careful study by Bedford and Yanagimachi (40) revealed that the sperm are deposited in the vagina but transported through the cervix into the uterus within a few minutes. At the time of vaginal deposition, the sperm are immotile or only weakly motile (40); thus, it is assumed that they are swept through a relaxed cervix into the uterus by muscular contractions of the female tract. In support of this assumption, Carballada and Esponda (41) detected several seminal plasma proteins in rat uterine fluid. As in the rat, the mouse uterine lumen is distended only minutes after coitus by a dense mass containing sperm and bacteria (3), indicating that mouse semen is rapidly swept through the cervix into the uterus as well.

Copulatory Plugs and Gels

Whereas most of the semen of murine rodents is rapidly transported into the uterine cavity, some remains in the vagina where it coagulates to form a

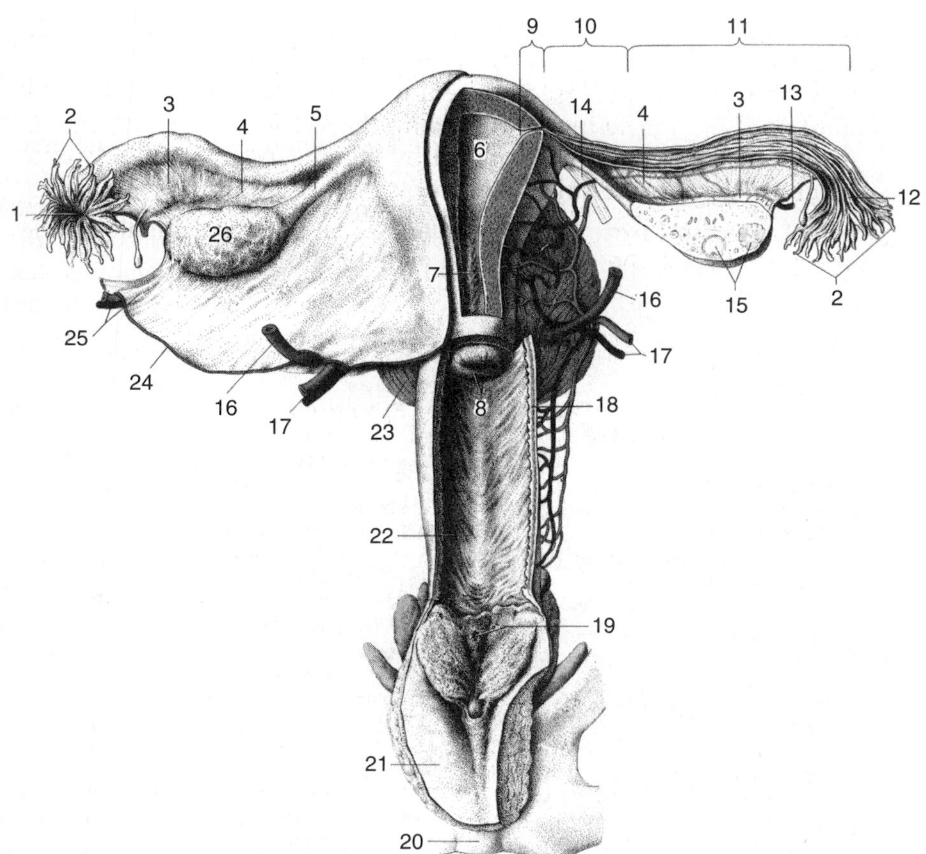

FIG. 1. Human female reproductive tract. (1) Ostium of oviduct (fallopian tube), (2) fimbria, (3) mesovarium, (4) mesosalpinx, (5) ovarian ligament, (6) uterus, (7) cervical canal, (8) external os of the cervix, (9) uterotubal junction, (10) oviductal isthmus, (11) oviductal ampulla, (12) infundibulum, (13) ovarian artery, (14) round ligament of the uterus, (15) preovulatory follicle and corpus luteum in ovary, (16) ureter, (17) uterine artery and vein, (18) vaginal wall, (19) external urethral opening, (20) pubic symphysis, (21) labium majus, (22) rugae in vagina, (23) urinary bladder, (24) cut margin of broad ligament of the uterus, (25) ovarian artery and vein in suspensory ligament of ovary, (26) ovary. [Adapted from (336).]

copulatory plug. The plug forms a cervical cap that promotes sperm transport into the uterus (42–44). Ligation of the vesicular and coagulating glands of rats prevented the formation of plugs and the transport of sperm into the uterus (45). The plugs formed by semen of guinea pigs and mice extend into the cervical canals and thus would seem to form a seal against retrograde sperm loss (42) (Fig. 2). The major protein components of the copulatory plugs of mice are the seminal vesicle-secreted proteins, SVS I and SVS II, the latter also known as semenoclotin (46,47). A transglutaminase secreted by the mouse coagulating gland polymerizes SVS I and SVS II and is thought to be responsible for plug formation in vivo (46,48). A third protein, SVS III, which is also cross-linked by transglutaminase, has been localized to the mouse copulatory plug (49).

Male mice deficient for the gene encoding the protease inhibitor, known as protease nexin-1 (PN-1), show a marked impairment in fertility (50). Vaginal plugs formed in females after mating with PN-1 null males were small, soft, and fibrous and did not lodge tightly in the dual cervical canals. When proteins extracted from plugs produced by mutant and wild-type males were compared by sodium dodecyl sulfate-polyacrylamide gel electrophoresis analysis, significantly fewer protein bands were detected in the mutant extracts, suggesting that, in the mutant, many of the proteins normally forming the plug had been degraded into peptides too small to be retained by the gel. No sperm could be found in the uterus 15 minutes after mating with PN-1 null males, demonstrating the importance of the plug for promoting transport of mouse sperm into the uterus (50).

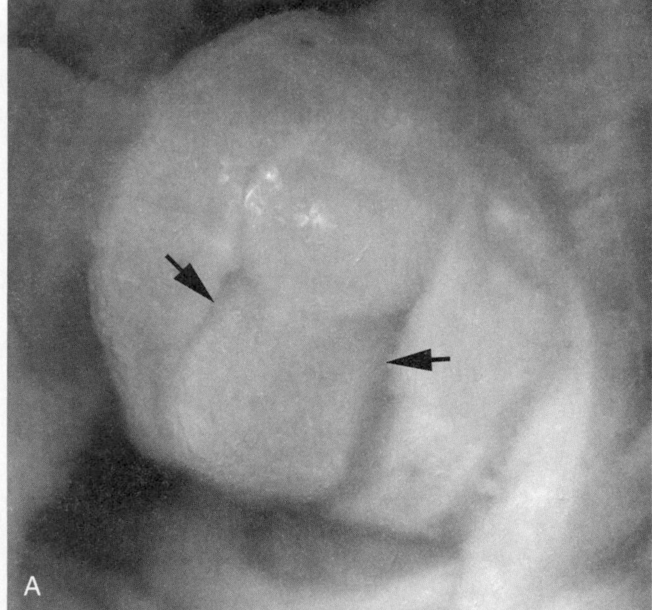

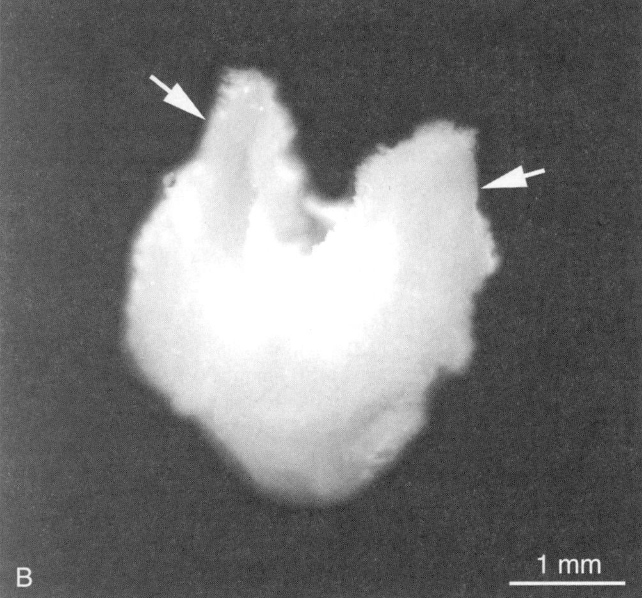

FIG. 2. Copulatory plugs. **A:** External os of mouse cervix. *Arrows* indicate openings. **B:** Anterior portion of copulatory plug removed from vagina. *Arrows* indicate projections that fit into cervical openings. (Photomicrographs by S. S. Suarez.)

The semen of humans also coagulates, although it forms a loose gel rather than the compact fibrous plug seen in rodents. The coagulate forms within about a minute of coitus and then is enzymatically degraded in 1/2 to 1 hour (51). The predominant structural proteins of the gel are the 50-kDa semenogelin I and the 63-kDa semenogelin II, as well as a glycosylated form of semenogelin II, all of which are secreted primarily by the seminal vesicles (52). The gel is degraded by prostate-specific antigen, a serine protease secreted by the prostate gland (53). It has been proposed that this coagulum serves to keep the sperm at the cervical os (9), that it protects sperm against the harsh environment of the vagina (54), and that it prevents semen of other males from entering the cervix (55).

Like humans, some primates produce semen that forms a soft gel. However, in chimpanzees (*Pan troglodytes*), a species in which females mate with more than one male in a brief time, the semen coagulates into a compact plug resembling that of rodents (55,56).

Some carnivores (e.g., domestic dogs, *Canis familiaris*) and some rat and mouse species of the family Cricetidae use the penis as a copulatory plug; that is, the mating pair remains joined together for a period of time after coitus (57). About a minute after a dog has mounted and locked with a bitch, he turns around, such that the two animals face opposite poles. They remain locked together in this position for 5 to 45 minutes. Close examination of the anatomy of the joined genitalia led Grandage (58) to propose that locking serves to "encourage uterine insemination."

Uterine Semen Deposition

Some species bypass the vagina altogether and deposit semen directly into the uterine cavity, where sperm may quickly gain access to the oviduct. In the pig (*Sus scrofa*), the penis is shaped like a corkscrew and the cervix contains complementary furrows. During copulation, the penis fits into the cervix and passes a large volume of semen (about 250 ml) into the uterine cavity (59–61).

Even those species that do not pass the penis through the cervix may use the penis to open the cervical canals. The glans of the rat penis contains projections that seem to be designed to push open flaps guarding the cervical canals (45).

Artificial Insemination

In dairy cattle (*Bos taurus*), bypassing the vagina and cervix has proven advantageous for artificial insemination. Whereas a bull normally deposits several billion sperm into the vagina (Table 1), artificial inseminators deposit 5–20 million frozen/thawed sperm directly into the body of the uterus (62–64). Experiments comparing the fertilization success of standard insemination into the uterine body with deep insemination into the uterine horns so far have yielded mixed results, and therefore deep insemination is not currently used in standard practice (63,65,66).

Artificial insemination into the peritoneal cavity, near the ovary, can result in pregnancy and has been used to treat infertility in humans and repeat breeding syndrome in dairy cows (67). Access to the peritoneal

TABLE 1. *Semen characteristics in domestic animals (61)*

	Bull	Stallion	Ram (Buck)	Boar	Dog	Cat
Volume, ml	4 (1–15)	70 (30–250)	1 (0.7–3.0)	250 (125–500)	10 (1.0–25.0)	0.04 (0.01–0.12)
Sperm concentration, millions/ml	1,200 (300–2,500)	120 (30–600)	3,000 (1,000–6,000)	150 (25–1,000)	125 (20–540)	1,730 (96–3,740)
pH	6.8 (6.2–7.5)	7.4 (7.0–7.8)	6.8 (6.2–7.0)	7.4 (7.0–7.8)	6.7 (6.0–6.8)	7.4
Total sperm per ejaculum, billions (approximate)	4.8	8.4	3.0	37.5	1.25	0.057

Values are means, with ranges in parentheses.

cavity is obtained through the abdominal wall or vaginal wall (67). Evidently, inseminated sperm are swept into the ostium or picked up by the cilia on the surface of the oviductal fimbriae and transported into the ampulla to fertilize oocytes. Oviductal pickup of sperm from the peritoneal cavity may also occur naturally. To demonstrate that sperm can exit an oviduct via the ostium of the ampulla, enter the peritoneal cavity, and then enter the contralateral oviduct through its ostium, Larsson (68) ligated and resected a 2- to 3-cm segment of isthmus of one oviduct of dairy heifers (virgin cows) and then deposited semen into the uterus at estrus. Sperm were recovered only 2 hours after insemination from the resected oviducts, cranial to the resected segment. Presumably, these sperm had reached the contralateral oviduct via the peritoneal cavity.

There are case reports of human tubal pregnancies that arose in spite of lack of access of sperm from the uterus into the oviduct on the side of ovulation (69–71). The only route available to the sperm in these cases was through the peritoneal cavity.

Vaginal Defenses against Sperm

The vagina is open to the exterior and thus to infection, especially at the time of mating; therefore, it is well equipped with antimicrobial defenses that can work against sperm as well as against pathogens. To enable fertilization to take place, both the female and the male have adopted mechanisms for protecting sperm from these defenses. As discussed above, some species bypass the vagina and deposit semen directly into the uterus, whereas vaginal deposition occurs right at the entrance to the cervix.

Vaginal pH

For vaginal inseminators, sperm must contend with the acidic pH of vaginal fluid. The vaginal pH of women is normally 5 or lower, which is microbicidal for many sexually transmitted disease pathogens. Evidence indicates that the acidity is maintained through lactic acid production by anaerobic lactobacilli that feed on glycogen present in shed vaginal epithelial cells (72). Lowering pH with lactic acid has been demonstrated to immobilize bull sperm (73,74). The pH of seminal plasma ranges from 6.7 to 7.4 in common domestic species (Table 1) (61) and has the potential to neutralize vaginal acid. Vaginal pH was measured by radiotelemetry in a fertile human couple during coitus. The pH rose from 4.3 to 7.2 within 8 seconds of the arrival of semen, whereas no change was detected when the partner used a condom (75). Vaginal washings of women with high levels of detectable seminal antigens had a median pH of 6.1, whereas the median pH of washings lacking detectable antigens was 3.7 (76).

Leukocytic Response in the Vagina

In addition to providing protective substances for sperm in seminal plasma, males may also overcome female defenses by inseminating millions or even billions of sperm. This strategy would be particularly effective in overcoming a cellular immune response. In the rabbit, deposition of semen results in an invasion of neutrophils into the vagina. This invasion takes time, however, to build to an effective level. Numerous leukocytes, many containing ingested sperm, were observed in the vaginas of rabbits 3–24 hours postcoitus (77,78). By that time, however, thousands of sperm have already reached the oviduct (79).

Sperm Transport through the Cervix

Cervical Patency

In some vaginal semen depositors, such as primates and dairy cattle, sperm presumably pass through the cervix by swimming through the cervical mucus. In others, such as mice (3), rats (40), and hamsters (40), sperm are swept through the cervix into the uterus within minutes of coitus, probably drawn through by uterine contractions.

The cervix may open during estrus in some species. Fluoroscopy and scintigraphy have been used in domestic dogs and cats (*Felis cattus*) to examine cervical patency. Opening of the cervix in these species has been correlated with estrus (80–82). Radiopaque fluid and also human serum albumin radiolabeled with technetium 99 could be seen rapidly passing through the cervix and filling the uterine lumen after deposition in the cranial vagina at estrus. In these studies, care was taken to avoid applying pressure in the vagina; thus, it is reasonable to conclude that the fluids and particles had been drawn through the cervix by actions of the female tract.

To investigate the role of cervical patency in transport of domestic dog sperm, pregnancy rates were compared after insemination into the cranial vagina or uterus, either before or after cervical closure. Of bitches inseminated into the vagina 24, 48, or 72 hours after cervical closure, none became pregnant; whereas, three of five, three of five, and one of five bitches became pregnant after insemination directly into the uterus at 24, 48, or 72 hours after cervical closure, respectively, with no effect on litter size. These findings indicate that cervical closure is a limiting factor for reproductive success in dogs (81).

Mucous Barrier

After deposition in the cranial vagina, sperm of humans and dairy cows enter the cervical canal rapidly, where they encounter large volumes of cervical mucus produced when systemic levels of estrogen are high. Estrous cervical mucus is highly hydrated, often exceeding 96% water in women (83). The extent of hydration is correlated with penetrability to sperm (84). Coitus on the day of maximal mucous hydration in women is more closely correlated with incidence of pregnancy than coitus timed with respect to ovulation detected using basal body temperature (85). Cervical mucus presents a greater barrier to abnormal sperm that cannot swim properly or that present a poor hydrodynamic profile than it does to morphologically normal vigorously motile sperm and is thus thought to be one means of sperm selection (83,86–88).

The greatest barrier to sperm penetration of cervical mucus is at its border, because here the structure is more compact (89). Components of seminal plasma may assist sperm in penetrating the mucous border. More human sperm were found to enter cervical mucus in vitro when an inseminate was diluted 1:1 with whole seminal plasma than when it was diluted with Tyrode's medium, even though the sperm swam faster in the Tyrode's dilution. It was concluded that components of seminal plasma facilitate penetration of sperm into cervical mucus (90a).

Immune Responses in the Cervix

The cervix is immunologically competent. In rabbits and humans, vaginal insemination stimulates the migration of leukocytes, particularly neutrophils and macrophages, into the cervix and into the vagina (91,92). Neutrophils migrate readily through midcycle human cervical mucus (93). In rabbits, neutrophils were found to heavily infiltrate cervices within a half hour of mating or artificial insemination; however, the invasion was confined to the posterior cervix at the external os (91). Interestingly, it was discovered that if female rabbits were mated to a second male during the neutrophilic infiltration induced by an earlier mating, sperm from the second male were still able to fertilize (94). Thus, although the cervix is capable of mounting a leukocytic response and neutrophils may migrate into cervical mucus, the leukocytes may not present a significant barrier to sperm, at least in the rabbit. It has been demonstrated that neutrophils bind to human sperm and ingest them only if serum that contains both serological complement and complement-fixing antisperm antibodies is present (95). This can happen in vivo if the female somehow becomes immunized against sperm antigens. In conclusion, the evidence indicates that leukocytic invasion serves to protect against microbes that accompany sperm and does not normally present a barrier to normal motile sperm, at least not shortly after coitus.

Immunoglobulins IgG and IgA have been detected in human cervical mucus. Secretory IgA is produced locally by plasma cells in subepithelial connective tissue. The amount secreted increases in the follicular phase and decreases at about the time of ovulation (96). The immunoglobulins provide greater protection from microbes at the time when the cervical mucus is highly hydrated and offers the least resistance to penetration. However, when there are antibodies present that recognize antigens on the surface of ejaculated sperm, infertility can result (7).

Complement proteins are also present in cervical mucus (97), along with regulators of complement activity (98). Thus, there is a potential for antibody-mediated destruction of sperm in the cervical mucus as well as leukocytic capture of sperm. Some antisperm antibodies are not complement activating; however, they can still interfere with movement of sperm through cervical mucus by physical obstruction (7,99).

Preferential Pathways through the Cervix for Sperm

An elegant three-dimensional reconstruction of serial sections of the bovine cervix produced by Mullins and Saacke (100) led them to conclude that mucosal folds in the cervix form channels leading to the uterine cavity (Fig. 3). Furthermore, based on histochemical staining characteristics of the mucus in the sections of cervix, they concluded that mucus deep in the channels is different in composition and less dense from that in the central portion of the cervix during the follicular phase. They proposed that sperm reach the uterine cavity by traveling through deep channels that originate at the external os and by avoiding the more viscous mucus in the center of the cervical lumen that serves to discharge uterine contents. Mattner (101) found that when he flushed the cervices of goats and cows 19–24 hours after mating at the onset of estrus, he recovered approximately 90% of the mucus and more than 90% of the luminal leukocytes but only about half of the sperm. The remaining half of the sperm were found deep in the mucosal grooves, presumably sheltered somewhat from the leukocytes. These observations also indicate that the cervix supports the passage of normal motile sperm while discouraging passage of microbes and sperm with abnormal form or motility. Normal, fresh, motile sperm can avoid the area most populated by neutrophils and they appear to be resistant to leukocytic phagocytosis.

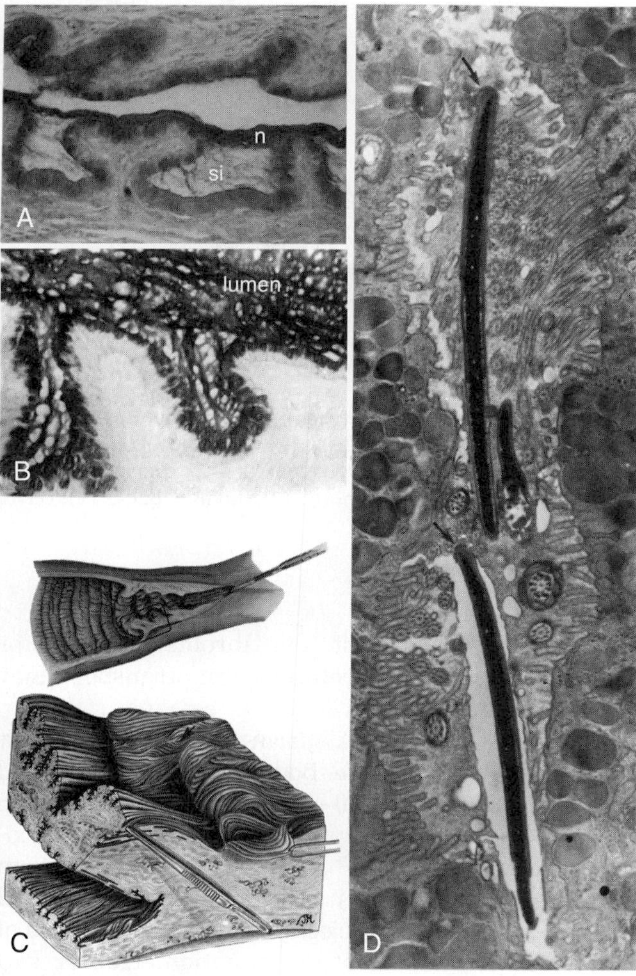

FIG. 3. The bovine cervix (*Bos taurus*). **A:** A section of a primary fold of cervical mucosa taken from a cow in the follicular phase of the estrous cycle. The tissue was fixed in Bouin's solution, dehydrated, embedded in paraffin, sectioned, and stained with Alcian blue (AB) and high-iron diamine (HID). AB-positive sialomucins (si) predominate in basal areas of minor grooves in the mucosa, underneath a layer of denser staining HID-positive neutral mucins and sulfomucins (n). This staining pattern demonstrates that sperm encounter a different type of mucus in the base of grooves (magnification × 100). **B:** In a similar section, taken from a cow in the luteal phase, the AB and HID staining reveal a loss in the layered organization seen in the estrous cow (magnification × 100). **C:** An illustration by K. J. Mullins of the three-dimensional structure of the folds of cervical mucosa, derived from stereomicroscopic examination of tissue stained on its mucosal surface and from three-dimensional reconstruction of serial sections. **D:** Transmission electron micrograph of cervical tissue showing sperm within grooves of cervical mucosa. *Arrows* indicate the rostral tips of the heads of sperm (magnification × 4,460). [All images adapted from (100).]

Sperm may also be guided through the cervix by the microarchitecture of the cervical mucus itself. Mucins, the chief glycoproteins comprising cervical mucus, are long, flexible, linear molecules (M_r of human mucins are approximately 10^7). The viscosity of mucus is due to the large size of mucins, whereas elasticity results from the entanglement of the molecules (102–105). It is thought that these long molecules become aligned by the secretory flow in mucosal grooves and thus serve to guide sperm. Human (106) and bull (107) sperm have been demonstrated to orient themselves along the long axis of threads of bovine cervical mucus. Human sperm swimming through cervical mucus swim in a straighter path than they do in seminal plasma or medium (Fig. 4) (108).

Are Sperm Stored in the Cervix?

Little is known about how long sperm spend traversing the cervix of mucus-laden species or whether sperm are stored there. About 1 day after natural mating, a few million sperm were recovered from the cervices of cows and goats and more than 60% of

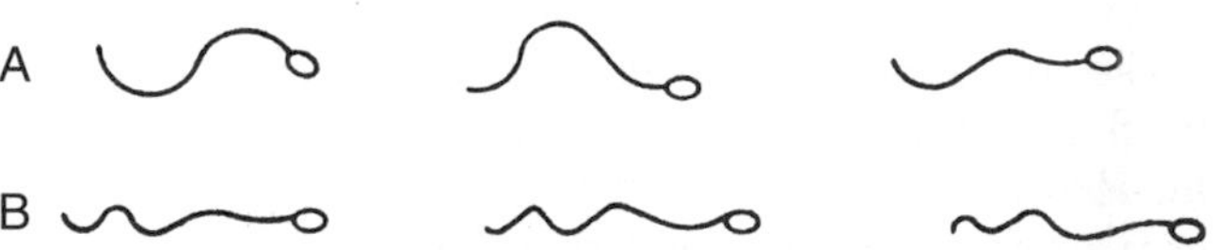

FIG. 4. Flagellar beating patterns of human sperm. **A:** Sperm swimming in seminal plasma or aqueous medium. **B:** Sperm swimming in cervical mucus. [Adapted from (108). Reproduced by permission of the Society for Reproduction and Fertility.]

these sperm were motile (101). Vigorously motile sperm have been recovered from the human cervix up to 5 days after insemination (109). Nevertheless, it is not known whether sperm collected from cervices this long after coitus would reach the oviducts and succeed in fertilizing, nor could it be known whether these sperm had reentered the cervix from the uterus. In vitro, human sperm swim through estrous cervical mucus at a rate of 2–3 mm/min (108). At this rate, the time for traversing the cervical canal would be about 10 minutes.

Sperm Transport through the Uterus

Rate of Transport

In uterine semen depositors, sperm may leave the uterus rapidly to enter the uterotubal junction. In pigs, for example, ligation of the uterotubal junction at various times after mating revealed that sperm in numbers sufficient to achieve fertilization reach the oviduct within 30 minutes of coitus (59). The large volume of porcine semen may ensure that sperm are washed up against the uterotubal junction shortly after coitus. In mice, rats, and hamsters, even though semen is deposited in the cranial vagina, enough fluid is soon transported into the uterine horns along with the sperm, causing visible distension (40,110). Mouse sperm reach the oviduct in substantial numbers within 30 minutes of mating (3).

At only a few centimeters in length, the human uterine cavity is relatively small and could be passed through in less than 10 minutes by sperm swimming at about 5 mm/min, which is the swimming speed of sperm in aqueous medium (111). The actual rate of passage of human sperm through the uterus is difficult to determine due to experimental limitations. Variation is high among women within a study and between studies (112). In one set of experiments, fertile women were inseminated into the cranial vagina shortly before surgical excision of both fallopian tubes. Sperm were recovered from the fimbrial segment of the ampulla in two women whose tubes were removed 5 minutes after insemination, even though they had been abstinent for at least 16 days. Sperm were recovered all along the tubes of two more women merely 10 minutes after insemination (113). Unfortunately, the motility of these sperm was not assessed; therefore, it could not be determined whether the sperm were capable of fertilizing. In another study (114), several motile sperm were recovered from fallopian tubes after hysterectomy 30 minutes after insemination in one patient and 1 hour after insemination in three of seven patients; however, these women underwent surgery for treatment of fibroids, polyps, or endometriosis and therefore sperm transport may have been abnormal.

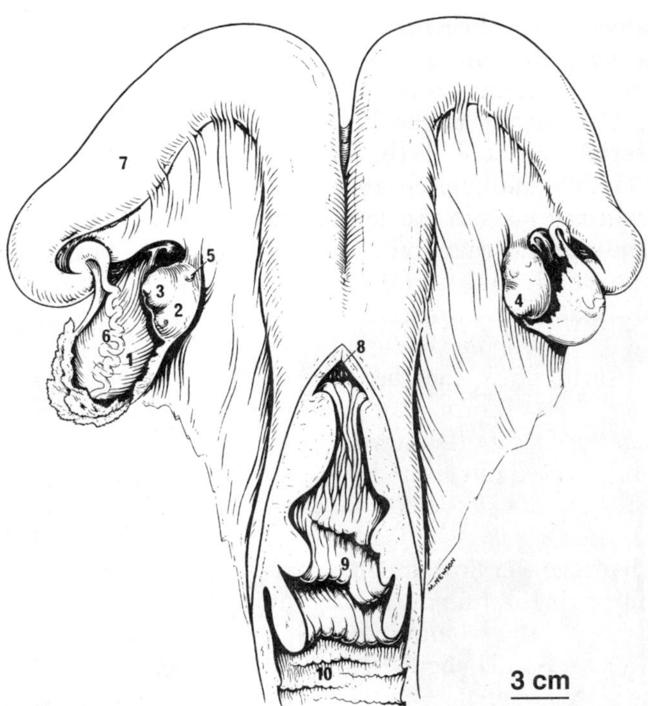

FIG. 5. The dorsal aspect of the female reproductive tract of a cow (*Bos taurus*). (1) ovarian bursa, (2) ovary, (3) corpus luteum, (4) follicle in ovary, (5) corpus albicans in ovary, (6) oviduct (uterine tube), (7) uterine horn, (8), uterine body, (9) cervix, (10) vagina. [From (61).]

In contrast to human sperm, bovine sperm must pass through a uterine body 2.5–4 cm long and uterine horns that are 20–40 cm long before reaching the uterotubal junction (Fig. 5) (61). At an approximate swimming speed of 7 mm/min, bull sperm would require about an hour to reach the uterotubal junction if they swam directly from the internal os of the cervix to the uterotubal junction. Hunter and Wilmut (115,116a,b) ligated the uterotubal junction at various times after mating and later flushed the oviduct to recover embryos. Their results indicate that about 8 hours are required for bovine sperm, in numbers sufficient to achieve fertilization, to reach the oviduct. A similar experiment in sheep also indicated 8–10 hours are required for sperm in sufficient numbers to reach the oviduct (117).

Uterine Peristalsis

Transport of sperm through the uterus is likely aided in some species by proovarian contractions of the myometrium. Ultrasonography of the human uterus revealed cranially directed waves of uterine smooth muscle contractions that increased in intensity during the late follicular phase (118,119). The uterine contractions occurring in women during the periovulatory period are limited to the layer of myometrium directly beneath the endometrium (118,120). This is in contrast to contractions occurring during menses, which involve all layers of the myometrium. In cows and ewes, electromyography indicated that strong contractile activity occurs during estrus, whereas contractions are weak and localized during the luteal phase (121).

In humans, contractile activity of uterine muscle may draw sperm and watery midcycle mucus from the cervix into the uterus. Fukuda and Fukuda (122) interpreted ultrasound images of the uteri of women in the late follicular phase to indicate that the uterine cavity is filled with mucus. They proposed that the cervical mucus assists sperm movement through the human uterine cavity. This is possible because the volume of uterine fluid in midcycle women is only about 100 μl (123) and cervical mucus is plentiful enough to fill the lumen.

Kunz and collaborators (119) deposited 5- to 40-μm albumen macrospheres radioactively tagged with technetium into the cranial vaginae of women to determine how such contractions might transport sperm. They found that spheres were rapidly and maximally transported into the uterine cavity and even into the oviductal isthmus during the late follicular phase. Interestingly, transport of the spheres was greater to the isthmus ipsilateral to the dominant follicle than to the contralateral isthmus. This preferential transport may result from signals passed via a vascular communication between the preovulatory follicle and the uterus and oviduct. An arterial anastomosis lies between the ovarian and uterine arteries (which also supply the oviduct) in the corneal region of the human uterus (Fig. 1). Doppler flow sonography revealed increased perfusion of the anastomosing vessels on the side of the preovulatory follicle (124). It is thought that these vessels can carry hormones from the dominant follicle directly to the uterus and oviduct without first passing through the systemic circulation, because the ovarian artery associates closely with the ovarian vein. This association is thought to enable a countercurrent transfer of ovarian hormones from the venous drainage of the ovary to the ovarian artery and then to the arterial supply of the uterus and oviduct (125,126). In addition, lymphatic drainage of the ovary might transfer hormones to the ovarian artery and then the oviductal vessels (126). Although these anatomical relationships appear to be widespread in mammals, the data demonstrating transfer of hormones or other signals along this route are still preliminary.

Studies of uterine contractions during estrus should be interpreted with caution if the females were not mated. Videolaparoscopic examination of mated and unmated rats revealed significant changes in contractile patterns of the uterine horns after mating. Unexpectedly, the change consisted of several-fold increases in both cranially and caudally propagating circular contractions (127). Caudally directed peristalsis would be expected to carry sperm away from the uterotubal junction. In estrous domestic cats, both ascending and descending contractions were observed by fluoroscopy (82). Perhaps the ebb and flow of contractions direct fresh waves of sperm to the junction.

Myometrial contractions may be stimulated by seminal components. When vasectomized male rats were mated with females, the incidence of strong uterine contractions declined, indicating that sperm or testicular or epididymal secretions have stimulatory activity (127). In addition, the products of male accessory sex glands can play a role. Removal of the seminal vesicles significantly reduced the pregnancy rate in mice (110). In boars, there is evidence that estrogens, which may reach 11.5 μg in an ejaculate, increase myometrial contraction frequency (128). Because boar semen is deposited directly into the uterine cavity, the uterus is exposed to the full amount of estrogens in the semen. There is evidence that the estrogens enhance contraction by stimulating secretion of prostaglandin $F_{2\alpha}$ (128).

Immunological Responses in the Uterus

Rapid transport of sperm through the uterus by myometrial contractions can enhance sperm survival by propelling them past the immunological defenses of the female. As is the case in the vagina and cervix, coitus induces a leukocytic infiltration of the uterine cavity, which reaches a peak several hours after mating in mice (1). The leukocytes are primarily neutrophils and have been observed phagocytizing uterine sperm in mice, rats, and rabbits (1,129). This phagocytosis was observed several hours after insemination and therefore might be directed primarily against damaged sperm. However, normal sperm may also be attacked, particularly in vaginal inseminators because their sperm have lost much of the protection afforded by seminal plasma. Seminal plasma inhibits immune responses and contains protective components that coat sperm (130,131).

When sperm first enter the uterus, they outnumber the leukocytes. As time passes, the leukocytes begin to outnumber the sperm, and the sperm probably lose some protective seminal plasma coating. At some point, even undamaged sperm may fall victim to the leukocytes. Thus, it is imperative that sperm pass through the uterine cavity before significant numbers of leukocytes arrive.

Transport through the Uterotubal Junction

The Junction as a Barrier to Sperm Migration

The uterotubal junction presents anatomical, physiological, and/or mucous barriers to sperm passage in most eutherian mammals. Anatomically, the lumen in species as distantly related as dairy cattle and mice is particularly tortuous and narrow (132–136). The narrowness of the lumen is especially apparent in living tissue (137) and in frozen sections (Fig. 6), in which tissue does not shrink as it does during standard preparation of paraffin-embedded sections (136).

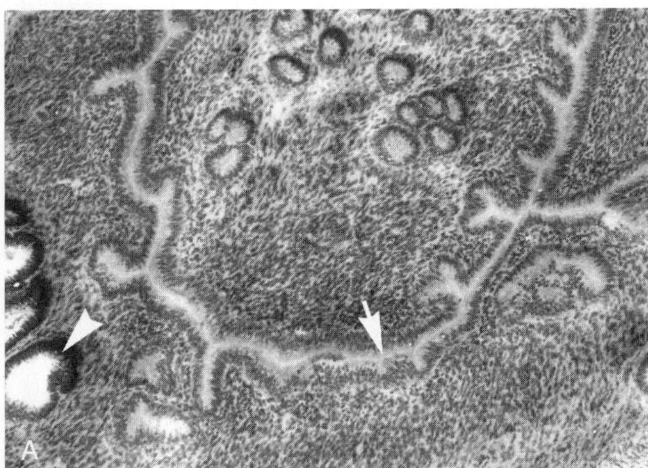

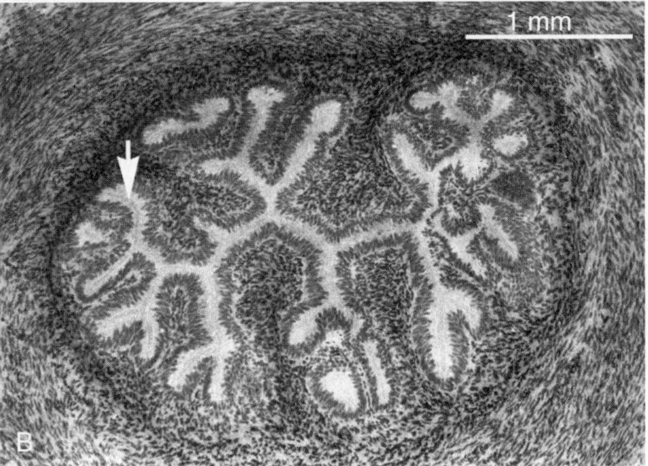

FIG. 6. Frozen sections of the bovine uterotubal junction and caudal isthmus, stained with periodic acid–Schiff reagent and counterstained with hematoxylin (136). **A:** A section of the uterotubal junction proximal to the uterus. *Arrow* indicates lumen of the uterotubal junction; *arrowhead* indicates a uterine gland. **B:** A section of caudal isthmus, proximal to the uterotubal junction. The narrow, mucus-filled lumen (*arrow*) may be distinguished as a slightly darker region bounded by the lightly stained apical cytoplasm of the mucosal epithelial cells. [Adapted from (337).]

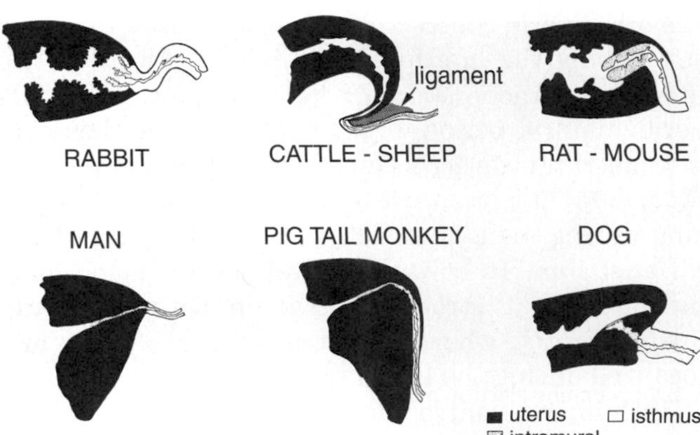

FIG. 7. Diagrammatic illustration (not drawn to scale) to show species differences in the morphology of the uterotubal junction. Various anatomical relationships between the oviduct and uterus can be seen: the conspicuous folds in the rabbit, the flexure of the junction and isthmus in cows, the colliculus in the rat and mouse, the funnel shape in humans and other primates, and the mound-like papilla in the dog. [From (133).]

The entrance to the junction is fairly simple in humans, whereas it is complicated by mucosal folds in cows, pigs, rabbits, and many other species (Fig. 7) (132–135). In mice and rats, the entrance forms a conical projection into the uterus, called a colliculus tubarius (Fig. 8) (3,137,138).

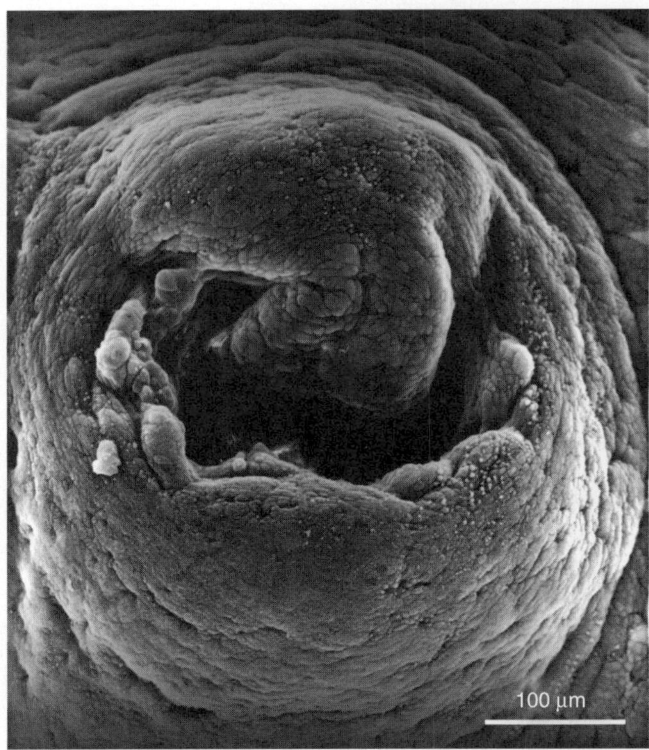

FIG. 8. Scanning electron micrograph of the colliculus tubarius, the uterine entrance to the uterotubal junction in the rat. (Provided by David M. Phillips.)

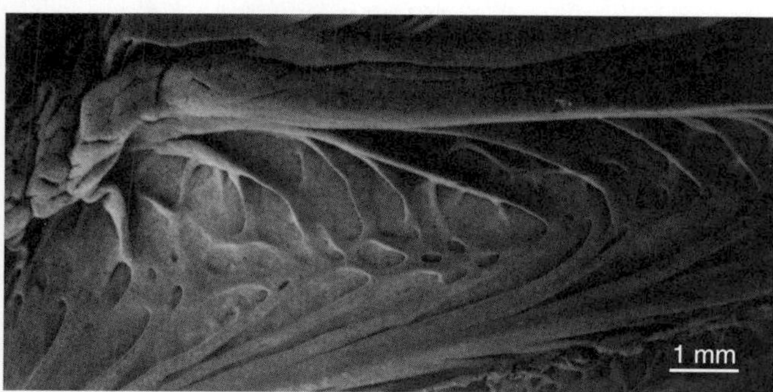

FIG. 9. Scanning electron micrograph of the mucosal (inner) surface of the bovine uterotubal junction, opened longitudinally. Between terminal primary folds are secondary folds that form culs-de-sac that open toward the uterus. [From (139).]

Within the lumen of the junction there are large and small folds in the mucosa, some of which create grooves that end blindly. In the cow, mucosal folds form culs-de-sac with openings that face back toward the uterus (135,139). This arrangement of folds seems designed to entrap sperm and prevent further ascent (Fig. 9).

A physiological valve may be created by a vascular plexus in the lamina propria/submucosal layer of the junction wall. When engorged, the plexus can compress the lumen. This plexus has been well described in cattle (135). The walls of the bovine junction and adjacent tubal isthmus also contain a thick muscular layer that could further constrict the lumen. The bovine uterotubal junction is sigmoidal in shape and supported by muscular ligaments that appear capable of increasing the flexure of the curve and thus compressing the lumen (132,133). The human junction traverses a thick muscular layer of uterine wall (133). In the mouse, the junction is reported to be patent shortly after coitus but to be tightly closed about an hour later (Fig. 10) (3,137).

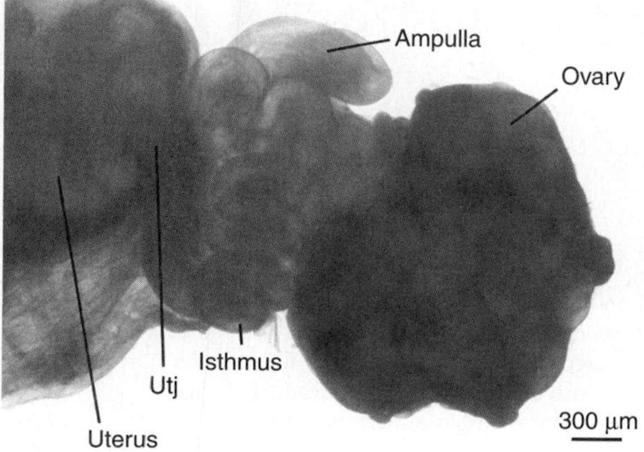

FIG. 10. Photomicrograph of a freshly dissected oviduct of a mouse, illustrating the long, sigmoidal uterotubal junction (Utj) and the highly coiled isthmus and ampulla. [From (337).]

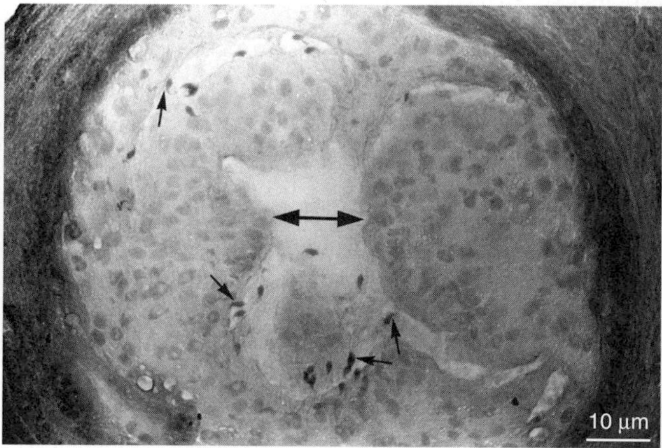

FIG. 11. Frozen section of sperm in the extramural segment of the mouse uterotubal junction, close to the isthmus, stained with periodic acid–Schiff reagent to show mucus and with hematoxylin as a nuclear counterstain (136). The *two-headed arrow* indicates the central part of the lumen, filled with mucus; *arrows* indicate heads of some of the sperm in the lumen. (Provided by Robert P. DeMott.)

The narrow lumen may be filled with a viscous mucus that can also impede the progress of sperm (Figs. 6 and 11). Mucus has been found in the uterotubal junction in rabbits (140,141), pigs (142a), dairy cattle (136,143), and humans (144).

The Junction as a Filter

For uterine semen depositors, the uterotubal junction, particularly when filled with mucus, may serve the filtration function served by the cervix in vaginal depositors. That is, it may filter out pathogens introduced with the semen and morphologically abnormal sperm or sperm with poor motility. In pigs (145), rats (138), and hamsters (146) motile sperm pass through the uterotubal junction much more successfully than immotile sperm. Sperm demonstrating linear progressive motility are more successful at passing through the uterotubal junction than are sperm swimming in nonlinear patterns, such as those circling because of an asymmetrical flagellar beating pattern (138,147). In species in which whole semen enters the uterus, the uterotubal junction may also serve to filter out seminal plasma. Seminal plasma components are left behind in the uterus and are not detected in the oviducts of rats (41).

Male mice that are null mutants for the genes encoding fertilin β (148), calmegin (149,150), or testis-specific angiotensin-converting enzyme (ACE) (151,152) are infertile because their sperm cannot pass through the uterotubal junction or bind to the zona pellucida. In these null mutants, both the motility and morphology of the sperm are normal when examined in vitro. Fertilin β is localized on the plasma membrane overlying the acrosome on mature sperm from wild-type males, whereas it is lacking in the null mutants (148). As for calmegin, sequence homology indicates that it is a chaperone protein, which would place it in the endoplasmic reticulum of spermatids, assisting in the proper folding of proteins destined for membranes. Both wild-type and null mutants lack calmegin in mature sperm; therefore, its effect on fertility is presumed to be due to the lack of proteins that rely on calmegin for proper placement in the sperm plasma membrane. In the case of *ACE* null mutants, antibody labeling indicates that testis-specific ACE is normally shed from mouse sperm when they enter the epididymis; therefore, it has been proposed that ACE has an enzymatic effect on the surface of maturing sperm that somehow enables them to pass through the female tract (153).

Figure 12 shows that sperm from mice that are null mutants for *calmegin* do not enter the oviduct. The role of calmegin in enabling sperm to pass into the oviduct was examined more closely using chimeric males that produced a mixture of germ cells with wild-type and disrupted *calmegin* genes. The question addressed was whether calmegin-chaperoned proteins are required by individual sperm to pass through the uterotubal junction or would the presence of wild-type sperm enable them to do so. Such would be the case, for example, if the proteins on the sperm surface assist passage by signaling the uterotubal junction to open. Chimeric males were created by fusing embryos from "wild-type" mice that had normal *calmegin* genes with those from a double transgenic line of mice that were homozygous null for *calmegin* and expressed enhanced green fluorescent protein in their acrosomes. The resulting chimeric XY/XY males produced a mixture of sperm, about half of which were mutant, as identified by the presence of the fluorescent acrosomes (as seen in Fig. 12). When these males were mated with wild-type females, only wild-type sperm could be found within the oviduct (154). This indicates that normal morphology and motility are not sufficient for enabling sperm to enter the oviduct. An additional factor, likely a sperm surface protein or proteins, is required by each sperm to pass through the junction.

Rapid Transit through the Junction

Sperm have been recovered in the cranial reaches of the ampulla only minutes after mating or insemination in humans (113) and several other species of mammals (39,121,155). This phenomenon has been

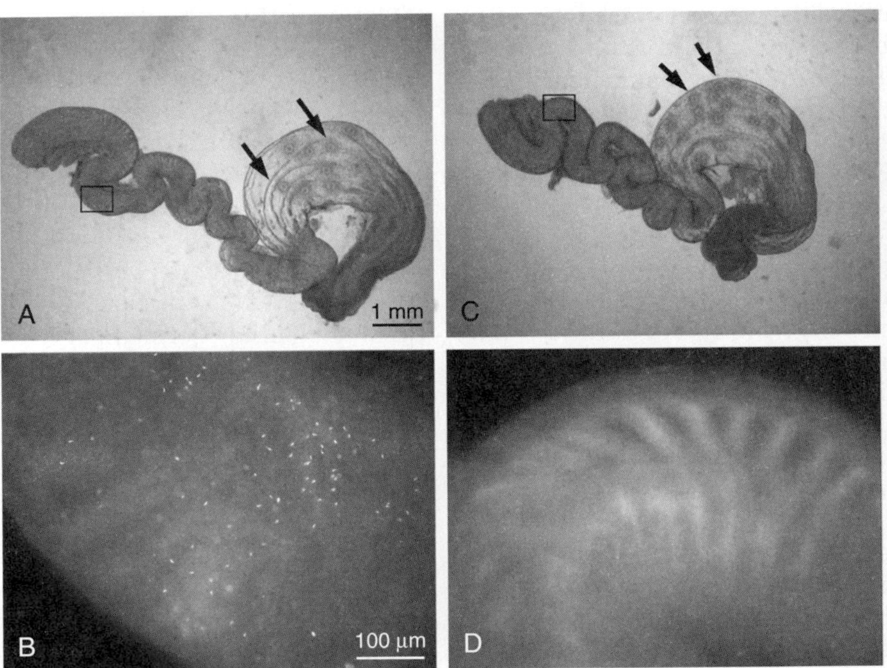

FIG. 12. Comparison of oviductal colonization by sperm from *calmegin* –/– and *calmegin* +/– male mice. Both homozygotes and heterozygotes produced sperm tagged with enhanced green fluorescent protein in their acrosomes, due to insertion of a transgene into the line. *Arrows* indicate oocytes within the oviductal ampullae. **A:** Whole mount of an oviduct from a female mated with a *calmegin* +/– male. The *box* indicates the isthmic region shown in B. **B:** Fluorescent illumination of the boxed area from A shows sperm within the oviductal isthmus. **C:** Whole mount of an oviduct from a female mated with a *calmegin* –/– male. The *box* indicates the isthmic region shown in D. **D:** Fluorescent illumination of the boxed area from C reveals no sperm in the isthmus. [From (154).]

termed rapid sperm transport or rapid transit (39). Rapid transport of sperm into the oviduct would seem to counter the proposed model of sperm swimming one by one through the uterotubal junction. However, when the condition of rabbit sperm recovered from the cranial ampulla shortly after mating was evaluated by Overstreet and Cooper (39), they found that most were immotile and damaged. They proposed that waves of contractions stimulated by insemination transport sperm rapidly to the site of fertilization, but these sperm are mortally damaged by the associated shear stress and do not fertilize. Later, motile sperm gradually pass through the uterotubal junction to establish an oviductal population capable of fertilizing. Rapid transport resulting in sperm damage has not been studied this carefully in other species, but it is likely to apply in cases where sperm are found near the ovary within a few minutes of coitus.

Sperm Transport through the Oviduct

Oviductal Sperm Reservoir

The Reservoir Stores Sperm and Prevents Polyspermy. Upon entering the oviduct, sperm become trapped and form a reservoir. The mammalian oviductal sperm reservoir may have been first described in hamsters by Yanagimachi and Chang (156) and has since been reported to exist in a variety of species, such as hamsters (157), rabbits (79,158), cows (116a), pigs (59), and sheep (159).

The oviduct provides a safe haven for sperm. Unlike the vagina, cervix, and uterus, the oviduct does not respond to insemination with an influx of leukocytes (160). However, the oviduct goes beyond providing a safe haven; it somehow acts to maintain the fertility of sperm between the onset of estrus and the time of ovulation. Sperm fertility and motility are maintained longer in vitro if the sperm are incubated with oviductal epithelium [bovine (161,162), porcine (143), equine (163), human (164), and canine (165)].

Sperm storage structures have developed in other groups of vertebrates. Sperm are stored in folds of ovarian tissue in several species of viviparous fishes (166). In several families of turtles, sperm storage tubules exist in the region of the oviduct that is homologous to the mammalian isthmus (167–169). Evidently, these storage tubules allow females to fertilize multiple clutches of eggs, sometimes years after mating. Several species of snakes and lizards are known to store sperm in specialized structures in the anterior vagina and infundibulum of the oviduct (170–173). Neither of these sites is homologous to the mammalian isthmus, however. Many birds are known to store sperm in tubular crypts at the uterovaginal junction (12,174–176), which enables them to lay multiple clutches of eggs after a single act of coitus.

The entrapment of mammalian sperm in the oviductal isthmus may also serve to prevent polyspermic fertilization by allowing only a few sperm at a time to reach the oocyte in the ampulla. Sperm numbers have been artificially increased at the site of fertilization in the pig by surgical insemination directly into the oviduct (177,178), by resecting the oviduct to bypass the reservoir (179), or by administering progesterone into the muscularis to inhibit smooth muscle constriction of the lumen (180,181). In each of these cases, the incidence of polyspermy increased.

Sperm Are Trapped Primarily by Binding to Epithelium. There is strong evidence from multiple species that the oviductal reservoir is created in eutherian mammals when sperm bind to oviductal epithelium. Motile sperm have been observed to bind by their heads to the apical surface of the oviductal epithelium in cattle (Fig. 13) (143), mice (137), hamsters (182), pigs (142), horses (183), and dogs (184). Entrapment may be enhanced by the mucus-filled narrow lumens of the uterotubal junction and isthmus, which would slow the sperm and increase their contact with the mucosal epithelium, thus facilitating binding.

Guinea pig sperm have been observed stacked in large rouleaux within mucosal pockets in the uterotubal junction 1–12 hours after copulation (19,20), indicating that sperm-to-sperm binding may also play a role in sperm storage.

Rabbit sperm flushed from the oviductal isthmus with oil showed only sluggish motility but could be revived by dilution in medium. This led to the proposal that motility suppression is a mechanism for trapping and storing sperm (90b,185,186). In hamsters (187) and mice (137), immotile sperm have been seen in the central part of the isthmic lumen. These sperm, however, may not be the ones to go on to fertilize, because highly motile sperm were found in the same isthmuses bound tightly to the epithelium (187).

Sperm Bind to Carbohydrates on Oviductal Epithelium. Sperm binding to oviductal epithelium involves carbohydrate recognition. Fetuin and its terminal sugar, sialic acid, were found to competitively inhibit binding of hamster sperm inseminated into oviducts, whereas desialylated fetuin did not (188). The sperm could be labeled over the rostral portion of the head by colloidal gold-tagged fetuin, indicating they bind fetuin in the same region in which they bind to epithelium. Gold-tagged fetuin also labeled certain protein bands on Western blots of membrane extracts from hamster sperm (188). These observations indicate that there is a

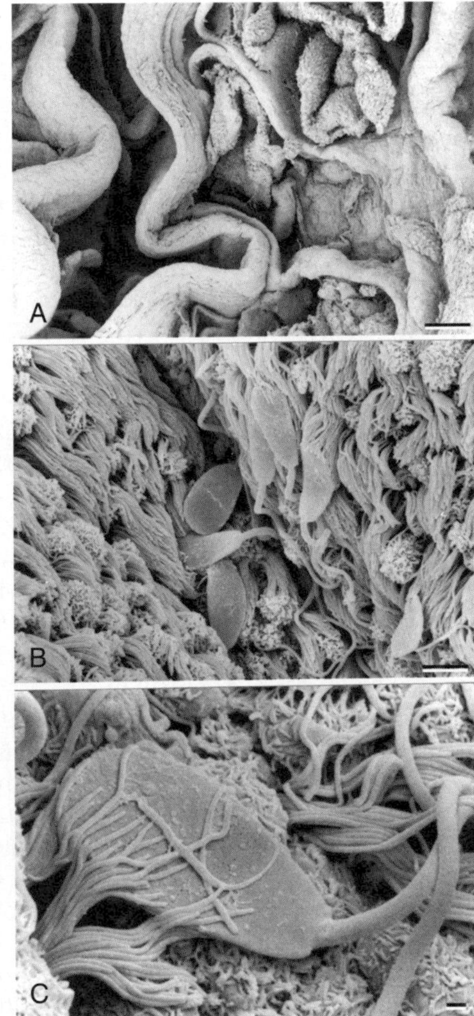

FIG. 13. Scanning electron micrographs of bovine sperm and mucosal epithelium of the oviductal isthmus. Sperm are located in grooves created by mucosal folds. **A:** A low magnification view of the isthmus (bar = 75 μm). **B:** A higher magnification of a mucosal groove (bar = 5 μm). **C:** A high magnification view of a sperm cell associated with the cilia of the epithelium (bar = 1 μm). [From (221).]

carbohydrate-binding molecule on the heads of hamster sperm that binds sialic acid and is responsible for attachment of sperm to the epithelium.

Binding of other species of sperm to conspecific epithelium could also be competitively inhibited by specific carbohydrates. Binding of stallion sperm to explants of oviductal epithelium was inhibited by asialofetuin and its terminal sugar, galactose (189,190a), whereas binding of boar sperm was blocked by mannose (191,192). Bull sperm binding was blocked by fucoidan and its component fucose (193). Pretreatment of bovine epithelium with fucosidase, but not galactosidase, also reduced sperm binding (193). Thus, carbohydrate involvement in sperm binding to epithelium appears to be a widespread phenomenon among eutherian mammals, although the particular carbohydrate moiety comprising the binding site varies according to species. In each of the species studied so far, a different sugar inhibited binding in vitro. These species differences may not seem so unusual when one considers that a single amino acid residue can determine the ligand specificity of a lectin (194,195) and that closely related animal lectins have different carbohydrate specificities (196).

Other forms of heterotypic binding between cells involve carbohydrate recognition. Examples are the selectins, which mediate leukocyte binding to endothelium (197), and glycolipid ligands on ciliated respiratory cells, which are recognized by mycoplasmas (198). Selectins mediate temporary binding between the two cell types, just as binding between sperm and epithelium is temporary. Carbohydrate recognition is also implicated in sperm–zona binding (199–201) and sperm–Sertoli cell binding (202,203). During the course of evolution, lectins with different specificities could have arisen to regulate sperm attachment to these different surfaces.

The binding of bull sperm to oviductal epithelium has been more fully characterized than that of other species. Fucose, in an alpha 1–4 linkage to *N*-acetylglucosamine, as in the trisaccharide Lewis-a, inhibited binding to oviductal epithelium more efficiently than fucose in monosaccharide form. Furthermore, Lewis-a tagged by conjugation to fluorescein-labeled polyacrylamide bound to the heads of live bull sperm (204).

Lewis-a was used to affinity purify a protein of approximately 16.5 kDa from extracts of bull sperm plasma membranes. This protein competitively inhibited sperm binding to explants of oviductal epithelium. The protein was identified by amino acid sequencing as PDC-109, a secretion of bovine seminal vesicles that is adsorbed onto sperm (Fig. 14) (205). PDC-109 purified from seminal plasma also inhibited sperm binding to epithelium. Bovine epididymal sperm lack PDC-109 on their membranes (Fig. 15) and, indeed, show only minimal binding to oviductal epithelium in vitro. However, when epididymal sperm were incubated with purified PDC-109, washed to remove unadsorbed protein, and then added to explants of oviductal epithelium, sperm binding was raised to the level of ejaculated sperm (206).

PDC-109 is the most abundant protein in bovine seminal plasma, present at concentrations of 15–20 mg/ml (207). It occurs in two forms: BSP-A1 and BSP-A2. BSP-A1 possesses a single trisaccharide, NeuNAc-Gal-GalNAc, that is *O*-linked via GalNAc to threonine residue number 11 (208), whereas BSP-A2 lacks the trisaccharide. The variably glycosylated N-terminal segment is followed by two tandem fibronectin type II domains (209). PDC-109 binds to choline phospholipids via short hydrophobic

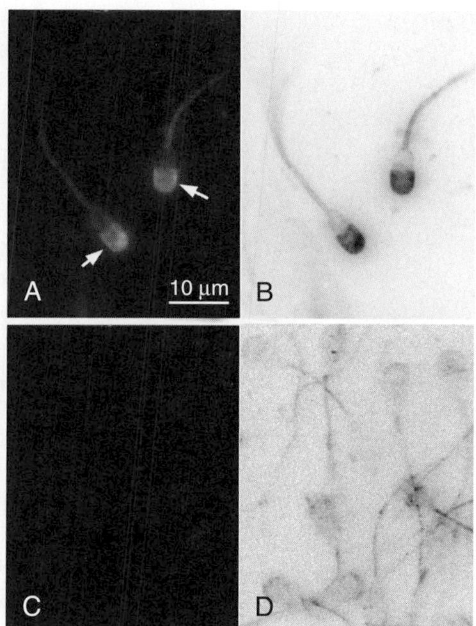

FIG. 14. Indirect immunofluorescent labeling of ejaculated bovine sperm with a polyclonal antibody to PDC-109. **A:** Anti-PDC-109 antiserum diluted 1:1,000. *Arrows* show labeling of acrosomal region. **B:** Inverted image of A. **C:** Preimmune serum diluted 1:50. **D:** Inverted image of C, enhanced to show all sperm. [From (206).]

sequences within its fibronectin domains (210). This is thought to be the mechanism by which it adsorbs onto epididymal sperm when they are exposed to seminal plasma (211,212).

Preserving Sperm Fertility during Storage. The oviductal mucosa protects sperm against aging damage during storage. Sperm incubated with oviductal epithelium in vitro remain viable longer than when they are incubated in medium alone [porcine (143), equine (163), human (164), canine (165)] or with tracheal epithelium [bovine (161)]. Viability can be extended by incubating sperm with vesicles prepared from the apical membranes of oviductal epithelium [rabbit (213), equine (214), human (215)], indicating that the epithelium can produce the effect by direct contact rather than by secretions. It was reported that equine sperm binding to epithelium or membrane vesicles maintained low levels of cytoplasmic Ca^{2+} compared with free-swimming sperm, sperm attached to Matrigel, or sperm incubated with vesicles made from kidney membranes (190b,214). Equine and human sperm incubated with oviduct membrane vesicles also capacitated more slowly than sperm incubated in capacitating medium alone, when capacitation was assessed by chlortetracycline fluorescence patterns (214,215). Possibly, viability is maintained by preventing capacitation and its concomitant rise in cytoplasmic Ca^{2+}. The mechanism for preventing rises of cytoplasmic Ca^{2+} in sperm are not known, but one suggestion is that catalase, which is present in the oviduct, serves to protect against peroxidative damage to the sperm membranes (216).

The oviductal binding protein on bull sperm, PDC-109, probably acts to stabilize sperm membranes. By using spin-labeled lipids, it was determined that PDC-109 reduces membrane fluidity and immobilizes cholesterol in phospholipid membranes, including those of epididymal sperm (217,218). PDC-109 can also contribute to membrane stability by inhibiting the activity of phospholipase A_2 (219,220). Thus, PDC-109 may play a role in preserving sperm fertility while they are stored in the reservoir.

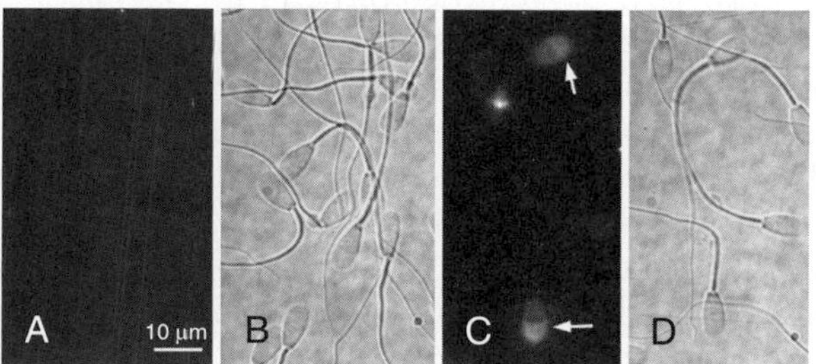

FIG. 15. Bovine epididymal sperm adsorbed PDC-109 over the acrosomal region of the head plasma membrane, shown using anti-PDC-109. **A:** Epididymal sperm without PDC-109 show no antibody label. **B:** Bright-field image of A. **C:** Epididymal sperm previously treated with PDC-109 are labeled over the acrosomal region (*arrows*). **D:** Bright-field image of C. [From (206).]

Release of Sperm from the Reservoir. Sperm could either be released from the reservoir via loss of binding sites on the epithelium or by loss of binding protein from sperm. Changes in the hormonal state of oviductal epithelium related to impending ovulation were not found to affect the density of binding sites for sperm (142a,183,221); therefore, it appears that the epithelium does not release sperm by reducing available binding sites. Instead, current evidence indicates that a change in the sperm brings about its release.

Sperm undergo two changes in preparation for fertilization: capacitation and hyperactivation. Capacitation involves changes in the plasma membrane, such as loss of proteins and cholesterol, which prepare sperm to undergo the acrosome reaction and fertilize oocytes (see Chapter 2). Loss or modification of proteins on the surface of the plasma membrane overlying the acrosome could reduce binding affinity for the oviductal epithelium. Hyperactivation is a change in flagellar beating that involves increased flagellar bend amplitude and, usually, increased asymmetry of the beat (Figs. 16 and 17) (222,223). Hyperactivation can provide the force necessary for overcoming the attraction between sperm and oviductal epithelium (223).

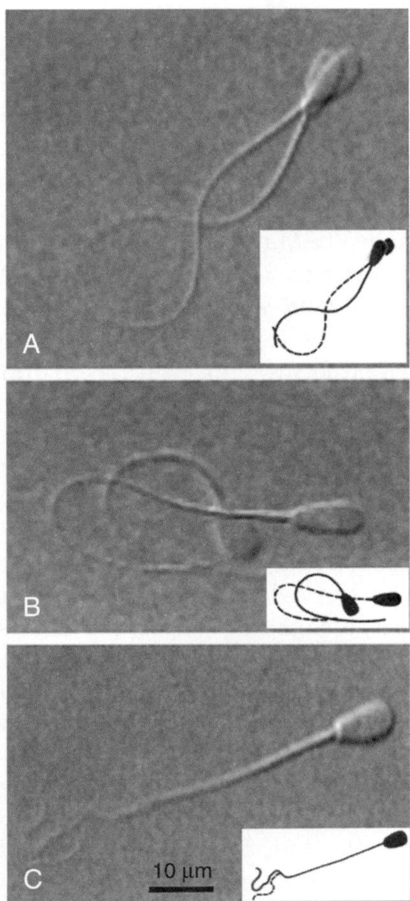

FIG. 16. Video images of swimming patterns of bull sperm on glass slides. Sperm were illuminated by flashes from a xenon stroboscope at 60 Hz and images were captured at 30 Hz. Traces of the images are shown in the insets. **A:** The symmetrical flagellar beat pattern of activated bull sperm in aqueous medium. **B:** The asymmetrical beat pattern of a hyperactivated sperm. **C:** The progressive swimming of a hyperactivated sperm through a thick viscoelastic solution of long-chain polyacrylamide. [From (222). Reproduced by permission of the Society for Reproduction and Fertility.]

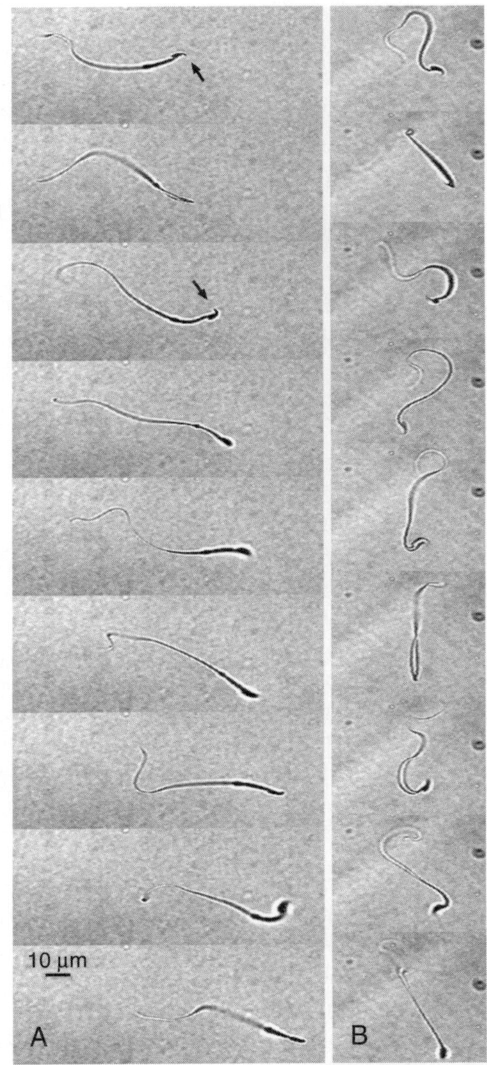

FIG. 17. Phase contrast sequential video images of (**A**) activated and (**B**) hyperactivated mouse sperm. Consecutive images were collected at 30 Hz. Each series begins at the top of the figure. Rotation of the head (rolling) can be envisioned by noting the position of the hook of the head (*arrows*). Note the increased flexing of the hyperactivated sperm. [From (222). Reproduced by permission of the Society for Reproduction and Fertility.]

Oviducts removed from mated mice can be transilluminated to examine the behavior of sperm within the reservoir. Under these conditions, it was noted that only hyperactivated sperm detached from epithelium (224). Hamster sperm that had undergone both capacitation and hyperactivation in vitro did not bind to epithelium when infused into hamster oviducts, whereas uncapacitated/unhyperactivated sperm did bind (182). When bull sperm were capacitated by incubation with heparin before being added to explants of oviductal epithelium, binding was significantly reduced (225). In this case, the sperm were capacitated but not hyperactivated; therefore, it was concluded that capacitation reduces binding affinity of sperm for epithelium. Taken together, these observations indicate that capacitation-induced changes in the sperm head surface are responsible for loss of binding affinity, although the pull produced by hyperactivation can enhance the ability of bound sperm to detach from the epithelium.

Hyperactivation of sperm can also assist their escape from the reservoir by propelling them through mucus in the oviductal lumen. Mucus that fills the uterotubal junction also extends into the isthmus in rabbits (140,141), pigs (142a), dairy cattle (136,143), and humans (144). Hyperactivated sperm penetrate artificial mucus, such as viscoelastic solutions of long-chain polyacrylamide or methyl cellulose, far more effectively than nonhyperactivated sperm (142b,226,227); therefore, they are better equipped to swim through oviductal mucus (Fig. 16). Even in the absence of mucus, hyperactivation can assist escape from the reservoir, because it endows sperm with greater flexibility for turning around in pockets of oviductal mucosa (Fig. 17) (222,223,228,229).

Although evidence is lacking for a release mechanism involving reduction in binding sites on the epithelium, the epithelium may still play a role in sperm release by secreting factors that affect sperm. For example, hormonal signals that induce ovulation or signals from the preovulatory follicle could stimulate the epithelium to secrete factors that stimulate sperm capacitation and hyperactivation, thereby bringing about sperm release. Soluble oviductal factors do enhance capacitation of bull sperm in vitro (230,231).

The carbohydrate-binding molecule responsible for binding sperm to the epithelium is lost or loses binding affinity as sperm become capacitated. Hamster sperm capacitated in vitro no longer labeled over the acrosomal region with fetuin (188), indicating they had lost the ability to bind to oviductal epithelium via sialic acid. Similarly, electrophoretically resolved proteins extracted from capacitated hamster sperm showed diminished fetuin labeling on blots, compared with proteins extracted from uncapacitated sperm (188).

Capacitated bull sperm also showed reduced binding to oviductal epithelium (225), as well as to fucose, as measured using fluorescein-labeled fucosylated bovine serum albumin (205,232). The loss of binding affinity for the oviductal epithelium can be accounted for by a loss of the adsorbed seminal plasma protein PDC-109 from the sperm head. When PDC-109 coats sperm by adsorbing to choline phospholipids in the plasma membrane, it forms homodimers with heparin-binding sites exposed (208,233). Heparin is used to capacitate bull sperm in vitro (234–236), and there is evidence that PDC-109 is lost from the sperm surface during heparin capacitation (237). Addition of heparin to cocultures of bovine sperm and oviductal epithelium enhances the release of sperm from epithelium (238). In vivo, as the time of ovulation approaches, an increase of heparin-like glycosaminoglycans in oviduct fluid could serve to release sperm. Oviduct fluid collected from cows in the periovulatory period showed a peak in heparin-like biochemical activity as well as a peak in capacitation-inducing activity (235).

Shedding of PDC-109 from sperm during capacitation somehow facilitates removal of cholesterol, which is a key step in the capacitation process (239,240). Thus, PDC-109 appears to play a dual role, one in building the sperm reservoir and the other in capacitating sperm.

Summary: Entrapment and Release of Sperm in the Reservoir. In summary of what is known about sperm movement in the oviduct, the following picture emerges. The sperm reservoir forms in the uterotubal junction and/or the caudal isthmus when a carbohydrate-binding molecule on the sperm head adheres to an oligosaccharide ligand on the oviductal epithelium. The narrowness of the oviductal lumen and mucus within the lumen can enhance sperm binding by slowing their progress and increasing their contact with the epithelial surface. Binding to the mucosal epithelium prolongs sperm survival and delays capacitation. As the time of ovulation approaches, capacitation is likely initiated by secretions in the oviduct fluid. The carbohydrate-binding molecule on the surface of sperm is shed during capacitation, thereby allowing sperm to release from the epithelium. As sperm gradually lose binding affinity for the epithelium, hyperactivation provides them with the force to pull away from its surface.

Oviductal Sperm Storage in Marsupials and Insectivores. In marsupial mammals (241,242), sperm are stored in special mucosal crypts in the oviduct (Figs. 18 and 19). However, sperm within

FIG. 18. Sperm in mucosal crypts in the isthmic region of an oviduct from an Australian marsupial, *Sminthopsis crassicaudata*. *Arrows* point to some sperm heads. Nomarski optics of a fresh specimen. Methods described in (243). (Provided by J. Michael Bedford.)

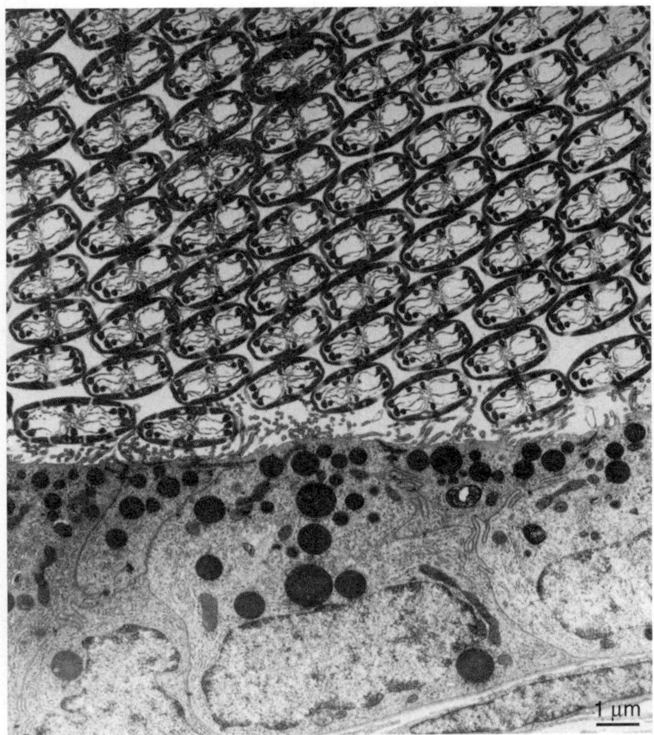

FIG. 19. Transmission electron micrograph of sperm in an isthmic crypt of a recently mated Australian marsupial, *Sminthopsis crassicaudata*. Cross-sections of tightly packed sperm tails are seen above sections of isthmic epithelium. [From (296), with permission of CSIRO Publishing, Melbourne, Australia.]

the crypts do not attach to the epithelium. Many of the sperm in the crypts of the marsupial *Sminthopsis crassicuadata* were observed to be immotile (243); thus, motility suppression may serve to keep sperm in the tubules until ovulation.

In primitive eutherian mammals in the Order Insectivora, some species of shrews and moles possess isthmic sperm storage crypts (244), whereas others store sperm in bubble-like outpocketings of the ampullar wall (Fig. 20) (245,246). Sperm found within these structures also do not bind to the epithelium. Perhaps the most complex oviductal sperm storage system is that of the white-toothed shrew (*Crocidure russula*) and the African shrew (*Myosorex varius*). In these insectivores, sperm inhabit crypts in the isthmus before ovulation and move up to bubble-shaped outpocketings in the ampulla at about the time of ovulation (Fig. 21). Sperm seen in the isthmic crypts have slowly beating flagella, whereas sperm in the ampullar structures have rapidly beating flagella (247,248).

Another insectivore, the African pygmy hedgehog (*Atelerix albiventris*), lacks distinctive oviductal sperm storage structures, but little is known about how they store sperm (249). More advanced eutherian mammals, such as mice, cows, and pigs, also lack distinctive storage structures. In these species, sperm bind to epithelium primarily in grooves created by folds of mucosa. It is curious that distinctive storage structures would be lost and sperm binding would evolve to replace them as a mechanism of sperm storage.

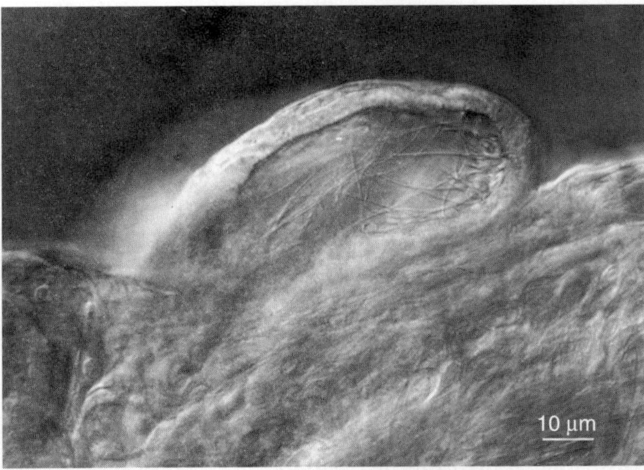

FIG. 20. A transilluminated bubble-shaped outpocketing of the wall of the oviductal ampulla of the insectivore *Cryptotis parva*. The sperm within were flagellating vigorously before they were exposed to 2% glutaraldehyde for the purpose of photography. [From (245).]

Sperm Storage in Women. So far, no conclusive evidence has emerged for a distinct oviductal sperm reservoir in humans (250). Although associations of human sperm with oviductal epithelium have

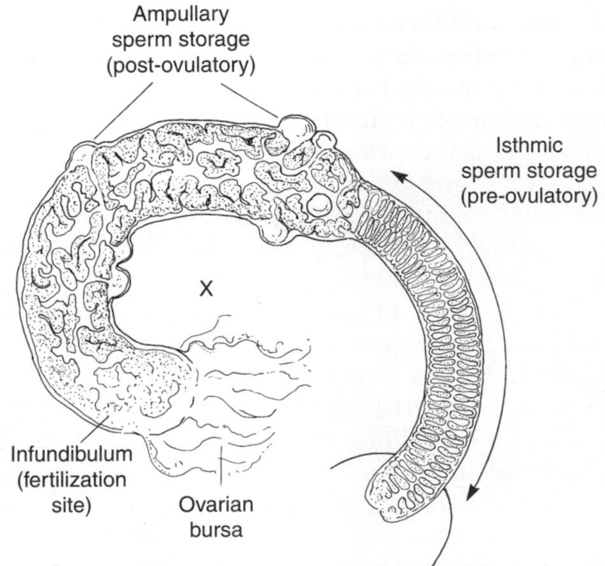

FIG. 21. Diagram of the main features of the oviduct of the shrew, *Crocidura russula* monacha. Crypts are seen in the wall of the isthmus, whereas bubble-shaped outpocketings are seen on the surface of the ampulla. The oviduct is approximately 5 mm long. X indicates the location of the ovary. [From (247).]

been observed in vitro, the sperm do not seem to bind as quickly or as tightly to the epithelium as those of nonprimate mammals (215,251–254). On the other hand, human sperm viability is maintained by incubation with oviductal epithelium (215), as it is in species in which there is immediate and strong binding of sperm to epithelium (161,162,165).

The human cervix has been designated a sperm reservoir. The cervical lumen is about 3 cm in length (255); therefore, it could store many thousands of sperm. The human uterus is rather small in proportion to body size, and theoretically, human sperm could travel through the lumen to reach the uterotubal junction in about 10 min. The entrance to the uterotubal junction in humans is shaped like a funnel and lacks the tall elaborate mucosal folds characteristic of nonprimate mammals (133,134). The short uterus and open uterotubal junction could enable human sperm to pass quickly into the oviduct and escape phagocytosis by uterine leukocytes.

The evidence that could be used to argue against a cervical reservoir is that very few sperm have been recovered from human uteri 24 hours after coitus (114,256). Furthermore, the leukocytic infiltration of the uterus, which becomes significant several hours after coitus (257), could present a barrier to passage of sperm that had been stored in the cervix. The leukocytes appear to outnumber human sperm in the uterus at 24 hours after coitus (257). Unless sperm are protected from phagocytosis (and they might be!), it is unlikely they could travel from the cervical reservoir to the oviduct 24 hours postcoitus.

Alternatively, human sperm could be stored for long periods of time in the oviduct, but not in a distinct reservoir and not as a result of binding tightly to the mucosal surface. The mucosal folds of the human oviductal lumen increase in height and complexity toward the ovary (Fig. 22), thus offering increasingly greater obstacles to the advancement of sperm. Sperm progress could be slowed by the mucus in the lumen (144) and by sperm sticking lightly to the epithelium (252,253). So, rather than having a distinct reservoir, human sperm advancement to the site of fertilization could be slowed in such a manner as to increase the chances that a few will be present at the site of fertilization when ovulation occurs. Muscular contractions and secretions at the time of ovulation could move or activate sperm and increase chances of encountering the oocyte. Furthermore, as discussed below, human sperm might be chemotactically drawn to the oocyte by follicular fluid introduced into the oviduct by the cumulus mass at ovulation (258).

In conclusion, data of sperm distribution in the tubes of women have not provided a clear picture of the events of sperm transport. Sperm recovered at various times in different regions of the human oviduct have varied so much in numbers that the data do not permit the construction of a model for the pattern of sperm transport (250). Perhaps fertilization is a relatively inefficient and unregulated process in humans, because the sexual act has taken on an additional role of promoting long-term pair bonding in addition to promoting fertilization success.

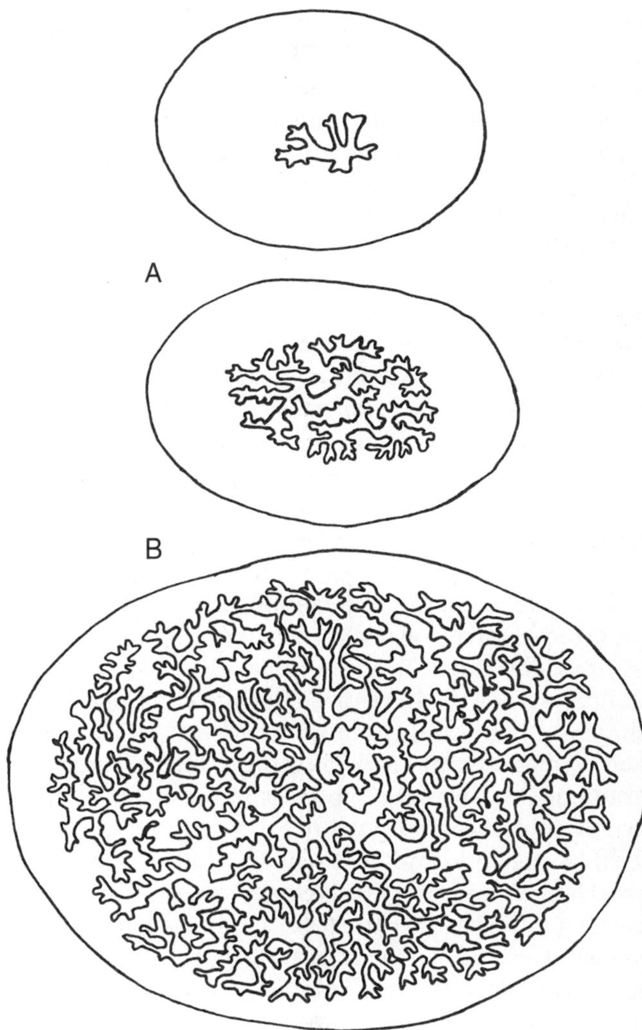

FIG. 22. Cross-sections showing the increasing complexity of the human oviductal lumen from the uterotubal junction (**A**), to the isthmus (**B**), to the ampulla (**C**). (Illustration by C. Rose Gottlieb.)

Sperm Migration into the Ampulla

When hyperactivated mouse sperm detach from oviductal epithelium, they bounce around in the lumen, reattaching and detaching several times (224). Covering space rapidly in this manner, they could eventually encounter the cumulus mass, which fills the ampullar lumen (Fig. 12). Mouse sperm are not driven back by ciliary movement in the ampulla, which sweeps the oocyte caudally as the sperm advance cranially. Hamster sperm observed through the ampullar wall glide rapidly over the mucosal surface (259).

Sperm from mice in which the gene for CatSper1 or CatSper2 has been disrupted do not reach the oocytes in the oviductal ampulla. Although they show normal vigorous progressive motility, they cannot hyperactivate and do not penetrate artificial mucus as well as wild-type sperm (227,260–262). Thus, hyperactivation is required for fertilization in vivo.

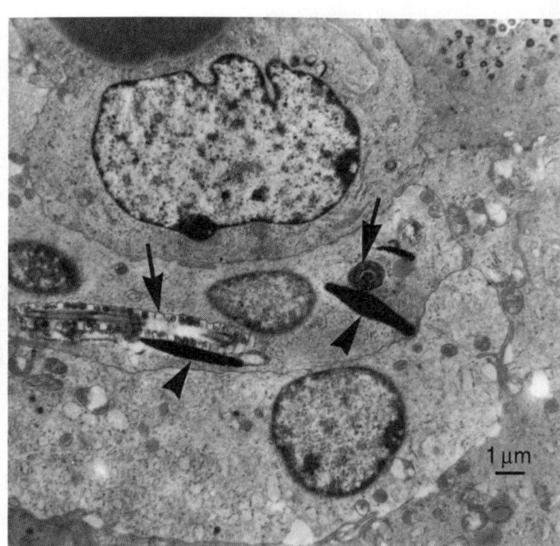

FIG. 23. Phagocytosis of surplus sperm by isthmic epithelium. Transmission electron micrograph of mouse isthmic epithelium, taken 15 to 22 hours after coitus. Flagella (*arrows*) and sperm heads (*arrowheads*) can be seen in the cytoplasm of the epithelial cells. Methods described in (272). (Provided by Joana Chakraborty.)

Chemotaxis

In vitro, human and rabbit sperm have been reported to turn toward or accumulate in a gradient of follicular fluid (258,263–266). Odorant receptors unique to sperm have been localized to a spot on the base of the flagellum of canine (267), rat (268), and human sperm (269). Placing human sperm in a gradient of the odorant bourgeonal caused them to orient into the gradient (269). Sperm are equipped with a mechanism for turning; that is, they can switch back and forth between symmetrical flagellar beating and the asymmetrical flagellar beating of hyperactivation. Hyperactivation is reversible (270), so sperm can alternate between turning and swimming straight ahead. However, the identity of the signal or signals that instigate switching between symmetrical and asymmetrical flagellar beating in vivo is not known.

Chemotactic guidance of mammalian sperm to oocytes, if it exists, is likely to be limited to short distances within the oviduct, because the coils of the oviduct, the mucosal folds, muscular contractions, and ciliary currents would disrupt a gradient over longer distances. Babcock (271) suggested that human sperm could be guided into the cumulus mass from nearby mucosal pockets in the oviductal ampulla. In other species, such as mice, large cumulus masses fill the width of the ampullar lumen (Fig. 12). In these species, chemoattraction could serve to guide sperm to the center of the cumulus mass, if a gradient of attractant were established within the cumulus.

Fate of Nonfertilizing Sperm in the Oviduct

After fertilization, mammalian sperm may be phagocytosed by isthmic epithelial cells (Fig. 23) (247,272,273) or may be eliminated into the peritoneal cavity (274) and then phagocytosed. Phagocytosis within the oviduct may be used by species, such as mice, which have an extensive ovarian bursa that would limit passage of sperm into the peritoneal cavity.

OOCYTE AND EMBRYO TRANSPORT IN THE OVIDUCT

In many mammalian species, oocytes progress from prophase of the first meiotic division to metaphase of the second meiotic division just before ovulation. Most then arrest in the second metaphase until fertilization activates completion of meiosis (275). Oocytes are usually fertilized soon after entering the oviduct. If mating or artificial insemination is delayed until the time of ovulation or after, the incidence of pregnancy is reduced and the occurrence of developmental anomalies is increased (276). The most common anomalies detected after delayed insemination are polyspermy (multiple sperm fertilizing a single oocyte), polygyny (failure to complete the second meiotic division and extrude the second polar body), and disintegration of the oocyte chromosomes (276,277). Aged unfertilized oocytes may undergo a spontaneous, though abnormal, form of activation, eventually leading to cell death (278,279). Aged fertilized oocytes exhibit developmental abnormalities as well. Female mice inseminated 10 hours after ovulation had shortened gestation length and decreased litter size. Their pups had a higher perinatal death rate, growth retardation, and other developmental delays (280). In the CF-1 mouse strain, oocyte abnormalities begin to appear about 4 hours after ovulation or about 16 hours after human chorionic gonadotropin administration (278), although the fertile life span of oocytes in Swiss Webster mice may be 15 hours or more (277). Early reports indicate that the fertile life span of oocytes is about 6 hours in the rabbit (281), 12 hours in the rat (282), 9 hours in the golden hamster (283), and 20 hours in the guinea pig (284). Ferret (*Mustela* sp.) oocytes may be fertilizable for 30–36 hours (285,286).

Oocyte Pickup

There are two steps involved in oocyte transport. The first is the picking up of the oocyte from the surface of the ovary or from the ovarian bursa by the fimbria. The second is transport of the oocyte through the oviductal ampulla.

Richard Blandau's pioneering films of ovulation and oocyte pickup in rabbits inspired an appreciation of the process. The films revealed that the mesosalpinx contracts rhythmically during ovulation, causing the fimbria to slide over the surface of the ovary (42). The mesosalpinx also moves the oviduct, mesovarium, and ovary to aid in the positioning of the fimbria over the ovary. In addition, contractions of the muscularis of the wall of the fimbria contribute to moving them over the ovarian surface.

In some species, such as rats and hamsters, the fimbriae are small and cannot sweep over the surface of the ovary. In these species, the space between the ovary and oviductal ostium is nearly completely enclosed by the mesovarium and mesosalpinx, which together form a bursa. The cumulus mass is actually ovulated into the bursa, where it is jostled by movement of the ovary and oviduct until it comes into contact with cilia on the fimbrial surface and is picked up (42,287,288). A movie of pickup of hamster oocytes by Talbot and collaborators can be viewed online (289). At ovulation in the hamster, follicular contents are extruded as a long sticky strand of cumulus, matrix, and oocyte. The strand soon makes contact with the cilia on the surface of the fimbriae, drawing the rest of the mass away from the surface of the ovary. The mass is then rapidly transported over the fimbrial surface toward the ostium (Fig. 24). Upon reaching the ostium, the mass churns in the entrance and becomes compacted before it enters the ampulla (288,289).

Ciliary currents alone cannot sweep an object the size of a cumulus mass or even an oocyte into the ostium of the ampulla; such currents can only move small particles, such as *Lycopodium* spores (290) or

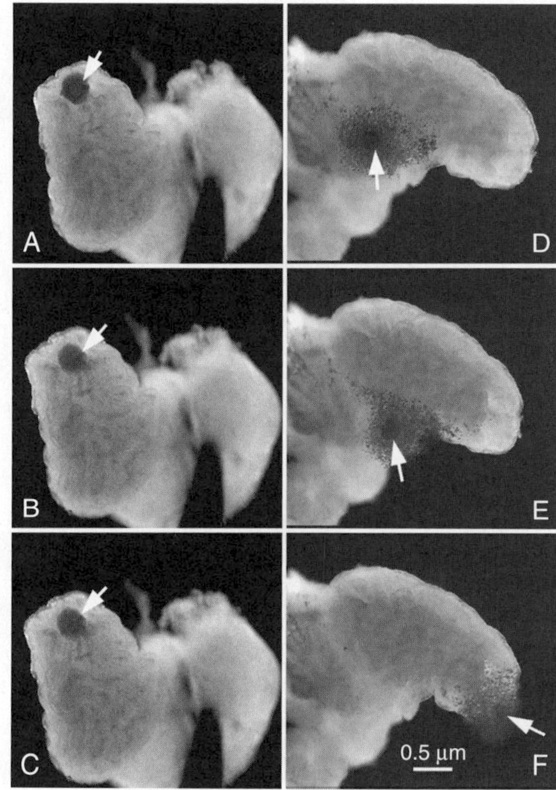

FIG. 24. Video images of unexpanded and expanded hamster oocyte cumulus complexes interacting with infundibula. **A–C:** The unexpanded oocyte–cumulus complex (shown 0, 3, and 6 minutes after being placed in the ostium) did not adhere to the surface of the infundibulum and did not undergo normal pickup. **D–F:** In contrast, the expanded oocyte–cumulus complex moved across the surface of the infundibulum to the entrance into the ampulla in about 10 seconds. [From (288).]

sperm (68–71). In the eutherian mammals studied so far, adhesion of the cumulus mass to the cilia is required for successful pickup. The sites of interaction on the surface of the fimbriae are the tips of its cilia (Fig. 25) (288,291).

Adhesion between the cilia and oocyte is mediated by the negative charge of sialic acid glycosyl moieties on the surface of the cilia. Polycationic molecules block oocyte pickup in the rabbit and hamster (287,290,291). Neuraminidase pretreatment of hamster fimbriae to remove sialic acid reduces pickup (287).

On the oocyte side of the adhesion, presence of cumulus is required for pickup. In rabbits and hamsters, if the cumulus and its matrix are removed from the oocyte, it is not picked up by the fimbria (42,287).

Hyaluronic acid is the chief component of the extracellular cumulus matrix; however, it is unlikely to be the molecule directly responsible for sticking the cumulus to the fimbrial cilia. When placed on the surface of hamster fimbriae, hyaluronate gel was not picked up. Furthermore, a solution of hyaluronate did not block pickup, and pretreatment

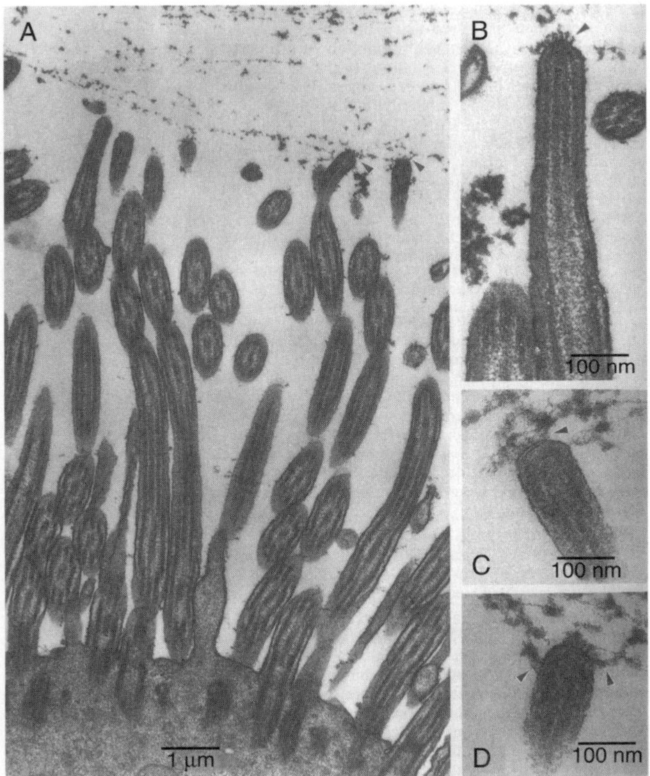

FIG. 25. Transmission electron micrographs showing the interaction of infundibular cilia with the matrix of the cumulus oophorus. **A:** Tips of cilia in contact with the cumulus matrix at its periphery (*arrowheads*). **B:** Higher magnification of the interaction, showing the glycocalyx of the tip of a cilium (*arrowhead*). **C:** Higher magnification showing granules and filaments of the matrix (*arrowhead*) associating with the tip of a cilium. **D:** Another high magnification view of matrix components associating with the tip of a cilium. [From (288).]

of the cumulus with hyaluronidase did not prevent pickup. Cumulus-free oocytes were not picked up, even if they were coated with Na-hyaluronate (287). In contrast, coating the cumulus-free oocytes with egg white restored pickup (287). The specific molecules involved in adhering cumulus to cilia have yet to be identified.

Fertility of female mice was severely reduced by targeted disruption of the bikunin gene (292). Bikunin is the light chain of proteins in the IαI family (inter-alpha-trypsin inhibitors). Heavy chains in the family, called SHAPs (serum-derived hyaluronan-associated proteins), bind directly to hyaluronic acid. IαI proteins formed from the light and heavy chains link hyaluronic acid and associated molecules together to form gels. In the mutant mice, cumulus formation was abnormal but could be significantly improved if IαI was infused into the peritoneal cavity (292). TNFAIP6 (hyaladherin tumor necrosis factor–α–induced protein 6) binds to hyaluronan and forms a complex with IαI proteins (293). It acts as a catalyst to covalently link the heavy chains of IαI to hyaluronan. Disruption of the gene encoding TNFAIP6 also severely reduced fertility of female mice. Fertility was significantly improved by injecting mouse recombinant TNFAIP6 protein into the peritoneal cavity at the time of ovulation induction by human chorionic gonadotropin (293). Similarly, disruption of the gene for PTX3 (pentraxin 3) resulted in abnormal cumulus formation and female subfertility. PTX3 binds to cumulus matrix protein TNFAIP6 (294,295). In vitro fertilization was normal in *Ptx3*$^{-/-}$ mice, indicating that although a normal matrix is not required in vitro, it is critical in vivo. PTX3 is also present in human cumulus matrix (295). The exact roles of these molecules in adhesion to the fimbriae are unknown. They may either interact directly with molecules of the ciliary plasma membranes, or they may assemble other cumulus matrix molecules into the proper molecular structure for adhesion to the cilia.

In hamsters, the adhesive attraction between cumulus mass and oviductal cilia increases 10- to 40-fold when the mass reaches the ostium. After compaction of the cumulus in the ostium, adhesive strength of the interaction decreases precipitously (288).

Although the cumulus is undoubtedly important for ovum pickup in the eutherian mammals studied, a cumulus oophorus has not been found in marsupial mammals. The granulosa cells do not accompany the oocyte at ovulation (241,296,297). In some shrews, which are considered primitive eutherian mammals, the cumulus does not have a visible matrix at the time of ovulation, although a matrix may be produced after fertilization, when the oocyte is in the oviduct (245,298).

Bedford (297) noted that the diameter of the central lumen of the oviduct matches the size of the ovulatory products in mammals. In most eutherian mammals, there is a large expanded cumulus mass surrounding the oocytes, which fills the relatively large central space of the ampulla or the entire ampullar lumen (Fig. 26). Cumulus expansion is accomplished by the secretion of hyaluronic acid and other matrix materials, followed by hydration. In marsupial mammals, there is no cumulus surrounding the ovulated egg and the ampulla is narrower than those of eutherian mammals (296). In shrews, an intermediate situation exists; that is, there is a cumulus around the ovulating oocytes, but it is not expanded (245,247,298). Correspondingly, the ampullar lumen of shrews is intermediate in diameter and fits closely around the compact cumulus mass (Fig. 26). So, in all cases, the oocyte and its vestments fit snugly in the central ampullar lumen. These observations raise an interesting question: Why did the vestments of the oocyte increase and expand? Bedford and Kim (299) proposed that the expanded cumulus evolved to serve to sequester sperm.

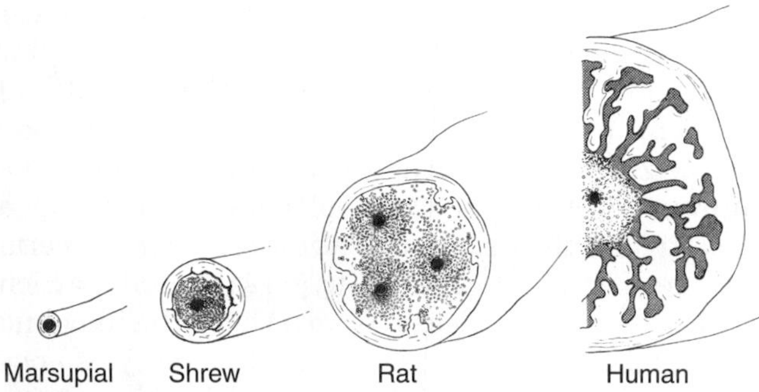

FIG. 26. Diagram of the spatial relationship between the egg or egg–cumulus complex and the site of fertilization in the oviduct of various representative mammals. Where the oviduct is much larger than the egg, the ability of the cumulus to fill the space is maximized by a variable degree of cumulus expansion. [From (297).]

Oocyte Transport in the Oviduct

In eutherian mammals, once the cumulus mass containing one or more oocytes enters the ampulla, it moves rapidly to the ampullary–isthmic junction. Potential effectors of this movement are the oviductal musculature and the cilia. When rabbit oviductal smooth muscle contractions were blocked by isoproterenol, the net rate of transport of cumulus–oocyte masses down the ampulla (about 0.12 mm/sec) was not affected (300). Isoproterenol was also tested in rats, where it inhibited muscle contractile activity without affecting ciliary activity or transport of surrogate ova (polystyrene microspheres suspended in egg white) (301). These observations indicate that cilia alone can move the cumulus–oocyte mass to its destination. When muscular contraction was allowed, back-and-forth motion of the mass was observed, whereas when it was blocked, the mass moved smoothly down the ampulla. Thus, although the overall rate of transport is not affected by inhibiting muscular action, the pattern of transport is affected. The back-and-forth movement could serve to enhance infiltration of the cumulus matrix with ovarian secretions or to initiate the process of cumulus removal.

Cilia lining the oviductal ampulla of dairy cows, sheep, guinea pigs, rats, and rabbits beat toward the uterus (302,303). When segments of rabbit ampullae were surgically reversed such that cilia in the segment beat toward the ovary, eggs were not transported across the segments. Controls, which were doubly transected but not reversed in orientation, transported eggs normally (303,304).

Some women diagnosed as having Kartagener syndrome (immotile cilia syndrome) are infertile (305,306), whereas others are fertile (307). This suggests that cilia are not absolutely necessary for oocyte transport. However, some women with Kartagener syndrome have some motile cilia (305,306), and this could explain why some with the syndrome are fertile.

Embryo Transport in the Oviduct

In most eutherian mammals, the cumulus disperses soon after fertilization, whereas the zona pellucida remains intact. The fertilized ovum usually pauses at the ampullary–isthmic junction before resuming transport toward the uterus (308). At the time of resumption of transport, the amplitude of isthmic contractions decreases in the rabbit, detected by insertion of a pressure-sensitive microballoon (309). The strong contractions that occur during estrus and in the periovulatory period could halt transport at the ampullary–isthmic junction, whereas subsequent weakening of contractions could enable transport to resume.

Oviductal transport of embryos differs from that of unfertilized oocytes in some species. In the mare, unfertilized oocytes remain in the oviduct for several estrous cycles, whereas embryos pass into the uterus 5–6 days after ovulation (310,311). In rats, unfertilized oocytes reach the uterus about 72 hours after ovulation, whereas fertilized oocytes take about 96 hours (312). These observations indicate that the early embryo interacts differently with the oviductal mucosa than the oocyte, perhaps initiating signaling pathways that modulate its transport to the uterus.

In horses, there is evidence that early embryos secrete prostaglandin E_2 (PGE_2), which acts to hasten the transport of embryos into the uterus. Infusion of PGE_2 into the oviducts of pregnant mares hastened the transport of both oocytes and embryos into the uterus (313,314).

Platelet-activating factor (PAF) bioactivity was detected in early hamster embryos, and PAF antagonists were found to prevent hastening of embryo transport into the uterus (315). PAF is also produced by human embryos (316), and receptors for PAF have been detected in the mucosal epithelium of human oviducts (317).

In the rat, the time of oviductal embryo transport is 72 to 94 hours from the approximate time of fertilization (318). Systemic administration of estradiol-17β at noon on day 1 of the estrous cycle in mated rats reduced the time of transport to 11–23 hours (318). Evidence indicates that the effects of estradiol are nongenomic, possibly through stimulation of protein kinases A and C (319–321).

The evidence supporting roles for estradiol, PAF, and PGE_2 indicate that smooth muscle contraction plays a role in transport of the early embryo toward the uterus. The isthmus of the oviduct is generally lined by far fewer ciliated cells than the ampulla (302) and possesses a thicker muscular tunic. In contrast to these findings, porcine ova fertilized by intrauterine insemination were not found to traverse the isthmus and reach the uterus any faster than unfertilized oocytes in uninseminated sows (322).

Rabbit embryos acquire a visible mucoid coat as they move down the oviduct (323). The oviductal epithelium of many species of eutherian mammals secretes an oviduct-specific group of glycoproteins with mucin-like domains that adhere to the zona pellucida of oocytes and early embryos (324–326). As marsupial embryos roll down the oviduct, they are wrapped in strands of mucin that form concentric layers (327). However, the role mucoid coats play in embryo transport is unknown.

Little is known about the passage of the embryo through the uterotubal junction. Relaxation of the smooth muscle is likely to open the lumen for passage of the embryos. In dairy cows, the sigmoid flexure at the junction must relax and the vascular plexus surrounding the junction must drain fluids out of the wall to open the lumen (132,135).

Cigarette Smoke Interferes with Oocyte and Embryo Transport

Epidemiological data have revealed a correlation between smoking and ectopic pregnancy in women (8,328). The soluble components of mainstream cigarette smoke inhibit ciliary activity in the oviduct and oocyte pickup in hamsters (329–331). Of the components of cigarette smoke, nicotine, pyridines, and pyrazine were found to interfere with oviductal functions. Pyridines and pyrazine and some of its derivatives were effective in picomolar or nanomolar doses at reducing oocyte pickup rate, ciliary beat frequency, and smooth muscle contractions in vitro in fresh preparations of hamster oviductal infundibula (332,333). Pyrazines are also added to foods in order to impart a roasted or smokey flavor and might reach the oviduct through this route (334).

Exposure of female hamsters to mainstream or sidestream cigarette smoke delayed the arrival of embryos into the uterus (335). Some of the delay could be attributed to impaired oocyte transport as well as embryo transport.

SUMMARY

Although much remains to be learned and some issues remain to be settled, the current picture of gamete transport in most eutherian mammals is as follows. Sperm are deposited at coitus into the uterus or anterior vagina. Those deposited in the vagina swim through the cervix or are drawn through by uterine contractions. Muscular contractions assist in moving sperm through the uterine cavity. A few thousand sperm swim through the uterotubal junction. While still within the junction or upon reaching the caudal isthmus, sperm are trapped by binding to the mucosal epithelium, forming a reservoir. As the time of ovulation approaches, sperm become capacitated and hyperactivated, which enables them to release from the epithelium and escape from the reservoir. Meanwhile, oocytes, surrounded by cumulus cells in a sticky viscoelastic matrix, are released from the ovary. The cumulus mass adheres lightly to cilia on the mucosal surface of the fimbriae and is transported into the oviductal ampulla and rapidly toward the ampullary–isthmic junction by ciliary action. During this time, a few sperm reach the cumulus mass. Fertilization occurs soon thereafter, as sperm penetrate the cumulus, reach and penetrate the zona pellucida, and finally fuse with the oocyte plasma membrane. After pausing at fertilization, the embryo resumes transport, probably by the actions of smooth muscle in the oviductal wall, and reaches the uterus in a few days.

ACKNOWLEDGMENTS

This chapter was inspired by its predecessor in the previous edition, which was written by Michael J. Harper (Harper, 1994). This work was supported by grants from the U.S. Department of Agriculture and the National Science Foundation. Thanks to George Ignotz, Becky Marquez, and Collin Wolff for reviewing the manuscript and to Deborah Grunder and Donna Wickham for editing the references.

REFERENCES

1. Austin, C. R. (1957). Fate of spermatozoa in the uterus of the mouse and rat. *J. Endocrinol.* 14, 335–342.
2. Braden, A. W. H., and Austin, C. R. (1954). Fertilization of the mouse egg and the effect of delayed coitus and of hot-shock treatment. *Aust. J. Biol. Sci.* 7, 522–565.
3. Zamboni, L. (1972). Fertilization in the mouse. In *Biology of Mammalian Fertilization and Implantation* (K. S. Moghissi and E. S. E. Hafez, Eds.), pp. 213–262. Charles C Thomas, Springfield, IL.
4. Hosken, D. J., O'Shea, J. E., and Blackberry, M. A. (1996). Blood plasma concentrations of progesterone, sperm storage, and sperm viability and fertility in Gould's wattled bat (*Chalinolobus gouldii*). *J. Reprod. Fertil.* 108, 171–177.
5. Bernard, R. T. F., Happold, D. C. D., and Happold, M. (1997). Sperm storage in a seasonally reproducing African vespertilionid, the banana bat (*Pipistrellus nanus*) from Malawi. *J. Zool. Lond.* 241, 161–174.
6. Bernard, R. T. F., and Cumming, G. S. (1997). African bats: evolution of reproductive patterns and delays. *Q. Rev. Biol.* 72, 253–274.

7. Menge, A. C., and Edwards, R. P. (1993). Mucosal immunity of the reproductive tract and infertility. In *Immunology of Reproduction* (R. K. Naz, Ed.), pp. 19–36. CRC Press, Boca Raton, FL.
8. Bouyer, J., Coste, J., Shojaei, T., Pouly, J. L., Fernandez, H., Gerbaud, L., and Job-Spira, N. (2003). Risk factors for ectopic pregnancy: a comprehensive analysis based on a large case-control, population-based study in France. *Am. J. Epidemiol.* 157, 185–194.
9. Harper, M. J. K. (1994). Gamete and zygote transport. In *The Physiology of Reproduction* (E. Knobil and J. D. Neill, Eds.), 2nd ed., pp. 123–187. Raven Press, New York.
10. Preston, B. T., Stevenson, I. R., Pemberton, J. M., Coltman, D. W., and Wilson, K. (2003). Overt and covert competition in a promiscuous mammal: the importance of weaponry and testes size to male reproductive success. *Proc. R. Soc. Lond. B Biol. Sci.* 270, 633–640.
11. Roldan, E. R. S., Gomendio, M., and Vitullo, A. D. (1992). The evolution of eutherian spermatozoa and underlying selective forces: female selection and sperm competition. *Biol. Rev.* 67, 551–593.
12. Birkhead, T. R., and Moeller, A. P. (1993). Sexual selection and the temporal separation of reproductive events: sperm storage data from reptiles, birds, and mammals. *Biol. J. Linn. Soc.* 50, 295–311.
13. Hosken, D. J. (1997). Sperm competition in bats. *Proc. R. Soc. Lond. B* 264, 385–392.
14. Manning, J. T., and Chamberlain, A. T. (1994). Sib competition and sperm competitiveness: an answer to "why so many sperms?" and the recombination/sperm number correlation. *Proc. R. Soc. Lond. B* 256, 177–182.
15. Moore, H., Dvorakova, K., Jenkins, N., and Breed, W. (2002). Exceptional sperm cooperation in the wood mouse. *Nature* 418, 174–177.
16. Rodger, J. C., and Bedford, J. M. (1982). Separation of sperm pairs and sperm-egg interaction in the opossum, *Didelphis virginiana*. *J. Reprod. Fertil.* 64, 171–179.
17. Moore, H. D. (1996). Gamete biology of the new world marsupial, the grey short-tailed opossum, *Monodelphis domestica*. *Reprod. Fertil. Dev.* 8, 605–615.
18. Moore, H. D., and Taggart, D. A. (1995). Sperm pairing in the opossum increases the efficiency of sperm movement in a viscous environment. *Biol. Reprod.* 52, 947–953.
19. Martan, J., and Shepherd, B. A. (1973). Spermatozoa in rouleaux in the female guinea pig genital tract. *Anat. Rec.* 175, 625–629.
20. Yanagimachi, R., and Mahi, C. A. (1976). The sperm acrosome reaction and fertilization in the guinea-pig: a study in vivo. *J. Reprod. Fertil.* 46, 49–54.
21. Flaherty, S. P., Swann, N. J., Primakoff, P., and Myles, D. G. (1993). A role for the WH-30 protein in sperm-sperm adhesion during rouleaux formation in the guinea pig. *Dev. Biol.* 156, 243–252.
22. Martan, J., and Hruban, Z. (1970). Unusual spermatozoan formations in the epididymis of the flying squirrel (*Glaucomys volans*). *J. Reprod. Fertil.* 21, 167–170.
23. Heath, E., Schaefferm N., Meritt, D. A. Jr., and Jeyendran, R. S. (1987). Rouleaux formation by spermatozoa in the naked-tail armadillo, *Cabassous unicinctus*. *J. Reprod. Fertil.* 79, 153–158.
24. Yanagimachi, R. (1982). In vitro sperm capacitation and fertilization of golden-hamster eggs in a chemically-defined medium. In *In Vitro Fertilization and Embryo Transfer* (E. S. E. Hafez and K. Semm, Eds.), pp. 65–76. MTP Press LTD, Lancaster, UK.
25. Yanagimachi, R., Kamiguchi, Y., Sugawara, S., and Mikamo, K. (1983). Gametes and fertilization in the Chinese hamster. *Gamete Res.* 8, 97–117.
26. Suarez, S. S. (1988). Hamster sperm motility transformation during development of hyperactivation in vitro and epididymal maturation. *Gamete Res.* 19, 51–65.
27. Daels, P. F., Hughes, J. P., and Stabenfeldt, G. H. (1991). Reproduction in horses. In *Reproduction in Domestic Animals* (P. T. Cupps, Ed.), 4th ed., pp. 414–444. Academic Press, San Diego.
28. Krutzsch, P. H., Crichton, E. G., and Nagle, R. B. (1982). Studies on prolonged spermatozoa survival in Chiroptera: a morphological examination of storage and clearance of intrauterine and caudal epididymal spermatozoa in the bats *Myotis lucifugus* and *M. velifer*. *Am. J. Anat.* 165, 421–434.
29. Overstreet, J. W., and Cooper, G. W. (1979). Effect of ovulation and sperm motility on the migration of rabbit spermatozoa to the site of fertilization. *J. Reprod. Fertil.* 55, 53–59.
30. Roth, T. L., O'Brien, J. K., McRae, M. A., Bellem, A. C., Romo, S. J., Kroll, J. L., and Brown, J. L. (2001). Ultrasound and endocrine evaluation of the ovarian cycle and early pregnancy in the Sumatran rhinoceros, *Dicerorhinus sumatrensis*. *Reproduction* 121, 139–149.
31. Milligan, S. R. (1982). Induced ovulation in mammals. In *Oxford Reviews of Reproductive Biology* (C. A. Finn, Ed.), Vol. 4, pp. 1–46. Clarendon Press, Oxford.
32. Brown, J. L., Graham, L. H., Wielebnowski, N., Swanson, W. F., Wildt, D. E., and Howard, J. G. (2001). Understanding the basic reproductive biology of wild felids by monitoring of faecal steroids. *J. Reprod. Fertil. Suppl.* 57, 71–82.
33. Skidmore, J. A. (2003). The main challenges facing camel reproduction research in the 21st century. *Reprod. Suppl.* 61, 37–47.
34. Boone, W. R., Keck, B. B., Catlin, J. C., Casey, K. J., Boone, E. T., Dye, P. S., Schuett, R. J., Tsubota, T., and Bahr, J. C. (2004). Evidence that bears are induced ovulators. *Theriogenology* 61, 1163–1169.
35. Addo, P., Dodoo, A., Adjei, S., Awumbila, B., and Awotwi, E. (2002). Determination of the ovulatory mechanism of the grasscutter (*Thryonomys swinderianus*). *Anim. Reprod. Sci.* 71, 125–137.
36. Kaneko, T., Iida, H., Bedford, J. M., Oda, S., and Mori, T. (2003). Mating-induced cumulus-oocyte maturation in the shrew, *Suncus murinus*. *Reproduction* 126, 817–826.
37. Dryden, C. L. (1969). Reproduction in *Suncus murinus*. *J. Reprod. Fertil. Suppl.* 117, 345–353.
38. Sobrero, A. J., and MacLeod, J. (1962). The immediate postcoital test. *Fertil. Steril.* 13, 184–189.
39. Overstreet, J. W., and Cooper, G. W. (1978). Sperm transport in the reproductive tract of the female rabbit. I. The rapid transit phase of transport. *Biol. Reprod.* 19, 101–114.
40. Bedford, J. M., and Yanagimachi, R. (1992). Initiation of sperm motility after mating in the rat and hamster. *J. Androl.* 13, 444–449.
41. Carballada, R., and Esponda, P. (1997). Fate and distribution of seminal plasma proteins in the genital tract of the female rat after natural mating. *J. Reprod. Fertil.* 109, 325–335.
42. Blandau, R. J. (1969). Gamete transport: comparative aspects. In *The Mammalian Oviduct*, pp. 129–162. The University of Chicago Press, Chicago.
43. Matthews, M. K. Jr., and Adler, N. T. (1978). Systematic interrelationship of mating, vaginal plug position, and sperm transport in the rat. *Physiol. Behav.* 20, 303–309.
44. Carballada, R., and Esponda, P. (1992). Role of fluid from seminal vesicles and coagulating glands in sperm transport into the uterus and fertility in rats. *J. Reprod. Fertil.* 95, 639–648.
45. Blandau, R. J. (1945). On the factors involved in sperm transport through the cervix uteri of the albino rat. *Am. J. Anat.* 77, 253–272.

46. Lundwall, A., Peter, A., Lovgren, J., Lilja, H., and Malm, J. (1997). Chemical characterization of the predominant proteins secreted by mouse seminal vesicles. *Eur. J. Biochem.* 249, 39–44.
47. Lundwall, A., Malm, J., Clauss, A., Valtonen-Andre, C., and Olsson, A. Y. (2003). Molecular cloning of complementary DNA encoding mouse seminal vesicle-secreted protein SVS I and demonstration of homology with copper amine oxidases. *Biol. Reprod.* 69, 1923–1930.
48. Williams-Ashman, H. G. (1984). Transglutaminases and the clotting of mammalian seminal fluids. *Mol. Cell. Biochem.* 58, 51–61.
49. Lin, H. J., Luo, C. W., and Chen, Y. H. (2002). Localization of the transglutaminase cross-linking site in SVS III, a novel glycoprotein secreted from mouse seminal vesicle. *J. Biol. Chem.* 277, 3632–3639.
50. Murer, V., Spetz, J. F., Hengst, U., Altrogge, L. M., de Agostini, A., and Monard, D. (2001). Male fertility defects in mice lacking the serine protease inhibitor protease nexin-1. *Proc. Natl. Acad. Sci. U S A* 98, 3029–3033.
51. Lilja, H., and Lundwall, A. (1992). Molecular cloning of epididymal and seminal vesicular transcripts encoding a semenogelin-related protein. *Proc. Natl. Acad. Sci. U S A* 89, 4559–4563.
52. Lilja, H. (1985). A kallikrein-like serine protease in prostatic fluid cleaves the predominant seminal vesicle protein. *J. Clin. Invest.* 76, 1899–1903.
53. Watt, K. W., Lee, P. J., M'Timkulu, T., Chan, W. P., and Loor, R. (1986). Human prostate-specific antigen: structural and functional similarity with serine proteases. *Proc. Natl. Acad. Sci. U S A* 10, 3166–3170.
54. Lundwall, A., Giwercman, A., Ruhayel, Y., Giwercman, Y., Lilja, H., Hallden, C., and Malm, J. (2003). A frequent allele codes for a truncated variant of semenogelin I, the major protein component of human semen coagulum. *Mol. Hum. Reprod.* 9, 345–350.
55. Jensen-Seaman, M. I., and Li, W. H. (2003). Evolution of the hominoid semenogelin genes, the major proteins of ejaculated semen. *J. Mol. Evol.* 57, 261–270.
56. Kingan, S. B., Tatar, M., and Rand, D. M. (2003). Reduced polymorphism in the chimpanzee semen coagulating protein, semenogelin I. *J. Mol. Evol.* 57, 159–169.
57. Dewsbury, D. A. (1975). Diversity and adaptation in rodent copulatory behavior. *Science* 190, 947–954.
58. Grandage, J. (1972). The erect dog penis: a paradox of flexible rigidity. *Vet. Rec.* 91, 141–147.
59. Hunter, R. H. F. (1981). Sperm transport and reservoirs in the pig oviduct in relation to the time of ovulation. *J. Reprod. Fertil.* 63, 109–117.
60. Hunter, R. H. F. (1982). *Reproduction of Farm Animals*. 149 pp. Longman, London.
61. Roberts, S. J. (1986). *Veterinary Obstetrics and Genital Diseases*, 3rd ed. Stephen Roberts, Woodstock, VT.
62. Foote, R. H., and Kaproth, M. T. (1997). Sperm numbers inseminated in dairy cattle and nonreturn rates revisited. *J. Dairy Sci.* 80, 3072–3076.
63. López-Gatius, F. (2000). Site of semen deposition in cattle: a review. *Theriogenology* 53, 1407–1414.
64. Vishwanath R. (2003). Artificial insemination: the state of the art. *Theriogenology* 59, 571–584.
65. Kurykin, J., Jaakma, U., Majas, L., Jalakas, M., Aidnik, M., Waldmann, A., and Padrik, P. (2003). Fixed time deep intracornual insemination of heifers at synchronized estrus. *Theriogenology* 60, 1261–1268.
66. Verberckmoes, S., Van Soom, A., De Pauw, I., Dewulf, J., Vervaet, C., and de Kruif, A. (2004). Assessment of a new utero-tubal junction insemination device in dairy cattle. *Theriogenology* 61, 103–115.
67. Yaniz, J. L., Lopez-Bejar, M., Santolaria, P., Rutllant, J., and Lopez-Gatius, F. (2002). Intraperitoneal insemination in mammals: a review. *Reprod. Domest. Anim.* 37, 75–80.
68. Larsson, B. (1986). Transperitoneal migration of spermatozoa in heifers. *Zentralbl. Veterinarmed. B* 33, 714–718.
69. Metz, K. G., and Mastroianni, L. Jr. (1979). Tubal pregnancy subsequent to transperitoneal migration of spermatozoa. *Obstet. Gynecol. Surv.* 34, 554–560.
70. Brown, C., La Vigne, W. E., and Padilla, S. L. (1987). Unruptured pregnancy in a heterotopic fallopian tube: evidence for transperitoneal sperm migration. *Am. J. Obstet. Gynecol.* 156, 88–90.
71. Ansari, A. H., and Miller, E. S. (1994). Sperm transmigration as a cause of ectopic pregnancy. *Arch. Androl.* 32, 1–4.
72. Boskey, E. R., Cone, R. A., Whaley, K. J., and Moench, T. R. (2001). Origins of vaginal acidity: high D/L lactate ratio is consistent with bacteria being the primary source. *Hum. Reprod.* 16, 1809–1813.
73. Acott, T. S., and Carr, D. W. (1984). Inhibition of bovine spermatozoa by caudal epididymal fluid. II. Interaction of pH and a quiescence factor. *Biol. Reprod.* 30, 926–935.
74. Carr, D. W., Usselman, M. C., and Acott, T. S. (1985). Effects of pH, lactate, and viscoelastic drag on sperm motility: a species comparison. *Biol. Reprod.* 33, 588–595.
75. Fox, C. A., Meldrum, S. J., Watson, B. W. (1973). Continuous measurement by radio-telemetry of vaginal pH during human coitus. *J. Reprod. Fertil.* 33, 69–75.
76. Bouvet, J. P., Gresenguet, G., and Belec, L. (1997). Vaginal pH neutralization by semen as a cofactor of HIV transmission. *Clin. Microbiol. Infect.* 3, 19–23.
77. Phillips, D. M., and Mahler, S. (1977). Phagocytosis of spermatozoa by the rabbit vagina. *Anat. Rec.* 189, 61–72.
78. Phillips, D. M., and Mahler, S. (1977). Leukocyte emigration and migration in the vagina following mating in the rabbit. *Anat. Rec.* 189, 45–60.
79. Overstreet, J. W., Cooper, G. W, and Katz, D. F. (1978). Sperm transport in the reproductive tract of the female rabbit. II. The sustained phase of transport. *Biol. Reprod.* 19, 115–132.
80. Silva, L. D. M., Onclin, K., and Verstegen, J. P. (1995). Cervical opening in relation to progesterone and oestradiol during heat in beagle bitches. *J. Reprod. Fertil.* 104, 85–90.
81. Verstegen, J. P., Silva, L. D. M., and Onclin, K. (2001). Determination of the role of cervical closure in fertility regulation after mating or artificial insemination in Beagle bitches. *J. Reprod. Fertil. Suppl.* 57, 31–34.
82. Chatdarong, K., Kampa, N., Axner, E., and Linde-Forsberg, C. (2002). Investigation of cervical patency and uterine appearance in domestic cats by fluoroscopy and scintigraphy. *Reprod. Dom. Anim.* 37, 275–281.
83. Katz, D. F., Slade, D. A., and Nakajima, S. T. (1997). Analysis of preovulatory changes in cervical mucus hydration and sperm penetrability. *Adv. Contracept.* 13, 143–151.
84. Morales, P., Roco, M., and Vigil, P. (1993). Human cervical mucus: relationship between biochemical characteristics and ability to allow migration of spermatozoa. *Hum. Reprod.* 8, 78–83.
85. Bigelow, J. L., Dunson, D. B., Stanford, J. B., Ecochard, R., Gnoth, C., and Colombo, B. (2004). Mucus observations in the fertile window: a better predictor of conception than timing of intercourse. *Hum. Reprod.* 19, 889–892.
86. Hanson, F. W., and Overstreet, J. W. (1981). The interaction of human spermatozoa with cervical mucus in vivo. *Am. J. Obstet. Gynecol.* 140, 173–178.
87. Barros, C., Vigil, P., Herrera, E., Arguello, B., and Walker, R. (1984). Selection of morphologically abnormal sperm by human cervical mucus. *Arch. Androl. Suppl.* 12, 95–107.

88. Katz, D. F., Morales, P., Samuels, S. J., and Overstreet, J. W. (1990). Mechanisms of filtration of morphologically abnormal human sperm by cervical mucus. *Fertil. Steril.* 54, 513–516.
89. Yudin, A. I., Hanson, F. W., and Katz, D. F. (1989). Human cervical mucus and its interaction with sperm: a fine-structural view. *Biol. Reprod.* 40, 661–671.
90a. Overstreet, J. W., Coats, C., Katz, D. F., and Hanson, F. W. (1980). The importance of seminal plasma for sperm penetration of cervical mucus. *Fertil. Steril.* 34, 569–572.
90b. Overstreet, J. W., Katz, D. F., and Johnson, L. L. (1980). Motility of rabbit spermatozoa in the secretions of the oviduct. *Biol. Reprod.* 22, 1083–1088.
91. Tyler, K. R. (1977). Histological changes in the cervix of the rabbit after coitus. *J. Reprod. Fertil.* 49, 341–345.
92. Pandya, I. J., and Cohen, J. (1985). The leukocytic reaction of the human uterine cervix to spermatozoa. *Fertil. Steril.* 43, 417–421.
93. Parkhurst, M. R., and Saltzman, W. M. (1994). Leukocytes migrate through three-dimensional gels of midcycle cervical mucus. *Cell. Immunol.* 156, 77–94.
94. Taylor, N. J. (1982). Investigation of sperm-induced cervical leukocytosis by a double mating study in rabbits. *J. Reprod. Fertil.* 66, 157–160.
95. D'Cruz, O. J., Wang, B.-L., and Haas, G. G. Jr. (1992). Phagocytosis of immunoglobulin G and C3-bound human sperm by human polymorphonuclear leukocytes is not associated with the release of oxidative radicals. *Biol. Reprod.* 46, 721–732.
96. Kutteh, W. H., Prince S. J., Hammond, K. R., Kutteh, C. C., and Mestecky, J. (1996). Variations in immunoglobulins and IgA subclasses of human uterine cervical secretions around the time of ovulation. *Clin. Exp. Immunol.* 104, 538–542.
97. Matthur, S., Rosenlund, C., Carlton, M., Caldwell, J., Barber, M., Rust, P. F., and Williamson, H. O. (1988). Studies on sperm survival and motility in the presence of cytotoxic sperm antibodies. *Am. J. Reprod. Immunol. Microbiol.* 17, 41–47.
98. Jensen, T. S., Bjorge, L., Wollen, A.-L., and Ulstein, M. (1995). Identification of the complement regulatory proteins CD46, CD55, and CD59 in human fallopian tube, endometrium, and cervical mucosa and secretion. *Am. J. Reprod. Immunol.* 34, 1–9.
99. Ulcova-Gallova, Z. (1997). Ten-year experience with anti-spermatozool activity in ovulatory cervical mucus and local hydrocortisone treatment. *Am. J. Reprod. Immunol.* 38, 231–234.
100. Mullins, K. J., and Saacke, R. G. (1989). Study of the functional anatomy of bovine cervical mucosa with special reference to mucus secretion and sperm transport. *Anat. Rec.* 226, 106–117.
101. Mattner, P. E. (1968). The distribution of spermatozoa and leukocytes in the female genital tract in goats and cattle. *J. Reprod. Fertil.* 17, 253–261.
102. Carlstedt, I., and Sheehan, J. K. (1984). Macromolecular architecture and hydrodynamic properties of human cervical mucins. *Biorheology* 21, 225–233.
103. Carlstedt, I., and Sheehan, J. K. (1989). Structure and macromolecular properties of cervical mucus glycoproteins. *Symp. Soc. Exp. Biol.* 43, 289–316.
104. Sheehan, J. K., and Carlstedt, I. (1984). Hydrodynamic properties of human cervical mucus glycoproteins in 6M guanidinium chloride. *Biochem. J.* 217, 93–101.
105. Sheehan, J. K., Oates, K., and Carlstedt, I. (1986). Electron microscopy of cervical, gastric and bronchial mucus glycoproteins. *Biochem. J.* 239, 147–153.
106. Chretien, F. C. (2003). Involvement of the glycoproteic meshwork of cervical mucus in the mechanism of sperm orientation. *Acta Obstet. Gynaecol. Scand.* 82, 449–461.
107. Tampion, D., and Gibons, R. A. (1962). Orientation of spermatozoa in mucus of the cervix uteri. *Nature* 194, 381.
108. Katz, D. F., Mills, R. N., and Pritchett, T. R. (1978). The movement of human spermatozoa in cervical mucus. *J. Reprod. Fertil.* 53, 259–265.
109. Gould, J. E., Overstreet, J. W., and Hanson, F. W. (1984). Assessment of human sperm function after recovery from the female reproductive tract. *Biol. Reprod.* 31, 888–894.
110. Peitz, B., and Olds-Clarke, P. (1986). Effects of seminal vesicles removal on fertility and uterine sperm motility in the house mouse. *Biol. Reprod.* 35, 608–617.
111. Mortimer, S. T., and Swan, M. A. (1995). Kinematics of capacitating human spermatozoa analysed at 60 Hz. *Hum. Reprod.* 10, 873–879.
112. Croxatto, H. B. (1996). Gamete transport. In *Reproductive Endocrinology, Surgery, Technology* (E. Y. Adashi, J. A. Rock, and Z. Rosenwaks, Eds.), pp. 385–402. Lippincott-Raven, Philadelphia.
113. Settlage, D. S. F., Motoshima, M., and Tredway, D. R. (1973). Sperm transport from the external cervical os to the fallopian tubes in women: a time and quantitation study. *Fertil. Steril.* 24, 655–661.
114. Rubenstein, B. B., Strauss, H., Lazarus, M. L., and Hankin, H. (1951). Sperm survival in women. *Fertil. Steril.* 2, 15–19.
115. Hunter, R. H. F., and Wilmut, I. (1982). The rate of functional sperm transport into the oviducts of mated cows. *Anim. Reprod. Sci.* 5, 167–173.
116. Wilmut, I., and Hunter, R. H. (1984). Sperm transport into the oviducts of heifers mated early in oestrus. *Reprod. Nutr. Dev.* 24, 461–468.
116a. Hunter, R. H. F., and Wilmut, I. (1984). Sperm transport in the cow: periovulatory redistribution of viable cells within the oviduct. *Reprod. Nutr. Dev.* 24, 597–608.
117. Hunter, R. H., Nichol, R., and Crabtree, S. M. (1980). Transport of spermatozoa in the ewe: timing of the establishment of a functional population in the oviduct. *Reprod. Nutr. Dev.* 20, 1869–1875.
118. Lyons, E. A., Taylor, P. J., Zheng, X. H., Ballard, G., Levi, C. S., and Kredentser, J. V. (1991). Characterization of subendometrial myometrial contractions throughout the menstrual cycle in normal fertile women. *Fertil. Steril.* 55, 771–774.
119. Kunz, G., Beil, D., Deininger, H., Wildt, L., and Leyendecker, G. (1996). The dynamics of rapid sperm transport through the female genital tract: evidence from vaginal sonography of uterine peristalsis and hysterosalpingoscintigraphy. *Hum. Reprod.* 11, 627–632.
120. De Ziegler, D., Bulletti, C., Fanchin, R., Epiney, M., and Brioschi, P. A. (2001). Contractility of the nonpregnant uterus: the follicular phase. *Ann. N. Y. Acad. Sci.* 943, 172–184.
121. Hawk, H. W. (1983). Transport and fate of spermatozoa after insemination of cattle. *J. Dairy Sci.* 70, 1487–1503.
122. Fukuda, M., and Fukuda, K. (1994). Uterine endometrial cavity movement and cervical mucus. *Hum. Reprod.* 9, 1013–1016.
123. Casslen, B. (1986). Uterine fluid volume: cyclic variations and possible extra uterine contributions. *J. Reprod. Med.* 31, 506–510.
124. Kunz, G. (1998). Sonographic evidence for the involvement of the utero-ovarian counter-current system in the ovarian control of directed uterine sperm transport. *Hum. Reprod. Update* 4, 667–672.
125. Hunter, R. H. F., Cook, B., and Poyser, N. L. (1983). Regulation of oviduct function in pigs by local transfer of

ovarian steroids and prostaglandins: a mechanism to influence sperm transport. *Eur. J. Obstet. Gynaecol. Reprod. Biol.* 14, 225–232.

126. Stefanczyk-Krzymowska, S., Grzegorzewski, W., Wasowska, B., Skipor, J., and Krzymowski, T. (1998). Local increase of ovarian steroid hormone concentration in blood supplying the oviduct and uterus during early pregnancy of sows. *Theriogenology* 50, 1071–1080.
127. Crane, L. H., and Martin, L. (1991). Postcopulatory myometrial activity in the rat as seen by video-laparoscopy. *Reprod. Fertil. Dev.* 3, 685–698.
128. Claus, R. (1990). Physiological role of seminal components in the reproductive tract of the female pig. *J. Reprod. Fertil. Suppl.* 40, 117–131.
129. Bedford, J. M. (1965). Effect of environment on phagocytosis of rabbit spermatozoa. *J. Reprod. Fertil.* 9, 249–256.
130. Suarez, S. S., and Oliphant, G. (1982). The interaction of rabbit spermatozoa and serum complement proteins. *Biol. Reprod.* 27, 473–483.
131. Dostal, J., Veselsky, L., Marounek, M., Zelezna, B., and Jonakova, V. (1997). Inhibition of bacterial and boar epididymal sperm immunogenicity by boar seminal immunosuppressive component in mice. *J. Reprod. Fertil.* 111, 135–141.
132. Hook, S. J., and Hafez, E. S. E. (1968). A comparative study of the mammalian uterotubal junction. *J. Morphol.* 125, 159–184.
133. Hafez, E. S. E., and Black, D. L. (1969). The mammalian uterotubal junction. In *The Mammalian Oviduct: Comparative Biology and Methodology* (E. S. E. Hafez and R. J. Blandau, Eds.), pp. 85–128. The University of Chicago Press, Chicago.
134. Beck, L. R., and Boots, L. R. (1974). The comparative anatomy, histology, and morphology of the mammalian oviduct. In *The Oviduct and Its Functions* (A. D. Johnson and C. W. Foley, Eds.), pp. 2–51. Academic Press, New York.
135. Wrobel, K.-H., Kujat, R., and Fehle, G. (1993). The bovine tubouterine junction: general organization and surface morphology. *Cell. Tissue Res.* 271, 227–239.
136. Suarez, S. S., Brockman, K., and Lefebvre, R. (1997). Distribution of mucus and sperm in bovine oviducts after artificial insemination. *Biol. Reprod.* 56, 447–453.
137. Suarez, S. S. (1987). Sperm transport and motility in the mouse oviduct: observations in situ. *Biol. Reprod.* 36, 203–210.
138. Gaddum-Rosse, P. (1981). Some observations on sperm transport through the uterotubal junction of the rat. *Am. J. Anat.* 160, 333–341.
139. Yániz, J. L., Lopez-Gatius, F., Santolaria, P., and Mullins, K. J. (2000). Study of the functional anatomy of bovine oviductal mucosa. *Anat. Rec.* 260, 268–278.
140. Jansen, R. P. S. (1978). Fallopian tube isthmic mucus and ovum transport. *Science* 201, 349–351.
141. Jansen, R. P. S., and Bajpai, V. K. (1982). Oviduct acid mucus glycoproteins in the estrous rabbit: ultrastructure and histochemistry. *Biol. Reprod.* 26, 155–168.

142a. Suarez, S. S., Redfern, K., Raynor, P., Martin, F., and Phillips, D. M. (1991). Attachment of boar sperm to mucosal explants of oviduct in vitro: possible role in formation of a sperm reservoir. *Biol. Reprod.* 44, 998–1004.

142b. Suarez, S. S., Katz, D. F., Owen, D. H., Andrew, J. B., and Powell, R. L. (1991). Evidence for the function of hyperactivated motility in mammalian sperm. *Biol. Reprod.* 44, 375–381.

143. Suarez, S. S., Drost, M., Redfern, K., and Gottlieb, W. (1990). Sperm motility in the oviduct. In *Fertilization in Mammals* (B. D. Bavister, J. Cummins, and E. R. S. Roldan, Eds.), pp. 111–124. Serono Symposia, Norwell.
144. Jansen, R. P. S. (1980). Cyclic changes on the human fallopian tubes isthmus and their functional importance. *Am. J. Obstet. Gynecol.* 136, 292–308.
145. Baker, R. D., and Degen A. A. (1972). Transport of live and dead boar spermatozoa within the reproductive tract of gilts. *J. Reprod. Fertil.* 28, 369–377.
146. Smith, T. T., Koyanagi, F., and Yanagimachi, R. (1988). Distribution and number of spermatozoa in the oviduct of the golden hamster after natural mating and artificial insemination. *Biol. Reprod.* 37, 225–234.
147. Shalgi, R., Smith, T. T., and Yanagimachi, R. (1992). A quantitative comparison of the passage of capacitated and uncapacitated hamster spermatozoa through the uterotubal junction. *Biol. Reprod.* 46, 419–424.
148. Cho, C., Bunch, D. O., Faure, J. E., Goulding, E. H., Eddy, E. M., Primakoff, P., and Myles, D. G. (1998). Fertilization defects in sperm from mice lacking fertilin beta. *Science* 281, 1857–1859.
149. Ikawa, M., Wada, I., Kominami, K., Watanabe, D., Toshimori, K., Nishimune, Y., and Okabe, M. (1997). The putative chaperone calmegin is required for sperm fertility. *Nature* 387, 607–611.
150. Yamagata, K., Nakanishi, T., Ikawa, M., Yamaguchi, R., Moss, S. B., and Okabe, M. (2002). Sperm from the calmegin-deficient mouse have normal abilities for binding and fusion to the egg plasma membrane. *Dev. Biol.* 250, 348–357.
151. Krege, J. H., John, S. W., Langenbach, L. L., Hodgin, J. B., Hagaman, J. R., Bachman, E. S., Jennette, J. C., O'Brien, D. A., and Smithies, O. (1995). Male–female differences in fertility and blood pressure in ACE-deficient mice. *Nature* 375, 146–148.
152. Hagaman, J. R., Moyer, J. S., Bachman, E. S., Sibony, M., Magyar, P. L., Welch, J. E., Smithies, O., Krege, J. H., and O'Brien, D. A. (1998). Angiotensin-converting enzyme and male fertility. *Proc. Natl. Acad. Sci. U S A* 95, 2552–2557.
153. Metayer, S., Dacheux, F., Dacheux, J. L., and Gatti, J. L. (2002). Germinal angiotensin I-converting enzyme is totally shed from the rodent sperm membrane during epididymal maturation. *Biol. Reprod.* 67, 1763–1767.
154. Nakanishi, T., Isotani, A., Yamaguchi, R., Ikawa, M., Baba, T., Suarez, S. S., and Okabe, M. (2004). Selective passage through the uterotubal junction of sperm from a mixed population produced by chimeras of calmegin-knockout and wild-type male mice. *Biol. Reprod.* 71, 959–965.
155. Hawk, H. W. (1987). Sperm survival and transport in the female reproductive tract. *J. Dairy Sci.* 66, 2645–2660.
156. Yanagimachi, R., and Chang, M. C. (1963). Sperm ascent through the oviduct of the hamster and rabbit in relation to the time of ovulation. *J. Reprod. Fertil.* 6, 413–420.
157. Smith, T. T., Koyanagi, F., and Yanagimachi, R. (1987). Quantitative comparison of the passage of homologous and heterologous spermatozoa through the uterotubal junction of the golden hamster. *Gamete Res.* 19, 227–234.
158. Harper, M. J. K. (1973). Relationship between sperm transport and penetration of eggs in the rabbit oviduct. *Biol. Reprod.* 8, 441–450.
159. Hunter, R. H. F., and Nichol, R. (1983). Transport of spermatozoa in the sheep oviduct: preovulatory sequestering of cells in the caudal isthmus. *J. Exp. Zool.* 228, 121–128.
160. Rodriguez-Martinez, H., Nicander, L., Viring, S., Einarsson, S., and Larsson, K. (1990). Ultrastructure of the uterotubal junction in preovulatory pigs. *Anat. Histol. Embryol.* 19, 16–36.
161. Pollard, J. W., Plante, C., King, W. A., Hansen, P. J., Betteridge, K. J., and Suarez, S. S. (1991). Fertilizing capacity of bovine sperm may be maintained by binding to oviductal epithelial cells. *Biol. Reprod.* 44, 102–107.

162. Chian, R.-C., and Sirard, M.-A. (1994). Fertilizing ability of bovine spermatozoa co-cultured with oviduct epithelial cells. *Biol. Reprod.* 52, 156–162.
163. Ellington, J. E., Ignotz, G. G., Varner, D. D., Marcucio, R. S., Mathison, P., and Ball, B. A. (1993). In vitro interaction between oviduct epithelia and equine sperm. *Arch. Androl.* 31, 79–86.
164. Kervancioglu, M. E., Djahanbakhch, O., and Aitken, R. J. (1994). Epithelial cell coculture and the induction of sperm capacitation. *Fertil. Steril.* 61, 1103–1108.
165. Kawakami, E., Kashiwagi, C., Hori, T., and Tsutsui, T. (2001). Effects of canine oviduct epithelial cells on movement and capacitation of homologous spermatozoa in vitro. *Anim. Reprod. Sci.* 68, 121–131.
166. Koya, Y., Munehara, H., and Takano, K. (1997). Sperm storage and degradation in the ovary of a marine copulating sculpin, *Alcichthys alcicornis* (Teleosti: scorpaeniformes) role of intercellular junctions between inner ovarian epithelial cells. *J. Morphol.* 233, 153–163.
167. Gist, D. H., and Jones, J. M. (1989). Sperm storage within the oviducts of turtles. *J. Morphol.* 199, 379–384.
168. Gist, D. H., and Congdon, J. D. (1998). Oviductal sperm storage as a reproductive tactic of turtles. *J. Exp. Zool.* 282, 526–534.
169. Sever, D. M., and Hamlett, W. C. (2002). Female sperm storage in reptiles. *J. Exp. Zool.* 292, 187–199.
170. Gist, D. H., and Jones, J. M. (1987). Storage of sperm in the reptilian oviduct. *Scann. Microsc.* 1, 1839–1849.
171. Srinivas, S. R., Shivanandappa, T., Hegde, S. N., and Sarkar, H. B. D. (1995). Sperm storage in the oviduct of the tropical rock lizard, *Psammophilus dorsalis. J. Morphol.* 224, 293–301.
172. Perkins, M. J., and Palmer, B. D. (1996). Histology and functional morphology of the oviduct of an oviparous snake, *Diadophis punctatus. J. Morphol.* 227, 67–79.
173. Murphy-Walker, S., and Haley, S. R. (1996). Functional sperm storage duration in female *Hemidactylus frenatus* (family gekkonidae). *Herpetologica* 52, 365–373.
174. Bakst, M. R. (1987). Anatomical basis of sperm-storage in the avian oviduct. *Scann. Microsc.* 1, 1257–1266.
175. Bakst, M. R. (1994). Oviductal sperm selection, transport, and storage in poultry. *Poultry Sci. Rev.* 5, 117–143.
176. Birkhead, T. R., Pellatt, E. J., and Fletcher, F. (1993). Selection and utilization of spermatozoa in the reproductive tract of the female zebra finch *Taeniopygia guttata. J. Reprod. Fertil.* 99, 593–600.
177. Polge, C., Salamon, S., and Wilmut, I. (1970). Fertilizing capacity of frozen boar semen following surgical insemination. *Vet. Rec.* 87, 424–428.
178. Hunter, R. H. F. (1973). Polyspermic fertilization in pigs after tubal deposition of excessive numbers of spermatozoa. *J. Exp. Zool.* 183, 57–64.
179. Hunter, R. H. F., and Leglise, P. C. (1971). Polyspermic fertilization following tubal surgery in pigs, with particular reference to the role of the isthmus. *J. Reprod. Fertil.* 24, 233–246.
180. Day, B. N., and Polge, C. (1968). Effects of progesterone on fertilization and egg transport in the pig. *J. Reprod. Fertil.* 17, 227–230.
181. Hunter, R. H. F. (1972). Local action of progesterone leading to polyspermic fertilization in pigs. *J. Reprod. Fertil.* 31, 433–444.
182. Smith, T. T., and Yanagimachi, R. (1991). Attachment and release of spermatozoa from the caudal isthmus of the hamster oviduct. *J. Reprod. Fertil.* 91, 567–573.
183. Thomas, P. G. A., Ball, B. A., and Brinsko, S. P. (1994). Interaction of equine spermatozoa with oviduct epithelial cell explants is affected by estrous cycle and anatomic origin of explant. *Biol. Reprod.* 51, 222–228.
184. Petrunkina, A. M., Simon, K., Gunzel-Apel, A. R., and Topfer-Petersen, E. (2004). Kinetics of protein tyrosine phosphorylation in sperm selected by binding to homologous and heterologous oviductal explants: how specific is the regulation by the oviduct? *Theriogenology* 61, 1617–1634.
185. Overstreet, J. W., and Cooper, G. W. (1975). Reduced sperm motility in the isthmus of the rabbit oviduct. *Nature* 258, 718–719.
186. Burkman, L. J., Overstreet, J. W., and Katz, D. F. (1984). A possible role for potassium and pyruvate in the modulation of sperm motility in the rabbit oviductal isthmus. *J. Reprod. Fertil.* 71, 367–376.
187. Smith, T. T., and Yanagimachi, R. (1990). The viability of hamster spermatozoa stored in the isthmus of the oviduct: the importance of sperm-epithelium contact for survival. *Biol. Reprod.* 42, 450–457.
188. DeMott, R. P., Lefebvre, R., and Suarez, S. S. (1995). Carbohydrates mediate the adherence of hamster sperm to oviductal epithelium. *Biol. Reprod.* 52, 1395–1403.
189. Lefebvre, R., DeMott, R. P., Suarez, S. S., and Samper, J. C. (1995). Specific inhibition of equine sperm binding to oviductal epithelium. Equine reproduction VI. *Biol. Reprod.* Mono 1, 689–696.
190a. Dobrinski, I., Ignotz, G. G., Thomas, P. G. A., and Ball, B. A. (1996). Role of carbohydrates in the attachment of equine spermatozoa to uterine tubal (oviductal) epithelial cells in vitro. *Am. J. Vet. Res.* 57, 1635–1639.
190b. Dobrinski, I., Suarez, S. S., and Ball, B. A. (1996). Intracellular calcium concentration in equine spermatozoa attached to oviductal epithelial cells in vitro. *Biol. Reprod.* 54, 783–788.
191. Green, C. E., Bredl, J., Holt, W. V., Watson, P. F., and Fazeli, A. (2001). Carbohydrate mediation of boar sperm binding to oviductal epithelial cells in vitro. *Reproduction* 122, 305–315.
192. Wagner, A., Ekhlasi-Hundrieser, M., Hettel, C., Petrunkina, A., Waberski, D., Nimtz, M., and Topfer-Petersen, E. (2002). Carbohydrate-based interactions of oviductal sperm reservoir formation-studies in the pig. *Mol. Reprod. Dev.* 61, 249–257.
193. Lefebvre, R., Lo, M. C., and Suarez, S. S. (1997). Bovine sperm binding to oviductal epithelium involves fucose recognition. *Biol. Reprod.* 56, 1198–1204.
194. Kogan, T. P., Revelle, B. M., Tapp, S., Scott, D., and Beck, P. J. (1995). A single amino acid residue can determine the ligand specificity of E-selectin. *J. Biol. Chem.* 270, 14047–14055.
195. Revelle, B. M., Scott, D., and Beck, P. J. (1996). Single amino acid residues in the E- and P-selectin epidermal growth factor domains can determine carbohydrate binding specificity. *J. Biol. Chem.* 271, 16160–16170.
196. Weiss, W. I. (1994). Recognition of cell surface carbohydrates by C-type animal lectins. In *Cellular Adhesion* (B. W. Metcalf, B. J. Dalton, and G. Poste, Eds.). Plenum Press, New York.
197. Varki, A. (1992). Selectins and other mammalian sialic acid-binding lectins. *Curr. Opin. Cell. Biol.* 4, 257–266.
198. Zhang, Q., Young, T. F., and Ross, R. F. (1994). Glycolipid receptors for attachment of *Mycoplasma hyopneumoniae* to porcine respiratory ciliated cells. *Infect. Immun.* 62, 4367–4373.
199. Yanagimachi, R. (1994). Mammalian fertilization. In *The Physiology of Reproduction* (E. Knobil and J. D. Neill, Eds.), pp. 189–317. Raven Press, New York.
200. Sinowatz, F., Topfer-Petersen, E., and Calvete, J. J. (1997). Glycobiology of fertilization. In *Glycosciences* (H.-J. Gabius and S. Gabius, Eds.), pp. 595–610. Chapman & Hall, Weinheim.

201. Kerr, C. L., Hanna, W. F., Shaper, J. H., and Wright, W. W. (2004). Lewis X-containing glycans are specific and potent competitive inhibitors of the binding of ZP3 to complementary sites on capacitated, acrosome-intact mouse sperm. *Biol. Reprod.* 71, 778–789.
202. Raychoudhury, S. S., and Millette, C. F. (1997). Multiple fucosyltransferases and their carbohydrate ligands are involved in spermatogenic cell-Sertoli cell adhesion in vitro in rats. *Biol. Reprod.* 56, 1268–1273.
203. Akama, T. O., Nakagawa, H., Sugihara, K., Narisawa, S., Ohyama, C., Nishimura, S., O'Brien, D. A., Moremen, K. W., Millan, J. L., and Fukuda, M. N. (2002). Germ cell survival through carbohydrate-mediated interaction with Sertoli cells. *Science* 295, 124–127.
204. Suarez, S. S., Revah, I., Lo, M., and Kölle, S. (1998). Bull sperm binding to oviductal epithelium is mediated by a Ca^{2+}-dependent lectin on sperm which recognizes Lewis-A trisaccharide. *Biol. Reprod.* 59, 39–44.
205. Ignotz, G. G., Lo, M., Perez, C., Gwathmey, T. M., and Suarez, S. S. (2001). Characterization of a fucose-binding protein from bull sperm and seminal plasma responsible for formation of the oviductal sperm reservoir. *Biol. Reprod.* 64, 1806–1811.
206. Gwathmey, T. M., Ignotz, G. G., and Suarez, S. S. (2003). PDC-109 (BSP-A1/A2) promotes bull sperm binding to oviductal epithelium in vitro and may be involved in forming the oviductal sperm reservoir. *Biol. Reprod.* 69, 809–815.
207. Calvete, J. J., Raida, M., Sanz, L., Wempe, F., Scheit, K.-H., Romero, A., and Töpfer-Petersen, E. (1994). Localization and structural characterization of an oligosaccharide *O*-linked to bovine PDC-109. *FEBS Lett.* 350, 203–206.
208. Calvete, J. J., Campanero-Rhodes, A., Raida, M., and Sanz, L. (1999). Characterization of the conformational and quaternary structure-dependent heparin-binding region of bovine seminal plasma protein PDC–109. *FEBS Lett.* 444, 260–264.
209. Romero, A., Varela, P. F., Topfer-Petersen, E., and Calvete, J. J. (1997). Crystallization and preliminary X-ray diffraction analysis of bovine seminal plasma PDC-109, a protein composed of two fibronectin type II domains. *Proteins* 28, 454–456.
210. Ramakrishnan, M., Anbazhagan, V., Pratap, T. V., Marsh, D., and Swamy, M. J. (2001). Membrane insertion and lipid-protein interactions of bovine seminal plasma protein PDC-109 investigated by spin-label electron spin resonance spectroscopy. *Biophys. J.* 81, 2215–2225.
211. Desnoyers, L., and Manjunath, P. (1992). Major proteins of bovine seminal plasma exhibit novel interactions with phospholipid. *Biol. Chem.* 267, 10149–10155.
212. Müller, P., Erlemann, K.-R., Müller, K., Calvete, J. J., Töpfer-Petersen, E., Marienfeld, K., and Herrman, A. (1998). Biophysical characterization of the interaction of bovine seminal plasma protein PDC-109 with phospholipid vesicles. *Eur. Biophys. J.* 27, 33–41.
213. Smith, T. T., and Nothnick, W. B. (1997). Role of direct contact between spermatozoa and oviductal epithelial cells in maintaining rabbit sperm viability. *Biol. Reprod.* 56, 83–89.
214. Dobrinski, I., Smith, T. T., Suarez, S. S., and Ball, B. A. (1997). Membrane contact with oviductal epithelium modulates the intracellular calcium concentration in equine spermatozoa in vitro. *Biol. Reprod.* 56, 861–869.
215. Murray, S. C., and Smith, T. T. (1997). Sperm interaction with fallopian tube apical plasma membrane enhances sperm motility and delays capacitation. *Fertil. Steril.* 68, 352–357.
216. Lapointe, S., Sullivan, R., and Sirard, M.-A. (1998). Binding of a bovine oviductal fluid catalase to mammalian spermatozoa. *Biol. Reprod.* 58, 747–753.
217. Greube, A., Müller, K., Töpfer-Petersen, E., Herrmann, A., and Müller, P. (2001). Influence of bovine seminal plasma protein PDC-109 on the physical state of membranes. *Biochemistry* 40, 8326–8334.
218. Müller, P., Greube, A., Töpfer-Petersen, E., and Herrmann, A. (2002). Influence of the bovine seminal plasma protein PDC-109 on cholesterol in the presence of phospholipids. *Eur. Biophys. J.* 31, 438–447.
219. Manjunath, P., Soubeyrand, S., Chandonnet, L., and Roberts, K. D. (1994). Major proteins of bovine seminal plasma inhibit phospholipase A_2. *Biochem. J.* 303, 121–128.
220. Soubeyrand, S., and Manjunath, P. (1997). Novel seminal phospholipase A_2 is inhibited by the major proteins of bovine seminal plasma. *Biochim. Biophys. Acta* 1341, 183–188.
221. Lefebvre, R., Chenoweth, P. J., Drost, M., LeClear, C. T., MacCubbin, M., Dutton, J. T., and Suarez, S. S. (1995). Characterization of the oviductal sperm reservoir in cattle. *Biol. Reprod.* 53, 1066–1074.
222. Ho, H. C., and Suarez, S. S. (2001). Hyperactivation of mammalian spermatozoa: function and regulation. *Reproduction* 122, 519–526.
223. Suarez, S. S., and Ho, H. C. (2003). Hyperactivated motility in sperm. *Reprod. Domest. Anim.* 38, 119–124.
224. DeMott, R. P., and Suarez, S. S. (1992). Hyperactivated sperm progress in the mouse oviduct. *Biol. Reprod.* 46, 779–785.
225. Lefebvre, R., and Suarez, S. S. (1996). Effect of capacitation on bull sperm binding to homologous oviductal epithelium. *Biol. Reprod.* 54, 575–582.
226. Suarez, S. S., and Dai, X. B. (1992). Hyperactivation enhances mouse sperm capacity for penetrating viscoelastic media. *Biol. Reprod.* 46, 686–691.
227. Quill, T. A., Sugden, S. A., Rossi, K. L., Doolittle, L. K., Hammer, R. E., and Garbers, D. L. (2003). Hyperactivated sperm motility driven by CatSper2 is required for fertilization. *Proc. Natl. Acad. Sci. U S A* 100, 14869–14874.
228. Suarez, S. S., Katz, D. F., and Overstreet, J. W. (1983). Movement characteristics and acrosomal status of rabbit spermatozoa recovered at the site and time of fertilization. *Biol. Reprod.* 29, 1277–1287.
229. Suarez, S. S., and Osman, R. A. (1987). Initiation of hyperactivated flagellar bending in mouse sperm within the female reproductive tract. *Biol. Reprod.* 36, 1191–1198.
230. Chian, R.-C, LaPointe, S, and Sirard, M. A. (1995). Capacitation in vitro of bovine spermatozoa by oviduct cell monolayer conditioned medium. *Mol. Reprod. Dev.* 42, 318–324.
231. Mahmoud, A. I., and Parrish, J. J. (1996). Oviduct fluid and heparin induce similar surface changes in bovine sperm during capacitation. *Mol. Reprod. Dev.* 43, 554–560.
232. Revah, I., Suarez, S. S., Flesch, F. M., Colenbrander, B., and Gadella, B. M. (2000). Physiological state of bull sperm affects fucose- and mannose-binding properties. *Biol. Reprod.* 62, 1010–1015.
233. Wah, D. A., Fernandez-Tornero, C., Sanz, L., Romero, A., and Calvete, J. J. (2002). Sperm coating mechanism from the 1.8 Å crystal structure of PDC-109-phosphorylcholine complex. *Structure* 10, 505–514.
234. Parrish, J. J., Susko-Parrish, J. L., Winer, M. A., and First, N. L. (1988). Capacitation of bovine sperm by heparin. *Biol. Reprod.* 38, 1171–1180.
235. Parrish, J. J., Susko-Parrish, J. L., Handrow, R. R., Sims, M. M., and First, N. L. (1989). Capacitation of bovine spermatozoa by oviduct fluid. *Biol. Reprod.* 40, 1020–1025.
236. Galantino-Homer, H. L., Visconti, P. E., and Kopf, G. S. (1997). Regulation of protein tyrosine phosphorylation during bovine sperm capacitation by a cyclic adenosine 3′5′-monophosphate-dependent pathway. *Biol. Reprod.* 56, 707–719.

237. Thérien, I., Bousquet, D., and Manjunath, P. (2001). Effect of seminal phospholipid-binding proteins and follicular fluid on bovine sperm capacitation. *Biol. Reprod.* 5, 41–51.
238. Bosch, P., de Avila, J. M., Ellington, J. E., and Wright, R. W. Jr. (2001). Heparin and Ca^{2+}-free medium can enhance release of bull sperm attached to oviductal epithelial cell monolayers. *Theriogenology* 56, 247–260.
239. Thérien, I., Bleau, G., and Manjunath, P. (1995). Phosphatidylcholine-binding proteins of bovine seminal plasma modulate capacitation of spermatozoa by heparin. *Biol. Reprod.* 52, 1372–1379.
240. Manjunath, P., and Thérien, I. (2002). Role of seminal plasma phospholipid-binding proteins in sperm membrane lipid modification that occurs during capacitation. *J. Reprod. Immunol.* 53, 101–119.
241. Bedford, J. M. (1991). The co-evolution of mammalian gametes. In *A Comparative Overview of Mammalian Fertilization* (B. S. Dunbar and M. G. O'Rand, Eds.), pp. 3–35. Plenum Press, New York.
242. Taggart, D. A. (1994). A comparison of sperm and embryo transport in the female reproductive tract of marsupial and eutherian mammals. *Reprod. Fertil. Dev.* 6, 451–472.
243. Bedford, J. M., and Breed, W. G. (1994). Regulated storage and subsequent transformation of spermatozoa in the fallopian tubes of an Australian marsupial. *Sminthopsis crassicaudata*. *Biol. Reprod.* 50, 845–854.
244. Bedford, J. M., Mori, T., and Oda, S. (1997). Ovulation induction and gamete transport in the female tract of the musk shrew, *Suncus murinus*. *J. Reprod. Fertil.* 110, 115–125.
245. Bedford, J. M., Mock, O. B., and Phillips, D. M. (1997). Unusual ampullary sperm crypts, and behavior and role of the cumulus oophorus, in the oviduct of the least shrew, *Cryptotis parva*. *Biol. Reprod.* 56, 1255–1267.
246. Bedford, J. M., Mock, O. B., Nagdas, S. K., Winfrey, E. P., and Olson, G. E. (1999). Reproductive features of the eastern mole (*Scalopus aquaticus*) and star-nosed mole (*Condylura cristata*). *J. Reprod. Fetil.* 117, 345–353.
247. Bedford, J. M., Phillips, D. M., and Mover-Lev, H. (1997). Novel sperm crypts and behavior of gametes in the fallopian tube of the white-toothed shrew, *Crocidura russula* Monacha. *J. Exp. Zool.* 277, 262–273.
248. Bedford, J. M., Bernard, R. T., and Baxter, R. M. (1998). The "hybrid" character of the gametes and reproductive tracts of the African shrew, *Mysorex varius*, supports its classification in the Crocidosoricinae. *J. Reprod. Fertil.* 112, 165–173.
249. Bedford, J. M., Mock, O. B., Nagdas, S. K., Winfrey, E. P., and Olson, G. E. (2000). Reproductive characteristics of the African pygmy hedgehog, *Atelerix albiventris*. *J. Reprod. Fertil.* 120, 143–150.
250. Williams, M., Hill, C. J., Scudamore, I., Dunphy, B. Cooke, I. D., and Barratt, C. L. R. (1993). Sperm numbers and distribution within the human fallopian tube around ovulation. *Hum. Reprod.* 8, 2019–2026.
251. Yeung, W. S. B., Ng, V. K. H., Lau, E. Y. L., and Ho, P. C. (1994). Human oviductal cells and their conditioned medium maintain the motility and hyperactivation of human spermatozoa *in vitro*. *Hum. Reprod.* 9, 656–660.
252. Pacey, A. A., Davies, N., Warren, M. A., Barratt, C. L., and Cooke, I. D. (1995). Hyperactivation may assist human spermatozoa to detach from intimate association with the endosalpinx. *Hum. Reprod.* 10, 2603–2609.
253. Pacey, A. A., Hill, C. J., Scudamore, I. W., Warren, M. A., Barratt, C. L. R., and Cooke, I. D. (1995). The interaction in vitro of human spermatozoa with epithelial cells from the human uterine (fallopian) tube. *Hum. Reprod.* 10, 360–366.
254. Baillie, H. S., Pacey, A. A., Warren, M. A., Scudamore, I. W., and Barratt, C. L. (1997). Greater numbers of human spermatozoa associate with endosalpingeal cells derived from the isthmus compared with those from the ampulla. *Hum. Reprod.* 12, 1985–1992.
255. Insler, V., Glezerman, M., Zeidel, L., Bernstein, D., and Misgav, N. (1980). Sperm storage in the human cervix: a quantitative study. *Fertil. Steril.* 33, 288–294.
256. Moyer, D. L., Rimdusit, S., and Mishell, D. R. Jr. (1970). Sperm distribution and degradation in the human female reproductive tract. *Obstet. Gynecol.* 35, 831–840.
257. Thompson, L. A., Barratt, C. L. R., Bolton, A. E., and Cooke, I. D. (1992). The leukocytic reaction of the human uterine cervix. *Am. J. Reprod. Immunol.* 28, 85–89.
258. Ralt, D., Goldenberg, M., Fetterolf, P., Thompson, D., Dor, J., Mashiachi, S., Garbers, D. L., and Eisenbach, M. (1991). Sperm attraction to a follicular fluid factor(s) correlates with human egg fertilizability. *Proc. Natl. Acad. Sci. U S A* 88, 2840–2844.
259. Katz, D. F., and Yanagimachi, R. (1980). Movement characteristics of hamster sperm within the oviduct. *Biol. Reprod.* 22, 759–764.
260. Ren, D., Navarro, B., Perez, G., Jackson, A. C., Hsu, S., Shi, Q., Tilly, J. L., and Clapham, D. E. (2001). A sperm ion channel required for sperm motility and male fertility. *Nature* 413, 603–609.
261. Carlson, A. E., Westenbroek, R. E., Quill, T., Ren, D., Clapham, D. E., Hille, B., Garbers, D. L., and Babcock, D. F. (2003). CatSper 1 required for evoked Ca2+ entry and control of flagellar function in sperm. *Proc. Natl. Acad. Sci. U S A* 100, 14864–14868.
262. Quill, T. A., Ren, D., Clapham, D. E., and Garbers, D. L. (2001). A voltage-gated ion channel expressed specifically in spermatozoa. *Proc. Natl. Acad. Sci. U S A* 98, 12527–12531.
263. Ralt, D., Manor, M., Cohen-Dayag, A., Tur-Kaspa, I., Ben-Shlomo, I., Makler, A., Yuli, I., Dor, J., Blumberg, S., Mashiach, S., and Eisenbach, M. (1994). Chemotaxis and chemokinesis of human spermatozoa to follicular factors. *Biol. Reprod.* 50, 774–785.
264. Cohen-Dayag, A., Ralt, D., Tur-Kaspa, I., Manor, M., Makler, A., Dor, J., Mashiach, S., and Eisenbach, M. (1994). Sequential acquisition of chemotactic responsiveness by human spermatozoa. *Biol. Reprod.* 50, 786–790.
265. Cohen-Dayag, A., Tur-Kaspa, I., Dor, J., Mashiach, S., and Eisenbach, M. (1995). Sperm capacitation in humans is transient and correlates with chemotactic responsiveness to follicular factors. *Proc. Natl. Acad. Sci. U S A* 92, 11039–11043.
266. Fabro, G., Rovasio, R. A., Civalero, S., Frenkel, A., Caplan, S. R., Eisenbach, M., and Giojalas, L. C. (2002). Chemotaxis of capacitated rabbit spermatozoa to follicular fluid revealed by a novel directionality-based assay. *Biol. Reprod.* 67, 1565–1571.
267. Vanderhaeghen, P., Schurmans, S., Vassart, G., and Parmentier, M. (1993). Olfactory receptors are displayed on dog mature sperm cells. *J. Cell. Biol.* 123, 1441–1452.
268. Walensky, L. D., Roskams, A. J., Lefkowitz, R. J., Snyder, S. H., and Ronnett, G. V. (1995). Odorant receptors and desensitization proteins colocalize in mammalian sperm. *Mol. Med.* 1, 130–141.
269. Spehr, M., Gisselmann, G., Poplawski, A., Riffell, J. A., Wetzel, C. H., Zimmer, R. K., and Hatt, H. (2003). Identification of a testicular odorant receptor mediating human sperm chemotaxis. *Science* 299, 2054–2058.
270. Suarez, S. S., Vincenti, L., and Ceglia, M. W. (1987). Hyperactivated motility induced in mouse sperm by calcium ionophore A23187 is reversible. *J. Exp. Zool.* 244, 331–336.
271. Babcock, D. F. (2003). Development. Smelling the roses? *Science* 299, 1993–1994.

272. Chakraborty, J., and Nelson, L. (1975). Fate of surplus sperm in the fallopian tube of the white mouse. *Biol. Reprod.* 12, 455–463.
273. Rasweiler, J. J. (1987). Prolonged receptivity to the male and the fate of spermatozoa in the female black mastiff bat, *Molossus ater. J. Reprod. Fertil.* 79, 643–654.
274. Mortimer, D., and Templeton, A. A. (1982). Sperm transport in the human female reproductive tract in relation to semen analysis characteristics and time of ovulation. *J. Reprod. Fertil.* 64, 401–408.
275. Alberts, B., Johnson, A. Lewis, J., Raf, M., Roberts, K., and Walter, P. (2002). *Molecular Biology of the Cell*, 4th ed. Garland Science, New York.
276. Austin, C. R. (1970). Ageing and reproduction: post-ovulatory deterioration of the egg. *J. Reprod. Fertil. Suppl.* 12, 39–53.
277. Marston, J. H., and Chang, M. C. (1964). The fertilizable life of ova and their morphology following delayed insemination in mature and immature mice. *J. Exp. Zool.* 155, 237–251.
278. Xu, Z., Abbott, A., Kopf, G., Schultz, R. M., and Ducibella, T. (1997). Spontaneous activation of ovulated mouse eggs: time-dependent effects on M-phase exit, cortical granule exocytosis, maternal messenger ribonucleic acid recruitment, and inositol 1,4,5-trisphosphate sensitivity. *Biol. Reprod.* 57, 743–750.
279. Gordo, A. C., Rodrigues, P., Kurokawa, M., Jellerette, T., and Exley, G. E. (2002). Intracellular calcium oscillations signal apoptosis rather than activation in in vitro aged mouse eggs. *Biol. Reprod.* 66, 1828–1837.
280. Tarin, J. J., Perez-Albala, S., Aguilar, A., Minarro, J., Hermenegildo, C., and Cano, A. (1999). Long-term effects of postovulatory aging of mouse oocytes on offspring: a two-generational study. *Biol. Reprod.* 61, 1347–1355.
281. Hammond, J. (1934). The fertilization of rabbit ova in relation to time. A method of controlling the litter size. The duration of pregnancy and the weight of young at birth. *J. Exp. Biol.* 11, 140–161.
282. Blandau, R. J. (1952). The female factor in fertility and infertility. I. Effects of delayed fertilization on the development of the pronuclei in rat ova. *Fertil. Steril.* 3, 349–365.
283. Yanagimachi, R., and Chang, M. C. (1961). Fertilizable life of golden hamster ova and their morphological changes at the time of losing fertilizability. *J. Exp. Zool.* 148, 185–204.
284. Blandau, R. J., and Young, W. C. (1939). The effects of delayed fertilization on the development of the guinea pig ovum. *Am. J. Anat.* 68, 275–291.
285. Hammond, J., and Walton, A. (1934). Notes on ovulation and fertilization in the ferret. *J. Exp. Biol.* 11, 307.
286. Chang, M. C., and Yanagimachi, R. (1963). Fertilization of ferret ova by deposition of epididymal sperm into the ovarian capsule, with special reference to the fertilizable life of ova and capacitation of sperm. *J. Exp. Zool.* 154, 175–187.
287. Mahi-Brown, C. A., and Yanagimachi, R. (1983). Parameters influencing ovum pickup by oviductal fimbria in the golden hamster. *Gamete Res.* 8, 1–10.
288. Lam, X., Gieseke, C., Knoll, M., and Talbot, P. (2000). Assay and importance of adhesive interaction between hamster (*Mesocricetus auratus*) oocyte-cumulus complexes and the oviductal epithelium. *Biol. Reprod.* 62, 579–588.
289. Talbot, P., Geiseke, C., and Knoll, M. (1999). Oocyte pickup by the mammalian oviduct. *Mol. Biol. Cell.* 10, 5–8.
290. Norwood, J. T., Hein, C. E., Halbert, S. A., and Anderson, R. G. (1978). Polycationic macromolecules inhibit cilia-mediated ovum transport in the rabbit oviduct. *Proc. Natl. Acad. Sci. U S A* 75, 4413–4416.
291. Norwood, J. T., and Anderson, R. G. (1980). Evidence that adhesive sites on the tips of oviduct cilia membranes are required for ovum pickup in situ. *Biol. Reprod.* 23, 788–791.
292. Zhuo, L., Yoneda, M., Zhao, M., Yingsung, W., Yoshida, N., Kitagawa, Y., Kawamura, K., Suzuki, T., and Kimata, K. (2001). Defect in SHAP-hyaluronan complex causes severe female infertility. A study by inactivation of the bikunin gene in mice. *J. Biol. Chem.* 276, 7693–7696.
293. Fulop, C., Szanto, S., Mukhopadhway, D., Bardos, T., Kamath, R. V., Rugg, M. S., Day, A. J., Salustri, A., Hascall, V. C., Glant, T. T., and Mikecz, K. (2003). Impaired cumulus mucification and female sterility in tumor necrosis factor-induced protein-6 deficient mice. *Development* 130, 2253–2261.
294. Varani, S., Elvin, J. A., Yan, C., DeMayo, J., DeMayo, F. J., Horton, H. F., Byrne, M. C., and Matzuk, M. M. (2002). Knockout of pentraxin 3, a downstream target of growth differentiation factor-9, causes female subfertility. *Mol. Endocrinol.* 16, 1154–1167.
295. Salustri, A., Garlanda, C., Hirsch, E., De Acetis, M., Maccagno, A., Bottazzi, B., Doni, A., Bastone, A., Mantovani, G., Peccoz, P. B., Salvatori, G., Mahoney, D. J., Day, A. J., Siracusa, G., Romani, L., and Mantovani, A. (2004). PTX3 plays a key role in the organization of the cumulus oophorus extracellular matrix and in in vivo fertilization. *Development* 131, 1577–1586.
296. Breed, W. G. (1994). How does sperm meet egg in a marsupial? *Reprod. Fertil. Dev.* 6, 485–506.
297. Bedford, J. M. (1996). What marsupial gametes disclose about gamete function in eutherian mammals. *Reprod. Fertil. Dev.* 8, 569–580.
298. Bedford, J. M., Cooper, G. W., Phillips, D. M., and Dryden, G. L. (1994). Distinctive features of the gametes and reproductive tracts of the Asian musk shrew, *Suncus murinus. Biol. Reprod.* 50, 820–834.
299. Bedford, J. M., and Kim, H. H. (1993). Cumulus oophorus as a sperm sequestering device, in vivo. *J. Exp. Zool.* 265, 321–328.
300. Halbert, S. A., Tam, P. Y., and Blandau, R. J. (1976). Egg transport in the rabbit oviduct: the roles of cilia and muscle. *Science* 191, 1052–1053.
301. Halbert, S. A., Becker, D. R., and Szal, S. E. (1989). Ovum transport in the rat oviductal ampulla in the absence of muscle contractility. *Biol. Reprod.* 40, 1131–1136.
302. Gaddum-Rosse, P., and Blandau, R. J. (1976). Comparative observations on ciliary currents in mammalian oviducts. *Biol. Reprod.* 14, 605–609.
303. Eddy, C. A., Flores, J. J., Archer, D. R., and Pauerstein, C. J. (1978). The role of cilia in fertility: an evaluation by selective microsurgical modification of the rabbit oviduct. *Am. J. Obstet. Gynecol.* 132, 814–821.
304. McComb, P. F., Halbert, S. A., and Gomel, V. (1980). Pregnancy, ciliary transport, and the reversed ampullary segment of the rabbit fallopian tube. *Fertil. Steril.* 34, 386–390.
305. McComb, P., Langley, L., Villalon, M., and Verdugo, P. (1986). The oviductal cilia and Kartagener's syndrome. *Fertil. Steril.* 46, 412–416.
306. Halbert, S. A., Patton, D. F. L., Zarutskie, P. W., and Soules, M. R. (1997). Function and structure of cilia in the Fallopian tube of an infertile woman with Kartegener's syndrome. *Hum. Reprod.* 12, 55–58.
307. Bleau, G., Richer, C.-L., and Bousquet, D. (1978). Absence of dynein arms in cilia of endocervical cells in a fertile woman. *Fertil. Steril.* 30, 362–363.
308. Halbert, S. A., Szal, S. E., and Broderson, S. H. (1988). Anatomical basis of a passive mechanism for ovum retention at the ampulloisthmic junction. *Anat. Rec.* 221, 841–845.
309. Spilman, C. H., Shaikh, A. A., and Harper, M. J. K. (1978). Oviductal motility amplitude and ovarian steroid secretion during egg transport in the rabbit. *Biol. Reprod.* 18, 409–417.

310. Betteridge, K. J., Eaglesome, M. D., Mitchell, D., Flood, P. F., and Beriault, R. (1982). Development of horse embryos up to twenty two days after ovulation: observations on fresh specimens. *J. Anat.* 135, 191–209.
311. Freeman, D. A., Woods, G. L., Vanderwall, D. K., and Weber, J. A. (1992). Embryo-initiated oviductal transport in mares. *J. Reprod. Fertil.* 95, 535–538.
312. Villalon, M., Ortiz, M. E., Aguayo, C., Munoz, J., and Croxatto, H. B. (1982). Differential transport of fertilized and unfertilized ova in the rat. *Biol. Reprod.* 26, 337–341.
313. Weber, J. A., Freeman, D. A., Vanderwall, D. K., and Woods, G. L. (1991). Prostaglandin E2 hastens oviductal transport of equine embryos. *Biol. Reprod.* 45, 544–546.
314. Weber, J. A., Freeman, D. A., Vanderwall, D. K., and Woods, G. L. (1991). Prostaglandin E2 secretion by oviductal transport-stage equine embryos. *Biol. Reprod.* 45, 540–543.
315. Velasquez, L. A., Aguilera, J. G., and Croxatto, H. B. (1995). Possible role of platelet-activating factor in embryonic signaling during oviductal transport in the hamster. *Biol. Reprod.* 52, 1302–1306.
316. Ammit, A. J., and O'Neill, C. (1991). Comparison of a radioimmunoassay and bioassay for embryo-derived platelet-activating factor. *Hum. Reprod.* 6, 872–878.
317. Velasquez, L. A., Maisey, K., Fernandez, R., Valdes, D., Cardenas, H., Imarai, M., Delgado, J., Aguilera, J., and Croxatto, H. B. (2001). PAF receptor and PAF acetylhydrolase expression in the endosalpinx of the human fallopian tube: possible role of embryo-derived PAF in the control of embryo transport to the uterus. *Hum. Reprod.* 16, 1583–1587.
318. Ortiz, M. E., Villalon, M., and Croxatto, H. B. (1979). Ovum transport and fertility following postovulatory treatment with estradiol in rats. *Biol. Reprod.* 21, 1163–1167.
319. Orihuela, P. A., and Croxatto, B. (2001). Acceleration of oviductal transport of oocytes induced by estradiol in cycling rats is mediated by nongenomic stimulation of protein phosphorylation in the oviduct. *Biol. Reprod.* 65, 1238–1245.
320. Orihuela, P. A., Ríos, M., and Croxatto, H. B. (2001). Disparate effects of estradiol on egg transport and oviductal protein synthesis in mated and cyclic rats. *Biol. Reprod.* 65, 1232–1237.
321. Orihuela, P. A., Parada-Bustamante, A., Cortés, P. P., Gatica, C., and Croxatto, H. B. (2003). Estrogen receptor, cyclic adenosine monophosphate, and protein kinase A are involved in the nongenomic pathway by which estradiol accelerates oviductal oocyte transport in cyclic rats. *Biol. Reprod.* 68, 1225–1231.
322. Mwanza, M., Razdan, P., Hulten, F., and Einarsson, S. (2002). Transport of fertilised and unfertilized ova in sows. *Anim. Reprod. Sci.* 74, 69–74.
323. Denker, H. W., and Gerdes, H. J. (1979). The dynamic structure of rabbit blastocysts coverings. I. Transformation during regular preimplantation development. *Anat. Embryol.* 157, 15–34.
324. Malette, B., Paquette, Y., Merlen, Y., and Bleau, G. (1995). Oviductins possess chitinase- and mucin-like domains: a lead in the search for the biological function of these oviduct-specific ZP-associating glycoproteins. *Mol. Reprod. Dev.* 41, 384–397.
325. Malette, B., Paquette, Y., and Bleau, G. (1995). Size variations in the mucin-type domain of hamster oviductin: identification of the polypeptide precursors and characterization of their biosynthetic maturation. *Biol. Reprod.* 53, 1311–1323.
326. Sendai, Y., Komiya, H., Suzuki, K., Onuma, T., Kikuchi, M., Hoshi, H., and Araki, Y. (1995). Molecular cloning and characterization of a mouse oviduct-specific glycoprotein. *Biol. Reprod.* 53, 285–294.
327. Selwood, L. (2000). Marsupial egg and embryo coats. *Cell. Tiss. Org.* 166, 208–219.
328. Stillman, R. J., Rosenberg, M. J., and Sachs, B. P. (1986). Smoking and reproduction. *Fertil. Steril.* 46, 545–566.
329. Knoll, M., Shaoulian, R., Magers, T., and Talbot, P. (1995). Ciliary beat frequency of hamster oviducts is decreased in vitro by exposure to solutions of mainstream and sidestream cigarette smoke. *Biol. Reprod.* 53, 29–37.
330. Knoll, M., and Talbot, P. (1998). Cigarette smoke inhibits oocyte cumulus complex pick-up by the oviduct in vitro independent of ciliary beat frequency. *Reprod. Toxicol.* 12, 57–68.
331. Talbot, P., DiCarlantonio, G., Knoll, M., and Gomez, C. (1998). Identification of cigarette smoke components that alter functioning of hamster (*Mesocricetus auratus*) oviducts in vitro. *Biol. Reprod.* 58, 1047–1053.
332. Riveles, K., Iv, M., Arey, J., and Talbot, P. (2003). Pyridines in cigarette smoke inhibit hamster oviductal functioning in picomolar doses. *Reprod. Toxicol.* 17, 191–202.
333. Riveles, K., Roza, R., Arey, J., and Talbot, P. (2004). Pyrazine derivatives in cigarette smoke inhibit hamster oviductal functioning. *Reprod. Biol. Endocrinol.* 2, 23–37.
334. Adams, T. B., Doull, J., Feron, V. J., Goodman, J. I., Marnett, L. J., Munro, I. C., Newberne, P. M., Portoghese, P. S., Smith, R. L., Waddell, W. J., and Wagner, B. M. (2002). The FEMA GRAS assessment of pyrazine derivatives used as flavor ingredients. Flavor and Extract Manufacturers Association. *Food Chem. Toxicol.* 40, 429–451.
335. DiCarlantonio, G., and Talbot, P. (1999). Inhalation of mainstream and sidestream cigarette smoke retards embryo transport and slows muscle contraction in oviducts of hamsters (*Mesocricetus auratus*). *Biol. Reprod.* 61, 651–656.
336. Frick, H., Leonhardt, H., and Starck, D. (1991). *Human Anatomy 2: Special Anatomy: Viscera and Nervous System, Classification of Muscles and Vessels, Organization of Lymphatics and Nerves*. Thieme Medical Publishers, New York.
337. Suarez, S. S. (2002). Gamete transport. In *Fertilization* (D. M. Hardy, Ed.), pp. 3–28. Academic Press, San Diego.

Knobil and Neill's Physiology of Reproduction, Third Edition
edited by Jimmy D. Neill,
Elsevier © 2006

CHAPTER 4

Implantation

Sudhansu K. Dey[1] and Hyunjung Lim[2]

Introduction, 147
Physiological and Morphological Aspects of Embryo–Uterine Interactions, 148
Three Stages of Implantation, 148
Types of Implantation, 150
Hormonal Requirements, 151
Endometrial Glands and Their Secretions, 153
Special Aspects of Embryo–Uterine Interactions, 153
Delayed Implantation (Diapause), 153
Window of Implantation, 156
Molecular Aspects of Embryo–Uterine Interactions, 157
Steroid Hormones, 157
Adhesion Molecules, 159
Vasoactive Factors, 162
Growth Factors, 164
Cytokines, 165
Transcription Factors, 167
Calcium-Associated Signaling Molecules, 168
Connexins, 169
Cell Cycle Regulation, 169
Matrix Remodeling and Angiogenesis, 170
Endocannabinoids, 172
Morphogens, 172
Future Directions: New Techniques, Novel Discovery, 173
Concluding Remarks, 174
Acknowledgments, 175
References, 175

INTRODUCTION

An essential task of procreation is to diversify and pass on superior genetic materials to the offspring. Asexual procreation in prokaryotes and in some eukaryotes is superior to sexual procreation in higher eukaryotes, especially in mammals, with respect to sheer number of progeny. Thus, reproduction in viviparous mammals has adapted to a more complex and highly regulated system. Fostering an offspring within the womb to produce a live birth is a demanding task, requiring safeguarded regulatory systems at multiple critical steps. The embarkment of a new life in eutherians first depends on the union between a sperm and an egg (ovum) that results in successful fertilization; failure to achieve this union leads to their demise. A one-cell fertilized egg, hence termed embryo, undergoes several mitotic cell divisions, ultimately forming a differentiated tissue called the blastocyst with two distinct cell populations, the inner cell mass (ICM) and a layer of trophectoderm cells surrounding the ICM (1). A reciprocal interaction between the blastocyst and the maternal uterus initiates the process of implantation, a process by which the vascular system of the embryo is brought into functional communication with the maternal circulation, leading to the establishment of functional placenta and pregnancy. During early days, the process of implantation was often referred to as nidation, the term originating from the word "nidus," meaning a nest or a breeding place. Maternal resources filtered across the placenta with selective barrier properties nourish and protect the conceptus. Placental types are classified into three categories: hemochorial (rodents, humans, and subhuman primates), epitheliochorial (horses, cow, sheep, and pigs), and endotheliochorial (most carnivores) (2).

Implantation is a prerequisite step for subsequent development of the embryo. A significant number of

[1]Departments of Pediatrics, Cell and Developmental Biology and Pharmacology, Vanderbilt University Medical Center, Nashville, Tennessee; [2]Department of Obstetrics and Gynecology, Washington University School of Medicine, St. Louis, Missouri.

pregnancy losses due to preimplantation embryonic death are common to many mammals and have been considered a selection process for the survival of superior embryos for implantation. However, dysregulated events before, during, or immediately after implantation are also often causes for poor pregnancy rates in eutherians. Therefore, a deeper understanding and unraveling the clues to preimplantation embryo development and implantation in the uterus have been a challenge to reproductive and developmental biologists with a mission of curing and improving infertility, ensuring birth of quality offspring, and/or developing novel contraceptive approaches to restrict rapid world population growth.

PHYSIOLOGICAL AND MORPHOLOGICAL ASPECTS OF EMBRYO–UTERINE INTERACTIONS

In 1947, Corner wrote, "…the uterine chamber is actually a less favorable place for early embryos than say, the anterior chamber of the eye, except when the hormones of the ovary act upon it and change it to a place of superior efficiency for its new functions." Subsequently, it was realized that the establishment of pregnancy results from the culmination of an intimate relationship between the developing embryo and the differentiating uterus. However, although conceptually accepted, the nature of such two-way interaction between the blastocyst and the uterus is still a challenging question. Embryo–uterine interaction leading to implantation is only initiated when embryonic development is synchronized with the preparation of the uterus to the receptive state. More specifically, the embryo must reach the blastocyst stage and gain implantation competency, and the uterus, through steroid hormone–dependent changes, must attain the receptive stage before successful implantation can occur. A lack of synchrony between the embryonic development and the preparation of the uterus results in implantation failure. The underlying mechanisms that coordinate blastocyst development to implantation competency with uterine receptivity are not yet fully appreciated.

Three Stages of Implantation

At the blastocyst stage, the trophectoderm of the developing embryo acquires competence to attach to the receptive uterine luminal epithelium that has been appropriately primed with the steroid hormones estrogen and progesterone (P_4). If this condition of synchrony is met, an implantation-initiating adhesion cascade begins upon engagement of cell adhesion molecules at the luminal epithelial and trophectoderm surfaces. Many of these adhesion molecules transduce cytoplasmic signals necessary to sustain embryonic and maternal contributions to the formation of a placenta that supports fetal development through the end of pregnancy.

Implantation begins when the blastocyst is placed into intimate physical and physiological contacts with the uterine endometrium. Enders and Schlafke (3,4) classified the process of implantation into three stages: apposition, adhesion and penetration. Apposition is the stage when the embryonic trophectoderm cells become closely apposed to the luminal epithelium. This is followed by the stage of adhesion when the association of the trophectoderm and the luminal epithelium is sufficiently intimate to resist dislocation of the blastocyst by flushing the uterine lumen. The stage of penetration involves the invasion of the luminal epithelium by the trophectoderm. Stromal cell differentiation to decidual cells (decidualization) is more extensive, and the loss of the luminal epithelium is evident at this stage. These three stages of implantation are a continuous process over a period of time.

Apposition

In rodents, a generalized stromal edema occurs before the beginning of apposition. This event leads to the closure of the uterine lumen and results in interdigitation of microvilli of the trophectoderm and luminal epithelia (apposition), followed by closer contact between them (adhesion or attachment reaction). The luminal closure occurs throughout the entire uterus during pregnancy or pseudopregnancy, and thus this event does not necessitate the presence of blastocysts. Priming of the uterus with P_4 alone appears to be sufficient for this event to occur; luminal closure does not occur in the absence of this hormone. Thus, although the events of luminal closure and apposition occur in P_4-treated delayed-implanting mice, the attachment reaction does not occur under this condition. The superimposition of estrogen treatment is essential for the latter event to occur.

Attachment

The attachment reaction coincides with the localized increased stromal vascular permeability at the site of the blastocyst as determined by intravenous injection of a macromolecular blue dye (uterine blue reaction) (5) (Fig. 1). The first sign of the attachment reaction in the process of implantation (day 1, spermatozoa in the vaginal smear) occurs in the mouse and rat on the evenings of days 4 and 5, respectively, and on day $6\frac{1}{2}$ in the rabbit (5–7). In primates, the

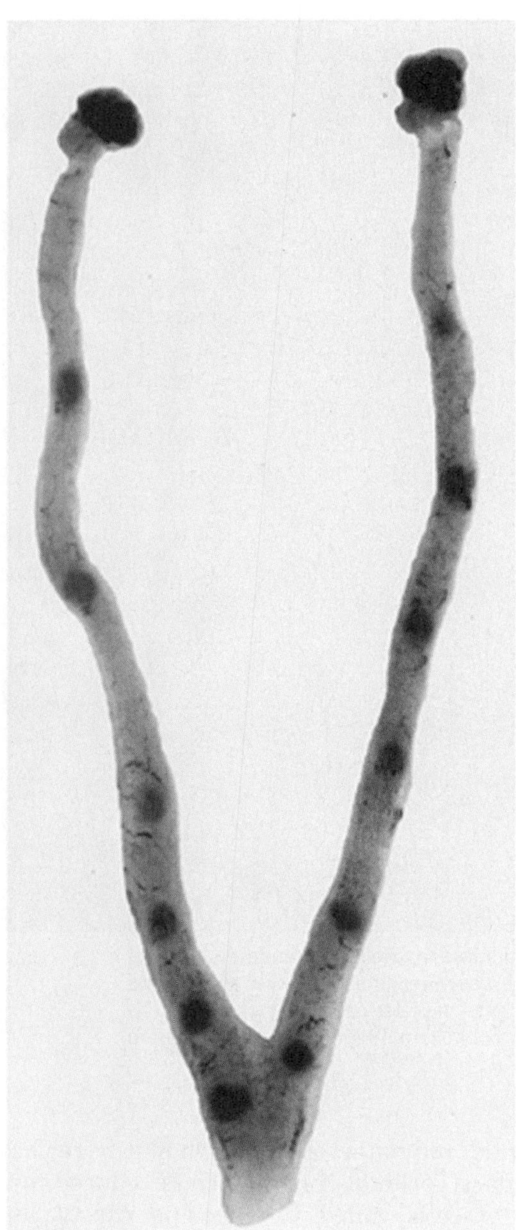

FIG. 1. Increased endometrial vascular permeability. Increased vascular permeability at the sites of blastocyst attachment with the uterine luminal epithelium was examined in a mouse at midnight (2400 h) of day 4 of pregnancy after an intravenous injection of a blue dye (Chicago Blue B6) solution. Distinct blue bands (dark bands in this picture) indicate that the attachment process has been initiated (see Fig. 2).

attachment reaction occurs approximately on day 8 in humans and baboons, on day 9 in macaques, and on day 11 in marmoset monkeys (8,9). In large domestic animals, the first signs of attachment occur on day 13 in pigs, day 20 in cows, day 16 in sheep, and day 19 in goats (10) (Table 1).

In mice, blastocysts are oriented with their ICMs directed mesometrially, whereas in humans the ICM is directed antimesometrially (Fig. 2). The mechanism by which the orientation of the blastocyst is achieved

TABLE 1. *Timing of implantation in various species*

Species	Day of implantation	Ovarian estrogen requirement	Decidualization
Mouse	4.5[a]	Yes	Yes
Rat	5.5[a]	Yes	Yes
Hamster	3.5[a]	No	Yes
Guinea pig	5[a]	No	Yes
Rabbit	6.5[b]	No	Yes
Pig	13–14[b]	No	No
Cow	19–20[b]	No	No
Sheep	15–16[b]	No	No
Human	8[c]	?	Yes
Baboon	8[c]	?	Yes
Rhesus monkey	9[c]	?	Yes

[a]Day 1, vaginal sperm.
[b]Day 0, mating at estrus.
[c]Day 0, Preovulatory estrogen or luteinizing hormone peak.

at the time of implantation remains elusive. There is evidence that in P_4-treated delayed-implanting mice, blastocysts are placed antimesometrially and interdigitation (apposition) of luminal epithelial cell microvilli occurs with those of the abembryonic or lateral trophectoderm cells of the blastocyst with its ICM oriented toward the uterine lumen. This observation led to the suggestion that upon initiation of the attachment reaction and subsequent implantation process by estrogen, blastocysts retain the orientation they adopted during delay. During normal implantation in mice with the onset of luminal closure, blastocysts are placed at the antimesometrial side of the lumen along the uterine axis. Shortly after the luminal closure, zona-encased blastocysts are located in implantation chambers with random orientation of their ICMs. However, with the beginning of the attachment reaction, blastocysts are correctly oriented with their ICMs directed at the mesometrial pole. This observation suggests that the trophectoderm of the entire blastocyst surface has the potential for attachment to the luminal epithelium and that attachment occurs randomly immediately after the loss of zona pellucida. Evidence was presented to suggest that free movement of the ICM directs the correct orientation of the blastocyst. However, this issue still remains unsettled and warrants further investigation. All these events from the luminal closure to the attachment reaction occur between about 86 and 92 hours postcoitum in mice (11,12).

Penetration and Decidualization

Penetration of embryos through the uterine epithelium and basal lamina into the stroma is required to establish a definitive vascular relationship with the mother. This process varies considerably from species to species with respect to timing and cytologic features (13). In rodents, penetration of

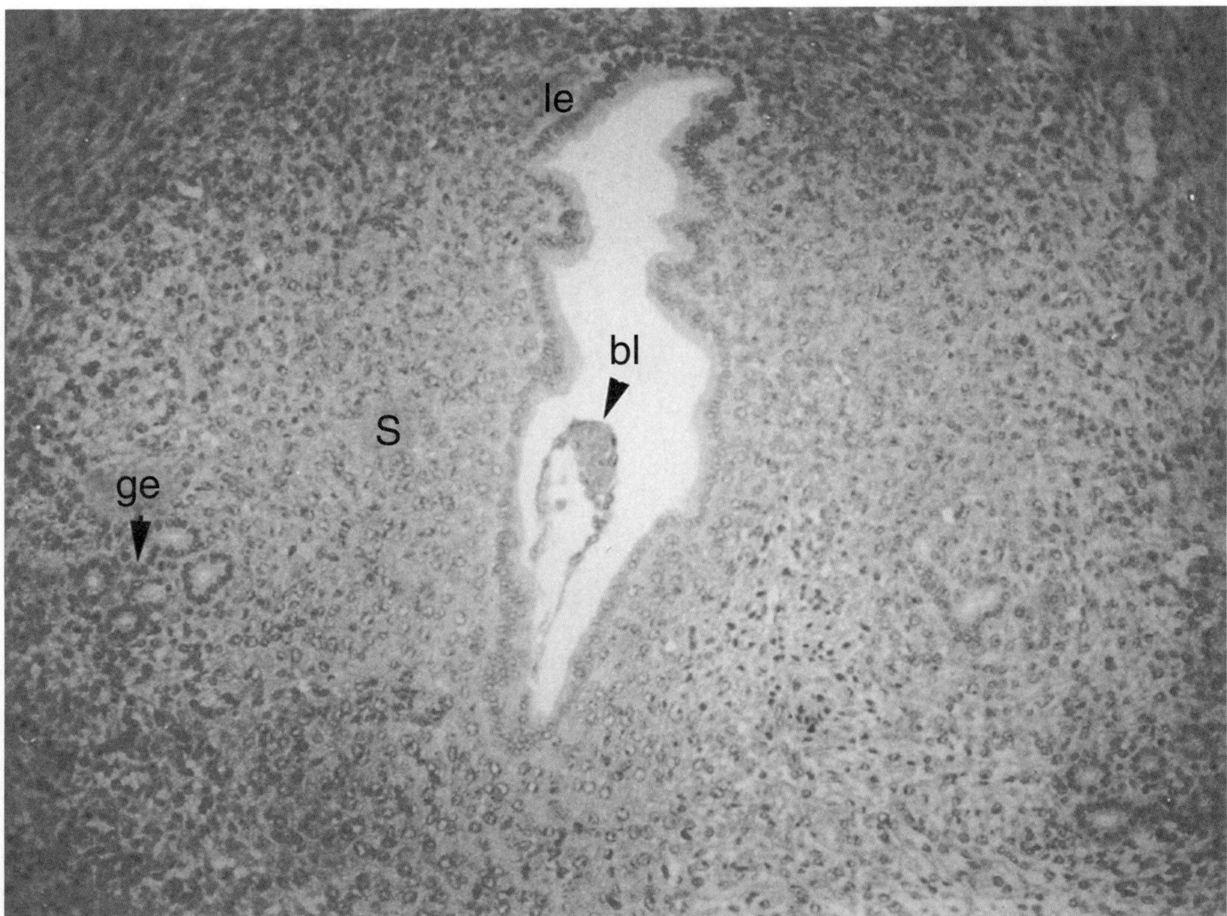

FIG. 2. Attachment reaction of the blastocyst with the uterine luminal epithelium. Around midnight of day 4 of pregnancy in mice, the attachment reaction occurs between the blastocyst and uterine luminal epithelium and is accompanied by localized endometrial vascular permeability (see Fig. 1). le, luminal epithelium; s, stroma; bl, blastocyst; ge, glandular epithelium. [Reprinted with permission from (199).]

trophoblasts is usually followed by decidualization of endometrial stromal cells surrounding the implanting blastocyst, eventually embedding the embryo into the stromal bed. Proliferation and differentiation of stromal cells to decidual cells occur in response to either the blastocyst or artificial stimuli. The decidual cell reaction is always preceded by increased endometrial vascular permeability (5,14). Pulse-labeling experiments with ^{3}H-thymidine suggested that decidual cells originate from undifferentiated stromal cells (15). The decidual cell reaction is first initiated at the antimesometrial sites where blastocysts implant. In mice, the differentiating stromal cells surrounding the blastocyst initially form the primary decidual zone on day 5. This zone is avascular and densely packed with decidual cells. By day 6, the secondary decidual zone is formed around the primary decidual zone. At this time DNA synthesis is high in the secondary decidual zone but low in the primary decidual zone (16). The primary decidual zone degenerates progressively up to day 8. After day 8, the placental and embryonic growth slowly replaces the secondary decidual zone, which is reduced to a thin layer of cells called the decidua capsularis. The mesometrial decidual cells ultimately form the decidua basalis (17). Presumed functions of decidua are to provide nutrition to the developing embryo, protect the embryo from immunological responses of the mother, and regulate trophoblast invasion into the uterine stroma.

Types of Implantation

Mechanisms of placentation vary between species and are broadly classified according to the degree of invasiveness and association of maternal and fetal structures. Placentation strategies range from the simplest superficial epitheliochorial placentas of domestic animals, in which the uterine luminal epithelium remains intact and the embryo and associated membranes are restricted to the uterine lumen throughout gestation, to the endotheliochorial and hemochorial placentas of rodents and humans in which the conceptus invades and/or erodes the uterine epithelium and embeds into the endometrial stroma (18). Despite the diversity of placentation strategies, the initial events of apposition, attachment, and adhesion between maternal uterine luminal epithelium and conceptus trophectoderm are shared among species.

Bonnet (19), on the basis of different types of blastocyst–uterine cell–cell interactions, classified implantation into three categories: central, eccentric, and interstitial. Central implantation occurs in mammals such as rabbits, ferrets, and some marsupials. In these species, blastocysts grow and extensively expand before implantation. In contrast, the blastocysts of mice, rats, and hamsters are small and show modest expansion. In these animals, an implantation chamber is formed by the invagination of the uterine epithelium, which is characteristic of eccentric implantation. In guinea pigs, chimpanzees, and humans, the implantation process is of the interstitial type, that is, blastocysts embed into the subepithelial stromal bed. Based on the results of ultrastructural studies, Schlafke and Enders (13) classified implantation into intrusive, displacement, and fusion types. In intrusive types of implantation, such as in humans and guinea pigs, trophoblast cells penetrate through the luminal epithelium, reaching and extending through the basal lamina. The displacement type of implantation occurs in rodents, in which the luminal epithelium is freed off the underlying basal lamina, facilitating the spread of trophoblasts through the epithelium. The fusion type of implantation, in which trophoblast cells make connection with the luminal epithelium by forming symplasma, occurs in the rabbit. In many rodents, including mice and rats, implantation always occurs at the antimesometrial side of the uterus, whereas in some bats implantation is mesometrial. The noninvasive type of implantation is observed in domestic animals such as the pig, sheep, cow, and horse and in the wallaby (2). Embryos in domestic animals maintain free-floating status longer than invasive concepti and become elongated up to 100 mm long by day 12. This increase is due to the growth of the extraembryonic tissue, enabling an efficient exchange of metabolites with uterine secretions until the attachment reaction occurs.

In sheep, the blastocyst enters the uterus on day 4 and hatches from the zona pellucida on day 9. After day 10, elongation of the blastocyst occurs, and it develops first into a tubular and then into a filamentous form (20). Starting on day 11, the spherical or slightly tubular blastocyst begins to elongate until it reaches a length of 25 cm or more by day 17 and resembles a long filament composed mainly of extraembryonic trophoblast (21). Elongation of the conceptus is critical for developmentally regulated production of interferon (IFN)-τ (22–24), a type I IFN that is the signal for maternal recognition of pregnancy and acts in a paracrine manner on the endometrial epithelia to inhibit the development of the luteolytic machinery (25). The cellular and molecular mechanism(s) regulating blastocyst elongation is not well understood. After day 14, the filamentous conceptus appears to be immobilized in the uterine lumen. The first changes in the endometrial luminal epithelium begin on day 14 in both uterine horns (26). In ruminants, distinct areas of projecting aglandular uterine mucosa, called caruncles, are exclusively involved in an attachment reaction. The caruncles become edematous with a folded and depressed surface. These modifications are progressive and do not occur simultaneously on all caruncles. Caruncular foldings are perhaps the first step in the formation of crypts that constitute the maternal side of the future placentomes, which receive hematotrophic nutrition for the conceptus (27). Dome-like cytoplasmic protrusions also appear on the caruncular epithelial cells. On day 16, the trophectoderm begins to firmly adhere to the endometrial luminal epithelium. The interdigitation of the trophoblast and endometrial luminal epithelium is observed in both the caruncular and intercaruncular areas of the endometrium. Adhesion of the trophectoderm to the endometrial epithelium progresses along the uterine horn and appears to be completed around day 22 (26,28). Unlike sheep, the attachment in horse and pig involves multiple sites covering most of the embryonic surface. In all these ruminants, a notable decidualization is absent.

Hormonal Requirements

In all eutherian mammals thus far studied, the uterus differentiates into an altered state when the blastocyst is capable of effective two-way interactions to initiate the process of implantation. This state is termed uterine receptivity for implantation, and this lasts for a limited period. At this stage the uterine environment is conducive to support blastocyst growth, attachment, and subsequent events of implantation (10,29–31). The major factors that specify uterine receptivity are the ovarian steroids, P_4 and/or estrogens. Ovarian P_4 and estrogen are crucial for implantation in mice and rats, but ovarian estrogen is not essential for implantation in several species, including pigs, guinea pigs, rabbits, and hamsters; P_4 alone can support implantation (29,32–36) (Table 1). The role of embryonic estrogen in implantation in these species has been implicated, and, in fact, rabbit, pig, and horse blastocysts have the capacity to synthesize estrogens. The mouse embryo lacks the enzymatic machinery for estrogen synthesis (37), whereas hamster blastocysts express the aromatase gene (unpublished results). Whether preimplantation estrogen secretion by the ovary or embryo plays a role in human implantation remains unknown.

The uterine tissue compartments are composed of heterogeneous cell types that respond uniquely to changing ovarian P_4 and estrogen secretion. In mice and rats, the coordinated actions of P_4 and estrogen regulating uterine cell proliferation and/or differentiation in a spatiotemporal manner establish the window of uterine receptivity for implantation (16). In the adult mouse or rat uterus, estrogen stimulates proliferation of luminal and glandular epithelial cells, whereas in the stroma these processes require both P_4 and estrogen (5,16,30). A similar steroid hormonal modulation of uterine cell-type specific proliferation and differentiation occurs in these species during the periimplantation period (38,39). For example, on the first day of pregnancy (vaginal plug) in mice, uterine epithelial cells undergo proliferation under the influence of the preovulatory estrogen secretion. In contrast, rising levels of P_4 secreted from freshly formed corpora lutea initiate stromal cell proliferation from day 3 onward. The stromal cell proliferation is further stimulated by a small amount of ovarian estrogen secreted on the morning of day 4 of pregnancy. These coordinated effects of P_4 and estrogen result in the cessation of uterine epithelial cell proliferation, initiating differentiation (16). This preimplantation secretion of ovarian estrogen is also necessary for the increased endometrial capillary permeability at the location of the blastocyst (a prerequisite event in the initiation of implantation) (5,30,38). In pseudopregnant mice, the steroid hormonal milieu within the uterus is similarly maintained due to the presence of newly formed corpora lutea. Thus, the sensitivity of the pseudopregnant uterus for implantation on days 1–4 is quite similar to normal pregnancy, and blastocyst transfer into the uterine lumen during the receptive phase provokes normal implantation reactions and subsequent decidualization. Although blastocysts are normal inducers of these events, various nonspecific stimuli, such as intraluminal infusion of oil, air, or trauma, can also simulate certain aspects of the decidual cell reaction (deciduoma) in pseudopregnant or steroid hormonally prepared uteri (10). However, there is evidence that the initial uterine reactions induced by nonspecific stimuli are different from those induced by blastocysts (40,41).

The baboon has been studied as the nonhuman primate model to understand human implantation. Uterine receptivity and implantation in this species can be categorized into three distinct phases. The first phase is regulated by estrogen and progesterone and is evident between days 8 and 10 postovulation of the normal menstrual cycle. The second phase of uterine receptivity is induced by blastocyst "signals" superimposed on the estrogen/P_4 primed receptive endometrium. This phase is associated with functional and morphological changes in the endometrium that are distinct from those observed between days 8 and 10 postovulation in a nonconceptual cycle. The third phase of uterine receptivity is initiated after attachment and implantation of the blastocyst. Several lines of evidence demonstrate that embryo-derived factors directly or indirectly influence endometrial receptivity and implantation in primates. Chorionic gonadotropin (CG), when infused in a manner that mimics blastocyst transit, has physiological effects on the three major cell types in the uterine endometrium (i.e., luminal and glandular epithelium and stromal fibroblasts) (42). The effects of CG on glandular transformation and stromal cell differentiation are direct and occur independent of the ovary. The glandular response to CG infusion is characterized by a marked increase in transcriptional and posttranslational modulation of glycodelin (42). Synthesis of glycodelin by the glandular epithelium parallels the rise and decline of CG in the peripheral circulation (43).

In pigs, estrogen is the major hormone produced by the placenta that acts on the endometrium. Pig concepti secrete estrogens between days 10 and 15 of pregnancy, which are considered essential for the establishment of pregnancy (44). Estrogens, directly or indirectly, alter the secretion of prostaglandin (PG)$F_{2\alpha}$ by the endometrium from an endocrine direction (toward the uterine vasculature) to an exocrine direction (toward the uterine lumen) (45). The $PGF_{2\alpha}$ sequestered in the uterine lumen is unavailable to exert a luteolytic effect on the corpus luteum. Additionally, an increase in selected histotroph components occurs in the uterine lumen immediately after the release of estrogens from the conceptus on day 11 of pregnancy (44,46). Placental estrogens also act on the endometrial epithelium in a paracrine manner to increase the expression of specific growth factors that, in turn, act on the trophectoderm to stimulate cell proliferation and development.

The impact of supraphysiological levels of estrogen on human endometrial receptivity, although controversial, has been investigated at the clinical level. In high responder patients to gonadotropins in in vitro fertilization (IVF) cycles, high serum estradiol levels (>3,000 pg/ml) on the day of hCG administration are detrimental to uterine receptivity (47), regardless of the number of oocytes retrieved or serum P_4 levels. In addition, an increase in serum estrogen levels during the preimplantation period has been observed in high responder patients compared with normal responder patients (48). Moreover, decreasing estrogen levels during the preimplantation period by a step-down follicle-stimulating hormone protocol increases implantation and pregnancy rates in high responder patients (49). Also, evidence shows that estradiol at concentrations of greater than 10^{-6} M in an in vitro culture induces a deleterious effect on embryonic adhesion to the substratum (50).

Endometrial Glands and Their Secretions

All mammalian uteri contain endometrial glands that synthesize, secrete, or transport a complex array of proteins and other factors, termed histotroph (27,51,52). The histotroph is a complex mixture of enzymes, growth factors, cytokines, lymphokines, hormones, transport proteins, and others. The idea that uterine secretions nourish the developing conceptus was discussed by Aristotle, in the third century B.C., and by William Harvey in the 17th century. In 1882, Bonnet (53) concluded that secretions of uterine glands are important for fetal well-being in ruminants. Evidence from primate and subprimate species during the last century supports an unequivocal role for secretions of endometrial glands as primary regulators of conceptus survival, development, production of pregnancy recognition signals, implantation, and placentation (54–58). Studies on the uterine gland knockout (UGKO) ewe model, generated by continuous administration of a synthetic nonmetabolizeable progestin in neonatal ewes from birth to postnatal day 56, reveal an essential role for endometrial glands and their secretions in normal estrous cycles and in periimplantation conceptus survival and growth (23,59,60). These ewes do not exhibit normal 17-day estrous cycles because of the inability of the uterus to produce sufficient luteolytic pulses of $PGF_{2\alpha}$. The lack of superficial or ductal glandular epithelium, coupled with an overall reduction in surface area of the luminal epithelium, reduces the numbers of oxytocin receptors that could respond to oxytocin (59,61).

Exogenous $PGF_{2\alpha}$ induces luteolysis in UGKO ewes, and they display normal estrus mating behavior (23,60,62). However, adult UGKO ewes are unable to establish pregnancy (23,59,63), and transfer of normal hatched blastocysts into the uteri of timed UGKO-recipient ewes fails to rescue this defect (63). Morphologically normal blastocysts are present in uterine flushings of bred UGKO ewes on day 6 or 9 postmating but not on day 14 (23,63). On day 14, uterine flushings of mated UGKO ewes contain either no concepti or a severely growth-retarded tubular concepti (63). The periimplantation period of pregnancy in sheep is marked by rapid elongation of the conceptus from a tubular to a filamentous form between days 11 and 13 and production of IFN-τ, the signal for maternal recognition of pregnancy (61). Although the growth-retarded concepti in mated UGKO ewes produce little or no IFN-τ, the endometrium of UGKO ewes nonetheless responds to intrauterine infusions of recombinant ovine IFN-τ with increased expression of IFN-τ–stimulated genes (23).

Initially, the nonadhesive property of the luminal epithelium appears to be partially due to apical expression of mucins, such as mucin glycoprotein 1 (MUC1), that sterically impairs interactions between the trophectoderm and adhesive glycoproteins, such as integrins, due to their extensive glycosylation and extended extracellular structure (64). MUC1 expression by the luminal epithelium decreases between days 9 and 17 of early pregnancy in normal (64) and UGKO (23) ewes. Extracellular matrix (ECM) and integrins are thought to be responsible for trophectoderm attachment and adhesion to the luminal epithelium (64,65). During the periimplantation period in ewes, integrin subunits av, a4, a5, β1, β3, and β5 are constitutively expressed on both the conceptus trophectoderm and the apical surface of the luminal epithelium, and these expression patterns are normal in UGKO ewes (64). Further, expression of receptors for estrogen (ERα), progesterone (PR), oxytocin (OTR), and several luminal epithelium–specific genes are also normal in the endometrium of UGKO ewes (23,59). Thus, by these measures the luminal epithelium does not appear to be defective in UGKO ewes.

Uterine flushings of UGKO ewes were analyzed for the presence of osteopontin (OPN) and glycosylated cell adhesion molecule 1 (GlyCAM-1) proteins, which are expressed by glandular epithelium of the ovine uterus (66,67) and are suggested to play a role in the regulation of implantation. Uterine flushings of day 14 bred UGKO ewes contained lower amounts of OPN and GlyCAM-1 compared with day 14 pregnant ewes (23). Genomics and proteomics are being used to identify specific components of histotroph that are absent or diminished in the UGKO ewe (68).

SPECIAL ASPECTS OF EMBRYO–UTERINE INTERACTIONS

Delayed Implantation (Diapause)

Delayed implantation (embryonic diapause) occurs when the embryo at the blastocyst stage achieves a state of suspended animation. During this period, blastocyst growth is very slow with minimal or no cell division. Nearly 100 mammals in seven different orders undergo delayed implantation (69). During delayed implantation, the uterus remains in a quiescent state, and in most cases embryos at the blastocyst stage undergo dormancy. But the underlying molecular mechanisms that direct this process among different species that have adopted this reproductive strategy are not clearly understood (70). In mice and rats, ovariectomy before the presumed "estrogen surge" in the morning of day 4 of pregnancy results in the failure of implantation and initiates a state of dormancy of the blastocyst within the uterine lumen (5,71). This condition, referred to as delayed implantation, can be maintained for many days by continued treatment with P_4. The process of implantation with blastocyst activation is rapidly initiated by a single injection of estrogen in the P_4-primed uterus (5,71,72). Although the mechanisms by which estrogen mediates the processes of blastocyst activation and implantation are poorly understood, several molecular markers have been identified, as shown in Fig. 3.

Two functionally distinct categories of embryonic diapause are recognized (69). Facultative diapause, best known in rodents and marsupials (wallabies and kangaroos), is the developmental arrest induced by environmental conditions, including the stress of lactation, insufficient diet, deprivation of water, or other forms of stress. In contrast, obligate diapause is present during every gestation of a species and is believed to be a mechanism for synchronizing parturition with favorable environmental conditions for neonatal survival (70). Obligate diapause is prevalent in mustelids, bears, seals, and some wallabies. Embryo quiescence appears most commonly at the blastocyst stage. In most mammals displaying discontinuity of development, the progression to the blastocyst stage and postimplantation development follow a preordained species-specific program. Notable exceptions are found in the bat family, where variation in the rate of postimplantation development has been documented (73). Species within the Mustelid family display periods of embryonic diapause that are variable between individuals but can be in excess of 350 days in the fisher (*Martes pennanti*) or can be as brief as 3 weeks in the mink (*Mustela vison*) (74). Delayed implantation does not occur in certain species, including hamsters, rabbits, guinea pigs, or pigs. Whether this phenomenon occurs in humans is not known.

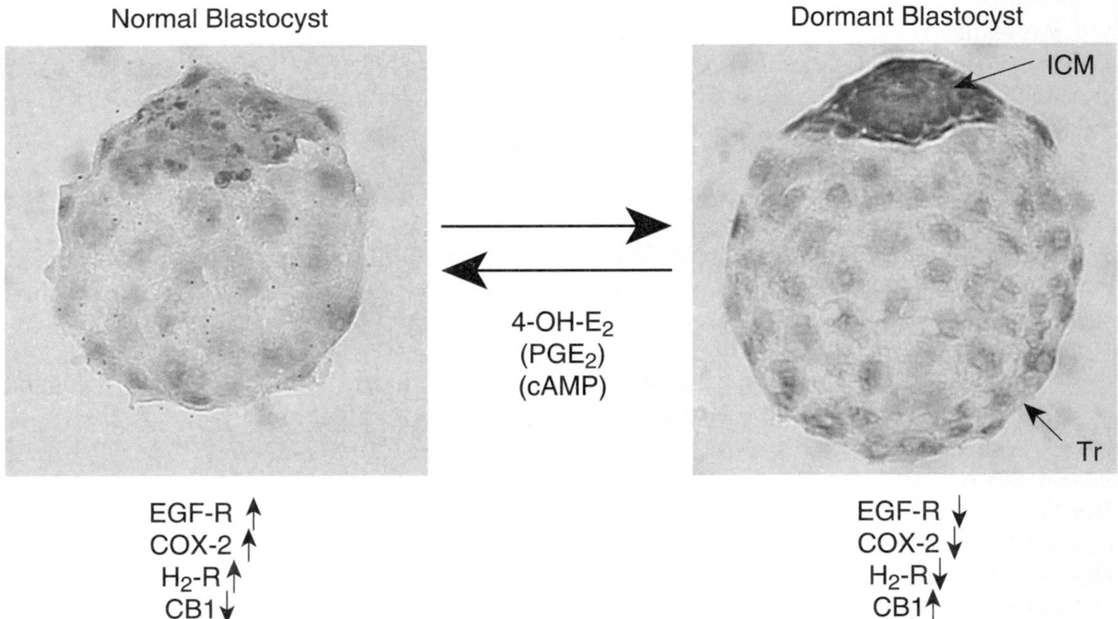

FIG. 3. Molecular markers for the blastocyst dormancy and activation. Although normal and dormant blastocysts apparently show morphological differences, several molecular markers are regulated reversibly by the blastocyst's state of activity. For example, 4-OH-E_2, PGE_2, or cAMP can activate dormant blastocysts in vitro and can either up- or down-regulate specific marker molecules as indicated. Likewise, if normal blastocysts are induced to undergo dormancy by the use of delayed implantation in the uterus before the attachment reaction, expression of these markers is shifted in the reverse direction. EGF-R, epidermal growth factor receptor; COX-2, cyclooxygenase-2; H_2-R, histamine type 2 receptor; CB1, brain-type cannabinoid receptor. [Reprinted with permission from (199).]

Environmental, Neural, and Pituitary Regulation of Embryonic Diapause

Seasonal cues synchronize reproductive events in mammals, and it was recognized early that photoperiod plays an important role in the termination of diapause and subsequent induction of implantation (74). The lengthening of days before and after the vernal equinox influences the timing of implantation in numerous species, including the spotted skunk (*Spilogale putorius*) (75) and the mink (76). Day length, or more precisely a regime of photoperiod in which minks are exposed to light during a critical period from 12 to 16 hours after dawn, provides a facultative signal that induces implantation (76). Melatonin produced by the pineal gland is considered the primary mediator of the photoperiodic regulation of diapause (76,77). The essential factor regulated by melatonin proved to be prolactin, because implantation could be advanced by treatment with prolactin (78) or dopamine antagonists (79) but further delayed by dopamine agonists (78). Indeed, prolactin alone induces implantation in hypophysectomized mink (80), and prolactin overcomes the effects of chronic melatonin treatment on the termination of diapause (81). Prolactin executes its effects directly on the mink corpus luteum (82). Although functional prolactin receptors are present in the mink uterus (83), it is not known whether prolactin has a role at the level of the endometrium.

Ovarian Regulation

After ovulation in mustelids that display an obligate diapause, the ovarian follicle collapses and forms the corpus luteum (84). The corpus luteum undergoes a remarkable structural reduction in size as diapause ensues, secreting low levels of P_4 (70,82). In contrast to the pattern of terminal differentiation that characterizes corpus luteum development in most species, the mink corpus luteum retains its mitotic potential during the period of diapause (85). In response to the pituitary prolactin signaling that terminates diapause, the corpus luteum is rejuvenated, a process characterized by a several-fold increase in corpus luteum volume and in P_4 output (82). As in other species, P_4 is essential for implantation and for the maintenance of gestation in mustelids. Nonetheless, numerous attempts to induce precocious implantation during diapause with P_4 or P_4 in combination with estrogens have been unsuccessful, suggesting luteal factors other than P_4 are required for successful implantation in this species (86). In the ferret (*Mustela putoris*), a Mustelid species that does not undergo delay, a luteal protein factor has been shown to be necessary for implantation to occur (87). Further studies have identified a secretory form of glucose-6-phosphoisomerase, also known as autocrine motility factor, as the luteal protein required for implantation in the ferret (88). Its role in other species with embryonic diapause has not yet been confirmed.

Uterine Factors and Embryonic Diapause

As the host of the embryo during diapause and an ultimate place for implantation, the uterus plays a significant role both in the initiation and termination of discontinuous development in obligate delay. Evidence that the uterus inhibits the renewal of embryonic development comes from reciprocal embryo transfer (ET) experiments in which it was shown that growth of nondiapausing ferret blastocysts is arrested when transferred to the mink uterus, whereas diapausing mink blastocysts reinitiates embryogenesis after transfer into the ferret uterus (89). In addition, mink embryos in diapause cocultured with somatic cell lines displayed the capacity to reinitiate embryonic development in vitro, providing further evidence that the uterus maintains diapause in this species (90). The factors associated with inhibition of renewed development are not known. Histochemical evidence demonstrates that the transition of the mink uterus from delay to implantation is accompanied by increases in the quantity and distribution of glycosaminoglycans (91). This concurs with observations of low levels of protein synthesis by the uterus during delay in the spotted skunk, followed by increased protein synthesis a few days before implantation (92). Studies of candidate gene expression during implantation in species with obligate delay have revealed patterns similar to those seen in the traditional rodent models. Epidermal growth factor (EGF) and its receptor (EGFR) are present in the spotted skunk uterus both during the delay and postimplantation phases of gestation, and EGF activity is significantly elevated during the implantation process (93). Leukemia inhibitory factor (LIF), a cytokine that is essential for implantation in the mouse (94), is expressed in the uterine glands just before and after implantation in the mink (95). LIF receptors (LIFRs) were found in uterine glands and in the invading trophoblast during implantation in the spotted skunk (96). Another gene essential for mouse implantation is cyclooxygenase-2 (COX-2) (94). It is not present in the uterus during diapause in either the mink (97) or the skunk (98) uterus but is expressed in the trophoblast and endometrium of both species during implantation.

The products of COX-2 activity that subserve the implantation process are not well known, but PGE and PGD are implicated in the up-regulation of transcription of the angiogenic factors associated with implantation (99,100).

The Embryo in Diapause

The carnivore blastocyst is large and comprises 200–400 cells (70). During diapause, embryos appear to undergo a slow expansion in the absence of mitotic proliferation (101,102). The reactivation of development comprises a rapid increase in volume, followed by renewal of cell proliferation within 48–72 hours (103). There is evidence, as in other species (104), to indicate that fibroblast growth factor-4 (FGF-4), produced by the ICM of the mink embryo, drives proliferation of the trophoblast cells before implantation (103). In the mouse, terminal differentiation of the trophoblast is associated with endoreduplication (105,106), and there are some intriguing indications that endocycles may occur in the mink trophoblast during its escape from diapause (102). Another peculiarity of the carnivore blastocyst is that it remains encapsulated in the zona pellucida, which is breached at multiple sites by trophoblast outgrowths only at the time of implantation (107). The progression of trophoblast into the endometrium does not differ greatly between species that exhibit diapause such as the spotted skunk (107,108).

A global gene expression analysis identified molecular pathways distinguishing blastocyst dormancy and activation in mice (109). The major functional categories of altered genes include the cell cycle, cell signaling, and energy metabolic pathways, particularly highlighting the importance of heparin-binding (HB)-EGF signaling in blastocyst–uterine cross-talk in implantation. The results provide evidence that the two different physiological states of the blastocyst, dormancy and activation, are molecularly distinguishable in a global perspective and underscore the importance of specific molecular pathways in these processes. This study has identified candidate genes that provide a scope for in-depth analysis of their functions and an opportunity for examining their relevance to blastocyst dormancy and activation in numerous other species for which microarray analysis is not available or possible due to very limited availability of blastocysts. Collectively, the delayed implantation models in mice and in other species could be exploited more extensively to better understand the molecular signaling that emanates from the embryo and influences uterine biology and vice versa.

Window of Implantation

Uterine sensitivity with respect to steroid hormonal requirements and implantation has been classified into prereceptive, receptive, and nonreceptive (refractory) phases (10,29). These phases of uterine sensitivity to implantation have been defined by using ET experiments in pseudopregnant mice. In the mouse, although the uterus is fully receptive on day 4, it is considered prereceptive on days 1–3 of pregnancy or pseudopregnancy, and by the afternoon of day 5 the uterus becomes refractory to blastocyst implantation. Early works on uterine receptivity in mice demonstrated that the mouse uterus is only rendered receptive for blastocyst implantation if exposed to estrogen 24–48 hours after P_4 priming (110). However, it is now known that the window of receptivity can be prolonged for an extended period by providing a low dose of estrogen in mice. For example, using a delayed-implanting mouse model, it has been shown that a low dose of estrogen (3 ng) postpones uterine refractoriness to implantation for at least 4 days, whereas a high dose of estrogen (10–25 ng) causes rapid closure of the receptive phase. Uterine nonreceptivity induced at high estrogen levels is accompanied by aberrant uterine expression of implantation-related genes (111). In a normal implantation model, this is not achievable, probably because of the presence of "preimplantation estrogen secretion" exclusively on day 4 of pregnancy. Thus, normal implantation of blastocysts occurs when the uterus is receptive but becomes faulty beyond this time. In fact, it was shown that a short delay in the initial attachment reaction in mice during pregnancy creates an adverse ripple effect, resulting in developmental anomalies during the subsequent course of pregnancy (112). These results suggest that careful regulation of estrogen levels is one of the important factors for improving implantation rates in IVF/ET programs.

Another critical factor determining the window of implantation is the blastocyst's state of activity. In experimentally induced delayed implantation in rodents, blastocysts undergo zona hatching, albeit at a slower pace, but they become dormant without initiating the attachment reaction even after P_4 priming of the uterus. However, a single injection of estrogen promptly initiates blastocyst activation with the initiation of implantation in the P_4-primed uterus. Blastocyst dormancy is molecularly and physiologically distinguishable from blastocyst activation. EGFR, COX-2, and histamine type 2 receptor (H_2), factors associated with blastocyst attachment reaction, are expressed in normal or activated blastocysts but are down-regulated in dormant blastocysts

(31,113–116). In contrast, the G-protein–coupled cannabinoid receptor, CB1, which responds to natural or endocannabinoids, is down-regulated in activated blastocysts but remains up-regulated in dormant blastocysts (117) (Fig. 3). Collectively, these findings suggest that a complex array of molecular networking regulates blastocyst activation and dormancy.

Although estrogen is essential for blastocyst activation and implantation in the P_4-primed mouse uterus, the mechanisms by which estrogen initiates these responses remain elusive. Evidence suggests that estrogen actions in uterine preparation and blastocyst activation for implantation are two distinct events. Indeed, ET experiments in delayed-implanting recipient mice provided evidence that although the primary estrogen, estradiol-17β, initiates uterine events for implantation, its catechol metabolite, 4-hydroxy-estradiol-17β (4-OH-E_2), participates in activation of dormant blastocysts (118). Blastocyst activation by 4-OH-E_2 involves COX-2–derived PGs and cAMP (118). Furthermore, the use of an ER antagonist (ICI-182, 780) has shown that whereas estradiol via its interaction with the nuclear ERs participates in the preparation of the P_4-primed uterus to the receptive state in an endocrine manner, 4-OH-E_2 produced from estradiol in the uterus mediates blastocyst activation in a paracrine manner that does not involve nuclear ERs. These results provide evidence that both primary and catecholestrogens are required for embryo–uterine interactions for successful implantation and that implantation occurs only when uterine receptivity coincides with the blastocyst's state of activity. Molecular pathways that are potentially involved in uterine receptivity and blastocyst activation are discussed below in more detail.

MOLECULAR ASPECTS OF EMBRYO–UTERINE INTERACTIONS

Despite experimental success in initiating embryonic development outside the womb and in identifying numerous signaling molecules involved in the embryo–uterine dialogue (57,94,119–121), there is still a significant knowledge gap in understanding the in vivo events of implantation. The successful implantation of an embryo is contingent upon cellular and molecular cross-talk between the uterus and the embryo. The coordination of the endocrine and cellular and molecular events via paracrine, autocrine, and/or juxtacrine factors in a dynamic manner produce within the uterus a favorable environment, the receptive state, to support implantation. The embryo also functions as an active unit with its own molecular program of cell growth and differentiation. Thus, deficiencies in uterine receptivity, embryo development, or the embryo–uterine dialogue compromise fertility. The current state of our knowledge of preimplantation and implantation physiology is the result of the accumulation of scientific observations gathered over many years. Implantation is a complex process involving spatiotemporally regulated endocrine, paracrine, autocrine, and juxtacrine modulators that span cell–cell and cell–matrix interactions. However, the precise sequence and details of the molecular interactions involved have not yet been fully defined. Furthermore, the implantation process varies among species, thus precluding the formulation of a unified theme. In addition, ethical restrictions and experimental difficulties prevent direct analysis of embryo–uterine interactions during implantation in humans. This section primarily focuses on the molecular basis of implantation in mice because more mechanistic information is now available for this species. However, an attempt has been made to indicate comparative analyses based on limited work in other species (Fig. 4).

Steroid Hormones

Lessons from Gene-Targeted Mouse Models

As mentioned earlier (see Hormonal Requirements), ovarian estrogen and P_4 are critical to the process of implantation. Estrogen and P_4 effects in the uterus are primarily regulated by nuclear estrogen receptors (ER-α and ER-β) and progesterone receptors (PR-A and PR-B). Differential uterine expression of ER and PR during the periimplantation period in mice suggests that coordinated effects of estrogen and P_4 in uterine events for implantation are mediated via these nuclear receptors (122). Furthermore, mouse models devoid of each receptor gene are informative as to how these receptors are involved in uterine biology. For example, *ERα(–/–)* mice exhibit infertility because of hyperstimulated ovaries and hypoplastic uteri (123). However, further experiments using *ERα(–/–)* mice showed that P_4 alone is sufficient to support decidualization (deciduoma) in response to artificial stimuli (124,125). These results suggest that *ERα(–/–)* mice have defective implantation perhaps due to the failure of the attachment reaction but not due to the failure of decidualization events (126). Microarray analysis of RNA from the *ERα(–/–)* uterus also demonstrates a minimal response to estrogen (127), suggesting that ER-α is essential for proliferative and genomic responses of the uterus. Female mice devoid of ER-β exhibit subfertility (128). As in *ERα(–/–)* or *ERα/ERβ(–/–)* double knockout

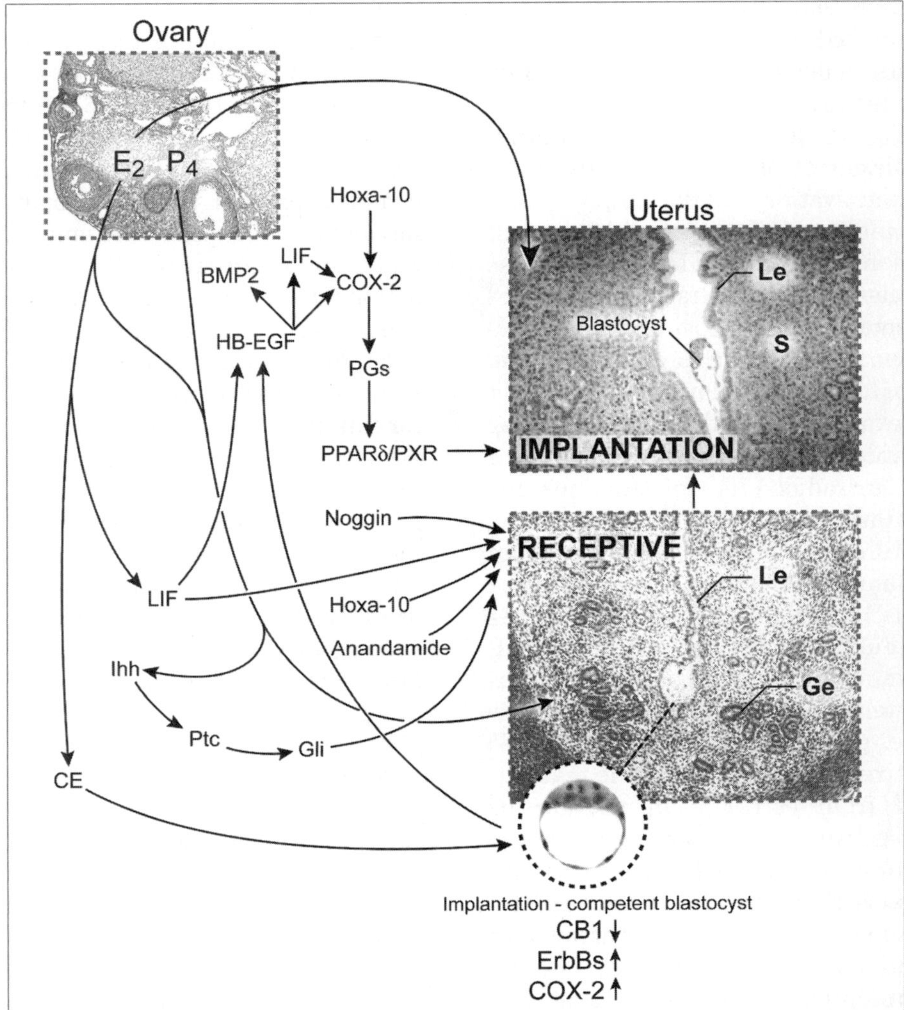

FIG. 4. A proposed signaling network in embryo–uterine communication during implantation in mice. Implantation in mammalians absolutely depends on synchronized development of the blastocyst to the stage when it is competent to implant and the uterus to the stage when it is receptive to blastocyst growth and implantation. Ovarian estrogen (E_2) and progesterone (P_4) are the primary effectors that direct the prereceptive uterus to a receptive state via a number of locally expressed growth factors, cytokines, transcription factors, morphogens, and vasoactive mediators in the uterus, whereas uterine-derived catecholestrogen and regulated levels of endocannabinoids activates the blastocyst to an implantation-competent state. During the attachment phase, signaling and adhesive events embracing the uterus and the blastocyst lead to implantation. Le, luminal epithelium; Ge, glandular epithelium; S, stroma. [Reprinted with permission from (445).]

mice, defects in the hypothalamic–pituitary–gonadal (ovarian) axis are apparent in *ERβ(–/–)* female mice (129). The uterus of the *ERβ(–/–)* mouse retains full biological function as demonstrated by uterine weight increase (128). Full genomic responsiveness to estrogen in these mice is also seen by microarray analysis (127,128). Additionally, although the *ERβ(–/–)* female mice exhibit subfertility, those that achieve pregnancy carry and deliver full-term pups, indicating that uterine function in terms of implantation, decidual response, and parturition are normal. These results point toward a minimal importance of ER-β in the adult uterus for estrogen response or pregnancy. *PR(–/–)* mice exhibit pleiotropic reproductive abnormalities, including impaired ovulation, uterine hyperplasia, and decidualization (130). Selective ablation of the PR-A isoform showed infertility with a milder phenotype, suggesting that PR-A and PR-B serve as functionally distinct mediators of P_4 action in vivo (131). Experiments using both *PR(–/–)* and *PR-A(–/–)* mice further reinforce a requirement of P_4 in decidualization (130,131).

Although both *ER(–/–)* and *PR(–/–)* mice show severe phenotypes of female reproductive failures, these mice have been used as model systems to study steroid hormonal regulation of several genes. Das and coworkers (126) reported a non-ER–mediated estrogen signaling pathway that is resistant to an ER antagonist (ICI-182, 780) in *ERα(–/–)* uteri. Lactoferrin is a well-known estrogen-responsive gene in mice, and many natural or synthetic estrogens can induce this gene in wild-type uteri (126). In contrast, although estradiol was not able to induce this gene in *ERα(–/–)* uteri, a catechol estrogen, 4-OH-E_2, was effective in this response. Because this induction by 4-OH-E_2 was resistant to an ER antagonist (ICI-182, 780), ER-β did not appear to mediate this effect. The possibility that an alternative estrogen signaling pathway independent of nuclear ERs exists in the mouse uterus was further explored (132). Using a differential display technique, it was found that several genes are up-regulated or down-regulated by estradiol and 4-OH-E_2 both in wild-type and *ERα(–/–)* uteri independent of nuclear ERs. The up-regulated genes were immunoglobulin heavy chain binding protein, calpactin I, calmodulin, and Sik-similar protein, whereas the gene that encodes secreted frizzled related protein-2 was readily down-regulated. An ER antagonist failed to neutralize these responses both in wild-type and *ERα(–/–)* mice, suggesting that these are ER-independent early responses. This provides compelling evidence that an alternative estrogen-signaling pathway is operative in the uterus. However, it is still unclear whether early estrogenic responses are mediated by a putative cell surface estrogen receptor (133) or by other nuclear receptors, such as ER-γ or ERRs (estrogen-receptor related receptors) (134,135). Alternatively, this could be explained by estrogen signaling via a membrane receptor of the G-protein–coupled receptor family. Although several putative membrane PRs have been cloned (136,137), whether similar types of membrane ERs are present in the uterus is not presently known.

Role of Catecholestrogen: An Activator of Dormant Blastocyst

Gene-targeting experiments have established the importance of both ER and PR in uterine preparation for implantation in mice. However, whether the preimplantation embryo is a direct target for steroid hormones remains unclear. There is evidence suggesting that preimplantation estrogen secretion on day 4 of pregnancy in mice has a dual role as a primary estrogen and as a catecholestrogen with distinct targets (118). Although primary estrogen acts via the uterine ER to prepare the uterus for implantation, a catecholestrogen formed locally in the uterus from the primary estrogen participates in blastocyst activation. Estradiol-17β undergoes hydroxylation to 4-OH-E_2 by a P-450–linked enzyme, CYP1B1 (138). This enzyme is present throughout the mouse uterus on day 4 but disappears from the implantation site on day 5 (118). Activation of dormant blastocysts appears to involve an "early response" to 4-OH-E_2, because dormant blastocysts transferred into delayed-implanting recipient uteri within 1 hour of estradiol administration of the recipients showed implantation, whereas similar blastocysts transferred beyond this 1-hour period failed to implant (118). These results suggested that a rapid response occurs in utero that is critical to implantation. In contrast, dormant blastocysts cultured in the presence of 4-OH-E_2, but not estradiol, gain implantation competency and upon transfer implant in pseudopregnant recipients well beyond the 1-hour "window" of estradiol treatment. Similar results were also obtained by culturing dormant blastocysts in the presence of PGE_2 or a permeable analogue of cAMP. This effect apparently involves the COX-2 signaling pathway (118). For example, coincubation of dormant blastocysts with a selective COX-2 inhibitor and 4-OH-E_2 efficiently blocks their activation and implantation upon transfer to suitable recipients. This effect of the COX-2 inhibitor was partially reversed by addition of PGE_2 to the culture media. The results strongly suggest that the action of 4-OH-E_2 on dormant blastocysts is mediated via the COX-2 signaling pathway, leading to an increase in intracellular cAMP levels. However, it is still a mystery how catecholestrogens mediate activation of blastocysts (118). Although nuclear ER-α is present in both the active and dormant blastocysts (139), dormant blastocysts do not respond to estradiol and fail to attain implantation competency in vitro. In contrast, dormant blastocysts do respond to a catecholestrogen 4-OH-E_2 and become implantation competent in vitro. The ER antagonist ICI-182, 780 fails to reverse this response, suggesting that nuclear ER signaling is not critical to blastocyst activation (118). These observations are surprising in the light of findings that *ER-α*, *ER-β*, and *efp* (estrogen responsive finger protein) mRNAs are expressed in the preimplantation embryos (140,141). Examining the direct roles of estrogens and/or P_4 in preimplantation embryo function and how steroid hormone signaling in the embryo and uterus are coordinated for implantation require further investigation.

Adhesion Molecules

Many glycoproteins and carbohydrate ligands and their receptors are expressed in the uterine luminal epithelium and blastocyst cell surfaces around the time of implantation (142,143). Primary adhesion molecules that are implicated in implantation are selectins, galectins, heparan sulfate proteoglycans, Muc-1, integrins, cadherins, and the trophinin–tastin–bystin complex.

MUC1

MUC1 acts as an antiadhesive masking molecule (144). *MUC1*, a stretch of long carbohydrate moieties, is expressed in the mouse uterine epithelium during the prereceptive period. The physical hindrance created by these branches is considered to prevent interaction between the embryo and the luminal epithelium of the uterus before the attachment reaction. This is consistent with timely down-regulation of *MUC1* from the luminal epithelium throughout the uterus before the attachment reaction on day 4 of pregnancy in mice (57). This seems paradoxical to the observation that overall Muc1 expression increases in the rabbit and human uterus during the receptive period. However, careful examination revealed a decrease in *MUC1* levels at the site of implantation in rabbits (145). In humans, the situation appears to be more complicated. During the apposition phase, the presence of an embryo increases the levels of Muc1 in the epithelium, but at the adhesion phase, the embryo induces cleavage of *MUC1* to clear this glycoprotein from the implantation site (146). Collectively, these findings suggest that Muc1 acts as an antiadhesive molecule that must be removed from the implantation site.

Integrins

Among the adhesion molecules, integrins have been studied more extensively in the human endometrium because of their cycle-dependent changes and the potential role in uterine receptivity. Members of the integrin family serve as receptors for various ECM ligands and modulate cell–cell adhesion and signal transduction cascades (147). Each integrin is comprised of two subunits, α and β, and each αβ combination has its own binding specificity and signaling properties. As membrane-associated receptors, integrins possess short cytoplasmic tails with no enzymatic activity. Signaling by integrins is mediated by associating adaptor proteins that bridge them to the cytoskeleton, cytoplasmic kinases, and transmembrane growth factor receptors (147). Several members of the integrin family, including αvβ3, are known to interact with the RGD (Arg-Gly-Asp) peptide sequence present in many ECM proteins, such as fibronectin, laminin, and entactin.

Integrins were first reported in the human uterus in 1992 with both constitutive and cycle-dependent patterns of expression (148,149). Characterization of integrin expression during the menstrual cycle further defined three specific integrin family members that were each expressed during a narrowly defined time between days 20 and 24 of a typical 28-day cycle in women (150–154). This period of integrin coexpression correlates well with a study that showed normal implantation in the human occurs 7 to 10 days after ovulation, corresponding to days 21 to 24 of the menstrual cycle (155). Implantation beyond this interval was associated with poor reproductive outcome. Although many integrin heterodimers show constitutive expression in the uterine epithelium or stroma, α1β, α3β1, α6β1, αvβ3, and αvβ1 heterodimers exhibit cycle-dependent changes (148,149,156–158). Among these, αvβ3 has been more convincingly shown to be localized to the luminal epithelium (158). Temporal and spatial expression of this integrin and its ligand, OPN, corresponds to the specialized surface modifications known as pinopodes (159,160) that are also expressed during implantation in other animal species (161–163). In mice, αvβ3 is expressed both in the uterine luminal epithelium and blastocyst during implantation. It has been shown that an intrauterine injection of the RGD peptide or a neutralizing antibody against αvβ3 reduces the number of implantation sites in mice and rabbits (161). The expression of αvβ3 has been well characterized in women with infertility and recurrent pregnancy loss. In women with histological delay (luteal phase defect), integrin expression is consistently delayed or absent (148,151). Other diagnoses associated with suspected implantation defects, such as polycystic ovarian syndrome (164), hydrosalpinges (165), endometriosis (166), and unexplained infertility, are also manifested by weak or absent αvβ3 integrin expression (167,168).

Among many subunits, α5β1, α6β1, and αvβ3 are expressed in the mouse embryo throughout the peri-implantation period, whereas several others exhibit stage-specific expression (169). Integrins are also expressed in the differentiating trophoblasts at later stages (169), suggesting their roles in trophoblast differentiation and adhesion. A role for fibronectin via integrin binding in blastocyst outgrowth was further confirmed in vitro using antibodies against αv, α5, β1, or β3 that inhibited adhesiveness of the outer surface of the trophoblast inducible by fibronectin (170). In addition, a gene-targeting experiment revealed that deletion of the β1 gene results in ICM defects and embryonic lethality (171). However, the mutant embryos form morphologically normal blastocysts and initiate implantation, but trophoblast invasion becomes defective (171). Adhesion-competent late-blastocyst-stage trophoblasts undergo intracellular signaling initiated upon ligation of α5β1 and αvβ3 by fibronectin (172). Integrin signaling mobilizes cytoplasmic Ca^{2+} and induces the trafficking of intracellular vesicles, resulting in stronger adhesion to fibronectin at the apical surface. Therefore, blastocyst adhesion to the endometrium during implantation is considered to be regulated by the endogenous developmental program as well as through interactions with ECM components in the local environment (173,174). Although there is evidence that the embryo is a site of action for integrin signaling, it is not yet clear if the uterus is also a site of action of this signaling. Results of gene targeting experiments of integrin subunits are not very informative in relation to their roles in implantation because of the complex phenotypes and apparent compensation by other subunits (175–178).

ECM Components

Invasive mouse trophoblasts adhere, spread, and migrate onto ECM substrates (179–182) and penetrate three-dimensional ECM structures (183,184). Several ECM components that are up-regulated in the periimplantation endometrium, including fibronectin, laminin, and collagen type IV (185–187), support trophoblast outgrowth in vitro (179,180). Trophoblast interactions with the ECM are mediated primarily by integrins (169,170,180–182,188). Hexapeptides containing the Arg-Gly-Asp (RGD) sequence recognized by integrins (189) block trophoblast outgrowth on fibronectin, collagen types II and IV, entactin, and vitronectin (180,182,188). However, trophoblast adhesion to type I laminin is independent of its RGD sequence and is primarily mediated through interaction of α7β1 with the E8 integrin-recognition domain of laminin (190,191).

OPN is an acidic phosphorylated glycoprotein component of the ECM detected in the epithelium and in secretions of many tissues, including the uterus (192). OPN binds to integrin heterodimers (αvβ1, αvβ3, αvβ5, αvβ6, αvβ8, α4β1, α5β1, and α8β1) via its RGD sequence and to α4β1 and α9β1 by other sequences to promote cell adhesion, spread, and migration (192). OPN is implicated in the process of implantation in ruminants and humans. In sheep, OPN expression is increased in pregnant sheep during the periimplantation period (days 11–17) when adherence and attachment of concepti to uterine luminal epithelium occur (66,193). Secreted OPN binds to integrin heterodimers expressed by trophectoderm and the uterus to stimulate changes in morphology of the conceptus extraembryonic placental membranes and to induce adhesion between the luminal epithelium and trophectoderm essential for implantation and placentation (192). OPN is also found in the porcine endometrium and is present in uterine luminal flushings from cyclic and pregnant gilts (194). In humans, the temporal and spatial distribution of OPN protein correlates with the expression pattern of αvβ3 integrin, appearing approximately 7 days after ovulation (195). In the mouse uterus, OPN is highly expressed in the metrial gland in the developing deciduum (196). However, OPN-deficient mice are fertile, suggesting OPN is not essential for fertility in mice (197).

Trophinin

Trophinin was identified by the cDNA library screening of a human trophoblastic cell line (198). This transmembrane protein can mediate homophilic interactions between two different cell types. For example, it mediates an interaction between human endometrial and trophoblastic cell lines, but this interaction is complex (198). Trophinin requires the presence of a cytoplasmic protein, tastin, to sustain adhesion between these two cell types. In addition, the presence of bystin, another cytoplasmic protein, is required for effective interaction between trophinin and tastin. This adhesion complex that is present in both trophoblastic teratocarcinomas and endometrial adenocarcinomas mediates adhesion between them. In humans and monkeys, trophinin is specifically expressed in cells involved in implantation. Furthermore, the trophinin complex was detected in both trophoblast and decidual cells at the human fetal–maternal interface as early as the sixth week of pregnancy (199). Trophinin, tastin, and bystin are all highly expressed in fallopian tube epithelial cells of women with tubal pregnancies, suggesting they may play a role in ectopic pregnancy, a condition unique to humans (200). In mice, trophinin expression is distinct from that in humans, and trophinin-deficient mice do not exhibit fertility defects, suggesting this adhesion molecule is not crucial to embryo implantation and placental development in mice (201).

E-Cadherin

E-cadherin, a calcium-dependent cell–cell adhesion molecule, participates in the formation of the epithelial adherens junctions in cooperation with α- and β-catenins (202,203). E-cadherin is a critical factor for blastocyst formation, because its targeted deletion leads to defective embryonic development, resulting in failure to form the trophectoderm (204,205). E-cadherin is implicated in uterine–embryo interactions because of its homotypic adhesive activity (206). Although the trophectoderm highly expresses E-cadherin, the components of the adherens junctional complex are also expressed in the uterine luminal epithelium at the time of the attachment reaction. The expression subsequently becomes evident in the subepithelial stroma surrounding the implanting blastocysts with apoptosis occurring in the luminal epithelium (206,207). This junctional complex, in turn, forms a barrier surrounding the embryo, perhaps restricting passage of injurious stimuli from the maternal circulation. The outcomes suggest that temporal and cell-specific expression of the adherence junction proteins in the uterus results in molecular guidance that is important for blastocyst attachment and subsequent invasion.

Selectins

The selectin adhesion system plays an important role in human implantation (208). This adhesion system, which is involved in leukocyte capture from the bloodstream, is also operative during implantation and placentation. On the maternal side, selectin oligosaccharide ligands are expressed in the receptive uterine epithelium and on the embryonic side trophoblast cells express l-selectin receptors. It was observed that beads coated with specific selectin ligands adhere to the trophoblast, suggesting that the trophoblast cell surface receptors are functional. This investigation suggests that trophoblast l-selectin mediates interactions with the uterus to establish an adhesion mechanism for implantation. This study shows that trophoblast cells share a system known to be active in the blood–vascular system. This is a very interesting finding in several respects. First, this study asserts that specific ligand–receptor signaling pathways between the embryo and uterus are critical for implantation and subsequent pregnancy establishment. Second, it shows that the same mechanism operative during implantation is also operative during the later phases of pregnancy. Third, it shows that trophoblast cells share a system known to be active in the blood–vascular system.

EMMPRIN

ECM metalloproteinase inducer (EMMPRIN) is a transmembrane glycosylated protein that is a member of the immunoglobulin superfamily. EMMPRIN, also known as basigin, CD147, or human leukocyte activation–associated M6 antigen, exhibits homophilic interactions. EMMPRIN is expressed by uterine epithelial and stromal cells in both mice and humans, and its expression is regulated by estrogen and P_4 (209,210). EMMPRIN is also expressed by developing embryos at the time of implantation, and EMMPRIN null mutant embryos are unable to implant successfully in uteri of their heterozygous mothers (211). These data suggest that EMMPRIN is one of the homophilic adhesion molecules that mediate embryo– uterine interactions during implantation.

Vasoactive Factors

It has long been speculated that vasoactive agents, such as histamine and PGs, are involved in many aspects of reproduction, including ovulation, fertilization, implantation, and decidualization.

Histamine

Histamine functions as a ubiquitous mediator of cell–cell signaling and is synthesized from L-histidine by histidine decarboxylase (HDC) both in peripheral tissues and in the nervous system (212). Histamine is not only a well-known neurotransmitter in the brain (213), it also participates in other physiological responses, including gastric acid secretion, regulation of allergic reactions, and vascular permeability (214). The process of implantation is considered a proinflammatory reaction, and increased vascular permeability at the site of blastocyst implantation is common to many species. Thus, it was suggested that histamine plays a role in implantation and decidualization (214). Earlier observations suggested that uterine mast cells are the possible source of histamine and its release from mast cells by estrogen is important for implantation (215,216). This suggestion was based on the observations that local histamine application stimulates uterine hyperemic and edematous responses (217) and that there is a reduction in uterine mast cell numbers and histamine content after estrogen treatment and during implantation (218). Furthermore, a histamine antagonist, pyrathiazine, or an inhibitor of HDC was shown to interfere with implantation when instilled into the uterine lumens of rats and rabbits (215,219).

Histamine works via at least four histamine receptor subtypes (H_1, H_2, H_3, and H_4) (220–222), and blocking of both H_1 and H_2 receptors was shown to interfere with implantation in rats (223). Subsequent studies also showed that histamine induces implantation in delayed-implanting rats when injected with a suboptimal dose of estrogen (224). However, successful implantation and birth of live offspring in mast cell–deficient mice and other evidence suggest that uterine mast cell histamine is not essential for implantation (183,225,226). Thus, if histamine is involved in implantation, it should be provided either by major uterine cell types or by embryonic cells. Although mouse blastocysts do not have the capacity for histamine synthesis (116), HDC is expressed in uterine epithelial cells on day 4 of pregnancy in mice before implantation but not in decidual cells (227). In addition, H_1, H_2, and H_3 receptor subtypes are not detectable in the uterus, but H_2 receptors are expressed in preimplantation mouse blastocysts. These observations as well as the inhibition of blastocyst zona-hatching and implantation by H_2 antagonists and an HDC inhibitor suggest that uterine histamine targets the blastocyst for implantation (116). However, apparently normal implantation occurs in mice lacking HDC or H_2 type histamine receptor genes, suggesting the possible involvement of other vasoactive agents with overlapping functions in this process (221,228).

Prostaglandins

PGs possess vasoactive, mitogenic, and differentiating properties (229) and are implicated in various female reproductive functions. Early studies showed that during artificially induced decidualization, uterine concentrations of PGs increase in a temporal pattern accompanied by increases in endometrial vascular permeability, an early event in the endometrial response to decidualization (230–235). These studies used pharmacological inhibitors of PG synthesis that substantially attenuated the increased endometrial vascular permeability changes (230,236, 237) and subsequent decidualization (233,238,239).

PGs are produced via the COX pathway. COX exists in two isoforms, COX-1 and COX-2, and is the rate-limiting enzyme in the biosynthesis of PGs. COX mediates the conversion of arachidonic acid into PGH_2, which is then converted to various PGs by specific synthases (229). The COX isoforms are encoded by two separate genes and exhibit distinct cell-specific expression, regulation, and subcellular localization, although they share similar structural and kinetic properties. COX-1 is considered a constitutive enzyme that mediates "housekeeping" functions. In contrast, COX-2 is an inducible enzyme and is induced in a variety of cell types by growth factors, cytokines, and inflammatory stimuli (229). PGs normally execute their functions by interacting with cell surface G-protein–coupled receptors, but they can also function as ligands for nuclear peroxisome proliferator-activated receptors (PPARs) (240–243). Because COX-2 is primarily responsible for increased PG production during inflammation, this isoform is the target for developing selective antiinflammatory drugs (244,245). COX-2 overexpression is also associated with tumorigenesis (246,247).

Because of "proinflammatory" characteristics of ovulation and implantation, participation of PGs in these processes has been speculated (248,249). For example, PGs are considered to participate in follicular rupture during ovulation (250). This is consistent with gonadotropin-mediated induction of COX-2 in ovarian follicles preceding ovulation (250,251). PGs are also implicated as important mediators of increased endometrial vascular permeability during implantation and decidualization (252). Both COX-1 and COX-2 exhibit a unique pattern of expression in the periimplantation mouse uterus (252). COX-1 is expressed in uterine luminal and glandular epithelial cells on the morning of day 4 of pregnancy, but its expression is down-regulated in the luminal epithelium by the time of the attachment reaction. In contrast, COX-2 is expressed in the luminal epithelium and underlying stromal cells solely at the site of blastocyst attachment. Using the delayed implantation model, this study also showed that the expression of COX-2 in the receptive uterus requires the presence of active blastocysts. Gene-targeting experiments further demonstrated that COX-2–derived PGs are essential for ovulation, fertilization, implantation, and decidualization (253–255). Experiments with *COX-1(–/–)* mice suggested that the loss of COX-1 is compensated by the expression of COX-2 for implantation (256). Among various PGs, the levels of prostacyclin (PGI_2) are highest at the implantation sites of wild-type mice and implantation defects are partially restored in *COX-2(–/–)* mice by administration of a more stable prostacyclin agonist, carbaprostacyclin (257).

The role of PGs is further illustrated by the reduced fertility of female mice lacking cytoplasmic phospholipase A_2 that is involved in the liberation of arachidonic acid from membrane phospholipids for PG synthesis by the COX system (258–260). The reduced fertility in these females is due to deferral of "on-time" implantation, leading to subsequent retarded fetoplacental development and reduced litter size (112). Collectively, these results indicate that the cytoplasmic phospholipase A_2–COX-2 axis is crucial to implantation.

COX-2 is also expressed either in the uterus, blastocyst, or both during implantation in a variety of species with different modes of implantation, including sheep, mink, skunk, baboon, pig, and hamster (97,98,207,261,262). In hamster, COX-2 and PGE synthase are coexpressed and produce PGE_2 as a major PG product in implantation sites (207). COX-2 expression in human endometrium has also been reported (263,264). These results suggest a conserved function of COX-2 in implantation in various species. It has also been shown that depending on the genetic background, COX-1 can improve female infertility in COX-2–deficient mice (265).

Membrane receptors for PGE_2, $PGF_{2\alpha}$, PGD_2, PGI_2, and thromboxanes have been named as EP_1–EP_4, FP, DP, IP, and TP, respectively; they belong to the G-protein–coupled family of cell surface receptors (266,267). Although PGE_2 synthase is expressed at the implantation sites with the presence of PGE_2 and EP receptors (267–269) and although PGE_2 has been shown to be associated with implantation and decidualization (233), gene-targeting experiments showed that three of the four EP receptor subtypes (EP_1–EP_3) are not critical to implantation. EP_4 deficiency mostly results in perinatal lethality, and thus its role in implantation has not yet been fully explored (267). Mice deficient in FP or IP show normal implantation. PGs can also exert their effects by utilizing PPARs that belong to a nuclear hormone receptor superfamily. The evidence that PG-mediated PPAR signaling is involved in implantation is discussed below in more detail (PPARδ).

PGs also appear to be important for embryonic functions relevant to preimplantation embryo development and implantation. Preimplantation embryos produce PGs, and inhibitors of PG synthesis have been shown to inhibit embryonic growth, functions, and zona hatching in vitro (208,270). Dormant mouse blastocysts can achieve implantation competence if cultured in the presence of PGE_2 or a permeable analogue of cAMP. This effect apparently involves the COX-2 signaling pathway (118). However, normal development of *COX-1(–/–)/COX-2(–/–)* double mutant embryos in the uterus suggests that PGs of embryonic origin are not essential for embryo development (271). However, compensation by maternal PGs in embryonic development cannot be ruled out.

Vascular endothelial growth factor (VEGF), also known as vascular permeability factor, is highly vasoactive in nature. It is a potent inducer of vasodilatation and angiogenesis. Its role in implantation is discussed below (see VEGF Signaling).

Growth Factors

The expression of various growth factors and their receptors in the uterus in a temporal and cell-specific manner during the periimplantation period suggests that these factors are important for implantation (57,94,119,120,272,273).

The EGF family of growth factors includes EGF itself, transforming growth factor (TGF)-α, HB-EGF, amphiregulin, betacellulin, epiregulin, and neuregulins (115,274). HB-EGF is the earliest molecular marker to be found in the uterus exclusively at the sites of active blastocysts, appearing several hours before the attachment reaction in mice (6). This induction is followed by the expression of betacellulin, epiregulin, neuregulin-1, and COX-2 around the time of the attachment reaction (115,252,274). In contrast, amphiregulin is expressed throughout the uterine epithelium on the morning of day 4 of pregnancy and is well characterized as a P_4-responsive gene in the uterus (275). Around the time of the attachment reaction, strong expression of amphiregulin in the luminal epithelium is only found around the implanting blastocysts, and this expression is absent by the morning of day 5. However, amphiregulin-deficient mice or compound knockout mice for EGF/TGF-α/amphiregulin apparently do not exhibit implantation defects (276,277). Because HB-EGF, betacellulin, epiregulin, neuregulin, and amphiregulin all show overlapping uterine expression patterns around the blastocyst at the time of attachment reaction (57,274), it is assumed that a compensatory mechanism rescues implantation in the absence of one or more members of the EGF family.

EGF-like growth factors interact with the receptor subtypes of the *erbB* gene family, which is comprised of four receptor tyrosine kinases: ErbB1 (EGFR), ErbB2, ErbB3, and ErbB4. They share common structural features but differ in their ligand specificity and kinase activity (278). The initial dimerization between coexpressed receptors upon ligand binding constitutes the classical mechanism of action of EGF-like ligands. Spatiotemporal expression patterns of EGF gene family members and ErbBs in the uterus during the periimplantation period suggest compartmentalized functions of EGF-like growth factors in implantation (115).

Numerous growth factors and their receptors are expressed in preimplantation embryos of several species, suggesting their roles in preimplantation mammalian development (279,280). ErbB1 (EGFR), ErbB2, and ErbB4, the receptor subtypes for the EGF family of growth factors, are expressed in the mouse blastocyst (113,281), and EGF or TGF-α has beneficial effects on embryonic development in vitro (282). There is evidence that embryonic ErbB1 and/or ErbB4 interact with uterine HB-EGF during blastocyst implantation (114,281). HB-EGF is expressed as soluble and transmembrane forms in the uterine luminal epithelium at the site of blastocysts before the attachment reaction (6,283). Both ErbB1 and ErbB4 are expressed in implantation-competent blastocysts but are down-regulated in dormant blastocysts during delayed implantation (113,281). Furthermore, although a recombinant soluble HB-EGF can promote blastocyst growth and differentiation (6), cells that express the transmembrane form of HB-EGF can adhere to active, but not dormant, blastocysts in vitro (114), suggesting HB-EGF's paracrine and juxtacrine effects. By directing an HB-EGF–toxin conjugate toward wild-type and *erbB1(–/–)* blastocysts, it was found that HB-EGF could also interact with embryonic ErbB4 and heparan sulfate proteoglycans (281). Collectively, these results suggest that an interaction between uterine HB-EGF and blastocyst ErbBs is important for the attachment reaction. However, the absolute necessity of HB-EGF in implantation requires genetic evidence. *HB-EGF* null mutant mice die early during postnatal life because of cardiac defects, precluding critical examination of the implantation phenotype (284). It should also be noted that early events of implantation do not appear to be affected by blastocysts deficient in either ErbB1 or ErbB4 (285,286), although the implantation-initiating efficiency of blastocysts deficient in more than one receptor type needs to be tested to delineate the functional redundancy among the receptor family.

Implantation-competent blastocysts can produce HB-EGF, and this growth factor induces its own expression in the uterus in a paracrine manner (109). The results elucidate an important role of HB-EGF on both sides of the blastocyst and uterus during implantation. These results provide evidence that one of the signaling molecules involved in establishing a hierarchy of events between the embryo and uterus for implantation is HB-EGF that originates first in implantation-competent blastocysts. This autoinduction loop of HB-EGF is perhaps the first example of a molecular cross-talk between these two different entities that leads to the initiation of the implantation process. In conclusion, detailed expression and gene targeting experiments with all of the ligands and receptors are required to define paracrine, autocrine, and/or juxtacrine roles of specific ligand or its receptors in implantation.

Among many growth factors that have been studied in humans, HB-EGF appears to play a role in implantation and embryonic development. Its expression is maximal during the late secretory phase (cycle days 20–24) when the endometrium becomes receptive for implantation (287,288), and cells expressing the transmembrane form of HB-EGF adhere to human blastocysts displaying cell surface ErbB4 (289). Furthermore, HB-EGF was shown to be one of the most potent growth factors for enhancing the development of human IVF-derived embryos to blastocysts and subsequent zona hatching (290). Thus, cumulative evidence suggests that HB-EGF has a significant role in preimplantation embryo development and implantation as a paracrine and/or juxtacrine factor in various species.

Temporal and cell-specific expression of TGF-β and receptor isoforms in the uterus during the periimplantation period suggests that this growth factor is also important for implantation (291–293). TGF-β is shown to be involved in the implantation process by modulating immunological responses. TGF-β derived from the seminal vesicle gland is identified as a major active constituent in the seminal plasma of several mammalian species, where it reaches concentrations of more than 200 ng/ml (294). In mice, exposure to semen at mating activates an inflammatory response in the uterine mucosa (295), and in women, sexual intercourse causes similar inflammatory events in the cervix (Robertson, personal communication). The response is activated when seminal TGF-β triggers synthesis of several proinflammatory cytokines and chemokines in female reproductive tract epithelial cells, notably granulocyte-macrophage colony-stimulating factor and interleukin (IL)-6 (296).

Cytokines

Expression of multiple cytokines and their receptors in the uterus and embryo during early pregnancy suggests cytokine signaling in various aspects of implantation (57,290,297,298). However, gene-targeting studies showed that mice lacking tumor necrosis factor-α, IL-1β, IL-1 receptor antagonist, IL-1 receptor type 1, IL-6, and granulocyte-macrophage colony-stimulating factor apparently did not show overt reproductive defects (297). These observations suggest that either these cytokines have minor roles in implantation or the loss of one cytokine is compensated by other cytokines with overlapping functions. In contrast, some cytokines are more important for normal female fertility (299–301). For example, female *op/op* mice with a naturally occurring mutation of the macrophage colony-stimulating factor gene have markedly impaired fertility (299), and mice with a null mutation of the *LIF* gene encoding leukemia inhibitory factor show complete failure of implantation, and blastocysts in these mutant mice stay dormant (300,302). Studies on *IL-11Rα* mutant mice have also shown that IL-11 is crucial to decidualization, but not for the attachment reaction (301). In humans, IL-1 was identified as one modulator of the communication between maternal endometrium and embryo (303). IL-1β is also implicated in decidualization in primates, because it induces COX-2 expression,

PGE_2 synthesis, and insulin-like growth factor binding protein-1 (IGFBP-1) expression in human and baboon stromal fibroblasts in the presence of steroid hormones (304).

Both LIF and IL-11 are members of the IL-6 family, which includes IL-6 itself, oncostatin M, ciliary neurotrophic factor, and cardiotrophin (305). LIF and IL-11 bind to ligand-specific receptors, LIFR and IL-11R, respectively, and share gp130 as a common signal transduction partner (305), suggesting that gp130 signaling is critical to implantation and decidualization. Although the mechanism underlying implantation and decidualization failures in the absence of LIF still remains to be elucidated, there is a loss or an aberrant expression of certain implantation-related genes in pregnant *LIF* mutant mice (306). For example, uterine expression of HB-EGF and epiregulin is absent and COX-2 expression is aberrant at the sites of blastocysts in *LIF* mutant mice during the anticipated time of implantation.

LIF and its receptors, LIFR, and gp130 exist in both soluble and membrane-bound forms, and soluble forms of these two receptors antagonize the actions of their ligands (307–309). LIF expression exhibits a biphasic pattern, first appearing in uterine glands on day 4 of pregnancy in mice (310) and then in stromal cells surrounding the blastocyst at the time of the attachment reaction (306). This suggests that LIF has dual roles: first in the preparation of the uterus and later in the attachment reaction. However, the molecular mechanism by which LIF executes its effects on implantation is not yet known. In this regard, it would be useful to define the complex ligand–receptor interactions and detailed expression patterns of LIF receptors that occur during the periimplantation period. However, inactivation of gp130 by deleting all signal transducer and activator of transcription binding sites results in implantation failure (311), reinforcing the importance of LIF signaling in implantation.

Reciprocal blastocyst transfer experiments revealed that the uterus in the absence of LIF is incapable to trigger implantation irrespective of the blastocyst genotypes (300,302). However, LIFR and gp130 are expressed at the blastocyst stage, and administration of LIF improves embryo viability and hatching in several species (312–314), implicating that LIF signaling is also important for preimplantation embryonic development.

LIF expression in the uterus is maximal around the time of implantation in most species examined, although the steroid hormonal requirements for the preparation of uterine receptivity and implantation differ depending on the species. Although uterine LIF expression in some species appears to be regulated by P_4 (315), estrogen regulates LIF expression in the mouse uterus. This is evident from LIF expression on day 1 of pregnancy and during the estrus stage of the cycle when the uterus is under the influence of estrogen stimulation (310,316,317). In addition, LIF is not expressed in the uterus during experimentally induced delayed implantation but is rapidly induced by an injection of estrogen (306,310). However, the mechanism by which estrogen induces LIF expression in the mouse uterus and the mechanism by which it is regulated by P_4 in other species are not known. In humans, LIF is expressed in the endometrium and at higher levels in the glandular epithelium of the secretory endometrium (318). Furthermore, LIF deficiency has been associated with unexplained recurrent abortions and infertility in women (319).

IFN-τ is a member of the type I IFN family that acts differentially on the endometrial epithelium and stroma to regulate expression of a number of IFN-stimulated genes, which are considered to play roles in endometrial differentiation and conceptus implantation in the sheep (320). The actions of IFN-τ to signal pregnancy recognition (321) and induce the expression of IFN-stimulated genes are dependent on the effects of P_4. The type I IFN receptor subunits, IFNAR1 and IFNAR2, are expressed in all endometrial cell types with highest expression in the luminal epithelium in the sheep (322). However, most IFN-stimulated genes are induced or increased in response to the conceptus or IFN-τ only in the endometrial stroma and middle to deep glandular epithelium of the ovine uterus (64,323–326). In the ovine uterus, the lack of IFN-stimulated gene induction in endometrial luminal and glandular epithelia by IFN-τ is apparently due to the expression of IFN regulatory factor 2, a potent repressor of gene transcription (323). In addition, IFN regulatory factor 2 appears to be involved in IFN-τ inhibition of ERα gene transcription in these cells (327). Interestingly, *Wnt7a* is one of the genes stimulated by IFN-τ in the uterine luminal epithelium (328). Therefore, the actions of IFN-τ may not be completely limited to inhibition of the endometrial luteolytic mechanism but may also be involved in the induction of genes important for conceptus survival and growth as well as uterine receptivity. Maternal recognition of pregnancy in ruminants (sheep, cattle, goats) requires that the conceptus elongates from a spherical to a tubular and then filamentous form to produce IFN-τ, which is the pregnancy recognition signal that prevents development of the endometrial luteolytic mechanism (61,329,330). This antiluteolytic effect of IFN-τ results in the maintenance of a functional corpus luteum and secretion of P_4 that is essential to maintain a uterine environment conducive to successful development of the conceptus to term.

The mononuclear cells of the conceptus trophectoderm synthesize and secrete IFN-τ between days 10 and 25 with maximal production on days 14 to 16 (329). IFN-τ appears to be the sole factor produced by the conceptus that prevents development of the endometrial luteolytic mechanism (331). IFN-τ acts in a paracrine fashion on uterine luminal and glandular epithelial cells to suppress transcription of ERα and OTR genes (327,332), thereby preventing development of the endometrial luteolytic mechanism. Indeed, increases in *ERα* and *OTR* gene expression detected in the epithelium on days 11 to 17 postestrus in cyclic sheep do not occur in pregnant sheep (333) or in cyclic sheep infused with IFN-τ (334). By inhibiting increases in *OTR* expression, IFN-τ prevents endometrial production of luteolytic pulses of PGF. However, IFN-τ does not inhibit basal production of PGF, which is higher in pregnant than in cyclic ewes, and the conceptus or IFN-τ does not affect COX-2 expression in the epithelium of early pregnant sheep (261,328). Thus, the antiluteolytic effects of IFN-τ are to prevent increases in epithelial *ERα*, *PR*, and *OTR* gene expression, which are all estrogen responsive, by directly inhibiting transcription of the *ERα* gene and maintaining secretion of P_4 by the corpus luteum (327).

Transcription Factors

Hoxa-10 and Hoxa-11

Hox genes are transcription factors that belong to a multigene family. They are developmentally regulated and share a common highly conserved sequence element, called the homeobox, encoding a 61-amino acid helix-turn-helix DNA-binding domain (335). Hox genes are organized in four clusters (A, B, C, and D) on four different chromosomes in mice and humans and follow a stringent pattern of spatial and temporal colinearity during embryogenesis (335). Several Hox genes at the 5′ end of each cluster are classified as AbdB-like Hox genes, because of their homology with the *Drosophila AbdB* gene. In vertebrates, *AbdB*-like Hox genes, similar to their *Drosophila* ortholog, are expressed in developing genitourinary systems (336). For example, Hoxa-10 and Hoxa-11 are highly expressed in developing genitourinary tracts and adult female reproductive tract, suggesting their roles in reproductive events (336–338). Hoxa-10 mutant mice exhibit oviductal transformation of the proximal one-third of the uterus. Furthermore, adult female mice deficient in Hoxa-10 show failures in blastocyst implantation and decidualization unrelated to the oviductal transformation (336). Subsequent studies revealed that uterine stromal cells in Hoxa-10–deficient female mice show reduced proliferation in response to P_4, leading to decidualization defects (336,339). The uterus in *Hoxa-11*–deficient mice is hypoplastic and devoid of uterine glands due to developmental defects (338). Defective proliferation of stromal cells in *Hoxa-10(–/–)* female mice suggests that Hoxa-10 is involved in the local events of cellular proliferation by regulating cell cycle molecules. Indeed, cyclin D3 is aberrantly expressed in *Hoxa-10* mutant uteri in response to a decidualizing stimulus (340). Because several P_4-responsive genes are dysregulated in the uterine stroma of *Hoxa-10* mutant mice (339), this transcription factor may convey P_4 responsiveness in the uterine stroma by regulating gene expression. A similar but more severe phenotype was also noted in *Hoxa-11*–deficient female mice (338).

A global gene expression study further revealed that the absence of Hoxa-10 in the uterus in response to P_4 is associated with two cellular disturbances (341). First, among many genes that were up-regulated in the *Hoxa-10*–deficient uteri, two cell cycle molecules, p15 and p57, were notable. These two genes are both cyclin-dependent kinase inhibitors (CKIs), suggesting that the previously observed defect in stromal cell proliferation in *Hoxa-10* mutant mice is also associated with the up-regulation of CKIs. Second, there is hyperproliferation of T lymphocytes in the *Hoxa-10*–deficient uterine stroma in response to P_4. These results suggest that an aberrant lymphocyte proliferation is one cause of implantation failure in *Hoxa-10*–deficient mice. In humans, both Hoxa-10 and Hoxa-11 genes are markedly up-regulated in the uterus during the midsecretory phase in a steroid hormone–dependent manner (342), suggesting their roles in human implantation.

During human pregnancy, decidualizing stromal cells express high levels of IGFBP-1 (343). Its proposed roles in reproductive physiology are numerous, and it is implicated in preeclampsia, fetal growth retardation, and polycystic ovarian syndrome (344). Multiple factors contribute to the regulation of *IGFBP-1* gene expression, including insulin, glucocorticoids, progesterone, cytokines, and hypoxia (345). As was demonstrated in the liver system, IGFBP-1 expression is also regulated in uterine stromal cells by binding of Hoxa-10 and FKHR (a member of the FOXO subfamily of forkhead/winged-helix family of transcription factors) to its promoter (346).

Other Homeobox Genes

The Hmx family of transcription factors belongs to a homeobox gene family, unrelated to other larger classes of homeobox genes. These genes show overlapping expression during development, but gene-targeting experiments have revealed a unique role for Hmx3 in female reproduction (347). *Hmx3* mutant female mice show normal fertilization and preimplantation embryo development to blastocysts. However, the blastocysts fail to implant in the uterus and subsequently die. Because Hmx3 is primarily expressed in the myometrium during early pregnancy, the mechanism of infertility in these mice is different from that of *Hoxa-10* or *Hoxa-11* mutant mice, yet to be explored. *Msx-1*, another homeobox gene formerly known as *Hox-7.1*, is also implicated in implantation in mice (348). This gene is differentially expressed in the mouse uterus during the peri-implantation period, suggesting its role in implantation. In addition, it was shown that uterine expression of *Msx-1* is aberrant in *LIF* mutant mice, further reinforcing the importance of these signaling pathways in implantation. The study provided evidence for a novel cytokine-homeotic-Wnt signaling network in implantation.

*PPAR*δ

The nuclear receptor superfamily of transcription factors modulates expression of target genes by binding to specific DNA elements. The members of this superfamily span from well-characterized steroid hormone receptors to orphan nuclear receptors with no known ligands. Steroid hormone receptor aside, the PPAR family of nuclear receptors has been implicated in female reproductive events. Three members of the PPAR family are PPARα, PPARγ, and PPARδ. To act as a transcriptional activator, PPARs must heterodimerize with a member of the retinoid X receptor (RXR) subfamily (242,257). The ligands for PPAR include natural and synthetic eicosanoids, fatty acids, and hypolipidemic and hypoglycemic drugs, and there is evidence that PPAR/RXR dimers mediate nuclear signaling of PGs (349).

PPARδ has been shown to participate in various physiological processes, including embryo implantation (257). As mentioned above (see Prostaglandins), PGs produced via the COX-2 pathway are essential for implantation. During early pregnancy in mice, PGI_2 is the most abundant PG in the uterus, and its levels are higher at the implantation sites than at the interimplantation sites (257). Consistent with the finding that COX-2 driven uterine PG production is crucial to implantation, COX-2 and PGI synthase are coexpressed at the implantation site, suggesting the availability of PGI_2 to uterine cells. Among known PGI_2 receptors, such as IP, PPARα, and PPARδ, PPARδ is colocalized with COX-2 and PGI synthase at similar regions of the implantation sites; the expression of IP and PPARα was very low to undetectable. PPARδ expression requires the presence of active blastocysts, because it is not detectable in uteri of delayed implanting mice (350). PPARδ is indeed functional as a PGI_2 receptor, because administration of cPGI or L-165041 (a selective PPARδ agonist) to COX-2–deficient mice improves implantation and decidualization (257). This work provides evidence for a role for PPARδ in embryo implantation (257,266). Three independent groups have reported diverse phenotypes of *PPARδ* knockout mice (351–353). Because of severe early developmental defects of PPARδ mutant embryos, it has been very difficult to use this model to directly address whether the absence of maternal PPARδ affects implantation as in *COX-2*–deficient mice. Therefore, the conditional knockout mouse model with uterine-specific deletion of PPARδ is necessary to address this issue. Uterine expression of PPARδ has also been reported in rats and humans (354,355).

Calcium-Associated Signaling Molecules

Calcitonin

Calcitonin, a 32-amino acid peptide hormone originally found to be secreted from the thyroid gland, is known to regulate calcium levels in bone and kidney cells (356). In the rat uterus, calcitonin is expressed in the epithelium at the time of implantation, and it is under P_4 regulation (357). By administering antisense oligonucleotides, it was shown that attenuation of calcitonin expression indeed affects embryo implantation (358). It was also shown that calcitonin expression in human endometrium is temporally restricted to the uterine epithelium during the midsecretory phase of the menstrual cycle, which closely overlaps with the putative window of uterine receptivity. Calcitonin expression in the human endometrium is also under P_4 regulation (359). Similar expression and regulation of calcitonin has also been observed in the baboon endometrium (360), suggesting a conserved function of this peptide hormone during implantation.

Calbindins

Calbindin (CaBP)-d9k and -d28k are members of a large family of over 250 intracellular calcium-binding proteins (361). They are thought to regulate intracytoplasmic concentration and transport of free Ca^{2+}, modulating its absorption. CaBP-d9k was first identified in an RNA differential display experiment as a unique gene that is highly expressed before implantation but decreases at the site of embryo attachment (362). CaBP-d28k also exhibits a similar expression pattern, being down-regulated at implantation sites after the attachment reaction (363). Deletion of both genes by using morpholino antisense against CaBP-d9k administered directly into the uterine lumen of *CaBP-d28k* null mice shows that deficiency of both proteins results in complete failure of implantation (363). CaBP-d9k is not expressed in the human and rhesus monkey endometrium, but CaBP-d28k has an expression pattern that suggests its role in implantation in the primate (364). However, the results of effects of direct infusion of agents in pregnant uterine lumens to influence implantation should be interpreted with caution.

Connexins

Connexins, the gap junction channels, direct intercellular communication and permit passage of small molecules between the cytoplasm of neighboring cells, thereby coupling cells electrically and metabolically (365,366). They comprise transmembrane proteins and belong to a multigene family with high sequence homology among various species (367). During implantation, a local induction of *Cx26* expression restricted to the luminal epithelium surrounding the implantation chamber is observed in rats and mice (368,369). Local increases in *Cx26* transcripts in response to embryo recognition, however, were not inhibited by antiestrogen. Thus, this blastocyst-mediated induction of this connexin apparently acts independently of the ER. In the delayed implantation model, it was shown that Cx26 is already induced in the presence of blastocysts without an implantation reaction. In addition, expression of Cx26 is significantly increased after induction of decidualization. In both cases, connexin expression cannot be abolished by the application of antiestrogens, illustrating again the ER independence of this response. Studies in *ERα* knockout mice confirmed this observation (370). Thus, endometrial connexin expression could be under regulation by two distinct signaling pathways: Cx26 can be induced by estrogen via ERα but uses an ER-independent signaling pathway during embryo implantation and decidualization. The physiological role of local communication properties of the uterine epithelium upon embryo recognition remains to be investigated.

Cell Cycle Regulation

Differentiation of the uterus to support embryo development and implantation is primarily directed by P_4 and estrogen (5,31). After blastocyst attachment with the luminal epithelium, decidualization is initiated at the antimesometrial site where blastocysts implant. This process, characterized by stromal cell proliferation and differentiation into specialized type of cells (decidual cells) with polyploidy, is critical to the establishment of pregnancy in many species. The mechanisms by which cell cycle events govern decidualization are poorly understood. The cell cycle is tightly controlled at two checkpoints, the G1-S and G2-M phases. Normally, the operation of these phases involves a complex interplay of cyclins, cyclin-dependent kinases (cdks) and CKIs. In rodents, stromal cells immediately surrounding the implanting blastocyst proliferate at the beginning of decidualization (16). In mice, stromal cells close to the embryo cease to proliferate, initiating the formation of the primary decidual zone later on day 5 of pregnancy; the primary decidual zone is fully established by day 6. However, stromal cell proliferation outside the primary decidual zone continues, eventually forming the secondary decidual zone (10). Normally, the stimulus for decidualization is the implanting blastocyst. However, a similar process (deciduoma) can be experimentally induced in the pseudopregnant or hormonally prepared rodent uterus by intraluminal infusion of various agents, including oil (10). The development of decidua or deciduoma in rodents is associated with the formation of multinucleated and giant cells (371–373). In mice, the decidual cells in the antimesometrial zone are characterized by polyploidy and endoreduplication, and most cells in this zone eventually enlarge, containing nuclei with as much as 64n DNA.

The well-known regulators of mammalian cell proliferation are the three D-type cyclins (D1, D2, and D3), also known as G1 cyclins (374). The D-type cyclins accumulate during the G1 phase, and their association with cdk4 or cdk6 is important for forming holoenzymes that facilitate cell entry into the S phase. The retinoblastoma protein (Rb) and its family members, p107 and p130, are negative regulators of the D-type cyclins. Inactivation of these regulators by phosphorylation is dependent on the cyclin/cdk complex activity and allows the cell cycle to proceed through the G1 phase (375). The action of cdks usually is constrained by at least two CKIs, p16 and p21. The p16 family includes p15, p16, p18, and p19, and they inhibit the catalytic partners of D-type cyclins, cdk4 and cdk6. The p21 family is comprised of p21, p27, and p57, and they inhibit cdks with a broader specificity. CKIs accumulate in quiescent cells but are down-regulated with the onset of proliferation. Thus, a critical balance between the positive and negative cell cycle regulators is a key decision-maker for cell division (376). The uterus is a unique and dynamic physiologically relevant model in which cellular proliferation, differentiation, polyploidization, and apoptosis occur in a spatiotemporal manner during the reproductive cycle and pregnancy. In the human uterus, various cyclins (A, B1, D1, and E) and cdks (cdk1, cdk2, and cdk4) and CKI (p27) are regulated during the menstrual cycle or after hormone treatments (377,378). These factors are expressed primarily in epithelial and stromal cells during the proliferative phase, suggesting their involvement in rhythmic proliferation of these cells (377). During the secretory phase or after P_4 administration, p27 expression correlates with P_4-induced growth suppression in endometrial glands and stromal basalis (378). In rodents, however, the uterine expression of D- and E-type cyclins and cdks is regulated by estrogen and/or P_4 in a temporal manner (379–381). P_4-dependent growth suppression of the endometrium is considered to be mediated by decreased cdk activity, which presumably occurs via decreased levels of cyclins and increased association of CKI (p27) with cdks (382). There is also evidence that P_4 inhibition of uterine epithelial cell proliferation is mediated by the inhibition of nuclear translocation of cyclin D1 and cdk4 in association with the activation of cyclins A– and E–dependent cdk2 activity (383). Cell cycle molecules are also involved in endoreduplication in trophoblast differentiation during placentation (106,384).

In mice, the expression of cyclin D3 is up-regulated in decidualizing stromal cells at the implantation site and is associated with cell proliferation (340,385,386). Furthermore, cyclin D3 is associated with the large polyploid cells that are defined as terminally differentiated stromal cells. Coordinate expression of cdk4 and cyclin D3 at the site of the embryo after the onset of implantation in mice on day 5 of pregnancy suggests that these regulators play roles in proliferation of stromal cells undergoing decidualization. However, the expression of p21 with concomitant down-regulation of cyclin D3 and cdk4 in the primary decidual zone at the implantation site in the afternoon of day 5 supports the view that cell proliferation activity of cdk4/cyclin D3 ceases with the development of the primary decidual zone. But the expression of cdk4/cyclin D3 in the decidualizing stroma outside the primary decidual zone is again consistent with their role in proliferation of the stroma at the secondary decidual zone. In contrast, down-regulation of cdk4 in the secondary decidual zone with persistent expression of p21 on day 6 of pregnancy perhaps directs differentiation of stromal cells in this zone. On this day of pregnancy, a switch from cdk4 to cdk6 with continued expression of cyclin D3 and p21 in stromal cells within the secondary decidual zone is noted in polyploid decidual cells. The presence of cyclin E, cyclin A, and cdk2 with concomitant down-regulation of cyclin B and cdk1 in these cells supports the view that these cells are entering the endocycle pathway.

The physiological significance of stromal cell polyploidy during decidualization is still unclear. The life span of decidual cells during pregnancy is limited, and their demise makes room for the rapidly growing embryo. Because most decidual cells become polyploid during their lifetime, it is speculated that polyploidy limits the life span of decidual cells. Furthermore, one of the many functions of the deciduum is to support embryonic growth that requires increased protein synthesis. Polyploidy thus may ensure increased synthetic capacity by increasing the number of gene copies for transcription. In conclusion, a tight coordination of cell cycle molecules appears to be critical for uterine cell proliferation and differentiation during implantation and decidualization.

Matrix Remodeling and Angiogenesis

Tissue remodeling and angiogenesis are two hallmark events during implantation and decidualization. The changing endocrine state of the female during the reproductive cycle and pregnancy results in extensive remodeling of the uterine tissue (186,387). For example, various basement membrane components, such as type IV collagen, laminin, fibronectin, and proteoglycans, in the human uterus undergo changes throughout the menstrual cycle and pregnancy (387). Likewise, the ECM components undergo remodeling during mouse uterine stromal cell decidualization (186).

Proteases

Matrix metalloproteinases and tissue inhibitors of matrix metalloproteinases are thought to be key mediators for matrix degradation during implantation and decidualization (272,387–389). There is evidence that a balance between a select set of matrix metalloproteinases and tissue inhibitors of matrix metalloproteinases is important for implantation. Mechanisms regulating the matrix metalloproteinase and tissue inhibitors of matrix metalloproteinase genes during the periimplantation period are not clear, although growth factors and cytokines, including the EGF and TGF-β family members and LIF, have been shown to modulate matrix metalloproteinases and tissue inhibitors of matrix metalloproteinases (387). The cathepsin family of cysteine proteases is also implicated in implantation. Cathepsin B and L are expressed in mature invasive trophoblast cells and injection of synthetic inhibitors into pregnant mice during the early phase of implantation resulted in implantation failure (390). Altered levels of certain matrix remodeling molecules are associated with unexplained infertility and recurrent miscarriages (391).

VEGF Signaling

Under physiological conditions, angiogenesis, a process by which new blood vessels develop from preexisting vessels, primarily occurs in the uterus and ovary of the adult during the reproductive cycle and pregnancy (392). Increased vascular permeability and angiogenesis are crucial to successful implantation, decidualization, and placentation. A number of studies provided indirect and descriptive evidence for the potential roles of estrogen and P_4 in these processes in various species (392–395). These studies primarily examined the changes in the whole uterus of the expression of a number of gene products known to regulate vascular permeability and angiogenesis, including VEGF and its receptors, without investigating the angiogenic status of the uterus. Thus, in vivo roles for estrogen and P_4 in uterine angiogenesis are not fully appreciated.

VEGF, originally discovered as a vascular permeability factor (393), is also a potent mitogen for endothelial cells and a key regulator of vasculogenesis and angiogenesis (396). Targeted disruption of even one allele of the *VEGF* gene results in embryonic death in utero during midgestation with aberrant blood vessel formation (397,398). Differential splicing of the *VEGF* gene generates several VEGF isoforms in both humans and mice; $VEGF_{121}$ and $VEGF_{165}$ are the predominant isoforms in humans, whereas $VEGF_{120}$ and $VEGF_{164}$ are the most abundant isoforms in mice (395,399). VEGF effects are primarily mediated by two tyrosine kinase receptors: FLT1 (VEGFR1) and FLK1/KDR (VEGFR2) (400–403). FLK1 is the major transducer of VEGF signals that induce chemotaxis, actin reorganization, and proliferation of endothelial cells (396,404,405). Another multifunctional VEGF receptor was identified as neuropilin-1 (NRP1). Although NRP1 functions as a receptor for at least five different ligands, it is expressed in human endothelial cells as a $VEGF_{165}$-specific receptor.

Murine VEGF isoforms and their receptors, FLT1, FLK1, and NRP1, are differentially expressed in the mouse uterus in a spatiotemporal manner during implantation, and the predominant $VEGF_{164}$ isoform interacts with FLK1 and NRP1 (394,395). These results suggest that the VEGF system is important for uterine vascular permeability and angiogenesis during implantation. Others have also shown the expression of VEGF and its receptors in the uterus during pregnancy and in response to steroid hormones (393). For example, estrogen rapidly induces uterine vascular permeability and VEGF expression transcriptionally via nuclear ER (393), and the *VEGF* gene contains estrogen response elements (406). P_4 also up-regulates uterine VEGF expression via activation of the nuclear PR, but at a slower rate (406). Because estrogen rapidly stimulates uterine vascular permeability and VEGF expression and because vascular permeability is considered a prerequisite for angiogenesis, it was widely believed that estrogen is a potent stimulator of uterine angiogenesis during normal reproductive processes in vivo. However, estrogen and P_4 seem to have different effects in vivo; estrogen promotes uterine vascular permeability but profoundly inhibits angiogenesis, whereas P_4 stimulates angiogenesis with little effect on vascular permeability. These effects of estrogen and P_4 are mediated by differential spatiotemporal expression of proangiogenic factors in the uterus (407). Cell type–specific expression of VEGF isoforms and their receptors during embryo implantation has also been reported in mink and primates (99, 408–410).

VEGF effects are complemented and coordinated by another class of angiogenic factors, the angiopoietins (411). VEGF acts during the early stages of vessel development (397,398,412), whereas angiopoietin-1 (Ang1) acts later to promote angiogenic remodeling, including vessel maturation, stabilization, and leakiness (413–415). In contrast to agonistic functions of Ang1, Ang2 behaves as an antagonist. They interact with an endothelial cell–specific tyrosine kinase receptor, Tie2 (416). There are two additional members of the angiopoietin family, Ang3 and Ang4, but definitive biological functions of Ang3 and Ang4 remain unclear. It is now shown that whereas VEGF and its receptor Flk1 are primarily important for uterine vascular permeability and angiogenesis before and during the attachment phase of the implantation process, VEGF together with the angiopoietins and their receptor Tie2 direct angiogenesis during decidualization after implantation (417).

PGs are also involved in uterine vascular permeability and angiogenesis during implantation and decidualization. Specifically, COX-2–derived PGs participate in these processes (417). Thus, one cause of failure of implantation and decidualization in *COX-2(–/–)* mice is the deregulated vascular events in the absence of COX-2. The attenuation of uterine angiogenesis in these mice is primarily due to defective VEGF signaling rather than to the angiopoietin system. Collectively, the results provide evidence that although ovarian steroid hormones influence uterine vascular permeability and angiogenesis during the preimplantation period, COX-2–derived PGs direct these events during implantation and decidualization by differentially regulating VEGF and angiopoietin signaling (407,417).

Hypoxia-inducible factors (HIFs) are intimately associated with vascular events and induce VEGF expression by binding to the hypoxia response element in the VEGF promoter. HIF-α isoforms function by forming heterodimers with the aryl hydrocarbon receptor nuclear translocator (ARNT) (HIF-β) family members. In the uterus, expression of HIFs and ARNTs does not correlate with VEGF expression during the preimplantation period (days 1–4) in mice. In contrast, their expression follows the localization of uterine VEGF expression with increasing angiogenesis during the postimplantation period (days 5–8). This disparate pattern of uterine HIFs, ARNTs, and VEGF expression on days 1–4 of pregnancy suggests HIFs have multiple roles in addition to the regulation of angiogenesis during the periimplantation period. Steroid hormones also differentially regulate HIFs, with P_4 primarily up-regulating uterine HIF-1α expression and estrogen transiently stimulating that of HIF-2α (418). The definitive role of HIFs in regulating uterine angiogenesis warrants further investigation.

Endocannabinoids

Psychoactive cannabinoids are active components of marijuana that work via activation of the G-protein–coupled cell surface receptors, CB1 and CB2 (419,420). The discovery of cannabinoid receptors led to the identification of endogenous cannabinoid ligands, arachidonoylethanolamine (anandamide) and 2-arachidonoylglycerol (421,422). The mouse uterus synthesizes anandamide, and the levels fluctuate in the uterus during early pregnancy, coincident with the window of uterine receptivity for implantation (423). Thus, anandamide levels were found to be lower in the receptive uterus and at the implantation sites but were higher in the nonreceptive uterus and at interimplantation sites (423). The P_4-treated delayed-implanting uterus also showed elevated levels of anandamide, but the levels were down-regulated with the termination of the delayed implantation by estrogen (117). Therefore, a correlation between the levels of anandamide and phases of uterine receptivity suggests that endocannabinoid ligand–receptor signaling is an important aspect in defining the window of uterine receptivity for implantation. Ligand–receptor signaling with cannabinoids and its receptor CB1 also directs preimplantation embryo development and implantation. CB1 is expressed in the embryo from the two-cell stage at the time of zygotic gene expression through the blastocyst stage (424). Embryonic CB1 is functional, because two-cell embryos cultured in the presence of natural, synthetic, and endocannabinoids fail to develop to the blastocyst stage, and this failure occurs between the eight-cell and blastocyst stages. This effect is reversed by a CB1-selective antagonist (425). Furthermore, the endocannabinoid anandamide at a low concentration stimulates blastocyst differentiation and trophoblast outgrowth, whereas at higher concentrations it inhibits these events via differential regulation of mitogen-activated protein kinase and Ca^{2+} signaling (426,427). These results suggest that a narrow range of cannabinoid concentrations regulates the embryonic developmental program.

A tightly regulated level of uterine anandamide and embryonic CB1 during early pregnancy is important for preimplantation embryonic development and implantation. Indeed, embryos develop asynchronously in *CB1* mutant mice during the preimplantation period. Moreover, implantation fails to occur in wild-type mice, but not in *CB1(–/–)/CB2(–/–)* double mutant mice, when they are maintained on experimentally induced sustained levels of exogenously administered cannabinoids (117). Collectively, the expression of cannabinoid receptors in the preimplantation mouse embryo, synthesis of anandamide in the uterus, and the dose- and stage-specific effects of anandamide on embryo development and implantation suggest that ligand–receptor signaling with endocannabinoids or cannabinoid agonists is important for these events. The observation that heightened levels of cannabinoids inhibit implantation in mice subsequently led to the discovery that elevated levels of anandamide due to its reduced metabolism induce spontaneous pregnancy losses in women (428,429). Thus, regulated cannabinoid signaling perhaps functions as a physiological surveillance system that ensures implantation of healthy, but not abnormal, embryos resulting from aberrant expression of CB1 or their exposure to aberrant levels of endogenous or exogenous cannabinoids.

Morphogens

The intricate cross-talk between the blastocyst and uterus during implantation has several features of the reciprocal epithelial–mesenchymal interactions during embryogenesis and involves evolutionarily conserved signaling pathways. Many of these evolutionary conserved genes, including those encoding fibroblast growth factors (FGFs), IGFs, bone morphogenetic proteins (BMPs), Wnts, Noggin, Indian hedgehog (IHH) proteins, and their receptors, are potential players in the process of implantation and embryo spacing in the uterus (41). These genes are expressed in the uterus in a spatiotemporal manner during the periimplantation period in mice. For example, the attachment reaction is associated with a localized stromal induction of genes encoding BMP-2, FGF-2, and Wnt4. A simple in vitro model of implantation is not yet available to examine either the hierarchy of the events elicited in the uterus by the embryo or the function of individual signaling proteins. These questions were addressed by selectively delivering factors via blastocyst-sized gelatin beads in the uterine lumen to provoke implantation-like reactions with correct gene expression similar to what is generated by living embryos during normal implantation. Beads soaked in HB-EGF or IGF-1, but not other proteins, induce many of the same discrete local responses elicited by the blastocyst, including increased localized vascular permeability, decidualization, and expression of *Bmp-2*, *COX-2*, and *HB-EGF* at the sites of the beads (94,109,430). Furthermore, beads containing BMP-2 do not themselves produce an implantation-like response but alter the spacing of implantation sites induced by blastocysts cotransferred with the beads (41).

Genes encoding the components of the hedgehog-signaling pathway, namely IHH, the multipass transmembrane HH-binding protein/receptor, PATCHED (PTC), and the transcription factors, GLI1, 2, and 3 (431–433) are expressed in a dynamic temporal and spatial pattern during the preparation of the uterus for implantation (430). The expression of Ihh increases in the luminal epithelium and glands from day 3, reaching high levels on day 4. During the same time, the expression of *Ptc*, *Gli1*, and *Gli2* is up-regulated in the underlying mesenchymal stroma. Transcription of Ihh in ovariectomized mice is induced by P_4 but not by estrogen. Lower induction of *Ihh*, *Ptc*, and *Hoxa-10* is also observed in response to P_4 treatment in the uteri of *PR* mutant mice lacking the nuclear PR. This finding suggests that this hormone regulates Ihh via both nuclear receptor dependent and independent pathways. Furthermore, in uterine explant cultures, recombinant N-SHH protein stimulates the proliferation of mesenchymal cells and the expression of noggin. These findings suggest that IHH generated by the epithelium functions as a paracrine growth factor for stromal cells during the early stages of pregnancy (430). P_4 regulation of Ihh expression in the mouse uterus was confirmed by another group of investigators (434).

FGF-7, also known as keratinocyte growth factor, is an established paracrine mediator of hormone-regulated epithelial growth and differentiation (435). In all organs studied, FGF-7 was uniquely expressed in cells of mesenchymal origin. Intriguingly, expression of FGF-7 in the porcine uterus is exclusively restricted to the endometrial luminal epithelium and is particularly abundant between days 12 and 15 of the estrous cycle and pregnancy (436). Endometrial *FGF-7* mRNA levels are highest on day 12 in pregnant gilts and day 15 in cyclic gilts and are greater on day 12 of pregnancy than on day 12 of the estrous cycle. FGF-7 protein was detected in the uterine flushings of both day 12 cyclic and pregnant gilts. The receptor for FGF-7, known as FGF receptor 2_{IIIb} (FGFR2 $_{IIIb}$) or keratinocyte growth factor receptor, is detected in both the endometrial epithelium and conceptus trophectoderm. Treatment of endometrial explants from day 9 cyclic gilts with estradiol increases FGF-7 expression (437). Further, treatment of porcine trophectoderm cells with recombinant rat FGF-7 increases their proliferation, phosphorylates FGFR2 $_{IIIb}$, activates the mitogen-activated protein kinase (MAPK or ERK1/2) cascade, and increases expression of urokinase-type plasminogen activator, a marker for trophectoderm cell differentiation (437). Collectively, these results indicate that estrogen, the pregnancy recognition signal from the pig conceptus, increases uterine epithelial FGF-7 expression and FGF-7, in turn, stimulates the proliferation and differentiation of the conceptus trophectoderm in pigs, which is the only species possessing a true epitheliochorial type of placentation except the camel (436,437).

FUTURE DIRECTIONS: NEW TECHNIQUES, NOVEL DISCOVERY

Recent conceptual and technological advances in molecular and genetic approaches have led to the discovery of numerous molecules involved in implantation. However, a precise hierarchical landscape of the signaling pathways has not yet been defined. It is also not known which pathways work in concert and which work independently. This chapter focuses on a select number of signaling molecules and pathways that are thought to be important for embryo–uterine interactions during implantation. A large number of other growth factors, cytokines, lipid mediators, vasoactive agents, and transcription factors that could well be involved in implantation are not addressed here because of space limitations.

The advent of microarray technology and sequencing of the mouse and human genomes allow us to analyze implantation-related genes in a global perspective. In mice, microarrays have been used to identify genes that show differential expression at implantation versus interimplantation sites (438), implantation versus postimplantation periods (439), or genes that are differentially regulated by changes in estrogen and P_4 signaling (341,434,440–444). Comparison of the array results of implantation versus interimplantation sites with those of P_4-treated uteri versus estrogen-stimulated uteri led to identification of a number of genes with previously recognized roles in implantation and a number of new genes related to implantation and steroid hormonal regulation (341,434,440,445). Closer examination of these genes and comparing results from similar microarray experiments may provide clues to the identity of important genes or gene families in the implantation process. In humans, microarray technology has been used to identify genes associated with the window of uterine receptivity (166,446). A comparison of the results shows differential regulation of a small number of genes around the time of uterine receptivity for implantation. Interestingly, a number of genes are also differentially expressed in mouse models of implantation and human uterine receptivity. Currently, emerging techniques in proteomics are also being applied to the understanding of uterine biology (447–449) and will likely identify known and unknown molecules with novel functions related to implantation.

CONCLUDING REMARKS

Infertility and rapid population growth are two pressing global reproductive health issues. The processes of preimplantation embryo development and uterine preparation for implantation are two major determinants for the reproductive success. Basic and clinical research to better understand these events will help alleviate problems of female infertility, improve fertility regulation in women, and lead to the development of new and improved contraceptive methods. Interactions between the major uterine and embryonic cell types with respect to endocrine, paracrine, juxtacrine, and autocrine factors during implantation are extremely complex (94,445). Thus, exploring and defining the molecular landscape during the critical time of implantation necessitate well-thought-out experimental designs in the context of both embryonic and uterine contributions to formulate a more meaningful blueprint. This mission is not easily achievable in humans due to experimental difficulties, ethical considerations, and current restrictions on research with human embryos. Therefore, animal models continue to be used for studying embryo–uterine interactions during implantation. In addition, efforts should be directed to establish reliable in vitro systems to study implantation, which are not currently available. However, experiments using endometrial biopsy samples to identify molecules associated with human uterine receptivity (window of implantation) during the menstrual cycle with changing estrogen and P_4 levels should continue to be pursued to better understand this process in humans.

Although considerable information regarding the roles of growth factors, cytokines, homeotic genes, transcription factors, and lipid mediators in implantation has been generated, their hierarchical blueprint in directing uterine and embryonic function during implantation remains to be deciphered. A difficult task before us is to unveil the intricate nature of the signaling pathways in implantation. Further investigation is required to understand whether these pathways function independently, in parallel, or converge to a common signaling pathway to execute reciprocal interactions between the embryo and uterus during implantation. Thus, our understanding of the implantation process is still far from complete. As Boving wrote, "The conceptus and uterus have mastered everything they need to know about implantation. We scientists have not" (540). For example, many of the genes, which are expressed in an implantation-specific manner and appear to be important for implantation, cannot be studied mechanistically because deletion of these genes results in embryonic lethality. Uterine- or embryo-specific conditional deletion of genes of interest is urgently needed to better understand the definitive roles of these genes in uterine biology and implantation. Our failure to identify suitable uterine cell-specific promoters is a hindrance to achieve this objective. There is also a difficulty in identifying the critical roles of signaling molecules within a gene family because of the redundant or compensatory functions of the gene products within the family.

Strategies comparing global gene expression profiles between the implantation and interimplantation sites have identified novel genes in the implantation process. Thus, a genome-wide screening approach coupled with functional assays will help elucidate these complex signaling pathways. In addition, experiments should be pursued to compare global gene expression patterns between wild-type and gene-deleted mouse uteri and blastocysts under defined physiological experimental conditions. The results obtained from these experiments may help uncover new signaling molecules and pathways not previously identified. The application of proteomics is also likely to provide information regarding interactions among various molecular pathways relevant to implantation. Another area of research that deserves particular attention is to identify embryonic signaling molecules that influence uterine functions for implantation. Although CGs in primates are well known for their role in pregnancy establishment, it is not yet clear whether they also function as implantation initiators. The situation is different in large animals in which embryo-derived IFNs and estrogens are known to function as important signaling molecules for pregnancy recognition (329,451,452). These studies have been possible because of the availability of relatively large amount of blastocyst tissues in these species. In other species, including rodents and humans, the most limiting factor is the availability of adequate amount of tissues for biochemical and molecular biology experiments. With the advent of microscale proteomics and genomic approaches, it is hoped that more information on embryonic signals is likely to be forthcoming.

Although the mechanics and cellular architecture of the implantation process vary, certain basic features are common to many species. For example, implantation occurs at the blastocyst stage, there is a defined window of uterine receptivity for implantation, a reciprocal interaction between the blastocyst and the uterus is essential for implantation, and a localized increase in uterine vascular permeability occurs at the site of the blastocyst during the attachment reaction. Identification and characterization of signaling pathways in these steps may elucidate a unifying scheme relevant to understanding the mechanism of human implantation.

ACKNOWLEDGMENTS

This chapter is dedicated to the late Professor Donald C. Johnson (Kansas University Medical Center, Kansas City) as a tribute to his life-long dedication to reproduction research and passion for scientific pursuits. We apologize for not citing numerous relevant work due to space constraints.

REFERENCES

1. Gardner, R. L., and Papaioannou, V. E. (1975). Differentiation in the trophectoderm and inner cell mass. In *The Early Development of Mammals* (M. Balls and A. E. Wild, Eds.), pp. 107–132. Cambridge University Press, London.
2. Renfree, M. B. (1982). Implantation and placentation. In *Reproduction in Mammals* (C. R. Austin and R. V. Short, Eds.), pp. 26–69. Cambridge University Press, Cambridge.
3. Enders, A. C., and Schlafke, S. (1967). A morphological analysis of the early implantation stages in the rat. *Am. J. Anat.* 120, 185–226.
4. Enders, A. C., and Schlafke, S. (1969). Cytological aspects of trophoblast-uterine interaction in early implantation. *Am. J. Anat.* 125, 1–29.
5. Psychoyos, A. (1973). Endocrine control of egg implantation. In *Handbook of Physiology* (R. O. Greep, E. G. Astwood, and S. R. Geiger, Eds.), pp. 187–215. American Physiology Society, Washington, DC.
6. Das, S. K., Wang, X. N., Paria, B. C., Damm, D., Abraham, J. A., Klagsbrun, M., Andrews, G. K., and Dey, S. K. (1994). Heparin-binding EGF-like growth factor gene is induced in the mouse uterus temporally by the blastocyst solely at the site of its apposition: a possible ligand for interaction with blastocyst EGF-receptor in implantation. *Development* 120, 1071–1083.
7. Hoos, P. C., and Hoffman, L. H. (1980). Temporal aspects of rabbit uterine vascular and decidual responses to blastocyst stimulation. *Biol. Reprod.* 23, 453–459.
8. Enders, A. C., and Schlafke, S. (1986). Implantation in nonhuman primates and in the human. *Comp. Prim. Biol.* 3, 453–459.
9. Enders, A. C., and Lopata, A. (1999). Implantation in the marmoset monkey: expansion of the early implantation site. *Anat. Rec.* 256, 279–299.
10. Dey, S. K. (1996). Implantation. In *Reproductive Endocrinology, Surgery, and Technology* (E. Y. Adashi, J. A. Rock, and Z. Rosenwaks, Eds.), pp. 421–434. Lippincott-Raven, New York.
11. Kirby, D. R. S. (1971). Blastocyst-uterine relationship before and during implantation. In *The Biology of Blastocysts* (R. J. Blandau, Ed.), pp. 393–410. The University of Chicago Press, Chicago.
12. Finn, C. A., and Porter, D. G. (1975). *The Uterus*. Publishing Sciences Group, Inc., Acton, MA.
13. Schlafke, S., and Enders, A. C. (1975). Cellular basis of interaction between trophoblast and uterus at implantation. *Biol. Reprod.* 12, 41–65.
14. Kennedy, T. G. (1986). Prostaglandins and uterine sensitization for the decidual cell reaction. *Ann. N. Y. Acad. Sci.* 476, 43–48.
15. Galassi, L. (1968). Autoradiographic study of the decidual cell reaction in the rat. *Dev. Biol.* 17, 75–84.
16. Huet, H., Andrews, G. K., and Dey, S. K. (1989). Cell type-specific localization of c-myc protein in the mouse uterus: modulation by steroid hormones and analysis of the periimplantation period. *Endocrinology* 125, 1683–1690.
17. Krehbiel, R. H. (1937). Cytological studies of the decidual reaction in the rat during early pregnancy and in the production of deciduoma. *Physiol. Zool.* 10, 212–238.
18. Burton, G. J. (1992). Human and animal models: limitations and comparisons. In *The First Twelve Weeks of Gestation* (E. R. Barnea, J. Hustin, and E. Janiaux, Eds.), pp. 469–485. Spinger-Verlag, Berlin.
19. Bonnet, R. (1884). Beitrage zur embryologie der wiederkauer, gewonnen am schafei. *Arch. Anat. Physiol.* 8, 170.
20. Wintenberger-Torres, S., and Flechon, J. E. (1974). Ultrastructural evolution of the trophoblast cells of the pre-implantation sheep blastocyst from day 8 to day 18. *J. Anat.* 118, 143–153.
21. Rowson, L. E., and Moor, R. M. (1966). Development of the sheep conceptus during the first fourteen days. *J. Anat.* 100, 777–785.
22. Farin, C. E., Imakawa, K., and Roberts, R. M. (1989). In situ localization of mRNA for the interferon, ovine trophoblast protein-1, during early embryonic development of the sheep. *Mol. Endocrinol.* 3, 1099–1107.
23. Gray, C. A., Burghardt, R. C., Johnson, G. A., Bazer, F. W., and Spencer, T. E. (2002). Evidence that absence of endometrial gland secretions in uterine gland knockout ewes compromises conceptus survival and elongation. *Reproduction* 124, 289–300.
24. Guillomot, M., Michel, C., Gaye, P., Charlier, N., Trojan, J., and Martal, J. (1990). Cellular localization of an embryonic interferon, ovine trophoblastin and its mRNA in sheep embryos during early pregnancy. *Biol. Cell.* 68, 205–211.
25. Bazer, F. W. (1992). Mediators of maternal recognition of pregnancy in mammals. *Proc. Soc. Exp. Biol. Med.* 199, 373–384.
26. Guillomot, M., Flechon, J. E., and Wintenberger-Torres, S. (1981). Conceptus attachment in the ewe: an ultrastructural study. *Placenta* 2, 169–182.
27. Wimsatt, W. A. (1950). New histological observations on the placenta of the sheep. *Am. J. Anat.* 87, 391–436.
28. Boshier, D. P. (1970). The pontamine blue reaction in pregnant sheep uteri. *J. Reprod. Fertil.* 22, 595–596.
29. Psychoyos, A. (1973). Hormonal control of ovoimplantation. *Vitam. Horm.* 31, 201–256.
30. Yoshinaga, K. (1988). Uterine receptivity for blastocyst implantation. *Ann. N. Y. Acad. Sci.* 541, 424–431.

31. Paria, B. C., Huet, H., and Dey, S. K. (1993). Blastocyst's state of activity determines the "window" of implantation in the receptive mouse uterus. *Proc. Natl. Acad. Sci. U S A* 90, 10159–10162.
32. Heap, R. B., and Deanesly, R. (1967). The increase in plasma progesterone levels in the pregnant guinea-pig and its possible significance. *J. Reprod. Fertil.* 14, 339–341.
33. Harper, M. J., Dowd, D., and Elliott, A. S. (1969). Implantation and embryonic development in the ovariectomized-adrenalectomized hamster. *Biol. Reprod.* 1, 253–257.
34. Kwun, J. K., and Emmens, C. W. (1974). Hormonal requirements for implantation and pregnancy in the ovariectomized rabbit. *Aust. J. Biol. Sci.* 27, 275–283.
35. McCormack, J. T., and Greenwald, G. S. (1974). Evidence for a preimplantation rise in oestradiol-17beta levels on day 4 of pregnancy in the mouse. *J. Reprod. Fertil.* 41, 297–301.
36. Heap, R. B., Flint, A. P., Hartmann, P. E., Gadsby, J. E., Staples, L. D., Ackland, N., and Hamon, M. (1981). Oestrogen production in early pregnancy. *J. Endocrinol.* 89 Suppl, 77P–94P.
37. Stromstedt, M., Keeney, D. S., Waterman, M. R., Paria, B. C., Conley, A. J., and Dey, S. K. (1996). Preimplantation mouse blastocysts fail to express CYP genes required for estrogen biosynthesis. *Mol. Reprod. Dev.* 43, 428–436.
38. Finn, C. A., and Martin, L. (1967). Patterns of cell division in the mouse uterus during early pregnancy. *J. Endocrinol.* 39, 593–597.
39. Huet-Hudson, Y. M., Andrews, G. K., and Dey, S. K. (1989). Cell type-specific localization of c-myc protein in the mouse uterus: modulation by steroid hormones and analysis of the periimplantation period. *Endocrinology* 125, 1683–1690.
40. Lundkvist, O., and Nilsson, B. O. (1982). Endometrial ultrastructure in the early uterine response to blastocysts and artificial deciduogenic stimuli in rats. *Cell. Tissue Res.* 225, 355–364.
41. Paria, B. C., Ma, W., Tan, J., Raja, S., Das, S. K., Dey, S. K., and Hogan, B. L. (2001). Cellular and molecular responses of the uterus to embryo implantation can be elicited by locally applied growth factors. *Proc. Natl. Acad. Sci. U S A* 98, 1047–1052.
42. Fazleabas, A. T., Donnelly, K. M., Srinivasan, S., Fortman, J. D., and Miller, J. B. (1999). Modulation of the baboon (*Papio anubis*) uterine endometrium by chorionic gonadotrophin during the period of uterine receptivity. *Proc. Natl. Acad. Sci. U S A* 96, 2543–2548.
43. Hausermann, H. M., Donnelly, K. M., Bell, S. C., Verhage, H. G., and Fazleabas, A. T. (1998). Regulation of the glycosylated beta-lactoglobulin homolog, glycodelin (placental protein 14:[PP14]) in the baboon (*Papio anubis*) uterus. *J. Clin. Endocrinol. Metab.* 83, 1226–1233.
44. Geisert, R. D., Renegar, R. H., Thatcher, W. W., Roberts, R. M., and Bazer, F. W. (1982). Establishment of pregnancy in the pig. I. Interrelationships between preimplantation development of the pig blastocyst and uterine endometrial secretions. *Biol. Reprod.* 27, 925–939.
45. Bazer, F. W., and Thatcher, W. W. (1977). Theory of maternal recognition of pregnancy in swine based on estrogen controlled endocrine versus exocrine secretion of prostaglandin F2alpha by the uterine endometrium. *Prostaglandins* 14, 397–400.
46. Fazleabas, A. T., Geisert, R. D., Bazer, F. W., and Roberts, R. M. (1983). Relationship between release of plasminogen activator and estrogen by blastocysts and secretion of plasmin inhibitor by uterine endometrium in the pregnant pig. *Biol. Reprod.* 29, 225–238.
47. Simon, C., Cano, F., Valbuena, D., Remohi, J., and Pellicer, A. (1995). Clinical evidence for a detrimental effect on uterine receptivity of high serum oestradiol concentrations in high and normal responder patients. *Hum. Reprod.* 10, 2432–2437.
48. Pellicer, A., Valbuena, D., Cano, F., Remohi, J., and Simon, C. (1996). Lower implantation rates in high responders: evidence for an altered endocrine milieu during the preimplantation period. *Fertil. Steril.* 65, 1190–1195.
49. Simon, C., Garcia Velasco, J. J., Valbuena, D., Peinado, J. A., Moreno, C., Remohi, J., and Pellicer, A. (1998). Increasing uterine receptivity by decreasing estradiol levels during the preimplantation period in high responders with the use of a follicle-stimulating hormone step-down regimen. *Fertil. Steril.* 70, 234–239.
50. Valbuena, D., Martin, J., de Pablo, J. L., Remohi, J., Pellicer, A., and Simon, C. (2001). Increasing levels of estradiol are deleterious to embryonic implantation because they directly affect the embryo. *Fertil. Steril.* 76, 962–968.
51. Amoroso, E. C. (1952). Placentation. In *Marshall's Physiology of Reproduction* (A. S. Parkes, Ed.), Vol. 2, pp. 127–311. Longmans Green, London.
52. Bazer, F. W. (1975). Uterine protein secretions: relationship to development of the conceptus. *J. Anim. Sci.* 41, 1376–1382.
53. Bonnet, R. (1882). Die Uterinmilch und irhe Bedeutung far die Fruct. In *Betrage zur Biologie als Fetgabe dem Anatomen und Physiologen*, pp. 221–263. Th. von Bischoff, Stuttgart.
54. Bazer, F. W., Roberts, R. M., and Thatcher, W. W. (1979). Actions of hormones on the uterus and effect on conceptus development. *J. Anim. Sci.* 49 Suppl 2, 35–45.
55. Roberts, R. M., and Bazer, F. W. (1988). The functions of uterine secretions. *J. Reprod. Fertil.* 82, 875–892.
56. Bartol, F. F., Wiley, A. A., Floyd, J. G., Ott, T. L., Bazer, F. W., Gray, C. A., and Spencer, T. E. (1999). Uterine differentiation as a foundation for subsequent fertility. *J. Reprod. Fertil.* 54 Suppl, 287–302.
57. Carson, D. D., Bagchi, I., Dey, S. K., Enders, A. C., Fazleabas, A. T., Lessey, B. A., and Yoshinaga, K. (2000). Embryo implantation. *Dev. Biol.* 223, 217–237.
58. Gray, C. A., Bartol, F. F., Tarleton, B. J., Wiley, A. A., Johnson, G. A., Bazer, F. W., and Spencer, T. E. (2001). Developmental biology of uterine glands. *Biol. Reprod.* 65, 1311–1323.
59. Allison, G., Bartol, F. F., Taylor, K. M., Wiley, A. A., Ramsey, W. S., Ott, T. L., Bazer, F. W., and Spencer, T. E. (2000). Ovine uterine gland knock-out model: effects of gland ablation on the estrous cycle. *Biol. Reprod.* 62, 448–456.
60. Gray, C. A., Bazer, F. W., and Spencer, T. E. (2001). Effects of neonatal progestin exposure on female reproductive tract structure and function in the adult ewe. *Biol. Reprod.* 64, 797–804.
61. Spencer, T. E., and Bazer, F. W. (2002). Biology of progesterone action during pregnancy recognition and maintenance of pregnancy. *Front. Biosci.* 7, d1879–d1898.
62. Gray, C. A., Taylor, K. M., Bazer, F. W., and Spencer, T. E. (2000). Mechanisms regulating norgestomet inhibition of endometrial gland morphogenesis in the neonatal ovine uterus. *Mol. Reprod. Dev.* 57, 67–78.
63. Gray, C. A., Taylor, K. M., Ramsey, W. S., Hill, J. R., Bazer, F. W., Bartol, F. F., and Spencer, T. E. (2001). Endometrial glands are required for preimplantation conceptus elongation and survival. *Biol. Reprod.* 64, 1608–1613.
64. Johnson, G. A., Stewart, M. D., Gray, C. A., Choi, Y., Burghardt, R. C., Yu, L., Bazer, F. W., and Spencer, T. E. (2001). Effects of the estrous cycle, pregnancy, and interferon tau on 2′,5′-oligoadenylate synthetase expression in the ovine uterus. *Biol. Reprod.* 64, 1392–1399.
65. Burghardt, R. C., Johnson, G. A., Jaeger, L. A., Ka, H., Garlow, J. E., Spencer, T. E., and Bazer, F. W. (2002). Integrins and extracellular matrix proteins at the maternal–fetal interface in domestic animals. *Cell. Tiss. Org.* 172, 202–217.
66. Johnson, G. A., Spencer, T. E., Burghardt, R. C., and Bazer, F. W. (1999). Ovine osteopontin. I. Cloning and expression of messenger ribonucleic acid in the uterus during the periimplantation period. *Biol. Reprod.* 61, 884–891.

67. Spencer, T. E., Bartol, F. F., Bazer, F. W., Johnson, G. A., and Joyce, M. M. (1999). Identification and characterization of glycosylation-dependent cell adhesion molecule 1-like protein expression in the ovine uterus. *Biol. Reprod.* 60, 241–250.
68. Spencer, T. E., Stagg, A. G., Joyce, M. M., Jenster, G., Wood, C. G., Bazer, F. W., Wiley, A. A., and Bartol, F. F. (1999). Discovery and characterization of endometrial epithelial messenger ribonucleic acids using the ovine uterine gland knockout model. *Endocrinology* 140, 4070–4080.
69. Renfree, M. B., and Shaw, G. (2000). Diapause. *Annu. Rev. Physiol.* 62, 353–75.
70. Mead, R. A. (1993). Embryonic diapause in vertebrates. *J. Exp. Zool.* 266, 629–641.
71. Yoshinaga, K., and Adams, C. E. (1966). Delayed implantation in the spayed, progesterone treated adult mouse. *J. Reprod. Fertil.* 12, 593–595.
72. McLaren, A. (1973). Blastocyst activation. In *The Regulation of Mammalian Reproduction* (S. J. Segal, R. Crozier, P. A. Corfman, and P. G. Condliffe, Eds.), pp. 321–328. Charles C Thomas, Springfield, IL.
73. Rasweiler, J. J. T., and Badwaik, N. K. (1997). Delayed development in the short-tailed fruit bat, *Carollia perspicillata*. *J. Reprod. Fertil.* 109, 7–20.
74. Pearson, O. P., and Enders, R. K. (1944). Duration of pregnancy in certain mustelids. *J. Exp. Zool.* 95, 21–35.
75. Mead, R. A. (1971). Effects of light and blinding upon delayed implantation in the spotted skunk. *Biol. Reprod.* 5, 214–20.
76. Murphy, B. D., and James, D. A. (1974). The effects of light and sympathetic innervation to the head on nidation in mink. *J. Exp. Zool.* 187, 267–276.
77. Bonnefond, C., Martinet, L., and Monnerie, R. (1990). Effects of timed melatonin infusions and lesions of the superchiasmatic nuclei on prolactin and progesterone secretion in pregnant and pseudopregnatn mink. *J. Neuroendocrinol.* 2, 583–591.
78. Papke, R. L., Concannon, P. W., Travis, H. F., and Hansel, W. (1980). Control of luteal function and implantation in the mink by prolactin. *J. Anim. Sci.* 50, 1102–1107.
79. Murphy, B. D. (1983). Precocious induction of luteal activation and termination of delayed implantation in mink with the dopamine antagonist pimozide. *Biol. Reprod.* 29, 658–662.
80. Murphy, B. D., Concannon, P. W., Travis, H. F., and Hansel, W. (1981). Prolactin: the hypophyseal factor that terminates embryonic diapause in mink. *Biol. Reprod.* 25, 487–491.
81. Murphy, B. D., DiGregorio, G. B., Douglas, D. A., and Gonzalez-Reyna, A. (1990). Interactions between melatonin and prolactin during gestation in mink (*Mustela vison*). *J. Reprod. Fertil.* 89, 423–429.
82. Murphy, B. D., Rajkumar, K., Gonzalez Reyna, A., and Silversides, D. W. (1993). Control of luteal function in the mink (*Mustela vison*). *J. Reprod. Fertil.* 47 Suppl, 181–188.
83. Rose, J., Stormshak, F., Adair, J., and Oldfield, J. E. (1983). Prolactin binding sites in the uterus of the mink. *Mol. Cell. Endocrinol.* 31, 131–139.
84. Hanssen, A. (1947). The physiology of reproduction in the mink (*Mustela vison* Schreb.) with special reference to delayed implantation. *Acta. Zool.* 28, 1–136.
85. Douglas, D. A., Song, J. H., Moreau, G. M., and Murphy, B. D. (1998). Differentiation of the corpus luteum of the mink (*Mustela vison*): mitogenic and steroidogenic potential of luteal cells from embryonic diapause and postimplantation gestation. *Biol. Reprod.* 58, 1163–1169.
86. Murphy, B. D., Mead, R. A., and McKibbin, P. E. (1983). Luteal contribution to the termination of preimplantation delay in mink. *Biol. Reprod.* 28, 497–503.
87. Mead, R. A., Joseph, M. M., Neirinckx, S., and Berria, M. (1988). Partial characterization of a luteal factor that induces implantation in the ferret. *Biol. Reprod.* 38, 798–803.
88. Schulz, L. C., and Bahr, J. M. (2003). Glucose-6-phosphate isomerase is necessary for embryo implantation in the domestic ferret. *Proc. Natl. Acad. Sci. U S A* 100, 8561–8566.
89. Chang, M. C. (1968). Reciprocal insemination and egg transfer between ferrets and mink. *J. Exp. Zool.* 168, 49–60.
90. Moreau, G. M., Arslan, A., Douglas, D. A., Song, J., Smith, L. C., and Murphy, B. D. (1995). Development of immortalized endometrial epithelial and stromal cell lines from the mink (*Mustela vison*) uterus and their effects on the survival in vitro of mink blastocysts in obligate diapause. *Biol. Reprod.* 53, 511–518.
91. Murphy, B. D., and James, D. A. (1974). Mucopolysaccharide histochemistry of the mink uterus during gestation. *Can. J. Zool.* 52, 687–693.
92. Mead, R. A., Rourke, A. W., and Swannack, A. (1979). Changes in uterine protein synthesis during delayed implantation in the western spotted skunk and its regulation by hormones. *Biol. Reprod.* 21, 39–46.
93. Paria, B. C., Das, S. K., Mead, R. A., and Dey, S. K. (1994). Expression of epidermal growth factor receptor in the preimplantation uterus and blastocyst of the western spotted skunk. *Biol. Reprod.* 51, 205–213.
94. Paria, B. C., Reese, J., Das, S. K., and Dey, S. K. (2002). Deciphering the cross-talk of implantation: advances and challenges. *Science* 296, 2185–2188.
95. Song, J. H., Houde, A., and Murphy, B. D. (1998). Cloning of leukemia inhibitory factor (LIF) and its expression in the uterus during embryonic diapause and implantation in the mink (*Mustela vison*). *Mol. Reprod. Dev.* 51, 13–21.
96. Passavant, C., Zhao, X., Das, S. K., Dey, S. K., and Mead, R. A. (2000). Changes in uterine expression of leukemia inhibitory factor receptor gene during pregnancy and its up-regulation by prolactin in the western spotted skunk. *Biol. Reprod.* 63, 301–307.
97. Song, J. H., Sirois, J., Houde, A., and Murphy, B. D. (1998). Cloning, developmental expression, and immunohistochemistry of cyclooxygenase 2 in the endometrium during embryo implantation and gestation in the mink (*Mustela vison*). *Endocrinology* 139, 3629–3636.
98. Das, S. K., Wang, J., Dey, S. K., and Mead, R. A. (1999). Spatiotemporal expression of cyclooxygenase 1 and cyclooxygenase 2 during delayed implantation and the periimplantation period in the western spotted skunk. *Biol. Reprod.* 60, 893–899.
99. Lopes, F. L., Desmarais, J., Gevry, N. Y., Ledoux, S., and Murphy, B. D. (2003). Expression of vascular endothelial growth factor isoforms and receptors Flt-1 and KDR during the peri-implantation period in the mink, *Mustela vison*. *Biol. Reprod.* 68, 1926–1933.
100. Hoozemans, D. A., Schats, R., Lambalk, C. B., and Hompes, P. G. (2004). Human embryo implantation: current knowledge and clinical implications in assisted reproductive technology. *Reprod. Biomed. Online* 9, 692–715.
101. Enders, A. C., Schlafke, S., Hubbard, N. E., and Mead, R. A. (1986). Morphological changes in the blastocyst of the western spotted skunk during activation from delayed implantation. *Biol. Reprod.* 34, 423–437.
102. Isakova, G., and Zhogoleva, N. N. (1997). Activity of the embryonal genome of mink during diapause (cytogenetic analysis): number of cells and cell nucleus size or various size and age. *Genetika* 33, 22–830 [In Russian].
103. Desmarais, J. A., Bordignon, V., Lopes, F. L., Smith, L. C., and Murphy, B. D. (2004). The escape of the mink embryo from obligate diapause. *Biol. Reprod.* 70, 662–670.
104. Tanaka, S., Kunath, T., Hadjantonakis, A. K., Nagy, A., and Rossant, J. (1998). Promotion of trophoblast stem cell proliferation by FGF4. *Science* 282, 2072–2075.

105. Cross, J. C. (2000). Genetic insights into trophoblast differentiation and placental morphogenesis. *Semin. Cell. Dev. Biol.* 11, 105–113.
106. MacAuley, A., Cross, J. C., and Werb, Z. (1998). Reprogramming the cell cycle for endoreduplication in rodent trophoblast cells. *Mol. Biol. Cell.* 9, 795–807.
107. Enders, A. C., and Mead, R. A. (1996). Progression of trophoblast into the endometrium during implantation in the western spotted skunk. *Anat. Rec.* 244, 297–315.
108. Enders, A. C., and Schlafke, S. (1972). Implantation in the ferret: epithelial penetration. *Am. J. Anat.* 133, 291–315.
109. Hamatani, T., Daikoku, T., Wang, H., Matsumoto, H., Carter, M. G., Ko, M. S., and Dey, S. K. (2004). Global gene expression analysis identifies molecular pathways distinguishing blastocyst dormancy and activation. *Proc. Natl. Acad. Sci. U S A* 101, 10326–10331.
110. Huet, H., and Dey, S. K. (1990). Requirement for progesterone priming and its long-term effects on implantation in the mouse. *Proc. Soc. Exp. Biol. Med.* 193, 259–263.
111. Ma, W. G., Song, H., Das, S. K., Paria, B. C., and Dey, S. K. (2003). Estrogen is a critical determinant that specifies the duration of the window of uterine receptivity for implantation. *Proc. Natl. Acad. Sci. U S A* 100, 2963–2968.
112. Song, H., Lim, H., Paria, B. C., Matsumoto, H., Swift, L. L., Morrow, J., Bonventre, J. V., and Dey, S. K. (2002). Cytosolic phospholipase A2alpha is crucial (correction of A2alpha deficiency is crucial) for "on-time" embryo implantation that directs subsequent development. *Development* 129, 2879–2889.
113. Paria, B. C., Das, S. K., Andrews, G. K., and Dey, S. K. (1993). Expression of the epidermal growth factor receptor gene is regulated in mouse blastocysts during delayed implantation. *Proc. Natl. Acad. Sci. U S A* 90, 55–59.
114. Raab, G., Kover, K., Paria, B. C., Dey, S. K., Ezzell, R. M., and Klagsbrun, M. (1996). Mouse preimplantation blastocysts adhere to cells expressing the transmembrane form of heparin-binding EGF-like growth factor. *Development* 122, 637–645.
115. Lim, H., Das, S. K., and Dey, S. K. (1998). erbB genes in the mouse uterus: cell-specific signaling by epidermal growth factor (EGF) family of growth factors during implantation. *Dev. Biol.* 204, 97–110.
116. Zhao, X., Ma, W., Das, S. K., Dey, S. K., and Paria, B. C. (2000). Blastocyst H(2) receptor is the target for uterine histamine in implantation in the mouse. *Development* 127, 2643–2651.
117. Paria, B. C., Song, H., Wang, X., Schmid, P. C., Krebsbach, R. J., Schmid, H. H., Bonner, T. I., Zimmer, A., and Dey, S. K. (2001). Dysregulated cannabinoid signaling disrupts uterine receptivity for embryo implantation. *J. Biol. Chem.* 276, 20523–20528.
118. Paria, B. C., Lim, H., Wang, X. N., Liehr, J., Das, S. K., and Dey, S. K. (1998). Coordination of differential effects of primary estrogen and catecholestrogen on two distinct targets mediates embryo implantation in the mouse. *Endocrinology* 139, 5235–5246.
119. Lim, H., Song, H., Paria, B. C., Reese, J., Das, S. K., and Dey, S. K. (2002). Molecules in blastocyst implantation: uterine and embryonic perspectives. *Vitam. Horm.* 64, 43–76.
120. Norwitz, E. R., Schust, D. J., and Fisher, S. J. (2001). Implantation and the survival of early pregnancy. *N. Engl. J. Med.* 345, 1400–1408.
121. Giudice, L. C. (1999). Potential biochemical markers of uterine receptivity. *Hum. Reprod.* 14 Suppl 2, 3–16.
122. Tan, J., Paria, B. C., Dey, S. K., and Das, S. K. (1999). Differential uterine expression of estrogen and progesterone receptors correlates with uterine preparation for implantation and decidualization in the mouse. *Endocrinology* 140, 5310–5321.
123. Lubahn, D. B., Moyer, J. S., Golding, T. S., Couse, J. F., Korach, K. S., and Smithies, O. (1993). Alteration of reproductive function but not prenatal sexual development after insertional disruption of the mouse estrogen receptor gene. *Proc. Natl. Acad. Sci. U S A* 90, 11162–11166.
124. Paria, B. C., Tan, J., Lubahn, D. B., Dey, S. K., and Das, S. K. (1999). Uterine decidual response occurs in estrogen receptor-alpha-deficient mice. *Endocrinology* 140, 2704–2710.
125. Curtis, S. W., Clark, J., Myers, P., and Korach, K. S. (1999). Disruption of estrogen signaling does not prevent progesterone action in the estrogen receptor alpha knockout mouse uterus. *Proc. Natl. Acad. Sci. U S A* 96, 3646–3651.
126. Das, S. K., Taylor, J. A., Korach, K. S., Paria, B. C., Dey, S. K., and Lubahn, D. B. (1997). Estrogenic responses in estrogen receptor-alpha deficient mice reveal a distinct estrogen signaling pathway. *Proc. Natl. Acad. Sci. U S A* 94, 12786–12791.
127. Hewitt, S. C., Deroo, B. J., Hansen, K., Collins, J., Grissom, S., Afshari, C. A., and Korach, K. S. (2003). Estrogen receptor-dependent genomic responses in the uterus mirror the biphasic physiological response to estrogen. *Mol. Endocrinol.* 17, 2070–2083.
128. Couse, J. F., and Korach, K. S. (1999). Estrogen receptor null mice: what have we learned and where will they lead us? *Endocr. Rev.* 20, 358–417.
129. Couse, J. F., Yates, M. M., Walker, V. R., and Korach, K. S. (2003). Characterization of the hypothalamic-pituitary-gonadal axis in estrogen receptor (ER) null mice reveals hypergonadism and endocrine sex reversal in females lacking ERalpha but not ERbeta. *Mol. Endocrinol.* 17, 1039–1053.
130. Lydon, J. P., DeMayo, F. J., Funk, C. R., Mani, S. K., Hughes, A. R., Montgomery, C. A., Shyamala, G., Conneely, O. M., and Malley, B. W. (1995). Mice lacking progesterone receptor exhibit pleiotropic reproductive abnormalities. *Genes. Dev.* 9, 2266–2278.
131. Mulac, J., Mullinax, R. A., DeMayo, F. J., Lydon, J. P., and Conneely, O. M. (2000). Subgroup of reproductive functions of progesterone mediated by progesterone receptor-B isoform. *Science* 289, 1751–1754.
132. Das, S. K., Tan, J., Raja, S., Halder, J., Paria, B. C., and Dey, S. K. (2000). Estrogen targets genes involved in protein processing, calcium homeostasis, and Wnt signaling in the mouse uterus independent of estrogen receptor-alpha and -beta. *J. Biol. Chem.* 275, 28834–28842.
133. Razandi, M., Pedram, A., Park, S. T., and Levin, E. R. (2003). Proximal events in signaling by plasma membrane estrogen receptors. *J. Biol. Chem.* 278, 2701–2712.
134. Shigeta, H., Zuo, W., Yang, N., DiAugustine, R., and Teng, C. T. (1997). The mouse estrogen receptor-related orphan receptor alpha 1: molecular cloning and estrogen responsiveness. *J. Mol. Endocrinol.* 19, 299–309.
135. Stefano, G. B., Cadet, P., Breton, C., Goumon, Y., Prevot, V., Dessaint, J. P., Beauvillain, J. C., Roumier, A. S., Welters, I., and Salzet, M. (2000). Estradiol-stimulated nitric oxide release in human granulocytes is dependent on intracellular calcium transients: evidence of a cell surface estrogen receptor. *Blood* 95, 3951–3958.
136. Zhu, Y., Rice, C. D., Pang, Y., Pace, M., and Thomas, P. (2003). Cloning, expression, and characterization of a membrane progestin receptor and evidence it is an intermediary in meiotic maturation of fish oocytes. *Proc. Natl. Acad. Sci. U S A* 100, 2231–2236.
137. Zhu, Y., Bond, J., and Thomas, P. (2003). Identification, classification, and partial characterization of genes in humans and other vertebrates homologous to a fish membrane progestin receptor. *Proc. Natl. Acad. Sci. U S A* 100, 2237–2242.
138. Shen, Z., Liu, J., Wells, R. L., and Elkind, M. M. (1994). cDNA cloning, sequence analysis, and induction by aryl

hydrocarbons of a murine cytochrome P450 gene, Cyp1b1. *DNA Cell. Biol.* 13, 763–769.

139. Hou, Q., Paria, B. C., Mui, C., Dey, S. K., and Gorski, J. (1996). Immunolocalization of estrogen receptor protein in the mouse blastocyst during normal and delayed implantation. *Proc. Natl. Acad. Sci. U S A* 93, 2376–2381.
140. Kowalski, A. A., Graddy, L. G., Vale, C., Choi, I., Katzenellenbogen, B. S., Simmen, F. A., and Simmen, R. C. (2002). Molecular cloning of porcine estrogen receptor-beta complementary DNAs and developmental expression in periimplantation embryos. *Biol. Reprod.* 66, 760–769.
141. Hiroi, H., Momoeda, M., Inoue, S., Tsuchiya, F., Matsumi, H., Tsutsumi, O., Muramatsu, M., and Taketani, Y. (1999). Stage-specific expression of estrogen receptor subtypes and estrogen responsive finger protein in preimplantational mouse embryos. *Endocr. J.* 46, 153–158.
142. Kimber, S. J., and Spanswick, C. (2000). Blastocyst implantation: the adhesion cascade. *Semin. Cell. Dev. Biol.* 11, 77–92.
143. Aplin, J. D. (1997). Adhesion molecules in implantation. *Rev. Reprod.* 2, 84–93.
144. Surveyor, G. A., Gendler, S. J., Pemberton, L., Das, S. K., Chakraborty, I., Julian, J., Pimental, R. A., Wegner, C. C., Dey, S. K., and Carson, D. D. (1995). Expression and steroid hormonal control of Muc-1 in the mouse uterus. *Endocrinology* 136, 3639–3647.
145. Hoffman, L. H., Olson, G. E., Carson, D. D., and Chilton, B. S. (1998). Progesterone and implanting blastocysts regulate Muc1 expression in rabbit uterine epithelium. *Endocrinology* 139, 266–271.
146. Meseguer, M., Aplin, J. D., Caballero, C., Connor, J. E., Martin, J. C., Remohi, J., Pellicer, A., and Simon, C. (2001). Human endometrial mucin MUC1 is up-regulated by progesterone and down-regulated in vitro by the human blastocyst. *Biol. Reprod.* 64, 590–601.
147. Giancotti, F. G., and Ruoslahti, E. (1999). Integrin signaling. *Science* 285, 1028–1032.
148. Lessey, B. A., Damjanovich, L., Coutifaris, C., Castelbaum, A., Albelda, S. M., and Buck, C. A. (1992). Integrin adhesion molecules in the human endometrium. Correlation with the normal and abnormal menstrual cycle. *J. Clin. Invest.* 90, 188–195.
149. Tabibzadeh, S. (1992). Patterns of expression of integrin molecules in human endometrium throughout the menstrual cycle. *Hum. Reprod.* 7, 876–882.
150. Lessey, B. A., Castelbaum, A. J., Buck, C. A., Lei, Y., Yowell, C. W., and Sun, J. (1994). Further characterization of endometrial integrins during the menstrual cycle and in pregnancy. *Fertil. Steril.* 62, 497–506.
151. Creus, M., Balasch, J., Ordi, J., Fabregues, F., Casamitjana, R., Quinto, L., Coutifaris, C., and Vanrell, J. A. (1998). Integrin expression in normal and out-of-phase endometria. *Hum. Reprod.* 13, 3460–3468.
152. Thomas, K., Thomson, A. J., Wood, S. J., Kingsland, C. R., Vince, G., and Lewis-Jones, D. I. (2003). Endometrial integrin expression in women undergoing IVF and ICSI: a comparison of the two groups and fertile controls. *Hum. Reprod.* 18, 364–369.
153. Nardo, L. G., Bartoloni, G., Di, M. S., and Nardo, F. (2002). Expression of alpha(v)beta3 and alpha4beta1 integrins throughout the putative window of implantation in a cohort of healthy fertile women. *Acta. Obstet. Gynaecol. Scand.* 81, 753–758.
154. Horne, A. W., White, J. O., and Lalani, E. (2002). Adhesion molecules and the normal endometrium. *Br. J. Obstet. Gynaecol.* 109, 610–617.
155. Wilcox, A. J., Baird, D. D., and Weinberg, C. R. (1999). Time of implantation of the conceptus and loss of pregnancy. *N. Engl. J. Med.* 340, 1796–1799.
156. Klentzeris, L. D., Bulmer, J. N., Trejdosiewicz, L. K., Morrison, L., and Cooke, I. D. (1993). Beta-1 integrin cell adhesion molecules in the endometrium of fertile and infertile women. *Hum. Reprod.* 8, 1223–1230.
157. Lessey, B. A. (1994). The use of integrins for the assessment of uterine receptivity. *Fertil. Steril.* 61, 812–814.
158. Lessey, B. A., Ilesanmi, A. O., Lessey, M. A., Riben, M., Harris, J. E., and Chwalisz, K. (1996). Luminal and glandular endometrial epithelium express integrins differentially throughout the menstrual cycle: implications for implantation, contraception, and infertility. *Am. J. Reprod. Immunol.* 35, 195–204.
159. Lessey, B. A. (2003). Two pathways of progesterone action in the human endometrium: implications for implantation and contraception. *Steroids* 68, 809–815.
160. Nardo, L. G., Nikas, G., Makrigiannakis, A., Sinatra, F., and Nardo, F. (2003). Synchronous expression of pinopodes and alpha v beta 3 and alpha 4 beta 1 integrins in the endometrial surface epithelium of normally menstruating women during the implantation window. *J. Reprod. Med.* 48, 355–361.
161. Illera, M. J., Cullinan, E., Gui, Y., Yuan, L., Beyler, S. A., and Lessey, B. A. (2000). Blockade of the alpha(v)beta(3) integrin adversely affects implantation in the mouse. *Biol. Reprod.* 62, 1285–1290.
162. Illera, M. J., Lorenzo, P. L., Gui, Y. T., Beyler, S. A., Apparao, K. B., and Lessey, B. A. (2003). A role for alphavbeta3 integrin during implantation in the rabbit model. *Biol. Reprod.* 68, 766–771.
163. Fazleabas, A. T., Bell, S. C., Fleming, S., Sun, J., and Lessey, B. A. (1997). Distribution of integrins and the extracellular matrix proteins in the baboon endometrium during the menstrual cycle and early pregnancy. *Biol. Reprod.* 56, 348–356.
164. Homburg, R. (1996). Polycystic ovary syndrome: induction of ovulation. *Baillieres Clin. Endocrinol. Metab.* 10, 281–292.
165. Strandell, A., Lindhard, A., Waldenstrom, U., Thorburn, J., Janson, P. O., and Hamberger, L. (1999). Hydrosalpinx and IVF outcome: a prospective, randomized multicentre trial in Scandinavia on salpingectomy prior to IVF. *Hum. Reprod.* 14, 2762–2769.
166. Kao, L. C., Tulac, S., Lobo, S., Imani, B., Yang, J. P., Germeyer, A., Osteen, K., Taylor, R. N., Lessey, B. A., and Giudice, L. C. (2002). Global gene profiling in human endometrium during the window of implantation. *Endocrinology* 143, 2119–2138.
167. Lessey, B. A., Castelbaum, A. J., Sawin, S. W., and Sun, J. (1995). Integrins as markers of uterine receptivity in women with primary unexplained infertility. *Fertil. Steril.* 63, 535–542.
168. Lessey, B. A., Castelbaum, A. J., Sawin, S. W., Buck, C. A., Schinnar, R., Bilker, W., and Strom, B. L. (1994). Aberrant integrin expression in the endometrium of women with endometriosis. *J. Clin. Endocrinol. Metab.* 79, 643–649.
169. Sutherland, A. E., Calarco, P. G., and Damsky, C. H. (1993). Developmental regulation of integrin expression at the time of implantation in the mouse embryo. *Development* 119, 1175–1186.
170. Schultz, J. F., and Armant, D. R. (1995). Beta 1- and beta 3-class integrins mediate fibronectin binding activity at the surface of developing mouse peri-implantation blastocysts. Regulation by ligand-induced mobilization of stored receptor. *J. Biol. Chem.* 270, 11522–11531.
171. Stephens, L. E., Sutherland, A. E., Klimanskaya, I. V., Andrieux, A., Meneses, J., Pedersen, R. A., and Damsky, C. H. (1995). Deletion of beta 1 integrins in mice results in inner cell mass failure and peri-implantation lethality. *Genes. Dev.* 9, 1883–1895.

172. Wang, J., Mayernik, L., and Armant, D. R. (2002). Integrin signaling regulates blastocyst adhesion to fibronectin at implantation: intracellular calcium transients and vesicle trafficking in primary trophoblast cells. *Dev. Biol.* 245, 270–279.
173. Armant, D. R., Wang, J., and Liu, Z. (2000). Intracellular signaling in the developing blastocyst as a consequence of the maternal-embryonic dialogue. *Semin. Reprod. Med.* 18, 273–287.
174. Wang, J., and Armant, D. R. (2002). Integrin-mediated adhesion and signaling during blastocyst implantation. *Cell. Tiss. Org.* 172, 190–201.
175. Fassler, R., and Meyer, M. (1995). Consequences of lack of beta 1 integrin gene expression in mice. *Genes Dev.* 9, 1896–1908.
176. Hynes, R. O. (1996). Targeted mutations in cell adhesion genes: what have we learned from them? *Dev. Biol.* 180, 402–412.
177. Kreidberg, J. A., Donovan, M. J., Goldstein, S. L., Rennke, H., Shepherd, K., Jones, R. C., and Jaenisch, R. (1996). Alpha 3 beta 1 integrin has a crucial role in kidney and lung organogenesis. *Development* 122, 3537–3547.
178. Gardner, H., Kreidberg, J., Koteliansky, V., and Jaenisch, R. (1996). Deletion of integrin alpha 1 by homologous recombination permits normal murine development but gives rise to a specific deficit in cell adhesion. *Dev. Biol.* 175, 301–313.
179. Armant, D. R., Kaplan, H. A., and Lennarz, W. J. (1986). Fibronectin and laminin promote in vitro attachment and outgrowth of mouse blastocysts. *Dev. Biol.* 116, 519–523.
180. Carson, D. D., Tang, J. P., and Gay, S. (1988). Collagens support embryo attachment and outgrowth in vitro: effects of the Arg-Gly-Asp sequence. *Dev. Biol.* 127, 368–375.
181. Sutherland, A. E., Calarco, P. G., and Damsky, C. H. (1988). Expression and function of cell surface extracellular matrix receptors in mouse blastocyst attachment and outgrowth. *J. Cell. Biol.* 106, 1331–1348.
182. Yelian, F. D., Edgeworth, N. A., Dong, L. J., Chung, A. E., and Armant, D. R. (1993). Recombinant entactin promotes mouse primary trophoblast cell adhesion and migration through the Arg-Gly-Asp (RGD) recognition sequence. *J. Cell. Biol.* 121, 923–929.
183. Wordinger, R. J., Jackson, F. L., and Morrill, A. (1986). Implantation, deciduoma formation and live births in mast cell-deficient mice (W/Wv). *J. Reprod. Fertil.* 77, 471–476.
184. Armant, D. R., and Kameda, S. (1994). Mouse trophoblast cell invasion of extracellular matrix purified from endometrial tissue: a model for peri-implantation development. *J. Exp. Zool.* 269, 146–156.
185. Wartiovaara, J., Leivo, I., and Vaheri, A. (1979). Expression of the cell surface-associated glycoprotein, fibronectin, in the early mouse embryo. *Dev. Biol.* 69, 247–257.
186. Wewer, U. M., Liotta, L. A., Jaye, M., Ricca, G. A., Drohan, W. N., Claysmith, A. P., Rao, C. N., Wirth, P., Coligan, J. E., and Albrechtsen, R. (1986). Altered levels of laminin receptor mRNA in various human carcinoma cells that have different abilities to bind laminin. *Proc. Natl. Acad. Sci. U S A* 83, 7137–7141.
187. Blankenship, T. N., Enders, A. C., and King, B. F. (1992). Distribution of laminin, type IV collagen, and fibronectin in the cell columns and trophoblastic shell of early macaque placentas. *Cell. Tiss. Res.* 270, 241–248.
188. Armant, D. R., Kaplan, H. A., Mover, H., and Lennarz, W. J. (1986). The effect of hexapeptides on attachment and outgrowth of mouse blastocysts cultured in vitro: evidence for the involvement of the cell recognition tripeptide Arg-Gly-Asp. *Proc. Natl. Acad. Sci. U S A* 83, 6751–6755.
189. Hynes, R. O. (1992). Specificity of cell adhesion in development: the cadherin superfamily. *Curr. Opin. Genet. Dev.* 2, 621–624.
190. Armant, D. R. (1991). Cell interactions with laminin and its proteolytic fragments during outgrowth of mouse primary trophoblast cells. *Biol. Reprod.* 45, 664–672.
191. Klaffky, E., Williams, R., Yao, C. C., Ziober, B., Kramer, R., and Sutherland, A. (2001). Trophoblast-specific expression and function of the integrin alpha 7 subunit in the peri-implantation mouse embryo. *Dev. Biol.* 239, 161–175.
192. Johnson, G. A., Burghardt, R. C., Bazer, F. W., and Spencer, T. E. (2003). Osteopontin: roles in implantation and placentation. *Biol. Reprod.* 69, 1458–1471.
193. Johnson, G. A., Burghardt, R. C., Spencer, T. E., Newton, G. R., Ott, T. L., and Bazer, F. W. (1999). Ovine osteopontin. II. Osteopontin and alpha(v)beta(3) integrin expression in the uterus and conceptus during the periimplantation period. *Biol. Reprod.* 61, 892–899.
194. Garlow, J. E., Ka, H., Johnson, G. A., Burghardt, R. C., Jaeger, L. A., and Bazer, F. W. (2002). Analysis of osteopontin at the maternal-placental interface in pigs. *Biol. Reprod.* 66, 718–725.
195. Apparao, K. B., Murray, M. J., Fritz, M. A., Meyer, W. R., Chambers, A. F., Truong, P. R., and Lessey, B. A. (2001). Osteopontin and its receptor alphavbeta(3) integrin are coexpressed in the human endometrium during the menstrual cycle but regulated differentially. *J. Clin. Endocrinol. Metab.* 86, 4991–5000.
196. Nomura, S., Wills, A. J., Edwards, D. R., Heath, J. K., and Hogan, B. L. (1988). Developmental expression of 2ar (osteopontin) and SPARC (osteonectin) RNA as revealed by in situ hybridization. *J. Cell. Biol.* 106, 441–450.
197. Liaw, L., Birk, D. E., Ballas, C. B., Whitsitt, J. S., Davidson, J. M., and Hogan, B. L. (1998). Altered wound healing in mice lacking a functional osteopontin gene (spp1). *J. Clin. Invest.* 101, 1468–1478.
198. Fukuda, M. N., Sato, T., Nakayama, J., Klier, G., Mikami, M., Aoki, D., and Nozawa, S. (1995). Trophinin and tastin, a novel cell adhesion molecule complex with potential involvement in embryo implantation. *Genes. Dev.* 9, 1199–1210.
199. Suzuki, N., Nakayama, J., Shih, I. M., Aoki, D., Nozawa, S., and Fukuda, M. N. (1999). Expression of trophinin, tastin, and bystin by trophoblast and endometrial cells in human placenta. *Biol. Reprod.* 60, 621–627.
200. Nakayama, J., Aoki, D., Suga, T., Akama, T. O., Ishizone, S., Yamaguchi, H., Imakawa, K., Nadano, D., Fazleabas, A. T., Katsuyama, T., Nozawa, S., and Fukuda, M. N. (2003). Implantation-dependent expression of trophinin by maternal fallopian tube epithelia during tubal pregnancies: possible role of human chorionic gonadotrophin on ectopic pregnancy. *Am. J. Pathol.* 163, 2211–2219.
201. Nadano, D., Sugihara, K., Paria, B. C., Saburi, S., Copeland, N. G., Gilbert, D. J., Jenkins, N. A., Nakayama, J., and Fukuda, M. N. (2002). Significant differences between mouse and human trophinins are revealed by their expression patterns and targeted disruption of mouse trophinin gene. *Biol. Reprod.* 66, 313–321.
202. Ozawa, M., Engel, J., and Kemler, R. (1990). Single amino acid substitutions in one Ca^{2+} binding site of uvomorulin abolish the adhesive function. *Cell* 63, 1033–1038.
203. Ozawa, M., and Kemler, R. (1992). Molecular organization of the uvomorulin-catenin complex. *J. Cell. Biol.* 116, 989–996.
204. Larue, L., Ohsugi, M., Hirchenhain, J., and Kemler, R. (1994). E-cadherin null mutant embryos fail to form a trophectoderm epithelium. *Proc. Natl. Acad. Sci. U S A* 91, 8263–8267.

205. Riethmacher, D., Brinkmann, V., and Birchmeier, C. (1995). A targeted mutation in the mouse E-cadherin gene results in defective preimplantation development. *Proc. Natl. Acad. Sci. U S A* 92, 855–859.
206. Paria, B. C., Zhao, X., Das, S. K., Dey, S. K., and Yoshinaga, K. (1999). Zonula occludens-1 and E-cadherin are coordinately expressed in the mouse uterus with the initiation of implantation and decidualization. *Dev. Biol.* 208, 488–501.
207. Wang, X., Matsumoto, H., Zhao, X., Das, S. K., and Paria, B. C. (2004). Embryonic signals direct formation of tight junctional permeability barrier in the decidualizing stroma during embryo implantation. *J. Cell. Sci.* 117, 53–62.
208. Genbacev, O. D., Prakobphol, A., Foulk, R. A., Krtolica, A. R., Ilic, D., Singer, M. S., Yang, Z. Q., Kiessling, L. L., Rosen, S. D., and Fisher, S. J. (2003). Trophoblast L-selectin-mediated adhesion at the maternal-fetal interface. *Science* 299, 405–408.
209. Noguchi, Y., Sato, T., Hirata, M., Hara, T., Ohama, K., and Ito, A. (2003). Identification and characterization of extracellular matrix metalloproteinase inducer in human endometrium during the menstrual cycle in vivo and in vitro. *J. Clin. Endocrinol. Metab.* 88, 6063–6072.
210. Xiao, L. J., Chang, H., Ding, N. Z., Ni, H., Kadomatsu, K., and Yang, Z. M. (2002). Basigin expression and hormonal regulation in mouse uterus during the peri-implantation period. *Mol. Reprod. Dev.* 63, 47–54.
211. Igakura, T., Kadomatsu, K., Kaname, T., Muramatsu, H., Fan, Q. W., Miyauchi, T., Toyama, Y., Kuno, N., Yuasa, S., Takahashi, M., Senda, T., Taguchi, O., Yamamura, K., Arimura, K., and Muramatsu, T. (1998). A null mutation in basigin, an immunoglobulin superfamily member, indicates its important roles in peri-implantation development and spermatogenesis. *Dev. Biol.* 194, 152–165.
212. Hakanson, R. (1967). Kinetic properties of mammalian histidine decarboxylase. *Eur. J. Pharmacol.* 1, 42–46.
213. Schwartz, J. C. (1977). Histaminergic mechanisms in brain. *Annu. Rev. Pharmacol. Toxicol.* 17, 325–339.
214. Marcus, G. J., and Shelesnyak, M. C. (1968). Studies on the mechanism of nidation. 33. Coital elevation of uterine histamine content. *Acta. Endocrinol. (Copenh.)* 57, 136–141.
215. Shelesnyak, M. C. (1957). Some experimental studies on the mechanism of ova-implantation in the rat. *Recent Prog. Hormone. Res.* 13, 269–322.
216. Padilla, L., Reinicke, K., Montesino, H., Villena, F., Asencio, H., Cruz, M., and Rudolph, M. I. (1990). Histamine content and mast cells distribution in mouse uterus: the effect of sexual hormones, gestation and labor. *Cell. Mol. Biol.* 36, 93–100.
217. Spaziani, E. (1963). Relationship between early vascular responses and growth in the rat uterus: stimulation of cell division by estradiol and vasodilating amines. *Endocrinology* 72, 180–188.
218. Marcus, G. J., Kracier, P. F., and Shelesnyak, M. C. (1963). Studies on the mechanism of decidualization. II. The histamine releasing of pyrathiazine. *J. Reprod. Fertil.* 5, 409–415.
219. Dey, S. K., Villanueva, C., Chien, S. M., and Crist, R. D. (1978). The role of histamine in implantation in the rabbit. *J. Reprod. Fertil.* 53, 23–26.
220. Inoue, I., Yanai, K., Kitamura, D., Taniuchi, I., Kobayashi, T., Niimura, K., Watanabe, T., and Watanabe, T. (1996). Impaired locomotor activity and exploratory behavior in mice lacking histamine H1 receptors. *Proc. Natl. Acad. Sci. U S A* 93, 13316–13320.
221. Kobayashi, T., Tonai, S., Ishihara, Y., Koga, R., Okabe, S., and Watanabe, T. (2000). Abnormal functional and morphological regulation of the gastric mucosa in histamine H2 receptor-deficient mice. *J. Clin. Invest.* 105, 1741–1749.
222. Toyota, H., Dugovic, C., Koehl, M., Laposky, A. D., Weber, C., Ngo, K., Wu, Y., Lee, D. H., Yanai, K., Sakurai, E., Watanabe, T., Liu, C., Chen, J., Barbier, A. J., Turek, F. W., Fung, L., and Lovenberg, T. W. (2002). Behavioral characterization of mice lacking histamine H(3) receptors. *Mol. Pharmacol.* 62, 389–397.
223. Brandon, J. M., and Wallis, R. M. (1977). Effect of mepyramine, a histamine H1-, and burimamide, a histamine H2-receptor antagonist, on ovum implantation in the rat. *J. Reprod. Fertil.* 50, 251–254.
224. Johnson, D. C., and Dey, S. K. (1980). Role of histamine in implantation: dexamethasone inhibits estradiol-induced implantation in the rat. *Biol. Reprod.* 22, 1136–1141.
225. Brandon, J. M., and Bibby, M. C. (1979). A study of changes in uterine mast cells during early pregnancy in the rat. *Biol. Reprod.* 20, 977–980.
226. Salamonsen, L. A., Jeziorska, M., Newlands, G. F., Dey, S. K., and Woolley, D. E. (1996). Evidence against a significant role for mast cells in blastocyst implantation in the rat and mouse. *Reprod. Fertil. Dev.* 8, 1157–1164.
227. Paria, B. C., Das, N., Das, S. K., Zhao, X., Dileepan, K. N., and Dey, S. K. (1998). Histidine decarboxylase gene in the mouse uterus is regulated by progesterone and correlates with uterine differentiation for blastocyst implantation. *Endocrinology* 139, 3958–3966.
228. Ohtsu, H., Tanaka, S., Terui, T., Hori, Y., Makabe, K., Pejler, G., Tchougounova, E., Hellman, L., Gertsenstein, M., Hirasawa, N., Sakurai, E., Buzas, E., Kovacs, P., Csaba, G., Kittel, A., Okada, M., Hara, M., Mar, L., Numayama, T., Ishigaki, S., Ohuchi, K., Ichikawa, A., Falus, A., Watanabe, T., and Nagy, A. (2001). Mice lacking histidine decarboxylase exhibit abnormal mast cells. *FEBS Lett.* 502, 53–56.
229. Smith, W. L., and Dewitt, D. L. (1996). Prostaglandin endoperoxide H synthases-1 and -2. *Adv. Immunol.* 62, 167–215.
230. Kennedy, T. G. (1979). Prostaglandins and increased endometrial vascular permeabiltiy resulting from the application of artificial stimulus to the uterus of the rat sensitized for the decidual cell reaction. *Biol. Reprod.* 20, 560–566.
231. Kennedy, T. G. (1980). Timing of uterine sensitivity for the decidual cell reaction: role of prostaglandins. *Biol. Reprod.* 22, 519–525.
232. Kennedy, T. G. (1980). Estrogen and uterine sensitization for the decidual cell reaction: role of prostaglandins. *Biol. Reprod.* 23, 955–962.
233. Kennedy, T. G. (1985). Evidence for the involvement of prostaglandins throughout the decidual cell reaction in the rat. *Biol. Reprod.* 33, 140–146.
234. Doktorcik, P. E., and Kennedy, T. G. (1986). 6-Keto-prostaglandin E1 and the decidual cell reaction in rats. *Prostaglandins* 32, 679–689.
235. Kennedy, T. G., and Doktorcik, P. E. (1988). Uterine decidualization in hypophysectomized-ovariectomized rats: effects of pituitary hormones. *Biol. Reprod.* 39, 318–328.
236. Hamilton, G. S., and Kennedy, T. G. (1994). Uterine vascular changes after unilateral intrauterine infusion of indomethacin and prostaglandin E2 to rats sensitized for the decidual cell reaction. *Biol. Reprod.* 50, 757–764.
237. Hamilton, G. S., and Kennedy, T. G. (1994). Uterine vascular permeability after uterine stimulation to rats differentially sensitized for the decidual cell reaction. *Can. J. Physiol. Pharmacol.* 72, 711–715.
238. Kennedy, T. G., and Lukash, L. A. (1982). Induction of decidualization in rats by the intrauterine infusion of prostaglandins. *Biol. Reprod.* 27, 253–260.
239. Keys, J. L., and Kennedy, T. G. (1990). Effect of indomethacin and prostaglandin E2 on structural differentiation of rat

endometrium during artificially induced decidualization. *Am. J. Anat.* 188, 148–162.
240. Forman, B. M., Chen, J., and Evans, R. M. (1997). Hypolipidemic drugs, polyunsaturated fatty acids, and eicosanoids are ligands for peroxisome proliferator-activated receptors alpha and delta. *Proc. Natl. Acad. Sci. U S A* 94, 4312–4317.
241. Forman, B. M., Tontonoz, P., Chen, J., Brun, R. P., Spiegelman, B. M., and Evans, R. M. (1995). 15-Deoxy-delta 12, 14-prostaglandin J2 is a ligand for the adipocyte determination factor PPAR gamma. *Cell* 83, 803–812.
242. Kliewer, S. A., Umesono, K., Noonan, D. J., Heyman, R. A., and Evans, R. M. (1992). Convergence of 9-cis retinoic acid and peroxisome proliferator signalling pathways through heterodimer formation of their receptors. *Nature* 358, 771–774.
243. Kliewer, S. A., Forman, B. M., Blumberg, B., Ong, E. S., Borgmeyer, U., Mangelsdorf, D. J., Umesono, K., and Evans, R. M. (1994). Differential expression and activation of a family of murine peroxisome proliferator-activated receptors. *Proc. Natl. Acad. Sci. U S A* 91, 7355–7359.
244. Copeland, R. A., Williams, J. M., Giannaras, J., Nurnberg, S., Covington, M., Pinto, D., Pick, S., and Trzaskos, J. M. (1994). Mechanism of selective inhibition of the inducible isoform of prostaglandin G/H synthase. *Proc. Natl. Acad. Sci. U S A* 91, 11202–11206.
245. Kurumbail, R. G., Stevens, A. M., Gierse, J. K., McDonald, J. J., Stegeman, R. A., Pak, J. Y., Gildehaus, D., Miyashiro, J. M., Penning, T. D., Seibert, K., Isakson, P. C., and Stallings, W. C. (1996). Structural basis for selective inhibition of cyclooxygenase-2 by anti-inflammatory agents. *Nature* 384, 644–648.
246. DuBois, R. N., Radhika, A., Reddy, B. S., and Entingh, A. J. (1996). Increased cyclooxygenase-2 levels in carcinogen-induced rat colonic tumors. *Gastroenterology* 110, 1259–1262.
247. Oshima, M., Dinchuk, J. E., Kargman, S. L., Oshima, H., Hancock, B., Kwong, E., Trzaskos, J. M., Evans, J. F., and Taketo, M. M. (1996). Suppression of intestinal polyposis in Apc delta716 knockout mice by inhibition of cyclooxygenase 2 (COX-2). *Cell* 87, 803–809.
248. Espey, L. L. (1994). Current status of the hypothesis that mammalian ovulation is comparable to an inflammatory reaction. *Biol. Reprod.* 50, 233–238.
249. McMaster, M. T., Dey, S. K., and Andrews, G. K. (1993). Association of monocytes and neutrophils with early events of blastocyst implantation in mice. *J. Reprod. Fertil.* 99, 561–569.
250. Sirois, J. (1994). Induction of prostaglandin endoperoxide synthase-2 by human chorionic gonadotropin in bovine preovulatory follicles in vivo. *Endocrinology* 135, 841–848.
251. Sirois, J., Simmons, D. L., and Richards, J. S. (1992). Hormonal regulation of messenger ribonucleic acid encoding a novel isoform of prostaglandin endoperoxide H synthase in rat preovulatory follicles. Induction in vivo and in vitro. *J. Biol. Chem.* 267, 11586–11592.
252. Chakraborty, I., Das, S. K., Wang, J., and Dey, S. K. (1996). Developmental expression of the cyclo-oxygenase-1 and cyclo-oxygenase-2 genes in the peri-implantation mouse uterus and their differential regulation by the blastocyst and ovarian steroids. *J. Mol. Endocrinol.* 16, 107–122.
253. Dinchuk, J. E., Car, B. D., Focht, R. J., Johnston, J. J., Jaffee, B. D., Covington, M. B., Contel, N. R., Eng, V. M., Collins, R. J., and Czerniak, P. M. (1995). Renal abnormalities and an altered inflammatory response in mice lacking cyclooxygenase II. *Nature* 378, 406–409.
254. Langenbach, R., Morham, S. G., Tiano, H. F., Loftin, C. D., Ghanayem, B. I., Chulada, P. C., Mahler, J. F., Lee, C. A., Goulding, E. H., and Kluckman, K. D. (1995). Prostaglandin synthase 1 gene disruption in mice reduces arachidonic acid-induced inflammation and indomethacin-induced gastric ulceration. *Cell* 83, 483–492.
255. Lim, H., Paria, B. C., Das, S. K., Dinchuk, J. E., Langenbach, R., Trzaskos, J. M., and Dey, S. K. (1997). Multiple female reproductive failures in cyclooxygenase 2-deficient mice. *Cell* 91, 197–208.
256. Reese, J., Brown, N., Paria, B. C., Morrow, J., and Dey, S. K. (1999). COX-2 compensation in the uterus of COX-1 deficient mice during the pre-implantation period. *Mol. Cell. Endocrinol.* 150, 23–31.
257. Lim, H., Gupta, R. A., Ma, W. G., Paria, B. C., Moller, D. E., Morrow, J. D., DuBois, R. N., Trzaskos, J. M., and Dey, S. K. (1999). Cyclo-oxygenase-2-derived prostacyclin mediates embryo implantation in the mouse via PPARdelta. *Genes Dev.* 13, 1561–1574.
258. Clark, J. D., Schievella, A. R., Nalefski, E. A., and Lin, L. L. (1995). Cytosolic phospholipase A2. *J. Lipid. Mediat. Cell. Signal.* 12, 83–117.
259. Bonventre, J. V., Huang, Z., Taheri, M. R., Leary, E., Li, E., Moskowitz, M. A., and Sapirstein, A. (1997). Reduced fertility and postischaemic brain injury in mice deficient in cytosolic phospholipase A2. *Nature* 390, 622–625.
260. Uozumi, N., Kume, K., Nagase, T., Nakatani, N., Ishii, S., Tashiro, F., Komagata, Y., Maki, K., Ikuta, K., Ouchi, Y., Miyazaki, J., and Shimizu, T. (1997). Role of cytosolic phospholipase A2 in allergic response and parturition. *Nature* 390, 618–622.
261. Charpigny, G., Reinaud, P., Tamby, J. P., Creminon, C., and Guillomot, M. (1997). Cyclooxygenase-2 unlike cyclooxygenase-1 is highly expressed in ovine embryos during the implantation period. *Biol. Reprod.* 57, 1032–1040.
262. Kim, J. J., Wang, J., Bambra, C., Das, S. K., Dey, S. K., and Fazleabas, A. T. (1999). Expression of cyclooxygenase-1 and -2 in the baboon endometrium during the menstrual cycle and pregnancy. *Endocrinology* 140, 2672–2678.
263. Critchley, H. O., Jones, R. L., Lea, R. G., Drudy, T. A., Kelly, R. W., Williams, A. R., and Baird, D. T. (1999). Role of inflammatory mediators in human endometrium during progesterone withdrawal and early pregnancy. *J. Clin. Endocrinol. Metab.* 84, 240–248.
264. Marions, L., and Danielsson, K. G. (1999). Expression of cyclo-oxygenase in human endometrium during the implantation period. *Mol. Hum. Reprod.* 5, 961–965.
265. Wang, H., Ma, W. G., Tejada, L., Zhang, H., Morrow, J. D., Das, S. K., and Dey, S. K. (2004). Rescue of female infertility from the loss of cyclooxygenase-2 by compensatory up-regulation of cyclooxygenase-1 is a function of genetic makeup. *J. Biol. Chem.* 279, 10649–10658.
266. Lim, H., and Dey, S. K. (2002). A novel pathway of prostacyclin signaling-hanging out with nuclear receptors. *Endocrinology* 143, 3207–3210.
267. Yang, Z. M., Das, S. K., Wang, J., Sugimoto, Y., Ichikawa, A., and Dey, S. K. (1997). Potential sites of prostaglandin actions in the periimplantation mouse uterus: differential expression and regulation of prostaglandin receptor genes. *Biol. Reprod.* 56, 368–379.
268. Ni, H., Sun, T., Ma, X. H., and Yang, Z. M. (2003). Expression and regulation of cytosolic prostaglandin e synthase in mouse uterus during the peri-implantation period. *Biol. Reprod.* 68, 744–750.
269. Ni, H., Sun, T., Ding, N. Z., Ma, X. H., and Yang, Z. M. (2002). Differential expression of microsomal prostaglandin e synthase at implantation sites and in decidual cells of mouse uterus. *Biol. Reprod.* 67, 351–358.
270. Racowsky, C., and Biggers, J. D. (1983). Are blastocyst prostaglandins produced endogenously? *Biol. Reprod.* 29, 379–388.

271. Reese, J., Paria, B. C., Brown, N., Zhao, X., Morrow, J. D., and Dey, S. K. (2000). Coordinated regulation of fetal and maternal prostaglandins directs successful birth and postnatal adaptation in the mouse. *Proc. Natl. Acad. Sci. U S A* 97, 9759–9764.
272. Cross, J. C., Werb, Z., and Fisher, S. J. (1994). Implantation and the placenta: key pieces of the development puzzle. *Science* 266, 1508–1518.
273. Tazuke, S. I., and Giudice, L. C. (1996). Growth factors and cytokines in endometrium, embryonic development, and maternal: embryonic interactions. *Semin. Reprod. Endocrinol.* 14, 231–245.
274. Das, S. K., Das, N., Wang, J., Lim, H., Schryver, B., Plowman, G. D., and Dey, S. K. (1997). Expression of betacellulin and epiregulin genes in the mouse uterus temporally by the blastocyst solely at the site of its apposition is coincident with the "window" of implantation. *Dev. Biol.* 190, 178–190.
275. Das, S. K., Chakraborty, I., Paria, B. C., Wang, X. N., Plowman, G., and Dey, S. K. (1995). Amphiregulin is an implantation-specific and progesterone-regulated gene in the mouse uterus. *Mol. Endocrinol.* 9, 691–705.
276. Luetteke, N. C., Qiu, T. H., Peiffer, R. L., Oliver, P., Smithies, O., and Lee, D. C. (1993). TGF alpha deficiency results in hair follicle and eye abnormalities in targeted and waved-1 mice. *Cell.* 73, 263–278.
277. Luetteke, N. C., Qiu, T. H., Fenton, S. E., Troyer, K. L., Riedel, R. F., Chang, A., and Lee, D. C. (1999). Targeted inactivation of the EGF and amphiregulin genes reveals distinct roles for EGF receptor ligands in mouse mammary gland development. *Development* 126, 2739–2750.
278. Olayioye, M. A., Neve, R. M., Lane, H. A., and Hynes, N. E. (2000). The ErbB signaling network: receptor heterodimerization in development and cancer. *EMBO J.* 19, 3159–3167.
279. Harvey, M. B., Leco, K. J., Arcellana, P., Zhang, X., Edwards, D. R., and Schultz, G. A. (1995). Roles of growth factors during peri-implantation development. *Hum. Reprod.* 10, 712–718.
280. Rappolee, D. A., Brenner, C. A., Schultz, R., Mark, D., and Werb, Z. (1988). Developmental expression of PDGF, TGF-alpha, and TGF-beta genes in preimplantation mouse embryos. *Science* 241, 1823–1825.
281. Paria, B. C., Elenius, K., Klagsbrun, M., and Dey, S. K. (1999). Heparin-binding EGF-like growth factor interacts with mouse blastocysts independently of ErbB1: a possible role for heparan sulfate proteoglycans and ErbB4 in blastocyst implantation. *Development* 126, 1997–2005.
282. Paria, B. C., and Dey, S. K. (1990). Preimplantation embryo development in vitro: cooperative interactions among embryos and role of growth factors. *Proc. Natl. Acad. Sci. U S A* 87, 4756–4760.
283. Das, S. K., Tsukamura, H., Paria, B. C., Andrews, G. K., and Dey, S. K. (1994). Differential expression of epidermal growth factor receptor (EGF-R) gene and regulation of EGF-R bioactivity by progesterone and estrogen in the adult mouse uterus. *Endocrinology* 134, 971–981.
284. Iwamoto, R., Yamazaki, S., Asakura, M., Takashima, S., Hasuwa, H., Miyado, K., Adachi, S., Kitakaze, M., Hashimoto, K., Raab, G., Nanba, D., Higashiyama, S., Hori, M., Klagsbrun, M., and Mekada, E. (2003). Heparin-binding EGF-like growth factor and ErbB signaling is essential for heart function. *Proc. Natl. Acad. Sci. U S A* 100, 3221–3226.
285. Gassmann, M., Casagranda, F., Orioli, D., Simon, H., Lai, C., Klein, R., and Lemke, G. (1995). Aberrant neural and cardiac development in mice lacking the ErbB4 neuregulin receptor. *Nature* 378, 390–394.
286. Threadgill, D. W., Dlugosz, A. A., Hansen, L. A., Tennenbaum, T., Lichti, U., Yee, D., LaMantia, C., Mourton, T., Herrup, K., and Harris, R. C. (1995). Targeted disruption of mouse EGF receptor: effect of genetic background on mutant phenotype. *Science* 269, 230–234.
287. Yoo, H. J., Barlow, D. H., and Mardon, H. J. (1997). Temporal and spatial regulation of expression of heparin-binding epidermal growth factor-like growth factor in the human endometrium: a possible role in blastocyst implantation. *Dev. Genet.* 21, 102–108.
288. Leach, R. E., Khalifa, R., Ramirez, N. D., Das, S. K., Wang, J., Dey, S. K., Romero, R., and Armant, D. R. (1999). Multiple roles for heparin-binding epidermal growth factor-like growth factor are suggested by its cell-specific expression during the human endometrial cycle and early placentation. *J. Clin. Endocrinol. Metab.* 84, 3355–3363.
289. Chobotova, K., Spyropoulou, I., Carver, J., Manek, S., Heath, J. K., Gullick, W. J., Barlow, D. H., Sargent, I. L., and Mardon, H. J. (2002). Heparin-binding epidermal growth factor and its receptor ErbB4 mediate implantation of the human blastocyst. *Mech. Dev.* 119, 137–144.
290. Martin, K. L., Barlow, D. H., and Sargent, I. L. (1998). Heparin-binding epidermal growth factor significantly improves human blastocyst development and hatching in serum-free medium. *Hum. Reprod.* 13, 1645–1652.
291. Tamada, H., McMaster, M. T., Flanders, K. C., Andrews, G. K., and Dey, S. K. (1990). Cell type-specific expression of transforming growth factor-beta 1 in the mouse uterus during the periimplantation period. *Mol. Endocrinol.* 4, 965–972.
292. Das, S. K., Flanders, K. C., Andrews, G. K., and Dey, S. K. (1992). Expression of transforming growth factor-beta isoforms (beta 2 and beta 3) in the mouse uterus: analysis of the periimplantation period and effects of ovarian steroids. *Endocrinology* 130, 3459–3466.
293. Das, S. K., Lim, H., Wang, J., Paria, B. C., BazDresch, M., and Dey, S. K. (1997). Inappropriate expression of human transforming growth factor (TGF)-alpha in the uterus of transgenic mouse causes downregulation of TGF-beta receptors and delays the blastocyst-attachment reaction. *J. Mol. Endocrinol.* 18, 243–257.
294. Robertson, S. A., Ingman, W. V., Leary, S., Sharkey, D. J., and Tremellen, K. P. Transforming growth factor beta—a mediator of immune deviation in seminal plasma. *J. Reprod. Immunol.* 57, 109–128.
295. Robertson, S. A., and Sharkey, D. J. (2001). The role of semen in induction of maternal immune tolerance to pregnancy. *Semin. Immunol.* 13, 243–254.
296. Tremellen, K. P., Seamark, R. F., and Robertson, S. A. (1998). Seminal transforming growth factor beta1 stimulates granulocyte-macrophage colony-stimulating factor production and inflammatory cell recruitment in the murine uterus. *Biol. Reprod.* 58, 1217–1225.
297. Stewart, C. L., and Cullinan, E. B. (1997). Preimplantation development of the mammalian embryo and its regulation by growth factors. *Dev. Genet.* 21, 91–101.
298. Chard, T. (1995). Cytokines in implantation. *Hum. Reprod. Update* 1, 385–396.
299. Pollard, J. W., Hunt, J. S., Wiktor, J., and Stanley, E. R. (1991). A pregnancy defect in the osteopetrotic (op/op) mouse demonstrates the requirement for CSF-1 in female fertility. *Dev. Biol.* 148, 273–283.
300. Stewart, C. L., Kaspar, P., Brunet, L. J., Bhatt, H., Gadi, I., Kontgen, F., and Abbondanzo, S. J. (1992). Blastocyst implantation depends on maternal expression of leukaemia inhibitory factor. *Nature* 359, 76–79.
301. Robb, L., Li, R., Hartley, L., Nandurkar, H. H., Koentgen, F., and Begley, C. G. (1998). Infertility in female mice lacking the receptor for interleukin 11 is due to a defective uterine response to implantation. *Nat. Med.* 4, 303–308.

302. Escary, J. L., Perreau, J., Dumenil, D., Ezine, S., and Brulet, P. (1993). Leukaemia inhibitory factor is necessary for maintenance of haematopoietic stem cells and thymocyte stimulation. *Nature* 363, 361–364.
303. Simon, C., Mercader, A., Gimeno, M. J., and Pellicer, A. (1997). The interleukin-1 system and human implantation. *Am. J. Reprod. Immunol.* 37, 64–72.
304. Strakova, Z., Srisuparp, S., and Fazleabas, A. T. (2000). Interleukin-1beta induces the expression of insulin-like growth factor binding protein-1 during decidualization in the primate. *Endocrinology* 141, 4664–4670.
305. Kishimoto, T., Taga, T., and Akira, S. (1994). Cytokine signal transduction. *Cell* 76, 253–262.
306. Song, H., Lim, H., Das, S. K., Paria, B. C., and Dey, S. K. (2000). Dysregulation of EGF family of growth factors and COX-2 in the uterus during the preattachment and attachment reactions of the blastocyst with the luminal epithelium correlates with implantation failure in LIF-deficient mice. *Mol. Endocrinol.* 14, 1147–1161.
307. Rathjen, P. D., Toth, S., Willis, A., Heath, J. K., and Smith, A. G. (1990). Differentiation inhibiting activity is produced in matrix-associated and diffusible forms that are generated by alternate promoter usage. *Cell* 62, 1105–1114.
308. Narazaki, M., Witthuhn, B. A., Yoshida, K., Silvennoinen, O., Yasukawa, K., Ihle, J. N., Kishimoto, T., and Taga, T. (1994). Activation of JAK2 kinase mediated by the interleukin 6 signal transducer gp130. *Proc. Natl. Acad. Sci. U S A* 91, 2285–2289.
309. Layton, M. J., Lock, P., Metcalf, D., and Nicola, N. A. (1994). Cross-species receptor binding characteristics of human and mouse leukemia inhibitory factor suggest a complex binding interaction. *J. Biol. Chem.* 269, 17048–17055.
310. Bhatt, H., Brunet, L. J., and Stewart, C. L. (1991). Uterine expression of leukemia inhibitory factor coincides with the onset of blastocyst implantation. *Proc. Natl. Acad. Sci. U S A* 88, 11408–11412.
311. Ernst, M., Inglese, M., Waring, P., Campbell, I. K., Bao, S., Clay, F. J., Alexander, W. S., Wicks, I. P., Tarlinton, D. M., Novak, U., Heath, J. K., and Dunn, A. R. (2001). Defective gp130-mediated signal transducer and activator of transcription (STAT) signaling results in degenerative joint disease, gastrointestinal ulceration, and failure of uterine implantation. *J. Exp. Med.* 194, 189–203.
312. Fry, R. C. (1992). The effect of leukaemia inhibitory factor (LIF) on embryogenesis. *Reprod. Fertil. Dev.* 4, 449–458.
313. Dunglison, G. F., Barlow, D. H., and Sargent, I. L. (1996). Leukaemia inhibitory factor significantly enhances the blastocyst formation rates of human embryos cultured in serum-free medium. *Hum. Reprod.* 11, 191–196.
314. Nichols, J., Davidson, D., Taga, T., Yoshida, K., Chambers, I., and Smith, A. (1996). Complementary tissue-specific expression of LIF and LIF-receptor mRNAs in early mouse embryogenesis. *Mech. Dev.* 57, 123–131.
315. Yoshida, K., Taga, T., Saito, M., Suematsu, S., Kumanogoh, A., Tanaka, T., Fujiwara, H., Hirata, M., Yamagami, T., Nakahata, T., Hirabayashi, T., Yoneda, Y., Tanaka, K., Wang, W. Z., Mori, C., Shiota, K., Yoshida, N., and Kishimoto, T. (1996). Targeted disruption of gp130, a common signal transducer for the interleukin 6 family of cytokines, leads to myocardial and hematological disorders. *Proc. Natl. Acad. Sci. U S A* 93, 407–411.
316. Shen, M. M., and Leder, P. (1992). Leukemia inhibitory factor is expressed by the preimplantation uterus and selectively blocks primitive ectoderm formation in vitro. *Proc. Natl. Acad. Sci. U S A* 89, 8240–8244.
317. Yang, Z. M., Chen, D. B., Le, S. P., and Harper, M. J. (1996). Differential hormonal regulation of leukemia inhibitory factor (LIF) in rabbit and mouse uterus. *Mol. Reprod. Dev.* 43, 470–476.
318. Arici, A., Engin, O., Attar, E., and Olive, D. L. (1995). Modulation of leukemia inhibitory factor gene expression and protein biosynthesis in human endometrium. *J. Clin. Endocrinol. Metab.* 80, 1908–1915.
319. Hambartsoumian, E. (1998). Endometrial leukemia inhibitory factor (LIF) as a possible cause of unexplained infertility and multiple failures of implantation. *Am. J. Reprod. Immunol.* 39, 137–143.
320. Hansen, T. R., Austin, K. J., Perry, D. J., Pru, J. K., Teixeira, M. G., and Johnson, G. A. (1999). Mechanism of action of interferon-tau in the uterus during early pregnancy. *J. Reprod. Fertil. Suppl.* 54, 329–339.
321. Ott, T. L., Mirando, M. A., Davis, M. A., and Bazer, F. W. (1992). Effects of ovine conceptus secretory proteins and progesterone on oxytocin-stimulated endometrial production of prostaglandin and turnover of inositol phosphate in ovariectomized ewes. *J. Reprod. Fertil.* 95, 19–29.
322. Rosenfeld, C. S., Han, C. S., Alexenko, A. P., Spencer, T. E., and Roberts, R. M. (2002). Expression of interferon receptor subunits, IFNAR1 and IFNAR2, in the ovine uterus. *Biol. Reprod.* 67, 847–853.
323. Choi, Y., Johnson, G. A., Burghardt, R. C., Berghman, L. R., Joyce, M. M., Taylor, K. M., Stewart, M. D., Bazer, F. W., and Spencer, T. E. (2001). Interferon regulatory factor-two restricts expression of interferon-stimulated genes to the endometrial stroma and glandular epithelium of the ovine uterus. *Biol. Reprod.* 65, 1038–1049.
324. Choi, Y., Johnson, G. A., Spencer, T. E., and Bazer, F. W. (2003). Pregnancy and interferon tau regulate major histocompatibility complex class I and beta2-microglobulin expression in the ovine uterus. *Biol. Reprod.* 68, 1703–1710.
325. Johnson, G. A., Spencer, T. E., Burghardt, R. C., Joyce, M. M., and Bazer, F. W. (2000). Interferon-tau and progesterone regulate ubiquitin cross-reactive protein expression in the ovine uterus. *Biol. Reprod.* 62, 622–627.
326. Kim, S., Choi, Y., Bazer, F. W., and Spencer, T. E. (2003). Identification of genes in the ovine endometrium regulated by interferon tau independent of signal transducer and activator of transcription 1. *Endocrinology* 144, 5203–5214.
327. Fleming, J. A., Choi, Y., Johnson, G. A., Spencer, T. E., and Bazer, F. W. (2001). Cloning of the ovine estrogen receptor-alpha promoter and functional regulation by ovine interferon-tau. *Endocrinology* 142, 2879–2887.
328. Kim, S., Choi, Y., Spencer, T. E., and Bazer, F. W. (2003). Effects of the estrous cycle, pregnancy and interferon tau on expression of cyclooxygenase two (COX-2) in ovine endometrium. *Reprod. Biol. Endocrinol.* 1, 58.
329. Roberts, R. M., Ealy, A. D., Alexenko, A. P., Han, C. S., and Ezashi, T. (1999). Trophoblast interferons. *Placenta* 20, 259–264.
330. Spencer, T. E., Ott, T. L., and Bazer, F. W. (1996). tau-Interferon: pregnancy recognition signal in ruminants. *Proc. Soc. Exp. Biol. Med.* 213, 215–229.
331. Vallet, J. L., Bazer, F. W., Fliss, M. F., and Thatcher, W. W. (1988). Effect of ovine conceptus secretory proteins and purified ovine trophoblast protein-1 on interoestrous interval and plasma concentrations of prostaglandins F-2 alpha and E and of 13,14-dihydro-15-keto prostaglandin F-2 alpha in cyclic ewes. *J. Reprod. Fertil.* 84, 493–504.
332. Spencer, T. E., and Bazer, F. W. (1996). Ovine interferon tau suppresses transcription of the estrogen receptor and oxytocin receptor genes in the ovine endometrium. *Endocrinology* 137, 1144–1147.
333. Spencer, T. E., and Bazer, F. W. (1995). Temporal and spatial alterations in uterine estrogen receptor and progesterone

receptor gene expression during the estrous cycle and early pregnancy in the ewe. *Biol. Reprod.* 53, 1527–1543.

334. Spencer, T. E., Becker, W. C., George, P., Mirando, M. A., Ogle, T. F., and Bazer, F. W. (1995). Ovine interferon-tau regulates expression of endometrial receptors for estrogen and oxytocin but not progesterone. *Biol. Reprod.* 53, 732–745.
335. Krumlauf, R. (1994). Hox genes in vertebrate development. *Cell* 78, 191–201.
336. Benson, G. V., Lim, H., Paria, B. C., Satokata, I., Dey, S. K., and Maas, R. L. (1996). Mechanisms of reduced fertility in Hoxa-10 mutant mice: uterine homeosis and loss of maternal Hoxa-10 expression. *Development* 122, 2687–2696.
337. Hsieh, L., Witte, D. P., Weinstein, M., Branford, W., Li, H., Small, K., and Potter, S. S. (1995). Hoxa 11 structure, extensive antisense transcription, and function in male and female fertility. *Development* 121, 1373–1385.
338. Gendron, R. L., Paradis, H., Hsieh, L., Lee, D. W., Potter, S. S., and Markoff, E. (1997). Abnormal uterine stromal and glandular function associated with maternal reproductive defects in Hoxa-11 null mice. *Biol. Reprod.* 56, 1097–1105.
339. Lim, H., Ma, L., Ma, W. G., Maas, R. L., and Dey, S. K. (1999). Hoxa-10 regulates uterine stromal cell responsiveness to progesterone during implantation and decidualization in the mouse. *Mol. Endocrinol.* 13, 1005–1017.
340. Das, S. K., Lim, H., Paria, B. C., and Dey, S. K. (1999). Cyclin D3 in the mouse uterus is associated with the decidualization process during early pregnancy. *J. Mol. Endocrinol.* 22, 91–101.
341. Yao, M. W., Lim, H., Schust, D. J., Choe, S. E., Farago, A., Ding, Y., Michaud, S., Church, G. M., and Maas, R. L. (2003). Gene expression profiling reveals progesterone-mediated cell cycle and immunoregulatory roles of hoxa-10 in the preimplantation uterus. *Mol. Endocrinol.* 17, 610–627.
342. Taylor, H. S., Arici, A., Olive, D., and Igarashi, P. (1998). HOXA10 is expressed in response to sex steroids at the time of implantation in the human endometrium. *J. Clin. Invest.* 101, 1379–1384.
343. Giudice, L. C. (1997). Multifaceted roles for IGFBP-1 in human endometrium during implantation and pregnancy. *Ann. N. Y. Acad. Sci.* 828, 146–156.
344. Lee, P. D., Giudice, L. C., Conover, C. A., and Powell, D. R. (1997). Insulin-like growth factor binding protein-1: recent findings and new directions. *Proc. Soc. Exp. Biol. Med.* 216, 319–357.
345. Fazleabas, A. T., Kim, J. J., and Strakova, Z. (2004). Implantation: embryonic signals and the modulation of the uterine environment—a review. *Placenta* 25 Suppl A, S26–S31.
346. Kim, J. J., Taylor, H. S., Akbas, G. E., Foucher, I., Trembleau, A., Jaffe, R. C., Fazleabas, A. T., and Unterman, T. G. (2003). Regulation of insulin-like growth factor binding protein-1 promoter activity by FKHR and HOXA10 in primate endometrial cells. *Biol. Reprod.* 68, 24–30.
347. Wang, W., Van, D., and Lufkin, T. (1998). Inner ear and maternal reproductive defects in mice lacking the Hmx3 homeobox gene. *Development* 125, 621–634.
348. Daikoku, T., Song, H., Guo, Y., Riesewijk, A., Mosselman, S., Das, S. K., and Dey, S. K. (2004). Uterine Msx-1 and Wnt4 signaling becomes aberrant in mice with the loss of leukemia inhibitory factor or Hoxa-10: evidence for a novel cytokine-homeobox-Wnt signaling in implantation. *Mol. Endocrinol.* 18, 1238–1250.
349. Desvergne, B., and Wahli, W. (1999). Peroxisome proliferator-activated receptors: nuclear control of metabolism. *Endocr. Rev.* 20, 649–688.
350. Ding, N. Z., Teng, C. B., Ma, H., Ni, H., Ma, X. H., Xu, L. B., and Yang, Z. M. (2003). Peroxisome proliferator-activated receptor delta expression and regulation in mouse uterus during embryo implantation and decidualization. *Mol. Reprod. Dev.* 66, 218–224.
351. Barak, Y., Liao, D., He, W., Ong, E. S., Nelson, M. C., Olefsky, J. M., Boland, R., and Evans, R. M. (2002). Effects of peroxisome proliferator-activated receptor delta on placentation, adiposity, and colorectal cancer. *Proc. Natl. Acad. Sci. U S A* 99, 303–308.
352. Peters, J. M., Lee, S. S., Li, W., Ward, J. M., Gavrilova, O., Everett, C., Reitman, M. L., Hudson, L. D., and Gonzalez, F. J. (2000). Growth, adipose, brain, and skin alterations resulting from targeted disruption of the mouse peroxisome proliferator-activated receptor beta(delta). *Mol. Cell. Biol.* 20, 5119–5128.
353. Michalik, L., Desvergne, B., Tan, N. S., Basu-Modak, S., Escher, P., Rieusset, J., Peters, J. M., Kaya, G., Gonzalez, F. J., Zakany, J., Metzger, D., Chambon, P., Duboule, D., and Wahli, W. (2001). Impaired skin wound healing in peroxisome proliferator-activated receptor (PPAR)alpha and PPARbeta mutant mice. *J. Cell. Biol.* 154, 799–814.
354. Ding, N. Z., Ma, X. H., Diao, H. L., Xu, L. B., and Yang, Z. M. (2003). Differential expression of peroxisome proliferator-activated receptor delta at implantation sites and in decidual cells of rat uterus. *Reproduction* 125, 817–825.
355. Tong, B. J., Tan, J., Tajeda, L., Das, S. K., Chapman, J. A., DuBois, R. N., and Dey, S. K. (2000). Heightened expression of cyclooxygenase-2 and peroxisome proliferator-activated receptor-delta in human endometrial adenocarcinoma. *Neoplasia* 2, 483–490.
356. Austin, L. A., and Heath, H., III (1981). Calcitonin: physiology and pathophysiology. *N. Engl. J. Med.* 304, 269–278.
357. Ding, Y. Q., Zhu, L. J., Bagchi, M. K., and Bagchi, I. C. (1994). Progesterone stimulates calcitonin gene expression in the uterus during implantation. *Endocrinology* 135, 2265–2274.
358. Zhu, L. J., Bagchi, M. K., and Bagchi, I. C. (1998). Attenuation of calcitonin gene expression in pregnant rat uterus leads to a block in embryonic implantation. *Endocrinology* 139, 330–339.
359. Kumar, S., Zhu, L. J., Polihronis, M., Cameron, S. T., Baird, D. T., Schatz, F., Dua, A., Ying, Y. K., Bagchi, M. K., and Bagchi, I. C. (1998). Progesterone induces calcitonin gene expression in human endometrium within the putative window of implantation. *J. Clin. Endocrinol. Metab.* 83, 4443–4450.
360. Kumar, S., Brudney, A., Cheon, Y. P., Fazleabas, A. T., and Bagchi, I. C. (2003). Progesterone induces calcitonin expression in the baboon endometrium within the window of uterine receptivity. *Biol. Reprod.* 68, 1318–1323.
361. Linse, S., Thulin, E., Gifford, L. K., Radzewsky, D., Hagan, J., Wilk, R. R., and Akerfeldt, K. S. (1997). Domain organization of calbindin D28k as determined from the association of six synthetic EF-hand fragments. *Protein Sci.* 6, 2385–2396.
362. Nie, G. Y., Li, Y., Wang, J., Minoura, H., Findlay, J. K., and Salamonsen, L. A. (2000). Complex regulation of calcium-binding protein D9k (calbindin-D[9k]) in the mouse uterus during early pregnancy and at the site of embryo implantation. *Biol. Reprod.* 62, 27–36.
363. Luu, K. C., Nie, G. Y., and Salamonsen, L. A. (2004). Endometrial calbindins are critical for embryo implantation: evidence from in vivo use of morpholino antisense oligonucleotides. *Proc. Natl. Acad. Sci. U S A* 101, 8028–8033.
364. Luu, K. C., Nie, G. Y., Hampton, A., Fu, G. Q., Liu, Y. X., and Salamonsen, L. A. (2004). Endometrial expression of calbindin-d28k but not -d9k in primates implies evolutionary changes and functional redundancy of calbindins at implantation. *Reproduction* 128, 433–441.
365. Bruzzone, R., White, T. W., and Paul, D. L. (1996). Connections with connexins: the molecular basis of direct intercellular signaling. *Eur. J. Biochem.* 238, 1–27.

366. Kumar, N. M., and Gilula, N. B. (1996). The gap junction communication channel. *Cell* 84, 381–388.
367. Willecke, K., Eiberger, J., Degen, J., Eckardt, D., Romualdi, A., Guldenagel, M., Deutsch, U., and Sohl, G. (2002). Structural and functional diversity of connexin genes in the mouse and human genome. *Biol. Chem.* 383, 725–737.
368. Winterhager, E., Grummer, R., Jahn, E., Willecke, K., and Traub, O. (1993). Spatial and temporal expression of connexin26 and connexin43 in rat endometrium during trophoblast invasion. *Dev. Biol.* 157, 399–409.
369. Grummer, R., Chwalisz, K., Mulholland, J., Traub, O., and Winterhager, E. (1994). Regulation of connexin26 and connexin43 expression in rat endometrium by ovarian steroid hormones. *Biol. Reprod.* 51, 1109–1116.
370. Grummer, R., Hewitt, S. W., Traub, O., Korach, K. S., and Winterhager, E. (2004). Different regulatory pathways of endometrial connexin expression: preimplantation hormonal-mediated pathway versus embryo implantation-initiated pathway. *Biol. Reprod.* 71, 273–281.
371. Sachs, L., and Shelesnyak, M. C. (1955). The development and suppression of polyploidy in the developing and suppressed deciduoma in the rat. *J. Endocrinol.* 12, 146–151.
372. Leroy, F., Bogaert, C., Van, H., and Delcroix, C. (1974). Cytophotometric and autoradiographic evaluation of cell kinetics in decidual growth in rats. *J. Reprod. Fertil.* 38, 441–449.
373. Ansell, J. D., Barlow, P. W., and McLaren, A. (1974). Binucleate and polyploid cells in the decidua of the mouse. *J. Embryol. Exp. Morphol.* 31, 223–227.
374. Sherr, C. J. (1993). Mammalian G1 cyclins. *Cell* 73, 1059–1065.
375. Riley, D. J., Lee, E. Y., and Lee, W. H. (1994). The retinoblastoma protein: more than a tumor suppressor. *Annu. Rev. Cell. Biol.* 10, 1–29.
376. Sherr, C. J., and Roberts, J. M. (1999). CDK inhibitors: positive and negative regulators of G1-phase progression. *Genes Dev.* 13, 1501–1512.
377. Shiozawa, T., Li, S. F., Nakayama, K., Nikaido, T., and Fujii, S. (1996). Relationship between the expression of cyclins/cyclin-dependent kinases and sex-steroid receptors/Ki67 in normal human endometrial glands and stroma during the menstrual cycle. *Mol. Hum. Reprod.* 2, 745–752.
378. Shiozawa, T., Nikaido, T., Nakayama, K., Lu, X., and Fujii, S. (1998). Involvement of cyclin-dependent kinase inhibitor p27Kip1 in growth inhibition of endometrium in the secretory phase and of hyperplastic endometrium treated with progesterone. *Mol. Hum. Reprod.* 4, 899–905.
379. Geum, D., Sun, W., Paik, S. K., Lee, C. C., and Kim, K. (1997). Estrogen-induced cyclin D1 and D3 gene expressions during mouse uterine cell proliferation in vivo: differential induction mechanism of cyclin D1 and D3. *Mol. Reprod. Dev.* 46, 450–458.
380. Altucci, L., Addeo, R., Cicatiello, L., Germano, D., Pacilio, C., Battista, T., Cancemi, M., Petrizzi, V. B., Bresciani, F., and Weisz, A. (1997). Estrogen induces early and timed activation of cyclin-dependent kinases 4, 5, and 6 and increases cyclin messenger ribonucleic acid expression in rat uterus. *Endocrinology* 138, 978–984.
381. Prall, O. W., Sarcevic, B., Musgrove, E. A., Watts, C. K., and Sutherland, R. L. (1997). Estrogen-induced activation of Cdk4 and Cdk2 during G1-S phase progression is accompanied by increased cyclin D1 expression and decreased cyclin-dependent kinase inhibitor association with cyclin E-Cdk2. *J. Biol. Chem.* 272, 10882–10894.
382. Musgrove, E. A., Swarbrick, A., Lee, C. S., Cornish, A. L., and Sutherland, R. L. (1998). Mechanisms of cyclin-dependent kinase inactivation by progestins. *Mol. Cell. Biol.* 18, 1812–1825.
383. Jones, S. R., Kimler, B. F., Justice, W. M., and Rider, V. (2000). Transit of normal rat uterine stromal cells through G1 phase of the cell cycle requires progesterone-growth factor interactions. *Endocrinology* 141, 637–648.
384. Bamberger, A., Sudahl, S., Bamberger, C. M., Schulte, H. M., and Loning, T. (1999). Expression patterns of the cell-cycle inhibitor p27 and the cell-cycle promoter cyclin E in the human placenta throughout gestation: implications for the control of proliferation. *Placenta* 20, 401–406.
385. Tan, J., Raja, S., Davis, M. K., Tawfik, O., Dey, S. K., and Das, S. K. (2002). Evidence for coordinated interaction of cyclin D3 with p21 and cdk6 in directing the development of uterine stromal cell decidualization and polyploidy during implantation. *Mech. Dev.* 111, 99–113.
386. Tan, Y., Li, M., Cox, S., Davis, M. K., Tawfik, O., Paria, B. C., and Das, S. K. (2004). HB-EGF directs stromal cell polyploidy and decidualization via cyclin D3 during implantation. *Dev. Biol.* 265, 181–195.
387. Aplin, J. D., Charlton, A. K., and Ayad, S. (1988). An immunohistochemical study of human endometrial extracellular matrix during the menstrual cycle and first trimester of pregnancy. *Cell. Tiss. Res.* 253, 231–240.
388. Vu, T. H., and Werb, Z. (2000). Matrix metalloproteinases: effectors of development and normal physiology. *Genes Dev.* 14, 2123–2133.
389. Das, S. K., Yano, S., Wang, J., Edwards, D. R., Nagase, H., and Dey, S. K. (1997). Expression of matrix metalloproteinases and tissue inhibitors of metalloproteinases in the mouse uterus during the peri-implantation period. *Dev. Genet.* 21, 44–54.
390. Afonso, S., Romagnano, L., and Babiarz, B. (1997). The expression and function of cystatin C and cathepsin B and cathepsin L during mouse embryo implantation and placentation. *Development* 124, 3415–3425.
391. Jokimaa, V., Oksjoki, S., Kujari, H., Vuorio, E., and Anttila, L. (2002). Altered expression of genes involved in the production and degradation of endometrial extracellular matrix in patients with unexplained infertility and recurrent miscarriages. *Mol. Hum. Reprod.* 8, 1111–1116.
392. Folkman, J. (1995). Angiogenesis in cancer, vascular, rheumatoid and other disease. *Nat. Med.* 1, 27–31.
393. Hyder, S. M., and Stancel, G. M. (1999). Regulation of angiogenic growth factors in the female reproductive tract by estrogens and progestins. *Mol. Endocrinol.* 13, 806–811.
394. Chakraborty, I., Das, S. K., and Dey, S. K. (1995). Differential expression of vascular endothelial growth factor and its receptor mRNAs in the mouse uterus around the time of implantation. *J. Endocrinol.* 147, 339–352.
395. Halder, J. B., Zhao, X., Soker, S., Paria, B. C., Klagsbrun, M., Das, S. K., and Dey, S. K. (2000). Differential expression of VEGF isoforms and VEGF(164)-specific receptor neuropilin-1 in the mouse uterus suggests a role for VEGF(164) in vascular permeability and angiogenesis during implantation. *Genesis* 26, 213–224.
396. Ferrara, N., and Davis, S. (1997). The biology of vascular endothelial growth factor. *Endocr. Rev.* 18, 4–25.
397. Carmeliet, P., Ferreira, V., Breier, G., Pollefeyt, S., Kieckens, L., Gertsenstein, M., Fahrig, M., Vandenhoeck, A., Harpal, K., Eberhardt, C., Declercq, C., Pawling, J., Moons, L., Collen, D., Risau, W., and Nagy, A. (1996). Abnormal blood vessel development and lethality in embryos lacking a single VEGF allele. *Nature* 380, 435–439.
398. Ferrara, N., Carver, M., Chen, H., Dowd, M., Lu, L., Shea, K. S., Powell, B., Hillan, K. J., and Moore, M. W. (1996). Heterozygous embryonic lethality induced by targeted inactivation of the VEGF gene. *Nature* 380, 439–442.

399. Tischer, E., Mitchell, R., Hartman, T., Silva, M., Gospodarowicz, D., Fiddes, J. C., and Abraham, J. A. (1991). The human gene for vascular endothelial growth factor. Multiple protein forms are encoded through alternative exon splicing. *J. Biol. Chem.* 266, 11947–11954.
400. Peters, K. G., De, V., and Williams, L. T. (1993). Vascular endothelial growth factor receptor expression during embryogenesis and tissue repair suggests a role in endothelial differentiation and blood vessel growth. *Proc. Natl. Acad. Sci. U S A* 90, 8915–8919.
401. Shibuya, M., Yamaguchi, S., Yamane, A., Ikeda, T., Tojo, A., Matsushime, H., and Sato, M. (1990). Nucleotide sequence and expression of a novel human receptor-type tyrosine kinase gene (flt) closely related to the fms family. *Oncogene* 5, 519–524.
402. Millauer, B., Wizigmann, V., Schnurch, H., Martinez, R., Moller, N. P., Risau, W., and Ullrich, A. (1993). High affinity VEGF binding and developmental expression suggest Flk-1 as a major regulator of vasculogenesis and angiogenesis. *Cell* 72, 835–846.
403. Quinn, T. P., Peters, K. G., De, V., Ferrara, N., and Williams, L. T. (1993). Fetal liver kinase 1 is a receptor for vascular endothelial growth factor and is selectively expressed in vascular endothelium. *Proc. Natl. Acad. Sci. U S A* 90, 7533–7537.
404. Waltenberger, J., Claesson, W., Siegbahn, A., Shibuya, M., and Heldin, C. H. (1994). Different signal transduction properties of KDR and Flt1, two receptors for vascular endothelial growth factor. *J. Biol. Chem.* 269, 26988–26995.
405. Yoshida, A., Anand, A., and Zetter, B. R. (1996). Differential endothelial migration and proliferation to basic fibroblast growth factor and vascular endothelial growth factor. *Growth Factors* 13, 57–64.
406. Hyder, S. M., Nawaz, Z., Chiappetta, C., and Stancel, G. M. (2000). Identification of functional estrogen response elements in the gene coding for the potent angiogenic factor vascular endothelial growth factor. *Cancer Res.* 60, 3183–3190.
407. Ma, W., Tan, J., Matsumoto, H., Robert, B., Abrahamson, D. R., Das, S. K., and Dey, S. K. (2001). Adult tissue angiogenesis: evidence for negative regulation by estrogen in the uterus. *Mol. Endocrinol.* 15, 1983–1992.
408. Ghosh, D., Sharkey, A. M., Charnock-Jones, D. S., Dhawan, L., Dhara, S., Smith, S. K., and Sengupta, J. (2000). Expression of vascular endothelial growth factor (VEGF) and placental growth factor (PlGF) in conceptus and endometrium during implantation in the rhesus monkey. *Mol. Hum. Reprod.* 6, 935–941.
409. Niklaus, A. L., Babischkin, J. S., Aberdeen, G. W., Pepe, G. J., and Albrecht, E. D. (2002). Expression of vascular endothelial growth/permeability factor by endometrial glandular epithelial and stromal cells in baboons during the menstrual cycle and after ovariectomy. *Endocrinology* 143, 4007–4017.
410. Rowe, A. J., Wulff, C., and Fraser, H. M. (2003). Localization of mRNA for vascular endothelial growth factor (VEGF), angiopoietins and their receptors during the peri-implantation period and early pregnancy in marmosets (*Callithrix jacchus*). *Reproduction* 126, 227–238.
411. Maisonpierre, P. C., Suri, C., Jones, P. F., Bartunkova, S., Wiegand, S. J., Radziejewski, C., Compton, D., McClain, J., Aldrich, T. H., Papadopoulos, N., Daly, T. J., Davis, S., Sato, T. N., and Yancopoulos, G. D. (1997). Angiopoietin-2, a natural antagonist for Tie2 that disrupts in vivo angiogenesis. *Science* 277, 55–60.
412. Shalaby, F., Rossant, J., Yamaguchi, T. P., Gertsenstein, M., Wu, X. F., Breitman, M. L., and Schuh, A. C. (1995). Failure of blood-island formation and vasculogenesis in Flk-1-deficient mice. *Nature* 376, 62–66.
413. Suri, C., Jones, P. F., Patan, S., Bartunkova, S., Maisonpierre, P. C., Davis, S., Sato, T. N., and Yancopoulos, G. D. (1996). Requisite role of angiopoietin-1, a ligand for the TIE2 receptor, during embryonic angiogenesis. *Cell* 87, 1171–1180.
414. Sato, T. N., Tozawa, Y., Deutsch, U., Wolburg, B., Fujiwara, Y., Gendron, M., Gridley, T., Wolburg, H., Risau, W., and Qin, Y. (1995). Distinct roles of the receptor tyrosine kinases Tie-1 and Tie-2 in blood vessel formation. *Nature* 376, 70–74.
415. Thurston, G., Suri, C., Smith, K., McClain, J., Sato, T. N., Yancopoulos, G. D., and McDonald, D. M. (1999). Leakage-resistant blood vessels in mice transgenically overexpressing angiopoietin-1. *Science* 286, 2511–2514.
416. Davis, S., and Yancopoulos, G. D. (1999). The angiopoietins: yin and yang in angiogenesis. *Curr. Top. Microbiol. Immunol.* 237, 173–185.
417. Matsumoto, H., Ma, W. G., Daikoku, T., Zhao, X., Paria, B. C., Das, S. K., Trzaskos, J. M., and Dey, S. K. (2002). Cyclooxygenase-2 differentially directs uterine angiogenesis during implantation in mice. *J. Biol. Chem.* 277, 29260–29267.
418. Daikoku, T., Matsumoto, H., Gupta, R. A., Das, S. K., Gassmann, M., DuBois, R. N., and Dey, S. K. (2003). Expression of hypoxia-inducible factors in the peri-implantation mouse uterus is regulated in a cell-specific and ovarian steroid hormone-dependent manner. Evidence for differential function of HIFs during early pregnancy. *J. Biol. Chem.* 278, 7683–7691.
419. Matsuda, L. A., Lolait, S. J., Brownstein, M. J., Young, A. C., and Bonner, T. I. (1990). Structure of a cannabinoid receptor and functional expression of the cloned cDNA. *Nature* 346, 561–564.
420. Munro, S., Thomas, K. L., and Abu, S. (1993). Molecular characterization of a peripheral receptor for cannabinoids. *Nature* 365, 61–65.
421. Devane, W. A., Hanus, L., Breuer, A., Pertwee, R. G., Stevenson, L. A., Griffin, G., Gibson, D., Mandelbaum, A., Etinger, A., and Mechoulam, R. (1992). Isolation and structure of a brain constituent that binds to the cannabinoid receptor. *Science* 258, 1946–1949.
422. Felder, C. C., Briley, E. M., Axelrod, J., Simpson, J. T., Mackie, K., and Devane, W. A. (1993). Anandamide, an endogenous cannabimimetic eicosanoid, binds to the cloned human cannabinoid receptor and stimulates receptor-mediated signal transduction. *Proc. Natl. Acad. Sci. U S A* 90, 7656–7660.
423. Schmid, P. C., Paria, B. C., Krebsbach, R. J., Schmid, H. H., and Dey, S. K. (1997). Changes in anandamide levels in mouse uterus are associated with uterine receptivity for embryo implantation. *Proc. Natl. Acad. Sci. U S A* 94, 4188–4192.
424. Paria, B. C., Das, S. K., and Dey, S. K. (1995). The preimplantation mouse embryo is a target for cannabinoid ligand-receptor signaling. *Proc. Natl. Acad. Sci. U S A* 92, 9460–9464.
425. Yang, Z. M., Paria, B. C., and Dey, S. K. (1996). Activation of brain-type cannabinoid receptors interferes with preimplantation mouse embryo development. *Biol. Reprod.* 55, 756–761.
426. Wang, J., Paria, B. C., Dey, S. K., and Armant, D. R. (1999). Stage-specific excitation of cannabinoid receptor exhibits differential effects on mouse embryonic development. *Biol. Reprod.* 60, 839–844.
427. Wang, H., Matsumoto, H., Guo, Y., Paria, B. C., Roberts, R. L., and Dey, S. K. (2003). Differential G protein-coupled cannabinoid receptor signaling by anandamide directs blastocyst activation for implantation. *Proc. Natl. Acad. Sci. U S A* 100, 14914–14919.

428. Maccarrone, M., Valensise, H., Bari, M., Lazzarin, N., Romanini, C., and Finazzi-Agro, A. (2000). Relation between decreased anandamide hydrolase concentrations in human lymphocytes and miscarriage. *Lancet* 355, 1326–1329.
429. Maccarrone, M., Bisogno, T., Valensise, H., Lazzarin, N., Fezza, F., Manna, C., Di, M., V, and Finazzi-Agro, A. (2002). Low fatty acid amide hydrolase and high anandamide levels are associated with failure to achieve an ongoing pregnancy after IVF and embryo transfer. *Mol. Hum. Reprod* 8, 188–195.
430. Matsumoto, H., Zhao, X., Das, S. K., Hogan, B. L., and Dey, S. K. (2002). Indian hedgehog as a progesterone-responsive factor mediating epithelial-mesenchymal interactions in the mouse uterus. *Dev. Biol.* 245, 280–290.
431. Ingham, P. W., and McMahon, A. P. (2001). Hedgehog signaling in animal development: paradigms and principles. *Genes Dev.* 15, 3059–3087.
432. Johnson, R. L., and Scott, M. P. (1998). New players and puzzles in the Hedgehog signaling pathway. *Curr. Opin. Genet. Dev.* 8, 450–456.
433. McMahon, A. P. (2000). More surprises in the Hedgehog signaling pathway. *Cell* 100, 185–188.
434. Takamoto, N., Zhao, B., Tsai, S. Y., and DeMayo, F. J. (2002). Identification of Indian hedgehog as a progesterone-responsive gene in the murine uterus. *Mol. Endocrinol.* 16, 2338–2348.
435. Rubin, J. S., Bottaro, D. P., Chedid, M., Miki, T., Ron, D., Cheon, G., Taylor, W. G., Fortney, E., Sakata, H., and Finch, P. W. (1995). Keratinocyte growth factor. *Cell. Biol. Int.* 19, 399–411.
436. Ka, H., Spencer, T. E., Johnson, G. A., and Bazer, F. W. (2000). Keratinocyte growth factor: expression by endometrial epithelia of the porcine uterus. *Biol. Reprod.* 62, 1772–1778.
437. Ka, H., Jaeger, L. A., Johnson, G. A., Spencer, T. E., and Bazer, F. W. (2001). Keratinocyte growth factor is up-regulated by estrogen in the porcine uterine endometrium and functions in trophectoderm cell proliferation and differentiation. *Endocrinology* 142, 2303–2310.
438. Reese, J., Das, S. K., Paria, B. C., Lim, H., Song, H., Matsumoto, H., Knudtson, K. L., DuBois, R. N., and Dey, S. K. (2001). Global gene expression analysis to identify molecular markers of uterine receptivity and embryo implantation. *J. Biol. Chem.* 276, 44137–44145.
439. Yoshioka, K., Matsuda, F., Takakura, K., Noda, Y., Imakawa, K., and Sakai, S. (2000). Determination of genes involved in the process of implantation: application of GeneChip to scan 6500 genes. *Biochem. Biophys. Res. Commun.* 272, 531–538.
440. Cheon, Y. P., Li, Q., Xu, X., DeMayo, F. J., Bagchi, I. C., and Bagchi, M. K. (2002). A genomic approach to identify novel progesterone receptor regulated pathways in the uterus during Implantation. *Mol. Endocrinol.* 16, 2853–2871.
441. Andrade, P. M., Silva, I. D., Borra, R. C., de, L., and Baracat, E. C. (2002). Estrogen regulation of uterine genes in vivo detected by complementary DNA array. *Horm. Metab. Res.* 34, 238–244.
442. Watanabe, H., Suzuki, A., Mizutani, T., Khono, S., Lubahn, D. B., Handa, H., and Iguchi, T. (2002). Genome-wide analysis of changes in early gene expression induced by oestrogen. *Genes. Cells.* 7, 497–507.
443. Lindberg, M. K., Weihua, Z., Andersson, N., Moverare, S., Gao, H., Vidal, O., Erlandsson, M., Windahl, S., Andersson, G., Lubahn, D. B., Carlsten, H., Dahlman, W., Gustafsson, J. A., and Ohlsson, C. (2002). Estrogen receptor specificity for the effects of estrogen in ovariectomized mice. *J. Endocrinol.* 174, 167–178.
444. Naciff, J. M., Jump, M. L., Torontali, S. M., Carr, G. J., Tiesman, J. P., Overmann, G. J., and Daston, G. P. (2002). Gene expression profile induced by 17alpha-ethynyl estradiol, bisphenol A, and genistein in the developing female reproductive system of the rat. *Toxicol. Sci.* 68, 184–199.
445. Dey, S. K., Lim, H., Das, S. K., Reese, J., Paria, B. C., Daikoku, T., and Wang, H. (2004). Molecular cues to implantation. *Endocr. Rev.* 25, 341–373.
446. Carson, D. D., Lagow, E., Thathiah, A., Al, S., Farach, C., Vernon, M., Yuan, L., Fritz, M. A., and Lessey, B. (2002). Changes in gene expression during the early to mid-luteal (receptive phase) transition in human endometrium detected by high-density microarray screening. *Mol. Hum. Reprod.* 8, 871–879.
447. Banks, R. E., Dunn, M. J., Forbes, M. A., Stanley, A., Pappin, D., Naven, T., Gough, M., Harnden, P., and Selby, P. J. The potential use of laser capture microdissection to selectively obtain distinct populations of cells for proteomic analysis: preliminary findings. *Electrophoresis* 20, 689–700.
448. Hoang, V. M., Foulk, R., Clauser, K., Burlingame, A., Gibson, B. W., and Fisher, S. J. (2001). Functional proteomics: examining the effects of hypoxia on the cytotrophoblast protein repertoire. *Biochemistry* 40, 4077–4086.
449. Cencic, A., Henry, C., Lefevre, F., Huet, J. C., Koren, S., and La, B. (2002). The porcine trophoblastic interferon-gamma, secreted by a polarized epithelium, has specific structural and biochemical properties. *Eur. J. Biochem.* 269, 2772–2781.
450. Boving, B. G., and Boving, R. L. (1978). Implantation perspective. In *Human Fertilization* (H. Ludwig and P. F. Tauber, Eds.). Georg Thieme Publishers, Stuttgart.
451. Bazer, F. W., Vallet, J. L., Ashworth, C. J., Anthony, R. V., and Roberts, R. M. (1987). The role of ovine conceptus secretory proteins in the establishment of pregnancy. *Adv. Exp. Med. Biol.* 230, 221–235.
452. Perry, J. S., Heap, R. B., and Amoroso, E. C. (1973). Steroid hormone production by pig blastocysts. *Nature* 245, 45–47.

Knobil and Neill's Physiology of Reproduction, Third Edition
edited by Jimmy D. Neill,

CHAPTER 5

Anatomy and Genesis of the Placenta

Graham J. Burton,[1] Peter Kaufmann,[2] and Berthold Huppertz[2]

Historical Survey, 189
Comparative Placentation, 193
Placental Shape, 194
Maternal–Fetal Interdigitation, 194
Maternal–Fetal Barrier, 195
Vascular Arrangement, 196
Placental Evolution, 197
Macroscopy of the Delivered Placenta, 197
Multiple Pregnancies, 199
Early Development, 199
Prelacunar Stage, 199
Lacunar Stage, 201
Early Villous Stage, 202
Basic Villous Structure, 203
Villous Trophoblast, 203
Cytotrophoblast (Langhans' Cells), 203
Syncytial Fusion, 205
Syncytiotrophoblast, 207
Trophoblast Turnover, 208
Fixed Connective Tissue, 209
Hofbauer Cells, 209
Fetal Vessels, 209
Integrity of the Placental Barrier, 212
Villous Trees, 213
Structure of Villous Types, 213
Development of Villous Trees, 214
Fetal Angioarchitecture, 215
Intervillous Space, 216
Oxygen as Regulator of Villous Development, 218
Stereological Data, 220
Nonvillous Parts, 222
Chorionic Plate, Secondary Yolk Sac, Umbilical Cord, and Membranes, 222
Basal Plate and Uteroplacental Vessels, 224
Placental Septa, 224
Cell Columns and Cell Islands, 224
Extravillous Trophoblast, 225
Uteroplacental Vessels, 225
Fibrinoid, 226
Composition and Origin, 227
Functional Aspects, 227
Genetic Control of Placental Development, 228
Mouse Placenta, 228
Genetic Control, 229
Imprinting and Placental Development, 229
Morphological Correlates of Noninvasive Diagnostic Methods, 230
Ultrasound, 230
Doppler Ultrasound, 231
Magnetic Resonance Imaging, 232
References, 232

HISTORICAL SURVEY

Although the placenta was venerated by the early Egyptians, DeWitt (1) credited the Greek physician Diogenes of Apollonia (c. 480 B.C.) with being the first to ascribe the function of fetal nutrition to the organ. Aristotle (384–322 B.C.) developed this idea further, realizing that the fetus is fully enclosed within membranes, which he termed the *chorion*. He thus challenged the previously held Hippocratic theories that the fetus is nourished by suckling from uterine paps or cotyledons. Instead, Aristotle believed that the vessels of the umbilical cord transmitted nutrients by communicating directly with those of the uterus, vein to vein and artery to artery.

Aristotle's teachings were reinforced by Galen (130–200 A.D.), who synthesized both Hippocratic and Aristotelian ideas on reproduction into a series of theoretic constructs. Thus, he believed that the embryo is

[1]Department of Anatomy, University of Cambridge, Cambridge, United Kingdom; [2]Department of Anatomy II, University Hospital Aachen, Germany.

formed from the menstrual blood that is no longer shed after conception and that whereas the allantois forms from the female semen, the chorion forms from the coagulation of the male sperm mass (1). Despite this rather teleological approach, Galen was a great experimenter, accurately describing the anatomy of the fetal membranes of ruminants and correctly deducing that the fetus excretes urine into the allantoic sac. These animal studies most likely colored his appreciation of the human situation, however, for he too believed that the vessels of the umbilical cord communicate directly with their maternal counterparts, opening on the surface of the uterine caruncles. Indeed, Amoroso (2) held the view that both Aristotle and Galen probably never actually dissected the human body.

This state of affairs persisted throughout the Middle Ages. For example, during this period it was commonly held that the human uterus was a seven-chambered structure similar to that of the pig, and it was Leonardo da Vinci (1452–1519) who first illustrated it correctly as a *uterus simplex*. Leonardo's manuscripts are justly famed for their beautiful illustrations, but often the accompanying notes indicate meticulous observations. In these, he asserted that the veins of the fetus are not directly connected to those of the uterus, thus challenging one of the most basic precepts of the ancients' teachings.

Despite these major contributions, da Vinci continued to propagate the misconception that the human placenta is cotyledonary in nature through his illustrations. He was not alone in his confusion, however, and in his defense it must be recognized that common domestic animals display a bewildering array of placental types. The next great anatomist, Andreas Vesalius (1514–1564), depicted, in the first edition of *De Humani Corporis Fabrica*, the human placenta as being of the zonary type characteristic of carnivores. He corrected this in the second edition, but the discoidal placenta then shown still bears many similarities to a single cotyledon from a ruminant, with large vessels running over the free chorion and the intimation of the presence of a large allantoic sac.

After nearly 1,300 years of stagnation since the time of Galen, scholastic activity suddenly blossomed during the Renaissance. Realdus Columbus (1516–1559) introduced the term *placenta* in 1559 (3), derived from the Latin root for a flat "cake." Julius Caesar Arantius (1530–1589) made great strides forward, not only in considering the placenta as a uterine liver involved in the "purification" of the fetal blood but also by establishing through many comparative dissections the independence of the maternal and fetal circulations. Despite da Vinci's earlier notes, which admittedly were not published as an anatomical text, Arantius is generally accredited with this fundamental discovery (4).

The idea was not readily accepted, however. Thus, Hieronymus Fabricius (1537–1619), a detailed observer of the placenta of a number of species, wavered for many years but was finally convinced by the weight of evidence in favor of the ancient teachings. His pupil, Adrianus Spigelius (1578–1625), adopted a compromise position, believing that maternal blood flows from the uterine vessels into the open mouths of the umbilical veins but that the umbilical arteries end in the placenta and do not communicate with their uterine equivalents. It is arguable that neither of these men really furthered knowledge of placental function, but what cannot be denied is the exquisite beauty of their illustrations (Fig. 1).

The next main advance in our understanding of the placenta was through the work of William Harvey (1578–1657) and his theories on the circulation of blood. He continued to see the main function of the organ as nutritive rather than respiratory and argued on the basis of his understanding of the circulation that the fetal and maternal bloodstreams must be separate. Nevertheless, Harvey could still not explain the vital step of how blood passed from arteries to veins, and

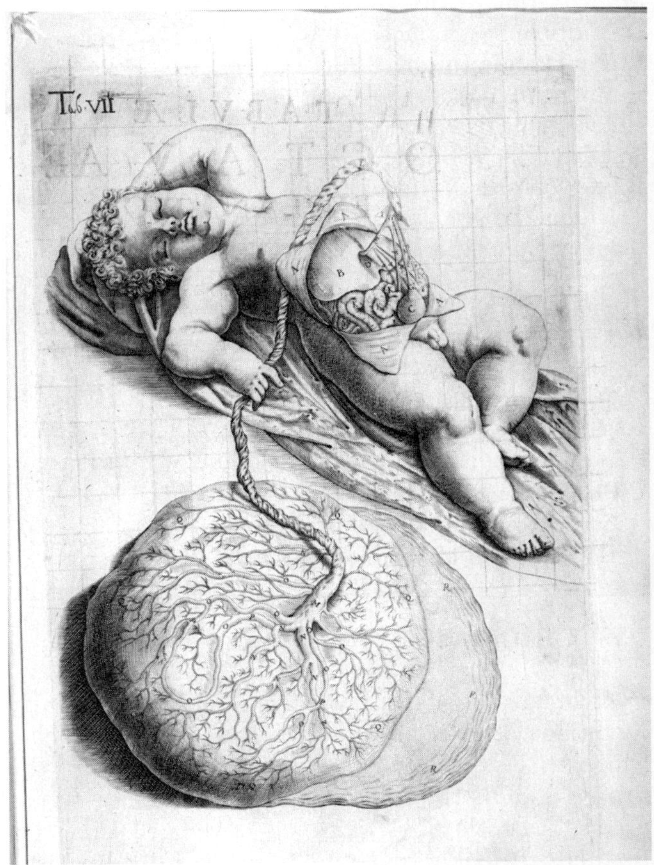

FIG. 1. One of the plates from De Formato Foetu of Spigelius (1626), demonstrating the exquisite beauty of illustrations from this period. As can be seen, the origin of the umbilical arteries from the pelvic vasculature and the termination of the umbilical vein in the liver are depicted correctly. Also, the urachus connecting the fetal bladder to the allantoic diverticulum in the proximal end of the umbilical cord is clearly shown.

it was not until after his death that Malpighi in 1660 provided the answer with his discovery of capillaries. It was also Malpighi who first suggested that the placenta is a substitute for the lungs while those of the fetus are at rest, and we come close to a correct appreciation of the organ's function when in 1668 John Mayow asserted that the maternal blood supplies the fetus with "nitro-aerial spirits," or oxygen. He proposed the replacement of Arantius's uterine liver concept with that of a uterine lung, but now at least two of the principal functions of the organ were clearly established.

No historical survey of the placenta, however brief, would be complete without due acknowledgment of the enormous contributions made by the brothers William (1718–1783) and John (1728–1792) Hunter. Through their renowned studies involving the injection of molten wax into the circulation of gravid uteri, graphically depicted by Corner (4), they succeeded in providing confirmatory experimental proof of the independence of the maternal and fetal circulations, first suggested by da Vinci almost 250 years earlier. Their writings suggest that the brothers had a clear appreciation of the way in which the uterine arteries open into what we now refer to as the intervillous space. For them, however, it was simply a large blood-filled space because the concept of placental villi was not realized until nearly halfway through the 19th century, when use of the microscope became more widespread.

Almost contemporaneously in the 1840s, Weber and Dalrymple published virtually identical drawings of isolated villi, illustrating within their core the fetal capillary network uniting umbilical artery to vein. Although these drawings accurately reflect the gross morphology of the villi, ideas concerning their histological nature were still confused. An epithelial covering to the villi was clearly depicted, but for many years this was believed to be maternal in origin. The illustrations of Ercolani from 1877 [see Fig. 9 in (5)] reflect contemporary opinion of how this was achieved, for in these the uterine vessels are shown opening through the basal plate and expanding into thin-walled sac-like structures that envelop the fetal villi like a glove. Only 5 years later in 1882, Theodor Langhans demonstrated that both layers of the epithelial covering, which are now referred to as the cytotrophoblast and the syncytiotrophoblast, were in fact fetal in origin, so establishing our current understanding of villous structure.

That such confusion should have existed is perhaps not surprising, for this was a time of intense debate on the issues of evolution and the diversity of animals after the publication of Darwin's *Origin of Species* in 1859. It saw the heyday of comparative anatomy, and the placental membranes certainly did not escape the attention of avid investigators intent on demonstrating phylogenetic links. Large collections of material were amassed by zoologists such as Hubrecht, and indeed it was he who first introduced the term trophoblast when describing the early development of the hedgehog in 1889 (6). In recognition of its direct nutritive significance to the embryo, Hubrecht conferred the term to the outer layer of the blastocyst that is in immediate contact with maternal tissues.

Necessary order was imposed on the broader zoological scene largely by Otto Grosser, who proposed a classification system based on the number of tissue layers that, under the light microscope, appear to separate the maternal and fetal bloodstreams (7). Hence, the terms *epitheliochorial*, *syndesmochorial*, *endotheliochorial*, and *hemochorial* were coined, and although this system had to be updated in the light of subsequent findings, and has on occasion been misconstrued, in general it has provided a useful framework for subsequent comparative studies. These have advanced our understanding of placental function enormously and mention must be made of the major contributions provided by, for example, the works of Mossman (8) and Amoroso (9). Grosser did not restrict himself to animal material, however, and his descriptions of the histological structure of the human placenta were also detailed and accurate.

Focusing on the human situation, it should now be apparent that the foundations of our present understanding of placental structure were in place by the start of the 20th century. Shortly after, in 1905, as a result of a series of astute observations, Halban (10) concluded that the organ must produce "active substances" and correctly pronounced that "during pregnancy the placenta usurps the protective function of the ovary and carries it to a higher degree." Although Halban obviously had no detailed knowledge of estrogen and progesterone, the third main function of the placenta, that of an endocrine organ, had clearly been recognized.

However, although an extensive body of knowledge concerning the structure of the mature placenta was beginning to be established at the beginning of that century, relatively little was known about earlier stages and, in particular, about implantation. The small amount of information available was based on very rare specimens recovered fortuitously during surgical or autopsy procedures, and many of these have since been interpreted as being pathologic. To rectify this situation, in the mid-1920s Streeter established a breeding colony of rhesus monkeys at the Carnegie Institute in Washington and so pioneered the systematic study of early embryological development. In conjunction with collaborators such as Hartman, Heuser, and Wislocki, a series of detailed descriptions of macaque development resulted (11). These not only yielded unique information on that species, but also provided a yardstick of development against which the more occasional human specimens could be compared. The success of these studies prompted attempts to recover early human embryos on a more systematic basis from hysterectomy material, and this led to the establishment of the renowned Hertig-Rock collection (12). The subsequent descriptions of human development based on this collection laid the foundations for much of modern embryology, although some of the conclusions regarding maternal blood flow have had to be reevaluated in the light of recent physiological measurements taken in vivo, as seen later.

Although the Carnegie Institute played a pivotal role, others [e.g., Stieve (13)] also made notable contributions through their meticulous observations of early specimens. Stieve is perhaps more widely remembered for his papers describing the configuration of the mature villous tree, which at the time were highly controversial. As stated, the general architecture of a placental villus had been known quite accurately since the illustrations of Weber and Dalrymple, but how these were arranged within the organ was another matter. The prevailing view was that the villi resembled a tree, with its "roots" attached to the chorionic plate and ever-diminishing branches extending toward the basal plate. A few were believed to attach to the uterine wall as "anchoring villi," but most were thought to hang freely in the intervillous space as terminal villi. Stieve (13), however, considered that fusion between the tips of terminal villi was commonplace and that this converted the villous tree into a lattice-type network through which the fetal blood flows. By contrast, Spanner (14) believed that the main trunk of the first tree extended to the basal plate and that the principal branches then recurved toward the chorionic plate before giving off terminal villi [see Figs. 171 to 173 in (3)]. Subsequent studies and three-dimensional reconstructions have now irrefutably confirmed that the traditional view is the correct picture, although as seen later, some fusion between villi does occur.

These differences in arrangement might at first sight appear unimportant and the subject of esoteric debates between morphologists, but when considering placental exchange it is critical to have an appreciation of the relative directions of the maternal and fetal blood flows. On the maternal side, one problem that had taxed investigators since the time of the Hunter brothers was how the direction of blood flow was regulated in the intervillous space. The problem they and their contemporaries rapidly perceived was why does the maternal blood, when delivered into a large space with no preformed channels, circulate to bathe all the villi rather than pass directly from the arterial to the nearest venous opening? One solution to the problem was offered by Bumm in 1893 [see Figs. 64 and 65 in (15)], who maintained that the openings of the arteries and veins were widely separated. Whereas the uterine veins open through the basal plate, he believed the mouths of the uterine arteries were to be found distributed over the surface of the placental septa, which he mistakenly considered to be of maternal origin.

Spanner (14) refuted these findings and instead proposed that although arterial openings are distributed at intervals over the basal plate, the venous openings are found exclusively at the margins of the placental disc. In his view, arterial blood flows toward a relatively villus-free area beneath the chorionic plate, the subchorial lake, and then disperses laterally into a marginal sinus before entering the uterine veins [see Fig. 66 in (15)]. Spanner formulated his theory on the basis of injection studies, and although it was enthusiastically received at that time, it is likely that he was misled by artifacts arising from premature hardening of his injection medium, celloidin.

Resolution of the problem of arteriovenous shunting in the intervillous space was finally provided by the introduction of cineradiography. This was pioneered in the human by Borell and coworkers (16), but the problem of radiation dosage limited systematic study. Again, the macaque colony at the Carnegie Institute played a vital role, and it was largely through the elegant work of Ramsey and her collaborator, Donner, that the "physiologic concept" was confirmed. In this, it is proposed that arterial blood is delivered from the maternal spiral arteries through openings in the basal plate and that the incoming force sends it in fountain-like spurts toward the chorionic plate [see Fig. 67 in (15)]. These are often referred to eponymously as *Borell jets*. As blood disperses laterally between the villi, it displaces the existing supply into the mouths of the uterine veins. Indeed, it appears that the force of this incoming blood that shapes the fetal villous tree into a series of lobules centered over the arterial openings (17), and these act as independent maternal–fetal exchange units.

The midpart of the 20th century saw other major advances in the correlation of structure and function, resultant again on the introduction of new technologies. During the 1940s, the elaboration and refinement of histochemical techniques, pioneered in their application to the placenta by workers such as Dempsey and Wislocki, enabled enzymatic activity and hormone synthesis to be localized to specific sites and tissues. Furthermore, it became possible to explore how patterns of activity may vary at different stages of gestation and under different antenatal conditions. Shortly after, in the next decade, the cellular organelles involved could be visualized through the advent of the transmission electron microscope.

With the superior resolution offered by this instrument, the detailed nature of, for example, the microvillous border, the vacuolar transport systems, the synthetic and degradatory organelles, and the relationship between the syncytium and the cytotrophoblast at once became apparent (18,19). The images obtained provided powerful evidence that the trophoblast does not act as a simple homogeneous semipermeable membrane and influenced physiologists greatly in their thinking on the mechanisms of transplacental transport.

Since these early studies, advances in microscope technology and in allied tissue preparative techniques have permitted more sophisticated structural studies to be performed. Membrane receptors, growth factors, extracellular matrix molecules, and other molecules regulating placental development can be localized and quantified by autoradiographic and immunolabeling procedures. Cytoskeletal proteins can now be mapped in three dimensions using fluorescent probes and confocal microscopy. Models of how placental growth and differentiation are regulated have been proposed on the basis of these findings and corroborated or refuted by subsequent in vitro studies. In situ hybridization permits messenger ribonucleic acids (mRNAs) to be localized, and so it is now possible to dissect placental structure, development, and function down to the molecular level. Most recently, the development of molecular techniques to manipulate the expression of individual genes within specific cell lineages of the mouse placenta is allowing the impact of particular growth factors in particular locations on placental structure and function to be elucidated (20,21). In the light of these new findings, it is necessary to update and reappraise our theories on placental development constantly.

The placenta is manifestly the result of maternal–fetal interactions, however, and mention must be made of the enormous contribution that immunology made during the second half of the 20th century. For many years it had been known that trophoblast extends beyond the placenta down the necks of the spiral arteries, in a process that destroys their media and is thought to be necessary to ensure an adequate uteroplacental blood supply. Trophoblast also migrates through the endometrium between the uterine glands, mingling with cells of the maternal immune system. The discovery of the major histocompatability complex led Medawar to draw comparisons between the relationship of the fetus to the mother and that of an allograft to its host. Despite much research, however, this self/non-self approach has not proved successful in explaining the maternal–fetal relationship, and it is now accepted that pregnancy cannot be viewed as simply acceptance or rejection akin to a transplant. More recently, advances in our understanding of histocompatibility antigens and of the effects of cytokines released through cellular interactions between these ligands and uterine natural killer cells and other leucocytes are beginning to clarify how the disparate genetic tissues may successfully coexist (22).

COMPARATIVE PLACENTATION

When one reviews the wide array of placenta types demonstrated by even the domestic mammals, it is little wonder they have been the source of such confusion over the centuries. Interspecies differences are marked, and it is essential they continue to be borne in mind when extrapolating physiological, endocrinological, immunological, or any other data from the animal to the human situation. All placentas function to supply nutrients to the fetus, but the pathways involved may be very different (23). This section highlights some of the more major differences, but for details on individual species, readers are directed to the specialist texts (8,24). Because the mouse placenta is increasingly being used as a model for the human, given the possibilities for genetic manipulation offered by transgenesis, special consideration is given to this species later in this chapter.

Placental types may be classified at several complementary levels, reflecting, for example, both gross and microscopic structure, the histological nature of the maternal–fetal interface, or the relative directions of the maternal and fetal blood flows. The most fundamental level relates to the origin of the fetal vessels that vascularize the chorion.

The chorion is intrinsically avascular, and so for it to take part in maternal–fetal exchange it must develop a functional circulation. This may be derived from vessels running in the extraembryonic mesodermal covering of either the yolk sac or the allantois. The former situation gives rise to a *choriovitelline* placenta, and this represents the main form of placentation in many marsupials. Among most mammals, however, the yolk sac is a very transient structure, functioning only for nutritive and respiratory exchange in the earliest stages of pregnancy. Notable exceptions to this general rule are the rodents and lagomorphs in which the yolk sac often persists throughout gestation, although it does not actually vascularize the chorion. In some species, such as the mouse and rabbit, the yolk sac is highly elaborated into the complete inverted type. Breakdown of overlying chorion and of the outer wall of the yolk sac exposes the endodermal lining to the secretions of the uterine glands. This represents an important pathway in the absorption of proteins and the transport of maternal immunoglobulins (25), which continues to function until term.

Nonetheless, in these species, as in most mammals, the definitive placenta is vascularized by vessels associated with the allantois. The allantois is a diverticulum of the hindgut, whose initial embryonic development is slightly later than that of the yolk sac. It may expand into the extraembryonic coelom as a large fluid-filled sac in species such as the pig, cow, and horse or remain a rudimentary structure penetrating only the proximal end of the connecting stalk, as in the human. In either case, development of its vasculature is prolific, and on making contact with the chorion it establishes the *chorioallantoic* placenta. Further description is limited to this form of placentation, in view of its paramount importance in mammalian maternal–fetal exchange.

Placental Shape

The second level of classification to be considered relates to the gross morphology of the placenta. Four main types are recognized, and the basis of the classification is whether physical interaction between the maternal and fetal tissues occurs over all the available surface of the chorionic sac or whether it is restricted to specialized regions. In the simplest situation, as in, for example, the pig, interaction takes place over virtually the entire sac, and so these placentas are described as *diffuse* (Fig. 2A). By contrast, interaction only occurs in ruminants opposite specialized nonglandular areas of the endometrium known as the caruncles. The number, shape, and size of the caruncles are species specific, ranging in number from 100 to 120 in the sheep to only 4 in some types of deer. Intervening areas of chorion are smooth and relatively avascular, and so the resultant placentas are termed *cotyledonary* (Fig. 2B). Among several unrelated orders, including the carnivores, the physical interaction is restricted to an equatorial belt encircling the chorionic sac, giving rise to a *zonary* placenta (Fig. 2C). In *discoidal* placentas, which may be single or double, interaction is confined to a roughly circular area. This situation is characteristic of the primates and many laboratory rodents (Fig. 2D).

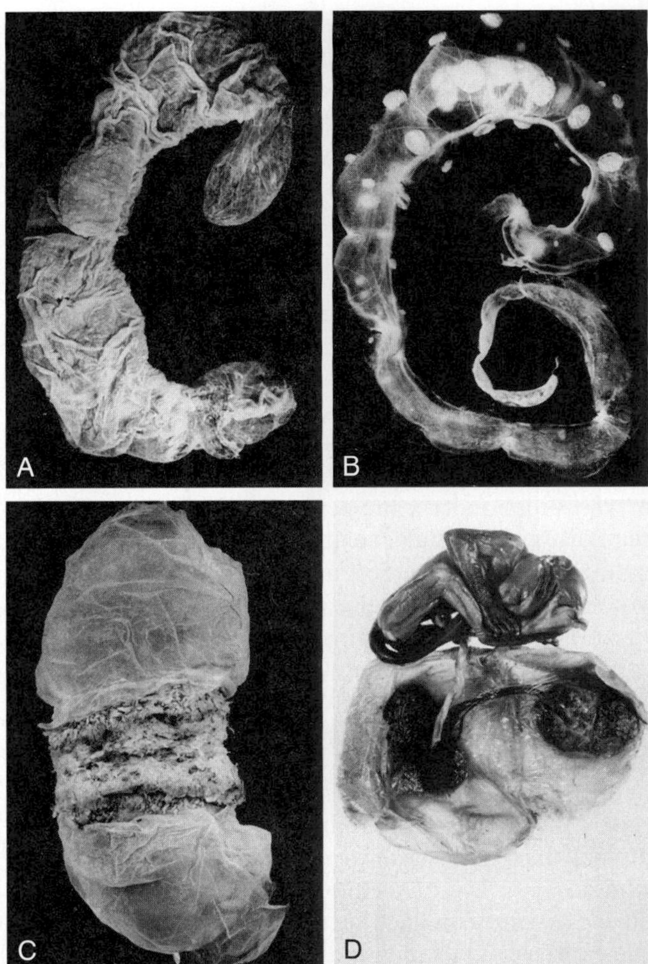

FIG. 2. Montage showing the (**A**) diffuse placenta of a domestic pig, (**B**) cotyledonary placenta of a cow, (**C**) zonary placenta of a dog, and (**D**) double discoid placenta of a leaf-eating monkey.

Maternal–Fetal Interdigitation

Within each of these gross forms, variations exist in the manner by which the maternal and fetal tissues interact. Again, the simplest situation, *folded*, is typified by the arrangement in the pig. The uterine surface is thrown into a series of undulations, referred to as primary folds, and the fetal chorion is initially draped over these (Fig. 3A). As gestation advances, a series of smaller ripples, the secondary folds, develop at the interface, so increasing the surface area of contact. A generally similar but more exaggerated type of folding is seen among the carnivores. Within the zonary placenta of the cat or dog, the maternal and fetal tissues interdigitate through an elaborate series of tall, slender, parallel folds or sheets. Because of their size and orientation, this arrangement is referred to as *lamellar* (Fig. 3B). In other species, for instance some of the higher primates, the parallel folds are not complete and so arrays of broad palmate branches are created (Fig. 3C). This pattern is described as *trabecular* and foreshadows the development of the *villous* configuration in which the fetal tissues form a three-dimensional tree-like structure, repeatedly branching into more slender units (Fig. 3D). Finally, the maternal and fetal tissues may each form a highly complex three-dimensional framework, which interlocks in a similar fashion to the substance and pores of a sponge (Fig. 3E). This *labyrinthine* arrangement is found in the placentas of rodents, lagomorphs, and insectivores.

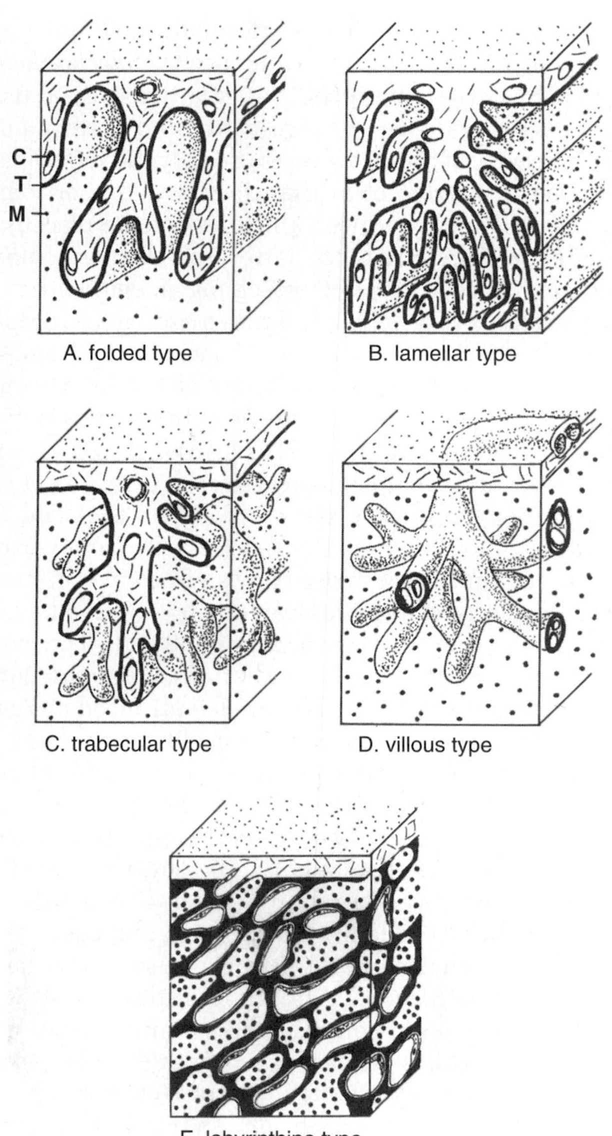

FIG. 3. Interdigitation between the maternal and fetal tissues at the gross morphological level may be of the (**A**) folded, (**B**) lamellar, (**C**) trabecular, (**D**) villous, and (**E**) labyrinthine type. M, maternal tissue or maternal blood (*larger dots*); fetal trophoblast (*black*); and C, fetal capillaries and fetal connective tissue.

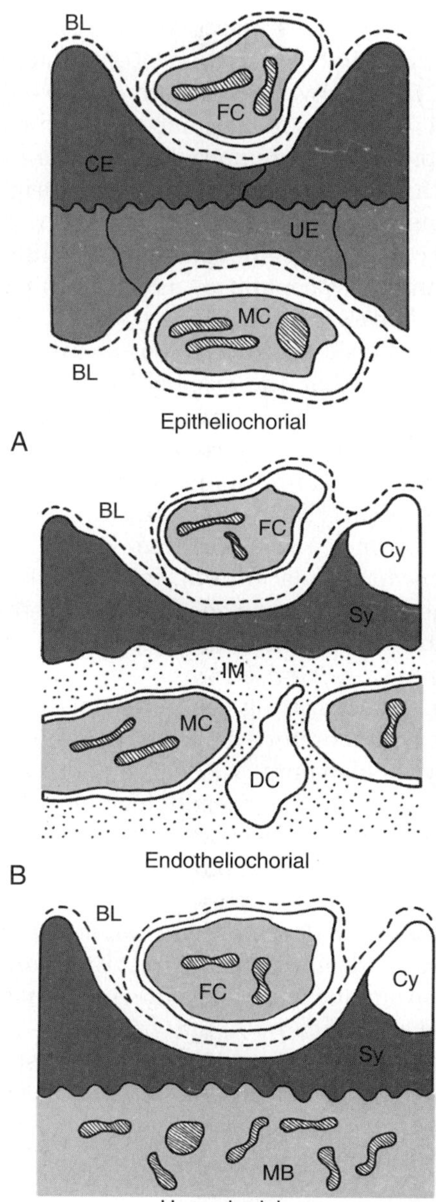

FIG. 4. Tissue layers of maternal-fetal barrier according to Grosser classification of (**A**) epitheliochorial, (**B**) endotheliochorial, and (**C**) hemochorial. MC, maternal capillary; MB, maternal blood; FC fetal capillary; BL, basal laminas; CE, chorionic epithelium (trophoblast); Sy, syncytiotrophoblast; Cy, cytotrophoblast; UE, uterine epithelium; and MI, maternal interstitium. The fetal components remain three layered, whereas the maternal components are reduced step by step until, in the hemochorial situation, the trophoblast comes into direct contact with the maternal blood.

Maternal–Fetal Barrier

Allied to these gross morphological and microscopical variants are differences in the fundamental relationship between the trophoblast and the uterine tissues. In most species, the trophoblast is simply apposed to the uterine epithelium, and there is no destruction or invasion of the maternal tissues. The conceptus therefore remains in the uterine lumen throughout gestation, and use of the term *implantation* is somewhat of a misnomer. Nonetheless, this situation is often referred to as "central implantation," and the histological relationship between the fetal and maternal tissues is classified as *epitheliochorial* (Fig. 4A). Although individual species variations exist in the detailed nature of the placental barrier, this arrangement is found in the pig and horse. In ruminants, the situation is more complex in that fetal binucleate cells migrate and fuse with the uterine epithelium, creating a syncytium of mixed maternal and fetal origin. The term *synepitheliochorial* was introduced to describe this relationship (24).

By contrast, invasion of the maternal tissues is seen in five apparently unrelated orders of mammals, namely, the carnivores, insectivores, rodents, bats, and primates. Limited destruction of the endometrium occurs in the carnivores, in many species of insectivores and bats, and among some lower primates. The uterine epithelium is removed, and in the carnivores this typically occurs opposite the equatorial zone of the chorionic sac. The trophoblast is now apposed to the maternal capillaries embedded in an acellular matrix of basal lamina-type material, a relationship classified as *endotheliochorial* (Fig. 4B).

Further invasion results in erosion into the maternal vessels, so that the trophoblast is now bathed directly by the mother's blood. This is termed *hemochorial* placentation (Fig. 4C), and is typical of the rodents and primates. Even within this category, variations occur. Depending on whether the trophoblast is one, two, or three layers thick, the *hemomonochorial* placenta of, for example, humans and guinea pig, the *hemodichorial* placenta of the rabbit, and the *hemotrichorial* placenta of the rat and mouse are recognized.

The layers separating the maternal and fetal circulations may therefore be a combination of cellular and syncytial elements, and so the detailed histological nature of the interhemal membrane is of key importance when considering the exchange properties of a particular placental type (23). Despite this seemingly aggressive behavior of the trophoblast, in most cases the depth of invasion is still relatively superficial and limited so that the bulk of the conceptus remains within the uterine lumen. This is known as *superficial implantation*, and it is only among the great apes that the blastocyst becomes completely embedded within the endometrium and *interstitial implantation* takes place.

Vascular Arrangement

A potential danger of classification systems is that once proposed, the information contained within them may be wrongly extrapolated to support hypotheses in fields quite different from those in which the scheme was originally devised. The history of the Grosser classification exemplifies this point.

When first put forward, the histological classification scheme provided a major advance by clarifying the different types of maternal–fetal interface observed among mammals. Assumptions were soon made, however, that the permeability of the placental barrier is dictated by the numbers of layers separating the two circulations. The epitheliochorial placenta was thus considered to be of lower efficiency and more "primitive" than its hemochorial counterpart, a supposition that dogged placental physiology for many decades.

As mentioned previously, although the number of tissue layers present may influence the *mechanisms* by which active transport of vesicular transfer occurs, there is no evidence to suggest it has any bearing on the efficiency of these processes. Equally, for diffusional transport, it is now clear that the rate is proportional to the area and the mean thickness of the placental barrier. With regard to the latter, it is a common phenomenon in many species possessing an epitheliochorial placenta for the fetal capillaries to invaginate deeply into the chorionic epithelium as gestation advances. As a result, the overall thickness of the interhemal barrier may be little different from that in the hemochorial situation, a fact that is lost when comparing placentas on their histological basis alone.

Another important determinant of the rate of diffusional exchange is the relative directions of the maternal and fetal blood flows. Several different arrangements can be envisaged, and these all have contrasting exchange characteristics (23,26). The most inefficient system would be that of *concurrent* flow, in which both the maternal and fetal bloodstreams run in the same direction. Perhaps for this reason, it does not seem to form the basis of any placental circulation. At the opposite extreme is the *countercurrent* system, in which the two bloodstreams run in opposite directions. This is highly efficient, theoretically allowing arteriovenous equilibration, and is seen in the placentas of rodents, lagomorphs, and equids.

In many ruminant species, there seems to be a mixture of both concurrent and countercurrent flow at different points along the villous interdigitations. As a result, the net effect is considered to be *cross-current*. A more clearly defined cross-current flow is seen within the lamellar placenta of carnivores. Finally, in those higher primates possessing a villous hemochorial placenta, the highly three-dimensional branching of the villous tree ensures an unpredictable and varying combination of all three patterns of relative blood flow. The term *multivillous* flow has been used to describe this situation.

These differences in the geometric arrangements of the fetal and maternal vascular networks and their influence on diffusional exchange are reflected in the fetoplacental weight ratio at term (pig, cross-current 9:1; rodents, countercurrent up to 20:1) (26). Again, on the basis of the Grosser classification, one might suppose that the villous hemochorial placenta of the human, providing such intimate contact between the two circulations, would be highly efficient. However, a combination of arteriovenous shunting through the intervillous space and multivillous blood flow conspire to ensure that the fetoplacental weight ratio remains at a lowly 6:1. Maintaining the maternal circulation in a geometrically defined network clearly confers some advantages.

Placental Evolution

The morphological arrangements of the placental membranes have in the past been used as a basis for assessing the phylogenetic relationships between mammalian species, in particular among the primates (8,27,28). In general, the primitive or ancestral form appears to be the choriovitelline placenta that forms the definitive placenta of marsupials (except bandicoots) and yet is only a transient structure in most eutherian mammals. The chorioallantoic placenta has assumed many of its functions, but is there an evolutionary link between the epitheliochorial, endotheliochorial, and hemochorial types? The advent of molecular analyses of large data sets has shed further light on phylogenetic relationships (29–31), and the data suggest that placental mammals fall into four superorders. The divergence of these ancestral groups was followed by periods of isolation caused by separation of the continental land masses, and this allowed for the convergent evolution of equivalent features, both between and within the different superorders. Of particular relevance, it is clear from all lines of evidence that hemochorial placentation arose independently in each of the four superorders, and at least four times within the superorder Glires and Euarchonta, to which the human belongs. Hence, although the placental interface may appear equivalent at the gross morphological level in various species, the fine structure and the developmental processes by which they are achieved may be quite different (31).

Analyses of the genetic control of placental development have also provided important insights into placental evolution. Thus, it is clear that most of the genes essential for normal placental development play key roles in the development of other organs, and the number of placental-specific genes is surprisingly small (20). Evolution of the placenta therefore did not rely on the emergence of a new set of genes, but rather existing genes were recycled to take on new functions.

MACROSCOPY OF THE DELIVERED PLACENTA

The full-term human placenta is usually a circular disc-like organ (Fig. 1). In approximately 10% of cases, however, it displays abnormal shapes, such as bilobate placenta, succenturiate placenta, duplex placenta, bidiscoidal placenta, zonary placenta, or placenta membranacea (32). The latter three types are comparable with placental shapes of other species (see Comparative Placentation) and may reflect implantation toward the margins of the uterus where the conceptus may impinge on both the anterior and posterior walls. At term, the delivered organ is, on average, 22 cm in diameter, 2.5 cm thick at the center, and weighs 470 g. These data show considerable individual variation and, moreover, are strongly dependent on the mode of delivery. Factors such as the time of clamping the umbilical cord are critical because loss of fetal and/or maternal blood can have a major impact on the dimensions of this highly vascularized organ (33).

The umbilical cord most frequently inserts in a slightly eccentric position on the fetal surface or *chorionic plate*. The avascular and glossy amnion covers the chorionic plate, including the chorionic vessels (Fig. 5). The latter are continuous with those of the umbilical cord and branch in a star-like pattern, their final branches supplying the villous trees. When arteries and veins cross, the arterial branches usually lie superficial. White opaque stripes often accompanying the larger chorionic vessels are due to locally increased number of collagen fibers. Corresponding opaque spots located between the chorionic vessels (bosselations) usually point to larger subchorionic deposits of fibrin-type fibrinoid, or Langhans' fibrinoid.

The uterine (maternal) surface of the placenta, or *basal plate*, is an artificial surface, originating from the separation of the organ from the uterine wall. It is composed of a heterogeneous mixture of trophoblastic and decidual cells, embedded in large amounts of extracellular debris, fibrinoid, and blood clot. An incomplete system of flat grooves or deeper clefts subdivides the basal surface of the placenta into 10 to 40 slightly elevated areas called lobes. Inside the placenta, the grooves correspond to the septa (Fig. 5). The grooves are the postpartal results of tearing at sites of minor mechanical resistance, because the basal central parts of the septa are often characterized by necrotic zones and pseudocysts. The septa should not be misunderstood as true separating structures that subdivide the intervillous space into chambers. Rather, they are irregular pillars or short sails that only trace the lobar borders.

The lobes show a fairly good correspondence with the position of the villous trees. From the chorionic plate at term, 60 to 70 villous trees (or fetal lobules) arise. Thus, according to Boyd and Hamilton (3) and Kaufmann (34), each lobe is occupied by one to four villous trees (Fig. 5). Small marginal lobes are likely to be occupied by only one single villous tree and thus correspond to what Schuhmann (35) described as a placentone. Near the placental margin, where the chorionic and basal plates join and form the smooth chorion or *chorion laeve* (Fig. 5), the transparency of the chorionic plate decreases, thus forming a largely

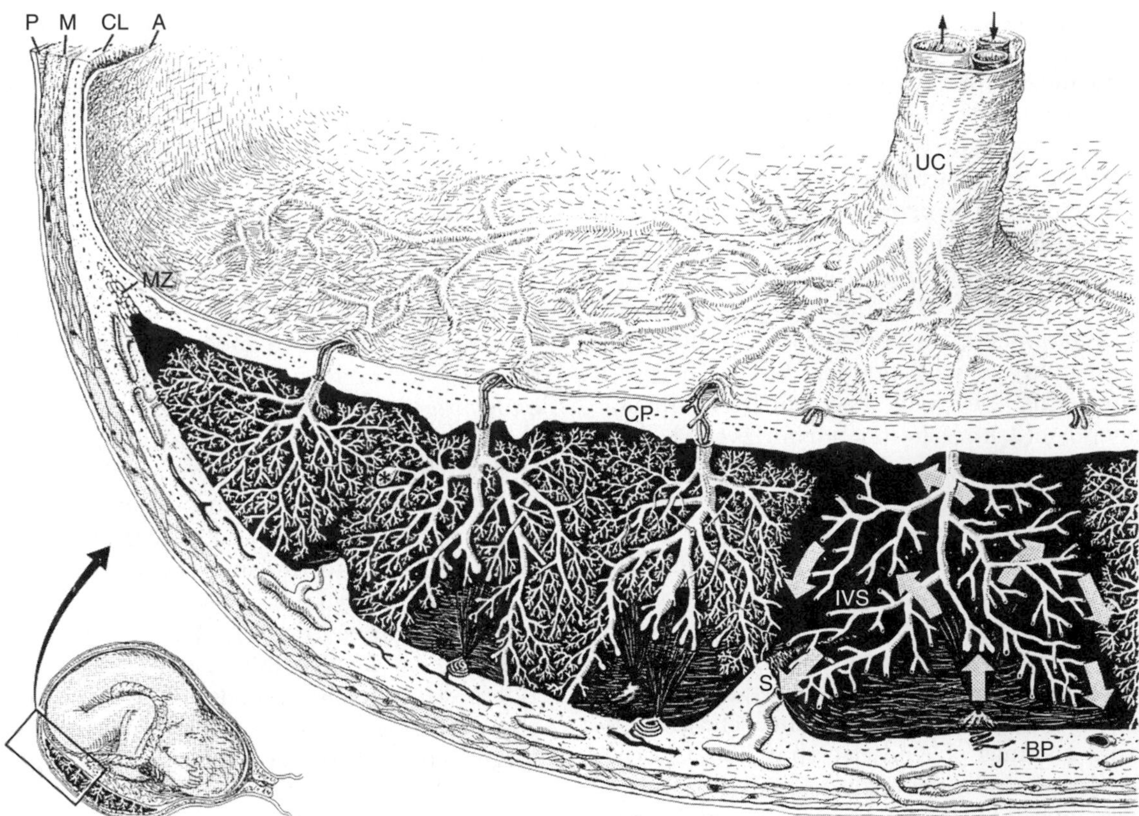

FIG. 5. Summarizing survey diagram of a nearly mature human placenta in situ. Loose centers of villous trees, arranged around maternal arterial inflow area, are frequent features. P, perimetrium; M, myometrium; CL, chorion laeve; A, amnion; MZ, marginal zone between placenta and fetal membranes with obliterated intervillous space and ghost villi; *, cell island; S, septum; J, junctional zone; BP, basal plate; CP, chorionic plate; IVS, intervillous space; UC, umbilical cord. [Reproduced from (389) with permission.]

incomplete, opaque, subchorial closing ring that connects the placenta with the rest of the fetal membranes.

Structurally impressive parameters such as placental shape and site of cord insertion are, with the exceptions described below, usually regarded as functionally unimportant. However, because they may reflect abnormal orientation of the blastocyst at the time of implantation or abnormal remodeling of the chorion frondosum into the definitive placenta, their potential association with some compromise of placental function cannot be entirely discounted. There are two abnormalities that are worthy of mention, because their presence is more directly associated with adverse pregnancy outcome. Velamentous insertion of the umbilical cord describes the situation when the cord is attached not to the placental disc but to the adjacent chorion laeve. It most likely reflects mal-orientation of the blastocyst at implantation and occurs in approximately 1% of pregnancies. The danger is that because the umbilical vessels ramify over the smooth chorion to reach the placenta, they are vulnerable to being torn when the membranes rupture at birth, resulting in antepartum hemorrhage.

The second condition is placenta circumvallata, in which the closing ring is peripherally undergrown by villous trees. In such cases the membranes are connected superficially to the chorionic plate, the margin of which is often fibrosed and inelastic. Because of the abnormal rigidity conferred, the condition is associated with a relatively high rate of premature rupture of the membranes, antepartum hemorrhage, and preterm onset of labor. For more detailed accounts of placental pathology, readers are directed to the specialist texts (36).

Terms such as *fetal placenta* for the chorionic plate, including the villous trees and intervillous space, or *maternal placenta* for the basal plate should be strictly avoided because they are inappropriate, misleading, and often cause misinterpretation (e.g., as soon as morphologically inexperienced scientists isolate respective parts of the organ and trusting in their putative judgment designate material to be solely of maternal or fetal origin). Both plates usually represent a colorful mixture of fetal and maternal tissues. Macroscopic dissection of placental structures can never guarantee tissues entirely of fetal or maternal composition. A corresponding warning is necessary regarding the placental bed, which is often thought to represent only maternal remains after separation of the placenta. Trophoblastic cells that invade the endometrium, even reaching the myometrium, remain in utero after delivery.

Multiple Pregnancies

In multiple pregnancies, a separate chorionic sac forms from each blastocyst. These sacs may develop independently of one another to form individual placentas, or if implantation occurs close together, the placentas may fuse and form a single mass. In the latter situation, the two apposed layers of chorion and their associated amniotic coverings always separate the fetuses, and so in the case of twins these are referred to as dichorionic diamniotic placentas. This arrangement occurs in approximately 50% of dizygotic twinning and in monozygotic twinning when the conceptus splits at the morula stage (Fig. 6). If splitting occurs at a later stage in monozygotic twinning, then two inner cell masses, and hence two amniotic sacs, are formed within a single chorionic sac. This leads to the formation of a monochorionic diamniotic placenta, where the fetuses are separated only by the apposed layers of amnion. More rarely, splitting occurs at the inner cell mass stage, and this results in the formation of a monochorionic monoamniotic placenta. This is associated with a high rate of intrauterine fetal death due to physical interactions between the fetuses and their umbilical cords becoming entwined and compressed.

Another complication of monochorionic placentation is that vascular anastomoses form between the placental territories of the two fetuses in 95% to 100% of cases. This can lead to a circulatory imbalance and the twin transfusion syndrome, in which one twin becomes the dominant recipient and the other becomes an emaciated donor. The recipient is often larger than normal, polycythemic, and associated with polyhydramnios, whereas the donor is growth retarded, anemic, and associated with oligohydramnios. In the most severe cases, the donor may die in utero and be delivered as a *fetus papyraceous* (3). The circulatory imbalances can be corrected by either drawing off excess amniotic fluid from the recipient sac or by laser ablation of the anastomoses. Vascular communications between the two placental territories are extremely rare in a dichorionic placenta, even though it may appear morphologically as one organ.

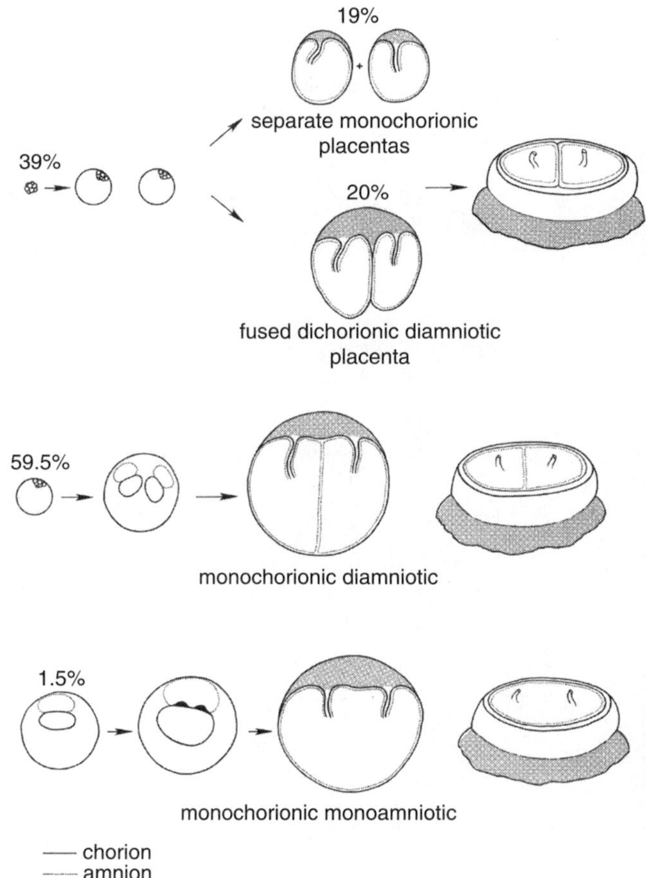

FIG. 6. Diagrammatic representation of the various arrangements of the placental membranes seen in monozygotic twins depending on the time of splitting of the conceptus. In monochorionic placentas, vascular anastomoses between the territories can lead to the twin transfusion syndrome. In dizygotic twins, two separate placentas or a single dichorionic placenta is formed with approximately equal frequency.

EARLY DEVELOPMENT

Prelacunar Stage

The trophoblast lineage is the first to differentiate during development, occurring at the transition between the morula and blastocyst stages, and the molecular events underlying this process are considered later (see Genetic Control of Placental Development). Morphologically, formation of the placenta begins as soon as the blastocyst has become attached to the endometrium, an event that occurs at about days 6–7 postconception. At this stage, the blastocyst has hatched from the zona pellucida and consists of an outer wall, comprising a single layer of uninucleate trophoblast cells, surrounding the blastocele and the inner cell mass (Fig. 7A). Cell lineage studies have now confirmed that the placenta and extraembryonic membranes are largely derived from

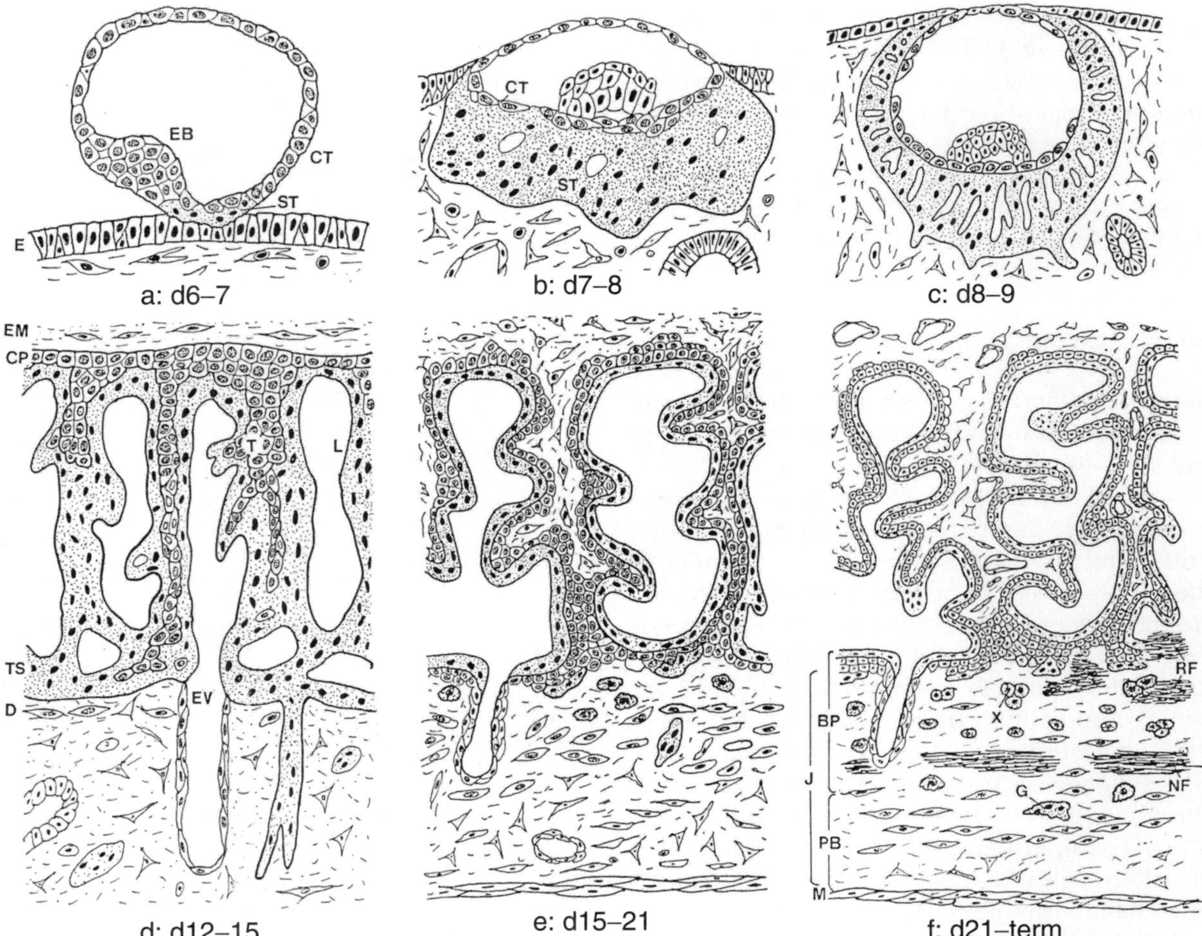

FIG. 7. Simplified drawings of typical stages of early placental development. **A, B:** Prelacunar stages; **C,** lacunar stage; **D,** transition from lacunar to primary villous stage; **E,** secondary villous stage; and **F,** tertiary villous stage. E, endometrial epithelium; EB, embryoblast; CT, cytotrophoblast; ST, syncytiotrophoblast; EM, extraembryonic mesoderm; CP, chronic plate; T, trabeculae and primary villi; L, maternal blood lacunae; TS, trophoblastic shell; EV, endometrial vessel; D, decidua; RF, Rohr's fibrinoid; NF, Nitabuch's fibrinoid; G, trophoblastic giant cell; x, X cells or extravillous cytotrophoblast; BP, basal plate; PB, placental bed; J, junctional zone.

the trophoblast, whereas the embryo and umbilical cord arise exclusively from the inner cell mass.

The inner cell mass confers an axis of symmetry on the blastocyst, and the overlying cells, referred to as polar trophoblast, are involved in making the first crucial attachments with the endometrial epithelium. Variations in the orientation of the blastocyst at the time of implantation almost certainly account for abnormalities in the site of insertion of the umbilical cord into the placental disc described previously (37). Implantation normally takes place in the upper part of the body of the uterus near the midsagittal plane and with almost equal frequency on the anterior and posterior walls. It is intriguing that a higher incidence of both abnormal placental shapes and eccentric insertions of the umbilical cord has been reported in pregnancies arising from in vitro fertilization techniques (38). One conclusion might be that the normal pattern of maternal–fetal interactions regulating implantation is disturbed in this group.

The earliest in vitro specimens available for study are those in the collection of Hertig et al. (12). The youngest of these, estimated to be at day 7 postconception, displays a blastocyst that is almost completely embedded within the endometrium (Fig. 7B). By this stage the trophoblast has differentiated into two fundamental subtypes. In contact with the maternal tissues and forming a mantle surrounding the conceptus is the multinucleated *syncytiotrophoblast*, whereas the trophoblast cells that formed the original wall of the blastocyst remain unicellular and are now referred to as *cytotrophoblast* cells. These act as a stem cell population, and their rapid division and subsequent fusion with the syncytiotrophoblast leads to the continual expansion of the mantle (39).

It has become clear over recent years that the maternal endometrial epithelium undergoes dramatic morphological changes before attachment of the blastocyst. The epithelial cells withdraw their microvillous brush border and replace it by flattened projections, termed pinopods or uterodomes (40). Only then are the trophoblast cells of the blastocyst able to firmly attach to the uterine epithelium. Exact knowledge of the process of invasion during implantation of the human blastocyst is still lacking. There is evidence from other mammals that display invasive implantation that it is not the uninucleate trophoblast cells that penetrate the epithelium but rather the multinucleated syncytiotrophoblast. Thus, syncytial fusion of uninucleate trophoblast cells seems to be a prerequisite for invasion of the blastocyst (3). Preventing syncytial fusion at that stage may therefore prevent successful implantation. But whether the first human syncytium that is generated here is comprised only of trophoblast or is a mixed syncytium of trophoblast and maternal epithelial cells, as found in the rabbit (41), still has to be determined. Syncytial fusion at the time of implantation is characterized by fusion of single uninucleate cells (trophoblast epithelium), whereas later during pregnancy syncytial fusion of the villous trophoblast of the placenta is characterized by fusion of uninucleate cytotrophoblast with the multinucleated syncytiotrophoblast (39).

Lacunar Stage

Toward the end of day 8 postconception, a series of fluid-filled spaces or vacuoles develop within the syncytiotrophoblastic mass (Fig. 7C). Although initially isolated, these vacuoles soon coalesce to form larger *lacunae*, which are separated by attenuated *trabeculae* of syncytiotrophoblast. This marks the start of the lacunar or trabecular stage of development, lasting from days 8 to 12 postconception.

At the end of this period, implantation may be considered to be complete. The conceptus is totally embedded within the uterine wall, and the endometrial epithelium has regrown over the site. The syncytiotrophoblastic mantle surrounds the conceptus but is thicker, and the lacunae are better developed beneath the embryonic pole. By this stage, the inner cell mass has transformed into the bilaminar germ disc, and extraembryonic mesoderm derived from the most caudal end of the future primitive streak (42) has begun to spread over the inner aspect of the trophoblast wall (Fig. 7D). The new combination of trophoblast and extraembryonic mesoderm is termed the *chorion*, and so the original wall of the blastocyst may now be referred to as the chorionic sac.

Formation of the lacunae subdivides the trophoblastic mantle into three fundamental zones: the primary chorionic plate facing the embryo, the lacunar system together with the trabeculae, and the trophoblastic shell abutting the endometrium. The lacunae are clearly the forerunners of the intervillous space of the mature placenta, whereas the primary chorionic plate and the trophoblastic shell become the chorionic and basal plates, respectively. In the mature placenta these are connected by anchoring villi, which represent the derivatives of the trabeculae as follows.

Commencing on day 12, cytotrophoblast cells originating from the primary chorionic plate penetrate into the trabeculae, extending as far as their tips by day 14 (Fig. 7D). At this point they extend laterally, establishing contacts with counterparts in adjacent trabeculae and so contribute to the trophoblastic shell. Indeed, many penetrate the syncytiotrophoblastic trabeculae completely and come to lie on the deepest side of the shell, in contact with the endometrial cells. Because of their location, these are referred to as *extravillous cytotrophoblast*, and as seen later, they represent a unique subset of trophoblast cells characterized by completely different functions as compared with the villous trophoblast. From this position, many migrate into the endometrial stroma, both between the uterine glands and around the spiral arteries, which they finally penetrate, as *interstitial trophoblast*, and down the lumens of the spiral arteries as *endovascular trophoblast* (43,44). One of their principal roles appears to be the transformation of the spiral arteries into dilated capacitance vessels, thus ensuring an adequate uteroplacental circulation later in gestation. This process involves the destruction of the arterial media and the replacement of the endothelial lining with trophoblast.

During the implantation process, the expanding syncytiotrophoblastic mantle comes into contact with the maternal capillary and venous plexus within the superficial endometrium. The proximity of the trophoblast induces these vessels to become sinusoidally dilated, and soon continuity is established between these vessels and the developing lacunae (12,45). As a result, maternal erythrocytes can be observed within the lacunae. This has frequently been equated with onset of the maternal intraplacental circulation, and the precocious onset of hemotrophic nutrition this allows has been considered an evolutionary advantage of the interstitial form of implantation displayed by humans. However, as commented by Hertig et al. (12), the erythrocytes are surprisingly sparse, and arterial connections with the lacunae cannot be traced at this stage of development (46). Any blood flow within the lacunae must therefore be, at best, a slow ebb of venous blood.

Early Villous Stage

Proliferation within the cytotrophoblastic cores of the trabeculae results in a considerable increase in their length. Around day 13, the trabeculae begin to develop side branches composed entirely of trophoblast that protrude into the lacunae (Fig. 7D). These are termed *primary villi*, and the lacunae can now justifiably be referred to as *intervillous spaces* (Fig. 7D).

Shortly after, extraembryonic mesoderm derived from the primary chorionic plate invades the trabeculae. It does not penetrate to their tips, and so the more distal parts of the trabeculae connecting with the trophoblastic shell remain composed of cytotrophoblast alone, often with the syncytiotrophoblastic covering being replaced by fibrinoid. These segments are referred to as the cytotrophoblastic cell columns and are the actively proliferating source of the extravillous cytotrophoblast. The mesoderm does penetrate the primary villi, however, and in doing so transforms them into *secondary villi* (Fig. 7E).

Within this villous mesoderm, hemangioblastic progenitor cells differentiate, and beginning around day 21, these give rise to the first fetal capillaries (47,48). The connections between the developing placental and fetal vessels remain very narrow at this stage, however, restricting the possibility of an effective umbilical circulation (49). Nonetheless, vascularization marks the formation of *tertiary villi* (Fig. 7F).

Hence, this form of villous classification reflects basic stages in the development of new villi, a process that continues throughout gestation. The first two categories represent transitory stages, and as tertiary villi accumulate they soon come to form the bulk of the placental tissue (50).

With further expansion and invasion, the extravillous trophoblast cells come into contact with the tips of the endometrial spiral arteries. However, free communication between the placenta and these vessels is not established until toward the end of the first trimester, for paradoxically the extent of the endovascular invasion is so great that aggregates of trophoblast cells plug their distal segments (51,52). Direct vision of the intervillous space by hysteroscopy reveals that for the first 12 weeks of pregnancy the space contains a clear fluid, free of maternal blood cells (53). This may arise as a plasma filtrate, percolating through the network of intercellular spaces within the plugs, but observations based on placenta-in situ specimens have revealed that the endometrial glands are another significant source of fluid (54). Glandular secretions are delivered into the intervillous space through openings in the trophoblastic shell until at least 10 weeks of pregnancy, and so the glands may represent an important source of potential nutrients, growth factors, and cytokines during the early stages of pregnancy.

One of the principal implications of the plugging of the arteries is that development of the conceptus at this stage takes place in a low oxygen environment. Measurements taken within the intervillous space in vivo confirm that the oxygen concentration is less than 20 mm Hg until 10 weeks of pregnancy (55,56), and metabolism is heavily anaerobic. The first trimester corresponds to the period of embryogenesis, when the principal organ systems are differentiating, and development at this stage is vulnerable to disruption by free radicals (57). Plugging of the spiral arteries may therefore serve to protect the fetus from free radical–mediated teratogenesis by restricting fetal exposure to oxygen at this critical phase of development (58,59).

This new interpretation of events during early pregnancy brings the human situation closer in line physiologically with that in most mammalian species. In many there is an extended period between arrival of the conceptus within the uterus and the establishment of the placenta, during which the fetus is more obviously supported by endometrial secretions. In view of their high lipid composition, these secretions have historically been referred to as "uterine milk" (9), and the growth factors they contain exert a powerful influence on placental development (60,61).

Toward the end of the first trimester, the trophoblastic plugs begin to dislocate, and evidence of maternal arterial inflow into the intervillous space is first observed. Doppler ultrasound studies have revealed that in normal pregnancies, arterial inflow starts in the periphery of the placenta and extends progressively toward the center (62) (Fig. 8). This finding correlates inversely with the depth of trophoblast invasion across the placental bed. Invasion is greatest in the center, where plugging might be expected to be most extensive (63). Onset of the maternal arterial circulation is associated with a threefold rise in oxygen concentration within the intervillous space (55,56). This may provide an important stimulus to trophoblast differentiation, as discussed later, but can also lead to placental oxidative stress and trophoblast damage, particularly in the syncytiotrophoblast (56,64).

Not surprisingly, morphological and immunohistochemical evidence of oxidative stress is greatest in the periphery of the placenta (62). Villi in this region are avascular, display a degenerate trophoblastic

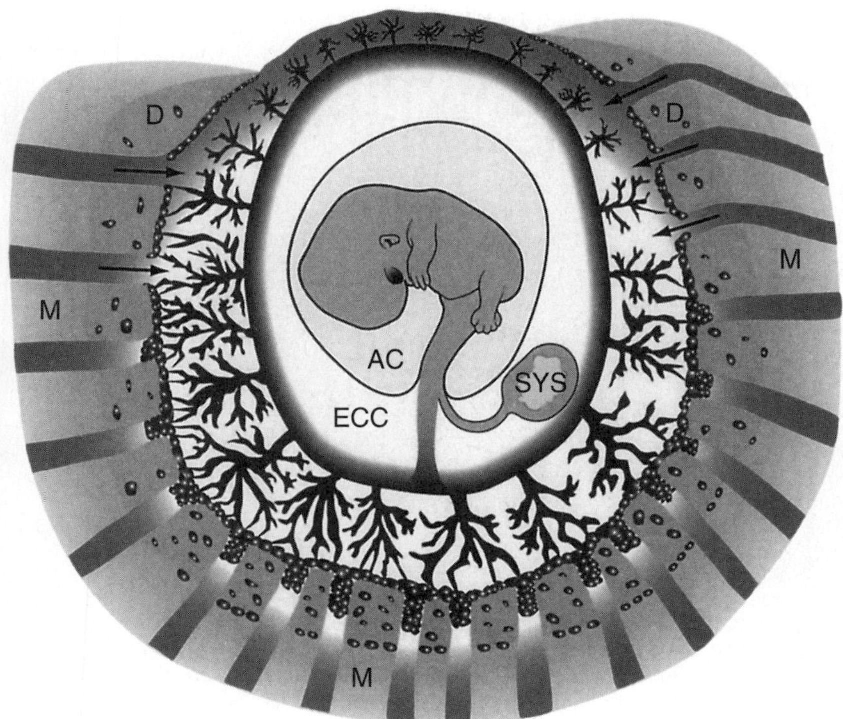

FIG. 8. Diagrammatic representation of the gestational sac at the end of the second month (8 to 9 weeks) showing the myometrium (M), the decidua (D), the developing placenta (P), the amniotic cavity (AC), and the secondary yolk sac (SYS) floating within the exocoelomic cavity (ECC). Onset of the maternal blood flow to the placenta (*arrows*) starts in the peripheral regions of the placenta, and the locally high concentration of oxygen may contribute to regression of villi over the abembryonic pole. [Reproduced from (308) with permission.]

covering, and gradually regress (Fig. 8). As a result, the chorionic sac becomes secondarily smooth except at the embryonic pole, forming the *chorion laeve* and the definitive discoidal placenta. This regression allows the placental membranes to rupture at birth without tearing the placental tissues and risking fetal hemorrhage but if incomplete can lead to the formation of succenturiate or accessory placental lobes described previously.

With onset of the maternal intraplacental circulation at the end of the first trimester, the placenta becomes truly hemochorial and able to support the rapid growth of the fetus during the second and third trimesters. Attention is now turned to the tissue components within the placenta that enable it to carry out its various functions.

BASIC VILLOUS STRUCTURE

Villous Trophoblast

Throughout gestation, the epithelium of the villi, the villous trophoblast, rests on a well-developed basement membrane, composed largely of type IV collagen, laminin, and fibronectin. The thickness of the basement membrane is very variable from point to point on the villous surface, but in general it increases toward term. Excessive thickening has been reported in pregnancies associated with maternal diabetes, hypertension, and cigarette smoking.

Cytotrophoblast (Langhans' Cells)

Lying deep to the syncytial layer always in direct contact to the basement membrane are the uninucleate villous cytotrophoblast cells (Fig. 9). These appear frequently in sectioned material from early gestation to the extent that for much of the first trimester the trophoblast is a two-layered epithelium. As gestation progresses, the cytotrophoblast cells are seen less often, and at term most of the villous trophoblast volume (about 85%) is attributable to the syncytiotrophoblast (65). This has led to the common misconception that the number of cytotrophoblast cells falls as pregnancy advances. In contrast, stereological studies have clearly shown that the total number of cells increases steadily until term (66). Mayhew et al. (65) quantified trophoblast nuclei throughout gestation and calculated the number of cytotrophoblast nuclei at 13–16 weeks of

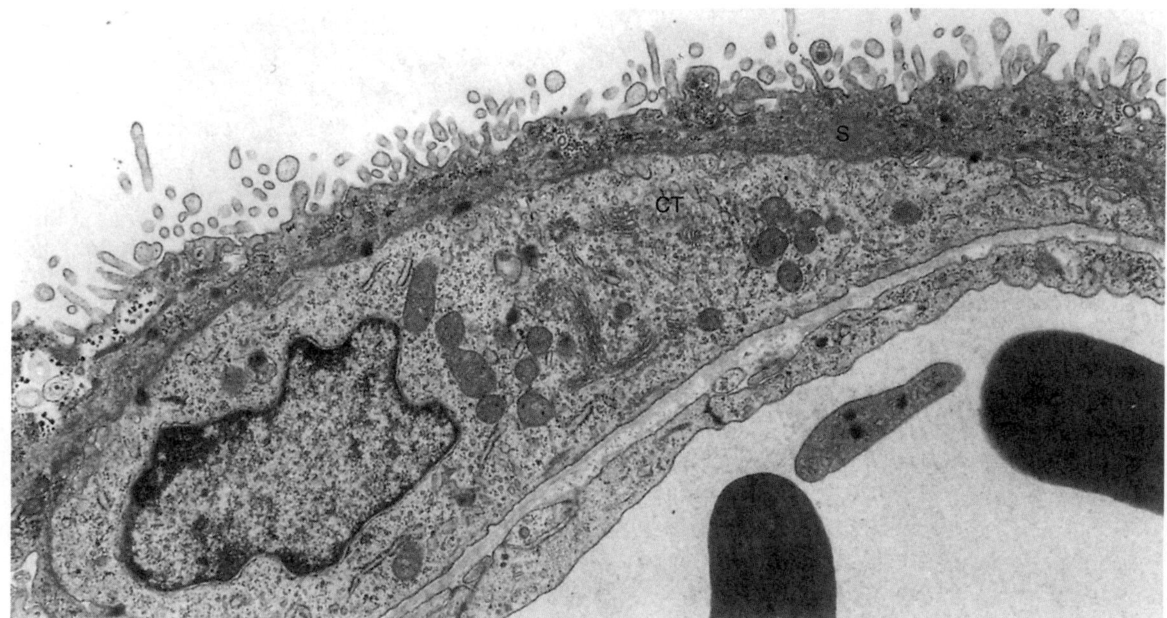

FIG. 9. Transmission electron micrograph of the placental barrier separating the maternal blood in the intervillous space (*above*) from fetal blood in the villous capillaries (*lower right*). The barrier consists of syncytiotrophoblast (S), cytotrophoblast cells (CT), the basal laminas and intervening connective tissue, and the fetal endothelium. (Magnification $\times$ 9,800.) [Reproduced from (390) with permission.]

gestation to be about 1×10^9, steadily increasing to 6×10^9 at 37–41 weeks of gestation. This implies continuous proliferation of the cytotrophoblast stem cells throughout pregnancy, which was confirmed using various proliferation markers (67). The volume of trophoblast associated with each cytotrophoblast nucleus appears to remain constant throughout gestation, indicating that trophoblastic growth is highly hyperplastic in nature (65,66).

The real situation is that as the villous surface rapidly expands, the cytotrophoblast cells become widely separated and so *appear* less numerous in sectioned material. This may have important consequences for maternal–fetal immunoglobulin transfer, for unlike the syncytiotrophoblast and endothelial cells, cytotrophoblast cells do not express Fcγ receptors (68). Transport of immunoglobulins may therefore only become possible in later pregnancy, when the cytotrophoblastic layer becomes incomplete.

Throughout pregnancy, villous cytotrophoblast cells do not come into direct contact with maternal blood. If the syncytiotrophoblast is damaged, fibrin-type fibrinoid is deposited on the surface as a result of maternal blood clotting and is used as a cover to separate maternal blood from these cells (69,70).

The control of cytotrophoblastic proliferation is uncertain, although it does seem to be heavily influenced by the prevailing oxygen tension. Culture in hypoxic conditions (71–73) or pregnancy at high altitude (74) leads to an increase in number. One factor that has been described as a trophoblast mitogen is hepatocyte growth factor, which is expressed by villous mesenchymal cells (75). The respective receptor of hepatocyte growth factor, the proto-oncogene protein product c-*met*, is expressed by cytotrophoblast (76). Deficiency of this potent stimulator of trophoblast proliferation leads to trophoblast maldevelopment (75) and intrauterine growth retardation (77).

Resting, or undifferentiated, cytotrophoblast cells are generally cuboidal in shape, and their cytoplasm displays a paucity of organelles (78,79). Secretory vesicles are infrequent, and the overall appearance is of a cell type that is metabolically quiescent. The expression pattern of cyclin D3 in villous cytotrophoblast led DeLoia et al. (80) to suggest that about half of the cells are in the cell cycle. McKenzie et al. (81) proposed that it is the interactions between a variety of proteins that are responsible for blockage of S phase and the cells' entrance into differentiation: (a) down-regulation of cyclin E; (b) inactivation of cyclin-dependent kinase-2 (Cdk2); (c) increased expression of Kip1, a cyclin-dependent kinase inhibitor; and (d) accumulation of hypophosphorylated and thus active retinoblastoma gene product (Rb).

Postproliferative differentiation leads to the formation of a smaller number of intermediate cells, which as their name implies are intermediate in morphological appearance between the resting state and syncytiotrophoblast (78,79). These cells contain large amounts of rough endoplasmic reticulum and numerous mitochondria and free ribosomes. These cells are more active, as evidenced by the display of enzymes for both aerobic and anaerobic glycolysis (36,82) and by the incorporation of high amounts of ^{3}H-uridine in vitro (83,84). The presence of high numbers of mRNA, including that for the human chorionic gonadotropin α subunit, has been demonstrated in intermediate cells (85,86).

The cascade of regulators of cytotrophoblast differentiation is still rather mysterious. Some distinct pathways with a variety of key molecules that regulate placental differentiation are known from mouse knockout models. The key molecules involve transcription factors, cytokines, matrix factors and other regulatory proteins (87–94). Other key molecules involved in the regulation of mitotic events, initiation of apoptosis, and control of cell death play active roles in differentiation and fusion of cytotrophoblast cells as well as the maintenance of the syncytiotrophoblast.

Differentiation and fusion of cytotrophoblast are orchestrated by a variety of transcription factors such as Hash-2 (the human homologue to the murine Mash-2), basic helix-loop-helix proteins, the Id family (inhibitor of DNA binding proteins), and Gcm-1 (glial cell missing-1) as well as by ligands of the nuclear receptor superfamily of proteins (90,94–98). The potential key regulators obviously differ depending on the trophoblast subtype, that is, villous cytotrophoblast or syncytiotrophoblast and extravillous trophoblast with interstitial or endovascular trophoblast (99). It was shown that species differences between mouse and human also exist in terms of trophoblast differentiation and the regulatory proteins involved. Differentiation of cytotrophoblast in both species shares specific regulatory proteins (e.g., Hash-2/ Mash-2), whereas other regulators are clearly different (Hand-1 is nearly undetectable in human cytotrophoblast) (93,100). Gcm-1 seems to be an important regulator for cytotrophoblast differentiation and fusion, not only in mice but also in humans. The human analogue of the *Drosophila melanogaster* neural transcription factor Gcm is expressed in differentiating cytotrophoblast and syncytiotrophoblast just before and after syncytial fusion (98,101,102).

The cytoplasm of highly differentiated cytotrophoblast cells comes to resemble that of the overlying syncytiotrophoblast, and finally the concentration of organelles clearly exceeds that of the covering syncytiotrophoblast (36), and fusion with the latter soon follows. The few fusion events that have been observed directly [see Fig. 6.15 in (36)] demonstrate that at least two characteristics are needed: a high degree of differentiation of the cytotrophoblast cells and at the same time a certain degree of regressive changes in the other fusion partner, the covering syncytiotrophoblast. The absolute increase in the number of cytotrophoblast cells throughout pregnancy implies that the rate of syncytial fusion is lower than the rate of proliferation and that both are tightly regulated processes.

Syncytial Fusion

Despite intensive research during the last decade, the molecular mechanisms of syncytial trophoblast fusion and its control are still poorly understood. Data suggest the involvement of several factors and proteins, but a clear cascade of events is still missing. An intimate contact with neighboring cells, maternal tissue, and extracellular matrix is fundamental for the generation and maintenance of the syncytiotrophoblast out of cytotrophoblast cells (103). Effectors such as epidermal growth factor, estradiol, glucocorticoids, and gonadotropic hormone have been identified to stimulate cytotrophoblast differentiation in vitro (89,104– 106).

Communication between cells depends on gap junctions enabling the flux of small signal molecules such as cAMP or Ca^{2+} from one cell to the other. The expression pattern of connexins, channels that cluster to build gap junctions, is not placenta specific. Nevertheless, it shows a pattern that is related to the stage of trophoblast differentiation (107–109). Different connexins are exposed between cytotrophoblast cells or between cytotrophoblast and syncytiotrophoblast (110). In villous trophoblast, connexins 40 and 43 are expressed in a characteristic temporal and spatial manner and seem to play important roles in differentiation and fusion of trophoblast (108,111–113). In particular, connexin 43 is one of the candidate channel proteins to be involved in syncytium formation, because it is expressed in certain trophoblast populations and is regulated by cAMP, human chorionic gonadotropin, and estrogen (107,108,112,114). Recent experimental work showed that direct involvement of connexin 43 in syncytial fusion is very likely (109,110).

In the rabbit uterine epithelium, the formation of an epithelial syncytium is introduced by the formation of gap junctions, as has been tested in pseudopregnant rabbits (115). The gap junctions establish an intercellular coupling of neighboring cells, finally inducing disintegration of the two separating plasma membranes. Firth et al. (116) described a similar role for gap junctions for syncytial fusion of trophoblast in the guinea pig. Desmosomes normally attach cytotrophoblast cells to the deep surface of the syncytium, and during recruitment gap junctions form in the intervening areas. Their formation appears to initiate fusion, and soon isolated remnants of the cell membranes, some still carrying desmosomes, are all that remain [for review, see (36)]. The circulation of these remnants within the syncytioplasm may result in the incorporation of redundant junctional complexes into the microvillous surface (117).

Accumulating data from the last few years make it very likely that mechanisms normally found in early stages of the apoptosis cascade are involved in the fusion process of villous trophoblast. The role of apoptotic processes and their regulation in syncytium formation has been characterized by Huppertz and coworkers, who suggested that syncytium formation is an effect of the initial stages of the apoptosis cascade, which are then inhibited and delayed within the syncytium (118–121).

Initiation of the apoptosis cascade within the cytotrophoblast and its control are still unclear. Placental macrophages (Hofbauer cells) secrete tumor necrosis factor (TNF) α (122), whereas cytotrophoblast cells express the respective receptor, TNF-R1 (86,123). In vitro experiments showed that addition of TNF-α to trophoblast cells or coculture of trophoblast cells with activated macrophages results in trophoblast apoptosis, and addition of TNF-R1 blocking antibodies leads to inhibition of trophoblast apoptosis in these systems (124,125). However, whether this system is active and involved in differentiation and fusion of villous trophoblast is still open and remains to be elucidated.

Early stages of the apoptosis cascade involve the activation of a subset of intracellular proteases, called initiator caspases. A variety of caspases is present as inactive proforms in the cytotrophoblast (126). The initiator caspase 8 is one of the few shown to be present in cytotrophoblast, not only as inactive proform but also as active enzyme (126). This enzyme seems to play a crucial role in cytotrophoblast differentiation and fusion because blockage of caspase 8 protein expression and enzyme activity leads to prevention of syncytial fusion and subsequently to the generation of a multilayer of uninucleate cytotrophoblast cells (121). The events taking place between caspase 8 activity and the direct process of syncytial fusion are still unclear. However, two classes of molecules have been shown to be involved in both caspase 8 cleavage and syncytial fusion:

1. Spectrin and its nonerythrocytic homologue, fodrin, are plasma membrane–associated cytoskeletal proteins that link cytoskeletal filaments to the plasma membrane. Fodrin is known to be associated with negatively charged phospholipids at the inner leaflet of the plasma membrane (127), and spectrin increases the resistance against fusion in red blood cells in vitro (128). Furthermore, fodrin is a substrate of initiator caspases such as caspase 8 and is typically cleaved in early stages of apoptosis (129,130). Immunoreactivity for fodrin is present in most of the cytotrophoblast cells but is absent in a subset of cytotrophoblast cells displaying morphological signs of high differentiation (86).
2. Mammalian cells actively sustain an asymmetrical distribution of phospholipids in their plasma membrane. Only the inner leaflet contains negatively charged aminophospholipids such as phosphatidylethanolamine and phosphatidylserine (131). In the outer leaflet, neutral phospholipids, including phosphatidylcholine, prevail. This asymmetrical distribution is achieved by an aminophospholipid translocase counteracting spontaneous flipping of phosphatidylserine and phosphatidylethanolamine to the outer leaflet (132). Inactivation or reversion of translocase activity and/or activation of counteracting enzymes such as floppases and scramblases results in the appearance of phosphatidylserine in the outer leaflet of the plasma membrane (phosphatidylserine-flip).

This phosphatidylserine-flip is characteristic for early stages of apoptosis (130,133,134); moreover, it is crucial for the process of syncytial fusion (135,136). Current views suggest that the phosphatidylserine-flip during apoptosis is an active process (133), that is, supported by the activity of scramblases. In contrast, the phosphatidylserine-flip during differentiation before syncytial fusion in the choriocarcinoma cell line BeWo is thought to be due to activation of the slower floppases (136,137).

In living placental villi ex vivo, a small subset of villous cytotrophoblast cells displays a phosphatidylserine-flip (118). As already depicted above, blockage of caspase 8 enzyme activity leads to a reduction in syncytial fusion and subsequently to the generation of a multilayer of uninucleate cytotrophoblast cells (121). Thus, mechanisms of the early apoptosis cascade seem to be involved in trophoblast differentiation and fusion. Taking into account both culture models, BeWo cell line studies and placental explant cultures, it is still open which of the enzymes (scramblase, floppase, translocase) is responsible for the phosphatidylserine-flip during differentiation and fusion of cytotrophoblast cells in vivo.

Cleavage of fodrin and flipping of phosphatidylserine seem to be basic mechanisms used as prerequisites for syncytial fusion. These changes in membrane composition and stability in the course of early apoptotic events contribute to the physicochemical properties needed for fusion. However, no apoptotic cell displaying a phosphatidylserine-flip normally fuses syncytially, and in most cases, only cells of the same type (myoblasts, macrophages, trophoblast) fuse syncytially with each other. Thus, more and other cell type–specific signals are needed as prerequisites for cell fusion.

Some years ago, the supporting roles of members of the ADAM family of proteins (membrane proteins with a disintegrin and a metalloprotease domain) for cell–cell fusion of myotubes and osteoclasts as well as for sperm–oocyte fusion were described [for review, see (138)]. The disintegrin domain provides specific binding properties to integrins of fusion partners. An integrated fusion peptide seems to be involved in the fusion process itself.

ADAM 12 is known to be involved in fusion of skeletal muscle, and its mRNA was detected in high levels in the human placenta (138,139). In vitro data from Martin et al. (140) suggested that the effects of the integrated fusion peptide are strongly enhanced by increasing the concentration of negatively charged lipids such as phosphatidylserine in the lipid bilayer. According to these authors, two factors seem to be required for syncytial fusion, the presence of members of the ADAM family and changes of electrostatic properties of the membranes.

It is intriguing that at the time of implantation in the marmoset, there is considerable retroviral expression at the interface between the syncytiotrophoblast layer and the cytotrophoblast cells, leading to speculation that retroviral proteins may be involved in the fusion process (141). A new player in the field of human trophoblast fusion is syncytin, the human endogenous retrovirus (HERV)–derived protein. Its necessity for human syncytiotrophoblast formation is supported by a number of observations. Syncytin mRNA expression is restricted to placenta apart from minor expression within the testis (142). Inside the placenta, it is expressed nearly exclusively in and near the syncytiotrophoblast. Moreover, syncytin has an intact open reading frame, whereas the open reading frames for other HERV-W genomic elements are defective (143), and a functional promoter is present in the syncytin gene (144). The syncytin promoter exhibits several binding sites for transcriptional regulation, which are involved in the control of proliferation and differentiation and in the expression of trophoblast-specific genes such as transcription factor activating proteins AP-1 and AP-2, specificity protein Sp1 and Oct, which binds to a DNA sequence termed the octamer motif (99,144–148). Syncytin interacts with a receptor protein that is known as the amino acid transporter ASCT2. After binding, the activated receptor exerts specific effects on cell differentiation and syncytialization in vitro (149,150). Although the signaling pathways after binding of syncytin to its receptor are still unclear, stimulation of trophoblast cells with cAMP agonists leads to an increase of syncytin mRNA and protein along with cell differentiation and fusion (106,150).

As one of the regulators of cytotrophoblast differentiation, Gcm-1 was described to be involved in regulating the gene expression of syncytin (151). Furthermore, Gcm-1 regulates syncytin gene expression in the choriocarcinoma cell lines BeWo and JEG3 and up-regulates syncytin-mediated cell fusion in vitro (151). The regulation of syncytin expression is highly important to fusing cells because overexpression of syncytin causes cell death in vitro after extensive cell fusion (151). Whether syncytin interacts with the membrane rearrangements during early apoptosis or whether it induces cell fusion like other exogenous retroviruses (152) are still open questions. Thus, the master genes, the route of molecular mechanisms, and the protein interactions in human syncytiotrophoblast formation still need to be elucidated.

Syncytiotrophoblast

In the human, only a few cell types differentiate into multinucleate syncytia via syncytial fusion: myoblasts to generate myotubes, macrophages to generate osteoclasts, and trophoblast to generate the syncytiotrophoblast. The syncytiotrophoblast maintains a polarity with an apical microvillous membrane adjoining the maternal bloodstream and a basal membrane adjacent to cytotrophoblast cells and a basement membrane.

The syncytiotrophoblast forms the outer covering of the villous tree (Figs. 9 and 10) and the surface of the chorionic and basal plate directed toward the intervillous space. It is a continuous multinucleated syncytial layer that contains variable concentrations of organelles (Fig. 9). Microvilli are found over the entire surface throughout gestation (Fig. 9). Although their density and morphology may vary according to the antenatal and fixation conditions (153), they provide an amplification factor of approximately sevenfold. Receptors for many factors, such as insulin, low-density lipoprotein, transferrin, and immunoglobulin (Ig) G, have been detected on the microvillous surface [for reviews, see (36,78)]. These receptors tend to cluster at the base of the microvilli where clathrin-coated pills are found (154). Proposed functions of the microvilli are that they present a large surface area containing receptors to the maternal blood and that movement of receptor and bound ligand into the region of a coated pit with subsequent internalization provides a means for concentration. The microvillous surface is also rich in different enzymes, particularly in alkaline phosphatase, which may exist in several forms. Other enzymes include 5′-nucleotidase, hexokinase, α-amylase, protein kinases, and galactosyl and sialyl transferases. Actin filaments provide a supporting framework for the microvilli and link with a dense meshwork of both microtubules and microfilaments lying just beneath the syncytial surface (155).

Within the cytoplasm are many pinocytotic vesicles, free ribosomes, mitochondria, lipid droplets, and different multivesicular and dense bodies (78,156). Endoplasmic reticulum, Golgi apparatus, and secretory droplets are also plentiful, confirming the wide range of metabolic activities in which the syncytiotrophoblast is involved. Mitosis has never been observed within syncytial nuclei, and transcription is reduced. Generation and maintenance of this layer therefore depend on the continual incorporation of cytotrophoblast cells into it (157).

In all stages of pregnancy, syncytial fusion of trophoblast takes place. The number of fusion events exceeds the needs for syncytiotrophoblast growth by about factor 6 (118), resulting in an excess amount of syncytial nuclei. This is balanced, after a few weeks of aging, by accumulation of old nuclei in so-called syncytial knots, which are shed from the syncytial surface into the maternal blood (65,118,158,159). The mean intrasyncytial survival time of a nucleus, between incorporation by fusion and shedding into the maternal blood, is about 3 to 4 weeks (118,120).

As has been described above, the early events of the apoptosis cascade take place inside the cytotrophoblast cells, providing tools for syncytial fusion. However, in most cells, execution of the apoptosis cascade takes less than 24 hours. Thus, there have to be mechanisms to prolong or even to stop the cascade within the syncytiotrophoblast for up to 3 weeks. One of the earliest reports of apoptosis-related proteins in the human placenta showed high concentrations of the apoptosis inhibitor Bcl-2 in the trophoblast (160). Other apoptosis inhibitors were detected in following years, such as Mcl-1 and XIAP (118,161).

Zhou et al. (162) showed that overexpression of Mcl-1 in hematopoietic cells delays apoptotic death for several days, and similar effects have been described for Bcl-2 [for review, see (163)]. Considerable amounts of the apoptosis inhibitors Mcl-1 and Bcl-2 are present in cytotrophoblast as mRNA and protein. During fusion, both forms are transferred into the syncytiotrophoblast by syncytial fusion (118,126). Here, both mitochondrial proteins may block furtherance of the apoptosis cascade in combination with XIAP, a potent inhibitor of active effector caspases (161). Using neuronal models, Marks et al. (164) showed that activation of effector caspases does not necessarily result in apoptotic death. If the caspases are inactivated again within a critical time interval, the cells survive and use caspase activity for neuronal plasticity. The time interval to rescue the cell varied between minutes and hours, depending on the size of the cell. Thus, multiple mechanisms seem to control and manipulate apoptosis in the syncytiotrophoblast.

Trophoblast Turnover

The necessity of a continuous turnover of syncytiotrophoblast with permanent syncytial fusion and permanent extrusion of syncytial knots can be explained by a general down-regulation of transcription in this tissue (36,84,157). Incorporation of tritiated uridine as a measure of transcriptional activity is hardly detectable and much less as compared with cytotrophoblast or other placental cells (84,157). Thus, the survival and functional activity of the syncytiotrophoblast depend on the transfer of mRNA, proteins, and organelles from the cytotrophoblast by syncytial fusion (36,165). Interestingly, down-regulation of transcription is a widespread phenomenon after starting the apoptosis cascade (166,167). The syncytiotrophoblast may be in a latent stage of apoptosis, and therefore transfer of mRNA and/or protein from cytotrophoblast to syncytiotrophoblast is necessary to maintain and feed the syncytiotrophoblast throughout pregnancy. If syncytial fusion is blocked in villous explant cultures, proteins that are normally expressed within the syncytiotrophoblast accumulate within the cytotrophoblast and do not reach the syncytiotrophoblast (73,121).

Aging of syncytiotrophoblast nuclei from fusion to extrusion into the maternal circulation is reflected by changes of nuclear shape. Freshly incorporated syncytial nuclei are large, ovoid, and rich in euchromatin. During the next 3 to 4 weeks, the nuclei become smaller and denser and finally display a dense aggregation of the chromatin beneath the nuclear membrane (65,70,118). Freshly incorporated nuclei are surrounded by high concentrations of rough endoplasmic reticulum, polysomes, and mitochondria. In parallel to nuclear aging, the endoplasmic reticulum degranulates into smooth endoplasmic reticulum (36). Enzyme histochemical studies revealed that aging and turnover inside the syncytiotrophoblast are accompanied by inactivation of mitochondrial and transport-related enzymes, whereas lysosomal enzymes and aminopeptidases are activated (82).

With increasing nuclear shrinkage and chromatin condensation, nuclear proteins that are known targets of execution caspases during late stages of apoptosis are cleaved and gradually lost. These proteins comprise lamin B, PARP, and topoisomerase IIα (118). Additionally, parts of the syncytial surface, mostly areas including syncytial knots, bind annexin V, thus suggesting a phosphatidylserine-flip (84,118). Those areas of the syncytiotrophoblast also show immunoreactivity for a caspase-dependent cytokeratin 18 cleavage product, indicating degradation of cytokeratin 18 by caspases (168). Degradation of cytoskeletal proteins is a general phenomenon during apoptosis and is thought to be responsible for the impaired anchorage of syncytial nuclei. The relatively freely moving old nuclei accumulate and appear to be sequestered into tightly packed clumps known as syncytial knots at the tips of villi. The driving force for nuclear accumulation may be provided by shear stress due to maternal blood flow in the neighboring intervillous space. This assumption is supported by the fact that nuclear accumulation does not take place at sites with arrested maternal circulation in vivo.

At all stages of gestation, it appears that syncytial knots break away, and the resulting syncytial globules are exported in the maternal blood, most becoming lodged in the pulmonary capillary bed. In the early 1960s it was estimated that up to 150,000 such globules enter the maternal circulation each day (169). More recent calculations (118,120) suggest that clustering and extrusion of aged parts of the syncytiotrophoblast into the maternal circulation involve apoptosis. About 50% of syncytial nuclei (320 nuclei per 10,000 μm^2 villous surface per 4 weeks during the first trimester of pregnancy) that entered the syncytiotrophoblast by syncytial fusion are subsequently lost by shedding into the maternal circulation. As revealed by TUNEL positivity and ultrastructural changes, these nuclei have executed the end stages of the apoptosis cascade (65,70, 118,170,171).

From the quantitative data above, it was calculated that the mean duration of the presence of a single nucleus inside the syncytiotrophoblast is 3–4 weeks (118,120). During that time, the nucleus becomes smaller and denser due to nuclear condensation (65). In the course of pregnancy, the rate of cytotrophoblast proliferation per villous surface unit declines (172); thus, by term only 13% to 15% of villous trophoblast volume is made up of uninucleate cytotrophoblast cells (65,173).

Syncytial knots with nuclear clusters inside detach from the apical membrane of the syncytiotrophoblast and can be found in uterine vein blood (174) as well as in pulmonary vessels of the mother (169,175), sometimes leading to trophoblast embolism inside the lung (176–178). It has to be stressed that these large apoptotic fragments of the syncytiotrophoblast can be observed in uterine venous, but not arterial, blood of pregnant women (174,179).

The multinucleate syncytial knots are products generated by apoptotic cleavage. As such, they are surrounded by a tightly sealed plasma membrane and do not release their content into the environment (i.e., maternal blood) and thus do not induce an inflammatory response of the mother. During placental pathologies such as preeclampsia, the normal apoptotic turnover of the syncytiotrophoblast seems to be altered, resulting in a more nonapoptotic release of trophoblast material (120,180). This necrotic or aponecrotic material might induce the endothelial damage and activation typical for preeclampsia.

Fixed Connective Tissue

One of the principal components of the stromal villous core is the population of fixed connective tissue cells. The appearance of these cells varies greatly, depending on the degree of maturity of the villi in question, and so they are described later in relation to the development of the villous tree.

Hofbauer Cells

Hofbauer cells are fetal macrophages that are found within the villous stroma at all stages of gestation. They possess a very characteristic morphology, for their surface is thrown into a complex system of lamellipodia and microplicae and their cytoplasm contains many vacuoles and lysosomes (Figs. 10 and 11). The origin of these cells has been the topic of much debate, but it now appears there may be different sources. In the first month of pregnancy, Hofbauer cells possibly originate from mesenchymal cells in the villous stroma (48,181), but this supply may be augmented by differentiation from fetal bone marrow–derived macrophages once the fetal circulation is established (36). Moreover, it has been shown that the macrophage population may increase through mitotic division (182).

Several diverse functions have been ascribed to Hofbauer cells, including the regulation of trophoblast differentiation and stromal fluid balance, modulation of vasculogenesis and stromal cell growth, and the absorption of immune complexes. Comparison of class II major histocompatibility complex antigen expression by the cells at different stages of pregnancy has shown they are progressively activated as gestation advances. They may thus also play an important role as antigen-presenting cells, acting as an active second line of defense to any infectious agent that succeeds in breaching the trophoblastic layer. In this regard, it should be noted that a subpopulation of Hofbauer cells are CD4 positive and so may act as portals of entry to the fetus or reservoirs for the human immunodeficiency virus [for reviews, see (36,183)].

Fetal Vessels

The fetal vascular network is lined by a nonfenestrated endothelium throughout pregnancy (Fig. 10). At least two populations of endothelial cells have been described, although the functional significance

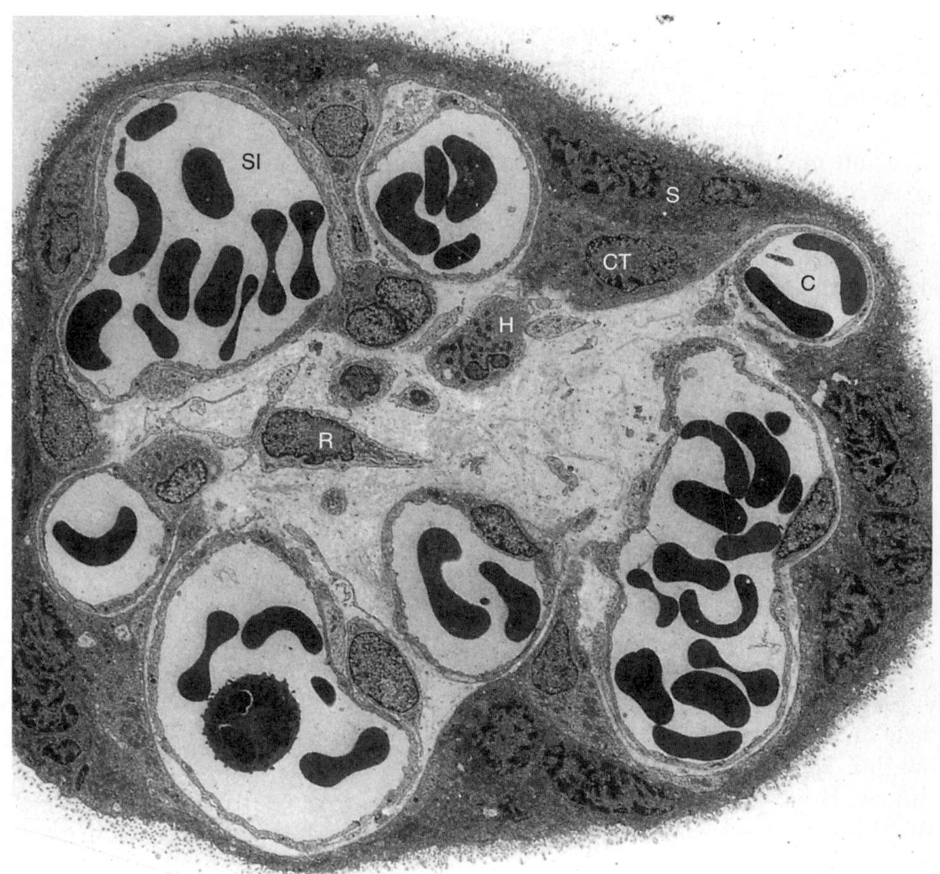

FIG. 10. Survey electron micrograph of a typical, well-preserved, terminal villus. It illustrates a high degree of fetal capillarization, with some of the capillaries (C) narrow and others dilated, forming sinusoids (SI). Sparse connective tissue is composed of macrophages (H), fixed connective tissue cells (R), and connective tissue fibers. The stroma is surrounded by a highly variable layer of syncytiotrophoblast (S) below which a few cytotrophoblastic cells (CT) can be seen. (Magnification × 2,000.) [Reproduced from (390) with permission.]

of this is not yet clear (184,185). One cell type is characterized by the presence of many microfilaments, which have been identified as predominantly vimentin and are therefore considered to have a structural rather than a contractile role (155). The second type possesses a well-developed rough endoplasmic reticulum and contains many secretory granules. Both cell types may be represented within a single cross-section of a capillary, and indeed the contrasting appearances may represent planes of section through central and peripheral parts of a single cell type (185). Therefore, it is unclear whether two cell types genuinely exist and, if so, whether there are regional variations in distribution throughout the villous tree.

Junctional complexes link adjacent cells and serve to both stabilize the endothelium and to limit the permeability of the paracellular transport route, particularly in respect to anionic molecules. Large molecules, including IgG, are transported across the endothelium by vesicles of the plasma membrane, which are present within both of the cell types described above (186).

The capillaries within the terminal villi are only surrounded by a basement membrane during the last third of pregnancy (48) (Fig. 10), and this contains type IV collagen, laminin, and fibronectin (187). It is secreted by both the endothelial cells and their associated pericytes. The pericytes may also contribute to angiogenesis within the villi (48). Pericytes are less frequently associated with capillaries during early gestation than toward term (188), suggesting that the vasculature is more plastic during early pregnancy and responsive to changes in angiogenic growth factors.

Further proximally in the villous tree, arteries and arterioles are found within the stem and intermediate villi (Figs. 11 and 12). These possess a smooth muscle coat, but elastic laminae are generally absent and the adventitia gradually blends with the fixed connective tissue of the stromal core. Because of the lack of neural innervation in the placenta, the caliber of the fetal vasculature must be controlled by paracrine and autocrine factors produced in response

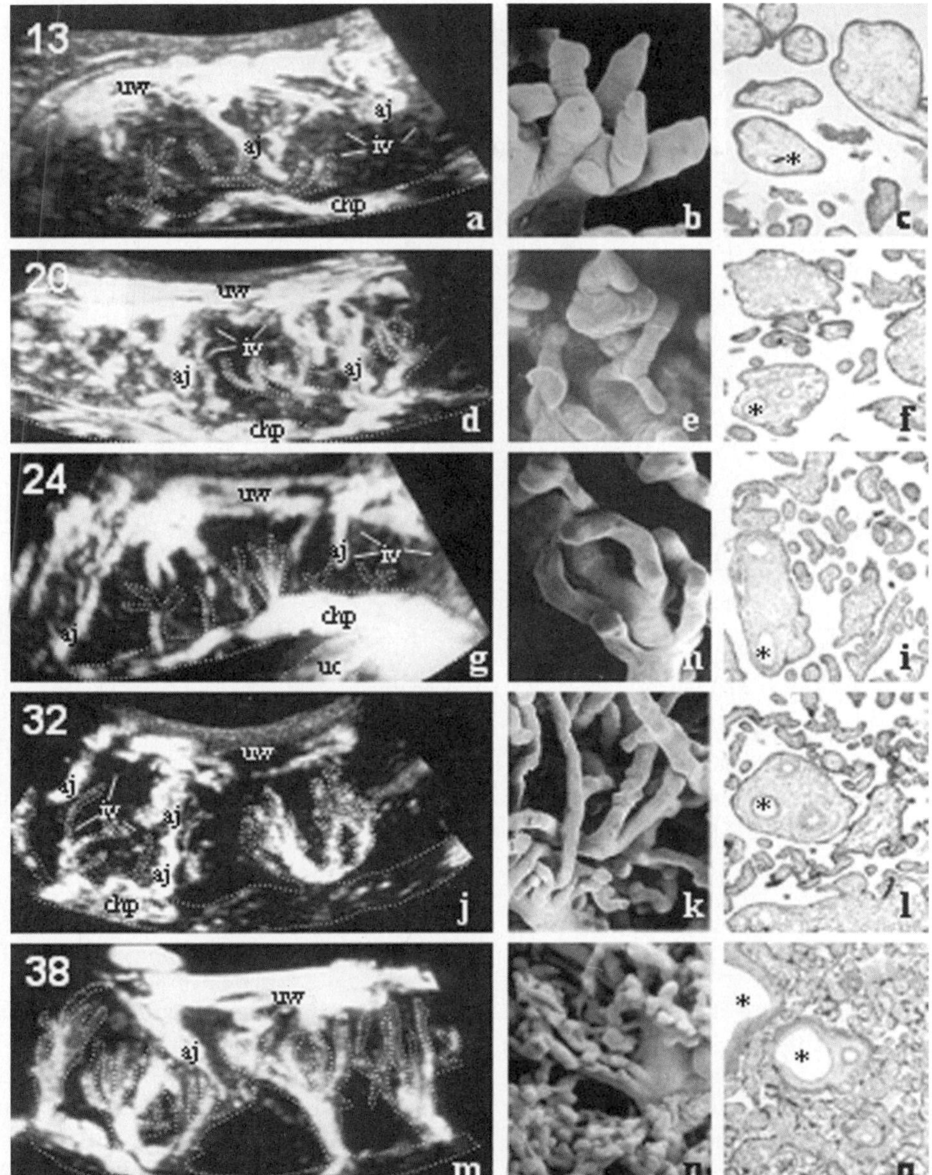

FIG. 11. Typical three-dimensional power Doppler images (*left vertical column*) from weeks 13 (**a–c**), 20 (**d–f**), 24 (**g–i**), 32 (**j–l**), and 38 (**m–o**), as compared with age-matched scanning electron micrographs of placental villi (*middle column*) and age-matched histological sections (*right column*). All images are from normal pregnancies. In the exchange area (*gray dots*) intervillous Doppler signals (iv) are reduced with advancing pregnancy paralleled by a decreased width of the intervillous space (*black space* in scanning electron micrograph; *white clefts* between villi in histology). Villous flow signals increase in extent, intensity, and width with increasing width of the villous stem vessels (*). aj, maternal arterial jets; chp, chorionic plate; uc, umbilical cord; uw, uterine wall. Magnification of scanning electron micrographs and paraffin sections. (Magnification × 65.) [Reproduced from (219) with permission.]

to physical stimuli such as transmural pressure or flow rate (189).

Studies from Graf and coworkers (190,191) clearly showed that large amounts of elastic fibers are present in the stroma of stem villi, in addition to the classical constituents of the stroma (i.e., reticular and collagen fibers). Elastin was found alone or in association with microfibrils, the latter forming long banding structures. This combination of structures, generally known to form elastic connective tissue, was present in close vicinity to extravascular smooth muscle cells of the stem villi. The latter cells belong to the perivascular contractile sheath of stem villi. The specific arrangement of extravascular smooth muscle cells together with elastic and collagen fibers points to a functional myofibroelastic unit within the perivascular

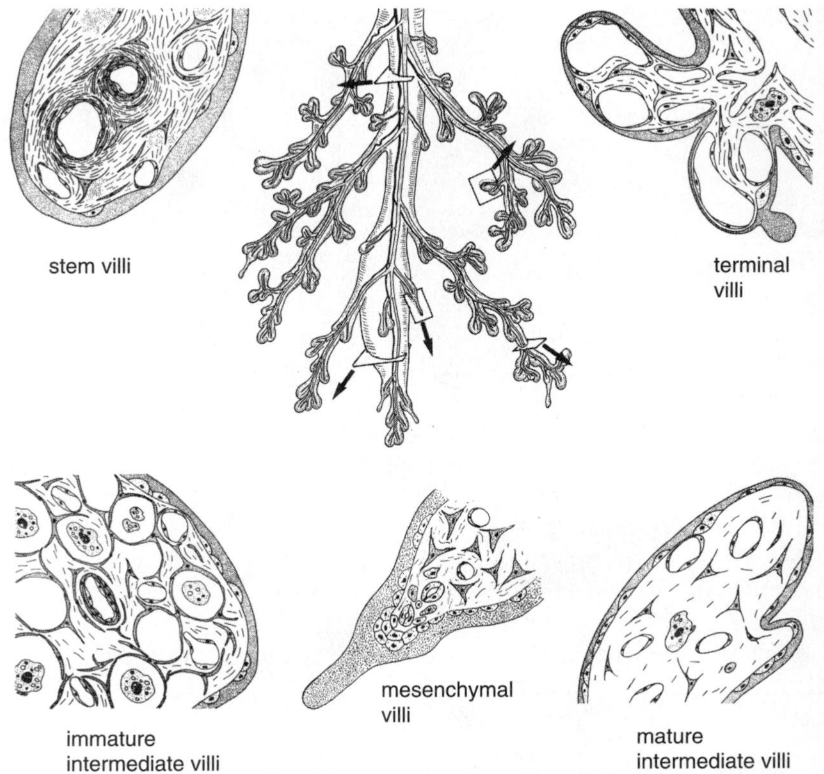

FIG. 12. Simplified representation of peripheral part of mature placental villous tree, together with typical cross-sections of different villous types. [Reproduced from (389) with permission.]

contractile sheath. Surrounding large fetal blood vessels inside the placenta, this system may contribute to elasticity and supporting tensile and/or contracting forces within stem villi. Moreover, an extensive system of signal transduction molecules (e.g., integrins and extracellular matrix proteins as well as proteins associated to the actin cytoskeleton) exists, providing a cross-talk between smooth muscle cells of the perivascular contractile sheath and the surrounding extracellular matrix.

Integrity of the Placental Barrier

There is physiological evidence for the existence of two different routes of transfer across the placental barrier, a transcellular and a paracellular route (192). The transcellular route involves transfer across the plasma membrane and the cytosol of the syncytiotrophoblast; the paracellular route is thought to be an extracellular water-filled pathway. The existence of the latter pathway has been proven with physiological in vitro and in vivo measurements. However, it is difficult to understand its morphological basis because the villous syncytiotrophoblast is a continuous uninterrupted layer, completely separating maternal and fetal circulations.

The syncytiotrophoblast extends over the surfaces of all villous trees (Figs. 10 and 11) as well as over parts of the inner surfaces of the chorionic and basal plates. It thus lines the intervillous space. Consequently, every substance passing from the maternal to the fetal circulation was thought to do so under the control of this trophoblastic layer. A few reports have pointed to the existence of accidental vertical cell membranes that represent the results of a syncytiotrophoblastic repair mechanism (36) after syncytial rupture (see below). Therefore, terms such as *syncytial cells* and *syncytiotrophoblasts*, widely used in experimental disciplines, are inappropriate and should be avoided.

Research has shown that the integrity and completeness of the syncytiotrophoblastic barrier is not perfect. Where the syncytiotrophoblast is interrupted by degeneration or by mechanical forces, the gap is filled by fibrin (so-called fibrin-type fibrinoid) as a result of blood clotting (70). Such areas are regular findings in every placenta throughout pregnancy (69) and account for approximately 7% of the villous surface of the normal term placenta (193). They may serve as paratrophoblastic routes for maternal–fetal macromolecule transfer bypassing the syncytiotrophoblast, because horseradish peroxidase (48,000 Da, 3.0-nm molecular radius) was shown to pass through the fibrin spots from the maternal circulation into the fetal villous stroma.

Transtrophoblastic channels can provide another paratrophoblastic transfer route for smaller molecules (194–196). The channels have luminal diameters of 15–25 nm (197) and pass through the syncytiotrophoblast as winding and branching channels from the apical to basal surface. They are likely to be routes for membrane recycling from the basal to the apical syncytiotrophoblastic plasma membrane because there are morphological hints of a basoapical membrane flux (196,197). Stulc and coworkers (198) revealed that the placental barrier in the rat contains pressure-dependent paracellular pathways connecting the maternal and fetal extracellular compartments.

Functionally, transtrophoblastic channels are possible sites of transfer for water-soluble lipid-insoluble molecules with an effective molecular diameter of about 1.5 nm (192,199). Under the conditions of fetal–maternal fluid shift caused by fetal venous pressure increase or fetal decrease of osmotic pressure, they may dilate to such an extent that even larger molecules may pass (194,200). Moreover, pressure-dependent dilation and closure of the channels may act as an important factor in fetal osmoregulation and water balance. Excessive fetal hydration causes an increase of fetal venous pressure and a decrease of osmotic pressure. Both factors have been experimentally proven to dilate initially narrow channels, thus allowing fetal–maternal fluid shift and equilibration of the surplus water.

VILLOUS TREES

Structure of Villous Types

The pattern of trabeculae seen during early development lays down the foundations of the arrangement of the villi in the mature organ. Hence, the villi are clustered together into a series of spherical units known as lobules or placentones (35). Each lobule arises from the chorionic plate by a thick villous trunk (Fig. 5), which is ultimately derived from a trabecula. The main trunk branches repeatedly, and although some subdivisions extend as far as the basal plate, the anchoring villi, most end freely in the intervillous space (Fig. 11). Five types of villi have been distinguished on the basis of their caliber, stromal characteristics, and vessel structure, and these types largely reflect differing subdivisions of this villous tree (3,173,201,202) (Fig. 12).

Stem Villi

Stem villi (3,201) (Fig. 12) represent the first 5 to 30 generations of unequal dichotomous branchings (diameters of 80 μm up to 3,000 μm) and serve to give mechanical support to the villous tree. They are characterized by a compact fibrous stroma containing centrally located arteries or larger arterioles and veins or venules. Capillaries are relatively infrequent, and so these villi probably play little part in placental exchange. Their physiological significance lies in the fact that with the perivascular contractile sheath surrounding the large vessels, inside a functional myofibroelastic unit is present. This system contributes to supporting tensile and/or contracting forces within the stem villi.

Mature Intermediate Villi

Repeated branching of stem villi results in the formation of long slender intermediate villi, ranging from 80 to 120 μm in diameter. These mature intermediate villi (202) are often gently curving, and terminal villi arise at intervals from all aspects of their surface (Figs. 11 and 12). Internally, they consist of a loose stromal cone, with scant fibers and cells, and occasional arterioles are embedded within this cone. Fetal vessels occupy at least half the villous volume, most in the form of capillaries. This increased degree of vascularization and the high level of enzyme activity within the syncytiotrophoblast indicate that this villous type may play a significant role in maternal–fetal exchange and hormone synthesis.

Terminal Villi

Terminal villi (3,201) (Figs. 10 to 12) are the final branches of the villous tree and are physiologically its most important component. They are short stubby protuberances up to 100 μm in length and 80 μm in diameter, although some possess a more narrow neck region. Most (95%) arise from the surface of mature intermediate villi, and one of their principal characterizing features is the high degree of capillarization (more than 50% of overall volume).

When viewing sections of terminal villi, it is immediately apparent that the thickness of the syncytiotrophoblast, and indeed of the villous membrane in general, is not uniform over the villous surface. Instead, there are areas where the syncytium is extremely attenuated and is devoid of nuclei or other large organelles. Underlying such an area is a dilated segment of a fetal capillary, known as a sinusoid, and hence the thickness of the villous membrane separating the maternal and fetal circulations may be as little as 0.5 to 2.0 μm (Fig. 10). These sites have been referred to as alpha zones, epithelial plates, and nephropneumoid regions, but now the term *vasculosyncytial membrane* is most widely used. Their morphology suggests they play an important role in diffusional exchange, and their formation is considered later in connection with the fetal angioarchitecture. At other points on the villous surface, the syncytiotrophoblastic layer is relatively thick (Fig. 10). Here the nuclei and other major organelles are found. The terms *beta zones* or *enteroid regions* have been used to denote these areas, and as the latter suggests, they are the most important sites of metabolic and endocrine activities.

Hence, different areas of the surface of terminal villi are responsible for different placental functions. Put in simple terms, this ensures that two of the placenta's main tasks, to act as a fetal lung and as a fetal liver, are arranged in parallel rather than in series. This is most certainly to the benefit of diffusional exchange, for the intervening barrier should be as thin as possible for maximum efficiency. To assist in this process, aged syncytiotrophoblastic nuclei are sequestered in clumps toward the tips of terminal villi, referred to as *syncytial knots*.

Immature Intermediate Villi

As their title suggests, immature intermediate villi (201) (Fig. 12) represent peripheral continuations of stem villi that are in the process of development. Thus, although common in immature placentas, their distribution in the mature organ is generally limited to the central regions of the lobules (Fig. 5), which are considered the principal germinal zones. Externally, these villi are rather bulbous, whereas in section they display highly characteristic profiles. The fixed stromal cells possess large sail-like processes that link together to form a series of fiber-free intercommunicating channels oriented parallel to the long axis of the villus (Fig. 12). Hofbauer cells appear morphologically to be able to move along and cross between these channels. Embedded among the stromal cells are arterioles, confirming that these villi are the forerunners of stem villi. It is important that their presence is recognized and that they are not incorrectly interpreted as edematous villi.

Mesenchymal Villi

Mesenchymal villi (103) (Fig. 12) are again a transient population, seen predominantly in the earliest stages of pregnancy in which they are the precursors of immature intermediate villi. In the more mature placenta, they are inconspicuous and may differentiate directly into mature intermediate villi. Structurally, they possess a stromal core that is not yet organized into stromal channels. Fetal vessels are poorly developed, but Hofbauer cells may be visible.

Development of Villous Trees

As mentioned previously, the development of the villous tree starts by the formation of side branches on the trabeculae. The earliest of these are composed of syncytiotrophoblast alone and so are termed *syncytial trophoblastic sprouts*. In early gestation, these are seen arising in an apparently random pattern from the surfaces of mesenchymal and immature intermediate villi (103,153) (Fig. 11) as well as along the surfaces of all villous types. In later pregnancy, similar structures can be observed, but then they are signs of syncytiotrophoblastic nuclear sequestration into syncytial knots as end stages of apoptosis (159).

Mesenchymal invasion into the proximal end of a true syncytial sprout is soon followed by the formation of capillaries, so rendering conversion into a tertiary villus complete. Up until the fifth week postconception, mesenchymal villi formed in this way progress to become primitive stem villi. However, after this point mesenchymal villi differentiate into immature intermediate villi (103). The latter continue to produce many new syncytial sprouts until they themselves are transformed into stem villi. Hence, growth of the more structural elements of the villous tree is very rapid during the first trimester.

The situation changes considerably during the third trimester. Placental growth tends to slow, and differentiation of the remaining villous types is seen. Mesenchymal villi no longer transform into immature intermediate villi but form mature intermediate villi instead. The remaining immature intermediate villi gradually differentiate into stem villi and so become scarce at term, normally being restricted to the centers of the lobules. Throughout the third trimester, terminal villi, the more functional components of the placenta, are elaborated from the surfaces of mature intermediate villi.

This switch in differentiation pathways at the start of the third trimester is of key importance in placental development (103,202). If it takes place too early, the organ stops growing too soon and differentiates prematurely. Consequently, the terminal villi are very densely packed because of a deficit of later generations of stem villi, and this may have an adverse effect on blood flow through the intervillous space. Alternatively, if the switch is delayed, the placenta is characterized by persisting immaturity and a relative lack of terminal villi.

Fetal Angioarchitecture

Vascularization of the human placenta is the result of local de novo formation of capillaries out of pluripotent mesenchymal precursor cells in villi, rather than sprouting of vessels from the embryo into the placenta. Vascularization of the placenta starts at day 21 postconception with a four somite embryo (48,203). At that time, progenitors of hemangiogenic cells differentiate within the core of secondary villi before the formation of first vessels. This process leads to the formation of capillaries within the mesenchyme and thus generates tertiary villi. The progenitor cells directly derive from mesenchymal cells (48,204) rather than originating from blood cells of the fetus. The early appearance of Hofbauer cells in the villous stroma suggests a paracrine role for these cells during the first stages of placental vasculogenesis (205,206).

During this period, vascular endothelial growth factor (VEGF) is strongly expressed in cytotrophoblast cells and subsequently also in Hofbauer cells. The respective receptors of VEGF, Flt-1 and Flk-1, are present on the vasculogenic and angiogenic precursor cells (207). Increased expression of VEGF and its receptors could systematically contribute to the altered differentiation and maturation of villous types associated with placental vascularization (208,209). At present, detailed information concerning the presence and function of VEGF, its two receptors, and their relation to angiogenesis and vasculogenesis in the very early stages of human placental vasculogenesis is developing (181,207,210,211).

In the term placenta, close to the point of insertion of the cord into the chorionic plate in most (96%) placentas, the umbilical arteries are linked by Hyrtl's anastomosis. This most likely ensures even pressure in the chorionic arteries as they ramify over the chorionic plate. At intervals, branches of these arteries penetrate the plate and take up a central location within stem villi. Such vessels are surrounded by a few layers of smooth muscle cells and by an adventitia, which is two to three times as thick as the media.

As the villous tree branches, so do the arteries. Arterioles are found within the smaller stem villi and both mature and immature intermediate villi (34,212) and are surrounded by an intercommunicating system of long slender capillaries, referred to as the paravascular network (153,212–214). This is orientated parallel to the longitudinal axis of the villi. Many functions have been ascribed to it in the past, the most popular of which has been the nutrition of deeper parts of the villous core for which diffusion from the intervillous space is arguably inadequate. Equally plausible, however, is the claim that the network represents a vestige of an earlier stage in development when the villi were better vascularized (3). What is clear as a result of exploring its connections through serial reconstructions is that there is no basis for suggestions that it regulates fetal blood flow resistance by acting as an arteriovenous shunting system (36).

Within mature intermediate villi, the arterioles and venules give way to long coiled capillary loops. Elongation of this capillary plexus is rapid during the third trimester and exceeds that of the containing villi. As a result, capillary coils are formed, which protrude from the surface, raising a blister of trophoblast before them (215). In this way, new terminal villi are formed, and the same capillary may run through several terminal villi in series (Fig. 13) before communicating with a venule.

The caliber of the capillary is not uniform within a terminal villus, because dilated regions, known as sinusoids, occur at intervals along its length (Figs. 10 and 13). At these points, the lumen may reach 40 μm in diameter, and it has been proposed that they serve to reduce blood flow resistance, so ensuring an even distribution of flow through peripheral parts of the villous tree. Because the resistance of a vessel is inversely proportional to the fourth power of the radius, their presence could have a significant impact on blood flow. Sinusoids are only seen in the final stages of placental development, and their formation coincides with increasing length of the fetal capillaries (212,215). Moreover, within the otherwise identical placentas of the capybara and guinea pig, sinusoids only occur in the capybara where the fetal capillaries are considerably longer (36).

In terms of their formation, it is notable that the sinusoids are invariably located on the outside wall of a tight capillary bend. Wall tension is greatest at this point, and consequently this is the preferential site of dilation under the influence of the high transmural pressure operating in later gestation. Dilation brings the outer wall of the capillary loop into close contact with the overlying syncytiotrophoblast, and pressure exerted on the latter may lead to remodeling, resulting in its local attenuation and the formation of a vasculosyncytial membrane (216). Experiments involving the perfusion of term placentas at different pressures have revealed that the pressure differential between the fetal capillaries and the intervillous space has a major impact on the dimensions of the capillary sinusoids (217). Normally, this is in the region of 10 mm Hg, but as the differential increases, the sinusoids dilate, and consequently the mean thickness of the villous membrane falls. Hence, there is always a very close and precise

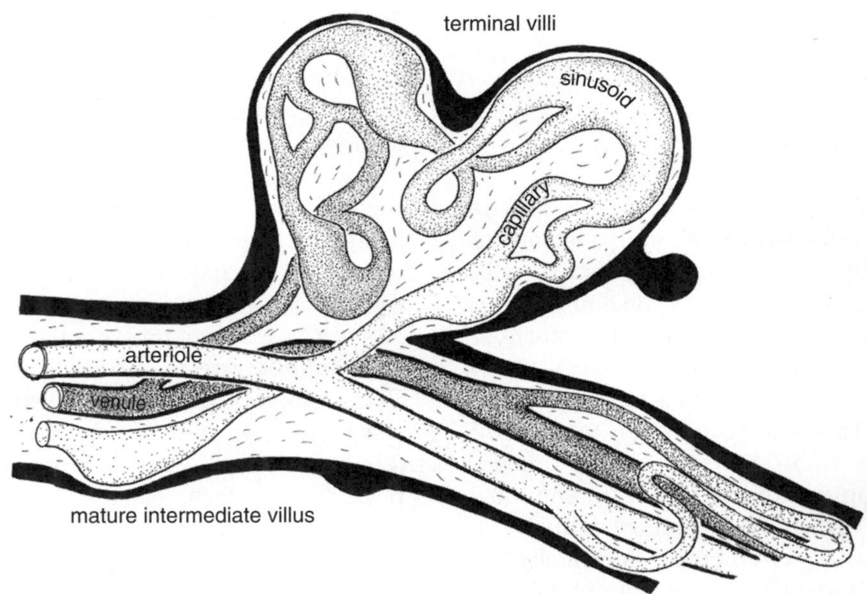

FIG. 13. Arrangement of fetal vessels in two terminal villi and one mature intermediate villus. Note highly complex loop formation of terminal fetal capillaries. Branching is usually followed shortly by refusing of two capillary branches, avoiding basal shortcuts. Local dilations, or sinusoids, reduce blood flow resistance, and lead to thinning of the barrier separating the two circulations. [Reproduced from (389) with permission.]

relationship between the volume of the fetal capillaries and mean membrane thickness.

The formation of terminal villi, capillary sinusoids, and vasculosyncytial membranes are therefore inextricably interlinked. Another benefit of sinusoids proposed in the past is that they serve to reduce the rate of blood flow opposite the attenuated regions of the trophoblast. This is only of physiological significance should the rate of diffusion across the membrane be a limiting factor. Alternatively, it could be argued that erythrocytes passing around the inner aspect of the capillary bend are disadvantaged by their presence, because they are further removed from the oxygen source. It seems likely, however, that given the marked changes in caliber, the pattern of blood flow through these regions is complex, although not turbulent. This is sufficient to ensure adequate mixing and more uniform oxygenation of the fetal erythrocytes.

Intervillous Space

The directional relationship between the maternal and fetal blood flow is of considerable physiological significance, as discussed previously in relation to comparative placental types. To understand the pattern of maternal blood flow in the human placenta, one needs knowledge of the configuration of the intervillous space.

As described previously, the fetal villi are arranged into 60 to 70 inverted cup-shaped lobules or placentones, each arising from the chorionic plate by a main stem villus. From the distribution of villous types and enzyme activity, it is clear that the central region of a lobule is the germinative zone (35). Here, the villi are loosely packed, and indeed as gestation advances, central areas free of stem villi can be detected ultrasonographically (218,219). A series of radiographic and injection studies confirmed that the arterial blood released from the openings of the maternal spiral arteries through the basal plate is delivered into these areas (220,221) (Figs. 5 and 11). This constant relationship suggests that the pattern of maternal arterial inflow plays a key role, either hemodynamically or by differing oxygen tensions, in the formation and location of the lobules (17).

The openings of the uterine veins are positioned opposite the more peripheral parts of the lobule (Fig. 5). As a result, the maternal blood disperses radially through the lobule in a centrifugal fashion. This led Wigglesworth (221) to distinguish three zones in the lobule: an arterial zone consisting of the stem villous-free central area, a capillary zone comprising the dense meshwork of villi, and a venous zone situated peripherally and under the chorionic plate. Although accurate measurements are difficult to obtain in vivo, it is likely that the oxygen tension in the intervillous space varies correspondingly, and recent data on the activity of the principal antioxidant enzymes support this assumption (222).

Because of the complex arrangement of villi described above, the interrelationship of the fetal and maternal blood flows cannot be assigned to one of the geometrically defined flow systems, such as concurrent, countercurrent, or cross-current. The situation in the human placenta comprises components of all these exchange systems (26) and corresponds to what Moll (223) defined as a multivillous flow system.

The rate of maternal blood flow through the lobule is dictated principally by two factors: the driving force, which is the pressure differential between the uterine arteries and veins, and the resistance, presented by the gaps between adjacent villi. As described in the historical introduction, our appreciation of the configuration of the intervillous space has changed over the years. Scanning electron microscopy has resolved many of the uncertainties, however, and it is now clear that within the bulk of the lobule, it comprises a series of irregular cleft-like spaces (153). These are generally of capillary dimensions only but are in free communication with one another. The width of the intervillous clefts is difficult to estimate, because not only are the walls very irregular but the spaces are influenced by several artificial phenomena. Using data for antepartum intervillous blood volume (23.3%–37.9% of the placental volume) and the villous surface area at term (11–13.3 m^2), the mean width has been calculated as ranging from 16.4 to 32.0 μm (36). One must bear in mind, however, that there is a considerable subchorial lake and large spaces in the centers of the lobules. Thus, the real "intervillous" volume is likely to be lower than that quoted above, and so intervillous clefts are even narrower. Occasional fusion does occur between neighboring villi (159,224), although not to the extent claimed by Stieve (13). The main significance of these points of fusion is that they can be disrupted (224), with the resulting syncytial injury providing a stimulus for fibrin deposition as mentioned previously.

Resin injection studies have demonstrated that the shape of the lobule and the disposition of the villi within it are heavily dependent on the perfusion pressure of the fetal vasculature (225). Equally, the volume of the intervillous space is approximately halved if clamping of the umbilical cord is delayed (33). It thus appears that the fetal vessels provide hydraulic support of the villous tree and, in doing so, may determine the shape and size of the intervillous clefts. During their cineradiographic studies of the rhesus monkey placenta, Ramsey and Donner (15) described the "villous pulse" whereby the lobule appeared to expand and contract in phase with the fetal heart beat. Equivalent studies are not possible in the human, but perfusion studies of delivered placentas at different pressure have confirmed that as the pressure rises over the physiological range, the gaps between the villi increase in size (226). It is envisaged that increasing turgidity of the fetal vasculature leads to straightening of the intermediate and smaller stem villi and their movement into the otherwise relatively villous-free interlobular and subchorionic regions. In this way, the pressures within the fetal circulation could influence the pattern of maternal blood flow.

The converse situation has also been proposed, whereby the pressures in the intervillous space could influence the fetal circulation. This concept was formulated into the *sluice-flow principal* (227), in which it was suggested that if pressure in the intervillous space exceeds umbilical venous pressure, the villous capillaries will be compressed. Doppler flow studies have demonstrated a rapidly reversible increase in the umbilical pulsatility index when women move from either the standing or the lateral position to the supine position (228,229). Because this increase in vascular resistance is independent of changes in the fetal heart rate, it is thought to be due to raised pressure in the intervillous space. This, in turn, results from impeded venous drainage, consequent on partial occlusion of the inferior vena cava by the gravid uterus (228). The fact that the size of the capillary sinusoids within terminal villi depends on the transmural pressure differential provides morphological evidence that the peripheral elements of the fetal vasculature are vulnerable to compression (217). It is therefore important that the pressure within the intervillous space is maintained at a low pressure to avoid compression of the villous vasculature, and this may be one of the most important consequences of the physiological conversion of the spiral arteries that occurs during the first trimester (230).

A further factor that can potentially influence the relationship between maternal and fetal blood flows is the deposition of fibrin within the intervillous space (see Fibrinoid). This occurs to a limited extent during the course of all normal pregnancies, principally in the relatively villous-free area immediately beneath the chorionic plate. Maternal blood flow through this region is particularly sluggish, and the fibrin enmeshes the roots of the main stem villi. Because these are not involved in maternal–fetal exchange, the deposition is of little physiological significance. Only if it is particularly excessive will it compromise placental surface area, but it may serve to remodel the intervillous space and reduce the volume of "redundant" regions.

Oxygen as Regulator of Villous Development

Because one of the principal functions of the placenta is gaseous exchange, it is not surprising on a teleological basis that development of the organ may be influenced by the prevailing oxygen concentration. During the last few years, major advances have been made in our understanding of how oxygen can modulate gene expression and cell behavior at the molecular level (231–233). Central to this process is the generation of oxygen free radicals, molecular species that contain one or more unpaired electrons (234). Once the preserve of chemists, free radicals are now increasingly recognized as key players in biological signaling and homeostatic mechanisms.

Under physiological conditions the most important intracellular site of free radical production is the mitochondria (234). The leakage of electrons from complexes I and III of the respiratory chain on to molecular oxygen results in the formation of the superoxide anion ($O_2 \cdot^-$) and occurs at a rate proportional to the oxygen concentration. Other potential sources include NAD(P)H oxidase, a membrane-associated enzyme that plays an important role in oxygen sensing in endothelial cells and myocytes. It is also present in the placenta (235–237) and has been implicated in the increased generation of $O_2 \cdot^-$ observed in preeclamptic placentas (238).

Under pathological conditions, however, additional mechanisms may come into play, and one potentially important source is the enzyme xanthine dehydrogenase/xanthine oxidase. In the dehydrogenase form, this enzyme converts hypoxanthine to xanthine and xanthine to uric acid, passing the electron released to NAD^+. During periods of hypoxia, the enzyme can be proteolytically cleaved by to the oxidase form that uses oxygen as the electron recipient, so generating $O_2 \cdot^-$. This conversion is responsible for the burst of free radical production that is associated with episodes of ischemia-reperfusion (239), and it is relevant that increased activity has been observed in placentas after vaginal delivery (240).

A number of redox-sensitive transcription factors have now been identified, including AP-1, CREB, HIF-1, NF-κB, p53, SP-1, and STAT3, that respond to changes in the homeostatic balance of free radicals (233). Although some of these modulate stress responses, others are involved in the regulation of cell proliferation and differentiation. The very reactivity that makes free radicals ideal for these purposes also poses a danger to cell integrity, however. If their concentrations rise beyond intracellular defenses, indiscriminate damage can occur to a wide range of biomolecules, including lipids, proteins, and DNA. Abnormal folding or strand breakage can ensue, leading to loss of function and even cell death. Hence, a spectrum of changes may be induced by oxygen, ranging from physiological adaptations to pathological insults.

In terms of the more physiological responses, these can be considered at the gross morphological and cellular levels. The topology of the villous tree is altered in different clinical conditions thought to reflect contrasting placental oxygenation (241), with branching of the villous tree being promoted by hypoxia. Under normal oxygen conditions, mature intermediate villi show repeated branching of terminal villi. In cases of hyperoxia, long slender intermediate villi with few terminal branches predominate, whereas in hypoxic situations the terminal villi are shorter and more clustered together. In this respect the question is how can hyperoxic conditions generate within the organ? The placenta is a unique organ in that one bloodstream (the maternal) provides oxygen and a second bloodstream (the fetal) extracts oxygen from the placenta to the fetus. Alterations in both blood supplies lead to alterations of the delicate balance of intraplacental oxygen. If there is less input from the maternal side (e.g., due to hypobaric hypoxia) with a steady output, the overall oxygen concentration within the placenta falls, leading to hypoxia inside the placenta. In contrast, if the extraction of oxygen is reduced (e.g., due to a fetal heart problem or increased resistance in the placental vasculature) with a steady input, the over all oxygen concentration within the placenta rises, leading to a relative hyperoxia inside the placenta (241).

As yet, there are no direct measurements of the oxygen concentration within the intervillous space to correlate with these findings. However, examination of placentas at high altitude, where the villous tree is undoubtedly subjected to hypobaric hypoxia (242), supports the hypothesis (243), as do experiments in animal models exposed to chronic hypoxia (244). The change in topology is believed to reflect a greater degree of vascularization and branching angiogenesis within the villous tree, induced by hypoxically driven changes in the expression of members of the VEGF family and the ratio of angiopoietins 1 and 2 (207,245). The increased vascularization also leads to greater elaboration of vasculosyncytial membranes, and hence thinning of the villous membrane, as discussed in Stereological Data. Similar, although less extensive, changes in villous morphology and maturity are observed across a placental lobule, where an oxygen gradient exists between the center and periphery due to the pattern of maternal blood flow (222,246).

At the cellular level, it has long been appreciated that the number of cytotrophoblast cell profiles seen in histological sections is increased under conditions of supposed intrauterine hypoxia (71,241), and again evidence from high altitude placentas supports the idea (74). The increase may reflect either greater proliferation of the cells or reduced fusion with the syncytiotrophoblast, and there is evidence to support both effects. Culture of villous explants under 2.5% oxygen favors proliferation of the cells (73), whereas 21% oxygen causes them to differentiate and to develop the invasive phenotype (247). On the other hand, culture under low oxygen conditions suppresses the expression of syncytin in trophoblast cell lines (248), and altering the redox potential of cytotrophoblast cells by manipulating the expression of antioxidant enzymes can also impair cell fusion (249).

Considerable attention has focused on the role of oxygenation in regulating cytotrophoblast differentiation along the extravillous pathway and the subsequent invasion of these cells into the endometrium. Most of this work has been performed in vitro with explant cultures or immortalized cell lines, and somewhat contradictory data have been obtained depending on the model and oxygen concentrations used. Often, comparisons have been drawn between the effects of 2.5% and 21% oxygen, and caution has to be exercised when extrapolating the results to the in vivo situation as ambient oxygen concentrations are clearly unphysiological [for a recent review of explant cultures, see (250)]. Nonetheless, measurements performed in vivo reveal that during early pregnancy, the extravillous trophoblast cells migrate from a low oxygen environment in the intervillous space to a higher concentration in the decidua (55,56). In the explant situation, 21% oxygen favors differentiation and migration of extravillous trophoblast cells compared with 2.5% oxygen, an effect that appears to be mediated via the HIF-1 pathway and transforming growth factor-β_3 (247,251,252). By contrast, cell lines cultured under 1% oxygen show increased invasiveness compared with controls cultured under 2.5% or 5% (253,254). Evidence suggests that this results from increased expression of the urokinase receptor via a novel mechanism involving suppression of nitric oxide production, and hence cyclic guanosine monophosphate, due to the limitation in oxygen supply. Oxygen and trophoblast redox status have also been implicated in the secretion of extracellular matrix components (255), endocrine function (256,257), and cytokine production by villous explants (258,259).

At the more pathological end of the spectrum, there is an increasing recognition of the role that placental oxidative stress plays in the pathophysiology of pregnancy disorders ranging from miscarriage to preeclampsia (260–264). It is clear from the physiological measurements performed in vivo during the first trimester that trophoblast is well adapted to a low oxygen environment (55,56). Indeed, it appears that trophoblast is more susceptible to elevated oxygen than to low oxygen, although if the concentration drops too low it may undergo necrosis (64,73). Thus, villi in the more highly oxygenated peripheral regions of the placenta display increased immunohistochemical evidence of oxidative stress during onset of the maternal circulation (62). They also become avascular, consistent with down-regulation of VEGF, and ultimately regress. Although these changes may be considered physiological in that regression of villi over the abembryonic pole to form the chorion laeve occurs in all normal pregnancies, equivalent changes are seen in cases of missed miscarriage (265). Instead of being confined to one region, however, they occur throughout the placenta, reflecting the disorganized onset of maternal blood flow observed in these cases (62). They are accompanied by reduced proliferation and increased apoptosis within the cytotrophoblast population, again consistent with exposure to a hyperoxic environment. Degeneration of the syncytiotrophoblast is extensive, but regeneration of a functional layer occurs through fusion of the remaining cytotrophoblast cells. This is hormonally active, producing human chorionic gonadotropin, and so pregnancy outcome in these cases depends on the balance of syncytial degeneration and regeneration (265).

In preeclampsia there is much evidence of increased placental oxidative stress (261,262,264,266), although the cause of this remains uncertain. It has long been considered that the placenta is hypoxic in these cases due to reduced perfusion secondary to deficient trophoblast invasion. As discussed in Stereological Data, there is no firm evidence to substantiate this claim, at least in cases of late-onset preeclampsia when placental development is normal. Recently, it was suggested that fluctuating oxygen concentrations within the intervillous space caused by intermittent perfusion from the spiral arteries may be the trigger (267), and work in vitro demonstrates that hypoxia–reoxygenation is indeed a potent stimulus (268). Preeclampsia is associated with increased apoptosis within the syncytiotrophoblast layer (269,270), but because of their apoptotic nature these syncytial knots should not lead to an inflammatory response of the mother. In contrast, deportation of nonapoptotic microvillous debris, most likely necrotic in nature, into the maternal circulation may cause activation of the maternal endothelial cells (120,261). Although the boundary between physiological and pathological changes induced by oxygen is indistinct, the clinical observation that maternal supplementation with antioxidant vitamins in pregnancies at high risk of preeclampsia significantly reduces the incidence of the disease raises hope for future treatment regimes (271).

Stereological Data

In the past, many attempts have been made to quantify different aspects of placental structure, with the aim of producing a measure of the organ's functional capabilities. Most have worked with sectioned material and have largely presented either planar morphometric data, such as profile areas and perimeter lengths, or stereological ratios, for example, component volume or surface area density per reference volume. Although such results are easily obtained and may be sufficient to draw comparisons between different regions or treatment groups, their physiological significance can be difficult to interpret. For example, the number of vasculosyncytial membranes per 100 villous profiles provides little indication of the overall diffusing capacity for several reasons. First, there are the problems of defining what actually constitutes a vasculosyncytial membrane and how one accounts for the influence of the plane of sectioning on apparent membrane thickness. Second, such an index does not take into account the areal extent of these specializations, because a few large vasculosyncytial membranes may have the same impact on gaseous exchange as many small ones. What is required is a descriptor that takes all these factors into account and provides a measure of membrane thickness that can be substituted directly into physiological diffusion equations (272). In this case, it is the harmonic mean thickness that is of key importance, because being based on the reciprocal of the local thicknesses, this index emphasizes the presence of the vasculosyncytial membranes where most gaseous exchange takes place. Whenever possible, it is therefore advisable to convert all planar data into component densities and true thicknesses, and this is the domain of stereology. The component densities should, in turn, be converted into absolute values (e.g., total lengths, surface areas, and numbers) by relating them to the reference volume (i.e., the total volume of the placenta). Doing such conversions enables an absolute value for each placental parameter to be calculated (Table 1). It is then easy to see whether changes in one compartment are mirrored or compensated for by alterations in another, so avoiding problems of misinterpretation (273).

Comparisons between studies can also be made difficult by differences in the way in which the placentas were delivered and subsequently sampled and processed. Because the placenta is such a highly vascular organ, the timing of the clamping of the cord has a profound impact on the values of many parameters (33). Similarly, placentas delivered by cesarean section are volumetrically different from those delivered vaginally (274). Furthermore, even material obtained after cesarean section differs considerably from that aspirated directly from the maternally perfused placenta in situ (275).

After delivery, the picture is complicated by leakage of fetal blood occurring from villous tissue torn during delivery and the effects of postdelivery ischemia. Thus, even in cesarean-delivered placentas in which the cord is clamped immediately, repeated sampling at intervals over a period of 20 min demonstrates major changes (20%–30%) in the size of the fetal vessels and the thickness of the villous membrane (276). Most of these changes take place in the first 5 minutes postdelivery. To overcome this problem, perfusion fixation under arterial pressure has been used (274,277), and this technique yields data that correspond closely to the in vivo situation.

Clearly, for a comparative study to be internally consistent, the collection of material must be from a uniform site throughout. Alternatively, if the data

TABLE 1. *Stereological data for normal term placentas*

	Aherne and Dunhill (391)	Laga et al. (392)	Bouw et al. (33)	Burton and Jauniaux (279)	Mayhew et al. (303)
Placental volume, cm^3	488	448	540	476	–
Total villous volume, cm^3	224	204	239	253	259
Volume of intervillous space, cm^3	144	110	210	180	276
Volume of fetal vessels, cm^3	45.0	36.0	74.9	36.8	48.4
Volume of trophoblast, cm^3	58.0	57.3	47.6	58.9	67.9
Villous surface area, m^2	11.0	16.7	13.3	12.6	11.0
Capillary surface area, m^3	12.2	12.4	15.8	14.7	6.94
Arithmetic mean thickness of villous membrane, μm	3.5	–	–	7.3	6.2[a]
Harmonic mean thickness of villous membrane, μm	–	10.0	–	5.7	–

[a]Trophoblast only.

are to reflect the structure of the whole organ, it is essential that an adequate number of samples are taken on a random basis (278). The combined use of paraffin embedded material for the estimation of large component volumes, such as the intervillous space or villous trees, and resin-embedded material for finer structural details, such as capillary volume or villous membrane thickness, provides the optimal balance between sample size and tissue resolution (279).

There have been many recent advances in stereological techniques, and these have been thoroughly reviewed by Gundersen and coworkers (280,281). The contributions of stereology to our broader understanding of placental morphology and to development of the fetal vasculature have been reviewed (245,272). Here we restrict ourselves to a consideration of how it has provided insight into the patterns of normal placental growth and maturation and the ways these can be influenced adversely in compromised pregnancies.

Placental mass and volume increase during pregnancy and correlate positively with fetal size. Most of this expansion occurs from 20 weeks onward and reflects the progressive elaboration of terminal villi and to a lesser extent intermediate villi (50). Hence, the surface area for maternal–fetal exchange increases, but there is also a close correlation between the surface area of trophoblast and maternal concentrations of several placental hormones (282). Enlargement of the trophoblastic compartment appears to be solely hyperplastic, with proliferation of cytotrophoblast cells and their subsequent fusion with the syncytiotrophoblast accounting for the increase in volume (66,283). There is no change in the ratio of cytotrophoblast cells to syncytial nuclei during gestation and equally no change in the tissue volume per nucleus. By contrast, growth of the fetal capillaries appears to be biphasic, with endothelial cell proliferation predominating during the first half of pregnancy and cell hypertrophy and remodeling occurring later (284,285).

Although villous surface area is an important determinant of placental diffusing capacity, the thickness of the villous membrane is considerably more influential (286). During gestation the thickness of the villous membrane gradually reduces (50,287), principally due to expansion of the fetal capillary bed (216).

If surface area and harmonic mean thickness are substituted into the Fick equation, the morphometric diffusing capacity of the villous membrane can be calculated (272). This provides an overall estimate of the maximum diffusing capacity of the membrane, although of course it cannot take into consideration the influence of the rates or relative directions of blood flow in the two circulations. Nonetheless, values estimated correlate closely with physiological measurements performed in animal species (21,244). Stereological studies in the human have confirmed that the placental morphometric diffusing capacity increases throughout gestation commensurate with the gain in fetal weight (288). They have also demonstrated a correlation between the diffusing capacity of the villous membrane and the maturity of the placenta as assessed ultrasonographically (279) and that there are no significant quantitative changes in the postmature placenta (277). Values of 3 to 5.5 ml/min/mm Hg have been reported at term (279,288).

Many studies have quantified structural changes in placentas from complicated pregnancies, often attributing these to reduced placental perfusion and hence chronic hypoxia. No physiological measurements of intervillous oxygen concentrations are available to confirm the latter assumption, however, except in the case of pregnancy at high altitude. At 4,300 m the maternal arterial partial pressure of oxygen is reduced by approximately 50% to 53 mm Hg (242), and so the placenta is exposed to significant hypobaric hypoxia. The effects of this challenge on placental morphology have recently been exhaustively reviewed (245,289), although comparisons between studies are difficult because of the different altitudinal levels, ethnic groups, and numbers of cases involved. For example, placental weight has been variously reported as increased, unchanged, or decreased at high altitude. By contrast, there is general consensus that the terminal villi display increased vascularization, thinning of the villous membrane, increased numbers of cytotrophoblast cells, and reduced perivillous fibrin deposition (74,242,290–292). As a result, the morphometric diffusing capacity has been reported to be maintained at sea level values (293) or to be doubled (292) at high altitude. It is tempting to assume that this represents a compensatory adaptation to facilitate oxygen transfer to the fetus, because there is no conclusive evidence at present that the fetus is hypoxic at high altitudes (289). Confounding factors of life at high altitude, such as differences in exercise levels and dietary intake, must also be considered when interpreting these data.

Maternal anemia potentially represents a different hypoxic challenge to the placenta, because the arterial blood entering the intervillous space has a reduced carrying capacity. By contrast to the situation at high altitude, increasing the diffusing capacity of the placenta in anemic hypoxia is likely to be of little benefit, as the partial pressure of oxygen in the maternal blood is normal. Hence, the transplacental oxygen gradient driving diffusion across the villous membrane is maintained. Stereological estimates indicate that the morphometric diffusing capacity is within the normal range, despite there being a reduced surface area of the villous tree (294). This is achieved through increased vascularization of the villi present and subsequent thinning of the villous membrane (295–297). Attributing these changes solely to hypoxia is potentially misleading, however, for the most common cause of maternal anemia, iron deficiency, is usually associated with generalized dietary insufficiency. Other effects may therefore be exerted on the placenta. In animal models, where dietary manipulations can be controlled to produce iron deficiency alone, analysis of the expression of hypoxically regulated genes provides no support for the contention that the placenta is hypoxic (298).

Similar assumptions of chronic hypoxia have been made in the past regarding the placenta in preeclampsia. Our appreciation of the syndrome has changed in recent years, however, and it is now appreciated that a spectrum of disorders are embraced by this term. The late-onset form is associated with normal birth weights and placental weights, and analysis of placental energetics provides no evidence that the organ is hypoxic (299). By contrast, the early-onset form is frequently associated with intrauterine growth restriction (IUGR) and atherotic changes in the spiral arteries, reducing placental perfusion. Recent ultrasonographic longitudinal studies of placental growth have revealed that placental volume follows different trajectories in the two forms and that growth of the placenta is reduced in cases of IUGR from as early as 10–12 weeks of pregnancy (300). At term, this is reflected in reduced elaboration of the villous tree (301). It is therefore important to make a distinction between the two forms of preeclampsia now recognized, because it is likely that they have different pathophysiologies (264,267). Stereological studies involving preeclamptic pregnancies not associated with IUGR have confirmed that placental development is essentially normal (302,303), whereas those associated with IUGR display severely reduced villous development and an increased thickness of the villous membrane (303,304). Both effects compromise placental diffusion. Importantly, comparison with placentas from pregnancies complicated by IUGR alone indicates that these changes are attributable to the IUGR process rather than to preeclampsia per se (303).

NONVILLOUS PARTS

Chorionic Plate, Secondary Yolk Sac, Umbilical Cord, and Membranes

The development of the chorionic plate, the umbilical cord, and the fetal membranes are closely related to that of the amnion. At day 13 postconception, the blastocystic cavity is occupied by a loose meshwork of cells, the extraembryonic mesoderm, that surrounds the embryoblast. The embryoblast is composed of two vesicles, the amniotic vesicle and the primary yolk sac. Where both vesicles are in contact with each other, they form the double-layered embryonic disc. From day 14 onward, the extraembryonic mesoderm cells are rearranged in such a way that they only line the inner surface of the trophoblastic vesicle (thus forming the chorion) and the surfaces of the two embryonic vesicles (making up the embryo). In between, the exocoelomic cavity forms. The exocoelom is bridged by a bundle of mesenchymal cells, which is referred to as the connecting stalk. It links the embryonic vesicles to the trophoblast and is the forerunner of the umbilical cord. During the same period, a duct-like extension of the hindgut originating in the caudal region of the embryo penetrates the proximal end of the connecting stalk. This structure is the allantois, the primitive extraembryonic urinary bladder.

In the course of the following weeks, one of the two embryonic vesicles, namely, the amniotic vesicle, enlarges considerably. It extends around the embryonic disc and the yolk sac and finally covers the connecting stalk (Fig. 8). The latter is thus transformed into the umbilical cord. Further fluid accumulation within the amniotic cavity causes its continued expansion until it completely occupies the former blastocystic cavity surrounding the fetus by 11–13 weeks of pregnancy. In the course of this process, the amniotic mesenchyme locally touches and finally fuses with the chorionic mesoderm lining the inner surface of the trophoblastic vesicle. This process starts near the insertion of the cord into the chorionic plate and continues until the middle of pregnancy, by which time the amniotic and chorionic mesenchyme are fused throughout. Unlike the situation in the cord in which the amnion fuses firmly with the underlying connective tissue, fusion of the amnion with chorionic plate or remaining membranes is never complete. Rather, amnion and chorion can always easily slide against each other. Histologically, they seem to be separated by a system of slender fluid-filled clefts, the intermediate layer of the chorionic plate, and the membranes.

The other vesicle, the yolk sac, floats freely in the exocoelomic fluid and attains its maximum diameter of approximately 6 mm at 10 weeks of gestation (Fig. 8). The outer mesothelial layer is covered with microvilli and displays the morphological characteristics of an active absorptive epithelium (305,306). The exocoelomic fluid contains a similar protein profile to maternal serum (307), and although it was suggested nearly three decades ago that the yolk sac may play a role in early maternal–fetal transport (305), firmer evidence has only recently become available. The protein glycodelin, secreted by the uterine glands, has been detected immunohistochemically within the mesothelial layer, and α-tocopherol transport protein, involved in the transport of vitamin E, has also been localized to this epithelium (54,308). These findings raise the possibility that during the earliest weeks of gestation the human placenta may function on a choriovitelline basis, although of course the yolk sac never makes physical contact with the chorion.

Already in the third week postconception, the allantois becomes supplied with fetal vessels. The human allantoic vessels, two allantoic arteries originating from the internal iliac arteries and one allantoic vein that enters into the hepatic vein, invade the chorionic plate and become connected to the villous vessels (for vascularization of the placenta, see Fetal Angioarchitecture). In recognition of the allantoic participation in vascularizing the organ, the human placenta falls into the category of "chorioallantoic" placentas previously described. Fusion of the allantoic vessels with the intravillous vessel system takes place within the chorionic plate and establishes a complete morphological fetoplacental circulation in the course of the fifth week postconception. Whether an effective physiological circulation occurs at that time is debatable, however, as most fetal erythrocytes remains nucleated until 10–12 weeks. Until then, the erythrocytes are relatively undeformable and so impose a high viscosity and resistance upon the fetal circulation.

In the same period, the developing cord has a length of between 0.5 and 1.0 cm. By the fourth month, it has grown to between 16 and 18 cm, by the sixth month to between 33 and 35 cm, and at term it reaches a final length of about 50 cm, with extreme values ranging from 18 to 122 cm (3). Even in early pregnancy, the cord is characterized by a spiral twisting, the number of turns increasing up to a maximum of 300 as pregnancy advances. In most cases, the twist is sinistral, or anticlockwise. The twists have been interpreted as the result of rotary movements of the fetus, caused by asymmetrical uterine contractions (36).

At day 14 postconception, the primary chorionic plate consists of three layers: syncytiotrophoblast, cytotrophoblast, and extraembryonic mesenchyme (Fig. 7D). They separate the intervillous space from the blastocystic cavity. Trophoblastic proliferation with subsequent degeneration and deposition of fibrin-type fibrinoid (Langhans' fibrinoid) causes continuous growth of the primary chorionic plate. Around the fourth and fifth week of pregnancy, allantoic blood vessels reach the primary chorionic plate through the connecting stalk and connect with the vessels formed independently inside the villi. As soon as the expanding amniotic sac comes into close contact with the mesenchymal surface of the chorionic plate (eighth to tenth week postconception), the definitive chorionic plate is formed. It is composed of the following layers: Langhans' fibrinoid lining the intervillous space and largely replacing the original syncytiotrophoblast; one or several layers of cytotrophoblast (becoming extravillous trophoblast); chorionic mesenchyme; a spongy layer with many clefts, which indicate the border between chorion and amnion; amniotic mesoderm; and amniotic epithelium, which lines the amniotic cavity [for further details, see (36,309)].

To understand the development of the fetal membranes, it is necessary to recollect that the villous developmental steps described above are valid only for the implantation pole (i.e., that part of the blastocystic circumference that becomes attached to the endometrium and that implants first). The other parts of the blastocystic circumference implant a few days later and undergo corresponding, although delayed, development (Fig. 7B). As early as the fourth week, they already show the first signs of regression; these parts are called the capsular chorion frondosum. The newly formed villi degenerate as described previously, and the surrounding intervillous space eventually obliterates. Finally, the chorionic plate, the obliterated intervillous space, villous remnants, and the basal plate all fuse, forming a multilayered compact lamella termed the chorion laeve. The first patches of the smooth chorion appear opposite to the implantation pole at the so-called abembryonic or antiimplantation pole. From there, they spread over about 70% of the surface of the chorionic sac until the fourth lunar month.

On the maternal surface, the smooth chorion is lined by capsular decidua. With complete implantation, the decidua closes again over the blastocyst, forming the capsular decidua. Between the fifteenth and twentieth week postconception, this layer fuses with the opposite wall of the uterus, thus obliterating the uterine cavity. From this point onward, the smooth chorion has contact over nearly its entire surface with the decidual surface of the uterine wall and may function as a paraplacental exchange organ. Because of the deficiency of a fetal vascularization of both the smooth chorion and the amnion, all paraplacental exchange between fetal membranes and fetus has to pass the amniotic fluid.

The mean thickness of the fetal membranes at term, after separation from the uterine wall, is about 200 to 300 μm. The membranes show some similarity with the structure of the chorionic plate (310). The innermost layer, the amniotic epithelium, encloses the amniotic fluid (Fig. 5). It is involved in the production of the latter and also in its resorption. Moreover, experimental results (311) indicate that the amniotic epithelium contains abundant carbonic anhydrase, an enzyme being involved in removal of carbon dioxide and in pH regulation. It was shown that the amniotic epithelium is a rich source of a variety of growth factors and their receptors (312). The epithelium rests on a thin layer of amniotic mesoderm.

The next layer, the chorionic mesoderm, is continuous with the connective tissue of the chorionic plate and is directly adherent to the outer cytotrophoblastic layer. The latter is of varying thickness. Near the placental margin, persisting ghost villi embedded in fibrinoid split the cytotrophoblast into two layers. These continue into the placenta and become confluent with the chorionic and basal plates, respectively. Attached to the outer surface of the cytotrophoblast is a decidual layer. The latter indicates that separation of the membranes, as in the placenta, does not take place along the maternal–fetal interface but instead it cleaves somewhat deeper in between maternal tissues.

Basal Plate and Uteroplacental Vessels

The basal plate and its forerunner, the trophoblastic shell, are formed at the base of the lacunar system at day 8 postconception (Fig. 7C,D). Initially, it is a purely trophoblastic layer. Already from day 13 postconception onward, uninucleate extravillous trophoblast cells detach from the shell and penetrate into the surrounding endometrium, thus forming the junctional zone. During invasion extravillous trophoblast cells further differentiate, displaying different phenotypes (44,313) such as the so-called X cells. A few invading trophoblast cells come into close contact with each other and fuse to form syncytial giant cells. Increasing numbers of endometrial stroma cells undergo considerable hypertrophy and become transformed into decidua cells.

Because of a combination of phagocytic and immunological activities, varying amounts of trophoblast, decidua, extracellular matrix, and intermingled fibrogen condense and form a heterogeneous extracellular material called fibrinoid (see Fibrinoid). Where the latter is in close contact with the intervillous space, it is a blood clot product (thus termed fibrin-type fibrinoid), formerly called Rohr's fibrinoid (Fig. 7F). More deeply positioned layers of fibrinoid, which surround groups of trophoblastic and/or decidua cells, are secretory products of extravillous trophoblast cells (thus termed matrix-type fibrinoid). The entire maternal–fetal "battlefield" stretches from the intervillous space down to the myometrium. It is described as the junctional zone (Fig. 7F). The superficial part of it, adhering to the placenta after placental separation, is the basal plate. It represents the bottom of the intervillous space and consists of trophoblastic and endometrial cells and much fibrinoid. On the lower surface of the basal plate again fibrin-type fibrinoid is found, formerly called Nitabuch's fibrinoid. Its intervillous surface may be partly lined by maternal endothelium derived from the maternal veins. Those parts of the junctional zone that remain in the uterus after delivery are called the placental bed. This consists mainly of intact and necrotic endometrial tissue, with trophoblastic cells intermingled (314).

Placental Septa

Placental septa are dome-shaped or sail-like extensions of the basal plate into the intervillous space, rather than real septa dividing the intervillous space into separately maternally perfused chambers (Fig. 5). Their tissue composition is identical with that of the basal plate, namely, a fibrinoid matrix with interspersed trophoblastic and decidual cells. Even maternal vessels, mostly veins, may extend into the septa. The septa are interpreted as dislocations of basal plate tissue into the intervillous space, caused by lateral movement and folding of the uterine wall and basal plate over each other (36).

Cell Columns and Cell Islands

Cell columns are the trophoblastic connections of larger stem villi, the so-called anchoring villi, to the basal plate. These are segments of the villous trees that persist in the primary villus stage, because mesenchymal invasion during formation of secondary villi does not reach the most basal segments of the anchoring villi (Fig. 7E,F). Because of continuous cytotrophoblastic proliferation at the stromal–trophoblastic interface (315,316), the cell columns serve as segments of longitudinal growth of the anchoring villi. From their distal ends, cytotrophoblast cells invade the basal plate, thus contributing to the growth of the latter, and invade the placental bed, finally reaching and transforming the spiral arteries. Because of this, cell columns serve as one of the richest sources for the so-called extravillous cytotrophoblast. Fibrin-type fibrinoid deposition on the lateral surfaces of the cell columns slowly "buries" them into the basal plate. As soon as they are completely incorporated into the latter, the cytotrophoblastic proliferation slows down. After partial degeneration of the cells and complete disintegration of their structure, cell columns largely disappear in the course of the last trimester and can only rarely be observed in the term placenta.

Cell islands obviously are largely comparable structures (316–318). They too are formed from villous tips that have not been opened up by connective tissue during the transition from primary to secondary villi. The only difference is that these villous tips are not connected to the basal plate, as are anchoring villi. Also, the cytotrophoblast of the cell islands proliferates and later becomes largely embedded into fibrinoid, which surrounds clusters and strings of surviving extravillous cytotrophoblast. Sometimes, central degeneration and liquefaction causes the development of fluid-filled cysts inside the cell islands. Cell biological studies concerning proliferative behavior of cytotrophoblast, the expression of growth factor receptors and of oncogene protein products, and the interactions with extracellular matrix have not revealed any differences between cell islands and cell columns (316–318). The only difference is that in cell columns there is a clear spatial distribution from the basement membrane of the anchoring villus toward the decidual tissue of the basal plate. In contrast, in cell islands this organization is missing.

Extravillous Trophoblast

Most cellular and syncytial trophoblast from the implanted blastocyst is consumed in the development of the placental villi. The remaining trophoblast, which is not used for villous formation, the extravillous trophoblast, is the basic material for the development of all other parts of the placenta (e.g., the smooth chorion, the chorionic plate, the basal plate including the cell columns, the septa, and the cell islands).

The nomenclature of the trophoblast residing outside of the villi is still confusing. Because some doubt existed in the past regarding its derivation, the first name used was *X cells*. Later, its trophoblastic origin was proven (319), and many authors proposed new designations: extravillous trophoblast, extravillous cytotrophoblast, nonvillous trophoblast, intermediate trophoblast, specialized trophoblast, interstitial trophoblast, intravascular trophoblast, intraarterial trophoblast, trophoblastic giant cells, trophocytes, spongiotrophoblast-like cells, and placental site giant cells. Regrettably, every one of these terms has a slightly different definition. We therefore propose the use of the term *extravillous trophoblast* as the most general heading for all types of trophoblast occurring outside of villi. When syncytial elements can be excluded, the name *extravillous cytotrophoblast* may be more appropriate.

The two main subpopulations of extravillous trophoblast are the endovascular trophoblast, located inside the media or lining the spiral artery lumina and partly occluding those, and the interstitial trophoblast, comprising all those extravillous trophoblast cells of the junctional zone that are not located inside vessel walls and lumina. In the last few years it has become evident that the endovascular trophoblast seems to use a side route of differentiation and has become a subpopulation of the interstitial trophoblast [for literature, see (36,44)].

The interstitial trophoblast has been shown to comprise different phenotypes of trophoblast cells (313). *Large polygonal trophoblast cells* are rare in early pregnancy, increase in number during pregnancy, and at term they are the prevailing phenotype of extravillous trophoblast. They show a large, polygonal, uninuclear phenotype with big, irregularly shaped, intensely staining nuclei. They are separated from each other by secretions of matrix-type fibrinoid (see Fibrinoid). *Small spindle-shaped trophoblast cells* are found from the distal part of cell columns until the first third of the myometrium and thus display the same spatial distribution pattern as the large polygonal trophoblast cells. They prevail throughout the invasive pathway in the first trimester and decrease in number until term. They display small ovoid nuclei and elongated partly filiform cell bodies mostly oriented radially to the uterine wall. Because the nuclei are often outside the sectional plane, they easily may escape observation. They seem to represent the really invasive phenotype of extravillous trophoblast, separated from each other by large polygonal trophoblast, decidual, or myometrial cells. *Multinucleated giant cells* prevail close to or even inside the myometrium where they tend to form a thin layer of giant cells. They comprise 2 to more than 10 irregularly shaped nuclei of varying size. Their generation via syncytial fusion might be a tool to stop invasion of trophoblast cells.

One of the crucial problems in histopathology and in experimental studies of the placenta for the inexperienced placentologist is the discrimination between extravillous cytotrophoblast and decidua. Reliable markers for decidual cells as opposed to trophoblastic cells are antiprolactin (320) and antivimentin (321). However, extravillous trophoblast as opposed to decidual cells can be identified by binding anticytokeratin (321) and antihuman placental lactogen (322,323). But it should be kept in mind that the different phenotypes of extravillous trophoblast described above do not stain equally well using anticytokeratin antibodies (313), even when using antibodies against cytokeratin isoforms 7 or 18 (168,324).

Uteroplacental Vessels

Because of the invasive capacities of the penetrating trophoblast, endometrial arteries (spiral arteries) and veins become eroded and thus connected to the intervillous space, as described previously. There is general agreement that the number of corresponding maternal vessels that supply the placenta, although originally high, is reduced considerably toward term by obliteration (3). The final number of spiral arteries given for the term placenta is about 100 and that for venous openings is 50 to 200.

Normal growth and development of the fetus firmly depends on the adequate remodeling of maternal uterine spiral arteries. The process of remodeling of uteroplacental arteries is called "physiological conversion of uteroplacental arteries."

In a first stage, changes to uteroplacental arteries involve a generalized perturbation of these arteries such as endothelial basophilia and vacuolation, disorganized vascular smooth muscle cells, and finally lumen dilation (325). These initial changes occur during very early pregnancy and are independent from direct trophoblast invasion. Rather, they are considered to involve maternal activation of local decidual artery renin-angiotensin systems (325). Craven and coworkers (325) presented evidence for trophoblast invasion–independent changes by demonstrating that during intrauterine pregnancies, spiral arteries from both implantation and nonimplantation regions display these "physiological" changes. Furthermore, endometrial spiral arteries undergo the same physiological vascular modifications in ectopic pregnancies.

In a second stage, the uteroplacental arteries within the implantation region are invaded by extravillous trophoblast cells. Extravillous trophoblast cells in direct vicinity of uteroplacental artery structures are associated with further vascular remodeling without a direct penetration of the arterial media. This finding, described in the guinea pig (326–328), shows reduction of the number of active media smooth muscle cells and deposition of fibrinoid material before infiltration of the media by trophoblast. Respective observations in the human have not yet been reported.

In a third stage, remodeling of uteroplacental arteries is characterized by infiltration of the arterial wall by endovascular trophoblast. Now the uteroplacental arteries undergo further dilation, reaching several times the original diameter of the lumen (36,329). Infiltration of trophoblast cells into the smooth muscle layer of the media coincides with loss of elastic fibers (330). It is still open whether smooth muscle cells undergo cell death and become replaced by infiltrating trophoblast (331) or display a temporary molecular and structural de-differentiation (328) during trophoblast invasion.

Trophoblast invasion into uteroplacental arteries is not a homogeneous process. The number of extravillous trophoblast cells and the depth of invasion of uteroplacental arteries are most pronounced in the central region of the placental bed. Both the density and depth of invasion of extravillous trophoblast as well as the degree of invasion of spiral arteries diminish toward the placental margin (63). This correlates with the pattern of onset of the maternal circulation, as described previously.

The anatomical pathway taken by extravillous trophoblast has been a matter of controversy. Recently, it became clear that it is the route of *intravasation* that leads to invasion of the spiral arteries (44). Immunohistochemical data as well as structural criteria place trophoblast cells within the wall of uteroplacental arteries at the end stage of differentiation of interstitial trophoblast derived from trophoblastic cell columns (332–334). As described above, a subpopulation of extravillous trophoblast cells takes a side route, invades the arterial walls from the surrounding junctional zone, and finally enters the arterial lumen. These *intravasated* endovascular trophoblast cells may migrate inside the arterial lumen along the arterial wall and may even locally reenter the arterial wall to extravasate the uteroplacental artery.

Extravillous trophoblast cells emanating from the cell columns provide the source for the interstitial route of endovascular trophoblast invasion. Cells from the interstitial route invade and thus *intravasate* uteroplacental arteries and contribute to the remodeling process of the arteries by replacing media, smooth muscle, and endothelium.

Pregnancy pathologies that are combined with impaired trophoblast invasion of the uteroplacental arteries, such as IUGR, indicate that only the precise sequence and summation of all three stages guarantees a degree of arterial dilation adequate to sufficiently perfuse the placenta throughout all stages of pregnancy. Complete absence of trophoblast invasion is frequently associated with spontaneous miscarriage due to premature and disorganized onset of the maternal circulation (62,335), whereas less severe deficiencies are associated with IUGR and preeclampsia (336–338). The picture emerging is that all three conditions are part of a spectrum of placental pathologies linked to oxidative stress secondary to impaired conversion of the spiral arteries (264).

Fibrinoid

In paraffin sections of normal and pathological placentas of all stages of development, one finds an acellular homogeneous material that preferable binds acid stains. This material was first described by Langhans as fibrin. Later, it was named *fibrinoid* by Hitschmann and Lindenthal (339) because there were some doubts as to the mere derivation from blood clotting.

Fibrinoid has been described in different localizations and under different names: as Langhans' stria or subchorial fibrinoid at the intervillous surface of the chorionic plate, as perivillous fibrin or perivillous fibrinoid encasing or partly covering placental villi, as intravillous fibrinoid or villous fibrinoid necrosis in the villous stroma, as Rohr's stria at the surface of the basal plate (Fig. 7F) facing the intervillous space, or as Nitabuch's stria or uteroplacental fibrinoid in the depth of the basal plate where maternal and fetal cells come in close contact to each other (Fig. 7F). Moreover, fibrinoid deposits can be found in placental septa and cells islands, in the walls of uteroplacental vessels replacing the usual constituents of vessels walls, and in the smooth chorion where it fills the former intervillous space. Studies of the last decade have provided new insight into the composition and origin of fibrinoid (69,340–342), leading to a division of fibrinoid into two subtypes, *fibrin-type fibrinoid* and *matrix-type fibrinoid*.

Composition and Origin

Fibrin-type fibrinoid has been shown to contain fibrinogen and fibrin (343,344), and ultrastructurally, it is characterized by a dense meshwork of fibrin fibers with a diameter of less than 10 nm [for review, see (36)]. Beside its immunoreactivity with antibodies that specifically bind to fibrin, fibrin-type fibrinoid binds to ulex europaeus lectin (341). This reaction indicates the presence of encased and disintegrated remnants of blood cells and endothelial cells. Moreover, plasma fibronectin can be detected in this type of fibrinoid, whereas cellular fibronectin and other matrix proteins such as collagens or laminins are absent. A typical feature of this type of fibrinoid is the absence of extravillous trophoblast cells. In most locations (see above), fibrin-type fibrinoid shows intense reaction with antifibrin and antifibrinogen antibodies because it is always related to the intervillous space and to maternal blood.

Such findings indicate that fibrin-type fibrinoid is derived from blood clotting but that also cellular degeneration may contribute to it. The typical location of fibrin-type fibrinoid lining the intervillous space points to its origin as a maternal blood clot product. Especially when lining degenerative breaks of the syncytiotrophoblast, its origin becomes obvious (70). However, because it may be detected inside villi, a contribution of fetal blood cannot be excluded. Thus, it seems to be a mixture of derivatives from maternal and fetal fibrinogen.

In deeper layers of fibrinoid not in direct contact with the intervillous space and maternal blood, an immunohistochemically different composition has been observed. At these sites matrix-type fibrinoid is found containing oncofetal fibronectin and tenascin, whereas fibrin is no longer detectable (69,342). Histologically, this type of fibrinoid is always related to single or clustered trophoblast cells. Matrix-type fibrinoid is characterized by the presence of basement membrane constituents such as collagen IV, laminin, and cellular fibronectins (340). Other matrix components found in this type of fibrinoid are heparin sulfate and vitronectin (342,345), fibrillin (346), and merosin (347,348). These findings indicate that matrix-type fibrinoid is derived from cellular secretion. The matrix components accumulate between the extravillous trophoblast cells and lead to separation and loss of contact of these cells. Finally, at the end of pregnancy the large polygonal extravillous trophoblast cells are found to be deeply embedded inside their own matrix.

Normally, matrix-type fibrinoid does not appear to have a direct contact to maternal blood. In those cases where it may face maternal blood, the surface is covered by fibrin-type fibrinoid. Thus, both types are clearly separate, although in most locations they are in direct vicinity to each other.

Functional Aspects

It is no longer justified to classify fibrinoid merely as being the result of a degenerative process, caused by placental aging or altered blood flow and nutrition (i.e., simply as an indicator for placental "degeneration"). Rather, most authors consider that fibrinoid is an unavoidable constituent of the normal placenta. Four different aspects are currently under discussion:

1. Hormann (349) pointed to fibrinoid as "constructive principal" of the placenta, serving for mechanical stability of the organ.
2. Additional importance may be attributed to fibrinoid as a regulator of the intervillous circulation (36). The intervillous space is an open cleft-like communicating and continuously growing system. Problems in perfusion are to be expected. One possibility to adapt the shape of the intervillous space to the maternal blood flow is to obstruct all poorly perfused areas by clotting of blood and fibrinoid deposits. Morphogenetically active extracellular matrix constituents such as tenascin (317) and oncofetal fibronectin (318) detected in such fibrinoid demonstrate high morphogenetic activities of the villous trees in those locations. Thus, fibrinoid may act as a limb in a complex process to adapt the shape of the villous trees to intervillous circulation.
3. The occurrence of fibrinoid at the maternal– fetal junction correlates with invasive placentation (318). Because of this, it has been discussed as

a barrier to limit the invasiveness of the trophoblast (350). Certainly, the extravillous trophoblast cells that secrete matrix-type fibrinoid (mostly the large polygonal cells) are of the type that no longer show an invasive phenotype (44,313), but it is not clear whether this is cause or effect.

4. Many authors discuss an immunologically protective function (351). An immunoprotective role of sialic acid as a normal constituent of fibrinoid has been stressed by Currie and Bagshawe (352). This molecule may mask fetal antigens and thus prevent their recognition by maternal cells; moreover, sialic acid is thought to protect fetal cells from already sensitized maternal lymphocytes. Also, the blood group antigen precursor "i," secreted by extravillous trophoblast cells (353), and heparan sulfate proteoglycan, secreted by decidual cells into the surrounding uteroplacental fibrinoid (354), have been suggested as molecules for immunoprotection.

Swinburne (355) proposed that the fibrinoid expresses target antigens that bind circulating maternal antibodies; the resulting immune complexes contribute to deposition of fibrinoid. The same concept was favored by Chaouat and coworkers (356) and by Hunziker and Wegmann (357).

In the noninvasive epitheliochorial placentation of Artiodactyla and Perissodactyla, which largely lack fibrinoid, the trophoblast is uninterrupted and serves with the trophoblastic glycocalix as a perfect immune barrier. By contrast, the trophoblast and its glycocalix are multiply interrupted in hemochorial placentation and may thus raise the demand for an additional focal immune barrier. Because fibrinoid is deposited in all sites where the trophoblast is discontinuous, the fibrinoid may serve as a substitute (36,193,318).

GENETIC CONTROL OF PLACENTAL DEVELOPMENT

In the past few years, significant advances have been made in our understanding of the genetic control of placental development, both at the level of the contribution of individual genes and the relative expression of maternal and paternal alleles at different sites. These advances are largely based on studies performed on the mouse, because this is the best-studied mammalian experimental genetic model system. In addition, the placenta of the mouse is discoid and hemochorial and therefore often considered analogous to the human organ. Important differences exist, however, and so the anatomy of the mouse placenta is briefly considered. For more detailed descriptions, readers are referred to several comprehensive reviews (93,358,359).

Mouse Placenta

The mouse blastocyst implants around day 4.5 postconception, and the polar trophectoderm cells overlying the inner cell mass proliferate to form a solid mass of cells comprising the ectoplacental cone distally and the extraembryonic ectoderm proximally. By contrast, the mural trophectoderm cells forming the remainder of the blastocyst wall stop proliferating, undergo endoreduplication, and form trophoblast giant cells. Over the next few days, the yolk sac enlarges rapidly and forms the inverted yolk sac placenta. This plays a key role in early fetal nutrition and the transport of large proteins up until term (25).

The allantois makes contact with the chorion beneath the ectoplacental cone around day 8.5 postconception, bringing with it fetal blood vessels. Soon after, the trophoblast is thrown into a series of simple folds and at the same time differentiates into two contrasting layers. Closest to the embryo is the labyrinth layer, derived most probably from the extraembryonic ectoderm, whereas against the maternal decidua is the spongiotrophoblast, derived most probably from the cells of the ectoplacental cone. A layer of trophoblast giant cells formed from the mural trophectoderm and the outer parts of the ectoplacental cone is interposed between the spongiotrophoblast and the decidua. Later, at around day 12.5, a second population of trophoblast cells is observed in close association with the spongiotrophoblast, the glycogen cells. As their name suggests, these cells are rich in glycogen and are thought to differentiate from that layer (360). The glycogen cells migrate deeply into the decidua after day 13, but there is disagreement as to whether they cluster around the central maternal artery (359,360).

With increasing elaboration of folding within the labyrinth, maternal blood spaces appear within the trophoblast layer. Traversing the decidua beneath the developing placenta are 5–10 maternal spiral-shaped arteries that converge to form 1–4 centrally located arterial channels that communicate with the maternal blood spaces at around day 10.5. The supplying arteries dilate over the next few days, losing their elastic lamina and smooth muscle cells. This process is not correlated spatially with decidual invasion of trophoblast cells but instead first occurs in regions rich in uterine natural killer cells (360).

On a superficial level the labyrinth zone can be equated with the villous tree of the human placenta, the spongiotrophoblast to the cytotrophoblastic cell columns, and the glycogen cells to the extravillous trophoblast. More detailed comparisons based on equivalent gene expression are provided by Cross et al. (20). There are important differences, however, that need to be borne in mind when extrapolating data between the two species (361). In the mouse the trophoblast lining the labyrinth is three layered, with two layers of syncytiotrophoblast and an outer interrupted layer of cytotrophoblast cells. The process of trophoblastic fusion may also be different from that in the human, because there is no murine equivalent of the retroviral protein syncytin. There are also significant endocrine differences, as the mouse trophoblast does not secrete steroid hormones or chorionic gonadotropins. Beyond the labyrinth, the trophoblast giant cells appear to have different origins in the two species, by fusion in the human and endoreduplication in the mouse. In view of these differences, it is pertinent to remember that despite the conservation of genes that typifies placental development, the hemochorial placentas of mice and humans are examples of convergent evolution.

Genetic Control

The separation of the trophoblast lineage from that of the inner cell mass is the first to occur developmentally, and recent work on four transcription factors (Oct4, Sox2, Nanog, and Cdx2) is beginning to shed light on this process in the mouse. Although the cues promoting this separation remain elusive, once it has occurred it appears to be supported by the maintenance of Cdx2 expression in the trophectoderm and the induction of Nanog and maintenance of Oct4 (or Oct4/Sox2) expression in the inner cell mass. Subsequent proliferation of the extraembryonic ectoderm is stimulated by fibroblast growth factor (Fgf4), secreted by cells of the epiblast layer of the germ disc, which acts via the mitogen-activated protein kinase pathway to activate downstream transcription factors such as Cdx2, Sox2, Eomes, and ERRß (93,362).

Differentiation of trophoblast giant cells from the mural trophectoderm and the ectoplacental cone appears to be dependent on the basic helix-loop-helix transcription factor Hand1. Conversely, another basic helix-loop-helix transcription factor, Mash2, suppresses giant cell formation and promotes differentiation into spongiotrophoblast. Further differentiation of spongiotrophoblast into glycogen cells is promoted by insulin-like growth factor 2 (20,93). Within the labyrinth there is strong evidence that indicates the initial folding of the trophoblast is regulated by the transcription factor, Gcm1. Expression of *Gcm1* is limited to the tips of the elongating folds, where the trophoblast cells fuse to form syncytium. Although *Gcm1* is not sufficient by itself to promote syncytial transformation, expression of the antisense transcript blocks that process. Although expression appears before allantoic fusion, the latter is required to maintain expression through subsequent development. Hence, *Gcm1* deficient mice or other mutants in which the allantois does not fuse with the chorion fail to form these primary folds (93). Equally, *Gcm1* mutants display abnormal shaping of the maternal blood spaces within the mature labyrinth (360). A number of other factors have been implicated in the regulation of branching morphogenesis within the labyrinth. Thus, several signaling pathways may be involved, including Fgf, hepatocyte growth factor, and wingless-related (Wnt).

Detailed reviews of the involvement of other transcription factors in mouse placental development are provided by Rossant and Cross (93) and Cross et al. (20). Comparison of expression patterns with the human placenta revealed many homologies, supporting the use of the mouse as a model system (99).

Imprinting and Placental Development

Contrary to earlier beliefs, the contributions of the maternal and paternal genetic material are not functionally equivalent during development. A class of genes has now been identified in placental mammals and angiosperms whose expression is dependent on their parental origin. They play an important role in regulating fetal and neonatal growth, and their evolution has been explained by the maternal–fetal "conflict" hypothesis (363,364). Briefly stated, the paternal genome attempts to extract more nutrients and resources from the mother, whereas her genome attempts to balance nutrient provision to the current fetus with her own needs and those of future fetuses. Most investigations on imprinted genes have been conducted on the mouse because of the ease of genetic manipulations, but because many of the genes are conserved in the human, it is likely they play similar roles in both species. Of the 60 or so imprinted genes currently identified, most are indeed involved in the regulation of fetal growth. Thus, knockouts of the paternally expressed genes *Igf2*, *Peg1*, *Peg3*, and *insulin* result in IUGR, whereas knockouts of the maternally expressed genes *H19* and *Igf2r* result in fetal hypertrophy (365).

All the imprinted genes studied so far, approximately half of those identified, have been found to be expressed in the placenta (366). Paternally expressed transcripts include *Igf2* in the glycogen cells, *Peg1* in the fetal labyrinthine vessels, and *Peg3* in the labyrinth trophoblast and spongiotrophoblast. As might be expected, knocking out each gene results in a reduction in placental size. In particular, deletion of just the trophoblast-specific promoter (P0) for *Igf2* results in a reduction in placental surface area and increased thickness of the interhemal membrane. Diffusion across the placenta is impaired and most likely contributes to the fetal growth retardation observed (21). For maternal transcripts *H19*, *Igf2r*, and *p57Kip2* are widely expressed, whereas *Ipl* is restricted to the labyrinth and *Mash2* to the labyrinth and spongiotrophoblast. Deletions of *H19*, *Igf2r*, *p57Kip2*, and *Ipl* result in the hypertrophy of all layers of the placenta, despite the restricted expression of *Ipl* (366). Although most attention has focused on the role of imprinted genes in regulating placental development, they may also play a role in controlling function through their effects on specific transporters and channels (366). For example, the *Ata3* gene that encodes a component of the system A amino acid transporter is paternally expressed.

Imprinted genes may therefore play various roles in the regulation of placental development and nutrient supply to the fetus, and so are pivotal to normal fetal development. Imprinting is controlled epigenetically by differential methylation of DNA and chromatin modifications, and the mechanism is highly susceptible to a variety of environmental insults. Disruption of imprinting can lead to abnormal development in cloned fetuses, and it will be of interest to see in the future whether less severe disturbances, caused for example through abnormal inheritance of specific transcripts or maternal dietary insufficiency, may underlie currently unexplained cases of human IUGR.

MORPHOLOGICAL CORRELATES OF NONINVASIVE DIAGNOSTIC METHODS

Ultrasound

Ultrasound has proven to be a simple, safe, and accurate way of imaging the placenta in vivo. From the outset, one of the great benefits it conferred was the ability to localize the placenta, but as the technology improves and with the introduction of three-dimensional ultrasound, it has become possible to monitor the organ's growth and assess its consistency in great detail. This is providing new information on the normal development and functioning of the placenta and informing clinical management of complicated pregnancies.

The development of the transvaginal probe has enabled events taking place during the first few weeks of pregnancy to be visualized. Only a few days after the expected menstrual period, it is possible to distinguish a gestational sac of 2–3 mm diameter buried within the uterine mucosa (367). The center of the sac appears as a fluid-filled echo-free cavity and is surrounded by an echo-rich peripheral zone corresponding to the proliferating trophoblastic mantle. The secondary yolk sac is the first structure to become visible ultrasonographically within the cavity during the sixth week of pregnancy. Because it is so delicate in structure, the amnion is not easily visualized until it has expanded more fully at around 8 weeks. By this time, it is therefore possible to distinguish multiple pregnancies and to diagnose whether these are mono- or dichorial and mono- or diamniotic by the thickness and shape of the septum separating the fetuses. Hence, one can be alerted to the possibility of the twin transfusion syndrome developing later in pregnancy.

At the end of the fifth week of pregnancy, the peripheral zone is 3- to 4-mm thick, and this thickness triples over the next 2 weeks. Regression of chorionic villi and formation of the chorion laeve is indicated during weeks 7–11 by an increasing disparity in the thickness of the peripheral zone at the embryonic and abembryonic poles. There is no difference in echogenicity at these two sites, however. That part destined to become the definitive placenta continues to expand, and uterine vessels become more prominent in the adjacent endometrium. Volumetric growth of normal and abnormal placentas was recently assessed in a longitudinal study from 12 weeks of pregnancy until term (300). Placental volume was found to be smaller in cases of IUGR at 12 weeks compared with normal subjects, although subsequent development followed the normal trajectory. Although this technique sheds important light on the timing of onset of the pathophysiology in complicated pregnancies, at present the data are too heterogeneous to allow it to be used as a diagnostic tool.

Textural changes within the placenta can also be detected ultrasonographically, and from 8 weeks of pregnancy onward it becomes possible to identify molar changes within the gestational sac. A complete mole typically presents as filling the uterine cavity with multiple sonolucent areas of varying size and shape ("snowstorm appearance") without any associated embryonic or fetal structures. In partial moles where there is focal swelling of villous tissue, the appearance is of a thickened placenta containing multicystic avascular sonolucent spaces ("Swiss cheese appearance"). Although complete moles can be diagnosed with an accuracy of 80% by 12 weeks of pregnancy, partial moles are more difficult to detect and 70% are missed by ultrasonographic examination alone (368–370). Sonolucent areas reflecting the development of villus-free placental lakes can also be visualized in the peripheral areas of most placentas toward term. Isolated examples are of little clinical importance, but they are particularly common in pregnancies complicated by IUGR and preeclampsia. The lakes are thought to be caused by the hemodynamic effects of the more jet-like inflow of maternal blood that occurs in these placentas due to deficient conversion of the spiral arteries (371).

In an attempt to assess placental development and to correlate this with fetal maturity, Grannum and coworkers (218) proposed a classification scheme based on ultrasonographic appearances. Within the scheme, four grades of placental appearances (0–3) are recognized, reflecting, for example, the development of placental septa, the central villus-free cavities of the lobules, and sites of calcification in the basal plate. The higher grades are therefore taken to indicate increasing maturity of the placenta, and stereological studies demonstrated that grade 3 placentas have a greater theoretical diffusing capacity (279). It is not unusual for individual cases to display a mixture of features, however, and not all placentas progress to a grade 3 configuration at term. The grade 1 configuration is generally seen at approximately 31 weeks of gestation, but in normal pregnancies 40% of cases remain grade 1 at term. Grade 2 placenta appears at approximately 36 weeks of gestation, and 45% of cases retain this configuration until term. The grade 3 placenta appears at 38 weeks and accounts for the remaining 15% of cases at term. In the original study of normal patients, there was a strong correlation between the gradings and the lecithin-to-sphingomyelin ratio in the amniotic fluid (218). These findings were confirmed by subsequent studies (372), but others have reported weaker correlations (373).

Other uses of ultrasound in pregnancy are to diagnose abnormalities of the placenta that may have clinical consequences at delivery, such as velamentous insertion of the cord, circumvallate placenta, and placenta praevia (374). It can also be used to assess the depth of villous invasion, because in cases of placenta accreta the decidual interface between placenta and myometrium is absent. Alerting the obstetrician in advance of delivery helps to reduce the higher risk of perinatal mortality associated with this condition.

Doppler Ultrasound

The application of the Doppler principle to ultrasound has opened up new possibilities for imaging fetoplacental unit, permitting blood flow velocities to be measured noninvasively. Pulsed Doppler has the advantage of being range selective and so allows velocities at specific points to be measured. It can be combined with conventional ultrasound imaging, and these points can be accurately located within the placenta or uterine wall. The refinement of color Doppler enables relative velocities to be more easily discriminated and blood flow to be determined in smaller vessels than is possible with traditional Doppler techniques.

The Doppler signals obtained represent the summation of multiple Doppler shift frequencies backscattered by erythrocytes traveling at different velocities. These velocities vary according to the phase of the cardiac cycle; therefore, in real-time imaging a flow velocity waveform is displayed. Analysis of this waveform can provide qualitative information such as the presence of direction of blood flow and also quantitative or semiquantitative data on velocity. The shape of the flow velocity waveform also gives an indication of the degree of resistance to flow offered by the vessel in question and its subsequent ramifications. In low resistance networks, end-diastolic velocities indicate that some blood flow occurs during all phases of the cardiac cycle. By contrast, in high-resistance vascular beds, flow is only observed during systole. In the most severe cases, reverse flow may even occur during diastole. Several indices, such as the resistance index and pulsality index, have therefore been devised and are all based on a comparison of the peak systolic and end-diastolic velocities. There is a strong correlation between these indices and umbilical blood gas measurements (375,376).

Doppler ultrasound can thus provide information on velocity and resistance, but of greater interest to physiologists is the volume of blood flow. This is determined by a combination of the mean velocity and the vessel's cross-sectional area. Measurements of the latter with ultrasound are technically demanding, and so initially quantitative measurements of blood flow proved to be poorly reproducible. As experience has grown, the results have become more consistent, and significantly reduced umbilical venous volume flow has been reported in cases of IUGR (377).

Using color Doppler techniques, it is possible to visualize blood flow in the umbilical vessel at 7 weeks of gestation. During the first trimester, the flow velocity waveform indicates a high vascular resistance in the placenta and no end-diastolic velocities. At the end of this period, between weeks 12 and 14 of gestation, end-diastolic velocities develop and are maintained throughout the remainder of pregnancy in normal cases. These changes coincide with the elaboration of terminal villi and the progressive enlargement of the fetal capillary bed (378). In addition, the decline in nucleated erythrocytes that occurs at the end of the first trimester has a major impact on the viscosity of the fetal blood. As a result, during the second half of pregnancy the placental vascular bed has the lowest resistance of any circulatory bed in the fetus. Independent of this change with gestational age, there is a progressive fall in resistance from the umbilical artery to the chorionic arteries and intraplacental arterioles (379). These findings correlate well with mathematical models that depict the placental vasculature as a tiered series of resistances, with the first level represented by the umbilical arteries and the second by the chorionic arteries, offering a larger number of smaller resistances acting in parallel (380).

There have been numerous attempts to correlate abnormal fetal velocity waveforms with changes in the placental vasculature in normal and complicated pregnancies, but this is a difficult task for several reasons. First, according to Poiseuille's law, resistance to flow is related to the fluid viscosity, to the length of the vascular system, and to the fourth power of the mean radius of the vessels. The latter is a particularly powerful influence, and yet vessel radius in vivo is difficult to measure accurately because the placental arteries collapse rapidly during and after delivery. Second, the vascular resistance depends on the total cross-sectional area of the resistance vessels in the tertiary stem villi presented to the umbilical arterial inflow. Hence, simple counts of vessel numbers provide little information, particularly if the vessels concerned are of widely differing caliber. Third, the fetal vasculature may be very heterogeneous across the placenta. Thus, it is possible to observe normal waveforms within intraplacental vessels despite the presence of extremely abnormal umbilical artery waveforms (381). Areas of the placenta may therefore receive normal or even increased blood flow being diverted from regions in which the stem villous arteries are totally or partially occluded. Despite these difficulties, there is general agreement that raised placental resistance is associated with a reduction in the elaboration of terminal villi and with a reduction in villous vascularization (301,382,383). Whether this represents primary pathology or a secondary response to deficiencies in the maternal circulation cannot be determined at this stage.

In addition to assessing vascular resistance, color Doppler can also be used to diagnose abnormalities of the fetal placental vasculature, such as chorioangiomas, from midgestation onward. These tumors may present various sonographic appearances, but if particularly vascular they can act as arteriovenous shunts within the placental circulation. Such cases are associated early features of fetal congestive heart failure, polyhydramnios, and premature onset of labor (384).

Color power angiography (CPA) is a new application in Doppler velocimetry (219). It is able to identify red blood cells in very small vessels and when combined with three-dimensional reconstruction is able to map out the vasculature within an organ such as the placenta (Fig. 11). The placental vascular patterns that can be observed using CPA reflect the development of the villous trees and its intervillous space in vivo. In normal placentas from 13 to 40 weeks of gestation, the patterns derived from the dynamic CPA images are very similar to those obtained by classical histology and scanning electron microscopy. Throughout pregnancy, placental vessels can be visualized by CPA as long as their luminal diameter exceeds 200 μm. Only the smaller stem villi with their terminal branches, which are equipped only with microcirculatory vessels, remain undetectable.

On the maternal side, the timing of the onset of the maternal circulation has now been clarified. In normal pregnancies this is a progressive phenomenon that starts toward the end of the first trimester in the peripheral regions of the placenta and then extends toward the center (62). Once established, there is a progressive reduction in resistance to blood flow in the uterine arteries that continues until approximately the twentieth week of gestation, from which point onward the resistance remains stable. This largely reflects hormonally induced vasodilation of the uterine and arcuate arteries, which can increase their radii by up to tenfold and so raise combined uterine artery blood flow from 45 ml/min during the follicular phase to 750 ml/min at term.

Magnetic Resonance Imaging

Magnetic resonance imaging is another safe and noninvasive imaging technique that has potential for monitoring the placenta in vivo. Pioneering studies have demonstrated that it can be used to monitor the volumetric growth of the organ sequentially during gestation and to assess the ratio between the vascular and nonvascular components across gestational age (385,386).

In an attempt to gain more functional data, radioisotopes have been injected into the maternal circulation during late pregnancy to measure placental blood flows. Transit times were reduced in pregnancies complicated by IUGR and hypertension (387), but these studies have not been repeated due to the danger of fetal exposure to radiation. Another approach has been to use perfusion-sensitive echo-planar imaging, a technique that allows variations in perfusion rate to be visualized across the placenta. Perfusion rates were found to be variable within the normal placenta, with areas of low perfusion being clustered together, but there were more areas of low perfusion in pregnancies complicated by IUGR (388). In these cases there was a significant correlation between areas of reduced perfusion and fetal size.

REFERENCES

1. DeWitt, F. (1958). An historical study of the theories of the placenta to 1900. *J. Hist. Med.* 14, 360–374.
2. Amoroso, E. C. (1971). Early theories of the placenta form: fancies to facts. The JY Simpson Oration. *In Royal College of Obstetricians and Gynaecologists, 42nd Annual Report,* pp. 76–91. RCOG Press, London.

3. Boyd, J. D., and Hamilton, W. J. (1970). *The Human Placenta*. Heffer and Sons, Cambridge.
4. Corner, G. W. (1963). Exploring the placental maze. The development of our knowledge of the relation between the blood-streams of mother and infant in utero. *Am. J. Obstet. Gynecol.* 86, 408–418.
5. Steven, D. H. (1983). Historical introduction: concepts of the trophoblast and its role in placentation. In *Biology of Trophoblast* (Y. W. Loke and A. Whyte, Eds.), pp. 3–21. Elsevier, Amsterdam.
6. Hubrecht, A. A. W. (1889). Studies in mammalian embryology. 1. The placentation of *Erinaceus europeaus*, with remarks on the phylogeney of the placenta. *J. Microsc. Sci.* 30, 283–404.
7. Grosser, O. (1927). *Frühentwicklung, Eihautbildung und Placentation des Menschen und der Säugetiere*. JF Bergmann, München.
8. Mossman, H. W. (1987). *Vertebrate fetal membranes: comparative ontogeny and morphology; evolution; phylogenetic significance; basic functions; research opportunities*. Macmillan, London.
9. Amoroso, E. C. (1952). Placentation. In *Marshall's Physiology of Reproduction* (A. S. Parkes, Ed.), pp. 127–311. Longmans, Green and Co., London.
10. Halban, J. (1905). Die innere Sekretion von Ovarium und placenta unde ihre Bedeutung für die Funktion der Milchdrüse. *Arch. Gynäkol.* 75, 353–441.
11. Wislocki, G. B., and Streeter, G. L. (1938). On the placentation in the macaque (*Macaca mulatta*) from the time of implantation until formation of the definitive placenta. *Carnegie Contrib. Embryol.* 27, 1–65.
12. Hertig, A. T., Rock, J., and Adams, E. C. (1956). A description of 34 human ova within the first 17 days of development. *Am. J. Anat.* 98, 435–494.
13. Stieve, H. (1941). Die Entwicklung und der Bau der menschlichen Placenta. Teil 2. Zotten: Zottenraumgitter und Gerfäße in der zweiten Hälfte der Schwangerschaft. *Z. Mikrosk. Anat. Forsch.* 50, 1–20.
14. Spanner, R. (1935). Mütterlicher und kindlicher Kreislauf der menschlichen Placenta und seine Strombahnen. *Z. Anat. Entwicklung.* 105, 163–242.
15. Ramsey, E. M., and Donner, M. W. (1980). *Placental Vasculature and Circulation. Anatomy, Physiology, Radiology, Clinical Aspects, Atlas and Textbook*. Georg Thieme, Stuttgart.
16. Borell, U., Fernström, I., and Westman, A. (1958). Eine arteriographische Studie des Plazentarkreislaufs. *Geburtsh. Frauenh.* 18, 1–9.
17. Reynolds, S. R. M. (1966). Formation of fetal cotyledons in the hemochorial placenta. A theoretical consideration of the functional implications of such an arrangement. *Am. J. Obstet. Gynecol.* 94, 425–439.
18. Boyd, J. D., and Hughes, A. W. (1954). Observations on human chorionic villi using the electron microscope. *J. Anat.* 88, 356–362.
19. Wislocki, G. B., and Dempsey, E. W. (1955). Electron microscopy of the human placenta. *Anat. Rec.* 123, 133–167.
20. Cross, J. C., Baczyk, D., Dobric, N., Hemberger, M., Hughes, M., Simmons, D. G., Yamamoto, H., and Kingdom, J. C. P. (2003). Genes, development and evolution of the placenta. *Placenta* 24, 123–130.
21. Sibley, C. P., Coan, P. M., Ferguson-Smith, A. C., Dean, W., Hughes, J., Smith, P., Reik, W., Burton, G. J., Fowden, A. L., and Constancia, M. (2004). Placental-specific insulin-like growth factor 2 (Igf2) regulates the diffusional exchange characteristics of the mouse placenta. *Proc. Natl. Acad. Sci. U S A* 101, 8204–8208.
22. Moffett, A. and Loke, Y. W. (2004). The immunological paradox of pregnancy: a reappraisal. *Placenta* 25, 1–8.
23. Enders, A. C., and Carter, A. M. (2004). What can comparative studies of placental structure tell us? A review. *Placenta* 25, Suppl. A, S3–S9.
24. Wooding, F. B. P., and Flint, A. P. F. (1994). Placentation. In *Marshall's Physiology of Reproduction* (G. E. Lamming, Ed.), pp. 233–460. Chapman & Hall, London.
25. Beckman, D. A., Brent, R. L., and Lloyd, J. B. (1996). Sources of amino acids for protein synthesis during early organogenesis in the rat. 4. Mechanisms before envelopment of the embryo by the yolk sac. *Placenta* 17, 635–641.
26. Dantzer, V., Leiser, R., Kaufmann, P., and Luckhardt, M. (1988). Comparative morphological aspects of placental vascularisation. *Trophoblast Res.* 3, 235–260.
27. Hill, J. P. (1932). The developemntal history of the primates. *Philos. Trans. R. Soc. B* 221, 45–178.
28. Luckett, W. P. (1976). Cladistic relationships among higher primate categories: evidence of the fetal membranes and placenta. *Folia Primatol.* 25, 245–276.
29. Madsen, O., Scally, M., Douady, C., Kao, D. J., DeBry, R. W., Adkins, R., Amrine, H. M., Stanhope, M. J., de Jong, W. W., and Springer, M. S. (2001). Parallel adaptive radiations in two major clades of placental mammals. *Nature* 409, 610–614.
30. Murphy, W. J., Eizirik, E., Johnson, W. E., Zhang, Y. P., Ryder, O. A., and O'Brien, S. J. (2001). Molecular phylogenetics and the origins of placental mammals. *Nature* 409, 614–618.
31. Carter, A. M. (2001). Evolution of the placenta and fetal membranes seen in the light of molecular phylogenetics. *Placenta* 22, 800–807.
32. Torpin, R. (1969). *The Human Placenta*. Charles C Thomas, Springfield, IL.
33. Bouw, G. M., Stolte, L. A. M., Baak, J. P. A., and Oort, J. (1976). Quantitative morphology of the placenta. I. Standardization of sampling. *Eur. J. Obstet. Gynaecol. Reprod. Biol.* 6, 325–331.
34. Kaufmann, P. (1985). Basic morphology of the fetal and maternal circuits in the human placenta. *Contrib. Gynecol. Obstet.* 13, 5–17.
35. Schuhmann, R. A. (1982). Placentone structure of the human placenta. *Biblthca. Anat.* 22, 46–57.
36. Benirschke, K., and Kaufmann, P. (2000). *Pathology of the Human Placenta*. Springer-Verlag, New York.
37. McLennan, J. E. (1968). Implications of eccentricity of the human umbilical cord. *Am. J. Obstet. Gynecol.* 101, 1124–1130.
38. Jauniaux, E., Englert, Y., Vanesse, M., Hidden, M., and Wilkin, P. (1990). Pathologic features of placentas from singleton pregnancies obtained by in vitro fertilization and embryo transfer. *Obstet. Gynecol.* 76, 61–64.
39. Potgens, A. J., Schmitz, U., Bose, P., Versmold, A., Kaufmann, P., and Frank, H. G. (2002). Mechanisms of syncytial fusion: a review. *Placenta* 23 Suppl. A, S107–S113.
40. Murphy, C. R. (2000). Understanding the apical surface markers of uterine receptivity: pinopods or uterodomes? *Hum. Reprod.* 15, 2451–2454.
41. Larsen, J. F. (1970). Electron microscopy of nidation in the rabbit and observations on the human trophoblastic invasion. In *Ovoimplantation, Human Gonadotropin and Prolactin* (P. O. Hubinot, C. R. Leroy, and P. Leleux, Eds.), pp. 38–51. Karger, Basel.
42. Luckett, W. P. (1978). Origin and differentiation of the yolk sac and extraembryonic mesoderm in presomite human and rhesus monkey embryos. *Am. J. Anat.* 152, 59–97.
43. Pijnenborg, R. (1990). Trophoblast invasion and placentation in the human: morphological aspects. *Trophoblast Res.* 4, 33–47.
44. Kaufmann, P., Black, S., and Huppertz, B. (2003). Endovascular trophoblast invasion: implications for the pathogenesis of intrauterine growth retardation and preeclampsia. *Biol. Reprod.* 69, 1–7.

45. Carter, A. M. (1997). When is the maternal placental circulation established in Man? *Placenta* 18, 83–87.
46. Hamilton, W. J., and Boyd, J. D. (1960). Development of the human placenta in the first three months of gestation. *J. Anat.* 94, 297–328.
47. Dempsey, E. W. (1972). The development of capillaries in the villi of early human placentas. *Am. J. Anat.* 134, 221–238.
48. Demir, R., Kaufmann, P., Castellucci, M., Erbengi, T., and Kotowski, A. (1989). Fetal vasculogenesis and angiogenesis in human placental villi. *Acta Anat.* 136, 190–203.
49. Corner, G. W. (1929). A well-preserved human embryo of 10 somites. *Carnegie Contrib. Embryol.* 20, 81–102.
50. Jackson, M. R., Mayhew, T. M., and Boyd, P. A. (1992). Quantitative description of the elaboration and maturation of villi from 10 weeks of gestation to term. *Placenta* 13, 357–370.
51. Hustin, J., and Schaaps, J. P. (1987). Echographic and anatomic studies of the maternotrophoblastic border during the first trimester of pregnancy. *Am. J. Obstet. Gynecol.* 157, 162–168.
52. Burton, G. J., Jauniaux, E., and Watson, A. L. (1999). Maternal arterial connections to the placental intervillous space during the first trimester of human pregnancy: the Boyd Collection revisited. *Am. J. Obstet. Gynecol.* 181, 718–724.
53. Schaaps, J. P., and Hustin, J. (1988). In vivo aspect of the maternal-trophoblastic border during the first trimester of gestation. *Trophoblast Res.* 3, 39–48.
54. Burton, G. J., Watson, A. L., Hempstock, J., Skepper, J. N., and Jauniaux, E. (2002). Uterine glands provide histiotrophic nutrition for the human fetus during the first trimester of pregnancy. *J. Clin. Endocrinol. Metab.* 87, 2954–2959.
55. Rodesch, F., Simon, P., Donner, C., and Jauniaux, E. (1992). Oxygen measurements in endometrial and trophoblastic tissues during early pregnancy. *Obstet. Gynecol.* 80, 283–285.
56. Jauniaux, E., Watson, A. L., Hempstock, J., Bao, Y.-P., Skepper, J. N., and Burton, G. J. (2000). Onset of maternal arterial bloodflow and placental oxidative stress: a possible factor in human early pregnancy failure. *Am. J. Pathol.* 157, 2111–2122.
57. Nicol, C. J., Zielenski, J., Tsui, L.-C., and Wells, P. G. (2000). An embryoprotective role for glucose-6-phosphate dehydrogenase in developmental oxidative stress and chemical teratogenesis. *FASEB J.* 14, 111–127.
58. Burton, G. J., Hempstock, J., and Jauniaux, E. (2003). Oxygen, early embryonic metabolism and free radical-mediated embryopathies. *Reprod. BioMed. Online* 6, 84–96.
59. Jauniaux, E., Gulbis, B., and Burton, G. J. (2003). The human first trimester gestational sac limits rather than facilitates oxygen transfer to the fetus-a review. *Placenta* 24, Suppl. A, S86–S93.
60. Lennard, S. N., Gerstenberg, C., Allen, W. R., and Stewart, F. (1998). Expression of epidermal growth factor and its receptor in equine placental tissues. *J. Reprod. Fertil.* 112, 49–57.
61. Gray, C. A., Taylor, K. M., Ramsey, W. S., Hill, J. R., Bazer, F. W., Bartol, F. F., and Spencer, T. E. (2001). Endometrial glands are required for preimplantation conceptus elongation and survival. *Biol. Reprod.* 64, 1608–1613.
62. Jauniaux, E., Hempstock, J., Greenwold, N., and Burton, G. J. (2003). Trophoblastic oxidative stress in relation to temporal and regional differences in maternal placental blood flow in normal and abnormal early pregnancies. *Am. J. Pathol.* 162, 115–125.
63. Pijnenborg, R., Bland, J. M., Robertson, W. B., Dixon, G., and Brosens, I. (1981). The pattern of interstitial trophoblastic invasion of the myometrium in early human pregnancy. *Placenta* 2, 303–316.
64. Watson, A. L., Skepper, J. N., Jauniaux, E., and Burton, G. J. (1998). Susceptibility of human placental syncytiotrophoblastic mitochondria to oxygen-mediated damage in relation to gestational age. *J. Clin. Endocrinol. Metab.* 83, 1697–1705.
65. Mayhew, T. M., Leach, L., McGee, R., Ismail, W. W., Myklebust, R. and Lammiman, M. J. (1999). Proliferation, differentiation and apoptosis in villous trophoblast at 13–41 weeks of gestation (including observations on annulate lamellae and nuclear pore complexes). *Placenta* 20, 407–422.
66. Simpson, R. A., Mayhew, T. M., and Barnes, P. R. (1992). From 13 weeks to term, the trophoblast of human placenta grows by the continuous recruitment of new proliferative units: a study of nuclear number using the disector. *Placenta* 13, 501–512.
67. Kosanke, G., Kadyrov, M., Korr, H., and Kaufmann, P. (1998). Maternal anemia results in increased proliferation in human placental villi. *Trophoblast Res.* 11, 339–357.
68. Bright, N. A., and Ockleford, C. D. (1994). Heterogeneity of Fc gamma receptor-bearing cells in human term amniochorion. *Placenta* 15, 247–255.
69. Kaufmann, P., Huppertz, B., and Frank, H.-G. (1996). The fibrinoids of the human placenta: origin, composition and functional relevance. *Ann. Anat.* 178, 485–501.
70. Nelson, D. M. (1996). Apoptotic changes occur in syncytiotrophoblast of human placental villi where fibrin type fibrinoid is deposited at discontinuities in the villous trophoblast. *Placenta* 17, 387–391.
71. Fox, H. (1964). The villous cytotrophoblast as an index of placental ischaemia. *J. Obstet. Gynaecol. Br. Commonw.* 71, 885–893.
72. Esterman, A., Greco, M. A., Mitani, Y., Finlay, T. H., Ismail-Beigi, F., and Dancis, J. (1997). The effect of hypoxia on human trophoblast in culture: morphology, glucose transport and metabolism. *Placenta* 18, 129–136.
73. Huppertz, B., Kingdom, J., Caniggia, I., Desoye, G., Black, S., Korr, H., and Kaufmann, P. (2003). Hypoxia favours necrotic versus apoptotic shedding of placental syncytiotrophoblast into the maternal circulation. *Placenta* 24, 181–190.
74. Ali, K. Z. M. (1997). Stereological study of the effect of altitude on the trophoblast cell populations of human term placental villi. *Placenta* 18, 447–450.
75. Uehara, Y., Minowa, O., Mori, C., Shiota, K., Kuno, J., Noda, T., and Kitamura, N. (1995). Placental defect and embryonic lethality in mice lacking hepatocyte growth factor/scatter factor. *Nature* 373, 702–705.
76. Furugori, K., Kurauchi, O., Itakura, A., Kanou, Y., Murata, Y., Mizutani, S., Seo, H., Tomoda, Y., and Nakamura, T. (1997). Levels of hepatocyte growth factor and its messenger ribonucleic acid in uncomplicated pregnancies and those complicated by preeclampsia. *J. Clin. Endocrinol. Metab.* 82, 2726–2730.
77. Somerset, D. A., Li, X. F., Afford, S., Strain, A. J., Ahmed, A., Sangha, R. K., Whittle, M. J., and Kilby, M. D. (1998). Ontogeny of hepatocyte growth factor (HGF) and its receptor (c-met) in human placenta: reduced HGF expression in intrauterine growth restriction. *Am. J. Pathol.* 153, 1139–1147.
78. Jones, C. J. P., and Fox, H. (1991). Ultrastructure of the normal human placenta. *Electron Microsc. Rev.* 4, 129–178.
79. Burton, G. J., Skepper, J. N., Hempstock, J., Cindrova, T., Jones, C. J. P., and Jauniaux, E. (2003). A reappraisal of the contrasting morphological appearances of villous cytotrophoblast cells during early human pregnancy; evidence for both apoptosis and primary necrosis. *Placenta* 24, 297–305.
80. DeLoia, J. A., Burlingame, J. M., and Krasnow, J. S. (1997). Differential expression of G1 cyclins during human placentogenesis. *Placenta* 18, 9–16.
81. McKenzie, P. P., Foster, J. S., House, S., Bukovsky, A., Caudle, M. R., and Wimalasena, J. (1998). Expression of G1 cyclins and cyclin-dependent kinase-2 activity during terminal differentiation of cultured human trophoblast. *Biol. Reprod.* 58, 1283–1289.

82. Kaufmann, P., and Stark, J. (1972). Enzyme-histochemical studies on mature human placental villi. I. Differentiation and degeneration of the trophoblast. *Histochemie* 29, 65–82.
83. Richart, R. (1961). Studies of placental morphogenesis. I. Radioautographic studies of human placenta utilizing tritiated thymidine. *Proc. Soc. Exp. Biol. Med.* 106, 829–831.
84. Huppertz, B., Frank, H.-G., Reister, F., Kingdom, J., Korr, H., and Kaufmann, P. (1999). Apoptosis cascade progresses during turnover of human trophoblast: analysis of villous cytotrophoblast and syncytial fragments in vitro. *Lab. Invest.* 79, 1687–1702.
85. Hoshina, M., Boothby, M., Hussa, R., Pattillo, R., Camel, H. M., and Boime, I. (1985). Linkage of human chorionic gonadotrophin and placental-lactogen biosynthesis to trophoblast differentiation and tumorigenesis. *Placenta* 6, 163–172.
86. Huppertz, B., Frank, H. G., and Kaufmann, P. (1999). The apoptosis cascade—morphological and immunohistochemical methods for its visualization. *Anat. Embryol.* 200, 1–18.
87. Copp, A. J. (1995). Death before birth: clues from gene knockouts and mutations. *Trends Genet.* 11, 87–93.
88. Rinkenberger, J. L., Cross, J. C., and Werb, Z. (1997). Molecular genetics of implantation in the mouse. *Dev. Genet.* 21, 6–20.
89. Morrish, D. W., Dakour, J., and Li, H. (1998). Functional regulation of human trophoblast differentiation. *J. Reprod. Immunol.* 39, 179–195.
90. Janatpour, M., Utset, M. F., Cross, J. C., Rossant, J., Dong, J., Israel, M. A., and Fisher, S. J. (1999). A repertoire of differentially expressed transcription factors that offers insight into mechansims of human cytotrophoblast differentiation. *Dev. Genet.* 25, 146–157.
91. Cross, J. C. (2000). Genetic insights into trophoblast differentiation and placental morphogenesis. *Semin. Cell. Dev. Biol.* 11, 105–113.
92. Morrish, D. W., Dakour, J., and Li, H. (2001). Life and death in the placenta: new peptides and genes regulating human syncytiotrophoblast and extravillous cytotrophoblast lineage formation and renewal. *Curr. Protein Pept. Sci.* 2, 245–259.
93. Rossant, J., and Cross, J. C. (2001). Placental development: lessons from mouse mutants. *Nat. Rev. Genet.* 2, 538–548.
94. Cross, J. C., Anson-Cartwright, L., and Scott, I. C. (2002). Transcription factors underlying the development and endocrine functions of the placenta. *Rec. Prog. Horm. Res.* 57, 221–234.
95. Janatpour, M. J., McMaster, M. T., Genbacev, O., Zhou, Y., Dong, J., Cross, J. C., Israel, M. A., and Fisher, S. J. (2000). Id-2 regulates critical aspects of human cytotrophoblast differentiation, invasion and migration. *Development* 127, 549–558.
96. Genbacev, O., McMaster, M. T., and Fisher, S. J. (2000). A repertoire of cell cycle regulators whose expression is coordinated with human cytotrophoblast differentiation. *Am. J. Pathol.* 157, 1337–1351.
97. Schild, R. L., Schaiff, W. T., Carlson, M. G., Cronbach, E. J., Nelson, D. M., and Sadovsky, Y. (2002). The activity of PPAR gamma in primary human trophoblasts is enhanced by oxidized lipids. *J. Clin. Endocrinol. Metab.* 87, 1105–1110.
98. Baczyk, D., Satkunaratnam, A., Nait-Oumesma, R. B., Huppertz, B., Cross, J. C., and Kingdom, J. C. (2004). Complex patterns of GCM1 mRNA and protein in villous and extravillous trophoblast cells of the human placenta. *Placenta* 25, 553–559.
99. Loregger, T., Pollheimer, J., and Knofler, M. (2003). Regulatory transcription factors controllong function and differentiation of human trophoblast: a review. *Placenta* 24 Suppl. A, S104–S110.
100. Scott, I. C., Anson-Cartwright, L., Riley, P., Reda, D., and Cross, J. C. (2000). The HAND1 basic helix-loop-helix transcription factor regulates trophoblast differentiation via multiple mechanisms. *Mol. Cell. Biol.* 20, 530–541.
101. Yamada, K., Ogawa, H., Tamiya, G., Ikeno, M., Morita, M., Asakawa, S., Shimizu, N., and Okazaki, T. (2000). Genomic organization, chromosomal localization, and the complete 22 kb DNA sequence of the human GCMa/GCM1, a placenta-specific transcription factor gene. *Biochem. Biophys. Res. Commun.* 278, 134–139.
102. Nait-Oumesmar, B., Copperman, A. B., and Lazzarini, R. A. (2000). Placental expression and chromosomal localization of the human Gcm 1 gene. *J. Histochem. Cytochem.* 48, 915–922.
103. Castellucci, M., Scheper, M., Scheffen, I., Celona, A., and Kaufmann, P. (1990). The development of the human placental villous tree. *Anat. Embryol.* 181, 117–128.
104. Dakour, J., Li, H., Chen, H., and Morrish, D. W. (1999). EGF promotes development of a differentiated trophoblast phenotype having c-myc and junB proto-oncogene activation. *Placenta* 20, 119–126.
105. Yang, M., Lei, Z. M., and Rao, C. V. (2003). The central role of human chorionic gonadotropin in the formation of human placental syncytium. *Endocrinology* 144, 1108–1120.
106. Frendo, J. L., Olivier, D., Cheynet, V., Blond, J. L., Bouton, O., Vidaud, M., Rabreau, M., Evain-Brion, D., and Mallet, F. (2003). Direct involvement of HERV-W Env glycoprotein in human trophoblast cell fusion and differentiation. *Mol. Cell. Biol.* 23, 3566–3574.
107. Cronier, L., Herve, J. C., Deleze, J., and Malassine, A. (1997). Regulation of gap junctional communication during human trophoblast differentiation. *Microsc. Res. Tech.* 38, 21–28.
108. Winterhager, E., Kaufmann, P., and Gruemmer, R. (2000). Cell-cell-communication during placental development and possible implications for trophoblast proliferation and differentiation. *Placenta* 21 Suppl. A, S61–S68.
109. Cronier, L., Frendo, J.-L., Defamie, N., Pidoux, G., Bertin, G., Guibourdenche, J., Pointis, G., and Malassine, A. (2003). Requirement of gap junctional intercellular communication for human villous trophoblast differentiation. *Biol. Reprod.* 69, 1472–1480.
110. Frendo, J. L., Cronier, L., Bertin, G., Guibourdenche, J., Vidaud, M., Evain-Brion, D., and Malassine, A. (2003). Involvement of connexin 43 in human trophoblast cell fusion and differentiation. *J. Cell Sci.* 116, 3413–3421.
111. Larsen, W. J., and Wert, S. E. (1988). Roles of cell junctions in gametogenesis and in early embryonic development. *Tissue Cell.* 20, 809–848.
112. Cronier, L., Bastide, B., Herve, J. C., Deleze, J., and Malassine, A. (1994). Gap junctional communication during human trophoblast differentiation: influence of human chorionic gonadotropin. *Endocrinology* 135, 402–408.
113. Winterhager, E., Von Ostau, C., Gerke, M., Gruemmer, R., Traub, O., and Kaufmann, P. (1999). Connexin expression patterns in human trophoblast cells during placental development. *Placenta* 20, 627–638.
114. Schreiber, J. R., Beckmann, M. W., Polacek, D., and Davies, P. F. (1993). Changes in gap junction connexin-43 messenger ribonucleic acid levels associated with rat ovarian follicular development as demonstrated by in situ hybridization. *Am. J. Obstet. Gynecol.* 168, 1094–1102.
115. Winterhager, E., Busch, L. C., and Kuhnel, W. (1984). Membrane events involved in fusion of uterine epithelial cells in pseudopregnant rabbits. *Cell Tissue Res.* 235, 357–363.
116. Firth, J. A., Farr, A., and Bauman, K. (1980). The role of gap junctions in trophoblastic cell fusion in the guinea-pig placenta. *Cell Tissue Res.* 205, 311–318.

117. Reale, E., Wang, T., Zaccheo, D., Maganza, C., and Pescetto, G. (1980). Junctions on the maternal blood surface of the human placental syncytium. *Placenta* 1, 245–258.
118. Huppertz, B., Frank, H.-G., Kingdom, J. C. P., Reister, F., and Kaufmann, P. (1998). Villous cytotrophoblastic regulation of the syncytial apoptotic cascade in the human placenta. *Histochem. Cell Biol.* 110, 495–508.
119. Huppertz, B., Tews, D. S., and Kaufmann, P. (2001). Apoptosis and syncytial fusion in human placental trophoblast and skeletal muscle. *Int. Rev. Cytol.* 205, 215–253.
120. Huppertz, B., Kaufmann, P., and Kingdom, J. (2002). Trophoblast turnover in health and disease. *Fetal Matern. Med. Rev.* 13, 103–118.
121. Black, S., Kadyrov, M., Kaufmann, P., Ugele, B., Emans, N., and Huppertz, B. (2004). Syncytial fusion of human trophoblast depends on caspase 8. *Cell Death Differ.* 11, 90–98.
122. Steinborn, A., von Gall, C., Hildenbrand, R., Stutte, H. J., and Kaufmann, M. (1998). Identification of placental cytokine-producing cells in term and preterm labor. *Obstet. Gynecol.* 91, 329–335.
123. Yui, J., Hemmings, D., Garcia-Lloret, M., and Guilbert, L. J. (1996). Expression of the human p55 and p75 tumor necrosis factor receptors in primary villous trophoblasts and their role in cytotoxic signal transduction. *Biol. Reprod.* 55, 400–409.
124. Yui, J., Garcia-Lloret, M., Wegmann, T. G., and Guilbert, L. J. (1994). Cytotoxicity of tumour necrosis factor-alpha and gamma-interferon against primary human placental trophoblasts. *Placenta* 15, 819–835.
125. Reister, F., Frank, H.-G., Kingdom, J. C. P., Heyl, W., Kaufmann, P., Rath, W., and Huppertz, B. (2001). Macrophage-induced apoptosis limits endovascular trophoblast invasion in the uterine wall of preeclamptic women. *Lab. Invest.* 81, 1143–1152.
126. Huppertz, B., and Kaufmann, P. (1999). The apoptosis cascade in human villous trophoblast. *Trophoblast Res.* 13, 215–242.
127. Diakowski, W., Prychidny, A., Swistak, M., Nietubyc, M., Bialkowska, K., Szopa, J., and Sikorski, A. F. (1999). Brain spectrin (fodrin) interacts with phospholipids as revealed by intrinsic fluorescence quenching and monolayer experiments. *Biochem. J.* 338, 83–90.
128. Baumann, M., and Sowers, A. E. (1996). Membrane skeleton involvement in cell fusion kinetics: a parameter that correlates with erythrocyte osmotic fragility. *Biophys. J.* 71, 336–340.
129. Greidinger, E. L., Miller, D. K., Yamin, T. T., Casciola-Rosen, L., and Rosen, A. (1996). Sequential activation of three distinct ICE-like activities in Fas-ligated Jurkat cells. *FEBS Lett.* 390, 299–303.
130. Vanags, D. M., Porn-Ares, M. I., Coppola, S., Burgess, D. H., and Orrenius, S. (1996). Protease involvement in fodrin cleavage and phosphatidylserine exposure in apoptosis. *J. Biol. Chem.* 271, 31075–31085.
131. Bevers, E. M., Comfurius, P., and Zwaal, R. F. (1996). Regulatory mechanisms in maintenance and modulation of transmembrane lipid asymmetry: pathophysiological implications. *Lupus* 5, 480–487.
132. Williamson, P., and Schlegel, R. A. (1994). Back and forth: the regulation and function of transbilayer phospholipid movement in eukaryotic cells. *Mol. Membr. Biol.* 11, 199–216.
133. Martin, S. J., Reutelingsperger, C. P., McGahon, A. J., Rader, J. A., van Schie, R. C., LaFace, D. M., and Green, D. R. (1995). Early redistribution of plasma membrane phosphatidylserine is a general feature of apoptosis regardless of the initiating stimulus: inhibition by overexpression of Bcl-2 and Abl. *J. Exp. Med.* 182, 1545–1556.
134. Martin, S. J., Finucane, D. M., Amarante-Mendes, G. P., O'Brien, G. A., and Green, D. R. (1996). Phosphatidylserine externalization during CD95-induced apoptosis of cells and cytoplasts requires ICE/CED-3 protease activity. *J. Biol. Chem.* 271, 28753–28756.
135. Lyden, T. W., Ng, A. K., and Rote, N. S. (1993). Modulation of phosphatidylserine epitope expression by BeWo cells during forskolin treatment. *Placenta* 14, 177–186.
136. Das, M., Xu, B., Lin, L., Chakrabarti, S., Shivaswamy, V., and Rote, N. S. (2004). Phosphatidylserine efflux and intercellular fusion in a BeWo model of human villous cytotrophoblast. *Placenta* 25, 396–407.
137. Huppertz, B., and Hunt, J. S. (2000). Trophoblast apoptosis and placental development—a workshop report. *Placenta* 21 Suppl. A, S74–S76.
138. Huovila, A. P., Almeida, E. A., and White, J. M. (1996). ADAMs and cell fusion. *Curr. Opin. Cell Biol.* 8, 692–699.
139. Gilpin, B. J., Loechel, F., Mattei, M. G., Engvall, E., Albrechtsen, R., and Wewer, U. M. (1998). A novel, secreted form of human ADAM 12 (meltrin alpha) provokes myogenesis in vivo. *J. Biol. Chem.* 273, 157–166.
140. Martin, I., Epand, R. M., and Ruysschaert, J. M. (1998). Structural properties of the putative fusion peptide of fertilin, a protein active in sperm-egg fusion, upon interaction with the lipid bilayer. *Biochemistry* 37, 17030–17039.
141. Smith, C. A., and Moore, H. D. (1988). Expression of C-type viral particles at implantation in the marmoset monkey. *Hum. Reprod.* 3, 395–398.
142. Mi, S., Lee, X., Li, X., Veldman, G. M., Finnerty, H., Racie, L., LaVallie, E., Tang, X. Y., Edouard, P., Howes, S., Keith, J. C., Jr., and McCoy, J. M. (2000). Syncytin is a captive retroviral envelope protein involved in human placental morphogenesis. *Nature* 403, 785–789.
143. Voisset, C., Bouton, O., Bedin, F., Duret, L., Mandrand, B., Mallet, F., and Paranhos-Baccala, G. (2000). Chromosomal distribution and coding capacity of the human endogenous retrovirus HERV-W family. *AIDS Res. Hum. Retrovir.* 16, 731–740.
144. Cheng, Y. H., Richardson, B. D., Hubert, M. A., and Handwerger, S. (2004). Isolation and characterization of the human syncytin gene promoter. *Biol. Reprod.* 70, 694–701.
145. Sharma, S. C., and Richards, J. S. (2000). Regulation of AP1 (Jun/Fos) factor expression and activation in ovarian granulosa cells. Relation of JunD and Fra2 to terminal differentiation. *J. Biol. Chem.* 275, 33718–33728.
146. Peters, T. J., Chapman, B. M., and Soares, M. J. (2000). Trophoblast differentiation. An in vitro model for trophoblast giant cell development. *Methods Mol. Biol.* 137, 301–311.
147. Vasicek, R., Meinhardt, G., Haidweger, E., Rotheneder, H., Husslein, P., and Knofler, M. (2003). Expression of the human Hand1 gene in trophoblastic cells is transcriptionally regulated by activating and repressing specificity protein (Sp)-elements. *Gene* 302, 115–127.
148. Wang, V. E., Schmidt, T., Chen, J., Sharp, P. A., and Tantin, D. (2004). Embryonic lethality, decreased erythropoiesis, and defective octamer-dependent promoter activation in Oct-1-deficient mice. *Mol. Cell Biol.* 24, 1022–1032.
149. Lavillette, D., Marin, M., Ruggieri, A., Mallet, F., Cosset, F. L., and Kabat, D. (2002). The envelope glycoprotein of human endogenous retrovirus type W uses a divergent family of amino acid transporters/cell surface receptors. *J. Virol.* 76, 6442–6452.
150. Kudo, Y., and Boyd, C. A. (2002). Changes in expression and function of syncytin and its receptor, amino acid transport system B(0) (ASCT2), in human placental choriocarcinoma BeWo cells during syncytialization. *Placenta* 23, 536–541.

151. Yu, C., Shen, K., Lin, M., Chen, P., Lin, C., Chang, G. D., and Chen, H. (2002). GCMa regulates the syncytin-mediated trophoblastic fusion. *J. Biol. Chem.* 277, 50062–50068.
152. LaBranche, C. C., Galasso, G., Moore, J. P., Bolognesi, D. P., Hirsch, M. S., and Hammer, S. M. (2001). HIV fusion and its inhibition. *Antiviral Res.* 50, 95–115.
153. Burton, G. J. (1987). The fine structure of the human placenta as revealed by scanning electron microscopy. *Scanning Microsc.* 1, 1811–1828.
154. Ockleford, C. D., and Whyte, A. (1977). Differeniated regions of human placental cell surface associated with exchange of materials between maternal and foetal blood: coated vesicles. *J. Cell. Sci.* 25, 293–312.
155. Ockleford, C. D., and Wakely, J. (1981). The skeleton of the placenta. In *Progress in Anatomy* (R. J. Harrison and R. L. Holmes, Eds.), pp. 19–48. Cambridge University Press, London.
156. Dearden, L., and Ockleford, C. D. (1983). Structure of human trophoblast: correlation with function. In *Biology of Trophoblast* (Y. W. Loke and A. Whyte, Eds.), pp. 69–110. Elsevier, Amsterdam.
157. Kaufmann, P., Nagl, W., and Fuhrmann, B. (1983). Die funktionelle Bedeutung der Langhanszellen der menschlichen Placenta. *Anat. Anz.* 77, 435–436.
158. Jones, C. J. P., and Fox, H. (1977). Syncytial knots and intervillous bridges in the human placenta: an ultrastructural study. *J. Anat.* 124, 275–286.
159. Cantle, S. J., Kaufmann, P., Luckhardt, M., and Schweikhart, G. (1987). Interpretation of syncytial sprouts and bridges in the human placenta. *Placenta* 8, 221–234.
160. LeBrun, D. P., Warnke, R. A., and Cleary, M. L. (1993). Expression of bcl-2 in fetal tissues suggests a role in morphogenesis. *Am. J. Pathol.* 142, 743–753.
161. Gruslin, A., Qiu, Q., and Tsang, B. K. (2001). X-linked inhibitor of apoptosis protein expression and the regulation of apoptosis during human placental development. *Biol. Reprod.* 64, 1264–1272.
162. Zhou, P., Qian, L., Kozopas, K. M., and Craig, R. W. (1997). Mcl-1, a Bcl-2 family member, delays the death of hematopoietic cells under a variety of apoptosis-inducing conditions. *Blood* 89, 630–643.
163. Hawkins, C. J., and Vaux, D. L. (1994). Analysis of the role of bcl-2 in apoptosis. *Immunol. Rev.* 142, 127–139.
164. Marks, N., Berg, M. J., Guidotti, A., and Saito, M. (1998). Activation of caspase-3 and apoptosis in cerebellar granule cells. *J. Neurosci. Res.* 52, 334–341.
165. Kaufmann, P., Gentzen, D. M., and Davidoff, M. (1977). The ultrastructure of Langhans cells in pathologic human placentas [author's transl]. *Arch. Gynakol.* 222, 319–332.
166. Owens, G. P., Hahn, W. E., and Cohen, J. J. (1991). Identification of mRNAs associated with programmed cell death in immature thymocytes. *Mol. Cell Biol.* 11, 4177–4188.
167. Kockx, M. M., Muhring, J., Knaapen, M. W., and de Meyer, G. R. (1998). RNA synthesis and splicing interferes with DNA in situ end labeling techniques used to detect apoptosis. *Am. J. Pathol.* 152, 885–888.
168. Kadyrov, M., Kaufmann, P., and Huppertz, B. (2001). Expression of a cytokeratin 18 neo-epitope is a specific marker for trophoblast apoptosis in human placenta. *Placenta* 22, 44–48.
169. Iklé, F. A. (1961). Trophoblastzellen im strömenden Blut. *Schweiz Med. Wochenschr.* 91, 934–945.
170. Smith, S. C., Leung, T. N., To, K. F., and Baker, P. N. (2000). Apoptosis is a rare event in first-trimester placental tissue. *Am. J. Obstet. Gynecol.* 183, 697–699.
171. Chan, C. C. W., Lao, T. T., and Cheung, A. N. Y. (1999). Apoptotic and proliferative activities in first trimester placentae. *Placenta* 20, 223–227.
172. Arnholdt, H., Meisel, F., Fandrey, K., and Lohrs, U. (1991). Proliferation of villous trophoblast of the human placenta in normal and abnormal pregnancies. *Virchows Arch. B Cell Pathol. Inc. Mol. Pathol.* 60, 365–372.
173. Sen, D. K., Kaufmann, P., and Schweikhart, G. (1979). Classification of placental villi. II. Morphometry. *Cell Tissue Res.* 200, 425–434.
174. Johansen, M., Redman, C. W., Wilkins, T., and Sargent, I. L. (1999). Trophoblast deportation in human pregnancy—its relevance for pre-eclampsia. *Placenta* 20, 531–539.
175. Lunetta, P., and Penttila, A. (1996). Immunohistochemical identification of syncytiotrophoblastic cells and megakaryocytes in pulmonary vessels in a fatal case of amniotic fluid embolism. *Int. J. Legal Med.* 108, 210–214.
176. Cohle, S. D., and Petty, C. S. (1985). Sudden death caused by embolization of trophoblast from hydatidiform mole. *J. Forensic Sci.* 30, 1279–1283.
177. Delmis, J., Pfeifer, D., Ivanisevic, M., Forko, J. I., and Hlupic, L. (2000). Sudden death from trophoblastic embolism in pregnancy. *Eur. J. Obstet. Gynaecol. Reprod. Biol.* 92, 225–227.
178. Kamoi, S., Ohaki, Y., Mori, O., Satomi, M., Takahashi, H., Kawamura, T., and Araki, T. (2003). Placental villotrophoblastic pulmonary emboli after elective abortion: immunohistochemical diagnosis and comparison with ten control cases. *Int. J. Gynecol. Pathol.* 22, 303–309.
179. Knight, M., Redman, C. W., Linton, E. A., and Sargent, I. L. (1998). Shedding of syncytiotrophoblast microvilli into the maternal circulation in pre-eclamptic pregnancies. *Br. J. Obstet. Gynaecol.* 105, 632–640.
180. Huppertz, B., and Kingdom, J. (2004). Apoptosis in the trophoblast: role of apoptosis in placental morphogenesis. *J. Soc. Gynecol. Invest.* 11, 353–362.
181. Demir, R., Kayisli, U. A., Seval, Y., Celik-Ozenci, C., Korgun, E. T., Demir-Weusten, A. Y., and Huppertz, B. (2004). Sequential expression of VEGF and its receptors in human placental villi during very early pregnancy: differences between placental vasculogenesis and angiogenesis. *Placenta* 25, 560–572.
182. Castellucci, M., Celona, A., Bartels, H., Steininger, B., Benedetto, V., and Kaufmann, P. (1987). Mitosis of the Hofbauer cell: possible implications for a fetal macrophage. *Placenta* 8, 65–76.
183. Burton, G. J., and Watson, A. L. (1997). The structure of the human placenta: implications for initiating and defending against viral infections. *Rev. Med. Virol.* 7, 219–228.
184. Nikolov, S. D., and Schiebler, T. H. (1973). Fetal blood vessel system of the human full-term placenta. *Z. Zellforsch. Mikrosk. Anat.* 139, 333–350.
185. Karimu, A. L., and Burton, G. J. (1995). Human term placental capillary endothelial cell specialization: a morphometric study. *Placenta* 16, 93–99.
186. Leach, L., Eaton, B. M., Firth, J. A., and Contractor, S. F. (1989). Immunogold localisation of endogenous immunoglobulin-G in ultrathin frozen sections of the human placenta. *Cell Tissue Res.* 257, 603–607.
187. Yamada, T., Isemura, M., Yamaguchi, Y., Munakata, H., Hayashi, N., and Kyogoku, M. (1987). Immunochemical localization of fibronectin in the human placentas at their different stages of maturation. *Histochemistry* 86, 579–584.
188. Zhang, E. C., Burton, G. J., Smith, S. K., and Charnock-Jones, D. S. (2002). Placental vessel adaptation during gestation and to high altitude: changes in diameter and perivascular cell coverage. *Placenta* 23, 751–762.
189. Myatt, L. (1992). Control of vascular resistance in the human placenta. *Placenta* 13, 329–341.

190. Graf, R., Neudeck, H., Gossrau, R., and Vetter, K. (1996). Elastic fibres are an essential component of human placental stem villous stroma and an integrated part of the perivascular contractile sheath. *Cell Tissue Res.* 283, 133–141.
191. Graf, R., Matejevic, D., Schuppan, D., Neudeck, H., Shakibaei, M., and Vetter, K. (1997). Molecular anatomy of the perivascular sheath in human placental stem villi: the contractile apparatus and its association to the extracellular matrix. *Cell Tissue Res.* 290, 601–607.
192. Stulc, J. (1989). Extracellular transport pathways in the haemochorial placenta. *Placenta* 10, 113–119.
193. Nelson, D. M., Crouch, E. C., Curran, E. M., and Farmer, D. R. (1990). Trophoblast interaction with fibrin matrix. Epithelialization of perivillous fibrin deposits as a mechanism for villous repair in the human placenta. *Am. J. Pathol.* 136, 855–865.
194. Kaufmann, P., Schroder, H., and Leichtweiss, H. P. (1982). Fluid shift across the placenta. II. Fetomaternal transfer of horseradish peroxidase in the guinea pig. *Placenta* 3, 339–348.
195. Kertschanska, S., Kosanke, G., and Kaufmann, P. (1994). Is there morphological evidence for the existence of transtrophoblastic channels in human placental villi? *Trophoblast Res.* 8, 581–596.
196. Kertschanska, S., Kosanke, G., and Kaufmann, P. (1997). Pressure dependence of so-called transtrophoblastic channels during fetal perfusion of human placental villi. *Microsc. Res. Tech.* 38, 52–62.
197. Kaufmann, P., Schröder, H., Leichtweiss, H. P., and Winterhager, E. (1987). Are there membrane-lined channels through the trophoblast? A study with lanthanum hydroxide. *Trophoblast Res.* 2, 557–571.
198. Kertschanska, S., Stulcova, B., Kaufmann, P., and Stulc, J. (2000). Distensible transtrophoblastic channels in the rat placenta. *Placenta* 21, 670–677.
199. Thornburg, K. L., and Faber, J. J. (1977). Transfer of hydrophilic molecules by placenta and yolk sac of the guinea pig. *Am. J. Physiol.* 233, C111–C124.
200. Schroder, H., Nelson, P., and Power, G. (1982). Fluid shift across the placenta. I. The effect of dextran T 40 in the isolated guinea-pig placenta. *Placenta* 3, 327–338.
201. Kaufmann, P., Sen, D. K., and Schweikhert, G. (1979). Classification of human placental villi. 1. Histology. *Cell Tissue Res.* 200, 409–423.
202. Kaufmann, P. (1982). Development and differentiation of the human placental villous tree. *Biblthca. Anat.* 22, 29–39.
203. Knoth, M. (1968). Ultrastructure of chorionic villi from a four-somite human embryo. *J. Ultrastruct. Res.* 25, 423–440.
204. Demir, R., Kaufmann, P., and Erbengi, T. (1992). Ultrastructural observations on angiogenetic cells formation and differentiation toward vascular endothelium in human placental villi. In *Vascular Endothelium: Physiological Basis of Clinical Problems* (J. D. Catravas, A. D. Callow, and U. S. Ryan, Eds.), pp. 240–242. Plenum Press, NATO Scientific Affairs Division I, New York.
205. Demir, R., and Erbengi, T. (1984). Some new findings about Hofbauer cells in the chorionic villi of the human placenta. *Acta. Anat.* 119, 18–26.
206. Demir, R., Kosanke, G., Kohnen, G., Kertschanska, S., and Kaufmann, P. (1997). Classification of human placental stem villi: review of structural and functional aspects. *Microsc. Res. Tech.* 38, 29–41.
207. Charnock Jones, D. S., Kaufmann, P., and Mayhew, T. M. (2004). Aspects of human fetoplacental vasculogenesis and angiogenesis. I. Molecular recognition. *Placenta* 25, 103–113.
208. Castellucci, M., Kosanke, G., Verdenelli, F., Huppertz, B., and Kaufmann, P. (2000). Villous sprouting: fundamental mechanisms of human placental development. *Hum. Reprod. Update* 6, 485–494.
209. Kingdom, J., Huppertz, B., Seaward, G., and Kaufmann, P. (2000). Development of the placental villous tree and its consequences for fetal growth. *Eur. J. Obstet. Gynaecol. Reprod. Biol.* 92, 35–43.
210. Challier, J. C., Carbillon, L., Kacemi, A., Vervelle, C., Bintein, T., Galtier, M., Espie, M. J., and Uzan, S. (2001). Characterization of first trimester human fetal placental vessels using immunocytochemical markers. *Cell. Mol. Biol.* 47 Online Pub, OL79–87.
211. Leach, L., Babawale, M. O., Anderson, M., and Lammiman, M. (2002). Vasculogenesis, angiogenesis and the molecular organisation of endothelial junctions in the early human placenta. *J. Vasc. Res.* 39, 246–259.
212. Kaufmann, P., Luckhardt, M., and Leiser, R. (1988). Three-dimensional representation of the fetal vessel system in the human placenta. *Trophoblast Res.* 3, 113–137.
213. Boe, F. (1953). Studies on the vascularization of the human placenta. *Acta Obstet. Gynaecol. Scand.* 32 Suppl., 1–92.
214. Arts, N. F. (1961). Investigations on the vascular system of the placenta. I. General introduction and the fetal vascular system. *Am. J. Obstet. Gynecol.* 82, 147–158.
215. Kaufmann, P., Bruns, U., Leiser, R., Luckhardt, M., and Winterhager, E. (1985). The fetal vascularisation of term placental villi. II. Intermediate and terminal villi. *Anat. Embryol.* 173, 203–214.
216. Burton, G. J., and Tham, S. W. (1992). The formation of vasculo-syncytial membranes in the human placenta. *J. Dev. Physiol.* 18, 43–47.
217. Karimu, A. L., and Burton, G. J. (1994). The effects of maternal vascular pressure on the dimensions of the placental capillaries. *Br. J. Obstet. Gynaecol.* 101, 57–63.
218. Grannum, P. A. T., Berkowitz, R. L., and Hobbins, J. C. (1979). The ultrasonic changes in the maturing placenta and their relation to fetal pulmonic maturity. *Am. J. Obstet. Gynecol.* 133, 915–922.
219. Konje, J. C., Huppertz, B., Bell, S. C., Taylor, D. J., and Kaufmann, P. (2003). 3-Dimensional colour power angiography for staging human placental development. *Lancet* 362, 1199–1201.
220. Nelson, J. H., Bernstein, R. L., Wilson-Houston, J., Garcia, N. A., and Gartenlaub, C. (1961). Percutaneous retrograde femoral arteriography in obstetrics and gynecology. *Obstet. Gynecol. Surv.* 16, 1–19.
221. Wigglesworth, J. S. (1969). Vascular anatomy of the human placenta and its significance for placental pathology. *J. Obstet. Gynaecol. Brit. Commonw.* 76, 979–989.
222. Hempstock, J., Bao, Y.-P., Bar-Issac, M., Segaren, N., Watson, A. L., Charnock Jones, D. S., Jauniaux, E., and Burton, G. J. (2003). Intralobular differences in antioxidant enzyme expression and activity reflect oxygen gradients within the human placenta. *Placenta* 24, 517–523.
223. Moll, W. (1972). Gas exchange in concurrent, countercurrent and cross-current flow systems. The concept of the fetoplacental unit. In *Respiratory Gas Exchange and Blood Flow in the Placenta* (L. D. Longo and H. Bartels, Eds.), pp. 281–296. U.S. Department of Health, Education and Welfare, Bethesda, MD.
224. Burton, G. J. (1986). Scanning electron microscopy of intervillous connections in the mature human placenta. *J. Anat.* 147, 245–254.
225. Freese, U. E. (1968). The uteroplacental vascular relationship in the human. *Am. J. Obstet. Gynecol.* 101, 8–16.
226. Karimu, A. L., and Burton, G. J. (1993). Star volume estimates of the intervillous gaps in the human placenta: how changes in umbilical arterial pressure might influence

the maternal placental circulation. *J. Dev. Physiol.* 19, 137–142.

227. Power, G. G., and Longo, L. D. (1973). Sluice flow in placenta: maternal vascular pressure effects on fetal circulation. *Am. J. Physiol. Cell. Physiol.* 225, 1490–1496.
228. Marx, G. F., Patel, S., Berman, J. A., Farmakides, G., and Schulman, H. (1986). Umbilical blood flow velocity waveforms in different maternal positions and with epidural analgesia. *Obstet. Gynecol.* 68, 61–64.
229. van Katwijk, C., and Wladimiroff, J. W. (1991). Effect of maternal posture on the umbilical artery flow velocity waveform. *Ultrasound Med. Biol.* 17, 683–685.
230. Moll, W., Künzel, W., and Herberger, J. (1975). Hemodynamic implications of hemochorial placentation. *Eur. J. Obstet. Gynaecol. Reprod. Biol.* 5, 67–74.
231. Sun, Y., and Oberley, L. W. (1996). Redox regulation of transcriptional activators. *Free Radic. Biol. Med.* 21, 335–348.
232. Droge, W. (2002). Free radicals in the physiological control of cell function. *Physiol. Rev.* 82, 47–95.
233. Chen, K., Thomas, S. R., and Keaney, J. F. (2003). Beyond LDL oxidation: ROS in vascular signal transduction. *Free Radic. Biol. Med.* 35, 117–132.
234. Halliwell, B., and Gutteridge, J. M. C. (1999). *Free Radicals in Biology and Medicine*. Oxford Science Publications, Oxford.
235. Matsubra, S., and Tamada, T. (1991). Ultracytochemical localisation of NAD(P)H oxidase activity in the human placenta. *Acta Obstet. Gynaecol. Jpn.* 43, 117–121.
236. Manes, C. (2001). Human placental NAD(P)H oxidase: solubilization and properties. *Placenta* 22, 58–63.
237. Raijmakers, M. T. M., Peters, W. H. M., Steegers, E. A. P., and Poston, L. (2004). NAD(P)H oxidase associated superoxide production in human placenta from normotensive and pre-eclamptic women. *Placenta* 25, Suppl. A, S85–S89.
238. Dechend, R., Viedt, C., Muller, D. N., Ugele, B., Brandes, R. P., Wallukat, G., Park, J.-K., Janke, J., Barta, P., Theuer, J., Fiebler, A., Homuth, V., Dietz, R., Haller, H., Kreuzer, J., and Luft, F. C. (2003). AT_1 receptor agonistic antibodies from preeclamptic patients stimulate NADPH oxidase. *Circulation* 107, 1632–1639.
239. Schachter, M., and Foulds, S. (1999). Free radicals and the xanthine oxidase pathway. In *Ischaemia-Reperfusion Injury* (P. A. Grace and R. T. Mathie, Eds.), pp. 137–147. Blackwell Science Ltd., Oxford.
240. Many, A., and Roberts, J. M. (1997). Increased xanthine oxidase during labour-implications for oxidative stress. *Placenta* 18, 725–726.
241. Kingdom, J. C. P., and Kaufmann, P. (1997). Oxygen and placental villous development: origins of fetal hypoxia. *Placenta* 18, 613–621.
242. Espinoza, J., Sebire, N. J., McAuliffe, F., Krampl, E., and Nicolaides, K. H. (2001). Placental villus morpholgy in relation to maternal hypoxia at high altitude. *Placenta* 22, 606–608.
243. Ali, K. Z. M., Burton, G. J., Morad, N., and Ali, M. E. (1996). Does hypercapillarization influence the branching pattern of terminal villi in the human placenta at high altitude? *Placenta* 17, 677–682.
244. Bacon, B. J., Gilbert, R. D., Kaufman, P., Dwight Smith, A., Trevino, F. T., and Longo, L. D. (1984). Placental anatomy and diffusing capacity in guinea pigs follwing long-term maternal hypoxia. *Placenta* 5, 475–488.
245. Mayhew, T. M., Charnock Jones, D. S., and Kaufmann, P. (2004). Aspects of human fetoplacental vasculogenesis and angiogenesis. III. Changes in complicated pregnancies. *Placenta* 25, 127–139.
246. Schuhmann, R., Stoz, F., and Maier, M. (1988). Histometric investigations in placentones (materno–fetal circulation units) of human placentae. *Trophoblast Res.* 3, 3–16.
247. Genbacev, O., Zhou, Y., Ludlow, J. W., and Fisher, S. J. (1997). Regulation of human placental development by oxygen tension. *Science* 277, 1669–1672.
248. Kudo, Y., Boyd, C. A., Sargent, I. L., and Redman, C. W. (2003). Hypoxia alters expression and function of syncytin and its receptor during trophoblast cell fusion of human placental BeWo cells: implications for impaired trophoblast syncytialisation in pre-eclampsia. *Biochim. Biophys. Acta* 1638, 63–71.
249. Frendo, J. L., Therond, P., Bird, T., Massin, N., Muller, F., Guibourdenche, J., Luton, D., Vidaud, M., Anderson, W. B., and Evain-Brion, D. (2001). Overexpression of copper zinc superoxide dismutase impairs human trophoblast cell fusion and differentiation. *Endocrinology* 142, 3638–3648.
250. Miller, R. K., Genbacev, O., Turner, M. A., Aplin, J., Caniggia, I., and Huppertz, B. (2005). Human placental explants in culture: approaches and assessments. *Placenta* 26, 439–448.
251. Caniggia, I., and Winter, J. L. (2002). Adriana and Luisa Castellucci Award Lecture 2001. Hypoxia inducible factor-1: oxygen regulation of trophoblast differentiation in normal and pre-eclamptic pregnancies—a review. *Placenta* 23 Suppl. A, S47–S57.
252. Schäffer, L., Scheid, A., Spielmann, P., Breymann, C., Zimmermann, R., Meuli, M., Gassmann, M., Marti, H. M., and Wenger, R. H. (2003). Oxygen-regulated expression of TGF-ß3, a growth factor involved in trophoblast differentiation. *Placenta* 24, 941–950.
253. Graham, C. H., Postovit, L. M., Park, H., Canning, M. T., and Fitzpatrick, T. E. (2000). Adriana and Luisa Castellucci Award Lecture 1999: role of oxygen in the regulation of trophoblast gene expression and invasion. *Placenta* 21, 443–450.
254. Postovit, L. M., Adams, M. A., Lash, G. E., Heaton, J. P., and Graham, C. H. (2002). Oxygen-mediated regulation of tumor cell invasiveness. Involvement of a nitric oxide signaling pathway. *J. Biol. Chem.* 277, 35730–35737.
255. Chen, C.-P., and Aplin, J. D. (2003). Placental extracellular matrix: gene expression, deposition by placental fibroblasts and the effect of oxygen. *Placenta* 24, 316–325.
256. Esterman, A., Finlay, T. H., and Dancis, J. (1996). The effect of hypoxia on term trophoblast: hormone synthesis and release. *Placenta* 17, 217–222.
257. Pidoux, G., Guibourdenche, J., Frendo, J.-L., Gerbaud, P., Conti, M., Luton, D., Muller, F., and Evain-Brion, D. (2004). Impact of trisomy 21 on human trophoblast behaviour and hormonal function. *Placenta* 25, Suppl. A, S79–S84.
258. Benyo, D. F., Miles, T. M., and Conrad, K. P. (1997). Hypoxia stimulates cytokine production by villous explants from the human placenta. *J. Clin. Endocrinol. Metab.* 82, 1582–1588.
259. Malek, A., Sager, R., and Schneider, H. (2001). Effect of hypoxia, oxidative stress and lipopolysaccharides on the release of prostaglandins and cytokines from human term placental explants. *Placenta* 22 Suppl. A, S45–S50.
260. Roberts, J. M., and Hubel, C. A. (1999). Is oxidative stress the link in the two-stage model of pre-eclampsia? *Lancet* 354, 788–789.
261. Redman, C. W. G., and Sargent, I. L. (2000). Placental debris, oxidative stress and pre-eclampsia. *Placenta* 21, 597–602.
262. Huppertz, B., Kaufmann, P., and Black, S. (2003). Trophoblast aponecrosis and maternal endothelial dysfunction. In *Preeclampsia. Proceedings of the 45th Study Group of the Royal College of Obstetricians and Gynaecologists* (H. O. D. Critchley, L. Poston, and R. J. J. Walke, Eds.), pp. 158–167. RCOG Press, London.
263. Poston, L., and Raijmakers, M. T. M. (2004). Trophoblast oxidative stress, antioxidants and pregnancy outcome: a review. *Placenta* 25, Suppl. A, S72–S78.

264. Burton, G. J., and Jauniaux, E. (2004). Placental oxidative stress; from miscarriage to preeclampsia. *J. Soc. Gynecol. Invest.* 11, 342–352.
265. Hempstock, J., Jauniaux, E., Greenwold, N., and Burton, G. J. (2003). The contribution of placental oxidative stress to early pregnancy failure. *Hum. Pathol.* 34, 1265–1275.
266. Hubel, C. A. (1999). Oxidative stress in the pathogenesis of preeclampsia. *Proc. Soc. Exp. Biol. Med.* 222, 222–235.
267. Burton, G. J., and Hung, T.-H. (2003). Hypoxia-reoxygenation; a potential source of placental oxidative stress in normal pregnancy and preeclampsia. *Fetal Matern. Med. Rev.* 14, 97–117.
268. Hung, T.-H., Skepper, J. N., Charnock-Jones, D. S., and Burton, G. J. (2002). Hypoxia/reoxygenation: a potent inducer of apoptotic changes in the human placenta and possible etiological factor in preeclampsia. *Circ. Res.* 90, 1274–1281.
269. Smith, S. C., Baker, P. N., and Symonds, E. M. (1997). Increased placental apoptosis in intrauterine growth restriction. *Am. J. Obstet. Gynecol.* 177, 1395–1401.
270. Allaire, A. D., Ballenger, K. A., Wells, S. R., McMahon, M. J., and Lessey, B. A. (2000). Placental apoptosis in preeclampsia. *Obstet. Gynecol.* 96, 271–276.
271. Chappell, L. C., Seed, P. T., Briley, A. L., Kelly, F. J., Lee, R., Hunt, B. J., Parmar, K., Bewley, S. J., Shennan, A. H., Steer, P. J., and Poston, L. (1999). Effect of antioxidants on the occurrence of pre-eclampsia in women at increased risk: a randomised trial. *Lancet* 354, 810–816.
272. Mayhew, T. M., and Burton, G. J. (1997). Stereology and its impact on our understanding of human placental functional morphology. *Microsc. Res. Tech.* 38, 195–205.
273. Mayhew, T. M., Huppertz, B., Kaufmann, P., and Kingdom, J. C. (2003). The "reference trap" revisited: examples of the dangers in using ratios to describe fetoplacental angiogenesis and trophoblast turnover. *Placenta* 24, 1–7.
274. Burton, G. J., Ingram, S. C., and Palmer, M. E. (1987). The influence of the mode of fixation on morphometrical data derived from terminal villi in the human placenta at term: a comparison of immersion and perfusion fixation. *Placenta* 8, 37–51.
275. Schweikhart, G., and Kaufmann, P. (1977). Zur Abgrenzung normaler, artefizieller und pathologischer Strukturen in reifen menschlichen Plazentazotten. I. Ultrastruktur des Syncytiotrophoblasten. *Arch. Gynäkol.* 222, 213–230.
276. Feneley, M. R., and Burton, G. J. (1991). Villous composition and membrane thickness in the human placenta at term: a stereological study using unbiased estimators and optimal fixation techniques. *Placenta* 12, 131–142.
277. Larsen, L. G., Clausen, H. V., Andersen, B., and Græm, N. (1995). A stereologic study of postmature placentas fixed by dual perfusion. *Am. J. Obstet. Gynecol.* 172, 500–507.
278. Mayhew, T. M., and Burton, G. J. (1988). Methodological problems in placental morphometry: apologia for the use of stereology based on sound sampling practice. *Placenta* 9, 565–581.
279. Burton, G. J., and Jauniaux, E. (1995). Sonographic, stereological and Doppler flow velocimetric assessments of placental maturity. *Br. J. Obstet. Gynaecol.* 102, 818–825.
280. Gundersen, H. J. G., Bendtsen, T. F., Korbo, L., Marcussen, N., Moller, A., Nielsen, K., Nyengaard, J. R., Pakkenberg, B., Sorensen, F. B., Vesterby, A., and West, M. J. (1988). Some new, simple and efficient stereological methods and their use in pathological research and diagnosis. *APMIS* 96, 379–394.
281. Gundersen, H. J. G., Bagger, P., Bendtsen, T. F., Evans, S. M., Korbo, L., Marcussen, N., Moller, A., Nielsen, K., Nyengaard, J. R., Pakkenberg, B., Sorensen, F. B., Vesterby, A., and West, M. J. (1988). The new stereological tools: disector, fractionator, nucleator and point sampled intercepts and their use in pathological research and diagnosis. *APMIS* 96, 857–881.
282. Vermeulen, R. C. W., Kurver, P. H. J., Arts, N. F. T., Van Kessel, H., Wilson, G. R., and Klopper, A. (1982). The relationship between the surface area of the trophoblast and some placental products. *Placenta* 3, 359–366.
283. Mayhew, T. M., and Simpson, R. A. (1994). Quantitative evidence of the spatial dispersal of trophoblast nuclei in human placental villi during gestation. *Placenta* 15, 837–844.
284. Mayhew, T. M. (2002). Fetoplacental angiogenesis during gestation is biphasic, longitudinal and occurs by proliferation and remodelling of vascular endothelial cells. *Placenta* 23, 742–750.
285. Kaufmann, P., Mayhew, T. M., and Charnock Jones, D. S. (2004). Aspects of human fetoplacental vasculogenesis and angiogenesis. II. Changes during normal pregnancy. *Placenta* 114–126.
286. Mayhew, T. M., Jackson, M. R., and Haas, J. D. (1986). Microscopical morphology of the human placenta and its effects on oxygen diffusion: a morphometric study. *Placenta* 7, 121–131.
287. Jauniaux, E., Burton, G. J., Moscosco, G. J., and Hustin, J. (1991). Development of the early placenta: a morphometric study. *Placenta* 12, 269–276.
288. Mayhew, T. M., Jackson, M. R., and Boyd, P. A. (1993). Changes in oxygen diffusive conductances of human placental during gestation (10–41 weeks) are commensurate with the gain in fetal weight. *Placenta* 14, 51–61.
289. Zamudio, S. (2003). The placenta at high altitude. *High Altitude Med. Biol.* 4, 171–191.
290. Jackson, M. R., Mayhew, T. M., and Haas, J. D. (1987). Morphometric studies on villi in human term placentae and the effects of altitude, ethnic grouping and sex of newborn. *Placenta* 8, 487–495.
291. Jackson, M. R., Mayhew, T. M., and Haas, J. D. (1988). On the factors which contribute to thinning of the villous membrane at high altitude. II. An increase in the degree of peripheralization of fetal capillaries. *Placenta* 9, 9–18.
292. Reshetnikova, O. S., Burton, G. J., and Milovanov, A. P. (1994). Effects of hypobaric hypoxia on the feto-placental unit; the morphometric diffusing capacity of the villous membrane at high altitude. *Am. J. Obstet. Gynecol.* 171, 1560–1565.
293. Mayhew, T. M., Jackson, M. R., and Haas, J. D. (1990). Oxygen diffusive conductances of human placentae from pregnancies at low and high altitudes. *Placenta* 11, 493–503.
294. Reshetnikova, O. S., Burton, G. J., and Teleshova, O. V. (1995). Placental histomorphometry and morphometric diffusing capacity of the villous membrane in pregnancies complicated by maternal iron-deficiency anemia. *Am. J. Obstet. Gynecol.* 173, 724–727.
295. Burton, G. J., Reshetnikova, O. S., Milovanov, A. P., and Teleshova, O. V. (1996). Stereological evaluation of vascular adaptations of human placental villi to differing forms of hypoxic stress. *Placenta* 17, 49–55.
296. Kadyrov, M., Kosanke, G., Kingdom, J., and Kaufmann, P. (1998). Increased fetoplacental angiogenesis during first trimester in anaemic women. *Lancet* 352, 1747–1749.
297. Huang, A., Zhang, R., and Yang, Z. (2001). Quantitative (stereological) study of placental structures in women with iron-deficiency anaemia. *Eur. J. Obstet. Gynaecol. Reprod. Biol.* 97, 59–64.
298. Lewis, R. M., James, L. A., Zhang, J., Bryne, C. D., and Hales, C. N. (2001). Effects of maternal iron restriction in the rat on hypoxia induced gene expression and fetal metabolite levels. *Br. J. Nutr.* 85, 193–201.

299. Bloxham, D. L., Bullen, B. E., Walters, B. N. J., and Lao, T. T. (1987). Placental glycolysis and energy metabolism in preeclampsia. *Am. J. Obstet. Gynecol.* 157, 97–101.
300. Hafner, E., Metzenbauer, M., Hofinger, D., Munkel, M., Gassner, R., Schuchter, K., Dillinger-Paller, B., and Philipp, K. (2003). Placental growth from the first to the second trimester of pregnancy in SGA-foetuses and pre-eclamptic pregnancies compared to normal foetuses. *Placenta* 24, 336–342.
301. Jackson, M. R., Walsh, A. J., Morrow, R. J., Mullen, J. B. M., Lye, S. J., and Ritchie, J. W. K. (1995). Reduced placental villous tree elaboration in small-for-gestational-age pregnancies: relationship with umbilical artery Doppler waveforms. *Am. J. Obstet. Gynecol.* 172, 518–525.
302. Teasdale, F. (1985). Histomorphometry of the human placenta in maternal preeclampsia. *Am. J. Obstet. Gynecol.* 152, 25–31.
303. Mayhew, T. M., Ohadike, C., Baker, P. N., Crocker, I. P., Mitchell, C., and Ong, S. S. (2003). Stereological investigation of placental morphology in pregnancies complicated by pre-eclampsia with and without intrauterine growth restriction. *Placenta* 24, 219–226.
304. Teasdale, F. (1987). Histomorphometry of the human placenta in pre-eclampsia associated with severe intra-uterine growth retardation. *Placenta* 8, 119–128.
305. Gonzalez-Crussi, F., and Roth, L. M. (1976). The human yolk sac and yolk sac carcinoma. An ultrastructural study. *Hum. Pathol.* 7, 675–691.
306. Jones, C. J. P. (1997). The life and death of the embryonic yolk sac. In *Embryonic Medicine and Therapy* (E. Jauniaux, E. R. Barnea, and R. G. Edwards, Eds.), pp. 180–196. Oxford University Press, Oxford.
307. Jauniaux, E., and Gulbis, B. (2000). Fluid compartments of the embryonic environment. *Hum. Reprod.* Update 6, 268–278.
308. Jauniaux, E., Cindrova-Davies, T., Johns, J., Dunster, C., Hempstock, J., Kelly, F. J., and Burton, G. J. (2004). Distribution and transfer pathways of antioxidant molecules inside the first trimester human gestational sac. *J. Clin. Endocrinol. Metab.* 89, 1452–1459.
309. Weser, H., and Kaufmann, P. (1978). Light microscopical and histochemical studies on the chorionic plate of the mature human placenta [author's transl]. *Arch. Gynakol.* 225, 15–30.
310. Bourne, G. L. (1962). *The Human Amnion and Chorion.* Lloyd-Luke Ltd., London.
311. Muhlhauser, J., Crescimanno, C., Rajaniemi, H., Castellucci, M., and Kaufmann, P. (1992). Localization of carbonic anhydrase isoenzymes in amnion and human placenta by immunofluorescence techniques. *Anat. Anz.* 174 Suppl., 128.
312. Koizumi, N. J., Inatomi, T. J., Sotozono, C. J., Fullwood, N. J., Quantock, A. J., and Kinoshita, S. (2000). Growth factor mRNA and protein in preserved human amniotic membrane. *Curr. Eye Res.* 20, 173–177.
313. Kemp, B., Kertschanska, S., Kadyrov, M., Rath, W., Kaufmann, P., and Huppertz, B. (2002). Invasive depth of extravillous trophoblast correlates with cellular phenotype: a comparison of intra- and extrauterine implantation sites. *Histochem. Cell Biol.* 117, 401–414.
314. Robertson, W. B., and Warner, B. (1974). The ultrastructure of the human placental bed. *J. Pathol.* 112, 203–211.
315. Okudaira, Y., Yoshiaki, M., and Kanoh, H. (1991). Morphological variability of human trophoblasts in normal and neoplastic conditions. In *Placenta: Basic Research for Clinical Application* (H. Soma, Ed.), pp. 176–187. Karger, Basel.
316. Mühlhauser, J., Crescimanno, C., Kaufmann, P., Höfler, H., Zaccheo, D., and Castellucci, M. (1993). Differentiation and proliferation patterns in human trophoblast revealed by c-erbB-2 oncogene product and EGF-R. J. *Histochem. Cytochem.* 41, 165–173.
317. Castellucci, M., Classen-Linke, I., Mühlhauser, J., Kaufmann, P., Zardi, L., and Chiquet-Ehrismann, R. (1991). The human placenta: a model for tenascin expression. *Histochemistry* 95, 449–458.
318. Castellucci, M., Crescimanno, C., Arezio, P., Muhlhauser, J., Cinti, S., and Kaufmann, P. (1991). Extracellular matrix molecules in the morphogenesis of the human placenta. *Placenta* 12, 376.
319. Faller, T., and Ferenci, P. (1973). The structure of the placental septa. A study using the quinacrine-staining of the Y-chromatin (author's transl). *Z. Anat. Entwicklung.* 142, 207–217.
320. Rosenberg, S. M., Maslar, I. A., and Riddick, D. H. (1980). Decidual production of prolactin in late gestation: further evidence for a decidual source of amniotic fluid prolactin. *Am. J. Obstet. Gynecol.* 138, 681–685.
321. Beham, A., Denk, H., and Desoye, G. (1988). The distribution of intermediate filament proteins, actin and desmoplakins in human placental tissue as revealed by polyclonal and monclonal antibodies. *Placenta* 9, 479–492.
322. Kurman, R. J., Main, C. S., and Chen, H. C. (1984). Intermediate trophoblast: a distinctive form of trophoblast with specific morphological, biochemical and functional features. *Placenta* 5, 349–369.
323. Beck, T., Schweikhart, G., and Stolz, E. (1986). Immunohistochemical location of HPL, SP1 and beta-HCG in normal placentas of varying gestational age. *Arch. Gynecol.* 239, 63–74.
324. Blaschitz, A., Weiss, U., Dohr, G., and Desoye, G. (2000). Antibody reaction patterns in first trimester placenta: implications for trophoblast isolation and purity screening. *Placenta* 21, 733–741.
325. Craven, C. M., Morgan, T., and Ward, K. (1998). Decidual spiral artery remodelling begins before cellular interaction with cytotrophoblasts. *Placenta* 19, 241–252.
326. Hees, H., Moll, W., Wrobel, K. H., and Hees, I. (1987). Pregnancy-induced structural changes and trophoblastic invasion in the segmental mesometrial arteries of the guinea pig (*Cavia porcellus* L.). *Placenta* 8, 609–626.
327. Moll, W., Nienartowicz, A., Hees, H., Wrobel, K.-H., and Lenz, A. (1988). Blood flow regulation in the uteroplacental arteries. *Trophoblast Res.* 3, 83–96.
328. Nanaev, A., Chwalisz, K., Frank, H. G., Kohnen, G., Hegele-Hartung, C., and Kaufmann, P. (1995). Physiological dilation of uteroplacental arteries in the guinea pig depends on nitric oxide synthase activity of extravillous trophoblast. *Cell Tissue Res.* 282, 407–421.
329. Hirano, H., Imai, Y., and Ito, H. (2002). Spiral artery of placenta: development and pathology-immunohistochemical, microscopical, and electron-microscopic study. *Kobe J. Med. Sci.* 48, 13–23.
330. Robertson, W. B. (1976). Uteroplacental vasculature. *J. Clin. Pathol.* 10 Suppl., 9–17.
331. De Wolf, F., De Wolf-Peeters, C., and Brosens, I. (1973). Ultrastructure of the spiral arteries in the human placental bed at the end of normal pregnancy. *Am. J. Obstet. Gynecol.* 117, 833–848.
332. Pijnenborg, R., Bland, J. M., Robertson, W. B., and Brosens, I. (1983). Uteroplacental arterial changes related to interstitial trophoblast migration in early human pregnancy. *Placenta* 4, 397–413.
333. Damsky, C. H., Fitzgerald, M. L., and Fisher, S. J. (1992). Distribution patterns of extracellular matrix components and adhesion receptors are intricately modulated during first trimester cytotrophoblast differentiation along the invasive pathway, in vivo. *J. Clin. Invest.* 89, 210–222.
334. Fisher, S. J., and Damsky, C. H. (1993). Human cytotrophoblast invasion. *Semin. Cell. Biol.* 4, 183–188.

335. Hustin, J., Jauniaux, E., and Schaaps, J. P. (1990). Histological study of the materno-embryonic interface in spontaneous abortion. *Placenta* 11, 477–486.
336. Brosens, I. A. (1988). The utero-placental vessels at term: the distribution and extent of physiological changes. *Trophoblast Res.* 3, 61–67.
337. Sheppard, B. L., and Bonnar, J. (1988). The maternal blood supply to the placenta in pregnancy complicated by intrauterine fetal growth retardation. *Trophoblast Res.* 3, 69–81.
338. Kadyrov, M., Schmitz, C., Black, S., Kaufmann, P., and Huppertz, B. (2003). Pre-eclampsia and maternal anaemia display reduced apoptosis and opposite invasive phenotypes of extravillous trophoblast. *Placenta* 24, 540–548.
339. Hitschmann, J., and Lindenthal, O. T. (1903). Der weisse Infarkt der Placenta. *Arch. Gynakol.* 69, 587–628.
340. Frank, H. G., Malekzadeh, F., Kertschanska, S., Crescimanno, C., Castellucci, M., Lang, I., Desoye, G., and Kaufmann, P. (1994). Immunohistochemistry of two different types of placental fibrinoid. *Acta Anat.* 150, 55–68.
341. Lang, I., Hartmann, M., Blaschitz, A., Dohr, G., Kaufmann, P., Frank, H. G., Hahn, T., Skofitsch, G., and Desoye, G. (1994). Differential lectin binding to the fibrinoid of human full-term placenta: correlation with a fibrin antibody and the PAF-Halmi method. *Acta Anat.* 150, 170–177.
342. Huppertz, B., Kertschanska, S., Frank, H. G., Gaus, G., Funayama, H., and Kaufmann, P. (1996). Extracellular matrix components of the placental extravillous trophoblast: immunocytochemistry and ultrastructural distribution. *Histochem. Cell Biol.* 106, 291–301.
343. Faulk, W. P. (1989). Placental fibrin. *Am. J. Reprod. Immunol.* 19, 132–135.
344. Sutcliffe, R. G., Davies, M., Hunter, J. B., Waters, J. J., and Parry, J. E. (1982). The protein composition of the fibrinoid material at the human uteroplacental interface. *Placenta* 3, 297–308.
345. Castellucci, M., Crescimanno, C., Schröter, C. A., and Kaufmann, P. (1993). Extravillous trophoblast: immunohistochemical localization of extracellular matrix molecules. In *Frontiers in Gynecologic and Obstetric Investigation* (A. R. Genazzani, F. Petraglia, and A. D. Genazzani, Eds.), pp. 19–25. Parthenon, New York.
346. King, B. F., and Blankenship, T. N. (1997). Immunohistochemical localization of fibrillin in developing macaque and term human placentas and fetal membranes. *Microsc. Res. Tech.* 38, 42–51.
347. Leivo, I., Laurila, P., Wahlstrom, T., and Engvall, E. (1989). Expression of merosin, a tissue-specific basement membrane protein, in the intermediate trophoblast cells of choriocarcinoma and placenta. *Lab. Invest.* 60, 783–790.
348. Ehrig, K., Leivo, I., Argraves, W. S., Ruoslahti, E., and Engvall, E. (1990). Merosin, a tissue-specific basement membrane protein, is a laminin-like protein. *Proc. Natl. Acad. Sci. U S A* 87, 3264–3268.
349. Hormann, G. (1965). Fibrinoidization of the chorion epithelium as a constructive principle of the human placenta. *Z. Geburtsh. Perinatol.* 164, 263–269.
350. Wynn, R. M. (1972). Cytotrophoblastic specializations: an ultrastructural study of the human placenta. *Am. J. Obstet. Gynecol.* 114, 339–355.
351. Faulk, P., Trenchev, P., Dorling, J., and Holborow, J. (1975). Antigens on post-implantation placentae. In *Immunobiology of Trophoblast* (R. G. Edwards, C. W. S. Howe, and M. H. Johnson, Eds.). University Press, Cambridge.
352. Currie, G. A., and Bagshawe, K. D. (1967). The masking of antigens on trophoblast and cancer cells. *Lancet* 1, 708–710.
353. Frank, H. G., Huppertz, B., Kertschanska, S., Blanchard, D., Roelcke, D., and Kaufmann, P. (1995). Anti-adhesive glycosylation of fibronectin-like molecules in human placental matrix-type fibrinoid. *Histochem. Cell Biol.* 104, 317–329.
354. Kisalus, L. L., and Herr, J. C. (1988). Immunocytochemical localization of heparan sulfate proteoglycan in human decidual cell secretory bodies and placental fibrinoid. *Biol. Reprod.* 39, 419–430.
355. Swinburne, L. M. (1970). Leucocyte antigens and placental sponge. *Lancet* 2, 592–594.
356. Chaouat, G., Kolb, J. P., and Wegmann, T. G. (1983). The murine placenta as an immunological barrier between the mother and the fetus. *Immunol. Rev.* 75, 31–60.
357. Hunziker, R. D., and Wegmann, T. G. (1986). Placental immunoregulation. *Crit. Rev. Immunol.* 6, 245–285.
358. Downs, K. M. (2002). Early placental development ontogeny in the mouse. *Placenta* 23, 116–131.
359. Georgiades, P., Ferguson-Smith, A. C., and Burton, G. J. (2002). Comparative developmental anatomy of the murine and human definitive placenta. *Placenta* 23, 3–19.
360. Adamson, S. L., Lu, Y., Whiteley, K. J., Holmyard, D., Hemberger, M., Pfarrer, C., and Cross, J. C. (2002). Interactions between trophoblast cells and the maternal and fetal circulation in the mouse placenta. *Dev. Biol.* 250, 358–373.
361. Malassiné, A., Frendo, J.-L., and Evain-Brion, D. (2003). A comparison of placental development and endocrine functions between the human and mouse model. *Hum. Reprod.* Update 9, 531–539.
362. Kunath, T., Strumpf, D., and Rossant, J. (2004). Early trophoblast determination and stem cell maintenance in the mouse: a review. *Placenta* 25, Suppl. A, S32–S38.
363. Moore, T., and Haig, D. (1991). Genomic imprinting in mammalian development: a parental tug-of-war. *Trends Genet.* 7, 45–49.
364. Moore, T. (2001). Genetic conflict, genomic imprinting and establishment of the epigenotype in relation to growth. *Reproduction* 122, 185–193.
365. Tycko, B., and Morison, I. M. (2002). Physiological functions of imprinted genes. *J. Cell. Physiol.* 192, 245–258.
366. Reik, W., Constancia, M., Fowden, A., Anderson, N., Dean, W., Ferguson-Smith, A., Tycko, B., and Sibley, C. (2003). Regulation of supply and demand for maternal nutrients in mammals by imprinted genes. *J. Physiol.* 547, 35–44.
367. Yeh, H., Godman, J., Carr, L., and Rabinowitz, J. (1986). Intradecidual sign: a US criterion of early intrauterine pregnancy. *Radiology* 161, 463–467.
368. Jauniaux, E., and Nicolaides, K. H. (1997). Early ultrasound diagnosis and follow-up of molar pregnancies. *Ultrasound Obstet. Gynecol.* 9, 17–21.
369. Lindholm, H., and Flam, F. (1999). The diagnosis of molar pregnancy by sonography and gross morphology. *Acta Obstet. Gynaecol. Scand.* 78, 6–9.
370. Benson, C. B., Genest, D. R., Bernstein, M. R., Soto-Wright, V., Goldstein, D. P., and Berkowitz, R. S. (2000). Sonographic appearance of first trimester complete hydatidiform moles. *Ultrasound Obstet. Gynecol.* 16, 188–191.
371. Jauniaux, E., and Nicolaides, K. H. (1996). Placental lakes, absent umbilical artery diastolic flow and poor fetal growth in early pregnancy. *Ultrasound Obstet. Gynecol.* 7, 141–144.
372. Tabsh, K. M. A. (1983). Correlation of real-time ultrasonic placental grading with amniotic lecithin/sphingomyelin ratio. *Am. J. Obstet. Gynecol.* 145, 504–508.
373. Harman, C. R., Manning, F. A., Stearns, E., and Morrison, I. (1982). The correlation of ultrasonicplacental grading and the fetal pulmonary maturation in five hundread and sixty-three pregnancies. *Am. J. Obstet. Gynecol.* 143, 941–943.
374. Jauniaux, E., and Campbell, S. (1990). Sonographic assessment of placental abnormalities. *Am. J. Obstet. Gynecol.* 163, 1650–1658.

375. Bilardo, C. M., Nicolaides, K. H., and Campbell, S. (1990). Doppler measurements of fetal and uteroplacental circulations: relationship with umbilical venous blood gases measured at cordocentesis. *Am. J. Obstet. Gynecol.* 162, 115–120.
376. Weiner, C. P. (1990). The relationship between the umbilical artery systolic/diastolic ratio and umbilical blood gas measurements in specimens obtained by cordcentesis. *Am. J. Obstet. Gynecol.* 162, 1198–1202.
377. Boito, S., Struijk, P. C., Ursem, N. T., Stijnen, T., and Wladimiroff, J. W. (2002). Umbilical venous volume flow in the normally developing and growth-restricted human fetus. *Ultrasound Obstet. Gynecol.* 19, 344–349.
378. Jauniaux, E., Jurkovic, D., Campbell, S., and Hustin, J. (1992). Doppler ultrasound features of the developing placental circulations: correlation with anatomic findings. *Am. J. Obstet. Gynecol.* 166, 585–587.
379. Jauniaux, E., Jurkovic, D., Campbell, S., Kurjak, A., and Hustin, J. (1991). Investigation of placental circulations by colour Doppler ultrasound. *Am. J. Obstet. Gynecol.* 164, 486–488.
380. Guiot, C., Piantà, P. G., and Todros, T. (1992). Modelling the feto-placental circulation. 1. A distributed network predicting umbilical haemodynamics throughout pregnancy. *Ultrasound Med. Biol.* 18, 535–544.
381. Rotmensch, S., Liberati, M., Luo, J.-S., Kliman, H. J., Gollin, Y., Bellati, U., Hobbins, J. C., and Copel, J. A. (1994). Color Doppler flow patterns and flow velocity waveforms of the intraplacental circulation in growth-retarded fetuses. *Am. J. Obstet. Gynecol.* 171, 1257–1264.
382. Hitschold, T., Weiss, E., Beck, T., Huntefering, H., and Berle, P. (1993). Low target birth weight or growth retardation? Umbilical Doppler flow velocity waveforms and histometric analysis of fetoplacental vascular tree. *Am. J. Obstet. Gynecol.* 168, 1260–1264.
383. Krebs, C., Macara, L. M., Leiser, R., Bowman, A. W., Greer, I. A., and Kingdom, J. C. P. (1996). Intrauterine growth restriction with absent end-diastolic flow velocity in the umbilical artery is associated with maldevelopment of the placental terminal villous tree. *Am. J. Obstet. Gynecol.* 175, 1534–1542.
384. Jauniaux, E., and Ogle, R. (2000). Color Doppler imaging in the diagnosis and management of chorioangiomas. *Ultrasound Obstet. Gynecol.* 15, 463–467.
385. Duncan, K. R., Sahota, D. S., Gowland, P. A., Moore, R., Chang, A., Baker, P. N., and Johnson, I. R. (2001). Multilevel modeling of fetal and placental growth using echo-planar magnetic resonance imaging. *J. Soc. Gynecol. Invest.* 8, 285–290.
386. Ong, S. S., Tyler, D. J., Moore, R. J., Gowland, P. A., Baker, P. N., Johnson, I. R., and Mayhew, T. M. (2004). Functional magnetic resonance imaging (magnetization transfer) and stereological analysis of human placentae in normal pregnancy and in pre-eclampsia and intrauterine growth restriction. *Placenta* 25, 408–412.
387. Bodis, J., Zambo, K., Nemessanyi, Z., Mate, E., and Csaba, I. F. (1985). Application of the parametric scan in the investigation of uteroplacental blood flow. *Eur. J. Nucl. Med.* 10, 286–287.
388. Francis, S. T., Duncan, K. R., Moore, R. J., Baker, P. N., and Johnson, I. R. (1998). Non-invasive mapping of placental perfusion. *Lancet* 351, 1397–1399.
389. Kaufmann, P., and Scheffen, I. (1992). Placental development. In *Fetal and Neonatal Physiology* (R. Polin and W. Fox, Eds.), pp. 47–56. WB Saunders, Philadelphia.
390. Schiebler, T. H., and Kaufmann, P. (1981). Reife Plazenta. In *Die Plazenta des Menschen* (V. Becker, T. H. Schiebler, and F. Kubli, Eds.), pp. 51–111. Thieme, Stuttgart.
391. Aherne, W., and Dunnill, M. S. (1966). Quantitative aspects of placental structure. *J. Pathol. Bacteriol.* 91, 123–139.
392. Laga, E. M., Driscoll, S. G., and Munro, H. N. (1973). Quantitative studies of human placenta. *I. Morphometry. Biol. Neonate* 23, 231–259.

Knobil and Neill's Physiology of Reproduction, Third Edition
edited by Jimmy D. Neill,
Elsevier © 2006

CHAPTER 6

Sex Determination and Differentiation

Amanda Swain

Introduction, 245
Chromosomal Sex, 245
Sex Chromosome Defects, 246
Gonadal Sex, 248
Sex Determination, 248
Genes Implicated in Sex Determination and Their Associated Defects, 249
Testis Development, 251
Phenotypic Sex, 254
Sexual Differentiation of Reproductive Organs, 254
Sexual Differentiation of the Nervous System, 255
Future, 256
Acknowledgments, 256
References, 256

INTRODUCTION

Sex determination is the choice to become male or female, a fundamental step in the life of an animal. Although this choice is common to most species, the mechanism that determines sexual dimorphism is highly diverse and includes a wide range of environmental and genetic sex-determining systems. In mammals, the primary sex determinant is genetic, and therefore the first step in sexual differentiation is the establishment of chromosomal sex at fertilization. For the most part, the initial stages of fetal development do not show any morphological differences between the sexes until the gonad develops. The second step in sexual differentiation is the choice of the bipotential gonad to become an ovary or a testis. The third step is the development of the sexual phenotype of the fetus, some aspects of which are controlled by diffusible substances produced by the gonads. Disorders of sexual differentiation can affect all three steps, and their analysis has provided great insight into the molecular and genetic pathways that control this process.

CHROMOSOMAL SEX

Mammalian sex determination is dependent on the Y chromosome. If the Y chromosome is present, the fetus develops as a male and if absent as a female. Furthermore, the only gene on the Y chromosome that is necessary and sufficient to determine male development is *SRY* (sex-determining region on the Y chromosome), which is present in most mammals. Therefore, in normal conditions, females have two X chromosomes, which they inherit from each parent, and males have an X chromosome, which they inherit from their mother, and a Y chromosome, which they inherit from their father.

Heteromorphic sex chromosomes like the X and Y are thought to have evolved from an identical pair of autosomes. Several evolutionary theories propose that the first step in the evolution of sex chromosomes is the appearance of an autosomal gene with two different alleles, where homozygosity leads to the development of one sex and heterozygosity to the other. Once this occurs, there is a suppression of recombination between the chromosomes, probably initiated by the reduced crossing over at the sex-determination locus. This leads to a degeneration of the sex-linked loci in the heterogametic sex (1–3) (Fig. 1A).

The properties of the human Y chromosome support this evolutionary model. It is far smaller than the X chromosome and consists of two main regions. One is a relatively short region, named the pseudoautosomal region, that is found at either end of the chromosome and is identical to the corresponding regions of the X chromosome. This is the only place where the X and Y chromosome still recombine

Gene Function and Regulation Section, The Institute of Cancer Research, London, United Kingdom.

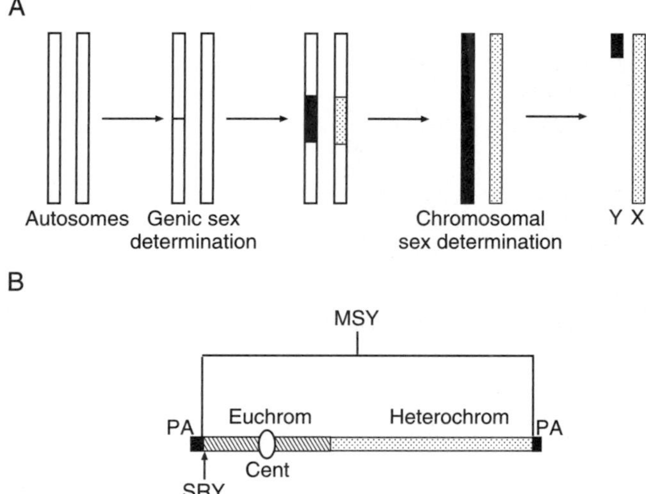

FIG. 1. A: Proposed mechanism for the evolution of sex chromosomes from a pair of autosomes. [Adapted from (2).] **B:** Structure of the human Y chromosome indicating the position of the *SRY* gene. PA, pseudoautosomal region; MSY, male-specific region of the Y; euchrom, euchromatic region; heterochrom, heterochromatic region; cent, centromere.

during male meiosis, ensuring sequence identity (4). The other region comprises 95% of the chromosome and is Y specific. This male-specific region of the Y chromosome (MSY) is a mosaic of heterochromatin, highly repetitive DNA that is thought to be nonfunctional, and euchromatin sequences (5). The determination of the physical map of the Y chromosome and the sequencing of the euchromatic portion of MSY provided evidence for the model that the human X and Y originated about 300 million years ago with the emergence of *SRY* as the primary sex determinant (5–7). X and Y recombination was then suppressed in a series of steps, probably through chromosomal inversions, which led to the differentiation of each chromosome. Selection then occurred on the Y chromosome for a group of testis-specific genes that enhanced male fertility, which are maintained by Y-Y gene conversion (8) (Fig. 1B).

The differentiation of the X and Y chromosome during evolution led to the problem of dosage compensation because females had a double dose of X-linked genes, whereas males only had one. To overcome this problem, mammals evolved a mechanism of silencing the transcription of genes from one X chromosome when two are present, a process called X-inactivation (9). A major component of this process is the noncoding *XIST* gene that maps to the X chromosome inactivation centre, Xic, and is preferentially expressed by the inactive X chromosome (10,11). The mechanism by which *XIST* triggers X-inactivation is not fully understood and is a subject of intense study. Not all genes are subject to X-inactivation, and reports show that up to 20% of X-linked genes in humans might escape this control (12,13). Some, but not all, of these genes have related Y-linked homologues, relics of the shared origin between the X and Y, leading to the proposal that the evolution of X-inactivation occurred on a gene-by-gene or cluster-by-cluster basis driven by the loss of the Y-linked homologues (5,14,15).

Several properties of this genetic sex determination system reflect the rapid rate of evolution of this process. SRY is found in most eutherian (placental) and metatheria (marsupial) mammals but has not been found in monotremes (egg-laying mammals), which diverged 170 million years ago (16). In marsupials, SRY controls most sex differentiation; however, the choice between a pouch or a scrotum is dependent on the number of X chromosomes (17,18). The extreme case is that of some, but not all, species of mole vole where no Y chromosome or *Sry* gene can be found, although sex determination does occur, giving rise to XX females and XO males (19,20). Another case is that of the wood lemming, where an X and Y chromosome are present but a variant X chromosome (X*) segregates at high frequency in the population and overrides the action of *Sry*, giving rise to X*Y females that are fertile (21).

Sex Chromosome Defects

SRY

Inactivating mutations in the testis-determining gene, *SRY*, are responsible for 46,XY female sex reversal in humans. In fact, this property was used to identify this gene in the first place. The study of DNA from XX male individuals identified small regions of the Y chromosome that had transferred onto the X chromosome by abnormal X–Y interchange during male meiosis. The *SRY* gene was found within the minimum region of Y-specific DNA required for male development (22). Confirmation of its role was found by the analysis of XY female patients with mutations within this gene (23,24).

The mouse has been an important working model to study the process of mammalian sex determination. As in the case of human patients, XX mice carrying the *Sxr* mutation develop as males, and it was shown that the X chromosome in these mice carried a small portion of the Y chromosome containing the *Sry* gene (25). Conversely, a mouse mutation that produced XY females was shown to have an 11-kb deletion on the Y chromosome that included the *Sry* gene (26). Direct evidence that *SRY* is the

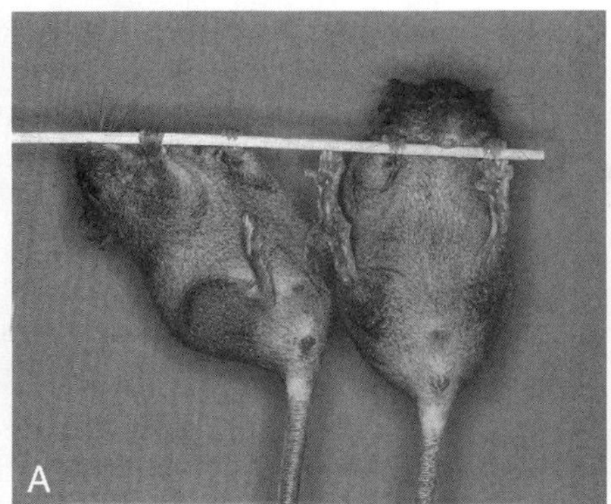

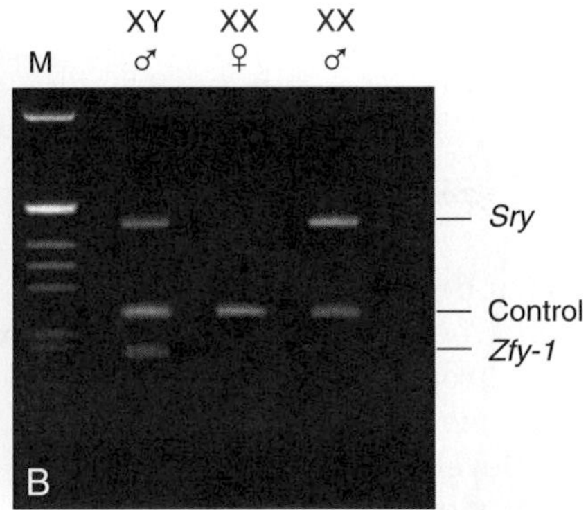

FIG. 2. XX male sex reversal with an *Sry* transgene. **A:** On the left is a control XY male and on the right is his litter mate, an XX transgenic carrying a 14-kb genomic fragment containing the mouse *Sry* gene. The external and internal genitalia are indistinguishable except for the testis, which is smaller in the transgenic animal due to the lack of germ cells. **B:** Polymerase chain reaction analysis of genomic DNA from these mice as well as an XX sibling control. Bands for *Sry* and the DNA loading control (myogenin) are present in the XX transgenic animal but the Y chromosome marker *Zfy-1* is missing. M, size marker. [Adapted from (27).]

only gene on the Y chromosome to be necessary and sufficient for establishing male development came from transgenic experiments in the mouse. XX mice carrying a 14-kb fragment of Y chromosome DNA that contains the mouse *Sry* gene and no other gene developed as males (27) (Fig. 2). These transgenic animals were infertile, however, because of a block in spermatogenesis due to the presence of two X chromosomes and the lack of other Y-linked genes.

Y-Specific Genes

Several genes identified in the MSY regions show testis-specific expression, suggesting they might have a role in male sexual function. Consistent with this, most 46,XX men are azoospermic and therefore infertile. Deletions in at least three regions of the long arm of the Y chromosome (azoospermia factor a, b, and c) have been associated with reduced (oligospermia) or lack (azoospermia) of sperm production in humans (28). Because of the nature of the sequence in this region, large repeats, and multicopy genes, it has been difficult to establish which genes are responsible for the spermatogenic failure in each case. Deletions of regions of the Y chromosome in the mouse also lead to a block in spermatogenesis. In this animal model system, it was possible to prove, in the case of the gene *Eif2s3y*, that it was responsible for the spermatogenic failure in the *Sxrb* strain of mice. Elegant transgenic studies where genes were introduced into mouse strains carrying the Y deletions have shown that *Eif2s3y* but not other genes in the interval rescued the spermatogenic phenotype observed in these mice (29).

X-Dosage

XO. In humans the loss of one X chromosome gives rise to Turner syndrome. In most cases XO fetuses are lost, but some survive with a frequency of 1 in 2,500–3,000 female births. The phenotypes of the affected individuals include short stature, a webbed neck, broad chest, and cognitive defects. Genes escaping X-inactivation have been proposed to be responsible for these defects because only one copy of these genes is expressed in these individuals. Consistent with this, fewer genes in the mouse escape X-inactivation, and XO mice have a less severe phenotype (30). Patients with Turner syndrome show gonadal dysgenesis, which is characterized by the loss of oocytes. XO mice also show a reduced number of oocytes at birth, and analysis of these mice suggests that this is due to the lack of chromosome pairing of the single X during meiosis (31,32).

XXY. Males with Klinefelter syndrome are characterized by an XXY genotype, increased stature, and infertility. Loss of germ cells in these individuals occurs before puberty, and a similar situation is seen in the mouse. It has been proposed that the loss of germ cells in these individuals is due to the presence of two X chromosomes. Before entry into meiosis, germ cells with two X chromosomes undergo X chromosome reactivation, giving rise to a double dose of X-linked genes. If this occurs in the testis, the X chromosome dosage acts to impair germ cell development (33).

GONADAL SEX

The action of SRY that determines male sex differentiation occurs within the developing gonad. The gonads are part of the urogenital system, which is derived from the intermediate mesoderm of the embryo. They initially develop as thickenings on the ventromedial surface of the mesonephros on either side of the midline at around 4 weeks in the human fetus and 9.5 days in the mouse. The somatic cells of the gonad are derived from the mesonephros and the coelomic epithelium that lines the coelomic cavity. The primordial germ cells originate at earlier stages of embryogenesis at the base of the allantois and migrate through the hindgut to join the gonad as it develops. This initial development is identical between males and females, but at early stages, around E10.5-11 in the mouse, SRY acts in the XY embryo to initiate testis development. Without SRY the gonad develops as an ovary; therefore, the mammalian embryo is programmed to develop as a female and the male-specific pathway is imposed on this basic body plan.

Sex Determination

Although SRY was identified in 1990, it is still not clear how it acts to initiate testis development. Its molecular properties suggest that it acts at the level of transcription. The SRY protein contains an HMG box type of DNA-binding domain found in a number of other transcription factors. This domain is thought to be the main functional part of the protein because almost all mutations found in human XY females cluster within this domain (24,34). In addition, species comparisons show that the HMG box domain is relatively conserved, unlike the rest of the protein, which is very different between all but closely related species (35,36). HMG box domains have been shown to bind DNA in the minor groove and to bend DNA at an acute angle, suggesting that SRY might act to modify the configuration of chromatin (37–39). However, there has been no in vivo evidence for this role.

Regions outside the HMG box domain are thought to be important in SRY function. This has been shown in transgenic mice where a construct containing a stop codon that led to the production of a truncated protein lacking the nonconserved carboxy terminal domain of mouse SRY was not able to induce male development in XX embryos (40). One explanation for this phenotype is that these regions are required for protein stability, and because various sequences can provide this function, they have been allowed to diverge. The human SRY gene was not able to sex reverse XX mice when the human locus was used to make transgenic mice. However, when regulatory elements from the mouse SRY gene were used to drive the human gene, sex reversal of XX transgenic mice did occur (41). Therefore, although the human and mouse genes differ outside the HMG box domain, they are functionally conserved. Differences between the expression pattern of human and mouse SRY could account for the failure of sex reversal when human regulatory sequences were used in the mouse (42).

Studies in the mouse have shown that *Sry* is expressed almost exclusively in the somatic cells of the gonad of XY embryos as it begins to develop. Transcript levels reach a peak around E11.5 and then decline to undetectable levels by E12.5 (43,44) (Fig. 3). This transient pattern of expression suggests that *SRY* acts as a trigger for testis differentiation but is not required for the subsequent

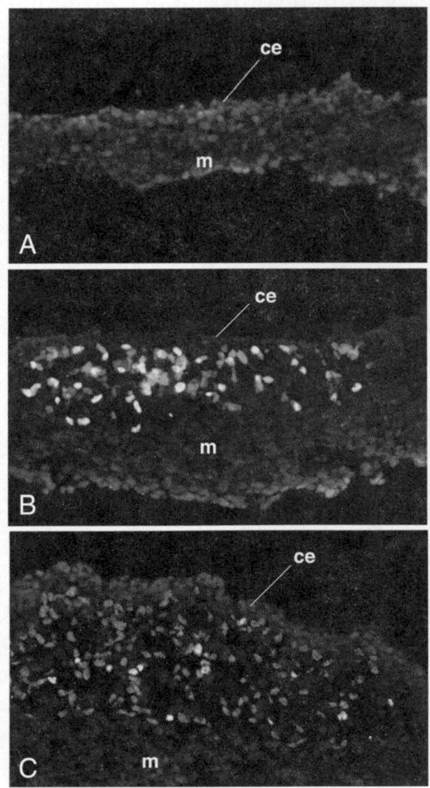

FIG. 3. Transient expression of *Sry* in the mouse embryonic gonad. Gonads from E11 to 12.5 (**A,** E11; **B,** E11.5; **C,** E12.5) transgenic embryos that produced SRY protein fused to a MYC epitope were stained with antibodies to SOX9 in green and MYC in red. SRY is initially found within the middle portion of the gonad, and as development proceeds it colocalizes with SOX9 (*yellow*) and then at later stages it is turned off whereas SOX9 marks the differentiating Sertoli cells. ce, coelomic epithelia; m, mesonephros. [Adapted from (46).] (See color insert.)

development of the organ or maintenance of testis-specific gene expression. Studies in chimeric mice derived from aggregations of early XX and XY embryos showed that Sertoli cells were the only cell type in the testis to have a bias toward an XY genotype. Consistent with this, cell fate studies in transgenic mice have shown that *Sry* is expressed in Sertoli cell precursors (45,46). These data show that *SRY* initiates the differentiation of the Sertoli cell lineage within the developing gonad and that these cells direct the differentiation of the rest of the cell lineages along the male pathway (47).

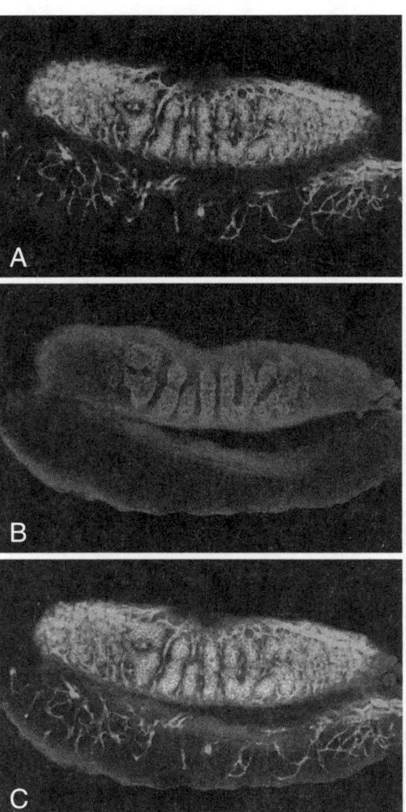

FIG. 4. Ovotestis from XY^{POS} mice on a C57Bl/6 background. Confocal images of gonads from E13.5 embryos that were stained with antibodies to PECAM, which marks endothelial cells and germ cells (**A** and **C**), and to AMH, which is produced by Sertoli cells (**B** and **C**). Testicular cords can be seen in the central portion of the gonad, whereas the ovarian portion is found at both poles. (Pictures courtesy of Kenn Albrecht and John Sullivan, Boston University. The AMH antibody was provided by Natalie Josso.) (See color insert.)

Genes Implicated in Sex Determination and Their Associated Defects

Studies in mice with different alleles of SRY have led to the proposal that there is a window of time during gonad development when SRY needs to achieve a certain level of expression to ensure testis differentiation. If this is not the case, testis development is delayed and ovary development occurs. In some cases this gives rise to the development of ovotestes where testicular cords are found in the middle portion of the gonad surrounded by ovarian tissue on either side (Fig. 4). This property of sex determination implies that genes that act to affect the levels of expression of SRY or interfere with SRY action, either directly or indirectly, are implicated in the process of sex determination.

WT1

The Wilms' tumor suppressor gene was first identified as one of the genes mutated in nephroblastomas found in newborns. This gene was also associated with defects in urogenital development in both mouse and humans. During embryo development, WT1 is expressed throughout the intermediate mesoderm at early stages of urogenital development, including the gonad (48,49). Mice with a disrupted WT1 gene show absence of gonad and adrenal and arrested kidney development (50). Studies on genetically modified mice have shown that not only is WT1 important in the development of the early genital ridge in both sexes, but it also has a role in sex determination in the XY gonad.

The WT1 locus encodes a series of variant transcripts due to alternative splicing, alternative transcriptional start sites, and RNA editing. These variants encode two main proteins containing a zinc finger domain that differ in the presence or absence of three amino acids (KTS) between two zinc fingers (designated +KTS or –KTS isoforms). Cell culture studies showed that these two proteins had different functions within the nucleus (51). The –KTS isoform was associated with transcriptional regulation, whereas the +KTS isoform colocalized with splicing factors and could bind RNA. Hammes et al. (52) generated two strains of mice with mutations in the WT1 gene that inhibit the splicing events that give rise to the different isoforms such that the modified mice only expressed one variant or the other. These two mouse strains showed different phenotypes in the gonad. The mice lacking the –KTS isoform showed a lack of proper gonad development in both sexes, whereas the mice lacking the +KTS isoform showed relatively normal gonad development but the XY gonad developed as an ovary. *Sry* expression in the gonad was significantly reduced in the latter case, accounting for the sex reversal phenotype. These data suggest that WT1 regulates the levels of *Sry* within the gonad, possibly posttranscriptionally, through the action of the +KTS isoform.

Human patients with Frasier syndrome carry a heterozygous mutation in the splice site of WT1 that prevents the production of the +KTS isoform, shifting the ratio of the two isoforms within cells (53). Consistent with the mouse studies, the XY Frasier individuals develop as females. However, the sex reversal phenotype observed in mice lacking the +KTS isoform is only observed when the mutation is present in a homozygous state. Differences between human and mice phenotypes have been found for mutations in various genes involved in early sexual development, suggesting that the sensitivities to the levels of different factors are particular to each species.

Genetic Background and tda Loci

A well-studied sex reversal system in mice is that of the *Sry* allele derived from a subspecies of *Mus musculus domesticus* from the Val Poschiavino in Switzerland. This allele (Sry^{POS}) is not able to trigger testis differentiation as efficiently as other alleles and causes a 14-hour delay in the activation of the testicular pathway within the gonad. On a particular inbred mouse strain background, C57BL/6, Sry^{POS} is unable to initiate proper testis development and the gonads of these animals develop as ovotestes, with both ovarian and testicular components, or as ovaries (54) (Fig. 4). Eicher et al. (55) mapped the C57Bl/6 loci involved in this effect to three regions on chromosome 4 (tda1), 2 (tda2), and 5 (tda3). The genes associated with these loci have not yet been identified, but it has been proposed they encode factors that are involved in the regulation of *Sry* expression and/or that promote ovary development (56).

These studies highlight the importance of the genetic background on the process of sex determination. Consistent with this, several mutant strains of mice, including those carrying mutations in *Fgf9* and *Dax1*, only show a sex reversal phenotype when the genetic background is derived from the C57Bl/6 background. This suggests that allelic variations that might have a small effect on available levels of active SRY could have a significant impact on phenotype due to the delicate balance that controls the decision to become male or female. Comparisons between human and mouse suggest that this balance is set up slightly differently between the two species such that mutations in the same gene can give different phenotypes, as mentioned in the case of WT1.

DAX-1 (Nr0b1)

The X-linked gene, *DAX-1*, has been implicated in sex determination and gonad and adrenal development, although its role is complex. XY mice with a targeted mutation in this gene showed defective testis cord development; however, in certain genetic backgrounds these mutant mice showed complete male-to-female sex reversal (57,58). *Sry* expression was not affected in these sex-reversed mice, suggesting that this gene is required for SRY action in the gonad. Mutations in *DAX-1* in XY human individuals gave rise to adrenal deficiencies and hypogonadotropic hypogonadism, where their testes were reduced in size but no sex reversal was observed (59,60).

DAX-1 is an unusual member of the nuclear hormone receptor family because it lacks the classic zinc finger DNA-binding domain but contains instead a novel domain with a repeated motif. Tissue culture studies showed that DAX-1 can act as a repressor of transcriptional activation by other classic nuclear hormone receptors (60–62). DAX-1 expression within the early gonad correlates with that of a classic nuclear hormone receptor, SF1 (Nr5a1) (63). *SF1* expression is specific to the early genital ridge and adrenal, and mice with a mutation in this gene lack gonads and adrenals (see Chapter 8 for a detailed analysis of early gonad development) (64,65). These properties suggest that the action of DAX-1 in sex determination is related to that of SF1 and that the levels of active SF1 protein are important to determine testis development. Because of the early gonadal phenotype of the *SF1* deficient mice, it has not been possible to address the role of this factor in sex determination and testis differentiation.

Human XY individuals with a duplication in the region of the X chromosome that contains DAX-1 show sex reversal, called the dosage sensitive sex reversal syndrome (66). Consistent with this, transgenic XY mice with extra levels of *Dax-1* show sex reversal, although only in certain genetic backgrounds (67). These studies show that although DAX-1 is required for testis development, at high levels it can act antagonistically to SRY and promote sex reversal.

Insulin Receptor Family

Targeted mutagenesis in the mouse has uncovered a role for the insulin family signaling pathway in sex determination. XY mice with mutations in all insulin receptors *Ir*, *Igf1r*, and *Irr* simultaneously show sex reversal (68). *Sry* expression during gonad development in these mice was found to be significantly reduced, accounting for the phenotype. However, cell proliferation and the size of the gonads in both sexes were reduced in these mutant animals. Cell proliferation has been shown to be important in sex determination because inhibition of proliferation at particular stages of gonad development can prevent testis determination (69). Therefore, the phenotype observed in the mutant mice was probably due to a defect in proliferation of cells of the gonad at the time when *Sry* is acting to determine testis development.

GATA4/FOG2

Members of the GATA family of transcription factors are important in many processes involved in embryo development. These factors contain a zinc finger DNA-binding domain and interact with FOG (Friend of GATA) proteins that contain multitype zinc finger domains. Expression analysis of GATA and FOG genes during urogenital development show that GATA4 and FOG2 are expressed in the somatic cells of the gonad (70,71). Mouse XY embryos with a mutation in *Fog2* show sex reversal (72). Analysis of the gonads of these mutant mice shows a reduction in *Sry* expression when compared with wild-type XY embryos. Ovary development was also seen in XY embryos with a mutation in *Gata4* that abrogates the interaction of GATA4 with FOG2 showing that both proteins are required for sex determination (72). Studies are underway to determine a direct role for GATA4 in the regulation of *Sry* transcription.

Chromatin Remodeling Factors

Two factors involved in chromatin domain organization, M33 and ATRX, have been implicated in sex determination. M33 is a member of the polycomb group gene family that is a general repressor of transcription, and XY mice mutants for this gene show sex reversal (73). However, the phenotype was variable, and gonad development was retarded; therefore, the role of this factor in sex determination has not been determined. Human XY individuals with mutations in the X-linked gene, *ATRX*, show sex reversal (74). The role of this factor in gonad development has not yet been studied in any detail. Interestingly, a homolog of the *ATRX* gene was found on the marsupial Y chromosome (75). This gene was found to have a testis-specific expression, in contrast to ATRX that is expressed in many tissues, suggesting that it has a role in marsupial testis development.

Testis Development

SRY acts in the gonad to determine testis development, and in this section I discuss the cellular and molecular pathways activated by this gene. A discussion on ovary development can be found in Chapter 8.

Sertoli Cell Differentiation

The action of SRY in the developing gonad initiates Sertoli cell differentiation. The transient nature of *Sry* expression in the gonad implies that it must activate the expression of genes required to define and maintain Sertoli cell identity. A candidate gene of this type is the transcription factor SOX9. Expression analysis in the mouse have shown that *Sox9* is one of the first genes up-regulated in the male gonad compared with the female, at around E11.5, and its expression is then associated with Sertoli cells throughout testis development (76,77). Mutations in SOX9 in human and mouse reveal the important role of this factor in testis differentiation. In humans, heterozygous mutations in *SOX9* give rise to XY female sex reversal associated with the severe dwarfism syndrome, camptomelic dysplasia (78,79). Mice that lack SOX9 in the developing testis showed defects in testis differentiation, giving rise to ovotestis-like gonads (80). These defects were shown to be more severe when these mice also lacked the related gene *Sox8*, which has been shown to be up-regulated in Sertoli cells of the developing gonad slightly after *Sox9* (81). This suggests that these genes have similar functions and therefore are part of a reinforced redundant system to ensure the differentiation of Sertoli cells.

Transgenic experiments in mice showed that *Sox9* can induce Sertoli cell differentiation and testis development in XX gonads (82). In one case, a chance integration of a transgenic construct 1.3 Mb from the *Sox9* locus resulted in the induction of the expression of this gene within the embryonic XX gonad (83). Consistent with this, duplications of the *SOX9* locus in humans are associated with XX male sex reversal (84). These studies suggest that the only role of SRY is to activate *SOX9* expression and that this gene is sufficient for testis differentiation. Studies are underway to determine the regulatory elements for *SOX9* expression in the developing gonad. Genetic studies have suggested a complex mode of regulation involving distant regulatory elements (85).

SOX9 protein is related to SRY in that it also contains an HMG box type of DNA-binding domain. Outside this domain, it differs from SRY in that it contains a strong transcriptional activation domain. Evolutionary studies have shown that the *SOX9* gene is highly conserved in vertebrates, and in many cases its expression is associated with early testis differentiation (76,77,86–88). This suggests a common pathway for ensuring Sertoli cell differentiation in vertebrates and that different sex determination mechanisms have evolved to control this process (89). Consistent with this model, conserved members of the DMRT family of proteins are found in the developing gonad of all vertebrates studied (90). The genes encoding at least two members of this family, *DMRT1* and *2*, map to regions of chromosome 9 in humans that are deleted in individuals with XY sex reversal, suggesting that these genes are involved in early testis development (91,92). However, mice deficient for *Dmrt1* only show testicular defects postnatally, suggesting that more than one member of the family is involved in this process (93). DMRT genes have been found to have a more central role in sex determination in the teleost fish Medaka, where a member of this family, *DMY*, maps to the Y chromosome and is most likely the testis determining gene in this species, although not in other fish species (94,95).

One of the earliest products of Sertoli cells during development is anti-müllerian hormone (AMH) that induces the regression of the müllerian duct in the male embryo. The regulation of *AMH* expression has been highly studied using a combination of tissue culture studies, transgenic mice to map embryonic specific regulatory elements, and genetically modified mice to define specific transcription binding sites in vivo. These studies identified several transcription factors implicated in *AMH* regulation: SOX9, SOX8, SF1, DAX-1, GATA4, and WT1 (81,96–100). Although some of these factors are found in many cell types of the embryo, the combination of factors is exclusive to Sertoli cells and therefore provides specificity. These studies also show that factors such as SF1 can have a role at different stages and in different cell types of the gonad during development.

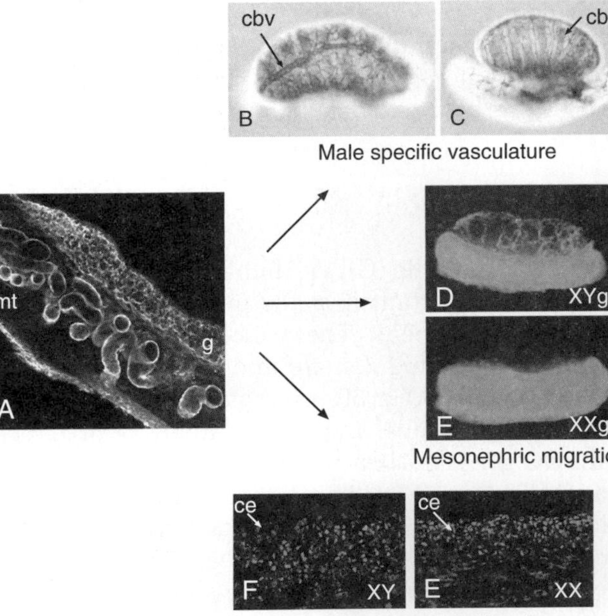

FIG. 5. Cellular changes in the gonad activated by SRY. **A:** Confocal image of a gonad and mesonephros stained with an antibody to laminin at a stage before morphological changes take place (E11.5). The action of *Sry* in the XY gonad induces three events: (1) The formation of male-specific vasculature as shown by a whole-mount antibody stain to PECAM, which marks endothelial cells, of gonad and mesonephros from an E14 embryo (**B** and **C** are pictures of the same gonad); (2) mesonephric cell migration is observed in the XY (**D**) but not the XX (**E**) gonad as shown by organ culture samples where the mesonephros but not the gonad is derived from transgenic mice that express GFP ubiquitously; (3) up-regulation of proliferation of the coelomic epithelial cells in the XY (**F**) but not the XX (**G**) gonad as shown by BrdU incorporation in red. mt, mesonephric tubules; g, gonad; cbv, coelomic blood vessel; ce, coelomic epithelium. [Adapted from (134).] (See color insert.)

Cellular Pathways Associated with Testis Development

Cell Proliferation. Once sex determination has occurred, the size of the male gonad increases relative to the female gonad. Studies in the mouse have shown this increase results from higher cell proliferation. The site of this male-specific proliferation is the coelomic epithelium, which has been shown to contribute cells to the Sertoli and interstitial lineage (101,102) (Fig. 5). *Sry* was shown to be expressed in the cells in the interior of the gonad and not in the coelomic epithelium (45,46). This suggests that although *Sry* induces proliferation in the male gonad, it does so non–cell autonomously. A possible mediator in this process is the fibroblast growth factor, FGF9, which is expressed in the gonad at early stages but not in a sex-specific manner. Mice lacking *Fgf9* show a reduction in cell proliferation in the XY gonad but not in the XX gonad (103). Testis development in the mutant mice was severely affected, and in certain genetic backgrounds complete sex reversal was observed, highlighting the important of this cell process in male sexual development (103,104). Other factors that affect proliferation in the developing testis are members of the platelet-derived growth factor (PDGF) family because mice lacking the PDGF receptor α (PDGFRα) show reduced male-specific proliferation in the coelomic epithelia (105).

Mesonephric Cell Migration. The characterization of the cellular processes involved in gonad development has been possible because of the development of in vitro organ culture systems. Tissues from genetically modified mice have been used in recombinants to determine the molecules involved in the different processes. This type of study has shown that cells from the mesonephros migrate into the gonad during development and that this only occurs in the male gonad (106,107) (Fig. 5). The migrating cells are mostly endothelial, but cells that resemble perivascular and peritubular myoid cells have also been implicated. Several growth factors, including

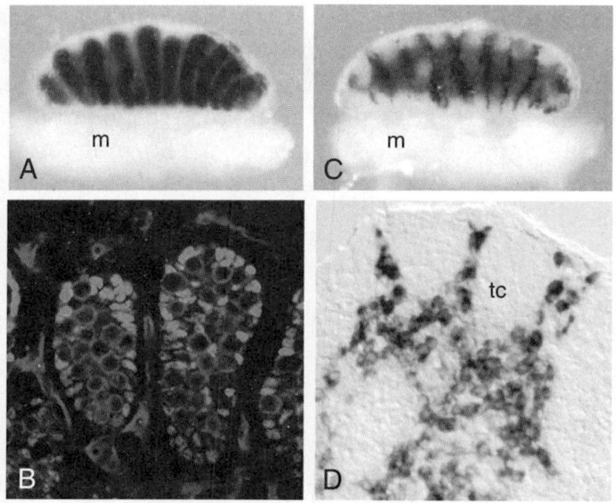

FIG. 6. Morphology of the embryonic testis. Expression of the different cell types of the testis within the E13.5 gonad are seen by whole-mount in situ hybridization for the expression of the Sertoli cell marker AMH (**A**) and the Leydig cell marker *Cyp11a1* (**C**). The structure of the testicular cords is seen in the confocal image of an E12.5 gonad that was stained with an antibody against SOX9 in green (**B**). Germ cells are found within the testicular cords as marked by the antibody stain for PECAM in red (B). In the interstitium, epithelial cells, as marked by the antibody stain for PECAM in red (B), and Leydig cells, as marked by the section of the whole-mount in situ hybridization on a gonad from an E14 embryo for the expression of *Cyp11a1* (**D**), are found. m, mesonephros. (Confocal image courtesy of Blanche Capel, Duke University.) (See color insert.)

FGF9 and PDGFs, were able to induce migration of mesonephric cells into an XX gonad in culture (104,105). However, evidence for a role of these factors in vivo has yet to be found.

Testicular Cord Formation. One of the most striking morphological changes in the XY gonad after the action of SRY is the organization of cells into testicular cords, which are visible by E12.5 in the mouse (Fig. 6). These cords are made up of Sertoli cells, forming an epithelial layer enclosing the germ cells. Surrounding the Sertoli cells are the peritubular myoid cells separated by a basal lamina layer. Outside the cords, in the interstitium, Leydig and endothelial cells are found. Mesonephric cell migration is thought to contribute to cord formation because blocking this process in organ cultures impairs the proper formation of cords (108). Germ cells are not essential for this process because testicular cords form in agametic gonads.

Male-Specific Vascular Formation. Testis cord formation in the XY gonad is accompanied by the formation of a prominent blood vessel running along the periphery of the testis, just under the coelomic epithelium. Although vascular formation is an active process in the gonad of both sexes during development, this coelomic blood vessel is only found in the male. The endothelial cells associated with this vessel were found to express arterial markers, suggesting that a male-specific arterial system that promotes the export of testosterone produced by Leydig cells needs to form to ensure the masculinization of the embryo during its development (109) (Fig. 5). The formation of the coelomic blood vessel is dependent on migration of endothelial cells from the mesonephros. Mice lacking PDGFRα show a defect in coelomic blood vessel assembly and branching in the testis (105). In addition, XY gonads from these mice were unable to induce mesonephric cell migration in organ cultures. *Pdgfrα* is expressed in the region of the testis where the coelomic blood vessel is forming, suggesting that PDGFs act within the gonad to ensure proper vascular formation but are not directly responsible for the induction of endothelial cell migration from the mesonephros.

The XX gonad does not normally form a coelomic vessel. Unexpectedly, it was found recently that an active repressive pathway is present in the ovary to prevent the migration of endothelial cells from the mesonephros and the formation of a coelomic blood vessel. Mice with mutations in the genes encoding either WNT4 or Follistatin, two cell signaling molecules, showed the presence of an ectopic coelomic blood vessel, similar to that found in the normal XY gonad (110,111). WNT4 was found to act upstream of Follistatin, and this latter factor has been shown to repress the activity of members of the transforming growth factor-β family of growth factors, such as activin. Whether a transforming growth factor-β family-dependent pathway is responsible for endothelial cell migration from the mesonephros into either the male or female gonad has yet to be determined.

Leydig Cell Differentiation

Leydig cells are the steroid-producing cell type of the testis. They are very active during development, and expression of genes involved in steroid biosynthesis can be first seen around E12.5-13 in the interstitial cells of the mouse testis. The origin of Leydig cells has not been established. Adrenal cortical cells, which also produce steroids during development, are thought to be derived from the same primordium as the gonad, called the adrenogenital primordium (112). These cells arise when the genital ridge first forms and separate into two populations as development proceeds. However, although the expression of genes involved in steroidogenesis is observed in the adrenal cells at early stages, it is not until after cord formation that they are seen in the gonad and only in the testis. These observations have led to the model that Leydig cell precursors are found within the gonad during testis development and that their differentiation is activated by Sertoli cells either directly or as a secondary effect of testis development.

A candidate factor involved in the activation of Leydig cell differentiation is the cell signaling molecule Desert Hedgehog (DHH). DHH is produced by Sertoli cells at early stages of testis development, and mice that lack this gene showed a severe reduction in Leydig cell numbers during development in certain genetic backgrounds (113). Expression analysis of *Patched1*, a downstream gene activated by the DHH pathway, showed that this factor acted on all interstitial cells, including Leydig cells and peritubular myoid cells. Cord formation was also affected in these mutant mice, and therefore this factor might have a more general role during testis development (114). Consistent with this, human individuals with mutations in DHH show variable gonadal dysgenesis phenotypes (115).

A similar phenotype is observed in mice and humans that carry mutations in the X-linked aristaless related homeobox (*Arx*) gene (116). This gene is expressed in interstitial and peritubular myoid cells of the developing testis, and Leydig cell differentiation is severely affected in the mutant mouse embryos. However, *Arx* is not expressed in Leydig cells, indicating that the differentiation and compartmentalization of the interstitial cells during development is important for the induction of steroid production. In humans, mutations in *ARX* are associated with the XLAG (X-linked lissencephaly with abnormal genitalia) syndrome, because this gene also has a function in forebrain development (116).

Transcription factors implicated in steroid production in Leydig cells during embryo development have not been identified, with the exception of SF1. Tissue culture studies have shown that SF1 can activate genes involved in steroid biosynthesis, and this factor was found to be present in Leydig cells as they differentiate (96,117). However, due to the early gonadal phenotype of *SF1* deficient mice, it has been difficult to establish the role of this factor in Leydig cells in vivo.

Germ Cells

Germ cells within the developing gonad follow a sexual dimorphic pathway. In the ovary they enter meiosis at around E13.5, but in the male they arrest in mitosis at about the same time. This difference is not dependent on the chromosomal sex of the germ cells but is dependent on the somatic environment of the gonad (118). Organ culture experiments have shown that germ cells enter meiosis unless they are grown with tissue derived from the embryonic testis (119). This indicates that entry into meiosis and the formation of oocytes is an intrinsic property of germ cells, and this process is prevented by products from the embryonic testis to ensure that spermatogenesis occurs. The testis-specific factor responsible for inducing mitotic arrest in germ cells has not yet been identified, but it has been shown to be produced during a brief period just after E11.5 (119).

PHENOTYPIC SEX

Sexual Differentiation of Reproductive Organs

Experiments done by Jost in the 1950s on rabbit fetuses showed that castrated male embryos develop as females. Jost postulated that secretions from the fetal testis induced the development of male sexual characteristics (120). Today we know of at least three diffusible factors produced by the fetal testis with these properties, AMH, insulin-like factor 3 (Insl3), and testosterone (Fig. 7). As mentioned above, AMH is produced by Sertoli cells, and its role is to activate the regression of the müllerian duct, which gives rise to the oviduct, uterus, and upper portion of the vagina in the female. Leydig cells produce Insl3 that, together with testosterone, regulate the descent of the testis from a location next to the kidney to the region where the scrotum is developing (121,122).

The development of most male-specific structures is dependent on testosterone produced by the Leydig cells of the embryonic testis. Testosterone regulates three main aspects of male phenotypic development: the conversion of the wolffian ducts into the epididymis, vas deferens, and seminal vesicles; the formation of the male urethra and prostate from the urogenital sinus; and the formation of the phallus and scrotum from the genital tubercle and urethral folds. In some of these target tissues, testosterone is not the acting hormone but is converted to a more potent androgen, dihydrotestosterone (DHT), through the action of the enzyme 5α-reductase found in these tissues. In marsupials it was found that the male-specific circulating androgen during the stage of development of male-specific organs was not DHT or testosterone but a 3α reduced derivative of DHT, 5α-androstane-3α, 17β-diol. However, 5α-androstane-3α, 17β-diol is considered a weak androgen, and therefore it is thought that it converts back to DHT to induce virilization (123).

Both testosterone and DHT act through the androgen receptor, an X-linked member of the nuclear hormone receptor family, with DHT having an approximate 10-fold higher affinity (124). Mutations in the androgen receptor gene and in genes encoding any of the enzymes required for DHT synthesis give

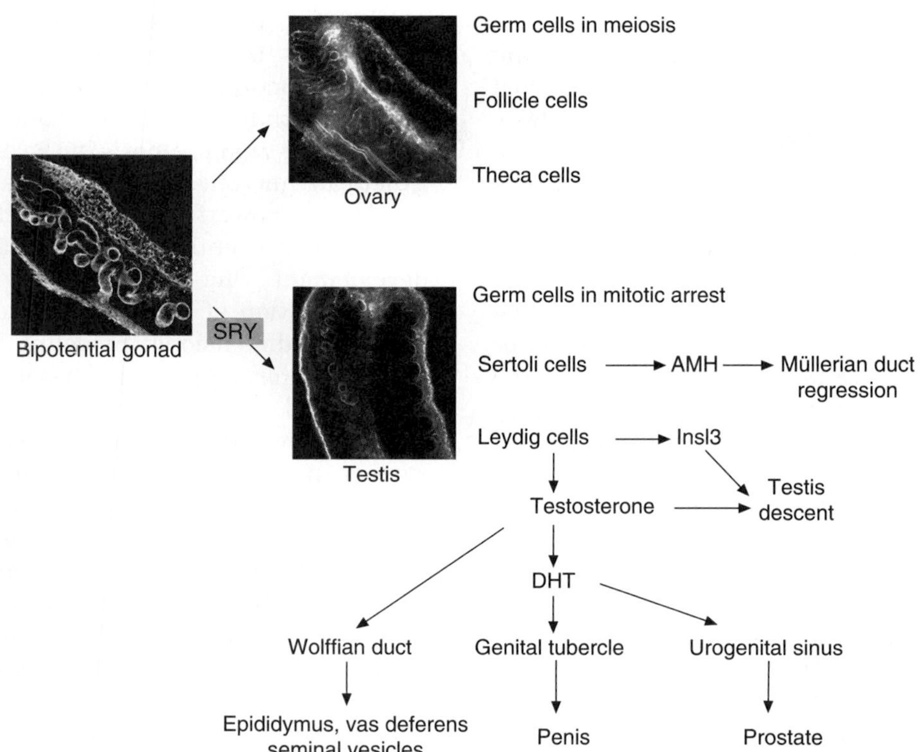

FIG. 7. Sexual differentiation of the internal and external genitalia. The bipotential gonad develops into an ovary in the absence of SRY or a testis if SRY is present. Three main sexually dimorphic cell types are found in the gonad: germ cells, supporting cells (Sertoli cells and follicle cells), and steroidogenic cells (Leydig and theca cells). In the male, products from the testis, AMH, Insl3, and testosterone, induce the differentiation of the male-specific internal and external genitalia. [Confocal images of laminin-stained gonads were adapted from (135).]

rise to defects in the development of all male-specific structures in both mouse and humans (125). These defects can be highly variable depending on the nature of the mutation and the importance of the gene involved in the process. Analysis of these mutations showed that not all male-specific structures require DHT for development as human individuals with 5α-reductase deficiencies show virilization of the wolffian duct but absence of other male-specific structures (125). In the mouse, males lacking both isoforms of the 5α-reductase enzyme showed proper internal and external genitalia, although the prostate and the seminal vesicles were smaller, showing that testosterone is able to induce most male sexual development in this species (126).

The development of sexual reproductive organs in the female is less dependent on diffusible factors made by the gonad. The exact role of steroids in female development has been hard to determine because although the ovary does not produce steroids during embryogenesis, the fetus is bathed in female hormones derived from the mother. Mice lacking both estrogen receptor α and β develop proper müllerian duct derivatives and ovary development is not affected initially. However, in adulthood these animals show uterine hypoplasia and testis-like cords within the ovary, showing that estrogen does play a role in late development of female reproductive organs (127). As expected from Jost's postulate, female embryos that have been treated with androgens show virilization of their reproductive organs. A similar situation occurs in XX human individuals with congenital adrenal hyperplasia. These individuals carry mutations in the genes encoding enzymes in the final steps of adrenal steroid synthesis, which increases the levels of production by the adrenal of steroidogenic intermediates that have androgenic properties.

Sexual Differentiation of the Nervous System

Androgens have been shown to have a role in the sexual differentiation of the nervous system. Several regions of the rodent brain, including the sexually dimorphic nucleus of the preoptic area, the anteroventral periventricular nucleus, and the spinal nucleus of the bulbocavernosus, show sexual differences that depend on the presence or absence of androgens during fetal or neonatal stages (128–130). In some cases it was found that androgens were not acting through the androgen receptor but were converted into estrogen in the brain by the enzyme aromatase, which was acting through the estrogen receptor. Consistent with these differences in brain structure between the sexes, behavior can depend on the level of androgens during fetal life. This was first demonstrated in guinea pigs where females were exposed to testosterone in utero, and this interfered with their female pattern of reproductive behavior in adulthood (131). Detailed descriptions of the role of steroids in different aspects of reproductive function and behavior are found in other chapters of this book.

Although many of the sex differences found in the brain are due to the effect of steroids, recent evidence has shown genetic contributions to these differences. Most of these differences are expected to be derived from the sex chromosomes because these contribute to gene differences between sexes. For example, females lack Y-linked genes; males may have lower doses of X-linked genes that escape X-inactivation; and because males only inherit their X chromosome from their mothers, they lack paternal imprinting effects. Experiments using genetically modified mice were designed to uncover the effects of sex chromosomes independently of the sex of the gonad (132). Using these mice, it was shown that the density of vasopressin innervation of the lateral septum was different between XX and XY animals independently of the phenotype of the gonad. A difference was also seen in the number of dopamine neurons in cultures from embryonic mesencephalic cells (133). Further characterization is needed to determine the importance of genetic effects in the development of brain structure and function and what effects these might have on behavior.

FUTURE

The study of sex determination and differentiation has provided great insight into different aspects of reproductive biology. The analysis of disorders of human sexual differentiation has been essential in the identification of genes regulating this process. In addition, the mouse has been an important model system with the advent of genetic technology such as targeted mutagenesis and transgenesis. Many more genes involved in this process have yet to be determined because the molecular basis for most sex reversed individuals is still not known. Future progress will capitalize on advances in genomics, proteomics, chromatin precipitation assays, and mutagenesis screens to identify other critical genes.

The particular nature of the gonad, a bipotential organ that is genetically triggered toward two sexually dimorphic fates, makes it an interesting system to study the molecular and cellular pathways required for organogenesis in the embryo. The use of genetically modified mice and organ culture systems has been a powerful tool in the characterization of the cellular events that take place during gonad development. One of the future goals in the field is the integration of cellular and molecular pathways into three-dimensional modeling systems with the aid of bioinformatics to understand how this process is regulated.

The poor conservation of the sex determining system in the animal kingdom makes comparative studies between different animals an important way of identifying pathways that are common and those that are not conserved and therefore provide the basis for reproductive functions that are unique to each species. The sequencing of the genomes of different animals will help identify novel genes in this process.

ACKNOWLEDGMENTS

I thank Pierre Val and Jaime Carvajal for help with the manuscript and figures.

REFERENCES

1. Bull, J. J. (1983). *Evolution of Sex Determining Mechanisms.* Benjamin Cummings, Menlo Park, CA.
2. Rice, W. R. (1996). Evolution of the Y sex chromosome in animals. *Bioscience* 46, 331–343.
3. Charlesworth, B. (2002). The evolution of chromosomal sex determination. *Novartis Found. Symp.* 244, 207–219.
4. Burgoyne, P. S. (1982). Genetic homology and crossing over in the X and Y chromosomes of mammals. *Hum. Genet.* 61, 85–90.
5. Skaletsky, H., Kuroda-Kawaguchi, T., Minx, P. J., Cordum, H. S., Hillier, L., Brown, L. G., Repping, S., Pyntikova, T., Ali, J., Bieri, T., Chinwalla, A., Delehaunty, A., Delehaunty, K., Du, H., Fewell, G., Fulton, L., Fulton, R., Graves, T., Hou, S. F., Latrielle, P., Leonard, S., Mardis, E., Maupin, R., McPherson, J., Miner, T., Nash, W., Nguyen, C., Ozersky, P., Pepin, K., Rock, S., Rohlfing, T., Scott, K., Schultz, B., Strong, C., Tin-Wollam, A., Yang, S. P., Waterston, R. H., Wilson, R. K., Rozen, S., and Page, D. C. (2003). The male-specific region of the human Y chromosome is a mosaic of discrete sequence classes. *Nature* 423, 825–837.
6. Tilford, C. A., Kuroda-Kawaguchi, T., Skaletsky, H., Rozen, S., Brown, L. G., Rosenberg, M., McPherson, J. D., Wylie, K., Sekhon, M., Kucaba, T. A., Waterston, R. H., and Page, D. C. (2001). A physical map of the human Y chromosome. *Nature* 409, 943–945.
7. Graves, J. A. (2002). Evolution of the testis-determining gene—the rise and fall of SRY. *Novartis Found. Symp.* 244, 86–97.
8. Rozen, S., Skaletsky, H., Marszalek, J. D., Minx, P. J., Cordum, H. S., Waterston, R. H., Wilson, R. K., and Page, D. C. (2003). Abundant gene conversion between arms of palindromes in human and ape Y chromosomes. *Nature* 423, 873–876.

9. Lyon, M. F. (1961). Gene action in the X-chromosome of the mouse (*Mus musculus* L.). *Naturwissenschaften* 190, 372–373.
10. Brockdorff, N., Ashworth, A., Kay, G. F., Cooper, P., Smith, S., McCabe, V. M., Norris, D. P., Penny, G. D., Patel, D., and Rastan, S. (1991). Conservation of position and exclusive expression of mouse Xist from the inactive X chromosome. *Nature* 351, 329–331.
11. Brown, C. J., Ballabio, A., Rupert, J. L., Lafreniere, R. G., Grompe, M., Tonlorenzi, R., and Willard, H. F. (1991). A gene from the region of the human X inactivation centre is expressed exclusively from the inactive X chromosome. *Nature* 349, 38–44.
12. Carrel, L., Cottle, A. A., Goglin, K. C., and Willard, H. F. (1999). A first-generation X-inactivation profile of the human X chromosome. *Proc. Natl. Acad. Sci. U S A* 96, 14440–14444.
13. Sudbrak, R., Wieczorek, G., Nuber, U. A., Mann, W., Kirchner, R., Erdogan, F., Brown, C. J., Wohrle, D., Sterk, P., Kalscheuer, V. M., Berger, W., Lehrach, H., and Ropers, H. H. (2001). X chromosome-specific cDNA arrays: identification of genes that escape from X-inactivation and other applications. *Hum. Mol. Genet.* 10, 77–83.
14. Graves, J. A. (1995). The evolution of mammalian sex chromosomes and the origin of sex determining genes. *Philos. Trans. R. Soc. Lond. B Biol. Sci.* 350, 305–311; discussion 311–312.
15. Lahn, B. T., and Page, D. C. (1999). Four evolutionary strata on the human X chromosome. *Science* 286, 964–967.
16. Foster, J. W., and Graves, J. A. (1994). An SRY-related sequence on the marsupial X chromosome: implications for the evolution of the mammalian testis-determining gene. *Proc. Natl. Acad. Sci. U S A* 91, 1927–1931.
17. Renfree, M. B., and Short, R. V. (1988). Sex determination in marsupials: evidence for a marsupial-eutherian dichotomy. *Philos. Trans. R. So.c Lond. B. Biol. Sci.* 322, 41–53.
18. Wai-Sum, O., Short, R. V., Renfree, M. B., and Shaw, G. (1988). Primary genetic control of somatic sexual differentiation in a mammal. *Nature* 331, 716–717.
19. Jimenez, R., Alarcon, F. J., Sanchez, A., Burgos, M., and De La Guardia, R. D. (1996). Ovotestis variability in young and adult females of the mole *Talpa occidentalis* (insectivora, mammalia). *J. Exp. Zool.* 274, 130–137.
20. Whitworth, D. J., Licht, P., Racey, P. A., and Glickman, S. E. (1999). Testis-like steroidogenesis in the ovotestis of the European mole, *Talpa europaea*. *Biol. Reprod.* 60, 413–418.
21. Fredga, K. (1983). Aberrant sex chromosome mechanisms in mammals. Evolutionary aspects. *Differentiation* 23, S23–S30.
22. Sinclair, A. H., Berta, P., Palmer, M. S., Hawkins, J. R., Griffiths, B. L., Smith, M. J., Foster, J. W., Frischauf, A. M., Lovell-Badge, R., and Goodfellow, P. N. (1990). A gene from the human sex-determining region encodes a protein with homology to a conserved DNA-binding motif. *Nature* 346, 240–244.
23. Berta, P., Hawkins, J. R., Sinclair, A. H., Taylor, A., Griffiths, B. L., Goodfellow, P. N., and Fellous, M. (1990). Genetic evidence equating SRY and the testis-determining factor. *Nature* 348, 448–450.
24. Goodfellow, P. N., and Lovell-Badge, R. (1993). SRY and sex determination in mammals. *Annu. Rev. Genet.* 27, 71–92.
25. Gubbay, J., Collignon, J., Koopman, P., Capel, B., Economou, A., Munsterberg, A., Vivian, N., Goodfellow, P., and Lovell-Badge, R. (1990). A gene mapping to the sex-determining region of the mouse Y chromosome is a member of a novel family of embryonically expressed genes. *Nature* 346, 245–250.
26. Gubbay, J., Vivian, N., Economou, A., Jackson, D., Goodfellow, P., and Lovell-Badge, R. (1992). Inverted repeat structure of the Sry locus in mice. *Proc. Natl. Acad. Sci. U S A* 89, 7953–7957.
27. Koopman, P., Gubbay, J., Vivian, N., Goodfellow, P., and Lovell-Badge, R. (1991). Male development of chromosomally female mice transgenic for Sry. *Nature* 351, 117–121.
28. Vogt, P. H., Edelmann, A., Kirsch, S., Henegariu, O., Hirschmann, P., Kiesewetter, F., Kohn, F. M., Schill, W. B., Farah, S., Ramos, C., Hartmann, M., Hartschuh, W., Meschede, D., Behre, H. M., Castel, A., Nieschlag, E., Weidner, W., Grone, H. J., Jung, A., Engel, W., and Haidl, G. (1996). Human Y chromosome azoospermia factors (AZF) mapped to different subregions in Yq11. *Hum. Mol. Genet.* 5, 933–943.
29. Mazeyrat, S., Saut, N., Grigoriev, V., Mahadevaiah, S. K., Ojarikre, O. A., Rattigan, A., Bishop, C., Eicher, E. M., Mitchell, M. J., and Burgoyne, P. S. (2001). A Y-encoded subunit of the translation initiation factor Eif2 is essential for mouse spermatogenesis. *Nat. Genet.* 29, 49–53.
30. Ashworth, A., Rastan, S., Lovell-Badge, R., and Kay, G. (1991). X-chromosome inactivation may explain the difference in viability of XO humans and mice. *Nature* 351, 406–408.
31. Burgoyne, P. S., and Baker, T. G. (1985). Perinatal oocyte loss in XO mice and its implications for the aetiology of gonadal dysgenesis in XO women. *J. Reprod. Fertil.* 75, 633–645.
32. Speed, R. M. (1986). Oocyte development in XO foetuses of man and mouse: the possible role of heterologous X-chromosome pairing in germ cell survival. *Chromosoma* 94, 115–124.
33. Mroz, K., Carrel, L., and Hunt, P. A. (1999). Germ cell development in the XXY mouse: evidence that X chromosome reactivation is independent of sexual differentiation. *Dev. Biol.* 207, 229–238.
34. Harley, V. R., Jackson, D. I., Hextall, P. J., Hawkins, J. R., Berkovitz, G. D., Sockanathan, S., Lovell-Badge, R., and Goodfellow, P. N. (1992). DNA binding activity of recombinant SRY from normal males and XY females. *Science* 255, 453–456.
35. Tucker, P. K., and Lundrigan, B. L. (1993). Rapid evolution of the sex determining locus in Old World mice and rats. *Nature* 364, 715–717.
36. Whitfield, L. S., Lovell-Badge, R., and Goodfellow, P. N. (1993). Rapid sequence evolution of the mammalian sex-determining gene SRY. *Nature* 364, 713–715.
37. Ferrari, S., Harley, V. R., Pontiggia, A., Goodfellow, P. N., Lovell-Badge, R., and Bianchi, M. E. (1992). SRY, like HMG1, recognizes sharp angles in DNA. *EMBO. J.* 11, 4497–4506.
38. Pontiggia, A., Rimini, R., Harley, V. R., Goodfellow, P. N., Lovell-Badge, R., and Bianchi, M. E. (1994). Sex-reversing mutations affect the architecture of SRY-DNA complexes. *EMBO. J.* 13, 6115–6124.
39. Rimini, R., Pontiggia, A., Spada, F., Ferrari, S., Harley, V. R., Goodfellow, P. N., and Bianchi, M. E. (1995). Interaction of normal and mutant SRY proteins with DNA. *Philos. Trans. R. Soc. Lond. B. Biol. Sci.* 350, 215–220.
40. Bowles, J., Cooper, L., Berkman, J., and Koopman, P. (1999). Sry requires a CAG repeat domain for male sex determination in *Mus musculus*. *Nat. Genet.* 22, 405–408.
41. Lovell-Badge, R., Canning, C., and Sekido, R. (2002). Sex-determining genes in mice: building pathways. *Novartis Found. Symp.* 244, 4–18.
42. Salas-Cortes, L., Jaubert, F., Barbaux, S., Nessmann, C., Bono, M. R., Fellous, M., McElreavey, K., and Rosemblatt, M. (1999). The human SRY protein is present in fetal and adult Sertoli cells and germ cells. *Int. J. Dev. Biol.* 43, 135–140.
43. Hacker, A., Capel, B., Goodfellow, P., and Lovell-Badge, R. (1995). Expression of Sry, the mouse sex determining gene. *Development* 121, 1603–1614.
44. Jeske, Y. W., Bowles, J., Greenfield, A., and Koopman, P. (1995). Expression of a linear Sry transcript in the mouse genital ridge. *Nat. Genet.* 10, 480–482.

45. Albrecht, K. H., and Eicher, E. M. (2001). Evidence that Sry is expressed in pre-Sertoli cells and Sertoli and granulosa cells have a common precursor. *Dev. Biol.* 240, 92–107.
46. Sekido, R., Bar, I., Narvaez, V., Penny, G., and Lovell-Badge, R. (2004). SOX9 is up-regulated by the transient expression of SRY specifically in Sertoli cell precursors. *Dev. Biol.* 274, 271–279.
47. Palmer, S. J., and Burgoyne, P. S. (1991). In situ analysis of fetal, prepuberal and adult XX–XY chimaeric mouse testes: Sertoli cells are predominantly, but not exclusively, XY. *Development* 112, 265–268.
48. Pelletier, J., Schalling, M., Buckler, A. J., Rogers, A., Haber, D. A., and Housman, D. (1991). Expression of the Wilms' tumor gene WT1 in the murine urogenital system. *Genes Dev.* 5, 1345–1356.
49. Armstrong, J. F., Pritchard-Jones, K., Bickmore, W. A., Hastie, N. D., and Bard, J. B. (1993). The expression of the Wilms' tumour gene, WT1, in the developing mammalian embryo. *Mech. Dev.* 40, 85–97.
50. Kreidberg, J. A., Sariola, H., Loring, J. M., Maeda, M., Pelletier, J., Housman, D., and Jaenisch, R. (1993). WT-1 is required for early kidney development. *Cell* 74, 679–691.
51. Larsson, S. H., Charlieu, J. P., Miyagawa, K., Engelkamp, D., Rassoulzadegan, M., Ross, A., Cuzin, F., van Heyningen, V., and Hastie, N. D. (1995). Subnuclear localization of WT1 in splicing or transcription factor domains is regulated by alternative splicing. *Cell* 81, 391–401.
52. Hammes, A., Guo, J. K., Lutsch, G., Leheste, J. R., Landrock, D., Ziegler, U., Gubler, M. C., and Schedl, A. (2001). Two splice variants of the Wilms' tumor 1 gene have distinct functions during sex determination and nephron formation. *Cell* 106, 319–329.
53. Barbaux, S., Niaudet, P., Gubler, M. C., Grunfeld, J. P., Jaubert, F., Kuttenn, F., Fekete, C. N., Souleyreau-Therville, N., Thibaud, E., Fellous, M., and McElreavey, K. (1997). Donor splice-site mutations in WT1 are responsible for Frasier syndrome. *Nat. Genet.* 17, 467–470.
54. Eicher, E. M., Washburn, L. L., Whitney, J. B. D., and Morrow, K. E. (1982). *Mus poschiavinus* Y chromosome in the C57BL/6J murine genome causes sex reversal. *Science* 217, 535–537.
55. Eicher, E. M., Washburn, L. L., Schork, N. J., Lee, B. K., Shown, E. P., Xu, X., Dredge, R. D., Pringle, M. J., and Page, D. C. (1996). Sex-determining genes on mouse autosomes identified by linkage analysis of C57BL/6J-YPOS sex reversal. *Nat. Genet.* 14, 206–209.
56. Albrecht, K. H., Young, M., Washburn, L. L., and Eicher, E. M. (2003). Sry expression level and protein isoform differences play a role in abnormal testis development in C57BL/6J mice carrying certain Sry alleles. *Genetics* 164, 277–288.
57. Meeks, J. J., Crawford, S. E., Russell, T. A., Morohashi, K., Weiss, J., and Jameson, J. L. (2003). Dax1 regulates testis cord organization during gonadal differentiation. *Development* 130, 1029–1036.
58. Meeks, J. J., Weiss, J., and Jameson, J. L. (2003). Dax1 is required for testis determination. *Nat. Genet.* 34, 32–33.
59. Muscatelli, F., Strom, T. M., Walker, A. P., Zanaria, E., Recan, D., Meindl, A., Bardoni, B., Guioli, S., Zehetner, G., Rabl, W., et al. (1994). Mutations in the DAX-1 gene give rise to both X-linked adrenal hypoplasia congenita and hypogonadotropic hypogonadism. *Nature* 372, 672–676.
60. Zanaria, E., Muscatelli, F., Bardoni, B., Strom, T. M., Guioli, S., Guo, W., Lalli, E., Moser, C., Walker, A. P., McCabe, E. R., et al. (1994). An unusual member of the nuclear hormone receptor superfamily responsible for X-linked adrenal hypoplasia congenita. *Nature* 372, 635–641.
61. Zazopoulos, E., Lalli, E., Stocco, D. M., and Sassone-Corsi, P. (1997). DNA binding and transcriptional repression by DAX-1 blocks steroidogenesis. *Nature* 390, 311–315.
62. Lalli, E., Melner, M. H., Stocco, D. M., and Sassone-Corsi, P. (1998). DAX-1 blocks steroid production at multiple levels. *Endocrinology* 139, 4237–4243.
63. Ikeda, Y., Swain, A., Weber, T. J., Hentges, K. E., Zanaria, E., Lalli, E., Tamai, K. T., Sassone-Corsi, P., Lovell-Badge, R., Camerino, G., and Parker, K. L. (1996). Steroidogenic factor 1 and Dax-1 colocalize in multiple cell lineages: potential links in endocrine development. *Mol. Endocrinol.* 10, 1261–1272.
64. Luo, X., Ikeda, Y., and Parker, K. L. (1994). A cell-specific nuclear receptor is essential for adrenal and gonadal development and sexual differentiation. *Cell* 77, 481–490.
65. Sadovsky, Y., Crawford, P. A., Woodson, K. G., Polish, J. A., Clements, M. A., Tourtellotte, L. M., Simburger, K., and Milbrandt, J. (1995). Mice deficient in the orphan receptor steroidogenic factor 1 lack adrenal glands and gonads but express P450 side-chain-cleavage enzyme in the placenta and have normal embryonic serum levels of corticosteroids. *Proc. Natl. Acad. Sci. U S A* 92, 10939–10943.
66. Bardoni, B., Zanaria, E., Guioli, S., Floridia, G., Worley, K. C., Tonini, G., Ferrante, E., Chiumello, G., McCabe, E. R., Fraccaro, M., and et al. (1994). A dosage sensitive locus at chromosome Xp21 is involved in male to female sex reversal. *Nat. Genet.* 7, 497–501.
67. Swain, A., Narvaez, V., Burgoyne, P., Camerino, G., and Lovell-Badge, R. (1998). Dax1 antagonizes Sry action in mammalian sex determination. *Nature* 391, 761–767.
68. Nef, S., Verma-Kurvari, S., Merenmies, J., Vassalli, J. D., Efstratiadis, A., Accili, D., and Parada, L. F. (2003). Testis determination requires insulin receptor family function in mice. *Nature* 426, 291–295.
69. Schmahl, J., and Capel, B. (2003). Cell proliferation is necessary for the determination of male fate in the gonad. *Dev. Biol.* 258, 264–276.
70. Viger, R. S., Mertineit, C., Trasler, J. M., and Nemer, M. (1998). Transcription factor GATA-4 is expressed in a sexually dimorphic pattern during mouse gonadal development and is a potent activator of the Müllerian inhibiting substance promoter. *Development* 125, 2665–2675.
71. Robert, N. M., Tremblay, J. J., and Viger, R. S. (2002). Friend of GATA (FOG)-1 and FOG-2 differentially repress the GATA-dependent activity of multiple gonadal promoters. *Endocrinology* 143, 3963–3973.
72. Tevosian, S. G., Albrecht, K. H., Crispino, J. D., Fujiwara, Y., Eicher, E. M., and Orkin, S. H. (2002). Gonadal differentiation, sex determination and normal Sry expression in mice require direct interaction between transcription partners GATA4 and FOG2. *Development* 129, 4627–4634.
73. Katoh-Fukui, Y., Tsuchiya, R., Shiroishi, T., Nakahara, Y., Hashimoto, N., Noguchi, K., and Higashinakagawa, T. (1998). Male-to-female sex reversal in M33 mutant mice. *Nature* 393, 688–692.
74. Ion, A., Telvi, L., Chaussain, J. L., Galacteros, F., Valayer, J., Fellous, M., and McElreavey, K. (1996). A novel mutation in the putative DNA helicase XH2 is responsible for male-to-female sex reversal associated with an atypical form of the ATR-X syndrome. *Am. J. Hum. Genet.* 58, 1185–1191.
75. Pask, A., Renfree, M. B., and Marshall Graves, J. A. (2000). The human sex-reversing ATRX gene has a homologue on the marsupial Y chromosome, ATRY: implications for the evolution of mammalian sex determination. *Proc. Natl. Acad. Sci. U S A* 97, 13198–13202.
76. Kent, J., Wheatley, S. C., Andrews, J. E., Sinclair, A. H., and Koopman, P. (1996). A male-specific role for SOX9 in vertebrate sex determination. *Development* 122, 2813–2822.
77. Morais da Silva, S., Hacker, A., Harley, V., Goodfellow, P., Swain, A., and Lovell-Badge, R. (1996). Sox9 expression during gonadal development implies a conserved role for the

gene in testis differentiation in mammals and birds. *Nat. Genet.* 14, 62–68.

78. Foster, J. W., Dominguez-Steglich, M. A., Guioli, S., Kowk, G., Weller, P. A., Stevanovic, M., Weissenbach, J., Mansour, S., Young, I. D., Goodfellow, P. N., et al. (1994). Camptomelic dysplasia and autosomal sex reversal caused by mutations in an SRY-related gene. *Nature* 372, 525–530.
79. Wagner, T., Wirth, J., Meyer, J., Zabel, B., Held, M., Zimmer, J., Pasantes, J., Bricarelli, F. D., Keutel, J., Hustert, E., et al. (1994). Autosomal sex reversal and camptomelic dysplasia are caused by mutations in and around the SRY-related gene SOX9. *Cell* 79, 1111–1120.
80. Chaboissier, M. C., Kobayashi, A., Vidal, V. I., Lutzkendorf, S., van de Kant, H. J., Wegner, M., de Rooij, D. G., Behringer, R. R., and Schedl, A. (2004). Functional analysis of Sox8 and Sox9 during sex determination in the mouse. *Development* 131, 1891–1901.
81. Schepers, G., Wilson, M., Wilhelm, D., and Koopman, P. (2003). SOX8 is expressed during testis differentiation in mice and synergizes with SF1 to activate the Amh promoter in vitro. *J. Biol. Chem.* 278, 28101–28108.
82. Vidal, V. P., Chaboissier, M. C., de Rooij, D. G., and Schedl, A. (2001). Sox9 induces testis development in XX transgenic mice. *Nat. Genet.* 28, 216–217.
83. Bishop, C. E., Whitworth, D. J., Qin, Y., Agoulnik, A. I., Agoulnik, I. U., Harrison, W. R., Behringer, R. R., and Overbeek, P. A. (2000). A transgenic insertion upstream of sox9 is associated with dominant XX sex reversal in the mouse. *Nat. Genet.* 26, 490–494.
84. Huang, B., Wang, S., Ning, Y., Lamb, A. N., and Bartley, J. (1999). Autosomal XX sex reversal caused by duplication of SOX9. *Am. J. Med. Genet.* 87, 349–353.
85. Pfeifer, D., Kist, R., Dewar, K., Devon, K., Lander, E. S., Birren, B., Korniszewski, L., Back, E., and Scherer, G. (1999). Camptomelic dysplasia translocation breakpoints are scattered over 1 Mb proximal to SOX9: evidence for an extended control region. *Am. J. Hum. Genet.* 65, 111–124.
86. Spotila, L. D., Spotila, J. R., and Hall, S. E. (1998). Sequence and expression analysis of WT1 and Sox9 in the red-eared slider turtle, *Trachemys scripta*. *J. Exp. Zool.* 281, 417–427.
87. Moreno-Mendoza, N., Harley, V. R., and Merchant-Larios, H. (1999). Differential expression of SOX9 in gonads of the sea turtle *Lepidochelys olivacea* at male- or female-promoting temperatures. *J. Exp. Zool.* 284, 705–710.
88. Western, P. S., Harry, J. L., Graves, J. A., and Sinclair, A. H. (1999). Temperature-dependent sex determination: upregulation of SOX9 expression after commitment to male development. *Dev. Dyn.* 214, 171–177.
89. Wilkins, A. S. (1995). Moving up the hierarchy: a hypothesis on the evolution of a genetic sex determination pathway. *Bioessays* 17, 71–77.
90. Raymond, C. S., Shamu, C. E., Shen, M. M., Seifert, K. J., Hirsch, B., Hodgkin, J., and Zarkower, D. (1998). Evidence for evolutionary conservation of sex-determining genes. *Nature* 391, 691–695.
91. Raymond, C. S., Parker, E. D., Kettlewell, J. R., Brown, L. G., Page, D. C., Kusz, K., Jaruzelska, J., Reinberg, Y., Flejter, W. L., Bardwell, V. J., Hirsch, B., and Zarkower, D. (1999). A region of human chromosome 9p required for testis development contains two genes related to known sexual regulators. *Hum. Mol. Genet.* 8, 989–996.
92. Ottolenghi, C., Veitia, R., Quintana-Murci, L., Torchard, D., Scapoli, L., Souleyreau-Therville, N., Beckmann, J., Fellous, M., and McElreavey, K. (2000). The region on 9p associated with 46,XY sex reversal contains several transcripts expressed in the urogenital system and a novel doublesex-related domain. *Genomics* 64, 170–178.
93. Raymond, C. S., Murphy, M. W., O'Sullivan, M. G., Bardwell, V. J., and Zarkower, D. (2000). Dmrt1, a gene related to worm and fly sexual regulators, is required for mammalian testis differentiation. *Genes Dev.* 14, 2587–2595.
94. Matsuda, M., Nagahama, Y., Shinomiya, A., Sato, T., Matsuda, C., Kobayashi, T., Morrey, C. E., Shibata, N., Asakawa, S., Shimizu, N., Hori, H., Hamaguchi, S., and Sakaizumi, M. (2002). DMY is a Y-specific DM-domain gene required for male development in the medaka fish. *Nature* 417, 559–563.
95. Kondo, M., Nanda, I., Hornung, U., Asakawa, S., Shimizu, N., Mitani, H., Schmid, M., Shima, A., and Schartl, M. (2003). Absence of the candidate male sex-determining gene dmrt1b(Y) of medaka from other fish species. *Curr. Biol.* 13, 416–420.
96. Shen, W. H., Moore, C. C., Ikeda, Y., Parker, K. L., and Ingraham, H. A. (1994). Nuclear receptor steroidogenic factor 1 regulates the müllerian inhibiting substance gene: a link to the sex determination cascade. *Cell* 77, 651–661.
97. De Santa Barbara, P., Bonneaud, N., Boizet, B., Desclozeaux, M., Moniot, B., Sudbeck, P., Scherer, G., Poulat, F., and Berta, P. (1998). Direct interaction of SRY-related protein SOX9 and steroidogenic factor 1 regulates transcription of the human anti-Müllerian hormone gene. *Mol. Cell. Biol.* 18, 6653–6665.
98. Nachtigal, M. W., Hirokawa, Y., Enyeart-VanHouten, D. L., Flanagan, J. N., Hammer, G. D., and Ingraham, H. A. (1998). Wilms' tumor 1 and Dax-1 modulate the orphan nuclear receptor SF-1 in sex-specific gene expression. *Cell* 93, 445–454.
99. Arango, N. A., Lovell-Badge, R., and Behringer, R. R. (1999). Targeted mutagenesis of the endogenous mouse Mis gene promoter: in vivo definition of genetic pathways of vertebrate sexual development. *Cell* 99, 409–419.
100. Tremblay, J. J., and Viger, R. S. (1999). Transcription factor GATA-4 enhances Müllerian inhibiting substance gene transcription through a direct interaction with the nuclear receptor SF-1. *Mol. Endocrinol.* 13, 1388–1401.
101. Karl, J., and Capel, B. (1998). Sertoli cells of the mouse testis originate from the coelomic epithelium. *Dev. Biol.* 203, 323–333.
102. Schmahl, J., Eicher, E. M., Washburn, L. L., and Capel, B. (2000). Sry induces cell proliferation in the mouse gonad. *Development* 127, 65–73.
103. Schmahl, J., Kim, Y., Colvin, J. S., Ornitz, D. M., and Capel, B. (2004). Fgf9 induces proliferation and nuclear localization of FGFR2 in Sertoli precursors during male sex determination. *Development* 131, 3627–3636.
104. Colvin, J. S., Green, R. P., Schmahl, J., Capel, B., and Ornitz, D. M. (2001). Male-to-female sex reversal in mice lacking fibroblast growth factor 9. *Cell* 104, 875–889.
105. Brennan, J., Tilmann, C., and Capel, B. (2003). Pdgfr-alpha mediates testis cord organization and fetal Leydig cell development in the XY gonad. *Genes Dev.* 17, 800–810.
106. Buehr, M., Gu, S., and McLaren, A. (1993). Mesonephric contribution to testis differentiation in the fetal mouse. [Published erratum appears in *Development* 1993 Aug; 118(4): following 1384.] *Development* 117, 273–281.
107. Martineau, J., Nordqvist, K., Tilmann, C., Lovell-Badge, R., and Capel, B. (1997). Male-specific cell migration into the developing gonad. *Curr. Biol.* 7, 958–968.
108. Tilmann, C., and Capel, B. (1999). Mesonephric cell migration induces testis cord formation and Sertoli cell differentiation in the mammalian gonad. *Development* 126, 2883–2890.
109. Brennan, J., Karl, J., and Capel, B. (2002). Divergent vascular mechanisms downstream of Sry establish the arterial system in the XY gonad. *Dev. Biol.* 244, 418–428.

110. Jeays-Ward, K., Hoyle, C., Brennan, J., Dandonneau, M., Alldus, G., Capel, B., and Swain, A. (2003). Endothelial and steroidogenic cell migration are regulated by WNT4 in the developing mammalian gonad. *Development* 130, 3663–3670.
111. Yao, H. H., Matzuk, M. M., Jorgez, C. J., Menke, D. B., Page, D. C., Swain, A., and Capel, B. (2004). Follistatin operates downstream of Wnt4 in mammalian ovary organogenesis. *Dev. Dyn.* 230, 210–215.
112. Hatano, O., Takakusu, A., Nomura, M., and Morohashi, K. (1996). Identical origin of adrenal cortex and gonad revealed by expression profiles of Ad4BP/SF-1. *Genes Cells* 1, 663–671.
113. Yao, H. H., Whoriskey, W., and Capel, B. (2002). Desert Hedgehog/Patched 1 signaling specifies fetal Leydig cell fate in testis organogenesis. *Genes Dev.* 16, 1433–1440.
114. Clark, A. M., Garland, K. K., and Russell, L. D. (2000). Desert hedgehog (Dhh) gene is required in the mouse testis for formation of adult-type Leydig cells and normal development of peritubular cells and seminiferous tubules. *Biol. Reprod.* 63, 1825–1838.
115. Canto, P., Soderlund, D., Reyes, E., and Mendez, J. P. (2004). Mutations in the desert hedgehog (DHH) gene in patients with 46,XY complete pure gonadal dysgenesis. *J. Clin. Endocrinol. Metab.* 89, 4480–4483.
116. Kitamura, K., Yanazawa, M., Sugiyama, N., Miura, H., Iizuka-Kogo, A., Kusaka, M., Omichi, K., Suzuki, R., Kato-Fukui, Y., Kamiirisa, K., Matsuo, M., Kamijo, S., Kasahara, M., Yoshioka, H., Ogata, T., Fukuda, T., Kondo, I., Kato, M., Dobyns, W. B., Yokoyama, M., and Morohashi, K. (2002). Mutation of ARX causes abnormal development of forebrain and testes in mice and X-linked lissencephaly with abnormal genitalia in humans. *Nat. Genet.* 32, 359–369.
117. Morohashi, K., Zanger, U. M., Honda, S., Hara, M., Waterman, M. R., and Omura, T. (1993). Activation of CYP11A and CYP11B gene promoters by the steroidogenic cell-specific transcription factor, Ad4BP. *Mol. Endocrinol.* 7, 1196–1204.
118. McLaren, A. (1995). Germ cells and germ cell sex. *Philos. Trans. R. Soc. Lond. B. Biol. Sci.* 350, 229–233.
119. McLaren, A., and Southee, D. (1997). Entry of mouse embryonic germ cells into meiosis. *Dev. Biol.* 187, 107–113.
120. Jost, A. (1953). Problems in fetal endocrinology; the gonadal and hypophtseal hormones. *Rec. Prog. Horm. Res.* 8, 379–418.
121. Nef, S., and Parada, L. F. (1999). Cryptorchidism in mice mutant for Insl3. *Nat. Genet.* 22, 295–299.
122. Zimmermann, S., Steding, G., Emmen, J. M., Brinkmann, A. O., Nayernia, K., Holstein, A. F., Engel, W., and Adham, I. M. (1999). Targeted disruption of the Insl3 gene causes bilateral cryptorchidism. *Mol. Endocrinol.* 13, 681–691.
123. Shaw, G., Renfree, M. B., Leihy, M. W., Shackleton, C. H., Roitman, E., and Wilson, J. D. (2000). Prostate formation in a marsupial is mediated by the testicular androgen 5 alpha-androstane-3 alpha,17 beta-diol. *Proc. Natl. Acad. Sci. U S A* 97, 12256–12259.
124. Deslypere, J. P., Young, M., Wilson, J. D., and McPhaul, M. J. (1992). Testosterone and 5 alpha-dihydrotestosterone interact differently with the androgen receptor to enhance transcription of the MMTV-CAT reporter gene. *Mol. Cell. Endocrinol.* 88, 15–22.
125. Griffin J. E, W. J. (2002). Disorders of the testes and the male reproductive tract. In *Williams Textbook of Endocrinology* (K. H. Larsen, S. Melmed, K. S. Polonsky KS, Eds.), Vol. 1, pp. 709–769. WB Saunders, Philadelphia.
126. Mahendroo, M. S., Cala, K. M., Hess, D. L., and Russell, D. W. (2001). Unexpected virilization in male mice lacking steroid 5 alpha-reductase enzymes. *Endocrinology* 142, 4652–4662.
127. Couse, J. F., Hewitt, S. C., Bunch, D. O., Sar, M., Walker, V. R., Davis, B. J., and Korach, K. S. (1999). Postnatal sex reversal of the ovaries in mice lacking estrogen receptors alpha and beta. *Science* 286, 2328–2331.
128. Breedlove, S. M., Jacobson, C. D., Gorski, R. A., and Arnold, A. P. (1982). Masculinization of the female rat spinal cord following a single neonatal injection of testosterone propionate but not estradiol benzoate. *Brain Res.* 237, 173–181.
129. Dohler, K. D., Coquelin, A., Davis, F., Hines, M., Shryne, J. E., and Gorski, R. A. (1984). Pre- and postnatal influence of testosterone propionate and diethylstilbestrol on differentiation of the sexually dimorphic nucleus of the preoptic area in male and female rats. *Brain Res.* 302, 291–295.
130. Simerly, R. B., Swanson, L. W., Handa, R. J., and Gorski, R. A. (1985). Influence of perinatal androgen on the sexually dimorphic distribution of tyrosine hydroxylase-immunoreactive cells and fibers in the anteroventral periventricular nucleus of the rat. *Neuroendocrinology* 40, 501–510.
131. Phoenix, C. H., Goy, R. W., Gerall, A. A., and Young, W. C. (1959). Organizing action of prenatally administered testosterone propionate on the tissues mediating mating behavior in the female guinea pig. *Endocrinology* 65, 369–382.
132. De Vries, G. J., Rissman, E. F., Simerly, R. B., Yang, L. Y., Scordalakes, E. M., Auger, C. J., Swain, A., Lovell-Badge, R., Burgoyne, P. S., and Arnold, A. P. (2002). A model system for study of sex chromosome effects on sexually dimorphic neural and behavioral traits. *J. Neurosci.* 22, 9005–9014.
133. Carruth, L. L., Reisert, I., and Arnold, A. P. (2002). Sex chromosome genes directly affect brain sexual differentiation. *Nat. Neurosci.* 5, 933–934.
134. Brennan, J., and Capel, B. (2004). One tissue, two fates: molecular genetic events that underlie testis versus ovary development. *Nat. Rev. Genet.* 5, 509–521.
135. Karl, J., and Capel, B. (1995). Three-dimensional structure of the developing mouse genital ridge. *Philos. Trans. R. Soc. Lond. B. Biol. Sci.* 350, 235–242.

Knobil and Neill's Physiology of Reproduction, Third Edition
edited by Jimmy D. Neill,
Elsevier © 2006

CHAPTER 7

Preimplantation Embryogenesis

Kathleen H. Burns[1] and Martin M. Matzuk[2]

Introduction, 261
Transgenic Mouse Technology, 263
Early Postfertilization Development and Maternal Effect Gene Products, 264
Maternal Effect Genes, 265
How Many Early Acting Maternal Effect Genes Exist?, 267
Translational Control of Maternal mRNAs, 267
Zygote Genome Activation, 268
Specific Products of Early Transcription, 270
Specific Transcription Factors, 271
Chromatin Remodeling and Epigenetics in Early Embryos, 272
Controlling ZGA: The Pandora Problem, 272
Genomic Methylation: Rewriting and Its Risks, 273
X-Chromosome Dosage Compensation, 276
The Destruction of Creation, 278
Degradation of RNA, 278
Ubiquitination of Proteins and Organelles, 279
Cellular Apoptosis, 280
Cell Cycle Control in Early Mitoses, 282
Cell Cycle Progression in One-Cell and Two-Cell Embryos, 282
Regulators of Later Cell Divisions, 284
Oviductal Factors and Extracellular Signals Influencing Early Development, 286
The Oviductal Microenvironment, 286
Extracellular Signaling Pathways in Early Development, 288
Morphological and Metabolic Changes in Early Cleavage Embryos, 288
Morula Compaction, 290
Glucose Import and the Relevance of Maternal Hyperglycemia and Hyperinsulinemia, 291
TE Sidedness and Blastocoel Development, 292
Preimplantation Polarity and Cell Fate Determination, 293
Polarity in Mammalian Embryos, 293
Totipotency Within the Embryo and Cell Fate Determination, 294
Ethics and the Embryo, 296
Acknowledgments, 297
References, 297

Who makes seed grow in women,
Who creates people from sperm,
Who feeds the son in his mother's womb,
Who soothes him to still his tears,
Who gives breath that it may vivify every part of his being.

THE GREAT HYMN TO ATEN, 1500 B.C.

INTRODUCTION

The beginnings of an individual's life have held intrigue throughout the ages. As its processes have been reconsidered in different times, perceptions of early development have taken shape in intimate interaction with societal and scientific convictions and controversies. Moral and religious positions delineating the sanctity of human life give context to current ethical debates surrounding embryo research. This reflects an innate reverence for early development that can be appreciated from the most ancient references to monotheism—as enumerated by the Egyptians, who believed that conception and gestation were conducted by divine presence of the sun god, Aten (Fig. 1).

Many scholars date the formal study of embryology and attempts to demystify development to Aristotle's treatise, *De Generatione Animalium*, important in part because of its comparative premise,

[1]Department of Pathology, The Johns Hopkins Hospital, Baltimore, Maryland; [2]Department of Pathology, Baylor College of Medicine, Houston, Texas.

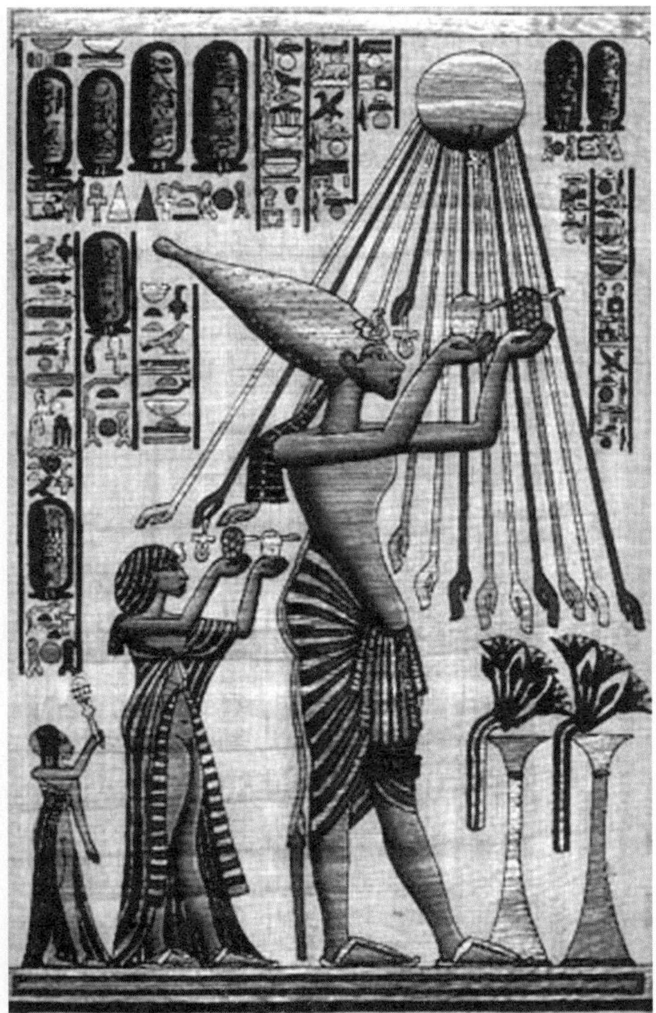

FIG. 1. An ancient Egyptian hieroglyphic of the Armana style depicting communication between the Pharaoh and the sun god, Aten. The image and the Great Hymn to Aten are among the works accomplished during the rule of Amenhotep IV, Akhenaten. The Hymn is inscribed at the Tomb of Aye at Amarna.

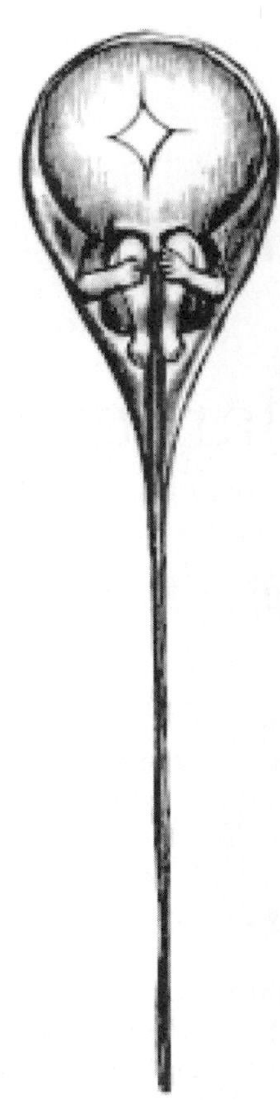

FIG. 2. The Hartsoeker illustration of an animalcule in the head of a spermatozoon. The sketch, made in 1694, is believed to be based on descriptions of Leeuwenhoek. The term *homunculus* was attached to such depictions retrospectively.

which maintained that observations made in several species could be used to infer universal patterns of development. During his lifetime, the question of how body structures came about was posed with two distinct answers later emerging—parties of the long-standing "epigenesis verses preformation" debate. Aristotle and Galen's early writings indicate that structures result de novo from "a shaping or formative process" acting on initially featureless antecedents. This concept of an invisible shaping force, or *vis essentialis* as it was termed by Caspar Wolff in 1759, was carried forward by epigenesis schools, whereas others contended that gestation represented growth of miniscule preexisting and wholly formed embryos until the close of the 18th century. Perhaps predictably, preformation schools were themselves divided into ovists and spermists, based on whether the small fetus or "homunculus" was thought to reside in the female egg or the head of the male sperm (Fig. 2).

The point was seriously deliberated not only for the promise of understanding, but also became emotionally charged with implications for biblical depictions of and social distinctions between the sexes (1). By the 1890s, the more influential epigenesis schools developed new ideas shaped by Charles Darwin's evolution concept, as articulated in Ernst Haeckel's proposal that "ontogeny recapitulates phylogeny." According to this historic concept, embryos progress through stages resembling the adult forms of more primitive species—each embryo's development thus ascended the *scala naturae*, recreating an accelerated evolution of its species. The earliest preimplantation embryos would be to these observers reminiscent of the first life forms on earth. Perhaps it would be the last most famous theory of embryology primarily based on descriptive works, which were soon to be complemented and challenged by forays into active experimentation. By the turn of the century, William Roux had pioneered such approaches with cleaving frog embryos, studying effects of cellular ablation, and Walter Heape had published the first experiments in mammalian systems, recovering preimplantation rabbit embryos and transferring them for development in foster mothers (2).

The field of early embryology today is being driven by advances in molecular biology, and we are heirs of the enormous challenge to decipher the molecular means of differentiation. How does total developmental potential arise, and what subsequently influences cellular "choices" at junctions of fate-determining algorithms? Mature gametes are highly specialized, postmitotic, haploid cells that have completed (spermatozoa) or have nearly completed (metaphase II oocytes) meiosis. After fertilization and cleavage to the two-cell embryo, the resultant cells (blastomeres) have assumed totipotency and remain totipotent through the morula stage. With the formation of a blastocoel (embryonic day 3.5 [E3.5] in the mouse), the embryo is partitioned into trophectoderm (TE) and inner cell mass (ICM). Cells of the ICM, from which embryonic stem (ES) cells are derived, are considered pluripotent because they can contribute to all embryonic tissues. During implantation and shortly thereafter, cell lineages become exponentially more numerous and pluripotency is lost, with the potential exception of lingering stem cells now being sought and studied with great excitement. During the years since the writing of the previous edition of *The Physiology of Reproduction*, we have enhanced our appreciation for factors crucial to this window of development. In this chapter, we review recent advances in our understanding of mammalian life between fertilization and implantation. We highlight insights gained from studies of transgenic mouse models, themselves a fruit of our expanding ability to manipulate the processes of early development.

TRANSGENIC MOUSE TECHNOLOGY

Transgenic mouse technology refers to the incorporation of engineered DNA into the heritable mouse genome. The genetic alteration can be accomplished through microinjection of a transgene into the male pronucleus of one-cell embryos, viral infection of early mouse embryos or ES cells, or targeted mutagenesis of ES cells. Oocytes microinjected with DNA are expected to incorporate concatamers of the engineered transgene and develop into founder mice with the potential to both pass on the transgene and express the gene product encoded. These F_0 mice and their mutant F_1 progeny are known as hemizygous mice because the transgene integration creates a new allele within the genome, random in its location and effects on any neighboring genes (Fig. 3). In contrast, mutagenized ES cells can be made to harbor targeted mutations; researchers can deliberately replace a portion of the genome with a "look-alike" construct and selection for nonrandom integration. The resulting heterozygote ES cells are maintained in pluripotent states in vitro and may be "reintroduced" into blastocyst-stage embryos by injection into the

Day 1: PMSG inject FVB strain oocyte donors

Day 3: hCG injection and mating
pseudopregnant foster mothers mated with vasectomized males

Day 4: oocyte collection from the oviducts of donor females and DNA
pronuclear microinjection
oocyte transfer to
pseudopregnant foster mothers

Day 23: F_0 founder generation is born

Day 37: tail DNA is recovered for Southern blot or PCR analysis

Day 65: each founder is crossed to wild-type mice to begin a unique transgenic
line 50% of the F_1 generation is expected to inherit an autosomal transgene

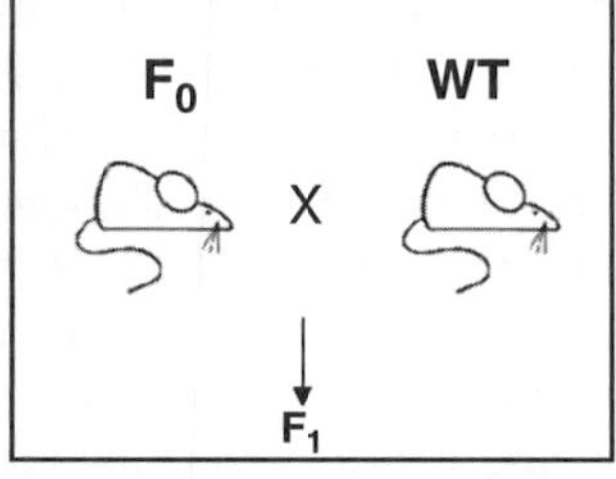

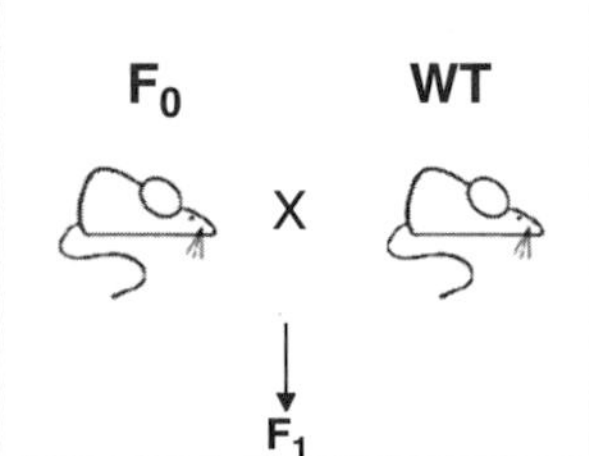

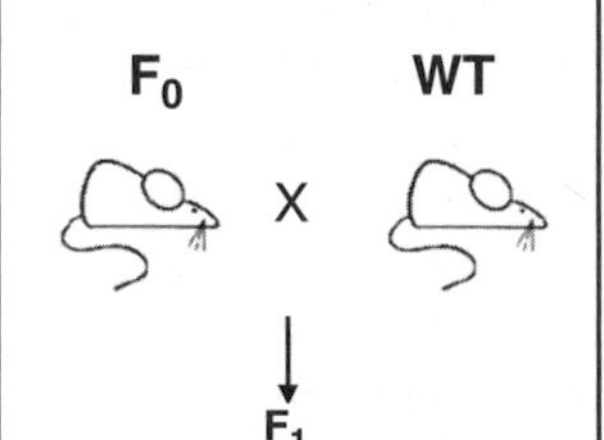

FIG. 3. Generation of transgenic mice by pronuclear microinjection. A transgene construct is injected into the pronucleus of a one-cell zygote and incorporates at a random site in the genome. The resulting mice become founders of the transgenic lines. It is expected that there is one integration site with each injection, though different founders will each have a unique site of insertion and unique numbers of transgenes that have concatamerized during the integration. Thus, different levels of transgene expression may be studied in different lines.

Day 1: mate C57 strain females for blastocyst donation

Day 2: check vaginal plugs to verify copulation
mate pseudopregnant F_1 foster mothers with vasectomized males

Day 3: check vaginal plugs to verify copulation

Day 5: collect E3.5 blastocysts from the uteri of donor females and inject 129 strain ES cells
transfer embryos to pseudopregnant foster mothers

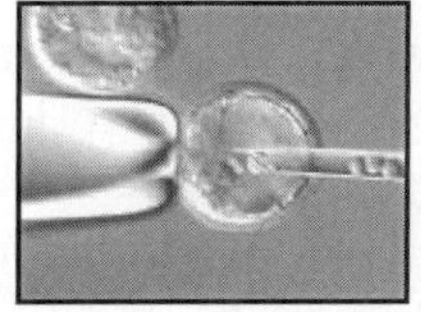

Day 22: C57/129 strain chimeras are born

Day 30: percent chimerism is estimated based on coat color mosaicism

Day 64: chimeras are mated to C57 strain (+/+) mice to generate heterozygotes
50% of mice inheriting the 129 strain agouti coat color will be heterozygotes (+/–) for an autosomal mutation
mating heterozygotes will yield homozygotes (–/–) at a 1:3 Mendelian ratio
if these mice are viable

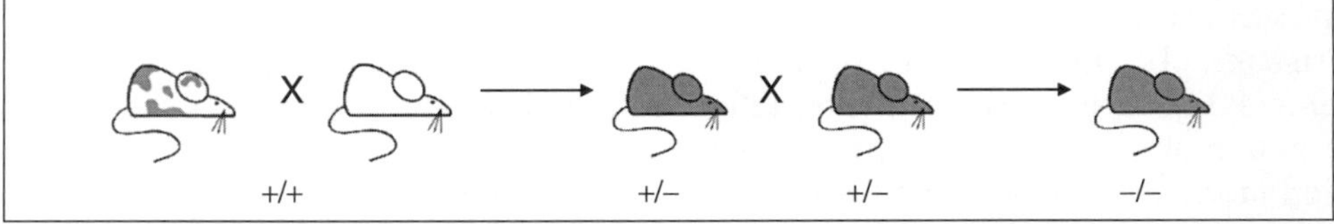

FIG. 4. Generation of targeted mutations in mice. Gene targeting is carried out and selected for in vitro in ES cells from one strain of mice (129Sv/Ev). Targeted clones are expanded and injected into the blastocoel of blastocysts from another strain (C57BL6). The targeted cells incorporate into the ICM and epiblast and contribute to lineages in the resulting chimeras. The extent to which the targeted cells contribute can be roughly assessed by noting the C57/129 coat color mosaicism. High percentage chimeras with germline contribution of the targeted cells have the potential to pass on the altered allele. Breeding the resulting heterozygote mice allows for disruption of both alleles of an autosomal locus.

blastocoel to give rise to chimeric mice, which represent a mixture of cell lineages derived from the manipulated ES cells and the original ICM. When the ES cell line contributes to gamete development in the resulting chimeras, this allows the establishment of mutant lines for use as research models [(3,4); reviewed in (5,6)] (Fig. 4).

Homozygosity for an autosomal mutant locus can be achieved by breeding heterozygote mice. When the goal of the experiment is to completely abrogate gene expression, the homozygote null, or "knockout," mice display in vivo effects of gene loss-of-function. These knockout models have led to major conceptual breakthroughs in many fields; several examples relevant to preimplantation development are described herein.

EARLY POSTFERTILIZATION DEVELOPMENT AND MATERNAL EFFECT GENE PRODUCTS

Transcription is silenced during gamete formation in both sexes. In the male, this occurs in spermiogenesis with the compact packaging of DNA and entails the exchange of histones for protamine proteins. In the female, the transition to a transcriptionally quiescent stage is associated with the assumption of a surrounded nucleolus (SN) DNA configuration in oocytes of multilayered ovarian follicles (7). Until reorganization of a diploid nucleus is completed during early cleavage stages, the zygote genome remains largely inactive. This varies between the late two-cell stage (E1.5) in mice to the four- and eight-cell stages in humans. Thus, fertilization and the earliest events of embryogenesis do not rely on de novo transcription.

Although very recent work indicates that sperm may contribute some functional messenger RNAs (mRNAs) during fertilization (8), their nongenomic contributions are regarded as relatively minor. Thus, many RNAs proteins, and small molecules needed for this developmental window are stored in the oocyte, having been encoded for by maternal effect genes or synthesized by oocyte-expressed proteins. Such effects of the maternal genotype can be appreciated in vitro; under standard culture conditions, embryos from some strains of mice cannot develop beyond the two-cell stage, and this "two-cell block" is determined by the maternal, but not paternal, strain (9).

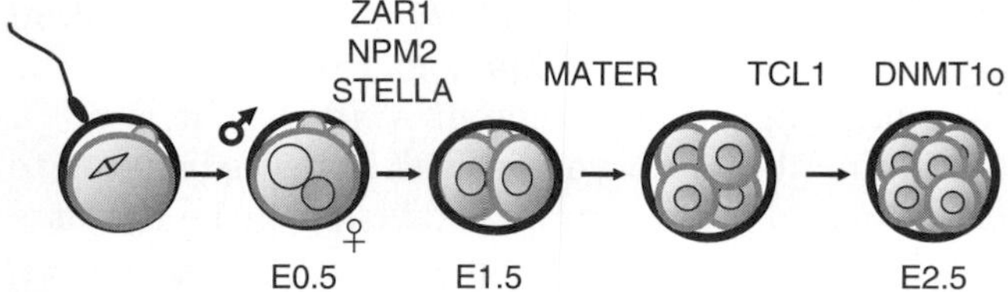

FIG. 5. Maternal effect gene products described in mice. Early embryonic development from fertilization to embryonic day 2.5 (E2.5) is depicted. Maternal effect proteins essential for each developmental transition are noted. This schematic is based on the phenotype of embryos developing in females that are null for the respective maternal effect alleles.

In vivo, zygotes can be lost at early stages when essential maternal effect genes are mutated in females (Fig. 5), resulting in female infertility or subfertility in homozygous mice, as we consider below.

It should be noted that oocytes remain functionally diploid (like somatic counterparts) throughout oogenesis, and so in the absence of bona fide haploinsufficiency, such loss-of-function effects are appreciated only in homozygous females (where each oocyte is affected), and heterozygotes are generally completely unaffected. We also consider briefly reports of mutations in maternal effect genes that have deleterious effects in embryo development after implantation when these are believed to be the long-term sequelae of a lack of the gene product in earlier development. The circumstance may also arise when insults are exclusively maternally influenced because of functional or actual hemizygosity—for example, imprinted silencing of the paternal allele or X-linked traits considered in male offspring are exclusively maternally influenced.

Maternal Effect Genes

Our laboratory recently described two early-acting maternal effect genes (10,11); both were initially recovered from a complementary DNA (cDNA) library enriched in oocyte-specific genes by subtractive hybridization (12). The first of these, zygote arrest 1 (*Zar1*), encodes an atypical plant homeodomain motif-containing protein essential for embryo development to the two-cell stage after fertilization (10). Although presence of this motif suggests that ZAR1 may function to regulate transcription or organize chromatin, the protein appears to be predominantly cytoplasmic throughout oogenesis and is not enriched in the nuclei of early zygotes. Nevertheless, it is crucial for development, and a *Zar1* null mutation results in complete infertility in female mice. *Zar1* knockout females have apparently normal ovarian follicle development and ovulate oocytes that are competent to undergo meiosis, fertilization, and pronuclear formation. Zygotes initiate genome activation but are lost shortly thereafter, so that only fragmenting remnants can be recovered on E1.5. The second gene, *Npm2*, encodes the mammalian nucleoplasmin 2 ortholog, which we first expected to be involved in decondensing the sperm head to form the male pronucleus, as suggested by in vitro studies with the *Xenopus laevis* protein (13–15). Mammalian NPM2 is expressed exclusively in oocytes and is a nuclear protein during oocyte growth and in early cleavage stage embryos through the blastocyst stage, consistent with this model. Targeted deletion of *Npm2* instead proved its importance in heterochromatin formation and nucleolar organization in both unfertilized oocyte nuclei and the pronuclei of fertilized eggs (11) (Fig. 6). NPM2 is not needed for pronuclear formation but is required for efficient progression through mitosis to the two-cell stage, and knockout females are markedly subfertile. Orthologs of both *Zar1* and *Npm2* have been described in a number of species, from invertebrates to humans, and sequence analyses and expression data suggest functional conservation (10,16,17). It will be interesting to pursue their mechanisms of action and how their roles may be modified in different organisms.

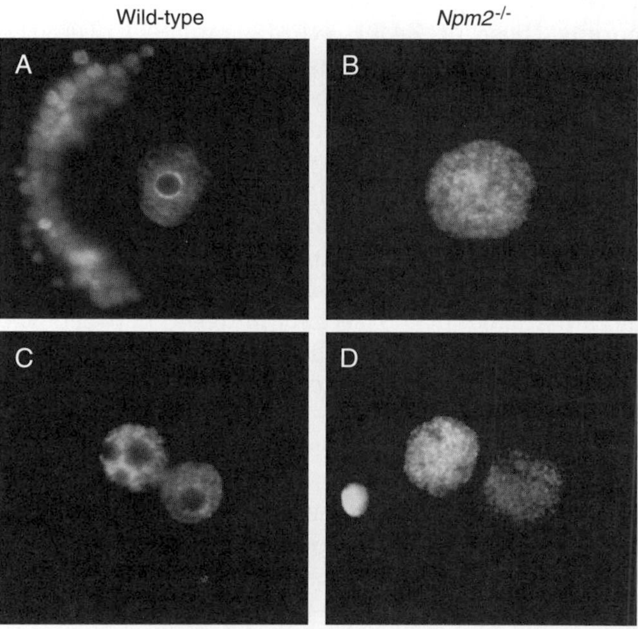

FIG. 6. The phenotype of NPM2-deficient GV oocytes and one-cell zygotes. Images are DAPI stained GV oocytes (**A, B**) and pronuclear stage one-cell zygotes (**C, D**) visualized by deconvolution microscopy. In the oocytes and zygotes of $Npm2^{-/-}$ females (**B, D**), there is no organization of the DNA around coalesced nucleolar structures, as compared with wild-type (**A, C**). Diminished quantities of hypoacetylated histone H3-marked heterochromatin were appreciable in the NPM2 deficient zygotes.

The parallel discoveries of *Zar1* and *Npm2* underscore that oocytes contain transcripts and proteins that are dispensable for oocyte viability, ovarian follicular development, and fertilization but are crucial to early embryos. This is reinforced by the recent description of *Stella* (also known as primordial germ cell 7 [*Pgc7*] and developmental pluripotency-associated 3 [*Dppa3*]), a maternal effect gene transcriptionally activated during germ cell specification in embryonic life that is also expressed in pluripotent cell lines and germ cell tumors (18,19). STELLA encodes a nuclear protein with SAP-DNA binding and splicing factor domains; the protein is expressed in oocytes and early embryos to the blastocyst stage. *Stella* knockout females exhibit infertility or subfertility and embryo loss during early development, with compromise evident at the transition to the two-cell stage (19). Unlike *Zar1* and *Npm2* knockout models, the fertility defect in these *Stella* null females is not exclusively a maternal effect. *Stella* null females mated to knockout males are completely infertile, whereas mating with *Stella* wild-type males demonstrates a partial rescue of fertility and production of STELLA protein from paternal transcription in the few surviving E3.5 embryos. Thus, though there is an absolute requirement for STELLA during early development, there seems to be a subset of embryos surviving zygote genome activation (ZGA) in the absence of maternal STELLA that gain a window for rescue by the paternal allele.

With essential functions closely following temporally those of ZAR1, NPM2, and STELLA, the MATER protein, *m*aternal *a*ntigen *t*hat *e*mbryos *r*equire (encoded by the *Nalp5* locus, *NA*CHT *l*eucine rich repeat and *P*YD-containing *5*), was the first oocyte-specific maternal effect product identified. MATER was originally discovered and named as a protein implicated in a mouse model of autoimmune oophoritis. MATER knockout mice (*Nalp5*$^{-/-}$) exhibit female infertility because of loss of MATER-deficient embryos around the second mitoses, between the two-cell and four-cell stages (20). The MATER protein contains a region of homology shared with ribonuclease inhibitors, and it is speculated that it may be involved in controlling a wave of RNA degradation that attends early development (21).

Although defects of the null embryos described above seem to largely precede or compromise robust transcriptional activation of the zygote genome, not all maternal effect gene mutations result in a phenotype so early in development. For example, T-cell leukemia/lymphoma 1 (*Tcl1*) heterozygote and knockout females are subfertile due to loss of zygotes beginning around the four- to eight-cell transition, the phenotype being more pronounced in *Tcl1*$^{-/-}$ mothers (22). Embryos from *Tcl1*$^{-/-}$ females survive activation of the zygote genome, and there is some evidence that the paternal *Tcl1* genetic contribution affects the number of blastomeres present at the time of compaction, although ultimately it is not a determinant of embryo survival. TCL1 is believed to function as a promoter of cell division, given that it is overexpressed in some hematological malignancies, and *Tcl1* deficient embryos have diminished numbers of blastomeres. Interestingly, TCL1 shuttles between the cell surface and nucleus of early embryos, being cortical in mitosis and G_1 phases and nuclear during S and G_2 (22).

Subcellular localization of another maternal effect protein, an oocyte-specific form of DNA methyltransferase 1, termed DNMT1o, provided a critical clue as to the timing of its function. DNMT1o is sequestered in the cytoplasm (despite having a nuclear localization sequence) during most stages of preimplantation development, the exception being the eight-cell stage morula, when the protein is transiently nuclear. At this time, DNMT1o seems to allow imprinted loci to retain their methylation patterns even in the context of the widespread demethylation occurring in cleavage stage embryos. Null mutations in DNMT1o result in late gestational demise in pups of homozygous females associated with loss of allele-specific methylation patterns at imprinted loci (23). Similarly, a maternal effect gene, encoding heat shock transcription factor 1 (HSF1), is important for gene regulation during early development and impacts viability at later time points. Deletion of maternal *Hsf1* results in embryo loss during both preimplantation and postimplantation development, and homozygous null females are infertile as a result (24).

Finally, a group of maternal effect factors critical in maintaining genomic integrity has been discovered. This includes two chromosome organizers involved in meiotic metaphases, formin 2 (FMN2) and synaptonemal complex protein (SCP3), and a DNA mismatch repair protein, postmeiotic segregation increased 2 (PMS2). *Fmn2* and *Scp3* knockout females have defects in chromosome positioning and segregation during meiosis and exhibit zygote loss after fertilization due to polyploid and aneuploid embryo formation (25,26). Females that are homozygous for *Pms2* deletion have pups that exhibit microsatellite instability. These acquired mutations are not only limited to the maternal genome where insults may have arisen during oogenesis, but also occur with mosaicism in the paternal genome, suggesting that maternal PMS2 normally prevents mutations during the first rounds of replication in early embryos (27). Thus, although appreciable numbers of *Tcl1*, *Dnmt1o*, *Hsf1*, *Fmn2*, *Scp3*, and *Pms2* deficient embryos survive ZGA, nothing fully compensates for the loss of the maternal product in regulating gene expression or maintaining genomic integrity early in development.

How Many Early Acting Maternal Effect Genes Exist?

It is estimated that there are between 35,000 and 40,000 protein-encoding gene loci in mammalian genomes, many capable of producing alternative transcripts. There appear to be over 10,000 different mRNA species expressed in oocytes and early embryos. What portion of these are functional and represent maternal effect genes, as opposed to those with their principal roles limited to oogenesis, remains a point of speculation. As we describe below, a portion of maternal effect mRNAs are marked with a signature sequence that delays their translation, and this may prove predictive of their importance during embryogenesis. In contrast, ZAR1, NPM2, and MATER proteins are all present in growing oocytes after ovarian follicle recruitment, and that their critical roles would manifest in embryo development could not be predicted solely on the basis of expression. Many candidate maternal effect gene products are widely expressed and may be essential for murine viability, limiting the utility of traditional knockout technologies. Even for genes exclusively expressed in oocytes, we may find ourselves challenged to bypass compromises in oocyte viability or ovarian folliculogenesis that arise from null mutations to perform loss-of-function experiments specifically in embryos.

Several alternative approaches hold promise in identifying maternal effect genes should a traditional null mutation result in lethality in homozygotes preceding sexual maturity. First, if null females survive to 3 weeks of age, they may be treated with exogenous gonadotropins to recover oocytes or embryos (depending on the feasibility of in vivo fertilization). In vitro conditions are established to allow studies of oocyte maturation, fertilization, and development to the blastocyst stage; implantation and gestation may be followed thereafter by embryo transfer to pseudopregnant foster mothers (28). Second, if null females die in the perinatal period, their ovaries can be transplanted to ovarian bursas of immunocompatible hosts. This approach readily enables evaluation of fertility and embryogenesis (29). Similarly, survival to the perinatal stage would permit oocyte/somatic cell exchange experiments if the gene of interest seemed to be important in ovarian somatic cells (30). Third, tissue-specific gene deletions can be performed to generate oocytes null for a gene while maintaining the functional allele in other cell types. This conditional knockout system is bipartite, comprised of a "floxed" allele in which critical sequence is flanked by *loxP* recombination sites and a transgene directing cell-specific expression of Cre recombinase. A *Zp3* (zona pellucida 3) promoter-driven Cre transgene (31,32) has been used for oocyte-specific deletion of genes such as phosphatidylinositol glycan A (*Piga*) (33), germ cell nuclear factor (Gcnf; nuclear receptor 6A1 [*Nr6a1*]) (34), and embryonic enhancer of zeste 2 (*Ezh2*) (35). Fourth, RNA interference (RNAi) could be used to "knock down" levels of oocyte mRNA in vitro or in vivo in transgenic mice (36,37). Similarly, morpholino antisense oligonucleotides may be used to preclude mRNA translation. These have been tested in mouse preimplantation embryos against E-cadherin and α catenin (38,39). Inherent in these third and fourth strategies is the potential to produce various degrees of hypomorphic effects, which may complicate interpretation, but should also provide important insights analogous to allelic series mutation experiments in other organisms. Finally, inhibitory agents or antibodies may be used to disable accumulated protein products in embryos in vitro, although the development and characterization of these can be expected to be arduous, as each must be met with individualized questions as to efficacy and specificity.

Translational Control of Maternal mRNAs

Biochemical studies have shown that with fertilization, there is a dramatic change in the profile of proteins synthesized by mouse oocytes (40,41). In lieu of reliance on transcriptional control, early zygotes seem to exercise extensive regulation over translation of maternally inherited mRNAs. These transcripts have *cis*-acting cytoplasmic polyadenylation element (CPE) sequences in their 3′-UTRs that allow binding with the dual-function CPE-binding (CPEB) protein. In *Xenopus* oocytes, unphosphorylated CPEB "masks" mRNAs by precluding initiation factor interaction with the 5′ mRNA cap, whereas CPEB phosphorylation causes recruitment of polyA polymerases and relaxes translation inhibition [reviewed in (42)]. In mouse oocytes, many CPE-containing mRNAs are regulated in a similar manner, during either oogenesis or early embryo development or both. In a study to identify mammalian homologs of cell cycle regulators expressed during the oocyte-to-embryo transition, many contained CPE or CPE-like elements that presumably allow such translational regulation (43) (Table 1). Direct experimental evidence for this regulation has been reported for *Mos* during oocyte maturation (44) and Cyclin A2 and Spindlin after fertilization (45,46).

Translational mRNA regulation is important in early oogenesis, and perhaps because of the deleterious potential of transcripts prematurely translated or a requirement for their regulated recruitment, CPEB is critical. Germ cells of *Cpeb* knockout females die around E16.5, a time point when CPEB

TABLE 1. *Select CPE-containing mRNAs*

Mouse gene	EST expression	Putative function	CPE
Ask activator of S phase	Egg, 2-cell	Damage checkpoint	UUUUUAU
Bub1 budding uninhibited by benzimidazole 1	2-cell	Spindle checkpoint	UAUUUAU
Ccnb1 cyclin B1	Egg, zygote, 2-cell	B-type cyclin	UUUUAU
Ccnb2 cyclin B2	Egg, zygote, 2-cell	B-type cyclin	AUUUUAUU
Cdc20 cell division cycle 20	Egg, zygote, 2-cell	Chromosome segregation	UUUUUAU
Cdc42 cell division cycle 42	Zygote, 2 cell	Cytoskeletal organization	UUUUUAAA
Cdc45l cell division cycle 45	Zygote, 2-cell	Replication	UUUUUAAU
Cdk2 cyclin-dependent kinase 2	2-cell	Cyclin dependent kinase	UUUUAU
Cenpa centromere autoantigen A	2-cell	Kinetochore protein	UUUUUAAA
Chd1 chromodomain helicase DNA binding protein 1	2-cell	Chromatin organization	UUUUUAU
Cse1l chromosome segregation 1-like	2-cell	Chromosome segregation	UUUUAUU
Cspg6 chondroitin sulfate proteoglycan 6	Zygote, 2-cell	Chromatid cohesion	AUUUUAU
Dncic1 dynein, cytoplasmic, intermediate chain 1	2-cell	Kinetochore protein	AUAUUUAU
Exo1 exonuclease 1	Egg, zygote, 2-cell	DNA repair	UUUUUAU
Hip2 Huntingtin-interacting protein 2	Zygote, 2-cell	Ubiquitin conjugation	UUUUAAA
Kif11 kinesin family member 11	Egg, zygote, 2-cell	Kinetochore protein	AUUUUAAU
Mapre1 microtubule-associated protein, RP/EB family, 1	Egg, zygote, 2-cell	Microtubule assembly	AUUUUAAU
Mcm3 minichromosome maintenance deficient 3	Zygote, 2-cell	Replication	UUUUUUAUU
Mcm7 minichromosome maintenance deficient 3	Zygote	Replication	UUUUUAUU
Psd3 proteasome 26S subunit, non-ATPase, 3	2-cell	Protein clearance	UUUUAAA
Rad51 RecA homolog	2-cell	DNA repair	UUUUAAA

protein normally is first phosphorylated by Aurora kinase (47,48). The early loss of *Cpeb* null oocytes has precluded the use of the knockout model to investigate CPEB loss-of-function in mammalian oocyte maturation and early embryonic development. Perhaps a postnatal oocyte-specific knockout of *Cpeb* will be helpful in defining the roles of CPEB in later oocyte functions and the oocyte to embryo transition. Current pharmacological evidence suggests that activation of CPE-containing transcripts is critical for achieving zygote transcription. Treating fertilized eggs with 3′-deoxyadenosine (preventing polyadenylation of maternal mRNAs) or cycloheximide (preventing protein synthesis) inhibits zygote transcription within the pronuclei (49,50). These data indicate that translational activation of stockpiled maternal transcripts is critical in setting the stage for ZGA.

ZYGOTE GENOME ACTIVATION

Within approximately 30 hours of fertilization, there is a transition from primary dependence on maternal effect factors to products of the zygote genome.

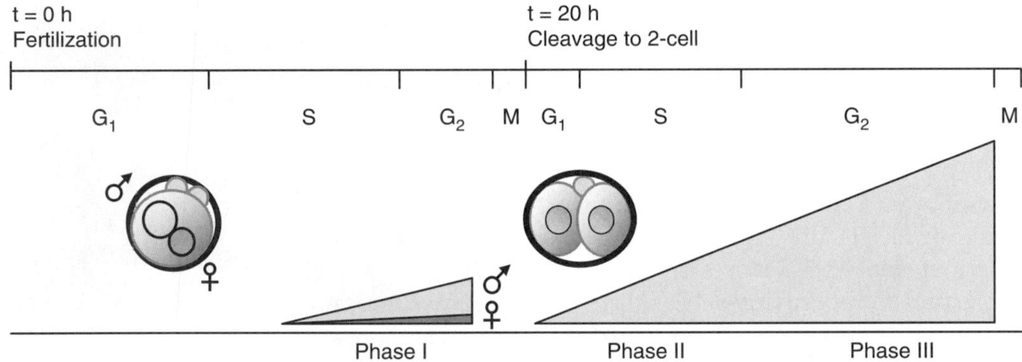

FIG. 7. An illustration of the timeline of zygote genome activation (ZGA). Detectable ZGA first takes place during the S/G_2 phase of the first cell cycle at approximately one-third of the levels seen in later embryos. This is known as phase I activation, and transcription in the male pronucleus predominates over that in the female pronucleus during this period. The diploid nucleus is being reorganized during phase II and III, and transcription becomes linked to translation and essential for continued development.

Transcriptional reactivation is essential for continued development as maternal stores are exhausted, and α-amanitin inhibition induces a block to murine embryo progression beyond the two-cell or four-cell stage, depending on the genetic strain (51).

Although widespread ZGA coupled to protein translation does not ensue before the two-cell stage, probe incorporation studies indicate some transcriptional activity in the pronuclei of one-cell zygotes (52,53). This early activation is known as phase I of ZGA (minor ZGA) (Fig. 7). Transcription is first appreciable during replication (S phase), and replication may play important roles in reorganizing genomic DNA to a transcriptionally permissive state (54,55). Although ZGA coincides with template modifications, factors independent from the chromatin state seem critical to attaining transcriptional ability, given that late one-cell zygotes (>20 hours postfertilization) transcribe microinjected reporter constructs and genes from transplanted two-cell embryo nuclei more so than younger one-cell embryos (56,57). During the two-cell stage in mouse development, phases II and III (major ZGA) occur, with increasing species and quantities of transcripts generated; these mRNAs have the potential to undergo direct translation. In some experimental systems, onset of transcription can be uncoupled from replication and cell division, leading to the proposition that a temporal regulation or "zygotic clock" allows for ZGA only after a predetermined period has elapsed postfertilization (58,59). Notably, this is unlike the scenario in the frog, *Xenopus laevis*, where several replication rounds and nuclear divisions appear necessary to achieve a given nuclear-to-cytoplasmic ratio and dilute maternal histone stores, prerequisites for transcription factor association with DNA (60).

What is the molecular mechanism for the "zygotic clock"? Although initiating ZGA may require the convergence of several regulatory pathways, RNA polymerase II (RNApolII) is implicated as a principal effecter. This is because earliest transcription is sensitive to α-amanitin inhibition and progression of the clock correlates well with presence of various forms of the catalytic subunit of RNApolII. The carboxyterminal domain (CTD) of RNApolII is dephosphorylated after fertilization in mammals, converting it from the oocyte hyperphosphorylated state (IIo) to a hypophosphorylated state (IIa). An embryonic form with intermediate electrophoretic mobility (IIe) becomes evident in preimplantation embryos, its appearance coinciding with minor ZGA. This is followed by the replacement of this intermediate form with the hyperphosphorylated form (IIo) beginning with major ZGA (61). Coincident with these changes, RNApolII accumulates in the nucleus, although its phosphorylation is separable from nuclear translocation. RNApolIIe and IIo states are lost with the addition of cyclohexamide postfertilization, a manipulation that precludes ZGA, although nuclear import is still conducted. Treatment with aphidicolin to inhibit replication does not prevent the appearance of RNApolIIe and IIo forms, their nuclear import, or the synthesis of a subset of α-amanitin sensitive gene products. Thus, three forms of RNApolII can be resolved by electrophoresis, and the appearance of slower migrating IIe and IIo after fertilization appear to be linked to minor and major ZGA, respectively.

Like ZGA, formation of RNApolIIe and IIo after fertilization depends on protein synthesis, though otherwise we have few insights into how their generation is regulated. In *Xenopus* oocytes, phosphorylation of the RNApolII CTD (IIa to IIo) has been attributed to microtubule-associated protein kinase activity (62). Whether this is true in mammalian embryos and whether RNApolIIe, which has only been described in mammals, represents the product of a distinct kinase remain unclear. RNApolIIe is not susceptible to 5,6-dichloro-1-β-D-ribofuranosylbenzimidazole, an inhibitor of the TFIIH CTD kinase that is required for maintenance of the IIo hyperphosphorylated form in cleavage embryos (61). Of note, it seems that CTD phosphorylation is critical for survival of early embryos; knockout mice lacking the MAT1 component of TFIIH die in the preimplantation period when maternal MAT1 stores are depleted (63). MAT1 is important for specifying CTD substrate affinity in vitro (64,65), and this was confirmed by the discovery of defects in CTD phosphorylation in *Mat1*$^{-/-}$ cells. However, no clear compromise of polymerase II–mediated transcription was appreciated in these cell lines, unlike in cells with disruptions in the yeast *TFB3* homolog. Together, these data suggest that redundancies of function allow for some compensation in mammalian cells with aberrations in CTD phosphorylation and raise the possibility that other mechanisms contribute to transcriptional regulation, at least after ZGA.

Although ZGA onset may be regulated by processes with no obvious preference for male or female pronuclei, phase I transcription levels in the two pronuclei are asymmetric, being greater in the male pronucleus as measured by reporter construct microinjection or BrDU incorporation (53,66). It has been proposed that steps in unpacking the paternal genome may allow for transient increases in transcription factor accessibility and their preferential association with the male pronucleus. Notably, there seems to be a competition between the two pronuclei for oocyte stores of transcription factors rather than something strictly repressive about the maternal pronucleus. In the absence of the paternal contribution, the maternal pronucleus can sequester increased levels of transcription factors and support high levels of transcription (67). Differential histone states between the two pronuclei may be chief determinants of their transcriptional status. For example, the male pronucleus has hyperacetylated histone H4 and hypomethylated histone H3 as compared with the female pronucleus; both are predicted to "loosen" nucleosomal structure and facilitate transcription of the paternal genome (Fig. 8) (35,68–70). Consistent with a prominent role of histone acetylation in affecting transcription, butyrate treatment (which inhibits histone deacetylases [HDACs]) allows for promoter derepression when reporter constructs are microinjected into female pronuclei, so that transcription becomes comparable with the higher levels observed without treatment in male pronuclei (71). The regulators responsible for pronuclear asymmetries in histone modifications include the EZH2/EED (enhancer of zeste 2/embryonic ectoderm development) protein complex, polycomb group (PcG) proteins important

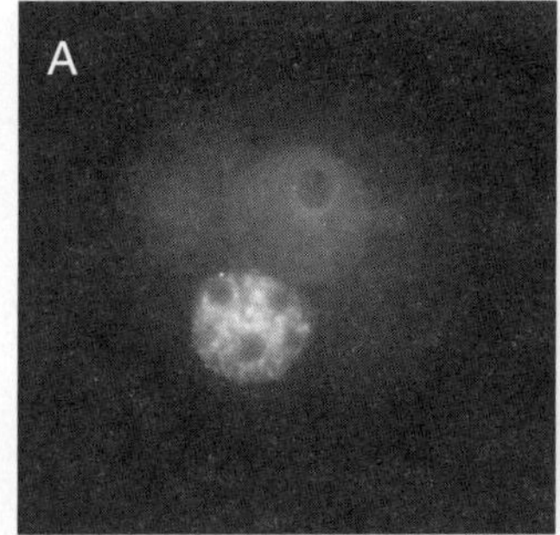

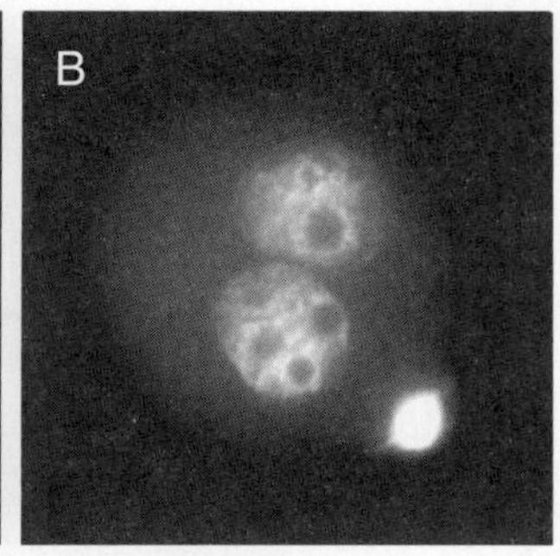

FIG. 8. Histone H3 methylation in one-cell zygotes. **A:** The two pronuclei have unequal quantities of histone H3 methylated on lysine 9 (MeK9), which is only detectable in the female pronucleus. The opposing male pronucleus is lightly counterstained with DAPI and demonstrates no histone H3 MeK9 immunoreactivity. **B:** Both maternal and paternal pronuclei stain equally brightly in detection of histone H3 methylated on lysine 4 (MeK4). Additionally, an extruded polar body is intensely stained. Antibodies used to detect these forms of histone H3 are from Upstate Biotechnology (#07-212 and #07-030) and were detected with a fluorescent anti-rabbit IgG secondary antibody and visualized by deconvolution microscopy.

in histone methylation. The complex is localized preferentially in the female pronucleus after fertilization, and methylation of histone H3 at lysines K9 and K27 in this period are compromised with oocyte-specific deletion of *Ezh2* (35). Thus, although the zygotic clock may govern the potential for gene expression by regulating *trans*-acting transcription factors, histone modifications and other alterations in chromatin structure are important mediators of template access for transcription. This theme persists throughout early development, and we consider it further in the sections to follow.

Specific Products of Early Transcription

Although the period surrounding ZGA is one that allows for some degree of promiscuous transcription, as we consider below, some regulation is imposed at both mRNA and protein synthesis levels to tailor the zygote transcriptome and proteome (72). Of the tens of thousands of genes transcribed before implantation, only a few hundred are activated during these early ZGA phases at the "oocyte-to-embryo" transition (73). As we gain an appreciation for this period of development, a list of genes is emerging that constitute the first to be expressed (Table 2). Although some transcripts may represent "noise" of immature chromatin, we infer that some hold important roles in early development.

Many of the earliest products of the zygote genome are involved in preparing for protein synthesis, which increases in both absolute quantity and diversity of proteins produced dramatically during early development. RNA processing proteins, such as

TABLE 2. *Early products of the zygote genome*

Function	Gene product	References
Cell cycle control	Cyclin A (*Ccna2*)	(45)
	Cyclin E (*Ccne1*)	(74)
	Prothymosin α (*Ptma*)	(74)
Cell recognition	Major histocompatibility complex MHC, class 1, H2D	(75)
Chromatin remodeling	Histone H_1	(76)
	Histone deacetylase 1 (*Hdac1*)	(77)
	High mobility group box 1 (Hmg1, *Hmgb1*)	(78)
Metabolism	Cytochrome C oxidase, subunit VIIc (*Cox7c*)	(50)
	Hypoxanthine guanine phosphoribosyl transferase (*Hprt*)	(50)
	Ornithine decarboxylase (*Odc*)	(50)
	Pyruvate dehydrogenase E1α1 (*Pdha1*)	(50)
	Phosphoribosyl pyrophosphate synthetase 1 (*Prps1*)	(50)
Transcription factors	OCT4; POU domain, class 5, transcription factor 1 (*Pou5f1*)	(79)
	TATA-box binding protein (*Tbp*)	(80)
	TEA domain family member 2 (*Tead2*)	(81)
	Trans-acting transcription factor 1 (*Sp1*)	(82)
RNA processing	Splicing factor, arginine/serine-rich 3 (*Sfrs3*; SRp20)	(83)
	U2 small nuclear ribonucleoprotein auxiliary factor 1, related sequence 1 (*U2af1-rs1*; U2afbp-rs)	(84)
Protein production	Eukaryotic translation initiation factor eIF-4c (*Eif1a*)	(54)
	Heat shock protein 25 (Hsp25; *Hspb1*)	(85)
	Heat shock protein 70.1 (*Hspa1b*)	(86)
	Ribosomal RNA	(87)
	Ribosomal proteins	(88)
	TRiC-P5 protein chaperonin (chaperonin subunit 3γ; *Cct3*)	(89)
X-chromosome inactivation	Inactive X specific transcript (*Xist*)	(90,91)
Y-chromosome products	Sex-related Y (*Sry*)	(92)
	Y-chromosome-linked zinc finger (*Zfy*)	(92)
Unknown	Murine endogenous retrovirus-like (MuERV-L; *Erv4*)	(93,94)
	Transcription requiring complex (TRC)	(95)

splicing factor SRp20 (83) and translation initiation factor eIF-4c (54); rRNA and protein components of ribosomes; and large heat shock proteins, like the Hsp70.1 chaperonin, are examples. It is known that preventing ZGA or protein synthesis precludes early cleavage development. Some early products of the zygote genome are individually essential to progression through this period; for example, mutant embryos lacking the *Sfrs3* gene encoding SRp20 fail to form blastocysts and die at the morula stage (83). Once processed for translation, mRNAs are recruited to ribosomes. High rates of ribosome assembly are accomplished by the two-cell stage to meet this demand, and transcription of rRNAs augments oocyte reserves early in development. The nucleolar organizer protein NOPP140 (nucleolar and coiled-body phosphoprotein1; NOLC1) and RNApolI are implicated in this activation of rDNA transcription (87). Stores of ribosomal component proteins sequestered in lattice-like structures are accumulated during oogenesis and become available thereafter (96); more are synthesized in a RNApolII-dependent manner after ZGA in late two-cell embryos in the mouse (88). AATF (apoptosis antagonizing transcription factor; also termed Traube [TRB] or Che-1) seems to be an important participant in ribosomal protein production (97). AATF is a leucine zipper protein localized to the nucleolus of preimplantation embryos in a cell cycle–dependent manner. It has been shown to interact with retinoblastoma (RB) and RNApolII subunits in other experimental systems (98). Knockout embryos arrest in development at the compacted morula stage and show depleted levels of ribosomes, polyribosomes, and rough endoplasmic reticulum (97). Similarly, knockout embryos lacking the S19 ribosomal protein encoding gene *Rps19* are lost before the blastocyst stage (99). Thus, although initial postfertilization translation is devoted to production of proteins encoded for by maternal RNAs and is not coupled with zygote transcription, early zygote genome products reflect the priority of continued protein synthesis, soon to fall exclusively under the direction of zygotic mRNAs.

Specific Transcription Factors

What determines the genes first activated in embryos? In addition to epigenetic regulation of chromatin structure, which we consider in the next section, specific transcription factors are responsible for regulating early gene expression, and there is an emerging list of maternal and embryonic transcription factors responsible for ZGA. Some, such as *trans*-acting transcription factor 1 (SP1) and TATA-binding protein (TBP) (82), function in both oocytes and embryos, whereas others are produced only after fertilization. Effects of oocyte depletion of SP1 and TBP have not been studied, but knockout work at each locus indicates important functions of the zygotic genome-encoded transcription factor. *Sp1*$^{-/-}$ embryos exhibit defects in postimplantation development (100), whereas *Tbp*$^{-/-}$ embryos illustrate defects before implantation. Maternally encoded TBP seems to be exhausted by the eight-cell stage, and zygotic expression of TBP is required by the blastocyst stage for viability and appropriate RNApolI and III activities (80). TEA domain family member 2 (TEAD2) is a transcription factor encoded by maternal mRNA stored in the oocyte; the *Tead2* mRNA is translated rapidly after fertilization (101). The onset of TEAD2 transcriptional activity thus correlates with ZGA, and *Tead2* mRNA is thereafter actively produced by the zygote genome, so that increasing concentrations of the mRNA are available from the two-cell to blastocyst stages (81). No loss-of-function experiments specifically disabling TEAD2 have been reported. Cooperation or redundancies of some transcription factors functional in early development are beginning to be investigated by double mutant studies (102). For example, activating transcription factor 1 (ATF1) and cAMP responsive element binding protein 1 (CREB1) are homologous transcription factors that seem to compensate for loss of one another during early development. Single knockout *Atf*$^{-/-}$ embryos have no phenotype (102), and *Creb1*$^{-/-}$ mice progress normally through preimplantation development to die in the perinatal period of respiratory distress (103). A unique phenotype is uncovered in double mutants lacking both transcription factors (*Atf1*$^{-/-}$, *Creb1*$^{-/-}$) as these embryos die during preimplantation development before blastocoel formation (102). Efforts are ongoing to identify both the transcription factors responsible for shaping the early transcriptome and their essential early targets.

CHROMATIN REMODELING AND EPIGENETICS IN EARLY EMBRYOS

Controlling ZGA: The Pandora Problem

Major ZGA represents a period of "permissiveness to transcription" characterized by many transiently up-regulated mRNA species that are repressed by the late two-cell and four-cell stages with attendant chromatin remodeling. This phenomenon has been described using divergent methodologies, including EST analysis and differential display polymerase chain reaction (104,105). Similarly, in rabbit embryos, transient expression of a subset of genes between the 8- and 16-cell stages is thought to be an observation that ZGA outpaces the acquisition of repressive chromatin structure (106).

Establishment of repressive nucleosomal configurations seems critical in controlling the initial promiscuity of transcription after ZGA. Interestingly, important participants in ZGA, including DNA replication complexes, may actually displace transcriptional proteins and make chromatin available for this reorganization in two-cell stage embryos. To support this, expression of the transcription requiring complex genes and the eukaryotic translation initiation factor (eIF-4c; *Eif1a*) gene requires the first round of DNA replication in one-cell zygotes but is then down-regulated in a manner dependent on the second round of replication (77). Similarly, total BrUTP incorporation reflecting transcription by two-cell blastomeres can be maintained at threefold post–S phase levels by treatment with aphidicolin to prevent DNA replication (53). The S phase of the two-cell embryo may participate in gene repression by allowing for activity of HDACs, because the addition of trapoxin to inhibit HDACs prevents transcription requiring complex genes and *Eif1a* down-regulation (54). Similarly, trichostatin A–mediated HDAC inhibition broadly prevents decreases in gene expression at this transition as assessed by differential display (105). Indeed, trichostatin A treatment inhibits development from the two-cell to four-cell stage, suggesting that HDAC activity and consequent transcriptional repression are required for this developmental window. In addition to changes in the acetylation state of histone proteins, increased availability of histones important for nucleosome condensation, including histone H1, H2A, and H2B, in two-cell embryos is likely to contribute to the chromatin maturation (76).

Cis-acting sequences (i.e., regional enhancers and gene-specific promoters) influence the expression of individual genes in the context of the local chromatin and thereby allow selective transcription in regions of facultative repression. Along with the developing suppression of transcriptional activity considered above, two-cell embryos acquire the ability to recognize enhancer sequences that relax repressed nucleosomal configurations (107,108). Onset of enhancer activity is related not only to the maturation of chromatin in these cells, but reflects the acquisition of a discrete coactivator function at the two-cell stage. Reporter construct studies demonstrate that the as yet unknown coactivator is coded for by the zygote genome, given that its presence requires ZGA and

TABLE 3. *Transcript persistence in early development*

Gene expression pattern	Number of genes	Adjusted
Constitutive or complex expression	327 (3.36%)	13.33%
Unfertilized egg	Stage-specific expression 291 (2.99%)	11.86%[a]
	Expression initiation 44 (0.45%)	1.79%
Fertilized zygote	Stage-specific expression 243 (2.50%)	9.91%[a]
	Expression initiation 25(0.26%)	1.02%
Two-cell embryo	Stage-specific expression 282 (2.90%)	11.50%[a]
	Expression initiation 22 (0.23%)	0.90%
Four-cell embryo	Stage-specific expression 248 (2.55%)	10.11%[a]
	Expression initiation 25 (0.26%)	1.02%
Eight-cell embryo	Stage-specific expression 347 (3.57%)	14.15%[a]
	Expression initiation 39 (0.40%)	1.59%
Morula	Stage-specific expression 387 (3.98%)	15.77%[a]
	Expression initiation 12 (0.12%)	0.49%
Blastocyst	Stage-specific expression 161 (1.66%)	6.56%[a]
Low-level/negligible expression	7265 (74.77%)	0.00%

Genes are categorized by their expression pattern in each stage of development. "Stage-specific expression" indicates gene expression is limited to a single time point in development; "Expression initiation" indicates that a transcript is first detected at this developmental stage and persists thereafter in subsequent stages.

[a]Stage-specific transcripts make up a total of nearly 80% of transcripts identified in unfertilized metaphase II eggs through blastocyst embryo stages, not considering genes with low level of expression (adjusted %).

can be generated prematurely in oocytes by supplementing their mRNAs with those of ES cells (109). Lack of this activity in earlier stages may prevent inappropriately broad transcriptional activation during early ZGA and the first stages of chromatin remodeling. Thus, early cleavage embryos attain the potential to use enhancers like more differentiated cells. However, use of particular promoter sequences remains distinct from that of more mature cells. Blastomeres between the two-cell and blastocyst stages have a unique potential to produce robust gene expression in the absence of TATA-box containing promoters. Preferential TATA-less promoter activity has been described at the endogenous eukaryotic translation initiation factor eIF-4c (*Eif1a*) locus, and TATA disruption does not compromise expression of reporter constructs with HSV-tk promoter elements during this window (110,111). Features of gene templates and their transactivators that contribute to this and how profound an effect this has on overall embryonic mRNA composition are unclear. Nevertheless, there seems to be potential for TATA-less gene promoters to be transcriptionally active in cleavage stage embryos, escaping effects of HDACs or other chromatin-condensing factors and shaping the transcriptome of these remarkable cells until differentiation.

Although emphasis has been placed on how ZGA is repressed in the transition from two-cell to four-cell stages in the mouse embryo, the wave of transcription being contained represents one of several following suit. Indeed, global analysis of ESTs from mouse preimplantation embryos conducted by Ko et al. (104) revealed that most genes expressed at appreciable levels in early development are present discretely in a stage-specific manner (Table 3). These stage-specific gene transcripts are rapidly made and degraded so that they are present in narrow well-defined intervals. Whether the disappearance of these mRNAs prevents or is coupled to their translation is largely unknown, although two-dimensional protein gel analyses provide evidence that the proteome of early embryos is very dynamic during this period (112,113). Equally intriguing about the report of Ko et al. (104) is the finding that stage-specific gene loci cluster physically within chromosomal regions, indirect evidence of the influences of chromatin configurations on gene expression. For example, groups of blastocyst-specific genes are found on chromosomes 1 and 8, and several morula-specific genes reside in close proximity on chromosome 11. Nearly half of the genes included in this analysis are novel, and most have not been studied in early embryos. Therefore, whether their stage-specific expression reflects unique functions in the embryo or is merely the byproduct of a systematic reorganization of the diploid nucleus remains a point of speculation.

Genomic Methylation: Rewriting and Its Risks

Methylation of CpG dinucleotides is an important determinant of gene transcription and is heritable in somatic cells because of methyltransferases that act on hemimethylated duplexes. During gametogenesis and early embryo development, there are two temporal windows when genome methylation marks are largely erased (Fig. 9). The first nidus in methylation

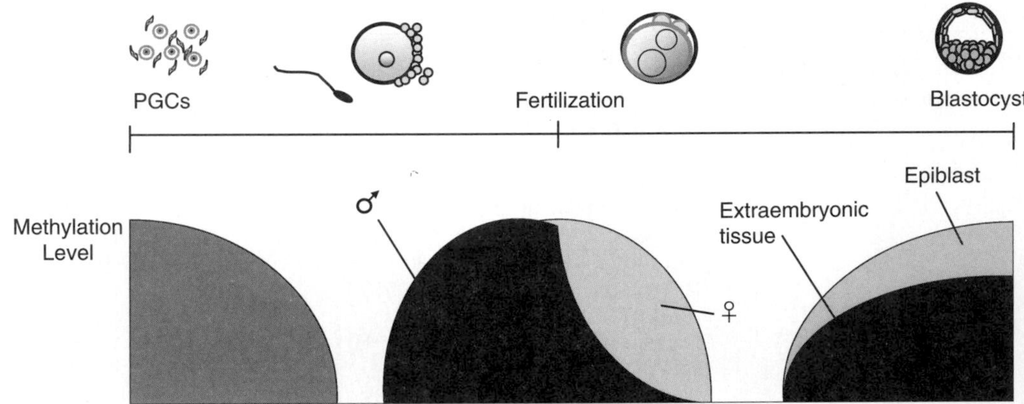

FIG. 9. Genome-wide methylation levels in germ cells, zygotes, and preimplantation embryos. Primordial germ cells (PGCs) undergo genome-wide demethylation after a window of mitosis and migration in the embryo. Methylation is regained during gametogenesis, occurring earlier in male gametes (beginning at the prospermatogonia stage in embryo development) than female gametes (largely accomplished during postnatal oocyte growth). Within a few hours of fertilization, active processes demethylate the male pronucleus, whereas the maternal genome undergoes a progressive loss of methylation with replication in the early cleavage embryo. Approaching the blastocyst stage, de novo methylation increases and is more pronounced in the epiblast as compared with extraembryonic tissues.

occurs in primordial germ cells by E13.5 during a period of postmigratory mitotic quiescence in the gonadal ridge (114,115). Its timing may follow primordial germ cell mitosis to minimize genomic instability (116). Many imprinted loci that display parent-of-origin–dependent methylation in somatic cells are demethylated at this time, resulting in biallelic expression of genes, including *H19*, *Igf2*, *Igf2r*, and *Snrpn* (Table 4) (117). Global DNA methylation then increases, beginning at the prospermatogonia stage in males and with postnatal oocyte growth in females, so that mature spermatozoa and GV stage oocytes have highly methylated genomes. Sex-specific methylation imprints are initiated at this time and presumably have roles in the parental "memory" these alleles maintain later in the development of a new individual. The timing of these methylation events is sexually dimorphic, occurring relatively early in spermatogenesis as compared with oogenesis; methylation-associated silencing of *Snrpn* and other processes essential for the up-regulation of maternal alleles of *Igf2r* and $p57^{Kip2}$ are acquired only as oocytes grow and mature in adult females (118,119).

The second wave of demethylation occurs shortly after fertilization, and this represents both an active process predominant in the paternal pronucleus and a passive or replication-dependent demethylation that occurs in both parental genomes (120). Most methylation sites established during gametogenesis are lost, though a subset of parental-specific marks secured at

TABLE 4. *Examples of loci with parent-of-origin specific imprinting*

Gene	Imprinting	References
Cyclin-dependent kinase inhibitor 1C (*Cdkn1c*; $p57^{Kip2}$)	Maternal expression enabled during oocyte maturation	(119)
H19 fetal liver mRNA (*H19*)	Paternal allele is silenced by methylation traced to male gametogenesis	(118)
Insulin-like growth factor 2 (*Igf2*)	Inverse expression as H19; maternal allele is silenced	(119)
Insulin-like growth factor 2 receptor (*Igf2r*)	Differential parental methylation; maternal expression enabled during oocyte maturation	(119,124)
Mesoderm-specific transcript (*Mest*)	Maternal allele is methylated in mature oocytes	(118)
Paternally-expressed gene 3 (*Peg3*)	Maternal allele is methylated in mature oocytes	(118)
Small nuclear ribonucleoprotein N (*Snrpn*)	Methylated in mature oocytes	(118)
U2 small nuclear ribonucleoprotein auxiliary factor related sequence (*U2afbp-rs1*)	Methylation in oocytes; independent methylation of the maternal allele in embryos	(122)
X inactive-specific transcript (*Xist*)	Methylated early in oocytes; paternal expression in sperm development, early embryos, and extraembryonic tissues	(125,126)

imprinted loci (including *H19*) are exceptions (121). Consequences of gamete imprinting at other loci are still unclear, however, the methylation marks being erased after fertilization (*Igf2*) (120) or proving dispensable for appropriate later embryonic imprinting in mutational analyses (U2afbp-rs) (122). Genome-wide de novo methylation begins in blastocysts, and methylation of the genome of ICM cells outpaces that of the TE (123). Thus, events of gametogenesis and early embryo development are attended by dramatic fluxes in genome-wide methylation, as well as deliberate modifications of methylation sites imparting parental-specific allele imprints.

Many methyltransferases, demethylation enzymes, and regulators are thought to mediate the changes in methylation patterns described above, with distinct participants involved during gametogenesis and embryogenesis (127). Several recent knockout models are providing our first insights into the in vivo functions of these factors (Table 5). In addition to knockout mouse models, other transgenic approaches have been used to study the roles of specific DNA binding proteins in protecting loci from methylation, including *trans*-acting transcription factor 1 (SP1) and CCCTC binding factor (CTCF). Transgenic mice with mutations of SP1 binding sites at the housekeeping gene encoding adenine phosphoribosyltransferase (*Aprt*) or the CTCF sites at the maternal *H19* locus demonstrate loss of protein binding, hypermethylation, and inappropriate gene silencing (128–131). Similarly, transgenic mice generating a hairpin RNA interfering with CTCF expression results in loss of maternal hypomethylation at *H19* (132).

It is thought that a primary purpose of CpG dinucleotide methylation is to silence the transcription of retrotransposons within the genome (138). These sequences are mobile genetic elements that transpose by reverse transcription of an RNA intermediate and then reintegration in the genome; they include long terminal repeats derived from retroviruses and long interspersed elements. Despite their potential to create transcriptional interference, produce insertional loss-of-function mutations, generate chimeric and antisense transcripts, and promote translocations and other insults to genomic integrity, there is a staggering tolerance for their presence. Retrotransposon sequences outnumber genes in mammalian genomes, and their representation on the X chromosome has suggested they may aid in propagating repressed chromatin configurations in *cis* (139,140). Though benign in most cell types, checks to their activity are relaxed with genomic hypomethylation in malignant cells where they contribute to genetic insults (141). In early embryos, low levels of genome methylation in combination with transcriptional permissiveness result in a high portion of these retrotransposable elements in the early transcriptome. Retrotransposon-related repetitive elements make up as much as 6% of two-cell embryo ESTs, and one long terminal repeat sequence, murine endogenous retrovirus-like (MuERV-L; *Erv4*), is represented among the earliest transcripts of the zygote genome (93,142). Moreover, reverse transcriptase activity is readily appreciable in two-cell embryos (142). Despite this seeming period of vulnerability, however, only 1 in 500 germline mutations are thought ascribable to retrotransposon presence (143). Interestingly, antisense oligonucleotide "knockdown" of MuERV-L causes embryo loss at the four-cell stage, although it is not clear if this is a nonspecific reaction to large amounts of double-stranded RNA (dsRNA) rather than reflecting important roles of the endogenous sequence (93). It is thought that active retrotransposable elements in early development may be processed to dsRNA that is in turn recognized by RNA-triggered gene silencing or RNAi participants (144,145). The expression of RNAi machinery and DICER (36), another dsRNAse termed ethanol

TABLE 5. *Mouse models of methylation and imprinting defects*

Mutated gene	Phenotype	References
DNA methyltransferase 1 (*Dnmt1*)	Global genome demethylation and dysregulation of imprinted genes	(133)
DNA methyltransferase 1 (*Dnmt1*); oocyte-specific variant (*Dnmt1o*)	Maternal effect phenotype; Dnmt1o$^{-/-}$ females are infertile, their embryos exhibit defects in maintaining imprinted methylations	(23)
DNA methyltransferase 3a (*Dnmt3a*)	Homozygotes survive development, but become runted and die around 4 weeks of age	(134)
DNA methyltransferase 3b (*Dnmt3b*)	Defective methylation of pericentric satellite DNA	(134)
DNA methyltransferases 3a and 3b (*Dnmt3a* and *Dnmt3b*)	Double knockout in ES cells blocks de novo methylation and prompts progressive methylation loss	(134,135)
DNA methyltransferase 3L (*Dnmt3l*)	Imprinted methylation lost in oocytes and embryos; *Dnmt3l*$^{-/-}$ males are azoospermic; pups of homozygous females die before midgestation	(136)
Embryonic ectoderm development (*Eed*)	Embryonic lethal; defects in maintenance of imprints and silencing at select loci	(137)

induced 2 (ETOHI2), and OAS family members (146) in oocytes and early embryos suggests that dsRNA surveillance pathways may be functional. Moreover, a recent study by Svoboda et al. (147) demonstrated a 50% increase in the quantities of MuERV-L and intracisternal A particle long terminal repeats in eight-cell morula after knockdown of *Dicer* mRNA. How these and as yet unknown mechanisms may protect the embryo from the dangerous potential of transposable elements and what evolutionary benefits the presence of these sequences may provide remain areas of conjecture.

X-Chromosome Dosage Compensation

Males inherit a single maternal X chromosome, whereas females inherit an X chromosome from each parent. The resulting differential dosage of X-linked genes in the two sexes is compensated for by different strategies in various taxonomic groups. In mammals, it is accomplished by female X-chromosome inactivation. This X-inactivation silences the majority of transcription from a single X chromosome, which becomes compacted to form a Barr body, while allowing gene expression from the other X chromosome. Importantly, the inactivated X chromosome is not completely inert but expresses a noncoding transcript, called inactive X specific transcript (*Xist*), from its X-inactivation center (XIC) locus (148). Copies of the 15-kb *Xist* transcript coat the X chromosome in *cis* and play a role in silencing other loci. On the active X chromosome, an antagonistic antisense transcript (*Tsix*) is synthesized at the XIC that seems to play a role in preventing *Xist* accumulation in *cis* (149).

The stage for *Xist* expression control in zygotes is set in gametogenesis (Fig. 10). During the first meiotic prophase in spermatogenesis, putative paternal

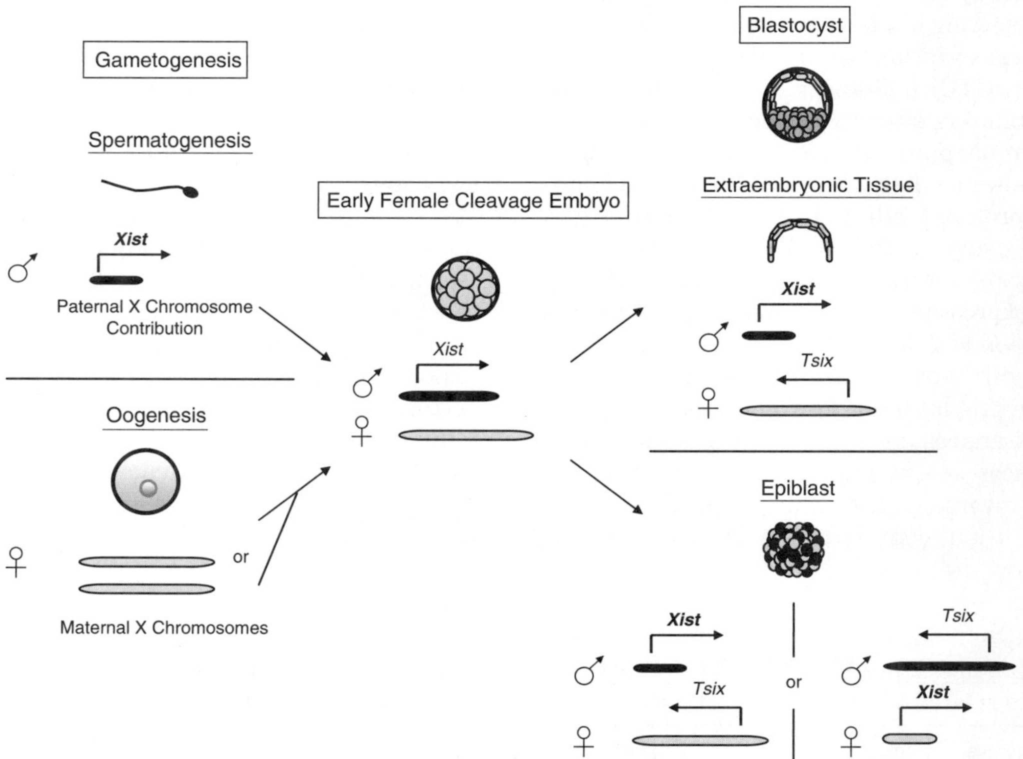

FIG. 10. X-chromosome regulation in gametes and early embryos. During spermatogenesis, the single X chromosome expresses *Xist* and is inactivated to form a tightly condensed structure in association with Y. In contrast, *Xist* expression is suppressed by methylation in the female germline, and immature oocytes are functionally diploid for the X chromosome. In early cleavage embryos, paternal and maternal *Xist* alleles continue to function similarly, in that only the paternal *Xist* is transcribed, although paternal X inactivation is partially relaxed or incomplete. With differentiation, the imprinted pattern of X-inactivation persists and becomes permanent in extraembryonic precursors, whereas it is randomized in the epiblast cells that will become the embryo proper. By this time, inactivation becomes accompanied by stabilization of *Xist* (bold) as well as *Tsix* up-regulation on the active X chromosome; "leakage" of transcription from inactive X chromosomes largely ceases and they undergo obvious chromatin compaction. Chromosome shade indicates the parent-of-origin; decreases in chromosome size reflect the extent of cytologic and epigenetic features of inactivation.

X chromosomes are inactivated, whereas in oogonia there is activation of both copies of the X chromosome (150,151). Like X-chromosome regulation in somatic tissues, inactivation of the X chromosome in the male germline is associated with *Xist* expression, whereas dual X activation in oocytes correlates with down-regulation of *Xist* (152). Regulation of *Xist* is associated with differential methylation of the XIC locus in the germline of each sex, modifications that persist during the genome-wide demethylation seen in early embryos (125,126). Methylation of CpG islands in the *Xist* promoter region and its 5′ transcribed portion preclude expression on active X chromosomes (153). These sites are methylated in the early germline (E12.5) and remain so in females, where dual X activation and cessation of *Xist* transcription are seen by E13.5 (125,126,152). The sites are demethylated in spermatogonia later in gestation (125,126,153). However, despite the sophisticated regulation of *Xist*, it does not prove essential for the process of germline X-inactivation (154,155). Functional silencing of the X chromosome, formation of the XY sex body, and spermatogenesis are all intact in males with a deletion of *Xist*. Why *Xist* is dispensable for inactivation, what roles *Tsix* may play in gametogenesis, and what, if any, specialized processes compensate for losses or gains of X chromosome activity in the germline are unknown.

Xist is actively transcribed by zygotes maintaining the paternal X-inactivation mark beginning at the two-cell stage; *Xist* RNA coats the paternal X chromosome by the four-cell stage and accumulates in increasing quantities until the morula stage (91,156,157). Despite early availability of *Xist*, many features of paternal X-chromosome inactivation in preimplantation embryos are delayed. In cleavage stage embryos, paternal X chromosomes do not demonstrate cytological compaction, lags in replication, or histone methylation and acetylation patterns associated with repression, and transcriptional inactivation is incomplete (156,158–160). Transcriptional silencing seems to be established progressively, so that by the morula stage expression from the paternal X chromosome is related to the distance of a locus from the XIC. Those genes less than 10 cM from the paternal XIC (such as α-thalassemia/mental retardation syndrome X-linked [*Atrx*] and phosphoglycerate kinase 1 [*Pgk1*]) show the most complete inactivation, whereas distant loci in *cis* (such as PCTAIRE-motif protein kinase [*Pctk1*] and pyruvate dehydrogenase E1α1 [*Pdha1*]) transiently escape silencing, and intermediately placed loci (such as hypoxanthine guanine phosphoribosyltransferase [*Hprt*] and methyl CpG binding protein 2 [*Mecp2*]) exhibit partial inactivation with respect to expression from the maternal X chromosome (161). Thus, female embryos inherit a paternal X chromosome silenced during male gametogenesis and induce its continued expression of *Xist* early in development, though through the morula stage epigenetic changes associated with silencing remain both incomplete and reversible.

With the partition of the early embryo into ICM and TE lineages, there is a relaxation of the imprinted X-inactivation in the inner mass that allows assumption of random X-inactivation in the epiblast. Random inactivation in these cells ultimately results in females that are mosaic for maternal and paternal X-inactivation, although the ratio can be skewed in some examples of disease. In contrast to random inactivation, the imprinted paternal *Xist* expression persists in extraembryonic tissues and leads to uniform paternal X-inactivation. Approaching implantation, *Xist* RNA is expressed in a more stable form, *Tsix* expression is activated on the opposite chromosome, and several epigenetic markers of the inactivated chromosome become evident; silencing thereafter is not reversed (156,162–164). Thus, during differentiation in blastocysts, X-inactivation remains imprinted in extraembryonic precursors and is randomized and subsequently fixed in the epiblast. Similar handling of X chromosome inactivation is recapitulated in cloned embryos, so that whichever X is inactivated in the somatic nucleus transferred to the oocyte is preserved as the silenced copy in extraembryonic lineages, whereas inactivation is randomly reassigned in the embryo proper (165).

Given the foregoing description of how sexually dimorphic X-chromosome handling during gamete development extends to influence X-chromosome usage in the embryo, it may be self-evident that parent-of-origin effects determine the phenotypes *Xist* and *Tsix* mutant mice. Mice with a maternally inherited *Xist* deletion are healthy, the females exhibiting a uniform paternal X-inactivation. However, *paternal* inheritance of the null allele causes early embryo loss with profound defects in the development of extraembryonic lineages (166,167); the phenotype is thought to arise from a loss of paternal X-inactivation and overexpression of X-linked products in XX TE cells. Conversely, deletions or mutations precluding expression of a *maternal Tsix* cause ectopic *Xist* expression and inappropriate X-inactivation. This results in loss of both male and female embryos and intrauterine growth retardation; there is no deleterious effect of paternal inheritance (168,169). Interestingly, the imprinted roles of the two X chromosomes in females can be reversed by a complementation of these mutations—early developmental defects in XX embryos can be rescued to some extent by inheritance of both a paternal *Xist* mutant and a maternal *Tsix* mutant (164). In addition to *Xist* and *Tsix* mutant models, a handful of other transgenics

TABLE 6. *Mouse models of X-inactivation*

Mutated gene	Phenotype	References
Breast cancer 1 (*Brca1*)	Embryonic lethal; *Brca1*$^{-/-}$ cell lines exhibit loss of *Xist* expression and epigenetic markers on the inactive X chromosome	(170)
DNA methyltransferase 1 (*Dnmt1*)	Embryonic lethal; extensive demethylation with unstable random X-inactivation in the epiblast	(171)
Embryonic ectoderm development (*Eed*)	Embryonic lethal; defects in maintenance of imprinted X-inactivation in extraembryonic tissue	(172)
Enhancer of zeste 2 (*Ezh2*)	Embryonic lethal; depletion of oocyte *Ezh2* results in loss of EED recruitment and histone H3 K9 and K27 methylation on the inactive X in trophectoderm cells	(35,173)
X inactive-specific transcript (*Xist*)	Embryonic lethal with paternal inheritance; loss of X-inactivation in trophectoderm cells	(166,167)
X inactive-specific transcript, antisense (*Tsix*)	Embryonic lethal with maternal inheritance; inappropriate X-inactivation	(168,169)

are emerging that illustrate defects in X-chromosome dosage compensation and thus provide insight into the effect of *trans*-acting factors (Table 6).

THE DESTRUCTION OF CREATION

Degradation of RNA

The oocyte-to-embryo transition involves rapid clearance of oocyte-specific transcripts so as to limit their functions in the embryo. A list of murine oocyte transcripts cleared in early cleavage embryos has come from polymerase chain reaction–based subtractive hybridization studies (146). It includes mRNAs encoding bone morphogenetic protein 15 (BMP15), growth and differentiation factor 9 (GDF9), an oocyte-specific H1 histone (H1oo), an RNA-binding Y box protein (MSY2), 2´-5´ oligoadenylate synthetase family members (OAS1c and OAS1e), oocyte-secreted protein-1 (OOSP1), tissue plasminogen activator (PLAT), ret-finger protein-like 4 (RFLP4), and zona pellucida proteins (ZP1 and ZP2). EST analyses have also added to this collection scores of unknown genes that are part of the maternal endowment cleared quickly during postfertilization development (104). Though their intended roles are presumably filled by early embryonic development, any deleterious effects of persistent mRNAs remain speculative. The fact that maternal mRNA degradation occurs early in development has been understood for many years (174–176), because it represents a major change in the overall zygote RNA constitution. Despite this insight, however, the molecular mechanisms and means of targeting specific transcripts for degradation remain elusive in mammals, although we consider some candidate participants below.

In addition to the unique task of degrading maternal RNAs, it should be stated that early embryos seem to have more generalized RNA clearance pathways, including those that mediate degradation of nonsense transcripts, mRNAs that have premature stop codons due to point mutations or frameshifts. One widely expressed protein, RENT1 (regulator of nonsense transcripts 1, the mammalian ortholog of *Saccharomyces* Upf1p), is produced in early development and contains a cysteine-rich zinc finger RNA binding domain and putative NTPase and helicase domains (177). Deletion of *Rent1* in knockout mice results in preimplantation lethality (178). *Rent1*$^{-/-}$ blastocysts demonstrate high levels of apoptosis and aberrant stabilization of nonsense β-glucuronidase (*Gus*) transcripts in combination with a naturally occurring *Gus* frameshift/premature stop mutation (*Gus*^{mps}; mucopolysaccharidosis VII) (178). Whether non-sense-mediated decay is conducted earlier by maternal factors and its importance in oocytes and earlier cleavage stage embryos are unknown. Nevertheless, these studies underscore that in considering RNA handling in embryos, degradation of RNA for purposes other than the transition between maternal and zygotic transciptomes may prove essential for cell viability in higher organisms.

Public EST collections (www.ncbi.nlm.nih.gov) indicate several other ribonuclease transcripts are expressed in eggs and early embryos, including diaphanous homolog 3 (*Diap3*), DIC-1 homolog (*Dicer1*), ethanol induced 2 (*Etohi2*), polyA-specific ribonuclease (*Parn*), as well as ribonucleases H1, P1, P2, and T2 (*Rnaseh1*, *Rnasep1*, *Rnasep2*, and *Rnaset2*). Of these, DICER1, a dsRNAse, and RNAse H1 are known to be critical for early embryo development in mice; *Dicer1*$^{-/-}$ and *Rnaseh1*$^{-/-}$ knockouts are severely compromised within approximately the first week of embryonic life (179,180). Two candidate negative regulators of RNA clearance have been suggested in the literature, MATER (encoded by *Nalp5*, NACHT leucine rich repeat, and PYD-containing 5) and MSY2 (encoded by *Ybx2*, Y box protein 2). As mentioned previously, MATER, a maternal protein essential for early embryo development, contains a C-terminal leucine-rich repeat sequence closely resembling ribonuclease inhibitors described in pigs and humans (20). Other ribonuclease enzyme inhibitors (such as RNH1, ribonuclease/angiogenin inhibitor 1) are expressed during preimplantation development and may share important roles in allowing persistence of key maternal transcripts or promoting the accumulation of zygotic transcripts. MSY2 is a germ cell RNA binding protein largely degraded during the transition from one-cell to two-cell stages, evidence that has prompted suggestion that loss of MSY2 leaves maternal transcripts vulnerable to RNAse activity (181,182). Of note, recent studies with *Drosophila* oocytes have identified two pathways, one maternal and one zygotic, that work in concert to clear mRNAs upon oocyte activation and after fertilization and ZGA, respectively (183). The maternal pathway has been found to rely on specific 3´-UTR "degradation elements" in target mRNAs, and lethal maternal effect mutations have been found at 13 loci crucial for operation of this pathway (183,184). However, many of these loci encode proteins that coordinate completion of meiosis—their functions likely more prerequisite than participatory in RNA destabilization.

Ubiquitination of Proteins and Organelles

Early embryos regulate protein presence not only by controlling transcript abundance and translation, but also by regulating protein clearance pathways. Three well-established classes of enzymes, termed E1, E2, and E3, covalently attach ubiquitin (Ub) moieties to potential substrates of proteolysis (Fig. 11). This process is highly regulated at the level of the E3 ligases that confer substrate specificity and is countered by the deubiquitylating and E3 inhibitory activities of Ub-specific proteases and *co*nstitutively *p*hotomorphogenic *s*ignalosome (COPS), an eight-subunit complex homologous to the proteasome lid named for its role in *Arabidopsis* development.

The process of Ub-mediated proteolysis has been described elegantly in oocyte maturation, wherein clearance of regulatory B-type cyclin subunits of maturation promoting factor (MPF) results in releases from cell cycle pauses at metaphase I and II (185). The early embryo seems to inherit ubiquitination machinery from the oocyte and may augment this with products of the zygote genome. Reflecting the presence of these pathways, mRNAs encoding F-box proteins (which link ubiquitination enzymes with specific substrates) and 19S and 20S proteosome components (which mediate degradation) are prevalent in both oocytes and two-cell embryos (142,143).

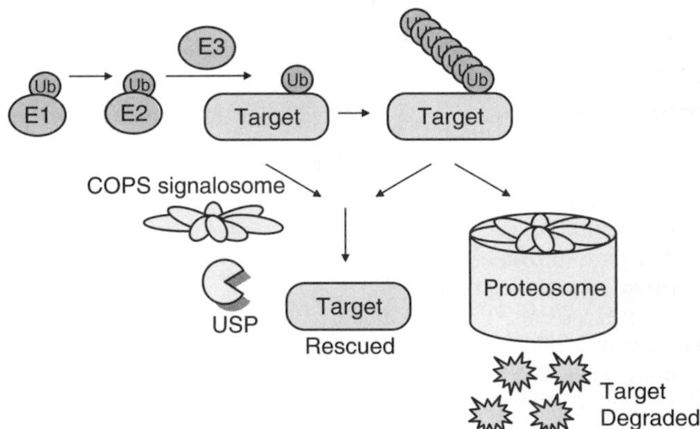

FIG. 11. Ubiquitination pathways regulating protein clearance. E1 enzymes are ATP-dependent activators of ubiquitin (Ub) by thiol-ester linkage; Ub is then transferred to an E2 conjugating enzyme. E3 enzymes are ligases that assist in the transfer of Ub from the E2 to a specific protein, targeting this substrate for degradation. Deubiquitylation mediated by Ub-specific proteases (UBPs) and the *co*nstitutively *p*hotomorphogenic *s*ignalosome (COPS) may protect targets from proteolysis.

Notably, some Ub-mediated degradation participants have expression profiles restricted to germ cells and early embryos, suggesting they play defined functions exclusively during meiosis and early development. These include the putative Ub ligase Ret Finger Protein-Like 4 (RFPL4), which is rapidly lost after the first embryonic mitosis and may control cyclin B1 clearance (186,187). Loss-of-function studies of several regulators of ubiquitination have led to defects in embryo development in the preimplantation period, many associated with the aberrant accumulation of cell cycle regulators known to be cleared by Ub-mediated proteolysis (Table 7).

Two analogous ubiquitination-like pathways exist for conjugating proteins with homologs of Ub, namely SUMO (*s*mall *U*b-related *mo*difier) and RUB (*r*elated to *Ub*) moieties. How sumoylation and rubylation tags direct protein fate remains an area of developing interest (197). In the case of RUB proteins, their only known targets are members of the Cullin family, and RUB conjugation seems to promote Cullin E3 activity. Although the extent to which Ub analogous pathways function in embryos remains to be pursued, at least one mutant model gives indication of their importance. Knockout mice lacking RUB1, also known as NEDD8 (neuronal precursor cell expressed, developmentally down-regulated gene 8), die in early development (before E7.5); their demise is preceded by defects in blastocyst formation and increased apoptosis of the ICM. Cyclin E, p57, and β-catenin are all overexpressed in the *Nedd8*$^{-/-}$ embryos, which is consistent with a role for NEDD8 in Cullin protein function (198).

Although ubiquitination is most often considered in the context of regulating individual protein clearance,

TABLE 7. *Models of defects in Ub-mediated proteolysis*

Gene targeted	Function	Phenotype	Notes	References
Tumor susceptibility gene (*Tsg101*)	E2 conjugating enzyme	*Tsg101*$^{-/-}$ embryos die around E6.5	Abnormal p53 accumulation	(188)
Cullin 1 (*Cul1*)	E3 ligase	*Cul1*$^{-/-}$ embryos die around E6.5	Cyclin E accumulation	(189,190)
Cullin 3 (*Cul3*)	E3 ligase	*Cul3*$^{-/-}$ embryos die around E7.5	Cyclin E accumulation	(191)
Cullin 4A (*Cul4a*)	E3 ligase	*Cul4a*$^{-/-}$ embryos die E4.5–E7.5	Putative targets unknown	(192)
Proteasome (prosome, macropain) 26S subunit, ATPase, 3 (*Psmc3*)	Proteasome component	*Psmc3*$^{-/-}$ embryos fail to form blastocysts	Broad inhibition of protein clearance expected	(193)
Proteasome (prosome, macropain) 26S subunit, ATPase, 4 (*Psmc4*)	Proteasome component	*Psmc4*$^{-/-}$ embryos fail to form blastocysts	Broad inhibition of protein clearance expected	(193)
Ub specific protease 9, X chromosome (*Usp9x*; FAM)	UBP	*Usp9x* antisense RNA prevents blastocyst development	Decreased β-catenin and AF-6	(194)
COPS, subunit 2 (*Cops2*, CSN2)	COPS subunit	*Cops2*$^{-/-}$ embryos die around implantation	Cyclin E, p21, and p53 accumulation	(195)
COPS, subunit 3 (*Cops3*, CSN3)	COPS subunit	*Cops3*$^{-/-}$ embryos die by E5.5-E7.5	Putative targets unknown	(196)

it also seems to be used in proteosomal clearance of paternal mitochondria (199,200). This destruction accounts for the exclusively maternal inheritance of these organelles and mitochondrial DNA, which can be compromised by the addition of proteosome inhibitors or microinjection of anti-Ub antibodies. Ooplasm donation has been used to replete presumed deficient or poor-functioning cytoplasmic factors in eggs of infertile women. The resulting mitochondrial heteroplasmy in their children emphasizes our potential to correct mitochondrial DNA disorders by this methodology as well as our need to better appreciate any importance of uniparental mitochondrial inheritance (201). Finally, in addition to regulating cell cycle progression and turnover of maternal protein stores and paternal mitochondria, Ub-mediated proteolysis may have roles in cellular differentiation in preimplantation stages. This is suggested in blastocysts by the differential expression of proteins such as COPS8 (which is predominantly in the ICM) and the preferential accumulation of Ub cross-reactive structures preferentially in the TE (202). Thus, there is both complex regulation of and a myriad of potential roles for proteolysis modulation in early development.

Cellular Apoptosis

Programmed cell death is carefully regulated in developing oocytes, serving as an important determinant of female reproductive life span as well as a means of surveillance to ensure oocyte quality (203). Similarly, oocytes can be lost shortly after ovulation or as early zygotes by an apoptotic process known as fragmentation, as can individual blastomeres in later cleavage stage embryos. In vitro, extent of apoptosis in mouse blastocysts can be measured by TUNEL (*t*erminal deoxynucleatidyl transferase-mediated d*U*TP *n*ick *e*nd *l*abeling) and is related to both genetic strain and culture conditions (204). One-cell zygote demise is mediated principally by maternal factors in response to derailed development, whereas by the two-cell stage both maternal and paternal genetic contributions affect the incidence of blastomere loss. Physiological cell death in preimplantation embryos does not occur to appreciable degrees until the blastocyst stage, when large numbers of ICM cells are eliminated (205).

In somatic cells, apoptotic pathways begin by extrinsic signaling (Fas-mediated) as well as by intracellular pathways (regulated by BCL2 family members) (Fig. 12). RNA expression studies and EST analyses indicate that key components of both of these pathways are expressed in preimplantation embryos, suggesting that they are in operation from early cleavage stages. Fas (*t*umor *n*ecrosis *f*actor *r*eceptor super*f*amily, member *6*; *Tnfrsf6*) and Fas ligand (FasL; *Tnfsf6*) mRNAs are induced in mouse morula (206), and monoclonal antibodies that recognize Fas inhibit two-cell embryo development to morula in vitro (207). Although *Tnfrsf6*$^{-/-}$ and *Tnfsf6*$^{-/-}$ knockout mice survive gestation to exhibit immunological defects, several downstream mediators of Fas-signaling pathways prove critical for postimplantation development. These include FADD (Fas-associated via death domain) and Casper (*C*aspase 8 and *F*add-*l*ike *a*poptosis *r*egulator; CFLAR), which seem to cooperate in vivo, as well as DAXX (Fas death domain-associated protein), which seems to counterinitiate the apoptotic pathway. Null embryos lacking any one of these three genes die around embryonic day 10 (208–210). Thus, mouse

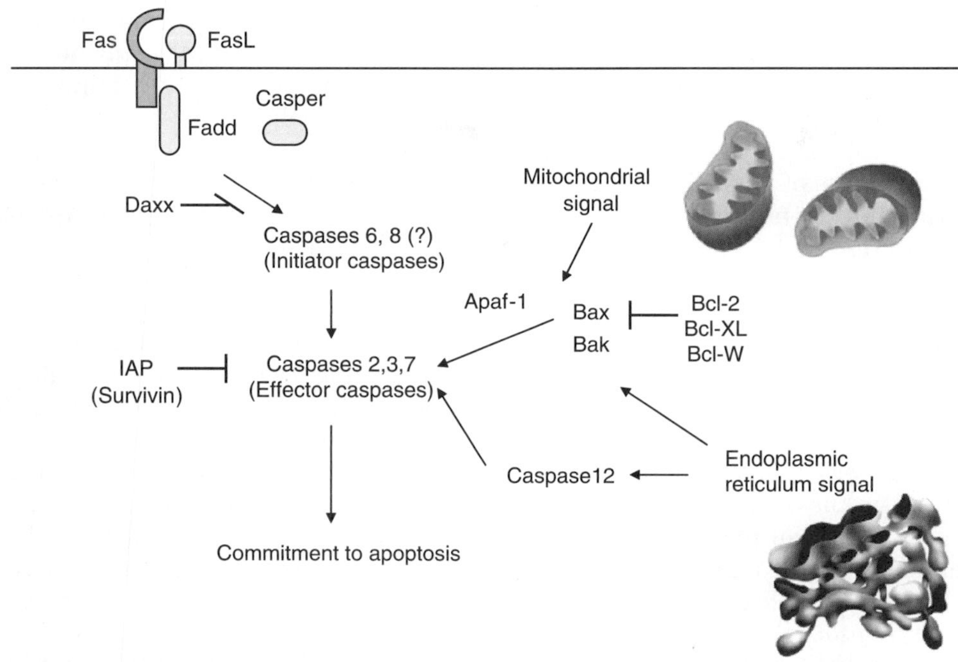

FIG. 12. Apoptotic pathways operative in early development. Apoptotic signals may be triggered by extracellular signaling or may originate in the mitochondria or endoplasmic reticulum. Initiator caspases in transmitting extracellular signals embryos are not clear. Expression has not been confirmed for known intiator caspases 8, 9, and 10; caspase 6 is expressed and resembles initiators in that it preferentially recognizes (L/V)EXD, the processing site for effector caspases, though evidence for an in vitro initiator activity is mixed (415). Caspase 12 is also present, suggesting the presence of a pathway hinging on ion fluxes at the endoplasmic reticulum (213). Several BCL family members that modify mitochondrial apoptotic pathways have been studied in early embryos, as has IAP (survivin), which may inhibit the effector caspase cascade.

embryos seem to have components of programmed cell death pathways important in conveying responsiveness to signals generated at the outer membrane and indeed cannot proceed through development without select examples. Other extracellular ligand–cell surface receptor interactions that participate in determining cellular viability in embryos (such as the transforming growth factor [TGF]-α survival signal) are considered further in a separate section. Intracellular apoptosis triggers involve specific stress signals and ion leakage from mitochondria and endoplasmic reticulum, signals that may be modified by BCL2 family members (namely, the proapoptotic BAX and BAK and the antiapoptotic BCL2, BCLXL, and BCLW). Although many other Bcl-2–like regulators of these pathways have been described, we focus on these five because their expression has been documented in preimplantation murine embryos (211). Notably, there seems to be sufficient redundancy of function among this group that single knockouts do not meet blocks in early development, although irreplaceable roles have emerged for several in oocyte development [reviewed in (212)].

Fas- and BCL2-mediated pathways converge to activate caspase enzyme cascades, culminating in the activation of effector caspases that then induce the morphological changes associated with cell death (i.e., DNA condensation and nuclear fragmentation, cell shrinkage, and blebbing or cellular fragmentation). In addition to transcripts of BCL2 family members, caspase gene sequences have also been sought in early embryos by reverse transcriptase polymerase chain reaction (211). The majority examined, including mRNAs encoding caspases 2, 3, 6, and 7, are detectable in preimplantation embryos from the two-cell stage onward. Interestingly, caspase 12, now implicated chiefly in endoplasmic reticulum–mediated pathways to programmed cell death, is unique in that its transcript is detectable in one-cell zygotes as well as later cleavage stages (211,213). The significance of the nadir of most (but not all) caspase transcripts in one-cell embryos, and whether it is reflected in caspase protein expression profiles remain to be explored. Apoptosis is suppressed at the level of the caspase cascade in oocytes and cleavage embryos in part because of the presence of survivin (baculoviral IAP repeat-containing 5, BIRC5), an *i*nhibitor of *a*poptosis *p*rotein (IAP) capable of binding and functionally sequestering caspases. Antisense oligonucleotides designed to knockdown survivin mRNA lead to arrest of morula or early blastocyst development and apoptosis that is responsive to caspase inhibitors (214). Similarly, targeted deletion of the survivin gene results in embryonic death by E4.5 (215,216), though this is ascribable not only to its role as a regulator of apoptosis, but also to loss of its mediation of chromosomal segregation during mitosis, as we discuss further in the next section.

CELL CYCLE CONTROL IN EARLY MITOSES

The somatic cell cycle is envisioned as being subdivided into four time periods, one for each mitosis (M phase) and DNA synthesis (S phase) as well as two intervening growth or gap phases (G_1 and G_2). Similarly, mammalian zygotes immediately have a recognizable four partite cell cycle organization, unlike early *Xenopus* and *Drosophila* cleavage embryos, which alternate seamlessly from mitosis into replication until the 12th and 14th cell cycles without appreciable G_1 and G_2 phases. In addition to basic cellular functions that occur during G_1 and G_2, "checkpoint" pathways exist at these junctures to ensure satisfactory cell status before commitment to replication or mitosis is made; this includes assessment of genomic integrity. As the intricacies of somatic cell cycle regulatory mechanisms are investigated, many are likely to be relevant to early embryos and the "oocyte-to-embryo" transition; however there are also unique aspects to the completion of meiosis and first mitotic divisions.

Cell Cycle Progression in One-Cell and Two-Cell Embryos

Few mammalian cell cycle regulators are known to be operative during the earliest cell divisions, and discovering them has been challenging for several reasons. The small size and limited numbers of readily collectable mammalian eggs and embryos makes direct manipulation difficult and biochemical analyses tedious. Also, redundancy of regulator functions, essential roles that some gene products hold in meiosis, and the presence of maternal protein and mRNA endowments in early zygotes leave us with a paucity of relevant mutant mouse models. Nevertheless, what is known alludes to a rewarding research area (Fig. 13).

Maturation or *M*-phase *p*romoting *f*actor (MPF; a complex composed of cyclin B1 and $p34^{Cdc2}$) is the most well-studied cell cycle driver important in the progression of meiosis and mitosis. MPF was originally described in *Xenopus* oocytes but has also been documented in mammalian species, peaking in activity levels during metaphase I and II of female meiosis. MPF activity peaks and plateaus at its maximum

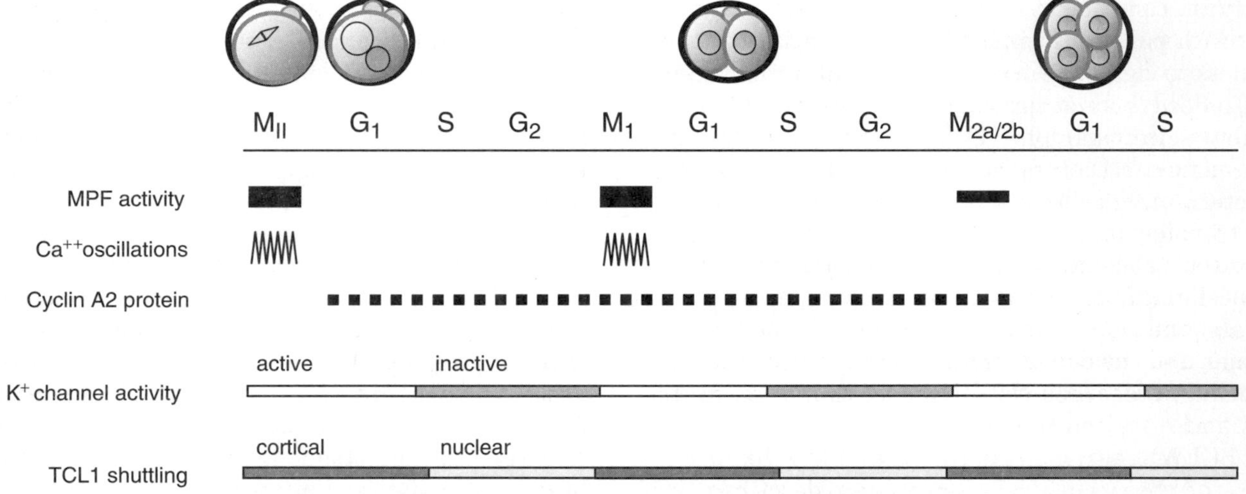

FIG. 13. Progression through the first cell cycles. A handful of candidate cell cycle regulators has been studied in the first two cleavages of mammalian embryos. The schematic shown represents a time line showing the completion of meiosis (metaphase II) and the first and second mitoses (M_1 and $M_{2a/2b}$). The *black bars* correspond to metaphase of meiosis II and the mitoses, points when MPF activity is high. Peak MPF activity during the second division is somewhat attenuates as compared with previous divisions, which is indicated by height of the black bar. Ca^{2+} ion fluxes (*M marks*) represent rapid oscillations and have been described in metaphase II and the first mitosis; their presence later in development is not clear. Cyclin A2 protein (*broken line*) is translated from maternal mRNAs and is present from the first G_1 through the first cell division to be cleared at the second $M_{2a/2b}$. Finally, the state of K^+ channels (active/inactive) and TCL1 location (cortical/nuclear) are depicted as alternating *shaded bars*.

when oocytes mature to arrest in metaphase II before fertilization, and during this time it is stabilized against Ub-mediated proteolysis by *cytostatic factor* (CSF; a microtubule-associated protein kinase kinase kinase termed Mos). Metaphase II ends with a calcium ion (Ca^{2+}) influx at fertilization that triggers calcium/calmodulin-dependent protein kinase II activation, inactivation of CSF, and degradation of both MPF and CSF [reviewed in (217)]. MPF activity is terminated with cyclin B1 protein clearance, which is ascribed to activation of a hypothetical E3 Ub–ligase complex, termed the anaphase promoting complex. Loss of MPF activity (which can be mimicked by the $p34^{Cdc2}$ kinase inhibitor roscovitine) seems to play a role in squelching the Ca^{2+} oscillations, whereas colcemid treatment to preserve MPF activity and metaphase II prolongs the fertilization/activation-induced Ca^{2+} ion fluxes (218).

Peaks of MPF activity recur during the first and second embryonic cell divisions (mitoses M_1 and $M_{2a/2b}$), and similar to egg activation at fertilization, Ca^{2+} transients occur during at least the first mitosis. Indeed, when the intracellular Ca^{2+} ion chelator 1, 2-bis(o-aminophenoxy)ethane-*N,N,N',N'*-tetraacetic acid is used, either metaphase II release or nuclear membrane breakdown and the start of M_1 can be blocked (219). At least one MPF target involved in cell cycle control, a downstream kinase termed polo-like kinase 1 (PLK1), is also activated in mouse oocytes, not only during progression through meiosis, but during an M_1-like mitosis after parthenogenetic activation (220). PLK1 subcellular localization indicates that it likely functions in microtubule assembly both during meiosis and mitosis (221). Together, these findings are indications that the first mitosis is regulated in part by the same mechanisms that control meiosis. During G_2 of the two-cell stage, abilities to destroy MPF and CSF activities are diminished, and a smaller increase in MPF activity is seen in this division (222). The precise extent that MPF is responsible for regulating the first cell divisions, the interplay between MPF and Ca^{2+} concentrations during mitoses, and the downstream and cooperating factors in operation as mitotic cell division regulators augment and replace meiotic ones are under investigation.

In addition to the cyclin B1 subunit of MPF, cyclin A2 is thought to be an important regulator of early mitoses. Cyclin A2 protein is thought to operate in somatic cells in a biphasic manner, at the onset of S phase and at late G_2, when it mediates commitment to DNA replication and stabilizes B-type cyclins during mitosis, respectively. Cyclin A2 is normally cleared from cells during mitosis to be replenished by de novo synthesis each cycle. Similarly, cyclin A2 protein is expressed in early one-cell zygotes by polyadenylation and translation of maternal mRNAs after fertilization so as to function during the first replication of the zygote genome (45). Interestingly, however, cyclin A2 escapes total degradation during the first mitosis (M_1) and is not cleared until the second mitotic divisions ($M_{2a/2b}$) at the transition from two to four cells (223). During the one-cell stage, cyclin A2 is located within both pronuclei, and in two-cell embryos it is recruited to the nuclei after G_1 and remains nuclear in S and G_2 (223). It has been shown in some systems that presence of an A-type cyclin resistant to degradation can prolong mitosis (224), and indeed the first mitosis M phase requires a lengthy 120 minutes, essentially twice the length of subsequent cleavages (225). Whether the duration of the first mitosis is in fact related to the persistence of cyclin A2 and what purpose such a delay system might have aside from a lesser need for cyclin A2 synthesis in early two-cell embryos is yet to be established.

During the G_2 phase of the second cell cycle, mRNAs encoding several cell cycle regulators (including cyclin D1, cyclin E, CDK2, CDK4, and p21) increase in abundance, consistent with broader activation of the zygote genome at this point and roles for the newly synthesized regulator proteins in the progression of subsequent cell cycles. In contrast, mRNAs encoding some regulators are diminished during the two-cell stage, including cyclin D2 and RB (226). For the latter this is reflected at the protein level so that RB persists at quantities detectable by Western blot only partially through the two-cell stage and is not appreciable again until the blastocyst stage. Moreover, inappropriate expression of RB is detrimental to preimplantation mouse development (227). Whether the down-regulation of RB midway through the two-cell stage represents a departure from the mechanisms governing the first cell cycle or whether RB does something else in one-cell and early two-cell zygotes remains a provocative question.

Independent from the fluctuations of cyclin proteins considered above, K^+ ion channel activities may direct aspects of cell cycle timing. Functioning K^+ channels (candidates include members of the EAG [*e*ther *a* *g*o-go] or ERG [*E*ag-*r*elated *g*ene] families) exist in all stages of preimplantation mammalian embryos, alternating between high activity levels during M and G_1 and inactivity or low activity in S and G_2 (228,229). Addition of puromycin to block protein synthesis interferes with cyclin production but does not disrupt the oscillations in K^+ conductance (228,229). This indicates that K^+ transit is mediated by channel activity that is separable from cyclin-CDK activities. Similarly, embryo fragmentation with exclusion of the nucleus does not preclude channel activity timing. Interestingly, however, inhibiting DNA synthesis (with aphidicolin) or inducing DNA damage (with mitomycin C) prevents the diminution of K^+ channel activity at S phase (230), indicating that K^+ conductance is influenced by genomic insult checkpoints. Along with the variations in MPF activity and cortical-nuclear translocations of the maternal effect T-cell leukemia/lymphoma 1 (TCL1) described previously (22), these active–inactive K^+ channel state switches represent a handful of molecular events known to correlate with cell cycle progression in the first mitotic divisions. Identifying the K^+ channels involved, origins of their fluctuating activity, downstream effects of K^+ ion fluxes, and potential interactions with other cell cycle regulators are intriguing areas of study.

We mentioned earlier that inhibiting DNA metabolism or inducing DNA damage precludes the S phase inactivation in K^+ ion channel activity, and evidence is emerging that checkpoints exist as early as the first mitosis to prevent cell cycle progression if defects of genomic content are detected. Compared with male meiosis, female meiosis during gametogenesis seems relatively resistant to the loss of gene products causing meiotic disruption (231), perhaps creating a necessity for stringent checks during the first embryonic cell cycle. Indeed, at least nearing the conclusion of meiosis, DNA damage (induced by doxorubicin or etoposide) does not interfere with appropriate release from metaphase II and completion of meiosis. The damaging insult does not induce fragmentation of metaphase II–arrested cells but triggers DNA damage recognition systems only after activation with strontium, which alone would release the meiosis block and allow progression of the parthenote to the two-cell stage (232).

Interesting evidence for a checkpoint before the first mitosis (M_1) was found using an experimental system fusing oocytes interrupted in meiosis after meiosis I in either a G_2-like postreplication stage or a G_1-like stage; the arrests were produced by kinase inhibitors 6-dimethylaminopurine and butyrolactone-1, respectively. Fusion of two G_2-paused oocytes leads directly to M_1 and the generation of a two-cell–like embryo within a few hours of culture in 6-dimethylaminopurine–free media. In contrast, similar fusion of a G_2-like cell with a cell containing an unreplicated G_1-like nucleus results in a stall in mitosis until replication in the latter was complete (233). This suggests the presence of a genome assessment before M_1-phase commitment, but interpretation is complicated by the interruption of meiosis and abnormal activation as well as the altered ploidy and exclusively maternal nature of the DNA.

Further insight into the regulation of S phase has come from studies of zygotes obtained by metaphase II oocyte fertilization with X-irradiated spermatozoa (234). Fertilization with spermatozoa exposed to 6 Gy is correlated to activation of p53-responsive reporter constructs in the female pronucleus of resulting one-cell zygotes and slows replication in both pronuclei in a manner dependent on functional p53 alleles. The p53 protein (also known as transformation related protein, TRP53) is an important checkpoint participant known to interfere with Cdk activity promoting S and M phases. Although p53 is normally dispensable for early development, *Trp53*$^{-/-}$ zygotes generated from irradiated spermatozoa rapidly complete S phase but undergo aberrant chromosomal segregation during mitoses and degenerate before implantation (235). In this system, a G_2/M phase checkpoint was not clearly demonstrated, given that mitosis ultimately progressed not only in *Trp53*$^{-/-}$ embryos with extensive genomic damage, but also in wild-type embryos that failed to complete replication during the prolonged S phase (235). However, it does seem that an "S phase checkpoint," as it has been termed, mediates a lengthened S phase in one-cell zygotes in response to DNA damage in a manner dependent on p53 and functioning symmetrically in the male and undamaged female pronuclei. The prolonged S phase damage response may be tied directly to the viability of these embryos during the period of implantation, and presumably this and other p53-mediated mechanisms are involved ultimately in facilitating repair (236–238). Clinically, research into checkpoints and repair pathways in early embryos is of great importance, with relevance to the study of medical conditions as diverse as molar pregnancies, aneuploidy syndromes, and the inheritance of more subtle mutations.

Regulators of Later Cell Divisions

As maternal factors are cleared, subsequent cell divisions are largely controlled by proteins encoded by the zygotic genome, and homozygous deletion of genes absolutely essential for cell viability and proliferation manifest themselves. Null mouse models provide us with both direct and indirect evidence for the importance of several cell cycle regulators during embryonic divisions (Fig. 14).

In the preceding section, we reviewed how cyclin A2 (CCNA2) protein accumulates after fertilization by translation of maternal mRNA stores as well as potential consequences of its persistence during the first embryonic mitosis. Maternal cyclin A2 stores are depleted at the second mitoses, and although *Ccna2*$^{-/-}$ mice die after implantation, they survive beyond the exhaustion of maternal stores (223,239). Thus, although expression of cyclin A2 suggests it plays important roles in preimplantation development, any necessary roles in the first cell cycles are met by maternally encoded protein. Embryos from

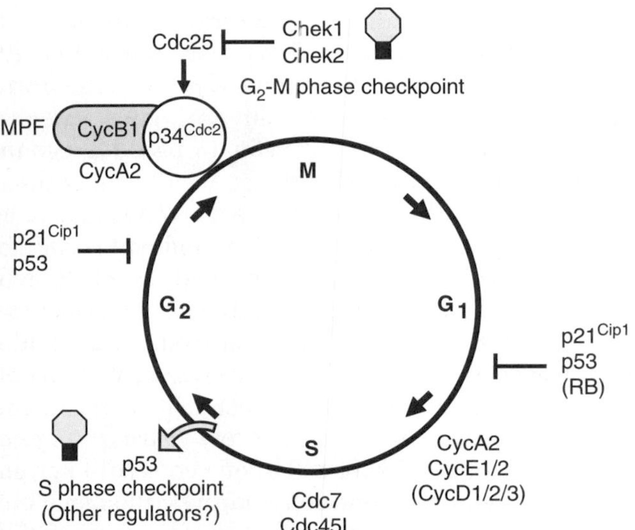

FIG. 14. Cell cycle regulation in cleavage stage embryos. The commitment to mitosis is made by cyclin/kinase complexes, like MPF. Cyclin A2 and E-type cyclins have also been implicated in either the commitment to mitosis or S phase in somatic cells and have been studied in early murine development. In somatic cells, D-type cyclins are involved in inactivation of retinoblastoma (RB) and facilitate S phase entry, although it seems that neither D-type cyclins nor RB are expressed in blastomeres; they are indicated in parentheses. Two checkpoints have emerged based on studies in early embryos, one regulating CDC25 phosphatase and MPF activation by CHEK proteins and one at the S phase commitment that seems to hinge on p53 during the early divisions, with a relaxed requirement for p53 in later blastomeres. S phase completion itself requires the DNA replication regulators CDC7 and CDC45L. In addition to p53, p21^{Cip1} has been studied in early development and may play roles in both S phase and M phase commitment.

the four-cell to blastocyst stages are capable of cell cycle progression with neither maternal nor embryonic cyclin A2. It has been postulated that cyclin A1 is instead critical to early development in mediating G_1-to-S and G_2-to-M transitions, although *Ccna1*$^{-/-}$ embryos have shown that cyclin A1 is only required for male gametogenesis (240). Similar to the early embryonic lethality of cyclin A2 knockouts, cyclin B1 (*Ccnb1*$^{-/-}$) knockout mice die before E10, although the exact timing of cell cycle disturbances and any correlation between their onset and the depletion of maternal stores has not been reported (241).

There seem to be large degrees of functional redundancy among other cyclin proteins in early development. Cyclin B2 (*Ccnb2*$^{-/-}$), cyclin E1 (*Ccne1*$^{-/-}$), and cyclin E2 (*Ccne2*$^{-/-}$) single knockout mice demonstrate no defects in development, though regulation of B- and E-type cyclin expression has been studied in early embryos in the context of translational control of maternal mRNAs, early targets of ZGA, and aberrant cell cycle regulation in *Cul1*, *Cul3*, and *Cops2* knockout models (241,242). Moreover, double knockouts lacking both cyclin E1 and cyclin E2 (*Ccne1*$^{-/-}$, *Ccne2*$^{-/-}$) and double knockouts lacking any combination of two of the three D-type cyclins (*Ccnd2*$^{-/-}$, *Ccnd3*$^{-/-}$; *Ccnd1*$^{-/-}$, *Ccnd3*$^{-/-}$; or *Ccnd1*$^{-/-}$, *Ccnd2*$^{-/-}$) survive early preimplantation development. *Ccne1*$^{-/-}$, *Ccne2*$^{-/-}$ double knockout mice die at midgestation from defects in placental development associated with failed endoreplication in trophoblast cells, whereas other D-type cyclin double knockouts live until late gestation (243,244). Other combinatorial mutant studies and further analyses of cyclin expression in existing knockout lines can be expected to reveal more about potential functional redundancies of cyclins in early cell divisions.

Two cell division cycle (Cdc) proteins, the orthologs for which were first described in *Saccharomyces cerevisiae*, have been found to be essential to cell cycle control in early mammalian development. In yeast genetic models, the first of these, Cdc7 kinase, has been implicated in initiation of DNA synthesis and functions in part to recruit Cdc45, an important factor for DNA polymerase α recruitment and elongation [reviewed in (245)]. *Cdc7*$^{-/-}$ and *Cdc45l*$^{-/-}$ knockout mice lacking alleles for either ortholog die around the implantation period between E3.5 and E6.5 (246,247). Somewhat surprisingly given mutant studies in yeast, mouse ES cells lacking CDC7 are able to initiate S phase, indicating that the mammalian protein is not essential for the G_1-to-S phase transition, but S phase is not completed and triggers a block to continued progression. This S phase checkpoint can be bypassed, partially rescuing cell cycle continuation in *Cdc7*$^{-/-}$, *Trp53*$^{-/-}$ double mutants (246).

Although evidence for a G_2-M phase checkpoint is somewhat equivocal in the earliest cell divisions, by the morula and blastocyst stages, a p53-independent post–S phase checkpoint becomes more readily appreciable. Compromise of this checkpoint is manifest in embryos lacking checkpoint kinase 1 homolog (CHEK1), first described in *Saccharomyces pombe*; *Chek1*$^{-/-}$ embryos die between E2.5 and E3.5, and *Chek1*$^{-/-}$ ES cells are not recoverable (248). The *Chek1*$^{-/-}$ phenotype ultimately leads to apoptosis and embryo loss, and *Chek1*$^{-/-}$, *Trp53*$^{-/-}$ double mutants indicate the phenotype is independent of p53-mediated pathways. Normally, irradiation of ES cells leads to a G_2 arrest response, which is lost in conditional (Cre-mediated) *Chek1*-deleted ES cells as well as in *Chek2*$^{-/-}$ cells lacking a related checkpoint activator (248,249). CHEK proteins are postulated to affect cell cycle arrest by an inhibitory phosphorylation of CDC25 phosphatase, precluding p34^{Cdc2} dephosphorylation and activation. In mice, CDC25 orthologs are generated by three homologous genes, *Cdc25a*, *Cdc25b*, and *Cdc25c*. CDC25B proves essential for MPF activation during meiosis, but roles during early mitotic divisions for the persisting protein are unclear (250). Interestingly, a switch to up-regulate the CDC25A homolog at the blastocyst stage may play a role in regulating $p34^{Cdc2}$ in later cell divisions and maturation of a G_2-to-M transition checkpoint under zygote control (251). It is also notable that in addition to a p53-independent G_2-M checkpoint, ES cells show a prolonged S phase after irradiation that, unlike in early embryos, is no longer strictly dependent on p53 (252). Thus, other checkpoints may be maturing and becoming regulated in multifactorial manners in later embryonic cell divisions.

The early checkpoints we considered link cell cycle progression to DNA damage surveillance and repair machinery, limiting cell division or inducing apoptosis in cells with inappropriate DNA content. A host of knockout models lacking gene products that regulate genomic integrity and DNA partitioning during mitosis show that these become crucial to development as checkpoints are established and DNA damage incurred because of a null mutation accumulates (Table 8). Also, within a single embryo, checkpoints make for selective proliferation of cells with appropriate genomic content, as evidenced by experiments using aggregation chimeras mixing cells of normal zygotes with those derived from irradiated sperm (253,254). How these pathways may skew genetic mosaicism postgestation and how this impacts presentation of human diseases is of continued interest.

OVIDUCTAL FACTORS AND EXTRACELLULAR SIGNALS INFLUENCING EARLY DEVELOPMENT

Embryos have long been considered independent of extraembryonic signals until the time of implantation, when intimate communication with the maternal environment becomes quite obvious. The fact that fertilization and preimplantation embryo development have been accomplished in vitro has supported this supposition, although we have some recent compelling indications that early development is optimized by the oviductal or Fallopian tube environment. Moreover, preimplantation embryos seem capable of responding to autocrine factors (generated within an individual embryo or by neighboring embryos) and paracrine or endocrine factors (elaborated by the female reproductive tract or distant maternal organs). In this section, we review highlights of a now rapidly expanding field.

The Oviductal Microenvironment

Evidence that the oviduct or Fallopian tube optimizes early embryo development is derived largely from in vitro coculture experiments. Mouse embryos cultured from the two-cell to morula stages with mouse oviduct cells have improved cleavage rates and viability in vitro as well as higher rates of fetal development after transfer to pseudopregnant females when compared with those cultured without oviduct cells (277,278). It seems that analogous relationships between the oviduct and embryo affect human development, in that addition of human oviductal cells in culture is capable of increasing the numbers of hatching human blastocysts by fivefold if coculture is started the first day after in vitro fertilization (during the one-cell pronuclear stage) and twofold if begun on the second day (during the two-cell stage) (279). Use of a 24-hour coculture system between the human one-cell and two-cell stages in a prospective randomized clinical in vitro fertilization setting increased the numbers of embryos recovered, raised both the pregnancy rate and numbers of viable fetuses, and decreased spontaneous abortion rates (280). The similarity between murine and primate mammalian species is further underscored by the finding that coculture with human oviductal cells has embryotrophic effects on mouse embryos; human oviduct cell-conditioned media and media fractions had dose-dependent effects on murine development that were abolished by heat or trypsin treatment (281). In addition to benefits appreciable at the level of embryo morphology, specific changes to the murine embryonic transcriptome induced by human oviductal coculture have been identified, affecting gene targets implicated in everything from transcriptional control to blastocoel formation (282).

Efforts to isolate and identify specific oviduct-derived embryotrophic factors are ongoing, and a number of candidate ligands have emerged, including those with known receptors on early embryos (as we consider below) and those (like complement protein iC3b) (283) for which candidate binding proteins are being studied. Other oviduct-secreted proteins, such as protease inhibitors (including secretory leukocyte protease inhibitor, placental protein 5, tissue inhibitor of metalloproteinase 1, and plasminogen activator inhibitor 1) (284–286) and immunoglobulins (IgG and IgA) (287), may not directly interact with gametes or embryos but likely have protective functions. A subset of oviduct-expressed genes is under hormonal control, being highest in expression at the postovulatory phase of the estrous cycle; these include a heterogeneous group of proteins termed oviductal secretory glycoproteins. Their regulation may reflect important roles in promoting embryo development, though the major identified oviductal secretory glycoprotein in mice (encoded for by the oviductal glycoprotein 1 gene [*Ovgp1*]) is dispensable for fertility in knockouts (288). Perhaps even more interesting than hormonal

TABLE 8. *Knockout models postulated to compromise genomic integrity or organization in early embryos*

Gene	Somatic cell functions	Loss of function phenotype	References
Amyloid β (A4) precursor like protein (*Aplp2*)	Centromere CDEI-element binding; important in segregation	Embryonic lethal before E3.5; abnormal DNA content	(255)
Ataxia telangietasia and Rad3 related (*Atr*)	DNA damage surveillance/Chk1 phosphorylation	Embryonic lethal between E3.5 and E7.5; chromosomal breaks precede apoptosis	(256,257)
Baculoviral IAP repeat-containing 5 (*Birc5*; Survivin)	Inhibitor of apoptosis protein; associates with spindle during mitosis	Null embryos die by E4.5; disrupted microtubule formation and polyploidy	(215)
Budding uninhibited by benzimidazoles 3 homolog (*Bub3*)	Component of the mitotic spindle assembly	Embryonic lethal by E7.5; abnormal nuclei/ mitotic errors accumulate from E4.5	(258)
Centromere autoantigen A (*Cenpa*)	Nucleosomal packaging of centromeric DNA	Embryonic lethal by E6.5 with abnormal nuclei; lost CENPC localization	(259)
Centromere autoantigen C (*Cenpc*)	DNA binding protein of the inner kinetochore plate	Lethal; abnormal nuclei and mitotic arrest in morula	(260)
Centromere autoantigen E (*Cenpe*)	Centromere "passenger protein"; microtubule capture at kinetochores	Lethal; *Cenpe*$^{-/-}$ embryos undergo early arrest	(252)
Erg-associated protein with SET domain (Eset; SET domain bifurcated 1, *Setdb1*)	Histone methyltransferase associated with pericentric heterochromatin	Embryonic lethality between E3.5 and E5.5; no *Setdb1*$^{-/-}$ ES cells recovered	(261)
Flap specific endonuclease 1 (*Fen1*)	Base excision repair and lagging-strand synthesis	Lethal; failed ICM development in blastocysts; defects evident in S phase of blastocyst outgrowths	(262)
Inner centromere protein (*Incenp*)	Centromere "passenger protein"; microtubule dynamics during segregation	*Incenp*$^{-/-}$ embryos exhibit failed blastocyst development; mitotic and nuclear abnormalities by E2.5	(263)
Nibrin (*Nbn*; homolog of Nijmegen breakage syndrome)	DNA damage surveillance/ checkpoint activation	Embryonic lethal from E3.5-E7.5; extensive apoptosis	(264)
Rad51 (mammalian homolog of *Escherichia coli recA* and *Saccharomyces cerevisiae RAD51*)	Homologous recombination and dsDNA break repairs	Embryonic lethal; reduced *Rad51*$^{-/-}$ embryos at four and eight-cell stages; Null ES cells are not viable	(265)
REV3-like, catalytic subunit of DNA polymerase ζ; RAD45 like (*Saccharomyces cerevisiae*; *Rev3l*)	Catalytic subunit of DNA polymerase ζ; involved in dsDNA break repair	*Rev3l*$^{-/-}$ embryos have poor development of the early ICM and die around E10.5; dsDNA breaks evident in mutant embryos	(266–269)
Structure specific recognition protein 1 (*Ssrp1*)	High mobility group box protein involved in chromatin remodeling, DNA replication and repair	*Ssrp*$^{-/-}$ embryos die around implantation; homozygous null ES cell lines were not recoverable	(270)
SWI/SNF related, matrix associated, actin dependent regulator of chromatin, subfamily a, member 4 (*Smarca4*; Brahma-related gene 1, Brg1)	ATP-dependent chromatin remodeling/nucleosome sliding	Embryonic lethality in the preimplantation period	(271)
SWI/SNF related, matrix associated, actin dependent regulator of chromatin, subfamily a, member 5 (*Smarca5*; Snf2h)	ATP-dependent chromatin remodeling/nucleosome sliding	Embryonic lethal between E5.5-E7.5; ES cells could not be recovered from *Smarca5*$^{-/-}$ blastocysts	(272)
SWI/SNF related, matrix associated, actin dependent regulator of chromatin, subfamily b, member 1 (*Smarcb1*; Snf5)	ATP-dependent chromatin remodeling/nucleosome sliding	Preimplantation embryonic lethal; tumor suppressor functions evident in heterozygotes	(273)
Telomeric DNA binding factor (*Trf1*)	Prevention of checkpoint activation by chromosome ends	Increased apoptosis in the blastocyst ICM; death between E5 and E6	(274)
Topoisomerase I (*Top1*)	Induces single strand breaks for relaxing DNA supercoiling	*Top1*$^{-/-}$ embryos die at the four-eight cell stage; homozygous ES cells could not be recovered	(275)
Topoisomerase IIα (*Top2a*)	Induces double strand breaks for relaxing DNA supercoiling	*Top2a*$^{-/-}$ embryos die at the four-eight cell stage	(276)

regulation of oviductal gene expression is the finding that developing embryos play a role in shaping the oviduct transcriptome, up-regulating genes like thymosin β4, which is involved in actin depolymerizaion (289). Thus, a dialogue between preimplantation embryos and the maternal oviduct takes place, perhaps as intricate and important as those between oocytes and ovarian somatic cells during folliculogenesis or those between embryos and the uterine environment around implantation (290,291). Our increasing understanding of these interactions in preimplantation development should continue to improve assisted reproductive techniques and may identify targets for novel contraceptives and etiologies of infertility and ectopic pregnancies.

Extracellular Signaling Pathways in Early Development

In addition to their responsiveness to paracrine factors derived from the oviduct, blastomeres of preimplantation embryos seem to have receptors capable of interpreting a multitude of extracellular signals (Table 9). Autocrine and paracrine factors may be derived from neighboring embryos or unfertilized eggs as well as from organizing signals from blastomeres within a single embryo (292). In contrast, embryos also express receptors for paracrine or endocrine factors, such as estrogens, produced in maternal tissues and not within the preimplantation embryos themselves (293,294). In several cases, there is potential for ligand production by multiple sources, such as both central maternal endocrine organs and the local maternal reproductive tract, or by both maternal and embryonic cells; in many, the location exerting a dominant signaling influence is still unclear.

Mouse models have been illustrative of the in vivo importance of several growth factor signaling cascades or their components in early development. For example, female knockout mice lacking leukemia inhibitory factor (LIF; *Lif*$^{-/-}$) demonstrate defects in the implantation of their blastocysts, and epidermal growth factor receptor null mice (*Egfr*$^{-/-}$) die in the preimplantation period on a CF1 genetic background with degeneration of their ICM (205,324). Notably however, loss of maternal LIF is not phenocopied by deletion of embryonic *Lifr* (332), and neither is the *Egfr*$^{-/-}$ embryonic phenotype recapitulated in offspring of *Egf* null mothers (333). Thus, complexities of these signaling pathways are being reviewed; explanations likely include as yet unappreciated related ligands and receptors, overlapping roles for maternal and fetal sources in their production, independent activities of free receptors, and other pathway participants that may function as modifiers. For example, other potential ligands for epidermal growth factor receptor include TGF-α and amphiregulin (AREG). TGF-α alone is dispensable for preimplantation embryo development, viability, and female fertility, though TGF-α homozygous mutant (*Tgfa*$^{-/-}$) embryos demonstrate increased levels of apoptosis within their ICM (334). Antisense *Areg* mRNA expression slows development to the blastocyst stage in vitro (300), though any preimplantation phenotype of *Areg*$^{-/-}$ mice has not been appreciated. Relatively subtle preimplantation phenotypes have been described in murine models of colony-stimulating factor and cannabinoid signaling pathways, though in many other potentially relevant models, such observations pertaining to early development have not been sought and reported.

Murine models have provided more information regarding signaling pathway structure in the cases of fibroblast growth factor (FGF) and TGF-β pathways. *Fgf4*$^{-/-}$ and *Fgfr2*$^{-/-}$ mice die shortly after implantation, the relatedness of these phenotypes bolstering the idea that this is an important ligand–receptor pair in vivo (303,304). TGF-β1 null (*Tgfb1*$^{-/-}$) embryos develop a block before morula formation with penetrance depending on the genetic background (335). Knowledge of this phenotype is helpful in interpreting findings of receptor loss-of-function and gain-of-function studies as well as effects of enhanced TGF-β signaling inhibitory pathways. The preimplantation block can be recreated by either expression of a dominant negative type II receptor (dnTβRII) (336) or by overexpression of an inhibitory SMAD, SMAD7 (337); it can be rescued by provision of a constitutively active type I receptor (TβR1) (329,336). Together, these studies provide compelling evidence that the basic organization of TGF-β superfamily signaling described in somatic cells (Fig. 15) is operational in early development. Moreover, cell-specific expression of TGF-β superfamily members in blastocysts (such as the ICM-specific expression of TGF-β1, TGF-β3, activin βA and activin βB, and the TE-specific expression of TGF-β2 and TβRII) as well as changes in expression of some TGF-βs that accompany implantation suggest that these proteins optimize later preimplantation development and implantation and may serve in intraembryonic signaling and differentiation pathways (329,338).

MORPHOLOGICAL AND METABOLIC CHANGES IN EARLY CLEAVAGE EMBRYOS

The first mitosis of mammalian embryos is completed with a meridional divide that bisects the zygote, and subsequent cleavage planes arise asynchronously and are rotational in their respective orientations.

TABLE 9. *Receptors and extracellular ligands potentially important during preimplantation development*

Receptor(s) expressed in preimplantation embryos	Potential extracellular ligand(s)	Notes	References
Calcitonin receptor (*Calcr*)	Calcitonin	*Calcr* mRNA is upregulated in early cleavage embryos; calcitonin accelerates blastocyst development in vitro and is expressed by the endometrium at implantation	(295)
Cannabinoid receptors (brain-type, *Cnr1* and macrophage-type, *Cnr2*)	Endocannabinoids (ex: anandamide)	*Cnr1* and *Cnr2* mRNAs are detected beyond the four-cell stage; uterine anandamide and blastocyst CNR1 are down-regulated around implantation; *Cnr1*$^{-/-}$ or *Cnr2*$^{-/-}$ mice exhibit delayed and asynchronous preimplantation development	[Reviewed in (296)]
Colony-stimulating factor 1 receptor (*Csf1r*)	Colony-stimulating factor 1, macrophage (CSF1)	Embryos express *Csf1r* mRNA during most stages of early development; CSF1 treatment in vitro speeds mouse blastocyst development and increases the number of cells in blastocysts	(297)
Epidermal growth factor receptor (*Egfr*)	Epidermal growth factor (EGF) (as well as transforming growth factor [TGF]-α and amphiregulin [AREG])	EGFR is expressed by blastocysts; EGF, TGF-α, and AREG enhance blastocoel formation; delayed implantation and blastocyst dormancy is associated with reversible down-regulation of *Egfr*	(298–301)
Estrogen receptors α and β (*Esr1* and *Esr2*)	Estrogen	*Esr1* and *Esr2* mRNAs are expressed during most of preimplantation development and estrogen-responsive gene expression can be appreciated; estrogen steroid is likely maternal rather than embryonic	(293,294)
Fibroblast growth factor receptor (*Fgfr2*)	Fibroblast growth factor 4 (FGF4)	*Fgfr2*,*Fgf4*, and other FGF ligands and receptors are differentially expressed in lineages of early embryos; *Fgfr2*$^{-/-}$ and *Fgf4*$^{-/-}$ mice die shortly after implantation, and expression of a dominant negative *Fgfr* transgene derails blastomere cell divisions	(302–305)
Frizzled homologs 2 and 4 (*Fzd2* and *Fzd4*)	Wingless-related MMTV integration site ligands (WNT3A, WNT7B)	mRNA microarray analyses demonstrate several WNT pathway ligands, receptors, and signal transducers are expressed during specific periods of early embryogenesis	(306)
Gonadotropin receptors (*Fshr* and *Lhr*)	Follicle-stimulating hormone (FSH) and luteinizing hormone (LH)	*Fshr* and *Lhr* mRNAs are present in all stages of murine preimplantation development and most stages of human preimplantation embryos	(307,308)
Gonadotropin-releasing hormone receptor (*Gnrhr*)	Gonadotropin-releasing hormone (GnRH)	GnRHR and GnRH proteins are expressed by mouse and human embryos and GnRH ligand is also produced in the human oviduct; GnRH agonists promote murine preimplantation development in vitro while antagonists block development	(309–311)
Growth hormone receptor (*Ghr*)	Growth hormone (GH)	GHR and GH immunoreactivity is demonstrable in preimplantation embryos; GH treatment improves mouse preimplantation development in vitro	(312,313)
Growth hormone secretagogue receptor (*Ghsr*)	Ghrelin (GHRL)	Embryos express *Ghsr* from the morula stage, and *Ghrl* mRNA is found in embryos and endometrium; GHRL protein is upregulated in the uterine fluid with fasting and inhibits preimplantation development in vitro	(314)
Insulin receptor (*Insr*)	Insulin (INS)	Embryos become responsive to maternal insulin during compaction; it affects their metabolism of glucose and amino acids as well as stimulates ICM cell proliferation	[Reviewed in (315)]
Insulin-like growth factor receptors (*Igf1r* and *Igf2r*)	Insulin-like growth factors (IGFI and IGFII)	Preimplantation embryos express both IGFI and IGFII, as well as their cognate receptors; antisense IGFII compromises in vitro development	(316,317)

(Continued)

TABLE 9. *Receptors and extracellular ligands potentially important during preimplantation development—Cont'd*

Receptor(s) expressed in preimplantation embryos	Potential extracellular ligand(s)	Notes	References
Interferon-γ receptor (*Ifngr*)	Interferon-γ (IFNG)	*Ifngr* mRNA and its encoded protein are present throughout preimplantation development	(318)
Kit oncogene (*Kit*)	Stem cell factor (SCF; also known as Steel and Kit Ligand)	Blastocysts express KIT receptor before implantation, and SCF promotes blastocyst outgrowth in vitro; both embryos and the maternal reproductive tract express KIT and SCF during implantation	(319)
Leptin receptor (*Lepr*)	Leptin (LEP)	Embryos express *Lepr* throughout preimplantation development; *Lep* mRNA is expressed by the oviduct, endometrium, and blastocysts. Leptin treatment optimizes blastocyst development in vitro, an effect that can be blocked with antibodies recognizing LEPR	(320,321)
Leukemia inhibitory factor receptor (*Lifr*)	Leukemia inhibitory factor (LIF)	Blastocysts express *Lifr* in the ICM and *Lif* in the trophectoderm; LIF supplementation promotes embryo development and ES pluripotency in vitro, and uterine LIF is crucial for implantation	(322–324)
Notch family receptors (*Notch2* and *Notch3*)	Delta and Jagged ligands (DLK1, DTX1, and JAG2)	mRNA microarray analyses demonstrate several Notch pathway ligands, receptors, and signal transducers are expressed during specific periods of early embryogenesis	(306)
Platelet-activating factor receptor (*Ptafr*)	Platelet-activating factor (PAF)	Both PAF and its receptor are expressed in early embryos; PAF supplementation improves in vitro blastocyst development; signaling pathways are being explored in *Ptafr*$^{-/-}$ embryos	(325–327)
Platelet-derived growth factor α receptor (*Pdgfra*)	Platelet-derived growth factor α (PDGFA)	Preimplantation embryos from the two-cell to blastocyst stages express both PDGF ligand and receptor; blastocyst receptors respond to exogenous PDGF	(328)
TGF-β/BMP receptors	TGFβ/BMP superfamily ligands	Several TGF-β/BMP superfamily ligands and receptors are expressed in early development; see text for both loss-of-function and gain-of-function studies in murine embryos	[Reviewed in (329)]
Tumor necrosis factor receptor superfamily, member 1a (*Tnfrsf1a*)	Tumor necrosis factor (TNF-α)	TNF-α treatment inhibits ICM cell proliferation and promotes apoptosis in vitro; antisense *Tnfrsf1a* is partially protective	(330,331)

These divisions are referred to as *indeterminate cleavages* in deuterostomes because each of the cells maintains totipotency. Cellular proliferation of preimplantation embryos takes place within the zona pellucida, and there is little increase in the total volume of the embryo until after hatching.

Morula Compaction

During the eight-cell morula stage, component blastomeres lose their distinct boundaries and express intercellular adhesive junctions, becoming *compacted* (Fig. 16) (339). Compaction is unique to mammalian embryos. The process is accompanied by establishment of cell polarity, with microvillar structures marking apical surfaces and nuclear localization basolaterally. In addition to normal development, perhaps the study of cell associations in early embryos has relevance to separation of blastomeres resulting in monozygotic (identical) twinning.

At the molecular level, cadherin 1 (CDH1, E-cadherin, uvomorulin) is an important participant in compaction. CDH1 is initially uniformly distributed on cell surfaces in early embryos and then clusters at intercellular contact sites during the eight-cell stage (340). Compaction is disrupted when CDH1 function is blocked by inhibitory antibodies (341), and inhibition also precludes premature compaction and relocalization of CDH1 induced by protein kinase C pathway modulators (342). Embryonic CDH1 is produced by transcription of maternal *Cdh1* alleles

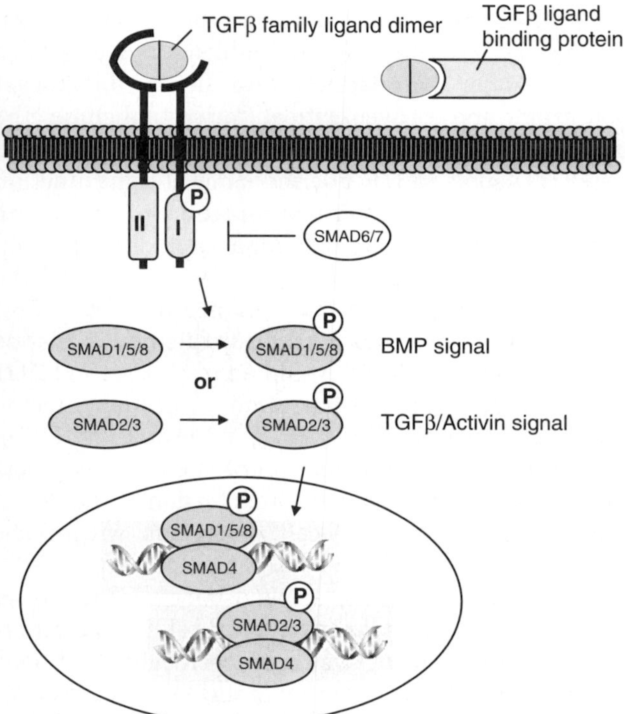

FIG. 15. TGF-β family signaling pathways in somatic cells. Extracellular ligand dimers may be bound to proteins or are available to interact with cell surface receptors. Ligand binding by type I and type II receptor pairs induces phosphorylation of the type I receptor and initiation of SMAD activation by phosphorylation. Bone morphogenetic protein (BMP) signals usually activate SMADs 1, 5, and 8, whereas TGF-β or activin ligands activate SMADs 2 and 3. Inhibitory SMADs attenuate this initiation. Activated SMAD proteins move into the nucleus with the common SMAD4 to affect gene transcription (416).

during oogenesis as well as by the zygotic genome beginning in the late two-cell stage, and its function has been studied by several approaches that preclude mRNA transcription or translation. Maternal stores of the protein can be depleted by oocyte-specific gene deletion, resulting in defects in blastomere adherence in eight-cell morula, after which point the phenotype is overcome by activation of the paternal *Cdh1* allele in the zygote (143). The delay in compaction does not seem to compromise later development. Cadherin function attributable to activation of the zygote genome has been tested by traditional knockout approaches. Early *Cdh1* knockout embryos retain stores of maternally encoded CDH1 and are able to undergo compaction, although cell associations are then loosened or "decompacted" by E4.5, and cavitation and trophoectoderm development fail (343,344). The knockout phenotype is recapitulated by RNAi in embryos injected with *Cdh1* dsRNA at the one-cell stage (345). Interestingly, injection of morpholino antisense oligonucleotides that block translation of *Cdh1* mRNAs but do not trigger RNAi degradation pathways results in an earlier phenotype. In these embryos, there is very early depletion of CDH1 protein and a complete block to further development at the two-cell stage (38). Perhaps the phenotype is so severe because of the extent or rapidity of RNA loss-of-function (as compared with RNAi) and the effect of the morpholino on both oocyte and zygotic *Cdh1* transcripts (as compared with the traditional knockout and the oocyte-specific knockout). The block appears specific to the loss of *Cdh1* function in that it is rescued by provision of a *Cdh1* mRNA lacking the 5′-UTR sequence targeted by the morpholino nucleotide. Thus, it seems that inhibition of *Cdh1* mRNA in oocytes and one-cell embryos could derail development even before compaction if not rescued by zygotic *Cdh1* activation, and the presence of one functioning zygotic *Cdh1* allele is required for the maintenance of compaction and cell specification in preimplantation development.

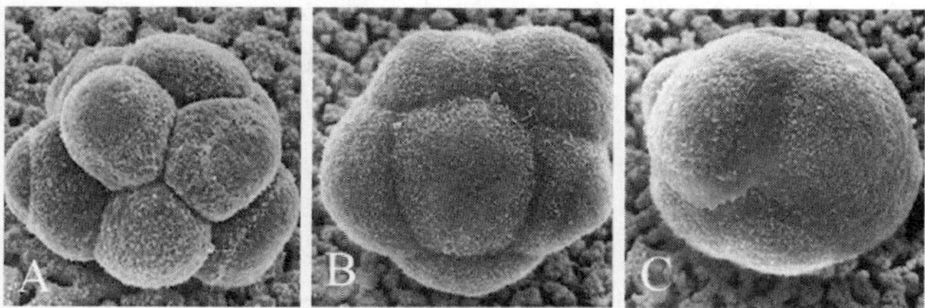

FIG. 16. Compaction of a mouse eight-cell embryo. The three panels represent a series of scanning electron micrograph images. [From (339) by copyright permission of The Rockefeller University Press.] **A:** An early eight-cell mouse morula beginning compaction. **B:** An eight-cell morula midway through compaction, with borders between cells becoming more obscured. **C:** An advanced eight-cell morula completing compaction where cell boundaries appreciable at this level have been essentially erased.

Glucose Import and the Relevance of Maternal Hyperglycemia and Hyperinsulinemia

In addition to the morphological changes of compaction, embryos in this period of development undergo a metabolic transition enabling the use of sugar substrates. Before this, primary sources of energy for the preimplantation embryo are pyruvate, lactate, and amino acids. Between the eight-cell and later morula stages, mRNAs encoding glucose transporters and several genes involved in glucose utilization are induced, regulated in part by maternal glucose, insulin, and insulin-like growth factor levels (346). Maternal hyperglycemia down-regulates expression of the embryonic facultative glucose transporter GLUT1, encoded for by the *Slc2a1* gene (solute carrier family 2, facilitated glucose transporter, member 1) as well as the related GLUT2 and GLUT3 transporters (347). Though at first counterintuitive, the reduced import of glucose in hyperglycemic conditions may itself compromise development, so that embryos are undergoing a self-imposed starvation (348). Others propose that glucose transporter down-regulation is actually protective. Whatever the nature of insult, a downstream mechanism for cell death has emerged wherein hyperglycemic embryotoxicity triggers Bax-mediated apoptosis (349,350). *Glut1*$^{-/-}$ cells are not viable, and loss-of-function of GLUT1 has instead been modeled by transgenic expression of antisense *Slc2a1* in mice; this leads to multiple developmental malformations (351). Interestingly, *Glut3*$^{-/-}$ embryos demonstrate failed blastocoel development, though glucose is not required for in vitro cavitation, suggesting other important functions for this transporter in development (352).

Preimplantation embryos also express the insulin-responsive GLUT8 transporter, which is translocated to the cell membrane in response to insulin and promotes blastocyst glucose utilization (353). Its transport to the cell membrane hinges on a vesicle–outer cell membrane fusion mediated by Syntaxin 4A (STX4A); *Stx4a*$^{-/-}$ embryos demonstrate impaired GLUT8 cell surface expression, reduced rates of insulin-stimulated glucose uptake at the blastocyst stage, and embryonic death by E7.5 (354). GLUT8 has also been "knocked-down" by incubation of preimplantation embryos in *Slc2a8* antisense mRNA, resulting in blastomere apoptosis in vitro and embryo resorption after transfer to pseudopregnant females (355). Given the dominant roles of the maternal environment in providing preimplantation embryos with glucose and insulin, early development may represent a period of vulnerability for embryos of diabetics, accounting for an aspect of fertility problems in these women and congenital malformations in their children. Also, studies in rats indicate that even in the absence of diabetes, dietary protein restriction during the few days of preimplantation development creates a transient period of hyperglycemia and reduces ICM cell number in blastocysts and impacts blood pressure and body weight in young adults weeks after gestation (356). Thus, the interplay between maternal and fetal metabolism during the preimplantation period seems to influence the establishment of pregnancy and may have sequelae long into development and adult life—bearing important implications for public health concerns, assisted reproductive technologies, and theories of the "fetal basis of adult disease."

TE Sidedness and Blastocoel Development

Between E2.5 and E3.5, mouse embryos move from the coiled oviduct into the uterus. The embryos enter the uterus at approximately the 16-cell morula stage and proceed to develop a blastocoel beginning at the 32-cell stage. Tight junction complexes between developing TE cells are assembled in a step-wise manner after compaction and before blastocyst cavitation and prove critical for establishing the sidedness of the TE as well as its selectivity in ion and water transport (357). The sequence of tight junction development includes (a) localization of punctuate and subsequently banded zona occludens protein (ZO1; tight junction protein 1 or TJP1) around the apicolateral aspect of the emerging TE (358), (b) accumulation and organization of cingulin plaque protein (359), and (c) development of a second ZO1 isoform band and colocalization of occludin transmembrane protein [(360,361), reviewed in (362)]. Coincident with the formation of these cell contacts within the TE and blastocoel expansion is the development of distinct cell–cell adhesions within the eccentrically located ICM (363).

Blastocoel development depends on cAMP-regulated transport of ions and water across the TE and requires numerous transporter channels specifically positioned with respect to the orientation of the TE cells (364,365) (Fig. 17). Evidence of an Na^+/K^+ exchange ATPase was first described by biochemical studies demonstrating that blastocoel development was inhibited by increasing extraembryonic concentrations of K^+ ions or by addition of ouabain (366), and immunolocalization experiments then indicated its presence on the basolateral TE membrane (367). The Na^+/K^+ ATPase isoform expressed in early development is thought to be comprised principally of two subunits, the catalytic $\alpha 1$ and its regulator $\beta 1$ (encoded for by *Atp1a1* and *Atp1b1*, respectively) (368). Temporal (beginning at the late morula) and spatial (confined to the TE) expression of the α protein is limited by posttranscriptional mechanisms (369). This controlled regulation of ATPase $\alpha 1$ expression is misleading, though, because *Atp1a1* knockout mice demonstrate

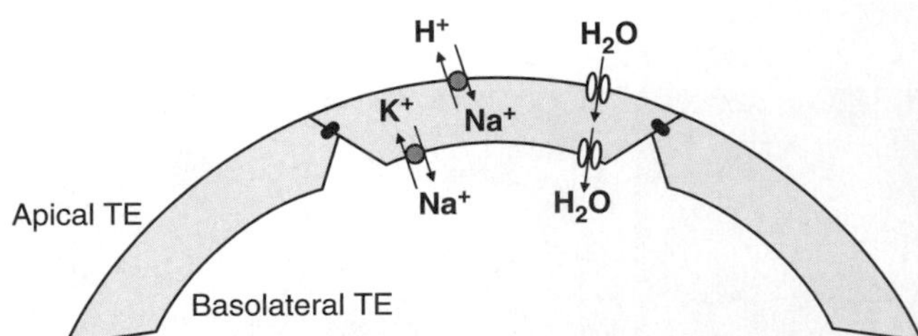

FIG. 17. A schematic of ion channels in trophectoderm (TE) cells. Basolateral Na^+/K^+ exchangers are known to actively transport sodium ions into the developing blastocoel cavity. Apical Na^+/H^+ exchangers and cotransporters provide intracellular sodium for continued basolateral transport. Aquaporin (AQP) proteins (AQP9 on the apical surface; APQ3 and APQ8 on the basolateral surface) allow for water to follow shifts in osmolarity resulting from the sodium gradient.

normal blastocyst development and cavitation. Other related Na^+/K^+ ATPases presumably compensate for the absence of the ATPase α1 protein in knockout morula and developing blastocysts, which lack immunodetectable protein. This effect is transient though, because *Atp1a1* null mice succumb to other problems later in development (370). A single-pass transmembrane γ protein associated with the Na^+/K^+ ATPases (encoded for by *Fxyd2*, named for FXYD amino acids in the family's invariant sequence) is also expressed in early embryos and is present on both the basolateral and apical membranes of the TE. Antisense experiments indicate it is important in Na^+/K^+ ATPase activity and blastocyst cavitation, though the precise roles of FXYD proteins are unclear (371). An apical Na^+/H^+ exchanger (NHE3; encoded for by solute carrier family 9 [sodium/hydrogen exchanger], member 3, *Slc9a3*) persistent from oogenesis is at least partially responsible for import of sodium ions from outside the embryo to supply the basolateral Na^+/K^+ ATPase, and this activity can be inhibited by treatment with 5-(*N*-ethyl-*N*-isopropyl) amiloride (365,372). Other avenues for sodium entry include Na^+/glucose and Na^+/amino acid cotransporters on the TE apical surface. Water follows sodium across the TE by means of aquaporin protein channels; different forms of aquaporins are present on each aspect of the TE and are known to facilitate rapid fluid transport given near-isosmotic gradients (373).

Throughout blastocoel development, blastocysts increase only subtly in size and remain within their surrounding zona pellucida until late in E3.5 or E4.5 in the mouse. There is regulated abembryonic expression at this time of a tryptase enzyme, termed strypsin or implantation serine protease 1 (ISP1), which has been implicated in blastocyst hatching from the zona pellucida. The enzyme activity is sensitive to trypsin inhibitors, and antisense *Isp1* mRNA precludes blastocyst hatching and outgrowth in vitro (374,375). Although defects in the hatching process have not been clearly described in humans as cause of failed implantation and establishment of pregnancy, blastocyst hatching is facilitated by inducing zona pellucida damage as a part of assisted reproductive techniques (376).

PREIMPLANTATION POLARITY AND CELL FATE DETERMINATION

Polarity in Mammalian Embryos

In *Drosophila*, *Xenopus*, and many other species that serve as models of development, it has long been understood that polarities of the egg are extrapolated to influence embryonic spatial patterning, though this has not been as readily appreciable in mammalian species. The difference in ideology arises because mammalian blastomeres up to the morula stage have been considered to be naive of cell fate determining influences and completely plastic or interchangeable experimentally with one another. Polarities in morula are indeed relatively subtle.

Recently, however, it has been suggested that during mammalian oogenesis, fertilization, and preimplantation development, a "default polarity" is established that is predictive of the cellular lineages to give rise to distinct blastocyst cell populations and ultimately later embryonic and extraembryonic structures. Oocyte meiotic cleavage positioning and the location of sperm penetration are obvious markers of early asymmetry and may provide the first cues for organization of the embryo. According to this model based on marking experiments, the first mitotic cleavage furrow is placed relative to these sites and extends from the stationary point of the first extruded polar body (termed by convention the animal pole) to pass through the sperm entry point; this plane later delineates embryonic and abembryonic components of the blastocyst (Fig. 18) (377,378). Others following development with time-lapse recording methods argue that the orientation of the first plane of cleavage is positioned with respect to the pronuclei (so as to form perpendicularly to a line between the pronuclei just before the first mitosis) and that the remnant polar body is actually mobile and frequently relocates to this plane (379). Interestingly, transplantation experiments designed to create ectopic animal poles interfere with chromosomal segregation during this division and compromise development (380), indicating a perturbation of morphogenetic determinants. These determinants, however, remain enigmatic. As the first cleavage approximates the sperm entry position, the entry position itself is relocated to the inner surface of one of the two blastomeres of the two-cell embryo. The blastomere acquiring the sperm entry position is more likely to divide shortly before its counterpart and contributes cells preferentially to the embryonic portion of the blastocyst (378,381).

Two populations of cells are established by blastocoel development, the ICM and the outer TE, and embryonic and abembryonic axes become clearly evident. Thereafter, two ICM lineages can be distinguished in the preimplantation embryo as the epiblast (primitive ectoderm) and hypoblast (primitive endoderm). Development of the former gives rise to embryonic constituents and the extraembryonic mesoderm, whereas the hypoblast cells differentiate into extraembryonic visceral and parietal endoderms. The TE is also subdivisible into two components, the mural TE and polar TE layers, that surround the blastocoel and the ICM and give rise to

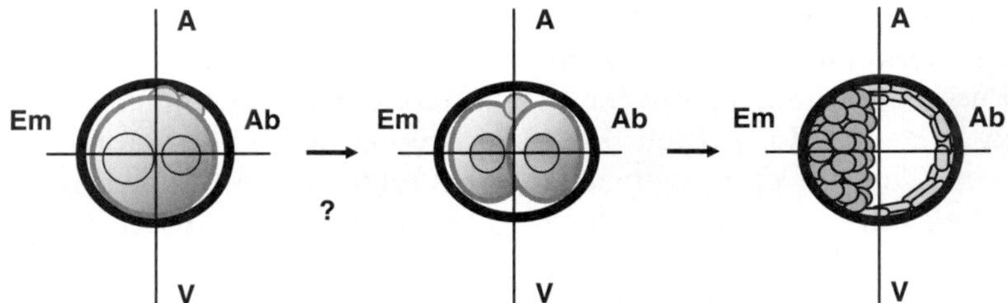

FIG. 18. Polarity of blastocysts seems traceable to zygote markers. The animal (A) and vegetal (V) poles are named by convention in mammalian embryos depending on the position of the second meiotic polar body; whether this can be considered relatively stationary is of some debate. Embryonic (Em) and abembryonic (Ab) progenitor regions are separated by the A–V axis running through the polar body and sperm entry point. The positioning of the first mitotic cleavage seems to be related to either the polar body position or the location of the two pronuclei, depending on the experimental system used for observation (*question mark*). The two cell blastomere to inherit the sperm entry site is predicted to be the first to divide and contributes preferentially to the ICM of the blastocyst and the epiblast and embryonic lineages in postimplantation development. Morphogenetic determinants of these processes are unknown, though some maternal proteins (including Leptin and STAT3) are asymmetrically distributed in mouse eggs and remains so during subdivisions of the blastomeres (417) and are interesting candidates.

the trophoblast giant cells and extraembryonic ectoderm, respectively. Cellular ontogeny in the early embryo can be appreciated by "tracing" the descendants of selected blastocyst cells, and such studies clearly demonstrate that the positioning of cells within the blastocyst ICM can be correlated to their assignment in the postimplantation egg cylinder (382).

Totipotency Within the Embryo and Cell Fate Determination

One of the most fascinating features of early embryonic cells is their pluripotency and resiliency—their ability to differentiate along many pathways to become any of the variety of cell types represented in an adult organism and to continue with successful development after disturbance, respectively. Pluripotent cells can also be derived from primordial germ cells (embryonic germ cells) and tumor tissues (embryonic carcinoma cells) in addition to preimplantation embryos, though cell lines from the latter (ES cells) seem to have greater adaptability when challenged to differentiate in both in vitro and in vivo experimental paradigms.

Currently, only a handful of gene products are known to be important in the acquisition and maintenance of pluripotency, though inquiries into their functions represent an emerging field that has piqued broad interest. If we are able to provide pluripotent cells with an environmental context that cues particular differentiation pathways while precluding others, medicinal applications hold promise to allow replacement of a defective differentiated cell type in disease or introduction of a transgene into an adult stem cell population for gene therapy. Exciting strides have been made as proof of the concept that we can direct differentiation. For example, in the field of reproductive biology, we have known from generating knockout mice that ES cells have the potential to contribute to the mammalian germline in vivo. Recent reports from several groups have documented that ES cells can be differentiated in vitro to give rise to both male and female germ cell-like lineages with the potential to generate blastocysts after fertilization or activation (Fig. 19) (383–385). With the development of cloning

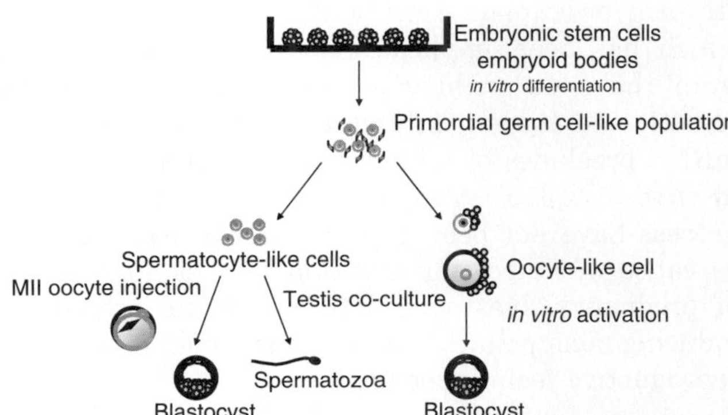

FIG. 19. In vitro differentiation of pluripotent ES cells to generate male and female gametes. Researchers have isolated primordial germ cell-like lineages from embryoid bodies. **Left:** Maturation of the primordial germ cells into haploid male gametes could be accomplished in embryoid bodies, and these are capable of fertilization and providing a chromosome complement to oocytes after intracytoplasmic sperm injection (ICSI) (383). These male gamete precursors could be brought through spermatogenesis to produce spermatozoa when transplanted into reconstituted testicular tubules (384). **Right:** Similarly, embryoid body cells can develop into oogonia that enter meiosis, recruit adjacent cells to form follicle-like structures, and activate to produce blastocysts (385).

techniques, wherein an oocyte nucleus is replaced with that of a diploid differentiated somatic cell, early embryos and stem cells with the genomic content of any preexisting individual seem to be attainable (386).

Attempting to identify candidate regulators of pluripotency, we have found that ICM cells and ES cells derived from them are characterized by unique transcriptomes. This can be readily appreciated by digital differential display comparisons of publicly available ES cell EST libraries with those derived from somatic tissues (387) as well as by ES cell microarray studies that have identified ES-specific gene products, most as yet uncharacterized in other tissues (388). Specific transcription factors (some interrelated) have emerged as critical determinants of ES cell viability, self-renewal, and pluripotency, as we now consider. These include signal transducer and activator of transcription 3 (STAT3), SMADs induced by BMPs, octamer binding transcription factor 3/4 (OCT3/4), SRY-box containing gene 2 (SOX2), fork-head box protein D3 (FOXD3), and NANOG (Fig. 20). In addition to the loss-of-function experiments demonstrating their importance, inappropriate expression of some of these and other transcription factors can cause loss of developmental potential and promote specific developmental programs (389,390).

STAT3 transactivation is induced by LIF signaling through its receptor, LIFR, and the LIFR-associated protein interleukin 6 signal transducer (IL6ST or gp130) [reviewed in (391)]. If mouse ES cells are not grown with fibroblast feeder cells, supplementation of their media with LIF is essential for maintaining pluripotency. STAT3 seems to be crucial to mouse ES cell pluripotency and self-renewal, as evidenced by the findings that (a) introduction of a dominant negative STAT3 results in ES cell differentiation (392), (b) an estrogen inducible STAT3 can be used to couple ES cell pluripotency to steroid administration (393), and (c) *Stat3* loss-of-function mutations are not normally tolerated (394). *Stat3* loss-of-function mutations in heterozygosity can be maintained in ES cells that have acquired less LIF dependency (394), though the LIF-mediated pathway is not solely responsible for STAT3 function in all early cell lineages in vivo.

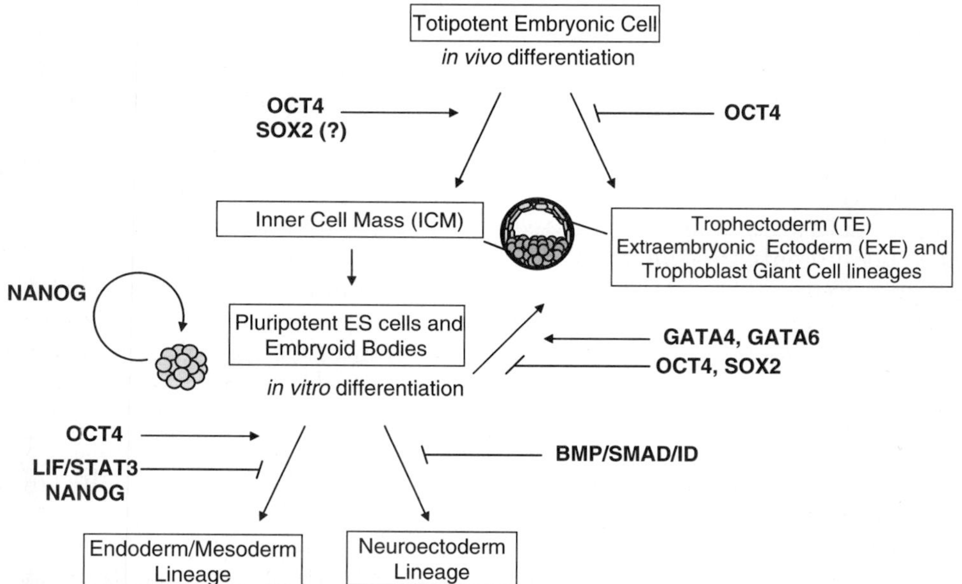

FIG. 20. Transcription factors influencing blastomere and mouse ES cell fate determinations in vivo and in vitro. OCT4 is important in the first differentiation step, being induced in blastomeres that are part of the ICM and perhaps directly repressing expression of transcription factors characteristic of the TE. Knockout mice lacking OCT4 have failed development of the ICM. It has been suggested that SOX2 also participates with OCT4 at this level, but it is difficult to assess in vivo due to persistence of maternal SOX2 in *Sox2*$^{-/-}$ embryos. ES cells established in vitro are studied for the propensity for self-renewal (*circular arrow*), differentiation into endoderm/mesoderm-like cells (*left*) or neuroectoderm cells (*right*), and "redifferentiation" into a TE-like lineage (*upgoing arrow, far right*). NANOG overexpression promotes continued renewal of the pluripotent population even in the absence of LIF/STAT3. It cooperates with LIF/STAT3 to preclude endoderm/mesoderm lineage assignment. Overexpression of OCT4 promotes this pathway. The BMP/SMAD/ID pathway seems to serve a similar function as LIF/STAT3 in inhibiting neuroectoderm-like differentiation. Absence of OCT4 or SOX2 or overexpression of GATA4 or GATA6 promotes the assumption of a TE-like morphology and up-regulation of TE markers.

Rather, STAT3 seems to have functions separable from LIF signaling in that STAT3 knockout mice are early embryonic lethal and defects in LIFR and gp130 knockouts are not manifest as early in development (332,395,396). Cooperating with LIF and STAT3 is a BMP–SMAD pathway. BMP4 augments LIF sustained self-renewal and pluripotency of ES cells in vitro and induces through SMADs inhibitor of differentiation (Id) genes, inhibitor of DNA binding 1 and 3 (*Ibd1* and *Ibd3* [also known as Id1 and Id3]). ID proteins in turn prevent the activity of helix-loop-helix transcription factors and progression of neuroectoderm lineage, and their expression can substitute for BMP supplementation (397,398). There is the suggestion that STAT3 and SMADs can complex to result in cooperative transactivation in neural cells (399), and STAT3 and phosphorylated SMAD1 can be coimmunoprecipitated from ES cells after simultaneous LIF and BMP stimulation (398), although what targets may exist in ES cells for such a complex remain unknown.

OCT3/4 is a POU family transcription factor with conserved functions in mouse and human development. It is expressed in the germline and early embryos; in blastocysts its expression is limited to the ICM (400). Homozygous targeted deletion of the murine OCT3/4-encoding gene (POU domain, class 5, transcription factor 1; *Pou5f1*) results in ES cell differentiation into trophoblasts, and *Pou5f1*$^{-/-}$ embryos die at implantation due to failure to form an ICM (401). Further, manipulation of *Pou5f1* expression levels can skew differentiation programs; overexpression results in ES cell differentiation toward primitive endoderm/mesoderm pathways and attenuated expression allows a "dedifferentiation" and assumption of a TE-like phenotype (402). OCT3/4 target genes include *Fgf4* and transcriptional regulators orthodenticle homolog 2 (*Otx2*), undifferentiated embryonic cell transcription factor 1 (*Utf1*), and zinc finger protein 42 (*Zfp42*); it may also serve to down-regulate TE-specific transcription factors such as caudal type homeo box 2 (*Cdx2*) and heart and neural crest derivatives expressed transcript 1 (*Hand1*) [reviewed in (403)]. These findings place OCT3/4 at the vertex of an array of downstream transcription factors shaping the transcriptome of pluripotent lineages, and understanding upstream events and factors that modulate OCT3/4 transactivation remain important areas of focus.

SOX2 is one potential factor cooperating with OCT3/4 in the epiblast and its derivatives. Crystal structures have illustrated their interaction (404), and biochemical approaches indicate that together SOX2 and OCT3/4 transcribe *Nanog*. SOX2 is coexpressed with OCT3/4 in the epiblast, though there is not complete overlap of their expression in other lineages. OCT3/4 is expressed without SOX2 in the hypoblast-derived extraembryonic endoderm, whereas SOX2 is expressed without OCT3/4 in the TE-derived extraembryonic ectoderm. SOX2 proves essential for the development of the lineages where it is expressed (407). *Sox2*$^{-/-}$ mice die with defects in epiblast and extraembryonic ectoderm development shortly after implantation when maternal SOX2 protein is exhausted, and *Sox2*$^{-/-}$ ICM cells are forced to differentiate to trophoblasts. Thus, though the presence of maternal SOX2 protein in the null model has slowed in vivo analysis of SOX2 roles in the initial ICM verses TE fate determination, it seems that zygotic SOX2 acts in concert with OCT3/4 in maintaining early epiblast development. Notably, in vitro evidence suggests that OCT3/4 can also interact with FOXD3, a winged-helix DNA-binding transcription factor necessary for postimplantation epiblast development and ES cell viability (408), though OCT3/4 seems to bind to the FOXD3 DNA-binding domain and act as a corepressor of FOXD3-mediated transactivation (409). Thus, OCT3/4 may control other transcription factors by modulating their expression, acting as a coactivator, or affecting repression. It is clear that the doses of these transcription factor activities and incremental modulation of their functions may have major impacts on cell lineage specification.

Finally, the most recently described transcription factor associated with pluripotency of ES cells, NANOG, is a homeobox transcription factor discovered essentially simultaneously by three groups and named after the land of the "ever young" or "ever living" in Celtic legend, Tir nan Og (387,410,411). *Nanog* mRNA is highly expressed specifically in ICM cells of blastocysts and primordial germ cells but is not maintained in mature gametes. Dependent on OCT3/4, its expression in ES cells is shut down when differentiation is induced in vitro by addition of retinoic acid or 3-methoxybenzamide. Overexpression of *Nanog* allows mouse ES cell colonies to remain morphologically undifferentiated after treatment with these factors or withdrawal of LIF or BMP, though NANOG seems to operate in a pathway that parallels rather than augments STAT3 activation (387,398,410). In contrast, the knockout of *Nanog* in ES cells results in slowed ES cell division and induction of visceral and parietal endoderm markers (387). In vivo, *Nanog* null embryos die between E3.5 and E5.5 of development, lacking epiblast cells and having disorganized extraembryonic tissues (387).

ETHICS AND THE EMBRYO

Like the biological basis of our inheritance, our ethical convictions regarding early development are shaped by many generations of human history and ideas both inspired and contrived. We find ourselves participants in the continuing efforts of society to place priority—and perhaps limitations—on aspects of this research and its applications. The discoveries described herein are the rewards of studies undertaken largely because of their potential to shed light on human infertility and developmental malformation. Availability of new tools, including the Primate Embryo Gene Expression Resource (PREGER) (412), will assist translational efforts to extend knowledge of murine biology to primate species and underscore the hope of bringing bench research to the bedside. We have seen staggering advances in the development of assisted reproductive technologies, though elements in society debate whether the benefit of providing otherwise infertile couples with children outweighs foreseeable risks (413) and the loss of discarded embryos. In the future, controlling genomic content of stem cells may become widespread by techniques initiating oocyte activation in the absence of fertilization, as may be preceded by somatic nuclear transfer. As a step toward this, epigenetic anomalies of parthegnotes and cloned embryos are beginning to be appreciated and manipulated (414,415). Recently, investigators have isolated ES cells from cloned human blastocysts, furthering groundwork for the tantalizing prospect of regenerative medicine (416). Though still in the future, potential applications for such cell lines have captivated the imagination and are being envisioned with both intense anticipation and trepidation. All of us with a scientific interest in these fields have a special perspective as our nations contemplate decisions that present themselves. Certainly, the ways in which these issues take shape—as much as the progress of our laboratory endeavors—have relevance to our understanding of the fundamentals of humanity.

ACKNOWLEDGMENTS

We thank Drs. Latham and Solter for critical review and helpful comments on this chapter and Ms. Shirley Baker for tireless help with manuscript formatting. Research in the Matzuk laboratory has been supported by National Institutes of Health (grants HD33438, HD32067, HD42500, and CA60651), the Specialized Cooperative Centers Program in Reproduction Research, and Wyeth Research. K. H. B. was a graduate of the M.S.T.P. program at Baylor College of Medicine, currently at the Johns Hopkins Hospital. We also gratefully acknowledge Dr. Ernie Knobil's support for our research program at Baylor College of Medicine (Fig. 21) and have confidence that he and his legacy will continue to inspire reproductive biologists at every level for years to come.

FIG. 21. Dr. Ernst Knobil congratulating Dr. Martin Matzuk at the inaugural Knobil Lectureship at the University of Pittsburgh in 1988.

REFERENCES

1. Pinto-Correia, C. (1997). *The Ovary of Eve: Egg and Sperm and Preformation*. The University of Chicago Press.
2. Biggers, J. D. (1991). Walter Heape, F.R.S.: a pioneer in reproductive biology. Centenary of his embryo transfer experiments. *J. Reprod. Fertil*. 93, 173–186.
3. Robertson, E., Bradley, A., Kuehn, M., and Evans, M. (1986). Germ-line transmission of genes introduced into cultured pluripotential cells by retroviral vector. *Nature* 323, 445–448.
4. Thomas, K. R., and Capecchi, M. R. (1987). Site-directed mutagenesis by gene targeting in mouse embryo-derived stem cells. *Cell* 51, 503–512.
5. Capecchi, M. R. (1994). Targeted gene replacement. *Sci. Am*. 270, 52–59.
6. Burns, K., DeMayo, F. J., and Matzuk, M. M. (2002). Transgenic technology, cloning and germ cell transplantation. In *Molecular Biology in Reproductive Medicine* (B. C. J. M. Fauser, Ed.), 2nd ed., pp. 169–195. Parthenon Publishing.
7. Bouniol-Baly, C., Hamraoui, L., Guibert, J., Beaujean, N., Szollosi, M. S., and Debey, P. (1999). Differential transcriptional activity associated with chromatin configuration in fully grown mouse germinal vesicle oocytes. *Biol. Reprod*. 60, 580–587.
8. Ostermeier, G. C., Miller, D., Huntriss, J. D., Diamond, M. P., and Krawetz, S. A. (2004). Reproductive biology: delivering spermatozoan RNA to the oocyte. *Nature* 429, 154.
9. Goddard, M. J., and Pratt, H. P. (1983). Control of events during early cleavage of the mouse embryo: an analysis of the "2-cell block." *J. Embryol. Exp. Morphol*. 73, 111–133.

10. Wu, X., Viveiros, M., Eppig, J., Bai, Y., Fitzpatrick, S., and Matzuk, M. M. (2003). Zygote arrest 1 (Zar1) is a novel maternal-effect gene critical for the oocyte-to-embryo transition. *Nat. Genet.* 33, 187–191.
11. Burns, K. H., Viveiros, M. M., Ren, Y., Wang, P., DeMayo, F. J., Frail, D. E., Eppig, J. J., and Matzuk, M. M. (2003). Roles of NPM2 in chromatin and nucleolar organization in oocytes and embryos. *Science* 300, 633–636.
12. Diatchenko, L., Lau, Y. F., Campbell, A. P., Chenchik, A., Moqadam, F., Huang, B., Lukyanov, S., Lukyanov, K., Gurskaya, N., Sverdlov, E. D., and Siebert, P. D. (1996). Suppression subtractive hybridization: a method for generating differentially regulated or tissue-specific cDNA probes and libraries. *Proc. Natl. Acad. Sci. U S A* 1996;93:6025–6030.
13. Philpott, A., and Leno, G. H. (1992). Nucleoplasmin remodels sperm chromatin in *Xenopus* egg extracts. *Cell* 69, 759–767.
14. Philpott, A., Leno, G. H., and Laskey, R. A. (1991). Sperm decondensation in *Xenopus* egg cytoplasm is mediated by nucleoplasmin. *Cell* 65, 569–578.
15. Earnshaw, W. C., Honda, B. M., Laskey, R. A., and Thomas, J. O. (1980). Assembly of nucleosomes: the reaction involving *X. laevis* nucleoplasmin. *Cell* 21, 373–383.
16. Wu, X., Wang, P., Brown, C. A., Zilinski, C. A., and Matzuk, M. M. (2003). Zygote arrest 1 (Zar1) is an evolutionarily conserved gene expressed in vertebrate ovaries. *Biol. Reprod.* 69, 861–867.
17. Crevel, G., Huikeshoven, H., Cotterill, S., Simon, M., Wall, J., Philpott, A., Laskey, R. A., McConnell, M., Fisher, P. A., and Berrios, M. (1997). Molecular and cellular characterization of CRP1, a *Drosophila* chromatin decondensation protein. *J. Struct. Biol.* 118, 9–22.
18. Saitou, M., Barton, S. C., and Surani, M. A. (2002). A molecular programme for the specification of germ cell fate in mice. *Nature* 418, 293–300.
19. Payer, B., Saitou, M., Barton, S. C., Thresher, R., Dixon, J. P., Zahn, D., Colledge, W. H., Carlton, M. B., Nakano, T., and Surani, M. A. (2003). Stella is a maternal effect gene required for normal early development in mice. *Curr. Biol.* 13, 2110–2117.
20. Tong, Z. B., Gold, L., Pfeifer, K. E., Dorward, H., Lee, E., Bondy, C. A., Dean, J., and Nelson, L. M. (2000). Mater, a maternal effect gene required for early embryonic development in mice. *Nat. Genet.* 26, 267–268.
21. Tong, Z. B., Nelson, L. M., and Dean, J. (2000). Mater encodes a maternal protein in mice with a leucine-rich repeat domain homologous to porcine ribonuclease inhibitor. *Mamm. Genome* 11, 281–287.
22. Narducci, M. G., Fiorenza, M. T., Kang, S. M., Bevilacqua, A., Di Giacomo, M., Remotti, D., Picchio, M. C., Fidanza, V., Cooper, M. D., Croce, C. M., Mangia, F., and Russo, G. TCL1 participates in early embryonic development and is overexpressed in human seminomas. *Proc. Natl. Acad. Sci. U S A* 99, 11712–11717.
23. Howell, C. Y., Bestor, T. H., Ding, F., Latham, K. E., Mertineit, C., Trasler, J. M., and Chaillet, J. R. (2001). Genomic imprinting disrupted by a maternal effect mutation in the Dnmt1 gene. *Cell* 104, 829–838.
24. Christians, E., Davis, A. A., Thomas, S. D., and Benjamin, I. J. (2000). Maternal effect of Hsf1 on reproductive success. *Nature* 407, 693–694.
25. Leader, B., Lim, H., Carabatsos, M. J., Harrington, A., Ecsedy, J., Pellman, D., Maas, R., and Leder, P. (2002). Formin-2, polyploidy, hypofertility and positioning of the meiotic spindle in mouse oocytes. *Nat. Cell. Biol.* 4, 921–928.
26. Yuan, L., Liu, J. G., Hoja, M. R., Wilbertz, J., Nordqvist, K., and Hoog, C. (2002). Female germ cell aneuploidy and embryo death in mice lacking the meiosis-specific protein SCP3. *Science* 296, 1115–1118.
27. Gurtu, V. E., Verma, S., Grossmann, A. H., Liskay, R. M., Skarnes, W. C., and Baker, S. M. (2002). Maternal effect for DNA mismatch repair in the mouse. *Genetics* 160, 271–277.
28. Hogan, B., Beddington, R., Costantini, F., and Lacy, E. (1994). *Manipulating the Mouse Embryo—A Laboratory Manual.* Cold Spring Harbor Laboratory Press, Plainview, NY.
29. Matzuk, M. M., Kumar, T. R., Shou, W., Coerver, K. A., Lau, A. L., Behringer, R. R., and Finegold, M. J. (1996). Transgenic models to study the roles of inhibins and activins in reproduction, oncogenesis, and development. *Rec. Progr. Horm. Res.* 51, 123–157.
30. Eppig, J. J., Wigglesworth, K., and Pendola, F. L. (2002). The mammalian oocyte orchestrates the rate of ovarian follicular development. *Proc. Natl. Acad. Sci. U S A* 99, 2890–2894.
31. Lewandoski, M., Wassarman, K. M., and Martin, G. R. (1997). Zp3-cre, a transgenic mouse line for the activation or inactivation of loxP-flanked target genes specifically in the female germ line. *Curr. Biol.* 7, 148–151.
32. de Vries, W. N., Binns, L. T., Fancher, K. S., Dean, J., Moore, R., Kemler, R., and Knowles, B. B. (2000). Expression of Cre recombinase in mouse oocytes: a means to study maternal effect genes. *Genesis* 26, 110–112.
33. Alfieri, J. A., Martin, A. D., Takeda, J., Kondoh, G., Myles, D. G., and Primakoff, P. (2003). Infertility in female mice with an oocyte-specific knockout of GPI-anchored proteins. *J. Cell. Sci.* 116, 2149–2155.
34. Lan, Z. J., Gu, P., Xu, X., Jackson, K. J., DeMayo, F. J., O'Malley, B. W., Cooney, A. J. (2003). GCNF-dependent repression of BMP-15 and GDF-9 mediates gamete regulation of female fertility. *EMBO J.* 22, 4070–4081.
35. Erhardt, S., Su, I. H., Schneider, R., Barton, S., Bannister, A. J., Perez-Burgos, L., Jenuwein, T., Kouzarides, T., Tarakhovsky, A., and Surani, M. A. (2003). Consequences of the depletion of zygotic and embryonic enhancer of zeste 2 during preimplantation mouse development. *Development* 130, 4235–4248.
36. Svoboda, P., Stein, P., Hayashi, H., and Schultz, R. M. (2000). Selective reduction of dormant maternal mRNAs in mouse oocytes by RNA interference. *Development* 127, 4147–4156.
37. Svoboda, P., Stein, P., and Schultz, R. M. (2001). RNAi in mouse oocytes and preimplantation embryos: effectiveness of hairpin dsRNA. *Biochem. Biophys. Res. Commun.* 287, 1099–1104.
38. Kanzler, B., Haas-Assenbaum, A., Haas, I., Morawiec, L., Huber, E., and Boehm, T. (2003). Morpholino oligonucleotide-triggered knockdown reveals a role for maternal E-cadherin during early mouse development. *Mech. Dev.* 120, 1423–1432.
39. Siddall, L. S., Barcroft, L. C., and Watson, A. J. (2002). Targeting gene expression in the preimplantation mouse embryo using morpholino antisense oligonucleotides. *Mol. Reprod. Dev.* 63, 413–421.
40. Schultz, R. M., and Wassarman, P. M. (1977). Biochemical studies of mammalian oogenesis: protein synthesis during oocyte growth and meiotic maturation in the mouse. *J. Cell. Sci.* 24, 167–194.
41. Cascio, S. M., and Wassarman, P. M. (1982). Program of early development in the mammal: post-transcriptional control of a class of proteins synthesized by mouse oocytes and early embryos. *Dev. Biol.* 89, 397–408.
42. Kuersten, S., and Goodwin, E. B. (2003). The power of the 3′UTR: translational control and development. *Nat. Rev. Genet.* 4, 626–637.
43. Hwang, S. Y., Oh, B., Knowles, B. B., Solter, D., and Lee, J. S. (2001). Expression of genes involved in mammalian meiosis during the transition from egg to embryo. *Mol. Reprod. Dev.* 59, 144–158.
44. Gebauer, F., Xu, W., Cooper, G. M., and Richter, J. D. (1994). Translational control by cytoplasmic polyadenylation of c-mos mRNA is necessary for oocyte maturation in the mouse. *EMBO J.* 13, 5712.

45. Fuchimoto, D., Mizukoshi, A., Schultz, R. M., Sakai, S., and Aoki, F. (2001). Posttranscriptional regulation of cyclin A1 and cyclin A2 during mouse oocyte meiotic maturation and preimplantation development. *Biol. Reprod.* 65, 986–993.
46. Oh, B., Hwang, S. Y., Solter, D., and Knowles, B. B. (1997). Spindlin, a major maternal transcript expressed in the mouse during the transition from oocyte to embryo. *Development* 124, 493–503.
47. Tay, J., Hodgman, R., Sarkissian, M., and Richter, J. D. (2003). Regulated CPEB phosphorylation during meiotic progression suggests a mechanism for temporal control of maternal mRNA translation. *Genes Dev.* 17, 1457–1462.
48. Tay, J., and Richter, J. D. (2001). Germ cell differentiation and synaptonemal complex formation are disrupted in CPEB knockout mice. *Dev. Cell.* 1, 201–213.
49. Aoki, F., Hara, K. T., and Schultz, R. M. (2003). Acquisition of transcriptional competence in the 1-cell mouse embryo: requirement for recruitment of maternal mRNAs. *Mol. Reprod. Dev.* 64, 270–274.
50. Wang, Q., and Latham, K. E. (1997). Requirement for protein synthesis during embryonic genome activation in mice. *Mol. Reprod. Dev.* 47, 265–270.
51. Rambhatla, L., and Latham, K .E. (1995). Strain-specific progression of alpha-amanitin-treated mouse embryos beyond the two-cell stage. *Mol. Reprod. Dev.* 41, 16–19.
52. Bouniol, C., Nguyen, E., and Debey, P. (1995). Endogenous transcription occurs at the 1-cell stage in the mouse embryo. *Exp. Cell. Res.* 218, 57–62.
53. Aoki, F., Worrad, D. M., and Schultz, R. M. (1997). Regulation of transcriptional activity during the first and second cell cycles in the preimplantation mouse embryo. *Dev. Biol.* 181, 296–307.
54. Davis, W. Jr., De Sousa, P. A., and Schultz, R. M. (1996). Transient expression of translation initiation factor eIF-4C during the 2-cell stage of the preimplantation mouse embryo: identification by mRNA differential display and the role of DNA replication in zygotic gene activation. *Dev. Biol.* 174, 190–201.
55. Forlani, S., Bonnerot, C., Capgras, S., and Nicolas, J. F. (1998). Relief of a repressed gene expression state in the mouse 1-cell embryo requires DNA replication. *Development* 125, 3153–3166.
56. Vernet, M., Bonnerot, C., Briand, P., and Nicolas, J. F. (1992). Changes in permissiveness for the expression of microinjected DNA during the first cleavages of mouse embryos. *Mech. Dev.* 36, 129–139.
57. Latham, K. E., Solter, D., and Schultz, R. M. (1992). Acquisition of a transcriptionally permissive state during the 1-cell stage of mouse embryogenesis. *Dev. Biol.* 149, 457–462.
58. Davis, W. Jr., and Schultz, R. M. (1997). Role of the first round of DNA replication in reprogramming gene expression in the preimplantation mouse embryo. *Mol. Reprod. Dev.* 47, 430–434.
59. Wiekowski, M., Miranda, M., and DePamphilis, M. L. (1991). Regulation of gene expression in preimplantation mouse embryos: effects of the zygotic clock and the first mitosis on promoter and enhancer activities. *Dev. Biol.* 147, 403–414.
60. Prioleau, M. N., Buckle, R. S., and Mechali, M. (1995). Programming of a repressed but committed chromatin structure during early development. *EMBO J.* 14, 5073–5084.
61. Bellier, S., Chastant, S., Adenot, P., Vincent, M., Renard, J. P., and Bensaude, O. (1997). Nuclear translocation and carboxyl-terminal domain phosphorylation of RNA polymerase II delineate the two phases of zygotic gene activation in mammalian embryos. *EMBO J.* 16, 6250–6262.
62. Bellier, S., Dubois, M. F., Nishida, E., Almouzni, G., and Bensaude, O. (1997). Phosphorylation of the RNA polymerase II largest subunit during *Xenopus laevis* oocyte maturation. *Mol. Cell. Biol.* 17, 1434–1440.
63. Rossi, D. J., Londesborough, A., Korsisaari, N., Pihlak, A., Lehtonen, E., Henkemeyer, M., and Makela, T. P. (2001). Inability to enter S phase and defective RNA polymerase II CTD phosphorylation in mice lacking Mat1. *EMBO J.* 20, 2844–2856.
64. Rossignol, M., Kolb-Cheynel, I., and Egly, J. M. (1997). Substrate specificity of the cdk-activating kinase (CAK) is altered upon association with TFIIH. *EMBO J.* 16, 1628–1637.
65. Yankulov, K. Y., and Bentley, D. L. Regulation of CDK7 substrate specificity by MAT1 and TFIIH. *EMBO J.* 16, 1638–1646.
66. Ram, P. T., and Schultz, R. M. (1993). Reporter gene expression in G2 of the 1-cell mouse embryo. *Dev. Biol.* 156, 552–556.
67. Henery, C. C., Miranda, M., Wiekowski, M., Wilmut, I., and DePamphilis, M. L. (1995). Repression of gene expression at the beginning of mouse development. *Dev. Biol.* 169, 448–460.
68. Adenot, P. G., Mercier, Y., Renard, J. P., and Thompson, E. M. (1997). Differential H4 acetylation of paternal and maternal chromatin precedes DNA replication and differential transcriptional activity in pronuclei of 1-cell mouse embryos. *Development* 124, 4615–4625.
69. Cowell, I. G., Aucott, R., Mahadevaiah, S. K., Burgoyne, P. S., Huskisson, N., Bongiorni, S., Prantera, G., Fanti, L., Pimpinelli, S., Wu, R., Gilbert, D. M., Shi, W., Fundele, R., Morrison, H., Jeppesen, P., and Singh, P. B. (2002). Heterochromatin, HP1 and methylation at lysine 9 of histone H3 in animals. *Chromosoma* 111, 22–36.
70. Jenuwein, T., and Allis, C. D. (2001). Translating the histone code. *Science* 293, 1074–1080.
71. Wiekowski, M., Miranda, M., and DePamphilis, M. L. (1993). Requirements for promoter activity in mouse oocytes and embryos distinguish paternal pronuclei from maternal and zygotic nuclei. *Dev. Biol.* 159, 366–378.
72. Nothias, J. Y., Miranda, M., and DePamphilis, M. L. (1996). Uncoupling of transcription and translation during zygotic gene activation in the mouse. *EMBO J.* 15, 5715–5725.
73. DePamphilis, M. L., Kaneko, K. J., and Vassilev, A. (2002). Activation of zygotic gene expression in mammals. In *Developmental Biology and Biochemistry* (M. DePamphilis, Ed.), Vol. 12, pp. 55–82.
74. Domashenko, A. D., Latham, K. E., and Hatton, K. S. (1997). Expression of myc-family, myc-interacting, and myc-target genes during preimplantation mouse development. *Mol. Reprod. Dev.* 47, 57–65.
75. Sprinks, M. T., Sellens, M. H., Dealtry, G. B., and Fernandez, N. (1993). Preimplantation mouse embryos express Mhc class I genes before the first cleavage division. *Immunogenetics* 38, 35–40.
76. Wiekowski, M., Miranda, M., Nothias, J. Y., and DePamphilis, M. L. (1997). Changes in histone synthesis and modification at the beginning of mouse development correlate with the establishment of chromatin mediated repression of transcription. *J. Cell. Sci.* 110(Pt 10), 1147–1158.
77. Schultz, R. M., Davis, W. Jr., Stein, P., and Svoboda, P. (1999). Reprogramming of gene expression during preimplantation development. *J. Exp. Zool.* 285, 276–282.
78. Spada, F., Brunet, A., Mercier, Y., Renard, J. P., Bianchi, M. E., and Thompson, E. M. (1998). High mobility group 1 (HMG1) protein in mouse preimplantation embryos. *Mech. Dev.* 76, 57–66.
79. Palmieri, S. L., Peter, W., Hess, H., and Scholer, H. R. (1994). Oct-4 transcription factor is differentially expressed in the mouse embryo during establishment of the first two extraembryonic cell lineages involved in implantation. *Dev. Biol.* 166, 259–267.
80. Martianov, I., Viville, S., and Davidson, I. (2002). RNA polymerase II transcription in murine cells lacking the TATA binding protein. *Science* 298, 1036–1039.

81. Kaneko, K. J., and DePamphilis, M. L. (1998). Regulation of gene expression at the beginning of mammalian development and the TEAD family of transcription factors. *Dev. Genet.* 22, 43–55.
82. Worrad, D. M., and Schultz, R. M. (1997). Regulation of gene expression in the preimplantation mouse embryo: temporal and spatial patterns of expression of the transcription factor Sp1. *Mol. Reprod. Dev.* 46, 268–277.
83. Jumaa, H., Wei, G., and Nielsen, P. J. (1999). Blastocyst formation is blocked in mouse embryos lacking the splicing factor SRp20. *Curr. Biol.* 9:899–902.
84. Latham, K. E., Rambhatla, L., Hayashizaki, Y., and Chapman, V. M. (1995). Stage-specific induction and regulation by genomic imprinting of the mouse U2afbp-rs gene during preimplantation development. *Dev. Biol.* 168, 670–676.
85. Kim, M., Geum, D., Khang, I., Park, Y. M., Kang, B. M., Lee, K. A., and Kim, K. (2002). Expression pattern of HSP25 in mouse preimplantation embryo: heat shock responses during oocyte maturation. *Mol. Reprod. Dev.* 61, 3–13.
86. Christians, E., Campion, E., Thompson, E. M., and Renard, J. P. (1995). Expression of the HSP 70.1 gene, a landmark of early zygotic activity in the mouse embryo, is restricted to the first burst of transcription. *Development* 121, 113–122.
87. Baran, V., Brochard, V., Renard, J. P., and Flechon, J. E. (2001). Nopp 140 involvement in nucleologenesis of mouse preimplantation embryos. *Mol. Reprod. Dev.* 59, 277–284.
88. Taylor, K. D., and Piko, L. (1992). Expression of ribosomal protein genes in mouse oocytes and early embryos. *Mol. Reprod. Dev.* 31, 182–188.
89. Sevigny, G., Kothary, R., Tremblay, E., De Repentigny, Y., Joly, E. C., and Bibor-Hardy, V. (1995). The cytosolic chaperonin subunit TRiC-P5 begins to be expressed at the two-cell stage in mouse embryos. *Biochem. Biophys. Res. Commun.* 216, 279–283.
90. Latham, K. E., and Rambhatla, L. (1995). Expression of X-linked genes in androgenetic, gynogenetic, and normal mouse preimplantation embryos. *Dev. Genet.* 17, 212–222.
91. Zuccotti, M., Boiani, M., Ponce, R., Guizzardi, S., Scandroglio, R., Garagna, S., and Redi, C. A. (2002). Mouse Xist expression begins at zygotic genome activation and is timed by a zygotic clock. *Mol. Reprod. Dev.* 61, 14–20.
92. Zwingman, T., Erickson, R. P., Boyer, T., and Ao, A. (1993). Transcription of the sex-determining region genes Sry and Zfy in the mouse preimplantation embryo. *Proc. Natl. Acad. Sci. U S A* 90, 814–817.
93. Kigami, D., Minami, N., Takayama, H., AND Imai, H. (2003). MuERV-L is one of the earliest transcribed genes in mouse one-cell embryos. *Biol. Reprod.* 68, 651–654.
94. Wang, Q., Chung, Y. G., deVries, W. N., Struwe, M., and Latham, K. E. (2001). Role of protein synthesis in the development of a transcriptionally permissive state in one-cell stage mouse embryos. *Biol. Reprod.* 65, 748–754.
95. Conover, J. C., Temeles, G. L., Zimmermann, J. W., Burke, B., and Schultz, R. M. (1991). Stage-specific expression of a family of proteins that are major products of zygotic gene activation in the mouse embryo. *Dev. Biol.* 144, 392–404.
96. Bachvarova, R., De Leon, V., and Spiegelman, I. (1981). Mouse egg ribosomes: evidence for storage in lattices. *J. Embryol. Exp. Morphol.* 62, 153–164.
97. Thomas, T., Voss, A. K., Petrou, P., and Gruss, P. (2000). The murine gene, Traube, is essential for the growth of preimplantation embryos. *Dev. Biol.* 227, 324–342.
98. Fanciulli, M., Bruno, T., Di Padova, M., De Angelis, R., Iezzi, S., Iacobini, C., Floridi, A., and Passananti, C. (2000). Identification of a novel partner of RNA polymerase II subunit 11, Che-1, which interacts with and affects the growth suppression function of Rb. *Faseb. J.* 14, 904–912.
99. Matsson, H., Davey, E. J., Draptchinskaia, N., Hamaguchi, I., Ooka, A., Leveen, P., Forsberg, E., Karlsson, S., and Dahl, N. (2004).Targeted disruption of the ribosomal protein S19 gene is lethal prior to implantation. *Mol. Cell. Biol.* 24, 4032–4037.
100. Marin, M., Karis, A., Visser, P., Grosveld, F., and Philipsen, S. (1997). Transcription factor Sp1 is essential for early embryonic development but dispensable for cell growth and differentiation. *Cell* 89, 619–628.
101. Wang, Q., and Latham, K. E. (2000). Translation of maternal messenger ribonucleic acids encoding transcription factors during genome activation in early mouse embryos. *Biol. Reprod.* 62, 969–978.
102. Bleckmann, S. C., Blendy, J. A., Rudolph, D., Monaghan, A. P., Schmid, W., and Schutz, G. (2002). Activating transcription factor 1 and CREB are important for cell survival during early mouse development. *Mol. Cell. Biol.* 22, 1919–1925.
103. Rudolph, D., Tafuri, A., Gass, P., Hammerling, G. J., Arnold, B., and Schutz, G. (1998). Impaired fetal T cell development and perinatal lethality in mice lacking the cAMP response element binding protein. *Proc. Natl. Acad. Sci. U S A* 95, 4481–4486.
104. Ko, M. S., Kitchen, J. R., Wang, X., Threat, T. A., Hasegawa, A., Sun, T., Grahovac, M. J., Kargul, G. J., Lim, M. K,, Cui, Y., Sano, Y., Tanaka, T., Liang, Y., Mason, S., Paonessa, P. D., Sauls, A. D., DePalma, G. E., Sharara, R., Rowe, L. B., Eppig, J., Morrell, C., and Doi, H. (2000). Large-scale cDNA analysis reveals phased gene expression patterns during preimplantation mouse development. *Development* 127, 1737–1749.
105. Ma, J., Svoboda, P., Schultz, R. M., and Stein, P. (2001). Regulation of zygotic gene activation in the preimplantation mouse embryo: global activation and repression of gene expression. *Biol. Reprod.* 64, 1713–1721.
106. Pacheco-Trigon, S., Hennequet-Antier, C., Oudin, J. F., Piumi, F., Renard, J. P., and Duranthon, V. (2002). Molecular characterization of genomic activities at the onset of zygotic transcription in mammals. *Biol. Reprod.* 67, 1907–1918.
107. Martinez-Salas, E., Linney, E., Hassell, J., and DePamphilis, M. L. (1989). The need for enhancers in gene expression first appears during mouse development with formation of the zygotic nucleus. *Genes Dev.* 3, 1493–1506.
108. Martinez-Salas, E., Cupo, D. Y., and DePamphilis, M. L. (1988). The need for enhancers is acquired upon formation of a diploid nucleus during early mouse development. *Genes Dev.* 2, 1115–1126.
109. Majumder, S., Zhao, Z., Kaneko, K., and DePamphilis, M. L. (1997). Developmental acquisition of enhancer function requires a unique coactivator activity. *EMBO J.* 16, 1721–1731.
110. Davis, W. Jr., and Schultz, R. M. (2000). Developmental change in TATA-box utilization during preimplantation mouse development. *Dev. Biol.* 218, 275–283.
111. Majumder, S., and DePamphilis, M. L. (1994). TATA-dependent enhancer stimulation of promoter activity in mice is developmentally acquired. *Mol. Cell. Biol.* 14, 4258–4268.
112. Latham, K. E., Garrels, J. I., Chang, C., and Solter, D. (1991). Quantitative analysis of protein synthesis in mouse embryos. I. Extensive reprogramming at the one- and two-cell stages. *Development* 112, 921–932.
113. Shi, C. Z., Collins, H. W., Garside, W. T., Buettger, C. W., Matschinsky, F. M., and Heyner, S. (1994). Protein databases for compacted eight-cell and blastocyst-stage mouse embryos. *Mol. Reprod. Dev.* 37, 34–47.
114. Kafri, T., Ariel, M., Brandeis, M., Shemer, R., Urven, L., McCarrey, J., Cedar, H., and Razin, A. (1992). Developmental

pattern of gene-specific DNA methylation in the mouse embryo and germ line. *Genes Dev.* 6, 705–714.
115. Hajkova, P., Erhardt, S., Lane, N., Haaf, T., El-Maarri, O., Reik, W., Walter, J., and Surani, MA. (1998). Epigenetic reprogramming in mouse primordial germ cells. *Mech. Dev.* 117, 15–23.
116. Chen, R. Z., Pettersson, U., Beard, C., Jackson-Grusby, L., and Jaenisch, R. (1998). DNA hypomethylation leads to elevated mutation rates. *Nature* 395, 89–93.
117. Szabo, P. E., and Mann, J. R. (1995). Biallelic expression of imprinted genes in the mouse germ line: implications for erasure, establishment, and mechanisms of genomic imprinting. *Genes Dev.* 9, 1857–1868.
118. Lucifero, D., Mertineit, C., Clarke, H. J., Bestor, T. H., and Trasler, J. M. (2002). Methylation dynamics of imprinted genes in mouse germ cells. *Genomics* 79, 530–538.
119. Obata, Y., Kaneko-Ishino, T., Koide, T., Takai, Y., Ueda, T., Domeki, I., Shiroishi, T., Ishino, F., and Kono, T. (1998). Disruption of primary imprinting during oocyte growth leads to the modified expression of imprinted genes during embryogenesis. *Development* 125, 1553–1560.
120. Oswald, J., Engemann, S., Lane, N., Mayer, W., Olek, A., Fundele, R., Dean, W., Reik, W., and Walter, J. (2000). Active demethylation of the paternal genome in the mouse zygote. *Curr. Biol.* 10, 475–478.
121. Olek, A., and Walter, J. (1997). The pre-implantation ontogeny of the H19 methylation imprint. *Nat. Genet.* 17, 275–276.
122. Sunahara, S., Nakamura, K., Nakao, K., Gondo, Y., Nagata, Y., and Katsuki, M. (2000). The oocyte-specific methylated region of the U2afbp-rs/U2af1-rs1 gene is dispensable for its imprinted methylation. *Biochem. Biophys. Res. Commun.* 268, 590–595.
123. Santos, F., Hendrich, B., Reik, W., and Dean, W. (2002). Dynamic reprogramming of DNA methylation in the early mouse embryo. *Dev. Biol.* 241, 172–182.
124. Stoger, R., Kubicka, P., Liu, C. G., Kafri, T., Razin, A., Cedar, H., and Barlow, D. P. (1993). Maternal-specific methylation of the imprinted mouse Igf2r locus identifies the expressed locus as carrying the imprinting signal. *Cell* 73, 61–71.
125. Ariel, M., Robinson, E., McCarrey, J. R., and Cedar, H. (1995). Gamete-specific methylation correlates with imprinting of the murine Xist gene. *Nat. Genet.* 9, 312–315.
126. Zuccotti, M., and Monk, M. (1995). Methylation of the mouse Xist gene in sperm and eggs correlates with imprinted Xist expression and paternal X-inactivation. *Nat. Genet.* 9, 316–320.
127. Tucker, K. L., Beard, C., Dausmann, J., Jackson-Grusby, L., Laird, P. W., Lei, H., Li, E., and Jaenisch, R. (1996). Germ-line passage is required for establishment of methylation and expression patterns of imprinted but not of nonimprinted genes. *Genes Dev.* 10, 1008–1020.
128. Brandeis, M., Frank, D., Keshet, I., Siegfried, Z., Mendelsohn, M., Nemes, A., Temper, V., Razin, A., and Cedar, H. (1994). Sp1 elements protect a CpG island from de novo methylation. *Nature* 371, 435–438.
129. Macleod, D., Charlton, J., Mullins, J., and Bird, A. P. (1994). Sp1 sites in the mouse aprt gene promoter are required to prevent methylation of the CpG island. *Genes Dev.* 8, 2282–2292.
130. Schoenherr, C. J., Levorse, J. M., and Tilghman, S. M. (2003). CTCF maintains differential methylation at the Igf2/H19 locus. *Nat. Genet.* 33, 66–69.
131. Pant, V., Mariano, P., Kanduri, C., Mattsson, A., Lobanenkov, V., Heuchel, R., and Ohlsson, R. (2003). The nucleotides responsible for the direct physical contact between the chromatin insulator protein CTCF and the H19 imprinting control region manifest parent of origin-specific long-distance insulation and methylation-free domains. *Genes Dev.* 17, 586–590.
132. Fedoriw, A. M., Stein, P., Svoboda, P., Schultz, R. M., and Bartolomei, M. S. (2004). Transgenic RNAi reveals essential function for CTCF in H19 gene imprinting. *Science* 303, 238–240.
133. Li, E., Bestor, T. H., and Jaenisch, R. (1992). Targeted mutation of the DNA methyltransferase gene results in embryonic lethality. *Cell* 69, 915–926.
134. Okano, M., Bell, D. W., Haber, D. A., and Li, E. (1999). DNA methyltransferases Dnmt3a and Dnmt3b are essential for de novo methylation and mammalian development. *Cell* 99, 247–257.
135. Chen, T., Ueda, Y., Dodge, J. E., Wang, Z., and Li, E. (2003). Establishment and maintenance of genomic methylation patterns in mouse embryonic stem cells by Dnmt3a and Dnmt3b. *Mol. Cell. Biol.* 23, 5594–5605.
136. Bourc'his, D., Xu, G. L., Lin, C. S., Bollman, B., and Bestor, T. H. (2001). Dnmt3L and the establishment of maternal genomic imprints. *Science* 294, 2536–2539.
137. Mager, J., Montgomery, N. D., de Villena, F. P., and Magnuson, T. (2003). Genome imprinting regulated by the mouse Polycomb group protein Eed. *Nat. Genet.* 33, 502–507.
138. Yoder, J. A., Walsh, C. P., and Bestor, T. H. (1997). Cytosine methylation and the ecology of intragenomic parasites. *Trends Genet.* 13, 335–340.
139. Fanning, T. G., and Singer, M. F. (1987). LINE-1: a mammalian transposable element. *Biochim. Biophys. Acta.* 910, 203–212.
140. Bailey, J. A., Carrel, L., Chakravarti, A., and Eichler, E. E. (2000). Molecular evidence for a relationship between LINE-1 elements and X chromosome inactivation: the Lyon repeat hypothesis. *Proc. Natl. Acad. Sci. U S A* 97, 6634–6639.
141. Lengauer, C., Kinzler, K. W., and Vogelstein, B. (1997). DNA methylation and genetic instability in colorectal cancer cells. *Proc. Natl. Acad. Sci. U S A* 94, 2545–2550.
142. Evsikov, A. V., de Vries, W. N., Peaston, A. E., Radford, E. E., Fancher, K. S., Chen, F. H., Blake, J. A., Bult, C. J., Latham, K. E., Solter, D., and Knowles, B. B. Systems biology of the two-cell mouse embryo. *Cytogenet. Genome Res.* 105, 240–250.
143. Knowles, B. B., Evsikov, A. V., de Vries, W. N. , Peaston, A. E., and Solter, D. (2003). Molecular control of the oocyte to embryo transition. *Philos. Trans. R. Soc. Lond. B. Biol. Sci.* 358, 1381–1387.
144. Stevenson, D. S., and Jarvis, P. (2003). Chromatin silencing: RNA in the driving seat. *Curr. Biol.* 13, R13–R15.
145. Baulcombe, D. C. (2000). Molecular biology: unwinding RNA silencing. *Science* 290, 1108–1109.
146. Zeng, F., and Schultz, R. M. (2003). Gene expression in mouse oocytes and preimplantation embryos: use of suppression subtractive hybridization to identify oocyte- and embryo-specific genes. *Biol. Reprod.* 68, 31–39.
147. Svoboda, P., Stein, P., Anger, M., Bernstein, E., Hannon, G. J., and Schultz, R. M. (2004). RNAi and expression of retrotransposons MuERV-L and IAP in preimplantation mouse embryos. *Dev. Biol.* 269, 276–285.
148. Brockdorff, N., Ashworth, A., Kay, G. F., McCabe, V. M., Norris, D. P., Cooper, P. J., Swift, S., and Rastan, S. (1992). The product of the mouse Xist gene is a 15 kb inactive X-specific transcript containing no conserved ORF and located in the nucleus. *Cell* 71, 515–526.
149. Lee, J. T., Davidow, L. S., and Warshawsky, D. (1999). Tsix, a gene antisense to Xist at the X-inactivation centre. *Nat. Genet.* 21, 400–404.
150. Lifschytz, E., and Lindsley, D. L. (1972). The role of X-chromosome inactivation during spermatogenesis (*Drosophila*-allocycly-chromosome evolution-male sterility-dosage compensation). *Proc. Natl. Acad. Sci. U S A* 69, 182–186.

151. Kratzer, P. G., and Chapman, V. M. (1981). X chromosome reactivation in oocytes of *Mus caroli*. *Proc. Natl. Acad. Sci. U S A* 78, 3093–3097.
152. McCarrey, J. R., and Dilworth, D. D. (1992). Expression of Xist in mouse germ cells correlates with X-chromosome inactivation. *Nat. Genet.* 2, 200–203.
153. Norris, D. P., Patel, D., Kay, G. F., Penny, G. D., Brockdorff, N., Sheardown, S. A., and Rastan S. (1994). Evidence that random and imprinted Xist expression is controlled by preemptive methylation. *Cell* 77, 41–51.
154. Turner, J. M., Mahadevaiah, S. K., Elliott, D. J., Garchon, H. J., Pehrson, J. R., Jaenisch, R., and Burgoyne, P. S. (2002). Meiotic sex chromosome inactivation in male mice with targeted disruptions of Xist. *J. Cell. Sci.* 115, 4097–4105.
155. McCarrey, J. R., Watson, C., Atencio, J., Ostermeier, G. C., Marahrens, Y., Jaenisch, R., and Krawetz, S. A. (2002). X-chromosome inactivation during spermatogenesis is regulated by an Xist/Tsix-independent mechanism in the mouse. *Genesis* 34, 257–266.
156. Okamoto, I., Otte, A. P., Allis, C. D., Reinberg, D., and Heard, E. (2003). Epigenetic dynamics of imprinted X inactivation during early mouse development. *Science* 303, 644–649.
157. Hartshorn, C., Rice, J. E., and Wangh, L. J. (2002). Developmentally-regulated changes of Xist RNA levels in single preimplantation mouse embryos, as revealed by quantitative real-time PCR. *Mol. Reprod. Dev.* 61, 425–436.
158. Takagi, N. (1974). Differentiation of X chromosomes in early female mouse embryos. *Exp. Cell. Res.* 86, 127–135.
159. Sugawara, O., Takagi, N., and Sasaki, M. (1985). Correlation between X-chromosome inactivation and cell differentiation in female preimplantation mouse embryos. *Cytogenet. Cell. Genet.* 39, 210–219.
160. Epstein, C. J., Smith, S., Travis, B., and Tucker, G. (1978). Both X chromosomes function before visible X-chromosome inactivation in female mouse embryos. *Nature* 274, 500–503.
161. Huynh, K. D., and Lee, J. T. (2003). Inheritance of a pre-inactivated paternal X chromosome in early mouse embryos. *Nature* 426, 857–862.
162. Sheardown, S. A., Duthie, S. M., Johnston, C. M., Newall, A. E., Formstone, E. J., Arkell, R. M., Nesterova, T. B., Alghisi, G. C., Rastan, S., and Brockdorff, N. (1997). Stabilization of Xist RNA mediates initiation of X chromosome inactivation. *Cell* 91, 99–107.
163. Panning, B., Dausman, J., and Jaenisch, R. (1997). X chromosome inactivation is mediated by Xist RNA stabilization. *Cell* 90, 907–916.
164. Sado, T., Wang, Z., Sasaki, H., and Li, E. (2001). Regulation of imprinted X-chromosome inactivation in mice by Tsix. *Development* 128, 1275–1286.
165. Eggan, K., Akutsu, H., Hochedlinger, K., Rideout, W. III, Yanagimachi, R., and Jaenisch, R. (2000). X-chromosome inactivation in cloned mouse embryos. *Science* 290, 1578–1581.
166. Penny, G. D., Kay, G. F., Sheardown, S. A., Rastan, S., and Brockdorff, N. (1996). Requirement for Xist in X chromosome inactivation. *Nature* 379, 131–137.
167. Marahrens, Y., Panning, B., Dausman, J., Strauss, W., and Jaenisch, R. (1997). Xist-deficient mice are defective in dosage compensation but not spermatogenesis. *Genes Dev.* 11, 156–166.
168. Lee, J. T., and Lu, N. (1999). Targeted mutagenesis of Tsix leads to nonrandom X inactivation. *Cell* 99, 47–57.
169. Lee, J. T. (2000). Disruption of imprinted X inactivation by parent-of-origin effects at Tsix. *Cell* 103, 17–27.
170. Ganesan, S., Silver, D. P., Greenberg, R. A., Avni, D., Drapkin, R., Miron, A., Mok, S. C., Randrianarison, V., Brodie, S., Salstrom, J., Rasmussen, T. P., Klimke, A., Marrese, C., Marahrens, Y., Deng, C. X., Feunteun, J., and Livingston, D. M. (2002). BRCA1 supports XIST RNA concentration on the inactive X chromosome. *Cell* 111, 393–405.
171. Sado, T., Fenner, M. H., Tan, S. S., Tam, P., Shioda, T., and Li, E. (2000). X inactivation in the mouse embryo deficient for Dnmt1: distinct effect of hypomethylation on imprinted and random X inactivation. *Dev. Biol.* 225, 294–303.
172. Wang, J., Mager, J., Chen, Y., Schneider, E., Cross, J. C., Nagy, A., and Magnuson, T. (2001). Imprinted X inactivation maintained by a mouse Polycomb group gene. *Nat. Genet.* 28, 371–375.
173. O'Carroll, D., Erhardt, S., Pagani, M., Barton, S. C., Surani, M. A., and Jenuwein, T. The polycomb-group gene Ezh2 is required for early mouse development. *Mol. Cell. Biol.* 21, 4330–4336.
174. Bachvarova, R., and De Leon, V. (1980). Polyadenylated RNA of mouse ova and loss of maternal RNA in early development. *Dev. Biol.* 74, 1–8.
175. Clegg, K. B., and Piko, L. (1983). Poly(A) length, cytoplasmic adenylation and synthesis of poly(A)+ RNA in early mouse embryos. *Dev. Biol.* 95, 331–341.
176. Paynton, B. V., Rempel, R., and Bachvarova, R. (1988). Changes in state of adenylation and time course of degradation of maternal mRNAs during oocyte maturation and early embryonic development in the mouse. *Dev. Biol.* 129, 304–314.
177. Perlick, H. A., Medghalchi, S. M., Spencer, F. A., Kendzior, R. J. Jr., and Dietz, H. C. (1996). Mammalian orthologues of a yeast regulator of nonsense transcript stability. *Proc. Natl. Acad. Sci. U S A* 93, 10928–10932.
178. Medghalchi, S. M., Frischmeyer, P. A., Mendell, J. T., Kelly, A. G., Lawler, A. M., and Dietz, H. C. (2001). Rent1, a trans-effector of nonsense-mediated mRNA decay, is essential for mammalian embryonic viability. *Hum. Mol. Genet.* 10, 99–105.
179. Bernstein, E., Kim, S. Y., Carmell, M. A., Murchison, E. P., Alcorn, H., Li, M. Z., Mills, A. A., Elledge, S. J., Anderson, K. V., and Hannon, G. J. (2003). Dicer is essential for mouse development. *Nat. Genet.* 35, 215–217.
180. Cerritelli, S. M., Frolova, E. G., Feng, C., Grinberg, A., Love, P. E., and Crouch, R. J. (2003). Failure to produce mitochondrial DNA results in embryonic lethality in Rnaseh1 null mice. *Mol. Cell.* 11, 807–815.
181. Gu, W., Tekur, S., Reinbold, R., Eppig, J. J., Choi, Y. C., Zheng, J. Z., Murray, M. T., and Hecht, N. B. (1998). Mammalian male and female germ cells express a germ cell-specific Y-Box protein, MSY2. *Biol. Reprod.* 59, 1266–1274.
182. Yu, J., Hecht, N. B., and Schultz, R. M. (2002). RNA-binding properties and translation repression in vitro by germ cell-specific MSY2 protein. *Biol. Reprod.* 67, 1093–1098.
183. Bashirullah, A., Halsell, S. R., Cooperstock, R. L., Kloc, M., Karaiskakis, A., Fisher, W. W., Fu, W., Hamilton, J. K., Etkin, L. D., and Lipshitz, H. D. (1999). Joint action of two RNA degradation pathways controls the timing of maternal transcript elimination at the midblastula transition in *Drosophila melanogaster*. *EMBO J.* 18, 2610–2620.
184. Tadros, W., Houston, S. A., Bashirullah, A., Cooperstock, R. L., Semotok, J. L., Reed, B. H., and Lipshitz, H. D. (2003). Regulation of maternal transcript destabilization during egg activation in *Drosophila*. *Genetics* 164, 989–1001.
185. Ledan, E., Polanski, Z., Terret, M. E., and Maro, B. (2001). Meiotic maturation of the mouse oocyte requires an equilibrium between cyclin B synthesis and degradation. *Dev. Biol.* 232, 400–413.
186. Suzumori, N., Burns, K. H., Yan, W., and Matzuk, M. M. (2003). RFPL4 interacts with oocyte proteins of the ubiquitin-proteasome degradation pathway. *Proc. Natl. Acad. Sci. U S A* 100, 550–555.

187. Rajkovic, A., Lee, J. H., Yan, C., and Matzuk, M. M. (2002). The ret finger protein-like 4 gene, Rfpl4, encodes a putative E3 ubiquitin-protein ligase expressed in adult germ cells. *Mech. Dev.* 112, 173–177.
188. Ruland, J., Sirard, C., Elia, A., MacPherson, D., Wakeham, A., Li, L., de la Pompa, J. L., Cohen, S. N., and Mak, T. W. (2001). p53 accumulation, defective cell proliferation, and early embryonic lethality in mice lacking tsg101. *Proc. Natl. Acad. Sci. U S A* 98, 1859–1864.
189. Wang, Y., Penfold, S., Tang, X., Hattori, N., Riley, P., Harper, J. W., Cross, J. C., and Tyers, M. (1999). Deletion of the Cul1 gene in mice causes arrest in early embryogenesis and accumulation of cyclin E. *Curr. Biol.* 9, 1191–1194.
190. Dealy, M. J., Nguyen, K. V., Lo, J., Gstaiger, M., Krek, W., Elson, D., Arbeit, J., Kipreos, E. T., and Johnson, R. S. (1999). Loss of Cul1 results in early embryonic lethality and dysregulation of cyclin E. *Nat. Genet.* 23, 245–248.
191. Singer, J. D., Gurian-West, M., Clurman, B., and Roberts, J. M. (1999). Cullin-3 targets cyclin E for ubiquitination and controls S phase in mammalian cells. *Genes Dev.* 13, 2375–2387.
192. Li, B., Ruiz, J. C., and Chun, K. T. (2002). CUL-4A is critical for early embryonic development. *Mol. Cell. Biol.* 22, 4997–5005.
193. Sakao, Y., Kawai, T., Takeuchi, O., Copeland, N. G., Gilbert, D. J., Jenkins, N. A., Takeda, K., and Akira, S. (2000). Mouse proteasomal ATPases Psmc3 and Psmc4: genomic organization and gene targeting. *Genomics* 67, 1–7.
194. Pantaleon, M., Kanai-Azuma, M., Mattick, J. S., Kaibuchi, K., Kaye, P. L., and Wood, S. A. (2001). FAM deubiquitylating enzyme is essential for preimplantation mouse embryo development. *Mech. Dev.* 109, 151–160.
195. Lykke-Andersen, K., Schaefer, L., Menon, S., Deng, X. W., Miller, J. B., and Wei, N. (2003). Disruption of the COP9 signalosome Csn2 subunit in mice causes deficient cell proliferation, accumulation of p53 and cyclin E, and early embryonic death. *Mol. Cell. Biol.* 23, 6790–6797.
196. Yan, J., Walz, K., Nakamura, H., Carattini-Rivera, S., Zhao, Q., Vogel, H., Wei, N., Justice, M. J., Bradley, A., and Lupski, J. R. (2003). COP9 signalosome subunit 3 is essential for maintenance of cell proliferation in the mouse embryonic epiblast. *Mol. Cell. Biol.* 23, 6798–6808.
197. Hochstrasser, M. (2000). Evolution and function of ubiquitin-like protein-conjugation systems. *Nat. Cell. Biol.* 2, E153–E157.
198. Tateishi, K., Omata, M., Tanaka, K., and Chiba, T. (2001). The NEDD8 system is essential for cell cycle progression and morphogenetic pathway in mice. *J. Cell. Biol.* 155, 571–579.
199. Sutovsky, P., Moreno, R. D., Ramalho-Santos, J., Dominko, T., Simerly, C., and Schatten, G. (1999). Ubiquitin tag for sperm mitochondria. *Nature* 402, 371–372.
200. Sutovsky, P., Van Leyen, K., McCauley, T., Day, B. N., and Sutovsky, M. (2004). Degradation of paternal mitochondria after fertilization: implications for heteroplasmy, assisted reproductive technologies and mtDNA inheritance. *Reprod. Biomed. Online.* 8, 24–33.
201. Brenner, C. A., Barritt, J. A., Willadsen, S., and Cohen, J. (2000). Mitochondrial DNA heteroplasmy after human ooplasmic transplantation. *Fertil. Steril.* 74, 573–578.
202. Sutovsky, P., Motlik, J., Neuber, E., Pavlok, A., Schatten, G., Palecek, J., Hyttel, P., Adebayo, O. T., Adwan, K., Alberio, R., Bagis, H., Bataineh, Z., Bjerregaard, B., Bodo, S., Bryja, V., Carrington, M., Couf, M., de la Fuente, R., Diblik, J., Esner, M., Forejt, J., Fulka, J. Jr., Geussova, G., Gjorret, J. O., Libik, M., Hampl, A., Hassane, M. S., Houshmand, M., Hozak, P., Jezova, M., Kania, G., Kanka, J., Kandil, O. M., Kishimoto, T., Klima, J., Kohoutek, J., Kopska, T., Kubelka, M., Lapathitis, G., Laurincik, J., Lefevre, B., Mihalik, J., Novakova, M., Oko, R., Omelka, R., Owiny, D., Pachernik, J., Pacholikova, J., Peknicova, J., Pesty, A., Ponya, Z., Preclikova, H., Sloskova, A., Svoboda, P., Strejcek, F., Toth, S., Tepla, O., Valdivia, M., Vodicka, P., and Zudova, D. (2001). Accumulation of the proteolytic marker peptide ubiquitin in the trophoblast of mammalian blastocysts. *Cloning Stem Cells* 3, 157–161.
203. Tilly, J. L. (2001). Commuting the death sentence: how oocytes strive to survive. *Nat. Rev. Mol. Cell. Biol.* 2, 838–848.
204. Kamjoo, M., Brison, D. R., Kimber, S. J. (2002). Apoptosis in the preimplantation mouse embryo: effect of strain difference and in vitro culture. *Mol. Reprod. Dev.* 61, 67–77.
205. Threadgill, D. W., Dlugosz, A. A., Hansen, L. A., Tennenbaum, T., Lichti, U., Yee, D., LaMantia, C., Mourton, T., Herrup, K., and Harris, R. C., et al. (1995). Targeted disruption of mouse EGF receptor: effect of genetic background on mutant phenotype. *Science* 269, 230–234.
206. Kelkar, R. L., Dharma, S. J., and Nandedkar, T. D. (2003). Expression of Fas and Fas ligand protein and mRNA in mouse oocytes and embryos. *Reproduction* 126, 791–799.
207. Zou, G. M., Reznikoff-Etievant, M. F., Leon, A., Verge, V., Hirsch, F., and Milliez, J. (2000). Fas-mediated apoptosis of mouse embryo stem cells: its role during embryonic development. *Am. J. Reprod. Immunol.* 43, 240–248.
208. Yeh, W. C., Pompa, J. L., McCurrach, M. E., Shu, H. B., Elia, A. J., Shahinian, A., Ng, M., Wakeham, A., Khoo, W., Mitchell, K., El-Deiry, W. S., Lowe, S. W., Goeddel, D. V., and Mak, T. W. (1998). FADD: essential for embryo development and signaling from some, but not all, inducers of apoptosis. *Science* 279, 1954–1958.
209. Yeh, W. C., Itie, A., Elia, A. J., Ng, M., Shu, H. B., Wakeham, A., Mirtsos, C., Suzuki, N., Bonnard, M., Goeddel, D. V., and Mak, T. W. (2000). Requirement for Casper (c-FLIP) in regulation of death receptor-induced apoptosis and embryonic development. *Immunity* 12, 633–642.
210. Michaelson, J. S., Bader, D., Kuo, F., Kozak, C., and Leder, P. (1999). Loss of Daxx, a promiscuously interacting protein, results in extensive apoptosis in early mouse development. *Genes Dev.* 13, 1918–1923.
211. Exley, G. E., Tang, C., McElhinny, A. S., and Warner, C. M. (1999). Expression of caspase and BCL-2 apoptotic family members in mouse preimplantation embryos. *Biol. Reprod.* 61, 231–239.
212. Burns, K. H., and Matzuk, M. M. (2004). The application of gene ablation and related technologies to the study of ovarian physiology. In *The Ovary* (P. C. K. Leung and E. Y. Adashi, Eds.). Academic Press, San Diego.
213. Szegezdi, E., Fitzgerald, U., and Samali, A. (2003). Caspase-12 and ER-stress-mediated apoptosis: the story so far. *Ann. N Y Acad. Sci.* 1010, 186–194
214. Kawamura, K., Sato, N., Fukuda, J., Kodama, H., Kumagai, J., Tanikawa, H., Shimizu, Y., and Tanaka, T. (2003). Survivin acts as an antiapoptotic factor during the development of mouse preimplantation embryos. *Dev. Biol.* 256, 331–341.
215. Uren, A. G., Wong, L., Pakusch, M., Fowler, K. J., Burrows, F. J., Vaux, D. L., and Choo, K. H. (2000). Survivin and the inner centromere protein INCENP show similar cell-cycle localization and gene knockout phenotype. *Curr. Biol.* 10, 1319–1328.
216. Conway, E. M., Pollefeyt, S., Steiner-Mosonyi, M., Luo, W., Devriese, A., Lupu, F., Bono, F., Leducq, N., Dol, F., Schaeffer, P., Collen, D., and Herbert, J. M. (2002). Deficiency of survivin in transgenic mice exacerbates Fas-induced apoptosis via mitochondrial pathways. *Gastroenterology* 123, 619–631.
217. Kubiak, J. Z., and Ciemerych, M. A. (2001). Cell cycle regulation in early mouse embryos. *Novartis Found. Symp.* 237, 79–89; discussion 89–99.

218. Deng, M. Q., and Shen, S. S. (2000). A specific inhibitor of p34(cdc2)/cyclin B suppresses fertilization-induced calcium oscillations in mouse eggs. *Biol. Reprod.* 62, 873–878.
219. Kono, T., Jones, K. T., Bos-Mikich, A., Whittingham, D. G., and Carroll, J. (1996). A cell cycle-associated change in Ca2+ releasing activity leads to the generation of Ca2+ transients in mouse embryos during the first mitotic division. *J. Cell. Biol.* 132, 915–923.
220. Pahlavan, G., Polanski, Z., Kalab, P., Golsteyn, R., Nigg, E. A., and Maro, B. (2000). Characterization of polo-like kinase 1 during meiotic maturation of the mouse oocyte. *Dev. Biol.* 220, 392–400.
221. Tong, C., Fan, H. Y., Lian, L., Li, S. W., Chen, D. Y., Schatten, H., and Sun, Q. Y. (2002). Polo-like kinase-1 is a pivotal regulator of microtubule assembly during mouse oocyte meiotic maturation, fertilization, and early embryonic mitosis. *Biol. Reprod.* 67, 546–554.
222. Zernicka-Goetz, M., Ciemerych, M. A., Kubiak, J. Z., Tarkowski, A. K., and Maro, B. (1995). Cytostatic factor inactivation is induced by a calcium-dependent mechanism present until the second cell cycle in fertilized but not in parthenogenetically activated mouse eggs. *J. Cell. Sci.* 108(Pt 2), 469–474.
223. Winston, N., Bourgain-Guglielmetti, F., Ciemerych, M. A., Kubiak, J. Z., Senamaud-Beaufort, C., Carrington, M., Brechot, C., and Sobczak-Thepot, J. (2000). Early development of mouse embryos null mutant for the cyclin A2 gene occurs in the absence of maternally derived cyclin A2 gene products. *Dev. Biol.* 223, 139–153.
224. den Elzen, N., and Pines, J. (2001). Cyclin A is destroyed in prometaphase and can delay chromosome alignment and anaphase. *J. Cell. Biol.* 153, 121–136.
225. Ciemerych, M. A., Maro, B., and Kubiak, J. Z. (1999). Control of duration of the first two mitoses in a mouse embryo. *Zygote* 7, 293–300.
226. Moore, G. D., Ayabe, T., Kopf, G. S., and Schultz, R. M. (1996). Temporal patterns of gene expression of G1-S cyclins and cdks during the first and second mitotic cell cycles in mouse embryos. *Mol. Reprod. Dev.* 45, 264–275.
227. Iwamori, N., Naito, K., Sugiura, K., and Tojo, H. (2002). Preimplantation-embryo-specific cell cycle regulation is attributed to the low expression level of retinoblastoma protein. *FEBS Lett.* 526, 119–123.
228. Day, M. L., Pickering, S. J., Johnson, M. H., and Cook, D. I. (1993). Cell-cycle control of a large-conductance K+ channel in mouse early embryos. *Nature* 365, 560–562.
229. Day, M. L., Winston, N., McConnell, J. L., Cook, D., and Johnson, M. H. (2001). tiK+ toK+: an embryonic clock? *Reprod. Fertil. Dev.* 13, 69–79.
230. Day, M. L., Johnson, M. H., and Cook, D. I. (1998). A cytoplasmic cell cycle controls the activity of a K+ channel in preimplantation mouse embryos. *EMBO J.* 17, 1952–1960.
231. Hunt, P. A., and Hassold, T. J. (2002). Sex matters in meiosis. *Science* 296, 2181–2183.
232. Liu, L., Trimarchi, J. R., Smith, P. J., and Keefe, D. L. (2002). Checkpoint for DNA integrity at the first mitosis after oocyte activation. *Mol. Reprod. Dev.* 62, 277–288.
233. Fulka, J. Jr., First, N. L., Fulka, J., and Moor, R. M. (1999). Checkpoint control of the G2/M phase transition during the first mitotic cycle in mammalian eggs. *Hum. Reprod.* 14, 1582–1587.
234. Shimura, T., Inoue, M., Taga, M., Shiraishi, K., Uematsu, N., Takei, N., Yuan, Z. M., Shinohara, T., and Niwa, O. (2002). p53-dependent S-phase damage checkpoint and pronuclear cross talk in mouse zygotes with X-irradiated sperm. *Mol. Cell. Biol.* 22, 2220–2228.
235. Shimura, T., Toyoshima, M., Taga, M., Shiraishi, K., Uematsu, N., Inoue, M., and Niwa, O. (2002). The novel surveillance mechanism of the Trp53-dependent s-phase checkpoint ensures chromosome damage repair and preimplantation-stage development of mouse embryos fertilized with x-irradiated sperm. *Radiat. Res.* 158, 735–742.
236. Brandriff, B., and Pedersen, R. A. (1981). Repair of the ultraviolet-irradiated male genome in fertilized mouse eggs. *Science* 211, 1431–1433.
237. Matsuda, Y., Seki, N., Utsugi-Takeuchi, T., and Tobari, I. (1989). X-ray- and mitomycin C (MMC)-induced chromosome aberrations in spermiogenic germ cells and the repair capacity of mouse eggs for the X-ray and MMC damage. *Mutat. Res.* 211, 65–75.
238. Matsuda, Y., and Tobari, I. (1989). Repair capacity of fertilized mouse eggs for X-ray damage induced in sperm and mature oocytes. *Mutat. Res.* 210, 35–47.
239. Murphy, M., Stinnakre, M. G., Senamaud-Beaufort, C., Winston, N. J., Sweeney, C., Kubelka, M., Carrington, M., Brechot, C., and Sobczak-Thepot, J. (1997). Delayed early embryonic lethality following disruption of the murine cyclin A2 gene. *Nat. Genet.* 15, 83–86.
240. Liu, D., Matzuk, M. M., Sung, W. K., Guo, Q., Wang, P., and Wolgemuth, D. J. (1998). Cyclin A1 is required for meiosis in the male mouse. *Nat. Genet.* 20, 377–380.
241. Brandeis, M., Rosewell, I., Carrington, M., Crompton, T., Jacobs, M. A., Kirk, J., Gannon, J., and Hunt, T. (1998). Cyclin B2-null mice develop normally and are fertile whereas cyclin B1-null mice die in utero. *Proc. Natl. Acad. Sci. U S A* 95, 4344–4349.
242. Geng, Y., Yu, Q., Sicinska, E., Das, M., Schneider, J. E., Bhattacharya, S., Rideout, W. M., Bronson, R. T., Gardner, H., and Sicinski, P. (2003). Cyclin E ablation in the mouse. *Cell* 114, 431–443.
243. Parisi, T., Beck, A. R., Rougier, N., McNeil, T., Lucian, L., Werb, Z., and Amati, B. (2003). Cyclins E1 and E2 are required for endoreplication in placental trophoblast giant cells. *EMBO J.* 22, 4794–4803.
244. Ciemerych, M. A., Kenney, A. M., Sicinska, E., Kalaszczynska, I., Bronson, R. T., Rowitch, D. H., Gardner, H., and Sicinski, P. (2002). Development of mice expressing a single D-type cyclin. *Genes Dev.* 16, 3277–3289.
245. Lei, M., and Tye, B. K. (2001). Initiating DNA synthesis: from recruiting to activating the MCM complex. *J. Cell. Sci.* 114, 1447–1454.
246. Kim, J. M., Nakao, K., Nakamura, K., Saito, I., Katsuki, M., Arai, K., and Masai, H. (2002). Inactivation of Cdc7 kinase in mouse ES cells results in S-phase arrest and p53-dependent cell death. *EMBO J.* 21, 2168–2179.
247. Yoshida, K., Kuo, F., George, E. L., Sharpe, A. H., Dutta, A. (2001). Requirement of CDC45 for postimplantation mouse development. *Mol. Cell. Biol.* 21, 4598–4603.
248. Liu, Q., Guntuku, S., Cui, X. S., Matsuoka, S., Cortez, D., Tamai, K., Luo, G., Carattini-Rivera, S., DeMayo, F., Bradley, A., Donehower, L. A., and Elledge, S. J. (2000). Chk1 is an essential kinase that is regulated by Atr and required for the G(2)/M DNA damage checkpoint. *Genes Dev.* 14, 1448–1459.
249. Hirao, A., Cheung, A., Duncan, G., Girard, P. M., Elia, A. J., Wakeham, A., Okada, H., Sarkissian, T., Wong, J. A., Sakai, T., De Stanchina, E., Bristow, R. G., Suda, T., Lowe, S. W., Jeggo, P. A., Elledge, S. J., and Mak, T. W. (2002). Chk2 is a tumor suppressor that regulates apoptosis in both an ataxia telangiectasia mutated (ATM)-dependent and an ATM-independent manner. *Mol. Cell. Biol.* 22, 6521–6532.
250. Lincoln, A. J., Wickramasinghe, D., Stein, P., Schultz, R. M., Palko, M. E., De Miguel, M. P., Tessarollo, L., and

Donovan, P. J. (2002). Cdc25b phosphatase is required for resumption of meiosis during oocyte maturation. *Nat. Genet.* 30, 446–449.

251. Wickramasinghe, D., Becker, S., Ernst, M. K., Resnick, J. L., Centanni, J. M., Tessarollo, L., Grabel, L. B., and Donovan, P. J. (1995). Two CDC25 homologues are differentially expressed during mouse development. *Development* 121, 2047–2056.
252. Putkey, F. R., Cramer, T., Morphew, M. K., Silk, A. D., Johnson, R. S., McIntosh, J. R., and Cleveland, D. W. (2002). Unstable kinetochore-microtubule capture and chromosomal instability following deletion of CENP-E. *Dev. Cell.* 3, 351–365.
253. Warner, P., Wiley, L. M., Oudiz, D. J., Overstreet, J. W., and Raabe, O. G. (1991). Paternally inherited effects of gamma radiation on mouse preimplantation development detected by the chimera assay. *Radiat. Res.* 128, 48–58.
254. Obasaju, M. F., Wiley, L. M., Oudiz, D. J., Raabe, O., and Overstreet, J. W. (1989). A chimera embryo assay reveals a decrease in embryonic cellular proliferation induced by sperm from X-irradiated male mice. *Radiat. Res.* 118, 246–256.
255. Rassoulzadegan, M., Yang, Y., and Cuzin, F. (1998). APLP2, a member of the Alzheimer precursor protein family, is required for correct genomic segregation in dividing mouse cells. *EMBO J.* 17, 4647–4656.
256. Brown, E. J., and Baltimore, D. (2000). ATR disruption leads to chromosomal fragmentation and early embryonic lethality. *Genes Dev.* 14, 397–402.
257. de Klein, A., Muijtjens, M., van Os, R., Verhoeven, Y., Smit, B., Carr, A. M., Lehmann, A. R., and Hoeijmakers, J. H. (2000). Targeted disruption of the cell-cycle checkpoint gene ATR leads to early embryonic lethality in mice. *Curr. Biol.* 10, 479–482.
258. Kalitsis, P., Earle, E., Fowler, K. J., and Choo, K. H. (2000). Bub3 gene disruption in mice reveals essential mitotic spindle checkpoint function during early embryogenesis. *Genes Dev.* 14, 2277–2282.
259. Howman, E. V., Fowler, K. J., Newson, A. J., Redward, S., MacDonald, A. C., Kalitsis, P., and Choo, K. H. (2000). Early disruption of centromeric chromatin organization in centromere protein A (Cenpa) null mice. *Proc. Natl. Acad. Sci. U S A* 97, 1148–1153.
260. Kalitsis, P., Fowler, K. J., Earle, E., Hill, J., and Choo, K. H. (1998). Targeted disruption of mouse centromere protein C gene leads to mitotic disarray and early embryo death. *Proc. Natl. Acad. Sci. U S A* 95, 1136–1141.
261. Dodge, J. E., Kang, Y. K., Beppu, H., Lei, H., and Li, E. (2004). Histone H3-K9 methyltransferase ESET is essential for early development. *Mol. Cell. Biol.* 24, 2478–2486.
262. Larsen, E., Gran, C., Saether, B. E., Seeberg, E., and Klungland, A. (2003). Proliferation failure and gamma radiation sensitivity of Fen1 null mutant mice at the blastocyst stage. *Mol. Cell. Biol.* 23, 5346–5353.
263. Cutts, S. M., Fowler, K. J., Kile, B. T., Hii, L. L., O'Dowd, R. A., Hudson, D. F., Saffery, R., Kalitsis, P., Earle, E., and Choo, K. H. (1999). Defective chromosome segregation, microtubule bundling and nuclear bridging in inner centromere protein gene (Incenp)-disrupted mice. *Hum. Mol. Genet.* 8, 1145–1155.
264. Dumon-Jones, V., Frappart, P. O., Tong, W. M., Sajithlal, G., Hulla, W., Schmid, G., Herceg, Z., Digweed, M., and Wang, Z. Q. (2003). Nbn heterozygosity renders mice susceptible to tumor formation and ionizing radiation-induced tumorigenesis. *Cancer Res.* 63, 7263–7269.
265. Tsuzuki, T., Fujii, Y., Sakumi, K., Tominaga, Y., Nakao, K., Sekiguchi, M., Matsushiro, A., Yoshimura, Y., and Morita, T. (1996). Targeted disruption of the Rad51 gene leads to lethality in embryonic mice. *Proc. Natl. Acad. Sci. U S A* 93, 6236–6240.
266. Bemark, M., Khamlichi, A. A., Davies, S. L., Neuberger, M. S. (2000). Disruption of mouse polymerase zeta (Rev3) leads to embryonic lethality and impairs blastocyst development in vitro. *Curr. Biol.* 10, 1213–1216.
267. Wittschieben, J., Shivji, M. K., Lalani, E., Jacobs, M. A., Marini, F., Gearhart, P. J., Rosewell, I., Stamp, G., and Wood, R. D. (2000). Disruption of the developmentally regulated Rev3l gene causes embryonic lethality. *Curr. Biol.* 10, 1217–1220.
268. Kajiwara, K., J, O. W., Sakurai, T., Yamashita, S., Tanaka, M., Sato, M., Tagawa, M., Sugaya, E., Nakamura, K., Nakao, K., Katsuki, M., and Kimura, M. (2001). Sez4 gene encoding an elongation subunit of DNA polymerase zeta is required for normal embryogenesis. *Genes Cells* 6, 99–106.
269. Van Sloun, P. P., Varlet, I., Sonneveld, E., Boei, J. J., Romeijn, R. J., Eeken, J. C., and De Wind, N. (2002). Involvement of mouse Rev3 in tolerance of endogenous and exogenous DNA damage. *Mol. Cell. Biol.* 22, 2159–2169.
270. Cao, S., Bendall, H., Hicks, G. G., Nashabi, A., Sakano, H., Shinkai, Y., Gariglio, M., Oltz, E. M., and Ruley, H. E. (2003). The high-mobility-group box protein SSRP1/T160 is essential for cell viability in day 3.5 mouse embryos. *Mol. Cell. Biol.* 23, 5301–5307.
271. Bultman, S., Gebuhr, T., Yee, D., La Mantia, C., Nicholson, J., Gilliam, A., Randazzo, F., Metzger, D., Chambon, P., Crabtree, G., and Magnuson, T. (2000). A Brg1 null mutation in the mouse reveals functional differences among mammalian SWI/SNF complexes. *Mol. Cell.* 6, 1287–1295
272. Stopka, T., and Skoultchi, A. I. (2003). The ISWI ATPase Snf2h is required for early mouse development. *Proc. Natl. Acad. Sci. U S A* 100, 14097–14102.
273. Klochendler-Yeivin, A., Fiette, L., Barra, J., Muchardt, C., Babinet, C., and Yaniv, M. (2000). The murine SNF5/INI1 chromatin remodeling factor is essential for embryonic development and tumor suppression. *EMBO Rep.* 1, 500–506.
274. Karlseder, J., Kachatrian, L., Takai, H., Mercer, K., Hingorani, S., Jacks, T., and de Lange, T. (2003). Targeted deletion reveals an essential function for the telomere length regulator Trf1. *Mol. Cell. Biol.* 23, 6533–6541.
275. Morham, S. G., Kluckman, K. D., Voulomanos, N., and Smithies, O. (1996). Targeted disruption of the mouse topoisomerase I gene by camptothecin selection. *Mol. Cell. Biol.* 16, 6804–6809.
276. Akimitsu, N., Adachi, N., Hirai, H., Hossain, M. S., Hamamoto, H., Kobayashi, M., Aratani, Y., Koyama, H., and Sekimizu, K. (2003). Enforced cytokinesis without complete nuclear division in embryonic cells depleting the activity of DNA topoisomerase IIalpha. *Genes Cells* 8, 393–402.
277. Sakkas, D., Trounson, A. O., and Kola, I. (1989). In vivo cleavage rates and viability obtained for early cleavage mouse embryos in co-culture with oviduct cells. *Reprod. Fertil. Dev.* 1, 127–136.
278. Sakkas, D., and Trounson, A. O. (1990). Co-culture of mouse embryos with oviduct and uterine cells prepared from mice at different days of pseudopregnancy. *J. Reprod. Fertil.* 90, 109–118.
279. Yeung, W. S., Ho, P. C., Lau, E. Y., and Chan, S. T. (1992). Improved development of human embryos in vitro by a human oviductal cell co-culture system. *Hum. Reprod.* 7, 1144–1149.
280. Yeung, W. S., Lau, E. Y., Chan, S. T., and Ho, P. C. (1996). Coculture with homologous oviductal cells improved the implantation of human embryos: a prospective randomized control trial. *J. Assist. Reprod. Genet.* 13, 762–767.

281. Liu, L. P., Chan, S. T., Ho, P. C., and Yeung, W. S. (1995). Human oviductal cells produce high molecular weight factor(s) that improves the development of mouse embryo. *Hum. Reprod.* 10, 2781–2786.
282. Lee, K. F., Chow, J. F., Xu, J. S., Chan, S. T., Ip, S. M., and Yeung, W. S. (2001). A comparative study of gene expression in murine embryos developed in vivo, cultured in vitro, and cocultured with human oviductal cells using messenger ribonucleic acid differential display. *Biol. Reprod.* 64, 910–917.
283. Lee, Y. L., Lee, K. F., Xu, J. S., He, Q. Y., Chiu, J. F., Lee, W. M., Luk, J. M., and Yeung, W. S. (2004). The embryotrophic activity of oviductal cell-derived complement C3b and iC3b, a novel function of complement protein in reproduction. *J. Biol. Chem.* 279, 12763–12768.
284. Buhi, W. C., Alvarez, I. M., and Kouba, A. J. (2000). Secreted proteins of the oviduct. *Cells Tiss. Organs* 166, 165–179.
285. Ota, Y., Shimoya, K., Zhang, Q., Moriyama, A., Chin, R., Tenma, K., Kimura, T., Koyama, M., Azuma, C., and Murata, Y. (2002). The expression of secretory leukocyte protease inhibitor (SLPI) in the fallopian tube: SLPI protects the acrosome reaction of sperm from inhibitory effects of elastase. *Hum. Reprod.* 17, 2517–2522.
286. Butzow, R. (1989). The human fallopian tube contains placental protein 5. *Hum. Reprod.* 4, 17–20.
287. Parr, M. B., and Parr, E. L. (1985). Immunohistochemical localization of immunoglobulins A, G and M in the mouse female genital tract. *J. Reprod. Fertil.* 74, 361–370.
288. Araki, Y., Nohara, M., Yoshida-Komiya, H., Kuramochi, T., Ito, M., Hoshi, H., Shinkai, Y., and Sendai, Y. (2003). Effect of a null mutation of the oviduct-specific glycoprotein gene on mouse fertilization. *Biochem. J.* 374, 551–557.
289. Lee, K. F., Yao, Y. Q., Kwok, K. L., Xu, J. S., and Yeung, W. S. (2002). Early developing embryos affect the gene expression patterns in the mouse oviduct. *Biochem. Biophys. Res. Commun.* 292, 564–570.
290. Matzuk, M. M., Burns, K. H., Viveiros, M. M., and Eppig, J. J. (2002). Intercellular communication in the mammalian ovary: oocytes carry the conversation. *Science* 296, 2178–2180.
291. Paria, B. C., Reese, J., Das, S. K., Dey, S. K. (2002). Deciphering the cross-talk of implantation: advances and challenges. *Science* 296, 2185–2188.
292. Salahuddin, S., Ookutsu, S., Goto, K., Nakanishi, Y., and Nagata, Y. (1995). Effects of embryo density and co-culture of unfertilized oocytes on embryonic development of in-vitro fertilized mouse embryos. *Hum. Reprod.* 10, 2382–2385.
293. Hiroi, H., Momoeda, M., Inoue, S., Tsuchiya, F., Matsumi, H., Tsutsumi, O., Muramatsu, M., and Taketani, Y. (1999). Stage-specific expression of estrogen receptor subtypes and estrogen responsive finger protein in preimplantational mouse embryos. *Endocr. J.* 46, 153–158.
294. Stromstedt, M., Keeney, D. S., Waterman, M. R., Paria, B. C., Conley, A. J., and Dey, S. K. (1996). Preimplantation mouse blastocysts fail to express CYP genes required for estrogen biosynthesis. *Mol. Reprod. Dev.* 43, 428–436.
295. Wang, J., Rout, U. K., Bagchi, I. C., and Armant, D. R. (1998). Expression of calcitonin receptors in mouse preimplantation embryos and their function in the regulation of blastocyst differentiation by calcitonin. *Development* 125, 4293–4302.
296. Paria, B. C., Wang, H., and Dey, S. K. (2002). Endocannabinoid signaling in synchronizing embryo development and uterine receptivity for implantation. *Chem. Phys. Lipids.* 121, 201–210.
297. Bhatnagar, P., Papaioannou, V. E., and Biggers, J. D. (1995). CSF-1 and mouse preimplantation development in vitro. *Development* 121, 1333–1339.
298. Dardik, A., and Schultz, R. M. (1991). Blastocoel expansion in the preimplantation mouse embryo: stimulatory effect of TGF-alpha and EGF. *Development* 113, 919–930.
299. Dardik, A., Smith, R. M., and Schultz, R. M. (1992). Colocalization of transforming growth factor-alpha and a functional epidermal growth factor receptor (EGFR) to the inner cell mass and preferential localization of the EGFR on the basolateral surface of the trophectoderm in the mouse blastocyst. *Dev. Biol.* 154, 396–409.
300. Tsark, E. C., Adamson, E. D., Withers, G. E. III, and Wiley, L. M. (1997). Expression and function of amphiregulin during murine preimplantation development. *Mol. Reprod. Dev.* 47, 271–283.
301. Paria, B. C., Das, S. K., Andrews, G. K., and Dey, S. K. (1993). Expression of the epidermal growth factor receptor gene is regulated in mouse blastocysts during delayed implantation. *Proc. Natl. Acad. Sci. U S A* 90, 55–59.
302. Rappolee, D. A., Basilico, C. , Patel, Y., and Werb, Z. (1994). Expression and function of FGF-4 in peri-implantation development in mouse embryos. *Development* 120, 2259–2269.
303. Feldman, B., Poueymirou, W., Papaioannou, V. E., DeChiara, T. M., and Goldfarb, M. (1995). Requirement of FGF-4 for postimplantation mouse development. *Science* 267, 246–249.
304. Arman, E., Haffner-Krausz, R., Chen, Y., Heath, J. K., and Lonai, P. (1998). Targeted disruption of fibroblast growth factor (FGF) receptor 2 suggests a role for FGF signaling in pregastrulation mammalian development. *Proc. Natl. Acad. Sci. U S A* 95, 5082–5087.
305. Chai, N., Patel, Y., Jacobson, K., McMahon, J., McMahon, A., and Rappolee, D. A. (1998). FGF is an essential regulator of the fifth cell division in preimplantation mouse embryos. *Dev. Biol.* 198, 105–115.
306. Wang, Q. T., Piotrowska, K., Ciemerych, M. A., Milenkovic, L., Scott, M. P., Davis, R. W., and Zernicka-Goetz, M. (2004). A genome-wide study of gene activity reveals developmental signaling pathways in the preimplantation mouse embryo. *Dev. Cell.* 6, 133–144.
307. Patsoula, E., Loutradis, D., Drakakis, P., Michalas, L., Bletsa, R., and Michalas, S. (2003). Messenger RNA expression for the follicle-stimulating hormone receptor and luteinizing hormone receptor in human oocytes and preimplantation-stage embryos. *Fertil. Steril.* 79, 1187–1193.
308. Patsoula, E., Loutradis, D., Drakakis, P., Kallianidis, K., Bletsa, R., and Michalas, S. (2001). Expression of mRNA for the LH and FSH receptors in mouse oocytes and preimplantation embryos. *Reproduction* 121, 455–461.
309. Casan, E. M., Raga, F., and Polan, M. L. (1999). GnRH mRNA and protein expression in human preimplantation embryos. *Mol. Hum. Reprod.* 5, 234–239.
310. Casan, E. M., Raga, F., Bonilla-Musoles, F., and Polan, M. L. (2000). Human oviductal gonadotropin-releasing hormone: possible implications in fertilization, early embryonic development, and implantation. *J. Clin. Endocrinol. Metab.* 85, 1377–1381.
311. Raga, F., Casan, E. M., Kruessel, J., Wen, Y., Bonilla-Musoles, F., and Polan, M. L. (1999). The role of gonadotropin-releasing hormone in murine preimplantation embryonic development. *Endocrinology* 140, 3705–3712.
312. Pantaleon, M., Whiteside, E. J., Harvey, M. B., Barnard, R. T., Waters, M. J., and Kaye, P. L. (1997). Functional growth hormone (GH) receptors and GH are expressed by preimplantation mouse embryos: a role for GH in early embryogenesis? *Proc. Natl. Acad. Sci. U S A* 94, 5125–5130.
313. Fukaya, T., Yamanaka, T., Terada, Y., Murakami, T., and Yajima, A. (1998). Growth hormone improves mouse embryo development in vitro, and the effect is neutralized by growth

hormone receptor antibody. *Tohoku J. Exp. Med.* 184, 113–122.

314. Kawamura, K., Sato, N., Fukuda, J., Kodama, H., Kumagai, J., Tanikawa, H., Nakamura, A., Honda, Y., Sato, T., and Tanaka, T. (2003). Ghrelin inhibits the development of mouse preimplantation embryos in vitro. *Endocrinology* 144, 2623–2633.
315. Kaye, P. L. (1997). Preimplantation growth factor physiology. *Rev. Reprod.* 2, 121–127.
316. Doherty, A. S., Temeles, G. L., and Schultz, R. M. (1994). Temporal pattern of IGF-I expression during mouse preimplantation embryogenesis. *Mol. Reprod. Dev.* 37, 21–26.
317. Rappolee, D. A., Sturm, K. S., Behrendtsen, O., Schultz, G. A., Pedersen, R. A., and Werb, Z. (1992). Insulin-like growth factor II acts through an endogenous growth pathway regulated by imprinting in early mouse embryos. *Genes Dev.* 6, 939–952.
318. Truchet, S., Wietzerbin, J., and Debey, P. (2001). Mouse oocytes and preimplantation embryos bear the two subunits of interferon-gamma receptor. *Mol. Reprod. Dev.* 60, 319–330.
319. Mitsunari, M., Harada, T., Tanikawa, M., Iwabe, T., Taniguchi, F., and Terakawa, N. (1999). The potential role of stem cell factor and its receptor c-kit in the mouse blastocyst implantation. *Mol. Hum. Reprod.* 5, 874–879.
320. Kawamura, K., Sato, N., Fukuda, J., Kodama, H., Kumagai, J., Tanikawa, H., Murata, M., and Tanaka, T. (2003). The role of leptin during the development of mouse preimplantation embryos. *Mol. Cell. Endocrinol.* 202, 185–189.
321. Kawamura, K., Sato, N., Fukuda, J., Kodama, H., Kumagai, J., Tanikawa, H., Nakamura, A., and Tanaka, T. (2002). Leptin promotes the development of mouse preimplantation embryos in vitro. *Endocrinology* 143, 1922–1931.
322. Nichols, J., Davidson, D., Taga, T., Yoshida, K., Chambers, I., and Smith, A. (1996). Complementary tissue-specific expression of LIF and LIF-receptor mRNAs in early mouse embryogenesis. *Mech. Dev.* 57, 123–131.
323. Lavranos, T. C., Rathjen, P. D., and Seamark, R. F. (1995). Trophic effects of myeloid leukaemia inhibitory factor (LIF) on mouse embryos. *J. Reprod. Fertil.* 105, 331–338.
324. Stewart, C. L., Kaspar, P., Brunet, L. J., Bhatt, H., Gadi, I., Kontgen, F., and Abbondanzo, S. J. (1992). Blastocyst implantation depends on maternal expression of leukaemia inhibitory factor. *Nature* 359, 76–79.
325. O'Neill, C. (1998). Autocrine mediators are required to act on the embryo by the 2-cell stage to promote normal development and survival of mouse preimplantation embryos in vitro. *Biol. Reprod.* 58, 1303–1309.
326. Lu, D. P., Li, Y., Bathgate, R., Day, M., and O'Neill, C. (2003). Ligand-activated signal transduction in the 2-cell embryo. *Biol. Reprod.* 69, 106–116.
327. Lu, D. P., Chandrakanthan, V., Cahana, A., Ishii, S., and O'Neill, C. (2004). Trophic signals acting via phosphatidylinositol-3 kinase are required for normal preimplantation mouse embryo development. *J. Cell. Sci.* 117, 1567–1576.
328. Palmieri, S. L., Payne, J., Stiles, C. D., Biggers, J. D., and Mercola, M. (1992). Expression of mouse PDGF-A and PDGF alpha-receptor genes during pre- and post-implantation development: evidence for a developmental shift from an autocrine to a paracrine mode of action. *Mech. Dev.* 39, 181–191.
329. Mummery, C. L. (2001). Transforming growth factor beta and mouse development. *Microsc. Res. Tech.* 52, 374–386.
330. Pampfer, S., Vanderheyden, I., Vesela, J., and De Hertogh, R. (1995). Neutralization of tumor necrosis factor alpha (TNF alpha) action on cell proliferation in rat blastocysts by antisense oligodeoxyribonucleotides directed against TNF alpha p60 receptor. *Biol. Reprod.* 52, 1316–1326.
331. Pampfer, S., Wuu, Y. D., Vanderheyden, I., and De Hertogh, R. (1994). Expression of tumor necrosis factor-alpha (TNF alpha) receptors and selective effect of TNF alpha on the inner cell mass in mouse blastocysts. *Endocrinology* 134, 206–212.
332. Ware, C. B., Horowitz, M. C., Renshaw, B. R., Hunt, J. S., Liggitt, D., Koblar, S. A., Gliniak, B. C., McKenna, H. J., Papayannopoulou, T., and Thoma, B., et al. (1995). Targeted disruption of the low-affinity leukemia inhibitory factor receptor gene causes placental, skeletal, neural and metabolic defects and results in perinatal death. *Development* 121, 1283–1299.
333. Luetteke, N. C., Qiu, T. H., Fenton, S. E., Troyer, K. L., Riedel, R. F., Chang, A., and Lee, D. C. (1999). Targeted inactivation of the EGF and amphiregulin genes reveals distinct roles for EGF receptor ligands in mouse mammary gland development. *Development* 126, 2739–2750.
334. Brison, D. R., and Schultz, R. M. (1998). Increased incidence of apoptosis in transforming growth factor alpha-deficient mouse blastocysts. *Biol. Reprod.* 59, 136–144.
335. Kallapur, S., Ormsby, I., and Doetschman, T. (1999). Strain dependency of TGFbeta1 function during embryogenesis. *Mol. Reprod. Dev.* 52, 341–349.
336. Roelen, B. A., Goumans, M. J., Zwijsen, A., and Mummery, C. L. (1998). Identification of two distinct functions for TGF-beta in early mouse development. *Differentiation* 64, 19–31.
337. Zwijsen, A., van Rooijen, M. A., Goumans, M. J., Dewulf, N., Bosman, E. A., ten Dijke, P., Mummery, C. L., and Huylebroeck, D. (2000). Expression of the inhibitory Smad7 in early mouse development and upregulation during embryonic vasculogenesis. *Dev. Dyn.* 218, 663–670.
338. Albano, R. M., Groome, N., and Smith, J. C. (1993). Activins are expressed in preimplantation mouse embryos and in ES and EC cells and are regulated on their differentiation. *Development* 117, 711–723.
339. Ducibella, T., Ukena, T., Karnovsky, M., and Anderson, E. (1977). Changes in cell surface and cortical cytoplasmic organization during early embryogenesis in the preimplantation mouse embryo. *J. Cell. Biol.* 74, 153–167.
340. Vestweber, D., Gossler, A., Boller, K., and Kemler, R. (1987). Expression and distribution of cell adhesion molecule uvomorulin in mouse preimplantation embryos. *Dev. Biol.* 124, 451–456.
341. Hyafil, F., Morello, D., Babinet, C., and Jacob, F. (1980). A cell surface glycoprotein involved in the compaction of embryonal carcinoma cells and cleavage stage embryos. *Cell* 21, 927–934.
342. Winkel, G. K., Ferguson, J. E., Takeichi, M., and Nuccitelli, R. (1990). Activation of protein kinase C triggers premature compaction in the four-cell stage mouse embryo. *Dev. Biol.* 138, 1–15.
343. Larue, L., Ohsugi, M., Hirchenhain, J., and Kemler, R. (1994). E-cadherin null mutant embryos fail to form a trophectoderm epithelium. *Proc. Natl. Acad. Sci. U S A* 91, 8263–8267.
344. Riethmacher, D., Brinkmann, V., and Birchmeier, C. (1995). A targeted mutation in the mouse E-cadherin gene results in defective preimplantation development. *Proc. Natl. Acad. Sci. U S A* 92, 855–859.
345. Wianny, F., and Zernicka-Goetz, M. (2000). Specific interference with gene function by double-stranded RNA in early mouse development. *Nat. Cell. Biol.* 2, 70–75.
346. Hamatani, T., Carter, M. G., Sharov, A. A., and Ko, M. S. (2004). Dynamics of global gene expression changes during mouse preimplantation development. *Dev. Cell.* 6, 117–131.

347. Moley, K. H., Chi, M. M., and Mueckler, M. M. (1998). Maternal hyperglycemia alters glucose transport and utilization in mouse preimplantation embryos. *Am. J. Physiol.* 275, E38–E47.
348. Moley, K. H. (1999). Diabetes and preimplantation events of embryogenesis. *Semin. Reprod. Endocrinol.* 17, 137–151.
349. Moley, K. H., Chi, M. M., Knudson, C. M., Korsmeyer, S. J., and Mueckler, M. M. (1998). Hyperglycemia induces apoptosis in pre-implantation embryos through cell death effector pathways. *Nat. Med.* 4, 1421–1424.
350. Chi, M. M., Pingsterhaus, J., Carayannopoulos, M., and Moley, K. H. (2000). Decreased glucose transporter expression triggers BAX-dependent apoptosis in the murine blastocyst. *J. Biol. Chem.* 275, 40252–40257.
351. Heilig, C. W., Saunders, T., Brosius, F. C. III, Moley, K., Heilig, K., Baggs, R., Guo, L., and Conner, D. (2003). Glucose transporter-1-deficient mice exhibit impaired development and deformities that are similar to diabetic embryopathy. *Proc. Natl. Acad. Sci. U S A* 100, 15613–15618.
352. Pantaleon, M., Harvey, M. B., Pascoe, W. S., James, D. E., and Kaye, P. L. (1997). Glucose transporter GLUT3: ontogeny, targeting, and role in the mouse blastocyst. *Proc. Natl. Acad. Sci. U S A* 94, 3795–3800.
353. Carayannopoulos, M. O., Chi, M. M., Cui, Y., Pingsterhaus, J. M., McKnight, R. A., Mueckler, M., Devaskar, S. U., and Moley, K. H. (2000). GLUT8 is a glucose transporter responsible for insulin-stimulated glucose uptake in the blastocyst. *Proc. Natl. Acad. Sci. U S A* 97, 7313–7318.
354. Wyman, A. H., Chi, M., Riley, J., Carayannopoulos, M. O., Yang, C., Coker, K. J., Pessin, J. E., and Moley, K. H. (2003). Syntaxin 4 expression affects glucose transporter 8 translocation and embryo survival. *Mol. Endocrinol.* 17, 2096–2102.
355. Pinto, A. B., Carayannopoulos, M. O., Hoehn, A., Dowd, L., and Moley, K. H. (2002). Glucose transporter 8 expression and translocation are critical for murine blastocyst survival. *Biol. Reprod.* 66, 1729–1733.
356. Kwong, W. Y., Wild, A. E., Roberts, P., Willis, A. C., and Fleming, T. P. (2000). Maternal undernutrition during the preimplantation period of rat development causes blastocyst abnormalities and programming of postnatal hypertension. *Development* 127, 4195–4202.
357. Kim, J., Gye, M. C., and Kim, M. K. (2004). Role of occludin, a tight junction protein, in blastocoel formation, and in the paracellular permeability and differentiation of trophectoderm in preimplantation mouse embryos. *Mol. Cells* 17, 248–254.
358. Fleming, T. P., McConnell, J., Johnson, M. H., and Stevenson, B. R. (1989). Development of tight junctions de novo in the mouse early embryo: control of assembly of the tight junction-specific protein, ZO-1. *J. Cell. Biol.* 108, 1407–1418.
359. Fleming, T. P., Hay, M., Javed, Q., and Citi, S. (1993). Localisation of tight junction protein cingulin is temporally and spatially regulated during early mouse development. *Development* 117, 1135–1144.
360. Sheth, B., Fesenko, I., Collins, J. E., Moran, B., Wild, A. E., Anderson, J. M., and Fleming, T. P. (1997). Tight junction assembly during mouse blastocyst formation is regulated by late expression of ZO-1 alpha+ isoform. *Development* 124, 2027–2037.
361. Sheth, B., Moran, B., Anderson, J. M., and Fleming, T. P. (2000). Post-translational control of occludin membrane assembly in mouse trophectoderm: a mechanism to regulate timing of tight junction biogenesis and blastocyst formation. *Development* 127, 831–840.
362. Fleming, T. P., Papenbrock, T., Fesenko, I., Hausen, P., and Sheth, B. (2000). Assembly of tight junctions during early vertebrate development. *Semin. Cell. Dev. Biol.* 11, 291–299.
363. Robson, P., Stein, P., Zhou, B., Schultz, R. M., and Baldwin, H. S. (2001). Inner cell mass-specific expression of a cell adhesion molecule (PECAM-1/CD31) in the mouse blastocyst. *Dev. Biol.* 234, 317–329.
364. Manejwala, F., Kaji, E., and Schultz, R. M. (1986). Development of activatable adenylate cyclase in the preimplantation mouse embryo and a role for cyclic AMP in blastocoel formation. *Cell* 46, 95–103.
365. Manejwala, F. M., Cragoe, E. J. Jr., and Schultz, R. M. (1989). Blastocoel expansion in the preimplantation mouse embryo: role of extracellular sodium and chloride and possible apical routes of their entry. *Dev. Biol.* 133, 210–220.
366. Wiley, L. M. (1984). Cavitation in the mouse preimplantation embryo: Na/K-ATPase and the origin of nascent blastocoele fluid. *Dev. Biol.* 105, 330–342.
367. Watson, A. J., and Kidder, G. M. (1988). Immunofluorescence assessment of the timing of appearance and cellular distribution of Na/K-ATPase during mouse embryogenesis. Dev. Biol. 126, 80–90.
368. MacPhee, D. J., Jones, D. H., Barr, K. J., Betts, D. H., Watson, A. J., and Kidder, G. M. (2000). Differential involvement of Na(+),K(+)-ATPase isozymes in preimplantation development of the mouse. *Dev. Biol.* 222, 486–498.
369. MacPhee, D. J., Barr, K. J., De Sousa, P. A., Todd, S. D., and Kidder, G. M. (1994). Regulation of Na+,K(+)-ATPase alpha subunit gene expression during mouse preimplantation development. *Dev. Biol.* 162, 259–266.
370. Barcroft, L. C., Moseley, A. E., Lingrel, J. B., and Watson, A. J. (2004). Deletion of the Na/K-ATPase alpha1-subunit gene (Atp1a1) does not prevent cavitation of the preimplantation mouse embryo. *Mech. Dev.* 121, 417–426.
371. Jones, D. H., Davies, T. C., and Kidder, G. M. (1997). Embryonic expression of the putative gamma subunit of the sodium pump is required for acquisition of fluid transport capacity during mouse blastocyst development. *J. Cell. Biol.* 139, 1545–1552.
372. Barr, K. J., Garrill, A., Jones, D. H., Orlowski, J., and Kidder, G. M. (1998). Contributions of Na+/H+ exchanger isoforms to preimplantation development of the mouse. *Mol. Reprod. Dev.* 50, 146–153.
373. Barcroft, L. C., Offenberg, H., Thomsen, P., and Watson, A. J. (2003). Aquaporin proteins in murine trophectoderm mediate transepithelial water movements during cavitation. *Dev. Biol.* 256, 342–354.
374. Perona, R. M., and Wassarman, P. M. (1986). Mouse blastocysts hatch in vitro by using a trypsin-like proteinase associated with cells of mural trophectoderm. *Dev. Biol.* 114, 42–52.
375. O'Sullivan, C. M., Rancourt, S. L., Liu, S. Y., and Rancourt, D. E. (2001). A novel murine tryptase involved in blastocyst hatching and outgrowth. *Reproduction* 122, 61–71.
376. Edi-Osagie, E., Hooper, L., and Seif, M. W. (2003). The impact of assisted hatching on live birth rates and outcomes of assisted conception: a systematic review. *Hum. Reprod.* 18, 1828–1835.
377. Gardner, R. L. (1997). The early blastocyst is bilaterally symmetrical and its axis of symmetry is aligned with the animal-vegetal axis of the zygote in the mouse. *Development* 124, 289–301.
378. Piotrowska, K., and Zernicka-Goetz, M. (2001). Role for sperm in spatial patterning of the early mouse embryo. *Nature* 409, 517–521.
379. Hiiragi, T., and Solter, D. (2004). First cleavage plane of the mouse egg is not predetermined but defined by the topology of the two apposing pronuclei. *Nature* 403, 360–364.

380. Plusa, B., Grabarek, J. B., Piotrowska, K., Glover, D. M., and Zernicka-Goetz, M. (2002). Site of the previous meiotic division defines cleavage orientation in the mouse embryo. *Nat. Cell. Biol.* 4, 811–815.
381. Piotrowska, K., Wianny, F., Pedersen, R. A., and Zernicka-Goetz, M. (2001). Blastomeres arising from the first cleavage division have distinguishable fates in normal mouse development. *Development* 128, 3739–3748.
382. Weber, R. J., Pedersen, R. A., Wianny, F., Evans, M. J., and Zernicka-Goetz, M. (1999). Polarity of the mouse embryo is anticipated before implantation. *Development* 126, 5591–5598.
383. Geijsen, N., Horoschak, M., Kim, K., Gribnau, J., Eggan, K., and Daley, G. Q. (2004). Derivation of embryonic germ cells and male gametes from embryonic stem cells. *Nature* 427, 148–154.
384. Toyooka, Y., Tsunekawa, N., Akasu, R., and Noce, T. (2003). Embryonic stem cells can form germ cells in vitro. *Proc. Natl. Acad. Sci. U S A* 100, 11457–11462.
385. Hubner, K., Fuhrmann, G., Christenson, L. K., Kehler, J., Reinbold, R., De La Fuente, R., Wood, J., Strauss, J. F. III, Boiani, M., and Scholer, H. R. (2003). Derivation of oocytes from mouse embryonic stem cells. *Science* 300, 1251–1256.
386. Latham, K. E. (2004). Cloning: questions answered and unsolved. *Differentiation* 72, 11–22.
387. Mitsui, K., Tokuzawa, Y., Itoh, H., Segawa, K., Murakami, M., Takahashi, K., Maruyama, M., Maeda, M., and Yamanaka, S. (2003). The homeoprotein Nanog is required for maintenance of pluripotency in mouse epiblast and ES cells. *Cell* 113, 631–642.
388. Tanaka, T. S., Kunath, T., Kimber, W. L., Jaradat, S. A., Stagg, C. A., Usuda, M., Yokota, T., Niwa, H., Rossant, J., and Ko, M. S. (2002). Gene expression profiling of embryo-derived stem cells reveals candidate genes associated with pluripotency and lineage specificity. *Genome Res.* 12, 1921–1928.
389. Fan, Y., Melhem, M. F., and Chaillet, J. R. (1999). Forced expression of the homeobox-containing gene Pem blocks differentiation of embryonic stem cells. *Dev. Biol.* 210, 481–496.
390. Fujikura, J., Yamato, E., Yonemura, S., Hosoda, K., Masui, S., Nakao, K., Miyazaki, Ji, J., and Niwa, H. (2002). Differentiation of embryonic stem cells is induced by GATA factors. *Genes Dev.* 16, 784–789.
391. Smith, A. G. (2001). Embryo-derived stem cells: of mice and men. *Annu. Rev. Cell. Dev. Biol.* 17, 435–462.
392. Niwa, H., Burdon, T., Chambers, I., and Smith, A. (1998). Self-renewal of pluripotent embryonic stem cells is mediated via activation of STAT3. *Genes Dev.* 12, 2048–2060.
393. Matsuda, T., Nakamura, T., Nakao, K., Arai, T., Katsuki, M., Heike, T., and Yokota, T. (1999). STAT3 activation is sufficient to maintain an undifferentiated state of mouse embryonic stem cells. *EMBO J.* 18, 4261–4269.
394. Raz, R., Lee, C. K., Cannizzaro, L. A., d'Eustachio, P., and Levy, D. E. (1999). Essential role of STAT3 for embryonic stem cell pluripotency. *Proc. Natl. Acad. Sci. U S A* 96, 2846–2851.
395. Takeda, K., Noguchi, K., Shi, W., Tanaka, T., Matsumoto, M., Yoshida, N., Kishimoto, T., and Akira, S. (1997). Targeted disruption of the mouse Stat3 gene leads to early embryonic lethality. *Proc. Natl. Acad. Sci. U S A* 94, 3801–3804.
396. Yoshida, K., Taga, T., Saito, M., Suematsu, S., Kumanogoh, A., Tanaka, T., Fujiwara, H., Hirata, M., Yamagami, T., Nakahata, T., Hirabayashi, T., Yoneda, Y., Tanaka, K., Wang, W. Z., Mori, C., Shiota, K., Yoshida, N., and Kishimoto, T. (1996). Targeted disruption of gp130, a common signal transducer for the interleukin 6 family of cytokines, leads to myocardial and hematological disorders. *Proc. Natl. Acad. Sci. U S A* 93, 407–411.
397. Benezra, R., Davis, R. L., Lockshon, D., Turner, D. L., and Weintraub, H. (1990). The protein Id: a negative regulator of helix-loop-helix DNA binding proteins. *Cell* 61, 49–59.
398. Ying, Q. L., Nichols, J., Chambers, I., and Smith, A. (2003). BMP induction of Id proteins suppresses differentiation and sustains embryonic stem cell self-renewal in collaboration with STAT3. *Cell* 115, 281–292.
399. Nakashima, K., Yanagisawa, M., Arakawa, H., Kimura, N., Hisatsune, T., Kawabata, M., Miyazono, K., and Taga, T. (1999). Synergistic signaling in fetal brain by STAT3-Smad1 complex bridged by p300. *Science* 284, 479–482.
400. Okamoto, K., Okazawa, H., Okuda, A., Sakai, M., Muramatsu, M., and Hamada, H. (1990). A novel octamer binding transcription factor is differentially expressed in mouse embryonic cells. *Cell* 60:461–472.
401. Nichols, J., Zevnik, B., Anastassiadis, K., Niwa, H., Klewe-Nebenius, D., Chambers, I., Scholer, H., and Smith, A. (1998). Formation of pluripotent stem cells in the mammalian embryo depends on the POU transcription factor Oct4. *Cell* 95, 379–391.
402. Niwa, H., Miyazaki, J., and Smith, A. G. (2000). Quantitative expression of Oct-3/4 defines differentiation, dedifferentiation or self-renewal of ES cells. *Nat. Genet.* 24, 372–376.
403. Niwa, H. (2001). Molecular mechanism to maintain stem cell renewal of ES cells. *Cell. Struct. Funct.* 26, 137–148.
404. Remenyi, A., Lins, K., Nissen, L. J., Reinbold, R., Scholer, H. R., and Wilmanns, M. (2003). Crystal structure of a POU/HMG/DNA ternary complex suggests differential assembly of Oct4 and Sox2 on two enhancers. *Genes Dev.* 17, 2048–2059.
405. Kuroda, T., Tada, M., Kubota, H., Kimura, H., Hatano, S. Y., Suemori, H., Nakatsuji, N., and Tada, T. (2005). Octamer and Sox elements are required for transcriptional cis regulation of Nanog gene expression. *Mol. Cell Biol.* 25(6), 2475–2485.
406. Rodda, D. J., Chew, J. L., Lim, L. H., Loh, Y. H., Wang, B., Ng, H. H., and Robson, P. (2005). Transcriptional regulation of nanog by OCT4 and SOX2. *J. Biol. Chem.* (epub).
407. Avilion, A. A., Nicolis, S. K., Pevny, L. H., Perez, L., Vivian, N., and Lovell-Badge, R. (2003). Multipotent cell lineages in early mouse development depend on SOX2 function. *Genes Dev.* 17, 126–140.
408. Hanna, L. A., Foreman, R. K., Tarasenko, I. A., Kessler, D. S., and Labosky, P. A. (2002). Requirement for Foxd3 in maintaining pluripotent cells of the early mouse embryo. *Genes Dev.* 16, 2650–2661.
409. Guo, Y., Costa, R., Ramsey, H., Starnes, T., Vance, G., Robertson, K., Kelley, M., Reinbold, R., Scholer, H., and Hromas, R. (2002). The embryonic stem cell transcription factors Oct-4 and FoxD3 interact to regulate endodermal-specific promoter expression. *Proc. Natl. Acad. Sci U S A* 99, 3663–3667.
410. Chambers, I., Colby, D., Robertson, M., Nichols, J., Lee, S., Tweedie, S., and Smith, A. (2003). Functional expression cloning of Nanog, a pluripotency sustaining factor in embryonic stem cells. *Cell* 113, 643–655.
411. Wang, S. H., Tsai, M. S., Chiang, M. F., and Li, H. (2003). A novel NK-type homeobox gene, ENK (early embryo specific NK), preferentially expressed in embryonic stem cells. *Gene Expr. Patterns.* 3, 99–103.
412. Zheng, P., Patel, B., McMenamin, M., Reddy, S. E., Paprocki, A. M., Schramm, R. D., and Latham, K. E. (2004). The primate embryo gene expression resource: a novel resource to facilitate rapid analysis of gene expression patterns in non-human primate oocytes and preimplantation stage embryos. *Biol. Reprod.*

413. De Rycke, M., Liebaers, I., and Van Steirteghem, A. (2002). Epigenetic risks related to assisted reproductive technologies: risk analysis and epigenetic inheritance. *Hum. Reprod.* 17, 2487–2494.
414. Humpherys, D., Eggan, K., Akutsu, H., Hochedlinger, K., Rideout, W. M. III, Biniszkiewicz, D., Yanagimachi, R., and Jaenisch, R. (2001). Epigenetic instability in ES cells and cloned mice. *Science* 293, 95–97.
415. Kono, T., Obata, Y., Wu, Q., Niwa, K., Ono, Y., Yamamoto, Y., Park, E. S., Seo, J. S., and Ogawa, H. (2004). Birth of parthenogenetic mice that can develop to adulthood. *Nature* 428, 860–864.
416. Hwang, W. S., Ryu, Y. J., Park, J. H., Park, E. S., Lee, E. G., Koo, J. M., Chun, H. Y., Lee, B. C., Kang, S. K., Kim, S. J., Ahn, C., Hwang, J. H., Park, K. Y., Cibelli, J. B., and Moon, S. Y. (2004). Evidence of a pluripotent human embryonic stem cell line derived from a cloned blastocyst. *Science*.
417. Chang, H. Y., Yang, X. (2000). Proteases for cell suicide: functions and regulation of caspases. *Microbiol. Mol. Biol. Rev.* 64, 821–846.
418. Chang, H., Brown, C. W., and Matzuk, M. M. (2002). Genetic analysis of the mammalian TGF-b superfamily. *Endocrine Reviews* 23, 787–823.
419. Antczak, M., and Van Blerkom, J. (1997). Oocyte influences on early development: the regulatory proteins leptin and STAT3 are polarized in mouse and human oocytes and differentially distributed within the cells of the preimplantation stage embryo. *Mol. Hum. Reprod.* 3, 1067–1086.

Female Reproductive System

Knobil and Neill's Physiology of Reproduction, Third Edition
edited by Jimmy D. Neill,
Elsevier © 2006

CHAPTER 8

Embryology and Genetics of the Mammalian Gonads and Ducts

Keith L. Parker[1] and Bernard P. Schimmer[2]

Formation of the Bipotential Gonad, 314
Anatomy and Cell Biology of Sexual Differentiation, 314
The Indifferent Stage, 314
Testes Formation, 315
Organization of Testis Cords and Vascular Development, 315
Ovary Formation, 316
Genes Essential for Development of the Bipotential Gonad, 317
Wilms' Tumor Related 1 (WT1), 317
Steroidogenic Factor 1 (SF-1), 318
LHX1, 319
LHX9, 319
EMX2, 320
Genes Essential for Testes Development, 320
SRY, 320
SOX9, 320
GATA4, 321
DMRT1, 321
M33, 321
DAX1, 322
ARX, 323
POD-1, 323
Genes Implicated in Ovarian Development, 323
DAX1, 323
WNT4, 324
Primordial Germ Cells, 325
Origin, 325
Migration, 326
Chromosomal Events, 326
Embryology of Ducts of the Reproductive Tract, 326
Overview, 326
Embryology, 327
Hormone Action on the Embryonic Ducts, 327
References, 331

The early embryo is sexually undifferentiated: it contains primitive ducts that give rise to the internal genitalia of both sexes, while its gonads cannot be distinguished as testes or ovaries. These indifferent, or bipotential, gonads arise in a region under the coelomic epithelium called the genital ridge and contain lineages derived from both somatic and germ cells. Their dimorphic differentiation into testes or ovaries is controlled by the presence or absence of the Y chromosome, which contains the *SRY* gene. In genetic males, SRY expression in the somatic compartment directs differentiation of the embryonic gonads into testes and thereby functions as the critical dominant determinant of testes development from which all subsequent events in sex differentiation follow (1). Whereas male-specific hormones produced by the testes direct the fetus to develop along the male pathway (2,3), hormone synthesis by the ovary is not required for the female pattern of prenatal sex differentiation. The sexually dimorphic development of the testes or ovaries from a common precursor provides a striking model system for studying organogenesis and differentiation, and gonadal development is perhaps the best understood area of human embryology. In addition, developmental defects characterized by varying degrees of sex reversal of the sex ducts, external genitalia, and other secondary sex characteristics represent an important group of

[1]Department of Internal Medicine, Division of Endocrinology and Metabolism, University of Texas Southwestern Medical Center, Dallas, Texas.
[2]Best Institute, Toronto, Ontario, Canada.

clinical diseases in human beings. This chapter reviews the events in early development of the indifferent gonads and ducts, the sexually dimorphic formation of ovaries or testes, and the mechanisms by which the fetal gonads direct either the male or female developmental pathways. A further discussion of specific events in male sex determination and differentiation is provided in Chapter 6.

FORMATION OF THE BIPOTENTIAL GONAD

The gonads develop adjacent to the primitive embryonic kidneys, or mesonephroi in the region called the urogenital ridge, extending from the forelimb bud to the hindlimb bud. Complex interactions between the gonad and mesonephros are critical for gonad development. Several events characterize the initial stages in gonad formation: migration of primordial germ cells (PGC) into the mesenchymal tissue adjacent to the mesonephros; proliferation and differentiation of the supporting cells immediately underlying the coelomic epithelium; and release and migration of cells derived from the mesonephros into the developing gonad.

Studies in a variety of animal models have led to our current understanding of the development of the gonads. In addition, more traditional analyses of human beings with defects in gonadal development have provided key insights into events and genes that mediate these processes. Most recently, transgenic and knockout mice have assumed an increasing role in studies of mammalian gonad differentiation based on the ability to examine specific effects of gene inactivation or mis-expression. Although a great deal remains to be learned, a rough outline of the processes of development of the bipotential gonad has emerged. Rather than providing an encyclopedic overview of studies carried out in many different experimental models, this chapter focuses on unifying mechanisms of gonadal differentiation derived from studies of human beings and mice. A detailed discussion of the historical aspects of comparative embryology in various species is provided in the previous version of this chapter (4).

ANATOMY AND CELL BIOLOGY OF SEXUAL DIFFERENTIATION

This section provides a brief overview of the origins of the embryonic anlagen that form the gonads and the genital structures and the series of events that direct sex differentiation. This overview is then followed by a more detailed discussion that focuses on the roles of specific genes that have been implicated as essential for these events.

The Indifferent Stage

During the indifferent stage, the ovaries and testes cannot be distinguished histologically, and therefore are termed bipotential or indifferent gonads [reviewed in (5)]. As shown in Fig. 1, the bipotential gonads arise from the urogenital ridges, structures derived from the intermediate mesoderm that contain cell precursors that contribute to the kidneys, gonads, and adrenal cortex. These indifferent gonads also arise in close proximity to the ductal structures that are destined to give rise to the male and female internal genitalia (Fig. 2), thereby permitting complex developmental interactions.

After sex determination, the testes and ovaries become histologically distinct as the testes organize into two distinct compartments: the testicular cords-precursors of the seminiferous tubules-and the interstitial region (Fig. 3). The testicular cords contain fetal Sertoli cells (6) and primordial germ cells, which originate outside of the urogenital ridge in the epiblast and migrate into the indifferent gonad (7,8). The interstitial region surrounding the testicular cords contains the steroidogenic Leydig cells, the peritubular myoid cells, and endothelial cell precursors. The ovaries, in contrast, maintain an amorphous, "ground-glass" appearance and exhibit little structural differentiation until considerably later in gestation.

The urogenital tracts also cannot be distinguished in male and female embryos initially (Fig. 4). During this indifferent stage, both male and female embryos have two sets of paired ducts: the Müllerian (paramesonephric) ducts and the Wolffian (mesonephric) ducts (5). Under the influence of the Y chromosome in males, testes develop and produce specific hormones

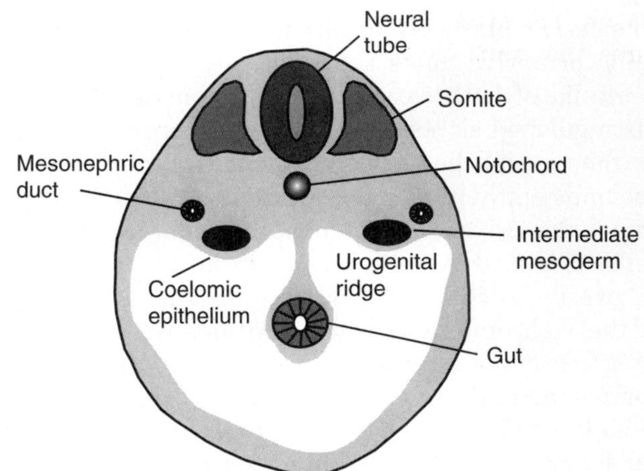

FIG. 1. The urogenital ridges derive from the intermediate mesoderm. Shown is a coronal section of a mouse embryo at embryonic day 9.5 (E9.5). At this indifferent stage, male and female embryos are indistinguishable. [Modified with permission from (5).]

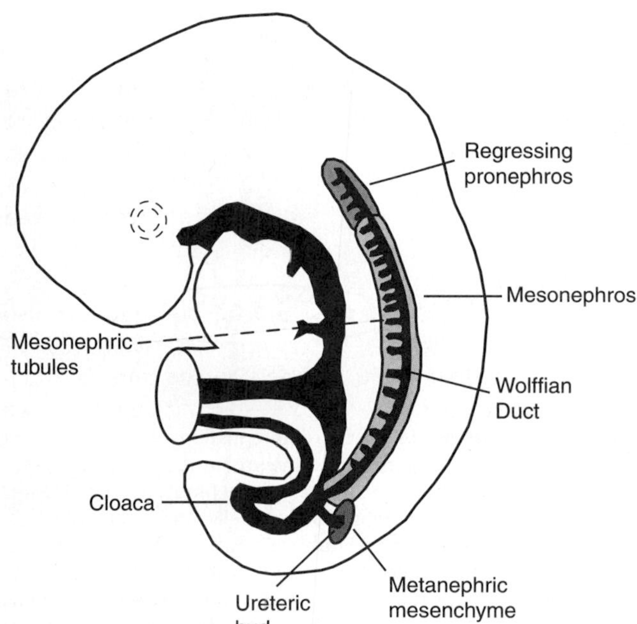

FIG. 2. Relative locations of the developing urogenital structures. Shown is a sagittal section of a mouse embryo at E11. At this stage, the pronephros is regressing, the mesonephros is still intact, as are the Wolffian and Müllerian ducts. [Modified with permission from (5).]

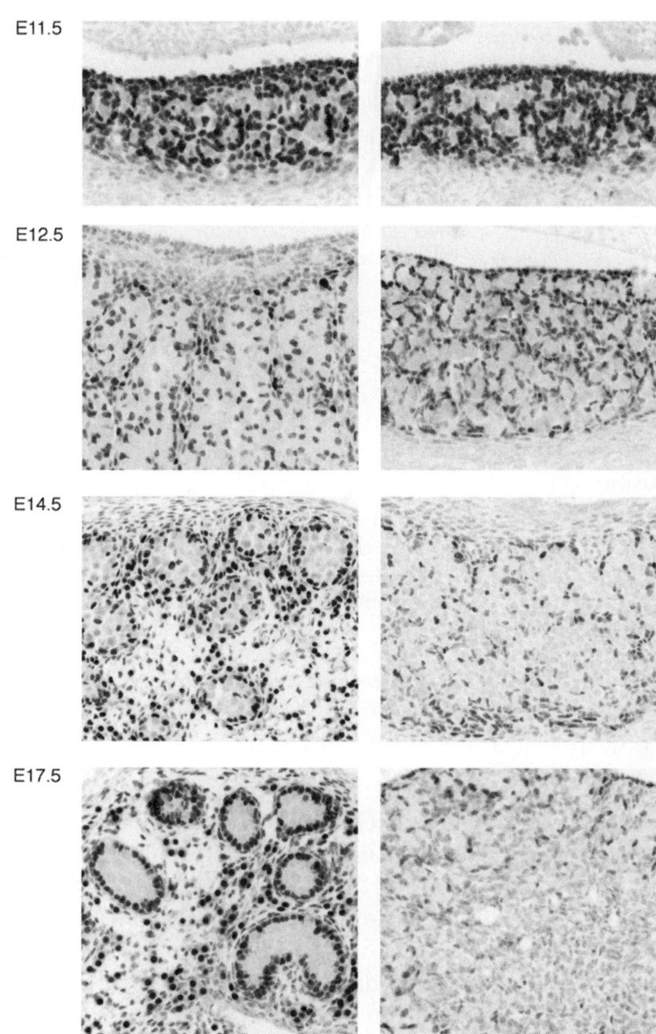

FIG. 3. Photomicrographs of the indifferent gonad at E11.5 and the testis (*left*) and ovary (*right*) after gonadal sex differentiation at E13.5. Note the well-formed testes cords in the testis versus the lack of histological organization in the ovary.

that trigger male sexual differentiation. The critical hormonal mediators of male sexual differentiation are testosterone, produced by the Leydig cells in the interstitial region, and anti-Müllerian hormone (AMH; also called Müllerian-inhibiting substance), produced by the Sertoli cells within the testicular cords (2,3). Testosterone induces the Wolffian ducts to differentiate into the seminal vesicles, epididymis, vas deferens, and ejaculatory ducts, while the Wolffian ducts regress in the absence of the testosterone. AMH induces regression of the Müllerian ducts, which otherwise would form the oviducts, uterus, and upper vagina. The external genitalia also develop from structures that initially are common to both sexes: the genital tubercle, urethral folds, urethral grooves and genital swelling that also are virilized by testicular androgens. For these structures, however, testosterone is not sufficient and must be converted to dihydrotestosterone (DHT) by the type 2 isozyme of 5α-reductase (9).

TESTES FORMATION

Organization of Testis Cords and Vascular Development

The testes first become distinct histologically when they form the testis cords, which are the precursors to the seminiferous tubules. In humans, this process begins by approximately the 7th week of gestation (Fig. 5). In mice, testis cords are first apparent at embryonic day 12.5 (E12.5). The organization of cords is initiated by pre-Sertoli cells, which express SRY at approximately 6 weeks of gestation in humans (10,11) or E10.5 in mice (12). Although it is proposed that the Sertoli cells arise from progenitors derived from the coelomic epithelium, cells that express Sry are not within the coelomic epithelium proper. Based on analyses of expression of a green fluorescent protein reporter gene driven by the Sry promoter, these Sertoli cells progenitors are found just interior to the coelomic epithelium and activate Sry expression with a wave that proceeds from the center of the gonad to the anterior and posterior poles (13). Following the onset of Sry expression, these cells undergo a rapid wave of proliferation and also exhibit changes in gene expression that denote the onset of the testis

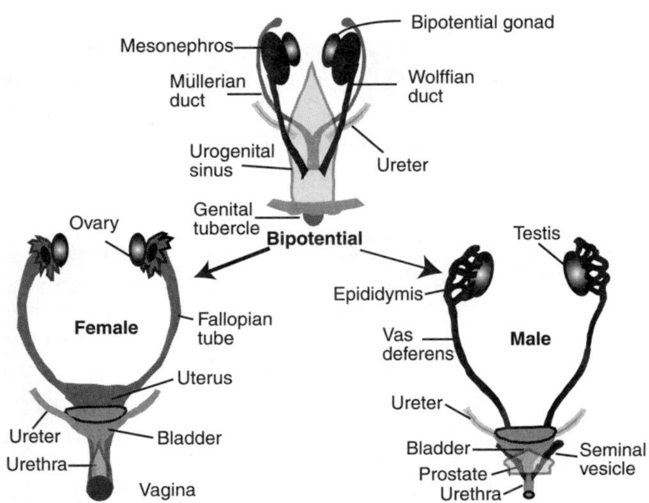

FIG. 4. Sexually dimorphic development of the Müllerian and Wolffian ducts to form the female (*left*) or male (*right*) internal genital structures. Under the influence of SRY, the bipotential gonad is induced to form a testis, which then produces hormones that cause the Wolffian ducts to form male internal genital structures and the Müllerian ducts to regress. If ovaries develop (or in the absence of testes hormones), the Wolffian ducts regress and the Müllerian ducts form the Fallopian tubes, uterus, and upper vagina.

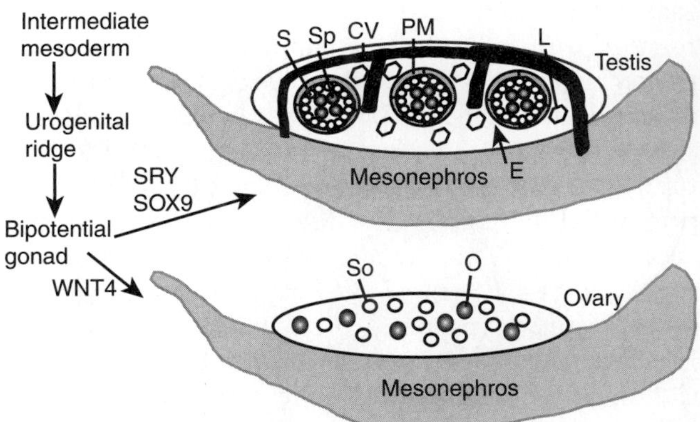

FIG. 6. Schematic model of the differentiation of the indifferent gonad into either a testis or ovary. Shown are the embryonic testis and ovary at ~E12.5 adjacent to the mesonephros. Note that testes cords have formed, containing Sertoli cells and spermatogonia, surrounded by a layer of peritubular myoid cells that separate the cords from the interstitial region, where the steroidogenic Leydig cells and vasculature are found. The ovary, in contrast, contains somatic cells and germ cells, but exhibits little histological organization and has not yet formed follicles. Genes that direct the bipotential gonad to form testes include *SRY* and *SOX9*, whereas *WNT4* is important for ovarian differentiation. S, Sertoli cell; Sp, spermatogonium; PM, peritubular myoid cell; L, Leydig cell; E, endothelial cell; CV, coelomic vessel; So, somatic cell; O, oogonium. [Modified with permission from (14).]

pathway [reviewed in (14)]. Foremost among these changes is the expression of the gene encoding the high mobility group (HMG) protein SOX9 (15,16). SOX9 was initially cloned based on its sequence homology to SRY, and its expression follows closely upon that of SRY. As discussed below, considerable information argues that SOX9 is both necessary and sufficient to direct the bipotential gonad along the testes pathway (17). Over the next several days, the Sertoli cells lay down a basement membrane that delineates the cords, which contain the Sertoli cells and the spermatogonia (Fig. 6). These spermatogonia are germ cells arrested in G1 of the mitotic cycle, presumably due to direct inhibitory effects of Sertoli cells [see below (18)]. The less-organized interstitial region contains the precursors to fetal Leydig cells, which are first visualized as steroidogenic cells at E13, although in situ hybridization analyses of genes that are specifically expressed by Leydig cells (e.g., the cholesterol side-chain cleavage enzyme, Cyp11a) suggest that the Leydig cell program of gene expression is initiated as early as E11.5 in mice (19). The onset of testosterone production, which stabilizes the Wolffian ducts, is not apparent until somewhat later at E14.

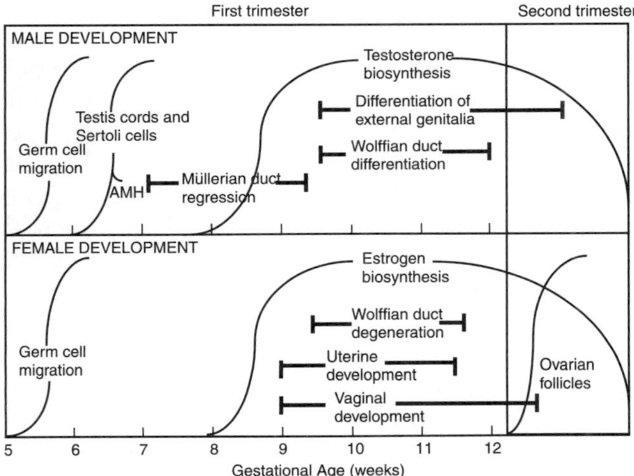

FIG. 5. Timing of different events in sex differentiation of human XX and XY embryos in utero. Shown are the gestational ages at which the various events occur that comprise prenatal sex differentiation. [Modified with permission from (213).]

The development of a vascular supply is another hallmark of testicular development. Before E11.5, vascular development in the gonads appears to be sexually indifferent. Thereafter, endothelial cell precursors are recruited from the neighboring region of the mesonephros and rapidly organize to form an extensively branched arterial blood vessel. The mechanisms underlying the reorganization of the testicular vasculature are unknown, but presumably involve SRY-dependent products of Sertoli cells (see below for further discussion).

OVARY FORMATION

As outlined above, the initial steps in ovarian development during the indifferent stage parallel those seen with the testes. The ovaries, however, retain a relatively undifferentiated appearance during the period when the male gonad begins its process of differentiation (Fig. 3). In humans, the ovaries do not exhibit positive histological evidence of their sex until approximately 16 weeks of gestation, when the primordial follicles are visualized as oocytes surrounded by a single layer of flattened granulosa cells within a basal lamina. At this stage, the oocytes are arrested in prophase of the first meiosis (18). In mice, histological evidence of ovarian differentiation is even further delayed, and the primordial follicles do not appear until shortly before birth at ~E18. By birth, a few primary follicles characterized by cuboidal granulosa cells surrounding the oocyte are present. At a molecular level, differences in gene expression in somatic cells and differences in the behavior of the germ cells provide a positive means to distinguish the ovaries (see below). All of these developmental processes in ovarian development are independent of gonadotropin stimulation, and the gonadotropin-dependent events do not occur until after birth.

GENES ESSENTIAL FOR DEVELOPMENT OF THE BIPOTENTIAL GONAD

Several genes are essential for the development of the bipotential gonad (Table 1). As such, mutations in these genes impair development of both the testes and ovaries; by nature of the mechanisms of sex differentiation discussed above, these mutations generally perturb the phenotypes in genetic males to a much greater degree than genetic females.

Wilms' Tumor Related 1 (WT1)

Wilms' tumor-related 1 (*WT1*) was identified initially through its association with embryonic kidney tumors (Wilms' tumors), which arise from abnormal proliferation of the metanephric blastema and occur almost exclusively during childhood [reviewed in (20)]. The genetics of Wilms' tumors are complex and multiple loci have been identified (21). Only *WT1* on chromosome 11q13 has been definitively linked to Wilms' tumors. The cloning of *WT1* was facilitated by the identification of patients with heterozygous deletions of human chromosome 11p13 that caused a contiguous gene deletion syndrome (WAGR) that included *W*ilms' tumors, *a*niridia, *g*enitourinary abnormalities (see below), and mental *r*etardation (22,23). Sequence analysis of WT1 identified four zinc fingers at the carboxyl-terminus, the last three of which show high homology with the transcription factors Sp1 and Egr1. Consistent with a possible role as a transcription factor, WT1 interacts with specific DNA recognition sequences upstream of a large number of genes, including genes that encode growth

TABLE 1. *Genes essential for gonadal development*

Gene	Actions	Other effects
Genes implicated in early gonadal development in both sexes		
WT1 (11p13)	Regulate SF-1, SRY, DAX1	Adrenal, renal, cardiac anomalies
Steroidogenic Factor 1 (9q33)	Regulate AMH, CYP11A, StAR	Adrenal, pituitary defects
LHX9 (1q31.3)	Regulate SF-1	
EMX2 (10q26)		Schizencephaly
M33 (17q25)		
Genes affecting later events		
Genes required for testis development		
SRY (Yp11.3)	Testes determination	
SOX9 (17q23)	Regulate AMH	Skeletal defects (campomelic dysplasia)
DAX1 (Xp21.3)	Repress SF-1 action	Adrenal hypoplasia congenita
GATA4 (8p23)	Regulate SF-1	Cardiac defects
FOG2 (8q23)		Cardiac defects
Desert hedgehog (12q13.1)	Leydig cell differentiation	Minifascicular polyneuropathy
FGF9 (13q11-12)		
Pod1 (6pterqter)		Renal defects
ARX (Xp22.13)		Lissencephaly
PDGFRα (4q12)		
Genes required for ovarian development		
WNT4 (1p35)		Müllerian agenesis
FIG1α		
BMP15 (Xp11.2)		
Follistatin (5q11.2)		

factors and their receptors, generally repressing their transcription [reviewed in (24)]. Although repression of growth-stimulating genes is a plausible explanation for tumor suppression, it is unlikely that this role can fully explain WT1's pleiotropic effects (see below).

WT1 is expressed from the very earliest stages of development of the urogenital ridge, subsequently localizing to both the kidney and the gonads (11,25–28). In the testes and ovaries, expression becomes restricted to the Sertoli and granulosa cells, respectively. In addition to the relatively mild genitourinary abnormalities seen in the WAGR syndrome (e.g., cryptorchidism and hypospadias in males and horseshoe kidneys in both sexes), WT1 mutations also cause two other syndromic disorders of genitourinary development: Denys-Drash syndrome and Frasier syndrome. Denys-Drash is an autosomal dominant syndrome that almost always results from point mutations in the zinc finger region of WT1 that abolish DNA binding; it is proposed that these mutated proteins inhibit wild-type WT1 by a dominant negative action (29). Denys-Drash patients exhibit gonadal and urogenital abnormalities in conjunction with diffuse mesangial sclerosis that appears in the first year of life and causes end-stage renal disease by age 3. Wilms' tumors also are common in Denys-Drash families (perhaps as frequently as 75% of the time). The gonadal abnormalities of these patients vary but generally correlate with the degree of inhibition of WT1 action. Gonadal abnormalities are more severe than those associated with the WAGR syndrome and range from streak gonads and sex-reversal of external and internal genitalia in 46, XY males to varying degrees of pseudohermaphroditism in males that are less severely affected.

WT1 mutations also have been identified in patients with Frasier syndrome (30–32). These mutations are in or adjacent to intron 9 of the *WT1* gene and selectively decrease the production of an isoform of the protein that contains the residues Lys-Thr-Ser (+KTS). Patients with Frasier syndrome rarely develop Wilms' tumors but rather exhibit gonadal dysgenesis, male pseudohermaphroditism, and a milder renal disease due to focal glomerular sclerosis. Renal insufficiency is not clinically apparent until childhood, typically after age 4, and patients typically maintain some renal function until adolescence. These findings suggest that the +KTS isoform of WT1 is essential for both kidney and gonadal development but is not needed to suppress the development of Wilms' tumors.

Transgenic and knockout (KO) mice have provided key insights into the functions of Wt1. KO mice lacking Wt1 exhibited renal and gonadal agenesis (33). The initial stages of gonadal development were intact, but the gonads did not increase in size and rather regressed by approximately E14 due to increased apoptosis. Additional insights into the distinct roles of the +KTS and –KTS isoforms have emerged from studies of KO mice that selectively lack one or the other of these isoforms (34). In these mice, the –KTS form was essential for survival of cell lineages that contributed to the bipotential gonad in both sexes. Specific mechanisms by which this effect might occur are discussed below. In contrast, the +KTS isoform stimulated the expression of Sry and thus contributed selectively to testis differentiation and onset of the male developmental pathway at a later stage, such that XY mice selectively lacking the +KTS isoform formed ovaries and exhibited female sex differentiation.

A goal of ongoing experiments is to define the target genes by which WT1 mediates gonadal development. Analyses of mice that are selectively impaired in the production of one isoform or the other suggest that SRY is a direct target (34), a finding also supported by cell transfection assays (35,36). WT1 binding sites also have been identified in the 5′-flanking region of SF-1, and the available evidence indicates that the –KTS isoform of Wt1 stimulates the expression of SF-1 in vivo (37). Given the key role of SF-1 itself in gonadal development, it is apparent that the lack of SF-1 could drastically impair gonadogenesis. Collectively, these findings support a hierarchical model in which WT1 contributes to gonadal development by inducing the expression of SF-1 in the urogenital ridge. As discussed further below, other proposed target genes of WT1 that play key roles in gonadal development are the atypical orphan nuclear receptor DAX1 (38) and the secreted signaling molecule Wnt4 (39). Cell transfection studies also suggest that WT1 and SF-1 interact synergistically via protein-protein interactions (40), probably in conjunction with other transcriptional regulators such as the LIM-only coactivator FHL2 (41), to regulate key genes in the sex differentiation cascade such as AMH.

Steroidogenic Factor 1 (SF-1)

The essential role for SF-1 in early gonadal development emerged from studies of SF-1 mutations that caused gonadal agenesis and impaired the formation of both testes and ovaries [reviewed in (42,43)]. SF-1 was identified initially as a transcription factor that regulates tissue-specific expression of the cytochrome P450 steroid hydroxylases—enzymes that catalyze most of the reactions required for the synthesis of steroid hormones (44,45). The sequence of the SF-1 cDNA showed that this factor belongs to the nuclear hormone receptor superfamily of transcription factors that mediate the actions of steroid hormones, thyroid hormone, vitamin D, and retinoids. In addition to the steroid hydroxylases, SF-1 also regulates adrenal and gonadal expression of many genes involved in steroidogenesis, including the type 2 isoform of 3β-hydroxysteroid dehydrogenase, the ACTH receptor, the gonadotropin receptors, scavenger receptor B-1, and the steroidogenic acute regulatory protein [reviewed in (42,43)]. Analyses of reporter genes driven by the AMH promoter region in transfected Sertoli cells and transgenic mice suggest that SF-1 also regulates expression of AMH (46–49). Thus, it appears that SF-1 controls the production of both hormones required for male phenotypic differentiation.

SF-1 is expressed in both male and female mouse embryos at E9 when the intermediate mesoderm first condenses to form the urogenital ridge (19). With the onset of testis cord formation, the levels of SF-1 transcripts increase in both functional compartments of the testes: the interstitial region, where Leydig cells produce androgens, and the testicular cords, where Sertoli cells synthesize AMH. Strikingly, although there is no major histological change, SF-1 transcripts in ovaries decrease coincident with sexual differentiation (19,50), suggesting that persistent SF-1 expression may impair female sexual differentiation. In addition to the gonads and adrenals, SF-1 transcripts also are detected in the anterior pituitary and hypothalamus.

Analyses of SF-1 KO mice confirmed essential roles of SF-1 at all three levels of the hypothalamic–pituitary–steroidogenic organ axis. SF-1 KO mice completely lack adrenal glands and gonads because of apoptosis in the primordial organs at very early developmental stages (51,52). Consistent with the degeneration of the testes before androgens and AMH are produced, SF-1 knockout mice also exhibit male-to-female sex reversal of their internal and external urogenital tracts. These mice also have impaired expression of the pituitary gonadotropins, which regulate gonadal steroidogenesis, and have marked structural abnormalities of the ventromedial hypothalamic nucleus, a cell group in the mediobasal hypothalamus linked to energy homeostasis and reproductive behavior (43).

The human *SF-1* gene on chromosome 9q33 shares extensive homology with its mouse counterpart. The identification of human subjects with clinical disorders caused by SF-1 mutations has revealed both similarities and differences in SF-1 functions between species. Although adrenal and gonadal agenesis in mice is seen only with two disrupted alleles, a greater degree of dose sensitivity is seen in human beings (see below). Also, gonadotrope function apparently is preserved in human subjects with SF-1 mutations.

In the initial report, a mutation in the zinc finger DNA binding domain of SF-1 (G35E) that abolished DNA binding and transcriptional activation was associated with adrenal insufficiency and 46, XY sex-reversal due to gonadal dysgenesis (53). Intriguingly, the other *SF-1* allele was apparently normal, suggesting that the clinical disorder resulted from haploinsufficiency of SF-1 function. A second subject with early-onset adrenal insufficiency harbored a missense mutation in the hinge region of SF-1 (R255L) that caused a loss of transcriptional activation (54). Apparently normal ovaries were seen on magnetic resonance imaging, suggesting that a single wild-type allele of *SF-1* suffices for human ovary development. A homozygous mutation (R92Q) in an accessory DNA binding domain that impaired but did not abolish DNA binding and transcriptional activity caused autosomal recessive adrenal insufficiency and 46, XY sex reversal (55). The residual function of the mutated protein likely accounts for the autosomal recessive pattern of inheritance. Most recently, two groups described 46, XY subjects that presented with ambiguous external genitalia and streak gonads but had normal adrenocortical function. The first subject had an 8 base-pair deletion that resulted in premature termination upstream of the activator function 2 (AF-2) transactivation domain (56), whereas the second subject had a missense mutation that truncated the SF-1 protein at the 15th residue (57).

LHX1

LHX1 (also called Lim1) encodes a transcription factor that contains a homeodomain DNA binding region and two Cys-rich LIM domains that participate in protein–protein interactions. LHX1 is expressed from the earliest stages of differentiation of the urogenital ridge. Knockout mice lacking Lhx1 have both renal and gonadal agenesis, although the phenotype and specifics of the abnormalities have not been described in detail (58).

LHX9

LIM-containing homeobox 9 (LHX9) is another member of the homeodomain family of transcription factors that is expressed in the bipotential gonad from early developmental stages (E9.5). Following the onset of sex differentiation, Lhx9 transcripts localize to the interstitial region of the testes and ovaries but do not precisely localize to any specific cell lineage (59). The generation of Lhx9 KO mice revealed its essential roles in gonadal development. In Lhx9 KO mice, germ cells migrated normally into the genital ridge but somatic cell proliferation was impaired and a discrete gonad never formed. SF-1 expression was markedly reduced in the genital ridges of Lhx9 KO mice, suggesting that SF-1 may lie downstream of Lhx9 in a developmental cascade; this conclusion is supported by experiments described above for WT1, which showed that LHX9, like WT1, activated the SF-1 promoter in cell transfection studies (37).

EMX2

The human *EMX2* gene located on chromosome 10q26, encodes a homeodomain transcription factor that is related to the *Drosophila empty spiracles* gene. It is expressed in the intermediate mesoderm by as early as E8.5; thereafter, it is expressed in the Wolffian and Müllerian ducts, the indifferent gonads, the developing kidney, and the developing brain, where it is proposed to regulate the expression of fibroblast growth factor 8. Both male and female KO mice lacking Emx2 exhibited gonadal regression at the time when sexual differentiation normally begins and also had renal agenesis (60). In human beings, mutations of *EMX2* are also associated with schizencephaly, a rare disorder associated with clefts within the cerebral cortex (61).

GENES ESSENTIAL FOR TESTES DEVELOPMENT

SRY

As discussed in Chapter 6, the indifferent gonad is induced down the testis pathway by a gene designated *SRY* (sex determining region, Y chromosome). *SRY* is located immediately adjacent to the pseudoautosomal region of the short arm of the Y chromosome (62). In human beings, the presence or absence of this gene correlated highly with the sex-reversal phenotype; *SRY* was abnormally present in most 46, XX subjects with a male phenotype and absent or mutated in a subset of 46, XY individuals with a female phenotype (63,64). Finally, an 11 kb transgene containing mouse *Sry* caused XX mice to develop as phenotypic males (65).

SRY encodes a protein with a DNA-binding motif, termed the high mobility group (HMG) box, which is found in a number of other proteins. Other HMG box proteins with greater than 60% homology with the *SRY* HMG box are termed *SOX* genes [reviewed in (66)]. As discussed below, *SOX9*, one of these genes, also plays critical roles in testes development. The large majority of *SRY* mutations that cause sex reversal are located in the HMG box—the only region tightly conserved through evolution—supporting an essential role for DNA binding in SRY action (67). Recent studies suggest that other *SRY* mutations may impair function by interfering with nuclear import of the protein (68).

Based on analyses of other HMG box proteins and NMR spectroscopic analyses, SRY interacts with DNA in its minor groove to induce a bend in the DNA double helix (69). Indeed, certain *SRY* mutations do not impair DNA binding, but rather impair its ability to bend DNA—either by disordering the packing of the HMG box or by impairing direct contacts with DNA (70). These observations suggest that SRY, like other HMG box proteins, regulates gene expression through “architectural” effects that render promoter regions more accessible, thereby facilitating the action of other transcriptional regulators [reviewed in (71)].

Coupled with the documented roles of SRY in activating the male developmental cascade, these findings have led to the hypothesis that *SRY* activates downstream genes, which in turn convert the bipotential gonad into a testis. Direct target genes of SRY, however, have not been isolated, and certain lines of evidence suggest that SRY does not positively activate testes development, but rather represses a negative regulator that normally inhibits this pathway (72,73).

At a cellular level, it is clear that SRY is expressed in the pre-Sertoli cells prior to the onset of overt sex differentiation. Within the genital ridge, SRY expression first is seen in the central region and then expands from the central region to the anterior and posterior poles. In contrast, Leydig cell differentiation is first seen in the anterior pole, with a subsequent progression to the posterior pole. These differences in patterns of expression suggest that distinct gradients of inducing factors may direct Sertoli and Leydig cell development during testis differentiation (74).

SOX9

SOX9 is another member of the HMG family that initially was isolated by virtue of its homology with SRY. SOX9 mutations cause campomelic dysplasia, an autosomal dominant form of dwarfism associated with 46, XY sex reversal in most but not all male offspring (75,76). SOX9 is expressed at low levels in the indifferent gonad of both sexes. Following the initiation of male sex determination by SRY, Sox9 expression is dramatically upregulated in the testes, but essentially disappears in the ovaries, suggesting that SOX9 plays key roles in the male developmental pathway (77,78).

In keeping with the skeletal phenotype seen in campomelic dysplasia, SOX9 is also expressed in developing chondrocytes and regulates the expression of a number of genes involved in skeletal development. For these genes, the ability of SOX9 to form homodimers on compound DNA elements is critical for its functions (79,80); homodimerization of SOX9, however, is not required for testes development. Instead, SOX9 regulates the expression of target genes involved in male development (e.g., AMH) through synergistic interactions with other transcription factors such as SF-1 (49,50). In addition to direct protein–protein interactions, SOX9 may reinforce SF-1 expression by interacting with a regulatory element in its 5′-flanking region (81). *SOX8*, another SOX family member, also is expressed in the gonads and, like SOX9, can interact with SF-1 to stimulate AMH promoter activity (82,83).

The sex reversal seen with *SOX9* mutations in campomelic dysplasia and in KO mice lacking Sox9 function (83) argues that SOX9 is essential for testes development. Other genetic studies argue that SOX9 expression is sufficient to induce testes formation and the male developmental pathway. Duplication of human chromosome 17q23-24, which includes *SOX9*, causes 46, XX sex reversal (84). Similar evidence in mice came from an insertional mutation called *oddsex*. In *oddsex* mice, insertion of a tyrosinase minigene transgene approximately 1 megabase upstream of the *Sox9* gene caused XX genetic females to develop as males, presumably by long-range effects of the minigene enhancer to activate Sox9 expression in the developing ovary, where it normally is silenced (85,86). Even more definitively, SF-1–driven transgenic expression of Sox9 in the developing gonads was sufficient to divert XX embryos to the male developmental pathway, inducing testes formation and male sex differentiation even in the absence of *Sry* (87). These studies have established SOX9 as a necessary and sufficient mediator of the male developmental pathway. A more severe gonadal phenotype is caused by the double knockout of SOX8 and SOX9 suggesting that SOX8 may reinforce SOX9 function (83).

GATA4

GATA4, a member of a family of zinc finger transcription factors named on the basis of their DNA recognition motif (GATA), is an important factor in gonadal function. GATA4 is expressed in both the ovary and testis during the indifferent gonad stage but exhibits sexually dimorphic expression after gonadal sex differentiation [testes > ovaries (88)]. Significantly, transfection analyses in cultured cell lines suggest that GATA4 activates the expression of a number of genes that mediate male sex differentiation, including AMH and several enzymes involved in steroidogenesis (89). Like WT1 and SOX9, GATA4 protein interacts with SF-1 to synergistically activate target genes, and DAX1 inhibits this synergy by binding SF-1 rather than by interacting with GATA4 (90). Another regulator of GATA4 is the coregulator FOG2 (Friends of GATA 2). FOG2 is co-expressed with GATA4 during gonadal development and, in cell transfection studies, inhibits GATA4-dependent transcriptional activation of a number of gonadal genes, including AMH, activin A, the steroidogenic acute regulatory protein, and aromatase. Furthermore, FOG2 KO mice or mice harboring mutations in Gata4 that disrupt its interaction with Fog2 have decreased levels of SRY and fail to develop testes cords (91). It is not clear at this time if SRY is a direct downstream target gene of GATA4.

DMRT1

Although there are striking differences in the roles of specific factors in sex differentiation in different species, some genes are conserved over a broad evolutionary scale. The doublesex, Mab-3 related transcription factor (DMRT1) is one such gene (92). DMRT1 is one of seven members of a family that share the DM DNA binding motif found in genes that mediate sex differentiation in *Drosophila* (doublesex) and *C. elegans* (Mab-3). In humans, DMRT1 maps to chromosome 9p32, in a region in which deletions are associated with impaired testes development and sex differentiation (92). In mouse embryos, Dmrt1 is expressed in the indifferent gonad of both sexes beginning at ~E10.5; expression then becomes sexually dimorphic (testes > ovaries) and localizes inside the developing testes cords (93). Knockout mice lacking Dmrt1 are not sex-reversed, but show defects in testis development and are infertile secondary to germ cell depletion (94). The ovaries in XX mice appear normal, and these mice are fertile. This relatively mild testis phenotype may reflect partial compensation by other members of the DMRT family that also are expressed in the embryonic testes.

M33

M33 (also called chromobox homolog-2, CBX-2) is a chromodomain protein that is related to the *Drosophila* polycomb proteins. In *Drosophila*, polycomb genes repress the expression of homeotic genes in a development-specific manner. Both male and female KO mice lacking M33 have gonadal agenesis (95), although an extensive description of the phenotype has not been reported.

DAX1

DAX1 (*d*osage-sensitive sex reversal, *a*drenal hypoplasia congenita, *X*-linked; officially designated NR0B1) was isolated by positional cloning of the Xp22 gene responsible for the X-linked disorder adrenal hypoplasia congenita [AHC (96)]. *DAX1* encodes a protein that contains the conserved ligand-binding domain of nuclear receptors but lacks the typical zinc finger DNA-binding motif. Instead, DAX1 contains three copies of a novel 67–69 amino acid repeat that reportedly binds DNA (97) or polyadenylated RNA (98). To date, missense mutations that impair DAX1 function in AHC patients map exclusively to the C-terminal, ligand-binding domain rather than to the novel N-terminal repeats (99). Many of these human mutations abolish the ability of DAX-1 to repress SF-1–mediated transcription (100–102), suggesting that a functional antagonism between these two orphan receptors may be important for endocrine development (100–102).

Additional functions of DAX-1 include regulation of gonadotropin production and testes determination. For example, AHC patients at the time of normal puberty exhibit impaired gonadotropin secretion due to a compound hypothalamic-pituitary defect, whereas female "carriers" of DAX1 mutations sometimes present with delayed puberty (103,104). In contrast to these presumptive loss-of-function mutations, duplication of the region of Xp22 that contains *DAX1* in humans or overexpression of a *Dax1* transgene in mice was associated with male-to-female sex reversal, a phenomenon termed "dosage-sensitive sex reversal" (105,106). These findings suggested that DAX1 might exert an anti-testis effect, and it was proposed that it might play important roles in ovarian development (see below).

Important new insights into DAX1 functions came from Dax1 KO mice (107). Unlike human subjects with AHC, Dax1 KO mice had relatively normal adrenal function but had a number of defects in testes function. Spermatogenesis was impaired despite apparently normal levels of gonadotropins, suggesting a distinct role for Dax1 in sperm development. Subsequent analyses have suggested that Dax-1 is also essential for testis differentiation and testis cord formation (108,109).

Genes that Impair Testes Development by Affecting Cell Proliferation or Migration

The fibroblast growth factors (FGFs) are secreted proteins that play key roles in the development of multiple tissues (110). The FGF family includes 24 members of secreted growth factors that act via membrane receptors with intrinsic tyrosine kinase activity and that couple to diverse signaling pathways, including the Ras-mitogen activated protein (MAP) kinase and inositol trisphosphate/Akt pathways. Gradients of FGF signaling have been implicated in mesoderm induction, gastrulation, neural induction, anterior-posterior patterning, and formation of the endoderm.

In KO mice lacking fgf9, ovary development was normal in genetic females, although the mice died shortly after birth due to lung defects. Genetic males, however, were sex-reversed and showed no histological evidence of testes cord formation (111). At a molecular level, these fgf9 KO gonads had impaired proliferation of Sertoli cell precursors between E11–E11.5, a critical feature for testes development and male sex differentiation, and decreased expression of SF-1. As a result of the failed development of Sertoli cells, testes did not form and mediators of male sex differentiation were not expressed. Subsequent studies have shown that exogenous FGF9 induces cell proliferation of a population of Sertoli cell precursors that is accompanied by translocation of the FGF receptor 2 (FGFR2) into the nucleus (112). Although FGF9 treatment also induced proliferation of gonads from XX embryos, fgfr2 did not translocate to the nucleus and organization of testes cords and expression of mediators of male sex differentiation did not occur. These findings imply that factors other than FGF9 are also required to support the male developmental pathway.

Another growth factor implicated in proliferation of the Sertoli cell precursors and subsequent testes development is platelet-derived growth factor [PDGF, reviewed in (113)]. KO mice lacking the alpha receptor for PDGF (pdgfra) exhibited impaired testes development and male sex differentiation in a manner that resembled the defects in the fgr9 KO mice. Specifically, pdgfa KO testes had impaired proliferation of Sertoli cell precursors, diminished migration of cells from the mesonephros, and impaired Leydig cell differentiation; they also failed to generate the male-specific vasculature (114).

Insulin/Insulin-Like Growth Factor Receptors

The insulin and insulin-like growth factors receptors play key roles in embryonic growth and development, signaling through membrane receptors with intrinsic tyrosine kinase activity. In XY compound KO mice deficient in the insulin receptor, insulin-like growth factor 1 receptor, and a related receptor, Insulin-related receptor (Irr), testes development was impaired and ovaries developed despite expression of Sry (115). Although the underlying mechanism is not known, this finding suggests that action of these growth factors via their cognate membrane receptors is essential for proliferation of the Sertoli cell precursors, and hence for testes development.

Hedgehog Genes

Cell–cell interactions mediated by secreted signaling molecules also play key roles in gonadal development. The hedgehogs (sonic hedgehog, desert hedgehog, and Indian hedgehog) comprise a family of cholesterol-modified proteins that signal via the Patched 12-transmembrane domain receptor to regulate development in multiple tissues (116). For example, a gradient of sonic hedgehog expression establishes dorsal ventral polarity in the developing embryo, while Indian hedgehog is an important mediator of bone development. The posttranslational transfer of cholesterol to the amino terminus of hedgehog proteins apparently is essential for hedgehog action on target tissues (117), and certain defects in male sex differentiation result from either genetic defects [e.g., the Smith-Lemli-Opitz syndrome (118)] or chemical inhibitors [e.g., cyclopamine (119)] that interfere with this cholesterol modification.

Desert hedgehog (DHH) is expressed in Sertoli cells of the testes, while the Patched receptor is expressed in the interstitial region. Dhh KO mice have male-to-female sex reversal associated with impaired differentiation of cells with the Leydig cell phenotype (120,121), suggesting that the Sertoli cells stimulate Leydig cell differentiation in a paracrine manner via secreted Dhh. A potential molecular basis for this defect is the observation that SF-1 expression was impaired in Leydig cells—but not Sertoli cells—of the Dhh KO testes. In human beings, heterozygous loss-of-function mutations of DHH also have been associated with ambiguous external genitalia (122,123). These findings support an evolutionarily conserved pathway of DHH in testes development and male sex differentiation. A separate role of DHH in neural development is supported by the observation that Dhh KO mice also had defects in the structure of peripheral nerves (124), while one study of human subjects with DHH mutations described a minifascicular polyneuropathy (122), a finding not seen in the second report (123). Finally, analyses in Dhh KO mice have revealed a key role of Dhh signaling in development of male germ cells in the testes, again presumably due to defective Sertoli cell function (125).

ARX

A defect in Leydig cell differentiation also is seen in KO mice lacking the homeobox transcription factor ARX (126). Although these Arx KO mice have impaired Leydig cell function, Arx is not expressed in the fetal Leydig cells; thus, the impairment of Leydig cell development and male sex differentiation probably result from the absence of paracrine signaling from other cells in the interstitial region. Support for an important role of ARX in testes development also is provided by human beings with *ARX* mutations, who present with a central nervous system developmental defect called lisencephaly and impaired male sex differentiation (127).

POD-1

POD-1 (also known as capsulin, epicardin, or Tcf21) is a member of the basic helix-loop-helix family of transcription factors that is essential for the normal development of both the ovary and testis (128). A role for Pod-1 as a determinant of gonadal cell fate emerged from studies of the phenotype of Pod-1 KO mice. These mice exhibited XY sex reversal, an expansion of the population of steroidogenic cells, an overexpression of the side-chain cleavage enzyme, and disruption of gonadal development in both males and females. POD-1 appears to negatively regulate SF1 function, thereby providing a possible mechanism for its action in gonadal development and sexual differentiation.

GENES IMPLICATED IN OVARIAN DEVELOPMENT

As discussed above, a number of genes have been shown to play essential roles in establishment of the bipotential gonad or in subsequent events in testes development and male sex differentiation. Our understanding of genes that mediate ovarian development is considerably more limited. This partly reflects the fact that the ovary is not required for female sexual differentiation in utero. Indeed, the diagnosis of ovarian dysgenesis often is not apparent until it is manifested by primary amenorrhea or sexual infantilism at the normal time of puberty. In addition, the ovary appears to differentiate considerably later than the testes, and, as a consequence, phenotypes associated with mutations in genes that interfere with ovarian development in mouse models therefore may be less apparent than those that affect the testes.

DAX1

One early candidate for an ovary-determining gene was *DAX1*. Following comparable expression in the indifferent gonad of both male and female mouse embryos, Dax1 expression becomes sexually dimorphic (ovary > testis) coincident with gonadal differentiation (129). In human beings, duplication of a 160 kb region of Xp22 that contains DAX1 was associated with male-to-female sex reversal and impaired testis development in human subjects (105). Finally, DAX1 interferes with the action of SF-1 in cell transfection assays (100–102) and of *SRY* in the male developmental pathway in vivo (106). Collectively, these studies suggest that *DAX1* is an anti-testis gene, leading to the model that *DAX1* was the first gene that specifically directed ovarian development.

As described above, the development of Dax1 KO mice mandated a re-evaluation of this model (107–109). Female Dax1 KO mice had normal ovaries and were fertile, arguing strongly that Dax1 is not an essential mediator of ovarian differentiation. Of equal significance, testis differentiation was impaired in Dax1 KO mice, arguing that Dax1 actually is required for normal testis development (see above). These surprising findings altered considerably our concepts of Dax1 action.

WNT4

Members of the WNT (wingless-type mammary tumor virus integration site) family of lipid-modified glycoproteins play critical roles in cell differentiation, intercellular signaling, and proliferation and apoptosis (130). The various WNTs signal through a family of G-protein–coupled receptors termed frizzled that couple to β-catenin, Ca^{2+}, and cGMP (131). An important advance in our understanding of ovarian development came from studies that examined the phenotype of KO mice lacking Wnt4. In mice, Wnt4 is expressed early in urogenital development in both males and females, but displays sexually dimorphic expression thereafter (ovary > testis). Wnt4 also is expressed in the mesonephros and in the mesenchyme that gives rise to the Müllerian ducts. Knockout mice lacking Wnt4 provided strong evidence that Wnt4 is a key mediator of normal sexual development and ovarian differentiation (132). These Wnt4 KO mice lacked Müllerian ducts; this presumably was a primary effect of the absence of Wnt4 expression in the mesenchymal precursors, because the Müllerian ducts regressed before AMH normally is made. In addition, the Wolffian ducts were stabilized in both males and females, suggesting that testosterone was produced. Consistent with this, the ovaries expressed the steroidogenic enzymes Cyp17 and 3β-hydroxysteroid dehydrogenase, which normally are made in testes but not ovaries. Although the primordial germ cells migrated normally into the genital ridges of Wnt4 KO mice, these germ cells were absent at birth. Subsequent studies showed that the germ cells died via apoptosis in much greater numbers than in wild-type females (133).

Further evidence for roles of WNT4 in female development came from studies of human beings. Duplication of the region of chromosome 1p that contains the *WNT4* gene is associated with male-to-female sex reversal (134), suggesting that too much WNT4 impairs testes development. More recently, a 46, XX human subject who presented with primary amenorrhea secondary to Müllerian agenesis and gonadal dysgenesis was found to be heterozygous for a *WNT4* mutation (135). Collectively, these studies define essential roles of WNT4 in ovarian develop and female sex differentiation.

These findings prompted efforts to define the mechanisms by which WNT4 contributes to ovarian development. In addition to containing ectopic cells that resembled Leydig cells and expressed the steroidogenic enzymes, ovaries from Wnt4 KO mice formed the coelomic vessel that normally is characteristic of the testis (136). Migration of endothelial cells from the mesonephros into the gonadal ridge normally contributes to this vessel, suggesting that WNT4 may repress the migration of these endothelial cell precursors. Studies have also sought to identify target genes whose expression is affected by WNT4 signaling. One such gene is follistatin, a protein that binds members of the bone morphogenetic protein (BMP) family of signaling molecules such as activin. Follistatin expression in the female gonad co-localizes with Wnt4 to cells several layers below the coelomic epithelium (137). Although Wnt4 expression is normal in follistatin KO mice, follistatin is not expressed in Wnt4 KO mice (138). These findings suggest that *WNT4* lies upstream of follistatin in a hierarchical pathway of ovarian differentiation. Similarly, the expression of bone morphogenetic protein 2 (Bmp-2) normally becomes sexually dimorphic as the ovary differentiates, and is dramatically diminished in Wnt4 KO mice. Unfortunately, Bmp-2 KO mice die during embryogenesis due to other defects, precluding further analyses of later stages of ovarian development.

Consistent with the model that follistatin is downstream of WNT4 in a common pathway, follistatin KO mice also exhibit a specific loss of germ cells at approximately E16.5 due to increased apoptosis. As a consequence, the germ cells signals that are required for normal follicle maturation are not expressed, and the follicle fails to develop normally. Another target of Wnt4 action in the ovary apparently is Dax1 (139), although the KO mice studies described above argue strongly that impaired Dax1 expression does not cause the observed defect in ovary development.

A number of investigators are now using various approaches to identify other genes that mediate ovary development. A couple of general statements should be borne in mind. First, ovarian development in utero lags considerably behind that of the testes. The human follicles develop only to the primary follicle state in utero, whereas mice, with their telescoped gestation, form only the primordial follicles. A second general principle in the ovary is that germ cells are absolutely required for formation of the primordial follicle and for normal follicle maturation (140). As described in detail in Chapter 7, several oocyte-expressed genes have now been identified that are critical for follicle development, providing a potential molecular basis for this essential role. These genes include the transcription factor FIG1α and the secreted transforming growth factor (TGF)β members GDF9, and BMP15. Knockout mice lacking the oocyte-specific basic helix-loop-helix transcription factor "Factor in germline 1 alpha" (FIG1α) have impaired formation of primordial follicles and their ovaries never develop beyond this stage and exhibit premature ovarian failure and sterility (141). This developmental defect likely reflects the failure to express FIG1α target genes that are essential for oocyte-granulosa cell interactions. Similarly, the oocyte-derived proteins GDF9 and BMP15 also are essential for follicle maturation [reviewed in (142)]. The winged helix transcription factor FOXL2, which is expressed in the somatic compartment, also has been shown to play critical roles in later stages of ovary development in both humans (143) and KO mice (144,145).

PRIMORDIAL GERM CELLS

The germ cells are those cells that give rise to the sperm or ova, respectively, in the testes and ovaries, and thus transmit genetic information to the next generation. The germ cells arise from a group of precursor cells, called primordial germ cells, or PGC. Studies in a number of mammalian and nonmammalian species revealed that the PGC do not initially arise from the genital ridges, but rather develop outside of the genital ridge and then migrate into the region (146–147). While these cells are readily identified within the gonads based on their characteristic morphology and position within the testes cords or follicles, defining their origin and migratory pathway proved to be very difficult. As detailed below, considerable progress in this regard has now emerged from studies in both mammalian systems (in particular mice) and in nonmammalian model organisms (e.g., zebrafish and *Drosophila*).

Origin

In certain model species (e.g., frogs, *Drosophila*, *C. elegans*), the germ cell precursors are segregated from the somatic compartment early in embryonic development and can be distinguished morphologically by a characteristic cytoplasm. In species such as mice, however, no such cell population can be identified during early development. Rather, PGC in the mouse traditionally were traced histochemically based on their expression of alkaline phosphatase. Using this assay, PGC could be visualized in mouse embryos at the base of the allantois at E8.5, and then could be followed as they moved through the hindgut to enter the region of the genital ridge at E9.5 to E11.5.

In an effort to extend developmental studies to earlier stages, single cells in the proximal epiblast were labeled to permit the tracing of their descendants. These studies revealed that the PGC were derived from proximal epiblast cells adjacent to the extraembryonic ectoderm (148). Other cell lineages also were derived from these epiblast cells, indicting that these cells were not restricted to the PGC fate at this time. These experiments led to the proposal that PGC arose at approximately E7 from a group of 45 or so progenitor cells.

Analyses of knockout mice defined a critical role for members of the BMP signaling family in PGC development. Mice lacking BMP4 lacked both the allantois and PGC, indicating an essential developmental role for this secreted signaling molecule (149). Subsequent studies have also implicated BMP8b in a coordinate, synergistic pathway of PGC development (150,151). It appears that BMP4 is expressed in the extraembryonic ectoderm and then acts on adjacent cells in the epiblast to induce differentiation along the PGC pathway. Thus, the PGC fate is not limited to a subpopulation of the epiblast cells, but rather reflects the inductive effect of signaling proteins on uncommitted precursors.

Shortly after the induction of the PGC phenotype, these precursor cells express two markers of PGC specification. Fragilis is a member of the interferon-induced family and is expressed in the epiblast, presumably activated by Bmp signaling (152). Shortly thereafter, stella, which encodes a protein proposed to function in chromosomal architecture and RNA processing, is expressed within the PGC precursors. It has been proposed that these genes play important roles in PGC specification (153), although others have proposed that stella is an important maternal factor during cleavage of the mouse embryo but is not required for germ cell specification (154).

New insights arising from molecular approaches to PGC development in other species have revealed some surprising parallels. Previous studies in *Drosophila* had identified a maternal effect gene that encodes a zinc finger protein, named nanos, which was essential for PGC migration into the gonad and germ cell development (155). Analyses of the human and mouse genomes revealed three nanos homologs, designated nanos1, nanos2, and nanos3. In nanos3 KO mice, the PGC developed normally in the epiblast but the germ cells subsequently were lost completely in both sexes (156). These findings suggest that nanos3 is required for PGC migration and/or proliferation. Other factors that are involved in germ cell proliferation and/or survival have been reviewed (157). They include a number of growth factors, including stem cell factor (also called kit ligand) and its membrane receptor C-kit and members of the basic fibroblast growth factor family.

Migration

Once they initially are formed in the epiblast, the PGC must migrate through the hindgut to reach the genital ridges. This process involves active migration by the PGC, apparently drawn by a chemoattractive gradient that is produced by cells in the genital ridge. The study of this migration has been advanced by the development of a green fluorescent protein (GFP) marker for cells that normally express Oct4, a specific marker of PGC (147).

In human beings, the germ cells migrate to the genital ridge by the 5th week of gestation and are referred to thereafter as gonocytes (either spermatogonia or oogonia). The oogonia enter a phase of rapid mitotic proliferation that continues until the 8th week of gestation. Between 8 and 13 weeks, some of the oogonia initiate meiosis and reach prophase I, the stage at which they are first called oocytes. Many of these oocytes form "germ cell clusters," which consist of multiple oocytes that are connected in a syncytium that may facilitate transfer of nutrients and organelles between cells. This precedes the first evidence of ovarian differentiation—the formation of primordial follicles containing individual oocytes surrounded by a layer of flattened granulosa cells—which begins at 16 weeks. The net effect of ongoing mitosis, entry into meiosis and atresia is such that the total number of germ cells in the ovary increases until it peaks at approximately 6 million at 20 weeks. Thereafter, the predominant fate is follicular atresia, such that only 1 to 2 million germ cells are present at birth. At around the time of birth, those oogonia that have not been incorporated into a primordial follicle and entered meiosis will die via apoptosis.

In mice, the PGC migrate into the genital ridges between E10.5 and E11.5, and—following a period of proliferation—initiate meiosis in the developing ovary at E13 to E13.5, whereas they again are arrested in mitosis as prospermatogonia in testes. Studies suggest that the Sertoli cells produce a paracrine signal that prevents the germ cells in the testes from entering meiosis (146). In the ovary, one marker of the initiation of meiosis is a down-regulation of the POU transcription factor Oct4 (158). The onset of meiosis also can be ascertained by examination of expression of synaptonemal complex protein 3, which encodes a component of the synaptonemal complex.

Cell–cell contact is essential, as revealed by knockout mice lacking β1-integrin, which do not migrate normally. As discussed above, the stem cell factor/c-kit signaling pathway is clearly critical, and PGC in mice lacking either ligand or receptor do not populate the genital ridges and rather die due to apoptosis. In experimental model systems, the chemokine SDF-1 (stromal-derived factor-1) and its receptor CXCR4, which is expressed on germ cells, also have emerged as critical mediators of PGC proliferation and/or migration (159,160).

Chromosomal Events

During the migratory period, PGC containing two X chromosomes randomly inactivate one of their X chromosomes. Shortly after they enter the genital ridge, the germ cells in males undergo several rounds of division, followed by arrest in the G0/G1 transition; development in male germ cells proceeds no further until sperm maturation resumes after birth. In females, however, the inactivated X chromosome is reactivated, after which the germ cells transit through the leptotene, zygotene, and pachytene stages before they are arrested in diplotene. Demethylation of differentially imprinted genes also occurs at around the time of arrival at the genital ridge, and it appears that new imprints are imposed at varying times for different genes.

EMBRYOLOGY OF DUCTS OF THE REPRODUCTIVE TRACT

Overview

The reproductive tracts of males and females develop from the Wolffian and Müllerian ducts respectively. These structures arise from mesodermal tissue during early embryogenesis and differentiate or regress as determined by hormones secreted from the committed male gonad, i.e., testosterone and AMH. Testosterone triggers the differentiation of the Wolffian ducts into the epididymis (cranial region), vas deferens (central region), and seminal vesicle and ejaculatory ducts (caudal region); while AMH promotes the apoptotic degeneration of the Müllerian ducts. In the absence of testosterone, the Wolffian ducts degenerate and the Müllerian ducts differentiate into the Fallopian tubes (also termed "oviducts"), uterus, cervix and the anterior portion of the vagina. The coexistence of the two ducts endows the early embryo with an inherent capability to develop the reproductive tracts of either the male or the female and is consistent with the hypothesis that the early embryo remains indifferent until the gonads commit to form testes or ovaries. Much like the commitment of the bipotential gonads to the ovarian pathway, the fates of the Müllerian and Wolffian ducts in the female embryo appear to be determined by intrinsic events rather than by an active hormonal influence [reviewed in (161–163)].

Embryology

Although the Wolffian and Müllerian ducts coexist in the embryo for a short period of time, the Wolffian ducts (also called the mesonephric ducts) first appear at 32 days after conception in humans or E9.5 in mice coincident with the formation of the genital ridge. The Wolffian ducts develop as the major excretory channels of the primitive kidney and contribute to the development both of the primitive kidney and the precursor to the permanent kidney, the metanephros. At approximately 54 days postconception (E13–14 in the mouse), the Wolffian ducts are induced by testosterone to become components of the male genital system. The upper portion of each duct proximal to the testes elongates significantly and ultimately differentiates to form the epididymis, the middle portion of the duct retains its tube-like structure and becomes the vas deferens, and the caudal portion of the duct dilates and forms the seminal vesicles (5).

The Müllerian ducts are formed 37 days after conception (E12 in the mouse), along with the appearance of the bipotential gonads, and regress in males around 51 days after conception in response to AMH. The Müllerian ducts are sensitive to AMH for only a limited time, and fail to respond either before or after this critical period. Regression begins in regions closest to the caudal poles of the testes and progresses both cranially and caudally, sparing only the tips at each end. AMH appears to target the neighboring mesenchyme leading to its condensation around the ductal epithelium (3); this, in turn, leads to increases in hyaluronidase (164) and protease (165) activities that destroy the extracellular matrix surrounding the ducts (166–168) and ultimately lead to the dissolution of the ducts through apoptotic mechanisms (169).

Hormone Action on the Embryonic Ducts

Testosterone and Wolffian Duct Differentiation. Testosterone per se, rather than its metabolite 5α-dihydrotestosterone (DHT), controls Wolffian duct differentiation. The direct role of testosterone is supported by studies of patients with 5α-reductase deficiency who cannot synthesize DHT (9) and of experimental animals treated with 5α-reductase inhibitors that prevent DHT synthesis 170); despite the inability to convert testosterone to DHT in these settings, the Wolffian ducts differentiate normally. The effects of testosterone are mediated by the androgen receptor (AR), an X chromosome-encoded transcription factor that belongs to the steroid/nuclear receptor superfamily. Mutations that disrupt the synthesis or activity of the AR cause the Wolffian ducts to degenerate despite the presence of testosterone, thereby establishing the importance of the AR in testosterone action (171). Interestingly, AR is expressed in the mesenchyme surrounding the Wolffian duct just prior to the onset of androgen biosynthesis, but is not expressed in the ductal epithelium until well after the onset of ductal differentiation. These observations suggest that the effects of testosterone on the Wolffian ducts during the initial period of differentiation are indirect and possibly mediated by paracrine influences originating in the mesenchyme (172). Very little is known about the molecular events underlying testosterone action. Studies of mouse Wolffian ducts in vivo and in organ culture suggest that epidermal growth factor (EGF) and prostaglandin E_2 (PGE2) pathways may be important downstream components of the testosterone-activated differentiation pathway. EGF receptor levels are higher in the developing reproductive tracts of males than females, while testosterone administration to XX mouse embryos increased EGF receptor levels in the ducts coincident with their virilization. Moreover, antisera against the EGF receptor inhibited Wolffian duct differentiation in organ explants, even in the presence of testosterone (173). Antisera specific for PGE2 (or inhibitors of PGE2 synthesis) similarly inhibited the effects of testosterone on Wolffian duct differentiation in organ culture (174), implicating PGE2 as another component of testosterone action on the duct. The specific mechanisms by which EGF and prostaglandin mediate these effects are unknown.

AMH and Müllerian Duct Regression. The existence of AMH as a distinct testicular factor that controlled Müllerian duct regression was first postulated by Alfred Jost [see (2) for a personal account of this work] and was definitively established with the cloning of the *AMH* gene (175). AMH is a dimer of 25 kDa derived from a larger precursor via cleavage by a kex2/subtilisin-like member of the prohormone convertase family of proteases (176,177). AMH is a member of the TGFβ family of growth factors and apparently acts through specific type I and type II receptor serine/threonine kinases linked to Smad pathways of signal transduction. In TGFβ signaling, the ligand binds as a dimer to both type I and type II receptors. The ligand-bound type II receptor recruits and phosphorylates its type I receptor partner; this, in turn, triggers a downstream signaling cascade that directly activates specific Smads. The available evidence suggests that AMH acts similarly. Significantly, the type I and type II receptors are located in the neighboring mesenchyme rather than the ductal epithelium itself (178,179), suggesting that the effects of AMH on the Müllerian ducts, like those of testosterone on the Wolffian ducts, are indirect and mediated by paracrine factors derived from the surrounding mesenchyme.

Genes Affecting Ductal Development and Differentiation

As with gonadal development, studies in KO mice and human subjects with abnormal development have provided some insights into the genes that play critical roles in the early formation and sexually dimorphic development of the internal genital structures [see (180) for a review focused on female development].

Homeobox Genes and Other Transcription Factors

Homeobox genes encode transcription factors with a conserved DNA binding domain that play key roles in embryonic development, some of which are discussed above in the context of gonadal development. Some of the homeobox genes (i.e., *HOX* genes) are found in gene clusters and show spatial and temporal gradients of expression that help to initiate patterning of the anterior–posterior axis [reviewed in (181)]. A number of these genes are expressed in the developing genitourinary tract and play important roles in reproduction, as revealed by analyses of knockout mice. Some of the *HOX* genes are also implicated in later events in implantation (182) (see also Chapter 4).

The abdominal B *HOX* genes reside at the 5′ ends of the four mammalian HOX clusters. Abdominal B members have been implicated as important, partially redundant regulators of development of the urogenital ducts with somewhat overlapping patterns of expression. As revealed by KO mice, Hoxa10 is required for complete development of seminal vesicles in males (183) and for proper development of the upper part of uterus and the secretory glands in females (184). Uterine receptivity to blastocyst implantation is also impaired, causing a fertility defect. The HOXA10 promoter contains an estrogen-responsive element that is differentially regulated by estradiol and diethylstilbestrol (185), providing a possible mechanism for the developmental anomalies associated with exposure in utero to the latter compound (186).

HOXA11 is expressed in the uterus and cervix and apparently is involved in uterine development. In female Hoxa11 KO mice, the uterus appears thin, and abnormally shaped, stromal tissue is lost, and endometrial glands do not develop; males have cryptorchid testes (187).

The congenital disorder Hand-Foot-Genital syndrome is an autosomal dominant disorder characterized by skeletal malformations and incomplete fusion of the Müllerian ducts that result in a variety of uterine malformations (188). Males in affected kindreds often exhibit hypospadias in association with the skeletal abnormalities. This syndrome is associated with heterozygous mutations or deletion (189) or polyalanine expansion (190) in HOXA13. In mice, spontaneously arising mutations are associated with similar developmental abnormalities (191,192), while Hoxa13 KO mice exhibit agenesis of the distal Müllerian ducts (193).

Hoxd13. Hoxd13 is broadly expressed in both the mesenchyme and epithelium of the lower genitourinary tract (of both Müllerian and Wolffian duct origin) in mouse embryos. Newborn Hoxd13 KO mice exhibit multiple abnormalities in the accessory sex organs, including diminished mesenchymal folding of the seminal vesicles, prostatic hypoplasia, and agenesis of the bulbourethral gland (193). In females, the mice again have multiple defects in development of the Müllerian-derived structures.

Pax-2 and Pax-8. *Pax-2* and *Pax-8* encode DNA-binding transcription factors of the paired box family that can function interchangeably in development (194). Pax-2 expression in the intermediate mesoderm is determined by a *cis*-acting enhancer element that targets Pax-2 to this region (195) and a nearby binding site for the transcription factor YY1 that acts cooperatively with the enhancer to maintain high levels of Pax-2 expression (196). By E10, Pax-2 can be detected in both the Wolffian ducts and surrounding mesenchyme (197) and by E13.5 *Pax-2* is evident in the Müllerian ducts as well. Pax-8 appears in the same region of the intermediate mesoderm as Pax-2 but at a somewhat earlier developmental stage. As demonstrated using Pax-2 KO mice (198), *Pax-2* plays an essential role in the differentiation and development of the kidneys and the reproductive tracts of males and females. In Pax-2 KO mice, the Wolffian ducts appear to develop normally up to E9.5. Thereafter, they fail to extend caudally and by E12.5 begin to degenerate. The Müllerian ducts also appear to develop normally until E13.5, when they begin to exhibit signs of impaired development; by

E16.5, they have degenerated completely. As a consequence of impaired Wolffian and Müllerian duct development, Pax-2 KO mice are devoid of ureters, kidneys and components of the genital tract. Males lack epididymis, vas deferens and seminal vesicles; females lack oviducts, uterus, and vagina. Gonadal development appears to be normal in *Pax-2* KO mice. Since disruption of *Pax-2* did not disrupt the very early stages of Wolffian and Müllerian duct development, consideration was given to the contributions of Pax-8 to this process. The KO of Pax-8 did not cause a phenotype in the early mouse embryo; however, disruption of Pax-8 together with Pax-2 prevented the differentiation of mesenchyme to epithelium, thereby blocking the early stages of duct formation in the intermediate mesoderm. The need to inactivate both Pax-2 and Pax-8 for this effect has been attributed to the redundant actions of the two closely related transcription factors (199). The findings outlined above position *Pax-2* and *Pax-8* at the apex of a gene hierarchy governing differentiation and development of the urogenital tract.

Emx2. *Emx2* is a transcription factor of the homeobox family whose disruption in mice prevents the formation of kidneys, ureters, gonads and the genital tracts in males and females; adrenal glands, which are derived from the same embryonic tissue, develop normally. In normal mice, Emx2 can be detected in the intermediate mesoderm at E8.5–9.0, the earliest stages of urogenital differentiation. Emx2 appears in the epithelia of the Wolffian duct by E9.5–10, in the gonadal primordia by E11.5, and in the Müllerian ducts at E12.5. Emx2 also is expressed in components of the developing renal system and thus appears to be a key regulator of urogenital differentiation (60,200). In Emx2 KO male mice, the Wolffian ducts appear normal until approximately E10.5; by E11.5, the Wolffian ducts exhibit signs of abnormal degeneration, and the vas deferens, seminal vesicles, and epididymis fail to develop. In Emx2 KO females, Müllerian ducts never form and, as a consequence, the oviducts, uterus and upper part of the vagina fail to develop (60). The KO of Emx2 does not affect Pax-2 expression at E10.5; however at E11.5, Pax-2 expression is extinguished in portions of the Wolffian duct that otherwise appear morphologically normal suggesting some form of reciprocal regulation.

LHX1. As discussed earlier, LHX1 encodes a transcription factor comprised of a homeodomain DNA binding region and two Cys-rich LIM domains that participate in protein–protein interactions. Lhx1 is expressed in both the Wolffian and Müllerian ducts of developing mouse embryos. Lhx1 is first detected in the intermediate mesoderm at E.7.5 and appears in epithelium of both the Wolffian and the Müllerian ducts coincident with their formation (201,202). In males, Lhx1 expression persists in the Wolffian ducts and its differentiated structures throughout development but disappears from the Müllerian ducts as they regress. In females, the Müllerian ducts continue to express Lhx1 until E15.5, after which expression declines and becomes restricted to the regions destined to become the oviducts. Gene disruption of Lhx1 in mice led to the loss of gonads and kidneys and to aberrant Wolffian duct development, which was evident as early as E9.5 (58,202). The Lhx1 KO mice died at E10, precluding an assessment of the role of Lhx1 in Müllerian duct development. Interestingly, Pax-2 expression was not extinguished in the Lhx1 KO mice, suggesting that *Pax2* functions upstream of *Lhx1* in the developmental cascade. The problem of embryo lethality associated with the absence of Lhx1 was partially solved by changing the genetic background of the Lhx1 KO mice (203). On the different genetic background, gonads appeared to develop normally in both males and females; however, females completely lacked the reproductive structures derived both from the epithelium and mesenchyme of the Müllerian ducts. Whereas Pax2 expression is independent of Lhx1, Lhx1 expression in Wolffian ducts seems to require Pax2; in contrast, Lim1 expression in the Müllerian ducts seemed to be *Pax2*-independent. Lim1 expression in the Müllerian ducts is similarly independent of *Wnt7a* (202). Taken together, these observations indicate that *Lhx1* is critical for the epithelium–mesenchyme interactions that lead to the formation of both the Wolffian and Müllerian ducts. Lhx1 expression appears to be autonomously regulated and, at least in the Müllerian ducts, appears to be independent of *Pax2* and *Wnt7a*, other genes implicated in Müllerian duct formation and differentiation. Nevertheless, there appear to be other genetic modifiers that influence LHX1 function as evidenced by the influence of genetic background on the Lhx1 KO phenotype.

Growth Factors and Signaling Molecules

AMH Type I and Type II Receptors. The gene encoding the AMH type II receptor has been cloned from several sources [e.g., (178,204)] and its role in AMH action has been established both in vitro and in vivo through studies of human patients with mutations in the receptor (205) and through studies of KO mice (206). Patients with mutations in the AMH type II receptor exhibit a rare form of male pseudohermaphroditism referred to as Persistent Müllerian Duct syndrome and characterized by the retention of oviducts and uterus along with a normal male reproductive tract. A similar phenotype also is seen in male mice carrying a null mutation of the AMH type II receptor. The identity of the AMH type I receptor has been more elusive. One strong candidate is the previously described type IA bone morphogenetic protein receptor (*Bmpr1a/Alk3*). Support for its involvement comes from studies using mice with a conditional null mutation in *Bmpr1a/Alk3*, targeted to mesenchyme of the Müllerian ducts. Conditional disruption of *Bmpr1a/Alk3* reproduced the phenotype of male pseudohermaphroditism seen with AMH type II receptor mutations (207). While these observations clearly demonstrate an essential role for *Bmpr1a/Alk3* in Müllerian duct regression, the failure to also demonstrate a direct interaction of this receptor with the AMH type II receptor leaves some doubt its identity as the AMH type I receptor (208). A second candidate for the AMH type I receptor is the Activin type I receptor (*Acvr1/Alk2*). *Acvr1/Alk2* antisense RNA disrupts AMH-stimulated regression of Müllerian ducts in organ culture, arguing that *Acvr1/Alk2* is essential for AMH action, at least in vitro (209). Consistent with this hypothesis, *Acvr1/Alk2* is co-expressed with the AMH type II receptor both temporally and spatially in the mesenchyme adjacent to the Müllerian ducts (209). Unfortunately, like *Bmpr1a/Alk3*, *Acvr1*/Alk2 does not interact directly with the AMH type II receptor and thus may not be the *bona-fide* MIS receptor. In addition, Acvr1/*Alk2*-null embryos die well before the Müllerian ducts normally form, so there is no in vivo evidence for the role of this receptor in AMH-stimulated regression of the Müllerian ducts (210).

WNT4 and WNT7a. As discussed above, members of the WNT family are secreted glycoproteins that function as paracrine factors regulating cell fates and other aspects of development. As reviewed (211), WNT4 and WNT7a play particularly important roles in the early stages of development of the female reproductive system.

Wnt4-Wnt4 can be found in the mesenchyme of the intermediate mesoderm at E9.5–10.5, in the indifferent gonads that develop at E11 and in the mesenchyme neighboring the Müllerian ducts at E12. Wnt4 is not expressed in the Wolffian ducts, but is essential for the formation of the Müllerian ducts, as evidenced by the phenotypes of Wnt4 KO mice (132) and a human XX subject with a *WNT4* mutation (135). In Wnt4 KO mice of both sexes, Müllerian ducts fail to develop whereas the Wolffian ducts persist and differentiate along the male pathway. As a consequence, the reproductive tracts of females are masculinized whereas those of males are unaffected. The gonads of Wnt4 null females have a testicular appearance and express enzymes associated with androgen biosynthesis, suggesting that they are the source of the androgenic influence on Wolffian duct development. In contrast to the internal genitalia, development of the external genitalia appears to be independent of *Wnt4* and not affected in *Wnt4* KO females. Wnt4 may have an additional role in oocyte maturation as suggested by the finding that the ovaries of *Wnt4* KO females contain a markedly reduced number of oocytes (see above for further discussion). A human female with a *WNT4* mutation presented a phenotype reminiscent of that of female Wnt4 KO mice. External features were normal; however, the patient lacked Müllerian-derived internal structures and exhibited slightly elevated levels of testosterone. The mutation in this patient was correlated with a single amino acid change in one *WNT4* allele (135). Interestingly, the phenotype of the patient harboring the *WNT4* mutation is similar to patients presenting with Müllerian agenesis, referred to as the Mayer-Rokitansky-Küster-Hauser syndrome. While the underlying cause of this syndrome is unknown, it has been speculated that the WNT4 signaling pathway may be involved.

Wnt7a-Wnt7a is expressed over the entire epithelium of the Müllerian ducts in both male and female mice from E12.5 to E14.5 (212). Whereas Wnt7a disappears in males as the Müllerian ducts regress, Wnt7a persists in females in both the Müllerian ducts and their derivative structures later in life. Wnt7a KO mice have improper development of the Müllerian ducts and both males and females are infertile. In males, the Müllerian ducts do not undergo normal regression; instead, they appear as thin undifferentiated ducts that run parallel to the epididymis and the vas deferens from the testis to the urogenital sinus. The Wolffian ducts and the gonads appear to develop normally in males; however, the persistence of the Müllerian ducts blocks sperm passage past the distal end of the vas deferens, accounting for the infertility. In females, the gonads appear to develop and function normally but Müllerian duct differentiation is impeded—the oviducts and the upper uterus are incompletely developed both in the fetus and in the adult mouse. Furthermore, the uterus exhibits deficiencies in epithelium-derived secretory glands and in the mesodermal-derived stromal layer between the musculature and the epithelium. These observations suggest that Wnt7a controls both epithelial and mesenchymal differentiation. In support of this hypothesis, Wnt7a KO mice exhibit loss of AMH type II receptors in the mesenchyme surrounding the Müllerian ducts of both male and female embryos. Since the disruption of Wnt7a produces a phenotype that is much different from that produced by disruption of the AMH receptor, Wnt7a may have a more global effect on Müllerian duct differentiation. Thus, Wnt7a appears to play an early role in signaling from the epithelium to the mesenchyme that ultimately determines the fate of the Müllerian ducts in both males and females.

REFERENCES

1. Swain, A., and Lovell-Badge, R. (1999). Mammalian sex determination: a molecular drama. *Genes Dev.* 13, 755–766.
2. Jost, A., Vigier, B., Prepin, J., and Perchellet, J. P. (1973). Studies on sex differentiation in mammals. *Rec. Prog. Horm. Res.* 29, 1–35.
3. MacLaughlin, D. T., and Donahoe, P. K. (2004). Sex determination and differentiation. *N. Engl. J. Med.* 350, 367–378.
4. Byskov, A. G., and Hoyer, P. E. (1994). Embryology of mammalian gonads and ducts. In *The Physiology of Reproduction* (E. Knobil and J. Neill, Eds.). Raven Press, New York.
5. Staack, A., Donjacour, A. A., Brody, J., Cunha, G. R., and Carroll, P. (2003). Mouse urogenital development: a practical approach. *Differentiation* 71, 402–413.
6. Cupp, A. S., Skinner, M. K. (2004). Embryonic Sertoli cell differentiation. In *Sertoli Cell Biology* (M. K. Skinner and M. D. Griswold, Eds.), pp. 43–70. Elsevier Science, New York.
7. McLaren, A. (2003). Primordial germ cells in the mouse. *Dev. Biol.* 262, 1–15.
8. Molyneaux, K., and Wylie, C. (2004). Primordial germ cell migration. *Int. J. Dev. Biol.* 48, 537–544.
9. Russell, D. W., and Wilson, J. D. (1994). Steroid 5 alpha-reductase: two genes/two enzymes. *Annu. Rev. Biochem.* 63, 25–61.
10. Hanley, N. A., Hagan, D. M., and Clement-Jones, M., et al. (2000). SRY, SOX9, and DAX1 expression patterns during human sex determination and gonadal development. *Mech. Dev.* 91, 403–407.
11. de Santa Barbara, P., Moniot, B., Poulat, F., and Berta P. (2000). Expression and subcellular localization of SF-1, SOX9, WT1, and AMH proteins during early human testicular development. *Dev Dyn.* 217, 293–298.
12. Hacker, A., Capel, B., Goodfellow, P., and Lovell-Badge, R. (1995). Expression of Sry, the mouse sex determining gene. *Development* 121, 1603–1614.
13. Albrect, K., and Eicher, E. (2001). Evidence that Sry is expressed in pre-Sertoli cells and Sertoli and granulosa cells have a common precursor. *Dev. Biol.* 240, 92–107.
14. Brennan, J., and Capel, B. (2004). One tissue, two fates: molecular genetic events that underlie testis versus ovary development. *Nat. Rev. Genet.* 5. 509–521.
15. Clarkson, M.J., and Harley, V.R. (2002). Sex with two SOX on: SRY and SOX9 in testis development. *Trends Endocrinol. Metab.* 13, 106–1011.
16. Koopman, P., Bullejos, M., and Bowles, J. (2001). Regulation of male sexual development by Sry and Sox9. *J. Exp. Zool.* 290, 463–474.
17. Vidal, V. P., Chaboissier, M. C., de Rooij, D. G., and Schedl, A. (2001). Sox9 induces testis development in XX transgenic mice. *Nat. Genet.* 28, 216–217.
18. McLaren, A. (1988). Somatic and germ cell sex in mammals. *Philos. Trans. R. Soc. Lond.* 322, 3–9.
19. Ikeda, Y., Shen, W. H., Ingraham, H. A., and Parker, K. L. (1994). Developmental expression of mouse steroidogenic factor-1, an essential regulator of the steroid hydroxylases. *Mol. Endocrinol.* 8, 654–662.
20. Wagner, K. D., Wagner, N., and Schedl, A. (2003). The complex life of WT1. *J. Cell Sci.* 116, 1653–1658.
21. Dome, J. S., and Coppes, M. J. (2002). Recent advances in Wilms tumor genetics. *Curr. Opin. Pediatr.* 14, 5–11.
22. Call, K. M., Glaser, T., and Ito, C. Y., et al. (1990). Isolation and characterization of a zinc finger polypeptide gene at the human chromosome 11 Wilms' tumor locus. *Cell* 60, 509–520.
23. Gessler, M., Poustka, A., Cavenee, W., Neve, R. L., Orkin, S. H., and Bruns, G. A. (1990). Homozygous deletion in Wilms tumors of a zinc-finger gene identified by chromosomal jumping. *Nature* 343, 774–778.
24. Discenza, M. T., and Pelletier, J. (2004). Insights into the physiological role of WT1 from studies of genetically modified mice. *Physiol. Genomics.* 16, 287–300.
25. Pelletier, J., Schalling, M., Buckler, A. J., Rogers, A., Haber, D. A., and Housman, D. (1991). Expression of the Wilms' tumor gene WT1 in the murine urogenital system. *Genes Dev.* 5, 1345–1356.
26. Rackley, R. R., Flenniken, A. M., Kuriyan, N. P., Kessler, P. M., Stoler, M. H., and Williams, B. R. (1993). Expression of the Wilms' tumor suppressor gene WT1 during mouse embryogenesis. *Cell Growth Differ.* 4, 1023–1031.
27. Armstrong, J. F., Pritchard-Jones, K., Bickmore, W. A., Hastie, N. D., and Bard, J. B. (1993). The expression of Wilms' tumour gene, WT1, in the developing mammalian embryos. *Mech. Dev.* 40, 85–97.
28. Hanley, N. A., Ball, S. G., Clement-Jones, M., Hagan, D. M., Strachan, T., Lindsay, S., Robson, S., Ostrer, H., Parker, K. L., and Wilson, D. I. (1999). Expression of steroidogenic factor 1 and Wilms' tumour 1 during early human gonadal development and sex determination. *Mech. Dev.* 87, 175–178.
29. Pelletier, J., Bruening, W., and Kashtan, C. E., et al. (1991). Germline mutations in the Wilms' tumor suppressor gene are associated with abnormal urogenital development in Denys-Drash syndrome. *Cell* 67, 437–447.
30. Barbaux, S., Niaudet, P., and Gubler, M. C., et al. (1997). Donor splice-site mutations in WT1 are responsible for Frasier syndrome. *Nat. Genet.* 17, 467–470.
31. Kikuchi, H., Takata, A., and Akasaka, Y., et al. (1998). Do intronic mutations affecting splicing of WT1 exon 9 cause Frasier syndrome? *J. Med. Genet.* 7:45–48.
32. Klamt, B., Koziell, A., Poulat, F., Wieacker, P., Scambler, P., Berta, P., and Gessler, M. (1998). Frasier syndrome is caused by defective alternative splicing of WT1 leading to an altered ratio of WT1 +/KTS splice isoforms. *Hum. Mol. Genet.* 7, 709–714.
33. Kreidberg, J. A., Sariola, H., Loring, J. M., Maeda, M., Pelletier, J., Housman, D., and Jaenisch, R. (1993). WT1 is required for early kidney development. *Cell* 74, 679–691.
34. Hammes, A., Guo, J. K., Lutsch, G., Leheste, J. R., Landrock, D., Ziegler, U., Gubler, M. C., and Schedl, A. (2001). Two splice variants of the Wilm's tumor 1 gene distinct functions during sex determination and nephron formation. *Cell* 106, 319–329.
35. Hossain, A., and Saunders, G. F. (2001). The human sex-determining gene SRY is a direct target of WT1. *J. Biol. Chem.* 276, 16817–16823.
36. Matsuzawa-Watanabe, Y., Inoue, J., and Semba, K. (2003). Transcriptional activity of testis-determining factor SRY is modulated by the Wilms' tumor 1 gene product, WT1. *Oncogene.* 22, 7900–7904.
37. Wilhelm, D., and Englert, C. (2002). The Wilms tumor suppressor WT1 regulates early gonad development by activation of Sf1. *Genes Dev.* 16, 1839–1851.
38. Kim, J., Prawitt, D., Bardeesy, N., Torban, E., Vicaner, C., Goodyer, P., Zabel, B., and Pelletier, J. (1999). The Wilms' tumor suppressor gene (wt1) product regulates Dax-1 gene expression during gonadal differentiation. *Mol. Cell. Biol.* 19, 2289–2299.
39. Sim, E. U., Smith, A., Szilagi, E., Rae, F., Ioannou, P., Lindsay, M. H., and Little, M. H. (2002). Wnt-4 regulation by the Wilms' tumour suppressor gene, WT1. *Oncogene.* 21, 2948–2960.
40. Nachtigal, M. W., Hirokawa, Y., Enyeart-VanHouten, D. L., Flanagan, J. N., Hammer, G. D., and Ingraham, H. A. (1998). Wilms' tumor 1 and Dax-1 modulate the orphan nuclear receptor SF1 in sex-specific gene expression. *Cell* 93, 445–454.

41. Du, X., Hublitz, P., Gunther, T., Wilhelm, D., Englert, C., and Schule, R. (2002). The LIM-only coactivator FHL2 modulates WT1 transcriptional activity during gonadal differentiation. *Biochim. Biophys. Acta.* 1577, 93–101.
42. Val, P., Lefrancois-Martinez, A. M., Veyssiere, G., and Martinez, A. (2003). SF-1 a key player in the development and differentiation of steroidogenic tissues. *Nucl. Recept.* 1, 1–23.
43. Parker, K. L., and Schimmer, B. P. (1997). Steroidogenic factor 1: a key determinant of endocrine development and function. *Endocrine Rev.* 18, 361–377.
44. Lala, D. S., Rice, D. A., and Parker, K. L. (1992). Steroidogenic factor I, a key regulator of steroidogenic enzyme expression, is the mouse homolog of fushi tarazu-factor I. *Mol. Endocrinol.* 6, 1249–1258.
45. Honda, S-I., Morohashi, K-I., Nomura, M., Takeya, H., Kitajima, M., and Omura, T. (1993). Ad4BP regulating steroidogenic P-450 gene is a member of steroid hormone receptor superfamily. *J. Biol. Chem.* 268, 7494–7502.
46. Shen, W-H., Moore, C. C. D., Ikeda, Y., Parker, K. L., and Ingraham, H. A. (1994). Nuclear receptor steroidogenic factor 1 regulates MIS gene expression: a link to the sex determination cascade. *Cell* 77, 651–661.
47. Giuili, G., Shen, W. H., and Ingraham, H. A. (1997). The nuclear receptor SF-1 mediates sexually dimorphic expression of Müllerian Inhibiting Substance, in vivo. *Development* 124, 1799–1807.
48. Arango, N., Lovell-Badge, R., and Behringer, R. R. (1999). Targeted mutagenesis of the endogenous mouse Mis gene promoter: in vivo definition of genetic pathways of vertebrate sexual development. *Cell* 99, 409–419.
49. De Santa Barbara, P., Bonneaud, N., Boizet, B., Desclozeaux, M., Moniot, B., Sudbeck, P., Scherer, G., Poulat, F., and Berta, P. (1998). Direct interaction of SRY-related protein SOX9 and steroidogenic factor 1 regulates transcription of the human anti-Müllerian hormone gene. *Mol. Cell. Biol.* 18, 6653–6665.
50. Hatano, O., Takayama, K., and Imai, T., et al. (1995). Sex-dependent expression of a transcription factor, Ad4BP, regulating steroidogenic P-450 genes in the gonads during prenatal and postnatal rat development. *Development* 120, 2787–2797.
51. Luo, X., Ikeda, Y., and Parker, K. L. (1994). A cell-specific nuclear receptor is essential for adrenal and gonadal development and for male sexual differentiation. *Cell* 77, 481–490.
52. Sadovsky, Y., Crawford, P. A., and Woodson, K. G., et al. (1995). Mice deficient in the orphan receptor steroidogenic factor 1 lack adrenal glands and gonads but express P450 side-chain-cleavage enzyme in the placenta and have normal embryonic serum levels of corticosteroids. *Proc. Natl. Acad. Sci. U S A* 92, 10939–10943.
53. Achermann, J. C., Ito, M., Ito, M., Hindmarsh, P. C., and Jameson, J. L. (1999). A mutation in the gene encoding steroidogenic factor 1 causes XY sex reversal and adrenal failure in humans. *Nat. Genet.* 22, 125–126.
54. Biason-Lauber, A., and Schoenle, E. J. (2000). Apparently normal ovarian differentiation in a prepubertal girl with transcriptionally inactive steroidogenic factor 1 (NR5A1/SF1) and adrenocortical insufficiency. *Am. J. Hum. Genet.* 67, 1563–1568.
55. Achermann, J. C., Ozisik, G., Ito, M., Orun, U. A., Harmancik, K., Gurakan, B., and Jameson, J. L. (2002). Gonadal determination and adrenal development are regulated by the orphan nuclear receptor steroidogenic factor-1, in a dose-dependent manner. *J. Clin. Endocrinol. Metab.* 87, 1829–1833.
56. Correa, R. V., Domenice, S., Bingham, N. R., Billerbeck, E. C., Rainey, W. E., Parker, K. L., and Mendonca, B. (2004). A microdeletion in the ligand binding domain of human steroidogenic factor 1 causes XY sex reversal without adrenal insufficiency. *J. Clin. Endocrinol. Metab.* 89, 1767–1772.
57. Mallet, D., Bretones, P., Michel-Calemard, L., Dijoud, F., David, M., and Morel, Y. (2004). Dysgenesis without adrenal insufficiency in a 46, XY patient heterozygous for the nonsense C16X mutation: a case of SF1 haploinsufficiency. *J. Clin. Endocrinol. Metab.* 89, 4829–4832.
58. Shawlot, W., and Behringer, R. R. (1995). Requirement for Lim1 in head-organizer function. *Nature* 374, 425–430.
59. Birk, O. S., Casiano, D. E., and Wassif, C. A., et al. (2000). The LIM homeobox gene *Lhx9* is essential for mouse gonad formation. *Nature* 403, 909–912.
60. Miyamoto, N., Yoshida, M., Kuratani, S., Matsuo, I., and Aizawa, S. (1997). Defects of urogenital development in mice lacking *Emx2*. *Development* 124, 1653–1664.
61. Cecchi, C., and Boncinelli, E. (2000). Emx homeogenes and mouse brain development. *Trends Neurosci.* 23, 347–352.
62. Sinclair, A. H., Berta, P., and Palmer, M. S., et al. (1990). A gene from the human sex-determining region encodes a protein with homology to a conserved DNA-binding motif. *Nature* 346, 240–244.
63. Berta, P., Hawkins, J. R., Sinclair, A. H., et al. (1990). Genetic evidence equating SRY and the testis-determining factor. *Nature* 348, 448–450.
64. Jager, R. J., Anvret, M., Hall, K., and Scherer, G. (1990). A human XY female with a frame shift mutation in the candidate tesis-determining gene SRY. *Nature* 348, 452–454.
65. Koopman, P., Gubbay, J., Vivian, N., Goodfellow, P., and Lovell-Badge, R. (1991). Male development of chromosomally female mice transgenic for Sry. *Nature* 351, 117–121.
66. Pevny, L. H., and Lovell-Badge, R. (1997). Sox genes find their feet. *Curr. Opin. Genet. Dev.* 7, 338–344.
67. Cameron, F. J., and Sinclair, A. H. (1997). Mutations in SRY and SOX9: testis determining genes. *Hum. Mutat.* 9, 388–395.
68. Harley, V. R., Layfield, S., and Mitchell, C. L., et al. (2003). Defective importin beta recognition and nuclear import of the sex-determining factor SRY are associated with XY sex-reversing mutations. *Proc. Natl. Acad. Sci. U S A* 100, 7045–7050.
69. Werner, M. H., Huth, J. R., Gronenborn, A. M., and Clore, G. M. (1995). Molecular basis of human 46, XY sex reversal revealed from the three-dimensional solution structure of the human SRY-DNA complex. *Cell* 81, 705–714.
70. Pontiggia, A., Rimini, R., Harley, V. R., Goodfellow, P. N., Lovell-Badge, R., and Bianchi, M. E. (1994). Sex reversing mutations affecting the architecture of SRY-DNA complexes. *EMBO J.* 13, 6115–6124.
71. Thomas, J. O., and Travers, A. A. (2001). HMG1 and 2, and related 'architectural' DNA-binding proteins. *Trends Biochem. Sci.* 26, 167–174.
72. Harley, V. R., Clarkson, M. J., and Argentaro, A. (2003). The molecular action and regulation of the testis-determining factors, SRY (sex-determining region on the Y chromosome) and SOX9 [SRY-related high-mobility group (HMG) box 9]. *Endocr. Rev.* 24, 466–487.
73. Canning, C. A., and Lovell-Badge, R. (2002). Sry and sex determination: how lazy can it be? *Trends Genet.* 18, 111–113.
74. Hiramatsu, R., Kanai, Y., and Mizukami, T., et al. (2003). Regionally distinct potencies of mouse XY genital ridge to initiate testis differentiation dependent on anteroposterior axis. *Dev. Dyn.* 228, 247–253.
75. Foster, J.W., Dominguez-Steglich, M.A., and Guioli, S., et al. (1994). Campomelic dysplasia and autosomal sex reversal caused by mutations in an SRY-related gene. *Nature* 372, 525–530.
76. Wagner, T., Wirth, J., and Meyer, J., et al. (1994). Autosomal sex reversal and campomelic dysplasia are caused by mutations in and around the SRY-related gene SOX9. *Cell* 79, 1111–1120.
77. Kent, J., Wheatley, S. C., Andrews, J. E., Sinclair, A. H., and Koopman, P. (1996). A male-specific role for SOX9 in vertebrate sex determination. *Development* 122, 2813–2822.

78. Morais da Silva, S., Hacker, A., Harley, V., Goodfellow, P., Swain, A., and Lovell-Badge, R. (1996). Sox9 expression during gonadal development implies a conserved role for the gene in testis determination in mammals and birds. *Nat. Genet.* 14, 62–68.
79. Bernard, P., Tang, P., Liu, S., Dewing, P., Harley, V. R., and Vilain, E. (2003). Dimerization of SOX9 is required for chondrogenesis, but not for sex determination. *Hum. Mol. Genet.* 12, 1755–1765.
80. Sock, E., Pagon, R. A., Keymolen, K., Lissens, W., Wegner, M., and Scherer, G. (2003). Loss of DNA-dependent dimerization of the transcription factor SOX9 as a cause for campomelic dysplasia. *Hum. Mol. Genet.* 12, 1439–1447.
81. Shen, J. H., and Ingraham, H. A. (2002). Regulation of the orphan nuclear receptor steroidogenic factor 1 by Sox proteins. *Mol. Endocrinol.* 16, 529–540.
82. Schepers, G., Wilson, M., Wilhelm, D., and Koopman, P. (2003). SOX8 is expressed during testis differentiation in mice and synergizes with SF-1 to activate the Amh promoter in vitro. *J. Biol. Chem.* 278, 28101–28108.
83. Chaboissier, M. C., Kobayashi, A., Vidal, V. I., Lutzkendorf, S., van de Kant, H. J., Wegner, M., de Rooij, D. G., Behringer, R. R., and Schedl, A. (2004). Functional analysis of Sox8 and Sox9 during sex determination in the mouse. *Development* 131, 1891–1990.
84. Huang, B., Wang, S., Ning, Y., Lamb, A. N., and Bartley, J. (1999). Autosomal sex reversal caused by duplication of SOX9. *Am. J. Med. Genet.* 87, 349–353.
85. Bishop, C. E., Whitworth, D. J., and Qin, Y., et al. (2000). A transgenic insertion upstream of sox9 is associated with dominant XX sex reversal in the mouse. *Nat. Genet.* 26, 490–494.
86. Qin, Y., Kong, L. K., Poirier, C., Truong, C., Overbeek, P. A., and Bishop, C. E. (2004). Long-range activation of Sox9 in Odd Sex (Ods) mice. *Hum. Mol. Genet.* 13, 1213–1218.
87. Vidal, V. P., Chaboissier, M. C., de Rooij, D. G., and Schedl, A. (2001). *Sox9* induces testis development in XX transgenic mice. *Nat. Genet.* 28, 216–217.
88. Viger, R. S., Mertineit, C., Trasler, J. M., and Nemer, M. (1998). Transcription factor GATA-4 is expressed in a sexually dimorphic pattern during mouse gonadal development and is a potent activator of the Müllerian inhibiting substance promoter. *Development* 125, 2665–2675.
89. Tremblay, J. J., and Viger, R. S. (2001). GATA factors differentially activate multiple gonadal promoters through conserved GATA regulatory elements. *Endocrinology* 142, 977–986.
90. Tremblay, J. J., and Viger, R. S. (2001). Nuclear receptor Dax-1 represses the transcriptional cooperation between GATA-4 and SF-1 in Sertoli cells. *Biol. Reprod.* 64, 1191–1199.
91. Tevosian, S. G., Albrecht, K. H., Crispino, J. D., Fujiwara, Y., Eicher, E. M., and Orkin, S. H. (2002). Gonadal differentiation, sex determination and normal Sry expression in mice require direct interaction between transcription partners GATA4 and FOG2. *Development* 129, 4627–4634.
92. Raymond, C. S., Shamu, C. E., Shen, M. M., Seifert, K. J., Hirsch, B., Hodgkin, J., and Zarkower, D. (1998). Evidence for evolutionary conservation of sex-determining genes. *Nature* 391, 691–695.
93. Raymond, C. S., Shamu, C. E., Shen, M. M., Seifert, K. J., Hirsch, B., Hodgkin, J., and Zarkower, D. (1999). Expression of Dmrt1 in the genital ridge of mouse and chicken embryos suggests a role in vertebrate sexual development. *Dev. Biol.* 215, 208–220.
94. Raymond, C. S., Murphy, M. W., O'Sullivan, M. G., Bardwell, V. J., and Zarkower, D. Dmrt1, a gene related to worm and fly sexual regulators, is required for mammalian testis differentiation. *Genes Dev.* 14, 2587–2595.
95. Katoh-Fukui, Y., Tsuchiya, R., Shiroishi, T., Nakahara, Y., Hashimoto, N., Noguchi, K., and Higashinakagawa, T. (1998). Male to female sex reversal in *M33* mutant mice. *Nature* 393, 688–692.
96. Zanaria, E., Muscatelli, F., and Bardoni, B., et al. (1994). An unusual member of the nuclear hormone receptor superfamily responsible for X-linked adrenal hypoplasia congenital. *Nature* 372, 635–641.
97. Zazopoulos, E., Lalli, E., Stocco, D.M., and Sassone-Corsi, P. (1997). DNA binding and transcriptional repression by DAX-1 blocks steroidogenesis. *Nature* 390, 311–315.
98. Lalli, E., Ohe, K., Hindelang, C., and Sassone-Corsi, P. (2000). Orphan receptor DAX-1 is a shuttling RNA binding protein associated with polyribosomes via mRNA. *Mol. Cell Biol.* 20, 4910–4921.
99. Iyer, A. K., and McCabe, E. R. (2004). Molecular mechanisms of DAX1 action. *Mol. Genet. Metab.* 83, 60–73.
100. Ito, M., Yu, R., and Jameson, J. L. (1997). DAX-1 inhibits SF-1-mediated transactivation via a carboxy-terminal domain that is deleted in adrenal hypoplasia congenita. *Mol. Cell. Biol.* 17, 1476–1483.
101. Crawford, P. A., Dorn, C., Sadovsky, Y., and Milbrandt, J. (1998). Nuclear receptor DAX-1 recruits nuclear receptor corepressor N-CoR to steroidogenic factor 1. *Mol. Cell. Biol.* 18, 2949–2956.
102. Nachtigal, M. W., Hirokawa, Y., Enyeart-VanHouten, D. L., Flanagan, J. N., Hammer, G. D., and Ingraham, H. A. (1998). Wilms' tumor 1 and DAX-1 modulate the orphan nuclear receptor SF-1 in sex-specific gene expression. *Cell* 93, 445–454.
103. Muscatelli, F., Strom, T. M., and Walker, A. P., et al. (1994). Mutations in the DAX-1 gene give rise to both X-linked adrenal hypoplasia congenita and hypogonadotrophic hypogonadism. *Nature* 372, 672–676.
104. Achermann, J. C., Meeks, J. J., and Jameson, J. L. (2001). Phenotypic spectrum of mutations in DAX-1 and SF-1. *Mol. Cell. Endocrinol.* 185, 17–25.
105. Bardoni, B., Zanaria, E., and Guioli, S., et al. (1994). A dosage sensitive locus at chromosome Xp21 is involved in male to female sex reversal. *Nat. Genet.* 7, 497–501.
106. Swain, A., Narvaez, S., Burgoyne, P., Camerino, G., and Lovell-Badge, R. (1998). Dax1 antagonizes Sry action in mammalian sex determination. *Nature* 391, 761–767.
107. Yu, R. N., Ito, M., Saunders, T. L., Camper, S. A., and Jameson, J. L. (1998). Role of AhCh in gonadal development and gametogenesis. *Nat. Genet.* 20, 353–357.
108. Meeks, J. J., Weiss, J., and Jameson, J. L. (2003). *Dax1* is required for testis determination. *Nat. Genet.* 34, 32–33.
109. Meeks, J. J., Crawford, S. E., Russell, T. A., Morohashi, K., Weiss, J., and Jameson, J. L. (2003). Dax1 regulates testis cord organization during gonadal differentiation. *Development* 30, 1029–1036.
110. Bottcher, R. T., and Niehrs, C. (2005). Fibroblast growth factor signalling during early vertebrate development. *Endocr. Rev.* 26, 63–77.
111. Colvin, J. S., Green, R. P., Schmahl, J., Capel, B., and Ornitz, D. M. (2001). Male-to-female sex reversal in mice lacking fibroblast growth factor 9. *Cell* 104, 875–889.
112. Schmahl, J., Kim, Y., Colvin, J. S., Ornitz, D. M., Capel, B. (2004). Fgf9 induces proliferation and nuclear localization of FGFR2 in Sertoli precursors during male sex differentiation. *Development* 131, 3627–3636.
113. Betsholtz, C., Karlsson, L., and Lindahlm, P. (2001). Developmental roles of platelet-derived growth factors. *Bioessays* 23, 494–507.
114. Brennan, J., Tilmann, C., and Capel, B. (2003). Pdgfr-alpha mediates testis cord organization and fetal Leydig cell development in the XY gonad. *Genes Dev.* 17, 800–810.

115. Nef, S., Verma-Kurvari, S., Merenmies, J., Vassalli, J. D., Efstratiadis, A., Accili, D., and Parada, L. (2003). Testis determination requires insulin receptor family function in mice. *Nature* 426, 291–295.
116. Ingham, P. W., McMahon, A. P. (2001). Hedgehog signaling in animal development: paradigms and principles. *Genes Dev.* 15, 3059–3087.
117. Jeong, J., and McMahon, A. P. (2002). Cholesterol modification of Hedgehog family proteins. *J. Clin. Invest.* 110, 591–596.
118. Neri, G., and Opitz, J. (1999). Syndromal (and nonsyndromal) forms of male pseudohermaphroditism. *Am. J. Med. Genet.* 89, 201–209.
119. Yao, H. H., and Capel, B. (2002). Disruption of testis cords by cyclopamine or forskolin reveals independent cellular pathways in testis organogenesis. *Dev. Biol.* 246, 356–365.
120. Clark, A., Garland, K., and Russell, L. (2000). Desert Hedgehog (Dhh) gene is required in the mouse testis for formation of adult-type Leydig cells and normal development of peritubular cells and seminiferous tubules. *Biol. Reprod.* 63, 1825–1838.
121. Yao, H. H., Whoriskey, W., and Capel, B. (2002). Desert hedgehog/Patched 1 signaling specifies fetal Leydig cell fate in testis organogenesis. *Genes Dev.* 16, 1433–1440.
122. Umehara, F., Tate, G., Itoh, K., Yamaguchi, N., Douchi, T., Mitsuya, T., and Osame, M. (2000). A novel mutation of desert hedgehog in a patient with 46, XY partial gonadal dysgenesis accompanied by minifascicular neuropathy. *Am. J. Hum. Genet.* 67, 1302–1305.
123. Canto, P., Soderlund, D., Reyes, E., and Mendez, J. P. (2004). Mutations in the desert hedgehog (DHH) gene in patients with 46,XY complete pure gonadal dysgenesis. *J. Clin. Endocrinol. Metab.* 89, 4480–4483.
124. Parmantier, E., Lynn, B., Lawson, D., Turmaine, M., Namini, S. S., Chakrabarti, L., McMahon, A. P., Jessen, K. R., and Mirsky, R. (1999). Schwann cell-derived Desert hedgehog controls the development of peripheral nerve sheaths. *Neuron.* 23, 713–724.
125. Bitgood, M. J., Shen, L., and McMahon, A. P. (1996). Sertoli cell signaling by Desert hedgehog regulates the male germline. *Curr. Biol.* 6, 298–304.
126. Kitamura, K., Yanazawa, M., Sugiyama, N., Miura, H., Iizuka-Kogo, A., Kusaka, M., Omichi, K., Suzuki, R., Kata-Fukui, Y., Kamiirisa, K., and Kamijo, S., et al. (2001). Mutation of ARX causes abnormal development of forebrain and testes and X-linked lissencephaly with abnormal genitalia in humans. *Nat. Genet.* 32, 359–369.
127. Sherr, E. H. (2003). The ARX story (epilepsy, mental retardation, autism, and cerebral malformations): one gene leads to many phenotypes. *Curr. Opin. Pediatr.* 15, 567–577.
128. Cui, S., Ross, A., Stallings, N., Parker, K. L., Capel, B., and Quaggin, S. E. (2004). Disrupted gonadogenesis and male-to-female sex reversal in Pod1 knockout mice. *Development* 131, 4095–4105.
129. Swain, A., Zanaria, E., Hacker, A., Lovell-Badge, R., and Camerino, G. (1996). Mouse Dax1 expression is consistent with a role in sex determination as well as in adrenal and hypothalamus function. *Nat Genet.* 12, 404–409.
130. Nelson, W. J., and Nusse, R. (2004). Convergence of Wnt, beta-catenin, and cadherin pathways. *Science* 303, 1483–1487.
131. Malbon, C. C. (2004). Frizzleds: new members of the superfamily of G-protein–coupled receptors. *Front. Biosci.* 9, 1048–1058.
132. Vainio, S., Heikkala, M., Kispert, A., Chin, N., and McMahon, A. P. (1999). Female development in mammals is regulated by Wnt4 signaling. *Nature* 397, 405–409.
133. Yao, H. H. C., Matzuk, M. M., Jorgez, C. J., Menke, D. B., Page, D. C., Swain, A., and Capel, B. (2004). Follistatin operates downstream of Wnt4 in mammalian ovary organogenesis. *Dev. Dyn.* 230, 210–215.
134. Jordan, B. K., Mohammed, M., Ching, S. T., Delot, E., Chen, X. N., Dewing, P., Swain, A., Rao, P. N., Elejaide, B. R., and Vilain, E. (2001). Up-regulation of WNT-4 signaling and dosage-sensitive sex reversal in humans. *Am. J. Hum. Genet.* 68, 1102–1109.
135. Biason-Lauber, A., Konrad, D., Navratil, F., and Schoenle, E. J. (2004). A WNT4 mutation associated with Müllerian-duct regression and virilization in a 46,XX woman. *N. Engl. J. Med.* 351, 792–798.
136. Jeays-Ward, K., Hoyle, C., Brennan, J., Dandonneau, M., Alldus, G., Capel, B., and Swain, A. (2003). Endothelial and steroidogenic cell migration are regulated by WNT4 in the developing mammalian gonad. *Development* 130, 3663–3670.
137. Menke, D. B., and Page, D. C. (2002). Sexually dimorphic gene expression in the developing mouse gonad. *Gene Expr. Patterns.* 2, 359–366.
138. Yao, H. H., Matzuk, M. M., Jorgez, C. J., Menke, D. B., Page, D. C., Swain, A., and Capel, B. (2004). Follistatin operates downstream of Wnt4 in mammalian ovary organogenesis. *Dev. Dyn.* 230, 210–215.
139. Mizusaki, H., Kawabe, K., Mukai, T., Ariyoshi, E., Kasahara, M., Yoshioka, H., Swain, A., and Morohashi, K. (2003). Dax-1 (dosage-sensitive sex reversal-adrenal hypoplasia congenita critical region on the X chromosome, gene 1) gene transcription is regulated by wnt4 in the female developing gonad. *Mol. Endocrinol.* 17, 507–519.
140. Matzuk, M. M., Burns, K., Viveiros, M. M., and Eppig, J. (2002). Intracellular communication in the ovary: oocytes carry the conversation. *Science* 296, 2178–2180.
141. Soyal, S. M., Amleh, A., and Dean, J. (2000). FIGalpha, a germ cell-specific transcription factor required for ovarian follicle formation. *Development* 127, 4645–4654.
142. Shimasaki, S., Moore, R. K., Otsuka, F., and Erickson, G. F. (2004). The bone morphogenetic protein system in mammalian reproduction. *Endocr. Rev.* 25, 72–101.
143. Crisponi, L., Deiana, M., Loi, A., Chiappe, F., Uda, M., Amati, P., Bisceglia, L., Zelante, L., Nagaraja, R., Porcu, S., Ristaldi, M. S., Marzella, R., Rocchi, M., Nicolino, M., Lienhardt-Roussie, A., Nivelon, A., Verloes, A., Schlessinger, D., Gasparini, P., Bonneau, D., Cao, A., and Pilia, G. (2001). The putative forkhead transcription factor FOXL2 is mutated in Blepharophimosis/ptosis/epicanthus inversus syndrome. *Nat. Genet.* 27, 159–166.
144. Uda, M., Ottolenghi, C., Crisponi, L., Garcia, J. E., Deina, M., Kimber, W., Forabosco, A., Cao, A., Schlessinger, D., and Pilia, G. (2004). Foxl2 disruption causes mouse ovarian failure by pervasive blockage of follicle development. *Hum. Mol. Genet.* 13, 1171–1181.
145. Schmidt, D., Ovitt, C. E., Anlag, K., Fehsenfeld, S., Gredsted, L., Treier, A. C., and Treier, M. (2004). The murine winged-helix transcription factor Foxl2 is required for granulosa cell differentiation and ovary maintenance. *Development* 131, 933–942.
146. McClaren, A. (2003). Primordial germ cells in the mouse. *Dev. Biol.* 262, 1–15.
147. Molyneaux, K., and Wylie, C. (2004). Primordial germ cell migration. *Int. J. Dev. Biol.* 48, 537–544.
148. Lawson, K. A., and Hage, W. J. (1994). Clonal analysis of the origin of primordial germ cells in the mouse. *Ciba Found. Symp.* 182, 68–91.
149. Lawson, K. A., Dunn, R. R., Roelen, B. A., Zeinstra, L. M., Davis, A. M., Wright, C. V., Korving, J. P., and Hogan, B. L. (1999). Bmp4 is required for the generation of primordial germ cells in the mouse embryo. *Genes Dev.* 13, 424–436.

150. Ying, Y., Liu, X. M., Marble, A., Lawson, K. A., and Zhao, G. O. (2000). Requirement of Bmp8b for the generation of primordial germ cells in the mouse. *Mol. Endocrinol.* 14, 1053–1063.
151. Ying, Y., Qi, X., and Zhao, G-Q. (2001). Induction of primordial germ cells from murine epiblasts by synergistic action of Bmp4 and Bmp8b signaling pathways. *Proc. Natl. Acad. Sci. U S A.* 98, 7858–7862.
152. Tanaka, S. S., and Matsui, Y. (2002). Developmentally regulated expression of mil-1 and mil-2, mouse interferon-induced transmembrane protein like genes, during formation and differentiation of primordial germ cells. *Gene Expr. Patterns* 2, 297–303.
153. Saitou, M., Barton, S. C., and Surani, M. A. (2002). A molecular program for the specification of germ cell fate in mice. *Nature* 418, 293–300.
154. Bortvin, A., Goodheart, M., Liao, M., and Page, D. C. (2004). Dppa3/Pgc7/stella is a maternal factor but is not required for germ cell specification in mice. *BMC Dev. Biol.* 23, 2.
155. Forbes, A., and Lehmann, R. (1998). Nanos and pumilio have critical roles in the development and function of Drosophila germline stem cells. *Development* 125, 679–690.
156. Tsuda, M., Sasaoka, Y., Kiso, M., Abe, K., Haraguchi, S., Kobayashi, S., and Saga, Y. (2003). Conserved role of nanos proteins in germ cell development. *Science* 301, 1239–1241.
157. Wylie, C. (1999). Germ cells. *Cell* 96, 165–174.
158. Scholer, H. R., Hatzopoulos, A. K., Balling, R., Suzuki, N., and Gruss, P. (1989). A family of octamer-specific proteins present during mouse embryogenesis: evidence for germline-specific expression of an Oct factor. *EMBO J.* 8, 2543–2550.
159. Ara, T., Nakamura, Y., Egawa, T., Sugiyama, T., Abe, K., Kishimoto, T., Matsui, Y., and Nagasawa, T. (2003). Impaired colonization of the gonads by primordial germ cells in mice lacking a chemokine, stromal cell-derived factor-1 (SDF-1). *Proc. Natl. Acad. Sci. U S A* 100, 5319–5323.
160. Molyneaux, K., Zinszner, H., Kunwar, P., Schaible, K., Stebler, J., Sunshine, M., O'Brien, W., Raz, E., Littman, D., and Wylie, C. (2003). The chemokine SDF1/CXCL12 and its receptor CXCR4 regulate mouse germ cell migration and survival. *Development* 130, 4279–4286.
161. Hunter, R. H. F. (1995). *Sex Determination, Differentiation and Intersexuality in Placental Mammals.* Cambridge University Press, London.
162. Josso, N., Rey, R., and Gonzalès, J. (2003). Sexual differentiation. New MI, Ed. Endotext.com/Pediatric Endocrinology, Chapter 7.
163. Lee, M. M., and Donahoe, P. K. (2003). Müllerian inhibiting substance: a gonadal hormone with multiple functions. *Endocrine Rev.* 14, 152–164.
164. Hayashi, A., Donahoe, P. K., Budzik, G. P., and Trelstad, R. L. (1982). Periductal and matrix glycosaminoglycans in rat Müllerian duct development and regression. *Dev. Biol.* 92, 16–26.
165. Roberts, L. M., Visser, J. A., and Ingraham, H. A. (2002). Involvement of a matrix metalloproteinase in MIS-induced cell death during urogenital development. *Development* 129, 1487–1496.
166. Trelstad, R. L., Hayashi, A., Hayashi, K., and Donahoe, P. K. (1982). The epithelial-mesenchymal interface of the male rate Müllerian duct: loss of basement membrane integrity and ductal regression. *Dev. Biol.* 92, 27–40.
167. Ikawa, H., Trelstad, R. L., Hutson, J. M., Manganaro, T. F., and Donahoe, P. K. (1984). Changing patterns of fibronectin, laminin, type IV collagen, and a basement membrane proteoglycan during rat Müllerian duct regression. *Dev. Biol.* 102, 260–263.
168. Paranko, J., Pelliniemi, L. J., and Foidart, J. M. (1984). Epithelio-mesenchymal interface and fibronectin in the differentiation of the rat mesonephric and paramesonephric ducts. *Differentiation* 27, 196–204.
169. Roberts, L. M., Hirokawa, Y., Nachtigal, M. W., and Ingraham, H. A. (1999). Paracrine-mediated apoptosis in reproductive tract development. *Dev. Biol.* 208, 110–122.
170. George, F. W., and Peterson, K. G. (1988). 5 alpha-dihydrotestosterone formation is necessary for embryogenesis of the rat prostate. *Endocrinology* 22, 1159–1164.
171. Gottlieb, B., Beitel, L. K., Wu, J. H., and Trifiro, M. (2004). The androgen receptor gene mutations database (ARDB): 2004 update. *Hum. Mutat.* 23, 527–533.
172. Cooke, P. S., Young, P., and Cunha, G. R. (1991). Androgen receptor expression in developing male reproductive organs. *Endocrinology* 128, 2867–2873.
173. Gupta, C. (1996). The role of epidermal growth factor receptor (EGFR) in male reproductive tract differentiation: stimulation of EGFR expression and inhibition of Wolffian duct differentiation with anti-EGFR antibody. *Endocrinology* 137, 905–910.
174. Gupta, C., and Bentlejewski, C. A. (1992). Role of prostaglandins in the testosterone-dependent wolffian duct differentiation of the fetal mouse. *Biol. Reprod.* 47, 1151–1160.
175. Cate, R. L., Mattaliano, R. J., Hession, C., Tizard, R., Farber, N. M., Cheung, A., Ninfa, E. G., Frey, A. Z., Gash, D. J., Chow, E. P., Fisher, R. A., Bertonis, J. M., Torres, G., Wallner, B. P., Ramachandran, K. L., Ragin, R. C., Manganaro, T. F., MacLaughlin, D. T., and Donahoe, P. K. (1986). Isolation of the bovine and human genes for Müllerian inhibiting substance and expression of the human gene in animal cells. *Cell* 45, 685–698.
176. Pepinsky, R. B., Sinclair, L. K., Chow, E. P., Mattaliano, R. J., Manganaro, T. F., Donahoe, P. K., and Cate, R. L. (1988) Proteolytic processing of Müllerian inhibiting substance produces a transforming growth factor-beta-like fragment. *J. Biol. Chem.* 263, 18961–18964.
177. Nachtigal, M. W., and Ingraham, H. A. (1996). Bioactivation of Müllerian inhibiting substance during gonadal development by a kex2/subtilisin-like endoprotease. *Proc. Natl. Acad. Sci. U S A* 93, 7711–7716.
178. Baarends, W. M., van Helmond, M. J., Post, M., van der Schoot, P. J., Hoogerbrugge, J. W., de Winter, J. P., Uilenbroek, J. T., Karels, B., Wilming, L.G., Meijers, J. H., Themmen, A. P. N., and Grootegoed, J. A. (1994). A novel member of the transmembrane serine/threonine kinase receptor family is specifically expressed in the gonads and in mesenchymal cells adjacent to the müllerian duct. *Development* 120, 189–197.
179. di Clemente, N., Wilson, C., Faure, E., Boussin, L., Carmillo, P., Tizard, R., Picard, J. Y., Vigier, B., Josso, N., and Cate, R. (1994). Cloning, expression, and alternative splicing of the receptor for anti-Müllerian hormone. *Mol. Endocrinol.* 8, 1006–1020.
180. Kobayashi, A., and Behringer, R. R. (2003). Developmental genetics of the female reproductive tract. *Nat. Rev. Genet.* 4, 969–980.
181. Kmita, M., and Deboule, D. (2003). Organizing axes in time and space; 25 years of collinear tinkering. *Science* 301, 331–333.
182. Daftary, G. S., and Taylor, H. S. (2000). Implantation in the human: the role of HOX genes. *Semin. Reprod. Med.* 18, 311–320.
183. Podlasek, C. A., Seo, R. M., Clemens, J. Q., Ma, L., Maas, R. L., and Bushman, W. (1999). Hoxa-10 deficient male mice exhibit abnormal development of the accessory sex organs. *Dev. Dyn.* 214, 1–12.

184. Benson, G. V., Lim, H., Paria, B. C., Satokata, I., Dey, S. K., and Maas, R. (1996). Mechanisms of femalea infertility in Hoxa-10 mutant mice: uterine homesis versus loss of maternal Hoxa-10 expression. *Development* 122, 2687–2696.
185. Akbas, G. E., Song, J., and Taylor, H. S. (2004). A HOXA10 estrogen response element (ERE) is differentially regulated by 17 beta-estradiol and diethylstilbestrol (DES). *J. Mol. Biol.* 340, 1013–1023.
186. Block, K., Kardana, A., Igarashi, P., and Taylor H. S. (2000). In utero diethylstilbestrol (DES) exposure alters Hox gene expression in the developing müllerian system. *FASEB J.* 14, 1101–1108.
187. Gendron, R. L., Paradis, H., Hsieh-Li, H. M., Lee, D. W., Potter, S. S., and Markoff, E. (1997). Abnormal uterine stromal and glandular function associated with maternal reproductive defects in Hoxa-11 null mice. *Biol. Reprod.* 56, 1097–1105.
188. Simpson, J. L. (1999). Genetics of the female reproductive tracts. *Am. J. Med. Genet.* 89, 224–239.
189. Mortlock, D. P., and Innis, J. W. (1997). Mutation of HOXA13 in hand-foot-genital syndrome. *Nat. Genet.* 15, 179–180.
190. Utsch, B., Becker, K., Brock, D., Lentze, M. J., Bidlingmaier, F., and Ludwig, M. (2002). A novel stable polyalanine [poly(A)] expansion in the HOXA13 gene associated with hand-foot-genital syndrome: proper function of poly(A)-harbouring transcription factors depends on a critical repeat length? *Hum. Genet.* 110, 488–494.
191. Johnson, K. R., Sweet, H., Donahue, L., Ward-Bailey, P., Bronson, R., and Davidson, M. (1998). A new spontaneous mouse mutation of Hoxd13 with a polyalanine expansion and phenotype similar to human synpolydactyly. *Hum. Mol. Genet.* 7, 1033–1038.
192. Mortlock, D. P., Post, L. C., and Innis, J. W. (1996). The molecular basis of hypodactyly (Hd); a deletion in Hoxa13 leads to arrest of digital arch formation. *Nat. Genet.* 13, 284–289.
193. Warot, X., Fromental-Ramain, C., Fraulob, V., Chambon, P., and Dolle, P. (1997). Gene dosage-dependent effects of the Hoxa-13 and Hoxd-13 mutations on morphogenesis of the terminal parts of the digestive and urogenital tracts. *Development* 124, 4781–4791.
194. Bouchard, M., Pfeffer, P., and Busslinger, M. (2000). Functional equivalence of the transcription factors Pax2 and Pax5 in mouse development. *Development* 127, 3703–3713.
195. Kuschert, S., Rowitch, D. H., Haenig, B., McMahon, A. P., and Kispert, A. (2001). Characterization of Pax-2 regulatory sequences that direct transgene expression in the Wolffian duct and its derivatives. *Dev. Biol.* 229, 128–140.
196. Patel, S. R., and Dressler, G. R. (2004). Expression of Pax2 in the intermediate mesoderm is regulated by YY1. *Dev. Biol.* 267, 505–516.
197. Dressler, G. R., and Douglass, E. C. (1992). Pax-2 is a DNA-binding protein expressed in embryonic kidney and Wilms tumor. *Proc. Natl. Acad. Sci. U S A* 89, 1179–1183.
198. Torres, M., Gomez-Pardo, E., Dressler, G. R., and Gruss, P. (1995). *Pax-2* controls multiple steps of urogenital development. *Development* 121, 4057–4065.
199. Bouchard, M., Souabni, A., Mandler, M., Neubüser, A., and Buslsinger, M. (2002). Nephric lineage specification by Pax2 and Pax8. *Genes Dev.* 16, 2958–2970.
200. Pellegrini, M., Pantano, S., Lucchini, F., Fumi, M., and Forabosco, A. (1997). Emx2 developmental expression in the primordia of the reproductive and excretory systems. *Anat. Embryol. (Berl).* 196, 427–433.
201. Barnes, J. D., Crosby, J. L., Jones, C. M., Wright, C. V., and Hogan, B. L. (1994). Embryonic expression of Lim-1, the mouse homolog of Xenopus Xlim-1, suggests a role in lateral mesoderm differentiation and neurogenesis. *Dev. Biol.* 161, 168–178.
202. Kobayashi, A., Shawlot, W., Kania, A., and Behringer, R. R. (2004). Requirement of Lim1 for female reproductive tract development. *Development* 131, 539–549.
203. Tsang, T. E., Shawlot, W., Kinder, S. J., Kobayashi, A., Kwan, K. M., Schughart, K., Kania, A., Jessell, T. M., Behringer, R. R., and Tam, P. P. (2000). Lim1 activity is required for intermediate mesoderm differentiation in the mouse embryo. *Dev. Biol.* 223, 77–90.
204. di Clemente, N., Josso, N., Gouédard, L., and Belville, C. (2003). Components of the anti-Müllerian hormone signaling pathway in gonads. *Mol. Cell. Endocrinol.* 211, 9–14.
205. Imbeaud, S., Belville, C., Messika-Zeitoun, L., Rey, R., di Clemente, N., Josso, N., and Picard, J. Y. (1996). A 27 base-pair deletion of the anti-müllerian type II receptor gene is the most common cause of the persistent müllerian duct syndrome. *Hum. Mol. Genet.* 5, 1269–1277.
206. Mishina, Y., Rey, R., Finegold, M. J., Matzuk, M. M., Josso, N., Cate, R. L., Behringer, R. R. (1996). Genetic analysis of the Müllerian-inhibiting substance signal transduction pathway in mammalian sexual differentiation. *Genes Dev.* 10, 2577–2587.
207. Jamin, S. P., Arango, N. A., Mishina, Y., Hanks, M. C., and Behringer, R. R. (2002). Requirement of Bmpr1a for Müllerian duct regression during male sexual development. *Nat. Genet.* 32, 408–410.
208. Gouedard, L., Chen, Y. G., Thevenet, L., Racine, C., Borie, S., Lamarre, I., Josso, N., Massague, J., and di Clemente, N. (200). Engagement of bone morphogenetic protein type IB receptor and Smad1 signaling by anti-Müllerian hormone and its type II receptor. *J. Biol. Chem.* 275, 27973–27978.
209. Visser, J. A., Olaso, R., Verhoef-Post, M., Dramer, P., and Themmen, A. P. (2001). The serine/threonine kinase transmembrane receptor ALK2 mediates Mulerian inhibiting substance binding. *Mol. Endocrinol.* 15, 936–945.
210. Mishina, Y., Crombie, R., Bradley, A., and Behringer, R. R. (1999). Multiple roles for activin-like kinase-2 signaling during mouse embryogenesis. *Dev. Biol.* 213, 314–326.
211. Heikkila, M., Peltoketo, H., and Vainio, S. (2001). *Wnts* and the female reproductive system. *J. Exptl. Zool.* 290, 616–623.
212. Parr, B. A., and McMahon, A. P. (1998). Sexually dimorphic development of the mammalian reproductive tract requires *Wnt-7a*. *Nature* 395, 707–710.
213. Parker, K. L. (2004). Sexual differentiation. In *Textbook of Endocrine Physiology* (J. E. Griffin and S. R. Ojeda, Eds.), pp. 167–185. Oxford University Press, New York.

Knobil and Neill's Physiology of Reproduction, Third Edition
edited by Jimmy D. Neill,
Elsevier © 2006

CHAPTER 9

Oviduct and Endometrium: Cyclic Changes in the Primate Oviduct and Endometrium

Alexandra P. Hess, Nihar R. Nayak, and Linda C. Giudice

Introduction, 337
Anatomy, Cell Constituents, and Embryology of the Oviduct and Endometrium, 338
The Oviduct, 340
Historical Perspective and Functions, 340
Cyclic Changes in the Primate Oviduct, 341
Estrogen and Progesterone Receptors, 343
Growth Factor and Cytokine Expression and Action in the Oviduct, 345
Oviductal Fluid and Oviductin, 347
The Endometrium, 349
Overview of Phases of the Menstrual Cycle and Endometrial Histology, 349
Steroid Hormone Receptors: Expression and Actions in Endometrium, 352
Molecular Mechanisms Underlying Endometrial Function, 354
Conclusion and Eye to the Future, 369
References, 369

INTRODUCTION

The oviduct and endometrium in primates have common embryologic origins. They are both highly sensitive to changes in circulating ovarian-derived steroid hormones, although the cyclicity of the two organs is distinct. In the oviduct, remarkable changes occur in epithelial ciliogenesis and secretory cell development primarily in response to increasing levels of estradiol in the proliferative phase of the menstrual cycle. In the peri- and postovulatory period, under the influence of progesterone, the oviductal epithelium atrophies, regresses, and becomes quiescent, primarily through cellular apoptosis. The endometrium also undergoes intense cellular proliferation under the influence of estradiol. However, in contrast to the oviduct, after ovulation and under the influence of progesterone, it does not regress. Rather, it undergoes morphological and biochemical changes, preparing for implantation of the conceptus. This is accompanied by differentiation of the epithelium, decidualization of the stroma, and the influx of immune cells. In the absence of implantation the endometrium undergoes apoptosis and tissue desquamation, and regeneration ensues by mechanisms likely involving stem cells in the basalis region of the tissue. Most information about the oviduct in primates derives from histological analysis and biochemical evaluation of steroid hormone receptors and select, secreted oviductal proteins and growth factors. Relatively little is known about dynamic biochemical changes in the cellular constituents of this tissue throughout the cycle—primarily because of its limited availability. In contrast, the endometrium has enjoyed extensive investigation, primarily because of relatively greater accessibility. Additionally, the endometrium can be visualized by ultrasound, and clinical correlates have been derived from these assessments. This is in contrast to the oviduct which can be visualized ultrasonographically primarily only when it is diseased, e.g., as a hydrosalpinx. This chapter describes what is currently known about the

Supported in part by German Research Foundation (Deutsche Forschungsgemeinschaft HE 3544/1 [APH]), NIH Building Interdisciplinary Research Careers in Women's Health (BIRCWH) HD 043452-03 (NRN), and the NIH Specialized Cooperative Centers Program in Reproduction Research HD #31398-08 (LCG).
Department of Obstetrics and Gynecology, Stanford University School of Medicine, Stanford, California.

cycle-dependent morphologic, biochemical, genomic, and proteomic changes and signaling events in human and nonhuman primate oviduct and endometrium, with translation of the biology to clinical medicine.

ANATOMY, CELL CONSTITUENTS, AND EMBRYOLOGY OF THE OVIDUCT AND ENDOMETRIUM

The oviducts are anatomically contiguous with the uterus (Fig. 1A) and develop in parallel with it (see below). Each oviduct is about 13 cm long and is divided into four regions (Fig. 1A) (1): the *intramural region* (passing through the uterine wall); the *isthmus* (narrow, proximal 1/3 of the tube, extending laterally from the intramural region to the ampulla); the *ampulla* (the expanded, distal 2/3 of the oviduct, contiguous with the isthmus); and the *infundibulum* (large, trumpet-shaped end of the tube, open to the peritoneal cavity by the abdominal osteum and containing multiple folds, the "fimbria" (2). The oviduct is composed of three distinct histological layers: an inner *mucosa*, middle *muscularis*, and the

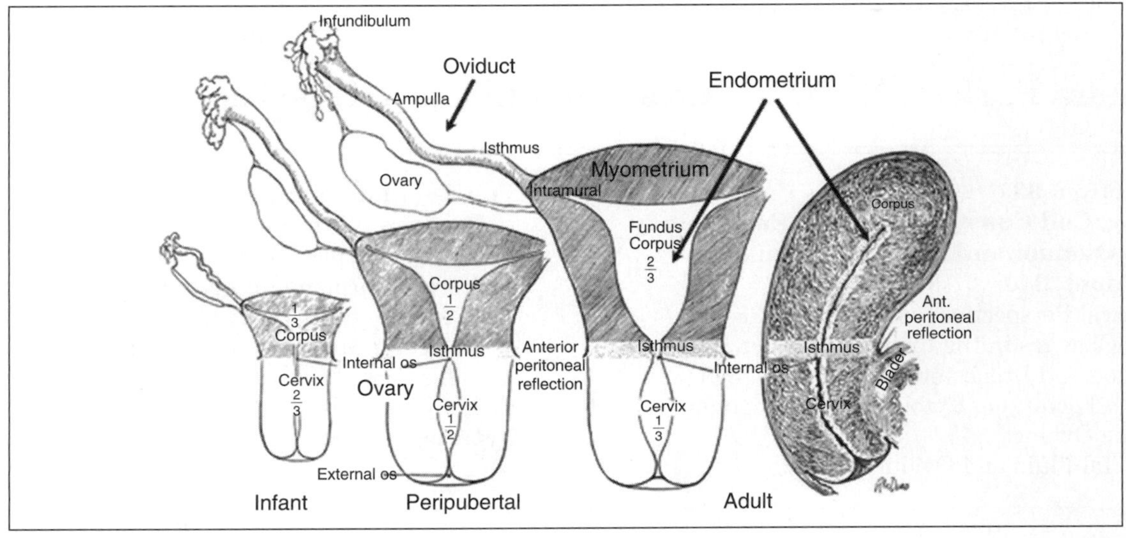

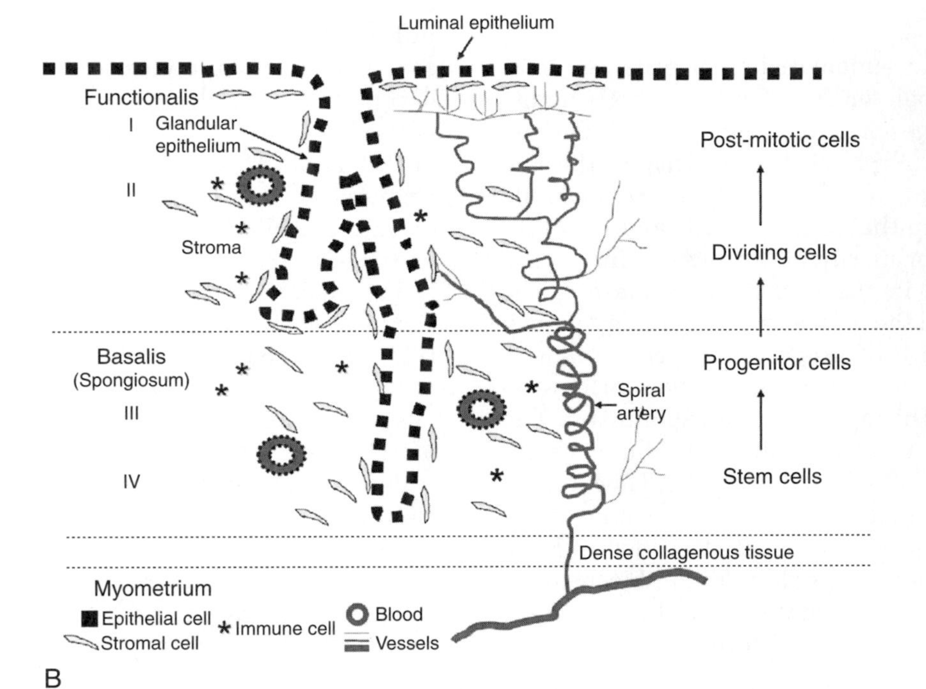

FIG. 1. Anatomy of the uterus and fallopian tube. **A:** Schematic of the uterus, cervix and oviducts, as a function of age. [Adapted from Ramsey (1), with permission.] **B:** Schematic diagram of the four zones of the endometrium and their cellular constituents. [Adapted with permission from Padykula et al. (427).] (See text.)

outer *serosa*. The mucosa forms an elaborate system of highly branched, interconnected, longitudinal folds in the ampulla, resembling a labyrinth, with less branching and folding along the tube towards the isthmus (2). A simple columnar epithelium lines the mucosa and is comprised of ciliated cells and nonciliated secretory cells. The lamina propria provides the support for the mucosa and is comprised of a thin vascular layer of loose connective tissue. The muscularis is comprised of a circular layer and an outer, longitudinal layer of smooth muscle and connective tissue. At the isthmus a third longitudinal layer is present, although it is not well delineated from the others. The serosa is the outer layer and is contiguous with the mesosalpinx (2).

The uterus is divided into three anatomical parts (Fig. 1A): the *corpus* (body), the *fundus* (top portion), and the *isthmus* (which is contiguous with the cervix). The endometrium lining the openings of the oviducts is designated as the peritubal-ostium or corneal region. The endometrium lining the isthmus is clinically referred to the lower uterine segment because histologically it does not undergo the classical hormone-dependent changes observed in the corpus. The endometrium lining the corpus is most profoundly affected by changes in circulating steroid hormones (see below).

The endometrium has a complex cellular constituency (Fig. 1B), including simple columnar epithelium (glandular and luminal) of secretory and, in women, ciliated cells, stromal fibroblasts, vascular endothelium and smooth muscle, and immune cells (3). The single layer of columnar epithelium, "luminal epithelium" is the interface between the uterine cavity and the rest of the endometrial mucosa (Fig. 1B). Glandular epithelium lines the tubular or branched glands that can extend as deep as the endometrial–myometrial junction. The stromal compartment ("lamina propria") is comprised of highly cellular connective tissue with an extracellular matrix that contains few connective tissue fibers and resembles embryonic mesenchyme. The endometrial–myometrial "junction" is indistinct, without an intervening submucosal membrane. Pockets of basal endometrium extend in the proximate regions of adjacent myometrium (4).

Primate (rhesus macaque) endometrium was characterized by Bartelmez (5) to be comprised of four histologically defined zones (Fig. 1B), and similar zonation is also described in human endometrium (6,7). These compartments are:

- Zone I, comprised of luminal epithelia and subjacent, densely packed stroma
- Zone II, the upper endometrium in which the straight region of the glands course and are widely separated by stroma
- Zone III, mid-regions of the glands, widely separated by stroma
- Zone IV, the deepest portion of the glands in a fibrous stroma, adjacent to the endometrial–myometrial junction

Zones I and II comprise the "functionalis" layer, and zones III and IV comprise the "basalis" layer (8). In humans and nonhuman primates, the functionalis responds cyclically to ovarian steroid hormones, is the site of implantation, and is shed with the menses in the absence of implantation. The basalis participates in regeneration of the tissue after the menses (9). Changes in the functionalis during the proliferative phase occur primarily in response to estradiol (E_2), and epithelial and stromal differentiation occur in the secretory phase in response to progesterone (P_4) (see below). In addition, growth factors and cytokines are involved in paracrine (or autocrine) actions to effect the histological and biochemical signatures of the endometrial functionalis throughout the cycle (3,10) (and see below).

The oviduct and the uterus develop in parallel during fetal development (11,12) and derive from coelomic epithelium and surrounding mesenchyme (4,13). Development of the uterus and oviducts depends on the caudal extensions of the paramesonephric (Müllerian) ducts, paired invaginations of the coelomic epithelium, first described by Johannes Müller in 1825 (4). The cranial portions open into the peritoneal cavity and develop into the oviducts, and the caudal portions migrate medially and fuse, forming the uterovaginal primordium. The lower portion of the latter contributes to the cervix and vagina, and the upper portions give rise to the epithelium and glands of the uterus. The endometrial stroma and myometrium derive from adjacent mesenchyme. Failure of fusion of the paramesonephric ducts results in uterine anomalies, ranging from a double or bicornuate uterus to a septate uterus. Such fusion abnormalities are relatively common, occurring in about 0.25% of women, including minor abnormalities, during postpartum examination (14).

The uterus changes shape and grows during development due to caudal growth of the uterovaginal primordium, rostral extension of the fused portion of the paramesonephric ducts, and a proliferation and thickening of the mesenchyme. By midgestation the caudal 2/3 of the uterus becomes the isthmus, and the cranial 1/3 is bulbous and becomes the body. Some layering of the mucosa is evident at this time (12), and by 20 weeks of gestation smooth muscle cells develop in the muscular layer (myometrium), growing during the second half of gestation and resulting in the pear-like shape seen in adults (Fig. 1A) (11,12). Before 20 weeks of gestation, the uterine cavity of the human fetus is lined by a single layer of columnar epithelium without glands. At midgestation the epithelium lining the cavity forms pouches of cells that grow into the underlying mesenchyme, becoming glands by the 7th month. At this time the endometrium responds to the high levels of steroid hormones from the placenta. Indeed, the proliferating glandular epithelium resembles estrogen-stimulated adult endometrium during the proliferative phase of the cycle (11,12) (and see below). By the 9th month, the fetal endometrium is hypertrophic, with edematous stroma, heavily congested blood vessels and actively secreting, coiled glands. In the newborn (15) the endometrium has features of proliferative and secretory endometrium (12). During infancy and childhood the uterus grows at a rate consistent with overall body growth, and the corpus remains smaller than the cervix (12,15) (Fig. 1A). The glands elongate and reach the basal portion of the endometrium (Fig. 1B). After puberty, uterine (endometrium and myometrium) growth is stimulated by circulating estradiol and progesterone, with the corpus becoming larger than the cervix, as in the adult (Fig. 1A).

Molecular mechanisms underlying Müllerian duct development have been elucidated in mice using homologous recombination approaches and are summarized by Cunha et al. (16). Gene knockout mice with deficiencies of Wolffian duct development also usually exhibit Müllerian duct defects and failure of uterine and vaginal development. Candidate genes include Pax-2, Emx-2, retinoic acid receptor-α and β, and vitamin A (16). Molecular pathways underlying anterior-posterior patterning in reproductive tract development have also been approached through select gene knockout studies. Candidate genes and families for these processes include homeobox (Hox) genes, Msx-1, and members of the Wnt family (Wnt 4, 5a, and 7a), important in patterning and establishing tissue boundaries in *Drosophila* (17,18). For example, Wnt 7a null mice have partial posteriorization of the female reproductive tract (the oviduct acquires characteristics of the uterus and the uterus acquires characteristics of the vagina), although both tissues retain some of their own characteristics (19). Interestingly, the homozygous Wnt 7a null (–/–) mouse lacks endometrial glands and has overgrown and poorly organized smooth muscle layers of the myometrium. In humans and nonhuman primates, there is no information on these or other factors involved in the *development* of the glands or other structures in the uterus (or oviduct), although the Wnt and Hox gene families (and perhaps others) are likely involved. Hox genes and members of the Wnt family are regulated during the menstrual cycle of adult women (20,21), and misexpression of Hox A-10 and Hox A-11 has been observed in women with infertility and endometriosis (22). In mouse models, in addition to their roles in reproductive tract development, it is likely that Hox genes, Msx1, and Wnt family members play an important role in adult uterine epithelial and stromal morphology and differentiation in response to cyclic changes of ovarian steroid hormones. Cooke et al. have demonstrated (23,24), using tissue recombinant technology of uterine specimens in which the epithelium, stroma, or both lack estrogen and progesterone receptors, that stromal–epithelial interactions play a crucial role in hormonal response of both the epithelium and stroma to steroid hormones (steroid hormone receptors section in The Endometrium, below). Given the cyclic changes in morphology and function of the adult uterus, it has been proposed that this tissue has conserved some developmental mechanisms into adulthood, perhaps involving Msx, Wnt, and Hox family members, in addition to several growth factor families (16,24). Whether this is true for primate endometrium (and oviduct) remains to be elucidated, but from an evolutionary perspective and conservation of mechanisms, this is teleologically reasonable.

THE OVIDUCT

Historical Perspective and Functions

The first anatomical description of the mammalian oviduct was made by Fallopius in 1561 (25), at which time the organ was considered to be a passive conduit for gametes. Appreciation of the cellular composition and fine structure of the oviduct and subsequently some of its molecular constituents have lead to the speculation that it is an active reproductive organ with transport and secretory functions, participating in transport of gametes, fertilization, and early embryonic development (26–33). The oviduct is comprised of a muscular layer lined by an epithelium that contains primarily two cell types—ciliated and secretory. From a teleological perspective, ciliation is an important cellular attribute for a conduit that facilitates sperm traveling from the uterus to meet an oocyte in its distal portion and to promote, in the opposite direction, embryo transport from the site of fertilization to the uterine cavity. Studies with human (34–37), rhesus monkey (38), mouse (39,40), and sheep (41) oviductal epithelial monolayers co-cultured with blastocysts of the same species, have demonstrated improved fertilization and cleavage rates, improved embryo quality of the inner cell mass and an increase in total embryonic cell number, compared to without co-culture, supporting the hypothesis that secreted products from oviductal epithelium are important in embryo development. However, the identity of these products, whether they derive from the secretory cells, ciliated cells, or stroma (or a combination of these), and the mechanisms whereby they facilitate embryo development and transport remain poorly understood. Herein, cyclic changes in the primate oviduct are reviewed.

Cyclic Changes in the Primate Oviduct

Human Oviduct

The anatomy of the oviduct can be highly variable from one species to another, but it generally consists (Fig. 1A) of a muscular layer associated with a secretory layer, divided into an infundibulary, ampullary, isthmic and intramural portion. In humans it is approximately four to five inches in length. The tubal epithelium is comprised of secretory and ciliated cells, interspersed with a cell type either representing an undifferentiated, morphologically altered, or functionally depleted form of secretory or ciliated cell (42). The ratio of ciliated to secretory cells varies along the length of the Fallopian tube. In the distal portion, the infundibulum, the majority of cells are ciliated (43,44). In the ampulla, the site of fertilization and early embryo cleavage, the mucosa is also densely ciliated, although less than in the infundibulum. The least ciliated portion of the oviduct is the isthmic region, the key region in regulating spermatozoa passage and the entrance of the cleaving blastocyst into the uterus.

Early studies on human oviduct by Novak and Everett (45) revealed specific changes in the development and growth of the secretory cells and ciliation of the ciliated cells during the menstrual cycle. Later studies found maximal growth (height) of the ciliated and secretory cells at mid-cycle, and subsequently a decrease in cell height in the luteal (secretory) phase, with minimal cell height in the premenstrual and menstrual phases. These observations are consistent with findings of Brenner et al., demonstrating that the ovarian-derived steroid hormones, estradiol (E_2) and progesterone (P), play a role in these cyclic changes—specifically, that E_2 stimulates cellular growth and P inhibits it (31). Over the next four decades numerous studies confirmed the observations of Novak and Everett regarding cyclic variations in the height of the secretory cells. However, conflicting information arose on cyclic changes in ciliation patterns throughout the cycle (46–49). Subsequently, Verhage et al. (50), and Donnez et al. (51) reported their observations on oviducts of normally cycling woman in the estrogen-dominant (follicular [proliferative] phase) and the progesterone-dominant (luteal [secretory]) phases of the menstrual cycle and also postpartum. These studies clearly demonstrated that the epithelial secretory cells reach their maximum height in the late follicular phase, when the highest degree of ciliation is observed in the fimbrial and ampullary portions of the oviduct. In the luteal phase, under the influence of progesterone, many cells decrease in height and lose some of their cilia. Re-ciliation and cellular hypertrophy in the early follicular phase are associated with moderate E_2 and low progesterone levels. Further atrophy and marked deciliation are found in pregnancy and postpartum. The theory that progesterone leads to atrophy and deciliation, whereas E_2 stimulates the ciliation and cell growth, is supported by several clinical studies. For example, in postmenopausal women, administration of estrogen results in induction of oviductal ciliogenesis (52). Also, increased ciliogenesis and cell height resulted when stilbestrol was administered to subjects in the postpartum period (5 mg/day for 5 to 9 days), beginning on the day of delivery. However, the stilbestrol effect vanished when it was administered in combination with progesterone (53). A study of primary cultures of oviductal epithelium, analyzed by scanning electron microscopy and immunocytochemistry, validated induction of a ciliated epithelial cell phenotype after administration of E_2. The cells adopted an immature, secretory-like phenotype in the absence of E_2 treatment (54).

Oviductal secretions in mammals, in general, are hormonally regulated by ovarian steroids, with estrogen stimulating and progesterone inhibiting the secretory process (30). The current of secretions is oriented towards the peritoneum at ovulation (facilitating sperm movement toward the ovulated oocyte) and for the following three days. Reversal of the current occurs subsequently, enabling embryo transport to the uterus (33).

Overall, these data on human oviduct support the view that estrogens stimulate differentiation and function of the oviductal epithelium and that progesterone suppresses these effects (31). Similar steroid regulation of nonhuman primate oviductal epithelium has been reported, but in more detail based on a larger number of anatomic specimens for analysis, compared to humans, and the ability to manipulate the hormonal environment in an animal model, which is not ethically permissible in women. These changes are reviewed below.

Nonhuman Primate Oviduct

Comparative anatomy of the primate oviduct was first summarized by Eckstein in the *Handbook of Primatology* (*Primatologia*) (55). Subsequently, several studies describing cyclic changes in the oviduct were reported in different nonhuman primate species including rhesus (56), baboon (57,58), cynomolgus (59), and pig-tailed macaques (60–62). Similar to human oviduct, estrogen promotes differentiation and reciliation of the oviductal epithelium in nonhuman primates; whereas, progesterone treatment leads to deciliation and atrophy of this epithelium (56,63–65). Ciliogenesis in nonhuman primate oviduct is more dramatically regulated than in human oviduct and is a good indicator of the effect ovarian steroid hormones on oviductal epithelium.

Cyclic changes in nonhuman primate oviduct have been extensively reviewed by Brenner and colleagues in the previous two editions of this book (4,31). In an elegant study, Brenner et al. correlated the histological changes in the oviductal epithelium with the changes in serum levels of E_2 and P and the histology of the ovaries and the endometrium (59). These authors and others have used several morphological characteristics to determine different stages of the oviductal cycle, including height of ciliated cells, percent of ciliation in different parts of the oviduct, size and roundness of cell nuclei, degree of mitotic activity, extent of ciliogenesis, content of glycogen and granules in secretory cells, degree of "pinching-off" of ciliated tips, presence of intraepithelial apoptotic bodies, and the presence of macrophages filled with nuclear and cellular fragments (28,56,59,66,67). Using these parameters, Brenner and colleagues have defined oviductal changes (Fig. 2) in the cynomolgus and rhesus macaques into eight specific stages throughout the menstrual cycle, and they named these stages in order of appearance of ciliated cells as follows:

1. Full regression (Fig. 2A): The epithelium is maximally atrophied and deciliated. Epithelial cell nuclei are maximally shriveled, and there is minimal secretory activity.
2. Preciliogenic (Fig. 2B): Hallmark is the onset of epithelial cell nuclei swelling, smoothing of the nuclear contours, cellular hypertrophy and mitotic activity. In general, light and dark cells and basal bodies are not apparent.
3. Ciliogenic (Fig. 2C): Mitosis and cellular hypertrophy continue, with round, smooth, and enlarged epithelial nuclei. Light and dark cells can be distinguished, and basal bodies are apparent in the apical cytoplasm of the light hypertrophied cells.
4. Ciliogenic-ciliated (Fig. 2D): Features of the ciliogenic phase persist, but numerous ciliated cells have now developed. Secretory dark cells are present, but are not prominent. Ciliogenic cells predominate over ciliated ones.
5. Ciliated-ciliogenic (Fig. 2E): In this phase, the majority of cells have become ciliated, but a few cells undergoing ciliogenesis can be found scattered throughout the epithelium. Moreover, mitotic activity has not ceased completely. Secretory cells have become more prominent and have developed bulbous tips filled with granules and glycogen. In this phase, ciliated cells predominate over ciliogenic ones.
6. Ciliated-secretory (Fig. 2F): Most epithelial cells are either ciliated or secretory, and ciliogenic cells are extremely rare. Secretory cells are fully developed with bulbous tips rich in granules, glycogen and some vacuoles, that extend beyond the basal body row in the fimbriae and past the cilia in the ampulla and isthmus. Epithelial cell nuclear contours are less smooth than during ciliogenesis.
7. Early regression (Fig. 2G): Apoptotic epithelial cells are scattered throughout the epithelium.

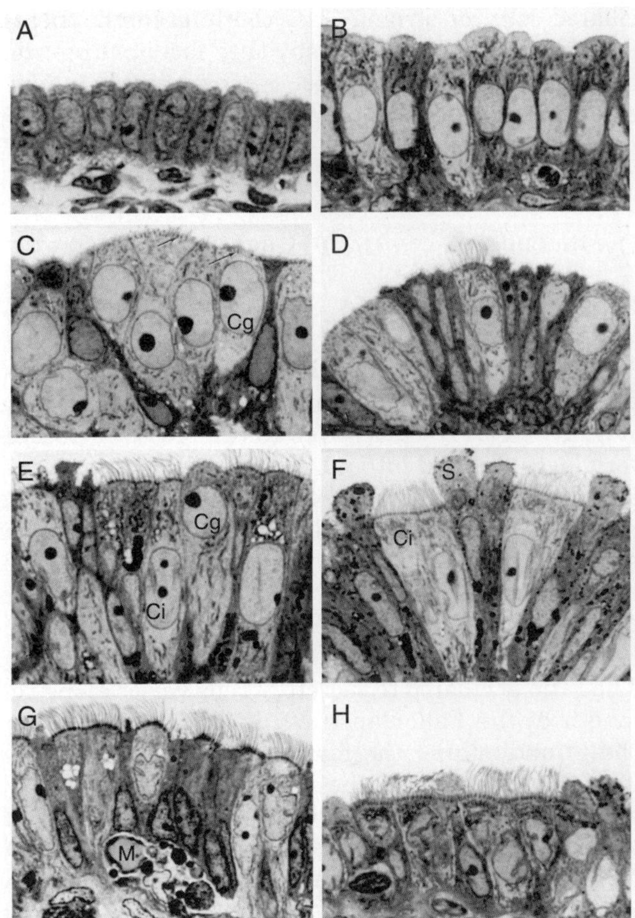

FIG. 2. Photomicrographs showing different histological changes in various stages of oviductal epithelium during the menstrual cycle in cynomolgus macaques. **A:** Full regression, fimbriae: atrophied and deciliated epithelium. **B:** Preciliogenic, ampulla: hypertrophied cells with swollen nuclei. **C:** Ciliogenic, fimbriae: light hypertrophied ciliogenic cells (Cg) with enlarged nucleoli and basal bodies in apical cytoplasm (*arrows*). **D:** Ciliogenic-ciliated, ampulla: mostly ciliogenic cells and few ciliated cells. **E:** Ciliated-ciliogenic, ampulla: mostly ciliated cells (Ci) and few ciliogenic cells (Cg). **F:** Ciliated-secretory, ampulla: mostly ciliated (Ci) and fully developed secretory (*dark*) cells with bulbous tips (S). **G:** Early regression, fimbriae: infiltration of large number of macrophages (M). **H:** Late regression, ampulla: atrophied ciliated and secretory cells, nuclei mostly shriveled. (Magnification × 1,050.) [Reproduced with permission from Brenner and Maslar (4).]

Macrophages have invaded the epithelium and are phagocytosing dead cells. There are no other differences from the ciliated-secretory phase.

8. Late regression (Fig. 2H): The epithelium is atrophied. Secretory activity has diminished, considerable deciliation has occurred and many ciliated cells are "pinching off" their tips. Dead cells and macrophages continue to be present. Epithelial cell nuclei appear shriveled.

In general, these cyclic phases are most evident in the fimbriae and ampulla and least evident in the isthmus. The exception is the secretory tip extension, which is least prominent in the fimbriae and ampulla and greatest in the isthmic region of the oviduct. Although these changes take place in a temporal sequence, some ciliogenic cells can be found in the ciliated-secretory phase where they are no longer supposed to be.

Estrogen and Progesterone Receptors

Menstrual cycle-dependent changes in human and nonhuman primate oviductal epithelial differentiation and ciliation highly suggest regulation of these processes by the ovarian-derived steroid hormones, estradiol and progesterone. Receptors for estrogen and progesterone have been extensively studied in the human and nonhuman primate oviduct with regard to binding characteristics, cellular localization, expression and regulation during the menstrual cycle. These findings are reviewed herein.

Binding Studies

Receptor binding capacities for estrogen and progesterone throughout the menstrual cycle were demonstrated, by autoradiographic grain counts, to be higher in the oviductal epithelium than in the underlying lamina propria (68). However, further measurements of both receptors in the cytosolic fraction revealed regional differences among the various anatomical segments of the Fallopian tubes (69,70). Highest estradiol and progesterone binding was detected in the ampullary region, and significantly lower binding was observed in the infundibulum and isthmus regions (69,70). In general, the number of cytoplasmic and nuclear binding sites for E_2 and progesterone vary dramatically across the menstrual cycle, with highest binding in the estrogen-dominant (follicular, proliferative) phase and lowest binding in progesterone-dominant (luteal, secretory) phase (71). Helm et al. conducted a study focused on progesterone binding in the cytosolic and nuclear fractions and found that, although nuclear progesterone binding was always lower than cytosolic, the overall binding pattern in the different anatomical regions of the oviduct was similar, with highest binding in both fractions in the ampulla. Furthermore, in the luteal phase of the cycle, cytoplasmic, but not nuclear, progesterone binding capacity was inversely correlated with plasma progesterone concentrations (72). The same pattern of inversely related tissue binding compared to serum levels was also observed with estradiol (68).

In humans, nonhuman primates, mice and chickens, E_2 stimulates the synthesis of the estrogen receptor (ER) and the progesterone receptor (PR) in female reproductive tract tissues, including the oviduct; whereas progesterone has the opposite effect. In all these species, the nuclear and cytosolic receptors are present at higher concentrations in the follicular phase, compared to the luteal phase. Steroid hormone regulation has been demonstrated by exogenous administration of E_2 and progesterone to nonhuman primates. For example, in rhesus monkey administered with E_2, there was an increase in oviductal epithelium ER, ciliation and decrease of epithelial cell height, followed by a decrease of receptor amount after administration of progesterone. The level of ER presence increased again after withdrawal of progesterone, clearly demonstrating regulation of ER by estradiol and progesterone. Significantly less binding of E_2 to its cognate receptor was also observed in the oviduct during the phase when progesterone was administered, even though the serum levels of E_2 remained constant (56). The same group also studied changes of oviductal ER throughout the natural menstrual cycle of cynomolgus macaques. Therein, they confirmed their previous findings in the rhesus monkey regarding a significant increase of nuclear and cytosolic ER combined with ongoing ciliogenesis and growth during the first part of the cycle, followed temporally by a marked decrease of ER, deciliogenesis, and cellular atrophy. Simultaneous measurements of serum steroid levels suggested that P suppresses the amount of ER present in the fallopian tube even though circulating E_2 levels remain stable (73). These observations further suggest that P likely facilitates its actions through antagonizing E_2 in increasing the ER and thus down-regulates ER levels below the minimum number of receptors required to fulfill adequate estrogen-dominated signaling in the oviduct.

Immunohistochemical Localization of Steroid Receptors in the Oviduct

Immunohistochemical localization of PR and ER in human Fallopian tubes obtained from subjects during the luteal phase of the cycle (day LH+4 to LH+6), revealed immunostaining in the cell nucleus in stromal as well as in epithelial cells. No uniform distribution was observed in glandular epithelial cells, and no staining at all was detected in perivascular cells. Also the intensity of staining for PR in the stromal cells was not evenly distributed throughout the oviduct, since there was significantly more staining detectable in the isthmus, compared to the ampulla. There were also no marked differences in the distribution of ER concentration along the oviduct (74,75).

Brenner et al. reported that most of the specific staining for the ER was present in the nuclei of the stromal and epithelial cells of oviducts of spayed monkeys. In this study E_2 treatment increased immunohistochemical nuclear staining in the stromal smooth muscle and secretory cells; whereas, progesterone administration decreased the amount. In contrast to earlier studies, Brenner and colleagues found immunoreactive ER only in the nuclear fraction, consistent with its role as a nuclear transcription factor (76,77). Furthermore ER was only detected in stromal cells, secretory cells and smooth muscle cells, but not in ciliated cells. PR immunostaining increased in response to E_2 treatment, equally in secretory, smooth muscle and stromal cells whereas in animals treated sequentially with E_2 and then P, PR staining was more suppressed in epithelium than stromal cells (31). Similar results in the baboon and human oviduct have reported (78). Taken together, the data suggest that long term, suppressive effects of P on the oviductal epithelium are likely mediated by the stroma, rather than a direct effect on the epithelium, because the stromal cells maintain their PR levels during treatment with P, while PR are down-regulated in the epithelium.

Progesterone and Estrogen Receptor Isoforms

Recent studies in the reproductive tract of humans and rodents have found that progesterone facilitates its physiological effects by interacting with two different isoforms of the PR: progesterone receptor A (PR-A, 94 kDa) and progesterone receptor B (PR-B, 116 kDa) (79–81). These two isoforms are encoded by a single gene under the control of separate promoters (82) and belong to the nuclear receptor superfamily of transcription factors (83–85). In the human, PR-B contains an additional 164 amino acid sequence in the N terminus, compared to PR-A (86,87). The expression of the two protein isoforms from one gene is highly conserved in several species, including human and rodents (88–92). The ratio of the individual isoforms varies among different reproductive tissues and depends on the developmental status (93) and hormonal environment (94,95). Although both of the isoforms are coexpressed in the same cells, they act as two functionally different transcription factors (96–99). Investigations with mice lacking both isoforms (PRKO) have shown that both PRs are essential for female fertility. A null mutation of both isoforms leads to a phenotype with various reproductive abnormalities, including the inability to ovulate, uterine hyperplasia and inflammation, defects in uterine implantation, impaired thymic immune adaption to pregnancy, and loss of sexual behavioral response to progesterone (99,100). Further analysis of mice lacking only PR-A (PRAKO) has provided evidence about the spatiotemporal expression and the contribution of PR-A and PR-B functions in a tissue-specific manner to facilitate different functions of progesterone action within the reproductive tract. Normal ovulatory and uterine responses require PR-A; whereas, in absence of PR-A, PR-B has revealed an unexpected capacity to support epithelial cell proliferation rather than to inhibit it (99). Focusing on the oviduct, Gava et al. observed in the mouse model, that PR-A is the predominant isoform present in all oviductal cells, except in the ampullary epithelium where PR-B is exclusively expressed (81). The predominance of PR-A in mouse oviduct suggests that this is the primary mediator of progesterone action in this tissue. Similar findings were observed in the rat oviduct (101).

Regarding the ER isoforms alpha (ER-α) and beta (ER-β) presence in the fallopian tube, little is known in humans; whereas, several comprehensive studies using the rat-model were performed (101–104). ER-α, as well as the more recently discovered ER-β, have been cloned and found in several species including rat (105), mouse (106), and human (107) showing a high homology among species. A study by Shah et al., performing immunohistochemistry, reported nuclear immunostaining for the ER in human adult fallopian tubes in sections from the early follicular phase and mid-cycle. Using Western blot analysis they found two isoforms of the ER in the adult oviduct—one, with a molecular weight of approximately 66 kDa, and a truncated second form of molecular weight approximately 49 kDa (108). Investigations of human fetal (13 and 20 weeks' gestation) tissues showed a very low to undetectable expression of ER-α mRNA; whereas, ER-β was detected in various fetal tissues at these gestational ages. Regarding the female reproductive tract in particular, immunoreactive ER-α and ER-β were both detected in the epithelium of the oviduct, leading the authors to conclude that estrogenic actions through both receptors may alter fetal oviductal development (109). However, in the adult human oviduct immunohistochemistry has revealed a pattern of ER-β with most of the expression restricted to the nucleus of epithelial cells and some diffuse staining in the cytoplasm of several epithelial cell types.

Specifically, ER-β was found in the nucleus and cytoplasm of the ciliated epithelium. The connective tissue showed restricted immunoreactivity for the nucleus. In contrast, ER-α was detected in all three cell types, but only in the nucleus (110). The differential expression of ER-α and ER-β in human oviduct suggests that estrogenic control mechanisms are more complex than envisioned in this tissue, and further investigation is required to identify clearly their effects on development and function of the human oviduct (and reproductive tract). To our knowledge no studies regarding the differential expression of ER-α and ER-β in the nonhuman primate oviduct have been reported to date, in comparison to investigations on rhesus ovary and endometrium that demonstrated mRNA expression for both types of the estrogen receptor (111).

Antiprogestin Effects on ER, PR, and Cellular Function

Progesterone receptor antagonists or "antiprogestins," such as mifepristone, block the biological effects of progesterone at the receptor level (112,113). The end effect depends on the time of treatment and the doses administered. Mifepristone's mechanisms of action are mediated in the human Fallopian tube by inhibiting the normal down regulation of PR and therefore increasing the amount of PR in both the ampulla and the isthmus. Studies in women given a single dose of 200 mg of mifepristone once a month, two days after the LH surge, have shown that this PR antagonist is an effective contraceptive (114,115), likely by interfering with gamete transport in the tube. Mifepristone has also been demonstrated to be an effective emergency contraceptive with low side effects, when taken postcoitally (116). A recent study, using immunohistochemistry and western blotting, elucidated the distribution of PR and ER in the Fallopian tube after treatment with mifepristone (74). Following mifepristone treatment, a significant increase in PR concentration was found in both epithelial and stromal cells, compared to the untreated control group. In untreated controls, immunostaining for the PR in stromal cells was less in the ampulla, compared to the isthmus; whereas, no differences in epithelial cells were observed. In the treatment group, however, the most pronounced and significant increase was in the ampulla, although marked but not significant changes occurred in the isthmic part of the oviduct. Furthermore, Western blot analysis revealed that both isoforms of the PR were up-regulated, but the predominant isoform was PR-B (74). Another study comparing PR-B expression in nontreated and mifepristone treated woman demonstrated that immunohistochemical staining for PR-B was present in epithelial and stromal cells, with no spatial difference in oviducts of untreated women. After mifepristone treatment, PR-B expression increased significantly in both cell types and both parts of the oviduct investigated (117). Regarding the ER, no differences in spatial expression throughout the oviduct was observed in the untreated control group. However, in the antiprogestin treated subjects, significantly increased ER immunostaining was observed in the epithelial cells, primarily in the isthmic region of the Fallopian tube; whereas, no changes were found in the stromal cells of treated, compared to untreated, subjects (74). Thus, it appears that PR-B and ER in human oviduct are regulated by progesterone.

That oviductal cellular differentiation and growth are E_2 and progesterone dependent has been further demonstrated by the use of mifepristone. In rhesus monkeys, e.g., when mifepristone was administered after two weeks of estrogen priming, it inhibited progesterone effects on oviductal cells, but permitted E_2 action in keeping cells in a secretory, ciliated state and maintaining the oviductal wet weight (31). Furthermore, when the antiprogestin ZK 1376 316 was administered to rhesus monkeys, it inhibited the antagonistic effects of progesterone on E_2-dependent differentiation of the oviduct in a dose-dependent manner (118).

Growth Factor and Cytokine Expression and Action in the Oviduct

Epidermal Growth Factor and Receptor

Cytokines are believed to play an important role in the communication between the oviduct and the developing embryo. Cytokine expression in human endometrium is cycle-dependent (see below), and cytokine expression in human Fallopian tubes is also cycle-dependent, suggesting hormonal regulation in both these tissues (119). The mitogenic effects of estradiol on tubal epithelial cells of different species, including human and nonhuman primates, are mediated by stimulating local biosynthesis of select growth factors and cytokines that may act in a autocrine or paracrine fashion (120–125). Epidermal growth factor (EGF) and transforming growth factor a (TGF-α) belong to the tubal secretions that are embryotrophic for the early developing blastocyst. EGF and TGF-α share considerable homology (126) and compete for binding to the epidermal growth factor receptor (EGFR), which has an extracellular ligand binding domain (127,128). They both have growth promoting actions in uterine tissues (122,129), and they also have been found in human tubal epithelium and early gestational embryonic membranes. Immunohistochemical investigations have demonstrated specific localization for EGF and TGF-α in the ampullary segment more than in the isthmus in the epithelium at late follicular, luteal and postpartum stages. In contrast, immunostaining in the follicular phase was weaker and was barely detectable in postmenopausal specimens (120). Interestingly, a similar pattern was observed in human specimens obtained 5 to 12 years after tubal ligation when compared to unligated samples from the same phase of the cycle (124). The ciliated and nonciliated tubal epithelial cells are the primary site of immunostaining of EGF, TGF-α, and EGFR, followed by stromal cells and to a lesser extent in smooth muscle cells, fibroblasts of serosal tissue and arterial endothelial and smooth muscle cells (124). The most intense immunostaining for TGF-α was predominantly in the apical regions of the mucosal epithelial cells, but was also evident to a lesser extent within the cytoplasm, nuclear envelope, the muscularis and serosa layers. Immunogold electron microscopy showed the receptor to be present predominantly in the basal regions of the epithelial cells, but also in the cilia, basal bodies (which control ciliary activity), endoplasmic reticulum, nuclear membranes and chromatin (120,121). In cumulus cells, intense staining for the EGF receptor appears in the cell membranes, in contact regions between the cumulus cells, leading to speculation of possible paracrine interactions between the oviduct and embryo via the EGF ligand-receptor. Overall, multiple studies have found that cycle-dependent expression of TGF-α, EGF, and EGFR, with high expression during late proliferative and early to mid secretory phases and significantly reduced expression during the postmenopausal period—highly suggesting hormonal regulation (123,124,130,131). To investigate a possible direct effect of estrogen on EGF, TGF-α, and EGFR, Adachi et al. investigated the tubal epithelium of menopausal women with and without estrogen replacement therapy. They found specific immunostaining for TGF-α, EGF, and EGFR in all subjects treated with estrogen, but no immunostaining in the untreated group, supporting the hypothesis that estrogen regulates these growth factors and their receptor (131).

In vitro studies with TGF-α and EGF have shown that they are potent stimulators of meiotic maturation in immature mouse oocytes (132,133). Co-culture experiments with 2-cell mouse embryos, cultured in the presence of human tubal epithelial cells, revealed a significant promotion of blastocyst formation. Furthermore, this effect was abolished in the presence of anti-TGF-α antibody (134). However in studies with later stages of mouse embryos, from the 4-cell stage to the blastocyst stage, EGF had an inhibitory influence on embryo development and spreading of the trophectoderm (135).

Cytokines and Effects of Progesterone Receptor Antagonist

The effects of mifepristone on cytokine expression in human oviduct was conducted by Li et al. (136). After administration of a single dose of 200 mg mifepristone on day LH+2, oviductal biopsies were taken on day LH+3 to LH+5 and investigated for tumor necrosis factor-α (TNF-α), transforming growth factor-β (TGF-β), leukemia inhibitory factor (LIF), and interleukin-8 (IL-8) expression. All cytokines were expressed in both stromal and epithelial cells, and the treatment did not affect the expression of TGF-β and LIF. However, the expression patterns of IL-8 and TGF-β were altered by antiprogestin treatment. IL-8, highly expressed in the ampullary region before treatment, was undetectable after treatment in this part of the tube, although it was increased in the isthmus. The expression of TGF-β in epithelial cells of the proximal oviduct was significantly increased after mifepristone treatment (136). TNF-α receptor is present in various developmental stages of human preimplantation embryos (137), and studies in mice have demonstrated that TNF-α binds to the inner cell mass (138) and reduces the average number of cells as well as the ability of the embryo to differentiate after implantation (139). It can thus be hypothesized, that increased expression of TGF-α in the oviduct could negatively influence embryo development, thereby suggesting a mechanism for mifepristone's contraceptive actions.

Angiogenic Growth Factors and Receptors

Several studies have addressed the issue of the highly orchestrated expression of angiogenic factors and their receptors in the maternal and/or embryonic compartments of the reproductive tract. However, very limited information is available about the presence and physiological actions of angiogenic factors and their receptors in the fallopian tube. One of the factors investigated is vascular endothelial growth factor (VEGF) and its two receptor (VEGF-R) tyrosine kinases, flt-1 (fms-like tyrosine kinase, VEGF-R1) and flk-1 or KDR (fetal liver kinase-1 or kinase domain region, VEGF-R2) (140,141). Flt-1 was shown to have a higher affinity for its VEGF ligand (142), compared to KDR. Human VEGF is a highly conserved homodimeric glycoprotein consisting of six isoforms with 121, 145, 165, 183, 189, and 206 amino acids, generated by alternative exon splicing of a single VEGF gene (143–145), with different heparin-binding properties that affect their solubility (146,147). VEGF promotes different actions throughout the human body. It is known to be a potent mitogen for vascular endothelial cells (147,148) and as a stimulator of vascular permeability (149,150). Both biological events play an important role in the chain of events that lead to angiogenesis, including during early ovarian follicle development or later in the reproductive cascade during the establishment of a connective vascular system between the decidua and the implanting embryo (151,152). However, VEGF is also abundantly found in human tissue that is not primarily involved in active angiogenesis, such as luminal organs like the digestive tract. It is thus considered to play an additional role in promoting secretory activity and increasing vascular permeability in epithelial cells of luminal organs (153,154). Therefore, it is teleologically reasonable that VEGF may influence oviductal fluid synthesis by oviductal epithelium, enhancing serum transudation of plasma-derived substances and proteins—one of the essential mechanisms for the formation of oviductal fluid (155,156).

Immunohistochemical studies of human fallopian tubes from regularly cycling parous women undergoing laparoscopic tubal ligation or hysterectomy for benign conditions revealed strong and specific expression of VEGF in the luminal epithelium, smooth muscle cells and pericytes lining small and large blood vessels within the oviduct (157,158). The cytoplasmic immunostaining was equally observed in all different parts of the oviduct and in all stages of the menstrual cycle.

Lam et al. (157) also investigated expression of VEGF mRNA throughout the oviduct, using semiquantitative reverse transcription polymerase chain reaction. The highest mRNA expression was observed during the periovulatory stage. Regarding the topological variation, the highest mRNA expression in the periovulatory stage was found in the ampullary and infundibular region compared to the isthmus (157). These findings support the theory that VEGF acts as a permeability enhancing factor that stimulates oviductal fluid production. Indeed, in several animal models oviductal secretions are temporally and spatially regulated, with the greatest fluid production noted in the pre-ovulatory phase, predominantly produced by the epithelium of the ampullary part of the oviduct (159,160).

Since the ampullary portion of the fallopian tube is where the fertilization of the blastocyst and early development of the embryo normally occur, VEGF-regulated transudation may be important in facilitating embryo transport to the uterus. The greatest fluid production in the pre-ovulatory phase may have a role in enhancing sperm transport during the periovulatory period.

To address the question whether VEGF expression is hormonally regulated in the oviduct (as it is in endometrium), serum samples of normally cycling subjects were investigated for FSH, LH, E_2, and progesterone concentrations. A positive correlation was found between VEGF mRNA expression and serum LH and FSH levels, peaking during the periovulatory stage. On the other hand, no correlation with the sex steroids was observed, as seen in endometrium (see The Oviduct). Therefore, VEGF mRNA expression in human oviduct is steroid-independent and the correlation with gonadotropins is more likely to be direct, rather than mediated by ovarian steroids. Furthermore, the independence from E_2 supports the notion that oviductal VEGF is predominantly involved in promoting permeability, rather than angiogenesis (157).

Studies on the VEGF receptors support additional evidence that KDR and flt-1 may promote different functions of VEGF. The KDR receptor appears to be involved in mitogenesis for endothelial cells (161,162); whereas, flt-1 is likely to be responsible for chemotaxis and permeability (163,164). Immunoreactive KDR and flt-1 have been found in the endothelial, epithelial and stromal cells of the oviduct (165). Although a similar distribution pattern for both receptors within the oviduct was demonstrated, a significant difference in their mRNA expression was observed. Although flt-1 mRNA expression was approximately the same in all parts of the oviduct, KDR mRNA was significantly higher in the ampullary and infundibular areas, compared to the isthmus. As observed with VEGF mRNA expression, steroid hormone independence was found. Peak expression of flt-1 is in the periovulatory stage and correlates with elevated serum FSH and LH concentrations (165). These findings additionally support the notion that VEGF acting through KDR is primarily responsible for angiogenesis occurring predominantly in the ampullary part of the oviduct, while its action mediated through flt-1 may increase oviductal secretion and permeability in the periovulatory phase. Increased angiogenesis may also enable delivery of macrophages for ultimate apoptosis of oviductal epithelia cells in the postovulatory period.

Oviductal Fluid and Oviductin

Oviductal fluid is a complex mixture of oviduct-specific secretory proteins, selective serum transudate and electrolytes (166–168). In humans, the oviductal epithelium synthesizes and subsequently secretes a rich glycocalyx, consisting of a range of different glycoproteins (169). Under the influence of periovulatory estrogen in the late follicular phase, an approximately 120–130 kDa oviduct-specific protein, oviductin (or oviductal glycoprotein [OGP]) is synthesized by the secretory cells of the human (65,170–172) baboon (155,173,174), and rhesus monkey (175) oviduct and secreted into the lumen (155,170). The cDNA sequence of OGP has been determined for a number of species including human (171), rhesus monkey (175), and baboons (173,176). The 3′ end of the human sequence exhibits 92% identity with the cDNA sequence of the baboon and additionally a 91% amino acid identity (176). This glycoprotein is synthesized and released by the secretory epithelium of all mammals, except rats (171) and mares (177). Oviductal glycoprotein (OGP) is found uniquely in the endoplasmic reticulum, Golgi apparatus, and putative secretory granules at apical regions of the secretory cells lining the oviductal lumen in the human (172,178) and baboon (57,174) and not in other parts of the oviduct or in the endometrium or cervix (170). Since there is a high percentage of conservation of the oviductin sequence throughout evolution and among different species, it may be that this glycoprotein plays a major role in prefertilization events during the reproductive process (179,180). Ovariectomy and consecutive hormone replacement therapy with estrogen restores expression of oviductin; whereas, addition of progesterone strongly antagonizes the E_2 effects, suppressing oviductin mRNA and protein expression.

Although a large body of evidence supports that the induction and expression of OGP is highly estrogen dependent (156,168,171,181), recent bovine studies indicate that LH and LH receptors have an additional effect (182). In vitro studies have demonstrated an increased half-life of oviductin when cells are treated with hCG, as a surrogate for LH (183). Furthermore, it has been postulated that the absence of E_2 and LH or the LH receptor in the standard oviduct mucosal cell culture system are responsible for the observed loss of oviductin expression in vitro (184). Another study by this group has revealed that hCG added to mucosal cells in culture results in increased oviductin expression (180). Although not particularly examined, it was suggested that LH/hCG may increase the transcription rate of the oviductin gene or mediate a decrease in the degradation of oviductin transcripts (183). When epithelial cells are derived from oviducts in the luteal phase, where no oviductin expression is detected initially, neither E_2 nor hCG can restore its expression. The strong correlation between serum E_2 concentrations and OGP and the inability of E_2 or hCG to stimulate OGP secretion in cultured cells suggests a block of the action by the hormone or that the cells have lost their responsiveness to E_2. Recent studies support the latter, because oviductal mucosal cells that lose polarity also lose the ability to respond to E_2 (54,180).

The expression of OGP mRNA in humans is highly correlated with serum E_2, luteinizing hormone (LH) and progesterone concentrations. Furthermore, since its expression is spatially regulated, being higher in the fimbrial-ampulla region than in the isthmus, functions for it, related to gametes and embryonic development have been proposed (184). Studies in humans and animal models have demonstrated that OGP binds to the zona pellucida and to sperm (168,172,185). It has also been detected in the perivitelline space of oocytes, in blastomeres and embryos of several species (174,186–189) except mouse, in which the association appears to be limited to the perivitelline space (190,191). However, it does not associate with oocytes directly collected from the ovary in different species (172,187,192–196). The association with the zona pellucida is likely due to specific binding sites for the glycoprotein in the zona pellucida, since other high molecular weight glycoproteins fail to compete in binding, even when vigorous washing steps and prolonged incubation times are applied (197). A recent study in the hamster and baboon, comparing the regional differences in OGP binding to the zona pellucida in vivo versus in vitro found that overall more binding sites for OGP are available in vivo than in vitro. Furthermore, the outer half of the zona binds significantly more OGP than the inner half in oviductal oocytes, whereas no regional differences have been observed with in vitro cultured oocytes. The functional significance is not yet known, but could be important for the process of sperm binding to the zona (198). Human OGP enhances human sperm binding to the blastocyst's outer surface of the zona pellucida by threefold in a hemizona assay, when included in the incubation media compared to the serum control (172). Furthermore, an antibody against the glycoprotein inhibits the enhancement of sperm binding. Although the homology of the OGP among different species and especially in primates is very high, substitution of heterologous baboon OGP for human OGP surprisingly results in preventing sperm binding to the zona pellucida, indicating the importance and selectivity of a homologous system (199). This could be a mechanism to prevent interspecies fertilization. In contrast to observations in the hamster, where OGP is bound in the acrosomal region of spermatozoa before and after capacitation (200), and in bovine sperm, where OGP is bound to the sperm head and midtail region (201), homologous binding of OGP on human spermatozoa has not been observed (172).

Contradictory results regarding OGP treatment of spermatozoa alone before fertilization have been reported in pig, cattle and bovine models. Preincubated bovine spermatozoa develop increased capacitation and fertilization rates (202) and pig spermatozoa show an increased penetration rate and polyspermy (203). However, another study in cattle did not support these findings (204). With regard to oocytes, positive effects of OGP have been observed. In pigs, for example, high penetration and fertilization rates are maintained and polyspermy reduced, comparable to the results of a combined pretreatment for both sperm and oocyte (203,205). Bovine oocytes, preincubated with OGP, result in a marked increase in fertilization, compared to controls (204). These findings suggest that the effects of OGP on fertilization are primarily mediated by interactions with the oocyte (178). Because oviductal glycoproteins within species have different molecular weights, different isoforms and, therefore, potentially different posttranslational modifications, further studies are necessary to determine whether OGPs associated with the zona pellucida, perivitelline and vitelline space or blastomere membrane consists of different OGP forms, each with a different target and activity (178).

Summary

The histological data strongly support a dynamic series of events at the cellular level that occur in the primate oviduct throughout the menstrual cycle. Epithelial cellular growth and proliferation, ciliogenesis and ciliation maximally occurring in the periovulatory period highly suggest a role for preparation of the oviduct for important epithelial cell functions during this period of time, including facilitating sperm entry and transport through the oviduct, fertilization, and perhaps facilitating early embryonic development. While select growth factors and cytokines and the oviduct-specific glycoprotein, oviductin, are likely candidates for these processes, precise knowledge about the molecular mechanisms operating within the oviduct have yet to be realized. Recent global gene profiling of mouse embryos at various stages of development (206) may give insight into the cross-talk between the embryo and its maternal conduit host. Expression of cholecystokinin and bradykinin, e.g., by the early developing mouse embryo suggests that these ligands may have receptors in the oviduct to facilitate their transport and/or development (206). A global approach to gene expression in oviductal epithelium across the menstrual cycle is an important complementary approach to establish the basis of molecular phenotyping, signaling pathways, and autocrine/paracrine/juxtacrine interactions occurring in the oviduct during the menstrual cycle and during gamete transport, fertilization, and early embryogenesis. It is likely that, in addition to destruction of the architecture of the epithelium in oviducts of women who have had reproductive tract infections and tubal infertility, factors that govern function are also severely impacted, leading to tubal dysfunction and ectopic pregnancy (207).

THE ENDOMETRIUM

Overview of Phases of the Menstrual Cycle and Endometrial Histology

The endometrium is the anatomic prerequisite for establishing and sustaining a pregnancy until delivery of the fetus. Nearly 100 years ago, Hitschmann and Adler first described histological changes in the endometrium (208). Over the following 50 years, endometrium from women and rhesus monkeys was studied in detail and revealed that the histological appearance of this tissue was directly correlated with changes in the ovary. Indeed, a central dogma in endometrial biology has been, and continues to be, that the ovarian hormones, estradiol and progesterone, control mitosis, differentiation, apoptosis, and function of the primate endometrium (209,210). This has lead to defining various histological stages of the functionalis layer of the endometrium, which is a dynamic tissue that responds on a cyclic basis to changes in the ovarian-derived steroid hormones, estradiol and progesterone (Fig. 3). Based on the analysis of nearly 8,000 endometrial biopsies, Noyes and colleagues (211) described different histological stages of the endometrium, based on an ideal 28-day cycle, with cycle day (CD) 1 being the first day of menstrual flow and CD 14 the day of ovulation. Today these phases (Fig. 3A) are best described as: the menstrual phase, the postmenstrual repair phase, the proliferative phase, interval (mid-cycle) endometrium, and the secretory phase. The proliferative phase can further be subdivided (Fig. 3A) into early (CD 4–7), mid (CD 8–10), and late (CD 11–14) stages, based chronologically during the cycle and the extent of cellular mitoses (see below). The secretory phase is also subdivided into early (CD 16–19), mid (CD 20–24), and late (CD 25–28) stages. During pregnancy, the endometrium has a special nomenclature designation as the "decidua" (Fig. 3B), and during hypoestrogenic states (such as the menopause), it eventually becomes atrophied. In normally cycling women and nonhuman primates, multiple events dynamically occur during the cycle.

In the proliferative phase (also known as the follicular phase, based on ovarian physiological changes) and under the influence of estradiol (E_2), cellular constituents of the endometrium undergo proliferation (and thus the name of the phase), as evidenced by high mitotic indices of these cells (212) and increasing

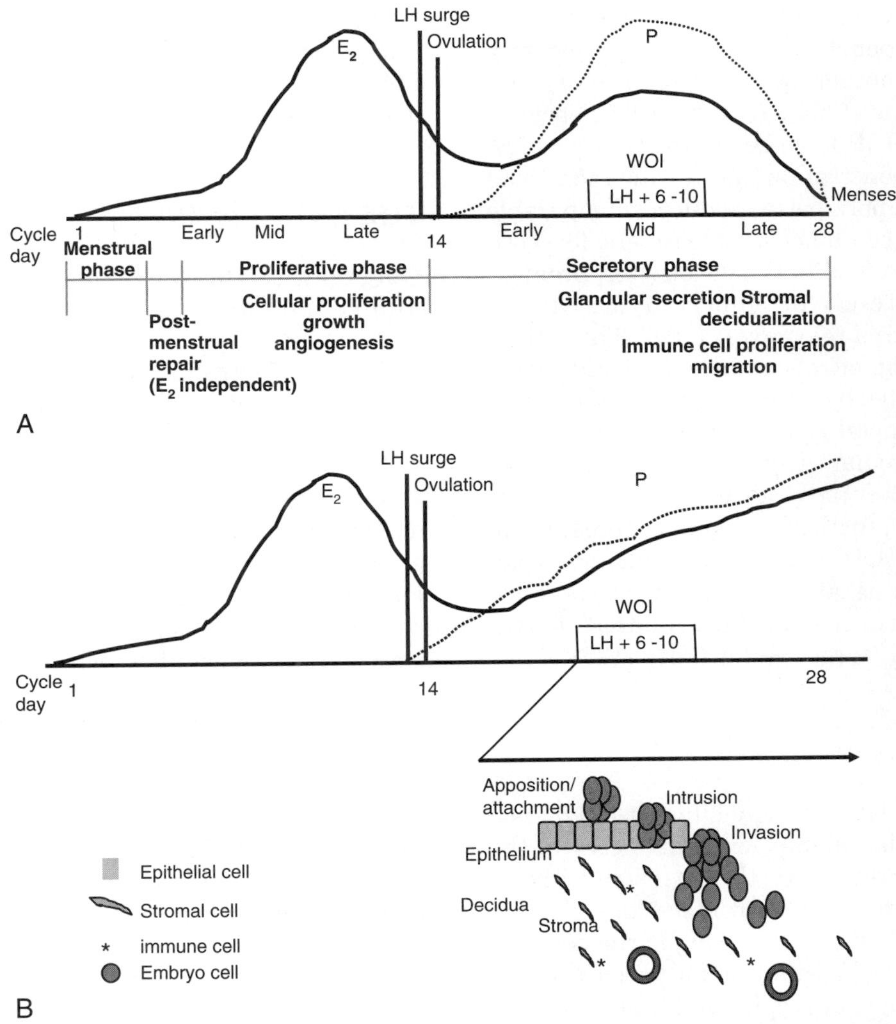

FIG. 3. Stages of the menstrual cycle and events in early pregnancy. **A:** Different stages of the menstrual cycle, with day 1 of menses taken as cycle day (CD) 1, ovulation on CD 14, and the end of the cycle as CD 28. **B:** Stages of the menstrual cycle and hormonal changes during a conception cycle. Relative levels of circulating estradiol (E2) and progesterone (P) are shown. WOI, window of implantation. (See text.)

height of the tissue (see below). DNA synthesis is increased, and mitoses are numerous in the epithelium, stroma, and vascular endothelium in the late proliferative phase (increased markedly over the mid-proliferative phase and nearly absent in the early proliferative phase) (Fig. 4), with the straight glands becoming more voluminous and tortuous as the phase proceeds reviewed in (213) (Fig. 4). This proliferation results in the development of a glandular network, an elaborate system of blood vessels, and stromal cell proliferation. Histological analysis reveals growth of the endometrium (Fig. 4), with an increase in thickness from about 2 mm in the postmenstrual repair phase to about 14 mm just before ovulation. Inappropriate growth of the endometrium can be associated with infertility, miscarriage, and pregnancy disorders (10,214).

After ovulation, in the secretory phase (also called the luteal phase, based on ovarian physiological changes), and under the influence of progesterone (P) (with prior E_2 priming and continued, relatively high circulating levels of E_2), glandular secretion and stromal cell differentiation (decidualization) occur. During the first 36–48 hours after ovulation the morphology of the endometrium is similar to the late proliferative phase (7,211). Thereafter, the effects of progesterone become evident, and endometrial histological hallmarks of ovulation and progesterone include (Fig. 4; Table 1) the following. On CD 16 glycogen begins to accumulate in the basalis region of the glandular epithelium, and the nuclei begin to adopt the appearance of being displaced and "pushed" upward by the glycogen granules. By CD 17 there are well-developed subnuclear glycogen "vacuoles" in

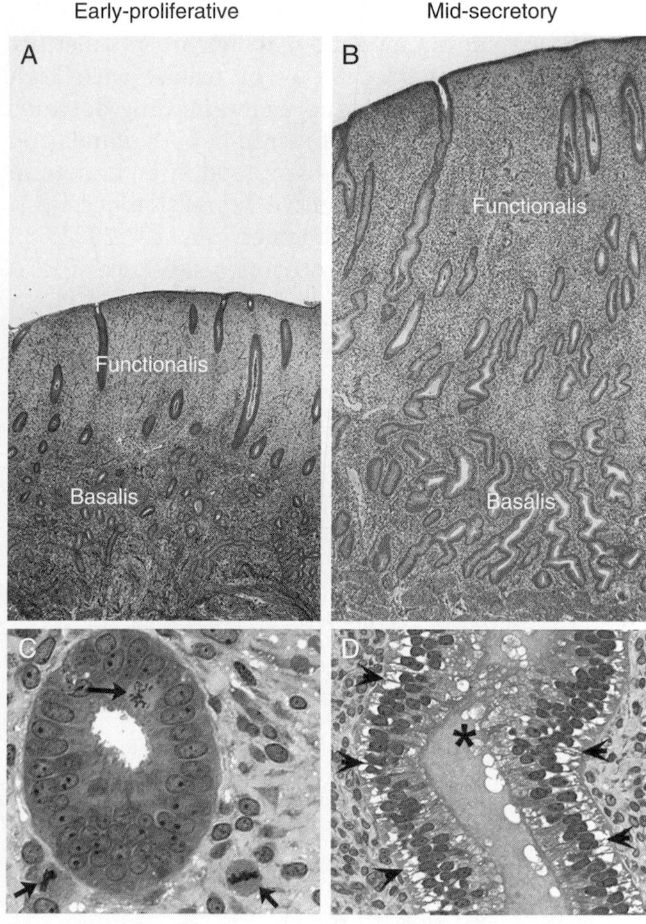

FIG. 4. Photomicrographs of GMA embedded hematoxylin-stained sections of rhesus monkey endometrium. The *upper panel* shows the functionalis and basalis zones in full-thickness endometrium during the early-proliferative (**A**) and mid-secretory (**B**) phases. The *lower panel* shows mitosis in glandular and stromal cells in the early-proliferative phase (**C**, *arrows*), and subnuclear vacuoles (**D**, *arrowheads*) and secretions in the lumen of glands (**D**, *asterisk*) during the mid-secretory phase. Note the marked increase in endometrial thickness, glandular sacculation, and stromal edema during the secretory phase (compare B with A).

TABLE 1. *Ultrastructural characteristics of secretory endometrium*

Cycle day	Process
17	Glandular cells:
	Subnuclear glycogen vacuoles and "palisading" of nuclei
	Giant mitochondria
	Nucleolar channel system (NCS)
19, 20	Glandular cells:
	Glandular secretions
	Cessation of DNA synthesis and mitoses
20–23	Stromal edema
22–25	Coiling of spiral arterioles
	Predecidualization of the stroma
24–25	Influx of leukocytes
26–28	Influx of PMNs

glandular cells and a "palisading" of gland cell nuclei (Fig. 4) (213). The "vacuoles" are pools of glycogen granules which, when the tissue is fixed in formalin, are solubilized, leaving morphological "vacuoles." The endometrium is unique in that the accumulation and synthesis occur in the absence of excessive glycogen intake (213). At the ultrastructural level, ovulation is heralded by an increase in mitochondrial size and an increased number of cristae (in response to an increased demand for energy for glycogen metabolism (215), as well as the appearance of the nucleolar channel system (NCS). The latter is only found in gland cells and represents an infolding of nuclear membranes into the nuclear or nucleolar substance. It is unique to women and can be elicited in response to administration of exogenous progesterone or progestins (7,213).

On days 19 and 20, few vacuoles remain, and the glycoprotein-rich supranuclear cytoplasmic products are expelled by apocrine-type secretion into the glandular lumen (Fig. 4). This uterine secretory fluid also contains plasma transudates derived from blood vessels in the endometrium. The nuclei now are near the basal portion of the epithelial cells, and the lumen is filled with proteinaceous material. The peak of glandular secretions coincides with the time of implantation of the blastocyst and cessation of DNA synthesis and cell division in glandular cells (7,213). The endometrium becomes receptive to implantation of a blastocyst only during a defined "window" that is temporally and spatially restricted, limited to 6–10 days after the LH surge (Fig. 3B) (see below) (216).

Beginning on day 20 and thereafter in the cycle, changes in the stroma become prominent, compared to in the glands prior to this day of the secretory phase. Stromal changes include increased capillary permeability, stromal edema, stromal mitoses, endothelial cell proliferation and coiling of the spiral arterioles, especially in the upper functionalis (211,213). Maximal capillary permeability and stromal edema occur on cycle day 22 and are prerequisites for predecidual transformation of endometrial stromal fibroblasts (7,211). Predecidualization refers to a unique morphological and secretory phenotype that results in conversion of spindle-shaped stromal cells into plump epithelial-like cells with enlarged nuclei and increased cytoplasm (217,218). Classical secretory products of predecidualized endometrial stromal cells include prolactin and IGF binding protein 1 (IGFBP-1) (219,220) (and see below). Predecidualized cells have ultrastructural and immunohistochemical features of epithelial and mesenchymal cells, with characteristics of the former being a laminin-rich basement membrane and intercellular gap–junction nexuses and of the latter, immunohistochemical staining for mesenchymal cell antigens, such as vimentin and desmin, but not epithelial membrane antigens or cytokeratins (7). Stromal mitoses are numerous on CD 24, and stromal predecidualization first occurs as "vascular cuffing," around the spiral arterioles and capillaries of the functionalis, then subepithelially, and by day 27 most of the stroma in the functionalis layer appears as a solid sheet of predecdualized cells. On CD 24–25, there is a marked increase in lymphocyte infiltration, and by CD 26–28 there is a marked influx of extravasated polymorphonuclear leukocytes (PMNs) (211,221). These cells are believed to function in protecting the foreign fetal allograft from rejection by the maternal immune system (see below), and in the absence of implantation may contribute to local tissue destruction. The histological changes in the secretory phase occur in humans in the absence of implantation of a blastocyst. Similar events occur in nonhuman primate endometrium, except stromal decidualization is not as marked in the rhesus monkey as in human endometrium in the absence of a conceptus, and in the baboon, decidualization begins only at implantation (4).

In the absence of implantation the endometrium undergoes cellular apoptosis, vascular basement membrane breakdown, tissue bleeding, vasospasm, and tissue desquamation, resulting in the menses. After three to four days of bleeding, the endometrium undergoes regeneration and healing, during the postmenstrual repair phase. The latter is independent of steroid hormones and is believed to occur as the result of a population of stem cells and/or their progenitors, derived from the deepest portions of the endometrium (see below).

Steroid Hormone Receptors: Expression and Actions in Endometrium

Steroid hormone receptors [estrogen (ER), progesterone (PR), and androgen (AR)] are expressed in human endometrium and are cyclically regulated in the functionalis layer (Fig. 4A). The ER and PR in the basalis are not steroid hormone- or cycle phase-dependent (222–224). Steroid hormone receptors are ligand-activated nuclear transcription factors.

Estrogen Receptors

Estradiol stimulates ER expression, and ER levels are highest during the proliferative phase (Fig. 4) (31,225–229). They decrease after ovulation, reflecting the suppressive effects of progesterone on the ER. Immunostaining has revealed that the ER is present in highest amounts in the glandular epithelium during the proliferative phase (226). Two forms of ER, ERα, and ERβ, are now appreciated, which are two distinct gene products encoded by eight exons, with high homology in their DNA and ligand-binding domains (105). ERa and ERβ are expressed in both glands and stroma, while ERβ only is expressed in endothelium (Fig. 5) (224,230,231). Relative expression of ERβ is low, compared to ERα, in endometrium (105,232,233), and ERβ does not compensate for the absence of ERα

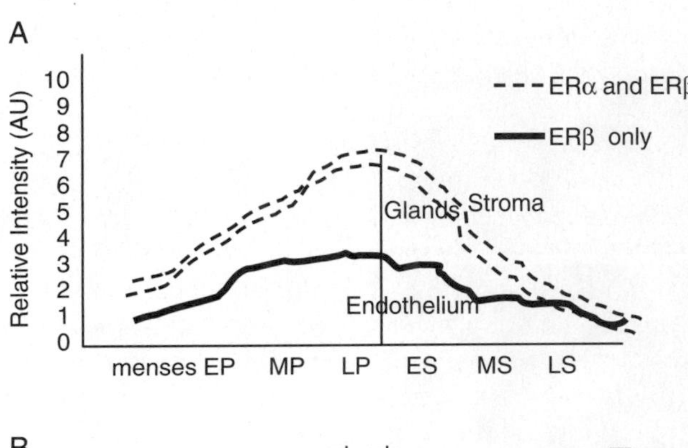

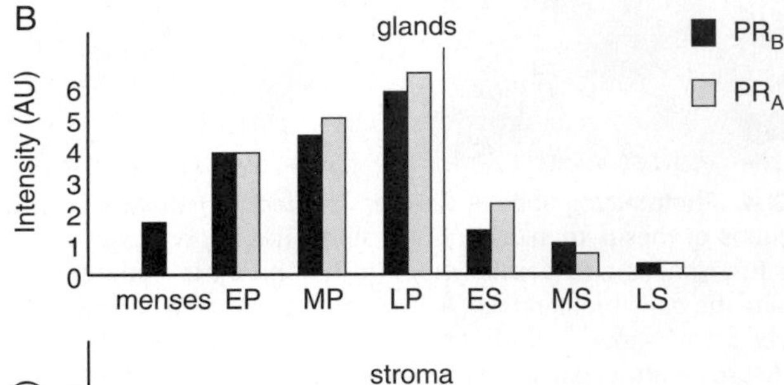

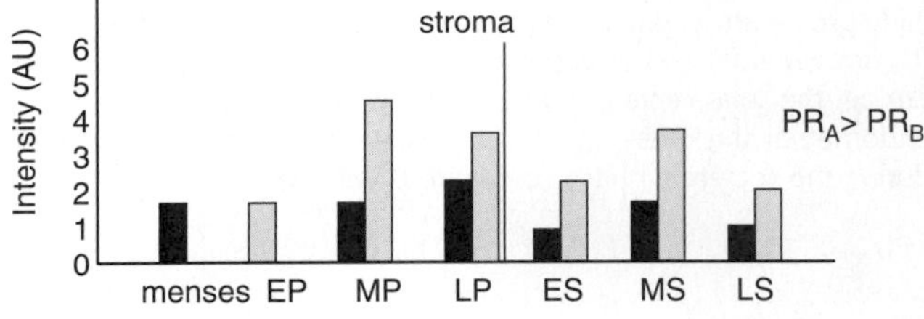

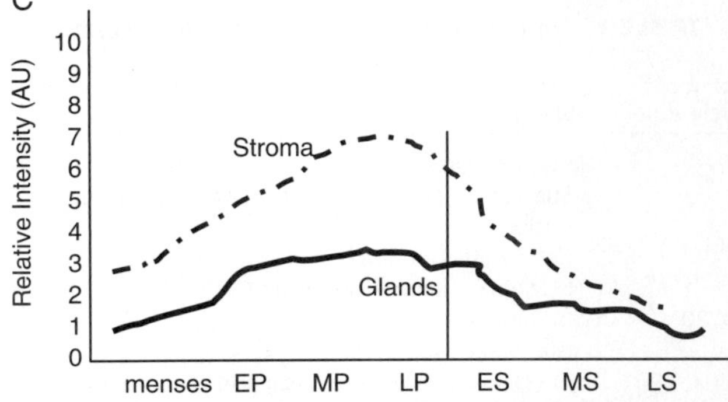

FIG. 5. Relative expression of steroid hormone receptors in human endometrium during the menstrual cycle. **A:** Estrogen receptors: ERα and ERβ. **B:** Progesterone receptors: PR-A and PR-B. [From Mote et al. (96), with permission.] **C:** Androgen receptors. E, early; M, mid; L, late; P, proliferative; S, secretory.

in the uterus of the ERα null mouse (234,235), suggesting that ERα is the predominant ER in endometrium. However, the finding that ERβ only is expressed in endothelium suggests that E_2 acts directly on endometrial blood vessels via this receptor isoform. ERα down-regulation in epithelium in the secretory phase is profound (Fig. 5). Indeed, the mid-secretory phase is characterized by a loss of nuclear immunostaining for ERα (and PR), a universal response in all mammalian species (236). In the initial menstrual phase, there is no detectable immunostaining for ERα (or PR) in the glandular epithelium, although both receptors continue to be expressed in stromal cells. During regeneration, ERα (and PR) are localized to all cell types throughout the endometrium (236).

E_2 action in the endometrium is direct, through its cognate receptors, and indirect, by induction of growth factors in either its target cells or by inducing growth factors as paracrine modulators (16,237,238) (and see below). E_2 is necessary for endometrial growth and for progesterone action. With regard to cell cycle regulation, E_2 complexed with ER leads to up-regulation of cell cycle proteins. In the endometrium in the proliferative phase, an increase in ERα expression is correlated with increased expression of cyclins D1, E, A, B1, and the cyclin-dependent kinases cdk4, cdk2, and cdc2 (239). The transcription factors c-fos and c-jun are mediators of E2 action, and both are more highly expressed in the proliferative phase, compared to the secretory phase (240). In addition, there is a correlation between ER and c-jun and c-fos transcription factors in stroma and epithelia, and PR expression correlates with c-jun in epithelia (240).

Progesterone Receptors

Progesterone binding to PR results in changes in gene expression, and it can also have nongenomic effects (241). Early studies demonstrated peak PR concentrations in human endometrium at the time of ovulation, reflecting induction of the PR by E_2 (225,226,229,233). PR is most prominent in glandular epithelium in the proliferative phase, but undetectable in the mid-secretory phase (226). Furthermore, stromal cells were demonstrated to have high levels of PR in the proliferative phase and throughout the secretory phase. Similar observations were made in nonhuman primate endometrium (31,227,228,242). Most of these studies were conducted before it was known that the human PR is expressed as PRA and PRB isoforms which are two functionally different transcription factors encoded by a single gene which differ functionally (241). While P action on target genes is conferred primarily by PRB homodimers, truncated PRA acts as a repressor of PRB function. By immunohistochemistry, both isoforms have been detected throughout the mid- to late proliferative phase of the menstrual cycle (Fig. 5). In endometrial glands PRA and PRB are expressed prior to subnuclear vacuole formation and glycogenolysis in the secretory phase (95,96,233,243), suggesting a role for both isoforms in these processes. In contrast, only PRB persists later, during the mid-secretory phase, suggesting a role for this isoform in glandular secretion. In endometrial stroma, PRA predominates throughout the cycle, suggesting that it is important in progesterone action on this cell type postovulation in the secretory phase (95,96,233,243). Overall, these results support the view that PRA and PRB mediate distinct pathways of progesterone action in the glandular epithelium and stroma throughout the menstrual cycle. It should be noted that the timely down-regulation of epithelial PR coincides with the opening of the window of implantation and uterine receptivity for embryonic implantation (vide infra) (244), and histological delay of the endometrium (a clinically abnormal state) is associated with a failure of such PR down-regulation.

Androgen Receptor

The androgen receptor (AR) is also expressed in human endometrium and follows similar regulation as the ERs and PRs (245,246) (Fig. 5). The functions of the AR remain unclear, although proliferation is likely to be mediated, in part, by this receptor. This is a clinically relevant issue in hyperandrogenic states, such as polycystic ovarian syndrome, in which there is a higher risk of endometrial hyperplasia in women with this disorder who have elevated circulating androgen (and insulin) levels and who have unique endometrial functioning (247).

Steroid Hormone Actions on Endometrial Cells

Lessons Learned from ERKO and PRKO Mice. Insight into the roles of E_2 and progesterone in reproductive processes has been greatly facilitated by the estrogen and progesterone receptor knockout mice (ERKO and PRKO, respectively). For example, studies in the ERα null (–/–) mouse demonstrate a lack of epithelial proliferation in response to estrogen and a lack of induction of mRNAs of classical estrogen-responsive genes, including lactorferrin, the progesterone receptor, and glucose-5-phosphate dehydrogenase (234,248). The PRKO mouse displays failure to ovulate (lack of follicular rupture), hyperplastic uteri, failure to respond to the decidual stimulus, and infertility (100,249). E_2 treatment of the PRAKO mouse results in endometrial epithelial hyperplasia, and progesterone treatment had no subsequent effect, suggesting that PR-A mediates progesterone inhibition of estradiol stimulated endometrial epithelial growth. In addition, in the PRAKO/PRBKO double knockout mouse, there is an influx of inflammatory leukocytes into the endometrium, and progesterone has no effect, demonstrating that *both* PRs inhibit influx of inflammatory leukocytes into the endometrium in mice (99,100).

The use of the progesterone receptor antagonist, mefipristone, has identified further progesterone-regulated events in endometrium. Mefipristone given early in the secretory phase in humans prevents progesterone-induced down-regulation of PR and ER (250,251). E_2 and progesterone action on rhesus monkey endometrium was studied by Slayden et al. (252) using a progesterone receptor antagonist. They found that the antagonist inhibited progesterone action, allowing estradiol to act in a normal fashion, and that the antagonist itself had anti-proliferative effects that opposed E_2 actions (252). In the mouse, after mefipristone is administered, global gene profiling of the uterus has revealed a multitude of up- and down-regulated transcription factor, cytoskeleton, extracellular matrix, signal transduction, growth factor, receptor, immune, and angiogenic families of genes, among others (253), which may give insight into progesterone-regulated genes in other mammalian species.

Epithelial–Stromal Interactions. It has classically been considered that the action of estradiol and progesterone on cells in the endometrium have been mediated by their respective receptors and that their effects are direct or the result of induction of growth factors. This is indeed true for many actions of E_2 and progesterone on endometrial cells. However, the absence of ER and PR in mid-secretory epithelium, e.g., raises the question of how these steroid hormones could effect the glandular changes that are so dramatic during this phase of the cycle and whether there is communication between the stroma and epithelium to achieve these effects. Using a unique recombination system with ERKO, PRKO, and wild type mice, Cooke and colleagues conducted an elegant set of experiments to study the mechanism of E_2 and progesterone action in female genital tract epithelia and stroma (23,24,238,254–256). This system involves enzymatically separating and recombining stroma and epithelium of uteri from these mice and is summarized in detail in Cooke et al., 2002 (24). They found that E_2 induces epithelial proliferation in the uterus *indirectly* through stromal ERa (238), that E_2 induction of epithelial secretory protein production in the endometrium requires both stromal and epithelial ERα, and that stromal ERα mediates the inhibitory effects of E_2 on uterine epithelial PR expression. In addition, they found that stromal PRs mediate the inhibitory effects of progesterone on estrogen-induced endometrial epithelial proliferation and epithelial PR expression (24,257).

Molecular Mechanisms Underlying Endometrial Function

Proliferation/Differentiation/ Epithelial–Stromal Cross-Talk

In pre-ovulatory endometrium there is morphological and biochemical evidence of proliferation of gland, stromal, and vascular endothelial cells. DNA synthesis is increased and mitoses are numerous in the late proliferative phase (213). Peak DNA synthesis is observed on cycle days 8–10 and corresponds to maximal mitoses, peak plasma estradiol levels and peak ERα and β levels. Several growth factor families have been identified in endometrium and have been postulated to be mediators of E_2-stimulated proliferation of cellular components. Extensive reviews on these growth factors in nonprimate models and mechanisms of action of these factors in uterine growth, epithelial, vascular endothelial and stromal proliferation have been reviewed by Tong and Pollard (258) and Cooke et al. (24). Some of these are summarized below.

Insulin-like Growth Factor-I. IGF-I is a member of the IGF family that is comprised of IGF-I, IGF-II, the type I IGF receptor (a tyrosine kinase signaling receptor), the type II IGF receptor (same as the mannose-6-phosphate receptor and important in IGF-II ligand turnover), and a family of six high affinity binding proteins (IGFBP-1 through -6) that mostly inhibit IGF actions (219). IGF-I and -II are mitogeneic to endometrial stromal and glandular epithelial cells in culture (259,260), acting via the type I receptor. In rat uterus, IGF-I expression is regulated by E_2 (261). In human endometrium, IGF-I is expressed primarily in proliferative and early secretory endometrium (262–264). IGF receptors are expressed in human endometrium, with greatest expression in the glands compared to stroma (265) and endometrial total IGF binding capacity to IGFBPs and IGF receptors is cycle- and steroid-hormone dependent (266,267). E_2 is a primary regulator of IGF-I mRNA and protein in the rodent uterus (268–271). In the IGF-I knockout mouse, administration of E_2 did not result in epithelial mitogenic response (272), mainly due to the inability of the epithelial cells to transit through G2 phase of the cell cycle. Richards and colleagues have found that E_2 induces phosphorylation of the type I IGF receptor and insulin receptor substrate-1 (273). In addition, E2 and progesterone interact and regulate insulin receptor signaling through IGF-I (274). In a transgenic mouse model in which IGFBP-1 is overexpressed in the stroma, there is marked glandular atrophy even in the face of estradiol administration (275), and the glandular epithelium is atrophic in women who use a progestin-containing intrauterine device (which amply stimulates IGFBP-1 in the stroma) (276). The transgenic study and the clinical observation both support inhibition of IGF-I action on the epithelium by stromal-derived IGFBP-1, even in the face of normal circulating E_2 levels, further supporting the hypothesis that IGF-I may mediate the mitotic actions of E_2 in this tissue. However, investigation into uterine growth and estrogen responsiveness in IGF-I null (–/–) and wild type mice has revealed that when IGF-I local expression is absent in the uterus, the organ grows normally when exposed to normal circulating IGF-I levels in its nude mouse host (277,278). The conclusion from this set of experiments is that IGF-I may be permissive for uterine growth and estrogen responsiveness, rather than a direct mediator of E_2's action (277,278).

IGF-II has been postulated to be a mediator of progesterone action in human endometrium, due to the differential expression of this member of the IGF family in stroma in the secretory phase of the cycle (263,265). Further studies are needed to elucidate the role(s) of IGF-II in endometrial function and a potential progestomedin (279).

Epidermal Growth Factor. EGF is a member of the TGF-β family, acting through its cognate EGF receptor. EGF is expressed in human endometrium and peaks in glandular epithelium during the late proliferative phase and increasing in stromal expression in the secretory phase (280–282). The EGF receptor in expressed in stromal and epithelial cells, although EGF does not induce proliferation of stromal cells, per se (282). The relationship between EGF and estradiol actions underscores, once again, the complexity of epithelial and stromal interactions. EGF induces endometrial epithelial growth and stimulates estrogen responsive genes (e.g., lactoferrin) (122), an effect that is blocked by the pure anti-estrogen, ICI 161384 and by anti-EGF antibodies (283). EGF requires the presence of ERα, as EGF treatment of αERKO mice does not induce DNA synthesis as with wild type mice (283), suggesting that EGF may use the ER and EREs in estrogen regulated genes to effect its mitogenic actions in the uterus. That EGF can phosphorylate and activate ER has furthered our understanding of this complex cross-talk between EGF and E_2 signaling in the endometrium (284). The EGF receptor is also in epithelial cells, which could allow for estradiol action on stroma and subsequent EGF action on the epithelium. In the EGFR-KO uterine transplant model to a nude host (EGFR KO is embryonic lethal), E_2 did not stimulate uterine stromal DNA synthesis and proliferation, although the endometrial epithelium responded equally well as the wild type. The data support the model that EGF receptor signaling is important for stroma, although not for estrogen-induced epithelial growth in the uterus (285).

Another EGFR ligand, HB-EGF, may also play a role in mediating mitogenic effects of E_2 on uterine epithelium and inhibition of this process by progesterone (286). Whether TGF-α, another EGFR ligand, is important in estrogen signaling remains uncertain at this time [for review see Cooke et al. (24)].

Fibroblast Growth Factor. Fibroblast growth factors comprise a family of nine members that signal through a group of homologous receptors. FGF-1, -2, -4, and -7 are expressed in human endometrium (287–289). FGF-2 mRNA and immunoreactive protein are most highly expressed in the proliferative phase (288). The FGF-2 receptor is expressed in stromal cells, and DNA synthesis is stimulated by FGF (288). With regard to FGF-7 [keratinocyte growth factor (KGF)], it is expressed in monkey and mouse uterus. In the former, it is stimulated by progesterone, and was suggested to be a mediator of progesterone action on epithelium (290). However, when administered during the luteal-follicular transition (when progesterone levels are declining), KGF had no effect on glandular cell proliferation in the endometrium of either juvenile or adult rhesus macaques (291). Interestingly, KGF had unanticipated effects on endometrium: strongly inhibiting apoptosis and increasing glandular sacculation in the basalis zone and being trophic to the spiral arteries (292). For a complete review of KGF in the macaque, the reader is referred to (293).

Hepatocyte Growth Factor and Others. Hepatocyte growth factor (HGF), also known as scatter factor, is a heparin-binding cytokine produced by mesenchymal cells in several tissue types. Its receptor, c-met, is expressed in epithelial and endothelial cells. Consistent with other systems, HGF is strongly expressed in stromal cells during the proliferative stage of baboon endometrium, and the expression is very low or undetectable in the secretory phase (219). However, c-met is constitutively expressed in glandular epithelium, luminal epithelium, and vascular endothelium throughout the menstrual cycle (219). HGF stimulates epithelial mitogenesis in cultured human endometrial epithelium (294), although a role for it in human endometrium during the cycle is not clear. It has been reported to be a morphogen for endometrial epithelial gland formation in vitro (295), although it may have other roles, as well in vivo. Many other growth factors and cytokines are differentially expressed in human endometrium, and the reader is referred to sections below and a recent review (296).

Angiogenesis

Endometrial Vasculature. The supply of oxygen and nutrients is an important aspect of the normal growth and differentiation of the endometrium which is achieved by the regulation of endometrial vasculature. Using an anti-angiogenic agent in mice, Klauber et al. (297) first demonstrated the critical role of endometrial vasculature in regulating endometrial development and differentiation. The endometrial blood vessels are unique, and unlike most other organs, they undergo dynamic changes during each menstrual cycle. Numerous recent studies on the mechanisms of angiogenesis, vessel growth and regression, coupled with development of different strategies for modulation of vessel growth, have opened new possibilities for treatment of a number of gynecological disorders involving the uterine vasculature.

The overall morphology and architecture of the endometrial vasculature has been well described by several elegant studies in both human and nonhuman primates (298–300). Understanding of the human endometrial vascular development and differentiation during the menstrual cycle has been particularly facilitated by the similarity of these processes among human, rhesus monkeys, and other primates (299,300). The uterine blood supply is primarily derived through the uterine and the iliac arteries. At about the middle third of the myometrium, the uterine artery branches and forms the arcuate arteries. The radial arteries arise at about right angles from the arcuate arteries in the myometrium and pass toward the uterine lumen. At the myometrial–endometrial junction (Fig. 1B) the radial arteries give off small branches, the basal arteries, which supply the basal endometrial zone, and continue towards the lumen as spiral arteries. Each spiral arteriole pursues a convoluted (coiled) course through the functional layer and supply blood to the upper endometrial zones. The coiling becomes extensive during the secretory phase of the menstrual cycle. The spiral arteries have basic structures similar to the parent radial and arcuate arteries, but the basal arteries are essentially muscular with little elastic and fibrous tissues. Capillaries from the spiral arterioles provide blood supply to the stroma and form an extensive network just below the luminal epithelium, called subepithelial capillary plexus. A large number of venules drains blood from the subepithelial plexus, and several smaller veins coalesce to form sinuses. From these sinuses, larger veins converge as they pass the endometrium into the myometrium and enter the uterine veins bilaterally. The vessels in the basal zone mostly remain unchanged, but the vessels in the functional zones of endometrium change dramatically throughout the menstrual cycle in response to circulating ovarian steroids (298,299,301).

Angiogenesis and Vascular Growth during the Menstrual Cycle. Angiogenesis is the formation of new blood vessels from preexisting vessels, and is a rare event in most adult organs under normal physiological conditions. However, it is a critical component of the endometrial regeneration during each menstrual cycle (302–304). Following menstruation, the deeper endometrial zones contain ruptured arterioles and venules. After day 5 of menses, the blood vessels have healed and the newly formed surface epithelium has spread and covered the ragged surface. The arterial stumps give rise to capillary sprouts which proliferate and grow towards the surface. As the endometrial thickness increases during the proliferative phase, there is a rapid growth of vasculature that supplies the upper layer of the regenerating endometrium. Although, there is considerable variability among reports on the pattern of human endometrial endothelial cell proliferation (302), studies in the human (212,305) and rhesus monkey (301) show a rapid burst of vascular proliferation in the mid-proliferative phase of the cycle which coincides with increased proliferation and growth of endometrial stroma and glands. Ferenczy et al. (212) have reported a second wave of endothelial cell proliferation during the mid-luteal phase of the cycle which is not evident in the rhesus macaque (301). It has been suggested that the differences among reports on human endometrial vascular proliferation may be due to variations in hormone levels at the time of endometrial sampling or to variations in the region biopsies (306). However, all studies are in agreement that most vascular proliferation occurs in the upper zones, not in the lower zones, of the primate endometrium (212,301,305).

The newly formed vessels gradually acquire a coat of vascular smooth muscle cells (VSMC) and elastic tissues around them converting them in to arterioles and arteries. The origin of endometrial VSMC and the mechanism of their recruitment by endothelial cells are currently unknown. Generally, it is believed that VSMC may differentiate from mesenchymal cells including stromal cells, or may transdifferentiate from endothelial cells or from bone marrow precursors and macrophages (307). After recruitment, the VSMCs proliferate and migrate along the capillaries and convert them into arterioles. Several factors including vascular endothelial growth factor (VEGF), angiopoietins, transforming growth factor beta (TGF-β), and platelet-derived growth factor (PDGF), have been implicated in these processes (307). Proliferation of the VSMC around the spiral arterioles significantly increases during the mid- to late-secretory phase (308). As they grow, the spiral arteries become highly convoluted in the secretory phase (299,308,309). In the absence of implantation, the corpus luteum regresses and the subsequent fall in progesterone leads to an intense vasoconstriction of the spiral arteries that precedes the bleeding and sloughing of the upper endometrial zones during menstruation (309).

Regulation of Endometrial Vessels. The sole ovarian steroid hormones required to induce the vascular growth and differentiation during the menstrual cycle are estrogen and progesterone. However, the basic mechanisms involved in regulating these processes are not clear. Some recent studies suggest expression of estrogen, particularly ERβ, and progesterone receptors in the vascular endothelium of primate endometrium (230,310). Numerous studies show both genomic and nongenomic effects of estrogen on endothelial cells in different systems (311), and there is emerging evidence that estrogen and progesterone can act on the vascular endothelium via their cognate endothelial receptors and/or by nongenomic mechanisms. However, the pattern of endothelial cell proliferation during the menstrual cycle in the rhesus macaque (301) and the human endometrium (305) do not suggest a direct role of estrogen and/or progesterone in this process.

Several angiogenic factors are expressed in different cell types of human endometrium and its resident population of macrophages and are presumed to play significant roles in mediating the action of estrogen and progesterone on endometrial vasculature. These factors include vascular endothelial growth factors (VEGF) (312), angiopoietins (313), platelet-derived growth factor (PDGF) (314), fibroblast growth factors (FGFs) (287), epidermal growth factor (EGF) (122), transforming growth factor-α (TGF-α) (315), transforming growth factor-β (TGF-β) (316), tumor necrosis factor-α (TNF-α) (317), acidic and basic fibroblast growth factor (316,318), interleukin (IL)-1, IL-6 (319), and IL-8 (320).

To date, most of the studies on endometrial angiogenic growth factors are focused on VEGF, a prime regulator of both physiological and pathological angiogenesis. VEGF is expressed in a wide variety of cells and tissues, including the endometrium (304,321). Different isoforms of VEGF are generated by alternative splicing from a single VEGF gene containing eight exons, separated by seven introns, resulting in the generation of four different molecular species having respectively 121, 165, 189, and 206 amino acids (321). Another variant having 145 amino acids has been reported in the human endometrium (312). $VEGF_{165}$ and $VEGF_{121}$ are the most abundant isoforms expressed in the human endometrium (322,323), and a recent study shows transient upregulation of $VEGF_{189}$ during the secretory phase (324). VEGF activity is mainly mediated by two high affinity tyrosine kinase receptors, VEGF receptor type 1 (fms-like tyrosine kinase receptor; Flt-1) and VEGF receptor type 2 (kinase insert domain-containing receptor; KDR/Flk-1) (321). These receptors are primarily expressed in the endometrial vascular endothelium throughout the cycle (304,325). Other receptors of VEGF include VEGFR-3 (Flt-4), expressed in endothelial lining of lymphatic vessels, and neuropilins, expressed in both endothelial and nonendothelial cells (326).

Numerous studies show cyclic changes in VEGF mRNA and protein expression in both human and nonhuman primate endometrium during the menstrual cycle, and suggest a role of VEGF in endometrial angiogenesis (304). Estrogen regulates VEGF mRNA expression in human endometrial adenocarcinoma cells (312), stromal cells (327,328), and in the rhesus and baboon endometrium in vivo (301,329). These are likely to be an estrogen receptor-mediated effects because the induction peaks within 2 hours of estradiol treatment (330,331), and is blocked by pure antiestrogens (332). Recently, it has been demonstrated that estradiol regulates VEGF gene transcription in endometrial cells through a variant estrogen response element located 1.5 kb upstream from the transcriptional start site (332). Although a progesterone response element has not yet been identified in the VEGF gene, several reports indicate that progestins alone or in combination with estrogens can stimulate VEGF expression in human endometrial stromal cells in vitro (327,328). However, most in vivo studies show an increase in glandular VEGF expression during the secretory and menstrual phases, while the stromal VEGF expression is downregulated in the secretory phase (301,312). These studies together do not depict a clear picture of how hormones regulate VEGF in the endometrium. It is likely that cell-specific regulatory factors, rather than the gene promoter, play critical roles in the regulation of VEGF expression in different cell types of endometrium. For example, hypoxia can stimulate several-fold increase in VEGF expression in endometrial gland cells compared to stromal cells (333) and upregulates VEGR expression in decidualized human endometrial stromal cells in culture (334).

Several other VEGF family growth factors have been identified in the endometrium that interact with the VEGF receptors (302,304). These include placental growth factor (PlGF), VEGF-B, VEGF-C, and VEGF-D. PlGF forms heterodimers with VEGF-A, and it binds to neuropilins and VEGFR-1, but not VEGFR-2 (335,336). VEGF-B binds to neuropilin-1, while VEGF-C binds to VEGFR-2 and -3 and is thought to play a major role in regulation of lymphatic endothelium. VEGF-D also signals through VEGFR-2 and -3; however, further research is required for better understanding of the specific role of each VEGF family members and their interactions in regulating the endometrial vasculature.

Another family of angiogenic factors, the angiopoietins, plays a critical role in angiogenesis particularly in stabilization of vessels by recruitment of pericytes around endothelial cells (307,337,338). Angiopoietins regulate endothelial cell functions by activating or inhibiting the activation of Tie2 tyrosine kinase receptor. Two members of this family of ligands, angiopoietin-1 (Ang-1) and angiopoietin-2 (Ang-2), have been studied extensively in several models of angiogenesis (338). In vivo, Ang-1 promotes angiogenesis, and is involved in vessel maturation and recruitment of VSMC. Ang-1 or Tie2 knockout in mice results in embryonic lethality because of severe vascular abnormalities characterized by disturbance in endothelial cells and pericyte interactions. Although Ang-1 and Ang-2 share 60% homology in protein structure, Ang-2 is a naturally occurring competitive antagonist of Ang-1 and is believed to play a role in destabilizing the vessels prior to remodeling. A number of recent studies demonstrate the expression of Ang-1, Ang-2, and Tie-2 in the human endometrium (337). Tie-2 is primarily expressed in the vascular endothelium, while the ligands, Ang-1 and Ang-2, are expressed in glands, stroma and endothelial cells. However, there is lack of agreement among different studies on cellular localization and hormonal regulation during the menstrual cycle, and it is difficult to draw any conclusion from these studies on specific role of angiopoietins in endometrial angiogenesis (337). Clearly, functional studies and more investigations on cellular localization of these molecules in full-thickness endometrial samples are required.

Mid-Secretory Endometrium and Embryonic Implantation

Evidence for the Window of Implantation in the Mid-Secretory Phase. Studies on early human pregnancies reveal distinct patterns of attachment of the blastocyst to the endometrial surface, invasion into and attachment to the underlying stroma, and replacement of the endothelial cells of the maternal spiral arteries with endothelial trophoblasts (339,340). These observations support a model of implantation in humans, shown schematically in Fig. 3B, which begins in a select period of the mid-secretory phase, the "window of implantation." Nearly 50 years ago, Adams et al. published a classic study on histological evaluation of 34 early pregnancy specimens obtained during the first 17 days of gestation (339), suggesting that the window of implantation in human endometrium resides in the mid-secretory phase. They based their conclusions on the findings that embryos identified in secretory phase hysterectomy specimens were free-floating when specimens were obtained before cycle day 20 and were attached or had invaded into the endometrial stroma when specimens were obtained after cycle day 20. Subsequent definition of the window of implantation was facilitated by identification of dome-like structures, "pinopodes" (Greek: *pino*, drink; *podia*, feet) that involve the luminal epithelium, are important for embryonic attachment, and participate in water and solute transport across the epithelium (341). Scanning electron microscopy of endometrial biopsies obtained from normally cycling women on postovulatory day (POD) 6 demonstrate 78% with pinopodes, compared to rare pinopodes observed on POD 2 or 9 (341–343). The structures persist for 1–2 days, with 5 days of variation in the days postovulation in which they occur, and their presence positively correlates with implantation sites (344). Furthermore, continuing pregnancy rates are high for embryos that implant between POD 8-10 (84%), compared to 18% when implantation occurs 11 days or more after ovulation (345). Thus, the data support a discrete time in the cycle between six to ten days after the LH surge (LH + 6 –LH +10 or POD 7-11) that defines the window of implantation.

Plasticity of the window has been demonstrated with fertility therapies and hormonal replacement. For example, it is advanced to POD 5–6 in clomiphene or gonadotropin-stimulated cycles (346,347) and is delayed to POD 9–10 in steroid hormone replacement cycles for donor-recipients (348). In assisted reproductive cycles pregnancy rates are 40.5% when embryo transfers are performed between cycle days 17–19, compared to no conceptions when embryo transfers occur after cycle day 20 (349). These data further demonstrate plasticity of the window under hormonal manipulation and underscore the narrow period of endometrial specialization that coincides with the window of implantation.

Gene Expression in the Mid-Secretory Phase. Much information on the molecular dialogue occurring between the endometrium and the implanting conceptus derives from transgenic and "knockout" experimental animal models, many of which have different mechanisms of implantation, compared to primates. These are reviewed in Chapter 4 and Dey et al. (186). Studies in the human derive mainly from in vitro studies with human endometrial and placental cells and explant cultures and investigation into expression of individual genes and proteins (or families thereof) in endometrium sampled during the implantation window. Recently, genomic approaches to global gene expression profiling in human endometrium during the mid-secretory phase, compared to other phases of the menstrual cycle, have provided vast amounts of information on gene transcriptional activity involved in endometrial cyclic changes preparing the tissue for embryonic implantation. Because the endometrium is highly responsive to steroid hormone actions and undergoes a continuum in histological and biochemical phenotypes across the menstrual cycle, several important aspects of investigations with human endometrial tissue warrant emphasis. Critical for studies designed to elucidate normal biology is the definition of the subject population with regard to, e.g., age, absence of infertility, absence of endometrial or uterine pathology, not on steroid hormones for at least 3 months of endometrial sampling, and obtaining fundal (and not lower uterine segment) endometrial tissue. Also, when in the cycle tissue samples are obtained is absolutely critical to interpretation of the data. Timing to the LH surge and/or endometrial histological evaluation, are essential for accurately assessing the stage of secretory phase samples for analysis. Recent molecular profiling in our laboratory reveals molecular signatures at each substage of the cycle (unpublished data), underscoring the importance of cycle stage definition prior to analysis. Using oligonucleotide microarray technology, several groups have investigated gene profiles in human endometrium during the mid-secretory versus late proliferative (350,351) or early secretory phases (352,353) (Fig. 6). Subtractive cDNA hybridization and differential display polymerase chain reaction (PCR) have provided additional information about

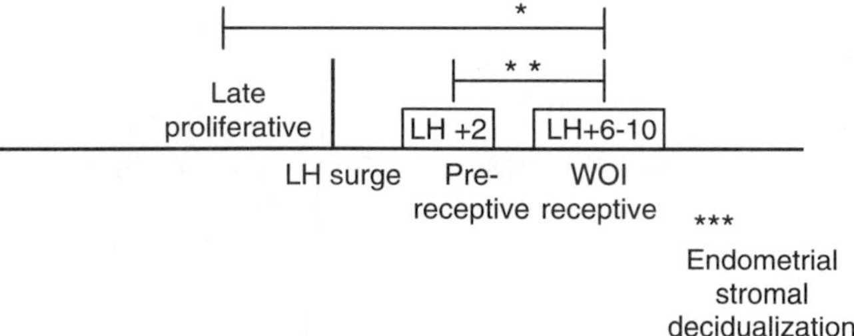

FIG. 6. Schematic of experimental designs comparing genes expressed at various times in the menstrual cycle in human and nonhuman primate endometrium. [Adapted from Giudice (10), with permission.]

comparative gene expression in endometrium during the secretory versus proliferative phases (354,355). Results from these approaches are discussed herein in the context of known and newly proposed players and mechanisms involved in endometrial receptivity to implantation.

Mid-Secretory Compared to Proliferative Phase. Kao et al. (350) and Borthwick et al. (351) investigated, using oligonucleotide microarrays and similar bioinformatics software analyses, genes expressed in mid-secretory versus late proliferative endometrium. The only difference in the approaches used by these investigators was how the samples were processed. The former group analyzed genes expressed in individual samples in both phases and compared mean levels between the phases, in contrast to the latter approach in which samples in each phase were pooled, analyzed, and then compared. Global gene profiling of 12,686 genes in endometrial biopsies obtained 8–10 days after the LH surge, compared to the late proliferative phase, in normally cycling, fertile subjects without endometriosis revealed mean levels of 156 genes that were up-regulated and 345 genes that were down regulated at least twofold (350). Up-regulated genes (Table 2) included those involved in cholesterol transport and trafficking, prostaglandin biosynthesis, carbohydrate and glycoprotein biosynthesis, secretory proteins, cell cycle regulators, proteases and peptidases, nitric oxide synthesis, extracellular matrix synthesis, cell adhesion molecules, neuromodulators and their receptors, immune modulators, cytokines and their receptors, detoxification, structural and cytoskeletal proteins, phospholipids binding proteins, cell surface proteins and receptors, transporters, and others. Secreted proteins, such as mammaglobin, osteopontin, and glycodelin, have been reported in human endometrium in response to progesterone. Osteopontin, e.g., is important in embryo attachment, and glycodelin, also found to be uniquely expressed in secretory endometrium by subtractive hybridization/PCR (354), is widely believed to participate in endometrial receptivity, due to its immunosuppressive properties (356). N-Acetyl-6-sulfotransferase is a member of a family of sulfotransferases important in synthesis, e.g., of the ligand for L-selectin which is expressed by the human blastocyst (340), and thus is important for embryonic adhesion. Other genes are members of the TGFβ and IGF families, known to be important in endometrial development for implantation (219,220). Importantly, a variety of molecules can now be investigated for their roles in endometrial–embryo interactions, including those with immune functions, members of the Wnt signaling pathway, water and ion transporters, transcription factors, apoptosis family members, and others. Maintaining an environment that is aseptic

TABLE 2. *Genes up-regulated in the implantation window versus the proliferative phase*

Families/GenBank accession no.	Fold up	*P*-value	Description (N = 156)
Cholesterol transport/trafficking			
M12529	100.0	0.013	Apolipoprotein-E
J02611	5.6	0.0013	Apolipoprotein-D
Prostaglandin biosynthesis			
M22430	18.2	0.0300	Phospholipase A2
U19487	3.6	0.0300	Prostaglandin E2 receptor
Carbohydrate/glycoprotein synthesis			
AB009598	15.6	0.03	Glucuronyltransferase I
AB014679	6.4	0.0066	N-acetylglucosamine-6-O-sulfotransferase (GlcNAc6ST)
Secretory proteins			
M61886	14.6	0.0272	Pregnancy-associated endometrial alpha2-globulin (glycodelin)
U33147	12.4	0.0255	Mammaglobin
AB020315	12.1	0.0057	Dickkopf-1 (hdkk-1)
M31452	7.0	0.0272	Proline-rich protein (PRP)
M57730	4.9	0.0057	B61
X16302	2.7	0.0130	Insulin-like growth factor binding protein (IGFBP-2)
AB000584	2.4	0.0057	TGF-beta superfamily protein
Cell cycle			
M69199	9.2	0.0184	G0S2 protein
M14752	6.4	0.0300	c-abl
M60974	3.9	0.0057	Growth arrest and DNA-damage-inducible protein (gadd45)
AF002697	2.2	0.0130	E1B 19K/Bcl-2-binding protein Nip3
U66469	2.0	0.0418	cell growth regulator CGR19
Proteases/peptidases			
M17016	9.0	0.0130	Serine protease-like protein
M30474	5.2	0.0343	Gamma-glutamyl transpeptidase type II
L12468	4.0	0.0279	Aminopeptidase A
AL008726	2.5	0.0013	Lysosomal protective protein precursor, cathepsin A, carboxypeptidase C
Nitric oxide synthesis			
U82256	8.3	0.0057	Arginase type II
Extracellular matrix/ cell adhesion molecules			
J04765	8.1	0.0013	Osteopontin
U17760	4.1	0.0017	Laminin S B3 chain
M61916	2.6	0.0184	Laminin B1 chain
Neuromodulators/synthesis/receptors			
M68840	7.5	0.0013	Monoamine oxidase A (MAOA)
U95367	2.6	0.0437	GABA-A receptor pi subunit
Immune modulators/cytokines			
L41268	7.2	0.0082	Natural killer-associated transcript 2 (NKAT2)
M84526	6.7	0.0272	Adipsin/complement factor D
M31516	5.9	0.0013	Decay-accelerating factor
AF031167	5.9	0.0013	Interleukin 15 precursor (IL-15)
D63789	4.5	0.0300	SCM-1 beta precursor (lymphotactin)
M85276	4.0	0.0437	NKG5 NK and T-cell specific gene
U14407	3.7	0.0300	Interleukin 15 (IL-15)
M34455	3.7	0.0049	Interferon-gamma-inducible indoleamine 2,3-dioxygenase (IDO)
U31628	3.3	0.0066	Interleukin-15 receptor alpha chain precursor (IL15RA)
L09708	2.1	0.0279	Complement component 2 (C2)
M14058	2.0	0.0130	Complement C1r
Detoxification			
J03910	5.9	0.0013	Metallothionein-IG (MTIG)
M10943	3.8	0.0049	Metallothionein-If
R93527	3.6	0.0049	Homo sapiens cDNA similar to metallothionein
M13485	3.5	0.0013	Metallothionein I-B
H68340	3.5	0.0049	Homo sapiens cDNA similar to metallothionein-If
K01383	3.0	0.0279	Metallothionein-I-A
X71973	2.9	0.0130	Phospholipid hydroperoxide glutathione peroxidase
M93311	2.4	0.0049	Metallothionein-III

Continued

TABLE 2. *Genes up-regulated in the implantation window versus the proliferative phase—cont'd.*

Families/GenBank accession no.	Fold up	*P*-value	Description (N = 156)
Structural/cytoskeletal proteins			
M88338	5.2	0.0418	Serum constituent protein (MSE55)
M34175	4.3	0.0212	Beta adaptin
M19267	3.7	0.0300	Tropomyosin
X06956	3.4	0.0082	Alpha-tubulin
Phospholipid binding proteins			
D28364	4.7	0.0300	Annexin II
M82809	2.2	0.0279	Annexin IV (ANX4)
Cell surface proteins/receptors			
L78207	4.3	0.0013	Sulfonylurea receptor (SUR1)(K^+-channel)
U11863	3.4	0.0418	HP-DAO2 diamine oxidase, copper/topa quinone containing mRNA
J03779	2.7	0.0184	Common acute lymphoblastic leukemia antigen (CALLA)
D50683	2.6	0.0437	TGF-beta II-R alpha
X97324	2.1	0.0272	Adipophilin
Transporters			
AB000712	3.9	0.0272	hCPE-R (*Clostridia perfringens enterotoxin* receptor-1)
U81800	3.4	0.0057	Monocarboxylate transporter (MCT3)
Other cellular functions			
X79882	3.2	0.0057	lrp
M62896	3.2	0.0057	Lipocortin (LIP) 2 pseudogene mRNA
EST/Unknown function	**N = 40**		

Modified from (350) with permission.

and permits detoxification is also likely to be important in enhancing embryonic implantation.

Down-regulated genes included (Table 3) those encoding secretory proteins, proteases, cell surface proteins, receptors, extracellular matrix proteins, cell adhesion molecules, immune modulators and receptors, vasoactive substances, transporters, ion binding proteins, steroid hormone action, and neuromodulators and receptors. These include, e.g., intestinal trefoil factor, tenascin C, matrilysin (matrix metallopteinase-7), frizzled related protein (FRP-HE, a Wnt inhibitor), and others. Validation of several up- and down-regulated genes was conducted and confirmed differential expression, and select up- and down-regulated ESTs were mined and revealed members of similar families, including aquaporins, important in water transport (Table 4) (10).

Interrogation of the dataset for absolute gene expression revealed over 200 genes commonly and abundantly expressed by proliferative and secretory endometrium, including ribosomal RNA, genes for structural proteins, and others (351). Electronic identification of progesterone response elements (PREs) and estrogen response elements (EREs) in the promoters of genes that were found to be up- or down-regulated in the window of implantation (peak circulating P) compared to the late proliferative phase (peak E_2) revealed candidate EREs in for example intestinal trefoil factor, a gene that is up-regulated in the proliferative phase, and candidate PREs in monoamine oxidase, upregulated in the secretory phase (350). Electronic identification of steroid hormone regulatory elements in some, but not all, genes expressed in this tissue at this time of the cycle has provided a powerful complementary approach to investigate steroid hormone-regulated genes. Indeed, many of the genes investigated could be considered progesterone-regulated, due to the times in the cycle in which their expression was investigated, and the absence of classical ERE or PRE does not preclude estradiol or progesterone regulation of genes (241).

In rhesus endometrium, Okulicz and Ace reported temporal regulation of gene expression during the expected window of receptivity (357). Using differential display reverse transcriptase-polymerase chain reaction, 12 cDNA fragments were isolated and sequenced whose mRNA levels were elevated in the time frame investigated. Two of these fragments were consistent with secretory leukocyte protease inhibitor (antileukoprotease), a neutrophil elastase inhibitor with antiinflammatory and anti-bacterial properties. The second fragment was syncytin, a fusogenic membrane glycoprotein that induces formation of syncytia and is important in placental development. In a subsequent study by this group, genes expressed on cycle days 21–23 were compared to genes expressed on day 13, using oligonucleotide microarrays (358). Numerous genes up- and down-regulated were identical to equivalent comparisons in human endometrium. For example, among the 39 up-regulated genes were uteroglobin, metallothionein IG, and secretory leukocyte protease inhibitor. Among the 69 down-regulated genes were TGFβ1, stromelysin 3, proenkephalin, collagen type VII

TABLE 3. *Genes down-regulated in the implantation window versus the proliferative phase*

Families/GenBank accession no.	Fold down	*P*-value	Description (N = 377)
Secretory proteins			
L08044	49.8	0.0418	Intestinal trefoil factor
AF026692	19.8	0.0017	Frizzled related protein frpHE
AF056087	6.3	0.0013	Secreted frizzled related protein FRP
AB000220	5.8	0.0047	Semaphorin E
X78947	2.9	0.0279	Connective tissue growth factor
U38276	2.6	0.0130	Semaphorin III family homolog
Proteases			
L22524	24.1	0.0082	Matrilysin
M96859	10.8	0.0213	Dipeptidyl aminopeptidase like protein
X51405	9.7	0.0117	Carboxypeptidase E
AF071748	3.1	0.0117	Cathepsin F (CATSF)
X02596	2.0	0.0013	bcr (breakpoint cluster region) gene in Philadelphia chromosome
Cell surface proteins/receptors			
AB011542	3.5	0.0177	MEGF9
M34641	3.4	0.0013	Fibroblast growth factor (FGF) receptor-1
M87770	3.2	0.0212	Fibroblast growth factor receptor (K-sam)
U09278	3.2	0.0130	Fibroblast activation protein
AB015633	3.0	0.0017	Type II membrane protein
X83425	2.7	0.0388	Lutheran blood group glycoprotein
Extracellular matrix/cell adhesion molecules			
M92642	11.2	0.0013	Alpha-1 type XVI collagen (COL16A1)
AL049946	10.1	0.0017	DKFZp564I1922
M34064	6.0	0.0013	Human N-cadherin
U69263	5.6	0.0117	Matrilin-2 precursor
J04599	4.2	0.0013	hPGI mRNA encoding bone small proteoglycan I (biglycan)
X78565	3.9	0.0130	Tenascin-C
U19718	3.0	0.0066	Microfibril-associated glycoprotein (MFAP2)
D13666	2.8	0.0117	Osteoblast specific factor-2 (OSF-2os)
X53002	2.4	0.0049	Integrin beta-5 subunit
X53586	2.3	0.0013	Integrin alpha 6
X17042	2.1	0.0279	Hematopoietic proteoglycan core protein
Apoptosis/inhibitors			
AF005775	2.3	0.0212	Caspase-like apoptosis regulatory protein 2 (clarp)
AF016266	2.2	0.0013	TRAIL receptor 2
M59465	2.0	0.0464	Tumor necrosis factor alpha inducible protein A20
Immune modulators/receptors			
M83664	4.7	0.0049	MHC class II lymphocyte antigen (HLA-DP) beta chain
M60028	4.6	0.0017	MHC class II HLA-DQ-beta (DQB1,DQw9)
X94232	3.5	0.0386	T-cell activation protein
J00194	2.9	0.0130	HLA-dr antigen alpha-chain
M24594	2.6	0.0213	Interferon-inducible 56Kd protein
Vasoactive substances			
J05081	4.7	0.0418	Endothelin 3 (EDN3)
AF022375	3.4	0.0279	Vascular endothelial growth factor
Transport proteins			
L04569	2.8	0.0213	L-type voltage-dependent calcium channel α1 subunit (hHT)
U83993	2.5	0.0057	P2X4 purinoreceptor
U07139	2.0	0.0130	Voltage-gated calcium channel beta subunit
Ion binding proteins			
X72964	2.5	0.0013	Caltractin
AF070616	2.1	0.0049	BDP-1 protein
Steroid hormone action			
Y12711	2.4	0.0057	Putative progesterone binding protein
AJ000882	2.1	0.0212	Steroid receptor coactivator 1e
Neuromodulators/receptors			
U29195	2.2	0.0418	Neuronal pentraxin II (NPTX2)
EST/Unknown function	**N = 153**		

Modified from (350) with permission.

TABLE 4. *Selected ESTs changing > threefold in the window of implantation versus proliferative phase in human endometrium*

Gene bank accession #	Gene	Fold-change	Description/Function
Up-regulated			
M22430	Phospholipase A2	18.2	Prostaglandin biosynthesis
N74607	Aquaporin 3	6.6	Water transport
M14752	v-abl oncogene homologue	6.4	Tyrosine kinase/cell differentiation, division, adhesion, stress response
AA420624	Amine oxidase	6.2	Integral mitochondrial membrane protein
AA976838	Apolipoprotein C-1	6.1	VLDL and HDL
R93527	Metallothionein 1H	3.6	Detoxification, metal binding
H68340	Metallothionein-1F	3.5	Detoxification, metal binding
AI309115	Beta-defensin	3.4	Bactericidal activity
AA056747	ATPase, H^+ transporter	3.3	Proton pump
AI887421	Retinoic acid receptor responder 1	3.2	Membrane protein induced by protein retinoic acid receptors
X79882	LRP	3.2	Drug resistance related protein/Wnt signaling
Down-regulated			
D10925	CCR1 chemokine receptor 1	11.3	Chemokine (C-C motif, RANTES. MCP-1) receptor
N95168	GADD45B	7.1	Growth arrest
A1810807	DNCI1	7.0	Intracellular organelle motion
D50920	TRAP 100 hormone receptor	6.7	Complexes with thyroid
AA149644	Junctional adhesion molecule 3	6.2	Localized in tight junctions between endothelial cells
AI688516	NADH-ubiquinone redox	6.1	Oxidoreductase
AB002305	Aryl hydrocarbon receptor nuclear translocator 2	5.9	Transcription factor family bHLH, PAS domain
AB007972	Protein phosphatase 1 regulatory subunit (inhibitor)	5.3	Phosphatase regulator
AA100961	Platelet endothelial cell adhesion molecule-1	5.2	Endothelial cell intercellular junctions
AI263885	Class I cytokine receptor	4.4	T-cell cytokine receptor GTP ase/cytokinesis/exocytosis
D50918	Septin 6	3.0	
AL049367	GNG 12	3.0	Guanine nucleotide binding protein
AB020647	FBXL7 F-box and leucine-rich repeat protein	3.0	Functions in ubiquitination
AF0308202	Syntaxin 6	3.0	Vesicle trafficking
AB029028	RAP140	3.0	Retinoblastoma-related

Data analyzed from database from (350) and (351).

alpha 1, secreted frizzled-related protein 4, PR membrane component 1, CXCL12, and biglycan. Studies like this underscore the validity of using the nonhuman primate model to investigate steroid hormone and other regulators' actions on the endometrium and provide in themselves insight into steroid hormone action on primate endometrium in general.

Mid-Secretory Compared to Early Secretory Phase. Changes in endometrial gene expression during the early secretory (pre-receptive) phase, compared to the mid-secretory (receptive) phase by Carson et al. (352) revealed 323 up-regulated, and 370 down-regulated genes greater than twofold during the transition. Many regulated genes were common to those reported in the transition from late proliferative to mid-secretory (350,351), including, but not limited to, the claudin-4 (CPE receptor), osteopontin, glycodelin, matrilysin, and FRP-HE. Riesewijk et al. (359), investigated genes in samples obtained from the same subjects at 2 and 7 days after the LH surge [compared to Carson et al. where pairwise analysis was not performed (352)], and precise timing was used. These investigators found 153 genes were up-regulated greater than threefold, and 58 were down-regulated, with numerous genes in common to all three studies described above. It is interesting that numerous genes are in common between the mid-secretory pahse and the early secretory and late proliferative phases. However, this is not too surprising, since there is not much histological difference in endometrial histology 36–48 hours after ovulation (LH+2) as compared to the late proliferative phase.

Model of Molecular Mechanisms Occurring in the Mid-Secretory Phase for Receptivity to Embryonic Implantation. Figure 7 represents a model of events occurring in the endometrium as it prepares for and participates in embryonic implantation. It has been postulated, by analogy to leukocyte migration from the vasculature into tissues, that before attaching to the endometrial epithelium, a blastocyst undergoes tethering and rolling on the endometrial surface, and finally bridges to cell surface carbohydrates and proteins (340). Mechanisms must be in place in the maternal endometrium for synthesis of molecules participating in these events. Once attachment occurs, a set of mechanisms is likely initiated to

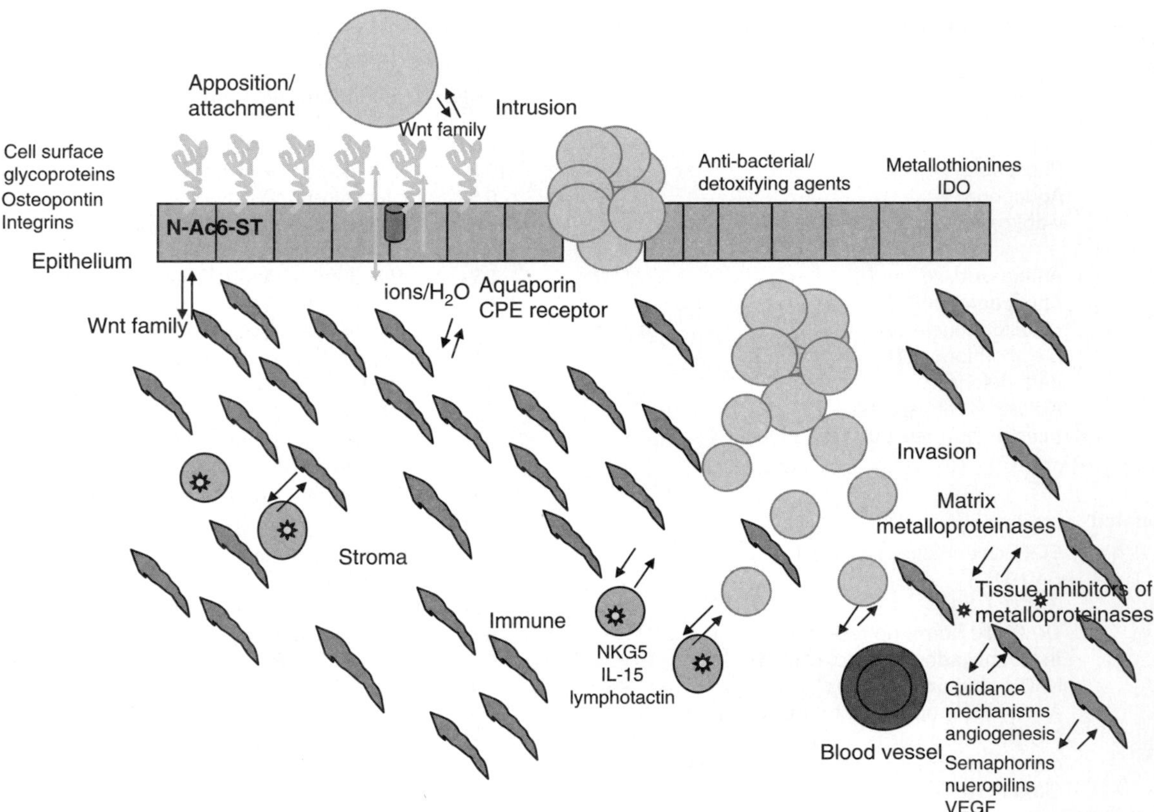

FIG. 7. Model of events occurring during the early stages of embryonic implantation. N-Ac6-St, N-acetylglucosamin-6-O-sulfotransferease; IDO, indolamine 2,3-dioxygenase; CPE, *Clostridia perfringens* enterotoxin; IL-15, interleukin 15; VEGF, vascular endothelial growth factor. [From Giudice (216), with permission.] (See color insert.)

permit intrusion of the trophoblast into the stromal compartment and guidance as the invading trophoblast migrates to the maternal spiral arteries, all the while maintaining integrity of the extracellular matrix and anticoagulation. It is possible that that these embryo–endometrial interactions involve ion transport and signaling through paracrine mechanisms via growth factors and cytokines, as well as adaptation of guidance mechanisms similar to those used in angiogenesis and, perhaps, neuronal migration (350). Tolerance of the implanting allograph, and other protective mechanisms (e.g., anti-bacterial, detoxification) are likely important to maximize viability of the implanting conceptus. Furthermore, tissue integrity is important, as are changes in extracellular matrix composition, to assure the conceptus is anchored to its maternal host. Many of these events likely occur simultaneously, some occur in microenvironments uniquely, and others occur in strict temporal, as well as spatial fashion. Also, the dialogue between conceptus and endometrium is likely dynamic, extending beyond the static dataset of molecules observed in the nonpregnant state. Nonetheless, these approaches provide insight into the molecular pathways, molecular signals, and physiological processing that await an embryo should implantation occur.

Stromal Predecidualization

Predecidualization of stromal cells is necessary for successful implantation and represents differentiation of the stromal cells to a distinct morphological appearance, as described above, and by a unique biosynthetic and secretory phenotype (7,20,218). Progesterone and cAMP promote this endometrial stromal differentiation process. In vitro predecidualization can be effected with progesterone, after estradiol priming, and also by cAMP and other ligands that activate the protein kinase A-dependent pathway (360–363). Numerous proteins and factors are up-regulated or induced or down-regulated during the predecidualization process. Predecidualization results in a unique extracellular matrix, rich in laminin, heparan proteoglycans, and type IV collagen (213,218). The precise molecular mechanisms underlying the process of decidualization are not completely understood.

Gene profiling of human endometrial stromal cells cultured in vitro and treated with progesterone after E_2 priming or with cAMP has revealed regulation of many known endometrial predecidualized stromal cell genes (e.g., IGFBP-1, prolactin, VEGF) as well as numerous genes not previously known to be expressed in endometrium (334). Included are specific cytokines, growth factors, nuclear transcription factors, members of the cyclin family, the insulin receptor, select neurotransmitter receptors, neuromodulators, the FSH receptor, inhibin/activin βA subunit, inhibin α, some members of the angiotensin/rennin family, and the TNF-related apoptosis inducing ligand (TRAIL) were observed. Delineating the functions of these newly found players in the decidualization process is a challenge for the future.

A time course of stromal predecidualization has been investigated on a more global scale by Brar et al. (217) and Tierney et al. (364). Brar and colleagues interrogated gene expression of human term pregnancy decidual fibroblasts in response to E_2, P and 8Br-cAMP after 0, 2, 4, 6, 9, 12, and 15 days of treatment. Tierney et al. investigated genes expressed in human endometrial stromal cells obtained from cycling (nonpregnant) subjects and decidualized in vitro in response to 8Br-cAMP for 0, 2, 4, 6, 12, 24, and 48 hours. Kinetic analyses of regulated genes revealed striking concomitance of gene expression in the cells obtained from different patient populations, treated with different protocols. During decidual fibroblast differentiation, marked changes were observed in genes encoding proteins involved in extracellular organization, cytoskeletal organization, and cell adhesion, transcription factors, members of the IGF family, prolactin, early induction of genes involved in MAP kinase signaling, the JAK/STAT pathway, as well as others, underscoring multiple processes occurring during the decidualization process. The most upregulated genes were preprosomatostatin, IGFBP-1, prolactin, NOT, and neuropeptides (217,364). Many genes were reprogrammed within specific functional groups and gene families, with simultaneous induction and down-regulation of sets of genes with related functions. The data suggest that reprogramming of gene expression within functional categories represents a fundamental component of cellular differentiation for the endometrial stromal cell. Figure 8 presents genes that are kinetically regulated and some of the processes with which they are associated during decidualization of endometrial stromal fibroblasts (217,364).

Animal models in which homologous recombination and gene knockout experiments have resulted in abnormal, incomplete, or absent decidualization phenotype have given insight into the importance of particular genes in the decidualization process (365).

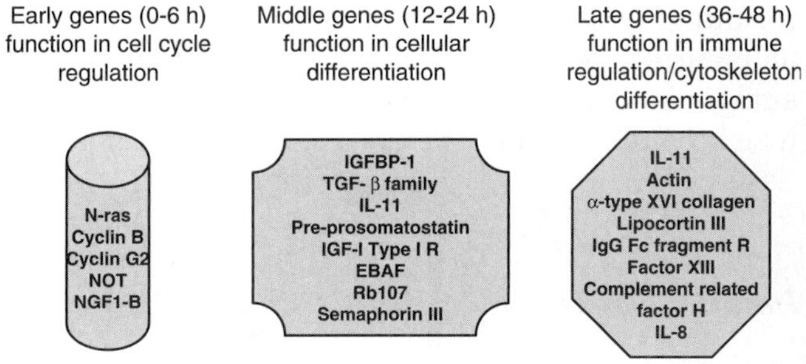

FIG. 8. Model of three phases of decidualization of human endometrial stromal cells in response to cAMP. Early response genes (0 to 6 hr) are in involved in cell cycle regulation, mitosis, endoreduplication, and subsequent cell cycle arrest. These set the stage for the next 12 to 48 hr in which the genes expressed are primarily involved in cellular differentiation, as the cell begins to adopt, both morphologically and functionally, the decidualized phenotype. This process involves middle regulated genes (12 to 24 hr) for cellular differentiation and adoption of the morphological and secretory phenotype, and late regulated genes (36 to 48 hr) that mediate more specialized functions of the stromal cell, including immune regulation, signal transduction, and enzymatic pathways. [From Tierney et al. (364), with permission.]

It should be noted that decidualization in mice occurs in the presence of an embryo or can be induced mechanically, and players promoting decidualization in humans may differ where decidualization occurs in the absence of an embryo cycle-to-cycle. However, increasing evidence suggests that some of the components found in mouse implantation may also play a role in decidualization in human endometrium. Among these are Hox A-10, Hox A-11, IL-11, and leukemia inhibitory factor (LIF) [extensively reviewed in Lim et al. (365) and Dey et al. (186)]. For example, Hox A-10 is upregulated by progesterone in uterine stroma in the mouse at day 3.5 postcoitum, 24 hours before implantation (366), and HoxA-10 deficient mice have severe infertility, a decidualization defect, and recurrent pregnancy loss due to abnormal implantation (367). Furthermore, progesterone treatment of HoxA-10 knock out mice, compared to wild-type, reveals a stromal cell proliferation defect accompanied by quantitative and spatial alterations in expression of two cyclin-dependent kinase inhibitor genes, p57 and p15, as well as a severe, local immunological disturbance with polyclonal T-cell proliferation (368). Thus, HoxA-10 is a mediator of progesterone-regulated stromal cell proliferation, decidualization and local immunosuppression. Interestingly, HoxA-10 expression in human endometrium is normally upregulated in mid-secretory phase endometrium, but is not up-regulated in infertile women with endometriosis (22). These clinical findings suggest an abnormality in the signaling for progesterone differentiation of the endometrium in women with this disorder. [The reader is referred to a recent review on the pathophysiology and abnormalities in women with endometriosis, including endometrial abnormalities (369).] Studies on mouse endometrium after administration of a PR antagonist to pregnant mice have provided additional insights into steroid-hormone regulated genes in this tissue (253).

Immune Environment in the Endometrium

Endometrial Immune Cells. Uterine immune cell populations undergo profound changes throughout the menstrual cycle (370,371), and their distribution within the female reproductive tract is regulated by E_2 and progesterone (372–374). Leukocytes account for 5% of stromal cells in the proliferative phase, 25% in the secretory phase, and 30% of decidual cells in early pregnancy (375,376). Between 5% to 15% of endometrial stromal cells are macrophages, found scattered throughout the stromal compartment and also in small aggregates (377–380). B-cells are present, but are few in numbers in human endometrium (379–383). They are believed to contribute to the mucosal immunity of the endometrium, in addition to the macrophages and T-cells (374). T-cells are the most abundant leukocyte type in endometrium and are found throughout the endometrium, scattered in the stroma or as intraepithelial lymphocytes, and as discrete lymphoid follicles (375,379,380,384,385). The largest leukocyte population in the endometrium consists of NK cells or large granulated lymphocytes (LGLs). These cells comprise over 70% of endometrial leukocytes at the time of implantation (375), and their increased numbers are attributable to proliferation in situ (386–388) and also to influx from the peripheral blood circulation (389,390). Endometrial NK cells differ both phenotypically and functionally from peripheral blood NK cells. They are $CD56^{bright}$ CD^{16null} and express the killer activatory and inhibitory receptors (KAR/KIR), in contrast to peripheral NK cells which are predominantly $CD56^{dim}$, CD^{16+} (371,391). They express early T cell markers, including CD2 and CD7, integrins such as CD11a and CD18, and the low affinity IL-2Rβ. In addition, in contrast to the majority of peripheral NK cells, endometrial NK cells express CD69, an early-activation marker (392). Furthermore, the endometrial NK cell phenotype changes during the normal menstrual cycle (371,376, 391). How some peripheral NK cells migrate to the endometrium is not certain, but uterine NK cells express the α_4 integrin and LFA-1, and decidual blood vessels in human uterus express the α_4 integrin ligand, VACM-1 (393–395). It has been suggested that the migration of peripheral NK cells to human endometrium may be similar to that of lymphocyte emigration using these integrin-ligand receptors (396). Furthermore, MIP-1α and β and several others endometrial stromal products have been suggested to be attractants for peripheral NK cells to emigrate to the endometrium [reviewed in (391)].

There are important functional differences between endometrial and peripheral NK cells [reviewed in (391)]. For example, uterine NK cells, in contrast to peripheral NK cells, exhibit weak lytic activity against the standard NK-sensitive target K56. However, they can be readily activated after IL-2 stimulation to kill effectively trophoblast cells, which are normally resistant to lysis. Also, endometrial NK cells express a different cytokine profile, compared to resting peripheral NK cells. Specifically, expression of G-CSF, M-CSF, GM-CSF, TNF-α, IFN-γ, TGF-β and LIF, has been found in decidual CD56+ cells; whereas, peripheral blood NK cells produce no cytokines in their resting state (391). Strominger and colleagues recently analyzed, by microarray technology, genes expressed in uterine NK cells in decidua versus peripheral NK cells (397). They found selective expression of CD9, galectin-1 and glycodelin and also up-regulation of tetraspanins, integrins, lectin-like receptors, and KIRs in uterine NKs, compared to peripheral NK cells. Functional studies will likely follow mining this important dataset.

The function of endometrial NK cells has not yet been delineated with certainty in humans. However, because of their increased numbers in early pregnancy, their hormonal dependence, and their physical proximity to the invading trophoblast in the implantation site, it has been suggested that they play an important role in regulating the maternal immune response to the fetal allograft and control of trophoblast growth and invasion (376). The essential function of uterine NK cells for normal pregnancy was established in a series of experiments with transgenic mice. Guimond et al. studied uterine NK cell function in TgE26 transgenic mice (398) which carry high copy numbers of the complete human CD3ε gene and exhibit NK and T-cell deficiencies from fetal life. TgE26 mice had no uterine NK cells present at implantation sites. Furthermore, TgE26 mice had normal rates of implantation and normal trophoblast differentiation, arguing against a role of NK cells in those two functions. However, starting on day 10 of gestation, TgE26 mice exhibited significantly increased fetal loss at midgestation, small placental size, small offspring size, and various histological abnormalities of the maternal vasculature in the decidua, with thickening of arterial walls. Reconstitution experiments, using bone marrow from B and T cell-deficient *scid/scid* donors, that have intact NK cell lineage, reversed the reproductive deficits with about 50% of uterine NK cells compared to normal mice, well developed metrial glands at implantation sites, larger placentae than nonengrafted transgenic mice, no major abnormalities in the vasculature of the placenta, and normal fetal viability (399). These experiments definitively demonstrated that there is an essential role of uterine NK cells in pregnancy success. Despite an essential role for NK cells in mouse pregnancy, their exact function in humans remains to be elucidated. The two main functions of peripheral NK cells include cytotoxicity and cytokine production. Such functions have been suggested for endometrial NK cells and could involve (a) cytotoxic activity against trophoblast invasion and (b) regulation of placental development. The reader is referred to a recent, extensive review on NK cells and clinical implications in pregnancy maintenance (391) and in mice (400).

In the basalis layer of the endometrium lymphoid follicles or aggregates have been described that are located between endometrial glands, approximately equidistant from the base of the glands in the basalis and often associated with blood vessels (385,401). They are comprised of a central B-cell core, surrounded by a mass of exclusively CD8+ T-cells, which is further surrounded by a halo of macrophages. Lymphoid aggregates develop during the proliferative phase of the cycle, are at their highest mass at mid-cycle, and then persist in size through the secretory phase. These observations and the fact that they are not present in postmenopausal women highly suggest steroid hormone regulation (402). While their precise functions are not known, it has been suggested that they may suppress cell mediated immune responses during the window of implantation to prevent immune responses against the conceptus, or they may play a major role in immune protection at the time of menses with the departure of stromal immune cells in the functionalis when it is shed and leaving the underlying basalis vulnerable to infection (374). Alternatively, the lymphoid aggregates may cluster T and B cells to enable re-seeding of the regenerated endometrium after the menstrual phase or serve as a source of growth factors and cytokines which may play a role in re-epithelialization of the endometrium after the menses (374).

Cytokines in the Endometrium. Numerous cytokines are expressed by endometrial epithelium, stroma, and immune cells, which have been shown in numerous studies to regulate endometrial epithelial and stromal function, vascular function, and leukocyte populations. Members of the CSF and interleukin families, interferon-γ (IFN-γ), LIF, TGF-β and TNF-α have been extensively studied (260,403–405). For example, CSF-1, derived from endometrial epithelial cells, uterine NK cells, lymphocytes, and stromal fibroblasts is stimulated by E_2 and progesterone and acts primarily on macrophages, the developing embryo and trophoblasts to regulate functions of these cells (406,407). IL-1 is believed to have a role in regulation of embryonic attachment and invasion, as IL-1 derived from human blastocysts can induce expression of the β_3 integrin in endometrial epithelium which is believed to be important for embryo attachment (408). Redundancy of cytokine production in the endometrium and embryo is the rule rather than the exception, and IL-1 is expressed in both human endometrial epithelium and stroma (as well as the blastocyst), and may contribute to the invasive phase of embryonic implantation, as well (409). INF-γ and TGF-β antagonize the proliferative effects of IGF-I and EGF and are believed to be involved in differentiation of the glandular and luminal epithelium to the secretory phenotype in the second half of the cycle (410). LIF is required for implantation in the mouse and is expressed in luminal epithelium at implantation (along with its co-receptor gp130) in mice and women (411–414). In LIF deficient mice, the embryos do not implant and the endometrium does not decidualize (415). Thus, LIF may influence epithelial cells by enabling them to transduce a decidualization signal to the underlying stroma, in addition to a potential role in embryonic attachment.

Numerous cytokines can affect stromal cell function. For example, IL-1, TNF-α, CSF-1, and GM-CSF stimulate secretion of prostaglandins, production of IL-6 and stem cell factor in endometrial stromal cells, and are important in the differentiation of stromal fibroblasts to the decidual phenotype (405). However, much redundancy must exist, as mice null for each of these cytokines do not have decidualization defects. However, IL-11 mice have implantation failure due to absence of decidualization of the stroma and dysregulated proliferation and invasion of the trophoblast (416). Thus, IL-11 (and LIF) are absolutely necessary for stromal decidualization.

During the menses, TNF-α and IL-1 are secreted by leukocytes that accumulate in the stromal compartment at the end of the cycle. These have been demonstrated to stimulate release of matrix degrading enzymes that contribute to breakdown of the vascular basement membrane and connective tissue integrity in the functionalis layer, thereby resulting in bleeding and tissue desquamation [see below and (417,418)]. TGF-β family members also play a role in tissue breakdown (419), and TNF-α may contribute to the apoptosis observed in human endometrium during the pre-menstrual and menstrual phases of the cycle. Cytokines are also believed to play a role in angiogenesis (see above) and perhaps vascular tone (420) and regulation of maternal leukocyte emigration during the secretory phase (405).

Tissue Desquamation and Menstruation

In the absence of implantation, E_2 and P levels begin to decline about cycle day 25 (Fig. 3A). This decline in steroid hormones heralds the activation of matrix degrading enzymes, which are normally suppressed by progesterone (421) and destabilization of lysosomal membrane integrity which is also a progesterone-dependent process (422). There is vasoconstriction of endometrial vessels and release of matrix degrading enzymes and lysosomal enzymes that result in destruction of glandular and stromal cells and the vascular endothelium. This then results in bleeding, but not clotting, because abundant plasmin, a fibrinolytic agent, is present in the menstrual effluent. Glandular cells also undergo apoptosis (423). Vascular luminal surface membrane injury results in platelet deposition and multiple small foci of ischemic tissue necrosis (424). PGF2α levels increase dramatically in the late secretory phase, which stimulate vasoconstriction of myometrial and basal arteries, and expulsion of the degenerated tissue is enhanced by PGF2-mediated myometrial contractions (425). The menstrual effluent is comprised of proteolyzed extracellular matrix, red blood cells, a heavy polymorphonuclear exudate, and high levels of proteolytic enzymes. Ultrastructural analysis reveals that the upper two thirds of late secretory endometrium gradually involutes, degenerates, and undergoes necrosis (426). On days 2–4 of the cycle, the functionalis gradually is detached from the basalis, and hysteroscopic evaluation reveals that the cleaved mucosa rolls on itself until it is detached from the underlying basalis and is shed from the uterine cavity. A normal menstrual period lasts about 4 days, with the highest degree of tissue desquamation and bleeding occurring in the 1st and 2nd days. Note that the isthmus and peritubal ostia remain intact during the menstrual period (426). At the biochemical level, matrix degrading enzymes are suppressed by progesterone, and in the late secretory phase, with the waning of progesterone production by the corpus luteum in nonconception cycles, effectively progesterone withdrawal occurs in the endometrium, stimulating multiple matrix metalloproteinases (418,421). Tissue inhibitors of MMPs (TIMPs) are expressed in human endometrium, with TIMP-1 and TIMP-2 not being cycle-dependent (3). TIMP-3 mRNA and protein, however, in decidualized endometrial stromal cells is upregulated by progesterone in vitro and in vivo (355), and thus with progesterone withdrawal, TIMP-3 declines, and MMPs can effectively proteolyze the basement membrane and connective tissue of the stroma. Coupled with vasoconstriction of the vasculature, these processes effectively initiate menstruation.

Tissue Regeneration

A remarkable feature of human and nonhuman primate endometrium in the absence of pregnancy is its regeneration on a cyclic basis, attributable, in part, to a unique stem cell population in the basalis. The functionalis is shed between days 1–3 of menses, leaving a denuded basalis layer (5,426). The basalis zone IV (Fig. 1B) is believed to contain the stem cells and progenitor cells for cyclic regeneration of endometrium in women and rhesus monkeys (8,427), based partially on the intense glandular epithelial proliferation that occurs in this zone *prior* to the onset (day 1) of menses (426–428). This event is independent of E_2, since circulating E_2 and ER and PR are unchanged from premenstrual values. On day 3 of menses, re-epithelialization occurs first by migration and then spreading of epithelial cells of the denuded basalis over underlying fibroblasts. On this day the epithelial glands in zone IV are now quiescent, whereas mitosis in glandular epithelium of zones I–III is dramatic (236,428). Subsequently, maximal proliferation of the surface epithelia on days 4–5 completes reconstruction of the luminal epithelial barrier, and stromal, vascular endothelial and smooth muscle proliferation occur, in zones I–III and continue throughout the proliferative phase of the cycle (429). It should be noted that stromal cells of the basalis proliferate and appear to replace shed endometrium and demonstrate immunohistochemical features of myofibroblasts. Both cell types are typical associated with normal wound healing, which is a mechanism that likely operational in the postmenstrual repair phase.

In human endometrium some of the upper basalis layer behaves as the functionalis, and some controversy exists whether glandular epithelial remnants of the functionalis remain and whether there are other cellular sources for re-epithelialization, such as the lower uterine segment and peritubal ostium that remain intact throughout the cycle. These possibilities notwithstanding, current data support the hypothesis (8,9) that endometrial regeneration is initiated and sustained by a small population of undifferentiated stem cells that divide slowly, are confined to the basalis layer, and give rise to a group of progenitor cells with a high proliferative capacity, with the postmitotic progeny of these progenitors cells continuing to differentiate to specific endometrial cell types within micro-environments of the tissue, resulting in its regeneration. A recent report on the clonogenic activity of human epithelial and stromal cells (392) supports the hypothesis of stem cells in the endometrium. Specifically, cultures of single cell types (epithelium, stroma, or leukocytes) were treated with various growth factors, and colony formation was observed. Cloning efficiencies were 0.14% for endometrial epithelium which needed feeder layers and 1.2% for stroma which was clonogenic in serum-free medium containing TGF-α, bFGF, EGF, or PDGF-BB. The authors concluded that in human endometrium 0.14% of epithelium and 1.2% of stromal cells have clonogenic activity, a characteristic of stem cells and progenitor cells. Also, a bone marrow population is in uterine endometrium (381,430,431), suggesting that the origin of endometrial stem cells is the bone marrow.

CONCLUSION AND EYE TO THE FUTURE

Cyclic changes in primate oviduct and endometrium are remarkable and occur in response to the ovarian-derived steroid horomes, E_2 and progesterone, and their downstream effectors. The multitude of functions of the oviduct and the endometrium and the numerous processes and players are profound. It is a marvel that primates have been able to reproduce with a minimum of difficulty and complications, although infections leading to tubal and endometrial damage are a testimony to the fragility of the reproductive process and its success. As we learn more about the biology of the endometrium from the perspective of the molecular mechanisms, we shall be able to understand the pathophysiology of disorders related to the endometrium (e.g., endometriosis, endometrial polyps, endometrial hyperplasia and cancer) and develop treatments that are based on physiology, biochemistry, and genetics and depend less on empirical therapies.

REFERENCES

1. Ramsey, E. M. (1994). Anatomy of the uterus. In *The Uterus* (T. Chard and J. G. Grudzinskas, Eds.), pp. 18–40. Cambridge University Press, Cambridge.
2. Dodd, E. E. (1979). *Atlas of Histology*. McGraw-Hill, New York.
3. Giudice, L. C. (2003). Implantation and endometrial function. In *Reproductive Medicine: Molecular, Cellular, and Genetic Fundamentals* (B. C. J. M. Fauser, Ed.), 2nd ed., pp. 439–465. Parthenon Pub. Group, Boca Raton, FL.
4. Brenner, R. M., and Maslar, I. A. (1988). The primate oviduct and endometrium. In *The Physiology of Reproduction* (E. Knobil and J. Neill, Eds.), Vol. 1. Raven Press, New York.
5. Bartelmez, G. W., Corner, G. W., and Hartman, C. G. (1951). Cyclic changes in the endometrium of the rhesus monkey (Macaca mulatta). In *Contributions to Embryology*. Carnegie Institution of Washington, Washington, DC.
6. Bartelmez, G. W. (1957). The phases of the menstrual cycle and their interpretation in terms of the pregnancy cycle. *Am. J. Obstet. Gynecol.* 74, 931–955.
7. Ferenczy, A., and Bergeron, C. (1991). Histology of the human endometrium: from birth to senescence. *Ann. N. Y. Acad. Sci.* 622, 6–27.
8. Padykula, H. A. (1991). Regeneration in the primate uterus: the role of stem cells. *Ann. N. Y. Acad. Sci.* 622, 47–56.
9. Okulicz, W. C. (2002). Regeneration. In *The Endometrium* (S. R. Glasser, J. R. Aplin, L. C. Giudice, S. Tabibzadeh S, Eds.), pp. 110–121. Taylor & Francis, London, New York.
10. Giudice, L. C. (2003). Elucidating endometrial function in the post-genomic era. *Hum. Reprod. Update* 9, 223–235.
11. Valdes-Dapena, M. (1973). Histologic and pathologic features of the fetal and neonatal lung. *Ann. Clin. Lab. Sci.* 3, 108–117.
12. Witschi, E. (1970). Development and differentiation of the uterus. *Perinatal life: Biological and clinical perspectives*. Proceedings of the 3rd Annual Symposium on the Physiology and Pathology of Human Reproduction. Wayne State University Press, Detroit.
13. Langman, J. (1979). *Medical Embryology*. Willliams & Wilkins, Baltimore.
14. Rock, J. A., and and Schlaff, W. D. (1985). The obstetric consequences of uterovaginal anomalies. *Fertil. Steril.* 43, 681–692.
15. O'Rahilly, R. W., Hertig, A. T., and Rock, J. (1973). The embryology and anatomy of the uterus. In *The Uterus* (H. J. Norris, A. T. Hertig, and M. R. Abell, Eds.). Williams & Wilkins, Baltimore.
16. Cunha, G. R., Kurita, T., Cooke, P.S., Sassoon, D., Miller, C., and Lubahn, D. B. (2002). The embryology of the uterus. In *The Endometrium* (S. R. Glasser, J. R. Aplin, L. C. Giudice, and S. S. Tabibzadeh, Eds.), pp. 9–10. Taylor & Francis, London, New York.
17. Baker, N. E. (1987). Molecular cloning of sequences from wingless a segment polarity gene in drosophila the spatial distribution of a transcript in embryos. *EMBO. J.* 6, 1765–1774.
18. Moon, R. T., Brown, J. D., and Torres, M. (1997). WNTs modulate cell fate and behavior during vertebrate development. *Trends Genet.* 13, 157–162.
19. Cygan, J. A., Johnson, R.L., and McMahon, A. P. (1997). Novel regulatory interactions revealed by studies of murine limb pattern in Wnt-7a and En-1 mutants. *Development* 124, 5021–5032.
20. Brar, A. K. (2002). Decidua as an endocrine organ. In *The Endometrium* (S. R. Glasser, J. D. Aplin, Giudice, L. C. and S. S. Tabibzadeh, Eds.). Taylor & Francis, London, New York.
21. Tulac, S., Nayak, N. R., Kao, L. C., Van Waes, M., Huang, J., Lobo, S., Germeyer, A., Lessey, B. A., Taylor, R. N., Suchanek, E., and Giudice, L. C. (2003). Identification, characterization, and regulation of the canonical Wnt signaling pathway in human endometrium. *J. Clin. Endocrinol. Metab.* 88, 3860–3866.
22. Taylor, H. S., Bagot, C., Kardana, A., Olive, D., and Arici, A. (1999). HOX gene expression is altered in the endometrium of women with endometriosis. *Hum. Reprod.* 14, 1328–1331.
23. Cooke, P. S., Buchanan, D. L., Lubahn, D.B., and Cunha, G. R. (1998). Mechanism of estrogen action: lessons from the estrogen receptor-alpha knockout mouse. *Biol. Reprod.* 59, 470–475.
24. Cooke, P. S., Buchanan, D. L., Lubahn, D. B., and Cunha, G. R. (2002). Role of stromal-epithelial interactions in hormonal responses of the uterus. In *The Endometrium* (S. R. Glasser, J. R. Aplin, L. C. Giudice, and S. S. Tabibzadeh, Eds.), pp. 151–166. Taylor & Francis, London, New York.
25. Beck, L. R., and Boots, L. R. (1974). The oviduct and its functions. In *The Oviduct and Its Functions* (A. D. Johnson, C. W. Foley, Eds.). Academic Press, London, New York.
26. Hafez, E. S. E., and Blandau, R. J. (1969). *The Mammalian Oviduct*. University of Chicago Press.

27. Blandau, J. R. (1973). Gamete transport in the female mammal. *Handbook of Physiology*, Sect. 7 Endocrinology, The Female Reproductive System, Vol. 2 (R. O. Greep and E. B. Astwood, Eds.) Williams & Wilkins, Baltimore.
28. Brenner, R. M., and Anderson, R. G. W. (1973). Endocrine control of ciliogenesis in the primate oviduct. *Handbook of Physiology*, Sect. 7 Endocrinology, The Female Reproductive System, Vol. 2 (R. O. Greep and E. B. Astwood, Eds.). Williams & Wilkins, Baltimore.
29. Johnson, A. D., and Foley, C. W. (1974). *The Oviduct and Its Functions*. Academic Press, New York.
30. Jansen, R. P. (1984). Endocrine response in the fallopian tube. *Endocr. Rev.* 5, 525–551.
31. Brenner, R. M., and Slayden, O. D. (1994). Cyclic changes in primate oviduct and endometrium. In *The Physiology of Reproduction* (E. Knobil and J. D. Neill, Eds.), Vol. 1, 2nd ed., pp. 541–569. Raven Press, New York.
32. Brenner, R. M., and Slayden, O. D. (1996). The fallopian tube cycle. In *Reproductive Endocrinology, Surgery, and Technology* (E. Y. Adashi, J. A. Rock, and Z. Rosenwaks, Eds.) Lippincott-Raven, New York.
33. Menezo, Y., and Guerin P. (1997). The mammalian oviduct: biochemistry and physiology. *Eur. J. Obstet. Gynecol. Reprod. Biol.* 73, 99–104.
34. Kervancioglu, M. E., Saridogan, E., Atasu, T., Camlibel, T., Demircan, A., Sarikamis, B., and Djahanbakhch, O. (1997). Human fallopian tube epithelial cell co-culture increases fertilization rates in male factor infertility but not in tubal or unexplained infertility. *Hum. Reprod.* 12, 1253–1258.
35. Bongso, T. A., Fong, C. Y., Ng, S. C., and Ratnam, S. S. (1992). Human ampullary co-cultures for blastocyst transfer in assisted reproduction. *Ann. Acad. Med. Singapore.* 21, 571–575.
36. Vlad, M., Walker, D., and Kennedy, R. C. (1996). Nuclei number in human embryos co-cultured with human ampullary cells. *Hum. Reprod.* 11, 1678–1686.
37. Yeung, W. S., Lau, E. Y., Chan, S. T., and Ho, P. C. (1996). Coculture with homologous oviductal cells improved the implantation of human embryos: a prospective randomized control trial. *J. Assist. Reprod. Genet.* 13, 762–767.
38. Goodeaux, L. L., Thibodeaux, J. K., Voelkel, S. A., Anzalone, C. A., Roussel, J. D., Cohen, J. C., and Menezo, Y. (1990). Collection, co-culture and transfer of rhesus preimplantation embryos. *ARTA* 1, 370–379.
39. Takeuchi, K., Nagata, Y., Sandow, B. A., and Hodgen, G. D. (1992). Primary culture of human fallopian tube epithelial cells and co-culture of early mouse pre-embryos. *Mol. Reprod. Dev.* 32, 236–242.
40. Fukaya, T., Chida, S., Murakami, T., and Yajima, A. (1996). Is direct cell-to-cell contact needed to improve embryonic development in co-culture? *Tohoku J. Exp. Med.* 180, 225–232.
41. Gandolfi, F., and Moor, R. M. (1987). Stimulation of early embryonic development in the sheep by co-culture with oviduct epithelial cells. *J. Reprod. Fertil.* 81, 23–28.
42. Eddy, C. A., and Pauerstein, C. J. (1980). Anatomy and physiology of the fallopian tube. *Clin. Obstet. Gynecol.* 23, 1177–1193.
43. Patek, E., Nilsson, L., and Johannisson, E. (1972). Scanning electron microscopic study of the human fallopian tube. Report I. The proliferative and secretory stages. *Fertil. Steril.* 23, 549–565.
44. Patek, E., Nilsson, L., and Johannisson, E. (1972). Scanning electron microscopic study of the human fallopian tube. Report II. Fetal life, reproductive life, and postmenopause. *Fertil. Steril.* 23, 719–733.
45. Novak, E., and Everett, H. S. (1928). Cyclical and other variations in the tubal epithelium. *Am. J. Obstet. Gynecol.* 16, 499.
46. Oberti, C., and Gomez-Rogers, C. (1972). "De novo" ciliogenesis in the human oviduct during the menstrual cycle. In *Symposium on Biology of Reproduction* (J. T. Velardo, B. A. Kasprow, Eds.), pp. 241–248. Pan American Congress of Anatomy, New Orleans.
47. Patek, E. (1974). The epithelium of the human Fallopian tube: a surface ultrastructural and cytochemical study. *Acta. Obstet. Gynecol. Scand. Suppl.* 31, 1–28.
48. Brosens, I. A., and Vasquez, G. (1976). Fimbrial microbiopsy. *J. Reprod. Med.* 16, 171–178.
49. Clyman, M. J. (1966). Electron microscopy of the human fallopian tube. *Fertil. Steril.* 17, 281–301.
50. Verhage, H. G., Bareither, M. L., Jaffe, R. C., and Akbar, M. (1979). Cyclic changes in ciliation, secretion and cell height of the oviductal epithelium in women. *Am. J. Anat.* 156, 505–521.
51. Donnez, J., Casanas-Roux, F., Caprasse, J., Ferin, J., and Thomas, K. (1985). Cyclic changes in ciliation, cell height, and mitotic activity in human tubal epithelium during reproductive life. *Fertil. Steril.* 43, 554–559.
52. Donnez, J., Casanas-Roux, F., Ferin, J., and Thomas, K. (1983). Changes in ciliation and cell height in human tubal epithelium in the fertile and post-fertile years. *Maturitas* 5, 39–45.
53. Andrews, M. C. (1951). Epithelial changes in the puerperal fallopian tube. *Am. J. Obstet. Gynecol.* 62, 28–37.
54. Comer, M. T., Leese, H.J., and Southgate, J. (1998). Induction of a differentiated ciliated cell phenotype in primary cultures of Fallopian tube epithelium. *Hum. Reprod.* 13, 3114–3120.
55. Eckstein, P. (1958). Internal reproductive organs. In *Primatologia* (H. Hofer, A. H. Schultz, D. Stark, Eds.). Karger, Basel.
56. Brenner, R. M., Resko, J. A., and West, N. B. (1974). Cyclic changes in oviductal morphology and residual cytoplasmic estradiol binding capacity induced by sequential estradiol-progesterone treatment of spayed Rhesus monkeys. *Endocrinology* 95, 1094–1104.
57. Verhage, H. G., Mavrogianis, P. A., Boice, M. L., Li, W., and Fazleabas, A. T. (1990). Oviductal epithelium of the baboon: hormonal control and the immuno-gold localization of oviduct-specific glycoproteins. *Am. J. Anat.* 187, 81–90.
58. Odor, D. L., and Augustine, J. R. (1995). Morphological study of changes in the baboon oviductal epithelium during the menstrual cycle. *Microsc. Res. Tech.* 32, 13–28.
59. Brenner, R. M., Carlisle, K. S., Hess, D. L., Sandow, B. A., and West, N. B. (1983). Morphology of the oviducts and endometria of cynomolgus macaques during the menstrual cycle. *Biol. Reprod.* 29, 1289–1302.
60. Rumery, R. E., Gaddum-Rosse, P., Blandau, R. J., and Odor, D. L. (1978). Cyclic changes in ciliation of the oviductal epithelium in the pig-tailed macaque (Macaca nemestrina). *Am. J. Anat.* 153, 345–351.
61. Odor, D. L., Gaddum-Rosse, P., Rumery, R. E., and Blandau, R. J. (1980). Cyclic variations in the oviuductal ciliated cells during the menstrual cycle and after estrogen treatment in the pig-tailed monkey, Macaca nemestrina. *Anat. Rec.* 198, 35–57.
62. Odor, D. L., Gaddum-Rosse, P., and Rumery, R. E. (1983). Secretory cells of the oviduct of the pig-tailed monkey, Macaca nemestrina, during the menstrual cycle and after estrogen treatment. *Am. J. Anat.* 166, 149–172.
63. Allen, E. (1928). Reactions of immature monkey (Macaca rhesus) to injections of ovarian hormone. *J. Morphol.* 46, 479–520.
64. Allen, E. (1928). Further experiments with an ovarian hormone in the ovariectomized adult monkey, Macaca rhesus, especially the degenerative phase of the experimental menstrual cycle. *Am. J. Anat.* 42, 467–487.
65. Verhage, H. G., Fazleabas, A. T., Mavrogianis, P. A., O'Day-Bowman, M. B., Schmidt, A., Arias, E. B., and Jaffe, R. C. (1997).

Characteristics of an oviductal glycoprotein and its potential role in fertility control. *J. Reprod. Fertil. Suppl.* 51, 217–226.

66. Kerr, J. F., Wyllie, A. H., and Currie, A. R. (1972). Apoptosis: a basic biological phenomenon with wide-ranging implications in tissue kinetics. *Br. J. Cancer* 26, 239–257.
67. Wyllie, A. H., Kerr, J. F., and Currie, A. R. (1980). Cell death: the significance of apoptosis. *Int. Rev. Cytol.* 68, 251–306.
68. Lindenbaum, E. S., Beach, D., and Peretz, B. A. (1987). Steroidal binding sites in the ampulla of the human fallopian tube-autoradiographic and biochemical study. *Eur. J. Obstet. Gynecol. Reprod. Biol.* 24, 201–209.
69. Pollow, K., Inthraphuvasak, J., Manz, B., Grill, H. J., and Pollow, B. (1981). A comparison of cytoplasmic and nuclear estradiol and progesterone receptors in human fallopian tube and endometrial tissue. *Fertil. Steril.* 36, 615–622.
70. Pollow, K., Inthraphuvasak, J., Grill, H. J., and Manz B. (1982). Estradiol and progesterone binding components in the cytosol of normal human fallopian tubes. *J. Steroid. Biochem.* 16, 429–435.
71. Pino, A. M., Devoto, L., Davila, M., and Soto, E. (1984). Changes during the menstrual cycle in cytosolic and nuclear concentrations of progestagen receptor in the human Fallopian tube. *J. Reprod. Fertil.* 70, 481–485.
72. Helm, G., Batra, S., and Owman, C. (1987). Cytoplasmic and nuclear progesterone receptors in human fallopian tube and their relationship to plasma steroids during the menstrual cycle. *Int. J. Fertil.* 32, 162–166.
73. West, N. B., and Brenner, R. M. (1983). Estrogen receptor levels in the oviducts and endometria of cynomolgus macaques during the menstrual cycle. *Biol. Reprod.* 29, 1303–1312.
74. Christow, A., Sun, X., and Gemzell-Danielsson, K. (2002). Effect of mifepristone and levonorgestrel on expression of steroid receptors in the human Fallopian tube. *Mol. Hum. Reprod.* 8, 333–340.
75. Amso, N. N., Crow, J., and Shaw, R. W. (1994). Comparative immunohistochemical study of oestrogen and progesterone receptors in the fallopian tube and uterus at different stages of the menstrual cycle and the menopause. *Hum. Reprod.* 9, 1027–1037.
76. Gasc, J. M., and Baulieu, E. E. (1986). Steroid hormone receptors: intracellular distribution. *Biol. Cell.* 56, 1–6.
77. Press, M. F., Nousek-Goebl, N. A., Bur, M., and Greene, G. L. (1986). Estrogen receptor localization in the female genital tract. *Am. J. Pathol.* 123, 280–292.
78. O'Day-Bowman, M. B., and Verhage, H. G. (1992). Immunocytochemical (ICC) localization and steroid regulation of estrogen (ER) and progestin (PR) receptors in baboon (Papio anubis) and human oviduct. 25th annual meeting of SSR, Raleigh, NC, July 12–15, 1992.
79. Horwitz, K. B., and Alexander, P. S. (1983). In situ photolinked nuclear progesterone receptors of human breast cancer cells: subunit molecular weights after transformation and translocation. *Endocrinology* 113, 2195–2201.
80. Savouret, J. F., Misrahi, M., and Milgrom, E. (1990). Molecular action of progesterone. *Int. J. Biochem.* 22, 579–594.
81. Gava, N., Clarke, C. L., Byth, K., Arnett-Mansfield, R. L., and deFazio, A. (2004). Expression of progesterone receptors A and B in the mouse ovary during the estrous cycle. *Endocrinology* 145, 3487–3494.
82. Kastner, P., Krust, A., Turcotte, B., Stropp, U., Tora, L., Gronemeyer, H., and Chambon, P. (1990). Two distinct estrogen-regulated promoters generate transcripts encoding the two functionally different human progesterone receptor forms A and B. *EMBO. J.* 9, 1603–1614.
83. Evans, R. M. (1988). The steroid and thyroid hormone receptor superfamily. *Science*. 240, 889–895.
84. Tsai, M. J., and O'Malley, B. W. (1994). Molecular mechanisms of action of steroid/thyroid receptor superfamily members. *Annu. Rev. Biochem.* 63, 451–486.
85. Graham, J. D., and Clarke, C. L. (1997). Physiological action of progesterone in target tissues. *Endocr. Rev.* 18, 502–519.
86. Sartorius, C. A., Melville, M. Y., Hovland, A. R., Tung, L., Takimoto, G. S., and Horwitz, K. B. (1994). A third transactivation function (AF3) of human progesterone receptors located in the unique N-terminal segment of the B-isoform. *Mol. Endocrinol.* 8, 1347–1360.
87. Wen, D. X., Xu, Y. F., Mais, D. E., Goldman, M. E., and McDonnell, D. P. (1994). The A and B isoforms of the human progesterone receptor operate through distinct signaling pathways within target cells. *Mol. Cell. Biol.* 14, 8356–8364.
88. Conneely, O. M., Kettelberger, D. M., Tsai, M. J., Schrader, W. T., and O'Malley, B. W. (1989). The chicken progesterone receptor A and B isoforms are products of an alternate translation initiation event. *J. Biol. Chem.* 264, 14062–14064.
89. Lessey, B. A., Alexander, P. S., and Horwitz, K. B. (1983). The subunit structure of human breast cancer progesterone receptors: characterization by chromatography and photoaffinity labeling. *Endocrinology* 112, 1267–1274.
90. Schott, D. R., Shyamala, G., Schneider, W., and Parry, G. (1991). Molecular cloning, sequence analyses, and expression of complementary DNA encoding murine progesterone receptor. *Biochemistry* 30, 7014–7020.
91. Bethea, C. L., and Widmann, A. A. (1998). Differential expression of progestin receptor isoforms in the hypothalamus, pituitary, and endometrium of rhesus macaques. *Endocrinology* 139, 677–687.
92. Kraus, W. L., Montano, M. M., and Katzenellenbogen, B. S. (1993). Cloning of the rat progesterone receptor gene 5′-region and identification of two functionally distinct promoters. *Mol. Endocrinol.* 7, 1603–1616.
93. Shyamala, G., Schneider, W., and Schott, D. (1990). Developmental regulation of murine mammary progesterone receptor gene expression. *Endocrinology* 126, 2882–2889.
94. Duffy, D. M., Wells, T. R., Haluska, G. J., and Stouffer, R. L. (1997). The ratio of progesterone receptor isoforms changes in the monkey corpus luteum during the luteal phase of the menstrual cycle. *Biol. Reprod.* 57, 693–699.
95. Mangal, R. K., Wiehle, R. D., Poindexter, A.N. III, and Weigel, N. L. (1997). Differential expression of uterine progesterone receptor forms A and B during the menstrual cycle. *J. Steroid. Biochem. Mol. Biol.* 63, 195–202.
96. Mote, P. A., Balleine, R. L., McGowan, E. M., and Clarke, C. L. (1999). Colocalization of progesterone receptors A and B by dual immunofluorescent histochemistry in human endometrium during the menstrual cycle. *J. Clin. Endocrinol. Metab.* 84, 2963–2971.
97. Hovland, A. R., Powell, R. L., Takimoto, G. S., Tung, L., and Horwitz, K. B. (1998). An N-terminal inhibitory function, IF, suppresses transcription by the A-isoform but not the B-isoform of human progesterone receptors. *J. Biol. Chem.* 273, 5455–5460.
98. Graham, J. D., and Clarke, C. L. (2002). Expression and transcriptional activity of progesterone receptor A and progesterone receptor B in mammalian cells. *Breast Cancer Res.* 4, 187–190.
99. Conneely, O. M., Mulac-Jericevic, B., Lydon, J. P., and De Mayo, F. J. (2001). Reproductive functions of the progesterone receptor isoforms: lessons from knock-out mice. *Mol. Cell. Endocrinol.* 179, 97–103.
100. Lydon, J. P., DeMayo, F. J., Funk, C. R., Mani, S. K., Hughes, A. R., Montgomery, C. A., Jr., Shyamala, G., Conneely, O. M., and O'Malley BW. (1995). Mice lacking progesterone receptor exhibit pleiotropic reproductive abnormalities. *Genes Dev.* 9, 2266–2278.

101. Okada, A., Ohta, Y., Inoue, S., Hiroi, H., Muramatsu, M., and Iguchi, T. (2003). Expression of estrogen, progesterone and androgen receptors in the oviduct of developing, cycling and pre-implantation rats. *J. Mol. Endocrinol.* 30, 301–315.
102. Okada, A., Ohta, Y., Buchanan, D. L., Sato, T., Inoue, S., Hiroi, H., Muramatsu, M., and Iguchi, T. (2002). Changes in ontogenetic expression of estrogen receptor alpha and not of estrogen receptor beta in the female rat reproductive tract. *J. Mol. Endocrinol.* 28, 87–97.
103. Okada, A., Ohta, Y., Brody, S. L., Watanabe, H., Krust, A., Chambon, P., and Iguchi, T. (2004). Role of foxj1 and estrogen receptor alpha in ciliated epithelial cell differentiation of the neonatal oviduct. *J. Mol. Endocrinol.* 32, 615–625.
104. Pelletier, G., Labrie, C., and Labrie, F. (2000). Localization of oestrogen receptor alpha, oestrogen receptor beta and androgen receptors in the rat reproductive organs. *J. Endocrinol.* 165, 359–370.
105. Kuiper, G. G., Enmark, E., Pelto-Huikko, M., Nilsson, S., and Gustafsson, J. A. (1996). Cloning of a novel receptor expressed in rat prostate and ovary. *Proc. Natl. Acad. Sci. U S A.* 93, 5925–5930.
106. Tremblay, G. B., Tremblay, A., Copeland, N. G., Gilbert, D. J., Jenkins, N. A., Labrie, F., and Giguere, V. (1997). Cloning, chromosomal localization, and functional analysis of the murine estrogen receptor beta. *Mol. Endocrinol.* 11, 353–365.
107. Mosselman, S., Polman, J., and Dijkema, R. (1996). ER beta: identification and characterization of a novel human estrogen receptor. *FEBS Lett.* 392, 49–53.
108. Shah, A., Nandedkar, T. D., Raghavan, V. P., Parulekar, S. V., and Natraj, U. (1999). Characterization and localization of estrogen and progesterone receptors of human fallopian tube. *Indian J. Exp. Biol.* 37, 893–899.
109. Takeyama, J., Suzuki, T., Inoue, S., Kaneko, C., Nagura, H., Harada, N., and Sasano, H. (2001). Expression and cellular localization of estrogen receptors alpha and beta in the human fetus. *J. Clin. Endocrinol. Metab.* 86, 2258–2262.
110. Taylor, A. H., (2000). Al-Azzawi, F. Immunolocalisation of oestrogen receptor beta in human tissues. *J. Mol. Endocrinol.* 24, 145–155.
111. Pau, C. Y., Pau, K. Y., and Spies, H. G. (1998). Putative estrogen receptor beta and alpha mRNA expression in male and female rhesus macaques. *Mol. Cell. Endocrinol.* 146, 59–68.
112. Spitz, I. M., Van Look, P. F., and Coelingh Bennink, H. J. (2000). The use of progesterone antagonists and progesterone receptor modulators in contraception. *Steroids* 65, 817–823.
113. Klein-Hitpass, L., Cato, A. C., Henderson, D., and Ryffel, G. U. (1991). Two types of antiprogestins identified by their differential action in transcriptionally active extracts from T47D cells. *Nucleic Acids Res.* 19, 1227–1234.
114. Gemzell-Danielsson, K., Swahn, M. L., Svalander, P., and Bygdeman, M. (1993). Early luteal phase treatment with mifepristone (RU 486) for fertility regulation. *Hum. Reprod.* 8, 870–873.
115. Bygdeman, M., Danielsson, K. G., Marions, L., and Swahn, M. L. (199). Contraceptive use of antiprogestin. *Eur. J. Contracept. Reprod. Health Care* 4, 103–107.
116. Cheng, L., Gulmezoglu, A. M., Ezcurra, E., and Van Look, P. F. (2000). Interventions for emergency contraception. *Cochrane Database Syst. Rev.* CD001324.
117. Sun, X., Christow, A., Marions, L., and Gemzell-Danielsson, K. (2003). Progesterone receptor isoform B in the human fallopian tube and endometrium following mifepristone. *Contraception* 67, 319–326.
118. Slayden, O. D., Zelinski-Wooten, M. B., Chwalisz, K., Stouffer, R. L., and Brenner, R. M. (1998). Chronic treatment of cycling rhesus monkeys with low doses of the antiprogestin ZK 137 316: morphometric assessment of the uterus and oviduct. *Hum. Reprod.* 13, 269–277.
119. Palter, S. F., Mulayim, N., Senturk, L., and Arici, A. (2001). Interleukin-8 in the human fallopian tube. *J. Clin. Endocrinol. Metab.* 86, 2660–2667.
120. Lei, Z. M., and Rao, C. V. (1992). Expression of epidermal growth factor (EGF) receptor and its ligands, EGF and transforming growth factor-alpha, in human fallopian tubes. *Endocrinology* 131, 947–957.
121. el-Danasouri, I., Frances, A., and Westphal, L. M. (1993). Immunocytochemical localization of transforming growth factor-alpha and epidermal growth factor receptor in human fallopian tubes and cumulus cells. *Am. J. Reprod. Immunol.* 30, 82–87.
122. Nelson, K. G., Takahashi, T., Bossert, N. L., Walmer, D. K., and McLachlan, J. A. (1991). Epidermal growth factor replaces estrogen in the stimulation of female genital-tract growth and differentiation. *Proc. Natl. Acad. Sci. U S A* 88, 21–25.
123. Kurachi, H., Morishige, K., Imai, T., Homma, H., Masumoto, N., Yoshimoto, Y., and Miyake, A. (1994). Expression of epidermal growth factor and transforming growth factor-alpha in fallopian tube epithelium and their role in embryogenesis. *Horm. Res.* 41 (Suppl 1), 48–54.
124. Chegini, N., Zhao, Y., and McLean, F. W. (1994). Expression of messenger ribonucleic acid and presence of immunoreactive proteins for epidermal growth factor (EGF), transforming growth factor alpha (TGF alpha) and EGF/TGF alpha receptors and 125I-EGF binding sites in human fallopian tube. *Biol. Reprod.* 50, 1049–1058.
125. Srivastava, M. D., Lippes, J., and Srivastava, B. I. (1996). Cytokines of the human reproductive tract. *Am. J. Reprod. Immunol.* 36, 157–166.
126. Burgess, A. W. (1989). Epidermal growth factor and transforming growth factor alpha. *Br. Med. Bull.* 45, 401–424.
127. Carpenter, G. (1987). Receptors for epidermal growth factor and other polypeptide mitogens. *Annu. Rev. Biochem.* 56, 881–914.
128. Carpenter, G., and Cohen, S. Epidermal growth factor. *J. Biol. Chem.* 265, 7709–7712.
129. Nelson, K. G., Takahashi, T., Lee, D. C., Luetteke, N. C., Bossert, N. L., Ross, K., Eitzman, B. E., and McLachlan, J. A. (1992). Transforming growth factor-alpha is a potential mediator of estrogen action in the mouse uterus. *Endocrinology* 131, 1657–1664.
130. Adachi, K., Kurachi, H., Adachi, H., Imai, T., Sakata, M., Homma, H., Higashiguchi, O., Yamamoto, T., and Miyake A. (1995). Menstrual cycle specific expression of epidermal growth factor receptors in human fallopian tube epithelium. *J. Endocrinol.* 147, 553–563.
131. Adachi, K., Kurachi, H., Homma, H., Adachi, H., Imai, T., Sakata, M., Higashiguchi, O., Yamaguchi, M., Morishige, K., and Sakoyama, Y., et al. (1995). Estrogen induces epidermal growth factor (EGF) receptor and its ligands in human fallopian tube: involvement of EGF but not transforming growth factor-alpha in estrogen-induced tubal cell growth in vitro. *Endocrinology* 136, 2110–2119.
132. Brucker, C., Alexander, N. J., Hodgen, G. D., and Sandow, B. A. (1991). Transforming growth factor-alpha augments meiotic maturation of cumulus cell-enclosed mouse oocytes. *Mol. Reprod. Dev.* 28, 94–98.
133. Downs, S. M., Daniel, S. A., and Eppig, J. J. (1988). Induction of maturation in cumulus cell-enclosed mouse oocytes by follicle-stimulating hormone and epidermal growth factor: evidence for a positive stimulus of somatic cell origin. *J. Exp. Zool.* 245, 86–96.
134. Morishige, K., Kurachi, H., Amemiya, K., Adachi, H., Adachi, K., Sakoyama, Y., Miyake, A., and Tanizawa, O. (1993). Menstrual stage-specific expression of epidermal growth factor and transforming growth factor-alpha in human

oviduct epithelium and their role in early embryogenesis. *Endocrinology* 133, 199–207.
135. Goldman, S., Dirnfeld, M., Koifman, M., Gonen, Y., Lissak, A., and Abramovici, H. (1993). The effect of epidermal growth factor on growth and differentiation of mouse preimplantation embryos in vitro. *Hum. Reprod.* 8, 1459–1462.
136. Li, H. Z., Sun, X., Stavreus-Evers, A., and Gemzell-Danielsson, K. (2004). Effect of mifepristone on the expression of cytokines in the human Fallopian tube. *Mol. Hum. Reprod.* 10, 489–493.
137. Sharkey, A. M., Dellow, K., Blayney, M., Macnamee, M., Charnock-Jones, S., and Smith, S. K. (1995). Stage-specific expression of cytokine and receptor messenger ribonucleic acids in human preimplantation embryos. *Biol. Reprod.* 53, 974–981.
138. Pampfer, S., Wuu, Y. D., Vanderheyden, I., and De Hertogh, R. (1994). Expression of tumor necrosis factor-alpha (TNF alpha) receptors and selective effect of TNF alpha on the inner cell mass in mouse blastocysts. *Endocrinology* 134, 206–212.
139. Wu, Y. D., Pampfer, S., Becquet, P., Vanderheyden, I., Lee, K. H., and De Hertogh, R. (1999). Tumor necrosis factor alpha decreases the viability of mouse blastocysts in vitro and in vivo. *Biol. Reprod.* 60, 479–483.
140. Terman, B. I., Carrion, M. E., Kovacs, E., Rasmussen, B. A., Eddy, R. L., and Shows, T. B. (1991). Identification of a new endothelial cell growth factor receptor tyrosine kinase. *Oncogene* 6, 1677–1683.
141. Shibuya, M., Yamaguchi, S., Yamane, A., Ikeda, T., Tojo, A., Matsushime, H., and Sato, M. (1990). Nucleotide sequence and expression of a novel human receptor-type tyrosine kinase gene (flt) closely related to the fms family. *Oncogene* 5, 519–524.
142. de Vries, C., Escobedo, J. A., Ueno, H., Houck, K., Ferrara, N., and Williams, L. T. (1992). The fms-like tyrosine kinase, a receptor for vascular endothelial growth factor. *Science* 255, 989–991.
143. Gabler, C., Einspanier, A., Schams, D., and Einspanier, R. (1999). Expression of vascular endothelial growth factor (VEGF) and its corresponding receptors (flt-1 and flk-1) in the bovine oviduct. *Mol. Reprod. Dev.* 53, 376–383.
144. Poltorak, Z., Cohen, T., Sivan, R., Kandelis, Y., Spira, G., Vlodavsky, I., Keshet, E., and Neufeld, G. (1997). VEGF145, a secreted vascular endothelial growth factor isoform that binds to extracellular matrix. *J. Biol. Chem.* 272, 7151–7158.
145. Lei, J., Jiang, A., and Pei, D. (1998). Identification and characterization of a new splicing variant of vascular endothelial growth factor: VEGF183. *Biochim. Biophys. Acta.* 1443, 400–406.
146. Tischer, E., Mitchell, R., Hartman, T., Silva, M., Gospodarowicz, D., Fiddes, J.C., and Abraham, J. A. (1991). The human gene for vascular endothelial growth factor. Multiple protein forms are encoded through alternative exon splicing. *J. Biol. Chem.* 266, 11947–11954.
147. Ferrara, N., Chen, H., Davis-Smyth, T., Gerber, H. P., Nguyen, T. N., Peers, D., Chisholm, V., Hillan, K. J., and Schwall, R. H. (1998). Vascular endothelial growth factor is essential for corpus luteum angiogenesis. *Nat. Med.* 4, 336–340.
148. Leung, D. W., Cachianes, G., Kuang, W. J., Goeddel, D. V., and Ferrara, N. (1989). Vascular endothelial growth factor is a secreted angiogenic mitogen. *Science* 246, 1306–1309.
149. Connolly, D. T. (1991). Vascular permeability factor: a unique regulator of blood vessel function. *J. Cell. Biochem.* 47, 219–223.
150. Senger, D. R., Van de Water, L., Brown, L. F., Nagy, J. A., Yeo, K. T., Yeo, T. K., Berse, B., Jackman, R. W., Dvorak, A. M., and Dvorak, H. F. (1993). Vascular permeability factor (VPF, VEGF) in tumor biology. *Cancer Metastasis Rev.* 12, 303–324.
151. Mattioli, M., Barboni, B., Turriani, M., Galeati, G., Zannoni, A., Castellani, G., Berardinelli, P., and Scapolo, P. A. (2001). Follicle activation involves vascular endothelial growth factor production and increased blood vessel extension. *Biol. Reprod.* 65, 1014–1019.
152. Koos, R. D. (1995). Increased expression of vascular endothelial growth/permeability factor in the rat ovary following an ovulatory gonadotropin stimulus: potential roles in follicle rupture. *Biol. Reprod.* 52, 1426–1435.
153. Shifren, J. L., Doldi, N., Ferrara, N., Mesiano, S., and Jaffe, R. B. (1994). In the human fetus, vascular endothelial growth factor is expressed in epithelial cells and myocytes, but not vascular endothelium: implications for mode of action. *J. Clin. Endocrinol. Metab.* 79, 316–322.
154. Witmer, A. N., Dai, J., Weich, H. A., Vrensen, G. F., and Schlingemann, R. O. (2002). Expression of vascular endothelial growth factor receptors 1, 2, and 3 in quiescent endothelia. *J. Histochem. Cytochem.* 50, 767–777.
155. Verhage, H. G., Fazleabas, A. T., and Donnelly, K. (1988). The in vitro synthesis and release of proteins by the human oviduct. *Endocrinology* 122, 1639–1645.
156. Boatman, D. E. (1997). Responses of gametes to the oviductal environment. *Hum. Reprod.* 12, 133–149.
157. Lam, P. M., Briton-Jones, C., Cheung, C. K., Lok, I. H., Yuen, P. M., Cheung, L. P., and Haines, C. (2003). Vascular endothelial growth factor in the human oviduct: localization and regulation of messenger RNA expression in vivo. *Biol. Reprod.* 68, 1870–1876.
158. Gordon, J. D., Mesiano, S., Zaloudek, C. J., and Jaffe, R. B. (1996). Vascular endothelial growth factor localization in human ovary and fallopian tubes: possible role in reproductive function and ovarian cyst formation. *J. Clin. Endocrinol. Metab.* 81, 353–359.
159. Stanke, D. F., Sikes, J. D., DeYoung, D. W., and Tumbleson, M. E. (1974). Proteins and amino acids in bovine oviducal fluid. *J. Reprod. Fertil.* 38, 493–496.
160. Leese, H. J. (1983). Studies on the movement of glucose, pyruvate and lactate into the ampulla and isthmus of the rabbit oviduct. *Q. J. Exp. Physiol.* 68, 89–96.
161. Gille, H., Kowalski, J., Li, B., LeCouter, J., Moffat, B., Zioncheck, T. F., Pelletier, N., and Ferrara, N. (2001). Analysis of biological effects and signaling properties of Flt-1 (VEGFR-1) and KDR (VEGFR-2). A reassessment using novel receptor-specific vascular endothelial growth factor mutants. *J. Biol. Chem.* 276, 3222–3230.
162. Keyt, B. A., Nguyen, H. V., Berleau, L. T., Duarte, C. M., Park, J., Chen, H., and Ferrara, N. (1996). Identification of vascular endothelial growth factor determinants for binding KDR and FLT-1 receptors. Generation of receptor-selective VEGF variants by site-directed mutagenesis. *J. Biol. Chem.* 271, 5638–5646.
163. Olofsson, B., Korpelainen, E., Pepper, M. S., Mandriota, S. J., Aase, K., Kumar, V., Gunji, Y., Jeltsch, M. M., Shibuya, M., Alitalo, K., and Eriksson, U. (1998). Vascular endothelial growth factor B (VEGF-B) binds to VEGF receptor-1 and regulates plasminogen activator activity in endothelial cells. *Proc. Natl. Acad. Sci. U S A* 95, 11709–11714.
164. Wang, H., and Keiser, J. A. (1998). Vascular endothelial growth factor upregulates the expression of matrix metalloproteinases in vascular smooth muscle cells: role of flt-1. *Circ. Res.* 83, 832–840.
165. Lam, P. M., Briton-Jones, C., Cheung, C. K., Leung, S. W., Cheung, L. P., and Haines, C. (2004). Increased messenger RNA expression of vascular endothelial growth factor and its receptors in the implantation site of the human oviduct with ectopic gestation. *Fertil. Steril.* 82, 686–690.
166. Leese, H. J. (1988). The formation and function of oviduct fluid. *J. Reprod. Fertil.* 82, 843–856.
167. Hunter, R. H. F. (1988). *The Fallopian Tubes: Their Role in Fertility and Infertility.* Springer-Verlag, Berlin.

168. Buhi, W. C., Alvarez, I. M., and Kouba, A. J. (2000). Secreted proteins of the oviduct. *Cells Tissues Organs* 166, 165–179.
169. Schulte, B. A., Rao, K. P., Kreutner, A., Thomopoulos, G. N., and Spicer, S. S. (1985). Histochemical examination of glycoconjugates of epithelial cells in the human fallopian tube. *Lab. Invest.* 52, 207–219.
170. Rapisarda, J. J., Mavrogianis, P. A., O'Day-Bowman, M. B., Fazleabas, A. T., and Verhage, H. G. (1993). Immunological characterization and immunocytochemical localization of an oviduct-specific glycoprotein in the human. *J. Clin. Endocrinol. Metab.* 76, 1483–1488.
171. Arias, E. B., Verhage, H. G., and Jaffe, R. C. (1994). Complementary deoxyribonucleic acid cloning and molecular characterization of an estrogen-dependent human oviductal glycoprotein. *Biol. Reprod.* 51, 685–694.
172. O'Day-Bowman, M. B., Mavrogianis, P. A., Reuter, L. M., Johnson, D. E., Fazleabas, A. T., and Verhage, H. G. (1996). Association of oviduct-specific glycoproteins with human and baboon (Papio anubis) ovarian oocytes and enhancement of human sperm binding to human hemizonae following in vitro incubation. *Biol. Reprod.* 54, 60–69.
173. Jaffe, R. C., Arias, E. B., O'Day-Bowman, M. B., Donnelly, K. M., Mavrogianis, P. A., and Verhage, H. G. (1996). Regional distribution and hormonal control of estrogen-dependent oviduct-specific glycoprotein messenger ribonucleic acid in the baboon (Papio anubis). *Biol. Reprod.* 55, 421–426.
174. Verhage, H. G., Boice, M. L., Mavrogianis, P., Donnelly, K., and Fazleabas, A. T. (1989). Immunological characterization and immunocytochemical localization of oviduct-specific glycoproteins in the baboon (Papio anubis). *Endocrinology* 124, 2464–2472.
175. Verhage, H. G., Mavrogianis, P. A., Boomsma, R. A., Schmidt, A., Brenner, R. M., Slayden, O. V., and Jaffe, R. C. (1997). Immunologic and molecular characterization of an estrogen-dependent glycoprotein in the rhesus (Macaca mulatta) oviduct. *Biol. Reprod.* 57, 525–531.
176. Donnelly, K. M., Fazleabas, A. T., Verhage, H. G., Mavrogianis, P. A., and Jaffe, R. C. (1991). Cloning of a recombinant complementary DNA to a baboon (Papio anubis) estradiol-dependent oviduct-specific glycoprotein. *Mol. Endocrinol.* 5, 356–364.
177. Buhi, W. C., Alvarez, I. M., Choi, I., Cleaver, B. D., Simmen, F. A. (1996). Molecular cloning and characterization of an estrogen-dependent porcine oviductal secretory glycoprotein. *Biol. Reprod.* 55, 1305–1314.
178. Buhi, W. C. (2002). Characterization and biological roles of oviduct-specific, oestrogen-dependent glycoprotein. *Reproduction* 123, 355–362.
179. Hirth, F., and Reichert, H. (1999). Conserved genetic programs in insect and mammalian brain development. *Bioessays* 21, 677–684.
180. Briton-Jones, C., Lok, I. H., Chiu, T. T., Cheung, L. P., and Haines, C. (2003). Human chorionic gonadotropin and 17-beta estradiol regulation of human oviductin/oviduct specific glycoprotein mRNA expression in vitro. *Fertil. Steril.* 80(Suppl 2), 720–726.
181. Briton-Jones, C., Lok, I. H., Yuen, P. M., Chiu, T. T., Cheung, L. P., and Haines, C. (2001). Regulation of human oviductin mRNA expression in vivo. *Fertil. Steril.* 75, 942–946.
182. Shemesh, M. (2001). Actions of gonadotrophins on the uterus. *Reproduction* 121, 835–842.
183. Sun, T., Lei, Z. M., and Rao, C. V. (1997). A novel regulation of the oviductal glycoprotein gene expression by luteinizing hormone in bovine tubal epithelial cells. *Mol. Cell. Endocrinol.* 131, 97–108.
184. Briton-Jones, C., Lok, I. H., Yuen, P. M., Chiu, T. T., Cheung, L. P., and Haines, C. (2002). Human oviductin mRNA expression is not maintained in oviduct mucosal cell culture. *Fertil. Steril.* 77, 576–580.
185. Anderson, S. H., and Killian, G. J. (1994). Effect of macromolecules from oviductal conditioned medium on bovine sperm motion and capacitation. *Biol. Reprod.* 51, 795–799.
186. Dey, S. K., Lim, H., Das, S. K., Reese, J., Paria, B. C., Daikoku, T., and Wang, H. (2004). Molecular cues to implantation. *Endocr. Rev.* 25, 341–373.
187. Boice, M. L., McCarthy, T. J., Mavrogianis, P. A., Fazlebas, A. T., and Verhage, H. G. (1990). Localization of oviductal glycoproteins within the zona pellucida and perivitelline space of ovulated ova and early embryos in baboons (Papio anubis). *Biol. Reprod.* 43, 340–346.
188. Wegner, C. C., and Killian, G. J. (1992). Origin of oestrus-associated glycoproteins in bovine oviductal fluid. *J. Reprod. Fertil.* 95, 841–854.
189. Kan, F. W., and Roux, E. (1995). Elaboration of an oviductin by the oviductal epithelium in relation to embryo development as visualized by immunocytochemistry. *Microsc. Res. Tech.* 31, 478–487.
190. Kapur, R. P., and Johnson, L. V. (1988). Ultrastructural evidence that specialized regions of the murine oviduct contribute a glycoprotein to the extracellular matrix of mouse oocytes. *Anat. Rec.* 221, 720–729.
191. Kim, H., Kim, S. R., Kim, M. K., and Schuetz, A. W. (1996). Oviductal protein produces fluorescence staining of the perivitelline space in mouse oocytes. *J. Exp. Zool.* 274, 351–357.
192. Araki, Y., Kurata, S., Oikawa, T., Yamashita, T., Hiroi, M., Naiki, M., and Sendo, F. (1987). A monoclonal antibody reacting with the zona pellucida of the oviductal egg but not with that of the ovarian egg of the golden hamster. *J. Reprod. Immunol.* 11, 193–208.
193. Leveille, M. C., Roberts, K. D., Chevalier, S., Chapdelaine, A., and Bleau, G. (1987). Uptake of an oviductal antigen by the hamster zona pellucida. *Biol. Reprod.* 36, 227–238.
194. Kan, F. W., Roux, E., St-Jacques, S., and Bleau, G. (1990). Demonstration by lectin-gold cytochemistry of transfer of glycoconjugates of oviductal origin to the zona pellucida of oocytes after ovulation in hamsters. *Anat. Rec.* 226, 37–47.
195. Wegner, C. C., and Killian, G. J. (1991). In vitro and in vivo association of an oviduct estrus-associated protein with bovine zona pellucida. *Mol. Reprod. Dev.* 29, 77–84.
196. Abe, H., and Abe, M. (1993). Immunological detection of an oviductal glycoprotein in the rat. *J. Exp. Zool.* 266, 328–335.
197. Reuter, L. M., O'Day-Bowman, M. B., Mavrogianis, P. A., Fazleabas, A. T., and Verhage, H. G. (1994). In vitro incubation of golden (Syrian) hamster ovarian oocytes and human sperm with a human oviduct specific glycoprotein. *Mol. Reprod. Dev.* 38, 160–169.
198. O'Day-Bowman, M. B., Mavrogianis, P. A., Minshall, R. D., and Verhage, H. G. (2002). In vivo versus in vitro oviductal glycoprotein (OGP) association with the zona pellucida (ZP) in the hamster and baboon. *Mol. Reprod. Dev.* 62, 248–256.
199. Schmidt, A., Mavrogianis, P. A., O'Day-Bowman, M. B., and Verhage, H. G. (1997). Species-specific effect of oviductal glycoproteins on hamster sperm binding to hamster oocytes. *Mol. Reprod. Dev.* 46, 201–207.
200. Boatman, D. E., and Magnoni, G. E. (1995). Identification of a sperm penetration factor in the oviduct of the golden hamster. *Biol. Reprod.* 52, 199–207.
201. King, R. S., and Killian, G. J. (1994). Purification of bovine estrus-associated protein and localization of binding on sperm. *Biol. Reprod.* 51, 34–42.
202. King, R. S., Anderson, S. H., and Killian, G. J. (1994). Effect of bovine oviductal estrus-associated protein on the ability of sperm to capacitate and fertilize oocytes. *J. Androl.* 15, 468–478.

203. McCauley, T. C., Buhi, W. C., Didion, B. A., and Day, B. N. (2001). Exposure of oocytes to porcine oviduct-specific glycoprotein reduces the incidence of polyspermic penetration in vitro. Sixth International Conference on Pig Reproduction 47.
204. Martus, N. S., Verhage, H. G., Mavrogianis, P. A., and Thibodeaux, J. K. (1998). Enhancement of bovine oocyte fertilization in vitro with a bovine oviductal specific glycoprotein. *J. Reprod. Fertil.* 113, 323–329.
205. Kouba, A. J., Abeydeera, L. R., Alvarez, I. M., Day, B. N., and Buhi, W. C. (2000). Effects of the porcine oviduct-specific glycoprotein on fertilization, polyspermy, and embryonic development in vitro. *Biol. Reprod.* 63, 242–250.
206. Wang, Q. T., Piotrowska, K., Ciemerych, M. A., Milenkovic, L., Scott, M. P., Davis, R. W., and Zernicka-Goetz, M. (2004). A genome-wide study of gene activity reveals developmental signaling pathways in the preimplantation mouse embryo. *Dev. Cell.* 6, 133–144.
207. Stovall, T. G., and McLord, M. L. (1996). Early pregnancy loss and ectopic pregnancy. In *Novak's Gynecology* (J. S. Berek and E. Y. Adashi, Eds.), 12th ed. Williams & Wilkins, Baltimore.
208. Hitschmann, F., and Adler, L. (1908). Der Bau der Uterusschleimhaut des geschlectsreifen Weibes mit besonderer Berucksightigung der Menstruation. *Monatschr. Geburtsch. Gynakol.* 27, 1–23.
209. Hodgen, G. D. (1983). Surrogate embryo transfer combined with estrogen-progesterone therapy in monkeys: implantation, gestation, and delivery without ovaries. *JAMA* 250, 2167–2171.
210. Good, R. G., and Moyer, D. L. (1968). Estrogen-progesterone relationships in the development of secretory endometrium. *Fertil. Steril.* 19, 37–49.
211. Noyes, R. W., Hertig, A. T., and Rock, J. (1950). Dating the endometrial biopsy. *Fertil. Steril.* 3, 25.
212. Ferenczy, A., Bertrand, G., and Gelfand, M. M. (1979). Proliferation kinetics of human endometrium during the normal menstrual cycle. *Am. J. Obstet. Gynecol.* 133, 859–867.
213. Ferenczy, A., and Giudice, L. C. (1996). The endometrial cycle: morphologic and biochemical events. In *Reproductive Endocrinology, Surgery, and Technology* (E. Y. Adashi, J. A. Rock, and Z. Rosenwaks, Eds.), pp. 171–194. Lippincott-Raven, Philadelphia.
214. Giudice, L. C. (2003). Implantation and endometrial function. In *Reproductive Medicine: Molecular, Cellular and Genetic Fundamentals* (B. C. J. M. Fauser, Ed.), pp. 439–465. Parthenon Publishers, New York.
215. Ferenczy, A., and Guralneck, M. (1983). Endometrial microstructure: structure-function relationships throughout the menstrual cycle. *Sem. Reprod. Endocrinol.* 1, 205–211.
216. Giudice, L. C. (2004). Microarray expression profiling reveals candidate genes for human uterine receptivity. *Am. J. Pharmacogenomics.* 4, 299–312.
217. Brar, A. K., Handwerger, S., Kessler, C. A., and Aronow, B. J. (2001). Gene induction and categorical reprogramming during in vitro human endometrial fibroblast decidualization. *Physiol. Genomics.* 7, 135–148.
218. Irwin, J. C., and Giudice, L. C. (1998). Decidua. In *Encyclopedia of Reproduction* (E. Knobil and J. D. Neill, Eds.), pp. 822–835. Academic Press, San Diego.
219. Nayak, N. R., and Giudice, L. C. (2003). Comparative biology of the IGF system in endometrium, decidua, and placenta, and clinical implications for foetal growth and implantation disorders. *Placenta* 24, 281–296.
220. Fazleabas, A. T., Kim, J. J., and Strakova, Z. (2004). Implantation: embryonic signals and the modulation of the uterine environment: a review. *Placenta* 25(Suppl A), S26–S31.
221. Daly, D. C., Tohan, N., Doney, T. J., Maslar, I. A., and Riddick, D. H. (1982). The significance of lymphocytic-leukocytic infiltrates in interpreting late luteal phase endometrial biopsies. *Fertil. Steril.* 37, 786–791.
222. Brosens, J. J., and Parker, M. G. (2003). Gene expression: Oestrogen receptor hijacked. *Nature* 423, 487–488.
223. Wang, G., Zheng, S., and Fan, Z. [The changes of sex steroid receptors and morphology in Norplant treated endometrium]. *Zhonghua. Fu. Chan. Ke. Za. Zhi.* 33, 493–497.
224. Garcia, M., Derocq, D., Freiss, G., and Rochefort, H. (1992). Activation of estrogen receptor transfected into a receptor-negative breast cancer cell line decreases the metastatic and invasive potential of the cells. *Proc. Natl. Acad. Sci. U S A* 89, 11538–11542.
225. Bergeron, C., Ferenczy, A., Toft, D. O., Schneider, W., and Shyamala, G. (1988). Immunocytochemical study of progesterone receptors in the human endometrium during the menstrual cycle. *Lab. Invest.* 59, 862–869.
226. Lessey, B. A., Killam, A. P., Metzger, D. A., Haney, A. F., Greene, G. L., and McCarty, K. S, Jr. (1988). Immunohistochemical analysis of human uterine estrogen and progesterone receptors throughout the menstrual cycle. *J. Clin. Endocrinol. Metab.* 67, 334–340.
227. Okulicz, W. C., Savasta, A. M., Hoberg, L. M., and Longcope, C. (1990). Biochemical and immunohistochemical analyses of estrogen and progesterone receptors in the rhesus monkey uterus during the proliferative and secretory phases of artificial menstrual cycles. *Fertil. Steril.* 53, 913–920.
228. Hild-Petito, S., Verhage, H. G., and Fazleabas, A. T. (1992). Immunocytochemical localization of estrogen and progestin receptors in the baboon (Papio anubis) uterus during implantation and pregnancy. *Endocrinology* 130, 2343–2353.
229. Bergqvist, A., and Ferno, M. (1993). Oestrogen and progesterone receptors in endometriotic tissue and endometrium: comparison of different cycle phases and ages. *Hum. Reprod.* 8, 2211–2217.
230. Critchley, H. O., Brenner, R. M., Henderson, T. A., Williams, K., Nayak, N. R., Slayden, O. D., Millar, M. R., and Saunders, P. T. (2001). Estrogen receptor beta, but not estrogen receptor alpha, is present in the vascular endothelium of the human and nonhuman primate endometrium. *J. Clin. Endocrinol. Metab.* 86, 1370–1378.
231. Snijders, M. P., de Goeij, A. F., Debets-Te Baerts, M. J., Rousch, M. J., Koudstaal, J., and Bosman, F. T. (1992). Immunocytochemical analysis of oestrogen receptors and progesterone receptors in the human uterus throughout the menstrual cycle and after the menopause. *J. Reprod. Fertil.* 94, 363–371.
232. Rey, J. M., Pujol, P., Dechaud, H., Edouard, E., Hedon, B., and Maudelonde, T. (1998). Expression of oestrogen receptor-alpha splicing variants and oestrogen receptor-beta in endometrium of infertile patients. *Mol. Hum. Reprod.* 4, 641–647.
233. Mylonas, I., Jeschke, U., Shabani, N., Kuhn, C., Balle, A., Kriegel, S., Kupka, M. S., and Friese, K. (2004). Immunohistochemical analysis of estrogen receptor alpha, estrogen receptor beta and progesterone receptor in normal human endometrium. *Acta. Histochem.* 106, 245–252.
234. Couse, J. F., Curtis, S. W., Washburn, T. F., Lindzey, J., Golding, T. S., Lubahn, D. B., Smithies, O., and Korach, K. S. (1995). Analysis of transcription and estrogen insensitivity in the female mouse after targeted disruption of the estrogen receptor gene. *Mol. Endocrinol.* 9, 1441–1454.
235. Couse, J. F., and Korach, K. S. (1998). Exploring the role of sex steroids through studies of receptor deficient mice. *J. Mol. Med.* 76, 497–511.

236. Okulicz, W. C., and Scarrell, R. (1998). Estrogen receptor alpha and progesterone receptor in the rhesus endometrium during the late secretory phase and menses. *Proc. Soc. Exp. Biol. Med.* 218, 316–321.
237. Cunha, G. R. (1991). Ontongeny of sex steroid hormone receptors in mammels. In *Nuclear Hormone Receptors: Molecular Mechanisms, Cellular Functions, Clinical Abnormalities* (M. G. Parker, Ed.), pp. 235–268. Academic Press, London.
238. Cooke, P. S., Buchanan, D L., Young, P., Setiawan, T., Brody, J., Korach, K. S., Taylor, J., Lubahn, D. B., and Cunha, G. R. (1997). Stromal estrogen receptors mediate mitogenic effects of estradiol on uterine epithelium. *Proc. Natl. Acad. Sci. U S A* 94, 6535–6540.
239. Shiozawa, T., Li, S. F., Nakayama, K., Nikaido, T., and Fujii, S. (1996). Relationship between the expression of cyclins/cyclin-dependent kinases and sex-steroid receptors/Ki67 in normal human endometrial glands and stroma during the menstrual cycle. *Mol. Hum. Reprod.* 2, 745–752.
240. Maldonado, V., Castilla, J. A., Martinez, L., Herruzo, A., Concha, A., Fontes, J., Mendoza, N., Garcia-Pena, M. L., Mendoza, J. L., Magan, R., Ortiz, A., and Gonzalez, E. (2003). Expression of transcription factors in endometrium during natural cycles. *J. Assist. Reprod. Genet.* 20, 474–481.
241. Giangrande, P. H., and McDonnell, D. P. The A and B isoforms of the human progesterone receptor: two functionally different transcription factors encoded by a single gene. *Recent Prog. Horm. Res.* 54, 291–313.
242. Brenner, R. M., McClellan, M. C., West, N. B., Novy, M. J., Haluska, G. J., and Sternfeld, M. D. (1991). Estrogen and progestin receptors in the macaque endometrium. *Ann. N Y Acad. Sci.* 622, 149–166.
243. Mote, P. A., Balleine, R. L., McGowan, E. M., and Clarke, C. L. (2000). Heterogeneity of progesterone receptors A and B expression in human endometrial glands and stroma. *Hum. Reprod.* 15(Suppl 3), 48–56.
244. Lessey, B. A., Yeh, I., Castelbaum, A. J., Fritz, M. A., Ilesanmi, A. O., Korzeniowski, P., Sun, J., and Chwalisz, K. (1996). Endometrial progesterone receptors and markers of uterine receptivity in the window of implantation. *Fertil. Steril.* 65, 477–483.
245. Mertens, H. J., Heineman, M. J., Theunissen, P. H., de Jong, F. H., Evers, J. L. (2001). Androgen, estrogen and progesterone receptor expression in the human uterus during the menstrual cycle. *Eur. J. Obstet. Gynecol. Reprod. Biol.* 98, 58–65.
246. Slayden, O. D., Nayak, N. R., Burton, K. A., Chwalisz, K., Cameron, S. T., Critchley, H. O., Baird, D. T., and Brenner, R. M. (2001). Progesterone antagonists increase androgen receptor expression in the rhesus macaque and human endometrium. *J. Clin. Endocrinol. Metab.* 86, 2668–2679.
247. Lathi, R., Bessina, M., Swiersz, L., Giudice, L. C. (2002). The endometrium in hyperinsulinemic states: Nestler J. *Curr. Opin. Endocrinol. Diabetes* 9, 480–485.
248. Lubahn, D. B., Moyer, J. S., Golding, T. S., Couse, J. F., Korach, K. S., and Smithies, O. (1993). Alteration of reproductive function but not prenatal sexual development after insertional disruption of the mouse estrogen receptor gene. *Proc. Natl. Acad. Sci. U S A* 90, 11162–11166.
249. Lydon, J. P., DeMayo, F. J., Conneely, O. M., and O'Malley, B. W. (1996). Reproductive phenotpes of the progesterone receptor null mutant mouse. *J. Steroid. Biochem. Mol. Biol.* 56, 67–77.
250. Cameron, S. T., Critchley, H. O., Thong, K. J., Buckley, C. H., Williams, A. R., and Baird, D. T. (1996). Effects of daily low dose mifepristone on endometrial maturation and proliferation. *Hum. Reprod.* 11, 2518–2526.
251. Maentausta, O., Svalander, P., Danielsson, K. G., Bygdeman, M., and Vihko, R. (1993). The effects of an antiprogestin, mifepristone, and an antiestrogen, tamoxifen, on endometrial 17 beta-hydroxysteroid dehydrogenase and progestin and estrogen receptors during the luteal phase of the menstrual cycle: an immunohistochemical study. *J. Clin. Endocrinol. Metab.* 77, 913–918.
252. Slayden, O. D., Hirst, J. J., and Brenner, R. M. (1993). Estrogen action in the reproductive tract of rhesus monkeys during antiprogestin treatment. *Endocrinology* 132, 1845–1856.
253. Cheon, Y. P., Li, Q., Xu, X., DeMayo, F. J., Bagchi, I. C., and Bagchi, M. K. (2002). A genomic approach to identify novel progesterone receptor regulated pathways in the uterus during implantation. *Mol. Endocrinol.* 16, 2853–2871.
254. Kurita, T., Young, P., Brody, J. R., Lydon, J. P., O'Malley, B. W., and Cunha, G. R. (1998). Stromal progesterone receptors mediate the inhibitory effects of progesterone on estrogen-induced uterine epithelial cell deoxyribonucleic acid synthesis. *Endocrinology* 139, 4708–4713.
255. Kurita, T., Lee, K. J., Cooke, P. S., Taylor, J. A., Lubahn, D. B., and Cunha, G. R. (2000). Paracrine regulation of epithelial progesterone receptor by estradiol in the mouse female reproductive tract. *Biol. Reprod.* 62, 821–830.
256. Kurita, T., Lee, K. J., Cooke, P. S., Lydon, J. P., and Cunha, G. R. (2000). Paracrine regulation of epithelial progesterone receptor and lactoferrin by progesterone in the mouse uterus. *Biol. Reprod.* 62, 831–838.
257. Kurita, T., Lee, K., Saunders, P. T., Cooke, P. S., Taylor, J. A., Lubahn, D. B., Zhao, C., Makela, S., Gustafsson, J. A., Dahiya, R., and Cunha, G. R. (2001). Regulation of progesterone receptors and decidualization in uterine stroma of the estrogen receptor-alpha knockout mouse. *Biol. Reprod.* 64, 272–283.
258. Tong, W., and Pollard, J. W. (2002). Female sex steroid regulation of cell proliferation in the endometrium. In *The Endometrium* (S. R. Glasser, L. C. Giudice, and S. S. Tabibzadeh, Eds.). Taylor & Francis, London, New York.
259. Irwin, J. C., de las Fuentes, L., Dsupin, B. A., and Giudice, L. C. (1993). Insulin-like growth factor regulation of human endometrial stromal cell function: coordinate effects on insulin-like growth factor binding protein-1, cell proliferation and prolactin secretion. *Regul. Pept.* 48, 165–177.
260. Giudice, L. C. (1994). Growth factors and growth modulators in human uterine endometrium: their potential relevance to reproductive medicine. *Fertil. Steril.* 61, 1–17.
261. Murphy, L. J., and Ghahary, A. (1990). Uterine insulin-like growth factor-1: regulation of expression and its role in estrogen-induced uterine proliferation. *Endocr. Rev.* 11, 443–453.
262. Boehm, K. D., Daimon, M., Gorodeski, I. G., Sheean, L. A., Utian, W. H., and Ilan, J. (1990). Expression of the insulin-like and platelet-derived growth factor genes in human uterine tissues. *Mol. Reprod. Dev.* 27, 93–101.
263. Giudice, L. C., Dsupin, B. A., Jin, I. H., Vu, T. H., and Hoffman, A. R. (1993). Differential expression of messenger ribonucleic acids encoding insulin-like growth factors and their receptors in human uterine endometrium and decidua. *J. Clin. Endocrinol. Metab.* 76, 1115–1122.
264. Roy, R. N., Cecutti, A., Gerulath, A. H., Steinberg, W. M., and Bhavnani, B. R. (1997). Endometrial transcripts of human insulin-like growth factors arise by differential promoter usage. *Mol. Cell. Endocrinol.* 135, 11–19.
265. Zhou, J., Dsupin, B. A., Giudice, L. C., and Bondy, C. A. (1994). Insulin-like growth factor system gene expression in human endometrium during the menstrual cycle. *J. Clin. Endocrinol. Metab.* 79, 1723–1734.

266. Rutanen, E. M., Pekonen, F., and Makinen, T. (1988). Soluble 34K binding protein inhibits the binding of insulin-like growth factor I to its cell receptors in human secretory phase endometrium: evidence for autocrine/paracrine regulation of growth factor action. *J. Clin. Endocrinol. Metab.* 66, 173–180.
267. Ritvos, O., Ranta, T., Jalkanen, J., Suikkari, A. M., Voutilainen, R., Bohn, H., and Rutanen, EM. (1988). Insulin-like growth factor (IGF) binding protein from human decidua inhibits the binding and biological action of IGF-I in cultured choriocarcinoma cells. *Endocrinology* 122, 2150–2157.
268. Ghahary, A., Chakrabarti, S., and Murphy, L. J. (1990). Localization of the sites of synthesis and action of insulin-like growth factor-I in the rat uterus. *Mol. Endocrinol.* 4, 191–195.
269. Kapur, S., Tamada, H., Dey, S. K., and Andrews, G. K. (1992). Expression of insulin-like growth factor-I (IGF-I) and its receptor in the peri-implantation mouse uterus, and cell-specific regulation of IGF-I gene expression by estradiol and progesterone. *Biol. Reprod.* 46, 208–219.
270. Bhattacharyya, N., Ramsammy, R., Eatman, E., Hollis, V. W., and Anderson WA. (1994). Protooncogene, growth factor, growth factor receptor, and estrogen and progesterone receptor gene expression in the immature rat uterus after treatment with estrogen and tamoxifen. *J. Submicrosc. Cytol. Pathol.* 26, 147–162.
271. Murphy, L. J., Murphy, L. C., and Friesen, H. G. (1987). Estrogen induces insulin-like growth factor-I expression in the rat uterus. *Mol. Endocrinol.* 1, 445–450.
272. Adesanya, O. O., Zhou, J., Samathanam, C., Powell-Braxton, L., and Bondy, C. A. (1999). Insulin-like growth factor 1 is required for G2 progression in the estradiol-induced mitotic cycle. *Proc. Natl. Acad. Sci. U S A* 96, 3287–3291.
273. Richards, R. G., DiAugustine, R. P., Petrusz, P., Clark, G. C., and Sebastian J. (1996). Estradiol stimulates tyrosine phosphorylation of the insulin-like growth factor-1 receptor and insulin receptor substrate-1 in the uterus. *Proc. Natl. Acad. Sci. U S A* 93, 12002–12007.
274. Richards, R. G., Klotz, D. M., Bush, M. R., Walmer, D. K., and DiAugustine, R. P. (2001). E2-induced degradation of uterine insulin receptor substrate-2: requirement for an IGF-I-stimulated, proteasome-dependent pathway. *Endocrinology* 142, 3842–3849.
275. Rajkumar, K., Dheen, T., Krsek, M., and Murphy, L. J. (1996). Impaired estrogen action in the uterus of insulin-like growth factor binding protein-1 transgenic mice. *Endocrinology* 137, 1258–1264.
276. Pekonen, F., Nyman, T., Lahteenmaki, P., Haukkamaa, M., Rutanen, E. M. (1992). Intrauterine progestin induces continuous insulin-like growth factor-binding protein-1 production in the human endometrium. *J. Clin. Endocrinol. Metab.* 75, 660–664.
277. Buchanan, D. L., Sato, T., Peterson, R. E., and Cooke, P. S. (2000). Antiestrogenic effects of 2,3,7,8-tetrachlorodibenzo-p-dioxin in mouse uterus: critical role of the aryl hydrocarbon receptor in stromal tissue. *Toxicol. Sci.* 57, 302–311.
278. Sato, T., Wang, G., Hardy, M. P., Kurita, T., Cunha, G. R., and Cooke, P. S. (2002). Role of systemic and local IGF-I in the effects of estrogen on growth and epithelial proliferation of mouse uterus. *Endocrinology* 143, 2673–2679.
279. Frost, R. A., Mazella, J., and Tseng, L. (1993). Insulin-like growth factor binding protein-1 inhibits the mitogenic effect of insulin-like growth factors and progestins in human endometrial stromal cells. *Biol. Reprod.* 49, 104–111.
280. Haining, R. E., Cameron, I. T., van Papendorp, C., Davenport, A. P., Prentice, A., Thomas, E. J., and Smith, S. K. (1991). Epidermal growth factor in human endometrium: proliferative effects in culture and immunocytochemical localization in normal and endometriotic tissues. *Hum. Reprod.* 6, 1200–1205.
281. Haining, R. E., Schofield, J. P., Jones, D. S., Rajput-Williams, J., and Smith, S. K. (1991). Identification of mRNA for epidermal growth factor and transforming growth factor-alpha present in low copy number in human endometrium and decidua using reverse transcriptase-polymerase chain reaction. *J. Mol. Endocrinol.* 6, 207–214.
282. Zhang, L., Rees, M. C., and Bicknell, R. (1995). The isolation and long-term culture of normal human endometrial epithelium and stroma: expression of mRNAs for angiogenic polypeptides basally and on oestrogen and progesterone challenges. *J. Cell. Sci.* 108 (Pt 1), 323–331.
283. Curtis, S. W., Washburn, T., Sewall, C., DiAugustine, R., Lindzey, J., Couse, J. F., and Korach, K. S. (1996). Physiological coupling of growth factor and steroid receptor signaling pathways: estrogen receptor knockout mice lack estrogen-like response to epidermal growth factor. *Proc. Natl. Acad. Sci. U S A* 93, 12626–12630.
284. Kato, S., Endoh, H., Masuhiro, Y., Kitamoto, T., Uchiyama, S., Sasaki, H., Masushige, S., Gotoh, Y., Nishida, E., and Kawashima, H., et al. (1995). Activation of the estrogen receptor through phosphorylation by mitogen-activated protein kinase. *Science* 270, 1491–1494.
285. Hom, Y. K., Young, P., Wiesen, J. F., Miettinen, P. J., Derynck, R., Werb, Z., and Cunha, G. R. (1998). Uterine and vaginal organ growth requires epidermal growth factor receptor signaling from stroma. *Endocrinology* 139, 913–921.
286. Zhang, Z., Laping, J., Glasser, S., Day, P., and Mulholland, J. (1998). Mediators of estradiol-stimulated mitosis in the rat uterine luminal epithelium. *Endocrinology* 139, 961–966.
287. Ferriani, R. A., Charnock-Jones, D. S., Prentice, A., Thomas, E. J., and Smith, S. K. (1993). Immunohistochemical localization of acidic and basic fibroblast growth factors in normal human endometrium and endometriosis and the detection of their mRNA by polymerase chain reaction. *Hum. Reprod.* 8, 11–16.
288. Sangha, R. K., Li, X. F., Shams, M., and Ahmed, A. (1997). Fibroblast growth factor receptor-1 is a critical component for endometrial remodeling: localization and expression of basic fibroblast growth factor and FGF-R1 in human endometrium during the menstrual cycle and decreased FGF-R1 expression in menorrhagia. *Lab. Invest.* 77, 389–402.
289. Siegfried, S., Pekonen, F., Nyman, T., Ammala, M., and Rutanen, E. M. (1997). Distinct patterns of expression of keratinocyte growth factor and its receptor in endometrial carcinoma. *Cancer*. 79, 1166–1171.
290. Koji, T., Chedid, M., Rubin, J. S., Slayden, O. D., Csaky, K. G., Aaronson, S. A., and Brenner, R. M. (1994). Progesterone-dependent expression of keratinocyte growth factor mRNA in stromal cells of the primate endometrium: keratinocyte growth factor as a progestomedin. *J. Cell. Biol.* 125, 393–401.
291. Slayden, O. D., Izumi, S. I., Rubin, J. A. S., Lacey, D. L., and Brenner, R. M. (1999). Keratinocyte growth factor in the nonhuman primate endometrium: regulation and action. In *Embryo Implantation: Molecular, Cellular and Clinical Aspects* (D. Carson, Ed.). Springer-Verlag, New York.
292. Slayden, O. D., Rubin, J. S., Lacey, D. L., and Brenner, R. M. (2000). Effects of keratinocyte growth factor in the endometrium of rhesus macaques during the luteal-follicular transition. *J. Clin. Endocrinol. Metab.* 85, 275–285.
293. Brenner, R. M., Slayden, O. D., Rubin, J. S., and Lacey, D. L. (2002). Keratinocyte growth factor has arteriotrophica and antiapoptotic effects in the primate endometrium. In *The Endometrium* (S.R. Glasser, J. D. Aplin, L. C. Giudice, and S. S. Tabibzadeh, Eds.). Taylor & Francis, London, New York.

294. Zarnegar, R., and Michalopoulos, G. K. (1995). The many faces of hepatocyte growth factor: from hepatopoiesis to hematopoiesis. *J. Cell. Biol.* 129, 1177–1180.
295. Sugawara, J., Fukaya, T., Murakami, T., Yoshida, H., and Yajima, A. (1997). Hepatocyte growth factor stimulated proliferation, migration, and lumen formation of human endometrial epithelial cells in vitro. *Biol. Reprod.* 57, 936–942.
296. Mulholland, J. (2002). Steroid regulated genes in the endometrium: a reference base. In *The Endometrium* (S. R. Glasser, J. D. Aplin, L. C. Giudice, and S. S. Tabibzadeh, Eds.). Taylor & Francis, London, New York.
297. Klauber, N., Rohan, R. M., Flynn, E., and D'Amato, R. J. (1997). Critical components of the female reproductive pathway are suppressed by the angiogenesis inhibitor AGM-1470. *Nat. Med.* 3, 443–446.
298. Rogers, P. A. (1996). Structure and function of endometrial blood vessels. *Hum. Reprod. Update* 2, 57–62.
299. Ramsey, E. (1982). Vascular anatomy. In *Biology of the Uterus* (R. Wynn, Ed.), 2nd ed. Plenum, New York.
300. Bartelmez, G. W. The form and functions of the uterine blood vessels in the rhesus monkey. In *Contributions to Embryology*, Carnegie Institution of Washington, 1951, pp. 155–181.
301. Nayak, N. R., and Brenner, R. M. (2002). Vascular proliferation and vascular endothelial growth factor expression in the rhesus macaque endometrium. *J. Clin. Endocrinol. Metab.* 87, 1845–1855.
302. Rogers, P. A., and Gargett, C. E. (1998). Endometrial angiogenesis. *Angiogenesis* 2, 287–294.
303. Reynolds, L. P., Killilea, S. D., and Redmer, D. A. (1992). Angiogenesis in the female reproductive system. *FASEB J.* 6, 886–892.
304. Smith, S. K. (1998). Angiogenesis, vascular endothelial growth factor and the endometrium. *Hum. Reprod. Update.* 4, 509–519.
305. Goodger, A. M., and Rogers, P. A. (1994). Endometrial endothelial cell proliferation during the menstrual cycle. *Hum. Reprod.* 9, 399–405.
306. Rogers, P. A., Lederman, F., and Taylor, N. (1998). Endometrial microvascular growth in normal and dysfunctional states. *Hum. Reprod. Update.* 4, 503–508.
307. Carmeliet, P. (2000). Mechanisms of angiogenesis and arteriogenesis. *Nat. Med.* 6, 389–395.
308. Abberton, K. M., Taylor, N. H., Healy, D. L., and Rogers, P. A. (1999). Vascular smooth muscle cell proliferation in arterioles of the human endometrium. *Hum. Reprod.* 14, 1072–1079.
309. Markee, J. (1940). Menstruation in intraocular endometrial transplants in the rhesus monkey. *Contrib. Embryol.* 28, 219–308.
310. Iruela-Arispe, M. L., Rodriguez-Manzaneque, J. C., and Abu-Jawdeh, G. (1999). Endometrial endothelial cells express estrogen and progesterone receptors and exhibit a tissue specific response to angiogenic growth factors. *Microcirculation* 6, 127–140.
311. Nasr, A., and Breckwoldt, M. (1998). Estrogen replacement therapy and cardiovascular protection: lipid mechanisms are the tip of an iceberg. *Gynecol. Endocrinol.* 12, 43–59.
312. Charnock-Jones, D. S., Sharkey, A. M., Rajput-Williams, J., Burch, D., Schofield, J. P., Fountain, S. A., Boocock, C. A., and Smith, S. K. (1993). Identification and localization of alternately spliced mRNAs for vascular endothelial growth factor in human uterus and estrogen regulation in endometrial carcinoma cell lines. *Biol. Reprod.* 48, 1120–1128.
313. Krikun, G., Schatz, F., Finlay, T., Kadner, S., Mesia, A., Gerrets, R., and Lockwood, C. J. (2000). Expression of angiopoietin-2 by human endometrial endothelial cells: regulation by hypoxia and inflammation. *Biochem. Biophys. Res. Commun.* 275, 159–163.
314. Chegini, N., Rossi, M. J., and Masterson, B. J. (1992). Platelet-derived growth factor (PDGF), epidermal growth factor (EGF), and EGF and PDGF beta-receptors in human endometrial tissue: localization and in vitro action. *Endocrinology* 130, 2373–2385.
315. Horowitz, G. M., Scott, R. T. Jr., Drews, M. R., Navot, D., and Hofmann, G. E. (1993). Immunohistochemical localization of transforming growth factor-alpha in human endometrium, decidua, and trophoblast. *J. Clin. Endocrinol. Metab.* 76, 786–792.
316. Gold, L. I., Saxena, B., Mittal, K. R., Marmor, M., Goswami, S., Nactigal, L., Korc, M., and Demopoulos, R. I. (1994). Increased expression of transforming growth factor beta isoforms and basic fibroblast growth factor in complex hyperplasia and adenocarcinoma of the endometrium: evidence for paracrine and autocrine action. *Cancer Res.* 54, 2347–2358.
317. Leibovich, S. J., Polverini, P. J., Shepard, H. M., Wiseman, D. M., Shively, V., and Nuseir, N. (1987). Macrophage-induced angiogenesis is mediated by tumour necrosis factor-alpha. *Nature* 329, 630–632.
318. Esch, F., Ueno, N., Baird, A., Hill, F., Denoroy, L., Ling, N., Gospodarowicz, D., and Guillemin, R. (1985). Primary structure of bovine brain acidic fibroblast growth factor (FGF). *Biochem. Biophys. Res. Commun.* 133, 554–562.
319. Tabibzadeh, S. (1991). Human endometrium: an active site of cytokine production and action. *Endocr. Rev.* 12, 272–290.
320. Critchley, H. O., Kelly, R. W., and Kooy, J. (1994). Perivascular location of a chemokine interleukin-8 in human endometrium: a preliminary report. *Hum. Reprod.* 9, 1406–1409.
321. Ferrara, N. (2000). Vascular endothelial growth factor and the regulation of angiogenesis. *Recent. Prog. Horm. Res.* 55, 15–35; discussion 35–36.
322. Torry, D. S., Holt, V. J., Keenan, J. A., Harris, G., Caudle, M. R., and Torry, R. J. (1996). Vascular endothelial growth factor expression in cycling human endometrium. *Fertil. Steril.* 66, 72–80.
323. Huang, J. C., Liu, D. Y., and Dawood, M. Y. (1998). The expression of vascular endothelial growth factor isoforms in cultured human endometrial stromal cells and its regulation by 17beta-oestradiol. *Mol. Hum. Reprod.* 4, 603–607.
324. Ancelin, M., Buteau-Lozano, H., Meduri, G., Osborne-Pellegrin, M., Sordello, S., Plouet, J., and Perrot-Applanat, M. (2002). A dynamic shift of VEGF isoforms with a transient and selective progesterone-induced expression of VEGF189 regulates angiogenesis and vascular permeability in human uterus. *Proc. Natl. Acad. Sci. U S A* 99, 6023–6028.
325. Nayak, N. R., Critchley, H. O., Slayden, O. D., Menrad, A., Chwalisz, K., Baird, D. T., and Brenner, R. M. (2000). Progesterone withdrawal up-regulates vascular endothelial growth factor receptor type 2 in the superficial zone stroma of the human and macaque endometrium: potential relevance to menstruation. *J. Clin. Endocrinol. Metab.* 85, 3442–3452.
326. Germeyer, A., Hamilton, A. E., Laughlin, L. S., Lasley, B. L., Brenner, R. M., Giudice, L. C., and Nayak, N. R. (2004). Cellular expression and hormonal regulation of neuropilin-1 and -2 mRNA in the human and rhesus macaque endometrium. *J. Clin. Endocrinol. Metab.* 90, 1783–1790.
327. Bausero, P., Cavaille, F., Meduri, G., Freitas, S., and Perrot-Applanat, M. (1998). Paracrine action of vascular endothelial growth factor in the human endometrium: production and target sites, and hormonal regulation. *Angiogenesis* 2, 167–182.
328. Shifren, J. L., Tseng, J. F., Zaloudek, C. J., Ryan, I. P., Meng, Y. G., Ferrara, N., Jaffe, R. B., and Taylor, R. N. (1996). Ovarian steroid regulation of vascular endothelial growth factor in the human endometrium: implications for angiogenesis during the menstrual cycle and in the pathogenesis of endometriosis. *J. Clin. Endocrinol. Metab.* 81, 3112–3118.

329. Albrecht, E. D., Aberdeen, G. W., Niklaus, A. L., Babischkin, J. S., Suresch, D. L., and Pepe, G. J. (2003). Acute temporal regulation of vascular endothelial growth/permeability factor expression and endothelial morphology in the baboon endometrium by ovarian steroids. *J. Clin. Endocrinol. Metab.* 88, 2844–2852.
330. Cullinan-Bove, K., and Koos, R. D. (1993). Vascular endothelial growth factor/vascular permeability factor expression in the rat uterus: rapid stimulation by estrogen correlates with estrogen-induced increases in uterine capillary permeability and growth. *Endocrinology* 133, 829–837.
331. Hyder, S. M., Stancel, G. M., Chiappetta, C., Murthy, L., Boettger-Tong, H. L., and Makela, S. (1996). Uterine expression of vascular endothelial growth factor is increased by estradiol and tamoxifen. *Cancer Res.* 56, 3954–3960.
332. Hyder, S. M., Chiappetta, C., Murthy, L., Stancel, G. M. (1997). Selective inhibition of estrogen-regulated gene expression in vivo by the pure antiestrogen ICI 182,780. *Cancer Res.* 57, 2547–2549.
333. Sharkey, A. M., Day, K., McPherson, A., Malik, S., Licence, D., Smith, S. K., and Charnock-Jones, D. S. (2000). Vascular endothelial growth factor expression in human endometrium is regulated by hypoxia. *J. Clin. Endocrinol. Metab.* 85, 402–409.
334. Popovici, R. M., Kao, L. C., and Giudice, L. C. (2000). Discovery of new inducible genes in in vitro decidualized human endometrial stromal cells using microarray technology. *Endocrinology* 141, 3510–3513.
335. Gluzman-Poltorak, Z., Cohen, T., Herzog, Y., and Neufeld, G. (2000). Neuropilin-2 is a receptor for the vascular endothelial growth factor (VEGF) forms VEGF-145 and VEGF-165 [corrected]. *J. Biol. Chem.* 275, 18040–18045.
336. Migdal, M., Huppertz, B., Tessler, S., Comforti, A., Shibuya, M., Reich, R., Baumann, H., and Neufeld, G. (1998). Neuropilin-1 is a placenta growth factor-2 receptor. *J. Biol. Chem.* 273, 22272–22278.
337. Rogers, P. A., and Abberton, K. M. (2003). Endometrial arteriogenesis: vascular smooth muscle cell proliferation and differentiation during the menstrual cycle and changes associated with endometrial bleeding disorders. *Microsc. Res. Tech.* 60, 412–419.
338. Yancopoulos, G. D., Davis, S., Gale, N. W., Rudge, J. S., Wiegand, S. J., and Holash, J. (2000). Vascular-specific growth factors and blood vessel formation. *Nature* 407, 242–248.
339. Adams, E. C., Hertig, A. T., and Rock, J. (1956). A description of 34 human ova within the first 17 days of development. *Am. J. Anat.* 98, 435–493.
340. Genbacev, O. D., Prakobphol, A., Foulk, R. A., Krtolica, A. R., Ilic, D., Singer, M. S., Yang, Z. Q., Kiessling, L. L., Rosen, S. D., and Fisher, S. J. (2003). Trophoblast L-selectin-mediated adhesion at the maternal-fetal interface. *Science* 299, 405–408.
341. Martel, D., Malet, C., Gautray, A., and Psychoyos, A. (1987). Surface changes of the luminal uterine epithelium during human menstrual cycle: a scanning electron microscopic study. In *The Endometrium: Hormonal Impacts* (J. De Brux, R. Mortel, and J. Gautray, Eds.), pp. 15–29. Plenum, New York.
342. Martel, D., Monier, M. N., Roche, D., and Psychoyos, A. Hormonal dependence of pinopode formation at the uterine luminal surface. *Hum. Reprod.* 6, 597–603.
343. Nikas, G. (1999). Cell-surface morphological events relevant to human implantation. *Hum. Reprod.* 14(Suppl 2), 37–44.
344. Nikas, G., Makrigiannakis, A., Hovatta, O., and Jones, H. W. Jr. (2000). Surface morphology of the human endometrium. Basic and clinical aspects. *Ann. N Y Acad. Sci.* 900, 316–324.
345. Wilcox, A. J., Baird, D. D., and Weinberg, C. R. (1999). Time of implantation of the conceptus and loss of pregnancy. *N. Engl. J. Med.* 340, 1796–1799.
346. Martel, D., Frydman, R., Glissant, M., Maggioni, C., Roche, D., and Psychoyos, A. (1987). Scanning electron microscopy of postovulatory human endometrium in spontaneous cycles and cycles stimulated by hormone treatment. *J. Endocrinol.* 114, 319–324.
347. Nikas, G., Develioglu, O. H., Toner, J. P., and Jones, H. W. Jr. (1999). Endometrial pinopodes indicate a shift in the window of receptivity in IVF cycles. *Hum. Reprod.* 14, 787–792.
348. Develioglu, O. H., Hsiu, J. G., Nikas, G., Toner, J. P., Oehninger, S., and Jones, H. W. Jr. (1999). Endometrial estrogen and progesterone receptor and pinopode expression in stimulated cycles of oocyte donors. *Fertil. Steril.* 71, 1040–1047.
349. Navot, D., Scott, R. T., Droesch, K., Veeck, L. L., Liu, H. C., and Rosenwaks, Z. (1991). The window of embryo transfer and the efficiency of human conception in vitro. *Fertil. Steril.* 55, 114–118.
350. Kao, L. C., Tulac, S., Lobo, S., Imani, B., Yang, J. P., Germeyer, A., Osteen, K., Taylor, R. N., Lessey, B. A., and Giudice, L. C. (2002). Global gene profiling in human endometrium during the window of implantation. *Endocrinology* 143, 2119–2138.
351. Borthwick, J. M., Charnock-Jones, D. S., Tom, B. D., Hull, M. L., Teirney, R., Phillips, S. C., and Smith, S. K. (2003). Determination of the transcript profile of human endometrium. *Mol. Hum. Reprod.* 9, 19–33.
352. Carson, D. D., Lagow, E., Thathiah, A., Al-Shami, R., Farach-Carson, M. C., Vernon, M., Yuan, L., Fritz, M. A., and Lessey, B. (2002). Changes in gene expression during the early to mid-luteal (receptive phase) transition in human endometrium detected by high-density microarray screening. *Mol. Hum. Reprod.* 8, 871–879.
353. Riesewijk, A., Martin, J., van Os, R., Horcajadas, J. A., Polman, J., Pellicer, A., Mosselman, S., and Simon, C. (2003). Gene expression profiling of human endometrial receptivity on days LH+2 versus LH+7 by microarray technology. *Mol. Hum. Reprod.* 9, 253–264.
354. Vaisse, C., Atger, M., Potier, B., and Milgrom, E. (1990). Human placental protein 14 gene: sequence and characterization of a short duplication. *DNA Cell. Biol.* 9, 401–413.
355. Higuchi, T., Kanzaki, H., Nakayama, H., Fujimoto, M., Hatayama, H., Kojima, K., Iwai, M., Mori, T., and Fujita, J. (1995). Induction of tissue inhibitor of metalloproteinase 3 gene expression during in vitro decidualization of human endometrial stromal cells. *Endocrinology* 136, 4973–4981.
356. Seppala, M., Taylor, R. N., Koistinen, H., Koistinen, R., and Milgrom, E. (2002). Glycodelin: a major lipocalin protein of the reproductive axis with diverse actions in cell recognition and differentiation. *Endocr. Rev.* 23, 401–430.
357. Okulicz, W. C., and Ace, C. I. (2003). Temporal regulation of gene expression during the expected window of receptivity in the rhesus monkey endometrium. *Biol. Reprod.* 69, 1593–1599.
358. Ace, C. I., and Okulicz, W. C. (2004). Microarray profiling of progesterone-regulated endometrial genes during the rhesus monkey secretory phase. *Reprod. Biol. Endocrinol.* 2, 54.
359. Riesewijk, A., Martin, J., Dominguez, F., van Os, R., Pellicer, A., Mosselman, S., and Simon, C. (2002). Genes expressed in receptive versus pre-receptive human endometrium. *J. Soc. Gynecol. Invest.* 9, 223A. P-489 (abstract).
360. Irwin, J. C., Utian, W. H., and Eckert, R. L. (1991). Sex steroids and growth factors differentially regulate the growth and differentiation of cultured human endometrial stromal cells. *Endocrinology* 129, 2385–2392.
361. Zhu, H. H., Huang, J. R., Mazela, J., Elias, J., and Tseng, L. (1992). Progestin stimulates the biosynthesis of fibronectin and accumulation of fibronectin mRNA in human endometrial stromal cells. *Hum. Reprod.* 7, 141–146.
362. Telgmann, R., Maronde, E., Tasken, K., and Gellersen, B. (1997). Activated protein kinase A is required for

differentiation-dependent transcription of the decidual prolactin gene in human endometrial stromal cells. *Endocrinology* 138, 929–937.
363. Brar, A. K., Frank, G. R., Kessler, C. A., Cedars, M. I., and Handwerger, S. (1997). Progesterone-dependent decidualization of the human endometrium is mediated by cAMP. *Endocrine* 6, 301–307.
364. Tierney, E. P., Tulac, S., Huang, S. T., and Giudice, L. C. (2003). Activation of the protein kinase A pathway in human endometrial stromal cells reveals sequential categorical gene regulation. *Physiol. Genomics* 16, 47–66.
365. Lim, H., Song, H., Paria, B. C., Reese, J., Das, S. K., and Dey S. K. (2002). Molecules in blastocyst implantation: uterine and embryonic perspectives. *Vitam. Horm.* 64, 43–76.
366. Ma, L., Benson, G. V., Lim, H., Dey, S. K., and Maas, R. L. (1998). Abdominal B (AbdB) Hoxa genes: regulation in adult uterus by estrogen and progesterone and repression in müllerian duct by the synthetic estrogen diethylstilbestrol (DES). *Dev. Biol.* 197, 141–154.
367. Benson, G. V., Lim, H., Paria, B. C., Satokata, I., Dey, S. K., Maas, R. L. (1996). Mechanisms of reduced fertility in Hoxa-10 mutant mice: uterine homeosis and loss of maternal Hoxa-10 expression. *Development* 122, 2687–2696.
368. Yao, M. W., Lim, H., Schust, D. J., Choe, S. E., Farago, A., Ding, Y., Michaud, S., Church, G. M., and Maas, R. L. (2003). Gene expression profiling reveals progesterone-mediated cell cycle and immunoregulatory roles of Hoxa-10 in the preimplantation uterus. *Mol. Endocrinol.* 17, 610–627.
369. Giudice, L. C., and Kao, L. C. (2004). Endometriosis. *Lancet* 364, 1789–1799.
370. Kodama, T., Hara, T., Okamoto, E., Kusunoki, Y., and Ohama, K. (1998). Characteristic changes of large granular lymphocytes that strongly express CD56 in endometrium during the menstrual cycle and early pregnancy. *Hum. Reprod.* 13, 1036–1043.
371. Hunt, J. S., Petroff, M. G., and Burnett, T. G. (2000). Uterine leukocytes: key players in pregnancy. *Semin. Cell. Dev. Biol.* 11, 127–137.
372. Givan, A. L., White, H. D., Stern, J. E., Colby, E., Gosselin, E. J., Guyre, P. M., and Wira, C. R. (1997). Flow cytometric analysis of leukocytes in the human female reproductive tract: comparison of fallopian tube, uterus, cervix, and vagina. *Am. J. Reprod. Immunol.* 38, 350–359.
373. Wira, C. R., and Sandoe, C. P. (1977). Sex steroid hormone regulation of IgA and IgG in rat uterine secretions. *Nature* 268, 534–536.
374. Wira, C. R., Fahey, J. V., White, H. D., Yeaman, G. R., Givan, A. L., and Howell, A.L. (2002). The mucosal immune system in the human reproductive tract: the influence of the menstrual cycle and menopause on mucosal immunity in the uterus. In *The Endometrium* (S. R. Glasser, J. D. Aplin, L. C. Giudice, and S. S. Tabibzadeh, Eds.). Taylor & Francis, London, New York.
375. Bulmer, J. N., Morrison, L., Longfellow, M., Ritson, A., and Pace, D. (1991). Granulated lymphocytes in human endometrium: histochemical and immunohistochemical studies. *Hum. Reprod.* 6, 791–798.
376. King, A. (2000). Uterine leukocytes and decidualization. *Hum. Reprod. Update.* 6, 28–36.
377. Bonatz, G., Hansmann, M. L., Buchholz, F., Mettler, L., Radzun, H. J., and Semm, K. (1992). Macrophage- and lymphocyte-subtypes in the endometrium during different phases of the ovarian cycle. *Int. J. Gynaecol. Obstet.* 37, 29–36.
378. Booker, S. S., Jayanetti, C., Karalak, S., Hsiu, J. G., and Archer, D. F. (1994). The effect of progesterone on the accumulation of leukocytes in the human endometrium. *Am. J. Obstet. Gynecol.* 171, 139–142.
379. Kamat, B. R., and Isaacson, P. G. (1987). The immunocytochemical distribution of leukocytic subpopulations in human endometrium. *Am. J. Pathol.* 127, 66–73.
380. Loke, Y. W., and King, A. (2000). Immunological aspects of human implantation. *J. Reprod. Fertil. Suppl.* 55, 83–90.
381. Fernandez-Shaw, S., Clarke, M. T., Hicks, B., Naish, C. E., Barlow, D. H., and Starkey, P. M. (1995). Bone marrow-derived cell populations in uterine and ectopic endometrium. *Hum. Reprod.* 10, 2285–2289.
382. King, A., and Loke, Y. W. (1988). Differential expression of blood-group-related carbohydrate antigens by trophoblast subpopulations. *Placenta* 9, 513–521.
383. Witz, C. A., Montoya, I. A., Dey, T. D., and Schenken, R. S. (1994). Characterization of lymphocyte subpopulations and T cell activation in endometriosis. *Am. J. Reprod. Immunol.* 32, 173–179.
384. Lachapelle, M. H., Miron, P., Hemmings, R., Roy, D. C. (1996). Endometrial T, B, and NK cells in patients with recurrent spontaneous abortion. Altered profile and pregnancy outcome. *J. Immunol.* 156, 4027–4034.
385. Tabibzadeh, S. (1990). Immunoreactivity of human endometrium: correlation with endometrial dating. *Fertil. Steril.* 54, 624–631.
386. Kammerer, U., Marzusch, K., Krober, S., Ruck, P., Handgretinger, R., and Dietl, J. (1999). A subset of CD56+ large granular lymphocytes in first-trimester human decidua are proliferating cells. *Fertil. Steril.* 71, 74–79.
387. Jones, R. K., Searle, R. F., Stewart, J. A., Turner, S., and Bulmer, J. N. (1998). Apoptosis, bcl-2 expression, and proliferative activity in human endometrial stroma and endometrial granulated lymphocytes. *Biol. Reprod.* 58, 995–1002.
388. Pace, D., Morrison, L., and Bulmer, J. N. (1989). Proliferative activity in endometrial stromal granulocytes throughout menstrual cycle and early pregnancy. *J. Clin. Pathol.* 42, 35–39.
389. Chantakru, S., Miller, C., Roach, L. E., Kuziel, W. A., Maeda, N., Wang, W. C., Evans, S. S., and Croy, B. A. (2002). Contributions from self-renewal and trafficking to the uterine NK cell population of early pregnancy. *J. Immunol.* 168, 22–28.
390. Jones, R. L., Kelly, R. W., and Critchley, H. O. (1997). Chemokine and cyclooxygenase-2 expression in human endometrium coincides with leukocyte accumulation. *Hum. Reprod.* 12, 1300–1306.
391. Dosiou, C., and Giudice, L. C. (2005). Natural killer cells in pregnancy and recurrent pregnancy loss: endocrine and immunologic perspectives. *Endocr. Rev.* 26, 44–62.
392. Chan, R. W., Schwab, K. E., and Gargett, C. E. (2004). Clonogenicity of human endometrial epithelial and stromal cells. *Biol. Reprod.* 70, 1738–1750.
393. Ruck, P., Marzusch, K., Kaiserling, E., Horny, H. P., Dietl, J., Geiselhart, A., Handgretinger, R., and Redman, C. W. (1994). Distribution of cell adhesion molecules in decidua of early human pregnancy. An immunohistochemical study. *Lab. Invest.* 71, 94–101.
394. Burrows, T. D., King, A., and Loke, Y. W. (1995). The role of integrins in adhesion of decidual NK cells to extracellular matrix and decidual stromal cells. *Cell. Immunol.* 166, 53–61.
395. Haynes, M. K., Wapner, R. L., Jackson, L. G., and Smith, J. B. (1997). Phenotypic analysis of adhesion molecules in first-trimester decidual tissue from chorion villus samples. *Am. J. Reprod. Immunol.* 38, 423–430.
396. Springer, T. A. (1995). Traffic signals on endothelium for lymphocyte recirculation and leukocyte emigration. *Annu. Rev. Physiol.* 57, 827–872.
397. Koopman, L. A., Kopcow, H. D., Rybalov, B., Boyson, J. E., Orange, J. S., Schatz, F., Masch, R., Lockwood, C. J., Schachter, A. D., Park, P. J., and Strominger, J. L. (2003).

Human decidual natural killer cells are a unique NK cell subset with immunomodulatory potential. *J. Exp. Med.* 198, 1201–1212.

398. Guimond, M. J., Luross, J. A., Wang, B., Terhorst, C., Danial, S., and Croy, B. A. (1997). Absence of natural killer cells during murine pregnancy is associated with reproductive compromise in TgE26 mice. *Biol. Reprod.* 56, 169–179.
399. Guimond, M. J., Wang, B., and Croy, B. A. (1998). Engraftment of bone marrow from severe combined immunodeficient (SCID) mice reverses the reproductive deficits in natural killer cell-deficient tg epsilon 26 mice. *J. Exp. Med.* 187, 217–223.
400. de Fougerolles, A. R., and Baines, M. G. (1987). Modulation of the natural killer cell activity in pregnant mice alters the spontaneous abortion rate. *J. Reprod. Immunol.* 11, 147–153.
401. Tabibzadeh, S. S., and Poubouridis, D. (1990). Expression of leukocyte adhesion molecules in human endometrium. *Am. J. Clin. Pathol.* 93, 183–189.
402. Yeaman, G. R., Guyre, P. M., Fanger, M. W., Collins, J. E., White, H. D., Rathbun, W., Orndorff, K. A., Gonzalez, J., Stern, J. E., and Wira, C. R. (1997). Unique CD8+ T cell-rich lymphoid aggregates in human uterine endometrium. *J. Leukoc. Biol.* 61, 42–435.
403. Tabibzadeh, S., and Sun, X. Z. (1992). Cytokine expression in human endometrium throughout the menstrual cycle. *Hum. Reprod.* 7, 1214–1221.
404. Kelly, R. W., King, A. E., and Critchley, H. O. (2001). Cytokine control in human endometrium. *Reproduction* 121, 3–19.
405. Robertson, S. A., and Hudson, S. N. (2002). Cytokines: pivotal regulators of endometrial immunobiology. In *The Endometrium* (S. R. Glasser, J. D. Aplin, L. C. Giudice, and S. S. Tabibzadeh, Eds.). Taylor & Francis, London, New York.
406. Arceci, R. J., Shanahan, F., Stanley, E. R., and Pollard, J. W. (1989). Temporal expression and location of colony-stimulating factor 1 (CSF-1) and its receptor in the female reproductive tract are consistent with CSF-1-regulated placental development. *Proc. Natl. Acad. Sci. U S A* 86, 8818–8822.
407. Daiter, E., Pampfer, S., Yeung, Y. G., Barad, D., Stanley, E. R., and Pollard, J. W. (1992). Expression of colony-stimulating factor-1 in the human uterus and placenta. *J. Clin. Endocrinol. Metab.* 74, 850–858.
408. Simon, C., Gimeno, M. J., Mercader, A., O'Connor, J. E., Remohi, J., Polan, M. L., and Pellicer, A. (1997). Embryonic regulation of integrins beta 3, alpha 4, and alpha 1 in human endometrial epithelial cells in vitro. *J. Clin. Endocrinol. Metab.* 82, 2607–2616.
409. Simon, C., Frances, A., Piquette, G. N., el Danasouri, I., Zurawski, G., Dang, W., and Polan, M. L. (1994). Embryonic implantation in mice is blocked by interleukin-1 receptor antagonist. *Endocrinology* 134, 521–528.
410. Tabibzadeh, S., Lessey, B., and Satyaswaroop, P. G. (1998). Temporal and site-specific expression of transforming growth factor-beta4 in human endometrium. *Mol. Hum. Reprod.* 4, 595–602.
411. Cullinan, E. B., Abbondanzo, S. J., Anderson, P. S., Pollard, J. W., Lessey, B. A., and Stewart, C. L. (1996). Leukemia inhibitory factor (LIF) and LIF receptor expression in human endometrium suggests a potential autocrine/paracrine function in regulating embryo implantation. *Proc. Natl. Acad. Sci. U S A* 93, 3115–3120.
412. Yang, Z. M., Le, S. P., Chen, D. B., Cota, J., Siero, V., Yasukawa, K., and Harper, M. J. (1995). Leukemia inhibitory factor, LIF receptor, and gp130 in the mouse uterus during early pregnancy. *Mol. Reprod. Dev.* 42, 407–414.
413. Bhatt, H., Brunet, L. J., and Stewart, C. L. (1991). Uterine expression of leukemia inhibitory factor coincides with the onset of blastocyst implantation. *Proc. Natl. Acad. Sci. U S A* 88, 11408–11412.
414. Charnock-Jones, D. S., Sharkey, A. M., Fenwick, P., and Smith, S. K. (1994). Leukaemia inhibitory factor mRNA concentration peaks in human endometrium at the time of implantation and the blastocyst contains mRNA for the receptor at this time. *J. Reprod. Fertil.* 101, 421–426.
415. Stewart, C. L., Kaspar, P., Brunet, L. J., Bhatt, H., Gadi, I., Kontgen, F., and Abbondanzo, S. J. (1992). Blastocyst implantation depends on maternal expression of leukaemia inhibitory factor. *Nature* 359, 76–79.
416. Robb, L., Li, R., Hartley, L., Nandurkar, H. H., Koentgen, F., and Begley, C. G. (1998). Infertility in female mice lacking the receptor for interleukin 11 is due to a defective uterine response to implantation. *Nat. Med.* 4, 303–308.
417. Salamonsen, L. A., and Woolley, D. E. (1996). Matrix metalloproteinases in normal menstruation. *Hum. Reprod.* 11(Suppl 2), 124–133.
418. Salamonsen, L. A. (2003). Tissue injury and repair in the female human reproductive tract. *Reproduction* 125, 301–311.
419. Moulton, B. C. (1994). Transforming growth factor-beta stimulates endometrial stromal apoptosis in vitro. *Endocrinology* 134, 1055–1060.
420. Mantovani, A., Bussolino, F., and Dejana, E. (1992). Cytokine regulation of endothelial cell function. *FASEB J.* 6, 2591–2599.
421. Rodgers, W. H., Matrisian, L. M., Giudice, L. C., Dsupin, B., Cannon, P., Svitek, C., Gorstein, F., and Osteen, K. G. (1994). Patterns of matrix metalloproteinase expression in cycling endometrium imply differential functions and regulation by steroid hormones. *J. Clin. Invest.* 94, 946–953.
422. Weissmann, G. (1964). Labilization and stabilization of lysosomes. *Fed. Proc.* 23, 1038–1044.
423. Tabibzadeh, S. (1996). The signals and molecular pathways involved in human menstruation, a unique process of tissue destruction and remodelling. *Mol. Hum. Reprod.* 2, 77–92.
424. Christiaens, G. C., Sixma, J. J., and Haspels, A. A. (1980). Morphology of haemostasis in menstrual endometrium. *Br. J. Obstet. Gynaecol.* 87, 425–439.
425. Fitzpatrick, R. L., and Liggins, G. C. (1980). Effects of prostaglandins on the cervix of pregnant women and sheep. In *Dilatation of the Uterine Cervix* (F. Naftolin, P. F. Stubblefield, Eds.). Raven Press, New York.
426. Ferenczy, A. (1980). Regeneration of the human endometrium. In *Progress in Surgical Pathology* (C. M. Fenoglio-Preiser, M. Wolff, and R. Lattes, Eds.), pp. 157–173. Masson, New York.
427. Padykula, H. A., Coles, L. G., Okulicz, W. C., Rapaport, S. I., McCracken, J. A., King, N. W., Jr., Longcope, C., and Kaiserman-Abramof, I. R. (1989). The basalis of the primate endometrium: a bifunctional germinal compartment. *Biol. Reprod.* 40, 681–690.
428. Okulicz, W. C., Ace, C. I., and Scarrell, R. (1997). Zonal changes in proliferation in the rhesus endometrium during the late secretory phase and menses. *Proc. Soc. Exp. Biol. Med.* 214, 132–138.
429. Ferenczy, A. (1989). Regeneration of the human endometrium. In *Progress in Surgical Pathology* (C. M. Fenoglio and M. Wolff, Eds.), pp. 157–173. Masson, New York.
430. Taylor, H. S. (2004). Endometrial cells derived from donor stem cells in bone marrow transplant recipients. *JAMA* 292, 81–85.
431. Lysiak, J. J., and Lala, P. K. (1992). In situ localization and characterization of bone marrow-derived cells in the decidua of normal murine pregnancy. *Biol. Reprod.* 47, 603–613.

Knobil and Neill's Physiology of Reproduction, Third Edition
edited by Jimmy D. Neill,

CHAPTER 10

Follicular Development: Mouse, Sheep, and Human Models

Aleksandar Rajkovic,[1] Stephanie A. Pangas,[2] and Martin M. Matzuk[2]

Introduction, 383
Morphology of Folliculogenesis, 384
Murine Folliculogenesis, 384
Human Folliculogenesis, 386
Primordial Follicle Formation, 387
Factor in the Germline (Figla), 387
The Primordial to Primary Follicle Transition, 387
Newborn Ovary Homeobox Gene (Nobox), 388
Kit Ligand (Kitl) and Kit Receptor (Kit), 389
Anti-Müllerian Hormone, 389
Nerve Growth Factor and Receptors, 389
Forkhead Box L2 (Foxl2), 390
Oocyte–Granulosa Cell Interactions, 391
GDF9, BMP15, and BMP Receptors in Preantral Follicle and Granulosa Cell Physiology, 391
Growth Differentiation Factor 9, 392
Bone Morphogenetic Protein 15 (Bmp15), 393
Bmp Receptor Type IB Receptor (Bmpr1b, Alk6), 394
Transcription Factors in Preantral Stage Folliculogenesis, 395
TAF 4b, 395
Foxo3a, 395
Antral Follicle Development, 395
Key Roles of the Hypothalamus and Pituitary, 395
Roles of Estrogen Signaling Pathways, 399
Gap Junction Membrane Channel Proteins and Connexins, 401
Metabolism and Folliculogenesis, 402
Leptin, 402
Growth Hormone, 403
Insulin-Like Growth Factor (IGF) Signaling System, 403
Additional Mouse Models for Studying TGFβ Superfamily Signaling in the Ovary, 404
Inhibins, 404
Activins, 408
Follistatin, 409
Conclusion, 410
Acknowledgments, 410
References, 410

INTRODUCTION

Folliculogenesis is the process by which the female germ cell develops within the somatic cells of the ovary and matures into a fertilizable egg (1,2). The development of the follicle is complex and involves the integration of signals from multiple organ systems. While early folliculogenesis appears to be directed by signals within the ovary, endocrine hormones from the pituitary are necessary for folliculogenesis to proceed beyond these early stages. In turn, the ovary produces a number of feedback and feedforward hormones that regulate pituitary physiology. Within the ovary, cross-talk between the oocyte and the somatic-derived granulosa and theca cells also must occur at every stage for normal development (3). Many of the advances in our understanding of the genetics and physiology of reproduction have accelerated due to the availability of multiple genetic and genomic technologies. The availability of nearly

[1]Departments of Obstetrics and Gynecology; [2]Department of Pathology, Baylor College of Medicine, Houston, Texas.

complete genome sequences for human, rat, mouse, chicken, cow, pig, dog, sheep, cat, and continuing sequencing of other genomes, as well as rapidly improving sequencing technologies, are allowing a much faster determination of naturally occurring mutations in humans and other animals. In addition, large scale sequencing of cDNA libraries generated from a variety of human and rodent tissues, including newborn and adult ovaries, and isolated oocytes, is leading to the unprecedented discovery of many novel genes that are critical for reproductive processes (4–7).

Genomic manipulation through engineered deletions, point mutations, duplications and genomic rearrangements in the mouse, has allowed investigators to generate new lines of mice lacking key reproductive genes that serve as in vivo models for human diseases including infertility. The technology to delete mouse genes and generate non-functional loci (gene "knockouts") or novel expression patterns (gene "knockins" and transgenic overexpression lines) has resulted in more than 200 mouse models with reproductive phenotypes (8). These mice are invaluable for various reasons: they allow us to study physiology in order to infer functional importance of a particular gene; they can give us insight into the compensatory mechanisms that occur when the particular gene is deleted; they can be used to attempt treatment of reproductive pathologies; and they can also be used to narrow the candidate genes for particular human disorders. One of the most instructive examples has been the mouse mutants for follicle stimulating hormone (FSH) and its receptor, FSH receptor (FSHR), which closely mimic humans with mutations in *FSHβ* and *FSHR* (9–12) (see below).

A number of searchable internet-based databases are available for investigators interested in genes that have a reproductive function. Jackson Laboratories (*http://www.jax.org*) maintains an extensive website with complete mouse genome informatics where information can be queried under individual genes and includes detailed information about the expression and phenotypes of spontaneous and targeted mutations as well as relevant references. The Ovarian Kaleidoscope database (*http://ovary.stanford.edu*) keeps up to date information on the expression, function and regulation of genes expressed in the ovary. The National Center for Biotechnology Information (*http://www.ncbi.nlm.nih.gov*) has databases of genomes, sequences of many transcriptomes, and invaluable tools for analyzing and discovering genes important in ovarian development. Other online resources include the Mammalian Reproductive Genetics database (*http://mrg.genetics.washington.edu*), a database of the genes and literature relating to reproduction that is searchable by phenotype or expression pattern, and the GermOnline database (*http://germonline.unibas.ch*), a database of microarray expression data for genes relevant to mitosis, meiosis, and gametogenesis.

This chapter focuses on the development and analysis of mouse models for the study of folliculogenesis. Although the emphasis will be placed on rodent models, relevant studies in humans, primates, and sheep will be included when data is available. Understanding reproduction through the manipulation of the mouse genome uncovers basic mechanisms of follicle development and the discovery of genes important in ovarian, follicular, and oocyte development. These models provide a foundation for addressing questions clinically relevant to reproductive disorders and disease in humans.

MORPHOLOGY OF FOLLICULOGENESIS

Murine Folliculogenesis

Folliculogenesis is a highly regulated developmental sequence resulting in the growth and differentiation of the oocyte and associated somatic cells. The most commonly used classification of oocytes and follicles is that of Pedersen and Peters (13,14). This classification is based on oocyte size, granulosa cell numbers, and follicular morphology (Fig. 1). The classification is useful as a means of uniformly describing blocks in follicular development. In addition, the names primordial, primary, secondary, preantral, and antral follicles are convenient for general purposes. Neither nomenclature is ideal, as not all follicles with the same oocyte size class share the same developmental fate. It would be useful to develop stage-specific molecular markers for each follicle class including transitional stages.

Newborn mouse ovaries are densely packed with oocytes, most of which are present in clusters with no evidence of surrounding granulosa cells (14) (Fig. 2A,D). These oocytes have various terms including "naked" oocytes, germ cell cysts, clusters, nests or syncytia and correspond to type 1 follicles in the Pedersen and Peters classification (14–17). The individual oocyte size within each cluster is smaller than 20 μm, and the number of oocytes in a single cluster frequently exceeds 7. Oocytes contained in germ cell clusters are connected by intercellular bridges that possibly play important roles in transferring metabolites and whole organelles between oocytes.

The vast majority of oocytes enter meiosis during embryonic life, and at birth, some oocytes are in the transitory stages of prophase (pachytene and early diplotene), while others have entered late diplotene and dictyate at which they will remain until meiosis is

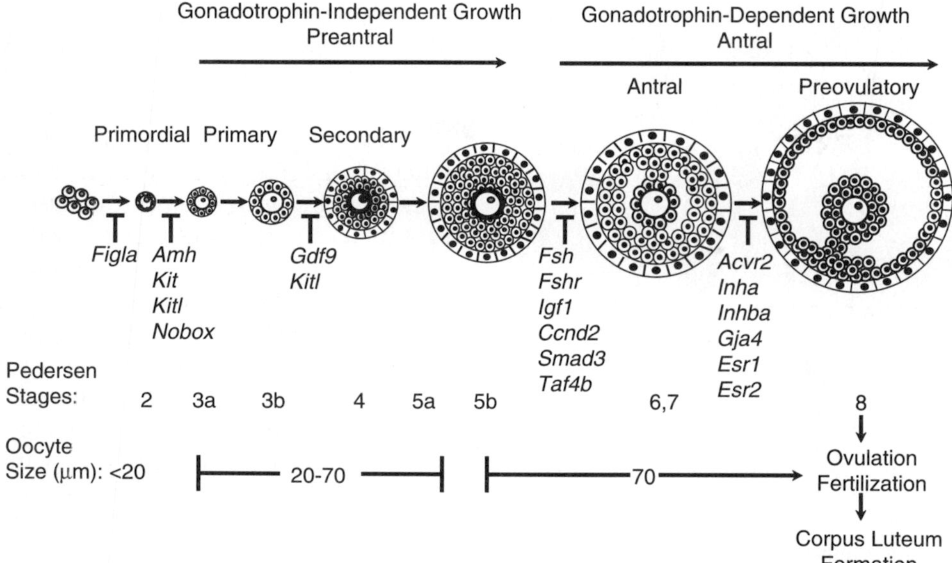

FIG. 1. Classification of mouse follicles. From left to right, postnatal follicle development begins at the primordial stage to the antral stage. Prior to antrum formation, follicle growth is independent of pituitary gonadotropins. Following ovulation, granulosa cell undergo luteinization to become the corpus luteum. Blocks in various knockout mouse models are indicated. Pedersen classification (types 2 to 8) is shown for each stage. Corresponding oocyte diameters for each type are also indicated.

again resumed shortly before ovulation. More centrally within the ovary are oocytes that have been surrounded by flat squamous pre-granulosa cells; these represent primordial follicles. Once formed, primordial follicles are thought to represent the only available source of oocytes during the entire reproductive lifespan of the female. Currently, primordial follicle physiology is under intense investigation as little is known about the mechanisms of primordial follicle arrest or activation (see below).

One to 2 days after birth, a number of primordial follicles appear while the number of germ cell clusters and the number of oocytes within the cluster declines. By postnatal day 3, a class of primary follicles is discernible by the presence of cuboidal granulosa cells and growth of the oocyte beyond 20 μm. These follicles correspond to type 3a follicles in the Pedersen classification (Figs. 1 and 2B,E). Numerous primordial follicles have now replaced the substantial number of germ cell cysts at the periphery of the ovary. At postnatal day 7, most of the germ cell cysts have disappeared and primordial follicles are the most abundant follicular type but primary and secondary follicles are present in the medullary region (Fig. 2). Secondary follicles (type 4) (Figs. 1 and 2B,E) are distinguished from primary follicles as having more than 1 layer of granulosa cells as well as acquiring an additional somatic cell layer, the theca. The thecal cell layer begins to form in growing follicles around the basement membrane and ultimately differentiates into theca interna and theca externa. Theca development also coincides with development of numerous small blood vessels presumably via angiogenesis.

A postnatal day 14 ovary looks similar to a day 7 ovary with multiple secondary follicles. Follicles are also degenerating at this stage, as lack of appropriate pituitary follicle stimulating hormone (FSH) levels does not support complete follicular development. By postnatal day 21, Type 6 follicles are also discernible. In Type 6 follicles, the oocyte reaches its final diameter of approximately 70 μm, and is surrounded by multiple layers of granulosa cells that contain scattered areas of interstitial fluid. Eventually, these scattered areas of fluid coalesce to form the antral cavity. This also divides the granulosa cells into two populations, mural and cumulus. Granulosa cells that surround the oocyte are called the cumulus granulosa cells while granulosa cells that line the basement membrane are called mural granulosa cells. Cumulus and mural granulosa cells differ not only in their location but also in their expression of different molecular markers.

Antral follicles, synonymous with the term Graafian follicle, are highly differentiated endocrine structures and correspond to types 6–8 follicles in the Pedersen classification (Figs. 1 and 2C,F). These follicles are responsive to gonadotropins and

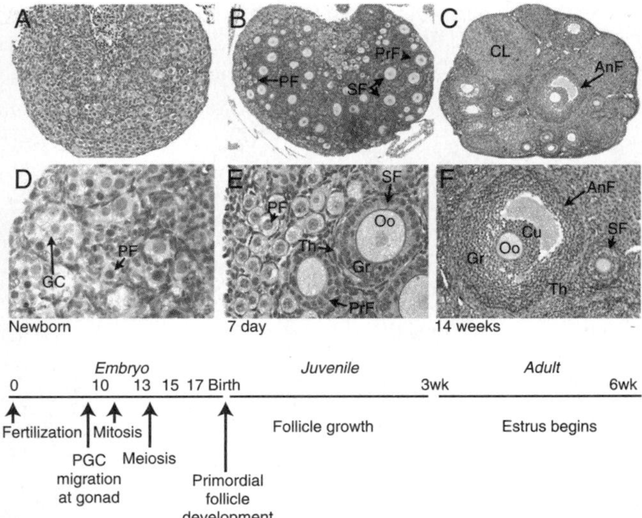

FIG. 2. Folliculogenesis in the mouse. Light micrographs depicting follicle stages are shown in panels **A–C** and magnified views are shown in panels **D–F** above the time line. Follicle development begins embryonically when primordial germ cells migrate to the developing gonad. By E11-13, the PGCs undergo mitotic division and by E13.5 enter meiosis. Oocytes will arrest in meiotic prophase I until ovulation. **Panels A, D:** Newborn ovaries contain oocytes in germ cell cysts (GC) and are beginning to form primordial follicles (PF). **Panels B, E:** By 7 days after birth, primary follicles (PF) and secondary follicles (SF) have developed. Primary follicles contain a growing oocyte and are surrounded by cuboidal granulosa cells. These further develop into secondary follicles by proliferation of granulosa cells and the acquisition of a thecal cell layer (Th). A 4- to 5-day estrous cycle begins at 4 weeks of age and pituitary gonadotropins are required to develop follicular stages beyond the secondary stage. **Panels C, F:** Adult ovaries contain multiple stage follicles including antral stage (AnF) and corpora lutea from previous ovulations (CL). Granulosa cells from antral follicles are divided into two types, mural granulosa cells (Gr) that line the follicle wall, and cumulus granulosa cells (Cu) that surround the oocyte (Oo). GC, germ cell cyst; PF, primordial follicle; PrF, primary follicle; SF, secondary follicle; AnF, antral follicle; CL, corpus luteum; Oo, oocyte; Gr, granulosa cell; Cu, cumulus granulosa cell; Th, thecal cell. Micrographs not shown to scale.

participate actively in the hypothalamic–pituitary–gonadal (HPG) axis. Unless rescued by FSH, granulosa cells of most early antral follicles undergo widespread apoptosis and ultimately follicular death (18,19). In dominant follicles, luteinizing hormone (LH) stimulates thecal cell androgen production, while FSH stimulates granulosa cell proliferation, aromatization of androgens to estrogens, and LH receptor expression (20–23). Follicular estrogens feedback on both the hypothalamus and pituitary to trigger the midcycle gonadotropin surge of LH that precedes ovulation. Inhibin, secreted by granulosa cells, also acts as an endocrine hormone and feeds back to the pituitary to inhibit FSH secretion (24). The complicated web of interactions between the oocyte, granulosa cells, thecal cells, pituitary and hypothalamus ultimately selects follicles for ovulation.

The LH surge triggers release of the oocyte from meiotic arrest, breakdown of the follicle wall and extrusion of the cumulus–oocyte complex. The oocyte completes meiosis I, gives off its first polar body, and progresses to the metaphase stage of meiosis II, where it arrests again until fertilization. Granulosa cells remaining in the postovulatory follicle undergo luteinization, a process that involves cell cycle exit, cellular hypertrophy, acquisition of steroidogenic morphology, and expression of cytochrome P450 cholesterol side chain cleavage. The resulting corpus luteum is a highly vascularized, transient endocrine organ that produces progesterone that is essential for uterine preparation and maintenance of pregnancy. In mice, the entire process of the primordial follicle to maturity takes about 3 weeks—considerably shorter than the several months thought to be required in humans.

Human Folliculogenesis

In the newborn human females, approximately 400,000 follicles remain out of 6–7 million that were present at midway (20 weeks) in gestation (25–27). The newborn ovary measures approximately 1.3 cm × 0.5 cm × 0.3 cm and weighs less than 0.3 grams (28,29). Follicular maturation and atresia occur prenatally and throughout childhood. Because ovulation does not occur, corpora lutea and corpora albicantia are absent in the prepubertal ovary, and all maturing follicles undergo atresia in a manner identical to that occurring in adults. By puberty, almost 90% of the initial pool is depleted—presumably by apoptosis. Beyond age 37, the number of follicles approach approximately 25,000 follicles and the rate of follicle loss increases so that by age 50, approximately 1,000 follicles remain (30,31). Thus, only approximately 400 follicles will ovulate during a woman's reproductive life with the vast majority never reaching the mature stage (32). Adult ovaries are ovoid, and measure approximately 4 cm × 2 cm × 1 cm and weigh approximately 7 grams while menopausal ovaries weigh half as much and contain few if any primordial follicles and a wide variety of stromal-like cells including luteinized stromal cells (33,34).

Primordial follicle formation in humans occurs in utero (35–37). Primordial follicles are observed in almost all ovaries after 20 weeks of gestation, whereas preantral and antral follicles are observed from 26 weeks and onwards. Three types of very small follicles can be defined in primates as well as other mammals: primordial follicles, follicles with mixed cuboidal and squamous granulosa cells (intermediate follicles), and primary follicles. It has been suggested that despite this histologic variation in granulosa cells of early follicles, human follicles with an oocyte nuclear diameter less than 19 μm can be considered resting (27,38). Unfortunately, no molecular markers are known that differentiate resting from early growing follicles.

PRIMORDIAL FOLLICLE FORMATION

Factor in the Germline (Figla)

The molecular events responsible for the breakdown of germ cell cysts and the concomitant formation of primordial follicles in the newborn ovary are poorly defined. No currently identified secreted hormone or growth factor has been shown to play a function in the breakdown of germ cell cysts and formation of primordial follicles. However, a mutation in the mouse helix-loop-helix transcription factor, factor in the germline alpha (*Figla*) (Table 1), causes a disruption of primordial follicle formation and the subsequent loss of most of the oocytes by postnatal day 2 (39). FIGLA is an oocyte-specific protein and was known to bind promoter elements (E-box) in the oocyte zona pellucida genes and regulate their expression in vitro (40). Zona pellucida proteins are a major component of the glycoprotein-rich matrix that surrounds the developing oocyte and is required for fertilization. It may also act as a barrier to the exchange of growth factors between the oocyte and the immediately adjacent granulosa cells (41). However, *Zp1*, *Zp2*, and *Zp3* knockout mice can form primordial through antral follicles, and therefore, individually, zona pellucida genes can not account for the observed phenotype of the *Figla* mutant mice (42–45). Of interest would be a triple knockout for all three zona pellucida genes and a comparison to the *Figla* knockout phenotype. Furthermore, it is likely that FIGLA regulates expression of other genes that are critical for follicle formation and oocyte survival. A human homolog of FIGLA is also restricted in its expression to germ cells, forms heterodimer with the E12 protein, and binds to the E-box of the human Zp2 promoter (46,47). Human *FIGLA* is expressed as early as 14 weeks gestational age with a dramatic increase in transcripts by midgestation (19 weeks gestational age), the time of primordial follicle formation in humans, suggesting a similar conserved function of the human and mouse FIGLA proteins.

THE PRIMORDIAL TO PRIMARY FOLLICLE TRANSITION

The regulation and activation of the nongrowing pool of primordial follicles is still one of the most poorly understood processes in ovarian reproductive biology. The transition from primordial to growing follicle occurs spontaneously in newborn mouse ovaries cultured in media in the presence of serum (48,49). After 8 days of culture of newborn ovaries, primary and secondary follicles form. Oocyte-granulosa cell complexes isolated from the 8-day organ cultures of newborn ovaries can be used to produce oocytes that can be fertilized and develop into embryos and live pups. Unlike mouse, follicular development in vitro occurs more rapidly in baboon ovaries, with many primary follicles observed by 2 days of culture and much diminished number of primordial follicles (49,50). Removal of ovaries from their natural milieu by culturing them may eliminate a potential inhibitor of follicular growth that may be present in the surrounding tissue, peritoneum, or blood. On the other hand, it is possible that current culture media provide a much richer source of nutrients and oxygen, leading to premature activation of follicular growth.

Numerous factors have been shown to accelerate development of follicles in vitro. Primordial follicles from 4-day-old rat newborn ovaries cultured in vitro develop faster when basic fibroblast growth factor, kit ligand, leukemia inhibitory factor, nerve growth factor, and bone morphogenetic protein 4 are introduced into the culture (51–55). Typically, without treatment with growth factors, 50% of follicles grow in vitro; this rises to 60% upon treatment with particular growth factors. The small increase in growing follicles is statistically significant and indicates that numerous factors can enhance follicular growth in vitro. The role of growth differentiation factor 9 (GDF9) in the transition from primordial to primary follicles is controversial; some authors report stimulation and other authors report no effect on primordial follicles (56–59). It is clear, however, that mice lacking GDF9 develop to the primary follicle stage (60).

In mice, a hypothesis was also advanced that the decline in progesterone after birth induces primordial follicle formation (61), but this is unlikely to be the case in humans since progesterone levels are high at 20 weeks of gestation (62–64). Pregnant baboons treated with the aromatase inhibitor CGS 20267 deliver fetuses with ovaries that contain 50% less primordial follicles and larger numbers of interfollicular nests (equivalent to germ cell cysts) as compared to female fetuses given CGS 20267 and estradiol benzoate (65,66). The aromatase inhibitor CGS 20267 reduces maternal and umbilical estradiol levels by 95% and the observed ovarian effects are attributed to decreased estrogen levels. Estrogen receptors alpha and beta (ERα and ERβ, respectively) are expressed during midgestation of baboon fetuses, in the mesenchymal-epithelial cells that surround oocytes in the germ cell cysts (67). The presence of estrogen receptors so early during folliculogenesis would argue that the estrogen effect on primordial follicle formation is direct. Further investigations are necessary to determine the role of hormones in the primordial to primary follicle transition.

Newborn Ovary Homeobox Gene (Nobox)

Nobox is an oocyte-specific homeobox gene discovered in screens to identify oocyte-specific genes using "in silico" subtraction of expressed sequence tag (EST) databases. Nobox was one of the first oocyte-specific homeobox genes to be discovered (68). *Nobox* RNA and protein are expressed both in germ cell cysts, primordial follicles and throughout folliculogenesis (68,69). A human homolog of *Nobox* exists, but little is known regarding its expression and function. Female mice lacking NOBOX cannot form growing follicles, arrest folliculogenesis at the transition from primordial to primary follicles and are infertile (69). Unlike *Gdf9* null ovaries, NOBOX deficient ovaries rapidly lose oocytes so that by 2 weeks after birth very few follicles are found (Fig. 3). Homeobox genes encode transcription factors that play critical functions during organogenesis and can be either ubiquitous or cell-type specific in their expression and function. Since *Nobox* expression is restricted to oocytes, it was hypothesized that NOBOX regulates expression of genes preferentially expressed in oocytes. Consistent with this hypothesis, the NOBOX deficiency disrupts expression of multiple oocyte-specific genes including *Gdf9*, *Bmp15*, *Oct4*, *Rfpl4*, and *Mos*. However, it is currently unclear whether NOBOX directly or indirectly regulates these genes. NOBOX does not affect expression of *Figla* and *Zp* genes, demonstrating that a number of transcription factors may be important in the transition from germ cell cysts to growing follicles. Importantly, it will be necessary to determine the mechanism of follicle loss in *Nobox* null ovaries, since it appears to be independent of known members of the apoptotic cascade (i.e., *Bcl2*, *Caspase2*, *Bax*, and *Bcl2l2*) that are still expressed in *Nobox* null ovaries. It is possible that other oocyte-specific genes are essential for oocyte survival and elucidation of their pathways in NOBOX deficient ovaries will be of great interest.

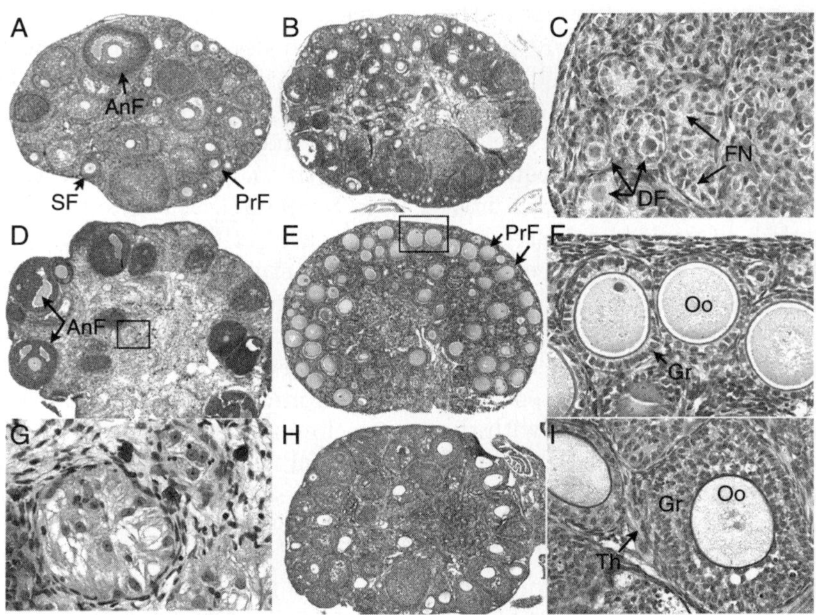

FIG. 3. Ovarian histology for various mouse knockout models. Light micrographs of adult ovarian sections from knockout models. **Panel A:** Adult wild-type ovaries demonstrate follicles from many stages. **Panel B:** Adult Fsh-/- mutant ovaries lack antral follicles. **Panel C:** 14-day old Nobox-/- ovaries contain only follicular nests and degenerating oocytes. **Panel D:** Erα-/-;Erβ-/- double mutant ovaries containing many advanced follicle types as well as structures that resemble male seminiferous tubules (boxed and shown at higher magnification in panel G). **Panel E:** Gdf9-/- null ovaries arrest at the primary follicle stage. **Panel F:** Magnification of boxed Gdf9-/- follicles shown in panel E. **Panel H:** Double mutant Inha-/-;Gdf9-/- ovary demonstrates partial rescue of the Gdf9-/- phenotype. **Panel I:** Higher magnification of the Inha-/-;Gdf9-/- showing multilaminar follicles and a thecal cell layer. PrF, primary follicle; SF, secondary follicle; AnF, antral follicle; Oo, oocyte; Gr, granulosa cell; Th, thecal cell; DF, degenerating follicle; FN, follicular nest. Micrographs not shown to scale.

Kit Ligand (Kitl) and Kit Receptor (Kit)

Kit encodes a transmembrane receptor with tyrosine kinase activity and belongs to a family that includes colony stimulating factor 1 (CSF1) and platelet-derived growth factors (PDGF) α and β. The receptor tyrosine kinase KIT and its ligand (KITL), are encoded by the *white spotting* (*W*) and *Steel* (*Sl*) loci in the mouse, respectively (70–73). KIT-KITL signaling plays important roles in hematopoiesis, melanogenesis and gametogenesis (74,75).

The first insight into the role of KITL and its receptor, KIT, came from the studies of mice spontaneously mutated at the *Kitl* and *Kit* loci (76,77). There are more than 60 spontaneous mutations in the *Kitl* and *Kit* loci and depending on the severity of functional compromise, these mice show mild to severe defects in hematopoiesis, pigmentation, and reproduction. The mutant alleles are semidominant: some are lethal when homozygous and others are viable. *Kit* expression can be detected as early as 7.5 days post conception and is high in proliferating germ cells (70). *Kit* ceases to be expressed when oogonia enter meiosis, around E13.5 and is re-expressed in newborn ovaries. A similar pattern of expression is observed for *Kitl* (71). *Kitl* is expressed in newborn ovaries in pre-granulosa cells and continues to be expressed in granulosa cells throughout folliculogenesis (78–80). *Kit* is expressed in primordial and other oocytes throughout follicular growth (78,81). KIT has also been detected in thecal cells and KITL–KIT interactions can stimulate thecal cell growth and androstenedione production in the absence of gonadotropins (82). Both in vivo and in vitro data provide evidence that KITL and KIT interactions play a significant role in the primordial to primary follicle transition (83,84). The Sl^d allele causes infertility due to a complete absence of germ cells in the ovary (85). The Sl^t allele, however, causes infertility due to the early growth arrest of ovarian follicles (84). Folliculogenesis from mice that carry the Sl^t allele is arrested early with many primordial follicles present and few growing follicles (84).

Monoclonal antibodies against KIT have been used to study the functions of KIT in early folliculogenesis (86). Mice injected with anti-KIT antibody (ACK2) blocked follicle development at the primordial follicle stage when mice were injected at birth. Incorporation of BrdU into granulosa cells was also diminished in these ovaries, arguing that DNA synthesis was also diminished. If antibodies were injected starting with day 2, when primordial follicles are already apparent in mouse ovaries, primary follicular development was only slightly perturbed, and these mice have a diminished number of antral follicles with decreased proliferation and DNA synthesis in granulosa cells. It therefore appears that functional KIT is important at the very early stages of folliculogenesis in order for it to act. Since primordial follicle formation was not disturbed in any of these experiments, the conclusion can be made that KIT is not essential for the initial step of germ cell cyst breakdown into individual primordial follicles, but is necessary for the primordial to primary transition and later during antral follicle formation.

Anti-Müllerian Hormone

The dimeric glycoprotein anti-Müllerian hormone (AMH) is a member of the transforming growth factor-beta (TGFβ) superfamily of growth and differentiation factors. Embryonically, AMH induces degeneration of Müllerian (female) ducts during male fetal sex differentiation (87). Postnatally in ovaries, AMH and its receptor AMHR2 are expressed in granulosa cells of primary and growing follicles in rodents and humans (88–91). AMH appears to be the only natural ligand of the AMHR2 receptor as *Amh* and *Amhr2* mutants phenocopy each other (92). Although *Amh* null mouse females are fertile with normal litter sizes (87), a careful histologic analysis of ovaries from 25-day-old, 4-month-old, and 13-month-old revealed significantly less primordial follicles in 4 and 13 month old *Amh* null ovaries (93). AMH therefore appears to inhibit growth of primordial follicles, and absence of AMH leads to a faster depletion of primordial follicles. It is possible that AMH attenuates the effects of FSH, and that absence of AMH potentiates the proliferative effects of FSH in the sexually mature ovary (4 and 13 month old ovary) leading to greater recruitment of primordial follicles (94). In vitro studies using 2-day postnatal mouse ovaries cultured in the presence and absence of AMH show that fewer growing follicles were found in ovaries treated with AMH (91). Transgenic mice that chronically overexpress AMH under the control of mouse metallothionein promoter are infertile, have less germ cells at birth, and lose germ cells in the span of two weeks after birth (95). These results are in contrast with observations from *Amh* null mice in which *Amh* inhibits growth of primordial follicles; the mechanism whereby overexpression of AMH accelerates germ cell loss is unclear. It is possible that AMH has dual roles in follicle development: one embryonically that results in the expected number of germ cells at birth, and one postnatally, that regulates the size of the growing follicle pool.

Nerve Growth Factor and Receptors

Nerve growth factor (NGF) belongs to a family of proteins known as neurotrophins (96). Neurotrophins are required for the survival and differentiation of neurons within both the central and peripheral nervous system. In the rat ovary, NGF and its two receptors, NGFR and NTRK1 are expressed before primordial follicle formation suggesting that NGF plays a role in early folliculogenesis (97–99). Indeed, analysis of mice lacking NGF revealed that *Ngf* null ovaries from 7-day mice had a drastic reduction in the number of primary and secondary follicles as compared to control mice (100,101). Although the number of primordial follicles was equivalent, the number of oocytes within the germ cell cysts was also increased in NGF null ovaries. NGF may play a role in differentiation of flat pre-granulosa cells into the cuboidal cells that characterize primary follicles. The reduction in the number of secondary follicles also indicates that NGF may functionally interact with GDF9 and kit ligand (KL) to induce proliferation and differentiation of somatic cells surrounding the oocyte. The number of PCNA positive cells was distinctly lower in *Ngf* null ovaries suggesting that somatic cells in *Ngf* null mice have reduced ovarian mitotic activity.

The lack of NGFR, one of the two receptors via which NGF acts, does not affect primordial, primary or secondary follicle formation, and mice are fertile (102). Unfortunately, NTRK1 deficiency causes death at or shortly after birth (103,104), and there are no reports of whether ovarian development is affected in these newborn mice. The NTRK2 receptor is also expressed in primordial and growing oocytes, while ligands that bind NTRK2, neurotrophin 5 (NT5) and brain derived neurotrophic factor (BDNF), are expressed in rodent and human primordial oocytes and granulosa cells (105,106). Although NT5 and BDNF are expressed in primordial follicles, *Ntrk2* deletion in mice does not affect the number of primordial follicles; however, secondary follicles are significantly lower in 7-day ovaries (106). Double mutants of *Nt5* and *Bdnf* also show significantly lower levels of secondary follicles (106). As measured by PCNA, granulosa cell proliferation is diminished in NTRK2 null animals, while apoptosis is not increased. Quantitative RT-PCR comparing *Ntrk2* null mice with wild type mice did not identify mRNA differences between the growth factor genes *Gdf9* and *Kitl*, or their receptors *Bmpr2* (a receptor for BMPs and GDFs) and *Kit* (the receptor for KITL). On the other hand, the amount of FSH receptor mRNA was significantly reduced in *Ntrk2* null ovaries as compared to wild type ovaries. It is possible that this reduction in FSH receptors leads to the decreased number of secondary follicles observed in NTRK2 null mice and diminished mitotic activity. FSH levels are somewhat diminished in NTRK2 null ovaries but not significantly different from heterozygous animals. *Ntkr2* null 4- to 5-day-old ovaries transplanted under the kidney capsule of adult female C57BL/6J mice rapidly lose oocytes and arrest follicular development. Therefore it appears that NTRK2 and its ligands NT5 and BDNF, act in similar fashion to KITL and KIT to regulate early survival of female germ cells. In vitro studies using K252a to block all NTRK receptors in the mouse newborn ovaries and human fetal ovaries show that NTRK receptors are important for germ cell survival (107). These data indicate that neurotrophins provide support to nonneuronal targets and play important role in intraovarian regulation of early folliculogenesis along with GDF9 and KITL.

Forkhead Box L2 (Foxl2)

Blepharophimosis ptosis and epicanthus inversus syndrome can occur with (BPES I) or without ovarian failure (BPES II). Nonsense mutations and duplications in *Foxl2*, a member of the forkhead/hepatocyte nuclear factor 3 (FKH/HNF3) gene family of transcription factors, cause BPES (108,109). The phenotype in women with ovarian failure varies from the presence of primordial follicles to ovarian streak ovaries (110,111). In situ hybridization of mouse ovaries indicates that *Foxl2* is expressed postnatally in the pre-granulosa cells that surround primordial follicles and beyond (108). Ovaries from *Foxl2*$^{-/-}$ mice contain oocytes surrounded mostly by flat granulosa cells that grow slowly compared to wild type ovaries (112,113). Expression of the oocyte genes *Figla*, *Gdf9*, and *Kit* was not affected by the lack of FOXL2, and ovarian morphology as well as oocyte count were also unaffected at birth. Therefore, the lack of FOXL2 caused granulosa cells in mutant ovaries to fail to complete the transition from squamous to cuboidal. Activin βA and anti-Müllerian hormone (AMH) expression is strongly diminished in these mutant ovaries. This is important as AMH has been suggested to be an inhibitor of primordial follicle recruitment (91,93, 114–118), and lack of AMH in *Foxl2*$^{-/-}$ ovaries would be expected to lead to greater number of more advanced follicles in adult mice. Clearly, other mechanisms must lead to the block in folliculogenesis or the role of AMH is not dominant in this mutation. Since FOXL2 is a transcription factor, identification of direct downstream targets would be useful for understanding genetic pathways that lead to the block in early folliculogenesis. The steroidogenic acute regulatory (*Star*) gene contains multiple putative forkhead DNA consensus sites, and FOXL2 binds these sites directly (119). FOXL2 most likely acts as a repressor of StAR, and dominant negative mutations within FOXL2 most likely interfere with repressor activity of the wild-type protein with subsequent derepression of StAR (119). *Star* is expressed in steroidogenic tissues including adrenal glands, granulosa and thecal cells of the ovary and is responsible for the increased mobilization and delivery of cholesterol precursors to the inner mitochondrial membrane. However, derepression of StAR activity in mice that lack FOXL2 is probably not the cause of the early follicular block at the primordial/primary follicular stage. FOXL2 downstream targets that are yet to be identified are most likely the cause of the early follicular block.

OOCYTE–GRANULOSA CELL INTERACTIONS

Oocyte–granulosa cell interactions are paramount for the growth of oocytes and proliferation and differentiation of granulosa cells (3). For many years, it was thought that the somatic cells that surround oocytes play a pivotal role in follicular development. There had been earlier reports suggesting an oocyte role in folliculogenesis, since removal of oocytes from the antral follicles resulted in increased circulation of progesterone (120,121). The authors concluded that oocytes secrete a factor that prevented premature granulosa cell luteinization. Epidermal growth factor (EGF), a potent stimulator of cumulus expansion in intact complexes, fails to stimulate cumulus expansion after oocytectomy (removal of oocytes from the follicles) (122). Oocytectomized complexes, however, respond to FSH and EGF when co-cultured with germinal vesicle stage denuded oocytes or culture medium conditioned by denuded oocytes. These experiments indicate that a factor or factors secreted by the oocytes plays an important role in cumulus cell expansion. Other studies showed that transforming growth factor beta 1 (TGFβ1) stimulated hyaluronic acid (HA) synthesis in both cumulus and mural granulosa cells (124). TGFβ1 stimulation was additive to the oocyte factor and neutralizing antibodies to TGFβ did not inhibit the response to the oocyte factor(s). These results indicated that the oocyte factor(s) and TGFβ are not the same and that they operate through different receptors in stimulating HA synthesis. Finally, subsequent experiments showed that GDF9 satisfies many of the criteria for being an oocyte-secreted factor important in granulosa cell proliferation and differentiation (58,80,125,126) (see below).

Currently, oocytes have been ascribed a dominant role to the somatic cells in directing early folliculogenesis. Eppig and colleagues have provided such evidence (127,128). The hypothesis that oocytes control the rate of follicular development was tested by isolating oocytes from the secondary follicles of 12-day-old mice and combining these oocytes with the somatic cells isolated from newborn mouse ovaries. When oocytes from a 12-day-old secondary follicle were removed, reaggregated with somatic cells from newborn ovary, and implanted beneath the renal capsules of bilaterally ovariectomized host females, antral follicles were visible in 9 days, indicating that the rate of follicular development was accelerated by almost twofold (128). As a control, primordial oocytes were re-aggregated with the same somatic cells and after 9 days, only secondary follicles were observed with absence of antral follicles. The granulosa cells from these reaggregated ovaries underwent cumulus expansion when exposed to FSH in vitro, and recovered oocytes were competent to undergo fertilization and embryonic development. These experiments indicate that a 12-day oocyte is able to accelerate somatic cell differentiation and proliferation with resulting accelerated production of antral follicles.

Oocyte to granulosa cell communication requires that oocytes either secrete factors that have an autocrine or paracrine function, and/or express receptors on their cell surface for detecting granulosa cell derived paracrine factors. Likely, there exist a number of yet undiscovered oocyte-produced proteins during follicle development. Strategies have been designed to identify genes encoding mouse oocyte secretory and transmembrane proteins using a signal sequence trap (129). This method exploits the yeast invertase gene (SUC2) that is necessary for the yeast to grow on sucrose medium (130). A screen was designed whereby a mutant SUC2 gene was engineered that required a cloned DNA fragment to contain a signal sequence in order for the SUC2 to be secreted (130). Using this method, a number of known and novel sequences have been identified (129). How these factors influence oocyte–granulosa cell communication and development is currently unknown.

GDF9, BMP15, AND BMP RECEPTORS IN PREANTRAL FOLLICLE AND GRANULOSA CELL PHYSIOLOGY

Preantral follicle growth is generally characterized by growth of both the oocyte and the granulosa cells. In addition, preantral follicles develop an additional somatic cell layer—the thecal cells. Preantral follicle growth corresponding to type 3b to type 5b follicles of Pedersen and Peters classification (Fig. 1) appears to be independent of the pituitary gonadotropins (see below). This independence from gonadotropins is supported by phenotypes in FSHβ subunit, FSH receptor (FSHR) and naturally occurring gonadotropin releasing hormone (GnRH) knockouts, all of which develop to type 5b follicles (10–12,131–133). Preantral follicle growth is regulated by a number of autocrine and paracrine factors, of which oocyte-secreted GDF9 and BMP15 play important roles.

Growth Differentiation Factor 9

Growth differentiation factor 9 (GDF9) is a member of the transforming growth factor-β superfamily (134). Like other members of the TGFβ superfamily, GDF9 is synthesized as a prepropeptide precursor that has a signal peptide, a pro region, and a bioactive mature domain. The pro region is cleaved by a specific protease in the intracellular or extracellular compartment to generate the bioactive mature domain. *Gdf9* mRNA is preferentially expressed in oocytes within the ovary as early as type 3a follicles, and its transcripts are also found abundantly in unfertilized eggs (134,135). To study the functions of GDF9, our group engineered *Gdf9* null mice (60). *Gdf9* null mice form primary follicles with large oocytes that are greater than 70 μm in diameter surrounded by a single layer of cuboidal granulosa cells (Fig. 3) (58,60,136). Histologically, growth of the oocyte is uncoupled from proliferation of granulosa cells, and *Gdf9* null mice showed convincingly for the first time that factors secreted by oocytes are critical for granulosa cell proliferation and follicular progression.

Ultrastructural studies showed that fully grown oocytes isolated from *Gdf9* null mice progress to advanced stages of differentiation equivalent to those found in antral follicles of control mice (136). In vitro maturation of oocytes from *Gdf9* null mice revealed that most of the oocytes are capable of resuming meiosis. Among the characteristic ultrastructural features of oocytes from *Gdf9* null mice are perinuclear organelle aggregation, unusual peripheral Golgi complexes, and a failure to form cortical granules (136).

The knockout model showed that GDF9 is not critical in the transition from primordial to primary follicles but plays a crucial role in the growth of follicle beyond primary follicular stage. Molecular analysis of the *Gdf9* knockout revealed a lack of thecal cell markers (17α-hydroxylase, LH receptor, and kit) (58). While thecal cell *kit* expression is absent in these mice, *kit* expression in the oocyte, however, is unchanged. kitl and inhibin α levels are up-regulated in follicles that lack GDF9, indicating that GDF9 may negatively regulate these growth factors. It is likely that up-regulation of kitl leads to greater signaling through oocyte-expressed KIT and thus the increase in oocyte size in the *Gdf9* null ovaries. In vitro studies corroborate the suppression of *Kitl* expression in granulosa cells of preantral and mural granulosa cells by recombinant GDF9 (80). However, some in vitro studies have shown conflicting data between species-specific forms of recombinant GDF9 (rat versus mouse) and tissue type (mouse, rat, and human) and whether GDF9 stimulates the growth of primordial follicles and upregulates the expression of inhibin α (57,59,137–139). Future studies are required to understand the differences between rat, mouse, and human GDF9. However, both the knockout data and the in vitro studies demonstrate the important role of GDF9 in early follicle growth and differentiation.

Differences in the process of programmed cell death (apoptosis) also appear to be part of the *Gdf9* null phenotype. Analysis of apoptosis with TUNEL (*t*erminal deoxynucleotidyl transferase-mediated d*U*TP *n*ick *e*nd *l*abeling) revealed that very few of the granulosa cells underwent apoptosis in *Gdf9* null ovaries. More than 50% of the granulosa cells in wild type type 3b follicles stained for the proliferating cell nuclear antigen (PCNA) while less than 10% of the type 3b follicles in the GDF9 null ovaries are positive. These data indicate that lack of GDF9 leads to a decrease in the granulosa cell proliferation as well as a decrease in their turnover (58,140). These data also indicates that granulosa cells of primary follicles do not die via an apoptotic mechanism.

Since the *Gdf9* knockout ovaries arrest very early in postnatal folliculogenesis, studies looking for effects of GDF9 at later stages have had to rely on in vitro studies. These studies showed that GDF9 signaling is also critical in antral follicle growth and ovulation (56,125,141,142). During ovulation, GDF9 induces cumulus cell expansion, an inflammation-like process that involves the mucification of the cells surrounding the oocyte. This process binds the cumulus cells to the oocyte, protects the oocyte during follicular extrusion and assists in sperm binding and fertilization. Cumulus cell expansion partly involves the production of hyaluronic acid (HA), prostaglandin E2 (PGE2), pentraxin 3 (PTX3), and tumor necrosis factor alpha induced protein 6 (TNFAIP6) (138,143,144), all of which have been demonstrated to be regulated by GDF9 in vitro (125,138,145). The role of GDF9 in preovulatory granulosa cells can be studied by isolating mural granulosa cells from large antral follicles, culturing them with and without recombinant GDF9 and comparing gene expression using oligonucleotide array hybridization with microarray chips. Several genes were noted to be highly induced in granulosa cells treated with recombinant GDF9, including *Ptx3*, *Tnfaip6*, and gremlin (138,146).

Ptx3 knockout female mice are subfertile and have defects in ovulation (138,147). Superovulated *Ptx3* null mice release fewer oocytes than controls. Surprisingly, many oocytes isolated from *Ptx3* null mice were denuded of cumulus granulosa cells, whereas oocytes from control mice retained their cumulus cells. Ovulation induction followed by fertilization revealed that only 1.3% of oocytes fertilized in *Ptx3* knockout, while 55% fertilized in heterozygous controls. It is therefore likely that disruption in the cumulus mass surrounding the oocytes causes reduced fertilization and the subfertility in *Ptx3* knockout mice.

The importance of cumulus cells expansion in fertility has also been demonstrated by the knockout mouse model for *Tnfaip6* (144). Female *Tnfaip6*$^{-/-}$ mice are infertile (144). The cumulus cell complexes from these mice fail to expand due to the lack of hyaluronan and heavy chain inter-α-trypsin inhibitor family complexes. Fertility can be restored by treating cumulus cell–oocyte complexes in vitro with recombinant TNFAIP6, which has been shown to catalzye the transfer of heavy chains to hyaluranon in cell-free systems (144). Thus, knockout mice for two of the genes demonstrated to be downstream of GDF9 signaling, *Ptx3* and *Tnfaip6*, have severe defects in cumulus expansion that render the mice sterile.

The in vitro data on GDF9 activity and cumulus cell gene expression suggest that GDF9, by many criteria, is the long-sought after oocyte-secreted growth factor involved in cumulus cell expansion. The following evidence firmly supports this view: (a) GDF9 stimulates hyaluronan synthase 2 mRNA expression; (b) GDF9 inhibits urokinase plasminogen activator; (c) GDF9 inhibits LH receptor mRNA; (d) GDF9 stimulates cumulus expansion; (e) GDF9 stimulates progesterone production via an increase in StAR; and (f) GDF9 stimulates cyclooxygenase 2, pentraxin 3, and Tnfaip6 mRNA synthesis (125,145). The vast amount of data regarding the roles of GDF9 in follicular development shows the importance of combining in vivo with in vitro models, both of which have been instrumental in delineating the key functions of the oocyte-specific protein GDF9.

Bone Morphogenetic Protein 15 (Bmp15)

BMP15 was cloned using degenerative oligonucleotides to identify novel members of the bone morphogenetic protein (BMP) family (148). BMPs are the largest subgroup of the TGFβ superfamily of growth and differentiation factors (149). BMP15 is expressed in the ovary, and by in situ hybridization, *Bmp15* appears to be exclusive to oocytes as early as type 3a but also within oocytes of more advanced follicular stages (135,148). Mouse and human BMP15 share 76% amino acid identity, and both map to the X chromosome.

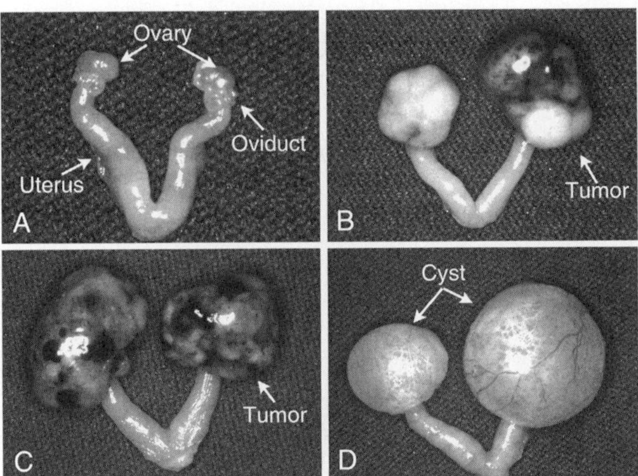

FIG. 4. Gross anatomy of ovarian tumor and cyst formation in several mouse models. **Panel A:** Reproductive tract of an adult female mouse. **Panel B:** Female Inha$^{-/-}$ mice develop ovarian tumors with 100% penetrance. **Panel C:** Double knockout of Inha$^{-/-}$; Acvr2$^{-/-}$ develop tumors similar to Inha$^{-/-}$ despite the lack of activin signaling through ACVR2. **Panel D:** Bilateral fluid filled cysts in a Gdf9$^{-/-}$; Bmp15$^{-/-}$ double knockout ovary.

Bmp15 null female mice are subfertile and have few ovarian histopathologies (150). Rather, these mice have defects in ovulation and fertilization rates. Since GDF9 and BMP15 expression patterns overlap, it is possible that GDF9 and BMP15 have redundant functions. Generation of double mutant mice support this hypothesis. *Gdf9*$^{+/-}$ *Bmp15*$^{-/-}$ ovaries have more defects than *Bmp15*$^{-/-}$ mutants alone, suggesting that there is some overlap in function between the two ligands. Double homozygous *Gdf9*$^{-/-}$ *Bmp15*$^{-/-}$ resemble *Gdf9*$^{-/-}$ mutants, and contain large fluid-filled cysts (Fig. 4).

The importance of BMP15 in ovarian function was initially shown in the study of Inverdale and Hanna sheep which carry naturally occurring X-linked mutations that cause increased ovulation in the heterozygotes but primary ovarian failure in homozygotes (151,152) (Table 1). Linkage mapping of the Inverdale mutation identified *BMP15* as one of the candidate genes. *BMP15* gene sequence comparison between Inverdale and wild type sheep identified a substitution of a highly conserved hydrophobic valine residue of the mature protein with aspartic acid (V31D) (153). Mutational analysis of the Hanna sheep identified a nonsense mutation in the *BMP15* coding region (153,154). Ovarian follicles in sheep homozygous for the *BMP15* mutations do not normally grow beyond the primary follicle stage (154–156). The ovarian pathology in the *BMP15* sheep mutants resembles the ovarian pathology observed in *Gdf9*

TABLE 1. *Sheep mutations causing ovarian defects*

Mutation (strain)	Gene	Type of mutation	Region of mutation	Possible mechanism of disruption (ref.)
FecX[I] (Inverdale)	*BMP15*	Missense V31D	Mature	Protein dimerization (153)
FecX[H] (Hanna)	*BMP15*	Nonsense Q23X	Mature	Truncated protein product (153)
FecG[H]	*GDF9*	Missense S77F	Mature	Possibly interrupts type I receptor binding (419)
FecX[B]	*BMP15*	Missense S99I		Residue may be involved in type II receptor binding (419)
FecX[G]	*BMP15*	Nonsense Q239X	Pre-pro	Premature termination in pre-pro region (419)
FecB (Booroola)	*BMPR1B*	Missense Q249R	Kinase domain	Kinase activity (164)

null mice. Interestingly, homozygous BMP15 null mice are phenotypically dissimilar from the homozygous sheep mutants (150). This may reflect the type of mutation (complete null versus amino acid substitution). Most likely, since the sheep mutation is not a null mutation, the mutated BMP15 could act in a dominant negative dosage-sensitive manner to block GDF9. This is supported by in vitro studies that use recombinant human BMP15 with an I31D substitution, which mimics the V31D substitution found in Inverdale sheep. When BMP15 I31D is co-expressed with GDF9, it severely impairs proteolytic processing and secretion of GDF9 (157,158). In the murine *Bmp15* knockout, this interaction would not exist because BMP15 is not produced and cannot interfere with GDF9 activity. Another alternative hypothesis is that the discrepancies between mouse and sheep models are reflective of the fundamental biological differences between polyovulatory (i.e., mouse and rat) and monoovulatory animals (i.e., sheep, human, and primate), and perhaps GDF9 signaling in mice is dominant over BMP15 as compared to sheep (150). The generation of mice with engineered BMP15 mutations that mimic the sheep mutations via knock-in transgenic strategies would answer these questions.

The dominant negative role of *Bmp15* mutations has also been hypothesized in humans, with the discovery of a family that had ovarian dysgenesis (ovaries characterized by lack of germ cells) (159). Two female siblings of an Italian family had primary amenorrhea, ovaries that lacked follicles, a normal 46 XX karyotype, and no evidence of autoimmune disease or consanguinity. A search for candidate genes revealed a mutation in the human *BMP15* gene that involved an A to G transition at nucleotide 704, changing a tyrosine at position 235 to a cysteine. The tyrosine residue is highly conserved in human, sheep, mouse, rat, and pig and is located in exon 2 in the propeptide region of the protein. The mutant Y235C protein was unable to stimulate incorporation of ^{3}H-thymidine into human granulosa cells, and it also inhibited wild-type BMP15's ability to stimulate ^{3}H-thymidine incorporation into granulosa cells. In addition, human granulosa cell proliferation was enhanced in the presence of wild-type BMP15, but was unaffected by mutant protein (159). These data support the hypothesis that secretion of unprocessed monomer or dimeric mutant products cause dominant negative interactions with the wild-type protein, thereby resulting in an early block in folliculogenesis. However, it is unclear whether the Y235C mutation affects human BMP15 interactions with GDF9 or with GDF9 processing. The father, who was hemizygous for the mutation, was fertile, demonstrating that BMP15 does not affect spermatogenesis, as in the BMP15 knockout mouse model. Clearly, BMP15 is essential in follicular growth in sheep and humans, and further studies in mice are necessary to establish whether BMP15 plays as critical role.

Bmp Receptor Type IB Receptor (Bmpr1b, Alk6)

Receptor complexes for the TGFβ superfamily members consist of two closely related transmembrane serine/threonine kinases, the "type I" and "type II" receptors. Seven type I and five type II receptors have been identified in mammals. Upon binding the dimeric ligands, the receptor complexes phosphorylate intracellular signaling molecules called SMADs which subsequently translocate to the nucleus and regulate transcription of target genes (149,160,161). Studies on BMP15 signaling using rat primary granulosa cells from early antral follicles and a human granulosa cell line COV434 indicate that BMP15 binds with the highest affinity to BMPR1B receptors on the surface of the granulosa cells. Modeled after signaling by other BMP ligands, it has been proposed the BMP15/BMPR1B complex then recruits the type II receptor BMPR2, resulting in signal transduction (SMAD1/5/8 phosphorylation) (162).

The mutation Q249R (glutamine→arginine) in the regulatory serine/threonine kinase domain of BMPR1B receptor was identified in Booroola ewes and causes increased fertility (154,163,164) (Table 1). Females with one copy of this mutation have an ovulation rate of 3 or 4, while those with two copies have an ovulation rate of between 5 and 14. Animals that carry both the Inverdale and Booroola mutation (i.e., the *BMP15* and BMPR1B mutation, respectively) have greater fertility, further corroborating an interaction between BMP15 and this receptor (154).

A mouse knockout of Bmpr1b has been generated, and these mice show infertility, with pathology both in the ovary and in the endometrium (165). In mice, BMPR1B is expressed in oocytes of maturing (type 6) follicles and in oocytes and granulosa cells of antral follicles. No Bmpr1b transcripts were detected in granulosa cells of resting, primordial, developing (type 1–5b), or atretic follicles, corporal lutea, or theca, as well as uterine endometrium (165). However, histological and transmission electron microscopy studies of Bmpr1b null ovaries demonstrate no obvious differences between null and wild type ovaries. Similar to *Bmp15* knockout mice, Bmpr1b knockout mice show defects in cumulus expansion and fertilization in vivo. The mechanism for defective cumulus expansion is not clear, but neither FSH receptor mRNA levels, nor prostaglandin E2 receptor mRNA levels are affected. Paradoxically, cyclooxygenase 2 (COX2) mRNA levels were elevated in Bmpr1b null mice, and this observation has not been explained. Just like BMP15 ligand, the Bmpr1b receptor knockout in mice has a phenotype quite different from that observed in sheep. However, it has not been demonstrated what type of mutation (inactivating or activating) the Booroola mutation represents. Again, it will be necessary to introduce identical mutations into mice to test the hypothesis whether polyovulatory animals have different signaling pathways, or whether effects of mutations in rodents and sheep and most likely humans, has the same effect.

TRANSCRIPTION FACTORS IN PREANTRAL STAGE FOLLICULOGENESIS

TAF4b

Only a few transcription factors have been shown to play key roles in folliculogenesis, and these have been oocyte-specific factors: *Figla* (necessary for the primordial follicle formation) (39) and *Nobox* (required for the primordial to primary transition) (69) (see above). It is unknown if other cell types in the ovary (granulosa and thecal cells) also maintain cell-specific transcription factors or if they simply utilize more ubiquitously expressed ones. One transcription factor that has been shown to have a critical role later in folliculogenesis is TAF4b. TAF4b is a cell-type specific subunit of the core promoter transcription factor, TFIID (166). TFIID is composed of the TATA-binding protein (TBP) and TAF_{II} subunits; TAF4b was identified in highly differentiated human B-cell line by binding to TBP (167). TAF4b expression is preferentially restricted to ovary and testes and within ovaries, localizes to granulosa cells. Mice that lack TAF4b are infertile and have small ovaries that contain preantral follicles (166). Transcripts for inhibin βB, aromatase, inhibin βA, cyclin D2, follistatin, inhibin α subunit, and 17-β hydroxysteroid dehydrogenase were down-regulated in mice lacking TAF4b. The difference in transcript level is likely due to decreased number of granulosa cell types in *Taf4b* null mice. It is also unclear whether TAF4b itself is hormonally regulated during folliculogenesis.

Foxo3a

The forkhead transcription factors (FOXO) have roles in cell cycle arrest and apoptosis, and as demonstrated by the embryonic lethality of a *Foxo1* knockout, these proteins may have important roles in embryogenesis (168). In contrast to *Foxo1*, the mouse knockout of *Foxo3a* is viable but shows some hematologic defects (169). In addition, an age-dependent infertility was discovered such that by 15 weeks of age, female *Foxo3a*$^{-/-}$ mice are infertile. The infertility is attributed to widespread activation of the follicle pool prior to sexual maturity followed by large-scale atresia of the oocytes. The loss of the ovarian reserve as well as elevated levels of serum FSH and LH is reminiscent of premature ovarian failure (POF) in women (169). Thus, understanding of the regulation of *Foxo3a*, as well as the identification of target genes for FOXO3a, will provide additional candidate genes for POF.

ANTRAL FOLLICLE DEVELOPMENT

Key Roles of the Hypothalamus and Pituitary

Formation of the antrum heralds the final phases of folliculogenesis and the transition from intraovarian to extraovarian regulation. The hypothalamic–pituitary–gonadal (HPG) axis (Fig. 4) in females coordinates follicle maturation with sexual behavior and the physiologic preparation for pregnancy. Prior to a mature HPG axis, follicles are recruited to grow from primordial follicles, but cannot form antral follicles without sufficient gonadotropin stimulation. Spontaneous and targeted mutations in mice as well as genetic disorders in humans affecting hypothalamic and pituitary production of various peptides and hormones, have provided important insights into the hormonal regulation of folliculogenesis.

Transcriptional Regulation of Gonadotropins Genes

A number of transcription factors that control pituitary development, survival of hypothalamic and pituitary neuroendocrine cells, and regulation of the gonadotrophin genes, have been identified. Mutations in these genes can lead to profound defects in reproduction (170,171). Homeobox genes such as LIM homeobox protein 3 (*Lhx3*), homeobox gene expressed in ES cells (*Hesx1*), paired-like homeodomain factor 1 (*Prop1*), and pituitary specific transcription factor 1 (*Pit1*) are required for the development of the anterior pituitary which contains gonadotrope cells along with corticotropes, thyrotropes, somatotropes, and lactotropes. Mutations in *LHX3*, *PROP1*, and *PIT1* have been associated with combined pituitary hormone deficiency (CPHD) in humans (170). Mouse mutants in another homeobox genes, *Otx1*, show a developmental-restricted period of dwarfism and hypogonadism due to decreased expression of GH, FSH, and LH (172,173). Additionally, knockout mice for the transcription factor nuclear receptor subfamily 5, group A (*Nr5a*) (also known as SF1) also have decreased expression of FSH, LH and the receptor for the gonadotropins releasing hormone (GnRH) (174,175).

Gonadotropin Releasing Hormone

Gonadotropin releasing hormone (GnRH) is a decapeptide secreted by hypothalamic neurons in a pulsatile manner into the capillary plexus of the median eminence, and affects the release of LH and FSH from gonadotropic cells of the anterior pituitary. Humans with Kallmann syndrome, characterized by the lack of GnRH, have hypogonadotropic (low levels of FSH and LH) hypogonadism (176,177). Kallmann syndrome is a genetically heterogeneous disorder with X-linked, autosomal dominant and autosomal recessive inheritance and occurs as a result of defective migration in GnRH producing neurons and olfactory neurons (178). One of the genetic loci responsible for Kallmann syndrome is located in the Xp22.3 region and encodes a KAL1 protein that shares functional domains with protease inhibitors and neural cell adhesion molecules. The absence of KAL1 prevents the embryonic migration and development of the neurons that are normally destined to secrete GnRH. However, since inheritance is X-linked, KAL mutations and deletions affect males more than females with 5:1 ratio, while carrier females are usually unaffected. The degree of GnRH deficiency is variable in women with Kallmann syndrome ranging from complete to partial deficiency in FSH and LH. Ovarian streak gonads were described in a 16-year-old woman with possibly autosomal dominant form of Kallmann syndrome (179). The cause of the ovarian streaks in this woman is unclear. A 32-year-old woman with Kallmann syndrome was also described as having primordial follicles on ovarian biopsy, but it is unclear whether more advanced follicular structures were present (180). Treatment with GnRH or FSH and LH can induce ovulation and fertility although response to the treatment is variable most likely due to the genetically heterogeneous nature of the disorder (180,181). Currently there are no published cases of women with mutations in the GnRH gene, therefore it is unclear what effect isolated deficiency of GnRH has on human folliculogenesis. A second GnRH peptide, GnRHII, and its receptor have been described in humans and primates although its role in ovarian folliculogenesis is not clear (182–185).

A spontaneous deletion in the hypogonadal (*hpg*) mouse removed 33.5 kb of the *Gnrh* locus leading to deletion of the last two exons but not the sequences encoding the signal sequence and the GnRH decapeptide (131). This partially deleted gene is transcriptionally active but GnRH cannot be detected by immunohistochemistry. *Hpg* mice have low or undetectable levels of GnRH and very low (but assayable) levels of FSH and LH. Ovaries from *hpg* mice at 4 weeks of age weigh 1/10th that of normal ovaries and lack antral follicles (133). In most cases, the oocyte is surrounded by two layers of granulosa cells (133). Further studies on follicular development revealed that at birth, *hpg* mice have 20% less follicles than wild type mice (132). Therefore, *hpg* mice are born with smaller endowment of follicles corresponding to types 1, 2, and 3a of the Pedersen and Peters scheme (13).

Gonadotropin releasing hormone receptor (GnRH-R) mutations in humans cause infertility (186). Several mutations have been described to date, and most are compound heterozygotes, with more severe phenotypes observed in homozygous mutations. Folliculogenesis is affected, but the degree of ovarian and follicle development is not clear due to difficulty in analyzing human ovaries. One affected woman had ovarian streaks at age 16 detected by laparoscopy (187). At age 22, the same woman underwent bilateral oophorectomy for benign ovarian tumors (seromucinous cystadenomas). Histologically, few primordial follicles were detected but no mention is made whether other follicular stages are present. Of interest is that ovarian tumors developed in this patient despite the low levels of gonadotropins. Since both *GnRH* and *Gnrh-R* are expressed in the human granulosa cells, disruption of signaling in the ovary may also play an important role in the phenotype (188). The phenotype in this particular woman also supports the thesis that gonadotropins may play a role in early folliculogenesis. No published *Gnrhr* knockout mouse model exists, and future studies are needed to assess contribution of GnRH autocrine signaling in ovarian folliculogenesis versus its central role in the hypothalamic–pituitary axis.

FSH

The pituitary glycoprotein hormone, FSH, plays an essential role in reproduction through interaction with gonadal FSH receptors. FSH is a dimeric glycoprotein composed of a unique β subunit complexed with a common α subunit that is shared with thyroid stimulating hormone, luteinizing hormone, and chorionic gonadotropin. FSH binds to the FSH receptor and activates cell signals that lead to germ cell maturation and follicular growth (189). To study the deficiency of FSH in mice, our group created an FSHβ (*Fshb*) knockout (10). FSH deficient males were fertile, with decreased sperm concentration, sperm motility, and testes size. In contrast, females lacking FSH were infertile, with small uteri and ovaries, elevated LH, and lack of estrous cycles (10) (Fig. 3). Ovarian histology at 6 weeks revealed the presence of primordial, primary and multilayered preantral follicles. However, no antral follicles are formed, and the knockout mice do not ovulate.

Multiple genes are abnormally regulated in the FSHβ knockout mice (190). FSHR is upregulated in null mice almost twofold as compared to wild type. LHR is expressed almost fivefold less in null ovaries, consistent with the lack of corpora lutea. Serum/glucocorticoid inducible kinase (SGK), which is induced in preovulatory granulosa cells by FSH and is thought to play a role in the transition of proliferative to terminal luteal stage (191–193), is also down-regulated in FSHβ knockout mice. As expected, *Cyplla* mRNA levels are down regulated almost sixfold, also likely due to the lack of corpora lutea. Aromatase, (CYP19), which converts androgens to estrogens (194,195), is decreased sixfold in the null mice.

FSH promotes granulosa cell proliferation, and FSHβ knockouts arrest at preantral stages of follicular development. *Cyclin D2* null (196) and *Igf1* null female mice (197,198) are also infertile and demonstrate an arrest at the same preantral follicle stages as FSHβ null ovaries. Interestingly, the lack of granulosa cell proliferation in FSHβ null mice is not due to the lack of cyclin D2 since it continues to be expressed (190), and the cyclin-dependent kinases, Cdk2 and Cdk4, RNA levels are not affected much in the FSHβ null ovaries. Although *Igf1* mRNA levels are not altered in FSHβ null ovaries, FSH receptor mRNA are depressed in *Igf1* null mice (198). Thus, one of the major defects in the *Igf1* null ovaries is due to defective FSH signaling.

The phenotype of FSHβ null mice can be rescued by transgenes expressing human FSHβ subunit gene under its own promoter, as well as constitutive and ectopic expression of the human FSHβ under the control of the metallothionein-I promoter (199). Complete restoration of fertility and folliculogenesis was accomplished when the human FSHβ gene was expressed, indicating that the regulation and functions of FSH in mammals are evolutionarily conserved.

In humans, inactivating mutations in the FSHβ gene exist and several women have been identified (200–202). A 27-year-old woman with primary amenorrhea, absent breast development, low FSH and high LH was found to be homozygous for a two base pair deletion at codon 61 (Val61X). This frameshift mutation was predicted to change amino acids 61 to 86 followed by a stop codon resulting in truncated FSHβ protein lacking amino acids 86-111 (200). A second patient was 15-years-old with absent breast development, primary amenorrhea, low FSH, undetectable estradiol, and elevated LH. She was a compound heterozygote for two different FSHβ gene mutations, Val61X, and a Cys51Gly missense mutation. Both the Val 61X and the Cys51Gly mutations resulted in unmeasurable immunoreactive and bioactive FSH levels (201). The mouse FSHβ mutant mimics well the phenotype observed in women with the FSHβ mutations and shows the essential role that FSH plays in fertility of mice and humans. Women with FSH mutations gave further insight into the roles of FSH in androgen production. Although LH levels were high in one patient, she had normal levels of testosterone, and no evidence of hirsutism. These findings argue that FSH is important in LH-induced androgen production in the theca, either by inducing inhibin, LHR receptors, and or through other growth factors (202). It is clear that by studying animals and humans with induced and naturally occurring mutations in particular genes yields useful information with regards to hormonal function during folliculogenesis.

FSH Receptor (FSHR)

FSH receptors are transmembrane G (guanine nucleotide binding) protein–coupled receptors. Members of this family have an extracellular hormone binding domain, a seven transmembrane spanning region, and an intracellular region that couples to G proteins. Two different groups have generated FSHR deficient mice by deleting the first exon and portions of the promoter and first intron (11,12). Female mice lacking FSHR receptor are infertile. The ovaries and uteri of mutant mice at 8 weeks of postnatal life are smaller and thinner in both studies, while in one study, vaginas were imperforate. A lack of biologically active estrogen is the probable cause for small uteri and imperforate vaginas. Primordial, primary, and secondary follicles are present but large antral follicles and corpora lutea are absent, which probably accounts for the smaller size of the mutant ovaries. Clearly, the lack of FSHR does not affect growth of early follicles, consistent with the FSHβ knockout (10) and the hypothesis that gonadotropins are not essential for the early stages of folliculogenesis. FSH and LH levels are higher in null mice as compared to control mice. Similar to FSHβ knockout mice (10), continuous exposure of ovarian stroma to elevated LH levels most likely accounts for the observed hypertrophy of the interstitial tissue as well as numerous small lipid droplets.

Studies on female mice older than 8 weeks revealed formation of tumor-like structures and cysts by 12 months of age in FSH receptor null mice. The histology of these tumors showed an equal admixture of tubular structures and vacuolated plump stromal cells reminiscent of Sertoli and Leydig cells, respectively. It is likely that elevated levels of gonadotropins are responsible for the tumor like defects similar to mice that overexpress FSH, LH, or hCG (203–206). However, there is no convincing evidence that inactivating mutations in FSHR cause tumors in primates and humans, and searches for such mutations in granulosa cell tumors have been unproductive.

The controversial role of FSHR in the primordial to primary follicle transition was also addressed using FSHR knockouts. FSHR mRNA is expressed in 1-day-old mice, a time when many germ cell cysts are present and primordial follicles form. Histomorphometric analysis which measured follicle numbers by types at days 2, 10, and 24, showed presence of primary and secondary follicles in 2-day-old knockout mice while secondary follicles were absent from wild-type controls with a subsequent decrease in the total number of primordial, primary and preantral follicles as compared with wild type. The differences were statistically significant and modest. Based on these data, the lack of FSHR appears to accelerate follicular maturation with a subsequent block at the preantral stage. Also, it is possible to conclude that FSHR plays a small but significant enough role in early follicular growth, and it is possible that the accelerated follicle recruitment in early ovaries leads to the premature ovarian failure observed in 1-year-old female mice.

Only a few mutations in the *FSHR* gene in humans have been identified (207). A Finnish population-based study identified patients having XX gonadal dysgenesis (46, XX women with primary or secondary amenorrhea and serum FSH > 40 mIU/mL) (9). These 75 cases included 57 sporadic cases and 18 cases having affected relatives (7 different families). Most cases were found in north central Finland, a sparsely populated part of the country. The overall frequency of the disorder in Finland was 1 per 8,300 live born females, a relatively high incidence attributed to a founder effect. An alanine to valine mutation at position 189 of FSHR segregated with the phenotype. The effect of this mutation on FSHR expression and activity has been studied in vitro, and it has been suggested that the mutant receptor is an inactivating mutation that has impaired trafficking and folding which results in the decreased ability of the receptor to bind ligand (207,208). Additional mutations in the FSHR gene include a set of compound heterozygous mutations group of French women who had primary or secondary amenorrhea and follicle development to the secondary stage (209). These inactivating mutations are in the extracellular domain and affect ligand binding and receptor trafficking (207).

LHβ Knockout and LHβ Mutations

Similar to FSH, LH is a glycoprotein hormone with a common α subunit but a unique hormone-specific β subunit and is synthesized in the pituitary gonadotropes. Recently, a mouse model for the loss of LHβ has been generated (210). LHβ (*LHb*) null mice are viable but males and females are infertile. Female mice are hypogonadal with decreased serum levels of estradiol and progesterone. Histologic analysis of the ovaries demonstrates that healthy large antral follicles and corpora lutea are absent. Primary and secondary follicles with intact thecal cell layers develop, and thecal cell markers, *Lhcgr*, *Bmp4*, and *Cyp17a1* are expressed, indicating that LH is not necessary for initial differentiation of the thecal cell layer. However, low serum steroid levels and the decreased expression of the steroidogenic enzymes *Cyp11a1*, *Cyp19a1*, and *Cyp17a1* demonstrate that LH is necessary for steroid production. Similar to the FSHβ knockout mouse model, the phenotype of LHβ null mice can be rescued by treatment with exogenous gonadotropins suggesting that the block in development is not irreversible.

Only one rare mutation that produces an inactive LHβ in humans has been identified and was found in a male with a family history of male infertility (211). Female members of that family do not appear to be affected.

LH Receptor (Lhcgr)

Like FSH receptors, LH receptors are single chain transmembrane glycoproteins that belong to the G-protein–coupled receptor family. Although LH and FSH are not thought to play major roles in early folliculogenesis, truncated LH receptor mRNA is expressed in newborn ovaries (212) at the time when germ cell cysts are present and primordial follicles are formed. Full-length transcripts are not detectable until postnatal day 5, but based on all LHβ and LH receptor mutations, and the role of these transcripts in early follicular formation is unclear. To study the effects of LH receptor deletions on ovarian development and folliculogenesis, two knockouts have been generated. In the first knockout, exon 1 and the upstream flanking region containing multiple transcription initiation sites and the promoter region were deleted (213). Both male and female null mice were sterile. At 60 days, null females had markedly smaller ovaries (more than 50% less than the weight of wild type ovaries) with pale and thin uteri and delayed vaginal opening. Preantral and antral follicles, but not preovulatory follicles and no corpora lutea, were present in females lacking LH receptor. As compared to FSH receptor knockouts, LH receptor knockouts have more advanced follicular structures, but ovaries in both knockouts continue to be quite small. Like the LHβ knockout, this argues that LH is not required for early gonadal and reproductive tract differentiation. Serum LH levels are high in LH knockouts which could be due to loss of estradiol negative feedback and/or loss of negative LH feedback on its own secretion through decreased hypothalamic GNRH levels in both sexes. Moderate elevation in serum FSH levels could be due to decreased gonadal inhibin secretion. FSH receptor and progesterone receptor mRNA levels were not significantly affected by the lack of LH receptor. Estrogen receptor α (ERα) and Star mRNA levels are decreased and estrogen receptor β (ERβ) mRNA levels are increased in both ovaries and testes. Since LH maintains ERα and *Star* and inhibits ERβ mRNA levels, these results in the knockout are consistent with previous studies. Hormone replacement therapy with estrogen and progesterone did not restore fertility nor reverse ovarian morphology which argues that other downstream targets are involved in stimulating follicular growth beyond the antral stage.

Another LH receptor knockout was generated by deleting exon 11 of the LH receptor which encodes the transmembrane and cytoplasmic receptor domains and part of the 3′ extracellular domain, thus preventing the formation of full-length functional LH receptor capable of anchoring to the plasma membrane and of signal transduction (214). These LHR knockout mice did not synthesize any mRNA encoding the full length functional LH receptor, and the phenotype was remarkably similar to the above-described LHR knockout in which promoter and first exon were removed.

In humans, mutations in the LH receptor cause gonadal dysgenesis. Sultan and Lumbroso (215) recently tabulated the 46,XY cases, 9 having female phenotype and 2 having micropenis. All 46,XX cases occurred in sibships in which an affected 46,XY male had Leydig cell hypoplasia. Latronico et al. (216) reported primary amenorrhea in a 22-year-old woman. In that family, 3 males and 1 female had a homozygous nonsense (stop) mutation at codon 554 (C554X). The newly produced stop codon resulted in a truncated protein having 5 rather than 7 transmembrane domains. The affected female had breast development but only a single episode of menstrual bleeding at age 20 years. The mutation reduced signal transduction activity of the LH receptor gene. In a 46,XX case reported by Toledo et al. (217), secondary amenorrhea occurred. The mutation was Ala593Pro. Activating LH receptor mutations seem to have little effect in females, although in males precocious puberty occurs (215). Females with activating LH receptor mutations fail to demonstrate precocious puberty.

Roles of Estrogen Signaling Pathways

Estrogen receptors α and β are expressed in many organs including the hypothalamus, anterior pituitary, mammary gland, uterus, and ovary (218). Estradiol stimulates granulosa cell growth, increases synthesis of insulin like growth factor 1 (IGF1), maintains FSH receptors, induces LH receptor, augments aromatase activity and subsequent estradiol production, and attenuates granulosa cell apoptosis. Estrogen receptors are members of the steroid receptor superfamily of ligand-activated transcription factors. Estrogen receptors and its ligand, 17β-estradiol, play critical roles in the development of female secondary sexual characteristics as well as development of the female reproductive tract, and maintenance of pregnancy.

Two receptors are known, ERα (*Esr1*) and ERβ (*Esr2*). ERβ is expressed at higher level in granulosa cells of growing follicles as compared to ERα while ERα is expressed in thecal cells and interstitial cells (219). The mRNAs corresponding to ERα and ERβ are first expressed around day 8 in rat ovaries, which at this age contain both primordial, primary and some secondary follicles. Although estrogen is thought to act mainly on antral follicles, earlier expression of mRNA of the receptors in growing follicles may indicate additional functions or possibly estrogen-independent activity of the receptors.

Initial observations in mice deleted for ERα receptor showed that ovaries from these mutant mice contained cystic and hemorrhagic follicles, and uteri were hypoplastic (220). The mouse model also indicated that another estrogen receptor existed, since estradiol binding in uterine preparation from mutant animals still occurred. This estrogen receptor turned out to be ERβ (221–223). ERβ knockout mice are viable and subfertile (224). Litter sizes in ERβ knockouts were 35% of wild type. Superovulation showed a dramatic reduction in the number of oocytes per female with only 6 oocytes per female in the knockouts as compared to 52 in the control mice. Other observed effects included decreases in the cellular mass of the oocyte cumulus, and an inadequate response to exogenous hormones. The endogenous gonadotropin levels were normal in these mice.

To study effects of complete lack of ERα and ERβ in reproduction, double knockouts were generated (225,226). The double mutant ovaries have phenotype that is distinct from either single receptor knockout. Unlike wild-type ovaries that showed mostly secondary follicles, prepubertal double knockout mice possessed adult-like follicles with defined antra and theca, characteristic of hypergonadotropic-precocious maturation of the ovary. Serum LH levels were higher in double knockouts as compared to ERα knockout alone, which suggest that both estrogen receptors are required for estradiol-mediated regulation of LH secretion in the hypothalamic–pituitary axis. In 10-week- to 7-month-old ovaries, primordial and growing follicles were visible, but corpora lutea were lacking. Interestingly, structures resembling seminiferous tubules of the testis were present in double knockout ovaries, some with degenerating oocytes and granulosa cells and Sertoli-like cells within the structures (Fig. 3). Sertoli cell differentiation markers, *Amh* and *Sox9*, were elevated in these structures. Female aromatase knockout mice (see below) also develop Sertoli and Leydig cells at puberty which argues that the lack of estrogens is detrimental to normal granulosa cell differentiation or estrogens are required for the maintenance of granulosa cell fate.

The aromatase knockout (Arko) mouse model has given us much insight into the effect of a complete lack of estrogen on follicular development (227). Aromatase is a cytochrome P450 enzyme which catalyzes the formation of aromatic C18 estrogens from C19 androgens. Since aromatase is required for the synthesis of estradiol, a knockout would render the mice estrogen deficient (Arko). Knockout mice were phenotypically normal but infertile. Testosterone levels were almost 10 times the wild-type levels. LH levels were elevated two- to tenfold, and FSH levels were also elevated three- to fourfold, over the wild-type levels. These findings corroborate previous studies that showed the importance of estrogen in the negative feedback regulation of gonadotropins and the high testosterone levels presumably reflect LH-stimulated androgen production by the thecal cells. Ovaries from 12- to 14-week-old mice showed presence of many antral follicles, with hypertrophied stromal cells and atretic follicles. No corpora lutea were visible, therefore no ovulation was occurring. Another different ArKo knockout was generated and analysis of these mice showed that at 8 months of age, ovaries were infiltrated with mast cells, had a reduced number of follicles and contained hemorrhagic cysts (228). Although the aromatase knockout mice could not be superovulated, oocytes recovered from *Arko* mice progress through in vitro maturation, fertilization and early embryo development (229). Therefore estrogen appears to play an important role in feedback regulation of ovulation and continued follicle growth but does not interfere with oocyte growth.

The role of estrogen metabolism is also important in humans. Ito et al. (230) reported a mutation in the CYP19 (P450 aromatase) gene in an 18-year-old 46,XX woman with primary amenorrhea and cystic ovaries. The patient was a compound heterozygote for two different point mutations in exon 10. The mutant protein had no activity. Most of the mutations in humans involve amino acid substitutions as well as premature stop codons. It is a rare disorder, with only six women reported.

Deficiency of the aromatase enzyme is more often associated with genital ambiguity. Shozu et al. (231) detected placental aromatase deficiency manifesting as maternal virilization during the third trimester. The 46,XX infant was born with genital ambiguity (female pseudohermaphroditism). Adrenal enzyme defects were not evident. The molecular basis of the mutation was an 87-bp insert in exon 6 of the aromatase gene, altering the splice junction site to produce a novel protein with 29 additional amino acids. An aromatase mutation in 46,XX female infants has been associated with genital ambiguity (230) or clitoromegaly (232). In the latter cases clitoral enlargement occurred at puberty, but breast development did not. Multiple ovarian follicular cysts were evident. FSH was elevated, estrone and estradiol low. Estrogen and progesterone therapy resulted in a growth spurt, decreased FSH, decreased androstenedione, decreased testosterone, breast development, menarche, and fewer follicular cysts.

Gap Junction Membrane Channel Proteins and Connexins

Connexins are a family of at least 15 proteins that form intercellular membrane channels of gap junctions and allow diffusional movement of ions, metabolites, and potential signaling molecules. These structural molecules play important roles in bi-directional communication between the oocyte and granulosa cells. Several connexins have been shown to be important during folliculogenesis. Connexin 37 protein is expressed in primary and growing follicles. Connexin 37 (*Gja4*) knockouts contained at 2 weeks of age many developing preantral follicles, but large Graafian follicles are not observed (233). Therefore, histologically, it appears that lack of Cx37 blocks preantral to antral follicle development. Electron microscopy studies showed that while the Cx37 knockouts contained adherens junctions, they lacked gap junctions present in wild type ovaries. Despite the lack of Graafian follicles, the null ovaries displayed smaller structures resembling corpora lutea that were 5–10 times more abundant. The premature luteinization observed in null animals suggests that junctional communication may be a major mechanism regulating corpus luteum formation. The connexin 37 knockout had smaller average oocyte size. While 52.8% of oocytes from *Cx37*$^{+/-}$ mice resumed meiosis with 26.8% of these proceeding to metaphase of meiosis I and 26% to metaphase of meiosis II, only 2.2% of *Cx37*$^{-/-}$ oocytes resumed meiosis in culture with the vast majority (86.3%) at the germinal vesical stage. All of these observations clearly indicate that connexin 37 is essential at the time from pre-antral to antral follicle transition, most likely involving in the transport of metabolites necessary for this transition.

Connexin 43 (*Gja1*) protein is the most abundant connexin in the ovary, and is expressed as early as E14.5. Connexin 43 gap junctions interconnect granulosa cells that surround the oocyte as well as oocytes and granulosa cells. Connexin 43 (*Gja1*) knockouts die at birth due to cardiovascular failure (234). Ovaries in null mice, retrieved at the time of birth are unusually small apparently due to a deficiency of germ cells (235). The germ cell deficiency was traced back as far as Day 11.5 of gestation, implying that it arises during early stages of germ line development. To study postnatal folliculogenesis in *Gja1*$^{-/-}$ mice, fetal and neonatal ovaries have been grafted into the right kidney capsule of an ovariectomized and immunocompromised prkdcscid/Prkdc^{scid} mouse (236). The grafts were removed after 1–3 weeks and ovarian histology examined. After 3 weeks of development in kidney grafts, a range of follicle stages from primordial through primary, secondary and tertiary could be seen in the wild-type ovaries. In contrast, *Gja1* null ovaries did not proceed beyond primary follicles. Examination of grafts after 1, 2, and 3 weeks showed that mutant granulosa cells were unable to proliferate. Oocyte growth was also slower in mutants as compared to wild type animals. Oocytes recovered from *Gja1* mutants could not be fertilized. Electron microscopy also showed that these oocytes are abnormal as compared to wild type.

Connexin 43 appears to be regulated by the gonadotropins, FSH and LH. Because Cx43 is more abundant in large antral follicles as compared to small antral and preantral follicles, some have hypothesized that FSH induces Cx43 synthesis. In contrast, the preovulatory surge of LH is followed by a drop in the level of mRNA encoding Cx43 (237,238). Further corroboration came from experiments utilizing exogenous administration of gonadotropins. Injection of FSH into hypophysectomized rats promoted an increase in the ovarian gap junctional membrane, while hCG, which induces ovulation, led to a significant reduction in the quantity of gap junctions in granulosa cells (239,240). In vitro administration of FSH and LH to granulosa cell lines corroborated in vivo observations (241,242). It has been suggested that effects of the gonadotropins on Cx43 is mediated through steroid hormones (243). Other signaling pathways may also regulate Cx43, including the BMP/Smad and the MAPK-Ras pathways which have been recently shown to regulate the Cx43 promoter in vitro (244, 245).

These results imply that gap junctional coupling mediated by Cx43 channels plays indispensable roles in both germ line development and postnatal folliculogenesis. The very different ovarian phenotypes resulting from the Cx37 and Cx43 knockouts indicate that gap-junctional coupling between the oocyte and its cumulus granulosa cells serves a distinct role from that between the granulosa cells themselves, possibly involving different molecules. Connexin 32, also expressed in the ovary, when knocked out does not appear to affect fertility (246), suggesting these are redundant gap junctions components that have yet to be identified.

Additional cellular connections may be important for antrum formation and granulosa cell–oocyte interactions. Some of the intercellular communication between granulosa cells and oocytes is thought to be mediated by transzonal projections (TZPs) between the oocyte and granulosa cells (41). Microtubule-TZPs (MT-TZP), in particular, appear to be involved in this communication (41). The MT-TZP connections between the oocyte and granulosa cells has been hypothesized to be under the control of FSH (247). FSH priming of wild type and *Fsh*$^{-/-}$ results in alterations in the MT-TZP such that the TZPs are retracted from the oocyte (247). This results in changes in oocyte development as measured by chromatin remodeling and acquisition of meiotic competence (247). From these experiments, it can be hypothesized that without FSH priming, granulosa cells have a stable interaction with the oocyte, allowing for the paracrine exchange of factors. Once MT-TZPs are retracted, oocyte development is accelerated. Whether oocyte-specific factor such as GDF9 and BMP15 or other unknown oocyte-factors also modulate the formation and/or relation of TZPs has still to be addressed.

METABOLISM AND FOLLICULOGENESIS

Globally, there are several peptide growth factors that may influence ovarian function due to their endocrine (metabolic) effects on the hypothalamic–pituitary–gonadal axis (171). In this section, we will briefly discuss leptin, growth hormone, and IGF1.

Leptin

Leptin, a protein hormone with ubiquitous expression, is encoded by the *ob* (obesity) gene. Leptin is mainly secreted by adipose tissue and much attention has been devoted to studying its effect on the central nervous system to diminish appetite and to increase energy consumption. In the murine ovary, leptin is expressed in oocytes, granulosa cells, and theca cells. Using immunohistochemistry, it appears that the highest expression of leptin is confined to the oocyte, while the leptin receptor has highest expression in theca cells (248). Leptin administration does not affect spontaneous or induced maturation of either isolated denuded oocytes or cumulus-oocyte complexes, but it does significantly increase the rate of meiotic resumption in preovulatory follicle-enclosed oocytes (248). This effect does not appear to be mediated by activation of phosphodiesterase 3B and subsequent cAMP reduction (248). It is possible that leptin induces secretion of a factor important in meiotic resumption (248). In humans, leptin immunolocalizes to oocytes in primordial and primary follicles as well as granulosa cells surrounding the oocytes (249). In preantral follicles, leptin is preferentially expressed in granulosa cells; and in antral follicles, leptin is preferentially expressed in the thecal layer. Rat studies also document expression of leptin and its receptor in granulosa, theca cells, and oocytes (250).

ob/ob mice, which are deficient in leptin, are morbidly obese and have a subnormal number of ovarian follicles (251). Weight loss in *ob/ob* mice does not restore fertility but *ob/ob* ovaries do produce viable eggs when transplanted into wild type mice, indicating that non-gonadal leptin can restore fertility (252). Ovaries from *ob/ob* mice contain antral follicles with pronounced apoptosis of granulosa cells. The infertility in these mice is partially due to altered pattern of GnRH secretion leading to decreased secretion of LH and FSH. Leptin replacement increases FSH and LH levels in *ob/ob* mice and restores fertility (252). In addition, administration of gonadotropins also restores fertility, indicating that the absence of leptin influences the hypothalamus and pituitary. Since FSH and LH levels are only partially suppressed in *ob/ob* mice, and some direct effects of leptin on the gonads is likely also important, it is unclear whether the increased apoptosis of granulosa cells that is observed in the *ob/ob* preantral and antral follicles is due to low FSH and LH levels, a result of absent leptin signaling in the ovary, or a combination of both.

Leptin shows an important link between metabolism and reproduction. Leptin may serve as a metabolic signal informing the reproductive system that sufficient fat stores are available to meet the caloric demands of reproduction. Metabolic stresses such as food restriction, and severe exercise will lead to low circulating levels of leptin and lower fertility. How much of this control is accomplished centrally or in the ovary is still unclear and remains a topic of study. Leptin deficiency in humans is rare and causes morbid obesity. Follicular development and function, as well as gonadotropin secretion in these children is unknown. Women with mutations in the leptin receptor are also obese and have primary amenorrhea (253). FSH and LH levels are low to normal, and estradiol concentration is low. Consistent with the *ob/ob* mouse studies, these findings reflect hypogonadism of central origin and is likely responsible for observed infertility. Since the leptin receptor is expressed in the ovary, the degree to which local versus central control of ovarian physiology contributes to infertility in humans and mice is unclear. In vitro studies lend support that leptin does have local effects in the ovary, although these reports are conflicting. Leptin can suppress ovarian steroid synthesis in primary rat granulosa cells and in rat and human granulosa cell lines (254). Other studies have shown that leptin inhibits IGF1 synergistic effects with FSH on estradiol production by rat granulosa cells as well as inhibits aromatase activity (255). In contrast, it has been reported that leptin increases aromatase mRNA and protein expression in human granulosa cells (256). The use of supraphysiologic amounts of leptin may account for different results and further studies may be helpful, such as conditional knockouts of leptin or its receptor in the ovary, to study local effects of leptin in ovarian physiology. In the rat, leptin mRNA increases threefold within 2 hours of hCG administration, and leptin receptor mRNA levels also increase dramatically 9 hours after hCG administration (250). These results suggest that expression of leptin and its receptor peak at the time of ovulation, but their specific functions are unknown. Lastly, leptin is also expressed in human vasculature as well as ovarian endothelial cells and promotes angiogenesis. Leptin-induced angiogenesis may be important in folliculogenesis (257).

Growth Hormone

Growth hormone (GH) is secreted by the anterior pituitary and acts via growth hormone receptor to effect growth and onset of sexual maturity (258). GH receptor is expressed in multiple tissues including the ovary. GH receptor RNA is located in the granulosa cells of antral follicles and corpus luteum but is not detected in preantral follicles, theca, oocytes, or stroma (259).

In horses and rodents, GH receptor RNA is detectable in oocytes, cumulus cells and granulosa cells throughout folliculogenesis (260,261). GH augments cumulus expansion and improves nuclear maturation in equine cumulus cell–oocyte complexes (260). In the cow, GH receptor protein and RNA are detected as early as primordial oocytes and granulosa cells of primary follicles and more advanced follicles (262). The discrepancy of expression between various species may be due to methods and material differences. Mice that lack GH receptor are subfertile dwarfs (263–265) with delayed age of conception, low plasma IGFI, and elevated plasma GH levels (266). A subset of GH receptor knockout females are infertile, although the reason for why some are subfertile while others are infertile is unknown, but it may have to do with epigenetic effects that lead to complete infertility (263). Treatment with recombinant human IGF1 normalizes the age of puberty (266). However, IGFI does not restore ovulation rate in GH receptor knockout mice, and therefore growth hormone may affect follicular growth independent of IGFI (267). More detailed analysis of ovaries from GH receptor knockouts revealed significantly smaller numbers of preovulatory follicles, corpora lutea, plasma estradiol levels, and increased number of apoptotic follicles (263). It is unclear whether effects of GH receptor deficiency affect the number and histology of primordial, primary, secondary, and preantral follicles and whether the initial pool of primordial follicles is equivalent in knockout animals. Women with mutations in *GHR* (Laron syndrome) can become pregnant despite low levels of IGF1 (268–270).

Female mice that overexpress bovine GH under the control of the strong phosphoenol pyruvate carboxykinase (PEPCK) promoter have a 60% rate of infertility, an increase in the numbers of preovulatory follicles and corpora lutea, and decreased granulosa cell apoptosis in developing follicles (271,272). The mechanism of GH suppression of apoptosis may involve production of local as well as systemic IGF1 (273,274) or poorly understood GH effects that are independent of the IGFI. Female mice that overexpress human GH using the metallothionein I (MT) promoter have a 20% rate of infertility. A similar rate of infertility is observed for bovine GH driven by the metallothionein I promoter (274). Infertility in these transgenic mice may be due to inadequate luteal function (275). Acromegaly, a condition that causes excessive growth due to excess GH secretion, often leads to infertility. Because acromegaly is often caused by adenomas (benign tumors of the pituitary), infertility may be due to hyperprolactinemia as well as gonadotropin dysregulation as the expanding tumor may affect other cell types in the pituitary. Analysis of 196 women with acromegaly revealed that 60% had amenorrhea, and of these 45% had elevated prolactin levels. A number of women therefore only had elevated GH levels without hyperprolactinemia. FSH and LH levels in these women were not reported, but it is possible that GH hypersecretion alone can lead to infertility (276). Unfortunately, little is known about folliculogenesis in these women, and further studies are necessary to assess whether excessive growth hormone production alone or other perturbations in the hypothalamic–pituitary axis cause follicular dysfunction in humans.

Insulin-Like Growth Factor (IGF) Signaling System

The IGF family has long been recognized as having an important role in female reproduction. The signaling components include two ligands (IGF1, IGF2), two receptors (IGF1R, IGF2R) and a number of binding proteins (IGFBP) that regulate IGF action (277). Of these, IGF1 is known to be important and required for sexual development and reproduction (278,279). *Igf1* is expressed in granulosa cells of growing follicles and one receptor, *Igf1r,* is expressed in both growing and atretic follicles (280,281). In mice, IGF1 levels are low at the primary follicle stage, but increase during preantral and antral follicle growth (281). The expression of *Igf1* may be under local paracrine control, as microarray analysis has shown that recombinant mouse GDF9 can upregulate expression of the *Igf1* gene (138,146). *Igf2* expression in the ovary is lower than *Igf1* (281) and the *Igf1* knockout mouse has reproductive defects, while the *Igf2* null mouse has normal fertility (197,282). Most *Igf1* null mice die perinatally, but those that survive to adulthood are infertile (197). Ovaries from *Igf1* knockout mice do not contain antral follicles and cannot be ovulated. This defect is partially due to decreased FSH receptor and aromatase expression (198) (see above; FSH).

ADDITIONAL MOUSE MODELS FOR STUDYING TGFB SUPERFAMILY SIGNALING IN THE OVARY

While the gonadotropins, FSH and LH, govern the final stages of follicular development and ovulation, the ovary participates in the endocrine control of gonadotrope function by the production of steroid and nonsteroid hormones. Some of the best-characterized nonsteroid hormones that regulate pituitary function are the activins and the inhibins (283). Inhibin was first described as a proteinaceous substance from the testes that could inhibit FSH secretion (284). Purification of inhibin lead to the discovery of activin, a protein that stimulates the release of pituitary FSH (285,286). In addition, the ovary produces follistatin, a negative regulator of activin and BMP function (287–289). Both inhibins and activins are members of the transforming growth factor-β superfamily. Members of this family are secreted dimeric proteins processed from larger prepro-precursor forms. Inhibin and activin share a common β subunit; activins are ββ dimers (β_A:β_A, β_B:β_B, and β_A:β_B) while inhibins are heterodimers of the β subunit with a distantly related α subunit (inhibin A [α:β_A] or inhibin B [α:β_B]). Genetic disruptions in inhibins, activins, and follistatin have demonstrated that these proteins have important paracrine as well as endocrine roles in the reproductive axis.

Inhibins

While inhibins are produced locally in many tissues including the brain, pituitary, placenta, and adrenal gland (290–295), the major site of circulating inhibin production are the gonads (296–298). Castration of male or female rats results in the loss of circulating inhibin and the subsequent rise in pituitary FSH (299,300). Serum inhibin is a major factor contributing to the rise in FSH as demonstrated by the increase in FSH during passive immunization of rats with anti-inhibin antisera (301–305). Inhibin was also hypothesized to have a paracrine role in the ovary since it was demonstrated that in vitro, inhibins modulate thecal cell androgen production, follicular growth, and oocyte maturation (306–310). In general, inhibin functions as an antagonist of activin, as no unique inhibin binding receptor or signal transduction pathway has ever been conclusively demonstrated (311,312). Recent data also have shown that inhibin can block BMP signaling (313) although inhibin suppression of BMP signaling in the ovary is less well defined than as an antagonist of activins.

In female rodents, inhibin A and inhibin B are produced from granulosa cells. In humans, inhibin α is also detected in thecal cells and corpora lutea (314,315). While very little is understood about the biological differences between inhibin A and B isoforms, it is known, at least in the rat and mouse, that their production varies with follicle size and cycle stage (58,299,316). Inhibin B is likely the product of small developing follicles (316), and in rats, inhibin B is inversely correlated with FSH levels throughout the cycle and increases in response to FSH stimulation, remaining high through midcycle and declining following the primary gonadotropin surge (299). In contrast, inhibin A levels are low following the primary gonadotropin surge, but expression is stimulated following the secondary FSH surge and slowly rises during early cycle stages (299). Inhibin A, unlike inhibin B, correlates with estradiol production and the development of preovulatory follicles (316). The highly regulated production of inhibin during the estrous cycle suggests that its regulation and function are critical to female fertility. It is not surprising then that genetic mutations in all three inhibin subunit genes (*Inha, Inhba*, and *Inhbb*) (317–320) show reproductive defects. In addition, the creation of an inhibin α (*Inha*) null mouse led to the unexpected discovery of inhibin α as a tumor suppressor gene (24).

Mouse Models to Study Inhibins

Inha was the first of the three inhibin/activin genes to be knocked-out in mouse by our group (24). Since the inhibin α subunit was targeted, activin (i.e., β_A and β_B subunit) expression is left "intact." While mice null for inhibin α are viable during embryonic and prepubertal development, all homozygous mice develop gonadal tumors by four weeks of age and die from a cancer cachexia-like wasting syndrome by 12–17 weeks of age (Fig. 4) (24,321). Tumors are mixed granulosa/Sertoli cell tumors. FSH levels are elevated two- to threefold, and activin is expressed in the serum is 20-fold higher in the null mice compared to wild-type mice. At 4 weeks of age, prior to gross tumor formation, granulosa cells appear abnormal and display hyperplasia. Germ cells seem to be unaffected prior to tumor formation, but when immature mice are superovulated, they produce fewer eggs than wild-type mice (322). Gonadectomy does not rescue the lethality of the mutation since these mice then develop adrenal tumors (321). Thus, the phenotype of *Inha* null mice is complex, and the contributions of elevated serum FSH and activins were further analyzed.

Activin is a potent growth factor, and rodents treated short term with recombinant activin exhibit a wasting syndrome (323,324). Thus, it was hypothesized that the cachexia was caused by superphysiologic levels of activin (321,325,326). Since activin partially signals through the activin type II receptor (ACVR2) (327), knockouts for the *Acvr2* gene (see below) (318) were crossed to *Inha* mutant mice to generate double homozygotes (*Inha*$^{-/-}$;*Acvr2*$^{-/-}$) (326). These mice would be deficient in activin signaling through ACVR2 as well as FSH synthesis, since mice null for *Acvr2* have reduced FSH levels due to the lack of activin signaling in the pituitary (318). The double homozygous mutant mice develop gonadal tumors similar to *Inha*$^{-/-}$ mice and reinforce the hypothesis that inhibin α is a tumor suppressor gene (Fig. 4). However, despite elevated peripheral activin levels, double homozygous mutant mice do not develop the wasting syndrome seen in the *Inha*$^{-/-}$ mice, demonstrating that early death, but not tumor formation, is due to activin signaling. However, another high affinity receptor for activin, ACVR2B, is expressed in granulosa cells (328), and it is unclear if additional activin signaling through this receptor also contributes to the tumor phenotype. Knockout mice for *Acvr2b* have been generated, but homozygous mice die shortly after birth due to cardiac defects (329), leaving questions of additional activin signaling in the ovary currently unanswered.

The contribution of elevated FSH to tumor formation in *Inha*$^{-/-}$ mice has also been examined. Elevated FSH by itself does not cause tumor development but results in the formation of hemorrhagic cysts in the ovaries, as demonstrated by overexpression of FSH in transgenic mice (203). Eventually, these mice die due to enlarged cystic ovaries and urinary bladder obstructions (203). Crosses of hypogonadal (*Gnrh*^{hpg}) mice, which lack hypothalamic GnRH, to *Inha* mice to generate *hpg*/*hpg*;*Inha*$^{-/-}$ mice, prevents the wasting syndrome, but in females, does not prevent premalignant ovarian tumors (330).

A clear demonstration that FSH modulates granulosa cell tumor progression, but not tumor formation, was shown by generating *Inha*$^{-/-}$;*Fshb*$^{-/-}$ mice (203). While these mice live longer than *Inha*$^{-/-}$ females, double homozygous female mice die by 39 weeks, in contrast to males, which survived to 1 year. Tumors from these mice are slow-growing and less invasive. In addition, they do not display the high levels of activin and estradiol as do the *Inha*$^{-/-}$ single mutant mice. Double mutants of inhibin α and both α and β forms of the estrogen receptor (or all three genes) in females does not affect tumor formation, suggesting that estrogen signaling does not contribute to the tumor pathogenesis (331). These data demonstrate that while FSH and activin signaling modulate the tumor phenotype, they do not directly contribute to the tumor formation. Additional and as of yet undescribed mechanisms for the tumor suppressor function of inhibin in sex-cord/stromal tumors must be operating.

Since tumor suppressor genes function through an antiproliferative signal, the cell cycle regulatory genes, p27 and cyclin D2 were examined in the *Inha*$^{-/-}$ background. Cyclins, cyclin-dependent kinases (Cdks), and cyclin-dependent kinase inhibitors (CKIs) regulate cell cycle progression. Overexpression of cyclins and Cdks, or underexpression of CKIs, results in uncontrolled cell proliferation. Cyclin D and p27 are involved in the G1-S phase transition; cyclin D2-CDK promotes S phase commitment, while p27 inhibits it (332). Cyclin D2 has been shown to be FSH responsive and cyclin D2 knockout mice are infertile due to a lack of granulosa cell proliferation (196), and cyclin D2 is upregulated in *Inha*$^{-/-}$ mice (333). Double knockout mice of cyclin D2 and inhibin α have delayed ovarian tumorigenesis but eventually develop tumors and die by 29 weeks of age, very similar to *Inha*$^{-/-}$;*Fshb*$^{-/-}$ mice (203,331). However, prior to advanced tumor stages, in the absence of inhibin, more advanced preantral follicles are seen than in the cyclin D2 knockout alone (331). p27 is a CKI that when mutated in mice results in increased body size, hyperplasia of multiple organs, cancer, and infertility in female mice due to granulosa cell defects (334–336). Double homozygous mutants of p27 and *Inha* have an enhanced *Inha*$^{-/-}$ phenotype and die by 8 weeks of age (333). It is clear from these studies that while modifiers of the *Inha*$^{-/-}$ phenotype have been discovered, the tumor-suppressing activity of inhibin is still largely a mystery.

Loss-of-function is one way to study gene function. Another is by gain-of-function, and two such models for inhibin have been generated. In one, a mifeprisone-inducible promoter was used to express the inhibin α and βA subunit in mice (337). This enabled the controlled expression of inhibin A from the liver and the possibility of "rescuing" the knockout inhibin

TABLE 2. *Female reproductive phenotypes in genetic mouse models**

Gene	Name	Mouse knockout model: female reproductive phenotype	Ref.
	Secreted Factors and Receptors		
Acvr2	Activin receptor IIA	Infertility; follicles arrest at early antral stage due to defects in pituitary function	(318)
Bmp15	Bone morphogenetic protein 15	Subfertility; defects in ovulation and fertilization; defects in combination with $Gdf9^{+/-}$	(150)
Bmpr1b	Bone morphogenetic protein receptor, type 1B	Infertility; defects in cumulus cell physiology and endometrium	(165)
Ccnd2	Cyclin D2	Infertility; granulosa cells do not proliferate in response to FSH	(196)
Cga	Glycoprotein hormones, alpha subunit	Infertility; FSH and LH deficiency	(421)
Fst	Follistatin	Complete loss of germ cells by birth; granulosa cell-specific knockout has early ovarian failure	(409,416)
Fshb	Follicle stimulating hormone beta subunit	Infertility; preantral follicle block	(10)
Fshr	Follicle stimulating hormone receptor	Infertility; block prior to antral follicle formation	(12)
Gdf9	Growth differentiation factor 9	Infertility; arrest of follicle growth at the primary stage	(60)
Gnrh	Gonadotropin releasing hormone	Infertility; hypogonadotropic hypogonadism	(131)
Ghr	Growth hormone receptor	Delayed puberty and prolonged pregnancy	(264)
Igf1	Insulin-like growth factor 1	Infertility; preantral follicle block	(197)
Inha	Inhibin alpha	Develop gonadal sex-cord tumors with 100% penetrance	(317)
Inhba	Inhibin/activin beta A	Perinatal lethality	(318)
Inhbb	Inhibin/activin beta B	Extended gestation; unable to nurse offspring	(422)
Lep	Leptin	Infertility in *ob/ob* mice; hypogonadotropic hypogonadism	(251)
Lhb	Luteinizing hormone beta	Infertility; hypogonadal; absence of large antral follicles and corpora lutea	(210)
Lhcgr	Luteinizing hormone/choriogonadotropin receptor	Infertility; folliculogenesis blocked at the antral stage	(213,214)
Ngfb	Nerve growth factor, beta	Reduced primary and secondary follicles	(100)
Ntrk2	Neurotrophic tyrosine kinase, receptor, type 2	Reduced number of secondary follicles and a decrease in granulosa cell proliferation	(106)
	Steroids and Receptors		
Cyp19a1	Cytochrome P450, family 19, subfamily a, polypeptide 1	Infertility; hypertrophied stromal cells; defects in ovulation; hemorrhagic cysts	(227)
Esr1	Estrogen receptor 1 (alpha)	Infertility; hemorrhagic ovarian cysts and uterine defects	(220)
Esr2	Estrogen receptor 2 (beta)	Subfertility; reduced litter sizes and granulosa cell defects	(224)
	Structural Proteins/Misc.		
Gja4	Gap junction membrane channel protein alpha 4	Infertility; no large antral follicles; defects in ovulation; oocyte defects	(233)
Gja1	Gap junction membrane channel protein alpha 1	Neonatal lethality; neonatal ovaries have germ cell deficiency	(235)
Ptx3	Pentraxin related 3	Subfertility; defects in ovulation and fertilization	(138)
Tnfaip6	Tumor necrosis factor alpha induced protein 6	Infertility; defects in cumulus cell physiology	(144)
	Transcription Factors		
Figla	Factor in the germline alpha	Infertility; oocyte loss by postnatal day 2	(39)
Foxl2	Forkhead box L2	Infertility; absence of secondary follicles and oocyte atresia; follicular depletion	(112,113)
Foxo3a	Forkhead box O3	Infertility by 15 weeks of age; widespread activation of follicle pool	(169)
Hesx1	Homeobox gene expressed in ES cells	Pituitary dysplasia	(423)
Lhx3	LIM homeobox protein 3	No development of anterior or intermediate lobes of pituitary gland	(424)
Nobox	Newborn ovary homeobox	Infertility; oocyte loss by postnatal day 14; disrupted primordial to primary transition	(69)
Nr5a1	Nuclear receptor subfamily 5 group A member 1 (SF-1)	Gonadal agenesis; fail to produce FSH and LH	(174,175)
Prop1	Paired-like homeodomain factor 1	Infertility; multiple anterior pituitary defects and hypogonadism	(425,426)
Pit1	Pituitary specific transcription factor 1	Infertility; multiple anterior pituitary defects and hypogonadism	(427)
Otx1	Orthodenticle homolog 1 (Drosophila)	Hypogonadism; age-specific defects in folliculogenesis and fertility	(173)
Taf4b	TAF_{II}105 RNA polymerase II, TATA box binding protein (TBP)-associated factor	Infertility; arrest of follicles at preantral stage	(166)

*The phenotypes of female homozygous null mice are described unless otherwise indicated in the text. Additional citations can be found in the text or in (140,149,171,420).

α phenotype. In this system, a mifepristone-activated chimeric nuclear receptor protein (GLVP) transgenic line was crossed to the target transgenic mouse containing the GLVP-responsive promoter upstream of the coding sequences for the α and β_A subunits to create an inhibin "bigenic" line. Overexpression of inhibin A in wild-type female mice results in a block in antral follicle development and no corpora lutea, due to suppressed levels of pituitary FSH similar to ACVR2 knockout mice. Overexpression of inhibin A was also able to suppress tumor formation in the males *Inha*$^{-/-}$ mice. Another transgenic line of mice has been developed that expresses the inhibin α subunit under the control of the mouse metallothionein-I (MT) promoter (338,339). In transgenic mice, inhibin α subunit is detected in multiple tissues including the ovary, liver, pituitary, testis, and kidney. Like the inhibin bigenic mice, overexpression of inhibin α alone results in suppressed pituitary FSH. However, MT-α ovaries are smaller than wild type and develop large, fluid filled cysts and contain polyovular follicles. The ovarian cysts remain despite pharmacological reduction in pituitary gonadotropins, and may be the result of excess androgen production. The etiology of the polyovular follicles is unknown.

All of the known inhibin knockout and transgenic lines highlight the difficultly in studying inhibin; that is, inhibin production is intrinsically linked to activin production. A large excess of inhibin α is required to produce dimeric inhibin over activin (340,341). Eliminating inhibin results in high activin levels while overexpression of inhibin α reduces activin levels. Therefore, it is difficult to interpret the distinct phenotypes without taking into account the effects on activin activity. Given that the primary role of gonadal inhibin is to regulate pituitary FSH, another strategy would be to knock out local signal transduction pathway components in the ovary instead of the secreted ligand. However, this would require a molecular understanding of how inhibins function, and currently this is unknown.

A direct role for inhibin in the pathogenesis of the *Gdf9* knockout phenotype has also been discovered. Initially, when ovaries of *Gdf9* null mice were examined, it was found that arrested primary follicles expressed abnormally high levels of inhibin α in granulosa cells (58). Inhibin α also was abnormally expressed in steroidogenic follicular nests and in follicles containing degenerating oocytes. Given the abnormal expression pattern and the proposed role of inhibin as a tumor-suppressor, it was possible that elevated levels of inhibin were partly responsible for the nonproliferative state of granulosa cells in *Gdf9* null ovaries. Crosses of *Inha* mutant mice to *Gdf9* mutant mice demonstrated that unlike *Gdf9*$^{-/-}$ ovaries, double mutant *Inha*$^{-/-}$;*Gdf9*$^{-/-}$ follicles become multilaminar and acquire a morphologically distinct thecal cell layer (Fig. 3) (139). Thus, by decreasing inhibin α expression, granulosa cell proliferation and thecal cell recruitment was recovered. However, the thecal cells appear nonfunctional and do not express markers associated with their differentiation including LH receptor, *Kit*, and the steroidogenic enzyme 17α-hydroxylase (*Cyp17*). One possibility is that the elimination of inhibin α in double null ovaries relieves the repression of granulosa cell growth and thecal cell recruitment by allowing the formation of activin dimers. Since activins are known regulators of preantral follicle growth and differentiation and have stage-specific effects on different classes of follicles, activins (or BMPs) may promote granulosa cell proliferation in the small preantral follicles (342–344). However, follicles in *Inha*$^{-/-}$;*Gdf9*$^{-/-}$ arrest again at the 5b stage, and thus additional proteins including GDF9, may be necessary again at this stage for subsequent development. Since inhibins and activins are secreted ligands and functional antagonists, it is likely that their production as well as access to the cell signaling machinery is under strict molecular regulation. Understanding this regulation will be key to deciphering their paracrine function in the ovary. As the genetic models have indicated, disruptions in the spatial and temporal expression patterns of the subunits may lead to disease states and infertility.

Inhibin Mutations and Human Disease

INHA has been hypothesized to have a role in premature ovarian failure (POF). Premature ovarian failure is characterized by amenorrhea, hypoestrogenism, and elevated levels of gonadotropins in women under the age of 40, and in the majority of cases, the disease is idiopathic. Since high levels of serum gonadotropins might indicate a disruption in the negative feedback from the gonads, inhibin represents a good candidate gene. Two studies of POF patients have identified mutations in the inhibin α gene. In one study of 43 POF patients from New Zealand and Slovenia, three women showed an G→A missense mutation at nucleotide 769 that would cause an alanine to threonine amino acid substitution in the coding region of inhibin α, while only 1 out of 150 control subjects were found to carry this same mutation (345). In a second study of a cohort of Italian women with ovarian failure, a significant increase was seen in the number of females with this same mutation in both POF (7/157) and primary amenorrhoea (3/12), but not in control (0/100) and early menopause (0/36) patients (346). While both of these studies show correlations between the mutation and POF, the functional significance of this mutation is unknown. Bioassays, such as the ability to inhibit production of FSH from primary pituitary cell cultures (341), will be necessary to determine if the mutant protein has any altered function. Additional patient analysis of the inhibin α gene may be warranted, as an additional mutation in the noncoding 5′-UTR of inhibin α has been shown to be more prevalent in POF (80.3%) patients than in the control group (66.7%) (346). Mutations in the 5′-UTR of genes can affect mRNA processing, stability and translation, leading to disease states as well (347).

Additional studies have looked for links between inhibin and female reproductive disease, but currently no other mutations in inhibin α have been described. Up-regulated inhibin A levels have been used as a marker for granulosa cell tumors and mucinous carcinomas (348–352), but unlike the mouse model, loss of inhibin α has not been associated with granulosa cell tumors. In contrast, absence of inhibin α expression and a loss of heterozygosity have been associated with some forms of prostate cancer (353,354).

Polycystic ovarian syndrome (PCOS) is another ovarian pathology of unknown origin. PCOS patients show infertility, disrupted menstrual cycles, hyperandrogenism, and hyperinsulinemia, and high levels of serum LH. Superficially, the phenotype of constitutive overexpression of inhibin α in mice resembles PCOS, but cyst morphology suggests that the mouse model and the human disease are different pathologies (338). In addition while some studies have shown higher levels of inhibin in cystic follicles of PCOS patients, other studies have not supported this (355–359). Clearly, the etiology of PCOS is complex and multifactorial, and multiple perturbations in the reproductive axis are present.

Activins

Activins are dimers of the inhibin β subunit. There are four isoforms of the mammalian β subunit: β_A, β_B, β_C, and β_E. Biological activity of activin A (β_A:β_A) and activin B (β_B:β_B) homodimers have been well described (360–362), but less is known about activin AB (β_A:β_B) heterodimers (363). The β_C and β_D isoforms are liver-specific, but knockout mice show no apparent defects, and therefore their biological significance is unclear (364,365). Activins signal through a heteromeric assembly of membrane bound serine-threonine kinase receptors that include a type II ligand binding receptor and a type I signaling receptor (327). Two type II receptors named ACVR2 and ACVR2B are the predominant activin binding receptors (366–368). The type II receptors complex with one of three type I receptors called activin receptor-like kinases (ALKs) ALK2, ALK4, or ALK7 (369–371) that results in the phosphorylation and activation of the transcription factors SMAD2 and SMAD3 (372). SMAD2 and SMAD3 then mediate gene transcription with interactions with transcriptional co-activators and co-repressors (373).

The roles of activin in the reproductive axis are diverse. A local feedback mechanism operates in the pituitary to activate synthesis and release of FSH (374–378). Within the ovary, activin has been proposed to act on multiple cell compartments. Activin subunits are expressed in granulosa cells in rats, mice and humans (296,379–381). In vitro, activin induces granulosa cell proliferation and FSH receptor expression in immature follicles, and increases FSH-dependent aromatase activity (342,343,382–384). In more mature follicles, activin can inhibit terminal differentiation by inhibiting premature luteinization through suppression of progesterone production (384–386). In thecal cells, activin opposes inhibin and decreases androgen production (387,388). Oocytes also respond to activin, and activin can promote oocyte maturation in primates and humans (309,389). Thus, the potential paracrine effects of activin in the ovary are wide-ranging.

To understand activin's effects in vivo, knockout mice for activin β_A (*Inhba*) and β_B (*Inhbb*) were

TABLE 3. *Human mutations causing reproductive axis defects in females*

Gene	Mutation	Type of mutation	Clinical presentation	Ref.
CYP19	Arg435Cys Cys437Tyr	Loss-of-function (compound)	Primary amenorrhea, cystic ovaries	(230)
FSHB	Val61fs Cys51Gly	Loss-of-function	Primary amenorrhea and infertility	(200,201)
FSHR	Ala189Val	Loss-of-function	Defects in follicle development	(9,428)
GNRHR	Gln106Arg Arg262Gln	Hypomorphic	Primary amenorrhea, low estrogens, low FSH and LH	(429)
LHR	Cys554X	Truncation	Primary amenorrhea	(216)
	Ala593Pro	Inactivating	Secondary amenorrhea	(217)
LEPR	Mutation in splice donor site exon 16	Truncation	Primary amenorrhea	(253)

generated (318,319). Mice homozygous for β_B are viable and give birth to live pups but are unable to nurse their offspring due to defects in milk production (319). In contrast, *Inhba* null mice die shortly after birth with craniofacial defects including cleft palate, loss of whiskers, upper and lower incisors, lower molars and defects in the retina (318,390–392). The perinatal lethality is attributed to the inability of homozygous pups to suckle. The perinatal lethality and craniofacial defects can be rescued by "knocking-in" the activin β_B subunit gene into the β_A locus suggesting that activin A and activin B have redundant functions in craniofacial development (320). However, the knockin allele (designated $Inhba^{BK}$) also functions as a hypomorphic β_A allele since hemizygous expression $Inhba^{BK}$ ($Inhba^{BK/-}$) results in decreased life expectancy, hypogonadism, delayed hair growth, and enlargement of external genitalia while $Inhba^{BK/+}$ and $Inhba^{+/-}$ mice are phenotypically normal. The knockin experiments also revealed a reproductive function for activin β_A. $Inhba^{BK/BK}$ mice have a severe reduction in fertility, and only one of six homozygous females was able to produce litters. Ovaries of homozygous knockin mice have fewer preovulatory follicles and hemizygous knockin mouse ovaries had fewer large preantral follicles. Thus, activin β_B is unable to compensate for activin β_A at all stages of follicle development. The latter result is interesting since it has been hypothesized that the β_A and β_B subunits are produced from different size follicles, [e.g., β_B from small preantral and β_A from large preantral and antral follicles (58,296,316, 381,393)]. The inhibin and activin content would therefore vary with follicle stage. Functionally, the knockin model suggests physiologic differences exist between activin A and activin B.

A null allele for one of the activin receptors, ACVR2, has also been generated (318). Twenty percent of the homozygous null mice die perinatally with a primary defect in the mandible and secondary defects in the palate. Surviving female homozygous *Acvr2* null mice display disrupted pituitary function related to the suppression of FSH. Ovaries from the surviving homozygous null females had small ovaries, with no corpora lutea, and minimal antral follicle development. Transplants of $Acvr2^{-/-}$ ovaries to ovariectomized immunocompatible donor females showed normal follicular development and restored fertility to the host mice (140). An additional activin receptor, ACVR2B, binds activin with a higher affinity than ACVR2 (368), and this receptor is expressed in mammalian ovaries (328,380,394). The contribution of ACVR2B to activin signaling in the ovary remains to be explored since these null mice mostly die after birth.

Activin signaling, like TGFβ signaling, results in the phosphorylation and activation of SMAD2 and SMAD3 (395,396). Null alleles for many of the Smad genes have been generated, but given the significance of TGFβ/activin and BMP signaling pathways in the formation of embryonic and extraembryonic tissues, many of the knockouts are embryonic lethal. *Smad2* null embryos have defective gastrulation mesoderm formation, and anteroposterior axis defects, and die at embryonic day E6.5–E8.5. Smad3 null mice, however, are viable (397,398), and one targeted allele shows reproductive defects when homozygous mice are generated (399,400). $Smad3^{-/-}$ mice have reduced fertility, reduced numbers of large preantral and antral follicles, and a larger primordial follicle pool during adulthood but not at birth (400). More atretic follicles are found in null ovaries than wild-type ovaries (399). Serum analysis in these mice shows higher levels of FSH and lower levels of estradiol, and ovarian extracts show reduced levels of inhibin α, cyclin D2, ERβ, but not FSH receptor. The low estradiol levels and lack of inhibin are consistent with the loss of antral follicles and the subsequent lack of negative feedback control in the pituitary. As both TGFβ and activin transmit signals through SMAD3, the challenge will be to understand the molecular output of TGFβ and/or activin signaling in normal ovarian physiology and how this is altered in the pathological state.

Follistatin

Follistatin was first identified from gonadal extracts as an FSH-modulatory factor (288,401,402). Follistatin is a single polypeptide protein not structurally related to the TGFβ superfamily, but which acts as a binding protein and functional antagonist of activin (394,403,404). In the ovary, granulosa cells and luteal cells are the main sites of production (381,405,406) but this is stage dependent. Follistatin is not expressed in primordial and primary follicles, but is expressed from antral stages (381,406).

Both loss-of-function and gain-of-function mutations have been generated for the follistatin gene (407,408). Follistatin null mice die within hours after birth, making it difficult to study the role of follistatin in ovarian function postnally. However, these mice will be a resource for understanding prenatal ovarian development since a recent study has shown that follistatin activity is required for embryonic ovarian development (409). *Fst* null ovaries, similar to *Wnt4* null XX gonads, lose 90% of their germ cells at E16.5 and have an almost compete loss of germ cells by birth. Epistasis analysis suggests that follistatin is downstream of the WNT4 pathway, but it is unclear how follistatin acts to prevent oocyte loss (409). The role of activin in prenatal ovary development is still unexplored, even though activin subunits are expressed embryonically (410). Follistatin also is known to block BMP signaling, particularly BMP7 and BMP15 (411–413), and the question of which of these pathways, activin or BMP, is implicated in the *Fst* null mouse model remains to be tested.

Transgenic mice overexpressing follistatin are viable but show reproductive defects (408). In transgenic lines with the highest expression of follistatin, females were infertile. In the transgenic lines with less follistatin expression, females developed infertility within several months. Histologically, ovaries appeared similar to FSHβ and *Acvr2* null mice with a follicle block prior to the antral stage and an increase in follicular atresia. The serum profiles confirmed that FSH was suppressed, and estradiol was undetectable. It is unclear whether all of the ovarian pathologies can be attributed to diminished pituitary function or whether there is a local abrogation of activin (and other TGFβ family) signaling in the ovary as well.

The embryonic and perinatal lethal phenotypes of ligand, receptor and SMAD components of the TGFβ superfamily highlight the need for tissue and cell-type specific conditional knockout models. Currently, one of the best conditional knockout strategies is the *Cre-loxP* system (414). All or portions of the gene of interest can be flanked with the bacteriophage P1 recombinase sites ("loxP" sites). Deletion of DNA *in cis* is facilitated by expression of Cre recombinase. Tissue-specific gene targeting can be achieved by utilizing tissue-specific promoters to express Cre, either by transgenic or knockin strategies. Crosses of the tissue-specific Cre line with the mouse line carrying the floxed allele produces a conditional allele only in offspring that express Cre. Several ovarian Cre-expressing lines have been generated. Zona pellucida 3 (*Zp3*)-Cre is specific for oocytes (415), and *Amhr2cre* is specific for granulosa cells (416). The latter line was used to generate a conditional knockout of follistatin (*Fst*-cKO) in the granulosa cells of the ovary (416). *Fst*-cKO mutant mice have fertility defects including reduced number and size of litters, fewer follicles, defects in ovulation and fertilization, increased serum FSH and LH, and reduced testosterone. In addition, the ovaries contain Sertoli-like structures also seen in inhibin α null, ERα and ERβ knockouts, and FSHβ, follistatin, and AMH transgenic mice (24,95,218,408). Granulosa cell function appears generally disrupted, possibly through the availability of free activin. In humans, it is known that serum generally contains high levels of follistatin-bound and therefore inactive, activin (417), and follicular fluid shows high levels of unbound follistatin (418). Disruptions in the balance of follistatin to activin would lead to more activin signaling, and potentially the disease state seen in follistatin-null ovaries.

CONCLUSION

Over the past decade, numerous mouse models have been generated that demonstrate the importance of multiple signaling systems in follicle development. These mice provide a basis for understanding human infertility and disease as well as revealing new candidate genes for mutational analysis. Newer technologies, such as conditional knockouts using the *Cre-lox* system, will allow specific cell-type deletions, thereby circumventing problems associated with global knockouts such as embryonic lethality and phenotypes associated with multiple tissue disruption. Combined, these models hold the promise of a comprehensive understanding of the physiology and pathology of human fertility.

ACKNOWLEDGMENTS

We thank Ms. Shirley Baker for expert assistance with manuscript formatting. Studies in the Rajkovic and Matzuk laboratories have been supported by National Institutes of Health grants HD00849, HD044858, HD047514, HD01426, HD32067, HD33438, HD42500, AAOGF, and CA60651; the Specialized Cooperative Centers Program in Reproduction Research (HD07495); and Wyeth Research. Additional funding was provided by a Basil O'Connor Starter Scholar Award from the March of Dimes Birth Defects Foundation to Dr. Rajkovic. Dr. Stephanie Pangas has been supported by a Postdoctoral Fellowship from the Center for Reproduction Training Grant HD007165 and a National Institutes of Health Ruth L. Kirschstein National Research Award (HD046335).

REFERENCES

1. Elvin, J. A., and Matzuk, M. M. (2001). Control of ovarian function. In *Transgenics in Endocrinology* (M. M. Matzuk, C. W. Brown, and T. R. Kumar, Eds.), pp. 61–89. Humana Press, Totowa, NJ.
2. Burns, K. H., and Matzuk, M. M. (2004). The application of gene ablation and related technologies to the study of ovarian function. In *The Application of Gene Ablation and Related Technologies to the Study of Ovarian Function* (P. C. Leung and E. Y. Adashi, Eds.), pp. 411–432. Elsevier Academic Press, London.
3. Matzuk, M. M., Burns, K., Viveiros, M. M., and Eppig, J. (2002). Intercellular communication in the mammalian ovary: oocytes carry the conversation. *Science* 296, 2178–2180.

4. Ko, M. S., Kitchen, J. R., Wang, X., Threat, T. A., Hasegawa, A., Sun, T., Grahovac, M. J., Kargul, G. J., Lim, M. K., Cui, Y., Sano, Y., Tanaka, T., Liang, Y., Mason, S., Paonessa, P. D., Sauls, A. D., DePalma, G. E., Sharara, R., Rowe, L. B., Eppig, J., Morrell, C., and Doi, H. (2000). Large-scale cDNA analysis reveals phased gene expression patterns during preimplantation mouse development. *Development* 127, 1737–1749.
5. Rajkovic, A., Yan, M. S. C., Klysik, M., and Matzuk, M. (2001). Discovery of germ cell-specific transcripts by expressed sequence tag database analysis. *Fertil. Steril.* 76, 550–554.
6. Stanton, J. L., and Green, D. P. (2001). A set of 840 mouse oocyte genes with well-matched human homologues. *Mol. Hum. Reprod.* 7, 521–543.
7. Ko, M. S. (2004). Embryogenomics of pre-implantation mammalian development: current status. *Reprod. Fertil. Dev.* 16, 79–85.
8. Matzuk, M. M., and Lamb, D. J. (2002). Genetic dissection of mammalian fertility pathways. *Nat. Cell. Biol.* 4, Suppl, s41–s49.
9. Aittomaki, K., Lucena, J. L., Pakarinen, P., Sistonen, P., Tapanainen, J., Gromoll, J., Kaskikari, R., Sankila, E. M., Lehvaslaiho, H., Engel, A. R., et al. (1995). Mutation in the follicle-stimulating hormone receptor gene causes hereditary hypergonadotropic ovarian failure. *Cell* 82, 959–968.
10. Kumar, T. R., Wang, Y., Lu, N., and Matzuk, M. M. (1997). Follicle stimulating hormone is required for ovarian follicle maturation but not male fertility. *Nat. Genet.* 15, 201–204.
11. Abel, M. H., Wootton, A. N., Wilkins, V., Huhtaniemi, I., Knight, P. G., and Charlton, H. M. (2000). The effect of a null mutation in the follicle-stimulating hormone receptor gene on mouse reproduction. *Endocrinology* 141, 1795–1803.
12. Dierich, A., Sairam, M. R., Monaco, L., Fimia, G. M., Gansmuller, A., LeMeur, M., and Sassone-Corsi, P. (1998). Impairing follicle-stimulating hormone (FSH) signaling in vivo: targeted disruption of the FSH receptor leads to aberrant gametogenesis and hormonal imbalance. *Proc. Natl. Acad. Sci. U S A* 95, 13612–13617.
13. Pedersen, T., and Peters, H. (1968). Proposal for a classification of oocytes and follicles in the mouse ovary. *J. Reprod. Fert.* 17, 555–557.
14. Peters, H. (1969). The development of the mouse ovary from birth to maturity. *Acta. Endocrinol. (Copenh).* 62, 98–116.
15. Pepling, M. E., and Spradling, A. C. (2001). Mouse ovarian germ cell cysts undergo programmed breakdown to form primordial follicles. *Dev. Biol.* 234, 339–351.
16. Spiegelman, M., and Bennett, D. (1973). A light- and electron-microscopic study of primordial germ cells in the early mouse embryo. *J. Embryol. Exp. Morphol.* 30, 97–118.
17. Ruby, J. R., Dyer, R. F., and Skalko, R. G. (1969). Continuities between mitochondria and endoplasmic reticulum in the mammalian ovary. *Z. Zellforsch. Mikrosk. Anat.* 97, 30–37.
18. Billig, H., Furuta, I., and Hsueh, A. J. (1994). Gonadotropin-releasing hormone directly induces apoptotic cell death in the rat ovary: biochemical and in situ detection of deoxyribonucleic acid fragmentation in granulosa cells. *Endocrinology* 134, 245–252.
19. Chun, S. Y., Eisenhauer, K. M., Minami, S., Billig, H., Perlas, E., and Hsueh, A. J. (1996). Hormonal regulation of apoptosis in early antral follicles: follicle-stimulating hormone as a major survival factor. *Endocrinology* 137, 1447–1456.
20. Zeleznik, A. J., Midgley, A. R. Jr., and Reichert, L. E. Jr. (1974). Granulosa cell maturation in the rat: increased binding of human chorionic gonadotropin following treatment with follicle-stimulating hormone in vivo. *Endocrinology* 95, 818–825.
21. Dorrington, J. H., Moon, Y. S., and Armstrong, D. T. (1975). Estradiol-17beta biosynthesis in cultured granulosa cells from hypophysectomized immature rats; stimulation by follicle-stimulating hormone. *Endocrinology* 97, 1328–1331.
22. Midgley, A. R. Jr. (1973). Autoradiographic analysis of gonadotropin binding to rat ovarian tissue sections. *Adv. Exp. Med. Biol.* 36, 365–378.
23. Richards, J. S. (1994). Hormonal control of gene expression in the ovary. *Endocr. Rev.* 15, 725–751.
24. Matzuk, M. M., Finegold, M. J., Su, J. G., Hsueh, A. J., and Bradley, A. (1992). Alpha-inhibin is a tumour-suppressor gene with gonadal specificity in mice. *Nature* 360, 313–319.
25. Forabosco, A., Sforza, C., De Pol, A., Vizzotto, L., Marzona, L., and Ferrario, V. F. (1991). Morphometric study of the human neonatal ovary. *Anat. Rec.* 231, 201–208.
26. Block, E. (1953). A quantitative morphological investigation of the follicular system in newborn female infants. *Acta. Anat. (Basel)* 17, 201–206.
27. Gougeon, A. (1996). Regulation of ovarian follicular development in primates: facts and hypotheses. *Endocr. Rev.* 17, 121–155.
28. Pryse-Davies, J. (1974). The development, structure and function of the female pelvic organs in childhood. *Clin. Obstet. Gynaecol.* 1, 483–508.
29. Valdes-Dapena, M. A. (1967). The normal ovary of childhood. *Ann. N Y Acad. Sci.* 142, 597–613.
30. Faddy, M. J., and Gosden, R. G. (1996). A model conforming the decline in follicle numbers to the age of menopause in women. *Hum. Reprod.* 11, 1484–1486.
31. Faddy, M. J., and Gosden, R. G. (1995). A mathematical model of follicle dynamics in the human ovary. *Hum. Reprod.* 10, 770–775.
32. Morita, Y., and Tilly, J. L. (1999). Oocyte apoptosis: like sand through an hourglass. *Dev. Biol.* 213, 1–17.
33. Snowden, J. A., Harkin, P. J., Thornton, J. G., and Wells, M. (1989). Morphometric assessment of ovarian stromal proliferation—a clinicopathological study. *Histopathology* 14, 369–379.
34. Boss, J. H., Scully, R. E., Wegner, K. H., and Cohen, R. B. (1965). Structural variations in the adult ovary: clinical significance. *Obstet. Gynecol.* 25, 747–764.
35. Baker, T. G. (1963). A quantitative and cytological study of germ cells in human ovaries. *Proc. R. Soc. Lond. B. Biol. Sci.* 158, 417–433.
36. Reynaud, K., Cortvrindt, R., Verlinde, F., De Schepper, J., Bourgain, C., and Smitz, J. (2004). Number of ovarian follicles in human fetuses with the 45,X karyotype. *Fertil. Steril.* 81, 1112–1119.
37. Kurilo, L. F. (1981). Oogenesis in antenatal development in man. *Hum. Genet.* 57, 86–92.
38. Gougeon, A., and Chainy, G. B. (1987). Morphometric studies of small follicles in ovaries of women at different ages. *J. Reprod. Fertil.* 81, 433–442.
39. Soyal, S. M., Amleh, A., and Dean, J. (2000). FIGalpha, a germ cell-specific transcription factor required for ovarian follicle formation. *Development* 127, 4645–4654.
40. Liang, L., Soyal, S. M., and Dean, J. (1997). FIGalpha, a germ cell specific transcription factor involved in the coordinate expression of the zona pellucida genes. *Development* 124, 4939–4947.
41. Albertini, D. F., Combelles, C. M., Benecchi, E., and Carabatsos, M. J. (2001). Cellular basis for paracrine regulation of ovarian follicle development. *Reproduction* 121, 647–653.
42. Rankin, T., Familari, M., Lee, E., Ginsberg, A., Dwyer, N., Blanchette-Mackie, J., Drago, J., Westphal, H., and Dean, J. (1996). Mice homozygous for an insertional mutation in the Zp3 gene lack a zona pellucida and are infertile. *Development* 122, 2903–2910.
43. Rankin, T. L., Tong, Z. B., Castle, P. E., Lee, E., Gore-Langton, R., Nelson, L. M., and Dean, J. (1998). Human ZP3

restores fertility in Zp3 null mice without affecting order-specific sperm binding. *Development* 125, 2415–2424.
44. Rankin, T., Talbot, P., Lee, E., and Dean, J. (1999). Abnormal zonae pellucidae in mice lacking ZP1 result in early embryonic loss. *Development* 126, 3847–3855.
45. Rankin, T. L., O'Brien, M., Lee, E., Wigglesworth, K., Eppig, J., and Dean, J. (2001). Defective zonae pellucidae in Zp2-null mice disrupt folliculogenesis, fertility and development. *Development* 128, 1119–1126.
46. Huntriss, J., Gosden, R., Hinkins, M., Oliver, B., Miller, D., Rutherford, A. J., and Picton, H.M. (2002). Isolation, characterization and expression of the human Factor In the Germline alpha (FIGLA) gene in ovarian follicles and oocytes. *Mol. Hum. Reprod.* 8, 1087–1095.
47. Bayne, R. A., Martins da Silva, S. J., and Anderson, R. A. (2004). Increased expression of the FIGLA transcription factor is associated with primordial follicle formation in the human fetal ovary. *Mol. Hum. Reprod.* 10, 373–381.
48. Eppig, J. J., and O'Brien, M. J. (1996). Development in vitro of mouse oocytes from primordial follicles. *Biol. Reprod.* 54, 197–207.
49. Fortune, J. E., Cushman, R. A., Wahl, C. M., and Kito, S. (2000). The primordial to primary follicle transition. *Mol. Cell. Endocrinol.* 163, 53–60.
50. Wandji, S. A., Srsen, V., Nathanielsz, P. W., Eppig, J. J., and Fortune, J. E. (1997). Initiation of growth of baboon primordial follicles in vitro. *Hum. Reprod.* 12, 1993–2001.
51. Driancourt, M. A., Reynaud, K., Cortvrindt, R., and Smitz, J. (2000). Roles of KIT and KIT LIGAND in ovarian function. *Rev. Reprod.* 5, 143–152.
52. Nilsson, E. E., and Skinner, M. K. (2003). Bone morphogenetic protein-4 acts as an ovarian follicle survival factor and promotes primordial follicle development. *Biol. Reprod.* 69, 1265–1272.
53. Parrott, J. A., and Skinner, M. K. (1999). Kit-ligand/stem cell factor induces primordial follicle development and initiates folliculogenesis. *Endocrinology* 140, 4262–4271.
54. Nilsson, E., Parrott, J. A., and Skinner, M. K. (2001). Basic fibroblast growth factor induces primordial follicle development and initiates folliculogenesis. *Mol. Cell. Endocrinol.* 175, 123–130.
55. Nilsson, E. E., Kezele, P., and Skinner, M. K. (2002). Leukemia inhibitory factor (LIF) promotes the primordial to primary follicle transition in rat ovaries. *Mol. Cell. Endocrinol.* 188, 65–73.
56. Vitt, U. A., and Hsueh, A. J. (2001). Stage-dependent role of growth differentiation factor-9 in ovarian follicle development. *Mol. Cell. Endocrinol.* 183, 171–177.
57. Vitt, U. A., McGee, E. A., Hayashi, M., and Hsueh, A. J. (2000). In vivo treatment with GDF-9 stimulates primordial and primary follicle progression and theca cell marker CYP17 in ovaries of immature rats. *Endocrinology* 141, 3814–3820.
58. Elvin, J. A., Yan, C., Wang, P., Nishimori, K., and Matzuk, M. M. (1999). Molecular characterization of the follicle defects in the growth differentiation factor 9-deficient ovary. *Mol. Endocrinol.* 13, 1018–1034.
59. Nilsson, E. E., and Skinner, M. K. (2002). Growth and differentiation factor-9 stimulates progression of early primary but not primordial rat ovarian follicle development. *Biol. Reprod.* 67, 1018–1024.
60. Dong, J., Albertini, D. F., Nishimori, K., Kumar, T. R., Lu, N., and Matzuk, M. M. (1996). Growth differentiation factor-9 is required during early ovarian folliculogenesis (see comments). *Nature* 383, 531–535.
61. Kezele, P., and Skinner, M. K. (2003). Regulation of ovarian primordial follicle assembly and development by estrogen and progesterone: endocrine model of follicle assembly. *Endocrinology* 144, 3329–3337.
62. Lindberg, B. S., Nilsson, B. A., and Johansson, E. D. (1974). Plasma progesterone levels in normal and abnormal pregnancies. *Acta. Obstet. Gynecol. Scand.* 53, 329–335.
63. Holmdahl, T. H., and Johansson, E. D. (1972). Peripheral plasma levels of 17 -hydroxyprogesterone during human pregnancy. *Acta. Endocrinol. (Copenh.)* 71, 765–772.
64. Johansson, E. D., and Jonasson, L. E. (1971). Progesterone levels in amniotic fluid and plasma from women. I. Levels during normal pregnancy. *Acta. Obstet. Gynecol. Scand.* 50, 339–343.
65. Zachos, N. C., Billiar, R. B., Albrecht, E. D., and Pepe, G. J. (2002). Developmental regulation of baboon fetal ovarian maturation by estrogen. *Biol. Reprod.* 67, 1148–1156.
66. Zachos, N. C., Billiar, R. B., Albrecht, E. D., and Pepe, G. J. (2004). Regulation of oocyte microvilli development in the baboon fetal ovary by estrogen. *Endocrinology* 145, 959–966.
67. Pepe, G. J., Billiar, R. B., Leavitt, M. G., Zachos, N. C., Gustafsson, J. A., and Albrecht, E. D. (2002). Expression of estrogen receptors alpha and beta in the baboon fetal ovary. *Biol. Reprod.* 66, 1054–1060.
68. Suzumori, N., Yan, C., Matzuk, M., and Rajkovic, A. (2001). Nobox is a homeobox-containing gene preferentially expressed in primordial and growing oocytes. *Mech. Dev.* 111, 137–141.
69. Rajkovic, A., Pangas, S. A., Ballow, D., Suzumori, N., and Matzuk, M. M. (2004). NOBOX deficiency disrupts early folliculogenesis and oocyte-specfic gene expression. *Science* 305, 1160–1163.
70. Manova, K., and Bachvarova, R. F. (1991). Expression of c-kit encoded at the W locus of mice in developing embryonic germ cells and presumptive melanoblasts. *Dev. Biol.* 146, 312–324.
71. Matsui, Y., Zsebo, K. M., and Hogan, B. L. (1990). Embryonic expression of a haematopoietic growth factor encoded by the Sl locus and the ligand for c-kit. *Nature* 347, 667–669.
72. Mintz, B., and Russell, E. S. (1957). Gene-induced embryological modifications of primordial germ cells in the mouse. *J. Exp. Zool.* 134, 207–237.
73. Geissler, E. N., McFarland, E. C., and Russell, E. S. (1981). Analysis of pleiotropism at the dominant white-spotting (W) locus of the house mouse: a description of ten new W alleles. *Genetics* 97, 337–361.
74. Besmer, P., Manova, K., Duttlinger, R., Huang, E. J., Packer, A., Gyssler, C., and Bachvarova, R. F. (1993). The kit-ligand (steel factor) and its receptor c-kit/W: pleiotropic roles in gametogenesis and melanogenesis. *Dev. Suppl.* 125–137.
75. Donovan, P. J., and de Miguel, M. P. (2001). The role of c-kit/kit ligand axis in mammalian gametogenesis. In *The Role of c-kit/kit Ligand Axis in Mammalian Gametogenesis* (M. M. Matzuk, C. W. Brown, and T. R. Kumar, Eds.), pp. 147–172. Humana Press, Totowa, NJ.
76. Silvers, W. K. (1979). *The Coat Colors of Mice: A Model for Mammalian Gene Action and Interactions.* Springer-Verlag, New York.
77. Russell, E. S. (1979). Hereditary anemias of the mouse: a review for geneticists. *Adv. Genet.* 20, 357–459.
78. Manova, K., Nocka, K., Besmer, P., and Bachvarova, R. F. (1990). Gonadal expression of c-kit encoded at the W locus of the mouse. *Development* 110, 1057–1069.
79. Manova, K., Huang, E. J., Angeles, M., De Leon, V., Sanchez, S., Pronovost, S. M., Besmer, P., and Bachvarova, R. F. (1993). The expression pattern of the c-kit ligand in gonads of mice supports a role for the c-kit receptor in oocyte growth and in proliferation of spermatogonia. *Dev. Biol.* 157, 85–99.
80. Joyce, I. M., Pendola, F. L., Wigglesworth, K., and Eppig, J. J. (1999). Oocyte regulation of kit ligand expression in mouse ovarian follicles. *Dev. Biol.* 214, 342–353.
81. Kang, J. S., Lee, C. J., Lee, J. M., Rha, J. Y., Song, K. W., and Park, M. H. (2003). Follicular expression of c-Kit/SCF

and inhibin-alpha in mouse ovary during development. *J. Histochem. Cytochem.* 51, 1447–1458.

82. Parrott, J. A., and Skinner, M. K. (1997). Direct actions of kit-ligand on theca cell growth and differentiation during follicle development. *Endocrinology* 138, 3819–3827.
83. Huang, E. J., Manova, K., Packer, A. I., Sanchez, S., Bachvarova, R. F., and Besmer, P. (1993). The murine steel panda mutation affects kit ligand expression and growth of early ovarian follicles. *Dev. Biol.* 157, 100–109.
84. Kuroda, H., Terada, N., Nakayama, H., Matsumoto, K., and Kitamura, Y. (1988). Infertility due to growth arrest of ovarian follicles in Sl/Slt mice. *Dev. Biol.* 126, 71–79.
85. McCoshen, J. A., and McCallion, D. J. (1975). A study of the primordial germ cells during their migratory phase in Steel mutant mice. *Experientia* 31, 589–590.
86. Yoshida, H., Takakura, N., Kataoka, H., Kunisada, T., Okamura, H., and Nishikawa, S. I. (1997). Stepwise requirement of c-kit tyrosine kinase in mouse ovarian follicle development. *Dev. Biol.* 184, 122–137.
87. Behringer, R. R., Finegold, M. J., and Cate, R. L. (1994). Müllerian-inhibiting substance function during mammalian sexual development. *Cell* 79, 415–425.
88. Rajpert-De Meyts, E., Jorgensen, N., Graem, N., Muller, J., Cate, R. L., and Skakkebaek, N. E. (1999). Expression of anti-Müllerian hormone during normal and pathological gonadal development: association with differentiation of Sertoli and granulosa cells. *J. Clin. Endocrinol. Metab.* 84, 3836–3844.
89. Weenen, C., Laven, J. S., Von Bergh, A. R., Cranfield, M., Groome, N. P., Visser, J. A., Kramer, P., Fauser, B. C., and Themmen, A. P. (2004). Anti-Müllerian hormone expression pattern in the human ovary: potential implications for initial and cyclic follicle recruitment. *Mol. Hum. Reprod.* 10, 77–83.
90. Baarends, W. M., Uilenbroek, J. T., Kramer, P., Hoogerbrugge, J. W., van Leeuwen, E. C., Themmen, A. P., and Grootegoed, J. A. (1995). Anti-müllerian hormone and anti-müllerian hormone type II receptor messenger ribonucleic acid expression in rat ovaries during postnatal development, the estrous cycle, and gonadotropin-induced follicle growth. *Endocrinology* 136, 4951–4962.
91. Durlinger, A. L., Gruijters, M. J., Kramer, P., Karels, B., Ingraham, H. A., Nachtigal, M. W., Uilenbroek, J. T., Grootegoed, J. A., and Themmen, A. P. (2002). Anti-Müllerian hormone inhibits initiation of primordial follicle growth in the mouse ovary. *Endocrinology* 143, 1076–1084.
92. Mishina, Y., Rey, R., Finegold, M. J., Matzuk, M. M., Josso, N., Cate, R. L., and Behringer, R. R. (1996). Genetic analysis of the Müllerian-inhibiting substance signal transduction pathway in mammalian sexual differentiation. *Genes Dev.* 10, 2577–2587.
93. Durlinger, A. L., Kramer, P., Karels, B., de Jong, F. H., Uilenbroek, J. T., Grootegoed, J. A., and Themmen, A. P. (1999). Control of primordial follicle recruitment by anti-Müllerian hormone in the mouse ovary. *Endocrinology* 140, 5789–5796.
94. Durlinger, A. L., Gruijters, M. J., Kramer, P., Karels, B., Kumar, T. R., Matzuk, M. M., Rose, U. M., de Jong, F. H., Uilenbroek, J. T., Grootegoed, J. A., and Themmen, A. P. (2001). Anti-Müllerian hormone attenuates the effects of FSH on follicle development in the mouse ovary. *Endocrinology* 142, 4891–4899.
95. Behringer, R. R., Cate, R. L., Froelick, G. J., Palmiter, R. D., and Brinster, R. L. (1990). Abnormal sexual development in transgenic mice chronically expressing Müllerian-inhibiting substance. *Nature* 345, 167–170.
96. Tessarollo, L. (1998). Pleiotropic functions of neurotrophins in development. *Cytokine Growth Factor Rev* 9, 125–137.
97. Lara, H. E., Hill, D. F., Katz, K. H., and Ojeda, S. R. (1990). The gene encoding nerve growth factor is expressed in the immature rat ovary: effect of denervation and hormonal treatment. *Endocrinology* 126, 357–363.
98. Dissen, G. A., Hill, D. F., Costa, M. E., Ma, Y. J., and Ojeda, S. R. (1991). Nerve growth factor receptors in the peripubertal rat ovary. *Mol. Endocrinol.* 5, 1642–1650.
99. Dissen, G. A., Romero, C., Hirshfield, A. N., and Ojeda, S. R. (2001). Nerve growth factor is required for early follicular development in the mammalian ovary. *Endocrinology* 142, 2078–2086.
100. Ojeda, S. R., Romero, C., Tapia, V., and Dissen, G. A. (2000). Neurotrophic and cell-cell dependent control of early follicular development. *Mol. Cell. Endocrinol.* 163, 67–71.
101. Dissen, G. A., Romero, C., Paredes, A., and Ojeda, S. R. (2002). Neurotrophic control of ovarian development. *Microsc. Res. Tech.* 59, 509–515.
102. Lee, K. F., Li, E., Huber, L. J., Landis, S. C., Sharpe, A. H., Chao, M. V., and Jaenisch, R. (1992). Targeted mutation of the gene encoding the low affinity NGF receptor p75 leads to deficits in the peripheral sensory nervous system. *Cell* 69, 737–749.
103. Smeyne, R. J., Klein, R., Schnapp, A., Long, L. K., Bryant, S., Lewin, A., Lira, S. A., and Barbacid, M. (1994). Severe sensory and sympathetic neuropathies in mice carrying a disrupted Trk/NGF receptor gene. *Nature* 368, 246–249.
104. Liebl, D. J., Klesse, L. J., Tessarollo, L., Wohlman, T., and Parada, L. F. (2000). Loss of brain-derived neurotrophic factor-dependent neural crest-derived sensory neurons in neurotrophin-4 mutant mice. *Proc. Natl. Acad. Sci. U S A* 97, 2297–2302.
105. Anderson, R. A., Robinson, L. L., Brooks, J., and Spears, N. (2002). Neurotropins and their receptors are expressed in the human fetal ovary. *J. Clin. Endocrinol. Metab.* 87, 890–897.
106. Paredes, A., Romero, C., Dissen, G. A., DeChiara, T. M., Reichardt, L., Cornea, A., Ojeda, S. R., and Xu, B. (2004). TrkB receptors are required for follicular growth and oocyte survival in the mammalian ovary. *Dev. Biol.* 267, 430–449.
107. Spears, N., Molinek, M. D., Robinson, L. L., Fulton, N., Cameron, H., Shimoda, K., Telfer, E. E., Anderson, R. A., and Price, D. J. (2003). The role of neurotrophin receptors in female germ-cell survival in mouse and human. *Development* 130, 5481–5491.
108. Crisponi, L., Deiana, M., Loi, A., Chiappe, F., Uda, M., Amati, P., Bisceglia, L., Zelante, L., Nagaraja, R., Porcu, S., Ristaldi, M. S., Marzella, R., Rocchi, M., Nicolino, M., Lienhardt-Roussie, A., Nivelon, A., Verloes, A., Schlessinger, D., Gasparini, P., Bonneau, D., Cao, A., and Pilia, G. (2001). The putative forkhead transcription factor FOXL2 is mutated in blepharophimosis/ptosis/epicanthus inversus syndrome. *Nat. Genet.* 27, 159–166.
109. Prueitt, R. L., and Zinn, A. R. (2001). A fork in the road to fertility. *Nat. Genet.* 27, 132–134.
110. Nicolino, M., Bost, M., David, M., and Chaussain, J. L. (1995). Familial blepharophimosis: an uncommon marker of ovarian dysgenesis. *J. Pediatr. Endocrinol. Metab.* 8, 127–133.
111. Fraser, I. S., Shearman, R. P., Smith, A., and Russell, P. (1988). An association among blepharophimosis, resistant ovary syndrome, and true premature menopause. *Fertil. Steril.* 50, 747–751.
112. Uda, M., Ottolenghi, C., Crisponi, L., Garcia, J. E., Deiana, M., Kimber, W., Forabosco, A., Cao, A., Schlessinger, D., and Pilia, G. (2004). Foxl2 disruption causes mouse ovarian failure by pervasive blockage of follicle development. *Hum. Mol. Genet.* 13, 1171–1181.
113. Schmidt, D., Ovitt, C. E., Anlag, K., Fehsenfeld, S., Gredsted, L., Treier, A. C., and Treier, M. (2004). The murine winged-helix transcription factor Foxl2 is required for granulosa cell

differentiation and ovary maintenance. *Development* 131, 933–942.
114. Salmon, N. A., Handyside, A. H., and Joyce, I. M. (2004). Oocyte regulation of anti-Müllerian hormone expression in granulosa cells during ovarian follicle development in mice. *Dev. Biol.* 266, 201–208.
115. Durlinger, A. L., Visser, J. A., and Themmen, A. P. (2002). Regulation of ovarian function: the role of anti-Müllerian hormone. *Reproduction* 124, 601–609.
116. Jiang, E. P., and Zhang, Y. Q. (2002). (The role of transforming growth factor beta superfamily in male germ cell development). *Zhonghua. Nan. Ke. Xue.* 8, 435–437.
117. Bath, L. E., Wallace, W. H., Shaw, M. P., Fitzpatrick, C., and Anderson, R. A. (2003). Depletion of ovarian reserve in young women after treatment for cancer in childhood: detection by anti-Müllerian hormone, inhibin B and ovarian ultrasound. *Hum. Reprod.* 18, 2368–2374.
118. Gruijters, M. J., Visser, J. A., Durlinger, A. L., and Themmen, A. P. (2003). Anti-Müllerian hormone and its role in ovarian function. *Mol. Cell. Endocrinol.* 211, 85–90.
119. Pisarska, M. D., Bae, J., Klein, C., and Hsueh, A. J. (2004). Forkhead l2 is expressed in the ovary and represses the promoter activity of the steroidogenic acute regulatory gene. *Endocrinology* 145, 3424–3433.
120. el-Fouly, M. A., Cook, B., Nekola, M., and Nalbandov, A. V. (1970). Role of the ovum in follicular luteinization. *Endocrinology* 87, 286–293.
121. Nekola, M. V., and Nalbandov, A. V. (1971). Morphological changes of rat follicular cells as influenced by oocytes. *Biol. Reprod.* 4, 154–160.
122. Buccione, R., Vanderhyden, B. C., Caron, P. J., and Eppig, J. J. (1990). FSH-induced expansion of the mouse cumulus oophorus in vitro is dependent upon a specific factor(s) secreted by the oocyte. *Dev. Biol.* 138, 16–25.
123. Salustri, A., Yanagishita, M., and Hascall, V. C. (1990). Mouse oocytes regulate hyaluronic acid synthesis and mucification by FSH-stimulated cumulus cells. *Dev. Biol.* 138, 26–32.
124. Salustri, A., Ulisse, S., Yanagishita, M., and Hascall, V. C. (1990). Hyaluronic acid synthesis by mural granulosa cells and cumulus cells in vitro is selectively stimulated by a factor produced by oocytes and by transforming growth factor-beta. *J. Biol. Chem.* 265, 19517–19523.
125. Elvin, J.A., Yan, C., and Matzuk, M. M. (2000). Growth differentiation factor-9 stimulates progesterone synthesis in granulosa cells via a prostaglandin E2/EP2 receptor pathway. *Proc. Natl. Acad. Sci. U S A* 97, 10288–10293.
126. Braw-Tal, R. (2002). The initiation of follicle growth: the oocyte or the somatic cells? *Mol. Cell. Endocrinol.* 187, 11–18.
127. Eppig, J. J. (2001). Oocyte control of ovarian follicular development and function in mammals. *Reproduction* 122, 829–838.
128. Eppig, J. J., Wigglesworth, K., and Pendola, F. L. (2002). The mammalian oocyte orchestrates the rate of ovarian follicular development. *Proc. Natl. Acad. Sci. U S A* 99, 2890–2894.
129. Taft, R. A., Denegre, J. M., Pendola, F. L., and Eppig, J. J. (2002). Identification of genes encoding mouse oocyte secretory and transmembrane proteins by a signal sequence trap. *Biol. Reprod.* 67, 953–960.
130. Klein, R. D., Gu, Q., Goddard, A., and Rosenthal, A. (1996). Selection for genes encoding secreted proteins and receptors. *Proc. Natl. Acad. Sci. U S A* 93, 7108–7113.
131. Mason, A. J., Hayflick, J. S., Zoeller, R. T., Young, W. S. III, Phillips, H. S., Nikolics, K., and Seeburg, P. H. (1986). A deletion truncating the gonadotropin-releasing hormone gene is responsible for hypogonadism in the hpg mouse. *Science* 234, 1366–1371.
132. Halpin, D. M., Charlton, H. M., and Faddy, M. J. (1986). Effects of gonadotrophin deficiency on follicular development in hypogonadal (hpg) mice. *J. Reprod. Fertil.* 78, 119–125.
133. Cattanach, B. M., Iddon, C. A., Charlton, H. M., Chiappa, S. A., and Fink, G. (1977). Gonadotrophin-releasing hormone deficiency in a mutant mouse with hypogonadism. *Nature* 269, 338–340.
134. McGrath, S. A., Esquela, A. F., and Lee, S. J. (1995). Oocyte-specific expression of growth/differentiation factor-9. *Mol. Endocrinol.* 9, 131–136.
135. Elvin, J. A., Yan, C., and Matzuk, M. M. (2000). Oocyte-expressed TGF-beta superfamily members in female fertility. *Mol. Cell. Endocrinol.* 159, 1–5.
136. Carabatsos, M. J., Elvin, J., Matzuk, M. M., and Albertini, D. F. (1998). Characterization of oocyte and follicle development in growth differentiation factor-9-deficient mice. *Dev. Biol.* 204, 373–384.
137. Hreinsson, J. G., Scott, J. E., Rasmussen, C., Swahn, M. L., Hsueh, A. J., and Hovatta, O. (2002). Growth differentiation factor-9 promotes the growth, development, and survival of human ovarian follicles in organ culture. *J. Clin. Endocrinol. Metab.* 87, 316–321.
138. Varani, S., Elvin, J. A., Yan, C., DeMayo, J., DeMayo, F. J., Horton, H. F., Byrne, M. C., and Matzuk, M. M. (2002). Knockout of pentraxin 3, a downstream target of growth differentiation factor-9, causes female subfertility. *Mol. Endocrinol.* 16, 1154–1167.
139. Wu, X., Chen, L., Brown, C. A., Yan, C., and Matzuk, M. M. (2004). Interrelationship of growth differentiation factor 9 and inhibin in early folliculogenesis and ovarian tumorigenesis in mice. *Mol. Endocrinol.* 18, 1509–1519.
140. Elvin, J. A., and Matzuk, M. M. (1998). Mouse models of ovarian failure. *Rev. Reprod.* 3, 183–195.
141. Matzuk, M. M. (2000). Revelations of ovarian follicle biology from gene knockout mice. *Mol. Cell. Endocrinol.* 163, 61–66.
142. Vitt, U. A., Hayashi, M., Klein, C., and Hsueh, A. J. (2000). Growth differentiation factor-9 stimulates proliferation but suppresses the follicle-stimulating hormone-induced differentiation of cultured granulosa cells from small antral and preovulatory rat follicles. *Biol. Reprod.* 62, 370–377.
143. Zhuo, L., and Kimata, K. (2001). Cumulus oophorus extracellular matrix: its construction and regulation. *Cell. Struct. Funct.* 26, 189–196.
144. Fulop, C., Szanto, S., Mukhopadhyay, D., Bardos, T., Kamath, R. V., Rugg, M. S., Day, A. J., Salustri, A., Hascall, V. C., Glant, T. T., and Mikecz, K. (2003). Impaired cumulus mucification and female sterility in tumor necrosis factor-induced protein-6 deficient mice. *Development* 130, 2253–2261.
145. Elvin, J. A., Clark, A. T., Wang, P., Wolfman, N. M., and Matzuk, M. M. (1999). Paracrine actions of growth differentiation factor-9 in the mammalian ovary. *Mol. Endocrinol.* 13, 1035–1048.
146. Pangas, S. A., Jorgez, C. J., and Matzuk, M. M. (2004). Growth differentiation factor 9 regulates expression of the bone morphogenetic protein antagonist, gremlin. *J. Biol. Chem.* 30, 32281–32286.
147. Salustri, A., Garlanda, C., Hirsch, E., De Acetis, M., Maccagno, A., Bottazzi, B., Doni, A., Bastone, A., Mantovani, G., Peccoz, P. B., Salvatori, G., Mahoney, D. J., Day, A.J., Siracusa, G., Romani, L., and Mantovani, A. (2004). PTX3 plays a key role in the organization of the cumulus oophorus extracellular matrix and in in vivo fertilization. *Development* 131, 1577–1586.
148. Dube, J. L., Wang, P., Elvin, J., Lyons, K. M., Celeste, A. J., and Matzuk, M. M. (1998). The bone morphogenetic protein

15 gene is X-linked and expressed in oocytes. *Mol. Endocrinol.* 12, 1809–1817.

149. Chang, H., Brown, C. W., and Matzuk, M. M. (2002). Genetic analysis of the mammalian transforming growth factor-beta superfamily. *Endocr. Rev.* 23, 787–823.
150. Yan, C., Wang, P., DeMayo, J., DeMayo, F. J., Elvin, J. A., Carino, C., Prasad, S. V., Skinner, S. S., Dunbar, B. S., Dube, J. L., Celeste, A. J., and Matzuk, M. (2001). Synergestic roles of bone morphogenetic protein 15 and growth differentiation factor 9 in ovarian function. *Mol. Endocrinol.* 15, 854–866.
151. Davis, G. H., McEwan, J. C., Fennessy, P. F., Dodds, K. G., and Farquhar, P. A. (1991). Evidence for the presence of a major gene influencing ovulation rate on the X chromosome of sheep. *Biol. Reprod.* 44, 620–624.
152. Davis, G. H., McEwan, J. C., Fennessy, P. F., Dodds, K. G., McNatty, K. P., and O, W. S. (1992). Infertility due to bilateral ovarian hypoplasia in sheep homozygous (FecXI FecXI) for the Inverdale prolificacy gene located on the X chromosome. *Biol. Reprod.* 46, 636–640.
153. Galloway, S. M., McNatty, K. P., Cambridge, L. M., Laitinen, M. P., Juengel, J. L., Jokiranta, T. S., McLaren, R. J., Luiro, K., Dodds, K. G., Montgomery, G. W., Beattie, A. E., Davis, G. H., and Ritvos, O. (2000). Mutations in an oocyte-derived growth factor gene (BMP15) cause increased ovulation rate and infertility in a dosage-sensitive manner. *Nat. Genet.* 25, 279–283.
154. McNatty, K. P., Juengel, J. L., Wilson, T., Galloway, S. M., and Davis, G. H. (2001). Genetic mutations influencing ovulation rate in sheep. *Reprod. Fertil. Dev.* 13, 549–555.
155. Braw-Tal, R., McNatty, K. P., Smith, P., Heath, D. A., Hudson, N. L., Phillips, D. J., McLeod, B. J., and Davis, G. H. (1993). Ovaries of ewes homozygous for the X-linked Inverdale gene (FecXI) are devoid of secondary and tertiary follicles but contain many abnormal structures. *Biol. Reprod.* 49, 895–907.
156. Smith, P., O, W. S., Corrigan, K. A., Smith, T., Lundy, T., Davis, G. H., and McNatty, K. P. (1997). Ovarian morphology and endocrine characteristics of female sheep fetuses that are heterozygous or homozygous for the inverdale prolificacy gene (fecX1). *Biol. Reprod.* 57, 1183–1192.
157. Liao, W. X., Moore, R. K., and Shimasaki, S. (2004). Functional and molecular characterization of naturally occurring mutations in the oocyte-secreted factors bone morphogenetic protein-15 and growth and differentiation factor-9. *J. Biol. Chem.* 279, 17391–17396.
158. Liao, W. X., Moore, R. K., Otsuka, F., and Shimasaki, S. (2003). Effect of intracellular interactions on the processing and secretion of bone morphogenetic protein-15 (BMP-15) and growth and differentiation factor-9. Implication of the aberrant ovarian phenotype of BMP-15 mutant sheep. *J. Biol. Chem.* 278, 3713–3719.
159. Di Pasquale, E., Beck-Peccoz, P., and Persani, L. (2004). Hypergonadotropic ovarian failure associated with an inherited mutation of human bone morphogenetic protein-15 (BMP15) gene. *Am. J. Hum. Genet.* 75, 106–111.
160. Kretzschmar, M., Liu, F., Hata, A., Doody, J., and Massague, J. (1997). The TGF-beta family mediator Smad1 is phosphorylated directly and activated functionally by the BMP receptor kinase. *Genes. Dev.* 11, 984–995.
161. Miyazono, K. (1999). Signal transduction by bone morphogenetic protein receptors: functional roles of Smad proteins. *Bone* 25, 91–93.
162. Moore, R. K., Otsuka, F., and Shimasaki, S. (2003). Molecular basis of bone morphogenetic protein-15 signaling in granulosa cells. *J. Biol. Chem.* 278, 304–310.
163. Davis, G. H., Dodds, K. G., and Bruce, G. D. (1999). Combined effect of the Inverdale and Booroola prolificacy genes on ovulation rate in sheep. *Proc. Assoc. Adv. Anim. Breed. Genet.* 13, 74–77.
164. Wilson, T., Wu, X. Y., Juengel, J. L., Ross, I. K., Lumsden, J. M., Lord, E. A., Dodds, K. G., Walling, G. A., McEwan, J. C., O'Connell, A. R., McNatty, K. P., and Montgomery, G. W. (2001). Highly prolific Booroola sheep have a mutation in the intracellular kinase domain of bone morphogenetic protein IB receptor (ALK-6) that is expressed in both oocytes and granulosa cells. *Biol. Reprod.* 64, 1225–1235.
165. Yi, S. E., LaPolt, P. S., Yoon, B. S., Chen, J. Y., Lu, J. K., and Lyons, K. M. (2001). The type I BMP receptor BmprIB is essential for female reproductive function. *Proc. Natl. Acad. Sci. U S A* 98, 7994–7999.
166. Freiman, R. N., Albright, S. R., Zheng, S., Sha, W. C., Hammer, R. E., and Tjian, R. (2001). Requirement of tissue-selective TBP-associated factor TAFII105 in ovarian development. *Science* 293, 2084–2087.
167. Dikstein, R., Zhou, S., and Tjian, R. (1996). Human TAFII 105 is a cell type-specific TFIID subunit related to hTAFII130. *Cell* 87, 137–146.
168. Hosaka, T., Biggs, W.H. III, Tieu, D., Boyer, A. D., Varki, N. M., Cavenee, W. K., and Arden, K. C. (2004). Disruption of forkhead transcription factor (FOXO) family members in mice reveals their functional diversification. *Proc. Natl. Acad. Sci. U S A* 101, 2975–2980.
169. Castrillon, D. H., Miao, L., Kollipara, R., Horner, J. W., and DePinho, R. A. (2003). Suppression of ovarian follicle activation in mice by the transcription factor Foxo3a. *Science* 301, 215–218.
170. Cushman, L. J., and Camper, S. A. (2001). Molecular basis of pituitary dysfunction in mouse and human. *Mamm. Genome.* 12, 485–494.
171. Burns, K. H., and Matzuk, M. M. (2002). Minireview: genetic models for the study of gonadotropin actions. *Endocrinology* 143, 2823–2835.
172. Acampora, D., Mazan, S., Avantaggiato, V., Barone, P., Tuorto, F., Lallemand, Y., Brulet, P., and Simeone, A. (1996). Epilepsy and brain abnormalities in mice lacking the Otx1 gene. *Nat. Genet.* 14, 218–222.
173. Acampora, D., Mazan, S., Tuorto, F., Avantaggiato, V., Tremblay, J. J., Lazzaro, D., di Carlo, A., Mariano, A., Macchia, P. E., Corte, G., Macchia, V., Drouin, J., Brulet, P., and Simeone, A. (1998). Transient dwarfism and hypogonadism in mice lacking Otx1 reveal prepubescent stage-specific control of pituitary levels of GH, FSH and LH. *Development* 125, 1229–1239.
174. Luo, X., Ikeda, Y., and Parker, K. L. (1994). A cell-specific nuclear receptor is essential for adrenal and gonadal development and sexual differentiation. *Cell* 77, 481–490.
175. Ingraham, H. A., Lala, D. S., Ikeda, Y., Luo, X., Shen, W. H., Nachtigal, M. W., Abbud, R., Nilson, J. H., and Parker, K. L. (1994). The nuclear receptor steroidogenic factor 1 acts at multiple levels of the reproductive axis. *Genes. Dev.* 8, 2302–2312.
176. Jones, J. R., and Kemmann, E. (1976). Olfacto-genital dysplasia in the female. *Obstet Gynecol. Annu.* 5, 443–466.
177. Rugarli, E. I., Lutz, B., Kuratani, S. C., Wawersik, S., Borsani, G., Ballabio, A., and Eichele, G. (1993). Expression pattern of the Kallmann syndrome gene in the olfactory system suggests a role in neuronal targeting. *Nat. Genet.* 4, 19–26.
178. McKusick, V. A. (1994). 308700 Kallmann syndrome. In *Mendelian Inheritance in Man* (V. A. McKusick, Ed.). Johns Hopkins University Press, Baltimore.
179. Persson, J. W., Humphrey, K., Watson, C., Taylor, P., Leigh, D., McDonald, B., and Fraser, I. S. (1999). Investigation

of a unique male and female sibship with Kallmann's syndrome and 46,XX gonadal dysgenesis with short stature. *Hum. Reprod.* 14, 1207–1212.
180. Battaglia, C., Salvatori, M., Regnani, G., Giulini, S., Primavera, M. R., and Volpe, A. (2000). Successful induction of ovulation using highly purified follicle-stimulating hormone in a woman with Kallmann's syndrome. *Fertil. Steril.* 73, 284–286.
181. Crowley, W. F. Jr., and McArthur, J. W. (1980). Simulation of the normal menstrual cycle in Kallman's syndrome by pulsatile administration of luteinizing hormone-releasing hormone (LHRH). *J. Clin. Endocrinol. Metab.* 51, 173–175.
182. Wang, A. F., Li, J. H., Maiti, K., Kim, W. P., Kang, H. M., Seong, J. Y., and Kwon, H. B. (2003). Preferential ligand selectivity of the monkey type-II gonadotropin-releasing hormone (GnRH) receptor for GnRH-2 and its analogs. *Mol. Cell. Endocrinol.* 209, 33–42.
183. White, R. B., Eisen, J. A., Kasten, T. L., and Fernald, R. D. (1998). Second gene for gonadotropin-releasing hormone in humans. *Proc. Natl. Acad. Sci. U S A* 95, 305–309.
184. Neill, J. D., Duck, L. W., Sellers, J. C., and Musgrove, L. C. (2001). A gonadotropin-releasing hormone (GnRH) receptor specific for GnRH II in primates. *Biochem. Biophys. Res. Commun.* 282, 1012–1018.
185. Kang, S. K., Tai, C. J., Nathwani, P. S., and Leung, P. C. (2001). Differential regulation of two forms of gonadotropin-releasing hormone messenger ribonucleic acid in human granulosa-luteal cells. *Endocrinology* 142, 182–192.
186. de Roux, N., and Milgrom, E. (2001). Inherited disorders of GnRH and gonadotropin receptors. *Mol. Cell. Endocrinol.* 179, 83–87.
187. Layman, L. C., McDonough, P. G., Cohen, D. P., Maddox, M., Tho, S. P., and Reindollar, R. H. (2001). Familial gonadotropin-releasing hormone resistance and hypogonadotropic hypogonadism in a family with multiple affected individuals. *Fertil. Steril.* 75, 1148–1155.
188. Minaretzis, D., Jakubowski, M., Mortola, J. F., and Pavlou, S. N. (1995). Gonadotropin-releasing hormone receptor gene expression in human ovary and granulosa-lutein cells. *J. Clin. Endocrinol. Metab.* 80, 430–434.
189. Findlay, J. K., and Drummond, A. E. (1999). Regulation of the FSH receptor in the ovary. *Trends Endocrinol. Metab.* 10, 183–188.
190. Burns, K. H., Yan, C., Kumar, T. R., and Matzuk, M. M. (2001). Analysis of ovarian gene expression in follicle-stimulating hormone beta knockout mice. *Endocrinology* 142, 2742–2751.
191. Gonzalez-Robayna, I. J., Falender, A. E., Ochsner, S., Firestone, G. L., and Richards, J. S. (2000). Follicle-stimulating hormone (FSH) stimulates phosphorylation and activation of protein kinase B (PKB/Akt) and serum and glucocorticoid-lnduced kinase (Sgk): evidence for A kinase-independent signaling by FSH in granulosa cells. *Mol. Endocrinol.* 14, 1283–1300.
192. Alliston, T. N., Gonzalez-Robayna, I. J., Buse, P., Firestone, G. L., and Richards, J. S. (2000). Expression and localization of serum/glucocorticoid-induced kinase in the rat ovary: relation to follicular growth and differentiation. *Endocrinology* 141, 385–395.
193. Alliston, T. N., Maiyar, A. C., Buse, P., Firestone, G. L., and Richards, J. S. (1997). Follicle stimulating hormone-regulated expression of serum/glucocorticoid-inducible kinase in rat ovarian granulosa cells: a functional role for the Sp1 family in promoter activity. *Mol. Endocrinol.* 11, 1934–1949.
194. Ishimura, K., Yoshinaga-Hirabayashi, T., Tsuri, H., Fujita, H., and Osawa, Y. (1989). Further immunocytochemical study on the localization of aromatase in the ovary of rats and mice. *Histochemistry* 90, 413–416.
195. Tamura, T., Kitawaki, J., Yamamoto, T., Osawa, Y., Kominami, S., Takemori, S., and Okada, H. (1992). Immunohistochemical localization of 17 alpha-hydroxylase/C17-20 lyase and aromatase cytochrome P-450 in the human ovary during the menstrual cycle. *J. Endocrinol.* 135, 589–595.
196. Sicinski, P., Donaher, J. L., Gene, Y., Parker, S. B., Gardner, H., Park, M. Y., Robker, R. L., Richard, J. S., McGinnis, L. K., Biggers, J. D., Eppig, J. J., Bronson, R. T., Elledge, S. J., and Weinberg, R. A. (1996). Cyclin D2 is an FSH-responsive gene involved in gonadal cell proliferation and oncogenesis. *Nature* 384, 470–474.
197. Baker, J., Hardy, M. P., Zhou, J., Bondy, C., Lupu, F., Bellve, A. R., and Efstratiadis, A. (1996). Effects of an Igf1 gene null mutation on mouse reproduction. *Mol. Endocrinol.* 10, 903–918.
198. Zhou, J., Kumar, T. R., Matzuk, M. M., and Bondy, C. (1997). Insulin-like growth factor I regulates gonadotropin responsiveness in the murine ovary. *Mol. Endocrinol.* 11, 1924–1933.
199. Kumar, T. R., Low, M. J., and Matzuk, M. M. (1998). Genetic rescue of follicle-stimulating hormone beta-deficient mice. *Endocrinology* 139, 3289–3295.
200. Matthews, C. H., Borgato, S., Beck-Peccoz, P., Adams, M., Tone, Y., Gambino, G., Casagrande, S., Tedeschini, G., Benedetti, A., and Chatterjee, V. K. (1993). Primary amenorrhoea and infertility due to a mutation in the beta- subunit of follicle-stimulating hormone. *Nat. Genet.* 5, 83–86.
201. Layman, L. C., Lee, E. J., Peak, D. B., Namnoum, A. B., Vu, K. V., van Lingen, B. L., Gray, M. R., McDonough, P. G., Reindollar, R. H., and Jameson, J. L. (1997). Delayed puberty and hypogonadism caused by mutations in the follicle- stimulating hormone beta-subunit gene. *N. Engl. J. Med.* 337, 607–611.
202. Layman, L. C. (2000). Mutations in the follicle-stimulating hormone-beta (FSH beta) and FSH receptor genes in mice and humans. *Semin. Reprod. Med.* 18, 5–10.
203. Kumar, T. R., Palapattu, G., Wang, P., Woodruff, T. K., Boime, I., Byrne, M. C., and Matzuk, M. M. (1999). Transgenic models to study gonadotropin function: the role of follicle-stimulating hormone in gonadal growth and tumorigenesis. *Molecular Endocrinology* 13, 851–865.
204. Rulli, S. B., Kuorelahti, A., Karaer, O., Pelliniemi, L. J., Poutanen, M., and Huhtaniemi, I. (2002). Reproductive disturbances, pituitary lactotrope adenomas, and mammary gland tumors in transgenic female mice producing high levels of human chorionic gonadotropin. *Endocrinology* 143, 4084–4095.
205. Risma, K. A., Clay, C. M., Nett, T. M., Wagner, T., Yun, J., and Nilson, J. H. (1995). Targeted overexpression of luteinizing hormone in transgenic mice leads to infertility, polycystic ovaries, and ovarian tumors. *Proc. Natl. Acad. Sci. U S A* 92, 1322–1326.
206. Matzuk, M. M., DeMayo, F. J., Hadsell, L. A., and Kumar, T. R. (2003). Overexpression of human chorionic gonadotropin causes multiple reproductive defects in transgenic mice. *Biol. Reprod.* 69, 338–346.
207. Themmen, A. P. N., and Huhtaniemi, I. T. (2000). Mutations of gonadotropins and gonadotropin receptors: elucidating the physiology and pathophysiology of pituitary-gonadal function. *Endocr. Rev.* 21, 551–583.
208. Davis, D., Liu, X., and Segaloff, D. L. (1995). Identification of the sites of N-linked glycosylation on the follicle-stimulating hormone (FSH) receptor and assessment of their role in FSH receptor function. *Mol. Endocrinol.* 9, 159–170.
209. Touraine, P., Beau, I., Gougeon, A., Meduri, G., Desroches, A., Pichard, C., Detoeuf, M., Paniel, B., Prieur, M., Zorn, J. R., Milgrom, E., Kuttenn, F., and Misrahi, M. (1999). New natural inactivating mutations of the follicle-stimulating hormone

receptor: correlations between receptor function and phenotype. *Mol. Endocrinol.* 13, 1844–1854.

210. Ma, X., Dong, Y., Matzuk, M. M., and Kumar, T. R. (2004). Targeted disruption of luteinizing hormone beta subunit leads to hypogonadism, defects in gonadal steroidogenesis and infertility. *Proc. Natl. Acad. Sci. U S A*, 101, 17294–17299.
211. Weiss, J., Axelrod, L., Whitcomb, R. W., Harris, P. E., Crowley, W. F., and Jameson, J. L. (1992). Hypogonadism caused by a single amino acid substitution in the beta subunit of luteinizing hormone. *N. Engl. J. Med.* 326, 179–183.
212. Sokka, T. A., Hamalainen, T. M., Kaipia, A., Warren, D. W., and Huhtaniemi, I. T. (1996). Development of luteinizing hormone action in the perinatal rat ovary. *Biol. Reprod.* 55, 663–670.
213. Lei, Z. M., Mishra, S., Zou, W., Xu, B., Foltz, M., Li, X., and Rao, C. V. (2001). Targeted disruption of luteinizing hormone/human chorionic gonadotropin receptor gene. *Mol. Endocrinol.* 15, 184–200.
214. Zhang, F. P., Poutanen, M., Wilbertz, J., and Huhtaniemi, I. (2001). Normal prenatal but arrested postnatal sexual development of luteinizing hormone receptor knockout (LuRKO) mice. *Mol. Endocrinol.* 15, 172–183.
215. Sultan, L. H., and Lumbroso, S. (1988). LH receptor defects. In *Fertility and Reproductive Medicine* (R. D. Kempers, J. Cohen, A. F. Haney, and J. B. Younger, Eds.), pp. 769–797). Elsevier Science, Amsterdam.
216. Latronico, A. C., Anasti, J., Arnhold, I. J., Rapaport, R., Mendonca, B. B., Bloise, W., Castro, M., Tsigos, C., and Chrousos, G. P. (1996). Brief report: testicular and ovarian resistance to luteinizing hormone caused by inactivating mutations of the luteinizing hormone-receptor gene. *N. Engl. J. Med.* 334, 507–512.
217. Toledo, S. P., Brunner, H. G., Kraaij, R., Post, M., Dahia, P. L., Hayashida, C. Y., and Kremer, H. T. A. P. (1996). An inactivating mutation of the luteinizing hormone receptor causes amenorrhea in a 46,XX female. *J. Clin. Endocrinol. Metab.* 81, 3850–3854.
218. Couse, J. F., and Korach, K. S. (1999). Estrogen receptor null mice: what have we learned and where will they lead us? *Endocr. Rev.* 20, 358–417.
219. Korach, K. S., Emmen, J. M., Walker, V. R., Hewitt, S. C., Yates, M., Hall, J. M., Swope, D. L., Harrell, J. C., and Couse, J. F. (2003). Update on animal models developed for analyses of estrogen receptor biological activity. *J. Steroid. Biochem. Mol. Biol.* 86, 387–391.
220. Lubahn, D. B., Moyer, J. S., Golding, T. S., Couse, J. F., Korach, K. S., and Smithies, O. (1993). Alteration of reproductive function but not prenatal sexual development after insertional disruption of the mouse estrogen receptor gene. *Proc. Natl. Acad. Sci. U S A* 90, 11162–11166.
221. Mosselman, S., Polman, J., and Dijkema, R. (1996). ER beta: identification and characterization of a novel human estrogen receptor. *FEBS Lett.* 392, 49–53.
222. Kuiper, G. G., Enmark, E., Pelto-Huikko, M., Nilsson, S., and Gustafsson, J. A. (1996). Cloning of a novel receptor expressed in rat prostate and ovary. *Proc. Natl. Acad. Sci. U S A* 93, 5925–5930.
223. Tremblay, G. B., Tremblay, A., Copeland, N. G., Gilbert, D. J., Jenkins, N. A., Labrie, F., and Giguere, V. (1997). Cloning, chromosomal localization, and functional analysis of the murine estrogen receptor beta. *Mol. Endocrinol.* 11, 353–365.
224. Krege, J. H., Hodgin, J. B., Couse, J. F., Enmark, E., Warner, M., Mahler, J. F., Sar, M., Korach, K. S., Gustafsson, J. A., and Smithies, O. (1998). Generation and reproductive phenotypes of mice lacking estrogen receptor beta. *Proc. Natl. Acad. Sci. U S A* 95, 15677–15682.
225. Couse, J. F., Hewitt, S. C., Bunch, D. O., Sar, M., Walker, V. R., Davis, B. J., and Korach, K. S. (1999). Postnatal sex reversal of the ovaries in mice lacking estrogen receptors a and b. *Science* 286, 2328–2331.
226. Dupont, S., Krust, A., Gansmuller, A., Dierich, A., Chambon, P., and Mark, M. (2000). Effect of single and compound knockouts of estrogen receptors alpha (ERalpha) and beta (ERbeta) on mouse reproductive phenotypes. *Development* 127, 4277–4291.
227. Fisher, C. R., Graves, K. H., Parlow, A. F., and Simpson, E. R. (1998). Characterization of mice deficient in aromatase (ArKO) because of targeted disruption of the cyp19 gene. *Proc. Natl. Acad. Sci. U S A* 95, 6965–6970.
228. Britt, K. L., Drummond, A. E., Cox, V. A., Dyson, M., Wreford, N. G., Jones, M. E., Simpson, E. R., and Findlay, J. K. (2000). An age-related ovarian phenotype in mice with targeted disruption of the Cyp 19 (aromatase) gene. *Endocrinology* 141, 2614–2623.
229. Huynh, K., Jones, G., Thouas, G., Britt, K. L., Simpson, E. R., and Jones, M. E. (2004). Estrogen is not directly required for oocyte developmental competence. *Biol. Reprod.* 70, 1263–1269.
230. Ito, Y., Fisher, C. R., Conte, F. A., Grumbach, M. M., and Simpson, E. R. (1993). Molecular basis of aromatase deficiency in an adult female with sexual infantilism and polycystic ovaries. *Proc. Natl. Acad. Sci. U S A* 90, 11673–11677.
231. Shozu, M., Akasofu, K., Harada, T., and Kubota, Y. (1991). A new cause of female pseudohermaphroditism: placental aromatase deficiency. *J. Clin. Endocrinol. Metab.* 72, 560–566.
232. Mullis, P. E., Yoshimura, N., Kuhlmann, B., Lippuner, K., Jaeger, P., and Harada, H. (1997). Aromatase deficiency in a female who is compound heterozygote for two new point mutations in the P450arom gene: impact of estrogens on hypergonadotropic hypogonadism, multicystic ovaries, and bone densitometry in childhood. *J. Clin. Endocrinol. Metab.* 82, 1739–1745.
233. Simon, A. M., Goodenough, D. A., Li, E., and Paul, D. L. (1997). Female infertility in mice lacking connexin 37. *Nature* 385, 525–529.
234. Reaume, A. G., de Sousa, P. A., Kulkarni, S., Langille, B. L., Zhu, D., Davies, T. C., Juneja, S. C., Kidder, G. M., and Rossant, J. (1995). Cardiac malformation in neonatal mice lacking connexin43. *Science* 267, 1831–1834.
235. Juneja, S. C., Barr, K. J., Enders, G. C., and Kidder, G. M. (1999). Defects in the germ line and gonads of mice lacking connexin43. *Biol. Reprod.* 60, 1263–1270.
236. Ackert, C. L., Gittens, J. E., O'Brien, M. J., Eppig, J. J., and Kidder, G. M. (2001). Intercellular communication via connexin43 gap junctions is required for ovarian folliculogenesis in the mouse. *Dev. Biol.* 233, 258–270.
237. Wiesen, J. F., and Midgley, A. R. Jr. (1993). Changes in expression of connexin 43 gap junction messenger ribonucleic acid and protein during ovarian follicular growth. *Endocrinology* 133, 741–746.
238. Granot, I., and Dekel, N. (1997). Developmental expression and regulation of the gap junction protein and transcript in rat ovaries. *Mol. Reprod. Dev.* 47, 231–239.
239. Burghardt, R. C., and Matheson, R. L. (1982). Gap junction amplification in rat ovarian granulosa cells. I. A direct response to follicle-stimulating hormone. *Dev. Biol.* 94, 206–215.
240. Larsen, W. J., Tung, H. N., and Polking, C. (1981). Response of granulosa cell gap junctions to human chorionic gonadotropin (hCG) at ovulation. *Biol. Reprod.* 25, 1119–1134.
241. Sommersberg, B., Bulling, A., Salzer, U., Frohlich, U., Garfield, R. E., Amsterdam, A., and Mayerhofer, A. (2000).

Gap junction communication and connexin 43 gene expression in a rat granulosa cell line: regulation by follicle-stimulating hormone. *Biol. Reprod.* 63, 1661–1668.

242. Granot, I., and Dekel, N. (1994). Phosphorylation and expression of connexin-43 ovarian gap junction protein are regulated by luteinizing hormone. *J. Biol. Chem.* 269, 30502–30509.
243. Granot, I., and Dekel, N. (2002). The ovarian gap junction protein connexin43: regulation by gonadotropins. *Trends Endocrinol. Metab.* 13, 310–313.
244. Chatterjee, B., Meyer, R. A., Loredo, G. A., Coleman, C. M., Tuan, R., and Lo, C. W. (2003). BMP regulation of the mouse connexin43 promoter in osteoblastic cells and embryos. *Cell. Commun. Adhes.* 10, 37–50.
245. Carystinos, G. D., Kandouz, M., Alaoui-Jamali, M. A., and Batist, G. (2003). Unexpected induction of the human connexin 43 promoter by the ras signaling pathway is mediated by a novel putative promoter sequence. *Mol. Pharmacol.* 63, 821–831.
246. Nelles, E., Butzler, C., Jung, D., Temme, A., Gabriel, H. D., Dahl, U., Traub, O., Stumpel, F., Jungermann, K., Zielasek, J., Toyka, K. V., Dermietzel, R., and Willecke, K. (1996). Defective propagation of signals generated by sympathetic nerve stimulation in the liver of connexin32-deficient mice. *Proc. Natl. Acad. Sci. U S A* 93, 9565–9570.
247. Combelles, C. M., Carabatsos, M. J., Kumar, T. R., Matzuk, M. M., and Albertini, D. F. (2004). Hormonal control of somatic cell oocyte interactions during ovarian follicle development. *Mol. Reprod. Dev.* 69, 347–355.
248. Ryan, N. K., Woodhouse, C. M., Van der Hoek, K. H., Gilchrist, R. B., Armstrong, D. T., and Norman, R. J. (2002). Expression of leptin and its receptor in the murine ovary: possible role in the regulation of oocyte maturation. *Biol. Reprod.* 66, 1548–1554.
249. Cioffi, J. A., Van Blerkom, J., Antczak, M., Shafer, A., Wittmer, S., and Snodgrass, H. R. (1997). The expression of leptin and its receptors in pre-ovulatory human follicles. *Mol. Hum. Reprod.* 3, 467–472.
250. Ryan, N. K., Van der Hoek, K. H., Robertson, S. A., and Norman, R. J. (2003). Leptin and leptin receptor expression in the rat ovary. *Endocrinology* 144, 5006–5013.
251. Zhang, Y., Proenca, R., Maffei, M., Barone, M., Leopold, L., and Friedman, J. M. (1994). Positional cloning of the mouse obese gene and its human homologue. *Nature* 372, 425–432.
252. Chehab, F. F., Lim, M. E., and Lu, R. (1996). Correction of the sterility defect in homozygous obese female mice by treatment with the human recombinant leptin. *Nat. Genet.* 12, 318–320.
253. Clement, K., Vaisse, C., Lahlou, N., Cabrol, S., Pelloux, V., Cassuto, D., Gourmelen, M., Dina, C., Chambaz, J., Lacorte, J. M., Basdevant, A., Bougneres, P., Lebouc, Y., Froguel, P., and Guy-Grand, B. (1998). A mutation in the human leptin receptor gene causes obesity and pituitary dysfunction. *Nature* 392, 398–401.
254. Barkan, D., Jia, H., Dantes, A., Vardimon, L., Amsterdam, A., and Rubinstein, M. (1999). Leptin modulates the glucocorticoid-induced ovarian steroidogenesis. *Endocrinology* 140, 1731–1738.
255. Zachow, R. J., and Magoffin, D. A. (1997). Direct intraovarian effects of leptin: impairment of the synergistic action of insulin-like growth factor-I on follicle-stimulating hormone-dependent estradiol-17 beta production by rat ovarian granulosa cells. *Endocrinology* 138, 847–850.
256. Kitawaki, J., Kusuki, I., Koshiba, H., Tsukamoto, K., and Honjo, H. (1999). Leptin directly stimulates aromatase activity in human luteinized granulosa cells. *Mol. Hum. Reprod.* 5, 708–713.
257. Sierra-Honigmann, M. R., Nath, A. K., Murakami, C., Garcia-Cardena, G., Papapetropoulos, A., Sessa, W. C., Madge, L. A., Schechner, J. S., Schwabb, M. B., Polverini, P. J., and Flores-Riveros, J. R. (1998). Biological action of leptin as an angiogenic factor. *Science* 281, 1683–1686.
258. Bartke, A., and Michalkiewicz, M. (2002). Overexpression and targeted disruption of genes involved in the control of growth, food intake, and obesity. In *Transgenics in Endocrinology* (M. M. Matzuk, C. W. Brown, and T. R. Kumar, Eds.), pp. 339–384. Humana Press, Totowa, NJ.
259. Sharara, F. I., and Nieman, L. K. (1994). Identification and cellular localization of growth hormone receptor gene expression in the human ovary. *J. Clin. Endocrinol. Metab.* 79, 670–672.
260. Marchal, R., Caillaud, M., Martoriati, A., Gerard, N., Mermillod, P., and Goudet, G. (2003). Effect of growth hormone (GH) on in vitro nuclear and cytoplasmic oocyte maturation, cumulus expansion, hyaluronan synthases, and connexins 32 and 43 expression, and GH receptor messenger RNA expression in equine and porcine species. *Biol. Reprod.* 69, 1013–1022.
261. Carlsson, B., Nilsson, A., Isaksson, O. G., and Billig, H. (1993). Growth hormone-receptor messenger RNA in the rat ovary: regulation and localization. *Mol. Cell. Endocrinol.* 95, 59–66.
262. Kolle, S., Sinowatz, F., Boie, G., and Lincoln, D. (1998). Developmental changes in the expression of the growth hormone receptor messenger ribonucleic acid and protein in the bovine ovary. *Biol. Reprod.* 59, 836–842.
263. Zaczek, D., Hammond, J., Suen, L., Wandji, S., Service, D., Bartke, A., Chandrashekar, V., Coschigano, K., and Kopchick, J. (2002). Impact of growth hormone resistance on female reproductive function: new insights from growth hormone receptor knockout mice. *Biol. Reprod.* 67, 1115–1124.
264. Zhou, Y., Xu, B. C., Maheshwari, H. G., He, L., Reed, M., Lozykowski, M., Okada, S., Cataldo, L., Coschigamo, K., Wagner, T. E., Baumann, G., and Kopchick, J. J. (1997). A mammalian model for Laron syndrome produced by targeted disruption of the mouse growth hormone receptor/binding protein gene (the Laron mouse). *Proc. Natl. Acad. Sci. U S A* 94, 13215–13220.
265. Lupu, F., Terwilliger, J. D., Lee, K., Segre, G. V., and Efstratiadis, A. (2001). Roles of growth hormone and insulin-like growth factor 1 in mouse postnatal growth. *Dev. Biol.* 229, 141–162.
266. Danilovich, N., Wernsing, D., Coschigano, K. T., Kopchick, J. J., and Bartke, A. (1999). Deficits in female reproductive function in GH-R-KO mice; role of IGF-I. *Endocrinology* 140, 2637–2640.
267. Bachelot, A., Monget, P., Imbert-Bollore, P., Coshigano, K., Kopchick, J. J., Kelly, P. A., and Binart, N. (2002). Growth hormone is required for ovarian follicular growth. *Endocrinology* 143, 4104–4112.
268. Dor, J., Ben-Shlomo, I., Lunenfeld, B., Pariente, C., Levran, D., Karasik, A., Seppala, M., and Mashiach, S. (1992). Insulin-like growth factor-I (IGF-I) may not be essential for ovarian follicular development: evidence from IGF-I deficiency. *J. Clin. Endocrinol. Metab.* 74, 539–542.
269. Amselem, S., Duquesnoy, P., Attree, O., Novelli, G., Bousnina, S., Postel-Vinay, M. C., and Goossens, M. (1989). Laron dwarfism and mutations of the growth hormone-receptor gene. *N. Engl. J. Med.* 321, 989–995.
270. Menashe, Y., Sack, J., and Mashiach, S. (1991). Spontaneous pregnancies in two women with Laron-type dwarfism: are growth hormone and circulating insulin-like growth factor mandatory for induction of ovulation? *Hum. Reprod.* 6, 670–671.
271. Mayerhofer, A., Weis, J., Bartke, A., Yun, J. S., and Wagner, T. E. (1990). Effects of transgenes for human and

bovine growth hormones on age-related changes in ovarian morphology in mice. *Anat. Rec.* 227, 175–186.

272. Danilovich, N. A., Bartke, A., and Winters, T. A. (2000). Ovarian follicle apoptosis in bovine growth hormone transgenic mice. *Biol. Reprod.* 62, 103–107.
273. Eisenhauer, K. M., Chun, S. Y., Billig, H., and Hsueh, A. J. (1995). Growth hormone suppression of apoptosis in preovulatory rat follicles and partial neutralization by insulin-like growth factor binding protein. *Biol. Reprod.* 53, 13–20.
274. Naar, E. M., Bartke, A., Majumdar, S. S., Buonomo, F. C., Yun, J. S., and Wagner, T. E. (1991). Fertility of transgenic female mice expressing bovine growth hormone or human growth hormone variant genes. *Biol. Reprod.* 45, 178–187.
275. Bartke, A., Steger, R. W., Hodges, S. L., Parkening, T. A., Collins, T. J., Yun, J. S., and Wagner, T. E. (1988). Infertility in transgenic female mice with human growth hormone expression: evidence for luteal failure. *J. Exp. Zool.* 248, 121–124.
276. Katznelson, L., Kleinberg, D., Vance, M. L., Stavrou, S., Pulaski, K. J., Schoenfeld, D. A., Hayden, D. L., Wright, M. E., Woodburn, C. J., Klibanski, A., and Stravou, S. (2001). Hypogonadism in patients with acromegaly: data from the multi-centre acromegaly registry pilot study. *Clin. Endocrinol. (Oxf).* 54, 183–188.
277. Monget, P., and Bondy, C. (2000). Importance of the IGF system in early folliculogenesis. *Mol. Cell. Endocrinol.* 163, 89–93.
278. Chandrashekar, V., Zaczek, D., and Bartke, A. (2004). The consequences of altered somatotropic system on reproduction. *Biol. Reprod.* 71, 17–27.
279. Bondy, C., and Accili, D. (2002). Insulin and insulin-like growth factors: targeted deletion of the ligands and receptors. In *Transgenics in Endocrinology* (M. M. Matzuk, C. W. Brown, and T. R. Kumar, Eds.), pp. 371–384. Humana Press, Totowa, NJ.
280. Adashi, E. Y., Resnick, C. E., Payne, D. W., Rosenfeld, R. G., Matsumoto, T., Hunter, M.K., Gargosky, S. E., Zhou, J., and Bondy, C. A. (1997). The mouse intraovarian insulin-like growth factor I system: departures from the rat paradigm. *Endocrinology* 138, 3881–3890.
281. Wandji, S. A., Wood, T. L., Crawford, J., Levison, S. W., and Hammond, J. M. (1998). Expression of mouse ovarian insulin growth factor system components during follicular development and atresia. *Endocrinology* 139, 5205–5214.
282. DeChiara, T. M., Efstratiadis, A., and Robertson, E. J. (1990). A growth-deficiency phenotype in heterozygous mice carrying an insulin-like growth factor II gene disrupted by targeting. *Nature* 345, 78–80.
283. Bernard, D. J., and Woodruff, T. K. (2002). Genetic approaches to the study of pituitary follicle-stimulating hormone regulation. In *Transgenics in Endocrinology* (M. M. Matzuk, C. W. Brown, and T. R. Kumar, Eds.), pp. 297–318. Humana Press, Totowa, NJ.
284. McCullagh, D. R. (1932). Dual endocrine activity of the testes. *Science* 76, 19–20.
285. Vale, W., Rivier, J., Vaughan, J., McClintock, R., Corrigan, A., Woo, W., Karr, D., and Spiess, J. (1986). Purification and characterization of an FSH releasing protein from porcine ovarian follicular fluid. *Nature* 321, 776–779.
286. Rivier, J., Spiess, J., McClintock, R., Vaughan, J., and Vale, W. (1985). Purification and partial characterization of inhibin from porcine follicular fluid. *Biochem. Biophys. Res. Commun.* 133, 120–127.
287. Otsuka, F., Yamamoto, S., Erickson, G. F., and Shimasaki, S. (2001). Bone morphogenetic protein-15 inhibits follicle-stimulating hormone (FSH) action by suppressing FSH receptor expression. *J. Biol. Chem.* 276, 11387–11392.
288. Ueno, N., Ling, N., Ying, S. Y., Esch, F., Shimasaki, S., and Guillemin, R. (1987). Isolation and partial characterization of follistatin: a single-chain Mr 35,000 monomeric protein that inhibits the release of follicle-stimulating hormone. *Proc. Natl. Acad. Sci. U S A* 84, 8282–8286.
289. Lin, H. J., Owens, T. R., Sinow, R. M., Fu, P.C. Jr., DeVito, A., Beall, M. H., and Lachman, R. S. (1998). Anomalous inferior and superior venae cavae with oculoauriculovertebral defect: review of Goldenhar complex and malformations of left-right asymmetry. *Am. J. Med. Genet.* 75, 88–94.
290. Fujimura, H., Ohsawa, K., Funaba, M., Murata, T., Murata, E., Takahashi, M., Abe, M., and Torii, K. (1999). Immunological localization and ontogenetic development of inhibin alpha subunit in rat brain. *J. Neuroendocrinol.* 11, 157–163.
291. Roberts, V., Meunier, H., Vaughan, J., Rivier, J., Rivier, C., Vale, W., and Sawchenko, P. (1989). Production and regulation of inhibin subunits in pituitary gonadotropes. *Endocrinology* 124, 552–554.
292. Meunier, H., Rivier, C., Evans, R. M., and Vale, W. (1988). Gonadal and extragonadal expression of inhibin alpha, beta A, and beta B subunits in various tissues predicts diverse functions. *Proc. Natl. Acad. Sci. U S A* 85, 247–251.
293. Petraglia, F., Garuti, G. C., Calza, L., Roberts, V., Giardino, L., Genazzani, A. R., Vale, W., and Meunier, H. (1991). Inhibin subunits in human placenta: localization and messenger ribonucleic acid levels during pregnancy. *Am. J. Obstet. Gynecol.* 165, 750–758.
294. Roberts, V. J., Peto, C. A., Vale, W., and Sawchenko, P. E. (1992). Inhibin/activin subunits are costored with FSH and LH in secretory granules of the rat anterior pituitary gland. *Neuroendocrinology* 56, 214–224.
295. Roberts, V. J., Bentley, C. A., Guo, Q., Matzuk, M. M., and Woodruff, T. K. (1996). Tissue-specific binding of radiolabeled activin A by activin receptors and follistatin in postimplantation rat and mouse embryos. *Endocrinology* 137, 4201–4209.
296. Woodruff, T. K., D'Agostino, J., Schwartz, N. B., and Mayo, K. E. (1988). Dynamic changes in inhibin messengerRNAs in rat ovarian follicles during the reproductive cycle. *Science* 239, 1296–1299.
297. Roberts, V., Meunier, H., Sawchenko, P. E., and Vale, W. (1989). Differential production and regulation of inhibin subunits in rat testicular cell types. *Endocrinology* 125, 2350–2359.
298. Schwall, R. H., Mason, A. J., Wilcox, J. N., Bassett, S. G., and Zeleznik, A. J. (1990). Localization of inhibin/activin subunit mRNAs within the primate ovary. *Mol. Endocrinol.* 4, 75–79.
299. Woodruff, T., Besecke, L., Groome, N., Draper, L., Schwartz, N., and Weiss, J. (1996). Inhibin A and inhibin B are inversely correlated to follicle-stimulating hormone, yet are discordant during the follicular phase of the rat estrous cycle, and inhibin A is expressed in a sexually dimorphic manner. *Endocrinology* 137, 5463–5467.
300. Burger, L. L., Dalkin, A. C., Aylor, K. W., Workman, L. J., Haisenleder, D. J., and Marshall, J. C. (2001). Regulation of gonadotropin subunit transcription after ovariectomy in the rat: measurement of subunit primary transcripts reveals differential roles of GnRH and inhibin. *Endocrinology* 142, 3435–3442.
301. Arai, K., Watanabe, G., Taya, K., and Sasamoto, S. (1996). Roles of inhibin and estradiol in the regulation of follicle-stimulating hormone and luteinizing hormone secretion during the estrous cycle of the rat. *Biol. Reprod.* 55, 127–133.
302. Arai, K., Komura, H., Akikusa, T., Iio, K., Kishi, H., Watanabe, G., and Taya, K. (1997). Contributions of endogenous inhibin and estradiol to the regulation of follicle-stimulating hormone and luteinizing hormone secretion in the pregnant rat. *Biol. Reprod.* 56, 1482–1489.
303. Attardi, B., Vaughan, J., and Vale, W. (1992). Regulation of FSH beta messenger ribonucleic acid levels in the rat by endogenous inhibin. *Endocrinology* 130, 557–559.

304. Rivier, C., Rivier, J., and Vale, W. (1986). Inhibin-mediated feedback control of follicle-stimulating hormone secretion in the female rat. *Science* 234, 205–208.
305. Rivier, C., and Vale, W. (1989). Immunoneutralization of endogenous inhibin modifies hormone secretion and ovulation rate in the rat. *Endocrinology* 125, 152–157.
306. Hillier, S. G., Yong, E. L., Illingworth, P. J., Baird, D. T., Schwall, R. H., and Mason, A. J. (1991). Effect of recombinant inhibin on androgen synthesis in cultured human thecal cells. *Mol. Cell. Endocrinol.* 75, R1–6.
307. Hsueh, A. J., Dahl, K. D., Vaughan, J., Tucker, E., Rivier, J., Bardin, C. W., and Vale, W. (1987). Heterodimers and homodimers of inhibin subunits have different paracrine action in the modulation of luteinizing hormone-stimulated androgen biosynthesis. *Proc. Natl. Acad. Sci. U S A* 84, 5082–5086.
308. Woodruff, T. K., Lyon, R. J., Hansen, S. E., Rice, G. C., and Mather, J. P. (1990). Inhibin and activin locally regulate rat ovarian folliculogenesis. *Endocrinology* 127, 3196–3205.
309. Alak, B. M., Smith, G. D., Woodruff, T. K., Stouffer, R. L., and Wolf, D. P. (1996). Enhancement of primate oocyte maturation and fertilization in vitro by inhibin A and activin A. *Fertil. Steril.* 66, 646–653.
310. O, W. S., Robertson, D. M., and de Kretser, D. M. (1989). Inhibin as an oocyte meiotic inhibitor. *Mol. Cell. Endocrinol.* 62, 307–311.
311. Bernard, D. J., Burns, K. H., Haupt, B., Matzuk, M. M., and Woodruff, T. K. (2003). Normal reproductive function in InhBP/p120-deficient mice. *Mol. Cell. Biol.* 23, 4882–4891.
312. Chapman, S. C., Bernard, D. J., Jelen, J., and Woodruff, T. K. (2002). Properties of inhibin binding to betaglycan, InhBP/p120 and the activin type II receptors. *Mol. Cell. Endocrinol.* 196, 79–93.
313. Wiater, E., and Vale, W. (2003). Inhibin is an antagonist of bone morphogenetic protein signaling. *J. Biol. Chem.* 278, 7934–7941.
314. Smith, K. B., Millar, M. R., McNeilly, A. S., Illingworth, P. J., Fraser, H. M., and Baird, D. T. (1991). Immunocytochemical localization of inhibin alpha-subunit in the human corpus luteum. *J. Endocrinol.* 129, 155–160.
315. Jaatinen, T. A., Penttila, T. L., Kaipia, A., Ekfors, T., Parvinen, M., and Toppari, J. (1994). Expression of inhibin alpha, beta A and beta B messenger ribonucleic acids in the normal human ovary and in polycystic ovarian syndrome. *J. Endocrinol.* 143, 127–137.
316. Woodruff, T., Krummen, L., McCray, G., and Mather, J. (1993). *In situ* ligand binding of recombinant human (^{125}I) activin-A and recombinant human (^{125}I) inhibin-A to the adult rat ovary. *Endocrinology* 133, 2998–3006.
317. Matzuk, M. M., Finegold, M. J., Su, J.-G. J., Hsueh, A. J. W., and Bradley, A. (1992). a-Inhibin is a tumor-suppressor gene with gonadal specificity in mice. *Nature* 360, 313-319.
318. Matzuk, M. M., Kumar, T. R., and Bradley, A. (1995). Different phenotypes for mice deficient in either activins or activin receptor type II. *Nature* 374, 356–360.
319. Vassalli, A., Matzuk, M. M., Gardner, H. A. R., Lee, K.-F., and Jaenisch, R. (1994). Activin/inhibin bB subunit gene disruption leads to defects in eyelid development and female reproduction. *Genes Dev.* 8, 414–427.
320. Brown, C. W., Houston-Hawkins, D. E., Woodruff, T. K., and Matzuk, M. M. (2000). Insertion of *Inhbb* into the *Inhba* locus rescues the *Inhba*-null phenotype and reveals new activin functions. *Nature Genetics* 25, 453–457.
321. Matzuk, M. M., Finegold, M. J., Mather, J. P., Krummen, L., Lu, H., and Bradley, A. (1994). Development of cancer cachexia-like syndrome and adrenal tumors in inhibin-deficient mice. *Proc. Natl. Acad. Sci. U S A* 91, 8817–8821.
322. Matzuk, M. M., Kumar, T. R., Shou, W., Coerver, K. A., Lau, A. L., Behringer, R. R., and Finegold, M. J. (1996). Transgenic models to study the roles of inhibins and activins in reproduction, oncogenesis, and development. *Recent Progress in Hormone Research* 51, 123–157.
323. Schwall, R. H., Robbins, K., Jardieu, P., Chang, L., Lai, C., and Terrell, T. G. (1993). Activin induces cell death in hepatocytes in vivo and in vitro. *Hepatology* 18, 347–356.
324. Hully, J. R., Chang, L., Schwall, R. H., Widmer, H. R., Terrell, T. G., and Gillett, N. A. (1994). Induction of apoptosis in the murine liver with recombinant human activin A. *Hepatology* 20, 854–862.
325. Coerver, K. A. (1996). Activin function in Cachexia-like syndrome and gonadal tumor development in inhibin-deficient mice. Ph.D. thesis, Baylor College of Medicine, Houston, TX.
326. Coerver, K. A., Woodruff, T. K., Finegold, M. J., Mather, J., Bradley, A., and Matzuk, M. M. (1996). Activin signaling through activin receptor type II causes the cachexia-like symptoms in inhibin-deficient mice. *Mol. Endocrinol.* 10, 534–543.
327. Mathews, L. S. (1994). Activin receptors and cellular signaling by the receptor serine kinase family. *Endocr. Rev.* 15, 310–325.
328. Cameron, V., Nishimura, E., Mathews, L., Lewis, K., Sawchenko, P., and Vale, W. (1994). Hybridization histochemical localization of activin receptor subtypes in rat brain, pituitary, ovary, and testis. *Endocrinology* 134, 799–808.
329. Oh, S. P., and Li, E. (1997). The signaling pathway mediated by the type IIB activin receptor controls axial patterning and lateral asymmetry in the mouse. *Genes Dev.* 11, 1812–1826.
330. Kumar, T. R., Wang, Y., and Matzuk, M. M. (1996). Gonadotropins are essential modifier factors for gonadal tumor development in inhibin-deficient mice. *Endocrinology* 137, 4210–4216.
331. Burns, K. H., Agno, J. E., Sicinski, P., and Matzuk, M. M. (2003). Cyclin D2 and p27 are tissue-specific regulators of tumorigenesis in inhibin alpha knockout mice. *Mol. Endocrinol.* 17, 2053–2069.
332. Sherr, C. J. (1995). D-type cyclins. *Trends Biochem. Sci.* 20, 187–190.
333. Cipriano, S., Chen, L., Burns, K., Koff, A., and Matzuk, M. (2001). Inhibin and p27 interact to regulate gonadal tumorigenesis. *Mol. Endocrinol.* 15, 985–996.
334. Kiyokawa, H., Kineman, R. D., Manova-Todorova, K. O., Soares, V. C., Hoffman, E. S., Ono, M., Khanam, D., Hayday, A. C., Frohman, L. A., and Koff, A. (1996). Enhanced growth of mice lacking the cyclin-dependent kinase inhibitor function of p27^{Kip1}. *Cell* 85, 721–732.
335. Tong, W., Kiyokawa, H., Soos, T. J., Park, M. S., Soares, V. C., Manova, K., Pollard, J. W., and Koff, A. (1998). The absence of p27Kip1, an inhibitor of G1 cyclin-dependent kinases, uncouples differentiation and growth arrest during the granulosa–luteal transition. *Cell Growth Differ.* 9, 787–794.
336. Fero, M. L., Rivkin, M., Tasch, M., Porter, P., Carow, C. E., Firpo, E., Polyak, K., Tsai, L. H., Broudy, V., Perlmutter, R. M., Kaushansky, K., and Roberts, J. M. (1996). A syndrome of multiorgan hyperplasia with features of gigantism, tumorigenesis, and female sterility in p27(Kip1)-deficient mice. *Cell* 85, 733–744.
337. Pierson, M. P., Wang, Y., DeMayo, F. J., Matzuk, M. M., Tsai, S. Y., and O'Malley, B. W. O. (2000). Regulable expression of inhibin A in wild-type and inhibin a null mice. *Mol. Endocrinol.* 14, 1075–1085.
338. McMullen, M. L., Cho, B. N., Yates, C. J., and Mayo, K. E. (2001). Gonadal pathologies in transgenic mice expressing the rat inhibin alpha-subunit. *Endocrinology* 142, 5005–5014.

339. Cho, B. N., McMullen, M. L., Pei, L., Yates, C. J., and Mayo, K. E. (2001). Reproductive deficiencies in transgenic mice expressing the rat inhibin alpha-subunit gene. *Endocrinology* 142, 4994–5004.
340. Vale, W., Bilezikjian, L. M., and Rivier, C. (1994). Reproductive and other roles of inhibins. In *The Physiology of Reproduction* (E. Knobil and J. D. Neill, Eds.), pp. 1861–1878. Raven Press, New York.
341. Pangas, S. A., and Woodruff, T. K. (2002). Production and purification of recombinant human inhibin and activin. *J. Endocrinol.* 172, 199–210.
342. Yokota, H., Yamada, K., Liu, X., Kobayashi, J., Abe, Y., Mizunuma, H., and Ibuki, Y. (1997). Paradoxical action of activin A on folliculogenesis in immature and adult mice. *Endocrinology* 138, 4572–4576.
343. Zhao, J., Taverne, M. A., van der Weijden, G. C., Bevers, M. M., and van den Hurk, R. (2001). Effect of activin A on in vitro development of rat preantral follicles and localization of activin A and activin receptor II. *Biol. Reprod.* 65, 967–977.
344. Miro, F., and Hillier, S. (1996). Modulation of granulosa cell deoxyribonucleic acid synthesis and differentiation by activin. *Endocrinology* 137, 464–468.
345. Shelling, A. N., Burton, K. A., Chand, A. L., van Ee, C. C., France, J. T., Farquhar, C. M., Milsom, S. R., Love, D. R., Gersak, K., Aittomaki, K., and Winship, I. M. (2000). Inhibin: a candidate gene for premature ovarian failure. *Hum. Reprod.* 15, 2644–2649.
346. Marozzi, A., Porta, C., Vegetti, W., Crosignani, P. G., Tibiletti, M. G., Dalpra, L., and Ginelli, E. (2002). Mutation analysis of the inhibin alpha gene in a cohort of Italian women affected by ovarian failure. *Hum. Reprod.* 17, 1741–1745.
347. Mendell, J. T., and Dietz, H. C. (2001). When the message goes awry: disease-producing mutations that influence mRNA content and performance. *Cell* 107, 411–414.
348. Boggess, J. F., Soules, M. R., Goff, B. A., Greer, B. E., Cain, J. M., and Tamimi, H. K. (1997). Serum inhibin and disease status in women with ovarian granulosa cell tumors. *Gynecol. Oncol.* 64, 64–69.
349. Petraglia, F., Luisi, S., Pautier, P., Sabourin, J. C., Rey, R., Lhomme, C., and Bidart, J. M. (1998). Inhibin B is the major form of inhibin/activin family secreted by granulosa cell tumors. *J. Clin. Endocrinol. Metab.* 83, 1029–1032.
350. Healy, D. L., Burger, H. G., Mamers, P., Jobling, T., Bangah, M., Quinn, M., Grant, P., Day, A. J., Rome, R., and Campbell, J. J. (1993). Elevated serum inhibin concentrations in postmenopausal women with ovarian tumors. *N. Engl. J. Med.* 329, 1539–1542.
351. Frias, A. E. Jr., Li, H., Keeney, G. L., Podratz, K. C., and Woodruff, T. K. (1999). Preoperative serum level of inhibin A is an independent prognostic factor for the survival of postmenopausal women with epithelial ovarian carcinoma. *Cancer* 85, 465–471.
352. Lappohn, R. E., Burger, H. G., Bouma, J., Bangah, M., and Krans, M. (1992). Inhibin as a marker for granulosa cell tumor. *Acta. Obstet. Gynecol. Scand. Suppl.* 155, 61–65.
353. Mellor, S. L., Richards, M. G., Pedersen, J. S., Robertson, D. M., and Risbridger, G. P. (1998). Loss of the expression and localization of inhibin alpha-subunit in high grade prostate cancer. *J. Clin. Endocrinol. Metab.* 83, 969–975.
354. Schmitt, J. F., Millar, D. S., Pedersen, J. S., Clark, S. L., Venter, D. J., Frydenberg, M., Molloy, P. L., and Risbridger, G. P. (2002). Hypermethylation of the inhibin alpha-subunit gene in prostate carcinoma. *Mol. Endocrinol.* 16, 213–220.
355. Mizunuma, H., Andoh, K., Obara, M., Yamaguchi, M., Kamijo, T., Hasegawa, Y., and Ibuki, Y. (1994). Serum immunoreactive inhibin levels in polycystic ovarian disease (PCOD) and hypogonadotropic amenorrhea. *Endocr. J.* 41, 409–414.
356. Anderson, R. A., Groome, N. P., and Baird, D. T. (1998). Inhibin A and inhibin B in women with polycystic ovarian syndrome during treatment with FSH to induce mono-ovulation. *Clin. Endocrinol. (Oxf).* 48, 577–584.
357. Lambert-Messerlian, G. M., Hall, J. E., Sluss, P. M., Taylor, A. E., Martin, K. A., Groome, N. P., Crowley, W. F. Jr., and Schneyer, A. L. (1994). Relatively low levels of dimeric inhibin circulate in men and women with polycystic ovarian syndrome using a specific two-site enzyme-linked immunosorbent assay. *J. Clin. Endocrinol. Metab.* 79, 45–50.
358. Magoffin, D. A., and Jakimiuk, A. J. (1998). Inhibin A, inhibin B and activin A concentrations in follicular fluid from women with polycystic ovary syndrome. *Hum. Reprod.* 13, 2693–2698.
359. Laven, J. S., Imani, B., Eijkemans, M. J., de Jong, F. H., and Fauser, B. C. (2001). Absent biologically relevant associations between serum inhibin B concentrations and characteristics of polycystic ovary syndrome in normogonadotrophic anovulatory infertility. *Hum. Reprod.* 16, 1359–1364.
360. Pangas, S. A., and Woodruff, T. K. (2000). Activin signal transduction pathways. *Trends Endocrinol. Metab.* 11, 309–314.
361. Welt, C., Sidis, Y., Keutmann, H., and Schneyer, A. (2002). Activins, inhibins, and follistatins: from endocrinology to signaling. A paradigm for the new millennium. *Exp. Biol. Med. (Maywood)* 227, 724–752.
362. Pierson, T. M., and Matzuk, M. M. (2001). Transgenic mouse models to study the inhibins and activins. In *Inhibins, Activins, and Follistatin in Human Reproductive Physiology* (S. Muttukrishna, Ed.). Imperial College Press, London.
363. Evans, L. W., Muttukrishna, S., Knight, P. G., and Groome, N. P. (1997). Development, validation and application of a two-site enzyme-linked immunosorbent assay for activin-AB. *J. Endocrinol.* 153, 221–230.
364. Lau, A. L., Kumar, T. R., Nishimori, K., Bonadio, J., and Matzuk, M. M. (2000). Activin bC and activin bE genes are not essential for mouse liver growth, differentiation, and regeneration. *Mol. Cell. Biol.* 20, 6127–6137.
365. Lau, A. L., Nishimori, K., and Matzuk, M. M. (1996). Structural analysis of the mouse activin beta C gene. *Biochim. Biophys. Acta.* 1307, 145–148.
366. Mathews, L. S., and Vale, W. W. (1991). Expression cloning of an activin receptor, a predicted transmembrane serine kinase. *Cell* 65, 973–982.
367. Matzuk, M. M., and Bradley, A. (1992). Structure of the mouse activin receptor type II gene. *Biochem. Biophys. Res. Commun.* 185, 404–413.
368. Attisano, L., Wrana, J. L., Cheifetz, S., and Massague, J. (1992). Novel activin receptors: distinct genes and alternative mRNA splicing generate a repertoire of serine/threonine kinase receptors. *Cell* 68, 97–108.
369. Massague, J., and Chen, Y. G. (2000). Controlling TGF-beta signaling. *Genes Dev.* 14, 627–644.
370. Tsuchida, K., Nakatani, M., Yamakawa, N., Hashimoto, O., Hasegawa, Y., and Sugino, H. (2004). Activin isoforms signal through type I receptor serine/threonine kinase ALK7. *Mol. Cell. Endocrinol.* 220, 59–65.
371. Zimmerman, C. M., and Mathews, L. S. (1996). Activin receptors: cellular signalling by receptor serine kinases. *Biochem. Soc. Symp.* 62, 25–38.
372. Moustakas, A., Souchelnytskyi, S., and Heldin, C. H. (2001). Smad regulation in TGF-beta signal transduction. *J. Cell. Sci.* 114, 4359–4369.
373. Attisano, L., and Wrana, J. L. (2000). Smads as transcriptional co-modulators. *Curr. Opin. Cell. Biol.* 12, 235–243.
374. Suszko, M. I., Lo, D. J., Suh, H., Camper, S. A., and Woodruff, T. K. (2003). Regulation of the rat follicle-stimulating

hormone beta-subunit promoter by activin. *Mol. Endocrinol.* 17, 318–332.

375. Weiss, J., Crowley, W. F. Jr., Halvorson, L. M., and Jameson, J. L. (1993). Perifusion of rat pituitary cells with gonadotropin-releasing hormone, activin, and inhibin reveals distinct effects on gonadotropin gene expression and secretion. *Endocrinology* 132, 2307–2311.
376. Weiss, J., Guendner, M. J., Halvorson, L. M., and Jameson, J. L. (1995). Transcriptional activation of the follicle-stimulating hormone beta-subunit gene by activin. *Endocrinology* 136, 1885–1891.
377. Besecke, L. M., Guendner, M. J., Schneyer, A. L., Bauer-Dantoin, A. C., Jameson, J. L., and Weiss, J. (1996). Gonadotropin-releasing hormone regulates follicle-stimulating hormone-beta gene expression through an activin/follistatin autocrine or paracrine loop. *Endocrinology* 137, 3667–3673.
378. DePaolo, L. V., Bald, L. N., and Fendly, B. M. (1992). Passive immunoneutralization with a monoclonal antibody reveals a role for endogenous activin-B in mediating FSH hypersecretion during estrus and following ovariectomy of hypophysectomized, pituitary-grafted rats. *Endocrinology* 130, 1741–1743.
379. Pangas, S. A., Rademaker, A. W., Fishman, D. A., and Woodruff, T. K. (2002). Localization of the activin signal transduction components in normal human ovarian follicles: implications for autocrine and paracrine signaling in the ovary. *J. Clin. Endocrinol. Metab.* 87, 2644–2657.
380. Drummond, A. E., Le, M. T., Ethier, J. F., Dyson, M., and Findlay, J. K. (2002). Expression and localization of activin receptors, Smads, and beta glycan to the postnatal rat ovary. *Endocrinology* 143, 1423–1433.
381. Roberts, V. J., Barth, S., el-Roeiy, A., and Yen, S. S. (1993). Expression of inhibin/activin subunits and follistatin messenger ribonucleic acids and proteins in ovarian follicles and the corpus luteum during the human menstrual cycle. *J. Clin. Endocrinol. Metab.* 77, 1402–1410.
382. Li, R., Phillips, D. M., and Mather, J. P. (*1995*). Activin promotes ovarian follicle development *in vitro*. *Endocrinology* 136, 849–856.
383. Nakamura, M., Nakamura, K., Igarashi, S., Tano, M., Miyamoto, K., Ibuki, Y., and Minegishi, T. (1995). Interaction between activin A and cAMP in the induction of FSH receptor in cultured rat granulosa cells. *J. Endocrinol.* 147, 103–110.
384. Miro, F., Smyth, C., and Hillier, S. (1991). Development-related effects of recombinant activin on steroid synthesis in rat granulosa cells. *Endocrinology* 129, 3388–3394.
385. Shukovski, L., Findlay, J. K., and Robertson, D. M. (1991). The effect of follicle-stimulating hormone-suppressing protein or follistatin on luteinizing bovine granulosa cells in vitro and its antagonistic effect on the action of activin. *Endocrinology* 129, 3395–3402.
386. Hillier, S. G., and Miro, F. (1993). Inhibin, activin, and follistatin. Potential roles in ovarian physiology. *Ann. N Y Acad. Sci.* 687, 29–38.
387. Sawada, H. (2002). Ascidian sperm lysin system. *Zoolog. Sci.* 19, 139–151.
388. Hillier, S. G., Yong, E. L., Illingworth, P. J., Baird, D. T., Schwall, R. H., and Mason, A. J. (1991). Effect of recombinant activin on androgen synthesis in cultured human thecal cells. *J. Clin. Endocrinol. Metab.* 72, 1206–1211.
389. Alak, B. M., Coskun, S., Friedman, C. I., Kennard, E. A., Kim, M. H., and Seifer, D. B. (1998). Activin A stimulates meiotic maturation of human oocytes and modulates granulosa cell steroidogenesis in vitro. *Fertil. Steril.* 70, 1126–1130.
390. Davis, A. A., Matzuk, M. M., and Reh, T. A. (2000). Activin A promotes progenitor differentiation into photoreceptors in rodent retina. *Mol. Cell. Neurosci.* 15, 11–21.
391. Ferguson, C. A., Tucker, A. S., Christensen, L., Lau, A. L., Matzuk, M. M., and Sharpe, P. T. (1998). Activin is an essential early mesenchymal signal in tooth development that is required for patterning of murine dentition. *Genes Dev.* 12, 2636–2649.
392. Jhaveri, S., Erzurumlu, R. S., Chiaia, N., Kumar, T. R., and Matzuk, M. M. (1998). Defective whisker follicles and altered brainstem patterns in activin and follistatin knockout mice. *Mol. Cell. Neurosci.* 12, 206–219.
393. Meunier, H., Cajander, S. B., Roberts, V. J., Rivier, C., Sawchenko, P. E., Hsueh, A. J. W., and Vale, W. (1988). Rapid changes in the expression of inhibin a-, bA-, and bB-subunits in ovarian cell types during the rat estrous cycle. *Mol. Endocrinol.* 2, 1352–1363.
394. Sidis, Y., Fujiwara, T., Leykin, L., Isaacson, K., Toth, T., and Schneyer, A. (1998). Characterization of Inhibin/Activin subunit, activin receptor, and follistatin messenger ribonucleic acid in human and mouse oocytes: evidence for Activin's paracrine signaling from granulosa cells to oocytes. *Biol. Reprod.* 59, 807–812.
395. Attisano, L., and Wrana, J.L. (2002). Signal transduction by the TGF-beta superfamily. *Science* 296, 1646–1647.
396. Shi, Y., and Massague, J. (2003). Mechanisms of TGF-beta signaling from cell membrane to the nucleus. *Cell* 113, 685–700.
397. Zhu, Y., Richardson, J. A., Parada, L. F., and Graff, J. M. (1998). Smad3 mutant mice develop metastatic colorectal cancer. *Cell* 94, 703–714.
398. Yang, X., Letterio, J. J., Lechleider, R. J., Chen, L., Hayman, R., Gu, H., Roberts, A. B., and Deng, C. (1999). Targeted disruption of SMAD3 results in impaired mucosal immunity and diminished T cell responsiveness to TGF-beta. *EMBO J.* 18, 1280–1291.
399. Tomic, D., Miller, K. P., Kenny, H. A., Woodruff, T. K., Hoyer, P., and Flaws, J. A. (2004). Ovarian follicle development requires Smad3. *Mol. Endocrinol.* 18, 2224–2240.
400. Tomic, D., Brodie, S. G., Deng, C., Hickey, R. J., Babus, J. K., Malkas, L. H., and Flaws, J.A. (2002). Smad 3 may regulate follicular growth in the mouse ovary. *Biol. Reprod.* 66, 917–923.
401. Robertson, D. M., Klein, R., de Vos, F. L., McLachlan, R. I., Wettenhall, R. E., Hearn, M. T., Burger, H. G., and de Kretser, D. M. (1987). The isolation of polypeptides with FSH suppressing activity from bovine follicular fluid which are structurally different to inhibin. *Biochem. Biophys. Res. Commun.* 149, 744–749.
402. Esch, F. S., Shimasaki, S., Mercado, M., Cooksey, K., Ling, N., Ying, S., Ueno, N., and Guillemin, R. (1987). Structural characterization of follistatin: a novel follicle-stimulating hormone release-inhibiting polypeptide from the gonad. *Mol. Endocrinol.* 1, 849–855.
403. Sidis, Y., Schneyer, A. L., Sluss, P. M., Johnson, L. N., and Keutmann, H. T. (2001). Follistatin: essential role for the N-terminal domain in activin binding and neutralization. *J. Biol. Chem.* 276, 17718–17726.
404. Nakamura, T., Takio, K., Eto, Y., Shibai, H., Titani, K., and Sugino, H. (1990). Activin-binding protein from rat ovary is follistatin. *Science* 247, 836–838.
405. Nakatani, A., Shimasaki, S., Depaolo, L. V., Erickson, G. F., and Ling, N. (1991). Cyclic changes in follistatin messenger ribonucleic acid and its protein in the rat ovary during the estrous cycle. *Endocrinology* 129, 603–611.
406. Kogawa, K., Ogawa, K., Hayashi, Y., Nakamura, T., Titani, K., and Sugino, H. (1991). Immunohistochemical localization of follistatin in rat tissues. *Endocrinol. Jpn.* 38, 383–391.
407. Matzuk, M. M., Lu, H., Vogel, H., Sellheyer, K., Roop, D. R., and Bradley, A. (1995). Multiple defects and perinatally death in mice deficient in follistatin. *Nature* 372, 360–363.

408. Guo, Q., Kumar, T. R., Woodruff, T., Hadsell, L. A., DeMayo, F. J., and Matzuk, M. M. (1998). Overexpression of mouse follistatin causes reproductive defects in transgenic mice. *Mol. Endocrinol.* 12, 96–106.
409. Yao, H. H., Matzuk, M. M., Jorgez, C. J., Menke, D. B., Page, D. C., Swain, A., and Capel, B. (2004). Follistatin operates downstream of Wnt4 in mammalian ovary organogenesis. *Dev. Dyn.* 230, 210–215.
410. Feijen, A., Goumans, M. J., and van den Eijnden-van Raaij, A. J. (1994). Expression of activin subunits, activin receptors and follistatin in postimplantation mouse embryos suggests specific developmental functions for different activins. *Development* 120, 3621–3637.
411. Otsuka, F., Moore, R. K., Iemura, S., Ueno, N., and Shimasaki, S. (2001). Follistatin inhibits the function of the oocyte-derived factor BMP-15. *Biochem. Biophys. Res. Commun.* 289, 961–966.
412. Glister, C., Kemp, C. F., and Knight, P. G. (2004). Bone morphogenetic protein (BMP) ligands and receptors in bovine ovarian follicle cells: actions of BMP-4, -6 and -7 on granulosa cells and differential modulation of Smad-1 phosphorylation by follistatin. *Reproduction* 127, 239–254.
413. Yamashita, H., ten Dijke, P., Huylebroeck, D., Sampath, T. K., Andries, M., Smith, J. C., Heldin, C. H., and Miyazono, K. (1995). Osteogenic protein-1 binds to activin type II receptors and induces certain activin-like effects. *J. Cell. Biol.* 130, 217–226.
414. Sauer, B., and Henderson, N. (1989). Cre-stimulated recombination at loxP-containing DNA sequences placed into the mammalian genome. *Nucleic Acids Res.* 17, 147–161.
415. Lewandoski, M., Wassarman, K. M., and Martin, G. R. (1997). Zp3-cre, a transgenic mouse line for the activation or inactivation of loxP-flanked target genes specifically in the female germ line. *Curr. Biol.* 7, 148–151.
416. Jorgez, C. J., Klysik, M., Jamin, S. P., Behringer, R. R., and Matzuk, M. M. (2004). Granulosa cell-specific inactivation of follistatin causes female fertility defects. *Mol. Endocrinol.* 18, 953–967.
417. Woodruff, T. K., Krummen, L., Baly, D., Wong, W. L., Garg, S., Sadick, M., Davis, G. H., Soules, M. R., and Mather, J. P. (1993). Inhibin and activin measured in human serum. In *Frontiers in Endocrinology: Inhibin and Inhibin-Related Proteins* (F. Petraglia, Ed.), pp. 55–68. Ares-Serono, Rome.
418. McConnell, D. S., Wang, Q., Sluss, P. M., Bolf, N., Khoury, R. H., Schneyer, A. L., Midgley, A. R. Jr., Reame, N. E., Crowley, W. F. Jr., and Padmanabhan, V. (1998). A two-site chemiluminescent assay for activin-free follistatin reveals that most follistatin circulating in men and normal cycling women is in an activin-bound state. *J. Clin. Endocrinol. Metab.* 83, 851–858.
419. Hanrahan, J. P., Gregan, S. M., Mulsant, P., Mullen, M., Davis, G. H., Powell, R., and Galloway, S. M. (2004). Mutations in the genes for oocyte-derived growth factors GDF9 and BMP15 are associated with both increased ovulation rate and sterility in Cambridge and Belclare sheep (Ovis aries). *Biol. Reprod.* 70, 900–909.
420. Pangas, S. A., and Matzuk, M. M. (2004). Genetic models for transforming growth factor b superfamily signaling in ovarian follicle development. *Mol. Cell. Endocrinol.* 225, 83–91.
421. Kendall, S. K., Samuelson, L. C., Saunders, T. L., Wood, R. I., and Camper, S. A. (1995). Targeted disruption of the pituitary glycoprotein hormone alpha-subunit produces hypogonadal and hypothyroid mice. *Genes Dev.* 9, 2007–2019.
422. Vassalli, A., Matzuk, M. M., Gardner, H. A., Lee, K. F., and Jaenisch, R. (1994). Activin/inhibin beta B subunit gene disruption leads to defects in eyelid development and female reproduction. *Genes Dev.* 8, 414–427.
423. Dattani, M. T., Martinez-Barbera, J. P., Thomas, P. Q., Brickman, J. M., Gupta, R., Martensson, I. L., Toresson, H., Fox, M., Wales, J. K., Hindmarsh, P. C., Krauss, S., Beddington, R. S., and Robinson, I. C. (1998). Mutations in the homeobox gene HESX1/Hesx1 associated with septo-optic dysplasia in human and mouse. *Nat. Genet.* 19, 125–133.
424. Sheng, H. Z., Zhadanov, A. B., Mosinger, B. Jr., Fujii, T., Bertuzzi, S., Grinberg, A., Lee, E. J., Huang, S. P., Mahon, K. A., and Westphal, H. (1996). Specification of pituitary cell lineages by the LIM homeobox gene Lhx3. *Science* 272, 1004–1007.
425. Sornson, M. W., Wu, W., Dasen, J. S., Flynn, S. E., Norman, D. J., O'Connell, S. M., Gukovsky, I., Carriere, C., Ryan, A. K., Miller, A. P., Zuo, L., Gleiberman, A. S., Andersen, B., Beamer, W. G., and Rosenfeld, M. G. (1996). Pituitary lineage determination by the Prophet of Pit-1 homeodomain factor defective in Ames dwarfism. *Nature* 384, 327–333.
426. Gage, P. J., Roller, M. L., Saunders, T. L., Scarlett, L. M., and Camper, S. A. (1996). Anterior pituitary cells defective in the cell-autonomous factor, df, undergo cell lineage specification but not expansion. *Development* 122, 151–160.
427. Li, S., Crenshaw, E. B. III, Rawson, E. J., Simmons, D. M., Swanson, L. W., and Rosenfeld, M. G. (1990). Dwarf locus mutants lacking three pituitary cell types result from mutations in the POU-domain gene pit-1. *Nature* 347, 528–533.
428. Tapanainen, J. S., Aittomaki, K., Min, J., Vaskivuo, T., and Huhtaniemi, I. T. (1997). Men homozygous for an inactivating mutation of the follicle-stimulating hormone (FSH) receptor gene present variable suppression of spermatogenesis and fertility. *Nat. Genet.* 15, 205–206.
429. de Roux, N., Young, J., Misrahi, M., Genet, R., Chanson, P., Schaison, G., and Milgrom, E. (1997). A family with hypogonadotropic hypogonadism and mutations in the gonadotropin-releasing hormone receptor. *N. Engl. J. Med.* 337, 1597–1602.

Knobil and Neill's Physiology of Reproduction, Third Edition
edited by Jimmy D. Neill,

CHAPTER 11

Ovulation

Lawrence L. Espey[1] and JoAnne S. Richards[2]

Introduction, 425
Morphological Features of Ovulation, 426
Follicular Apex at the Beginning of the Ovulatory Process, 427
Follicular Apex at Less than One Hour before Rupture, 430
Follicular Apex at Less than Five Minutes before Rupture, 430
Biochemical Reactions in Ovulation, 431
Progesterone and Ovulation, 432
Prostaglandins and Ovulation, 435
Proteases and Ovulation, 438
Ovarian Gene Expression during Ovulation, 440
Complexity of Integrating Gene Expression Data, 440
Catalog of Current Genes, 443
Additional Comments about Ovarian Gene Expression, 455
Concluding Summary, 458
Acknowledgments, 458
References, 458

INTRODUCTION

Mammalian ovaries have two principal functions. They produce sex steroids to prepare the adult female for reproduction, and they release eggs at appropriate intervals during the fertile years of the organism (1). The latter function is commonly referred to as ovulation, which is the topic of this chapter. The ovulatory process in mammals is a distinct biological phenomenon that begins when gonadotropic hormones stimulate mature ovarian follicles and it ends when the follicles rupture and release eggs into the oviduct (Fig. 1). The main focus of this chapter is on the molecular events that lead to the release of eggs, i.e., on the biochemical changes that occur in the ovary as the result of gene expression in response to gonadotropic hormones that excite specific ovarian cells and cause ovulation.

It is generally thought that the underlying mechanisms of hormone action that cause ovulation are homologous in all mammals. Initially, it was assumed that the process was a relatively simple phenomenon, involving a small number of regulatory factors that were activated by a surge in the secretion of pituitary gonadotropins. For more than 100 years, reproductive biologists "stressed the role of increasing intrafollicular pressure as the cause of rupture,...and the change in pressure was ascribed to the action of smooth muscle in the ovarian stroma" (2). However, in recent decades, it has become clear that intrafollicular pressure does not increase significantly during the hours preceding rupture (3), and the ovary is not endowed with any functional smooth muscle tissue (4). Instead, the current hypothesis is that the ovulatory surge in gonadotropins induces an acute inflammatory reaction that involves the stratum granulosum and the thecal layers of mature follicles (5–7). The inflammatory process generates protease activity in the granulosa and/or thecal layers of the follicles and this proteolytic activity degrades extracellular matrices within the connective tissue in the ovary. The degraded elements of the follicle wall dissociate and rupture under the force of a steady intrafollicular pressure, resulting in release of the cumulus cell-enclosed oocyte complex (COC). The current chapter is primarily a review of the cascade of ovarian gene expression that induces local inflammation that results in follicle rupture via proteolytic disintegration

[1]Biology Department, Trinity University, San Antonio, Texas; [2]Department of Molecular and Cell Biology, Baylor College of Medicine, Houston, Texas.

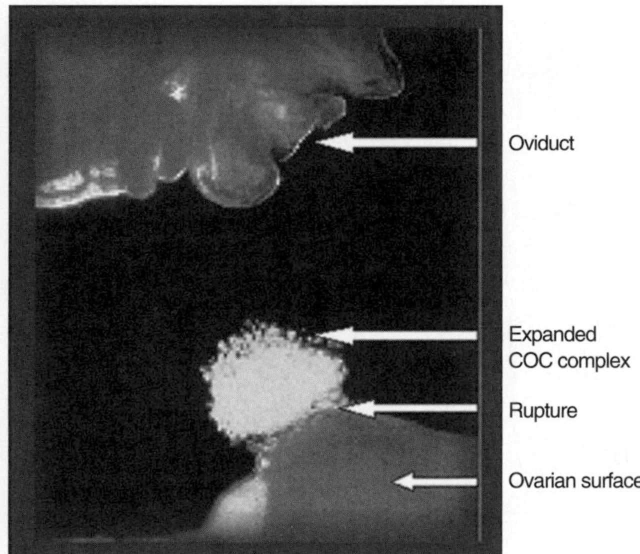

FIG. 1. Moment of rupture of a rabbit follicle. Ovulation of a fertilizable oocyte involves rupture of the ovarian surface and release of the expanded cumulus–oocyte complex (COC), followed by retrieval of the COC by the ciliated cells of the oviduct and its transport down the oviduct. (See text.) (See color insert.)

of the connective tissue elements of the follicle wall and the expansion of the COC associated with production of a specialized extracellular matrix. Each of these processes (rupture and COC expansion) is critical for successful release of fertilizable oocytes, hence successful ovulation. However, the chapter will initially summarize the basic morphology of an ovarian follicle and it will update contemporary views on the roles of certain biochemical agents such as progesterone and the prostaglandins in ovulation.

Reproductive biologists who have an enduring interest in ovulation research are encouraged to examine the exceptional review by Carl Hartman in 1932 (8). His account provides many interesting insights into the lineage of earlier research on this subject. Details about ovarian innervation and vascularity can be found in the chapter on ovulation in the initial edition of *The Physiology of Reproduction* (9). A more comprehensive review of the literature on ovulation is available in the second edition of these volumes (10). The present review will concentrate on information that has become available during the decade since publication of that second edition.

Innumerable reviews on topics related to ovulation have been written during the past 10–12 years. These include reviews of a generally allied nature (11–13), on prostaglandins (14–16), on progesterone (17–19), on proteases (20–29), on angiogenesis (30) and especially on cytokines (31–37) and leukocytes (38–40). In addition, there are a number of reviews that itemize the various genes that are uniquely expressed in the ovary in association with the ovulatory process (7,41–47). Some of the reviews emphasize the *inflammatory* nature of the ovulatory process (6,7,46). The contents of this chapter are based in part on information from the above assortment of reviews and in part from a number of different primary research papers. Because of the sheer volume of information it has been impossible to include all relevant primary literature.

An initial note on terminology is also in order. It is common to use the expressions *preovulatory*, *ovulatory*, and *ovulation* all in reference to the entire gonadotropin-induced process. However, in this chapter, the adjective *preovulatory* will be used in reference to mature follicles that have not yet been stimulated by the ovulatory surge of gonadotropins, whereas the term *ovulatory* will be applied to the entire process. The term *ovulation* may indicate either the process or the moment of egg release, but in instances where clear delineation of the latter phenomenon is important, the term *follicular rupture* will be used for clarity.

MORPHOLOGICAL FEATURES OF OVULATION

Mammalian ovulation is an exceptional biological phenomenon in that it requires the physical disintegration of healthy tissue at the surface of the ovary. The disturbance begins when a mature ovarian follicle is stimulated by an ovulatory surge in pituitary gonadotropins that react with G-protein–coupled receptors on the plasma membranes of the two innermost layers of cells in the follicle, namely the granulosa cells and the theca interna cells (1,10,11,41–43,47) (Fig. 2). There is a variety of evidence to indicate that the molecular response to gonadotropic stimulation

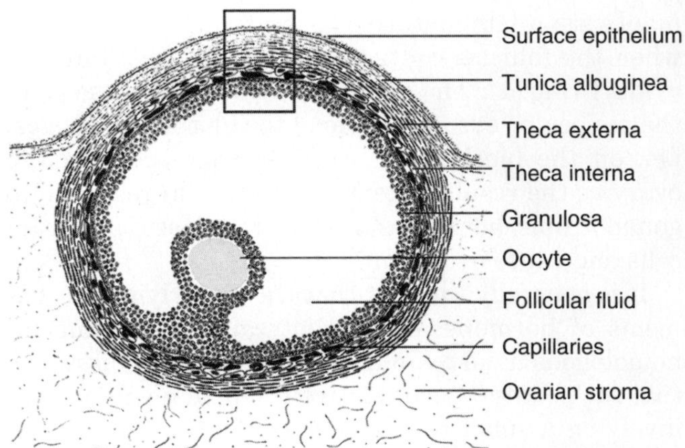

FIG. 2. Morphology of a typical ovarian follicle. Sketch is based on a rabbit follicle. Rectangle at the apex of the follicle indicates the region where rupture usually occurs, and the area represented in Figs. 2, 7, and 8. [From (10), with permission.]

begins rapidly, and these biochemical changes occur rather evenly throughout the follicle wall, i.e., rather uniformly around the entire circumference of the follicle (7,41–47). It is generally assumed that the site of rupture is in the apical most region of the follicle because this area happens to be, morphologically, the weakest portion of the follicle wall (1,10). The present section of this chapter summarizes the ultrastructure of the apical region of a rabbit follicle before it has been stimulated by an ovulatory dose of gonadotropin and at several stages near the time of follicular rupture.

The ovulatory process in rabbits requires approximately 10 hours. During the first several hours after a mature rabbit follicle has been stimulated by an ovulatory surge in pituitary gonadotropins, there is no conspicuous change in the macroscopic appearance of the follicle. However, by 4 hours into the process, a follicle will begin to blush (1). There is clear evidence that the capillaries in the follicle wall have dilated, and the tissue becomes hyperemic (48). There is no other morphological indication of pending rupture until 1–2 hours before the follicle wall will actually burst. As the time of rupture nears, the apex of a mature follicle protrudes more and more above the surface of the ovary and the follicle wall itself gradually becomes thinner. Eventually, in the last minutes before rupture, the apical most portion of the follicle becomes translucent and rapidly balloons above the normal curvature of the ovary to form a stigma. A follicle usually ruptures within several minutes after the stigma forms. Ovulation is complete when the expanded cumulus–oocyte complex or (COC) is discharged, usually within 1–2 minutes after the follicle wall bursts. Although the release of the oocyte is assumed to be passive, the formation of the hyaluronan COC-derived matrix is critical for egg release. Therefore, it is perhaps worth considering the notion that the COC and its surrounding matrix contribute along with the mural granulosa to follicular rupture via delivery of proteases attached to the matrix or via the matrix molecules themselves (49). This matrix also contributes to the viability of the COC in the oviduct (Fig. 3).

The anatomical description that follows will highlight the apical most area where rupture normally occurs in a follicle. Specifically, this analysis will examine morphological aspects of the five main layers of the follicle wall before initiation of the ovulatory process (i.e., at 10 hours before rupture of a rabbit follicle), at approximately 1 hour before rupture, and at less than 5 minutes before rupture.

Follicular Apex at the Beginning of the Ovulatory Process

Surface Epithelium

The surface epithelium is a single layer of cuboidal epithelial cells that cover the entire surface of the ovary (Fig. 4). They are loosely attached to a thin basal lamina at the surface of the collagenous connective tissue that comprises the tunica albuginea, a layer that also surrounds the ovary (Fig. 5). A dominant feature of the cells of the surface epithelium is the large granules that are normally present in their cytoplasm. The composition of these dense granules is unknown. Another distinct characteristic of the

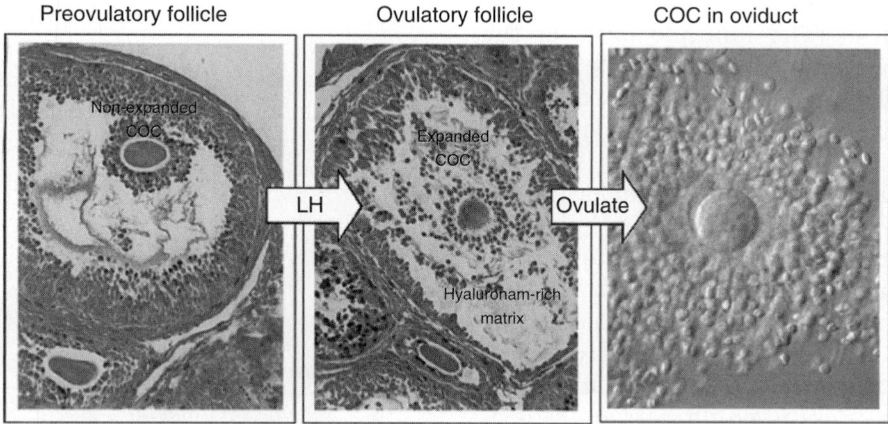

FIG. 3. COC expansion prior to and following ovulation. Cumulus granulosa cells within the preovulatory follicle are compacted around the oocyte (nonexpanded COC). Following the surge of gonadotropins the cumulus cells produce an extracellular matrix along which they migrate outwardly from the oocyte (expanded COC). The expanded COC is released at the time of ovulation and is retained during transport within the oviduct for approximately 24 hours. Specific molecular and biochemical events occur to ensure production and stabilization of the matrix prior to and following ovulation. (See text and Figs. 13, 15, and 20 for further details and discussion.) (See color insert.)

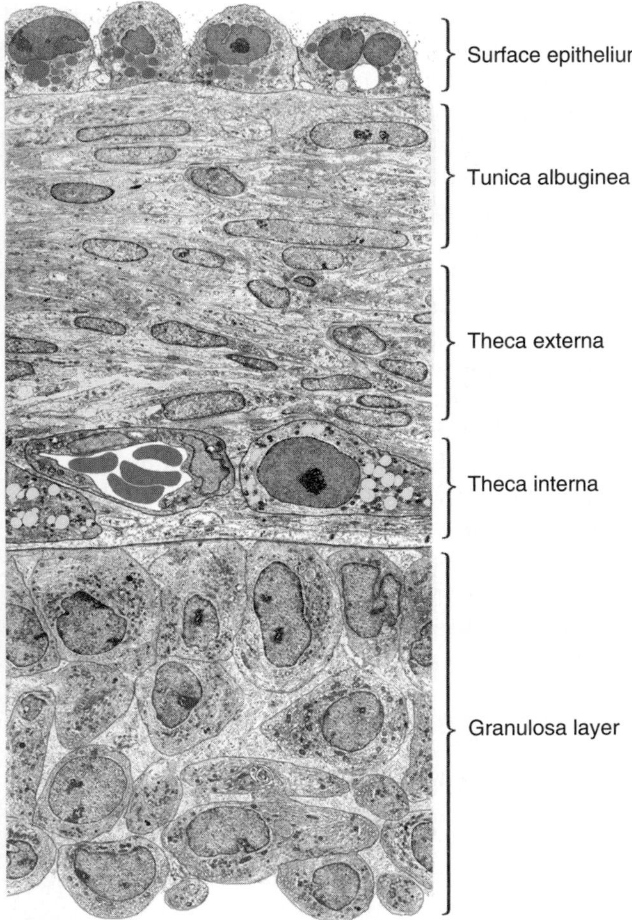

FIG. 4. Ultrastructure of the apex of a rabbit follicle 10 hours before rupture, i.e., before treating the animal with gonadotropin to induce ovulation. Cytoplasmic spheres in theca interna cells indicate steroidogenic activity is primarily in this layer of the follicle wall at the beginning of the ovulatory process. [From (1), with permission.]

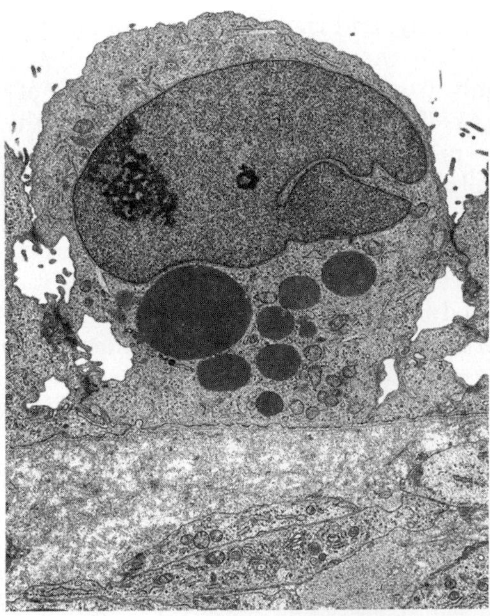

FIG. 5. A cross-sectional view of a typical cell in the surface epithelium of a rabbit at 10 hours before rupture. Dense granules on the basal side of the nucleus are of unknown composition, and there is no evidence that the content of these granules is secreted during the ovulatory process.

cells of the surface epithelium is the polymorphous nature of their nuclei. The multiple lobes of the nuclei are more conspicuous when the tissue sections for transmission electron microscopy are cut on a plane that is tangential to the surface of the ovary (Fig. 6).

Three decades ago, Bjersing and Cajander (50,51) concluded that the surface epithelium is the source of hydrolytic enzymes that cause ovulation. More recently, Murdoch and others (23,25,28,52,53) have reported that the surface epithelium produces plasminogen activator (PA) to catalyze the activation of proteolysis of follicular connective tissue and cause ovulation. However, there is no ultrastructural evidence that the surface epithelium has any role in ovulation. In fact, this outer layer of cells is usually sloughed from the stigma region of an ovulatory follicle before it ruptures (54–56). Furthermore, the surface epithelium can be scraped from the apical area of mature follicles, yet the follicles can still ovulate following stimulation by gonadotropin (57). Therefore, considering the evidence that ovulation can occur in follicles that lack the surface epithelium, it is doubtful this outer layer has an essential role in the process.

Tunica Albuginea

The whitish tunic of collagenous connective tissue known as the tunica albuginea forms an outer layer of protection for the ovary (Fig. 4). This tenacious tissue is the component in the apical area of the follicle wall that represents the stratum of greatest resistance to rupture. It is composed almost entirely of fibroblasts and associated collagen fibrils that are imbedded in the extracellular matrix; the thickness

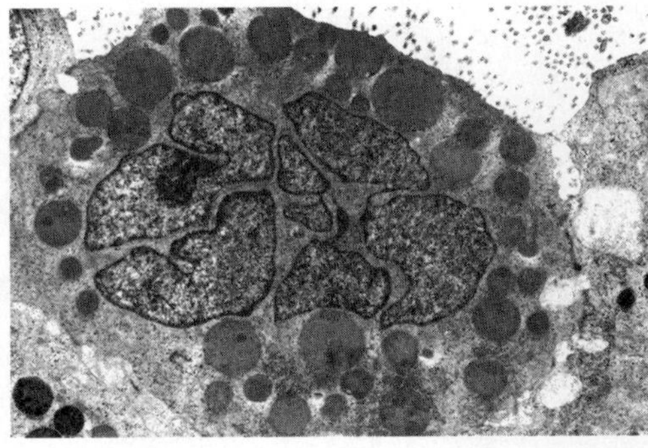

FIG. 6. Appearance of a rabbit surface epithelial cell that was sectioned on a plane tangential to the surface of the ovary. Cells that are sectioned on this plane typically display a polymorphous nucleus.

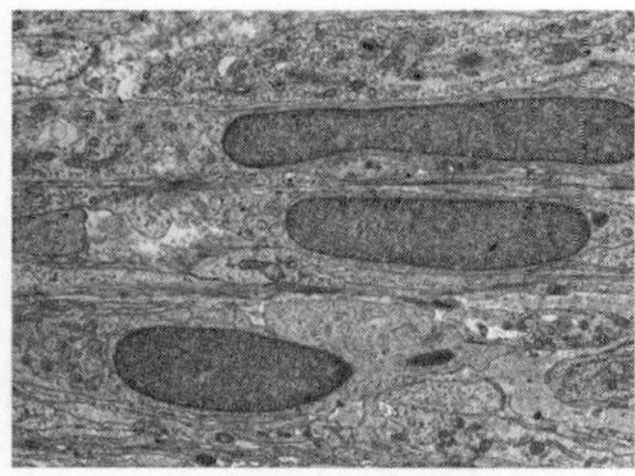
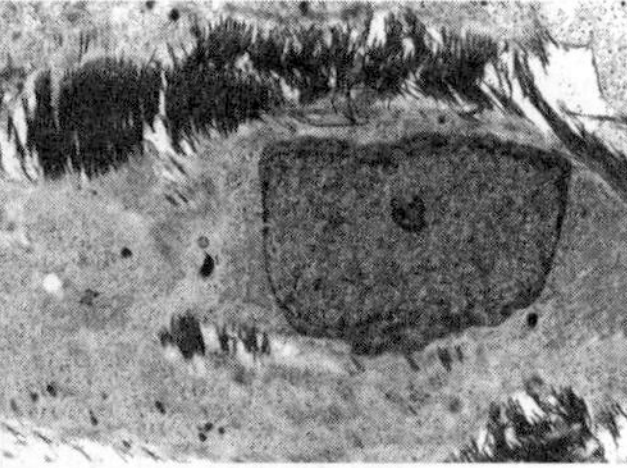

FIG. 7. Appearance of fibroblasts in the tunica albuginea at the apex of a rabbit follicle at 10 hours from rupture. *Left plate* is a cross-sectional view and *right plate* is a micrograph of tissue that was sectioned tangentially to the surface of the ovary. Taken together, these two micrographs reveal that the fibroblasts are platter-shaped, which is characteristic of layers of thecal connective tissue. Note that the secretion of collagen from a fibroblast is more conspicuous in the tangentially sectioned tissue.

of which is species dependent. For many years, the fibroblasts were considered to be spindle-shaped smooth muscle cells that might play an active role in the mechanism of ovulation (4), but the era of electron microscopy has made it evident that the cells are fibroblasts. If the follicular tissue is cut on a plane that is tangential to the surface of the ovary (instead of the usual cross-sectional cut), the cells in the tunica albuginea appear oval or round (Fig. 7). Upon considering this third dimension of the cells, it becomes clear that they are platter-shaped, rather than spindle-shaped. Such a flat structure is characteristic of fibroblasts in layers of thecal connective tissue (1,4,58). Also, it is easy to locate follicular fibroblasts that appear to be actively secreting tropocollagen, a molecule that rapidly polymerizes into distinct collagen fibrils when it is exposed to the ionic components of the extracellular environment (Fig. 8). Thus, the predominant cells in the tunica albuginea are fibroblasts and the depth of this layer varies among species, e.g., there is more in the human and rabbit ovary than in the rodent.

FIG. 8. A tangential view of a fibroblast in the tunica albuginea of a follicle before rupture. Only a small portion of the nucleus is visible in the *upper right corner* of this cell. Notice the extensive polymerization of collagen fibrils from the soluble tropocollagen that is secreted by this cell.

Theca Externa

The theca externa is another layer of collagenous connective tissue that is somewhat similar in composition to the tunica albuginea (Fig. 4). However, this tissue is limited to the outer layer of a mature follicle and is contiguous with the tunica albuginea only at the apical most segment of the follicle wall. One distinction between these two layers of connective tissue is that the amount of collagen in the theca externa is usually less than in the tunica albuginea (54). This difference might be due, at least in part, to the fact the collagenous tissue of a follicle that has just reached maturity is less established than the connective tissue that surrounds the ovary.

Theca Interna

The theca interna is a thin layer of steroid-secreting cells that are supplied nutrients and oxygen by numerous large capillaries that collectively receive most of the blood from the ovarian arterial supply (Fig. 4). Fibroblasts and collagen are sparse in this thin layer just inside the theca externa (10). The cytoplasm of these cells is dominated by lipid droplets, numerous mitochondria and Golgi networks that are distributed throughout their smooth endoplasmic reticulum. In regard to the theca interna, it is noteworthy that it has become common practice to refer to "theca interstitial cells" (59–62), but it is not always clear whether this terminology is being applied specifically to the secretory cells of the theca interna, to cells of both the theca interna and theca externa, to interstitial cells in the ovarian stroma, or to indefinite cells in all of these areas. In any case, since ultrastructural studies have established that the theca interna consists primarily of distinct, steroid-secreting cells (along with capillary endothelial cells), this chapter will use the terminology *theca interna cells* in reference to the steroid-secreting cells that are characteristic of this layer.

Stratum Granulosum

The innermost layer of follicular cells is the stratum granulosum, which extends inward from a basement membrane (i.e., the membrana propria) at the inner margin of the theca interna (Fig. 4). The first layer of granulosa cells adjacent to the membrana propria are more columnar in shape, while the remaining cells that extend toward the follicular antrum are cuboidal.

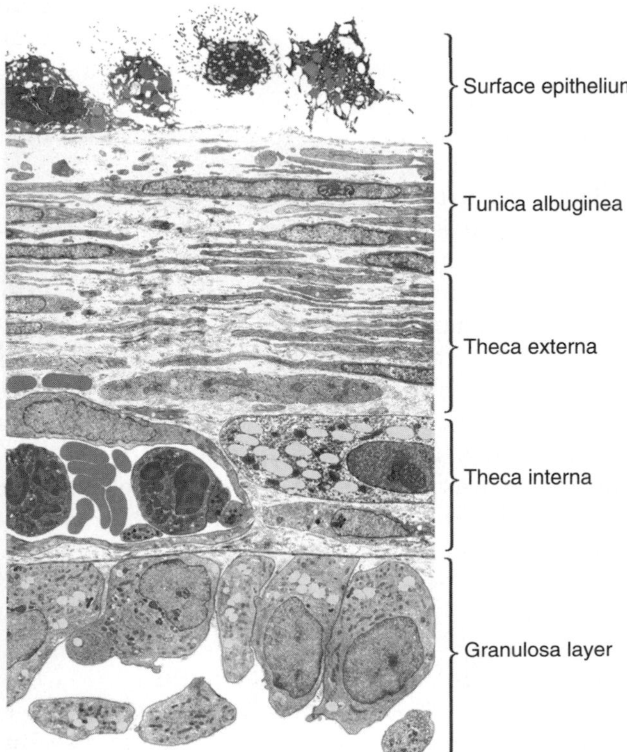

FIG. 9. Ultrastructure of the apex of a rabbit follicle less than 1 hour before rupture. Notice that the ovarian surface epithelial cells have been traumatized by the degradative events in the ovulatory follicle. The fibroblasts in the tunica albuginea and theca externa are more elongated than comparable cells at 10 hours from rupture (see Fig. 4). Cells in the theca interna continue to exhibit lipid droplets throughout their cytoplasm, and adjacent capillaries contain leukocytes and blood platelets. Notice that granulosa cells now exhibit lipid droplets, probably in association with the ovulatory increase in progesterone synthesis by these cells. [From (1), with permission.]

Collectively, the cells of the stratum granulosum are metabolically integrated by an extensive labyrinth of gap junctions that couple this layer into a syncytium (Fig. 9). It has been suggested that the ovulatory surge in gonadotropins that react with the abundance of G-protein–coupled receptors on the granulosa cells might initiate action potentials that are propagated via the gap junctions throughout this layer (63). In any case, the granulosa cells appear to be the primary site of onset of the ovulatory process.

Cumulus Granulosa Cells and Oocyte Microenvironment

The cumulus granulosa cells (i.e., those surrounding and directly in contact with the oocyte) comprise the innermost microenvironment of the follicle. Cumulus cells have specific functions and exhibit specific patterns of gene expression in the ovulating follicle (46,47). In particular, cumulus cell expression of the hyaluronon rich matrix and hyaluronan-binding molecules and inflammation-related products during COC expansion is essential for release of the oocyte (Fig. 3) and will be discussed in more detail below.

Follicular Apex at Less than One Hour before Rupture

Within 1 hour from rupture, there are conspicuous changes in the ultrastructure of an ovulatory follicle (Fig. 10). The cells of the surface epithelium develop necrotic-like vacuoles in their cytoplasm, and they appear to be less firmly attached to the tunica albuginea. The fibroblasts in the tunica albuginea and theca externa are more elongate and appear to have transformed from quiescent to motile cells. In the theca interna, the secretory cells look basically the same as earlier, but the adjacent capillaries contain more leukocytes and the lumenal surface of the capillary endothelial cells have blood platelets adhering to them (Fig. 9). As the apical region of the follicle wall balloons and the tissue begins to dissociate, the wall becomes thinner. This stretching of the apex of the follicle results in sloughing of some of the granulosa cells into the follicular antrum. A striking new development in the cytoplasm of the granulosa cells is the formation of lipid droplets, which suggests that the cells have become more active steroidogenically during the hours preceding follicular rupture.

Follicular Apex at Less than Five Minutes before Rupture

Shortly before a follicle ruptures, only traces of the surface epithelium remain clinging to the disintegrated

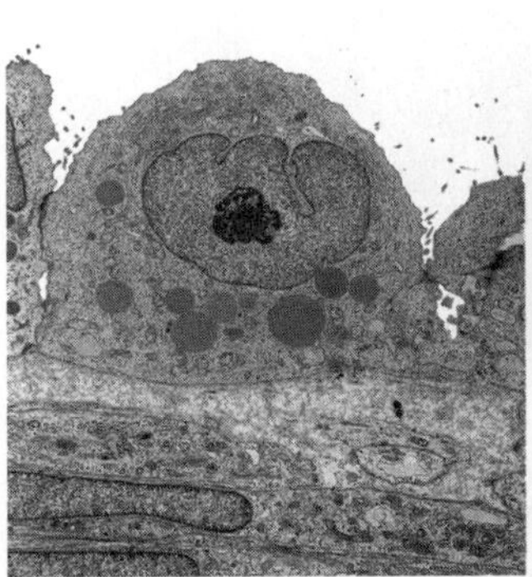
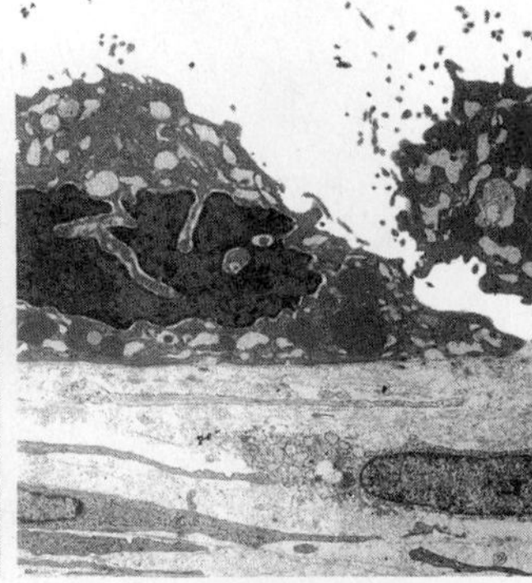

FIG. 10. Comparison of appearance of surface epithelial cells at 10 hours before rupture (*left plate*) and 1 hour before rupture (*right plate*). Notice in particular the change in the texture of the nuclear material and the high degree of vacuolization in the cytoplasm of the necrotic-looking cells that are nearer to the time of follicular rupture. [From (1), with permission.]

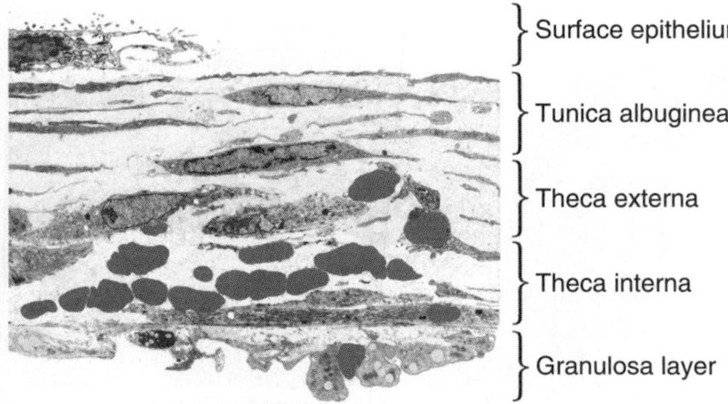

FIG. 11. Ultrastructure of the apex of a rabbit follicle within only several minutes of rupture. The follicle wall is significantly thinner and most of the epithelial cells have sloughed from the surface of the ovary. The extracellular matrix in the thecal layers is quite sparse, and the fibroblasts have dissociated from one another. Oftentimes, the capillaries in the theca interna have ruptured, and extravasated red blood cells are randomly distributed in the region of the theca interna. At the very apex of a follicle, where the ovulatory stigma forms, most of the granulosa cells have dislodged and floated into the follicular antrum. [From (1), with permission.]

tunica albuginea at the apex of the follicle (Fig. 11). The extracellular matrix of the tunica albuginea and theca externa is sparse, and the fibroblasts in these layers are quite dissociated from one another. The fibroblasts appear to be more motile and occasionally they exhibit amoeboid-like movement around extravascular red blood cells (Fig. 12). The capillaries in the theca interna usually have either ruptured or contain thrombi that impair blood flow in the local area. By this stage, essentially all of the granulosa cells have sloughed into the follicular fluid, or have retracted toward the base of the stigma that oftentimes forms during this final phase of the ovulatory process. Rupture ultimately occurs at the apical most area of the follicle simply because this is, morphologically, the thinnest point along the segment where a mature follicle is contiguous with the tunica albuginea at the ovarian surface.

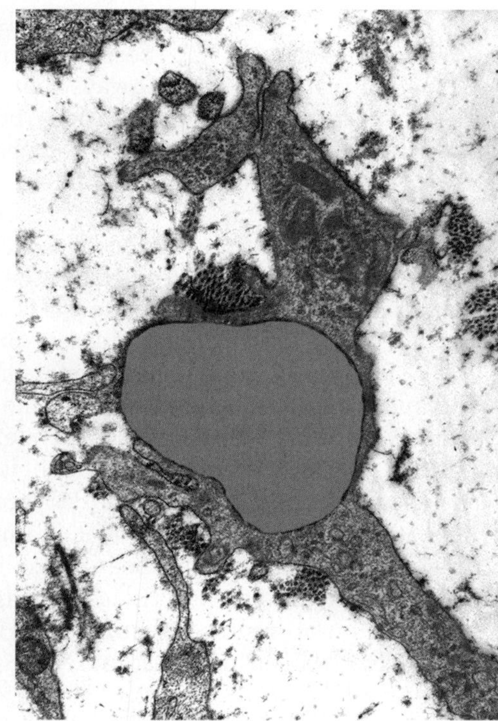

FIG. 12. Close-up view of a fibroblast-like cell surrounding an extravasated red blood cell in the theca interna several minutes before rupture. It is not clear whether this cell is a fibroblast or a macrophage, but it is evident this cell is assuming a phagocytic posture around the blood cell.

BIOCHEMICAL REACTIONS IN OVULATION

A wide variety of biochemical studies related to ovulation have been conducted during the past several decades. Many of these investigations have explored the possibility that mediators of inflammatory reactions are components of the ovulatory process (5,6,10). While this approach has provided useful information, it has not been as unbiased or as definitive as the recent molecular methods that have detected precise changes in gene expression and identified protein products of ovulation-related genes. However, the current status of the knowledge that has been gained from modern molecular biology will be discussed in detail later in this chapter. This section of the chapter on ovulation will summarize information about the three groups of biochemicals that have received the most attention during the past one-half century, namely progesterone, prostaglandins, and proteases (and related hydrolytic enzymes) (Fig. 13).

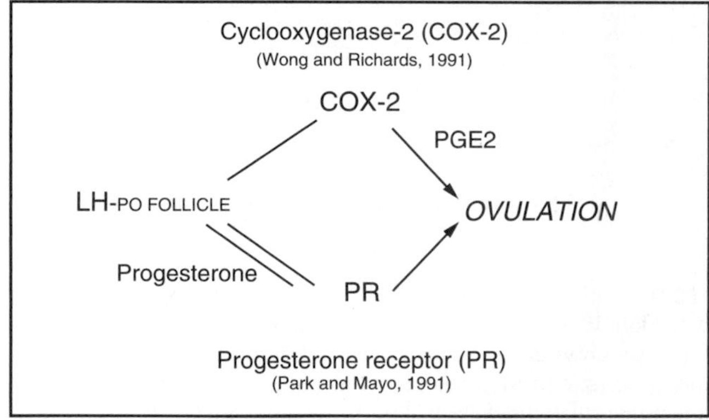

FIG. 13. Schematic to illustrate two critical events in LH mediated ovulation: induction of COX-2 and PR. The events that control the induction of these factors as well as the impact of prostaglandins and PR on ovulation-related genes are among the major topics discussed in this chapter.

Progesterone and Ovulation

Progesterone was initially associated with the corpus luteum because of the enormous production of this steroid in lutein tissue after ovulation. However, it is now clear that the ovary begins to produce a significant amount of progesterone within only a few hours after the ovulatory process has been initiated by gonadotropin (Fig. 14) (64–68). In essence, what this means is that the luteinization process also begins from the time a mature follicle is stimulated by *luteinizing* hormone (or its equivalent). Thus, in a certain sense, the ovulatory process can be considered as an initial phase of the protracted luteinization process (1,7). However, if the luteinization events *precede* those of ovulation or if the timing is altered, luteinization can proceed without ovulation and result in corpora lutea with entrapped oocytes. Several mutant mouse models have this type of anovulatory phenotype; including mice null for the progesterone receptor (PR) (69), cyclooxygenase-2 (COX-2) (70), the prostaglandin E_2 receptor subtype EP2 (71), phosphodiesterase 4 (PDE4) (72), CAAT enhancer binding protein beta (C/EBPbeta) (73) as well as cyclin D2 (74) (Fig. 15). That is to say, follicular rupture is a biologically programmed morphological phenomenon that must occur at a precise interval during the transformation of an ovarian follicle into a progesterone-producing corpus luteum.

The sites of ovulation-related progesterone synthesis are the theca interna and the granulosa layer (75–77) where most of the LH/hCG receptors have been identified (78–83). Additionally, at the time of ovulation, cumulus cells also make progesterone, in response to either external or COC-derived factors (84,85). Predictably, in the granulosa and theca cells, the activated G-protein–coupled LH receptor initiates cyclic AMP signal transduction pathways that mediate progesterone synthesis and ovulation (86–91). This increase in progesterone synthesis requires the transport of free cholesterol to mitochondria where $P450_{scc}$ and 3β-HSD sequentially convert the cholesterol to pregnenolone and progesterone, respectively (92–94). The shuttling of cholesterol from the cytoplasm into mitochondria for steroidogenesis is highly dependent on steroidogenic acute regulatory protein (StAR) (93,95–98). In addition, the transfer of electrons to the mitochondrial forms of cytochrome P450 during steroid hormone synthesis is dependent on the iron-sulfur protein adrenodoxin, which is normally expressed in conjunction with P450scc and StAR (76,99,100). By a mechanism that is unclear, once progesterone synthesis has been established, persistent steroidogenesis during the early stages of luteinization does not require sustained LH/hCG stimulation of the follicle or constant cyclic AMP generation in the follicle (77,101,102). Hence, spontaneously luteinized granulosa cells in culture can be maintained for weeks without adding LH or cAMP derivatives. However, the lengthy luteal phase of ovarian function in vivo is maintained by factors and

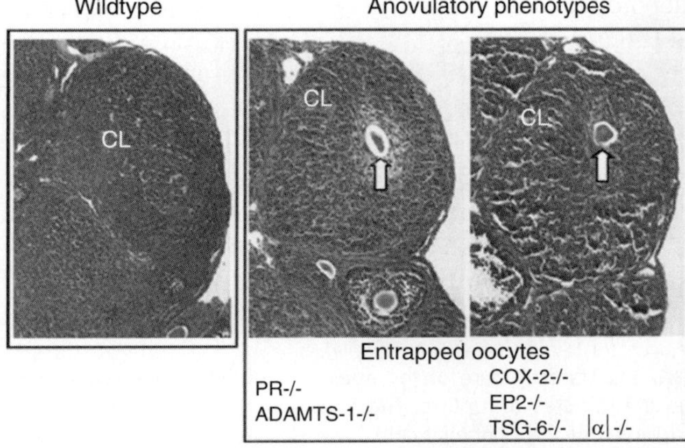

FIG. 15. Anovulatory ovarian phenotypes of mice null for the progesterone receptor (PR) and cyclooxygenase-2 (COX-2) and related downstream targets all exhibit functional corpora lutea with entrapped oocytes. The underlying causes for ovulation failure in these mutant mice are just beginning to be understood. One PR-regulated gene ADAMTS-1 appears to be important for matrix function and proteolysis of the hyaluronan binding proteoglycan versican. COX-2 produces PGE2 that binds the PGE2 receptor subtype EP2 and induces TSG-6. TSG-6 is another hyaluronan binding factor that in conjunction with IαI, a serum-derived, hyaluronan-binding factor is essential for matrix stabilization. [Modified from (44) with permission. See Figs. 20 and 24 and text for further discussion.] (See color insert.)

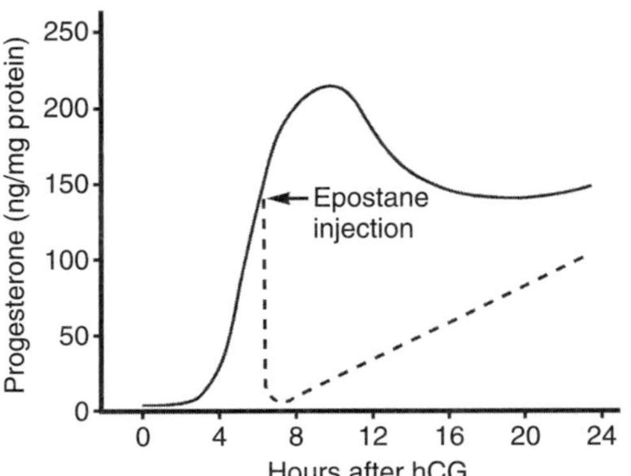

FIG. 14. Temporal pattern of ovarian progesterone synthesis during ovulation in the immature rat model. Notice that progesterone synthesis slows down during the most traumatic hours of the ovulatory process, and then increases substantially during post-ovulatory luteinization of the follicle. Epostane, an inhibitor of 3β-hydroxysteroid dehydrogenase, transiently inhibits ovarian progesterone synthesis (note *dotted line*) when administered within 7 hours after inducing ovulation with hCG, and such treatment effectively blocks ovulation. [Figure based on data from (67,68)].

hormones that are species specific. Although LH is the luteotropin in many mammals, rodents such as the mouse and rat utilize prolactin or prolactin-like molecules and steroid receptors to maintain the corpus luteum and pregnancy (see Chapter 12).

Besides initiating ovarian progesterone synthesis, the LH surge concomitantly induces the expression of mRNA for progesterone receptor (PR), a nuclear receptor transcription factor (7,103–106). PR mRNA and protein expression appears to be upregulated specifically in the granulosa cells of mature follicles in response to LH or hCG (103–107) (Fig. 16). The gonadotropin activates Sp1/Sp3 binding sites within the mouse PR proximal promoter, but the molecular mechanisms by which this activation occurs is unknown (108). Nonetheless, it is clear that the ovulatory effects of the gonadotropin surge are mediated at least in part by induction of the PR, because mice lacking PR fail to ovulate (105) (Fig. 16).

During the first several decades after its discovery, progesterone was not considered as a component of the ovulatory process. Then, in 1961, Frederick Hisaw (109) stressed during the proceedings of a conference that the accumulation of progesterone over estrogen "might take a responsible part in the process of ovulation." Eventually, the experiment that firmly established a role for this steroid was conducted by Snyder et al. (110) who clearly demonstrated that the inhibition of follicular steroidogenesis and ovulation by epostane could be overcome by treatment of the experimental animals with exogenous progesterone. Subsequently, it was reported that the optimum time to administer epostane is during the hour preceding the ovulatory increase in progesterone synthesis, that this inhibitory agent severely impairs all steroidogenic activity in the ovary within a few minutes after its injection, that its anti-steroidogenic effect is transient, and that ovarian progesterone synthesis rises back almost to normal levels by the time follicular rupture would have normally occurred in the animals that received an ovulation-inhibiting dose of epostane (68). Thus, the evidence reveals that there is a relatively narrow *temporal window* in the early stages of the ovulatory process (i.e., at approximately the time when the progesterone levels begin to rise) during which it is vital for follicular steroidogenesis to proceed unabated. However, in spite of this obvious participation of progesterone in the mechanism of ovulation, treatment of animals or follicles with progesterone, alone, does not induce follicular rupture (111,112). That is to say, the ovulatory surge in LH (or the injection of an analog of LH) induces other metabolic changes (e.g., the induction of PR) that are just as obligatory for ultimate rupture of the follicle (69,103–105)

The specific action of progesterone in ovulation had been rather uncertain until recently. It is now clear that the LH surge induces transcription and translation of the ovarian gene for a disintegrin and metalloproteinase with thrombospondin motifs (ADAMTS-1), the expression of which is dependent not only on LH but also on increases in both progesterone (113) and PR (105,106) (Fig. 17). For example, the inhibition of ovarian progesterone synthesis and ovulation by epostane also involves inhibition of ovarian ADAMTS-1 expression, and this interference can be overcome by treating the animals with exogenous progesterone (113). Furthermore, genetically deficient mice that lack PR fail to ovulate (105) (Fig. 13). ADAMTS-1 was first discovered in association with inflamed intestinal tissue (114), but a non-steroidal anti-inflammatory drug such as indomethacin (which can

LH/hCG induces the progesterone receptor (PR) in granulosa cells

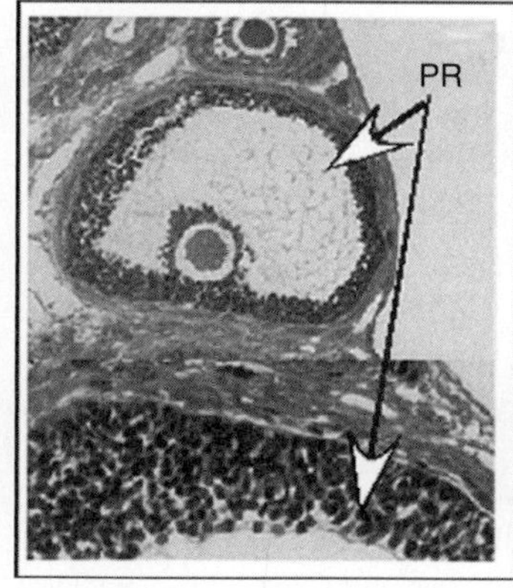

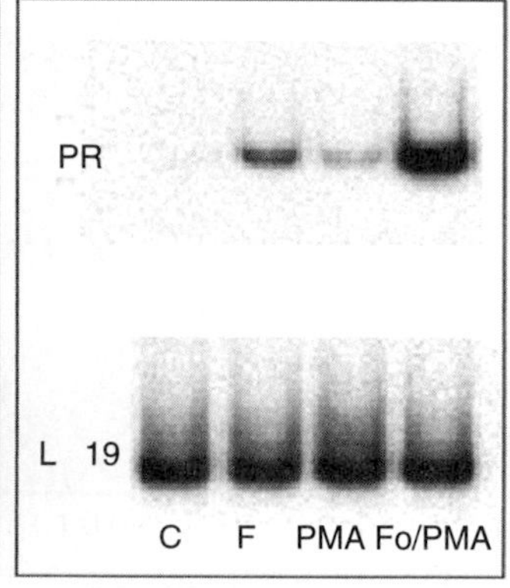

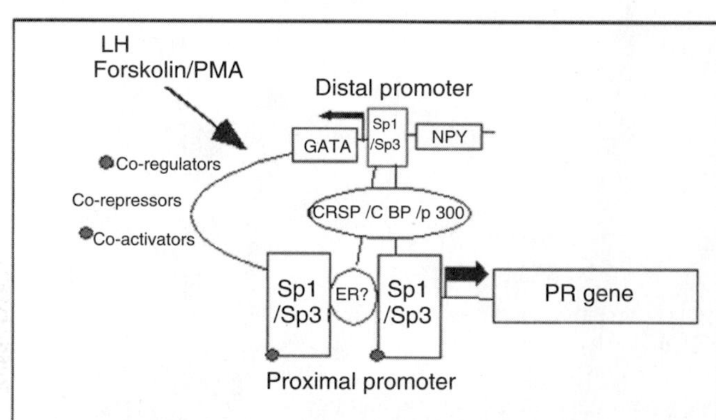

FIG. 16. LH induces PR in granulosa cells of preovulatory follicles. Agonists that mimic LH action (e.g., forskolin and PMA) likewise induce PR in granulosa cells in culture. LH and LH agonists mediate induction of PR via activation of Sp1/Sp3 binding sites within the proximal and distal regions of the murine PR promoter. [From (108), with permission.]

LH/hCG and PR induce ADAMTS-1 in granulosa cells

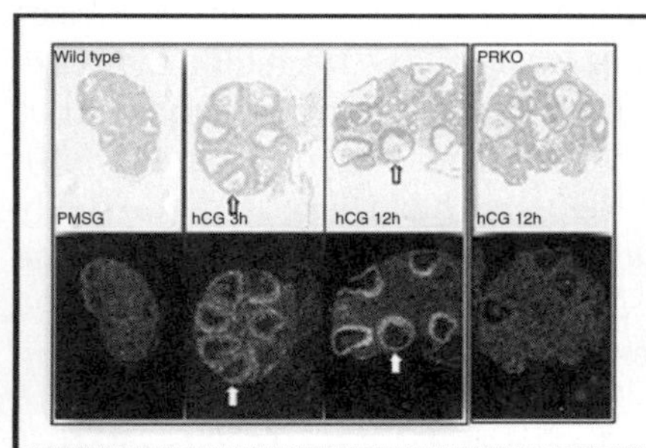

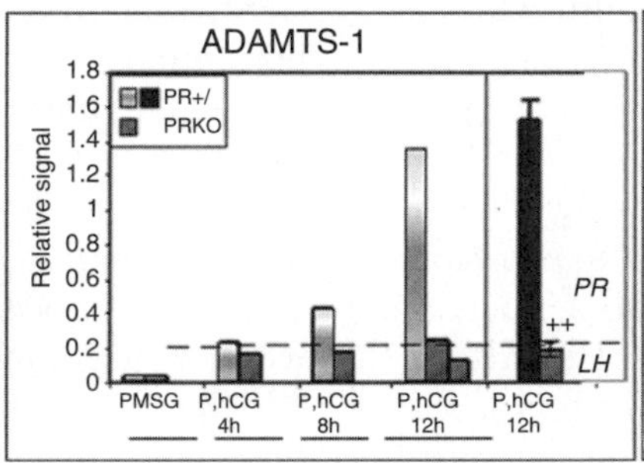

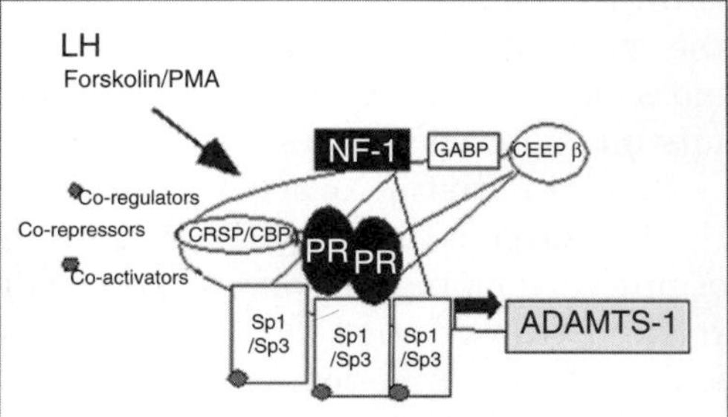

FIG. 17. LH and PR mediate the induction of ADAMTS-1 in granulosa cells of preovulatory follicles. ADAMTS-1 mRNA is selectively decreased in granulosa cells of PR knockout (PRKO) mice compared to wildtype and heterozygote mice. LH and PR mediate the induction of ADAMTS-1 via multiple regulatory domains within the promoter, including sites that bind Sp1/Sp3, NF-1, and C/EBPβ. Thus, ADAMTS-1, like PR, is dependent on activation of Sp1/Sp3 sites, but this metalloproteinase also requires PR interactions with these Sp1/Sp3 sites. [From (106), with permission.]

inhibit ovulation) does not block expression of the ADAMTS-1 gene during ovulation (113). Nevertheless, ADAMTS-1 appears to be an important enzyme for ovulation, and it is currently being studied more extensively (115–118). One known substrate of ADAMTS-1 is versican, a hyaluronan (HA) binding proteoglycan, that like HA is induced in granulosa cells of preovulatory follicles by LH (116). The exact function of these inflammation-related matrix components in the ovulation process and the obligatory requirement for ADAMTS-1 remain to be determined. However, it is clear that the ovulation cone (stigma) contains vascular elements (Fig. 18) and exhibits intense localization of ADAMTS-1 and ADAMTS-4 (Fig. 19). These results indicate that these two ADAMTSs might be involved in regulating angiogenic activity (49). In addition, since ADAMTS-1 localizes to the expanded COC matrix, it may contribute to a protective shield around the oocyte (Fig. 20). In addition, recent observations in the porcine ovary indicate that PR antagonists can disrupt expression of ADAMTS-1 and cumulus expansion in cultured COCs (119). As noted below, several structurally related ADAMTS proteases are expressed in the ovary and may have either specific or redundant functions (49,115). In addition to progesterone action on ADAMTS-1

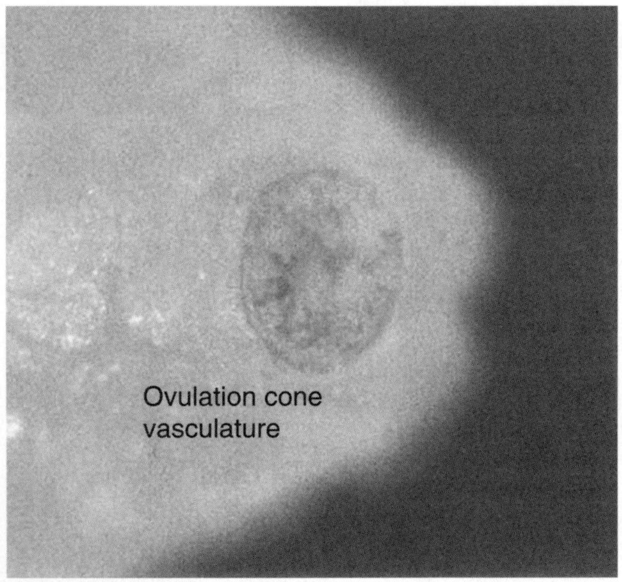

FIG. 18. An ovulation cone (stigma) with obvious vascular elements. This photomicrograph depicts the surface of an ovulated follicle 24 hours after the injection of hCG to induce ovulation in the PMSG-primed mouse. Note the enriched and complex vascular network in this region. [From (49); see text for details.] (See color insert.)

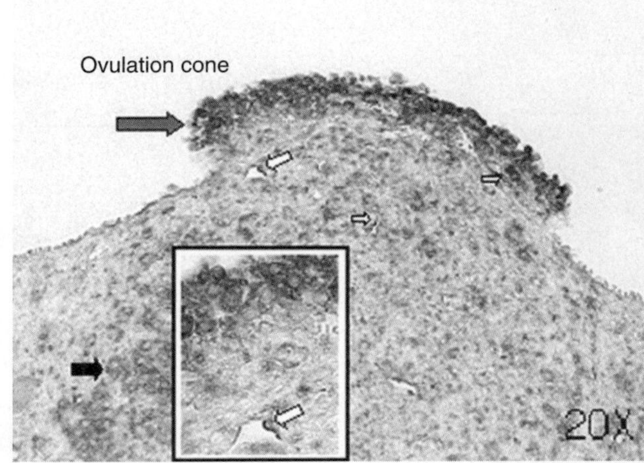

FIG. 19. Histological image of the ovulation stigma. This section of a paraffin-embedded, PMSG-hCG 24h mouse ovary depicts the immunolocalization of ADAMTS-1 in the ovulation cone (*gray arrow*) and to endothelial cells of the newly forming corpus luteum (*inset, white arrows*). ADAMTS-4 gives a similar immunolocalization pattern. [From (49); see text for details.] (See color insert.)

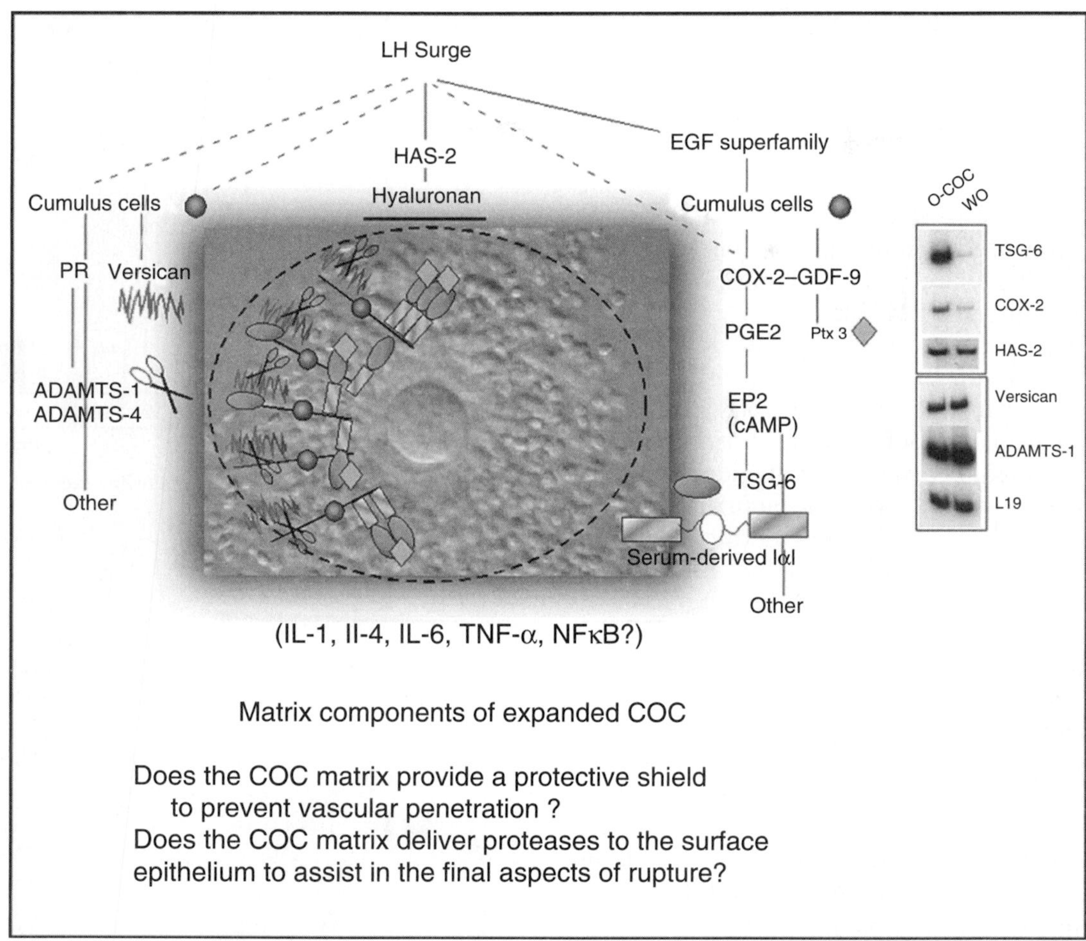

FIG. 20. Schematic of LH induced regulatory pathways that impact matrix formation and stabilization as well as ovulation. Ovulation is a complex process. Based on gene induction patterns and the phenotypes of mutant mice, it is clear that PR- and PGE2-mediated events are distinct phenomena, yet both are essential for ovulation. PR impacts the function and formation of the COC matrix by inducing ADAMTS-1, a protease that binds to and cleaves versican, another LH-induced HA-binding factor. The functions of ADAMTS-1 and versican remain to be established. However, based on the anti-angiogenic functions of ADAMTS-1, one might hypothesize that the HA superstructure provides a protective shield to prevent vascular penetration of the COC that would otherwise impede (physically) oocyte release. The attachment of proteases to the matrix may also provide a highly concentrated delivery system by which the COC assists in the final aspects of rupture. LH also induces transcription of the HAS-2 gene that encodes hyaluronan (HA), the major structural backbone of the COC matrix (*solid black bar*). LH induction of COX-2 results in the production of PGE2 that binds EP2 to generate cAMP and causes the eventual transcription of TSG-6 (*oval*). TSG-6 and the heavy chain (*rectangle*) of the serum-derived factor IαI form a complex that allows IαI to bind covalently to HA, thereby stabilizing HA. The cytokines Il-1β, IL-6, Il-4, TNF-α and downstream signaling cascades involving NFkB appear to participate in these events as well. [From (46,47,49,116); see text for details.] (See color insert.)

expression, this important steroid is involved in regulating ovarian expression of cathepsin L (cat L), which is an elastinolytic cysteine protease (43,105,120), and the action of this enzyme presumably contributes in some way to proteolytic degradation of the follicle wall. Also, ovarian expression of rat pituitary adenylate cyclase activating polypeptide (PACAP), which is structurally and functionally related to vasoactive intestinal peptide (VIP), is dependent on the ovulatory increase in progesterone (44,121,122). However, the function of PACAP in ovulation has not been well defined. Most likely, other genes regulated by PR will be identified in future studies.

Prostaglandins and Ovulation

Prostaglandins are produced from arachidonic acid, which is formed from membrane phospholipids in response to virtually any type of environmental condition that disturbs the plasma membrane (123). These molecules are classically associated with sites of inflammation and thus their presence in ovulating follicles lends credence to the idea that ovulation is an inflammatory-like process. The first indication that prostaglandins might be generated in the ovary in response to an ovulatory surge in gonadotropin arose circuitously from five simultaneous reports that indomethacin, a nonsteroidal antiinflammatory agent that is commonly used to inhibit prostaglandin synthesis, could prevent ovulation when administered to various laboratory animals (124–128). Shortly thereafter, LeMaire, Marsh and coworkers (129,130) measured a marked increase in prostaglandins E_2 and $F_{2\alpha}$ in rabbit and rat follicles during the ovulatory process, and their observations have been confirmed many times. In the immature rat model, ovarian prostaglandin levels begin to increase sharply at 3–4 hours after the administration of hCG to induce ovulation, and then they start to decline even before the follicles actually rupture (66,131). Characteristically, prostaglandin of the E-type is about twice as concentrated as the F-type in the ovary during ovulation.

The prostanoid pathway by which arachidonic acid is metabolized to bioactive prostaglandins is regulated primarily by cyclooxygenase (COX) enzymes (also known as prostaglandin H_2 synthases) (132). COX-1 is constitutively present in many tissues, including the ovary, but it is *not* upregulated in response to the LH/hCG surge (133–136). In contrast, COX-2 is an inducible form of the enzyme, and in the ovary it is expressed in ovarian granulosa cells and cumulus cells after the LH/hCG surge in both rat and mouse models (133–139) (Figs. 13 and 20). The enzyme is translated from COX-2 mRNA, which is undetectable in mature follicles that have not yet been stimulated to ovulate. However, by 4 hours after the administration of an ovulatory dose of hCG to immature rats and mice, the expression of mRNA for COX-2 reaches a peak and then decreases dramatically in granulosa cells by 6 hours (137,139). However, the peak in ovarian COX-2 protein that enzymatically converts arachidonic acid to the various prostaglandins lags about 1–3 hours behind COX-2 mRNA expression, and the enzyme is still present at elevated levels at 12–14 hours after hCG at the time when follicles begin to rupture, but also at a time when COX-2 mRNA in the granulosa cells is low (67,131,139). However, COX-2 is highly expressed in the COCs immediately prior to ovulation (12 hours post-hCG) as well as in ovulated COC within the oviduct (16 hours post-hCG) as well as in COC exposed to hormone in culture (Fig. 21) (137,138,140). Collectively, these observations indicate that newly translated COX-2 enzyme persists in an active form in the ovary for several hours after the COX-2 mRNA has been down-regulated in granulosa cells and that other mechanisms in conjunction with the LH surge control the expression of COX-2 mRNA and protein in the cumulus cells prior to and after ovulation. Most recently, the EGF-related factors amphiregulin, epiregulin and betacellulin have been shown to regulate COX-2 message in cumulus cells (72). Of relevance, the expression of COX-2 in the COCs closely mirrors that of HAS-2 in the mouse (140) and likely in other species (141–143).

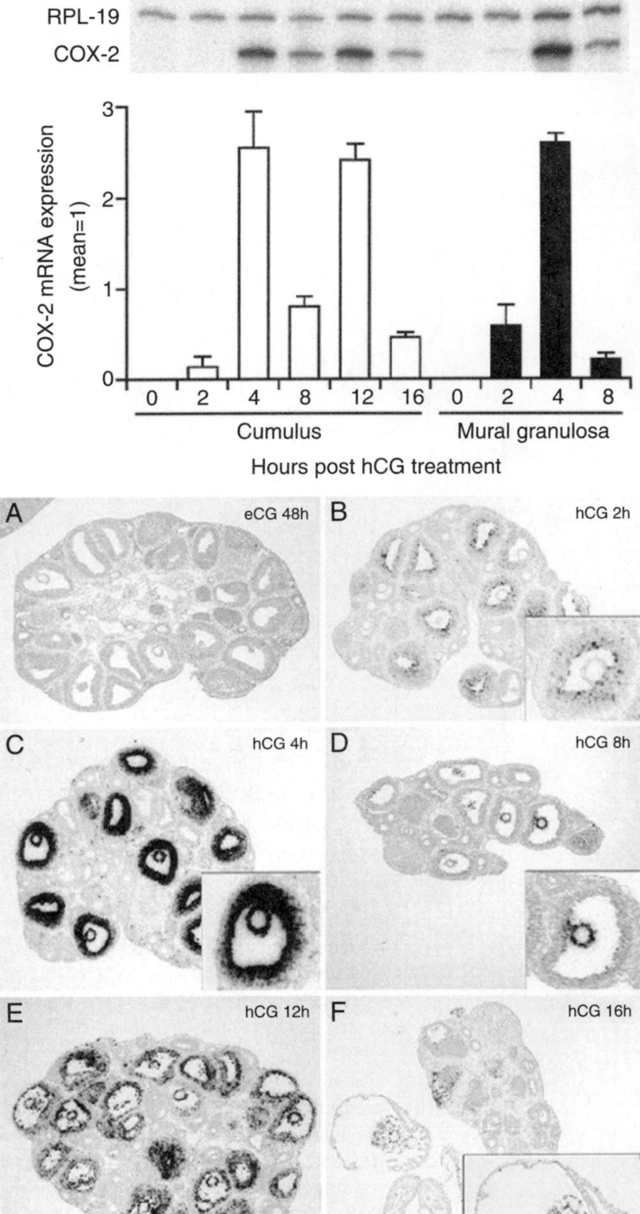

FIG. 21. Spatial and temporal induction of COX-2 mRNA in granulosa cells and cumulus cells during the hours preceding ovulation. Note the marked and specific induction of COX-2 at 12 hours. The EGF-like factors amphiregulin, epiregulin, and betacellulin may participate in this latter phase response. [From (137) with permission.]

On a related note, it has been hypothesized that there is a possible correlation between the time of peak expression of COX-2 and the species-specific duration of the ovulatory process (143–145). Limited data from the rat, cow, mare and monkey indicate that in all of these species there is approximately a 10-hour interval between the onset of ovulation-related ovarian prostaglandin synthesis and the time of follicular rupture, even though the length of the ovulatory process varies from a low of 12–14 hours in the rat to a high of 32–40 hours in the monkey. That is to say, the duration of the ovulatory process in a given species might be a function of the interval between the LH/hCG surge and the time of onset of ovarian COX-2 mRNA expression.

Some of the components of the complex signaling pathway that begins at the LH/hCG receptor on the surface of granulosa cells and ends with the expression of COX-2 to generate bioactive prostaglandins have been determined. The LH/hCG receptor is coupled to a G_s-protein that naturally activates adenylyl cyclase, which catalyzes the formation of cyclic AMP and leads to transient activation of protein kinase A (134,144,146). Other elements of the signaling pathway have not been deciphered, but it is known that the promoter of the rat COX-2 gene contains a CAAT/enhancer binding protein (C/EBP) consensus site, and CEBPβ mRNA and protein are induced rapidly in granulosa cells in vivo following an ovulatory dose of hCG (144,146,147). Moreover C/EBPβ KO mice are infertile and exhibit impaired ovulation (73).

It is generally accepted that prostaglandins have a fundamental role in the mechanism of ovulation (10,15,43,47,146). However, even though ovarian prostaglandins have been studied almost as extensively as progesterone, their functions in the rupture of a follicle are not as well established as is their critical role in COC expansion (46,47,138,140,150). The expansion process is essential for release of the oocyte from an ovulatory follicle as confirmed by results in four mutant mouse models, namely COX-2 (70), EP2 (71), IαI (148,149), and TSG-6 (150). Suggested roles for the prostaglandins have included wide-ranging ideas such as activation of follicular proteolytic enzymes (151), stimulation of follicular angiogenesis (152) and promotion of nitric oxide synthase activity (153). A promising new experimental approach that might help elucidate the role of prostanoids in ovulation is the assessment of specific prostaglandin receptors. It has been reported recently that, 3 hours after the injection of hCG into mice, ovarian expression of mRNAs for the two G_s-coupled prostaglandin E_2 receptors (i.e., EP2 and EP4) increased significantly in both granulosa cells and cumulus cells (154). Furthermore, mice that are null for either COX-2 or EP2 exhibit impaired ovulation (70,71) and have reduced expression of mRNA for tumor necrosis factor-stimulated gene-6 (TSG-6) (138). TSG-6 is a hyaluronan-binding protein that is expressed temporally and spatially in granulosa cells and cumulus cells with hyaluronan synthase-2 (HAS-2) (140) (Fig. 22). In the follicle, the long hyaluronan polymers provide the structural backbone of the extracellular matrix surrounding the oocyte and appear to be stabilized by various hyaluronan binding proteins (Fig. 20). TSG-6 is one of these factors. TSG-6 is also essential for delivery and covalent linkage of the heavy chains of the serum-derived protein inter-alpha trypsin-inhibitor (IαI) to hyaluronan. Therefore, it has been proposed that one possible function of prostaglandin E_2 is the induction of ovarian TSG-6 expression during the ovulatory process and that TSG-6 along with IαI stabilizes the COC matrix. Proof of this hypothesis has been obtained by several approaches. Mice null for TSG-6 or IαI exhibit impaired ovulation that can be reversed by providing exogenous TSG-6 or IαI, respectively (148–150). To support these in vivo data, FSH-induced expansion of COCs in culture can be disrupted by peptide specific antibodies to TSG-6 (138). Collectively, these data indicate that the formation and stabilization of the expanded matrix is critical for oocyte release. It is possible that the formation of expanded matrix with all of the attached HA binding proteins is essential to provide a protective shield around the oocyte. Stabilization of this hyaluronan COC-derived matrix is also essential following ovulation and is dependent on the oocyte-derived factors GDF-9/BMP15 and their induction of pentraxin-3, an inflammation-related factor that binds specifically to TSG-6 (155,156). The effects that the matrix exerts on the surface epithelium or

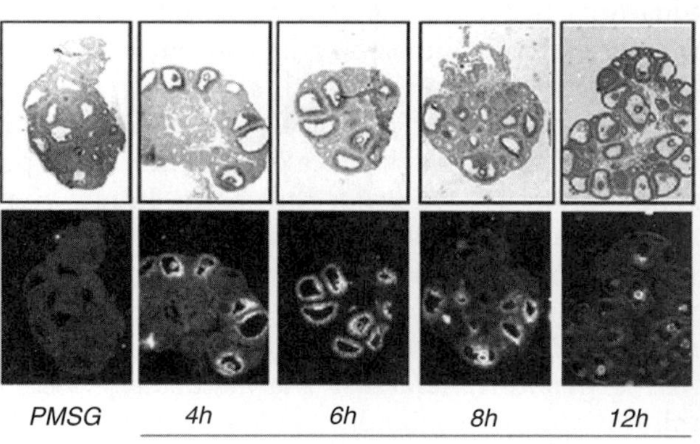

FIG. 22. Spatial and temporal induction of TSG-6 in granulosa cells and cumulus cells in the hours preceding ovulation. Note that the induction of TSG-6 mirrors that of COX-2 (e.g., see Fig. 21). [From (140), with permission.]

other components of the rupture process remain to be determined. However, an inviting hypothesis is that the matrix around the cumulus-enclosed oocyte plays a significant role in the final steps of ovulation, perhaps by delivering proteases such as ADAMTS-1 and ADAMTS-4 to the surface of the ovary or by providing a protective shield for the oocyte. In any event, further information about other downstream responses to activation of the prostaglandin receptor will improve the chances of deciphering other basic role(s) of prostaglandins in the mechanism of ovulation.

Experiments that assess the metabolic effects of inhibitors of prostaglandin synthesis also have the potential of providing information about the function(s) of prostaglandins in ovulation. For example, it has been reported that indomethacin significantly decreases the ability of interleukin-1β (IL-1β) to upregulate ovarian IL-6 transcripts in dispersed cells taken from whole ovaries of rats, and this inhibitory action could be overcome by addition of prostaglandin E_2 to the cultured cells (157). Furthermore, recent experiments with indomethacin continue to support the earlier evidence (10) that indomethacin can inhibit ovarian prostaglandin synthesis and impede ovulation without obstructing ovarian progesterone synthesis and luteal development (158–160). Thus, it is clear that the syntheses of prostaglandin and progesterone in ovulatory follicles are two distinctly independent phenomena that activate different genetic programs that converge on the COC matrix formation and function (Figs. 13, 15, and 20).

Like the inhibitory action of epostane on ovarian progesterone synthesis, the inhibitory effect of indomethacin on ovarian prostaglandin synthesis has served as an interesting subject for ovulation-related research. The drug has a very rapid effect on ovarian prostaglandin metabolism. When indomethacin is administered intravenously during the ovulatory process, the normally elevated levels of prostaglandins E_2 and $F_{2\alpha}$ decline to almost naught within only a few minutes (161). However, in spite of the powerful inhibitory action of indomethacin on prostaglandin synthesis, it is intriguing that excessively high doses of this non-steroidal antiinflammatory drug still permits the rupture of a limited number of follicles (161–163). Even more puzzling is the observation that relatively low doses of indomethacin that significantly inhibit the normal increase in ovarian prostaglandins during ovulation have no significant effect on the ovulation rate in rabbits and rats (161–163). This anomaly might be explained by the fact that induction of COX-2 mRNA and protein in the COCs is sustained at 12–16 hours post hCG and may not be inhibited easily by entrance of the drug into the follicle (137). However, it is relevant to note that indomethacin is not as specific of an inhibitor of COX-2 as originally thought. There is evidence that this drug also affects the activity of lipoxygenase enzymes that convert arachidonic acid into biologically important leukotrienes and lipoxins (163–165). In view of this information, a more thorough analysis of the potential roles of ovarian lipoxygenase products in the ovulatory process is warranted.

Proteases and Ovulation

In 1916, Irving Hardesty first suggested to Schochet (166) that the ovarian follicular fluid might "exert some special digestive action on the resisting tissues" of the follicle wall. However, after a number of unsuccessful attempts during the next three decades to demonstrate a role for proteases in ovulation, it was concluded that the "enzyme theory" did not fit the facts and that the immediate cause of ovulation was a mystery (167). During the next two decades, several additional attempts to detect ovarian proteolytic activity did not yield convincing results (10). Then, in the mid 1960s, it was demonstrated that small quantities of collagenolytic enzymes could be injected into rabbit follicles and cause rupture (but not necessarily oocyte release) (168), and that the collagenous tissue in the follicle wall of sows became weaker at the time of ovulation (169). Since that time, most of the work on the so-called "enzyme theory" of ovulation has focused on assessments of the potential roles of PA, matrix metalloproteinases (MMPs), and ADAMTS enzymes.

Plasminogen Activators

PA is a serine protease that has been implicated in many types of tissue degradation (170), and it is secreted from fibroblasts in association with collagenase activity (171). PA of the tissue type (tPA) was first studied in relation to ovulation in 1975 by Beers et al. (172). However, in the first two decades of research on the potential role of this protease in ovulation, there was considerable incongruity among the various studies (10). It has been concluded that this enzyme is mainly in the granulosa, in significant amounts in the theca interna, or especially abundant in the fibroblasts in the apical follicle wall. There were contradictory reports that granulosa cell production of tPA is stimulated chiefly by prostaglandins of the E type or chiefly by prostaglandins of the F type. Also, there were conflicting reports that indomethacin treatment of experimental animals suppresses the secretion of tPA, or that indomethacin has no apparent effect. There were differing reports on whether progesterone has a minor role or a significant role in the expression of ovarian tPA activity. Thus, there has been considerable confusion about ovarian tPA in the past, and the most recent decade of research on this enzyme has not established the significance of this enzyme in ovulation. The data on ovarian expression of tPA inhibitor type-1 (PAI-1) is equally confounding. It has been reported both that PAI-1 *is expressed* in the rat ovary during the ovulatory process (173) and that it *declines* dramatically in the mouse and monkey ovary as the time of ovulation approaches (62,174).

It has been reported that the urokinase type of PA (uPA) from ovarian surface epithelial cells is important for ovulation (28), and TNF-α and IL-1β have been found to drastically increase the secretion of this enzyme from ovarian surface epithelial cells (175). However, as stated earlier in this chapter, during the ovulatory process the surface epithelium sloughs from the apex of the follicle and is not attached to the ovary at the site of rupture. Therefore, this thin layer of cells does not appear to be essential for ovulation. Nevertheless, there continue to be reports that the tissue type variety of PA is generated in the granulosa cells (and surrounding connective tissues) of rats (173) and monkeys (61), and that such activity is important for follicular rupture. Yet, mice lacking tPA and/or uPA gene function exhibit little or no impairment of ovulation efficiency (176,177), and plasmin does not appear to be required for follicular rupture in mice (178). Thus, the issue of whether PAs are important for ovulation would seem to be immaterial except, perhaps, for the evidence in other experimental models that such enzymes have a central role in activation of the four classes of inflammatory MMPs discussed in the following paragraph (179).

Matrix Metalloproteinases

Most of the original work on ovarian MMPs and tissue inhibitors of MMPs (TIMPs) has been carried out by Curry et al. (26,29,61,180–182). The four categories of MMPs that have been studied in ovarian tissues are the collagenases, gelatinases, membrane-type MMPs, and stromelysins. The granulosa cells of ovulatory follicles appear mainly to transcribe and translate MMP-2 (182,183) and MMP-19 (184), whereas thecal and stromal tissues are associated with gonadotropin-induced elevations in collagenase-3 (185), MMP-2 (182,183), MMP-9 (26,182,183), and MMP-19 (184). However, a recent study of MMPs in equine follicles failed to detect any increase in MMP-2 or MMP-9 in follicular fluid or in explant cultures of stromal tissue at the time of ovulation (186). The conclusion from this latter report was that MMP-2 and MMP-9 are not key regulators for the changes in follicular shape immediately prior to ovulation. Another recent study of MMP-2 and MMP-9 in human follicular fluid yielded too much variability among the samples to clarify whether these proteases change significantly in the ovary during the ovulatory process (187). A novel approach to the assessment of MMPs in ovulation is the recent study of membrane type-1-MMP (MT1-MMP; MMP14) in ovulatory follicles (61). MT1-MMP hydrolyzes type I collagen, and it cleaves pro-MMP-2 in order to render it active (188,189). In view of this action, it is relevant that high expression of MT1-MMP mRNA occurs in the "theca-interstitial layer" of rat follicles near the time of ovulation (61). However, on a cautious note, there is a recent report that the phenotype of MT1-MMP deficient mice does not entirely mimic that of mice deficient in MMP-2 (190), suggesting that MT1-MMP might have functions other than the activation of pro-MMP-2. Nevertheless, it appears likely that MMPs have a significant function in the degradation of the collagenous connective tissue in the follicle wall. That MMP2 and MMP9 null mice are fertile and ovulate may indicate that these two proteases have some overlapping and redundant functions in the ovulating follicle (191). It is also worth noting that the expression patterns and activation of MMP2 and MMP9 are normal in PRKO null mice that fail to ovulate (105).

ADAMTS Enzymes

ADAMTS-1 and other members of this novel family of metalloproteinases differ from the usual MMPs by the fact that they are readily secreted and thereafter either bind extracellular matrix components, or attach to the cell surface via specific regulatory mechanisms (114,192). By means of the differential display method for gene discovery, ADAMTS-1 mRNA was first detected in rat ovarian tissue by Espey et al. (113). This initial report revealed that ADAMTS-1 mRNA is expressed primarily in the granulosa cells of ovulatory follicles, that transcription of the mRNA is significantly impaired by epostane treatment of the animals, and that injection of progesterone to the experimental animals overcomes the inhibitory action of epostane. Shortly after this demonstration that ADAMTS-1 gene expression is dependent on progesterone, it was shown that ovulatory expression of ADAMTS-1 mRNA is also dependent on PR, and that expression of ADAMTS-1 mRNA and ovulation is diminished significantly in PR knockout mice (44,105,106,116) (Figs. 15 and 17). It should be noted that, in addition to the substantial expression of ADAMTS-1 mRNA in the stratum granulosum of ovulatory follicles, ADAMTS-1 is expressed in varying degrees in cells in the thecal connective tissue and ovarian stroma (115,117,253). In granulosa cells, the ADAMTS-1 protein normally increases greater than tenfold during the ovulatory process, but less in PR knockout mice (116). It appears that one function of this enzyme in ovulation might be to cleave versican in the expanded cumulus–oocyte complex. ADAMTS-1 may also provide antiangiogenic effects to prevent vascularization of the expanded COC matrix that would impede release of this complex. A more comprehensive study of the ADAMTS family of metalloproteinases in the bovine ovary has revealed that more than one ADAMTS subtype might be involved in the ovulatory process (49,115). In the dominant follicle of the cow ovary, the ovulatory surge in gonadotropins induces an increase in ADAMTS-1, -2, and -5 mRNA levels in the granulosa cells and ADAMTS-1, -3, and -9 mRNA levels in thecal cells. In the mouse and rat ADAMTS-4 and ADAMTS-9 also increase in granulosa cells and theca cells of preovulatory follicles following administration of hCG, whereas ADAMTS-5 does not (49). ADAMTS-19 is highly expressed in the embryonic gonad (193). Thus, the regulation of ADAMTS expression in the ovary is complex, and this family of novel enzymes will undoubtedly be investigated in greater detail in future studies on ovulation.

OVARIAN GENE EXPRESSION DURING OVULATION

Over the years, there have been innumerable attempts to identify the basic biochemical components of the ovulatory process. Most of the efforts to differentiate the principal mediators of this process have been conducted under the assumption that ovulation might be a relatively straightforward series of chemical reactions-involving a reasonably small number of regulatory factors that are activated by an ovulatory surge in gonadotropin. However, this simplified notion was dispelled by the hypothesis that the biochemical events of ovulation are comparable to an inflammatory process (5). Since the introduction of that provocative idea 25 years ago, there have been a number of discoveries of ovarian agents (e.g., various cytokines that are well known components of acute inflammatory reactions), which are now considered to be important factors in ovulation (7,36,46). Still, the supposition that has driven much of the research since 1980 has been that appropriate experiments would elucidate uncomplicated signaling pathways starting with G-protein–coupled receptors for LH/hCG and leading to activation of kinases that could promote transcription and translation of one or more collagenolytic enzymes to degrade the follicle wall and cause rupture. The greatest experimental challenges appeared to be merely the identification of one or more ovulation-specific collagenases and the delineation of the roles that ovarian prostaglandins and progesterone play in the expression of such enzymes. Although a number of studies have provided useful information about the signaling pathways and the collagenolytic enzymes that likely play a critical role in ovulation, progress toward a comprehensive picture of the ovulatory process has been slow.

The recent development of novel molecular techniques to detect tissue-specific gene expression has promoted greater expectations of gaining a more precise understanding of the biochemical changes that occur in the ovary at the time of ovulation. However, in spite of an exponential growth in the number of ovulation-related genes that have been discovered during the past decade, the contemporary image of the ovulatory process seems to have become more convoluted rather than clearer. Therefore, the aims of the present section of this chapter are to identify some of the reasons for this unexpected complexity, to catalog most of the ovulation-related genes that have been reported to date, and to briefly summarize the state of our current knowledge about gene expression pathways (i.e., gene cascades) in the ovary at the time of ovulation. The genes listed in Table 1 include those that are induced by the ovulatory surge of LH/hCG. It excludes the genes that are expressed in the proliferative phase of follicular growth or are turned off as a consequence of the ovulatory surge in gonadotropin. Although the down-regulation of certain genes may be important for the transition of an ovulatory follicle into a viable corpus luteum, for reasons of simplicity these genes have not been included herein.

Complexity of Integrating Gene Expression Data

The scope and diversity of genes that have been described as participants in the ovulatory process were quite unpredictable. To date, there are at least 88 ovulation-related genes that have been reported (Table 1), and this compilation probably excludes some genes that were inadvertently overlooked within the voluminous literature. (The term *ovulation-related* is meant to indicate there is direct or indirect evidence to suggest that the gonadotropin surge induces expression of a given gene that is considered to be a contributing factor to the interwoven processes of ovulation and luteinization.) Ideally, it would be nice if each of the newly discovered ovulation-related genes could be assigned to a known position in one of

TABLE 1. *Ovulation-specific genes*

Gene name	Abbreviation	Tissue	Animal	References
A disintegrin and metalloproteinase with thrombospondin motifs-1	ADAMTS-1	Gr, TI, COC	rat, mou, hor, cow	(45–47,105,106,113,115–119)
A disintegrin and metalloproteinase with thrombospondin motifs-2	ADAMTS-2	Gr	cow	(115)
A disintegrin and metalloproteinase with thrombospondin motifs-3	ADAMTS-3	TI	cow	(115)
A disintegrin and metalloproteinase with thrombospondin motifs-4	ADAMTS-4	Gr, TI	cow, mou, pig	(49,115,116,119)
A disintegrin and metalloproteinase with thrombospondin motifs-5	ADAMTS-5	Gr	mou, cow	(49,115)
A disintegrin and metalloproteinase with thrombospondin motifs-9	ADAMTS-9	TI	mou, cow	(49,115)
Adrenodoxin	ADX	Gr	rat	(45)
5-Aminolevulinate synthase	ALAS	Gr	rat	(45)
Amphiregulin	AR	Gr, COC	mou	(72)
Apolipoprotein-E	apoE	Gr	rat	(195)
Aryl hydrocarbon receptor	AHR	Gr	rat, mon	(196,197)
Betacellulin	BTC	Gr, COC	mou	(72)
cAMP-response element binding protein	CREB	Gr	rat	(47,199)
cAMP-response element modulator	CREM	Gr	rat	(200,201)
Carbonyl reductase	CBR	TE, St	rat	(202)
Cathepsin L	catL	Gr, COC	rat, mou	(46,105,120)
CCAAT/enhancer-binding protein-β	C/EBP-β	Gr	Rat, mou	(139,205,206)
CD63 (cell surface antigen)	CD63	Gr	rat	(45)
Corticotropin-releasing hormone receptor-1	CRHR1	TE, St	rat, mou	(207–209)
Cutaneous fatty acid binding protein	C-FABP	TE	rat	(210)
Cyclooxygenase-2	COX-2	Gr, COC	rat, mou, hor, cow	(21,45,46,70,137,140,143,211)
Cytochrome P450 side chain cleavage (CYP11A1)	P450scc	Gr, Theca, ci	rat, hor	(91,93,95,212–215)
Cytokine-induced neutrophil chemoattractant	CINC	Gr, TE	rat, hum	(216–218)
Early growth response protein-1	Egr-1	Gr	rat, mou	(220,221)
Epiregulin	Epi	Gr, COC	rat, mou	(45,72)
FOS-like antigen-2	Fra-2	Gr, CL	rat	(224)
Frizzled G-protein–coupled receptor-1	Fz-1	Gr, TI	rat, mou	(225)
Frizzled G-protein–coupled receptor-4	Fz-4	CL	rat, mou	(225)
Frizzled-related protein-4 (secreted)	sFRP-4	Gr, CL	rat,mou	(226,227)
γ-Glutamylcysteine synthetase	γ-GCS	Gr	rat	(45)
Glutathione S-transferase	GST	CL	rat	(45)
G-protein–coupled receptor-54	GPR54	Gr	rat	(45)
Hyaluronan synthase-2	(HAS-2)	Gr, COC	mouse, cow, pig	(140–142)
Hyaluronan synthase-3	(HAS-3)	oocyte	(pig)	
3α-Hydroxysteroid dehydrogenase	3α-HSD	Gr	rat	(228)
3β-Hydroxysteroid dehydrogenase	3β-HSD	Gr, COC	rat, hor, mon	(91,212–215)
11β-Hydroxysteroid dehydrogenase	11β-HSD	Gr, SE	rat, hum	(229,230)
17β-Hydroxysteroid dehydrogenase, type 4	17β-HSD4	Gr,TI	rat, hor	(76,231)
IGF-binding protein-1	IGFBP-1	TE, St	mon	(59)
IGF-binding protein-2	IGFBP-2	Gr (?)	mon	(59)
IGF-binding protein-4	IGFBP-4	Gr,TI	mon	(232)
IGF-binding protein-5	IGFBP-5	Gr, St	mon	(59)
IGF-binding protein-6	IGFBP-6	SE(?)	mon	(59)
Interleukin-1β	IL-1β	Gr, TI	rat	(233–236)
Interleukin-6	IL-6	Gr, TI	rat	(233–236)
Interleukin-4 receptor-α	IL-4R-α	TE	rat	(219)
JunD proto-oncogene	JunD	Gr, CL	rat	(224)
Leptin	Ob	(?)	rat, mou	(237)
Leptin receptor (short form)	Ob-Ra	(?)	rat	(238,239)
Leptin receptor (long form)	Ob-Rb	Oo, TI, En, CL	rat	(237–239)
Liver receptor homolog-1	LRH-1	Gr, CL	rat, mou, hum	(240–243)
Macrophage migration inhibitory factor	MIF	Gr(?)	hum	(249)
Matrix metalloproteinase-1	MMP-1	Gr	mon	(250)
Matrix metalloproteinase-2	MMP-2	Gr, TE, St	rat	(61,181–184)
Matrix metalloproteinase-9	MMP-9	TE, St	rat	(26,181–184)
Matrix metalloproteinase-13	MMP-13	TE, St	rat	(185)

Continued

TABLE 1. *Ovulation-specific genes–cont'd*

Gene name	Abbreviation	Tissue	Animal	References
Matrix metalloproteinase-14	MMP-14	TI	cow	(251)
Matrix metalloproteinase-19	MMP-19	Gr, TE, St	mou	(184)
Membrane type-1 matrix metalloproteinase	MT1-MMP	TE, St	rat, mou	(61,251,252)
Metallothionein-1	Mt-1	Gr, CL	rat	(45,253)
Mitogen-activated protein kinase	MAPK	Gr	rat,hum,pig	(254–256)
Myeloid cell leukemia-1	MCL-1	Gr, TE	rat	(258)
Nerve growth factor	NGF	TE, St	rat	(259,260)
Growth factor tyrosine kinase receptor	TrkA	Gr, TE, St	rat	(259,260)
Nitric oxide synthase (endothelial)	eNOS	TE, St, CL	rat, mou	(261–265)
(Inducible)	eNOS	TE, St, CL	rat, mou	(261–265)
Oxytocin receptor	OTR	Gr	cow	(223–224)
p53 tumor suppressor	p53	Gr	mou	(25,266)
Pancreatitis associated protein-III	PAP-III	En	rat	(45,267)
(cAMP-specific) phosphodiesterase	PDE4	Gr, TE, St	rat, mou	(45,269,270)
Pituitary adenylate cyclase activating polypeptide	PACAP	Gr	rat	(45,121,270)
(Tissue) plasminogen activator	tPA	Gr,TE, St	rat, mou	(62,173–178)
Plasminogen activator inhibitor-1	PAI-1	Gr	rat	(173)
Prepronociceptin	ppN	TE	rat	(210)
Progesterone receptor	PR	Gr, TI, COC	rat, mou, hor, pig	(103–108,117,119,271,272)
Prolactin receptor	PRL-R	Gr, CL	rat, mou, hum, pig	(273–275)
Receptor interacting protein-140	RIP-140	Gr, TI	mou	(276–278)
Regulator of G-protein signaling protein-2	RGS2	Gr, COC	rat	(279)
Steroidogenic acute regulatory protein-1	StAR	Gr, TI, COC	rat, mou hor, mon	(45,95,281,282)
(Manganese) superoxide dismutase	Mn-SOD	Gr, TI	rat, mou	(283–286)
Tissue inhibitor of metalloproteinase-1	TIMP-1	Gr, TE, St	rat, mou	(182,191,288–291)
Tissue inhibitor of metalloproteinase-2	TIMP-2	Gr, TE, St	rat,cow	(61,251)
Tissue inhibitor of metalloproteinase-3	TIMP-3	Gr, TE, St	rat	(182,191,287–289)
Tumor necrosis factor-α	TNF-α	Gr	rat	(292–294)
Tumor necrosis factor-stimulated gene-6	TSG-6	Gr, coc	rat	(138,140,150,295)
Vascular endothelial growth factor	VEGF	Gr, TE, St	rat, cow	(65,296,297)
Versican (proteoglycan M)	PG-M	Gr, COC	rat, mou	(46,47,116,298)
Wingless-type MMTV integration site family member-4	Wnt-4	Gr, CL	rat, mou	(225,226)

Gr, granulosa; TI, theca interna; mou, mouse; mon, monkey; TE, theca externa; St, stroma; hor, horse; CL, corpus luteum; SE, surface epithelium; hum, human; Oo, oocyte; En, endothelial cell; COC, cumulus–oocyte complex.

the established cascades of gene expression in some complex process such as an acute inflammatory reaction. In some cases, as with the ILs and TNF-α, this has been feasible. But, for most of the genes that have been discovered, it has not been possible to relate them to a common pathway. Thus, the mounting challenge in the field of ovulation research is the task of delineating meaningful relationships among the numerous genes and their protein products.

The magnitude of the mission of integrating all of the data is not merely a matter of managing the ever-expanding number of genes that have been linked to ovulation. There are certain problems inherent in the multiplicity of experimental designs that are currently being used to identify ovulation-related genes. First, it is difficult to place the individual genes into a chronological order of expression because the diverse experimental organisms have included rats, mice, horses, cows, monkeys and humans, which have as much as a threefold variation in the lengths of their ovulatory processes—ranging from 12 hours (in the rodent; i.e., rat and mouse) to 36 hours (in the human). Second, a number of the studies have been based on a very limited number of time-points during the ovulatory process (e.g., a 0-hour control group and only a 12-hour experimental group in rat experiments) that detected gene expression, but did not reveal the time of onset of expression of the given gene. While the detection of any previously unidentified gene is certainly important, a single time-point does not allow assignment of a precise time of onset of expression of the gene in any kind of meaningful chronological order in relation to other ovulation-expressed genes. Third, some of the work has been conducted in vitro on cultured granulosa cells, or on cells of the theca interna, or on other cell types in the ovary. While this sort of single-cell study can provide certain useful information, the utility of the information is limited by the fact that the data does not reveal whether the expression of a given gene extends beyond the individual cell type. Furthermore, the spatial demarcation of ovarian gene expression is difficult to decipher from some reports because of inconsistent use of terminology, e.g., the use of "ovarian interstitial tissue" versus "ovarian stroma," or merely referring to "thecal cells" rather than "theca interna cells" or "theca externa cells." Thus, the typical diversity of experimental designs in different research laboratories is a certain impediment to integrating the information from different reports on assorted genes.

One of the original objectives of this chapter was to integrate as much of the data as possible in Table 1 into one grand schematic that would cover all of the principal pathway(s) of ovarian gene expression during the ovulatory process. However, that endeavor was reduced to the schematic (Fig. 20) presented above on the progesterone and prostaglandin related cascades in part because the problems identified in the previous paragraph quickly became apparent, and in part because too many of the genes simply do not fit into any of the established *cascades* of gene expression. Before other investigators attempt to integrate the information in Table 1 into a comprehensive flow chart, they are encouraged to examine (as starting points) several preliminary attempts by Espey et al. (7,45) to categorize at least some of the genes in Table 1, and a number of efforts by Richards et al. (41–44,46,47,74,300) to schematically depict the sequential and spatial relationships among small clusters of these genes. In such efforts to assemble the growing information into more meaningful graphics, a paradoxical question that arises is whether sufficient numbers of the full compliment of genes have been discovered to complete the picture. The current data from random methods of gene discovery, such as the differential display procedure (45), tends to suggest there might be hundreds of additional ovulation-related genes that remain to be identified. This deduction is based on the awareness that the last 20 genes that have been detected by the differential display procedure have less than a 25% chance of having been discovered previously (unpublished observation). That is to say, as many as 75% of the genes that are actually involved in ovulation might still remain to be discovered. Therefore, in metaphorical terms, the question is whether enough of the pieces of the jigsaw puzzle have been placed on the assembly table to begin a serious effort to unite those pieces. The optimum time to assemble a comprehensive flowchart might not have arrived, yet. Such an endeavor would be facilitated by a more complete list of the relevant genes, by a greater understanding of the functional relationships among those genes, and by more experience with the new computer software that is just becoming available to assist in literature searches that can better establishment connections among the diverse genes that have been detected so far. In the meantime, it is the aim of this chapter to summarize pertinent information about most of the genes that have been discovered to date.

Catalog of Current Genes

To conserve space, some of the closely related genes that are listed in Table 1, are considered together in this section.

ADAMTS Enzymes

The ADAMTS family of enzymes consists of at least 20 secreted proteins that have multiple domains with disintegrin and metalloproteinase activities that are involved in inflammation, angiogenesis, development and coagulation (115,301–303). The diverse members of this family act on variable substrates, including aggrecan, brevican, versican, procollagen I and procollagen II (115,116,118,302). During the ovulatory process, a number of the ADAMTS genes are expressed in different areas of the ovary and the ovarian follicle, but the most abundant transcription appears to be in the stratum granulosum (Table 1) (105,106,113,117). It is clear that at least ADAMTS-1 is dependent on the ovulatory increase in LH, ovarian progesterone and PR (105,106,113). Moreover, mice null for ADAMTS-1 exhibit impaired ovulation (118).

Adrenodoxin

ADX is a soluble, cAMP-regulated ferredoxin that transports electrons from NADPH-dependent adrenodoxin reductase to P450scc as well as to several other mitochondrial cytochromes that are involved in the steroid biosynthetic pathway (99,304–306). The expression of this electron carrier is upregulated mainly in the granulosa layer of rat follicles that have been stimulated with hCG (45). Ovarian ADX follows a temporal and spatial pattern of expression that is parallel to the expression of StAR (another steroidogenic regulatory molecule) during the hours of the ovulatory process; however, after a follicle ruptures, ADX transcription down-regulates back to its 0-hour control level while StAR continues to increase during the luteinization process (45).

5-Aminolevulinate Synthase

ALAS catalyzes the production of 5-aminolevulinate, which also functions in the process of transporting electrons to cytochrome P450 enzymes that are involved in steroidogenesis (307,308). Transcription of the ALAS gene in granulosa cells is one of the earliest responses to hCG stimulation of rat ovaries (45). ALAS expression has been associated with the *acute phase response*, which is a cascade of gene expression that is initiated in parallel with an acute inflammatory reaction and might function as a delay mechanism to moderate the degradative events of inflammation (309).

Amphiregulin

AR is a member of the EGF family of growth factors that are initially expressed as transmembrane precursor molecules that are cleaved by proteases to form active growth factors in the extracellular domain (310). This particular EGF family member has been associated with the inflammatory-like reaction that occurs in the endometrial stroma at the time of embryonic attachment (311,312). Thus, it is interesting to note that AR is transiently expressed in granulosa cells of the mouse ovary during follicular remodeling in response to an ovulatory dose of LH (72). Also, of note, AR null mice are fertile (194), perhaps because of redundant functions with epiregulin and betacellulin.

Apolipoprotein-E

ApoE is a 34 kDa segment of various lipoproteins that serves as the ligand for a cellular lipoprotein receptor engaged in the regulation of cholesterol transport and steroid metabolism (313–316). The promotion of steroidogenesis by apoE appears to diminish the intensity and duration of the acute inflammatory response (314,316). However, the significance of apoE in ovulation is not clear, because one report claims that apoE mRNA increases in granulosa cells of rats during ovulation (195), while another study concludes that apoE mRNA is not detectable in the granulosa during ovulation in the same animal (313).

Aryl Hydrocarbon Receptor

AHR has been described as a relatively *promiscuous* receptor because of a structural design that allows it to be bound and activated by many divergent chemicals ranging from numerous naturally occurring endogenous ligands to a variety of manmade environmental toxins such as xenobiotics (317–319). Once AHR is activated by a ligand, it can operate in conjunction with NF-kB as a transcription factor with the capacity to induce an array of genes associated with inflammation and/or an oxidative stress response that can either intensify or diminish the local disturbance (320,321). AHR is expressed in granulosa cells and the oocyte of growing follicles in the mouse (196). In regards to ovulation, there is a significant increase in AHR mRNA in primate granulosa cells by 12 hours after the administration of hCG (197) and AHR might also have a role in rodent ovulation, although this possibility has not been explored extensively (198).

Betacellulin

BTC was initially analyzed in association with its unusually high expression by the beta cells of the pancreas (322). Similar to other members of the EGF family of growth factors, it is proteolytically processed from a larger membrane-anchored precursor and functions as a potent mitogen when it binds to ErbB receptors that are present on a wide variety of cell types. Like AR and EPI, BTC is expressed in ovarian follicles that have been stimulated to ovulate, and it may play a key role in cumulus expansion (72).

cAMP-Response Element Binding Protein

CREB is one of three ubiquitous genes (including CREM and ATF-1) that yield binding proteins that bond with cAMP response elements (CRE) in the promoters of various genes to stimulate transcription (323,324). As a component of the protein kinase A pathway, and in conjunction with glucocorticoid-induced kinase (Sgk), it appears that a transient increase in a phospho-CREB can be induced in granulosa cells under in vitro conditions that mimic the ovulatory transition to lutein tissue (199). It is possible that the ovarian expression of CREB is involved in the regulation of Egr-1 expression in granulosa cells during the ovulatory process (325). This of course is just one of many sites at which CREB is likely to act.

cAMP-Response Element Modulator

As mentioned in the previous paragraph, CREM is a gene that is commonly expressed concurrently with CREB in response to cAMP elevation in a cell (326). And, like CREB, CREM is activated by PKA and binds to CRE in the promoter region of target genes. However, in contrast to CREB, CREM is an autoregulatory gene that encodes the inducible cAMP early repressor (ICER), and this product of the CREM gene governs the down-regulation of its own expression, along with down-regulation of the expression of CREB and other early response genes (323,327). Thus, the gonadotropin-induced expression of ovarian CREM that encodes ICER in the granulosa cells of ovulatory follicles of rats probably represents a negative feedback mechanism to modulate the intense metabolic changes that cause the follicle to rupture (200). The repressor action of ICER presumably serves to reduce CREB to basal levels and terminate the signaling cascade initiated by the LH/hCG-receptor-mediated elevation in local cAMP. However, despite the expression of CREM and ICER in the ovary, female CREM knockout mice are fertile whereas males are infertile (201).

Carbonyl Reductase

CBR is an aldoketo reductase with broad specificity for converting carbonyl compounds into alcohols (328). In the immature rat ovary, CBR mRNA is significantly elevated in ovarian thecal and stromal tissue (but not in the granulosa cells) beginning 4 hours after initiating the ovulatory process with hCG, and this increase persists through the time that follicles begin to rupture at 12 hours after hCG (202). Another aldoketo reductase, mouse vas deferens protein (MVDP) is also induced in granulosa cells of preovulatory follicles by LH/hCG (203). It has been suggested that ovarian CBR and MVDP could function to reduce the local toxicity that might develop from the aldo-keto functional groups on the progesterone and prostaglandin molecules that are copiously generated in the ovary at the time of ovulation (202–203).

Cathepsin L

CatL is important for degradation of elastic tissue in the extracellular matrix during an inflammatory response (329,330). It is also involved in alteration of the extracellular matrix at the invasive margins of tumors, and its expression in tumors is in conjunction with a number of protease-related genes that have been associated with ovulation, including MMP-2, MMP-9, MT1-MMP, PAI-1, and several members of the ADAMTS family (331). As mentioned earlier, ovarian expression of catL during ovulation is dependent in part on progesterone and PR as well as Sp1/Sp3 and CREB (44,105,120). However, unlike ADAMTS-1 null mice, catL null mice are fertile (204).

CCAAT/Enhancer-Binding Protein-β

C/EBP-β is a bZIP transcription factor that seems to be involved in inflammatory responses to injury, as well as to target tissue responses to certain hormonal stimulation (332,333). Interestingly, many of the AP-1 genes that require the inflammatory cytokines IL-1 and IL-6 for their induction have adjacent C/EBP and NF-κB motifs in their promoter regions, suggesting cooperation between these two families of transcription factors (332). C/EBP-β is rapidly induced in the granulosa cells of rats after they have been injected with an ovulatory dose of hCG (147,205,206), but there are conflicting reports as to whether expression of this transcription factor does (147), or does not (206), have an important role in regulating the induction of pro-inflammatory COX-2 expression in granulosa cells during the ovulatory process. Importantly, mice null for C/EBPβ are infertile and exhibit corpora lutea with entrapped oocytes (73).

CD63 (Cell Surface Antigen)

CD63 is a member of the tetraspanin superfamily of activation-linked cell surface antigens that is known for its abnormally high levels on the surface of activated basophils (334), on proliferating mast cells (335,336), and on the surface of endothelial cells in inflamed tissue (336). This antigen has also been identified as one of the two principal proteins in the membranes of so-called Weibel-Palade bodies, which are lysosome-related secretory organelles associated with inflamed endothelial cells (337,338). Before the induction of ovulation in immature rats, CD63 is expressed constitutively in follicular thecal cells and stromal interstitial cells, whereas expression in the granulosa layer must be induced by an ovulatory dose of hCG (45). This unusual spatial pattern of expression of an ovulation-related gene might be associated in some way to the fact that the stratum granulosum is avascular prior to ovulation and then experiences endothelial cell infiltration during luteal angiogenesis.

Corticotropin-Releasing Hormone Receptor

CRHR was first characterized as the target for CRH, which serves as the principal regulator of the hypothalamic–pituitary–adrenal stress response axis. However, CRHR is now recognized as a member of a growing family of CRH ligands, CRH receptors, and CRH-binding proteins that are expressed in peripheral inflammatory sites, especially in female reproductive tissues (207,339–343). It has been more than a decade since CRHR1 was detected in ovarian stromal cells and in the thecal cells surrounding ovulatory follicles in the rat (207,208). The potential role of this proinflammatory factor in ovulation is worthy of further investigation, but future efforts will likely require conditional knockout approaches since mice null for CRHR1 die after birth due to lung dysplasia and adrenal abnormalities (209).

Cutaneous Fatty Acid Binding Protein

C-FABP was named because of its induction in epidermal tissue following skin irritation and cutaneous inflammation, and because of its roles in the synthesis and transport of fatty acids (344,345). C-FABP mRNA is expressed in the thecal layer (i.e., in the "skin") of follicles in the ovaries of gonadotropin-primed rats that have been injected with hCG (210), and the significance of expression of this binding protein during the ovulatory process also deserves further evaluation.

Cyclooxygenase-2

COX-2 is the readily inducible, rate-limiting enzyme for the conversion of arachidonic acid to prostanoids in most cells and tissues that have been activated by inflammatory cytokines, trophic hormones, or tumor promoters (346–349). Over the past three decades, ovarian COX-2, along with PGE_2 and $PGF_{2\alpha}$, has received enormous attention in a wide variety of ovulation studies (10,41–43,47,67,68,70,131,136–140,147). In many different species of mammals, it has been demonstrated consistently that an ovulatory dose of LH/hCG stimulates a very marked but very transient increase in COX-2 mRNA expression in granulosa cells (45), whereas the synthesis of ovarian PGE_2 and $PGF_{2\alpha}$ persists through the time of follicular rupture (68,131). The specific inflammation-mediating action(s) of the ovarian prostanoid products of this enzyme are beginning to emerge. Based on recent evidence it is now clear that COX-2 and EP2 are essential for expression of the hyaluronan binding protein TSG-6 that is known to be essential for expansion of the COC and ovulation (138,140,150). Ovarian prostaglandins may also be involved in the ovarian hyperemia, although this has not been definitively shown (48). Mice null for COX-2 exhibit other abnormal reproductive functions (70,211).

Cytochrome P450 Side Chain Cleavage

Mitochondrial P450scc enzyme carries out the first step of steroidogenesis by converting cholesterol to pregnenolone, which is then metabolized to progesterone by the action of 3β-HSD (93,95,350). During the ovulatory process there is an enormous increase in ovarian progesterone synthesis that is associated with expression of P450scc in the granulosa cells in most mammals that have been studied including rat, horse, and monkey (84,91,101,212,213,351). Importantly, there is also an increase in P450scc expression in cumulus cells of ovulating COCs (84,85), thus providing a ligand for PR expressed in these same cells and appears in the pig to regulate ADAMTS-1 (49,119).

Cytokine-Induced Neutrophil Chemoattractant

CINC is a member of the CXC chemokine family that is induced by the inflammatory cytokines IL-1β, IL-6, and TNF-α to attract neutrophils into sites of inflammation (352–354). Although the data on ovarian expression of CINC during ovulation is not very clear, based on preliminary studies with rat and human tissues it appears that CINC is expressed primarily in granulosa cells, and possibly in thecal tissue as early as 6 hours after the ovulatory process has been induced by hCG (216–218,355). As a side note, it is somewhat interesting that a Japanese herbal medicine *Unkei-to*, which promotes ovulation, can stimulate ovarian steroidogenesis and other ovulation-like events by inducing the secretion of CINC with IL-1β and TNF-α in vitro (356). The role of the immune cells in ovulation is just beginning to unravel and no doubt other factors will soon be identified that impact ovulation. For example, T cells have been found in cumulus cells and produce IL-4 (219).

Early Growth Response Protein-1

Egr-1 is an inducible zinc-finger transcription factor that binds to specific GC-rich enhancer elements on an estimated 80–100 other genes that comprise the Egr-1-induced cascade that promotes acute inflammation, vascular hyper-permeability, and angiogenesis (220,357–360). Two of the principal targets for transcription are interleukin-1β (IL-1β) (359) and tumor necrosis factor-α (TNF-α) (360), which are central components of the inflammatory cascade (291,361). Of all the ovulation-specific genes that have been discovered to date, mRNAs for Egr-1 and ALAS are the two earliest transcripts to be upregulated in the ovary following onset of the ovulatory process (45,360). Like ALAS, Egr-1 is expressed in significantly greater amounts in the granulosa layer within 30 min after the ovulatory process has been initiated by hCG (221). Egr-1 null mice are infertile most likely due to the key role of this transcription factor in the pituitary gonadotropes where it is essential for transcription of the LH beta subunit gene (220–223). Altered ovarian function may be a consequence of chronically low levels of LH that would alter the growth of preouvlatory follicles by reducing/preventing theca cell differentiation and the production of thecal cell androgen precursors for aromatase. Alternatively, loss of Egr-1 may also alter the regulation of genes critical in the ovulation process.

Epiregulin

EPI is another member of the EGF family of growth factors that are characterized by a six-cysteine consensus motif that forms three intramolecular disulfide bonds crucial for binding of the growth factor to an ErbB receptor (322,362). While it has cytokinetic-promoting properties like the other growth factors, it is interesting to note that EPI has been associated with other genes such as Egr-1 (363) and NF-κB (364) that are common components of acute inflammatory reactions. An increase in EPI gene expression in ovulatory follicles has been reported in both the rat (45) and the mouse (72), and there is speculation that there might be some relationship between EPI and COX-2 during COC expansion and ovulation (45,72).

FOS-Like Antigen-2

Fra-2 is related to the family of FOS genes that encode proteins that dimerize with proteins of the JUN family to form a group of transcription factors known as activator proteins (e.g., AP-1) (365,366). Such FOS-JUN dimers function as local regulators of gene expression for cell proliferation, differentiation, and transformation (224). AP-1 dimers that include Fra-2 as one of the proteins are abundant in transformed fibroblasts of chickens and rodents (365). However, in ovaries that have been stimulated by LH, Fra-2 and JunD are rapidly induced in granulosa cells that are differentiating into luteal cells, rather than in the thecal fibroblasts in the outer layers of the follicle (224).

Frizzled G-Protein–Coupled Receptors

Fzs are true members of the G-protein–coupled receptor family that have 7-transmembrane segments and associate with heterotrimeric G-proteins to propagate intracellular signaling (367,368). Upon stimulation by the Wnt family of ligands, Fz receptors signal well-known effector responses that include the canonical pathway in which beta-catenin is activated as a transcription factor as well as a pathway in which intracellular Ca^{++} is mobilized. In the ovaries of PMSG/hCG-treated mice, Fz-1 mRNA increases first in the cells of the theca interna and then in the granulosa cells of ovulating follicles (225). However, during transformation of a ruptured follicle into a corpus luteum, Fz-1 declines and is replaced by Fz-4 (225) (Fig. 23). The downstream signaling targets of Fz-1 and Fz-4 remain to be clearly defined but it is most likely that Fz-1 activates the beta-catenin pathway whereas Fz-4 may activate other pathways. Interestingly, female Fz-4 null mice are infertile but the underlying causes are not yet entirely clear (Richards, unpublished observations).

Frizzled-Related Protein-4 (Secreted)

sFRPs are a group of antagonists (or in some instances agonists) of the Wnt signaling pathways that act extracellularly by binding both the Wnt ligands and the Fz membrane receptors of those ligands (369–371). This binding action might promote apoptosis by modulating the cell survival signals that are usually transduced by Wnt-stimulated Fzs. The expression of ovarian sFRP-4 mRNA and protein during ovulation reportedly increases in the granulosa cells of mouse follicles (226), but it increases in the thecal layers of ovulatory follicles of the rat (227). The ovarian increase in this Wnt signaling regulator is presumably important for modulating Wnt-frizzled signaling that occurs during and after ovulation, especially in the process of luteinization (226,227) (Fig. 23).

γ-Glutamylcysteine Synthetase

γ-GCS is a zinc metalloprotein enzyme that synthesizes glutathione from its three constituent amino acids glycine, glutamate, and arginine (372–374). A principal redox function of glutathione is to remove the excessive amounts of toxic peroxides and reduce the oxidative stress that can be harmful to cells during times of tissue inflammation or other forms of metabolic stress (375–377). During the inflammatory conditions of the ovulatory process, γ-GCS mRNA is expressed in large follicles from 2–8 hours after treating the animals with hCG (45). It is expressed in an irregular pattern, mostly in the theca externa layer, but also in the stratum granulosum.

Glutathione S-Transferase

GST is a member of a family of enzymes that function to detoxify hydrophobic electrophiles, i.e., compounds that contribute to oxidative stress because

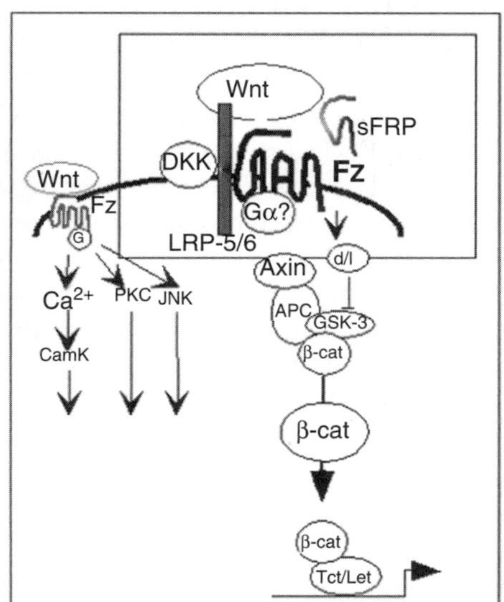

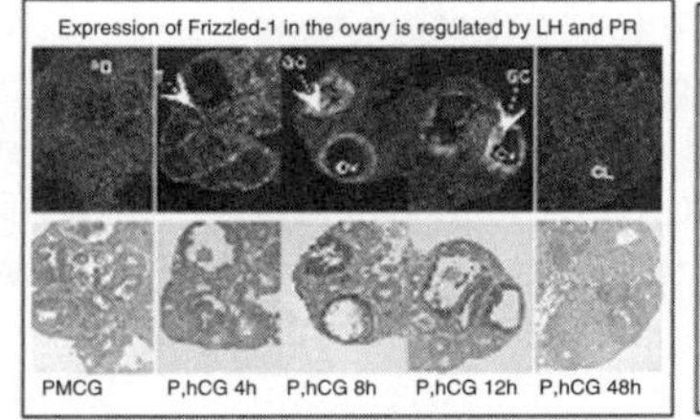

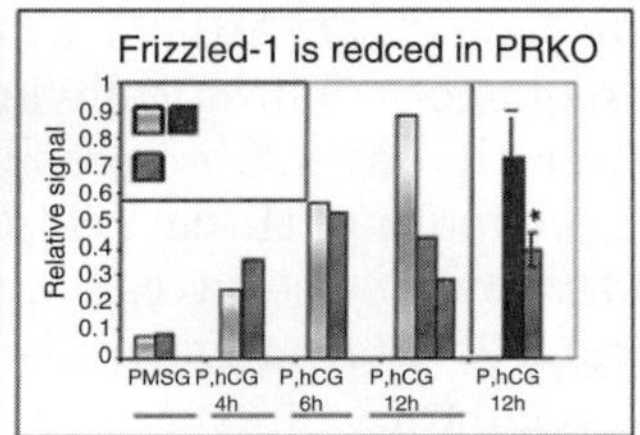

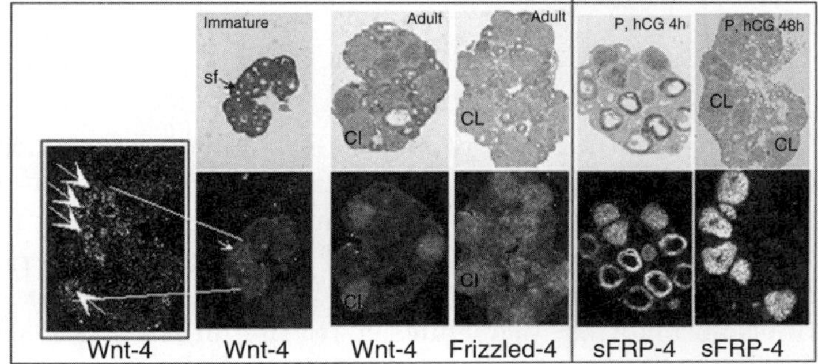

FIG. 23. The Wnt/Frizzled pathway of signaling molecules is hormonally regulated in the ovary and may impact ovulation and luteinization. Fz-1 is induced by LH in theca cells and granulosa cells of ovulating follicles, and its induction is regulated, in part, by PR. A potential downstream mediator of Fz-1 activation is beta-catenin. Wnt-4, Fz-4, and sFRP-4 all appear to be upregulated in luteal cells. Mice null for Fz-4 are infertile and appear to have a defect in their luteinization process. [From (226,227), with permission. See text for details.]

they contain electron deficient atoms such as are found in the unsaturated carbonyl groups in some steroids and eicosanoids (378–381). Specifically, GST functions by catalyzing the conjugation of glutathione (generated by γ-GCS action) to any of a wide variety of endogenous electrophilic compounds (382). In view of the functional relationships among GST, glutathione and γ-GCS, one would predict that the pattern of expression of GST would parallel that of γ-GCS. However, as indicated in the previous paragraph, γ-GCS is expressed transiently during the middle of the ovulatory process, whereas GST begins to be expressed only after γ-GCS mRNA levels have returned to 0-hour control values, i.e., after the follicles have ruptured (45). Nevertheless, GST is included in this catalog of ovulation-specific genes because it is induced as a consequence of the original ovulatory surge in gonadotropin.

G-Protein–Coupled Receptor-54

The GPR54 gene encodes a protein that is a galanin-like G-protein–coupled receptor that binds metastin, a 54-amino-acid peptide encoded by the metastasis suppressor gene KISS1 (382,383). Activation of the GPR54 receptor stimulates signaling events that result in Ca^{++} mobilization, phosphatidyl inositol diphosphate hydrolysis, arachidonic acid release, and ERK and p38 MAPK phosphorylation (383). GPR54 mRNA has been detected in placental, hypophyseal, pancreatic, and spinal tissue (383), and it is also highly expressed in the theca externa and stromal tissue of the rat ovary at 4–12 hours after initiation of the ovulatory process with hCG (45). It might be noted that the gene has been proposed as a regulator of puberty because GPR54-deficient mice exhibit hypogonadotropic hypogonadism, with delayed vaginal opening and an absence of folliculogenesis—abnormalities that can be overcome by exogenous gonadotropins (384,385). In view of its expression as an ovulation-related gene, it would be interesting to know whether gonadotropin-treated GPR54-deficient animals exhibit a normal ovulation rate.

3α-Hydroxysteroid Dehydrogenase

3α-HSD belongs to the aldoketo reductase superfamily that has, among its functions, the responsibility of inactivating metabolically disturbing aldehyde and ketone functional groups that are components of most bioactive steroids and eicosanoids (386,387). 3α-HSD mRNA increases in the granulosa layer of the gonadotropin-primed immature rat ovary as early as 2 hours after hCG administration, and it remains elevated for the duration of the ovulatory process (228). Since non-steroidal antiinflammatory drugs like indomethacin have been shown to be potent inhibitors of mammalian 3α-HSD (386–389), it would be useful to know whether part of the anti-ovulatory action of indomethacin is due to the inhibition of ovulation-specific 3α-HSD activity.

3β-Hydroxysteroid Dehydrogenase

It is well established that 3β-HSD catalyzes the formation of progesterone from pregnenolone. Thus, it is one of the two principal *gateway* enzymes that regulate the shuttle of cholesterol into the wide array of bioactive mineralocorticoids, glucocorticoids, and sex steroids. 3β-HSD mRNA reportedly increases in the granulosa cells of the rat (91), monkey (215), and equine ovary (212) during ovulation. It would be useful to have a more comprehensive description of the temporal and spatial patterns of expression of this important gene in one of the common rodent models for ovulation studies.

11β-Hydroxysteroid Dehydrogenase

11β-HSD is a reductase that is thought to suppress inflammatory reactions as a result of its ability to elevate local levels of the antiinflammatory glucocorticoid cortisol by regenerating this bioactive steroid from cortisone (229,390,391). Since the ovulatory process has been likened to an acute inflammatory reaction, it is interesting that elevated 11β-HSD mRNA has been detected in rat granulosa cells (230) and in human surface epithelial cells (229) that have been stimulated with either LH or with the inflammation-promoting ILs. In any case, it is noteworthy that 11β-HSD, 3α-HSD, and CBR proteins are all antiinflammatory reductases that are up-regulated in different parts of the ovary at the time of ovulation (45,229).

17β-Hydroxysteroid Dehydrogenase, Type 4

17β-HSDs exist as a number of different isoforms in a wide variety of tissues (392). Each isoform has selective substrate affinity that can lead to either reductive catalysis of some steroids, or oxidative catalysis (requiring NAD^+ as a cofactor) of other steroid substrates (392,393). Ovarian 17β-HSD4 is an oxidase that is induced by LH/hCG action on cultured theca interna cells and especially on granulosa cells (231). The expression of this particular HSD in the equine ovary appears to function in reducing the circulating levels of 17β-estradiol during follicular luteinization.

IGF-Binding Proteins

IGFBPs function as modulators of IGF-mediated cellular growth, survival and differentiation by binding to IGF-I and -II and thereby antagonizing the coupling of these growth factors to their respective receptors (348,394,395). It has been suggested that elevated levels of ovarian IGFBP-4 might inhibit ovulation by interfering with the actions of IGF-I and -II (232). Although the precise role(s) of these binding agents in ovulation remain to be determined, they are readily detected in virtually all regions of the ovary, with IGFBP-4 (232) and IGFBP-5 (59) being especially abundant in ovulatory follicles of the primate.

Interleukins

IL-1β and IL-6 are primarily proinflammatory cytokines, with their most salient and relevant properties being their ability to initiate expression of COX-2, phospholipase A_2, eicosanoids, iNOS, tPA, MMPs, NF-κB-inducing kinase, receptor activator of NF-κB (RANKL), and a myriad of other mediators of inflammatory reactions (37,396–400). IL-1β and IL-6 gene expression in the ovary has been studied extensively, and transcripts of these genes increase in the granulosa and surrounding theca interna within 4 hours after stimulation of the ovulatory process in rats (36,157,233–236,397). IL-4 has recently been shown to be expressed by T cells localized in cumulus cells and thus may play a role in COC expansion (219).

Interleukin-4 Receptor-α

IL-4Rα has been detected in the plasma membrane of a wide range of cells, and it is especially common in fibroblasts (401). IL-4Rα is the receptor for IL-4 which is distinctly different from the other IL cytokines because when it couples with IL-4Rα it initiates a signaling process that suppresses inflammation (402–404). IL-4Rα mRNA is expressed in the thecal fibroblasts (or T cells) of follicles in the ovaries of gonadotropin-primed rats that have been injected with hCG (210). Although its specific function in ovulation has not been established, IL-4Rα might be involved in terminating the ovarian inflammatory response and setting the stage for T-cell–mediated wound healing after rupture of the ovarian surface.

JunD Proto-Oncogene

As mentioned above, JunD dimerizes with proteins of the FOS family (e.g., Fra-2) to form the group of transcription factors known as activator proteins (e.g., AP-1) (365,366). Such FOS-JUN dimers function as local regulators of gene expression for cell proliferation, differentiation, and transformation (224). Like Fra-2, ovarian JunD is induced by LH and remains elevated through ovulation and luteinization (224). It is not clear whether this concurrent expression of ovarian Fra-2 and JunD might be induced by transforming growth factor-β—as occurs in epithelial and carcinoma cells (405).

Leptin and Leptin Receptors

Ob (i.e., leptin), a common product of the so-called *obesity* gene in adipocytes, is now considered to be a marker for the inflammatory response, and it is regulated by inflammatory cytokines such as IL-1 and TNF-α (406–410). Ob mRNA increases markedly in whole ovaries of immature rats within 2 hours after the animals have been injected with an ovulatory dose of hCG (237). Several hours later, when Ob mRNA expression is declining, there are substantial increases in mRNA for both the short form (Ob-Ra) and the long form (Ob-Rb) of the leptin receptor. The OB-Rb protein is especially high in oocytes, endothelial cells and thecal cells at the time of ovulation (238). These observations suggest that Ob could have a significant role in the ovarian inflammatory reaction during the ovulatory process. Of note, Ob null mice are subfertile (239).

Liver Receptor Homolog-1

LRH-1 is a member of a nuclear receptor superfamily that participates in the regulation of steroid metabolism. It reportedly might function as the principal transcriptional regulator of ovarian P450arom (240), or of ovarian StAR gene expression (241). In any case, LRH-1 mRNA and protein are significantly upregulated in rat and human granulosa cells after gonadotropic stimulation (240–242). Although the specific ovarian role of LRH-1 versus SF-1 remains to be clearly established, a conditional knockout of SF-1 in granulosa cells renders females infertile (243). These results indicate that LRH-1 is not redundant to SF-1 in the female gonad, an observation that lends support to a report that LRH-1 protein is much lower in granulosa cells than is SF-1 (242).

α2-Macroglobulin

α2-M is a nonspecific proteinase inhibitor that is abundant in the plasma and body fluids of vertebrates (411,412). It functions as a broad-spectrum protease-binding protein that inactivates MMPs, growth factors, cytokines, and other mediators of inflammation and tissue remodeling (244–246,413). α2-M mRNA and protein are detectable in granulosa cells of rat follicles at the time of ovulation, and this proteinase inhibitor probably functions to control damage in the ovary during the time when an ovulatory follicle is remodeled into a corpus luteum (244–247). Mice null for α2-M appear to be fertile (248).

Macrophage Migration Inhibitory Factor

MIF was named after its initial identification as a T-cell cytokine, but then it was rediscovered as a protein that is released by pituitary cells when they are exposed to endotoxins (414). Now, MIF is further established as a powerful proinflammatory cytokine that acts in concert with glucocorticoids to control both the *set point* and the magnitude of the inflammatory response (414–417). MIF might be a component of the proinflammatory cascade of ovarian gene expression during ovulation, because hCG can induce MIF expression in cultured granulosa cells (249).

Matrix Metalloproteinases

MMPs comprise a large family of zinc-dependent proteinases that degrade numerous proteins of the extracellular matrix during the tissue remodeling that occurs in a wide variety of physiological and pathological processes including angiogenesis, tumorigenesis, inflammation, and wound healing (26,418,419). Various reports in the literature use confusing and inconsistent nomenclature for the MMPs, but a convenient index of the most common terminology is available in a recent review by Curry and Osteen (26). Since the ovulatory process has characteristics similar to an inflammatory reaction, and since luteinization requires substantial vascular proliferation into the lutein granulosa, it is not surprising that there is considerable expression of MMPs in ovulatory follicles. Ovarian MMP-2, along with MMP-9 and -19, appear to be the most elevated in granulosa cells as well as in thecal and stromal tissues in response to LH/hCG (26,182–185). The specific functions of each protease that might be critical to ovulation remain to be delineated, because mice that are null to individual MMPs are fertile (191).

Membrane Type-1 Matrix Metalloproteinase

MT-1-MMP is an integral membrane proteinase that functions to regulate degradation of the local extracellular matrix and promote cell migration and invasion (420). One of its specific functions is to activate other cell surface proteinases such as proMMP-2, the active form of which can degrade type IV collagen (421). Thus, it is relevant that MT1-MMP RNA increases in the thecal connective tissue of rat ovarian follicles beginning 4 hours after hCG, and that pro-MMP-2 increases by 12 hours after hCG, when the follicles are just beginning to rupture (251). MT1-MMP null mice have severe developmental defects (252).

Metallothionein-1

Met-1 is a small, cysteine-rich protein that is a member of the single most abundant group of intracellular zinc-binding proteins that are conserved evolutionarily in virtually all animals, eukaryotic plants, bacteria and fungi (422–424). The hallmark of Met genes is the fact they are expressed in an unusually broad assortment of cells and tissues in response to an extraordinarily wide variety of chemical and physical stimuli, including agents that are known to mediate inflammation (422–426). The metabolic significance of the Mets has not been firmly established, but the three most common hypotheses are that they control zinc availability to proteins that require zinc to function, they protect cells from oxygen-free radicals, and/or they absorb accumulating heavy metals that might otherwise cause local toxicity (422–426). Ovarian Met-1 gene expression is a latent response to LH/hCG stimulation, because mRNA transcripts are not detectable in any significant amount in the stratum granulosum until after the follicles have ruptured and the luteal tissue is flourishing (45,253). Nevertheless, even though Met-1 is expressed only after rupture, it appears to be a unique marker of progesterone-secreting corpora lutea, and clarification of the role of this zinc-binding protein in luteal tissue might help to elucidate the full significance of progesterone as both a mediator of the ovulatory process and a persistence attribute of the corpus luteum.

Mitogen Activated Protein Kinase

MAPK consists of three major pathways of kinases, namely extracellular signal-regulated kinase (ERK), c-JUN N-terminal kinase (JNK), and p38 MAP kinase, that function in regulating gene expression related to cell growth, proliferation, and differentiation (426). Activation of these MAPK pathways by cytokines IL-1 and TNF-α induces the expression of transcription factors that promote the transcription of genes in the inflammatory cascade (427). In the ovary, gonadotropins operate through cAMP/PKA signaling mechanisms to activate MAPK pathways (254–257,4278). One function of ovarian MAPKs appears to be the regulation of StAR expression and progesterone synthesis (255,428,429).

Myeloid Cell Leukemia-1

Mcl-1 is an anti-apoptotic member of the Bcl-2 family of about 20 homologues of pro- or anti-apoptotic agents that regulate programmed cell death by either disrupting or by preserving, respectively, the permeability of mitochondrial membranes (430,431). Mcl-1 expression can be upregulated through activation of the transcription factor NF-κB, a central component of inflammatory cascades (432). Once the Mcl-1 protein is translated, it can exert its anti-apoptotic effect by complexing with pro-apoptotic members of the Bcl-2 family (e.g., Bak) and rendering them inoperative (431,433). During the ovulatory process in PMSG/hCG-primed rats, Mcl-1 expression increases in granulosa and thecal tissue (257) presumably for the purpose of moderating the pro-apoptotic momentum of the acute inflammatory reaction that traumatizes the follicle and causes it to rupture.

Nerve Growth Factor

NGF, the founding member of the neurotrophin family of growth factors, was initially a focus of intense investigation because of its ability to prevent apoptosis in peripheral neurons (434–436). Now, NGF is also recognized for its associations with cytokines (437–440), its involvement in inflammation and tissue repair (441–443), its expression by and activation of fibroblasts (437,438,444), and its contributions to endothelial cell survival and angiogenesis (439,441,445). Taking into account the reports that NGF is expressed in fibroblasts and is involved in inflammatory reactions, it is not totally surprising that the site of NGF mRNA elevation in ovulatory rat follicles is in the fibroblast-containing tissues of the thecal and stromal areas of the ovary and that this expression is intensified by the inflammatory cytokine IL-1β (259).

Nerve Growth Factor Tyrosine Kinase Receptor

NGF induces cellular responses by binding two principal receptors, namely high affinity TrkA and low affinity p75 neurotrophin receptor (p75NTR), which is a receptor that resembles members of the tumor necrosis factor receptor family (434,439,446,447). During ovulation, TrkA expression is detectable not only in the thecal and stromal areas of rat ovaries, but also in the stratum granulosum (260).

Nitric Oxide Synthases

NOSs consist mainly of endothelial eNOS and inducible iNOS, which both function to oxidize L-arginine and produce nitric oxide (448), a free radical that causes vasodilatation and edema and promotes other characteristics of inflammation (449–451). Based mainly on immunohistochemical studies that have detected eNOS and iNOS in rat and mouse ovaries, eNOS expression appears to increase mainly in the theca externa, ovarian stroma, and luteal tissue, as well as in the granulosa and the oocyte (261–265), whereas iNOS is predominantly in the granulosa and the oocyte (262,264,265).

p53 Tumor Suppressor

p53 has emerged as a sequence-specific DNA-binding protein that induces the expression of target genes that functions not only to mediate tumor suppression but also to facilitate the repair and survival of damaged cells (452–455). The expression of this transcription factor has been associated with the moderation of oxidative stress arising from inflammatory conditions (453,456–458), as well as with the promotion of angiogenesis (459). p53 mRNA expression reportedly increases significantly in granulosa cells of the mouse ovary near the time of ovulation (266), and its possible function is to promote gene expression that will facilitate the survival of follicular tissue that undergoes severe cellular stress during the inflammatory events of the ovulatory process.

Pancreatitis Associated Protein-III

PAP-III, a so-called secretory stress protein, was first detected because of its over-expression in inflamed pancreatic tissue—with highest expression localized in pancreatic acini (460,461). Although the function of PAP proteins has not been firmly established, the current hypothesis is that they perform in some way as endogenous protective agents in inflamed tissues (461,462). During the inflammatory-like conditions of ovulation, the ovarian increase in PAP-III mRNA is confined spatially to the endothelial lining of blood vessels and to small secondary follicles that are distributed randomly throughout the ovarian stroma, rather than to the ovulatory follicles (45,267). In view of the hypothetical role of PAP proteins as constituents of a defense mechanism in inflamed tissues, it is possible that the intense expression of PAP-III in small ovarian follicles is for the purpose of protecting future crops of ovulatory follicles from excessive oxidative stress at the time of ovulation.

(cAMP-Specific) Phosphodiesterase

PDE4 belongs to 1 of 11 major families of phosphodiesterases that hydrolyze intracellular cAMP and cGMP to their inactive 5′ metabolites. PDE4 is specific for cAMP (463), has been related to inflammatory conditions (463–465) and appears to be directly or indirectly involved in the activation of NF-κB, a key transcription factor in inflammatory cascades (464,466,467). During the past decade, PDE4 has been linked to ovarian thecal cells (268) and to granulosa cells (269). mRNA transcripts of this phosphodiesterase increase in the ovary within 2 hours after the administration of hCG to rats, and they remain elevated for the duration of the ovulatory process (45). Therefore, since there is evidence that disruption of expression of the PDE4 gene significantly impairs ovulation (269,468), it would be interesting to know whether a PDE4 inhibitor, such as rolipram (466,467) would suppress ovulation. It is also worth noting a recent finding that disruption of PDE4 gene transcription markedly reduces the expression of ovarian COX-2, PR, and other ovulation-related genes, and this impairment of PDE4 gene expression results in the inhibition of follicular rupture without blocking ovarian progesterone synthesis and luteinization, i.e., it leads to the formation of so-called *luteinized unruptured follicles* in which luteinization occurs in the absence of release of the oocyte (72,269,468).

Pituitary Adenylate Cyclase Activating Polypeptide

PACAP is actually a neuropeptide that is classified in the VIP/secretin/glucagon family of peptides (469). However, over the past decade this polypeptide has been firmly established as a potent antiinflammatory factor that exerts its attenuating effect on inflammation in a variety of tissues by stimulating the production of antiinflammatory mediators while inhibiting proinflammatory mediators (469–471). During the inflammatory events of the ovulatory process, PACAP mRNA is expressed quite transiently as a progesterone-dependent message in the stratum granulosum from approximately 4 to 8 hours after administration of hCG to rats (45,121,122). Somewhat unexpectedly, ovarian PACAP mRNA returns to the 0-hour control level by 12 hours after hCG, when rat follicles first begin to rupture and when the ovary would presumably be experiencing the most intense inflammation. Of interest, PACAP has also been shown to induce tPA (270).

(Tissue) Plasminogen Activator

tPA, along with urokinase-PA (uPA), enzymatically converts plasminogen into the protease plasmin, which then contributes to inflammatory reactions by activating MMPs (179). Only tPA is included in this catalog of ovulation-related genes because it appears to be more significant in the mechanism of ovulation. The roles of PAs in ovulation have already been described in this chapter and will not be detailed again, here. In brief, tPA activity increases throughout most of the follicular tissue in rats (172,173,472) and monkeys (62), reaching a peak near the time of follicular rupture.

Plasminogen Activator Inhibitor-1

PAI-1, which is the main inhibitor of fibrinolytic enzymes such as tPA and uPA (472), plays an essential role in tissue remodeling by suppressing inflammation and blood clotting in irritated and damaged tissues (473–477). Information on the ovarian expression of this potentially important regulator of inflammatory reactions is not very consistent, but it appears that PAI-1 declines to minimal levels in the rat (472) and monkey (62) at the time when follicles are actually rupturing.

Prepronociceptin

The ppN gene (also known as orphanin FQ), which encodes the precursor form of an endogenous agonist that ligates to opioid receptor-like-1 (478,479), contains a promoter region that appears to bind Sp1 transcription factor (480) and has two cAMP response elements (CRE) near the start site of transcription (481). ppN expression is induced in acutely inflamed tissues and probably mediates pro-inflammatory features of inflammatory reactions (480,482–484). ppN mRNA is expressed in the thecal layer of follicles in the ovaries of gonadotropin-primed rats that have been injected with hCG (210).

Progesterone Receptor(s)

There are actually two isoforms of PR, i.e., PR-A and PR-B, that vary mainly in the length of the processed protein and appear to have differing functional activities in a wide variety of biological processes (103,105,270,271,485). Based on studies of the mouse ovary, both isoforms of PR are highly expressed in granulosa cells of preovulatory follicles in response to an ovulatory surge in gonadotropins, and they may also be present in thecal cells (107,271). The importance of the PR receptor in ovulation has been demonstrated by the anovulatory state of mice that lack a functional PR gene (105). An extensive review of the PRs has been published recently (485).

Prolactin Receptor

PRL-R is expressed in a variety of cells in and out of the reproductive system, and it might be an important link in the defense reaction known as *the acute phase response* that helps re-establish homeostatic conditions in insulted tissues that become acutely inflamed (486–488). The coupling of PRL to this receptor initiates various signaling cascades involved in regulating cell differentiation, proliferation, and survival (489). In inflamed ovarian follicles, PRL-R mRNA increases significantly in the granulosa cells as they differentiate into luteal cells (273–275). PRL receptor null mice exhibit severe defects in luteinization (275).

Receptor Interacting Protein-140

RIP-140 is reportedly a nuclear receptor corepressor that suppresses the expression of several genes, but its regulation of nuclear receptor action is currently under debate (490,491). Among its array of functions, RIP-140 reportedly controls steroidogenic factor-1-dependent transcription (491), AP-1-mediated transcription (492), and vitamin A-regulated transcription (493), as well as appearing to have an indispensable role in interacting with the TNF-R1 receptor to induce NF-κB activation during inflammatory reactions (494–496). Ovarian expression of RIP-140 is reportedly essential for ovulation, even though mice that are null for this protein can nevertheless develop follicles that undergo luteinization without ever having ruptured (276–278). This phenotype is similar to the characteristics of mice null for PR and COC-2.

Regulator of G-Protein Signaling Protein-2

RGS proteins are GTPase activating proteins (GAPs) that enhance the intrinsic rate at which the α-subunits of certain heterotrimeric G-proteins hydrolyze GTP to GDP and thereby limit the time-span that α-subunits incite the activity of downstream effectors (497–499). In simpler terms, RGS2 regulates the duration of signaling by G-protein–coupled receptors. During ovulation in the immature rat model, RGS2 mRNA begins to increase significantly in the granulosa layer within 2–4 hours after the animals have received hCG, but it is already declining by the time the follicles actually begin to rupture (279). RGS-2 null mice appear to be fertile (280).

Steroidogenic Acute Regulatory Protein-1

StAR has a critical role at the onset of steroid hormone synthesis because it is responsible for the transfer of cholesterol from the outer mitochondrial membrane to the inner membrane where this primary substrate can be converted to pregnenolone by P450scc (93,95,96,281,282,350,500). Therefore, predictably, ovarian StAR expression is upregulated during the peri-ovulatory period in a temporal and spatial pattern that correlates with the known pattern of steroid synthesis. It is mainly present in the steroid-secreting cells of the theca interna at the beginning of the ovulatory process, it increases significantly in the granulosa layer when ovulatory follicles begin producing substantial amounts of progesterone, and it dominates the steroidogenically intense lutein granulosa after ovulation (41,45,96,281). It was noted in an earlier section of this chapter that LRH-1, which is also upregulated in the ovary at the time of ovulation, can significantly induce StAR promoter activity in a dose-dependent manner (242). Mice null for StAR exhibit severe adrenal abnormalities and eventual premature ovarian failure (282).

(Manganese) Superoxide Dismutase

Mn-SOD exists as a tetramer that is initially synthesized with a leader peptide that restricts this enzyme exclusively to the mitochondrial matrix where it functions to efficiently catalyze the dismutation of superoxide anions (501,502). Expression of this superoxide scavenger can be induced by the cytokines IL-6 and TNF-α—presumably in anticipation of the oxidative stress that is generated by these two mediators of acute inflammatory reactions (502–506). During the ovulatory process in the rat, Mn-SOD reportedly is upregulated in the granulosa layer (283–285,507) and in the theca interna (284,295). SOD knockout mice die perinatally due to mitochodrial injury and several cardiac and neurological abnormalities that preclude any chance of performing analyses on the mature ovary (286).

Tissue Inhibitors of Metalloproteinases

TIMPs are a group of four related endogenous inhibitors that form high affinity complexes with the active sites of MMPs in a 1:1 stoichiometry, but their bonding to different MMPs is not particularly selective (508). The integrity of the extracellular matrix at any given time in any given tissue is determined by the local ratio between TIMPs and MMPs (508–510). During ovulatory degradation of the extracellular matrix of the follicle wall, it is mainly TIMP-1 and TIMP-3 mRNAs that increase in the thecal connective tissue, but TIMP-1 also increases somewhat in the granulosa layer (45,182,287–289). TIMP null mice have a relatively mild ovarian phenotype, suggesting possible redundant functions among some of the inhibitors (290,291).

Tumor Necrosis Factor-α

TNF-α is a homotrimer consisting of 157 amino acid subunits that collectively activate two distinct cell surface receptors, namely TNF-R1 and TNF-R2, that trigger signaling pathways leading to activation of major transcription factors such as NF-κB, AP-1, and c-Jun, all of which regulate the inflammatory cascade (395,511,512). Considering the frequency with which cytokines such as TNF-α are mentioned in relation to the inflammatory events of the ovulatory process, and considering the common use of TNF-α, IL-1β and IL-6 as experimental probes in studies on ovulation, it is surprising how difficult it is to find specific information in the literature to clarify the temporal and spatial patterns of expression of ovarian TNF-α in any of the common experimental models in ovulation studies. The literature research for this chapter located one review that contained limited information (292) and one primary research report that focused on the effect of TNF-α on expression of uPA and MMP-9 in the ovarian surface epithelium (175). TNFα null animals exhibit premature ovarian failure (293).

Tumor Necrosis Factor-Stimulated Gene-6

TSG-6 was originally cloned from diploid human fibroblasts that were stimulated with TNF-α (513). TSG-6 expression is upregulated in many cell types in response to a variety of proinflammatory stimuli, but its actual function might be antiinflammatory since it appears to be a component of a negative feedback loop that down-regulates the inflammatory response (513,514). Specifically, TSG-6 binds to hyaluronan in inflamed tissues via its link module. TSG-6 also forms a complex with the heavy chains of inter-α-trypsin inhibitor (IαI), and in this manner "delivers" IαI heavy chains to hyaluronan. IαI then forms a covalent bond with hyaluronan to help stabilize it. Although the interaction of TSG-6 and IαI results in the release of bikunin, a protease inhibitor, evidence for bikunin in the ovary has not been reported (138,140,510,515,516). During ovulation in the rat, TSG-6 mRNA is upregulated throughout the granulosa layer, and, especially in the cumulus mass around the oocyte, and the translated protein appears to play a critical role in stabilizing the complex hyaluronan matrix that surrounds the oocyte (138,140,295,517). As indicated above, mice null for either TSG-6 or IαI are infertile, a phenotype that can be reversed by restoring TSG-6 or IαI, respectively (126–128). Furthermore, TSG-6-blocking antibodies disrupt the normal pattern of COC expansion in culture (138).

Vascular Endothelial Growth Factor

VEGF was originally detected as a result of its ability to induce vascular permeability and stimulate mitosis of vascular endothelial cells. However, it is now recognized as a ubiquitous growth factor that promotes neovascularization during any morphological remodeling of tissues, including the changes that occur during the growth of tumors and the trauma of acute inflammation (518–522). In the context of this review, it is relevant that the inflammatory cytokines TNF-α and IL-1β can induce VEGF expression (236,523). During the transformation of an ovulatory follicle into a corpus luteum, VEGF transcripts are expressed primarily in ovarian granulosa cells, and to a lesser extent in thecal and stromal cells (296,297).

Versican (Proteoglycan M)

Versican is a large extracellular matrix proteoglycan that readily binds with hyaluronan to provide strength and elasticity to tissues and to influence the adhesion and migration of cells within tissues (288,524). In rat and mouse ovaries, versican mRNA increases as much as tenfold in granulosa cells and cumulus cells after hCG treatment of the animals (49,116,298). It is relevant to ovulation that versican appears to be a substrate for several members of the ADAMTS family (116). Versican null mice are embryonic lethal (116).

Wingless-Type MMTV Integration Site Family Member-4

Wnts consist of a structurally related family of secreted glycoprotein ligands that have essential roles in cell adhesion, migration, proliferation, and differentiation during morphological movement and development of tissues (367,368,525). Members of this family of signaling proteins bind to the frizzled family of serpentine receptors (e.g., Fz-1 and Fz-4) that initiate a number of intracellular signaling pathways. Wnt-4 is best known for its critical role in specifying embryonic development of the ovary (299). Wnt-4 null mice exhibit sex reversal of the gonad at birth; i.e., genotypic XX null mice have gonads that are like a testis, only lacking germ cells. Wnt-4 mRNA is also detectable in ovulatory follicles and increases significantly during the morphological transformation of a ruptured follicle into luteal tissue (226). The role of Wnt-4 in luteal tissue has not yet been clearly defined.

Additional Comments about Ovarian Gene Expression

The above list of ovulation-related genes represents a wide range of agents that have been associated with a number of different gene expression cascades. Most of the genes are components of pathways that have already been identified as parts of the ovulatory process. These include genes that are required for steroidogenesis and for inflammation, as well as genes that are known to minimize the side effects of acute inflammatory reactions, such as oxidative stress (7). On the other hand, some of the genes such as ALAS do not seem to fit into any of the established cascades of gene expression, while other factors such as NF-κB that appear to be predictable components of the ovulatory process have not yet been studied. Therefore, this closing section provides several additional comments that might be useful during the two principal challenges of the next decade of ovulation research—namely the discovery of other essential genes in the ovulatory process and the integration of as much of the information as possible into a comprehensive cascade of ovulation-specific gene expression and action.

Potential Role of Nuclear Factor-κB

Based on the literature search for this review, many of the genes listed in Table 1 are related in some way to NF-κB. Therefore, this ubiquitous transcription factor merits special attention. NF-κB is responsible for the expression of a wide variety of genes that control the inflammatory response (512,526,527). This transcription factor normally exists in essentially all cells as a dimer that is restrained in an inactive state by bondage to an inhibitory subunit known as I-κB (528–530). The bioactive form of NF-κB can be rapidly released from I-κB when cell signaling from any of a number of proinflammatory stimuli produce I-κB kinase (IKK), which phosphorylates the inhibitory subunit I-κB and releases NF-κB (527,529–532). Thus, an elevation in active NF-κB is not brought about by de novo transcription and translation, but by release of the preexisting NF-κB dimer from its bound state with I-κB. Although the term *NF-κB* is generally used in reference to any of the possible combinations of the dimer subunits, the p50-p65 dimer is by far the most common in most cell types (529).

Inflammatory cytokines such as IL-1β and TNF-α, along with a number of other inflammatory agents such as NOS, activate MAPK and lead to the activation of NF-κB, AP-1, and other transcription factors (511,512,526,527,533,534). It is now recognized that "NF-κB is at the heart of the acute inflammatory response" (528). Once the NF-κB dimer is released from I-κB, it operates through a positive feedback mechanism to promote further expression of proinflammatory cytokines, including IL-1β, IL-6, and TNF-α (529,531,535). Furthermore, the NF-κB pathway leads to production of COX-2, MMPs, cell adhesion molecules, acute phase proteins, and other inflammation-related agents (527,529,535–537). Useful diagrams of the signaling pathways related to NF-κB are available in the literature (529,531,535).

It is also well known that antiinflammatory steroids act by inhibiting the transactivation of NF-κB-dependent genes. Therefore, it is relevant to note that 13 years ago, it was suggested that the ovulatory increase in ovarian progesterone might be a part of an antiinflammatory response (63). Now, progesterone has been firmly established as an inhibitor of NF-κB, and the action of this sex steroid appears to be either by stimulating synthesis of I-κB, or by competing with NF-κB for recognition sites on the target genes for this transcription factor (536). It is also worthy of mention that ovulation-related PACAP (along with vasoactive intestinal peptide) might exert its antiinflammatory action by interfering with NF-κB action (538,539).

In summary, it appears that NF-κB has a central role in the cascade of ovarian gene expression that is induced by an ovulatory surge in gonadotropin, and this transcription factor (along with AP-1 and other inflammation-related transcription factors) merit more attention in future studies on ovulation. The potential role of this agent in ovulation was suggested *recently* (7), and one interesting report has appeared in the literature (540). It was shown that the gene for serum amyloid A3 (SAA3), a principal protein in the acute phase response to inflammatory stimuli, can be induced in mouse granulosa cells by TNFα. It was further demonstrated that two SAA3 promoters that are responsive to TNFα are also responsive to p65, i.e., to one of the common subunits of the NF-κB dimer. Thus, based on the literature search for the present review, this is the first evidence (albeit indirect) that NF-κB might be activated in ovarian granulosa cells during ovulation. In conclusion, it would appear to be worthwhile in the future to use techniques based on immunohistochemistry with NF-κB-specific antibodies, or based on incubation of tissue sections with labeled probes containing κB-specific binding sites (539), to further analyze the temporal and spatial patterns of ovarian NF-κB activation during the ovulatory process.

Thoughts about Organizing Ovarian Gene Expression into Pathways

As mentioned earlier, it will be a challenging task to integrate all of the ovulation-related genes into meaningful pathways. Nevertheless, a significant number of genes have already been discovered, and several preliminary attempts to commensurate the relationships among those genes have already begun. In any such efforts, it is necessary to initially organize the different genes into practical groupings that can facilitate the construction of pathways of gene interaction. Therefore, four possible ways to group the genes are identified, below.

Categorizing by Temporal Pattern of Expression. Ovulation is a process that begins at the moment a surge in gonadotropin(s) couples with the LH/hCG receptors on the plasma membranes of cells in the granulosa layer and the theca interna. This process has an equally distinct ending when the stimulated follicles rupture and release ova into the oviduct. Based on the current information, some genes such as ALAS and Egr-1 begin to be expressed as early as 30 minutes after the start of the ovulatory process, while other genes such as ADAMTS-1 and TIMP-1 begin to be expressed several hours later. Yet, still other genes such as Mt-1 and γ-GCS, do not begin to be expressed until after the follicles rupture (45) and probably represent genes that are only indirectly related to the gonadotropin surge that initiates the ovulatory process. Therefore, any effort to establish pathways of ovulation-related gene expression would be less problematical if all of the genes listed in Table 1 had been studied in the same experimental animal and if all of the time points of ovarian extraction were at a number of consistent intervals. Obviously, most of the work to date has been carried out using the gonadotropin-primed immature rat model, and data from such studies will likely be the most useful in establishing pathways of gene expression. However, even with this common model, there is only limited value to studies that measure gene expression at one control time and only one experimental time somewhere within the 12-hour ovulatory process of the rat. The most useful data comes from studies that have examined ovarian gene expression at either 0, 3, 6, 9, 12, and 24 hours after the animals received hCG to initiate ovulation, or at 0, 2, 4, 8, 12, and 24 hours after hCG. The 0/3/6/9/12/24-hour protocol has the advantage of even distribution throughout the principal 12 hours of the ovulatory process, but it has the disadvantage of possibly missing the earliest gene expression (e.g., it would have missed ovarian ALAS expression that peaks at 1 hour after hCG and is declining by 2 hours) (45). The 0/2/4/8/12/24-hour protocol allows a little more scrutiny of the first four hours of the ovulatory process, i.e., during the interval when progesterone and the prostaglandins are first beginning to increase in the ovary.

Categorizing by Spatial Distribution of Expression. In situ hybridization and immunocytochemistry have provided useful information about the intraovarian location of each gene that is expressed during the ovulatory process. At least 72% of the 88 genes listed in Table 1 are expressed in the granulosa layer of the follicle, whereas approximately 20% have also been reported in cumulus cells. Obviously, these are the sites of the greatest response to the ovulatory surge in gonadotropin. The magnitude of the response at this site in the follicle is to be expected since it is the location of most of the LH/hCG receptors. Granulosa cells are also the location of the two gene transcripts (namely, ALAS and Egr-1) that are first detectable during the ovulatory process. Thus, until genes that are expressed earlier are identified it would seem acceptable to predict that the ovulatory process begins in the granulosa layer of the follicle. One can also predict that the ovulatory process continues in the COC where specific functions related to production of the ECM are critical. Therefore, any graphical illustration of the ovulatory cascade of gene expression must begin with the granulosa cells and continue to cumulus cells, and only those genes that are detected in these areas of the follicle should be considered for integration into granulosa/cumulus-specific pathways. In some cases, it is difficult to decipher from the published data whether a given gene is expressed only in the granulosa, or also in the adjacent, thin layer of cells belonging to the theca interna. Also, for genes that were detected in vitro (usually in cultured granulosa cells), it would be useful to know whether such expression occurs exclusively in the granulosa cells, or also in cumulus cells, in theca interna cells, and/or in other ovarian cells. In any case, it is reasonable to conclude that the inflammatory-like events of the ovulatory process begin primarily in the granulosa layer.

The response of the fibroblasts of the theca externa and the surrounding ovarian stromal tissue appears to be secondary to the gene expression in the granulosa cells. Therefore, another challenge for the next decade is to determine whether certain protein products of the ovulation-specific genes in the granulosa are having a direct effect on the thecal fibroblasts, or whether arousal of these quiescent fibroblasts is a result of action by serum proteins that have exuded from hyperpermeable vessels in the vicinity of the theca interna of ovulatory follicles.

Categorizing by Nature/Function of the Protein Product. It is also possible to categorize most of the genes based on the characteristics of the protein products that are generated by transcription and translation. From the ovulation/luteinization-related genes listed in Table 1, there is generated a diverse array of *ligands* (e.g., ApoE, ppN, Wnts, and PACAP), and *membrane receptors* (e.g., CRHR1, Fz-1, Fz-4, GPR54, IL-4Rα, LH, PRL-R, and TrkA) that are not distinctly related to one another. There are a number of G-protein–coupled receptor (*GPCR*)-*signaling regulators* (e.g., CREM [i.e., ICER], PDE4, PACAP, RGS2, and sFRP4) that might regulate the duration and intensity of the signaling cascade that originates from the activation of the LH/hCG receptor; but, it is also possible that these diverse signaling regulators might be affecting signaling pathways associated with some of the other GPCRs (mentioned above) that are upregulated during the ovulatory process. Some of the genes produce *growth factors* (e.g., AR, BTC, EPI, NGF, and VEGF), and various types of *binding proteins* (e.g., C-FABP, CREB, IGFBPs, α2-M, versican, and TSG-6) that interact with many different target molecules in order to carry out their metabolic functions. Another important class of ovulation-related agents is a group of *transcription factors* (e.g., AHR, CREB, LRH-1, C/EBP-B, Egr-1, Fra-2, JunD, p53, PR), along with several factors that significantly influence transcription (e.g., CBP, MAPK, p53, and RIP140). Central to the ovulation cascade of gene expression are several common *cytokines* (e.g., IL-1β, IL-6, TNF-α, and MIF), along with at least one *chemokine* (e.g., CINC). A number of the downstream products of the expression cascades include a variety of *proteolytic enzymes* (e.g., ADAMTSs, MMPs, tPA, catL, and MT-1-MMP) and *protease inhibitors* (e.g., the versatile α2-M, along with PAI-1, and the TIMPs). Ovarian steroid metabolism involves several *steroidogenic enzymes* (e.g., P450scc, 3β-HSD, 17β-HSD, and StAR) and several *electron transport proteins* (e.g., ADX and the ALAS product 5-aminolevulinate) that function in steroidogenic activity. Also, several genes for *aldo-keto reductases* (e.g., 3α-HSD, 11β-HSD, and 17-β-HSD), along with several genes for other *enzymes* (e.g., COX-2, γ-GCS, GST, MAPK, and NOS) with highly diverse functions are expressed in the ovary during ovulation. Finally, there are still other *proteins* (e.g., Met-1, PAP-III, Mn-SOD) that do not fit readily into any of the above categories of gene products.

Categorizing by Relevance to Functional Processes/Pathways. A fourth possibility is to classify the genes in Table 1 into categories of biological processes with which they have been commonly associated. In attempting to sort in this manner, one is immediately faced with the fact that many of the ovulation-related genes (e.g., the cytokines) have been associated with multiple gene expression cascades and/or signaling pathways. On the other hand, if one groups the genes according to their associations with several established biological processes that are known to occur in the ovary during ovulation, it turns out that the vast majority have been linked to the *inflammatory response* (e.g., ADAMTSs, AR, ApoE, AHR, C/EBP-B, CD63, CRHR, C-FABP, COX-2, Egr-1, EPI, Fra-2, Fz-1, IGFBPs, IL-1β, IL-6, IL-4Rα, Ob, α2-M, MIF, MMPs, MT-1-MMP, Met-1, MAPK, Mcl-1, NGF, NOS, p53, PAP-III, PDE4, PACAP, tPA, PAI-1, ppN, PRL-R, PR, RIP140, TIMPs, TNF-α, TSG-6, and VEGF) with some of the proteins exerting pro-inflammatory actions, and others having anti-inflammatory effects (7). Some of the other genes in Table 1 fall into less congested functional groups such as *oxidative stress* (e.g., AHR, CBR, γ-GCS, GST, 3α-HSD, 11β-HSD, and Mn-SOD), *steroidogenesis* (e.g., ADX, ALAS, P450scc, 3β-HSD, LRH-1, Met-1, p53, PR, and StAR), and *angiogenesis* (e.g., ADAMTS-1, ADAMTS-4, Egr-1, MMPs, NGF, and VEGF). Thus, the challenge of the future is to integrate this *omnium gatherum* of genes (as well as other novel genes that remain to be discovered) into a meaningful conglomeration of gene expression that might be entitled, simply, the "ovulation cascade."

CONCLUDING SUMMARY

Ovulation is a complicated cascade of molecular events that is initiated when LH (or an analogous gonadotropin) couples with LH/hCG receptors located in the membranes of granulosa cells as well as theca cells of mature ovarian follicles. These G-protein–coupled receptors operate through cAMP/PKA signaling pathways to induce the expression of a repertoire of genes in granulosa cells that are associated with the up-regulation of inflammatory cytokines such as IL-1β, IL-6, and TNF-α. Within this inflammatory *blitz* of up-regulated gene expression is the inducible form of cyclooxygenase, namely COX-2, which catalyzes the conversion of arachidonic acid into several prostanoids including the vasodilatory prostaglandin E_2. One known, critical site of PGE_2 action (based on the phenotypes of COX-2 and COX-1 null mice) occurs during COC expansion where this prostaglandin induces TSG-6, a factor that is obligatory for stabilizing the extracellular hyaluronan-rich COC matrix (Fig. 20). A possible additional consequence of the biological action of prostaglandins (combined possibly with the cooperation of other vasoactive agents) is the hyperemic response that occurs during the first hours of the ovulatory process. As a result either of the exudation of serum proteins into the thecal connective tissue surrounding ovulatory follicles, or of the diffusion of unknown signaling agents from the inflamed granulosa layer of the follicles, the fibroblasts in the ovary undergo a transformation from a quiescent state to a more active, proliferating state. This arousal of the fibroblasts induces gene expression for tPA, catL, MT-1-MMP, and a number of MMPs that weaken the extracellular matrix in the connective tissue surrounding the follicle. This proteolytic degradation of the collagenous tissue in the follicle wall may be augmented by the action of the progesterone-regulated ADAMTS-1 enzyme. In addition, ADAMTS-1 that is secreted by the granulosa cells and cumulus cells and localizes to extracellular hyaluronan-rich matrices of the expanded COCs may act to help provide a protective shield around the oocyte during its release and transport within the oviduct. Eventually, the integrity of the follicle is destabilized by proteolytic events and, under the force of a steady intrafollicular pressure, the apex of the follicle wall begins to thin out and protrude to the point where it ruptures. Thus, the ultimate cause of ovulation is a diverse array of proteolytic activity that originates from several different ovarian tissues in and around a mature follicle.

ACKNOWLEDGMENTS

This effort was supported in part by NSF Grant #0234358 (L.L.E.), in part by an endowment to Trinity University from Mrs. Ruth C. and Dr. Andrew G. Cowles, and in part by NIH Grants HD-16229, HD-07495 (Project III) (J.S.R.). The assistance of Karla Moncada and Rebecca Garcia with certain organizational aspects of this manuscript is greatly appreciated. We also thank Dr. Derek Boerboom and Dr. Haruhiro Kondo for reading and commenting on the text and figures.

REFERENCES

1. Espey, L. L. (1999). Ovulation. In *Encyclopedia of Reproduction* (E. Knobil and J. D. Neill, Eds.), Vol. III, pp. 605–614. Academic Press, San Diego.
2. Asdell, S. A. (1962). Mechanism of ovulation. In *The Ovary* (S. Zuckerman, Ed.), pp. 435–449. Academic Press, London.

3. Espey, L. L., and Lipner, H. (1963). Measurements of intrafollicular pressures in the rabbit ovary. *Am. J. Physiol.* 205, 1067–1072.
4. Espey, L. L. (1978). Ovarian contractility and its relationship to ovulation: a review. *Biol. Reprod.* 19, 540–551.
5. Espey, L. L. (1980). Ovulation as an inflammatory reaction—a hypothesis. *Biol. Reprod.* 22, 73–106.
6. Espey, L. L. (1994). Current status of the hypothesis that mammalian ovulation is comparable to an inflammatory reaction. *Biol. Reprod.* 50, 233–238.
7. Espey, L. L., Bellinger, A., and Healy, J. A. (2004). Ovulation: an inflammatory cascade of gene expression. In *The Ovary* (P. C. K. Leung and E. Y. Adashi, Eds.), pp. 145–165. Academic Press, New York.
8. Hartman, C. G. (1932). Ovulation and the transport and viability of ova and sperm in the female genital tract. In *Sex and Internal Secretions* (E. Allen, Ed.), pp. 647–688. Williams & Wilkins, Baltimore.
9. Lipner, H. (1988). Mechanism of mammalian ovulation. In *The Physiology of Reproduction* (E. Knobil and J. D. Neill, Eds.), pp. 447–488. Raven Press, New York.
10. Espey, L. L., and Lipner, H. (1994). Ovulation. In *The Physiology of Reproduction* (E. Knobil and J. D. Neill, Eds.), pp. 725–780. Raven Press, New York.
11. Davis, J. S. (1994). Mechanisms of hormone action: luteinizing hormone receptors and second-messenger pathways. *Curr. Opin. Obstet. Gynecol.* 6, 254–261.
12. Monniaux, D., Huet, C., Besnard, N., Clement, F., Bosc, M., Pisselet, C., Monget, P., and Mariana, J. C. (1997). Follicular growth and ovarian dynamics in mammals. *J. Reprod. Fertil. Suppl.* 51, 3–23.
13. Hillier, S. G. (2001). Gonadotropic control of ovarian follicular growth and development. *Mol. Cell. Endocrinol.* 179, 39–46.
14. Gelety, T. J., and Chaudhuri, G. (1992). Prostaglandins in the ovary and fallopian tube. *Baillieres Clin. Obstet. Gynaecol.* 6, 707–729.
15. Priddy, A. R., and Killick, S. R. (1993). Eicosanoids and ovulation. *Prostaglandins Leukot. Essent. Fatty Acids* 49, 827–831.
16. Olofsson, J. I., and Leung, P. C. (1996). Prostaglandins and their receptors: Implications for ovarian physiology. *Biol. Signals* 5, 90–100.
17. Zalanyi, S. (2001). Progesterone and ovulation. *Eur. J. Obstet. Gynecol. Reprod. Biol.* 98, 152–159.
18. Chaffin, C. L., and Stouffer, R. L. (2002). Local role of progesterone in the ovary during the periovulatory interval. *Rev. Endocr. Metab. Disord.* 3, 65–72.
19. Stouffer, R. L. (2003). Progesterone as a mediator of gonadotrophin action in the corpus luteum: beyond steroidogenesis. *Hum. Reprod. Update* 9, 99–117.
20. Tsafriri, A. (1995). Ovulation as a tissue remodeling process. Proteolysis and cumulus expansion. *Adv. Exp. Med. Biol.* 377, 121–140.
21 Tsafriri, A., and Reich, R. (1999). Molecular aspects of mammalian ovulation. *Exp. Clin. Endocrinol. Diabetes* 107, 1–11.
22. Smith, M. F., McIntush, E. W., Ricke, W. A., Kojima, F. N., and Smith, G. W. (1999). Regulation of ovarian extracellular matrix remodeling by metalloproteinases and their tissue inhibitors: Effects on follicular development, ovulation and luteal function. *J. Reprod. Fertil. Suppl.* 54, 367–381.
23. Murdoch, W. J. (1999). Plasmin-tumour necrosis factor interaction in the ovulatory process. *J. Reprod. Fertil. Suppl.* 54, 353–358.
24. Liu, Y. X. (1999). Regulation of the plasminogen activator system in the ovary. *Biol. Signals Recept.* 8, 160–177.
25. Murdoch, W. J. (2000). Proteolytic and cellular death mechanisms in ovulatory ovarian rupture. *Biol. Signals Recept.* 9, 102–114.
26. Curry, T. E. Jr., and Osteen, K. G. (2001). Cyclic changes in the matrix metalloproteinase system in the ovary and uterus. *Biol. Reprod.* 64, 1285–1296.
27. Richards, J. S. (2002). Delivery of the oocyte from the follicle to the oviduct: a time of vulnerability. *Ernst Schering Res. Found. Workshop* 43–62.
28. Murdoch, W. J., Gottsch, M. L. (2003). Proteolytic mechanisms in the ovulatory folliculo-luteal transformation. *Connect. Tissue Res.* 44, 50–57.
29. Curry, T. E. Jr., and Osteen, K. G. (2003). The matrix metalloproteinase system: changes, regulation, and impact throughout the ovarian and uterine reproductive cycle. *Endocr. Rev.* 24, 428–465.
30. Stouffer, R. L., Martinez-Chequer, J. C., Molskness, T. A., Xu, F., and Hazzard, T. M. (2001). Regulation and action of angiogenic factors in the primate ovary. *Arch. Med. Res.* 32, 567–575.
31. Vinatier, D., Dufour, P., Tordjeman-Rizzi, N., Prolongeau, J. F., Depret-Moser, S., and Monnier, J. C. (1995). Immunological aspects of ovarian function: role of the cytokines. *Eur. J. Obstet. Gynecol. Reprod. Biol.* 63, 155–168.
32. Adashi, E. Y. (1996). Immune modulators in the context of the ovulatory process: a role for interleukin-1. *Am. J. Reprod. Immunol.* 35, 190–194.
33. Norman, R. J., and Brannstrom, M. (1996). Cytokines in the ovary: pathophysiology and potential for pharmacological intervention. *Pharmacol. Ther.* 69, 219–236.
34. Machelon, V., and Emilie, D. (1997). Production of ovarian cytokines and their role in ovulation in the mammalian ovary. *Eur. Cytokine Netw.* 8, 137–143.
35. Adashi, E. Y. (1997). The potential role of IL-1 in the ovulatory process: an evolving hypothesis. *J. Reprod. Immunol.* 35, 1–9.
36. Adashi, E. Y. (1998). The potential role of interleukin-1 in the ovulatory process: an evolving hypothesis. *Mol. Cell. Endocrinol.* 140, 77–81.
37. Gerard, N., Caillaud, M., Martoriati, A., Goudet, G., and Lalmanach, A. C. (2004). The interleukin-1 system and female reproduction. *J. Endocrinol.* 180, 203–212.
38. Brannstrom, M., and Norman, R. J. (1993). Involvement of leukocytes and cytokines in the ovulatory process and corpus luteum function. *Hum. Reprod.* 8, 1762–1775.
39. Bukulmez, O., and Arici, A. (2000). Leukocytes in ovarian function. *Hum. Reprod. Update* 6, 1–15.
40. Brannstrom, M., and Enskog, A. (2002). Leukocyte networks and ovulation. *J. Reprod. Immunol.* 57, 47–60.
41. Richards, J. S. (1994). Hormonal control of gene expression in the ovary. *Endocr. Rev.* 15, 725–751.
42. Richards, J. S., Fitzpatrick, S. L., Clemens, J. W., Morris, J. K., Alliston, T., and Sirois, J. (1995). Ovarian cell differentiation: a cascade of multiple hormones, cellular signals, and regulated genes. *Recent Prog. Horm. Res.* 50, 223–254.
43. Richards, J. S., Russell, D. L., Robker, R. L., Dajee, M., and Alliston, T. N. (1998). Molecular mechanisms of ovulation and luteinization. *Mol. Cell. Endocrinol.* 145, 47–54.
44. Robker, R. L., Russell, D. L., Yoshioka, S., Sharma, S. C., Lydon, J. P., O'Malley, B. W., Espey, L. L., and Richards, J. S. (2000). Ovulation: a multi-gene, multi-step process. *Steroids* 65, 559–570.
45. Espey, L. L., and Richards, J. S. (2002). Temporal and spatial patterns of ovarian gene transcription following an ovulatory dose of gonadotropin in the rat. *Biol. Reprod.* 67, 1662–1670.
46. Richards, J. S., Russell, D. L., Ochsner, S., and Espey, L. L. (2002). Ovulation: new dimensions and new regulators of the inflammatory-like response. *Annu. Rev. Physiol.* 64, 69–92.
47. Richards, J. S., Russell, D.L., Ochsner, S., Hsieh, M., Doyle, K. H., Falender, A. E., Lo, Y. K., and Sharma, S. C. (2002). Novel signaling pathways that control ovarian follicular development, ovulation and luteinization. *Recent Prog. Horm. Res.* 57, 195–220.

48. Tanaka, N., Espey, L. L., and Okamura, H. (1989). Increase in ovarian blood volume during ovulation in the gonadotropin-primed immature rat. *Biol. Reprod.* 40, 762–768.
49. Teuling, E., Lo, Y., Thompson, V., Falender, A. E., Sandy, J. D., and Richards, J. S. (2004). Regulation expression of ADAMTS family members during follicular development and ovulation: evidence for specific and redundant functions. Endocrinology (submitted).
50. Bjersing, L., and Cajander, S. (1974). Ovulation and the mechanism of follicle rupture. III. Transmission electron microscopy of rabbit germinal epithelium prior to induced ovulation. *Cell. Tissue Res.* 149, 313–327.
51. Cajander, S. (1976). Structural alterations of rabbit ovarian follicles after mating with special reference to the overlying surface epithelium. *Cell. Tissue Res.* 173, 437–449.
52. Murdoch, W. J., and McDonnel, A. C. (2002). Roles of the ovarian surface epithelium in ovulation and carcinogenesis. *Reproduction* 123, 743–750.
53. Colgin, D. C., and Murdoch, W. J. (1997). Evidence for a role of the ovarian surface epithelium in the ovulatory mechanism of the sheep: secretion of urokinase-type plasminogen activator. *Anim. Reprod. Sci.* 47, 197–204.
54. Espey, L. L. (1967). Ultrastructure of the apex of the rabbit graafian follicle during the ovulatory process. *Endocrinology* 81, 267–276.
55. Parr, E. L. (1975). Rupture of ovarian follicles at ovulation. *J. Reprod. Fertil.* 22 (suppl.), 1–22.
56. Motta, P., and Van Blerkom, J. (1975). A scanning electron microscopic study of the luteo-follicular complex. II. Events leading to ovulation. *Am. J. Anat.* 143, 241–263.
57. Rawson, J. M., and Espey, L. L. (1977). Concentration of electron dense granules in the rabbit ovarian surface epithelium during ovulation. *Biol. Reprod.* 17, 561–566.
58. Espey, L. L. (1991). Ultrastructure of the ovulatory process. In *Ultrastructure of the Ovary* (G. Familiari, S. Makabe, and P. M. Motta, Eds.), pp. 143–159. Kluwer Academic Publishers, Boston.
59. Arraztoa, J. A., Monget, P., Bondy, C., and Zhou, J. (2002). Expression patterns of insulin-like growth factor-binding proteins 1, 2, 3, 5, and 6 in the mid-cycle monkey ovary. *J. Clin. Endocrinol. Metab.* 87, 5220–5228.
60. Erickson, G. F., and Shimasaki, S. (2003). The spatiotemporal expression pattern of the bone morphogenetic protein family in rat ovary cell types during the estrous cycle. *Reprod. Biol. Endocrinol.* 1, 9.
61. Jo, M., Thomas, L. E., Wheeler, S. E., and Curry, T. E. Jr. (2004). Membrane type 1-matrix metalloproteinase (MMP)-associated MMP-2 activation increases in the rat ovary in response to an ovulatory dose of human chorionic gonadotropin. *Biol. Reprod.* 70, 1024–1032.
62. Liu, Y. X., Liu, K., Feng, Q., Hu, Z. Y., Liu, H. Z., Fu, G. Q., Li, Y. C., Zou, R. J., and Ny, T. (2004). Tissue-type plasminogen activator and its inhibitor plasminogen activator inhibitor type 1 are coordinately expressed during ovulation in the rhesus monkey. *Endocrinology* 145, 1767–1775.
63. Espey, L. L. (1992). A review of factors that could influence membrane potentials of ovarian follicular cells during mammalian ovulation. *Acta Endocrinologica*, Suppl. 2 126, 1–31.
64. Bahr, J. M. (1978). Simultaneous measurements of steroids in follicular fluid and ovarian venous blood in the rabbit. *Biol. Reprod.* 18, 193–197.
65. Goff, A. K., and Henderson, K. M. (1979). Changes in follicular fluid and serum concentrations of steroids in PMS treated immature rats following LH administration. *Biol. Reprod.* 20, 1153–1157.
66. Hubbard, C. J., and Greenwald, G. S. (1982). Cyclic nucleotides, DNA, and steroid levels in ovarian follicles and corpora lutea of the cyclic hamster. *Biol. Reprod.* 26, 230–240.
67. Espey, L. L., Norris, C., Forman, J., and Siler-Khodr, T. (1989). Effect of indomethacin, cycloheximide, and aminoglutethimide on ovarian steroid and prostanoid levels during ovulation in the gonadotropin-primed immature rat. *Prostaglandins* 38, 531–539.
68. Espey, L. L., Adams, R. F., Tanaka, N., and Okamura, H. (1990). Effects of epostane on ovarian levels of progesterone, 17β-estradiol, prostaglandin E_2, and prostaglandin $F_{2\alpha}$ during ovulation in the gonadotropin-primed immature rat. *Endocrinology* 127, 259–263.
69. Conneely, O. M., Mulac-Jericevic, B., DeMayo, F., Lydon, J. P., and O'Malley, B. W. (2002) Reproductive functions of progesterone receptors. *Recent Prog. Horm. Res.* 57, 339–355.
70. Davis, B. J., Lennard, D. E., Lee, C. A., Tiano H. F., Morham, S. G., Wetsel, W. C., and Langenbach, R. (1999). Anovulation in cyclooxygenase-2 deficient mice is restored byu prostglandin E2 and interleukin-1b. *Endocrinology* 142, 3187–3197.
71. Hizaki, H., Segi, E., Sugimoto, Y., Hirose, M., Saji, T., Ushikubi, F., Matsuoka, T., Noda, Y., Tanaka, T., Yoshida, N., Narumiya, S., and Ichikawa, A. (1999). Abortive expansion of the cumulus and impaired fertility in mice lacking the prostaglandin E receptor subtype EP(2). *Proc. Natl. Acad. Sci. U S A* 96, 10501–10506.
72. Park, J. Y., Su, Y. Q., Ariga, M., Law, E., Jin, S. L., and Conti, M. (2004). EGF-like growth factors as mediators of LH action in the ovulatory follicle. *Science* 303, 682–684.
73. Sterneck, E., Tassarollo, L., and Johnson, P. F. (1997). An essential role for C/EBPβ in female reproduction. *Genes Dev.* 11, 2153–2162.
74. Robker, R. L., and Richards, J. S. (1998). Hormonal control of the cell cycle in ovarian cells: proliferation versus differentiation. *Biol. Reprod.* 59, 476–482.
75. Richards, J. S., Hedin, L., and Caston, L. (1986). Differentiation of rat ovarian thecal cells: evidence for functional luteinization. *Endocrinology* 118, 1660–1668.
76. Hedin, L., Rodgers, R. J., Simpson, E. R., and Richards, J. S. (1987). Changes in content of cytochrome $P450_{17\alpha}$, cytochrome $P450_{scc}$ and 3-hydroxy-3-methylglutaryl CoA reductase in developing rat ovarian follicles and corpora lutea: correlation with theca cell steroidogenesis. *Biol. Reprod.* 37, 211–223.
77. Goldring, N. B., Durica, J. M., Lifka, J., Hedin, L., Ratoosh, S. L., Miller, W. L., Orly, J., and Richards, J. S. (1987). Cholesterol side-chain cleavage P450 messenger ribonucleic acid: evidence for hormonal regulation in rat ovarian follicles and constitutive expression in corpora lutea. *Endocrinology* 120, 1942–1950.
78. Amsterdam, A., Koch, Y., Lieberman, M. E., and Lindner, H. R. (1975). Distribution of binding sites for human chorionic gonadotropin in the preovulatory follicle of the rat. *J. Cell. Biol.* 67, 894–900.
79. Uilenbroek, J. T., and Richards, J. S. (1979). Ovarian follicular development during the rat estrous cycle: gonadotropin receptors and follicular responsiveness. *Biol. Reprod.* 20, 1159–1165.
80. Richards, J. S. (1980). Maturation of ovarian follicles: actions and interactions of pituitary and ovarian hormones on follicular cell differentiation. *Physiol. Rev.* 60, 51–89.
81. Bogovich, K., Richards, J. S., and Reichert, L. E. Jr. (1981). Obligatory role of luteinizing hormone (LH) in the initiation of preovulatory follicular growth in the pregnant rat: specific effects of human chorionic gonadotropin and follicle-stimulating hormone on LH receptors and steroidogenesis in theca, granulosa and luteal cells. *Endocrinology* 109, 860–867.
82. Richards, J. S., and Bogovich, K. (1982). Effects of human chorionic gonadotropin and progesterone on follicular development in the immature rat. *Endocrinology* 111, 1429–1438.
83. Webb, R., and England, B. G. (1982). Identification of the ovulatory follicle in the ewe: associated changes in follicular size, thecal and granulosa cell luteinizing hormone receptors,

antral fluid steroids, and circulating hormones during the preovulatory period. *Endocrinology* 110, 873–881.
84. Orly, J. (1989). Orchestrated expression of steroidogenic side-chain cleavage cytochrome P-450 during follicular development in the rat ovary. *J. Reprod. Fert.* 37, 155–162.
85. Goldschmidt, D., Kraicer, P., and Orly, J. (1989). Periovulatory expression of cholesterol side-chain cleavage cytochrome P-450 in cumulus cells. *Endocrinology* 124, 369–378.
86. Marsh, J. M., Mills, T. M., and LeMaire, W. J. (1972). Cyclic AMP synthesis in rabbit graafian follicles and the effect of luteinizing hormone. *Biochim. Biophys. Acta* 273, 389–394.
87. Marsh, J. M., Mills, T. M., and LeMaire, W. J. (1973). Preovulatory changes in the synthesis of cyclic AMP by rabbit graafian follicles. *Biochim. Biophys. Acta* 304, 197–202.
88. Mills, T. M. (1975). Effect of luteinizing hormone and cyclic adenosine 3′,5′-monophosphate on steroidogenesis in the ovarian follicle of the rabbit. *Endocrinology* 96, 440–445.
89. Mason, N. R., and Marsh, R. (1975). The effect of LH on cyclic AMP and progesterone in rat ovaries in vivo. *Endocr. Res. Commun.* 2, 167–177.
90. Hedin, L., McKnight, G. S., Lifka, J., Durica, J. M., and Richards, J. S. (1987). Tissue distribution and hormonal regulation of mRNA for regulatory and catalytic subunits of adenosine 3′,5′-monophosphate-dependent protein kinases during ovarian follicular development and luteinization in the rat. *Endocrinology* 120, 1928–1935.
91. Eimerl, S., and Orly, J. (2002). Regulation of steroidogenic genes by insulin-like growth factor-1 and follicle stimulating hormone: differential responses of P450 side-chain cleavage, steroidogenic acute regulatory protein and 3β-hydroxysteroid dehydrogenase/isomerase in rat granulosa cells. *Biol. Reprod.* 67, 900–910.
92. Richards, J. S., and Hedin, L. (1988). Molecular aspects of hormone action in ovarian follicular development, ovulation, and luteinization. *Annu. Rev. Physiol.* 50, 441–463.
93. Niswender, G. D. (2002). Molecular control of luteal secretion of progesterone. *Reproduction* 123, 333–339.
94. Herrmann, M., Scholmerich, J., and Straub, R. H. (2002). Influence of cytokines and growth factors on distinct steroidogenic enzymes in vitro: a short tabular data collection. *Ann. N Y Acad. Sci.* 996, 166-168.
95. Stocco, D. M. (2001). StAR protein and the regulation of steroid hormone biosynthesis. *Annu. Rev. Physiol.* 63, 193–213.
96. Ronen-Fuhrmann, T., Timberg, R., King, S. R., Hales, K. H., Hales, D. B., Stocco, D. M., and Orly, J. (1998). Spatio-temporal expression patterns of steroidogenic acute regulatory protein (StAR) during follicular development in the rat ovary. *Endocrinology* 139, 303–315.
97. Granot, Z., Geiss-Friedlander, R., Melamed-Book, N., Eimerl, S., Timberg, R., Weiss, A. M., Hales, K. H., Hales, D. B., Stocco, D. M., and Orly, J. (2003). Proteolysis of normal and mutated steroidogenic acute regulatory proteins in the mitochondria: fate of unwanted proteins. *Mol. Endocrinol.* 17, 2461–2476.
98. Watson, E. D., Bae, S. E., Steele, M., Thomassen, R., Pedersen, H. G., Bramley, T., Hogg, C. O., and Armstrong, D. G. (2004). Expression of messenger ribonucleic acid encoding steroidogenic acute regulatory protein and enzymes, luteinizing hormone receptor during the spring transitional season in equine follicles. *Domest. Anim. Endocrinol.* 26, 215–230.
99. Grinberg, A. V., Hannemann, F., Schiffler, B., Muller, J., Heinemann, U., and Bernhardt, R. (2000). Adrenodoxin: structure, stability, and electron transfer properties. *Proteins* 40, 590–612.
100. Selvaraj, N., Dantes, A., and Amsterdam, A. (2000). Establishment and characterization of steroidogenic granulosa cells expressing β_2-adrenergic receptor: regulation of adrenodoxin and steroidogenic acute regulatory protein by adrenergic agents. *Mol. Cell. Endocrinol.* 168, 53–63.
101. Oonk, R. B., Krasnow, J. S., Beattie, W. G., and Richards, J. S. (1989). Cyclic AMP-dependent and -independent regulation of cholesterol side chain cleavage cytochrome P-450 (P-450_{scc}) in rat ovarian granulosa cells and corpora lutea. cDNA and deduced amino acid sequence of rat P-450_{scc}. *J. Biol. Chem.* 264, 21934–21942.
102. Oonk, R. B., Parker, K. L., Gibson, J. L., and Richards, J. S. (1990). Rat cholesterol side-chain cleavage cytochrome P-450 (P-450_{scc}) gene. Structure and regulation by cAMP in vitro. *J. Biol. Chem.* 265, 22392–22401.
103. Park, O.K., and Mayo, K.E. (1991). Transient expression of progesterone receptor messenger RNA in ovarian granulosa cells after the preovulatory luteinizing hormone surge. *Mol. Endocrinol.* 5, 967–978.
104. Natraj, U., and Richards, J. S. (1993). Hormonal regulation, localization, and functional activity of the progesterone receptor in granulosa cells of rat preovulatory follicles. *Endocrinology* 133, 761–769.
105. Robker, R. L., Russell, D. L., Espey, L. L., Lydon, J. P., O'Malley, B. W., and Richards, J. S. (2000). Progesterone-regulated genes in the ovulation process: ADAMTS-1 and cathepsin L proteases. *Proc. Natl. Acad. Sci. U S A* 97, 4689–4694.
106. Doyle, K. M., Russell, D. L., Sriraman, V., and Richards, J. S. (2004). Coordinate transcription of the ADAMTS-1 gene by LH and PR. *Mol. Endocrinol.* 18, 2463–2478.
107. Cassar, C. A., Dow, M. P., Pursley, J. R., and Smith, G. W. (2002). Effect of the preovulatory LH surge on bovine follicular progesterone receptor mRNA expression. *Domest. Anim. Endocrinol.* 22, 179–187.
108. Sriraman, V., Sharma, S. C., and Richards, J. S. (2003). Transactivation of the progesterone receptor gene in granulosa cells: evidence that Sp1/Sp3 binding sites in the proximal promoter play a key role in luteinizing hormone inducibility. *Mol. Endocrinol.* 17, 436–449.
109. Hisaw, F. L. (1961). Endocrines and the evolution of viviparity among the vertebrates. In *Physiology of Reproduction*, pp. 119–138. Proceedings of the Twenty-second Biology Colloquium. Oregon State University Press, Corvallis.
110. Snyder, B. W., Beecham, G. D., and Schane, H. P. (1984). Inhibition of ovulation in rats with epostane, an inhibitor of 3-beta-hydroxysteroid dehydrogenase. *Proc. Soc. Exp. Biol. Med.* 176, 238–242.
111. Rondell, P. (1974). Role of steroid synthesis in the process of ovulation. *Biol. Reprod.* 10, 199–215.
112. Stouffer, R. L. (2002). Pre-ovulatory events in the rhesus monkey follicle during ovulation induction. *Reprod. Biomed.* 3 (Suppl), 1–4.
113. Espey, L. L., Yoshioka, S., Russell, D. L., Robker, R. L., Fujii, S., and Richards, J. S. (2000). Ovarian expression of a disintegrin and metalloproteinase with thrombospondin motifs during ovulation in the gonadotropin-primed immature rat. *Biol. Reprod.* 62, 1090–1095.
114. Kuno, K., Kanada, N., Nakashima, E., Jujiki, F., Ichimura, F., and Matsushima, K. (1997). Molecular cloning of a gene encoding a new type of metalloproteinase-disintegrin family protein with thrombospondin motifs as an inflammation associated gene. *J. Biol. Chem.* 272, 556–562.
115. Madan, P., Bridges, P. J., Komar, C. M., Beristain, A. G., Rajamahendran, R., Fortune, J. E., and MacCalman, C. D. (2003). Expression of messenger RNA for ADAMTS subtypes changes in the periovulatory follicle after the gonadotropin surge and during luteal development and regression in cattle. *Biol. Reprod.* 69, 1506–1514.
116. Russell, D. L., Doyle, K. M., Ochsner, S. A., Sandy, J. D., and Richards, J. S. (2003). Processing and localization of

ADAMTS-1 and proteolytic cleavage of versican during cumulus matrix expansion and ovulation. *J. Biol. Chem.* 278, 42330–42339.

117. Boerboom, D., Russell, D. L., Richards, J. S., and Sirois, J. (2003). Regulation of transcripts encoding ADAMTS-1 (a disintegrin and metalloproteinase with thrombospondin-like motifs-1) and progesterone receptor by human chorionic gonadotropin in equine preovulatory follicles. *J. Mol. Endocrinol.* 31, 473–485.
117. Mittaz, L., Russell, D. L., Wilson, T., Brasted, M., Tkalcevic, J., Salamonsen, L. A., Hertzog, P. J., and Pritchard, M. A. (2004). Adamts-1 is essential for the development and function of the urogenital system. *Biol. Reprod.* 70, 1096–1105.
119. Shimada, M., Nishibori, M., Yamashita, Y., Ito, J., Mori, T., and Richards, J. S. (2004). Down-regulated expression of ADAMTS-1 by progesterone receptor antagonists is associated with impaired expansion of porcine cumulus-oocyte complexes. *Endocrinology* 145, 4603–4614.
120. Sriraman, V., and Richards, J. S. (2004). Cathepsin L gene expression and promoter activation in rodent granulosa cells. *Endocrinology* 145, 582–591.
121. Ko, C., In, Y. H., and Park-Sarge, O. K. (1999). Role of progesterone receptor activation in pituitary adenylate cyclase-activating polypeptide gene expression in rat ovary. *Endocrinology* 140, 5185–5194.
122. Park, J. I., Kim, W. J., Wang, L., Park, H. J., Lee, J., Park, J. H., Kwon, H. B., Tsafriri, A., and Chun, S. Y. (2000). Involvement of progesterone in gonadotrophin-induced pituitary adenylate cyclase-activating polypeptide gene expression in pre-ovulatory follicles of rat ovary. *Mol. Hum. Reprod.* 6, 238–245.
123. O'Flaherty, J. T. (1987). Phospholipid metabolism and stimulus-response coupling. *Biochem. Pharmacol.* 36, 407–412.
124. Orczyk, G. P., and Behrman, H. R. (1972). Ovulation blockade by aspirin or indomethacin-*in vivo* evidence for a role of prostaglandin in gonadotrophin secretion. *Prostaglandins* 1, 3-20.
125. Armstrong, D. T., and Grinwich, D. L. (1972). Blockade of spontaneous and LH-induced ovulation in rats by indomethacin, an inhibitor of prostaglandin biosynthesis. *Prostaglandins* 1, 21–28.
126. O'Grady, J. P., Caldwell, B. V., Auletta, F. J., and Speroff, L. (1972). The effects of an inhibitor of prostaglandin synthesis (indomethacin) on ovulation, pregnancy, and pseudopregnancy in the rabbit. *Prostaglandins* 1, 97–106.
127. Behrman, H. R., Orczyk, G. P., and Greep, R. O. (1972). Effect of synthetic gonadotrophin-releasing hormone (Gn-RH) on ovulation blockade by aspirin and indomethacin. *Prostaglandins* 1, 245–258.
128. Tsafriri, A., Lindner, H. R., Zor, U., and Lamprecht, S. A. (1972). Physiological role of prostaglandins in the induction of ovulation. *Prostaglandins* 2, 1–10.
129. LeMaire, W. J., Yang, N. S., Behrman, H. H., and Marsh, J. M. (1973). Preovulatory changes in the concentration of prostaglandins in rabbit graafian follicles. *Prostaglandins* 3, 367–376.
130. LeMaire, W. J., Leidner, R., and Marsh, J. M. (1975). Pre and post ovulatory changes in the concentration of prostaglandins in rat graafian follicles. *Prostaglandins* 9, 221–229.
131. Espey, L. L., Tanaka, N., Adams, R. F., and Okamura, H. (1991). Ovarian hydroxyeicosatetraenoic acids compared with prostanoids and steroids during ovulation in rats. *Am. J. Physiol.* 260, E163–E169.
132. Nelson, D. L., and Cox, M. M. (2005). *Lehninger Principles of Biochemistry*, p. 800. W. H. Freeman, New York.
133. Hedin, L., Gaddy-Kurten, D., Kurten, R., DeWitt, D. L., Smith, W. L., and Richards, J. S. (1987). Prostaglandin endoperoxide synthase in rat ovarian follicles: content, cellular distribution, and evidence for hormonal induction preceding ovulation. *Endocrinology* 121, 722–731.
134. Wong, W. Y., DeWitt, D. L., Smith, W. L., and Richards, J. S. (1989). Rapid induction of prostaglandin endoperoxide synthase in rat preovulatory follicles by luteinizing hormone and cAMP is blocked by inhibitors of transcription and translation. *Mol. Endocrinol.* 3, 1714–1723.
135. Wong, W. Y., and Richards, J. S. (1991). Evidence for two antigenically distinct molecular weight variants of prostaglandin H synthase in the rat ovary. *Mol. Endocrinol.* 5, 1269–1279.
136. Sirois, J., and Richards, J. S. (1992). Purification and characterization of a novel, distinct isoform of prostaglandin endoperoxide synthase induced by human chorionic gonadotropin in granulosa cells of rat preovulatory follicles. *J. Biol. Chem.* 267, 6382–6388.
137. Joyce, I. M., Pendola, F. L., O'Brien, M., and Eppig, J. J. (2001). Regulation of prostaglandin-endoperoxide synthase 2 messenger ribonucleic acid in mouse granulosa cells during ovulation. *Endocrinology* 142, 3187–3197.
138. Ochsner, S. A., Day, A. J., Rugg, M. S., Breyer, R. M., Gomer, R. H., and Richards, J. S. (2003). Disrupted function of tumor necrosis factor-a-stimulated gene 6 blockes cumulus cell-oocyte complex expansion. *Endocrinology* 144, 4376–4384.
139. Sirois, J., Simmons, D. L., and Richards, J. S. (1992). Hormonal regulation of messenger ribonucleic acid encoding a novel isoform of prostaglandin endoperoxide H synthase in rat preovulatory follicles. Induction in vivo and in vitro. *J. Biol. Chem.* 267, 11586–11592.
140. Ochsner, S. A., Russell, D. L., Day, A. J., Breyer, R. M., and Richards, J. S. (2003). Decreased expression of tumor necrosis factor-α-stimulated gene 6 in cumulus cells of the cyclooxygenase-2 and EP2 null mice. *Endocrinology* 144, 1008–1019.
141. Schoenfelder, M., and Einspanier, R. (2003). Expression of hyaluronan synthases and corresponding hyaluronan receptors is differentially regulated during oocyte maturation in cattle. *Biol. Reprod.* 69, 269–277.
142. Kimura, N., Konno, Y., Miyoshi K., Matsumoto, H., and Sato, E. (2002). Expression of hyaluronan synthases and CD44 messenger RNAs in porcine cumulus-oocyte complexes during in vitro maturation. *Biol. Reprod.* 66, 707–717.
143. Sirois, J., and Dore, M. (1997). The late induction of prostaglandin G/H synthase-2 in equine preovulatory follicles supports its role as a determinant of the ovulatory process. *Endocrinology* 138, 4427–4434.
144. Richards, J. S. (1997). Editorial: sounding the alarm—does induction of prostaglandin endoperoxide synthase-2 control the mammalian ovulatory clock? *Endocrinology* 138, 4047–4048.
145. Duffy, D. M., and Stouffer, R. L. (2001). The ovulatory gonadotrophin surge stimulates cyclooxygenase expression and prostaglandin production by the monkey follicle. *Mol. Hum. Reprod.* 7, 731–739.
146. Morris, J. K., and Richards, J. S. (1996). An E-box region within the prostaglandin endoperoxide synthase-2 (PGS-2) promoter is required for transcription in rat ovarian granulosa cells. *J. Biol. Chem.* 271, 16633–16643.
147. Sirois, J., and Richards, J. S. (1993). Transcriptional regulation of the rat prostaglandin endoperoxide synthase 2 gene in granulosa cells. Evidence for the role of a cis-acting C/EBPβ promoter element. *J. Biol. Chem.* 268, 21931–21938.
148. Sato, H., Kajikawa, S., Kuroda, S., Horisawa, Y., Nakamura, N., Kaga, N., Kakinuma, C., Kato, K., Morishita, H., Niwa, H., Miyazaki, J. (2001). Impaired infertility in female mice lacking urinary trypsin inhibitor. *Biochem. Biophys. Res. Commun.* 281, 1154–1160.
149. Zhuo, L., Yoneda, M., Yingsung, W., Yoshida, N., Kitagawa, Y., Kawamura, K., Suzuki, T., and Kimata, K. (2001). Defect in SHAP-hyaluronan complex causes severe female infertility: a

study by inactivation of the bikunin gene in mice. *J. Biol. Chem.* 276, 7693–7696.

150. Fulop, C., Szanto, S., Mukhopadhyay, D., Bardos, T., Kamath, R. V., Rugg, M. S., Day, A. J., Salustri, A., Hascall, V. C., Glant, T. T., and Mikecz, K. (2003). Impaired cumulus cell mucification and female sterility in tumor necrosis factor induced protein-6 deficient mice. *Development* 130, 2253–2261.
151. Miyazaki, T., Dharmarajan, A. M., Atlas, S. J., Katz, E., and Wallach, E. E. (1991). Do prostaglandins lead to ovulation in the rabbit by stimulating proteolytic enzyme activity? *Fertil. Steril.* 55, 1183–1188.
152. Sakurai, T., Tamura, K., Okamoto, S., Hara, T., and Kogo, H. (2003). Possible role of cyclooxygenase II in the acquisition of ovarian luteal function in rodents. *Biol. Reprod.* 69, 835–842.
153. Yamauchi, J., Miyazaki, T., Iwasaki, S., Kishi, I., Kuroshima, M., Tei, C., and Yoshimura, Y. (1997). Effects of nitric oxide on ovulation and ovarian steroidogenesis and prostaglandin production in the rabbit. *Endocrinology* 138, 3630–3637.
154. Segi, E., Haraguchi, K., Sugimoto, Y., Tsuji, M., Tsunekawa, H., Tamba, S., Tsuboi, K., Tanaka, S., and Ichikawa, A. (2003). Expression of messenger RNA for prostaglandin E receptor subtypes EP4/EP2 and cyclooxygenase isozymes in mouse periovulatory follicles and oviducts during superovulation. *Biol. Reprod.* 68, 804–811.
155. Salustri, A., Garlanda, C., Hirsch, E., DeAcetis, M., Maccagno, A., Bottazzi, B., Doni, A., Bastone, A., Mantovani, G., Peccoz, P. B., Savatori, G., Mahoney, D. J., Day, A. J., Siracusa, G., Romani, L., and Mantovani, A. (2004). PTX3 plays a key role in the organization of the cumulus oophorus extracellular matrix and in in vivo fertilization. *Development* 131, 1577–1586.
156. Varani, S., Elvin, J. A., Yan, C., DeMayo, J., DeMayo, F. J., Horton, H. F., Byrne, M. C., and Matzuk, M. M. (2002). Knockout of pentraxin 3, a downstream target of growth differentiation factor-9, causes female subfertility. *Mol. Endocrinol.* 16, 1154–1167.
157. Chung, K. W., Ando, M., and Adashi, E. Y. (2000). Periovulatory and interleukin (IL)-1-dependent regulation of IL-6 in the immature rat ovary: a specific IL-1 receptor-mediated eicosanoid-dependent effect. *J. Soc. Gynecol. Investig.* 7, 301–308.
158. Mikuni, M., Pall, M., Peterson, C. M., Peterson, C. A., Hellberg, P., Brannstrom, M., Richards, J. S., and Hedin, L. (1998). The selective prostaglandin endoperoxide synthase-2 inhibitor, NS-398, reduces prostaglandin production and ovulation in vivo and in vitro in the rat. *Biol. Reprod.* 59, 1077–1083.
159. Rose, U. M., Hanssen, R. G., and Kloosterboer, H. J. (1999). Development and characterization of an in vitro ovulation model using mouse ovarian follicles. *Biol. Reprod.* 61, 503–511.
160. Duffy, D. M., and Stouffer, R. L. (2002). Follicular administration of a cyclooxygenase inhibitor can prevent oocyte release without alteration of normal luteal function in rhesus monkeys. *Human Reprod.* 17, 2825–2831.
161. Espey, L. L., Norris, C., and Saphire, D. (1986). Effect of time and dose of indomethacin on follicular prostaglandins and ovulation in the rabbit. *Endocrinology* 119, 746–754.
162. Espey, L. L., Tanaka, N., and Okamura, H. (1989). Increase in ovarian leukotrienes during hormonally induced ovulation in the rat. *Am. J. Physiol.* 256, E753–E759.
163. Tanaka, N., Espey, L. L., and Okamura, H. (1989). Increase in 15-hydroxyeicosatetraenoic acid during ovulation in the gonadotropin-primed immature rat. *Endocrinology* 125, 1373–1377.
164. Siegel, M. I., McConnell, R. T., Porter, N. A., Selph, J. L., Truax, J. F., Vinegar, R., and Cuatrecasas, P. (1980). Aspirin-like drugs inhibit arachidonic acid metabolism via lipoxygenase and cyclo-oxygenase in rat neutrophils from carrageenan pleural exudates. *Biochem. Biophys. Res. Commun.* 92, 688–695.
165. Randall, R. W., Eakins, K. E., Higgs, G. A., Salmon, J. A., and Tateson, J. E. (1980). Inhibition of arachidonic acid cyclo-oxygenase and lipoxygenase activities of leukocytes by indomethacin and compound BW755c. *Agents Actions* 10, 553–555.
166. Schochet, S. S. (1916). A suggestion as to the process of ovulation and ovarian cyst formation. *Anat. Rec.* 10, 447–457.
167. Kraus, S. D. (1947). Observations on the mechanism of ovulation in the frog, hen and rabbit. *West. J. Surg. Obstet. Gynecol.* 55, 424–437.
168. Espey, L. L., and Lipner, H. (1965). Enzyme-induced rupture of rabbit Graafian follicle. *Am. J. Physiol.* 208, 208–213.
169. Espey, L. L. (1967). Tenacity of porcine Graafian follicle as it approaches ovulation. *Am. J. Physiol.* 212, 1397–1401.
170. Campbell, E. J., Senior, R. M., and Welgus, H. G. (1987). Extracellular matrix injury during lung inflammation. *Chest* 92, 161–167.
171. Werb, Z., and Aggeler, J. (1978). Proteases induce secretion of collagenase and plasminogen activator by fibroblasts. *Proc. Natl. Acad. Sci. U S A* 75, 1839–1843.
172. Beers, W. J. (1975). Follicular plasminogen and plasminogen activator and the effect of plasmin on ovarian follicle wall. *Cell* 6, 376–386.
173. Shen, X., Minoura, H., Yoshida, T., and Toyoda, N. (1997). Changes in ovarian expression of tissue-type plasminogen activator and plasminogen activator inhibitor type-1 messenger ribonucleic acids during ovulation in rat. *Endocr. J.* 44, 341–348.
174. Hagglund, A. C., Ny, A., Liu, K., and Ny, T. (1996). Coordinated and cell-specific induction of both physiological plasminogen activators creates functionally redundant mechanisms for plasmin formation during ovulation. *Endocrinology* 137, 5671–5677.
175. Yang, W. L., Godwin, A. K., and Xu, X. X. (2004). Tumor necrosis factor-α-induced matrix proteolytic enzyme production and basement membrane remodeling by human ovarian surface epithelial cells: molecular basis linking ovulation and cancer risk. *Cancer Res.* 64, 1534–1540.
176. Leonardsson, G., Peng, X. R., Liu, K., Nordstrom, L., Carmeliet, P., Mulligan, R., Collen, D., and Ny, T. (1995). Ovulation efficiency is reduced in mice that lack plasminogen activator gene function: functional redundancy among physiological plasminogen activator. *Proc. Natl. Acad. Sci. U S A* 92, 12446–12450.
177. Ny, A., Nordstrom, L., Carmeliet, P., and Ny, T. (1997). Studies of mice lacking plasminogen activator gene function suggest that plasmin production prior to ovulation exceeds the amount needed for optimal ovulation efficiency. *Eur. J. Biochem.* 244, 487–493.
178. Ny, A., Leonardsson, G., Hagglund, A. C., Hagglof, P., Ploplis, V. A., Carmeliet, P., and Ny, T. (1999). Ovulation in plasminogen-deficient mice. *Endocrinology* 140, 5030–5035.
179. Cuzner, M. L., and Opdenakker, G. (1999). Plasminogen activators and matrix metalloproteases, mediators of extracellular proteolysis in inflammatory demyelination of the central nervous system. *J. Neuroimmunol.* 94, 1–14.
180. Mann, J. S., Kindy, M. S., Edwards, D. R., and Curry, T. E. Jr. (1991). Hormonal regulation of matrix metalloproteinase inhibitors in rat granulosa cells and ovaries. *Endocrinology* 128, 1825–1832.
181. Curry, T. E. Jr., Mann, J. S., Huang, M. H., and Keeble, S. C. (1992). Gelatinase and proteoglycanase activity during the periovulatory period in the rat. *Biol. Reprod.* 46, 256–264.
182. Curry, T. E. Jr., Song, L., and Wheeler, S. E. (2001). Cellular localization of gelatinases and tissue inhibitors of

metalloproteinases during follicular growth, ovulation, and early luteal formation in the rat. *Biol. Reprod.* 65, 855–865.
183. Bagavandoss, P. (1998). Differential distribution of gelatinases and tissue inhibitor of metalloproteinase-1 in the rat ovary. *J. Endocrinol.* 158, 221–228.
184. Hagglund, A. C., Ny, A., Leonardsson, G., and Ny, T. (1999). Regulation and localization of matrix metalloproteinases and tissue inhibitors of metalloproteinases in the mouse ovary during gonadotropin-induced ovulation. *Endocrinology* 140, 4351–4358.
185. Balbin, M., Fueyo, A., Lopez, J. M., Diez-Itza, I., Velasco, G., and Lopez-Otin, C. (1996). Expression of collagenase-3 in the rat ovary during the ovulatory process. *J. Endocrinol.* 149, 405–415.
186. Riley, S. C., Thomassen, R., Bae, S. E., Leask, R., Pedersen, H. G., and Watson, E. D. (2004). Matrix metalloproteinase-2 and -9 secretion by the equine ovary during follicular growth and prior to ovulation. *Anim. Reprod. Sci.* 81, 329–339.
187. Nikolettos, N., Asimakopoulos, B., Tentes, L., Schopper, B., and al-Hasani, S. (2003). Matrix metalloproteinases 2 and 9 in follicular fluids of patients undergoing controlled ovarian stimulation for ICSI/ET. *In Vivo* 17, 201–204.
188. Yamakawa, S., Asai, T., Uchida, T., Matsukawa, M., Akizawa, T., and Oku, N. (2004). (-)-Epigallocatechin gallate inhibits membrane-type 1 matrix metalloproteinase, MT1-MMP, and tumor angiogenesis. *Cancer Lett.* 210, 47–55.
189. Atkinson, S. J., English, J. L., Holway, N., and Murphy, G. (2004). Cellular cholesterol regulates MT1 MMP dependent activation of MMP2 via MEK-1 in HT1080 fibrosarcoma cells. *FEBS Lett.* 566, 65–70.
190. Holmbeck, K., Bianco, P., Yamada, S., and Birkedal-Hansen, H. (2004). MT1-MMP: a tethered collagenase. *J. Cell. Physiol.* 200, 11–19.
191. Vu, T. H., and Werb, Z. (2000). Matrix metalloproteinases: effectors of development and normal physiology. *Genes Dev.* 14, 2123–2133.
192. Tang, B.L. (2001). ADAMTS: a novel family of extracellular matrix proteases. *Int. J. Biochem. Cell. Biol.* 33, 33–44.
193. Menke, D. B., and Page, D. C. (2002). Sexually dimorphic gene expression in the developing mouse gonad. *Gene Expr. Patterns* 2, 359–367.
194. Luetteke, N. C., Qiu, T. H., Fenton, S. E., Tryoer, K. L., Riedel, R. F., Chang, A., and Lee, D. C. (1999). Targeted inactivation of the EGF and amphiregulin genes reveals distinct roles for the EGF receptor ligands in mouse mammary gland development. *Development* 126, 2739–2750.
195. Polacek, D., Beckmann, M. W., and Schreiber, J. R. (1992). Rat ovarian apolipoprotein E: localization and gonadotropic control of messenger RNA. *Biol. Reprod.* 46, 65–72.
196. Robles, R., Morita, Y., Mann, K. K., Perez, G. I., Yang, S., Matikainen, T., Sherr, D. H., and Tilly, J. L. (2000). The aryl hydrocarbon receptor, a basic helix-loop-helix transcription factor of the PAS gene family, is required for normal ovarian germ cell dynamics in the mouse. *Endocrinology* 141, 450–453.
197. Chaffin, C. L., Stouffer, R. L., and Duffy, D. M. (1999). Gonadotropin and steroid regulation of steroid receptor and aryl hydrocarbon receptor messenger ribonucleic acid in macaque granulosa cells during the periovulatory interval. *Endocrinology* 140, 4753–4760.
198. Mizuyachi, K., Son, D. S., Rozman, K. K., and Terranova, P. F. (2002). Alteration in ovarian gene expression in response to 2,3,7,8-tetrachlorodibenzo-p-dioxin: reduction of cyclooxygenase-2 in the blockage of ovulation. *Reprod. Toxicol.* 16, 299–307.
199. Gonzalez-Robayna, I. J., Alliston, T. N., Buse, P., Firestone, G. L., and Richards, J. S. (1999). Functional and subcellular changes in the A-kinase-signaling pathway: relation to aromatase and Sgk expression during the transition of granulose cells to luteal cells. *Mol. Endocrinol.* 13, 1318–1337.
200. Mukherjee, A., Urban, J., Sassone-Corsi, P., and Mayo, K. E. (1998). Gonadotropin regulate inducible cyclic adenosine 3′,5′-monophosphate early repressor in the rat ovary: implications for inhibit-α subunit gene expression. *Mol. Endocrinol.* 12, 785–800.
201. Sassone-Corsi, P. (2000). CREM: a master switch regulating the balance between differentiation and apoptosis in male germ cells. *Mol. Reprod. Dev.* 56, 228–229.
202. Espey, L. L., Yoshioka, S., Russell, D., Ujioka, T., Vladu, B., Skelsey, M., Fujii, S., Okamura, H., and Richards, J. S. (2000). Characterization of ovarian carbonyl reductase gene expression during ovulation in the gonadotropin-primed immature rat. *Biol. Reprod.* 62, 390–397.
203. Brockstedt, E., Peters-Kottig, M., Badock, V., Hegele-Hartung, C., and Lessl, M. (2000). Luteinizing hormone induces mouse vas deferens protein expression in the murine ovary. *Endocrinology* 141, 2574–2581.
204. Benevides, F., Starost, M.F., Flores, M., Gimenez-Conti, I.B., Guenet, J.L., and Conti, C.J. (2002). Impaired hair follicle morphogenesis and cycling with abnormal epidermal differentiation in nackt mice, a cathepsin L-deficient mutation. *Am. J. Pathol.* 161, 693–703.
205. Piontkewitz, Y., Enerback, S., and Hedin, L. (1996). Expression of CCAAT enhancer binding protein-α (C/EBPα) in the rat ovary: implications for follicular development and ovulation. *Dev. Biol.* 179, 288–296.
206. Pall, M., Hellberg, P., Brannstrom, M., Mikuni, M., Peterson, C. M., Sundfeldt, K., Norden, B., Hedin, L., and Enerback, S. (1997). The transcription factor C/EBP-β and its role in ovarian function; evidence for direct involvement in the ovulatory process. *EMBO J.* 16, 5273–5279.
207. Mastorakos, G., Webster, E. L., Friedman, T. C., and Chrousos, G. P. (1993). Immunoreactive corticotropin-releasing hormone and its binding sites in the rat ovary. *J. Clin. Invest.* 92, 961–968.
208. Nappi, R. E., and Rivest, S. (1995). Stress-induced genetic expression of a selective corticotropin-releasing factor-receptor subtype within the rat ovaries, an effect dependent on the ovulatory cycle. *Biol. Reprod.* 53, 1417–1428.
209. Smith, G. W., Aubry, J. M., Dellu, F., Contarino, A., Bilezikjian, L. M., Gold, L. H., Chen, R., Marchuk, Y., Hause, C., Bentley, C. A., Sawchenko, P. E., Koob, G. F., Vale, W., and Lee, K. F. (1998). Corticotropin releasing factor receptor 1-deficient mice display decreased anxiety, impaired stress response and aberrant neuroendocine development. *Neuron.* 20, 1093–1102.
210. Leo, C. P., Pisarska, M. D., and Hsueh, A. J. (2001). DNA array analysis of changes in preovulatory gene expression in the rat ovary. *Biol. Reprod.* 65, 269–276.
211. Lim, H., Paria, B. C., Das, S. K., Dinchuk, J. E., Langenbach, R., Trzaskos, J. M., and Dey, S. K. (1997). Multiple female reproductive failures in cyclooxygenase 2-deficient mice. *Cell* 91, 197–208.
212. Boerboom, D., and Sirois, J. (2001). Equine P450 cholesterol side-chain cleavage and 3β-hydroxysteroid dehydrogenase/Δ(5)-Δ(4)-isomerase molecular cloning and regulation of their messenger ribonucleic acids in equine follicles during the ovulatory process. *Biol. Reprod.* 64, 206–215.
213. Komar, C. M., and Curry, T. E. Jr. (2003). Inverse relationship between the expression of messenger ribonucleic acid for peroxisome proliferator-activated receptor γ and P450 side chain cleavage in the rat ovary. *Biol. Reprod.* 69, 549–555.
214. Kaynard, A. H., Periman, L. M., Simard, J., and Melner, M. H. (1992). Ovarian 3β-hydroxysteroid dehydrogenase and sulfated glycoprotein-2 gene expression are differentially

regulated by the induction of ovulation, pseudopregnancy, and luteolysis in the immature rat. *Endocrinology* 130, 2192–2200.

215. Chaffin, C. L., Dissen, G. A., and Stouffer, R. L. (2000). Hormonal regulation of steroidogenic enzyme expression in granulosa cells during the peri-ovulatory interval in monkeys. *Mol. Hum. Reprod.* 6, 11–18.
216. Ushigoe, K., Irahara, M., Fukumochi, M., Kamada, M., and Aono, T. (2000). Production and regulation of cytokine-induced neutrophil chemoattractant in rat ovulation. *Biol. Reprod.* 63, 121–126.
217. Oral, E., Seli, E., Bahtiyar, M. O., Jones, E. E., and Arici, A. (1997). Growth-regulated-α expression in human preovulatory follicles and ovarian cells. *Am. J. Reprod. Immunol.* 38, 19–25.
218. Karstrom-Encrantz, L., Runesson, E., Bostrom, E. K., and Brannstrom, M. (1998). Selective presence of the chemokine growth-regulated oncogene-α (GRO-α) in the human follicle and secretion from cultured granulosa-lutein cells at ovulation. *Mol. Hum. Reprod.* 4, 1077–1083.
219. Piccinni, M. P., Scaletti, C., Mavilia, C., Lazzeri, E., Romagnani, P., Natali, I., Pellegrini, S., Livi, C., Romagnani, S., and Maggi, E. (2001). Production of IL-4 and leukemia inhibitory factor by T cell of the cumulus oophorus: a favorable microenvironement for pre-implantation embryo development. *Eur. J. Immunol.* 31, 2431–2437.
220. Russell, D. L., Doyle, K. M., Gonzales-Robayna, I., Pipaon, C., and Richards, J. S. (2003). Egr-1 induction in rat granulosa cells by follicle-stimulating hormone and luteinizing hormone: combinatorial regulation by transcription factors cyclic adenosine 3´,5´-monophosphate regulatory element binding protein, serum response factor, sp1, and early growth response factor-1. *Mol. Endocrinol.* 17, 520–533.
221. Espey, L. L., Ujioka, T., Russell, D., Skelsey, M., Vladu, B., Robker, R. L., Okamura, H., and Richards, J. S. (2000). Induction of early growth response protein-1 gene expression in the rat ovary in response to an ovulatory dose of human chorionic gonadotropin. *Endocrinology* 141, 2385–2391.
222. Mouillet, J. F., Sonnenberg-Hirche, C., Yan, X., and Sadovsky, Y. (2004). P300 regulates the synergy of steroidogenic factor-1 and early growth response-1 in activating luteinizing hormone beta subunit gene. *J. Biol. Chem.* 279, 7832–7839.
223. Horton, C. D., and Halvorson, L. M. (2004). The cAMP signaling system regulates LHbeta gene expression: roles of early growth response protein-1, Sp1, and steroidogenic factor-1. *J. Mol. Endocrinol.* 32, 291–306.
224. Sharma, S. C., and Richards, J. S. (2000). Regulation of AP1 (Jun/Fos) factor expression and activation in ovarian granulosa cells. Relation of JunD and Fra2 to terminal differentiation. *J. Biol. Chem.* 275, 33718–33728.
225. Hsieh, M., Johnson, M. A., Greenberg, N. M., and Richards, J. S. (2002). Regulated expression of Wnts and Frizzleds at specific stages of follicular development in the rodent ovary. *Endocrinology* 143, 898–908.
226. Hsieh, M., Mulders, S. M., Friis, R. R., Dharmarajan, A., and Richards, J. S. (2003). Expression and localization of secreted frizzled-related protein-4 in the rodent ovary: evidence for selective up-regulation in luteinized granulosa cells. *Endocrinology* 144, 4597–4606.
227. Drake, J. M., Friis, R. R., and Dharmarajan, A. M. (2003). The role of sFRP4, a secreted frizzled-related protein, in ovulation. *Apoptosis* 8, 389–397.
228. Espey, L. L., Yoshioka, S., Ujioka, T., Fujii, S., and Richards, J. S. (2001). 3α-hydroxysteroid dehydrogenase messenger RNA transcription in the immature rat ovary in response to an ovulatory dose of gonadotropin. *Biol. Reprod.* 65, 72–78.
229. Yong, P. Y., Harlow, C., Thong, K. J., and Hillier, S. G. (2002). Regulation of 11β-hydroxysteroid dehydrogenase type 1 gene expression in human ovarian surface epithelial cells by interleukin-1. *Hum. Reprod.* 17, 2300–2306.
230. Tetsuka, M., Haines, L. C., Milne, M., Simpson, G. E., and Hillier, S. G. (1999). Regulation of 11β-hydroxysteroid dehydrogenase type 1 gene expression by LH and interleukin-1β in cultured rat granulosa cells. *J. Endocrinol.* 163, 417–423.
231. Brown, K. A., Boerboom, D., Bouchard, N., Lussier, J. G., and Sirois, J. (2004). Human chorionic gonadotropin-dependent regulation of 17β-hydroxysteroid dehydrogenase type 4 in preovulatory follicles and its potential role in follicular luteinization. *Endocrinology* 145, 1906–1915.
232. Zhou, J., Wang, J., Penny, D., Monget, P., Arraztoa, J. A., Fogelson, L. J., and Bondy, C. A. (2003). Insulin-like growth factor binding protein 4 expression parallels luteinizing hormone receptor expression and follicular luteinization in the primate ovary. *Biol. Reprod.* 69, 22–29.
233. Kol, S., Kehat, I., and Adashi, E. Y. (2001). Ovarian interleukin-1-induced gene expression: privileged genes threshold theory. *Med. Hypotheses* 58, 6–8.
234. Ando, M., Kol, S., Kokia, E., Ruutiainen-Altman, K., Sirois, J., Rohan, R. M., Payne, D. W., and Adashi, E. Y. (1998). Rat ovarian prostaglandin endoperoxide synthase-1 and -2: periovulatory expression of granulosa cell-based interleukin-1-dependent enzymes. *Endocrinology* 139, 2501–2508.
235. Kol, S., Ruutianinen-Altman, K., Scherzer, W. J., Ben-Shlomo, I., Ando, M., Rohan, R. M., and Adashi, E. Y. (1999). The rat intraovarian interleukin (IL)-1 system: cellular localization, cyclic variation and hormonal regulation of IL-1β and of the type I and type II IL-1 receptors. *Mol. Cell Endocrinol.* 149, 115–128.
236. Levitas, E., Chamoun, D., Udoff, L. C., Ando, M., Resnick, C. E., and Adashi, E. Y. (2000). Periovulatory and interleukin-1β-dependent up-regulation of intraovarian vascular endothelial growth factor (VEGF) in the rat: potential role for VEGF in the promotion of periovulatory angiogenesis and vascular permeability. *J. Soc. Gynecol. Investig.* 7, 51–60.
237. Ryan, N. K., Van der Hoek, K. H., Robertson, S. A., and Norman, R. J. (2003). Leptin and leptin receptor expression in the ovary. *Endocrinology* 144, 5006–5013.
238. Duggal, P. S., Weitsman, S. R., Magoffin, D. A., and Norman, R. J. (2002). Expression of the long (OB-RB) and short (OB-RA) forms of the leptin receptor throughout the oestrous cycle in the mature rat ovary. *Reproduction* 123, 899–905.
239. Chehab, F. F., Lim, M. E., and Lu, R. (1996). Correction of the sterility defect in homozygous obese female mice by treatment with the human recombinant leptin. *Nat. Genet.* 12, 318–320.
240. Liu, D. L., Liu, W. Z., Li, Q. L., Wang, H. M., Qian, D., Treuter, E., and Zhu, C. (2003). Expression and functional analysis of liver receptor homologue 1 as a potential steroidogenic factor in rat ovary. *Biol. Reprod.* 69, 508–517.
241. Kim, J. W., Peng, N., Rainey, W. E., Carr, B. R., and Attia, G. R. (2004). Liver receptor homolog-1 regulates the expression of steroidogenic acute regulatory protein in human granulosa cells. *J. Clin. Endocinol. Metab.* 89, 3042–3047.
242. Falender, A. E., Lanz, R., Malenfant, D., Belanger, L., and Richards, J. S. (2003). Differential expression of steroidogenic factor 1 and FTF/LRH-1 in the rodent ovary. *Endocrinology* 144, 3598–3610.
243. Jeyasuria, P., Ikeda, Y., Jamin, S. P., Zhao, L., de Rooij, D. G., Themmen, A. P. N., Behringer, R. R., and Parker, K. L. (2004). Cell-specific knockout of steroidogenic factor 1 reveals its essential role in gonadal function. *Mol. Endocrinol.* 18, 1610–1619.

244. Gaddy-Kurten, D., Hickey, G. J., Fey, G. H., Gauldie, J., and Richards, J. S. (1989). Hormonal regulation and tissue-specific localization of a2-macroglobulin in rat ovarian follicles and corpora lutea. *Endocrinolgy* 125, 2985–2995.

245. Gaddy-Kurten, D., and Richards, J. S. (1991). Regulation of a2-macroglobulin by luteinizing hormone and prolactin during cell differentiation in the rat ovary. *Mol. Endocrinol.* 5, 1280–1291.

246. Dajee, M., Kazansky, A. V., Raught, B., Hocke, G. M., Fey G. H., and Richards, J. S. (1996). Prolactin induction of the α2-macroglobulin gene in rat ovarian granulose cells: Stat 5 activation and binding to the interleukin-6 response element. *Mol. Endocrinol.* 10, 171–184.

247. Zhu, C., and Woessner, J. F. Jr. (1991). A tissue inhibitor of metalloproteinases and α-macroglobulins in the ovulating rat ovary: possible regulators of collagen matrix breakdown. *Biol. Reprod.* 45, 334–342.

248. Hochepied, T., Van Leuven, F., and Libert, C. (2002). Mice lacking alpha2-macroglobulin shown an increased host defense against Gram-negative bacterial sepsis, but are more susceptible to endotoxic shock. *European Cytokine Network* 1, 86–91.

249. Wada, S., Kudo, T., Kudo, M., Sakuragi, N., Hareyama, H., Nishihira, J., and Fujimoto, S. (1999). Induction of macrophage migration inhibitory factor in human ovary by human chorionic gonadotrophin. *Hum. Reprod.* 14, 395–399.

250. Chaffin, C. L., and Stouffer, R. L. (1999). Expression of matrix metalloproteinases and their tissue inhibitor messenger ribonucleic acids in macaque periovulatory granulosa cells: time course and steroid regulation. *Biol. Reprod.* 61, 14–21.

251. Bakke, L. J., Dow, M. P., Cassar C. A., Peters, M. W., Pursley, J. R., and Smith, G. W. (2002). Effect of the preovulatory gonadotropin surge on matrix metalloproteinases (MMP)-14, MMP-2, and tissue inhibitor of metalloproteinases-2 expression within bovine periovulatory follicular and luteal tissue. *Biol. Reprod.* 66, 1627–1634.

252. Holmbeck, K., Bianco, P., Caterina, J., Yamada, S., and Kromer, M. (1999). MT1-MMP deficient mice develop dwarfism, osteopenia, arthritis, and connective tissue disease due to inadequate collagen turnover. *Cell* 99, 81–91.

253. Espey, L. L., Ujioka, T., Okamura, H., and Richards, J. S. (2003). Metallothionein-1 messenger RNA transcription in steroid-secreting cells of the rat ovary during the periovulatory period. *Biol. Reprod.* 68, 1895–1902.

254. Cameron, M. R., Foster, J. S., Bukovsky, A., and Wimalasena, J. (1996). Activation of mitogen-activated protein kinases by gonadotropin and cyclic adenosine 5′-monophosphates in porcine granulosa cells. *Biol. Reprod.* 55, 111–119.

255. Tajima, K., Dantes, A., Yao, Z., Sorokina, K., Kotsuji, F., Seger, R., and Amsterdam, A. (2003). Down-regulation of steroidogenic response to gonadotropin in human and rat preovulatory granulosa cells involves mitogen-activated protein kinase activation and modulation of DAX-1 and steroidogenic factor-1. *J. Clin. Endocrinol. Metab.* 88, 2288–2299.

256. Maizels, E. Y., Mukherjee, A., Sithanandam, G., Peters, C. A., Cottom, J., Mayo, K. E., and Hunzicker-Dunn, M. (2001). Developmental regulation of mitogen-activated protein kinase-activated kinases-2 and -3 (MAPKAPK-2/-3) in vivo during corpus luteum formation in the rat. *Mol. Endocrinol.* 15, 716–733.

257. Seto-Young, D., Zajac, J., Liu, H. C., Rosenwaks, Z., and Poretsky, L. (2003). The role of mitogen-activated protein kinase in insulin and insulin-like growth factor I (IGF-I) signaling cascades for progesterone and IGF-binding protein-1 production in human granulosa cells. *J. Clin. Endocrinol. Metab.* 88, 3385–3391.

258. Leo, C. P., Hsu, S. Y., Chun, S. Y., Bae, H. W., and Hsueh, A. J. (1999). Characterization of the antiapoptotic Bcl-2 family member myeloid cell leukemia-1 (Mcl-1) and the stimulation of its message by gonadotropin in the rat ovary. *Endocrinology* 140, 5465–5468.

259. Dissen, G. A., Hill, D. F., Costa, M. E., Les Dees, C. W., Lara, H. E., and Ojeda, S. R. (1996). A role for trkA nerve growth factor receptors in mammalian ovulation. *Endocrinology* 137, 198–209.

260. Dissen, G. A., Parrott, J. A., Skinner, M. K., Hill, D. F., Costa, M. E., and Ojeda, S. R. (2000). Direct effects of nerve growth factor on thecal cells from antral ovarian follicles. *Endocrinology* 141, 4736–4750.

261. Yamagata, Y., Nakamura, Y., Sugino, N., Harada, A., Takayama, H., Kashida, S., and Kato, H. (2002). Alterations in nitrate/nitrite and nitric oxide synthase in preovulatory follicles in gonadotropin-primed immature rats. *Endocr. J.* 49, 219–226.

262. Nakamura, Y., Kashida, S., Nakata, M., Takiguchi, S., Yamagata, Y., Takayama, H., Sugino, N., and Kato, H. (1999). Changes in nitric oxide synthase activity in the ovary of gonadotropin treated rats: the role of nitric oxide during ovulation. *Endocr. J.* 46, 529–538.

263. Jablonka-Shariff, A., and Olson, L. M. (1997). Hormonal regulation of nitric oxide synthases and their cell-specific expression during follicular development in the rat ovary. *Endocrinology* 138, 460–468.

264. Faletti, A. G., Mohn, C., Farina, M., Lomniczi, A., and Rettori, V. (2003). Interaction among β-endorphin, nitric oxide and protaglandins during ovulation in rats. *Reproduction* 125, 469–477.

265. Mitchell, L. M., Kennedy, C. R., and Hartshorne, G. M. (2004). Expression of nitric oxide synthase effect of substrate manipulation of the nitric oxide pathway in mouse ovarian follicles. *Hum. Reprod.* 19, 30–40.

266. Yaron, Y., Schwartz, D., Evans, M. I., Aloni, R., Kapon, A., and Rotter, V. (1999). p53 tumor suppressor gene expression in the mouse ovary during an artificially induced ovulatory cycle. *J. Reprod. Med.* 44, 107–114.

267. Yoshioka, S., Fujii, S., Richards, J. S., and Espey, L. L. (2002). Gonadotropin-induced expression of pancreatitis-associated protein-III mRNA in the rat ovary at the time of ovulation. *J. Endocrinol.* 174, 485–492.

268. Taylor, C. C., Limback, D., and Terranova, P. F. (1997). Src tyrosine kinase activity in rat thecal-interstitial cells and mouse TM3 Leydig cells is positively associated with cAMP-specific phosphodiesterase activity. *Mol. Cell. Endocrinol.* 126, 91–100.

269. Park, J. Y., Richard, F., Chun, S. Y., Park, J. H., Law, E., Horner, K., Jin, S. L., and Conti, M. (2003). Phosphodiesterase regulation is critical for the differentiation and pattern of gene expression in granulosa cells of the ovarian follicle. *Mol. Endocrinol.* 17, 1117–1130.

270. Apa, R., Lanzone, A., Miceli, F., Vaccari, S., Macchione, E., Stefanini, M., and Canipari, R. (2002). Pituitary adenylate cyclase-activating polypeptide modulates plasminogen activator expression in rat granulosa cell. *Biol. Reprod.* 66, 830–835.

271. Jo, M., Komar, C. M., and Fortune, J. E. (2002). Gonadotropin surge induces two separate increases in messenger RNA for progesterone receptor in bovine preovulatory follicles. *Biol. Reprod.* 67, 1981–1988.

272. Gava, N., Clarke, C. L., Byth, K., Arnett-Mansfield, R. L., and deFazio, A. (2004). Expression of progesterone receptors A and B in the mouse ovary during the estrous cycle. *Endocrinology* 145, 3487–3494.

272. Russell, D. L., and Richards, J. S. (1999). Differentiation-dependent prolactin responsiveness and Stat (signal

transducers and activators of transcription) signaling in rat ovarian cells. *Mol. Endocrinol.* 13, 2049–2064.

274. Vlahos, N. P., Bugg, E. M., Shamblott, M. J., Phelps, J. Y., Gearhart, J. D., and Zacur, H. A. (2001). Prolactin receptor gene expression and immunolocalization of the prolactin receptor in human luteinized granulosa cells. *Mol. Hum. Reprod.* 7, 1033–1038.
275. Grosdemouge, I., Bachelot, A., Lucas, A., Baran, N., Kelly, P. A., and Binart, N. (2003). Effects of deleting the prolactin receptor on ovarain gene expression. *Reprod. Biol. Endocrinol.* 6, 1–12.
276. Parker, M., Leonardsson, G., White, R., Steel, J., and Milligan, S. (2003). Identification of RIP140 as a nuclear receptor cofactor with a role in female reproduction. *FEBS Lett.* 546, 149–153.
277. White, R., Leonardsson, G., Rosewell, I., Ann Jacobs, M., Milligan, S., and Parker, M. (2000). The nuclear receptor co-repressor nrip1 (RIP140) is essential for female fertility. *Nat. Med.* 6, 1368–1374.
278. Leonardsson, G., Jacobs, M. A., White, R., Jeffery, R., Poulsom, R., Milligan, S., and Parker, M. (2002). Embryo transfer experiments and ovarian transplantation identify the ovary as the only site in which nuclear receptor interacting protein 1/RIP140 action is crucial for female fertility. *Endocrinology* 143, 700–707.
279. Ujioka, T., Russell, D. L., Okamura, H., Richards, J. S., and Espey, L. L. (2000). Expression of regulator of G-protein signaling protein-2 gene in the rat ovary at the time of ovulation. *Biol. Reprod.* 63, 1513–1517.
280. Doggrell, S. A. (2004). Is RGS-2 a new drug development target in cardiovascular disease? *Expert. Opin. Ther. Targets.* 8, 355–358.
281. Kerban, A., Boerboom, D., and Sirois, J. (1999). Human chorionic gonadotropin induces an inverse regulation of steroidogenic acute regulatory protein messenger ribonucleic acid in theca interna and granulosa cells of equine preovulatory follicles. *Endocrinology* 140, 667–674.
282. Hasegawa, T., Zhao, L., Caron, K. M., Majdic, G., Suzuki, T., Shizawa, S., Sasano, H., and Parker, K. L. (2000). Developmental roles of the steroidogenic actue regulatory protein (StAR) as revealed by StAR knockout mice. *Mol. Endocrinol.* 14, 1462–1471.
283. Sato, E. F., Kobuchi, H., Edashige, K., Takahashi, M., Yoshioka, T., Utsumi, K., and Inoue, M. (1992). Dynamic aspects of ovarian superoxide dismutase isozymes during the ovulatory process in the rat. *FEBS Lett.* 303, 121–125.
284. Sasaki, J., Sato, E. F., Nomura, T., Mori, H., Watanabe, S., Kanda, S., Watanabe, H., Utsumi, K., and Inoue, M. (1994). Detection of manganese superoxide dismutase mRNA in the theca interna cells of rat ovary during the ovulatory process by in situ hybridization. *Histochemistry* 102, 173–176.
285. Nomura, T., Sasaki, J., Mori, H., Sato, E. F., Watanabe, S., Kanda, S., Matsuura, J., Watanabe, H., and Inoue, M. (1996). Expression of manganese superoxide dismutase mRNA in reproductive organs during the ovulatory process and the estrous cycle of the rat. *Histochem. Cell. Biol.* 105, 1–6.
286. Lebovitz, R. M., Zhang, H., Vogel, H., Cartwright, J. Jr., Dionne, L., Lu, N., Huang, S., and Matzuk, M. M. (1996). Neurodegeneration, myocardial injury, and perinatal death in mitochondrial superoxide dismutase-deficient mice. *Proc. Natl. Acad. Sci. U S A* 93, 9782–9787.
287. Simpson, K. S., Byers, M. J., and Curry, T. E. Jr. (2001). Spatiotemporal messenger ribonucleic acid expression of ovarian tissue inhibitors of metalloproteinases throughout the rat estrous cycle. *Endocrinology* 142, 2058–2069.
288. Komar, C. M., Matousek, M., Mitsube, K., Mikuni, M., Brannstrom, M., and Curry, T. E. Jr. (2001). Effects of genistein on the periovulatory expression of messenger ribonucleic acid for matrix metalloproteinases and tissue inhibitors of metalloproteinases in the rat ovary. *Reproduction* 121, 259–265.
289. Curry, T. E. Jr., and Wheeler, S. E. (2002). Cellular localization of tissue inhibitors of metalloproteinase in the rat ovary throughout pseudopregnancy. *Biol. Reprod.* 67, 1943–1951.
290. Northnick, W. B. (2001). Reduction in reproductive lifespan of tissue inhibitor of metalloproteinase 1 (TIMP-1)-deficient female mice. *Reproduction* 122, 923–927.
291. Northnick, W. B. (2002). Tissue inhibitor of metalloproteinase-1 (TIMP-1) deficient mice display reduced serum progesterone levels during corpus luteum development. *Endocrinology* 144, 5–8.
292. Terranova, P. F., and Rice, V. M. (1997). Review: cytokine involvement in ovarian processes. *Am. J. Reprod. Immunol.* 37, 50–63.
293. Roby, K. F., Son, D. S., and Terranova, P. F. (1999). Alterations of events related to ovarian functions in tumor necrosis factor receptor type I knockout mice. *Biol. Reprod.* 61, 1616–1621.
294. Chen, Y. J., Feng, Q., and Liu, Y. X. (1999). Expression of the steroidogenic acute regulatory protein and luteinizing hormone receptor and their regulation by tumor necrosis factor-α in rat corpora lutea. *Biol. Reprod.* 60, 419–427.
295. Yoshioka, S., Ochsner, S., Russell, D. L., Ujioka, T., Fujii, S., Richards, J. S., and Espey, L. L. (2000). Expression of tumor necrosis factor-stimulated gene-6 in the rat ovary in response to an ovulatory dose of gonadotropin. *Endocrinology* 141, 4114–4119.
296. Garrido, C., Saule, S., and Gospodarowicz, D. (1993). Transcriptional regulation of vascular endothelial growth factor gene expression in ovarian bovine granulosa cells. *Growth Factors* 8, 109–117.
297. Koos, R. D. (1995). Increased expression of vascular endothelial growth/permeability factor in the rat ovary following an ovulatory gonadotropin stimulus: potential roles in follicular rupture. *Biol. Reprod.* 52, 1426–1435.
298. Russell, D. L., Ochsner, S. A., Hseih, M., Mulders, S., and Richards, J. S. (2003). Hormone-regulated expression and localization of versican in the rodent ovary. *Endocrinology* 144, 1020–1031.
299. Vainio, S., Heikkila, M., Kispert, A., Chin, N., and McMahon, A. P. (1999). Female development in mammals is regulated by Wnt-4 signalling. *Nature* 397, 405–409.
300. Richards, J. S. (2001). Perspective: the ovarian follicle—a perspective in 2001. *Endocrinology* 142, 2184–2193.
301. Thai, S. N., and Iruela-Arispe, M. L. (2002). Expression of ADAMTS1 during murine development. *Mech. Dev.* 115, 181–185.
302. Sugiura, T. (2003). Baculoviral expression of correctly processed ADAMTS proteins fused with the human IgG-Fc region. *J. Biotechnol.* 100, 193–201.
303. Rodriguez-Manzaneque, J. C., Westling, J., Thai, S. N., Luque, A., Knauper, V., Murphy, G., Sandy, J. D., and Iruela-Arispe, M. L. (2002). ADAMTS1 cleaves aggrecan at multiple sites and is differentially inhibited by metalloproteinase inhibitors. *Biochem. Biophys. Res. Commun.* 293, 501–508.
304. Sewer, M. B., and Waterman, M. R. (2003). ACTH modulation of transcription factors responsible for steroid hydroxylase gene expression in the adrenal cortex. *Microsc. Res. Tech.* 61, 300–307.
305. Rodgers, R. J., Lavranos, T. C., Rodgers, H. F., Young, F. M., and Vella, C. A. (1995). The physiology of the ovary: maturation of ovarian granulosa cells and a novel role for antioxidants in the corpus luteum. *J. Steroid Biochem.* 53, 241–246.

306. Hum, D. W., and Miller, W. L. (1993). Transcriptional regulation of human genes for steroidogenic enzymes. *Clin. Chem.* 39, 333–340.
307. Oh-hama, T. (1997). Evolutionary consideration on 5-aminolevulinate synthase in nature. *Origins Life Evol. Biosph.* 27, 405–412.
308. Thunell, S. (2000). Porphyrins, porphyrin metabolism and porphyrias. I. Update. *Scand. J. Clin. Lab. Invest.* 60, 509–540.
309. Suffredini, A. F., Fantuzzi, G., Badolato, R., Oppenheim, J. J., and O'Grady, N. P. (1999). New insights into the biology of the acute phase response. *J. Clin. Immunol.* 19, 203–214.
310. Lee, D. C., Sunnarborg, S. W., Hinkle, C. L., Myers, T. J., Stevenson, M. Y., Russell, W. E., Castner, B. J., Gerhart, M. J., Paxton, R. J., Black, R. A., Chang, A., and Jackson, L. F. (2003). TACE/ADAM17 processing of EGFR ligands indicates a role as a physiological convertase. *Ann. N Y Acad. Sci.* 995, 22–38.
311. Giudice, L. C. (1999). Genes associated with embryonic attachment and implantation and the role of progesterone. *J. Reprod. Med.* 44 (2 Suppl), 165–171.
312. Rice, A., and Chard, T. (1998). Cytokines in implantation. *Cytokine Growth Factor Rev.* 9, 287–296.
313. Nicosia, M., Moger, W. H., Dyer, C. A., Prack, M. M., and Williams, D. L. (1992). Apolipoprotein-E messenger RNA in rat ovary is expressed in theca and interstitial cells and presumptive macrophage, but not in granulosa cells. *Mol. Endocrinol.* 6, 978–988.
314. Larkin, L., Khachigian, L. M., and Jessup, W. (2000). Regulation of apolipoprotein E production in macrophages (review). *Int. J. Mol. Med.* 6, 253–258.
315. Ashavaid, T. F., Todur, S. P., and Nair, K. G. (2003). Apolipoprotein E polymorphism and coronary heart disease. *J. Assoc. Physicians India* 51, 784–788.
316. LaDu, M. J., Shah, J. A., Reardon, C. A., Getz, G. S., Bu, G., Hu, J., Guo, L., and Van Eldik, L. J. (2001). Apolipoprotein E and apolipoprotein E receptors modulate A β-induced glial neuroinflammatory responses. *Neurochem. Int.* 39, 427–434.
317. Denison, M. S., and Nagy, S. R. (2003). Activation of the aryl hydrocarbon receptor by structurally diverse exogenous and endogenous chemicals. *Annu. Rev. Pharmacol. Toxicol.* 43, 309–334.
318. Benedict, J. C., Miller, K. P., Lin, T. M., Greenfeld, C., Babus, J. K., Peterson, R. E., and Flaws, J. A. (2003). Aryl hydrocarbon receptor regulates growth, but not atresia, of mouse preantral and antral follicles. *Biol. Reprod.* 68, 1511–1517.
319. Eflerink, C. J. (2003). Aryl hydrocarbon receptor-mediated cell cycle control. *Prog. Cell. Cycle Res.* 5, 261–267.
320. Dalton, T. P., Puga, A., and Shertzer, H. G. (2002). Induction of cellular oxidative stress by aryl hydrocarbon receptor activation. *Chem. Biol. Interact.* 141, 44–95.
321. Tian, Y., Rabson, A. B., and Gallo, M. A. (2002). Ah receptor and NF-kappaB interactions: mechanisms and physiological implications. *Chem. Biol. Interact.* 141, 97–115.
322. Dunbar, A. J., and Goddard, C. (2000). Structure–function and biological role of betacellulin. *Int. J. Biochem. Cell. Biol.* 32, 805–815.
323. Rosenberg, D., Groussin, L., Jullian, E., Perlemoine, K., Bertagna, X., and Bertherat, J. (2002). Role of the PKA-regulated transcription factor CREB in development and tumorigenesis of endocrine tissues. *Ann. N. Y. Acad. Sci.* 968, 65–74.
324. Quinn, P. G. (2002). Mechanisms of basal and kinase-inducible transcription activation by CREB. *Prog. Nucleic Acid Res. Mol. Biol.* 72, 269–305.
325. Russell, D. L., Doyle, K. M., Gonzales-Robayna, I., Pipaon, C., and Richards, J. S. (2003). Egr-1 induction in rat granulosa cells by follicle-stimulating hormone and luteinizing hormone: combinatorial regulation by transcription factors cyclic adenosine 3′,5′-monophosphate regulatory element binding protein, serum response factor, sp1, and early growth response factor-1. *Mol. Endocrinol.* 17, 520–533.
326. Servillo, G., Della Fazia, M. A., and Sassone-Corsi, P. (2002). Coupling cAMP signaling to transcription in the liver: pivotal role of CREB and CREM. *Exp. Cell. Res.* 275, 143–154.
327. Don, J., and Stelzer, G. (2002). The expanding family of CREB/CREM transcription factors that are involved with spermatogenesis. *Mol. Cell. Endocrinol.* 187, 115–124.
328. Forrest, G. L., and Gonzalez, B. (2000). Carbonyl reductase. *Chem. Biol. Interact.* 129, 21–40.
329. Dong, Z., Katar, M., Linebaugh, B. E., Sloane, B. F., and Berk, R. S. (2001). Expression of cathepsins B, D and L in mouse corneas infected with Pseudomonas aeruginosa. *Eur. J. Biochem.* 268, 6408–6416.
330. Fiebiger, E., Maehhr, R., Villadangos, J., Weber, E., Erickson, A., Bikoff, E., Ploegh, H. L., and Lennon-Dumenil, A. M. (2002). Invariant chain controls the activity of extracellular cathepsin L. *J. Exp. Med.* 196, 1263–1269.
331. Levicar, N., Nuttall, R. K., and Lah, T. T. (2003). Proteases in brain tumour progression. *Acta Neurochir (Wien)* 145, 825–838.
332. Poli, V. (1998). The role of C/EBP isoforms in the control of inflammatory and native immunity functions. *J. Biol. Chem.* 273, 29279–29282.
333. Cassel, T. N., and Nord, M. (2003). C/EBP transcription factors in the lung epithelium. *Am. J. Physiol. Lung Cell. Mol. Physiol.* 285, 773–781.
334. Ebo, D. G., Hagendorens, M. M., Bridts, C. H., Schuerwegh, A. J., Clerck, L. S., and Stevens, W. J. (2004). In vitro allergy diagnosis: should we flow? *Clin. Exp. Allergy.* 34, 332–339.
335. Escribano, L., Diaz-Agustin, B., Nunes, R., Prados, A., Rodriguez, R., and Orfao, A. (2002). Abnormal expression of CD antigens in mastocytosis. *Int. Arch. Allergy Immunol.* 127, 127–132.
336. Castells, M. C. (2004). Mastocytosis: classification, diagnosis, and clinical presentation. *Allergy Asthma Proc.* 25, 33–36.
337. Vischer, U. M., and Wagner, D. D. (1993). CD63 is a component of Weibel-Palade bodies of human endothelial cells. *Blood* 82, 1184–1191.
338. Hannah, M. J., Williams, R., Kaur, J., Hewlett, L. J., and Cutler, D. F. (2002). Biogenesis of Weibel-Palade bodies. *Semin. Cell. Biol.* 13, 313–324.
339. Schafer, M., Mousa, S. A., and Stein, C. (1997). Corticotropin-releasing factor in antinociception and inflammation. *Eur. J. Pharmacol.* 323, 1–10.
340. Baigent, S. M. (2001). Peripheral corticotropin-releasing hormone and urocortin in the control of the immune response. *Peptides* 22, 809–820.
341. Mastorkos, G., and Ilias, I. (2003). Maternal and fetal hypothalamic-pituitary-adrenal axes during pregnancy and postpartum. *Ann. N. Y. Sci.* 997, 136–149.
342. Kalantaridou, S. N., Makrigiannakis, A., Mastorakos, G., and Chrousos, G. P. (2003). Roles of reproductive corticotropin-releasing hormone. *Ann. N. Y. Acad. Sci.* 997, 129–135.
343. Bale, T. L., and Vale, W. W. (2004). CRF and CRF receptors: role in stress responsivity and other behaviors. *Annu. Rev. Pharmacol. Toxicol.* 44, 525–527.
344. Watanabe, R., Fujii, H., Yamamoto, A., Yamaguchi, H., Takenouchi, T., Kameda, K., Ito, M., and Ono, T. (1996). Expression of cutaneous fatty acid-binding protein and its mRNA in rat skin. *Arch. Dermatol. Res.* 289, 184.
345. Schurer, N. Y. (2002). Implementation of fatty acid carriers to skin irritation and the epidermal barrier. *Contact Dermatitis* 47, 199–205.

346. Gately, S. (2000). The contributions of cyclooxygenase-2 to tumor angiogenesis. *Cancer Metastasis Rev.* 19, 19–27.
347. Crofford, L. J. (1997). COX-1 and COX-2 tissue expression: implications and predictions. *J. Rheumatol.* 49 (suppl.), 15–19.
348. Ristimaki, A. (2004). Cyclooxygenase 2: from inflammation to caricinogenesis. *Novartis Found. Symp.* 256, 215–221.
349. Bunn, R. C., and Fowlkes, J. L. (2003). Insulin-like growth factor binding protein proteolysis. *Trends Endocrinol. Metab.* 14, 176–181.
350. Thomson, M. (2003). Does cholesterol use the mitochondrial contact site as a conduit to the steroidogenic pathway? *Bioessays* 25, 252–258.
351. Hanukoglu, I., Suh, B. S., Himmelhoch, S., and Amsterdam, A. (1990). Induction and mitochondrial localization of cytochrome P450scc system enzymes in normal and transformed ovarian granulosa cells. *J. Cell. Biol.* 111, 1373–1381.
352. Sahnoun Z., Jamoussi, K., Zeghal, K. M. (1998). Free radicals and antioxidants: physiology, human pathology and therapeutic aspects (part II). *Therapie* 53, 315–339.
353. Suzuki, H., Mori, M., Seto, K., Shibata, F., Nagahashi, S., Kawaguchi, C., Suzuki, M., Matsui, H., Watanabe, K., Miura, S., and Ishii, H. (2000). Rat CXC chemokine GRO/CINC-1 paradoxically stimulate the growth of gastric epithelial cells. *Aliment. Pharmacol. Ther.* 14 (suppl 1), 94–100.
354. Shibata, F. (2002). The role of rat cytokine-induced neutrophil chemoattractants (CINCs) in inflammation. *Yakugaku Zasshi* 122, 263–268.
355. Irahara, M., Yasui, T., Tezuka, M., Ushigoe, K., Yamano, S., Kamada, M., and Aono, T. (2000). Evidence that Toki-shakuyaku-san and its ingredients enhance the secretion of a cytokine-induced neutrophil chemoattractant (CINC/gro) in the ovulatory process. *Methods Find. Exp. Clin. Pharmacol.* 22, 725–730.
356. Yasui, T., Matsuzaki, T., Ogata, R., Kiyokawa, M., Ushigoe, K., Uemura, H., Kuwahara, A., Ikawa, H., Maegawa, M., Furumoto, H., Aono, T., and Irahara, M. (2003). The herbal medicine Unkei-to stimulates the secretion of a cytokine-induced neutrophil chemoattractant, CINC/gro, in the rat ovarian cell culture. *Am. J. Reprod. Immunol.* 49, 14–20.
357. Gashler, A., and Sukhatme, V. P. (1995). Early growth response protein 1 (Egr-1): prototype of a zinc-finger family of transcription factors. *Prog. Nucleic Acids Res. Mol. Biol.* 50, 191–224.
358. Kachigian, L. M., and Collins, T. (1998). Early growth response factor 1: a pleiotropic mediator of inducible gene expression. *J. Mol. Med.* 76, 613–616.
359. Okada, M., Fujita, T., Olson, K. E., Collins, T., Stern, D. M., Yan, S. F., and Pinsky, D. J. (2001). Extinguishing Egr-1-dependent inflammatory and thrombotic cascades after lung transplantation. *FASEB J.* 15, 2757–2759.
360. Silverman, E. S., DeSanctis, G. T., Boyce, J., Maclean, J. A., Jiao, A., Green, F. H., Grasemann, H., Faunce, D., Fitzmaurice, G., Shi, G. P., Stein-Streilein, J., Milbrandt, J., Collins, T., and Drazen, J. M. (2001). The transcription factor early growth-response factor 1 modulates tumor necrosis factor-α, immunoglobulin E, and airway responsiveness in mice. *Am. J. Respir. Crit. Care Med.* 163, 778–785.
361. Adashi, E. Y. (1990). The potential relevance of cytokines to ovarian physiology: the emerging role of resident ovarian cells of the white blood cell series. *Endocr. Rev.* 11, 454–464.
362. Shirakata, Y., Komurasaki, T., Toyoda, H., Hanakawa, Y., Yamasaki, K., Tokumaru, S., Sayama, K., and Hashimoto, K. (2000). Epiregulin, a novel member of the epidermal growth factor family, is an autocrine growth factor in normal human keratinocytes. *J. Biol. Chem.* 275, 5748–5753.
363. Robert, C., Gagne, D., Bousquet, D., Barnes, F. L., and Sirard, M. A. (2001). Differential display and suppressive subtractive hybridization used to identify granulosa cell messenger RNA associated with bovine oocyte developmental competence. *Biol. Reprod.* 64, 1812–1820.
364. Li, X., Massa, P. E., Hanidu, A., Peet, G. W., Aro, P., Savitt, A., Mische, S., Li, J., and Marcu, K. B. (2002). IKKα, IKKβ, and NEMO/IKKγ are each required for the NF-kB-mediated inflammatory response program. *J. Biol. Chem.* 227, 45129–45140.
365. Tulchinsky, E. (2000). Fos family members: regulation, structure and role in oncogenic transformation. *Histol. Histopathol.* 15, 921–928.
366. Reddy, S. P., and Mossman, B. T. (2002). Role and regulation of activator protein-1 in toxicant-induced responses of the lung. *Am. J. Physiol. Lung. Cell. Mol. Physiol.* 283, L1161–L1178.
367. Malbon, C. C. (2004). Frizzleds: new members of the superfamily of G-protein-coupled receptors. *Front. Biosci.* 9, 1048–1058.
368. Hsieh, J. C. (2004). Specificity of WNT-receptor interactions. *Front. Biosci.* 9, 1333–1338.
369. Goodwin, A. M., and D'Amore, P. A. (2002). Wnt signaling in the vasculature. *Angiogenesis* 5, 1–9.
370. Jones, S. E., and Jomary, C. (2002). Secreted Frizzled-related proteins: searching for relationships and patterns. *Bioessays* 24, 811–820.
371. Kawano, Y., and Kypta, R. (2003). Secreted antagonists of the Wnt signaling pathway. *J. Cell. Sci.* 116, 2627–2634.
372. Rahman, I. (1999). Inflammation and the regulation of glutathione level in lung epithelial cells. *Antioxid. Redox Signal* 1, 425–447.
373. Soltaninassab, S. R., Sekhar, K. R., Meredity, M. J., and Freeman, M. L. (2000). Multifaceted regulation of γ-glutamylcysteine synthetase. *J. Cell. Physiol.* 182, 163–170.
374. Wild, A. C., and Mulcahy, R. T. (2000). Regulation of γ-glutamylcysteine synthetase subunit gene expression: insights into transcriptional control of antioxidant defenses. *Free Radic. Res.* 31, 281–301.
375. Anderson, M. E. (1998). Glutathione: an overview of biosynthesis and modulation. *Chem. Biol. Interact.* 111–112, 1–14.
376. Rahman, I., and MacNee, W. (2000). Oxidative stress and regulation of glutathione in lung inflammation. *Eur. Respir. J.* 16, 534–554.
377. Griffith, O. W. (1999). Biologic and pharmacologic regulation of mammalian glutathione synthesis. *Free Radic. Biol. Med.* 27, 922–935.
378. Hayes, J. D., and Strange, R. C. (2000). Glutathione S-transferase polymorphisms and their biological consequences. *Pharmacology* 61, 154–166.
379. Rahman, Q., Abidi, P., Afaq, F., Schiffmann, D., Mossman, B. T., Kamp, D. W., and Athar, M. (1999). Glutathione redox system in oxidative lung injury. *Crit. Rev. Toxicol.* 29, 543–568.
380. Adler, V., Yin, Z., Tew, K. D., and Ronai, Z. (1999). Role of redox potential and reactive oxygen species in stress signaling. *Oncogene* 18, 6104–6111.
381. Spiteri, M. A., Biancoo, A., Strange, R. C., and Fryer, A. A. (2000). Polymorphisms at the glutathione S-transferase, GSTP1 locus: a novel mechanism for susceptibility and development of atopic airway inflammation. *Allergy* 55 (suppl 66), 15–20.
382. Townsend, P. A., and Tew, K. D. (2003). The role of glutathione-S-transferase in anti-cancer drug resistance. *Oncogene* 22, 7369–7375.
383. Kotani, M., Detheux, M., Vandenbogaerde, A., Communi, D., Vanderwinden, J. M., Le Poul, E., Brezillon, S., Tyldesley, R., Suarez-Huerta, N., Vandeput, F., Blanpain, C., Schiffmann, S. N., Vassart, G., and Parmentier, M. (2001). The metastasis suppressor gene KiSS-1 encodes kisspeptins,

the natural ligands of the orphan G protein-coupled receptor GPR54. *J. Biol. Chem.* 276, 34631–34636.

384. Ohtaki, T., Shintani, Y., Honda, S., Matsumoto, H., Hori, A., Kanehashi, K., Terao, Y., Kumano, S., Takatsu, Y., Masuda, Y., Ishibashi, Y., Watanabe, T., Asada, M., Yamada, T., Suenaga, M., Kitada, C., Usuki, S., Kurokawa, T., Ond, H., Nishimura, O., and Fujino, M. (2001). Metastasis suppressor gene KiSS-1 encodes peptide ligand of a G-protein-coupled receptor. *Nature* 411, 613–617.
385. Seminara, S. B., Messager, S., Chatzidaki, E. E., Thresher, R. R., Acierno, J. S. Jr., Shagoury, J. K., Bo-Abbas, Y., Kuohung, W., Schwinof K. M., Hendrick, A. G., Zahn, D., Dixon, J., Kaiser, U. B., Slaugenhaupt, S. A., Gusella, J. F., O'Rahilly, S., Carlton, M. B., Crowley, W. F. Jr., Aparicio, S. A., and Colledge, W. H. (2003). The GPR54 gene as a regulator of puberty. *N. Engl. J. Med.* 349, 1614–1627.
386. Penning, T. M., Pawlowski, J. E., Schlegel, B. P., Jez, J. M., Lin, H. K., Hoog, S. S., Bennett, M. J., and Lewis, M. (1996). Mammalian 3α-hydroxysteroid dehydrogenases. *Steroids* 61, 508–523.
387. Ma, H., and Penning, T. M. (1999). Conversion of mammalian 3α-hydroxysteroid dehydrogenase to 20α-hydroxysteroid dehydrogenase using loop chimeras: changing specificity from androgens to progestins. *Proc. Natl. Acad. Sci. U S A* 96, 11161–11166.
388. Yamamoto, T., Matsuura, K., Shintani, S., Hara, A., Miyabe, Y., Sugiyama, T., and Katagiri, Y. (1998). Dual effects of anti-inflammatory 2-arylpropionic acid derivatives on a major isoform of human liver 3α-hydroxysteroid dehydrogenase. *Biol. Pharm. Bull.* 21, 1148–1153.
389. Matsuura, K., Shiraishi, H., Hara, A., Sato, K., Deyashiki, Y., Ninomiya, M., and Sakai, S. (1998). Identification of a principal mRNA species for human 3α-hydroxysteroid dehydrogenase isoform (AKR1C3) that exhibits high prostaglandin D2 11-ketoreductase active. *J. Biochem.* 124, 940–946.
390. Blum, A., and Maser, E. (2003). Enzymology and molecular biology of glucocorticoid metabolism in humans. *Prog. Nucleic Acid Res. Mol. Biol.* 75, 173–216.
391. Holmes, M. C., Yau, J. L., Kotelevtsev, Y., Mullins, J. J., and Seckl, J. R. (2003). 11β-hydroxysteroid dehydrogenases in the brain: two enzymes two roles. *Ann. N. Y. Acad. Sci.* 1007, 357–366.
392. Poirier, D. (2003). Inhibitors of 17β-hydroxysteroid dehydrogenases. *Curr. Med. Chem.* 10, 453–477.
393. Baker, M. E. (2001). Evolution of 17β-hydroxysteroid dehydrogenases and their role in androgen, estrogen and retinoid action. *Mol. Cell. Endocrinol.* 171, 211–215.
394. Mazerbourg, S., Bondy, C. A., Zhou, J., and Monget, P. (2003). The insulin-like growth factor system: a key determinant role in the growth and selection of ovarian follicles? A comparative species study. *Reprod. Domest. Anim.* 38, 247–258.
395. Zhou, R., Diehl, D., Hoeflich, A., Lahm, H., and Wolf, E. (2003). IGF-binding protein-4: biochemical characteristics and functional consequences. *J. Endocrinol.* 178, 177–193.
396. Stylianou, E., and Saklatvala, J. (1998). Interleukin-1. *Int. J. Biochem. Cell. Biol.* 30, 1075–1079.
397. Saliba, E., and Henrot, A. (2001). Inflammatory mediators and neonatal brain damage. *Biol. Neonate* 79, 224–227.
398. Dinarello, C. A. (2002). The IL-1 family and inflammatory diseases. *Clin. Exp. Rheumatol.* 20, S1–S13.
399. Dayer, J. M. (2002). Evidence for the biological modulation of IL-1 activity: the role of IL-1Ra. *Clin. Exp. Rheumatol.* 20, S14–S20.
400. Fernandes, J. C., Martel-Pelletier, J. P., and Pelletier, J. P. (2002). The role of cytokines in osteoarthritis pathophysiology. *Biorheology* 39, 237–246.
401. Doucet, C., Brouty-Boye, D., Pottin-Clemenceau, C., Canonica, G. W., Jasmin, C., and Azzarone, B. (1998). Interleukin (IL) 4 and IL-13 act on human lung fibroblasts. Implication in asthma. *J. Clin. Invest.* 101, 2129–2139.
402. Bonder, C. S., Dickensheets, H. L., Finlay-Jones, J. J., Donnelly, R. P., and Hart, P. H. (1998). Involvement of the IL-2 receptor γ-chain (γc) in the control by IL-4 of human monocyte and macrophage proinflammatory mediator production. *J. Immunol.* 160, 4048–4056.
403. Mozo, L., Gayo, A., Suarez, A., Rivas, D., Zamorano, J., and Gutierrez, C. (1998). Glucocorticoids inhibit IL-4 and mitogen-induced IL-4Rα chain expression by different posttranscriptional mechanisms. *J. Allergy Clin. Immunol.* 102, 968–976.
404. Hart, P. H., Bonder, C. S., Balogh, J., Dickensheet, H. L., Donnelly, R. P., and Finlay-Jones, J. J. (1999). Differential responses of human monocytes and macrophages to IL-4 and IL-13. *J. Leukoc. Biol.* 66, 575–578.
405. Mulder, K. M. (2000). Role of Ras and Mapks in TGFβ signaling. *Cytokine Growth Factor Rev.* 11, 23–35.
406. Finck, B. N., and Johnson, R. W. (2000). Tumor necrosis factor-α regulates secretion of the adipocyte-derived cytokine, leptin. *Microsc. Res. Tech.* 50, 209–215.
407. Fantuzzi, G., and Faggioni, R. (2000). Leptin in the regulation of immunity, inflammation, and hematopoiesis. *J. Leukoc. Biol.* 68, 437–446.
408. Das, U. N. (2001). Is obesity an inflammatory condition? *Nutrition.* 17, 953–966.
409. Gonzalez, R. R., Simon, C., Caballero-Campo, P., Norman, R., Chardonnens, D., Devoto L., and Bischof, P. (2000). Leptin and reproduction. *Hum. Reprod. Update* 6, 290–300.
410. Grimble, R. F. (2002). Inflammatory status and insulin resistance. *Curr. Opin. Clin. Nutr. Metab. Care* 5, 551–559.
411. Hibbetts, K., Hines, B., and Williams, D. (1999). An overview of proteinase inhibitors. *J. Vet. Intern. Med.* 13, 302–308.
412. Armstrong, P. B., and Quigley, J. P. (1999). α2-macroglobulin: an evolutionarily conserved arm of the innate immune system. *Dev. Comp. Immunol.* 23, 375–390.
413. Tchetverikov, I., Verzijl, N., Huizinga, T. W., TeKoppele, J. M., Hanemaaijer, R., and DeGroot, J. (2003). Active MMPs captured by α2-macroglobulin as a marker of disease activity in rheumatoid arthritis. *Clin. Exp. Rheumatol.* 21, 711–718.
414. Calandra, T. (2003). Macophage migration inhibitory factor and host innate immune responses in microbes. *Scand. J. Infect. Dis.* 35, 573–576.
415. Mitchell, R. A., and Bucala, R. (2000). Tumor growth-promoting properties of macrophage migration inhibitory factor (MIF). *Semin. Cancer Biol.* 10, 359–366.
416. Baugh, J. A., and Bucala, R. (2002). Macrophage migration inhibitory factor. *Crit. Care Med.* 30, S27–S35.
417. Baugh, J. A., and Donnelly, S. C. (2003). Macrophage migration inhibitory factor: a neuroendocrine modulator of chronic inflammation. *J. Endocrionl.* 179, 15–23.
418. Tsuruda, T., Costello-Boerrigter, L. C., and Burnett, J. C. Jr. (2004). Matrix metalloproteinases: pathways of induction by bioactive molecules. *Heart Fail. Rev.* 9, 53–61.
419. Ramnath, N., and Creaven, P. J. (2004). Matrix metalloproteinase inhibitors. *Curr. Oncol. Rep.* 6, 96–102.
420. Seiki, M., Mori, H., Kajita, M., Uekita, T., and Itoh, Y. (2003). Membrane-type 1 matrix metalloproteinase and cell migration. *Biochem. Soc. Symp.* 70, 253–262.
421. Seiki, M. (2003). Membrane-type 1 matrix metalloproteinase: a key enzyme for tumor invasion. *Cancer Lett.* 194, 1–11.
422. Chung, R. S., and West, A. K. (2004). A role for extracellular metallothioneins in CNS injury and repair. *Neuroscience* 123, 595–599.
423. Coyle, P., Philcox, J. C., Carey, L. C., and Rofe, A. M. (2002). Metallothionein: the multipurpose protein. *Cell. Mol. Life Sci.* 59, 627–647.

424. Haq, F., Mahoney, M., and Koropatnick, J. (2003). Signaling events for metallothionein induction. *Mutat. Res.* 533, 211–226.
425. Miles, A. T., Hawksworth, G. M., Beattie, J. H., and Rodilla, V. (2000). Induction, regulation, degradation, and biological significance of mammalian metallothioneins. *Crit. Rev. Biochem. Mol. Biol.* 35, 35–70.
426. Kyosseva, S. V. (2004). Mitogen-activated protein kinase signaling. *Int. Rev. Neurobiol.* 59, 201–220.
427. Saklatvala, J., Dean, J., and Clark, A. (2003). Control of the expression of inflammatory response genes. *Biochem. Soc. Symp.* 70, 95–106.
428. Seger, R., Hanoch, T., Rosenberg, R., Dantes, A., Merz, W. E., Strauss, J. F. 3rd, and Amsterdam, A. (2001). The ERK signaling cascade inhibits gonadotropin-stimulated steroidogenesis. *J. Biol. Chem.* 276, 13957–13964.
429. Amsterdam, A., Tajima, K., Frajese, V., and Seger, R. (2003). Analysis of signal transduction stimulated by gonadotropin in granulosa cells. *Mol. Cell. Endocrinol.* 202, 77–80.
430. Kirkin, V., Joos, S., and Zornig, M. (2004). The role of Bcl-2 family members in tumorigenesis. *Biochem. Biophys. Acta.* 1644, 229–249.
431. Sharpe, J. C., Arnoult, D., and Youle, R. J. (2004). Control of mitochondrial permeability by Bcl-2 family members. *Biochem. Biophys. Acta.* 1644, 107–113.
432. Henson, E. S., Gibson, E. M., Villanueva, J., Bristow, N. A., Haney, N., and Gibson, S. B. (2003). Increased expression of Mcl-1 is responsible for the blockage of TRAIL-induced apoptosis meditated by EGF/ErbB1 signaling pathway. *J. Cell. Biochem.* 89, 1177–1192.
433. Cuconati, A., Mukherjee, C., Perez, D., and White, E. (2003). DNA damage response and MCL-1 destruction initiate apoptosis in adenovirus-infected cells. *Genes Dev.* 17, 2922–2932.
434. Frade, J. M., and Barde, Y. A. (1998). Nerve growth factor: two receptors, multiple functions. *Bioessays* 20, 137–145.
435. Saragovi, H. U., and Zaccaro, M. C. (2002). Small molecule peptidomimetic ligands of neurtrophin receptor, identifying binding sites, activation sites and regulatory sites. *Curr. Pharm. Des.* 8, 2201–2216.
436. Lad, S. P., Neet, K. E., and Mufson, E. J. (2003). Nerve growth factor: structure, function and therapeutic implications for Alzheimer's disease. *Curr. Drug Targets CNS Neurol. Disord.* 2, 315–334.
437. Bonini, S., Rasi, G., Bracci-Laudiero, M. L., Procoli, A., and Aloe, L. (2003). Nerve growth factor: neurotrophin or cytokine. *Int. Arch. Allergy Immunol.* 131, 80–84.
438. Manni, L., Lundeberg, T., Fiorito, S., Bonini, S., Vigenti, E., and Aloe, L. (2003). Nerve growth factor release by human synovial fibroblasts prior to and following exposure to tumor necrosis factor-α, interleukin 1-β and cholecystokinin-8: the possible role of NGF in the inflammatory response. *Clin. Exp. Rheumatol.* 21, 617–624.
439. Tanaka, A., Wakita, U., Kambe, N., Iwasaki, T., and Matsuda, H. (2004). An autocrine function of nerve growth factor for cell cycle regulation of vascular endothelial cells. *Biochem. Biophys. Res. Commun.* 313, 1009–1014.
440. Villoslada, P., and Genain, C. P. (2004). Role of nerve growth factor and other trophic factors in brain inflammation. *Prog. Brain Res.* 146, 403–414.
441. Cantarella, G., Lempereur, L., Presta, M., Ribatti, D., Lombardo, G., Lazarovici, P., Zappala, G., Pafumi, C., and Bernardini, R. (2002). Nerve growth factor-endothelial cell interaction leads to angiogenesis in vitro and in vivo. *FASEB J.* 16, 1307–1309.
442. Freund, V., and Frossard, N. (2004). Expression of nerve growth factor in the airways and its possible roles in asthma. *Prog. Brain Res.* 146, 335–346.
443. Kawamoto, K., and Matsuda, H. (2004). Nerve growth factor and wound healing. *Prog. Brain Res.* 146, 369–384.
444. Micera, A., Vigneti, E., Pickholtz, D., Reich, R., Pappo, O., Bonini, S., Maquart, F. X., Aloe, L., and Levi-Schaffer, F. (2001). Nerve growth factor displays stimulatory effects on human skin and lung fibroblasts, demonstrating a direct role for this factor in tissue repair. *Proc. Natl. Acad. Sci. U S A* 98, 6162–6167.
445. Kraemer, R., and Hempstead, B. L. (2003). Neurotrophins: novel mediators of angiogenesis. *Front. Biosci.* 8, S1181–S1186.
446. Bull, H. A., Leslie, T. A., Chopra, S., and Dowd, P. M. (1998). Expression of nerve growth factor in cutaneous inflammation. *Br. J. Dermatol.* 139, 776–783.
447. Lee, F. S., Rajagopal, R., and Chao, M. V. (2002). Distinctive features of Trk neurotrophin receptor transactivation by G protein-coupled receptors. *Cytokine Growth Factor Rev.* 13, 11–17.
448. Weller, R. (2003). Nitric oxide: a key mediator in cutaneous physiology. *Clin. Exp. Dermatol.* 28, 511–514.
449. Abramson, S. B., Amin, A. R., Clancy, R. M., and Attur, M. (2001). The role of nitric oxide in tissue destruction. *Best Pract. Res. Clin. Rheumatol.* 15, 831–845.
450. Coleman, J. W. (2002). Nitric oxide: a regulator of mast cell activation and mast cell mediated inflammation. *Clin. Exp. Immunol.* 129, 4–10.
451. Blantz, R. C., and Munger, K. (2002). Role of nitric oxide in inflammatory conditions. *Nephron* 90, 373–378.
452. Balint, E. E., and Vousden, K. H. (2001). Activation and activities of the p53 tumour suppression protein. *Br. J. Cancer* 85, 1813–1823.
453. Xu, H., and el-Gewely, M. R. (2001). p53-responsive genes and the potential for cancer diagnostics and therapeutics development. *Biotechnol. Annu. Rev.* 7, 131–164.
454. Harms, K., Nozell, S., and Chen, X. (2004). The common and distinct target genes of the p53 family transcription factors. *Cell. Mol. Life Sci.* 61, 822–842.
455. Hofseth, L. J., Hussain, S. P., and Harris, C. C. (2004). p53: 25 years after its discovery. *Trends Pharmacol. Sci.* 25, 177–181.
456. Matuoka, K., and Chen, K. Y. (2002). Transcriptional regulation of cellular ageing by the CCAAT box-binding factor CBF/NF-Y. *Ageing Res. Rev.* 1, 639–651.
457. Yodoi, J., Nakamura, H., and Masutani, H. (2002). Redox regulation of stress signals: possible roles of dendritic stellate TRX producer cells (DST cell types). *Biol. Chem.* 383, 585–590.
458. Fitzpatrick, F. A. (2001). Inflammation, carcinogenesis and cancer. *Int. Immunopharmacol.* 1, 1651–1667.
459. Klafter, R., and Arbiser, J. L. (2000). Regulation of angiogenesis and tumorigenesis by signal transduction cascades: lessons from benign and malignant endothelial tumors. *J. Investig. Dermatol. Symp. Proc.* 5, 79–82.
460. Folch-Puy, E., Garcia-Movtero, A., Iovanna, J. L., Dagorn, J. C., Prats, N., Vaccaro, M. I., and Closa, D. (2003). The pancreatitis-associated protein induces lung inflammation in the rat through activation of TNFα expression in hepatocytes. *J. Pathol.* 199, 398–408.
461. Bimmler, D., Schiesser, M., Perren, A., Scheele, G., Angst, E., Meili, S., Ammann, R., Graf, R. (2004). Coordinate regulation of PSP/reg and PAP isoforms as a family of secretory stress proteins in an animal model of chronic pancreatitis. *J. Surg. Res.* 118, 122–135.
462. Heller, A., Fiedler, F., Schmeck, J., Luck, V., Iovanna, J. L., and Koch, T. (1999). Pancreatitis-associated protein protects the lung from leukocyte-induced injury. *Anesthesiology* 91, 1408–1414.
463. Burnouf, C., and Pruniaux, M. P. (2002). Recent advances in PDE4 inhibitors as immunoregulators and anti-inflammatory drugs. *Curr. Pharm. Des.* 8, 1255–1296.

464. Jin, S. L., and Conti, M. (2002). Induction of the cyclic nucleotide phosphodiesterase PDE4B is essential for LPS-activated TNF-α responses. *Proc. Natl. Acad. Sci. U S A* 99, 7628–7633.
465. Souness, J. E., and Rao, S. (1997). Proposal for pharmacologically distinct conformers of PDE cyclic AMP phosphodiesterases. *Cell. Signal.* 9, 227–236.
466. Combes, P., and Dickenson, J. M. (2001). Inhibition of NF-κB-mediated gene transcription by human A2B adenosine receptor in Chinese hamster ovary cells. *J. Pharm. Pharmacol.* 53, 1153–1156.
467. Kasyapa, C. S., Stentz, C. L., Davey, M. P., and Carr, D. W. (1999). Regulation of IL-15-stimulated TNF-α production by rolipram. *J. Immunol.* 163, 2836–2843.
468. Jin, S. L., Richard, F. J., Kuo, W. P., D'Ercole, A. J., and Conti, M. (1999). Impaired growth and fertility of cAMP-specific phosphodiesterase PDE4D-deficient mice. *Proc. Natl. Acad. Sci. U S A* 96, 11998–12003.
469. Delgado, M., Abad, C., Martinez, C., Juarranz, M. G., Leceta, J., Ganea, D., and Gomariz, R. P. (2003). PACAP in immunity and inflammation. *Ann. N. Y. Acad. Sci.* 992, 141–157.
470. Ganea, D., and Delgado, M. (2002). Vasoactive intestinal peptide (VIP) and pituitary adenylate cyclase-activating polypeptide (PACAP) as modulators of both innate and adaptive immunity. *Crit. Rev. Oral Biol. Med.* 13, 229–237.
471. Martinez, C., Abad, C., Delgado, M., Arranz, A., Juarranz, M. G., Rodriguez-Hench, N., Brabet, P., Leceta, J., and Gomariz, R. P. (2002). Anti-inflammatory role in septic shock of pituitary adenylate cyclase-activating polypeptide receptor. *Proc. Natl. Acad. Sci. U S A* 99, 1053–1058.
472. Shimada, H., Mori, T., Takada, A., Takada, Y., Noda, Y., Takai, I., Kohda, H., and Nishimura, T. (1981). Use of chromogenic substrate S-2251 for determination of plasminogen activator in rat ovaries. *Thromb. Haemost.* 46, 507–510.
473. Cho, S. H., Ryu, C. H., and Oh, C. K. (2004). Plasminogen activator inhibitor-1 in the pathogenesis of asthma. *Exp. Biol. Med.* 229, 138–146.
474. Mbebi, C., Hantai, D., Jandrot-Perrus, M., Doyennette, M. A., and Verdiere-Sahuque, M. (1999). Protease nexin I expression is up-regulated in human skeletal muscle by injury-related factors. *J. Cell. Physiol.* 179, 305–314.
475. Wiedermann, C. J., and Romisch, J. (2002). The anti-inflammatory actions of antithrombin—a review. *Acta Med. Austriaca* 29, 89–92.
476. Opal, S. M., Kessler, C. M., Roemisch, J., and Knaub, S. (2002). Antithrombin, heparin, and heparan sulfate. *Crit. Care Med.* 30 (suppl 5), S325–S331.
477. Hasan, S., Hosseini, G., Princivalle, M., Dong, J. C., Birsan, D., Cagide, C., and de Agostini, A. I. (2002). Coordinate expression of anticoagulant heparan sulfate proteoglycans and serine protease inhibitors in the rat ovary: a potent system of proteolysis control. *Biol. Reprod.* 66, 144–158.
478. Mika, J., Li, Y., Weihe, E., and Schafer, M. K. (2003). Relationship of pronociceptin/orphanin FQ and the nociceptin receptor ORL1 with substance P and calcitonin gene-related peptide expression in dorsal root ganglion of the rat. *Neurosci. Lett.* 348, 190–194.
479. Zeilhofer, H. U., and Calo, G. (2003). Nociceptin/orphanin FQ and its receptor-potential targets for pain therapy? *J. Pharmacol. Exp. Ther.* 306, 423–429.
480. Zaveri, N. T., Green, C. J., Polgar, W. E., Huynh, N., and Toll, L. (2002). Regulation of transcription of the human prepronociceptin gene by Sp1. *Gene* 290, 45–52.
481. Zaveri, N. T., Green, C. J., and Toll, L. (2000). Transcriptional regulation of the human prepronociceptin gene. *Biochem. Biophys. Res. Commun.* 276, 710–717.
482. Nakagawa, T., Kaneko, M., Inamura, S., and Satoh, M. (1999). Intracerebroventricular administration of nocistatin reduces inflammatory hyperalgesia in rats. *Neurosci. Lett.* 265, 64–66.
483. Itoh, M., Takasaki, I., Andoh, T., Nojima, H., Tominaga, M., and Kuraishi, Y. (2001). Induction by carrageenan inflammation of prepronociceptin mRNA in VR1-immunoreactive neurons in rat dorsal root ganglia. *Neurosci. Res.* 40, 227–233.
484. Xie, G. X., Ito, E., Maruyama, K., Suzuki, Y., Sugano. S., Sharma, M., Pietruck, C., and Palmer, P. P. (1999). The promoter region of human prepro-nociceptin gene and its regulation by cyclic AMP and steroid hormones. *Gene* 238, 427–436.
485. Li, X., and O'Malley, B. W. (2003). Unfolding the action of progesterone receptors. *J. Biol. Chem.* 278, 39261–39264.
486. Corbacho, A. M., Macotela, Y., Nava, G., Eiserich, J. P., Cross, C. E., Martinez de la Escalera, G., and Clapp, C. (2003). Cytokine induction of prolactin receptors mediates prolactin inhibition of nitric oxide synthesis in pulmonary fibroblasts. *FEBS Lett.* 544, 171–175.
487. Corbacho, A. M., Valacchi, G., Kubala, L., Olano-Martin, E., Schock, B. C., Kenny, T. P., and Cross, C. E. (2004). Tissue-specific gene expression of prolactin receptor in the acute phase response induced by lipopolysaccharides. *Am. J. Physiol. Endocrinol. Metab.* Epub, June 8.
488. Sun, R., Li, A. L., Wei, H. M., and Tian, Z. G. (2004). Expression of prolactin receptor and response to prolactin stimulation of human NK cell lines. *Cell. Res.* 14, 67–73.
489. Li, Y., Kumar, K. G., Tang, W., Spiegelman, V. S., and Fuchs, S. Y. (2004). Negative regulation of prolactin receptor stability and signaling mediated by SCF(β-TrCP) E3 ubiquitin ligase. *Mol. Cell. Biol.* 24, 4038–4048.
490. Christian, M., Tullet, J. M., and Parker, M. G. (2004). Characterization of four autonomous repression domains in the corepressor receptor interacting protein 140. *J. Biol. Chem.* 279, 15645–15651.
491. Mellgren, G., Borud, B., Hoang, T., Yri, O. E., Fladeby, C., Lien, E. A., and Lund, J. (2003). Characterization of receptor-interacting protein RIP140 in the regulation of SF-1 responsive target genes. *Mol. Cell. Endocrinol.* 203, 91–103.
492. Teyssier, C., Belguise, K., Galtier, F., Cavailles, V., and Chalbos, D. (2003). Receptor-interacting protein 140 binds c-Jun and inhibits estradiol-induced activator protein-1 activity by reversing glucocorticoid receptor-interacting protein 1 effect. *Mol. Endocrinol.* 17, 287–299.
493. Wei, L. N. (2004). Retinoids and receptor interacting protein 140 (RIP140) in gene regulation. *Curr. Med. Chem.* 11, 1527–1532.
494. Jung, Y., Isaacs, J. S., Lee, S., Trepel, J., Liu, Z. G., and Neckers, L. (2002). Hypoxia-inducible factor induction by tumor necrosis factor in normoxic cells requires receptor-interacting protein-dependent nuclear factor κ B activation. *Biochem. J.* 370, 1011–1017.
495. Chen, G., and Goeddel, D. V. (2002). TNF-R1 signaling: a beautiful pathway. *Science* 296, 1634–1635.
496. Fiorentino, L., Stehlik, C., Oliveira, V., Ariza, M. E., Godzik, A., and Reed, J. C. (2002). A novel PAAD-containing protein that modulates NF-κB induction by cytokines tumor necrosis factor-α and interleukin-1β. *J. Biol. Chem.* 277, 35333–35340.
497. Kehrl, J. H., and Sinnarajah, S. (2002). RGS2: a multifunctional regulator of G-protein signaling. *Int. J. Biochem. Cell. Biol.* 34, 432–438.
498. Berman, D. M., and Gilman, A. G. (1998). Mammalian RGS proteins: barbarians at the gate. *J. Biol. Chem.* 273, 1269–1272.

499. Hepler, J. R. (1999). Emerging roles for RGS proteins in cell signaling. *Trends Pharmacol. Sci.* 20, 376–382.
500. Tajima, K., Babich, S., Yoshida, Y., Dantes, A., Strauss, J. F. III, and Amsterdam, A. (2001). The proteasome inhibitor MG132 promotes accumulation of the steroidogenic acute regulatory protein (StAR) and steroidogenesis. *FEBS Lett.* 490, 59–64.
501. Zelko, I. N., Mariani, T. J., and Folz, R. J. (2002). Superoxide dismutase multigene family: a comparison of the CuZn-SOD (SOD1), Mn-SOD (SOD2) and EC-SOD (SOD3) gene structures, evolution and expression. *Free Radic. Biol. Med.* 33, 337–349.
502. Fujii, J., Suzuki, K., and Taniguchi, N. (1995). Physiological significance of superoxide dismutase isozymes. *Nippon Rinsho.* 53, 1227–1231.
503. Macmillan-Crow, L. A., and Cruthirds, D. L. (2001). Invited review: manganese superoxide dismutase in disease. *Free Radic. Res.* 34, 325–336.
504. Maier, C. M., and Chan, P. H. (2002). Role of superoxide dismutases in oxidative damage and neurodegenerative disorders. *Neuroscientist* 8, 323–334.
505. Hoshida, S., Yamashita, N., Otsu, K., and Hori, M. (2002). The importance of manganese superoxide dismutase in delayed preconditioning: involvement of reactive oxygen species and cytokines. *Cardiovasc. Res.* 55, 495–505.
506. Kinnula, V. L., and Crapo, J. D. (2003). Superoxide dismutases in the lung and human lung diseases. *Am. J. Respir. Crit. Care Med.* 167, 1600–1619.
507. Laloraya, M., Kumar, G. P., and Laloraya, M. M. (1989). Histochemical study of superoxide dismutase in the ovary of the rat during the oestrous cycle. *J. Reprod. Fertil.* 86, 583–587.
508. Nagase, H., and Brew, K. (2003). Designing TIMP (tissue inhibitor of metalloproteinases) variants that are selective metalloproteinase inhibitors. *Biochem. Soc. Symp.* 201–212.
509. Reynolds, J. J. (1996). Collagenases and tissue inhibitors of metalloproteinases: a functional balance in tissue degradation. *Oral Dis.* 2, 70–76.
510. Opdenakker, G., Van den Steen, P. E., Dubois, B., Nelissen, I., Van Coillie, E., Masure, S., Proost, P., and Van Damme, J. (2001). Gelatinase B functions as regulator and effector in leukocyte biology. *J. Leukoc. Biol.* 69, 851–859.
511. Leong, K. G., and Karsan, A. (2000). Signaling pathways mediated by tumor necrosis factor α. *Histol. Histopathol.* 15, 1303–1325.
512. Siwik, D. A., and Colucci, W. S. (2004). Regulation of matrix metalloproteinases by cytokines and reactive oxygen/nitrogen species in the myocardium. *Heart Fail. Rev.* 9, 43–51.
513. Wisniewski, H. G., and Vilcek, J. (2004). Cytokine-induced gene expression at the crossroads of innate immunity, inflammation and fertility: TSG-6 and PTX3/TSG-14. *Cytokine Growth Factor Rev.* 15, 129–146.
514. Milner, C. M., and Day, A. J. (2003). TSG-6: a multifunctional protein associated with inflammation. *J. Cell. Sci.* 116, 1863–1873.
515. Bardos, T., Kamath, R. V., Mikecz, K., and Glant, T. T. (2001). Anti-inflammatory and chondroprotective effect of TSG-6 (tumor necrosis factor-α-stimulated gene-6) in murine models of experimental arthritis. *Am. J. Pathol.* 159, 1711–1721.
516. Glant, T. T., Kamath, R. V., Bardos, T., Gal, I., Szanto, S., Murad, Y. M., Sandy, J. D., Mort, J. S., Roughley, P. J., and Mikecz, K. (2002). Cartilage-specific constitutive expression of TSG-6 protein (product of tumor necrosis factor α-stimulated gene 6) provides a chondroprotective, but not anti-inflammatory, effect in antigen-induced arthritis. *Arthritis Rheum.* 46, 2207–2218.
517. Carrette, O., Nemade, R. V., Day, A. J., Brickner, A., and Larsen, W. J. (2001). TSG-6 is concentrated in the extracellular matrix of mouse cumulus oocyte complexes through hyaluronan and inter-α-inhibitor binding. *Biol. Reprod.* 65, 301-308.
518. Verheul, H. M., and Pinedo, H. M. (2003). Vascular endothelial growth factor and its inhibitors. *Drugs Today (Barc)* 39 Suppl C, 81–93.
519. Dvorak, H. F., Nagy, J. A., Feng, D., Brown, L. F., and Dvorak, A. M. (1999). Vascular permeability factor/vascular endothelial growth factor and the significance of microvascular hyperpermeability in angiogenesis. *Curr. Top. Microbiol. Immunol.* 237, 97–132.
520. van Hinsbergh, V. W., Collen, A., and Koolwijk, P. (2001). Role of fibrin matrix in angiogenesis. *Ann. N. Y. Acad. Sci.* 936, 426–437.
521. Thurston, G. (2002). Complementary actions of VGF and angiopoietin-1 on blood vessel growth and leakage. *J. Anat.* 200, 575–580.
522. Bouma-ter Steege, J. C., Mayo, K. H., and Griffioen, A. W. (2001). Angiostatic proteins and peptides. *Crit. Rev. Eukaryot. Gene Expr.* 11, 319–334.
523. Wilson, K. T. (2002). Angiogenic markers, neovascularization and malignant deformation of Barrett's esophagus. *Dis. Esophagus* 15, 16–21.
524. Wight, T. N. (2002). Versican: a versatile extracellular matrix proteoglycan in cell biology. *Curr. Opin. Cell. Biol.* 14, 617–623.
525. Wang, H. Y. (2004). WNT-frizzled signaling via cyclic GMP. *Front. Biosci.* 9, 1043–1047.
526. Yamamoto, Y., and Gaynor, R. B. (2004). IκB kinases: key regulators of the NF-κB pathway. *Trends Biochem. Sci.* 29, 72–79.
527. Valen, G., Yan, Z. Q., and Hansson, G. K. (2001). Nuclear factor kappa-B and the heart. *J. Am. Coll. Cardiol.* 38, 307–314.
528. Lentsch, A. B., and Ward, P. A. (2000). The NFκB /IκB system in acute inflammation. *Arch. Immunol. Ther. Exp. (Warsz)* 48, 59–63.
529. Guijarro, C., and Egido, J. (2001). Transcription factor-κB (NF-κB) and renal disease. *Kidney Int.* 59, 415–424.
530. Karin, M., and Delhase, M. (2000). The IκB kinase (IKK) and NF-κB: key elements of proinflammatory signaling. *Semin. Immunol.* 12, 85–98.
531. Chen, F., Castranova, V., and Shi, X. (2001). New insights into the role of nuclear factor-κB in cell growth regulation. *Am. J. Pathol.* 159, 387–397.
532. Perkins, N. D. (2000). The Rel/NF-κB family: friend and foe. *Trends Biochem. Sci.* 9, 434–440.
533. Renard, P., and Raes, M. (1999). The proinflammatory transcription factor NF-κB: a potential target for novel therapeutical strategies. *Cell. Biol. Toxicol.* 15, 341–344.
534. Zhang, H. G., Wang, J., Yang, X., Hsu, H. C., and Mountz, J. D. (2004). Regulation of apoptosis proteins in cancer cells by ubiquitin. *Oncogene* 23, 2009–2015.
535. Foxwell, B. M., Bonderson, J., Brennan, F., and Feldmann, M. (2000). Adenoviral transgene delivery provides an approach to identifying important molecular processes in inflammation: evidence for heterogenicity in the requirement for NFκB in tumour necrosis factor production. *Ann. Rheum. Dis.* 59 Suppl 1, 154–159.
536. Kelly, R. W., King, A. E., and Critchley, H. O. (2001). Cytokine control in human endometrium. *Reproduction* 121, 3–19.
537. van den Berg, R., Haenen, G. R., van den Berg, H., and Bast, A. (2001). Transcription factor NF-κB as a potential biomarker for oxidative stress. *Br. J. Nutr.* 86 (Suppl 1), S121–S127.

538. Said, S. I., and Dickman, K. G. (2000). Pathways of inflammation and cell death in the lung: modulation by vasoactive intestinal peptide. *Regul. Pept.* 93, 21–29.
539. Ganea, D., and Delgado, M. (2001). Neuropeptides as modulators of macrophage functions. Regulation of cytokine production and antigen presentation by VIP and PACAP. *Arch. Immunol. Ther. Exp. (Warsz)* 49, 101–110.
540. Son, D. S., Roby, K. F., and Terranova, P. F. (2004). Tumor necrosis factor-α (TNF) induces serum amyloid A3 in mouse granulosa cells. *Endocrinology* 145, 2245–2252.

Knobil and Neill's Physiology of Reproduction, Third Edition
edited by Jimmy D. Neill,
Elsevier © 2006

CHAPTER 12

Structure, Function, and Regulation of the Corpus Luteum

Richard L. Stouffer

Introduction, 475
Historical Perspective, 476
Classification of Types of Corpora Lutea, 478
Long-Lived Corpora Lutea, 478
Short-Lived Corpora Lutea, 478
Ultrashort-Lived Corpora Lutea, 479
Luteinization, 480
Morphological Changes, 481
Regulation and Role of Luteal Angiogenesis, 482
Differentiation of Steroid-Secreting Cells, 484
Luteolysis, 489
Regulation of Luteal Structure–Function, 491
Luteotropic Factors, 491
Luteolytic Factors, 496
Extension of Luteal Function in Early Pregnancy, 502
Domestic Animals, 502
Primates, 503
Luteal Cell Types and Their Regulation, 505
Small versus Large Luteal Cells, 505
Function and Regulation of Luteal Cell Types, 506
Summary and Future Perspectives, 508
Acknowledgments, 509
References, 509

Although much is known, much still remains to be discovered, and much of what has been discovered needs to be uncovered.

–R. V. Short (1)

INTRODUCTION

The corpus luteum is an endocrine gland in the adult ovary that forms from the follicle wall after ovulation (2,3). It is associated with four unique features. The first is its ephemeral nature: the corpus luteum exists for a limited life span that in many species depends on the fate of the oocyte released by the antecedent ovulatory follicle. Second, its primary, if not sole essential function is to synthesize and secrete the steroid hormone, progesterone. Progesterone has numerous biological effects on target tissues (3), but particularly acts in the reproductive tract to permit implantation of the early embryo in the uterine endometrium and to support a maternal environment that sustains intrauterine pregnancy. Thus, the third feature of the corpus luteum is its physiological connection to viviparity (2). All mammals, many reptiles, and a few fishes that produce live young exhibit corpora lutea, although the essential connection between the corpus luteum and viviparity is less obvious in nonmammalian species. In some of these species, luteal function may be related to final oocyte maturation, egg movement in the oviduct, or oviposition, because the corpus luteum exists in some oviparous reptiles, amphibians, and birds as well (4). The latter point serves as a preface to and reminder of the fourth key feature of the corpus luteum, namely, the tremendous species diversity in the mechanisms that evolved to control the structure and function of the corpus luteum. Although the mammalian corpus luteum seems similar to the steroidogenic adrenal cortex (2), and develops from the steroidogenic ovarian follicle, the latter endocrine tissues display relatively minor differences in regulation of development and function across mammalian species. In contrast, the

Division of Reproductive Sciences, Oregon National Primate Research Center, Oregon Health and Science University, Beaverton, Oregon.

differences in endocrine and local control of the development, function, and life span of the corpus luteum are remarkable. Limitations of space prevent addressing many individual species in this chapter, but the reader is encouraged to peruse other comparative reviews to marvel at the challenges facing ovarian biologists in their attempts to promote or control fertility in domestic and wild animals, endangered species, and human beings. This review focuses on selected species of the traditional laboratory (rodents, rabbits) and domesticated (sheep, cow, horse) animals, plus primates (Old World macaques, humans), where the majority of research activity has occurred since publication of the earlier chapter on the corpus luteum by Niswender and Nett (3), in the second edition of *The Physiology of Reproduction.*

HISTORICAL PERSPECTIVE

An interesting reference on the discovery and study of the corpus luteum was provided by R. V. Short in 1977 (1). Existence of the corpus luteum as a structure in the ovary was noted by several renowned anatomists in the 16th and 17th centuries, such as Fabricius (5,6), who provided the first illustrated accounts of the comparative anatomy of the female reproductive tract, including ovaries in the pregnant pig containing "numerous little glands." Although Regnier de Graaf is chiefly remembered for his description of the antral graafian follicle, he provided the first detailed descriptions of the "globular bodies" (corpora lutea; Fig. 1) that formed in place of the "ova" (follicles) and which gave an indication of the number of fetuses present [(7), as translated by Jocelyn and Setchell (8)]. His appreciation of the diversity of colors (pink, red, purple, yellow) associated with the gland in various species did not lead him to name the structure the *corpus luteum* ("yellow body"), which is attributed to Malpighi's observations of cow ovaries. Nevertheless, Malpighi speculated correctly on the corpus luteum's "glandular nature" and stated that its structure closely resembled the suprarenal (adrenal) gland (9). During this era, the ovaries and their structures were identified and described, but their functions and regulation were not appreciated.

Although de Graaf and the Scottish anatomists William and John Hunter recognized the positive correlation between the number of corpora lutea and the number of offspring arising from pregnancy (10), and clinicians such as Pott (11) recognized that ovariectomy resulted in cessation of menstruation in women, the importance of these relationships remained unclear (12). Many scientists remained thoroughly indoctrinated in the prevailing views of

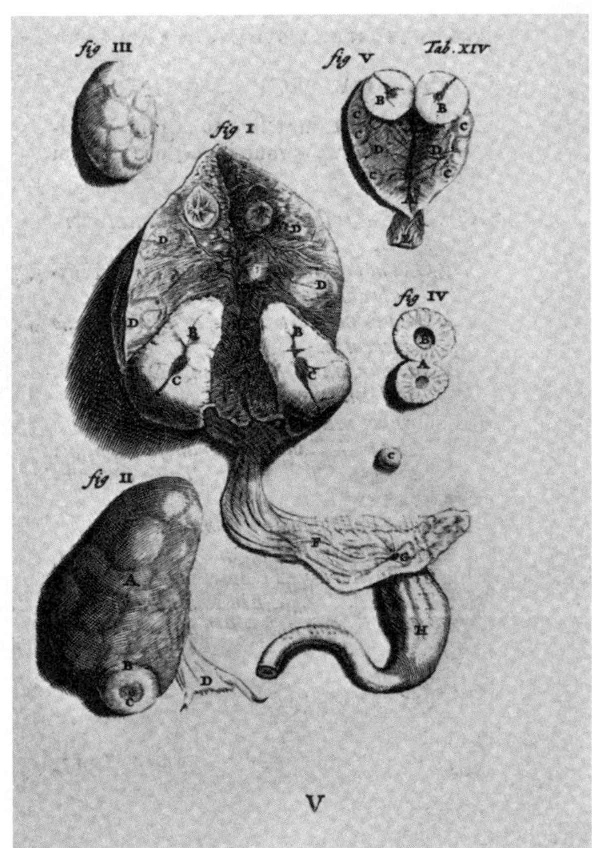

FIG. 1. Early (1672 A.D.) drawing of sheep ovaries by Regnier de Graaf. The ovary (*center, top right*) is bisected to display the presumptive corpus luteum, as well as other tissue compartments. [From de Graaf (7), as reported in Short, R. V. (1977). The discovery of the ovaries. In *The Ovary* (H. Zuckerman and B. J. Weir, Eds.), pp. 1–39. Academic Press, New York, with permission.]

Aristotle's "seed and soil" concept that the male provided the "seed" and the female played a passive role by providing the "soil" in which the seed could grow [as recounted by Thompson, 1910 (13)]. The soil was the uterine fluids, and women menstruated because their nature resulted in periodic repletion with excess blood that was removed by menstruation. Although there was some belief that the ovarian follicle was the egg, it was not until the 1800s that the egg was first described in the mammalian follicle (14) and the process of ovulation and formation of the corpus luteum from the follicle wall was identified (15). Nevertheless, the general opinion was that corpora lutea were just a form of scar tissue formed after ovulation. Only after Knauer (16) definitively proved that the ovary was a gland of "internal secretion" through classic ablation–replacement studies in various mammals, did investigators such as Beard (17) and Prenant (18) hypothesize that the corpus luteum had specific functions in uterine development, gestation, and lactation owing to its secretion of one or more products into the bloodstream. Studies by Fraenkel (19) and Magnus (20) confirmed the endocrine actions of the corpus luteum, but the search began in earnest to isolate the luteal hormone(s) only after suitable bioassays, such as the maternal deciduoma response (21), had been devised. In 1929, Corner and Allen (22) prepared an alcoholic extract from pig corpora lutea that mimicked luteal actions in the uterus and maintained pregnancy in ovariectomized animals. Hisaw et al. (23) also established that similar luteal extracts could produce the full secretory phase development of uterine endometrium in a primate, the rhesus monkey. By then it was recognized that menstruation in primates, including women, did not correspond to the period of "sexual excitement" or estrus in other species, that is, it was not the time of ovulation and optimal fertility (24). Rather, it marked the end of the cycle and the absence of ovulation—a wiser basis for the "rhythm method" of contraception.

The modern era of research began in 1934 with the isolation and characterization of the steroid progesterone from luteal extracts by four independent groups around the world (25–28). Investigators quickly established that this progestin [see Parkes (29) for an explanation of how the name was agreed upon] mimicked many of the roles of the corpus luteum. With the development of reliable radioimmunoassays (30) to replace bioassays, attention turned to understanding the processes controlling the functional life span of the corpus luteum in various species. After the discovery that hypophysectomy eliminated cyclic ovarian function in adult mammals, particular attention focused on identifying the pituitary gonadotropins (31) and established the vital roles of luteinizing hormone (LH) and prolactin (PRL) as "luteotropic hormones" in the development and steroidogenic function of the corpus luteum in various species (32). Also, the report that hysterectomy would prolong the life span of the corpus luteum led to investigations into the role of the uterus (33), and the subsequent discovery that the arachidonate metabolite prostaglandin (PG) $F_{2\alpha}$ is the uterine "luteolytic hormone" terminating the life span of the corpus luteum in nonfecund ovarian cycles of various species (34).

The discovery of the luteal hormone, progesterone, also had pronounced effects on clinical approaches to ensure sufficient luteal support of pregnancy, to control fertility, and to prevent undesirable effects of menopause in women. Although major ovarian pathologic processes (e.g., ovarian cancers, polycystic ovarian diseases) do not emanate from the corpus luteum, there are situations of deficient luteal function in natural ovarian cycles (short or insufficient luteal phases) or controlled ovarian stimulation cycles (short luteal phases) that may be associated with subfertility (35). Therefore, a synthetic progestin or, more recently, progesterone supplements can be administered to replace or assist luteal functions during the menstrual cycle (36). Also, with the discovery that administration of progesterone could inhibit ovulation in mammals (37), researchers began to formulate the idea that administration of a combination of estrogen and progestational agents would block ovulation and thus prevent conception in women while preserving the semblance of normal menstrual cycles in women. This work was facilitated by the success of chemists [e.g., R. Marker; see Barber (38)] in generating inexpensive, abundant supplies of steroid analogs from plant sapogenins, which are steroids that structurally resemble cholesterol. The basic (39) and clinical testing by Pincus, Chang, Rock, and Rice-Way proved that the combinational steroid "pill" provided virtually complete protection from pregnancy. The oral contraceptive pill introduced in 1960 had a profound impact on every aspect of society, and for the first time gave women the opportunity to control their reproductive status. Likewise, the discovery of molecules with antiprogestin activity (e.g., RU486 at Roussel Uclaf) demonstrated the molecular basis of luteal ablation on uterine function and pregnancy, resulting in the availability of a nonsurgical protocol for pregnancy termination or contragestation (40). The use of progestins, antiprogestins, and, more recently, selective progestin receptor modulators [which may have agonist or antagonist actions at specific target tissues (41)] continues to be of great interest for such purposes as birth control, disorders of the reproductive tract (e.g., persistent or irregular menstruation), and cancer treatment.

Since publication of the last edition (3), advances in cell and molecular biology have contributed greatly to our understanding of the mechanisms of action of luteotropic and luteolytic factors, the intracellular pathways controlling steroidogenesis, and the cell-to-cell interactions that are critical for the development, function, and normal life span of the corpus luteum. Unlike in earlier eras or even up to 1970, investigators of the corpus luteum are now faced with a tremendous increase in scientific information on a daily basis. A PubMed search of the National Library of Medicine database identified over 3,000 articles per decade on the corpus luteum since the 1970s. In this century of information overload and short information "sound bites" or "abstracts," the scholar is urged to rediscover the major breakthroughs, prevailing controversies, and dogmas of earlier eras that provided the groundwork for current concepts that have yet to stand the test of time. This review attempts to build on prior chapters and coalesce ideas from recent reviews of leaders in this field as referenced in the following sections and cited in the Acknowledgments.

CLASSIFICATION OF TYPES OF CORPORA LUTEA

Because of its transient nature, one can typically discern a developing phase, a functional phase, and a regressing phase in the life span of the corpus luteum. These phases are usually associated with increasing levels, elevated levels, and then declining levels of progesterone in the circulating blood (2). As adapted from Rothchild and others (2,42), mammals can be divided into three major categories (Fig. 2) based on how long the corpus luteum functions after ovulation in the nonfecund ovarian cycle and how the luteal life span is affected by pregnancy onset and gestation.

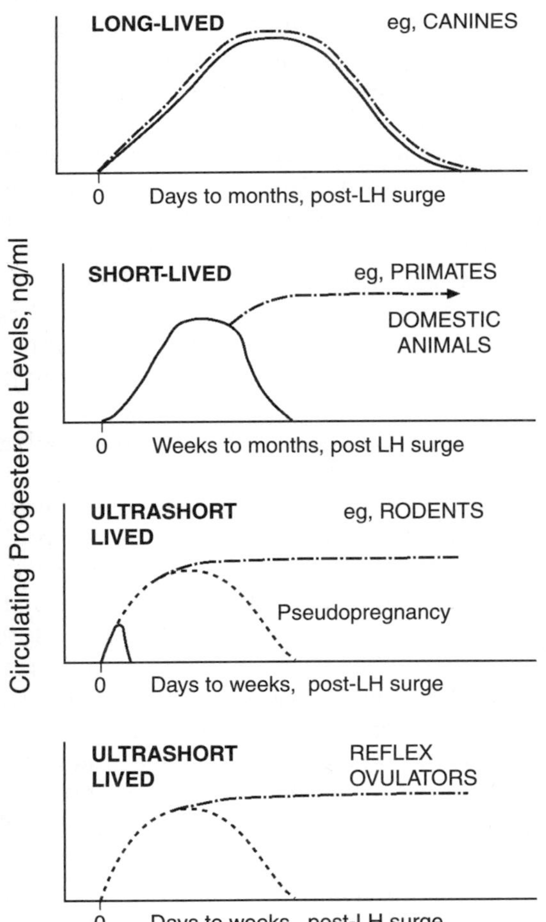

FIG. 2. Types of mammalian corpora lutea based on their functional life span (long, short, or ultrashort) during the nonfecund ovarian cycle (*solid lines*). The effects of nonfertile mating on ultrashort-lived corpora lutea (pseudopregnancy; *dotted line*) and of pregnancy in all three categories (*dashed lines*) are demonstrated. Reflex ovulators, such as rabbits and ferrets, can also be categorized as having ultrashort-lived corpora lutea because they have no luteal phase owing to the absence of a luteinizing hormone surge unless mating occurs.

Long-Lived Corpora Lutea

There are some monestrous species (i.e., animals with estrous cycles that occur only at long intervals, such as every 6 to 12 months) in which a fully functional corpus luteum develops after ovulation and its functional life span does not differ remarkably regardless of whether the corpus luteum originates during a fertile or nonfertile cycle (2). This category, which includes mammalian carnivores [e.g., dogs, wolves, foxes, cats, ferrets, skunks (43,44)] and other animals such as the roe deer, armadillo, and marsupials (45–47), contains corpora lutea that last from less than 2 weeks to as long as 6 months. In the domesticated dog (48,49), for example, the length of the luteal phase in pregnant and nonpregnant animals is 64 to 66 days. At the end of pregnancy, which typically occurs in each cycle of monestrous animals, there is rapid luteolysis. However, if pregnancy does not occur, the loss of luteal structure–function is very protracted, the mechanisms of which are unclear. Many marsupials can be included in this category because, even though the luteal life span may last only approximately 14 days, it is not altered by the occurrence or lack of pregnancy. However, this category of corpora lutea illustrates the evolution of mechanisms whereby the corpus luteum's life span can be modified by the addition of a *quiescent* interval after ovulation before luteal function begins. This can be related to seasonal cues [e.g., skunk, badger (50)] or lactation and season [e.g., wallaby (51)].

Short-Lived Corpora Lutea

In many polyestrous species (i.e., animals with frequent ovarian cycles), the corpus luteum develops and functions for a finite interval during the ovarian cycle, but the luteal life span is markedly increased if pregnancy ensues. This category includes the primate species of monkeys, great apes, and humans (52–54), wherein a functional corpus luteum forms rapidly after ovulation and functions for a sufficient interval (approximately 2 weeks) to permit timely movement of the early embryo through the oviduct and to prepare the uterus for implantation. If pregnancy does not occur, the corpus luteum regresses after 2 weeks. However, if pregnancy occurs, the functional life span of the corpus luteum is extended for a finite interval (55) until the developing placenta assumes its critical activities (i.e., progesterone production). The time of the luteal-placental shift, after which the primate corpus luteum is no longer essential for pregnancy, varies between species [3 and 7 weeks of gestation in rhesus macaques and women, respectively (56)]. It is not clear if the processes of luteal regression around the luteal-placental shift are comparable with those during luteolysis in the nonfecund cycle. However, there is intriguing evidence that after a quiescent interval at mid-gestation, the corpus luteum in primates rejuvenates by term pregnancy and is again a progesterone-secreting gland (57)—but its importance is unclear at least as an endocrine gland in view of the massive steroidogenic activity of the term placenta.

Many of the ungulates, including the domesticated farm animals, also display short-lived corpora lutea. For example, in sheep and cattle, the corpus luteum forms rapidly after ovulation and functions until days 17 and 21 postestrus, respectively (58,59). Again, if pregnancy does not occur, the corpus luteum undergoes rapid luteolysis; however, if intrauterine gestation begins, the functional corpus luteum is sustained for several months until the luteal-placental shift or, in the pig (60), even until the end of pregnancy. However, there are important differences between primates and ungulates with short-lived corpora lutea in the activities and processes controlling the length of the ovarian cycle and the luteal phase, as well as in the maternal-ovarian recognition of pregnancy. First, the ungulate corpus luteum permits waves of antral follicle growth during the luteal phase (61), and hence the interval of the follicular phase after luteolysis to development of the next ovulatory follicle is very short (2 to 3 days). In contrast, in many primates, the ensuing follicular phase is relatively long (approximately 2 weeks) because of the "dominance" of the corpus luteum, which prevents antral follicle development until after luteolysis (62,63). Second, luteolysis in ungulates requires the presence of a nonfecund uterus, leading to the concept that in the absence of implantation, a uterine luteolytic signal promotes the regression of the corpus luteum of the cycle (64). In contrast, hysterectomy does not alter the functional life span of the corpus luteum in primates exhibiting menstrual cycles [e.g., Old World macaques, women (65,66)], indicating the absence of a uterine luteolytic signal. Third, whereas ungulates have evolved a mechanism that apparently blocks the synthesis or secretion of the uterine luteolytic signal in early pregnancy, higher primates use an embryonic hormone that acts directly on the corpus luteum to rescue it from impending regression and promote luteal structure–function in early pregnancy (56). Again, there are exceptions or additions to these concepts. For example, some primates [e.g., New World monkeys, prosimians (67,68)] do not menstruate and may produce a uterine luteolytic signal (69). Likewise, the horse also produces a fetal hormone after the first month of gestation that not only stimulates the function of existing corpora lutea, but induces ovulation of large antral follicles and formation of "secondary" corpora lutea that assist in progesterone secretion for 5 months until the luteal–placental shift (3).

Ultrashort-Lived Corpora Lutea

Species with this feature (2) do not form a truly functional corpus luteum (polyestrus rodents, perhaps some insectivores) or indeed any corpus luteum during the ovarian cycle unless mating results in ovulation [reflex ovulators, e.g., rabbit (70)] or pseudopregnancy or pregnancy (e.g., rat, mouse). In rodents such as the laboratory rat or mouse (71,72), the estrous cycle lasts 4 to 5 days and lacks a true luteal phase. After ovulation, the ruptured follicle takes on some of the characteristics of a corpus luteum, but the luteal gland is not well developed and is considered nonfunctional because it does not secrete sufficient progesterone to stimulate a uterine response (i.e., decidualization) to implantation cues. Progesterone produced within the corpus luteum is rapidly metabolized to 20α-hydroxyprogesterone, an inactive progestin that will not support pregnancy. Mating or a facsimile (cervical stimulation) during estrus that does not result in pregnancy results in development of functional corpora lutea that secrete progesterone for 12 to 14 days. This interval of luteal function, known as *pseudopregnancy*, is an unusual artifact because mating in the natural environment typically results in pregnancy. However, pseudopregnancy is a valuable means of distinguishing mechanisms initiated by the neuroendocrine reflex arc in response to cervical stimulation that promote luteal development in animals with ultrashort-lived corpora lutea. If mating initiates pregnancy, continued progesterone secretion by the corpus luteum is required throughout gestation. Continued luteal structure–function is achieved by a combination of maternal (uterine decidual) and fetal (trophoblast) hormones or hormonal precursors that act on or are converted to essential factors by the rodent corpus luteum of pregnancy (73,74). Regression of the corpus luteum at the end of gestation, and probably at the end of pseudopregnancy, is controlled by a uterine luteolytic factor (71).

Thus, mammalian species have developed diverse means (Fig. 2) to ensure that ovarian cyclicity returns in the event of a nonfecund cycle (i.e., to restore fertility potential), yet provide adequate progesterone support if pregnancy ensues (i.e., to support intrauterine gestation until delivery of offspring). The simplest case is in animals with a "long-lived" corpus luteum, where gestation lasts no longer than the normal life span of the corpus luteum in the ovarian cycle. However, many animals have evolved means for keeping the functional life span of the corpus luteum "short" (i.e., less than necessary for successful pregnancy) in the ovarian cycle to increase reproductive efficiency. Therefore, processes evolved for preventing development of a functional corpus luteum ("ultrashort" life span), for uterine control of luteolysis in the nonfecund cycle ("short" life span), and for uterine-placental extension of luteal function in the event of pregnancy initiation. Detailed discussions of these processes and the factors that control luteal structure–function are included in the following sections.

LUTEINIZATION

The corpus luteum forms from the cells comprising the wall of the ovulatory follicle. Luteal formation (i.e., luteinization) begins before ovulation but proceeds in earnest after follicle rupture and continues for a species-specific number of days until the fully formed corpus luteum is present. As reviewed by Murphy (75), luteinization is a remarkable event involving cell proliferation, cell differentiation, and tissue remodeling that is unparalleled in the adult mammal. It comprises two major processes: (a) the terminated proliferation plus rapid hypertrophy and differentiation of the steroidogenic cells of follicle into the luteal cells of the corpus luteum; and (b) the rapid growth of blood vessels (i.e., angiogenesis), and in some species the lymphatics (i.e., lymphangiogenesis) (76), into the previously avascular granulosa layer of the follicle to form the extensive microvascular bed of the corpus luteum (Fig. 3). Although luteinization is often referred to as "terminal differentiation" of the follicle, in many species, from rodents to primates, further differentiation may occur in luteal tissue in response to pregnancy. The angiogenic process in the corpus luteum is remarkable, even in comparison with the rapid vascular development in tumors or the placenta, and offers a unique model for studying microvessel development, maintenance, and regression in normal adult tissue. As adapted from Murphy (75) and others (3,77), this section discusses the processes and regulation of luteinization resulting in a fully developed, progesterone-secreting corpus luteum.

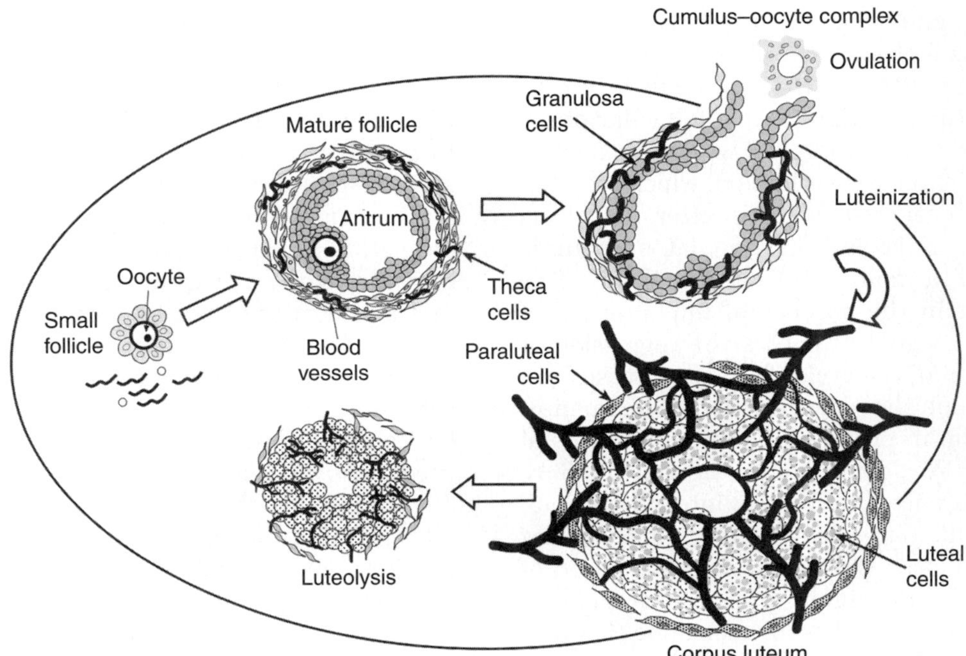

FIG. 3. Development of the corpus luteum from the mature (graafian) follicle. Folliculogenesis results in a large antral follicle in which vascularization (black blood vessels) is limited to the outer somatic (theca) cell layers. A basement membrane separates the theca from the avascular somatic (granulosa) cell layer, which encloses the gametogenic (oocyte) cell and a fluid-filled antrum. Near the time of ovulation, the basement membrane breaks down and angiogenesis occurs in the granulosa layer. Granulosa cells hypertrophy, and in many species luteinized granulosa and theca cells intermingle within the luteal tissue. In other species (e.g., primates), some compartmentalization remains because theca-derived paraluteal cells are found around the periphery and along infoldings in the corpus luteum. At luteolysis (see section on Luteolysis), both the luteal cells and microvasculature degenerate. (From Stouffer, R. L. [2003]. Corpus luteum in primates. In *Encyclopedia of Hormones* [E. R. Simpson, H. L. Henry, and A. W. Norman, Eds.], pp. 288–297. Elsevier, San Diego, with permission.)

Morphological Changes

A common event during collapse of the follicle antrum at ovulation is the infolding of the follicle wall, with breakdown of the basement membrane separating the granulosa and theca layers of the follicle. This is followed by invasion of microvascular cells into the avascular granulosa layer, as well as other cells associated with the vasculature [e.g., perivasculature support cells, or pericytes (78); migrating cells of the reticuloendothelial system]. In a few species, such as the dog (79), this process starts well before ovulation, but in most mammals it follows rupture of the follicle. The follicular infolding and vascular invasion also transport theca cells toward the center of the follicle cavity. In many species, including rodents and various domestic animals, theca-derived luteal cells appear to disperse throughout the parenchyma of the corpus luteum (80). Thus, unlike in the antecedent follicle, there is no apparent compartmentalization of theca- and granulosa-derived cells in the corpus luteum of these species. However, in some primate species, such as Old World macaques and humans, areas of the theca-derived compartment remain discrete from the remainder (including granulosa-derived) of the luteal tissue throughout its life span, even into pregnancy (81). The fate of the theca interna versus theca externa is controversial; they may degenerate or be absorbed into the surrounding stroma, become incorporated into the luteal parenchyma as theca-luteal cells, or remain as paraluteal cells clustered along the periphery or along infoldings of the luteal gland. The fate of thecal cells may depend on their differentiated state or proximity to luteinizing granulosa cells at ovulation. Certainly, their destruction is documented in some species, such as the horse (82), and there is speculation that they degenerate and are absent from the corpus luteum of a New World monkey, the marmoset (83).

In addition to the dramatic increase in vascularity, the large increase in mass of the corpus luteum is due to hypertrophy of the luteinizing granulosa cells (84). The granulosa cell with a high nuclear-to-cytoplasmic ratio is converted into a luteal cell with a low nuclear-to-cytoplasmic ratio, at a volume increase of up to tenfold (85). Even before ovulation, there are significant cellular changes as the nuclear chromatin disperses and the nucleolus forms concomitant with the increase in number of polyribosomes (84,86,87). Intracellular connections (e.g., gap junctions) between granulosa cells (as well as theca cells) disappear, presumably as a prelude to tissue remodeling that includes angiogenesis and cell hypertrophy (80,88,89). The amount of smooth endoplasmic reticulum increases and the characteristics of the mitochondria change, as they become rounded and their cristae convert from lamellar to primarily tubular form. The luteal cells may also contain well-developed Golgi complexes, as well as lipid droplets and protein-secretory granules. These changes are consistent with a cell that is very active in steroidogenesis, as well as peptide/protein secretion. However, again, the characteristics (e.g., lipid droplets, protein-secretory granules) and the stage at which they develop [e.g., early after ovulation in domestic animals or primates versus mid-pregnancy in rodents (84)] vary between species.

It appears that marked changes in the cytoskeleton play an important role in optimizing cell size as well as steroidogenic function (75). Limited research suggests that luteinizing granulosa cells can transiently alter their expression of microtubule components (e.g., tubulin) around ovulation, as well as acquire the ability to express smooth muscle actin, cytokeratin, vimentin, and desmin (90,91). Again, there are likely differences between granulosa- and theca-derived luteal cells because theca-derived cells express only large amounts of vimentin, at least in some species (90).

Also, it is likely that changes in the extracellular matrix (ECM) and cell–ECM interactions are critical in the tissue remodeling and function in the developing corpus luteum. For example, in the bovine corpus luteum, the type IV collagen in the basement membrane of the follicle is replaced by the fibrillar (type I) collagen serving as a major component in the luteal ECM (92). Remarkably, type I collagen comprises one sixth of the mass of the mature bovine corpus luteum (92). The dynamics of ECM components and extracellular surface glycoproteins, such as the integrins, is receiving recent attention. There is speculation that removal of integrin-mediated inhibition after ovulation, such as integrin $\alpha6\beta1$ in the follicle, may permit functional luteinization (93). However, restoration or expansion of other cell–ECM interactions may be critical as well. For example, integrin $\beta3$ expression is upregulated in the mouse corpus luteum (94). Further characterization of the follicular/luteal ECM, its regulation, and its interactions with cells of the corpus luteum is an important research area. A number of proteases are implicated in luteal remodeling in various species, including serine proteases [e.g., the plasminogen–plasmin system (95)], the matrix metalloproteinases (MMPs) and their endogenous tissue inhibitors [tissue inhibitors of metalloproteinases (TIMPs); for review, see Curry and Osteen (96)] and recently a novel family of proteases called ADAMTS [a disintegrin and metalloproteinase with thrombospondin-type repeats (97)]. There are reports of marked differences in protease expression between cell types comprising the corpus luteum and between species (96), but there is general consensus over elevated enzyme expression/activity during the extensive ECM remodeling associated with luteinization. An important challenge is to unravel the roles of and interactions between various proteases in the angiogenesis and luteal differentiation occurring during development of the corpus luteum. There is one intriguing report (98) that TIMP-1$^{-/-}$ null mutant mice display reduced serum progesterone levels during luteal development. In vitro models for luteinization of granulosa and theca cells (99) can be valuable for examining changes during the shift to producing large quantities of progestin (e.g., accumulation of lipid droplets for cholesterol storage) and cellular hypertrophy (e.g., microtubule enlargement). However, creative approaches using in vivo models will remain essential for dissecting the roles and interactions between cell populations and the ECM during the life span of the corpus luteum.

Regulation and Role of Luteal Angiogenesis

The changes in vascularity associated with the development of a functional corpus luteum are remarkable [for review, see Hazzard and Stouffer (100)]. Before ovulation, the primary follicular layer destined to form the corpus luteum (i.e., granulosum) is avascular, whereas after luteal development each steroidogenic cell is adjacent to a microvascular element (101) and cells associated with the microvasculature comprise approximately 50% of the total cell population (102). However, the precise mechanisms [e.g., sprouting, intussusception (103,104)] whereby angiogenesis forms the dense network of capillaries in the developing corpus luteum are not well defined. There is evidence of early migration of microvascular cells into the luteinizing granulosa cell layer (78), but much of the capillary expansion appears caused by endothelial cell proliferation. Cell proliferation is greatest in the corpus luteum during early development, and evidence suggests that the vast majority (up to 95%) of the dividing cells are vascular endothelial cells (101). The interval of angiogenesis varies between species, but in primates the capillaries reach the central cavity by day 4 and venules appear along the cavity border that lead back through the luteal tissue to large veins outside the corpus luteum by day 6 of the luteal phase (81). Nevertheless, there is evidence for some continued microvascular cell proliferation after luteal development. For example, the vascular space within the corpus luteum of domestic animals and primates increases from early to mid-luteal phase, and the number of nonsteroidogenic cells (i.e., primarily associated with the microvasculature) increases to late luteal phase (102). This may be associated with further maturation of the microvessels [as they acquire endothelial support cells, or pericytes (78,105)], but this important feature has received little attention to date.

Since the discovery that follicular fluid and ovarian extracts contained "angiogenic activity" (106,107), investigators have striven to identify the factors that control vascular events (e.g., endothelial cell proliferation) in the developing corpus luteum. Many of the earlier candidates, such as basic fibroblast growth factor, were pleiotropic factors that stimulated proliferation or migration of various cell types. Although some of these factors, including basic fibroblast growth factor, may have a role in luteal angiogenesis (107), major progress occurred when researchers determined that newly discovered endothelial-specific growth factors (108) associated with vasculogenesis in the developing embryo and angiogenesis in pathological conditions (e.g., tumor growth) were also present in the ovary. Ongoing studies are testing the hypothesis that such factors play key roles in physiological angiogenesis in the corpus luteum, that their regulation may be different from that observed in other tissues, and that cell–cell interactions through these factors are critical for the normal development, function, and perhaps life span of the corpus luteum.

Vascular Endothelial Growth Factor

Vascular endothelial growth factor (VEGF) (108), also discovered as a vascular permeability factor, specifically acts on vascular endothelial cells to promote proliferation, migration, tube formation, and vessel permeability. The five molecular forms of VEGF (121, 145, 165, 189, and 206 amino acids) can be produced by alternative splicing and they differ in solubility and heparin-binding ability (108). Evidence in other tissues indicates that the isoforms may have different physiological functions (77). Also, the VEGF family continues to expand to include newly discovered, structurally related proteins that share homology with VEGF (now called VEGF-A)—including VEGF-B through D (108), which may well be expressed in the ovary (109). In addition, researchers have identified high-affinity receptors (VEGF-R1, R2, R3) and proposed coreceptors (neuropilin-1 and 2) on endothelial cells that mediate or facilitate VEGF action (110,111).

Evidence suggests that all the necessary components of the VEGF-R system are present and dynamically expressed in the corpus luteum of several species [for reviews, see Hazzard and Stouffer (100) and Fraser and Lunn (112)]. Although primarily based on messenger RNA (mRNA) data, luteinizing granulosa cells appear to express high levels of VEGF-A (109), with VEGF-A expression remaining high in the corpus luteum during luteal development (113–115) and in the developed corpus luteum. However, expression may decline during luteal regression in primates and domestic animals. Limited experiments localized VEGF mRNA and protein to the steroidogenic luteal cells (116) and, in one report, to endothelial support cells (78). Also, VEGF-R1 and R2 and neuropilin-1 and 2 were localized to microvascular cells and, intriguingly in a few reports, to luteal cells (117–119). However, there are few studies to date on specific cell types to evaluate directly the production and regulation of VEGF-A isoforms, VEGF types, or VEGF receptors and coreceptors in the corpus luteum.

Limited studies suggest that regulation of VEGF expression in the ovulatory follicle and corpus luteum differs from that in many tissues. It is generally believed that a decline in local oxygen concentrations (hypoxia) is a primary stimulator of VEGF production and angiogenesis in normal and pathological conditions (120,121). Indeed, based on limited data, a similar scenario was proposed for the follicle (122). However, Martinez-Chequer and colleagues (123) observed that hypoxic conditions did not increase VEGF-A production by nonluteinizing granulosa cells from the preovulatory follicle or luteinizing granulosa cells from the periovulatory follicle in Old World monkeys. Rather, evidence from several species indicates that gonadotropin, notably the mid-cycle LH surge, promotes VEGF production (124) through either transcriptional or post-transcriptional (125) actions in the luteinizing granulosa layer and developing corpus luteum. In addition, local factors such as insulin-like growth factors (IGF)-1 and 2, may modulate or synergize with LH to control VEGF expression in luteinizing granulosa cells (123). However, the control of VEGF-A expression in cells of the corpus luteum has not been examined in detail. Limited reports suggest that LH action on luteinized tissue is less or not important in controlling VEGF production per se [except as a general luteotropin for maintaining luteal structure–function (113)], and that local regulators, including hypoxia, may predominate (126,127).

A critical role for VEGF-A in luteal angiogenesis and corpus luteum development and function was recently established. Because homozygous and heterozygous null mutations for VEGF-A were embryonic lethal in mice due to vascular malformations (128,129), more classic ablation approaches were required, including protocols that prevented potential antecedent effects during follicle development. Administration of VEGF-A antagonists (e.g., VEGF antibody, soluble VEGF-R chimeras) to mice or monkeys either at the time of ovulation (i.e., during luteinization) or at mid-luteal phase (i.e., after luteal development) markedly inhibited subsequent luteal structure–function (130–132). The suppression of luteal endothelial cell proliferation, endothelial cell area, and circulating progesterone levels, plus evidence that direct injection of VEGF antagonist into the periovulatory follicle blocks its rupture and impairs luteal architecture (133), suggest that VEGF (a) acts primarily through local effects in the follicle and corpus luteum, not through indirect actions on other tissues controlling ovarian function [e.g., through the hypothalamic–pituitary axis to alter gonadotropin secretion (134)]; (b) is essential for ovulation of the mature follicle and normal development of the corpus luteum; and (c) remains important in the developed corpus luteum to maintain luteal structure–function. Nevertheless, there are few studies to date on the mechanisms of action or specific effects of VEGF-A, or other VEGF family members, on ovarian cells. There is one report that VEGF stimulates proliferation of endothelial cells isolated from the monkey corpus luteum (135).

Angiopoietins

The angiopoietins (Angs), unlike VEGFs, are not endothelial cell mitogens, but nevertheless are essential for normal vascular development (136). Four different Angs were identified in the mouse or human, and they all bind to the Tie-2 receptor as either an agonist (Ang-1, plus Ang-4 in the mouse) or antagonist (Ang-2, plus Ang-3 in humans). An evolving model proposes that interactions between VEGF and Ang agonist/antagonists are critical for vessel development and maturation during embryogenesis and tumorigenesis (137). Whereas VEGF promotes angiogenesis to the stage of capillary tube formation, the ratio of Ang-1 to Ang-2 appears critical for capillary growth, maturation, or degeneration. Ang-2 may loosen cells from the ECM, allowing VEGF to stimulate endothelial cell events, whereas Ang-1 recruits periendothelial cells (i.e., pericytes) to stabilize and mature capillaries. In the relative absence of VEGF or other angiogenic factors, Ang-2 may destabilize vessels and cause capillary degeneration, especially of those not supported by pericytes.

It is tempting to hypothesize that a similar scenario is important for vascular events occurring during development, maintenance, and regression of the corpus luteum. Both Ang-1 and 2, plus Tie-2 mRNA or protein, were detected in luteal and endothelial cells, respectively, of several species. Limited reports on (a) the different temporal expression patterns for Ang-1/2 and VEGF during the luteal life span (116,138), and (b) the preponderance of immature vessels (not supported by pericytes) in the ovine corpus luteum (105), are consistent with this model. However, further research is needed on the cellular sites of Ang-Tie expression and action in the corpus luteum. Preliminary evidence that intrafollicular injection of the antagonist Ang-2, but not the agonist Ang-1, destroys the periovulatory follicle and resets the ovarian cycle in monkeys portends a vital role for Angs in follicular-luteal structure–function (139). Likewise, little is known about the regulation of Ang expression, except that the ovulatory gonadotropin surge stimulates Ang-1 (but not Ang-2) expression in the follicle (125), and chorionic gonadotropin (CG) exposure mimicking pregnancy initiation promotes Ang-1 and 2 expression in the corpus luteum (138) of primate species. However, given the in vivo approach of these studies and the delayed response (36 hours) of the ovulatory follicle (125), it remains to be determined if these are direct or indirect effects of gonadotropin.

Advances described in this section are adding cellular and molecular details to the concept that luteinization of the ovulatory follicle involves critical events whereby luteinizing cells (e.g., granulosa-derived luteal cells), and possibly microvascular cells (78), produce angiogenic factors that promote neovascularization of the developing corpus luteum (Fig. 4). To date, VEGF-A has received the most attention. But future research must resolve the role of specific isoforms of VEGF-A (140), other VEGF (B through D) family members (109), more promiscuous growth factors [e.g., basic fibroblast growth factor (107)], as well as emerging tissue-specific endothelial growth factors [e.g., endocrine gland (EG)-VEGF (141)] that may synergize with or complement VEGF actions. EG-VEGF is particularly intriguing because it is (a) structurally distinct but has similar biological actions to VEGF (141); and (b) highly expressed in steroidogenic tissues, with an expression pattern in the primate corpus luteum that may differ from that of VEGF (142,143). The importance of EG-VEGF and its relevance to nonprimate species (144) await investigation. Likewise, the proposed role(s) of Ang agonists and antagonists in regulating vessel stability, function, and degeneration and their relationship to control of luteal function or life span warrant rigorous comparative study.

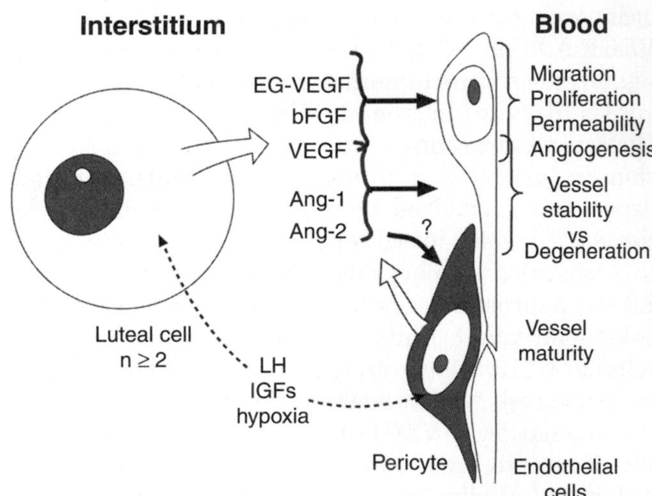

FIG. 4. Schematic of the proposed interactions between luteal cells, capillary endothelial cells, and (in some species) periendothelial support cells (pericytes) for the production and action of angiogenic factors that control the development, maintenance, and regression of the microvasculature in the corpus luteum during the ovarian cycle. *White arrows* denote synthesis and release of vascular endothelial growth factors (VEGF), endocrine gland (EG)-VEGF, basic fibroblast growth factor (bFGF), and the angiopoietins (Ang-1, Ang-2). *Dotted arrows* indicate regulators of VEGF/EG-VEGF/Ang expression. *Black arrows* denote sites of action leading to indicated effects on the microvasculature. (See text for further details. Adapted from Stouffer, R. L. [2004]. The functions and regulation of cell populations comprising the corpus luteum during the ovarian cycle. In *The Ovary* [P. C. K. Leung and E. Y. Adashi, Eds.], pp. 169–184. Academic Press, San Diego.)

Differentiation of Steroid-Secreting Cells

Suppression of Steroidogenic Cell Proliferation

It is proposed that further differentiation of follicular steroidogenic (i.e., theca and granulosa) cells into primarily progesterone-secreting (i.e., luteal) cells is associated with cessation of proliferation (75). Based on limited studies in a few species, primarily during superovulation protocols wherein the LH surge is replaced with a bolus of the LH-like hormone human CG (hCG), this appears to be the case. In rodents, cell mitotic activity generally ceased by 7 hours after administration of the hCG bolus, and before the occurrence of ovulation at 12 hours (145). In a primate model (146), the percentage of dividing granulosa cells declined markedly within the earliest time interval examined (i.e., 12 hours after the hCG bolus), but a small percentage (10%) of cells continued cell cycle activity up to the time of ovulation (36 hours post-hCG). Indeed, there is evidence that proliferation of granulosa/luteal cells persists through ovulation (80), and in the developed corpus luteum (101,147,148). However, the percentage of dividing cells is small, may decline with advancing age of the corpus luteum, and remains to be shown to have biological relevance. For example, in primates there is little evidence of increased proliferation of steroidogenic, or even non-steroidogenic [microvascular (149)] cells in species where the luteal life span is prolonged in actual or simulated pregnancy (101,150,151).

Some information has accrued in understanding how follicle/luteal cell progression through the cell cycle, or its cessation, is controlled by the intracellular proteins called cyclins (152), their interaction with specific cyclin-dependent kinases [CDK, e.g., CDK4 (153)], and inhibition of cyclin-CDK activity by CDK inhibitors (152). Studies in rodents indicate that cyclins D2 and E, which promote cell passage through the G1 and G1 to S transition phases of the cell cycle, are downregulated in granulosa cells after the ovulatory hCG bolus (154,155). Cyclin D2 is rapidly suppressed, whereas cyclin E does not change until 48 hours post-hCG. Although the mechanisms regulating cyclin expression require further investigation, there is some evidence for dephosphorylation of proteins [i.e., retinoblastoma proteins (156)] during luteinization that could suppress cyclin D and E activity. The importance of the cyclin D system in controlling follicle growth is supported by evidence that granulosa cells from cyclin $D^{-/-}$ null mutant mice are unable to proliferate, but they do luteinize after a gonadotropin bolus (157). However, again there may be important species differences at this control point because Chaffin et al. (146) did not observe a decrease in cyclin D2 or E mRNA expression in granulosa cells of monkey preovulatory follicles during the decline in cell proliferation initiated by an hCG bolus. Rather, the levels of cyclin B1 mRNA are reduced in macaque granulosa cells within 12 hours of exposure to a gonadotropin bolus. Also, phosphorylated retinoblastoma proteins are present in the developing corpus luteum in primates (158).

Unlike cyclin expression or activity, evidence suggests that the expression of CDK inhibitors (CKI) is more consistent among species during luteinization. A similar profile of a transient increase in CKIp21 inhibitor, followed by a later rise in CKIp27, occurs in granulosa cells of rodents (154) and primates (146) after an ovulatory hCG bolus. The early rise in CKIp21 inhibitor is associated with the marked decline in cell mitosis, but the later rise in p27 may be critical for sustained cell cycle arrest. Studies in $p27^{-/-}$ null mutant mice indicate that follicles grow normally, but luteal cells do not stop proliferating after an ovulatory stimulus (159). Further studies are needed to understand the dynamics of cyclin-CDK-CKI pathways in the follicle and corpus luteum, and their roles in controlling cell proliferation. It is also likely that other pathways, such as the Myc/Max/Mad family of transcription factors and p53-regulated Wips, also regulate cell proliferation (160) through or independent of the cyclin system.

Little is known about the gonadotropin actions that inhibit follicle cell proliferation. Gonadotropin surge levels could act directly or through LH/CG-stimulated intermediates, such as local growth factors or steroids [e.g., progesterone (161)], to control early and late (periovulatory) events in the cell cycle. Likewise, there is little information on proliferation or its control in luteinizing theca cells (75).

Enhancement of Progesterone Production

One of the hallmarks of luteinization is the switch from the principal steroid product of the maturing follicle (i.e., estrogens) to that of the developing and mature corpus luteum, progesterone. This is both a qualitative and quantitative change because the mammalian corpus luteum produces up to 100-fold greater amounts of steroid than the follicle. Thus, circulating levels of estradiol in the presence of the preovulatory follicle reach the picomolar range, whereas progesterone levels produced by the corpus luteum can reach the nanomolar range. The human corpus luteum, for example, is estimated to secrete up to 40 mg of progesterone per day during the menstrual cycle (162). To achieve such steroidogenic activity, it appears that at least two major pathways are enhanced in luteal cells: (a) cell-specific amplification of steps in the steroidogenic pathway converting cholesterol to progesterone, and (b) increased ability to sequester blood-borne sources of cholesterol and to store cholesterol-rich lipid deposits (163).

In some species, such as primates, circulating progesterone levels increase within 30 minutes after initial exposure to a gonadotropin surge/bolus (164). This suggests that a basal level of LH-sensitive progestogenic activity can be in place, but is relatively inactive until the LH surge. Indeed, nonluteinized granulosa cells from monkey preovulatory follicles appear to possess substantial levels of the enzyme 3β-hydroxysteroid dehydrogenase (3β-HSD) because they readily convert pregnenolone to progesterone (164). However, these cells cannot readily convert cholesterol analogs or cholesterol sources such as low-density lipoprotein (LDL) into progesterone. Evidence from some species suggests that luteinization is associated with a marked increase in expression or activity of the cytochrome P450 side-chain cleavage enzyme ($P450_{scc}$ or CYP-11A) responsible for converting cholesterol to pregnenolone in the inner mitochondrial membrane, but there may be a divergence in CYP-11A expression and enzyme activity because there are reports that CYP-11A mRNA initially declines after the onset of the LH surge (165). Because CYP-11A is required for androgen synthesis in the theca layer of the follicle (166), its enhancement during luteinization may be primarily in the granulosa cells. Moreover, a further increase in 3β-HSD expression, and a decrease in 17α-hydroxylase,17–20 lyase ($P450_{17\alpha}$ or CYP-17, which catalyzes androgen production from progesterone) and aromatase ($P450_{arom}$ or CYP-19, which converts androgen to estrogen) may facilitate the cellular emphasis on progestin production. However, there may be a divergence between mRNA expression and steroidogenic enzyme activity (165).

Because initial steroidogenesis (i.e., CYP-11A activity) occurs inside the mitochondria, the mechanism by which cholesterol is transported from outside to the inner mitochondrial membrane has been an area of important investigation (163,167). Although recognized for years as the rate-limiting step in steroidogenesis, and the site of acute regulation through synthesis of a short-lived protein, it was not until the mid-1990s that this protein was identified, characterized, and named *steroidogenic acute regulatory protein* [StAR (168)]. Mutations of the StAR gene in mice (169) or those occurring spontaneously in humans (163) are characterized by the inability to synthesize adequate amounts of steroids. However, placental steroidogenesis in some species (e.g., humans) does not appear to involve StAR [but possibly a related protein, MLN64 (170)]. There may be a family of related proteins that have StAR-related lipid-transfer (START) domains; indeed, there is limited evidence for MLN64 in ovarian cells (163). The critical role of StAR in cholesterol transport in adrenal and gonadal cells is established, but the mechanism of action awaits resolution. Although targeted to the mitochondrial membrane by specific amino acid sequences, StAR does not appear to require a receptor-type molecule for cholesterol transfer once it reaches the mitochondria. StAR has been postulated to act as an intermitochondrial shuttle (171), to interact with other proteins such as the peripheral benzodiazepine receptor (172,173) on the outer mitochondrial membrane, and to form a molten globule into the membrane to facilitate cholesterol transfer (174). Notably, the expression of StAR is markedly upregulated in the granulosa layer during luteinization in many species (175,176).

In several species, there is a transient peak in StAR expression occurring early after the LH/CG bolus, followed by a decline (175–177), before returning to high levels in the corpus luteum. Biphasic patterns were also reported for expression of steroidogenic enzymes [e.g., CYP-11A (176)]. These patterns could be related to transient downregulation of LH/CG receptors or desensitization of the receptor-stimulated pathways (3), which is known to occur in rodents at ovulation (178), but await investigation in other species. Alternatively, the renewed expression could reflect the end of tissue remodeling activities essential for follicle rupture and the emergence of steroidogenic activities in the developing corpus luteum. Finally, the renewal of expression of other steroidogenic enzymes [e.g., CYP-17 (178,179) or CYP-19 (180,181)] in some species likely reflects the unique ability of their corpus luteum also to synthesize androgens or estrogens. It is intriguing that the compartmentalization of the luteal tissue in Old World monkeys, baboons, and women retains the distribution of androgen- and estrogen-producing enzymes observed in the follicle. Immunocytochemical localization of CYP-17 to the paraluteal/theca-luteal cells and CYP-19 to the granulosa luteal cells implies the continuation of a two-cell model for estrogen production in the corpus luteum of higher primates (182,183).

To produce large quantities of progesterone, a major increase in availability of the steroid precursor, cholesterol, is required. The process of de novo synthesis of cholesterol by cells is well established (163). However, de novo synthesis by luteal cells appears to be a minor source for cholesterol steroidogenesis because expression of 3-hydroxy-3-methylglutaryl coenzyme A (HMG-CoA) reductase (the enzyme catalyzing the rate limiting in cholesterol synthesis from two carbon precursors) appears constant in the corpus luteum (184). Moreover, pharmacological blockage of this enzyme does alter circulating progesterone levels in women (185) or progesterone secretion by human luteinized granulosa cells (186). Rather, most species use blood-borne sources of cholesterol esters, either in the form of high-density lipoprotein (HDL) or LDL, for luteal steroidogenesis. The relative contribution of the two lipoproteins varies between species, with the rodent corpus luteum relying primarily on HDL and the primate corpus luteum requiring LDL (187). Notably, marked increases in the expression of receptors for and binding of HDL [SR-B1 (188,189)] and LDL [LDL-R (190,191)] occur during luteinization, that may actually precede cell exposure to circulating substrate after ovulation. Other components of cholesterol movement in luteal cells are not as well characterized, but there is evidence that apolipoprotein A-1 binds to SR-B1 and facilitates transfer from HDL-derived cholesterol ester (192). Also, the Niemann-Pick-C1 protein, which promotes transfer of LDL-derived cholesterol from the endosome to the Golgi apparatus, is also upregulated during luteinization (193). Deficiencies in any of these components in rodents [e.g., SR-B1 or apolipoprotein A-1 (194,195)] or women [LDL-R, Niemann-Pick-C1 (196–198)] will diminish circulating progesterone levels and impair fertility. Moreover, lack of tissue reserves of cholesterol in these conditions points to the importance of these pathways in building intracellular supplies of cholesterol esters in lipid droplets. Cholesterol esters derived from LDL or HDL, by the actions of lysosomal acid lipase or hormone-sensitive lipase, respectively, contribute to the intracellular pool of free cholesterol, and may be stored in lipid droplets that form from the action of acyl coenzyme A–cholesterol acyl transferase (ACAT-1) in the rough endoplasmic reticulum (163). Thus, another mechanism whereby luteal cells can acutely increase the availability of substrate for steroidogenesis is by hydrolysis of intracellular stores of cholesterol esterase in lipid droplets. The enzyme catalyzing this process is also hormone-sensitive lipase or neutral cholesteryl ester hydrolase, and its expression is enhanced in the corpus luteum of some species at luteinization (199,200). Free cholesterol is then directed intracellularly through sterol carrier proteins [e.g., SCP-2 in rodents (201,202)], apparently with involvement of cytoskeletal elements which, as noted earlier, change markedly during luteinization. Figure 5 provides a schematic of progesterone biosynthesis in a "generic luteal cell," as adapted from Niswender, Ho, and colleagues (163,167). However, there are many molecular and cellular events in this complex pathway, plus their regulation, which remain poorly understood and display species diversity.

Steroid- and Peptide-Secreting Cells

Although the primary hormone produced by the corpus luteum is the steroid progesterone, it is clear that luteal cells synthesize and secrete many peptides/proteins that can serve in a species-dependent manner as local (paracrine/autocrine) or endocrine factors. Three types of hormones have been identified: oxytocin, relaxin, and inhibin/activin.

As reviewed by Sherwood and Fields (203) and Ivell (204), after discovery of a "contractin factor" in cow luteal extracts that stimulated uterine contractions, 70 years passed before it was identified as molecularly homologous to oxytocin in the posterior

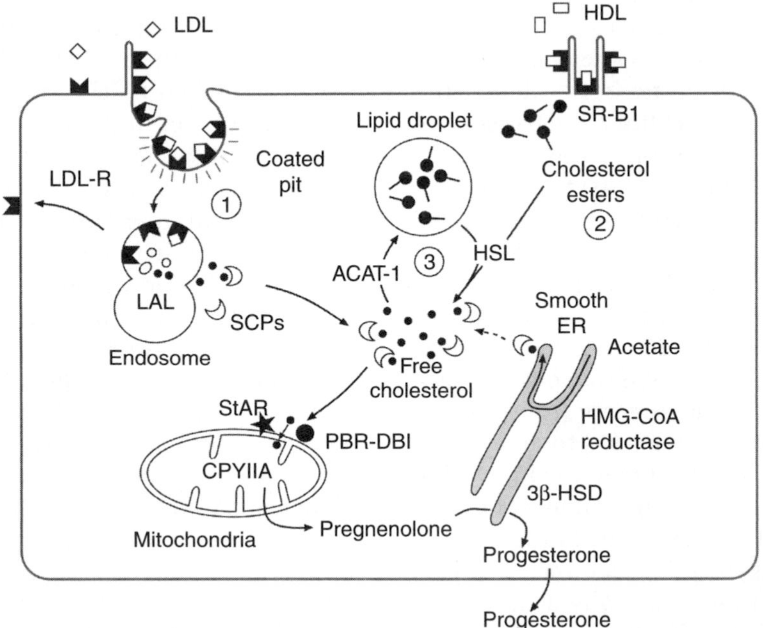

FIG. 5. Simplified illustration of the three major sources of intracellular, unesterified (i.e., "free") cholesterol for progesterone synthesis in luteal cells. **1:** The low-density lipoprotein (LDL) receptor–mediated pathway leading to internalization of coated pits, fusion with lysosomes to form endosomes, where lysosomal acid lipase (LAL), generates free cholesterol (*closed circle*). **2:** The high-density lipoprotein (HDL) receptor (SR-B1) pathway, wherein hormone-sensitive lipase (HSL) generates free cholesterol. **3:** The storage and turnover of free cholesterol in lipid droplets through acyl coenzyme A-cholesterol acyl transferase (ACAT-1)–catalyzed cholesterol ester formation and HSL-catalyzed ester hydrolysis. Intracellular transport of hydrophobic, free cholesterol appears actively directed by various proteins, including NPC1 protein and sterol carrier proteins (e.g., SCP2). Cholesterol transport to the inner mitochondrial membrane involves the steroidogenic acute regulatory (StAR) protein, possibly in close association with the peripheral-type benzodiazepine receptor (PBR) and an endogenous ligand (e.g., diazepam-binding inhibitor [DBI]). Cholesterol is enzymatically catalyzed to pregnenolone by the CPY-11A complex. After moving to the smooth endoplasmic reticulum (ER), pregnenolone is converted to progesterone by 3β-hydroxysteroid dehydrogenase-Δ^{5-4} isomerase (3β-HSD). Progesterone then exits the cell. Except under certain pathological conditions where it may serve as a compensatory pathway (163), de novo synthesis of cholesterol from acetyl CoA/acetate by 3-hydroxy-3-methylglutaryl CoA (HMG-CoA) synthase/reductase (*dotted line*) appears minimal in luteal cells.

pituitary (205). Subsequent studies confirmed that oxytocin is synthesized in large luteal cells in much the same way as pituitary cells: first as a precursor protein that contains the nonapeptide sequence for oxytocin plus its carrier protein neurophysin I, followed by packaging of the prohormone into secretory granules (206), where it undergoes a series of enzymatic steps to release oxytocin (204). The factors promoting luteal oxytocin mRNA expression, protein synthesis, and secretion are not well understood (203,204). In the cow, oxytocin–neurophysin mRNA increases markedly in luteinizing granulosa cells after ovulation, but production of luteal oxytocin is markedly delayed and does not peak until well into the luteal phase (day 8–12). The delay appears due to a lag in the availability of peptidyl glycine α-amidating monooxygenase, which is the terminal enzyme in post-translational processing. Moreover, oxytocin secretion and action in the late luteal phase may occur when its mRNA levels are reduced or virtually absent; this paradox can be explained by post-transcriptional regulation and oxytocin stores in granules that are released in response to a secretagogue (e.g., a $PGF_{2\alpha}$ pulse). In domestic animals, such as cows, sheep, goats, and pigs, there is evidence that luteal oxytocin plays a role in luteal regression at the end of the nonfertile cycle (42) (see section on Prostaglandin $F_{2\alpha}$). The release of oxytocin at midcycle may upregulate its own receptors in the uterus, and the subsequent pulses of Ca^{2+}-dependent oxytocin release provide a positive feedback loop for promoting $PGF_{2\alpha}$ release from the uterus at luteolysis.

Luteal oxytocin is detectable in other species, including primates, but at 1,000-fold lower levels; more sensitive techniques (e.g., reverse transcriptase polymerase chain reaction) are needed to detect oxytocin mRNA, and oxytocin concentrations in ovarian venous blood are typically not high enough to have an endocrine effect on nonovarian tissues (204). A physiological role for oxytocin in such species is perhaps limited to a local action, especially at luteinization. In many species (207), including primates (208), there is evidence that the LH surge promotes expression and production of oxytocin, and its receptor (204,208), in luteinizing cells of the ovulatory follicle. In general, data from a variety of species support a proluteinization action of oxytocin [e.g., to promote progesterone production and secretion (208,209)]. Ivell (204) speculates that the paracrine luteotropic action of oxytocin during development of the corpus luteum occurs in various species, only to be usurped by an endocrine role as a luteolytic factor during regression of the corpus luteum in domestic animals. Further studies are needed to establish the physiological importance of oxytocin in periovulatory events and luteinization.

Although reports (210,211) suggest that relaxin or a relaxin-like factor (RLF; insulin-like peptide-3 [INSL-3]) is expressed by the ovulatory, luteinizing follicle, and could play a role in ovulation (212), relaxin has generally been considered a hormone of pregnancy and parturition (203). Although discovered by Hisaw in 1926 (213), until recently the molecular nature of relaxin family members and their receptors proved elusive [for review, see Sherwood (214)]. Three genes encoding relaxins were identified in humans that belong to the superfamily of structurally related hormones, including insulin, IGFs, INSLs, and relaxin-like factor (RLF or INSL-3). Relaxin 1 is typically produced and secreted by rodents (rat, mouse) and domestic animals (pig), whereas relaxin 2 is found in the circulation of primates (human). Relaxin 3 is expressed in the brain, but its physiological significance remains undefined. Orphan receptors belonging to the leucine-rich guanine nucleotide–binding (G-protein) coupled receptors, LGR-7 and 8, have been identified as relaxin 1/2 and RLF/INSL-3 receptors, respectively (215).

The corpus luteum of many species produces and secretes relaxin in a stage-dependent manner (214). In species such as the rat, mouse, pig, and human, the corpus luteum is the sole source of relaxin throughout pregnancy. However, the dynamics of relaxin production and resultant levels and patterns in the circulation during pregnancy vary between species (214). In the pig, relaxin mRNA and protein increase at mid-luteal phase, particularly in the fertile cycle. The protein then progressively accumulates in membrane-bound cytoplasmic granules in large luteal cells throughout gestation until approximately 2 days before delivery, when degranulation occurs, luteal relaxin levels markedly decline, and circulating relaxin levels surge 12 to 24 hours before delivery. A similar prepartum surge occurs during late gestation in rats and coincides with the decline in progesterone levels that is required for onset of parturition. In contrast, in primates, including women, the levels of relaxin are detectable but much lower in the corpus luteum, and it does not appear that this protein accumulates in membrane-bound granules. Also, circulating relaxin levels are higher during early pregnancy compared with the second or third trimester, and there is no evidence of a prepartum surge (216,217). In nonprimates (214), data suggest that relaxin serves as a hormone during pregnancy to promote growth of the lower reproductive tract (cervix, vagina), facilitate rapid, successful delivery (cervical softening), and contribute to development of mammary gland alveoli (pig) or nipples (rat). In primates, limited studies have not demonstrated a physiological role for relaxin, but abnormal patterns of circulating relaxin have been associated with early pregnancy loss in women (218). Nevertheless, there is reason to suspect that it does not play a critical role in women because sole use of sequential estrogen and progesterone treatment in assisted reproductive technology protocols to mimic ovarian function results in normal gestations with timely delivery of offspring and successful lactation (56). This does not, however, rule out a local ovarian action (219) or a hormonal role in nonhuman primates. Clearly, the parturitional and lactational activities of relaxin vary between species because effects of relaxin deficiency are more dramatic in rats or pigs (214) than mice (relaxin $1^{-/-}$ or $LGR7^{-/-}$ null mutants (220,221). An extreme example is the sheep corpus luteum, which may not produce relaxin (222).

Inhibin and activin are members of the transforming growth factor-β superfamily of ligands initially identified based on their ability respectively to inhibit and stimulate pituitary follicle-stimulating hormone (FSH) (223). The sequential expression of inhibin B (αβB) and inhibin A (αβA) as antral follicle growth proceeds is well documented (224). Although inhibin production typically ends after the LH surge in many species, the corpus luteum of Old World monkeys (225) and women (226) continues to synthesize and secrete significant amounts of inhibin A and its α subunit. Investigators have localized the inhibin subunits primarily to granulosa luteal, but not theca luteal cells (227,228); these may also produce the binding proteins follistatin and follistatin-related gene products (229) that neutralize inhibin/activin activity. However, follistatin is highly expressed in luteal tissue of other species [e.g., rat (230)] that do not produce inhibin/activin. Thus, these proteins may bind other factors, such as bone morphogenic proteins, which could have local actions during luteinization and luteolysis (230). Although the major source of inhibins and activins after early pregnancy in primates is the fetal-placental unit (231,232), the corpus luteum secretes appreciable amounts of inhibin A in early gestation (233). However, an endocrine role for luteal inhibin or activin remains unclear. Immunoneutralization studies using inhibin antisera suggest that testicular inhibin B plays a critical role in regulating circulating FSH levels in male primates (234), but such studies failed to establish a similar role for ovarian inhibin A in females during the luteal phase of the menstrual cycle. Proposed local actions to modulate luteal structure–function await delineation (235).

LUTEOLYSIS

The corpus luteum typically ceases its function and structurally involutes at one of three stages: (a) near the end of a nonfecund ovarian cycle; (b) at the point in gestation where its progestogenic function is no longer needed to maintain intrauterine pregnancy; or, if the corpus luteum is required throughout pregnancy, then (c) at term delivery of the offspring. This loss of luteal structure–function is often referred to as *luteolysis*. As elegantly reviewed by Davis and Rueda (236), the area is not without controversy because some investigators have used the terminology of functional luteolysis versus structural luteolysis and debated whether these events are independent, dependent, or interdependent. This issue is again related to species differences, such that in some species [e.g., the hamster (237)] the corpus luteum undergoes loss of function and essentially involutes within one ovarian cycle, whereas in others (e.g., the mouse and rat) the corpus luteum no longer functions but three or more generations of corpora lutea may be evident in the ovary during the cycle (238). Likewise, the corpus luteum of primates may cease function 3 to 4 days before menstruation at the end of the cycle (52), but residual luteal mass is evident into the next follicular phase and can even respond in a limited manner to exogenous gonadotropin (LH/CG) treatment (239). Another level of complexity may be added when comparing luteolysis at the end of the ovarian cycle with that in pregnancy. There are reports that structural involution can occur at a slower rate after pregnancy compared with the end of the ovarian cycle (240). Such differences may be due to exposure to different milieus of luteotropic or luteolytic factors (see section on Regulation of Luteal Structure–Function) or reliance on different cell processes for destruction (74,241). For example, although there is little evidence for a significant role of PRL promoting the structure–function of the primate corpus luteum during the menstrual cycle, this hormone may delay the luteolytic process in the "rejuvenated corpus luteum" after parturition, especially during lactation (52). To date, for various reasons, many researchers investigated the process of luteolysis near the end of the nonfecund cycle, and often used the model of $PGF_{2\alpha}$-induced luteolysis once this PG was established as the uterine luteolytic factor in rodent and domestic animals.

The hallmark of early luteal regression is declining progesterone production, clearly demarcated by the fall of circulating blood levels of this steroid hormone (242). The next section discusses evidence for mechanisms whereby the luteolytic factor $PGF_{2\alpha}$ may directly or indirectly reduce steroid production by luteal tissue. Nevertheless, the mechanisms and cellular processes involved in loss of luteal structure–function remain poorly understood, particularly in species not using a uterine luteolytic ($PGF_{2\alpha}$) process. However, with the considerable interest in recent years in the processes that control cell death during tissue development and homeostasis, investigators began to apply advances in this field to studies on germ cell depletion, follicle atresia, and luteolysis in the ovary. As reviewed by Tilly and colleagues (243), at least six different forms of cell death have been described, but ovarian biologists have focused primarily on one type of programmed cell death termed *apoptosis* (244). Apoptosis describes the coordinated collapse of the cell, protein degradation, and DNA fragmentation followed by rapid loss of dead cell fragments (e.g., through engulfment by immune cells). Key characteristics include cell shrinkage (not swelling as noted in necrosis) and a "ladder" of 180–base pair–incremental sizes of genomic DNA after gel electrophoresis.

Increasing evidence supports the theory that apoptosis is an important mechanism controlling involution of the corpus luteum. Apoptotic indices, either morphological or biochemical, have been reported during structural involution of the corpus luteum in rodents (245,246), domestic animals (247–250), and primates (251,252). However, numerous issues remain to be resolved. For example, in primates there is some controversy as to whether cells are dying by apoptosis (148,251) or by autophagocytosis (253,254). It seems possible that further research may identify more than one form of cell death, and indeed diverse forms of apoptosis, during luteal regression. There is evidence the pharmacological methods of induced luteolysis (e.g., gonadotropin-releasing hormone [GnRH] antagonist–induced LH deprivation versus $PGF_{2\alpha}$-induced luteolysis) can result in different patterns of cell death (253). Studies are needed to compare processes of cell death during luteolysis in the ovarian cycle, pseudopregnancy, and pregnancy. Likewise, the physiological stimuli and cellular pathways that result in cell death in the corpus luteum await clear definition. Extensive research in invertebrate and vertebrate systems established that deprivation of critical survival-promoting "growth factors" or expression of "death ligands" is used in normal tissue turnover to eliminate cells by apoptosis (243). One can hypothesize that removal of or reduced sensitivity to luteotropic factors, or addition or increased sensitivity to luteolytic factors promotes apoptosis in the corpus luteum. However, whether these factors act directly or indirectly [e.g., through regulation of local death ligands (243)] and how cellular processes are temporarily activated are not clear.

There is considerable evidence that members of the B-cell leukemia/lymphoma-2 (Bcl-2) family of apoptosis-regulatory proteins are present in the mammalian ovary, including the corpus luteum. It is possible the prevailing theory (243)—that the balance between proapoptotic (e.g., Bax, Bak) and antiapoptotic (e.g., Bcl-2, Bcl-X) members, perhaps modulated by stimulus-sensitive BH3-only proteins (e.g., Nip3, Bad), controls cell death—applies to follicle and corpus luteum turnover. The evidence is strongest for follicle atresia, where Bax protein levels are elevated (255,256) and Bax-deficient mice accumulate abnormal follicles as their granulosa cells atrophy, but presumably fail to undergo apoptosis during attempted atresia (257). However, there is less evidence on the dynamics and role of Bcl-like proteins in the corpus luteum; indeed, whether the ratio of proapoptotic (e.g., Bax) to antiapoptotic (e.g., Bcl-2) proteins changes in association with apoptosis in the primate corpus luteum during the menstrual cycle (258–260) or early pregnancy (261) is controversial. There is increasing evidence that members of the death receptor family and their ligands are involved in luteal regression, including early disruption of progesterone production and structural involution. Notably, luteal expression of Fas ligand (FasL) and its receptor Fas is associated with luteolysis in several species (262–264), and exposure to FasL or Fas-activating antibody causes apoptosis of luteal cells in several species (263,265,266), as well as luteinized granulosa cells from women (267). The functional importance of the FasL/Fas system during luteolysis in the mouse is supported by evidence that (a) Fas-activating antibody causes premature luteal regression (262), whereas (b) mutants lacking FasL ($^{gld/gld}$–mice) or Fas ($^{lpr/lpr}$–mice) have irregular ovarian cycles attributed to defects including aberrant luteolysis (262). Tumor necrosis factor-α (TNF-α) has also been implicated in luteolysis (see section on Luteolytic Factors). This cytokine can cause death of endothelial cells derived from the corpus luteum (268,269) and augment interferon (IFN)-γ–induced death of luteal cells (270) from domestic animals. Evidence for a possible role in luteolysis is derived from TNF-α receptor-1–deficient mice (271), which fail to progress through the luteal (diestrous) phase of the cycle.

Although some progress has occurred in evaluating local death ligands and apoptotic regulatory proteins in the corpus luteum, there is little information on the molecular pathways that execute apoptosis and degrade cellular targets. In mammals, at least 14 different *c*ysteine *as*partate–specific prote*ases* [caspases 1 through 14 (243)] have been identified that serve either as "initiator" caspases (i.e., activate other caspases or initiate early mitochondrial changes) or "effector" caspases (hydrolyze a broad spectrum of proteins, including protein kinases, cytoskeletal components, and inhibitors of DNAses). Initial reports suggest several caspases are present in luteal tissue (272,273), and limited evidence supports a key role for the effector caspase-3 in luteolysis. Notably, the uterine luteolytic factor, $PGF_{2\alpha}$, activates caspase-3 during luteal regression in rodents (274) and sheep (275). Also, caspase-3 null$^{(-/-)}$ mice exhibit delayed onset of luteal involution in gonadotropin-stimulated ovarian cycles (276). However, there may be caspase-3–independent mechanisms that depend on the stage of the corpus luteum (277). Further research is needed to evaluate the regulation and processes of apoptosis, as well as other forms of cell death, in the corpus luteum and to discern their role(s) in functional as well as structural luteolysis.

REGULATION OF LUTEAL STRUCTURE–FUNCTION

It is well established that the ovulatory LH surge initiates many of the early events in luteinization; indeed, the LH-induced early rise in progesterone synthesis appears essential for a number of hormonal as well as local actions of progesterone, ranging from controlling the strength and duration of the LH surge (278,279) to ovulation (280), respectively. However, the luteinizing ovulating follicle rapidly becomes dependent on additional luteotropic support for development and maintenance of luteal structure–function, as evidenced by (a) the undeveloped corpus luteum and ultrashort luteal phase after the LH surge in unmated, nonfecund rodents; and (b) the ability of luteotropin antagonists quickly to suppress luteal development and function within 24 hours after the LH surge in species with a short luteal phase, such as monkeys and humans (281,282). A widely accepted concept is that the functional life span of the corpus luteum after the LH surge depends on the balance of and often sequential exposure to factors that either promote (luteotropins) or inhibit (luteolysins) luteal structure–function (283). Again, species have evolved various mechanisms for controlling the corpus luteum. Thus, luteotropic or luteolytic factors may originate from within the ovary/corpus luteum, as well as from classic or nonclassic endocrine glands, such as the pituitary, uterus, and placenta.

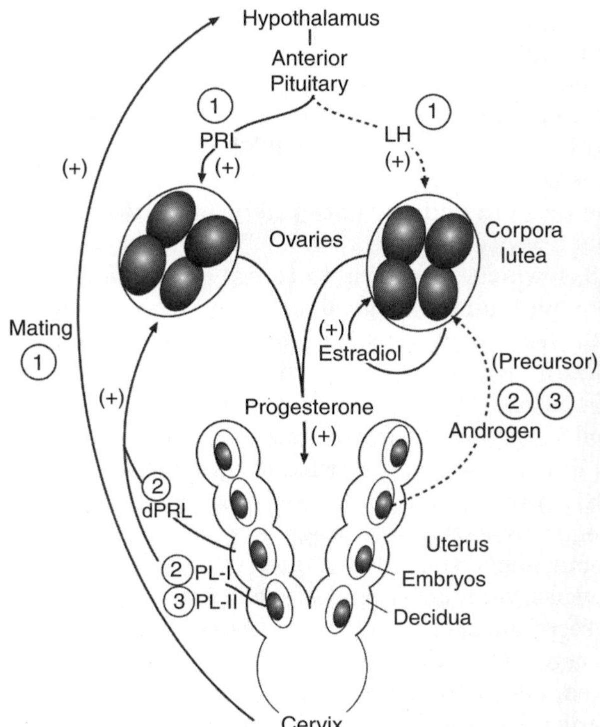

FIG. 6. Diagram of the neuroendocrine, endocrine, and local mechanisms promoting luteal structure–function during early (1), middle (2), and late (3) pregnancy in the rat. Three different sources of prolactin (PRL)-like hormones provide luteotropic support as gestation proceeds: (a) a neuroendocrine reflex arc activated by mating results in PRL release from the anterior pituitary; (b) by the second week of pregnancy, decidual PRL (dPRL) can provide support; and (c) by middle to late pregnancy, placental lactogens (PL-I, then PL-II) are the primary luteotropins. Also, in early pregnancy, luteinizing hormone (LH) promotes production of the local luteotropin, estradiol. By middle to late pregnancy, LH is no longer required because androgen from the placenta serves as a precursor for luteal estrogen production. PRL-like hormones and estradiol combine to maintain the functional corpus luteum throughout pregnancy, which is essential because the rat placenta is not a significant source of progesterone. (From Stouffer, R. L. [1999]. Corpus luteum of pregnancy. In *Encyclopedia of Reproduction* [E. Knobil and J. D. Neill, Eds.], pp. 709–717. Academic Press, San Diego, with permission.)

Luteotropic Factors

The primary luteotropic hormones vary between species, but generally include PRL- (284) or LH-like (285) hormones from the pituitary or placental-uterine unit. However, vital complementary roles may be played by local factors such as ovary- or luteal-derived steroids or growth factors.

Prolactin-Like Luteotropic Hormones

As reviewed by Bowen-Shauver and Gibori (74), either PRL secreted by the anterior pituitary or PRL-like hormones from the uterine decidua and placenta are the essential luteotropins in rodents, such as the rat, depending on the stage of the corpus luteum after ovulation (Fig. 6). Their critical role is evident from $PRL^{-/-}$ or PRL $receptor^{-/-}$ null mutant mice, in which ovulation and early luteinization occur, but progesterone levels are subnormal and the corpora lutea regress rapidly, resulting in sterility (286,287). However, the defect in uterine decidualization, implantation, and early embryonic development can be overcome by exogenous progesterone. Progesterone replacement does not, however, prevent fetal death at mid-pregnancy, suggesting additional vital roles for PRL in the uterus-placenta. Mating at estrus (i.e., stimulation of the cervix) activates a neuroendocrine loop that results in secretion of PRL from the pituitary in two daily surges (one nocturnal, one diurnal) for 10 to 12 days (72,288). It is experimentally established that pituitary PRL secreted during the first week of pregnancy or pseudopregnancy is solely capable of supporting luteal development and sustaining normal levels of progesterone secretion (289). However, by day 7, pituitary PRL is no longer essential because of the expression of PRL-like luteotropins by the decidua and placenta (290). From days 7 to 11, this activity appears to be a "decidual luteotropin" (291), which, according to recent data, is identical in sequence and biological activity to pituitary PRL (292). Although capable of compensating for pituitary PRL in experimental settings, its role is probably comparable for decidua-derived PRL in primates, that is, as a local "growth factor" (293). Nevertheless, by day 12, as pituitary PRL surges end and decidual PRL production declines, the pituitary is no longer essential for pregnancy maintenance (74). Remarkably, the corpora lutea of pregnancy increase in mass, vascularity, and steroidogenic capacity owing to secretion of other PRL-like factors, termed placental lactogen (PL)-I and II (294). Rat PL-I is a glycosylated, 36- to 40-kDa protein produced by trophoblast giant cells that is detectable in the circulation only at mid-pregnancy. Rat PL-II is a nonglycosylated, 25-kDa protein with a much shorter half-life that is also produced by the trophoblast giant cells. Onset of PL-II precedes the decline in PL-I expression in these cells, and circulating PL-II levels rise from mid-pregnancy until parturition. It is now believed that both rat and mouse PL-II, as well as PL-I, bind to the PRL receptor (295,296) and are sequentially responsible for maintenance of luteal structure–function in late pregnancy [as reviewed by Bowen-Shauver and Gibori (74)].

Thus, in rodents, such as the rat and mouse, PRL or PL binding to PRL receptors promotes luteal structure–function. Experiments in hypophysectomized rats suggest that between days 6 and 11 of pregnancy, LH also acts in concert with PRL to maintain progesterone secretion (297). However, subsequent studies suggest that the actual luteotropin is locally synthesized estradiol (298), which results from LH-stimulated production of androgen by the ovary/corpus luteum that is aromatized to estradiol (73). This absolute requirement for estradiol develops after the first week of pregnancy and it remains until parturition. But by mid-pregnancy, pituitary LH is no longer required because placental androgen production by trophoblast cells increases markedly and serves as a substrate for estrogen production by the ovary (299). Although the rodent decidua and placenta produce PRL-like hormones, much as in primates (293), there is no conclusive evidence that the former produces LH-like hormones, analogous to CG in monkeys or women.

The mechanisms of PRL action in the rodent corpus luteum are an area of active investigation. The expression of both isoforms of the PRL receptor (i.e., long and short forms) is induced by luteinization (300,301), and PRL itself may stimulate its own receptor's expression (301). PRL also activates a number of signal transduction pathways, but investigators to date have focused on the pathway through the Jak2/Stat5 family of transcription factors activated by the long form of PRL receptor [for review, see Risk and Gibori (302)]. The importance of this pathway was established from dual Stat5a$^{-/-}$ and 5b$^{-/-}$ null mutant mice that exhibit inadequate luteal function and infertility (303). Several kinase pathways activated by PRL in other tissues warrant further evaluation, such as inositol-1,4,5 triphosphate (IP3) kinase/AKT, mitogen-activated protein (MAP) kinase, and C-Src (302), but especially protein kinase C (PKC) δ, which appears essential for PRL-stimulated relaxin expression in the rat corpus luteum (304).

PRL directly stimulates progesterone production by luteal cells by several mechanisms [for review, see Risk and Gibori (302)]. First, PRL promotes steroidogenesis by upregulating enzymes involved in progesterone [CYP-11A, 3β-HSD (305,306)] and estradiol [CYP-19, 17β-HSD (305,307)] synthesis, as well as preventing progesterone degradation by inhibiting 20α-HSD expression (308,309). Because the enzymes involved in progesterone production (i.e., CYP-11A, 3β-HSD) are generally expressed in excess and are not rate-limiting, a primary action of PRL appears to be total suppression of 20α-HSD expression and activity, thereby preventing progesterone catabolism to the inactive progestin, 20α-dihydroprogesterone. Another enzyme that may be regulated, at least by stabilization (310), is 17β-HSD, which is involved in estradiol synthesis. This moiety was originally identified in the rodent corpus luteum at middle and term pregnancy as PRL receptor–associated protein (PRAP) because of its association with the short form of the PRL receptor (311). A second, and probably lesser mechanism of PRL action is to promote cholesterol availability from extracellular and intracellular stores, while promoting intraluteal cholesterol stores (312). PRL increases the expression of HDL receptor [SR-B1 (313,314)], hormone-sensitive lipase (314), and SCP-2 (314), and many of these effects appear mediated at the transcriptional level.

Less is known of the mechanisms by which PRL regulates other luteal activities, thereby maintaining luteal structure–function. Because PRL promotes luteal expression of LH receptors (315,316), enzymes involved in estradiol synthesis (305,307), and both estrogen receptors (ERs) α and β (317), some of the effects are likely indirect and mediated by LH (e.g., StAR expression) or estradiol (e.g., angiogenesis) (74). Alternatively, some PRL effects may be direct, such as stimulation of proteases or their endogenous inhibitors [e.g., α_2-macroglobulin (318)] or enzymes that scavenge oxygen free radicals [e.g., superoxide dismutase (314,319)]. Such actions may be important for stage-specific tissue remodeling and stability during luteal development and function, respectively. PRL may also influence cell hypertrophy either directly [e.g., induction of elongation factor-2 (EF-2), which promotes protein translation events (320)] or indirectly [e.g., through local synthesis and action of growth factors, such as IGF-1, on large luteal cells (321,322)]. However, much remains to be understood regarding the actions of PRL and PL in the corpus luteum at the physiologically relevant stages during the luteal life span of rodent species, as well as in other species in which the corpus luteum is PRL dependent or sensitive.

Luteinizing Hormone–Like Luteotropins

Despite considerable controversy that extended into the 1980s [e.g., in primates, see Zeleznik and Benyo (323); in domestic animals, see Niswender and Nett (3)], it is now generally accepted that after ovulation, the corpus luteum of many species, including domestic animals (e.g., sheep, cow, horse) and primates (monkeys to humans), is primarily dependent on LH from the anterior pituitary for development and maintenance of luteal structure–function. Hypophysectomy (324), administration of LH antisera (325), or suppression of endogenous LH secretion (281,282) inhibits the corpus luteum, whereas coadministration of LH with such treatments can restore the normal functional life span of the corpus luteum during the nonfecund ovarian cycle (326). LH depletion/neutralization beginning within 1 to 2 days of ovulation can prevent the normal development of the corpus luteum, whereas treatment at mid-luteal phase causes premature regression of the developed corpus luteum [for review, see Stouffer (327)].

Nevertheless, there are likely differences in LH support of the corpus luteum between primate and nonprimate species, including acute regulation of progesterone secretion. In monkeys and women, pulses of secreted LH are entrained to increased progesterone secretion by mid-luteal phase, with steroid levels at baseline in the lengthening intervals between LH pulses by late luteal phase of the nonfecund cycle (328). This, plus evidence that GnRH antagonist treatment rapidly and completely suppresses circulating progesterone levels [and causes early menstruation (281,282)], demonstrates the absolute requirement of primate luteal cells for LH exposure to maintain and stimulate progesterone secretion. In contrast, Diaz and colleagues (329) review evidence from cows and sheep that LH pulses are not entrained to progesterone secretion, and GnRH antagonist treatment only modestly (50%) decreases circulating progesterone levels in a stage-specific manner. Such findings suggest that a portion of luteal function in these domestic animals is not dependent on LH, and is consistent with the concept that the developed corpus luteum contains a cell type (i.e., large luteal cells) that constitutively produces progesterone. The purported role of "constitutive" versus "LH-dependent" stimulation of luteal progesterone production between cell types during the luteal life span and between species warrants further evaluation (see section on Small versus Large Luteal Cells).

Despite this "constitutive activity," the corpus luteum in many domestic animals requires continual LH support until luteal function is no longer required to maintain pregnancy (3). This interval in gestation may be relatively short (30 days, sheep), long (200 days, cow), or until term delivery (115 days, pig). In contrast, in many primate species, pituitary LH secretion is required only until late in the luteal phase of the fecund cycle. Thereafter, an LH-like hormone, CG, is secreted by trophoblast cells of the developing placenta, which extends the functional life span of the corpus luteum in early pregnancy (55). CG is synthesized for a species-specific interval during gestation in primates, including throughout pregnancy in women (56). The primary, although perhaps not sole [e.g., local uterine-placental (330)], action of CG is to sustain luteal structure–function for several days to weeks in early pregnancy until progesterone production is pronounced in the placenta. After this time (i.e., the luteal-placental shift), the ovary and corpus luteum are not required to maintain pregnancy in primates. Domestic animals do not typically produce a placental LH-like hormone, but the horse is a unique exception. Although not likely the initial sustainer of luteal function is early pregnancy (see section on Other Luteotropic Factors), the mare placenta begins to secrete large quantities of an LH-like hormone, originally termed pregnant mare serum gonadotropin and now referred to as equine CG (eCG) (3). The eCG stimulates function of the primary corpus luteum, but also induces ovulation of additional follicles that grow in intervals during early pregnancy and promotes the function of these secondary corpora lutea (331,332). The corpora lutea secrete progesterone until eCG disappears from the circulation at 150 days of gestation, but lutea-derived progesterone is essential only until approximately day 70. Thereafter, progesterone support from the fetal–placental unit is sufficient to sustain pregnancy (333).

Pituitary LH, placental CG, and eCG are members of the glycoprotein family of hormones that includes pituitary FSH and thyroid-stimulating hormone (TSH). Containing noncovalently linked α and β subunits with distinct glycosylation patterns [for review, see Bousfield (285)], the luteotropic LH-CG moieties bind with high affinity to one receptor, the LH-CG receptor, but not the FSH or TSH receptors. However, there is species specificity of the receptor, because primate LH-CG receptors (like primate growth hormone [GH] receptors) have a much higher affinity for human LH or hCG than nonprimate gonadotropins (334). Since the cloning of the LH receptor in 1989 (335), considerable information has accrued on this member of the seven-transmembrane domain, rhodopsin-like, G-protein–coupled receptor family (336,337). It is apparent that LH-CG receptor expression and hormone binding correlate with luteal function and circulating progesterone levels because activity increases during luteal development and decreases during luteal regression in various species (3). However, whether LH-CG receptor levels directly influence luteal structure–function is less clear, in part because LH-CG receptor$^{-/-}$ null mutant mice (338,339), as well as inactivating LH-CG receptor gene mutations in women (336), are associated with follicular defects and anovulation, which preclude analysis of luteal function. Also, the observation that few of the LH-CG receptors need to be occupied for maximal steroidogenesis in the corpus luteum led to the concept of "spare receptors" for this luteotropin [see Niswender and Nett (3) for review]. However, it seems unlikely that these receptors are really "spare" because these receptors are coupled to signal transduction (e.g., cyclic adenosine monophosphate [cAMP] production) that continues to increase even though acute steroid production is maximal. There is speculation that this level of receptor occupancy and cAMP signaling is important for other aspects of luteal structure–function, especially in conditions of higher luteotropic support such as CG exposure in pregnancy.

G-protein–coupled signaling though the LH-CG receptor involves activation of adenylate cyclase, production of cAMP, and activation of cAMP-dependent protein kinases (C-kinases), which can have profound acute (steroidogenic) and chronic (structure–function) effects on the corpus luteum [see Hunzicker-Dunn and Mayo (337) for review of signal transduction pathways]. Although LH-CG can activate cAMP-independent pathways [e.g., phospholipase C–mediated synthesis of IP3 and diacylglycerol (DAG), leading to activation of protein kinase C (PKC) and intracellular calcium fluxes (340,341)], many of the effects of LH on steroidogenesis can be mimicked by exogenous cAMP and appear similar to those described earlier for PRL in rodent luteal cells, namely, enhancement of LDL receptors and uptake, metabolism of stored cholesterol esters to free cholesterol, and promotion or maintenance of key enzymes such as CYP-11A and 3β-HSD (74,167). However, there are two major differences between PRL-PL and LH-CG action in luteal cells of relevant species: (a) LH-CG does not regulate circulating progesterone levels by inhibiting its rapid metabolism by 20α-HSD; rather, (b) LH-CG action acutely to promote progesterone secretion involves rapid increases in StAR activity (through protein kinase A–mediated phosphorylation), followed by a transcriptional effect to promote StAR expression. This mechanism, which remains an area of intense investigation, may be a critical site of LH-CG action in luteal cells of various species, including rodents before mid-pregnancy (74). Notably, StAR expression, as well as that of key steroidogenic enzymes, is sustained during prolongation of primate luteal function by CG treatment mimicking pregnancy initiation (342).

Much less is known regarding the physiological actions of LH through cAMP-independent pathways (337) or the nonsteroidogenic actions of LH to maintain luteal structure–function. LH may regulate proteases and their inhibitors [including MMPs, ADAMTS, and TIMPs (96,97,343)], cell–cell communication by gap junctions (344), and expression of other local [e.g., angiogenic (108)] or hormonal [e.g., relaxin (345,346)] factors. Investigators using differential display techniques or cDNA microarrays are discovering novel or underappreciated LH-regulated pathways, such as the corticotropin receptor–binding protein (CRF/urocortin-R_1/R_2-BP) system in the monkey and human corpus luteum (347,348), that are LH-regulated and dynamically expressed during the luteal life span. Rigorous investigations on these genes/gene families and their proteins in the corpus luteum will increase our understanding of the complex systems controlling luteal structure, function, and life span.

Other Luteotropic Factors

There is growing evidence that the primary luteotropic hormones, PRL and LH, exert their actions, at least in part, through LH-stimulated synthesis and secretion of tropic factors. These factors include steroids, peptide/proteins, and arachidonate metabolites.

Estradiol is an essential luteotropin for the maintenance of luteal structure–function in the rabbit [for review, see Keyes et al. (349)], and also at a specific interval during pregnancy (days 7 to 11) in the rat (74). Indeed, the corpus luteum of many species contains the classic nuclear ER (ERα) or the recently discovered ERβ form. (317,350). Experiments in gene-mutant mice suggest that estradiol can act through either ERα or ERβ to influence luteal structure–function, but one cannot rule out effects predetermined by loss of estrogen action in the antecedent follicle (351). Also, one cannot rule out the local actions of high estrogen levels on putative membrane receptors (352,353). Estrogen's actions in the corpus luteum are not well defined, but there are limited reports that it enhances StAR expression and mitochondrial cholesterol transport in rabbit luteal tissue (354,355), and SR-B1 expression, sterol content, and cholesterol transport (SCP-2) capacity in rat luteal tissue (356). One can hypothesize that estrogen synergizes with LH or PRL action to promote luteal structure–function. For example, LH-stimulated estrogen increases EF-2 expression in rat luteal cells; EF-2 is then activated by PRL, leading to a significant increase in luteal cell protein content (320,357). Estrogen may also promote luteal angiogenesis, notably at mid-pregnancy in the rat (74). However, evidence for ER expression in endothelial cells and estrogen regulation of angiogenic factors, such as that found in other vascular beds, including the uterus, has not been reported for the corpus luteum.

Since Rothchild (2) first proposed that progesterone itself is a local luteotropin in the mammalian corpus luteum, evidence [for review, see Stouffer (327)] has accumulated for LH-induced transient expression of progesterone receptors (PRA and B) in the ovulatory, luteinizing follicle of many species, and continued PR expression in the developed corpus luteum of several species with short and long luteal phases (from dogs to cows and primates). Studies in null mutant mice established an essential role of PRA, but not PRB, in ovulation, but progesterone's role during luteinization appeared minimal (358). Nevertheless, there are reports of progestin/antiprogestin effects in the rodent corpus luteum despite the absence of nuclear PR (359). Progesterone inhibited the expression of 20α-HSD (and hence suppressed progesterone metabolism) in rat luteal cells, and evidence was presented for this action through progestin binding to the glucocorticoid receptor (360). Progestins may also have actions through plasma membrane receptors proposed to exist in granulosa and luteal cells (361,362). However, in species other than rodents, there are data suggesting that progesterone acts as a local luteotropin to promote luteal development and sustain luteal structure–function during the ovarian cycle and perhaps in pregnancy [for review, see Stouffer (327)]. In vitro and in vivo studies using antiprogestins [e.g., RU486 (363)] or progesterone synthesis inhibition and replacement [e.g., a 3β-HSD inhibitor with and within a progestin, R5020 (364)] demonstrate a critical role for progesterone in controlling luteal structure–function as well as ovulation. Specific progesterone actions to control tissue remodeling [e.g., protease expression (365,366)], health [suppression of apoptosis (363)], and sensitivity to other local factors [e.g., through ER expression (350)] have been reported. Again, one can hypothesize that LH actions that are independent of progestin action and those that are mediated by LH-stimulated progesterone action synergize to control the development or structure–function of the corpus luteum in a species-dependent manner (Fig. 7). For example, there is evidence for LH-dependent, progesterone-independent versus progesterone-dependent control of several components of the MMP-TIMP family in the primate corpus luteum (343,365). Further studies are needed to discern the role of progesterone and its potential receptor–signal transduction pathways in controlling the development and functional life span of the corpus luteum in species with short versus ultrashort luteal phases.

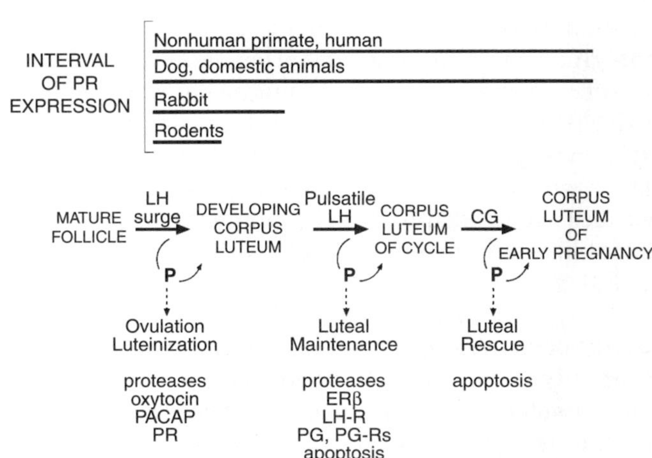

FIG. 7. Schematic of the duration of "classic" progesterone receptor (PR) expression in the corpus luteum of various species during its functional life span in the ovarian cycle and early pregnancy (*top*), with the hypothesis that luteinizing hormone (LH)/chorionic gonadotropin (CG)–stimulated synthesis of progesterone (P) generates a critical luteotropic factor essential for ovulation (all species examined to date), development of the corpus luteum (except rodents), maintenance of the corpus luteum during the ovarian cycle (cows and primates), and extension of luteal structure–function in pregnancy (cows and primates). Some of the gene products and processes identified as P-regulated pathways are listed. Growing evidence supports the concept of P as a local luteotropin (2). Proteases include ADAMTS-1, cathepsin L (rodent), and matrix metalloproteinase-1 (MMP-1; nonhuman primate). PACAP, pituitary adenylate cyclase activating peptide. [See text and other reviews (280,327) for further details.]

Direct effects of GH on the ovarian follicle and corpus luteum were reported [for review, see Hull and Harvey (367)], and there is some evidence that GH may be required for optimal luteal development. For example, both LH and GH were needed to obtain normal luteal weight and development in hypophysectomized ewes, although LH alone supported normal progesterone levels in the circulation (368). Although there is some evidence for receptors for the metabolic hormones GH, insulin, and thyroid hormone in luteal tissues, there is speculation that many of the first two hormones' observed effects pertain to GH-induced elevations in circulating IGF-1 or 2, or local IGF production in response to luteotropic hormones (74). There is evidence that IGF-1 or 2 stimulates progesterone secretion by luteinizing granulosa cells or luteal cells from various species (123,322,369,370), but the mechanism of IGF action in luteal cells has received little attention. In the human, LH stimulation of StAR expression is enhanced by IGFs (371). There are a few reports of IGF stimulating peptidergic activity [oxytocin production (370)], production of angiogenic factors [VEGF (123)], or expression of luteotropin receptors [e.g., ERβ in rat luteal cells (321)]—often in synergy with LH or PRL action. Since the mid-1990s, significant advances have been made in our understanding of gonadotropin control and dynamic expression of components of the IGF receptor–binding protein (IGF-R-BP) system in the follicle, and its integral role in influencing steroidogenesis (372) and growth versus atresia of the maturing antral follicle (373). It is possible that similar dynamics of the IGF system play a role in the development and functional life span of the corpus luteum (374,375). Further studies are needed to discern similarities and differences in the regulation and role of the species-specific IGF (e.g., IGF-1 in rodents, IGF-2 in primates)-R-BP system in the corpus luteum (376).

Although investigations on eicosanoids have focused primarily on the luteolytic action of $PGF_{2\alpha}$ (see section on Luteolytic Factors), there is growing evidence that other PGs and arachidonate metabolites can have luteotropic or luteoprotective effects on the corpus luteum of various species. Notably, PGs of the E, I, and D series stimulate cAMP and progesterone by luteal preparations in vitro (377–379), and can prevent spontaneous or $PGF_{2\alpha}$-induced luteolysis during local intrauterine or intraluteal administration in monkeys (380) and domestic animals [see Arosh et al. (381)]. The concept of a balance between luteotropic versus luteolytic PGs controlling luteal development, structure–function, and life span (283) is supported by recent evidence from the bovine corpus luteum that components in PGE synthesis transport, and signaling (e.g., PGE synthase, PG transporter, PGE receptors EP2 and EP3) are activated during luteal development and maintenance, whereas the PGF system (cyclooxygenase-2, PGF receptor PF) is activated at luteal regression (381). These data may explain the puzzling observations of why infusions of a PG synthesis inhibitor can decrease circulating progesterone levels (382,383). Further studies are needed on the putative luteotropic role of PGs [e.g., PGI (384), as well as PGE], the role of other arachidonate metabolites (385–389), and their relationship to the luteolysin $PGF_{2\alpha}$ in controlling luteal structure–function. To date, the best evidence for a physiological role of luteotropic PGs is the purported role of PGE_2 during early pregnancy in domestic animals [after IFN-τ declines and luteal resistance to $PGF_{2\alpha}$ is lost (390); see section on Domestic Animals].

Luteolytic Factors

As elegantly reviewed by Niswender and Nett (3) and McCracken and colleagues (42), the concept of physiologically important luteolytic factors (i.e., those that promote the loss of function and structural involution of the corpus luteum) began with the observation by Loeb in 1923 (33) that removal of the uterus (i.e., hysterectomy) lengthened the life span of the corpus luteum in guinea pigs. Subsequently, investigators reported that the timely demise of the corpus luteum did not occur after hysterectomy in many species during pseudopregnancy [e.g., rat (391); mice (392); hamster (393); rabbit (394)], or in other species that develop a functional corpus luteum during the nonfecund ovarian cycle [e.g., cow and sheep (395); pig (396); horse (397)]. There is variation among such species in the length of time that corpora lutea survive after hysterectomy, but luteal tissue usually retains both structure and function for at least the duration of normal pregnancy. Nevertheless, the extended luteal function may be at a lower level than that achieved earlier in pseudopregnancy [rabbit (398)] or the luteal phase of the estrus cycle [horse (397)]. In many, but not all (3) such species, the luteolytic effect of the uterus appears to be exerted locally. Removal of one of two uterine horns in many domestic (399,400) and laboratory animals (401,402) does not prevent timely luteal regression in the ovary adjacent to the intact horn, but prevents luteolysis on the other ovary. Therefore, it appears unlikely that the luteolytic factor is transported from the uterus to the ovary through the systemic circulation. Studies by Ginther (403) and others (42) suggest that the uterine luteolytic factor passes directly from the uterine vein into the ovarian artery in sheep and cattle, not through direct connections but by a countercurrent mechanism that permits transfer between regions of extensive contact between these two vessels. Throughout the broad ligament, the vessels have thinner walls where they are in contact and share a common tunica adventitia, and the ovarian artery is very tortuous on the uterine vein—all features consistent with venoarterial exchange (403).

However, in some species such as primates [Old World monkeys (65); humans (66)], the dog (404), and ferret (405), hysterectomy does not alter the normal life span of the corpus luteum. In such species, as well as in species where the uterine luteolytic effect is not exerted locally, there is less contact between the uterine vein and ovarian artery (406,407). Nevertheless, once the concept developed of a uterine luteolytic signal near the end of the nonfecund ovarian cycle, pseudopregnancy, or parturition in many species, investigators found it very difficult to elucidate the bioactive agent. It was not until the late 1970s, after considerable controversy, that there was general agreement that the uterine luteolytic factor was not a protein, peptide, or steroid, but rather $PGF_{2\alpha}$ (34).

Prostaglandin $F_{2\alpha}$

As reviewed by Sirois et al. (408), PGs are lipid molecules that belong to the eicosanoid family, which also includes prostacyclin, thromboxanes, leukotrienes, and lipoxins. They are derived from 20-carbon (C), polyunsaturated fatty acids, typically arachidonic acid. Various classes of PGs can be synthesized through the intermediate PGH_2, and PGF compounds have hydroxyl groups on C-9 and C-11. As summarized by Niswender and Nett (3), a major breakthrough occurred with the discovery that $PGF_{2\alpha}$, but not other PGs, caused premature luteolysis in pseudopregnant rats (409). This observation was followed by evidence that (a) exogenous $PGF_{2\alpha}$ had luteolytic effects in every species tested that depended on uterine control of the corpus luteum, except only to a limited extent in pigs (410); and (b) $PGF_{2\alpha}$ synthesis and secretion is elevated in endometrial tissue during luteal regression (411–413), plus pulses of $PGF_{2\alpha}$ release into the uterine vein (and often detectable in the circulation) increase in frequency and amplitude during the onset of luteolysis in several species (414–416). Critical evidence for a causal role of endogenous $PGF_{2\alpha}$ was derived from studies wherein nonsteroidal antiinflammatory drugs such as indomethacin, which inhibit PG synthesis, or active/passive immunization to $PGF_{2\alpha}$, blocked spontaneous luteolysis during the estrous cycle and prolonged pseudopregnancy and pregnancy in domestic animals and rodents (417–420). With increases in our understanding of the enzymatic pathways of PG synthesis and metabolism, and the development of more selective inhibitors of PG synthesis [see Sirois et al. (408)], further studies are warranted. However, evidence that null mutant mice lacking the $PGF_{2\alpha}$ receptor [$FP^{-/-}$ (421,422)] do not display a timely preparturitional decline in circulating progesterone and fail to deliver, and that this can be overcome by ovariectomy, supports a critical role for $PGF_{2\alpha}$ in signaling luteolysis. Similarly, defects in mice lacking critical genes in arachidonate mobilization from membrane phospholipids [phospholipase A2, or $PLA2^{-/-}$ (423)] or PG synthesis [$PGHS\text{-}1^{-/-}$ 424)] support this concept. However, because of possible roles of other PGs in the antecedent follicle [notably, PGE in cumulus–oocyte expansion and ovulation (425–427)] and corpus luteum (see section on Other Luteotropic Factors), it is more difficult to assess the effects of the latter defects in null mutant mouse models.

The mechanisms controlling the synthesis and release of uterine $PGF_{2\alpha}$ during luteolysis are an area of intense interest (Fig. 8). McCracken and colleagues (42) introduced a model where changes in steroid sensitivity toward the end of the luteal phase in sheep (a) stimulate the hypothalamic oxytocin pulse generator to increase the frequency of oxytocin release from the posterior pituitary and (b) upregulate oxytocin receptors in the uterine endometrium. These combined effects lead to the release of low (subluteolytic) pulses of $PGF_{2\alpha}$ from the uterus, which are sufficient to cause release of oxytocin stores from the corpus luteum through $PGF_{2\alpha}$-FP signal transduction. The supplemental release of luteal oxytocin amplifies the release of uterine $PGF_{2\alpha}$, resulting in large (luteolytic) pulses of $PGF_{2\alpha}$ that cause luteal regression. Moreover, Wiltbank and colleagues (329) have proposed another amplification system in the corpus luteum, wherein small amounts of uterine $PGF_{2\alpha}$ stimulate intraluteal production of $PGF_{2\alpha}$ through induction of PG synthase 2 (PGHS-2). The positive feedback loops between the uterus and corpus luteum, and within the corpus luteum, lead to sufficient $PGF_{2\alpha}$ action causing luteolysis. Whether such pathways occur or are physiologically relevant in other species (including those without appreciable luteal stores of oxytocin, e.g., rodents) awaits investigation. However, it is noteworthy that luteal tissue in domestic animals acquires the ability for $PGF_{2\alpha}$-induced PGF synthesis at the stage when the corpus luteum first develops the capacity for luteal regression [day 7 in cows; day 13 in pigs (329)]. Because many other substances, including gonadotropins, progesterone, and cytokines, are proposed modulators of luteal or uterine $PGF_{2\alpha}$ production (42,329), the precise manner in which the luteolytic signal is controlled remains obscure (390).

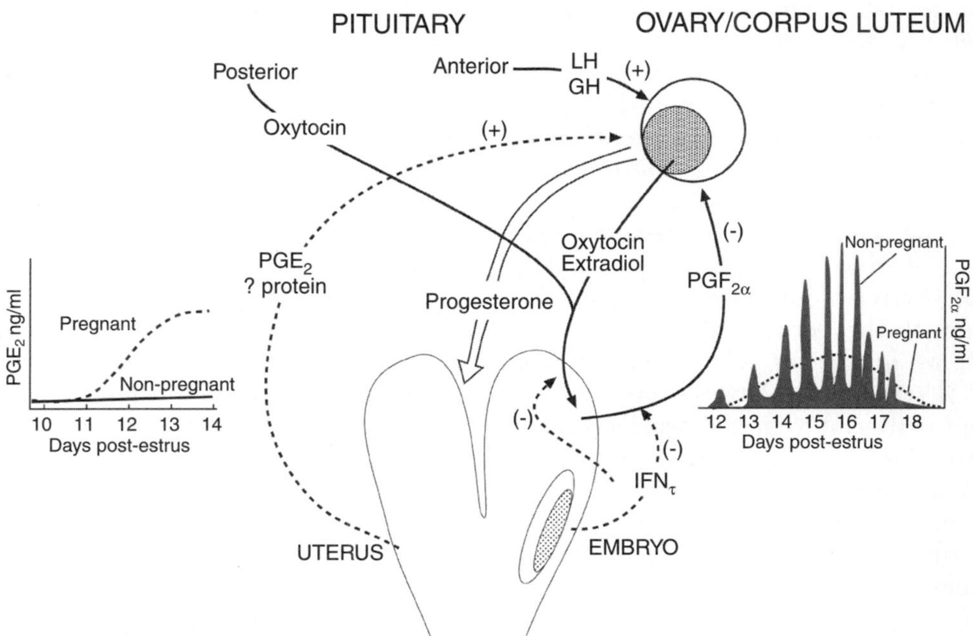

FIG. 8. Model of the neuroendocrine, endocrine, and local mechanisms controlling the functional life span of the corpus luteum during the estrous cycle (*solid lines*) and during early pregnancy (*dotted lines*) in ruminants, such as sheep. During the cycle, the corpus luteum is controlled by luteotropic factors (luteinizing hormone [LH]; growth hormone [GH]) from the anterior pituitary and a uterine luteolytic factor (prostaglandin [PG] $F_{2\alpha}$) whose timely pulsatile release (*right inset*) is influenced by pituitary (oxytocin) and ovarian (steroids, oxytocin) hormones. After implantation, embryo-derived interferon-τ (IFN-τ) prevents the pulsatile release of $PGF_{2\alpha}$, thereby preventing luteolysis in the fertile cycle. Also, secretion of PGE_2 (*left inset*) or unknown proteins from the uterus may promote luteal function during pregnancy. Hence, the corpus luteum continues to produce progesterone, which sustains intrauterine pregnancy. (Adapted from Niswender, G. D., and Nett, T. M. [1994]. Corpus luteum and its control in infraprimate species. In *The Physiology of Reproduction* [E. Knobil and J. D. Neill, Eds.], pp. 781–816. Raven Press, New York, and Stouffer, R. L. [1999]. Corpus luteum of pregnancy. In *Encyclopedia of Reproduction* [E. Knobil and J. D. Neill, Eds.], pp. 709–717. Academic Press, San Diego.)

Likewise, the mechanisms of $PGF_{2\alpha}$ action in luteolysis remain unclear, with several avenues proposed that range from reducing luteal blood flow, blocking the action of luteotropic factors, inhibiting steroidogenesis, and promoting cell destruction and tissue dissolution. However, again, the effects may vary between species and differ between experimental systems in vivo (e.g., intact tissue) and in vitro (dispersed, cultured cells). Based on its potent vasoconstrictor activity in various vascular beds, and evidence that exogenous $PGF_{2\alpha}$ can markedly decrease luteal blood flow in sheep (428), it was originally proposed that one of the early, important actions of the luteolytic $PGF_{2\alpha}$ signal was to limit influx of blood-borne nutrients and the like to the luteal tissue and efflux of metabolic byproducts, thereby causing tissue damage and dissolution. However, other investigators did not observe early changes in total or regional (luteal) ovarian blood flow during spontaneous or $PGF_{2\alpha}$-induced luteolysis in rodents (429), rabbits (349), or domestic animals (42) when progesterone levels were first declining (up to 50%). The effects of $PGF_{2\alpha}$ are dose dependent, and it was proposed that more physiological levels elicit little change in ovarian or luteal blood flow (42). However, rigorous studies correlating the effect of pulsatile exposure to either exogenous or endogenous $PGF_{2\alpha}$ on luteal blood flow throughout the duration of luteolysis have not been performed. There is little doubt that later decreases in blood flow are associated with continued reduction in luteal function and tissue destruction. However, in vivo and in vitro evidence that $PGF_{2\alpha}$ can suppress progesterone production and antagonize luteotropic hormone action independent of blood flow points to other major actions in luteal tissue.

Current evidence suggests that $PGF_{2\alpha}$ can rapidly decrease luteal progesterone synthesis by inhibiting the intracellular transport [e.g., by decreasing the mRNA levels and activity of SCP-2 in rodents (430)] and entrance of cholesterol into mitochondria [e.g., by decreasing the activity and expression of StAR in domestic animals and rodents (431,432)]. It now appears that previous observations that $PGF_{2\alpha}$ inhibited lipoprotein-supported progesterone production by luteal cells (329) were not due primarily to lesions in cholesterol uptake or storage (433), but rather to a lesion or lesions in cholesterol delivery to CYP-11A (329). Reports that transport-independent hydroxylated cholesterol substrates enhance steroid production by luteal tissue during spontaneous or $PGF_{2\alpha}$-induced regression support this concept. Nevertheless, there is evidence that $PGF_{2\alpha}$ causes a similar reduction in mRNA expression of some steroidogenic enzymes [e.g., 3β-HSD in the corpus luteum of sheep (434)]. In addition, although the loss of PRL/PL stimulation may be essential for the rapid expression of 20α-HSD in the rat corpus luteum during luteolysis at parturition, a key role for $PGF_{2\alpha}$ in regulating the transcription of this enzyme was reported in rodent luteal tissue (74). Further studies, such as that by Gibori and colleagues (314) examining global changes in gene expression by cDNA microarrays in $PGF_{2\alpha}$-treated corpora lutea, are needed to understand the temporal aspects of events initiated during luteolysis, and their regulation by luteolytic factors in the corpus luteum of various species. The elucidation of direct versus indirect [e.g., through other local factors (435)] actions of $PGF_{2\alpha}$ in subsequent changes in luteal structure–function, such as induction of enzymes [e.g., caspases (274,275)] involved in tissue dissolution, awaits investigation.

Receptors for $PGF_{2\alpha}$ comprise the FP subtype of PG receptors, which are G-protein–coupled, cell surface receptors with seven transmembrane domains (408). They are part of the group 2 contractile PG receptors, which also includes some of the PGE (EP1) and thromboxane A_2 receptors, thought to signal through $G_{\alpha q}$ and the phospholipase C pathway to increase intracellular Ca^{2+} levels. Although many studies in the 1970s reported specific PGF ligand binding to luteal tissue preparations (42), there have been relatively few studies on the expression and activity of PG receptor subtypes in luteal tissue since their discovery. There is evidence that the FP receptor is expressed in the corpus luteum of many species (436–438), including those lacking a uterine luteolytic mechanism (439), and that FP receptor expression increases in luteal cells during development (236), and thus can mediate $PGF_{2\alpha}$-induced luteolysis late in the ovarian cycle. A growing number of reports link the Ca^{2+}–PKC pathway to $PGF_{2\alpha}$ action in the corpus luteum, particularly to antisteroidogenic effects and cell death (236,440). However, it is likely that other signaling pathways contribute to $PGF_{2\alpha}$ action (236). In the rat corpus luteum, $PGF_{2\alpha}$-stimulated rises in intracellular Ca^{2+} lead to an induction of the transcription factor nur77, which promotes 20α-HSD expression (441). Other reports suggest that Ca^{2+}-dependent $PGF_{2\alpha}$ signaling through the extracellular signal–regulated kinase (Erk, a member of the MAP kinases) interferes with gonadotropin upregulation of StAR expression and progesterone secretion (236). There is also evidence that $PGF_{2\alpha}$ increases expression of SOCS-3, a member of the suppressors of cytokine signaling family, which could suppress PRL/PL action through the Jak2/Stat5 pathway (442).

It is also likely that $PGF_{2\alpha}$ acts through various cell types to promote the production of local luteolytic factors that contribute to the destruction of the corpus luteum. Although the role of $PGF_{2\alpha}$ as the luteolytic signal and cause of early loss of luteal function in many species is generally accepted, its role in loss of luteal structure and cell death is less clear. The separation of actions is obvious from in vitro studies, where addition of $PGF_{2\alpha}$ suppresses steroidogenesis without reducing cell viability (443). Morphometric and biochemical studies of cellular constituents, especially in the bovine and ovine corpus luteum, suggest that cell–cell interactions are important in luteal regression (444). For example, although evidence suggests that the "large" luteal cells are the predominant target for $PGF_{2\alpha}$ (445), there are reports that the numbers of "small" luteal cells and microvascular cells decrease before any loss of large cells at luteolysis (446). Evidence of apoptosis in small luteal cells and endothelial cells before large cells (64,440,447,448) supports this concept. Cell–cell interaction may involve a number of cytokines or chemokines from cells within (e.g., microvascular cells) or attracted (e.g., immune cells) to the corpus luteum as the luteal tissue ages or regresses. Three examples that received recent attention are endothelin-1, TNF-α, and monocyte chemotactic protein-1.

Endothelin-1

Endothelin-1 (ET-1) is a member of the peptide family containing ET-2, ET-3, and sarafotoxins that was originally defined by their cardiovascular actions, notably vasoconstriction (449). However, ET-1 has also emerged as a local modulator in the endocrine and reproductive system, including the ovary (450). Research by Milvae (451) and Meidan and Levy (452) suggests that ET-1 is an important factor controlling luteolysis in domestic animals. Several reports indicate that the ET-1 receptor system is present in the corpus luteum. There is evidence that endothelial cells in luteal tissue are the primary site of ET-1 mRNA expression. However, ET-1 is transcribed as a "prepro ET-1" molecule and the active peptide is ultimately formed from the intermediate "big ET-1" by the action of endothelin-converting enzyme (ECE-1) on endothelial cells and luteal cells. In addition, at least one of the ET receptors (ET-R, type A, or ETA) is expressed on both endothelial and luteal cells (452). Thus, the microvasculature synthesizes precursor/big ET-1, but both luteal and endothelial cells can produce bioactive ET-1 and transduce ET-1 signals through receptor-mediated pathways. There are reports that addition of ET-1 inhibits basal or LH-stimulated progesterone production by mixed or purified preparations of cells from the corpus luteum. Notably, preincubation with an ETA antagonist not only prevented the effects of exogenous ET-1, it blocked the antisteroidogenic effect of $PGF_{2\alpha}$ in luteal tissue slices (453). The latter effect is supported by evidence that intraluteal injection of an ETA antagonist delayed somewhat the luteolytic effect of exogenous $PGF_{2\alpha}$, and that combined treatment of ET-1 plus a subluteolytic dose of $PGF_{2\alpha}$ caused complete luteolysis in sheep (454–456). Finally, data on the dynamic expression and regulation of the ET-1 synthetic (ECE)-receptor (type A) system led Meidan and Levy (452) to propose that such changes could explain the insensitivity of the corpus luteum to the luteolytic actions of $PGF_{2\alpha}$ during the first half of the luteal phase.

Collectively, the data suggest that the ET-1 system (Fig. 9), through its regulation by luteotropic and luteolytic hormones, plays a local inhibitory role in the corpus luteum and mediates, at least in part, $PGF_{2\alpha}$ actions during luteal regression (452,457). However, key experiments remain to be performed, including investigations discerning whether modulating endogenous ET-1 synthesis or action alters the function or life span of the corpus luteum during the natural cycle. Although limited evidence suggests that the ET-1 system is present in other species, including rodents and primates (458,459), further studies need to address its importance beyond domestic animals. Finally, clarification is needed on the relative importance of $PGF_{2\alpha}$ action on luteal versus endothelial cells, and on the local amplification of $PGF_{2\alpha}$ (329) versus ET-1 (452) in the cascade of events leading to loss of luteal structure–function during luteolysis (Fig. 9).

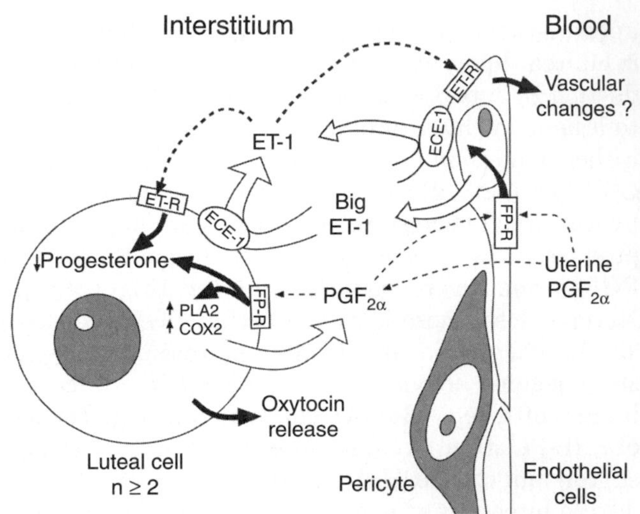

FIG. 9. Diagram of proposed interactions between luteal cells and microvascular cells for the production and actions of local endothelin (ET-1) and prostaglandin (PG) $F_{2\alpha}$ in response to the uterine luteolysin $PGF_{2\alpha}$. It is hypothesized that the initial $PGF_{2\alpha}$ signal is amplified or mediated by positive feedback loops to increase intraluteal levels of $PGF_{2\alpha}$ (329) and ET-1 (452), which then inhibit luteal cell function and cause tissue regression. *Dotted lines*, $PGF_{2\alpha}$ and ET-1 action on cell types; *white arrows*, $PGF_{2\alpha}$ and ET synthesis; *black arrows*, possible effects. ECE-1, endothelin-converting enzyme; ET-R, endothelin receptors; FP-R, $PGF_{2\alpha}$ receptors; COX-2, cyclooxygenase-2; PLA2, phospholipase A2. (From Stouffer, R. L. [2004]. The functions and regulation of cell populations comprising the corpus luteum during the ovarian cycle. In *The Ovary* [P. C. K. Leung and E. Y. Adashi, Eds.], pp. 169–184. Academic Press, San Diego, with permission.)

Tumor Necrosis Factor-α

Increasing attention has been directed at the cells of the immune system that, either residing throughout or attracted at specific stages, comprise a significant portion of the luteal tissue [for reviews, see Brannstrom and Friden (460) and Bukulmez and Arici (461)]. All types of leukocytes appear present, with their relative numbers changing during the life span of the corpus luteum, sometimes in a species-specific manner. For example, neutrophilic granulocytes are at their highest density during luteal development in the rat (i.e., early pregnancy or pseudopregnancy), whereas considerable numbers were also observed at luteal regression (462). In contrast, few neutrophils were detected in the pig corpus luteum (463), and although present in the human corpus luteum, no differences in neutrophil density were noted at luteolysis (464). Eosinophilic granulocytes are relatively sparse in the rat corpus luteum (462), yet infiltrate the ovine corpus luteum in large numbers after ovulation and are suspected to have a luteolytic role in sheep (460). Likewise, the numbers of lymphocytic subtypes in luteal tissue, such as cytotoxic T cells, helper T (T_H) cells, and natural killer cells, vary markedly between species. In the human corpus luteum, B lymphocytes and natural killer cells were sparse to nondetectable (464,465), although T_H cells were abundant, especially at luteolysis in several species (462,466). Macrophages are the cell type most consistently detected in the corpus luteum; these cells reportedly comprise more than 10% of all cells in human luteal tissue (465). Moreover, an increase in macrophage density was detected in a number of species during luteal regression (460).

In general, the changes in leukocyte cell density during the early and late luteal phase correlate with the tissue reorganization occurring during development and dissolution of the corpus luteum, respectively. Investigators have related these events to an acute inflammatory response/tissue repair during ovulation and luteinization (467), and an immune reaction/tissue elimination at luteolysis (468–470). Indeed, the expression of major histocompatibility complex (MHC) class II proteins on the surface of bovine (471) and primate, but not rodent (472) luteal cells, and their enhancement by $PGF_{2\alpha}$ (471), is consistent with an immune response. However, limited evidence to date supports the physiological relevance of local immune cells in controlling luteal function or life span; and MHC II molecules could serve in alternate roles [e.g., as receptors triggering apoptosis (473)]. This is due in part to difficulty in specifically modulating leukocyte numbers or action in the corpus luteum in vivo. One early approach to suppress macrophages and T lymphocytes with glucocorticoid administration yielded mixed results in rabbits and rats (474,475). Such approaches may be confounded by direct actions of glucocorticoids in luteal tissue through receptor-mediated pathways (476). In an opposite approach, Bagavant and colleagues (477) noted that an exaggerated inflammatory response and elevated cytokine levels in mouse ovaries, mediated by T_H1 cells, were compatible with normal ovarian cyclicity, ovulation, and fertility.

Nevertheless, it is well established that leukocytes express a cell-specific array of cytokines, degrading enzymes, and reactive oxygen species that promote the destruction of cells and ECM during tissue reorganization or destruction (460,461). Alternatively, cytokines may act to stimulate or inhibit steroid synthesis, at least as demonstrated in numerous in vitro experiments (478,479). Also, macrophages reportedly serve to eliminate dying luteal cells or cell fragments (apoptotic bodies) by phagocytosis (480). Brannstrom and Friden (460) provide a useful schematic of potential luteotropic versus luteolytic actions of cytokines on cell types of the corpus luteum.

Numerous studies established that TNF-α, a 17-kDa polypeptide, is produced in the ovary (481) and can have luteotropic or luteolytic effects depending on the species, stage of the luteal phase, and local milieu (236). Although perhaps produced by nonimmune cells, including endothelial and luteal cells, detailed studies in the pig (482) indicated that the luteal macrophage population secretes up to 100-fold greater quantities of bioactive TNF-α than other luteal cell types. Reports suggest that TNF-α inhibits luteal cell response to luteotropins, as judged from progesterone production in vitro, by middle to late luteal phase in rodents, domestic animals, and primates (236). Moreover, TNF-α reportedly induces the death of microvascular endothelial cells from the bovine corpus luteum in vitro (269), induces regression of blood vessels in the rabbit corpus luteum (483), and augments IFN-γ–induced death of luteal cells (270). Evidence suggests that TNF receptors (TNF-R) are present on luteal steroidogenic and endothelial cells (236). However, studies delineating a physiological role for the TNF-R system in the corpus luteum have not been reported. Recent generation of $TNF\text{-}R^{-/-}$ null mutant mice (271) suggests that TNF-RII (generally thought to mediate TNF's tropic functions, e.g., cell proliferation) deletion does not markedly alter ovarian function or reproductive capacity. However, TNF-RI (the cell death receptor) deletion caused numerous female reproductive defects, including early onset of puberty and premature cessation of ovarian cyclicity, with those that continued to cycle having prolonged diestrus intervals. The prolonged diestrus may indicate a role of the TNF-RII system in luteolysis, but other defects in the antecedent follicle or hypothalamic–pituitary function cannot be ruled out. Nevertheless, evidence that exogenous TNF-α can cause functional as well as structural luteolysis (483,484) led investigators to propose that TNF-α interacts with $PGF_{2\alpha}$, and perhaps with ET-1 (457), to promote regression of the corpus luteum. Also, TNF-α's signal transduction pathways in the corpus luteum remain to be resolved (236,340), but the TNF-RII system is implicated in the TNF suppression of luteotropic action and steroidogenesis (271).

Monocyte Chemotactic Protein-1

After the discovery of chemokines, which are defined as a family of small cytokines with two adjacent cysteine residues (C-C family) or one intervening amino acid (C-X-C family) that exhibit selective chemoattractant and activating properties on leukocytes (461), several reports suggested that members were either present (interleukin-8 [IL-8], monocyte chemotactic protein-1 [MCP-1]) or absent (regulated upon activation, normal T-cell expressed and secreted [RANTES]) from the ovary of certain species. Evidence suggests that IL-8 promotes neutrophil accumulation and activation [e.g., elastase activity (485)] and plays a critical role in gonadotropin-induced ovulation (486) and perhaps luteal development [e.g., angiogenesis (486)]. The chemokine receiving the most attention to date is MCP-1, which attracts and activates T-lymphocyte subtypes and monocytes/macrophages (487). Increased MCP-1 mRNA expression occurs in the corpus luteum of rodents, domestic animals, and primates during luteolysis (488–490). Also, luteolytic agents such as $PGF_{2\alpha}$ increase MCP-1 expression in luteal tissue either directly or indirectly through production of other cytokines (487). The cellular sources of MCP-1 are unclear, and there are conflicting reports regarding the ability of luteal cells to synthesize MCP-1 (491,492). Nevertheless, the data are consistent with the hypothesis that MCP-1 promotes the influx of macrophages that occurs in aging luteal tissue and that these cells then play a role in the structural, if not earlier functional, demise of the corpus luteum. As proposed by Penny (487), chemokines such as MCP-1 may prove to be a paracrine link between hormonal and cellular events in the corpus luteum involving steroidogenic luteal, microvascular endothelial, and immune cell types.

EXTENSION OF LUTEAL FUNCTION IN EARLY PREGNANCY

Mammals have developed diverse means for ensuring that ovarian cyclicity returns in the event of a nonfecund cycle, yet providing adequate progesterone support if pregnancy ensues. In the simplest case (e.g., many carnivores and marsupials), gestation lasts as long as or is shorter than the normal life span of the corpus luteum in the ovarian cycle (see section on Long-Lived Corpora Lutea). In these animals with "long-lived corpora lutea," there is no need for the conceptus or mother to alter luteal function because the developed corpus luteum is destined to provide the necessary progestational support. However, most animals have evolved a means for keeping the functional life span of the corpus luteum "short" (i.e., less than necessary for successful pregnancy) in the nonfecund cycle to increase reproductive efficiency. The most extreme cases (see section on Ultrashort-Lived Corpora Lutea) are the reflex ovulators (e.g., the rabbit), which do not even ovulate unless mating occurs, and the spontaneous ovulators that do not form fully developed, functional corpora lutea unless mating occurs (e.g., mouse, rat). Earlier in this review, emphasis was placed on the complex and sequential roles of maternal and decidual PRL, placental PL, and ovarian-placental steroids (i.e., estrogen) in promoting the functional life span of the rodent corpus luteum during pseudopregnancy or pregnancy. However, other animals that develop a functional corpus luteum during the ovarian cycle evoked processes for extending luteal function if pregnancy ensues, either by circumventing the luteolytic signal at the end of the cycle or providing additional luteotropic support to maintain luteal structure–function. The mechanism whereby the developing conceptus (implanting blastocyst) extends luteal function is the critical event in "maternal recognition of pregnancy" (55,56). The corpus luteum may be a critical source of progesterone throughout pregnancy, or become unnecessary after the "luteal-placental shift" when the placenta replaces the corpus luteum as the primary source of progesterone (55). Domestic animals (notably the ruminants, such as sheep and cattle) and primates with short-lived corpora lutea provide excellent models for illustrating different approaches to extending luteal function in early pregnancy.

Domestic Animals

It is generally thought that uterine exposure to progesterone during the 10 to 12 days after ovulation not only prepares the endometrium for implantation, but primes the mechanisms for production of the uterine luteolytic signal, $PGF_{2\alpha}$, if pregnancy does not occur. The proposed interaction between the posterior pituitary, uterus, and ovary to control the release and action of oxytocin and $PGF_{2\alpha}$ was alluded to earlier (see section on Prostaglandin $F_{2\alpha}$). It is now well established that extension of the functional life span of the corpus luteum in the fertile cycle is due to production of an antiluteolytic factor or factors from the developing conceptus. In sheep, embryos transferred to synchronized recipients as late as day 12 postestrus will prevent timely luteolysis (493). Hence, the critical period for maternal recognition of pregnancy in the sheep is days 12 to 13 of the cycle (3). Evidence indicates that the conceptus secretes antiluteolytic activity, which is not an LH- or PRL-like luteotropin (494), beginning around day 10, increasing through day 16, and becoming nondetectable by days 21 to 22. An important advance was the discovery by Bazer, Roberts, and colleagues (495,496) that the antiluteolytic substance in ewes is IFN-τ produced by mononuclear cells of the embryonic trophectoderm.

Ovine IFN-τ was originally called trophoblastin (497) or trophoblast protein-1 because it is the first major protein secreted by trophectoderm of the peri-implantational conceptus. This protein was subsequently identified as a type 1 IFN with potent antiviral, antiproliferative, and immunomodulatory activities (496). There is a high sequence homology of IFN-τ across ruminant species, where it apparently has a unique role in pregnancy initiation. There are multiple IFN-τ isoforms that arise from multiple genes; however, their functional significance remains unclear because all IFN-τ isoforms tested to date extend luteal function when injected into the uterine lumen at the appropriate time in ewes, cows, and goats.

The mechanisms of paracrine action of IFN-τ to prolong the function of the corpus luteum are an area of intense research (Fig. 8). It is well established that the high-amplitude pulses of luteolytic $PGF_{2\alpha}$ coming from the uterus in nonpregnant ewes are absent in fertile cycles (42). IFN-τ appears to act locally to prevent or alter the pattern of $PGF_{2\alpha}$ secretion (498–500). Although basal secretion of uterine $PGF_{2\alpha}$ is higher in early pregnancy, the high-amplitude, episodic pulses of $PGF_{2\alpha}$ release do not occur. There is also evidence from some species that $PGF_{2\alpha}$ secretion may be directed into the uterine lumen rather than the bloodstream. IFN-τ appears to interact with type 1 IFN receptors in the endometrium to block expression of estrogen and oxytocin receptors, but whether this action prevents estrogen and oxytocin response and is the primary event preventing uterine-induced luteolysis awaits further study (496). Also, its intracellular pathways of signal transduction remain poorly defined, but IFN-τ appears to target primarily the stroma and deep glandular epithelium to activate several STATs (signal transducer and activator of transcription, i.e., STAT 1 and 2) and to increase expression of several genes, including IFN regulatory factor-1 and 2, in the ruminant uterus (496,501).

Because intrauterine infusions of IFN-τ alone can extend the life span of the ruminant corpus luteum (502), some investigators believe that this is the only pregnancy recognition factor produced by the conceptus. However, others noted that the sensitivity of the corpus luteum to the luteolytic action of $PGF_{2\alpha}$ is reduced in early pregnancy (503). Moreover, luteal insensitivity to $PGF_{2\alpha}$ is transient and coincides with the time of maternal recognition of pregnancy. Because the uterine endometrium secretes another PG, PGE_2, in greater amounts during early pregnancy (504), and PGE_2 induces delays in luteolysis (505), it was suggested that this PG suppresses the luteolytic actions of $PGF_{2\alpha}$ (390). However, a physiological role for PGE_2 or an embryonic protein as a luteotropin that reduces the luteolytic effect of $PGF_{2\alpha}$ remains to be established. Again, species differences in maternal recognition of pregnancy exist even between domestic animals. Although the critical role for IFN-τ appears valid for sheep, cows, and goats, the mechanisms are clearly different in pigs and horses (3,496). The corpus luteum in the pig is required throughout pregnancy (60), and it is hypothesized that estrogens (not IFN-τ) secreted by the porcine conceptus prevent the luteolytic actions of uterine $PGF_{2\alpha}$ (506). A similar mechanism may occur in the mare, except the fetal trophoblast also begins to secrete an additional luteotropin (eCG; see section on Luteinizing Hormone–like Luteotropins) that sustains the function of the primary and secondary corpora lutea until the time of the luteal-placental shift around day 150 of gestation (333).

Primates

In primates, such as Old World monkeys and women, it is clear that the demise of the corpus luteum near the end of the nonfecund menstrual cycle is not due to a uterine luteolytic factor. Nevertheless, the factors responsible for luteolysis in primates remain an enigma. As reviewed by Zeleznik and Benyo (323), research has focused on either a relative loss of luteotropic support (LH) or intraovarian luteolytic factors (estrogen or $PGF_{2\alpha}$). Although the frequency of LH pulses declines by mid-luteal phase (328), subsequent studies demonstrated that this apparent reduction in luteotropic support does not cause luteolysis because the length of the luteal life span was unaltered if pulse frequency was not allowed to decline [by giving exogenous GnRH pulses (507)] or if pulse frequency was prematurely reduced (508). Nevertheless, considerable evidence suggests that luteal cell sensitivity to LH, including LH receptors, LH-stimulated cAMP production, and cAMP signaling, declines as the life span of the primate corpus luteum progresses (509–512), and incremental increases in LH are needed to sustain the normal functional life span of the corpus luteum in the menstrual cycle (513). However, the intracellular mechanisms controlling LH sensitivity and its role in onset of luteolysis are unclear.

The lack of a uterine or pituitary role in signaling luteolysis led investigators to propose "self-destruct" mechanisms (52) whereby local factors synthesized in the ovary or corpus luteum lead to loss of luteal structure–function. Based on the ability of the primate corpus luteum to synthesize estrogen, plus reports that systemic or local delivery of estradiol induced premature luteolysis, it was proposed that estrogen was the local luteolytic factor (52). However, further studies investigating endogenous (514,515) and exogenous (516) estrogen actions did not support its physiological or local luteolytic role. Difficulty in detecting appreciable levels of the classic ER, ERα, in primate luteal tissue (517,518) further reduced enthusiasm for this scenario. However, the discovery of the β isoform of ER in primate luteal tissue (519,520), plus evidence that progesterone downregulates luteal ERβ expression, which is highest at luteolysis in the monkey (350), suggests that the issue of local estrogen action in the primate corpus luteum warrants reevaluation. Because primate luteal tissue synthesizes PGs (521,522), a self-destruct scenario involving $PGF_{2\alpha}$ was also envisioned. Early in vivo studies yielded variable results, in part because of rapid metabolism, stage-specific sensitivity of the corpus luteum, or high levels of endogenous PGs in luteal tissue. However, intraluteal administration of $PGF_{2\alpha}$ analogs induces premature luteolysis in monkeys and women (380,523). Evidence of $PGF_{2\alpha}$ binding and FP receptors in luteal tissue, plus antigonadotropic effects of $PGF_{2\alpha}$ on primate luteal cells in vitro (524), support a direct action of this PG in the corpus luteum of the menstrual cycle. Nevertheless, it has been difficult to substantiate a physiological role for endogenous $PGF_{2\alpha}$ in primate luteolysis. With advances in our understanding of PG metabolism and action (408), and the production of more selective inhibitors of PG synthesis and PG receptors, further studies are warranted to define the putative roles of luteotropic (PGE, PGI) versus luteolytic (PGF) PGs, and other arachidonate products, in the primate corpus luteum.

Although the causes of luteolysis in primates remain unclear, it is generally believed that secretion of another LH-like hormone, CG, by the implanting blastocyst and the syncytiotrophoblast cells of the placenta is responsible for "rescuing" the primate corpus luteum near the end of the fertile cycle and extending luteal function in early pregnancy (55,56). Administration of hCG to mimic the patterns and levels of endogenous CG elicits changes in the corpus luteum that are characteristic of maternal recognition of pregnancy: (a) the functional luteal life span is prolonged and timely menstruation does not occur; (b) circulating levels of progesterone increase transiently and then decline as the time of the luteal-placental shift approaches; and (c) nonsteroidogenic activities, including relaxin and inhibin A production, also increase. Notably, passive or active immunization of nonhuman primates and women against CG produces infertility despite normal ovulatory menstrual cycle (525,526). Thus, CG independent of any other embryonic factors appears capable of rescuing the corpus luteum of the menstrual cycle and converting it into that of early pregnancy.

Although CG secretion appears characteristic of pregnancy in all primates, the patterns and peak levels vary among humans, apes, Old World monkeys, and New World monkeys (56). Peak CG concentrations circulating in women are 100- to 1,000-fold higher than in baboons, macaques, or marmosets. Although CG is first detected in utero-ovarian venous and then peripheral blood around the time of implantation in all primates, the duration of placental production varies. In humans and apes, CG levels peak in the first trimester, then decline but remain at substantial levels throughout gestation. However, in baboons and macaques, CG levels decline to near or below the limits of detection by mid-pregnancy. In marmosets and squirrel monkeys, CG levels do not peak until mid-pregnancy and decline just before parturition. In addition, the forms of CG secreted by the implanting blastocyst and developing placenta may vary. There is evidence from assisted reproductive technology protocols in women that free hCGβ was first detectable approximately 2 days before intact hCG (527), and that by the later stages of the first trimester and again at term pregnancy there is a relative excess of free hCGα subunit when CG levels are declining. Research on the control of CG expression and secretion is limited by difficulty of obtaining early primate embryos or placental tissue (55). However, such studies are needed to understand the control of pregnancy initiation and early pregnancy loss in women.

CG is an LH-like glycoprotein hormone that shares the same α subunit with pituitary LH [and FSH and TSH (285)]. However, the CGβ gene(s) is distinct from the LHβ gene, and it is proposed that the primate CGβ gene evolved from an ancestral LHβ-like gene through a single base pair deletion in exon III and a two–base pair insertion that extended the translated region of the C-terminal region. Thus, human LHβ and CGβ demonstrate remarkable similarity (85% homology) in the first 114 amino acids, but CGβ is unique from other glycoprotein hormone β subunits because of its 24- to 31-amino acid extension at the C-terminus. Also, their glycosylation with N-linked complex carbohydrates and sulfation varies between pituitary gonadotropes and placental trophoblast cells. Consequently, hCGα and β (but not LH subunits) contain terminal sialic acid residues, and are not sulfated like oligosaccharide termini on LH. These differences markedly influence the molecule's size and charge, as well as circulating half-life and bioactivity.

Again, for ethical and practical reasons, there have been few studies on the cellular and molecular changes during CG rescue of the corpus luteum in early pregnancy. However, in every primate species studied (55), evidence supports the concept that pituitary LH and placental CG share a common receptor that regulates luteal cells at least in part through stimulation of adenylate cyclase, to form cAMP and activate cAMP-dependent pathways (e.g., C-kinases). Models simulating early pregnancy in monkeys and women through exogenous CG exposure provide some insight (55,528). Morphologically (529), there are rapid changes in granulosa luteal cells suggestive of rapid use of steroid precursors (fewer lipid droplets) and increased production of secretory proteins (more membrane-bound granules). In contrast, the theca luteal cells display changes suggestive of increased steroid precursor accumulation (more lipid droplets) and steroidogenesis (more agranular endoplasmic reticulum). However, the increased mass and steroidogenic (both progesterone and estrogen) activity are not apparently accompanied by a renewal of angiogenesis in the corpus luteum. There is little evidence that the marked decline in microvascular cell proliferation that occurs as the corpus luteum ages in the menstrual cycle is altered by CG exposure (101,149–151). However, this does not rule out CG effects on the luteal vasculature either directly or through vascular-specific factors [e.g., Ang-1 or 2 (138)] to influence vessel structure–function (e.g., vascular permeability).

CG exposure results in divergent changes in luteal expression of mRNAs and proteins involved in steroid and peptide hormone production. In the monkey corpus luteum (530), 7 days of CG exposure transiently stimulates progesterone production by enhancing the uptake and availability of cholesterol stores (e.g., LDL) for steroidogenesis, not by stimulating key enzymatic steps (e.g., CYP11 or 3β-HSD). However, early time points (1 to 3 days of CG exposure) are needed to quantitate acute effects on mRNA and protein [immunostaining for these enzymes appears to increase (182)]. In contrast, CG markedly increases the expression or enzyme activity for androgen ($P450_{c17}$) and estrogen ($P450_{arom}$) production in primate luteal tissue (530,531). Remarkably, although relaxin secretion by the corpus luteum increases in early pregnancy, it remains unclear whether CG directly or acutely exerts transcriptional or post-transcriptional control over relaxin expression from the H2 (not H1) gene in the primate corpus luteum (55). Consideration of the direct versus indirect effects of CG action must take into account the dynamics of LH/CG receptor–effector systems in primate luteal tissue. Unlike in rodent systems, where a large bolus of LH or CG results in “downregulation” of gonadotropin receptors, the response in primate luteal tissue to high, rising levels of CG appears different. Although many of the LH/CG receptors became occupied, there was a remarkable constancy in the total receptor population as the duration and level of CG exposure was increased (55,532). Nevertheless, homologous desensitization of adenylate cyclase became apparent, as cAMP production in response to LH or CG (but not PGE_2 or forskolin) became severely impaired (533,534). It is possible that uncoupling of LH/CG receptor–signal transduction is important for the transient progestational response of the corpus luteum to continued CG exposure in early pregnancy. However, the continued response to local luteotropic factors, notably PGE_2 (533) or progesterone (535), may promote luteal structure–function.

Given that LH continues to circulate in the late luteal phase of the menstrual cycle, it remains unclear why or how another LH-like hormone (i.e., CG) rescues the primate corpus luteum in early pregnancy [for review, see Stouffer and Hearn (55)]. Based on studies on cells and cell membranes from nonprimate species, it was initially proposed that there were differences in receptor movement or turnover after binding of human CG versus ovine LH that prolong CG action (e.g., progesterone production) by luteal cells. However, further studies in primate cell systems (536) comparing hLH with hCG failed to support this concept. That LH/CG receptors in monkeys and women demonstrate a unique specificity for primate gonadotropins (334) emphasizes the need for further studies in relevant species. More recently, Zeleznik (537) presented data suggesting that administration of increasing amounts of recombinant hLH, as well as hCG, around the expected time of implantation increased progesterone levels and prevented timely luteolysis during the menstrual cycle in macaque monkeys. These data are consistent with the concept that CG's ability to rescue the primate corpus luteum is due not to a unique activity in the CG versus the LH molecule, but to qualitative and quantitative differences in the circulating gonadotropin support between LH (intermittent pulses with intervals of gonadotropin deprivation) and CG (continuous, increasing levels) near the end of the nonfecund versus fertile cycle. However, further studies are needed to discern how CG prevents timely luteolysis in the fertile cycle (261) and whether luteal regression after the luteal-placental shift involves similar processes as in the nonfecund cycle.

LUTEAL CELL TYPES AND THEIR REGULATION

Small versus Large Luteal Cells

As early as 1919, investigators first proposed that the corpus luteum consisted of at least two types of luteal cells based on differences in size, morphology, and staining properties (538). In domestic animals, the “large” luteal cells contain more smooth and rough endoplasmic reticulum, mitochondria, extensive Golgi apparatuses, and secretory granules than “small” luteal cells—features that are consistent with greater steroid and peptide/protein hormone production by large cells [for review, see Niswender and Nett (3)]. The often-stated opinion is that the large cells are derived from the granulosa layer, whereas the small cells originate from the theca layer of the luteinizing follicle. This concept has not been definitively tested, although Meidan and colleagues (99) report that pharmacologically induced luteinization of granulosa and theca cells in vitro results in many of the characteristics of large and small luteal cells, respectively. However, limited evidence suggests that this concept is an oversimplification. Using monoclonal antibodies (Abs) to granulosa and theca cell surface antigens, Alila and Hansel (539) determined that small luteal cells bound theca-specific

Abs, whereas large luteal cells bound either granulosa-specific or theca-specific Abs. Moreover, the percentage of large luteal cells that bound granulosa-specific Abs declined, whereas those that bound theca-specific Abs increased, as the bovine corpus luteum aged. This evidence implies that small luteal cells become large luteal cells as the life span of the corpus luteum progresses. But whether this scenario (a) implies further cell differentiation as well as hypertrophy, with discreet large cell subtypes, or (b) is conceptually relevant to other species, including primates with compartmentalized theca-luteal and granulosa-luteal cells in luteal tissue, is not clear.

Studies in several species indicate that the number and size of luteal cells change during the life span of the corpus luteum. In general, the numbers and sizes of steroidogenic luteal cells increase as the corpus luteum develops (e.g., during the early to mid-luteal phase in domestic animals and primates [for review, see Stouffer and Brannian (540)], whereas the number and size of cells decline during luteal regression near the end of the cycle. It is likely that increases in luteal cell number and size along with vascular development (increased endothelial cells; see section on Luteinization) influence corpus luteum activity after ovulation, because these events parallel the rise in circulating progesterone levels. Moreover, removal of granulosa cells from the preovulatory follicle markedly reduced luteal mass and serum progesterone levels (but not luteal life span) in domestic animals (541) and primates (542). These studies established the critical importance of granulosa-derived luteal cells in the progestogenic function of the corpus luteum during the ovarian cycle.

However, it appears that marked changes in function occur in luteal cells before changes in cell number or size at luteolysis. Moreover, morphometric evaluations of cellular constituents in a limited number of species suggest that cellular destruction does not occur simultaneously in all cell types during luteal regression. For example, small luteal cells and endothelial cells were the first cell types to decrease in the sheep corpus luteum, followed by the loss of large luteal cells (446). However, because luteolysis also involves a decline in cell size (e.g., as a result of loss of lipid droplets or apoptosis), it is difficult to establish whether large luteal cells actually disappear or no longer meet the size criterion. Again, differences appear evident in species with compartmentalized luteal tissue. In primates, notably the human, it was suggested that luteolytic changes first occur in the granulosa-derived luteal tissue. Sasano and Suzuki (543) observed that the number of granulosa luteal cells decreased, whereas the paraluteal region of theca-luteal cells remain prominent and steroidogenically active in the corpus luteum near the end of the menstrual cycle.

Function and Regulation of Luteal Cell Types

Despite the caveats of identifying luteal cells on the basis of size, which clearly varies between species and during the life span of the corpus luteum [for review, see Stouffer and Brannian (540)], considerable progress in characterizing luteal cell types was made by Niswender and Juengel (446), Hansel and colleagues (544), and others (329,452) when they separated cells from luteal tissue of domestic animals based on differences in cell size and density. Consistent with fine structural analyses of cellular organelles, the large luteal cells ($\geq$20 μm and 30 μm diameter, in sheep and cows, respectively) secrete ten- to 30-fold more progesterone than small luteal cells (<20 μm and 17 μm diameter), but the presence of cholesterol precursor (e.g., LDL) is required to sustain this difference during longer-term incubation in vitro. The higher progestogenic activity of large luteal cells in sheep is associated with greater expression of $P450_{scc}$ and 3β-HSD enzymes, and mitochondrial cholesterol transport is elevated. However, the mechanisms responsible for maintaining this high steroidogenic activity are not well defined. Other reports indicate that the large luteal cells of various species are the primary, although not necessarily exclusive, producer of relaxin or oxytocin (203). This research suggests that large luteal cells can produce more than one hormone (e.g., dual secretor of steroid and peptide/protein hormones). Taylor and colleagues (545) identified four subpopulations of large luteal cells in the pig by combining chemical and hemolytic plaque assays to detect steroid and relaxin production.

Small and large luteal cell preparations that differ in steroidogenic activity were also prepared from the corpus luteum of other species [e.g., rodents, rabbits, primates; for review, see Stouffer and Brannian (540)]. In monkeys, the 30- to 40-fold greater level of basal progesterone production by large luteal cells compared with small luteal cells is similar to that observed in other species. However, small three- to fourfold differences were generally observed between small and large cells from the human corpus luteum (546). A critical unanswered question is how these large and small luteal cell preparations in primates compare with the compartmentalized luteal and paraluteal cells observed in situ (182). Evidence indicates that basal estrogen production by all luteal cells is low, with only the large cells having the capacity to convert exogenous androgen to estrogen. Thus, the large cells contain aromatase (CYP-19), which is also localized to the true luteal cells in the tissue parenchyma. One might expect the small luteal cells, which in other species are thought to be of thecal origin, to correspond to the exteriorized paraluteal cells that contain $P450_{C17}$ (CYP-17) for androgen synthesis. However, androgen synthesis was tenfold greater in large versus small cells from the monkey corpus luteum (547), with reports on human luteal cells yielding conflicting data (548,549). These data and histological observations (529) suggest that size alone cannot distinguish between luteal and paraluteal cells in the primate corpus luteum. Rather, the large cell population may contain highly steroidogenic (both progestogenic and either androgenic or estrogenic) cells of both granulosa and theca origin [as proposed from the original work of Alila and Hansen in the sheep (539)]. The origin and ultimate fate (e.g., to differentiate further into large cells as the corpus luteum ages) of small luteal cells in the primate corpus luteum remain speculative.

There are also marked differences between small and large luteal cells in their responsiveness and receptor levels for luteotropic and luteolytic factors. For example, ovine large cells are minimally responsive to gonadotropin (LH), cAMP, or other activators of the protein kinase A pathway, whereas small cells respond with up to a tenfold increase in progesterone production (329). Related observations and calculations led to the proposal that in sheep the chronically elevated steroidogenic activity of large cells is "constitutive" (i.e., minimally regulated by ongoing hormone exposure) and responsible for 80% or more of the secreted progesterone. LH exposure may increase progesterone levels somewhat through acute actions on small luteal cells, but may also sustain levels through the conversion of small "less active" to large "high steroidogenic" luteal cells (550). However, $PGF_{2\alpha}$ appears to have early luteolytic actions through the large luteal cells. In domestic animals, such as sheep and cows, large luteal cells express much greater levels of FP receptors than small luteal cells (329). It appears that $PGF_{2\alpha}$ activates the phosphatidyl inositol (PI) pathway, with both branches of PI signaling, that is, elevated intracellular Ca^{2+} levels and PKC activation occurring primarily in large cells (329). Hoyer (440) summarizes evidence suggesting that $PGF_{2\alpha}$ activation of PKC mediates the loss of steroidogenic function, whereas the Ca^{2+} (IP3) pathway promotes the loss (through apoptosis) of large cells at luteolysis.

Although the sheep and cow are excellent large animal models for preparations of luteal cells, important species differences are likely with rodents and primates. For example, although small luteal cells typically respond to LH-like hormones, there are reports that large luteal cells also respond [e.g., rat (551); women (552)] or are more responsive [e.g., rhesus monkey (511)] than small luteal cells. This issue requires further investigation. Some differences may be technique related; for example, the human corpus luteum is difficult to disperse into cell preparations, such that the prevalent use of harsh proteases (e.g., trypsin) likely alters cell responsiveness or causes cell damage. This may explain why steroid production in vitro by large versus small cells from women is reportedly much less than that of other species. However, Brannian and colleagues (553) used similar techniques to isolate cell types from the sheep and monkey corpus luteum and confirmed species differences. As expected, small cells from the ovine corpus luteum were much more responsive to LH than large cells; however, the opposite was observed for the monkey—large luteal cells were more responsive to LH. This difference, if validated (detailed studies on LH/CG receptors and gonadotropin-activated signaling pathways in primate luteal cell types are limited), could relate to the apparent differences in the luteotropic roles of LH-like hormones in ruminants versus primates. Although elimination of endogenous LH support (using a GnRH antagonist) may only modestly reduce (up to 50%) circulating progesterone levels in cows (554) or sheep (555) depending on the stage of the luteal phase, it causes complete and rapid suppression of progesterone levels and early luteolysis at all stages of the luteal phase in primates (513,556). Likewise, the entrainment of LH pulses to progesterone secretion by mid-luteal phase of the menstrual cycle, with steroid levels at baseline in the lengthening intervals between LH pulses by middle to late luteal phase (328), demonstrates (a) the continued absolute requirement of primate luteal cells for LH-like hormone, and (b) the absence of any luteal cell type that constitutively produces progesterone. These findings would suggest that both small and large luteal cells in the primate corpus luteum respond to LH and CG during the menstrual cycle and at recognition of early pregnancy, respectively.

The effects of other luteotropic or luteolytic factors, and the possible dynamics of cell responsiveness or receptor signaling during the luteal life span, have received less attention. It was proposed that LH-stimulated estradiol production in the rabbit corpus luteum leads to estrogen action on both small and large cells to promote progesterone production (557). There are limited reports of different steroidogenic responses of small and large luteal cells from rodent corpora lutea to PRL (558). Large luteal cells in sheep have greater numbers of GH receptors than small cells (329), whereas small, but not large, luteal cells from women responded to GH with enhanced progesterone production (546). There is also evidence that luteal cell responsiveness to LH, but not the local luteotropin PGE_2, decreases in large (as well as small) cells as the monkey corpus luteum ages (511). Luteal cell changes in responsiveness or actions of the luteolytic factors $PGF_{2\alpha}$ were alluded to earlier. Whether or how such changes assist in regulating the functional life span of the corpus luteum in various species requires further investigation. After considerable effort to isolate and characterize luteal cell types in the 1980s to the mid-1990s, this research area has received less attention over the past decade in part because of (a) vagaries in arbitrarily defining cell types on the basis of a variable (i.e., size or diameter) that changes during the life span of the corpus luteum and in response to luteotropic or luteolytic factors; and (b) the lack of established markers (e.g., cell surface proteins) to identify and separate luteal cell types. Nevertheless, new information [e.g., differences in the peripheral benzodiazepine receptor–ligand system, but not StAR protein, between ovine small and large luteal cells (167)] continues to increase our understanding of the heterogeneity of luteal cells in the corpus luteum. As global gene/protein expression and novel tissue micromanipulative techniques (e.g., laser capture microdissection) are applied to the corpus luteum, a better perspective will be obtained on the origins, dynamics, and interactions between luteal cell subpopulations that are important for the functional life span of the corpus luteum during the ovarian cycle and pregnancy (Fig. 10).

SUMMARY AND FUTURE PERSPECTIVES

Since the mid-1990s, there have been significant increases in our knowledge of the processes that promote luteal development, the cellular components and control of steroidogenesis, and the mechanisms involved in extending luteal function in early pregnancy (Fig. 10). Likewise, information has accrued on the cellular actions of luteotropic hormones and luteolytic $PGF_{2\alpha}$, and the possible role of local factors in controlling luteal structure–function. Nevertheless, many issues need to be resolved at the systemic, cellular, and molecular levels to understand the activities of the corpus luteum and its regulation by endocrine and local factors. The identification of genes that are exclusively or preferentially expressed in ovarian compartments at specific stages of the reproductive cycle (559), and the initial results from

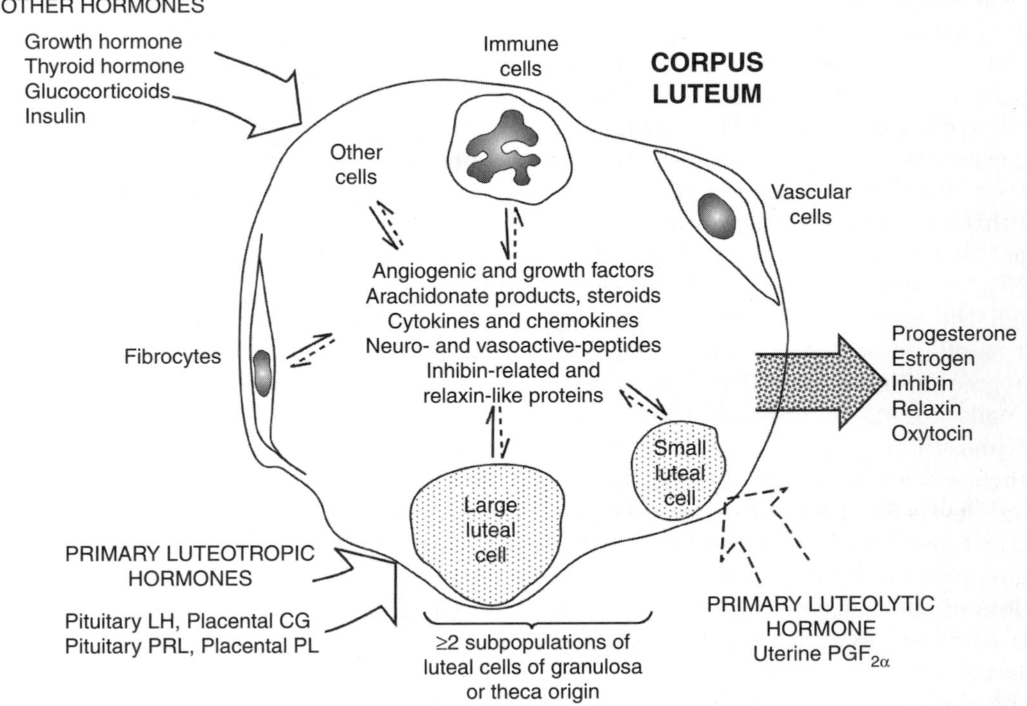

FIG. 10. Conceptual summary of the multiple cell types in the corpus luteum, the luteotropic and luteolytic hormones regulating luteal structure–function, and the categories of products that may serve as autocrine or paracrine regulators of the corpus luteum as well as secreted hormones. *Solid open arrows* denote hormones from the pituitary, placenta, and other endocrine glands that promote the development, function, or life span of the corpus luteum, whereas the *dotted open arrow* depicts the uterine luteolysin prostaglandin (PG) $F_{2\alpha}$. *Line arrows* denote secretion (*solid*) by or action (*dotted*) on cell types in the corpus luteum. The *gray open arrow* denotes hormone secretion from the corpus luteum. As evident from the text, this composite corpus luteum does not exist in nature because species evolved mechanisms relying on various hormones and local factors for their specific niche in reproduction. (Adapted from Stouffer, R. L. [2003]. Corpus luteum in primates. In *Encyclopedia of Hormones* [E. R. Simpson, H. L. Henry, and A. W. Norman, Eds.], pp. 288–297. Elsevier Science, San Diego.)

limited gene microarray analyses (314,347), portend rapid advances in identifying novel or underappreciated processes in the corpus luteum. Because of the corpus luteum's origin from the follicle, traditional gene deletion techniques may be difficult to interpret; however, advances in acute, conditional, and tissue-specific gene knockout approaches, as well as "knockdown" approaches using antisense oligonucleotides or small interfering (si) RNAs, should provide valuable tools for analyzing the roles of genes and gene products in the corpus luteum. Likewise, the advent of basic and clinical proteomics will affect ovarian biology in the near future. Recognizing the myriad of species differences in the control of the corpus luteum, a comparative database will be essential for future applications to improve or control fertility for such purposes as preserving endangered species, controlling species with burgeoning populations, improving the efficiency or quality of food (meat production), and treating ovarian disorders in women's health. The ovarian biologist will continue to find a plethora of intriguing problems and hypotheses for testing to understand this unique gland, the corpus luteum, that is essential for intrauterine pregnancy.

ACKNOWLEDGMENTS

Appreciation is expressed to those who kindled the author's interest in this field: Drs. David Schomberg, Gary Hodgen, Harold Behrman, Ernst Knobil, Gordon Niswender, and Irv Rothchild. Also, the input of other outstanding scientists, especially through their recent, timely reviews as cited in the text, is noted, including Drs. Roger Short, John McCracken, Geula Gibori, Bruce Murphy, Rena Meidan, Jerome Strauss III, David Sherwood, Richard Ivell, Bo Rueda, Jonathan Tilly, Tony Zeleznik, Milo Wiltbank, Mats Brannstrom, and Fuller Bazer. Owing to space limitations, many appropriate references could not be included; the reader is encouraged to read other timely and historical reviews, plus the relevant literature for in-depth appreciation of the advances and issues in this field. The scientific and intellectual contributions of the students, fellows, and collaborators to the author's program is gratefully acknowledged. Program support for the author during preparation of this review was provided by the Oregon National Primate Research Center (RR00163), the Specialized Cooperative Centers Program in Reproductive Research (HD18185), and individual research grants (R01 HD20869 and WHO/Rockefeller Foundation, RF96020). A special thanks to Ms. Carol Gibbins and Mr. Joel Ito for their assistance in preparing this manuscript.

REFERENCES

1. Short, R. V. (1977). The discovery of the ovaries. In *The Ovary* (H. Zuckerman and B. J. Weir, Eds.), pp. 1–39. Academic Press, New York.
2. Rothchild, I. (1981). The regulation of the mammalian corpus luteum. *Rec. Prog. Horm. Res.* 37, 183–283.
3. Niswender, G. D., and Nett, T. M. (1994). Corpus luteum and its control in infraprimate species. In *The Physiology of Reproduction* (E. Knobil and J. D. Neill, Eds.), pp. 781–816. Raven Press, New York.
4. Browning, H. C. (1973). The evolutionary history of the corpus luteum. *Biol. Reprod.* 8, 128–157.
5. Fabricius, H. (1604). *De Formato Foetu*. Patavii; L. Pasquati, Padua, Italy.
6. Adelmann, H. B. (1942). *The Embryological Treatises of Hieronymus Fabricius of Aquapendente*. Cornell University Press, Ithaca, NY.
7. de Graaf, R. (1672). *De Mulierum Organis Generationi Inservientibus Tractatus Novus*. Lugduni Batav., Ex Officina Hackiana.
8. Jocelyn, H. D., and Setchell, B. P. (1972). Regnier de Graaf on the human reproductive organs. *J. Reprod. Fertil. Suppl.* 17, 1–222.
9. Adelmann, H. B. (1966). *Marcello Malpighi and the Evolution of Embryology*. Cornell University Press, Ithaca, NY.
10. Hunter, W., and Baillie, M. (1794). *An Anatomical Description of the Human Gravid Uterus and Its Contents*. J. Johnson, London.
11. Pott, P. (1775). *Chirurgical Observations Relative to the Cataract, the Polypus of the Nose, the Cancer of the Scrotum, the Different Kinds of Ruptures, and the Mortification of the Toes and Feet*. T. J. Carnegy, London.
12. Corner, G. W. (1950). The relation of the ovary to the menstrual cycle: notes on the history of a belated discovery. *An. Fac. Med. Univ. Repub. Montevideo* 35, 758–766.
13. Thompson, D. W. (1910). The works of Aristotle translated into English. In *Historia Animalium* (J. A. Smith and W. D. Ross, Eds.), p. 633. Oxford University Press (Clarendon), London.
14. von Baer, K. E. (1827). *De Ovi Mammalium et Hominis Genesi*. Lipsiae, Sumptibus Vossii.
15. Prevost, J. L., and Dumas, J. A. B. (1824). De la generation dans les mammiferes, et des premiers indices du developpement de l'embryon. *Ann. Sci. Nat.* 3, 113–138.
16. Knauer, E. (1986). Einige Versuche über Ovarientransplantation bei Kaninchen. *Zentralbl. Gynaekol.* 20, 524–528.
17. Beard, J. (1897). *The Span of Gestation and the Cause of Birth*. Fischer, Jena, Germany.
18. Prenant, A. (1898). La valeur morphologique du corps jaune: Son action physiologique et therapeutique possible. *Rev. Gen. Sci. Pure Appl.* 9, 646–650.
19. Fraenkel, L. (1903). Die Function des Corpus Luteum. *Arch. Gynaekol.* 20, 461–466.
20. Magnus, V. (1901). Ovariets betydning for svangerskabet med saerligt hensyn til corpus luteum. *Nor. Mag. Laegevidensk.* 62, 1138–1145.
21. Loeb, L. (1908). The experimental production of the maternal part of the placenta in the rabbit. *Proc. Soc. Exp. Biol. Med.* 5, 102–104.

22. Corner, G. W., and Allen, W. M. (1929). Physiology of corpus luteum; production of special uterine reaction (progestational proliferation) by extracts of corpus luteum. *Am. J. Physiol.* 88, 326–339.
23. Hisaw, F. L., Fevold, H. L., and Meyer, R. K. (1930). The corpus luteum hormone: II. Methods of extraction. *Physiol. Zool.* 3, 135–144.
24. Pouchet, F.-A. (1847). *Theorie Positive de l'Ovulation Spontanee et de la Fecondation des Mammiferes et de l'Espece Humaine*. Bailliere et Fils, Paris.
25. Allen, W. M., and Wintersteiner, O. (1934). Crystalline progestin. *Science* 80, 190–191.
26. Butenandt, A., Westphal, U., and Cobler, H. (1934). Über einen Abbau des Stigmasterins zu corpus-luteum-worksamen Stoffen; ein Beitrag zur Konstitution des Corpus-luteum-Hormons (Vorlauf Mitteil). *Ber. Dtsch. Chem. Ges. B* 67, 1611–1616.
27. Hartmann, M., and Wettstein, A. (1934). Ein krystallisiertes Hormon aus Corpus Luteum. *Helv. Chim. Acta.* 17, 878–882.
28. Slotta, K. H., Ruschig, H., and Fels, E. (1934). Reindarstellung der Hormone aus dem Corpus Luteum (II. Mitteil). *Ber. Dtsch. Chem. Ges. A* 67, 1624–1626.
29. Parkes, A. S. (1962). Prospect and retrospect in the physiology of reproduction. *Br. Med. J.* 2, 71–75.
30. Niswender, G. D., and Midgley, A. R. Jr. (1970). Hapten-radioimmunoassay for steroid hormones. In *Immunologic Methods in Steroid Determination* (F. G. Peron and B. V. Caldwell, Eds.), pp. 149–173. Appleton-Century-Crofts, New York.
31. Schwartz, N. B. (1974). The role of FSH and LH and of their antibodies on follicle growth and on ovulation. *Biol. Reprod.* 10, 236–272.
32. Hutchinson, J. S. M., and Sharp, P. J. (1977). Hypothalamus-pituitary control of the ovary. In *The Ovary* (L. Zuckerman and B. J. Weir, Eds.), pp. 227–303. Academic Press, New York.
33. Loeb, L. (1923). The effect of extirpation of the uterus on the life and function of the corpus luteum in the guinea pig. *Proc. Soc. Exp. Biol. Med.* 20, 441–443.
34. Horton, E. W., and Poyser, N. L. (1976). Physiological reviews. Uterine luteolytic hormone: a physiological role for prostaglandin F2α. *Am. Physiol. Soc.* 56, 651.
35. Usadi, R. S., and Fritz, M. A. (2003). Luteal dysfunction. In *The Ovary* (P. C. K. Leung and E. Y. Adashi, Eds.), pp. 523–540. Elsevier Academic Press, San Diego.
36. Check, J. H., Chase, J. S., Wu, C.-H., Adelson, H. G., Teichman, M., and Rankin, A. (1987). The efficacy of progesterone in achieving successful pregnancy: I. Prophylactic use during luteal phase in anovulatory women. *Int. J. Fertil.* 32, 135–138.
37. Makepeace, A. W., Weinstein, G. L., and Friedman, M. H. (1937). Effect of progestin and progesterone on ovulation in rabbit. *Am. J. Physiol.* 119, 512–516.
38. Barber, H. R. K. (1989). The ovary—then and now. *Int. J. Fertil.* 34, 173–182.
39. Pincus, G., and Chang, M. C. (1953). Effects of progesterone and related compounds on ovulation and early development in rabbit. *Acta. Physiol. Lat. Am.* 3, 177–183.
40. Garfield, R. E., and Baulieu, E. E. (1987). The antiprogesterone steroid RU486: a short pharmacological and clinical review, with emphasis on the interruption of pregnancy. *Baillieres Clin. Endocrinol. Metab.* 1, 207–221.
41. Giannoukos, G., Szapary, D., Smith, C. L., Meeker, J. E. W., and Simons, S. S. Jr. (2001). New antiprogestins with partial agonist activity: Potential selective progesterone receptor modulators (SPRMs) and probes for receptor- and coregulator-induced changes in progesterone receptor induction properties. *Mol. Endocrinol.* 15, 255–270.
42. McCracken, J. A., Custer, E. E., and Lamsa, J. C. (1999). Luteolysis: a neuroendocrine-mediated event. *Physiol. Rev.* 79, 263–323.
43. Eckstein, P., and Zuckerman, S. (1956). The oestrous cycle in the mammalia. In *Marshall's Physiology of Reproduction* (A. S. Parkes, Ed.), pp. 226–396. Longmans, Green, New York.
44. Asdell, S. A. (1964). *Patterns of Mammalian Reproduction.* Cornell University Press, Ithaca, NY.
45. Hoffmann, B., Barth, D., and Karg, H. (1978). Progesterone and estrogen levels in peripheral plasma of the pregnant and nonpregnant roe deer (*Capreolus capreolus*). *Biol. Reprod.* 19, 931–935.
46. Tyndale-Biscoe, C. H. (1973). *The Life of Marsupials.* Academic Press, New York.
47. Sharman, G. B. (1976). Evolution of viviparity in mammals. In *Reproduction in Mammals. Book 6. The Evolution of Reproduction.* (C. A. Austin and R. V. Short, Eds.), p 32–70. Cambridge University Press, London.
48. Concannon, P. W. (1986). Canine physiology of reproduction. In *Small Animal Reproduction and Infertility* (E. T. Burke, Ed.), pp. 23–77. Lea & Febiger, Philadelphia.
49. Olson, P. N., Nett, T. M., Bowen, R. A., Sawyer, H. R., and Niswender, G. D. (1989). Endocrine regulation of the corpus luteum of the bitch as a potential target for altering fertility. *J. Reprod. Fertil.* 39 (Suppl), 27–40.
50. Mead, R. A. (1993). Embryonic diapause in vertebrates. *J. Exp. Zool.* 266, 629–641.
51. Tyndale-Biscoe, C. H. (1986). Embryonic diapause in a marsupial: roles of the corpus luteum and pituitary in its control. In *Comparative Endocrinology: Developments and Directions* (C. L. Ralph, Ed.), pp. 137–155. Alan R. Liss, New York.
52. Knobil, E. (1973). On the regulation of the primate corpus luteum. *Biol. Reprod.* 8, 246–258.
53. Vande Wiele, R. L., Bogumil, J., Dyrenfurth, I., Ferin, M., Jewelewicz, R., Warren, M., Rizkallah, T., and Mikhail, G. (1970). Mechanisms regulating the menstrual cycle in women. *Recent. Prog. Horm. Res.* 26, 63–103.
54. Buffet, N. C., Djakoure, C., Maitre, S. C., and Bouchard, P. (1998). Regulation of the human menstrual cycle. *Front. Neuroendocrinol.* 19, 151–186.
55. Stouffer, R. L., and Hearn, J. P. (1998). Endocrinology of the transition from menstrual cyclicity to establishment of pregnancy in primates. In *Endocrinology of Pregnancy* (F. W. Bazer, Ed.), pp. 35–57. Humana Press, Totowa, NJ.
56. Hodgen, G. D., and Itskovitz, J. (1988). Recognition and maintenance of pregnancy. In *The Physiology of Reproduction* (E. Knobil and J. Neill, Eds.), pp. 1995–2021. Raven Press, New York.
57. Treloar, O. L., Wolf, R. C., and Meyer, R. K. (1972). The corpus luteum of the rhesus monkey during late pregnancy. *Endocrinology* 91, 665–668.
58. Hansel, W., Concannon, P. W., and Lukaszewska, J. H. (1973). Corpora lutea of the large domestic animals. *Biol. Reprod.* 8, 222–245.
59. Goodman, R. L., and Inskeep, E. K. (2006). Neuroendocrine control of the ovarian cycle of the sheep. In *Knobil and Neill's Physiology of Reproduction* (J. D. Neill, Ed.), pp. 2389–2447. Academic Press, St. Louis.
60. Ellicott, A. R., and Dziuk, P. J. (1973). Minimum daily dose of progesterone and plasma concentration for maintenance of pregnancy in ovariectomized gilts. *Biol. Reprod.* 8, 300–304.
61. Fortune, J. E. (1994). Ovarian follicular growth and development in mammals. *Biol. Reprod.* 50, 225–232.
62. di Zerega, G. S., and Hodgen, G. D. (1981). Folliculogenesis in the primate ovarian cycle. *Endocr. Rev.* 2, 27–49.

63. Zeleznik, A. J. (2001). Follicle selection in primates: "Many are called but few are chosen." *Biol. Reprod.* 65, 655–659.
64. Knickerbocker, J. J., Wiltbank, M. C., and Niswender, G. D. (1988). Mechanisms of luteolysis in domestic livestock. *Domest. Anim. Endocrinol.* 5, 91–107.
65. Castracane, V. D., Moore, G. T., and Shaikh, A. A. (1979). Ovarian function in hysterectomized *Macaca fascicularis*. *Biol. Reprod.* 20, 462–472.
66. Ranney, B. and Abu-Ghazaleh, S. (1977). The future function and fortune of ovarian tissue which is retained in vivo during hysterectomy. *Am. J. Obstet. Gynecol.* 128, 626–634.
67. Spies, H. G., and Chappel, S. C. (1984). Mammals: Nonhuman primates. In *Marshall's Physiology of Reproduction: Reproductive Cycles of Vertebrates* (G. E. Lamming, Ed.), pp. 659–712. Churchill Livingstone, Edinburgh.
68. Hearn, J. P., Abbott, D. H., Chambers, P. D., Hodges, J. K., and Lunn, S. F. (1978). Use of the common marmoset, *Callithrix jacchus*, in reproductive research. *Primates Med.* 10, 40–49.
69. Summers, P. M., Wennink, C. J., and Hodges, J. K. (1985). Cloprostenol-induced luteolysis in the marmoset monkey (*Callithrix jacchus*). *J. Reprod. Fertil.* 73, 133–138.
70. Kauffman, A. S., and Rissman, E. (2006). Neuroendocrine control of mating-induced ovulation. In *Knobil and Neill's Physiology of Reproduction* (J. D. Neill, Ed.), pp. 2283–2326. Academic Press, St. Louis.
71. Hilliard, J. (1973). Corpus luteum function in guinea pigs, hamsters, rats, mice and rabbits. *Biol. Reprod.* 8, 203–221.
72. Freeman, M. E. (2006). Neuroendocrine control of the ovarian cycle of the rat. In *Knobil and Neill's Physiology of Reproduction* (J. D. Neill, Ed.), pp. 2327–2388. Academic Press, St. Louis.
73. Gibori, G., Khan, I., Warshaw, M. L., Mclean, M. P., Puryear, T. K., Nelson, S., Durkee, T. J., Azhar, S., Steinschneider, A., and Rao, M. S. (1988). Placental-derived regulators and the complex control of luteal cell function. *Recent Prog. Horm. Res.* 44, 377–429.
74. Bowen-Shauver, J. M., and Gibori, G. (2004). The corpus luteum of pregnancy. In *The Ovary* (P. C. K. Leung and E. Y. Adashi, Eds.), pp. 201–230. Elsevier Academic Press, San Diego.
75. Murphy, B. D. (2004). Luteinization. In *The Ovary* (P. C. K. Leung and E. Y. Adashi, Eds.), pp. 185–199. Elsevier Academic Press, San Diego.
76. Ichikawa, S., Uchino, S., and Hirata, Y. (1987). Lymphatic and blood vasculature of the forming corpus luteum. *Lymphology* 20, 73–83.
77. Stouffer, R. L. (2004). The functions and regulation of cell populations comprising the corpus luteum during the ovarian cycle. In *The Ovary* (P. C. K. Leung and E. Y. Adashi, Eds.), pp. 169–184. Elsevier Academic Press, San Diego.
78. Redmer, D. A., Doraiswamy, V., Bortnem, B. J., Fisher, K., Jablonka-Shariff, A., Grazul-Bilska, A. T., and Reynolds, L. P. (2001). Evidence for a role of capillary pericytes in vascular growth of the developing ovine corpus luteum. *Biol. Reprod.* 65, 879–889.
79. Concannon, P., Hansel, W., and McEntee, K. (1977). Changes in LH, progesterone and sexual behavior associated with preovulatory luteinization in the bitch. *Biol. Reprod.* 17, 604–613.
80. Murphy, B. D., Gevry, N., Ruis-Cortes, T., Cote, F., Downey, B. R., and Sirois, J. (2001). Formation and early development of the corpus luteum in pigs. *Reproduction* 58 (Suppl), 47–63.
81. Corner, G. Jr. (1956). The histological dating of the human corpus luteum of menstruation. *Am. J. Anat.* 98, 377–392.
82. Van Niekerk, C. H., Morgenthal, J. C., and Gerneke, W. H. (1975). Relationship between morphology of and progesterone production by the corpus luteum of the mare. *J. Reprod. Fertil.* 23 (Suppl), 171–175.
83. Webley, G. E., Richardson, M. C., Smith, C. A., Masson, G. M., and Hearn, J. P. (1990). Size distribution of luteal cells from pregnant and non-pregnant marmoset monkeys and a comparison of the morphology of marmoset luteal cells with those from the human corpus luteum. *J. Reprod. Fertil.* 90, 427–437.
84. Enders, A. C. (1973). Cytology of the corpus luteum. *Biol. Reprod.* 8, 158–182.
85. Smith, M. F., McIntush, E. W., and Smith, G. W. (1994). Mechanisms associated with corpus luteum development. *J. Anim. Sci.* 72, 1857–1872.
86. Anderson, E., and Little, B. (1985). The ontogeny of the rat granulosa cell. In *Proceedings of the Fifth Ovarian Workshop* (D. O. Toft and R. J. Ryan, Eds.), pp. 203–225. Ovarian Workshops, Champaign, IL.
87. McClellan, M. C., Diekman, M. A., Abel, J. H. Jr., and Niswender, G. D. (1975). Luteinizing hormone, progesterone and the morphological development of normal and superovulated corpora lutea in sheep. *Cell. Tissue Res.* 164, 291–307.
88. Kerban, A., Dore, M., and Sirois, J. (1999). Characterization of cellular and vascular changes in equine follicles during hCG-induced ovulation. *J. Reprod. Fertil.* 117, 115–123.
89. Mayerhofer, A., Dissen, G. A., Parrott, J. A., Hill, D. F., Mayerhofer, D., Garfield, R. E., Costa, M. E., Skinner, M. K., and Ojeda, S. R. (1996). Involvement of nerve growth factor in the ovulatory cascade: *trk*A receptor activation inhibits gap junctional communication between thecal cells. *Endocrinology* 137, 5662–5670.
90. Khan-Dawood, F. S., Dawood, M. Y., and Tabibzadeh, S. (1996). Immunohistochemical analysis of the microanatomy of primate ovary. *Biol. Reprod.* 54, 734–742.
91. Murdoch, W. J. (1996). Microtubular dynamics in granulosa cells of periovulatory follicles and granulosa-derived (large) lutein cells of sheep: relationships to the steroidogenic folliculo-luteal shift and functional luteolysis. *Biol. Reprod.* 54, 1135–1140.
92. Luck, M. R., and Zhao, Y. (1993). Identification and measurement of collagen in the bovine corpus luteum and its relationship with ascorbic acid and tissue development. *J. Reprod. Fertil.* 99, 647–652.
93. Fujiwara, H., Kataoka, N., Honda, T., Ueda, M., Yamada, S., Nakamura, K., Suginami, H., Mori, T., and Maeda, M. (1998). Physiological roles of integrin alpha 6 beta 1 in ovarian functions. *Horm. Res.* 50 (Suppl 2), 25–29.
94. Burns, K. H., Owens, G. E., Fernandez, J. M., Nilson, J. H., and Matzuk, M. M. (2002). Characterization of integrin expression in the mouse ovary. *Biol. Reprod.* 67, 743–751.
95. Liu, K., Liu, Y. X., Hu, Z. Y., Zou, R. Y., Chen, Y. J., Mu, X. M., and Ny, T. (1997). Temporal expression of urokinase type plasminogen activator, tissue type plasminogen activator, plasminogen activator inhibitor type 1 in rhesus monkey corpus luteum during the luteal maintenance and regression. *Mol. Cell. Endocrinol.* 133, 109–116.
96. Curry, T. E. Jr., and Osteen, K. G. (2003). The matrix metalloproteinase system: Changes, regulation, and impact throughout the ovarian and uterine reproductive cycle. *Endocr. Rev.* 24, 428–465.
97. Young, K. A., Bumlinson, B., and Stouffer, R. L. (2004). ADAMTS-1/METH-1 and TIMP-3 expression in the primate corpus luteum: Divergent patterns and stage-dependent regulation during the natural menstrual cycle. *Mol. Hum. Reprod.* 10, 559–565.
98. Nothnick, W. B. (2003). Tissue inhibitor of metalloproteinase-1 (TIMP-1) deficient mice display reduced serum progesterone levels during corpus luteum development. Endocrinology 144, 5–8.

99. Meidan, R., Girsh, E., Blum, O., and Aberdam, E. (1990). In vitro differentiation of bovine theca and granulosa cells into small and large luteal-like cells: morphological and functional characteristics. *Biol. Reprod.* 43, 913–921.
100. Hazzard, T. M., and Stouffer, R. L. (2000). Angiogenesis in ovarian follicular and luteal development. *Baillieres Best. Pract. Res. Clin. Obstet. Gynaecol.* 14, 883–900.
101. Christenson, L. K., and Stouffer, R. L. (1996). Proliferation of microvascular endothelial cells in the primate corpus luteum during the menstrual cycle and simulated early pregnancy. *Endocrinology* 137, 367–374.
102. Lei, Z. M., Chegini, N., and Rao, C. V. (1991). Quantitative cell composition of human and bovine corpora lutea from various reproductive states. *Biol. Reprod.* 44, 1148–1156.
103. Carmeliet, P. (2000). Mechanisms of angiogenesis and arteriogenesis. *Nat. Med.* 6, 389–395.
104. Rossant, J., and Howard, L. (2002). Signaling pathways in vascular development. *Annu. Rev. Cell. Dev. Biol.* 18, 541–573.
105. Goede, V., Schmidt, T., Kimmina, S., Kozian, D., and Augustin, H. G. (1998). Analysis of blood vessel maturation processes during cyclic ovarian angiogenesis. *Lab. Invest.* 78, 1385–1394.
106. Koos, R. D. (1993). Ovarian angiogenesis. In *The Ovary* (E. Y. Adashi and P. C. K. Leung, Eds.), pp. 433–453. Raven Press, New York.
107. Reynolds, L. P., and Redmer, D. A. (1998). Expression of the angiogenic factors, basic fibroblast growth factor and vascular endothelial growth factor, in the ovary. *J. Anim. Sci.* 76, 1671–1681.
108. Ferrara, N., and Davis-Smyth, T. (1997). The biology of vascular endothelial growth factor. *Endocr. Rev.* 18, 4–25.
109. Laitinen, M., Ristimaki, A., Honkasalo, M., Narko, K., Paavonen, K., and Ritvos, O. (1997). Differential hormonal regulation of vascular endothelial growth factors VEGF, VEGF-B, and VEGF-C messenger ribonucleic acid levels in cultured human granulosa-luteal cells. *Endocrinology* 138, 4748–4756.
110. Cross, M. J., Dixelius, J., Matsumoto, T., and Claesson-Welsh, L. (2003). VEGF-receptor signal transduction. *Trends Biochem. Sci.* 28, 488–494.
111. Neufeld, G., Cohen, T., Shraga, N., Lange, T., Kessler, O., and Herzog, Y. (2002). The neuropilins: Multifunctional semaphorin and VEGF receptors that modulate axon guidance and angiogenesis. *Trends Cardiovasc. Med.* 12, 13–19.
112. Fraser, H. M., and Lunn, S. F. (2001). Regulation and manipulation of angiogenesis in the primate corpus luteum. *Reproduction* 121, 355–362.
113. Ravindranath, N., Little-Ihrig, L., Phillips, H. S., Ferrara, N., and Zeleznik, A. J. (1992). Vascular endothelial growth factor messenger ribonucleic acid expression in the primate ovary. *Endocrinology* 131, 254–260.
114. Kamat, B. R., Brown, L. F., Manseau, E. J., Senger, D. R., and Dvorak, H. F. (1995). Expression of vascular permeability factor/vascular endothelial growth factor by human granulosa and theca lutein cells. *Am. J. Pathol.* 146, 157–165.
115. Yamamoto, S., Konishi, I., Tsuruta, Y., Nanbu, K., Mandai, M., Kuroda, H., Matsushita, K., Hamid, A. A., Yura, Y., and Mori, T. (1997). Expression of vascular endothelial growth factor (VEGF) during folliculogenesis and corpus luteum formation in the human ovary. *Gynecol. Endocrinol.* 11, 371–381.
116. Hazzard, T. M., Christenson, L. K., and Stouffer, R. L. (2000). Changes in expression of vascular endothelial growth factor and angiopoietin-1 and -2 in the macaque corpus luteum during the menstrual cycle. *Mol. Hum. Reprod.* 6, 993–998.
117. Sugino, N., Kashida, S., Takiguchi, S., Karube-Harada, A., and Kato, H. (2001). Expression of vascular endothelial growth factor (VEGF) receptors in rat corpus luteum: regulation by oestradiol during mid-pregnancy. *Reproduction* 122, 875–881.
118. Sugino, N., Kashida, S., Takiguchi, S., Karube, A., and Kato, H. (2000). Expression of vascular endothelial growth factor and its receptors in the human corpus luteum during the menstrual cycle and in early pregnancy. *J. Clin. Endocrinol. Metab.* 85, 3919–3924.
119. Xu, F., Hazzard, T. M., Scheffler, L. J., and Stouffer, R. L. (2002). Neuropilin-1 and -2 expression in the monkey corpus luteum during the menstrual cycle. *Biol. Reprod.* 66 (Suppl 1), 181–182.
120. Shweiki, D., Itin, A., Soffer, D., and Keshet, E. (1992). Vascular endothelial growth factor induced by hypoxia may mediate hypoxia-initiated angiogenesis. *Nature* 359, 843–845.
121. Sharkey, A. M., Day, K., McPherson, A., Malik, S., Licence, D., Smith, S. K., and Charnock-Jones, D. S. (2000). Vascular endothelial growth factor expression in human endometrium is regulated by hypoxia. *J. Clin. Endocrinol. Metab.* 85, 402–409.
122. Neeman, M., Abramovitch, R., Schiffenbauer, Y. S., and Tempel, C. (1997). Regulation of angiogenesis by hypoxic stress: From solid tumours to the ovarian follicle. *Int. J. Exp. Pathol.* 78, 57–70.
123. Martinez-Chequer, J. C., Stouffer, R. L., Hazzard, T. M., Patton, P. E., and Molskness, T. A. (2003). Insulin-like growth factor (IGF)-1 and -2, but not hypoxia, synergize with gonadotropin hormone to promote vascular endothelial growth factor (VEGF)-A secretion by monkey granulosa cells from preovulatory follicles. *Biol. Reprod.* 68, 1112–1118.
124. Christenson, L. K., and Stouffer, R. L. (1997). Follicle-stimulating hormone and luteinizing hormone/chorionic gonadotropin stimulation of vascular endothelial growth factor production by macaque granulosa cells from pre- and periovulatory follicles. *J. Clin. Endocrinol. Metab.* 82, 2135–2142.
125. Hazzard, T. M., Molskness, T. A., Chaffin, C. L., and Stouffer, R. L. (1999). Vascular endothelial growth factor (VEGF) and angiopoietin regulation by gonadotrophin and steroids in macaque granulosa cells during the peri-ovulatory interval. *Mol. Hum. Reprod.* 5, 1115–1121.
126. Molskness, T. A., and Stouffer, R. L. (2002). Hypoxia, but not gonadotropin, stimulates vascular endothelial growth factor production by primate luteal cells in vitro. *Biol. Reprod.* 66 (Suppl 1), 283–284.
127. Friedman, C. I., Danforth, D. R., Herbosa-Encarnacion, C., Arbogast, L., Alak, B. M., and Seifer, D. B. (1997). Follicular fluid vascular endothelial growth factor concentrations are elevated in women of advanced reproductive age undergoing ovulation induction. *Fertil. Steril.* 68, 607–612.
128. Carmeliet, P., Ferreira, V., Breier, G., Pollefeyt, S., Kieckens, L., Gertsenstein, M., Fahrig, M., Vandenhoeck, A., Harpal, K., Eberhardt, C., Declercq, C., Pawling, J., Moons, L., Collen, D., Risau, W., and Nagy, A. (1996). Abnormal blood vessel development and lethality in embryos lacking a single VEGF allele. *Nature* 380, 435–439.
129. Ferrara, N., Carver-Moore, K., Chen, H., Dowd, M., Lu, L., O'Shea, K. S., Powell-Braxton, L., Hillan, K. J., and Moore, M. W. (1996). Heterozygous embryonic lethality induced by targeted inactivation of the VEGF gene. *Nature* 380, 439–442.
130. Zimmermann, R. C., Hartman, T., Bohlen, P., Sauer, M. V., and Kitajewski, J. (2001). Preovulatory treatment of mice

with anti-VEGF receptor 2 antibody inhibits angiogenesis in corpora lutea. *Microvasc. Res.* 62, 15–25.

131. Fraser, H. M., Dickson, S. E., Lunn, S. F., Wulff, C., Morris, K. D., Carroll, V. A., and Bicknell, R. (2000). Suppression of luteal angiogenesis in the primate after neutralization of vascular endothelial growth factor. *Endocrinology* 141, 995–1000.
132. Dickson, S. E., Bicknell, R., and Fraser, H. M. (2001). Mid-luteal angiogenesis and function in the primate is dependent on vascular endothelial growth factor. *J. Endocrinol.* 168, 409–416.
133. Hazzard, T. M., Xu, F., and Stouffer, R. L. (2002). Injection of soluble vascular endothelial growth factor receptor 1 into the preovulatory follicle disrupts ovulation and subsequent luteal function in rhesus monkeys. *Biol. Reprod.* 67, 1305–1312.
134. Zimmermann, R. C., Xiao, E., Bohlen, P., and Ferin, M. (2002). Administration of antivascular endothelial growth factor receptor 2 antibody in the early follicular phase delays follicular selection and development in the rhesus monkey. *Endocrinology* 143, 2496–2502.
135. Christenson, L. K. and Stouffer, R. L. (1996). Isolation and culture of microvascular endothelial cells from the primate corpus luteum. *Biol. Reprod.* 55, 1397–1404.
136. Ward, N. L., and Dumont, D. J. (2002). The angiopoietins and Tie2/Tek: adding to the complexity of cardiovascular development. *Cell. Dev. Biol.* 13, 19–27.
137. Hanahan, D. (1997). Signaling vascular morphogenesis and maintenance. *Science* 277, 48–50.
138. Wulff, C., Wilson, H., Largue, P., Duncan, W. C., Armstrong, D. G., and Fraser, H. M. (2000). Angiogenesis in the human corpus luteum: Localization and changes in angiopoietins, Tie-2, and vascular endothelial growth factor messenger ribonucleic acid. *J. Clin. Endocrinol. Metab.* 85, 4302–4309.
139. Xu, F. and Stouffer, R. L. (2005). Local delivery of angiopoietin-2 into the preovulatory follicle terminates the menstrual cycle in rhesus monkeys. *Biol. Reprod.* 72, 1352–1358.
140. Koos, R. D. (1995). Increased expression of vascular endothelial growth/permeability factor in the rat ovary following an ovulatory gonadotropin stimulus: potential roles in follicle rupture. *Biol. Reprod.* 52, 1426–1435.
141. LeCouter, J., Kowalski, J., Foster, J., Hass, P., Zhang, Z., Dillard-Telm, L., Frantz, G., Rangell, L., DeGuzman, L., Keller, G. A., Peale, F., Gurney, A., Hillan, K. J., and Ferrara, N. (2001). Identification of an angiogenic mitogen selective for endocrine gland endothelium. *Nature* 412, 877–884.
142. Ferrara, N., Frantz, G., LeCouter, J., Dillard-Telm, L., Pham, T., Draksharapu, A., Giordano, T., and Peale, F. (2003). Differential expression of the angiogenic factor genes vascular endothelial growth factor (VEGF) and endocrine gland-derived VEGF in normal and polycystic human ovaries. *Am. J. Pathol.* 162, 1881–1893.
143. Fraser, H. M., Bell, J., Wilson, H., Taylor, P. D., Morgan, K., Anderson, R. A., and Duncan, W. C. (2005). Localization and quantification of cyclic changes in the expression of endocrine gland vascular endothelial growth factor in the human corpus luteum. *J. Clin. Endocrinol. Metab.* 90, 427–434.
144. Kisliouk, T., Levy, N., Hurwitz, A., and Meidan, R. (2003). Presence and regulation of endocrine gland vascular endothelial growth factor/prokineticin-1 and its receptors in ovarian cells. *J. Clin. Endocrinol. Metab.* 88, 3700–3707.
145. Richards, J. S., Russell, D. L., Robker, R. L., Dajee, M., and Alliston, T. N. (1998). Molecular mechanisms of ovulation and luteinization. *Mol. Cell. Endocrinol.* 145, 47–54.
146. Chaffin, C. L., Schwinof, K. M., and Stouffer, R. L. (2001). Gonadotropin and steroid control of granulosa cell proliferation during the periovulatory interval in monkeys. *Biol. Reprod.* 65, 755–762.
147. Ricke, W. A., Redmer, D. A., and Reynolds, L. P. (1999). Growth and cellular proliferation of pig corpora lutea throughout the estrous cycle. *J. Reprod. Fertil.* 117, 369–377.
148. Gaytan, F., Morales, C., Garcia-Pardo, L., Reymundo, C., Bellido, C., and Sanchez-Criado, J. E. (1998). Macrophages, cell proliferation, and cell death in the human menstrual corpus luteum. *Biol. Reprod.* 59, 417–425.
149. Wulff, C., Dickson, S. E., Duncan, W. C., and Fraser, H. M. (2001). Angiogenesis in the human corpus luteum: simulated early pregnancy by hCG treatment is associated with both angiogenesis and vessel stabilization. *Hum. Reprod.* 16, 2515–2524.
150. Rowe, A. J., Morris, K. D., Bicknell, R., and Fraser, H. M. (2002). Angiogenesis in the corpus luteum of early pregnancy in the marmoset and the effects of vascular endothelial growth factor immunoneutralization on establishment of pregnancy. *Biol. Reprod.* 67, 1180–1188.
151. Rodger, F. E., Young, F. M., Fraser, H. M., and Illingworth, P. J. (1997). Endothelial cell proliferation follows the mid-cycle luteinizing hormone surge, but not human chorionic gonadotrophin rescue, in the human corpus luteum. *Hum. Reprod.* 12, 1723–1729.
152. Johnson, D. G., and Walker, C. L. (1999). Cyclins and cell cycle checkpoints. *Annu. Rev. Pharmacol. Toxicol.* 39, 295–312.
153. Moons, D. S., Jirawatnotai, S., Tsutsui, T., Franks, R., Parlow, A. F., Hales, D. B., Gibori, G., Fazleabas, A. T., and Kiyokawa, H. (2002). Intact follicular maturation and defective luteal function in mice deficient for cyclin-dependent kinase-4. *Endocrinology* 143, 647–654.
154. Robker, R. L., and Richards, J. S. (1996). Hormone-induced proliferation and differentiation of granulosa cells: A coordinated balance of the cell cycle regulators cyclin D2 and p27Kip1. *Mol. Endocrinol.* 12, 924–940.
155. Robker, R. L., and Richards, J. S. (1996). Hormone control of the cell cycle in ovarian cells: proliferation versus differentiation. *Biol. Reprod.* 59, 476–482.
156. Hampl, A., Pachernik, J., and Dvorak, P. (2000). Levels and interactions of p27, cyclin D3, and CDK4 during the formation and maintenance of the corpus luteum in mice. *Biol. Reprod.* 62, 1393–1401.
157. Sicinski, P., Donaher, J. L., Geng, Y., Parker, S. B., Gardner, H., Park, M. Y., Robker, R. L., Richards, J. S., Mcginnis, L. K., Biggers, J. D., Eppig, J. J., Bronson, R. T., Elledge, S. J., and Weinberg, R. A. (1996). Cyclin D2 is an FSH-responsive gene involved in gonadal cell proliferation and oncogenesis. *Nature* 384, 470–474.
158. Bukovsky, A., Caudle, M. R., Keenan, J. A., Wimalasena, J., Foster, J. S., and Van Meter, S. E. (1995). Quantitative evaluation of the cell cycle-related retinoblastoma protein and localization of Thy-1 differentiation protein and macrophages during follicular development and atresia, and in human corpora lutea. *Biol. Reprod.* 52, 776–792.
159. Tong, W., Kiyokawa, H., Soos, T. J., Park, M. S., Soares, V. C., Manova, K., Pollard, J. W., and Koff, A. (1998). The absence of p27Kip1, an inhibitor of G1 cyclin-dependent kinases, uncouples differentiation and growth arrest during the granulosa→luteal transition. *Cell. Growth Differ.* 9, 787–794.

160. Chaffin, C. L., Brogan, R. S., Stouffer, R. L., and VandeVoort, C. A. (2003). Dynamics of Myc/Max/Mad expression during luteinization of primate granulosa cells in vitro: association with periovulatory proliferation. *Endocrinology* 144, 1249–1256.
161. Chaffkin, L. M., Luciano, A. A., and Peluso, J. J. (1993). The role of progesterone in regulating human granulosa cell proliferation and differentiation *in vitro*. *J. Clin. Endocrinol. Metab.* 76, 696–700.
162. Lipsett, M. B. (1978). Steroid hormones. In *Reproductive Endocrinology* (S. C. Yen and R. B. Jaffe, Eds.). WB Saunders, Philadelphia.
163. Ho, C. K. M., Christenson, L. K., and Strauss, J. F. III (2004). Intracellular cholesterol dynamics in steroidogenic cells. In *The Ovary* (P. C. K. Leung and E. Y. Adashi, Eds.), pp. 93–110. Elsevier Academic Press, San Diego.
164. Chaffin, C. L., Hess, D. L., and Stouffer, R. L. (1999). Dynamics of periovulatory steroidogenesis in the rhesus monkey follicle after ovarian stimulation. *Hum. Reprod.* 14, 642–649.
165. Voss, A. K., and Fortune, J. E. (1993). Levels of messenger ribonucleic acid for cholesterol side-chain cleavage cytochrome P-450 and 3β-hydroxysteroid dehydrogenase in bovine preovulatory follicles decrease after the luteinizing hormone surge. *Endocrinology* 132, 888–894.
166. Ronen-Fuhrmann, T., Timberg, R., King, S. R., Hales, K. H., Hales, D. B., Stocco, D. M., and Orly, J. (1998). Spatio-temporal expression patterns of steroidogenic acute regulatory protein (StAR) during follicular development in the rat ovary. *Endocrinology* 139, 303–315.
167. Niswender, G. D. (2002). Molecular control of luteal secretion of progesterone. *Reproduction* 123, 333–339.
168. Stocco, D. M. (1999). Steroidogenic acute regulatory protein. *Vitam. Horm.* 55, 399–441.
169. Caron, K. M., Soo, S. C., Wetsel, W. C., Stocco, D. M., Clark, B. J., and Parker, K. L. (1997). Targeted disruption of the mouse gene encoding steroidogenic acute regulatory protein provides insights into congenital lipoid adrenal hyperplasia. *Proc. Natl. Acad. Sci. U S A* 94, 11540–11545.
170. Tuckey, R. C., Bose, H. S., Czerwionka, I., and Miller, W. L. (2004). Molten globule structure and steroidogenic activity of N-218 MLN64 in human placental mitochondria. *Endocrinology* 145, 1700–1707.
171. Kallen, C. B., Billheimer, J. G., Summers, S. A., Stayrook, S. E., Lewis, M., and Strauss, J. F. III (1998). Steroidogenic acute regulatory protein (StAR) is a sterol transfer protein. *J. Biol. Chem.* 273, 26285–26288.
172. West, A., Horvat, R. D., Roess, D. A., Barisas, B. G., Juengel, J. L., and Niswender, G. D. (2001). Steroidogenic acute regulatory protein and peripheral-type benzodiazepine receptor associate at the mitochondrial membrane. *Endocrinology* 142, 502–505.
173. Lacapere, J.-J., and Papadopoulos, V. (2003). Peripheral-type benzodiazepine receptor: Structure and function of a cholesterol-binding protein in steroid and bile acid biosynthesis. *Steroids* 68, 569–585.
174. Christinsen, K., Bose, H. S., Harris, F. M., Miller, W. L., and Bell, J. D. (2001). Binding of steroidogenic acute regulatory protein to synthetic membranes suggests an active molten globule. *J. Biol. Chem.* 276, 17044–17051.
175. Pescador, N., Soumano, K., Stocco, D. M., Price, C. A., and Murphy, B. D. (1996). Steroidogenic acute regulatory protein in bovine corpora lutea. *Biol. Reprod.* 55, 485–491.
176. Chaffin, C. L., Dissen, G. A., and Stouffer, R. L. (2000). Hormonal regulation of steroidogenic enzyme expression in granulosa cells during the periovulatory interval in monkeys. *Mol. Hum. Reprod.* 6, 11–18.
177. Espey, L. L., and Richards, J. S. (2002). Temporal and spatial patterns of ovarian gene transcription following an ovulatory dose of gonadotropin in the rat. *Biol. Reprod.* 67, 1662–1670.
178. Bogovich, K., Richards, J. S., and Reichert, L. E. Jr. (1981). Obligatory role of luteinizing hormone (LH) in the initiation of preovulatory follicular growth in the pregnant rat: specific effects of human chorionic gonadotropin and follicle-stimulating hormone on LH receptors and steroidogenesis in theca, granulosa and luteal cells. *Endocrinology* 109, 860–867.
179. Khan, I., Sridaran, R., Johnson, D. C., and Gibori, G. (1987). Selective stimulation of luteal androgen biosynthesis by luteinizing hormone: comparison of hormonal regulation of P45017alpha activity in corpora lutea and follicles. *Endocrinology* 121, 1312–1319.
180. Gibori, G., Sridaran, R., and Basuray, R. (1982). Control of aromatase activity in luteal and ovarian nonluteal tissue of pregnant rats. *Endocrinology* 111, 781–788.
181. Lephart, E. D., Simpson, E. R., and McPhaul, M. J. (1992). Ovarian aromatase cytochrome P450 mRNA levels correlate with enzyme activity and serum estradiol levels in anestrous, pregnant and lactating rats. *Mol. Cell. Endocrinol.* 85, 205–214.
182. Sanders, S. L., and Stouffer, R. L. (1996). Localization of steroidogenic enzymes in macaque luteal tissue during the menstrual cycle and simulated early pregnancy: Immunohistochemical evidence supporting the two-cell model for estrogen production in the primate corpus luteum. *Biol. Reprod.* 56, 1077–1087.
183. Sasano, H., Okamoto, M., Mason, J. I., Simpson, E. R., Mendelson, C. R., Sasano, N., and Silverberg, S. G. (1989). Immunolocalization of aromatase, 17alpha-hydroxylase and side-chain-cleavage cytochromes P450 in the human ovary. *J. Reprod. Fertil.* 85, 163–165.
184. LaVoie, H. A., Benoit, A. M., Garmey, J. C., Dailey, R. A., Wright, D. J., and Veldhuis, J. D. (1997). Coordinate developmental expression of genes regulating sterol economy and cholesterol side-chain cleavage in the porcine ovary. *Biol. Reprod.* 57, 402–407.
185. Plotkin, D., Miller, S., Nakajima, S., Peskin, E., Burkman, R., Richardson, D., Mitchel, Y., Waldstreicher, J., Liu, M., Shapiro, D., and Santoro, N. (2002). Lowering low-density lipoprotein cholesterol with simvastatin, a hydroxy-3-methylglutaryl-coenzyme alpha reductase inhibitor, does not affect luteal function in premenopausal women. *J. Clin. Endocrinol. Metab.* 87, 3155–3161.
186. Tureck, R. W., and Strauss, J. F. III (1982). Progesterone synthesis by luteinized human granulosa cells in culture: the role of de novo sterol synthesis and lipoprotein-carried esterol. *J. Clin. Endocrinol. Metab.* 54, 367–373.
187. Murphy, B. D., and Silavin, S. L. (1989). Luteotropic agents and steroid substrate utilization. *Oxford Rev. Reprod. Biol.* 11, 180–223.
188. Li, X., Peegel, H., and Menon, K. M. J. (1998). In situ hybridization of high density lipoprotein (scavenger, type 1) receptor messenger ribonucleic acid (mRNA) during folliculogenesis and luteinization: evidence for mRNA expression and induction by human chorionic gonadotropin specifically in cell types that use cholesterol for steroidogenesis. *Endocrinology* 139, 3043–3049.
189. Rajapaksha, W. R., McBride, M., Robertson, L., and O'Shaughnessy, P. J. (1997). Sequence of the bovine HDL-receptor (SR-BI) cDNA and changes in receptor mRNA expression during granulosa cell luteinization in vivo and in vitro. *Mol. Cell. Endocrinol.* 134, 708–716.
190. Golos, T. G., and Strauss, J. F. III (1987). Regulation of low density lipoprotein receptor gene expression in cultured

human granulosa cells: roles of human chorionic gonadotrophin, 8-bromo-3′,5′-cyclic adenosine monophosphate, and protein synthesis. *Mol. Endocrinol.* 1, 321–326.
191. Brannian, J. D., Shiigi, S. M., and Stouffer, R. L. (1992). Gonadotropin surge increases fluorescent-tagged low-density lipoprotein uptake by macaque granulosa cells from preovulatory follicles. *Biol. Reprod.* 47, 355–360.
192. Glass, C., Pittman, R. C., Civen, M., and Steinberg, D. (1985). Uptake of high-density lipoprotein-associated apoprotein A-I and cholesterol esters by 16 tissues of the rat in vivo and by adrenal cells and hepatocytes in vitro. *J. Biol. Chem.* 260, 744–750.
193. Gevry, N., Lacroix, D., Song, J. H., Pescador, N., Dobias, M., and Murphy, B. D. (2002). Porcine Niemann Pick-C1 protein is expressed in steroidogenic tissue and modulated by cAMP. *Endocrinology* 143, 708–716.
194. Rigotti, A., Trigatti, B. L., Penman, M., Rayburn, H., Herz, J., and Krieger, M. (1997). A targeted mutation in the murine gene encoding the high density lipoprotein (HDL) receptor scavenger receptor class B type I reveals its key role in HDL metabolism. *Proc. Natl. Acad. Sci. U S A* 94, 12610–12615.
195. Plump, A. S., Erickson, S. K., Weng, W., Partin, J. S., Breslow, J. L., and Williams, D. L. (1996). Apolipoprotein A-I is required for cholesteryl ester accumulation in steroidogenic cells and for normal adrenal steroid production. *J. Clin. Invest.* 97, 2660–2671.
196. Illingworth, D. R., Corbin, D. K., Kemp, E. D., and Keenan, E. J. (1982). Hormone changes during the menstrual cycle in abetalipoproteinemia: reduced luteal phase progesterone in a patient with homozygous hypobetalipoproteinemia. *Proc. Natl. Acad. Sci. U S A* 79, 6685–6689.
197. Parker, C. R. Jr., Illingworth, D. R., Bissonnette, J., and Carr, B. R. (1986). Endocrine changes during pregnancy in a patient with homozygous familial hypobetalipoproteinemia. *N. Engl. J. Med.* 314, 557–560.
198. Patterson, M. C., Vanier, M. T., Suzuki, K., Morris, J. A., Carstea, E., Neufeld, E. B., Blanchette-Mackie, J. E., and Pentchev, P. (2000). Niemann-Pick disease type C: A lipid trafficking disorder. In *The Metabolic and Molecular Basis of Inherited Diseases* (C. R. S. W. Scriver, B. Childs, A. L. Beaudet, D. Valle, K. W. Kinzler, and B. Vogelstein, Eds.), pp. 3611–3634. McGraw-Hill, New York.
199. Behrman, H. R., and Armstrong, D. T. (1969). Cholesterol esterase stimulation by luteinizing hormone in luteinized rat ovaries. *Endocrinology* 85, 474–480.
200. Aten, R. F., Kolodecik, T. R., MacDonald, G. J., and Behrman, H. R. (1995). Modulation of cholesteryl ester hydrolase messenger ribonucleic acid levels, protein levels, and activity in the rat corpus luteum. *Biol. Reprod.* 53, 1110–1117.
201. Rennert, H., Amsterdam, A., Billheimer, J. T., and Strauss, J. F. III (1991). Regulated expression of sterol carrier protein 2 in the ovary: a key role for cyclic AMP. *Biochemistry* 30, 11280–11285.
202. Mclean, M. P., Puryear, T. K., Khan, I., Azhar, S., Beillheimer, J. T., Orly, J., and Gibori, G. (1989). Estradiol regulation of sterol carrier protein 2 independent of cytochrome P450 side-chain cleavage expression in the rat corpus luteum. *Endocrinology* 125, 1337–1344.
203. Sherwood, O. D., and Fields, P. A. (1999). Corpus luteum peptides. In *Encyclopedia of Reproduction* (E. Knobil and J. D. Neill, Eds.), pp. 718–729. Academic Press, San Diego.
204. Ivell, R. (1999). The physiology of ovarian oxytocin. *Reprod. Med. Rev.* 7, 11–25.
205. Wathes, D. C., and Swann, R. W. (1982). Is oxytocin an ovarian hormone. *Nature* 297, 225–227.
206. Sawyer, H. R., Moeller, C. L., and Kozlowski, G. P. (1986). Immunocytochemical localization of neurophysin and oxytocin in ovine corpora lutea. *Biol. Reprod.* 34, 543–548.
207. Voss, A. K., and Fortune, J. E. (1992). Oxytocin/neurophysin-I messenger ribonucleic acid in bovine granulosa cells increases after the luteinizing hormone (LH) surge is stimulated by LH *in vitro*. *Endocrinology* 131, 2755–2762.
208. Einspanier, A., Jurdzinski, A., and Hodges, J. K. (1997). A local oxytocin system is part of the luteinization process in the preovulatory follicle of the marmoset monkey (*Callithrix jacchus*). *Biol. Reprod.* 57, 16–26.
209. Mayerhofer, A., Sterzik, K., Link, H., Wiemann, M., and Gratzl, M. (1993). Effect of oxytocin on free intracellular Ca^{2+} levels and progesterone release by human granulosa-lutein cells. *J. Clin. Endocrinol. Metab.* 77, 1209–1214.
210. Bagnell, C. A., Zhang, Q., Ohleth, K., Connor, M. L., Downey, B. R., Tsang, B. K., and Ainsworth, L. (1993). Developmental expression of the relaxin gene in the porcine corpus luteum. *J. Mol. Endocrinol.* 10, 87–97.
211. Ivell, R., and Bathgate, R. A. (2002). Reproductive biology of the relaxin-like factor (RLF/INSL3). *Biol. Reprod.* 67, 699–705.
212. Brannstrom, M., and MacLennan, A. H. (1993). Relaxin induces ovulations in the in vitro perfused rat ovary. *Hum. Reprod.* 8, 1011–1014.
213. Hisaw, F. L. (1926). Experimental relaxation of the pubic ligament of the guinea pig. *Proc. Soc. Exp. Biol. Med.* 23, 661–663.
214. Sherwood, O. D. (2004). Relaxin's physiological roles and other diverse actions. *Endocr. Rev.* 25, 205–234.
215. Hsu, Y.-H., Nakabayashi, K., Nishi, S., Kumagai, J., Kudo, M., Sherwood, O. D., and Hseuh, A. J. W. (2002). Activation of orphan receptors by the hormone relaxin. *Science* 295, 671–674.
216. Bell, R. J., Eddie, L. W., Lester, A. R., Wood, E. C., Johnston, P. D., and Niall, H. D. (1987). Relaxin in human pregnancy serum measured with a homologous radioimmunoassay. *Obstet. Gynecol.* 69, 585–589.
217. Einspanier, A., Nubbemeyer, R., Scholte, S., Schumacher, M., Ivell, R., Fuhrmann, K., and Marten, A. (1999). Relaxin in the marmoset monkey: secretion pattern in the ovarian cycle and early pregnancy. *Biol. Reprod.* 61, 512–520.
218. Stewart, D. R., Celniker, A. C., Taylor, C. A., Jr., Cragun, J. R., Overstreet, J. W., and Lasley, B. L. (1990). Relaxin in the peri-implantation period. *J. Clin. Endocrinol. Metab.* 70, 1771–1773.
219. Mayerhofer, A., Engling, R., Stecher, B., Ecker, A., Sterzik, K., and Gratzl, M. (1995). Relaxin triggers calcium transients in human granulosa-lutein cells. *Eur. J. Endocrinol.* 132, 513.
220. Zhao, L., Roche, P. J., Gunnersen, J. M., Hammond, V. E., Tregear, G. W., Wintour, E. M., and Beck, F. (1999). Mice without a functional relaxin gene are unable to deliver milk to their pups. *Endocrinology* 140, 445–453.
221. Krajnc-Franken, M. A. M., van Disseldorp, J. M., Koenders, J. E., Mosselman, S., van Duin, M., and Gossen, J. A. (2004). Impaired nipple development and parturition in LGR7 knockout mice. *Mol. Cell. Biol.* 24, 687–696.
222. Roche, P. J., Crawford, R. J., and Tregear, G. W. (1993). A single-copy relaxin-like gene sequence is present in the sheep. *Mol. Cell. Endocrinol.* 91, 21–28.
223. Chapman, S. C., Kenny, H. A., and Woodruff, T. K. (2004). Activin, inhibin, and follistatin in ovarian physiology. In *The Ovary* (P. C. K. Leung and E. Y. Adashi, Eds.), pp. 273–287. Elsevier Academic Press, San Diego.
224. Woodruff, T. K., and Mather, J. P. (1995). Inhibin, activin and the female reproductive axis. *Annu. Rev. Physiol.* 57, 219–244.

225. Fraser, H. M., Groome, N. P., and McNeilly, A. S. (1999). Follicle-stimulating hormone-inhibin B interactions during the follicular phase of the primate menstrual cycle revealed by gonadotropin-releasing hormone antagonist and antiestrogen treatment. *J. Clin. Endocrinol. Metab.* 84, 1365–1369.
226. Groome, N. P., Illingworth, P. J., O'Brien, M., Pai, R., Rodger, F. E., Mather, J. P., and McNeilly, A. S. (1996). Measurement of dimeric inhibin B throughout the human menstrual cycle. *J. Clin. Endocrinol. Metab.* 81, 1401–1405.
227. Yamoto, M., Minami, S., and Nakano, R. (1991). Immunohistochemical localization of inhibin subunits in human corpora lutea during menstrual cycle and pregnancy. *J. Clin. Endocrinol. Metab.* 73, 470–477.
228. Smith, K. B., Millar, M. R., McNeilly, A. S., Illingworth, P. J., Fraser, H. M., and Baird, D. T. (1991). Immunocytochemical localization of inhibin a-subunit in the human corpus luteum. *J. Endocrinol.* 129, 155–160.
229. Wada, M., Shintani, Y., Kosaka, M., Sano, T., Hizawa, K., and Saito, S. (1996). Immunohistochemical localization of activin A and follistatin in human tissues. *Endocr. J.* 43, 375–385.
230. Erickson, G. F., and Shimasaki, S. (2003). The spatiotemporal expression pattern of the bone morphogenetic protein family in rat ovary cell types during the estrous cycle. *Reprod. Biol. Endocrinol.* 1, 9. Available at www.rbej.com/content/1/1/9.
231. Muttukrishna, S., George, L., Fowler, P. A., Groome, N. P., and Knight, P. G. (1995). Measurement of serum concentrations of inhibin-A (α-β_A dimer) during human pregnancy. *Clin. Endocrinol.* 42, 391–397.
232. Petraglia, F., Garg, S., Florio, P., Sadick, M., Gallinelli, A., Wong, W.-L., Krummen, L., Comitini, G., Mather, J., and Woodruff, T. K. (1993). Activin A and activin B measured in maternal serum, cord blood serum and amniotic fluid during human pregnancy. *Endocr. J.* 1, 323–327.
233. Lockwood, G. M., Ledger, W. L., Barlow, D. H., Groome, N. P., and Muttukrishna, S. (1998). Identification of the source of inhibins at the time of conception provides a diagnostic role for them in very early pregnancy. *Am. J. Reprod. Immunol.* 40, 303–308.
234. Medhamurthy, R., Culler, M. D., Gay, V. L., Negro-Vilar, A., and Plant, T. M. (1991). Evidence that inhibin plays a major role in the regulation of follicle-stimulating hormone secretion in the fully adult male rhesus monkey (*Macaca mulatta*). *Endocrinology* 129, 389–395.
235. Fraser, H. M., and Lunn, S. F. (1993). Does inhibin have an endocrine function during the menstrual cycle? *Trends Endocrinol. Metab.* 4, 187–194.
236. Davis, J. S., and Rueda, B. R. (2002). The corpus luteum: An ovarian structure with maternal instincts and suicidal tendencies. *Front Biosci.* 7, 1949–1978.
237. McCormack, J. T., Friederichs, M. G., Limback, S. D., and Greenwald, G. S. (1998). Apoptosis during spontaneous luteolysis in the cyclic golden hamster: biochemical and morphological evidence. *Biol. Reprod.* 58, 255–260.
238. Allen, E. (1922). The oestrous cycle of the mouse. *Am. J. Anat.* 30, 297.
239. Castracane, V. D., Stevens, V., Knickerbocker, J., Powell, J., Randolph, M., and Gimpel, T. (1998). Late luteal rescue in the baboon (*Papio cynocephalus*). *Hum. Reprod. Update* 4, 383–388.
240. O'Shea, J. D., and Wright, P. J. (1985). Regression of the corpus luteum of pregnancy following parturition in the ewe. *Acta. Anat.* 122, 69–76.
241. Orlicky, D. J., Fisher, L., Dunscomb, N., and Miller, G. J. (1992). Immunohistochemical localization of PGF2 alpha receptor in the rat ovary. *Prostaglandins Leukot. Essent. Fatty Acids* 46, 223–229.
242. Niswender, G. D., Juengel, J. L., McGuire, W. J., Belfiore, C. J., and Wiltbank, M. C. (1994). Luteal function: the estrous cycle and early pregnancy. *Biol. Reprod.* 50, 239–247.
243. Tilly, J. L., Pru, J. K., and Rueda, B. R. (2004). Apoptosis in ovarian development, function, and failure. In *The Ovary* (P. C. K. Leung and E. Y. Adashi, Eds.), pp. 321–352. Elsevier Academic Press, San Diego.
244. Kerr, J. F., Wyllie, A. H., and Currie, A. R. (1972). Apoptosis: a basic biological phenomenon with wide-ranging implications in tissue kinetics. *Br. J. Cancer* 26, 239–257.
245. Bowen, J. M., Towns, R., Warren, J. S., and Keyes, P. L. (1999). Luteal regression in the normally cycling rat: Apoptosis, monocyte chemoattractant protein-1, and inflammatory cell involvement. *Biol. Reprod.* 60, 740–746.
246. Hasumoto, K., Sugimoto, Y., Yamasaki, A., Morimoto, K., Kakizuka, A., Negishi, M., and Ichikawa, A. (1997). Association of expression of mRNA encoding the PGF2 alpha receptor with luteal cell apoptosis in ovaries of pseudopregnant mice. *J. Reprod. Fertil.* 109, 45–51.
247. O'Shea, J. D., Nightingale, M. G., and Chamley, W. A. (1977). Changes in small blood vessels during cyclical luteal regression in sheep. *Biol. Reprod.* 17, 162–177.
248. Murdoch, W. J. (1995). Temporal relationships between stress protein induction, progesterone withdrawal, and apoptosis in corpora lutea of ewes treated with $PGF_{2\alpha}$. *J. Anim. Sci.* 73, 1789–1792.
249. Rueda, B. R., Wegner, J. A., Marion, S. L., Wahlen, D. D., and Hoyer, P. B. (1995). Internucleosomal DNA fragmentation in ovine luteal tissue associated with luteolysis: in vivo and in vitro analyses. *Biol. Reprod.* 52, 305–312.
250. Bacci, M. L., Barazzoni, A. M., Forni, M., and Costerbosa, G. L. (1996). In situ detection of apoptosis in regressing corpus luteum of pregnant sow: evidence of an early presence of DNA fragmentation. *Domest. Anim. Endocrinol.* 13, 361–372.
251. Shikone, T., Yamoto, M., Kokawa, K., Yamashita, K., Nishimori, K., and Nakano, R. (1996). Apoptosis of human corpora lutea during cyclic luteal regression and early pregnancy. *J. Clin. Endocrinol. Metab.* 81, 2376–2380.
252. Young, F. M., Illingworth, P. J., Lunn, S. F., Harrison, D. J., and Fraser, H. M. (1997). Cell death during luteal regression in the marmoset monkey (*Callithrix jacchus*). *J. Reprod. Fertil.* 111, 109–119.
253. Fraser, H. M., Lunn, S. F., Harrison, D. J., and Kerr, J. B. (1999). Luteal regression in the primate: different forms of cell death during natural and gonadotropin-releasing hormone antagonist or prostaglandin analogue-induced luteolysis. *Biol. Reprod.* 61, 1468–1479.
254. Morales, C., Garcia-Pardo, L., Reymundo, C., Bellido, C., Sanchez-Criado, J. E., and Gaytan, F. (2000). Different patterns of structural luteolysis in the human corpus luteum of menstruation. *Hum. Reprod.* 15, 2119–2128.
255. Kugu, K., Ratts, V. S., Piquette, G. N., Tilly, K. I., Tao, X. J., Martimbeau, S., Aberdeen, G. W., Krajewski, S., Reed, J. C., Pepe, G. J., Albrecht, E. D., and Tilly, J. L. (1998). Analysis of apoptosis and expression of *bcl-2* gene family members in the human and baboon ovary. *Cell. Death Differ.* 5, 67–76.
256. Van Nassauw, L., Tao, L., and Harrisson, F. (1998). Distribution of apoptosis-related proteins in the quail ovary during folliculogenesis: BCL-2, BAX and CPP32. *Acta. Histochem.* 101, 103–112.
257. Knudson, C. M., Tung, K. S., Tourtellotte, W. G., Brown, G. A., and Korsmeyer, S. J. (1995). Bax-deficient mice with lymphoid hyperplasia and male germ cell death. *Science* 270, 96–99.
258. Rodger, F. E., Fraser, H. M., Krajewski, S., and Illingworth, P. J. (1998). Production of the proto-oncogene *BAX* does not vary with changing luteal function in women. *Mol. Hum. Reprod.* 4, 27–32.

259. Rodger, F. E., Fraser, H. M., Duncan, W. C., and Illingworth, P. J. (1995). Immunolocalization of Bcl-2 in the human corpus luteum. *Hum. Reprod.* 10, 1566–1570.
260. Devoto, L., Vega, M., Kohen, P., Castro, O., Carvallo, P., and Palomino, A. (2002). Molecular regulation of progesterone secretion by the human corpus luteum throughout the menstrual cycle. *J. Reprod. Immunol.* 55, 11–20.
261. Sugino, N., Suzuki, T., Kashida, S., Karube, A., Takiguchi, S., and Kato, H. (2000). Expression of Bcl-2 and Bax in the human corpus luteum during the menstrual cycle and in early pregnancy: regulation by human chorionic gonadotropin. *J. Clin. Endocrinol. Metab.* 85, 4379–4386.
262. Sakamaki, K., Yoshida, H., Nishimura, Y., Nishikawa, S., Manabe, N., and Yonehara, S. (1997). Involvement of Fas antigen in ovarian follicular atresia and luteolysis. *Mol. Reprod. Dev.* 47, 11–18.
263. Roughton, S. A., Lareu, R. R., Bittles, A. H., and Dharmarajan, A. M. (1999). Fas and Fas ligand messenger ribonucleic acid and protein expression in the rat corpus luteum during apoptosis-mediated luteolysis. *Biol. Reprod.* 60, 797–804.
264. Taniguchi, H., Yokomizo, Y., and Okuda, K. (2002). Fas-Fas ligand system mediates luteal cell death in bovine corpus luteum. *Biol. Reprod.* 66, 754–759.
265. Quirk, S. M., Harman, R. M., Huber, S. C., and Cowan, R. G. (2000). Responsiveness of mouse corpora luteal cells to Fas antigen (CD95)-mediated apoptosis. *Biol. Reprod.* 63, 49–56.
266. Kuranaga, E., Kanuka, H., Furuhata, Y., Yonezawa, T., Suzuki, M., Nishihara, M., and Takahashi, M. (2000). Requirement of the Fas ligand-expressing luteal immune cells for regression of corpus luteum. *FEBS Lett.* 472, 137–142.
267. Quirk, S. M., Cowan, R. G., Joshi, S. G., and Henrikson, K. P. (1995). Fas antigen-mediated apoptosis in human granulosa/luteal cells. *Biol. Reprod.* 52, 279–287.
268. Pru, J. K., Lynch, M. P., Davis, J. S., and Rueda, B. R. (2003). Signaling mechanisms in tumor necrosis factor alpha-induced death of microvascular endothelial cells of the corpus luteum. *Reprod. Biol. Endocrinol.* 1, 17.
269. Friedman, A., Weiss, S., Levy, N., and Meidan, R. (2000). Role of tumor necrosis factor alpha and its type 1 receptor in luteal regression: induction of programmed cell death in bovine corpus luteum-derived endothelial cells. *Biol. Reprod.* 63, 1905–1912.
270. Benyo, D. F., and Pate, J. L. (1992). Tumor necrosis factor-α alters bovine luteal cell synthetic capacity and viability. *Endocrinology* 130, 854–860.
271. Roby, K. F., Son, D. S., and Terranova, P. F. (1999). Alterations of events related to ovarian function in tumor necrosis factor receptor type I knockout mice. *Biol. Reprod.* 61, 1616–1621.
272. Khan, S. M., Dauffenbach, L. M., and Yeh, J. (2000). Mitochondria and caspases in induced apoptosis in human luteinized granulosa cells. *Biochem. Biophys. Res. Commun.* 269, 542–545.
273. Peluffo, M. C., Young, K. A., and Stouffer, R. L. (2005). Dynamic expression of caspases-2, -3, -8, and -9 proteins and enzyme activity, but not mRNA, in the monkey corpus luteum during the menstrual cycle. *J. Clin. Endocrinol. Metab.* 90, 2327–2335.
274. Boone, D. L., and Tsang, B. K. (1998). Caspase-3 in the rat ovary: localization and possible role in follicular atresia and luteal regression. *Biol. Reprod.* 58, 1533–1539.
275. Rueda, B. R., Hendry, I. S., Tilly, J. L., and Hamernik, D. L. (1999). Accumulation of caspase-3 messenger ribonucleic acid and induction of caspase activity in the ovine corpus luteum following prostaglandin F2α treatment in vivo. *Biol. Reprod.* 60, 1087–1092.
276. Carambula, S. F., Matikainen, T., Lynch, M. P., Flavell, R. A., Goncalves, P. B. D., Tilly, J. L., and Rueda, B. R. (2002). Caspase-3 is a pivotal mediator of apoptosis during regression of the ovarian corpus luteum. *Endocrinology* 143, 1495–1501.
277. Takiguchi, S., Sugino, N., Esato, K., Karube-Harada, A., Sakata, A., Nakamura, Y., Ishikawa, H., and Kato, H. (2004). Differential regulation of apoptosis in the corpus luteum of pregnancy and newly formed corpus luteum after parturition in rats. *Biol. Reprod.* 70, 313–318.
278. Chappell, P. E., Lydon, J. P., Conneely, O. M., O'Malley, B. W., and Levine, J. E. (1997). Endocrine defects in mice carrying a null mutation for the progesterone receptor gene. *Endocrinology* 138, 4147–4152.
279. Chappell, P. E., Schneider, J. S., Kim, P., Xu, M., Lydon, J. P., O'Malley, B. W., and Levine, J. E. (1999). Absence of gonadotropin surges and gonadotropin-releasing hormone self-priming in ovariectomized (OVX), estrogen (E_2)-treated, progesterone receptor knockout (PRKO) mice. *Endocrinology* 140, 3653–3658.
280. Chaffin, C. L., and Stouffer, R. L. (2002). Local role of progesterone in the ovary during the periovulatory interval. *Rev. Endocr. Metab. Disord.* 3, 65–72.
281. Fraser, H. M., Nestor, J. J. Jr., and Vickery, B. H. (1987). Suppression of luteal function by a luteinizing hormone-releasing hormone antagonist during the early luteal phase in the stumptailed macaque monkey and the effects of subsequent administration of human chorionic gonadotropin. *Endocrinology* 121, 612–618.
282. Dubourdieu, S., Charbonnel, B., Massai, M. R., Marraoui, J., Spitz, I., and Bouchard, P. (1991). Suppression of corpus luteum function by the gonadotropin-releasing hormone antagonist Nal-Glu: Effect of the dose and timing of human chorionic gonadotropin administration. *Fertil. Steril.* 56, 440–445.
283. Hamberger, L., Hahlin, M., Hillensjö, T., Johanson, C., and Sjögren, A. (1988). Luteotropic and luteolytic factors regulating human corpus luteum function. *Ann. N Y Acad. Sci.* 541, 485–497.
284. Gregerson, K. A. (2006). Prolactin: structure, function, and regulation of secretion. In *Knobil and Neill's Physiology of Reproduction* (J. D. Neill, Ed.), 1703–1726. Academic Press, St. Louis.
285. Bousfield, G. R., Jia, L., and Ward, D. N. (2006). Gonadotropins: chemistry and biosynthesis. In *Knobil and Neill's Physiology of Reproduction* (J. D. Neill, Ed.), pp. 1581–1634. Academic Press, St. Louis.
286. Binart, N., Helloco, C., Ormandy, C. J., Barra, J., Clement-Lacroix, P., Baran, N., and Kelly, P. A. (2000). Rescue of preimplantatory egg development and embryo implantation in PRL receptor-deficient mice after progesterone administration. *Endocrinology* 141, 2691–2697.
287. Reese, J., Binart, N., Brown, N., Ma, W. G., Paria, B. C., Das, S. K., Kelly, P. A., and Dey, S. K. (2000). Implantation and decidualization defects in PRL receptor (PRLR)-deficient mice are mediated by ovarian but not uterine PRLR. *Endocrinology* 141, 1872–1881.
288. Butcher, R. L., Fugo, N. W., and Collins, W. D. (1972). Semicircadian rhythm in plasma levels of PRL during early gestation in the rat. *Endocrinology* 90, 1125–1127.
289. Kraicer, P. F., and Shelesnyak, M. C. (1964). Studies on the mechanism of nidation: IX. Analysis of the response to ergocornine, an inhibitor of nidation. *J. Reprod. Fertil.* 8, 225–233.
290. Morishige, W. K., and Rothchild, I. (1974). Temporal aspects of the regulation of corpus luteum function by luteinizing hormone, PRL and placental luteotrophin during the first half of pregnancy in the rat. *Endocrinology* 95, 260–274.

291. Gibori, G., Rothchild, I., Pepe, G. J., Morishige, W. K., and Lam, P. (1974). Luteotropic action of decidual tissue in the rat. *Endocrinology* 95, 1113–1118.
292. Prigent-Tessier, A., Tessier, C., Hirosawa-Takamori, M., Boyer, C., Ferguson-Gottschall, S., and Gibori, G. (1999). Rat decidual PRL: Identification, molecular cloning, and characterization. *J. Biol. Chem.* 274, 37982–37989.
293. Jabbour, H. N., and Critchley, H. O. D. (2001). Potential roles of decidual prolactin in early pregnancy. *Reproduction* 121, 197–205.
294. Faria, T. N., Deb, S., Kwok, S. C. M., Talamantes, F., and Soares, M. J. (1990). Ontogeny of placental lactogen-I and placental lactogen-II expression in the developing rat placenta. *Dev. Biol.* 141, 279–291.
295. Glaser, L. A., Kelly, P. A., and Gibori, G. (1984). Differential action and secretion of rat placental lactogens. *Endocrinology* 115, 969–976.
296. MacLeod, K. R., Smith, W. C., Ogren, L., and Talamantes, F. (1989). Recombinant mouse placental lactogen-I binds to lactogen receptors in mouse liver and ovary: partial characterization of the ovarian receptor. *Endocrinology* 125, 2258–2266.
297. Jayatilak, P. G., Glaser, L. A., Warshaw, M. L., Herz, Z., Gruber, J. R., and Gibori, G. (1984). Relationship between LH and decidual luteotropin in the maintenance of luteal steroidogenesis. *Biol. Reprod.* 31, 556–564.
298. Gibori, G., Basuray, R., and McReynolds, B. (1981). Luteotropic role of the decidual tissue in the rat: dependency on intraluteal estradiol. *Endocrinology* 108, 2060–2066.
299. Jackson, J. A., and Albrecht, E. D. (1985). The development of placental androstenedione and testosterone production and their utilization by the ovary for aromatization to estrogen during rat pregnancy. *Biol. Reprod.* 33, 451–457.
300. Russell, D. L., and Richards, J. S. (1999). Differentiation-dependent PRL responsiveness and stat (signal transducers and activators of transcription) signaling in rat ovarian cells. *Mol. Endocrinol.* 13, 2049–2064.
301. Telleria, C. M., Parmer, T. G., Zhong, L., Clarke, D. L., Albarracin, C. T., Duan, W. R., Linzer, D. I., and Gibori, G. (1997). The different forms of the PRL receptor in the rat corpus luteum: Developmental expression and hormonal regulation in pregnancy. *Endocrinology* 138, 4812–4820.
302. Risk, M., and Gibori, G. (2001). Mechanisms of luteal cell regulation by prolactin. In *Prolactin* (N. D. Horseman, Ed.), pp. 265–295. Kluwer Academic, Boston.
303. Teglund, S., McKay, C., Schuetz, E., van Deursen, J. M., Stravopodis, D., Wang, D., Brown, M., Bodner, S., Grosveld, G., and Ihle, J. N. (1998). Stat5a and Stat5b proteins have essential and nonessential, or redundant, roles in cytokine responses. *Cell* 93, 841–850.
304. Peters, C. A., Maizels, E. T., Robertson, M. C., Shiu, R. P. C., Soloff, M. S., and Hunzicker-Dunn, M. (2000). Induction of relaxin messenger RNA expression in response to prolactin receptor activation requires protein kinase C δ signaling. *Mol. Endocrinol.* 14, 576–590.
305. Hickey, G. J., Oonk, R. B., Hall, P. F., and Richards, J. S. (1989). Aromatase cytochrome P450 and cholesterol side-chain cleavage cytochrome P450 in corpora lutea of pregnant rats: Diverse regulation by peptide and steroid hormones. *Endocrinology* 125, 1673–1682.
306. Martel, C., Gagne, D., Couet, J., Labrie, Y., Simard, J., and Labrie, F. (1994). Rapid modulation of ovarian 3-β-hydroxysteroid dehydrogenase\D5–D4 isomerase gene expression by prolactin and human chorionic gonadotropin in the hypophysectomized rat. *Mol. Cell. Endocrinol.* 99, 63–71.
307. Nokelainen, P., Peltoketo, H., Vihko, R., and Vihko, P. (1998). Expression cloning of a novel estrogenic mouse 17β-hydroxysteroid dehydrogenase/17-ketosteroid reductase (m17HSD7), previously described as a prolactin receptor-associated protein (PRAP) in rat. *Mol. Endocrinol.* 12, 1048–1059.
308. Akinola, L. A., Poutanen, M., Vihko, R., and Vihko, P. (1997). Expression of 17β-hydroxysteroid dehydrogenase type 1 and type 2, P450 aromatase, and 20α-hydroxysteroid dehydrogenase enzymes in immature, mature, and pregnant rats. *Endocrinology* 138, 2886–2892.
309. Albarracin, C. T., Parmer, T. G., Duan, W. R., Nelson, S. E., and Gibori, G. (1994). Identification of a major PRL-regulated protein as 20α-hydroxysteroid dehydrogenase: Coordinate regulation of its activity, protein content, and messenger ribonucleic acid expression. *Endocrinology* 134, 2453–2460.
310. Duan, W. R., Parmer, T. G., Albarracin, C. T., Zhong, L., and Gibori, G. (1997). PRAP, a prolactin receptor associated protein: Its gene expression and regulation in the corpus luteum. *Endocrinology* 138, 3216–3221.
311. Duan, W. R., Linzer, D. I. H., and Gibori, G. (1996). Cloning and characterization of an ovarian-specific protein that associates with the short form of the prolactin receptor. *J. Biol. Chem.* 271, 15602–15607.
312. Armstrong, D. T., Knudsen, K. A., and Miller, L. S. (1970). Effects of prolactin upon cholesterol metabolism and progesterone biosynthesis in corpora lutea of rats hypophysectomized during pseudopregnancy. *Endocrinology* 86, 634–641.
313. Rajkumar, K., Couture, R. L., and Murphy, B. D. (1985). Binding of high-density lipoproteins to luteal membranes: the role of prolactin, luteinizing hormone, and circulating lipoproteins. *Biol. Reprod.* 32, 546–555.
314. Stocco, C., Callegari, E., and Gibori, G. (2001). Opposite effect of PRL and prostaglandin F2α on the expression of luteal genes as revealed by rat cDNA expression array. *Endocrinology* 149, 4158–4161.
315. Gafvels, M., Bjurulf, E., and Selstam, G. (1992). PRL stimulates the expression of luteinizing hormone/chorionic gonadotropin receptor messenger ribonucleic acid in the rat corpus luteum and rescues early pregnancy from bromocriptine-induced abortion. *Biol. Reprod.* 47, 534–540.
316. Bjurulf, E., Selstam, G., and Olofsson, J. I. (1994). Increased LH receptor mRNA and extended corpus luteum function induced by PRL and indomethacin treatment in vivo in hysterectomized pseudopregnant rats. *J. Reprod. Fertil.* 102, 139–145.
317. Telleria, C. M., Zhong, L., Deb, S., Srivastava, R. K., Park, K. S., Sugino, N., Park-Sarge, O.-K., and Gibori, G. (1998). Differential expression of the estrogen receptors α and β in the rat corpus luteum of pregnancy: regulation by prolactin and placental lactogens. *Endocrinology* 139, 2432–2442.
318. Gaddy-Kurten, D., Hickey, G. J., Fey, G. H., Gauldie, J., and Richards, J. S. (1989). Hormonal regulation and tissue-specific localization of α_2-macroglobulin in rat ovarian follicles and corpora lutea. *Endocrinology* 125, 2985–2995.
319. Sugino, N., Hirosawa-Takamori, M., Zhong, L., Telleria, C. M., Shiota, K., and Gibori, G. (1998). Hormonal regulation of copper-zinc superoxide dismutase and manganese superoxide dismutase in the rat corpus luteum: induction by PRL and placental lactogens. *Biol. Reprod.* 59, 599–605.
320. Albarracin, C. G., Palfrey, H. C., Duan, W. R., Rao, M. C., and Gibori, G. (1994). PRL regulation of the calmodulin-dependent protein kinase III elongation factor-2 system in the rat corpus luteum. *J. Biol. Chem.* 269, 7772–7776.

321. Sugino, N., Telleria, C. M., Tessier, C., and Gibori, G. (1999). Regulation and role of the insulin-like growth factor I system in rat luteal cells. *J. Reprod. Fertil.* 115, 349–355.
322. Parmer, T. G., Roberts, C. T., Jr., LeRoith, D., Adashi, E. Y., Khan, I., Solan, N., Nelson, S., Zilberstein, M., and Gibori, G. (1991). Expression, action, and steroidal regulation of insulin-like growth factor-I (IGF-I) and IGF-I receptor in the rat corpus luteum: their differential role in the two cell populations forming the corpus luteum. *Endocrinology* 129, 2924–2932.
323. Zeleznik, A. J., and Benyo, D. F. (1994). Control of follicular development, corpus luteum function, and the recognition of pregnancy in higher primates. In *The Physiology of Reproduction* (E. Knobil and J. D. Neill, Eds.), pp. 751–782. Raven Press, New York.
324. Denamur, R., Martinet, J., and Short, R. V. (1973). Pituitary control of the ovine corpus luteum. *J. Reprod. Fertil.* 32, 207–220.
325. Moudgal, N. R., MacDonald, G. J., and Greep, R. O. (1972). Role of endogenous primate LH in maintaining corpus luteum function of the monkey. *J. Clin. Endocrinol. Metab.* 35, 113–116.
326. Farin, C. E., Nett, T. M., and Niswender, G. D. (1990). Effects of luteinizing hormone on luteal cell populations in hypophysectomized ewes. *J. Reprod. Fertil.* 88, 61–70.
327. Stouffer, R. L. (2003). Progesterone as a mediator of gonadotropin action in the primate corpus luteum: beyond steroidogenesis. *Hum. Reprod. Update* 9, 99–117.
328. Ellinwood, W. E., Norman, R. L., and Spies, H. G. (1984). Changing frequency of pulsatile luteinizing hormone and progesterone secretion during the luteal phase of the menstrual cycle of rhesus monkeys. *Biol. Reprod.* 31, 714–722.
329. Diaz, F. J., Anderson, L. E., Wu, Y. L., Rabot, A., Tsai, S. J., and Wiltbank, M. C. (2002). Regulation of progesterone and prostaglandin F2α production in the CL. *Mol. Cell. Endocrinol.* 191, 65–80.
330. Srisuparp, S., Strakova, Z., Brudney, A., Mukherjee, S., Reierstad, S., Hunzicker-Dunn, M., and Fazleabas, A. T. (2003). Signal transduction pathways activated by chorionic gonadotropin in the primate endometrial epithelial cells. *Biol. Reprod.* 68, 457–464.
331. Cole, H. H., Howell, C. E., and Hart, G. H. (1931). Changes occurring in the ovary of the mare during pregnancy. *Anat. Rec.* 49, 199–210.
332. Squires, E. L., Stevens, W. B., Pickett, B. W., and Nett, T. M. (1979). Role of pregnant mare serum gonadotropin in luteal function of pregnant mares. *Am. J. Vet. Res.* 40, 889–891.
333. Moss, G. E., Estergreen, V. L., Becker, S. R., and Grant, B. D. (1979). The source of 5α-pregnanes that occur during gestation in the mare. *J. Reprod. Fertil.* 27, 511–519.
334. Cameron, J. L., and Stouffer, R. L. (1981). Comparisons of the species specificity of gonadotropin binding to primate and nonprimate corpora lutea. *Biol. Reprod.* 25, 568–572.
335. McFarland, K. C., Sprengel, R., Phillips, H. S., Kohler, M., Rosemblit, N., Nikolics, K., Segaloff, D. L., and Seeburg, P. H. (1989). Lutropin-choriogonadotropin receptor: an unusual member of the G protein-coupled receptor family. *Science* 245, 494–499.
336. Huhtaniemi, I. (2004). Functional consequences of mutations and polymorphisms in gonadotropin and gonadotropin receptor genes. In *The Ovary* (P. C. K. Leung and E. Y. Adashi, Eds.), pp. 55–78. Elsevier Academic Press, San Diego.
337. Hunzicker-Dunn, M., and Mayo, K. (2006). Gonadotropin signaling in the ovary. In *Knobil and Neill's Physiology of Reproduction* (J. D. Neill, Ed.), pp. 547–592. Academic Press, St. Louis.
338. Zhang, F.-P., Poutanen, M., Wilbertz, J., and Huhtaniemi, I. (2001). Normal prenatal but arrested postnatal sexual development of luteinizing hormone receptor knockout (LuRKO) mice. *Mol. Endocrinol.* 15, 172–183.
339. Lei, Z. M., Mishra, S., Zou, W., Xu, B., Foltz, M., Li, X., and Rao, Ch. V. (2001). Targeted disruption of luteinizing hormone/human chorionic gonadotropin receptor gene. *Mol. Endocrinol.* 15, 184–200.
340. Wood, J. R., and Strauss, J. F. III (2002). Multiple signal transduction pathways regulate ovarian steroidogenesis. *Rev. Endocrinol. Metab. Disord.* 3, 33–46.
341. Houmard, B. S., Guan, Z., Stokes, B. T., and Ottobre, J. S. (1994). The effects of gonadotropin on the phosphatidylinositol pathway in the primate corpus luteum. *Mol. Cell. Endocrinol.* 104, 113–120.
342. Duncan, W. C., Cowen, G. M., and Illingworth, P. J. (1999). Steroidogenic enzyme expression in human corpora lutea in the presence and absence of exogenous human chorionic gonadotrophin (HCG). *Mol. Hum. Reprod.* 5, 291–298.
343. Young, K. A., and Stouffer, R. L. (2004). Gonadotropin and steroid regulation of matrix metalloproteinases and their endogenous tissue inhibitors in the developed corpus luteum of the rhesus monkey during the menstrual cycle. *Biol. Reprod.* 70, 244–252.
344. Grazul-Bilska, A. T., Reynolds, L. P., Bilski, J. J., and Redmer, D. A. (2001). Effects of second messengers on gap junctional intercellular communication of ovine luteal cells throughout the estrous cycle. *Biol. Reprod.* 65, 777–783.
345. Quagliarello, J., Goldsmith, L., Steinetz, B., Lustig, D. S., and Weiss, G. (1980). Induction of relaxin secretion in nonpregnant women by human chorionic gonadotropin. *J. Clin. Endocrinol. Metab.* 51, 74–77.
346. Gagliardi, C. L., Goldsmith, L. T., Saketos, M., Weiss, G., and Schmidt, C. L. (1992). Human chorionic gonadotropin stimulation of relaxin secretion by luteinized human granulosa cells. *Fertil. Steril.* 58, 314–320.
347. Xu, J., Stouffer, R. L., and Hennebold, J. D. (2005). Discovery of luteinizing hormone (LH)-regulated genes in the primate corpus luteum. *Mol. Hum. Reprod.* 11, 151–159.
348. Muramatsu, Y., Sugino, N., Suzuki, T., Totsune, K., Takahashi, K., Tashiro, A., Hong, M., Oki, Y., and Sasano, H. (2001). Urocortin and corticotropin-releasing factor receptor expression in normal cycling human ovaries. *J. Clin. Endocrinol. Metab.* 86, 1362–1369.
349. Keyes, P. L., Gadsby, J. E., Yuh, K.-C. M., and Bill, C. H. III (1983). The corpus luteum. *Int. Rev. Physiol.* 27, 57–97.
350. Duffy, D. M., Chaffin, C. L., and Stouffer, R. L. (2000). Expression of estrogen receptor α and β in the rhesus monkey corpus luteum during the menstrual cycle: regulation by luteinizing hormone and progesterone. *Endocrinology* 141, 1711–1717.
351. Rosenfeld, C. S., Roberts, R. M., and Lubahn, D. B. (2001). Estrogen receptor- and aromatase-deficient mice provide insight into the roles of estrogen within the ovary and uterus. *Mol. Reprod. Dev.* 59, 336–346.
352. Morley, P., Whitfield, J. F., Vanderhyden, B. C., Tsang, B. K., and Schwartz, J.-L. (1992). A new, nongenomic estrogen action: the rapid release of intracellular calcium. *Endocrinology* 131, 1305–1312.
353. Levin, E. R. (1999). Cellular functions of the plasma membrane estrogen receptor. *Trends Endocrinol. Metab.* 10, 374–377.
354. Townson, D. H., Wang, X. J., and Keyes, P. L. (1996). Expression of the steroidogenic acute regulatory protein in

the corpus luteum of the rabbit: dependence upon the luteotropic hormone, estradiol-17β. *Biol. Reprod.* 55, 868–874.
355. Cok, S. J., Hay, R. V., and Holt, J. A. (1997). Estrogen-mediated mitochondrial cholesterol transport and metabolism to pregnenolone in the rabbit luteinized ovary. *Biol. Reprod.* 57, 360–366.
356. Lopez, D., Sanchez, M. D., Shea-Eaton, W., and Mclean, M. P. (2002). Estrogen activates the high-density lipoprotein receptor gene via binding to estrogen response elements and interaction with sterol regulatory element binding protein-1A. *Endocrinology* 143, 2155–2168.
357. Mclean, M. P., Khan, I., Puryear, T. K., and Gibori, G. (1990). Induction and repression of specific estradiol sensitive protein in the rat corpus luteum. *Chin. J. Physiol.* 33, 353–366.
358. Conneely, O. M., Mulac-Jericevic, B., Lydon, J. P., and De Mayo, F. J. (2001). Reproductive functions of the progesterone receptor isoforms: lessons from knock-out mice. *Mol. Cell. Endocrinol.* 179, 97–103.
359. Park-Sarge, O.-K., Parmer, T. G., Gu, Y., and Gibori, G. (1995). Does the rat corpus luteum express the progesterone receptor gene? *Endocrinology* 136, 1537–1543.
360. Sugino, N., Telleria, C. M., and Gibori, G. (1997). Progesterone inhibits 20α-hydroxysteroid dehydrogenase expression in the rat corpus luteum through the glucocorticoid receptor. *Endocrinology* 138, 4497–4500.
361. Peluso, J. J., Fernandez, G., Pappalardo, A., and White, B. A. (2001). Characterization of a putative membrane receptor for progesterone in rat granulosa cells. *Biol. Reprod.* 65, 94–101.
362. Bramley, T. A., Menzies, G. S., Rae, M. T., and Scobie, G. (2002). Non-genomic steroid receptors in the bovine ovary. *Domest. Anim. Endocrinol.* 23, 3–12.
363. Rueda, B. R., Hendry, I. R., Hendry, W. J. III, Stormshak, F., Slayden, O. D., and Davis, J. S. (2000). Decreased progesterone levels and progesterone receptor antagonists promote apoptotic cell death in bovine luteal cells. *Biol. Reprod.* 62, 269–276.
364. Hibbert, M. L., Stouffer, R. L., Wolf, D. P., and Zelinski-Wooten, M. F. (1996). Midcycle administration of a progesterone synthesis inhibitor prevents ovulation in primates. *Proc. Natl. Acad. Sci. U S A* 93, 1897–1901.
365. Chaffin, C. L., and Stouffer, R. L. (1999). Expression of matrix metalloproteinases and their tissue inhibitor messenger ribonucleic acids in macaque periovulatory granulosa cells: time course and steroid regulation. *Biol. Reprod.* 61, 14–21.
366. Robker, R. L., Russell, D. L., Espey, L. L., Lydon, J. P., O'Malley, B. W., and Richards, J. S. (2000). Progesterone-regulated genes in the ovulation process: ADAMTS-1 and cathepsin L proteases. *Proc. Natl. Acad. Sci. U S A* 97, 4689–4694.
367. Hull, K. L., and Harvey, S. (2001). Growth hormone: Roles in female reproduction. *J. Endocrinol.* 168, 1–23.
368. Niswender, G. D., Juengel, J. L., Silva, P. J., Rollyson, M. K., and McIntush, E. W. (2000). Mechanisms controlling the function and life span of the corpus luteum. *Physiol. Rev.* 80, 1–29.
369. Constantino, C. X., Keyes, P. L., and Kostyo, J. L. (1991). Insulin-like growth factor-I stimulates steroidogenesis in rabbit luteal cells. *Endocrinology* 128, 1702–1708.
370. McArdle, C. A., and Holtorf, A. P. (1989). Oxytocin and progesterone release from bovine corpus luteal cells in culture: effects of insulin-like growth factor I, insulin, and prostaglandins. *Endocrinology* 124, 1278–1286.
371. Devoto, L., Christenson, L. K., McAllister, J. M., Makrigiannakis, A., and Strauss, J. F. III (1999). Insulin and insulin-like growth factor-I and II modulate human granulosa-lutein cell steroidogenesis: enhancement of steroidogenic acute regulatory protein (StAR) expression. *Mol. Hum. Reprod.* 5, 1003–1010.
372. Natesampillai, S., and Veldhuis, J. D. (2004). Actions of insulin and insulin-like growth factor-1 on sterol-metabolizing gene expression in ovarian cells. In *The Ovary* (P. C. K. Leung and E. Y. Adashi, Eds.), pp. 249–259. Elsevier Academic Press, San Diego.
373. Spicer, L. J. (2004). Proteolytic degradation of insulin-like growth factor binding proteins by ovarian follicles: a control mechanism for selection of dominant follicles. *Biol. Reprod.* 70, 1223–1230.
374. Neuvians, T. P., Pfaffl, M. W., Berisha, B., and Schams, D. (2003). The mRNA expression of the members of the IGF-system in bovine corpus luteum during induced luteolysis. *Domest. Anim. Endocrinol.* 25, 359–372.
375. Neuvians, T. P., Pfaffl, M. W., Berisha, B., and Schams, D. (2003). The mRNA expression of insulin receptor isoforms (IR-A and IR-B) and IGFR-2 in the bovine corpus luteum during the estrous cycle, pregnancy, and induced luteolysis. *Endocrine* 22, 93–99.
376. Bondy, C. A., Zhou, J., and Arraztoa, J. A. (2006). Growth hormone, insulin-like growth factors, and the ovary. In *Knobil and Neill's Physiology of Reproduction* (J. D. Neill, Ed.), pp. 527–546. Academic Press, St. Louis.
377. Molskness, T. A., VandeVoort, C. A., and Stouffer, R. L. (1987). Stimulatory and inhibitory effects of prostaglandins on the gonadotropin-sensitive adenylate cyclase in the monkey corpus luteum. *Prostaglandins* 34, 279–290.
378. Hahlin, M., Dennefors, B., Johanson, C., and Hamberger, L. (1988). Luteotropic effects of prostaglandin E2 on the human corpus luteum of the menstrual cycle and early pregnancy. *J. Clin. Endocrinol. Metab.* 66, 909–914.
379. Boiti, C., Zampini, D., Zerani, M., Guelfi, G., and Gobbetti, A. (2001). Prostaglandin receptors and role of G protein-activated pathways on corpora lutea of pseudopregnant rabbit in vitro. *J. Endocrinol.* 168, 141–151.
380. Zelinski-Wooten, M. B., and Stouffer, R. L. (1990). Intraluteal infusions of prostaglandins of the E, D, I, and A series prevent $PGF_{2\alpha}$-induced, but not spontaneous, luteal regression in rhesus monkeys. *Biol. Reprod.* 43, 507–516.
381. Arosh, J. A., Banu, S. K., Chapdelaine, P., Madore, E., Sirois, J., and Fortier, M. A. (2004). Prostaglandin biosynthesis, transport, and signaling in corpus luteum: a basis for autoregulation of luteal function. *Endocrinology* 145, 2551–2560.
382. Sargent, E. L., Baughman, W. L., Novy, M. J., and Stouffer, R. L. (1988). Intraluteal infusion of a prostaglandin synthesis inhibitor, sodium meclofenamate, causes premature luteolysis in rhesus monkeys. *Endocrinology* 123, 2261–2269.
383. Kim, L., Weems, Y. S., Bridges, P. J., LeaMaster, B. R., Ching, L., Vincent, D. L., and Weems, C. W. (2001). Effects of indomethacin, luteinizing hormone (LH), prostaglandin E2 (PGE2), trilostane, mifepristone, ethamoxytriphetol (MER-25) on secretion of prostaglandin E (PGE), prostaglandin F2α (PGF2α) and progesterone by ovine corpora lutea of pregnancy or the estrous cycle. *Prostaglandins Other Lipid Mediat.* 63, 189–203.
384. Kawate, N., Tsuji, M., Tamada, H., Inaba, T., and Sawada, T. (2004). Enhancement of prostacyclin synthesis at the beginning of formation of caprine corpora lutea. *Mol. Reprod. Dev.* 67, 308–312.
385. Milvae, R. A., Alila, H. W., and Hansel, W. (1986). Involvement of lipoxygenase products of arachidonic acid metabolism in bovine luteal function. *Biol. Reprod.* 35, 1210–1215.

386. Ichikawa, F., Yoshimura, Y., and Oda, T. (1990). The effects of lipoxygenase products on progesterone and prostaglandin production by human corpora lutea. *J. Clin. Endocrinol. Metab.* 70, 849.
387. Yoshimura, Y., Nakamura, Y., Ichikawa, F., Oda, T., Jinno, M., Ando, M., Koyama, N., and Shiokawa, S. (1991). Possible involvement of leukotrienes in human luteal function. *Acta Endocrinol.* 127, 246–251.
388. Blair, R. M., Saatman, R., Liou, S. S., Fortune, J. E., and Hansel, W. (1997). Roles of leukotrienes in bovine corpus luteum regression: An in vivo microdialysis study. *Proc Soc Exp Biol Med* 216, 72–80.
389. van Voorhis, B. J., Dunn, M. S., Falck, J. R., Bhatt, R. K., VanRollins, M., and Snyder, G. D. (1993). Metabolism of arachidonic acid to epoxyeicosatrienoic acids by human granulosa cells may mediate steroidogenesis. *J. Clin. Endocrinol. Metab.* 76, 1555–1559.
390. Pate, J. L. (2003). Lives in the balance: responsiveness of the corpus luteum to uterine and embryonic signals. *Reprod. Suppl.* 61, 207–217.
391. Bradbury, J. T., Brown, W. E., and Gray, L. A. (1950). Maintenance of the corpus luteum and physiologic actions of progesterone. *Recent Prog. Horm. Res.* 5, 151–194.
392. Critser, E. S., Rutledge, J. J., and French, R. L. (1980). Role of the uterus and the conceptus in regulating luteal lifespan in the mouse. *Biol. Reprod.* 23, 558–563.
393. Duby, R. T., McDaniel, J. W., Spilman, C. H., and Black, D. L. (1969). Utero-ovarian relationships in the gold hamster: I. Ovarian periodicity following hysterectomy. *Acta Endocrinol.* 60, 595–602.
394. Asdell, S. A., and Hammond, J. (1933). The effect of prolonging the life of the corpus luteum in the rabbit by hysterectomy. *Am. J. Physiol.* 103, 600–605.
395. Wiltbank, J. N., and Casida, L. E. (1956). Alteration of ovarian activity by hysterectomy. *J. Anim. Sci.* 15, 134–140.
396. Spies, H. G., Zimmerman, D. R., Self, H. L., and Casida, L. E. (1960). Effect of exogenous progesterone on the corpora lutea of hysterectomized gilts. *J. Anim. Sci.* 19, 108.
397. Squires, E. L., Wentworth, B. C., and Ginther, O. J. (1975). Progesterone concentration in blood of mares during the estrous cycle, pregnancy and after hysterectomy. *J. Anim. Sci.* 39, 759–767.
398. Miller, J. B., and Keyes, P. L. (1976). A mechanism for regression of the rabbit corpus luteum: uterine induced loss of luteal responsiveness to 17β-estradiol. *Biol. Reprod.* 15, 511–518.
399. Inskeep, E. K., and Butcher, R. I. (1966). Local component of utero-ovarian relationships in the ewe. *J. Anim. Sci.* 25, 1164–1168.
400. Ginther, O. J., Woody, C. O., Mahajan, S., Janakiraman, K., and Casida, L. E. (1967). Effect of oxytocin administration on the oestrous cycle of unilaterally hysterectomized heifers. *J. Reprod. Fertil.* 14, 225–229.
401. Butcher, R. L., Barley, D. A., and Inskeep, E. K. (1969). Local relationship between the ovary and uterus of rats and guinea pigs. *Endocrinology* 84, 476–481.
402. Duby, R. T., McDaniel, J. W., Spilman, C. H., and Black, D. L. (1969). Utero-ovarian relationships in the golden hamster: II. Quantitative and local influences of the uterus on ovarian function. *Acta. Endocrinol.* 60, 603–610.
403. Ginther, O. J. (1974). Internal regulation of physiological processes through venoarterial pathways: a review. *J. Anim. Sci.* 39, 550–564.
404. Hadley, J. C. (1975). The effect of serial uterine biopsies and hysterectomy on peripheral blood levels of total unconjugated oestrogen and progesterone in the bitch. *J. Reprod. Fertil.* 45, 389–393.
405. Deanesly, R., and Parkes, A. S. (1927). The effect of hysterectomy on the oestrous cycle of the ferret. *Am. J. Physiol.* 82, 14–18.
406. Del Campo, C. H., and Ginther, O. J. (1973). Vascular anatomy of the uterus and ovaries and the unilateral luteolytic effect of the uterus: horses, sheep and swine. *Am. J. Vet. Res.* 34, 305–316.
407. Del Campo, C. H., and Ginther, O. J. (1973). Vascular anatomy of the uterus and ovaries and the unilateral luteolytic effect of the uterus: Guinea pigs, rats, hamsters and rabbits. *Am. J. Vet. Res.* 33, 2561–2578.
408. Sirois, J., Boerboom, D., and Sayasith, K. (2004). Prostaglandin biosynthesis and action in the ovary. In *The Ovary* (P. C. K. Leung and E. Y. Adashi, Eds.), pp. 233–247. Elsevier Academic Press, San Diego.
409. Pharriss, B. B., and Wyngarden, L. (1969). The effect of prostaglandin F2 alpha on the progestogen content of ovaries from pseudopregnant rats. *Proc. Soc. Exp. Biol. Med.* 130, 92–94.
410. Gleeson, A. R. (1974). Luteal function of the cyclic sow after infusion of prostaglandin F2 alpha through a uterine vein. *J. Reprod. Fertil.* 36, 518–522.
411. Shemesh, M., and Hansel, W. (1975). Levels of prostaglandin F in bovine endometrium, uterine venous, ovarian arterial and jugular plasma during the estrous cycle. *Proc. Soc. Exp. Biol. Med.* 148, 123–126.
412. Vernon, M. W., Zavy, M. G., Aisquith, R. L., and Sharp, D. C. (1981). Prostaglandin F2 alpha in the equine endometrium: steroid production and production capacities during the estrous cycle and early pregnancy. *Biol. Reprod.* 25, 581–589.
413. Doebler, J. A., Wickersham, E. W., and Anthony, A. (1981). Uterine prostaglandin F2α content and 20-α-hydroxysteroid dehydrogenase activity in individual ovarian compartments during pseudopregnancy in the rat. *Biol. Reprod.* 24, 871–878.
414. McCracken, J. A., Barcikowski, B., Carlson, J. C., Green, K., and Samuelsson, B. (1973). The physiological role of prostaglandin F2 alpha in corpus luteum regression. *Adv. Biosci.* 9, 599–624.
415. Baird, D. T., Land, R. B., Scaramuzzi, R. J., and Wheeler, A. G. (1976). Endocrine changes associated with luteal regression in the ewe: the secretion of ovarian oestradiol, progesterone and androstenedione and uterine prostaglandin F2α throughout the oestrous cycle. *J. Endocrinol.* 69, 275–286.
416. Thorburn, G. D., Cox, R. I., Currie, W. B., Restall, B. J., and Schneider, W. (1973). Prostaglandin F and progesterone concentrations in the utero-ovarian venous plasma of the ewe during the oestrous cycle and early pregnancy. *J. Reprod. Fertil.* 18 (Suppl), 151–158.
417. Scaramuzzi, R. J., and Baird, D. T. (1976). The oestrous cycle of the ewe after active immunization against prostaglandin F2 alpha. *J. Reprod. Fertil.* 46, 39–47.
418. Fairclough, R. J., Smith, J. F., and McGowan, L. T. (1981). Prolongation of the oestrous cycle in cows and ewes after passive immunization with PGR antibodies. *J. Reprod. Fertil.* 62, 213–219.
419. Lau, I. F., Saksena, S. K., and Chang, M. C. (1975). Effects of indomethacin, an inhibitor of prostaglandin biosynthesis on the length of pseudo-pregnancy in rats and hamsters. *Acta Endocrinol.* 78, 343–348.
420. Critser, E. S., Rutledge, J. J., and French, L. R. (1981). Effect of indomethacin on the interestrous interval of intact and hysterectomized pseudopregnant mice. *Biol. Reprod.* 24, 1000–1005.
421. Sugimoto, Y., Yamasaki, A., Segi, E., Tsuboi, K., Aze, Y., Nishimura, T., Oida, H., Yhosida, N., Tanaka, T.,

Katsuyama, M., Hasumoto, K., Murata, T., Hirata, M., Ushikubi, F., Negishi, M., Ichikawa, A., and Narumiya, S. (1997). Failure of parturition in mice lacking the prostaglandin F receptor. *Science* 277, 681–683.
422. Sugimoto, Y., Segi, E., Tsuboi, K., Ichikawa, A., and Narumiya, S. (1998). Female reproduction in mice lacking the prostaglandin F receptor: Roles of prostaglandin and oxytocin receptors in parturition. *Adv. Exp. Med. Biol.* 449, 317–321.
423. Uozumi, N., Kume, K., Nagase, T., Nakatani, N., Ishii, S., Tashiro, F., Komagata, Y., Maki, K., Ikuta, K., Ouchi, Y., Miyazaki, J., and Shimizu, T. (1997). Role of cytosolic phospholipase A2 in allergic response and parturition. *Nature* 390, 618–622.
424. Langenbach, R., Morham, S. G., Tiano, H. F., Loftin, C. D., Ghanayem, B. I., Chulada, P. C., Mahler, J. F., Lee, C. A., Goulding, E. H., and Kluckman, K. D. (1995). Prostaglandin synthase 1 gene disruption in mice reduces arachidonic acid-induced inflammation and indomethacin-induced gastric ulceration. *Cell* 83, 483–492.
425. Salustri, A., Fulop, C., Camaioni, A., and Hascall, V. C. (2004). Oocyte-granulosa cell interactions. In *The Ovary* (P. C. K. Leung and E. Y. Adashi, Eds.), pp. 131–143. Elsevier Academic Press, San Diego.
426. Tilley, S. L., Audoly, L. P., Hicks, E. H., Kim, H. S., Flannery, P. J., Coffman, T. M., and Koller, B. H. (2004). Reproductive failure and reduced blood pressure in mice lacking the EP2 prostaglandin E2 receptor. *J. Clin. Invest.* 103, 1539–1545.
427. Duffy, D. M., and Stouffer, R. L. (2002). Follicular administration of a cyclooxygenase inhibitor can prevent oocyte release without alteration of normal luteal function in rhesus monkeys. *Hum. Reprod.* 17, 2825–2831.
428. Niswender, G. D., Reimers, T. J., Diekman, M. A., and Nett, T. M. (1976). Blood flow: a mediator of ovarian function. *Biol. Reprod.* 14, 64–81.
429. Behrman, H. R., Luborsky-Moore, J. L., Pang, C. Y., Wright, K., and Dorflinger, L. J. (1979). Mechanisms of PGF2 alpha action in functional luteolysis. *Adv. Exp. Med. Biol.* 112, 557–571.
430. Mclean, M. P., Billheimer, J. T., Warden, K. J., and Irby, R. B. (1995). Prostaglandin F2α mediates ovarian sterol carrier protein-2 expression during luteolysis. *Endocrinology* 136, 4963–4972.
431. Juengel, J. L., Meberg, B. M., Turzillo, A. M., Nett, T. M., and Niswender, G. D. (1995). Hormonal regulation of messenger ribonucleic acid encoding steroidogenic acute regulatory protein in ovine corpora lutea. *Endocrinology* 136, 5423–5429.
432. Fiedler, E. P., Plouffe, L., Jr., Hales, D. B., Hales, K. H., and Khan, I. (1999). Prostaglandin F2α induces a rapid decline in progesterone production and steroidogenic acute regulatory protein expression in isolated rat corpus luteum without altering messenger ribonucleic acid expression. *Biol. Reprod.* 61, 643–650.
433. Tandeski, T. R., Juengel, J. L., Nett, T. M., and Niswender, G. D. (1996). Regulation of mRNA encoding low density lipoprotein receptor and high density lipoprotein-binding protein in ovine corpora lutea. *Reprod. Fertil. Dev.* 8, 1107–1114.
434. McGuire, W. J., Juengel, J. L., and Niswender, G. D. (1994). Protein kinase C second messenger system mediates the antisteroidogenic effects of prostaglandin F2α in the ovine corpus luteum in vivo. *Biol. Reprod.* 51, 800–806.
435. Sarvas, K., Angervo, M., Koistinen, R., Tiitinen, A., and Seppala, M. (1994). Prostaglandin $F_{2\alpha}$ stimulates release of insulin-like growth factor binding protein-3 from cultured human granulosa-luteal cells. *Hum. Reprod.* 9, 1643–1646.
436. Sugimoto, Y., Hasumoto, K.-Y., Namba, T., Irie, A., and Katsukawa, M. (1994). Cloning and expression of a cDNA for mouse prostaglandin F receptor. *J. Biol. Chem.* 269, 1356–1360.
437. Kitanaka, J., Hashimoto, H., Sugimoto, Y., Negishi, M., Aino, H., Gotoh, M., Ichikawa, A., and Baba, A. (1994). Cloning and expression of cDNA for rat prostaglandin F2 alpha receptor. *Prostaglandins* 48, 31–41.
438. Graves, P. E., Pierce, K. L., Bailey, T. J., Rueda, B. R., Gil, D. W., Woodward, F. F., Yool, A. J., Hoyer, P. B., and Regan, J. W. (1995). Cloning of receptor for prostaglandin F2 alpha from the ovine corpus luteum. *Endocrinology* 136, 3430–3426.
439. Lake, S., Gullberg, H., Wahlqvist, J., Sjogren, A.-M., Kinhult, A., Lind, P., Hellstrom-Lindahl, E., and Stjernschantz, J. (1994). Cloning of the rat and human prostaglandin F2α receptors and the expression of the rat prostaglandin F2α receptor. *FEBS Lett.* 355, 317–325.
440. Hoyer, P. B. (1998). Regulation of luteal regression: the ewe as a model. *J. Soc. Gynecol. Invest.* 5, 49–57.
441. Stocco, C. O., Lau, L. F., and Gibori, G. (2002). A calcium/calmodulin-dependent activation of ERK1/2 mediates JunD phosphorylation and induction of nur77 and 20α HSD genes by prostaglandin F2α in ovarian cells. *J. Biol. Chem.* 277, 3293–3302.
442. Curlewis, J. D., Tam, S. P., Lau, P., Kusters, D. H., Barclay, J. L., Anderson, S. T., and Waters, M. J. (2002). A prostaglandin F2α analog induces suppressors of cytokine signaling-3 expression in the corpus luteum of the pregnant rat: a potential new mechanism in luteolysis. *Endocrinology* 143, 3984–3993.
443. Pate, J. L., and Condon, W. A. (1984). Effects of prostaglandin F2 alpha on agonist-induced progesterone production in cultured bovine luteal cells. *Biol. Reprod.* 31, 427–435.
444. Girsh, E., Greber, Y., and Meidan, R. (1995). Luteotrophic and luteolytic interactions between bovine small and large luteal-like cells and endothelial cells. *Biol. Reprod.* 52, 954–962.
445. Fitz, T. A., Mock, E. J., Mayan, M. H., and Niswender, G. D. (1984). Interactions of prostaglandins with subpopulations of ovine luteal cells: II. Inhibitory effects of PGF2 alpha and protection by PGE2. *Prostaglandins* 28, 127–138.
446. Juengel, J. L., and Niswender, G. D. (1999). Molecular regulation of luteal progesterone synthesis in domestic ruminants. *J. Reprod. Fertil. Suppl.* 54, 193–205.
447. Gaytan, F., Morales, C., Garcia-Pardo, L., Reymundo, C., Bellido, C., and Sanchez-Criado, J. E. (1999). A quantitative study of changes in the human corpus luteum microvasculature during the menstrual cycle. *Biol. Reprod.* 60, 914–919.
448. Sawyer, H. R., Niswender, K. D., Braden, T. D., and Niswender, G. D. (1990). Nuclear changes in ovine luteal cells in response to $PGF_{2\alpha}$. *Domest. Anim. Endocrinol.* 7, 229–238.
449. Grossman, J. D., and Morgan, J. P. (1997). Cardiovascular effects of endothelin. *News Physiol. Sci.* 12, 113–117.
450. Nussdorfer, G. G., Rossi, G. P., Malendowicz, L. K., and Mazzocchi, G. (1999). Autocrine-paracrine endothelin system in the physiology and pathology of steroid-secreting tissues. *Pharmacol. Rev.* 51, 406–437.
451. Milvae, R. A. (2000). Inter-relationships between endothelin and prostaglandin F2α in corpus luteum function. *Rev. Reprod.* 5, 1–5.

452. Meidan, R., and Levy, N. (2002). Endothelin-1 receptors and biosynthesis in the corpus luteum: molecular and physiological implications. *Domest. Anim. Endocrinol.* 5340, 1–12.
453. Girsh, E., Milvae, R. A., Wang, W., and Meidan, R. (1996). Effect of endothelin-1 on bovine luteal cell function: role in PGF2α-induced anti-steroidogenic action. *Endocrinology* 137, 1306–1312.
454. Hinckley, S. T., and Milvae, R. A. (2001). Endothelin-1 mediates prostaglandin $F_{2\alpha}$-induced luteal regression in the ewe. *Biol. Reprod.* 64, 1619–1623.
455. Meidan, R., Milvae, R. A., Weiss, S., Levy, N., and Friedman, A. (1999). Intraovarian regulation of luteolysis. *J. Reprod. Fertil. Suppl.* 54, 217–228.
456. Miyamoto, A., Naoko, O., and Masayuki, O. (2001). Effect of endothelin-1 on functional luteolysis in the cow. *Biol. Reprod.* 64, 232 (abstract).
457. Ohtani, M., Takase, S., Wijayagunawardane, M. P. B., Tetsuka, M., and Miyamoto, A. (2004). Local interaction of prostaglandin F2α with endothelin-1 and tumor necrosis factor-α on the release of progesterone and oxytocin in ovine corpora lutea in vivo: a possible implication for a luteolytic cascade. *Reproduction* 127, 117–124.
458. Usuki, S., Suzuki, N., Matsumoto, H., Yanagisawa, M., and Masaki, T. (1991). Endothelin-1 in luteal tissue. *Mol. Cell. Endocrinol.* 80, 147–151.
459. Miceli, F., Minici, F., Pardo, M. G., Navarra, P., Proto, C., Mancuso, S., Lanzone, A., and Apa, R. (2001). Endothelins enhance prostaglandin (PGE_2 and $PGF_{2\alpha}$) biosynthesis and release by human luteal cells: evidence of a new paracrine/autocrine regulation of luteal function. *J. Clin. Endocrinol. Metab.* 86, 811–817.
460. Brannstrom, M., and Friden, B. (1997). Immune regulation of corpus luteum function. *Semin. Reprod. Endocrinol.* 15, 363–370.
461. Bukulmez, O., and Arici, A. (2000). Leukocytes in ovarian function. *Hum. Reprod. Update* 6, 1–15.
462. Brannstrom, M., Giesecke, L., van den Heuvel, C. J., Moore, I. C., and Robertson, S. A. (1994). Leukocyte subpopulations in the rat corpus luteum during pregnancy and pseudopregnancy. *Biol. Reprod.* 50, 1161–1167.
463. Standaert, F. S., Zamor, C. S., and Chew, B. P. (1991). Quantitative and qualitative changes in blood leukocytes in the porcine ovary. *Am. J. Reprod. Immunol.* 215, 163–168.
464. Brannstrom, M., Pascoe, V., Norman, R. J., and McClure, N. (1994). Localization of leukocyte subsets in the follicle wall and in the corpus luteum throughout the human menstrual cycle. *Fertil Steril* 61, 488–495.
465. Wang, L. J., Pascoe, V., Petrucco, O. M., and Norman, R. J. (1992). Distribution of leukocyte subpopulations in the human corpus luteum. *Hum. Reprod.* 7, 197–202.
466. Bagavandoss, P., Wiggins, R. C., Kunkel, S. L., Remick, D. G., and Keyes, P. L. (1990). Tumor necrosis factor production and accumulation of inflammatory cells in the corpus luteum of pseudopregnancy and pregnancy in rabbits. *Biol. Reprod.* 42, 367–376.
467. Espey, L. L., Bellinger, A. S., and Healy, J. A. (2004). Ovulation: an inflammatory cascade of gene expression. In *The Ovary* (P. C. K. Leung and E. Y. Adashi, Eds.), pp. 145–165. Elsevier Academic Press, San Diego.
468. Murdoch, W. J., Steadman, L. E., and Belden, E. L. (1988). Immunoregulation of luteolysis. *Med. Hypotheses* 27, 197–199.
469. Bukovsky, A., Caudle, M. R., Keenan, J. A., Wimalasena, J., Upadhyaya, N. B., and Van Meter, S. E. (1995). Is corpus luteum regression an immune-mediated event? Localization of immune system components and luteinizing hormone receptor in human corpora lutea. *Biol. Reprod.* 53, 1373–1384.
470. Pate, J. L., and Keyes, P. L. (2001). Immune cells in the corpus luteum: friends or foes. *Reproduction* 122, 665–676.
471. Petroff, M. G., Coggeshall, K. M., Jones, L. S., and Pate, J. L. (1997). Bovine luteal cells elicit major histocompatability complex class II-dependent T-cell proliferation. *Biol. Reprod.* 57, 887–893.
472. Khoury, E. L., and Marshall, L. A. (1990). Luteinization of human granulosa cells in vivo is associated with expression of MHC class II antigens. *Cell Tissue Res.* 262, 217–224.
473. Nagy, Z. A., and Mooney, N. A. (2003). A novel, alternative pathway of apoptosis triggered through class II major histocompatibility complex molecules. *J. Mol. Med.* 81, 757–765.
474. Seiner, S. J., Schramm, W., and Keyes, P. L. (1992). Effect of treatment with methylprednisolone on duration of pseudopregnancy and on macrophages and T lymphocytes in rabbit corpora lutea. *J. Reprod. Fertil.* 96, 347–353.
475. Wang, F., Riley, J. C. M., and Behrman, H. R. (1993). Immunosuppressive levels of glucocorticoid block extrauterine luteolysins in the rat. *Biol. Reprod.* 49, 66–73.
476. Gaytan, F., Morales, C., Bellido, C., and Sanchez-Criado, J. E. (2002). Selective apoptosis of luteal endothelial cells in dexamethasone-treated rats leads to ischemic necrosis of luteal tissue. *Biol. Reprod.* 66, 232–240.
477. Bagavant, H., Adams, S., Terranova, P., Chang, A., Kraemer, F. W., Lou, Y., Kasai, K., Luo, A. M., and Tung, K. S. K. (1999). Autoimmune ovarian inflammation triggered by proinflammatory (Th1) T cells is compatible with normal ovarian function in mice. *Biol. Reprod.* 61, 635–642.
478. Castro, A., Castro, O., Troncoso, J. L., Kohen, P., Simon, C., Vega, M., and Devoto, L. (1998). Luteal leukocytes are modulators of the steroidogenic process of human mid-luteal cells. *Hum. Reprod.* 13, 1584–1589.
479. Hashii, K., Fujiwara, H., Yoshioka, S., Kataoka, N., Yamada, S., Hirano, T., Mori, T., Fujii, S., and Maeda, M. (1998). Peripheral blood mononuclear cells stimulate progesterone production by luteal cells derived from pregnant and non-pregnant women: possible involvement of interleukin-4 and interleukin-10 in corpus luteum function and differentiation. *Hum. Reprod.* 13, 2738–2744.
480. Paavola, L. G. (1979). The corpus luteum of the guinea pig: IV. Fine structure of macrophages during pregnancy and postpartum luteolysis and the phagocytosis of luteal cells. *Am. J. Anat.* 154, 337–364.
481. Terranova, P. F., and Rice, V. M. (1997). Review: Cytokine involvement in ovarian processes. *Am. J. Reprod. Immunol.* 37, 50–63.
482. Zhao, Y., Burbach, J. A., Roby, K. F., Terranova, P. F., and Brannian, J. D. (1998). Macrophages are the major source of tumor necrosis factor α in the porcine corpus luteum. *Biol. Reprod.* 59, 1385–1391.
483. Nariai, K., Kanayama, K., Endo, T., and Tsukise, A. (1995). Effects of TNF-α injection into the ovarian parenchyma on luteal blood vessels in rabbits. *Endocr. J.* 42, 761–766.
484. Wuttke, W., Spiess, S., Knoke, I., Pitzel, L., Leonhardt, S., and Jarry, H. (1998). Synergistic effects of prostaglandin $F_{2\alpha}$ and tumor necrosis factor to induce luteolysis in the pig. *Biol. Reprod.* 58, 1310–1315.
485. Ujioka, T., Matsukawa, A., Tanaka, N., Matsuura, K., Yoshinaga, M., and Okamura, H. (1998). Interleukin-8 as an essential factor in the human chorionic gonadotropin-induced rabbit ovulatory process: interleukin-8 induces neutrophil accumulation and activation in ovulation. *Biol. Reprod.* 58, 526–530.

486. Arici, A., Oral, E., Bukulmez, O., Buradagunta, S., Engin, O., and Olive, D. L. (1996). Interleukin-8 expression and modulation in human preovulatory follicles and ovarian cells. *Endocrinology* 137, 3762–3769.
487. Penny, L. A. (2000). Monocyte chemoattractant protein 1 in luteolysis. *Rev. Reprod.* 5, 63–66.
488. Townson, D. H., Warren, J. S., Flory, C. M., and Keyes, P. L. (1996). Expression of monocyte chemoattractant protein-1 in the corpus luteum in the rat. *Biol. Reprod.* 54, 513–520.
489. Penny, L. A., Armstrong, D. G., Baxter, G., Hogg, C., Kindahl, H., Bramley, T., Watson, E. D., and Webb, R. (1998). Expression of monocyte chemoattractant protein-1 in the bovine corpus luteum around the time of natural luteolysis. *Biol. Reprod.* 59, 1464–1469.
490. Senturk, L. M., Seli, E., and Gutierrez, L. S. (1999). Monocyte chemotactic protein-1 expression in human corpus luteum. *Mol. Hum. Reprod.* 5, 697–702.
491. Haworth, J. D., Rollyson, M. K., Silva, P., McIntush, E. W., and Niswender, G. D. (1998). Messenger ribonucleic acid encoding monocyte chemoattractant protein 1 is expressed by the ovine corpus luteum in response to prostaglandin F2α. *Biol. Reprod.* 58, 169–174.
492. Hosang, K., Knoke, I., Klaudiny, J., Wempe, F., Wuttke, W., and Scheit, K. H. (1994). Porcine luteal cells express monocyte chemoattractant protein 1 (MCP-1): analysis by polymerase chain reaction and cDNA cloning. *Biochem. Biophys. Res. Commun.* 199, 962–968.
493. Moor, R. M., and Rowson, L. E. A. (1966). The corpus luteum of the sheep: functional relationship between the embryo and corpus luteum. *J. Endocrinol.* 34, 233–239.
494. Ellinwood, W. E., Nett, T. M., and Niswender, G. D. (1979). Maintenance of the corpus luteum of early pregnancy in the ewe: I. Luteotropic properties of embryonic homogenates. *Biol. Reprod.* 21, 281–288.
495. Roberts, R. M., Xie, S., and Mathialagan, N. (1996). Maternal recognition of pregnancy. *Biol. Reprod.* 54, 294–302.
496. Bazer, F. W., Ott, T. L., and Spencer, T. E. (1998). Endocrinology of the transition from recurring estrous cycles to establishment of pregnancy in subprimate mammals. In *Endocrinology of Pregnancy* (F. W. Bazer, Ed.), pp. 1–34. Humana Press, Clifton, NJ.
497. Martal, J., Lacroix, M.-C., Loudes, C., and Saunier, M. (1979). Trophoblastin, an antiluteolytic protein present in early pregnancy in sheep. *J. Reprod. Fertil.* 56, 63–73.
498. Godkin, J. D., Bazer, F. W., Thatcher, W. W., and Roberts, R. M. (1984). Proteins released by cultured day 15–16 conceptuses prolong luteal maintenance when introduced into the uterine lumen of cyclic ewes. *J. Reprod. Fertil.* 71, 57–64.
499. Godkin, J. D., Bazer, F. W., Moffatt, J., Sessions, F., and Roberts, R. M. (1982). Purification and properties of a major, low molecular weight protein released by the trophoblast of sheep blastocysts at day 13–21. *J. Reprod. Fertil.* 65, 141–150.
500. Ott, T. L., Fleming, J. G., Spencer, T. E., Joyce, M. M., Chen, P., Green, C. N., Zhu, D., Welsh, T. H., Jr., Harms, P. G., and Bazer, F. W. (1997). Effects of exogenous recombinant ovine interferon tau on circulating concentrations of progesterone, cortisol, luteinizing hormone, and antiviral activity; interestrous interval; rectal temperature, and uterine response to oxytocin in cyclic ewes. *Biol. Reprod.* 57, 621–629.
501. Stewart, M. D., Choi, Y., Johnson, G. A., Yu-Lee, L. Y., Bazer, F. W., and Spencer, T. E. (2002). Roles of Stat1, Stat2, and interferon regulatory factor-9 (IRF-9) in interferon tau regulation of IRF-1. *Biol. Reprod.* 66, 393–400.
502. Spencer, T. E., Gray, A., Johnson, G. A., Taylor, K. M., Gertler, A., Gootwine, E., Ott, T. L., and Bazer, F. W. (1999). Effects of recombinant ovine interferon tau, placental lactogen, and growth hormone on the ovine uterus. *Biol. Reprod.* 61, 1409–1418.
503. Inskeep, E. K., Smutny, W. J., Butcher, R. L., and Pexton, J. E. (1975). Effects of intrafollicular injections of prostaglandins in nonpregnant and pregnant ewes. *J. Anim. Sci.* 41, 1098–1104.
504. LaCroix, M. C., and Kann, G. (1982). Comparative studies of prostaglandins F2α and E2 in late cyclic and early pregnant sheep: In vitro synthesis by endometrium and conceptus; effect of in vivo indomethacin treatment on establishment of pregnancy. *Prostaglandins* 23, 507–526.
505. Magness, R. R., Huie, J. M., and Weems, C. W. (1981). Effect of chronic ipsilateral or contralateral intrauterine infusion of prostaglandin E2 (PGE2) on luteal function of unilaterally ovariectomized ewes. *Prostaglandins Med.* 6, 389–401.
506. Guthrie, H. D., and Rexroad, C. E. Jr. (1981). Endometrial prostaglandin F release in vitro and plasma 13,14-dihydro-15-keto-prostaglandin F2 alpha in pigs with luteolysis blocked by pregnancy, estradiol benzoate or human chorionic gonadotropin. *J. Anim. Sci.* 52, 330–339.
507. Knobil, E., Plant, T. M., Wildt, L., Belchetz, P. E., and Marshall, G. (1980). Control of the rhesus monkey menstrual cycle: permissive role of hypothalamic gonadotropin releasing hormone. *Science* 207, 1371–1373.
508. Zeleznik, A. J., and Little-Ihrig, L. L. (1990). Effect of reduced luteinizing hormone concentrations on corpus luteum function during the menstrual cycle of rhesus monkeys. *Endocrinology* 126, 2237–2244.
509. Cameron, J. L., and Stouffer, R. L. (1982). Gonadotropin receptors of the primate corpus luteum: II. Changes in available LH and CG binding sites in macaque luteal membranes during the nonfertile menstrual cycle. *Endocrinology* 110, 2068–2073.
510. Eyster, K. M., Ottobre, J. S., and Stouffer, R. L. (1985). Adenylate cyclase in the corpus luteum of the rhesus monkey: III. Changes in basal and gonadotropin-sensitive activities during the luteal phase of the menstrual cycle. *Endocrinology* 117, 1571–1577.
511. Brannian, J. D., and Stouffer, R. L. (1991). Progesterone production by monkey luteal cell subpopulations at different stages of the menstrual cycle: changes in agonist responsiveness. *Biol. Reprod.* 44, 141–149.
512. Benyo, D. F., and Zeleznik, A. J. (1997). Cyclic adenosine monophosphate signaling in the primate corpus luteum: maintenance of protein kinase A activity throughout the luteal phase of the menstrual cycle. *Endocrinology* 138, 3452–3458.
513. Duffy, D. M., Stewart, D. R., and Stouffer, R. L. (1999). Titrating LH replacement to sustain the structure and function of the corpus luteum after GnRH antagonist treatment in rhesus monkeys. *J. Clin. Endocrinol. Metab.* 84, 342–349.
514. Ellinwood, W. E., and Resko, J. A. (1983). Effect of inhibition of estrogen synthesis during the luteal phase on function of the corpus luteum in rhesus monkeys. *Biol. Reprod.* 28, 636–644.
515. Albrecht, E. D., Haskins, A. L., Hodgen, G. D., and Pepe, G. J. (1981). Luteal function in baboons with administration of the antiestrogen ethamoxytriphetol (MER-25) throughout the luteal phase of the menstrual cycle. *Biol. Reprod.* 25, 451–457.
516. Schoonmaker, J. N., Bergman, K. S., Steiner, R. A., and Karsch, F. J. (1982). Estradiol-induced luteal regression in the rhesus monkey: evidence for an extraovarian site of action. *Endocrinology* 110, 1708–1715.
517. Hild-Petito, S., Stouffer, R. L., and Brenner, R. M. (1988). Immunocytochemical localization of estradiol and progesterone

receptors in the monkey ovary throughout the menstrual cycle. *Endocrinology* 123, 2896–2905.

518. Aladin Chandrasekher, Y., Melner, M. H., Nagalla, S. R., and Stouffer, R. L. (1994). Progesterone receptor, but not estradiol receptor, messenger ribonucleic acid is expressed in luteinizing granulosa cells and the corpus luteum in rhesus monkeys. *Endocrinology* 135, 307–314.
519. Enmark, E., Pelto-Huikko, M., Grandien, K., Lagercrantz, S., Lagercrantz, J., Fried, G., Nordenskjold, M., and Gustafsson, J.-A. (1997). Human estrogen receptor β-gene structure, chromosomal localization, and expression pattern. *J. Clin. Endocrinol. Metab.* 82, 4258–4265.
520. Pau, C. Y., Pau, K.-Y. F., and Spies, H. G. (1998). Putative estrogen receptor β and α mRNA expression in male and female rhesus macaques. *Mol. Cell. Endocrinol.* 146, 59–68.
521. Challis, J. R. G., Calder, A. A., and Dilley, S. E. (1976). Production of prostaglandins E and F by corpora lutea, corpora albicantes and stroma from the human ovary. *J. Endocrinol.* 68, 401–408.
522. Houmard, B. S., and Ottobre, J. S. (1989). Progesterone and prostaglandin production by primate luteal cells collected at various stages of the luteal phase: modulation by calcium ionophore. *Biol. Reprod.* 41, 401–408.
523. Auletta, F. J., and Flint, A. P. (1988). Mechanisms controlling corpus luteum function in sheep, cows, nonhuman primates, and women especially in relation to the time of luteolysis. *Endocr. Rev.* 9, 88–105.
524. Patwardhan, V. V., and Lanthier, A. (1984). Effect of prostaglandin F2 alpha on the hCG-stimulated progesterone production by human corpora lutea. *Prostaglandins* 27, 465–473.
525. Talwar, G. P., Sharma, N. C., and Dubey, S. K. (1976). Isoimmunization against hCG with conjugates of processed β-subunit of the hormone and tetanus toxoid. *Proc. Natl. Acad. Sci. U S A* 73, 218–222.
526. Thau, R. B., and Sundaram, K. (1980). The mechanisms of action of an antifertility vaccine in the rhesus monkey: reversal of the effects of antisera to β-oLH by medroxyprogesterone acetate. *Fertil. Steril.* 33, 317–320.
527. Hay, D. L. (1985). Discordant and variable production of human chorionic gonadotropin and its free α- and β-subunits in early pregnancy. *J. Clin. Endocrinol. Metab.* 61, 1195–1200.
528. Illingworth, P. J., Reddi, K., Smith, K., and Baird, D. T. (1990). Pharmacological "rescue" of the corpus luteum results in increased inhibin production. *Clin. Endocrinol.* 33, 323–332.
529. Booher, C., Enders, A. C., Hendrickx, A. G., and Hess, D. L. (1981). Structural characteristics of the corpus luteum during implantation in the rhesus monkey (*Macaca mulatta*). *Am. J. Anat.* 160, 17–36.
530. Benyo, D. F., Little-Ihrig, L., and Zeleznik, A. J. (1993). Noncoordinated expression of luteal cell messenger ribonucleic acids during human chorionic gonadotropin stimulation of the primate corpus luteum. *Endocrinology* 133, 699–704.
531. Ellinwood, W. E., Stanczyk, F. Z., Lazur, J. J., and Novy, M. J. (1989). Dynamics of steroid biosynthesis during the luteal-placental shift in rhesus monkeys. *J. Clin. Endocrinol. Metab.* 69, 348–355.
532. Duncan, W. C., McNeilly, A. S., Fraser, H. M., and Illingworth, P. J. (1996). Luteinizing hormone receptor in the human corpus luteum: lack of down-regulation during maternal recognition of pregnancy. *Hum. Reprod.* 11, 2291–2297.
533. VandeVoort, C. A., Molskness, T. A., and Stouffer, R. L. (1988). Adenylate cyclase in the primate corpus luteum during chorionic gonadotropin treatment simulating early pregnancy: homologous versus heterologous desensitization. *Endocrinology* 122, 734–740.
534. Rojas, F. J., Moretti-Rojas, I., Balmaceda, J. P., and Asch, R. H. (1991). The role of the adenylyl cyclase system in the regulation of corpus luteum function in the human and in nonhuman primates. *Steroids* 56, 252–257.
535. Duffy, D. M., and Stouffer, R. L. (1997). Gonadotropin versus steroid regulation of the corpus luteum of the rhesus monkey during simulated early pregnancy. *Biol. Reprod.* 57, 1451–1460.
536. Molskness, T. A., Zelinski-Wooten, M. B., Hild-Petito, S. A., and Stouffer, R. L. (1991). Comparison of the steroidogenic response of luteinized granulosa cells from rhesus monkeys to luteinizing hormone and chorionic gonadotropin. *Biol. Reprod.* 45, 273–281.
537. Zeleznik, A. J. (1998). In vivo responses of the primate corpus luteum to luteinizing hormone and chorionic gonadotropin. *Proc. Natl. Acad. Sci. U S A* 95, 11002–11007.
538. Corner, G. W. (1919). On the origin of the corpus luteum of the sow from both granulosa and theca interna. *Am. J. Anat.* 26, 117–183.
539. Alila, H. W., and Hansel, W. (1984). Origin of different cell types in the bovine corpus luteum as characterized by specific monoclonal antibodies. *Biol. Reprod.* 31, 1015–1025.
540. Stouffer, R. L., and Brannian, J. D. (1993). The function and regulation of cell populations composing the corpus luteum of the ovarian cycle. In *The Ovary* (E. Y. Adashi and P. C. K. Leung, Eds.), pp. 245–259. Raven Press, New York.
541. Milvae, R. A., Alila, H. W., Bushmich, S. L., and Hansel, W. (1991). Bovine corpus luteum function after removal of granulosa cells from the preovulatory follicle. *Domest. Anim. Endocrinol.* 8, 439–443.
542. Marut, E. L., Huang, S.-C., and Hodgen, G. D. (1983). Distinguishing the steroidogenic roles of granulosa and theca cells of the dominant ovarian follicle and corpus luteum. *J. Clin. Endocrinol. Metab.* 57, 925–930.
543. Sasano, H., and Suzuki, T. (1997). Localization of steroidogenesis and steroid receptors in human corpus luteum. *Semin. Reprod. Endocrinol.* 15, 345–351.
544. Hansel, W., Alila, H. W., Dowd, J. P., and Milvae, R. A. (1991). Differential origin and control mechanisms in small and large bovine luteal cells. *J. Reprod. Fertil. Suppl.* 43, 77–89.
545. Taylor, M. J., Clark, C. L., and Frawley, L. S. (1987). Evidence for the existence of a luteal cell type that is steroidogenic and releases relaxin. *Proc. Soc. Exp. Biol. Med.* 185, 469–473.
546. Di Simone, N., Castellani, R., Lanzone, A., Caruso, A., and Mancuso, S. (1993). Human growth hormone enhances progesterone production by human luteal cells in vitro: II. Evidence of a distinct effect on two luteal cell types. *Fertil. Steril.* 60, 47–52.
547. Sanders, S. L., Stouffer, R. L., and Brannian, J. D. (1996). Androgen production by monkey luteal cell subpopulations at different stages of the menstrual cycle. *J. Clin. Endocrinol. Metab.* 81, 591–596.
548. Ohara, A., Mori, T., Taii, S., Ban, C., and Narimoto, K. (1987). Functional differentiation in steroidogenesis of two types of luteal cells isolated from mature human corpora lutea of menstrual cycle. *J. Clin. Endocrinol. Metab.* 65, 1192–1200.
549. Retamales, I., Carrasco, I., Troncoso, J. L., Las Heras, J., Devoto, L., and Vega, M. (1994). Morpho-functional study of human luteal cell subpopulations. *Hum. Reprod.* 9, 591–596.
550. Farin, C. E., Moeller, C. L., Mayan, H., Gamboni, F., Sawyer, H. R., and Niswender, G. D. (1988). Effect of luteinizing hormone and human chorionic gonadotropin on cell populations in the ovine corpus luteum. *Biol. Reprod.* 38, 413–421.

551. Smith, C. J., Greer, T. B., Banks, T. W., and Sridaran, R. (1989). The response of large and small luteal cells from the pregnant rat to substrates and secretagogues. *Biol. Reprod.* 41, 1123–1132.
552. Friden, B. E., Hagstrom, H.-G., Lindblom, B., Sjoblom, P., Wallin, A., Brannstrom, M., and Hahlin, M. (1999). Cell characteristics and functions of two enriched fractions of human luteal cells during prolonged culture. *Mol. Hum. Reprod.* 5, 714–719.
553. Brannian, J. D., Stouffer, R. L., Shiigi, S. M., and Hoyer, P. B. (1993). Isolation of ovine luteal cell subpopulations by flow cytometry. *Biol. Reprod.* 48, 495–502.
554. Peters, K. E., Bergfeld, E. G., Cupp, A. S., Kojima, F. N., Mariscal, V., Sanchez, T., Wehrman, M. E., Grotjan, H. E., Hamernik, D. L., Kittok, R. J., and Kinder, J. E. (1994). Luteinizing hormone has a role in development of fully functional corpora lutea (CL) but is not required to maintain CL function in heifers. *Biol. Reprod.* 51, 1248–1254.
555. McNeilly, A. S., Crow, W. J., and Fraser, H. M. (1992). Suppression of pulsatile luteinizing hormone secretion by gonadotrophin-releasing hormone antagonist does not affect episodic progesterone secretion or corpus luteum function in ewes. *J. Reprod. Fertil.* 96, 865–874.
556. Reissmann, Th., Felberbaum, R., Diedrich, K., Engel, J., Comaru-Schally, A. M., and Schally, A. V. (1995). Development and applications of luteinizing hormone-releasing hormone antagonists in the treatment of infertility: an overview. *Hum. Reprod.* 10, 1974–1981.
557. Arioua, R. K., Benhaim, A., Feral, C., and Leymarie, P. (1997). Luteotrophic factors in hyperstimulated pseudopregnant rabbit: II. High sensitivity to hCG of luteal tissue and small luteal cells. *J. Endocrinol.* 154, 259–265.
558. Gregoraszczuk, E. L. (1990). Different response of porcine large and small luteal cells to PRL in terms of progesterone and estradiol secretion in vitro. *Exp. Clin. Endocrinol.* 96, 234–237.
559. Hennebold, J. D. (2003). Characterization of the ovarian transcriptome through the use of differential analysis of gene expression methodologies. *Hum. Reprod. Update.* 10, 227–239.

CHAPTER 13

Growth Hormone, Insulin-Like Growth Factors, and the Ovary

Carolyn A. Bondy, Jian Zhou, and Jose A. Arraztoa

Introduction, 527
Growth Hormone, 528
Growth Hormone and Growth Hormone Receptor, 528
Growth Hormone and the Ovary in Rodents and Farm Animals, 528
Growth Hormone and Human Fertility, 529
Insulin-Like Growth Factor System, 529
Insulin-Like Growth Factor Structure–Function, 530
Insulin-Like Growth Factor Receptor, 530
Insulin-Like Growth Factors and Ovary, 531
Insulin-Like Growth Factors, Insulin, and Ovarian Hyperandrogenism, 535
Insulin-Like Growth Factors and Ovarian Cancer, 537
Insulin-Like Growth Factor–Binding Proteins and the Ovary, 537
Insulin-Like Growth Factor–Binding Protein Proteolysis, Follicular Selection, and Luteolysis, 539
Summary of Growth Hormone/Insulin-Like Growth Factors and Ovarian Function, 539
References, 540

INTRODUCTION

Growth hormone (GH) and insulin-like growth factors (IGFs) function at many different levels to promote coordinated somatic growth, sexual maturation, and reproductive function. This chapter focuses specifically on the actions of GH and IGFs at the level of the ovary. We review information on the role of the GH and IGF systems in the development and functioning of the ovarian follicle and corpus luteum (CL) in diverse species, seeking common themes that may illuminate general principles.

GH and the IGF system affect all aspects of female reproduction, including ovarian, uterine, placental, and mammary growth and function as well as the neuroendocrinological integration of nutritional status and reproductive competency. This review concentrates primarily on the role of GH and the IGF system in the mammalian ovary, comparing observations in rodents, large farm animals and primates. Rodent models have been very informative because of the power of transgenics and targeted deletions, although their reproductive programs may be rather specialized and differ substantially from larger species, especially humans. There is also abundant information on the ovarian GH-IGF system in swine, cattle, and sheep, largely because of agribusiness support for research promoting fertility and fecundity. Finally, we review somewhat more limited data available for the human and nonhuman primates, and address some clinical aspects of the GH-IGF system and reproduction (e.g., the use of GH to enhance ovulatory response in assisted reproduction protocols) and the role of insulin/IGFs in polycystic ovary syndrome. Although we focus on mammals, it is interesting to note that a connection between insulin-like signaling and reproductive capacity is observed in evolutionarily distant species such as *Caenorhabditis elegans* and *Drosophila melanogaster*,

Developmental Endocrinology Branch, National Institute of Child Health and Human Development, Bethesda, Maryland.

in which interruptions in insulin-like signaling pathways may result in infertility (1).

GROWTH HORMONE

Growth Hormone and Growth Hormone Receptor

GH, also known as *somatotropin*, is best known for its role in postnatal somatic growth. GH is released from the anterior pituitary into the circulation and stimulates the synthesis and release of IGF-1 from the liver and some local tissues, with GH-stimulated IGF-1 a major mediator of statural growth during childhood (2). GH may also have additional, direct effects on growth, for example by increasing the stock of epiphysial growth plate chondrocyte precursors, or stem cells (3–5). Likewise with respect to its role in reproduction, it appears that GH may influence ovarian follicle development and function both through enhancing circulating and possibly local ovarian IGF-1, and by direct, GH receptor (GHR)–mediated effects in the ovary. The GHR is a member of the cytokine/hematopoietin receptor superfamily, consisting of a single polypeptide chain containing modular extracellular, transmembrane, and intracellular domains [for review, see Herrington and Carter-Su (6)]. GH binding leads to receptor dimerization and activation, with signaling mediated by phosphorylation of Janus kinase (Jak) 2; subsequent interactions may be with STAT (signal transducers and activators of transcription) transcription factors as well as a number of other, nongenomic pathways (6). GH's differential effects on somatic growth, intermediary metabolism, and regulation of adipogenesis appear to be determined by activation of distinct signaling pathways; there is, however, little information on specific molecular pathways implicated in GH effects in the ovary. The GHR is expressed in the ovary of many different species, from teleost (7) and chicken (8), to rats (9,10), farm animals, including swine, cattle, and sheep (11–15), and the human ovary (16). The GHR is concentrated in the membrana granulosa of primary and small- to medium-sized antral follicles and in CL. In the CL, GHR messenger RNA (mRNA) and immunoreactivity are concentrated in the large luteal cells responsible for high-level progesterone production (15,17,18).

Growth Hormone and the Ovary in Rodents and Farm Animals

Observations in GHR knockout mice indicate an important role for GH in murine ovarian function. These mice are dwarfs and have profoundly reduced circulating IGF-1 levels (19). Sexual maturation is delayed (20), as in GH- or GHR-deficient humans (21). These mice are fertile, although litter size is reduced. The ovarian phenotype in these mutant mice has been closely investigated (20,22,23). The number of follicles per ovary is markedly reduced in GHR null mice, though all stages of follicle development are represented. Atretic follicles are not increased (23). These mice show a reduced ovulatory response to exogenous gonadotropin treatment (22), suggesting a primary defect at the level of the ovary. There are, however, neuroendocrine abnormalities in these mice that have been reviewed (24). Two weeks of IGF-1 infusion did not rescue fertility or ability to respond to gonadotropin in GHR null mice (22). These observations suggest that endogenous GH enhances the recruitment or survival of follicles in the murine ovary—perhaps independent of IGF-1 effects—thus enhancing reproductive potential. Supporting this view, transgenic GH overexpression as well as GH injections produce an increased ovulation rate in mice (25). In vitro studies also show that GH enhances the development of murine follicles (26) and prevents granulosal apoptosis in cultured preovulatory rat follicles (27). GH does not seem to have an essential role in CL development or function in mice, however, because implantation and gestation occur normally in the GHR null mice.

Chronic administration of GH increases the number of small, estrogen-dominant antral follicles and the number of CL in mature cattle (28), but impairs ovarian development and reproductive function in prepubertal swine (29). GH's role in luteal function has been extensively investigated in cattle (30,31) and sheep (32,33). The CL fails to develop in hypophysectomized ewes, but replacement with luteinizing hormone (LH) and GH restores normal CL development; thus, GH is regarded as a true "luteotropic hormone" in this species. GH promotes progesterone and oxytocin secretion from the microdialyzed bovine CL within such a short period that direct, GHR-mediated action appears likely (34). GH stimulates prostaglandin F_2 ($PGF_{2\alpha}$) and progesterone production in the early bovine CL more effectively than does LH (35). Finally, both GH and IGF-1 stimulate progesterone secretion and suppress apoptosis in cultured porcine luteal cells (36). GH is a member of a polypeptide hormone family that includes prolactin and placental lactogen, both of which have roles in female reproduction (37). The cognate receptors are also related (37). Although prolactin's role as a lactogenic hormone is conserved across most mammalian species, this hormone, rather than GH, has a luteotrophic function in some species, including mice (38) and perhaps dogs (39).

A major question that remains unresolved is whether GH stimulates local IGF-1 production by granulosa or luteal cells. Conflicting results from different studies may be due to in vivo versus in vitro models and species differences. Hypophysectomy does not reduce ovarian IGF-1 content in the rat (40,41), suggesting that neither GH nor gonadotropins regulate follicular IGF-1 expression in this species. It is unclear, however, whether IGF-1 levels are altered in the GHR null mouse ovary (22,23). GH increases IGF-1 production (42–45) and estradiol synthesis by porcine granulosa cells (46). Gonadotropins also increase IGF-1 production by porcine follicles in vivo (47) and in vitro (43). GH treatment enhances ovarian IGF-1 mRNA level in hypophysectomized ewes (32), but does not increase IGF-1 production by cultured ovine luteal cells (48). Systemic GH treatment does not increase IGF-1 expression in the bovine (49) or human ovary (50). Although there has been substantial progress in elucidation of GH's signaling pathways in other tissues, little is known about GH signaling in the ovary. One study, however, reports that protein kinase A is involved in GH's stimulation of IGF-1 and prostaglandin production by porcine granulosa cells (45).

Growth Hormone and Human Fertility

GH stimulates steroidogenesis in cultured human granulosa and luteal cells (51–53); there is some evidence that this is due to enhanced IGF-1 or 2 production rather than direct GH effects (53). As noted previously, however, systemic administration of GH does not increase IGF-1 expression in the human ovary (50). Women treated with GH for a diagnosis of GH deficiency during childhood frequently have impaired fertility as adults (54). It is not known whether impaired reproductive function in these women is due primarily to GH deficiency affecting the ovary, or to additional pituitary defects such as gonadotropin or thyrotropin deficiency. Women with Laron's syndrome, or GH insensitivity due to GHR gene defects, have profoundly reduced IGF-1 levels, but spontaneous pregnancies are reported (55,56). These women may have reduced ovarian reserves or limited reproductive potential, similar to the GHR null mouse, but there is no information on their ovarian phenotype or age at menopause. GH does not seem to have an essential role in CL development or function in mice or humans, given normal term pregnancies in individuals with GHR deletions.

Given observations of GH's facilitatory effects on female reproduction in laboratory and farm animals, it was of interest to determine whether GH treatment or amplification of endogenous GH production by administration of GH-releasing factor (GRF) might help women with impaired fertility. There have been many small clinical studies, but here we discuss only the larger studies or meta-analyses. One multicenter study showed that use of GH together with human menopausal gonadotropin (hMG) in hypogonadotropic patients allowed a decrease in effective doses of hMG; however, the pregnancy rate was significantly higher in the placebo group, and the incidence of "side effects" higher in the GH-treated group (57). Another relatively large study showed that treatment of "poor responders" with GRF in addition to gonadotropins produced an increase in GH and IGF-1 levels but did not improve any parameter of ovarian response (58) or pregnancy rates (59). A systematic review of clinical trials involving the use of GH or GRF to improve fertility was provided by the Cochrane Menstrual Disorders and Subfertility Study Group (60). This review assessed the effects of GH and GRF on live births in patients on ovulation induction before in vitro fertilization, with and without history of poor ovarian response to standard protocols. They concluded that there is not enough evidence to support the routine use of GH in ovulation induction protocols. They note that among poor responders there is a suggestion of an increase in live births, but because of the marginal statistical significance (95% confidence interval 1.06–18.01), the reviewers recommend the use of GH only in the context of clinical research (60). GH overexpression in some transgenic mice models and in women with GH-producing pituitary tumors may be associated with impaired reproduction (61,62). Infertility associated with GH excess is attributed to systemic alterations in energy balance and disruption of the hypothalamic–pituitary–gonadal axis (63).

INSULIN-LIKE GROWTH FACTOR SYSTEM

IGF-1 and IGF-2 were originally described in the 1950s as sulfation factor/somatomedin-C (IGF-1) and nonsuppressible insulin-like activity (NSILA) and multiplication stimulating activity (MSA) for IGF-2 (64). IGF-1 was identified as the prime mediator of GH's growth-promoting effects on cartilage in the original "somatomedin hypothesis" (65). This hypothesis has been modified over the years as it has been appreciated that IGF-1 is synthesized locally in many different tissues (e.g., ovary, uterus, and mammary gland), where it is not consistently under regulation by GH, so that there are GH-independent effects by IGF-1 on growth and local tissue homeostasis (2). In addition, as mentioned earlier, there is substantial evidence for some IGF-1–independent actions by GH on growth plate (3–5) and potentially other sites, such as the ovary. IGF-2, in contrast to IGF-1, does not appear to be a mediator of GH's effects on somatic growth. In rodents, IGF-2 is responsible for proportionate somatic growth during fetal development by regulating placental size (66). Also in rodents, IGF-2 levels are high during fetal development but are greatly reduced postnatally. In primates, in contrast, circulating IGF-2 levels are high throughout development and adulthood, indeed several-fold greater than those of IGF-1, and IGF-2 mRNA and peptide are abundant in bone, ovary, and many other human tissues (67). The physiological role of IGF-2 in primates remains unclear. Pathologically high IGF-2 levels, as may be seen in patients with sarcomatous tumors, cause hypoglycemia because of this peptide's insulin-like effects at both IGF-1 and insulin receptors (68). The regulation of IGF-2 gene expression is complicated by genomic imprinting; targeted *IGF2* gene deletion studies in the mouse showed that the paternal allele is operative during fetal development (69). Although IGF-2 is also genomically imprinted in humans, its regulation is quite complex, with tissue- and developmental stage–specific patterns of imprinting (70). For example, expression is biallelic in the liver (71), which is the major source of circulating IGF-2. Loss of IGF-2 imprinting has been associated with tumor development or progression in a variety of tissues, including colon, adrenal, renal, and gynecological tumors (72,73).

Insulin-Like Growth Factor Structure–Function

IGF-1, IGF-2, and insulin belong to a family of anabolic peptides sharing a common evolutionary origin. It appears that an ancestral gene encoding an insulin-like peptide gave rise to multiple genes encoding more specialized peptides about the time gastroenteric and central nervous systems differentiated in early chordate evolution (74). From that time, insulin became progressively more specialized in terms of undergoing secondary processing (proteolytic excision of the C peptide and joining of the A and B peptides by disulfide bonds) and packaging in acidic secretory granules. Insulin expression is largely restricted to pancreatic beta cells, where its synthesis and secretion are tightly coupled to ingested substrates. IGF-1 and 2, in contrast, never acquired such extensive post-translational processing, and have continued to be widely expressed in many cell types demonstrating constitutive secretion. The A and B domains in insulin, IGF-1, and IGF-2 share approximately 50% sequence identity, and the predicted tertiary structures of these peptides are quite similar (67). The IGF-1 gene is located on chromosome 12 (12q22–q24.1), the IGF-2 gene on chromosome 11 (11p15), and the insulin gene nearby at 11p15.5 (75).

Insulin-Like Growth Factor Receptor

The insulin and IGF-1 receptors also arose from a common ancestor, in parallel with the ligands' evolution (76,77). These receptors consist of ligand-binding α subunits and membrane-spanning, tyrosine kinase β subunits (Fig. 1). The α and β sequences are encoded by a single gene and derived from post-translational cleavage followed by disulfide linking (78). The receptors are inserted in the plasma membrane as disulfide-linked dimers, which on binding a single ligand molecule undergo conformational changes resulting in kinase activation and autophosphorylation. The ligand-binding domains of these receptors have diverged enough to confer relative selectivity for binding of the cognate ligands, but crossover, particularly for IGF-2 at the insulin receptor, may be significant. The IGF-1 receptor, however, mediates most of the biological effects of both IGF-1 and IGF-2 (79). The IGF-2/mannose-6-phosphate receptor, which is structurally unrelated to the tyrosine kinase receptor family, promotes IGF-2 degradation and does not interact with IGF-1 (80).

The insulin and IGF-1 receptors demonstrate approximately 85% conservation of the tyrosine

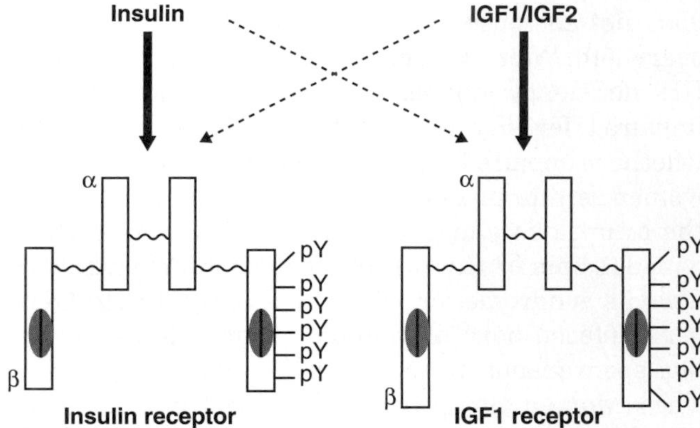

FIG. 1. Schematic diagram illustrating homologies between insulin and insulin-like growth factor (IGF)-1 receptors. Both IGF-1 and IGF-2 bind with high affinity to the IGF-1 receptor, and with ten- to 1,000-fold less affinity to the insulin receptor. The same is the case for insulin with respect to its cognate receptor and the IGF-1 receptor. The disulfide linked α subunits span the plasma membrane, and the tyrosine kinase–containing β subunits are entirely intracellular. Tyrosine phosphorylation sites are indicated as pY. The IGF-1 receptor has a carboxy-terminal extension with an additional pY not found in the insulin receptor.

kinase domain (78) and interact with the same set of intracellular docking proteins (insulin receptor substrate [IRS]-1 through 4, Shc, Gab1) and downstream signaling molecules, including phosphatidylinositol 3-kinase (PI3K), Akt/protein kinase B (PKB), protein tyrosine phosphatases, and Grb2/Sos (79,81). Despite the apparent convergence in signaling pathways, insulin and IGF-1 and 2 have distinct physiological roles, with insulin responsible for maintaining glucose and lipid homeostasis in the circulation, whereas IGF-1 and 2 are primarily responsible for promoting somatic growth. Much effort has been devoted to trying to delineate metabolic signaling pathways specific to the insulin receptor and growth or mitogenic pathways unique to the IGF-1 receptor. *Mitogenic* is probably not the most accurate term because although IGFs promote progression through the mitotic cycle by proliferating cells, they do not stimulate entry into mitosis. In any case, both insulin and IGFs acting through their cognate receptors promote anabolic effects such as glucose, amino acid, and lipid utilization. It seems likely that diverse functional outcomes of various in vitro assay systems reflect more on cell type, differentiation, culture, and experimental conditions than on intrinsic molecular properties of these ligand–receptor interactions. No obvious differences in hard-wired signaling pathways unique to the insulin or IGF-1 receptor have been identified, although subtle differences in the degree of kinase activation are discerned in some genetically manipulated in vitro systems (82,83).

The different physiological roles served by insulin and IGFs may be explained by differences in tissue- and developmental stage–specific bioavailability of the ligands rather than by intrinsic differences in ligand–receptor interactions. High levels of the growth factors are maintained in the circulation and tissue compartments by more or less continuous, constitutive secretion from liver and other sites and also by protection from clearance or proteolysis through high-affinity binding to IGF-binding proteins (IGFBPs) (84–87). This is in marked contrast to insulin availability. Insulin secretion is tightly coupled to ingested substrates; fasting insulin levels are normally quite low, then rise dramatically in response to a meal, followed by rapid clearance from the circulation. Moreover, insulin is not bound to IGFBPs or other specific binding proteins. Thus, peripheral tissues' exposure to insulin and to the IGFs is quite different. It seems likely that the IGFs' growth-promoting effects are due at the most fundamental level to insulin-like actions, directing the allocation of resources (e.g., glucose, lipids, amino acids) to cells and tissues with extraordinary anabolic needs during development and tissue repair processes (88–90).

Insulin-Like Growth Factors and Ovary

Rodents

IGFs appear to be important regulators of ovarian follicular growth and selection and luteal steroidogenesis in the rat and mouse (91). These rodents have served as important experimental models for elucidation of IGF-1's role in folliculogenesis (92). The cellular and developmental patterns of IGF ligand and receptor expression are very similar in the rat and mouse (41,93,94). The IGF-1 receptor is highly expressed in the oocyte, in granulosa cells of follicles at all stages, from small primary to mature preovulatory follicles, and CL in both rat and mouse (41,93,95). IGF-1 expression is more selective, however, being limited to the granulosa of healthy, growing, and selected follicles (41,94,96,97). Ovarian IGF-1 expression is selectively localized in follicle-stimulating hormone (FSH) receptor–expressing follicles but does not appear to be regulated by gonadotropin or GH, whereas IGF-1 receptor expression is dramatically increased by gonadotropin, as determined by in vivo observations in hypophysectomized animals (41,98). Local IGF-1 expression is significantly correlated with granulosa cell proliferation (97), whereas IGF-1 receptor expression is correlated with high-level metabolic activity, reflected by glucose utilization (Fig. 2). It seems likely that extraordinary energy requirements, associated with rapid follicular growth in the case of preovulatory follicles, and rapid growth and steroidogenesis in the case of CL, may be facilitated by "insulin-like" IGF-1 effects promoting nutrient acquisition and cellular anabolic processes at the follicular level.

The cellular patterns of IGF-1 and IGF-1 receptor expression in the normal murine ovary correlate well

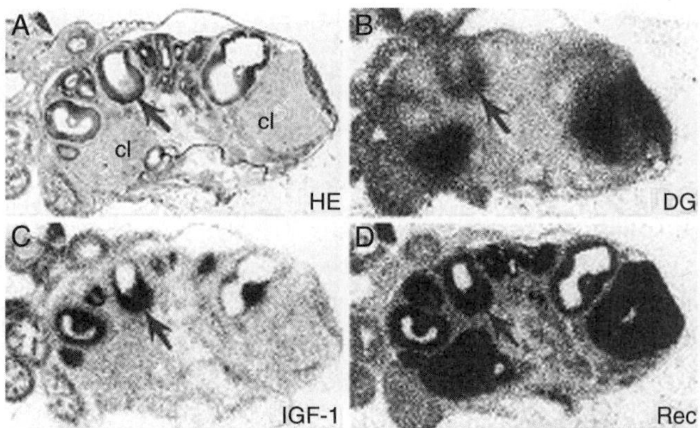

FIG. 2. Ovarian insulin-like growth factor (IGF)-1 receptor expression reflects metabolic activity. **A:** Hematoxylin and eosin–stained section from a mature rat ovary. **B–D:** Film autoradiographs from sequential ovary sections. **B:** ^{14}C-Labeled 2-deoxyglucose uptake. **C:** IGF-1. **D:** IGF-1 receptor mRNA by in situ hybridization.

with in vitro studies showing that IGF-1 promotes proliferation in granulosa cells from the smallest follicles and steroid production in more differentiated luteinized granulosa cells (92,99). If IGF-1's fundamental effects are to promote nutrient utilization, then IGF-1 should effectively enhance whatever cellular processes are in progress. In the case of the murine ovary, local, granulosal IGF-1 production promotes granulosa proliferation and hence follicular growth, whereas local (95) or circulating IGF-1 may act on the CL to promote steroidogenesis (Fig. 2). IGF-2 is minimally expressed in the murine ovary, primarily in perifollicular blood vessels (100), and *IGF2* deletion does not impair fertility in the mouse. IGF-2 is much less abundant in the circulation and tissues postnatally in murine species than in other mammals, where IGF-2 is significantly more abundant than IGF-1 (67).

The large literature on IGF-1 actions on murine granulosa and theca cells in vitro has been extensively reviewed (92,101,102). This chapter focuses on more recent in vivo findings illuminated by targeted gene deletion mice. *IGF1* gene deletion mice have small ovaries, consistent with their dwarf size; these ovaries appear anatomically normal in the prepubertal mouse, with a full complement of primordial follicles, as determined for independent deletions in different mouse lines (98,103). Thus, IGF-1 apparently is not essential for the normal development or differentiation of the ovary. This finding is in contrast to the observation of significantly reduced follicle numbers in the GHR null mouse (22,23). IGF-1 null mice are infertile owing to impaired follicular growth, because although primordial follicle numbers are normal, few antral follicles are evident, no pregnancies occur, and no CL are seen (98,103). Moreover, IGF-1 null mice do not ovulate in response to gonadotropin hyperstimulation (103). Granulosa cell proliferation (104) and expression of glucose transporter, FSH receptor, and aromatase gene are all significantly reduced in the IGF-1 null ovary (98,105), and administration of IGF-1 systemically restored their expression to wild-type levels, suggesting that the reduced gonadotropin responsiveness of these mice may be at least partially explained by reduced FSH receptor expression. In fact, the IGF-1 null ovarian phenotype appears very similar to that of FSH receptor null mice (106), with a block at the late primary or early antral stage of development. Based on these observations, it has been proposed that local IGF-1 production and reception in the murine ovarian follicle promotes an amplification cycle, as illustrated in Fig. 3. Local IGF-1 produced by granulosa cells in small growing follicles enhances cell metabolic function, proliferation, and FSH receptor expression and thus follicle response to FSH. FSH may enhance IGF-1 receptor expression directly, or

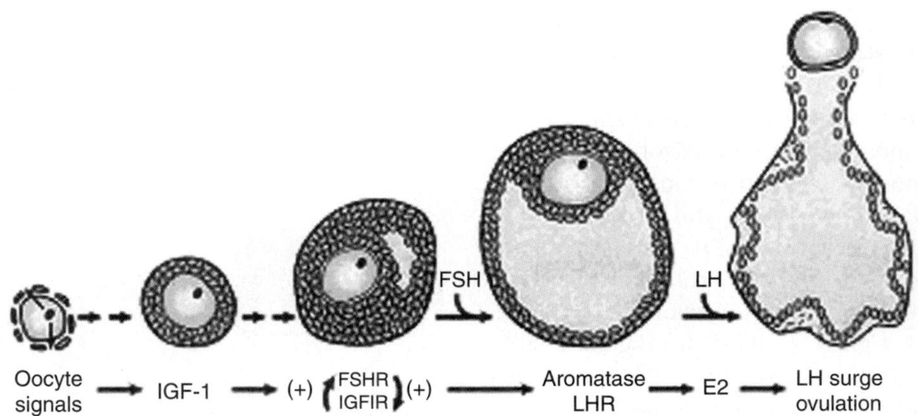

FIG. 3. Schematic illustrating insulin-like growth factor (IGF)-1's proposed role in follicular development. Oocyte autonomous signals are hypothesized to initiate granulosa cell proliferation and IGF-1 production. IGF-1 augments follicle-stimulating hormone receptor (FSHR) expression and FSH augments both IGF-1 receptor (IGFIR) and FSHR expression. This mutual complementary positive feedback loop in the follicle is hypothesized to be critical for the amplification of FSH action to induce the formation of multiple mature graafian follicles. In the absence of IGF-1 or FSH, follicles arrest at a prenatal/early antral stage of development. Maximal FSH action leads to mature antrum formation and granulosa cell aromatase and luteinizing hormone receptor (LHR) expression. The resulting peak in follicular estradiol (E2) synthesis stimulates an LH surge, which in turn stimulates ovulation. For the sake of simplicity, thecal layer development, which is also impaired in the IGF-1 knockout mouse, has been omitted from the diagram. (From Zhou, J., Kumar, T. R., Matzuk, M. M., and Bondy, C. [1997]. Insulin-like growth factor I regulates gonadotropin responsiveness in the murine ovary. *Mol. Endocrinol.* 11, 1924–1933.)

indirectly through enhanced estradiol production (107), creating a positive feedback loop allowing rapid follicular growth. Unfortunately, it is not possible to elucidate IGF-1's role in luteal function in the IGF-1 null mouse because in the absence of ovulation, no CL are present.

There has been substantial progress in elucidation of the specific molecular pathways engaged by IGF-1 and FSH receptors in the murine granulosa cell (107,108). Richards and coworkers have identified major signaling components of IGF-1 receptor action in the ovary, including the PI3K cascade, Akt/PKB, the serum- and glucocorticoid-inducible kinase SGK1, and members of the forkhead transcription factor family. These specific kinases are all involved in insulin- and IGF-induced anabolic effects, including glucose and amino acid transport and protein synthesis. They have also demonstrated that estradiol participates in the growth amplification cycle we outlined previously by augmenting IGF-1 receptor expression. Because IGF-1 deletion blocks follicle development at the preovulatory stage, the IGF-1 null mouse is not informative about IGF-1's role in luteal function. As noted earlier, the high level IGF-1 receptor expression in steroid-synthesizing luteal cells correlated with high levels of glucose metabolism (Fig. 2), suggesting that IGF-1 receptor activation may support the bioenergetic requirements of steroidogenesis, perhaps by enhancing cellular stores of adenosine triphosphate or reducing equivalents.

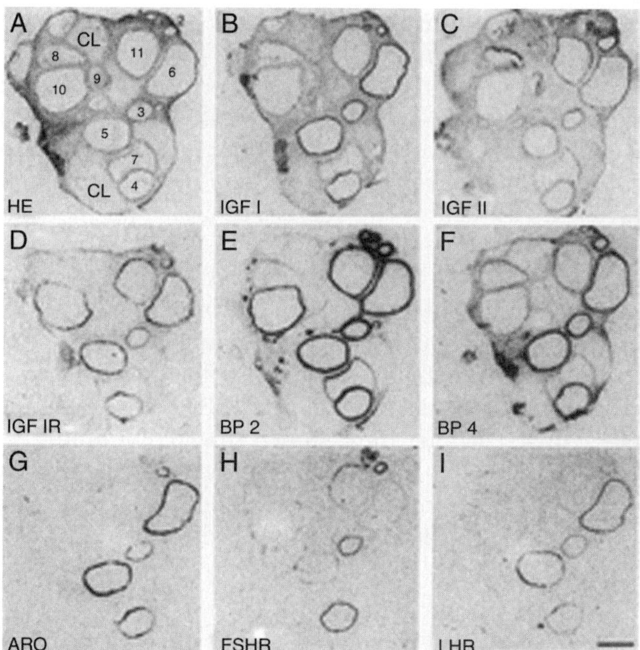

FIG. 4. Insulin-like growth factor (IGF) system gene expression in the mid-follicular (day 16) sow ovary. **A:** Hematoxylin and eosin–stained section. **B–I:** Film autoradiographs of sequential sections hybridized to antisense RNA probes for the indicated mRNAs. Follicles 1 to 6 are healthy growing follicles of various sizes. Follicles 7 to 9 are clearly atretic as demonstrated by shredding of the membrana granulosa and by loss of the follicle's oval shape. Follicles 10 and 11 appear healthy by morphological criteria but are presumed to be destined for atresia based on absence of gonadotropin receptor or aromatase mRNA. IGFIR, IGF-1 receptor; ARO, aromatase; FSHR, FSH receptor; LHR, LH receptor; BP, binding protein; HE, hematoxylin and eosin. Scale bar=2.5 mm. (From Zhou, J., Adesanya, O. O., Vatzias, G., Hammond, J. M., and Bondy, C. A. [1996]. Selective expression of insulin-like growth factor system components during porcine ovary follicular selection. *Endocrinology* 137, 4893–4901.)

Farm Animals

The IGF system in ovarian follicular development and selection has been extensively investigated in domestic animals, as reviewed be Mazerbourg and colleagues (109). Interestingly, the sow is the only large animal that selectively expresses IGF-1 in growing follicles, similar to the pattern in rats and mice (110,111). Swine are similar to the murine species in being polytocous, in contrast to cattle, sheep, and primates. IGF-1 and IGF-1 receptor are selectively concentrated in healthy graafian follicles, characterized by expression of aromatase and the gonadotropin receptors, in the sow ovary (Fig. 4). GH and gonadotropins are reported to increase follicular IGF-1 in the sow (29,112,113), although this effect may be due to transudation of circulating IGF-1 into the follicular fluid. However, treatment of cultured porcine granulosa cells with GH or FSH was found to increase IGF-1 mRNA levels (43). IGF-1 expression does not characterize the emergence of ovulatory follicles in any of the other species that have been examined to date, although increasing IGF-1 content, presumably derived from the circulation, is described for dominant follicle formation in cattle (114). IGF-2 is expressed in the theca-interstitial compartment of porcine, bovine, ovine, and primate ovaries and is abundant in the fluid obtained from large follicles of these species, as is IGF-1, apparently derived from the systemic circulation (109,115). Considerable evidence suggests that circulating IGF-1 plays a role in follicular growth in domestic animals, because reduction in GH secretion and hence IGF-1 levels leads to reduced follicular development, whereas increased IGF-1 levels enhance follicular growth (14,116,117). This observation may be relevant to the nutritional regulation of reproduction in these animals, in which nutritional deficiencies are associated with reduced circulating IGF-1 levels and reduced follicular maturation/ovulation (117,118).

The role of the IGF-1 system in CL development and function in livestock has also been investigated. In murine species, the IGF-1 receptor is highly expressed in the CL, but little IGF-1 or 2 is detected.

In the swine, however, both IGF-1 and its cognate receptor are highly expressed (111,119), suggesting that IGF-1 may play a paracrine role in porcine luteal function. Indeed, several lines of evidence suggest that the IGF system has a luteotrophic role in cattle and sheep as well as pigs (30,120–122). In general, IGF-1 and the IGF-1 receptor appear most abundant during the first half of the luteal phase, suggesting a role in CL growth and steroidogenesis. In vitro studies suggest that IGF-1 promotes steroidogenesis most robustly during early stages of swine luteal development, and that its effects on steroidogenesis require activation of PI3K (123). Further mechanistic studies have shown that IGF-1 amplifies gonadotropin-induced steroidogenic acute regulatory protein (StAR) and low-density lipoprotein receptor expression in luteinized granulosa cells in (124,125).

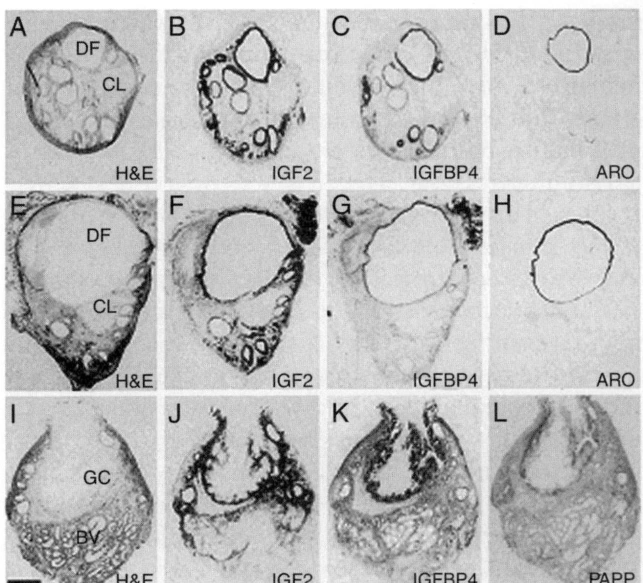

FIG. 5. Insulin-like growth factor (IGF)-2 and IGF-binding protein 4 (IGFBP4) gene expression in normal mid-cycle (days 13 to 14) rhesus monkey ovaries. On the extreme left of each row are hematoxylin and eosin–stained ovary sections, and the subsequent panels show film autoradiographs of sequential sections hybridized to cRNA probes for IGF-2, IGFBP4, and aromatase, or in **L**, the IGFBP4 protease, pregnancy-associated protein-A (PAPP-A). A single dominant follicle (DF), defined by maximum cross-sectional diameter >2 mm and aromatase expression, is apparent in each ovary. **A–D:** The emerging DF has the highest aromatase expression, but other, smaller follicles still have discernible aromatase and IGFBP4 expression (**C**, **D**). **E–H:** In the more advanced DF, aromatase and IGFBP4 expression are restricted to the DF, whereas IGF-2 mRNA is still evident in numerous smaller follicles. **I:** The DF is in the process of ovulating. **J–L:** IGF-2, IGFBP4, and PAPP-A mRNAs are concentrated in the exfoliating membrana granulosa. This ovary is cut through the ovarian pedicle and the profusion of engorged blood vessels (BV) is quite impressive. CL, corpus luteum. Scale bar = 0.8 mm. (From Zhou, J., Wang, J., Penny, D., Monget, P., Arraztoa, J. A., Fogelson, L. J., and Bondy, C. A. [2003]. Insulin-like growth factor binding protein 4 expression parallels luteinizing hormone receptor expression and follicular luteinization in the primate ovary. *Biol. Reprod.* 69, 22–29.)

Primates

IGF system expression is similar in the nonhuman primate (rhesus macaque) and human ovary (100,126–130). As reported for monotocous farm animals, IGF-1 expression is not normally detected in growing follicles in the primate ovary, whereas IGF-2 is expressed in the theca-interstitial compartment without apparent selectivity for follicle or cycle stage (Fig. 5B) (127–129). IGF-2 is also expressed in the membrana granulosa of large preovulatory and ovulatory follicles in the monkey (Fig. 5J). It is not clear whether this IGF-2 is expressed by granulosa cells or by invading stromal elements in the periovulatory follicle, however. As in all other species described, the IGF-1 receptor is abundantly expressed in oocytes, the membrana granulosa of graafian follicles and CL, and to a lesser extent thecal-interstitial compartments (127,128,131).

The human ovary is much more difficult to study because specimens are not available from healthy, naturally cycling women. Surgically obtained specimens are from women with gynecological disorders, with most having been on ovarian suppressive regimens before surgery. Autopsy material has the limitation that the cycle normality and stage are usually unknown. However, data pooled from both types of subjects suggest that IGF-1 mRNA is very scarce in the human ovary, perhaps localized in some stromal interstitial cells (127,128), but not in the steroidogenic apparatus. As in other species, IGF-1 receptor expression is abundant in human ovary oocytes, granulosa cells, and CL cells. IGF-1 receptor abundance in oocytes is interesting, because apparently IGF receptor activation promotes oocyte maturation (132). IGF-2 mRNA is abundant in the theca-interstitial compartment of most follicles, most likely primarily in the vascular tissue (127,128,133). As in the monkey, IGF-2 is found in the granulosa of preovulatory follicles (127,128), but also in the granulosa of some apparently atretic follicles (128). IGF-2 mRNA is detected in granulosa and granulosa luteal cells harvested from women undergoing gonadotropin stimulation for assisted reproduction (134), and IGF-2 peptide is secreted from these cells as well (134,135).

The regulation of follicular IGF-2 expression is not well understood. One study found that gonadotropins and prolactin increased IGF-2 peptide release (134), whereas another found that gonadotropins but not GH or prolactin increased IGF-2 mRNA in cultured granulosa cells (135). With the caveat that these observations are from cells cultured for 18 to 20 generations in an artificial medium and absent the normal follicular microstructure, it seems that granulosal IGF-2 may be regulated by gonadotropins in the human. Consistent with this notion, IGF-2 concentration is highest in the follicular fluid from "estrogen-dominant" and larger preovulatory follicles (136), suggesting that healthy, preovulatory follicles actively synthesize IGF-2 in vivo.

The effects of IGF-1 and IGF-2 on human ovarian cells in vitro have been comprehensively reviewed (137). In general, these effects are similar to those described for other species, with effects on proliferation and steroidogenesis dependent on degree of cell differentiation and culture conditions. There has been limited study of signaling pathways used by IGFs in human ovary cells. In vitro studies on granulosa lutein cells obtained from women after ovarian hyperstimulation show that IGF-1 amplifies StAR expression in (138) and promotes progesterone production through a mitogen-activated protein kinase (MAPK)–independent pathway (139). Studies on CL slices show that in addition to enhancing steroid production, IGF-1 also inhibits apoptosis (140).

Insulin-Like Growth Factors, Insulin, and Ovarian Hyperandrogenism

A connection between insulin and ovarian hyperandrogenism was first described in young women with insulin receptor defects or insulin receptor autoantibodies producing extreme hyperinsulinism and marked virilization (141,142). The association of hyperinsulinism and virilization led to the hypothesis that insulin, at high levels, might have direct effects on ovarian steroidogenesis, either through insulin receptors, IGF-1 receptors, or hybrid insulin–IGF receptors (143,144). Supporting this view, the insulin receptor is expressed in the human ovary in a pattern similar to the IGF-1 receptor (128,145), and insulin is able to stimulate steroidogenesis in human ovarian tissue (146). There have been many investigations of insulin's effects on the ovary, especially as related to the polycystic ovary syndrome (PCOS), resulting in a large body of literature pertaining to this clinically important subject. This work has been comprehensively reviewed (137) and is outside the scope of this chapter, which maintains a focus on IGFs and the ovary.

Many different theories have been proposed to explain the development of PCOS, or "chronic ovarian hyperandrogenism" (147–152). There is a basic controversy over whether the primary defect is metabolic [i.e., hyperinsulinism secondary to insulin resistance causing ovarian hyperandrogenism (150)], or whether the primary defect is overproduction of androgen by adrenal gland or ovary, leading secondarily to visceral adiposity, insulin resistance, and so forth (151,153,154). There is solid evidence supporting both these etiologies, and it seems quite likely that, in fact, there is more than one path to this common and heterogeneous disorder. In the case of the woman with extreme insulin resistance, the metabolic defect is clearly primary. Presumably the elevated insulin levels act on the ovary IGF-1 receptor (because their insulin receptors are nonfunctional or ab-blocked) to promote excess androgen production, which in turn interrupts normal hypothalamic–pituitary–ovarian regulation and ovulation. Also, in the more typical, obesity-related insulin resistance syndromes, weight loss or insulin-reducing medications such as metformin are able to reduce androgen levels in many women (155). Evidence for a primary role for excess androgen production is found in cases where androgen-producing tumors, congenital adrenal hyperplasia, and exogenous androgen treatment are all associated with the development of the hypertrophic PCOS phenotype (156–159).

The mechanism whereby excess androgen exposure leads to chronic ovarian hyperandrogenism is unclear. Because LH secretion is abnormal in chronic hyperandrogenism, a primary effect on the hypothalamic–pituitary axis has been invoked (153,159). Exposure of primates and sheep to large androgen excess prenatally is thought to "reprogram" the hypothalamic–pituitary unit in such a manner as to lead to chronic anovulation (153,160), although both species appear to have hypertrophic ovary tissue that is hyperresponsive to gonadotropin stimulation, similar to women with PCOS (161).

Given the well-known hyperresponsiveness of the hypertrophic polycystic ovary to gonadotropin stimulation (161) and the effective treatment of the syndrome in many cases by ovarian wedge resection or cautery (162), it seems possible that androgens may promote ovarian growth and function. We investigated the effects of androgen treatment in vivo to determine if androgens stimulate ovarian growth factor production or reception, and hence growth (163). Although IGF-1 is usually below detection limits in the normal primate ovary, increased IGF-1 and IGF-1 receptor expression are found in ovaries from androgen-treated monkeys (164). Small preantral and antral follicles were significantly and progressively increased in number and thecal layer thickness in testosterone- and dihydrotestosterone-treated monkeys over 3 to 10 days of treatment. Granulosa and thecal cell proliferation are significantly increased in these follicles. Androgens appear to boost the recruitment of resting, primordial follicles into the actively growing pool because the number of primary follicles significantly increased over time during androgen treatment (165). Androgen treatment resulted in a threefold increase in IGF-1 and a fivefold increase in IGF-1 receptor mRNA in primordial follicle oocytes

(165), suggesting that androgens may promote initiation of primordial follicle growth by augmenting IGF-1 activity in the primordial follicle. In addition, androgen treatment results in significant, three- to fourfold increases in IGF-1 and IGF-1 receptor expression in granulosa, thecal, and interstitial compartments (Fig. 6). Interestingly, FSH receptor expression increased in parallel with IGF-1 receptor in these androgen-treated animals (166) (Fig. 7), suggesting that in primates as in rodents, IGFs promote FSH receptor expression and thus FSH responsiveness. Follicular atresia was not increased, and there were actually significantly fewer apoptotic granulosa cells in the androgen-treated monkeys (163). These findings suggest that androgens enhance IGF-1 and IGF-1 receptor expression along with follicular growth and survival in the primate ovary, implicating IGF-1 in androgen-induced ovarian follicular and thecal-interstitial growth.

The mechanism whereby androgens promote primordial follicle development may involve activation of oocyte IGF-1 signaling. The IGF-1 receptor is expressed in oocytes from all species investigated (41,93,111,128), and IGF-1 stimulates oocyte metabolic activity and maturation in vitro (167,168). Thus, enhanced IGF-1 signaling through increased expression of both the peptide and its receptor in oocytes from androgen-treated monkeys may trigger oocyte activation and initiation of follicle growth. These observations provide a plausible explanation for the pathogenesis of "polycystic" ovaries in hyperandrogenism (Fig. 8). Pathological studies show increased numbers of growing follicles, including primary follicles, in ovaries from hyperandrogenic women with PCOS (169). Increased follicle numbers are also seen in testosterone-treated women (157,170), showing that androgens, whether derived from ovarian, adrenal, or exogenous sources, may stimulate excessive ovarian follicular growth. Further support for this notion is provided by the observation that treatment of women with PCOS with androgen receptor blockade results in diminution of follicle numbers (171).

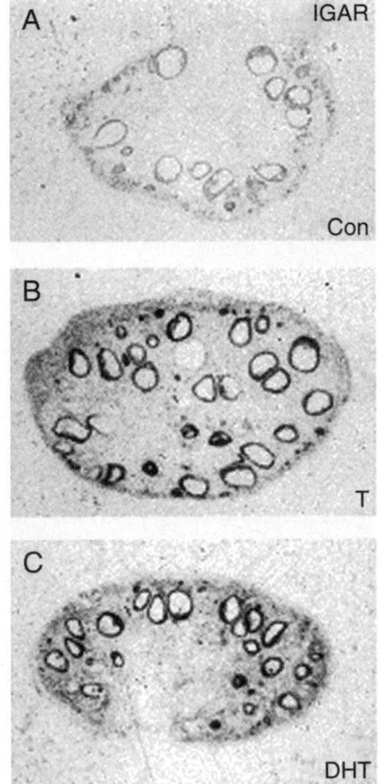

FIG. 6. Androgens increase insulin-like growth factor (IGF)-1 receptor expression in the monkey ovary. Representative film autoradiographs illustrate in situ hybridization results from control, placebo-treated (**A**), testosterone-treated (**B**), and dihydrotestosterone (DHT)-treated (**C**) monkeys. Androgen treatment increases the number of small follicles and increases IGF-1 receptor mRNA levels in the membrana granulosa of these follicles and in the stroma. (From Vendola, K., Zhou, J., Wang, J., and Bondy, C. A. [1999]. Androgens promote insulin-like growth factor-I and insulin-like growth factor-I receptor gene expression in the primate ovary. *Hum. Reprod.* 14, 2328–2332.)

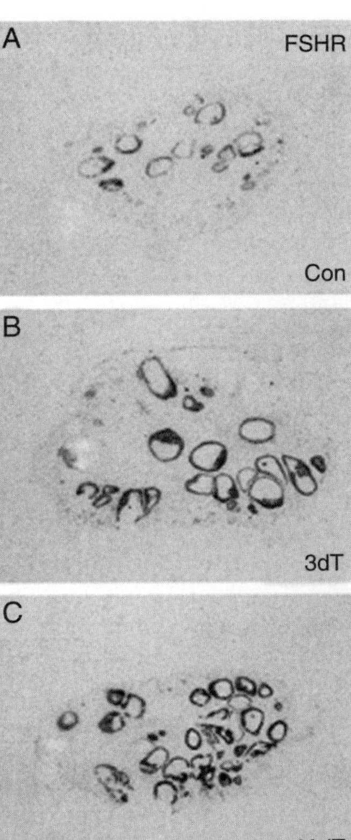

FIG. 7. Androgen treatment increases follicle-stimulating hormone (FSH) receptor gene expression in the primate ovary. In parallel with increased insulin-like growth factor (IGF)-1 and IGF-1 receptor expression, FSH receptor mRNA is increased in the follicles from androgen-treated monkeys. Shown are representative film autoradiographs from placebo-treated (**A**), 3-day testosterone (**B**), and 10-day testosterone treated animals (**C**). (From Weil, S., Vendola, K., Zhou, J., and Bondy, C. A. [1999]. Androgen and follicle-stimulating hormone interactions in primate ovarian follicle development. *J. Clin. Endocrinol. Metab.* 84, 2951–2956.)

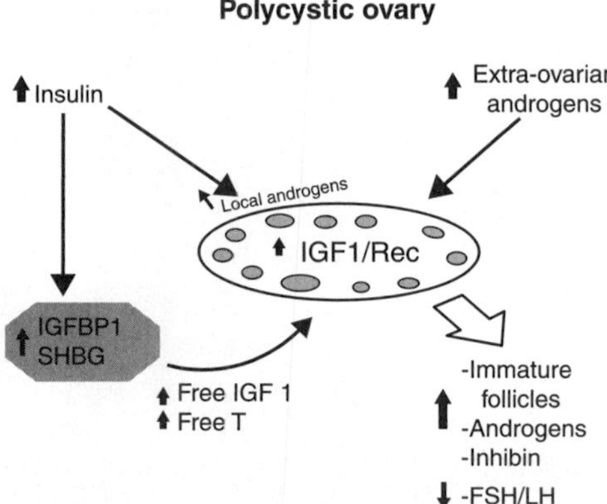

FIG. 8. Potential routes to the polycystic ovary phenotype. Increased circulating insulin secondary to adiposity or "extreme" insulin resistance may have direct effects on the ovary to promote follicular growth and androgen production, and also suppresses hepatic production of sex hormone binding globulin (SHBG) and insulin-like growth factor–binding protein 1 (IGFBP1), resulting in increased free androgen and IGF-1 levels in the circulation, which may then act on the ovary to promote follicular growth and androgen production. High circulating androgen derived from adrenal or exogenous sources stimulates the ovary directly, promoting early follicle growth and stromal hypertrophy through enhanced local IGF-1/IGF-1 receptor expression. By either pathway, increased numbers of immature follicles produce androgen and inhibin, interfering with gonadotropin dynamics.

In summary, although very little IGF-1 is detected in the normal primate ovary, both IGF-1 and IGF-1 receptor expression are increased by exposure to elevated androgen levels. In the model for generation of PCOS illustrated in Fig. 7, the polycystic ovarian phenotype may result from at least two different pathways. Excess systemic androgen levels, whether from genetically determined androgen overproduction by the ovary or adrenal, or from tumors or exogenous administration, may stimulate increased follicle numbers and stromal growth, presumably acting through augmented IGF-1 and IGF-1 receptor expression. An alternative pathway to PCOS involves hyperinsulinism. High insulin levels may act directly on the ovary (172) to increase androgen production. Moreover, high insulin levels increase bioavailable IGF-1 and testosterone by reducing the hepatic production of IGFBP1 and sex hormone–binding globulin, so that increases in circulating free testosterone and free IGF-1 may also stimulate ovarian growth (173–176). Increased follicle numbers and stromal tissue result in increased production of inhibin and androgen, suppressing gonadotropins and impairing cyclic function. This scenario may explain why some women with PCOS respond to gonadotropins with hyperstimulation, because they have increased numbers of growing follicles and enhanced FSH responsiveness. It also helps explain why wedge resection of the ovary may normalize cycles in women with this ovarian phenotype (177)—presumably reduction in follicle number and stromal volume reduces excessive inhibin and androgen production, allowing resumption of normal gonadotropin entrainment of the ovary. These findings are also consistent with the dynamic studies of gonadotropin responsiveness in women with PCOS (154,161).

Insulin-Like Growth Factors and Ovarian Cancer

Both IGF-1 and 2 and the IGF-1 receptor have been implicated in many types of neoplasia (72). Both IGFs and their cognate receptor are detected in ovarian tumor tissue and cell lines (178). IGF-1 levels were found to be significantly higher in the fluid obtained from malignant compared with benign ovarian cysts (179). In a large, nested, case–control, multicenter, prospective study, circulating IGF-1 levels were directly associated with ovarian cancer risk in women younger than 55 years of age (odds ratio, 4.97; 95% confidence interval, 1.22–20.2) (180). Finally, several small studies have reported loss of imprinting for IGF-2 gene expression in ovarian tumors (181–183). Given that androgens stimulate ovarian IGF-1 and IGF-1 receptor expression and ovarian growth, it might be expected that hyperandrogenism as in PCOS would be associated with increased rates of ovarian cancer, and at least one epidemiological study supports such an association (184).

Insulin-Like Growth Factor–Binding Proteins and the Ovary

Complex interactions with a large family of IGFBPs (IGFBP1 to 6) may modulate local IGF effects (84–87). These proteins bind IGF-1 and 2 with high affinity, regulating their transport and presentation to the receptor and prolonging their half-life in the circulation and tissues. There have been many studies on the expression patterns, in vitro secretion, proteolysis, regulation, and follicular fluid concentrations of these IGFBPs in humans and many of the larger species that have been extensively and thoroughly reviewed elsewhere (92,99,137,185–188). Although the IGFs and IGF-1 receptor display similar ovarian expression patterns in closely related species such as the mouse and rat, IGFBP2 through 5 are expressed in diverse patterns in the rat and mouse ovary (93,94,189). For example, IGFBP2 expression is confined to the ovarian stroma in the rat (93,189), but is expressed by granulosa cells in healthy follicles of mice (93,94). IGFBP4 is found in atretic follicles in both rat and mouse (94,189), whereas IGFBP5 is reportedly localized in atretic follicles in rats (189,190) but not in mice (94). Based on in vitro studies showing that IGFBPs inhibit IGF-1's effects, it was suggested that follicular IGFBP expression leads to atresia. However, mice with targeted gene deletions for IGFBP2 through 5 display no obvious reproductive phenotype (191). Given the variability in expression patterns and apparent insensitivity to deletion, it seems most likely that IGFBPs play dispensable, modulatory roles in murine ovarian function that are compensated by other IGFBPs or that are too subtle to be easily detected in the knockout models.

In larger animals and humans, there is also considerable variability in the ovarian synthesis as well as circulating levels of IGFBPs. For example, equine granulosa cells produce only IGFBP2 and 5 (192), whereas porcine granulosa produce IGFBP1 through 5 (193). Furthermore, physiological regulation of ovarian IGFBP expression may be species specific (115). IGFBP levels have been intensively investigated in follicular fluid, mainly by ligand-binding studies. IGFBP levels generally are reported to decrease in dominant follicles and increase in atretic follicles, although IGFBP4 levels increase markedly in ovulatory follicles in the sow (110). IGFBP4 is concentrated in the theca interna of healthy, medium to large, growing follicles, characterized by IGF-1, aromatase, and gonadotropin receptor expression, but is not detected in atretic follicles in the sow ovary (Fig. 4). IGFBP4 appears in granulosa cells of periovulatory follicles showing evidence of luteinization and LH receptor expression, consistent with the increase in IGFBP4 levels found in follicular fluid of ovulatory follicles (110). IGFBP4 is also abundant in the CL of the sow (111) and most other animals. IGFBP3 is also present in the porcine granulosa cells, and appears to be regulated by FSH in a biphasic manner (194).

In the rhesus monkey, IGFBP4 expression is selective for LH receptor–expressing, steroidogenic thecal and granulosa cells of healthy follicles (Fig. 4) and enhanced by the LH analog human chorionic gonadotropin (hCG; Fig. 9). LH/hCG also increases IGFBP4 expression in the bovine and murine ovary (94,128,195). The observation that IGFBP4 gene expression is selective for healthy follicles is supported by demonstration of thecal and granulosa cell proliferation (129), in addition to selective LH and FSH receptor and aromatase expression in these follicles (129) (Fig. 4). IGFBP4 is also expressed by healthy theca interna and CL in human ovaries (127,128). Thus, LH/hCG stimulates ovarian IGFBP4 expression in rodents, large animals, and primates, and IGFBP4 expression closely parallels LHR expression in healthy developing follicles and CL. These observations suggest a role for IGFBP4 in LH-induced steroidogenesis or luteinization.

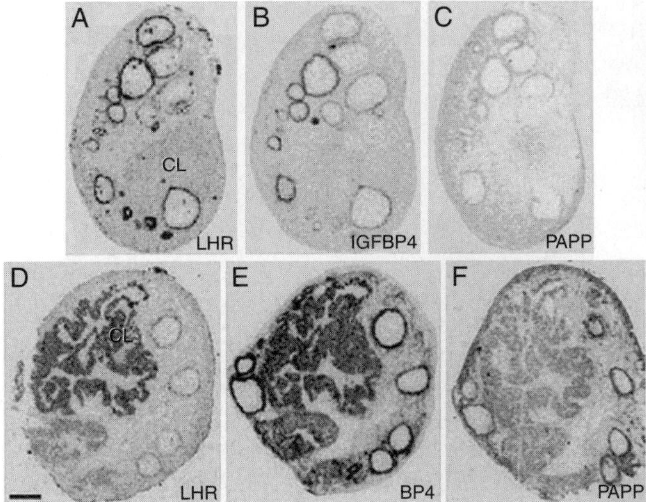

FIG. 9. Human chorionic gonadotropin (hCG) increases insulin-like growth factor–binding protein 4 (IGFBP4) gene expression in the primate ovary. Representative film autoradiographs comparing luteinizing hormone receptor (LHR), IGFBP4, and pregnancy-associated protein-A (PAPP-A) mRNA localization in sequential sections from an untreated mid-cycle monkey ovary (**A–C**) and from an hCG-treated mid-cycle monkey ovary (**D–F**). Both are nondominant ovaries with corpora lutea (CL) from the previous cycle. IGFBP4 and PAPP-A mRNAs are selectively localized in LHR-expressing follicles, but little IGFBP4 or PAPP-A mRNA is detected in the CL of the untreated monkey ovary (as also seen in Fig. 1C). After hCG treatment, however, IGFBP4 and PAPP-A expression are abundant in the CL, and increased also in the antral follicles (**E, F**). There are two small scars of old CL at the lower pole of the ovary that also demonstrate heightened IGFBP4 and PAPP-A expression. Bar = 1 mm. (From Zhou, J., Wang, J., Penny, D., Monget, P., Arraztoa, J. A., Fogelson, L. J., and Bondy, C. A. [2003]. Insulin-like growth factor binding protein 4 expression parallels luteinizing hormone receptor expression and follicular luteinization in the primate ovary. *Biol. Reprod.* 69, 22–29.)

The monkey and human ovary are replete with IGFBPs in addition to IGFBP4, although none of the others demonstrates an obvious connection with follicular development or sensitivity to gonadotropin regulation (127,128,130). IGFBP1 is detected in theca-interstitial cells and at low levels in granulosa cells of atretic follicles in the rhesus ovary (130). IGFBP2 is abundant in the ovarian surface epithelium and in granulosa cells of all antral follicles, including obviously atretic as well as dominant follicles (130). IGFBP3 is localized in oocytes, in the ovarian vascular endothelium, and the superficial cortical stroma—where it is distinctly more abundant in the nondominant ovary (130). IGFBP5 is selectively expressed by granulosa cells of mature ovulatory follicles in the monkey ovary (130). IGFBP5 is also widely expressed in the ovarian stroma, where, in contrast to IGFBP3, it is distinctly more abundant in the dominant than in the nondominant ovary. IGFBP6 mRNA is detected at low levels in the ovary interstitium and theca externa, and in the ovary surface epithelium.

Although most intensive study has been directed at IGFBPs in developing follicles, these proteins also are abundant in the CL of most species (30,33,120, 129,130,196–199).

Insulin-Like Growth Factor–Binding Protein Proteolysis, Follicular Selection, and Luteolysis

A widely held view on IGF effects on follicular selection involves the idea that selective IGFBP proteolysis may promote formation of ovulatory follicles by increasing “free” or bioavailable IGFs (101,186,195). This theory is based in large part on findings of IGFBP fragments and increased free IGF-1 in the follicular fluid of dominant follicles of cattle. IGFBP4 has been intensely studied in this context. One enzyme implicated in IGFBP4 proteolysis in follicular fluid is pregnancy-associated protein-A (PAPP-A) (195,200). This enzyme displays proteolytic activity against IGFBP4 only in the presence of IGF-1 or 2 and is present in the circulation of pregnant women bound to a protein known as the proform of the major eosinophil basic protein (proMBP), which inhibits its proteolytic activity. The source and function of this protein complex in the circulation is unclear. The PAPP-A in ovarian follicular fluid is produced by granulosa cells and may be associated with proMBP (201). A complex series of inhibitory interactions has been proposed to explain the role of IGFBPs, in particular IGFBP4, in folliculogenesis (186,202,203). IGF activation of the IGF-1 receptor is thought to be inhibited by IGFBP4—until such time as an IGFBP4 protease such as PAPP-A, is unleashed from *its* inhibitor, allowing binding protein degradation. Although PAPP-A has received much attention, there are many other proteases and protease inhibitors found in follicular fluid, particularly around the time of ovulation. Whereas IGFBP4 is generally considered an exclusively inhibitory factor (204), IGFBP5 may promote IGF actions but it appears that this IGFBP, as well as IGFBP2, may also be targeted by PAPP-A follicular fluid (195,205).

This view of IGFBPs as inhibitors of follicular selection is derived primarily from studies of IGF and IGFBP dynamics in follicular fluid; however, interactions between IGFs and their receptor take place at the plasma membrane. In the developing follicle, IGF target cells exist in complex, interlocking microenvironments in the membrana granulosa and theca interna. IGFBPs produced by granulosa and thecal cells decorate cell membranes and extracellular matrix within these highly specialized compartments. Thus, the scene of action for IGFs produced by granulosa cells or entering the follicle from the bloodstream will be in the membrana granulosa or theca interna, where binding proteins may reside to protect IGFs from proteolysis, provide local storage sites, present the peptides to the receptor, or selectively inhibit their interaction with the receptor. It is not clear that the contents of the follicular fluid accurately reflect these interactions between IGF system components and cells of the functional follicular compartments. The increased proteolysis of IGFBPs noted in periovulatory follicles may simply reflect a general increase in protease activity as follicles prepare for rupture and transformation to CL. Moreover, it seems unlikely that IGFBP4's highly selective expression by steroidogenic cells in healthy follicles and CL (at least in swine and monkeys), and regulation by LH, signify a role in follicular demise. Nevertheless, given the convincing correlations between accumulation of free IGF and loss of intact IGFBPs in dominant follicle fluid from well-studied species such as cattle, it seems that there are important species differences in the mechanisms used for emergence of ovulatory follicles, and that species that depend more on circulating than locally produced IGF for follicular growth may control emergence of a dominant follicle through local regulation of IGFBP and protease production.

The role of IGFBPs in luteolysis has received less study, but the hypothesis is similar to that for follicular selection. The notion is that IGF-1 or 2 promotes the rapid CL growth characteristic of early luteal development (typically a greater than 15-fold increase in size) and steroidogenesis during later development (33). Thus, presence of inhibitory binding proteins may promote luteolysis by suppressing IGF's anabolic activity. Increased expression of IGFBP1 was noted during induced luteolysis in the bovine ovary (120,197); IGFBP5 mRNA levels also increased, whereas IGFBPs 3 and 4 were decreased in this model of luteolysis (120). At this point, further study is required to clarify the role of IGFs and IGFBPs in luteal formation, function, and dissolution.

SUMMARY OF GROWTH HORMONE/INSULIN-LIKE GROWTH FACTORS AND OVARIAN FUNCTION

Targeted gene deletion mice have been very helpful to understanding the roles of GH and IGF-1 in murine ovarian function. The GHR null mouse is fertile, but ovarian follicle numbers, ovulation rate, and litter size are reduced (20,22,23). It is not known whether the reduced follicle number is due to decreased recruitment of new follicles or decreased survival of young follicles, although there appears to be no evidence of increased follicular atresia (23), suggesting the former mechanism. The possibility that GH increases the follicle endowment is intriguing in light of recent findings suggesting there is a dynamic stem cell population in the ovary apparently capable of supplying new primordial follicles in mature mice (206). Such an action in the ovary would be consistent with GH's apparent amplification of the supply of epiphysial growth plate stem cells (3–5). Normal development and ovulation of a smaller number of follicles in the GHR null mouse may be supported by normal follicular IGF-1 expression (22). In contrast to the GHR null phenotype, IGF-1 deletion results in sterility despite a normal endowment of ovarian follicles (98,103). IGF-1 null follicles do not progress past the late primary stage and are resistant to gonadotropin treatment (103). IGF-1 null granulosa cells exhibit reduced proliferation and decreased glucose transporter, FSH receptor, and aromatase expression (98,104,105). The comparison of GHR null and IGF-1 null phenotypes in the ovary poses an interesting analogy with the epiphysial growth plate. In the GHR null growth plate, there are reduced numbers of chondrocyte precursors and chondrocytes (4), whereas in the IGF-1 null growth plate, chondrocyte numbers are preserved but their growth rate is severely compromised (90).

Although the elucidation of roles for GH and IGF-1 in murine ovarian function has been very interesting, it is not certain that the murine findings can be extrapolated to other species. The mechanisms of reproduction have diverged radically during evolution, particularly in intensively bred laboratory mice, so caution is essential in applying observations from the mouse to different species. Women with GHR deletion, or Laron's syndrome, may be fertile (21,24), but it is possible that the pool of ovarian follicles is reduced in these women. Thus, the role of GH in regulation of ovarian follicle supply in species other than the rodent deserves further study. Observations from murine models on the role of IGF-1 in follicular development may not apply to species, such as the human, that do not normally express IGF-1 during follicle development. Although local sources of IGF peptides and IGFBP expression patterns in the ovary vary considerably between species, IGF-1 receptor expression is quite constant. This receptor is most highly expressed in growing follicles and in CL, in parallel with the level of metabolic activity in these structures, as reflected in glucose utilization. So whatever the source of the IGF ligand, and however the ligands are affected by the IGFBPs, the ultimate effect of IGF-1 receptor activation is to promote anabolic processes required to support rapid growth and steroid biosynthesis of graafian follicles and CL. Interestingly, IGF-2 serves a similar role in placental and fetal development, controlling the placental supply of nutrients available for fetal growth, regardless of the maternal nutritional state. Thus, the IGF family appears to be important in nutrient partitioning, supporting the extraordinary, energy-expensive processes of follicular growth, steroidogenesis, uterine hyperplasia and remodeling, and fetal growth.

REFERENCES

1. Tatar, M., Bartke, A., and Antebi, A. (2003). The endocrine regulation of aging by insulin-like signals. *Science* 299, 1346–1351.
2. Le Roith, D., Bondy, C., Yakar, S., Liu, J. L., and Butler, A. (2001). The somatomedin hypothesis: 2001. *Endocr. Rev.* 22, 53–74.
3. Isaksson, O. G., Ohlsson, C., Nilsson, A., Isgaard, J., and Lindahl, A. (1991). Regulation of cartilage growth by growth hormone and insulin-like growth factor I. *Pediatr. Nephrol.* 5, 451–453.
4. Wang, J., Zhou, J., Cheng, C. M., Kopchick, J. J., and Bondy, C. A. (2004). Evidence supporting dual, IGF-I-independent and IGF-I-dependent, roles for GH in promoting longitudinal bone growth. *J. Endocrinol.* 180, 247–255.
5. Hunziker, E. B., Wagner, J., and Zapf, J. (1994). Differential effects of insulin-like growth factor I and growth hormone on developmental stages of rat growth plate chondrocytes in vivo. *J. Clin. Invest.* 93, 1078–1086.
6. Herrington, J., and Carter-Su, C. (2001). Signaling pathways activated by the growth hormone receptor. *Trends. Endocrinol. Metab.* 12, 252–257.
7. Kajimura, S., Kawaguchi, N., Kaneko, T., Kawazoe, I., Hirano, T., Visitacion, N., Grau, E. G., and Aida, K. (2004). Identification of the growth hormone receptor in an advanced teleost, the tilapia (*Oreochromis mossambicus*) with special reference to its distinct expression pattern in the ovary. *J. Endocrinol.* 181, 65–76.
8. Heck, A., Metayer, S., Onagbesan, O. M., and Williams, J. (2003). mRNA expression of components of the IGF system and of GH and insulin receptors in ovaries of broiler breeder hens fed ad libitum or restricted from 4 to 16 weeks of age. *Domest. Anim. Endocrinol.* 25, 287–294.
9. Zhao, J., Taverne, M. A., van der Weijden, G. C., Bevers, M. M., and van den Hurk, R. (2002). Immunohistochemical localisation of growth hormone (GH), GH receptor (GHR), insulin-like growth factor I (IGF-I) and type I IGF-I receptor, and gene expression of GH and GHR in rat pre-antral follicles. *Zygote* 10, 85–94.
10. Carlsson, B., Nilsson, A., Isaksson, O. G., and Billig, H. (1993). Growth hormone-receptor messenger RNA in the rat ovary: regulation and localization. *Mol. Cell. Endocrinol.* 95, 59–66.
11. Kolle, S., Sinowatz, F., Boie, G., Lincoln, D., and Waters, M. J. (1997). Differential expression of the growth hormone receptor and its transcript in bovine uterus and placenta. *Mol. Cell. Endocrinol.* 131, 127–136.
12. Quesnel, H. (1999). Localization of binding sites for IGF-I, insulin and GH in the sow ovary. *J. Endocrinol.* 163, 363–372.
13. Eckery, D. C., Moeller, C. L., Nett, T. M., and Sawyer, H. R. (1997). Localization and quantification of binding sites for follicle-stimulating hormone, luteinizing hormone, growth hormone, and insulin-like growth factor I in sheep ovarian follicles. *Biol. Reprod.* 57, 507–513.
14. Cohick, W. S., Armstrong, J. D., Whitacre, M. D., Lucy, M. C., Harvey, R. W., and Campbell, R. M. (1996). Ovarian expression of insulin-like growth factor-I (IGF-I), IGF binding proteins,

and growth hormone (GH) receptor in heifers actively immunized against GH-releasing factors. *Endocrinology* 137, 1670–1677.

15. Lucy, M. C., Collier, R. J., Kitchell, M. L., Dibner, J. J., Hauser, S. D., and Krivi, G. G. (1993). Immunohistochemical and nucleic acid analysis of somatotropin receptor populations in the bovine ovary. *Biol. Reprod.* 48, 1219–1227.
16. Sharara, F. I., and Nieman, L. K. (1994). Identification and cellular localization of growth hormone receptor gene expression in the human ovary. *J. Clin. Endocrinol. Metab.* 79, 670–672.
17. Yuan, W., and Lucy, M. C. (1996). Effects of growth hormone, prolactin, insulin-like growth factors, and gonadotropins on progesterone secretion by porcine luteal cells. *J. Anim. Sci.* 74, 866–872.
18. Kolle, S., Sinowatz, F., Boie, G., and Lincoln, D. (1998). Developmental changes in the expression of the growth hormone receptor messenger ribonucleic acid and protein in the bovine ovary. *Biol. Reprod.* 59, 836–842.
19. Zhou, Y., Xu, B. C., Maheshwari, H. G., He, L., Reed, M., Lozykowski, M., Okada, S., Cataldo, L., Coschigamo, K., Wagner, T. E., Baumann, G., and Kopchick, J. J. (1997). A mammalian model for Laron syndrome produced by targeted disruption of the mouse growth hormone receptor/binding protein gene (the Laron mouse). *Proc. Natl. Acad. Sci. U S A* 94, 13215–13220.
20. Danilovich, N., Wernsing, D., Coschigano, K. T., Kopchick, J. J., and Bartke, A. (1999). Deficits in female reproductive function in GH-R-KO mice: role of IGF-I. *Endocrinology* 140, 2637–2640.
21. Laron, Z. (1999). Natural history of the classical form of primary growth hormone (GH) resistance (Laron syndrome). *J. Pediatr. Endocrinol. Metab.* 12[Suppl 1], 231–249.
22. Bachelot, A., Monget, P., Imbert-Bollore, P., Coshigano, K., Kopchick, J. J., Kelly, P. A., and Binart, N. (2002). Growth hormone is required for ovarian follicular growth. *Endocrinology* 143, 4104–4112.
23. Zaczek, D., Hammond, J., Suen, L., Wandji, S., Service, D., Bartke, A., Chandrashekar, V., Coschigano, K., and Kopchick, J. (2002). Impact of growth hormone resistance on female reproductive function: new insights from growth hormone receptor knockout mice. *Biol. Reprod.* 67, 1115–1124.
24. Chandrashekar, V., Zaczek, D., and Bartke, A. (2004). The consequences of altered somatotropic system on reproduction. *Biol. Reprod.* 71, 17–27.
25. Cecim, M., Kerr, J., and Bartke, A. (1995). Effects of bovine growth hormone (bGH) transgene expression or bGH treatment on reproductive functions in female mice. *Biol. Reprod.* 52, 1144–1148.
26. Liu, X., Andoh, K., Yokota, H., Kobayashi, J., Abe, Y., Yamada, K., Mizunuma, H., and Ibuki, Y. (1998). Effects of growth hormone, activin, and follistatin on the development of preantral follicle from immature female mice. *Endocrinology* 139, 2342–2347.
27. Eisenhauer, K. M., Chun, S. Y., Billig, H., and Hsueh, A. J. (1995). Growth hormone suppression of apoptosis in preovulatory rat follicles and partial neutralization by insulin-like growth factor binding protein. *Biol. Reprod.* 53, 13–20.
28. Jimenez-Krassel, F., Binelli, M., Tucker, H. A., and Ireland, J. J. (1999). Effect of long-term infusion with recombinant growth hormone-releasing factor and recombinant bovine somatotropin on development and function of dominant follicles and corpora lutea in Holstein cows. *J. Dairy Sci.* 82, 1917–1926.
29. Bryan, K. A., Hammond, J. M., Canning, S., Mondschein, J., Carbaugh, D. E., Clark, A. M., and Hagen, D. R. (1989). Reproductive and growth responses of gilts to exogenous porcine pituitary growth hormone. *J. Anim. Sci.* 67, 196–205.
30. Schams, D., and Berisha, B. (2004). Regulation of corpus luteum function in cattle: an overview. *Reprod. Domest. Anim.* 39, 241–251.
31. Lucy, M. C., Byatt, J. C., Curran, T. L., Curran, D. F., and Collier, R. J. (1994). Placental lactogen and somatotropin: hormone binding to the corpus luteum and effects on the growth and functions of the ovary in heifers. *Biol. Reprod.* 50, 1136–1144.
32. Juengel, J., Nett, T., Anthony, R., and Niswender, G. (1997). Effects of luteotrophic and luteolytic hormones on expression of mRNA encoding insulin-like growth factor I and growth hormone receptor in the ovine corpus luteum. *J. Reprod. Fertil.* 110, 291–298.
33. Niswender, G. D., Juengel, J. L., Silva, P. J., Rollyson, M. K., and McIntush, E. W. (2000). Mechanisms controlling the function and life span of the corpus luteum. *Physiol. Rev.* 80, 1–29.
34. Liebermann, J., and Schams, D. (1994). Actions of somatotrophin on oxytocin and progesterone release from the microdialysed bovine corpus luteum in vitro. *J. Endocrinol.* 143, 243–250.
35. Kobayashi, S., Miyamoto, A., Berisha, B., and Schams, D. (2001). Growth hormone, but not luteinizing hormone, acts with luteal peptides on prostaglandin F2α and progesterone secretion by bovine corpora lutea in vitro. *Prostaglandins Other Lipid Mediat.* 63, 79–92.
36. Ptak, A., Kajta, M., and Gregoraszczuk, E. L. (2004). Effect of growth hormone and insulin-like growth factor-I on spontaneous apoptosis in cultured luteal cells collected from early, mature, and regressing porcine corpora lutea. *Anim. Reprod. Sci.* 80, 267–279.
37. Forsyth, I. A., and Wallis, M. (2002). Growth hormone and prolactin: molecular and functional evolution. *J. Mammary Gland Biol. Neoplasia* 7, 291–312.
38. Grosdemouge, I., Bachelot, A., Lucas, A., Baran, N., Kelly, P., and Binart, N. (2003). Effects of deletion of the prolactin receptor on ovarian gene expression. *Reprod. Biol. Endocrinol.* 1, 12.
39. Hoffmann, B., Busges, F., Engel, E., Kowalewski, M. P., and Papa, P. (2004). Regulation of corpus luteum-function in the bitch. *Reprod. Domest. Anim.* 39, 232–240.
40. Hernandez, E. R., Roberts, C. T. Jr., LeRoith, D., and Adashi, E. Y. (1989). Rat ovarian insulin-like growth factor I (IGF-I) gene expression is granulosa cell-selective: 5′-untranslated mRNA variant representation and hormonal regulation. *Endocrinology* 125, 572–574.
41. Zhou, J., Chin, E., and Bondy, C. (1991). Cellular pattern of insulin-like growth factor-I (IGF-I) and IGF-I receptor gene expression in the developing and mature ovarian follicle. *Endocrinology* 129, 3281–3288.
42. Hsu, C. J., and Hammond, J. M. (1987). Concomitant effects of growth hormone on secretion of insulin-like growth factor I and progesterone by cultured porcine granulosa cells. *Endocrinology* 121, 1343–1348.
43. Samaras, S. E., Canning, S. F., Barber, J. A., Simmen, F. A., and Hammond, J. M. (1996). Regulation of insulin-like growth factor I biosynthesis in porcine granulosa cells. *Endocrinology* 137, 4657–4664.
44. Sirotkin, A. V., Makarevich, A. V., Kwon, H. B., Kotwica, J., Bulla, J., and Hetenyi, L. (2001). Do GH, IGF-I and oxytocin interact by regulating the secretory activity of porcine ovarian cells? *J. Endocrinol.* 171, 475–480.
45. Sirotkin, A. V., and Makarevich, A. V. (2002). Growth hormone can regulate functions of porcine ovarian granulosa cells through the cAMP/protein kinase A system. *Anim. Reprod. Sci.* 70, 111–126.
46. Kolodziejczyk, J., Gregoraszczuk, E. L., Leibovich, H., and Gertler, A. (2001). Different action of ovine GH on porcine

theca and granulosa cells proliferation and insulin-like growth factors I- and II-stimulated estradiol production. *Reprod. Biol.* 1, 33–41.
47. Hammond, J. M., Hsu, C. J., Klindt, J., Tsang, B. K., and Downey, B. R. (1988). Gonadotropins increase concentrations of immunoreactive insulin-like growth factor-I in porcine follicular fluid in vivo. *Biol. Reprod.* 38, 304–308.
48. Wathes, D. C., Perks, C. M., Davis, A. J., and Denning-Kendall, P. A. (1995). Regulation of insulin-like growth factor-I and progesterone synthesis by insulin and growth hormone in the ovine ovary. *Biol. Reprod.* 53, 882–889.
49. Kirby, C. J., Thatcher, W. W., Collier, R. J., Simmen, F. A., and Lucy, M. C. (1996). Effects of growth hormone and pregnancy on expression of growth hormone receptor, insulin-like growth factor-I, and insulin-like growth factor binding protein-2 and -3 genes in bovine uterus, ovary, and oviduct. *Biol. Reprod.* 55, 996–1002.
50. Penarrubia, J., Balasch, J., Garcia-Bermudez, M., Casamitjana, R., Vanrell, J. A., and Hernandez, E. R. (2000). Growth hormone does not increase the expression of insulin-like growth factors and their receptor genes in the pre-menopausal human ovary. *Hum. Reprod.* 15, 1241–1246.
51. Mason, H. D., Martikainen, H., Beard, R. W., Anyaoku, V., and Franks, S. (1990). Direct gonadotrophic effect of growth hormone on oestradiol production by human granulosa cells in vitro. *J. Endocrinol.* 126, R1-R4.
52. Lanzone, A., Di Simone, N., Castellani, R., Fulghesu, A. M., Caruso, A., and Mancuso, S. (1992). Human growth hormone enhances progesterone production by human luteal cells in vitro: evidence of a synergistic effect with human chorionic gonadotropin. *Fertil. Steril.* 57, 92–96.
53. Barreca, A., Artini, P. G., Del Monte, P., Ponzani, P., Pasquini, P., Cariola, G., Volpe, A., Genazzani, A. R., Giordano, G., and Minuto, F. (1993). In vivo and in vitro effect of growth hormone on estradiol secretion by human granulosa cells. *J. Clin. Endocrinol. Metab.* 77, 61–67.
54. de Boer, J. A., Schoemaker, J., van der Veen, E. A. (1997). Impaired reproductive function in women treated for growth hormone deficiency during childhood. *Clin. Endocrinol. (Oxf)* 46, 681–689.
55. Menashe, Y., Sack, J., and Mashiach, S. (1991). Spontaneous pregnancies in two women with Laron-type dwarfism: are growth hormone and circulating insulin-like growth factor mandatory for induction of ovulation? *Hum. Reprod.* 6, 670–671.
56. Dor, J., Ben-Shlomo, I., Lunenfeld, B., Pariente, C., Levran, D., Karasik, A., Seppala, M., and Mashiach, S. (1992). Insulin-like growth factor-I (IGF-I) may not be essential for ovarian follicular development: evidence from IGF-I deficiency. *J. Clin. Endocrinol. Metab.* 74, 539–542.
57. Cotreatment with growth hormone and gonadotropin for ovulation induction in hypogonadotropic patients: a prospective, randomized, placebo-controlled, dose-response study. European and Australian Multicenter Study. *Fertil. Steril.* 64, 917–23, 1995.
58. Howles, C. M., Loumaye, E., Germond, M., Yates, R., Brinsden, P., Healy, D., Bonaventura, L. M., and Strowitzki, T. (1999). Does growth hormone-releasing factor assist follicular development in poor responder patients undergoing ovarian stimulation for in-vitro fertilization? *Hum. Reprod.* 14, 1939–1943.
59. Busacca, M., Fusi, F. M., Brigante, C., Bonzi, V., Gonfiantini, C., Vignali, M., and Ferrari, A. (1996). Use of growth hormone-releasing factor in ovulation induction in poor responders. *J. Reprod. Med.* 41, 699–703.
60. Harper, K., Proctor, M., and Hughes, E. (2003). Growth hormone for in vitro fertilization. *Cochrane Database Syst. Rev.* CD000099.
61. Naar, E. M., Bartke, A., Majumdar, S. S., Buonomo, F. C., Yun, J. S., and Wagner, T. E. (1991). Fertility of transgenic female mice expressing bovine growth hormone or human growth hormone variant genes. *Biol. Reprod.* 45, 178–187.
62. Kalro, B. N. (2003). Impaired fertility caused by endocrine dysfunction in women. *Endocrinol. Metab. Clin. North Am.* 32, 573–592.
63. Bartke, A., Chandrashekar, V., Turyn, D., Steger, R. W., Debeljuk, L., Winters, T. A., Mattison, J. A., Danilovich, N. A., Croson, W., Wernsing, D. R., and Kopchick, J. J. (1999). Effects of growth hormone overexpression and growth hormone resistance on neuroendocrine and reproductive functions in transgenic and knock-out mice. *Proc. Soc. Exp. Biol. Med.* 222, 113–123.
64. Van den Brande, J. L. (1999). A personal view on the early history of the insulin-like growth factors. *Horm. Res.* 51[Suppl 3], 149–175.
65. Salmon, W. D. Jr., and Daughaday, W. H. (1990). A hormonally controlled serum factor which stimulates sulfate incorporation by cartilage in vitro. 1956. *J. Lab. Clin. Med.* 116, 408–419.
66. Reik, W., Constancia, M., Fowden, A., Anderson, N., Dean, W., Ferguson-Smith, A., Tycko, B., and Sibley, C. (2003). Regulation of supply and demand for maternal nutrients in mammals by imprinted genes. *J. Physiol. (Lond)* 547, 35–44.
67. Daughaday, W. H., and Rotwein, P. (1989). Insulin-like growth factors I and II: peptide, messenger ribonucleic acid and gene structures, serum, and tissue concentrations. *Endocr. Rev.* 10, 68–91.
68. Daughaday, W. H. (1989). Hypoglycemia in patients with non-islet cell tumors. *Endocrinol. Metab. Clin. North Am.* 18, 91–101.
69. DeChiara, T. M., Efstratiadis, A., and Robertson, E. J. (1990). A growth-deficiency phenotype in heterozygous mice carrying an insulin-like growth factor II gene disrupted by targeting. *Nature* 345, 78–80.
70. O'Dell, S. D., and Day, I. N. (1998). Insulin-like growth factor II (IGF-II). *Int. J. Biochem. Cell. Biol.* 30, 767–771.
71. Vu, T. H., and Hoffman, A. R. (1994). Promoter-specific imprinting of the human insulin-like growth factor-II gene. *Nature* 371, 714–717.
72. Pollak, M. N., Schernhammer, E. S., and Hankinson, S. E. (2004). Insulin-like growth factors and cancer. *Nat. Rev. Cancer* 4, 505–518.
73. Pavelic, K., Bukovic, D., and Pavelic, J. (2002). The role of insulin-like growth factor 2 and its receptors in human tumors. *Mol. Med.* 8, 771–780.
74. Reinecke, M., and Collet, C. (1998). The phylogeny of the insulin-like growth factors. *Int. Rev. Cytol.* 183, 1–94.
75. Bell, G. I., Gerhard, D. S., Fong, N. M., Sanchez-Pescador, R., and Rall, L. B. (1985). Isolation of the human insulin-like growth factor genes: Insulin-like growth factor II and insulin genes are contiguous. *Proc. Natl. Acad. Sci. U S A* 82, 6450–6454.
76. Navarro, I., Leibush, B., Moon, T. W., Plisetskaya, E. M., Banos, N., Mendez, E., Planas, J. V., and Gutierrez, J. (1999). Insulin, insulin-like growth factor-I (IGF-I) and glucagon: the evolution of their receptors. *Comp. Biochem. Physiol. B Biochem. Mol. Biol.* 122, 137–153.
77. Leibush, B. N., Lappova, Y. L., Bondareva, V. M., Chistyacova, O. V., Gutierrez, J., and Plisetskaya, E. M. (1998). Insulin-family peptide-receptor interaction at the early stage of vertebrate evolution. *Comp. Biochem. Physiol. B Biochem. Mol. Biol.* 121, 57–63.

78. Yarden, Y., and Ullrich, A. (1988). Growth factor receptor tyrosine kinases. *Annu. Rev. Biochem.* 57, 443–478.
79. Werner, H., Adamo, M., Roberts, C. T. Jr., and LeRoith, D. (1994). Molecular and cellular aspects of insulin-like growth factor action. *Vitam. Horm.* 48, 1–58.
80. Braulke, T. (1999). Type-2 IGF receptor: a multi-ligand binding protein. *Horm. Metab. Res.* 31, 242–246.
81. Barbieri, M., Bonafe, M., Franceschi, C., and Paolisso, G. (2003). Insulin/IGF-I-signaling pathway: an evolutionarily conserved mechanism of longevity from yeast to humans. *Am. J. Physiol.* 285, E1064–E1071.
82. Nakae, J., Barr, V., and Accili, D. (2000). Differential regulation of gene expression by insulin and IGF-1 receptors correlates with phosphorylation of a single amino acid residue in the forkhead transcription factor FKHR. *EMBO J.* 19, 989–996.
83. Kim, J. J., and Accili, D. (2002). Signalling through IGF-I and insulin receptors: where is the specificity? *Growth Horm. IGF Res.* 12, 84–90.
84. Duan, C. (2002). Specifying the cellular responses to IGF signals: roles of IGF-binding proteins. *J. Endocrinol.* 175, 41-54.
85. Monzavi, R., and Cohen, P. (2002). IGFs and IGFBPs: role in health and disease. *Best Pract. Res. Clin. Endocrinol. Metab.* 16, 433–447.
86. Firth, S. M., and Baxter, R. C. (2002). Cellular actions of the insulin-like growth factor binding proteins. *Endocr. Rev.* 23, 824–854.
87. Mohan, S., and Baylink, D. J. (2002). IGF-binding proteins are multifunctional and act via IGF-dependent and -independent mechanisms. *J. Endocrinol.* 175, 19–31.
88. Bondy, C. A., and Cheng, C. M. (2002). Insulin-like growth factor-1 promotes neuronal glucose utilization during brain development and repair processes. *Int. Rev. Neurobiol.* 51, 189–217.
89. Cheng, C. M., Reinhardt, R. R., Lee, W. H., Joncas, G., Patel, S. C., and Bondy, C. A. (2000). Insulin-like growth factor 1 regulates developing brain glucose metabolism. *Proc. Natl. Acad. Sci. U S A* 97, 10236–10241.
90. Wang, J., Zhou, J., Bondy, C. A. (1999). Igf1 promotes longitudinal bone growth by insulin-like actions augmenting chondrocyte hypertrophy. *FASEB J.* 13, 1985–1990.
91. Adashi, E., Resnick, C., D'Ercole, A., Svoboda, M., and Van Wyk, J. (1985). Insulin-like growth factors as intraovarian regulators of granulosa cell growth and function. *Endocr. Rev.* 6, 400–420.
92. Adashi, E. Y. (1998). The IGF family and folliculogenesis. *J. Reprod. Immunol.* 39, 13–19.
93. Adashi, E. Y., Resnick, C. E., Payne, D. W., Rosenfeld, R. G., Matsumoto, T., Hunter, M. K., Gargosky, S. E., Zhou, J., and Bondy, C. A. (1997). The mouse intraovarian insulin-like growth factor I system: departures from the rat paradigm. *Endocrinology* 138, 3881–3890.
94. Wandji, S. A., Wood, T. L., Crawford, J., Levison, S. W., and Hammond, J. M. (1998). Expression of mouse ovarian insulin growth factor system components during follicular development and atresia. *Endocrinology* 139, 5205–5214.
95. Parmer, T. G., Roberts, C. T. Jr., LeRoith, D., Adashi, E. Y., Khan, I., Solan, N., Nelson, S., Zilberstein, M., and Gibori, G. (1991). Expression, action, and steroidal regulation of insulin-like growth factor-I (IGF-I) and IGF-I receptor in the rat corpus luteum: their differential role in the two cell populations forming the corpus luteum. *Endocrinology* 129, 2924–2932.
96. Oliver, J. E., Aitman, T. J., Powell, J. F., Wilson, C. A., and Clayton, R. N. (1989). Insulin-like growth factor I gene expression in the rat ovary is confined to the granulosa cells of developing follicles. *Endocrinology* 124, 2671–2679.
97. Zhou, J., Refuerzo, J., and Bondy, C. (1995). Granulosa cell DNA synthesis is strictly correlated with the presence of insulin-like growth factor I and absence of c-fos/c-jun expression. *Mol. Endocrinol.* 9, 924–931.
98. Zhou, J., Kumar, T. R., Matzuk, M. M., and Bondy, C. (1997). Insulin-like growth factor I regulates gonadotropin responsiveness in the murine ovary. *Mol. Endocrinol.* 11, 1924–1933.
99. Monget, P., and Bondy, C. (2000). Importance of the IGF system in early folliculogenesis. *Mol. Cell. Endocrinol.* 163, 89–93.
100. Bondy, C. A., Chin, E., and Zhou, J. (1993). Significant species differences in local IGF-I and -II gene expression. *Adv. Exp. Med. Biol.* 343, 73–77.
101. Giudice, L. C. (2001). Insulin-like growth factor family in graafian follicle development and function. *J. Soc. Gynecol. Investig.* 8, S26–S29.
102. Erickson, G. F., and Danforth, D. R. (1995). Ovarian control of follicle development. *Am. J. Obstet. Gynecol.* 172, 736–747.
103. Baker, J., Hardy, M. P., Zhou, J., Bondy, C., Lupu, F., Bellve, A. R., and Efstratiadis, A. (1996). Effects of an Igf1 gene null mutation on mouse reproduction. *Mol. Endocrinol.* 10, 903–918.
104. Kadakia, R., Arraztoa, J. A., Bondy, C., and Zhou, J. (2001). Granulosa cell proliferation is impaired in the Igf1 null ovary. *Growth Horm. IGF Res.* 11, 220–224.
105. Zhou, J., Bievre, M., and Bondy, C. A. (2000). Reduced GLUT1 expression in Igf1-/- null oocytes and follicles. *Growth Horm. IGF Res.* 10, 111–117.
106. Kumar, T. R., Wang, Y., Lu, N., Matzuk, M. M. (1997). Follicle stimulating hormone is required for ovarian follicle maturation but not male fertility. Nat. Genet. 15, 201–204.
107. Richards, J. S., Sharma, S. C., Falender, A. E., and Lo, Y. H. (2002). Expression of FKHR, FKHRL1, and AFX genes in the rodent ovary: evidence for regulation by IGF-I, estrogen, and the gonadotropins. *Mol. Endocrinol.* 16, 580–599.
108. Richards, J. S., Russell, D. L., Ochsner, S., Hsieh, M., Doyle, K. H., Falender, A. E., Lo, Y. K., and Sharma, S. C. (2002). Novel signaling pathways that control ovarian follicular development, ovulation, and luteinization. *Recent Prog. Horm. Res.* 57, 195–220.
109. Mazerbourg, S., Bondy, C., Zhou, J., and Monget, P. (2003). The insulin-like growth factor system: a key determinant role in the growth and selection of ovarian follicles? A comparative species study. *Reprod. Domest. Anim.* 38, 247–258.
110. Liu, J., Koenigsfeld, A. T., Cantley, T. C., Boyd, C. K., Kobayashi, Y., and Lucy, M. C. (2000). Growth and the initiation of steroidogenesis in porcine follicles are associated with unique patterns of gene expression for individual components of the ovarian insulin-like growth factor system. *Biol. Reprod.* 63, 942–952.
111. Zhou, J., Adesanya, O. O., Vatzias, G., Hammond, J. M., and Bondy, C. A. (1996). Selective expression of insulin-like growth factor system components during porcine ovary follicular selection. *Endocrinology* 137, 4893–4901.
112. Samaras, S. E., Hagen, D. R., Bryan, K. A., Mondschein, J. S., Canning, S. F., and Hammond, J. M. (1994). Effects of growth hormone and gonadotropin on the insulin-like growth factor system in the porcine ovary. *Biol. Reprod.* 50, 178–186.
113. Bryan, K. A., Clark, A. M., Hammond, J. M., and Hagen, D. R. (1991). Effect of constant versus adjusted dose of exogenous porcine growth hormone (pGH) on growth and reproductive characteristics of gilts. *J. Anim. Sci.* 69, 2980–2987.
114. Ginther, O. J., Beg, M. A., Bergfelt, D. R., Donadeu, F. X., and Kot, K. (2001). Follicle selection in monovular species. *Biol. Reprod.* 65, 638–647.

115. Spicer, L. J. (2004). Proteolytic degradation of insulin-like growth factor binding proteins by ovarian follicles: a control mechanism for selection of dominant follicles. *Biol. Reprod.* 70, 1223–1230.
116. Lucy, M. C., Bilby, C. R., Kirby, C. J., Yuan, W., and Boyd, C. K. (1999). Role of growth hormone in development and maintenance of follicles and corpora lutea. *J. Reprod. Fertil. Suppl.* 54, 49–59.
117. Armstrong, D. G., Gong, J. G., and Webb, R. (2003). Interactions between nutrition and ovarian activity in cattle: physiological, cellular and molecular mechanisms. *Reprod. Suppl.* 61, 403–414.
118. Zulu, V. C., Nakao, T., and Sawamukai, Y. (2002). Insulin-like growth factor-I as a possible hormonal mediator of nutritional regulation of reproduction in cattle. *J. Vet. Med. Sci.* 64, 657–665.
119. Ge, Z., Nicholson, W. E., Plotner, D. M., Farin, C. E., and Gadsby, J. E. (2000). Insulin-like growth factor I receptor mRNA and protein expression in pig corpora lutea. *J. Reprod. Fertil.* 120, 109–114.
120. Neuvians, T. P., Pfaffl, M. W., Berisha, B., and Schams, D. (2003). The mRNA expression of the members of the IGF-system in bovine corpus luteum during induced luteolysis. *Domest. Anim. Endocrinol.* 25, 359–372.
121. Webb, R., Woad, K. J., and Armstrong, D. G. (2002). Corpus luteum (CL) function: local control mechanisms. *Domest. Anim. Endocrinol.* 23, 277–285.
122. Woad, K. J., Baxter, G., Hogg, C. O., Bramley, T. A., Webb, R., and Armstrong, D. G. (2000). Expression of mRNA encoding insulin-like growth factors I and II and the type 1 IGF receptor in the bovine corpus luteum at defined stages of the oestrous cycle. *J. Reprod. Fertil.* 120, 293–302.
123. Miller, E. A., Ge, Z., Hedgpeth, V., and Gadsby, J. E. (2003). Steroidogenic responses of pig corpora lutea to insulin-like growth factor I (IGF-I) throughout the oestrous cycle. *Reproduction* 125, 241–249.
124. Sekar, N., Lavoie, H. A., and Veldhuis, J. D. (2000). Concerted regulation of steroidogenic acute regulatory gene expression by luteinizing hormone and insulin (or insulin-like growth factor I) in primary cultures of porcine granulosa-luteal cells. *Endocrinology* 141, 3983–3992.
125. Sekar, N., and Veldhuis, J. D. (2001). Concerted transcriptional activation of the low density lipoprotein receptor gene by insulin and luteinizing hormone in cultured porcine granulosa-luteal cells: possible convergence of protein kinase A, phosphatidylinositol 3-kinase, and mitogen-activated protein kinase signaling pathways. *Endocrinology* 142, 2921–2928.
126. Bondy, C., and Zhou, J. (1993). Insulin-like growth factor system gene expression in the postpubertal human ovary. *Ann. N Y Acad. Sci.* 687, 65–76.
127. el-Roeiy, A., Chen, X., Roberts, V. J., Shimasakai, S., Ling, N., LeRoith, D., Roberts, C. T. Jr., and Yen, S. S. (1994). Expression of the genes encoding the insulin-like growth factors (IGF-I and II), the IGF and insulin receptors, and IGF-binding proteins-1-6 and the localization of their gene products in normal and polycystic ovary syndrome ovaries. *J. Clin. Endocrinol. Metab.* 78, 1488–1496.
128. Zhou, J., and Bondy, C. (1993). Anatomy of the human ovarian insulin-like growth factor system. *Biol. Reprod.* 48, 467–482.
129. Zhou, J., Wang, J., Penny, D., Monget, P., Arraztoa, J. A., Fogelson, L. J., and Bondy, C. A. (2003). Insulin-like growth factor binding protein 4 expression parallels luteinizing hormone receptor expression and follicular luteinization in the primate ovary. *Biol. Reprod.* 69, 22–29.
130. Arraztoa, J. A., Monget, P., Bondy, C., and Zhou, J. (2002). Expression patterns of insulin-like growth factor-binding proteins 1, 2, 3, 5, and 6 in the mid-cycle monkey ovary. *J. Clin. Endocrinol. Metab.* 87, 5220–5228.
131. Poretsky, L., Bhargava, G., and Levitan, E. (1990). Type I insulin-like growth factor receptors in human ovarian stroma. *Horm. Res.* 33, 22–26.
132. Gomez, E., Tarin, J. J., and Pellicer, A. (1993). Oocyte maturation in humans: the role of gonadotropins and growth factors. *Fertil. Steril.* 60, 40–46.
133. Geisthovel, F., Moretti-Rojas, I., Asch, R. H., and Rojas, F. J. (1989). Expression of insulin-like growth factor-II (IGF-II) messenger ribonucleic acid (mRNA), but not IGF-I mRNA, in human preovulatory granulosa cells. *Hum. Reprod.* 4, 899–902.
134. Voutilainen, R., and Miller, W. L. (1987). Coordinate tropic hormone regulation of mRNAs for insulin-like growth factor II and the cholesterol side-chain-cleavage enzyme, P450scc (corrected), in human steroidogenic tissues. *Proc. Natl. Acad. Sci. U S A* 84, 1590–1594.
135. Ramasharma, K., and Li, C. H. (1987). Human pituitary and placental hormones control human insulin-like growth factor II secretion in human granulosa cells. *Proc. Natl. Acad. Sci. U S A* 84, 2643–2647.
136. Thierry van Dessel, H. J., Chandrasekher, Y., Yap, O. W., Lee, P. D., Hintz, R. L., Faessen, G. H., Braat, D. D., Fauser, B. C., and Giudice, L. C. (1996). Serum and follicular fluid levels of insulin-like growth factor I (IGF-I), IGF-II, and IGF-binding protein-1 and -3 during the normal menstrual cycle. *J. Clin. Endocrinol. Metab.* 81, 1224–1231.
137. Poretsky, L., Cataldo, N. A., Rosenwaks, Z., and Giudice, L. C. (1999). The insulin-related ovarian regulatory system in health and disease. *Endocr. Rev.* 20, 535–582.
138. Devoto, L., Christenson, L. K., McAllister, J. M., Makrigiannakis, A., and Strauss, J. F. III. (1999). Insulin and insulin-like growth factor-I and -II modulate human granulosa-lutein cell steroidogenesis: enhancement of steroidogenic acute regulatory protein (StAR) expression. *Mol. Hum. Reprod.* 5, 1003–1010.
139. Seto-Young, D., Zajac, J., Liu, H. C., Rosenwaks, Z., and Poretsky, L. (2003). The role of mitogen-activated protein kinase in insulin and insulin-like growth factor I (IGF-I) signaling cascades for progesterone and IGF-binding protein-1 production in human granulosa cells. *J. Clin. Endocrinol. Metab.* 88, 3385–3391.
140. Villavicencio, A., Iniguez, G., Johnson, M. C., Gabler, F., Palomino, A., and Vega, M. (2002). Regulation of steroid synthesis and apoptosis by insulin-like growth factor I and insulin-like growth factor binding protein 3 in human corpus luteum during the midluteal phase. *Reproduction* 124, 501–508.
141. Kahn, C. R., Flier, J. S., Bar, R. S., Archer, J. A., Gorden, P., Martin, M. M., and Roth, J. (1976). The syndromes of insulin resistance and acanthosis nigricans. Insulin-receptor disorders in man. *N. Engl. J. Med.* 294, 739–745.
142. Taylor, S. I., Dons, R. F., Hernandez, E., Roth, J., and Gorden, P. (1982). Insulin resistance associated with androgen excess in women with autoantibodies to the insulin receptor. *Ann. Intern. Med.* 97, 851–855.
143. Barbieri, R. L., and Ryan, K. J. (1983). Hyperandrogenism, insulin resistance, and acanthosis nigricans syndrome: a common endocrinopathy with distinct pathophysiologic features. *Am. J. Obstet. Gynecol.* 147, 90–101.
144. Poretsky, L. (1991). On the paradox of insulin-induced hyperandrogenism in insulin-resistant states. *Endocr. Rev.* 12, 3–13.
145. Poretsky, L., Smith, D., Seibel, M., Pazianos, A., Moses, A. C., and Flier, J. S. (1984). Specific insulin binding sites in human ovary. *J. Clin. Endocrinol. Metab.* 59, 809–811.

146. Barbieri, R. L., Makris, A., and Ryan, K. J. (1984). Insulin stimulates androgen accumulation in incubations of human ovarian stroma and theca. *Obstet. Gynecol.* 64, 73S–80S.
147. Franks, S. (1995). Polycystic ovary syndrome. *N. Engl. J. Med.* 333, 853–861.
148. Tsilchorozidou, T., Overton, C., and Conway, G. S. (2004). The pathophysiology of polycystic ovary syndrome. *Clin. Endocrinol. (Oxf)* 60, 1–17.
149. Strauss, J. F. III. (2003). Some new thoughts on the pathophysiology and genetics of polycystic ovary syndrome. *Ann. N Y Acad. Sci.* 997, 42–48.
150. Sam, S., and Dunaif, A. (2003). Polycystic ovary syndrome: syndrome XX? *Trends Endocrinol. Metab.* 14, 365–370.
151. Azziz, R. (2003). Androgen excess is the key element in polycystic ovary syndrome. *Fertil. Steril.* 80, 252–254.
152. Barnes, R., and Rosenfield, R. L. (1989). The polycystic ovary syndrome: pathogenesis and treatment. *Ann. Intern. Med.* 110, 386–399.
153. Abbott, D. H., Dumesic, D. A., and Franks, S. (2002). Developmental origin of polycystic ovary syndrome: a hypothesis. *J. Endocrinol.* 174, 1–5.
154. Ehrmann, D. A., Barnes, R. B., and Rosenfield, R. L. (1995). Polycystic ovary syndrome as a form of functional ovarian hyperandrogenism due to dysregulation of androgen secretion. *Endocr. Rev.* 16, 322–353.
155. Lord, J. M., Flight, I. H., and Norman, R. J. (2003). Metformin in polycystic ovary syndrome: systematic review and meta-analysis. *B M J* 327, 951–953.
156. Kase, N., Kowal, J., Perloff, W., and Soffer, L. J. (1963). In vitro production of androgens by a virilizing adrenal adenoma and associated polycystic ovaries. *Acta. Endocrinol. (Copenh)* 44, 15–19.
157. Futterweit, W., and Deligdisch, L. (1986). Histopathological effects of exogenously administered testosterone in 19 female to male transsexuals. *J. Clin. Endocrinol. Metab.* 62, 16–21.
158. Carmina, E., and Lobo, R. A. (1998). Adrenal hyperandrogenism in the pathophysiology of polycystic ovary syndrome. *J. Endocrinol. Invest.* 21, 580–588.
159. Barnes, R. B., Rosenfield, R. L., Ehrmann, D. A., Cara, J. F., Cuttler, L., Levitsky, L. L., and Rosenthal, I. M. (1994). Ovarian hyperandrogynism as a result of congenital adrenal virilizing disorders: evidence for perinatal masculinization of neuroendocrine function in women. *J. Clin. Endocrinol. Metab.* 79, 1328–1333.
160. Robinson, J. E., Birch, R. A., Foster, D. L., and Padmanabhan, V. (2002). Prenatal exposure of the ovine fetus to androgens sexually differentiates the steroid feedback mechanisms that control gonadotropin releasing hormone secretion and disrupts ovarian cycles. *Arch. Sex. Behav.* 31, 35–41.
161. Barnes, R. B. (1998). The pathogenesis of polycystic ovary syndrome: lessons from ovarian stimulation studies. *J. Endocrinol. Invest.* 21, 567–579.
162. Donesky, B. W., and Adashi, E. Y. (1996). Surgical ovulation induction: the role of ovarian diathermy in polycystic ovary syndrome. *Baillieres Clin. Endocrinol. Metab.* 10, 293–309.
163. Vendola, K. A., Zhou, J., Adesanya, O. O., Weil, S. J., and Bondy, C. A. (1998). Androgens stimulate early stages of follicular growth in the primate ovary. *J. Clin. Invest.* 101, 2622–2629.
164. Vendola, K., Zhou, J., Wang, J., and Bondy, C. A. (1999). Androgens promote insulin-like growth factor-I and insulin-like growth factor-I receptor gene expression in the primate ovary. *Hum. Reprod.* 14, 2328–2332.
165. Vendola, K., Zhou, J., Wang, J., Famuyiwa, O. A., Bievre, M., and Bondy, C. A. (1999). Androgens promote oocyte insulin-like growth factor I expression and initiation of follicle development in the primate ovary. *Biol. Reprod.* 61, 353-357.
166. Weil, S., Vendola, K., Zhou, J., and Bondy, C. A. (1999). Androgen and follicle-stimulating hormone interactions in primate ovarian follicle development. *J. Clin. Endocrinol. Metab.* 84, 2951–2956.
167. Yoshimura, Y., Ando, M., Nagamatsu, S., Iwashita, M., Adachi, T., Sueoka, K., Miyazaki, T., Kuji, N., and Tanaka, M. (1996). Effects of insulin-like growth factor-I on follicle growth, oocyte maturation, and ovarian steroidogenesis and plasminogen activator activity in the rabbit. *Biol. Reprod.* 55, 152–160.
168. Thomson, F. J., Jess, T. J., Moyes, C., Plevin, R., and Gould, G. W. (1997). Characterization of the intracellular signalling pathways that underlie growth-factor-stimulated glucose transport in *Xenopus* oocytes: evidence for ras- and rho-dependent pathways of phosphatidylinositol 3-kinase activation. *Biochem. J.* 325(Pt 3), 637–643.
169. Hughesdon, P. E. Morphology and morphogenesis of the Stein-Leventhal ovary and of so-called "hyperthecosis." *Obstet. Gynecol. Surv.* 37, 59–77.
170. Pache, T. D., Chadha, S., Gooren, L. J., Hop, W. C., Jaarsma, K. W., Dommerholt, H. B., and Fauser, B. C. Ovarian morphology in long-term androgen-treated female to male transsexuals: a human model for the study of polycystic ovarian syndrome? *Histopathology* 19, 445–452.
171. De Leo, V., Lanzetta, D., D'Antona, D., la Marca, A., and Morgante, G. (1998). Hormonal effects of flutamide in young women with polycystic ovary syndrome. *J. Clin. Endocrinol. Metab.* 83, 99–102.
172. Phy, J. L., Conover, C. A., Abbott, D. H., Zschunke, M. A., Walker, D. L., Session, D. R., Tummon, I. S., Thornhill, A. R., Lesnick, T. G., and Dumesic, D. A. Insulin and messenger ribonucleic acid expression of insulin receptor isoforms in ovarian follicles from nonhirsute ovulatory women and polycystic ovary syndrome patients. *J. Clin. Endocrinol. Metab.* 89, 3561–3566.
173. Thierry van Dessel, H. J., Lee, P. D., Faessen, G., Fauser, B. C., and Giudice, L. C. (1999). Elevated serum levels of free insulin-like growth factor I in polycystic ovary syndrome. *J. Clin. Endocrinol. Metab.* 84, 3030–3035.
174. Franks, S., Gilling-Smith, C., Watson, H., and Willis, D. (1999). Insulin action in the normal and polycystic ovary. *Endocrinol. Metab. Clin. North Am.* 28, 361–378.
175. Giudice, L. C. (1999). Growth factor action on ovarian function in polycystic ovary syndrome. *Endocrinol. Metab. Clin. North Am.* 28, 325–339, vi.
176. Udoff, L. C., and Adashi, E. Y. (1996). Polycystic ovarian disease: current insights into an old problem. *J. Pediatr. Adolesc. Gynecol.* 9, 3–8.
177. Donesky, B. W., and Adashi, E. Y. Surgically induced ovulation in the polycystic ovary syndrome: wedge resection revisited in the age of laparoscopy. *Fertil. Steril.* 63, 439–463.
178. Yee, D., Morales, F. R., Hamilton, T. C., and Von Hoff, D. D. (1991). Expression of insulin-like growth factor I, its binding proteins, and its receptor in ovarian cancer. *Cancer Res.* 51, 5107–5112.
179. Karasik, A., Menczer, J., Pariente, C., and Kanety, H. (1994). Insulin-like growth factor-I (IGF-I) and IGF-binding protein-2 are increased in cyst fluids of epithelial ovarian cancer. *J. Clin. Endocrinol. Metab.* 78, 271–276.
180. Lukanova, A., Lundin, E., Toniolo, P., Micheli, A., Akhmedkhanov, A., Rinaldi, S., Muti, P., Lenner, P., Biessy, C., Krogh, V., Zeleniuch-Jacquotte, A., Berrino, F., Hallmans, G., Riboli, E., and Kaaks, R. (2002). Circulating levels of insulin-like growth factor-I and risk of ovarian cancer. *Int. J. Cancer* 101, 549–554.

181. Chen, C. L., Ip, S. M., Cheng, D., Wong, L. C., and Ngan, H. Y. (2000). Loss of imprinting of the IGF-II and H19 genes in epithelial ovarian cancer. *Clin. Cancer Res.* 6, 474–479.
182. Kim, H. T., Choi, B. H., Niikawa, N., Lee, T. S., and Chang, S. I. (1998). Frequent loss of imprinting of the H19 and IGF-II genes in ovarian tumors. *Am. J. Med. Genet.* 80, 391–395.
183. Yaginuma, Y., Nishiwaki, K., Kitamura, S., Hayashi, H., Sengoku, K., and Ishikawa, M. (1997). Relaxation of insulin-like growth factor-II gene imprinting in human gynecologic tumors. *Oncology* 54, 502–507.
184. Schildkraut, J. M., Schwingl, P. J., Bastos, E., Evanoff, A., and Hughes, C. (1996). Epithelial ovarian cancer risk among women with polycystic ovary syndrome. *Obstet. Gynecol.* 88, 554–559.
185. Monget, P., and Martin, G. B. (1997). Involvement of insulin-like growth factors in the interactions between nutrition and reproduction in female mammals. *Hum. Reprod.* 12[Suppl 1], 33–52.
186. Monget, P., Besnard, N., Huet, C., Pisselet, C., and Monniaux, D. (1996). Insulin-like growth factor-binding proteins and ovarian folliculogenesis. *Horm. Res.* 45, 211–217.
187. Wang, H. S., and Chard, T. (1999). IGFs and IGF-binding proteins in the regulation of human ovarian and endometrial function. *J. Endocrinol.* 161, 1–13.
188. Giudice, L. C., van Dessel, H. J., Cataldo, N. A., Chandrasekher, Y. A., Yap, O. W., and Fauser, B. C. (1995). Circulating and ovarian IGF binding proteins: potential roles in normo-ovulatory cycles and in polycystic ovarian syndrome. *Prog. Growth Factor Res.* 6, 397–408.
189. Nakatani, A., Shimasaki, S., Erickson, G. F., and Ling, N. (1991). Tissue-specific expression of four insulin-like growth factor-binding proteins (1, 2, 3, and 4) in the rat ovary. *Endocrinology* 129, 1521–1529.
190. Erickson, G. F., Nakatani, A., Ling, N., and Shimasaki, S. (1992). Localization of insulin-like growth factor-binding protein-5 messenger ribonucleic acid in rat ovaries during the estrous cycle. *Endocrinology* 130, 1867–1878.
191. Wood, T. L., Rogler, L. E., Czick, M. E., Schuller, A. G., and Pintar, J. E. (2000). Selective alterations in organ sizes in mice with a targeted disruption of the insulin-like growth factor binding protein-2 gene. *Mol. Endocrinol.* 14, 1472–1482.
192. Davidson, T. R., Chamberlain, C. S., Bridges, T. S., and Spicer, L. J. (2002). Effect of follicle size on in vitro production of steroids and insulin-like growth factor (IGF)-I, IGF-II, and the IGF-binding proteins by equine ovarian granulosa cells. *Biol. Reprod.* 66, 1640–1648.
193. Leighton, J. K., Grimes, R. W., Canning, S. F., and Hammond, J. M. (1994). IGF-binding proteins are differentially regulated in an ovarian granulosa cell line. *Mol. Cell. Endocrinol.* 106, 75–80.
194. Ongeri, E. M., Zhu, Q., Verderame, M. F., and Hammond, J. M. (2004). Insulin-like growth factor-binding protein-3 in porcine ovarian granulosa cells: gene cloning, promoter mapping, and follicle-stimulating hormone regulation. *Endocrinology* 145, 1776–1785.
195. Mazerbourg, S., Overgaard, M. T., Oxvig, C., Christiansen, M., Conover, C. A., Laurendeau, I., Vidaud, M., Tosser-Klopp, G., Zapf, J., and Monget, P. (2001). Pregnancy-associated plasma protein-A (PAPP-A) in ovine, bovine, porcine, and equine ovarian follicles: involvement in IGF binding protein-4 proteolytic degradation and mRNA expression during follicular development. *Endocrinology* 142, 5243-5253.
196. Schams, D., Berisha, B., Kosmann, M., and Amselgruber, W. M. (2002). Expression and localization of IGF family members in bovine antral follicles during final growth and in luteal tissue during different stages of estrous cycle and pregnancy. *Domest. Anim. Endocrinol.* 22, 51–72.
197. Sayre, B. L., Taft, R., Inskeep, E. K., and Killefer, J. (2000). Increased expression of insulin-like growth factor binding protein-1 during induced regression of bovine corpora lutea. *Biol. Reprod.* 63, 21–29.
198. Ge, Z., Miller, E., Nicholson, W., Hedgpeth, V., and Gadsby, J. (2003). Insulin-like growth factor (IGF)-I and IGF binding proteins-2, -3, -4, -5 in porcine corpora lutea during the estrous cycle: evidence for inhibitory actions of IGFBP-3. *Domest. Anim. Endocrinol.* 25, 183–197.
199. Gadsby, J. E., Lovdal, J. A., Samaras, S., Barber, J. S, and Hammond, J. M. (1996). Expression of the messenger ribonucleic acids for insulin-like growth factor-I and insulin-like growth factor binding proteins in porcine corpora lutea. *Biol. Reprod.* 54, 339–346.
200. Conover, C. A., Oxvig, C., Overgaard, M. T., Christiansen, M., and Giudice, L. C. (1999). Evidence that the insulin-like growth factor binding protein-4 protease in human ovarian follicular fluid is pregnancy associated plasma protein-A. *J. Clin. Endocrinol. Metab.* 84, 4742–4745.
201. Rhoton-Vlasak, A., Gleich, G. J., Bischof, P., and Chegini, N. (2003). Localization and cellular distribution of pregnancy-associated plasma protein-A and major basic protein in human ovary and corpora lutea throughout the menstrual cycle. *Fertil. Steril.* 79, 1149–1153.
202. Hourvitz, A., Kuwahara, A., Hennebold, J. D., Tavares, A. B., Negishi, H., Lee, T. H., Erickson, G. F., and Adashi, E. Y. (2002). The regulated expression of the pregnancy-associated plasma protein-A in the rodent ovary: a proposed role in the development of dominant follicles and of corpora lutea. *Endocrinology* 143, 1833–1844.
203. Giudice, L. C., Conover, C. A., Bale, L., Faessen, G. H., Ilg, K., Sun, I., Imani, B., Suen, L. F., Irwin, J. C., Christiansen, M., Overgaard, M. T., and Oxvig, C. (2002). Identification and regulation of the IGFBP-4 protease and its physiological inhibitor in human trophoblasts and endometrial stroma: evidence for paracrine regulation of IGF-II bioavailability in the placental bed during human implantation. *J. Clin. Endocrinol. Metab.* 87, 2359–2366.
204. Zhou, R., Diehl, D., Hoeflich, A., Lahm, H., and Wolf, E. (2003). IGF-binding protein-4: biochemical characteristics and functional consequences. *J. Endocrinol.* 178, 177–193.
205. Rivera, G. M., and Fortune, J. E. (2003). Selection of the dominant follicle and insulin-like growth factor (IGF)-binding proteins: evidence that pregnancy-associated plasma protein A contributes to proteolysis of IGF-binding protein 5 in bovine follicular fluid. *Endocrinology* 144, 437–446.
206. Johnson, J., Canning, J., Kaneko, T., Pru, J. K., and Tilly, J. L. (2004). Germline stem cells and follicular renewal in the postnatal mammalian ovary. *Nature* 428, 145–150.

Knobil and Neill's Physiology of Reproduction, Third Edition
edited by Jimmy D. Neill,
Elsevier © 2006

CHAPTER 14

Gonadotropin Signaling in the Ovary

Mary Hunzicker-Dunn[1] and Kelly Mayo[2]

Overview, 547
FSH Signals Promote Follicular Maturation, 547
Follicle Maturation, 547
FSH and the FSH Receptor, 548
FSH Signaling Pathways that Stimulate Granulosa Cell Differentiation: cAMP and PKA-Dependent Signaling, 550
Cross-Talk among FSH-Regulated Signaling Pathways Regulates Target Gene Expression, 566
LH Signals Promote Ovulation and Corpus Luteum Formation, 568
Ovulation and Luteinization, 568
LH and the LH Receptor, 570
LH Signaling Pathways that Stimulate Ovulation and Luteinization, 572
Molecular Basis for the Differential Regulation of Target Gene Expression by FSH in Preantral and LH in Preovulatory Granulosa Cells, 575
Concluding Remarks, 576
Acknowledgments, 577
References, 577

OVERVIEW

The gonadotropic hormones follicle-stimulating hormone (FSH) and luteinizing hormone (LH) initiate signaling events in ovarian target cells that govern the reproductive cycle and thus the continuation of a species. The ovarian follicle is the key structure that houses the oocyte. Upon appropriate stimulation by FSH and LH, follicles produce hormones that promote the development of secondary sex characteristics and that regulate the hypothalamic–pituitary axis as well as uterine receptivity. Additionally, the follicle must extrude the oocyte at an appropriate time to allow for fertilization, and remaining cells within the follicle must differentiate into cells of the corpus luteum that produce hormones necessary to sustain pregnancy. Although there is a great deal of species variability, LH is uniformly required to initiate progesterone production by the corpus luteum, which is necessary for implantation and the maintenance of pregnancy. In the absence of FSH follicles do not develop beyond the preantral stage, and animals are infertile. Similarly, in the absence of LH ovulation does not ensue, corpora lutea do not form, and animals are infertile. Both gonadotropic hormones initiate their activities by binding to surface protein receptors. Through complex signaling pathways, FSH and LH initiate distinct highly coordinated programs of gene expression that are only now beginning to be unraveled.

In this chapter we review current knowledge about the signaling pathways by which FSH promotes follicle maturation and LH promotes ovulation and corpus luteum formation. We refer the reader to excellent recent reviews of the FSH (1–3) and LH (4–7) receptors. We focus on FSH and LH signaling in granulosa cells of immature preantral follicles and mature preovulatory follicles and in thecal cells primarily of the rodent, although examples from other species are also discussed. We do not discuss LH signaling pathways and the regulation of LH target gene expression in the corpus luteum; readers are referred to recent reviews on this subject (8–10).

[1]Department of Cell and Molecular Biology, Center for Reproductive Science, Northwestern Medical School, Chicago; [2]Department of Biochemistry, Cellular and Molecular Biology, Center for Reproductive Science, Northwestern University Medical School, Evanston, Illinois.

FSH SIGNALS PROMOTE FOLLICULAR MATURATION

Follicle Maturation

Initial formation and early growth of ovarian follicles occur in a gonadotropin-independent fashion and rely on a number of characterized locally acting regulatory proteins that signal between the oocyte and the somatic cells of the follicle (11–13). Once the follicle acquires functional FSH receptors, it can respond to this pituitary gonadotropin to undergo further maturation in preparation for ovulation. Only a small subset of ovarian follicles survives to full maturity, with most being lost along the way to a programmed cell death process termed atresia (14,15). Maturation of ovarian follicles to a preovulatory phenotype encompasses not only an explosive increase in the proliferation of granulosa cells contained with in the follicle, resulting in rapid follicle growth, but also granulosa cell differentiation. As a consequence, the follicle develops a fluid-filled antrum. Although markers for antrum formation are not well-established, well-known responses that are often used as markers for FSH-dependent granulosa cell differentiation include the expression of the α subunit of the hormone inhibin (16), which feeds back to the anterior pituitary to repress production of FSH; increased estrogen production resulting largely from expression of the rate-limiting enzyme in estrogen biosynthesis that converts testosterone to estrogen, P-450 aromatase (aromatase); increased progesterone production as a result of increased expression of the rate-limiting enzyme in progesterone production that converts cholesterol to pregnenolone, P-450 cholesterol side chain cleavage (SCC); increased expression of 3β-hydroxysteroid dehydrogenase, which converts pregnenolone to progesterone; and increased expression of membrane receptors for LH and epidermal growth factor (EGF), both of which are required for the ovulatory response (17) and have been previously reviewed (18–20).

FSH also stimulates the transcription of genes that encode for intracellular signaling molecules, such as the regulatory (R) IIβ subunit of protein kinase A (PKA) (21), the A-kinase anchoring protein (AKAP), microtubule-associated protein 2D (MAP2D) (22), and serum glucocorticoid kinase (SGK) (23). FSH has also been shown to induce expression of the EGF receptor agonist epiregulin (24,25); the transcription factors early growth response factor-1 (Egr-1) (26), gonadotropin inducible ovarian transcription factor-1 (GIOT-1) (27), and liver receptor homolog (LRH)-1 (28); the follicular-fluid associated protein pregnancy-associated plasma protein-A (29), which is recognized to be the insulin-like growth factor (IGF) binding protein-4 protease (30); phosphodiesterases (PDE) 4D1 and 4D2 (31) to degrade cAMP; and vascular endothelial growth factor (VEGF), which has been shown to be critical for antrum formation, granulosa cell proliferation, and estrogen production and for increasing the vascularity of the thecal cell layer (32,33). FSH also inhibits the expression of calmodulin kinase IV (CaMK IV) (34) and *3,5,3′-triiodothyronine binding protein* messenger RNA (mRNA) (35).

FSH enhances expression at the protein level of immediate early genes such as *JunB*, c-*Jun*, c-*Fos*, *Fra* (36), and c-*Myc* (37,38), although these increases could result in part from posttranslational modifications to enhance protein stability (39) rather than strictly transcriptional regulation. The predominant marker for granulosa cell proliferation is increased expression of cyclin D2 (40,41). Cyclin D2 deficient mice are infertile as a result of the inability of granulosa cells to proliferate in response to FSH (40). It is interesting, however, that granulosa cells of cyclin D2 null mice differentiate to a preovulatory phenotype, and upon stimulation with an ovulatory concentration of LH, follicles do not ovulate but granulosa cells luteinize in follicles with entrapped ova (40). Moreover, recent microarray and differential display data reveal that FSH promotes increased expression of a large number of genes, including ones involved in reorganization of the microtubule and actin cytoskeleton, such as β-tubulin, the heavy chain of kinesin, and tropomysin-4 (33,42). These results indicate that the response of granulosa cells to FSH involves a complex and coordinated program of gene expression that comprises probably hundreds of genes (42,43). The FSH stimulus to promote granulosa cell differentiation requires relatively low but constant levels of FSH; withdrawal of FSH or its mediator cAMP during the 48-hour time course of granulosa cell differentiation (of the 4-day rodent estrous cycle) results in incomplete induction of differentiation markers (44,45).

FSH and the FSH Receptor

FSH, also known as follitropin, is a glycoprotein hormone consisting of two distinct subunits, an α subunit shared with the related glycoprotein hormones LH, human chorionic gonadotropin (hCG), and thyroid-stimulating hormone, and a unique β subunit (Fig. 1). FSH is produced in gonadotrope cells of the anterior pituitary, and its synthesis and secretion is highly regulated by neural, pituitary, and gonadal factors (46). The FSH receptor is a seven membrane-spanning G-protein–coupled receptor (GPCR) that belongs to the rhodopsin/β-adrenergic family of GPCRs (47). GPCRs are characterized by the presence of intracellular and extracellular loops linked by putative transmembrane (TM) α-helices

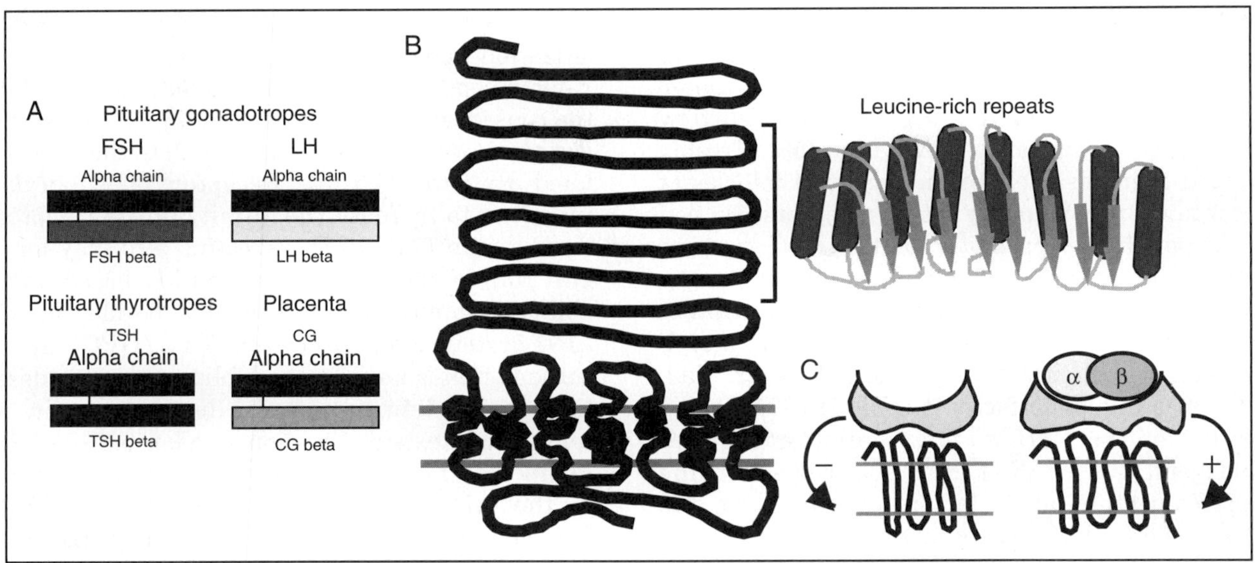

FIG. 1. The gonadotropins and gonadotropin receptors. **A:** A schematic of the family of dimeric glycoproteins hormones, which includes the pituitary gonadotropins FSH and LH, pituitary thyroid-stimulating hormone, and in some species placental chorionic gonadotropin. The α subunit is shared by all four hormones. **B:** Generic schematic of the glycoprotein hormone receptors, indicating the large extracellular domain and the seven membrane-spanning domains characteristic of GPCRs. A significant portion of the extracellular is composed of leucine-rich repeats, each composed of a β-strand followed by an α-helix, and these are proposed to form a horseshoe-shaped domain as shown that likely plays an important role in ligand binding. **C:** A model showing the view that the extracellular domain of the receptor plays a repressive role in the absence of ligand binding and that ligand interaction and ensuing conformational changes relieve this repression, allowing G protein activation. [Modeled after (502).]

and by their ability to couple to one or more guanine (G) nucleotide binding proteins. The FSH receptor belongs to the subgroup of class A GPCRs that contain Leu-rich-repeat motifs in their extracellular domain as well as the NSxxNPxxY motif in TM 7 and the DRY motif at the border of TM 3 and inner loop 2 (48). The extracellular domain contains at least eight Leu-rich-repeat motifs of approximately 24 residues (3) that are believed to provide structural integrity to the receptor (2). FSH binds to the large extracellular N-terminal domain of the receptor (49), although a recent report stated that the interior of the receptor is also important for both hormone binding and signal generation (50). Upon binding of FSH to this receptor, the receptor becomes activated. Activation of GPCRs is believed to represent a conformational change that alters the orientation of the TM domains (51). Recent circular dichroism studies detected a difference in the secondary structure of the receptor when FSH was bound compared with the unbound state, consistent with the generation of an altered conformation upon hormone binding (52). Very recently, the structure of partially deglycosylated human FSH bound to the extracellular domain of the FSH receptor was resolved at 2.9 Å (53). Results showed that binding specificity is mediated by both the FSH α and β subunits and that upon hormone binding, FSH undergoes a conformational change to a more rigid structure (53). Additionally, Fan and Hendrickson (53) provided the first structural evidence for the existence of gonadotropin receptor dimerization. Additional studies are required to understand how binding of FSH alters the conformation of the complete FSH receptor.

The predominant signal generated upon FSH receptor activation is cAMP, consistent with receptor coupling primarily to the stimulatory G protein (Gs) (3). Although there are reports that the FSH receptor can also couple to a pertussis toxin–sensitive G protein to modulate cAMP production, at least in a Chinese hamster ovarian cell line (53a), and to activate phospholipase C (PLC) in human embryonic kidney cells (54), the physiological relevance of these results has not been established.

The FSH GPCR is encoded by a single rather large gene that consists of 10 exons and 9 introns (2,3,55). The extracellular domain is encoded by the first 9 exons; the C-terminal part of the extracellular domain, TM, and intracellular C-terminal domains are encoded by exon 10 (55). This receptor is expressed in females only on follicular granulosa cells (3); expression is absent from ovaries of postmenopausal women (3). FSH receptor null mice are infertile, and follicles do not develop beyond the preantral stage (56,57), consistent with the view that this receptor is necessary and sufficient to mediate the effects of FSH. FSH receptor expression is not detected in primordial follicles but has been detected on granulosa cells of preantral follicles from postnatal day 3 in rats; therefore, it is expressed at the very early stages of follicular development (58-60). FSH receptor expression is increased by FSH or cAMP (61,62), activin (61–63), transforming-growth factor β (59), nerve growth factor (64), and IGF-1 (65) and is abolished by the LH surge that promotes ovulation (58). Freshly harvested granulosa cells from preantral follicles of rats contain approximately 4,500 receptors per cell (66). Although the promoter of the *FSH receptor* gene contains an E-box, which is conserved among species and binds basic helix-loop-helix transcription factors such as upstream stimulatory factors 1 and 2, c-Myc and its dimerization partners Max or Mad, and hypoxia-inducible factor 1 (HIF-1) (67,68), and activation of the *FSH receptor* promoter requires steroidogenic factor (SF)-1 (67,69) as well as binding of upstream stimulatory factors to the E-box, the mechanism for selective expression of the FSH receptor only in granulosa cells of female mammals has eluded identification. Exhaustive studies by Heckert and colleagues (68) revealed that elements required for cell-selective expression of the receptor are not present within the –5,000 to +123 gene region surroundings the transcription start site.

Several naturally occurring inactivating mutations of the *FSH receptor* gene have been described that result in arrested follicular development and hypergonadotropism. These include the following amino acid changes: I143T, A172V, and D207V in the extracellular domain, L584V in the extracellular loop 3, and R556C in cytoplasmic loop 3 (2). These mutations reduced or blocked FSH-stimulated cAMP generation, either as a result of poor hormone binding to the receptor and/or poor coupling to Gs. Although these results point to cytoplasmic loop 3 as containing critical sites for coupling to Gs, the specific Gs coupling region of the FSH receptor has not been identified. The only naturally occurring activating mutation identified for the human FSH receptor is D550G, which is located in cytoplasmic loop 3 (71). The FSH receptor with this mutation exhibits a modest increase in basal cAMP levels upon transfection into heterologous cells, but the patient with this mutation exhibited normal testis function despite being hypophysectomized (because of a pituitary tumor) (3,71). It is surprising that more activating mutations of this receptor have not been identified.

Like most other GPCRs, the FSH receptor appears to exhibit desensitization or reduced FSH-stimulated cAMP production (54) upon exposure to saturating agonist, although granulosa cells never see saturating concentrations of FSH under physiological conditions. The mechanism of GPCR desensitization has been particularly well studied for the β-adrenergic receptor. Desensitization of the β-adrenergic receptor occurs rapidly after receptor activation and is characterized by both a right-hand shift in the agonist dose–response curve and a decrease in maximal stimulation of adenylyl cyclase (72). β-Adrenergic receptor desensitization requires the phosphorylation of the activated receptor by a G protein-regulated kinase (73) and the consequent high affinity binding of arrestin2 (β-arrestin1) or arrestin3 (β-arrestin2) to the phosphorylated receptor (74,75). Receptor phosphorylation functions to increase the affinity of the receptor for the arrestin (74,76,77). Arrestins also play an integral role in receptor internalization based on their ability to bind both clathrin and β_2-adaptin of the AP-2 protein complex (78–80). Thus, GPCR internalization often is linked to and follows receptor desensitization. Arrestins have also been shown in some cellular contexts to function as adaptors to bind signaling intermediates, such as upstream components in the mitogen-activated protein kinase (MAPK)/ extracellular regulated protein kinase (ERK) and c-Jun NH2-terminal kinase (JNK) pathways, to redirect receptor signals into another pathway (81).

The FSH receptor, upon over-expression (by approximately tenfold) in a heterologous cell model, becomes phosphorylated in response to FSH predominately on Ser/Thr residues in intracellular loops 1 and 3 (82) in a G-protein–regulated kinase 2-dependent manner (83,84). Although a link between FSH receptor desensitization and receptor phosphorylation has not been reported, mutation of the phosphorylatable residues in intracellular loop 1 of the FSH receptor leads to reduced receptor internalization (82). Most of the internalized ligand-bound receptor recycles back to the cell surface, where intact hormone then dissociates (85). However, neither the rate nor the extent of FSH receptor internalization or FSH receptor desensitization has been studied under physiological conditions in granulosa cells.

There is also a report of an alternatively spliced FSH receptor that contains the first eight exons of the classic FSH GPCR plus a unique C-terminal extension that is hypothesized to traverse the plasma membrane a single time and to function more like a typical growth factor receptor (86). This receptor, upon expression in human embryonic kidney cells, localizes to the plasma membrane and binds FSH in a specific and high affinity manner (87). This alternative FSH receptor is reported to be expressed in mouse ovaries (88) and to activate the ERK pathway in immortalized granulosa cells (89). The physiological significance of this alternative FSH receptor to granulosa cell function remains to be elucidated.

FSH Signaling Pathways that Stimulate Granulosa Cell Differentiation: cAMP and PKA-Dependent Signaling

The predominant intracellular signal generated by FSH-dependent signaling through its GPCR

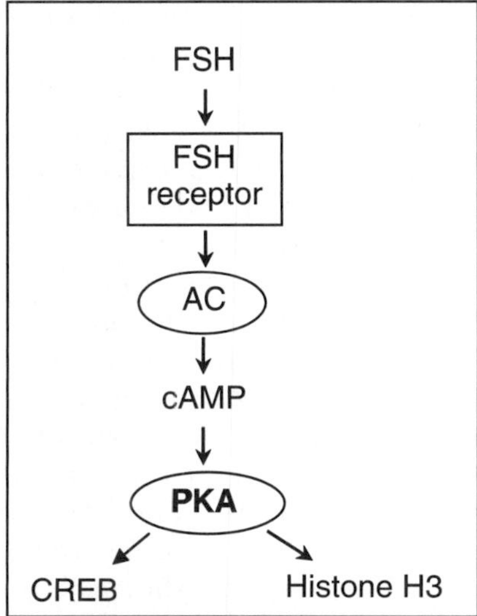

FIG. 2. FSH signals via cAMP to activate PKA. Two PKA substrates in immature granulosa cells are CREB and histone H3.

is cAMP (Fig. 2), based on the ability of forskolin to mimic FSH's ability to induce such differentiation markers as progesterone synthesis and LH receptors (44,45,90,91) [reviewed in (18)]. cAMP signals predominately by activating PKA (92). PKA is a tetrameric holoenzyme that consists of two regulatory (R) subunits and two catalytic (C) subunits. PKA is activated upon binding of cAMP to the R subunits, resulting in the dissociation of the C subunits that function as active kinases to phosphorylate substrates (93). Although representatives of an alternative family of cAMP effectors, consisting of cAMP-activated guanine nucleotide exchange factors also known as exchange proteins activated by cAMP (Epacs) (94–96), have been identified in rodent and human granulosa cells (97), the Epac target Rap1 is already activated in granulosa cells and FSH does not appear to further activate Rap1 (98). Moreover, an Epac-selective cAMP analogue does not promote induction of aromatase (99), a well-known marker of granulosa cell differentiation. Rather, in granulosa cells FSH activates PKA downstream of cAMP (100). The cell-permeable selective PKA C subunit inhibitor peptide myristoylated-(Myr-) PKI, the active portion of the ubiquitous PKA inhibitor protein, inhibits the induction of a number of proteins induced by FSH, including MAP2D, inhibin-α, and progesterone producing enzymes (98,101). PKI functions by binding to the substrate binding site of the C subunits of PKA, thus preventing other PKA substrates from binding, and preventing PKA from phosphorylating substrates (102). Additionally, competitive ATP antagonists that are established inhibitors of PKA, such as H89 and KT5720 (103,104), also inhibit many of these responses initiated by FSH, including the induction of progesterone synthesis, aromatase, LH receptor, RIIβ, Egr-1, SGK, and MAP2D (26,98,105). These results suggest a prominent role for PKA in FSH-stimulated granulosa cell differentiation.

Two classes of PKA holoenzymes, PKA I and PKA II, exist based on the association of two possible RI subunits (RIα and RIβ) or two possible RII subunits (RIIα and RIIβ) with four possible C subunits (α, β1, β2, and γ) (93). PKA phosphorylates a large number of substrates in various cells (106). Well-documented PKA substrates include CREB (107), PDE4D3 (108), Src (109), histone H3 (110), glycogen synthase kinase (GSK) α and β (111,112), nuclear factor (NF) κB (113), and L-type Ca^{2+} channels (114).

Most PKA in preantral granulosa cells consists of a PKA IIβ holoenzyme (115), yet RIIβ knockout mice are fertile (116). PKA Iα comprises less than 5% of the PKA activity in preantral granulosa cells, although there is a large amount of C subunit–free RI (115). The presence of substantial C subunit–free RI is surprising because RIα not bound to C is reported to be rapidly degraded (117). Granulosa cells do not express RIβ (118). Taken together, these results suggest that the actions of FSH could be mediated by PKA Iα rather than by the seemingly more abundant PKA IIβ holoenzyme. Because RIα null mice died at embryonic day 10.5 (119), proof that RIα is crucial to granulosa cell differentiation requires the generation of mice with an RIα deletion targeted to granulosa cells.

AKAPs Target PKA to Specific Subcellular Locations

The specificity of PKA action is accomplished by the targeting of PKA as well as its substrates to specific cellular locales by virtue of the binding of PKA to a large family of AKAPs. It has been estimated that more that 75% of PKA holoenzymes are targeted to specific intracellular sites via association of PKA regulatory subunits with AKAPs (93). RII subunits of PKA bind with nM affinity to AKAPs (93,120). The domain on the AKAP responsible for RII binding comprises an amphipathic helix that binds the N-termini of the PKA-RII dimer (121). AKAPs direct PKA to such locations as the mitochondria (122,123), Golgi apparatus (124), centrosome (125,126), nuclear envelope (127), and actin (128–130) and microtubule (131) cytoskeletons by the presence of specific subcellular targeting domains on each AKAP (132–134), as depicted in Fig. 3. Localization of PKA and one or more substrates to distinct regions within the cell is thought to promote both specific and efficient substrate phosphorylation in response to a stimulus (134). AKAPs not only bind PKA but also can function as platforms to coordinate signaling cascades by binding additional signaling proteins such as CaM, PDEs, protein phosphatases

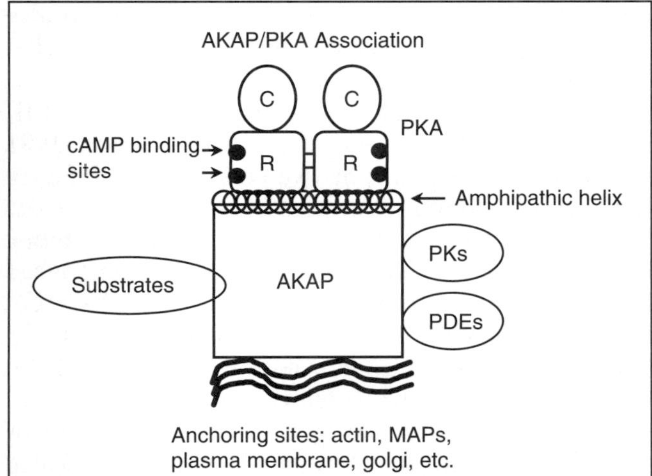

FIG. 3. Schematic model of the association of AKAPs with PKA, other protein kinases (PKs), PDEs, substrates, and anchoring sites in cells.

such as PP1 and PP2A, and protein kinases such as the tyrosine kinase Abl and the Ser/Thr protein kinase C (PKC) enzymes (132–134). By confining PKA and specific substrates to isolated cAMP gradients at discrete cellular locations, not only are signaling cascades optimized but also their fidelity is maintained, thereby preventing inappropriate cross-talk among pathways (133).

Most known AKAPs anchor PKA II holoenzymes and exhibit at least a 100-fold lower affinity for PKA I holoenzymes (133). Although there are a growing number of "dual" AKAPs that readily bind PKA I holoenzymes (135–139), these AKAPs still exhibit a ten- to 25-fold preference for PKA II (132). However, fibrous sheath protein 1 (137), the *Caenorhabditis elegans* $AKAP_{CE}$ (140), the peripheral benzodiazepine receptor-associated protein PAP7 (141), and the neurofibromatosis 2 tumor-suppresser protein merlin (142) appear to preferentially bind PKA I and thus are putative PKA I AKAPs.

Preantral granulosa cells express a large number of AKAPs, including AKAP-KL, AKAP79, Ezrin, AKAP-149, AKAP95, AKAP220, and AKAP100, based on RII overlay and Western blotting results (115). However, neither the expected selective link of each AKAP to individual PKA substrates and signaling pathways nor the cellular location of the granulosa cell AKAPs has been investigated. Thus, the cellular mechanism(s) by which each of the AKAPs present in granulosa cells modulates cellular functions is not known.

PKA Targets in Immature Granulosa Cells: CREB and Histones H3 and H1

Defining the signaling pathways that are activated by FSH in granulosa cells and determining how PKA coordinates signaling into these pathways remains a central challenge in ovarian biology. Although details of the signaling pathways activated by FSH are just beginning to be elucidated, it is clear that FSH stimulates the PKA-dependent phosphorylation of the established PKA substrates CREB on Ser133 (143,144) and histone H3 on Ser10 (100,101). Both FSH-stimulated CREB and histone H3 phosphorylations are abrogated by the PKA inhibitor Myr-PKI, suggesting that PKA is the predominant kinase that phosphorylates these proteins (Fig. 2). Both CREB and histone H3 are direct PKA substrates (101,107). It is interesting that in other cellular models, both CREB and histone H3 are phosphorylated on the same sites by protein kinases other than PKA. For example, CREB kinases in other cells include the phosphatidylinositol-3 kinase (PI-3K) substrate Akt (also known as protein kinase B) (145), CaMKs, the ERK substrates p90 ribosomal S6 protein kinases (RSK) and the mitogen- and stress-activated protein kinases 1 and 2 (MSK1 and -2), and the p38 MAPK substrates MAPK-activated protein kinases 2 and 3 (MK-2 and -3) (146,147); however, inhibitors of these kinases do not reduce FSH-stimulated CREB phosphorylation in granulosa cells (98,101).

Similarly, in other cells histone H3 is commonly phosphorylated by the ERK substrate RSK2 and by the ERK and p38 MAPK substrate MSK-1 (148–151). However, inhibitors of ERK activation and of activated p38 MAPK do not reduce histone H3 phosphorylation in granulosa cells (101). These data support the notion that PKA plays a unique role in granulosa cells to regulate signaling pathways.

FSH enhances the phosphorylation of histone H3 on Ser10 as well as the acetylation of Lys14, although it is not clear whether these modifications occur simultaneously or in a sequential manner (101). The histone code hypothesis put forth by Strahl and Allis (152) stated that covalent modifications of core histone tails result in the remodeling of chromatin to affect downstream events. The best-known core histone modifications consist of the covalent addition of an acetyl group to Lys, the addition of one or more methyl groups to Lys or Arg, and addition of a phospho group to Ser or Thr. These additions occur primarily on the N-terminal tails of histones H3 and H4. Acetylation neutralizes the positive charge of the histone and phosphorylation adds a negative charge, thereby decreasing the affinity of histone for DNA (153). The predicted result of histone acetylation and/or phosphorylation is the loosening of chromatin structure, resulting in increased access of select promoter regions to transcription factors and coactivators (152,154). Methylation can correlate either with gene activation or gene inhibition and can occur on the same Lys residues that are acetylated, and

Lys residues can be mono-, di-, or trimethylated, whereas Arg residues can be mono- or dimethylated. Arg methylation of histone H3 generally correlates with gene activation (155). For histone H3, methylation of Lys4 correlates with gene activation, whereas methylation of Lys9 and Lys27 correlates with gene repression (156,157). Histone acetylation is catalyzed by histone acetyltransferases, such as CREB binding protein (CBP), histone deacetylation is catalyzed by histone deacetylases, and histone methylation is catalyzed by methyltransferases.

Consistent with the histone code hypothesis, increased *c-Fos*, *SGK*, and *inhibin-α* promoter DNA is detected in the dual-phosphorylated and -acetylated histone H3-chromatin pools in response to FSH treatment of granulosa cells in chromatin immunoprecipitation assays (101). Because the *c-Fos* promoter contains a functional cAMP response element (158), signaling to activate c-Fos in granulosa cells probably reflects signaling to the *c-Fos* gene via CREB. In this instance, the C subunit of PKA would function not only to phosphorylate CREB to promote binding of CBP (159,160) and promoter activation but also to phosphorylate histone H3, whereas CBP perhaps promotes the acetylation of histone H3 as well as recruitment of other coactivators and the basal transcription machinery. Activation of the *SGK* gene in response to FSH/PKA requires an Sp1/Sp3 binding site (23). Because the phosphorylation and acetylation of histone H3 are linked to the rapid activation of the *SGK* gene, perhaps coactivators with histone acetyltransferases activity as well as the C subunit of PKA complex with Sp1/Sp3 at the *SGK* promoter. It is possible that the stimulatory effect of PKA on SGK transcription reflects H3 phosphorylation and chromatin reorganization rather than, or in addition to, direct phosphorylation of transcription factors.

Although PKA enhances the transcriptional activity of Sp1/Sp3 (161,162), it is not known whether Sp1/Sp3 are PKA targets. Histone H3 phosphorylation and acetylation is also linked to activation of *inhibin-α* subunit gene expression. Activation of the *inhibin-α* promoter requires synergism among PKA-phosphorylated CREB, SF-1 or LRH-1, and the coactivators CBP and SRC-1 (163). SF-1 and CREB constitutively bind to and interact (directly or indirectly) on the *inhibin-α* promoter (163), and CBP is recruited by and binds to phosphorylated CREB (159,160) and to SF-1 (164). The dependence on PKA for activation of the *inhibin-α* gene has historically been assumed to be a consequence of the phosphorylation of CREB (107), leading to CBP recruitment (160,163). It is also well established that PKA potentiates SF-1 transactivation activity. Although SF-1 can be phosphorylated by PKA in vitro, there is no convincing evidence for its in vivo phosphorylation (165), and mutation of its consensus PKA phosphorylation site (Ser430Ala) does not affect the stimulatory effects of PKA on SF-1 transactivational activity (166). There is also a report that PKA can decrease the degradation of SF-1, leading to an increase in SF-1 protein levels (166). However, recent studies suggest that PKA in an unknown manner weakens the interaction between the inhibitory dosage sensitive sex reversal protein-1 and SF-1, leading to recovery of SF-1 transcriptional activity (167). It is conceivable that activation of SF-1 target genes such as *inhibin-α* by PKA additionally and/or preferentially reflects the requirement for histone H3 phosphorylation by PKA.

It is likely that the rapid phosphorylation of histone H3 on Ser10 by PKA and acetylation on Lys14 by CBP or other histone acetyltransferases constitute a necessary step in the transcriptional activation of many FSH responsive genes, leading to granulosa cell differentiation. FSH is also predicted to promote alterations in the methylation patterns of histone H3, although this H3 modification has not yet been reported. It is possible that histone H3, in its phosphorylated, acetylated, and possibly methylated conformation, functions as a scaffold to mediate the assembly of the multiprotein complex of transcription factors, coactivators, and basal transcription factors, which leads to transcription (150,152). Based on results in a human embryonic kidney cell line (163), it is also expected that histone H4 is covalently modified in granulosa cells in response to FSH. Using the chromatin immunoprecipitation assay and immunoprecipitating with an antibody to acetylated histone H4, it was shown that PKA increased histone H4 acetylation associated with the *inhibin-α* promoter (163). The increased histone acetyltransferase activity could result either from the recruitment of CBP (168,169) or the recruitment by CBP of PCAF (170,171).

Histone H1 is also phosphorylated in granulosa cells in response to FSH, although with a slower time course than that of histone H3 (100). Histone H1 binds to the outer surface of the DNA that surrounds the core histones and to the stretches of linker DNA that connect nucleosomes (172). Histone H1 is known to be an in vitro PKA substrate, with phosphorylation on Ser37 (110,173), and is phosphorylated on Ser37 in vivo in liver in response to glucagon treatment (110) and in response to forskolin addition to N18 neuroblastoma cells (173). It is predicted that histone H1 phosphorylation contributes to chromatin remodeling, although less is known about the role of this histone in regulating transcriptional activation.

Because most of the known differentiation responses to FSH are mediated by cAMP/PKA, it was initially assumed that the genes activated by FSH would all contain CREB-binding sites. However, cAMP response elements have been identified only in the promoters of a subset of FSH-regulated genes, including *inhibin-α* (174), *aromatase* (144), *GIOT-1* (27), *Egr-1* (26), and *c-Fos* (158). Therefore, FSH is expected to activate additional signaling pathways. Moreover, activation of a number of FSH target genes, such as inhibin-α and GIOT-1, require not only CREB but also the ERK target SF-1 (27,163). Indeed, as discussed below, FSH also stimulates the activation of the ERKs in a PKA-dependent manner, based on the ability of both Myr-PKI and H89 to prevent FSH-stimulated ERK activation (98,175,176).

PKA Signals to Activate the ERKs in Granulosa Cells

The ubiquitous ERKs, which belong to the MAPK family, are activated by a variety of receptor agonists. These kinases are classically activated by receptor tyrosine kinases such as the insulin, EGF, or IGF-1 receptors. Upon activation of these receptors, their consequent autophosphorylation creates specific binding sites for Src homology 2 containing proteins such as Grb2 (growth factor receptor binding protein 2) (177). Grb2, which is complexed with Sos, binds to the receptor tyrosine kinase, and the guanine nucleotide exchange factor Sos promotes activation of Ras by stimulating its GDP release. Active Ras then promotes activation of the Ser/Thr kinase Raf-1, which in turn phosphorylates/activates the ERK kinase MEK. MEK then phosphorylates ERK on Thr and Tyr residues, resulting in ERK activation. GPCRs also activate ERKs generally by promoting the transactivation of a receptor tyrosine kinase (178), resulting in its phosphorylation and consequent activation of the receptor tyrosine kinase pathway to ERK, although the mechanisms of receptor tyrosine kinase transactivation vary for GPCRs and are incompletely understood (179–182). Although GPCR-generated cAMP is most commonly reported to inhibit ERK activation, especially in nonendocrine cells (183), cAMP also activates ERK either via PKA by stimulating the phosphorylation of the small G protein Rap1, leading to the activation of B-Raf, MEK, and ERK (184,185), or by directly binding to and activating the Rap1 guanine nucleotide exchange factor Epac, leading to activation of B-Raf, MEK, and ERK (94–96).

In preantral granulosa cells, FSH does not promote the phosphorylation/activation of MEK, yet ERK phosphorylation is rapidly enhanced by FSH (98). FSH was shown to regulate the association of a 100-kDa protein tyrosine phosphatase with ERK in a PKA-dependent manner (98). In the absence of FSH, a tonic stimulatory pathway promotes activation of the components of the ERK pathway upstream of ERK,

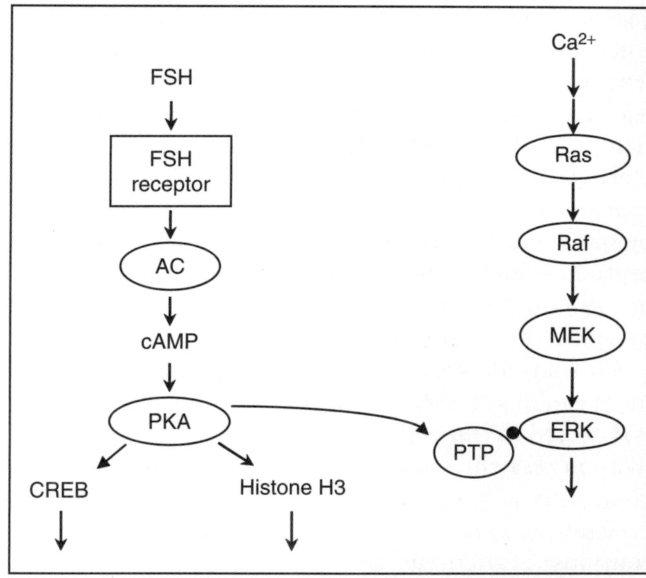

FIG. 4. FSH via cAMP/PKA activates ERK by stimulating the phosphorylation and consequent dissociation of an inhibitory protein tyrosine phosphatase (PTP).

including MEK; however, ERK activity is blunted by its association with a protein tyrosine phosphatase that inactivates ERK (Fig. 4). Thus, although addition of the MEK inhibitor PD98059 blocks FSH-stimulated ERK phosphorylation, the upstream kinase MEK is already activated in untreated granulosa cells and is not further activated by FSH (98). Upon FSH stimulation of granulosa cells, PKA catalyzes the phosphorylation of a 100-kDa protein tyrosine phosphatase that is complexed with ERK, resulting in dissociation of the phosphatase from ERK (98). In the absence of the associated phosphatase, ERK is relieved from inhibition and becomes activated by the tonic stimulatory pathway (98). This granulosa cell phosphatase is recognized by an antibody that recognizes the ERK-associated phosphatase, protein tyrosine phosphatase-SL, but based on its larger size and lack of recognition by other protein tyrosine phosphatase-SL antibodies, the ERK-associated phosphatase in granulosa cells appears to be distinct from protein tyrosine phosphatase-SL (98).

Regulation of the tonic pathway that promotes activation of MEK in granulosa cells is poorly understood. Both extracellular Ca^{2+} and the EGF receptor appear to be necessary for FSH-stimulated ERK activation, based on the abilities of Ca^{2+} chelation with ethyleneglycotetraacetic acid and the recognized EGF receptor inhibitor AG1478 (186) to abrogate this response (98). Moreover, in a primary culture model, Ca^{2+} entry appears to be mediated by L-type Ca^{2+} channels, based on the ability of the L-type Ca^{2+} channel inhibitor nifedipine (98) to block FSH-stimulated ERK activation. Consistent with Ca^{2+} entry via L-type

Ca^{2+} channels, the α1c subunit of the class C L-type Ca^{2+} channel appears to be phosphorylated on the PKA site, Ser1928 (114), in vehicle-treated granulosa cells (98), consistent with Ca^{2+} entry. Ca^{2+} entry also appears to be upstream of the EGF receptor, because the EGF receptor inhibitor AG1478 blocks the ability of the Ca^{2+} ionophore A23187 to stimulate ERK phosphorylation (98). However, the stimulus that promotes phosphorylation of the α1c subunit of the L-type Ca^{2+} channel leading to Ca^{2+} entry in rat granulosa cells has not been identified, and EGF receptor phosphorylation is not detected under conditions where the EGF receptor appears to be active (98).

Despite an incomplete understanding of the tonic pathway that directs activation of the ERK pathway in granulosa cells, FSH-stimulated ERK activation appears to be necessary, but not sufficient, for induction of the MAP2D (98) as well as for RIIβ, VEGF, and the LH receptor (unpublished data). These conclusions are based on the expression of these proteins or their promoter-reporter activities in the absence and presence of the MEK inhibitor, PD98059 (187). Similarly, this MEK inhibitor has also been shown to inhibit the FSH-dependent induction of the immediate early gene *Egr-1* in granulosa cells (26), suggesting that induction of *Egr-1* by FSH requires MEK-dependent ERK activation.

FSH has also been shown to increase protein expression for a number of immediate early genes that comprise the AP-1 family, including *JunB*, c-*Jun*, c-*Fos*, and *Fra2* (36) as well as c-*Myc* (37,188). *JunB* has been shown in granulosa cells to regulate the expression of the *inhibin-β* promoter (189). A recent report suggested ERK-dependent phosphorylation of these proteins stabilizes them by inhibiting their degradation; however, this response requires persistent ERK activation over a time course consistent with the expression of these immediate early genes (39). Based on the transient activation of ERK in granulosa cells [which is generally undetectable by 1 hour post-FSH (98)], it is unlikely that the mechanism by which FSH increases the expression of these proteins rests with their ERK-dependent phosphorylation.

ERK can also phosphorylate or direct the phosphorylation of a number of transcription factors and coactivators that have been identified as participants in the induction of FSH target genes. Figure 5 reviews ERK targets that have been established in various cellular models (147). For example, ERK is established to phosphorylate SF-1 on Ser203, resulting in recruitment of coactivators as well as driving the formation of a compact structure that results in formation of an active conformation (190,191). ERK also phosphorylates Sp1/Sp3 on Thr266, thereby enhancing its DNA binding activity (192). ERK is believed to phosphorylate CBP on Ser436 (193), and this phosphorylation has been suggested to be necessary for its recruitment of the AP-1 complex (194). It is interesting that p300 lacks this phosphorylation site (194). However, phosphorylation of CBP by ERK in vivo has not yet been reported, to our knowledge. ERK also phosphorylates a number of kinases, including RSKs 1–4, MSK1 and -2, and MAPK-interacting kinases (MNK1 and -2) (147). These kinases in turn phosphorylate a number of transcription factors, in a cell specific manner, as reviewed in Fig. 5. Only ERK-catalyzed RSK phosphorylation has been positively identified in granulosa cells treated with FSH (101), although a number of additional ERK substrates are expected to be phosphorylated in granulosa cells in response to FSH. Although RSK can phosphorylate CREB and histone

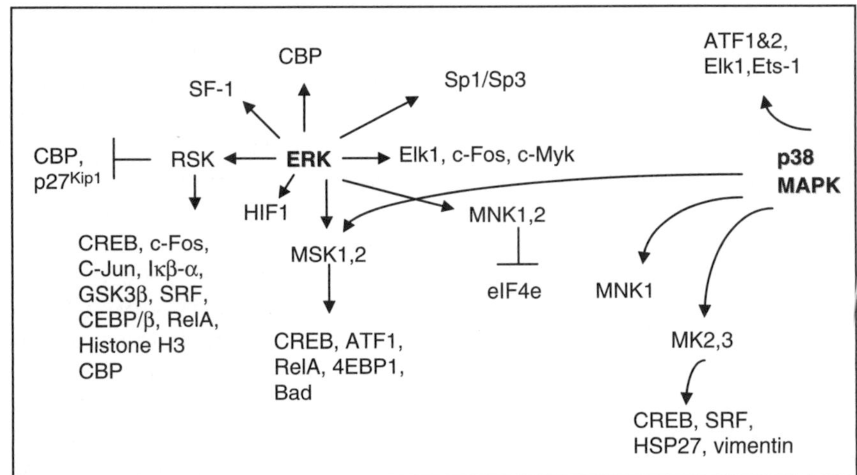

FIG. 5. Summary of identified ERK and p38 MAPK substrates. *Arrows* indicate stimulation; *perpendicular lines* indicate inhibition of activity.

H3 in some cells, FSH-stimulated phosphorylation of these proteins is not inhibited by the MEK inhibitor PD98059 (101), indicating that neither CREB nor histone H3 is an RSK substrate in granulosa cells. Unphosphorylated RSK has been reported to sequester CBP and inhibit its histone acetyltransferases activity (195,196), although this might be cell specific because RSK has also been reported to enhance coactivator activity of CBP (197). An association between RSK and CBP has not been investigated in granulosa cells, but based on the number of transcription factors that require CBP in FSH-regulated signaling pathways in granulosa cells, including CREB (159,198), HIF (199), and FOXO1 (200), as described below, or that bind that CBP, including SF-1 (164) and Smad (201,202), regulation of the availability of CBP could be an important regulatory mechanism by which FSH enhances target gene expression. Thus, ERK signaling likely participates at least indirectly in the regulation of the expression of a number of FSH target genes by regulating the phosphorylation of a number of transcription factors and/or coactivators.

Mammalian cells contain two additional MAPKs, the p38 MAPKs and the JNKs. Although activation of the JNKs by FSH to our knowledge has not been reported, the p38 MAPKs are activated by FSH in rat granulosa cells (203–205) (Fig. 6). The effect of FSH is mimicked by forskolin (205) and appears to be PKA dependent, based on the ability of H89 to inhibit p38 MAPK phosphorylation (205), although this result is controversial (204). A downstream p38 MAPK target whose phosphorylation is increased by FSH is the small heat shock protein HSP27 (205) (Fig. 5).

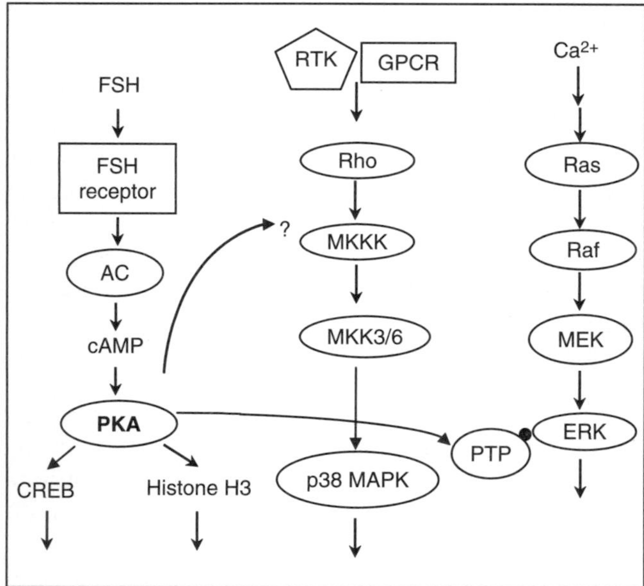

FIG. 6. FSH via cAMP/PKA signals to activate the p38 MAPK pathway through an unknown mechanism.

Although the significance of HSP27 phosphorylation in granulosa cells has not been delineated, upon phosphorylation HSP27 loses its ability to form oligomers and to block actin polymerization (206,207). HSP27 is also recognized to inhibit apoptosis in various cellular models (208). Rounding of granulosa cells, a characteristic response to FSH stimulation (209), is inhibited by the p38 MAPK inhibitor SB203580, suggesting that the p38 MAPK pathway contributes to granulosa cell cytoskeletal changes (205). Although FSH activates the upstream kinases MKK3 and MKK6 (203), the cellular mechanism by which FSH, presumably via PKA, signals into this pathway is not known (Fig. 6). Often, this pathway is activated by the small GTPases of the Rho family, leading to the activation of a number of possible kinases that direct activation of MKK3/6 (147). The potential ability of FSH to signal into a Rho family member has not been investigated to our knowledge.

p38 MAPK also regulates the phosphorylation of a number of transcription factors in various cells, as reviewed in Fig. 5, and also promotes the phosphorylation of MNK1 (but not MNK2) and MK2 and -3 (147). HSP27 is actually phosphorylated by MK2 and -3 and not directly by p38 MAPK. CREB can also be a target of MK2 and -3 but is not likely to be phosphorylated by these kinases in granulosa cells in response to FSH, based on the inability of the p38 MAPK inhibitor SB203580 to affect FSH-stimulated CREB phosphorylation (101). Other than HSP27, p38 MAPK targets in immature granulosa cells have not been identified.

FSH Increases Intracellular Ca^{2+}

FSH is reported to increase intracellular Ca^{2+} in a cAMP-dependent manner in porcine granulosa cells, and the Ca^{2+} appears to be derived largely from extracellular sources (210–212). Ca^{2+} entry is inhibited with verapamil, an L-type Ca^{2+} channel blocker, suggesting that plasma membrane Ca^{2+} channels are opened by cAMP or PKA (211). Moreover, Ca^{2+} is required for FSH to activate the *SCC* promoter-reporter in porcine granulosa cells (210), consistent with the participation of Ca^{2+} in a signaling pathway leading to activation of the *SCC* promoter. Although cAMP analogues that selectively activate PKA mimic FSH to raise intracellular Ca^{2+} levels in Sertoli cells (213), equivalent studies have not been performed in granulosa cells. However, in contrast to these data, there are two reports that FSH itself does not mobilize intracellular Ca^{2+}, one in human embryonic kidney cells heterologously expressing the human FSH receptor (214) and another in a rat ovarian granulosa cell line (215). Moreover, as discussed above, in primary rat granulosa cells in culture, L-type Ca^{2+} channels appear to be open and Ca^{2+} appears to signal at least into the ERK pathway, leading to activation of MEK (98), in the absence of FSH. Carnegie and Tsang (216) reported that in rat granulosa cells, ethyleneglycotetraacetic acid and the Ca^{2+} channel blocker verapamil inhibit FSH-stimulated progesterone production and that the Ca^{2+} ionophore A23187 stimulates progesterone production. Taken together, these results suggest that whereas FSH appears to promote the entry of Ca^{2+} into porcine granulosa cells in a cAMP dependent manner, FSH does not appear to stimulate a rise in intracellular Ca^{2+} concentrations in rat granulosa cells. Rather, in rat granulosa cells intracellular Ca^{2+} is elevated by a pathway that is independent of FSH.

Potential Ca^{2+}-dependent signaling pathways have also been studied in granulosa cells. CaMKs II and IV are present in immature rat and porcine granulosa cells (34,217), and FSH promotes a decline in CaMK IV expression to undetectable levels by 24 hours in rat granulosa cells (34). Mice deficient in CaMK IV exhibit impaired follicular development and ovulation (34). Cotransfection of active CaMK IV in addition to an *SCC* promoter-luciferase construct into porcine granulosa/luteal cells resulted in a significant increase in basal transcription of this gene that was blocked by a dominant negative CREB mutant (217). Cotransfection with CaMK II was ineffective (217). This result suggests that a possible CaMK IV target in porcine granulosa/luteal cells could be CREB. Ser133 on CREB is a recognized CaMK IV target in neuronal cells (218). Although these results suggest that CaMK IV plays a role in the basal transcription of SCC in mature granulosa cells, the role of CaMKs in immature cells is less clear. An early report by Conti and collaborators (219) showed that FSH treatment of rat granulosa cells activated a cAMP-PDE that required Ca^{2+}/CaM. Additionally, FSH-stimulated progesterone production in rat granulosa cells is reported to be inhibited by the CaM inhibitor R24571 (216), although FSH-stimulated ERK activation (Hunzicker-Dunn, unpublished data) is not inhibited by the CaMK II inhibitor Kn62 (220). It is not known whether FSH-stimulated SCC expression is regulated by the ERK pathway in granulosa cells. Thus, additional studies are required to clarify the role of Ca^{2+} and its regulated kinases in FSH actions.

FSH Signals to Activate the PI-3K Pathway

There is abundant evidence that FSH promotes the activation of PI-3K, leading to activation of Akt (protein kinase B) and downstream targets in granulosa cells (98,99,204,221–223). The PI-3K inhibitors wortmannin and/or LY294002 block FSH-stimulated induction of target gene products that characterize differentiated granulosa cells, including the LH receptor, MAP2D, VEGF, inhibin-α, and RIIβ (222). In addition, constitutively active Akt enhances and dominant negative Akt blocks FSH-stimulated induction of the LH receptor, aromatase, inhibin-α, and 3β-hydroxysteroid dehydrogenase (3βHSD) (99). These data suggest that FSH-dependent activation of the PI-3K pathway is obligatory for the induction of critical genes that characterize the differentiated granulosa cell phenotype. However, the cellular mechanism by which FSH directs activation of PI-3K is poorly understood.

PI-3K. PI-3K consists of a 100-kDa catalytic subunit and an 85-kDa regulatory subunit (224). Reports suggest the 85-kDa regulatory subunit contains a domain that inhibits activation of the catalytic subunit and a binding domain for small GTPase proteins like Ras (225). Binding of the 85-kDa subunit to tyrosine-phosphorylated residues and consequent phosphorylation of Tyr688 on this subunit is sufficient to relieve its inhibitory domain, resulting in activation of the catalytic subunit of PI-3K (225). However, full activation of the catalytic subunit also requires binding of active Ras to the 85-kDa regulatory subunit, based on evidence that a dominant negative Ras reduces the activation of the catalytic subunit (225).

PI-3K is best characterized as being activated downstream of receptor tyrosine kinases such as the insulin receptor or the IGF-1 receptor, where binding of the 85-kD regulatory subunit to the tyrosine-phosphorylated adaptor proteins, the insulin receptor substrates, has been demonstrated. However, PI-3K can also be activated downstream of other receptor tyrosine kinases, such as the EGF receptor, by binding to phosphorylated tyrosine residues on the receptor, or downstream of nonreceptor tyrosine kinases such as Src and the Janus kinases (224,226). PI-3K phosphorylates the 3′-OH position of the inositol ring of phosphatidylinositol 4,5-bis phosphate, generating phosphatidylinositol 3,4,5-trisphosphate. By a pathway that is incompletely understood, activation of PI-3K stimulates the recruitment of Akt from the cytosol to the plasma membrane. There is obligatory binding of phosphatidylinositol 3,4,5-trisphosphate to pleckstrin homology domains of both Akt and 3′-phosphoinositide-dependent kinase-1 (224). 3′-Phosphoinositide-dependent kinase-1 then phosphorylates Akt on Thr308; Akt is subsequently phosphorylated on Ser473 by an unidentified protein kinase, resulting in the activation of Akt and its translocation to the nucleus and likely to other cellular locations. 3′-Phosphoinositide-dependent kinase-1 is thought to be constitutively active and to reside predominately at the plasma membrane (224).

Consistent with the classic manner in which the PI-3K pathway is activated, in granulosa cells IGF-1 activates the PI-3K pathway via the IGF-1 receptor tyrosine kinase (159,204,222,227). That IGF-1 is important for follicular function is evidenced by the fact that IGF-1 null mice are infertile; follicles of IGF-1 null mice do not progress beyond the preantral stage of development (228). Although granulosa cells of preantral and antral follicles are the primary site of IGF-1 synthesis (229), FSH does not promote IGF-1 expression by rat granulosa cells (65,230). Rather, IGF-1 appears to be necessary for the augmented expression of the FSH receptor in preantral and early antral follicles, and this reduced FSH receptor expression in IGF-1 null mice has been suggested to be the basis for their infertility (65). It is well established that FSH and IGF-1 strongly synergize to activate a number of FSH target genes, including those for *SCC*, *aromatase*, and the *LH receptor* (230–234). However, IGF-1 alone promotes only a minimal activation of the target genes (230,234), indicating that additional FSH targets/responses are required to induce the full differentiation response.

Although FSH also activates the PI-3K pathway, the cellular mechanism is poorly understood. Signaling by FSH to activate the PI-3K pathway leading to the phosphorylation of Akt in granulosa cells is mimicked by cell permeable cAMP analogues or the adenylyl cyclase activator forskolin (99,204,222). However, the site at which cAMP or its effector intersects the PI-3K signaling pathway has not been identified but clearly must be upstream of PI-3K itself (Fig. 7). Although FSH appears to activate this pathway in Sertoli cells by stimulating the expression of IGF-1 (235), there is no evidence that the rapid activation of Akt, which is detected within 10–30 minutes of FSH addition to rat or porcine granulosa cells (204,221,222), is mediated by the autocrine production of IGF-1.

Akt Substrates. General downstream targets of Akt can include the transcription factor CREB, the proapoptotic protein Bad, PDE3β, nitric oxide synthase (NOS), the translation antagonist tuberin, GSK3β, the transcription factor NF-κβ, select members of the FOXO transcription factor family, D-type cyclins, the coactivator CBP, the cell cycle inhibitor $p27^{Kip1}$, the translation elongation factor E2F, and others (145,226) (Fig. 8). Thus, Akt can signal to regulate transcription, translation, metabolism, cell survival, and differentiation, but most likely this occurs in a cell-specific and perhaps agonist-dependent manner.

CREB. In rat granulosa cells, FSH-stimulates CREB phosphorylation on Ser133. However, CREB phosphorylation is mediated predominately by PKA, as reviewed above, and not by Akt based on the inability of the PI-3K inhibitor wortmannin to reduce

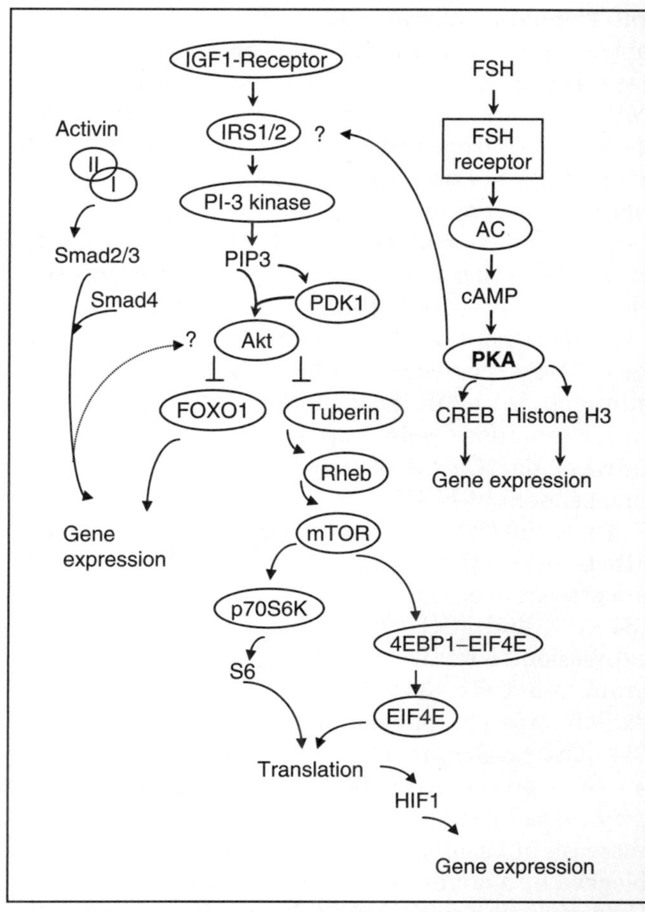

FIG. 7. FSH via cAMP/PKA signals via an unknown pathway to activate PI-3K, leading to mTOR-stimulated translation. FSH and activin synergize to promote gene expression. The mechanism by which activin in the presence of FSH promotes prolonged Akt phosphorylation is not known.

CREB phosphorylation (98) and on the inability of CREB to be further phosphorylated in granulosa cells infected with constitutively active Akt (99).

Tuberin. FSH also stimulates the phosphorylation of tuberin, a product of the *tuberous sclerosis* (TSC)

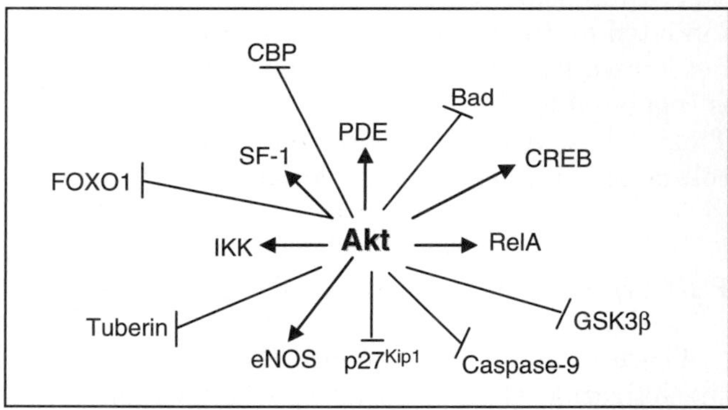

FIG. 8. Summary of identified Akt substrates. *Arrows* indicate stimulation; *perpendicular lines* indicate inhibition.

tumor suppressor gene (222). In its unphosphorylated state, tuberin (TSC2) in a complex with TSC1 (Hamartin) is active and inhibits translation (236). The tuberin–TSC1 complex has recently been shown to function as a GTPase activating protein for the protein ras homolog enriched in brain (Rheb) (237,238). This TSC complex stimulates the hydrolysis of the GTP bound to Rheb, converting $Rheb^{GTP}$ to $Rheb^{GDP}$, resulting in the inactivation of Rheb. $Rheb^{GDP}$ does not stimulate translation. Phosphorylation of tuberin on Thr1462 by Akt (236) results in an inhibition of the GTPase activating protein activity of the TSC1–TSC2 complex (237). As a result, the hydrolysis of the $Rheb^{GTP}$ is inhibited, maintaining Rheb in its active conformation. By a poorly understood mechanism, $Rheb^{GTP}$ promotes activation of the mammalian target of rapamycin (mTOR) protein kinase. mTOR signals to at least two substrate proteins to stimulate translation, p70 ribosomal S6 protein kinase ($p70^{S6K}$), and a binding protein that sequesters the eukaryotic translation initiation factor eIF4E (eIF4E-BP1) (239–243). Phosphorylation of $p70^{S6K}$ on Thr389 leads to the phosphorylation of its substrate protein, the ribosomal S6 protein. Phosphorylation of S6 protein, one of 30 ribosomal proteins that along with the 18S rRNA comprises the 40S ribosomal subunit complex, stimulates the translation of mRNAs containing a 5′-oligopyrimidine tract (244). Phosphorylation of eIF4E-BP1 on Ser65 and other sites releases eIF4E (245,246). Freed eIF4E, which binds to the 5′-methyl cap of mRNAs (244,247,248), now stimulates cap-dependent translation and is the rate-limiting protein in translation initiation (247).

FSH stimulates the phosphorylation of tuberin on Thr1462 within 10 minutes of its addition to rat granulosa cells, leading to the activation of mTOR and phosphorylation of its substrates $p70^{S6K}$ and downstream S6 protein as well as 4E-BP1 (222). That mTOR activation is crucial for activation of a subset of FSH target genes is evidenced by the ability of the mTOR inhibitor rapamycin (239) to inhibit the induction of MAP2D and RIIβ protein expression as well as promoter-reporter activities for the *LH receptor*, *VEGF*, and *inhibin-α* (222). These results suggest that a very rapid effect of FSH in granulosa cells is to stimulate translation. Indeed, it has been recently suggested that activation of the PI-3K and ERK pathways preferentially promotes translation rather than transcription (249). mTOR-dependent targets of translation whose expression is increased in various cellular models include the cyclins, resulting in increased cell proliferation (250,251). Certainly a future area of research should be the identification of FSH-targeted translation products, something that should be facilitated by the advancing field of proteomics.

It is quite interesting that IGF-1 also stimulates the phosphorylation of Akt, tuberin, $p70^{S6K}$, and S6 protein in rat granulosa cells (222), but as already indicated, IGF-1 does not stimulate expression of FSH target genes (230,234), at least to levels seen with FSH. This result suggests that additional pathways activated by FSH but not by IGF-1 are required for FSH target gene expression.

One of the translation targets that FSH stimulates in granulosa cells is the transcription factor HIF-1 (222). HIF-1 is a heterodimer, consisting of HIF-1α and HIF-1β, the latter also known as the aryl hydrocarbon receptor nuclear transporter. HIF-1 is a member of the basic helix-loop-helix/Per/aryl hydrocarbon receptor nuclear transporter/Sim family of transcription factors that bind to a modified E-box on DNA promoters (252). HIF-1β is constitutively expressed; HIF-1α protein levels are regulated. HIF-1 is best known for its regulation of cellular responses to hypoxia.

Under normoxic conditions, HIF-1α is hydroxylated on two critical proline residues by oxygen-dependent prolyl hydroxylases, triggering HIF-1α degradation [reviewed in (252,253)]. Hydroxylation of the proline residues creates a binding site for the protein product of the *von Hippel-Lindau* tumor suppressor gene, VHL. VHL is an E3-ubiquitin ligase and promotes the polyubiquitination of proline hydroxylated forms of HIF-1α, leading to its degradation by the proteosomal system. Cobalt has been shown to inhibit the interaction of HIF-1α and VHL (254) and is often used experimentally to stabilize HIF-1α protein levels. HIF-1 activity is also regulated by a second oxygen-dependent mechanism. Under normoxic conditions, a protein called "factor inhibiting HIF-1" catalyzes the hydroxylation of an asparagine residue in the C-terminal transactivation domain of HIF-1α, preventing the binding of its coactivator p300/CBP. Under hypoxic conditions, the activity of the oxygen-dependent prolyl hydroxylases is reduced, and HIF-1α protein is stabilized. The activity of the asparaginyl hydroxylase is also reduced under hypoxic conditions, thereby allowing the binding of p300/CBP to stimulate HIF-1 activity. The prolyl hydroxylases require iron as a cofactor, so agents that chelate iron such as desferrioxamine inhibit the propyl hydroxylases, thereby stabilizing HIF-1α protein levels.

HIF-1α protein can also be regulated at the level of translation upon activation of the PI-3K pathway by insulin or IGF-1 (255–257). In these studies, expression of HIF-1α protein correlated with activation of HIF-1 DNA binding activity and/or HIF-1–dependent reporter activities. Inhibitors that abrogate the ability of these growth factors to stimulate the phosphorylation of $p70^{S6K}$ and 4E-BP1 block the increase in HIF-1α protein (256,257).

In rat granulosa cells, FSH promotes an increase in HIF-1α protein (Fig. 7), which is detected in the presence either of a proteosomal inhibitor or $CoCl_2$ (222). FSH also promotes an increase in HIF-1 activity, detected using a minimal hypoxia response element linked to the luciferase reporter in rat granulosa cells. Moreover, coexpression of a HIF-1 dominant negative construct, lacking its DNA binding and transactivation domains (258), reduces the ability of FSH to activate promoters for the *LH receptor*, *VEGF* and *inhibin-α*, as detected using promoter-reporter assays (222). These results suggest that HIF-1 activity is required for FSH to activate the *LH receptor*, *VEGF*, and *inhibin-α* genes. However, although each of these genes contains apparent HIF response elements and *VEGF* is an established HIF target, it is not known in granulosa cells if HIF directly regulates the activities of these genes or if the action of HIF is indirect.

Transcriptional activity of HIF-1α requires binding of CBP/p300 to the C-terminal transactivation domain (253). The binding of CBP/p300 is negatively regulated in an oxygen-dependent manner by the hydroxylation of Asn803 by "factor inhibiting HIF-1" [reviewed in (253)]. There is also evidence, albeit controversial, that HIF-1α must be phosphorylated by ERK to achieve full transcriptional activity [reviewed in (199)]. It is interesting that activation by FSH of promoter-reporter activities or induction of protein for a number of apparent HIF targets in granulosa cells, such as VEGF, RIIβ, and MAP2D, are inhibited by the MEK inhibitor PD98059 (unpublished data). Moreover, activation of VEGF transcription in other cells requires cooperation between HIF and an AP-1 complex [reviewed in (199)]. This apparent ERK effect could reflect phosphorylation of HIF-1α, CBP, or members of the AP-1 complex, as reviewed in Fig. 5.

FOXO1. A second Akt target in rat and porcine granulosa cells is FOXO1 (97,221,259,260). FOXO1 is a member of the forkhead O class of transcription factors that were originally identified by their disruption in the pediatric tumor rhabdomyosarcoma (261). Three of the members of this family are Akt targets: FOXO1, FOXO3a, and FOXO4 [reviewed in (262,263)]. FOXO-regulated genes control the cell cycle, cellular metabolism, and cell death. These transcription factors bind to DNA as monomers at the insulin response element consensus sequence TT(A/G)TT(T/G)(A/G)(T/C) on target genes. In the absence of signaling down the PI-3K pathway, the Akt-regulated FOXO transcription factors are bound to DNA and are active. They can function as transcriptional activators to stimulate such target genes as the cell cycle inhibitor *p27^{Kip1}* (264–266), glucose-6-phosphatase activity (267), *IGF binding protein-1* (200), as well as transcriptional repressors to inactivate such target genes as *cyclin D1* and *D2* (268). Upon phosphorylation by Akt, FOXO1, -04, and -03a exit the nucleus and are targeted for degradation. Three serine or threonine residues on FOXO proteins are phosphorylated by Akt. For FOXO1, phosphorylation of Ser256 diminishes DNA binding and is necessary for the phosphorylation of Thr24 and Ser319 (269). Phosphorylation of Thr24 disrupts association of FOXO1 with CBP and stimulates binding to 14-3-3 proteins, and phosphorylation of Ser319 stimulates nuclear export (263,270,271). Based on a report that Akt phosphorylates CBP on Ser1814, suppressing transcriptional activation by CAATT enhancer binding protein β (CEBP/β) by disrupting the association of CBP with CEBP/β (272), perhaps Akt-catalyzed CBP phosphorylation in granulosa cells contributes to FOXO1 inactivation. The phosphorylation status of CBP in granulosa cells has not been reported.

FOXO factors also bind nuclear receptors, including those for estrogen (preferentially ER-α over ER-β), progesterone, androgen, thyroid hormone, glucocorticoid, and retinoid acid, presumably via the conserved LxxLL domain located in the C-terminal region of the protein (263). Although the actions of the FOXO factors can require binding to DNA, there is also evidence that these factors can affect transcription of target genes independent of their DNA binding activity, possibly via their association with coactivators such as CBP or with steroid receptors (273). FOXO factors can also be acetylated, possibly by CBP, a modification that is thought to activate FOXO transcriptional activity, although the effect of acetylation is controversial [reviewed in (263)].

There is accumulating evidence from a number of cellular models that FOXO factors promote cell cycle arrest (263). Cell cycle arrest can occur as a result of increased expression of the cyclin dependent kinase (Cdk) inhibitors p27^{Kip1}, p21^{Cip1}, and p130 and/or by decreased expression of the Cdk or their binding partners cyclin D or cyclin E (274,275). In general, progression from G_1 to the S phase of the cell cycle is mediated by increased expression of cyclin D, which partners with Cdk 4/6, followed by increased expression of the Cdk 2 partner cyclin E (274,275). Hyperphosphorylation of the retinoblastoma protein catalyzed by Cdk 4/6 and Cdk 2 relieves inhibition of the transcription factor E2F by retinoblastoma, allowing for the expression of genes necessary for DNA replication and S phase entry.

FOXO transcription factors have been reported to inhibit cell cycle progression by enhancing transcription of p27^{Kip1} and other Cdk inhibitors (264–266, 276,277), by repressing cyclin D transcription (268,278,279), and/or by increasing transcription of the unconventional cyclin G2 which inhibits the cell cycle (280). Porcine granulosa cells undergo proliferation upon stimulation with FSH and insulin (259).

Recent results show that infection of porcine granulosa cells with a dominant negative FOXO1 truncation mutant, which lacks its transactivation domain, increases entry of these cells into the S phase of the cell cycle by redirecting the compartmentalization of $p27^{Kip1}$ from the nucleus to the cytosol and decreasing $p27^{Kip1}$ content (259). Although $p27^{Kip1}$ is known to be phosphorylated on Thr157 by Akt, resulting in its redistribution from the nucleus to the cytosol (281–285), the mechanism by which FOXO1 directs relocation of $p27^{Kip1}$ to the cytosol followed by its apparent degradation is not known.

In contrast to porcine granulosa cells, proliferation of rat granulosa cells and the obligatory induction of cyclin D2 (41) and the S-phase marker proliferating cell nuclear antigen under primary serum-free culture conditions does not occur in the presence of FSH alone or FSH plus IGF-1 (40,234,286). However, in the presence of activin, *cyclin D2* mRNA expression is strongly enhanced by FSH under serum-free culture conditions (234,286). These results suggest that FSH, directly or indirectly, stimulates the expression of activin or a similar transforming growth factor β-family ligand and that both of these ligands converge to stimulate cyclin D2 expression and consequent granulosa cell proliferation. Rat granulosa cells are reported to produce activin subunits throughout early follicular development (287). However, activin levels produced under culture conditions are apparently too low to synergize with FSH, possibly due to contributions from the activin inhibitor follistatin and/or to utilization of activin subunits by inhibin-α, as these cells also produce follistatin and inhibin (288).

Recent results have shown that FOXO1 functions to repress activation not only of the *cyclin D2* gene but also to repress activation of a subset of FSH targets that characterize the preovulatory phenotype, including those for SF-1 and LRH-1 (260). Phosphorylation of FOXO1 by Akt is necessary but not sufficient to achieve the induction of cyclin D2, SF-1, and LRH-1, because FSH alone does not promote these responses (260). Rather, FSH plus activin is required to induce cyclin D2, SF-1, and LRH-1. The ability of FOXO1 to repress the activation of these FSH target genes is inferred from studies in which granulosa cells were infected with an adenoviral vector encoding a constitutively active FOXO1 mutant in which the three Akt phosphorylation sites were mutated to alanine (A3-FOXO1). Direct binding of FOXO1 to the *cyclin D2* promoter was evidenced by chromatin immunoprecipitation assays, this association was abrogated upon stimulation of cells with FSH plus activin, and the association was enhanced and not regulated in cells infected with the constitutively active A3-FOXO1 mutant (260). However, it is not known if other FSH target genes regulated by inactivation of FOXO1 are repressed directly or indirectly by FOXO1.

The induction of cyclin D2, SF-1, and LRH-1 was also abrogated in cells infected with a dominant negative Smad2/3 mutant (260). These results suggest that induction of these proliferation and differentiation responses requires not only relief from repression by FOXO1, but also the contribution of the Smad2/3 transcription factors (Fig. 7). However, the mechanism by which Smad2/3 cooperates with the Akt pathway leading to FOXO1 phosphorylation, resulting in expression not only of FSH target genes that promote granulosa cell differentiation but also those leading to cellular proliferation, has not been elucidated. It is interesting that active (unphosphorylated) FOXO1 can complex with phospho-Smad to promote expression of the cell cycle inhibitor $p21^{Cip1}$ in neuroepithelial and glioblastoma cell lines (289). In granulosa cells, however, phospho-Smad2/3 functions upon inactivation of FOXO1 to drive proliferation. Thus, there is cross-talk between these two pathways but the outcome in granulosa cells is cell specific.

It is not known if FSH plus activin treatment also promotes a redistribution of $p27^{Kip1}$ to the cytosolic fraction in rat granulosa cells upon FOXO1 phosphorylation. However, $p27^{Kip1}$ does not show any evidence of down-regulation, and interestingly, the phosphorylatable Thr157 residue believed to be responsible for its exit from the nucleus is not conserved in rodents (281,282).

A similar repressive function in follicular maturation has been attributed to FOXO3a, but at a much earlier stage of follicular development. Unlike FOXO1 null mice, which die at embryonic day 10.5 (290), FOXO3a null mice show global activation of primordial follicles, resulting in oocyte death and depletion of follicles as a result of increased granulosa cell mitotic activity (291). These results suggest that FOXO3a functions to suppress maturation of primordial follicles.

GSK3β. An additional Akt target in many cells is the Ser/Thr protein kinase GSK3β (Fig. 8). GSK3β is active when the PI-3K pathway is silent. Active Akt phosphorylates GSK3β on Ser9, resulting in the inactivation of GSK3β (224). One of the GSK3β targets in a number of cellular models is the D-type cyclins. In the absence of signaling down the PI-3 kinase pathway, GSK3β phosphorylates the D-type cyclins, leading to their ubiquitination and degradation (292–295). Activation of the PI-3K pathway promotes Akt-catalyzed phosphorylation of GSK3β, thereby rescuing D-type cyclins from degradation. Although there is evidence that FSH promotes the phosphorylation of GSK3β in rat granulosa cells (204), it is not known if the stability of cyclin D2 in granulosa cells is also regulated by GSK3β. However, based on evidence that FSH stimulates only a modest two- to threefold increase in *cyclin D2* promoter-reporter activity in transiently transfected granulosa cells compared with the robust increase in cyclin D2 protein expression (260), it is likely that FSH also stabilizes cyclin D2, possibly via inactivation of GSK3β.

GSK3β can also be phosphorylated by a number of other serine/threonine kinases, including PKA, PKC, and the ERK substrate RSK, on Ser9 to reduce its activity (296,297). Although FSH activates PKA, Akt, and RSK in granulosa cells (98–101,204,221,222), FSH-stimulated GSK3β phosphorylation is strongly inhibited by PI-3 kinase inhibitors, suggesting that GSK3β is predominately an Akt substrate in granulosa cells.

In addition to regulating the stability of the D-type cyclins, GSK3β phosphorylates a number of transcription factors, including β-catenin, CREB, HIF-1, NF-κB, and AP-1, as well as structural proteins and metabolic and signaling proteins such as MAP2, kinesin light chains, glycogen synthase, eIF2B, PKA RIIβ subunit, and protein phosphatase 1 (296,297). Thus, GSK3β can regulate glucose production, translation, transcription, cell survival, and cell motility (296,297). Generally, GSK3β substrates are initially phosphorylated by another protein kinase, priming the substrate for phosphorylation by GSK3β.

Substrate phosphorylation by GSK3β is regulated not only by the phosphorylation status/activity of GSK3β but also by the association of GSK3β with specific protein complexes in which GSK3β is often colocalized with its substrate and the obligatory priming protein kinase(s). One of the best known pathways to regulate GSK3β phosphorylation of substrates is that for Wnt (296,297). In absence of the Wnt ligand, GSK3β is complexed with a scaffold protein (axin), casein kinase 1, adenomatous polyposis coli, and β-catenin, as depicted in Fig. 9. In this complex, β-catenin is targeted for ubiquitination and proteosomal degradation as a result of a priming phosphorylation by casein kinase 1 followed by phosphorylation by GSK3β(297). Activation of the plasma membrane Frizzled receptor by Wnt leads to disruption of this complex and inhibition of GSK3β activity by a poorly understood mechanism that requires Dishevelled and the GSK3β binding protein. As a result, β-catenin is not phosphorylated and escapes from degradation. Unphosphorylated β-catenin translocates to the nucleus where it functions as a coactivator to bind T-cell factor/lymphoid enhancer factor and to activate T-cell factor/lymphoid enhancer factor–regulated genes (297).

Although signaling by the Wnt/β-catenin pathway in preantral granulosa cells has not been investigated, these immature granulosa cells express *Wnt-2*, *Wnt-4*, *Frizzled-1*, *Frizzled-4*, and *Dishevelled-2* mRNA (298,299), suggesting that this pathway is operating in these cells. Moreover, it is likely that FSH signals impact this pathway. It is also possible that FSH via the PI-3K/Akt-stimulated phosphorylation/inactivation of GSK3β stimulates an accumulation of dephosphorylated β-catenin in the nucleus independently of Wnt signaling, as shown in Fig. 9. Future studies are required to determine whether β-catenin, stabilized via the Wnt or FSH-stimulated PI-3K pathways, participates in FSH-dependent target gene activation in granulosa cells.

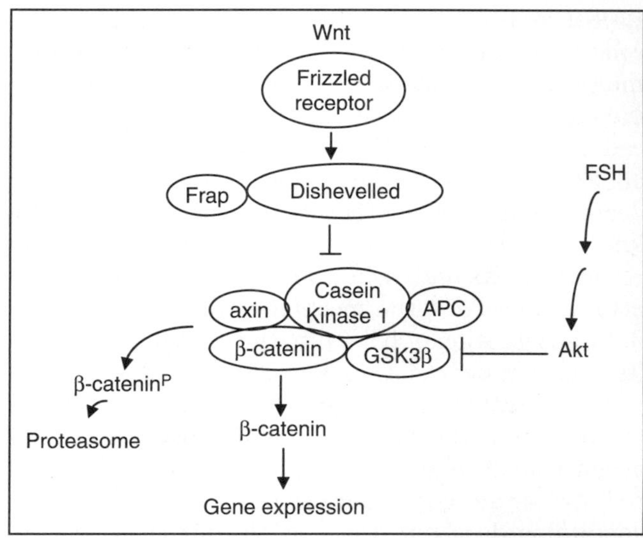

FIG. 9. Schematic diagram of the rescue of hypophosphorylated β-catenin from proteosomal degradation by either Wnt or signaling via Akt.

SGK. FSH and forskolin also induce the immediate early gene *SGK* (101,105,204) and stimulate its apparent phosphorylation, based on an upward shift in mobility on sodium dodecyl phosphate polyacrylamide gel electrophoresis (204). SGK is quite homologous to Akt, as it is activated in a PI-3K–dependent manner by 3′-phosphoinositide-dependent kinase-1 (300) and phosphorylates many of the same substrates phosphorylated by Akt, including members of the FOXO family (301). However, recent evidence suggests that the primary function of this kinase is to regulate the abundance of Na^+, K^+, and Cl^- ion channels by preventing their proteosomal degradation (302,303) by phosphorylating and inactivating a ubiquitin ligase (304). SGK is induced in granulosa cells in a PKA-dependent manner, and its induction is prevented by the PKA inhibitor H89 (105). SGK is not induced in granulosa cells by IGF-1 (204). In contrast to the signaling pathway that regulates the FSH-stimulated induction of SGK in granulosa cells, FSH-stimulated phosphorylation of SGK in granulosa cells is abolished by PI-3K inhibitors wortmannin and LY294002, consistent with FSH signaling to stimulate phosphorylation/activation of SGK via the PI-3K pathway (204). However, neither the substrates nor the function of SGK in granulosa cells is known.

Apoptosis-related PI-3K Substrates. The PI-3K pathway is also known to regulate cell survival versus apoptotic pathways. Indeed, addition of the PI-3K inhibitor LY294002 to rat granulosa cells induces apoptosis (305). Up to 99% of the ovarian follicles undergo apoptosis primarily at the early antral stage of development (in rats) (306). Follicles survive and proceed to final stages of maturation only if they are rescued from apoptosis, predominately by FSH. It is not surprising that apoptosis is a highly regulated event in the ovary and that a number of proteins in the apoptotic pathway that have been identified as being highly expressed in ovarian cells (306).

There are two major pathways that lead to apoptosis: one directed by the death receptors of the tumor necrosis family (TNF) and one directed by the release of cytochrome c (and other proteins) from mitochondria as a result of various cellular stresses such as heat shock, oxidative stress, and DNA damage (307–309). In both of these pathways, cysteine aspartic acid–specific proteases (caspases) become activated and cleave proteins that are crucial to cellular function, resulting in programmed cell death. Initiator caspases are activated in response to a proapoptotic signal and they in turn activate effector caspases, which catalyze substrate proteolysis, leading to cell death.

The gonadotropins appear to regulate primarily the pathway controlling the release of cytochrome c from mitochondria, as depicted in Fig. 10. Three groups of proteins are involved in this pathway in the ovary: (a) those that suppress apoptosis by inhibiting the release of cytochrome c from the mitochondria, including the Bcl-2 (B-cell leukemia/lymphoma 2) protein family members Mcl-1 (myeloid cell lymphoma-1) and Diva/Boo (310,311); (b) those that actively stimulate the release of cytochrome c, such as Bok (Bcl-2–related ovarian killer); and (c) those that do not stimulate cell death by themselves but must heterodimerize with selective antiapoptotic Bcl-2 proteins (present in group a) or with proapoptotic proteins (group b proteins) to stimulate cell death, including Bad (Bcl-xL/Bcl-2–associated death promoter) and Bod (Bcl-2-related ovarian death agonist)/Bim (306,309). Group a proteins contain four Bcl-2 homology (BH) domains, BH1–4; group b proteins do not contain the BH4 domain; group c proteins contain only the BH3 domain.

Upon release of cytochrome c from the mitochondria, cytochrome c complexes with Apaf-1 (apoptotic protease activating factor-1), resulting in the oligomerization of Apaf-1 and recruitment of procaspase-9 (307). Caspase-9 becomes activated as a result of an autocatalytic event, forming the active apoptosome, leading to the activation of effector caspases-3

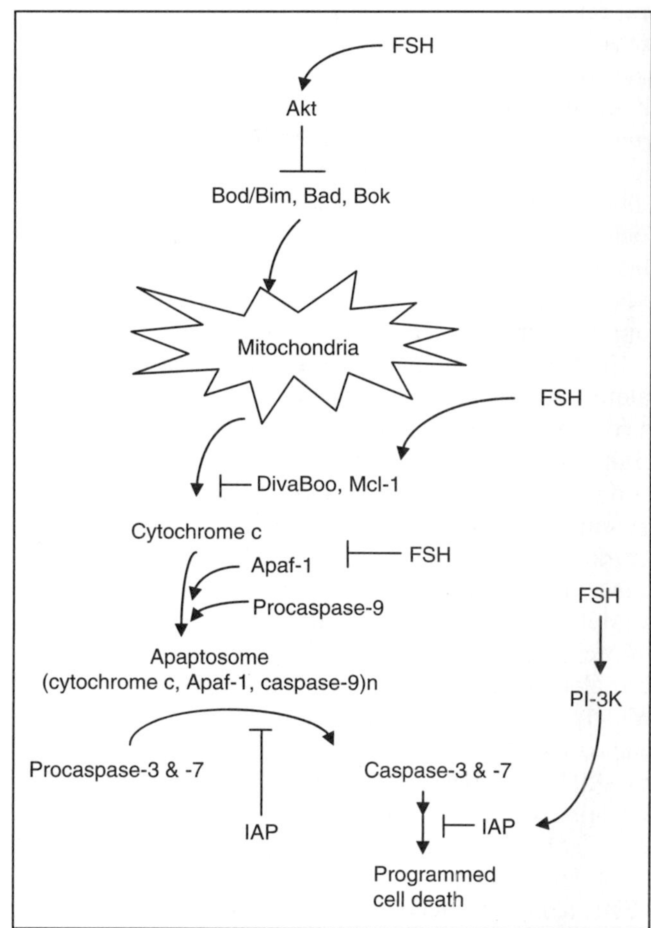

FIG. 10. Schematic diagram of the mitochondrial-dependent apoptotic pathway in ovarian cells and sites at which FSH has been reported to inhibit (*perpendicular lines*) or stimulate (*arrows*).

and -7, which promote the proteolysis of a large number of proteins crucial to cell survival and consequent cell death (307). PI-3K-stimulated Akt has been reported to phosphorylate and inactivate caspase-9 (312); however, the Akt phosphorylation site found in human caspase-9 is not conserved in rat, mouse, or monkey caspase-9 (313). Thus, regulation of caspase-9 phosphorylation/activity at least by Akt does not appear to be a widespread mechanism by which PI-3K signaling promotes cell survival. Although IGF-1 has been shown in porcine granulosa cells to inhibit apoptosis (227), a definitive role for IGF-1 in rat granulosa cell apoptosis, to our knowledge, has not been reported.

The fate of a cell appears to be determined by the balance between survival and apoptotic signals. Transgenic mice overexpressing the antiapoptotic protein Bcl-2 in the ovary show decreased follicular apoptosis and increased folliculogenesis; however, Bcl-2 does not appear to be expressed in the ovary (314). Rather, ovaries appear to express predominately the antiapoptotic Bcl-2 family protein Mcl-1, and the expression of Mcl-1 is increased by treatment of rats with FSH containing pregnant mare's serum gonadotropin (315). Increased expression of Mcl-1 likely contributes to the ability of FSH and pregnant mare's serum gonadotropin to promote cell survival (306). Ovaries also express high levels of the antiapoptotic protein Boo/Diva (310,311). These antiapoptotic Bcl-2 proteins not only inhibit cytochrome c release from the mitochondria but also are believed to sequester Apaf-1, thereby preventing it from activating downstream caspases-9 (306,309).

Bad is a member of the group c proteins, which alone do not trigger apoptosis but rather interact with select antiapoptotic Bcl-2 proteins to initiate apoptosis. Bad, like other members of this group of proteins, is thought to integrate positive and negative apoptotic signals into the apoptotic pathway. Yeast-two hybrid studies with an ovarian cDNA library have shown that one of the proteins that Bad interacts with is Mcl-1, thereby diminishing the antiapoptotic effect of Mcl-1 (315). The phosphorylation of Bad on one of two Ser residues prevents the interaction of Bad with Mcl-1, thereby allowing Mcl-1 to inhibit apoptosis (224). Upon phosphorylation of Bad on Ser137, Bad binds 14-3-3 proteins (316). Although Bad can be phosphorylated by a number of kinases on both Ser137 and Ser113, including PKA, RSK, and MSK (147,224, 306), Bad is a well-known target of Akt (145,224) (Figs. 5 and 8). Based on the ability of FSH to activate Akt, it is likely that the ability of FSH to promote granulosa cell survival requires Akt-dependent Bad phosphorylation. However, FSH-stimulated Bad phosphorylation has not been formally demonstrated in granulosa cells. Activation of Akt by IGF-1 via the PI-3K pathway in granulosa cells is also expected to promote the phosphorylation of Bad, leading to granulosa cell survival.

Antiapoptotic Mcl-1 has also been shown to interact with Bok and Bod/Bim, both of which are proapoptotic Bcl-2 family members and are highly expressed in the ovary (317,318), presumably leading to cell death. Bok, Bod/Bim, and Bad are expected to antagonize the survival effect of Mcl-1, in part by releasing Apaf-1, and Bok additionally directly triggers release of cytochrome c to promote apoptosis (309,318). The high expression of Bok and Bod/Bim in the ovary suggests that these proteins function as specific mediators of ovarian cell death (306,317).

Granulosa cells express Apaf-1 and in the absence of gonadotropin support, and there is mitochondrial release of cytochrome c and increased activation of caspase-3 (319). There is also evidence that gonadotropins decrease Apaf-1 expression (306) (Fig. 10), a response also expected to be antiapoptotic. However, although granulosa cells of caspase-3 null mice show attenuated apoptosis, these cells do undergo apoptosis, suggesting that other effector caspases can compensate for caspase-3 in granulosa cells to direct apoptosis (320).

Additional regulation of the apoptotic pathway occurs via a family of inhibitors of apoptosis (IAPs) that act as substrate inhibitors of caspases and block the activation of the both effector caspases and, at higher concentrations, initiator caspases (309). Of these IAPs, the X-linked IAP (XIAP) is reported to be the most potent (309). FSH has been shown to increase expression of XIAP in rat granulosa cells (223,305) (Fig. 10). Based on evidence that the addition of XIAP antisense oligonucleotides to granulosa cells enhances apoptosis (305), it is likely that the FSH-dependent regulation of XIAP expression is critical to the antiapoptotic effect of FSH (321). There is also a recent report that Akt interacts with and phosphorylates XIAP, protecting it from ubiquitin-directed degradation (322). These results suggest that the regulation of XIAP is crucial to the ability of FSH to rescue granulosa cells from programmed cell death and that FSH potentially regulates XIAP protein at multiple levels. However, FSH-stimulated XIAP phosphorylation via Akt has not been demonstrated in granulosa cells. A second group of regulators of the apoptotic pathway function as inhibitors primarily of the initiator caspases. One of the members of this group of proteins is FLICE inhibitory protein, or FLIP (309). Although FSH has not been shown to regulate the expression of FLIP in granulosa cells, the prosurvival effects of TNF-α in rat granulosa cells are accompanied by increased expression of both XIAP and FLIP (323,324).

NF-κB. The PI-3K pathway can also signal to regulate cell survival by promoting the nuclear translocation of the ubiquitous transcription factor NF-κB. Seven proteins comprise the NF-κB family: p105 and its proteolytic product p50, p100 and its proteolytic product p52, RelA (p65), RelB, and c-Rel (325). These transcription factors function as heterodimers and bind to κB response elements on target genes. More than 200 genes have been reported to be regulated by NF-κB (326). The prototypical NF-κB transcription factor complex consists of p50 and RelA (325). p50/RelA exists in the cytosol as a latent transcription factor complexed to an IκB inhibitor, as shown schematically in Fig. 11. The family of IκB inhibitors include IκB–α, the prototypical IκB inhibitor, as well as IκB-β, γ, and ε (325). In response to growth factor or cytokine stimulation, the IκB kinase (IKK2) (which exists in the cytosol in a complex with IKK1 and the scaffold protein Nemo) is activated and phosphorylates the IκB inhibitor, targeting IκB for ubiquinitation and proteosome-mediated degradation (325). With the dissociation of IκB, the nuclear localization signal of RelA becomes exposed

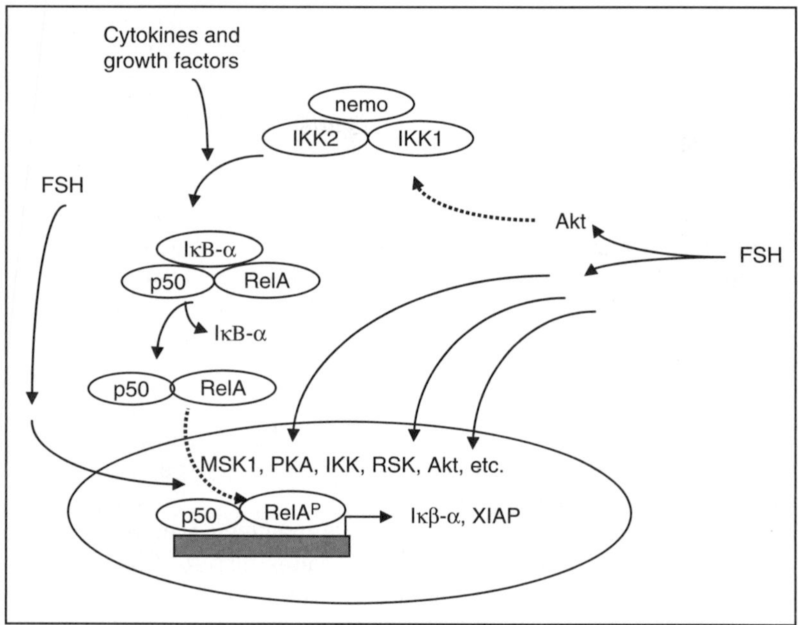

FIG. 11. Schematic diagram of the pathway leading to NF-κB–dependent activation of target genes. Potential sites of FSH stimulation of this pathway are indicated.

and p50/RelA is translocated into the nucleus. Transcriptional activation of this complex, however, additionally requires the phosphorylation of RelA in a cell-specific manner (325). RelA kinases include PKA, MSK1, RSK, Akt, PKCζ, casein kinase 2, IKK, and NF-κB–activating kinase (325,327). Phosphorylation of RelA generally increases the binding of coactivators, like CBP, and/or DNA binding activity (325).

FSH has been shown to promote the translocation of RelA from the cytoplasm to the nucleus of rat granulosa cells (as shown in Fig. 11) and to increase NF-κB–DNA binding activity, detected by formation of a protein–DNA complex using the consensus κB binding site and granulosa cell nuclear extracts in mobility shift assays (223). Additionally, the DNA–protein complex could be super shifted by antibodies to p50 and RelA, consistent with the presence of these proteins in the protein–DNA complex (223). Inhibition of the nuclear translocation of p50/RelA with SN50, a drug that binds to and blocks the nuclear localization signal of NF-κB, blocked the ability of FSH to induce the antiapoptotic protein XIAP and suppressed NF-κB–DNA binding activity (223). These results suggest that the caspase inhibitor XIAP may be an NF-κB target in granulosa cells, as depicted in Fig. 11. However, these authors did not detect the phosphorylation of IκB–α or its degradation during the first 30 minutes after treatment with FSH, in contrast to results with TNF-α treatment of granulosa cells (223). Moreover, modest (2.5-fold) over-expression of a dominant negative IκB–α (which had its IKK phosphorylation sites mutated from Ser to Ala) in granulosa cells did not inhibit the ability of FSH to increase p50/RelA DNA binding activity or XIAP expression but did inhibit the ability of TNF-α to increase p50/RelA DNA binding activity (223). One interpretation of these results is that FSH regulates the nuclear translocation of p50/RelA by an alternative pathway independent of IκB phosphorylation.

Although a noncanonical pathway to activate NF-κB has been reported that involves the proteolytic processing of p100, this pathway generates primarily p52/RelB and not p50/RelA (325). Pretreatment of granulosa cells with the PI-3K inhibitors LY292004 or wortmannin inhibited FSH-stimulated activation of p50/Rel DNA binding activity as well as XIAP expression, suggesting a role for PI-3K in this pathway (223). However, these authors did not determine whether the PI-3K inhibitors blocked translocation of NF-κB or affected its transactivation activity. Because Akt has been reported to enhance IKK activity (224) to phosphorylate IκB and to promote transactivation of RelA (327), either/or both sites of action for the PI-3K inhibitors is possible. The inability to detect FSH-stimulated IκB–α phosphorylation in granulosa cells (223) suggests either that the phosphorylation of IκB–α by IKK2 is below the detection level or that the FSH-sensitive p50/RelA complex is bound to another IκB inhibitor, such as IκB-β, γ, or ε. Because the transactivation of RelA can also be enhanced by PKA (325), it is tempting to speculate that FSH-stimulated PKA also participates in RelA transcriptional activation, as shown in Fig. 11. Additional studies are thus required to clarify the mechanism by which FSH enhances NF-κB activity and to identify additional NF-κB target genes in granulosa cells.

NOS. An additional Akt substrate is endothelial nitric oxide synthase (eNOS). NOS catalyzes the oxidation of L-arginine to nitric oxide (NO) and L-citrulline. NO regulates a number of biological functions including apoptosis, cell migration, and angiogenesis (328). Phosphorylation of eNOS on Ser1177 by Akt results in its activation (329). A major target of NO is the soluble guanylyl cyclase, the activation of which results in the production of cGMP.

Ovarian cells have been shown to express both eNOS and the inducible NOS (iNOS) in both a gonadotropin-dependent and -independent manner. Both eNOS and iNOS appear to be expressed in theca and stroma of rat follicles in the absence of gonadotropin stimulation (330), and the injection of pregnant mare's serum gonadotropin to immature rats or addition of FSH to porcine granulosa cells induces expression of eNOS in granulosa cells (330,331). There is evidence that iNOS is selectively expressed in granulosa cells only of healthy follicles and not in atretic follicles, where atresia was identified by TUNEL staining (332). eNOS null mice exhibit fewer preovulatory follicles and reduced ovulatory efficiency (333). iNOS null mice do not exhibit a striking reproductive phenotype (333).

Although expression of NOS has been evaluated in ovarian cells, the signaling pathways that regulate eNOS activation have not been studied, although regulation by the PI-3K pathway is likely, especially in view of the role of NOS in angiogenesis. FSH has been shown to produce only a minimal increase in cGMP levels (334), and direct activation of guanylyl cyclase in rat granulosa cells results in reduced FSH-stimulated cAMP production, estrogen, and inhibin-α synthesis (334,335). These latter results suggest that it is unlikely that FSH signals by activating guanylyl cyclase activity in granulosa cells. However, FSH-dependent NO signaling to regulate pathways other than guanylyl cyclase cannot be excluded. A role for eNOS downstream of the preovulatory surge of LH in ovulation, however, has been suggested and is discussed below.

Cross-Talk among FSH-Regulated Signaling Pathways Regulates Target Gene Expression

As reviewed in the proceeding sections, signaling cascades initiated by FSH activate target genes that inhibit apoptosis, drive proliferation, stimulate steroidogenesis, and promote a complex differentiation response that enables granulosa cells to respond appropriately to the LH surge. FSH accomplishes this array of functions by stimulating translation and transcription of specific target genes and by regulating posttranslational modifications not only of proteins in the canonical signaling pathways but also of transcription factors and coactivators. Figure 12 reviews some of the FSH target genes that are regulated downstream of FSH-stimulated Akt, CREB, and ERK. In many cases, it is not known if these genes are direct targets, for example, of FOXO1 or HIF-1, or are regulated indirectly by one or more immediate early genes whose transcription or translation is activated upon relief from FOXO1 repression

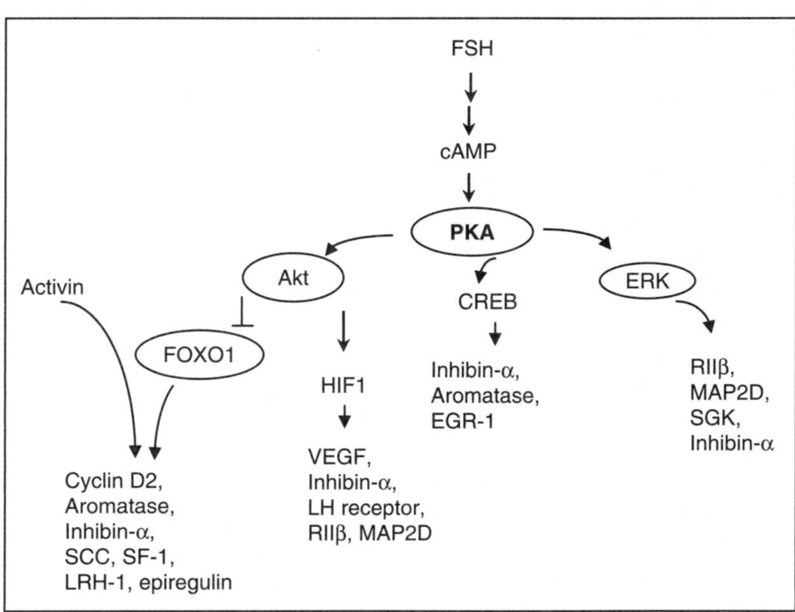

FIG. 12. Regulation of FSH-responsive gene targets.

and/or HIF-1 activation. Based on the observation that inhibin-α, for example, is regulated not only by CREB activation but also by HIF-1 activation and by relief from FOX01 repression, it is clear that activation of a single transcription factor is not sufficient to regulate target gene expression; rather, target gene expression requires a combination of transcription factors and coactivators.

Of the FSH target genes, *inhibin-α* has probably been one of the most extensively studied, along with aromatase, regarding the combinations of transcription factors and coactivators that are required to promote its transcription. SF-1 and cAMP/PKA synergize to activate the *inhibin-α* promoter (163). The SF-1 and CREB sites are adjacent to each other on the promoter, and although SF-1 and CREB can interact in extracts, it is not known if this interaction is direct (163). Although SF-1 is a key regulator in the basal state, upon FSH/cAMP stimulation it is likely that the related nuclear receptor LRH-1 synergizes with CREB to activate this gene (336). Additionally, CBP and SRC-1 further increase *inhibin-α* transcriptional activity, and CBP and SRC-1 interact with both SF-1 and CREB on the *inhibin-α* promoter (336). The ability of cAMP to activate the *inhibin-α* promoter likely reflects actions at a number of sites. cAMP via PKA phosphorylates CREB on Ser133, promoting recruitment of CBP (159,198). PKA also promotes the phosphorylation on Ser10 and acetylation on Lys14 of histone H3, as previously discussed, thus remodeling chromatin in the region of the *inhibin-α* promoter (101). PKA is also recognized to potentiate SF-1 transcriptional activity, although the mechanism by which PKA elicits this effect is controversial. Inhibin-α expression is also regulated, either directly or indirectly, by HIF-1 based on the ability of a dominant negative HIF-1α to reduce significantly the ability of FSH to activate *inhibin-α* promoter-reporter activity (222). Additionally, expression of constitutively FOXO1 in granulosa cells reduces the ability of FSH plus activin to activate inhibin-α expression (260). Although this result could reflect direct repression of the *inhibin-α* gene by FOXO1, this result could also be explained by increased expression of the SF-1 in response to FSH plus activin, because SF-1 is obligatory for activation of the *inhibin-α* gene (163). Finally, preliminary results suggest that inhibin-α expression is inhibited by the MEK inhibitor PD98059 (unpublished data). The most probable site of action of ERK is the phosphorylation of SF-1, resulting in the formation of an active SF-1 conformation (190,191), although additional ERK sites of action, such as CBP, cannot be ruled out.

There is also accumulating evidence that one or more of the GATA transcription factors could regulate *inhibin-α* gene expression in granulosa cells. *Inhibin-α* gene expression is activated by GATA-1 in testicular cell lines (337,338), GATA-4 has been reported to enhance an *inhibin-α* promoter-reporter construct in CV-1 cells (339), and GATA-4 is reported to be phosphorylated by PKA after FSH treatment of granulosa cells (340). However, to our knowledge, FSH regulation of *inhibin-α* gene expression by GATA-4 in granulosa cells has not been reported. Another transcription factor likely to be important for regulation of *inhibin-α* gene expression in granulosa cells is β-catenin downstream of Wnt, based on the ability of β-catenin to synergize with SF-1 to regulate *inhibin-α* gene expression in the adrenal (341). Additionally, at least two known transcription factors, the induced cAMP early repressor (ICER) and C/EBPβ, are involved in the transcriptional repression of the *inhibin-α* gene that occurs in response to LH stimulation (342,343). Thus, multiple signaling pathways and transcription factors are involved in a complex way in the regulation of this single gene.

FSH-dependent transcription factor requirements for the expression of aromatase, Egr-1, epiregulin, LH receptor, SCC, GIOT-1, and SGK have also been investigated. FSH-stimulated expression of aromatase requires CREB (344) and SF-1 (144). As with inhibin-α, expression of constitutively active FOXO1 in granulosa cells reduces the ability of FSH plus activin to induce aromatase expression (260). This result suggests either direct or indirect repression of the *aromatase* gene by FOXO1, as discussed for *inhibin-α*. Although there is also a report that GATA-4 can activate an *aromatase* promoter-reporter in a heterologous cell model (339), GATA-4 regulation of the *aromatase* gene in FSH-stimulated granulosa cells has not been reported. Expression of the immediate early gene *Egr-1* requires Sp1/Sp3 as well as CREB and a second CREB-like factor binding to a cAMP response element (26). Egr-1 is a zinc-finger transcription factor that binds to GC-rich enhancer elements (345). Egr-1 null mice are sterile (346). In addition to a defect in pituitary LH-β expression, follicles of Egr-1 null mice do not acquire LH receptors and thus cannot ovulate (346). Consistent with this result, LH receptor expression in rat granulosa cells has been shown to require Egr-1 (347). LH receptor expression has also been shown to require HIF-1 (222) and Sp1/Sp3 (348).

SGK is also an immediate early gene that responds to FSH whose promoter exhibits chromatin remodeling resulting from FSH-stimulated histone H3 phosphorylation and acetylation (101). Binding of Sp1/Sp3 to a GC-rich region of the *SGK* promoter accounts for the PKA-dependent activation of this gene (23). Despite the fact that FSH additionally promotes the activation of this kinase, based on an upshift in its migration on sodium dodecyl sulfate polyacrylamide gel electrophoresis that likely reflects phosphorylation (204), the function of this kinase in granulosa cells remains elusive. GIOT-1 is a member of the C_2-H_2-type zinc finger family of transcription factors and contains a kruppel-associated box-A domain (349) that exhibits transcriptional repressor activity (27). FSH-stimulated expression of GIOT-1 in granulosa cells is synergistically activated by SF-1 and PKA. The promoter contains SF-1 and cAMP response elements that bind SF-1 and CREB or a CREB family member, respectively (27). Regulation of *GIOT-1* promoter activity by SF-1 was abolished when cells were transfected with a Dax-1 expression vector, and FSH reduces *Dax-1* mRNA levels as well as protein (27). These authors concluded that activation of the *GIOT-1* promoter by SF-1 requires the FSH-dependent down-regulation of Dax-1 protein (27), consistent with recent evidence that the PKA-dependence of SF-1's activity is attributed to its dissociation from Dax-1 (167).

Expression of the *aromatase* gene has been shown to require CREB and SF-1 (144,344), to be inhibited by Dax (351), and to require derepression from FOXO1 (260). *Epiregulin* expression requires Sp1/Sp3 (24) and derepression from FOXO1 (260), and *SCC* expression requires LRH1 (352), CBP binding to SF-1 (164,353), and derepression from FOXO1. It is interesting that a number of FSH target genes require Sp1/Sp3 binding. Sp1/Sp3 bind to "GC" boxes (GGGGCGGGG) and are recognized to functionally interact with Egr-1 [reviewed in (161)]. Moreover, the DNA binding activity of Sp1/Sp3 is recognized to be enhanced upon phosphorylation at Thr266 by ERK (192) as well as to be regulated in a positive manner by PKA (162), although the mechanism of PKA regulation is poorly understood (161). It is thus possible that Sp1/Sp3 are ERK targets in granulosa cells and that PKA-regulated ERK phosphorylation of Sp1/Sp3 contributes to the dependence of FSH-target gene expression on ERK activation, as evidenced by results using the MEK inhibitor PD98059. Consistent with this idea, FSH-stimulated Sp1/Sp3 binding to the *Egr-1* promoter is reduced by pretreatment of granulosa cells with PD98059 (26). However, it has not formally been demonstrated in granulosa cells that ERK or one of their substrate kinases phosphorylates Sp1/Sp3 and that this phosphorylation is required for the ability of Sp1/Sp3 to activate the *Egr-1* promoter.

Taken together, FSH-stimulated target gene activation appears to require input from a number of pathways, including those for Akt and ERK in addition to the well-studied PKA-dependent pathway. Inhibitors of both PKA and PI-3K abrogate the ability of FSH to activate target genes, consistent with a requirement of both these kinases in the actions of FSH. The requirement for both PKA-phosphorylated CREB as well as SF-1 or Sp1/Sp3, both of which are PKA-dependent ERK targets, for the transcriptional activation of a number of FSH target genes coupled with their dependence on the PI-3K pathway explains why activation of the PI-3K pathway by IGF-1 is not sufficient to induce FSH target genes. Although some of the transcription factors and coactivators required to activate some of the FSH target genes have been identified, much needs to be learned regarding the combinatorial requirements for transcription factors and coactivators and how the activity of these factors is regulated. Moreover, it is likely that additional signaling pathways are regulated by FSH. Finally, we know very little about how FSH-dependent pathways converge with the many signaling systems involved in initial formation and early growth of the follicle. Thus, there is a great deal remaining to be elucidated regarding how FSH signals to promote follicular maturation.

LH SIGNALS PROMOTE OVULATION AND CORPUS LUTEUM FORMATION

Ovulation and Luteinization

Follicles are exposed to relatively low levels of LH as follicle maturation proceeds and then to a single high-level surge of LH driven by hypothalamic release of gonadotropin-releasing hormone, which induces ovulation and formation of the corpus luteum. Low levels of LH induce 17α-hydroxylase/C17-20-lyase (CYP17), the rate-limiting enzyme in androgen biosynthesis in theca cells (354–356) and the low- (357) or high- (358) density lipoprotein receptor, depending on the species, to provide cholesterol for steroid biosynthesis. The resulting increased production of androgens provides the substrate for aromatase in granulosa cells to synthesize estrogen. In addition to facilitating the actions of FSH, these increasing levels of estrogen act in a positive feedback fashion on the brain to induce the preovulatory LH surge.

The preovulatory surge of LH promotes ovulation and differentiation of granulosa and theca cells into luteal cells (luteinization) and triggers final maturation of the oocyte of preovulatory follicles. These responses to LH in preovulatory follicles involve not only the induction of a number of genes that are required for each of these events, as discussed below, but also termination of the expression of select FSH target genes that were required for follicular maturation. For example, the surge of LH terminates the expression in granulosa cells of the FSH receptor (58), estrogen receptor (ER)-β (359,360), cyclin D2 (41,361), aromatase (362), LH receptor (363,364), inhibin-α (365), and FOXO1 (97) but does not reduce the expression of a number of other FSH targets, including MAP2D (22), SCC (366), RIIβ (personal observation), or SGK (97). Many of these responses are variable dependent on species; for example, inhibin-α is repressed by the LH surge in the rodent (365), but inhibin A is actively secreted during the luteal phase in the primate reproductive cycle (367).

The surge of LH induces the transient expression of a number of genes that are required for ovulation, such as the cyclic nucleotide degrading enzyme *PDE4D* (31), the *progesterone receptor* (368–370), the progesterone receptor target gene *PACAP* (pituitary adenylate cyclase activating polypeptide) (371–373), the lysosomal protease cathepsin L (370), the matrix metalloproteinase *ADAMTS-1* (a disintegrin and metalloproteinase with thrombospondin-like repeats) (370,374), and cyclooxygenase-2 (COX-2), the rate-limiting enzyme in prostaglandin synthesis (375). As is the case with FSH regulation, a significant early effect of LH is to up-regulate transcription factors involved in the regulation of those genes playing functional roles in ovulation and luteinization. Examples include Egr-1 (373,376), a transcription factor that in other cellular models regulates expression of *cathepsin L*, metalloproteinases (MMPs), and *ADAMTS-1,* all of which are LH target genes [reviewed in (26)]; CEBP/β (377), a transcription factor that is required for induction of the luteinization-associated steroidogenic acute regulatory protein (StAR) (378–380) and participates in (381) but is not obligatory for (377) the induction of COX-2; ICER, which is a repressor of CREB-regulated genes including *inhibin-α* (342); and the immediate-early orphan nuclear receptors Nur77/NGF-1B, Nurr1, and Nor1 (382,383), which have been implicated in negative regulation of the *aromatase* gene after the LH surge.

Mice that are null for PDE4D (31), the progesterone receptor (384), COX-2 (385), Egr-1 (386), and CEBP/β (377,387) do not ovulate in response to the preovulatory LH surge and are thus infertile. After the injection of an ovulatory concentration of hCG, PDE4D null mice exhibit SCC and serum progesterone levels that are equivalent to those of wild-type mice but show significantly blunted induction of the progesterone receptor, cathepsin L, and PACAP (31). As discussed below, the basis for reduced induction of these ovulatory genes likely results from the significantly reduced levels of cAMP synthesized by granulosa cells of PDE4D null mice in response to hCG (31). The basis for the anovulation of COX-2 null mice is perplexing because mice that are null for either the prostaglandin (PG) $F_{2\alpha}$ receptor (388) or PGE_2 receptor (389) exhibit apparently normal ovulation, although PGE_2 receptor null mice exhibit reduced implantation (389). However, because administration of PGE_2 overcomes the ovulation block induced by the COX inhibitor indomethacin (390), apparently the PGE_2 requirement for ovulation is mediated via another prostanoid receptor or a PGE_2 metabolite. The LH surge also induces a number of MMPs that are believed to be critical for remodeling the follicle wall to accommodate ovulation (391,392), including MMP2 in theca cells (393). Although mice that are null for MMP2 and tissue inhibitors of metalloproteinases 1 and 2 are fertile (394), the large number of MMPs present in follicular cells suggests considerable redundant functions for these proteins.

The LH surge induces the transient expression by granulosa cells of genes associated with cumulus expansion and oocyte maturation, including the EGF family members amphiregulin, epiregulin, and betacellulin, which were shown to trigger cumulus expansion and oocyte maturation via the EGF receptor located on cumulus cells and oocytes (which contain few LH receptors) (17); COX-2 (375), as discussed above; hyaluronic acid, a glycosaminoglycan required for cumulus expansion (394) and hyaluronic acid synthase-2, the rate-limiting enzyme in hyaluronic acid synthesis (373); and the PG target gene *TNF-stimulated gene-6* (395). COX-2 null mice, in addition to being anovulatory, exhibit abnormal cumulus expansion and stigmata formation as well as reduced production of hyaluronic acid but exhibit normal oocyte maturation (396). Ovulation and cumulus expansion can be restored in COX-2 null mice by the administration of either PGE_2 or interleukin (IL)-1β, suggesting that IL-1β is either downstream of COX-2 or mediates a redundant pathway leading to ovulation (396). Although the LH surge induces IL-1β and IL-1β receptor antagonists inhibit ovulation (397), IL-1β null mice are fertile (398), suggesting that IL-1β is not an obligatory intermediate in the ovulatory response.

Luteinization is associated with a cessation of granulosa and theca cell proliferation and is characterized by hypertrophy of both granulosa and theca cells of preovulatory follicles (399) as well as by increased steroid hormone synthesis and secretion. The preovulatory surge of LH accordingly induces the persistent and elevated expression of genes that characterize luteinized granulosa and thecal cells, such as the cell cycle inhibitors $p21^{Cip1}$ and $p27^{Kip1}$ (41,400,402), the rate-limiting enzyme in progesterone synthesis SCC (366), the cholesterol-mobilizing protein StAR (373), and immediate early genes associated with the AP-1 complex *Fra2* and *JunD* (36). Luteinization can occur in the absence of ovulation based on the ability of follicle cells in progesterone receptor null mice to form a corpus luteum, in response to the injection of an ovulatory concentration of hCG, which is indistinguishable from that of wild-type mice except for the retention of the oocyte (370).

In preovulatory follicles of progesterone receptor null mice, aromatase is down-regulated, COX-2 is induced in response to LH receptor activation, and luteal cells express elevated levels of SCC (370). Similarly, in cyclin D2 null mice, granulosa cells differentiate in response to FSH to a preovulatory phenotype, and upon subsequent stimulation with an ovulatory concentration of the LH receptor agonist hCG, follicles do not ovulate but granulosa and thecal cells still luteinize in follicles with entrapped ova (40).

Also in PDE4D null mice, in response to hCG injection, granulosa cells of preovulatory follicles exhibit increased SCC and steroidogenesis, granulosa cell hypertrophy, and increased expression of $p27^{Kip1}$, characteristic of luteinization, but do not exhibit increased progesterone receptor, COX-2, or cathepsin L expression; consequently follicles do not ovulate (31). Likewise, in COX-2 null mice, normal corpora lutea form, albeit with entrapped oocytes (396). However, although luteinization is associated with cell cycle arrest, luteinization and exit from the cell cycle are independent responses to the surge of LH. Thus, in both $p27^{Kip1}$ null (402,403) and $p27^{Kip1}/p21^{Cip1}$ double-null mice (402), preovulatory follicles undergo ovulation and luteinization to form a corpus luteum in response to the LH surge even when granulosa cells continue to proliferate, as evidenced by retention of cyclin D2 expression and high bromodeoxyuridine incorporation and proliferative cell nuclear antigen expression in cells expressing SCC. The LH surge also induces the expression of P11 in granulosa and theca cells, a protein that interacts with the proapoptotic protein Bad (404), perhaps to promote survival of differentiating granulosa and theca cells.

LH and the LH Receptor

The pituitary gonadotropin LH shares a common alpha subunit with FSH, but its synthesis and secretion from pituitary gonadotropes is under distinct regulation. For many of the studies discussed in this chapter, the related placental gonadotropin hCG (Fig. 1) is used as an agonist of the LH receptor. The LH receptor, like the FSH GPCR, contains nine Leu-rich-repeat motifs in its extracellular domain and is classified in subgroup A of the rhodopsin/β-adrenergic family of GPCRs (7,405) (Fig. 1). These Leu-rich-repeats contribute to hormone binding and structural integrity of this receptor (406). The LH receptor is encoded by a gene with 11 exons and 10 introns (7). Exons 1–7 encode the regions of the large extracellular domain that are essential for high affinity binding of LH receptor agonists and the placental-derived hCG (407–409). Exons 9 and 10 also encode the extracellular domain but appear to be less important for agonist binding (407,409). Exon 11 encodes the seven TM domains and the C-terminal tail. The LH receptor in female mammals is expressed in the ovary in theca cells of ovarian follicles, in granulosa cells of mature preovulatory follicles, and in corpora lutea. LH receptor null mice are infertile; their ovaries contain only early antral follicles and lack preovulatory follicles or corpora lutea (410). The absence of follicular development to an apparent preovulatory stage likely reflects, at least in part, the reduced levels of theca-derived androgens and thus granulosa-derived estrogens in these animals (410) and the apparent requirement of these steroid hormones for final follicular maturation. Granulosa cells also express androgen receptors (411), and androgen receptor null mice exhibit increased granulosa cell apoptosis and defective cumulus cell expansion in response to the LH surge (412), suggesting a role for androgens in granulosa cells distinct from their role as an aromatase substrate. Although granulosa cells express primarily ER-β and theca cells express ER-α, mice null for ER-α or ER-β ovulate in response to exogenous gonadotropins, although ovulation efficiency is reduced, suggesting that estrogen facilitates ovulation (412a). Incomplete follicular maturation in the LH receptor null mice could also result from effects of basal LH on granulosa cells distinct from those described for the surge of LH.

In the absence of agonist, GPCRs are believed to exist in the membrane primarily in an inactive conformation that is in equilibrium with an active conformation (51). Agonist binding is thought to stabilize the receptor in the active conformation. Consistent with this notion, and unlike the FSH receptor, a number of mutations in the LH receptor have been identified that stabilize the receptor in an active conformation in the absence of agonist (413). The first human mutation of the LH receptor identified was Asp578 in TM 6 in patients with familial male-linked precocious puberty (414). This mutation results in a constitutively active LH receptor, which in males leads to precocious puberty. For reasons that are not clear, females with this LH receptor mutation do not exhibit abnormal fertility (413). This may reflect a stringent requirement for postpubertal levels of FSH to induce LH receptor expression in the female ovary. The absence of a phenotype in women with constitutively active LH receptors is especially unexpected in light of the clear phenotype in mice over-expressing a longer lived LH mutant protein, which is characterized by precocious puberty, anovulation, and resulting infertility (415). TM 6 and its extension into the cytoplasm as the third intracellular loop have been defined as hot spots for LH receptor mutations that promote a constitutively active receptor, because more than 16 mutations have been identified in this region (4,413). Additionally, Ser431 in TM 3 (416), Arg442 in the second intracellular loop (417), Asp397 in exoloop 1 (418), and Lys583 in exoloop 3 (419) of the rat LH receptor appear to contribute to the stabilized agonist-dependent active conformation of the LH receptor to generate cAMP (without affecting agonist binding), whereas Lys583 in exoloop 3 also appears to stabilize the agonist-stimulated active state of the receptor to generate inositol triphosphate (IP_3) (420). A large number of inactivating mutations of the LH receptor in women have been reported (413,421). These women exhibit amenorrhea and no preovulatory follicles or corpora lutea, consistent with the established functions for LH, and do not respond to exogenous hCG.

The LH receptor in granulosa cells is induced by FSH (422); LH receptor synthesis then ceases in response to the surge of LH, as discussed below, and is reinitiated with the formation of the corpus luteum. Hormonal regulation of the induction of LH receptors in corpora lutea is species dependent (423,424), consistent with the high degree of species specificity in the factors that regulate formation, maintenance, and regression of the corpus luteum. Theca cells appear to express LH receptors coincident with the morphological formation of this cell layer at very early stages of follicular development (58,425); however, it is not known if LH receptor expression is constitutive in these cells or is regulated by a paracrine factor produced either by the oocyte or by enclosed granulosa cells. FSH has been shown to promote up-regulation of *LH receptor* mRNA in granulosa cells via a transcriptional mechanism (426) that requires activation of Egr-1 (347), Sp1/3 (348,427), and HIF-1 (222) response elements on the *LH receptor* promoter. LH receptor expression also requires the inhibition of histone deacetylase activity at specific regions of the promoter (428).

In response to the surge of LH, LH receptors are down-regulated. Receptor down-regulation is relatively rapid when these receptors are expressed in heterologous cells: $T_{1/2}$ for internalization of the rat LH receptor in human embryonic kidney cells is 138 minutes; that for the human LH receptor is 17 minutes (429). However, in rat and porcine granulosa cells, LH receptor down-regulation in response to the LH surge has a $T_{1/2}$ of at least 8.5 hours (430–432). LH receptor down-regulation in rat granulosa cells is mimicked by cell permeable cAMP analogues (433) and primarily attributable to a reduced rate of synthesis rather than increased degradation (434). The reduced rate of LH receptor synthesis is not due to decreased transcription but rather to reduced *LH receptor* mRNA half-life (435). A surge of LH promotes increased *LH receptor* mRNA degradation via its interaction with an LH receptor mRNA binding protein (436) recently identified as mevalonate kinase (437).

The LH receptor couples to activate Gs (438,439), uniformly resulting, both in physiological contexts and heterologous expression models, in activation of adenylyl cyclase to increase synthesis of cAMP (440–442). The LH receptor has also been shown to activate Gi (438,439,443), Gq/11 and G13, based on the ability of LH to stimulate binding of a GTP photoaffinity analogue to these Gα proteins (439). Although LH receptor activation can stimulate Gi to inhibit adenylyl cyclase activity upon expression in a heterologous cell model (444), LH-stimulated adenylyl cyclase activity in porcine follicular membranes is not increased by pretreatment of membranes with pertussis toxin, which uncouples GPCRs from Gi (443). Thus, the physiological significance of LH receptor coupling to Gi is not known. Constitutively active LH receptors expressed in heterologous cells also activate PLC, resulting in elevated levels of IP_3 and Ca^{2+} (440–442). Similarly, when LH receptors are expressed at sufficiently high concentrations, receptor activation raises intracellular Ca^{2+} and activates PLC (445–447). This response is pertussis toxin insensitive (445) and thus most likely mediated via activation of Gq/11. In rat granulosa cells, which express approximately 10,000 receptors per cell (448), LH raises intracellular levels of IP_3 presumably via activation of PLC (449). Thus, signaling through cAMP appears to be the predominant but not the only second messenger through which the LH receptor can modulate responses.

Although the exact residues that bind Gs on the LH receptor have not been identified, the C-terminal portion of the third intracellular loop extending into TM 6 domain is important, based on the ability of a synthetic peptide corresponding to this region to activate adenylyl cyclase in the absence of LH receptor expression (450). Truncation of the C-terminal region of the rat LH receptor at residue 653 does not affect agonist-stimulated adenylyl cyclase activation but hampers receptor-stimulated PLC activation, suggesting that the G protein that activates PLC couples to a distinct region of the receptor from that which signals to Gs (451).

The LH receptor also exhibits desensitization and, in heterologous cell models, relatively rapid internalization in response to saturating levels of agonist. However, the preovulatory LH surge does not promote rapid internalization of the LH receptor in granulosa cells, as previously discussed. Rather, cell surface receptor numbers around the time of ovulation decrease with a $T_{1/2}$ of approximately 8.5 hours (430–432). Although this receptor can be phosphorylated on up to five serine residues located in the C-terminal tail (452,453) in a G protein–regulated kinase-dependent manner (454,455), phosphorylation of the human and porcine receptors does not contribute to either uncoupling from Gs/adenylyl cyclase (452,453,456,457) or receptor internalization (458). Similarly, for the rat LH receptor, mutation of the serine residues in the C-terminal tail retards the rate of desensitization but does not affect the extent of LH receptor desensitization (451,452,454). These results are consistent with evidence that truncation of the C-terminal tail of the LH receptor does not prevent LH receptor desensitization (459). A protein that interacts with the C-terminus of the human LH receptor, the type I PDZ domain protein GIPC (460), appears to regulate levels of cell surface LH receptor during hormone internalization and to play a role in ligand recycling.

Thus, unlike the β-adrenergic and rhodopsin GPCRs, in which high affinity binding of arrestin to the receptor is facilitated by receptor phosphorylation (75,461), binding of arrestin to the LH receptor and consequent uncoupling from Gs appears to depend on receptor activation rather than phosphorylation (453,462–464). Indeed, arrestin2 binds with pM affinity to a synthetic peptide corresponding to the third intracellular loop of the porcine LH receptor to promote desensitization (463,464). The accessibility of arrestin2 to the LH receptor in porcine membranes and in human embryonic kidney cells stably transfected with the murine LH receptor is regulated by the association of arrestin2 with the small G protein ADP-ribosylation factor (Arf)6 (464,465). LH receptor activation promotes activation of membrane-delimited Arf6 (466) by ARNO (Arf-nucleotide binding site opener) or a similar Arf6 activator (467), releasing arrestin2 from its membrane docking site (468). The binding of arrestin to GPCRs has been shown to sterically hinder the ability of the receptor to bind Gs (76). Consistent with this model, both Gs binding and arrestin2 binding appear to require the C-terminal portion of the 3i loop of the LH receptor (450,463). Although this pathway for Arf6-dependent desensitization has not been tested in other cellular models, over-expressed ARNO is reported to stimulate arrestin-dependent β-adrenergic receptor internalization, and an inactive Arf6 that cannot bind GTP inhibits β-adrenergic receptor internalization (469). These results suggest that Arf6 might play a more universal role in regulating arrestin availability to GPCRs.

Coupling of the LH receptor to Gs is also regulated in a poorly understood manner by PDE4D (31). Unexpectedly, in granulosa cells of PDE4D null mice, the immediate (detected by 1 hour) and the sustained (over 6 hours) cAMP synthetic response to an ovulatory concentration of hCG (both in vivo and in vitro) is significantly *blunted* (and not increased), whereas the response to the adenylyl cyclase activator forskolin (in vitro) is equivalent compared with that of wild-type mice (31). This result suggests that PDE4D in some manner facilitates coupling between the LH receptor and Gs to activate adenylyl cyclase and that in the absence of PDE4D, coupling between LH receptor and Gs is impaired. LH receptor–Gs coupling is likely regulated by the sustained levels of PDE4D induced in granulosa cells by FSH rather than that induced by LH (31). How PDE4D enhances LH receptor/Gs coupling remains to be elucidated.

LH Signaling Pathways that Stimulate Ovulation and Luteinization

cAMP- and PKA-Dependent Signaling in Preovulatory Cells

The LH receptor is believed to signal predominately via Gs to activate adenylyl cyclase and raise intracellular levels of cAMP. Consistent with this notion, the adenylyl cyclase activator forskolin mimics LH to stimulate expression of a number of LH target genes, including the progesterone receptor (470,471), Egr-1 (26), COX-2 (381,472), ICER (342), and the low-density lipoprotein receptor (357). The primary cAMP target in cells is PKA. Indeed, ovulatory concentrations of LH/hCG in preovulatory follicles lead to activation of PKA (473), and PKA inhibitors H89 or PKI inhibit the induction of a number of LH targets or responses, including COX-2 (470), progesterone biosynthesis (470), Egr-1 (26), and the low-density lipoprotein receptor (357).

As in preantral granulosa cells, hCG or forskolin treatment of preovulatory granulosa cells stimulates the phosphorylation of the PKA target CREB on Ser133 (143,474), as depicted in Fig. 13. Although LH-stimulated CREB phosphorylation is undetectable when granulosa cells are pretreated with the PKA inhibitor H89 (474), formal evidence that other pathways downstream of PKA do not also contribute to CREB phosphorylation is missing. LH targets that contain cAMP response elements and have been shown to bind CREB include Egr-1 (26), ICER (475), and StAR (380).

LH has also been shown to signal via PKA to activate ERK, based on the ability of H89 and Myr-PKI to strongly inhibit LH-stimulated ERK activity (474). ERK phosphorylation peaks at 10 minutes post-hCG and is barely detected by 1 hour in granulosa cells cultured under serum-free conditions (474). ERK phosphorylation is also readily detected in pregnant mare's serum gonadotropin-primed immature rats in response to a subcutaneous injection of hCG (203) and in isolated follicles in response to LH (476). However, in contrast to regulating the association of a phosphotyrosine phosphatase with ERK as is observed in preantral granulosa cells (98), PKA signals to regulate ERK activity in preovulatory granulosa cells at a step upstream of ERK. Thus, although FSH does not promote MEK phosphorylation in preantral

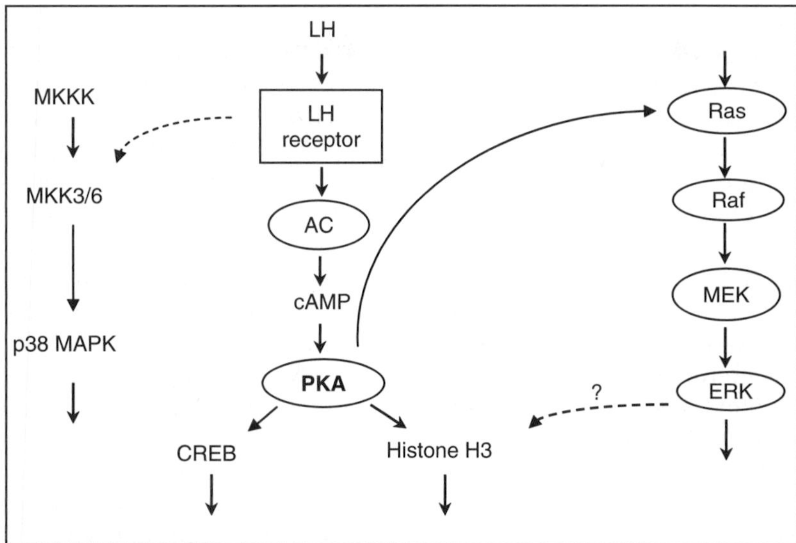

FIG. 13. LH via cAMP/PKA signals to targets such as ERK, CREB, and histone H3, whereas the LH signaling pathway to activate p38 MAPK is not known.

granulosa cells as discussed above (98), LH does stimulate MEK phosphorylation in preovulatory cells (474). LH-stimulated MEK phosphorylation is inhibited by H89, consistent with PKA dependence of this pathway. Although the site of regulation by PKA in the ERK signaling pathway has not been elucidated in preovulatory granulosa cells, it has been shown in MA-10 and primary Leydig cells that LH signals via PKA into the ERK pathway to activate Ras (477). LH activation of this pathway is inhibited by Myr-PKI and is not mimicked by a cAMP analogue that selectively activates cAMP-regulated guanine nucleotide exchange factors like Epac, consistent with regulation by PKA (477). However, the relevant direct PKA substrate in either of these cell types has yet to be identified. It is likely that in preovulatory granulosa cells, as in Leydig cells, PKA signals via an unidentified mechanism to regulate the activity of Ras, thus activating ERK, as depicted in Fig. 13. Downstream ERK substrates have not been reported in preovulatory granulosa cells, and thus the physiological significance of ERK signaling in these cells is not known. However, the MEK inhibitors PD98059 or U0126 have been shown to inhibit LH-stimulated *low-density lipoprotein receptor* promoter reporter activity (357), consistent with a role for ERK in activation of this LH target gene, as well as oocyte maturation (476). PD98059 has also been shown to inhibit binding of Sp1/Sp3 to the *Egr-1* promoter and inhibit activation of *Egr-1* promoter-reporter activity by LH (26), consistent with the evidence in another cellular model that ERK can phosphorylate and enhance the DNA binding activity of Sp1/Sp3 (192). Sp1/Sp3 binding to the *StAR* gene promoter also contributes to the cAMP activation of this gene (478). Another possible ERK target in preovulatory granulosa cells is SF-1. As discussed earlier, phosphorylation of SF-1 by ERK stabilizes the active conformation of this transcription factor (190,191). A number of LH target genes in preovulatory cells require binding of SF-1, including *StAR* (380,478,479) and *SCC* (164,353).

hCG also signals to phosphorylate histone H3 on Ser10 in preovulatory granulosa cells, and this response is inhibited by H89, consistent with an upstream PKA signal (474). However, it is not known whether other protein kinases downstream of PKA contribute to histone H3 phosphorylation in preovulatory cells or if H3 is directly phosphorylated by the catalytic subunit of PKA. Target LH genes whose promoters are associated with phosphorylated histone H3 (as detected by chromatin immunoprecipitation assays) have also not been reported.

Additional Signaling Pathways Activated by LH

LH Activates p38 MAPK. hCG signals to phosphorylate p38 MAPK in preovulatory granulosa cells (474) as well as its upstream kinases MKK3 and MKK6 (203) (Fig. 13). p38 MAPK phosphorylation in granulosa cells is detected 10 minutes post-hCG treatment (474). However, signaling into the p38 MAPK pathway appears to be independent of PKA, based on the inability of H89 to modulate hCG-stimulated p38 MAPK phosphorylation in cells in which hCG-stimulate CREB phosphorylation was blocked by H89 (474). Although signaling into the p38 MAPK may occur via cAMP-regulated Epac or a related guanine nucleotide exchange factor (94–96), the cAMP dependence of this pathway in preovulatory cells has not been established. p38 MAPK targets have not been reported in granulosa cells. An established p38 MAPK target in other cells is COX-2 (480,481). p38 MAPK activity is often regulated by IL-1β (480–482), and hCG increases the expression of IL-1β (483,484) and COX-2 (375). Although IL-1β has been shown to increase COX-2 expression and hyaluronic acid synthesis as well as to activate NOS in granulosa cells [reviewed in (485)], IL-1β has also been reported to act downstream of COX-2 to restore ovulation in COX-2 null mice (396). Moreover, IL-1β null mice are fertile (398). Similarly, inhibitors of NOS reduce the number of ovulations (335,486), but NOS null mice (333) are fertile. Because COX-2 null mice are infertile (385), these results indirectly suggest that neither p38 MAPK nor IL-1β are required for the induction of COX-2 by the LH surge.

LH Increases Intracellular Ca^{2+} but Does Not Appear to Activate PKC. It is well established that the LH receptor upon expression in a heterologous cell activates PLC to raise IP_3 levels and/or intracellular Ca^{2+} (440–442,445–447). The rise in intracellular Ca^{2+} in one report was shown to be independent of the concomitant rise in cAMP and thus likely dependent on IP_3 (446). In rat granulosa cells, LH is reported to increase IP_3 (487). In porcine thecal cells, LH has been shown to generate an oscillatory Ca^{2+} signal that is not inhibited by pertussis toxin (thus independent of Gi) but is inhibited by the Ca^{2+} chelator ethyleneglycotetraacetic acid and by the PLC inhibitor U73211 (487a). However, the function(s) of Ca^{2+} as a second messenger in granulosa or theca cells has not been elucidated. The rise in intracellular Ca^{2+} likely activates a CaMK, and LH-stimulated progesterone production is reduced by 50% by the CaMK inhibitor Kn93 (470), although the induction of COX-2 is not affected by this inhibitor. It is also possible that LH activates one or more PKC isoforms.

PKC consists of a number of isoforms (488): the conventional PKCs (α, β, and γ) require Ca^{2+}, diacylglycerol, and phosphatidylserine for activation; the novel PKCs (δ, ε, η, θ) do not require Ca^{2+}; and the atypical PKCs (ζ, ι/λ), require only phosphatidylserine for activation. LH-stimulated COX-2 induction is partially inhibited by the PKC inhibitor calphostin C (470) (which inhibits conventional and novel PKCs); subovulatory concentrations of LH plus 200 nM PMA (phorbol myristate acetate), which binds to the diacylglycerol binding site on conventional and novel forms of PKC, mimic the effect of an ovulatory concentration of LH to promote progesterone production, COX-2 induction, and luteinization of granulosa cells (489); 200 nM PMA mimics the surge of LH to induce the progesterone receptor (368) and Egr-1 (26). However, hCG treatment of granulosa cells does not promote the activation of the conventional (α, β) or novel (δ, ε) isoforms of PKC present in granulosa cells, based on (a) the inability of hCG to stimulate the translocation of these PKC isoforms to the membrane fraction; (b) the inability of the PKC inhibitor GF109203X to inhibit the hCG-stimulated phosphorylation of proteins that are downstream of PKC in other cells, namely CREB and ERK; or (c) the inability of hCG to stimulate the autophosphorylation of PKCδ on an activation-related site in contrast to the PMA positive control (474). The ability of LH to activate the atypical forms of PKC has not been thoroughly investigated, but LH is reported to stimulate the membrane translocation of the atypical PKC isoform PKCζ in preovulatory granulosa cells, and this pathway is proposed to mediate LH-stimulated induction of the immediate early transcription factor NGF-1B in these cells (382), and NGF-1B is in turn implicated in *aromatase* gene repression (383).

LH and Insulin/IGF-1 Synergize to Activate Akt and LH Target Genes. There is abundant evidence in preovulatory granulosa cells and in thecal cells that LH/hCG synergizes with insulin/IGF-1 to promote target gene expression. This synergism is observed in granulosa cells with respect to stimulation of progesterone production (490,491), *StAR* mRNA expression or promoter-reporter activity (490,491), *low-density lipoprotein receptor* promoter-reporter activity (357), and, surprisingly, cAMP production (490,491). Synergism is reported in thecal cells for the *CYP17* promoter-reporter (354,355) and high-density lipoprotein receptor (358). Cotransfection with the PKA inhibitor PKI plasmid inhibited LH-stimulation as well as the LH/hCG synergism with insulin/IGF-1 in regulation of *StAR* (491) and *low-density lipoprotein receptor* (357) promoter-reporter activities, consistent with LH signaling to these targets via PKA. Synergism between these pathways to stimulate *low-density lipoprotein receptor* promoter-reporter activity was also inhibited with the PI-3K inhibitors wortmannin and LY24002 (357), consistent with a role of the PI-3K pathway leading to Akt activation in synergism.

In an in vivo model in which LH is infused into the vena cava of rats, LH has been shown to stimulate Akt phosphorylation (492). These authors also showed, primarily after immunoprecipitation of indicated targets from ovarian extracts, that LH stimulated the tyrosine phosphorylation of Jak2, the association and tyrosine phosphorylation of Jak2 with signal transducer and activator of transcription (Stat)-1 and -5b, the tyrosine phosphorylation of insulin receptor substrates-1 and its association with PI-3K, and Akt phosphorylation, as shown in Fig. 14. They also demonstrated that LH stimulated the tyrosine phosphorylation of Shc (492), an adaptor protein that binds to phosphorylated tyrosine residues and associates with Grb2/SOS to activate Ras (493) and leads to

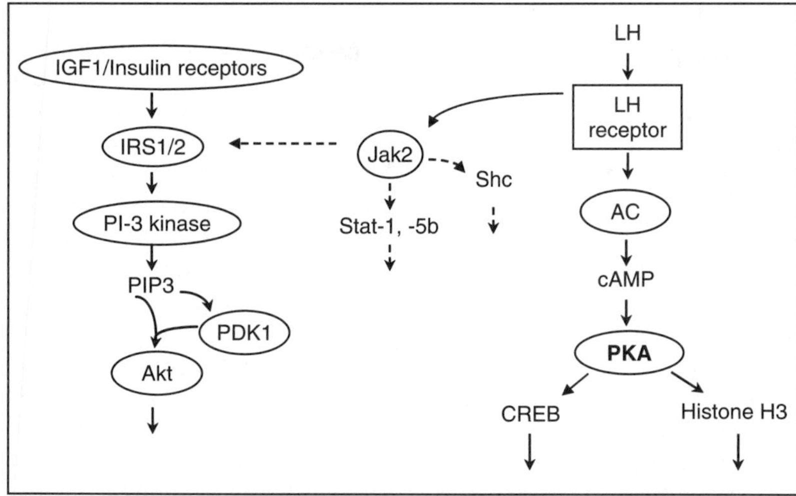

FIG. 14. LH signals into the Jak/Stat and PI-3K pathways.

an activation of ERK (492). Moreover, these authors reported that LH and insulin synergized to increase all these responses except for ERK activation and tyrosine phosphorylation of Stat-1 (492). The cellular pathway leading from the LH receptor to Jak2 activation is not known. Moreover, no link between Jak2 phosphorylation and PI-3K activation or ERK activation has been established.

Molecular Basis for the Differential Regulation of Target Gene Expression by FSH in Preantral and LH in Preovulatory Granulosa Cells

As can be seen from the proceeding review, there is currently more information regarding the cellular signaling pathways by which FSH regulates target gene expression compared with LH. It is also clear that cAMP/PKA plays a major role in the actions of FSH and at least in many of the responses to LH in granulosa cells. Despite the fact that these two gonadotropins are structurally related, act on similar receptors, and signal predominantly through cAMP-dependent mechanisms, FSH and LH exert very different responses in preantral versus preovulatory granulosa cells. For example, FSH acts via cAMP/PKA to activate transcription of target genes such as *aromatase* and *inhibin-α* in preantral granulosa cells but does not stimulate expression of the progesterone receptor or COX-2, whereas LH has the opposite effects in preovulatory granulosa cells and stimulates the down-regulation of aromatase and inhibin-α and activates transcription of the *progesterone receptor* and *COX-2* genes. Thus, a centrally important question in reproductive biology is how the same second messenger/effector exerts such different responses in these two stages of granulosa cell differentiation. Are the different responses attributable to the activation of distinct signaling pathways, such as only PKA for FSH and PKA plus PKC for LH? Or, do the different responses of the two cell types reflect the fact that preovulatory granulosa cells have differentiated and contain a different proteome, including distinct AKAPs, kinases, phosphatases, transcription factors, and other regulators important in generating their unique response?

To address the question of different signaling pathways for FSH versus LH, Zeleznik's group (494) cleverly investigated target gene expression after infection of preantral granulosa cells with constitutively active LH receptors. Results from these experiments showed that the LH receptor expressed in preantral cells partially mimicked responses of FSH and induced inhibin-α and 3β-hydroxysteroid dehydrogenase but interestingly did not induce aromatase or the LH receptor. These results suggest that these two GPCRs not only couple to overlapping signaling pathways but also appear to couple, perhaps more efficiently, to different signaling pathways to regulate distinct cellular events in preantral versus preovulatory granulosa cells.

Alternatively, rather than there being different second messengers generated by FSH versus LH receptor activation in preantral versus preovulatory granulosa cells, differences in the amount, duration, or cellular location of cAMP may distinguish FSH signaling from that of LH signaling. It is well known that sustained FSH at relatively low concentrations is necessary to stimulate granulosa cell differentiation (90), whereas a bolus of LH is required to stimulate ovulation. Indeed, it has been proposed that the actions LH and FSH are distinct in mature versus immature granulosa cells as a result of the ability of

LH receptor activation in mature cells to generate greater levels of cAMP (495). Consistent with this view is the observation that high doses of recombinant FSH can substitute for LH/hCG in promoting ovulation and luteinization in the rat (496). FSH and LH receptor activation also regulate the expression of the cAMP degrading PDEs (497). PDEs certainly participate in LH receptor signaling because PDE4D null mice exhibit an ovulation defect (497). However, the contribution of the PDEs to the distinct signaling by FSH and LH receptors in immature and mature granulosa cells is not fully understood. There is also abundant evidence for the cellular compartmentalization of cAMP in various cellular models (133,498). Consistent with the conceivably restricted production of cAMP in select cellular sites, PKA is recognized to be anchored to select cellular locations by the growing family of AKAPs (132,499,500). Potentially, then, the distinct expression of one or more AKAPs between immature and mature granulosa cells could also localize PKA and distinct PKA substrates to different cellular locals and account for distinct signaling in immature versus mature cells by FSH versus LH receptors. That PKA regulates ERK activity by distinct mechanisms in preantral versus preovulatory granulosa cells indicates that these two cells are indeed quite distinct and that preovulatory cells are not simply preantral granulosa cells that express, for example, LH receptors. The biochemical basis for the distinct signaling by PKA in preantral versus preovulatory granulosa cells remains to be elucidated but certainly could relate, in part, to expression of distinct AKAPs or other scaffolding proteins between the two cells that could direct pools of cAMP/PKA to different substrates.

CONCLUDING REMARKS

The existence of the gonadotropins was postulated based on the pioneering physiological studies of Long and Evans in 1922, and this work was quickly followed by the separation of pituitary extracts into the distinct fractions that would become known as FSH and LH by Fevold, Hisaw, and Leonard in 1931 (501). As approaches for the purification of the gonadotropins from various species were perfected in the 1970s, it became possible to begin detailed studies of their actions, leading to our fundamental understanding of the critical roles they play in both male and female reproduction.

As reviewed in this chapter, tremendous progress has been made in understanding mechanisms of gonadotropin action in regulating ovarian follicle maturation, ovulation, and luteinization. The structures of FSH and hCG are now known to atomic level resolution: There is a fundamental understanding of the gonadotropin receptors and how they selectively bind FSH and LH; it is clear that cAMP serves as the key second messenger for nearly all gonadotropin actions in the ovary; substantial in-roads have been made in defining the signaling pathways activated by gonadotropins; and many of the target genes activated by FSH or LH have been identified and their physiologic functions explored. In addition, genetic studies in mice and naturally occurring mutations in women have verified the fundamental importance of the gonadotropins and their receptors to ovarian function and allowed new insight into the precise stages and processes in which they act. Finally, some insight into what differentiates FSH and LH action in the ovary is now being obtained.

Despite this wealth of information, much remains to be learned about gonadotropin signaling pathways in the ovary. Structures of the gonadotropin receptors, which will likely begin with their extracellular domains and extend to the full-length proteins, are needed to gain further insight into hormone-binding selectivity and into the conformational changes that underlie receptor activation and G protein coupling. Studies to more fully understand the tissue- and cell-specific expression and hormonal regulation of the gonadotropin receptors are also important. Although some of the major signaling pathways downstream of gonadotropin-regulated cAMP generation have been explored in depth, as discussed in this chapter, others are just beginning to be explored and more detailed analysis is required. New advances in proteomics are likely to be important here, particularly methods to carefully examine the phosphorylation status of kinase target proteins in ovarian cells and to more fully characterize the kinases and phosphatases involved in particular regulatory events. As discussed, limited data also point to activation of additional signaling via cAMP-independent pathways in ovarian cells, and this is an area worthy of further investigation. A wealth of new information is accumulating from gene expression profiling and other high throughput analyses with respect to potential target genes activated by gonadotropin signaling, and verifying these targets, establishing mechanisms of their regulation, and exploring their functions in the ovary promises to be a significant focus of future work.

Cross-talk between signaling pathways is an often discussed but poorly understood aspect of signaling that is likely to be increasingly important as we begin to understand the molecular details of individual pathways. It will be particularly valuable to understand how gonadotropin signaling integrates with the many other signaling pathways important for follicle formation and maintenance. For example, transforming growth factor β family proteins play numerous roles in cell communication in the follicle, but little is known regarding how these pathways impact gonadotropin signaling and vice versa. Similarly, understanding how other signaling pathways influence the fundamental processes of cell proliferation, cell differentiation, and cell death in the ovary and how gonadotropin action might modify these events remains a key challenge.

We raised the issue of spatially regulated signaling as it relates to the AKAPs and their roles in determining cell-specific responses to cAMP, but the broader issue of spatially restricted signaling domains is likely to be an important one that applies to many pathways and contributes in a significant way to tissue- and cell-selective responses. Further characterization of such spatial domains in ovarian cells is warranted. This also reinforces the view that signaling principles learned in generic cell systems are not always applicable to granulosa or thecal cells and must be tested in the ovary. The temporal aspects of regulation are likely to be just as important. We broadly define for the purposes of this discussion preantral and preovulatory granulosa cells as distinct stages of differentiation in the follicle that respond differentially to the gonadotropins, yet the reality is that there is a continuum of differentiation states that define the precise nature of the response to gonadotropins or other stimuli.

Gene disruption technologies have been an integral component of studies of gonadotropin signaling pathways. A significant number of gene knockouts yield reproductive phenotypes, and as discussed throughout the chapter, these mice have been invaluable in establishing roles for particular signaling proteins and pathways in the ovary. However, many gene disruptions are not particularly informative with respect to reproduction because of early lethality or broad spread phenotype that impact reproduction in an indirect fashion. Thus conditional disruption approaches, including both ovary-specific and temporally regulated knockouts, are likely to be very informative for deciphering details of protein functions in gonadotropin-dependent signaling. In the same vein, recent advances in RNA interference technologies that include in vivo delivery hold much promise for down-regulating selective gene products in targets like the ovary to establish their functional importance.

Finally, studies of gonadotropin action in the ovary continue to be informed by clinical investigation and genetic analysis of reproductive disorders, and this is an area that will surely receive much more attention in the future. Naturally occurring mutations in the gonadotropins and their receptors provide a rich context in which to learn about interaction with and activation of the gonadotropin receptors. An increasing number of mutations in signaling molecules and transcription factors exhibit some reproductive phenotype, and these are tremendously informative in inferring the functional importance of particular genes and pathways. Even complex diseases that impact ovarian reserve, leading to premature ovarian failure, or that impact follicle maturation, leading to polycystic ovarian syndrome, are now being explored in genetic detail, and aspects of gonadotropin signaling will likely come into play in these as well. Thus, the future holds much promise for better understanding the molecular details of how FSH and LH function as master regulators in the maturation, ovulation, and luteinization of the ovarian follicle.

ACKNOWLEDGMENTS

We express our deepest gratitude to Evelyn Maizels, Hena Alam, and Tehnaz Parakh for critically reading this review. Supported by HD P01 21921 (to M. H. D. and K. M.), HD R01 38060 (to M. H. D.), and HD U54 41857 (to K. M.).

REFERENCES

1. Heckert, L. L., and Griswold, M. D. (2002). The expression of the follicle-stimulating hormone receptor in spermatogenesis. *Rec. Prog. Horm. Res.* 57, 129–148.
2. Dias, J. A., Cohen, B. D., Lindau-Shepard, B., et al. (2002). Molecular, structural, and cellular biology of follitropin and follitropin receptor. *Vitam. Horm.* 64, 249–322.
3. Simoni, M., Gromoll, J., and Nieschlag, E. (1997). The follicle-stimulating hormone receptor: biochemistry, molecular biology, physiology, and pathophysiology. *Endocr. Rev.* 18, 739–773.
4. Fanelli, F., Themmen, A. P., and Puett, D. (2001). Lutropin receptor function: insights from natural, engineered, and computer-simulated mutations. *IUBMB Life* 51, 149–155.
5. Ji, T. H., Grossmann, M., and Ji, I. (1998). G protein-coupled receptors. I. Diversity of receptor-ligand interactions. *J. Biol. Chem.* 273, 17299–17302.
6. Tsai, S-C., Adamik, R., Tsuchiya, M., et al. (1991). Differential expression during development of ADP-ribosylation factors, 20-kDa guanine nucleotide-binding protein activators of cholera toxin. *J. Biol. Chem.* 268, 8213–8219.
7. Ascoli, M., Fanelli, F., and Segaloff, D. L. (2002). The lutropin/choriogonadotropin receptor, a 2002 perspective. *Endocr. Rev.* 23, 141–174.
8. Davis, J. S. (1994). Mechanisms of hormone action: luteinizing hormone receptors and second-messenger pathways. *Curr. Opin. Obstet. Gynecol.* 6, 254–261.
9. Zeleznik, A. J. (2001). Modifications in gonadotropin signaling: a key to understanding cyclic ovarian function. *J. Soc. Gynecol. Invest.* 8, S24–S25.
10. Diaz, F. J., Anderson, L. E., Wu Y. L., et al. (2002). Regulation of progesterone and prostaglandin F2alpha production in the CL. *Mol. Cell. Endocrinol.* 191, 65–80.
11. Epifano, O., and Dean, J. (2002). Genetic control of early folliculogenesis in mice. *Trends Endocrinol. Metab.* 13, 169–173.
12. Eppig, J. J. (2001). Oocyte control of ovarian follicular development and function in mammals. *Reproduction* 122, 829–838.

13. Matzuk, M. M. (2000). Revelations of ovarian follicle biology from gene knockout mice. *Mol. Cell. Endocrinol.* 163, 61–66.
14. Johnson, A. L. (2003). Intracellular mechanisms regulating cell survival in ovarian follicles. *Anim. Reprod. Sci.* 78, 185–201.
15. Pru, J. K., and Tilly, J. L. (2001). Programmed cell death in the ovary: insights and future prospects using genetic technologies. *Mol. Endocrinol.* 15, 845–853.
16. Woodruff, T. K., Meunier, H., Jones, P. B., Hsueh, A. J., and Mayo, K. E. (1987). Rat inhibin: molecular cloning of alpha- and beta-subunit complementary deoxyribonucleic acids and expression in the ovary. *Mol. Endocrinol.* 1, 561–568.
17. Park, J. Y., Su, Y. Q., Ariga, M., et al. (2004). EGF-like growth factors as mediators of LH action in the ovulatory follicle. *Science* 303, 682–684.
18. Hsueh, A. J. W., Adashi, E. Y., Jones, P. B. C., and Welsh, T. H. Jr. (1984). Hormonal regulation of the differentiation of cultured ovarian granulosa cells. *Endocr. Rev.* 5, 76–110.
19. Richards, J. S. (1994). Hormonal control of gene expression in the ovary. *Endocr. Rev.* 15, 725–751.
20. McGee, E. A., and Hsueh, A. J. (2000). Initial and cyclic recruitment of ovarian follicles. *Endocr. Rev.* 21, 200–214.
21. Ratoosh, S. L., Lifka, J., Hedin, L., Jahnsen, T., and Richards, J. S. (1987). Hormonal regulation of the synthesis and mRNA content of the regulatory subunit of cyclic AMP-dependent protein kinase type II in cultured rat ovarian granulosa cells. *J. Biol. Chem.* 262, 7306–7313.
22. Salvador, L. M., Flynn, M. P., Avila, J., et al. (2004). Neuronal microtubule-associated protein 2D is a dual A-kinase anchoring protein expressed in rat ovarian granulosa cells. *J. Biol. Chem.* 279, 27621–27632.
23. Alliston, T. N., Maiyar, A. C., Buse, P., Firestone, G. L., and Richards, J. S. (1997). Follicle stimulating hormone-regulated expression of serum/glucocorticoid-inducible kinase in rat ovarian granulosa cells: a functional role for the Sp1 family in promoter activity. *Mol. Endocrinol.* 11, 1934–1949.
24. Sekiguchi, T., Mizutani, T., Yamada, K., et al. (2002). Transcriptional regulation of the epiregulin gene in the rat ovary. *Endocrinology* 143, 4718–4729.
25. Sekiguchi, T., Mizutani, T., Yamada, K., et al. (2004). Expression of epiregulin and amphiregulin in the rat ovary. *J. Mol. Endocrinol.* 33, 281–291.
26. Russell, D. L., Doyle, K. M., Gonzales-Robayna, I., Pipaon, C., and Richards, J. S. (2003). Egr-1 induction in rat granulosa cells by follicle-stimulating hormone and luteinizing hormone: combinatorial regulation by transcription factors cyclic adenosine 3′,5′-monophosphate regulatory element binding protein, serum response factor, sp1, and early growth response factor-1. *Mol. Endocrinol.* 17, 520–533.
27. Yazawa, T., Mizutani, T., Yamada, K., et al. (2003). Involvement of cyclic adenosine 5′-monophosphate response element-binding protein, steroidogenic factor 1, and Dax-1 in the regulation of gonadotropin-inducible ovarian transcription factor 1 gene expression by follicle-stimulating hormone in ovarian granulosa cells. *Endo.* 144, 1920–1930.
28. Falender, A. E., Lanz, R., Malenfant, D., Belanger, L., and Richards, J. S. (2003). Differential expression of steroidogenic factor-1 and FTF/LRH-1 in the rodent ovary. *Endocrinology* 144, 3598–3610.
29. Matsui, M., Sonntag, B., Hwang, S. S., et al. (2004). Pregnancy-associated plasma protein-a production in rat granulosa cells: stimulation by follicle-stimulating hormone and inhibition by the oocyte-derived bone morphogenetic protein-15. *Endocrinology* 145, 3686–3695.
30. Lawrence, J. B., Oxvig, C., Overgaard, M. T., et al. (1999). The insulin-like growth factor (IGF)-dependent IGF binding protein-4 protease secreted by human fibroblasts is pregnancy-associated plasma protein-A. *Proc. Natl. Acad. Sci. U S A* 96, 3149–3153.
31. Park, J. Y., Richard, F., Chun, S. Y., et al. (2003). Phosphodiesterase regulation is critical for the differentiation and pattern of gene expression in granulosa cells of the ovarian follicle. *Mol. Endocrinol.* 17, 1117–1130.
32. Zimmermann, R. C., Hartman, T., Kavic, S., et al. (2003). Vascular endothelial growth factor receptor 2-mediated angiogenesis is essential for gonadotropin-dependent follicle development. *J. Clin. Invest.* 112, 659–669.
33. Sasson, R., Dantes, A., Tajima, K., and Amsterdam, A. (2003). Novel genes modulated by FSH in normal and immortalized FSH-responsive cells: new insights into the mechanism of FSH action. *FASEB J.* 17, 1256–1266.
34. Wu, J. Y., Gonzalez-Robayna, I. J., Richards, J. S., and Means, A. R. (2000). Female fertility is reduced in mice lacking Ca2+/calmodulin-dependent protein kinase IV. *Endocrinology* 141, 4777–4783.
35. Ko, C., Grieshaber, N. A., Ji, I., and Ji, T. H. (2003). Follicle-stimulating hormone suppresses cytosolic 3,5,3′-triiodothyronine-binding protein messenger ribonucleic acid expression in rat granulosa cells. *Endocrinology* 144, 2360–2367.
36. Sharma, S. C., and Richards, J. S. (2000). Regulation of AP1 (Jun/Fos) factor expression and activation in ovarian granulosa cells. Relation of JunD AND Fra2 to terminal differentiation. *J. Biol. Chem.* 275, 33718–33728.
37. Piontkewitz, Y., Sundfeldt, K., and Hedin, L. (1997). The expression of c-myc during follicular growth and luteal formation in the rat ovary in vivo. *J. Endocrinol.* 152, 395–406.
38. Delidow, B. C., Lynch, J. P., White, B. A., and Peluso, J. J. (1992). Regulation of proto-oncogene expression and deoxyribonucleic acid synthesis in granulosa cells of perifused immature rat ovaries. *Biol. Repro.* 47, 428–435.
39. Murphy, L. O., MacKeigan, J. P., and Blenis, J. (2004). A network of immediate early gene products propagates subtle differences in mitogen-activated protein kinase signal amplitude and duration. *Mol. Cell. Biol.* 24, 144–153.
40. Sicinski, P., Donaher, J. L., Geng, Y., et al. (1996). Cyclin D2 is an FSH-responsive gene involved in gonadal cell proliferation and oncogenesis. *Nature* 384, 470–474.
41. Robker, R. L., and Richards, J. S. (1998). Hormone-induced proliferation and differentiation of granulosa cells: a coordinated balance of the cell cycle regulators cyclin D2 and $p27^{Kip1}$. *Mol. Endocrinol.* 12, 924–940.
42. Grieshaber, N. A., Ko, C., Grieshaber, S. S., Ji, I., and Ji, T. H. (2003). Follicle-stimulating hormone-responsive cytoskeletal genes in rat granulosa cells: class I beta-tubulin, tropomyosin-4, and kinesin heavy chain. *Endocrinology* 144, 29–39.
43. Hennebold, J. D., Tanaka, M., Saito, J., Hanson, B. R., and Adashi, E. Y. (2000). Ovary-selective genes I: the generation and characterization of an ovary-selective complementary deoxyribonucleic acid library. *Endocrinology* 141, 2725–2734.
44. Ranta, T., Knecht, M., Darbon, J., Baukal, A. J., and Catt, K. J. (1984). Induction of granulosa cell differentiation by forskolin: stimulation of adenosine 3′,5′-monophosphate production, progesterone synthesis, and luteinizing hormone receptor expression. *Endocrinology* 114, 845–852.
45. Knecht, M., and Catt, K. J. (1982). Induction of luteinizing hormone receptors by adenosine 3′,5′ monophosphate in cultured granulosa cells. *Endocrinology* 111, 1192–1200.
46. Padmanabhan, V., Karsch, F. J., and Lee, J. S. (2002). Hypothalamic, pituitary and gonadal regulation of FSH. *Reprod. Suppl.* 59, 67–82.
47. Kolakowski, L. F. Jr. (1994). GCRDb: a G-protein-coupled receptor database. *Receptors Channels* 2, 1–7.
48. Fredriksson, R., Lagerstrom, M. C., Lundin, L. G., and Schioth, H. B. (2003). The G-protein-coupled receptors in the human genome form five main families. Phylogenetic analysis, paralogon groups, and fingerprints. *Mol. Pharmacol.* 63, 1256–1272.

49. Ryu, K., Gilchrist, R. L., Tung, C. S., Ji, I., and Ji, T. H. (1998). High affinity hormone binding to the extracellular N-terminal exodomain of the follicle-stimulating hormone receptor is critically modulated by exoloop 3. *J. Biol. Chem.* 273, 28953–28958.
50. Sohn, J., Youn, H., Jeoung, M., et al. (2003). Orientation of follicle-stimulating hormone (FSH) subunits complexed with the FSH receptor. Beta subunit toward the N terminus of exodomain and alpha subunit to exoloop 3. *J. Biol. Chem.* 278, 47868–47876.
51. Gether, U. (2000). Uncovering molecular mechanisms involved in activation of G protein-coupled receptors. *Endocr. Rev.* 21, 90–113.
52. Schmidt, A., MacColl, R., Lindau-Shepard, B., Buckler, D. R., and Dias, J. A. (2001). Hormone-induced conformational change of the purified soluble hormone binding domain of follitropin receptor complexed with single chain follitropin. *J. Biol. Chem.* 276, 23373–23381.
53. Fan, Q. R., and Hendrickson, W. A. (2005). Structure of human follicle-stimulating hormone in complex with its receptor. *Nature* 433, 269–277.
53a. Arey, B. J., Stevis, P. E., Deecher, D. C., Shen, E. S., Frail, D. E., Negro-Vilar, A., and Lopez, F. J. (1997). Induction of promiscuous G protein coupling of the follicle-stimulating hormone (FSH) receptor: a novel mechanism for transducing pleiotropic actions of FSH isoforms. *Mol. Endocrinol.* 11, 517–526.
54. Quintana, J., Hipkin, R. W., Sanchez-Yague, J., and Ascoli, M. (1994). Follitropin (FSH) and a phorbol ester stimulate the phosphorylation of the FSH receptor in intact cells. *J. Biol. Chem.* 269, 8772–8779.
55. Sprengel, R., Braun, T., Nikolics, K., Segaloff, D. L., and Seeburg, P. H. (2000). The testicular receptor for follicle stimulating hormone: structure and functional expression of cloned cDNA. *Mol. Endocrinol.* 4, 525–530.
56. Dietrich, A., Sairam, M. R., Monaco, L., et al. (1998). Impairing follicle-stimulating hormone (FSH) signaling in vivo: targeted disruption of the FSH receptor leads to aberrant gametogenesis and hormonal imbalance. *Proc. Natl. Acad. Sci. U S A* 95, 13612–13617.
57. Abel, M. H., Wootton, A. N., Wilkins, V., et al. (2000). The effect of a null mutation in the follicle-stimulating hormone receptor gene on mouse reproduction. *Endocrinology* 141, 1795–1803.
58. Camp, T. A., Rahal, J. O., and Mayo, K. E. (1991). Cellular localization and hormonal regulation of follicle-stimulating hormone and luteinizing hormone receptor messenger RNAs in the rat ovary. *Mol. Endocrinol.* 5, 1405–1417.
59. Dunkel, L., Tilly, J. L., Shikone, T., Nishimori, K., and Hsueh, A. J. (1994). Follicle-stimulating hormone receptor expression in the rat ovary: increases during prepubertal development and regulation by the opposing actions of transforming growth factors beta and alpha. *Biol. Reprod.* 50, 940–948.
60. Sokka, T., and Huhtaniemi, I. (1990). Ontogeny of gonadotrophin receptors and gonadotrophin-stimulated cyclic AMP production in the neonatal rat ovary. *J. Endocrinol.* 127, 297–303.
61. Minegishi, T., Hirakawa, T., Kishi, H., et al. (2000). A role of insulin-like growth factor I for follicle-stimulating hormone receptor expression in rat granulosa cells. *Biol. Reprod.* 62, 325–333.
62. Nakamura, M., Nakamura, K., Igarashi, S., et al. (1995). Interaction between activin A and cAMP in the induction of FSH receptor in cultured rat granulosa cells. *J. Endocrinol.* 147, 103–110.
63. Findlay, J. K. (1993). An update on the roles of inhibin, activin, and follistatin as local regulators of folliculogenesis. *Biol. Reprod.* 48, 15–23.
64. Romero, C., Paredes, A., Dissen, G. A., and Ojeda, S. R. (2002). Nerve growth factor induces the expression of functional FSH receptors in newly formed follicles of the rat ovary. *Endocrinology* 143, 1485–1494.
65. Zhou, J., Kumar, T. R., Matzuk, M. M., and Bondy, C. (1997). Insulin-linked growth factor 1 regulates gonadotropin responsiveness in the murine ovary. *Mol. Endocrinol.* 11, 1924–1933.
66. Sanford, J. C., and Batten, B. E. (1989). Endocytosis of follicle-stimulating hormone by ovarian granulosa cells: analysis of hormone processing and receptor dynamics. *J. Cell. Physiol.* 138, 154–164.
67. Heckert, L. L. (2001). Activation of the rat follicle-stimulating hormone receptor promoter by steroidogenic factor 1 is blocked by protein kinase A and requires upstream stimulatory factor binding to a proximal E box element. *Mol. Endocrinol.* 15, 704–715.
68. Heckert, L. L., Sawadogo, M., Daggett, M. A., and Chen, J. K. (2000). The USF proteins regulate transcription of the follicle-stimulating hormone receptor but are insufficient for cell-specific expression. *Mol. Endocrinol.* 14, 1836–1848.
69. Levallet, J., Koskimies, P., Rahman, N., and Huhtaniemi, I. (2001). The promoter of murine follicle-stimulating hormone receptor: functional characterization and regulation by transcription factor steroidogenic factor 1. *Mol. Endocrinol.* 15, 80–92.
70. Deleted by author.
71. Gromoll, J., Simoni, M., and Nieschlag, E. (1996). An activating mutation of the follicle-stimulating hormone receptor autonomously sustains spermatogenesis in a hypophysectomized man. *J. Clin. Endocrinol. Metab.* 81, 1367–1370.
72. Hausdorff, W. P., Caron, M. G., and Lefkowitz, R. J. (1990). Turning off the signal: desensitization of beta-adrenergic receptor function. *FASEB J.* 4, 2881–2889.
73. Lefkowitz, R. J. (1993). G protein-coupled receptor kinases. *Cell* 74, 409–412.
74. Lohse, M. J., Andexinger, S., Pitcher, J., et al. (1992). Receptor-specific desensitization with purified proteins. *J. Biol. Chem.* 267, 8558–8564.
75. Gurevich, V. V., Dion, S. B., Onorato, J. J., et al. (1995). Arrestin interactions with G protein-coupled receptors. *J. Biol. Chem.* 270, 720–731.
76. Krupnick, J. G., Gurevich, V. V., and Benovic, J. L. (1998). Mechanism of quenching of phototransduction. *J. Biol. Chem.* 272, 18125–18131.
77. Barak, L. S., Ferguson, S. S., Zhang, J., and Caron, M. G. (1997). A beta-arrestin/green fluorescent protein biosensor for detecting a G protein-coupled receptor activation. *J. Biol. Chem.* 272, 27497–27500.
78. Laporte, S. A., Oakley, R. H., Holt, J. A., Barak, L. S., and Caron, M. G. (2000). The interaction of β-arrestin with the AP-2 adaptor is required for clustering of β2-adrenergic receptor into clathrin-coated pits. *J. Biol. Chem.* 275, 23120–23126.
79. Laporte, S. A., Oakley, R., Zhang, J., et al. (1999). The β2-adrenergic receptor/b arrestin complex recruits the clathrin adaptor AP-2 during endocytosis. *Proc. Natl. Acad. Sci. U S A* 96, 3712–3717.
80. Goodman, O. B., Krupnick, J. G., Santini, F., et al. (1996). β-arrestin acts as a clathrin adaptor in endocytosis of the β2-adrenergic receptor. *Nature* 383, 447–450.
81. Pierce, K. L., Lefkowitz, R. J., and Lefkowitz, R. J. (2001). Classical and new roles of β-arrestins in the regulation of G-protein coupled receptors. *Nat. Rev. Neurosci.* 2, 727–733.
82. Nakamura, K., Hipkin, R. W., and Ascoli, M. (1998). The agonist-induced phosphorylation of the rat follitropin receptor maps to the first and third intracellular loops. *Mol. Endocrinol.* 12, 580–591.
83. Lazari, M. F., Liu, X., Nakamura, K., Benovic, J. L., and Ascoli, M. (1999). Role of G protein-coupled receptor kinase

on the agonist-induced phosphorylation and internalization of the follitropin receptor. *Mol. Endocrinol.* 13, 866–878.
84. Krishnamurthy, H. Galet, C., and Ascoli, M. (2003). The association of arrestin-3 with the follitropin receptor depends on receptor activation and phosphorylation. *Mol. Cell Endocrinol.* 204, 126–140.
85. Krishnamurthy, H., Kishi, H., Shi, M., et al. (2003). Postendocytotic trafficking of the follicle-stimulating hormone (FSH)-FSH receptor complex. *Mol. Endocrinol.* 17, 2162–2176.
86. Yarney, T. A., Jiang, L., Khan, H., et al. (1997). Molecular cloning, structure, and expression of a testicular follitropin receptor with selective alteration in the carboxy terminus that affects signaling function. *Mol. Reprod. Dev.* 48, 458–470.
87. Sairam, M. R., Jiang, L. G., Yarney, T. A., and Khan, H. (1997). Alternative splicing converts the G-protein coupled follitropin receptor gene into a growth factor type I receptor: implications for pleiotropic actions of the hormone. *Mol. Reprod. Dev.* 48, 471–479.
88. Babu, P. S., Danilovich, N., and Sairam, M. R. (2001). Hormone-induced receptor gene splicing: enhanced expression of the growth factor type I follicle-stimulating hormone receptor motif in the developing mouse ovary as a new paradigm in growth regulation. *Endocrinology* 142, 381–389.
89. Babu, P. S., Krishnamurthy, H., Chedrese, P. J., and Sairam, M. R. (2000). Activation of extracellular-regulated kinase pathways in ovarian granulosa cells by the novel growth factor type I follicle-stimulating hormone receptor. *J. Biol. Chem.* 275, 27615–27626.
90. Knecht, M., Ranta, T., and Catt, K. J. (1983). Granulosa cell differentiation in vitro: induction and maintenance of follicle-stimulating hormone receptors by adenosine 3′,5′-monophosphate. *Endocrinology* 113, 949–956.
91. Knecht, M., Amsterdam, A., and Catt, K. (1981). The regulatory role of cyclic AMP in hormone-induced granulosa cell differentiation. *J. Biol. Chem.* 256, 10628–10633.
92. Taylor, S. S. (1989). cAMP-dependent protein kinase. *J. Biol. Chem.* 264, 8443–8446.
93. Dell'Acqua, M. L., and Scott, J. D. (1997). Protein kinase A anchoring. *J. Biol. Chem.* 272, 12881–12884.
94. de Rooij, J., Zwartkruis, F. J., Verheijen, M. H., et al. (1998). Epac is a Rap1 guanine-nucleotide-exchange factor directly activated by cyclic AMP. *Nature* 396, 474–477.
95. de Rooij, J., Zwartkruis, F. J. T., Verheijen, M. H. G., et al. (2000). Epac is a Rap1 guanine-nucleotide-exchange factor directly activated by cyclic AMP. *Nature* 396, 474–477.
96. de Rooij, J., Rehmann, H., van Triest, M., et al. (2000). Mechanism of regulation of the Epac family of cAMP-dependent RapGEFs. *J. Biol. Chem.* 275, 20829–20836.
97. Richards, J. S., Sharma, S. C., Falender, A. E., and Lo, Y. H. (2002). Expression of FKHR, FKHRL1, and AFX genes in the rodent ovary: evidence for regulation by IGF-I, estrogen, and the gonadotropins. *Mol. Endocrinol.* 16, 580–599.
98. Cottom, J., Salvador, L. M., Maizels, E. T., et al. (2003). Follicle stimulating hormone activates extracellular signal regulated kinase kinases by not extracellular regulated signal kinase through a 100 kDa phosphotyrosine phosphatase. *J. Biol. Chem.* 278, 7167–7179.
99. Zeleznik, A. J., Saxena, D., and Little-Ihrig, L. (2003). Protein kinase B is obligatory for follicle-stimulating hormone-induced granulosa cell differentiation. *Endocrinology* 144, 3985–3994.
100. DeManno, D. A., Cottom, J. E., Kline, M. P., et al. (1999). Follicle stimulating hormone promotes histone H3 phosphorylation on serine-10. *Mol. Endocrinol.* 13, 91–105.
101. Salvador, L. M., Park, Y., Cottom, J., et al. (2001). Follicle stimulating hormone stimulates protein kinase A mediated histone H3 phosphorylation and acetylation leading to select gene activation in ovarian granulosa cells. *J. Biol. Chem.* 276, 40146–40155.
102. Walsh, D. A., Ashby, C. D., Gonzalez, C., et al. (1971). Purification and characterization of a protein kinase inhibitor of adenosine 3′, 5′-monophosphate-dependent protein kinases. *J. Biol. Chem.* 246, 1977–1985.
103. Hidaka, H., Watanabe, M., and Kobayashi, R. (1991). Properties and use of H-series compounds as protein kinase inhibitors. *Methods Enzymol.* 201, 328–339.
104. Engh, R. A., Girod, A., Kinzel, V., Huber, R., and Bossemeyer, D. (1996). Crystal structures of catalytic subunit of cAMP-dependent protein kinase in complex with isoquinolinesulfonyl protein kinase inhibitors H7, H8 and H89. *J. Biol. Chem.* 271, 26157–26164.
105. Gonzalez-Robayna, I., Alliston, T. N., Buse, P., Firestone, G. L., and Richards, J. S. (1999). Functional and subcellular changes in the A-kinase-signaling pathway: relation to aromatase and Sgk expression during the transition of granulosa cells to luteal cells. *Mol. Endocrinol.* 13, 1318–1337.
106. Shabb, J. B. (2001). Physiological substrates of cAMP-dependent protein kinase. *Chem. Rev.* 101, 2381-2411.
107. Montminy, M. (2000). Transcriptional regulation by cyclic AMP. *Annu. Rev. Biochem.* 66, 807–822.
108. Lim, J., Pahlke, G., and Conti, M. (1999). Activation of the cAMP-specific phosphodiesterase PDE4D3 by phosphorylation. Identification and function of an inhibitory domain. *J. Biol. Chem.* 274, 19677–19685.
109. Schmitt, J. M., and Stork, P. J. (2002). PKA phosphorylation of Src mediates cAMP's inhibition of cell growth via Rap1. *Mol. Cell.* 9, 85–94.
110. Taylor, S. S. (1982). The in vitro phosphorylation of chromatin by the catalytic subunit of cAMP-dependent protein kinase. *J. Biol. Chem.* 257, 6056–6063.
111. Fang, X., Yu, S. X., Lu, Y., et al. (2000). Phosphorylation and inactivation of glycogen synthase kinase 3 by protein kinase A. *Proc. Natl. Acad. Sci. U S A* 97, 11960–11965.
112. Tanji, C., Yamamoto, H., Yorioka, N., et al. (2002). A-kinase anchoring protein AKAP220 binds to glycogen synthase kinase-3beta (GSK-3beta) and mediates protein kinase A-dependent inhibition of GSK-3beta. *J. Biol. Chem.* 277, 36955–36961.
113. Zhong, H., Yang, H. S., Erdjement-Bromage, H., Tempst, P., and Ghosh, S. (1997). The transcriptional activity of NF-kappaB is regulated by I kappaB-associated PKAC subunit through a cyclic AMP-independent mechanism. *Cell* 89, 413–424.
114. Davare, M. A., Dong, F., Rubin, C. S., and Hell, J. W. (1999). The A-kinase anchor protein MAP2B and cAMP-dependent protein kinase are associated with class C L-type calcium channels in neurons. *J. Biol. Chem.* 274, 30280–30287.
115. Carr, D. W., Cutler, R. E. Jr., Cottom, J. E., et al. (1999). Identification of cAMP-dependent protein kinase holoenzymes in preantral- and preovulatory-follicle-enriched ovaries, and their association with A-kinase-anchoring proteins. *Biochem. J.* 344, 613–623.
116. Cummings, D. E., Brandon, E. P., Planas, J. V., et al. (1996). Genetically lean mice result from targeted disruption of the RII beta subunit of protein kinase A. *Nature* 382, 622–626.
117. Amieux, P. S., Cummings, D. E., Motamed, K., et al. (1997). Compensatory regulation of RI alpha protein levels in protein kinase A mutant mice. *J. Biol. Chem.* 272, 3993–3998.
118. DeManno, D. A., Jackiw, V., Brooks, E. J., and Hunzicker-Dunn, M. (1994). Characterization of recombinant RI beta and evaluation of the presence of RI beta protein in rat brain and testicular extracts. *Biochem. Biophys. Acta.* 1222, 501–510.
119. Amieux, P. S., and McKnight, G. S. (2002). The essential role of RI alpha in the maintenance of regulated PKA activity. *Ann. N Y Acad. Sci.* 968, 75–95.

120. Carr, D. W., Hausken, Z. E., Fraser, I. D. C., Stofko-Hahn, R. E., and Scott, J. D. (1992). Association of the type II cAMP-dependent protein kinase with a human thyroid RII-anchoring protein. *J. Biol. Chem.* 267, 13376–13382.
121. Carr, D. W., Stofko-Hahn, R. E., Fraser, I. D. C., et al. (1991). Interaction of the regulatory subunit (RII) of cAMP-dependent protein kinase with RII-anchoring proteins occurs through an amphipathic helix binding motif. *J. Biol. Chem.* 266, 14188–14192.
122. Lin, R-Y., Moss, S. R., and Rubin, C. S. (1995). Characterization of S-AKAP84, a novel developmentally regulated A kinase anchor protein of male germ cells. *J. Biol. Chem.* 270, 27804–27811.
123. Ginsberg, M. D., Feliciello, A., Jones, J. K., Avvedimento, E. V., and Gottesman, M. E. (2003). PKA-dependent binding of mRNA to the mitochondrial AKAP121 protein. *J. Mol. Biol.* 327, 885–897.
124. Keryer, G., Rios, R. M., Landmark, B. F., et al. (1993). A high-affinity binding protein for the regulatory subunit of cAMP-dependent protein kinase II in the centrosome of human cells. *Exp. Cell. Res.* 204, 230–240.
125. Diviani, D., Langeberg, L. K., Doxsey, S. J., and Scott, J. D. (2000). Pericentrin anchors protein kinase A at the centrosome through a newly identified RII-binding domain. *Curr. Biol.* 10, 417–420.
126. Witczak, O., Skalhegg, B. S., Keryer, G., et al. (1999). Cloning and characterization of a cDNA encoding an A-kinase anchoring protein located in the centrosome, AKAP450. *EMBO J.* 18, 1858–1868.
127. Steen, R. L., Beullens, M., Landsverk, H. B., Bollen, M., and Collas, P. (2003). AKAP149 is a novel PP1 specifier required to maintain nuclear envelope integrity in G1 phase. *J. Cell. Sci.* 116, 2237–2246.
128. Li, Y., Ndubuka, C., and Rubin, C. S. (1996). A kinase anchor protein 75 targets regulatory (RII) subunits of cAMP-dependent protein kinase II to the cortical actin cytoskeleton in non-neuronal cells. *J. Biol. Chem.* 271, 16862–16869.
129. Dong, F., Feldmesser, M., Casadevall, A., and Rubin, C. S. (1998). Molecular characterization of a cDNA that encodes six isoforms of a novel murine A kinase anchor protein. *J. Biol. Chem.* 273, 6533–6541.
130. Westphal, R. S., Soderling, S. H., Alto, N. M., Langeberg, L. K., and Scott, J. D. (2000). Scar/WAVE-1, a Wiscott-Aldrich syndrome protein, assembles an actin-associated multi-kinase scaffold. *EMBO J.* 19, 4589–4600.
131. Obar, R. A., Dingus, J., Bayley, H., and Vallee, R. B. (1989). The RII subunit of cAMP-dependent protein kinase binds to a common amino-terminal domain in microtubule-associated proteins 2A, 2B, and 2C. *Neuron* 3, 639–645.
132. Diviani, D., and Scott, J. D. (2001). AKAP signaling complexes at the cytoskeleton. *J. Cell. Sci.* 114, 1431–1437.
133. Michel, J. J., and Scott, J. D. (2002). AKAP mediated signal transduction. *Annu. Rev. Pharmacol. Toxicol.* 42, 235–257.
134. Pawson, T., and Scott, J. D. (1997). Signaling through scaffold, anchoring, and adaptor proteins. *Science* 278, 2075–2080.
135. Reinton, N., Collas, P., Haugen, T. B., et al. (2000). Localization of a novel human A-kinase-anchoring protein, hAKAP220, during spermatogenesis. *Dev. Biol.* 223, 194–204.
136. Li, H., Adamik, R., Pacheco-Rodriguez, G., Moss, J., and Vaughan, M. (2003). Protein kinase A-anchoring (AKAP) domains in brefeldin A-inhibited guanine nucleotide-exchange protein 2 (BIG2). *Proc. Natl. Acad. Sci. U S A* 100, 1627–1632.
137. Miki, K., and Eddy, E. M. (1998). Identification of tethering domains for protein kinase A type I alpha regulatory subunits on sperm fibrous sheath protein FSC1. *J. Biol. Chem.* 273, 34384–34390.
138. Huang, L. J., Wang, L., Ma, Y., et al. (1999). NH2-Terminal targeting motifs direct dual specificity A-kinase-anchoring protein 1 (D-AKAP1) to either mitochondria or endoplasmic reticulum. *J. Cell. Biol.* 145, 951–959.
139. Huang, L. J., Durick, K., Weiner, J. A., Chun, J., and Taylor, S. S. (1997). D-AKAP2, a novel protein kinase A anchoring protein with a putative RGS domain. *Proc. Natl. Acad. Sci. U S A* 94, 11184–11189.
140. Angelo, R., and Rubin, C. S. (1998). Molecular characterization of an anchor protein ($AKAP_{CE}$) of type I protein kinase A from *Caenorhabditis elegans*. *J. Biol. Chem.* 273, 14633–14643.
141. Li, H., Degenhardt, B., Tobin, D., et al. (2001). Identification, localization, and function in steroidogenesis of PAP7: a peripheral-type benzodiazepine receptor- and PKA (RIalpha)-associated protein. *Mol. Endocrinol.* 15, 2211–2228.
142. Gronholm, M., Vossebein, L., Carlson, C. R., et al. (2003). Merlin links to the cAMP neuronal signaling pathway by anchoring the RI beta subunit of protein kinase A. *J. Biol. Chem.* 278, 41167–41172.
143. Mukherjee, A., Park-Sarge, O. K., and Mayo, K. E. (1996). Gonadotropins induce rapid phosphorylation of the 3′,5′-cyclic adenosine monophosphate response element binding protein in ovarian granulosa cells. *Endocrinology* 137, 3234–3245.
144. Carlone, D. L., and Richards, J. S. (1997). Functional interactions, phosphorylation, and levels of 3′,5′-cyclic adenosine monophosphate-regulatory element binding protein and steroidogenic factor-1 mediate hormone-regulated and constitutive expression of aromatase in gonadal cells. *Mol. Endocrinol.* 11, 292–304.
145. Toker, A. (2000). Protein kinases as mediators of phosphoinositide 3-kinase signaling. *Mol. Pharmacol.* 57, 652–658.
146. Shaywitz, A. J., and Greenberg, M. E. (1999). CREB: a stimulus-induced transcription factor activated by a diverse array of extracellular signals. *Annu. Rev. Biochem.* 68, 821–861.
147. Roux, P. P., and Blenis, J. (2004). ERK and p38 MAPK-activated protein kinases: a family of protein kinases with diverse biological functions. *Microbiol. Mol. Biol. Rev.* 68, 320–344.
148. Frodin, M., and Gammeltoft, S. (1999). Role and regulation of 90 kDa ribosomal S6 kinase (RSK) in signal transduction. *Mol. Cell. Endocrinol.* 151, 65–77.
149. Sassone-Corsi, P., Mizzen, C. A., and Cheung, P., et al. (1999). Requirement of Rsk-2 for epidermal growth factor-activated phosphorylation of histone H3. *Science* 295, 886–891.
150. Thomson, S., Clayton, A. L., and Hazzaslin, C. A., et al. (1999). The nucleosomal response associated with immediate-early gene induction is mediated via alternative MAP kinase cascades; MSK1 as a potential histone H3/HMG-14 kinase. *EMBO J.* 18, 4779–4793.
151. Frodin, M., Jensen, C. J., Merienne, K., and Gameltoft, S. (2000). A phosphoserine-regulated docking site in the protein kinase RSK2 that recruits and activates PDK1. *EMBO J.* 19, 2924–2934.
152. Strahl, B. D., and Allis, D. (2000). The language of covalent histone modifications. *Nature* 403, 41–45.
153. Struhl, K. (1998). Histone acetylation and transcriptional regulatory mechanisms. *Genes Dev.* 12, 599–606.
154. Wolffe, A. P., and Pruss, D. (1996). Chromatin: hanging on to histones. *Curr. Biol.* 6, 234–237.
155. Davie, J. K., and Dent, S. Y. (2002). Transcriptional control: an activating role for arginine methylation. *Curr. Biol.* 12, R59–R61.
156. Bannister, A. J., Schneider, R., and Kouzarides, T. (2002). Histone methylation: dynamic or static? *Cell* 109, 801–806.
157. Fischle, W., Wang, Y., and Allis, C. D. (2003). Binary switches and modification cassettes in histone biology and beyond. *Nature* 425, 475–479.

158. Sassone-Corsi, P., Visvader, J., Ferland, L., Mennon, P. L., and Verma, I. M. (1988). Induction of proto-oncogene fos transcription through the adenylate cyclase pathway: characterization of a cAMP-responsive element. *Genes Dev.* 2, 1529–1538.
159. Chrivia, J. C., Kwok, R. P., Lamb, N., et al. (1993). Phosphorylated CREB binds specifically to the nuclear protein CBP. *Nature* 365, 855–859.
160. Kwok, R. P. S., Lundblad, J. R., Chrivia, J. C., et al. (1994). Nuclear protein CBP is a coactivator for the transcription factor CREB. *Nature* 370, 223–225.
161. Samson, S. L., and Wong, N. C. (2002). Role of Sp1 in insulin regulation of gene expression. *J. Mol. Endocrinol.* 29, 265–279.
162. Ahlgren, R., Suske, G., Waterman, M. R., and Lund, J. (1999). Role of Sp1 in cAMP-dependent transcriptional regulation of the bovine CYP11A gene. *J. Biol. Chem.* 274, 19422–19428.
163. Ito, M., Park, Y., Weck, J., Mayo, K. E., and Jameson, J. L. (2000). Synergistic activation of the inhibin alpha-promoter by steroidogenic factor-1 and cyclic adenosine 3′,5′-monophosphate. *Mol. Endocrinol.* 14, 66–81.
164. Monte, D., DeWitte, F., and Hum, D. W. (1998). Regulation of the human P450scc gene by steroidogenic factor 1 is mediated by CBP/p300. *J. Biol. Chem.* 273, 4585–4591.
165. Lund, J., Bakke, M., Mellgren, G., Morohashi, K., and Doskeland, S. O. (1997). Transcriptional regulation of the bovine CYP17 gene by cAMP. *Steroids* 62, 43–45.
166. Aesoy, R., Mellgren, G., Morohashi, K., and Lund, J. (2002). Activation of cAMP-dependent protein kinase increases the protein level of steroidogenic factor-1. *Endocrinology* 143, 295–303.
167. Fan, W., Yanase, T., Wu, Y., et al. (2004). Protein kinase A potentiates adrenal 4 binding protein/steroidogenic factor 1 transactivation by reintegrating the subcellular dynamic interactions of the nuclear receptor with its cofactors, general control nonderepressed-5/transformation/transcription domain-associated protein, and suppressor, dosage-sensitive sex reversal-1: a laser confocal imaging study in living KGN cells. *Mol. Endocrinol.* 18, 127–141.
168. Ogryzko, V. W., Schiltz, R. L., Russanov, V., Howard, B. H., and Nakatani, Y. (1996). The transcriptional coactivators p300 and CBP are histone acetyltransferases. *Cell* 87, 953–959.
169. Bannister, A. J., and Kouzarides, T. (1996). The CBP co-activator is a histone acetyltransferase. *Nature* 384, 641–643.
170. Yang, X. J., Ogryzko, W., Nishikawa, J., Howard, B. H., and Nakatani, Y. (1996). A p300/CBP-associated factor that competes with the adenoviral oncoprotein E1A. *Nature* 382, 319–324.
171. Blanco, J. C., Minucci, S., Lu, J., et al. (1998). The histone acetylase PCAF is a nuclear receptor coactivator. *Genes Dev.* 12, 1638–1651.
172. Wolffe, A. P. (1992). New insights into chromatin function in transcriptional control. *FASEB J.* 6, 3354–3361.
173. Ajiro, K., Shibata, K., and Nishikawa, Y. (1990). Subtype-specific cyclic AMP-dependent histone H1 phosphorylation at the differentiation of mouse neuroblastoma cells. *J. Biol. Chem.* 265, 6494–6500.
174. Pei, L., Dodson, R., Schoderbek, W. E., Maurer, R. A., and Mayo, K. E. (1991). Regulation of the alpha inhibin gene by cyclic adenosine 3′,5′-monophosphate after transfection into rat granulosa cells. *Mol. Endocrinol.* 5, 521–534.
175. Das, S., Maizels, E. T., DeManno, D., et al. (1996). A stimulatory role of cyclic adenosine 3′,5′-monophosphate in follicle-stimulating hormone-activated mitogen-activated protein kinase signaling pathway in rat ovarian granulosa cells. *Endocrinology* 137, 967–974.
176. Seger, R., Hanoch, T., Rosenberg, R., et al. (2001). The ERK signaling cascade inhibits gonadotropin-stimulated steroidogenesis. *J. Biol. Chem.* 276, 13957–13964.
177. Cobb, M. H., and Goldsmith, E. J. (1995). How MAP kinases are regulated. *J. Biol. Chem.* 270, 14843–14846.
178. Luttrell, L. M., Daaka, Y., and Lefkowitz, R. J. (1999). Regulation of tyrosine kinase cascades by G-protein-coupled receptors. *Curr. Opin. Cell. Biol.* 11, 177–183.
179. Andreev, J., Galisteo, M. L., Kranenburg, O., et al. (2001). Src and Pyk2 mediate G-protein-coupled receptor activation of epidermal growth factor receptor (EGFR) but are not required for coupling to the mitogen-activated protein (MAP) kinase signaling cascade. *J. Biol. Chem.* 276, 20130–20135.
180. Prenzel, N., Zwick, E., Daub, H., et al. (1999). EGF receptor transactivation by G-protein-coupled receptors requires metalloproteinase cleavage of proHB-EGF. *Nature* 402, 884–888.
181. Herrlich, A., Daub, H., Knebel, A., Herrlich, P., Ulrich, A., Schultz G., and Gudermann, T. (1998). Ligand-independent activation of platelet-derived growth factor receptor is a necessary intermediate in lysophosphatidic, acid-stimulated mitogenic activity in L cells. *Proc. Natl. Acad. Sci. USA* 95, 8985–8990.
182. Della Rocca, G. L., van Biesen, T., Daaka, Y., Luttrell, D. K., Luttrell, L. M., and Lefkowitz, R. J. (1997). Ras-dependent mitogen-activated protein kinase activation by G protein-coupled receptors. Convergence of Gi- and Gq-mediated pathways on calcium/calmodulin, Pyk2, and Src kinase. *J. Biol. Chem.* 272, 19125–19132.
183. Burgering, B. M. T., and Bos, J. L. (1995). Regulation of Ras-mediated signalling. *Trends Biochem. Sci.* 20, 18–22.
184. Grewal, S. S., Fass, D. M., Yao, H., et al. (2000). Calcium and cAMP signals differentially regulate cAMP-responsive element-binding protein function via a Rap1-extracellular signal-regulated kinase pathway. *J. Biol. Chem.* 275, 34433–34441.
185. Schmitt, J. M., and Stork, P. J. S. (2000). β_2-Adrenergic receptor activates extracellular signal-regulated kinases (ERKs) via the small G protein Rap1 and the serine/threonine kinase B-raf. *J. Biol. Chem.* 275, 25342–25350.
186. Levitski, A., and Gazit, A. (1995). Tyrosine kinase inhibition: an approach to drug development. *Science* 267, 1782–1788.
187. Alessi, D., Cuenda, A., Cohen, P., Dudley, D. T., and Saltiel, A. A. (2000). PD 098059 is a specific inhibitor of the activation of mitogen-activated protein kinase kinase in vitro and in vivo. *J. Biol. Chem.* 270, 27489–27494.
188. Lim, K., and Hwang, B. D. (1995). Follicle-stimulating hormone transiently induces expression of protooncogene c-myc in primary Sertoli cell cultures of early pubertal and prepubertal rat. *Mol. Cell. Endocrinol.* 111, 51–56.
189. Ardekani, A. M., Romanelli, J. C., and Mayo, K. E. (1998). Structure of the rat inhibin and activin betaA-subunit gene and regulation in an ovarian granulosa cell line. *Endocrinology* 139, 3271–3279.
190. Hammer, G. D., Krylova, I., Zhang, Y., et al. (1999). Phosphorylation of the nuclear receptor SF-1 modulates cofactor recruitment: integration of hormone signaling in reproduction and stress. *Mol. Cell.* 3, 521–526.
191. Desclozeaux, M., Krylova, I. N., Horn, F., Fletterick, R. J., and Ingraham, H. A. (2002). Phosphorylation and intramolecular stabilization of the ligand binding domain in the nuclear receptor steroidogenic factor 1. *Mol. Cell. Biol.* 22, 7193–7203.
192. Merchant, J. L., Du, M., and Todisco, A. (1999). Sp1 phosphorylation by Erk 2 stimulates DNA binding. *Biochem. Biophys. Res. Commun.* 254, 454–461.
193. Liu, Y. Z., Chrivia, J. C., and Latchman, D. S. (1998). Nerve growth factor up-regulates the transcriptional activity of CBP through activation of the p42/p44(MAPK) cascade. *J. Biol. Chem.* 273, 32400–32407.
194. Zanger, K., Radovick, S., and Wondisford, F. E. (2001). CREB binding protein recruitment to the transcription

complex requires growth factor-dependent phosphorylation of its GF box. *Mol. Cell.* 7, 551–558.

195. Merienne, K., Pannetier, S., Harel-Bellan, A., and Sassone-Corsi, P. (2001). Mitogen-regulated RSK2-CBP interaction controls their kinase and acetylase activities. *Mol. Cell. Biol.* 21, 7089–7096.
196. Wang, Z., Zhang, B., Wang, M., and Carr, B. I. (2003). Persistent ERK phosphorylation negatively regulates cAMP response element-binding protein (CREB) activity via recruitment of CREB-binding protein to pp90RSK. *J. Biol. Chem.* 278, 11138–11144.
197. Nakajima, T., Fukamizu, A., Takahashi, J., et al. (1996). The signal-dependent coactivator CBP is a nuclear target for pp90RSK. *Cell* 86, 465–474.
198. Du, K., and Montminy, M. (1998). CREB is a regulatory target for the protein kinase Akt/PKB. *J. Biol. Chem.* 273, 32377–32379.
199. Michiels, C., Minet, E., Michel, G., et al. (2001). HIF-1 and AP-1 cooperate to increase gene expression in hypoxia: role of MAP kinases. *IUBMB Life* 52, 49–53.
200. Nasrin, N., Ogg, S., Cahill, C. M., et al. (2000). DAF-16 recruits the CREB-binding protein coactivator complex to the insulin-like growth factor binding protein 1 promoter in HepG2 cells. *Proc. Natl. Acad. Sci. U S A* 97, 10412–10417.
201. Janknecht, R., Wells, N. J., and Hunter T. (1998). TGF-beta-stimulated cooperation of smad proteins with the coactivators CBP/p300. *Genes Dev.* 12, 2114–2119.
202. Feng, X. H., Zhang, Y., Wu, R. Y., and Derynck, R. (1998). The tumor suppressor Smad4/DPC4 and transcriptional adaptor CBP/p300 are coactivators for smad3 in TGF-beta-induced transcriptional activation. *Genes Dev.* 12, 2153–2163.
203. Maizels, E. T., Mukherjee, A., Sithanandam, G., et al. (2001). Developmental regulation of mitogen-activated protein kinase-activated kinases-2 and -3 (MAPKAPK-2/-3) in vivo during corpus luteum formation in the rat. *Mol. Endocrinol.* 15, 716–733.
204. Gonzalez-Robayna, I. J., Falender, A. E., Ochsner, S., Firestone, G. L., and Richards, J. S. (2000). Follicle-stimulating hormone (FSH) stimulates phosphorylation and activation of protein kinase B (PKB/AKt) and serum glycocorticoid-induced kinase (Sgk): evidence for A kinase-independent signaling by FSH in granulosa cells. *Mol. Endocrinol.* 14, 1283–1300.
205. Maizels, E. T., Cottom, J., Jones, J. R., and Hunzicker-Dunn, M. (1998). Follicle stimulating hormone (FSH) activates the p38 mitogen-activated kinase pathway, induces small heat shock protein phosphorylation and cell rounding in immature rat ovarian granulosa cells. *Endocrinology* 139, 3353–3356.
206. Lambert, H., Charette, S. J., Bernier, A. F., Guimond, A., and Landry, J. (1999). HSP27 multimerization mediated by phosphorylation-sensitive intermolecular interactions at the amino terminus. *J. Biol. Chem.* 274, 9378–9385.
207. Rogalla, T., Ehrnsperger, M., Preville, X., et al. (1999). Regulation of Hsp27 oligomerization, chaperone function, and protective activity against oxidative stress/tumor necrosis factor alpha by phosphorylation. *J. Biol. Chem.* 274, 18947–18956.
208. Concannon, C. G., Gorman, A. M., and Samali, A. (2003). On the role of Hsp27 in regulating apoptosis. *Apoptosis* 8, 61–70.
209. Amsterdam, A., and Rotmensch, S. (1987). Structure-function relationships during granulosa cell differentiation. *Endocr. Rev.* 8, 309–337.
210. Jayes, F. C. L., Day, R. N., Garmey, J. C., et al. (2000). Calcium ions positively modulate follicle-stimulating hormone- and exogenous cyclic 3′,5′-adenosine monophosphate-driven transcription of the P450scc gene in porcine granulosa cells. *Endocrinology* 141, 2377–2384.
211. Flores, J. A., Veldhuis, J. D., and Leong, D. A. (1990). Follicle-stimulating hormone evokes an increase in intracellular free calcium ion concentrations in single ovarian (granulosa) cells. *Endocrinology* 127, 3172–3179.
212. Flores, J. A., Aguirre, C., Sharma, O. P., and Veldhuis, J. D. (1998). Luteinizing hormone (LH) stimulates both intracellular calcium ion ([Ca2+]i) mobilization and transmembrane cation influx in single ovarian (granulosa) cells: recruitment as a cellular mechanism of LH-[Ca2+]i dose response. *Endocrinology* 139, 3606–3612.
213. Sharma, O. P., Flores, J. A., Leong, D. A., and Veldhuis, J. D. (1994). Cellular basis for follicle-stimulating hormone-stimulated calcium signaling in single rat sertoli cells: possible dissociation from effects of adenosine 3′,5′-monosphosphate. *Endocrinology* 134, 1915–1923.
214. Lee, P. S., Buchan, A. M., Hsueh, A. J., Yuen, B. H., and Leung, P. C. (2002). Intracellular calcium mobilization in response to the activation of human wild-type and chimeric gonadotropin receptors. *Endocrinology* 143, 1732–1740.
215. Grieshaber, N. A., Boitano, S., Ji, I., Mather, J. P., and Ji, T. H. (2000). Differentiation of granulosa cell line: follicle-stimulating hormone induces formation of lamellipodia and filopodia via the adenylyl cyclase/cyclic adenosine monophosphate signal. *Endocrinology* 141, 3461–3470.
216. Carnegie, J. A., and Tsang, B. K. (1984). The calcium-calmodulin system: participation in the regulation of steroidogenesis at different stages of granulosa cell differentiation. *Biol. Reprod.* 30, 515–522.
217. Seals, R. C., Urban, R. J., Sekar, N., and Veldhuis, J. D. (2004). Up-regulation of basal transcriptional activity of the cytochrome P450 cholesterol side-chain cleavage (CYP11A) gene by isoform-specific calcium-calmodulin–dependent protein kinase in primary cultures of ovarian granulosa cells. *Endocrinology* 145, 5612–5622.
218. Finkbeiner, S., Tavazoie, S. F., Maloratsky, A., et al. (1997). CREB: a major mediator of neuronal neurotrophin responses. *Neuron* 19, 1031–1047.
219. Conti, M., Kasson, B. G., and Hsueh, A. J. (1984). Hormonal regulation of 3′,5′-adenosine monophosphate phosphodiesterases in cultured rat granulosa cells. *Endocrinology* 114, 2361–2368.
220. Marley, P. D., and Thomson, K. A. (1996). The Ca++/calmodulin-dependent protein kinase II inhibitors KN62 and KN93, and their inactive analogues KN04 and KN92, inhibit nicotinic activation of tyrosine hydroxylase in bovine chromaffin cells. *Biochem. Biophys. Res. Commun.* 221, 15–18.
221. Cunningham, M. A., Zhu, Q., Unterman, T. G., and Hammond, J. M. (2003). Follicle-stimulating hormone promotes nuclear exclusion of the forkhead transcription factor FoxO1a via phosphatidylinositol 3-kinase in porcine granulosa cells. *Endocrinology* 144, 5585–5594.
222. Alam, H., Maizels, E. T., Park, Y., et al. (2004). Follicle-stimulating hormone activation of hypoxia-inducible factor-1 by the phosphatidylinositol 3-kinase/AKT/Ras homolog enriched in brain (Rheb)/mammalian target of rapamycin (mTOR) pathway is necessary for induction of select protein markers of follicular differentiation. *J. Biol. Chem.* 279, 19431–19440.
223. Wang, Y., Chan, S., and Tsang, B. K. (2002). Involvement of inhibitory nuclear factor-kappaB (NFkappaB)-independent NFkappaB activation in the gonadotropic regulation of X-linked inhibitor of apoptosis expression during ovarian follicular development in vitro. *Endocrinology* 143, 2732–2740.
224. Vanhaesebroeck, B., and Alessi, D. R. (2000). The PI3K-PDK1 connection: more than just a road to PKB. *Biochem. J.* 346, 561–576.
225. Chan, T. O., Rodeck, U., Chan, A. M., et al. (2002). Small GTPases and tyrosine kinases coregulate a molecular switch

in the phosphoinositide 3-kinase regulatory subunit. *Cancer Cell.* 1, 181–191.
226. Vanhaesebroeck, B., and Waterfield, M. D. (1999). Signaling by distinct classes of phosphoinositide 3-kinases. *Exp. Cell. Res.* 253, 239–254.
227. Westfall, S. D., Hendry, I. R., Obholz, K. L., Rueda, B. R., and Davis, J. S. (2000). Putative role of the phosphatidylinositol 3-kinase-Akt signaling pathway in the survival of granulosa cells. *Endocrine* 12, 315–321.
228. Baker, J., Hardy, M. P., Zhou, J., et al. (1996). Effects of an Igf1 gene null mutation on mouse reproduction. *Mol. Endocrinol.* 10, 903–918.
229. Oliver, J. E., Aitman, T. J., Powell, J. F., Wilson, C. A., and Clayton, R. N. (1989). Insulin-like growth factor I gene expression in the rat ovary is confined to the granulosa cells of developing follicles. *Endocrinology* 124, 2671–2679.
230. Adashi, E. Y., Resnick, C. E., D'Ercole, J., Svoboda, M. E., and Van Wyk, J. J. (1985). Insulin-like growth factors as intraovarian regulators of granulosa cell growth and function. *Endocr. Rev.* 6, 400–420.
231. Adashi, E. Y., Resnick, C. E., Hernandez, E. R., et al. (1988). Insulin-like growth factor-I as an amplifier of follicle-stimulating hormone action: studies on mechanism(s) and site(s) of action in cultured rat granulosa cells. *Endocrinology* 122, 1583–1591.
232. Adashi, E. Y., Resnick, C. E., Brodie, A. M. H., Svoboda, M. E., and Van Wyk, J. J. (1985). Somatomedin-C-mediated potentiation of follicle stimulating hormone-induced aromatase activity of cultured rat granulosa cells. *Endocrinology* 117, 2313–2320.
233. Hammond, J. M., Mondschein, J. S., Samaras, S. E., and Canning, S. F. (1991). The ovarian insulin-like growth factors, a local amplification mechanism for steroidogenesis and hormone action. *J. Steroid Biochem. Mol. Biol.* 40, 411–416.
234. El Hefnawy, T., and Zeleznik, A. J. (2001). Synergism between FSH and activin in the regulation of proliferating cell nuclear antigen (PCNA) and cyclin D2 expression in rat granulosa cells. *Endocrinology* 142, 4357–4362.
235. Khan, S. A., Ndjountche, L., Pratchard, L., Spicer, L. J., and Davis, J. S. (2002). Follicle-stimulating hormone amplifies insulin-like growth factor I-mediated activation of AKT/protein kinase B signaling in immature rat Sertoli cells. *Endocrinology*143, 2259–2267.
236. Li, Y., Corradetti, M. N., Inoki, K., and Guan, K-L. (2004). TSC2: filling the gap in the mTOP signaling pathway. *Trends Biochem. Sci.* 29, 32–38.
237. Tee, A. R., Manning, B. D., Roux, P. P., Cantley, L. C., and Blenis, J. (2003). Tuberous sclerosis complex gene products, Tuberin and Hamartin, control mTOR signaling by acting as a GTPase-activating protein complex toward Rheb. *Curr. Biol.* 13, 1259–1268.
238. Fingar, D. C., and Blenis, J. (2004). Target of rapamycin (TOR): an integrator of nutrient and growth factor signals and coordinator of cell growth and cell cycle progression. *Oncogene* 23, 3151–3171.
239. Brown, E. J., Beal, P. A., Keith, C. T., et al. (1995). Control of p70 s6 kinase by kinase activity of FRAP in vivo. *Nature* 377, 441–446.
240. Weng, Q. P., Kozlowski, M., Belham, C., et al. (1998). Regulation of the p70 S6 kinase by phosphorylation in vivo. Analysis using site-specific anti-phosphopeptide antibodies. *J. Biol. Chem.* 273, 16621–16629.
241. Burnett, P. E., Barrow, R. K., Cohen, N. A., Snyder, S. H., and Sabatini, D. M. (1998). RAFT1 phosphorylation of the translational regulators p70 S6 kinase and 4E-BP1. *Proc. Natl. Acad. Sci. U S A* 95, 1432–1437.
242. Hay, N., and Sonenberg, N. (2004). Upstream and downstream of mTOR. *Genes Dev.* 18, 1926–1945.
243. Schalm, S. S., Fingar, D. C., Sabatini, D. M., and Blenis, J. (2003). TOS motif-mediated raptor binding regulates 4E-BP1 multisite phosphorylation and function. *Curr. Biol.* 13, 797–806.
244. Dufner, A., and Thomas, G. (1999). Ribosomal S6 kinase signaling and the control of translation. *Exp. Cell. Res.* 253, 100–109.
245. Hara, K., Yonezawa, K., Weng, Q. P., et al. (1998). Amino acid sufficiency and mTOR regulate p70 S6 kinase and eIF-4E BP1 through a common effector mechanism. *J. Biol. Chem.* 273, 14484–14494.
246. Schalm, S. S., and Blenis, J. (2002). Identification of a conserved motif required for mTOR signaling. *Curr. Biol.* 12, 632–639.
247. Pyronnet, S. (2000). Phosphorylation of the cap-binding protein eIF4E by the MAPK-activated protein kinase Mnk1. *Biochem. Pharmacol.* 60, 1237–1243.
248. Pyronnet, S., and Sonenberg, N. (2001). Cell-cycle-dependent translational control. *Curr. Opin. Genet. Dev.* 11, 13–18.
249. Rajasekhar, V. K., Viale, A., Socci, N. D., et al. (2003). Oncogenic Ras and Akt signaling contribute to glioblastoma formation by differential recruitment of existing mRNAs to polysomes. *Mol. Cell.* 12, 889–901.
250. Muise-Helmericks, R. C., Grimes, H. L., Bellacosa, A., et al. (1998). Cyclin D expression is controlled post-transcriptionally via a phosphatidylinositol 3-kinase/Akt-dependent pathway. *J. Biol. Chem.* 273, 29864–29872.
251. Tapon, N., Ito, N., Dickson, B. J., Treisman, J. E., and Hariharan, I. K. (2001). The *Drosophila* tuberous sclerosis complex gene homologs restrict cell growth and cell proliferation. *Cell* 105, 345–355.
252. Mazure, N. M., Brahimi-Horn, M. C., Berta, M. A., et al. (2004). HIF-1: master and commander of the hypoxic world. A pharmacological approach to its regulation by siRNAs. *Biochem. Pharmacol.* 68, 971–980.
253. Semenza, G. L. (2002). Physiology meets biophysics: visualizing the interaction of hypoxia-inducible factor 1 alpha with p300 and CBP. *Proc. Natl. Acad. Sci. U S A* 99, 11570–11572.
254. Yuan, Y., Hilliard, G., Ferguson, T., and Millhorn, D. E. (2003). Cobalt inhibits the interaction between hypoxia-inducible factor-alpha and von Hippel-Lindau protein by direct binding to hypoxia-inducible factor-alpha. *J. Biol. Chem.* 278, 15911–15916.
255. Chavez, J. C., and LaManna, J. C. (2002). Activation of hypoxia-inducible factor-1 in the rat cerebral cortex after transient global ischemia: potential role of insulin-like growth factor-1. *J. Neurosci.* 22, 8922–8931.
256. Treins, C., Giorgetti-Peraldi, S., Murdaca, J., Semenza, G. L., and Van Obberghen, E. (2002). Insulin stimulates hypoxia-inducible factor 1 through a phosphatidylinositol 3-kinase/target of rapamycin-dependent signaling pathway. *J. Biol. Chem.* 277, 27975–27981.
257. Fukuda, R., Hirota, K., Fan, F., et al. (2002). Insulin-like growth factor 1 induces hypoxia-inducible factor 1-mediated vascular endothelial growth factor expression, which is dependent on map kinase and phosphatidylinositol 3-kinase signaling in colon cancer cells. *J. Biol. Chem.* 277, 38205–38211.
258. Forsythe, J. A., Jiang, B. H., Iyer, N. V., et al. (1996). Activation of vascular endothelial growth factor gene transcription by hypoxia-inducible factor 1. *Mol. Cell. Biol.* 16, 4604–4613.
259. Cunningham, M. A., Zhu, Q., and Hammond, J. M. (2004). FOX01a can alter cell cycle progression by regulating the nuclear localization of p27kip in granulosa cells. *Mol. Endocrinol.* 18, 1756–1767.
260. Park, Y., Maizels, E. T., Feiger, Z. J., Alam, H., Peters, C. A., Woodruff, T. K., Unterman, T. G., Lee, E. J., Jameson, L. J., and Hunzicker-Dunn, M. (2005). Induction of cyclin D2 in

rat granulosa cells requires FSH-dependent relief from FOXO1 repression coupled with positive signals from Smad. *J. Biol. Chem.* 280, 9135–9148.

261. Tang, E. D., Nunez, G., Barr, F. G., and Guan, K. L. (1999). Negative regulation of the forkhead transcription factor FKHR by Akt. *J. Biol. Chem.* 274, 16741–16746.
262. Burgering, B. M., and Kops, G. J. (2002). Cell cycle and death control: long live Forkheads. *Trends Biochem. Sci.* 27, 352–360.
263. Van Der Heide, L. P., Hoekman, M. F., and Smidt, M. P. (2004). The ins and outs of FoxO shuttling: mechanisms of FoxO translocation and transcriptional regulation. *Biochem. J.* 380, 297–309.
264. Medema, R. H., Kops, G. J., Bos, J. L., and Burgering, B. M. (2000). AFX-like Forkhead transcription factors mediate cell-cycle regulation by Ras and PKB through p27kip1. *Nature* 404, 782–787.
265. Graff, J. R., Konicek, B. W., McNulty, A. M., et al. (2000). Increased AKT activity contributes to prostate cancer progression by dramatically accelerating prostate tumor growth and diminishing p27Kip1 expression. *J. Biol. Chem.* 275, 24500–24505.
266. Collado, M., Medema, R. H., Garcia-Cao, I., et al. (2000). Inhibition of the phosphoinositide 3-kinase pathway induces a senescence-like arrest mediated by p27Kip1. *J. Biol. Chem.* 275, 21960–21968.
267. Schmoll, D., Walker, K. S., Alessi, D. R., et al. (2000). Regulation of glucose-6-phosphatase gene expression by protein kinase Balpha and the forkhead transcription factor FKHR. Evidence for insulin response unit-dependent and -independent effects of insulin on promoter activity. *J. Biol. Chem.* 275, 36324–36333.
268. Schmidt, M., de Mattos, S. F., van der, H. A., et al. (2002). Cell cycle inhibition by FoxO forkhead transcription factors involves downregulation of cyclin D. *Mol. Cell. Biol.* 22, 7842–7852.
269. Zhang, X., Gan, L., Pan, H., et al. (2002). Phosphorylation of serine 256 suppresses transactivation by FKHR (FOXO1) by multiple mechanisms. Direct and indirect effects on nuclear/cytoplasmic shuttling and DNA binding. *J. Biol. Chem.* 277, 45276–45284.
270. Zhao, X., Gan, L., Pan, H., et al. (2004). Multiple elements regulate nuclear/cytoplasmic shuttling of FOXO1: characterization of phosphorylation- and 14-3-3-dependent and -independent mechanisms. *Biochem. J.* 378, 839–849.
271. Rena, G., Prescott, A. R., Guo, S., Cohen, P., and Unterman, T. G. (2001). Roles of the forkhead in rhabdomyosarcoma (FKHR) phosphorylation sites in regulating 14-3-3 binding, transactivation and nuclear targetting. *Biochem. J.* 354, 605–612.
272. Guo, S., Cichy, S. B., He, X., et al. (2001). Insulin suppresses transactivation by CAAT/enhancer-binding proteins beta (C/EBPbeta). Signaling to p300/CREB-binding protein by protein kinase B disrupts interaction with the major activation domain of C/EBPbeta. *J. Biol. Chem.* 276, 8516–8523.
273. Hirota, K., Daitoku, H., Matsuzaki, H., et al. (2003). Hepatocyte nuclear factor-4 is a novel downstream target of insulin via FKHR as a signal-regulated transcriptional inhibitor. *J. Biol. Chem.* 278, 13056–13060.
274. Coqueret, O. (2002). Linking cyclins to transcriptional control. *Gene* 299, 35–55.
275. Sherr, C. J. (1994). G1 phase progression: cycling on cue. *Cell* 79, 551–555.
276. Kops, G. J., Medema, R. H., Glassford, J., et al. (2002). Control of cell cycle exit and entry by protein kinase B-regulated forkhead transcription factors. *Mol. Cell. Biol.* 22, 2025–2036.
277. Dijkers, P. F., Medema, R. H., Pals, C., et al. (2000). Forkhead transcription factor FKHR-L1 modulates cytokine-dependent transcriptional regulation of p27(KIP1). *Mol. Cell. Biol.* 20, 9138–9148.
278. Furukawa-Hibi, Y., Yoshida-Araki, K., Ohta, T., Ikeda, K., and Motoyama, N. (2002). FOXO forkhead transcription factors induce G(2)-M checkpoint in response to oxidative stress. *J. Biol. Chem.* 277, 26729–26732.
279. Ramaswamy, S., Nakamura, N., Sansal, I., Bergeron, L., and Sellers, W. R. (2002). A novel mechanism of gene regulation and tumor suppression by the transcription factor FKHR. *Cancer Cell* 2, 81–91.
280. Martinez-Gac, L., Marques, M., Garcia, Z., Campanero, M. R., and Carrera, A. C. (2004). Control of cyclin G2 mRNA expression by forkhead transcription factors: novel mechanism for cell cycle control by phosphoinositide 3-kinase and forkhead. *Mol. Cell. Biol.* 24, 2181–2189.
281. Brazil, D. P., Yang, Z. Z., and Hemmings, B. A. (2004). Advances in protein kinase B signalling: AKTion on multiple fronts. *Trends Biochem. Sci.* 29, 233–242.
282. Sekimoto, T., Fukumoto, M., and Yoneda, Y. (2004). 14-3-3 suppresses the nuclear localization of threonine 157-phosphorylated p27(Kip1). *EMBO J.* 23, 1934–1942.
283. Viglietto, G., Motti, M. L., Bruni, P., et al. (2002). Cytoplasmic relocalization and inhibition of the cyclin-dependent kinase inhibitor p27(Kip1) by PKB/Akt-mediated phosphorylation in breast cancer. *Nat. Med.* 8, 1136–1144.
284. Liang, J., Zubovitz, J., Petrocelli, T., et al. (2002). PKB/Akt phosphorylates p27, impairs nuclear import of p27 and opposes p27-mediated G1 arrest. *Nat. Med.* 8, 1153–1160.
285. Shin, I., Yakes, F. M., Rojo, F., et al. (2002). PKB/Akt mediates cell-cycle progression by phosphorylation of p27(Kip1) at threonine 157 and modulation of its cellular localization. *Nat. Med.* 8, 1145–1152.
286. Ogawa, T., Yogo, K., Ishida, N., and Takeya, T. (2003). Synergistic effects of activin and FSH on hyperphosphorylation of Rb and G1/S transition in rat primary granulosa cells. *Mol. Cell. Endocrinol.* 210, 31–38.
287. Kishi, H., Minegishi, T., Tano, M., et al. (1998). The effect of activin and FSH on the differentiation of rat granulosa cells. *FEBS Lett.* 422, 274–278.
288. Miyanaga, K., Erickson, G. F., DePaolo, L. V., Ling, N., and Shimasaki, S. (1993). Differential control of activin, inhibin and follistatin proteins in cultured rat granulosa cells. *Biochem. Biophys. Res. Commun.* 194, 253–258.
289. Seoane, J., Le, H. V., Shen, L., Anderson, S. A., and Massague, J. (2004). Integration of Smad and forkhead pathways in the control of neuroepithelial and glioblastoma cell proliferation. *Cell* 117, 211–223.
290. Hosaka, T., Biggs, W. H. III, Tieu, D., et al. (2004). Disruption of forkhead transcription factor (FOXO) family members in mice reveals their functional diversification. *Proc. Natl. Acad. Sci. U S A* 101, 2975–2980.
291. Castrillon, D. H., Miao, L., Kollipara, R., Horner, J. W., and DePinho, R. A. (2003). Suppression of ovarian follicle activation in mice by the transcription factor Foxo3a. *Science* 301, 215–218.
292. Bai, J., Nakamura, H., Ueda, S., et al. (2004). Proteasome-dependent degradation of cyclin D1 in 1-methyl-4-phenylpyridinium ion (MPP+)-induced cell cycle arrest. *J. Biol. Chem.* 279, 38710–38714.
293. Takahashi-Yanaga, F., Taba, Y., Miwa, Y., et al. (2003). Dictyostelium differentiation-inducing factor-3 activates glycogen synthase kinase-3beta and degrades cyclin D1 in mammalian cells. *J. Biol. Chem.* 278, 9663–9670.
294. Klarlund, J., Tsiaras, W., Holik, J. J., Chawla, A., and Czech, M. P. (2000). Distinct polyphosphoinositide binding selectivities for PH domains of GRP1-like proteins based on diglycine versus triglycine motifs. *J. Biol. Chem.* 275, 32816–32821.

295. Diehl, J. A., Cheng, M., Roussel, M. F., and Sherr, C. J. (1998). Glycogen synthase kinase-3beta regulates cyclin D1 proteolysis and subcellular localization. *Genes Dev.* 12, 3499–3511.
296. Jope, R. S., and Johnson, G. V. (2004). The glamour and gloom of glycogen synthase kinase-3. *Trends Biochem. Sci.* 29, 95–102.
297. Doble, B. W., and Woodgett, J. R. (2003). GSK-3: tricks of the trade for a multi-tasking kinase. *J. Cell. Sci.* 116, 1175–1186.
298. Ricken, A., Lochhead, P., Kontogiannea, M., and Farookhi, R. (2002). Wnt signaling in the ovary: identification and compartmentalized expression of wnt-2, wnt-2b, and frizzled-4 mRNAs. *Endocrinology* 143, 2741–2749.
299. Hsieh, M., Johnson, M. A., Greenberg, N. M., and Richards, J. S. (2002). Regulated expression of wnts and frizzleds at specific stages of follicular development in the rodent ovary. *Endocrinology* 143, 898–908.
300. Park, J., Leong, M. L., Buse, P., et al. (1999). Serum and glucocorticoid-inducible kinase (SGK) is a target of the PI 3-kinase-stimulated signaling pathway. *EMBO J.* 18, 3024–3033.
301. Brunet, A., Park, J., Tran, H., et al. (2001). Protein kinase SGK mediates survival signals by phosphorylating the forkhead transcription factor FKHRL1 (FOXO3a). *Mol. Cell. Biol.* 21, 952–965.
302. Lang, F., Henke, G., Embark, H. M., et al. (2003). Regulation of channels by the serum and glucocorticoid-inducible kinase-implications for transport, excitability and cell proliferation. *Cell. Physiol. Biochem.* 2003;13:41–50.
303. Lang, F., and Cohen, P. (2001). Regulation and physiological roles of serum- and glucocorticoid-induced protein kinase isoforms. *Sci. STKE.* 108, RE17.
304. Snyder, P. M., Olson, D. R., Kabra, R., Zhou, R., and Steines, J. C. (2004). cAMP and SGK regulate ENaC through convergent phosphorylation of Nedd4-2. *J. Biol. Chem.* 279, 45753–45758.
305. Asselin, E., Wang, Y., and Tsang, B. K. (2001). X-linked inhibitor of apoptosis protein activates the phosphatidylinositol 3-kinase/Akt pathway in rat granulosa cells during follicular development. *Endocrinology* 142, 2451–2457.
306. Hsu, S. Y., and Hsueh, A. J. (2000). Tissue-specific Bcl-2 protein partners in apoptosis: an ovarian paradigm. *Physiol. Rev.* 80, 593–614.
307. Adrain, C., and Martin, S. J. (2001). The mitochondrial apoptosome: a killer unleashed by the cytochrome seas. *Trends Biochem. Sci.* 26, 390–397.
308. Kumar, S., and Cakouros, D. (2004). Transcriptional control of the core cell-death machinery. *Trends Biochem. Sci.* 29, 193–199.
309. Kaufmann, S. H., and Hengartner, M. O. (2001). Programmed cell death: alive and well in the new millennium. *Trends Cell. Biol.* 11, 526–534.
310. Song, Q., Kuang, Y., Dixit, V. M., and Vincenz, C. (1999). Boo, a novel negative regulator of cell death, interacts with Apaf-1. *EMBO J.* 18, 167–178.
311. Inohara, N., Gourley, T. S., Carrio, R., et al. (1998). Diva, a Bcl-2 homologue that binds directly to Apaf-1 and induces BH3-independent cell death. *J. Biol. Chem.* 273, 32479–32486.
312. Cardone, M. H., Roy, N., Stennicke, H. R., et al. (1998). Regulation of cell death protease caspase-9 by phosphorylation. *Science* 282, 1318–1321.
313. Fujita, E., Jinbo, A., Matuzaki, H., et al. (1999). Akt phosphorylation site found in human caspase-9 is absent in mouse caspase-9. *Biochem. Biophys. Res. Commun.* 264, 550–555.
314. Hsu, S. Y., Lai, R. J., Finegold, M., and Hsueh, A. J. (1996). Targeted overexpression of Bcl-2 in ovaries of transgenic mice leads to decreased follicle apoptosis, enhanced folliculogenesis, and increased germ cell tumorigenesis. *Endocrinology* 137, 4837–4843.
315. Leo, C. P., Hsu, S. Y., Chun, S. Y., Bae, H. W., and Hsueh, A. J. (1999). Characterization of the antiapoptotic Bcl-2 family member myeloid cell leukemia-1 (Mcl-1) and the stimulation of its message by gonadotropins in the rat ovary. *Endocrinology* 140, 5469–5477.
316. Hsu, S. Y., Kaipia, A., Zhu, L., and Hsueh, A. J. (1997). Interference of BAD (Bcl-xL/Bcl-2-associated death promoter)-induced apoptosis in mammalian cells by 14-3-3 isoforms and P11. *Mol. Endocrinol.* 11, 1858–1867.
317. Hsu, S. Y., Kaipia, A., McGee, E., Lomeli, M., and Hsueh, A. J. (1997). Bok is a pro-apoptotic Bcl-2 protein with restricted expression in reproductive tissues and heterodimerizes with selective anti-apoptotic Bcl-2 family members. *Proc. Natl. Acad. Sci. U S A* 94, 12401–12406.
318. Hsu, S. Y., Lin, P., Hsueh, A. J. (1998). BOD (Bcl-2-related ovarian death gene) is an ovarian BH3 domain-containing proapoptotic Bcl-2 protein capable of dimerization with diverse antiapoptotic Bcl-2 members. *Mol. Endocrinol.* 12, 1432–1440.
319. Robles, R., Tao, X. J., Trbovich, A. M., et al. (1999). Localization, regulation and possible consequences of apoptotic protease-activating factor-1 (Apaf-1) expression in granulosa cells of the mouse ovary. *Endocrinology* 140, 2641–2644.
320. Matikainen, T., Perez, G. I., Zheng, T. S., et al. (2001). Caspase-3 gene knockout defines cell lineage specificity for programmed cell death signaling in the ovary. *Endocrinology* 142, 2468–2480.
321. Jiang, J. Y., Cheung, C. K., Wang, Y., and Tsang, B. K. (2003). Regulation of cell death and cell survival gene expression during ovarian follicular development and atresia. *Front. Biosci.* 8, 222–237.
322. Dan, H. C., Sun, M., Kaneko, S., et al. (2004). Akt phosphorylation and stabilization of X-linked inhibitor of apoptosis protein (XIAP). *J. Biol. Chem.* 279, 5405–5412.
323. Xiao, C. W., Ash, K., and Tsang, B. K. (2001). Nuclear factor-kappaB-mediated X-linked inhibitor of apoptosis protein expression prevents rat granulosa cells from tumor necrosis factor alpha-induced apoptosis. *Endocrinology* 142, 557–563.
324. Xiao, C. W., Asselin, E., and Tsang, B. K. (2002). Nuclear factor kappaB-mediated induction of Flice-like inhibitory protein prevents tumor necrosis factor alpha-induced apoptosis in rat granulosa cells. *Biol. Reprod.* 67, 436–441.
325. Chen, L. F., and Greene, W. C. (2004). Shaping the nuclear action of NF-kappaB. *Nat. Rev. Mol. Cell. Biol.* 5, 392–401.
326. Shishodia, S., and Aggarwal, B. B. (2004). Nuclear factor-kappaB: a friend or a foe in cancer? *Biochem. Pharmacol.* 68, 1071–1080.
327. Mayo, M. W., Madrid, L. V., Westerheide, S. D., et al. (2002). PTEN blocks tumor necrosis factor-induced NF-kappa B-dependent transcription by inhibiting the transactivation potential of the p65 subunit. *J. Biol. Chem.* 277, 11116–11125.
328. Kawasaki, K., Smith, R. S. Jr., Hsieh, C. M., et al. (2003). Activation of the phosphatidylinositol 3-kinase/protein kinase Akt pathway mediates nitric oxide-induced endothelial cell migration and angiogenesis. *Mol. Cell. Biol.* 23, 5726–5737.
329. Fulton, D., Gratton, J. P., McCabe, T. J., et al. (1999). Regulation of endothelium-derived nitric oxide production by the protein kinase Akt. *Nature* 399, 597–601.
330. Jablonka-Shariff, A., and Olson, L. M. (1997). Hormonal regulation of nitric oxide synthases and their cell-specific expression during follicular development in the rat ovary. *Endocrinology* 138, 460–468.

331. Takesue, K., Hattori, M. A., Nishida, N., Kato, Y., and Fujihara, N. (2001). Expression of endothelial nitric oxide synthase gene in cultured porcine granulosa cells after FSH stimulation. *J. Mol. Endocrinol.* 26, 259–265.
332. Matsumi, H., Koji, T., Yano, T., et al. (1998). Evidence for an inverse relationship between apoptosis and inducible nitric oxide synthase expression in rat granulosa cells: a possible role of nitric oxide in ovarian follicle atresia. *Endocr. J.* 45, 745–751.
333. Jablonka-Shariff, A., Ravi, S., Beltsos, A. N., Murphy, L. L., and Olson, L. M. (1999). Abnormal estrous cyclicity after disruption of endothelial and inducible nitric oxide synthase in mice. *Biol. Reprod.* 61, 171–177.
334. Tafoya, M. A., Chen, J. Y., Stewart, R. L. Jr., and Lapolt, P. S. (2004). Activation of soluble guanylyl cyclase inhibits estradiol production and cyclic AMP accumulation from cultured rat granulosa cells. *Fertil. Steril.* 82 (Suppl 3), 1154–1159.
335. Chen, Y. H., Tafoya, M., Ngo, A., and Lapolt, P. S. (2003). Effects of nitric oxide and cGMP on inhibin A and inhibin subunit mRNA levels from cultured rat granulosa cells. *Fertil. Steril.* 79 (Suppl 1), 687–693.
336. Weck, J., and Mayo, K. E. (2004). Regulated recruitment of LRH-1 to the inhibin-alpha subunit promoter. Presented at the Endocrine Society 86th Meeting, New Orleans, Abstract # P1–257.
337. Feng, Z. M., Wu, A. Z., Zhang, Z., and Chen, C. L. (2000). GATA-1 and GATA-4 transactivate inhibin/activin beta-B-subunit gene transcription in testicular cells. *Mol. Endocrinol.* 14, 1820–1835.
338. Feng, Z. M., Wu, A. Z., and Chen, C. L. (1998). Testicular GATA-1 factor up-regulates the promoter activity of rat inhibin alpha-subunit gene in MA-10 Leydig tumor cells. *Mol. Endocrinol.* 12, 378–390.
339. Tremblay, J. J., and Viger, R. S. (2001). GATA factors differentially activate multiple gonadal promoters through conserved GATA regulatory elements. *Endocrinology* 142, 977–986.
340. Lavoie, H. A., Singh, D., and Hui, Y. Y. (2004). Concerted regulation of the porcine steroidogenic acute regulatory protein gene promoter activity by follicle-stimulating hormone and insulin-like growth factor I in granulosa cells involves GATA-4 and CCAAT/enhancer binding protein beta. *Endocrinology* 145, 3122–3134.
341. Gummow, B. M., Winnay, J. N., and Hammer, G. D. (2003). Convergence of Wnt signaling and steroidogenic factor-1 (SF-1) on transcription of the rat inhibin alpha gene. *J. Biol. Chem.* 278, 26572–26579.
342. Mukherjee, A., Urban, J., Sassone-Corsi, P., and Mayo, K. E. (1998). Gonadotropins regulate inducible cyclic adenosine 3′,5′-monophosphate early repressor in the rat ovary: implications for inhibin a subunit gene expression. *Mol. Endocrinol.* 12, 785–800.
343. Burkart, A. D., Mukherjee, A., Sterneck, E., Johnson, P. F., and Mayo, K. E. (2005). Repression of the inhibin-alpha subunit gene by transcription factor C/EBP beta. *Endocrinology* 146, 1909–1921.
344. Fitzpatrick, S. L., and Richards, J. S. (1994). Identification of a cyclic adenosine 3′,5′-monophosphate-response element in the rat aromatase promoter that is required for transcriptional activation in rat granulosa cells and R2C Leydig cells. *Mol. Endocrinol.* 8, 1309–1319.
345. Christy, B., and Nathans, D. (1989). DNA binding site of the growth factor-inducible protein Zif268. *Proc. Natl. Acad. Sci. U S A* 86, 8737–8741.
346. Topilko, P., Schneider-Maunoury, S., Levi, G., et al. (1998). Multiple pituitary and ovarian defects in Krox-24 (NGFI-A, Egr-1)-targeted mice. *Mol. Endocrinol.* 12, 107–122.
347. Yoshino, M., Mizutani, T., Yamada, K., et al. (2002). Early growth response gene-1 regulates the expression of the rat luteinizing hormone receptor gene. *Biol. Reprod.* 66, 1813–1819.
348. Zhang, Y., and Dufau, M. L. (2003). Dual mechanisms of regulation of transcription of luteinizing hormone receptor gene by nuclear orphan receptors and histone deacetylase complexes. *J. Steroid Biochem. Mol. Biol.* 85, 401–414.
349. Bellefroid, E. J., Poncelet, D. A., Lecocq, P. J., Revelant, O., and Martial, J. A. (1991). The evolutionarily conserved Kruppel-associated box domain defines a subfamily of eukaryotic multifingered proteins. *Proc. Natl. Acad. Sci. U S A* 88, 3608–3612.
350. Brannian, J. D., Christianson, H., Flynn, S., and Kurz, S. G. (1995). Loss of low-density lipoprotein utilization by regressing porcine luteal cells: effects of protein kinase C activation. *Biol. Reprod.* 52, 793–797.
351. Gurates, B., Amsterdam, A., Tamura, M., et al. (2003). WT1 and DAX-1 regulate SF-1-mediated human P450arom gene expression in gonadal cells. *Mol. Cell. Endocrinol.* 208, 61–75.
352. Saxena, D., Safi, R., Little-Ihrig, L., and Zeleznik, A. J. (2004). Liver receptor homolog-1 (LRH-1) stimulates the progesterone biosynthetic pathway during follicle stimulating hormone (FSH)-induced granulosa cell differentiation. *Endocrinology* 145, 3821–3829.
353. Clemens, J. W., Lala, D. S., Parker, K. L., and Richards, J. S. (1994). Steroidogenic factor-1 binding and transcriptional activity of the cholesterol side-chain cleavage promoter in rat granulosa cells. *Endocrinology* 134, 1499–1508.
354. Zhang, G., and Veldhuis, J. D. (2004). Requirement for proximal putative Sp1 and AP-2 cis-deoxyribonucleic acid elements in mediating basal and luteinizing hormone- and insulin-dependent in vitro transcriptional activation of the CYP17 gene in porcine theca cells. *Endocrinology* 145, 2760–2766.
355. Zhang, G., and Veldhuis, J. D. (2004). Insulin drives transcriptional activity of the CYP17 gene in primary cultures of swine theca cells. *Biol. Reprod.* 70, 1600–1605.
356. Magoffin, D. A., Kurtz, K. M., and Erickson, G. F. (1990). Insulin-like growth factor-I selectively stimulates cholesterol side-chain cleavage expression in ovarian theca-interstitial cells. *Mol. Endocrinol.* 4, 489–496.
357. Sekar, N., and Veldhuis, J. D. (2001). Concerted transcriptional activation of the low density lipoprotein receptor gene by insulin and luteinizing hormone in cultured porcine granulosa-luteal cells: possible convergence of protein kinase a, phosphatidylinositol 3-kinase, and mitogen-activated protein kinase signaling pathways. *Endocrinology* 142, 2921–2928.
358. Li, X., Peegel, H., and Menon, K. M. (2001). Regulation of high density lipoprotein receptor messenger ribonucleic acid expression and cholesterol transport in theca-interstitial cells by insulin and human chorionic gonadotropin. *Endocrinology* 142, 174–181.
359. Byers, M., Kuiper, G. G., Gustafsson, J-A., and Park-Sarge, O. K. (1997). Estrogen receptor-beta mRNA expression in rat ovary: down-regulation by gonadotropins. *Mol. Endocrinol.* 11, 172–182.
360. Sharma, S. C., Clemens, J. W., Pisarska, M. D., and Richards, J. S. (1999). Expression and function of estrogen receptor subtypes in granulosa cells: regulation by estradiol and forskolin. *Endocrinology* 140, 4320–4334.
361. Robker, R. L., and Richards, J. S. (1998). Hormonal control of the cell cycle in ovarian cells: proliferation versus differentiation. *Biol. Reprod.* 59, 476–482.
362. Hickey, G. J., Chen, S., Besman, M. J., et al. (1988). Hormonal regulation, tissue distribution, and content of

aromatase cytochrome P450 messenger ribonucleic acid and enzyme in rat ovarian follicles and corpora lutea: relationship to estradiol biosynthesis. *Endocrinology* 122, 1426–1436.

363. Segaloff, D. L., Wang, H., and Richards, J. S. (1990). Hormonal regulation of luteinizing hormone/choionic gonadotropin receptor mRNA in rat ovarian cells during follicular development and luteinization. *Mol. Endocrinol.* 4, 1856–1865.
364. Lapolt, P. S., Oikawa, M., Jia, X-C., Dargan, C., and Hsueh, A. J. W. (1990). Gonadotropin-induced up- and down-regulation of rat ovarian LH receptor message levels during follicular growth, ovulation, and luteinization. *Endocrinology* 126, 3277–3279.
365. Woodruff, T. K., D'Agostino, J. B., Schwartz, N. B., and Mayo, K. E. (1989). Decreased inhibin gene expression in preovulatory follicles requires primary gonadotropin surge. *Endocrinology* 124, 2193–2199.
366. Goldring, N. B., Durica, J. M., Lifka, J., et al. (1987). Cholesterol side-chain cleavage P450 messenger RNA: evidence for hormonal regulation in rat ovarian follicles and constitutive expression in corpora lutea. *Endocrinology* 120, 1942–1950.
367. Sehested, A., Juul, A. A., Andersson, A. M., et al. (2000). Serum inhibin A and inhibin B in healthy prepubertal, pubertal, and adolescent girls and adult women: relation to age, stage of puberty, menstrual cycle, follicle-stimulating hormone, luteinizing hormone, and estradiol levels. *J. Clin. Endocrinol. Metab.* 85, 1634–1640.
368. Natraj, U., and Richards, J. S. (1993). Hormonal regulation, localization, and functional activity of the progesterone receptor in granulosa cells of rat preovulatory follicles. *Endocrinology* 133, 761–769.
369. Park, O-K., and Mayo, K. E. (1991). Transient expression of progesterone receptor messenger RNA in ovarian granulosa cells after the preovulatory hormone surge. *Mol. Endocrinol.* 5, 967–978.
370. Robker, R. L., Russell, D. L., Espey, L. L., et al. (2000). Progesterone-regulated genes in the ovulation process: ADAMTS-1 and cathepsin L proteases. *Proc. Natl. Acad. Sci. U S A* 97, 4689–4694.
371. Ko, C., and Park-Sarge, O. K. (2000). Progesterone receptor activation mediates LH-induced type-I pituitary adenylate cyclase activating polypeptide receptor (PAC(1)) gene expression in rat granulosa cells. *Biochem. Biophys. Res. Commun.* 277, 270–279.
372. Park, J. I., Kim, W. J., Wang, L., et al. (2000). Involvement of progesterone in gonadotrophin-induced pituitary adenylate cyclase-activating polypeptide gene expression in pre-ovulatory follicles of rat ovary. *Mol. Hum. Reprod.* 6, 238–245.
373. Espey, L. L., and Richards, J. S. (2002). Temporal and spatial patterns of ovarian gene transcription following an ovulatory dose of gonadotropin in the rat. *Biol. Reprod.* 67, 1662–1670.
374. Espey, L. L., Yoshioka, S., Russell, D. L., et al. (2000). Ovarian expression of a disintegrin and metalloproteinase with thrombospondin motifs during ovulation in the gonadotropin-primed immature rat. *Biol. Reprod.* 62, 1090–1095.
375. Sirois, J., Simmons, D. L., and Richards, J. S. (1992). Hormonal regulation of messenger ribonucleic acid encoding a novel isoform of prostaglandin endoperoxide H synthase in rat preovulatory follicles. Induction in vivo and in vitro. *J. Biol. Chem.* 267, 11586–11592.
376. Espey, L. L., Ujioka, T., Russell, D. L., et al. (2000). Induction of early growth response protein-1 gene expression in the rat ovary in response to an ovulatory dose of human chorionic gonadotropin. *Endocrinology* 141, 2385–2391.
377. Sterneck, E., Tessarollo, L., and Johnson, P. F. (1997). An essential role for C/EBPbeta in female reproduction. *Genes Dev.* 11, 2153–2162.
378. Christenson, L. K., Johnson, P. F., McAllister, J. M., and Strauss, J. F. III. (1999). CCAAT/enhancer-binding proteins regulate expression of the human steroidogenic acute regulatory protein (StAR) gene. *J. Biol. Chem.* 274, 26591–26598.
379. Silverman, E., Eimerl, S., and Orly, J. (1999). CCAAT enhancer-binding protein beta and GATA-4 binding regions within the promoter of the steroidogenic acute regulatory protein (StAR) gene are required for transcription in rat ovarian cells. *J. Biol. Chem.* 274, 17987–17996.
380. Manna, P. R., Dyson, M. T., Eubank, D. W., et al. (2002). Regulation of steroidogenesis and the steroidogenic acute regulatory protein by a member of the cAMP response-element binding protein family. *Mol. Endocrinol.* 16, 184–199.
381. Sirois, J., and Richards, J. S. (1993). Transcriptional regulation of the rat prostaglandin endoperoxide synthase 2 gene in granulosa cells. *J. Biol. Chem.* 268, 21931–21938.
382. Park, J. I., Park, H. J., Lee, Y. I., Seo, Y. M., Chun, S. Y. (2003). Regulation of NGFI-B expression during the ovulatory process. *Mol. Cell. Endocrinol.* 202, 25–29.
383. Wu, Y., Ghosh, S., Nishi, Y., et al. (2005). The orphan nuclear receptors NURR1 and NGFI-B modulate aromatase gene expression in ovarian granulosa cells: a possible mechanism for repression of aromatase expression upon luteinizing hormone surge. *Endocrinology* 146, 237–246.
384. Lydon, J. P., DeMayo, F. J., Funk, C. R., et al. (1995). Mice lacking progesterone receptor exhibit pleiotropic reproductive abnormalities. *Genes Dev.* 9, 2266–2278.
385. Lim, H., Paria, B. C., Das, S. K., et al. (1997). Multiple female reproductive failures in cyclooxygenase 2-deficient mice. *Cell* 91, 197–208.
386. Lee, S. L., Sadovsky, Y., Swirnoff, A. H., et al. (1996). Luteinizing hormone deficiency and female infertility in mice lacking the transcription factor NGFI-A (Egr-1). *Science* 273, 1219–1221.
387. Pall, M., Hellberg, P., Brannstrom, M., et al. (1997). The transcription factor C/EBP-beta and its role in ovarian function; evidence for direct involvement in the ovulatory process. *EMBO J.* 16, 5273–5279.
388. Sugimoto, Y., Segi, E., Tsuboi, K., Ichikawa, A., and Narumiya, S. (1998). Female reproduction in mice lacking the prostaglandin F receptor. Roles of prostaglandin and oxytocin receptors in parturition. *Adv. Exp. Med. Biol.* 449, 317–321.
389. Tilley, S. L., Audoly, L. P., Hicks, E. H., et al. (1999). Reproductive failure and reduced blood pressure in mice lacking the EP2 prostaglandin E2 receptor. *J. Clin. Invest.* 103, 1539–1545.
390. Tsafriri, A. (1995). Ovulation as a tissue remodelling process. Proteolysis and cumulus expansion. *Adv. Exp. Med. Biol.* 377, 121–140.
391. Curry, T. E. Jr., and Osteen, K. G. (2003). The matrix metalloproteinase system: changes, regulation, and impact throughout the ovarian and uterine reproductive cycle. *Endocr. Rev.* 24, 428–465.
392. Murdoch, W. J., and Gottsch, M. L. (2003). Proteolytic mechanisms in the ovulatory folliculo-luteal transformation. *Connect. Tiss. Res.* 44, 50–57.
393. Curry, T. E. Jr., Song, L., and Wheeler, S. E. (2001). Cellular localization of gelatinases and tissue inhibitors of metalloproteinases during follicular growth, ovulation, and early luteal formation in the rat. *Biol. Reprod.* 65, 855–865.
394. Richards, J. S., Russell, D. L., Ochsner, S., and Espey, L. L. (2002). Ovulation: new dimensions and new regulators of the inflammatory-like response. *Annu. Rev. Physiol.* 64, 69–92.

395. Ochsner, S. A., Day, A. J., Rugg, M. S., et al. (2003). Disrupted function of tumor necrosis factor-alpha-stimulated gene 6 blocks cumulus cell-oocyte complex expansion. *Endocrinology* 144, 4376–4384.
396. Davis, B. J., Lennard, D. E., Lee, C. A., et al. (1999). Anovulation in cyclooxygenase-2-deficient mice is restored by prostaglandin E2 and interleukin-1beta. *Endocrinology* 140, 2685–2695.
397. Simon, C., Tsafriri, A., Chun, S. Y., et al. (1994). Interleukin-1 receptor antagonist suppresses human chorionic gonadotropin-induced ovulation in the rat. *Biol. Reprod*. 51, 662–667.
398. Kozak, W., Zheng, H., Conn, C. A., et al. (1995). Thermal and behavioral effects of lipopolysaccharide and influenza in interleukin-1 beta-deficient mice. *Am. J. Physiol.* 269, R969–R977.
399. Murphy, B. D. (2000). Models of luteinization. *Biol. Reprod.* 63, 2–11.
400. Nakayama, K., Ishida, N., Shirane, M., et al. (1996). Mice lacking p27(Kip1) display increased body size, multiple organ hyperplasia, retinal dysplasia, and pituitary tumors. *Cell* 85, 707–720.
401. Deleted by author.
402. Jirawatnotai, S., Moons, D. S., Stocco, C. O., et al. (2003). The cyclin-dependent kinase inhibitors p27Kip1 and p21Cip1 cooperate to restrict proliferative life span in differentiating ovarian cells. *J. Biol. Chem.* 278, 17021–17027.
403. Tong, W., Kiyokawa, H., Soos, T. J., et al. (1998). The absence of p27Kip1, an inhibitor of G1 cyclin-dependent kinases, uncouples differentiation and growth arrest during the granulosa->luteal transition. *Cell. Growth Differ.* 9, 787–794.
404. Chun, S. Y., Bae, H. W., Kim, W. J., et al. (2001). Expression of messenger ribonucleic acid for the antiapoptosis gene P11 in the rat ovary: gonadotropin stimulation in granulosa cells of preovulatory follicles. *Endocrinology* 142, 2311–2317.
405. Ji, T. H., Ryu, K-S., Gilchrist, R., and Ji, I. (1997). Interaction, signal generation, signal divergence, and signal transduction of LH/CG and the receptor. *Rec. Prog. Horm. Res.* 52, 431–454.
406. Song, Y. S., Ji, I., Beauchamp, J., Isaacs, N. W., and Ji, T. H. (2001). Hormone interactions to Leu-rich repeats in the gonadotropin receptors. II. Analysis of Leu-rich repeat 4 of human luteinizing hormone/chorionic gonadotropin receptor. *J. Biol. Chem.* 276, 3436–3442.
407. Braun, T., Schofield, P. R., and Sprengel, R. (1991). Amino-terminal leucine-rich repeats in gonadotropin receptors determine hormone selectivity. *EMBO J.* 10, 1885–1890.
408. Thomas, D., Rozell, T. G., Liu, X., and Segaloff, D. L. (1996). Mutational analyses of the extracellular domain of the full-length lutropin/choriogonadotropin receptor suggest leucine-rich repeats 1-6 are involved in hormone binding. *Mol. Endocrinol.* 10, 760–768.
409. Hong, S., Phang, T., Ji, I., and Ji, T. H. (1998). The amino-terminal region of the luteinizing hormone/choriogonadotropin receptor contacts both subunits of human choriogonadotropin. I. Mutational analysis. *J. Biol. Chem.* 273, 13835–13840.
410. Huhtaniemi, I., Zhang, F. P., Kero, J., Hamalainen, T., and Poutanen, M. (2002). Transgenic and knockout mouse models for the study of luteinizing hormone and luteinizing hormone receptor function. *Mol. Cell. Endocrinol.* 187, 49–56.
411. Tetsuka, M., and Hillier, S. G. (1996). Androgen receptor gene expression in rat granulosa cells: the role of follicle-stimulating hormone and steroid hormones. *Endocrinology* 137, 4392–4397.
412. Hu, Y. C., Wang, P. H., Yeh, S., et al. (2004). Subfertility and defective folliculogenesis in female mice lacking androgen receptor. *Proc. Natl. Acad. Sci. U S A* 101, 11209–11214.
412a. Couse, J. F. and Korach, K. S. (2001). Contrasting phenotypes in reproductive tissues of female estrogen receptor null mice. *Ann. NY Acad. Sci.* 948, 1–8.
413. Themmen, A. P. N., and Huhtaniemi, I. T. (2000). Mutations of gonadotropins and gonadotropin receptors: elucidating the physiology and pathophysiology of pituitary-gonadal function. *Endocr. Rev.* 21, 551–583.
414. Shenker, A., Laue, L., Kosugi, S., et al. (1993). A constitutively activating mutation of the luteinizing hormone receptor in familial male precocious puberty. *Nature* 365, 652–654.
415. Mann, R. J., Keri, R. A., and Nilson, J. H. (2003). Consequences of elevated luteinizing hormone on diverse physiological systems: use of the LHbetaCTP transgenic mouse as a model of ovarian hyperstimulation-induced pathophysiology. *Rec. Prog. Horm. Res.* 58, 343–375.
416. Munshi, U. M., Pogozheva, I. D., and Menon, K. M. (2003). Highly conserved serine in the third transmembrane helix of the luteinizing hormone/human chorionic gonadotropin receptor regulates receptor activation. *Biochemistry* 42, 3708–3715.
417. Dhanwada, K. R., Vijapurkar, U., and Ascoli, M. (1996). Two mutations of the lutropin/choriogonadotropin receptor that impair signal transduction also interfere with receptor-mediated endocytosis. *Mol. Endocrinol.* 10, 544–554.
418. Ji, I., and Ji, T. H. (1993). Receptor activation is distinct from hormone binding in intact lutropin-choriogonadotropin receptors and Asp^{397} is important for receptor activation. *J. Biol. Chem.* 268, 20851–20854.
419. Ryu, K. S., Gilchrist, R. L., Ji, I., Kim, S. J., and Ji, T. H. (1996). Exoloop 3 of the luteinizing hormone/choriogonadotropin receptor. Lys583 is essential and irreplaceable for human choriogonadotropin (hCG)-dependent receptor activation but not for high affinity hCG binding. *J. Biol. Chem.* 271, 7301–7304.
420. Gilchrist, R. L., Ryu, K-S., Ji, I., and Ji, T. H. (1996). The luteinizing hormone/chorionic gonadotropin receptor has distinct transmembrane conductors for cAMP and inositol phosphate signals. *J. Biol. Chem.* 271, 19283–19287.
421. Latronico, A. C. (2000). Naturally occurring mutations of the luteinizing hormone receptor gene affecting reproduction. *Semin. Reprod. Med.* 18, 17–20.
422. Zeleznik, A. J., Midgley, A. R. Jr., and Reichert, L. E. Jr. (1974). Granulosa cell maturation in the rat: increased binding of human chorionic gonadotropin following treatment with follicle-stimulating hormone in vivo. *Endocrinology* 95, 818–825.
423. Webb, R., Woad, K. J., and Armstrong, D. G. (2002). Corpus luteum (CL) function: local control mechanisms. *Domest. Anim. Endocrinol.* 23, 277–285.
424. Gibori, G. (1993). The corpus luteum of pregnancy. In *The Ovary* (E. Y. Adashi and P. C. K. Leung, Eds.), pp. 261–317. Raven Press, New York.
425. Gelety, T. J., and Magoffin, D. A. (1997). Ontogeny of steroidogenic enzyme gene expression in ovarian theca-interstitial cells in the rat: regulation by a paracrine theca-differentiating factor prior to achieving luteinizing hormone responsiveness. *Biol. Reprod.* 56, 938–945.
426. Shi, H., and Segaloff, D. L. (1995). A role for increased lutropin/choriogonadotropin receptor (LHR) gene transcription in the follitropin-stimulated induction of the LHR in granulosa cells. *Mol. Endocrinol.* 9, 734–744.
427. Chen, S., Shi, H., Liu, X., and Segaloff, D. L. (1999). Multiple elements and protein factors coordinate the basal and cyclic adenosine 3′,5′-monophosphate-induced transcription of the lutropin receptor gene in rat granulosa cells. *Endocrinology* 140, 2100–2109.
428. Zhang, Y., and Dufau, M. L. (2002). Silencing of transcription of the human luteinizing hormone receptor gene by

histone deacetylase-mSin3A complex. *J. Biol. Chem.* 277, 33431–33438.
429. Nakamura, K., Liu, X., and Ascoli, M. (2000). Seven non-contiguous intracellular residues of the lutropin/choriogonadotropin receptor dictate the rate of agonist-induced internalization and its sensitivity to non-visual arrestins. *J. Biol. Chem.* 275, 241–247.
430. Meduri, G., Vu Hai, M. T., Takemori, S., et al. (1996). Comparison of cellular distribution of LH receptors and steroidogenic enzymes in the porcine ovary. *J. Endocr.* 148, 435–446.
431. Robinson, M. S., Rhodes, J. A., and Albertini, D. F. (1983). Slow internalization of human chorionic gonadotropin by cultured granulosa cells. *J. Cell. Physiol.* 117, 43–50.
432. Amsterdam, A., Berkowitz, A., Nimrod, A., and Kohen, F. (1980). Aggregation of luteinizing hormone receptors in granulosa cells: a possible mechanism of desensitization to the hormone. *Proc. Natl. Acad. Sci. U S A* 77, 3440–3444.
433. Schwall, R. H., and Erickson, G. F. (1983). A central role for cyclic AMP, but not progesterone, in luteinizing hormone receptor down-regulation in the granulosa cell. *J. Biol. Chem.* 258, 13199–13204.
434. Schwall, R. H., and Erickson, G. F. (1984). Inhibition of synthesis of luteinizing hormone (LH) receptors by a down-regulating dose of LH. *Endocrinology* 114, 1114–1123.
435. Lu, D. L., Peegel, H., Mosier, S. M., and Menon, K. M. J. (1993). Loss of lutropin/human choriogonadotropin receptor messenger ribonucleic acid during ligand-induced down-regulation occurs post transcriptionally. *Endocrinology* 132, 235–240.
436. Nair, A. K., Kash, J. C., Peegel, H., and Menon, K. M. (2002). Post-transcriptional regulation of luteinizing hormone receptor mRNA in the ovary by a novel mRNA-binding protein. *J. Biol. Chem.* 277, 21468–21473.
437. Nair, A. K., and Menon, K. M. (2004). Isolation and characterization of a novel trans-factor for luteinizing hormone receptor mRNA from ovary. *J. Biol. Chem.* 279, 14937–14944.
438. Rajagopalan-Gupta, R. M., Rasenick, M. M., and Hunzicker-Dunn, M. (1997). LH/choriogonadotropin-dependent, cholera toxin-catalyzed adenosine 5′-diphosphate (ADP)-ribosylation of the long and short forms of Gsalpha and pertussis toxin-catalyzed ADP-ribosylation of Gialpha. *Mol. Endocrinol.* 11, 538–549.
439. Rajagopalan-Gupta, R. M., Lamm, M. L., Mukherjee, S., Rasenick, M. M., and Hunzicker-Dunn, M. (1998). Luteinizing hormone/choriogonadotropin receptor-mediated activation of heterotrimeric guanine nucleotide binding proteins in ovarian follicular membranes. *Endocrinology* 139, 4547–4555.
440. Kosugi, S., Mori, T., and Shenker, A. (1998). An anionic residue at position 564 is important for maintaining the inactive conformation of the human lutropin/choriogonadotropin receptor. *Mol. Pharm.* 53, 894–901.
441. Kosugi, S., Mori, T., and Shenker, A. (1996). The role of Asp^{578} in maintaining the inactive conformation of the human lutropin/choriogonadotropin receptor. *J. Biol. Chem.* 271, 31813–31817.
442. Kosugi, S., Van Dop, C., Geffner, M. E., et al. (1995). Characterization of heterogeneous mutations causing constitutive activation of the luteinizing hormone receptor in familial male precocious puberty. *Hum. Mol. Genet.* 4, 183–188.
443. Rajagopalan-Gupta, R. M., Mukherjee, S., Zhu, X., et al. (1999). Roles of Gi and Gq/11 in mediating desensitization of the luteinizing hormone/choriogonadotropin receptor in porcine ovarian follicular membranes. *Endocrinology* 140, 1612–1621.
444. Herrlich, A., Kuhn, B., Grosse, R., et al. (1996). Involvement of G_s and G_i proteins in dual coupling of the luteinizing hormone receptor to adenylyl cyclase and phospholipase C. *J. Biol. Chem.* 271, 16764–16772.
445. Gudermann, T., Nichols, C., Levy, F. O., Birnbaumer, M., and Birnbaumer, L. (1992). Calcium mobilization by the LH receptor expressed in Xenopus oocytes independent of 3′,5′-cyclic adenosine monophosphate formation: evidence for parallel activation of two signaling pathways. *Mol. Endocrinol.* 6, 272–278.
446. Gudermann, T., Birnbaumer, M., and Birnbaumer, L. (1992). Evidence for dual coupling of the murine luteinizing hormone receptor to adenylyl cyclase and phosphoinositide breakdown and Ca^{2+} mobilization. *J. Biol. Chem.* 267, 4479–4488.
447. Zhu, X., Gilbert, S., Birnbaumer, M., and Birnbaumer, L. (1994). Dual signaling potential is common among Gs-coupled receptors and dependent on receptor density. *Mol. Pharm.* 46, 460–469.
448. Rao, M. C., Richards, J. S., Midgley, A. R., and Reichert, L. E. (1977). Regulation of gonadotropin receptors by luteinizing hormone in granulosa cells. *Endocrinology* 101, 512–523.
449. Davis, J. S., Weakland, L. L., Coffey, R. G., and West, L. A. (1989). Acute effects of phorbol esters on receptor-mediated IP3, cAMP, and progesterone levels in rat granulosa cells. *Am. J. Physiol.* 256, E368–E374.
450. Abell, A. N., and Segaloff, D. L. (1997). Evidence for the direct involvement of transmembrane region 6 and the lutropin/choriogonadotropin receptor in activating G_s. *J. Biol. Chem.* 272, 14586–14591.
451. Wang, Z., Hipkin, R. W., and Ascoli, M. (1996). Progressive cytoplasmic tail truncations of the lutropin-choriogonadotropin receptor prevent agonist- or phorbol ester-induced phosphorylation, impair agonist- or phorbol ester-induced desensitization, and enhance agonist-induced receptor down-regulation. *Mol. Endocrinol.* 10, 748–759.
452. Wang, Z., Liu, X., and Ascoli, M. (1997). Phosphorylation of the lutropin/choriogonadotropin receptor facilitates uncoupling of the receptor from adenylyl cyclase and endocytosis of the bound hormone. *Mol. Endocrinol.* 11, 183–192.
453. Min, L., Galet, C., and Ascoli, M. (2001). The association of arrestin-3 with the human lutropin/choriogonadotropin receptor depends mostly on receptor activation rather than on receptor phosphorylation. *J. Biol. Chem.* 277, 702–710.
454. Lazari, M. F. M., Bertrand, J. E., Nakamura, K., et al. (1998). Mutation of individual serine residues in the C-terminal tail of the lutropin/choriogonadotropin receptor reveal distinct structural requirements for agonist-induced internalization. *J. Biol. Chem.* 273, 18316–18324.
455. Nakamura, K., Lazari, M. F. M., Li, S., Korgaonkar, C., and Ascoli, M. (1999). Role of the rate of internalization of the agonist-receptor complex on the agonist-induced down-regulation of the lutropin/choriogonadotropin receptor. *Mol. Endocrinol.* 13, 1295–1304.
456. Lamm, M. L. G., and Hunzicker-Dunn, M. (1994). Phosphorylation-independent desensitization of the luteinizing hormone/chorionic gonadotropin receptor in porcine follicular membranes. *Mol. Endocrinol.* 8, 1537–1546.
457. Ekstrom, R. C., and Hunzicker-Dunn, M. (1989). Homologous desensitization of ovarian luteinizing hormone/human chorionic gonadotropin-responsive adenylyl cyclase is dependent upon GTP. *Endocrinology* 124, 956–963.
458. Li, S., Liu, X., Min, L., and Ascoli, M. (2001). Mutations of the second extracellular loop of the human lutropin receptor emphasize the importance of receptor activation and de-emphasize the importance of receptor phosphorylation in agonist-induced internalization. *J. Biol. Chem.* 276, 7968–7973.

459. Zhu, X., Gudermann, T., Birnbaumer, M., and Birnbaumer, L. (1993). A luteinizing hormone receptor with a severely truncated cytoplasmic tail (LHR-ct628) desensitizes to the same degree as the full-length receptor. *J. Biol. Chem.* 268, 1723–1728.
460. Hirakawa, T., Galet, C., Kishi, M., and Ascoli, M. (2003). GIPC binds to the human lutropin receptor (hLHR) through an unusual PDZ domain binding motif, and it regulates the sorting of the internalized human choriogonadotropin and the density of cell surface hLHR. *J. Biol. Chem.* 278, 49348–49357.
461. Lohse, M. J., Benovic, J. L., Codina, J., Caron, M. G., and Lefkowitz, R. J. (1990). b-arrestin: a protein that regulates b-adrenergic receptor function. *Science* 248, 1547–1550.
462. Mukherjee, S., Palczewski, K., Gurevich, V. V., et al. (1999). A direct role for arrestins in desensitization of the luteinizing hormone/choriogonadotropin receptor in porcine ovarian follicular membranes. *Proc. Natl. Acad. Sci. U S A* 96, 493–498.
463. Mukherjee, S., Palczewski, K., Gurevich, V. V., and Hunzicker-Dunn, M. (1999). b arrestin-dependent desensitization of luteinizing hormone/choriogonadotropin receptor is prevented by a synthetic peptide corresponding to the third intracellular loop of the receptor. *J. Biol. Chem.* 274, 12984–12989.
464. Mukherjee, S., Gurevich, V. V., Preninger, A., et al. (2002). Aspartic acid 564 in the third cytoplasmic loop of luteinizing hormone/choriogonadotropin receptor is crucial for phosphorylation-independent interaction with arrestin2. *J. Biol. Chem.* 277, 17916–17927.
465. Hunzicker-Dunn, M., Gurevich, V. V., Casanova, J. E., and Mukherjee, S. (2002). ARF6: A newly appreciated player in G protein coupled receptor desensitization. *FEBS Lett.* 521, 3–8.
466. Salvador, L. M., Mukherjee, S., Kahn, R. A., et al. (2001). Activation of the luteinizing hormone/choriogonadotropin hormone receptor promotes ADP ribosylation factor (ARF) 6 activation in porcine ovarian follicular membranes. *J. Biol. Chem.* 276, 33773–33781.
467. Mukherjee, S., Casanova, J. E., and Hunzicker-Dunn, M. (2001). Desensitization of the luteinizing hormone/choriogonadotropin receptor in ovarian follicular membranes is inhibited by catalytically inactive ARNO. *J. Biol. Chem.* 276, 6524–6528.
468. Mukherjee, S., Gurevich, V. V., Jones, J. C. R., et al. (2000). The ADP ribosylation factor nucleotide exchange factor ARNO promotes b-arrestin release necessary for luteinizing hormone/choriogonadotropin receptor desensitization. *Proc. Natl. Acad. Sci. U S A* 97, 5901–5906.
469. Claing, A., Chen, W., Miller, W. E., et al. (2001). b arrestin-mediated ARF6 activation and b2-adrenergic receptor endocytosis. *J. Biol. Chem.* 276, 42509–42513.
470. Morris, J. K., and Richards, J. S. (1995). Luteinizing hormone induces prostaglandin endoperoxide synthase-2 and luteinization in vitro by A-kinase and C-kinase pathways. *Endocrinology* 136, 1549–1558.
471. Park-Sarge, O-K., and Mayo, K. E. (1994). Regulation of the progesterone receptor gene by gonadotropins and cyclic adenosine 3′,5′-monophosphate in rat granulosa cells. *Endocrinology* 134, 709–718.
472. Sayasith, K., Bouchard, N., Sawadogo, M., Lussier, J. G., and Sirois, J. (2004). Molecular characterization and role of bovine upstream stimulatory factor 1 and 2in the regulation of the prostaglandin G/H synthase-2 promoter in granulosa cells. *J. Biol. Chem.* 279, 6327–6336.
473. Hunzicker-Dunn, M. (1981). Selective activation of rabbit ovarian protein kinase isozymes in rabbit ovarian follicles and corpora lutea. *J. Biol. Chem.* 256, 12185–12193.
474. Salvador, L. M., Maizels, E., Hales, D. B., et al. (2002). Acute signaling by the LH receptor is independent of protein kinase C activation. *Endocrinology* 143, 2986–2994.
475. Krueger, D. A., Mao, D., Warner, E. A., and Dowd, D. R. (1999). Functional analysis of the mouse ICER (inducible cAMP early repressor) promoter: evidence for a protein that blocks calcium responsiveness of the CAREs (cAMP autoregulatory elements). *Mol. Endocrinol.* 13, 1207–1217.
476. Sela-Abramovich, S., Chorev, E., Galiani, D., and Dekel, N. (2005). MAPK mediates LH-induced breakdown of communication and oocyte maturation in rat ovarian follicles. *Endocrinology* 146, 1236–1244.
477. Hirakawa, T., and Ascoli, M. (2003). The lutropin/choriogonadotropin receptor-induced phosphorylation of the extracellular signal-regulated kinases in leydig cells is mediated by a protein kinase a-dependent activation of ras. *Mol. Endocrinol.* 17, 2189–2200.
478. Sugawara, T., Saito, M., and Fujimoto, S. (2000). Sp1 and SF-1 interact and cooperate in the regulation of human steroidogenic acute regulatory protein gene expression. *Endocrinology* 141, 2895–2903.
479. Sugawara, T., Holt, J. A., Kiriakidou, M., and Strauss, J. F. Jr. (1996). Steroidogenic factor 1-dependent promoter activity of the human steroidogenic acute regulatory protein (StAR) gene. *Biochemistry* 35, 9052–9059.
480. Guan, Z., Buckman, S. Y., Miller, B. W., Springer, L. D., and Morrison, A. R. (1998). Interleukin-1beta-induced cyclooxygenase-2 expression requires activation of both c-Jun NH2-terminal kinase and p38 MAPK signal pathways in rat renal mesangial cells. *J. Biol. Chem.* 273, 28670–28676.
481. Dean, J. L., Brook, M., Clark, A. R., and Saklatvala, J. (1999). p38 mitogen-activated protein kinase regulates cyclooxygenase-2 mRNA stability and transcription in lipopolysaccharide-treated human monocytes. *J. Biol. Chem.* 274, 264–269.
482. O'Neill, L. A., and Greene, C. (1998). Signal transduction pathways activated by the IL-1 receptor family: ancient signaling machinery in mammals, insects, and plants. *J. Leuk. Biol.* 63, 650–657.
483. Adashi, E. Y. (1996). Immune modulators in the context of the ovulatory process: a role for interleukin-1. *Am. J. Reprod. Immunol.* 35, 190–194.
484. Hurwitz, A., Ricciarelli, E., Botero, L., et al. (1991). Endocrine- and autocrine-mediated regulation of rat ovarian (theca-interstitial) interleukin-1 beta gene expression: gonadotropin-dependent preovulatory acquisition. *Endocrinology* 129, 3427–3429.
485. Kol, S., Donesky, B. W., Ruutiainen-Altman, K., et al. (1999). Ovarian interleukin-1 receptor antagonist in rats: gene expression, cellular localization, cyclic variation, and hormonal regulation of a potential determinant of interleukin-1 action. *Biol. Reprod.* 61, 274–282.
486. Shukovski, L., and Tsafriri, A. (1994). The involvement of nitric oxide in the ovulatory process in the rat. *Endocrinology* 135, 2287–2290.
487. Davis, J. S., Weakland, L. L., West, L. A., and Farese, R. V. (1986). Luteinizing hormone stimulates the formation of inositol triphosphate and cyclic AMP in rat granulosa cells. *Biochem. J.* 238, 597–604.
487a. Aguirre, C., Jayes, F. C., and Veldhuis, J. D. (2000). Luteinizing hormone (LH) drives diverse intracellular calcium second messenger signals in isolated porcine ovarian thecal cells: preferential recruitment of intracellular Ca^{2+} oscillatory cells by higher concentrations of LH. *Endocrinology* 141, 2220–2228.
488. Newton, A. C. (2003). Regulation of the ABC kinases by phosphorylation: protein kinase C as a paradigm. *Biochem. J.* 370, 361–371.

489. Morris, J. K., and Richards, J. S. (1993). Hormone induction of luteinization and prostaglandin endoperoxide synthase-2 involves multiple cellular signalling pathways. *Endocrinology* 133, 770–779.
490. Sekar, N., Garmey, J. C., and Veldhuis, J. D. (2000). Mechanisms underlying the steroidogenic synergy of insulin and luteinizing hormone in porcine granulosa cells: joint amplification of pivotal sterol-regulatory genes encoding the low-density lipoprotein (LDL) receptor, steroidogenic acute regulatory (stAR) protein and cytochrome P450 side-chain cleavage (P450scc) enzyme. *Mol. Cell. Endocrinol.* 159, 25–35.
491. Sekar, N., Lavoie, H. A., and Veldhuis, J. D. (2000). Concerted regulation of steroidogenic acute regulatory gene expression by luteinizing hormone and insulin (or insulin-like growth factor I) in primary cultures of porcine granulosa-luteal cells. *Endocrinology* 141, 3983–3992.
492. Carvalho, C. R., Carvalheira, J. B., Lima, M. H., et al. (2003). Novel signal transduction pathway for luteinizing hormone and its interaction with insulin: activation of janus kinase/signal transducer and activator of transcription and phosphoinositol 3-kinase/akt pathways. *Endocrinology* 144, 638–647.
493. Ihle, J. N., and Kerr, I. M. (1995). Jaks and Stats in signaling by the cytokine receptor superfamily. *Trends Genet.* 11, 69–74.
494. Bebia, Z., Somers, J. P., Liu, G., et al. (2001). Adenovirus-directed expression of functional LH receptors in undifferentiated rat granulosa cells: evidence for differential signaling through FSH and LH receptors. *Endocrinology* 142, 2252–2259.
495. Yong, E. L., Hillier, S. G., Turner, M., et al. (1994). Differential regulation of cholesterol side-chain cleavage (P450scc) and aromatase (P450arom) enzyme mRNA expression by gonadotrophins and cyclic AMP in human granulosa cells. *J. Mol. Endocrinol.* 12, 239–249.
496. Tapanainen, J. S., Lapolt, P. S., Perlas, E., and Hsueh, A. J. (1993). Induction of ovarian follicle luteinization by recombinant follicle-stimulating hormone. *Endocrinology* 133, 2875–2880.
497. Conti, M. (2002). Specificity of the cyclic adenosine 3′,5′-monophosphate signal in granulosa cell function. *Biol. Reprod.* 67, 1653–1661.
498. Zaccolo, M., and Pozzan, T. (2002). Discrete microdomains with high concentration of cAMP in stimulated rat neonatal cardiac myocytes. *Science* 295, 1711–1715.
499. Bauman, A. L., and Scott, J. D. (2002). Kinase- and phosphatase-anchoring proteins: harnessing the dynamic duo. *Nat. Cell. Biol.* 4, E203–E206.
500. Dodge, K. L., Khouangsathiene, S., Kapiloff, M. S., et al. (2001). mAKAP assembles a protein kinase A/PDE4 phosphodiesterase cAMP signaling module. *EMBO J.* 20, 1921–1930.
501. Pirece, J. G. (1988). Gonadotropins: chemistry and biosynthesis. In *The Physiology of Reproduction* (E. Knobil and J. D. Neill, Eds.), 1335–1348. Raven Press, New York.
502. Vassart, G., Pardo, L., and Costagliola, S. (2004). A molecular dissection of the glycoprotein hormone receptors. *Trends Biochem. Sci.* 29, 119–126.

Knobil and Neill's Physiology of Reproduction, Third Edition
edited by Jimmy D. Neill,
Elsevier © 2006

CHAPTER 15

Steroid Receptors in the Ovary and Uterus

John F. Couse, Sylvia C. Hewitt, and Kenneth S. Korach

Introduction, 594
The Steroid Receptors, 594
Genes, mRNA, and Regulation, 594
Receptor Structure, 597
Receptor Mechanisms of Action, 602
Steroid Signaling Null Mice, 607
Mice Null for PR Signaling, 607
Mice Null for AR Signaling, 607
Mice Null for ER Signaling, 607
Sex Steroid Receptors in Uterine Function, 608
Estrogen Receptor Signaling in Uterine Function, 608
ER Expression in the Uterus, 608
Uterine Phenotypes in Murine Models of Disrupted Estrogen Signaling, 610
Biphasic Uterine Response to Estradiol, 612
Estrogen–Growth Factor Cross-Talk in the Uterus, 614
Maintenance of Selective Estrogen Actions in the Absence of ER-α, 615
Maintenance of Progesterone Actions in the Absence of ER-α, 616
Progesterone Receptor Signaling in the Uterus, 616
PR Expression in the Uterus, 616
PR Actions in the Rodent Uterus: PR-Null Mice, 617
Androgen Receptor Signaling in Uterine Function, 619
Glucocorticoid Receptor Signaling in Uterine Function, 619
Sex Steroid Receptors in Ovarian Function, 619
Estrogen Receptor Signaling in Ovarian Function, 619
ER Expression in the Ovary, 619
Ovarian Phenotypes in Murine Models of Disrupted Estrogen Signaling, 627
Intraovarian Roles of Estradiol in Ovarian Function, 633
ER-α–Mediated Repression of Leydig Cell Development in the Ovary, 642
Disrupted Estrogen Signaling in Humans, 643
Progesterone Receptor Signaling in Ovarian Function, 644
PR Expression in the Ovary, 644
Ovarian Phenotypes in Murine Models of Disrupted Progesterone Signaling, 646
Intraovarian Role of Progesterone in Ovarian Function, 646
Androgen Receptor Signaling in Ovarian Function, 648
AR Expression in the Ovary, 648
Ovarian Phenotypes in Murine Models of Disrupted Androgen Signaling, 649
Intraovarian Role of Androgen in Ovarian Function, 650
Extraovarian Roles of Sex Steroids in Ovarian Function, 653
Negative Feedback Control of Gonadotropin Secretion, 654
Positive Feedback Control of Gonadotropin Secretion, 655
Estrogen Regulation of Prolactin Synthesis, 656
Summary, 657
References, 657

National Institute of Environmental Health Sciences, NIH, Research Triangle Park, North Carolina.

INTRODUCTION

Receptor mediated actions of progestins, androgens, and estrogens are central to reproduction. Perhaps unique among the various categories of signaling pathways, the sex steroid hormones and their cognate receptors are surely mentioned if not discussed in detail in every chapter of this volume. The sex steroid hormones are integrated into every aspect of mammalian reproductive physiology in both sexes, including sexual development, gametogenesis, hypothalamic–pituitary control of gonadal function, sexual and maternal behavior, pregnancy, and lactation. Disruption of the signaling pathways for any one of the three gonadal steroids leads to reduced fecundity if not infertility due to aberrations in multiple organ systems. Therefore, a better understanding of the sex steroid receptors in terms of their expression in reproductive and endocrine tissues, their mechanism of action, their role in reproductive processes, and their interaction with nonsteroidal signaling pathways is instrumental to our ability to manage infertility and reproductive disease.

This chapter principally covers the signaling pathways for progestins, androgens, and estrogens and the role of each in ovarian and uterine function in mammals. Because of the importance of the sex steroids and the likelihood that each is discussed in several other chapters of this volume, it is appropriate to first discuss the structure, mechanism of action, and signaling pathways for each of the sex steroid receptors. Sections then follow on the role of sex steroids in ovarian and uterine function. We chose to divide each of these sections according to the sex steroid receptors: the estrogen receptors (ERs), progesterone receptors (PRs), and androgen receptors (ARs). In the section on Sex Steroid Receptors in Ovarian Function, we discuss what is currently known concerning each particular sex steroid signaling pathway in terms of its expression pattern and distribution within the ovary; its intraovarian role in granulosa cell proliferation and differentiation, thecal cell function, and ovulation; and its extraovarian role in modulating gonadotropin secretion from the hypothalamic–pituitary axis. Because a separate chapter (Chapter 12) is dedicated to the corpus luteum, the role of the sex steroid receptors in luteal function is not covered in detail. In the section on Sex Steroid Receptors in Uterine Function, we discuss what is currently known concerning each sex steroid signaling pathway in terms of its expression pattern and distribution within the uterus and its role in uterine cell proliferation and differentiation. New since the second edition of this volume is the development of null mouse models for each of the sex steroids. The study of each of these models has already and will continue to make enormous contributions to our understanding of the reproductive role of each receptor signaling pathway in toto. Therefore, we include a detailed description of the ovarian and uterine phenotypes for each of these models in their respective sections.

THE STEROID RECEPTORS

The steroid receptors are part of the family of nuclear receptors (NRs) that is ubiquitous throughout the animal kingdom. The members of the NR family fulfill a plethora of functions and are integral to the development and maintenance of multiple physiological systems. Continued discovery of new NRs has led to a considerable expansion of the NR family and demanded a system of categorization. An early scheme used commonalities in receptor dimerization and DNA binding properties to divide the NR family into four distinct classes (1). The PR, AR, and ER, casually referred to as the sex steroid receptors, form class I in this scheme and are defined as ligand-induced homodimers that bind to DNA half-sites organized as inverted repeats (1). A more recent scheme was devised by a Nuclear Receptors Nomenclature Committee in 1999 (2) and is based on an earlier system for classification of the cytochrome P-450 family (3). The Nuclear Receptors Nomenclature Committee scheme is based largely on sequence homology and evolutionary comparisons and divides the over 60 NRs into seven subfamilies and a varied number of groups within each subfamily (2). The ERs (ER-α and ER-β) are the sole members of the NR3A subgroup, ER-α and ER-β being NR3A1 and NR3A2, respectively (2). The PR and AR are the number 3 and 4 members of the NR3C subgroup that also includes the glucocorticoid receptor (GR) and the mineralocorticoid receptor (2). A detailed description of the structure and multitude of functions among the NR family members is beyond the scope of this chapter but may be found in several excellent reviews (Table 1). The discussion below describes the general properties of the sex steroid receptor family and those mechanisms of receptor action that are most relevant to ovarian and reproductive tract function.

Genes, mRNA, and Regulation

The genes encoding the sex steroid receptors exhibit a highly conserved structural organization. The human PR (*PGR* or *NR3C3*), AR (*AR* or *NR3C4*), ER-α (*ESR1* or *NR3A1*), and ER-β (*ESR2* or *NR3A2*) genes are each composed of eight coding exons and vary in length from 40 kb (ER-β) to >140 kb (ER-α) (4–7).

TABLE 1. *Recent literature reviews of nuclear receptor function*

Aranda, A., and Pascual, A. (2001). Nuclear hormone receptors and gene expression. *Physiol. Rev.* 81, 1269–1304.
Coleman, K. M., and Smith, C. L. (2001). Intracellular signaling pathways: nongenomic actions of estrogens and ligand-independent activation of estrogen receptors. *Front. Biosci.* 6, 1379–1391.
Driggers, P. H., and Segars, J. H. (2002). Estrogen action and cytoplasmic signaling cascades. Part II: the role of growth factors and phosphorylation in estrogen signaling. *Trends Endocrinol. Metab.* 13, 422–427.
Gelman, E. P. (2002). Molecular biology of the androgen receptor. *J. Clin. Oncol.* 20, 3001–3015.
Giangrande, P. H., and McDonnell, D. P. (1999). The A and B isoforms of the human progesterone receptor: two functionally different transcription factors encoded by a single gene. *Rec. Prog. Horm. Res.* 54, 291–314.
Gobinet, J., Poujol, N., and Sultan, Ch. (2002). Molecular actions of androgens. *Mol. Cell. Endocrinol.* 198, 15–24.
Heinlein, C. A., and Chang, C. (2002). The roles of androgen receptors and androgen-binding proteins in nongenomic androgen actions. *Mol. Endocrinol.* 16, 2181–2187.
Katzenellenbogen, B. S., Montano, M. M., Ediger, T. R., Sun, J., Ekena, K., Lazennec, G., Martini, P. G., McInerney, E. M., Delage-Mourroux, R., Weis, K., and Katzenellenbogen, J. A. (2000). Estrogen receptors: selective-ligands, partners, and distinct pharmacology. *Rec. Prog. Horm. Res.* 55, 163–195.
Kushner, P. J., Agard, D. A., Greene, G. L., Scanlan, T. S., Shiau, A. K., Uht, R. M., and Webb, P. 2000. Estrogen receptor pathways to AP-1. *J. Steroid Biochem. Mol. Biol.* 74, 311–317.
Laudet, V., and Gronemeyer, H. (2002). *The Nuclear Receptor Facts Book.* Academic Press, London.
Lösel, R. M. (2004). Nongenomic steroid action: controversies, questions, and answers. *Physiol. Rev.* 83, 965–1016.
McEwan, I. J. (2004). Molecular mechanisms of androgen receptor-mediated gene regulation: structure-function analysis of the AF-1 domain. *Endocr. Rel. Cancer* 11, 281–293.
McKenna, N. J., Lanz, R. B., and O'Malley, B. W. (1999). Nuclear receptor coregulators: cellular and molecular biology. *Endocr. Rev.* 20, 321–344.
Segars, J. H., and Driggers, P. H. (2002). Estrogen action and cytoplasmic signaling cascades. Part I: membrane-associated signaling complexes. *Trends Endocrinol. Metab.* 13, 349–354.

In each case, the N-terminal domain (NTD) of the receptor is usually encoded by a single exon, the two zinc fingers of the DNA-binding domain, each encoded by separate exons (2 and 3), and the ligand-binding domain (LBD), encoded by exons 4–8.

Progesterone Receptor

The rabbit and chicken *Pgr* cDNAs were first cloned in 1986 (8–10) and have since been followed by similar clones from human (11), rat (12), mouse (13), and sheep (14). Two distinct promoters in the *PGR* gene provide for the generation of two major PR isoforms, PR-B (114 kDa) and PR-A (94 kDa), in most species except rabbit, which possess PR-B only. The two PR forms are identical in all regions except the NTD, which is truncated by 128–164 amino acids depending on the species (Fig. 1). Both PR isoforms are expressed at relatively equal levels in tissues, although differences in the ratio have been noted in certain tissues. A third isoform, termed PR-C, was recently discovered in humans and is an N-terminally truncated form that lacks the full A/B domain and first zinc finger of the C-domain but is able to bind ligand and enhance the actions of PR-A and PR-B (15,16). PR expression is directly induced by ER-mediated estradiol in certain but not all tissues, and this is especially true in the female reproductive tract (17–19). Several naturally existing *PGR* variant transcripts harboring exon deletions or distinct 5′-untranslated sequences have been described but are not well characterized in terms of expression or functionality (20,21).

Androgen Receptor

The human *AR* cDNA was first cloned in 1988 (22,23) and has since been described in multiple species, including monkey (24), mouse (25), rat (26), rabbit (27), and fish (28,29). The *AR* gene is located on the X chromosome, and therefore genetic males possess only a single copy (6,30,31). Two distinct start sites in the *AR* gene are used to produce two isoforms, AR-B (110 kDa) and AR-A (87 kDa), which differ only in the N-terminus (Fig. 1) (14). However, in contrast to the PR, the two known AR isoforms exhibit only subtle functional differences (14). A unique feature of the *AR* gene relative to its sex steroid receptor counterparts is the presence of polymorphic repeats of glutamine and glycine in the NTD, the former of which has been linked to certain chronic diseases and cancer in humans (32). There are binding sites for a wide array of transcription factors in the *AR* promoter, including Sp1, cAMP-response element binding (CREB) protein, and c-*myc*, suggesting complex tissue-specific regulation (31). Direct androgen-mediated autoregulation of *AR* expression has also been shown (32).

Estrogen Receptor

The human *ESR1* (ER-α) cDNA was first cloned in 1985 (33,34) and has since been isolated from over 20 additional species (14), including mouse (35), rat (36), pig (37), chicken (38), *Xenopus laevis* (39), and fish (40–42). Surprisingly, a second ER gene, termed

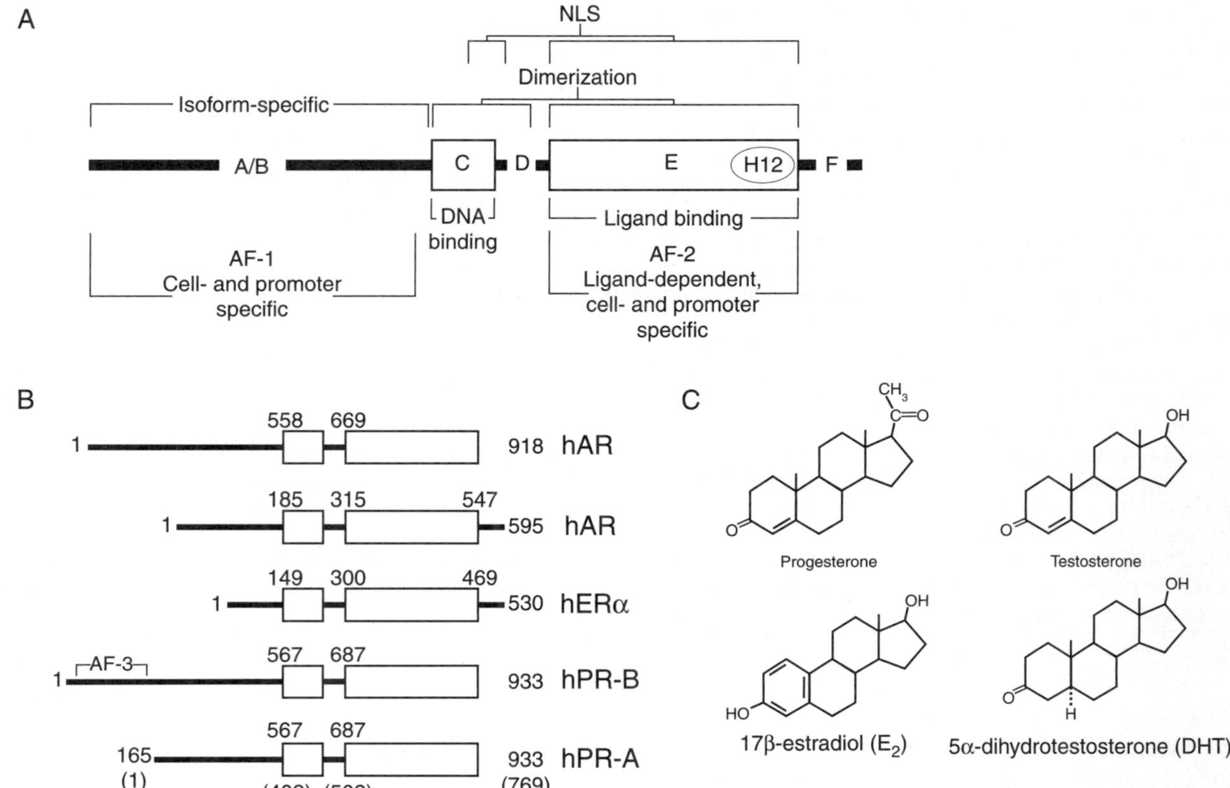

FIG. 1. The sex steroid family of receptors and endogenous ligands. **A:** Schematic illustration of the structural organization of the sex steroid nuclear receptors. The more highly conserved C and E domains are depicted as *open boxes* and the less well-conserved A/B, D, and F domains as *filled bars*. The F domain is unique to the estrogen receptors (ER). The functions of the modular domains are indicated as are the two known transcriptional activation function domains, AF-1 and AF-2. The AF-1 domain is harbored in the A/B region and exhibits constitutive activity in vitro, whereas the AF-2 domain lies with the ligand-binding (E) domain and is critical to ligand-induced receptor activation. Both domains synergize during ligand-activated receptor actions. NLS, nuclear localization signal. **B:** Structural comparison of the sex steroid receptors, the androgen receptor (AR), estrogen receptors-α (ERα) and -β (ERβ), and progesterone receptor-B (PR-B) and -A (PR-A). The number of amino acids that compose each module and whole receptor of the human forms are indicated. For PR-A, the numbers above refer to the amino acid residues in reference to PR-B; the numbers in parenthesis indicate the primary amino acid sequence for PR-A. An AF-3 region has been identified in the PR-B receptor. **C:** Chemical structure of the endogenous sex steroids that act as activating ligands for the respective receptors. [Parts of this figure were modified from (14).]

ESR2 (ER-β), was discovered in 1996 in rat (43) and human (44) and has since been cloned in almost 20 species (14), including mouse (45), monkey (46), cow (47), and fish (48,49). Unlike the PR and AR, ER-α and ER-β are not isoforms but rather distinct receptor forms encoded by separate genes.

The encoded ER-α proteins are 595 and 599 amino acids in length in human and mice, respectively, with an approximate molecular weight of 66 kDa (Fig. 1) (33–35). Multiple promoter and regulatory regions in the 5′-untranslated sequences of the human and rat *ESR1* gene have been described yet all possess the same single open reading frame (50–53). Numerous naturally occurring variants of the *ESR1* mRNA in normal and neoplastic tissues of several species have been described, but the existence of corresponding proteins remains controversial (54–56).

The promoter region of the *ESR1* gene has been relatively well characterized and indicates a complex regulation of expression (51,52,57–59). Cicatiello et al. (60) demonstrated that the proximal 0.4 kb of the mouse *Esr1* promoter is sufficient to provide for widespread but specific expression of a reporter construct in vivo. Regulatory elements that may provide for AP-1, Sp1, and ER autoregulation of the human *ESR1* promoter have been described (61,62). Decreased *ESR1* expression is linked to receptor-mediated actions of vitamin D (63) and increased intracellular cAMP or mitogen-activated protein kinase activity (61). Increased methylation of the

ESR1 promoter is implicated in reducing ER levels, especially in tumorigenic tissues (61).

The *ESR2* genes of multiple species yield numerous transcripts that range from 1 to >9 kb in length, in contrast to the single predominant transcript of ~7 kb transcribed from the *ESR1* gene. Initial descriptions of human and rodent ER-β projected a protein of 485 amino acids. However, it is now apparent that translation of the *ESR2* mRNA initiates upstream of these original open reading frames and yields a receptor of 549 amino acids in rodents and 530 amino acids in humans, each with an approximate molecular weight of 60–63 kDa (Fig. 1). Therefore, ER-β is slightly smaller than ER-α, and most of this difference lies within the N-terminus. A number of variant transcripts of the *ESR2* gene have been described; however, unlike ER-α there is growing evidence that some of these variants may coexist with the wild-type receptor form in certain tissues (Fig. 2). To date there is little known about the regulatory sequences of the *ESR2* gene. Portions of the mouse *Esr2* gene have been characterized and possess potential binding sites for numerous transcription factors (64) and the initiation of transcription from at least two distinct untranslated exons (20).

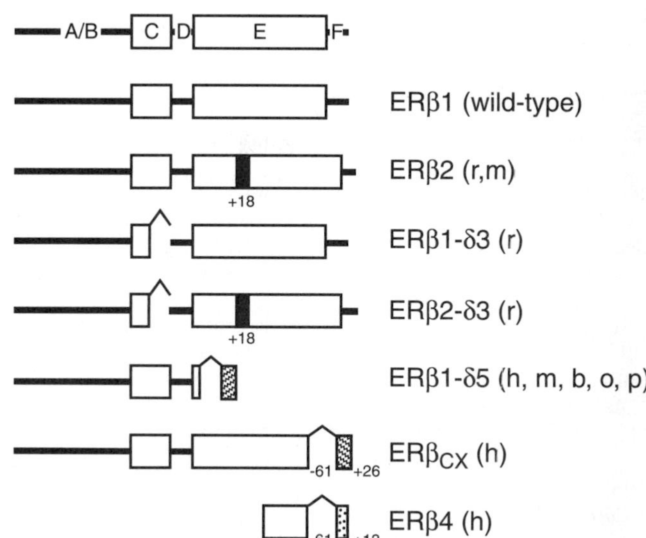

FIG. 2. Common estrogen receptor-β variant transcripts found in mammals. **Top:** Shown is the general modular structure of steroid hormone receptors illustrating the N′-terminal A/B domain, DNA binding (C) domain, hinge (D) region, ligand binding (E) domain, and C′-terminal F domain (see Fig. 1 for details). **Bottom:** Shown are the modular structures of full-length wild-type ER-β (or ER-β1), followed by several putative ER-β isoforms encoded by variant transcripts that have been described in mammals. ER-β2 is described in rats (r) and mice (m) and possesses a 54-bp insert that codes for an additional 18 amino acids in the ligand-binding (E) domain (347,364). ER-β2 exhibits an approximate 35-fold decreased affinity for estradiol relative to wild-type ER-β and a significantly reduced transactivational activity in vitro (347). ER-β1-δ3 is described in rats (r) and is a splice variant lacking exon 3, which encodes for the C′-terminal zinc finger of the DNA (C) binding domain (347). ER-β1-δ3 exhibits an affinity for estradiol similar to that of wild-type ERβ but is unable to bind DNA and therefore possesses no transactivational activity via a "classic" mechanism of receptor action in vitro (347). ER-β2-δ3 is a variant that exhibits both the 18 amino acid insertion of ER-β2 and the zinc finger deletion of ER-β1-δ3 and is described in rat (r) (347). ER-β1-δ5 is described in humans (h) (364), mice (m) (364), cows (b) (359), sheep (o) (343), and pigs (p) (344) and is an exon 5 deletion that results in a frame-shift, leading to a putative ER-β isoform that is truncated 15 amino acids into the E domain and terminated with 4–9 heterologous residues (depending on species). ER-β1-δ5 possesses some ligand-independent transactivational activity in vitro (344). ER-β_{cx} is described in humans (h) (361) and is a splicing variant that encodes a putative receptor isoform in which the C-terminal 61 amino acids are replaced by 26 amino acids encoded by an alternative downstream exon. ER-β_{cx} is unable to bind estradiol but acts as a dominant negative receptor when heterodimerized with ER-α in vitro (361). ER-β4 is described in humans (h) (361) and is a severe N- and C′-terminal truncated isoform that also possesses 13 heterologous amino acids in the C′-terminus due to a frame-shift.

Receptor Structure

Common to all members of the NR family is a modular structure of domains, each of which harbors an autonomous function that is critical to total receptor action (1,14,65). The sex steroid receptors are composed of five functional modules, an NTD or A/B domain, the DNA-binding (C) domain, a hinge (D) region, and an LBD (E) (Fig. 1). The ERs also possess a unique C-terminal F domain of unknown function. Our understanding of the NR functional domains and their importance to overall receptor activity is largely derived from the in vitro study of artificially generated mutant receptors and more recently from x-ray crystallography studies.

NTD or A/B Domain

The NTD or A/B domains of the NR family members greatly vary in length and share little homology among the steroid receptors, although some structural features are conserved (Fig. 1). Crystallography studies of the steroid receptor NTD have been largely unsuccessful due to the unstructured nature of this portion of the receptor in aqueous solutions. However, evidence suggests that intramolecular interactions between the A/B and other receptor domains are likely to induce a more structured NTD (31). The NTD of each of the sex steroid receptors harbors the transcriptional activation function-1 (AF-1) domain and provides for cell and promoter-specific activity of the receptor as well as a site for coreceptor protein interaction (Table 2). The AF-1 domain alone can confer constitutive transcriptional (i.e., ligand-independent) activity when linked to a heterologous NR DNA binding domain in vitro, but

TABLE 2. *Nuclear receptor coregulators*

Cofactor	Alternative designations	Comments
ERAP-160	GRIP-170 p160	Bind ER in a ligand-dependent manner.
RIP-140		Interacts with and coactivates or corepresses ER activity depending on the cell and promoter context.
TBP/TAF_{II}s		Interacts with and coactivates nuclear receptors.
SRC-1	hSRC-1 NcoA-1/mSRC-1 p160	Interacts with and coactivates nuclear receptors; interacts with CBP/p300, basal transcription factors; targeted deletion causes partial hormone insensitivity in mice.
ARA-70		Interacts with and coactivates AR.
CBP		Interacts with and coactivates multiple nuclear receptors and other activators; similar to p300.
p300		Interacts with and coactivates multiple nuclear receptors and other activators; similar to CBP.
PCAF	hGCN5	Interacts with PR and SRC-1.
TIF2/hSRC-2	GRIP-1/mSRC-2 NCoA-2, p160	Interacts with and coactivates nuclear receptors; interacts with CBP.
L7/SPA		Interacts with RU 486-bound PR and enhances partial agonist activity of RU 486 with PR.
p/CIP/mSRC-3	ACTR/hSRC-3 RAC3/hSRC-3 AIB-1/hSRC TRAM-1/hSRC-3 p160, SRC-3	P/CIP coactivates CBP-mediated signaling pathways; interacts with CBP and p300. AIB-1 interacts with and coactivates ER. SRC-3 preferentially coactivates ER-α over ER-β.
E6-AP		E6-AP interacts with and coactivates AR, ER, PR, and GR.
BRG-1	SWI2/SNF2	Interacts with GR and ER.
HMG-1		Coactivation specific for steroid receptors by promoting DNA binding.
SRA		Functions as an RNA transcript; selectively coactivates AF-1 of steroid receptors; part of SRC-1 complex.
SNURF		Co-activates AR, PR, GR, Sp1, and AP-1 activity.
ARIP3		Interacts with and modestly coactivates AR.
NCoR	RIP-13	Reduces RU 486-bound PR partial agonist activity.
SMRT	TRAC3	Reduces tamoxifen-bound ER and RU 486-bound partial agonist activity.

From McKenna et al. (1999). Nuclear receptor coregulators: cellular and molecular biology. *Endocr. Rev.* 20, 321–344.

this function is largely overcome in the context of the whole receptor (14). Posttranslational modifications to the A/B domain can dramatically affect the overall behavior of the receptor and are thought to be an important mechanism for the modulation of AF-1 functions. Phosphorylation of the A/B domain is the most well-characterized posttranslational modification and occurs in all three sex steroid receptors via the actions of multiple intracellular signaling pathways, including mitogen-activated protein kinase pathways, the cAMP/protein kinase A (PKA) pathway, and cyclin-dependent kinases (14,31,32,65,66).

The extended N-terminal sequences that are unique to PR-B (Fig. 1) provide a third transcriptional AF-3 domain in this isoform that is lacking in PR-A and may allow for PR-B specific expression of certain progesterone regulated genes (15,67). In general, PR-B is a stronger activator of transcription versus PR-A when acting on a hormone response element (HRE)-driven gene construct in vitro (67). Furthermore, antagonist-bound PR-B acts as a strong transcriptional activator in certain cell and promoter contexts, whereas antagonist-bound PR-A is inactive (15).

The NTD of the AR is especially autologous in vitro and is likely more important to overall AR transactivational activity than the C-terminal AF-2 domain (described below) (31,32). Furthermore, the first 30 residues of the AR NTF are highly conserved and important for interactions with the LBD that provide for agonist-induced stabilization of the receptor (31). Unique to the AR among the sex steroid receptors are a series of amino acid repeat sequences in the NTD, most notably the polymorphic tracts of glutamine and glycine, of which certain lengths are thought to be linked to various human diseases or cancer (31).

The greatest structural disparities between ER-α and ER-β lie within the A/B domain, which is approximately 30 amino acids shorter in ER-β and exhibits only >20% homology (68). This divergence likely accounts for the many functional differences that have been revealed from comparative studies of the two ER forms (45,68–72). In general, ER-β tends to be a less effective transcriptional activator compared with ER-α when acting in a classic estrogen response element (ERE)-driven mechanism in vitro (44,45,68, 73,74). An ER-α/ER-β chimera in which the A/B domain of ER-β has been replaced with that of ER-α is able to activate transcription much better than the native ER-β (68). Furthermore, certain antagonists (e.g., tamoxifen) that exhibit some agonist-like properties when bound to ER-α exhibit no such activity with ER-β (68,71).

DNA-Binding or C Domain

The C domain of the NR family members is that portion of the receptor that specifically functions to recognize and bind to the *cis*-acting enhancer sequences, or HREs, that are located within the regulatory regions of target genes (Fig. 3). It is the most highly conserved (55%–80%) region among the NR family members (1,75). A comparison of C domains among the human sex steroid receptor forms indicates 79% homology between the AR and PR and 56% homology between the AR and ER-α (32). The C domains of ER-α and ER-β are practically identical (>95% homology) in most species and are therefore expected to exhibit a similar affinity for the same HREs (7,43,76). The functionality of the C domain is provided by a motif of two zinc fingers, each composed of four cysteine residues that chelate a single Zn^{2+} ion, and are always encoded by separate exons within the gene (Fig. 3). Crystallography studies indicate a

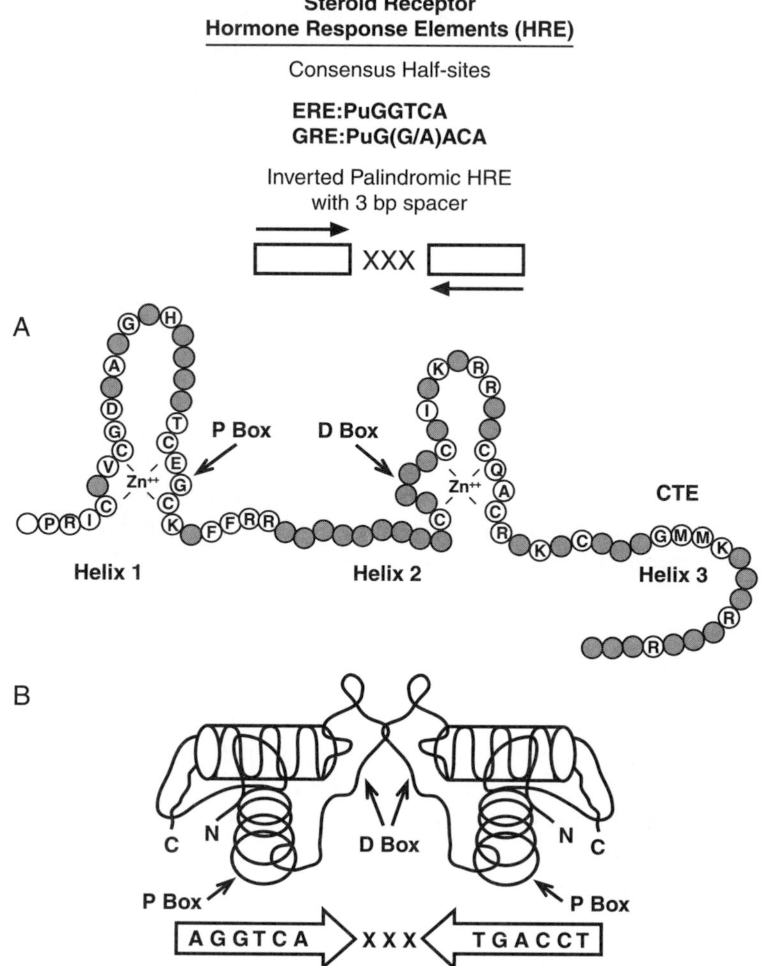

FIG. 3. The nuclear receptor response elements and primary structure of the DNA-binding domain. **Top:** Shown is the canonical core hormone response element (HRE) for the estrogen receptors (ERE) and androgen and progesterone receptors (GRE). These sequences are located in the regulatory regions of sex steroid target genes and provide a site for receptor binding and transactivational activity. A full HRE consists of two core palindromic sequences arranged as an inverted repeat, always separated by 3 bp. Note that a single change from T to A at position 4 of the core sequence modifies the ERE to a GRE. **Bottom: A:** The DNA-binding (C) domain of the sex steroid receptors is highly conserved and composed of two zinc fingers and a C′-terminal extension (CTE). Each zinc finger is composed of four conserved cysteine residues that coordinate to chelate a single zinc ion. Other conserved residues are shown and designated by the corresponding letter. Helix 1 of the first zinc finger contains the P-box residues involved in the discrimination of the HRE. Residues in the second zinc finger (helix 2) form the D-box that provides a dimerization interface. The CTE contains the T and A boxes critical for monomeric DNA binding. **B:** Shown is a ribbon diagram of the DNA-binding (C) domain illustrating the putative arrangement of helix 1 and helix 2 crossing at right angles to form the core of the DNA-binding domain that recognizes a hemi-site of the response element. [A reproduced with permission from (65). B reproduced with permission from Glass, C. K. (1994). Differential recognition of target genes by nuclear receptor monomers, dimers and heterodimers. *Endocr. Rev.* 15, 391–407.]

highly conserved structure consisting of dual α-helices positioned perpendicular to each other (77) (Fig. 3). Amino acids in the C-terminal base of the first zinc finger form the "P-box" (proximal box) of the DNA binding domain and confer sequence specificity to the receptor; hence, the proximal zinc finger is often referred as forming the "recognition helix" (Fig. 3) (14). Amino acids at the N-terminal base of the second zinc finger form the "D-box" (distal box) and are more specifically involved in differentiating the "spacer" sequence within the HRE as well as providing an interface for receptor dimerization (Fig. 3) (14).

The consensus HREs for the sex steroid receptors are composed of two 6-base pair (bp) palindromic sequences arranged as an inverted repeat and always separated by a 3-bp spacer (Fig. 3). Because the P-box residues are identical among the AR, PR, and GR, these receptors bind a common consensus HRE (or GRE) consisting of two 6-bp palindromic half-sites [5′-PuG(G/A)ACA] arranged as an inverted repeat and always separated by a 3-bp spacer (Fig. 3) (14). The consensus ERE bears the same arrangement but is composed of a unique palindromic half-site (5′-PuGGTCA) (Fig. 3) (14). The inverted-repeat arrangement of the palindromic sequences is unique to sex steroid receptor HREs and dictates that the receptors homodimerize in a "head-to-head" position when bound to DNA (Fig. 4) (65). Evidence indicates that the AR can uniquely dimerize in a "head-to-tail" fashion on an HRE of the monomer 5′-GGTTCT arranged as a direct repeat, providing a possible mechanism for androgen-specific regulation of target genes (77–80). Furthermore, the existence of multiple PR and ER forms also provides for heterodimerization in the form of PR-B/PR-A or ER-α/ER-β complexes (65), which exhibit distinct transcriptional activities relative to the respective homodimers (74,81–83) (Fig. 4). A consensus HRE within the promoter region of a known NR target gene is quite rare because most HREs described to date are imperfect. There is considerable evidence that multiple HREs within a target gene promoter can provide synergism and increased transcriptional activation by the respective steroid receptor; and the degree of activation may be further influenced by the distance of the HRE from the core promoter (14). In addition, ER-α has been shown to induce gene expression when bound as a monomer to an ERE half-site in the vicinity of a GC-rich region or Sp1 binding site (84–86).

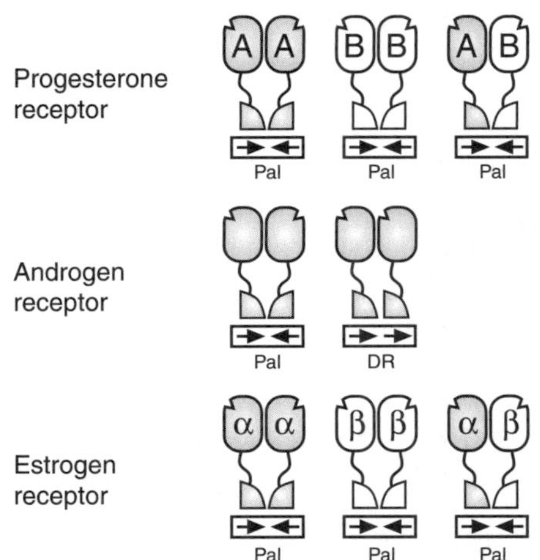

FIG. 4. Binding arrangements of the sex steroid receptors to HREs. The sex steroid receptors can bind as homodimers or heterodimers within each receptor family to palindromic (Pal) HREs spaced by three nucleotides. Dimerization is mediated by a strong dimerization interface present in the ligand-binding (E) domain and the receptors bind in a "head-to-tail" arrangement. Multiple studies indicate differential transactivational activity of homodimers vs. heterodimers. Recent studies indicate that AR may also bind to a direct repeat (DR) HRE in a "head-to-head" arrangment. Also, the sex steroid receptors may also bind HRE half-sites as monomers (not shown). [Reproduced with permission from (65).]

Hinge Region or D Domain

The D domain primarily serves to connect the more highly conserved C and E domains of the receptor (14). Commonly referred to as the "hinge" region, the D domain is postulated to dampen any steric hindrance that may occur between the C and E domains, thereby allowing the DNA-binding and ligand-binding regions to arrange in a multitude of combinations that may differentially affect receptor action (14). The D domain also harbors a nuclear localization signal that influences cellular compartmentalization of the receptor. Evidence also indicates that certain lysine residues with the hinge region of ER-α (87) and AR (31,88,89) are subject to acetylation by coactivator proteins (e.g., p300).

LBD or E Domain

The LBD or E domain of the steroid receptors is a highly structured multifunctional region that primarily serves to specifically bind ligand and provide for ligand-dependent transcriptional activity (14). An AF-2 domain located in the C-terminus of the E domain (14) mediates this latter function. The AF-2 domain is subject to posttranslational modifications (14) and is an especially strong activator of transcription in the ER and PR but is markedly weaker in the AR, where it is more involved in interactions with residues in the NTD (31). Also harbored within the E domain is a strong receptor dimerization interface, sites for interaction with heat shock proteins, and nuclear localization signals (14). Although there is minimal homology in the primary sequence of the

LBD for the sex steroid receptors, comparative studies of the crystal structures of liganded and unliganded LBDs indicate a highly conserved structural arrangement (14,90,91). These studies indicate that the LBD is composed of 12 α-helices (H1 through H12) arranged in a three-layer α-helical sandwich to create a hydrophobic ligand-binding pocket near the C-terminus of the receptor (14). The resulting shape and volume of the ligand binding pocket is larger than necessary to accommodate the corresponding ligand, suggesting that key interactions between the ligand and specific receptor residues are more critical to conferring ligand specificity (68,92). Receptor binding to an agonist ligand leads to rearrangement of the LBD such that H11 is repositioned and H12 swings back toward the core of domain to form a "lid" over the binding pocket. This agonist-induced repositioning of H12 leads to the formation of a hydrophobic cleft, or "NR box", by helices 3, 4, and 5 on the receptor surface, which serves to recruit coactivators (Table 2) to the receptor complex. In contrast, receptor antagonists are unable to induce a similar repositioning of H12, leading to a receptor formation that is incompatible with coactivator recruitment and is therefore less likely to activate transcription.

The LBDs of PR-B and PR-A are identical and provide for high affinity binding to progesterone (14). The LBD of the AR in humans, rats, and mice is identical and provides for high affinity binding of two endogenous androgens, testosterone and 5α-hydroxytestosterone (DHT), the latter of which binds with much greater affinity (93). Numerous high-affinity synthetic ligands for PR and AR have been developed, including the agonists R5020 and R1181, respectively (Table 3) (14). The LBDs of ER-α and ER-β exhibit less than 60% homology but bind the endogenous ligand, estradiol, with similar affinity (ER-α, 0.1 nM; ER-β, 0.4 nM) (14). Both ER forms also bind the synthetic estrogen, diethylstilbestrol (DES), with relatively equal affinity (Table 3) (94). However, given the divergence in homology, it is not surprising that ER-α and ER-β exhibit measurable differences in their affinity for other endogenous steroids and xenoestrogens (94–96). For example, ER-β tends to exhibit a stronger affinity for certain phytoestrogens (e.g., genestein and coumestrol) (94,95) as well as for the endogenous androgen metabolite, 5α-androstane-3β,17β-diol (97). Numerous synthetic steroidal and nonsteroidal ER antagonists have been developed over the years, but none proved to be specific for one or the other ER subtype (Table 3) (98). However, more recent advances have allowed for the generation of ER-selective nonsteroidal ligands (68,96,99) that exploit as well as illustrate differences between the LBDs of ER-α and ER-β and provide for powerful pharmacological tools to discern the overall function of each ER (Table 3).

TABLE 3. *Endogenous and exogenous ligands for the sex steroid receptors*

	Receptor Agonists		Receptor Antagonists
	Endogenous	Synthetic	
Estrogen	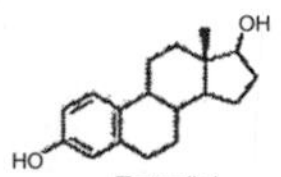 Estradiol	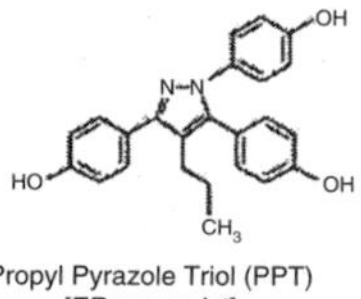 Propyl Pyrazole Triol (PPT) [ER-α agonist]; 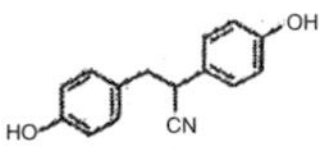 Diarylpropionitrile (DPN) [ER-β agonist]	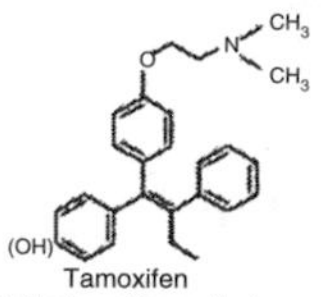 Tamoxifen (4-Hydroxy Tamoxifen); 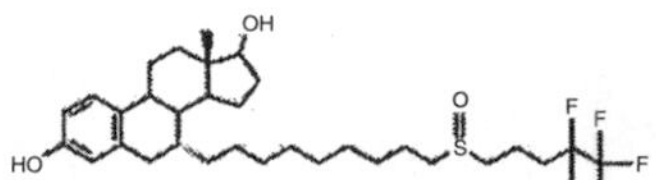 ICI 182.780
Progesterone	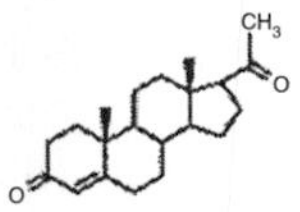 Progesterone	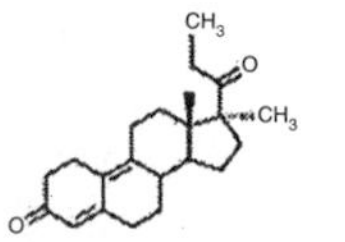 Promegestone (R5020)	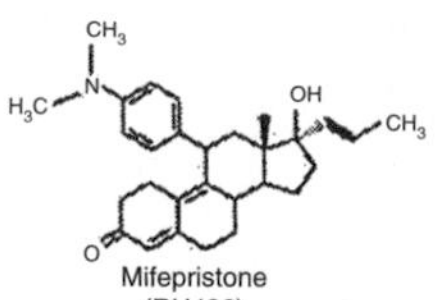 Mifepristone (RU486); 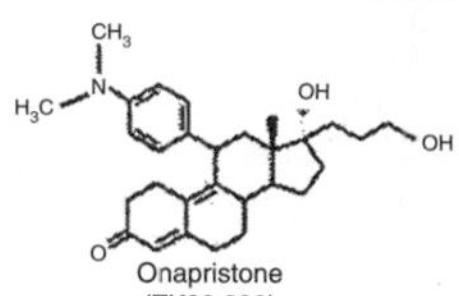 Onapristone (ZK98.299)
Androgen	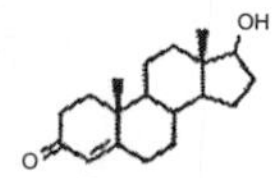 Testosterone; 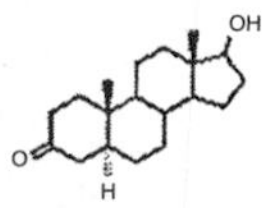 5α-Dihydrotestosterone	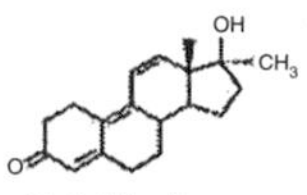 Methyltrienolone (R1881)	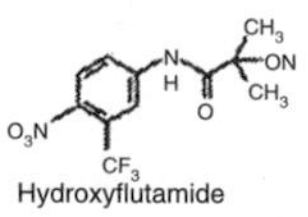 Hydroxyflutamide; 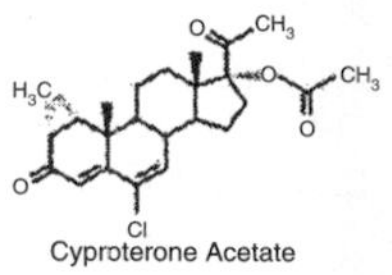 Cyproterone Acetate

F Domain

Among the sex steroid receptors, only ERs possess a well-defined F domain. This region is relatively unstructured and harbors little known function, although some data indicate a role in coactivator recruitment and receptor stability (68,100).

Receptor Mechanisms of Action

Our understanding of the mechanisms by which steroid hormones and their cognate intracellular receptors influence cell function and behavior has expanded profoundly since 1988 when Clark and Markaverich (101) reviewed the field for the inaugural edition of this volume. Much of that earlier chapter remains contemporary in reference to general receptor biochemistry and ligand-dependent activation, which is now referred to as the "classic" model of receptor function (Fig. 5). In the years since, numerous discoveries have been made that illuminate the complexity of sex steroid receptor signaling, such as the discovery of additional receptor forms and variants (e.g., ER-β) and high-resolution crystallography of receptor domains. In addition, several alternative receptor signaling mechanisms that diverge from the classic model have become apparent, including "cross-talk" with intracellular and second messenger systems that provide for sex steroid receptor activation in the absence of the cognate steroid ligand (Fig. 6); "tethering" of the sex steroid receptor to heterologous DNA-bound transcription factors to provide for steroid regulation of genes that lack HRE sequences within their promoter (Fig. 7); and plasma membrane steroid signaling, often referred to as "nongenomic" steroid actions. The existence of multiple receptor-mediated signaling pathways likely accounts for the refined control and plasticity of tissue responses to sex steroids. These modes of sex steroid receptor action as currently understood are discussed below.

Classic (Ligand/HRE Dependent)

The classic model of steroid receptor action states that the receptor resides in the nucleus or cytoplasm but is sequestered in a multiprotein inhibitory complex in the absence of hormone (Fig. 5). The lipophilic steroid ligands are able to freely diffuse across the plasma and nuclear membranes and bind their cognate receptor. The binding of ligand results in a conformational change in the receptor, transforming it to an "activated" state that is now available for homodimerization, increased phosphorylation, and binding to an HRE within target gene promoters.

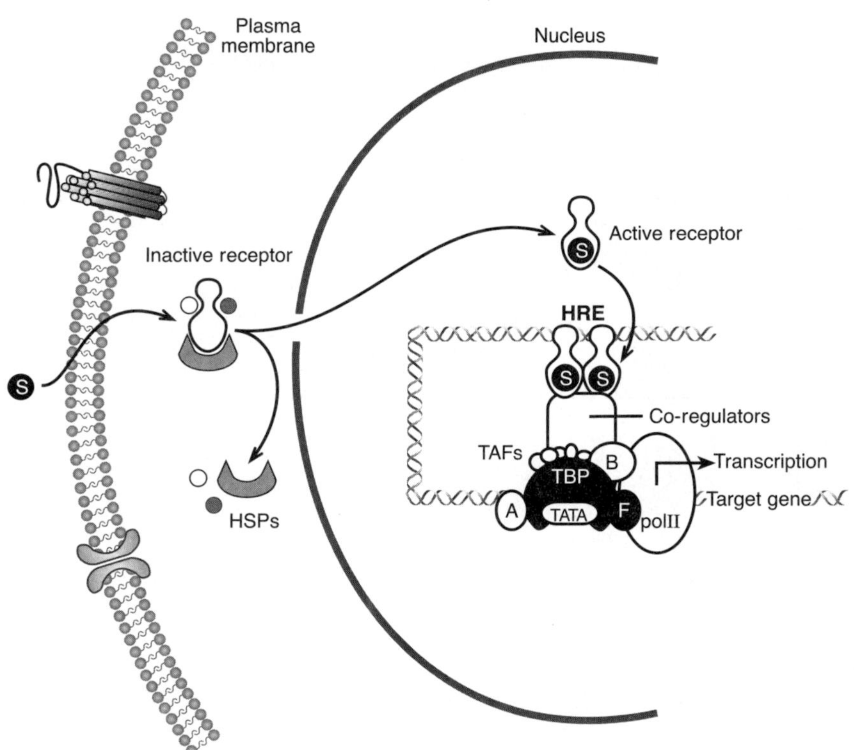

FIG. 5. The "classic" model of sex steroid receptor action. See text for description. S, steroid; HRE, hormone response element; HSPs, heat shock proteins; TBP, TATA binding protein; TAFs, TBP-associated factors; poll II, RNA polymerase II. [Elements of this drawing first appeared in (68).]

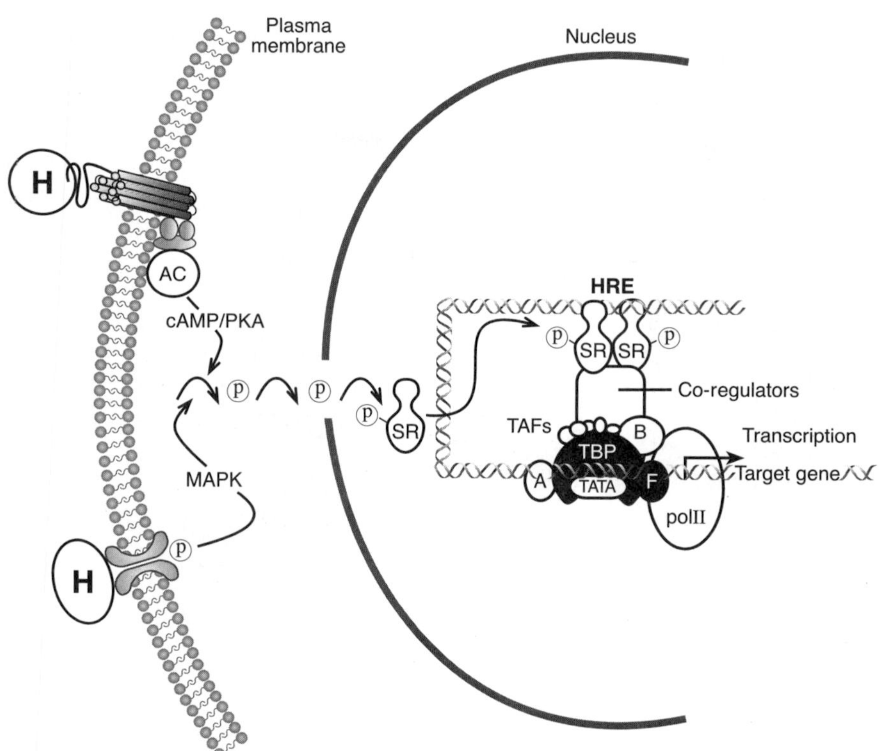

FIG. 6. Ligand-independent or "cross-talk" mechanism of sex steroid receptor action. Hormone stimulation of membrane receptor activated kinase signaling cascades may activate the sex steroid receptors via phosphorylation in the absence of steroid ligand. See text for more detailed description. H, trophic hormone; AC, adenylyl cyclase; cAMP, cyclic adenosine monophosphase; PKA, protein kinase A; P, phosphorylation; HRE, hormone response element; SR, steroid receptor; CoA, coactivator; TBP, TATA binding protein; TAFs, TBP-associated factors; poll II, RNA polymerase II. [Elements of this drawing first appeared in (68).]

The ligand–HRE-bound receptor complex interacts with the general transcription apparatus either directly or indirectly via cofactor proteins (Table 2) to promote transcription of the target gene (Fig. 5). This classic steroid receptor mechanism is dependent on the functions of both AF-1 and AF-2 domains of the receptor, which synergize via the recruitment of coactivator proteins, most notably the p160 family members. It is generally believed that the DNA-bound receptor–coactivator complex facilitates disruption of the chromatin and formation of a stable transcription preinitiation complex. Depending on the cell and promoter context, the DNA-bound receptor complex may positively or negatively affect expression of the downstream target gene; the latter effect has been shown for AR only.

Several sex steroid target genes are regulated via the classic ligand/HRE-dependent mechanism, for example, progesterone/PR regulation of the tyrosine amino transferase and chicken lysozyme genes (102,103); androgen/AR regulation of prostatic specific antigen, probasin, and keratinocyte growth factor (31); and estradiol/ER regulation of prolactin, vitellogenin, and lactoferrin (84,104). These early findings have been corroborated and the overall extent of ligand-dependent sex steroid receptor transactivation realized by the recent use of differential display and microarray techniques to generate gene expression profiles in various cell types and tissues after hormonal stimulation. Several recent studies have indeed demonstrated differential regulation of progesterone target genes by PR-A and PR-B (105–110). Similarly, differential display, microarray, and proteomics have been used to identify a large number of androgen/AR-mediated (111–115) and estradiol/ER-mediated (116–124) genes.

In general, PR-B can better or uniquely activate transcription versus PR-A when acting on progesterone response element–regulated genes in various cell types in vitro (17,67,125–127). Furthermore, PR-A can repress the transactivational activities of PR-B in certain cell and promoter contexts (127), suggesting that PR-A may modulate progesterone responsiveness in certain tissues. Interestingly, PR-A can also repress the actions of other heterologous NRs, including the AR, ER, and GR (14,67). The more recently discovered variant, PR-C, which lacks the A/B domain and the 5′ portion of the C domain and is primarily cytosolic,

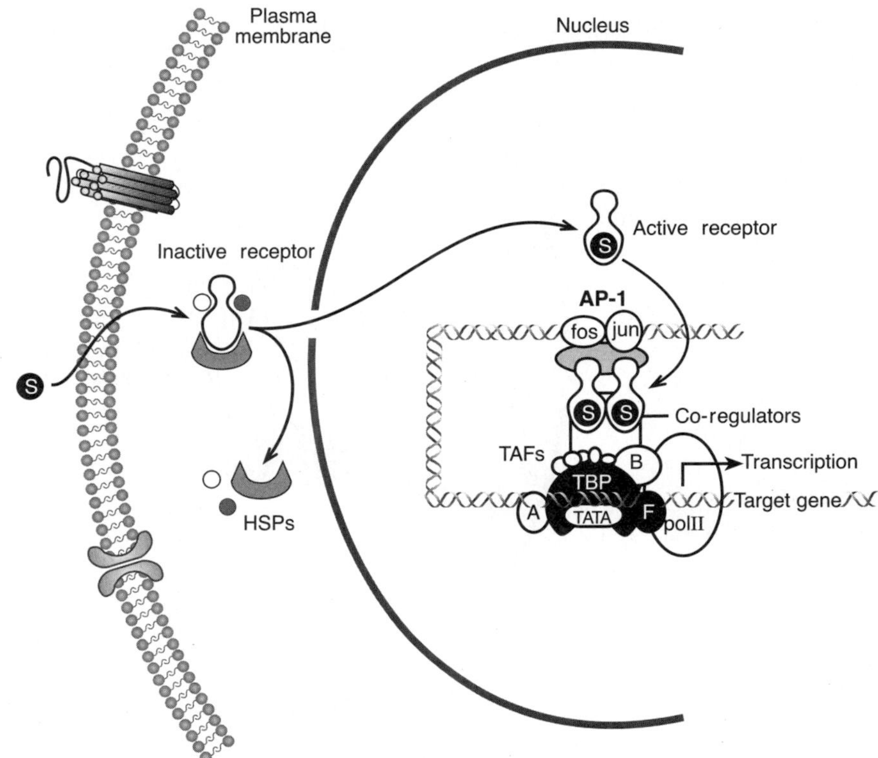

FIG. 7. Sex steroid receptor regulation of genes that lack HREs via "tethering." Shown is the putative mechanism by which ligand-activated sex steroid receptors may modulate the expression of genes that do not possess an HRE within their promoter regions. The ligand-activated receptors are thought to interact with other transcription factors, e.g., FOS/JUN, that is bound to its respective DNA response element (AP-1). This complex is then able to recruit coregulator proteins and induce gene expression in response to the respective sex steroid. See text for more detailed description. S, steroid; HRE, hormone response element; HSPs, heat shock proteins; TBP, TATA binding protein; TAFs, TBP-associated factors; poll II, RNA polymerase II. [Elements of this drawing first appeared in (68).]

has been shown to bind progesterone and enhance the actions of PR-B and PR-A (16).

When acting via a classic ERE-driven mechanism in vitro, ER-α homodimers and ER-α/ER-β heterodimers tend to be stronger activators of transcription compared with ER-β homodimers (44,45,68,73,74). Corroborating in vivo evidence of differential regulation and heterodimer formation by the ER subtypes has been difficult to generate. However, a microarray study by Lindberg et al. (118) found that ER-β generally inhibits ER-α–mediated gene expression, but in the absence of ER-α, ER-β can partially provide some estradiol-stimulated gene expression.

Cross-Talk (Ligand Independent)

We now have ample evidence that the sex steroid receptors can be activated via intracellular second messenger and signaling pathways, allowing for the induction of sex steroid target genes in the absence of steroid ligand (Fig. 6) (15,128,129). Polypeptide growth factors are able to activate ER-α–mediated gene expression via mitogen-activated protein kinase activation of ER-α in the absence of estradiol (Fig. 6) (128). Similarly, interleukin-6 stimulation of cells leads to increased AR-mediated gene expression (130–132), and cyclin-dependent kinases have been shown to activate PR-regulated gene expression (129,133). The intracellular signaling molecule cAMP, a common second messenger of G-protein–coupled receptors and an activator of the PKA pathway, can stimulate increased PR-, AR-, and ER-mediated transcription of target genes in the absence of steroid ligand (128,134–136). A well-characterized example of ligand-independent cAMP activation of PR occurs after dopamine stimulation of dopaminergic neurons (137,138) and is the mechanism by which sexual behavior is induced in female rodents (139–141). Likewise, growth factors are able to mimic the effects of estradiol in the rodent uterus via estradiol-independent activation of ER-α (142–144).

Ligand-independent activation of sex steroid receptors is believed to rely largely on cellular kinase pathways that alter the phosphorylation state of the receptor and/or its associated proteins (e.g., coactivators, heat shock proteins) (Fig. 6). As in the classic ligand-dependent mechanism described above, specific receptor domains are critical to ligand-independent activation as well. ER activation by peptide growth factor signaling pathways appears to be more dependent on AF-1 functions, whereas the effects of increased intracellular cAMP are postulated to depend on the AF-2 domain and do not require a functional AF-1 (128).

Tethered (HRE Independent)

Ligand-activated steroid receptors can stimulate the expression of genes that lack a conspicuous HRE within their promoter. This has been especially demonstrated for ER-α (86,145) but is also known to occur for PR (146). This mechanism of HRE-independent steroid receptor activation is postulated to involve a "tethering" of the ligand-activated receptor to transcription factors that are directly bound to DNA via their respective response elements (Fig. 7) (86,145). Therefore, sex steroid receptors acting in this fashion may be better defined as coregulators rather than a direct acting transcription factor. Estradiol/ER-α regulation of several genes, including ovalbumin, collagenase, insulin-like growth factor-1 (IGF-1), and cyclin D, is believed to occur via a tethering of the receptor to a DNA-bound AP-1 (Fos/Jun) complex with the gene promoter (145). The intricacies of this mechanism remain unclear but are postulated to involve a mediator component (e.g., p160) between ER and AP-1 versus direct interaction (145). Interestingly, ER-β is unable to enhance the actions of a DNA-bound AP-1 complex when bound to estradiol but can do so in the presence of certain selective ER modulators (SERMs) such as tamoxifen, raloxifene, and ICI 164,384 (147). A similar HRE-independent mechanism of sex steroid receptor regulation has been documented for genes that possess a GC-rich region or Sp1 binding site within the promoter, upon which the actions of a bound Sp1 complex can be enhanced by ER-α (86,146).

Plasma Membrane Associated Steroid Signaling

All models of sex steroid receptor action described thus far influence cellular phenotypes by acting in the nucleus and modulating target gene expression. The required gene transcription and mRNA translation is a relatively long process and is not likely to provide measurable effects within the cell or tissue until hours or even days after steroid stimulation. Therefore, the numerous examples described to date in which steroids affect cellular dynamics within seconds to minutes of exposure have remained difficult to explain within the context of the above models. Rapid cellular changes such as increased intracellular calcium or cAMP or activation of kinase signaling cascades have been attributed to sex steroid exposure of varied cell types and tissues (148–150). Because these steroid effects do not involve direct steroid receptor activation of gene transcription, they are often collectively referred to as representing "nongenomic" pathways of steroid action (148–150). Indeed, several of the documented effects lack a conspicuous nuclear component or have even been demonstrated to occur in enucleated cells, yet others do ultimately affect gene expression (148). Therefore, rather than nongenomic, some have proposed these types of steroid actions to be loosely categorized as plasma membrane associated steroid signaling events (151,152). Although there are numerous examples of putative membrane associated steroid signaling effects in reproductive tissues and processes, a thorough description of these mechanisms is beyond the scope of this chapter. Lösel et al. (148) thoroughly reviewed the current literature on the nongenomic actions of all the endogenous steroids. A series of reviews by Segars and Driggers (151,152) specifically concentrated on the membrane-associated effects of estradiol, and Heinlein and Chang (150) reviewed the nongenomic actions of androgens (Table 1).

An obstacle to our better understanding of membrane-associated effects of steroids is the lack of data on the nature of the cell-surface steroid "receptors." Questions remain concerning whether the membrane-associated receptors are identical or variant forms of the sex steroid NR or instead distinct receptors altogether (148). The nongenomic effects of progesterone are perhaps best characterized relative to the other sex steroids (148,153). Progesterone is known to rapidly induce maturation in *Xenopus* oocytes that are arrested in the G_2 phase of meiosis I (148), and this process does not require protein synthesis (154). Most evidence indicates this process begins at plasma membrane progesterone binding sites that are unique from the classic nuclear forms of PR (148), although cDNA homologs of the PR were cloned from *Xenopus* and found to be associated with progesterone-induced oocyte maturation (155,156). In other studies, progesterone is able to inhibit apoptosis of rat granulosa cells in vitro via a membrane-bound receptor that is immunoreactive to antisera directed against the LBD of the classic PR (157–159).

Rapid effects of estradiol have been described in a vast array of tissues, including a rapid activation of endothelial nitric oxide synthase in endothelial cells, potentiation of kainite-induced currents in hippocampal neurons, and influx of calcium in uterine endometrial cells (148). The evidence of an association between ER-α and plasma membrane bound components is becoming increasingly convincing and may indeed account for certain rapid effects of estradiol (151,152). In this sense, the various nongenomic actions of estradiol are often categorized according to their susceptibility to inhibition by classical ER antagonists (e.g., the ICI compounds) (151). Estradiol-induced activation of membrane ion channels, endothelial nitric oxide synthase, and mitogen-activated protein kinase kinase are all inhibited by ER antagonists and therefore likely involve a membrane-associated ER form that is similar to nuclear ER (151). In contrast, examples of estradiol-induced PKA and protein kinase C (PKC) activation are not inhibited by ICI compounds and are therefore not likely to involve the nuclear ER form (151).

An especially intriguing discovery made in the past few years is the existence of a multiprotein complex of extracellular signal related kinases, the inner plasma membrane component of tyrosine kinase receptors (c-Src), and the nuclear sex steroid receptors (151,160–163). There is also speculation that androgens or estradiol may act as extracellular signaling hormones via binding to sex hormone binding globulin that is already bound to the cell membrane via a G-protein–coupled-like sex hormone binding globulin cell-surface receptor (150).

Receptor Antagonists

Recent structural analyses of the sex steroid receptors provide insight into the agonist/antagonist actions of endogenous or synthetic ligands (Table 3). Certain ligands possess "mixed" agonist/antagonist activity that is dependent on the receptor, cell, and promoter context, leading to the more descriptive terms of SERMs, selective PR modulator, and selective AR modulator. As described in the section on Steroid Receptors: Receptor Mechanisms of Action, sex steroid receptor agonists interact with the LBD to generate an ordered array of 12 α-helices that is most conducive to receptor interaction with coactivator proteins and/or promotes disassociation of corepressor proteins. In turn, antagonist ligands generally fail to generate the necessary changes in receptor structure that induce full transcriptional activity (164,165).

Antiprogestins have a number of applications in reproductive medicine, including use as contraceptives due to their putative ability to induce abortion, prevent implantation, or possibly block ovulation (166). The currently available PR antagonists are all competitive inhibitors of progesterone binding to the PR LBD (166). Type I antiprogestins, such as ZK 98299 (onapristone, Table 3), bind PR but fail to induce receptor phosphorylation and induce weak receptor binding to HREs (166,167). In contrast, type II antiprogestins, such as RU 486 (mifepristone, Table 3), bind PR and promote receptor phosphorylation, homodimerization, and binding of the receptor complex to an HRE but are unable to activate transcription (166,167). Crystallography data indicate that RU 486 binding to PR induces an arrangement of helix 12 that is not conducive to binding of steroid receptor coactivators and may even promote recruitment of corepressors such as NCoR (167). Type II antiprogestins do possess some PR agonist activity when acting in the context of the cAMP/PKA signaling pathway in certain cell types (168–170).

The field of antiestrogens and selective ER modulators has been an area of intense research in the past decade because the potential clinical applications include treatments for fertility, cardiovascular disease, osteoporosis, cognitive function, and breast and gynecological cancers (98). The most well-characterized antiestrogens are competitive inhibitors of estradiol binding to ER and have been classified into two major groupings. The type I class of ER antagonists includes the triphenylethlylene compounds, hydroxy-tamoxifen (Table 3), and raloxifene and are characterized by mixed agonist/antagonist activity depending on the receptor, cell, and promoter context (98). Upon binding the ER, these compounds selectively inhibit AF-2 function but leave AF-1–mediated receptor functions intact, thus explaining the selective agonist activity (98,164,165,171,172). In contrast, the type II compounds are considered pure antagonists, the most well-characterized being the ICI compounds, ICI 164,384 and ICI 182,780 (Table 3), both of which are 7α-substituted derivatives of estradiol (98). These compounds bind the ER but possess extended side chains that prevent coactivator association with the NR box of helix 12 and may even promote interaction with corepressors as well as increased receptor degradation (98,173). More recent advances have allowed the generation of ER-selective nonsteroidal ligands (68,96,99) that exploit differences between the LBDs of ER-α and ER-β and provide powerful pharmacological tools to discern the overall function of each ER (Table 3).

Clinical use of antiandrogens has focused largely on the treatment of prostate cancer (166). The most well-characterized AR antagonists include the nonsteroidal compounds hydroxyflutamide (174) (Table 3), casodex (175), and nilutamide (176) and the steroidal compound cyproterone acetate (177) (Table 3). Hydroxyflutamide is considered a pure AR antagonist that competitively binds AR and reduces AR transcriptional activity by preventing interaction with coactivators (174,178). In classic bioassays, cyproterone acetate is less potent than flutamide and may even possess some agonist activity in certain tissues (177).

STEROID SIGNALING NULL MICE

A significant advance toward our understanding of steroid receptor functions in the female reproductive tract is the development of gene-targeted or "knockout" murine models. Descriptions of the uterine and ovarian phenotypes that follow the loss of each of the sex steroid receptor forms have provided tremendous insight into the role of steroid signaling and the contribution of each receptor to female reproduction. Furthermore, these models now serve as platforms for in vivo experimentation that were not previously feasible. We have largely used the numerous studies on steroid receptor null mice as a foundation of this chapter and to highlight the recent advances made in the field of steroid actions in uterine and ovarian function.

Mice Null for PR Signaling

There are currently three reported lines of PR-null mice. Mice lacking both nuclear PR forms (PRKO) were first described by Lydon et al. in 1995 and were generated via targeted disruption of exon 1 of the murine *Pgr* gene (179). Mice lacking only PR-A (PRAKO) (180) or PR-B (PRBKO) (181) were generated via a *Cre*/loxP-based gene targeting strategy that mutated the respective start codons for each isoform while preserving the translational reading frame of the other. Comparative studies in all three models have greatly furthered our understanding of the divergent functions for the two PR isoforms.

Mice Null for AR Signaling

There are three reported lines of AR-null mice, the testicular feminized male (*Tfm*) and two lines recently generated via homologous recombination in embryonic stem cells. *Tfm* mice were first described in 1970 and are a naturally existing androgen-resistant mutant (182) due to an inactivating mutation of the *Ar* gene (183,184). Comparable inactivation mutations of the *AR* gene and resulting phenotypes are well described in rats and humans (185). Because the *Ar* gene is located on the X chromosome and *Tfm* males are infertile, it is impossible to breed for XX female mice that are homozygous for the mutation. Lyon and Glenister (186) overcame this challenge more than 25 years ago by using embryo aggregation to generate a limited number of *Tfm* chimeric males that were fertile and carried germ cells harboring the *Ar* mutation. More recently, AR-null female mice were generated via a *Cre*/loxP targeting scheme that allows for tissue and temporal specific disruption of the *Ar* gene and therefore generation of fertile male carriers of the targeted *Ar* allele (187–189). Exon 2 of the murine *Ar* gene was targeted for deletion in both lines (187–189) and fertility and ovarian phenotypes for one have been described in the literature (189,190). The AR-null animals are the most recently generated among the sex steroid receptor knockouts, and only limited descriptions exist in the literature. Studies to date indicate that AR-null females exhibit reproductive phenotypes that are similar to those originally described in *Tfm*/*Tfm* females.

Mice Null for ER Signaling

Mice Lacking ER-α

The ER-α–null mice are the first steroid receptor null animal generated and preceded the discovery of ER-β. Many hypothesized roles of estrogen in the female reproductive tract have since been confirmed, rejected, or reevaluated since the generation of ER-α–null mice. The two reported lines of ER-α–null mice are the αERKO, first described by Lubahn et al. in 1993 (191), and the ERαKO, described by Dupont et al. in 2000 (192). Homologous recombination was used in both to either disrupt (αERKO) or completely excise (ERαKO) exon 3 (192) of the murine *Esr1* (ER-α) gene. As described to date, the reproductive, endocrine, and ovarian phenotypes in both lines are indistinguishable.

Mice Lacking ER-β

The first description of ER-β–null mice followed the discovery of ER-β by only 2 years. Because so little is known about the role of ER-β before the generation of the null model, it is difficult to predict possible phenotypes. Therefore, the ER-β–null mice have served a tremendous role in providing insight into the importance of the newly discovered ER form to female fertility, and studies to date indicate ER-β plays a particularly important role in ovarian function. The two reported lines of ER-β-null mice are the βERKO, first described by Krege et al. in 1998 (193), and the ERβKO, described by Dupont et al. in 2000 (192). Homologous recombination was used in both lines to disrupt exon 3 (192) of the murine *Esr2* (ER-β) gene. As described to date, the reproductive, endocrine, and ovarian phenotypes in both lines are indistinguishable.

Mice Lacking ER-α and ER-β

Much insight into the physiological roles of estradiol has been gained from the study of mice lacking one or the other known ER forms. Still, conclusions drawn from these models are confounded by possible compensatory mechanisms provided by the opposite ER in each respective ERKO model. Furthermore, mice lacking only one ER form do not provide for study of possible ER-α/ER-β heterodimer mediated actions (74,83). Therefore, compound ER-null mice (i.e., αβERKO) represent an opportunity to elucidate potential compensatory and cooperative actions of ER-α and ER-β. The two reported lines of compound ER-null mice are the αβERKO, described by Couse et al. in 1999 (194), and the ERαβKO, described by Dupont et al. in 2000 (192). Both were generated by cross-breeding the respective individual ER-null mice and, as described to date, exhibit comparable reproductive, endocrine, and ovarian phenotypes.

Mice Lacking Estradiol Synthesis

Estrogen signaling fundamentally requires ligand (i.e., estradiol) and the cognate nuclear ERs. Much has been gained from the study of mice lacking one or both ERs. Complementary to these models are the Cyp19-null mice that lack the capacity to synthesize estradiol. The Cyp19-null mice may prove especially valuable to the study of ligand-independent ER actions (Fig. 6) as well as the development and study of ER-selective agonists or antagonists. The two reported lines of Cyp19-null mice are those described by Fisher et al. in 1998 (195) and those described by Toda et al. in 2001 (196). In both, a similar region of exon 9 of the mouse *Cyp19* gene was targeted and removed. Most data indicate that females from the two independent Cyp19-null lines exhibit comparable reproductive, endocrine, and ovarian phenotypes. Certain caveats when making phenotypic comparisons among Cyp19-null and ER-null mice warrant consideration. For example, maternally derived estrogens may provide a more normal gestational environment for development in Cyp19-null mice versus ER-null models. Similarly, dietary estrogens during adulthood are able to abate the ovarian and uterine phenotypes in Cyp19-null mice (197).

SEX STEROID RECEPTORS IN UTERINE FUNCTION

The ovarian-derived sex steroid hormones dictate the uterine estrous or menstrual cycle in mammals and are therefore essential to the establishment and maintenance of pregnancy. The uterine response to the preovulatory rise in circulating estradiol is required to prepare the tissue for the forthcoming rise in progesterone that accompanies ovulation and is critical to embryo implantation. This physiological coordination between the ovary and uterus is common to the large majority of mammals studied. It is generally believed that the spectrum of actions and effects of the sex steroids in uterine tissue are mediated by their cognate NRs. Although the dramatic physiological effects of estrogens on the mammalian uterus have been appreciated and well described for more than five decades, only recently have investigations begun to elucidate the precise mechanisms and signaling pathways involved. Our understanding of steroid hormone action in the uterus has been greatly advanced by the generation of receptor null models in mice and the advent of gene array methods that allow for global mapping of the genomic response to steroids in uterine tissues. An especially intriguing outcome from these studies is the remarkable overlap in gene expression patterns that occurs in the uterus when exposed to progestins, androgens, or estrogens despite the divergent uterine phenotypes elicited by each steroid hormone. The sections below discuss what is currently known and hypothesized about sex steroid receptor actions in the mammalian uterus.

ESTROGEN RECEPTOR SIGNALING IN UTERINE FUNCTION

ER Expression in the Uterus

Rodents

ER-α is present in the female reproductive tract of rodents throughout late fetal and neonatal development, puberty, and adulthood (198,199). In mice, ER-α immunoreactivity first appears in the mesenchymal cells of the fetal uterus as early as gestational day 15 (198,199), whereas epithelial expression is delayed until the late fetal period but increases substantially on postnatal day 4 and peaks on day 16 (200,201). The level of ER-α transcripts in the murine uterus during neonatal development closely mirrors the patterns indicated by earlier immunohistochemical studies (202). There is reportedly no ER-β immunoreactivity in the neonatal mouse uterus, although transcripts are detected at a modest level (202).

In adult mice, ER-α transcripts and immunoreactivity are quite high in uterine epithelial cells before ovulation (203) but decrease during pregnancy, whereas expression in the stroma continues (204).

In contrast, ER-β expression is relatively low in uteri of virgin and pregnant mice (202,205,206). Adult rats exhibit a similar uterine ER expression pattern such that ER-α transcripts and immunoreactivity are easily detectable in the epithelium, stoma, and myometrium and levels peak during the proestrous period of increased circulating estradiol levels (207). The evidence of ER-β expression in the rat uterus is conflicting; whereas some report expression in approximately 40% of epithelial and stromal cells regardless of the estrous cycle stage (207), others report a total lack of detectable ER-β (208). In ovariectomized mice, ER-α and ER-β protein was detected in the epithelial, stromal, and myometrial cells (196,209), indicating that removal of ovarian hormones alters the ER expression pattern. There is overwhelming functional evidence that ER-α is responsible for mediating the many effects of estrogens in the uterus. As described below, only ER-α–null female mice exhibit a severely attenuated response to acute and chronic estrogen (e.g., estradiol or DES) treatment in terms of uterine growth and altered gene expression, as well as exhibit impaired embryo implantation (104,210,211). Furthermore, categorical studies have shown that the toxic effects of neonatal estrogen exposure in the neonatal female mouse uterus are clearly mediated by ER-α (212,213).

Domestic Animals

In the neonatal ovine uterus, ER-α mRNA and protein levels are highest in the developing glandular structures; detectable but lower in the luminal epithelia, stroma, and vascular endothelial cells; and absent in the myometrium (214,215). In adult ewe, ER-α expression is detected in all uterine cell types and is highest on day 1 of the 16- to 18-day cycle, concurrent with peak plasma estradiol levels (214,216). During days 1–6 when progestin levels rise and estradiol declines, ER-α expression decreases accordingly and does not begin to rise again until the end of the cycle nears on days 11–15 (214,216). A similar pattern of uterine ER-α expression is reported during the 21-day bovine cycle (217). In situ hybridization and immunohistochemical analyses indicate maximum ER-α levels on days 1–3, especially in the glandular and stromal cells; these levels generally decrease by day 6, although cells comprising the deep glands maintain ER-α expression (218). The observed cyclical changes in ER-α expression in ovine and bovine uteri are consistent with steroid regulation of receptor levels. Indeed, estradiol treatment of ovariectomized cows leads to increased ER-α expression in the glands, stroma, and luminal epithelia, whereas progesterone alone or in combination with estradiol elicits a decrease in ER-α levels (218).

Primates

Early investigations in human uterine tissue found that levels of ER immunoreactivity or high-affinity estradiol binding vary during the menstrual cycle such that peak expression occurs in the mid to late proliferative phase and then decreases at ovulation and commencement of the secretory phase (219–221). This peak in uterine ER levels at the time of rising plasma estradiol levels is similar to several other species and suggests that expression is autoregulated. However, maintenance of uterine ER expression in postmenopausal women suggests that nonsteroidal regulatory factors are also involved (221). The ovulatory decline in ER-α levels among all uterine cell types is concurrent with the rise in circulating progesterone levels and suggests that PR-mediated progesterone actions may act to downregulate ER expression. A distinct gradient of ER levels among the functional components of the human uterus was also reported such that levels are highest in the fundus and progressively decrease in those tissues closer to the cervix (222,223).

The discovery of ER-β forced a reevaluation of uterine ER expression because the early studies discussed above did not use methods that adequately distinguished between the two ER forms. More recent studies definitively show that ER-α is the predominant form in the human uterus and accounts for the increased receptor levels that occur during the proliferative phase (224). ER-α transcripts and protein are especially high in glandular epithelial and stromal cells during the proliferative phase and exhibit a dramatic decrease among uterine tissues upon entering the secretory phase of the cycle (224). The levels of ER-α transcripts and immunoreactivity in the uterine myometrium are much lower and exhibit little change during the menstrual cycle (224). Still, comparatively higher ER-α levels are described in the subendometrial myometrium during the proliferative phase and are postulated to be involved in uterine peristaltic activity (224).

ER-β expression in the human uterus is much lower relative to ER-α but is generally present in the same cell types and exhibits comparative changes during the menstrual cycle (224,225). There is one report of increased ER-β immunoreactivity in the uterine stroma during the secretory phase (225). ER-β immunoreactivity is especially detectable in vascular smooth muscle cells and increases therein during the secretory phase, suggesting a role for this receptor form in vascularization (225). The human specific ER-β isoform, ER-β_{cx} (Fig. 2), is also detected in the uterus and is present along with full-length ER-β (ER-β1) in the functional and basal layer endometrial glands (226). Interestingly, ER-β_{cx} immunoreactivity decreases in the basal layer glandular cells during the mid-secretory phase, whereas little change occurs in the level of full-length ER-β (226), suggesting that differential expression of ER-β isoforms over the course of the menstrual cycle may modify cellular responsivity to estradiol.

Nonhuman primates exhibit uterine ER expression patterns similar to those described in humans. ER-α immunoreactivity in baboon uteri is highest in endometrial glandular, stroma, and myometrial smooth muscle cells during the proliferative phase and decreases in all cell types upon commencement of the secretory phase (227). A similar ER-α expression pattern in epithelial, stromal, and myometrial cells is reported in the cynomolgus monkey uterus (228). ER-α immunoreactivity is also reported in the vasculature, including the spinal arteries, of the baboon uterus (227). Ovariectomy followed by exogenous estradiol or estradiol plus progestin treatments in baboons leads to changes in uterine ER expression that mirror the menstrual cycle, supporting the direct role of ovarian steroids in the regulation of uterine ER levels (229). Considerable levels of ER-β transcripts are also detected in the epithelial, stromal, and myometrial cells of cynomolgus monkey uteri (230). In contrast, ER-β immunoreactivity is reportedly low in normal uterine tissue from baboons, although higher levels are observed in diseased glands associated with endometriosis (231).

Uterine Phenotypes in Murine Models of Disrupted Estrogen Signaling

The murine models of disrupted estrogen signaling have proven invaluable to experimental investigation of estrogen actions in the uterus and the contribution of each ER form to these functions. In addition to the ER-null models are two independently derived lines of mice that lack the capacity to synthesize estradiol due to disruption of the *Cyp19* gene (195,196). Females within each respective model exhibit a similar phenotypic syndrome. Female mice lacking ER-α or Cyp19 are infertile due to dysfunction of numerous physiological systems, including the ovary (see section on Estrogen Receptor Signaling in Ovarian Function) and uterus, whereas ER-β–null females exhibit reduced fecundity that is largely attributable to ovarian dysfunction (see section on Estrogen Receptor Signaling in Ovarian Function). A level of caution is warranted when making phenotypic comparisons between the ER-null and Cyp19-null models because sensitivity to maternally derived estrogens may provide a more normal developmental environment during gestation in Cyp19-null mice and sensitivity to dietary estrogens during adulthood is able to abate several phenotypes in Cyp19-null mice (197). The reported uterine phenotypes are summarized in Table 4.

TABLE 4. *Uterine phenotypes in mice null for sex steroid signaling*

ER-α–null females
Adult uteri exhibit normal development and architecture but are immature and hypoplastic.
Adult uteri are insensitive to the proliferative and differentiating effects of endogenous and exogenous estrogens.
Adult uteri exhibit a decidualization response when induced by exogenous steroid treatments but do not provide for embryo implantation.
ER-β–null females
Exhibit grossly normal uterine development and function.
Cyp19-null females
Adult uteri exhibit normal development and architecture but are immature and hypoplastic.
Adult uteri are responsive to the proliferative effects of exogenous estrogen treatment.
PR-null females
Exhibit grossly normal uterine development.
Adult uteri lack a normal decidual response and do not provide for embryo implantation.
Adult uteri exhibit epithelial hyperplasia in response to estrogens.
PR-A–null females
Exhibit grossly normal uterine development.
Adult uteri lack a normal decidual response and do not provide for embryo implantation.
Adult uteri exhibit an abnormal epithelial proliferation in response to exogenous progesterone.
PR-B–null female
Exhibit grossly normal uterine development.
Adult uteri exhibit a normal decidual response and provide for embryo implantation.
Adult uteri do not exhibit epithelial hyperplasia in response to estrogens.
AR-null females
Adult uteri exhibit grossly normal uterine development, function and maintenance of pregnancy.
Adult uteri exhibit slightly reduced size at estrous or following gonadotropin-treatments, suggesting reduced ovarian synthesis of estradiol.

See text for references.

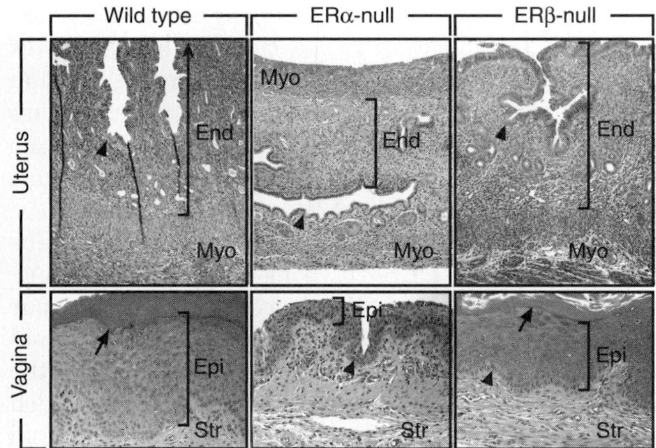

FIG. 8. Histology of uterine and vaginal tissue of wild-type, ER-α–null, and ER-β–null females. **Top:** Shown are cross-sections of uterine tissue from representative adult wild-type, ER-α–null, and ER-β–null adult females illustrating the presence of a normal uterine architecture in all (magnification × 33). The representative wild-type uterine section illustrates a normal myometrium (Myo), endometrial stroma (End), and luminal epithelium (*arrowhead*). The representative ER-α–null uterine section illustrates the characteristic hypoplasia of each anatomical compartment, a slightly disorganized endometrial stroma, and lack of estrogenization of the luminal (*arrowhead*) and glandular epithelium. Note the dramatically smaller diameter of the ER-α–null uterus is indicated by the ability to fit the whole transverse section of the tissue in the field of view vs. just one-half of the uterine cross-section for wild-type and ER-β–null tissues. The representative ER-β–null uterine section is indistinguishable from that of the wild-type, including the presence of estrogen-stimulated luminal epithelium (*arrowhead*). **Bottom:** Shown are cross-sections of vaginal tissue from representative adult wild-type, ER-α–null, and ER-β–null female mice (magnification × 66). The representative wild-type and ER-β–null vaginal tissues illustrate a normal stroma (Str) and hypertrophied epithelium (Epi) showing estrogen-induced stratification and cornification (*arrow*). In contrast, the ER-α–null vaginal tissue indicates complete estrogen insensitivity, as illustrated by a thin epithelium (Epi) and total lack of cornification. [Reproduced from (104).]

All three lines of ER-null females exhibit uteri that possess the expected tissue compartments, myometrium, endometrial stroma, and epithelium (104,194) (Fig. 8). However, in females lacking functional ER-α or Cyp19, uteri are overtly hypoplastic and exhibit severely reduced weights relative to wild-type littermates (194–196,232), whereas ER-β–null females possess uteri that are grossly normal and responsive to ovarian-derived steroids (104,194) (Fig. 8). The uterine endometrium of ER-α–null females is severely hypotrophic, poorly organized, and possesses a paucity of glandular structures (233). The luminal and glandular epithelial cells in ER-α–null uteri are healthy but consistently exhibit a cuboidal morphology, versus the tall columnar morphology and basal location of the nucleus of an "estrogenized" epithelium (Fig. 8). Therefore, fetal, neonatal and perinatal development of the female reproductive tract in mice is largely independent of ER-α– and ER-β–mediated actions, but estrogen responsiveness and sexual maturation of the adult uterus are ablated after the loss of functional ER-α. The totality of the ER-α–null phenotype and lack of any overt uterine abnormalities in ER-β–null females suggest that the latter receptor has little function in mediating estrogen actions in the uterus. However, although ER-α–null uteri lack the overt indications of "estrogenization" despite the enormously high levels of circulating estradiol in these females, a further loss in uterine weight is observed after ovariectomy (234), suggesting that ER-β–mediated estradiol actions may provide some compensatory role in the ER-α–null uterus. In further support of this argument, a more severely decreased uterine weight is observed in compound ER-null (194) and Cyp19-null females (195,196,232). ER-β–null females are reported to exhibit a slightly aberrant uterine growth response after ovariectomy and estrogen replacement (235); still, these animals are able to establish and sustain pregnancies to term (193).

The vaginal mucosa is composed of an epithelium and underlying stroma and possesses significant levels of ER (236) and is highly sensitive to estrogens (237). Endogenous or exogenous estrogens induce stromal differentiation, epithelial proliferation, and epithelial keratin synthesis to produce the stratified layer of cornified cells that lines the vaginal lumen (237–239). These changes in the vaginal mucosa are used experimentally to estimate the level of circulating gonadal steroids and approximate the estrous cycle stage. Despite exposure to elevated endogenous estradiol levels, the vaginal tissue of ER-α–null females lacks any indications of estrogenization (Fig. 8) (104). Exogenous administration of estradiol or DES also elicits no discernible vaginal response in ER-α–null females (233). In contrast, the vaginal mucosa of ER-β–null females undergoes the normal cyclic changes as dictated by ovarian sex steroids (Fig. 8). An effect on the vaginal mucosa similar to that observed in ER-α–null females is produced in rodents after prolonged ovariectomy or exposure to ER antagonists (240–242). Therefore, estrogenization of the vaginal mucosa in rodents, which is critical to mating, is ER-α dependent.

Interestingly, females that are heterozygous for a mutant ER that lacks DNA-binding function are infertile and exhibit a uterine abnormality of enlarged hyperplastic endometrial glands despite possessing normal levels of circulating sex steroids (209). The importance of ER coactivator proteins to estrogen action is illustrated by the measurably diminished uterine response reported in mice lacking sufficient SRC-1 (243).

Biphasic Uterine Response to Estradiol

Changes in Physiology

Estrogens stimulate a complex process of epithelial proliferation and differentiation in the sexually mature uterus that leads to the formation of a multilayered secretory endometrium. This effect is critical to providing a suitable intrauterine environment for the establishment and maintenance of pregnancy. Early studies in neonatal and prepubertal rodents found that both the uterine stroma and epithelium proliferate in response to estrogens (244); however, estrogen-induced mitogenesis in uteri of sexually mature rodents is limited to the epithelium (Fig. 9) (245,246). Therefore, sexual maturation of the rodent uterus is not simply marked by the presence of ER or an estrogen response but is rather the acquired capacity to undergo synchronized phases of proliferation and differentiation as dictated by the ovarian-derived sex steroids. The lack of estrogen-induced uterine epithelial proliferation in ER-α–null uteri indicates the essential role of ER-α in this process. Although ER-α is present in both epithelial and stromal compartments, tissue recombination studies indicate the proliferative epithelial response to estrogens is indirect and dependent on stromal ER-α actions (247).

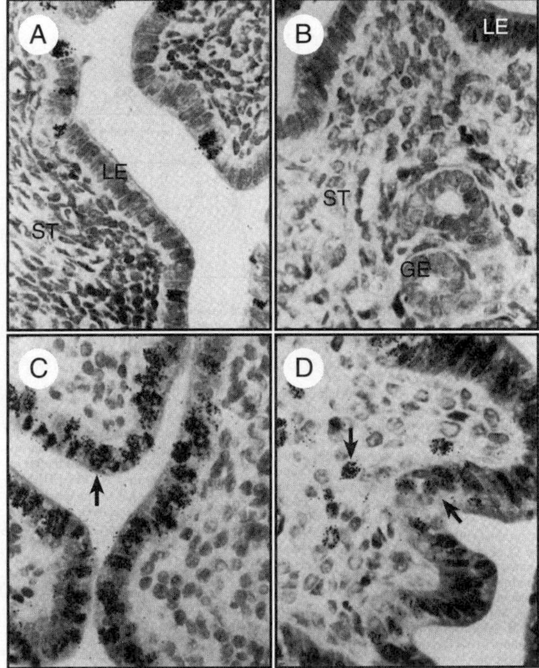

FIG. 9. Comparison of estrogenic response in prepubertal vs. adult uteri in mice. Light microscopic autoradiographs (magnification × 680) from adult and immature uteri 16 hours after treatment with estradiol (C and D) or saline (A and B). Mice were ovariectomized adults (A and C) or 21-day-old animals (B and D). All animals received an injection of [^{3}H]-thymidine (1 μCi/g body weight) 4 hours before death. Tissue was sectioned at 4 μm, exposed for 10 days, and then stained with hematoxylin and eosin. Examples of labeled nuclei are marked with arrows in the appropriate tissue compartments. Estradiol treatment in ovariectomized adult females leads to increased DNA synthesis (*arrows*) in the luminal and glandular epithelium of the uterus only and not in the stroma. In contrast, similar estradiol treatments of intact immature (21-day-old) females leads to increased DNA synthesis in both the luminal/glandular epithelium and stroma of the uterus. [Reproduced from (246).]

Treatment of ovariectomized mice with estrogens (e.g., estradiol or DES) has long served as an experimental model to mimic the uterine events that occur during the estrous phase of the rodent cycle or immediately after the preovulatory estradiol surge. In the first edition of this volume, Clark and Markaverich (101) thoroughly reviewed the morphological and biochemical changes that occur in the rodent uterus after estrogen stimulation and established the biphasic scheme that categorized the various changes according to their temporal pattern (Table 5). Estrogen-stimulated changes in the rodent uterus that occur early, within the first 6 hours after treatment, include increases in nuclear ER occupancy, water imbibition, vascular permeability and hyperemia, prostaglandin release, glucose metabolism, eosinophil infiltration, gene expression (e.g., c-*fos*), lipid and protein synthesis, and increased glucose-6-phosphate dehydrogenase (Table 5) (101). The above processes are then accompanied by a delayed response that peaks after 24–72 hours and includes

TABLE 5. *Biphasic response of the rodent uterus to estrogens*

Early uterotropic responses (within 6 hours)
Nuclear localization of estrogen receptor
Activation of receptor-tyrosine kinase pathways
Changes in gene expression (induction/repression of "early" genes)
Increased protein synthesis
Water imbibition
Hyperemia
Eosinophil infiltration
Albumin accumulation
Increased electrolytes
Lysozyme labilization
Increased cyclic nucleotides, prostaglandins and associated enzyme activity
Increased glucose metabolism and associated enzyme activity
Calcium influx
Increased lipid synthesis
Late uterotropic responses (within 24 hours)
Second peak of nuclear localization of estrogen receptor
Changes in gene expression (induction/repression of "late" genes)
Increased protein synthesis
Increased DNA synthesis
Epithelial proliferation in "waves"
Cellular hypertrophy
Overall increase in uterine dry weight

Updated from Clark, J. H., and Markaverich, B. M. (1988). Actions of ovarian steroid hormones. In *The Physiology of Reproduction* (E. Knobil and J. D. Neill, Eds.), pp. 675–724. Raven Press, New York.

dramatic increases in RNA and DNA synthesis, epithelial proliferation, and differentiation toward a more columnar secretory phenotype, dramatic increases in uterine weight, and continued gene expression (e.g., lactoferrin) (Table 5) (101). Ovariectomized wild-type female mice exhibit a three- to fourfold increase in uterine weight after three daily treatments with estradiol or DES, whereas no such response is observed in the uteri of ER-α–null females (191,249) (Fig. 10). The early phase effects of water imbibition and hyperemia as well as the late-phase effects of increased DNA synthesis and epithelial proliferation are absent in ER-α–null uteri (233,250) (Fig. 10). Interestingly, females that are heterozygous for the *Esr1* gene disruption possess approximately one-half the normal level of ER-α in the uterus, but their response to estrogen treatment is comparable with wild-type females. The uteri of ER-α–null females are also unresponsive to tamoxifen (233), a SERM that is known to be an ER agonist in the murine uterus (98). It was initially postulated that the biphasic uterine response to estrogens is mediated by at least two types of acceptor estradiol receptor complexes (251). However, the total lack of response to estrogens in ER-α–null uteri provides strong evidence that ER-α is required to mediate the full uterine response to estrogens. In turn, ER-β–null mice reportedly exhibit an exaggerated response to exogenous estrogen treatment, suggesting that this receptor form may act as a negative modulator of estrogen actions in the uterus (235).

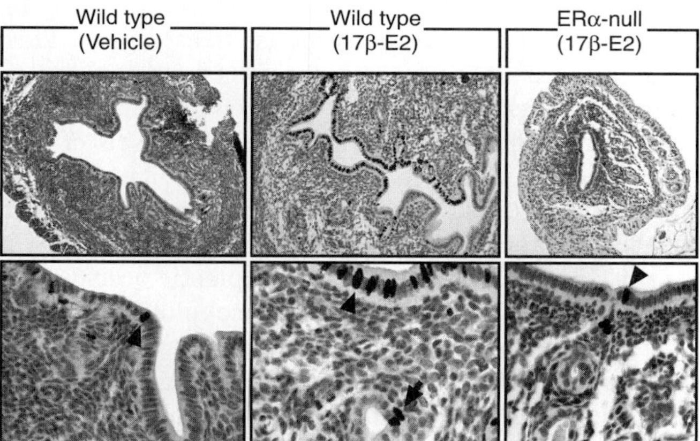

FIG. 10. Estrogen-induced epithelial proliferation in the uterus of adult mice is dependent on ER-α. Shown are cross-sections of uterine tissues from ovariectomized wild type or ER-α–null mice treated for 16 hours with vehicle or estradiol (17β-E_2). All animals were injected with bromodeoxyuridine (BrDU) 2 hours before death. Basal DNA synthesis as indicated by nuclear BrDU incorporation is detected in the luminal epithelia (*arrowheads*) of vehicle treated wild-type animals, and this is dramatically increased after estradiol treatment. In contrast, ER-α–null uteri exhibit a basal level of DNA synthesis (as indicated by BrDU incorporation) that is unchanged after estradiol treatment, indicating that estrogen-induced proliferation of the uterine epithelium is dependent on functional ER-α.

Interestingly, Cyp19-null mice are reported to exhibit a less than maximal uterine response to exogenous estrogen treatment (S. C. Hewitt, E. Simpson, and M. Jones, unpublished data). This phenotype may be similar to the partial uterine growth response characteristic of neonatal mice (244) and suggests that ER signaling may be important during postnatal uterine development.

Changes in Gene Expression

The dramatic physiological changes that occur in the rodent uterus in response to estrogens are presumably the ultimate effects of equally dramatic changes in gene expression among the uterine compartments. It is unlikely that the estradiol–ER complex is directly involved in mediating the whole genomic response in the uterus but more plausibly serves to stimulate a cascade of downstream signaling pathways that act to amplify the estrogen action. However, early investigations of the genomic response to estrogens in the rodent uterus discovered a handful of genes that are directly regulated via the classic ER mode of action, including *Pgr* (progesterone receptor) (17,19) and *Ltf* (lactoferrin) (252). The uteri of ER-α–null females fail to exhibit estrogen-induced increases in *Pgr* and *Ltf* expression (250), indicating the importance of ER-α to this response. In contrast, estrogens are reported to cause increased stromal *Pgr* expression in αERKO uteri, perhaps due to the actions of ER-β (253). As the previous study examined transcripts from a whole tissue preparation, the elevated constitutive epithelial *Pgr* transcript may have masked the stromal estrogen increase in the αERKO. Furthermore, estrogen-stimulated increases in PR in the rodent uterus are localized to the stromal and myometrial compartments, whereas increased lactoferrin is limited to luminal and glandular epithelium (254), indicating that ER-α functions are critical to induced gene expression in multiple uterine cell types. Further complexity of estrogen action in the uterus is illustrated by the simultaneous induction of PR expression in the myometrium and stroma while eliciting a decrease in PR levels in the luminal epithelium (254). Similarly, the c-*fos* gene is rapidly induced by estrogen in the uterus, and this is ER-α dependent (117).

The introduction of microarray analysis of gene expression has significantly advanced our abilities to define the genomic response of the rodent uterus to estradiol. Several studies have used microarray techniques to map the global gene expression patterns after estrogen exposure in the uterus and largely demonstrate that the biphasic uterine response to estrogens, so well characterized by physiological

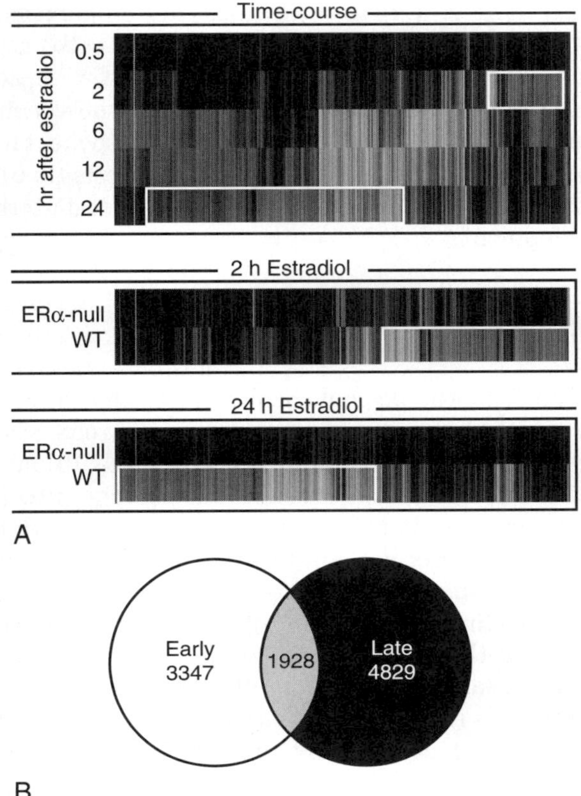

FIG. 11. Microarray analysis of estrogen-induced gene regulation in the mouse uterus. Adult ovariectomized wild-type or ER-α–null mice were treated with estradiol for the indicated time points. Uterine tissues were then collected and the gene expression profile determined by microarray analysis. **A** (*Top*): Cluster diagram showing significant changes in gene expression within 2 hours of estradiol treatment that continues out to 24 hours. Each *horizontal row* represents data from simultaneous analysis of vehicle and estradiol-treated uterine RNA at the indicated time. Each *vertical line* represents a single gene. Genes increased relative to vehicle control are shown in shades of gray, and those unchanged relative to vehicle control in black. Comparison of responses illustrates clusters of genes that are characteristic of early or late responses (indicated by *boxes*) as well as those that occur throughout the time course. Also shown are cluster diagrams illustrating that most estrogen-induced changes in gene expression 2 hours (*middle*) and 24 hours (*bottom*) after treatment in the mouse uterus are lacking in ER-α–null female mice. **B:** Venn diagram showing the numbers of significantly regulated genes characteristic of the early and late response to estrogen treatment.

indicators above, is mirrored by the global changes in gene expression (Fig. 11) (116,122,255–257). The clearly defined patterns of early and late response genes found in wild-type mouse uterine tissues are completely lacking in ER-α–null uteri (Fig. 11). The identified genes fall into functional groupings, including signal transduction, gene transcription, metabolism, protein synthesis and processing, immune function, and cell cycle (116,256–258). Surprisingly, the expression levels of a striking number of genes are repressed by estrogen in the mouse uterus, and these effects were either absent in ER-α–null uteri or relieved by cotreatment with ER antagonists, indicating that ER-α is also actively involved in this process (116). Comparative gene array analyses have also been conducted on human endometrial tissues during the proliferative and secretory phases of the menstrual cycle and have revealed gene expression patterns that are similar to those described in rodents (259,260). Others have used microarray techniques to assess the genomic response in uterine tissues after prolonged estrogen exposure during the final 10 days of gestation (261) or 6 weeks of exposure in adults (262). Although informative, these data are more applicable to evaluating the developmental and toxic effects of estrogens in the uterus.

The above microarray studies revealed a precise regulation of several cell-cycle modulating factors in the rodent uterus after estrogen exposure, including cyclin G1, cyclin E1, cdc2, and p21 (116). Follow-up studies have confirmed that transcript and protein levels of p21, which inhibits the G_1-S phase transition of the cell cycle, are maximally induced 12 hours after estrogen treatment and are localized to the nuclei of the epithelia (116). These data suggest that increased p21 expression is an important early genomic response to estrogen and may act to synchronize the uterine epithelium and provide for a coordinated proliferative response.

Estrogen–Growth Factor Cross-Talk in the Uterus

The autocrine and paracrine actions of polypeptide growth factors are believed to be an integral component of the uterine response to estrogens. This hypothesis is based on observations from numerous animal studies:

- Estradiol up-regulates the uterine levels of epidermal growth factor (EGF) and its receptor (EGFR) (263,264), transforming growth factor-α (265), and IGF-1 (144,266).
- EGFR-null mice exhibit hypoplastic uteri (267).
- EGF or IGF-1 mimic the morphological, proliferative, and genomic effects of estradiol or DES in uteri of ovariectomized mice (Fig. 12) (268).
- Anti-EGF antiserum is able to inhibit the uterine response to estrogens in vivo (268).
- ER antagonists inhibit the uterine response to EGF (269).

In addition, IGF-1 or EGF treatment of ovariectomized wild-type mice elicits a pattern of global gene expression similar to that induced by estradiol, although some estrogen-specific genes are revealed (117). IGF-1 treatment was shown to increase the expression of an ERE-driven luciferase reporter gene in transgenic mice, providing the first in vivo evidence of estradiol-independent ER activation by growth

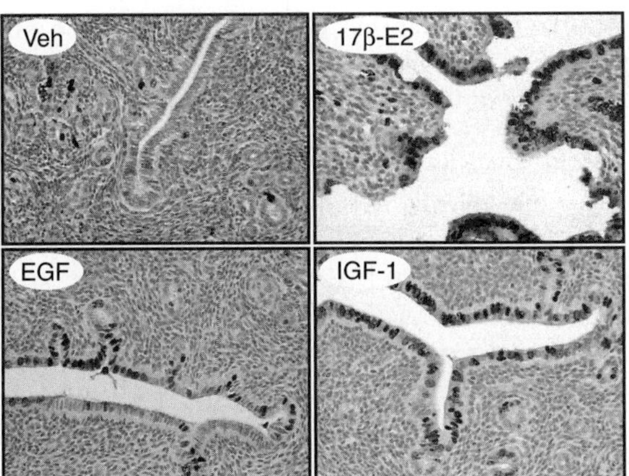

FIG. 12. Growth factors are able to mimic the proliferative effects of estradiol in the mouse uterus. Shown are uterine cross-sections from ovariectomized wild-type mice treated for 24 hours with vehicle (Veh), estradiol (17β-E_2), epidermal growth factor (EGF), or insulin-like growth factor 1 (IGF-1). Uterine tissue was then subjected to immunohistochemistry for the antigen Ki67, a marker of cell proliferation. As expected, estradiol treatment induces a dramatic increase in Ki67 immunoreactivity in uterine luminal and glandular epithelium and EGF and IGF-1 mimic this effect, in the absence of endogenous estradiol.

factors (143). The culmination of these and other studies has led to the proposed model that the mitogenic actions of estradiol in the rodent uterus are at least partially mediated by locally produced growth factors, but these actions require functional ER-α (Fig. 13).

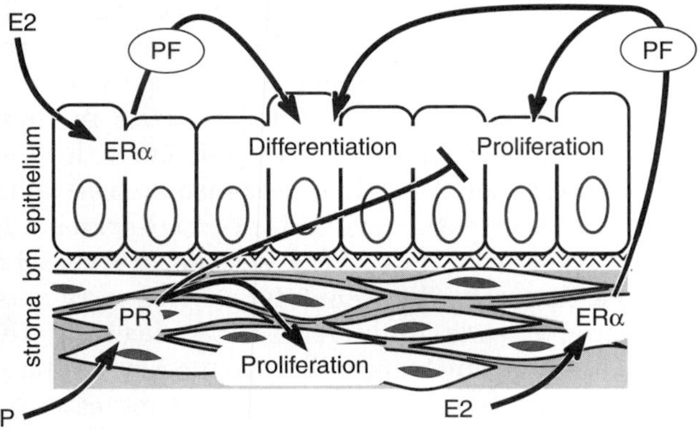

FIG. 13. Putative paracrine and autocrine mechanisms of uterine proliferation and differentiation in response to estrogens. Estradiol (E_2) stimulation of proliferation of the uterine epithelium requires the presence of functional ER-α in the underlying stroma but not the epithelium, indicating that estradiol/ER-α actions in the stroma induce the secretion of a paracrine factors (PF) that then act on the epithelium to stimulate proliferation. In contrast, differentiation of the uterine epithelium as indicated by secretory products (not shown) requires functional ER-α in the epithelium and may be a direct effect via ER-α or a paracrine/autocrine effect via the synthesis of secreted factors (PF). Progesterone (P) acts through stromal PRs to inhibit the proliferative response of the epithelium to estradiol, while inducing proliferation of the underlying stroma. This model is based on the findings from studies using ER-α–null uterine tissues (247, 271). bm, basement membrane.

ER-null mice provide an excellent in vivo model for the study of ER–growth factor "cross-talk" mechanisms in the uterus. ER-α–null uteri possess wild-type levels of functional EGF and EGFR but are unresponsive to the mitogenic actions of EGF, confirming the interaction of these two signaling systems (142). However, not all EGF responses are lacking in the uteri of αERKO females, as up-regulation of the *c-fos* gene remains intact (142). Similar studies have demonstrated that the uterine response to IGF-1 is compromised in ER-α–null females (143). Cunha and colleagues (199) used the ER-α–null mice in series of tissue recombination experiments to further demonstrate interaction between ER and growth factor signaling in the murine reproductive tract. In these studies, uterine stoma and epithelium are enzymatically disassociated and recombined with corresponding tissue from animals of different treatments or genotypes. The resulting chimeric stromal–epithelial unit is then implanted under the kidney capsule of an ovariectomized nude mouse that is then treated with various hormonal combinations (199). A caveat to these studies is that neonatal tissues are used and may not accurately reflect the uterine physiology of sexually mature females (199). Nonetheless, these methods have been effectively used to demonstrate that estrogen-induced proliferation of uterine and vaginal epithelium requires functional ER-α in the underlying stroma only (247,270), whereas estrogen-induced increases in secretory products (e.g., lactoferrin, complement C3, keratins) in the uterine or vaginal epithelium requires functional ER-α in both uterine compartments (247,270,271). These studies strongly support a paracrine mechanism of estrogen-mediated epithelial proliferation that requires ER-α in the stroma (Fig. 13). Interestingly, recent microarray studies indicate that the growth-factor induced genomic response is intact in ER-α–null uteri, although epithelial proliferation is absent (117), suggesting a greater complexity to the estrogen–growth factor signaling "cross-talk" than was originally perceived.

Maintenance of Selective Estrogen Actions in the Absence of ER-α

A distinct advantage of null receptor models, whether naturally existing or experimentally generated, is their use as an in vivo tool for discerning alternate pathways of hormone action. Studies toward these ends using ER-α–null mice indicate the preservation of a potential distinct estrogen signaling pathway in the uterus (272). Das et al. (272) reported that two consecutive treatments (over a period of 12 hours) with the catecholestrogen 4-hydroxyestradiol at 10 μg/kg body weight leads to significant increases in water imbibition and lactoferrin expression in the uteri of ovariectomized ER-α–null but not in wild-type mice. Furthermore, these responses are not inhibited by cotreatment with pure ER antagonists (ICI compounds), indicating the possibility of a non–ER-α–mediated signaling pathway for certain compounds exhibiting estrogenic activity (272). The temporal pattern of lactoferrin expression elicited by 4-hydroxyestradiol in ER-α–null uteri is distinct from that induced by estradiol in wild-type uteri (273). The low to absent expression of ER-β in the mouse uterus, including ER-α–null females (205), as well as documentation that ICI compounds block ER-β activities likely excludes this receptor in mediating the 4-hydroxyestradiol effects observed. Catecholestrogens are locally synthesized in tissues and are proposed to play a role in steroid regulated functions of the hypothalamus, pituitary, and ovary. The discovery of local synthesis of catecholestrogens in mammary tissue has led to postulations of their involvement in breast cancer (274). Therefore, further investigations into the alternate mechanisms by which these compounds may activate nuclear processes in the absence of ER-α are warranted.

Maintenance of Progesterone Actions in the Absence of ER-α

The *Pgr* gene is a well-described target of estrogen-induced expression via the classic model of ER action, especially in the uterus (18,19). The lack of estradiol-induced increases in *Pgr* expression in ER-α–null uteri confirms the regulatory dependence on ER-α action (250). Therefore, it was hypothesized that disruption of the ER-α gene may subsequently result in abnormally low levels of PR in ER-α–null uteri and thereby render this tissue refractory to progesterone as well. However, assays for *Pgr* expression and progestin binding in ER-α–null uteri indicate PR levels are only reduced by approximately half (210,250). Furthermore, a greater proportion of PR is nuclear localized in ER-α–null uteri (≈25%) relative to wild-type (≈5%) (210). Western blots indicate no difference in the relative levels of PR-A and PR-B between genotypes, with PR-A consistently present in greater amounts in both (210). Therefore, a loss of ER-α action in the uterus has not led to a complete lack of PR or to altered and preferential transcription from one of the two *Pgr* gene promoters. Furthermore, several progesterone-dependent actions are preserved in ER-α–null uteri, including progestin-induced expression of amphiregulin and calcitonin and the uterine decidual response (210).

The postovulatory nadir in circulating estradiol levels has long been known to be critical to uterine decidualization in mice (275). Studies have shown that the ER antagonist ICI 182,780 prevents artificially induced uterine decidualization in wild-type mice, indicating involvement of ER signaling (210). Therefore, it is surprising that ER-α–null mice exhibit a uterine response when exposed to an artificial model of hormonal and mechanical model of uterine decidualization (210,276). These findings suggest that an altered "organization" of the uterine tissue or compensatory pathways provide for hormonally driven uterine decidualization in ER-α–null mice. Alternatively, some genes detected at the time of implantation, including the gap junction protein connexin 26 (277) and the cytokine leukemia inhibitory factor (211), are regulated by dual pathways. One pathway involves estrogen-stimulated ER-α, and the ability to induce via this mechanism is lost in the αERKO. The second pathway is initiated by decidualization associated signals and is ER-α independent and retained in the αERKO. This indicates a redundancy of regulatory mechanisms that allows retention of gene regulation in the αERKO and accounts for the ER-α–independent uterine decidualization response observed in ER-α–null mice.

PROGESTERONE RECEPTOR SIGNALING IN THE UTERUS

PR Expression in the Uterus

In the murine uterus, PR transcript and protein levels are low on days 1–2 of pregnancy with the highest expression occurring in the epithelia and subepithelial stroma (204). PR levels rise in these same tissues on days 3 and 4 of pregnancy, leading up to implantation (204). Receptor-mediated estradiol actions are a primary regulator of PR expression in the mouse uterus, but these actions are complex and specific to the different uterine compartments. Ovariectomy leads to increased PR levels in the uterine luminal epithelium that can be reduced upon exogenous estradiol treatments, indicating estradiol acts to repress PR expression in this uterine compartment (206,254). Recombination experiments of stromal and epithelial tissues similar to those described above indicate this to be a paracrine-mediated effect that requires functional ER-α in the stroma (278) (Fig. 13). Simultaneous with causing decreased PR expression in the epithelium, estradiol treatments lead to increased PR levels in the uterine stroma of ovariectomized mice (206,254). These observations have been largely confirmed in studies of a recently generated PR-lacZ reporter mouse that represents *Pgr* gene promoter activity as β-galactosidase activity in situ (279). In these studies, β-galactosidase activity representative of PR expression is observed in luminal and glandular epithelium after ovariectomy and decreased after estrogen treatment (279). Progesterone treatments also lead to decreased PR-lacZ expression in all cell types (279).

In the adult ovine and bovine uterus, PR expression is highest during the early proliferative phase and present in all uterine cell types but is most abundant in the stroma and myometrium (214,216–218). PR levels decline as the cycle progresses to the secretory phase, initially in the stroma and myometrial cells (214,216–218).

In the human uterus, PR immunoreactivity is increased in all cell types during the proliferative phase but measurably lower than ER levels (220,221). During the secretory phase, PR levels are maintained in the stroma and myometrium but decrease in the glandular epithelium such that levels are eventually undetectable (220,221). More recent investigations have delineated PR-A and PR-B specific expression in the human uterus (280). The decrease in PR levels that occurs in the glandular epithelium during the mid-secretory phase is largely representative of PR-A, leading to an increased PR-B/PR-A ratio in these cells (280). In stroma, where fewer cells expressed PR relative to glandular epithelium, PR-A is more abundant (280). During the late secretory phase, PR-A levels in the stroma decrease but remain detectable, whereas stromal PR-B expression decreases occur earlier and are significantly diminished by the late secretory phase (280). The predominance of PR-A in the glandular epithelium during the early secretory phase may serve to inhibit further estrogen-mediated proliferation, as the uteri in PR-A–null mice exhibit an exaggerated proliferative response to estrogens (281). As progesterone levels rise during the mid-secretory phase, PR-B is concentrated in nuclear foci, suggesting active transcriptional activity (282). A similar finding of focal-concentrated PR-B is reported in endometrial cancer cells in post-menopausal women, suggesting that inappropriate PR-B activity may be associated with uterine cancer (282). The pattern of PR expression in uteri of nonhuman primates is similar to that described in humans. In female baboons, PR levels are increased during the proliferative phase and then decrease accordingly during the secretory phase (227).

PR Actions in the Rodent Uterus: PR-Null Mice

PR-null mice exhibit normally developed uteri that possess the expected uterine architecture (Table 4) (181,281). During the secretory phase of the murine estrous cycle, decreasing circulating levels of estradiol concurrent with rising progesterone cause a shift in uterine proliferation from the epithelial to stroma cells (283). Therefore, PR-mediated progesterone actions are thought to negatively modulate estradiol-induced proliferation of the uterine epithelium (Fig. 13) (283). Indeed, ovariectomized PR-null females treated with daily injections of estradiol and progesterone for 3 weeks exhibit abnormally large fluid-filled uteri that are characterized by a thickened uterine wall due to extensive extracellular edema, extensive disorganized proliferation of the glandular epithelia, and acute inflammation of the endometrium (179) (Fig. 14). The hyperestrogenic response in

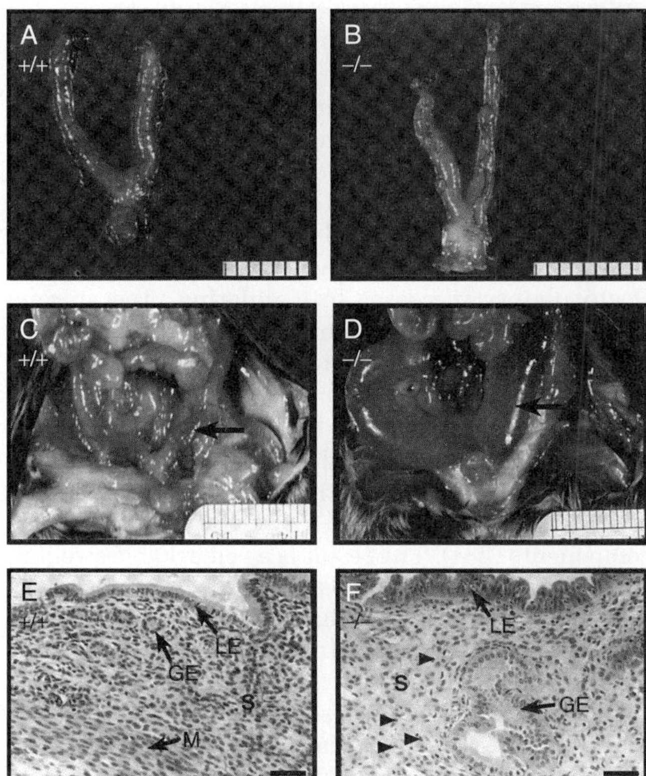

FIG. 14. Uterine response to estrogen and progesterone treatment in PR-null female mice. Comparative gross anatomy of uteri isolated from hormonally untreated ovariectomized wild-type (**A**) and PR-null (**B**) female mice does not reveal a significant difference in morphology. All mice were 6 weeks old. In situ gross anatomy of wild-type (**C**) and PR-null (**D**) uteri, after estrogen and progesterone treatment, is indicated by arrows. Note the marked enlarged fluid-filled uterus in the PR-null female. **E:** Histological analysis of a representative cross-section of the uterine wall of hormonally treated wild-type mouse shows the presence of a normal uterine architecture, luminal epithelium (LE), glandular epithelium (GE), stromal cell layer (S), and myometrium (M). Scale bar = 50 μm. **F:** Morphological analysis of a typical transverse section of the uterine wall of a PR-null female after estrogen and progesterone treatment reveals an abnormal uterine structure. Note the hyperplastic luminal epithelium (LE) and hypertrophic glandular epithelium (GE), loosely arranged stromal layer (S), and presence of polymorphonuclear leucocytes (indicated by *arrowheads*). Scale bar = 50 μm. [Reproduced with permission from (179).]

PR-null uteri is similar to that observed in ovariectomized mice exposed to prolonged estrogens only (246) and provides strong support to the modulatory actions of PR in the uterus (Fig. 14) (179). Similar experiments in isoform-specific PR-null mice indicate this phenotype is reproduced in PR-A–null mice only, whereas PR-B–null mice exhibit a normal uterine response, indicating that PR-A is primarily responsible for negatively modulating the uterine response to estradiol (180,181,281). Recombination experiments using uterine tissues from PR-null mice demonstrate that the antiproliferative functions of progestins in the uterine epithelium are paracrine mediated and require PR-A in the stroma (279) (Fig. 13).

PR-mediated progesterone actions in the uterus are critical to preparing the uterine endometrium for pregnancy. Embryo implantation is a highly complex process that requires synchronized cooperation between the blastocyst and uterine endometrium (see Chapter 4). Circulating estradiol levels peak at ovulation and elicit a cascade of proliferation and differentiation in the luminal and glandular epithelium of the uterus, including induction of PR expression in the endometrial stroma and myometrium (254,284). Postovulatory increases in circulating progesterone then cause decidualization, a complex process that involves massive proliferation and differentiation of the endometrial stroma along with localized increases in vascular permeability and edema (284). This process involves the synthesis and interaction of numerous hormones and signaling pathways, including prolactin, cytokines, prostaglandins, and extracellular matrix components (284). The result is a remarkable swelling of the uterine stroma that is thought to be necessary for implantation by forcing uterus to close down on the blastocyst (284). The final stage of apposition is a grasping of the blastocyst and ultimate attachment to the uterine wall, a process thought to be dependent on secondary rises in ovarian-derived estradiol (284). Therefore, preparation of the uterus for blastocyst implantation is dependent on the multifunctional and sometimes opposing effects of estradiol and progesterone. PR-null female mice, more specifically those lacking only PR-A, fail to exhibit a uterine decidual response, illustrating the importance of PR-A (179,180). Numerous progesterone/PR target genes have been identified using traditional methods (Table 6). Microarray analyses have also been used to map the uterine response to progesterone and have identified numerous target genes, including those involved in immune function, metabolism, growth factor regulation, signal transduction, and extra- and intracellular structure (285,286). These observations have led to the identification of the 12/15 lipoxygenase signaling pathway as a target of *Pgr* regulation and demonstrated its importance in implantation (287).

TABLE 6. *Progesterone regulated genes in the mammalian uterus*

Gene name	Species	Detection	Reference
Amphiregulin (*Areg*)	mouse	transcript	(675)
Calcitonin (*Calca*)	rat	transcript	(676)
Calcyclin	mouse	transcript	(285)
c-myc (*Myc*)	rat	protein	(677)
Connexin 26 (*Gjb2*)	rat	transcript	(678)
Cyclin G1 (*Ccng1*)	mouse	transcript	(679)
Fibronectin (*FN1*)	human	protein	(680)
Heparin-binding epidermal growth factor-like growth factor (*Hbegf*)	rat	transcript	(681)
Histidine decarboxylase (*Hdc*)	mouse	transcript, protein	(682)
Homeobox-a9 (*Hoxa9*)	mouse	transcript	(683)
Homeobox-a10 (*Hoxa10*)	mouse	transcript	(683,684)
Homeobox-a11 (*Hoxa11*)	mouse	transcript	(683)
Hypoxia induced factor 1α	mouse	transcript	(685)
Immune-responsive gene 1	mouse	transcript	(285)
Implantation serine proteinases-1 and -2	mouse	transcript, protein	(686)
Indian hedgehog (*Ihh*)	mouse	transcript, protein	(110)
Insulin like growth factor I (*Igf1*)	mouse	transcript	(687)
Insulin-like growth factor binding protein-1 (*IGFBP1*)	human	transcript	(680)
Interleukin-1 (*IL1*)	human	protein	(680)
L-12/15 lipoxygenase	mouse	transcript	(285)
Mucin 1 (*Muc1*)	rabbit	transcript	(688)
Placental 17β-hydroxysteroid dehydrogenase	human	transcript	(689)
Prostaglandin E receptor 2 (subtype EP2) (*Ptger2*)	mouse	transcript	(690)
Tissue inhibitors of metallo-proteinase-3 (*TIMP3*)	human	transcript	(680)
Transforming growth factor-β (*TGFB*)	human	protein	(680)
Uteroferrin	porcine	transcript	(691)
Uteroglobin (*Scgb1a1*)	rabbit	transcript	(692)
Vascular endothelial growth factor (*Vegf*)	mouse	transcript	(685)

The inflammatory response observed in PR-null uteri after prolonged estrogen/progesterone treatment is extensive and consists of "marked" infiltration of polymorphonuclear leukocytes into the endometrial stroma, mucosal epithelium, and uterine luminal fluid (179) (Fig. 14). In vitro studies have shown that progesterone inhibits the expression of chemotactic cytokines for neutrophil and lymphocyte infiltration (288,289) and may reduce prostaglandin E levels in uterine decidual and chorionic tissue during early pregnancy (290). Therefore, PR-mediated progesterone actions may be critical to suppressing the uterine inflammatory response that may accompany embryo implantation (179,291).

ANDROGEN RECEPTOR SIGNALING IN UTERINE FUNCTION

ARs are present in uterine tissues of multiple species (208,292–295), although the function of androgen signaling in the uterus remains unclear. In rodents, ARs are present in all uterine cell types but most highly expressed in the myometium, where expression may be positively regulated by estrogens (235,293). In humans, ARs are also detected in the myometrial and endometrial uterine tissues and levels increase during the proliferative phase (296).

Treatment of hypophysectomized rats with DHT is known to cause increased uterine weight in rodents, indicating that ligand-dependent AR actions can affect uterine phenotypes (297). Elevated androgens are also postulated to provide for some uterine maintenance in ER-α–null females (104,234). More recent studies have demonstrated that AR agonists cause increased myometrial thickness in the uteri of ovariectomized rats as well as inhibit estrogen-induced epithelial proliferation (298). AR-null female mice exhibit relatively normal uteri that are somewhat hypoplastic, although this phenotype may be more representative of decreased estradiol synthesis in the ovaries rather than a role for AR in maintaining uterine weight (Table 4) (190). Furthermore, AR-null females are able to establish and maintain pregnancies to term (190). A recent microarray study has indicated that DHT leads to a pattern of uterine gene regulation that is remarkably similar to that elicited by estrogens (298), with 86% of the DHT response consisting of a subset of estrogen responses. The fold-response of the overlapping genes was in general more robust after estrogen treatment versus DHT. Thus, despite quite divergent biological outcomes, the global genomic patterns elicited by estrogens and nonaromatizable androgens largely overlap. Genes noted included those involved in metabolism, tissue growth and remodeling, transcription, protein synthesis and processing, and signal transduction (298).

GLUCOCORTICOID RECEPTOR SIGNALING IN UTERINE FUNCTION

The uterus is not considered a "classic" target tissue of glucocorticoid action, although GRs are present throughout the cell types of the rodent uterus (299). The limited experimental data available indicate that GR-mediated glucocorticoid actions are present in the uterus (299). Surprisingly, stimulation of the GR signaling pathway by dexamethasone treatment in ovariectomized rats elicits a pattern of gene expression that is remarkably similar to estrogen (299).

SEX STEROID RECEPTORS IN OVARIAN FUNCTION

The functions of the sex steroids and their cognate receptors in ovarian function are especially complex and multifaceted. Progesterone, testosterone, and estradiol are all synthesized and secreted by the ovary during folliculogenesis and act via both extraovarian (i.e., endocrine) and intraovarian (i.e., paracrine/autocrine) pathways to profoundly influence all aspects of ovarian function. The endocrine actions of the three sex steroids in the hypothalamic–pituitary axis are critical to the regulation of gonadotropin secretion and the ovarian cycle. In turn, our appreciation of the extent and importance of the paracrine/autocrine actions of sex steroids within the ovary has increased substantially over the past years. Accompanying chapters in this volume discuss the interactions between sex steroid signaling and other key pathways that influence ovarian function, for example, gonadotropin secretion (Chapters 31, 44, and 45), gonadotropin and growth factor signaling (Chapters 31, 44, and 45), ovulation (Chapter 11), and luteal function (Chapter 12). In this chapter, we specifically review the literature concerning the expression of the sex steroid receptors in the mammalian ovary and use the newly described null mouse models as a platform to focus on recent revelations concerning the intraovarian and extraovarian roles of steroid signaling in ovarian function.

ESTROGEN RECEPTOR SIGNALING IN OVARIAN FUNCTION

ER Expression in the Ovary

Unlike a decade ago, there is currently ample evidence of ER expression among the somatic cell types of the mammalian ovary. This is largely due to a resurgence of interest in estrogen actions in the ovary brought about by the discovery of ER-β. Still, as early as 1969 Stumpf (300) demonstrated the specific uptake of [^{3}H]-estradiol in rat ovaries using dry-mount autoradiography. Later, Richards (301) and colleagues (302) demonstrated that the predominance of high-affinity estradiol binding sites in the rat ovary are in the granulosa cell fraction and that these levels are regulated by estrogen and gonadotropins. Today, successful cloning of the individual ER genes from multiple species and continued development of better ER-specific immunoglobulins allow more detailed characterization of ER-α and ER-β expression and regulation in the mammalian ovary.

ER expression patterns in the ovary are well conserved among mammalian species, although distinct differences are apparent in late-stage follicles. A high level of ER-β expression in granulosa cells of growing follicles is conserved among rodents (Fig. 15), large animals, and primates. ER-α expression is exclusive to thecal/interstitial cells in rats and mice (Fig. 15), but this compartmental expression pattern does not hold true in hamsters, domestic animals, monkeys, or humans (Fig. 15). In fact, current data indicate that granulosa cells of preovulatory follicles in large animals and primates express comparable levels of both ER-α and ER-β, and in humans the former may predominate (303). Similarly, thecal cells in large animal and human ovaries posses both ER forms. These data force us to consider a greater potential role for ER-α/ER-β heterodimers in the ovaries of large animals and primates that may not occur in rodents. Still, the more restricted expression pattern of ER-α and ER-β in rodent ovaries does not exclude possible cooperative action between the two receptors, as the unique ovarian phenotypes in compound ER-null mice suggest otherwise (192,194). Although much has been learned concerning ER localization over the past 10 years, it remains difficult to delineate which of the incongruent findings among species are truly representative of divergent expression patterns or are more attributable to disparities in techniques and immunoreagents. This is further confounded by evidence that levels of ER mRNA do not always correspond with the levels of immunoreactive protein within the ovarian compartments (303). The following section reviews the current literature concerning the expression patterns of ER-α and ER-β in the ovaries of rodents, domestic animals, and primates and the changes in expression patterns that occur during folliculogenesis.

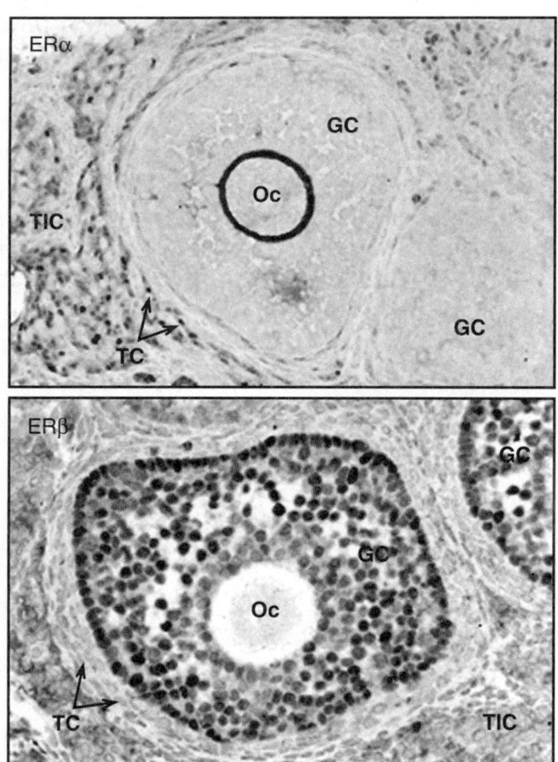

FIG. 15. Immunohistochemistry for ER-α and ER-β in adult mouse ovary. Immunohistochemistry for ER-α (*top*) indicates specific nuclear immunoreactivity in thecal/interstitial cells (TIC) and thecal cells (TC) in and around a preantral follicle. Granulosa cells (GC) lack any measurable immunoreactivity for ER-α. The staining around the outer surface of the oocyte (Oc) is nonspecific and not representative of ERα immunoreactivity. Immunohistochemistry for ER-β (*bottom*) indicates specific nuclear immunoreactivity in the granulosa cells (GC) throughout a preantral follicle. Thecal cells (TC) lack any measurable immunoreactivity for ER-β, but the surrounding thecal/interstitial cells (TIC) exhibit some cytoplasmic staining. Immunohistochemistry was done by Dr. Madhabananda Sar [see (304) for protocol].

Estrogen Receptor-α

Rodents. In the adult ovary, ER-α is localized to the ovarian interstitial/stroma, thecal cells of growing follicles, and surface epithelium in rat (207,208,294, 304–310) and mouse (202,311) ovaries. This expression pattern is not as evident in rabbits (312) or hamsters (313) where both receptors are detectable to varied degrees throughout the different somatic cell types of the ovary.

In the fetal rat, ER-α expression is detectable shortly after the first indications of gonadal differentiation (314). Postnatal rat and mouse ovaries exhibit ER-α mRNA shortly after birth but levels remain relatively constant, whereas ER-β levels rise substantially during this period (202,315). Immunoreactivity for ER-α in neonatal rat ovaries indicates ER-α expression is limited to thecal/interstitial cells and the ovarian surface epithelia and absent in granulosa cells and oocytes (304,314). A similar pattern of ER-α expression occurs in fetal hamster ovaries, which exhibit immunoreactivity as early as gestational day 14 and significant increases thereafter during the neonatal period (316). Yang et al. (316) reported that a noticeable increase in ER-α immunoreactivity occurs in the granulosa cells and oocytes of primordial follicles during neonatal days 8–15.

In adult rats and mice, ER-α expression continues to be exclusively localized to thecal cells of growing follicles, interstitial/stromal cells, and the ovarian surface epithelium (207,208,306,307,309,310,317,318) (Fig. 15). Electrophoresis mobility shift assays of nuclear extracts from rat granulosa cells indicate that virtually all specific ERE-bound complexes are supershifted by anti–ER-β but not anti–ER-α immunoglobulin, providing further support that rat granulosa cells possess very little ER-α (308,309). Still, Western blot analyses of whole cell or nuclear extracts from isolated rat granulosa cells do indicate a low level of ER-α protein of 61 kDa (308). Some studies using in situ hybridization, although generally congruent with the immunohistochemical findings of a paucity of ER-α in granulosa cells (314,319), describe scant and scattered signals over the granulosa population of larger follicles (207,320). Unlike ER-β, ER-α expression remains relatively constant throughout the rat estrous cycle (321). Furthermore, no change in ER-α levels is detected in granulosa cells cultured in follicle-stimulating hormone (FSH)/testosterone over a period of 72 hours; however, a 24-hour treatment with forskolin induces a marked shift in nuclear localization of ER-α immunoreactivity (308).

ER-α localization in the adult hamster ovary is somewhat divergent from that in rats and mice. Yang et al. (313) reported moderate but clear ER-α immunoreactivity in thecal/interstitial cells but also appreciable immunoreactivity in granulosa cells of small preantral follicles. In the granulosa cells of antral follicles, ER-α immunoreactivity is strongest among those cells in close proximity to the antrum (313). FSH treatment (2x/d × 2d) or a single injection of estradiol elicits a significant induction of ER-α expression in both thecal and granulosa cells in hypophysectomized hamsters (313). Still, Western blot analyses of whole ovarian homogenates from hamsters indicate an ER-β/ER-α ratio of 14:1 during the follicular phase, indicating that ER-β predominates. However, the rapid decline in ER-β and concurrent increase in ER-α that occurs just before and shortly after the gonadotropin surge adjusts this ratio to 2:1, suggesting that ER-β plays a more predominant role in follicular growth and ER-α is more important during luteinization (313).

Domestic Animals. In bovine ovaries, ER-α immunoreactivity is relatively high in thecal cells of secondary and tertiary follicles, substantially weaker in granulosa cells of tertiary follicles, and totally absent in primordial, primary, and secondary follicles (322). In agreement, ER-α transcripts are detectable in both granulosa and thecal cell fractions, but levels are substantially higher in the latter cell type (323). Furthermore, thecal ER-α expression appears to be differentially regulated because it is almost fourfold higher in follicles of 20–180 mm versus those <0.5 mm, whereas ER-β expression in thecal cells remains constant among follicles of different sizes (323). ER-α expression in granulosa cells, although much lower relative to theca, is increased with advanced follicle size (323). In porcine ovaries, ER-α immunoreactivity is detected in both theca interna and granulosa cells, but in large follicles only; even then ER-α immunoreactivity is considerably moderate relative to ER-β (324). In ovine ovaries, ER-α mRNA and immunoreactivity is detected in both cell types but predominates in granulosa cells and is particularly high in cumulus oophorus cells of small antral follicles (325).

Primates. The first reported study of ER-α immunoreactivity in rhesus or cynomolgus monkey ovaries found that ER-α was undetectable in all ovarian cell types except the surface epithelia, regardless of menstrual stage (326). However, a later study in baboon ovaries using the same antibody found nuclear ER-α immunoreactivity in 30%–40% of the granulosa cells of healthy antral follicles and detectable but less intense staining in granulosa cells of preantral follicles (327). This same study found the stroma, interstitial, and thecal cells to be largely unlabeled (327). Pau et al. (328) produced similar findings of ER-α expression in the granulosa cells of rhesus monkey ovaries by in situ hybridization.

High-affinity estradiol binding sites in normal human ovarian tissue were first described many years ago (329–331), but these techniques do not differentiate the two ER forms. Iwai et al. (332) used anti–ER-α antisera similar to that in the nonhuman primate studies to demonstrate immunoreactivity in the granulosa cells of antral and preovulatory follicles in human ovary but a total absence in primordial, preantral follicles and atretic follicles, and thecal/interstitial cells. A more recent study by Pelletier and El-Alfy (333) using a different anti-human ER-α antiserum produced results in contrast to those of Iwai et al. (332): a total lack of ER-α immunoreactivity in granulosa cells but clear nuclear staining of theca interna cells, interstitial gland cells, and ovarian surface epithelia. A possible explanation for this discrepancy may be a lack of large antral follicles in the samples evaluated by Pelleteir and El-Alfy because both studies agree that ER-α is not detectable in the granulosa cells of less mature follicles. In support of this explanation is a report by Taylor and Al-Azzawi (334) in which ER-α immunoreactivity is illustrated in the granulosa cells of what is clearly a large antral follicle when using antisera raised against the bovine ER-α. Indeed, Saunders et al. (335) found ER-α immunoreactivity is limited to granulosa cells of large antral follicles and undetectable in smaller follicles in human ovary. Also, in agreement with Pelletier and El-Alfy (333), Saunders et al. (335) found ER-α immunoreactivity in thecal cells of preantral and antral follicles and ovarian surface epithelia.

In agreement with the immunohistochemical findings described above, ER-α transcripts are detectable in human nonluteinized granulosa cells (336) and granulosa–luteal cells collected at the time of oocyte retrieval during in vitro fertilization procedures (336,337). In fact, Jakimiuk et al. (303) found ER-α mRNA and protein levels are two- to fourfold higher in granulosa versus thecal cells of small antral follicles in human ovaries. Furthermore, ER-α protein levels remain elevated in granulosa cells but decrease in thecal cells of dominant follicles (303). In luteinized human granulosa cells, Chiang et al. (338) found ER-α levels to remain relatively constant over a period of 10 days in culture but always much lower than ER-β mRNA levels. However, ER-α regulation in these cells was essentially the same as that found for ER-β in that activation of the luteinizing hormone (LH) or gonadotropin-releasing hormone (GnRH) signaling pathways leads to a 50% decrease within 12 hours via the PKA and PKC pathways, respectively (338).

Estrogen Receptor-β

In nonpregnant ovaries of most mammalian species, ER-β is clearly the predominant ER form and is most often exclusively localized to granulosa cells of follicles from the primary to preovulatory stage. This expression pattern is documented in the ovaries of rats (43,208,304,307–310,319,321,339,340), mice (202,311,318,340), hamsters (313), cow (47,323, 341), sheep (342,343), pigs (340,344), nonhuman primates (294,328), and humans (7,333,334,340,345) but may not be true in the rabbit (312).

Rodents. In developing rat ovaries, ER-β expression is detectable shortly after the first indications of gonadal differentiation on gestational day 14 and levels increase thereafter during prenatal development (314). A second more robust increase in ER-β expression occurs during days 10–15 of neonatal development in rat ovaries (315), and immunoreactivity is largely localized to granulosa cells of growing follicles and notably absent in primordial follicles and oocytes (304,314). Mouse ovaries exhibit a comparable developmental pattern of ER-β expression (202). In contrast, ER-α is limited to thecal and interstitial cells (304,314) and exhibits little change in levels during postnatal development rat and mouse ovaries (202,315). The hamster ovary exhibits a fairly similar ontogeny as ER-β is detectable as early as gestational day 13 and increases thereafter to peak on postnatal day 10, throughout which ER-β immunoreactivity is predominantly localized to granulosa cells but detectable in some interstitial cells and oocytes (316).

In the ovaries of prepubertal and adult rats and mice, ER-β immunoreactivity is nearly exclusive to the nuclei of granulosa cells of healthy follicles at all advanced stages of folliculogenesis (202,207,208,304–307,309,313,340) and is notably decreased or absent in atretic follicles (207,304) (Fig. 15). Primordial follicles, thecal cells, oocytes, ovarian surface epithelium, and luteal cells lack ER-β immunoreactivity in adult rat and mouse ovaries (207,208,304–306,309, 313,340), although some cytoplasmic staining is reported when using certain antisera (207,304). Granulosa cell specific expression of ER-β in the rat ovary was reproduced using three separate anti-rat ER-β antisera, two raised against residues 467–485 and a third raised against residues 54–71 (306). Furthermore, independent studies using in situ hybridization produced results congruent with the immunohistochemical findings of strict localization of ER-β mRNA to granulosa cells of healthy growing follicles (43,207,310,339), although Bao et al. (319) reported scattered but specific hybridization for ER-β mRNA in thecal cells of healthy medium and large follicles in the rat ovary.

A study by Saunders et al. (307) stands in contrast to those above. In this study, strong nuclear immunoreactivity is described throughout the different somatic cell types of the rat ovary, including thecal, interstitial, and luteal cells, as well granulosa cells of maturing follicles when using anti-rat ER-β antisera raised against residues 196–213. Yang et al. (313) reported similar findings of ER-β immunoreactivity in thecal/interstitial cell preparations from hamster ovary using different ER-β–specific antisera. Still, the results of Saunders et al. are problematic because they include reports of substantial ER-β immunoreactivity in other reproductive tissues of the rat, such as the oviduct and uterus (307), that are otherwise thought to possess relatively low or null levels of ER-β mRNA and immunoreactivity. Two obvious differences between the immunohistochemical study of Saunders et al. (307) and others are the aforementioned use of different antisera as well as different tissue fixative.

The above discrepant findings notwithstanding, ER-β is clearly the predominant ER form present in granulosa cells of healthy growing follicles in the rodent ovaries. Any specific distribution of ER-β throughout the subpopulations of granulosa cells of growing follicles has been difficult to ascertain. In general, ER-β immunoreactivity is continuous and strong among the mural cells and relatively uniform among those layers closer to the antrum and oocyte (304,306,309). However, not all granulosa cells of healthy growing follicles in the rodent ovary expressing ER-β as negative cells are found randomly distributed throughout the follicle (304,309).

Data from Western blots of nuclear extracts from whole ovaries or isolated granulosa cells support the immunohistochemical data that ER-β is the predominant receptor form in rodent granulosa cells (308,309). Electrophoresis mobility shift assays of whole cell extracts from ovaries or isolated granulosa cells of rats provide further evidence of the predominance of ER-β (308,309,346). Sharma et al. (308) found that antisera raised against residues 54–71 of rat ER-β produced the optimum results on Western blot and detected immunoreactive bands of 58/52 to

46/44 kDa in size, the 58-kDa protein presumably a full-length ER-β (ER-β1 in Fig. 2) (308). Antisera raised against the AF-2 (residues 182–485) domain of rat ER-β detected a single specific protein of 60 kDa in whole cell extracts from both rat and mouse ovaries that comigrates with the putative full-length ER-β (309). Choi et al. (340) and Hiroi et al. (305) produced comparable findings of immunoreactive proteins at 60 and 55 kDa in rat and mouse ovarian extracts when using either a monoclonal antibody mapped to residues 272–285 of human ER-β or antisera raised against C-terminal residues 467–485 of rat ER-β, respectively. In the hamster, a single ER-β immunoreactive protein of 54 kDa was detected (313).

Several different ER-β isoforms are present in rodent ovaries, notably ER-β2, ER-β1-δ3, and ER-β2-δ3, the latter being a compound form of the two former variants (Fig. 2). ER-β1-δ3 is an exon 3 deletion that results in a receptor lacking the C-terminal zinc finger of the DNA binding domain and is therefore unable to bind an ERE (83). ER-β2 is an especially interesting variant because it possesses an insert of 18 amino acids in the N-terminal region of the LBD (Fig. 2) that causes a 35-fold reduction in affinity for estradiol and a 1,000-fold decrease in estradiol-induced transactivational activity in vitro (83). Pettersson et al. (83) speculated that the reduced affinity of ER-β2 for estradiol may make the isoform especially attune to mediating estradiol actions within healthy growing follicles where intrafollicular estradiol levels greatly exceed that required to activate wild-type ER forms. Differential reverse transcriptase polymerase chain reaction indicates a 1:1 ratio of ER-β1/ER-β2 mRNA in the ovary of adult (308,346,347) and neonatal rats (315), whereas transcripts encoding ER-β1-δ3 or ER-β2-δ3 variants are detectable but at substantially lower levels (346,347). FLAG epitope-tagged clones of ER-β1 and ER-β2 expressed in 293T cells comigrate at approximately 60 kDa (347), suggesting that resolution of the two isoforms by Western blot may be difficult. However, several studies illustrated two distinct ER-β–specific bands on Western blots from rat (305,308,309,346) and mouse (340) ovarian extracts. Furthermore, O'Brien et al. (346) fractionated rat granulosa cell preparations in a 7.5% SDS-PAGE from which immunoreactive protein bands presumed to be ER-β1 (60 kDa) and ER-β2 (62 kDa) were resolved and further remarked that ER-β1 is present in much greater amounts. In retrospect, past [^{3}H]-estradiol binding data also support a predominance of ER-β1 versus ER-β2 in rat granulosa cells. Because recombinant ER-β1 and ER-β2 differentially bind estradiol with affinities of 0.14 and 5.1 nM, respectively (347), one would expect differentiation of the two forms when assessed by Scatchard plot analysis of estradiol binding data. Therefore, the earlier studies using [^{3}H]-estradiol (348) and more recent studies using [^{125}I]-17α-iodovinyl-11β-methoxyestradiol (309) that report a single, saturable, high-affinity binding factor with a K_d = 0.4 nM in rat ovarian or granulosa cell extracts are congruent with ER-β1 as the predominant form present. Similar data of a single estradiol binding component with a K_d = 1–1.4 nM in hamster ovaries are also reported (349,350).

Most studies on the regulation of ER-β expression during folliculogenesis focus on the rat ovary. Although some report high but relatively constant levels of ER-β mRNA leading up to proestrous (321), more detailed studies indicate a substantial increase in ER-β expression that peaks in medium size (275–450 mm) follicles possessing clearly defined antra (319). Most studies agree on the precipitous decline in ER-β levels that occur shortly after the ovulatory gonadotropin surge and the low levels that remain during estrous in rats (321) and hamsters (Fig. 16A) (313). In retrospect, it evident that the decreasing effect of the gonadotropin surge on ER-β levels may have been first reported by Richards in 1975 (301), in which a 75% decrease in high-affinity estradiol binding sites in granulosa cells of hypophysectomized FSH/estradiol primed rats was found to occur after a single LH treatment. This effect of LH on ER-β expression was reproduced in vivo in gonadotropin-primed rats (309,321) and mice (351) in which a single human chorionic gonadotropin (hCG) injection led to a rapid decrease in ER-β mRNA and protein levels of more than 50% within 6–9 hours (Fig. 16B). Similarly, granulosa cells isolated from pregnant mares' serum gonadotropin (PMSG)-stimulated rats and exposed to hCG in vitro exhibited a comparable loss of ER-β protein when assessed by Western blot (309) or electrophoresis mobility shift assay (308).

Therefore, induction of the LH-signaling pathway in differentiated granulosa cells is the primary stimulus for rapidly decreased ER-β expression. Evidence of a direct role of LH comes from findings that only preovulatory follicles that coexpress LH receptor exhibit a decline in ER-β expression, whereas smaller LH-receptor-negative follicles maintain ER-β expression as late as 24 hours after hCG exposure (Fig. 16C) (309,321). The inhibitory effect of LH on ER-β expression can be reproduced by exposing differentiated granulosa cells to either forskolin or 12-O-tetradeconylphorbol-13-acetate (TPA) (321,346), which are thought to mimic the effects of LH stimulation (352). Furthermore, forskolin or TPA treatment reproduce the same temporal pattern of decreased ER-β expression that follows LH or hCG exposure (353), suggesting a common mechanism. Guo et al. (353) demonstrated that both activators and mimics of the LH signaling pathway elicit a rapid decline in granulosa cell ER-β

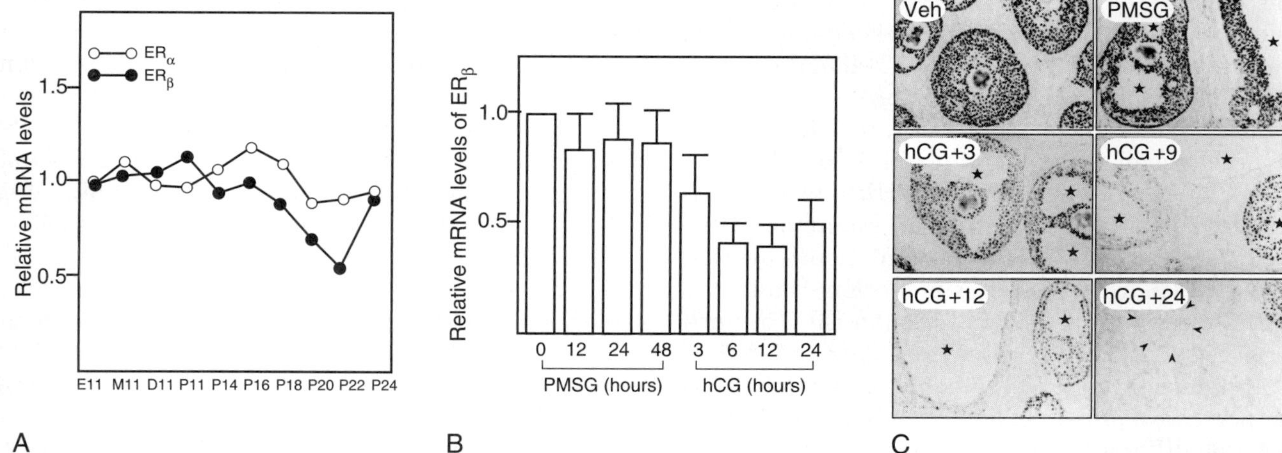

FIG. 16. Regulation of ER-β expression in adult rat ovaries during the estrous cycle or after exogenous gonadotropin treatments. **A:** Quantification of gene expression from Northern blot analysis (not shown) for ER-α and ER-β in adult rat ovaries during the estrous cycle. Band intensities were measured on a phosphorimager and normalized to an S16 internal control for each time point. mRNA levels are shown relative to the level at E1100 (set to 1.0). Sera from animals were used to determine LH concentrations; the onset of the LH surge was observed at 1600 hours of proestrous, and the peak was observed at 1800 hours of proestrous. Annotation at bottom indicates the estrous cycle stage and hour of tissue collection: E, estrous; M, metestrous, D, diestrous, P, proestrous. [Reproduced with permission from (321).] **B:** Reverse transcriptase polymerase chain reaction was used to quantify ER-β mRNA levels in the ovaries of immature rats after treatment with PMSG alone or PMSG for 48 hours followed by hCG. The ratio of ER-β/S16 of control rats with no hormonal treatment was set to 1.0. Shown are the mean ± SE (n = 4). [Reproduced with permission from (321).] **C:** Effect of hCG on ER-β immunoreactivity in rat ovaries treated with vehicle (Veh), PMSG for 48 hours, or PMSG for 48 hours followed by hCG. ER-β immunoreactivity is detected in granulosa cells of small and large antral follicles in vehicle-treated (Veh) and PMSG-treated (PMSG) animals. *Stars* indicate the location of antra in antral follicles. PMSG-treated rats were injected with an ovulatory dose of hCG and ovaries isolated after 3, 9, 12, or 24 hours. The expression of ER-β protein in granulosa cells 9 hours after hCG is reduced in large antral follicles (*left*), greatly reduced in preovulatory follicles (center), but do not change in small antral follicles (*right*). Similar expression is observed 12 hours after hCG administration. One day (24 hours) after hCG treatment, ER-β expression is not detected in corpora lutea (*small arrowheads*) but is highly expressed in preantral (*right*) and small antral follicles. Magnification × 250. [Reproduced with permission from (309).]

levels by decreasing the stability of ER-β transcripts rather than via a repression of gene expression.

In contrast to the periovulatory decrease in ER-β expression that occurs, there is little known about the positive regulation of ER-β expression in rat granulosa cells. Induction of folliculogenesis by PMSG or FSH in hypophysectomized rats leads to a significant rise in estradiol binding among granulosa cells, presumably representative of ER-β (301,354), but this is likely due more to an increased granulosa cell population rather than direct gonadotropin-stimulated ER-β expression. The presence of ER-β in the granulosa cells of primary follicles, which are considered to be insensitive to direct gonadotropin stimulation, supports a minor role for FSH in ER-β expression (304,314). Furthermore, ER-β levels exhibit little change in immature rat or mouse ovaries 48 hours after a single injection with PMSG (Fig. 16B) (309, 321,351), and no reports exist of altered ER-β expression in the ovaries of mice null for FSH signaling (355–357). Sharma et al. (308) found that ER-β protein levels in gonadotropin-primed rat granulosa cells are highest if assessed after isolation and decrease steadily to undetectable levels within 72 hours in culture, even in the presence of FSH and testosterone. However, this pattern was not totally reproduced at the level of ER-β mRNA, which declines by only 55% when cultured for the first 24 hours in the absence of hormones but stabilizes upon the addition of FSH and testosterone (308), suggesting that FSH may be more important for maintenance rather than stimulation of granulosa cell ER-β expression. In contrast to the above findings, preovulatory rat granulosa cells allowed to lose ER-β expression after long-term (6-day) culture exhibit a significant rise in levels 48 hours after exposure to forskolin, a direct PKA activator and believed to mimic FSH signaling (308). Interestingly, FSH may have a greater positive influence on ER-β expression in the hamster ovary where a substantial increase in expression in preantral and antral follicles is observed in hypophysectomized females after 1–2 days of treatment with ovine FSH (313) and in ovaries of neonatal hamsters in which prenatal FSH action is inhibited and exhibit significant decreases in ER-β expression (316).

Evidence of estradiol regulation of ER-β expression is conflicting. Treatment of hypophysectomized rats with estradiol for 3–4 days results in steady and dramatic increases in estradiol binding sites in granulosa cells (301,358), but once again this is more likely due to an increased granulosa cell population. However, Tonetta et al. (354) demonstrated that FSH-induced increases in estradiol binding sites in granulosa cells of hypophysectomized rats are blocked by coadministration of an ER antagonist, suggesting an autoregulatory element for ER expression. More recently, Drummond et al. (315) reported that 1–4 days of treatment with DES has no discernable effect on the levels of ER-β1 or ER-β2 transcripts in immature rat ovaries. Kim and Greenwald (349) found little change in the number of estradiol binding sites in hamster ovaries after estradiol or DES treatment, although animals were exposed for only 1–2 days and therefore significant gains in granulosa cell number may not have been achieved. Still, Yang et al. (313) found that hypophysectomized hamsters exhibit an almost sixfold increase in ER-β expression as detected by immunohistochemistry 24 hours after a single injection of 0.1 mg estradiol-valerate. Furthermore, rat granulosa cells isolated from estradiol-primed versus untreated animals exhibit a four- to sixfold higher basal estrogenic activity on an ERE-luciferase reporter construct, suggesting that individual cellular levels of ER-β are increased by estradiol treatment (308). Also, isolated rat granulosa cells exhibit an increased nuclear intensity for ER-β immunoreactivity 1.5 hours after estradiol exposure that is maintained for 24 hours but totally lost by 48 hours (308). Still, the failure of granulosa cells to totally sustain ER-β expression when maintained in FSH and testosterone, and therefore capable of synthesizing endogenous estradiol, is puzzling. It is plausible that estradiol regulation of ER-β expression in granulosa cells may require the actions of ER-α in thecal cells, which cannot be mimicked when granulosa cells are isolated in culture. Preservation of ER-β expression in preantral follicles of ER-α–null ovaries argues against this hypothesis (205,311). Therefore, although the signaling factors that may be most important to inducing ER-β expression remain unknown, it is clear that the level of ER-β is highly dependent on the state of granulosa cell differentiation as cells from primary, growing, and preovulatory follicles exhibit divergent expression patterns and mechanisms of regulation.

Domestic Animals. Descriptions of ER-β expression in the ovaries of domestic animals are limited but generally indicate patterns that are congruent with rodent ovaries. In bovine ovaries, Rosenfeld et al. (47) found that ER-β mRNA and immunoreactivity is restricted to granulosa cells of small and large antral follicles with no detectable levels in thecal cells. A follow-up study further demonstrated substantial ER-β immunoreactivity in cumulus oophorus and antral granulosa cells relative to mural granulosa and mentioned significant ER-β expression in oocytes (341), the latter finding being common only to hamsters (313,316). In contrast to the immunohistochemical data, Walther et al. (359) and Berisha et al. (323) found ER-β transcripts are detectable in both granulosa and thecal cells of bovine ovaries. In ovine ovaries, ER-β is highly expressed in granulosa cells of growing follicles (342,343), and levels exhibit little change over the course of the estrous cycle (343). However, Jansen et al. (342) reported via in situ hybridization that ER-β mRNA is greatest in follicles of 3 mm and declines in larger follicles during the early follicular phase). Both reports described low but detectable ER-β expression in thecal cells of growing follicles in the sheep ovary (342,343) as well as in ovarian surface epithelia (343). In porcine ovaries, ER-β is almost equally expressed among the granulosa and theca interna of medium and large follicles, as well as appreciable detection in the oocyte and ovarian surface epithelia (324). Like the rodent, atretic follicles in porcine ovaries exhibit reduced ER-β expression (324). LaVoie et al. (344) found that ER-β mRNA levels in whole ovarian lysates exhibit little change during the porcine estrous cycle and are relatively equal when comparing follicles of 1–5 mm in size.

An interesting commonality among bovine, ovine, and porcine ovaries is the presence of a truncated ER-β isoform lacking most of the ligand binding domain, termed ER-βΔLBD in the cow (359) and ER-β-δ5 in sheep and pig (343,344). In all three species, ER-β-δ5 is a deletion of exon 5 and encodes a receptor isoform that is truncated 15 residues into the LBD and terminates with four to nine heterologous residues depending on the species (343,344,359) (Fig. 2). LaVoie et al. (344) and Choi et al. (340) produced Western blot evidence that ER-β-δ5 may indeed be translated as a 35-kDa protein in porcine ovaries. Given that ER-β-δ5 lacks the LBD, it predictably exhibits little estrogen-induced transactivational activity when overexpressed in vitro but does possess ligand-independent transactivational activity that is threefold lower than full-length ER-β, suggesting preservation of AF-1 functions (344). When coexpressed, ER-β-δ5 is also able to reduce the in vitro activity of full-length ER-β by approximately 40% when present in excess (344,359).

Primates. In ovaries of known fertile cynomolgus monkeys, Pelletier et al. (23) found ER-β expression via in situ hybridization in granulosa cells of follicles at all different stages of development and substantial labeling in theca interna and ovarian surface epithelia. Similar findings are described by Pau et al. (328) in granulosa cells of large follicles in rhesus monkey ovary. In the baboon ovary, Pepe et al. (360) found ER-β immunoreactivity to be abundant in granulosa cells of follicles at all stages, including large antral follicles but notably lacking in thecal cells. Western blots of baboon ovarian homogenates indicate two specific ER-β protein bands at 55 and 63 kDa (360). An extensive immunohistochemical study in marmoset ovaries using antisera raised against human ER-β found extensive immunoreactivity in granulosa cells of primary, mid, and late follicular stage follicles, including those with a large antrum, but a clear lack of staining in atretic follicles (335). No specific pattern of ER-β immunoreactivity was apparent among subpopulations of granulosa cells within any one follicle (335). In contrast to baboon ovaries but in agreement with cynomolgus monkey ovaries, thecal cells of large growing follicles as well as the ovarian surface epithelia in marmoset ovaries possess appreciable levels of ER-β immunoreactivity (335).

High-affinity estradiol binding sites in normal human ovarian tissue were described many years ago (329–331). Al-Timimi et al. (330) found 45 of 89 normal ovaries from premenopausal women to possess detectable estradiol binding sites; whereas all ($n = 10$) postmenopausal ovaries assayed were devoid of detectable binding. Vierikko et al. (331) reported high-affinity estradiol binding in a similar percentage of premenopausal ovaries but did find 67% of postmenopausal samples were also positive, although the levels of estradiol binding in both are notably lower than that for progesterone. The advent of better reagents has more recently allowed for differentiation of the two ER forms within human ovaries. ER-β expression was first detected in the human ovary by Northern blot analysis (7,44,361,362). Follow-up studies using RNase-protection assays or reverse transcriptase polymerase chain reaction indicate a relatively equal ratio of ERα/ERβ transcripts in mixed cell homogenates from ovaries of pre- and postmenopausal women but exclusively ER-β transcripts in luteinized granulosa cells isolated from women undergoing in vitro fertilization (7,362). Hillier et al. (336) reported equal levels of ER-α and ER-β mRNA in both nonluteinized and luteinized granulosa cells from normal human ovaries, although levels of both ER forms are reduced in the latter. Another study comparing ER-β mRNA and protein levels within isolated thecal and granulosa cells of small antral follicles found mRNA levels are twofold higher in thecal cells, but protein levels are similar in both cell types (303).

Three reports describing immunohistochemical localization of ER-β in human ovaries produced comparable findings of specific immunoreactivity among granulosa cells of follicles from primary to late antral stage and specific but considerably lower levels in thecal cells of preantral and antral follicles (333–335). Furthermore, all three demonstrate substantial ER-β immunoreactivity in human ovarian surface epithelia (333–335). Hillier et al. (336) also found ER-β transcripts are detectable in primary cultures of human ovarian surface epithelial cells; however, similar evaluations in cell lines derived from human ovarian surface epithelium indicate no detectable ER-β mRNA (362).

There is little known about ER-β regulation human ovaries. Jakimiuk et al. (303) found ER-β expression is significantly decreased in thecal and granulosa cells of dominant follicles relative to those at earlier stages, yet ER-β protein levels did not mirror this difference. In primary cultures of human granulosa-luteal cells, ER-β is more highly expressed relative to ER-α and levels of the former increase 1.6-fold over a 10-day period during which spontaneous luteinization is believed to occur, whereas ER-α levels remain constant throughout (338). Treatment of human granulosa-luteal cells maintained in culture for 7 days with hCG reduces ER-β and ER-α expression by almost 50% within 24 hours, and this effect is mimicked by activators of PKA (forskolin) or PKC (TPA) (338), suggesting that ER-β is reduced by LH in human granulosa cells in a manner similar to that found in rodent ovaries.

Studies to date indicate that human tissues express a unique C-terminal variant of ER-β termed ER-β_{cx} (361) or human ER-β2 (345,363) (Fig. 2). ER-β_{cx} is a 495 amino acid long isoform in which the 61 C-terminal residues are replaced by a heterologous string of 26 amino acids encoded by an alternative downstream exon (361,363). Transcripts encoding ER-β_{cx} are detectable in multiple human tissues but most especially in ovary, testis, thymus, and spleen and are often at levels that are equal to or greater than wild-type ERβ (361,363). Immunohistochemistry using ER-β_{cx}–specific antisera indicate the variant is nuclear localized in granulosa cells of early follicles (345). The physiological function of ER-β_{cx} remains unclear. It is unable to bind estradiol, but in vitro studies indicate the isoform may preferentially heterodimerize with ER-α and act as a dominant negative modulator of ER-α action. Rodents also possess an ER-β variant (ER-β2) that poorly binds estradiol, but this receptor form does not inhibit wild-type ER-α or ER-β activity (83), and categorical studies have shown that the rodent ER-β2 variant is not detected in human tissues (345,364).

Ovarian Phenotypes in Murine Models of Disrupted Estrogen Signaling

Mice Lacking ER-α

Neonatal and prepubertal ER-α–null mice exhibit relatively normal ovaries except for a sporadic presence of large antral follicles, indicative of premature folliculogenesis (192,311). Adult ER-α–null females are anovulatory and hence infertile and exhibit ovaries that possess normal preantral and small antral stage follicles, multiple hemorrhagic cysts, absence of corpora lutea, reduced number of interstitial glandular cells, and mast cell infiltration in the interstitium (104,192,311,365) (Fig. 17 and Table 7). The cystic and hemorrhagic follicles likely originate from antral follicles that fail to ovulate due to acyclicity (i.e., lack of an LH surge) and therefore become atretic and accumulate within the ovary. They are characterized by a mural granulosum of one to several cells thick surrounding an enormous fluid-filled antrum containing blood and immune cells, a degenerating ovum if visible at all, elevated FSH-receptor and LH-receptor expression in the granulosum, and a hypertrophied theca that also possesses elevated LH-receptor levels (311,351) (Fig. 17). Therefore, ER-α is not required for the recruitment and early growth of follicles or the induction of gonadotropin receptors in thecal and granulosa cells but is vital to maintaining the health of late stage follicles in the mouse ovary (192,311).

ER-α–null (αERKO) females exhibit a severely disrupted reproductive hormonal milieu, which in toto represents the cause and effect of the overt ovarian phenotypes (351,366) (Table 7). Plasma LH levels in αERKO females are elevated three- to eightfold relative to their wild-type littermates, whereas FSH levels remain normal (366). Hence, the ovarian phenotypes of thecal hypertrophy (311), elevated steroidogenic enzyme expression, and increased sex steroid synthesis in αERKO females are congruent with hypergonadotropic-hypergonadism (366). A synopsis of evidence that supports acyclicity and chronically elevated LH as a primary cause of the ovarian phenotypes in ER-α–null females is as follows:

1. The cystic and hemorrhagic follicles appear after the onset of puberty (approximately 40 days of age) (311).
2. The cystic follicles and elevated steroidogenesis are prevented when plasma LH levels are reduced to normal via treatments with a GnRH antagonist (351,367) (Fig. 17).
3. Transgenic mice possessing elevated LH but functional ER-α exhibit a comparable ovarian phenotype (368–370) that is rescued by periodic ovulatory doses of exogenous hCG to induce luteinization (371,372).
4. Immature αERKO females successfully ovulate and form corpora lutea when treated with exogenous gonadotropins before the onset of the cystic phenotype (351,365,367).

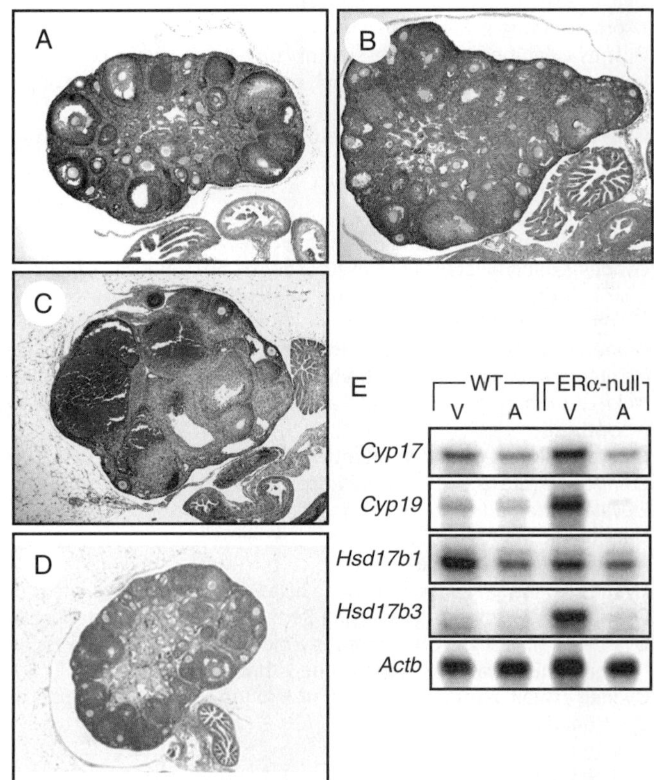

FIG. 17. Ovarian phenotypes in ER-null mice. **A–C:** Shown are cross-sections from representative adult ovaries of wild-type (**A**), ER-β–null (**B**), and ER-α–null (**C**) female mice. Wild-type and ER-β–null ovaries each exhibit all stages of folliculogenesis except corpora lutea, although large antral follicles are sparse in the ER-β–null ovary. In contrast, ER-α–null ovaries are characterized by several large, hemorrhagic, and cystic follicles; a sparse number of follicles at the early stages of proliferation; and a lack of corpora lutea. **D:** Cross-section of a representative ER-α–null ovary after prolonged treatment with a GnRH antagonist. Circulating LH levels were reduced in ER-α–null females by treatment with a GnRH antagonist (60 μg Antide) every 48 hours from the age of 28–53 days. Ovaries were collected within 24 hours of the final treatment. The characteristic ovarian phenotypes of ER-α–null mice (shown in C) are prevented by reducing circulating LH levels, indicating these more dramatic phenotypes are secondary to the loss of ER-α actions in the hypothalamic-pituitary axis that are necessary to maintain proper LH levels and not due to the loss of ER-α within the ovary. [Reproduced from (351).] **E:** Ribonuclease protection assays for the steroidogenic enzymes in adult wild-type (WT) and ER-α–null ovaries treated either with vehicle (V) or a GnRH antagonist (A) every 48 hours for 12 days (as described above). ER-α–null ovaries exhibit increased expression of *Cyp17* and *Cyp19*, both of which are reduced to normal after reduction of gonadotropin levels. ER-α–null ovaries also uniquely express *Hsd17b3*, a Leydig cell specific enzyme, and this expression is dependent on gonadotropin stimulation as it is ablated after the reduction of circulating LH levels with GnRH-antagonist treatments. All samples are normalized to β-actin (*Actb*) mRNA levels. [Reproduced from (366).]

TABLE 7. *Endocrine and ovarian phenotypes in mice null for estrogen signaling*

ER-α–null females
Elevated plasma LH.
Normal plasma FSH.
Anovulatory and infertile.
Ovaries are enlarged, possess multiple hemorrhagic, cystic follicles, lack corpora lutea, and exhibit mast cell infiltration.
Ovaries exhibit increased expression of gonadotropin receptor mRNAs.
Ovaries exhibit increased steroidogenic enzyme expression and secrete abnormally high levels of androstenedione, testosterone and estradiol; progesterone is low to normal.
Ovaries exhibit ectopic expression of *Hsd17b3*, a Leydig cell specific enzyme for the reduction of androstenedione to testosterone.
Immature females exhibit normal ovulatory response to exogenous gonadotropins and harvested oocytes behave normally in in vitro fertilization assays.
Isolated follicles exhibit normal rates of growth and ovulation in vitro but continue to produce elevated levels of sex steroids.
ER-β–null females
Normal plasma LH.
Normal plasma FSH.
Oligoovulatory and subfertile (some animals infertile).
Ovaries exhibit all stages of folliculogenesis but a paucity of corpora lutea.
Ovaries exhibit normal expression of steroidogenic enzymes and secrete relatively normal levels of progesterone, androstenedione, testosterone, and estradiol.
Granulosa cells of preovulatory follicles exhibit increased androgen receptor immunoreactivity.
Immature animals respond poorly to gonadotropin-induced ovulation.
Isolated primary granulosa cells and in vitro propagated follicles exhibit measurable decreases in *Cyp19* and *Lhcgr* expression and estradiol synthesis.
Isolated follicles exhibit an attenuated rate of growth and ovulation in vitro.
Compound (ER-α/ER-β)-null females
Elevated plasma LH.
Normal plasma FSH.
Anovulatory and infertile.
Ovaries exhibit few large healthy follicles and lack corpora lutea.
Ovaries exhibit follicles possessing Sertoli-like cells and ectopic expression of *Sox9*, a Sertoli cell specific gene.
Ovaries exhibit slightly increased steroidogenic enzyme expression and secrete increased testosterone, slightly higher levels of androstenedione and estradiol; progesterone is low to normal.
Ovaries exhibit ectopic expression of *Hsd17b3*, a Leydig cell specific enzyme for the reduction of androstenedione to testosterone.
Cyp19-null females
Elevated plasma LH.
Elevated plasma FSH.
Anovulatory and infertile.
Ovaries lack corpora lutea, are enlarged, and possess multiple hemorrhagic cystic follicles that worsen with age.
Ovaries secrete abnormally high levels of androstenedione and testosterone, but undetectable estradiol.
Ovaries exhibit ectopic expression of *Hsd17b3*, a Leydig cell specific enzyme for the reduction of androstenedione to testosterone.
Ovaries exhibit follicles possessing Sertoli-like cells and ectopic expression of *Sox9*, a Sertoli cell specific gene.
Immature animals respond poorly to gonadotropin-induced ovulation but harvested oocytes behave normally in in vitro fertilization.

See text for references.

Others propose that aberrantly high intraovarian histamine levels due to infiltration of mast cells in the interstitium may also contribute to cyst formation in ER-α–null ovaries (192). The above findings strongly indicate that a primary role of ER-α in murine ovarian function is extraovarian, that is, as an essential mediator of the endocrine actions of estradiol in the hypothalamic–pituitary axis that are critical to gonadotropin regulation (see section on Extraovarian Roles of Sex Steroids in Ovarian Function). Indeed, prolonged treatment of female rodents with antiestrogens that cross the blood–brain barrier (e.g., ZM-189, 154, EM-800) produces an αERKO-like gonadotropin profile and similar ovarian phenotype, whereas treatments with tamoxifen, a receptor antagonist that does not alter gonadotropin secretion, generates no such effects in the ovary (240–242).

A prominent phenotype in ER-α–null (αERKO) ovaries that may be indicative of an intraovarian role for ER-α is their elevated capacity to synthesize androgens (366). Relative to their wild-type littermates, αERKO females possess plasma levels of androstenedione and testosterone that are increased three- and 40-fold, respectively (366). Indeed, LH is the primary stimulus of thecal cell androgen synthesis (373), and plasma LH and thecal cell LH-receptor levels are both significantly increased in αERKO females. Therefore, it is not totally unexpected that αERKO ovaries exhibit increased expression of the enzymes necessary for androgen synthesis, most notably a remarkable increase of *Cyp17* expression (Fig. 17) (366), the enzyme necessary for the final step of androstenedione synthesis (373). However, the elevated plasma androgens found in αERKO females is likely not due solely to LH-mediated hyperstimulation of the theca. Estradiol from granulosa cells is proposed to mediate an intraovarian short feedback loop upon thecal cells to negatively modulate androgen synthesis during the later stages of folliculogenesis by primarily targeting CYP17 activity (374). Given that ER-α is the dominant ER form expressed in rodent thecal cells, the elevated *Cyp17* expression found in αERKO ovaries suggests ER-α mediates this effect of estradiol on androgen synthesis. Additional support for a specific ER-α–mediated effect on thecal cell steroidogenesis comes from our findings that chronically elevated LH in wild-type and ER-β–null females leads to a much more moderate increase in *Cyp17* expression and androgen synthesis relative to αERKO females, presumably due to the presence of ER-α (375). Furthermore, in vitro cultured ER-α–null follicles secrete substantially more androgens relative to similarly propagated wild-type follicles even though the level of gonadotropin stimulation is held constant (376). Therefore, it may be concluded that ER-α is paramount to maintaining proper androgen synthesis in rodent females via endocrine actions in the hypothalamic–pituitary axis that negatively modulate LH secretion and intraovarian actions on thecal cells to negatively modulate *Cyp17* expression.

The granulosa cell specific enzymes necessary for estradiol synthesis, *Cyp19* and *Hsd17b1*, are also expressed at elevated levels in αERKO ovaries (Fig. 17) (366). As a result, plasma estradiol levels are typically increased almost tenfold relative to wild-type littermates (366). It is likely that the plasma androstenedione levels discussed above would be even greater in αERKO females if ovarian aromatase activity was not also equally elevated and providing for efficient conversion (366). Although these findings in αERKO females suggest that ER-α may negatively modulate estradiol synthesis in granulosa cells, there is little precedent for this hypothesis. The chronically elevated plasma LH in αERKO females is not likely to positively influence granulosa cell estradiol synthesis as most evidence indicates that LH stimulation of granulosa cells leads to decreased aromatase activity (377–379). Furthermore, individually cultured ER-α–null follicles in vitro continue to exhibit heightened estradiol synthesis relative to similarly propagated wild-type follicles in an environment of controlled FSH and LH stimulation (376). Instead, increased aromatase activity in ER-α–null ovaries is likely due to ER-β–mediated positive actions of estrogens on FSH-induced granulosa cell steroidogenesis (378,380–383). Androgens have also been shown to augment FSH-induction of estradiol synthesis (378,384,385), suggesting that elevated androstenedione and testosterone levels in αERKO females may contribute to increased estradiol synthesis in granulosa cells.

The several attempts to induce ovulation ER-α–null females collectively indicate an age-dependent effect of the loss of ER-α on ovulatory efficiency. Schomberg et al. (311) reported that 4-month-old αERKO females do not successfully ovulate after exogenous treatments with PMSG and hCG, although this conclusion was based solely on the absence of corpora lutea rather than actual oocyte numbers. Two later studies focusing on younger αERKO females (3–5 weeks) produced comparable results of successful ovulation (oocytes in the oviduct) and formation of functional corpora lutea, although the oocyte yield was reduced compared with age-matched wild-type females (351,365). Oocytes harvested from ER-α–null females successfully undergo in vitro fertilization, indicating that ER-α may not be important to oocyte function (351). In contrast, Dupont et al. (192) reported that the other line of ER-α–null females (ERαKO) fail to ovulate or exhibit corpora lutea after exogenous gonadotropin treatments even at 21–25 days of age. Repeated studies in immature αERKO females of pure C57BL6 background (versus the mixed background of earlier studies) found these mice fully respond to gonadotropin-induced ovulation and exhibit an average oocyte yield comparable with that of similarly treated wild-type littermates (367). This discrepancy in in vivo ovulatory success between the αERKO and ERαKO mice may be due to differences in genetic strain, the considerable disparity in the doses of gonadotropin used, or the small sample size in the Dupont et al. study (n = 3). In further support of a minor role for ER-α in ovulation, Emmen et al. (376) demonstrated that individual ER-α–null follicles propagated and induced to ovulate in vitro did not respond differently from similarly cultured wild-type follicles.

Mice Lacking ER-β

Several facets of ovarian physiology thought to be dependent on the paracrine actions of estradiol are clearly maintained in ER-α–null ovaries, including granulosa cell proliferation, LH-receptor and aromatase expression, antrum formation, and the attenuation of atresia. The preservation of these estrogen actions in ER-α–null ovaries strongly suggests their dependence on ER-β, which continues to be expressed in αERKO granulosa cells (311,366). Neonatal ovaries from ER-β–null mice exhibit no gross abnormalities (104,192,193,386). Adult ER-β–null ovaries possess all stages of folliculogenesis (192,193,376,386), a slight but perceptible increase in atretic follicles (193,376,386), and a paucity of corpora lutea (Table 7) (193,376). Cheng et al. (386) found that among 5-month-old ER-β–null females, less than 30%

exhibit corpora lutea compared with 100% incidence in wild-type littermates (386). Furthermore, when corpora lutea are present in ER-β–null ovaries, there is rarely more than two per ovary versus upward of seven per ovary in wild-type females (386). Continuous mating studies with ER-β–null females indicate a severe impairment in fecundity, but there is discernible variability in this phenotype among age-matched animals (192,193). Consistent among βERKO pregnancies is reduced litter size of approximately one-third relative to wild-type (192,193). However, whereas some βERKO females exhibit the expected number of pregnancies over a 4-month period, others become pregnant only once, and still others exhibit total infertility (193). This variability in fecundity among ER-β–null females remains puzzling but is reported in both independently generated lines (192,193). Given that the ER-β–null lines represent distinct targeting schemes for the *Esr2* gene on different genetic backgrounds, the observed variability in fertility is likely related to the loss of ER-β functions. A similar variability in oocyte yield is found when immature ER-β–null females are induced to ovulate by exogenous gonadotropin treatments (192,193,367). Therefore, the observed subfertility/infertility in ER-β–null females appears to originate from disrupted ovarian function that is best characterized as infrequent and inefficient spontaneous ovulation.

Histological evaluation of ER-β–null ovaries after gonadotropin-induced ovulation reveals multiple preovulatory follicles possessing underdeveloped antra and minimal cumulus expansion (192,193,367). Krege et al. (193) reported obvious corpora lutea in immature βERKO ovaries 20 hours after hCG treatment, but Dupont et al. (192) remarked that no luteal cells were observed in similarly treated ERβKO ovaries that did not ovulate. We have since shown that immature βERKO females do indeed exhibit functional corpora lutea after induced ovulation, as indicated by histological evaluation of the ovaries and a substantial rise in plasma progesterone levels (367). Luteinization and increased progesterone synthesis is also observed when individually propagated ER-β–null follicles are exposed to an hCG bolus in vitro (376). Furthermore, luteinization occurs in unruptured preovulatory follicles both in vivo and in vitro (104,193,376), resulting in "trapped" oocytes similar to those observed in other null mouse models (179,387).

ER-β–null females exhibit a relatively normal reproductive endocrine milieu (366). Basal gonadotropin levels are within the range of control values (366), although one report remarked that plasma LH levels are slightly increased in βERKO females (388). This notwithstanding, ER-β–null females do not exhibit the ovarian phenotypes that are associated with chronically elevated LH (351,370). Therefore, ER-α is clearly the principal receptor involved in mediating the negative feedback effects of estradiol on gonadotropin secretion from the hypothalamic–pituitary axis. Sex steroid hormone levels in ER-β–null females are also comparable to wild-type littermates (366). Additionally, the overtly estrogenized uterine and vaginal tissues observed in ER-β–null females indicate sufficient ovarian estrogen synthesis in the absence of functional ER-β (104). Preservation of normal estradiol levels in βERKO females is surprising given the evidence that estrogens are necessary to maximize FSH stimulation of aromatase activity in granulosa cells (378,380–383). As mentioned below, androgens also augment FSH induction of estradiol synthesis (378,384,385) and provide for compensatory actions in ER-β–null granulosa cells. Still, βERKO females clearly exhibit an abnormal ovarian cycle, as indicated by persistent estrous and a paucity of corpora lutea in the ovary, suggesting that the preovulatory rise in estradiol necessary to induce the gonadotropin surge from the hypothalamic–pituitary axis may be insufficient.

Mice Lacking ER-α and ER-β

Congruent with the loss of ER-α, αβERKO females are infertile and do not spontaneously ovulate (192,194). Adult αβERKO ovaries possess structures appropriate to the normal ovary, including primordial and growing follicles, although the latter possess an underdeveloped antrum, reduced granulosa number, and a thin, poorly structured theca (Table 7 and Fig. 18) (192,194). Cystic follicles that are somewhat characteristic of those found in ER-α–null ovaries are present but are not as large or as hemorrhagic in αβERKO ovaries and are more apt to exhibit a thinner layer of granulosa cells (192,194). Hemorrhagic and cystic follicles in the mouse ovary are due to ovarian acyclicity compounded by chronically elevated plasma LH in the presence of normal FSH levels, as illustrated by ER-α–null females (351) and transgenic mice overexpressing an *Lhb* gene (370,372). Therefore, the absence of hemorrhagic and cystic follicles in αβERKO ovaries indicates an important intraovarian role for ER-β in this pathology, which is confirmed by the absence of such diseased follicles in transgenic ER-β–null mice that possess elevated LH (375).

A remarkable feature of adult αβERKO ovaries is the presence of seminiferous tubule-like structures that often occupy large portions of the gonad but are conspicuously absent in ER-α– or ER-β–null ovaries (192,194) (Fig. 18). Morphological observations indicate these structures are the "ghosts" of atretic follicles because they possess an intact basal lamina,

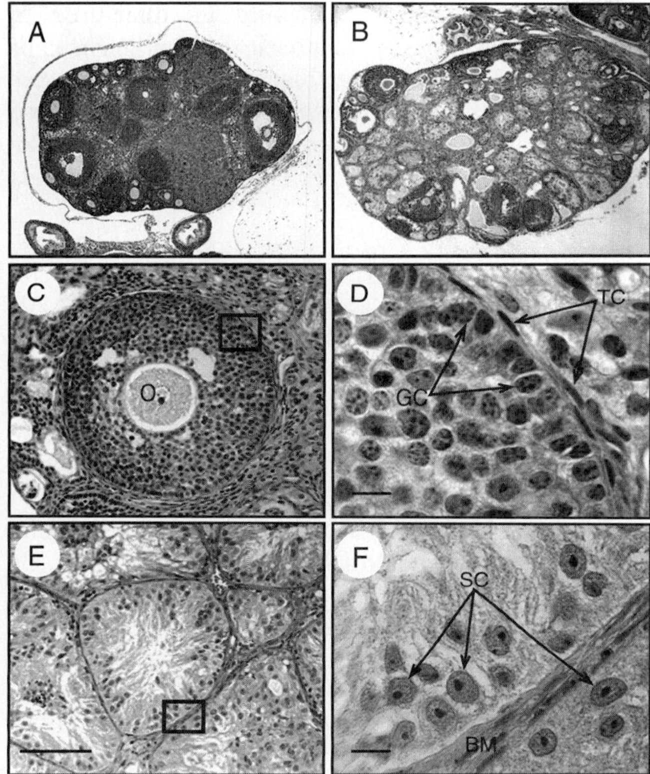

FIG. 18. Ovarian phenotype in adult mice lacking both ER-α and ER-β. **A, B:** Low-power magnification of representative adult wild-type (A) and ER-αβ–null (B) ovaries. The wild-type ovary exhibits all stages of folliculogenesis, including a preovulatory follicle (*right*) and corpus luteum (*right, bottom*). The ER-αβ–null ovary exhibits a few relatively healthy maturing follicles but is overwhelmed by multiple "sex-reversed" follicles. **C–F:** High-power magnification of representative follicles from adult ER-αβ–null ovaries (C, 66×; D, 330×). A healthy follicle shows a single oocyte (O), several layers of granulosa cells (GC), an intact basal lamina, and thecal cells (TC) (E, 66×; F, 330×). A representative "sex-reversed" follicle in which there is no longer evidence of an oocyte and the somatic cells have undergone re-differentiation to a Sertoli cell (SC) phenotype. Preservation of the basement membrane (BM) of the follicles provides for the "tubular" appearance that assimilates testicular cords. The Sertoli-like cells (SC) possess the characteristic tripartite nucleolus and veil-like cytoplasmic extensions (F). Scale bars: E = 100 μm; D, F = 10 μm. [Reproduced from (194).]

partial layers of granulosa cells, and an invariably degenerating oocyte (194) (Fig. 18). Furthermore, these structures are postpubertal because no such structures are detected in αβERKO ovaries at 10 days of age (194) and or in ERαβKO ovaries as late as 23 days of age (192). The defining characteristic of the ghost follicles that led to their initial description as testis-like seminiferous tubules is the overt presence of Sertoli-like cells in the lumen (192,194). Several morphological features of these cells in both αβERKO and ERαβKO ovaries are strongly indicative of a Sertoli cell phenotype (Fig. 18), including alignment with the basal lamina of the follicle wall, a tripartite nucleolus, numerous veil-like cytoplasmic processes extending inward toward the lumen, presence of ectoplasmic specializations that are unique to Sertoli cells in prepubertal testis, and immunoreactivity for Müllerian-inhibiting substance and sulfated glycoprotein-2 (SGP-2).

Granulosa and Sertoli cells fulfill analogous gametogenic roles as "nurse" cells in the ovary and testis, respectively, and are postulated to derive from a common embryological precursor cell during early gonadal differentiation (389). Thus, the origin of the Sertoli-like cells in αβERKO ovaries remains a perplexing question. Do they originate from a bipotent precursor population present in the αβERKO ovary at birth or are they the result of granulosa cell transdifferentiation after oocyte death and follicle atresia? Reports of extrafollicular Sertoli cells in the ERαβKO ovary (192,390) support the existence of an embryological precursor cell population, yet similar interstitial cell populations are not found in ovaries from the other αβERKO line (194). The following findings common to both lines support a pathway of granulosa cell transdifferentiation: the Sertoli-like cells are not apparent before 30 days of age, the spherical shape of the "tubules" indicated by two-dimensional histology suggests they originate from a once healthy follicle in which only the basement membrane remains, and the appearance of Sertoli cells is strongly correlated with oocyte death and follicle atresia (192,194). Furthermore, *Sox9* is highly expressed in both αβERKO and ERαβKO ovaries (194) and localized to the Sertoli-like cells (390) (Fig. 18). SOX9 is an *Sry*-related transcription factor that is critical to normal Sertoli cell differentiation during testis development in rodents and humans (391,392). Transgenic overexpression of *Sox9* leads to phenotypic male gonadal development in XX mice and therefore acts "downstream" of the Y-linked male-determining gene, *Sry* (393,394). In strong favor of a pathway of granulosa cell transdifferentiation, *Sox9* expression in ERαβKO ovaries is restricted to the granulosa cells of atretic follicles, precedes the overt appearance of Sertoli-like cells, and is not detectable in the granulosa cells of primordial follicles (390). Furthermore, those cells expressing *Sox9* are initially present within the confines of the follicle ghosts and appear only in the intrafollicular spaces of the ovary after the phenotype is progressed to an advanced stage (390), supporting a follicular (i.e., granulosa) origin.

Similar phenotypes of "sex reversal" are reported in fetal rodent ovaries after transplantation to an adult host (395,396), in vitro exposure to purified Müllerian inhibiting substance (397), transgenic Müllerian inhibiting substance overexpression in vivo (398), transgenic *Sox9* overexpression in vivo (393,394), or targeted disruption of the *Wnt4* gene in mice (399). Although the αβERKO ovarian phenotype shares a number of morphological similarities with the above findings, including aberrant expression of the Müllerian inhibiting substance, SGP-2, and *Sox9* genes, a remarkable distinction in the αβERKO ovary is the postnatal onset of the phenotype in the αβERKO ovary. The above descriptions of ovarian sex reversal are reported to occur or even require *fetal* ovarian tissue (397,400). Therefore, the observed sex-reversal adult αβERKO ovaries are the first to be described in the postnatal mouse gonad, indicating that the potential of female ovarian somatic cells to redifferentiate into Sertoli-like cells is present throughout life in the mouse.

A causal link between the loss of ER functions and postnatal ovarian sex reversal is unclear. Dupont et al. (192) reported that female mice possessing only one functional ER-α allele but lacking functional ER-β (i.e., *Esr1*$^{+/-}$ *Esr2*$^{-/-}$) also exhibit the αβERKO ovarian phenotype, suggesting a gene-dosage effect for ER-α (192). Similar phenotypes in the mammalian ovary are reported to follow massive germ cell loss (401). The remarkable oocyte attrition that occurs in the αβERKO ovaries suggests a similar mechanism occurs in this model, yet it remains unclear whether the progressive loss of germ cells in αβERKO ovaries illustrates the loss of critical intraovarian estrogen signaling that is dependent on ERα/ERβ cooperation. A role of direct estrogen/ER actions in gonadal differentiation is supported by reports of sex reversal in turtles and whiptail lizards after developmental exposure to aromatase inhibitors (402,403). The mouse *Sox9* promoter lacks an obvious ERE but does possess two consensus binding sites for GATA-1 (392), a transcription factor expressed in Sertoli cells during differentiation of mouse testis but repressed by germ cells in adult testis (404). Furthermore, ER-α can repress the transcriptional activities of GATA-1 (405), suggesting that aberrant *Sox9* expression leading to Sertoli-cell differentiation in αβERKO ovaries may be due to increased GATA-1 activity after the combined loss of ER-α and oocyte-derived inhibitory factors.

αβERKO females exhibit a gonadotropin profile that is similar to αERKO females, although plasma LH levels are noticeably higher. Therefore, the hormonal milieu in αβERKO females may impact the ovarian sex reversal observed (194,366). Transgenic mice possessing chronically elevated plasma LH also display a progressive loss of oocytes but lack any evidence of ovarian sex reversal (370,375,406). Interestingly, αβERKO ovaries do not exhibit αERKO-like increases in ovarian aromatase or estradiol synthesis (366). This finding provides further support that elevated LH leads to increased precursor substrates, which combined with the positive actions of ER-β lead to enhanced estradiol synthesis in ER-α–null ovaries, a scenario that would be disrupted in αβERKO ovaries. The dramatic loss of germ cells in the αβERKO ovary may also impact gonadotropin responsiveness of granulosa cells as oocyte-derived factors are known to positively influence steroidogenesis (407).

Mice Lacking Estradiol Synthesis

Initial reports of Cyp19-null ovaries describe all stages of folliculogenesis, including multiple large antral follicles, but no corpora lutea at 10–14 weeks of age (195,196,408). At 21 weeks of age, Cyp19-null ovaries continue to exhibit all stages of folliculogenesis, but larger follicles possess a reduced number of granulosa cells and become enlarged, cystic, and hemorrhagic, similar to ER-α–null follicles (408). By 1 year of age, Cyp19-null ovaries are characterized by a paucity of normal follicles, an increased number of cystic and hemorrhagic follicles, significant collagen deposition, and massive macrophage infiltration into the interstitium (Table 7) (408). Several findings indicate infertility among Cyp19-null females due to an inability to spontaneously ovulate, including absence of corpora lutea at all ages (195,196,408),) lack of pregnancies during continuous mating (196), and total failure to ovulate after exogenous gonadotropin treatments (196) (Table 7).

Surprisingly, initial descriptions of the ovarian phenotypes in Cyp19-null females did not include remarks of Sertoli-like cells similar to those found in αβERKO females. However, Britt et al. (197) later demonstrated that mutant females maintained on a diet free of phytoestrogens exhibit a clearly exacerbated ovarian phenotype by 6 weeks of age that included hemorrhagic and cystic follicles, seminiferous tubule-like structures possessing Sertoli-like cells, and an almost 40-fold increase in *Sox9* expression (Table 7). By 16 weeks of age, the tubule-like structures and Sertoli cells occupy up to 80% of the ovary and are strikingly similar to those found in αβERKO ovaries in both morphology and gene expression (197). Additional supporting evidence of Sertoli-like cells in Cyp19-null ovaries is documentation of Sertoli cell-specific ectoplasmic specializations and immunoreactivity for Espin, an important component of Sertoli cell junctions in testis (197). A similar phenotype of sex reversal has not yet been described in the ovaries of a second line of Cyp19-null females (196).

As expected, the reproductive hormonal milieu in female Cyp19-null mice is severely disrupted. Plasma estradiol levels (196) and ovarian aromatase activity (195) are below the level of detection, supporting the existence of a single aromatase-encoding gene in mice. In turn, plasma testosterone levels in Cyp19-null females are tenfold that found in wild-type littermates (195,196). Increased plasma androgens in Cyp19-null females are likely the combined effect of an accumulation of androgen precursors that occurs in the absence of aromatization to estradiol and increased synthesis due to hyperstimulation of the ovarian theca by fivefold increased plasma LH (195,408). Thus, increased plasma LH and androgens are characteristic of mice lacking ER-α–mediated estrogen actions after the loss of receptor (ER-α–null animals) or ligand (Cyp19-null animals). However, one striking difference between Cyp19-null and ER-α–null females is the four- to sixfold increase in plasma FSH that is unique to the former (408). Interestingly, Cyp19-null females exhibit the above abnormal hormone milieu regardless of phytoestrogen content in the diet (197), suggesting that dietary phytoestrogens may attenuate some of the phenotypes in the reproductive tract but not in the hypothalamic–pituitary axis.

More recent studies have indeed shown that Cyp19-null mice still are responsive to exogenous estrogens (409). Tonic estradiol replacement therapy in Cyp19-null females over a period of 21 days leads to restored gonadotropin homeostasis, improved follicular development in the ovary, a reduced presence of Sertoli-like cells and *Sox9* expression, and restoration of ovulation and corpora lutea formation (409,410). In a similar study, Toda et al. (196) reported that estradiol administration to Cyp19-null females every fourth day for 4 weeks ameliorated the ovarian phenotypes but did not restore ovulation.

Mice Possessing Mutant ER-α

As discussed in the section on The Steroid Receptors: Receptor Mechanisms of Action, the ERs may act as transcription factors on certain genes in the absence of a classic ERE within the regulatory sequences (Fig. 7). This alternative mechanism of ER action, often referred to as "tethering," is thought to occur via interaction with transcription factors that are bound directly to DNA (Fig. 7). To develop an in vivo model for the study of ERα signaling via ERE-independent pathways, Jakacka et al. (209) generated mice that only possess a mutant form of ER-α that lacks the N-terminal zinc finger of the DNA-binding domain. Because this mutation is incorporated into the endogenous *Esr1* gene (i.e., a "knockin"), transcriptional expression of the mutant ER-α is presumably no different from that of the wild-type *Esr1* allele. Animals that are heterozygous for the ER-α mutation, termed NERKI$^{+/-}$ mice, are infertile and exhibit distinct ovarian phenotypes characterized by follicles of all stages of growth but no corpora lutea and lipid-filled cells throughout the ovarian stroma (209). The latter phenotype is similar to that of female mice lacking functional steroid acute regulatory protein (STAR) and correlates with reduced *StAR* expression in NERKI$^{+/-}$ ovaries (209). Upon stimulation with exogenous gonadotropins, NERKI$^{+/-}$ ovaries exhibit multiple large hemorrhagic follicles similar to those found in ER-α–null ovaries (209). NERKI$^{+/-}$ females exhibit normal plasma gonadotropin and estradiol levels and proper expression of steroidogenic enzymes in the ovary (209). Still, induced ovulation in NERKI$^{+/-}$ females is only partially successful in causing oocytes to be released and the formation of corpora lutea, as several preovulatory follicles fail to rupture and become hemorrhagic (209).

Without further data it is difficult to determine the exact cause of the NERKI$^{+/-}$ ovarian phenotypes. Jakacka et al. (209) postulated that the mutant ER-α may act as a dominant-negative form of ER; however, the mutant ER-α does exhibit this property in vitro. Furthermore, prolactin (*Prl*) is a known ERE-regulated gene and exhibits no change in expression in NERKI$^{+/-}$ females (209). An imbalance between ERE-dependent and -independent pathways or tissue-specific inhibitory effects of the mutant ER-α may explain the some of the resulting phenotypes (209).

Intraovarian Roles of Estradiol in Ovarian Function

The extraovarian or endocrine roles of estrogen signaling in ovarian function have long been recognized and are the precedent for steroid-based oral contraceptives (411,412). In contrast, the significance of estrogen signaling within the ovary is not well understood due to inherent difficulties associated with the study of a particular hormone's action within the very tissue from which it is synthesized. Past in vivo investigations using ER antagonists (240–242,413) or aromatase inhibitors (414–416) to inhibit receptor actions within the ovary have provided informative but often equivocal findings (240–242) due to the following limitations:

1. The enormous levels of endogenous estradiol in the ovary are difficult to overcome by pharmacological administration of a receptor antagonist;
2. The enormous levels of aromatase activity in the ovary are difficult to overcome by pharmacological administration of an enzyme inhibitor;
3. It is difficult to discern which effects of a pharmacological ER antagonist are due to actions directly in the ovary versus actions in the hypothalamus or pituitary;
4. The well-characterized agonists (e.g., estradiol and DES) comparably activate both ER-α and ER-β;

5. Early antagonists were not selective for the two known ER isoforms
6. The two ER isoforms may respond differently to agonist or antagonist ligands.

The generation of ER and Cyp19-null mice and development of improved ER-selective ligands or SERMs (Table 3) (68,99,417) have largely overcome these limitations in recent years.

The abundance of data on the localization of ER-α and ER-β within rodent follicles allows for a simplified model of estrogen actions within the ovary. For example, given that ER-α predominates in thecal cells, does it mediate the inhibitory effects of estradiol on LH-stimulated androgen synthesis? In turn, although ER-β is clearly the predominant receptor in granulosa cells, is it responsible for mediating estrogen augmentation of FSH actions? This section uses the contributions from studies of the pertinent null mouse models as a platform to discuss the current view on intraovarian estrogen/ER signaling.

Granulosa Cell Proliferation

In the second edition of this volume, Greenwald and Roy (418) scrupulously reviewed the role of estrogens in granulosa cell proliferation, and their discussion remains contemporary. A number of studies in immature hypophysectomized rodents demonstrated that DES or estradiol treatments over 2–4 days leads to a marked increase in ovarian weight (419,420), a significant rise in the number of medium-sized preantral follicles, and increased DNA synthesis in both thecal and granulosa cells (358,421–423), indicating a direct and gonadotropin-independent effect of estrogens on the ovary. Similar data have been reported after DES treatment of intact immature rats (424). There is equally convincing data that sequential treatment of immature hypophysectomized rodents with DES followed by FSH results in a synergistic response in terms of increased ovarian weight and DNA synthesis and elicits several major indices of granulosa cell differentiation, that is, antrum formation and acquisition of LH-receptor expression that are not observed with either hormone alone (302,421,422,425). Furthermore, coadministration of an antiestrogen (*cis*-clomiphene) inhibits FSH-induced increases in ovarian weight and follicular growth in hypophysectomized rats, suggesting that ER-mediated effects are necessary for a full response to gonadotropin (426). The synergistic actions of estradiol on FSH-induced granulosa cell proliferation have been replicated in vitro (427), although others report that FSH-mediated growth of isolated murine follicles is not affected by antiestradiol antisera or ER antagonist (ICI 182,780) (428). The capacity of estradiol to enhance the granulosa cell response to FSH is not believed to rely on estrogen-induced increases in FSH-receptor levels, although some evidence indicates such an effect may play a greater role in mice versus rats (302,423,425). Still, pretreatment of hypophysectomized rats with antiestrogen (CI628) inhibits FSH-induced increases in granulosa cell FSH-receptor levels (354).

It is generally agreed that the synergistic response of granulosa cells to estrogens and FSH is via mutual coordination of the respective signaling systems (425). The discovery of ER-β and its substantial expression in the granulosa cells of growing follicles and absence in atretic or luteinized follicles strongly suggest that the trophic actions of estrogens are direct and ER-β mediated. Supporting data of a greater importance of ER-β versus ER-α in estrogen-induced granulosa cell proliferation comes from a report of Hegele-Hartung et al. (417), in which the ER-β–selective agonist, 8-vinylestra-1,3,5-(10)-triene-3,17β-diol, induces increased ovarian weight and granulosa cell proliferation to levels comparable with that elicited by estradiol in hypophysectomized rats, whereas the ER-α–selective agonist, 3,17-dihydroxy-19-nor-17α-pregna-1,3,5 (10)-triene-21,16α-lactone, has little effect. Although these data indicate that estradiol/ER-β actions may indeed enhance FSH-induced granulosa cell proliferation, several findings question the overall importance of this. First, ER-β–null ovaries exhibit a relatively normal number of growing preantral follicles with no obvious reduction in the granulosa cell population (192,193,429). Furthermore, immature ER-β–null females treated with PMSG exhibit an expected increase in the number of small and large preantral follicles (192,376). In addition, granulosa cell tumors due to the loss of inhibin-α (*Inha*) in mice continue to proliferate regardless of the presence or absence of functional ER-β (430). It is possible that ER-α may provide some compensatory actions in ER-β–null ovaries because the ovaries of mice null for both ER forms exhibit follicles that are markedly deficient in granulosa cells. However, this phenotype in compound ER-null ovaries may be due to increased atresia rather than lack of cell proliferation (192,194). Furthermore, Cyp19-null mice are presumably devoid of estradiol-dependent ER action but exhibit ovaries possessing all stages of follicle growth, each with a normal complement of granulosa cells (408). Therefore, the sum of these data indicates that although estrogen-induced granulosa cell proliferation is likely mediated by ER-β, it is not obligatory to gonadotropin-induced follicle growth in rodent ovaries.

ER-β may facilitate gonadotropin-induced granulosa cell proliferation via the induction of cyclin D2 (*Ccnd2*), which mediates progression of the cell cycle from the G_0 to S phase. Targeted disruption of the

Ccnd2 gene in mice indicates this cyclin protein is obligatory to granulosa cell proliferation because Ccnd2-null ovaries lack follicles beyond the small preantral stage (431). Furthermore, the arrest in folliculogenesis observed in Ccnd2-null mice is not rescued by exogenous FSH treatment in vivo or in vitro, indicating the critical downstream role of cyclin D2 in gonadotropin-mediated granulosa cell proliferation (431). In rats, *Ccnd2* expression is localized to the granulosa cells of early growing follicles and is significantly induced by PMSG or FSH treatment (432). Estradiol also induces *Ccnd2* expression in rat granulosa cells, and although the response is less rapid relative to that elicited by FSH, the overall induction by estradiol is higher and better sustained over 24 hours (432). Furthermore, estradiol elicits a concomitant decrease in the expression of *Cdkn1b* (p27), a cyclin-dependent kinase inhibitor (432). Estradiol induction of *Ccnd2* in undifferentiated rat granulosa cells in vitro is inhibited by ER antagonist (ICI 164,384), indicating this to be an ER-mediated event and presumably ER-β (432). Therefore, parallel pathways of FSH and estradiol converge to induce and maintain *Ccnd2* expression in granulosa cells, forcing cell-cycle progression and subsequent proliferation (432). Interestingly, adult ER-β–null ovaries possess normal granulosa cell expression of *Ccnd2* (386) and exhibit the expected induction after PMSG exposure (367), further supporting the hypothesis that FSH and estradiol work in parallel. Therefore, these observations and the absence of any gross deficiency in granulosa cell number in ER-β–null ovaries suggests that FSH may play a greater role in inducing *Ccnd2* expression and granulosa cell proliferation. Interestingly, estradiol specifically induces *Ccnd2* expression in keratinocytes via the cAMP/PKA signaling pathway that leads to activation of CREB (433), the same intracellular pathway used by FSH signaling in granulosa cells. These findings suggest that estradiol may also function within the FSH signaling pathway to induce *Ccnd2* in granulosa.

A second mechanism by which estradiol may stimulate granulosa cell proliferation is via interactions with the ovarian IGF-1 axis. Like Ccnd2-null nice, Igf1-null mice exhibit a total failure of follicles to progress beyond the small preantral stage (434,435). This arrest in folliculogenesis in Igf1-null ovaries is primarily due to an >50% reduction in FSH-receptor levels in granulosa cells, severely hampering their response to FSH (435). However, Kadakia et al. (436) demonstrated that Igf1-null granulosa cells are also refractory to estradiol-induced proliferation and exhibit a 50% lower rate of DNA synthesis and decreased *Ccnd2* induction after estradiol treatment in vivo. The poor response of Igf1-null granulosa cells to estradiol may be due to a similar reduction in ER-β levels because IGF-1 is reportedly necessary for the maintenance of ER-β expression in granulosa cells in vitro (437). There are currently no reports describing the levels of ER-β expression in the ovaries of IGF-1–null mice. Interestingly, Richards et al. (437) demonstrated that estradiol significantly up-regulates the expression of several components of the IGF-1 signaling pathway in rat granulosa cells, including the IGF-1 receptor β subunit; the glucose transporter, Glut-1; and the Forkhead family member, FKHR (Foxo1). These findings suggest that estradiol may enhance IGF-1 actions in granulosa cells by regulating targets important for cellular energy flow, glucose metabolism, and cell survival (437). IGF-1 actions are clearly necessary to provide for sufficient FSH signaling in granulosa cells, which in turn is necessary for the induction of aromatase activity and estradiol synthesis. Therefore, IGF-1 actions may provide ligand for ER-β to further enhance the effects of IGF-1, hence forming an autocrine regulatory pathway to promote granulosa cell proliferation (437). The extent to which these findings may translate to the human ovary remains to be determined because IGF-II, rather than IGF-1, is the predominant and gonadotropin-regulated IGF form in the human ovary (438).

The above discussion provides a compelling argument that the mitogenic actions of estradiol on granulosa cells are direct and ER-β mediated. However, thecal cell ER actions may also be involved. For example, ER-α in the thecal cells may respond to granulosa cell-derived estradiol by inducing the synthesis and secretion of paracrine-acting growth factors that then become the principal stimuli for granulosa cell proliferation. This type of mesenchymal–epithelial interface for estrogen-induced proliferation is known to occur in the uterus and mammary gland where ER-α in the underlying stroma is essential to estrogen-induced epithelial proliferation (Fig. 13) (247,270,439). Indeed, Dorrington and colleagues proposed that transforming growth factor-β is estrogen induced in thecal cells and stimulates granulosa cell proliferation (440,441). Similar examples of growth factor–mediated cooperative actions between theca and granulosa cells are postulated to regulate follicle steroid production (442,443). Still, ER-α–null follicles do not exhibit any gross deficits in granulosa cell population and possess a normal number of preovulatory follicles after gonadotropin stimulation (351). It is plausible that a dramatic effect on granulosa cell proliferation may only become apparent after the loss of both putative estrogen pathways, that is, the direct (ER-β mediated) and indirect (ER-α mediated) actions, explaining the unique somatic cell phenotypes observed in the ovaries of compound ER-null (192,194) and older Cyp19-null mice (408).

Granulosa Cell Differentiation

The process of granulosa cell differentiation that occurs during progression from a preantral to preovulatory follicle has been an area of intense research. Fully differentiated preovulatory follicles in mammalian ovaries are distinctly characterized by a large fluid-filled antrum, acquisition of LH responsiveness, significant increases in aromatase (CYP19) activity, and increased inhibin synthesis (see Chapter 10). Although FSH is the primary stimulus for both granulosa cell proliferation and differentiation, these two processes appear to be mutually exclusive as the undersized follicles in Ccnd2-null ovaries still develop an antrum, possess sufficient aromatase activity, and are responsive to exogenous LH stimulation (432). Therefore, granulosa cell proliferation and differentiation are mutually exclusive gonadotropin-dependent events during folliculogenesis.

The absolute requirement of FSH signaling in the process of granulosa cell differentiation is illustrated by a lack of antral follicles, aromatase and estradiol synthesis, and granulosa cell LH-receptor in mice null for FSH action (356,444,445). Igf1-null follicles exhibit a similar failure to differentiate, presumably because of a lack of sufficient FSH-receptor expression (434). Recent evidence also indicates that activins strongly potentiate these FSH actions on granulosa cells (446). However, numerous studies in rat ovaries have demonstrated that FSH induction of antrum formation (421), aromatase expression and activity (378,380–383), and LH-receptor expression (413,416, 447) requires estradiol for maximum effect. FSH acts via a classic heterotrimeric G-protein–coupled receptor pathway that may activate multiple intracellular second messenger systems, the most well characterized being activation of adenylyl cyclase, leading to intracellular accumulation of cAMP and activation of PKA (352,425). The long-recognized synergism between estradiol and FSH on granulosa cell differentiation is believed to be due to estradiol's capacity to both amplify the level of FSH-stimulated intracellular cAMP (302,377,425,447,448) and augment the downstream actions of cAMP itself (425,447). The mechanisms involved remain poorly understood but are unlikely to involve estrogen-induced increases in FSH-receptor levels, but rather via estradiol's capacity to positively modulate the granulosa cell adenylyl cyclase system (377,425). ER-mediated estrogen activation of the adenylyl cyclase system and subsequent cAMP-regulated gene expression has been demonstrated in uterine and breast cancer cells (449). Furthermore, in vitro and in vivo studies indicate that ER-α or ER-β can activate ERE-dependent gene transcription via the cAMP/PKA signaling pathway in the absence of estradiol (128,152,450) (Fig. 6). Furthermore, this pathway often involves the CREB protein (451), which is also known to be critical to FSH induction of gene expression (378,452,453). Therefore, given the existing evidence that estradiol is required to maximize the FSH response combined with the known cooperative activity between the ER and PKA signaling pathways, it may be reasonable to expect that a loss of ER action in the ovary would cause severe deficits in granulosa cell differentiation. Indeed, the ovarian phenotypes in ER-β–null females in terms of antrum formation, aromatase expression, and LH responsiveness collectively represent an attenuated response to FSH-induced granulosa cell differentiation in the absence of ER-β. Further supporting evidence comes from a report that only the ER-β–specific agonist (8-vinylestra-1,3,5-(10)-triene-3,17β-diol) but not the ER-α–specific agonist (3,17-dihydroxy-19-nor-17α-pregna-1,3,5 (10)-triene-21,16α-lactone) can effectively mimic the effects of exogenous estradiol pretreatment that are necessary for gonadotropin-induced ovulation in hypophysectomized rats (417).

Antrum Formation. Goldenberg et al. (421) first demonstrated that both FSH and estrogen are required for full antrum formation in preovulatory follicles of hypophysectomized rat ovaries. Wang and Greenwald (423) produced similar data in hypophysectomized mice, reporting that FSH (4 μg, two times a day for 4 days) plus estradiol (250 μg/day for 4 days) leads to an average of >70 large antral follicles per ovary versus an average of 6 or none when treated with FSH or estradiol alone, respectively. The requirement of FSH plus estradiol for antrum formation holds true when rat preantral follicles are grown in vitro (454). Others have shown that exposure to aromatase inhibitors during gonadotropin-induced folliculogenesis also reduces the number of healthy antral follicles in rats (414,415), although contradictory findings are reported when follicles are similarly exposed in vitro (455). Independent evaluations of ovaries from two separate lines of ER-β–null mice indicate a paucity of follicles with fully developed antra, even following standard gonadotropin stimulation (192,376). These data strongly suggest that the vital role of estradiol in FSH-induced antrum formation is mediated by ER-β. A role for ER-α in antrum formation cannot be excluded but is unlikely given the hallmark phenotype in ER-α–null ovaries is the invariable presence of severely oversized antral follicles, lending to their description as "cystic" (191,192,311,351). One may hypothesize that the exaggerated antrum or "cyst" formation among follicles in ER-α–null females is due to hyperstimulation of ER-β by the elevated estradiol levels that are characteristic of ER-null females. This hypothesis is supported by the lack of similar cystic follicles in the ovaries of mice lacking both ER forms (αβERKO) (192,194) or in transgenic βERKO females that possess αERKO-like levels of elevated LH (375).

As mentioned, FSH alone induces the formation of small antra in growing follicles, whereas estradiol has no such effect in vivo and in vitro; but both hormones are required for full antrum development (421,423, 454). Compared with the total failure of antrum formation among follicles in the murine models of disrupted FSH signaling (356,434,444,445,456), similarly sized follicles in ER-β–null ovaries exhibit an initiation of antrum formation but fail to develop to a full preovulatory stage (192). Quantitative analyses of immature ER-β–null ovaries 48 hours after PMSG stimulation indicate an increased number of small antral follicles but fewer large (preovulatory) follicles relative to wild-type females, suggesting an arrest in antrum formation (376). Likewise, a large percentage of ER-β–null follicles fail to reach maximum size when grown in culture with FSH and estradiol (376). These data support the role of FSH actions in the initiation of antrum formation, but in the absence of ER-β–mediated estradiol augmentation, maximum antrum development is prohibited.

Because so little is known about the physiology of antrum formation, the role of ER-β–mediated estrogen actions is speculative. It has long been proposed that antrum development follows gonadotropin-stimulated increases in granulosa cell proteogylcan secretion, which acts to increase osmotic pressure within the follicle and causes the influx of water from the ovarian vasculature (457,458). McConnell et al. (459) reported that water movement into mouse follicles in vitro occurs via a transcellular pathway and is likely mediated by aquaporins–7, –8, or –9 on the surface of granulosa cells (459). Although there is currently no evidence of estrogen regulation of aquaporin expression in the ovary, estradiol is known to influence fluid transport in the uterus via regulation of aquaporins–1 (460), –2, and –3 (461). Furthermore, ER-α is principally involved in the regulation of Na^+ and water transport across the efferent ductules in the rodent testis (462), suggesting ER-β may play a similar role in follicles. Antrum formation may also require cell–cell interactions among the granulosum. Gore-Langton and Daniel (454) demonstrated in vitro that rat preantral follicles lacking an intact basement membrane and therefore unable to provide a sealed environment exhibit a reorganization of granulosa cells that is unmistakably antrum-like in response to FSH plus estradiol, supporting a role for cell–cell interaction during antrum formation. A lack of antral follicles in the ovaries of mice lacking connexin-37 (*Cx37*) further supports a need for cell–cell interactions among granulosa cells (463). Estradiol is known to dramatically increase the number of gap junctions between granulosa cells (464) and to regulate ovarian expression of certain component proteins of gap junctions (465–467). Future studies of a potential regulatory role for estradiol in the expression of the aquaporin and connexin proteins may reveal a precise role for ER-β in antrum formation.

Aromatase and Estradiol Synthesis. Acquisition of aromatase activity and estradiol synthesis is a hallmark of healthy preovulatory follicles in mammalian ovaries. In monotocous species, 90% of the circulating estradiol is estimated to originate from the dominant follicle in the ovary (373). Aromatase (*CYP19*) expression in rat, bovine, marmoset, and human ovaries is exclusive to and highest in the granulosa cells of fully differentiated healthy follicles (319,468–470). The endocrine actions of ovarian-derived estradiol are critical to preparing the female reproductive tract for a forthcoming pregnancy (see section on Estrogen Receptor Signaling in Uterine Function) and priming the hypothalamic–pituitary axis for generation of the gonadotropin surge (see Chapters 43 and 44). However, there is equally abundant evidence that estradiol acts in a paracrine or autocrine manner to augment FSH stimulation of aromatase activity in granulosa cells and thereby positively influence its own rate of synthesis (Fig. 19). This pattern of *CYP19* expression described above strongly resembles the profile of ER-β during folliculogenesis, suggesting that the estradiol-induced enhancement of aromatase activity in granulosa cells is ER-β mediated.

FSH is clearly the principal stimulus for the acquisition of aromatase activity in granulosa cells (373). Once again, female mice lacking FSH (*Fshb*-null mice) (355,471) or FSH receptor (*Fshr*-null mice) (357,445) exhibit phenotypes that are consistent with a severe reduction in circulating estrogens, including uterine hypoplasia and elevated plasma levels of LH. Furthermore, the ovaries of *Fshb*-null mice exhibit a sixfold reduction in *Cyp19* transcripts (355), although these data are not corroborated in *Fshr*-null mice (357). FSH regulation of *Cyp19* expression in granulosa cells is believed to require the actions of multiple transcriptions factors, including steroidogenic factor-1 (SF-1), CREB protein, GATA-4, and CBP, the first three of which possess cognate response elements within the rodent and human *CYP19* promoter II (377,452,453,472). Female mice lacking functional SF-1 in the granulosa cells exhibit phenotypes consistent with insufficient estradiol production, similar to the FSH-null mice described above (473).

The mechanism by which estradiol augments FSH stimulation of aromatase activity or *Cyp19* expression in granulosa cells is unclear but known to occur in isolated granulosa cell cultures (378,380–383), ruling out the influence of a thecal cell–derived factor and supporting a direct role for ER-β. Coadministration of FSH and an ER antagonist (tamoxifen) to granulosa cells in vivo or in vitro inhibits the expected induction of aromatase activity (474), although others report conflicting data using a

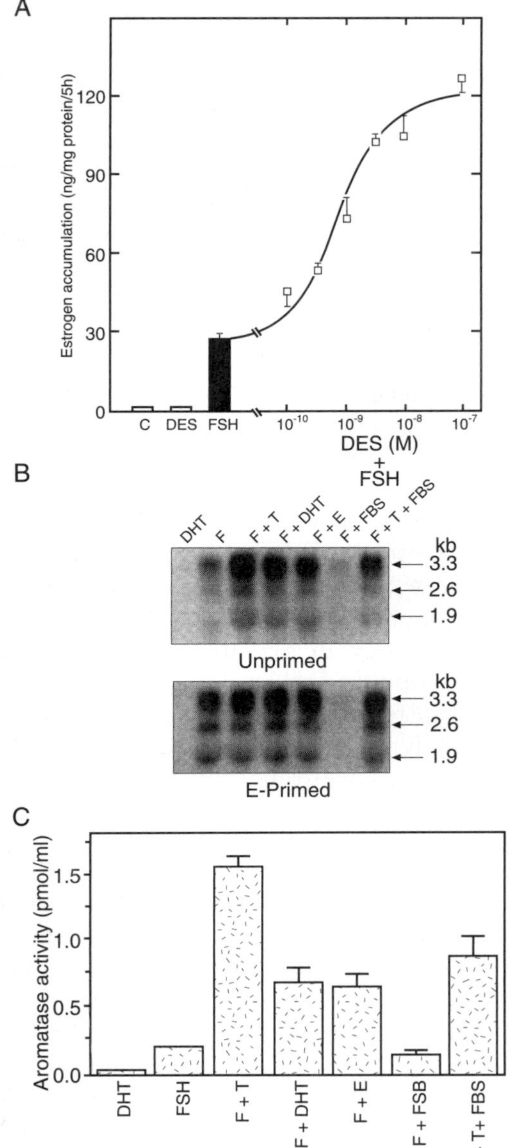

FIG. 19. Synergistic action of sex steroids and FSH in the induction of *Cyp19* expression and aromatase activity in rodent granulosa cells. **A:** Shown is the effect of in vitro treatment with the synthetic estrogen, DES, on FSH-induced aromatase activity in granulosa cells isolated from immature hypophysectomized rats. Granulosa cells were cultured for 3 days in the presence or absence of FSH (10 ng/ml) with or without increasing concentrations of DES. Three days later, the cells were washed with medium and reincubated for a 5-hour test interval in medium supplemented with androstenedione. The accumulation of estrogen (as measured by radioimmunoassay) during this test period is taken as a measure of the level of aromatase activity. C, controls. [Reproduced with permission from (381).] **B, C:** Shown is the effect of FSH and sex steroids on *Cyp19* expression and aromatase activity in cultured granulosa cells. Granulosa cells were isolated from 26-day-old rats that had been untreated (Unprimed) or treated with estradiol (1.5 mg/day) for 3 days (E-primed). Cells were incubated in serum-free medium with 5α-dihydrotestosterone (DHT) at 20 nM, FSH (50 ng/ml), or FSH in combination with DHT (20 nM), testosterone (T, 20 nM), or estradiol (E, 20 nM). Cells were also incubated in medium containing 5% fetal bovine serum (FBS), with FSH alone (50 ng/ml), or in combination with T (20 nM). After 48 hours in culture, cells were collected for RNA extraction and evaluation of *Cyp19* expression by Northern Blot analysis (B) or for measurement of aromatase activity (C). [Reproduced with permission from (378).]

different antagonist (ICI 182,780) in whole follicle culture experiments (428). More recently, Emmen et al. (376) found that ER-β–null follicles grown in vitro secrete significantly less amounts of estradiol relative to wild-type follicles. The absence of a consensus ERE within the rat or human *Cyp19* promoter II makes it unlikely that the positive effect of estradiol is via an ER-β–mediated "classic" mechanism of receptor action (Fig. 5). However, ER-β may interact with and enhance the actions of the cAMP/PKA pathway, SF-1, CREB, GATA-4, CBP, or other unknown transcription factors that act to promote *Cyp19* expression. Indeed, ER-β has been shown to associate with both CREB (451) and CBP (70,475) in heterologous in vitro systems. In fact, GnRH-induced phosphorylation of CREB, which also occurs via a G-protein–coupled receptor cAMP/PKA pathway, was recently demonstrated to require functional ER-β in GnRH-secreting neurons (388). The ERs are also known to influence GATA-directed gene expression via direct interactions with multiple members of the GATA family of transcription factors (405,476). Therefore, there is ample evidence to support ER interaction with the multitude of transcription factors known to influence *Cyp19* expression, but the exact nature by which ER-β is involved requires further study.

Given the above evidence that estradiol is required for maximum FSH induction of aromatase and estradiol synthesis, it is somewhat puzzling to find that adult ER-β–null ovaries exhibit relatively normal *Cyp19* expression and relatively normal plasma estradiol levels (104,366). More recent investigations have revealed a reduced capacity of individual ER-β follicles to synthesize estradiol in vitro (376) as well as a less than maximal induction of *Cyp19* expression in ovaries or isolated granulosa cells of PMSG-treated immature ER-β–null females (367). Still, these data indicate that ER-β in granulosa cells is not obligatory to *Cyp19* expression and aromatase activity in the mouse ovary. Potential compensatory pathways within the ovary include AR-mediated actions of testosterone and dihydrotestosterone, a nonaromatizable androgen, both of which also enhance FSH-induced expression of *Cyp19* in rat granulosa cells (378,477,478). AR-mediated androgen actions may provide for sufficient enhancement of FSH-induced aromatase activity in the granulosa cells in the absence of ER-β. This is especially plausible given a report of abnormally high *Ar* expression in granulosa cells of ER-β–null ovaries (386).

Acquisition of LH Receptor. In contrast to the constitutive expression of LH receptor in thecal cells, granulosa cells express LH receptor only in follicles in the late preovulatory stage. This limited expression of LH receptor to the granulosa cells of healthy preovulatory follicles provides for an intraovarian mechanism that ensures only those follicles that are suitable for ovulation acquire the capacity to respond to the LH surge. As with the other markers of follicle differentiation discussed above, FSH is the primary stimulus of LH-receptor (*Lhcgr*) expression in preovulatory granulosa cells. However, FSH exposure alone leads to a minimal increase of LH-receptor levels in the ovaries of hypophysectomized rats, and then only after 4 days of treatment, whereas pretreatment of animals with estradiol for 4 days before FSH exposure leads to an enormous induction of LH-receptor levels that are limited to the granulosa cells of preovulatory follicles (Fig. 20) (302). This requirement for FSH plus estradiol and the temporal pattern of LH-receptor expression has been reproduced in rat granulosa cells in vitro, with maximum levels reached after 48–72 hours (447,479,480). *Lhcgr* mRNA levels mirror the LH-receptor protein levels and exhibit a comparable FSH/estradiol regulation in rat granulosa cells (481). Further confirmation that estrogens are required to augment FSH-induced LH-receptor expression in granulosa cells comes from studies that have collectively shown the following:

1. Only estrogens (e.g., estradiol, DES) or substrates for aromatization (e.g., androstenedione, testosterone, estrone) are effective (447,479,480).
2. Nonaromatizable androgens (e.g., DHT) and progestins have little effect (447,480).
3. Cotreatment with aromatase inhibitors, ER antagonists, or 17β-estradiol–specific antisera block FSH induction of LH-receptor (Fig. 20) (413,416,479) and hCG-induced ovulation (415).
4. *Lhcgr* levels are decreased fivefold in the ovaries of Cyp19-null mice (409).

The poor ovulatory response of ER-β–null ovaries to an ovulatory dose of the LH analogue, hCG (193,367), strongly suggests that ER-β mediates the synergistic actions of estradiol during FSH induction of the *Lhcgr* gene in granulosa cells (104,193). In contrast, the granulosa cells of large antral follicles in ER-α–null ovaries exhibit abnormally high levels of LH-receptor, even after becoming enlarged and cystic (104,311).

The complexity of the *Lhcgr* promoter is illustrated by the differential regulation required between thecal and preovulatory granulosa cells. Not surprisingly, FSH induction of LH-receptor expression in granulosa cells occurs via the cAMP/PKA pathway (482–485). However, the proximal promoter of the *Lhcgr* gene in both humans and rats lacks a consensus cAMP response element (CRE) similar to those that confer FSH regulation of the *Cyp19* promoter II (486,487). Instead, recent evidence indicates that a GC-rich region possessing a cluster of Sp1 binding sites and an ERE half site contributes to FSH and 8-bromo-cAMP induction of the rat *Lhcgr* promoter in vitro (488–490). Furthermore, multiple nonconsensus cAMP response elements within this GC-rich region of the rat *Lhcgr*

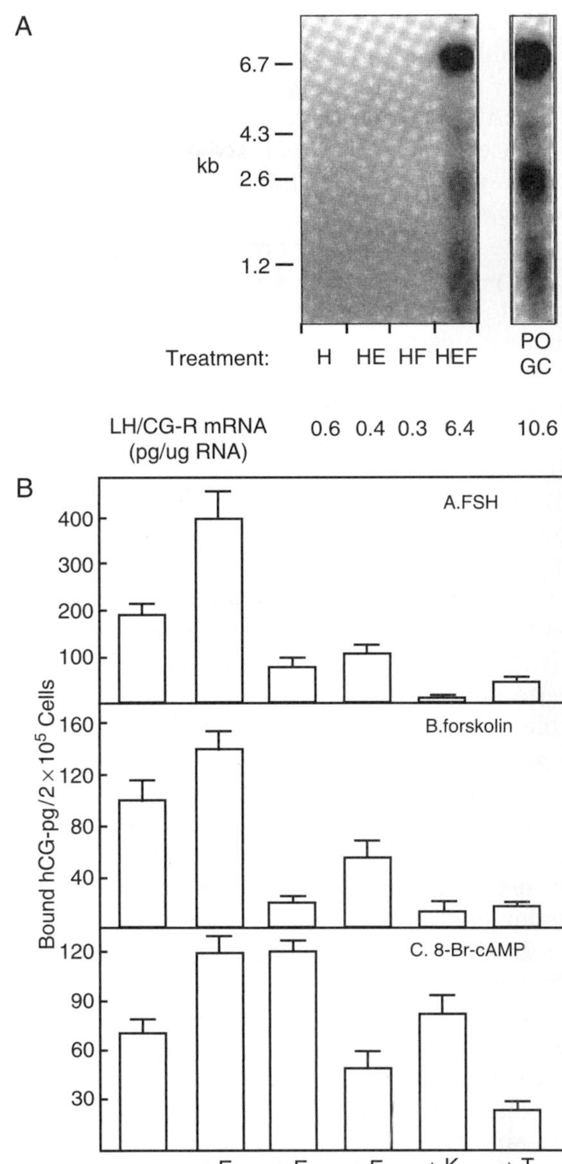

FIG. 20. Synergistic action of estradiol and FSH in the induction of LH receptor in rodent granulosa cells. **A:** Induction of *Lhcgr* (LH/CG-R) mRNA in granulosa cells of hypophysectomized rats after treatment with FSH and/or estradiol. Animals were untreated (H) or injected with 1.5 mg estradiol/day for 3 days (HE), 1 mg FSH twice daily for 2 days (HF), or a sequential combination of E then FSH (HEF). Animals were killed after treatment and granulosa cells isolated from the ovaries for total RNA extraction. *Top*: Northern blot of 20 μg RNA/lane probed for *Lhcgr* (LH/CG-R) mRNA. *Bottom*: Results of RNase-protection assay for *Lhcgr* (LH/CG-R) mRNA expressed as pg LH/CG-R mRNA per μg total RNA assayed. [Reproduced with permission from (481).] **B:** Effect of antiestrogens on the induction of LH-receptor levels by FSH, forskolin, or 8-bromo-cAMP. Granulosa cells were isolated from immature rats implanted with DES pellets for 4 days. 2×10^5 cells were cultured for 48 hours with FSH (100 ng), forskolin (10 μM), or 8-bromo-cAMP (5 nM), with or without estradiol (E_2, 10^{-8} M) in the absence or presence of antiestrogens, keoxifene (K, 1 μM) or tamoxifen (T, 1 μM). Media were then removed and cells were analyzed for LH-receptor content by radiolabeled binding assay. The *first bar* on the left side of the figure is FSH (A), forskolin (B), or 8-bromo-cAMP (C) alone. Control cells bound less than 10 pg hCG/2×10^5 cells, and the data are not shown. [Reproduced with permission from (413).]

promoter have been discovered and provide a platform for assembling an undefined complex of nuclear proteins present within granulosa cell extracts that confers cAMP induction (487,491). With such limited data concerning *Lhcgr* regulation, it is difficult to speculate where the actions of ER-β may impact LH responsiveness in granulosa cells. The ERE half site in the proximal *Lhcgr* promoter appears to be an inhibitory element in the context of a heterologous construct in vitro (489). Still, ERE-less GC-rich promoters possessing Sp1 sites are known to be responsive to estradiol via "tethering" of ER (Fig. 7), although studies to date indicate this capacity may be limited to ER-α and may not be present in ER-β (85,86).

Therefore, the mechanism by which estradiol/ER-β actions augment FSH stimulation of the *Lhcgr* gene is unclear. As discussed above, estradiol enhances the level and action of FSH-stimulated increases in intracellular cAMP to maximize *Cyp19* expression, and it is likely that a similar mechanism is used for induction of LH-receptor levels. However, there are notable differences in the estradiol/FSH regulation of *Cyp19* versus *Lhcgr* expression. FSH induction of *Lhcgr* expression in preovulatory granulosa cells is augmented by estrogens only (447,479,480), whereas nonaromatizable androgens or estrogens act through their respective receptor pathways to enhance FSH induction of *Cyp19* expression (378). This divergence in regulation is critical because androgen augmentation of FSH on *Cyp19* expression allows FSH/AR-mediated initiation of estradiol synthesis in preantral follicles shortly after thecal cells begin synthesizing aromatizable precursors, whereas the specificity for ER/estradiol actions in the FSH induction of the *Lhcgr* gene provides that LH-receptors are acquired only by those follicles that have first acquired sufficient estradiol synthesis and are hence healthy and suitable for ovulation. Also, Farookhi and Desjardins (480) demonstrated that FSH induction of *Lhcgr* but not *Cyp19* expression is lost when estrogen-pretreated granulosa cells are dispersed in culture by EGTA disruption of cell–cell contacts. A requirement for intact cell–cell contacts for estradiol/FSH-mediated *Lhcgr* expression may ensure that atretic follicles, which often exhibit a fragmented and less compact granulosa cell population, do not acquire the capacity to respond to an LH surge and ovulate an unhealthy oocyte.

Estradiol-Mediated Negative Feedback on Thecal Cell Function

Like granulosa cells, thecal cells also undergo a process of differentiation during progression from a preantral to preovulatory follicle (492). A principal element of the *two-cell, two-gonadotropin* paradigm of steroidogenesis in maturing follicles is that differentiated thecal cells exclusively possess the 17α-hydroxylase: $C_{17,20}$-lyase (CYP17) activity necessary for synthesizing androgens, forcing the granulosum to be dependent on the theca as a source of aromatizable precursors (373). Thecal cell synthesis of androstenedione is considered the rate-limiting step in estradiol production by preovulatory follicles and involves the sequential enzymatic actions of $P450_{scc}$ (CYP11), $P450_{17\alpha/17,20\text{-lyase}}$ (CYP17), and Δ^5-3β-hydroxysteroid dehydrogenase (3β-HSD 1) (373). It is generally accepted that LH regulates thecal cell steroidogenesis via the induction of CYP17 and CYP11 activity (492), whereas 3β-HSD 1 is constitutive (373). The requirement of LH stimulation to induce CYP17 activity in thecal cells is effectively illustrated by the 12-fold reduction in *Cyp17* expression in ovaries of LH-receptor (*Lhcgr*) null mice (493) as well as the fivefold increase in *Cyp17* expression in the ovaries of transgenic mice that possess chronically elevated LH levels (375). Likewise, chronically elevated plasma LH leads to a comparable increase in *Cyp17* expression and androstenedione synthesis in both ER-α–null and ER-β–null ovaries, indicating that neither ER is obligatory to the positive regulation of thecal cell steroidogenesis (366,375).

Therefore, androgens are necessary to serve as both a stimulus of and substrate for aromatase activity in granulosa cells. However, a proper androgen-to-estrogen ratio within the follicle is critical because increased androgens are strongly associated with atresia (494). Therefore, an intraovarian mechanism must exist to negatively modulate thecal cell androgen synthesis so that levels do not surpass the aromatase capacity of the granulosum. Defects in this intraovarian mechanism are postulated to be an underlying cause of the excess thecal cell androgen synthesis that is a hallmark of polycystic ovarian syndrome in women. There is ample evidence that granulosa cell–derived estradiol may mediate a feedback loop on thecal cells to decrease further androgen synthesis in thecal cells (495); however, the mechanism remains unclear. Magoffin (492) posed the question several years ago as to whether estradiol inhibition of androgen synthesis occurs at the level of steroidogenic enzyme expression or via direct inhibition of enzymatic activity. Indeed, estrogens have been shown to inhibit CYP17 activity in both thecal and Leydig cell preparations while having no effect on activities of the upstream steroidogenic enzymes (495–500). Early studies found that estradiol can compete with androstenedione for CYP17 enzymatic activity (496); however, this effect could not be reproduced in ovarian lysates in vitro (501). Additional studies in dispersed rat ovarian cells found that estradiol has no effect on the number of LH receptors or level of hCG-stimulated cAMP synthesis, further supporting a direct estradiol inhibition of CYP17 activity as a mechanism for regulating androgen synthesis (499). However, more recent evidence supports a role for ER-mediated regulation of *Cyp17* expression at the transcriptional level, including demonstrations that an antiestrogen blocks the expected drop in CYP17 activity that occurs just before ovulation in rats (502) and that estradiol significantly reduces *Cyp17* expression in the testes of rats (503) and fish (504). Our findings that ER-α–null ovaries exhibit increased *Cyp17* expression and enzymatic activity despite a milieu of significantly elevated estradiol also supports a mechanism of ER-α–mediated suppression of *Cyp17* transcription in thecal cells versus direct inhibition of CYP17 activity (Fig. 17) (366). Furthermore, wild-type and ER-β–null ovaries exposed to chronically elevated LH also exhibit increased *Cyp17* expression, but levels are modest relative to those found in ER-α–null ovaries (375). Therefore, the excess androgen synthesis observed in ER-α–null ovaries is likely the combined effect of LH hyperstimulation of the theca and loss of estradiol-mediated suppression of *Cyp17* expression, indicating that ER-α mediates both extra- and intraovarian feedback actions of estradiol to maintain homeostasis of androgen levels in female mice.

Ovulation

Estradiol clearly facilitates the acquisition of several gonadotropin-dependent effects during folliculogenesis that are conducive to ovulation, including the proper cellular organization of the follicle (e.g., antrum and cumulus oocyte complex), the necessary enzymatic activity (e.g., estradiol synthesis), and the essential receptor signaling pathways (e.g., LH-receptor). All these characteristics define a healthy differentiated follicle and are necessary for a proper response to the gonadotropin surge and rupture of a competent oocyte. However, a need for estrogen signaling in the ovulating follicle or after the postgonadotropin surge remains unclear. Selvaraj et al. (415) found that coadministration of immature rats with PMSG and an aromatase inhibitor (Fadrazole) leads to a reduced number of healthy antral follicles that culminates with a severely reduced ovulatory response to a bolus of hCG. These data support the intrafollicular role of estradiol in preparing competent preovulatory follicles for ovulation. However, in an additional experiment, Selvaraj et al. (415) once again treated immature rats with PMSG but delayed administration of the aromatase inhibitor by 40 hours such that exposure occurred just 6 hours before hCG-induced ovulation. These animals exhibited an equally poor response in terms of follicular rupture and recoverable oocytes from the oviduct, and this effect was abated by coadministration of estradiol just before induced ovulation. These data in turn suggest that estradiol does facilitate the ovulatory process. A caveat to this experimental scheme is that inhibition of estradiol synthesis is known to cause severe intraovarian (415) and intrafollicular (455) accumulation of progesterone and androgens, which could also lead to a poor ovulatory response. In contrast to the above in vivo findings, Hu et al. (455) reported that isolated mouse follicles exposed to an aromatase inhibitor in vitro exhibit a normal ovulatory response to hCG, including mucification of the cumulus–oocyte complex. Furthermore, detrimental effects of aromatase inhibition such as those reported in the rat ovary are not observed in similarly treated hamsters, rabbits, or monkeys (414).

Multiple investigations of the ovulatory capacity of ER-null and CYP19-null mice have been conducted. As described above, ER-α–null females are anovulatory throughout life and gonadotropin-induced ovulation is unsuccessful in ER-α–null females at 4 months of age when the ovarian cystic phenotype is advanced (311). However, immature (age 28 days) ER-α–null (αERKO) females do respond to induced ovulation and yield recoverable oocytes in the oviduct at an average yield of 14.5 oocytes/ER-α–null female versus 40.6 oocytes/wild-type female (351). Similar data have been reported in ER-α–null females as old as 5 weeks of age (351,365). Interestingly, Dupont et al. (192) reported that immature females from their ER-α–null (ERαKO) line exhibit a total failure to ovulate in response to exogenous gonadotropin treatments, although this study involved only three females.

The cause of the reduced ovulatory response to exogenous gonadotropins in ER-α–null females is unclear and suggests a facilitatory role for ER-α in follicular rupture However, the expected induction of several genes that are critical to follicle rupture, such as PR and prostaglandin-synthase-2 (505), occurs in gonadotropin-primed ER-α–null ovaries shortly after hCG treatment (351,367). Furthermore, ER-α–null ovaries form functional corpora lutea after induced ovulation, indicating that granulosa cells of ER-α–null preovulatory follicles possess the capacity to respond to LH (351,365). Therefore, the reduced ovulatory response in ER-α–null ovaries is more likely attributable to premature increases in endogenous plasma LH that lead to abnormally high ovarian androgen synthesis (351). Although immature ER-α–null ovaries do not yet manifest the overt morphological effects of LH hyperstimulation (e.g., cystic follicles), they do exhibit elevated LH-receptor levels (351). Cumulative damage in the ovary from chronic LH hyperstimulation likely causes the age-related reduction in gonadotropin-induced ovulatory capacity in ER-α–null females discussed above. Indeed, Armstrong and colleagues demonstrated in rats that treatment with PMSG preparations possessing increased LH activity lead to a reduced ovulatory response to subsequent hCG stimulation (506–508) and that this effect is likely due to inappropriately high stimulation of androgen synthesis during follicle maturation (508). Recent evidence that ER-α–null follicles allowed to differentiate in vitro under controlled hormonal stimulation exhibit a rate of hCG-induced rupture that is comparable with similarly cultured wild-type follicles further indicates that any role for ER-α in the ovulatory process is minimal (376).

Although less extensive studies on the ovulatory capacity of ER-β–null mice have been reported, the loss of functional ER-β clearly leads to ovarian defects that are inhibitory to ovulation. Both lines of ER-β–null females exhibit severely reduced fertility due to ovarian defects; however, unlike ER-α–null females, ER-β–null animals do spontaneously ovulate (192,193). Krege et al. (193) found that immature ER-β–null (βERKO) females exhibit a severely reduced oocyte yield of 6 oocytes/female versus 34 oocytes/female among age-matched wild-type females after gonadotropin-induced ovulation (193). Dupont et al. (192) reported a similar average yield in immature ERβKO females (17.6 versus 37 oocytes/female), even when excluding those females that do not ovulate at all. Once again, however, the question remains as to whether ER-β functions are required only in preparing a competent preovulatory follicle for ovulation or whether ER-β is also directly involved in the ovulatory process that follows the gonadotropin surge. The precipitous decrease in ER-β expression that occurs in granulosa cells of preovulatory follicles soon after the gonadotropin surge is highly conserved among species (308,309,313,321,336,351) and argues against a direct role for the receptor in follicle rupture.

More recent studies are beginning to elucidate the cause of the ovulatory defect in ER-β–null mice. Immature ER-β–null ovaries collected 20 hours after hCG-induced ovulation invariably reveal multiple preovulatory follicles that retain their oocyte and fail to undergo expansion of the cumulus–oocyte complex (192,193,367). Furthermore, hCG induction of prostaglandin synthase 2 and PR, two events that are critical to follicular rupture, is significantly compromised in immature βERKO ovaries (367). ER-β–null follicles induced to ovulate in vitro exhibit an equally poor rate of rupture and gene induction after hCG exposure (376). The sum of in vivo and in vitro data from ER-β–null mice strongly indicates an inability to fully respond to an endogenous LH surge or an exogenous bolus of hCG. Recent data indicate this is likely due to insufficient acquisition of LH-receptor expression among granulosa cells of ER-β–null preovulatory follicles (367,376), congruent with a role of ER-β in mediating the synergistic effects of estradiol on FSH-induced granulosa cell differentiation (see above section). Indeed, Clemens et al. (509) found that an ER -antagonist (ICI 164,384) inhibits FSH-induced differentiation of rat granulosa cells in vitro and causes a severely compromised induction of PR expression after stimulation of the cAMP/PKA signaling pathway. However, if exposure to the ER antagonist is delayed until the time of ovulatory stimulation, little effect is observed (509), supporting a role for ER-β in granulosa cell differentiation and LH-receptor expression rather than during ovulation.

Evidence of any cooperative action between ER-α and ER-β in ovulation may be derived from studies of compound ER-null and CYP19-null mice. The single report of induced ovulation in immature ER-α/ER-β–null (ERβKO) females describes a total failure of ovulation (192). Not surprisingly, ovaries from these females exhibited underdeveloped preovulatory follicles that failed to exhibit cumulus–oocyte complex expansion or luteinization (192), indicating an attenuated response to hCG similar to that in ER-β–null ovaries. Investigations in both lines of Cyp19-null females produced comparable results of a total failure to ovulate and luteinize (196,510). Interestingly, Huynh et al. (510) remarked that cumulus–oocyte complexes lacking oocytes were recovered from the oviducts of 4-week-old Cyp19-null females after induced ovulation, indicating that some elements of follicle rupture may occur after degradation of the oocyte. The more severe ovulatory phenotypes in compound ER-null and Cyp19-null females suggest that the intraovarian functions of both ER-α and ER-β are necessary for ovulation but not for the process of follicular rupture.

ER-α–Mediated Repression of Leydig Cell Development in the Ovary

As discussed in earlier sections, female mice lacking ER-α action due either to the loss of receptor (αERKO and αβERKO) or ligand (Cyp19-null) exhibit abnormally high levels of plasma testosterone. For example, the average plasma testosterone level in adult αERKO and αβERKO females is 5.5 ng/ml and 2.1 ng/ml, respectively, versus 0.14 ng/ml in wild-type females (366). In females of the two existing Cyp19-null lines, Fisher et al. (195) reported average plasma testosterone levels in mutant females to be 2.3 ng/ml versus 0.24 ng/ml in wild-type (195); Toda et al. (196) reported 1.4 ng/ml in the mutant females versus 0.13 ng/ml in wild-type (196). These levels of plasma testosterone in ER-α–null and Cyp19-null females surpass the lower limits characteristic of normal male mice (2–10 ng/ml) (195,196,511) and are obviously very uncharacteristic of normal females. The mouse ovary possesses the enzymatic machinery necessary for testosterone synthesis, although it is not a major product. The *two-cell, two-gonadotropin* paradigm of ovarian steroidogenesis postulates that thecal cell–derived androstenedione serves as a substrate for two sequential reactions in granulosa cells, CYP19-mediated aromatization to estrone followed by a reduction to estradiol by 17β-hydroxysteroid dehydrogenase type I (17β-HSD1) (373). The murine form of 17β-HSD 1 is reported to efficiently reduce androstenedione to testosterone as well as estrone to estradiol, whereas the human form may not possess this activity (512). Therefore, the accumulation of androstenedione that occurs in ER-α–null and Cyp19-null ovaries due to LH hyperstimulation of the theca in both, and the absence of aromatization in the latter (195,366), provides sufficient substrate for 17β-HSD 1 conversion to testosterone. However, further investigations have discovered that ER-α–null (366) and Cyp19-null (409) ovaries uniquely possess substantial levels of 17β-hydroxysteroid dehydrogenase type III (17β-HSD3) (Fig. 17), a Leydig cell–specific enzyme in testes that specifically functions to reduce androstenedione to testosterone (513–517). Remarkably, the level of *Hsd17b3* expression in αERKO, αβERKO (366), and Cyp19-null (409) ovaries is comparable with that found in testes of wild-type males. Therefore, the abnormally high capacity of ER-α–null and Cyp19-null ovaries to synthesize testosterone may be attributed to the accumulation of androstenedione and its conversion to testosterone by 17β-HSD 3 rather than 17β-HSD 1.

Hsd17b3 expression in Leydig cells is primarily LH regulated (515). Hence, we have shown that the ectopic expression in ER-α–null ovaries is also dependent on LH hyperstimulation as expression is eradicated after prolonged treatment of animals with a GnRH antagonist (Fig. 17) (366). In ER-α–null females, *Hsd17b3* expression is limited to the ovarian interstitium and possibly the theca (J. F. Couse and K. S. Korach, unpublished data). However, it is unclear whether these cells are Leydig-like and suggestive of an "organizational" defect in ER-α–null ovaries or whether they are thecal/interstitial type cells within which a lack of ER-α leads to loss of *Hsd17b3* repression suggestive of an "activational" defect. Evidence to support the latter hypothesis includes the following: testes from ER-α–null males also exhibit elevated *Hsd17b3* expression and testosterone synthesis (518), female mice possessing chronically elevated LH but intact ER-α signaling do not exhibit ovarian *Hsd17b3* expression (375), and prolonged estradiol therapy in Cyp19-null females eradicates *Hsd17b3* expression (409). In turn, supporting evidence exists for the former hypothesis that a defect during ovarian differentiation may occur in the absence of ER-α. Thecal and Leydig cells are generally believed to derive from a common bipotent embryological precursor cell during gonadal differentiation. In Cyp19-null ovaries, Britt et al. (197) described the presence of cells resembling mature Leydig cells of testes in that they possessed a tubulovesicular arrangement of mitochondrial cristae, an annular nucleolus, and an abundance of smooth endoplasmic reticulum in whorl-like formations. These Leydig-like cells in Cyp19-null ovaries are located in the interstitial regions and often proximal to the basement membrane of atretic follicles that exhibit Sertoli-like cells in the follicle lumen (197). Indeed, estradiol is a potent inhibitor of Leydig cell proliferation in vitro and in vivo (519), and estradiol replacement in Cyp19-null females reduces the presence of Leydig-like cells in the ovaries (409). Therefore, ER-α may be critical to promoting thecal cell differentiation and inhibiting the Leydig cell phenotype during ovarian differentiation. A similar phenotype of Leydig cell differentiation, *Hsd17b3* expression, and "masculinization" of the ovary and reproductive tract is reported in female mice lacking *Wnt4* function (399).

Disrupted Estrogen Signaling in Humans

There are no documented cases of a human female who is autosomal recessive for an inactivating mutation of the *ESR1* (ER-α) or *ESR2* (ER-β) genes. There are, however, six reported cases of human females who are autosomal recessive for inactivating mutations of the *CYP19* gene and therefore lack the ability to synthesize estradiol (520–525). These females present with androgen-induced pseudohermaphrodism at birth (520–522) and amenarche, polycystic ovaries, and an absence of female secondary sex characteristics at puberty (520,522,523). Three reports describe elevated plasma FSH and LH levels, leading to hypergonadism that manifests as increased plasma androgens and virilization (520,522,523). The ovarian pathology exhibited is similar to that observed in ER-α–null females and compatible with a diagnosis of polycystic ovarian syndrome (522). To date, there are no indications of "sex reversal" in terms of male cell types or gene expression in the ovaries of CYP19-null human females. Estrogen and progesterone replacement therapy effectively alleviate all the above phenotypes (520,522,523).

Polymorphisms of the human *ESR1* gene have been linked to breast cancer, cardiovascular disease, osteoporosis cognitive function, and multiple reproductive anomalies (526–531). A polymorphic $(TA)_n$ repeat within the *ESR1* promoter region has been linked to premature ovarian failure (532) but has no effect on plasma levels of estradiol, sex hormone binding globulin, testosterone, or FSH in premenopausal women during the follicular phase (533). Two *PvuII* single nucleotide polymorphisms (SNP) in the *ERS1* gene have been described to date. The first of these is a glutamine to glycine transformation at amino acid 400 (534) that has received little investigative attention. The second is an anonymous SNP located in intron 1,400 bp upstream of exon 2 (535) and is found in approximately 30% of women (536–540). Although the intronic *ESR1 PvuII* SNP does not affect ER-α receptor levels (535), it may be associated with poor performance in women undergoing in vitro fertilization (536,540), increased plasma estradiol and androstenedione levels (538,540), and late onset of menarche and/or menopause (537,539,541,542). Two SNP of the *ESR2* gene have been described to date: an *RsaI* SNP leads to a silent nucleotide change in exon 5 and an *AluI* SNP occurs in the 3′-untranslated region of exon 8 (543). Sundarrajan et al. (544) found that Chinese women who are homozygous for either the *RsaI* or *AluI* SNP of the *ESR2* gene are more prone to exhibit ovulatory defects, menstrual disorders, and elevated LH, whereas women who are homozygous for both exhibit severe idiopathic ovulatory defects.

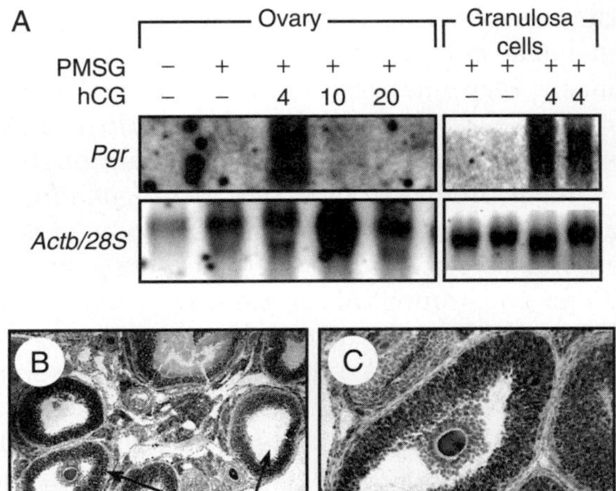

FIG. 21. Regulation of PR expression in mouse ovaries and granulosa cells during gonadotropin-induced ovulation. Immature mice were left untreated or were injected with PMSG followed 48 hours later with hCG. Animals were killed 48 hours after PMSG (+) and 4, 10, or 20 hours after subsequent hCG treatment. Ovaries were collected for RNA extraction, granulosa cell isolation followed by RNA extraction, or immunohistochemistry. **A:** Northern blot for progesterone receptor (*Pgr*) mRNA in whole ovaries or isolated granulosa cells during the indicated time points and treatments. *Pgr* expression is dramatically increased in whole ovaries or isolated granulosa cells within 4 hours after hCG treatment in PMSG-primed mice and is reduced to basal levels within 10 hours. Levels of β-actin (*Actb*) or 28S are shown for normalization. **B, C:** Immunohistochemistry for PR in the ovary of a PMSG-primed mouse 4 hours after hCG treatment, illustrating the dramatic induction of PR immunoreactivity in the granulosa cells of preovulatory follicles (POF) but not surrounding preantral follicles. (C, higher magnification of B.)

PROGESTERONE RECEPTOR SIGNALING IN OVARIAN FUNCTION

PR Expression in the Ovary

The first report of saturable high-affinity progestin binding sites in the rat ovary is that of Schreiber and colleagues in 1979 (545,546) and has since been followed by similar descriptions in human (329–331,547) and cow (548,549) ovaries. Later immunohistochemical studies primarily localized ovarian PR expression to the theca of large preovulatory follicles, the surface epithelia, and stromal/interstitium and found this expression pattern to be well conserved among species, including mice (550,551), dogs (552), rabbits (553,554), cows (555), pig (556), monkeys (326,557), and humans (332,558,559). Numerous studies in species ranging from rodents, large domestic animals, and primates agree that granulosa cells possess minimal PR expression throughout folliculogenesis except for the period 2–4 hours after the preovulatory gonadotropin surge, whereupon an enormous induction of PR expression occurs in preovulatory granulosa cells and peaks just before ovulation (326,332,351, 550,551,553,556,557,560–562) (Fig. 21). This dramatic increase in PR expression in the granulosa cells of ovulatory follicles is absolutely essential to follicle rupture (181,563). This induction of PR expression in ovulating follicles is quite transient and lasts less than 12 hours in rodent ovaries (351,550,560) but is maintained throughout luteinization and corpus luteum formation in rabbit (553), porcine (556), and primate (326,332,557,564) ovaries.

PR expression before the preovulatory surge is poorly characterized. PMSG treatment of immature rats leads to a fourfold increase in cytosolic progestin binding sites (565) and a modest twofold increase in *Pgr* mRNA levels (560) after 48 hours, indicating that FSH or subsequent estradiol signaling may stimulate basal levels of PR expression. In contrast, the mechanisms underlying the dramatic induction of PR expression in preovulatory granulosa after the gonadotropin surge are relatively well characterized. Over the course of the estrous cycle in rats, *Pgr* mRNA levels are increased >30-fold during a 4-hour period (1800–2200 hours) on the evening of proestrous and then return to baseline shortly after (2400 hours) (560). This increased expression is almost exclusive to the mural granulosa cells of healthy preovulatory follicles in rats (560,566), indicating that only fully differentiated granulosa cells possess the capacity for such rapid induction of the *Pgr* gene. Although PR expression climaxes within hours after plasma estradiol and LH levels climax, the following evidence indicates the latter hormone is most directly involved:

1. Expression peaks just 2 hours after plasma LH levels climax versus >8 hours after peak plasma estradiol levels (560).
2. hCG treatment 48 hours after PMSG exposure leads to a comparable increase in PR expression, whereas PMSG and concomitant increases in estradiol have a minimal effect (351,550,560).
3. Treatment of hypophysectomized or immature rats or differentiated rat granulosa cells with estradiol has no marked effect on ovarian PR expression (12,358).
4. Inhibition of the LH surge by pentobarbital abates the increase in PR expression despite having no effect on the rise in plasma estradiol levels (560).
5. LH but not estradiol elicits a dramatic PR induction in differentiated granulosa cells from rat (12,567) or porcine (561) ovaries in vitro.
6. An ER antagonist (ICI 164,3384) does not inhibit forskolin-induced PR expression in differentiated rat granulosa cells (509).
7. The putative ERE-like region of the rat *Pgr* promoter does not bind ER-α or ER-β in ovarian extracts (509).
8. ER (mainly ER-β) levels are rapidly decreasing at the time of rising PR expression in preovulatory granulosa cells (308,309,321,351).

The above data provide a convincing argument that the rapid and transient induction of PR expression in preovulatory granulosa cells is directly mediated by the LH surge. Clearly, only granulosa cells of preovulatory follicles in vivo or fully differentiated granulosa cells in vitro possess the capacity to respond to the LH surge and increase PR expression, presumably due to their unique acquisition of LH receptor (12,509,560). Indeed, granulosa cells isolated from estrogen-primed immature rats require a 48-hour exposure to physiological concentrations of FSH and testosterone to acquire LH induction of PR expression (509). LH-mediated induction of the intracellular cAMP/PKA pathway is likely the principal stimulus for increased PR expression in preovulatory granulosa cells as large doses of FSH, forskolin, or $(Bu)_2$cAMP also elicit a comparable response in rat ovaries in vivo (12) and differentiated rat (12,567) or porcine (561) granulosa cells in vitro. However, GnRH, PMA, or EGF exposure of differentiated rat granulosa cells in vitro elicits a comparable induction and temporal pattern of PR expression, suggesting that stimulation of PKC or tyrosine kinase signaling cascades may also be involved (567). In fact, Sriraman et al. (568) reported that forskolin and PMA synergize to induce increased PR expression in undifferentiated primary rat granulosa cells. However, studies in fully differentiated rat granulosa cells demonstrate that forskolin-induction of PR expression is entirely blocked by cotreatment with a PKA inhibitor (H89) but only minimally affected by a PKC inhibitor (Cal-C) (509), indicating a greater importance for the former signaling pathway.

The mechanisms by which LH activation of the PKA pathway leads to increased PR expression in preovulatory granulosa cells remain unclear. The *Pgr* promoter is complex, consisting of distal and proximal regions, and able to give rise to multiple transcripts (17,19). Initial studies concluded that an ERE3 within the proximal rat *Pgr* promoter is important to LH-induction (509), but later studies that better modeled the structure of the rat *Pgr* promoter indicated these sequences are dispensable (568). Furthermore, studies indicating that a GC and CCAAT box within the distal promoter may mediate cAMP activation (569) are now being questioned as more recent studies demonstrate that the whole *Pgr* distal promoter may not be required at all for LH-mediated induction (568). However, Sriraman et al. (568) recently demonstrated that two Sp1/Sp3 binding sites within the proximal promoter of the mouse *Pgr* gene bind Sp1/Sp3 and are essential to forskolin/PMA-induced expression.

Differential expression of the two PR isoforms, PR-A and PR-B, among thecal and granulosa cells further illustrates the complexity of PR expression in the ovary. Western blot analyses after LH induction indicate a PR-A/PR-B ratio of 3:1 in differentiated granulosa cells from rats (567) and a 2:1 ratio in whole ovaries from mice (550). In rat ovaries, PR-A immunoreactivity is restricted to the thecal cells of preantral and antral follicles before the gonadotropin surge, whereas PR-B immunoreactivity is detected in both thecal and mural granulosa cells throughout the stages of folliculogenesis and shows less variation over the course of the estrous cycle (551). The enormous induction of PR expression that occurs in preovulatory granulosa cells on the evening of proestrous is primarily representative of PR-A, resulting in a 2:1 ratio of PR-A/PR-B in these cells (551). In turn, the more moderate LH-induced increases in PR expression in thecal cells is predominantly PR-B (551). During the metestrous stage and rising progesterone synthesis that follow ovulation, PR-B is the sole isoform in thecal cells of preantral and antral follicles (551).

TABLE 8. *Endocrine and ovarian phenotypes in mice null for progesterone signaling*

PR-null females
Slightly elevated plasma LH.
Normal plasma FSH.
Exhibit a blunted gonadotropin surge.
Anovulatory and infertile.
Adult ovaries are grossly normal.
Immature females exhibit normal development of preovulatory follicles in response to PMSG but are unresponsive to subsequent hCG-induced ovulation and instead exhibit multiple unruptured follicles that eventually luteinize to form corpora lutea.
Immature female ovaries exhibit poor induction of *Adamts-1* and cathepsin-L (*Ctsl*) by hCG during gonadotropin-induced ovulation.
PR-A–null females
Anovulatory and infertile.
Adult ovaries are grossly normal.
Immature females exhibit normal development of preovulatory follicles in response to PMSG but respond poorly to subsequent hCG-induced ovulation and exhibit multiple unruptured follicles similar to PR-null females.
PR-B–null females
Fertile.
Adult ovaries are grossly normal.
Immature females exhibit normal response to gonadotropin-induced ovulation.

See text for references.

Ovarian Phenotypes in Murine Models of Disrupted Progesterone Signaling

Mice lacking both PR forms exhibit grossly normal ovaries but are anovulatory and therefore infertile (Table 8) (179). Furthermore, gonadotropin-induced ovulation of 6-week-old PR-null females is unable to rescue the anovulatory phenotype because no oocytes are recovered from the oviducts or upper uterine horns after treatment (Fig. 22) (179). Instead, PR-null ovaries after induced ovulation exhibit multiple, large, preovulatory follicles, each possessing a healthy competent oocyte but a total failure to rupture (179) (Fig. 22). The follicular cells of the unruptured preovulatory follicles undergo luteinization and form functional corpora lutea within the expected time frame, suggesting that some elements of an LH response are intact (181,566) (Fig. 22). Similar experiments in isoform-specific PR-null females indicate a more severe phenotype after the loss of PR-A relative to PR-B, as females from the former line exhibit an average of 9 oocytes/female versus 32 oocytes/female among wild-type littermates, whereas PR-B females exhibit no deficits in oocyte yield (180,181) (Fig. 22). The finding that the PR-A isoform is more critical to ovulation (180) is congruent with the studies of Gava et al. (551) that demonstrate this isoform is much more dramatically induced by the LH surge in preovulatory granulosa cells. Still, the absolute failure of ovulation even after gonadotropin induction in females lacking both PR forms indicates a level of cooperation between PR-A and PR-B that is vital to follicular rupture (180).

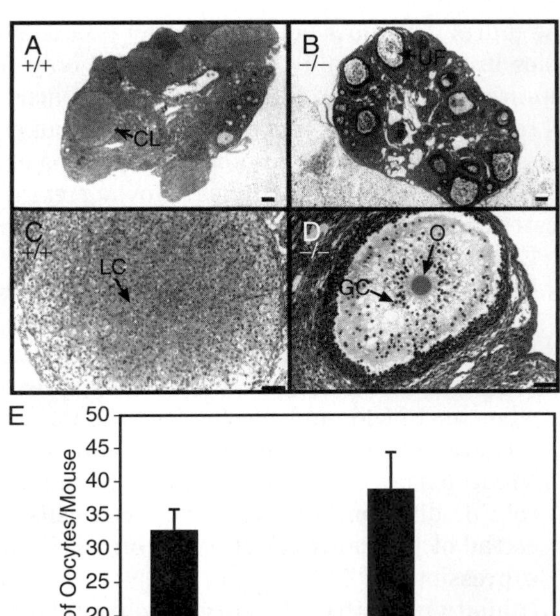

FIG. 22. Ovarian response to gonadotropin-induced ovulation in PR-null mice. **A:** Transverse section of a typical ovary isolated from a 6-week-old wild-type mouse (+/+) treated with an intraperitoneal injection of 5 IU of PMSG, followed 48 hours later with 5 IU of hCG, and killed 24 hours later. Note the presence of numerous corpora lutea (CL). Scale bar = 100 μm. **B:** A representative cross-section of an ovary from an age-matched PR-null (–/–) female that was hormonally treated exactly as the wild-type described above. Note the unusual presence of several unruptured follicles (UF) as indicated. Scale bar = 100 μm. **C:** High magnification of a corpus luteum in the wild-type ovary, exhibiting the characteristic hypertrophied luteal cells (LC). Scale bar = 50 μm. **D:** High magnification of an unruptured follicle present in a PR-null ovary, exhibiting an intact oocyte (O) with a zona pellucida and granulosa cells (GC) that have undergone cumulus expansion. Note the lack of luteinization among the GCs in the PR-null follicle compared to those in the wild-type (C). Scale bar = 50 μm. [A–D reproduced with permission from (179).] **E:** Average number of oocytes (± SEM) released per mouse after gonadotropin-induced ovulation in wild-type (WT), PRA-null (PRAKO(–/–), PRB-null (PRBKO–/–), and PR-null (PRKO(–/–) mice; *n* = 8 per test group. [Reproduced with permission from (181).]

Intraovarian Role of Progesterone in Ovarian Function

Follicle Growth and Differentiation

Although recent immunohistochemical studies indicate that both PR isoforms are expressed at basal levels throughout thecal and granulosa cells (551), there is little evidence that progesterone action profoundly influences follicle growth and differentiation in the rodent ovary. Administration of progesterone or synthetic progestins to immature or hypophysectomized rats has no obvious effect in the ovary in terms of ER levels (301,358) or gonadotropin/estrogen-induced follicle growth and maturation (301,570). In addition, progesterone has no measurable influence on FSH-induced granulosa cell proliferation (427) or growth and differentiation of whole follicles in vitro (428). Several studies suggest that progesterone inhibits estradiol synthesis in rodent ovaries, but these data in whole are conflicting (373). Perhaps most indicative of a minor role for progesterone signaling during folliculogenesis before ovulation is the lack of any overt phenotype in follicle growth, maturation, or steroidogenesis in the ovaries of PR-null mice (179,571). Indeed, Robker et al. (566) demonstrated that PR-null ovaries exhibit the expected FSH- or PMSG-induced increases in *Cyp19* and *Lhcgr* expression.

Ovulation

PR expression in the ovary is relatively low throughout folliculogenesis except during the period immediately after the ovulatory gonadotropin surge, during which an enormous induction of PR expression occurs in granulosa cells of ovulating follicles (12,560,567) and is synchronized with an equally acute increase in ovarian progesterone synthesis (572). These phenomena alone are indicative of an important role for PR-mediated progesterone signaling during ovulation. Experimental evidence of a critical role for PR in ovulation comes from studies in which ovulation is blocked when periovulatory ovaries or follicles are exposed to anti-progesterone antisera (573,574), inhibitors of progesterone synthesis (575–577), or PR antagonists (550,578,579). Finally, the anovulatory phenotype of PR-null mice discussed above provides definitive evidence that PR actions are required in the mammalian ovary during the period just before ovulation (179,180). PR-null females exhibit multiple, large, preovulatory but unruptured follicles after induced ovulation (179) (Fig. 22). Therefore, follicle growth and differentiation are unaffected by the loss of PR functions as PR-null follicles exhibit fully formed antra and increased *Cyp19* and *Lhcgr* expression after FSH or PMSG treatment (179,566). Furthermore, although follicle rupture is impaired, PR-null ovaries exhibit several indications that other LH- or hCG-induced responses are intact, including rapidly decreased *Cyp19* expression (566), dramatic induction of *Ptgs2* expression (566), expansion of the cumulus–oocyte complex (179), and luteinization and formation of functional corpora lutea (Fig. 22) (181,566). These findings in PR-null females are congruent with earlier studies using progesterone synthesis inhibitors (576,580) or PR antagonists (578) that also report that changes in the ovulatory levels of prostaglandins and steroids are largely unaffected by inhibition of progesterone signaling.

The above data clearly indicate that PR actions are highly specific and critical to follicle rupture. Recent studies have discovered that PR-null ovaries fail to exhibit LH-induced expression of two proteases, ADAMTS-1 and cathepsin-L, that are postulated to be involved in degradation of the follicle wall and extracellular matrix and facilitate oocyte extrusion (Fig. 23). ADAMTS-1 (*Adamts-1*) is a member of the A disintegrin and metalloproteinase family of proteases and is dramatically increased in granulosa cells of preovulatory follicles 12 hours after hCG treatment, approximately 4 hours after peak PR expression and coinciding with period of follicular rupture (Fig. 23) (577). A similar pattern of periovulatory *Adamts-1* expression is documented in preovulatory follicles of porcine (581), equine (562), and primate (582) ovaries. Cathepsin-L (*Ctsl*) is also significantly increased after hCG exposure in granulosa cells of preovulatory follicles and peaks 12 hours after treatment (566,583). PR-null females (566) as well wild-type female rats exposed to a progesterone synthesis inhibitor near the time of hCG treatment

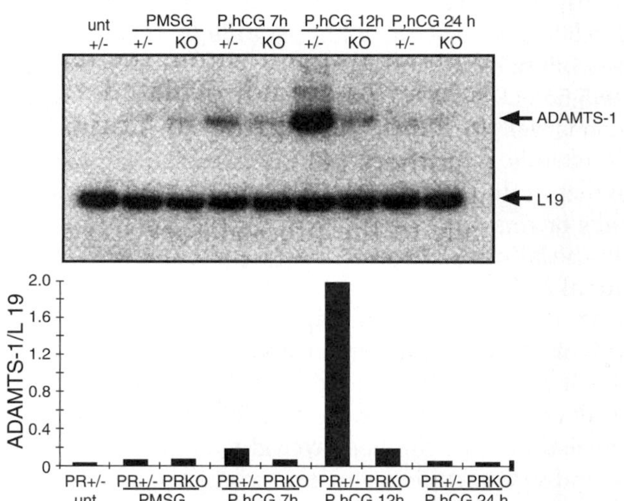

FIG. 23. Ovarian expression of ADAMTS-1 in PR-null ovaries during gonadotropin-induced ovulation. Immature heterozygous (PR+/–) and homozygous PR-null (PRKO) mice were left untreated or injected with PMSG and killed 46 hours later or were treated with PMSG for 46 hours followed by hCG and killed 7, 12, and 24 hours later. Reverse transcriptase polymerase chain reaction analysis of total RNA from whole ovaries of indicates that ADAMTS-1 expression is dramatically increased by hCG in PMSG-primed mice, but this induction is lacking in PR-null ovaries, indicating dependence on PR function. ADAMTS-1 expression was normalized to expression of ribosomal protein L19. [Reproduced with permission from (566).]

(577,584) failed to exhibit hCG-stimulated increases in *Adamts-1* expression (Fig. 23), indicating this effect of LH is primarily dependent on PR-mediated progesterone action. Support for a role of ADAMTS-1 in degradation of the follicle wall comes from a recent report that ADAMTS-1–null female mice exhibit severely compromised ovulation and a phenotype of preovulatory but unruptured follicles after gonadotropin-induced ovulation (585). Interestingly, PR-A–null mice exhibit a severe ovulatory phenotype similar to total PR-null females but possess normal LH-stimulated induction of *Adamts-1* and *Ctsl* during induced ovulation (181), suggesting possible compensatory actions by PR-B in gene induction but also questioning the importance of ADAMTS-1 and cathepsin-L in follicle rupture. Doyle et al. (584) recently demonstrated that both LH and PR signaling are able to induce *Adamts-1* expression in differentiated rodent granulosa cells via parallel pathways that ultimately coordinate to provide the proper temporal pattern of ADAMTS-1 activity in ovulating follicles.

ANDROGEN RECEPTOR SIGNALING IN OVARIAN FUNCTION

AR Expression in the Ovary

AR expression is documented in the ovaries of multiple species, including mouse (311,386), rat (208, 478,586–588), rabbit (229), pig (589,590), sheep (591), cow (592), monkey (229,294,335,593–595), and human (335,558,596–599). The ovarian pattern of *AR* expression is well conserved among species, with levels being detectable throughout the stages of folliculogenesis except in the primordial stage, highest in the granulosa cells of small preantral follicles, detectable but lower in thecal/interstitial cells, and generally absent in luteal cells.

In the ovaries of multiple species, *AR* expression among growing follicles is inversely correlated with the degree of granulosa differentiation. A mechanism to decrease AR levels during follicle maturation is consistent with the need to reduce follicle sensitivity to androgens, which continue to rise to provide levels sufficient for aromatization to estradiol (600). High-affinity androgen binding activity, AR immunoreactivity, and *Ar* transcripts are highest in small preantral follicles and considerably lower in large antral follicles (386,478,586,587,589,590,601). In the ovaries of gonadotropin-stimulated rats, Testsuka and Hillier (478) found that large antral follicles (>400 μm) exhibit a 2.75-fold higher level of *Cyp19* expression relative to small follicles (<200 μm) but a 51% decrease in *Ar* expression (Fig. 24). In rat ovaries, Szoltys and

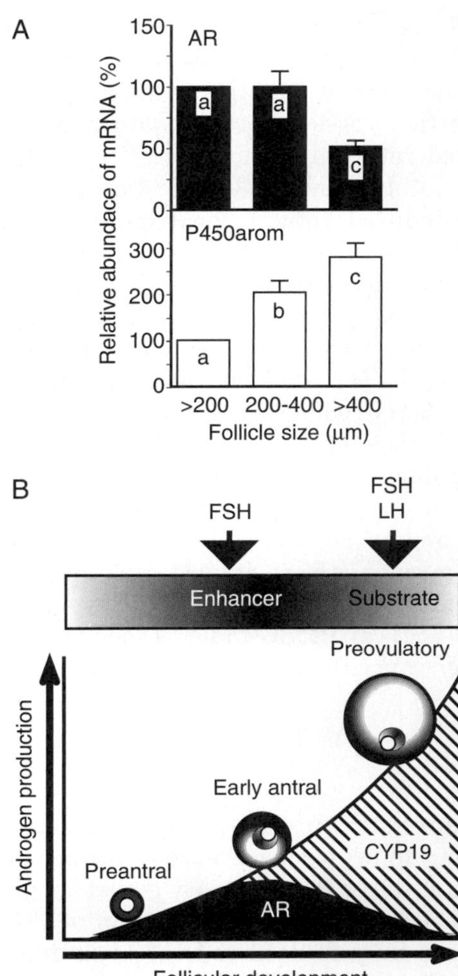

FIG. 24. Relationship between expression of androgen receptor and aromatase (CYP19) during folliculogenesis. **A:** Expression of androgen receptor (*Ar*) and aromatase (*Cyp19*) expression in granulosa cells of small (~200 μm), medium (200–400 μm), and large (>400 μm) follicles from immature rats after treatment with PMSG. Expression was quantified by RNase-protection assay and normalized to 18S rRNA (not shown); data are expressed as percentage of control values (SEM of 3 separate trials, each consisting of 8–10 animals). a, b: $P < 0.05$; a, c: $P < 0.01$ (by ANOVA with the Newman-Keuls test). [Reproduced with permission from (478).] **B:** Hypothetical model of androgen utilization during folliculogenesis. As follicular development progresses, thecal androgen production gradually increases. During the early stages of follicular differentiation, androgens act via androgen receptor (AR) to enhance FSH-induced differentiation, including the stimulation of *Cyp19* expression. During the final stages of follicular development, androgens primarily serve as a substrate for CYP19-mediated estradiol synthesis under stimulation by FSH and LH. This differential regulation of AR and CYP19 may be important in shifting androgen utilization from action to metabolism, thereby ensuring a healthy transition of a follicle from the early maturation to full maturation stage. [Reproduced with permission from (477).]

Slomczynska (587) demonstrated that decreasing AR immunoreactivity during follicle differentiation is first apparent in mural granulosa cells and then progresses in those cells closest to the antrum (Fig. 25). Interestingly, cells composing the cumulus–oocyte

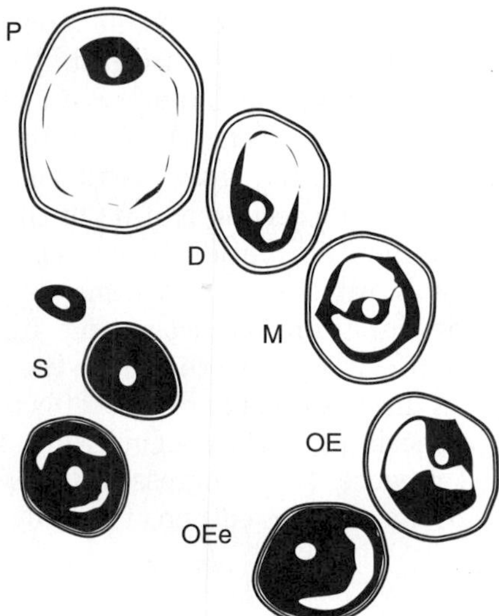

FIG. 25. Diagrammatic representation of AR distribution in the granulosa cells of developing and maturing rat follicles. S: preantral and early antral follicles, not exhibiting pseudostratification and present in the ovaries at all stages of the estrous cycle; AR expression in all granulosa cells. OEe: early antral follicle at estrous; AR expression in cumulus, antral, and mural cells, with exception of pseudostratified cells. OE: late estrous follicle; AR expression in cumulus, antral, and a few mural cells. D: diestrous follicle; AR expression in cumulus, some antral (mainly adjacent to the antrum), and a few mural cells. P: proestrous follicle; AR expression in cumulus and a few bordering the antrum cells. [Reproduced with permission from (587).]

complex of preovulatory follicles are believed to be the last to differentiate (602) and maintain *Ar* expression even in late-stage follicles (587).

In primate and human ovaries, *AR* expression is highest in the granulosa cells of small growing follicles, yet evidence of an inverse correlation with follicle differentiation are conflicting (229,558,594,595,597, 603). Hillier et al. (594) found AR immunoreactivity in the marmoset ovary is most abundant in granulosa cells of healthy preantral to small antral follicles and low or absent in preovulatory follicles of late follicular stage. In contrast, minimal differences in granulosa cell *AR* expression between preantral and large antral follicles are reported in rhesus monkey ovaries (595). Furthermore, healthy follicles at late stages of maturation in human ovaries are described to possess significant AR immunoreactivity (558,597,599).

The regulatory factors for *AR* expression in the ovary are poorly understood. Small preantral follicles in rats maintain high AR levels after hypophysectomy, indicating that gonadotropins are not required to induce *AR* expression (586). In fact, most evidence indicates that FSH- or PMSG-induced differentiation of granulosa cells is the primary cause of reduced *AR* expression. Campo et al. (601) demonstrated that PMSG treatment of rats leads to the replacement of high-affinity low-capacity androgen binding sites by nonsaturable low-affinity binding sites in the ovary. Similarly, immature rats treated with recombinant FSH over 48 hours exhibit a 65% reduction in ovarian *Ar* mRNA levels and a further decrease when LH is included (586). However, neither FSH nor 8-bromo-cAMP affects *Ar* mRNA levels in cultured rat granulosa cells in vitro (478). Interestingly, DHT elicits a 20% reduction in *Ar* expression in rat granulosa cells in vitro that is prevented by cotreatment with FSH (478). In contrast, testosterone reportedly has little effect on *AR* expression in rhesus monkey ovaries (595). There is increasing evidence that estradiol may play a role in decreasing granulosa cell AR levels during follicle differentiation. Like DHT, estradiol exposure of rat granulosa cells leads to a 20% reduction in *Ar* transcripts, but this effect is unabated by FSH (478). Furthermore, granulosa cells retrieved from untreated hypophysectomized rats respond well to DHT plus FSH and exhibit increased *Cyp19* expression accordingly, but the effect of DHT is lost in granulosa cells isolated from estradiol-primed hypophysectomized rats, suggesting that estrogen pretreatment decreases *Ar* expression and responsiveness (378). Indeed, mature follicles in ER-β–null ovaries are reported to possess aberrantly high *Ar* expression relative to similarly staged follicles in wild-type ovaries (386).

Ovarian Phenotypes in Murine Models of Disrupted Androgen Signaling

Tfm mice were first described in 1970 and are a naturally existing androgen-resistant mutant (182) due to an inactivating mutation of the *Ar* gene (183,184). Comparable inactivation mutations of the *AR* gene and resulting phenotypes have been described in rats and humans (185). *Tfm*/*Tfm* female mice are fertile but exhibit a noticeably shortened reproductive life span that becomes apparent after approximately four litters compared with control (*Tfm*/X or X/X) females; however, individual litter sizes are no different (Table 9) (186). Histological evaluation of

TABLE 9. *Endocrine and ovarian phenotypes in mice null for androgen signaling*

Tfm/Tfm mice
Fertile but exhibit reduced fecundity with age and a shortened reproductive life span.
AR-null females
Exhibit reduced fertility and a shortened reproductive life span.
Adults exhibit grossly normal ovaries but fewer corpora lutea.
Abnormal estrous cycle.
Immature females exhibit a poor response to gonadotropin-induced ovulation that is illustrated by reduced expression of FSH receptor, indications of reduced estradiol synthesis, below normal ovulation rate, and attenuated granulosa cell luteinization.

See text for references.

ovaries from the *Tfm*/*Tfm* breeding females indicated a sparse number of healthy follicles and a hypertrophied interstitium, both illustrative of premature ovarian failure (186). The difficulties in generating *Tfm*/*Tfm* female mice limited more extensive study.

More recently, AR-null female mice were generated via a *Cre*/loxP targeting scheme that allows for tissue and temporal specific disruption of the *Ar* gene and therefore the generation of fertile male carriers of the targeted *Ar* allele (187–189). AR-null females exhibit normal fecundity only during the first 12 weeks of continuous breeding, after which they produce a reduced number of litters per female (2.3 versus 3.3 in wild-type) and less offspring per litter (4.5 versus 9.8 in wild-type), whereas others become totally infertile (190) (Table 9). This apparent premature ovarian failure in AR-null females is remarkably similar to that described above in *Tfm*/*Tfm* females (186). Adult AR-null females possess ovaries exhibiting a normal number of growing and antral follicles but notably fewer corpora lutea (190) and reduced circulating progesterone (604), suggesting that folliculogenesis may be preserved but ovulation and luteinization are greatly disrupted. Gonadotropin-induced ovulation of immature AR-null females leads to a severely reduced ovulatory yield and oocytes that often exhibit a less condensed cumulus–oocyte complex (190). This latter finding is especially interesting in light of the preserved *Ar* expression in the cumulus oophorus cells of preovulatory follicles. Additional evidence that AR-null ovaries fail to fully respond to an ovulatory dose of hCG is an inadequate induction of progesterone receptor (*Pgr*), hyaluronyl synthetase-2 (*Has2*), and tumor necrosis factor-α–stimulated gene 6 (*Tnfaip6*) (190), all factors that are critical to expansion of the cumulus–oocyte complex and follicle rupture. Furthermore, granulosa cells of ovulatory follicles in AR-null ovaries fail to terminally differentiate and luteinize after hCG-induced ovulation, as indicated by decreased expression of cyclin-dependent kinase inhibitor 1A (*Cdkn1a*) and cytochrome P-450_{scc} (*Cyp11*) (190). There are currently no reported data on the reproductive hormone levels in AR-null females; however, adult females exhibit hypoplastic uteri 48 hours after PMSG stimulation, suggesting that gonadotropin-induced estradiol synthesis in the ovary may be impaired (190).

Intraovarian Role of Androgen in Ovarian Function

The intraovarian roles of androgens may be categorized into three distinct functions: as substrates for estradiol synthesis, as an enhancer of follicle differentiation, and as a stage-specific inhibitor of follicle growth (477). The role of thecal-derived androgens, most notably androstenedione, as substrates for estradiol synthesis in granulosa cells is critical given the profound importance of the latter hormone to reproductive function. Gore-Langton and Armstrong (373) thoroughly reviewed ovarian steroidogenesis in the second edition of this volume, and therefore this information is not repeated here. Both additional intraovarian actions of androgen involve their function as activating ligands for AR-mediated effects and are therefore the subject of this section. A discussion of androgen effects on progesterone synthesis in follicles of multiple species is not included because little progress has been made in this area since the previous edition of this volume (373).

Granulosa Cell Proliferation

The follicular response to androgens is dependent on the stage of growth and differentiation. In preantral undifferentiated follicles that have not yet acquired the capacity to synthesize estradiol, androgens promote gonadotropin-induced granulosa cell proliferation and maturation. In large differentiated follicles that have acquired aromatase activity, estradiol assumes the role of enhancing FSH-induced granulosa cell proliferation and differentiation, and the role of androgens is shifted to that as a substrate for conversion to estradiol. At this late stage, excess androgens are detrimental to follicle health and lead to atresia. This paradox of androgen action during follicle growth and differentiation explains the contradictory results when comparing experimental studies. Several studies in hypophysectomized rats have shown that coadministration of testosterone inhibits estrogen or gonadotropin-induced granulosa cell proliferation and increased ovarian weight and promotes degeneration in more mature follicles (358,420,605–607). In contrast, Armstrong and Papkoff (297) found that aromatizable androgens (e.g., testosterone or androstenedione) enhance the promotional effects of FSH in the ovaries of hypophysectomized rats but that the nonaromatizable androgen, DHT, is inhibitory. Similar treatment with hCG instead of testosterone has a comparable effect on gonadotropin or estradiol-induced follicle growth in hypophysectomized rats but is prevented by AR antagonists, indicating the detrimental effects of hCG are due to stimulation of thecal cell androgen synthesis (606,608,609). However, Bley et al. (427) found that granulosa cells from preantral follicles of estrogen-primed rats proliferate at comparable rates in response to FSH plus estradiol or DHT in vitro and that the promotional effects of the latter steroid are blocked by the antiandrogen, hydroxy-flutamide. Investigations of the mechanism by which androgens inhibit granulosa cell proliferation indicate that DHT exposure to estrogen-primed rats 24 hours before granulosa cell isolation causes a blunted response to forskolin-induced cyclin D2 (*Ccnd2*) expression in vitro, leading to cell cycle arrest and reduced proliferation (610).

The stage-specific effects of androgens during folliculogenesis described above may be more apparent in rat ovaries relative to other species. In hypophysectomized mice, Wang and Greenwald (423) found that testosterone or DHT, in combination with FSH, induces greater levels of DNA synthesis in antral follicles than FSH alone; a similar response was not observed in small and medium-sized follicles. In preantral mouse follicles grown in vitro, anti-androgen antisera reduce FSH-stimulated growth and DNA synthesis; however, these data must be interpreted with caution because anti-androgen antisera may also remove androgens from the pool of substrates for estradiol synthesis. Additional evidence that androgens may promote granulosa cell proliferation in mice comes from the findings of Burns et al. (430) that the AR-antagonist flutamide reduces the growth rate of granulosa cell tumors in animals lacking functional inhibins. In cumulus cells of large antral follicles from porcine ovaries, DHT augments FSH or IGF-1–induced proliferation, and this effect is inhibited by flutamide (611). Rhesus monkey ovaries exhibit a strong correlation ($r = 0.91$) between *AR* expression and cell proliferation (595) and exhibit an increased number of follicles of all stages except large antral after testosterone or DHT treatments (612).

Granulosa Cell Differentiation

Fully differentiated preovulatory follicles in mammalian ovaries are distinctly characterized by a large fluid-filled antrum, significantly increased aromatase (CYP19) activity, and acquisition of LH responsiveness. The absolute requirement of FSH signaling in this process is illustrated by the absence of all three phenotypes in mice null for FSH action (356,444,445). As discussed in the section on Estrogen Receptor Signaling in Ovarian Function, estradiol is also required for full FSH-induced follicle differentiation. However, there is substantial evidence that androgens may be as equally efficient as estradiol in augmenting certain responses to FSH, most especially the induction of aromatase activity and antrum formation (477). Both testosterone and DHT enhance FSH induction of aromatase activity in hypophysectomized rat ovaries or isolated granulosa cells in vitro in a dose-dependent fashion (297,613). Furthermore, pharmacological inhibition of ovarian androgen synthesis in rats simultaneously reduces aromatase activity that can be restored by exogenous testosterone (614). Fitzpatrick and Richards (378) found that testosterone, and DHT to a lesser extent, significantly enhances FSH induction of *Cyp19* expression and aromatase activity in granulosa cells from untreated hypophysectomized rats (Fig. 19). Interestingly, this effect is lost in granulosa cells isolated from estrogen-primed hypophysectomized rats (378), supporting the hypothesis that early growing follicles are more sensitive to androgen action because estrogens down-regulate *AR* expression as part of the process of follicle differentiation. Similar data have been produced in mice. For example, AR antagonists are able to inhibit FSH-induced estradiol synthesis in individually cultured murine follicles if included at the beginning of the culture period before any indications of differentiation (615). Similarly, estradiol output by large-antral murine follicles in vitro is unaffected by DHT (423). In marmoset ovaries, granulosa cells isolated from small (0.5–1 mm) follicles are highly responsive to the enhancing effects of testosterone or DHT on FSH-induced aromatase activity, but this response is lost in granulosa cells from larger (>2 mm) follicles whereupon DHT exposure becomes inhibitory (616,617). In contrast, 3 or 10 days of testosterone treatment in rhesus monkeys had no effect on ovarian *CYP19* expression (603).

Most data indicate that healthy preantral follicles are responsive to testosterone and perhaps require the steroid for the initiation of *Cyp19* expression by FSH in granulosa cells. Because androgen synthesis in thecal cells precedes the acquisition of aromatase activity in the accompanying granulosum, this mechanism of androgen/FSH synergism allows for more efficient induction of *Cyp19* expression than would occur with FSH alone. Once the granulosa cells acquire sufficient aromatase activity, thecal-derived androgens become more important as substrates for estradiol synthesis, allowing ER-mediated estradiol actions to continue the role of synergizing with FSH to promote follicle differentiation (Fig. 24). Hence, the hallmark of a healthy preovulatory follicle in rodent ovaries may be the presence of substantial aromatase activity and estradiol output and decreased *AR* expression and androgen sensitivity.

The mechanism by which androgens augment FSH action on granulosa cells is unclear but may differ from that hypothesized for estradiol (see Estrogen Receptor Signaling in Ovarian Function). Whereas ER-mediated estradiol actions are believed to enhance the amount and effectiveness of FSH-stimulated intracellular cAMP without obvious changes in FSH-receptor levels, evidence suggests that androgens may act at a site upstream of adenylyl cyclase. For example, neither testosterone nor DHT synergize with 8-bromo-cAMP to induce aromatase activity in rat granulosa cells (618), whereas estradiol does (378,619). Instead, AR-mediated androgen actions may largely act by increasing the level of FSH receptor in granulosa cells of preantral follicles. Immature rat ovaries exposed to DHT (620) or isolated granulosa cells treated with testosterone (447) exhibit increased FSH binding. Testosterone can also restore the losses in FSH receptor and responsiveness that occur in rat granulosa cells cultured without gonadotropins, and this effect is inhibited by AR antagonists (618). More recent evidence that androgens increase granulosa FSH-receptor levels comes from reports that ovaries of 10-day-old and prepubertal AR-null females exhibit significantly reduced *Fshr* mRNA levels, even 48 hours after PMSG treatment in the latter age group (190). Similarly, Weil et al. (603) found that 3 or 10 days of testosterone treatment in adult rhesus monkeys leads to increased FSH-receptor expression in follicles of all stages except primary. In addition to inducing or maintaining FSH-receptor levels in preantral granulosa cells, there is also evidence that androgens may enhance FSH action by decreasing the rate of intracellular cAMP turnover (621).

AR-mediated androgen actions may also be involved in antrum formation. Murray et al. (615) found DHT potentiates suboptimal doses of FSH to stimulate increased follicle diameter of murine follicles in culture, and this promotional effect is inhibited by AR antagonist. Testosterone or DHT are also reported to be almost as effective as estradiol in augmenting FSH-induced increases in the number of antral follicles in the ovaries of hypophysectomized mice (423). However, the ovaries of AR-null mice exhibit no difference in the number of antral follicles at 4 or 16 weeks of age (190). Therefore, FSH-dependent antrum formation may be enhanced by either receptor-mediated androgen or estrogen actions, congruent with the process being part of the transition period during which follicles are sensitive to both androgens and estrogens.

Not all FSH-dependent processes are enhanced by androgens. As described in Chapters 10 and 11, LH-receptor expression and LH responsiveness by granulosa cells occurs only in preovulatory follicles and during the final stages of folliculogenesis, just before the LH surge. Unlike the above processes of follicle differentiation (e.g., antrum formation and aromatase expression), FSH stimulation of LH-receptor expression in preovulatory granulosa cells is facilitated by estrogen or aromatizable androgens only (447,479, 480), whereas DHT has no effect or may even be inhibitory (447,480). These data are consistent with the mutually exclusive pattern of AR and LH receptor in preovulatory follicles of rodent ovaries, because *AR* expression is limited to the cumulus cells (587), whereas LH receptor is predominantly localized to mural and antral granulosa cells (622–624). In addition, in vivo studies demonstrating that AR antagonists have no effect on LH-receptor expression in the ovaries of FSH-stimulated DES-primed hypophysectomized rats suggest that AR-mediated actions are not required to promote or maintain LH-receptor expression in preovulatory follicles (625). The specific requirement of ER-mediated estradiol actions to augment FSH-stimulated LH-receptor expression in preovulatory granulosa cells provides a mechanism by which only those follicles that acquire sufficient aromatase activity and are suitable to ovulate have the capacity to respond to the LH surge.

Thecal Cell Steroidogenesis

There are few studies on the role of androgen signaling in thecal cell function and steroidogenesis despite these cells being the primary source of androgen synthesis. Several nonsteroidal paracrine factors are known to modulate thecal cell steroidogenesis and have been more thoroughly studied (407,495). The limited available evidence suggests that AR-mediated androgen actions may negatively regulate thecal cell steroidogenesis, thereby forming an autocrine regulatory feedback loop. Mahesh and colleagues (626) showed in enriched thecal/interstitial cell cultures from rats that AR activation by nonaromatizable agonists attenuates hCG or hCG/IGF-1–stimulated increases in androstenedione synthesis by 32% and 40%, respectively, via selective inhibition of *Cyp17* expression. Similar experiments demonstrated that AR antagonists enhance hCG-induced androstenedione synthesis, further indicating a receptor-mediated negative effect of androgens on thecal cell steroidogenesis (626). Therefore, ER-mediated estradiol actions (see section on Estrogen Receptor Signaling in Ovarian Function) and AR-mediated androgen actions may both contribute to maintaining homeostasis of androgen synthesis in the thecal cells of developing follicles.

Ovulation

Over the past several years there have been few studies aimed at determining a role for androgen signaling in follicle rupture. This is especially surprising since Mori et al. (627) demonstrated in 1977 that inhibition of androgen action at the time of ovulation is detrimental to follicle rupture. These studies used anti-testosterone and anti-progesterone antisera during induced ovulation in immature or hypophysectomized rats to demonstrate that acute treatment with anti-testosterone antisera at the time or shortly after hCG-induced ovulation leads to a dose-dependent reduction in the number of ovulated oocytes and that inhibition of ovulation by acute treatment with an anti-progesterone antiserum at the time of hCG treatment is rescued by concurrent treatment with testosterone or DHT but not estradiol (627). In a similar study, Peluso et al. (628) demonstrated that treatment of immature rats with the AR antagonists, cyproterone acetate or flutamide, 3 hours before hCG-induced ovulation drastically reduces the number of ovulated oocytes and prevents expansion of the cumulus–oocyte complex. Similar findings are reported in mice (629). Still, a search of more recent literature indicates no further categorical studies of the mechanisms by which androgen signaling may facilitate ovulation via activities after the gonadotropin surge. This area of research is likely to witness a resurgence given the ovulatory defects recently described in AR-null mice (190). A defect in spontaneous ovulation among AR-null mice is suggested by the presence of a normal number of preovulatory follicles but few corpora lutea (190) and reduced circulating progesterone levels (604). Furthermore, gonadotropin-induced ovulation of immature AR-null females indicates a severely reduced response in terms of recoverable oocytes in the oviducts and may be attributable to their failure to exhibit hCG-induced expression of PR, hyaluronyl synthetase-2, and tumor necrosis factor-α–stimulated gene 6 (190).

Follicle Atresia

Follicle atresia in the mammalian ovary has been thoroughly described in several reviews (494,600,630), including an excellent detailed description by Greenwald and Roy in the second edition of this volume (418). Follicular atresia is a complex hormonally driven process involving both putative "survival" factors (e.g., IGF-1, EGF, basic fibroblast growth factor, estradiol, and activin) and putative atretogenic factors (e.g., tumor necrosis factor-α, GnRH, and androgens) (494,600,630). The respective action of each of these factors on atresia is dependent on the follicular stage (630).

Circumstantial evidence of the atretogenic actions of androgen includes numerous observations that atretic follicles consistently exhibit an increased androgen-to-estrogen ratio within the intrafollicular fluid (494). Supporting experimental evidence includes reports that low doses of hCG (to induce endogenous androgen production), testosterone, or DHT after 4 days of estradiol treatment in hypophysectomized rats led to an increased numbers of atretic follicles within 24 hours (609,631). The detrimental effects of hCG or exogenous androgens are inhibited by cotreatment with anti-testosterone antisera or AR antagonists, indicating the involvement of AR in this process (609,631). Interestingly, large (>150 μm) follicles are more susceptible to the atretogenic effects of androgens than small (<150 μm) follicles, indicating that androgens are largely atretogenic in late-stage follicles (609,631). In similar experiments on hypophysectomized rats, Billig et al. (632) demonstrated that the follicular atresia that follows the withdrawal of 2 days of continuous estrogen (DES) treatment is enhanced by subsequent testosterone exposure but prevented by estradiol.

EXTRAOVARIAN ROLES OF SEX STEROIDS IN OVARIAN FUNCTION

The levels circulating of sex steroids are considered the most influential physiological determinant of serum gonadotropin levels in animals and humans (633). GnRH secreted from the hypothalamus stimulates the gonadotropes of the anterior pituitary to synthesize and secrete FSH and LH. The subsequent gonadotropin stimulation of the ovaries leads to gametogenesis and concomitant synthesis of sex steroids and endocrine peptides (e.g., inhibins and activins), which then feed back upon the hypothalamus and pituitary to regulate FSH and LH secretion. Hypothalamic secretion of GnRH is pulsatile, and this pulsatility is mirrored in the release of gonadotropins from the anterior pituitary and is obligatory to proper ovarian function and reproductive success (634,635). Females are uniquely endowed with an ability to produce an enormous gonadotropin surge that induces ovulation and is the hallmark of the female reproductive cycle. A discussion of the mechanisms of how estrogen acts on the hypothalamic–pituitary axis to induce the gonadotropin surge is beyond the scope of this chapter and can be found in more appropriate chapters in this volume. However, to gain insight into the role of estradiol in ovarian function via the study of the gene-targeted models, we must remain cognizant of those ovarian phenotypes that may be more attributable to abnormalities in estrogen signaling in the hypothalamic–pituitary axis.

Like the reproductive tract, the neuroendocrine system undergoes a developmental process that is profoundly influenced by steroid receptor signaling. Sexual *differentiation* of the neuroendocrine system is best illustrated by the permanent "organizational" changes that provide the sexually dimorphic ability of the female hypothalamus to produce an LH surge (636–638). Sexual *maturation* of the neuroendocrine system may be defined as the acquisition of pituitary responsiveness to hypothalamic factors and ovarian steroids and the onset of steroid "activated" sexual behaviors. The imprinting and activational effects of the sex steroids on the hypothalamic–pituitary axis are believed to be receptor mediated (639) as numerous studies demonstrate a broad expression pattern of the steroid receptors throughout the brain and pituitary (640–645).

Negative Feedback Control of Gonadotropin Secretion

The abundance of experimental evidence demonstrating that estradiol suppresses gonadotropin secretion from the anterior pituitary in multiple species is thoroughly reviewed in the second edition of this volume (646). Ovariectomy of female rodents rapidly leads to increased expression of the gonadotropin subunit genes in the pituitary and correlating increases in circulating FSH and LH, all of which are restored to presurgery levels by administration of exogenous estradiol at physiological levels (646). The precise mechanisms by which estrogens exert their effects are unclear because evidence exists to support both direct (ER mediated) and indirect pathways (647). In the rodent, it is generally believed that the loss of circulating estradiol that follows castration leads to increased frequency of GnRH pulses from the hypothalamus and hence elevated tonic levels of gonadotropins (646,648). Studies in mice bearing reporter transgenes under the control of the gonadotropin subunit gene promoters support estrogen regulation of hypothalamic GnRH secretion versus direct pituitary actions as the primary pathway of gonadotropin regulation in mice (649,650).

Comparative studies in ER-null female mice indicate the following in reference to estradiol-mediated negative feedback of gonadotropin secretion: (a) negative feedback depends on ER-α and ER-β is dispensable, (b) ER-α–mediated negative feedback principally regulates LH synthesis and secretion and not FSH, (c) negative feedback occurs primarily at the level of GnRH secretion from the hypothalamus, and (d) the loss of ER-α–mediated negative feedback leads to hypersecretion of LH and severe ovarian dysfunction (351,366,651). Plasma LH levels in ER-α–null (αERKO, αβERKO) females are three- and sixfold higher than their wild-type littermates and more comparable with those observed in ovariectomized wild-type females, whereas ER-β–null females exhibit wild-type like levels (366). In contrast, plasma FSH levels are normal in all three ER-null lines, whereas ovariectomy of wild-type littermates leads to a more than sixfold increase (366). Prolonged treatment of ER-α–null females with a GnRH-antagonist restores plasma LH levels to normal, suggesting the primary defect is dysregulated hypothalamic GnRH secretion (366). In addition, in situ hybridization studies reveal increased GnRH expression in the hypothalamus of ER-α–null females versus wild-type and ER-β–null mice (651). Slightly higher plasma LH in αβERKO versus αERKO females suggests that preserved ER-β functions in the hypothalamic–pituitary axis of the latter may provide some degree of negative feedback (366,651). Female PR-null mice also exhibit increased plasma LH levels (571) but not nearly to the extent observed in female ER-α–null mice. Furthermore, ovariectomy and the loss of ovarian-derived estrogens in PR-null females leads to a substantial increase in plasma LH (571), but no such exacerbation is observed in ER-α–null females (651, 652). Still, Levine et al. (644) reported that exogenous progesterone treatments abate the ovariectomized-induced increases in plasma LH, and this effect is lost in PR-null females, indicating that ligand-dependent PR actions may also contribute to maintaining LH homeostasis in females. Collectively, these data highlight the vital role of ER-α–mediated estrogen signaling in the maintenance of LH secretion in female mice. The disturbances that result from the loss of ER-α signaling, either due to the loss of receptor or ligand, are most dramatically manifested in the ovary. Restoration of plasma LH levels in the ER-α–null females dramatically abates the characteristic phenotype of hemorrhagic/cystic follicles and elevated steroidogenesis (Fig. 17) (351,366). Therefore, to a large extent these ovarian phenotypes in ER-α–null females may be considered "secondary" to the loss of estrogen signaling in the hypothalamic–pituitary axis.

Several possibilities may explain the preservation of normal FSH regulation in αERKO and αβERKO females, which is surprising given the dramatic increase that follows ovariectomy in wild-type females. Plasma FSH levels are also normal in PR-null females (571). FSH synthesis and secretion is primarily regulated by the inhibin/activin family of hormones and is less dependent on steroid-mediated feedback (653). This is effectively illustrated in the activin receptor type II null mice, which exhibit low plasma FSH levels due to unopposed suppression by inhibins (654). All three ERKO models exhibit normal expression of the three inhibin/activin subunit genes in the ovary and pituitary, and therefore inhibin-mediated suppression of FSH secretion is presumably intact. Furthermore, ovariectomy of the αERKO females results in elevated FSH levels similar to those of ovariectomized wild-type females, presumably due to the loss of ovarian inhibin secretion (652). Alternatively, we observed modifications in the level of GnRH-receptor expression in ER-α–null female pituitaries (366) that may favor LH but not FSH synthesis and secretion (655).

The preservation of properly regulated FSH secretion in all three ER-null females also contradicts a report of elevated plasma FSH in Cyp19-null female mice (197). This divergence in phenotypes between the ER-null and Cyp19-null models may illustrate possible "organizational" differences in the hypothalamic–pituitary axis due to the latter models' sensitivity to maternal estrogens during gestational development. Differences may also exist in the ovarian secretion of inhibin/activin between ER-α–null and Cyp19-null mice as these data are not yet reported. However, the ovaries of *Fshb*-null females exhibit an expected decrease in *Cyp19* expression (<20% versus control) and a more substantial decrease in the expression of the inhibin subunit genes (each <5% versus control) (355). These data suggest that estradiol may be required for ovarian inhibin synthesis; however, ovarian expression of all three inhibin subunit genes is not altered in ER-null females (366). Therefore, further study is necessary to reconcile this discrepancy between the ER-null and Cyp19-null models in terms of maintaining FSH homeostasis.

Positive Feedback Control of Gonadotropin Secretion

In paradox to the negative feedback effects of sex steroids on gonadotropin secretion are the positive actions of estrogen and progesterone exposure on the hypothalamic–pituitary axis that are essential to stimulating the preovulatory gonadotropin surge (638,656). The duration and dose of estradiol exposure in the hypothalamic–pituitary axis is believed to be critical to this switch from an inhibitory to stimulatory response to estradiol (656). It is generally believed that external and internal cues transduced from the brain integrate with the positive feedback of gonadal hormones to produce a combined effect of highly synchronized patterns of GnRH secretion upon an anterior pituitary that has been primed to be transiently hypersensitive to GnRH (656). Again, the precise site within the hypothalamic–pituitary axis that is principal to the positive feedback of estradiol remains unclear. Destruction of the GnRH-secreting neurons in the monkey can be overcome by pulsatile administration of exogenous GnRH, whereupon a gonadotropin surge can be superimposed by treatments with exogenous estradiol, indicating the pituitary as the predominant site of estrogen actions necessary to launch the gonadotropin surge (657). However, similar experiments in rodents are unsuccessful, suggesting a greater degree of cooperation between the hypothalamus and pituitary in these species (658).

Anovulation in ER-α–null females appears to be predominantly due to chronically high plasma LH levels and the absence of a periodic gonadotropin surge, because ovulation can be induced by exogenous gonadotropins. As a result, ER-α–null ovaries accumulate large differentiated follicles that eventually become atretic and hemorrhagic but continue to synthesize large amounts of estradiol. Interestingly, the αERKO-like phenotype in transgenic mice due to chronically elevated LH is abated after regularly administered exogenous gonadotropins to mimic the gonadotropin surge (371,659). Several recent findings in ER-β–null females may also reveal a possible primary defect in the hypothalamic–pituitary functions that leads to infertility. ER-β–null females exhibit a much lower than expected frequency of pregnancies during the course of a continuous mating study (192,193), and evaluations of the ovarian cycle via vaginal cytology indicate that a sizeable cohort of ER-β–null females exhibit persistent estrous (J. M. A. Emmen, personal communication). These observations suggest that the ovarian dysfunction and reduced fecundity in ER-β–null females may be due to their inability to generate a properly timed and sufficient gonadotropin surge. Therefore, preliminary evidence suggests that a loss of ER signaling during development may lead to "defeminization" of those elements in the neuroendocrine axis necessary for the gonadotropin surge. Interestingly, intracerebroventricular administration of ER-antisense oligos in rats immediately leads to persistent estrous (660). Furthermore, prenatal exposure of female mice to either an ER-α or ER-β selective agonist also leads to persistent estrous during adulthood (661).

There is ample evidence that progesterone also amplifies the effects of estradiol in the induction of the gonadotropin surge (572,662,663). PR may act at both the hypothalamus and anterior pituitary during generation of the preovulatory surge (644,664). Support for an important role for PR comes from studies demonstrating that PR-null females are acyclic and fail to exhibit a gonadotropin surge after exposure to mature male bedding (571). In addition, Chappell et al. (665) further demonstrated that ovariectomized PR-null females fail to produce an LH surge after a surge-inducing regiment of estradiol treatments, indicating that the hypothalamic–pituitary axis of PR-null females is refractory to the positive actions of estradiol. The ability of estradiol to elicit an LH surge in ovariectomized wild-type but not PR-null females indicates that functional PRs in the hypothalamic–pituitary axis are required but progesterone is not, suggesting ligand-independent PR actions (665); subsequent studies suggest PR activation by the cAMP/PKA signaling pathway (644,666). Therefore, estradiol/ER induction of hypothalamic PR expression, likely in the anteroventral periventricular nucleus, is obligatory to estradiol stimulation of the LH surge on the evening of proestrous (644,665, 667). Furthermore, additional studies in PR-null females also indicate that PR expression in the gonadotropes of the anterior pituitary are obligatory to GnRH self-priming of the gonadotrope, a phenomenon that allows for increased magnitude of LH release after sequential GnRH stimulation (644,665).

Estrogen Regulation of Prolactin Synthesis

Prolactin is a multifunctional hormone with various physiological roles, including actions as a growth factor, neurotransmitter, and immunoregulator (668) (see Chapter 33). Relevant to this discussion are the well-known effects of prolactin as a luteotrophic hormone in the ovary and in promoting blastocyst implantation in the uterus (668,669). Estradiol is critical to the expression and synthesis of prolactin in the anterior pituitary (642,670). Most evidence indicates that ER-α may be the predominant receptor form in lactotrophs, although some species exhibit detectable levels of ER-β as well (642,671,672). Congruent with these findings, female mice lacking functional ER-α exhibit decreased *Prl* gene expression of >80% (366,673), whereas ER-β–null females exhibit wild-type like levels (366). Interestingly, reduced *Prl* expression was not fully reflected in

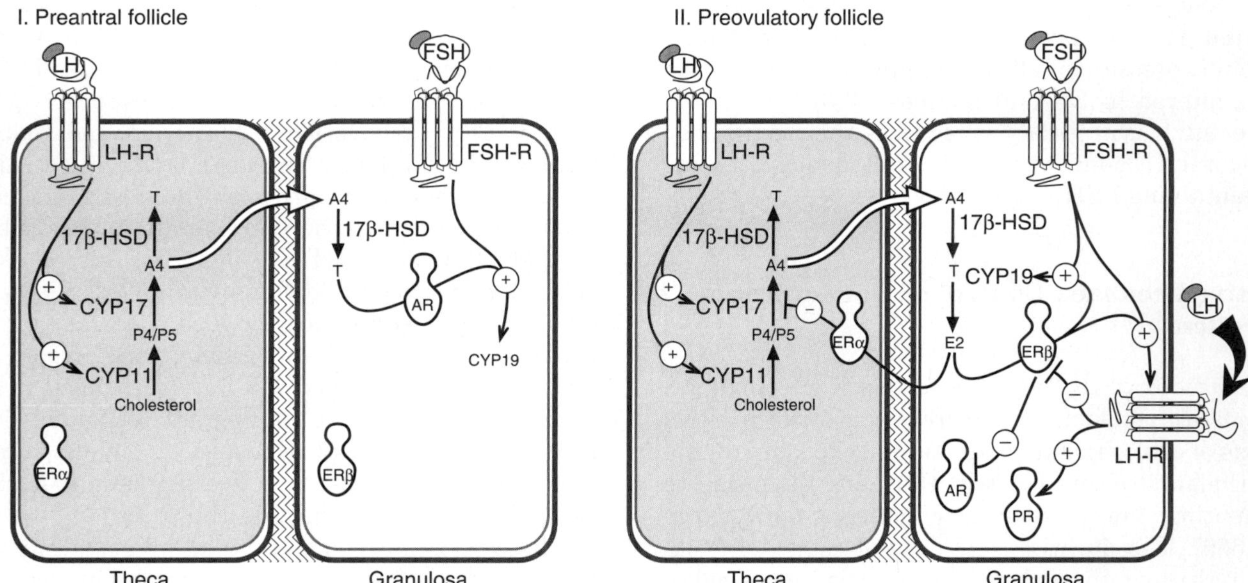

FIG. 26. Model of sex steroid receptor actions in the ovary during follicular differentiation. The process of follicle differentiation from preantral to preovulatory follicle is marked in granulosa cells by the induction of aromatase (CYP19) activity for the synthesis of estradiol and LH-receptor (LH-R) expression. This process is principally dependent on stimulation of thecal cells by LH and granulosa cells by FSH. However, there is substantial evidence of a modulatory role for the sex steroid receptors in this process. The above model illustrates the putative expression pattern and actions of the sex steroid receptors during follicle differentiation. Some aspects of the above model are supported by experimental data, whereas others are more speculation at this time (see text for details). (I) Preantral Follicle: In undifferentiated preantral follicles, LH stimulation of thecal cells (which constitutively express LH-R) leads to de novo synthesis of androstenedione (A4) and testosterone (T). A4 and T diffuse across the basement membrane of the follicle and into the granulosa cells. A4 may be converted to T in granulosa cells. Thecal-cell derived or locally synthesized T activates the androgen receptor (AR) and this action synergizes with FSH to induce aromatase (CYP19) expression in the granulosa cells. The induction of estradiol synthesis commences the process of follicle differentiation. (II) Preovulatory Follicle: Increasing estradiol synthesis in the granulosa cells leads to activation of ER-β and induces differentiation by synergizing with FSH to further increase CYP19 expression and estradiol synthesis and induce LH-R expression in the granulosa cells. Ligand-dependent ER-β actions may also lead to decreased AR expression, allowing the role of A4 and T to shift from that of a ligand for AR-mediated actions to that as substrates for estradiol synthesis. Increasing estradiol levels eventually activate ER-α in thecal cells to reduce further androgen synthesis by decreasing CYP17 activity, forming a negative feedback loop to ensure the proper estrogen-to-androgen ratio in the follicle. LH-Surge: Climaxing estradiol levels in the circulation facilitate the release of the LH surge from the pituitary, which stimulates ovulation and terminal differentiation of the granulosa cells. Only those follicles that possess properly differentiated granulosa cells (i.e., have acquired sufficient LH-R) are able to respond to the LH surge, which causes a dramatic rise in PR expression that occurs in the preovulatory granulosa cells within hours of the LH surge, whereas ER-β and CYP19 levels are both dramatically decreased.

plasma prolactin levels, which are only slightly reduced in ER-α–null females (366,674). Nonetheless, the potential impact of insufficient plasma prolactin levels on the αERKO or αβERKO ovarian phenotypes must be considered.

SUMMARY

The salient aspects of the above discussion are summarized in a model of sex steroid and steroid receptor action during folliculogenesis (Fig. 26). Although certain postulated mechanisms and pathways portrayed in this model are supported by substantial evidence, other aspects are more speculative and remain to be demonstrated experimentally. Regardless, the evidence is strong that receptor-mediated steroid pathways exist in the ovary and their intraovarian functions are critical to female fertility.

REFERENCES

1. Mangelsdorf, D. J., Thummel, C., Beato, M., Herrlich, P., Schutz, G., Umesono, K., Blumberg, B., Kastner, P., Mark, M., Chambon, P., and Evans, R. M. (1995). The nuclear receptor superfamily: the second decade. *Cell* 83, 835–839.
2. Nuclear Receptors Nomenclature Committee. (1999). A unified nomenclature system for the nuclear receptor superfamily. *Cell* 97, 161–163.
3. Nebert, D. W., Adesnik, M., Coon, M. J., Estabrook, R. W., Gonzalez, F. J., Guengerich, F. P., Gunsalus, I. C., Johnson, E. F., Kemper, B., Levin, W., et al. (1987). The P450 gene superfamily: recommended nomenclature. *DNA* 6, 1–11.
4. Ponglikitmongkol, M., Green, S., and Chambon, P. (1988). Genomic organization of the human oestrogen receptor gene. *EMBO J.* 7, 3385–3388.
5. Misrahi, M., Venencie, P. Y., Saugier-Veber, P., Sar, S., Dessen, P., and Milgrom, E. (1993). Structure of the human progesterone receptor gene. *Biochim. Biophys. Acta.* 1216, 289–292.
6. Kuiper, G. G., Faber, P. W., van Rooij, H. C., van der Korput, J. A., Ris-Stalpers, C., Klaassen, P., Trapman, J., and Brinkmann, A. O. (1989). Structural organization of the human androgen receptor gene. *J. Mol. Endocrinol.* 2, R1–R4.
7. Enmark, E., Pelto-Huikko, M., Grandien, K., Lagercrantz, S., Lagercrantz, J., Fried, G., Nordenskjold, M., and Gustafsson, J. A. (1997). Human estrogen receptor-β gene structure, chromosomal localization, and expression pattern. *J. Clin. Endocrinol. Metab.* 82, 4258–4265.
8. Loosfelt, H., Atger, M., Misrahi, M., Guiochon-Mantel, A., Meriel, C., Logeat, F., Benarous, R., and Milgrom, E. (1986). Cloning and sequence analysis of rabbit progesterone-receptor complementary DNA. *Proc. Natl. Acad. Sci. U S A* 83, 9045–9049.
9. Jeltsch, J. M., Krozowski, Z., Quirin-Stricker, C., Gronemeyer, H., Simpson, R. J., Garnier, J. M., Krust, A., Jacob, F., and Chambon, P. (1986). Cloning of the chicken progesterone receptor. *Proc. Natl. Acad. Sci. U S A* 83, 5424–5428.
10. Conneely, O. M., Sullivan, W. P., Toft, D. O., Birnbaumer, M., Cook, R. G., Maxwell, B. L., Zarucki-Schulz, T., Greene, G. L., Schrader, W. T., and O'Malley, B. W. (1986). Molecular cloning of the chicken progesterone receptor. *Science* 233. 767–770.
11. Misrahi, M., Atger, M., d'Auriol, L., Loosfelt, H., Meriel, C., Fridlansky, F., Guiochon-Mantel, A., Galibert, F., and Milgrom, E. (1987). Complete amino acid sequence of the human progesterone receptor deduced from cloned cDNA. *Biochem. Biophys. Res. Commun.* 143, 740–748.
12. Park-Sarge, O.-K., and Mayo, K. E. (1994). Regulation of the progesterone receptor gene by gonadotropins and cyclic adenosine 3′,5′-monophosphate in rat granulosa cells. *Endocrinology* 134, 709–718.
13. Schott, D. R., Shyamala, G., Schneider, W., and Parry, G. (1991). Molecular cloning, sequence analyses, and expression of complementary DNA encoding murine progesterone receptor. *Biochemistry* 30, 7014–7020.
14. Laudet, V., and Gronemeyer, H. (2002). *The Nuclear Receptor: Factsbook*. Academic Press, San Diego.
15. Li, X., and O'Malley, B. W. (2003). Unfolding the action of progesterone receptors. *J. Biol. Chem.* 278, 39261–39264.
16. Ogle, T. F. (2002). Progesterone-action in the decidual mesometrium of pregnancy. *Steroids* 67, 1–14.
17. Kastner, P., Krust, A., Turcotte, B., Stropp, U., Tora, L., Gronemeyer, H., and Chambon, P. (1990). Two distinct estrogen-regulated promoters generate transcripts encoding the two functionally different human progesterone receptor forms A and B. *EMBO J.* 9, 1603–1614.
18. Kraus, W. L., and Katzenellenbogen, B. S. (1993). Regulation of progesterone receptor gene expression and growth in the rat uterus: modulation of estrogen actions by progesterone and sex steroid hormone antagonists. *Endocrinology* 132, 2371–2379.
19. Kraus, W. L., Montano, M. M., and Katzenellenbogen, B. S. (1993). Cloning of the rat progesterone receptor gene 5′-region and identification of two functionally distinct promoters. *Mol. Endocrinol.* 7, 1603–1616.
20. Hirata, S., Shoda, T., Kato, J., and Hoshi, K. (2003). Isoform/variant mRNAs for sex steroid hormone receptors in humans. *Trends Endocrinol. Metab.* 14, 124–129.
21. Richer, J. K., Lange, C. A., Wierman, A. M., Brooks, K. M., Tung, L., Takimoto, G. S., and Horwitz, K. B. (1998). Progesterone receptor variants found in breast cells repress transcription by wild-type receptors. *Breast Cancer Res. Treat.* 48, 231–241.
22. Chang, C. S., Kokontis, J., and Liao, S. T. (1988). Molecular cloning of human and rat complementary DNA encoding androgen receptors. *Science* 240, 324–326.
23. Lubahn, D. B., Joseph, D. R., Sullivan, P. M., Willard, H. F., French, F. S., and Wilson, E. M. (1988). Cloning of human androgen receptor complementary DNA and localization to the X chromosome. *Science* 240, 327–330.
24. Choong, C. S., Kemppainen, J. A., and Wilson, E. M. (1998). Evolution of the primate androgen receptor: a structural basis for disease. *J. Mol. Evol.* 47, 334–342.
25. He, W. W., Fischer, L. M., Sun, S., Bilhartz, D. L., Zhu, X. P., Young, C. Y., Kelley, D. B., and Tindall, D. J. (1990). Molecular cloning of androgen receptors from divergent species with a polymerase chain reaction technique: complete cDNA sequence of the mouse androgen receptor and isolation of androgen receptor cDNA probes from dog, guinea pig and clawed frog. *Biochem. Biophys. Res. Commun.* 171, 697–704.
26. Tan, J. A., Joseph, D. R., Quarmby, V. E., Lubahn, D. B., Sar, M., French, F. S., and Wilson, E. M. (1988). The rat androgen receptor: primary structure, autoregulation of its messenger ribonucleic acid, and immunocytochemical localization of the receptor protein. *Mol. Endocrinol.* 2, 1276–1285.
27. Krongrad, A., Wilson, J. D., and McPhaul, M. J. (1995). Cloning and partial sequence of the rabbit androgen receptor: expression in fetal urogenital tissues. *J. Androl.* 16, 209–212.

28. Takeo, J., and Yamashita, S. (1999). Two distinct isoforms of cDNA encoding rainbow trout androgen receptors. *J. Biol. Chem.* 274, 5674–5680.
29. Touhata, K., Kinoshita, M., Tokuda, Y., Toyohara, H., Sakaguchi, M., Yokoyama, Y., and Yamashita, S. (1999). Sequence and expression of a cDNA encoding the red seabream androgen receptor. *Biochim. Biophys. Acta.* 1450, 481–485.
30. Lubahn, D. B., Brown, T. R., Simental, J. A., Higgs, H. N., Migeon, C. J., Wilson, E. M., and French, F. S. (1989). Sequence of the intron/exon junctions of the coding region of the human androgen receptor gene and identification of a point mutation in a family with complete androgen insensitivity. *Proc. Natl. Acad. Sci. U S A* 86, 9534–9538.
31. McEwan, I. J. (2004). Molecular mechanisms of androgen receptor-mediated gene regulation: structure-function analysis of the AF-1 domain. *Endocr. Relat. Cancer* 11, 281–293.
32. Gelmann, E. P. (2002). Molecular biology of the androgen receptor. *J. Clin. Oncol.* 20, 3001–3015.
33. Walter, P., Green, S., Greene, G., Krust, A., Bornert, J.-M., Jeltsch, J.-M., Staub, A., Jensen, E., Scrace, G., Waterfield, M., and Chambon, P. (1985). Cloning of the human estrogen receptor cDNA. *Proc. Natl. Acad. Sci. U S A* 82, 7889–7893.
34. Green, S., Walter, P., Kumar, V., Krust, A., Bornert, J.-M., Argos, P., and Chambon, P. (1986). Human oestrogen receptor cDNA: sequence, expression and homology to *v-erb-A*. *Nature* 320, 134–139.
35. White, R., Lees, J. A., Needham, M., Ham, J., and Parker, M. (1987). Structural organization and expression of the mouse estrogen receptor. *Mol. Endocrinol.* 1, 735–744.
36. Koike, S., Sakai, M., and Muramatsu, M. (1987). Molecular cloning and characterization of rat estrogen receptor cDNA. *Nucleic Acids Res.* 15, 2499–2513.
37. Bokenkamp, D., Jungblut, P. W., and Thole, H. H. (1994). The C-terminal half of the porcine estradiol receptor contains no post-translational modification: determination of the primary structure. *Mol. Cell. Endocrinol.* 104, 163–172.
38. Krust, A., Green, S., Argos, P., Kumar, V., Walter, P., Bornert, J.-M., and Chambon, P. (1986). The chicken oestrogen receptor sequence: homology with v-erb-A and the human oestrogen and glucocorticoid receptors. *EMBO J.* 5, 891–897.
39. Weiler, I. J., Lew, D., and Shapiro, D. J. (1987). The *Xenopus laevis* estrogen receptor: sequence homology with human and avian receptors and identification of multiple estrogen receptor messenger ribonucleic acids. *Mol. Endocrinol.* 1, 355–362.
40. Pakdel, F., Le Guellec, C., Vaillant, C., Le Roux, M. G., and Valtaire, Y. (1989). Identification and estrogen induction of two estrogen receptor (ER) messenger ribonucleic acids in rainbow trout liver: sequence homology with other ERs. *Mol. Endocrinol.* 3, 44–51.
41. Munoz-Cueto, J. A., Burzawa-Gerard, E., Kah, O., Valotaire, Y., and Pakdel, F. (1999). Cloning and sequencing of the gilthead sea bream estrogen receptor cDNA. *DNA Seq.* 10, 75–84.
42. Tan, N. S., Lam, T. J., and Ding, J. L. (1996). The first contiguous estrogen receptor gene from a fish, *Oreochromis aureus*: evidence for multiple transcripts. *Mol. Cell. Endocrinol.* 120, 177–192.
43. Kuiper, G. G. J. M., Enmark, E., Pelto-Huikko, M., Nilsson, S., and Gustafsson, J. A. (1996). Cloning of a novel receptor expressed in rat prostate and ovary. *Proc. Natl. Acad. Sci. U S A* 93, 5925–5930.
44. Mosselman, S., Polman, J., and Dijkema, R. (1996). ERβ: identification and characterization of a novel human estrogen receptor. *FEBS Lett.* 392, 49–53.
45. Tremblay, G. B., Tremblay, A., Copeland, N. G., Gilbert, D. J., Jenkins, N. A., Labrie, F., and Giguere, V. (1997). Cloning, chromosomal localization, and functional analysis of the murine estrogen receptor beta. *Mol. Endocrinol.* 11, 353–365.
46. Wu, W. X., Ma, X. H., Smith, G. C., and Nathanielsz, P. W. (2000). Differential distribution of ERalpha and ERbeta mRNA in intrauterine tissues of the pregnant rhesus monkey. *Am. J. Physiol. Cell. Physiol.* 278, C190–C198.
47. Rosenfeld, C. S., Yuan, X., Manikkam, M., Calder, M. D., Garverick, H. A., and Lubahn, D. B. (1999). Cloning, sequencing, and localization of bovine estrogen receptor-β within the ovarian follicle. *Biol. Reprod.* 60, 691–697.
48. Tchoudakova, A., Pathak, S., and Callard, G. V. (1999). Molecular cloning of an estrogen receptor beta subtype from the goldfish, *Carassius auratus*. *Gen. Comp. Endocrinol.* 113, 388–400.
49. Hawkins, M. B., Thornton, J. W., Crews, D., Skipper, J. K., Dotte, A., and Thomas, P. (2000). Identification of a third distinct estrogen receptor and reclassification of estrogen receptors in teleosts. *Proc. Natl. Acad. Sci. U S A* 97, 10751–10756.
50. Keaveney, M., Klug, J., Dawson, M. T., Neator, P. V., Neilan, J. G., Forde, R. C., and Gannon, F. (1991). Evidence for a previously unidentified upstream exon in the human oestrogen receptor gene. *J. Endocrinol.* 6, 111–115.
51. Keaveney, M., Klug, J., and Gannon, F. (1992). Sequence analysis of the 5_ flanking region of the human estrogen receptor gene. *DNA Seq.* 2, 347–358.
52. Grandien, K., Berkenstam, A., and Gustafsson, J.-A. (1997). The estrogen receptor gene: promoter organization and expression. *Int. J. Biochem. Cell. Biol.* 29, 1343–1369.
53. Fasco, M. J. (1998). Estrogen receptor mRNA splice variants produced from the distal and proximal promoter transcripts. *Mol. Cell. Endocrinol.* 138, 51–59.
54. Murphy, L. C., Dotzlaw, H., Leygue, E., Douglas, D., Coutts, A., and Watson, P. H. (1997). Estrogen receptor variants and mutations. *J. Steroid Biochem. Mol. Biol.* 62, 363–372.
55. Miksicek, R. J. (1994). Steroid receptor variants and their potential role in cancer. *Sem. Cancer Biol.* 5, 369–379.
56. Sluyser, M. (1995). Mutations in the estrogen receptor gene. *Hum. Mutat.* 6, 97–103.
57. Kos, M., Denger, S., Reid, G., and Gannon, F. (2002). Upstream open reading frames regulate the translation of the multiple mRNA variants of the estrogen receptor alpha. *J. Biol. Chem.* 277, 37131–37138.
58. Kos, M., Reid, G., Denger, S., and Gannon, F. (2001). Minireview: genomic organization of the human ERalpha gene promoter region. *Mol. Endocrinol.* 15, 2057–2063.
59. Flouriot, G., Griffin, C., Kenealy, M., Sonntag-Buck, V., and Gannon, F. (1998). Differentially expressed messenger RNA isoforms of the human estrogen receptor-alpha gene are generated by alternative splicing and promoter usage. *Mol. Endocrinol.* 12, 1939–1954.
60. Cicatiello, L., Cobellis, G., Addeo, R., Papa, M., Altucci, L., Sica, V., Bresciani, F., LeMeur, M., Kumar, V. L., Chambon, P., et al. (1995). In vivo functional analysis of the mouse estrogen receptor gene promoter: a transgenic mouse model to study tissue-specific and developmental regulation of estrogen receptor gene transcription. *Mol. Endocrinol.* 9, 1077–1090.
61. Pinzone, J. J., Stevenson, H., Strobl, J. S., and Berg, P. E. (2004). Molecular and cellular determinants of estrogen receptor alpha expression. *Mol. Cell. Biol.* 24, 4605–4612.
62. Donaghue, C., Westley, B. R., and May, F. E. (1999). Selective promoter usage of the human estrogen receptor-alpha gene and its regulation by estrogen. *Mol. Endocrinol.* 13, 1934–1950.
63. Stoica, A., Saceda, M., Fakhro, A., Solomon, H. B., Fenster, B. D., and Martin, M. B. (1999). Regulation of estrogen receptor-alpha gene expression by 1, 25-dihydroxyvitamin D in MCF-7 cells. *J. Cell. Biochem.* 75, 640–651.

64. Ishibashi, O., and Kawashima, H. (2001). Cloning and characterization of the functional promoter of mouse estrogen receptor beta gene. *Biochim. Biophys. Acta*. 1519, 223–229.
65. Aranda, A., and Pascual, A. (2001). Nuclear hormone receptors and gene expression. *Physiol. Rev*. 81, 1269–1304.
66. Zhang, Y., Beck, C. A., Poletti, A., Edwards, D. P., and Weigel, N. L. (1995). Identification of a group of Ser-Pro motif hormone-inducible phosphorylation sites in the human progesterone receptor. *Mol. Endocrinol*. 9, 1029–1040.
67. Giangrande, P. H., and McDonnell, D. P. (1999). The A and B isoforms of the human progesterone receptor: two functionally different transcription factors encoded by a single gene. *Recent Prog. Horm. Res*. 54, 291–313; discussion 313–294.
68. Katzenellenbogen, B. S., Montano, M. M., Ediger, T. R., Sun, J., Ekena, K., Lazennec, G., Martini, P. G., McInerney, E. M., Delage-Mourroux, R., Weis, K., and Katzenellenbogen, J. A. (2000). Estrogen receptors: selective ligands, partners, and distinctive pharmacology. *Recent Prog. Horm. Res*. 55, 163–193; discussion 194–165.
69. Delaunay, F., Pettersson, K., Tujague, M., and Gustafsson, J. A. (2000). Functional differences between the amino-terminal domains of estrogen receptors alpha and beta. *Mol. Pharmacol*. 58, 584–590.
70. Tremblay, A., Tremblay, G. B., Labrie, F., and Giguere, V. (1999). Ligand-independent recruitment of SRC-1 to estrogen receptor beta through phosphorylation of activation function AF-1. *Mol. Cell*. 3, 513–519.
71. Watanabe, T., Inoue, S., Ogawa, S., Ishii, Y., Hiroi, H., Ikeda, K., Orimo, A., and Muramatsu, M. (1997). Agonist effect of tamoxifen is dependent on cell type, ERE-promoter context, and estrogen receptor subtype: functional difference between estrogen receptors α and β. *Biochem. Biophys. Res. Commun*. 236, 140–145.
72. Hall, J. M., and McDonnell, D. P. (1999). The estrogen receptor beta-isoform (ERbeta) of the human estrogen receptor modulates ERalpha transcriptional activity and is a key regulator of the cellular response to estrogens and antiestrogens. *Endocrinology*. 140, 5566–5578.
73. Cowley, S. M., and Parker, M. G. (1999). A comparison of transcriptional activation by ER alpha and ER beta. *J. Steroid Biochem. Mol. Biol*. 69, 165–175.
74. Cowley, S. M., Hoare, S., Mosselman, S., and Parker, M. G. (1997). Estrogen receptors alpha and beta form heterodimers on DNA. *J. Biol. Chem*. 272, 19858–19862.
75. Tsai, M.-J., and O'Malley, B. W. (1994). Molecular mechanisms of action of steroid/thyroid receptor superfamily members. *Annu. Rev. Biochem*. 63, 451–486.
76. Kuiper, G. G. J. M., and Gustafsson, J.-A. (1997). The novel estrogen receptor-β subtype: potential role in the cell- and promoter-specific actions of estrogens and anti-estrogens. *FEBS Lett*. 410, 87–90.
77. Verrijdt, G., Haelens, A., and Claessens, F. (2003). Selective DNA recognition by the androgen receptor as a mechanism for hormone-specific regulation of gene expression. *Mol. Genet. Metab*. 78, 175–185.
78. Claessens, F., Alen, P., Devos, A., Peeters, B., Verhoeven, G., and Rombauts, W. (1996). The androgen-specific probasin response element 2 interacts differentially with androgen and glucocorticoid receptors. *J. Biol. Chem*. 271, 19013–19016.
79. Schoenmakers, E., Verrijdt, G., Peeters, B., Verhoeven, G., Rombauts, W., and Claessens, F. (2000). Differences in DNA binding characteristics of the androgen and glucocorticoid receptors can determine hormone-specific responses. *J. Biol. Chem*. 275, 12290–12297.
80. Verrijdt, G., Schoenmakers, E., Haelens, A., Peeters, B., Verhoeven, G., Rombauts, W., and Claessens, F. (2000). Change of specificity mutations in androgen-selective enhancers. Evidence for a role of differential DNA binding by the androgen receptor. *J. Biol. Chem*. 275, 12298–12305.
81. DeMarzo, A. M., Onate, S. A., Nordeen, S. K., and Edwards, D. P. (1992). Effects of the steroid antagonist RU486 on dimerization of the human progesterone receptor. *Biochemistry* 31, 10491–10501.
82. DeMarzo, A. M., Beck, C. A., Onate, S. A., and Edwards, D. P. (1991). Dimerization of mammalian progesterone receptors occurs in the absence of DNA and is related to the release of the 90-kDa heat shock protein. *Proc. Natl. Acad. Sci U S A* 88, 72–76.
83. Pettersson, K., Grandien, K., Kuiper, G. G. J. M., and Gustafsson, J.-A. (1997). Mouse estrogen receptor β forms estrogen response element binding heterodimers with estrogen receptor α. *Mol. Endocrinol*. 11, 1486–1496.
84. O'Lone, R., Frith, M. C., Karlsson, E. K., and Hansen, U. (2004). Genomic targets of nuclear estrogen receptors. *Mol. Endocrinol*. 18, 1859–1875.
85. Kim, K., Thu, N., Saville, B., and Safe, S. (2003). Domains of estrogen receptor alpha (ERalpha) required for ERalpha/Sp1-mediated activation of GC-rich promoters by estrogens and antiestrogens in breast cancer cells. *Mol. Endocrinol*. 17, 804–817.
86. Safe, S., and Kim, K. (2004). Nuclear receptor-mediated transactivation through interaction with Sp proteins. *Prog. Nucleic Acid Res. Mol. Biol*. 77, 1–36.
87. Wang, C., Fu, M., Angeletti, R. H., Siconolfi-Baez, L., Reutens, A. T., Albanese, C., Lisanti, M. P., Katzenellenbogen, B. S., Kato, S., Hopp, T., Fuqua, S. A., Lopez, G. N., Kushner, P. J., and Pestell, R. G. (2001). Direct acetylation of the estrogen receptor alpha hinge region by p300 regulates transactivation and hormone sensitivity. *J. Biol. Chem*. 276, 18375–18383.
88. Fu, M., Rao, M., Wang, C., Sakamaki, T., Wang, J., Di Vizio, D., Zhang, X., Albanese, C., Balk, S., Chang, C., Fan, S., Rosen, E., Palvimo, J. J., Janne, O. A., Muratoglu, S., Avantaggiati, M. L., and Pestell, R. G. (2003). Acetylation of androgen receptor enhances coactivator binding and promotes prostate cancer cell growth. *Mol. Cell. Biol*. 23, 8563–8575.
89. Fu, M., Wang, C., Reutens, A. T., Wang, J., Angeletti, R. H., Siconolfi-Baez, L., Ogryzko, V., Avantaggiati, M. L., and Pestell, R. G. (2000). p300 and p300/cAMP-response element-binding protein-associated factor acetylate the androgen receptor at sites governing hormone-dependent transactivation. *J. Biol. Chem*. 275, 20853–20860.
90. Matias, P. M., Donner, P., Coelho, R., Thomaz, M., Peixoto, C., Macedo, S., Otto, N., Joschko, S., Scholz, P., Wegg, A., Basler, S., Schafer, M., Egner, U., and Carrondo, M. A. (2000). Structural evidence for ligand specificity in the binding domain of the human androgen receptor. Implications for pathogenic gene mutations. *J. Biol. Chem*. 275, 26164–26171.
91. Sack, J. S., Kish, K. F., Wang, C., Attar, R. M., Kiefer, S. E., An, Y., Wu, G. Y., Scheffler, J. E., Salvati, M. E., Krystek, S. R., Jr., Weinmann, R., and Einspahr, H. M. (2001). Crystallographic structures of the ligand-binding domains of the androgen receptor and its T877A mutant complexed with the natural agonist dihydrotestosterone. *Proc. Natl. Acad. Sci. U S A* 98, 4904–4909.
92. Brzozowski, A. M., Pike, A. C. W., Dauter, Z., Hubbard, R. E., Bonn, T., Engstrom, O., Ohman, L., Greene, G. L., Gustafsson, J.-A., and Carlquist, M. (1997). Molecular basis of agonism and antagonism in the oestrogen receptor. *Nature* 389, 753–758.
93. Hiipakka, R. A., and Liao, S. (1998). Molecular mechanisms of androgen action. *Trends Endocrinol. Metab*. 9, 317–324.

94. Kuiper, G. G., Carlsson, B., Grandien, K., Enmark, E., Haggblad, J., Nilsson, S., and Gustafsson, J. A. (1997). Comparison of the ligand binding specificity and transcript tissue distribution of estrogen receptors alpha and beta. *Endocrinology* 138, 863–870.
95. Kuiper, G. G., Lemmen, J. G., Carlsson, B., Corton, J. C., Safe, S. H., van der Saag, P. T., van der Burg, B., and Gustafsson, J. A. (1998). Interaction of estrogenic chemicals and phytoestrogens with estrogen receptor beta. *Endocrinology* 139, 4252–4263.
96. Harris, H. A., Bapat, A. R., Gonder, D. S., and Frail, D. E. (2002). The ligand binding profiles of estrogen receptors alpha and beta are species dependent. *Steroids* 67, 379–384.
97. Weihua, Z., Lathe, R., Warner, M., and Gustafsson, J. A. (2002). An endocrine pathway in the prostate, ERbeta, AR, 5a-androstane-3b,17b-diol, and CYP7B1, regulates prostate growth. *Proc. Natl. Acad. Sci. U S A* 99, 13589–13594.
98. Macgregor, J. I., and Jordan, V. G. (1998). Basic guide to the mechanisms of antiestrogen action. *Pharmacol. Rev.* 50, 151–196.
99. Harris, H. A., Katzenellenbogen, J. A., and Katzenellenbogen, B. S. (2002). Characterization of the biological roles of the estrogen receptors, ERalpha and ERbeta, in estrogen target tissues in vivo through the use of an ERalpha-selective ligand. *Endocrinology* 143, 4172–4177.
100. Montano, M. M., Muller, V., Trobaugh, A., and Katzenellenbogen, B. S. (1995). The carboxy-terminal F domain of the human estrogen receptor: role in the transcriptional activity of the receptor and the effectiveness of antiestrogens as estrogen antagonists. *Mol. Endocrinol.* 9, 814–825.
101. Clark, J. H., and Markaverich, B. M. (1988). Actions of ovarian steroid hormones. In *The Physiology of Reproduction* (E. Knobil and J. D. Neill, Eds.), pp. 675–724. Raven Press, New York.
102. Beato, M., Herrlich, P., and Schutz, G. (1995). Steroid hormone receptors: many actors in search of a plot. *Cell* 83, 851–857.
103. Hecht, A., Berkenstam, A., Stromstedt, P. E., Gustafsson, J. A., and Sippel, A. E. (1988). A progesterone responsive element maps to the far upstream steroid dependent DNase hypersensitive site of chicken lysozyme chromatin. *EMBO J.* 7, 2063–2073.
104. Couse, J. F., and Korach, K. S. (1999). Estrogen receptor null mice: what have we learned and where will they lead us? *Endocr. Rev.* 20, 358–417.
105. Kester, H. A., van der Leede, B. M., van der Saag, P. T., and van der Burg, B. (1997). Novel progesterone target genes identified by an improved differential display technique suggest that progestin-induced growth inhibition of breast cancer cells coincides with enhancement of differentiation. *J. Biol. Chem.* 272, 16637–16643.
106. Richer, J. K., Jacobsen, B. M., Manning, N. G., Abel, M. G., Wolf, D. M., and Horwitz, K. B. (2002). Differential gene regulation by the two progesterone receptor isoforms in human breast cancer cells. *J. Biol. Chem.* 277, 5209–5218.
107. Jacobsen, B. M., Richer, J. K., Sartorius, C. A., and Horwitz, K. B. (2003). Expression profiling of human breast cancers and gene regulation by progesterone receptors. *J. Mamm. Gland Biol. Neoplasia.* 8, 257–268.
108. Jacobsen, B. M., Schittone, S. A., Richer, J. K., and Horwitz, K. B. (2005). Progesterone-independent effects of human progesterone receptors (PRs) in estrogen receptor-positive breast cancer: PR isoform-specific gene regulation and tumor biology. *Mol. Endocrinol.* 19, 574–587.
109. Smid-Koopman, E., Blok, L. J., Kuhne, L. C., Burger, C. W., Helmerhorst, T. J., Brinkmann, A. O., and Huikeshoven, F. J. (2003). Distinct functional differences of human progesterone receptors A and B on gene expression and growth regulation in two endometrial carcinoma cell lines. *J. Soc. Gynecol. Invest.* 10, 49–57.
110. Takamoto, N., Zhao, B., Tsai, S. Y., and DeMayo, F. J. (2002). Identification of Indian hedgehog as a progesterone-responsive gene in the murine uterus. *Mol. Endocrinol.* 16, 2338–2348.
111. Nelson, P. S., Clegg, N., Arnold, H., Ferguson, C., Bonham, M., White, J., Hood, L., and Lin, B. (2002). The program of androgen-responsive genes in neoplastic prostate epithelium. *Proc. Natl. Acad. Sci. U S A* 99, 11890–11895.
112. Eder, I. E., Haag, P., Basik, M., Mousses, S., Bektic, J., Bartsch, G., and Klocker, H. (2003). Gene expression changes following androgen receptor elimination in LNCaP prostate cancer cells. *Mol. Carcinog.* 37, 181–191.
113. Jiang, F., and Wang, Z. (2003). Identification of androgen-responsive genes in the rat ventral prostate by complementary deoxyribonucleic acid subtraction and microarray. *Endocrinology* 144, 1257–1265.
114. Umar, A., Ooms, M. P., Luider, T. M., Grootegoed, J. A., and Brinkmann, A. O. (2003). Proteomic profiling of epididymis and vas deferens: identification of proteins regulated during rat genital tract development. *Endocrinology* 144, 4637–4647.
115. Umar, A., Luider, T. M., Berrevoets, C. A., Grootegoed, J. A., and Brinkmann, A. O. (2003). Proteomic analysis of androgen-regulated protein expression in a mouse fetal vas deferens cell line. *Endocrinology* 144, 1147–1154.
116. Hewitt, S. C., Deroo, B. J., Hansen, K., Collins, J., Grissom, S., Afshari, C. A., and Korach, K. S. (2003). Estrogen receptor-dependent genomic responses in the uterus mirror the biphasic physiological response to estrogen. *Mol. Endocrinol.* 17, 2070–2083.
117. Hewitt, S. C., Collins, J., Grissom, S., Deroo, B., and Korach, K. S. (2005). Global uterine genomics in vivo: microarray evaluation of the estrogen receptor alpha-growth factor cross-talk mechanism. *Mol. Endocrinol.* 19, 657–668.
118. Lindberg, M. K., Moverare, S., Skrtic, S., Gao, H., Dahlman-Wright, K., Gustafsson, J. A., and Ohlsson, C. (2003). Estrogen receptor (ER)-beta reduces ERalpha-regulated gene transcription, supporting a "ying yang" relationship between ERalpha and ERbeta in mice. *Mol. Endocrinol.* 17, 203–208.
119. Fujimoto, N., Igarashi, K., Kanno, J., Honda, H., and Inoue, T. (2004). Identification of estrogen-responsive genes in the GH3 cell line by cDNA microarray analysis. *J. Steroid Biochem. Mol. Biol.* 91, 121–129.
120. Kato, N., Shibutani, M., Takagi, H., Uneyama, C., Lee, K. Y., Takigami, S., Mashima, K., and Hirose, M. (2004). Gene expression profile in the livers of rats orally administered ethinylestradiol for 28 days using a microarray technique. *Toxicology* 200, 179–192.
121. Terasaka, S., Aita, Y., Inoue, A., Hayashi, S., Nishigaki, M., Aoyagi, K., Sasaki, H., Wada-Kiyama, Y., Sakuma, Y., Akaba, S., Tanaka, J., Sone, H., Yonemoto, J., Tanji, M., and Kiyama, R. (2004). Using a customized DNA microarray for expression profiling of the estrogen-responsive genes to evaluate estrogen activity among natural estrogens and industrial chemicals. *Environ. Health Perspect.* 112, 773–781.
122. Ho Hong, S., Young Nah, H., Yoon Lee, J., Chan Gye, M., Hoon Kim, C., and Kyoo Kim, M. (2004). Analysis of estrogen-regulated genes in mouse uterus using cDNA microarray and laser capture microdissection. *J. Endocrinol.* 181, 157–167.
123. Wang, D. Y., Fulthorpe, R., Liss, S. N., and Edwards, E. A. (2004). Identification of estrogen-responsive genes by complementary deoxyribonucleic acid microarray and characterization of a novel early estrogen-induced gene: EEIG1. *Mol. Endocrinol.* 18, 402–411.

124. Inoue, A., Yoshida, N., Omoto, Y., Oguchi, S., Yamori, T., Kiyama, R., and Hayashi, S. (2002). Development of cDNA microarray for expression profiling of estrogen-responsive genes. *J. Mol. Endocrinol.* 29, 175–192.
125. Tora, L., Gronemeyer, H., Turcotte, B., Gaub, M. P., and Chambon, P. (1988). The N-terminal region of the chicken progesterone receptor specifies target gene activation. *Nature* 333, 185–188.
126. Meyer, M. E., Quirin-Stricker, C., Lerouge, T., Bocquel, M. T., and Gronemeyer, H. (1992). A limiting factor mediates the differential activation of promoters by the human progesterone receptor isoforms. *J. Biol. Chem.* 267, 10882–10887.
127. Vegeto, E., Shahbaz, M. M., Wen, D. X., Goldman, M. E., O'Malley, B. W., and McDonnell, D. P. (1993). Human progesterone receptor A form is a cell- and promoter-specific repressor of human progesterone receptor B function. *Mol. Endocrinol.* 7, 1244–1255.
128. Coleman, K. M., and Smith, C. L. (2001). Intracellular signaling pathways: nongenomic actions of estrogens and ligand-independent activation of estrogen receptors. *Front Biosci.* 6, D1379–D1391.
129. Weigel, N. L., and Zhang, Y. (1998). Ligand-independent activation of steroid hormone receptors. *J. Mol. Med.* 76, 469–479.
130. Ueda, T., Bruchovsky, N., and Sadar, M. D. (2002). Activation of the androgen receptor N-terminal domain by interleukin-6 via MAPK and STAT3 signal transduction pathways. *J. Biol. Chem.* 277, 7076–7085.
131. Chen, T., Wang, L. H., and Farrar, W. L. (2000). Interleukin 6 activates androgen receptor-mediated gene expression through a signal transducer and activator of transcription 3-dependent pathway in LNCaP prostate cancer cells. *Cancer Res.* 60, 2132–2135.
132. Culig, Z. (2004). Androgen receptor cross-talk with cell signalling pathways. *Growth Factors* 22, 179–184.
133. Zhang, Y., Beck, C. A., Poletti, A., Clement, J. P. T., Prendergast, P., Yip, T. T., Hutchens, T. W., Edwards, D. P., and Weigel, N. L. (1997). Phosphorylation of human progesterone receptor by cyclin-dependent kinase 2 on three sites that are authentic basal phosphorylation sites in vivo. *Mol. Endocrinol.* 11, 823–832.
134. Sadar, M. D. (1999). Androgen-independent induction of prostate-specific antigen gene expression via cross-talk between the androgen receptor and protein kinase A signal transduction pathways. *J. Biol. Chem.* 274, 7777–7783.
135. Nazareth, L. V., and Weigel, N. L. (1996). Activation of the human androgen receptor through a protein kinase A signaling pathway. *J. Biol. Chem.* 271, 19900–19907.
136. Denner, L. A., Weigel, N. L., Maxwell, B. L., Schrader, W. T., and O'Malley, B. W. (1990). Regulation of progesterone receptor-mediated transcription by phosphorylation. *Science* 250, 1740–1743.
137. Power, R. F., Mani, S. K., Codina, J., Conneely, O. M., and O'Malley, B. W. (1991). Dopaminergic and ligand-independent activation of steroid hormone receptors. *Science* 254, 1636–1639.
138. Power, R. F., Conneely, O. M., and O'Malley, B. W. (1992). New insights into activation of the steroid hormone receptor superfamily. *Trends Pharmacol. Sci.* 13, 318–323.
139. Mani, S. K., Allen, J. M. C., Lydon, J. P., Mulac-Jericevic, B., Blaustein, J. D., DeMayo, F. J., Conneely, O., and O'Malley, B. W. (1996). Dopamine requires the unoccupied progesterone receptor to induce sexual behavior in mice. *Mol. Endocrinol.* 10, 1728–1737.
140. Mani, S., Allen, J. M. C., Clark, J. H., Blaustein, J. D., and O'Malley, B. W. (1994). Convergent pathways for steroid hormone- and neurotransmitter-induced rat sexual behavior. *Science* 265, 1246–1249.
141. Mani, S. K., Fienberg, A. A., O'Callaghan, J. P., Snyder, G. L., Allen, P. B., Dash, P. K., Moore, A. N., Mitchell, A. J., Bibb, J., Greengard, P., and O'Malley, B. W. (2000). Requirement for DARPP-32 in progesterone-facilitated sexual receptivity in female rats and mice. *Science* 287, 1053–1056.
142. Curtis, S. W., Washburn, T., Sewall, C., DiAugustine, R., Lindzey, J., Couse, J. F., and Korach, K. S. (1996). Physiological coupling of growth factor and steroid receptor signaling pathways: estrogen receptor knockout mice lack estrogen-like response to epidermal growth factor. *Proc. Natl. Acad. Sci. U S A* 93, 12626–12630.
143. Klotz, D. M., Hewitt, S. C., Ciana, P., Raviscioni, M., Lindzey, J. K., Foley, J., Maggi, A., DiAugustine, R. P., and Korach, K. S. (2002). Requirement of estrogen receptor-alpha in insulin-like growth factor-1 (IGF-1)-induced uterine responses and in vivo evidence for IGF-1/estrogen receptor cross-talk. *J. Biol. Chem.* 277, 8531–8537.
144. Klotz, D. M., Hewitt, S. C., Korach, K. S., and Diaugustine, R. P. (2000). Activation of a uterine insulin-like growth factor I signaling pathway by clinical and environmental estrogens: requirement of estrogen receptor-alpha. *Endocrinology* 141, 3430–3439.
145. Kushner, P. J., Agard, D. A., Greene, G. L., Scanlan, T. S., Shiau, A. K., Uht, R. M., and Webb, P. (2000). Estrogen receptor pathways to AP-1. *J. Steroid Biochem. Mol. Biol.* 74, 311–317.
146. Owen, G. I., Richer, J. K., Tung, L., Takimoto, G., and Horwitz, K. B. (1998). Progesterone regulates transcription of the p21(WAF1) cyclin-dependent kinase inhibitor gene through Sp1 and CBP/p300. *J. Biol. Chem.* 273, 10696–10701.
147. Paech, K., Webb, P., Kuiper, G. G., Nilsson, S., Gustafsson, J., Kushner, P. J., and Scanlan, T. S. (1997). Differential ligand activation of estrogen receptors ERa and ERb at AP1 sites. *Science* 277, 1508–1510.
148. Losel, R. M., Falkenstein, E., Feuring, M., Schultz, A., Tillmann, H. C., Rossol-Haseroth, K., and Wehling, M. (2003). Nongenomic steroid action: controversies, questions, and answers. *Physiol. Rev.* 83, 965–1016.
149. Schmidt, B. M., Gerdes, D., Feuring, M., Falkenstein, E., Christ, M., and Wehling, M. (2000). Rapid, nongenomic steroid actions: a new age? *Front Neuroendocrinol.* 21, 57–94.
150. Heinlein, C. A., and Chang, C. (2002). The roles of androgen receptors and androgen-binding proteins in nongenomic androgen actions. *Mol. Endocrinol.* 16, 2181–2187.
151. Segars, J. H., and Driggers, P. H. (2002). Estrogen action and cytoplasmic signaling cascades. Part I: membrane-associated signaling complexes. *Trends Endocrinol. Metab.* 13, 349–354.
152. Driggers, P. H., and Segars, J. H. (2002). Estrogen action and cytoplasmic signaling pathways. Part II: the role of growth factors and phosphorylation in estrogen signaling. *Trends Endocrinol. Metab.* 13, 422–427.
153. Bramley, T. (2003). Non-genomic progesterone receptors in the mammalian ovary: some unresolved issues. *Reproduction* 125, 3–15.
154. Masui, Y., and Markert, C. L. (1971). Cytoplasmic control of nuclear behavior during meiotic maturation of frog oocytes. *J. Exp. Zool.* 177, 129–145.
155. Bayaa, M., Booth, R. A., Sheng, Y., and Liu, X. J. (2000). The classical progesterone receptor mediates *Xenopus* oocyte maturation through a nongenomic mechanism. *Proc. Natl. Acad. Sci. U S A* 97, 12607–12612.
156. Tian, J., Kim, S., Heilig, E., and Ruderman, J. V. (2000). Identification of XPR-1, a progesterone receptor required for *Xenopus* oocyte activation. *Proc. Natl. Acad. Sci. U S A* 97, 14358–14363.

157. Peluso, J. J., Fernandez, G., Pappalardo, A., and White, B. A. (2001). Characterization of a putative membrane receptor for progesterone in rat granulosa cells. *Biol. Reprod.* 65, 94–101.
158. Peluso, J. J., Bremner, T., Fernandez, G., Pappalardo, A., and White, B. A. (2003). Expression pattern and role of a 60-kilodalton progesterone binding protein in regulating granulosa cell apoptosis: involvement of the mitogen-activated protein kinase cascade. *Biol. Reprod.* 68, 122–128.
159. Peluso, J. J., and Pappalardo, A. (2004). Progesterone regulates granulosa cell viability through a protein kinase G-dependent mechanism that may involve 14-3-3sigma. *Biol. Reprod.* 71, 1870–1878.
160. Migliaccio, A., Castoria, G., Di Domenico, M., de Falco, A., Bilancio, A., Lombardi, M., Barone, M. V., Ametrano, D., Zannini, M. S., Abbondanza, C., and Auricchio, F. (2000). Steroid-induced androgen receptor-oestradiol receptor beta-Src complex triggers prostate cancer cell proliferation. *EMBO J.* 19, 5406–5417.
161. Kousteni, S., Bellido, T., Plotkin, L. I., O'Brien, C. A., Bodenner, D. L., Han, L., Han, K., DiGregorio, G. B., Katzenellenbogen, J. A., Katzenellenbogen, B. S., Roberson, P. K., Weinstein, R. S., Jilka, R. L., and Manolagas, S. C. (2001). Nongenotropic, sex-nonspecific signaling through the estrogen or androgen receptors: dissociation from transcriptional activity. *Cell* 104, 719–730.
162. Boonyaratanakornkit, V., Scott, M. P., Ribon, V., Sherman, L., Anderson, S. M., Maller, J. L., Miller, W. T., and Edwards, D. P. (2001). Progesterone receptor contains a proline-rich motif that directly interacts with SH3 domains and activates c-Src family tyrosine kinases. *Mol. Cell.* 8, 269–280.
163. Boonyaratanakornkit, V., and Edwards, D. P. (2004). Receptor mechanisms of rapid extranuclear signalling initiated by steroid hormones. *Essays Biochem.* 40, 105–120.
164. Nettles, K. W., and Greene, G. L. (2005). Ligand control of coregulator recruitment to nuclear receptors. *Annu. Rev. Physiol.* 67, 309–333.
165. Renaud, J. P., and Moras, D. (2000). Structural studies on nuclear receptors. *Cell. Mol. Life Sci.* 57, 1748–1769.
166. Fuhrmann, U., Parczyk, K., Klotzbucher, M., Klocker, H., and Cato, A. C. (1998). Recent developments in molecular action of antihormones. *J. Mol. Med.* 76, 512–524.
167. Leonhardt, S. A., and Edwards, D. P. (2002). Mechanism of action of progesterone antagonists. *Exp. Biol. Med. (Maywood).* 227, 969–980.
168. Beck, C. A., Weigel, N. L., Moyer, M. L., Nordeen, S. K., and Edwards, D. P. (1993). The progesterone antagonist RU486 acquires agonist activity upon stimulation of cAMP signaling pathways. *Proc. Natl. Acad. Sci. U S A* 90, 4441–4445.
169. Sartorius, C. A., Groshong, S. D., Miller, L. A., Powell, R. L., Tung, L., Takimoto, G. S., and Horwitz, K. B. (1994). New T47D breast cancer cell lines for the independent study of progesterone B- and A-receptors: only antiprogestin-occupied B-receptors are switched to transcriptional agonists by cAMP. *Cancer Res.* 54, 3868–3877.
170. Sartorius, C. A., Tung, L., Takimoto, G. S., and Horwitz, K. B. (1993). Antagonist-occupied human progesterone receptors bound to DNA are functionally switched to transcriptional agonists by cAMP. *J. Biol. Chem.* 268, 9262–9266.
171. Shao, W., and Brown, M. (2004). Advances in estrogen receptor biology: prospects for improvements in targeted breast cancer therapy. *Breast Cancer Res.* 6, 39–52.
172. McDonnell, D. P., Wijayaratne, A., Chang, C. Y., and Norris, J. D. (2002). Elucidation of the molecular mechanism of action of selective estrogen receptor modulators. *Am. J. Cardiol.* 90, 35F–43F.
173. Gibson, M. K., Nemmers, L. A., Beckman, W. C., Jr., Davis, V. L., Curtis, S. W., and Korach, K. S. (1991). The mechanism of ICI 164,384 antiestrogenicity involves rapid loss of estrogen receptor in uterine tissue. *Endocrinology* 129, 2000–2010.
174. Labrie, F. (1993). Mechanism of action and pure antiandrogenic properties of flutamide. *Cancer* 72, 3816–3827.
175. Blackledge, G. (1993). Casodex—mechanisms of action and opportunities for usage. *Cancer* 72, 3830–3833.
176. Gaillard-Moguilewsky, M., de Gery, A., and Ulmann, A. (1993). Pharmacology of nilutamide. *Cancer* 72, 3828–3829.
177. Schroder, F. H. (1993). Cyproterone acetate—mechanism of action and clinical effectiveness in prostate cancer treatment. *Cancer* 72, 3810–3815.
178. Gobinet, J., Poujol, N., and Sultan, C. (2002). Molecular action of androgens. *Mol. Cell. Endocrinol.* 198, 15–24.
179. Lydon, J. P., DeMayo, F. J., Funk, C. R., Mani, S. K., Hughes, A. R., Montgomery, C. A. J., Shyamala, G., Conneely, O. M., and O'Malley, B. W. (1995). Mice lacking progesterone receptor exhibit pleiotropic reproductive abnormalities. *Genes Dev.* 9, 2266–2278.
180. Mulac-Jericevic, B., Mullinax, R. A., DeMayo, F. J., Lydon, J. P., and Conneely, O. M. (2000). Subgroup of reproductive functions of progesterone mediated by progesterone receptor-B isoform. *Science.* 289, 1751–1754.
181. Conneely, O. M., Mulac-Jericevic, B., DeMayo, F., Lydon, J. P., and O'Malley, B. W. (2002). Reproductive functions of progesterone receptors. *Recent Prog. Horm. Res.* 57, 339–355.
182. Lyon, M. F., and Hawkes, S. G. (1970). X-linked gene for testicular feminization. *Nature* 227, 1217–1219.
183. Young, C. Y. F., Johnson, M. P., Prescott, J. L., and Tindall, D. J. (1989). The androgen receptor of the testicular-feminized (*Tfm*) mutant mouse is smaller than the wild-type receptor. *Endocrinology* 124, 771–775.
184. Gaspar, M.-L., Meo, T., Bourgarel, P., Guenet, J.-L., and Tosi, M. (1991). A single base deletion in the *Tfm* androgen receptor gene creates a short-lived messenger RNA that directs internal translation initiation. *Proc. Natl. Acad. Sci. U S A* 88, 8606–8610.
185. Patterson, M. N., McPhaul, M. J., and Hughes, I. A. (1994). Androgen insensitivity syndrome. In *Bailliere's Clinical Endocrinology and Metabolism: Hormones, Enzymes and Receptors* (M. C. Sheppard, P. M. Stewart, Eds.), pp. 379–404. Bailliere Tindall, London.
186. Lyon, M. F., and Glenister, P. H. (1980). Reduced reproductive performance in androgen-resistant *Tfm/Tfm* female mice. *Proc. R. Soc. Lond. B. Biol. Sci.* 208, 1–12.
187. Kato, S., Matsumoto, T., Kawano, H., Sato, T., and Takeyama, K. (2004). Function of androgen receptor in gene regulations. *J. Steroid Biochem. Mol. Biol.* 89–90, 627–633.
188. Sato, T., Kawano, H., and Kato, S. (2002). Study of androgen action in bone by analysis of androgen-receptor deficient mice. *J. Bone Miner. Metab.* 20, 326–330.
189. Yeh, S., Tsai, M. Y., Xu, Q., Mu, X. M., Lardy, H., Huang, K. E., Lin, H., Yeh, S. D., Altuwaijri, S., Zhou, X., Xing, L., Boyce, B. F., Hung, M. C., Zhang, S., Gan, L., and Chang, C. (2002). Generation and characterization of androgen receptor knockout (ARKO) mice: an in vivo model for the study of androgen functions in selective tissues. *Proc. Natl. Acad. Sci. U S A* 99, 13498–13503.
190. Hu, Y. C., Wang, P. H., Yeh, S., Wang, R. S., Xie, C., Xu, Q., Zhou, X., Chao, H. T., Tsai, M. Y., and Chang, C. (2004). Subfertility and defective folliculogenesis in female mice lacking androgen receptor. *Proc. Natl. Acad. Sci. U S A* 101, 11209–11214.
191. Lubahn, D. B., Moyer, J. S., Golding, T. S., Couse, J. F., Korach, K. S., and Smithies, O. (1993). Alteration of reproductive function but not prenatal sexual development after

insertional disruption of the mouse estrogen receptor gene. *Proc. Natl. Acad. Sci. U S A* 90, 11162–11166.

192. Dupont, S., Krust, A., Gansmuller, A., Dierich, A., Chambon, P., and Mark, M. (2000). Effect of single and compound knockouts of estrogen receptors alpha (ERalpha) and beta (ERbeta) on mouse reproductive phenotypes. *Development* 127, 4277–4291.
193. Krege, J. H., Hodgin, J. B., Couse, J. F., Enmark, E., Warner, M., Mahler, J. F., Sar, M., Korach, K. S., Gustafsson, J. A., and Smithies, O. (1998). Generation and reproductive phenotypes of mice lacking estrogen receptor beta. *Proc. Natl. Acad. Sci. U S A* 95, 15677–15682.
194. Couse, J. F., Hewitt, S. C., Bunch, D. O., Sar, M., Walker, V. R., Davis, B. J., and Korach, K. S. (1999). Postnatal sex reversal of the ovaries in mice lacking estrogen receptors alpha and beta. *Science* 286, 2328–2331.
195. Fisher, C. R., Graves, K. H., Parlow, A. F., and Simpson, E. R. (1998). Characterization of mice deficient in aromatase (ArKO) because of targeted disruption of the cyp19 gene. *Proc. Natl. Acad. Sci. U S A* 95, 6965–6970.
196. Toda, K., Takeda, K., Okada, T., Akira, S., Saibara, T., Kaname, T., Yamamura, K., Onishi, S., and Shizuta, Y. (2001). Targeted disruption of the aromatase P450 gene (Cyp19) in mice and their ovarian and uterine responses to 17beta-oestradiol. *J. Endocrinol.* 170, 99–111.
197. Britt, K. L., Kerr, J., O'Donnell, L., Jones, M. E. E., Drummond, A. E., Davis, S. R., Simpson, E. R., and Findley, J. K. (2002). Estrogen regulates development of the somatic cell phenotype in the eutherian ovary. *FASEB J.* 16, 1389–1397.
198. Greco, T. L., Duello, T. M., and Gorski, J. (1993). Estrogen receptors, estradiol, and diethylstilbestrol in early development: the mouse as a model for the study of estrogen receptors and estrogen sensitivity in embryonic development of male and female reproductive tracts. *Endocr. Rev.* 14, 59–71.
199. Cunha, G. R., Cooke, P. S., Bigsby, R., and Brody, J. R. (1991). Ontogeny of sex steroid receptors in mammals. In *Nuclear Hormone Receptors: Molecular Mechanisms, Cellular Functions, Clinical Abnormalities* (M. G. Parker, Ed.), pp. 235–268. Academic Press, London.
200. Korach, K. S., Horigome, T., Tomooka, Y., Yamashita, S., Newbold, R. R., and McLachlan, J. A. (1988). Immunodetection of estrogen receptor in epithelial and stromal tissues of neonatal mouse uterus. *Proc. Natl. Acad. Sci. U S A* 85, 3334–3337.
201. Yamashita, S., Newbold, R. R., McLachlan, J. A., and Korach, K. S. (1990). The role of the estrogen receptor in uterine epithelial proliferation and cytodifferentiation in neonatal mice. *Endocrinology* 127, 2456–2463.
202. Jefferson, W. N., Couse, J. F., Banks, E. P., Korach, K. S., and Newbold, R. R. (2000). Expression of estrogen receptor beta is developmentally regulated in reproductive tissues of male and female mice. *Biol. Reprod.* 62, 310–317.
203. Kurita, T., Lee, K. J., Cooke, P. S., Lydon, J. P., and Cunha, G. R. (2000). Paracrine regulation of epithelial progesterone receptor and lactoferrin by progesterone in the mouse uterus. *Biol. Reprod.* 62, 831–838.
204. Tan, J., Paria, B. C., Dey, S. K., and Das, S. K. (1999). Differential uterine expression of estrogen and progesterone receptors correlates with uterine preparation for implantation and decidualization in the mouse. *Endocrinology* 140, 5310–5321.
205. Couse, J. F., Lindzey, J., Grandien, K., Gustafsson, J. A., and Korach, K. S. (1997). Tissue distribution and quantitative analysis of estrogen receptor-α (ERα) and estrogen receptor-β (ERβ) messenger ribonucleic acid in the wild-type and ERα-knockout mouse. *Endocrinology* 138, 4613–4621.
206. Kurita, T., Lee, K., Saunders, P. T., Cooke, P. S., Taylor, J. A., Lubahn, D. B., Zhao, C., Makela, S., Gustafsson, J. A., Dahiya, R., and Cunha, G. R. (2001). Regulation of progesterone receptors and decidualization in uterine stroma of the estrogen receptor-alpha knockout mouse. *Biol. Reprod.* 64, 272–283.
207. Wang, H., Eriksson, H., and Sahlin, L. (2000). Estrogen receptors alpha and beta in the female reproductive tract of the rat during the estrous cycle. *Biol. Reprod.* 63, 1331–1340.
208. Pelletier, G., Labrie, C., and Labrie, F. (2000). Localization of oestrogen receptor alpha, oestrogen receptor beta and androgen receptors in the rat reproductive organs. *J. Endocrinol.* 165, 359–370.
209. Jakacka, M., Ito, M., Martinson, F., Ishikawa, T., Lee, E. J., and Jameson, J. L. (2002). An estrogen receptor (ER)alpha deoxyribonucleic acid-binding domain knock-in mutation provides evidence for nonclassical ER pathway signaling in vivo. *Mol. Endocrinol.* 16, 2188–2201.
210. Curtis, S. W., Clark, J., Myers, P., and Korach, K. S. (1999). Disruption of estrogen signaling does not prevent progesterone action in the estrogen receptor alpha knockout mouse uterus. *Proc. Natl. Acad. Sci. U S A* 96, 3646–3651.
211. Curtis Hewitt, S., Goulding, E. H., Eddy, E. M., and Korach, K. S. (2002). Studies using the estrogen receptor alpha knockout uterus demonstrate that implantation but not decidualization-associated signaling is estrogen dependent. *Biol. Reprod.* 67, 1268–1277.
212. Couse, J. F., Dixon, D., Yates, M., Moore, A. B., Ma, L., Maas, R., and Korach, K. S. (2001). Estrogen receptor-alpha knockout mice exhibit resistance to the developmental effects of neonatal diethylstilbestrol exposure on the female reproductive tract. *Dev. Biol.* 238, 224–238.
213. Couse, J. F., and Korach, K. S. (2004). Estrogen receptor-alpha mediates the detrimental effects of neonatal diethylstilbestrol (DES) exposure in the murine reproductive tract. *Toxicology* 205, 55–63.
214. Spencer, T. E., and Bazer, F. W. (2002). Biology of progesterone action during pregnancy recognition and maintenance of pregnancy. *Front Biosci.* 7, 1879–1898.
215. Taylor, K. M., Gray, C. A., Joyce, M. M., Stewart, M. D., Bazer, F. W., and Spencer, T. E. (2000). Neonatal ovine uterine development involves alterations in expression of receptors for estrogen, progesterone, and prolactin. *Biol. Reprod.* 63, 1192–1204.
216. Spencer, T. E., and Bazer, F. W. (1995). Temporal and spatial alterations in uterine estrogen receptor and progesterone receptor gene expression during the estrous cycle and early pregnancy in the ewe. *Biol. Reprod.* 53, 1527–1543.
217. Meikle, A., Sahlin, L., Ferraris, A., Masironi, B., Blanc, J. E., Rodriguez-Irazoqui, M., Rodriguez-Pinon, M., Kindahl, H., and Forsberg, M. (2001). Endometrial mRNA expression of oestrogen receptor alpha, progesterone receptor and insulin-like growth factor-I (IGF-I) throughout the bovine oestrous cycle. *Anim. Reprod. Sci.* 68, 45–56.
218. Kimmins, S., and MacLaren, L. A. (2001). Oestrous cycle and pregnancy effects on the distribution of oestrogen and progesterone receptors in bovine endometrium. *Placenta* 22, 742–748.
219. Lessey, B. A., Metzger, D. A., Haney, A. F., and McCarty, K. S., Jr. (1989). Immunohistochemical analysis of estrogen and progesterone receptors in endometriosis: comparison with normal endometrium during the menstrual cycle and the effect of medical therapy. *Fertil. Steril.* 51, 409–415.
220. Lessey, B. A., Killam, A. P., Metzger, D. A., Haney, A. F., Greene, G. L., and McCarty, K. S. Jr. (1988). Immunohistochemical analysis of human uterine estrogen and

progesterone receptors throughout the menstrual cycle. *J. Clin. Endocrinol. Metab.* 67, 334–340.

221. Noe, M., Kunz, G., Herbertz, M., Mall, G., and Leyendecker, G. (1999). The cyclic pattern of the immunocytochemical expression of oestrogen and progesterone receptors in human myometrial and endometrial layers: characterization of the endometrial-subendometrial unit. *Hum. Reprod.* 14, 190–197.
222. Tsibris, J. C., Fort, F. L., Cazenave, C. R., Cantor, B., Bardawil, W. A., Notelovitz, M., and Spellacy, W. N. (1981). The uneven distribution of estrogen and progesterone receptors in human endometrium. *J. Steroid Biochem.* 14, 997–1003.
223. Kauppila, A., Janne, O., Stenback, F., and Vihko, R. (1982). Cytosolic estrogen and progestin receptors in human endometrium from different regions of the uterus. *Gynecol. Oncol.* 14, 225–229.
224. Matsuzaki, S., Fukaya, T., Suzuki, T., Murakami, T., Sasano, H., and Yajima, A. (1999). Oestrogen receptor alpha and beta mRNA expression in human endometrium throughout the menstrual cycle. *Mol. Hum. Reprod.* 5, 559–564.
225. Lecce, G., Meduri, G., Ancelin, M., Bergeron, C., and Perrot-Applanat, M. (2001). Presence of estrogen receptor beta in the human endometrium through the cycle: expression in glandular, stromal, and vascular cells. *J. Clin. Endocrinol. Metab.* 86, 1379–1386.
226. Saunders, P. T., Millar, M. R., Macpherson, S., Irvine, D. S., Groome, N. P., Evans, L. R., Sharpe, R. M., and Scobie, G. A. (2002). ERbeta1 and the ERbeta2 splice variant (ERbetacx/beta2) are expressed in distinct cell populations in the adult human testis. *J. Clin. Endocrinol. Metab.* 87, 2706–2715.
227. Hild-Petito, S., Verhage, H. G., and Fazleabas, A. T. (1992). Immunocytochemical localization of estrogen and progestin receptors in the baboon (*Papio anubis*) uterus during implantation and pregnancy. *Endocrinology* 130, 2343–2353.
228. Koji, T., and Brenner, R. M. (1993). Localization of estrogen receptor messenger ribonucleic acid in rhesus monkey uterus by nonradioactive in situ hybridization with digoxigenin-labeled oligodeoxynucleotides. *Endocrinology* 132, 382–392.
229. Hild-Petito, S., West, N. B., Brenner, R. M., and Stouffer, R. L. (1991). Localization of androgen receptor in the follicle and corpus luteum of the primate ovary during the menstrual cycle. *Biol. Reprod.* 44, 561–568.
230. Pelletier, G., Luu-The, V., Charbonneau, A., and Labrie, F. (1999). Cellular localization of estrogen receptor beta messenger ribonucleic acid in cynomolgus monkey reproductive organs. *Biol. Reprod.* 61, 1249–1255.
231. Fazleabas, A. T., Brudney, A., Chai, D., Langoi, D., and Bulun, S. E. (2003). Steroid receptor and aromatase expression in baboon endometriotic lesions. *Fertil. Steril.* 80 Suppl 2, 820–827.
232. Simpson, E. R., Clyne, C., Rubin, G., Boon, W. C., Robertson, K., Britt, K., Speed, C., and Jones, M. (2002). Aromatase—a brief overview. *Annu. Rev. Physiol.* 64, 93–127.
233. Korach, K., Couse, J., Curtis, S., Washburn, T., Lindzey, J., Kimbro, K., Eddy, E., Migliaccio, S., Snedeker, S., Lubahn, D., Schomberg, D., and Smith, E. (1996). Estrogen receptor gene disruption: molecular characterization and experimental and clinical phenotypes. *Recent Prog. Horm. Res.* 51, 159–188.
234. Lindzey, J., Curtis, S. W., Washburn, T. F., and Korach, K. S. (1996). *Uterotropic Effects of Dihydrotestosterone in Estrogen Receptor Knockout and Wild Type Mice.* Endocrine Society Abstracts, Washington, DC, p. 77.
235. Weihua, Z., Saji, S., Makinen, S., Cheng, G., Jensen, E. V., Warner, M., and Gustafsson, J. A. (2000). Estrogen receptor (ER) beta, a modulator of ERalpha in the uterus. *Proc. Natl. Acad. Sci. U S A* 97, 5936–5941.
236. Stumpf, W., and Sar, M. (1976). Autoradiographic localization of estrogen, androgen, progestin, and glucocorticoid in "target tissues" and "non-target" tissues. In *Receptors and Mechanisms of Action of Steroid Hormones* (J. Pasqualini, Ed.), pp. 41–84. Marcel Dekker, New York.
237. Galand, P., Leroy, F., and Chretien, J. (1971). Effect of oestradiol on cell proliferation and histological changes in the uterus and vagina of mice. *J. Endocrinol.* 49, 243–252.
238. Kronenberg, M. S., and Clark, J. H. (1985). Changes in keratin expression during the estrogen-mediated differentiation of rat vaginal epithelium. *Endocrinology* 117, 1480–1489.
239. Gimenez-Conti, I. B., Lynch, M., Roop, D., Bhowmik, S., Majeski, P., and Conti, C. J. (1994). Expression of keratins in mouse vaginal epithelium. *Differentiation* 56, 143–151.
240. Sourla, A., Luo, S., Labrie, C., Belanger, A., and Labrie, F. (1997). Morphological changes induced by 6-month treatment of intact and ovariectomized mice with tamoxifen and the pure antiestrogen EM-800. *Endocrinology* 138, 5605–5617.
241. Luo, S., Martel, C., Sourla, A., Gauthier, S., Merand, Y., Belanger, A., Labrie, C., and Labrie, F. (1997). Comparative effects of 28-day treatment with the new anti-estrogen EM-800 and tamoxifen on estrogen-sensitive parameters in intact mice. *Int. J. Cancer* 73, 381–391.
242. Dukes, M., Chester, R., Yarwood, L., and Wakeling, A. E. (1994). Effects of a non-steroidal pure antioestrogen, ZM 189,154, on oestrogen target organs of the rat including bones. *J. Endocrinol.* 141, 335–341.
243. Xu, J., Qiu, Y., DeMayo, F. J., Tsai, S. Y., Tsai, M.-J., and O'Malley, B. W. (1998). Partial hormone resistance in mice with disruption of the steroid receptor coactivator-1 (SRC-1) gene. *Science* 279, 1922–1925.
244. Katzenellenbogen, B. S., and Gregor, N. G. (1974). Ontogeny of uterine responsiveness to estrogen during early development in the rat. *Mol. Cell. Endocrinol.* 2, 31–42.
245. Martin, L. (1980). Estrogens, anti-estrogens and the regulation of cell proliferation in the female reproductive tract in vivo. In *Estrogens in the Environment* (J. A. McLachlan, Ed.), pp. 103–130. Elsevier North Holland, New York.
246. Quarmby, V. E., and Korach, K. S. (1984). The influence of 17b-estradiol on patterns of cell division in the uterus. *Endocrinology* 114, 694–702.
247. Cooke, P. S., Buchanan, D. L., Young, P., Setiawan, T., Brody, J., Korach, K. S., Taylor, J., Lubahn, D. B., and Cunha, G. R. (1997). Stromal estrogen receptors mediate mitogenic effects of estradiol on uterine epithelium. *Proc. Natl. Acad. Sci. U S A* 94, 6535–6540.
248. Clarke, R. B., Howell, A., Potten, C. S., and Anderson, E. (1997). Dissociation between steroid receptor expression and cell proliferation in the human breast. *Cancer Res.*57, 4987–4991.
249. Korach, K. S. (1994). Insights from the study of animals lacking functional estrogen receptor. *Science* 266, 1524–1527.
250. Couse, J. F., Curtis, S. W., Washburn, T. F., Lindzey, J., Golding, T. S., Lubahn, D. B., Smithies, O., and Korach, K. S. (1995). Analysis of transcription and estrogen insensitivity in the female mouse after targeted disruption of the estrogen receptor gene. *Mol. Endocrinol.* 9, 1441–1454.
251. Stancel, G. M., Blatti, S. P., Benson, R. H., Gardner, R. M., Kirkland, J. L., Ireland, J. S., Sidikaro, J., and Weil, P. A. (1979). Hormonal control of uterine growth and regulation of responsiveness to estrogen. In *Ontogeny of Receptors and*

Reproductive Hormone Action (T. H. Hamilton, J. H. Clark, and W. A. Sadler, Eds.). Raven Press, New York.

252. Liu, Y., and Teng, C. (1992). Estrogen response module of the mouse lactoferrin gene contains overlapping chicken ovalbumin upstream promoter transcription factor and estrogen receptor-binding elements. *Mol. Endocrinol.* 6, 355–364.
253. Kurita, T., Lee, K. J., Zhao, C., Cooke, P. S., Taylor, J. A., Lubahn, D. B., and Cunha, G. R. (1998). Estrogen induces progesterone receptors in uterine stroma of estrogen receptor alpha knockout mouse. *Mol. Biol. Cell.* 9, 238A.
254. Tibbetts, T. A., Mendoza-Meneses, M., O'Malley, B. W., and Conneely, O. M. (1998). Mutual and intercompartmental regulation of estrogen receptor and progesterone receptor expression in the mouse uterus. *Biol. Reprod.* 59, 1143–1152.
255. Watanabe, H., Suzuki, A., Mizutani, T., Khono, S., Lubahn, D. B., Handa, H., and Iguchi, T. (2002). Genome-wide analysis of changes in early gene expression induced by oestrogen. *Genes Cells.* 7, 497–507.
256. Fertuck, K. C., Eckel, J. E., Gennings, C., and Zacharewski, T. R. (2003). Identification of temporal patterns of gene expression in the uteri of immature, ovariectomized mice following exposure to ethynylestradiol. *Physiol. Genom.* 15, 127–141.
257. Andrade, P. M., Silva, I. D., Borra, R. C., de Lima, G. R., and Baracat, E. C. (2002). Estrogen regulation of uterine genes in vivo detected by complementary DNA array. *Horm. Metab. Res.* 34, 238–244.
258. Watanabe, H., Suzuki, A., Kobayashi, M., Takahashi, E., Itamoto, M., Lubahn, D. B., Handa, H., and Iguchi, T. (2003). Analysis of temporal changes in the expression of estrogen-regulated genes in the uterus. *J. Mol. Endocrinol.* 30, 347–358.
259. Kao, L. C., Tulac, S., Lobo, S., Imani, B., Yang, J. P., Germeyer, A., Osteen, K., Taylor, R. N., Lessey, B. A., and Giudice, L. C. (2002). Global gene profiling in human endometrium during the window of implantation. *Endocrinology* 143, 2119–2138.
260. Borthwick, J. M., Charnock-Jones, D. S., Tom, B. D., Hull, M. L., Teirney, R., Phillips, S. C., and Smith, S. K. (2003). Determination of the transcript profile of human endometrium. *Mol. Hum. Reprod.* 9, 19–33.
261. Naciff, J. M., Jump, M. L., Torontali, S. M., Carr, G. J., Tiesman, J. P., Overmann, G. J., and Daston, G. P. (2002). Gene expression profile induced by 17alpha-ethynyl estradiol, bisphenol A, and genistein in the developing female reproductive system of the rat. *Toxicol. Sci.* 68, 184–199.
262. Jelinsky, S. A., Harris, H. A., Brown, E. L., Flanagan, K., Zhang, X., Tunkey, C., Lai, K., Lane, M. V., Simcoe, D. K., and Evans, M. J. (2003). Global transcription profiling of estrogen activity: estrogen receptor alpha regulates gene expression in the kidney. *Endocrinology* 144, 701–710.
263. DiAugustine, R. P., Petruez, P., Bell, G. I., Brown, C. F., Korach, K. S., McLachlan, J. A., and Teng, C. T. (1988). Influence of estrogens on mouse uterine epidermal growth factor percursor protein and messenger ribonucleic acid. *Endocrinology* 122, 2355–2363.
264. Huet-Hudson, Y. N., Chakraborty, C., Das, S. K., Suzuki, Y., Andrews, G. K., and Dey, S. K. (1990). Estrogen regulates the synthesis of epidermal growth factor in mouse uterine epithelial cells. *Mol. Endocrinol.* 4, 510–523.
265. Nelson, K. G., Takahshi, T., Lee, D. C., Luetteke, N. C., Bossert, N. L., Ross, K., Eitzman, B. E., and McLachlan, J. A. (1992). Transforming growth factor-α is a potential mediator of estrogen action in the mouse uterus. *Endocrinology* 131, 1657–1664.
266. Murphy, L. J., and Ghahary, A. (1990). Uterine insulin-like growth factor-1: regulation of expression and its role in estrogen-induced uterine proliferation. *Endocr. Rev.* 11, 443–453.
267. Hom, Y. K., Young, P., Wiesen, J. F., Miettinen, P. J., Derynck, R., Werb, Z., and Cunha, G. R. (1998). Uterine and vaginal organ growth requires epidermal growth factor receptor signaling from stroma. *Endocrinology* 139, 913–921.
268. Nelson, K. G., Takahashi, T., Bossert, N. L., Walmer, D. K., and McLachlan, J. A. (1991). Epidermal growth factor replaces estrogen in the stimulation of female genital-tract growth and differentiation. *Proc. Natl. Acad. Sci. U S A* 88, 21–25.
269. Ignar-Trowbridge, D. M., Nelson, K. G., Bidwell, M. C., Curtis, S. W., Washburn, T. F., McLachlan, J. A., and Korach, K. S. (1992). Coupling of dual signalling pathways: epidermal growth factor action involves the estrogen receptor. *Proc. Natl. Acad. Sci. U S A* 89, 4658–4662.
270. Buchanan, D. L., Kurita, T., Taylor, J. A., Lubahn, D. B., Cunha, G. R., and Cooke, P. S. (1998). Role of stromal and epithelial estrogen receptors in vaginal epithelial proliferation, stratification, and cornification. *Endocrinology* 139, 4345–4352.
271. Buchanan, D. L., Setiawan, T., Lubahn, D. B., Taylor, J. A., Kurita, T., Cunha, G. R., and Cooke, P. S. (1999). Tissue compartment-specific estrogen receptor-alpha participation in the mouse uterine epithelial secretory response. *Endocrinology* 140, 484–491.
272. Das, S. K., Taylor, J. A., Korach, K. S., Paria, B. C., Dey, S. K., and Lubahn, D. B. (1997). Estrogenic responses in estrogen receptor-alpha deficient mice reveal a distinct estrogen signaling pathway. *Proc. Natl. Acad. Sci. U S A* 94, 12786–12791.
273. Teng, C. T., Pentecost, B. T., Chen, Y. H., Newbold, R. R., Eddy, E. M., and McLachlan, J. A. (1989). Lactotransferrin gene expression in the mouse uterus and mammary gland. *Endocrinology* 124, 992–999.
274. Liehr, J. G. (2000). Is estradiol a genotoxic mutagenic carcinogen? *Endocr. Rev.* 21, 40–54.
275. Finn, C. A. (1965). Oestrogen and the decidual cell reaction of implantation in mice. *J. Endocrinol.* 32, 223–229.
276. Paria, B. C., Tan, J., Lubahn, D. B., Dey, S. K., and Das, S. K. (1999). Uterine decidual response occurs in estrogen receptor-alpha-deficient mice. *Endocrinology* 140, 2704–2710.
277. Grummer, R., Hewitt, S. W., Traub, O., Korach, K. S., and Winterhager, E. (2004). Different regulatory pathways of endometrial connexin expression: preimplantation hormonal-mediated pathway versus embryo implantation-initiated pathway. *Biol. Reprod.* 71, 273–281.
278. Kurita, T., Lee, K. J., Cooke, P. S., Taylor, J. A., Lubahn, D. B., and Cunha, G. R. (2000). Paracrine regulation of epithelial progesterone receptor by estradiol in the mouse female reproductive tract. *Biol. Reprod.* 62, 821–830.
279. Ismail, P. M., Li, J., DeMayo, F. J., O'Malley, B. W., and Lydon, J. P. (2002). A novel LacZ reporter mouse reveals complex regulation of the progesterone receptor promoter during mammary gland development. *Mol. Endocrinol.* 16, 2475–2489.
280. Mote, P. A., Johnston, J. F., Manninen, T., Tuohimaa, P., and Clarke, C. L. (2001). Detection of progesterone receptor forms A and B by immunohistochemical analysis. *J. Clin. Pathol.* 54, 624–630.
281. Conneely, O. M., Mulac-Jericevic, B., and Lydon, J. P. (2003). Progesterone-dependent regulation of female reproductive activity by two distinct progesterone receptor isoforms. *Steroids* 68, 771–778.
282. Arnett-Mansfield, R. L., DeFazio, A., Mote, P. A., and Clarke, C. L. (2004). Subnuclear distribution of progesterone

receptors A and B in normal and malignant endometrium. *J. Clin. Endocrinol. Metab.* 89, 1429–1442.

283. Tong, W., and Pollard, J. W. (1999). Progesterone inhibits estrogen-induced cyclin D1 and cdk4 nuclear translocation, cyclin E- and cyclin A-cdk2 kinase activation, and cell proliferation in uterine epithelial cells in mice. *Mol. Cell. Biol.* 19, 2251–2264.
284. Weitlauf, H. M. (1988). Biology of implantation. In *The Physiology of Reproduction* (E. Knobil and J. D. Neill, Eds.), pp. 231–262. Raven Press, New York.
285. Cheon, Y. P., Li, Q., Xu, X., DeMayo, F. J., Bagchi, I. C., and Bagchi, M. K. (2002). A genomic approach to identify novel progesterone receptor regulated pathways in the uterus during implantation. *Mol. Endocrinol.* 16, 2853–2871.
286. Yao, M. W., Lim, H., Schust, D. J., Choe, S. E., Farago, A., Ding, Y., Michaud, S., Church, G. M., and Maas, R. L. (2003). Gene expression profiling reveals progesterone-mediated cell cycle and immunoregulatory roles of Hoxa-10 in the preimplantation uterus. *Mol. Endocrinol.* 17, 610–627.
287. Li, Q., Cheon, Y. P., Kannan, A., Shanker, S., Bagchi, I. C., and Bagchi, M. K. (2004). A novel pathway involving progesterone receptor, 12/15-lipoxygenase-derived eicosanoids, and peroxisome proliferator-activated receptor gamma regulates implantation in mice. *J. Biol. Chem.* 279, 11570–11581.
288. Kelly, R. W., Leask, R., and Calder, A. A. (1992). Choriodecidual production of interleukin-8 and mechanism of parturition. *Lancet* 339, 776–777.
289. Ito, A., Imada, K., Takashi, S., Kubo, T., Matsushima, K., and Mori, Y. (1994). Suppression of interleukin 8 production by progesterone in rabbit uterine cervix. *Biochem. J.* 301, 183–186.
290. Cheung, L., Kelly, R. W., Thong, K. J., Hume, R., and Baird, D. T. (1993). The effect of mifepristone (RU486) on the immunohistochemical distribution of prostagland E and its metabolite in decidual and chorionic tissue in early pregnancy. *J. Clin. Endocrinol. Metab.* 77, 873–877.
291. Lydon, J. P., DeMayo, F. J., Conneely, O. M., and O'Malley, B. W. (1996). Reproductive phenotypes of the progesterone receptor null mutant mouse. *J. Steroid Biochem. Mol. Biol.* 56, 67–77.
292. Hirai, M., Hirata, S., Osada, T., Hagihara, K., and Kato, J. (1994). Androgen receptor mRNA in the rat ovary and uterus. *J. Steroid Biochem. Mol. Biol.* 49, 1–7.
293. Pelletier, G., Luu-The, V., Li, S., and Labrie, F. (2004). Localization and estrogenic regulation of androgen receptor mRNA expression in the mouse uterus and vagina. *J. Endocrinol.* 180, 77–85.
294. Pelletier, G. (2000). Localization of androgen and estrogen receptors in rat and primate tissues. *Histol. Histopathol.* 15, 1261–1270.
295. Kimura, N., Mizokami, A., Oonuma, T., Sasano, H., and Nagura, H. (1993). Immunocytochemical localization of androgen receptor with polyclonal antibody in paraffin-embedded human tissues. *J. Histochem. Cytochem.* 41, 671–678.
296. Mertens, H. J., Heineman, M. J., Theunissen, P. H., de Jong, F. H., and Evers, J. L. (2001). Androgen, estrogen and progesterone receptor expression in the human uterus during the menstrual cycle. *Eur. J. Obstet. Gynaecol. Reprod. Biol.* 98, 58–65.
297. Armstrong, D. T., and Papkoff, H. (1976). Stimulation of aromatization of exogenous and endogenous androgens in ovaries of hypophysectomized rats in vivo by follicle-stimulating hormone. *Endocrinology* 99, 1144–1151.
298. Nantermet, P. V., Masarachia, P., Gentile, M. A., Pennypacker, B., Xu, J., Holder, D., Gerhold, D., Towler, D., Schmidt, A., Kimmel, D. B., Freedman, L. P., Harada, S., and Ray, W. J. (2005). Androgenic induction of growth and differentiation in the rodent uterus involves the modulation of estrogen-regulated genetic pathways. *Endocrinology* 146, 564–578.
299. Rhen, T., Grissom, S., Afshari, C., and Cidlowski, J. A. (2003). Dexamethasone blocks the rapid biological effects of 17beta-estradiol in the rat uterus without antagonizing its global genomic actions. *FASEB J.* 17, 1849–1870.
300. Stumpf, W. E. (1969). Nuclear concentration of 3H-estradiol in target tissues. Dry-mount autoradiography of vagina, oviduct, ovary, testis, mammary tumor, liver and adrenal. *Endocrinology* 85, 31–37.
301. Richards, J. S. (1975). Estradiol receptor content in rat granulosa cells during follicular development: modification by estradiol and gonadotropins. *Endocrinology* 97, 1174–1184.
302. Richards, J. S., Ireland, J. J., Rao, M. C., Bernath, G. A., Midgley, A. R., Jr., and Reichert, L. E. Jr. (1976). Ovarian follicular development in the rat: hormone receptor regulation by estradiol, follicle stimulating hormone and luteinizing hormone. *Endocrinology* 99, 1562–1570.
303. Jakimiuk, A. J., Weitsman, S. R., Yen, H. W., Bogusiewicz, M., and Magoffin, D. A. (2002). Estrogen receptor alpha and beta expression in theca and granulosa cells from women with polycystic ovary syndrome. *J. Clin. Endocrinol. Metab.* 87, 5532–5538.
304. Sar, M., and Welsch, F. (1999). Differential expression of estrogen receptor-β and estrogen receptor-α in the rat ovary. *Endocrinology* 140, 963–971.
305. Hiroi, H., Inoue, S., Watanabe, T., Goto, W., Orimo, A., Momoeda, M., Tsutsumi, O., Taketani, Y., and Muramatsu, M. (1999). Differential immunolocalization of estrogen receptor a and b in rat ovary and uterus. *J. Mol. Endocrinol.* 22, 37–44.
306. Okada, A., Ohta, Y., Buchanan, D. L., Sato, T., Inoue, S., Hiroi, H., Muramatsu, M., and Iguchi, T. (2002). Changes in ontogenetic expression of estrogen receptor alpha and not of estrogen receptor beta in the female rat reproductive tract. *J. Mol. Endocrinol.* 28, 87–97.
307. Saunders, P. T. K., Maguire, S. M., Gaughan, J., and Millar, M. R. (1997). Expression of oestrogen receptor beta (ERb) in multiple rat tissues visualized by immunohistochemistry. *J. Endocrinol.* 154, R13–R16.
308. Sharma, S. C., Clemens, J. W., Pisarska, M. D., and Richards, J. S. (1999). Expression and function of estrogen receptor subtypes in granulosa cells: regulation by estradiol and forskolin. *Endocrinology* 140, 4320–4334.
309. Fitzpatrick, S. L., Funkhouser, J. M., Sindoni, D. M., Stevis, P. E., Deecher, D. C., Bapat, A. R., Merchenthaler, I., and Frail, D. E. (1999). Expression of estrogen receptor-beta protein in rodent ovary. *Endocrinology* 140, 2581–2591.
310. Mowa, C. N., and Iwanaga, T. (2000). Differential distribution of oestrogen receptor-alpha and -beta mRNAs in the female reproductive organ of rats as revealed by in situ hybridization. *J. Endocrinol.* 165, 59–66.
311. Schomberg, D. W., Couse, J. F., Mukherjee, A., Lubahn, D. B., Sar, M., Mayo, K. E., and Korach, K. S. (1999). Targeted disruption of the estrogen receptor-alpha gene in female mice: characterization of ovarian responses and phenotype in the adult. *Endocrinology* 140, 2733–2744.
312. Monje, P., and Boland, R. (2001). Subcellular distribution of native estrogen receptor alpha and beta isoforms in rabbit uterus and ovary. *J. Cell. Biochem.* 82, 467–479.
313. Yang, P., Kriatchko, A., and Roy, S. K. (2002). Expression of ER-alpha and ER-beta in the hamster ovary: differential regulation by gonadotropins and ovarian steroid hormones. *Endocrinology* 143, 2385–2398.
314. Mowa, C. N., and Iwanaga, T. (2000). Developmental changes of the oestrogen receptor-alpha and -beta mRNAs in

the female reproductive organ of the rat—an analysis by in situ hybridization. *J. Endocrinol.* 167, 363–369.

315. Drummond, A. E., Baillie, A. J., and Findlay, J. K. (1999). Ovarian estrogen receptor alpha and beta mRNA expression: impact of development and estrogen. *Mol. Cell. Endocrinol.* 149, 153–161.
316. Yang, P., Wang, J., Shen, Y., and Roy, S. K. (2004). Developmental expression of estrogen receptor α (ERα) and ERβ in the hamster ovary: regulation by follicle-stimulating hormone (FSH). *Endocrinology* 145, 5757–5766.
317. Sar, M., and Parikh, I. (1986). Immunohistochemical localization of estrogen receptor in the brain, pituitary and uterus with monoclonal antibodies. *J. Steroid Biochem.* 24, 497–503.
318. Hishikawa, Y., Damavandi, E., Izumi, S., and Koji, T. (2003). Molecular histochemical analysis of estrogen receptor alpha and beta expressions in the mouse ovary: in situ hybridization and Southwestern histochemistry. *Med. Electron Microsc.* 36:67–73.
319. Bao, B., Kumar, N., Karp, R. M., Garverick, H. A., and Sundaram, K. (2000). Estrogen receptor-beta expression in relation to the expression of luteinizing hormone receptor and cytochrome P450 enzymes in rat ovarian follicles. *Biol. Reprod.* 63, 1747–1755.
320. Shughrue, P. J., Lane, M. V., and Merchenthaler, I. (1997). Comparative distribution of estrogen receptor-alpha and -beta mRNA in the rat central nervous system. *J. Comp. Neurol.* 388, 507–525.
321. Byers, M., Kuiper, G. G., Gustafsson, J. A., and Park-Sarge, O. K. (1997). Estrogen receptor-β mRNA expression in rat ovary: down-regulation by gonadotropins. *Mol. Endocrinol.* 11, 172–182.
322. Van Den Broeck, W., Coryn, M., Simoens, P., and Lauwers, H. (2002). Cell-specific distribution of oestrogen receptor-alpha in the bovine ovary. *Reprod. Domest. Anim.* 37, 291–293.
323. Berisha, B., Pfaffl, M. W., and Schams, D. (2002). Expression of estrogen and progesterone receptors in the bovine ovary during estrous cycle and pregnancy. *Endocrine* 17, 207–214.
324. Slomczynska, M., and Wozniak, J. (2001). Differential distribution of estrogen receptor-beta and estrogen receptor-alpha in the porcine ovary. *Exp. Clin. Endocrinol. Diabetes* 109, 238–244.
325. Tomanek, M., Pisselet, C., Monget, P., Madigou, T., Thieulant, M. L., and Monniaux, D. (1997). Estrogen receptor protein and mRNA expression in the ovary of sheep. *Mol. Reprod. Dev.* 48, 53–62.
326. Hild-Petito, S., Stouffer, R. L., and Brenner, R. M. (1988). Immunocytochemical localization of estradiol and progesterone receptors in the monkey ovary throughout the menstrual cycle. *Endocrinology* 123, 2896–2905.
327. Billiar, R. B., Loukides, J. A., and Miller, M. M. (1992). Evidence for the presence of the estrogen receptor in the ovary of the baboon (*Papio anubis*). *J. Clin. Endocrinol. Metab.* 75, 1159–1165.
328. Pau, C. Y., Pau, K.-Y. F., and Spies, H. G. (1998). Putative estrogen receptor b and a mRNA expression in male and female rhesus maques. *Mol. Cell. Endocrinol.* 146, 59–68.
329. Lantta, M. (1984). Estradiol and progesterone receptors in normal ovary and ovarian tumors. *Acta. Obstet. Gynaecol. Scand.* 63, 497–503.
330. Al-Timimi, A., Buckley, C. H., and Fox, H. (1985). An immunohistochemical study of the incidence and significance of sex steroid hormone binding sites in normal and neoplastic human ovarian tissue. *Int. J. Gynecol. Pathol.* 4, 24–41.
331. Vierikko, P., Kauppila, A., and Vihko, R. (1983). Cytosol and nuclear estrogen and progestin receptors and 17 beta-hydroxysteroid dehydrogenase activity in non-diseased tissue and in benign and malignant tumors of the human ovary. *Int. J. Cancer.* 32, 413–422.
332. Iwai, T., Nanbu, Y., Iwai, M., Taii, S., Fujii, S., and Mori, T. (1990). Immunohistochemical localization of oestrogen receptors and progesterone receptors in human ovary throughout the menstrual cycle. *Virchow. Arch. A.* 417, 369–375.
333. Pelletier, G., and El-Alfy, M. (2000). Immunocytochemical localization of estrogen receptors alpha and beta in the human reproductive organs. *J. Clin. Endocrinol. Metab.* 85, 4835–4840.
334. Taylor, A. H., and Al-Azzawi, F. (2000). Immunolocalisation of oestrogen receptor beta in human tissues. *J. Mol. Endocrinol.* 24, 145–155.
335. Saunders, P. T., Millar, M. R., Williams, K., Macpherson, S., Harkiss, D., Anderson, R. A., Orr, B., Groome, N. P., Scobie, G., and Fraser, H. M. (2000). Differential expression of estrogen receptor-alpha and -beta and androgen receptor in the ovaries of marmosets and humans. *Biol. Reprod.* 63, 1098–1105.
336. Hillier, S. G., Anderson, R. A., Williams, A. R., and Tetsuka, M. (1998). Expression of oestrogen receptor alpha and beta in cultured human ovarian surface epithelial cells. *Mol. Hum. Reprod.* 4, 811–815.
337. Hurst, B. S., Zilberstein, M., Chou, J. Y., Litman, B., Stephens, J., and Leslie, K. K. (1995). Estrogen receptors are present in human granulosa cells. *J. Clin. Endocrinol. Metab.* 80, 229–232.
338. Chiang, C. H., Cheng, K. W., Igarashi, S., Nathwani, P. S., and Leung, P. C. (2000). Hormonal regulation of estrogen receptor alpha and beta gene expression in human granulosa-luteal cells in vitro. *J. Clin. Endocrinol. Metab.* 85, 3828–3839.
339. Shughrue, P. J., Lane, M. V., Scrimo, P. J., and Merchenthaler, I. (1998). Comparative distribution of estrogen receptor-alpha (ER-alpha) and beta (ER-beta) mRNA in the rat pituitary, gonad, and reproductive tract. *Steroids* 63, 498–504.
340. Choi, I., Ko, C., Park-Sarge, O. K., Nie, R., Hess, R. A., Graves, C., and Katzenellenbogen, B. S. (2001). Human estrogen receptor beta-specific monoclonal antibodies: characterization and use in studies of estrogen receptor beta protein expression in reproductive tissues. *Mol. Cell. Endocrinol.* 181, 139–150.
341. Manikkam, M., Bao, B., Rosenfeld, C. S., Yuan, X., Salfen, B. E., Calder, M. D., Youngquist, R. S., Keisler, D. H., Lubahn, D. B., and Garverick, H. A. (2001). Expression of the bovine oestrogen receptor-beta (bERbeta) messenger ribonucleic acid (mRNA) during the first ovarian follicular wave and lack of change in the expression of bERbeta mRNA of second wave follicles after LH infusion into cows. *Anim. Reprod. Sci.* 67, 159–169.
342. Jansen, H. T., West, C., Lehman, M. N., and Padmanabhan, V. (2001). Ovarian estrogen receptor-beta (ERbeta) regulation. I. Changes in ERbeta messenger RNA expression prior to ovulation in the ewe. *Biol. Reprod.* 65, 866–872.
343. Cardenas, H., Burke, K. A., Bigsby, R. M., Pope, W. F., and Nephew, K. P. (2001). Estrogen receptor beta in the sheep ovary during the estrous cycle and early pregnancy. *Biol. Reprod.* 65, 128–134.
344. LaVoie, H. A., DeSimone, D. C., Gillio-Meina, C., and Hui, Y. Y. (2002). Cloning and characterization of porcine ovarian estrogen receptor beta isoforms. *Biol. Reprod.* 66, 616–623.
345. Scobie, G. A., Macpherson, S., Millar, M. R., Groome, N. P., Romana, P. G., and Saunders, P. T. (2002). Human oestrogen receptors: differential expression of ER alpha and beta and the identification of ER beta variants. *Steroids* 67, 985–992.

346. O'Brien, M. L., Park, K., In, Y., and Park-Sarge, O. K. (1999). Characterization of estrogen receptor-beta (ERbeta) messenger ribonucleic acid and protein expression in rat granulosa cells. *Endocrinology* 140, 4530–4541.
347. Petersen, D. N., Tkalcevic, G. T., Koza-Taylor, P. H., Turi, T. G., and Brown, T. A. (1998). Identification of estrogen receptor β2, a functional variant of estrogen receptor β expressed in normal rat tissues. *Endocrinology* 139, 1082–1092.
348. Saiduddin, S., and Zassenhaus, H. P. (1977). Estradiol-17beta receptors in the immature rat ovary. *Steroids* 29, 197–213.
349. Kim, I., and Greenwald, G. S. (1987). Estrogen receptors in ovary and uterus of immature hamster and rat: effects of estrogens. *Endocrinol. Jpn.* 34, 45–53.
350. Kawashima, M., and Greenwald, G. S. (1993). Comparison of follicular estrogen receptors in rat, hamster, and pig. *Biol. Reprod.* 48, 172–179.
351. Couse, J. F., Bunch, D. O., Lindzey, J., Schomberg, D. W., and Korach, K. S. (1999). Prevention of the polycystic ovarian phenotype and characterization of ovulatory capacity in the estrogen receptor-alpha knockout mouse [In Process Citation]. *Endocrinology* 140, 5855–5865.
352. Conti, M. (2002). Specificity of the cyclic adenosine 3′,5′-monophosphate signal in granulosa cell function. *Biol. Reprod.* 67, 1653–1661.
353. Guo, C., Savage, L., Sarge, K. D., and Park-Sarge, O. K. (2001). Gonadotropins decrease estrogen receptor-beta messenger ribonucleic acid stability in rat granulosa cells. *Endocrinology* 142, 2230–2237.
354. Tonetta, S. A., Spicer, L. J., and Ireland, J. J. (1985). CI628 inhibits follicle-stimulating hormone (FSH)-induced increases in FSH receptors of the rat ovary: requirement of estradiol for FSH action. *Endocrinology* 116, 715–722.
355. Burns, K. H., Yan, C., Kumar, T. R., and Matzuk, M. M. (2001). Analysis of ovarian gene expression in follicle-stimulating hormone beta knockout mice. *Endocrinology* 142, 2742–2751.
356. Dierich, A., Sairam, M. R., Monaco, L., Fimia, G. M., Gansmuller, A., LeMeur, M., and Sassone-Corsi, P. (1998). Impairing follicle-stimulating hormone (FSH) signaling in vivo: targeted disruption of the FSH receptor leads to aberrant gametogenesis and hormonal imbalance. *Proc. Natl. Acad. Sci. U S A* 95, 13612–13617.
357. Danilovich, N., Babu, P. S., Xing, W., Gerdes, M., Krishnamurthy, H., and Sairam, M. R. (2000). Estrogen deficiency, obesity, and skeletal abnormalities in follicle-stimulating hormone receptor knockout (FORKO) female mice. *Endocrinology* 141, 4295–4308.
358. Saiduddin, S., and Zassenhaus, H. P. (1978). Effect of testosterone and progesterone on the estradiol receptor in the immature rat ovary. *Endocrinology* 102, 1069–1076.
359. Walther, N., Lioutas, C., Tillmann, G., and Ivell, R. (1999). Cloning of bovine estrogen receptor beta (ERbeta). expression of novel deleted isoforms in reproductive tissues. *Mol. Cell. Endocrinol.* 152, 37–45.
360. Pepe, G. J., Billiar, R. B., Leavitt, M. G., Zachos, N. C., Gustafsson, J. A., and Albrecht, E. D. (2002). Expression of estrogen receptors alpha and beta in the baboon fetal ovary. *Biol. Reprod.* 66, 1054–1060.
361. Ogawa, S., Inoue, S., Watanabe, T., Orimo, A., Hosoi, T., Ouchi, Y., and Muramatsu, M. (1998). Molecular cloning and characterization of human estrogen receptor betacx: a potential inhibitor ofestrogen action in human. *Nucleic Acids Res.* 26, 3505–3512.
362. Brandenberger, A. W., Tee, M. K., and Jaffe, R. B. (1998). Estrogen receptor alpha (ER-α) and beta (ER-β) mRNAs in normal ovary, ovarian serous cystadenocarcinoma and ovarian cancer cell lines: down-regulation of ER-β in neoplastic tissues. *J. Clin. Endocrinol. Metab.* 83, 1025–1028.
363. Moore, J. T., McKee, D. D., Slentz-Kesler, K., Moore, L. B., Jones, S. A., Horne, E. L., Su, J. L., Kliewer, S. A., Lehmann, J. M., and Willson, T. M. (1998). Cloning and characterization of human estrogen receptor beta isoforms. *Biochem. Biophys. Res. Commun.* 247, 75–78.
364. Lu, B., Leygue, E., Dotzlaw, H., Murphy, L. J., Murphy, L. C., and Watson, P. H. (1998). Estrogen receptor-b mRNA variants in human and murine tissues. *Mol. Cell. Endocrinol.* 138, 199–203.
365. Rosenfeld, C. S., Murray, A. A., Simmer, G., Hufford, M. G., Smith, M. F., Spears, N., and Lubahn, D. B. (2000). Gonadotropin induction of ovulation and corpus luteum formation in young estrogen receptor-alpha knockout mice. *Biol. Reprod.* 62, 599–605.
366. Couse, J. F., Yates, M. M., Walker, V. R., and Korach, K. S. (2003). Characterization of the hypothalamic-pituitary-gonadal axis in estrogen receptor (ER) Null mice reveals hypergonadism and endocrine sex reversal in females lacking ERalpha but not ERbeta. *Mol. Endocrinol.* 17, 1039–1053.
367. Couse, J. F., Yates, M. M., Deroo, B., and Korach, K. S. (2005). Estrogen receptor-beta is critical to granulosa cell differentiation and the ovulatory response to gonadotropins. *Endocrinology* 146, 3247–3262.
368. Risma, K. A., Hirshfield, A. N., and Nilson, J. H. (1997). Elevated luteinizing hormone in prepubertal transgenic mice causes hyperandrogenemia, precocious puberty, and substantial ovarian pathology. *Endocrinology* 138, 3540–3547.
369. Risma, K. A., Clay, C. M., Nett, T. M., Wagner, T., Yun, J., and Nilson, J. H. (1995). Targeted overexpression of luteinizing hormone in transgenic mice leads to infertility, polycystic ovaries, and ovarian tumors. *Proc. Natl. Acad. Sci. U S A* 92, 1322–1326.
370. Nilson, J. H., Abbud, R. A., Keri, R. A., and Quirk, C. C. (2000). Chronic hypersecretion of luteinizing hormone in transgenic mice disrupts both ovarian and pituitary function, with some effects modified by the genetic background. *Recent Prog. Horm. Res.* 55, 69–89; discussion 89–91.
371. Owens, G. E., Keri, R. A., and Nilson, J. H. (2002). Ovulatory surges of human CG prevent hormone-induced granulosa cell tumor formation leading to the identification of tumor-associated changes in the transcriptome. *Mol. Endocrinol.* 16, 1230–1242.
372. Mann, R. J., Keri, R. A., and Nilson, J. H. (2003). Consequences of elevated luteinizing hormone on diverse physiological systems: use of the LHbetaCTP transgenic mouse as a model of ovarian hyperstimulation-induced pathophysiology. *Recent Prog. Horm. Res.* 58, 343–375.
373. Gore-Langton, R. E., and Armstrong, D. T. (1994). Follicular steroidogenesis and its control. In *The Physiology of Reproduction* (E. Knobil and J. D. Neill, Eds.), pp. 571–627. Raven Press, New York.
374. Magoffin, D. A., and Weitsman, S. R. (1993). Differentiation of ovarian theca-interstitial cells in vitro: regulation of 17 alpha-hydroxylase messenger ribonucleic acid expression by luteinizing hormone and insulin-like growth factor-I. *Endocrinology* 132, 1945–1951.
375. Couse, J. F., Yates, M. M., Sanford, R., Nyska, A., Nilson, J. H., and Korach, K. S. (2004). Formation of cystic ovarian follicles associated with elevated luteinizing hormone requires estrogen receptor-beta. *Endocrinology* 145, 4693–4702.
376. Emmen, J. M., Couse, J. F., Elmore, S. A., Yates, M. M., Kissling, G. E., and Korach, K. S. (2005). In vitro growth and ovulation of follicles from ovaries of estrogen receptor (ER)-alpha and ERbeta null mice indicate a role for ERbeta in follicular maturation. *Endocrinology* 146, 2817–2826.
377. Richards, J. S. (1994). Hormonal control of gene expression in the ovary. *Endocr. Rev.* 15, 725–751.

378. Fitzpatrick, S. L., and Richards, J. S. (1991). Regulation of cytochrome P450 aromatase messenger ribonucleic acid and activity of steroids and gonadotropins in rat granulosa cells. *Endocrinology* 129, 1452–1462.
379. Fitzpatrick, S. L., Carlone, D. L., Robker, R. L., and Richards, J. S. (1997). Expression of aromatase in the ovary: down-regulation of mRNA by the ovulatory luteinizing hormone surge. *Steroids* 62, 197–206.
380. Zhuang, L. Z., Adashi, E. Y., and Hsuch, A. J. (1982). Direct enhancement of gonadotropin-stimulated ovarian estrogen biosynthesis by estrogen and clomiphene citrate. *Endocrinology* 110, 2219–2221.
381. Adashi, E. Y., and Hsueh, A. J. (1982). Estrogens augment the stimulation of ovarian aromatase activity by follicle-stimulating hormone in cultured rat granulosa cells. *J. Biol. Chem.* 257, 6077–6083.
382. Daniel, S. A., and Armstrong, D. T. (1983). Involvement of estrogens in the regulation of granulosa cell aromatase activity. *Can. J. Physiol. Pharmacol.* 61, 507–511.
383. Welsh, T. H., Jr., Jia, X. C., Jones, P. B., Zhuang, L. Z., and Hsueh, A. J. (1984). Disparate effects of triphenylethylene antiestrogens on estrogen and progestin biosyntheses by cultured rat granulosa cells. *Endocrinology* 115, 1275–1282.
384. Ghersevich, S., Nokelainen, P., Poutanen, M., Orava, M., Autio-Harmainen, H., Rajaniemi, H., and Vihko, R. (1994). Rat 17 beta-hydroxysteroid dehydrogenase type 1: primary structure and regulation of enzyme expression in rat ovary by diethylstilbestrol and gonadotropins in vivo. *Endocrinology* 135, 1477–1487.
385. Ghersevich, S., Poutanen, M., Tapanainen, J., and Vihko, R. (1994). Hormonal regulation of rat 17 beta-hydroxysteroid dehydrogenase type 1 in cultured rat granulosa cells: effects of recombinant follicle- stimulating hormone, estrogens, androgens, and epidermal growth factor. *Endocrinology* 135, 1963–1971.
386. Cheng, G., Weihua, Z., Makinen, S., Makela, S., Saji, S., Warner, M., Gustafsson, J. A., and Hovatta, O. (2002). A role for the androgen receptor in follicular atresia of estrogen receptor beta knockout mouse ovary. *Biol. Reprod.* 66, 77–84.
387. Lim, H., Paria, B. C., Das, S. K., Dinchuk, J. E., Langenbach, R., Trzaskos, J. M., and Dey, S. K. (1997). Multiple female reproductive failures in cyclooxygenase 2-deficient mice. *Cell* 91, 197–208.
388. Abraham, I. M., Han, S. K., Todman, M. G., Korach, K. S., and Herbison, A. E. (2003). Estrogen receptor beta mediates rapid estrogen actions on gonadotropin-releasing hormone neurons in vivo. *J. Neurosci.* 23, 5771–5777.
389. Swain, A., and Lovell-Badge, R. (1999). Mammalian sex determination: a molecular drama. *Genes Dev.* 13, 755–767.
390. Dupont, S., Dennefeld, C., Krust, A., Chambon, P., and Mark, M. (2003). Expression of Sox9 in granulosa cells lacking the estrogen receptors, ERalpha and ERbeta. *Dev. Dyn.* 226, 103–106.
391. Morais da Silva, S., Hacker, A., Harley, V., Goodfellow, P., Swain, A., and Lovell-Badge, R. (1996). Sox9 expression during gonadal development implies a conserved role for the gene in testis differentiation in mammals and birds. *Nat. Genet.* 14, 62–68.
392. Kanai, Y., and Koopman, P. (1999). Structural and functional characterization of the mouse Sox9 promoter: implications for campomelic dysplasia. *Hum. Mol. Genet.* 8, 691–696.
393. Vidal, V. P., Chaboissier, M. C., de Rooij, D. G., and Schedl, A. (2001). Sox9 induces testis development in XX transgenic mice. *Nat. Genet.* 28, 216–217.
394. Bishop, C. E., Whitworth, D. J., Qin, Y., Agoulnik, A. I., Agoulnik, I. U., Harrison, W. R., Behringer, R. R., and Overbeek, P. A. (2000). A transgenic insertion upstream of sox9 is associated with dominant XX sex reversal in the mouse. *Nat. Genet.* 26, 490–494.
395. Taketo-Hosotani, T. (1987). Factors involved in the testicular development from fetal mouse ovaries following transplantation. *J. Exp. Zool.* 241, 95–100.
396. Whitworth, D. J., Shaw, G., and Renfree, M. B. (1996). Gonadal sex reversal of the developing marsupial ovary in vivo and in vitro. *Development* 122, 4057–4063.
397. Vigier, B., Forest, M. G., Eychenne, B., Bezard, J., Garrigou, O., Robel, P., and Josso, N. (1989). Anti-Mullerian hormone produces endocrine sex reversal of fetal ovaries. *Proc. Natl. Acad. Sci. U S A* 86, 3684–3688.
398. Behringer, R. R., Cate, R. L., Froelick, G. J., Palmiter, R. D., and Brinster, R. L. (1990). Abnormal sexual development in transgenic mice chronically expressing mullerian inhibiting substance. *Nature* 345, 167–170.
399. Vainio, S., Heikkila, M., Kispert, A., Chin, N., and McMahon, A. P. (1999). Female development in mammals is regulated by Wnt-4 signalling. *Nature* 397, 405–409.
400. Hashimoto, N., Kubokawa, R., Yamazaki, K., Noguchi, M., and Kato, Y. (1990). Germ cell deficiency causes testis cord differentiation in reconstituted mouse fetal ovaries. *J. Exp. Zool.* 253, 61–70.
401. Whitworth, D. J. (1998). XX germ cells: the difference between an ovary and a testis. *Trends Endocrinol. Metab.* 9, 2–6.
402. Richard-Mercier, N., Dorizzi, M., Desvages, G., Girondot, M., and Pieau, C. (1995). Endocrine sex reversal of gonads by the aromatase inhibitor Letrozole (CGS 20267) in Emys orbicularis, a turtle with temperature-dependent sex determination. *Gen. Comp. Endocrinol.* 100, 314–326.
403. Wennstrom, K. L., and Crews, D. (1995). Making males from females: the effects of aromatase inhibitors on a parthenogenetic species of whiptail lizard. *Gen. Comp. Endocrinol.* 99, 316–322.
404. Yomogida, K., Ohtani, H., Harigae, H., Ito, E., Nishimune, Y., Engel, J. D., and Yamamoto, M. (1994). Developmental stage- and spermatogenic cycle-specific expression of transcription factor GATA-1 in mouse Sertoli cells. *Development* 120, 1759–1766.
405. Blobel, G. A., Sieff, C. A., and Orkin, S. H. (1995). Ligand-dependent repression of the erythroid transcription factor GATA- 1 by the estrogen receptor. *Mol. Cell. Biol.* 15, 3147–3153.
406. Flaws, J. A., Abbud, R., Mann, R. J., Nilson, J. H., and Hirshfield, A. N. (1997). Chronically elevated luteinizing hormone depletes primordial follicles in the mouse ovary. *Biol. Reprod.* 57, 1233–1237.
407. Magoffin, D. A. (2002). The ovarian androgen-producing cells: a 2001 perspective. *Rev Endocr. Metab. Disord.* 3, 47–53.
408. Britt, K. L., Drummond, A. E., Cox, V. A., Dyson, M., Wreford, N. G., Jones, M. E., Simpson, E. R., and Findlay, J. K. (2000). An age-related ovarian phenotype in mice with targeted disruption of the Cyp 19 (aromatase) gene. *Endocrinology* 141, 2614–2623.
409. Britt, K. L., Stanton, P. G., Misso, M., Simpson, E. R., and Findlay, J. K. (2004). The effects of estrogen on the expression of genes underlying the differentiation of somatic cells in the murine gonad. *Endocrinology* 145, 3950–3960.
410. Britt, K. L., Saunders, P. K., McPherson, S. J., Misso, M. L., Simpson, E. R., and Findlay, J. K. (2004). Estrogen actions on follicle formation and early follicle development. *Biol. Reprod.* 71, 1712–1723.
411. Pincus, G. (1966). Control of conception by hormonal steroids. *Science* 153, 493–500.

412. Pincus, G. (1962). Reproduction. *Annu. Rev. Physiol.* 24, 57–84.
413. Knecht, M., Tsai-Morris, C. H., and Catt, K. J. (1985). Estrogen dependence of luteinizing hormone receptor expression in cultured rat granulosa cells. Inhibition of granulosa cell development by the antiestrogens tamoxifen and keoxifene. *Endocrinology* 116, 1771–1777.
414. Moudgal, N. R., Shetty, G., Selvaraj, N., and Bhatnagar, A. S. (1996). Use of a specific aromatase inhibitor for determining whether there is a role for oestrogen in follicle/oocyte maturation, ovulation and preimplantation embryo development. *J. Reprod. Fertil.* 50 Suppl, 69–81.
415. Selvaraj, N., Shetty, G., Vijayalakshmi, K., Bhatnagar, A. S., and Moudgal, N. R. (1994). Effect of blocking oestrogen synthesis with a new generation aromatase inhibitor CGS 16949A on follicular maturation induced by pregnant mare serum gonadotrophin in the immature rat. *J. Endocrinol.* 142, 563–570.
416. Knecht, M., Brodie, A. M., and Catt, K. J. (1985). Aromatase inhibitors prevent granulosa cell differentiation: an obligatory role for estrogens in luteinizing hormone receptor expression. *Endocrinology* 117, 1156–1161.
417. Hegele-Hartung, C., Siebel, P., Peters, O., Kosemund, D., Muller, G., Hillisch, A., Walter, A., Kraetzschmar, J., and Fritzemeier, K. H. (2004). Impact of isotype-selective estrogen receptor agonists on ovarian function. *Proc. Natl. Acad. Sci. U S A* 101, 5129–5134.
418. Greenwald, G. S., and Roy, S. K. (1994). Follicular development and its control. In *The Physiology of Reproduction* (E. Knobil and J. D. Neill, Eds.), pp. 629–724. Raven Press, New York.
419. Williams, P. C. (1940). Effect of stilbestrol on the ovaries of the hypophysectomized rat. *Nature* 145, 388–389.
420. Pencharz, R. I. (1940). Effect of estrogens and androgens alone and in combination with chorionic gonadotropin on the ovary of the hypophysectomized rat. *Science* 91, 554–555.
421. Goldenberg, R. L., Vaitukaitis, J. L., and Ross, G. T. (1972). Estrogen and follicle stimulation hormone interactions on follicle growth in rats. *Endocrinology* 90, 1492–1498.
422. Rao, M. C., Midgley, A. R. J., and Richards, J. S. (1978). Hormonal regulation of ovarian cellular proliferation. *Cell* 14, 71–78.
423. Wang, X. N., and Greenwald, G. S. (1993). Synergistic effects of steroids with FSH on folliculogenesis, steroidogenesis and FSH- and hCG-receptors in hypophysectomized mice. *J. Reprod. Fertil.* 99, 403–413.
424. Chakravorty, A., Mahesh, V. B., and Mills, T. M. (1991). Regulation of follicular development by diethylstilboestrol in ovaries of immature rats. *J. Reprod. Fertil.* 92, 307–321.
425. Richards, J. S. (1980). Maturation of ovarian follicles: actions and interactions of pituitary and ovarian hormones on follicular cell differentiation. *Physiol. Rev.* 60, 51–89.
426. Nakano, R., Nakayama, T., and Iwao, M. (1982). Inhibition of ovarian follicle growth by a chemical antiestrogen. *Horm. Res.* 16, 230–236.
427. Bley, M. A., Saragueta, P. E., and Baranao, J. L. (1997). Concerted stimulation of rat granulosa cell deoxyribonucleic acid synthesis by sex steroids and follicle-stimulating hormone. *J. Steroid Biochem. Mol. Biol.* 62, 11–19.
428. Spears, N., Murray, A. A., Allison, V., Boland, N. I., and Gosden, R. G. (1998). Role of gonadotrophins and ovarian steroids in the development of mouse follicles in vitro. *J. Reprod. Fertil.* 113, 19–26.
429. Couse, J. F., and Korach, K. S. (1998). Exploring the role of sex steroids through studies of receptor deficient mice. *J. Mol. Med.* 76, 497–511.
430. Burns, K. H., Agno, J. E., Chen, L., Haupt, B., Ogbonna, S. C., Korach, K. S., and Matzuk, M. M. (2003). Sexually dimorphic roles of steroid hormone receptor signaling in gonadal tumorigenesis. *Mol. Endocrinol.* 17, 2039–2052.
431. Sicinski, P., Donaher, J. L., Geng, Y., Parker, S. B., Gardner, H., Park, M. Y., Robker, R. L., Richards, J. S., McGinnis, L. K., Biggers, J. D., Eppig, J. J., Bronson, R. T., Elledge, S. J., and Weinberg, R. A. (1996). Cyclin D2 is an FSH-responsive gene involved in gonadal cell proliferation and oncogenesis. *Nature* 384, 470–474.
432. Robker, R. L., and Richards, J. S. (1998). Hormone-induced proliferation and differentiation of granulosa cells: a coordinated balance of the cell cycle regulators cyclin D2 and $p27^{kip1}$. *Mol. Endocrinol.* 12, 924–940.
433. Kanda, N., and Watanabe, S. (2004). 17Beta-estradiol stimulates the growth of human keratinocytes by inducing cyclin D2 expression. *J. Invest. Dermatol.* 123, 319–328.
434. Baker, J., Hardy, M. P., Zhou, J., Bondy, C., Lupu, F., Bellve, A. R., and Efstatiadis, A. (1996). Effects of an Igf1 gene null mutation on mouse reproduction. *Mol. Endocrinol.* 10, 903–918.
435. Zhou, J., Kumar, R., Matzuk, M. M., and Bondy, C. (1997). Insulin-like growth factor I regulates gonadotropin responsiveness in the murine ovary. *Mol. Endocrinol.* 11, 1924–1933.
436. Kadakia, R., Arraztoa, J. A., Bondy, C., and Zhou, J. (2001). Granulosa cell proliferation is impaired in the Igf1 null ovary. *Growth Horm. IGF Res.* 11, 220–224.
437. Richards, J. S., Sharma, S. C., Falender, A. E., and Lo, Y. H. (2002). Expression of FKHR, FKHRL1, and AFX genes in the rodent ovary: evidence for regulation by IGF-I, estrogen, and the gonadotropins. *Mol. Endocrinol.* 16, 580–599.
438. Poretsky, L., Cataldo, N. A., Rosenwaks, Z., and Giudice, L. C. (1999). The insulin-related ovarian regulatory system in health and disease. *Endocr. Rev.* 20, 535–582.
439. Cunha, G. R., Young, P., Hom, Y. K., Cooke, P. S., Taylor, J. A., and Lubahn, D. B. (1997). Elucidation of a role of stromal steroid hormone receptors in mammary gland growth and development by tissue recombination experiments. *J. Mamm. Gland. Biol. Neopl.* 2, 393–402.
440. Bendell, J. J., and Dorrington, J. (1988). Rat thecal/interstitial cells secrete a transforming growth factor-beta-like factor that promotes growth and differentiation in rat granulosa cells. *Endocrinology* 123, 941–948.
441. Dorrington, J. H., Bendell, J. J., and Khan, S. A. (1993). Interactions between FSH, estradiol-17 beta and transforming growth factor-beta regulate growth and differentiation in the rat gonad. *J. Steroid Biochem. Mol. Biol.* 44, 441–447.
442. Roberts, A. J., and Skinner, M. K. (1990). Estrogen regulation of thecal cell steroidogenesis and differentiation: thecal cell-granulosa cell interactions. *Endocrinology* 127, 2918–2929.
443. Parrott, J. A., and Skinner, M. K. (1998). Developmental and hormonal regulation of keratinocyte growth factor expression and action in the ovarian follicle. *Endocrinology* 139, 228–235.
444. Kumar, T. R., Wang, Y., Lu, N., and Matzuk, M. M. (1997). Follicle stimulating hormone is required for ovarian follicle maturation but not male fertility. *Nat. Genet.* 15, 201–204.
445. Danilovich, N., Roy, I., and Sairam, M. R. (2001). Ovarian pathology and high incidence of sex cord tumors in follitropin receptor knockout (FORKO) mice. *Endocrinology* 142, 3673–3684.
446. Park, Y., Maizels, E. T., Feiger, Z. J., Alam, H., Peters, C. A., Woodruff, T. K., Unterman, T. G., Lee, E. J., Jameson, J. L., and Hunzicker-Dunn, M. (2005). Induction of cyclin D2 in rat granulosa cells requires FSH-dependent relief from FOXO1 repression coupled with positive signals from Smad. *J. Biol. Chem.* 280, 9135–9148.

447. Knecht, M., Darbon, J. M., Ranta, T., Baukal, A. J., and Catt, K. J. (1984). Estrogens enhance the adenosine 3′,5′-monophosphate-mediated induction of follicle-stimulating hormone and luteinizing hormone receptors in rat granulosa cells. *Endocrinology* 115, 41–49.
448. Richards, J. S., Jonassen, J. A., Rolfes, A. I., Kersey, K., and Reichert Jr. L. E. (1979). Adenosine 3′,5′-monophosphate, lutienizing hormone receptor, and progesterone during granulosa cell differentiation: effects of estradiol and follicle-stimulating hormone. *Endocrinology* 104, 765–773.
449. Aronica, S. M., Kraus, W. L., and Katzenellenbogen, B. S. (1994). Estrogen action via the cAMP signaling pathway: stimulation of adenylate cyclase and cAMP-regulated gene transcription. *Proc. Natl. Acad. Sci. U S A* 91, 8517–8521.
450. Coleman, K. M., Dutertre, M., El-Gharbawy, A., Rowan, B. G., Weigel, N. L., and Smith, C. L. (2003). Mechanistic differences in the activation of estrogen receptor-alpha (ER alpha)- and ER beta-dependent gene expression by cAMP signaling pathway(s). *J. Biol. Chem.* 278, 12834–12845.
451. Lazennec, G., Thomas, J. A., and Katzenellenbogen, B. S. (2001). Involvement of cyclic AMP response element binding protein (CREB) and estrogen receptor phosphorylation in the synergistic activation of the estrogen receptor by estradiol and protein kinase activators. *J. Steroid Biochem. Mol. Biol.* 77, 193–203.
452. Carlone, D. L., and Richards, J. S. (1997). Evidence that functional interactions of CREB and SF-1 mediate hormone regulated expression of the aromatase gene in granulosa cells and constitutive expression in R2C cells. *J. Steroid Biochem. Mol. Biol.* 61, 223–231.
453. Young, M., and McPhaul, M. J. (1997). Definition of the elements required for the activity of the rat aromatase promoter in steroidogenic cell lines. *J. Steroid Biochem. Mol. Biol.* 61, 341–348.
454. Gore-Langton, R. E., and Daniel, S. A. (1990). Follicle-stimulating hormone and estradiol regulate antrum-like reorganization of granulosa cells in rat preantral follicle cultures. *Biol. Reprod.* 43, 65–72.
455. Hu, Y., Cortvrindt, R., and Smitz, J. (2002). Effects of aromatase inhibition on in vitro follicle and oocyte development analyzed by early preantral mouse follicle culture. *Mol. Reprod. Dev.* 61, 549–559.
456. Mason, A. J., Hayflick, J. S., Zoeller, R. T., Young, W. S., 3rd, Phillips, H. S., Nikolics, K., and Seeburg, P. H. (1986). A deletion truncating the gonadotropin-releasing hormone gene is responsible for hypogonadism in the hpg mouse. *Science* 234, 1366–1371.
457. Guraya, S. S. (1985). *Biology of Ovarian Follicles in Mammals*. Springer-Verlag, Berlin, New York.
458. Edwards, R. G. (1974). Follicular fluid. *J. Reprod. Fertil.* 37, 189–219.
459. McConnell, N. A., Yunus, R. S., Gross, S. A., Bost, K. L., Clemens, M. G., and Hughes, F. M., Jr. (2002). Water permeability of an ovarian antral follicle is predominantly transcellular and mediated by aquaporins. *Endocrinology* 143, 2905–2912.
460. Richard, C., Gao, J., Brown, N., and Reese, J. (2003). Aquaporin water channel genes are differentially expressed and regulated by ovarian steroids during the periimplantation period in the mouse. *Endocrinology* 144, 1533–1541.
461. Jablonski, E. M., McConnell, N. A., Hughes, F. M., Jr., and Huet-Hudson, Y. M. (2003). Estrogen regulation of aquaporins in the mouse uterus: potential roles in uterine water movement. *Biol. Reprod.* 69, 1481–1487.
462. Hess, R. A. (2003). Estrogen in the adult male reproductive tract: a review. *Reprod. Biol. Endocrinol.* 1, 52.
463. Simon, A. M., Goodenough, D. A., Li, E., and Paul, D. L. (1997). Female infertility in mice lacking connexin 37. *Nature* 385, 525–529.
464. Burghardt, R. C., and Anderson, E. (1981). Hormonal modulation of gap junctions in rat ovarian follicles. *Cell. Tissue Res.* 214, 181–193.
465. Kidder, G. M., and Mhawi, A. A. (2002). Gap junctions and ovarian folliculogenesis. *Reproduction* 123, 613–620.
466. Wright, C. S., Becker, D. L., Lin, J. S., Warner, A. E., and Hardy, K. (2001). Stage-specific and differential expression of gap junctions in the mouse ovary: connexin-specific roles in follicular regulation. *Reproduction* 121, 77–88.
467. Risek, B., Klier, F. G., Phillips, A., Hahn, D. W., and Gilula, N. B. (1995). Gap junction regulation in the uterus and ovaries of immature rats by estrogen and progesterone. *J. Cell. Sci.* 108(Pt 3), 1017–1032.
468. Turner, K. J., Macpherson, S., Millar, M. R., McNeilly, A. S., Williams, K., Cranfield, M., Groome, N. P., Sharpe, R. M., Fraser, H. M., and Saunders, P. T. (2002). Development and validation of a new monoclonal antibody to mammalian aromatase. *J. Endocrinol.* 172, 21–30.
469. Garrett, W. M., and Guthrie, H. D. (1997). Steroidogenic enzyme expression during preovulatory follicle maturation in pigs. *Biol. Reprod.* 56, 1424–1431.
470. Bao, B., and Garverick, H. A. (1998). Expression of steroidogenic enzyme and gonadotropin receptor genes in bovine follicles during ovarian follicular waves: a review. *J. Anim. Sci.* 76, 1903–1921.
471. Kumar, T. R., Kelly, M., Mortrud, M., Low, M. J., and Matzuk, M. M. (1995). Cloning of the mouse gonadotropin β-subunit-encoding genes, I. Structure of the follicle-stimulating hormone β-subunit-encoding gene. *Gene* 166, 333–334.
472. Tremblay, J. J., and Viger, R. S. (2001). GATA factors differentially activate multiple gonadal promoters through conserved GATA regulatory elements. *Endocrinology* 142, 977–986.
473. Jeyasuria, P., Ikeda, Y., Jamin, S. P., Zhao, L., De Rooij, D. G., Themmen, A. P., Behringer, R. R., and Parker, K. L. (2004). Cell-specific knockout of steroidogenic factor 1 reveals its essential roles in gonadal function. *Mol. Endocrinol.* 18, 1610–1619.
474. Watson, J., and Howson, J. W. (1977). Inhibition of Tamoxifen of the stimulatory action of FSH on oestradiol-17beta synthesis by rat ovaries in vitro. *J. Reprod. Fertil.* 49, 375–376.
475. Tremblay, A., and Giguere, V. (2001). Contribution of steroid receptor coactivator-1 and CREB binding protein in ligand-independent activity of estrogen receptor beta. *J. Steroid Biochem. Mol. Biol.* 77, 19–27.
476. Blobel, G. A., and Orkin, S. H. (1996). Estrogen-induced apoptosis by inhibition of the erythroid transcription factor GATA-1. *Mol. Cell. Biol.* 16, 1687–1694.
477. Tetsuka, M., and Hillier, S. G. (1997). Differential regulation of aromatase and androgen receptor in granulosa cells. *J. Steroid Biochem. Mol. Biol.* 61, 233–239.
478. Tetsuka, M., and Hillier, S. G. (1996). Androgen receptor gene expression in rat granulosa cells: the role of follicle-stimulating hormone and steroid hormones. *Endocrinology* 137, 4392–4397.
479. Kessel, B., Liu, Y. X., Jia, X. C., and Hsueh, A. J. (1985). Autocrine role of estrogens in the augmentation of luteinizing hormone receptor formation in cultured rat granulosa cells. *Biol. Reprod.* 32, 1038–1050.
480. Farookhi, R., and Desjardins, J. (1986). Luteinizing hormone receptor induction in dispersed granulosa cells requires estrogen. *Mol. Cell. Endocrinol.* 47, 13–24.

481. Segaloff, D. L., Wang, H. Y., and Richards, J. S. (1990). Hormonal regulation of luteinizing hormone/chorionic gonadotropin receptor mRNA in rat ovarian cells during follicular development and luteinization. *Mol. Endocrinol.* 4, 1856–1865.
482. Knecht, M., and Catt, K. J. (1982). Induction of luteinizing hormone receptors by adenosine 3′,5′-monophosphate in cultured granulosa cells. *Endocrinology* 111, 1192–1200.
483. Erickson, G. F., Wang, C., Casper, R., Mattson, G., and Hofeditz, C. (1982). Studies on the mechanisms of LH receptor control by FSH. *Mol. Cell. Endocrinol.* 27, 17–30.
484. Nimrod, A. (1981). The induction of ovarian LH-receptors by FSH is mediated by cyclic AMP. *FEBS Lett.* 131, 31–33.
485. Segaloff, D. L., and Limbird, L. E. (1983). Luteinizing hormone receptor appearance in cultured porcine granulosa cells requires continual presence of follicle-stimulating hormone. *Proc. Natl. Acad. Sci. U S A* 80, 5631–5635.
486. Wang, H., Nelson, S., Ascoli, M., and Segaloff, D. L. (1992). The 5′-flanking region of the rat luteinizing hormone/chorionic gonadotropin receptor gene confers Leydig cell expression and negative regulation of gene transcription by 3′,5′-cyclic adenosine monophosphate. *Mol. Endocrinol.* 6, 320–326.
487. Chen, S., Liu, X., and Segaloff, D. L. (2000). A novel cyclic adenosine 3′,5′-monophosphate-responsive element involved in the transcriptional regulation of the lutropin receptor gene in granulosa cells. *Mol. Endocrinol.* 14, 1498–1508.
488. Chen, S., Shi, H., Liu, X., and Segaloff, D. L. (1999). Multiple elements and protein factors coordinate the basal and cyclic adenosine 3′,5′-monophosphate-induced transcription of the lutropin receptor gene in rat granulosa cells. *Endocrinology* 140, 2100–2109.
489. Geng, Y., Tsai-Morris, C. H., Zhang, Y., and Dufau, M. L. (1999). The human luteinizing hormone receptor gene promoter: activation by Sp1 and Sp3 and inhibitory regulation. *Biochem. Biophys. Res. Commun.* 263, 366–371.
490. Zhang, Y., and Dufau, M. L. (2003). Dual mechanisms of regulation of transcription of luteinizing hormone receptor gene by nuclear orphan receptors and histone deacetylase complexes. *J. Steroid Biochem. Mol. Biol.* 85, 401–414.
491. Chen, S., Liu, X., and Segaloff, D. L. (2001). Identification of an SAS (Sp1c adjacent site)-like element in the distal 5′-flanking region of the rat lutropin receptor gene essential for cyclic adenosine 3′,5′-monophosphate responsiveness. *Endocrinology* 142, 2013–2021.
492. Magoffin, D. A. (1991). Regulation of differentiated functions in ovarian theca cells. *Sem. Reprod. Endocrinol.* 9, 321–331.
493. Zhang, F. P., Poutanen, M., Wilbertz, J., and Huhtaniemi, I. (2001). Normal prenatal but arrested postnatal sexual development of luteinizing hormone receptor knockout (LuRKO) mice. *Mol. Endocrinol.* 15, 172–183.
494. Hsueh, A. J. W., Billig, H., and Tsafriri, A. (1994). Ovarian follicle atresia: a hormonally controlled apoptotic process. *Endocr. Rev.* 15, 707–724.
495. Erickson, G. F., Magoffin, D. A., Dyer, C. A., and Hofeditz, C. (1985). The ovarian androgen producing cells: a review of structure/function relationships. *Endocr. Rev.* 6, 371–399.
496. Onoda, M., and Hall, P. F. (1981). Inhibition of testicular microsomal cytochrome P-450 (17 alpha- hydroxylase/C-17,20-lyase) by estrogens. *Endocrinology* 109, 763–767.
497. Samuels, L. T., Bussmann, L., Matsumoto, K., and Huseby, R. A. (1975). Organization of androgen biosynthesis in the testis. *J. Steroid Biochem.* 6, 291–296.
498. Samuels, L. T., Uchikawa, T., Zain-ul-Abedin, M., and Huseby, R. A. (1969). Effect of diethylstilbestrol on enzymes of cryptochid mouse testes of Balb-c mice. *Endocrinology* 85, 96–102.
499. Magoffin, D. A., and Erickson, G. F. (1981). Mechanism by which 17 beta-estradiol inhibits ovarian androgen production in the rat. *Endocrinology* 108, 962–969.
500. Magoffin, D. A., and Erickson, G. F. (1982). Direct inhibitory effect of estrogen on LH-stimulated androgen synthesis by ovarian cells cultured in defined medium. *Mol. Cell. Endocrinol.* 28, 81–89.
501. Johnson, D. C., Martin, H., and Tsai-Morris, C. H. (1984). The in vitro and in vivo effect of estradiol upon the 17 alpha-hydroxylase and C17,20-lyase activity in the ovaries of immature hypophysectomized rats. *Mol. Cell. Endocrinol.* 35, 199–204.
502. Banks, P. K., Meyer, K., and Brodie, A. M. (1991). Regulation of ovarian steroid biosynthesis by estrogen during proestrus in the rat. *Endocrinology*. 129, 1295–1304.
503. Sakaue, M., Ishimura, R., Kurosawa, S., Fukuzawa, N. H., Kurohmaru, M., Hayashi, Y., Tohyama, C., and Ohsako, S. (2002). Administration of estradiol-3-benzoate down-regulates the expression of testicular steroidogenic enzyme genes for testosterone production in the adult rat. *J. Vet. Med. Sci.* 64, 107–113.
504. Govoroun, M., McMeel, O. M., Mecherouki, H., Smith, T. J., and Guiguen, Y. (2001). 17Beta-estradiol treatment decreases steroidogenic enzyme messenger ribonucleic acid levels in the rainbow trout testis. *Endocrinology* 142, 1841–1848.
505. Richards, J. S., Russell, D. L., Ochsner, S., and Espey, L. L. (2002). Ovulation: new dimensions and new regulators of the inflammatory-like response. *Annu. Rev. Physiol.* 64, 69–92.
506. Murphy, B. D., Mapletoft, R. J., Manns, J., and Humphrey, W. D. (1984). Variability in gonadotrophin preparations as a factor in the superovulatory response. *Theriogenology* 21, 117–125.
507. Opavsky, M. A., and Armstrong, D. T. (1989). Effects of luteinizing hormone on superovulatory and steroidogenic responses of rat ovaries to infusion with follicle-stimulating hormone. *Biol. Reprod.* 41, 15–25.
508. Armstrong, D. T., Siuda, A., Opavsky, M. A., and Chandrasekhar, Y. (1989). Bimodal effects of luteinizing hormone and role of androgens in modifying superovulatory responses of rats to infusion with purified porcine follicle-stimulating hormone. *Biol. Reprod.* 41, 54–62.
509. Clemens, J. W., Robker, R. L., Kraus, W. L., Katzenellenbogen, B. S., and Richards, J. S. (1998). Hormone induction of progesterone receptor (PR) messenger ribonucleic acid and activation of PR promoter regions in ovarian granulosa cells: evidence for a role of cyclic adenosine 3′,5′-monophosphate but not estradiol. *Mol. Endocrinol.* 12, 1201–1214.
510. Huynh, K., Jones, G., Thouas, G., Britt, K. L., Simpson, E. R., and Jones, M. E. (2004). Estrogen is not directly required for oocyte developmental competence. *Biol. Reprod.* 70, 1263–1269.
511. Eddy, E. M., Washburn, T. F., Bunch, D. O., Goulding, E. H., Gladen, B. C., Lubahn, D. B., and Korach, K. S. (1996). Targeted disruption of the estrogen receptor gene in male mice causes alteration of spermatogenesis and infertility. *Endocrinology* 137, 4796–4805.
512. Nokelainen, P., Puranen, T., Peltoketo, H., Orava, M., Vihko, P., and Vihko, R. (1996). Molecular cloning of mouse 17 beta-hydroxysteroid dehydrogenase type 1 and characterization of enzyme activity. *Eur. J. Biochem.* 236, 482–490.
513. Andersson, S., and Moghrabi, N. (1997). Physiology and molecular genetics of 17 beta-hydroxysteroid dehydrogenases. *Steroids* 62, 143–147.
514. Sha, J., Baker, P., and O'Shaughnessy, P. J. (1996). Both reductive forms of 17 beta-hydroxysteroid dehydrogenase (types 1 and 3) are expressed during development in the mouse testis. *Biochem. Biophys. Res. Commun.* 222, 90–94.

515. Baker, P. J., Sha, J. H., and O'Shaughnessy, P. J. (1997). Localisation and regulation of 17beta-hydroxysteroid dehydrogenase type 3 mRNA during development in the mouse testis. *Mol. Cell. Endocrinol.* 133, 127–133.
516. Mustonen, M. V., Poutanen, M. H., Isomaa, V. V., Vihko, P. T., and Vihko, R. K. (1997). Cloning of mouse 17beta-hydroxysteroid dehydrogenase type 2, and analysing expression of the mRNAs for types 1, 2, 3, 4 and 5 in mouse embryos and adult tissues. *Biochem. J.* 325, 199–205.
517. Tsai-Morris, C. H., Khanum, A., Tang, P. Z., and Dufau, M. L. (1999). The rat 17beta-hydroxysteroid dehydrogenase type III: molecular cloning and gonadotropin regulation. *Endocrinology* 140, 3534–3542.
518. Akingbemi, B. T., Ge, R., Rosenfeld, C. S., Newton, L. G., Hardy, D. O., Catterall, J. F., Lubahn, D. B., Korach, K. S., and Hardy, M. P. (2003). Estrogen receptor-alpha gene deficiency enhances androgen biosynthesis in the mouse Leydig cell. *Endocrinology* 144, 84–93.
519. Abney, T. O. (1999). The potential roles of estrogens in regulating Leydig cell development and function: a review. *Steroids* 64, 610–617.
520. Conte, F. A., Grumbach, M. M., Ito, Y., Fisher, C. R., and Simpson, E. R. (1994). A syndrome of female pseudohermaphrodism, hypergonadotropic hypogonadism, and multicystic ovaries associated with missense mutations in the gene encoding aromatase (P450arom). *J. Clin. Endocrinol. Metab.* 78, 1287–1292.
521. Shozu, M., Akasofu, K., Harada, T., and Kubota, T. (1991). A new cause of female pseudohermaphroditism: placental aromatase deficiency. *J. Clin. Endocrinol. Metab.* 72, 560–566.
522. Morishima, A., Grumbach, M. M., Simpson, E. R., Fisher, C., and Qin, K. (1995). Aromatase deficiency in male and female siblings caused by a novel mutation and the physiological role of estrogens. *J. Clin. Endocrinol. Metab.* 80, 3689–3698.
523. Mullis, P. E., Yoshimura, N., Kuhlmann, B., Lippuner, K., Jaeger, P., and Harada, H. (1997). Aromatase deficiency in a female who is compound heterozygote for two new point mutations in the P450arom gene: impact of estrogens on hypergonadotropic hypogonadism, multicystic ovaries, and bone densitometery in childhood. *J. Clin. Endocrinol. Metab.* 82, 1739–1745.
524. Ludwig, M., Beck, A., Wickert, L., Bolkenius, U., Tittel, B., Hinkel, K., and Bidlingmaier, F. (1998). Female pseudohermaphroditism associated with a novel homozygous G-to-A (V370-to-M) substitution in the P-450 aromatase gene. *J. Pediatr. Endocrinol. Metab.* 11, 657–664.
525. Grumbach, M. M., and Auchus, R. J. (1999). Estrogen: consequences and implications of human mutations in synthesis and action. *J. Clin. Endocrinol. Metab.* 84, 4677–4694.
526. Ongphiphadhanakul, B. (2003). Genetic polymorphisms of estrogen receptor-alpha: possible implications for targeted osteoporosis therapy. *Am. J. Pharmacogenom.* 3, 5–9.
527. Andersen, T. I., Heimdal, K. R., Skrede, M., Tveit, K., Berg, K., and Borresen, A. L. (1994). Oestrogen receptor (ESR) polymorphisms and breast cancer susceptibility. *Hum. Genet.* 94, 665–670.
528. Herrington, D. M. (2003). Role of estrogen receptor-alpha in pharmacogenetics of estrogen action. *Curr. Opin. Lipidol.* 14, 145–150.
529. Herrington, D. M., and Howard, T. D. (2003). ER-alpha variants and the cardiovascular effects of hormone replacement therapy. *Pharmacogenomics* 4, 269–277.
530. Liu, Y. Z., Liu, Y. J., Recker, R. R., and Deng, H. W. (2003). Molecular studies of identification of genes for osteoporosis: the 2002 update. *J. Endocrinol.* 177, 147–196.
531. Tempfer, C. B., Schneeberger, C., and Huber, J. C. (2004). Applications of polymorphisms and pharmacogenomics in obstetrics and gynecology. *Pharmacogenomics* 5, 57–65.
532. Syrrou, M., Georgiou, I., Patsalis, P. C., Bouba, I., Adonakis, G., and Pagoulatos, G. N. (1999). Fragile X premutations and (TA)n estrogen receptor polymorphism in women with ovarian dysfunction. *Am. J. Med. Genet.* 84, 306–308.
533. Westberg, L., Baghaei, F., Rosmond, R., Hellstrand, M., Landen, M., Jansson, M., Holm, G., Bjorntorp, P., and Eriksson, E. (2001). Polymorphisms of the androgen receptor gene and the estrogen receptor beta gene are associated with androgen levels in women. *J. Clin. Endocrinol. Metab.* 86, 2562–2568.
534. Wang, Y., and Miksicek, R. J. (1994). Characterization of estrogen receptor cDNAs from human uterus: identification of a novel PvuII polymorphism. *Mol. Cell. Endocrinol.* 101, 101–110.
535. Yaich, L., Dupont, W. D., Cavener, D. R., and Parl, F. F. (1992). Analysis of the PvuII restriction fragment-length polymorphism and exon structure of the estrogen receptor gene in breast cancer and peripheral blood. *Cancer Res.* 52, 77–83.
536. Georgiou, I., Konstantelli, M., Syrrou, M., Messinis, I. E., and Lolis, D. E. (1997). Oestrogen receptor gene polymorphisms and ovarian stimulation for in-vitro fertilization. *Hum. Reprod.* 12, 1430–1433.
537. Weel, A. E., Uitterlinden, A. G., Westendorp, I. C., Burger, H., Schuit, S. C., Hofman, A., Helmerhorst, T. J., van Leeuwen, J. P., and Pols, H. A. (1999). Estrogen receptor polymorphism predicts the onset of natural and surgical menopause. *J. Clin. Endocrinol. Metab.* 84, 3146–3150.
538. Zofkova, I., Zajickova, K., and Hill, M. (2002). The estrogen receptor alpha gene determines serum androstenedione levels in postmenopausal women. *Steroids* 67, 815–819.
539. Gorai, I., Tanaka, K., Inada, M., Morinaga, H., Uchiyama, Y., Kikuchi, R., Chaki, O., and Hirahara, F. (2003). Estrogen-metabolizing gene polymorphisms, but not estrogen receptor-alpha gene polymorphisms, are associated with the onset of menarche in healthy postmenopausal Japanese women. *J. Clin. Endocrinol. Metab.* 88, 799–803.
540. Sundarrajan, C., Liao, W., Roy, A. C., and Ng, S. C. (1999). Association of oestrogen receptor gene polymorphisms with outcome of ovarian stimulation in patients undergoing IVF. *Mol. Hum. Reprod.* 5, 797–802.
541. Stavrou, I., Zois, C., Ioannidis, J. P., and Tsatsoulis, A. (2002). Association of polymorphisms of the oestrogen receptor alpha gene with the age of menarche. *Hum. Reprod.* 17, 1101–1105.
542. Kok, H. S., Onland-Moret, N. C., van Asselt, K. M., van Gils, C. H., van der Schouw, Y. T., Grobbee, D. E., and Peeters, P. H. (2004). No association of estrogen receptor alpha and cytochrome P450c17alpha polymorphisms with age at menopause in a Dutch cohort. *Hum. Reprod.* 20, 536–542.
543. Rosenkranz, K., Hinney, A., Ziegler, A., Hermann, H., Fichter, M., Mayer, H., Siegfried, W., Young, J. K., Remschmidt, H., and Hebebrand, J. (1998). Systematic mutation screening of the estrogen receptor beta gene in probands of different weight extremes: identification of several genetic variants. *J. Clin. Endocrinol. Metab.* 83, 4524–4527.
544. Sundarrajan, C., Liao, W. X., Roy, A. C., and Ng, S. C. (2001). Association between estrogen receptor-beta gene polymorphisms and ovulatory dysfunctions in patients with menstrual disorders. *J. Clin. Endocrinol. Metab.* 86, 135–139.
545. Schreiber, J. R., and Erickson, G. F. (1979). Progesterone receptor in the rat ovary: further characterization and localization in the granulosa cell. *Steroids* 34, 459–469.

546. Schreiber, J. R., and Hsueh, J. W. (1979). Progesterone "receptor" in rat ovary. *Endocrinology* 105, 915–919.
547. Jacobs, B. R., Suchocki, S., and Smith, R. G. (1980). Evidence for a human ovarian progesterone receptor. *Am. J. Obstet. Gynecol.* 138, 332–336.
548. Jacobs, B. R., and Smith, R. G. (1980). Evidence for a receptor-like protein for progesterone in bovine ovarian cytosol. *Endocrinology* 106, 1276–1282.
549. Jacobs, B. R., and Smith, R. G. (1981). A comparison of progesterone and R5020 binding in endometrium, ovary, pituitary, and hypothalamus. *Fertil. Steril.* 35, 438–441.
550. Shao, R., Markstrom, E., Friberg, P. A., Johansson, M., and Billig, H. (2003). Expression of progesterone receptor (PR) A and B isoforms in mouse granulosa cells: stage-dependent PR-mediated regulation of apoptosis and cell proliferation. *Biol. Reprod.* 68, 914–921.
551. Gava, N., Clarke, C. L., Byth, K., Arnett-Mansfield, R. L., and deFazio, A. (2004). Expression of progesterone receptors A and B in the mouse ovary during the estrous cycle. *Endocrinology* 145, 3487–3494.
552. Vermeirsch, H., Simoens, P., Coryn, M., and Van den Broeck, W. (2001). Immunolocalization of progesterone receptors in the canine ovary and their relation to sex steroid hormone concentrations. *Reproduction* 122, 73–83.
553. Iwai, T., Fujii, S., Nanbu, Y., Nonogaki, H., Konishi, I., Mori, T., and Okamura, H. (1991). Effect of human chorionic gonadotropin on the expression of progesterone receptors and estrogen receptors in rabbit ovarian granulosa cells and the uterus. *Endocrinology* 129, 1840–1848.
554. Korte, J. M., and Isola, J. J. (1988). An immunocytochemical study of the progesterone receptor in rabbit ovary. *Mol. Cell. Endocrinol.* 58, 93–101.
555. Van den Broeck, W., D'Haeseleer, M., Coryn, M., and Simoens, P. (2002). Cell-specific distribution of progesterone receptors in the bovine ovary. *Reprod. Domest. Anim.* 37, 314–320.
556. Slomczynska, M., Krok, M., and Pierscinski, A. (2000). Localization of the progesterone receptor in the porcine ovary. *Acta. Histochem.* 102, 183–191.
557. Chandrasekher, Y. A., Melner, M. H., Nagalla, S. R., and Stouffer, R. L. (1994). Progesterone receptor, but not estradiol receptor, messenger ribonucleic acid is expressed in luteinizing granulosa cells and the corpus luteum in rhesus monkeys. *Endocrinology* 135, 307–314.
558. Suzuki, T., Sasano, H., Kimura, N., Tamura, M., Fukaya, T., Yajima, A., and Nagura, H. (1994). Immunohistochemical distribution of progesterone, androgen and oestrogen receptors in the human ovary during the menstrual cycle: relationship to expression of steroidogenic enzymes. *Hum. Reprod.* 9, 1589–1595.
559. Revelli, A., Pacchioni, D., Cassoni, P., Bussolati, G., and Massobrio, M. (1996). In situ hybridization study of messenger RNA for estrogen receptor and immunohistochemical detection of estrogen and progesterone receptors in the human ovary. *Gynecol. Endocrinol.* 10, 177–186.
560. Park, O.-K., and Mayo, K. E. (1991). Transient expression of progesterone receptor messenger RNA in ovarian granulosa cells after the preovulatory luteinizing hormone surge. *Mol. Endocrinol.* 5, 967–978.
561. Iwai, M., Yasuda, K., Fukuoka, M., Iwai, T., Takakura, K., Taii, S., Nakanishi, S., and Mori, T. (1991). Luteinizing hormone induces progesterone receptor gene expression in cultured porcine granulosa cells. *Endocrinology* 129, 1621–1627.
562. Boerboom, D., Russell, D. L., Richards, J. S., and Sirois, J. (2003). Regulation of transcripts encoding ADAMTS-1 (a disintegrin and metalloproteinase with thrombospondin-like motifs-1) and progesterone receptor by human chorionic gonadotropin in equine preovulatory follicles. *J. Mol. Endocrinol.* 31, 473–485.
563. Conneely, O. M., and Lydon, J. P. (2000). Progesterone receptors in reproduction: functional impact of the A and B isoforms. *Steroids* 65, 571–577.
564. Duffy, D. M., Molskness, T. A., and Stouffer, R. L. (1996). Progesterone receptor messenger ribonucleic acid and protein in luteinized granulosa cells of rhesus monkeys are regulated in vitro by gonadotropins and steroids. *Biol. Reprod.* 54, 888–895.
565. Arakawa, S., Iyo, M., Ohkawa, R., Kambegawa, A., Okinaga, S., and Arai, K. (1989). Steroid hormone receptors in the uterus and ovary of immature rats treated with gonadotropins. *Endocrinol. Jpn.* 36, 219–228.
566. Robker, R. L., Russell, D. L., Espey, L. L., Lydon, J. P., O'Malley, B. W., and Richards, J. S. (2000). Progesterone-regulated genes in the ovulation process: ADAMTS-1 and cathepsin L proteases. *Proc. Natl. Acad. Sci. U S A* 97, 4689–4694.
567. Natraj, U., and Richards, J. S. (1993). Hormonal regulation, localization, and functional activity of the progesterone receptor in granulosa cells of rat preovulatory follicles. *Endocrinology* 133, 761–769.
568. Sriraman, V., Sharma, S. C., and Richards, J. S. (2003). Transactivation of the progesterone receptor gene in granulosa cells: evidence that Sp1/Sp3 binding sites in the proximal promoter play a key role in luteinizing hormone inducibility. *Mol. Endocrinol.* 17, 436–449.
569. Park-Sarge, O. K., and Sarge, K. D. (1995). Cis-regulatory elements conferring cyclic 3′,5′-adenosine monophosphate responsiveness of the progesterone receptor gene in transfected rat granulosa cells. *Endocrinology* 136, 5430–5437.
570. Smith, B. D., and Bradbury, J. T. (1966). Influence of progestins on ovarian responses to estrogen and gonadotrophins in immature rats. *Endocrinology* 78, 297–301.
571. Chappell, P. E., Lydon, J. P., Conneely, O. M., O'Malley, B. W., and Levine, J. E. (1997). Endocrine effects in mice carrying null mutation for the progesterone receptor gene. *Endocrinology* 138, 4147–4152.
572. Freeman, M. E. (1994). The ovarian cycle of the rat. In *The Physiology of Reproduction* (E. Knobil and J. D. Neill, Eds.), pp. 1893–1928. Raven Press, New York.
573. Mori, T., Suzuki, A., Nishimura, T., and Kambegawa, A. (1977). Inhibition of ovulation in immature rats by anti-progesterone antiserum. *J. Endocrinol.* 73, 185–186.
574. Lipner, H. (1994). Mechanism of mammalian ovulation. In *The Physiology of Reproduction* (E. Knobil and J. D. Neill, Eds.), pp. 447–488. Raven Press, New York.
575. Brannstrom, M., and Janson, P. O. (1989). Progesterone is a mediator in the ovulatory process of the in vitro-perfused rat ovary. *Biol. Reprod.* 40, 1170–1178.
576. Tanaka, N., Espey, L. L., Kawano, T., and Okamura, H. (1991). Comparison of inhibitory actions of indomethacin and epostane on ovulation in rats. *Am. J. Physiol.* 260, E170–E174.
577. Espey, L. L., Yoshioka, S., Russell, D. L., Robker, R. L., Fujii, S., and Richards, J. S. (2000). Ovarian expression of a disintegrin and metalloproteinase with thrombospondin motifs during ovulation in the gonadotropin-primed immature rat. *Biol. Reprod.* 62, 1090–1095.
578. Brannstrom, M. (1993). Inhibitory effect of mifepristone (RU 486) on ovulation in the isolated perfused rat ovary. *Contraception* 48, 393–402.
579. Loutradis, D., Bletsa, R., Aravantinos, L., Kallianidis, K., Michalas, S., and Psychoyos, A. (1991). Preovulatory effects of the progesterone antagonist mifepristone (RU486) in mice. *Hum. Reprod.* 6, 1238–1240.

580. Espey, L. L., Adams, R. F., Tanaka, N., and Okamura, H. (1990). Effects of epostane on ovarian levels of progesterone, 17 beta-estradiol, prostaglandin E2, and prostaglandin F2 alpha during ovulation in the gonadotropin-primed immature rat. *Endocrinology* 127, 259–263.
581. Shimada, M., Nishibori, M., Yamashita, Y., Ito, J., Mori, T., and Richards, J. S. (2004). Down-regulated expression of A disintegrin and metalloproteinase with thrombospondin-like repeats-1 by progesterone receptor antagonist is associated with impaired expansion of porcine cumulus-oocyte complexes. *Endocrinology* 145, 4603–4614.
582. Young, K. A., Tumlinson, B., and Stouffer, R. L. (2004). ADAMTS-1/METH-1 and TIMP-3 expression in the primate corpus luteum: divergent patterns and stage-dependent regulation during the natural menstrual cycle. *Mol. Hum. Reprod.* 10, 559–565.
583. Sriraman, V., and Richards, J. S. (2004). Cathepsin L gene expression and promoter activation in rodent granulosa cells. *Endocrinology* 145, 582–591.
584. Doyle, K. M., Russell, D. L., Sriraman, V., and Richards, J. S. (2004). Coordinate transcription of the ADAMTS-1 gene by luteinizing hormone and progesterone receptor. *Mol. Endocrinol.* 18, 2463–2478.
585. Mittaz, L., Russell, D. L., Wilson, T., Brasted, M., Tkalcevic, J., Salamonsen, L. A., Hertzog, P. J., and Pritchard, M. A. (2004). Adamts-1 is essential for the development and function of the urogenital system. *Biol. Reprod.* 70, 1096–1105.
586. Tetsuka, M., Whitelaw, P. F., Bremner, W. J., Millar, M. R., Smyth, C. D., and Hillier, S. G. (1995). Developmental regulation of androgen receptor in rat ovary. *J. Endocrinol.* 145, 535–543.
587. Szoltys, M., and Slomczynska, M. (2000). Changes in distribution of androgen receptor during maturation of rat ovarian follicles. *Exp. Clin. Endocrinol. Diabetes.* 108, 228–234.
588. Slomczynska, M., and Szoltys, M. (1997). Immunohistochemical localization of androgen receptor (AR) in rat ovary. *Folia. Histochem. Cytobiol.* 35, 101–102.
589. Garrett, W. M., and Guthrie, H. D. (1996). Expression of androgen receptors and steroidogenic enzymes in relation to follicular growth and atresia following ovulation in pigs. *Biol. Reprod.* 55, 949–955.
590. Cardenas, H., and Pope, W. F. (2002). Androgen receptor and follicle-stimulating hormone receptor in the pig ovary during the follicular phase of the estrous cycle. *Mol. Reprod. Dev.* 62, 92–98.
591. Campo, S. M., Carson, R. S., and Findlay, J. K. (1985). Distribution of specific androgen binding sites within the ovine ovarian follicle. *Mol. Cell. Endocrinol.* 39, 255–265.
592. Hampton, J. H., Manikkam, M., Lubahn, D. B., Smith, M. F., and Garverick, H. A. (2004). Androgen receptor mRNA expression in the bovine ovary. *Domest. Anim. Endocrinol.* 27, 81–88.
593. Duffy, D. M., Abdelgadir, S. E., Stott, K. R., Resko, J. A., Stouffer, R. L., and Zelinski-Wooten, M. B. (1999). Androgen receptor mRNA expression in the rhesus monkey ovary. *Endocrine* 11, 23–30.
594. Hillier, S. G., Tetsuka, M., and Fraser, H. M. (1997). Location and developmental regulation of androgen receptor in primate ovary. *Hum. Reprod.* 12, 107–111.
595. Weil, S. J., Vendola, K., Zhou, J., Adesanya, O. O., Wang, J., Okafor, J., and Bondy, C. A. (1998). Androgen receptor gene expression in the primate ovary: cellular localization, regulation, and functional correlations. *J. Clin. Endocrinol. Metab.* 83, 2479–2485.
596. Milwidsky, A., Younes, M. A., Besch, N. F., Besch, P. K., and Kaufman, R. H. (1980). Receptor-like binding proteins for testosterone and progesterone in the human ovary. *Am. J. Obstet. Gynecol.* 138, 93–98.
597. Horie, K., Takadura, K., Fujiwara, H., Suginami, H., Liao, S., and Mori, T. (1992). Immunohistochemical localization of androgen receptor in the human ovary throughout the menstrual cycle in relation to oestrogen and progesterone receptor expression. *Human Reproduction.* 7, 284–190.
598. Chadha, S., Pache, T. D., Huikeshoven, J. M., Brinkmann, A. O., and van der Kwast, T. H. (1994). Androgen receptor expression in human ovarian and uterine tissue of long-term androgen-treated transsexual women. *Hum. Pathol.* 25, 1198–1204.
599. Takayama, K., Fukaya, T., Sasano, H., Funayama, Y., Suzuki, T., Takaya, R., Wada, Y., and Yajima, A. (1996). Immunohistochemical study of steroidogenesis and cell proliferation in polycystic ovarian syndrome. *Hum. Reprod.* 11, 1387–1392.
600. Kaipia, A., and Hsueh, A. J. (1997). Regulation of ovarian follicle atresia. *Annu. Rev. Physiol.* 59, 349–363.
601. Campo, S., Carson, R. S., and Findlay, J. K. (1992). Acute effect of PMSG on ovarian androgen-binding sites in the intact immature female rat. *Reprod. Fertil. Dev.* 4, 55–65.
602. Hirshfield, A. N. (1986). Patterns of [^{3}H] thymidine incorporation differ in immature rats and mature, cycling rats. *Biol. Reprod.* 34, 229–235.
603. Weil, S., Vendola, K., Zhou, J., and Bondy, C. A. (1999). Androgen and follicle-stimulating hormone interactions in primate ovarian follicle development. *J. Clin. Endocrinol. Metab.* 84, 2951–2956.
604. Yeh, S., Hu, Y. C., Wang, P. H., Xie, C., Xu, Q., Tsai, M. Y., Dong, Z., Wang, R. S., Lee, T. H., and Chang, C. (2003). Abnormal mammary gland development and growth retardation in female mice and MCF7 breast cancer cells lacking androgen receptor. *J. Exp. Med.* 198, 1899–1908.
605. Payne, R. W., Hellbaum, A. A., and Owens, J. N., Jr. (1956). The effect of androgens on the ovaries and uterus of the estrogen treated hypophysectomized immature rat. *Endocrinology* 59, 306–316.
606. Payne, R. W., and Runser, R. H. (1958). The influence of estrogen and androgen on the ovarian response of hypophysectomized immature rats to gonadotropins. *Endocrinology* 62, 313–321.
607. Conway, B. A., Mahesh, V. B., and Mills, T. M. (1990). Effect of dihydrotestosterone on the growth and function of ovarian follicles in intact immature female rats primed with PMSG. *J. Reprod. Fertil.* 90, 267–277.
608. Hillier, S. G., and Ross, G. T. (1979). Effects of exogenous testosterone on ovarian weight, follicular morphology and intraovarian progesterone concentration in estrogen-primed hypophysectomized immature female rats. *Biol. Reprod.* 20, 261–268.
609. Louvet, J. P., Harman, S. M., Schrieber, J. R., and Ross, G. T. (1975). Evidence of a role of androgens in follicular maturation. *Endocrinology* 97, 366–372.
610. Pradeep, P. K., Li, X., Peegel, H., and Menon, K. M. (2002). Dihydrotestosterone inhibits granulosa cell proliferation by decreasing the cyclin D2 mRNA expression and cell cycle arrest at G1 phase. *Endocrinology* 143, 2930–2935.
611. Hickey, T. E., Marrocco, D. L., Gilchrist, R. B., Norman, R. J., and Armstrong, D. T. (2004). Interactions between androgen and growth factors in granulosa cell subtypes of porcine antral follicles. *Biol. Reprod.* 71, 45–52.
612. Vendola, K. A., Zhou, J., Adesanya, O. O., Weil, S. J., and Bondy, C. A. (1998). Androgens stimulate early stages of follicular growth in the primate ovary. *J. Clin. Invest.* 101, 2622–2629.
613. Daniel, S. A., and Armstrong, D. T. (1980). Enhancement of follicle-stimulating hormone-induced aromatase activity by androgens in cultured rat granulosa cells. *Endocrinology* 107, 1027–1033.

614. Katz, Y., Leung, P. C., and Armstrong, D. T. (1979). Testosterone restores ovarian aromatase activity in rats treated with a 17,20-lyase inhibitor. *Mol. Cell. Endocrinol.* 14, 37–44.
615. Murray, A. A., Gosden, R. G., Allison, V., and Spears, N. (1998). Effect of androgens on the development of mouse follicles growing in vitro. *J. Reprod. Fertil.* 113, 27–33.
616. Harlow, C. R., Hillier, S. G., and Hodges, J. K. (1986). Androgen modulation of follicle-stimulating hormone-induced granulosa cell steroidogenesis in the primate ovary. *Endocrinology* 119, 1403–1405.
617. Harlow, C. R., Shaw, H. J., Hillier, S. G., and Hodges, J. K. (1988). Factors influencing follicle-stimulating hormone-responsive steroidogenesis in marmoset granulosa cells: effects of androgens and the stage of follicular maturity. *Endocrinology* 122, 2780–2787.
618. Daniel, S. A., and Armstrong, D. T. (1984). Site of action of androgens on follicle-stimulating hormone-induced aromatase activity in cultured rat granulosa cells. *Endocrinology* 114, 1975–1982.
619. Fitzpatrick, S. L., and Richards, J. S. (1994). Identification of a cyclic adenosine 3′,5′-monophosphate-response element in the rat aromatase promoter that is required for transcriptional activation in rat granulosa cells and R2C leydig cells. *Mol. Endocrinol.* 8, 1309–1319.
620. Farookhi, R. (1980). Effects of androgen on induction of gonadotropin receptors and gonadotropin-stimulated adenosine 3′,5′- monophosphate production in rat ovarian granulosa cells. *Endocrinology* 106, 1216–1223.
621. Hillier, S. G., and de Zwart, F. A. (1982). Androgen/antiandrogen modulation of cyclic AMP-induced steroidogenesis during granulosa cell differentiation in tissue culture. *Mol. Cell. Endocrinol.* 28, 347–361.
622. Wang, X.-N., and Greenwald, G. S. (1993). Hypophysectomy of the cyclic mouse. I. Effects on folliculogenesis, oocyte growth, and follicle-stimulating hormone and human chorionic gonadotropin receptors. *Biol. Reprod.* 48, 585–594.
623. Wang, X. N., and Greenwald, G. S. (1993). Hypophysectomy of the cyclic mouse. II. Effects of follicle-stimulating hormone (FSH) and luteinizing hormone on folliculogenesis, FSH and human chorionic gonadotropin receptors, and steroidogenesis. *Biol. Reprod.* 48, 595–605.
624. Whitelaw, P. F., Smyth, C. D., Howles, C. M., and Hillier, S. G. (1992). Cell-specific expression of aromatase and LH receptor mRNAs in rat ovary. *J. Mol. Endocrinol.* 9, 309–312.
625. Zeleznik, A. J., Hillier, S. G., and Ross, G. T. (1979). Follicle stimulating hormone-induced follicular development: an examination of the role of androgens. *Biol. Reprod.* 21, 673–681.
626. Simone, D. A., Chorich, L. P., and Mahesh, V. B. (1993). Mechanisms of action for an androgen-mediated autoregulatory process in rat thecal-interstitial cells. *Biol. Reprod.* 49, 1190–1201.
627. Mori, T., Suzuki, A., Nishimura, T., and Kambegawa, A. (1977). Evidence for androgen participation in induced ovulation in immature rats. *Endocrinology* 101, 623–626.
628. Peluso, J. J., Stude, D., and Steger, R. W. (1980). Role of androgens in hCG-induced ovulation in PMSG-primed immature rats. *Acta. Endocrinol. (Copenh).* 93, 505–512.
629. Ware, V. C. (1982). The role of androgens in follicular development in the ovary. I. A quantitative analysis of oocyte ovulation. *J. Exp. Zool.* 222, 155–167.
630. Markstrom, E., Svensson, E., Shao, R., Svanberg, B., and Billig, H. (2002). Survival factors regulating ovarian apoptosis—dependence on follicle differentiation. *Reproduction* 123, 23–30.
631. Azzolin, G. C., and Saiduddin, S. (1983). Effect of androgens on the ovarian morphology of the hypophysectomized rat. *Proc. Soc. Exp. Biol. Med.* 172, 70–73.
632. Billig, H., Furuta, I., and Hsueh, A. J. W. (1993). Estrogens inhibit and androgens enhance ovarian granulosa cell apoptosis. *Endocrinology* 133, 2204–2212.
633. Shupnik, M. A. (1996). Gonadal hormone feedback on pituitary gonadotropin genes. *TEM* 7, 272–276.
634. Belchetz, P. E., Plant, T. M., Nakai, Y., Keogh, E. J., and Knobil, E. (1978). Hypophysial responses to continuous and intermittent delivery of hypothalamic gonadotropin-releasing hormone. *Science* 202, 631–633.
635. Veldhuis, J. D. (1990). The hypothalamic pulse generator: the reproductive core. *Clin. Obstet. Gynecol.* 33, 538–550.
636. MacLusky, N. J., Leiberburg, I., and McEwen, B. S. (1979). Development of steroid receptor systems in the rodent brain. In *Ontogeny of Receptors and Reproductive Hormone Action* (T. H. Hamilton, J. H. Clark, and W. A. Sadler, Eds.), pp. 393–402. Raven Press, New York.
637. McEwen, B. S. (1992). Effects of the steroid/thyroid hormone family on neural and behavioral plasticity. In *Neuroendocrinology* (C. B. Nemeroff, Ed.), pp. 333–351. CRC Press, Boca Raton, FL.
638. Mahesh, V. B., and Brann, D. W. (1998). Regulation of the preovulatory gonadotropin surge by endogenous steroids. *Steroids* 63, 616–629.
639. McEwen, B. S. (1992). Steroid hormones: effect on brain development and function. *Hormone Res.* 37 Suppl 3, 1–10.
640. Shughrue, P., Scrimo, P., Lane, M., Askew, R., and Merchenthaler, I. (1997). The distribution of estrogen receptor-β mRNA in forebrain regions of the estrogen receptor-α knockout mouse. *Endocrinology* 138, 5649–5652.
641. Laflamme, N., Nappi, R. E., Drolet, G., Labrie, C., and Rivest, S. (1998). Expression and neuropeptidergic characterization of estrogen receptors (ERalpha and ERbeta) throughout the rat brain: anatomical evidence of distinct roles of each subtype. *J. Neurobiol.* 36, 357–378.
642. Stefaneanu, L. (1997). Pituitary sex steroid receptors: localization and function. *Endocrine Pathol.* 8, 91–108.
643. Register, T. C., Shively, C. A., and Lewis, C. E. (1998). Expression of estrogen receptor a and b transcripts in female monkey hippocampus and hypothalamus. *Brain Res.* 788, 320–322.
644. Levine, J. E., Chappell, P. E., Schneider, J. S., Sleiter, N. C., and Szabo, M. (2001). Progesterone receptors as neuroendocrine integrators. *Front Neuroendocrinol.* 22, 69–106.
645. MacLusky, N. J., and McEwen, B. S. (1980). Progestin receptors in rat brain: distribution and properties of cytoplasmic progestin-binding sites. *Endocrinology* 106, 192–202.
646. Haisenleder, D. J., Dalkin, A. C., and Marshall, J. C. (1994). Regulation of gonadotropin gene expression. In *The Physiology of Reproduction* (E. Knobil and J. D. Neill, Eds.), pp. 1793–1813. Raven Press, New York.
647. Petersen, S. L., Ottem, E. N., and Carpenter, C. D. (2003). Direct and indirect regulation of gonadotropin-releasing hormone neurons by estradiol. *Biol. Reprod.* 69, 1771–1778.
648. Gharib, S. D., Wierman, M. E., Shupnik, M. A., and Chin, W. W. (1990). Molecular biology of the pituitary gonadotropins. *Endocr. Rev.* 11, 177–199.
649. Keri, R. A., Andersen, B., Kennedy, G. C., Hamernik, D. L., Clay, C. M., Brace, A. D., Nett, T. M., Notides, A. C., and Nilson, J. H. (1991). Estradiol inhibits transcription of the human glycoprotein hormone α-subunit gene despite the absence of a high affinity binding site for estrogen receptor. *Mol. Endocrinol.* 5, 725–733.

650. Keri, R. A., Wolfe, M. W., Sauders, T. L., Anderson, I., Kendall, S. K., Wagner, T., Yeung, J., Gorski, J., Nett, T. M., Camper, S. A., and Nilson, J. H. (1994). The proximal promoter of the bovine luteinizing hormone β-subunit gene confers gonadotrope-specific expression and regulation by gonadotropin-releasing hormone, testosterone, and 17β-estradiol in transgenic mice. *Mol. Endocrinol.* 8, 1807–1816.
651. Dorling, A. A., Todman, M. G., Korach, K. S., and Herbison, A. E. (2003). Critical role for estrogen receptor alpha in negative feedback regulation of gonadotropin-releasing hormone mRNA expression in the female mouse. *Neuroendocrinology* 78, 204–209.
652. Lindzey, J., Couse, J. F., Stoker, T., Wetsel, W. C., Cooper, R., and Korach, K. S. (1998). Steroid regulation of gonadotrope function in female wild-type (WT) and estrogen receptor-α knockout (ERKO) mice. Endocrine Society, New Orleans, p. 112.
653. Woodruff, T. K., and Mather, J. P. (1995). Inhibin, activin and the female reproductive axis. *Annu. Rev. Physiol.* 57, 219–244.
654. Matzuk, M. M., Kumar, T. R., and Bradley, A. (1995). Different phenotypes for mice deficient in either activins or activin receptor type II. *Nature* 374, 356–360.
655. Kaiser, U. B., Sabbagh, E., Katzenellenbogen, B. S., Conn, P. M., and Chin, W. W. (1995). A mechanism for the differential regulation of gonadotropin subunit gene expression by gonadotropin-releasing hormone. *Proc. Natl. Acad. Sci. U S A* 92, 12280–12284.
656. Fink, G. (1988). Gonadotropin secretion and its control. In *The Physiology of Reproduction* (E. Knobil and J. D. Neill, Eds.), pp. 1349–1377. Raven Press, New York.
657. Nakai, Y., Plant, T. M., Hess, D. L., Keogh, E. J., and Knobil, E. (1978). On the sites of negative and positive feedback actions of estradiol in the control of gonadotropin secretion in the rhesus monkey. *Endocrinology* 102, 1008–1014.
658. Karsch, F. J. (1984). The hypothalamus and anterior pituitary gland. In *Reproduction in Mammals: Hormonal Control of Reproduction* (C. R. Austin and R. V. Short, Eds.), pp. 1–20. Cambridge University Press, Cambridge.
659. Mann, R. J., Keri, R. A., and Nilson, J. H. (1999). Transgenic mice with chronically elevated luteinizing hormone are infertile due to anovulation, defects in uterine receptivity, and midgestation pregnancy failure. *Endocrinology* 140, 2592–2601.
660. Orikasa, C., and Sakuma, Y. (2003). Possible involvement of preoptic estrogen receptor beta positive cells in luteinizing hormone surge in the rat. *Domest. Anim. Endocrinol.* 25, 83–92.
661. Patchev, A. V., Gotz, F., and Rohde, W. (2004). Differential role of estrogen receptor isoforms in sex-specific brain organization. *FASEB J.* 18, 1568–1570.
662. Goodman, R. L. (1994). Neuroendocrine control of the ovine estrous cycle. In *The Physiology of Reproduction* (E. Knobil and J. D. Neill, Eds.), pp. 1919–1969. Raven Press, New York.
663. Knobil, E., and Hotchkiss, J. (1994). The menstrual cycle and its neuroendocrine control. In *The Physiology of Reproduction* (E. Knobil and J. D. Neill, Eds.), pp. 1971–2021. Raven Press, New York.
664. Levine, J. E. (1997). New concepts of the neuroendocrine regulation of gonadotropin surges in rats. *Biol. Reprod.* 56, 293–302.
665. Chappell, P. E., Schneider, J. S., Kim, P., Xu, M., Lydon, J. P., O'Malley, B. W., and Levine, J. E. (1999). Absence of gonadotropin surges and gonadotropin-releasing hormone self-priming in ovariectomized (OVX), estrogen (E2)-treated, progesterone receptor knockout (PRKO) mice. *Endocrinology* 140, 3653–3658.
666. Chappell, P. E., Lee, J., and Levine, J. E. (2000). Stimulation of gonadotropin-releasing hormone surges by estrogen. II. Role of cyclic adenosine 3′5′-monophosphate. *Endocrinology* 141, 1486–1492.
667. Chappell, P. E., and Levine, J. E. (2000). Stimulation of gonadotropin-releasing hormone surges by estrogen. I. Role of hypothalamic progesterone receptors. *Endocrinology* 141, 1477–1485.
668. Bole-Feysot, C., Goffin, V., Edery, M., Binart, N., and Kelly, P. A. (1998). Prolactin (PRL) and its receptor: actions, signal transduction pathways and phenotypes observed in PRL receptor knockout mice. *Endocr. Rev.* 19, 225–268.
669. Richards, J. S., Fitzpatrick, S. L., Clemens, J. W., Morris, J. K., Alliston, T., and Sirois, J. (1995). Ovarian cell differentiation: a cascade of multiple hormones, cellular signals, and regulated genes. *Recent Prog. Horm. Res.* 50, 223–254.
670. Maurer, R. A., Kim, K. E., Day, R. N., and Notides, A. C. (1990). Regulation of prolactin gene expression by estradiol. *Prog. Clin. Biol. Res.* 322, 159–169.
671. Wilson, M. E., Price Jr., R. H., and Handa, R. J. (1998). Estrogen receptor-b messenger ribonucleic acid expression in the pituitary gland. *Endocrinology* 139, 5151–5156.
672. Mitchner, N. A., Garlick, C., and Ben-Jonathan, N. (1998). Cellular distribution and gene regulation of estrogen receptors a and b in the rat pituitary gland. *Endocrinology* 139, 3976–3983.
673. Scully, K. M., Gleiberman, A. S., Lindzey, J., Lubahn, D. B., Korach, K. S., and Rosenfeld, M. G. (1997). Role of estrogen receptor alpha in the anterior pituitary gland. *Mol. Endocrinol.* 11, 674–681.
674. Bocchinfuso, W. P., Lindzey, J. K., Hewitt, S. C., Clark, J. A., Myers, P. H., Cooper, R., and Korach, K. S. (2000). Induction of mammary gland development in estrogen receptor-alpha knockout mice. *Endocrinology* 141, 2982–2994.
675. Das, S. K., Chakraborty, I., Paria, B. C., Wang, X. N., Plowman, G., and Dey, S. K. (1995). Amphiregulin is an implantation-specific and progesterone-regulated gene in the mouse uterus. *Mol. Endocrinol.* 9, 691–705.
676. Ding, Y. Q., Zhu, L. J., Bagchi, M. K., and Bagchi, I. C. (1994). Progesterone stimulates calcitonin gene expression in the uterus during implantation. *Endocrinology* 135, 2265–2274.
677. Huet-Hudson, Y. M., Andrews, G. K., and Dey, S. K. (1989). Cell type-specific localization of c-myc protein in the mouse uterus: modulation by steroid hormones and analysis of the periimplantation period. *Endocrinology* 125, 1683–1690.
678. Orsino, A., Taylor, C. V., and Lye, S. J. (1996). Connexin-26 and connexin-43 are differentially expressed and regulated in the rat myometrium throughout late pregnancy and with the onset of labor. *Endocrinology* 137, 1545–1553.
679. Yue, L., Daikoku, T., Hou, X., Li, M., Wang, H., Noji, H., Dey, S. K., and Das, S. K. (2005). Cyclin G1 and cyclin G2 are expressed in the periimplantation mouse uterus in a cell-specific and progesterone-dependent manner: Evidence for aberrant regulation with Hoxa-10 deficiency. *Endocrinology* 146, 2424–2433.
680. Giudice, L. C. (1999). Genes associated with embryonic attachment and implantation and the role of progesterone. *J. Reprod. Med.* 44, 165–171.
681. Zhang, Z., Funk, C., Glasser, S. R., and Mulholland, J. (1994). Progesterone regulation of heparin-binding epidermal growth factor-like growth factor gene expression during sensitization and decidualization in the rat uterus: effects of the antiprogestin, ZK 98.299. *Endocrinology* 135, 1256–1263.

682. Paria, B. C., Das, N., Das, S. K., Zhao, X., Dileepan, K. N., and Dey, S. K. (1998). Histidine decarboxylase gene in the mouse uterus is regulated by progesterone and correlates with uterine differentiation for blastocyst implantation. *Endocrinology* 139, 3958–3966.
683. Ma, L., Benson, G. V., Lim, H., Dey, S. K., and Maas, R. L. (1998). Abdominal B (AbdB) Hoxa genes: regulation in adult uterus by estrogen and progesterone and repression in mullerian duct by the synthetic estrogen diethylstilbestrol (DES). *Dev. Biol.* 197, 141–154.
684. Taylor, H. S., Arici, A., Olive, D., and Igarashi, P. (1998). HOXA10 is expressed in response to sex steroids at the time of implantation in the human endometrium. *J. Clin. Invest.* 101, 1379–1384.
685. Daikoku, T., Matsumoto, H., Gupta, R. A., Das, S. K., Gassmann, M., Dubois, R. N., and Dey, S. K. (2003). Expression of hypoxia-inducible factors in the peri-implantation mouse uterus is regulated in a cell-specific and ovarian steroid hormone-dependent manner—evidence for differential function of HIFs during early pregnancy. *J. Biol. Chem.* 278, 7683–7691.
686. O'Sullivan, C. M., Ungarian, J. L., Singh, K., Liu, S., Hance, J., and Rancourt, D. E. (2004). Uterine secretion of ISP1 & 2 tryptases is regulated by progesterone and estrogen during pregnancy and the endometrial cycle. *Mol. Reprod. Dev.* 69, 252–259.
687. Kapur, S., Tamada, H., Dey, S. K., and Andrews, G. K. (1992). Expression of insulin-like growth factor-I (IGF-I) and its receptor in the peri-implantation mouse uterus, and cell-specific regulation of IGF- I gene expression by estradiol and progesterone. *Biol. Reprod.* 46, 208–219.
688. Hewetson, A., and Chilton, B. S. (1997). Molecular cloning and hormone-dependent expression of rabbit Muc1 in the cervix and uterus. *Biol. Reprod.* 57, 468–477.
689. Isomaa, V. V., Ghersevich, S. A., Maentausta, O. K., Peltoketo, E. H., Poutanen, M. H., and Vihko, R. K. (1993). Steroid biosynthetic enzymes: 17 beta-hydroxysteroid dehydrogenase. *Ann. Med.* 25, 91–97.
690. Lim, H., and Dey, S. K. (1997). Prostaglandin E2 receptor subtype EP2 gene expression in the mouse uterus coincides with differentiation of the luminal epithelium for implantation. *Endocrinology* 138, 4599–4606.
691. Simmen, R. C., Simmen, F. A., and Bazer, F. W. (1991). Regulation of synthesis of uterine secretory proteins: evidence for differential induction of porcine uteroferrin and antileukoproteinase gene expression. *Biol. Reprod.* 44, 191–200.
692. Chandra, T., Bullock, D. W., and Woo, S. L. (1981). Hormonally regulated mammalian gene expression: steady-state level and nucleotide sequence of rabbit uteroglobin mRNA. *DNA* 1, 19–26.

Knobil and Neill's Physiology of Reproduction, Third Edition
edited by Jimmy D. Neill,
Elsevier © 2006

CHAPTER 16

Physiology and Molecular Biology of the Relaxin Peptide Family

Ross A. D. Bathgate,[1] Aaron J. W. Hsueh,[2] and O. David Sherwood[3]

Introduction, 680
Discovery of Relaxin, INSL3, and Other Relaxin Family Hormones, 681
Original Isolation of the Relaxin Peptide, 681
Cloning of the Relaxin Genes, 684
Discovery and Cloning of Novel Relaxin Family Genes, 687
Evolution of Relaxin Peptide Family, 691
Chromosomal Locations, 692
INSL4 and H1 Relaxin-Equivalent Genes are Found Only in Primates, 693
Evolution of the Relaxin Genes, 693
Chemistry of Relaxin Peptides, 694
Biosynthesis and Processing of Relaxin Peptides, 694
Structure of Relaxin Peptides, 696
Chemical and Recombinant Synthesis, 696
Structure–Activity Relationships, 698
Relaxin Binding and Signaling, 698
Relaxin Binding Sites, 698
Binding Sites for Other Relaxin Family Peptides, 700
Relaxin Signaling, 700
Signaling by Other Relaxin Family Peptides, 702
Relaxin Family Peptide Receptors, 702
Sources and Secretion of Relaxin in the Female, 705
Pig, 706
Rat, 707
Human, 708
Primates, 710
Mouse, 711
Horse, 711
Ruminants, 712
Other Species, 713
Physiological Roles of Relaxin in Female and Male Reproductive Processes, 714
Pregnancy and Parturition, 714
Lactation, 726
Implantation, 728
Intraovarian Effects of Relaxin, 730
Male, 730
Physiological Roles of Relaxin in Nonreproductive Processes, 731
Fibrosis, 731
Wound Healing, 734
Cardiac Protection, 734
Other Actions, 735
Physiological Roles of INSL3, 735
Expression of the *INSL3* Gene, 735
Testis INSL3 Expression and Secretion, 738
Role of INSL3 in Testis Descent, 738
Gonadal Functions, 741
Physiological Roles of Other Relaxin Family Peptides, 741
Relaxin-3, a Putative Neuropeptide?, 741
INSLs 4 to 6, 742
Conclusion, 744
Acknowledgments, 745
References, 745

[1]The University of Melbourne, Howard Florey Institute, Victoria, Australia;
[2]Stanford University School of Medicine, Department of Obstetrics and Gynecology, Division of Reproductive Biology, Stanford, California;
[3]University of Illinois Urbana-Champaign, Department of Molecular and Integrative Physiology, Urbana, Illinois.

ABSTRACT

Relaxin was first identified in 1926 as a substance influencing the reproductive tract and was subsequently found to be a peptide hormone with a two-chain structure similar to insulin. It is now accepted that relaxin is a member of a family of peptide hormones that includes seven members in humans—three relaxin peptides (H1, H2, and H3) and insulin-like peptides (INSL) 3, 4, 5, and 6. All these genes evolved from an ancestral relaxin-3 gene in (lower) vertebrates. The reproductive hormone relaxin is the product of one gene in most species, *RLN1*. Among these genes, the physiological roles of relaxin and INSL3 have been more extensively studied. Relaxin-3 is likely an important neuropeptide but has no known roles in reproduction. In contrast, relaxin has vital physiological roles during pregnancy, although its actions vary between species. Hence, relaxin has essential actions on the cervix, pubic symphysis, vagina, uterus, and mammary apparatus during pregnancy. It is also a mediator of the important cardiovascular changes that occur during pregnancy in many species. Relaxin may also play a crucial role in implantation, especially in primates. In addition, relaxin has nonreproductive actions in wound healing, cardiac protection, and as an antifibrotic agent. INSL3 is essential for testis descent by promoting gubernacular development in the fetus. It also plays an important role in female and male germ cell maturation and survival, respectively. With the recent identification of relaxin family peptide receptors, future investigations should unravel the diverse actions of relaxin and relaxin family peptides and lead to novel therapeutic applications.

INTRODUCTION

F. L. Hisaw's interest in modifications of the pelvic girdle that many mammalian species undergo to facilitate giving birth to their young led to the discovery of relaxin. In 1926, Hisaw reported that the injection of serum from pregnant guinea pigs or rabbits into virgin guinea pigs shortly after estrus promoted a noticeable relaxation of the pubic ligament (1). The following year he found the relaxative substance in sow corpora lutea and rabbit placentas (2). In 1930, Hisaw and coworkers obtained a crude aqueous extract of the relaxative hormone from sow corpora lutea, and the hormone was named relaxin (3).

During the 1930s and most of the 1940s, there was little interest in relaxin and not much progress was made toward its isolation. Then, from the late 1940s through the 1950s, research on relaxin surged. Although researchers used impure porcine relaxin, they made pioneering discoveries concerning the biological effects of relaxin on the female reproductive tract of nonpregnant animals. Relaxin was found to promote elongation of the interpubic ligament in estrogen-primed mice (4,5), inhibit spontaneous contractions of the uterine myometrium in estrogen-primed guinea pigs (6), and promote cervical softening in estrogen-primed cattle (7). These biological effects, which have since been confirmed with highly purified relaxin preparations, providing valuable insight into probable physiological roles of relaxin during pregnancy and at parturition in several species.

Interest in relaxin lagged again during the 1960s and until the mid-1970s, when investigators began to describe straightforward techniques for isolating high-purity relaxin preparations. Hence, purified relaxin was isolated from numerous species and the availability of these relaxin preparations led to the first detailed studies on the chemistry and biology of relaxin. Highly purified relaxin was used to (a) determine the primary structure of the relaxin peptide from multiple species; (b) develop specific and sensitive homologous and heterologous radioimmunoassays which were used to determine blood levels and tissue sources of relaxin in many species; (c) determine target tissues and biological actions of relaxin in the female and male of many species; (d) develop specific monoclonal antibodies for multiple purposes, including studies to determine the biological effects of endogenous rat relaxin; and (e) investigate the mechanism of relaxin's action on myometrial cells and endometrial cells. Furthermore, highly purified porcine relaxin was used for the first clinical trials in humans.

In addition, information obtained from the primary structure of purified relaxin preparations was used with recombinant DNA techniques to clone the first relaxin genes. Cloning of the rat (8) and porcine (9) relaxin genes confirmed the peptide structures and highlighted that the peptides were produced as prohormones with a B-, C-, and A-chain structure like insulin. This was followed by the cloning of the first human relaxin gene, *RLN1* (10). Subsequently, another human relaxin gene was discovered, *RLN2* (11), which was demonstrated to be the gene encoding the relaxin peptide produced by the corpus luteum and circulating in the blood during pregnancy. This peptide is referred to as H2 relaxin, whereas the *RLN1* gene product is H1 relaxin. Because H2 relaxin is the circulating form of relaxin in humans, it is the functional ortholog of relaxins from other species. Relaxin genes were subsequently cloned from multiple species, confirming the peptide structures and demonstrating that only higher primates have two relaxin genes.

The discovery of the relaxin genes, the development of the relaxin knockout (KO) mouse, and the availability of recombinant H2 relaxin led to an acceleration in relaxin research and clinical development in the 1990s. Furthermore, the advent of molecular techniques as well as the availability of the expressed sequence tag (EST) and genome databases led to the discovery of five novel genes with high homology to relaxin. Hence, insulin-like peptide 3 (INSL3) (12) and INSL4 (13) were discovered by differential cloning projects, whereas INSL5 (14,15) and INSL6 (15–17) were identified by searching the EST databases. These novel peptide products, although not having a relaxin binding motif and unable to mimic relaxin's actions, have clearly coevolved with relaxin and are members of an extended peptide family. The subsequent discovery of an additional relaxin gene, *RLN3* (18), in the genome databases, and the discovery that this gene was the ancestor of the other known genes (19,20), firmly established the concept of the relaxin peptide family. The product of this third relaxin gene, relaxin-3, is highly expressed in the brain and is unlikely to be a circulating hormone. Hence, although relaxin-3 activates the relaxin receptor, it is unlikely to be involved with the reproductive functions of relaxin outlined in detail in this chapter. Throughout this chapter, *relaxin* refers to the product of the *RLN1* gene or, in the case of the human, the *RLN2* gene, as described previously.

Although relaxin was discovered in 1926, it was not until 2002 that the elusive relaxin receptor was finally identified. Thus, the leucine-rich, repeat-containing, G-protein–coupled receptor 7 (LGR7) was shown to be the relaxin receptor (21). A related receptor, LGR8, was demonstrated to be the receptor for INSL3 (22). Subsequently, two unrelated G-protein–coupled receptors (GPCR), GPCR135 and GPCR142, were shown to be the receptors for relaxin-3 (23) and INSL5 (24), respectively. Hence, there has been coevolution of ligand–receptor pairings in the relaxin peptide family and their receptors. In addition, there are some peptide- and species-specific interactions between ligands and receptors that may complicate the biology of the relaxin peptide family in some species.

Although relaxin-3 is the oldest member of the relaxin peptide family and likely has important functions in the brain, it has no known reproductive actions. Likewise, although INSL4, INSL5, and INSL6 are expressed in reproductive tissues, there is little known as to their potential reproductive roles. Therefore, this review, while outlining the discovery and evolution of the relaxin peptide family, focuses on the important reproductive actions of both relaxin and INSL3. No other polypeptide hormone has been demonstrated to be as rich as relaxin in the diversity of both its chemistry and physiology among species. This diversity, which includes relaxin's structure, source, secretory profile, and physiological effects during pregnancy, has influenced the organization and emphasis of this review.

The review first outlines the isolation of relaxin and the cloning of the relaxin and relaxin peptide family genes. This is followed by the chemistry of relaxin and INSL3 to highlight the differences in these peptides and the important variations in the structure of relaxin between species. We then outline early studies on relaxin binding sites that led to the identification and characterization of the relaxin peptide family receptors. Importantly, in view of the marked differences that exist among species, relaxin's sources and secretion are then described for individual species. This is followed by a detailed description of the physiological roles of relaxin in female and male reproductive processes. Some description of relaxin's important actions in nonreproductive processes is also included. The crucial roles of INSL3 in reproductive biology are then described, followed by a discussion on the other relaxin family peptides and their potential physiological roles. The review is concluded with comments on the state of the relaxin field and the outlook for the future of this intriguing peptide family.

For additional descriptions of the chemistry and biology of both relaxin and INSL3, the reader is referred to the previous version of this chapter (25), recent reviews on relaxin (26–39) and INSL3 (30,37,40–45), and proceedings books from international congresses on relaxin in 1994 (46), and relaxin and related peptides in 2000 (47) and 2004 (48).

DISCOVERY OF RELAXIN, INSL3, AND OTHER RELAXIN FAMILY HORMONES

Original Isolation of the Relaxin Peptide

Early Efforts to Isolate Relaxin (Before 1974)

In 1930, Fevold et al. (3) reported the first effort to isolate and characterize porcine relaxin. Corpora lutea from unselected sows were ground and extracted with hydrochloric acid (HCl) in alcohol, and the relaxin was enriched by pH adjustment and fractionations with alcohol, acetone, and ether. Although the hormone preparation was impure, these workers discovered that relaxin was probably a peptide: it was soluble in aqueous media, amphoteric, and inactivated by digestion with trypsin (3,49).

For nearly 50 years, essentially all efforts to isolate relaxin used ovaries from pregnant pigs because this source has a high content of relaxin bioactivity (50) and is relatively easy to acquire in large quantities. Until the late 1950s, efforts to isolate relaxin were hindered by the limited techniques available for isolating proteins (50–53). A second major obstacle was the lack of physicochemical techniques for determining the degree of purity of protein preparations. Instead, investigators relied heavily on the specific biological activity of relaxin preparations to assess their degree of purity. The first quantitative bioassay for relaxin used ovariectomized, estrogen-primed guinea pigs and defined a "guinea pig unit" (GPU) as the dose of relaxin required to induce unmistakable mobility of the pubic symphysis of two thirds of a group of 12 animals (51). This guinea pig pubic symphysis palpation bioassay had limited utility for determining potency estimates because it was cumbersome and based on the subjective and imprecise assessment of the degree of relaxin-induced mobility of the pubic symphysis.

During the late 1950s, investigators developed improved quantitative bioassays for relaxin. In 1959, Kroc et al. (54) described a bioassay based on the ability of relaxin to inhibit spontaneous contractions of the mouse uterus in vitro. In 1960, Steinetz et al. (55) developed an objective bioassay in which the interpubic ligament formed in estrogen-primed female mice in response to relaxin is transilluminated and precisely measured with a dissecting microscope fitted with an ocular micrometer. Most efforts to isolate porcine relaxin after 1960 used one of these two mouse bioassays to locate relaxin-containing fractions and to determine the specific bioactivity of relaxin preparations.

During the 1960s, numerous laboratories using new protein isolation techniques were able to obtain preparations of porcine relaxin that contained high specific biological activity (56–59). Although they did not precisely describe the physicochemical properties of their preparations, they determined correctly that porcine relaxin is a protein with a molecular weight between 4,000 and 10,000 Da (56,58,59), has a basic isoelectric point (58,59), and contains disulfide bonds essential for biological activity (58,60).

Subsequent Isolation and Characterization of Relaxin

Since the original research outlined previously, relaxin has been isolated and fully or partially characterized from the ovaries, corpora lutea, or placentas of numerous species as well as from the seminal plasma of humans. Yields of relaxin per kilogram equivalent of tissue source differ greatly among species. The concentrations of relaxin in pig and rat corpora lutea are much higher than they are in tissue sources in other species from which relaxin has been isolated. Therefore, with nearly all species, investigators developed isolation procedures applicable to kilogram quantities of the tissue source to obtain sufficient relaxin for chemical or physiological studies. Subsequent characterization of highly purified relaxin preparations revealed that the superficial features of relaxin, such as its size and two-chain composition, are similar among species, but the primary structure differed markedly. Distinct variation was found in the isoelectric points, amino acid contents, and specific bioactivities of relaxin preparations from different species.

This section describes the isolation and characterization of highly purified relaxin from various species, with particular emphasis on the pig, rat, and horse because much of the original research on the physiological actions of relaxin was conducted using preparations of these peptides.

Isolation of Porcine Relaxin. In 1974, Sherwood and O'Byrne (61) described a procedure for isolating relaxin in high yields from the domestic pig (*Sus scrofa*). They extracted relaxin from 1 kg of frozen ovaries obtained from sows containing fetuses with a crown–rump length of 10 cm or greater with an acid-acetone method (0.15 N HCl, 70% acetone). After gel filtration of the extract on Sephadex G-50 (Pharmacia; Piscataway, NJ), relaxin bioactivity was adsorbed to carboxymethyl cellulose and then eluted as three contiguous peaks designated CMB, Cma, and Cma′ by the addition of a linear salt gradient. The yields of each of the three relaxin preparations were approximately 35 mg/kg equivalent of ovarian tissue (61). Characterization studies demonstrated that the three porcine relaxin preparations were essentially homogeneous and that their structures were nearly identical—that is, they were microheterogeneous (61). The specific bioactivities of CMB, Cma, and Cma′ were high and equipotent. The amino acid compositions of CMB, Cma, and Cma′ were also in close agreement, and they contained no histidine, proline, or tyrosine. Sedimentation equilibrium analysis indicated that the molecular weights of the three relaxin preparations were approximately 6,000 Da, and gel filtration of reduced and carboxymethylated CMB, Cma, and Cma′ showed they consisted of two chains of similar size (designated A and B) linked by disulfide bonds. Complete amino acid sequence analysis of CMB, Cma, and Cma′ demonstrated that the microheterogeneity among these three preparations was attributable to slight differences in the length of C-termini of the their B chains (62) (Fig. 1).

Subsequently, many groups developed additional procedures for isolating highly purified porcine relaxin (63,64). Walsh and Niall (65) were able to limit suspected proteolysis during extraction protocols by use of a strongly acidic extraction and nearly eliminated the multiplicity of forms of purified porcine relaxin.

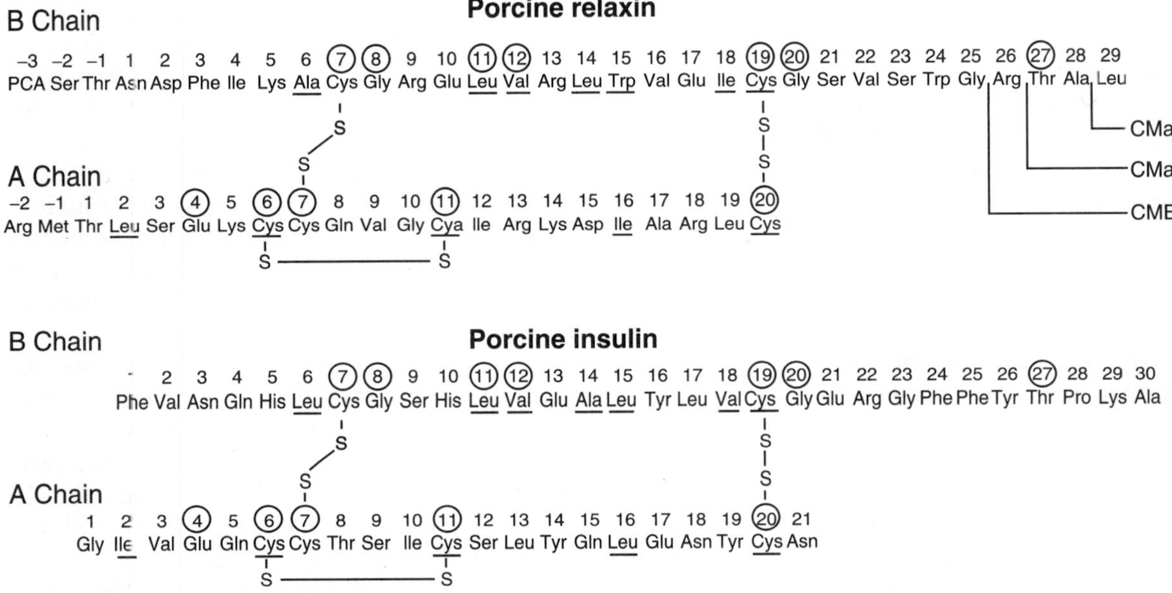

FIG. 1. Covalent structure of porcine relaxin and porcine insulin. Residues are numbered with respect to the insulin sequence to facilitate comparison of the two hormones. CMB, Cma, and Cma′ forms of porcine relaxin are attributable to small differences in the lengths of their B chains, as shown by *vertical lines*. The numbers of amino acids that are identical in relaxin and insulin are *circled*. Residues that contribute to the hydrophobic core of insulin and those in comparable positions in relaxin are *underlined*.

However, this procedure was not easily applied to the isolation of relaxin from kilogram batches of pig ovaries. There is reason to suspect that porcine relaxin is not normally processed to a major stored form in the corpora lutea. Büllesbach and Schwabe (66) concluded that in vivo conversion of porcine prorelaxin to relaxin may be much less stringently controlled than the conversion of other prohormones to active hormones. Nearly all studies conducted with highly purified porcine relaxin used relaxin obtained by the procedures of Sherwood and O'Byrne (61) or Büllesbach and Schwabe (66), or modifications of those procedures.

Isolation of Rat Relaxin. In 1979, a method for the isolation of relaxin from the rat (*Rattus norvegicus*) was first described (67). In contrast to the purification of porcine relaxin, two major forms of rat relaxin were isolated, designated CM1 and CM2. The yields of both CM1 and CM2 were approximately 140 mg/kg equivalent of ovarian tissue; therefore, the combined yield of rat relaxin was approximately 280 mg/kg equivalent of ovarian tissue. Characterization studies indicated that the two rat relaxin preparations were nearly homogeneous and microheterogeneous. No difference in biological activity between CM1 and CM2 was found with the mouse interpubic ligament bioassay, and the amino acid compositions of the two preparations were similar. Superficial structural features of rat relaxin were similar to those of porcine relaxin. Sedimentation equilibrium analysis indicated the molecular weights of CM1 and CM2 to be approximately 6,000 Da, and slab gel electrophoresis of reduced rat relaxin showed that it consisted of two chains of similar size linked by disulfide bonds. However, the amino acid composition of rat relaxin differed distinctly from that of porcine relaxin, and it was less potent than porcine relaxin in the mouse interpubic ligament bioassay (67,68).

The multiplicity of forms of rat relaxin appears to be attributable to limited proteolysis of a major stored form during the isolation procedure. As for porcine relaxin, extraction with strongly acidic solvent resulted in the isolation of a single form of rat relaxin (65). In view of the minimal microheterogeneity associated with relaxin obtained by the procedure of Walsh and Niall (65), and the similar yields obtained with this method, it is the method of choice for the isolation of rat relaxin.

Isolation of Equine Relaxin. In 1986, relaxin was isolated from the domestic horse (*Equus caballus*) (69). Relaxin was extracted from placentas obtained at term and resulted in the purification of one major (designated R1) and two minor products (designated R2 and R3). The yield of the predominant equine relaxin preparation R1 was only approximately 1.5 mg/kg equivalent of placentas, but it was sufficient for characterization studies. Slab gel electrophoresis of unreduced and reduced R1 indicated that its molecular weight was approximately 6,000 Da and that it consisted of two chains of similar size linked by disulfide bonds. The specific bioactivity of R1 as determined in the mouse interpubic ligament bioassay was far lower than that of porcine relaxin. The amino acid composition of equine relaxin differed from that of porcine, rat, and shark relaxin; it lacked phenylalanine and methionine. Stewart et al. (70) generated a monoclonal antibody (5F-7A) against equine relaxin preparation R1. The monoclonal antibody was used to produce an affinity column that was used for rapid and efficient isolation of equine relaxin. Four major and two minor isoforms of equine relaxin were obtained and mass spectrometry indicated that the molecular weights of the isoforms ranged from 5,170 to 5,570 Da, with the five predominant isoforms displaying biological activity in the mouse interpubic ligament bioassay. Sequence analysis demonstrated that the isoforms of equine relaxin are attributable to heterogeneity of the C-termini of the B chains, and two of the isoforms (AP-R1 and AP-R4) were used to determine the amino acid sequence of horse relaxin.

Isolation of Human Relaxin. Four source tissues were used over several years in efforts to isolate relaxin from the human (*Homo sapiens*). Attempts were made during the early to mid-1980s to purify relaxin from human corpora lutea (71), placentas (72–75), decidua (76), and seminal plasma (77). Although none of the human relaxin preparations was demonstrated to be highly purified, all displayed characteristics similar to those of porcine relaxin. Because of the low quantities of relaxin in the corpus luteum of pregnancy (71,78), isolation and characterization of human relaxin did not occur until the late 1980s and early 1990s.

In the late 1980s, Drolet et al. (79) and Winslow et al. (80) isolated nanogram quantities of human relaxin from corpora lutea surgically removed from women with ectopic pregnancies. Both groups were able to take advantage of the knowledge of the predicted amino acid sequence of human relaxin that was determined earlier by nucleotide sequence analysis (10,11). A synthetic human relaxin peptide based on these sequences was used to develop a monoclonal antibody that was used for affinity purification. In 1992, Winslow et al. (81) isolated nanogram quantities of human relaxin from seminal plasma using a similar protocol. Yields of human luteal and seminal relaxin were sufficient for amino acid sequence analysis, which demonstrated that the luteal- and seminal-derived products were identical.

Other Species. In 1987, Büllesbach et al. (82) isolated relaxin from the ovary of the skate (*Raja erinacea*). The skate relaxin has the longest B chain (36 amino acids) of all isolated relaxins. A relaxin peptide isolated from the placenta of the domestic dog (*Canis familiaris*) (83) also demonstrates a longer B chain than other relaxin peptides. Relaxin has been isolated from the ovaries of two species of shark, the dogfish shark (*Squalus acanthias*) (84) and the sand tiger shark (*Odontaspis taurus*) (85). Both relaxin sequences differed distinctly from that of porcine and rat relaxin, and dogfish shark relaxin displayed very poor bioactivity in the mouse interpubic ligament assay (84). Hamster relaxin and prorelaxin peptides have been isolated from late pregnant hamster placenta (86,87). The native peptide is chemically identical to a synthetic hamster peptide (87), and a partial B chain sequence was characterized from the native peptide and shown to be consistent with relaxin B chains from other species and demonstrated highest homology to rat relaxin (86). The relaxin prohormone was N-terminal sequenced and demonstrated identical sequence to the B chain, indicating that it was an authentic prorelaxin.

Relaxin peptides have been isolated and characterized from corpora lutea of the minke whale (*Balaenoptera acutorostrata*) and Bryde's whale (*Balaenoptera edeni*) (88). Paradoxically, the amino acid sequence of both whale relaxins is almost identical to that of porcine relaxin. A relaxin peptide extracted from the corpus luteum of the porpoise (*Phocaena phocaena*) (89) was also demonstrated to differ in only one residue from porcine relaxin.

All the relaxin peptide sequences outlined previously are aligned with relaxin sequences derived from complementary DNA (cDNA) or genomic sequence (see later) in Fig. 2.

Cloning of the Relaxin Genes

Rat and Porcine Relaxin

Niall and coworkers (8,9) conducted detailed structural analyses of putative relaxin precursors in the early 1980s, and their findings are consistent with Gast's findings (see section on Relaxin). These workers determined the complete amino acid sequences of porcine (9) and rat (8) preprorelaxin by cloning of relaxin cDNA. In general, this work involved (a) amino acid sequence analysis of porcine and rat relaxin, (b) synthesis of oligoribonucleotide primers complementary to the putative messenger RNA (mRNA) sequences predicted from the amino acid sequences of porcine and rat relaxin, (c) use of these synthetic DNA primers with mRNA isolated from pig and rat corpora lutea of pregnancy for the production of radiolabeled relaxin-specific cDNA probes, and (d) the identification of relaxin-specific clones in cDNA libraries constructed from total pregnancy-derived corpus luteum mRNA (90). These studies were the first to demonstrate that relaxin, like insulin, is

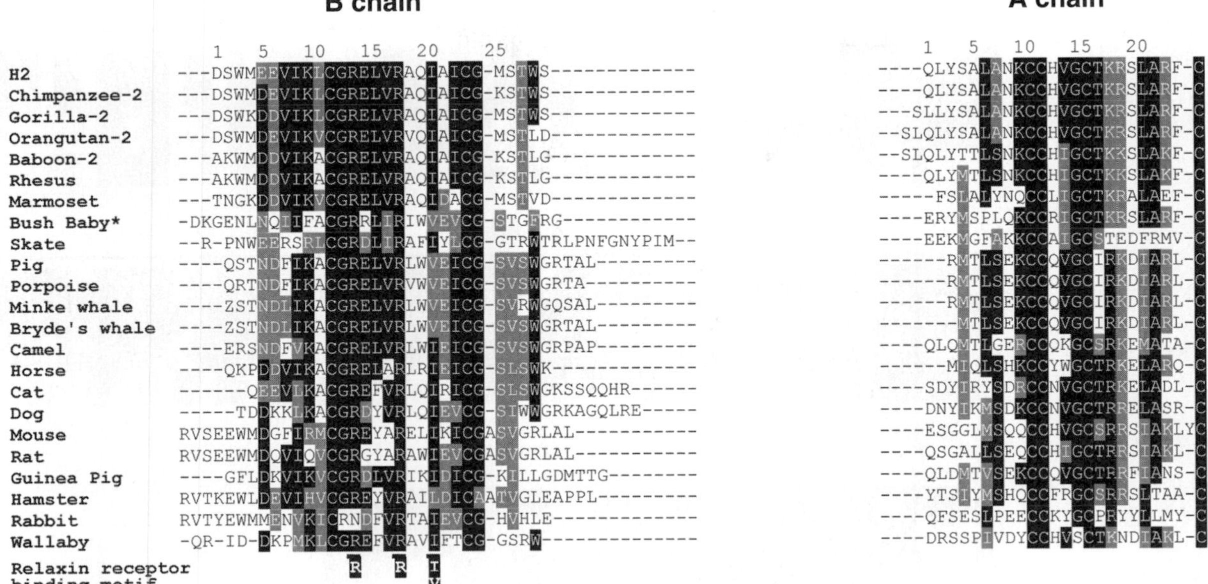

FIG. 2. Alignment of A- and B-chain sequences of primate relaxin-2 (human [H2], chimpanzee-2, gorilla-2, orangutan-2) and the equivalent relaxin peptides from all other known species. Sequences are derived from both peptide and cDNA sequencing, as outlined in the text. Conserved amino acid residues are *boxed in black*, and conservative amino acid substitutions are *boxed and shaded*. The chains are numbered according to the H2 relaxin sequence. The residues that make up the conserved relaxin receptor binding motif, Arg-13, Arg-17, Ile/Val-20, are marked. Bush baby, African lorisiform (*Galago crasicaudatus*), *identical to the Malagasy lemur (*Varecia variegata*) relaxin sequence; Wallaby, tammar (*Macropus eugenii*).

initially synthesized as a single-chain preprohormone with this overall structure: signal peptide/B chain/connecting peptide/A chain, as shown in Fig. 3.

The rat (91) and porcine (92) relaxin genes have been subsequently cloned from genomic libraries. The sequences are identical to those determined from the cDNA sequence other than a potential single allelic variation shown in the porcine sequence (92), and the gene structure conforms to that shown for other relaxin genes (outlined in the following sections).

Human

The use of recombinant DNA technology was essential in establishing the sequences of human relaxin owing to the lack of primary amino acid sequence data. This was due to the difficulties in obtaining sufficient amounts of human ovarian tissue for the isolation of relaxin and the insensitivity of peptide sequencing technologies at the time. Knowledge of the structures of the C peptides of porcine and rat relaxin formed the basis of the strategy used by Hudson et al. in 1983 (10) to determine the structure of one form of human relaxin. Human relaxin had not been isolated, and its primary structure was not known; consequently, Hudson et al. (10) used a probe based on the porcine relaxin cDNA sequence to screen a human genomic library. A clone was isolated, and the complete coding sequence of the encoded gene was determined (10). This gene was designated *RLN1* and the subsequent peptide product would be human relaxin-1 or H1 relaxin. In 1984, Hudson et al. (11) identified a clone for a second form of human relaxin by screening a cDNA library prepared from a human corpus luteum of pregnancy with radiolabeled relaxin-specific cDNA probes corresponding to exon I and exon II of the human *RLN1* gene (11). This second gene was designated *RLN2*, with the subsequent peptide product being called human relaxin-2 or H2 relaxin. This derived peptide sequence is identical to the relaxin peptide that was isolated from the corpus luteum (79,80). Hence, H2 relaxin is a luteal cell product and the circulating form of relaxin in humans, and therefore the human *RLN2* gene is the functional equivalent of the rat and porcine *RLN1* genes as discussed in section on *INSL4* and H1 Relaxin-Equivalent Genes Are Found Only in Primates. The derived amino acid sequences of human preprorelaxin 1 and preprorelaxin 2 aligned to the rat and porcine sequences are shown in Fig. 3. Although the four to

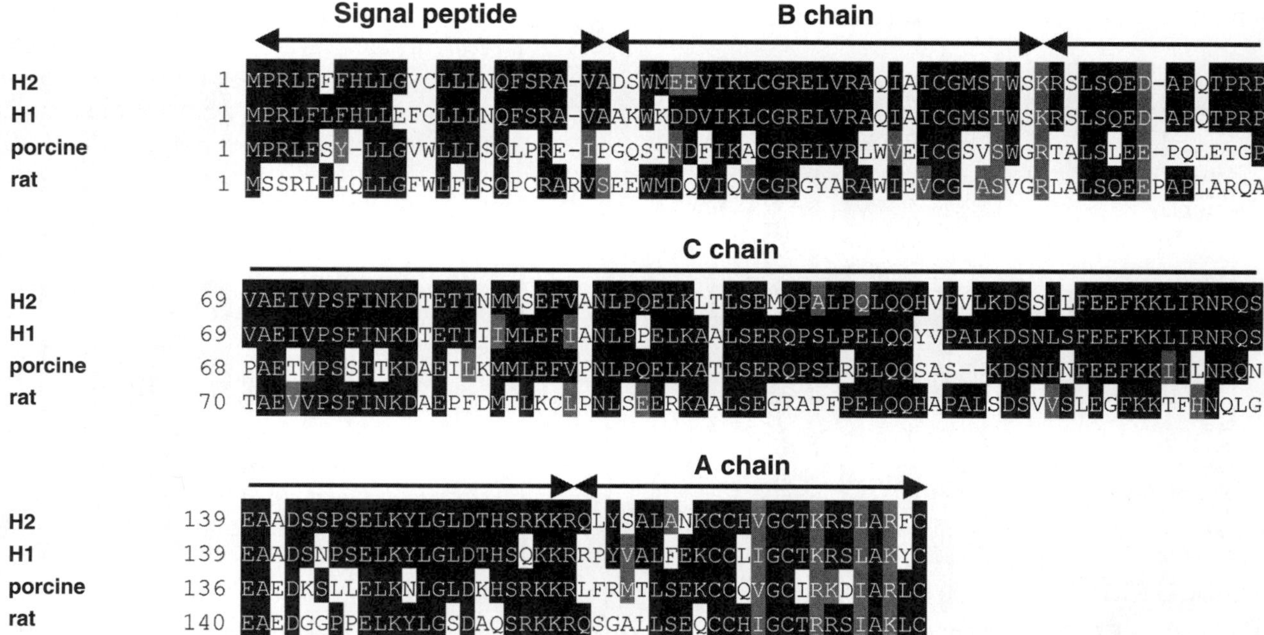

FIG. 3. Alignment of the amino acid sequences of human gene-2 (H2), human gene-1 (H1), porcine, and rat preprorelaxin. Conserved amino acid residues are *boxed in black* and conservative amino acid substitutions are *boxed and shaded*. The domains of the preprorelaxin sequences, based on H2 relaxin, are marked: Signal peptide, B chain, C chain, and A chain.

six amino acid residues on the N-termini of the B and A chains of putative H1 and H2 relaxin differ rather markedly, both chemically synthesized forms of the hormone had relaxin-like biological activity, as judged by their ability to inhibit contractions of rat uterine strips in vitro (10,11).

Primates

In 1989, clones for rhesus monkey relaxin were identified by screening a cDNA library prepared from corpora lutea of pregnancy with radiolabeled human relaxin-specific cDNA probes (93). The derived amino acid sequence of rhesus monkey preprorelaxin showed highest homology with H2 preprorelaxin. Southern blot analysis of genomic DNA indicated that there was only a single copy of the relaxin gene in the rhesus monkey. Subsequent studies using reverse transcriptase-polymerase chain reaction (RT-PCR) on genomic DNA have demonstrated that whereas Old World monkeys (rhesus monkey and baboon) have only one relaxin gene, all great apes (chimpanzee, pygmy chimpanzee, gorilla, and orangutan) have two relaxin genes (94,95). Relaxin-2 orthologous genes show high sequence similarity at the derived amino acid level to H2 relaxin. Although this is also the case for relaxin-1 orthologs compared with H1 relaxin, the derived relaxin-1 peptide in both the orangutan and gorilla is likely to be nonfunctional (95). This is discussed in more detail in section on *INSL4* and H1 Relaxin-Equivalent Genes Are Found Only in Primates. More recently, full-length relaxin cDNAs were cloned from marmoset relaxin ovarian mRNA (96) and African lorisiform (*Galago crasicaudatus*) and Malagasy lemur (*Varecia variegata*) placental mRNA (97). In both these studies the authors demonstrated that the species contained only one relaxin gene, and the derived amino acid sequences showed greatest homology to H2 preprorelaxin (Fig. 3).

Other Species

The availability of the rat, pig, and human relaxin gene sequences and the advent of recombinant DNA technology had led to the cloning of relaxin genes from numerous species. In 1992, Lee et al. (98) used a two-part strategy to clone the coding region of the guinea pig relaxin gene. Initially, they used RT-PCR to amplify part of the cDNA sequence using first-strand cDNA pooled from the endometrium of pregnant guinea pigs and heterologous primers based on the porcine cDNA sequence. The remaining sequence minus the signal peptide was amplified using rapid amplification of cDNA ends (RACE)-PCR and subcloning using homologous primers. The complete cDNA sequence of guinea pig relaxin remains to be determined.

A relaxin-like cDNA sequence was cloned from rabbit tracheobronchial epithelial cells in 1992 and was initially named *SQ10* (99). The derived amino acid sequence demonstrated 48% identity with H2 preprorelaxin, and the protein was implicated in squamous cell differentiation. Subsequently the *SQ10* mRNA was demonstrated to be highly expressed in the rabbit placenta and a partial amino acid sequence of a peptide isolated from the placenta was shown to be identical to the *SQ10*-derived amino acid sequence (100). The peptide isolated from the placenta demonstrated relaxin activity and the authors suggest that it is likely rabbit relaxin.

The sequence of the full coding region of the mouse relaxin cDNA was determined by Evans et al. in 1993 (101). An RT-PCR strategy was once again used because the amino acid sequence of the mouse relaxin peptide was not known. However, in the same study the authors were also able to isolate the native mouse relaxin peptide and confirm the sequence derived from the cDNA sequence, including the presence of an extra tyrosine residue in the A chain near the C-terminal cysteine that is not present in any other relaxin peptide (Fig. 2).

A partial hamster B-chain relaxin peptide sequence (86) was used to design oligonucleotide primers to prime a 3′RACE-PCR to isolate a hamster relaxin cDNA (102). The subsequent product was used to screen a hamster placental cDNA library, and a full-length hamster relaxin cDNA was isolated (102). The derived amino acid sequence of this clone showed highest homology to rat prorelaxin (51.8%).

A relaxin cDNA sequence has been isolated from the tammar wallaby (*Macropus eugenii*) corpus luteum (103). The derived amino acid sequence shows considerable variation from that of other mammals, although the basic preprohormone structure is retained, as are all the residues known to be necessary for biological activity. A relaxin peptide was subsequently isolated from the wallaby corpus luteum that corresponded to the cDNA sequence and demonstrated relaxin activity in vitro (104).

A partial relaxin cDNA sequence has been cloned from the horse placenta (105) using degenerate RT-PCR primers based on the known equine relaxin peptide sequence (70). The derived amino acid sequence of this clone was identical to the known horse relaxin A- and B-chain sequences (70).

Additional relaxin cDNA sequences have been cloned using RT- and RACE-PCR strategies from the dog (106) and cat (107) placenta and the camel ovary (108). Once again, the derived amino acid sequences demonstrated relatively low homology to preprorelaxin sequences from other species while retaining the amino acids demonstrated in other species to be essential for relaxin activity. The discovery of a relaxin gene in the camel is interesting because attempts to clone relaxin genes from other ruminant species have been unsuccessful (see section on Evolution of the Relaxin Genes).

Relaxin Gene Structure

Partial or complete analysis of the genomic DNA from human, rhesus monkey, porcine, rat, and mouse relaxin genomic clones indicated that their general structures are consistent with that of the human *RLN1* and *RLN2* genes shown in Fig. 4 (90). An intron interrupts the coding region at Glu in position 46 in the C peptide (10,91–93,101) in all relaxin genes so far studied. The position of this intron corresponds closely to that of one of the two introns found in insulin genes (109). There is no evidence that relaxin genes have a second intron corresponding to the position of the second insulin intron, which is located in the 5′ untranslated region preceding the signal peptide. There are, however, characterized splice variants of the human *RLN1* and *RLN2* genes that contain an extra exon within the known intron region (110). This splice variant contains a frame shift that would result in the expression of a truncated protein missing the A-chain sequence. It is therefore unlikely that it produces a functional relaxin peptide in vivo.

Discovery and Cloning of Novel Relaxin Family Genes

In stark contrast to the relaxin gene, which was finally cloned many years after the relaxin peptide was discovered, novel members of the relaxin gene family were first discovered as cDNA clones or as sequences from searches of the EST or genomic databases. All members of the relaxin gene family with their gene and peptide names are outlined in Table 1.

The INSL3 Gene

In the early 1990s, two groups independently discovered a novel cDNA clone that was differentially expressed in both the porcine (12) and mouse (111) testis. The deduced amino acid sequences of these cDNAs encoded a novel protein that demonstrated the characteristic structure of the insulin–relaxin peptide family with a signal peptide, B-, C-, A-chain structure. The disposition of the cysteine residues

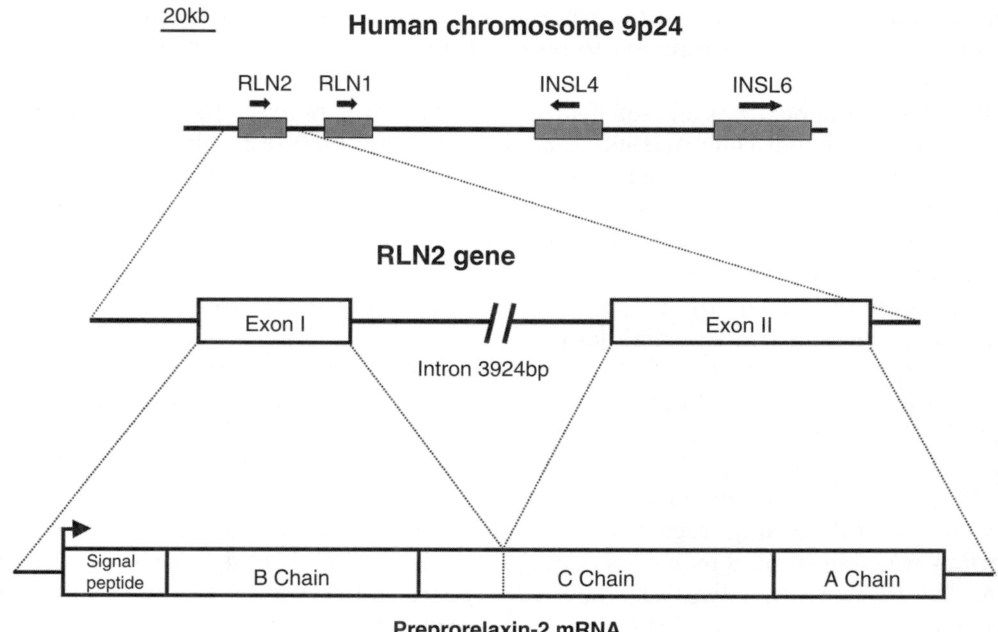

FIG. 4. Schematic representation of the transcription of the human *RLN2* gene. The gene is localized with the *INSL6*, *INSL4*, and *RLN1* genes on chromosome 9 at 9p24. The *RLN2* gene consists of two exons and is transcribed to give preprorelaxin-2 mRNA. Exon 1 encodes the signal peptide, B chain, and part of the C chain, and exon II encodes the rest of the C chain and the A chain of H2 relaxin. *Arrows* on the chromosome locations indicate the orientation of the genes.

was also identical to insulin and relaxin, indicating that it was likely that a mature peptide would be produced that would be similar in structure to these peptides. The cDNA was demonstrated to be highly expressed in the testicular Leydig cells and was initially given the name Leydig cell insulin-like peptide (Ley-I-L) (12). The gene for this peptide was subsequently cloned from the human, porcine (112), mouse (113,114), and rat (115) and shown to be a single-copy gene with a similar structure to the relaxin gene, with two exons interrupted by a single intron. This intron is located in the middle of the C-peptide coding region in a very similar position to intron in the relaxin gene. Splice variants of the *INSL3* gene containing an extra exon in the single intron, similar to those characterized for relaxin, have been identified (116). However, like the relaxin variants, these sequences contain a frame shift resulting in a premature stop codon after the B chain. Subsequently, the truncated peptide products of these variants are unlikely to encode functional INSL3 peptides.

TABLE 1. *Members of the relaxin peptide and insulin-like peptide gene families*

Peptide names	Abbreviations	Gene name
Insulin	INS	*INS*
Insulin-like growth factor-1	IGF-1	*IGF1*
Insulin-like growth factor-2	IGF-2	*IGF2*
Relaxin-1	RLX1 (human H1)	*RLN1*
Relaxin-2	RLX2 (human H2)	*RLN2*[a]
Relaxin-3	RLX3, INSL7	*RLN3*
Insulin-like peptide 3	INSL3	
Leydig–insulin-like peptide	Ley-I-L	*INSL3*
Relaxin-like factor	RLF	
Placentin	INSL4	*INSL4*
Early placental insulin-like factor	EPIL	
Insulin-like peptide 5	INSL5	*INSL5*
Relaxin–insulin-like factor 2	RIF2	
Insulin-like peptide 6	INSL6	*INSL6*
Relaxin–insulin-like factor 1	RIF1	

[a]The human *RLN2* gene is the equivalent of the *RLN1* gene in nonprimate species. Other species have only *RLN1* and *RLN3* genes.

An interesting feature of the *INSL3* gene was the discovery that the mouse gene is colinear within the 3′ end of the gene encoding Janus kinase 3 (*JAK3*) (113). This was subsequently confirmed also for the other species studied (42). Accordingly, the presumptive promoter region of the relaxin-like factor (RLF)/*INSL3* gene is restricted within intron 23 of the *JAK3* gene. Certainly, a short region of only approximately 500 to 700 base pairs from within this intron appears to be sufficient to govern specific *INSL3* gene expression when transfected into mouse tumor Leydig cells (113,114).

INSL3 has subsequently been cloned in the bovine (117), sheep (118), marmoset monkey (116), goat (119), dog (120), deer (121), and a subprimate species (97). In all species so far studied, the major site of expression is in the testicular Leydig cells.

An interesting feature of the INSL3 amino acid sequences is the high homology between species, especially in the B and A chains (Fig. 5). This is in contrast to the relatively low homology seen between species for the relaxin peptides (Fig. 2). The INSL3 peptide sequence shows higher homology to relaxin than insulin, especially in the B chain. Subsequent work using a synthetic peptide based on the INSL3 cDNA sequences and putative cleavage sites for the C peptide deduced on the relaxin sequence demonstrated that the peptide had some relaxin activity,

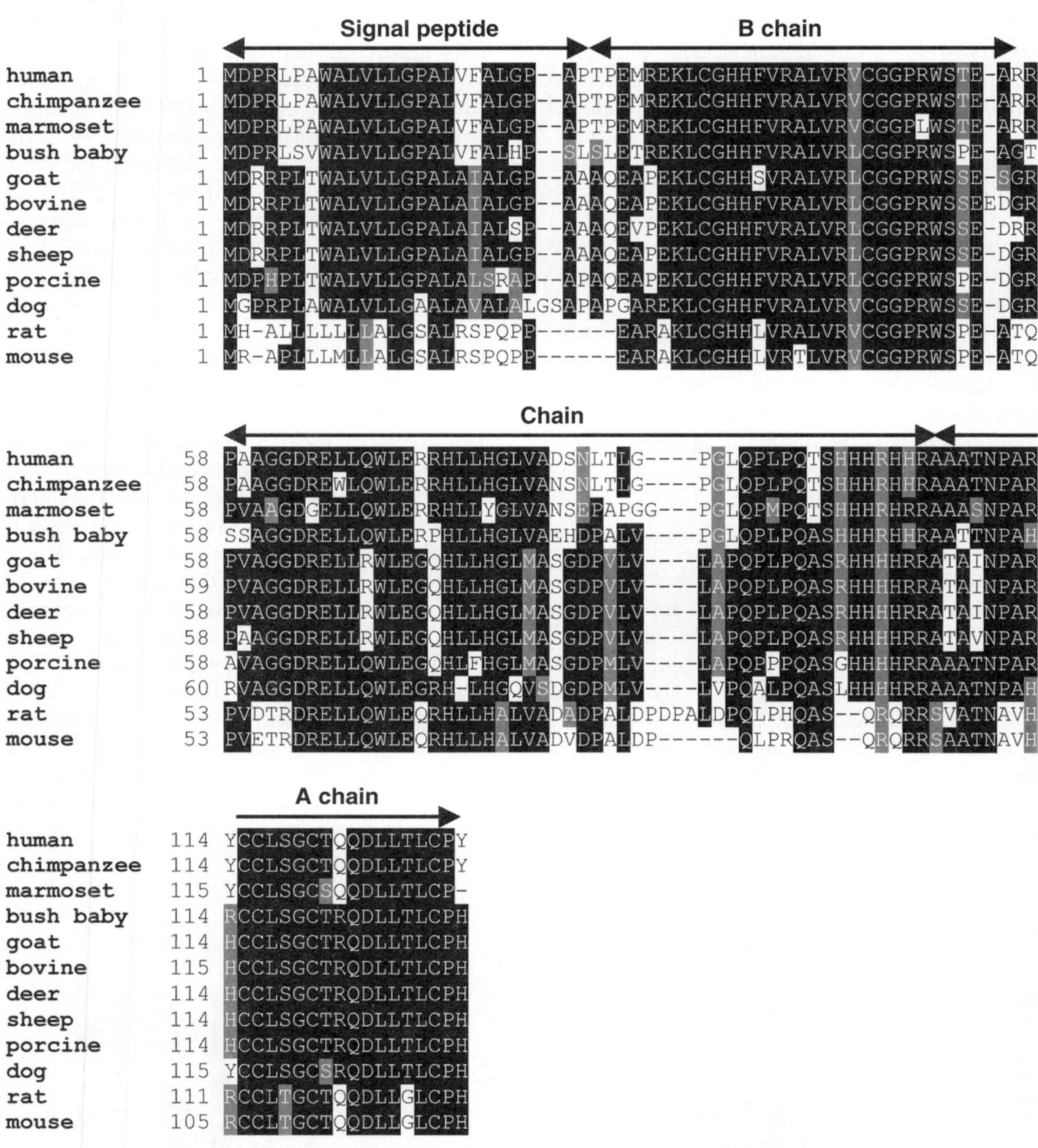

FIG. 5. Alignment of derived prepro–insulin-like peptide 3 (INSL3) amino acid sequences from all known species (see text for details). Chimpanzee prepro-INSL3 was determined from genomic sequence (sequence ID: BK005155). Conserved amino acid residues are *boxed in black* and conservative amino acid substitutions are *boxed and shaded*. The domains of the predicted prepro-INSL3 sequences are marked: Signal peptide, B chain, C chain, and A chain. Bush baby, African lorisiform (*Galago crasicaudatus*).

and the name relaxin-like factor or RLF was proposed (122). The gene was subsequently called *insulin-like 3* (*INSL3*) because it was the third insulin-like gene to be discovered, *IGF1* and *IGF2* being the first two. The protein product of this gene is hence referred to as insulin-like peptide 3 or INSL3.

Only in 2002 was the structure of an in vivo synthesized INSL3 peptide elucidated (123). Although the extraction procedure adopted probably prevented an analysis of potential prohormone forms of the protein, sufficient mature peptide could be extracted from bovine testis to show that in this tissue, at least, a mature A-B heterodimer INSL3 peptide is produced (123). The A chain at 26 amino acids was as predicted and of similar length to relaxin peptides. Interestingly, the B chain of the bovine INSL3 peptide is longer than predicted (40 amino acids compared with 31 amino acids) and considerably longer than relaxin peptides (usually 32 amino acids). However, a synthetic peptide based on a B chain of 31 amino acids demonstrated slightly higher bioactivity to the longer B-chain form. It remains to be seen if the INSL3 peptide is further processed in the blood or in its target tissues. A human INSL3 peptide sequence based on the predicted sequence is aligned to other members of the relaxin/insulin peptide family in Fig. 6.

The INSL4 Gene

Shortly after the discovery of INSL3, a cDNA clone for another new member of the relaxin peptide family was identified during screening of a subtracted cDNA library of first-trimester human placenta (13). The clone was tentatively named early placenta insulin-like peptide (EPIL). Comparison of the deduced amino acid sequence of EPIL demonstrated significant overall and structural homologies with members of the insulin/relaxin hormone superfamily. The signal peptide, B-, C-, A-chain prohormone structure was identified, as were the positions of the six cysteine residues, suggesting the native peptide would have a similar primary structure to relaxin and insulin. The gene for this novel cDNA was also cloned in the original study and demonstrated to be composed of two exons and one intron, similar to the insulin and relaxin genes (13). It was named the *insulin-like 4* (*INSL4*) gene based on the nomenclature used for the *INSL3* gene. At the same time, another group cloned the same cDNA from a cDNA library of first-trimester placentas and termed the clone placentin (124). This group further demonstrated that the cDNA, when inserted into a mammalian expression construct and transfected into 293 cells, resulted in the production of putative pro- and mature forms of the deduced protein. However the exact nature of these proteins is not known, and the native INSL4 protein has not been isolated. A putative human INSL4 peptide sequence is aligned to other members of the relaxin–insulin peptide family in Fig. 6.

The INSL5 and INSL6 Genes

Searching the EST databases for novel insulin–relaxin genes using the conserved cysteine pattern of this peptide family resulted in the independent discovery by three groups of two new members of this peptide family, INSL5 (14,15) and INSL6 (15–17). INSL5 was identified from a single human EST clone generated from a colon cDNA library (14). The complete human sequence was subsequently cloned and then used to identify a mouse EST from a colon cDNA library. The full-length mouse INSL5 cDNA was also subsequently cloned. Mouse INSL5 was independently identified in the EST databases and then cloned using RT-PCR and RACE-PCR from mouse ovary (15). An additional mouse clone was identified that had a longer 5′ end encoding an additional 10 amino acids. The clones were named

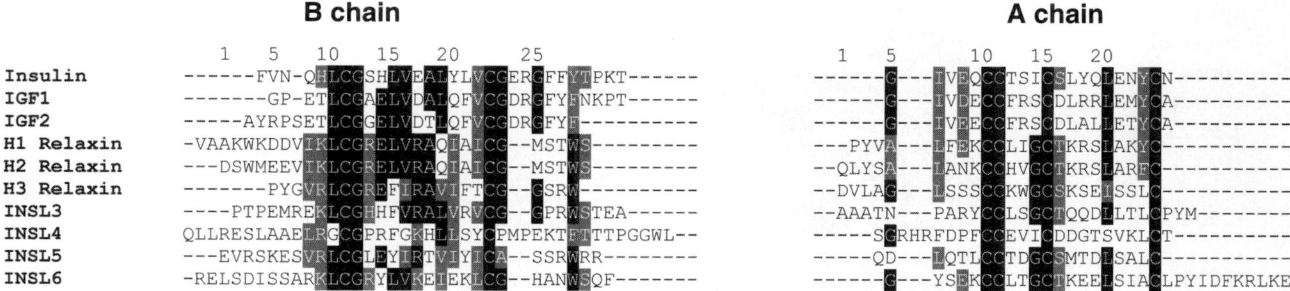

FIG. 6. Alignment of A- and B-chain sequences of all human insulin- and relaxin-like peptides. Conserved amino acid residues are *boxed in black* and conservative amino acid substitutions are *boxed and shaded*. Gaps have been introduced to aid the alignment where necessary. The chains are numbered according to the H2 relaxin sequence.

relaxin/insulin-like factor 2 (RIF2) long and short (15), the short form being identical to mouse INSL5 (14). Human and mouse INSL5 derived amino acid sequences contain the conserved relaxin cysteine motif as well as the classic signal peptide, B-, C-, A-chain structure of the relaxin family. The mouse and human sequences demonstrate 71% identity in the B- and A-chain regions, but display poor homology to other relaxin peptides in this region. The native INSL5 peptide has not been isolated, although recombinant INSL5 proteins have been expressed in mammalian cell lines using the cloned cDNA sequence (125).

At the same time RIF2/INSL5 was discovered from the EST databases, additional EST clones were discovered derived from testis cDNA libraries that encoded mouse and human orthologs of another novel relaxin family member (15). The full-length cDNAs of these sequences were subsequently amplified using RT- and RACE-PCR from testis cDNA. These clones were named RIF1. Subsequently, two other groups independently reported the discovery of the same EST sequences from the mouse (16) as well as the human (17). These groups termed the new clones INSL6, based on the nomenclature used for other family members. Lok et al. (17) identified the human EST and cloned the full-length human and rat cDNA clones using screening testis cDNA libraries. Although the derived amino acid sequences of the rodent clones show relatively low homology to the human, they are clearly all INSL6 orthologs (16). Human, mouse, and rat INSL6 show low homology to the RIF2/INSL5 derived amino acid sequences, but contain the conserved insulin–relaxin cysteine motif (from which they were identified) as well as the classic signal peptide, B-, C-, A-chain structure of the insulin–relaxin family.

The predicted human INSL5 and INSL6 B- and A-chain sequences are aligned with other relaxin family peptides in Fig. 6.

Discovery of Relaxin-3

Until recently, and as outlined in detail over preceding pages, it was generally accepted that there was only one relaxin gene in most mammals, with the exception of the human and higher primates, which have two genes. However, searches of the completed genome databases of the human and the mouse using B-chain sequences of relaxin identified a third human relaxin gene (*RLN3*) and its equivalent in the mouse (18). These genes showed a similar two-exon structure to the human *RLN1* and *RLN2* genes, with the characteristic signal peptide, B-, C-, and A-chain structure in the derived amino acid sequence. Interestingly, prorelaxin-3 displays a much shorter C chain than prorelaxin-1 and 2. Although the derived amino acid sequence showed highest homology to INSL5, the presence of an amino acid motif in the B chain of the peptide, which has been demonstrated to be essential for relaxin binding (126), indicated that the gene codes for a relaxin-, not an insulin-related peptide (Fig. 6). Subsequent synthesis of a peptide based on this sequence demonstrated it to bind and activate THP-1 cells (a human monocyte cell line), which express the relaxin receptor (18). Hence the new gene was called *RLN3*, the human ortholog being human relaxin-3 or H3 relaxin. The derived amino acid sequence of the mouse ortholog of this gene showed very high homology to H3 relaxin peptide and has also been designated as relaxin-3, even though there is no *RLN2* gene in mammals lower than higher primates. The rat version of this gene has been cloned from brain cDNA and the sequence confirmed from the completed genome databases (127). The structure of the rat relaxin-3 gene was identical to that of the mouse and human relaxin-3 genes. The porcine relaxin-3 sequence was determined using RT-PCR with oligonucleotide primers based on the human relaxin-3 sequence and porcine genomic DNA (128). Further searching of the genome and EST databases has identified numerous relaxin-3 genes in zebrafish (129) and fugu (pufferfish) genomes (19). It now seems likely that relaxin-3 is the ancestral relaxin gene. The striking feature of the relaxin-3 peptide family is the high homology in amino acid sequences in the A and B chains across species, which is in stark contrast to the other relaxin peptides (Fig. 7). This may be related to relaxin-3's putative role as a neuropeptide as opposed to an endocrine/paracrine hormone like the other relaxin peptides (34). In 2003, the endogenous relaxin-3 peptide was isolated from pig brain and demonstrated to be identical to the predicted mature peptide sequence (23).

EVOLUTION OF RELAXIN PEPTIDE FAMILY

Since the cloning of the first relaxin gene (*RLN1*) in 1981 (8), six additional human relaxin-like genes have been discovered, *RLN2*, *RLN3*, *INSL3*, *INSL4*, *INSL5*, and *INSL6*. Because the human and mouse genomes are now completely sequenced and exhaustive searches of the databases have failed to identify additional members (19,129,130), it is likely that there are no other members of this peptide family. All these genes and their predicted protein products show greater similarity to relaxin than to insulin or insulin-like growth factor (IGF)-1 or 2, and phylogenetic analysis indicates that these genes evolved

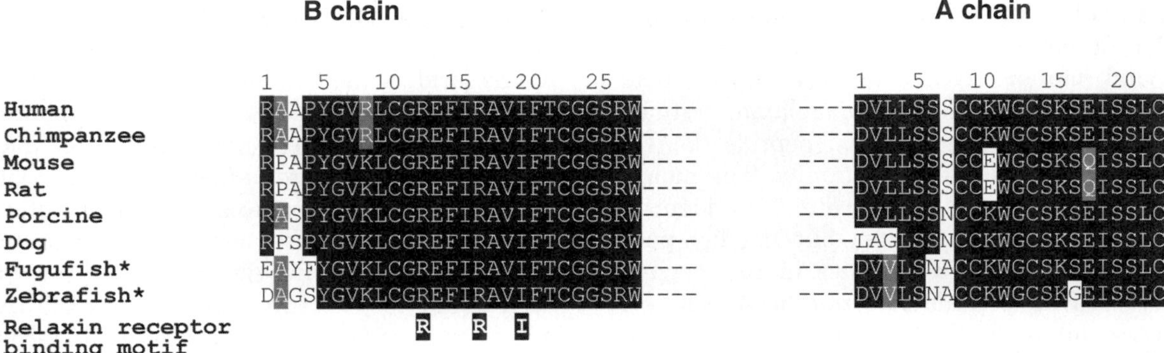

FIG. 7. Alignment of A- and B-chain sequences of relaxin-3 from various species. Sequences are derived from various sources, as outlined in the text. Chimpanzee relaxin-3 was determined from genomic sequence (sequence ID: BK005156). Conserved amino acid residues are *boxed in black* and conservative amino acid substitutions are *boxed and shaded.* The residues that make up the conserved relaxin receptor binding motif, Arg-12, Arg-16, Ile/Val-19, are marked. *There are multiple relaxin-3 genes in the fugu or pufferfish (*Takifugu rubripes*) and zebrafish (*Danio rerio*; see text for details); the sequences with highest homology have been used for the alignment. Note the remarkable homology in the sequence of the relaxin-3 A- and B-chain sequences from fish to humans.

from a common relaxin ancestor (19). These considerations have led to the concept of the relaxin peptide family.

Chromosomal Locations

The chromosomal locations of the 10 human members of the insulin and relaxin family are shown in Fig. 8. The relaxin peptide family genes are in different locations than the insulin and IGF-1 and 2 genes. The human *RLN1* and *RLN2* genes map in a tight cluster with the *INSL4* (131) and *INSL6* (17) genes on chromosome 9 at 9p24. The *RLN3* gene is localized on chromosome 19 at 19p13.3 in close proximity to the *INSL3* gene at 19p13.2 (18). The *INSL5* gene, however, on human chromosome 1 at 1p31.1, does not colocalize with any other relaxin-like genes (14).

In contrast to humans, the mouse genome contains only five relaxin family genes, although they are

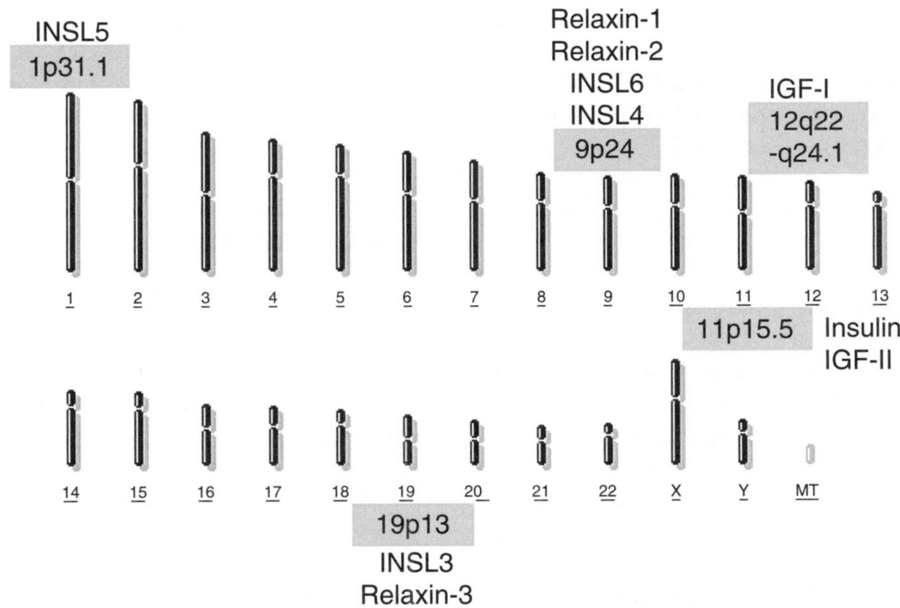

FIG. 8. Chromosomal localizations of the human relaxin family peptide and insulin-like peptide genes. (See the text for details.)

localized in a similar cluster arrangement to the human. Hence, mouse *RLN1* is adjacent to *INSL6* on chromosome 19 at 19C3 (15), *INSL3* is in close proximity to *RLN3* on chromosome 8 at 8C2 (18), and *INSL5* is located on chromosome 4 at 4C5–C7 (15).

INSL4 and H1 Relaxin-Equivalent Genes are Found Only in Primates

Exhaustive searching of the genome databases has failed to identify an *INSL4* or a human *RLN1*-equivalent gene in the completed mouse and rat genomes or any other nonprimate EST or genome databases (19,129,130). It was previously shown that the *RLN1* gene was found only in great apes and not in Old World or New World monkeys (95). Hence, H1 relaxin peptide orthologs are not found in other species. The H2 relaxin peptide is the functional equivalent of the relaxin gene in other species. The relaxin peptide produced by the human ovary during pregnancy is H2 relaxin (11,79,80), and ovarian expression of the H1 relaxin gene has not been detected (11,132). Winslow et al. (81) determined that the relaxin found in human seminal plasma is also derived from the *RLN2* gene. Although a synthetic peptide based on the deduced H1 relaxin amino acid sequence is bioactive (10), showing only slightly lower activity than H2 relaxin (133), a native H1 relaxin peptide has not been isolated, and hence there is currently no evidence that a peptide is actually produced by the H1 relaxin gene. Interestingly, there is a stop codon in the orangutan relaxin-1 gene at the start of the A chain, and the gorilla-derived B-chain sequence contains an extra cysteine residue (95). Therefore, in both of these species a bioactive H1 relaxin-equivalent peptide cannot be produced.

In the human, *INSL6*, *INSL4*, *RLN2*, and *RLN1* are all in close proximity on chromosome 9, whereas in the mouse only *INSL6* and *RLN1* are in close proximity on chromosome 19. It is therefore likely that human *INSL4* and *RLN1* are the result of a gene duplication that occurred only in humans and higher primates. Although an equivalent *INSL4* gene has not been found in nonprimate species, an *INSL4* gene has been identified in the rhesus monkey as well as the African green monkey genomes (130). It therefore is likely that the *INSL4* gene emerged before the divergence of New and Old World monkeys (130), predating the emergence of the *RLN1* gene. The higher homology between *INSL4* and *RLN1* gene sequences compared with *INSL6* and *RLN1*, and the discovery of *INSL4* splice variants with an extra exon similar to relaxin variants (110), is further evidence that the *INSL4* gene arose through duplication of the *RLN1* gene (130).

Evolution of the Relaxin Genes

The evolution of the relaxin peptide family has been a subject of much study, debate, and conjecture for many years. Relaxin peptides are well known for their high sequence variability between closely related species, although startling similarities have also been observed between peptide-derived sequences of very distant species such as pigs and whales (88). Other studies have reported the existence of an invertebrate relaxin (134,135). A cDNA and peptide sequence with almost 100% similarity to porcine relaxin was isolated from *Ciona intestinalis* (136). However, this sequence is not present in the recently completed *C. intestinalis* genome (20), and there are no relaxin-like sequences present in any of the complete or incomplete invertebrate genomes (19,20). Peptides with sequences almost identical to porcine relaxin were identified from Bryde's whale (*B. edeni*) and the minke whale (*B. acutorostrata*) (88). Based on the evolutionary distances between whales and pigs, it is highly unlikely that these sequences are correct.

Another oddity in the evolution of this family has been the inability of researchers to find a relaxin gene in cows and other ruminants, although there is a clear relaxin physiology in these species (118,137). In cows and sheep typical softening and dilatation of the cervix can be achieved with injections of a porcine relaxin preparation (138). Purified porcine relaxin, when injected intravenously into late pregnant sheep, induces a typical hormone-specific quiescence in myometrial activity (139). Peptide fractions that demonstrate characteristic relaxin-like activity have been partially purified from cows (140,141) and sheep (142), but these peptides were never purified to homogeneity. A possible explanation for the absence of a relaxin gene in ruminants was provided by the discovery that sheep have a nonfunctional relaxin-like gene (118). A relaxin-like cDNA fragment with 72% identity to exon II of pig relaxin (corresponding to the C peptide and A chain) and two genomic clones from sheep placenta were isolated, but sequencing revealed several stop codons in the region encoding the putative C chain, which would prevent translation of the A chain (118). Further analysis of sheep mRNA using 5′ RACE-PCR was unable to detect any relaxin-like sequences with similarity to the relaxin exon I, which codes for the B chain (118). The authors therefore concluded that sheep have a single-copy, nonfunctional relaxin-like gene. Exhaustive studies have failed to identify a relaxin gene sequence or nonfunctional sequences similar to those in the sheep in the cow or the goat (137). However, full genome sequences from these species will be required to confirm the presence of a nonfunctional relaxin gene sequence similar to that observed in the ovine.

Interestingly, a relaxin cDNA sequence has been identified in the camel (108) and relaxin immunoreactivity is found in the pregnant serum of the closely related llama and alpaca (143). Although classified as ruminants, Camelidae have a unique reproductive anatomy and physiology that more closely resembles that of pigs, which have a well-characterized relaxin gene.

Insulin and IGF orthologs have been identified in invertebrate genomes (144,145). In contrast, searches of the genomes of lower organisms have not identified any relaxin genes in species lower than fish (19,129). Hence, coupled with the structural similarities between the relaxin and insulin peptides, relaxin genes could have evolved from an insulin-like ancestor. Current evidence suggests that the relaxin ligand family likely underwent major expansion only during recent mammalian evolution. A cluster of five relaxin genes has been identified in the pufferfish genome (19), but all of these sequences show highest homology to relaxin-3. The high primary sequence identity of these sequences suggests that these genes all evolved from the same relaxin-like ancestor and that the relaxin-like genes in mammals also evolved from this ancestor. It therefore has been postulated the relaxin-like genes in mammals evolved from a single relaxin-3–like ancestral gene during mammalian evolution. However, without full genome sequences from all organisms, it is not possible to exclude the possibility that some relaxin family members evolved earlier in vertebrate evolution. A phylogenetic tree of the likely evolution of the human relaxin peptide family is shown in Fig. 9. The tree highlights that the relaxin peptides diverged from insulin-like peptides and have branched into two distinct relaxin peptide subfamilies. Subfamily A includes INSL3, INSL4, INSL5, INSL6, and H1 and H2 relaxin, and subfamily B includes INSL5 and H3 relaxin (20).

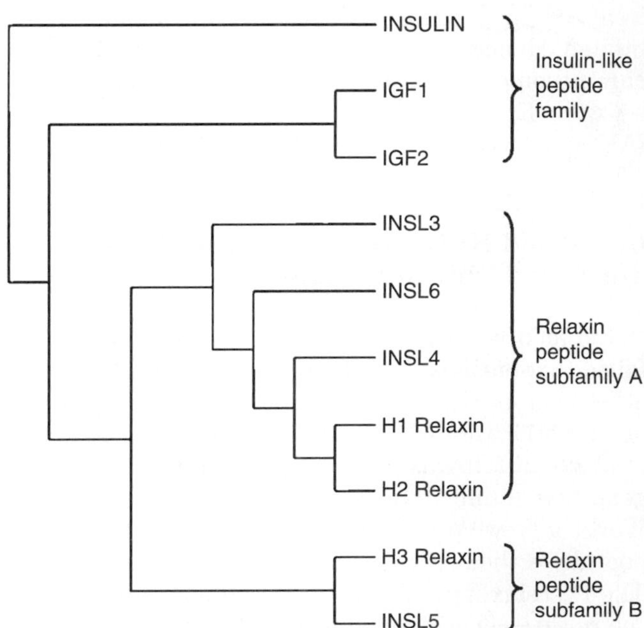

FIG. 9. Phylogenetic tree of the human relaxin peptide family showing the relationship to insulin, insulin-like growth factor (IGF)-1, and IGF-2. (See text for details.) Note that the relaxin peptide family has diverged into two subfamilies; subfamily A includes insulin-like peptide (INSL) 3, INSL4, INSL5, INSL6, and H1 and H2 relaxin, and subfamily B includes INSL5 and H3 relaxin. The tree is based on the prohormone amino acid sequences. (Adapted from Wilkinson, T. N., Speed, T. P., Tregear, G. W., and Bathgate, R. A. D. [2005]. Evolution of the relaxin-like peptide family. *BMC Evol. Biol.* 5, 14.)

CHEMISTRY OF RELAXIN PEPTIDES

Biosynthesis and Processing of Relaxin Peptides

Relaxin

Gel filtration of sow ovarian extracts provided the first indication that relaxin is synthesized originally in precursor forms and that the putative relaxin precursors are larger than preproinsulin and proinsulin (146,147). Subsequently, Kwok et al. (147) reported that a small portion of the relaxin in an acid-acetone extract of porcine ovaries was associated with a biologically active fraction that had an apparent molecular weight of 19,000 Da. In both studies, treatment of the large, relaxin-like fraction with the proteolytic enzyme trypsin appeared to convert the putative precursor to 6,000-Da relaxin. Gast and coworkers (148,149) postulated that 23,000-Da preprorelaxin is the primary translation product and that the first step in relaxin biosynthesis is the membrane-dependent cleavage of a 3,000-Da signal peptide to form prorelaxin (148,149) (Fig. 10). Subsequent cloning of the rat, porcine, and human relaxin genes (8–11) confirmed this preliminary work and demonstrated that relaxin, like insulin, is initially synthesized as a single-chained preprohormone with the following overall structure: signal peptide/B chain/connecting peptide/A chain, as shown diagrammatically in Fig. 10. The size difference between relaxin and insulin precursors is largely attributable to marked differences in the lengths of the connecting C peptides. The C peptides in porcine (9) and rat relaxin (8), for example, contain 104 or 105 residues, respectively; in porcine and rat proinsulin, however, they contain only approximately 30 residues (150). The function of the C peptide in relaxin is unknown. One function presumably is to direct the folding of the precursors so the correct disulfide bonds are formed between the B and A chains. However, studies expressing a mutant

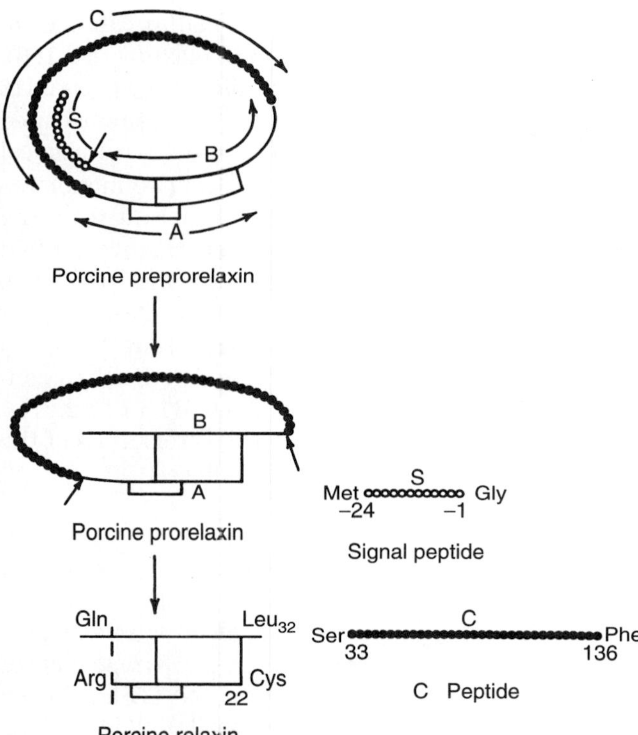

FIG. 10. Schematic summary of the processing of porcine preprorelaxin to relaxin by the successive removal of the signal peptide (S) and the connecting peptide (C) by proteolytic digestion. (Modified from Kemp, B. E., and Niall, H. D. [1984]. Relaxin. *Vitam. Horm.* 41, 79–115, with permission.)

human relaxin in yeast (151) or *Escherichia coli* (152) with connecting peptides of only 6 or 13 amino acids, respectively, have been demonstrated to produce folded bioactive relaxin peptides. It has been postulated that the C peptide may contain peptide sequences with hormonal activities (8,90), but there is no experimental evidence to support this.

Although prorelaxin proteins have been demonstrated to be bioactive (153–156), and do circulate in the plasma in some species [e.g., humans (152)], it is not known if they have a biological function. Generation of mature relaxin peptides requires cleavage of the C peptide, and this process is less well defined for prorelaxin and appears to be less stringently controlled than with proinsulin. Many relaxins contain dibasic residues at either the B/C or C/A junctions that likely direct cleavage, although others do not. Some relaxins actually contain a motif at their B/C or C/A chain junctions that is a consensus motif for the proprotein convertase furin (R-X-K/R-R) (157). In contrast, cleavage at the C terminus of the B chain in porcine and rat relaxin appears to require an enzyme with chymotrypsin-like specificity that recognizes the neutral aliphatic side chains of leucine at position 32 (32 Leu) in prorelaxin (9). If this is the case with porcine relaxin, subsequent cleavage of three amino acids must occur to give the predominant B29 form (65,66). Hudson et al. (11) originally proposed that H2 relaxin was also processed by a chymotrypsin-like enzyme at 32 Leu, but H2 relaxin isolated from the corpus luteum of pregnancy is a B29 form (158). Subsequent studies coexpressing H2 relaxin with murine prohormone convertase 1 (PC1) demonstrated production of authentic B29 relaxin, whereas expression without PC1 resulted only in prohormone production (159). The authors postulated that a similar enzyme might exist in human luteal cells and that the resultant basic amino acids Lys-Arg on the C terminus of the B chain may be cleaved subsequently by a carboxypeptidase enzyme in the luteal cells. Thus, it is likely that the cleavage of the prorelaxin C peptide is different between species.

Other Relaxin Family Peptides

All of the other relaxin family peptides have been initially characterized from their cDNA or genomic sequences. Their derived amino acid structures are identical to relaxin, with a signal peptide/B-, C-, A-chain structure. It is likely that these peptides are processed in a similar fashion to relaxin. However, as with relaxin peptides, there is a wide variation in the presence of putative prohormone cleavage sites between peptides and even between species for a particular peptide. INSL3 has been isolated from the bovine testis and has a B chain that is nine amino acids longer than predicted (123). Interestingly, synthetic peptides based on the predicted B31 sequence actually demonstrate higher bioactivity than the native form. It remains to be seen whether the native peptide is further cleaved by proteases in the blood or target tissues to produce the shorter form of the peptide, which shows higher bioactivity.

In contrast, relaxin-3 sequences for all species so far identified have conserved furin motifs at B/C and C/A chain junctions (127,128), and the native form of relaxin-3 isolated from pig brain conforms to the predicted cleaved mature peptide (23).

Very little is known of the processing of INSLs 4 through 6, and no native peptide has been isolated. Studies using a human INSL5 cDNA expressed in a rat insulinoma cell line that is capable of processing insulin demonstrate that only a small amount of mature INSL5 peptide is produced (125). Similarly, when INSL4 was expressed in 293 cells, seemingly both pro- and mature forms of the deduced protein were produced (124), although the exact nature of these proteins was not determined.

Structure of Relaxin Peptides

Relaxin

The first information on the structure of the mature relaxin peptide came from studies that determined the amino acid sequences of the A and B chains of various relaxins by peptide sequence analysis of native hormone preparations (see section on Discovery of Relaxin, INSL3, and Other Relaxin Family Hormones, for details). Nucleotide sequence analysis not only confirmed the amino acid sequences of many of these native relaxin sequences but predicted the amino acid sequences of H1 and H2 relaxin (see section on Discovery of Relaxin, INSL3, and Other Relaxin Family Hormones, for details). Conformation of the predicted disulfide linkages has been determined for porcine relaxin (160) and H2 relaxin (161), with an intrachain bond in the A chain between A10 and A15 and two interchain bonds between A11 and B11, and A24 and B23 (Fig. 2). In all species where the amino acid sequences of relaxin are known, with the exception of mouse relaxin (101), the half-cysteine residues are found in positions comparable with those in porcine and human relaxin (Fig. 2). The structure of relaxin has diverged considerably among species during evolution. In general, only 30% to 60% amino acid sequence identity exists between species, and the invariant positions in relaxin from all species are largely confined to the cysteine residues and adjacent glycine residues as well as the relaxin binding motif (Arg-X-X-X-Arg-X-X-Ile/Val; Fig. 2). The extensive differences in amino acid residues, as well as A- and B-chain lengths, between relaxin peptides from different species form a structural basis for the observations that relaxin preparations display species-specific differences in immunological cross-reactivity in radioimmunoassays (162). Furthermore, although the presence of the relaxin receptor binding cassette generally means that relaxin peptides from different species are bioactive across species, there are marked differences in biological activities among species. Importantly, relaxin does not interact with insulin receptors, show any insulin-like activity, or cross-react with any insulin immunoassays [see Sherwood (25) for more details]. In fact, the relaxin receptor is a GPCR, completely distinct from the insulin and IGF receptors, which are tyrosine kinases (see section on Relaxin Family Peptide Receptors, for details).

Direct analysis of the tertiary structure of a relaxin peptide was accomplished by Eigenbrot and coworkers in 1991 (163), and this work generally confirmed the predicted structural homology of relaxin to insulin first deduced from molecular modeling techniques. These modeling studies demonstrated that the crystallographic structure of insulin can accommodate the sequence of porcine relaxin (164–166). Eigenbrot and colleagues crystallized synthetic H2 relaxin, and x-ray structure analysis demonstrated that H2 relaxin, like insulin, forms an asymmetrical dimer. Whereas the orientation of the two monomers in the relaxin dimer is completely different from the two monomers in the insulin dimer, the overall fold of relaxin is similar to that of insulin (Fig. 11). Principal features of the molecule are a 24-residue A chain with two α-helices extending from A3 to A9 and from A13 to A20, and a 29-residue B chain containing one α-helix extending from B7 to B22. The relaxin model does not include the C terminus of the B chain, which appears to be disordered.

Other Relaxin Family Peptides

Unlike relaxin, the structure of all other relaxin peptides was first determined from their cDNA sequences (see section on Discovery of Relaxin, INSL3, and Other Relaxin Family Hormones, for details). The derived amino acid sequences of these gene sequences all show the characteristic relaxin/insulin structure. The actual native peptides have been isolated only for bovine INSL3 (123) and porcine relaxin-3 (23). The disulfide linkages characteristic of relaxin and insulin were confirmed for the bovine INSL3 peptide. The native peptides of INSLs 4 through 6 have not been isolated. There is currently no other structural information on any other relaxin peptide, although circular dichroism analysis of INSL3 indicates that it has a tertiary structure similar to relaxin (122, 167,168).

Chemical and Recombinant Synthesis

Much of the knowledge of the physiological actions of relaxin has been gained through the use of native relaxin preparations, especially porcine relaxin. The subsequent discovery of relaxin peptides from various species and the low amounts of the native peptide able to be extracted, their marked amino acid differences, and the variability in their biological activities between species has led to the need for species-specific relaxin peptides to be produced by alternative means. Solid-phase peptide synthesis has been the method of choice for producing relaxin, although recombinant production has also been used, especially to "scale up" production for clinical trials.

One of the greatest challenges in solid-phase synthesis of relaxin peptides is the synthesis and purification of the individual A and B chains. The B chain

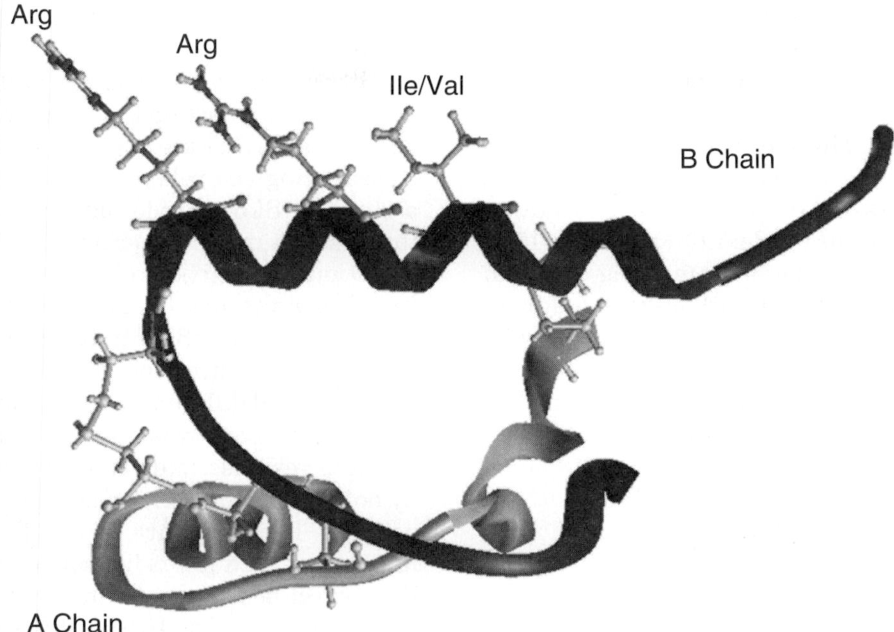

FIG. 11. Representation of the structure of H2 relaxin based on the crystal structure (163). The conserved residues incorporating the relaxin receptor binding motif (Arg-X-X-X-Arg-X-X-Ile/Val) are indicated, as are the separate A and B chains.

in particular is difficult to handle because of its poor solubility and tendency to aggregate (169). Usually, the choice of synthetic strategies and chain combination methods depends on the individual sequence, and is treated on a case-by-case basis. There are two main methods used to achieve the combination of the A and B chain chemically. The combination of the two chains in solution at high pH has been successfully used to synthesize relaxins from many species (10,170,171), but is ineffective for some relaxin peptides, proving completely unsuccessful for the synthesis of relaxin-3 (18) and INSL4 (172). This method also works quite effectively for synthesis of INSL3, in fact proceeding more efficiently than for relaxin synthesis (167–169). The use of novel cysteine side chain protecting groups allows the assembly of relaxin peptides through stepwise formation of the disulfide bonds. A variation of this method has been used successfully to synthesize relaxins from a variety of species (173–175) as well as INSL3 (122) and INSL4 (172,176). More recently, this method was also used to synthesize the first relaxin-3 peptide (18). Both of these methods have been extensively used to synthesize relaxin peptides from multiple species, as well as truncated and mutated peptides that have been invaluable in determining the key residues responsible for biological activity.

Many researchers have demonstrated the capability of recombinant systems to produce bioactive prorelaxin protein. Porcine prorelaxin has been produced in *E. coli* (153,177) and Chinese hamster ovarian cells (154). H2 prorelaxin has been expressed in yeast (151), *E. coli* (158,178), and mammalian cells through adenoviral infection (179) or by standard transfection (23). An insect cell expression system has been used to produce bioactive marmoset prorelaxin (156).

The need to produce large quantities of mature H2 relaxin peptide on a commercial scale led to the development of methods for large-scale recombinant production in *E. coli* (152). Initially, the A and B chains were produced separately in *E. coli* and the chains subsequently purified and combined to produce H2 relaxin (161). This material was used in the phase I and II clinical trials for cervical ripening (180). This method proved inappropriate for large-scale production of relaxin, and subsequently a one-chain process was developed. A human prorelaxin expression construct was designed with a mini–C peptide that would allow folding of the prorelaxin and enzymatic cleavage at the signal peptide/B, B/C, and C/A chain junctions to allow removal of the C chain and leader peptide. Expression of the mini–C peptide H2 prorelaxin in *E. coli* resulted in large quantities of insoluble protein in inclusion bodies. The protein was subsequently solubilized, purified, refolded, proteolytically processed, and purified again to produce mature H2 relaxin peptide that was identical to the circulating form (152). This material has been used in subsequent clinical trials (181) as well as in numerous in vitro and in vivo studies.

Other relaxin family peptides have also been produced recombinantly. Human prorelaxin-3 has been produced in mammalian cells with transfection of a PC1- and PC2-expressing cell line (AtT20), resulting in production of mature relaxin-3 peptide (128). A mammalian cell expression system using a human relaxin-3 expression construct with engineered furin cleavage sites and cotransfection with furin enzyme directs the production of mature human relaxin-3 peptide (23). An insect cell expression system has been used to produce marmoset pro-INSL3 (116).

Structure–Activity Relationships

Relaxin

Comparison of relaxin sequences from many species as well as testing of synthetic peptide analogs has defined many of the key structural features of relaxin. Relaxin A and B chains on their own do not display any biological activity (133). In addition, the intrachain molecular bond in the A chain is necessary for activity. Although the amino acids in the N-terminal regions of the A and B chains are not directly involved in activity, retaining secondary structure in this region of the chains is essential (182–184). The C-terminal region of the B chain does not appear to be important; up to eight amino acids can be deleted from this region of pig relaxin without loss of activity (182). One of the invariant amino acids across all relaxin peptides, the glycine residue at A14 adjacent to the cysteine residue, is absolutely essential for maintaining chain flexibility (185). Hence, the overall structure of the A chain seems to be more important than the individual amino acids in the chain. It would seem that the main role of the A chain is to act as a scaffold to hold the B chain in the correct formation. In contrast, individual amino acids in the B chain are important for receptor binding. The absolutely conserved arginine B13 and B17 residues (Fig. 3) are essential for interaction with the relaxin receptor (186). More recently, it has been demonstrated that the B20 isoleucine residue together with the two arginines is essential for relaxin binding (126). Furthermore, this residue can be substituted to valine without loss of activity, and virtually all native relaxins have either a valine or isoleucine in this position. Therefore, these amino acid residues, Arg-X-X-X-Arg-X-X-Ile/Val (Fig. 3), constitute what is now regarded as the "relaxin receptor binding motif." These residues are also conserved in the relaxin-3 peptide and are the basis for its interaction with relaxin receptors (18) (Fig. 7; see section on Relaxin Family Peptide Receptors).

Although these residues are undoubtedly the key to relaxin-specific actions, there are other structural determinants that are essential for full relaxin activity. Substituting the relaxin receptor binding motif into a sheep INSL3 peptide did not result in a dramatic increase in relaxin activity (187,188). Similarly, insertion of a partial relaxin receptor binding motif into a porcine insulin analog did not result in full relaxin activity (189). It is unlikely that there are other key residues for receptor interaction owing to the marked differences in relaxin structure between species; however, what is clear is that a combination of these residues and tertiary structure determined by both the A and B chain is necessary for full relaxin activity. The structure of the B chain is especially important, with full α-helical structure necessary for the correct positioning of the relaxin receptor binding motif (173,186). The INSL3 analogs with a full relaxin receptor binding motif demonstrate significantly reduced α-helicity compared with H2 relaxin (187). Interestingly, surface regions on the relaxin structure involved in receptor binding and that comprise the relaxin receptor binding motif were accurately predicted from molecular modeling studies more than 20 years ago (190).

INSL3

Although all INSL3 peptides contain a partial "relaxin-like" binding motif with two arginine residues in a pattern like relaxin (Fig. 12), these are displaced towards the C terminus by four residues, and INSL3 peptides only show minimal relaxin activity (122,167,187). Furthermore, INSL3 peptides only show limited ability to interact with the relaxin receptor (LGR7) (191) and are the cognate ligands for a related receptor (LGR8) (22) (see section on Relaxin Family Peptide Receptors). The determinants for INSL3 binding to its receptor are located in the N terminus of the B chain (192), in particular a tryptophan residue at B27 (Fig. 12).

RELAXIN BINDING AND SIGNALING

Relaxin Binding Sites

Because the identity of the relaxin receptor was unknown until recently, most of the research investigating the target tissues for relaxin has used labeled relaxin peptides to determine its sites of action. Relaxin binding sites have been identified in numerous reproductive tissues as well as some nonreproductive tissues.

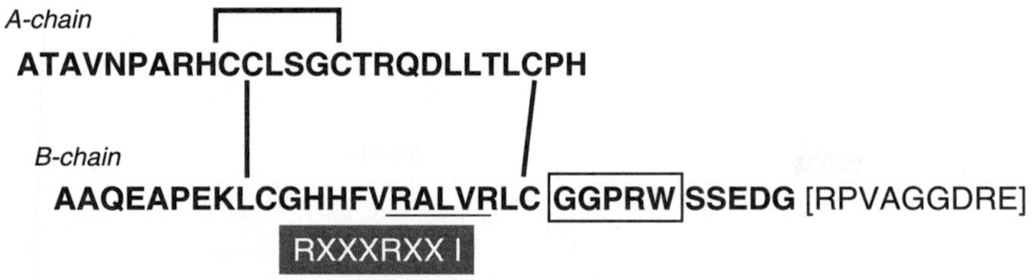

A-chain
ATAVNPARHCCLSGCTRQDLLTLCPH

B-chain
AAQEAPEKLCGHHFVRALVRLC GGPRW SSEDG [RPVAGGDRE]

RXXXRXX I

FIG. 12. Primary structure of synthetic sheep insulin-like peptide 3 (INSL3). The extra residues at the C-terminal end of the B chain (*light font*) have been shown to be present in the native INSL3 extracted from bovine testis (123). The residues implicated in binding of INSL3 to its receptor (192) are *boxed*, and the displaced relaxin-like receptor binding motif is *underlined*. The position of the relaxin receptor binding motif included in some INSL3 analogs (187) (see text) is indicated below the B-chain sequence. (From Ivell, R., and Bathgate, R. A. [2002]. Reproductive biology of the relaxin-like factor [RLF/INSL3]. *Biol. Reprod.* 67, 699–705, with permission.)

Early studies used purified porcine relaxin labeled by two alternative methods: (a) iodination of multiple tyrosine residues added to the N terminus of porcine relaxin to produce [^{125}I]-polytyrosyl-relaxin (193); or (b) incorporation of a ^{125}I group on the N terminus using the Bolton-Hunter reagent (194) to produce [^{125}I]-porcine relaxin (195). Both methods produced a product that retained its bioactivity and hence ability to bind to relaxin receptors, although the final product was not always completely pure and was often a mix of multiply iodinated forms. These ligands were subsequently used to perform the first characterization of the localizations and concentrations of relaxin receptors in multiple tissues from various species. Importantly, specific binding was not displaced by insulin, IGF-1, or IGF-2. Relaxin binding sites were identified in the uterus of rats (195–197), mice (195,198), and pigs (199); the pubic symphysis of mice (195,198) and guinea pigs (195); the cervix of pigs (199), rats (200), and guinea pigs (195); the rat mammary gland (195); human fetal membranes (201); and in fibroblasts from the mouse pubic symphysis and human skin (202).

H2 relaxin contains a tyrosine residue that can be labeled by conventional chloramine-T iodination (203). Hence, [^{125}I]-H2 relaxin has been demonstrated to bind to epithelial, stromal, and smooth muscle cells in human fallopian tube and smooth muscle cells in the arterioles (203). It has been used to demonstrate specific binding to fibroblasts isolated from human lower uterine segments (204). The authors demonstrated that stimulation of these cells with H2 relaxin resulted in the tyrosine phosphorylation of a 220-kDa protein, consistent with the relaxin receptor being a tyrosine kinase. This ligand has also been used to characterize the binding activity of relaxin analogs (126).

H2 relaxin has also been used for relaxin binding studies through phosphorylation of a long form (B33) of H2 relaxin with ^{32}P (205) and subsequently ^{33}P (206). High-pressure liquid chromatography purification of the phosphorylated product results in a specific high-activity radioligand monolabeled at the B32 serine residue (205,206). [^{32}P]-H2 relaxin was initially demonstrated to bind specifically to rat uterus, cervix, and brain (205). Specific binding was not displaced by insulin or IGF-1. The ligand has subsequently been used to demonstrate specific relaxin binding sites in the rat heart atrium (207), rat atrial cardiomyocytes (208), and numerous regions of the rat brain (209,210). In addition, binding sites have been identified in human uterine cells (208), human fetal membranes (211), and in the human monocytic cell line THP-1 (212). More recently, [^{33}P]-H2 relaxin has been used to demonstrate relaxin binding sites in the rat uterus, atria, and brain (206) (Fig. 13), and for numerous structure–function studies (18,104,191).

Cross-linking of [^{32}P]-H2 relaxin to human uterine cells and rat atrial cardiomyocytes demonstrated binding to a protein of at least 200 kDa that was postulated (208) to be a tyrosine kinase. These data are consistent with the 220-kDa protein that was tyrosine phosphorylated after relaxin stimulation in human lower uterine segment fibroblasts (204). These data, together with other studies outlined later, were strong evidence for the relaxin receptor being a tyrosine kinase and were probably part of the reason that the relaxin receptor defied discovery for many years.

The methodology developed to iodinate porcine relaxin was also modified to biotinylate porcine relaxin (213). Biotinylated relaxin has subsequently been used to localize relaxin binding sites in the cervix, mammary gland, and nipple of rats (214); the cervix, mammary gland, nipple, small intestine, skin (215), ovary, and testis (216) of pigs; the uterus, vagina, cervix, mammary gland, nipple, and placenta of pregnant women (217); and the uterine

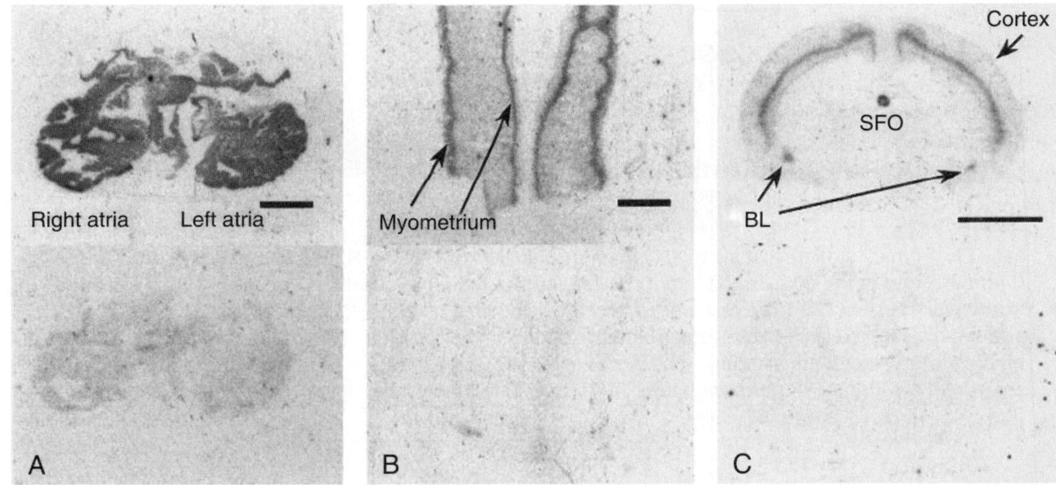

FIG. 13. Autoradiographic localization of [^{33}P]-H2 relaxin (B33) binding to slide-mounted sections of (**A**) rat atria with nonspecific binding (NSB) in lower panel (bar = 2 mm); (**B**) uterus with NSB in lower panel (bar = 2 mm); and (**C**) brain with NSB in lower panel (bar = 5 mm). Note that the specific binding was evenly distributed in atria but was confined to myometrium and specific areas in the brain, such as the fifth layer of cortex, subfornical organ (SFO), and basolateral amygdaloid nucleus (BL). (From Tan, Y. Y., Wade, J. D., Tregear, G. W., and Summers, R. J. [1999]. Quantitative autoradiographic studies of relaxin binding in rat atria, uterus and cerebral cortex: characterization and effects of oestrogen treatment. *Br. J. Pharmacol.* 127, 91–98. Copyright 1999 with permission from Macmillan Publishers.)

endometrium and myometrium of the marmoset monkey (218).

A monobiotinylated rat relaxin peptide has been produced by peptide synthesis and was demonstrated to be biologically active (219). This peptide has also been immobilized on the surface of a biosensor to study its interaction with THP-1 cell relaxin receptors (220).

Binding Sites for Other Relaxin Family Peptides

Very little is known of specific binding sites for other relaxin family peptides. An iodinated human INSL3 peptide has been prepared using the tyrosine residues in the sequence (122). [^{125}I]-INSL3 was used to demonstrate specific binding sites for INSL3 in the mouse uterus and brain (122). These binding sites were specific for INSL3 and had a 1,000-fold lower affinity for H2 relaxin. In addition, [^{125}I]-INSL3 has been chemically cross-linked to binding sites in the mouse uterus and demonstrated an approximately 200-kDa band (221) similar to that shown for relaxin receptor cross-linking (208).

H3 relaxin has also been labeled using chloramine-T to produce specific mono-[^{125}I]-H3 relaxin (23). This label was able to bind specifically to GPCR135 and GPCR142 receptors expressed in mammalian cells (see section on Relaxin Family Peptide Receptors). Because H3 relaxin also binds to the relaxin receptor (LGR7) with high affinity (18,191) (see section on Relaxin Family Peptide Receptors), determining specific receptor binding sites for H3 relaxin distinct from other relaxin binding sites is a challenge.

Relaxin Signaling

The specific signaling pathways whereby relaxin mediates its actions on numerous cell types are unknown. Some of these potential pathways are outlined in this section and are also detailed in specific sections in the context of relaxin physiology (see section on Physiological Roles of Relaxin in Female and Male Reproductive Processes).

Some of the earliest studies on the biochemical effects of relaxin have demonstrated its association with increased levels of cyclic adenosine monophosphate (cAMP) in target tissues (196,222,223). Relaxin increases intracellular cAMP in the mouse pubic symphysis (222), in rat uterus (196,223,224), and in cultures of rat (225) and rhesus monkey myometrium (226), human endometrium (227,228), rat anterior pituitary cell (229), and human THP-1 monocytes (212). These relaxin-stimulated increases in cAMP are important components in relaxin-induced myometrial inhibition (230) and decidualization of endometrial stromal cells (231,232).

Although relaxin stimulation is clearly linked to cAMP increases in many tissues, it is not clear exactly how relaxin activates cAMP. There seems to be at least some G-protein component of this increase in cAMP, which is consistent with the relaxin receptor LGR7 being a GPCR (21) (see section on Relaxin Binding and Signaling). Studies have demonstrated that cAMP elevation in THP-1 and human endometrial stromal cells can be blocked by tyrosine kinase or mitogen-activated protein kinase (MAPK) inhibitors (233), and relaxin activates MAPK signaling in these cells (234). It is therefore possible that relaxin is using a MAPK cascade to inhibit a cell-specific phosphodiesterase to induce sustained increases in cAMP (235) (Fig. 14). A separate study in THP-1 cells suggests that the sustained increases in cAMP are mediated by phosphoinositide 3-kinase and not by a phosphodiesterase (236). Although it is still likely that relaxin links to G proteins and activates adenylate cyclase in these cells, the sustained increases in cAMP are induced through these alternate pathways. In fact, these alternate mechanisms for producing a sustained increased in cAMP are an important aspect of relaxin's action in decidualization of endometrial cells and inhibition of myometrial quiescence (233).

Relaxin's actions in other cell types are not associated with sustained increases in cAMP but may still involve some component of G-protein–coupled cAMP signaling. A recent (2004) study indicated that in rat ventricular fibroblasts relaxin-mediated effects were associated with a weak, transient increase in cAMP (237). Earlier studies in human lower uterine segment fibroblasts did not demonstrate a link with cAMP, but did show that relaxin stimulation was linked to MAPK activation (204,238). Hence, in fibroblasts, primary G-protein coupling to cAMP release is not the predominant signaling mechanism, but it may link to MAPK signaling, which in turn exerts the biological responses.

There is also evidence that relaxin's effects on increasing heart rate may be mediated by cAMP (239). Rabbit sinoatrial node cells respond to relaxin, and its effects appear to be mediated by increases in intracellular cAMP and activation of cAMP-dependent protein kinase (239). Relaxin also stimulates the secretion of atrial natriuretic peptide from isolated perfused rat hearts (240). The authors demonstrated that these effects were mediated through activation of protein kinase C, although treatment with a cAMP-dependent protein kinase inhibitor also blocked the atrial natriuretic peptide secretion as well as the chronotropic response to relaxin, suggesting an involvement of cAMP. Indeed, one study has demonstrated that relaxin directly stimulates cAMP production in rat atria, although the authors indicated that there was only a weak coupling to cAMP (241).

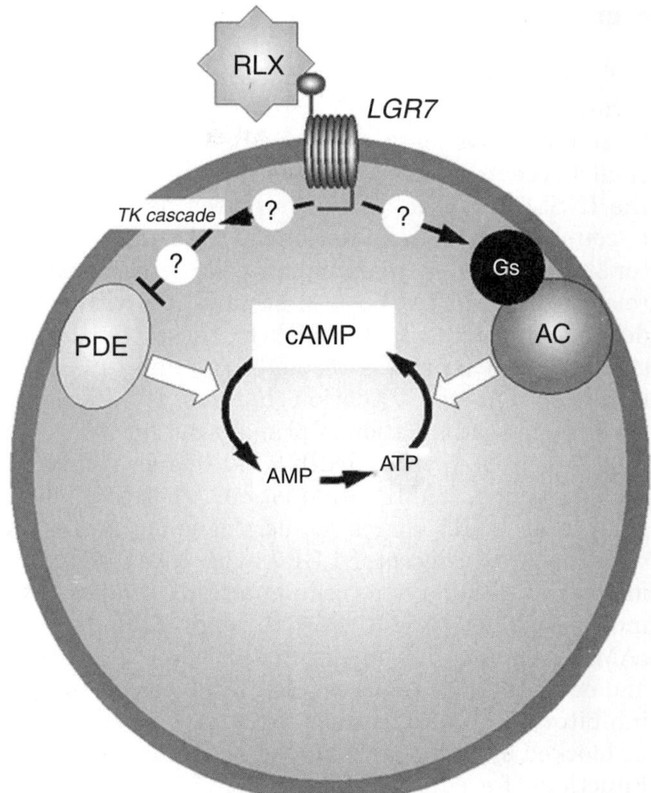

FIG. 14. Model of possible signal transduction mechanisms generating cyclic adenosine monophosphate (cAMP) in relaxin-stimulated cells. Conventionally, the relaxin receptor, LGR7, would be expected to stimulate adenylyl cyclase (AC) through a G_s-protein intermediate. However, considerable pharmacological evidence supports an alternative pathway, involving a tyrosine kinase (TK) cascade (probably using a mitogen-activated protein kinase) to inhibit a specific phosphodiesterase (PDE), thereby causing an effective upregulation of cAMP. (See text for details.) (Reprinted from Ivell, R., and Einspanier, A. [2002]. Relaxin peptides are new global players. *Trends Endocrinol. Metab.* 13, 343–348. Copyright 2002, with permission from Elsevier.)

Numerous studies have demonstrated that relaxin can exert its actions in some tissues, including the heart and blood vessels, by increasing the expression or activity of nitric oxide synthase isoenzymes, thereby promoting the generation of nitric oxide (29,242–245). However, the mechanisms by which relaxin exerts these actions are unknown.

Now that the identity of the relaxin receptor is known, it is likely that there will be an acceleration in the understanding of how relaxin mediates its cellular actions.

Signaling by Other Relaxin Family Peptides

Very little is known about the mechanisms involved in signaling of other relaxin family peptides. Most data have been gleaned from studies on the cloned receptors in transfected cell systems. Hence, the INSL3 receptor, LGR8, has been demonstrated to couple to cAMP stimulation (21). When the receptor in expressed in mammalian cells, it responds to relaxin (21) and INSL3 (22) stimulation with dose-dependent increases in cAMP. In these transfected cells, the mechanism is through coupling to G_s protein (21), but the situation in vivo is unknown. Nevertheless, stimulation of primary rat gubernacular cells in culture by both INSL3 and relaxin results in dose-dependent cAMP stimulation. Therefore, it is likely that LGR8 shares similar signaling pathways with the relaxin receptor LGR7, with cAMP being an important mediator of its cellular actions. Studies have indicated that INSL3 acts through decreases in cAMP to induce its effects in testicular germ cells and oocytes (246). In germ cells, INSL3 activates the inhibitory G protein because the effects of INSL3 can be blocked by pertussis toxin (see section on Gonadal Functions, for details).

H3 relaxin also stimulates cAMP accumulation in LGR7-transfected mammalian cells (191) and THP-1 cells (18). Hence, signaling by relaxin-3 through the LGR7 receptor is likely to be the same as that by relaxin. However, the result of relaxin-3 signaling through the GPCR135 and GPCR142 receptors is inhibition of cAMP (see section on Relaxin Family Peptide Receptors). These actions can be blocked by pertussis toxin, indicating that the receptor is coupled to G_I, G_o, or G_z (23,247). The potential signaling mechanisms for relaxin-3 through GPCR132 GPCR142 in vivo are unknown.

RELAXIN FAMILY PEPTIDE RECEPTORS

Because of their two-chain structure, relaxin and INSL3 genes have traditionally been thought to belong to the insulin ligand family, and several of the relaxin paralogs have been designated INSLs 3 to 7 based on their order of discovery (see section on Discovery and Cloning of Novel Relaxin Family Genes). Based on the hypothesized coevolution of polypeptide ligands and their receptors, the receptors for relaxin and INSL3 were believed to be related to the known insulin receptors with tyrosine kinase activity. However, attempts to purify the relaxin receptors were hampered by the high levels of nonspecific binding of the tracer and the low levels of relaxin and INSL3 binding sites in target tissues.

Advances in genome sequencing allowed identification of novel genes based on their sequence relatedness to known genes in the hormonal signaling pathway (248). Searches for paralogs of the known gonadotropin and thyrotropin receptors led to the identification of a group of GPCRs called LGRs. LGRs, similar to glycoprotein hormone receptors (249,250), are mosaic proteins that contain an extracellular domain with multiple leucine-rich repeats (LRRs) important in ligand binding, and a GPCR transmembrane domain. Studies of LGRs from different species suggest that three LGR subtypes (A, B, and C) evolved during the early evolution of metazoans (Fig. 15), and that each subtype of LGR shares a similar LRR domain and a unique hinge region between the LRR and the transmembrane region. The type A LGRs include the follicle-stimulating hormone receptor (FSHR), the luteinizing hormone receptor (LHR), and the thyroid-stimulating hormone receptor (TSHR), important for signaling of the heterodimeric glycoprotein hormones FSH, LH, and TSH, respectively. In mammals, the type B LGR comprises three members, LGRs 4 to 6, which remain orphan GPCRs without known ligands at present. By contrast, type C LGRs have only two members, LGR7 and LGR8. Because type A LGRs and the coevolved genes encoding glycoprotein hormone subunits could be traced to both nematodes and insects, it was concluded that the three LGR subtypes evolved before the emergence of vertebrates and nematodes (250). Therefore, the type C LGR signaling pathway represents one of the earliest forms of

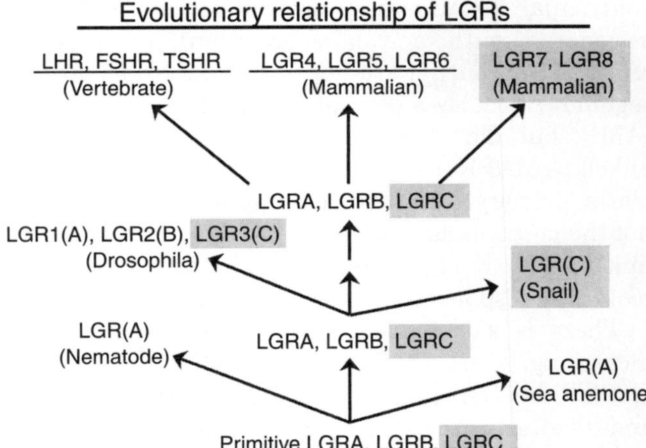

FIG. 15. Hypothetical model for the evolution of three subgroups of leucine-rich repeats containing G-protein–coupled receptors (LGRs) in the metazoan. The three types of LGR, A, B, and C, appear to occur early during metazoan evolution. The type C LGRs are orthologs of mammalian LGR7 and LGR8, and are *shaded*. Although *Caenorhabditis elegans* lacks a type B and type C LGR, they appear to have been lost in the nematode lineage after the divergence of nematodes from other metazoans. Similar to vertebrates, *Drosophila* encodes three subtypes of LGRs homologous to the mammalian types A, B, and C LGRs. FSHR, follicle-stimulating hormone receptor; LHR, luteinizing hormone receptor; TSHR, thyroid-stimulating hormone receptor.

GPCR signaling. A diagram of LGR7 is shown in Fig. 16. The ectodomain consists of a low-density lipoprotein cysteine-rich region, followed by an alternative splicing region and the LRRs. The ectodomain is connected to the seven transmembrane spanning region, which is followed by the C-terminal tail. Based on the differential splicing in the ectodomain, different isoforms of the receptors can be generated.

Based on the comparison of phenotypes of mice deficient in INSL3 (251,252) and mice lacking a 550-kilobase region of chromosome 3 that contained the LGR8 gene (253), it was hypothesized that the relaxin family peptides could function as cognate ligands for type C LGRs (21). Indeed, functional characterization established that porcine relaxin activates both LGR7 and LGR8, with increases in cAMP (21) (Fig. 17), and subsequent studies demonstrated they were also activated by H2 relaxin (191). LGR7 and LGR8 have more than 700 residues, share approximately 60% amino acid sequence identity, and contain 10 LRRs in their extracellular domain. LGR7 transcripts have been identified in reproductive tissues, as well as nonreproductive tissues such as the brain, kidney, heart, and lung, where actions of relaxin have been reported (see section on Physiological Roles of Relaxin in Female and Male Reproductive Processes).

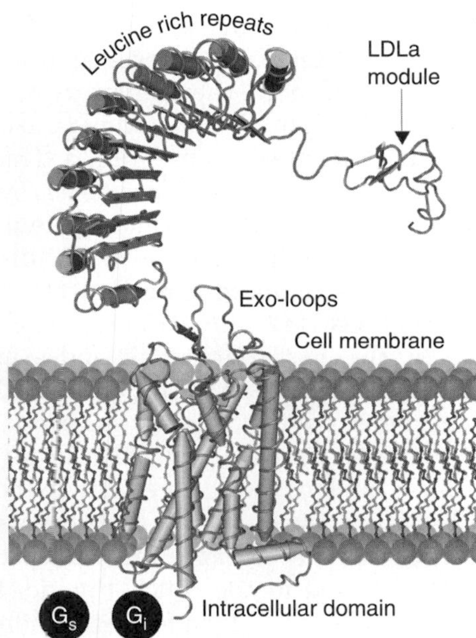

FIG. 16. Schematic representation of the putative structure of the relaxin receptor, LGR7. The major structural features include seven transmembrane spanning domains, intracellular and extracellular (exoloops) loops, extracellular leucine rich repeats (LRRs), and low-density lipoprotein class A (LDLa) module. The model was constructed based on the structures of bovine rhodopsin (805) for the transmembrane domains, porcine ribonuclease inhibitor (806) for the LRRs, and complement-like repeat CR8 from the LDL receptor–related protein (807) for the LDLa module.

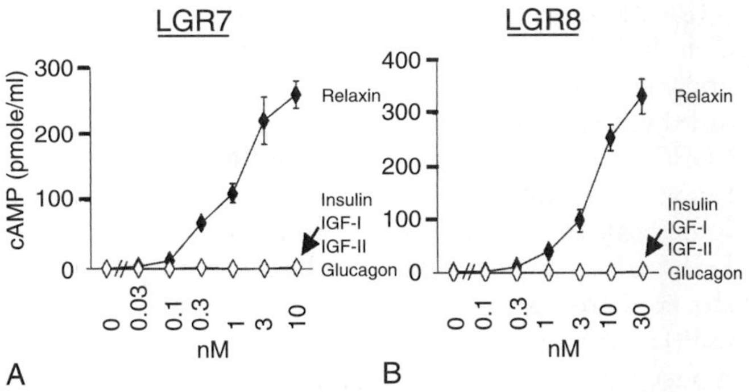

FIG. 17. The initial discovery that LGR7 and LGR8 respond to relaxin. Porcine relaxin stimulated dose-dependent cyclic adenosine monophosphate (cAMP) production in transfected 293T cells (10^5 cells per culture) expressing LGR7 (**A**) or LGR8 (**B**) using the pcDNA3.1-Zeo expression vector (21). In contrast, treatment with insulin, insulin-like growth factor (IGF)-1, IGF-2, or glucagon (a known G_s-protein–coupled receptor activator) had no effect on cAMP production. (From Hsu, S. Y., Nakabayashi, K., Nishi, S., Kumagai, J., Kudo, M., Sherwood, O. D., and Hsueh, A. J. [2002]. Activation of orphan receptors by the hormone relaxin. *Science* 295, 671–674, with permission.)

After the identification of LGR7 and LGR8 as relaxin receptors, the closely related relaxin-3 and INSL3 have been shown to function as selective agonists for LGR7 and LGR8, respectively (22,191). In addition, the ectodomains of these LGRs were shown to be important for ligand binding, similar to that found for type A LGRs (254). When the soluble ligand-binding portion of the human relaxin receptor LGR7 was administered subcutaneously to antagonize endogenous circulating relaxin during the last 4 days of mouse pregnancy, delivery was delayed by 27 hours and nipple development was retarded (21). Although the observed nipple phenotype is consistent with earlier work in relaxin (255) and LGR7 (256) null mice, the observed delay in parturition seems inconsistent. However, delivery of pups is prolonged in rats treated with a monoclonal antibody against relaxin (257), and some relaxin KO mice were observed to have difficulties in giving birth (255).

The rat and mouse orthologs of LGR7 have been cloned (258). At the protein level, mouse LGR7 and rat LGR7 are 85.2% and 85.7% identical to human LGR7, respectively. Expression of the receptor mRNA was seen in the cerebral cortex, ventricle, atrium, lung, nipple, endometrium, myometrium, uterus, cervix/vagina, and placenta, all known sites of relaxin action and relaxin binding. Both rat and mouse LGR7 transfected in mammalian cells bound H2 relaxin with high affinity, resulting in cAMP release. Both receptors had a higher affinity for rat relaxin than the human receptors. Both receptors also bind H3 relaxin with high affinity (R. A. D. Bathgate, unpublished data). Hence, these rodent LGR7 orthologs demonstrate all the characteristics of relaxin receptors.

Based on the preceding information and the ability of individual ligands to activate each receptor, it is apparent that relaxin is the cognate ligand for LGR7, although relaxin peptides from some species activate LGR8 at higher concentrations. Importantly, rat relaxin does not activate human, rat, or mouse LGR8, indicating that in rodents relaxin is not a ligand of LGR8 (R. A. D. Bathgate, unpublished data). INSL3 is the cognate ligand for LGR8; the LGR8 KO mouse (259) shows an identical phenotype to the INSL3 KO mouse (251,252). INSL3's actions on day 17 fetal rat gubernaculum demonstrated using an organ culture system (260), and cultured fetal gubernaculum cells in vitro (22) were mediated through LGR8, firmly establishing INSL3 as the endogenous ligand for LGR8. Additional studies using cross-breeds of INSL3-overexpressing mice and LGR8 and LGR7 KO mice have firmly established that LGR8 is the only receptor for INSL3 and that there is no interaction between the INSL3/LGR8 and relaxin/LGR7 signaling systems in vivo (261,262). These data have led to a better understanding of relaxin and INSL3 physiology and provide a basis for the future characterization of these signaling systems.

The findings that two orphan GPCRs are the receptors for relaxin and INSL3, respectively, are surprising. Thus, in spite of their structural similarity, relaxin and insulin family peptides act through two independent signaling pathways: the relaxin group activates the GPCRs, whereas the insulin group activates tyrosine kinase receptors (Fig. 18). In contrast, type A LGRs are activated by glycoprotein hormones, whereas insulin and related ligands activate insulin and IGF receptors. Phylogenetic analysis of LGRs and coevolved relaxin family peptides from different metazoans suggests that, whereas the number of relaxin receptors remained constant during vertebrate evolution, the ancestor gene for relaxin duplicated multiple times in a vertebrate branch-specific manner (19) (Fig. 9; see section on Evolution of the Relaxin Genes). Therefore, relaxin family peptides from different branches of vertebrates might have adapted to distinct physiological roles through a limited number of receptor genes. Although preliminary studies with synthetic INSL4 (172) and INSL6 (261) peptides indicate that they do not interact with LGR7 or LGR8, future studies with native forms of these peptides may show definitively whether these additional relaxin family ligands can interact with LGR7 or LGR8.

In contrast to the observed high-affinity interactions between relaxin and LGR7, and INSL3 and LGR8, relaxin-3 has a lower affinity for LGR7 than relaxin (18,191). A recent study indicates that relaxin-3 is a ligand for two orphan receptors, GPCR135 (also known as the somatostatin- and angiotensin-like

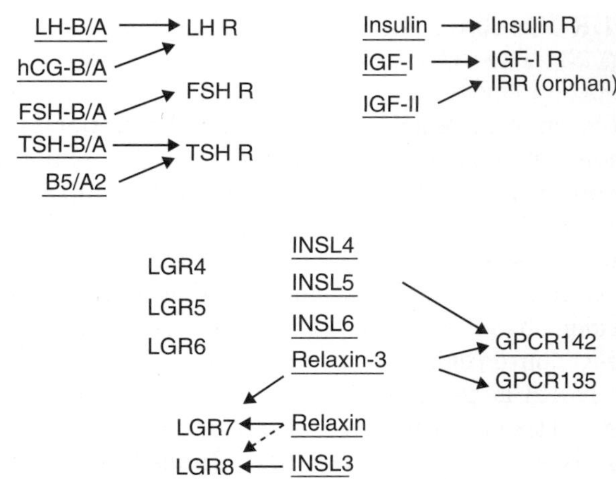

FIG. 18. Activation of leucine-rich repeat–containing G-protein–coupled receptor (LGR) subtypes by different polypeptide ligand families and the distinct signaling pathways activated by relaxin and insulin family ligands. The type A LGRs, luteinizing hormone receptor (LH R), follicle-stimulating hormone receptor (FSH R), and thyroid-stimulating hormone receptor (TSH R), are activated by glycoprotein hormones and the newly discovered $\alpha 2/\beta 5$ heterodimer (808), whereas the type C LGRs, LGR7 and LGR8, are activated by relaxin, insulin-like peptide (INSL) 3, and relaxin-3. In contrast, relaxin-3 is capable of activating GPR135 and GPR142, and INSL5 can activate GPCR142.

peptide receptor [SALPR]) and GPCR142 (23,247). These putative relaxin-3 receptors differ structurally and functionally from LGR7 and LGR8. They have relatively short N-terminal extracellular domains, and they are coupled to the inhibitory G protein important for cAMP inhibition. Studies using both native relaxin-3 purified from brain extracts and recombinant human relaxin-3 produced by transfected cells indicate that this hormone potently stimulated guanosine triphosphate-γS binding and inhibited cAMP accumulation in cells overexpressing GPCR135 and GPCR142. GPCR135 is predominantly expressed in the paraventricular and supraoptic nuclei of the hypothalamus in the rat brain (23), regions with connections to the nucleus incertus, where relaxin-3 is also expressed (18,127). These are regions where relaxin binding sites were also found (209). GPCR142 binds to relaxin-3 with slightly lower affinity than GPCR142 and is expressed less specifically, with mRNA detected in the colon, thyroid, salivary gland, prostate, placenta, thymus, testis, kidney, and brain (23). Although relaxin, INSL3, and INSL6 were shown not to interact with GPCR135 and GPCR142, one study has demonstrated that INSL5 is a specific ligand for GPCR 142, but not GPCR135 (24). The authors demonstrated that the expression of INSL5 and GPCR142 overlaps extensively, indicating that GPCR142 is likely the endogenous receptor for INSL5.

SOURCES AND SECRETION OF RELAXIN IN THE FEMALE

This section describes the sources and secretion of relaxin in various mammalian species. The sources and secretion of other relaxin family peptides are not discussed here and can be found in the section on Discovery and Cloning of Novel Relaxin Family Genes; the section on Physiological Roles of INSL3; and the section on Physiological Roles of Other Relaxin Family Peptides.

Relaxin is produced in highest levels by female reproductive tract tissues during pregnancy. Primary sources are the corpus luteum, placenta, and uterus. The tissue that is the primary source of the relaxin circulating in the peripheral blood varies among species. Table 2 contains a summary of known sources of relaxin in several mammalian species. No clear pattern of pregnancy-related physiological factors that might predict the source of relaxin within species has emerged. With some species, such as the pig, rat, and mouse, where luteal progesterone production is required throughout pregnancy, the corpus luteum is the primary source of relaxin. However, in the dog, rabbit, and golden hamster, where luteal progesterone also is required throughout pregnancy, the placenta is the apparent primary source of relaxin. The horse, guinea pig, and human, species in which the corpora lutea are not the sole source of progesterone throughout pregnancy, produce most if not all of their circulating relaxin in the placenta, uterus, and corpus luteum, respectively.

The cloning of the relaxin genes and advent of RT-PCR and in situ hybridization have enabled the identification of secondary tissues in some species that appear to produce small amounts of relaxin mRNA. The relaxin produced in these so-called secondary sources may have local physiological functions, but appears to be produced in quantities unlikely to contribute appreciably to circulating relaxin levels. In this section, for each species, the tissue thought to be the source of secreted relaxin and blood levels of the hormone is described before the putative secondary tissue source or sources. Regulation of relaxin *synthesis* varies extensively among species [reviewed in Sherwood (25)]. For example, rat ovarian relaxin levels remain low unless the animal is pregnant, whereas pig ovarian relaxin can increase to high levels in the absence of pregnancy. Regulation of relaxin *secretion* also varies among species; consequently, the profiles of relaxin levels in the peripheral blood throughout pregnancy differ strikingly among species. Hence, the sources and secretion of relaxin are described in the following for individual domestic species, laboratory species, and primates. Emphasis is on the three species studied most extensively—the pig, rat, and human.

TABLE 2. *Sources of relaxin in female mammals*

Species	Ovary needed throughout pregnancy	Primary source of relaxin	Secondary source(s) of relaxin
Pig *(Sus scrofa)*	Yes	Corpus luteum (273,276)	Uterus[a] (296) Theca interna (679)
Horse *(Equus caballus)*	No	Placenta (69,105)	Ovary (408)
Camel *(Camelus dromedarius)*	Yes	Corpus luteum (108)	Uterus (108)
Cat *(Felis catus)*	No	Placenta (107,429)	
Dog *(Canis familiaris)*	Yes	Placenta (106,430,431)	Ovaries[a] (430)
Rat *(Rattus norvegicus)*	Yes	Corpus luteum (305,307)	Uterus[a] (306) Heart, brain (210) Placenta[a], uterus[a], kidney[a], pancreas[a] mammary gland[a], brain, heart (299)
Mouse *(Mus musculus)*	Yes	Corpus luteum (101,347)	Brain, thymus[a], heart[a], kidney[a], lung[a], liver[a], spleen[a], skin[a], mammary gland[a] (18)
Guinea pig *(Cavia porcellus)*	No	Uterus (439,440)	Mammary gland[a] (445)
Rabbit *(Oryctolagus cuniculus)*	Yes	Placenta (418,803)	Uterus (418,419)
Golden hamster *(Mesocricetus auratus)*	Yes	Placenta (102,426)	
Hyena *(Crocuta crocuta)*	?	Placenta (446)	
Tammar wallaby *(Macropus eugenii)*	No	Corpus luteum (103,104)	Placenta, endometrium, follicle (103)
Human *(Homo sapiens)*	No	Corpus luteum (332)	Decidua, placenta (667,804) Mammary gland[a] (386, 387) Fallopian tube (203) Heart (390)

[a]Putative secondary sources of relaxin that are not well established.

Pig

Estrous Cycle

Relaxin is produced in preovulatory follicles as well as in corpora lutea during the approximately 21-day estrous cycle in the pig (*S. scrofa*). The theca interna is the source of relaxin in the preovulatory follicle as demonstrated by relaxin secretion from cultured thecal tissue (263), immunohistochemistry (264–266), Northern blotting (92), and in situ hybridization (267). After ovulation and with differentiation of the theca cells and granulosa cells to form the corpus luteum, both cell types appear to produce relaxin (265,268). Relaxin production in the corpus luteum appears to be highest during the luteal phase of the estrous cycle (days 5 to 16) and declines to low levels after luteolysis. Relaxin bioactivity in ovarian extracts (50,269,270), immunoreactivity in corpus luteal sections (271) and in luteal extracts (266), and relaxin mRNA (272) are all highest during the luteal phase. However, the content of relaxin in the corpora lutea during the cycle (273), and consequently blood levels (274), are much lower during the estrous cycle than in pregnancy.

Pregnancy

The corpora lutea are the principal, and perhaps the sole, source of the relaxin secreted into the peripheral circulation during the approximately 114-day gestation period. Using immunohistochemical techniques, relaxin immunoactivity was found associated with the cytoplasm of cells from the corpora lutea but not other ovarian components (266,271,275–277) (Fig. 19A). Bioactive relaxin levels in the corpora lutea increase steadily from approximately day 20 of pregnancy until approximately day 110, and then decline rapidly within 16 hours of birth (273,278). Consistent with these findings, ovarian relaxin mRNA levels were reported to be approximately 50-fold greater from day 40 to 90 of pregnancy than on day 13 of the estrous cycle, and to decline approximately 1,000-fold by day 2 of lactation (272).

Large luteal cells, but not small luteal cells, show relaxin immunostaining (265,266,277–279). The large luteal cells have the organelles associated with both steroid and protein synthesis (278,280), and it seems likely that both progesterone and relaxin are produced by these cells. Dense membrane-limited cytoplasmic granules (200 to 600 nm in diameter) are observed in the luteal cells of pregnant pigs (Fig. 19B), and relaxin immunoactivity is associated with these cytoplasmic granules (276,279) (Fig. 19C). The amount of relaxin stored in each granule may increase during pregnancy (276).

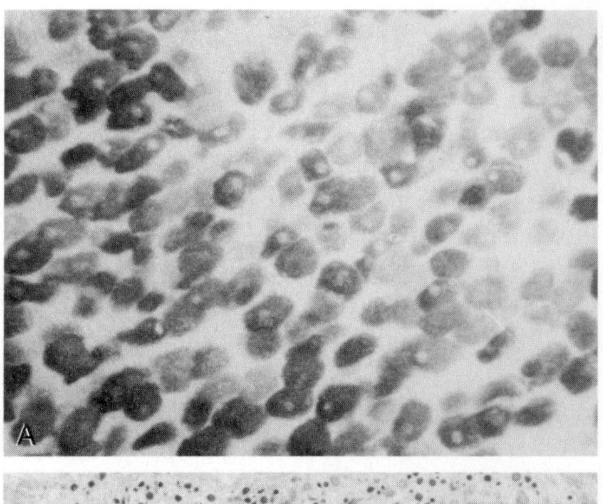

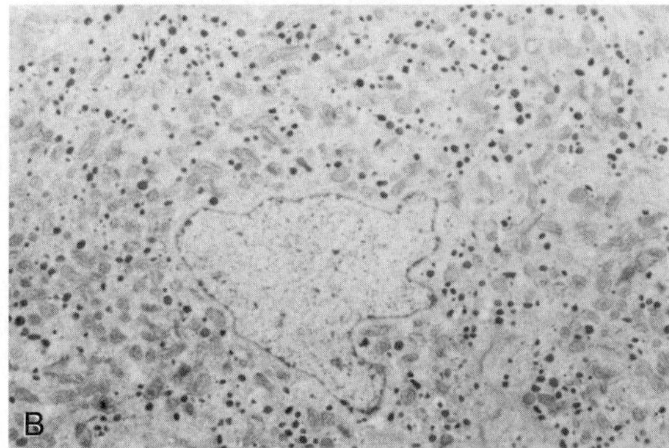

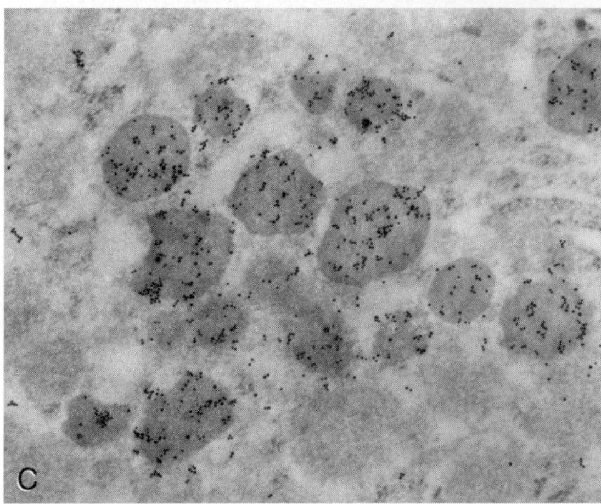

FIG. 19. A: Light microscopy immunohistochemical localization of relaxin in cytoplasm of pig luteal cells on day 106 of pregnancy using rabbit anti-porcine relaxin serum and peroxidase-antiperoxidase as marker. Note that nuclei do not stain. **B:** Electron microscopy of a portion of pig luteal cells on day 110 of pregnancy when granule content is maximal. (Original magnification × 2,800.) **C:** Electron microscopy immunocytochemical localization of relaxin in pig luteal granules (200 to 600 nm diameter) on day 106 of pregnancy using rabbit anti-porcine relaxin serum and goat anti-rabbit immunoglobulin G-colloidal gold (10 nm) as marker. (Original magnification × 37,400.) (**A** and **C** from Fields, P. A., and Fields, M. J. [1985]. Ultrastructural localization of relaxin in the corpus luteum of the nonpregnant, pseudopregnant, and pregnant pig. *Biol. Reprod.* 32, 1169–1179, with permission; **B** from Belt, W. D., Anderson, L. L., Cavazos, L. F., and Melampy, R. M. [1971]. Cytoplasmic granules and relaxin levels in porcine corpora lutea. *Endocrinology* 89, 1–10, with permission. Copyright 1971, The Endocrine Society.)

The profile of relaxin levels in the peripheral blood throughout pregnancy is consistent with the view that relaxin accumulates in corpora lutea during most of pregnancy and is released during the rapid degranulation that occurs during the 2 days before birth (278). Numerous studies have determined levels of relaxin immunoactivity in peripheral blood during pregnancy (281–292). Plasma relaxin levels remain below 2 ng/mL until approximately day 100 and then increase gradually to approximately 10 ng/mL 3 days before delivery (281,286). During the 2 days before birth, relaxin levels increase markedly and attain maximal levels, which usually range from 50 to 250 ng/mL (282–295) (Fig. 20). Relaxin levels decline to less than 1 ng/mL by 24 to 48 hours after delivery (283,286,287,294).

Other Sources of Relaxin

There is limited evidence that small amounts of relaxin may also be produced by extraovarian sites. Relaxin immunoactivity and mRNA expression have been detected during early pregnancy in pigs (296). There is some evidence that relaxin may also be produced by the pituitary (272,297).

Rat

Estrous Cycle and Pseudopregnancy

Both relaxin immunoactivity (270) and relaxin mRNA (298,299) are extremely low, but detectable, in ovarian extracts obtained during the estrous cycle in rats (*R. norvegicus*). They are both maximal at estrus (270,298). In pseudopregnant rats, in which luteal function and elevated progesterone levels are maintained approximately 18 days, ovarian relaxin levels are approximately 20-fold greater than those in cycling rats (270), but are extremely low compared with those in the ovaries of rats after day 10 of pregnancy (270,300,301).

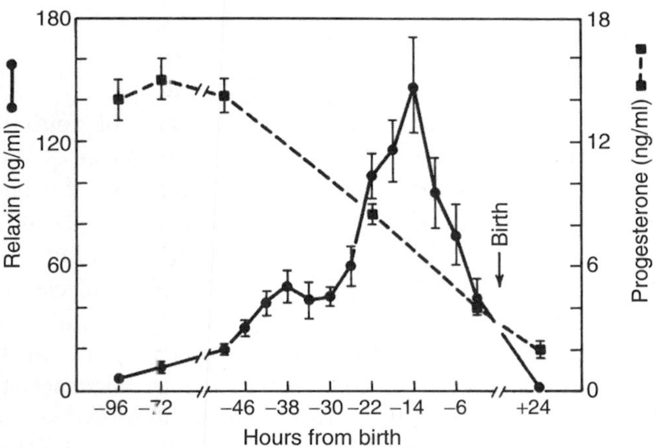

FIG. 20. Mean relaxin and progesterone immunoreactivity levels (±SE) in peripheral blood plasma of pigs from 96 hours before to 24 hours after birth. (From Sherwood, O. D., Chang, C. C., Bevier, G. W., and Dziuk, P. J. [1975]. Radioimmunoassay of plasma relaxin levels throughout pregnancy and at parturition in the pig. *Endocrinology* 97, 834–837, and Sherwood, O. D., Nara, B. S., Crnekovic, V. E., and First, N. L. [1979]. Relaxin concentrations in pig plasma after the administration of indomethacin and prostaglandin F2 alpha during late pregnancy. *Endocrinology* 104, 1716–1721, with permission. Copyright 1975 and 1979, The Endocrine Society.)

Pregnancy

As in the pig, the corpora lutea in rats are the source of both the relaxin (8,302–307) and progesterone (308,309) released into the peripheral blood during the approximate 23-day gestation period. Ovarian levels of relaxin bioactivity, relaxin immunoactivity, prorelaxin immunoactivity, and relaxin mRNA increase from approximately day 10 of pregnancy to maximal levels 2 or 3 days before birth, and then decline rapidly to low levels by 2 days postpartum (8,298,300,301,310) (Fig. 21). Immunohistochemical localization of relaxin with an antiserum to rat relaxin (306,307) and hybridization–histochemical localization of rat relaxin mRNA (8) identified the corpora lutea, but not other ovarian elements, as the ovarian source of relaxin during the second half of pregnancy. Unlike the pig, relaxin immunostaining was associated with both small (20 to 25 μm in diameter) and large (>25 μm in diameter) luteal cells (306,307). As in the pig, both relaxin and progesterone are probably produced by the same cell because the luteal cells have organelles associated with both protein and steroid synthesis (311). Relaxin in these cells is stored in small, membrane-bound granules (100 to 270 nm in diameter) (304,306).

Evaluation of the profile of both relaxin and progesterone immunoactivity levels in the peripheral serum throughout pregnancy led Sherwood and coworkers to postulate that the regulation of relaxin release from day 10 to day 20 differs from its regulation during the 3 days immediately before birth, which are designated the antepartum period (312,313). During the first period, relaxin immunoactivity becomes detectable in the serum by day 10, increases rapidly to 40 to 80 ng/mL by day 14, and remains relatively constant until day 20 (Fig. 21). Serum progesterone levels also remain high, at greater than 60 ng/mL, throughout this period (312,314). During the antepartum period, there is an elevation of serum relaxin immunoactivity to maximal levels, which usually range from 120 to 220 ng/mL, and this surge in relaxin levels is followed by a rapid decline throughout the approximately 24 hours before birth (301,313,315,316). Delivery in the rat normally lasts for approximately 90 minutes. During delivery, relaxin immunoactivity

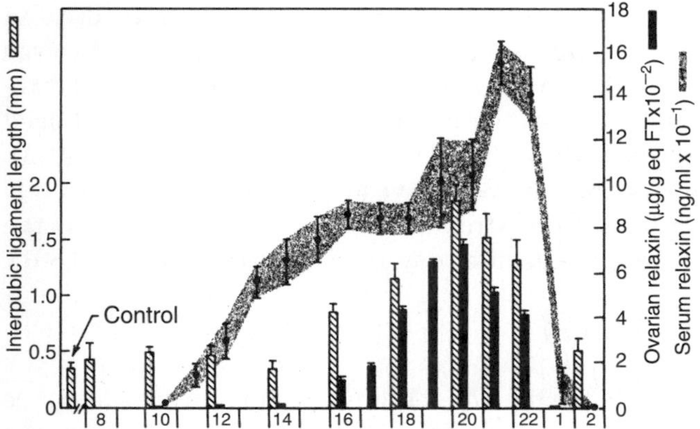

FIG. 21. Mean relaxin bioactivity and immunoreactivity levels (±SE) in extracts of ovaries and mean relaxin immunoreactivity levels (±SE) in peripheral blood sera obtained from rats from each of the days of pregnancy and lactation shown. (From Sherwood, O. D., Crnekovic, V. E., Gordon, W. L., and Rutherford, J. E. [1980]. Radioimmunoassay of relaxin throughout pregnancy and during parturition in the rat. *Endocrinology* 107, 691–698, with permission. Copyright 1980, The Endocrine Society.)

in the peripheral serum surges to high levels (>100 ng/mL) for 1 hour or more (301).

Other Sources of Relaxin

There is some evidence that relaxin immunoreactivity is present in the uterus during pregnancy in granulated metrial gland cells (317) and endometrial epithelial cells (306), but not in placental tissue (317). Relaxin-like immunoreactivity (299,318) and relaxin mRNA (210,299,318) have also been localized in numerous regions of the rat brain. Relaxin mRNA expression has also been demonstrated in the pregnant uterus, kidney, heart, placenta, pancreas, and mammary gland using RT-PCR (299). Levels of relaxin mRNA in all of these tissues were considerably lower than levels in the corpus luteum of pregnancy.

Human

Menstrual Cycle

Early studies indicated that bioactive relaxin may be produced in extraluteal ovarian sites during the follicular phase of the menstrual cycle (319,320). There is not total agreement on the cells that produce relaxin during the follicular phase of the menstrual cycle; some studies have suggested that granulosa cells are a potential source (321–323), whereas another study has demonstrated that the theca interna contains immunoreactive relaxin and prorelaxin (324). Although extracts of the corpus luteum of the menstrual cycle contain insufficient relaxin to be readily detected with in vivo relaxin bioassays (325,326), there is strong evidence that the corpus luteum of the cycle produces relaxin. Relaxin immunoactivity was found in luteal extracts (325), luteal cyst fluid (327), and culture medium bathing freshly dispersed luteal cells obtained during the menstrual cycle (328). Relaxin immunoreactivity has been found to be associated with luteal cells (329–332). Luteal relaxin is likely H2 relaxin because H2 relaxin mRNA, but not H1 relaxin mRNA, is expressed in the corpus luteum of the menstrual cycle in women (132,333).

There is now good evidence that the corpus luteum of the menstrual cycle releases relaxin into the circulation. Relaxin immunoactivity is detected as a small but consistent peak in peripheral plasma approximately 9 to 10 days after ovulation in nonconceptive cycles (334) (Fig. 22). Eddie et al. (335) also detected low levels of relaxin in the peripheral sera of nonpregnant women. A more recent study indicated that there is no difference in serum relaxin levels in conceptive and nonconceptive cycles (336).

The endometrium also appears to synthesize relaxin during the menstrual cycle. Relaxin immunoactivity has been detected in endometrial gland epithelial cells during the luteal phase of the menstrual cycle (330,337). Relaxin-specific mRNA is detected in human endometrial stromal and glandular epithelial cells, and relaxin protein is secreted into the medium taken from primary cultures of these cell types (338). Both H1 and H2 relaxin mRNA have been identified by RT-PCR and real-time PCR in the human uterus from the menstrual phase of the cycle (339).

Pregnancy

The corpus luteum is the major site of relaxin production during pregnancy and is the source of the circulating hormone. Extracts of corpora lutea of pregnancy contain relaxin immunoactivity (325,340,341) and bioactivity (325,326). Venous blood draining the ovary containing the corpus luteum has plasma relaxin levels three to four times higher during week 6 of gestation than during the luteal phase of the cycle (342). Human corpus luteal cells produce immunoreactive relaxin in vitro (328,343). Immunocytochemical studies at the ultrastructural level indicated that, like the pig, most of the relaxin immunoactivity was associated with secretory granules (332). The relaxin peptide produced by the pregnant corpus luteum and circulating in the blood has been isolated and demonstrated to be H2 relaxin (79,80). In addition, H2 relaxin mRNA, but not H1 relaxin mRNA, is produced in pregnant corpora lutea (333). Two laboratories reported that serum H2 relaxin is not detectable in patients with premature ovarian failure who become pregnant with egg donation and do not have a corpus

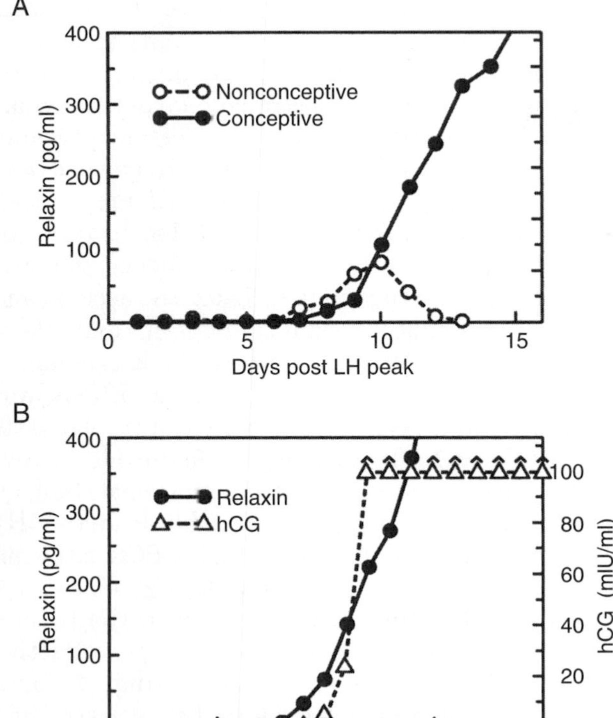

FIG. 22. A: Peripheral plasma relaxin concentrations in a nonconceptive cycle followed by a conceptive cycle in the same subject. **B:** Peripheral plasma relaxin and human chorionic gonadotropin (hCG) concentrations from one pregnant subject. Values of hCG greater than 100 mIU/mL were off scale and are indicated by ^. (From Stewart, D. R., Celniker, A. C., Taylor, C. A., Jr., Cragun, J. R., Overstreet, J. W., and Lasley, B. L. [1990]. Relaxin in the periimplantation period. *J. Clin. Endocrinol. Metab.* 70, 1771–1773, with permission. Copyright 1990, The Endocrine Society.)

luteum (344,345). Consistent with these findings, H2 relaxin fell to undetectable levels after ovariectomy in molar pregnancies (346).

Available evidence indicates that the levels of H2 relaxin in the corpus luteum of pregnancy in women (325) are much lower than those in the corpora lutea of pigs (61) and rats (67). There is no evidence that the membrane-bound granules that store H2 relaxin in the cytoplasm of human luteal cells accumulate during late pregnancy (332), as they do in pigs (280), rats (304,305), and mice (347).

H2 relaxin levels are elevated in the peripheral blood throughout nearly all of gestation in the human (325,348–359). The earliest indications that H2 relaxin levels might be higher during early pregnancy than late pregnancy were obtained in the 1930s when Pommerenke (354) and Abramson et al. (348) reported that with advancing pregnancy, fewer and fewer human serum samples were capable of inducing pelvic relaxation in estrogen-primed guinea pigs. Consistent with these reports, subsequent radioimmunoassays (325,327,353,356–358,360) or enzyme-linked immunosorbent assays (ELISAs) (349–352) demonstrated that H2 relaxin immunoactivity levels are higher during the first trimester than during the second or third trimester (Fig. 23). Importantly, the highest levels of relaxin in the blood (<1 ng/mL; Fig. 23) are much lower than those seen in the pig, rat, and other species during pregnancy. Also unlike the pig and rat, there is no antepartum surge in relaxin levels in women (78,355,358,361,362). Perhaps this is a reflection of the apparent low level of H2 relaxin that is stored in secretory granules in the corpus luteum of women during late pregnancy. There is limited evidence that H2 relaxin concentrations change relatively rapidly and are somewhat elevated during labor (353).

Clinical Implications of Serum Relaxin Levels

Because serum levels of H2 relaxin reflect both effective luteotropic stimulation and the amount of luteal tissue, investigators have examined the possibility that H2 relaxin levels can be used for clinical diagnostic purposes. Luteal phase defect, which is characterized by a short luteal phase and inadequate action of progesterone on the endometrium, is a cause of infertility and pregnancy loss. Stewart et al. (363) demonstrated that there is a marked reduction in both the duration of secretion and levels of H2 relaxin during the luteal phase in women with luteal phase defect. After in vitro fertilization and embryo transfer, Eddie et al. (335) initially detected H2 relaxin, but not human chorionic gonadotropin

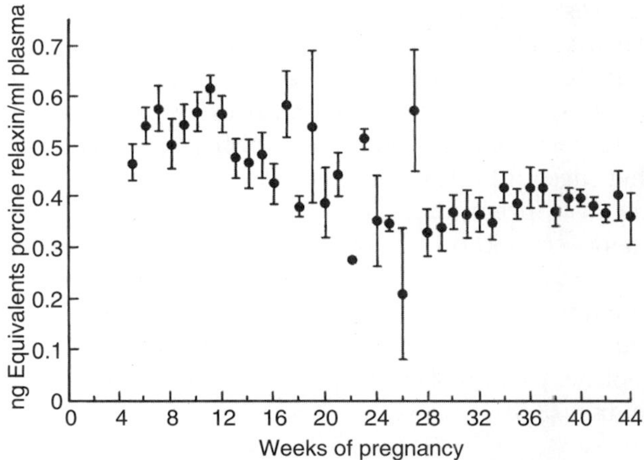

FIG. 23. Relaxin immunoreactivity levels in peripheral blood of women throughout pregnancy. Relaxin levels (±SE) were determined with a homologous relaxin radioimmunoassay that used rabbit anti-porcine relaxin serum R6, and hence levels are expressed as nanogram porcine relaxin equivalents. (From O'Byrne, E. M., Carriere, B. T., Sorensen, L., Segaloff, A., Schwabe, C., and Steinetz, B. G. [1978]. Plasma immunoreactive relaxin levels in pregnant and nonpregnant women. *J. Clin. Endocrinol. Metab.* 47, 1106–1110, with permission. Copyright 1978, The Endocrine Society.)

(hCG), in the peripheral serum of several nonpregnant women on days 6 to 8 postlaparoscopy. In contrast, both H2 relaxin and hCG were first detected on days 10 to 12 postlaparoscopy in pregnant women. Because the maintenance of early pregnancy requires synchrony of factors from the embryo, endometrium, and corpus luteum, these workers (335) postulated that early secretion of H2 relaxin in nonpregnant women could indicate asynchronous luteal function that results in loss of the embryo. Investigators have explored the possibility that atypically low peripheral serum H2 relaxin levels can be used to predict abortions during the first trimester, when maintenance of pregnancy depends on luteal support. Although all reports but one (364) indicated that serum H2 relaxin levels were frequently low or declined before spontaneous abortion (350,351,365,366), serum levels of H2 relaxin offer no advantage over serum levels of hCG or progesterone in predicting early abortion.

It has also been postulated that atypically high serum H2 relaxin levels during pregnancy may have clinical implications. There is limited evidence that elevated serum H2 relaxin levels might be useful in predicting preterm labor (367). Both prematurity risk and premature delivery are significantly associated with elevated serum H2 relaxin in patients with ovarian hyperstimulation (368). One study suggests that serum H2 relaxin may be an independent predictor of risk of very preterm delivery when serum is sampled in the 18th gestational week (369). MacLennan et al. (370) reported that serum H2 relaxin levels were higher in patients with pelvic pain and joint laxity than in control subjects. Further studies suggested that the serum H2 relaxin level was a factor that may best explain differences in sacral pain prevalence during pregnancy (371). Steinetz et al. (372) reported that serum relaxin levels are significantly higher in all three trimesters in diabetic women than in nondiabetic control subjects. Neither the mechanism that accounts for the elevated relaxin levels in diabetic pregnancy nor the physiological consequences are known. More recently, the same authors have suggested that part of the increased serum relaxin immunoreactivity in diabetic pregnancy could be due to the presence of another relaxin-like peptide, perhaps H1 or H3 relaxin, in addition to H2 relaxin (373).

Other Sources of Relaxin in the Human

Although reports are not entirely consistent, it seems likely that relaxin is produced in the decidua and placenta. Relaxin bioactivity has been demonstrated in extracts of term decidual tissue (76,374) and placental tissue (74,75). However, two studies found little (341) or no (327) relaxin immunoactivity in human placental extracts obtained at term. Several studies have demonstrated relaxin immunoactivity in decidual tissue (76,330,375–377) and placental tissue (74,329,339,375–378). Bryant-Greenwood and coworkers (132,377,379–381) and MacLennan et al. (340) used immunohistochemical localization or Northern blot analysis to obtain evidence that the small amounts of relaxin identified in the decidua and placenta are produced locally and not attributable to bound circulating relaxin. Interestingly, expression of H1 relaxin mRNA, as well as H2 relaxin mRNA, was detected in both decidua and placental trophoblasts using RT-PCR (132), providing the first evidence that the *RLN1* gene may be expressed in the female human. However, another group were unable to detect H1 relaxin mRNA in the placenta using RT-PCR, although they detected H2 relaxin mRNA (382).

Evidence that relaxin is produced in the human mammary gland regardless of the reproductive status of the woman is accumulating. Relaxin immunoactivity was detected in breast cyst fluid (383), milk (384), and mammary parenchyma in nonpregnant women (385). Bryant-Greenwood and coworkers localized relaxin in normal lobular and ductal epithelial cells as well as myoepithelial cells in mammary tissue obtained from prepubertal, cyclic, gestational, lactational, and postmenopausal women (386). Strong relaxin immunostaining was also observed in ductal epithelial cells of both benign and atypical hyperplasia as well as epithelial and myoepithelial cells of fibroadenomas (386). Both H2 and H1 relaxin mRNA have been detected in extracts of benign and neoplastic breast tissue by RT-PCR (387).

There were brief and isolated reports that relaxin immunoactivity is present in the myometrium (385), cervix (385), and cervicovaginal secretions (388,389). Both mRNA and protein for H1 and H2 relaxin have been detected in the heart (390).

Primates

Menstrual/Estrous Cycle

As is the case in the human, circulating relaxin levels rise in the late luteal phase of the menstrual cycle in rhesus monkeys (391), chimpanzees (392), and macaque monkeys (391). The high levels of relaxin in luteal tissue and ovarian venous blood of macaques suggested that the corpus luteum was the source of the circulating relaxin. Relaxin levels also rise in the luteal phase in the marmoset and were higher in conceptive cycles (393). The authors suggested that relaxin may be a reliable indicator of early pregnancy status in the common marmoset. Relaxin mRNA is expressed in the ovary in both the corpus luteum and thecal cells during the estrous cycle of the marmoset (96). Relaxin gene expression was also identified at a low level in the uterus and placenta. In addition, relaxin-like immunoreactivity was detected in the endometrial epithelium of the late proliferative phase, and levels increased through the secretory phase (96).

Pregnancy

During pregnancy, the major source of relaxin; the profile of relaxin levels in the blood; and the control of relaxin secretion in monkeys (391,394–396), baboons (*Papio ursinus*) (397), and chimpanzees (392) appear similar to those in the human. The corpus luteum of pregnancy has been demonstrated to be the major source of relaxin in monkeys and baboons. After bilateral ovariectomy of the rhesus monkey (*Macaca mulatta*) during the first trimester (396), luteectomy of the cynomolgus monkey (*Macaca fascicularis*) during the first trimester (391) or on day 150 of pregnancy (394), and luteectomy of baboons on day 30 of pregnancy (397), serum relaxin immunoactivity levels fell abruptly. The chimpanzee corpus luteum of pregnancy expresses relaxin-2 mRNA (94). Hence, in higher primates, as in humans, it seems the *RLN2* gene is the source of relaxin in the corpus luteum (see section on INSL4 and H1 Relaxin Equivalent Genes Are Found Only in Primates). Serum relaxin immunoactivity levels remain relatively constant in monkeys, baboons, and chimpanzees throughout most of pregnancy, with no evidence of a prelabor surge (392,394,396,397).

In contrast, the pattern of relaxin secretion during pregnancy in the New World monkey, the common marmoset, more closely resembles that of nonprimates (398), although plasma levels are highest in the second trimester and decrease toward basal levels in the weeks before birth (96). Relaxin mRNA expression in the corpus luteum matches the levels of relaxin in the blood (96).

There is some evidence that low levels of relaxin may be produced in nonluteal tissues in monkeys and baboons. On about day 140 of the approximately 175-day gestation period, low levels of relaxin immunoactivity were found in the decidual, placental, and myometrial tissue of baboons in which the corpus luteum had been removed on day 30 (397). Because pregnancy proceeds normally after the first trimester in the absence of the corpus luteum in humans and baboons, Castracane and coworkers (397) suggested that relaxin produced in nonluteal tissue may have a physiological role during late pregnancy and at parturition. The chimpanzee placenta expresses relaxin-2, but not relaxin-1 mRNA (94). Relaxin mRNA is expressed in the uterus and placenta of pregnant marmoset monkeys (96), although the levels are lower than those in the corpus luteum.

Mouse

In 1959, Steinetz et al. (399) detected relaxin bioactivity in the ovaries of pregnant mice (*Mus musculus*). Studies by Anderson and coworkers (347,400) indicate that relaxin synthesis in the mouse is similar to that in the rat; relaxin immunoreactivity was demonstrated in corpora lutea obtained during the second half of pregnancy, and no immunostaining was demonstrated in ovaries obtained throughout the estrous cycle. Relaxin immunostaining has been localized to the perinuclear region of luteal cells; it was first detected on day 11.5, increased progressively until day 18, dropped dramatically a few hours before delivery at day 18.5, and reached background levels shortly after delivery (401). The appearance and disappearance of membrane-bound cytoplasmic granules (200 nm in diameter) closely parallels that of relaxin immunostaining of luteal cells (347). Relaxin mRNA has been detected in the ovaries of late pregnant mice, with no expression being found in the uterus, placenta, or fetuses from the same animals (101). O'Byrne and Steinetz (402) determined the general profile of relaxin immunoactivity in the peripheral blood serum of mice during the 20-day gestation period with a porcine relaxin radioimmunoassay. Relaxin immunoactivity levels rose from approximately day 12 to maximal levels 1 or 2 days before delivery, and declined to low levels by 2 days postpartum (402).

Relaxin mRNA was found in highest abundance in late pregnant ovaries (101) but was also detected in numerous regions of the brain, thymus, heart, kidney, lung, liver, spleen, skin, mammary gland, and nonpregnant ovary by RT-PCR (18) (Fig. 24). Because relaxin mRNA expression was not observed using Northern blotting, it is likely that relaxin levels are very low in these tissues.

Horse

The placenta is the major, and perhaps the sole, source of relaxin in the horse (*E. caballus*) (69,403). Relaxin mRNA has been detected in the horse placenta using both Northern blotting and in situ hybridization, and these transcripts were localized to the fetal trophoblasts (105). Unlike corpora lutea of the pregnant pig, the placenta of the pregnant horse does not store much relaxin during pregnancy. The profiles of plasma relaxin immunoactivity throughout the approximately 350-day gestation period of the

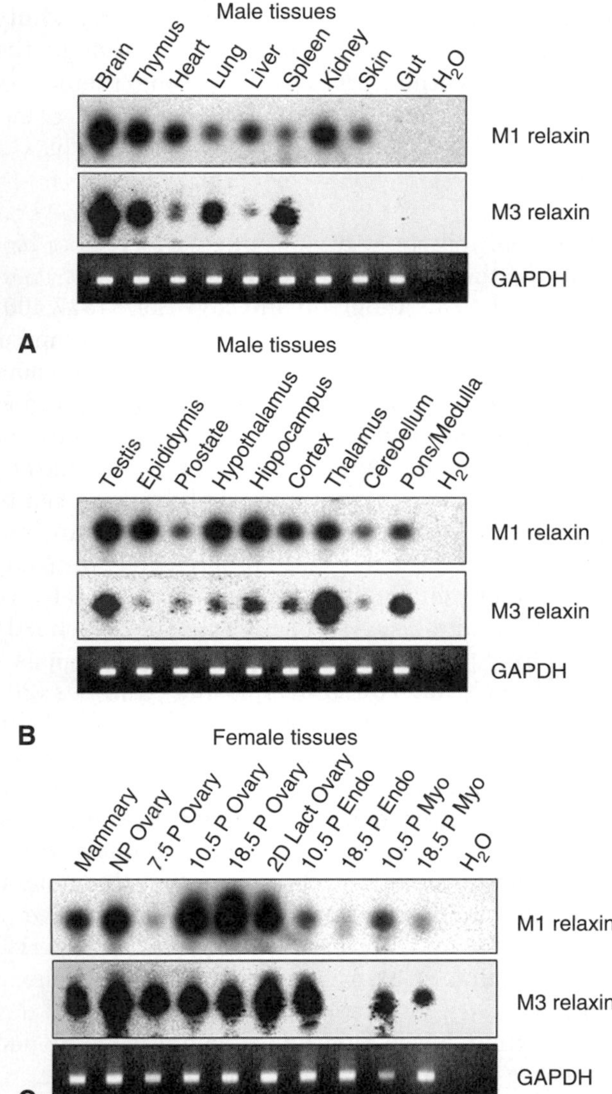

FIG. 24. Reverse transcriptase-polymerase chain reaction (RT-PCR) and Southern blot analysis of mouse relaxin (M1) and relaxin-3 (M3) gene expression in the mouse. Southern blots were conducted on RT-PCR products (from various male and female mouse tissues) using specific internal oligonucleotide primers to mouse relaxin and relaxin-3 cDNA, respectively. Ultraviolet photographs of glyceraldehyde-6-phosphate dehydrogenase (GAPDH) PCR products are shown as controls for quality, and equivalent loading, of the cDNA. Water (H_2O) replaced cDNA in negative control reactions for each PCR. Relaxin mRNA expression was determined in a number of nonreproductive male tissues (**A**), in reproductive tissues and specific brain regions of the male (**B**), and in female reproductive tissues at different stages of pregnancy (**C**). (From Bathgate, R. A., Samuel, C. S., Burazin, T. C., Layfield, S., Claasz, A. A., Reytomas, I. G., Dawson, N. F., Zhao, C., Bond, C., Summers, R. J., Parry, L. J., Wade, J. D., and Tregear, G. W. [2002]. Human relaxin gene 3 [H3] and the equivalent mouse relaxin [M3] gene: novel members of the relaxin peptide family. *J. Biol. Chem.* 277, 1148–1157, with permission.)

horse show marked among-breed differences in both amount and form (404–406). Relaxin immunoactivity is detectable by approximately day 80 in standardbreds, thoroughbreds, and ponies, but concentrations rise from baseline earlier in standardbreds than in thoroughbreds or ponies. Peak concentrations on about day 175 are approximately 100 ng/mL in standardbreds, and they are higher than levels in thoroughbreds. Thoroughbred mares exhibit a nadir in relaxin concentrations around day 225 of gestation, but standardbred mares do not. In thoroughbred mares, relaxin immunoactivity levels increase throughout approximately the last 125 days of gestation, whereas relaxin levels decline gradually from approximately day 175 to foaling in standardbreds. Pony mares have much lower levels of relaxin immunoactivity than both standardbreds and thoroughbreds, and unlike these two breeds, relaxin immunoactivity levels in ponies increase gradually from day 115 to foaling. Plasma relaxin levels were low in each of three standardbred mares with abnormal termination of pregnancy, and it was postulated that relaxin may, as an indicator of placental function, be used to assess at-risk pregnancies in the mare (406). Furthermore, plasma relaxin has been postulated to be a valuable clinical tool for both diagnosing placental insufficiency and monitoring treatment efficacy in mares (407). Relaxin immunoreactivity has also been localized in the ovary of the mare during the estrous cycle, in theca and granulosa cells, as well as in the corpus luteum (408). Northern blotting and in situ hybridization confirmed the presence of relaxin mRNA in the corpus luteum but not in the follicles (408).

Ruminants

As discussed in the section on Evolution of the Relaxin Genes, a relaxin gene has never been discovered in ruminants and a relaxin peptide has not been isolated, although numerous studies have documented the presence of relaxin bioactivity in the reproductive tract and plasma of ruminants.

Low levels of relaxin bioactivity have been demonstrated in extracts of corpora lutea obtained from pregnant cows (140,409–411). The placenta in cows also produces little, if any, relaxin (140,409,410). Anderson and coworkers reported that relaxin plasma levels usually peaked at concentrations ranging from 1 to 2 ng equivalents of porcine relaxin/mL between days 5 and 1 prepartum (409,412).

Porter and coworkers demonstrated that the sheep uterus is responsive to relaxin (139). Nevertheless, a source of relaxin during the approximately 150-day gestation period in sheep remains to be established. Relaxin-specific immunofluorescence has been observed in luteal cells of pregnant ewes (413), and both crude ovarian extracts (139,142) and partially purified ovarian extracts (142,414) from pregnant ewes contained apparent low levels of a substance that cross-reacted in porcine relaxin radioimmunoassays. In contrast, relaxin immunoactivity was not detected in ovarian tissue or serum by alternate anti-porcine relaxin antisera (414). Also, extracts of ovarian tissue obtained from ewes during late pregnancy demonstrated no relaxin bioactivity (139,414).

Low amounts of relaxin have been reported to be produced in the placenta and endometrium in sheep (142,414). However, other investigators failed to identify relaxin bioactivity in extracts of placentomes and endometrium (139,414). Whereas one group reported surges in relaxin immunoactivity during birth and suckling (415), a second group found little or no relaxin immunoactivity in the serum during middle or late gestation (414).

Although a pseudogene for relaxin was found in sheep (118), it is possible that the investigators were detecting the presence of another relaxin-like peptide. INSL3 mRNA is expressed in large amounts in the sheep (416) and cow (117) corpus luteum and is also produced in the cow uterus and placenta (117). It is not known if INSL3 circulates in the plasma of the sheep and cow; however, it has been postulated that INSL3 may substitute for relaxin in ruminants (417).

Other Species

The presence of relaxin bioactivity in the plasma and placental extracts of the rabbit was first reported by Hisaw in 1926 (1). Subsequently relaxin immunoreactivity was demonstrated in the multinucleated syncytiotrophoblast of placentas obtained on days 11, 16, 23, and 30 (418,419). Furthermore, Eldridge and Fields (418) obtained evidence at the ultrastructural level that rabbit relaxin is stored in cytoplasmic membrane-bound granules (150 to 400 nm in diameter) within the syncytiotrophoblast that migrate to the cell membrane and are released by exocytosis. The rabbit relaxin gene (*SQ10*) (99) (see section on Other Species) is highly expressed in the placenta and uterus, but not the ovary (420). Relaxin bioactivity has been demonstrated in uterine extracts (399,421) and relaxin immunoreactivity has been localized to the apical region of mononuclear endometrial gland cells or multinuclear luminal epithelial cells from day 4 of pregnancy until day 5 of lactation (418,419,422). During pregnancy, rabbit relaxin is first detected in the plasma during the preimplantation period (days 4 to 9), rises to maximal levels by day 15, remains elevated until delivery on day 32, and then declines (423). Relaxin was not detected in pseudopregnant, cycling, or male rabbits.

The placenta synthesizes relaxin during the second half of the 16-day gestation period in the golden (Syrian) hamster (*Mesocricetus auratus*). Relaxin bioactivity was demonstrated in placental homogenates obtained on days 11, 14, and 15 of pregnancy (424) and relaxin immunoreactivity was demonstrated in the placenta from days 8 to 15 (425–427). Most of the relaxin immunoactivity was in fetal primary and secondary giant trophoblast cells found in the decidua capsularis and trophospongium, respectively. Consistent with these findings, relaxin mRNA was highly expressed in the hamster placenta in the second half of pregnancy and was not detected in the ovary (102). O'Byrne and Steinetz (402) determined the general profile of relaxin immunoactivity in the blood of hamsters. Relaxin immunoactivity in peripheral serum appeared on day 8, rose to maximal levels by day 15, and dropped precipitously at parturition on day 16. These findings were confirmed using a homologous hamster radioimmunoassay, and the authors confirmed that the ovary was not a site of relaxin production (87).

Relaxin levels have been determined in the plasma of cats throughout pregnancy (428). Immunoreactive relaxin levels become detectable on approximately day 20, rise from approximately day 25 to maximal levels by approximately day 35, remain stable until just before parturition, and then decline sharply to undetectable levels by 24 hours after delivery. The primary source of relaxin in the cat appears to be the placenta (107,429).

The placenta also appears to be the major source of relaxin during the approximately 60-day pregnancy in dogs (*C. familiaris*) (83,106,430,431), but the ovary may also be a source (430). Steinetz and coworkers (430,432,433) and Tsutsui and Stewart (431) used porcine relaxin radioimmunoassays to demonstrate that relaxin immunoactivity becomes detectable in the peripheral serum during the fourth week of pregnancy and that relaxin levels remain elevated throughout the remainder of gestation.

Unlike other species, the primary source of relaxin in the guinea pig (*Cavia porcellus*) is the uterus. The first evidence for this was obtained in the 1940s (421,434,435), and subsequently relaxin immunoreactivity was demonstrated in the endometrial glands during the second half of pregnancy (436–440). Ultrastructural studies demonstrated that relaxin immunoactivity was located over large (0.5 to 1.5 μm in diameter) granules located between the nucleus and luminal surface (apical region) of the endometrial gland cells (436,437,439,441,442), and that these granules were released, at least in part, into the lumen of the uterus (441). Levels of relaxin mRNA measured by Northern blotting in guinea pig endometrium matched the levels of immunoreactive relaxin (440,443). A clear pattern of plasma levels of relaxin during pregnancy has not been established, although bioactive (435) and immunoreactive (402) plasma levels of relaxin seem to be highest in late pregnancy in the guinea pig. Low levels of relaxin immunoactivity have been reported in uterine extracts and peripheral blood plasma obtained during the estrous cycle (444) and in the mammary gland (445).

Relaxin immunoreactivity has been measured in the plasma and placental extracts of the spotted hyena (*Crocuta crocuta*) (446). The placenta is the source of the relaxin immunoreactivity and plasma levels are highest in the 2 weeks preceding parturition. Relaxin was not detected in the plasma of nonpregnant females. Relaxin has been detected in the plasma of pregnant llamas and alpacas (143) as well as Alpine marmots (*Marmota marmota*) (447). In both studies, the authors suggested that relaxin may be a good early indicator of pregnancy in these species. Studies in Asian elephants (*Elephas maximas*) suggest that relaxin may be important for the maintenance of pregnancy in this species (448).

PHYSIOLOGICAL ROLES OF RELAXIN IN FEMALE AND MALE REPRODUCTIVE PROCESSES

The multitude of information outlined in the previous sections demonstrates that relaxin is likely to have important roles in reproductive and nonreproductive processes in both males and females. The high levels of circulating relaxin in the blood during pregnancy in multiple species particularly highlight important roles for relaxin in pregnancy and parturition. The availability of purified porcine relaxin and recombinant H2 relaxin has enabled researchers to establish better the biological effects of relaxin. Furthermore, the recent availability of relaxin and LGR7 KO mice has allowed the determination of the roles of relaxin in reproductive and nonreproductive processes. This section emphasizes relaxin's effects on well-established target tissues: the pubic symphysis, cervix, uterus, nipples and mammary glands. In addition, it describes putative biological effects of relaxin on other tissues. It should be recognized that some biological effects of relaxin occur in certain species, but not in others.

Pregnancy and Parturition

Pubic Symphysis

In many species, modifications of the pelvic girdle occur to enable safe delivery of young. Growth of the interpubic ligament can enable marked increases in the size of the birth canal. Transformation of the pubic joint cartilage to a flexible and elastic interpubic ligament occurs during pregnancy in several species, including guinea pigs, mice, bats, and humans (449,450). It does not occur in species such as rats (451) and sheep (452). Although the development of the pubic symphysis into an interpubic ligament does not occur in rats (451), there is a decrease in collagen content of the pubic symphysis during pregnancy (453). Treatment with H2 relaxin in ovariectomized pregnant rats induced this decreased collagen content normally seen in pregnant rats (453). It is relaxin's capacity to promote transformation of the interpubic ligament that led to the hormone's discovery in guinea pigs (1) and subsequently to the development of the widely used mouse interpubic ligament bioassay (55). As with the rodent reproductive tract, relaxin's effects on the interpubic ligament in mice and guinea pigs are estrogen dependent (25). Relaxin is essential for the development of the pubic ligament in mice because the interpubic ligament fails to develop in the relaxin KO mouse (255) (Fig. 25). Interpubic ligament length, wet weight, dry weight, and water content are all significantly lower than in wild-type mice at term pregnancy (255,454). The pubic symphysis has not been examined in the LGR7 KO mouse. The mechanisms whereby relaxin brings about its effects on the interpubic ligament in mice, guinea pigs, and other species are not well understood (25). However, as with the cervix and vagina (discussed later), endogenous relaxin reduces the density of collagen fiber bundles in the interpubic ligament (454). The physiological significance of relaxin-induced transformation of the interpubic ligament during pregnancy likely varies from species to species.

Surprisingly, the more than fourfold increase in length of the interpubic ligament that occurs in pregnant mice is not essential at delivery. Relaxin KO mice (255), as well as mice that had their pubic bones tied together (455), delivered their young as rapidly as did wild-type controls. It is not known if endogenous relaxin contributes to the modest separation of the pubic symphysis that occurs during pregnancy in humans or other primates (370). In addition, there is limited evidence that severe pelvic pain and excessive joint laxity that occur in some women during late pregnancy is attributable, at least in some cases, to relaxin levels significantly higher than those in control subjects (456).

Cervix

During the first half of pregnancy in mammals, the cervix is a relatively firm and inextensible structure, which not only protects the conceptuses from the

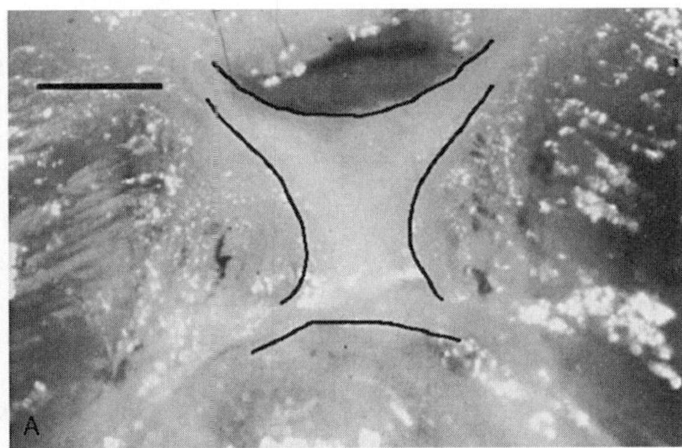

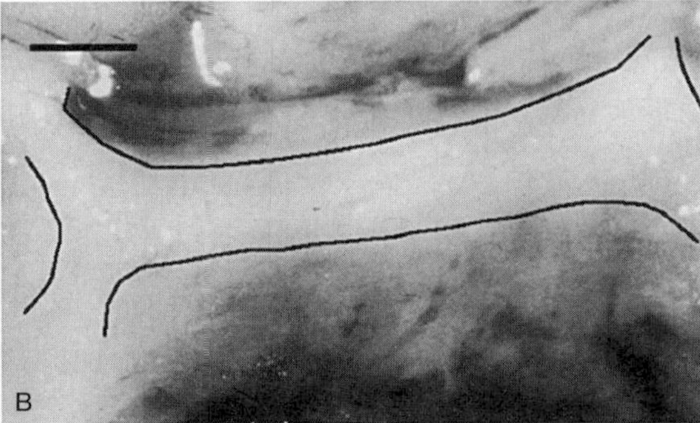

FIG. 25. Pubic symphysis of pregnant relaxin knockout (KO) (**A**) and relaxin wild-type (**B**) mice at 18.5 days' gestation. The interpubic ligament is relaxed in the wild-type mouse, but remains short in the relaxin KO mouse. Bar = 1 mm. (From Zhao, L., Roche, P. J., Gunnersen, J. M., Hammond, V. E., Tregear, G. W., Wintour, E. M., and Beck, F. [1999]. Mice without a functional relaxin gene are unable to deliver milk to their pups. *Endocrinology* 140, 445–453, with permission. Copyright 1999, The Endocrine Society.)

external environment but impedes premature delivery of the fetus. During the second half of pregnancy, however, the cervix increases in size and its consistency changes strikingly. The term *softens* is commonly used to refer to the changes in the tensile properties of the cervix that occur during late pregnancy. The tensile properties of the cervix are largely attributable to the connective tissue that predominates (457,458). In the firm cervix, collagen fibers are highly organized and arranged in dense parallel bands with little amorphous ground substance, which is rich in proteoglycans, glycosaminoglycans, and water, separating the collagen bundles. In the soft cervix, however, the collagen fibers are dispersed and randomly oriented in a considerably increased matrix (459). Relaxin plays a major role in promoting the cervical modifications that occur during late pregnancy in several mammalian species. Although there is considerable evidence that estrogen, prostaglandins, and, in some cases, progesterone also influence the tensile properties, morphology, and biochemical composition of the cervix (460), little is known concerning the mechanisms whereby relaxin interacts with these hormones to promote cervical softening. However, it is clear that the influence of relaxin on cervical softening appears to be a general phenomenon among mammalian species.

Relaxin plays a major role in bringing about the dramatic growth of the cervix that occurs in rats (461–464), mice (454), and pigs (465,466) during pregnancy. Both the wet weight of the cervix and the circumference of the cervical lumen at term in control rats and pigs are approximately twofold greater than those in relaxin-deficient animals (462,466) (Fig. 26). In anti-relaxin monoclonal antibody (MCA1)–treated rats, where endogenous relaxin is neutralized, there are far fewer cervical lumen involutions than in control rats, indicating that control rats would have greater expansion of cervical lumens than is possible in relaxin-deficient rats (467). Similarly, the cervix in relaxin KO mice is poorly developed and the cervical epithelium is thinner than in control mice (454). Studies have demonstrated a potential mechanism for the action of relaxin on the growth of the rat cervix. Relaxin promotes the accumulation of both epithelial cells and stromal cells (461,462) by not only stimulating cell proliferation (468) but by inhibiting apoptosis (469). Within the stroma, relaxin's effects on cell proliferation (468) and apoptosis are primarily on fibroblasts and not on smooth muscle cells (469).

Endogenous relaxin also promotes a dramatic increase in the extensibility (softening) of the cervix in pregnant rats and pigs (463,466,470,471) (Fig. 26). It is the extracellular matrix components, including, types I, III, and IV collagen, elastin, proteoglycans, and glycosaminoglycans, that are thought largely responsible for the tensile properties of the cervix. Analysis of cervices of late pregnant, relaxin-deficient rats and pigs, and relaxin KO mice reveals that the collagen fiber bundles fail to disperse or to become as disorganized as do those in cervices of control animals (454,467,472,473). The relaxin-induced dispersion of collagen fiber bundles may allow the fibers to be readily pulled past one another as the fetus enters the cervix. Indeed, elastin fiber–like structures are more prevalent and longer in relaxin-deficient pregnant rats than in controls (467). Endogenous relaxin also plays a major role in promoting changes in the biochemical composition of the cervix that occur during late pregnancy in rats and pigs (470,474). Hence, relaxin increases cervical hydration, dry weight, and glycosaminoglycan (dermatan sulfate, heparan sulfate, hyaluronic acid) content. The mechanisms associated with relaxin-induced hydration of the cervix are not well understood, but enlargement of the cervical arteries (467) may be a factor. Because it is the extracellular matrix that is largely responsible for the tensile properties of the cervix, it is postulated that degradation of collagen and other extracellular components is the key to cervical softening. Recombinant H2 relaxin upregulates the activity of matrix metalloproteinase-1 (MMP-1; collagenase)

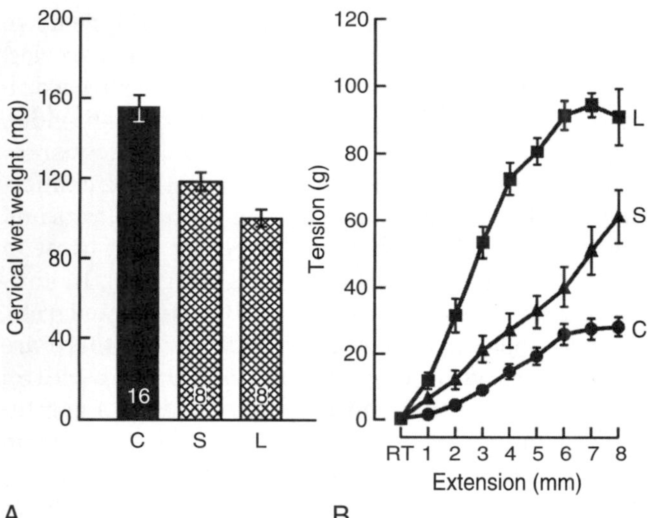

FIG. 26. Influence of neutralization of exogenous relaxin on weight and tensile properties of rat cervix on day 22 of pregnancy. **A:** Mean wet weight (±SE) of cervices obtained from phosphate-buffered saline–treated controls (C; n = 16) and rats treated with anti-relaxin monoclonal antibody MCA1 either throughout the second half of pregnancy (long treatment, L; n = 8) or during the 3-day antepartum period (short period, S; n = 8). **B:** Mean tension (±SE) at extension of cervices in the three groups for eight successive 1-mm extensions at 30-min intervals. RT, resting tension. (Modified from Hwang, J. J., and Sherwood, O. D. [1988]. Monoclonal antibodies specific for rat relaxin: III. Passive immunization with monoclonal antibodies throughout the second half of pregnancy reduces cervical growth and extensibility in intact rats. *Endocrinology* 123, 2486–2490, and Hwang, J. J., Shanks, R. D., and Sherwood, O. D. [1989]. Monoclonal antibodies specific for rat relaxin: IV. Passive immunization with monoclonal antibodies during the antepartum period reduces cervical growth and extensibility, disrupts birth, and reduces pup survival in intact rats. *Endocrinology* 125, 260–266, with permission. Copyright 1988 and 1989, The Endocrine Society.)

in primary cultures of cervical cells from pregnant guinea pigs (475), and increases levels of both mRNA and protein for proMMP-1, proMMP-2 (gelatinase), and proMMP-3 (stromelysin) in human lower uterine fibroblasts (238). Porcine relaxin has been demonstrated to stimulate the secretion of several metalloproteinases in primary cultures of human cervix fibroblast-like cells (476). However, relaxin seems to have little effect on collagen content in the cervix during pregnancy, although the ratio of glycosaminoglycans to collagen increases in relaxin-replete rats and pigs during late pregnancy (470,474). Hence, the direct degradation of collagen may not be a major mechanism of relaxin-induced cervical softening.

Relaxin clearly has actions on the cervix and numerous studies have demonstrated that cervical cells contain relaxin receptors. Biotinylated porcine relaxin has been used to demonstrate binding sites for relaxin in epithelial cells, circular and longitudinal smooth muscle cells, and cells associated with blood vessels in rat (214), pig (215), and human (217) cervices. Human lower uterine segment fibroblasts contain relaxin binding sites and LGR7 mRNA (R. A. D. Bathgate, unpublished data), and induce cellular signaling in response to relaxin (204,238). LGR7 immunoreactivity has also been demonstrated in rat cervical smooth muscle cells but not in epithelial cells (21).

Another potential mechanism by which relaxin may act in the cervix is through stimulation of prostaglandin synthesis. Prostaglandin E_2 administration promotes cervical softening in rats, humans, and other species (477,478) and is used frequently to promote ripening of the human cervix before delivery (479). Investigators used the cyclooxygenase inhibitor indomethacin to block prostaglandin synthesis during relaxin treatment. However, the inhibitor failed to block the effects of porcine relaxin on either cervical wet weight or softening in ovariectomized nonpregnant rats (480). Another potential mediator of relaxin's action on the cervix is nitric oxide. Blocking the synthesis of nitric oxide with the nitric oxide synthase inhibitor NG-nitro-L-arginine methyl ester (L-NAME) restrained cervical softening, and local application of nitric oxide donors promoted cervical softening in pregnant rats (477,481). Studies aimed at elucidating the role of nitric oxide in relaxin-induced cervical softening in rats provided inconclusive results (477,482). However, there is limited evidence that nitric oxide may contribute to relaxin's effects on cervical wet weight (482).

The potential contributions of estrogen and progesterone to relaxin's effects on the cervix have been investigated in pigs and rats. Relaxin's effects on growth and softening of the cervix are estrogen dependent in rats (54,483), but not in pigs (484,485). The administration of neither estrogen nor progesterone independently promotes marked cervical softening in nonpregnant rats and pigs (54,483–485), although estrogen alone promotes growth of the cervix. Moreover, in the presence of physiological levels of both estrogen and progesterone during late pregnancy in relaxin-deficient rats and pigs, the cervix fails to soften (463,466,471).

It is not known whether relaxin plays a major role in cervical ripening in the human. Plasma relaxin is elevated during the period of cervical ripening in humans (see section on Pregnancy), but levels are much lower than in the rat and pig. However, cervical dilatation occurs in an ovum donation pregnancy in which serum relaxin levels are not detectable (486). It is possible that relaxin produced by the decidua or placenta contributes to cervical softening in the human (see section on Other Sources of Relaxin in the Human).

Regardless of whether endogenous H2 relaxin is required for cervical softening during pregnancy in the human, there is evidence that the hormone may be administered at term to enhance cervical softening. Two early studies conducted with impure preparations of porcine relaxin demonstrated that relaxin could ripen the cervix (487,488), although these studies were compromised because of the poor purity of the preparation. Later clinical studies used highly purified porcine relaxin administered either in the vagina (489,490) or in the cervix (491,492) to prepare the cervix for labor induction. Both trials demonstrated that porcine relaxin could ripen the cervix. Because of the structural differences between porcine and human relaxin and the potential clinical complications inherent in using porcine relaxin, further trials used recombinant H2 relaxin for cervical ripening. In these new clinical trials, recombinant H2 relaxin was placed in methyl cellulose gel and then deposited in the vagina (180,493). It was expected that the hormone would effectively pass through the squamous epithelium, enter the blood, and thereby reach relaxin receptors in the cervix. Unfortunately, this did not appear to occur and the trials were unsuccessful. When circulating blood levels of H2 relaxin were measured, it was demonstrated that no relaxin entered the blood. It was postulated that the lack of adsorption of relaxin may be the reason the treatment failed to influence cervical ripening. The early trials with porcine relaxin treatment indicate that relaxin may still have clinical application as a cervical ripening agent.

Uterus

The uterus is one of earliest known target organs of relaxin. The effect of relaxin on uterine contractility was discovered in 1950, when Krantz et al. (6) found that intravenous administration of a crude extract of porcine relaxin to estrogen-primed guinea pigs diminished spontaneous uterine activity in situ. A short time later, Steinetz and colleagues demonstrated that relaxin causes several rapid changes in the biochemical composition of rat and mouse uteri (494). The actions of relaxin on myometrial activity and growth and development of the uterus are outlined in the following sections.

Myometrial Contractile Activity. The myometrium is relatively inactive throughout most of pregnancy, maintaining the developing fetus within a tranquil uterine cavity. Normally, at term the uterine myometrium contracts rhythmically and forcefully, and this highly coordinated pattern of contractions aids in dilating the cervix and expelling the fetus and placenta. The changes in the contractile properties of the uterus during pregnancy are regulated, at least in part, by changing levels of (a) hormones that stimulate or inhibit uterine contractility, (b) hormone receptors on myometrial cells, and (c) gap junctions between adjacent myometrial cells. This section largely consists of a description of the influence of relaxin on uterine contractility in rats and pigs because these are the species on which most of the research has been conducted. Unlike some other actions of relaxin, its effects on myometrial activity are highly species specific.

Numerous in vivo or in vitro pharmacological studies in nonpregnant animals from several species have demonstrated that either porcine relaxin or recombinant H2 relaxin reduces the frequency of myometrial contractions. Hence, partially purified porcine relaxin was reported to reduce the frequency or amplitude of uterine contractions in the rat (59,495–508), mouse (54,506,507), guinea pig (6,509,510), and hamster (511). Furthermore, studies with highly purified porcine relaxin that examined uterine contractility either in vivo or in vitro provided confirmation that relaxin renders the uterus quiescent in nonpregnant rats (223,512–533). Moreover, highly purified porcine relaxin inhibited contractions of the pig uterus (526,534,535). However, there are species where the pregnant myometrium seems unresponsive to relaxin. Although experiments using crude extracts of porcine relaxin injected intravenously into ovariectomized sheep induced typical hormone-specific quiescence in myometrial activity (139), other studies with highly purified porcine relaxin (536) or recombinant H2 relaxin (417) have demonstrated that the ovine myometrium is unresponsive to relaxin. In humans it is also likely that relaxin has a minor role, if any, in inhibiting spontaneous myometrial activity (526,537).

In rats and pigs, myometrial activity is low when serum relaxin levels are elevated, and then it increases markedly at delivery (538,539). Studies in rats that were ovariectomized on day 9 of pregnancy and given hormone replacement therapy with physiological levels of progesterone and estrogen throughout the remainder of pregnancy demonstrated that the frequency of intrauterine pressure cycles remained well above that in intact controls. When this hormone replacement therapy was supplemented with porcine relaxin, the frequency of intrauterine pressure cycles declined to levels that did not differ from those in controls (538) (Fig. 27).

During pregnancy, other hormones, including progesterone, estrogen, oxytocin, and prostaglandins, also have direct actions on the myometrium. The possible interaction of relaxin with the actions of these hormones is poorly understood. Progesterone is essential for maintaining pregnancy by providing hormonal support for the endometrium and preventing strong, highly coordinated uterine contractions. There is in vitro evidence that progesterone increases the sensitivity of the myometrium to the quiescent effect of relaxin in rats and pigs (495,502,526). Studies in ovariectomized nonpregnant rats demonstrated that estrogen also markedly increases the sensitivity of the myometrium to the quiescent effect of relaxin (518,540). This action is potentially mediated by induction of relaxin receptor expression because estrogen treatment increases the density of relaxin binding sites in the myometrium of ovariectomized rats (206,541). In late pregnant rats, mice, and pigs there was an increase in serum estrogen levels and decline in progesterone levels associated with an increase in the production of uterine components, such as gap junctions, oxytocin, oxytocin receptors, and prostaglandins, that increase the myometrium's capacity for highly coordinated contractions (25). Numerous studies provide evidence that pharmacological doses of relaxin inhibit myometrial contractions induced by either oxytocin or prostaglandin in rats (500,501,514,515,517,522,528,529,542) or pigs (535). Clearly, the inhibitory actions of relaxin on the myometrium need to be lost in late pregnancy to allow the contractions triggered by oxytocin or prostaglandins to induce birth. It is not clear how this is mediated because relaxin levels in the blood are still relatively high at birth (see section on Sources and Secretion of Relaxin in the Female).

The actions of relaxin are mediated by relaxin receptors on myometrial cells. Relaxin binding sites have been demonstrated in the rat (195,197), guinea pig (543), pig (199), and mouse uterus (198). These sites have been localized to the muscular layers of the rat myometrium by receptor autoradiography (200, 205,206). LGR7 mRNA is localized in the myometrium in the mouse (256,262) and rat (258). Isolated rat myometrial cells respond to relaxin with cAMP production (see section on Relaxin Signaling). The response to relaxin in these cells is mediated by LGR7 because it was blocked by coincubation with the soluble ectodomain of the receptor (21). In contrast, the human myometrium of the menstrual cycle does not express LGR7 mRNA (339,544), and does not contain relaxin binding sites (339) or demonstrate LGR7 immunoreactivity (544). However, immortalized human myometrial cells do respond to relaxin with cAMP release (230) (see later).

Contraction of smooth muscle occurs as a result of actomyosin interaction in a process that involves the hydrolysis of adenosine triphosphate (ATP). The actin-activated myosin ATPase activity requires phosphorylation of the 20,000-Da light chains that are associated with the globular head of myosin. Activation of the enzyme myosin light-chain kinase (MLCK), which catalyzes myosin light-chain (MLC) phosphorylation, is a sequential process requiring an elevation of the intracellular calcium levels. Calcium first binds to the calcium-regulatory protein calmodulin, which is found in high concentrations and is not

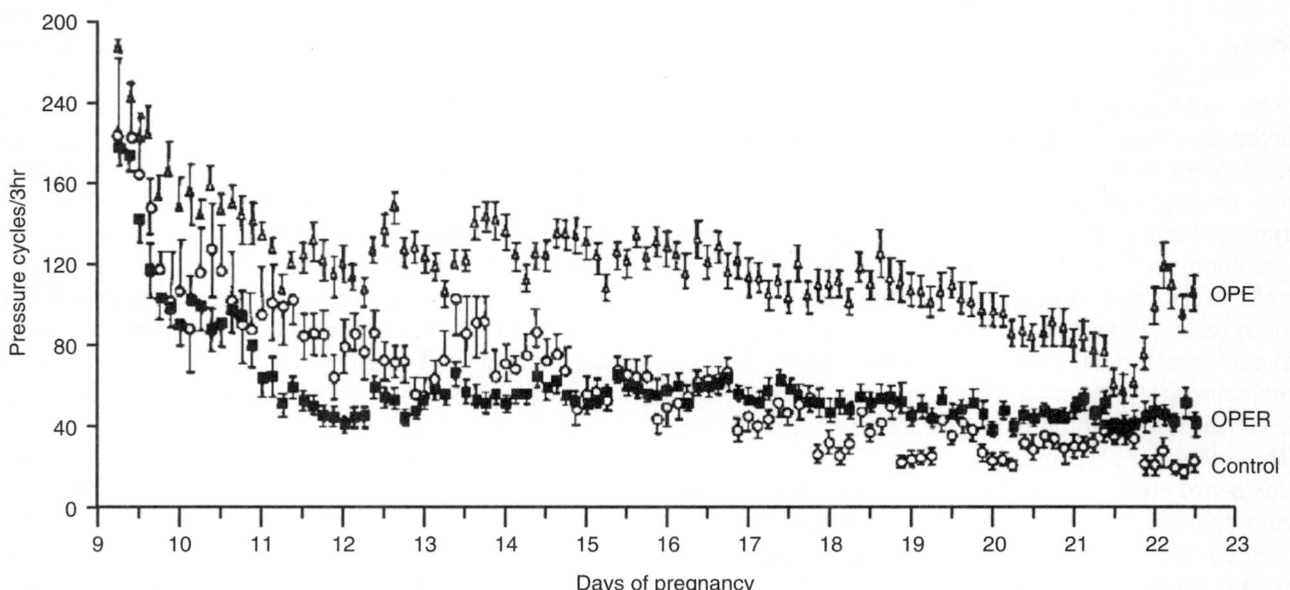

FIG. 27. Influence of endogenous relaxin on uterine contractility in pregnant rats. Mean frequency (±SEM, n ≥ 25) of intrauterine pressure cycles from day 9 until day 23 in intact pregnant rats (control), ovariectomized pregnant rats treated with progesterone and estrogen (OPE), and ovariectomized pregnant rats treated with progesterone, estrogen, and porcine relaxin (OPER). (Reprinted from Downing, S. J., and Sherwood, O. D. [1985]. The physiological role of relaxin in the pregnant rat: II. The influence of relaxin on uterine contractile activity. *Endocrinology* 116, 1206–1214, with permission. Copyright 1985, The Endocrine Society.)

limiting in the uterus, to form a Ca^{2+} • calmodulin complex. The Ca^{2+} • calmodulin complex then binds to the inactive MLCK to form the active holoenzyme complex Ca^{2+} • calmodulin • MLCK. Phosphorylation of MLCK, which is catalyzed by cAMP-dependent protein kinase (protein kinase A [PKA]), reduces the capacity of the MLCK to combine with Ca^{2+} • calmodulin. Thus, phosphorylation of MLCK inhibits formation of the active Ca^{2+} • calmodulin • MLCK complex (545,546). Relaxin induces its actions on myometrial inhibition by blocking the aforementioned cascade of events. Hence, relaxin treatment of the rat myometrium increases cAMP and PKA activity, decreases the affinity of MLCK for the Ca^{2+} • calmodulin complex , and decreases MLCK activity, MLC phosphorylation, and actomysin ATPase activity, with the result that uterine contractility diminishes (540,545–547). It has not been definitively established how relaxin controls the ratios of phosphorylated-MLC/MLC and phosphorylated-MLCK/MLCK. Relaxin's actions on myometrial cells are only partially mediated through activation of PKA (540,545,547). There is limited evidence that relaxin also acts by reducing intracellular Ca^{2+} levels by promoting increased Ca^{2+} efflux and inhibiting mobilization of Ca^{2+} from intracellular microsomal stores (540,545,546). A potential mechanism for promoting hyperpolarization and repolarization that can influence the Ca^{2+} transient and affect uterine contractility is the opening of K^+ channels (540,545,546). Sanborn and coworkers demonstrated that relaxin stimulates myometrial Ca^{2+}-activated K^+ channel activity and does so through PKA in a human myometrial cell line (548). There is also limited, but not consistent, evidence that relaxin may stimulate the opening of ATP-dependent K^+ channels in isolated rat uterus and myometrium (540,547). Also inconsistent are reports that relaxin may upregulate the L-arginine–nitric oxide pathway to increase the second messenger cyclic guanosine monophosphate, thereby inhibiting uterine smooth muscle contractility (243,547). Studies in immortalized human myometrial cells demonstrate that relaxin attenuates the contractile effects of oxytocin on these cells through activation of PKA. After stimulation with porcine relaxin, the regulatory subunit of PKA is anchored to the myometrial plasma membrane through association with A kinase anchoring proteins, and this obligatory anchoring is required for inactivation of G_q/phospholipase C (PLC) coupling of the oxytocin receptor (230). There is evidence that G_q/PLC inactivation is attributable to phosphorylation of $PLC\beta_3$ and not G_q (549).

Uterine Growth and Development. Early studies with partially purified porcine relaxin demonstrated that relaxin influenced metabolism in the rat and mouse uterus (54,550–554). More recent studies have demonstrated that relaxin plays an important role in uterine growth and development. Hence, porcine relaxin administration promotes growth of the uterine myometrium and endometrium in nonpregnant rodents and pigs (555–558). Recombinant H2 relaxin promotes growth of the uterus in a model of early pregnancy in the rhesus monkey (559) and in ovariectomized marmoset monkeys (560). The uterotropic effects of relaxin in nonpregnant rats or pigs were reported to be associated with dilation of small arteries or veins (561) and increased content of water, protein, collagen, glycogen, DNA, IGFs, IGF-binding proteins (IGFBPs), connexins, E-cadherin, vascular endothelial growth factor (VEGF), and tissue inhibitors of MMPs (558,562–565). The effects of relaxin in both the rhesus and marmoset monkey uterus were associated with increased endometrial angiogenesis (559,560). The uterotropic effects of relaxin are markedly influenced by estrogen. Whereas the administration of porcine relaxin alone over relatively short periods ranging from 6 to 54 hours (acute period) promotes growth of the uterus in immature or ovariectomized rats and pigs (555,558, 566,567), greater growth is obtained when rats are primed with estrogen before relaxin treatment (566). Moreover, prolonged treatment of ovariectomized pigs over 10 and 14 days (chronic period) with porcine relaxin has no effect on uterine growth in the absence of estrogen priming (556,557). Although relaxin-induced growth of the marmoset uterus did not require estrogen, the effects of relaxin were enhanced by cotreatment with estrogen (560). The mechanisms whereby relaxin and estrogen act in combination to promote uterine growth remain poorly understood. Estrogen may affect relaxin receptor expression in the endometrium (see section on Implantation), similar to actions on the myometrium outlined previously. There is also evidence that relaxin's acute effects on the rat uterus are mediated through ligand-independent activation of the estrogen receptor (565). In addition, estrogen and relaxin inhibit the expression of estrogen receptor β in the rat uterus, and it has been postulated that downregulation of estrogen receptor β might be necessary for estrogen or other estrogen receptor activators to exert their full trophic effects on the uterus (568). Progesterone also influences relaxin's effects on uterine growth in rats and pigs, but the influence of progesterone may differ in the two species. Whereas progesterone inhibited acute relaxin-induced increases in uterine wet weight and collagen content in ovariectomized prepubertal rats (566), progesterone augmented both the acute (555) and chronic (557) uterotropic effects of relaxin in ovariectomized gilts.

The actions of relaxin on the uterus outlined in this section have led researchers to hypothesize that relaxin plays an important role in uterine development during pregnancy. Studies in relaxin-deficient rats and mice do not support this hypothesis, whereas experiments in pigs do support a role. The uterus in relaxin-deficient rats was as large at term as in control animals (463). Consistent with this finding, uterine wet and dry weights increased as dramatically during pregnancy in relaxin KO mice as they did in wild-type controls (454). In contrast, when gilts were ovariectomized on day 40 and given hormone replacement therapy with progesterone only, the wet weight of the uterus at term was approximately 30% lower than in intact controls (465). Treatment with porcine relaxin, however, resulted in uterine wet weights similar to those in normal pregnant animals (465) (Fig. 28). Although relaxin may not have an important role to play in uterine growth during pregnancy in some species, it is postulated that the actions of relaxin on the uterus may be associated with implantation. This potential action is outlined in the section on Implantation.

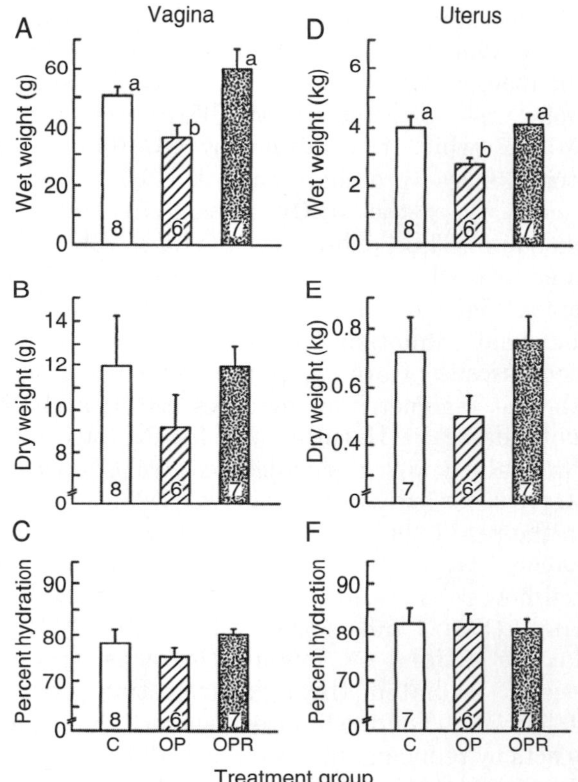

FIG. 28. Relaxin increases the size of the uterus and vagina in ovariectomized pregnant pigs. Mean (±SE) wet weight (**A, D**), dry weight (**B, E**), and percentage hydration (**C, F**) of the vagina and uterus of day 110 of pregnancy in control pigs (**C**), pigs treated with estrogen and progesterone (OP), and pigs treated with estrogen, progesterone and porcine relaxin (OPR). *Bars* with different superscripts differ significantly ($p < 0.05$). The number of pigs per group is indicated at the base of each bar. [See Min et al. (465) for more details.] (From Min, G., Hartzog, M. G., Jennings, R. L., Winn, R. J., and Sherwood, O. D. [1997]. Evidence that endogenous relaxin promotes growth of the vagina and uterus during pregnancy in gilts. *Endocrinology* 138, 560–565, with permission. Copyright 1997, The Endocrine Society.)

Vagina

Early studies demonstrated that relaxin treatment induced actions on the vagina in mice (551,569) and guinea pigs (570). Indeed, relaxin brings about marked growth of the vagina in rats (461,469,571) and pigs (465) (Fig. 28) during pregnancy. Studies of the vagina in pregnant rats indicate that relaxin promotes increased wet weight, dry weight, arterial cross-sectional area, epithelial cells, and stromal cells (461,571). The growth of the vagina in the relaxin KO mouse is severely compromised, indicating that relaxin is essential for vaginal growth during pregnancy in mice (454); the vaginal wet weight, dry weight, and collagen content were similar to virgin mice in late pregnant relaxin KO mice. Furthermore, the normal pregnancy-induced proliferation of the vaginal epithelium does not occur in relaxin KO mice (454). In the rat, relaxin acts to increase vaginal epithelial cells, at least in part, by reducing the rate of apoptosis (469). Relaxin also increased the extensibility of the vagina in rats (572) and, consistent with this finding, collagen fiber bundles in vaginas of relaxin-deficient rats and relaxin KO mice failed to disperse or to become as disorganized as do those in vaginas from control animals (454). Biotinylated porcine relaxin has been used to localize relaxin binding sites in the luminal epithelial cells as well as circular and longitudinal smooth muscle cells in the rat (571) and human (217). In contrast, LGR7 immunoreactivity was reported in rat vaginal smooth muscle cells and not in epithelial cells (21). It is clear that relaxin plays an important role in the development of the vagina during pregnancy. It is possible that relaxin-induced growth and softening of the vagina facilitates delivery.

Timing and Duration of Parturition

Although relaxin has important actions on the reproductive tract during pregnancy, it does not appear to be essential for the duration of pregnancy. The time of onset of delivery in relaxin-deficient rats (257,464) and in relaxin (255) or LGR7 (256,262) KO mice did not differ from that in controls. In contrast, when LGR7 function was blocked during the last

4 days of mouse pregnancy by treatment with the soluble ectodomain of the receptor, delivery was delayed by 27 hours (21).

There is limited evidence that there may be a central relaxin system involving the subfornical organ (SFO) that influences the time of onset of birth in rats (573,574). Both relaxin mRNA and immunoreactivity have been localized in the rat and mouse brain (18,210,299,318). Daily injections of a monoclonal antibody against rat relaxin into the right lateral ventricle throughout the second half of pregnancy advanced the onset of both luteolysis and delivery by approximately 24 hours (575). It was postulated that central relaxin may influence the time of delivery by acting on the SFO to inhibit oxytocin secretion (573,575). In contrast, intravenous administration of porcine relaxin increased the secretion of oxytocin in unanesthetized nonpregnant and pregnant rats (576,577). Moreover, neither the time of onset nor the duration of active labor in oxytocin KO mice differed from those in wild-type controls (578).

Although relaxin does not appear to be involved in the initiation of parturition, there is considerable evidence that it is important for the duration of parturition. When primiparous rats and pigs were bilaterally ovariectomized during the second half of pregnancy and given hormone replacement with physiological amounts of progesterone plus estrogen (rats) or progesterone only (pigs), the duration of delivery was several times longer and the incidence of live births was far lower than in intact controls (461,577,579,580). However, when hormone replacement was supplemented with porcine relaxin, birth parameters were similar to those of controls (461,579,580). Comparable findings were obtained in both rats (257,464, 581) and pigs (582) when circulating relaxin was immunoneutralized.

Studies in relaxin and LGR7 KO mice demonstrate that relaxin is essential for normal delivery in mice. In relaxin KO mice, two of eight pregnant females were unable to deliver their pups normally. One animal had a protracted labor and one was unable to deliver her young; all pups were either stillborn or died in utero during parturition (255). Similarly, in pregnant LGR7 KO mice, 25 of 162 pups (distributed among 9 of 21 litters) were found dead on the morning of delivery (256) (Table 3). An independent study with a different strain of LGR7 KO mice demonstrated comparable results in which 9% of pregnant females were unable to deliver pups. In some animals the dead pups were entrapped in the birth canal, indicating a problem with parturition (262). However, relaxin deficiency does not disrupt delivery as dramatically in mice as it does in rats and pigs because most of the relaxin and LGR7 KO mice delivered their young normally (255,256,262).

As discussed in the section on Cervix, there is no evidence that endogenous relaxin is necessary for normal birth in humans. However it has been postulated that H1 and H2 relaxin produced by the decidua and placenta may act through local autocrine/paracrine signaling to increase the expression of MMPs in fetal membranes and thereby bring about their rupture and the induction of delivery (583,584).

Available evidence supports the view that the primary means whereby relaxin facilitates birth in rats, mice, and pigs is by promoting dramatic growth and remodeling of the cervix. Relaxin's effects on two other portions of the birth canal—the vagina and the interpubic ligament—may also play roles at delivery in some species.

Plasma Osmolality Regulation

Plasma osmolality is maintained within a narrow range by brain mechanisms that control water intake, the neurohypophysial release of arginine vasopressin

TABLE 3. *Fertility data and pup survival of wild-type and homozygous male and female LGR7 mice*

Mating	Generation/age	Pregnancy rate	Litter size	No. of pups found dead/alive
WT M × WT F (1:1)	F3 / 11 wks	6/6	9±1	0/54
WT M × WT F (1:1)	F3 / 4.5 mo	6/6	7±2	0/41
KO M × WT F (1:2)	F2 / 14 wks	8/24	8±2	0/63
KO M × WT F (1:1)	F2 / 7 mo	7/11	9±3	0/62
KO M × WT F (1:1)	F3 / 4.5 mo	4/6	10±1	0/38
KO M × WT F (1:1)	F3 / 11 wks	5/8	9±1	11/41[a]
WT M × KO F (1:1)	F2 / 14 wks	8/8	10±2	6/76[a]
WT M × KO F (1:1)	F3 / 11 wks	5/5	9±1	7/45[a]

WT, wild-type mice; KO, LGR7 knockout mice; M, male; F, female.

[a]Pups died within 24 to 48 hours; no milk was found in their stomach.

From Krajnc-Franken, M. A., van Disseldorps A. J., Koenders, J. E., Mosselman, S., van Duin, M., and Gossen, J. A. (2004). Impaired nipple development and parturition in LGR7 knockout mice. *Mol. Cell. Biol.* 24, 687–696, with permission.

(AVP), which controls the excretion of free water by the kidney, and solute excretion. Studies in humans and rats demonstrated that during pregnancy there is an approximately 10-mosmol decline in plasma osmolality that occurs without a change in plasma AVP concentrations, and this adaptation of pregnancy is attributable to a reduced osmotic threshold for both thirst and AVP secretion (585). There is considerable evidence that relaxin is responsible for the decline in plasma osmolality that occurs during pregnancy in rats and mice. Infusion of either recombinant H2 relaxin or porcine relaxin for 5 or 6 days results in a decrease in plasma osmolality of approximately 10 mosmol in non-pregnant rats (586–588). This action of relaxin is associated with a resetting of the osmotic threshold for AVP release in ovariectomized rats (587). There is also evidence that relaxin treatment altered the thirst threshold (587); hence, both these actions of relaxin mimic the changes seen in pregnancy. In pregnant rats, the normal reduction in plasma osmolality coincides with elevated serum relaxin levels during the second half of pregnancy (301,585), and in pregnant relaxin-deficient rats this normal reduction in plasma osmolality does not occur (586). In addition, plasma osmolarity does not decrease in relaxin KO mice during late pregnancy, whereas it decreases by approximately 10 mosmol in wild-type controls (255). In humans, although plasma osmolarity is also reduced during pregnancy, this reduction may not be due to relaxin. Hence, in women with singleton pregnancies after ovum donation who have undetectable serum relaxin levels, plasma osmolarity levels decreased to a degree similar to that in normal women (345,589).

Water consumption increases markedly during the second half of rat pregnancy (575,590,591), and relaxin is a potent stimulator of water drinking in rats. Both intracerebroventricular and intravenous administration of relaxin promote drinking within minutes in nonpregnant rats (575,592–596). Intravenously infused relaxin and angiotensin II (which also increases in plasma during pregnancy) had a synergistic action on water drinking in rats (594). Furthermore, removal of relaxin from either the peripheral circulation (597) or cerebrospinal fluid (575) throughout the second half of rat pregnancy using passive immunization with monoclonal antibodies against rat relaxin resulted in a reduction in water consumption. The effects on reduction of water consumption were greatest when relaxin was neutralized in the central nervous system (575).

The actions of relaxin on drinking and resetting the threshold of AVP release in rats are mediated through actions on the SFO and organum vasculosum of the lamina terminalis (OVLT), two circumventricular organs located on the anterior wall of the third cerebral ventricle that lack a blood–brain barrier and are accessible to circulating relaxin (Fig. 29). Both high-affinity relaxin binding sites (206,209) and LGR7 mRNA expression (598) have been demonstrated in the SFO and OVLT. Electrophysiological studies demonstrate that H2 relaxin is able directly to activate neurons in the SFO (596). In addition, after intravenous administration of either porcine or H2 relaxin, the expression of *c-fos* increased in groups of neurons located in the more peripheral and dorsal parts of the SFO and in the dorsal cap region of the OVLT as well as in the supraoptic and paraventricular nuclei of the hypothalamus (595,596,599,600) (Fig. 29). Electrolytic ablation of the SFO blocked relaxin's actions on drinking, but ablation of the OVLT did not (596), suggesting that circulating relaxin acts on neurons in the outer shell of the SFO to mediate relaxin-induced drinking. These workers postulated the OVLT may mediate relaxin's action on AVP release because osmosensitive neurons reside in the dorsal cap of the OVLT, where enhanced *c-fos* expression is observed after relaxin administration. Because relaxin-sensitive neurons in the dorsal part of the OVLT send direct projections to the hypothalamic supraoptic and paraventricular nuclei (the sites of AVP-containing magnocellular neurons that project to the posterior pituitary) (595) (Fig. 29), they could be the neuroanatomical site where relaxin resets the osmostat during pregnancy (587,601). Another study demonstrated that lesions of the SFO on day 12 of pregnancy reduced the rate of increase in water consumption during the second half of pregnancy (573). There is evidence that angiotensin functions as a neurotransmitter acting through angiotensin AT1 receptors by pathways originating in the SFO and OVLT that mediate the effects of relaxin on drinking and vasopressin secretion (593,594,602–604).

Cardiovascular Effects

The potential role of relaxin in cardiovascular function was first postulated by Frederick Hisaw, who noted that administration of relaxin to castrated monkeys resulted in marked morphological changes in the endothelial cells of the endometrium consistent with hypertrophy and hyperplasia (605). Since this initial discovery, a considerable amount of evidence has accumulated suggesting that relaxin can affect blood vessel structure and function (38). Furthermore, these effects, together with effects on the kidney and heart, demonstrate that relaxin may contribute to the cardiovascular adaptations that occur in pregnancy (38). During human pregnancy, cardiovascular adaptations that are observed by 5 to 8 weeks include not only increased plasma volume, cardiac output, and heart rate but decreased blood pressure and vascular

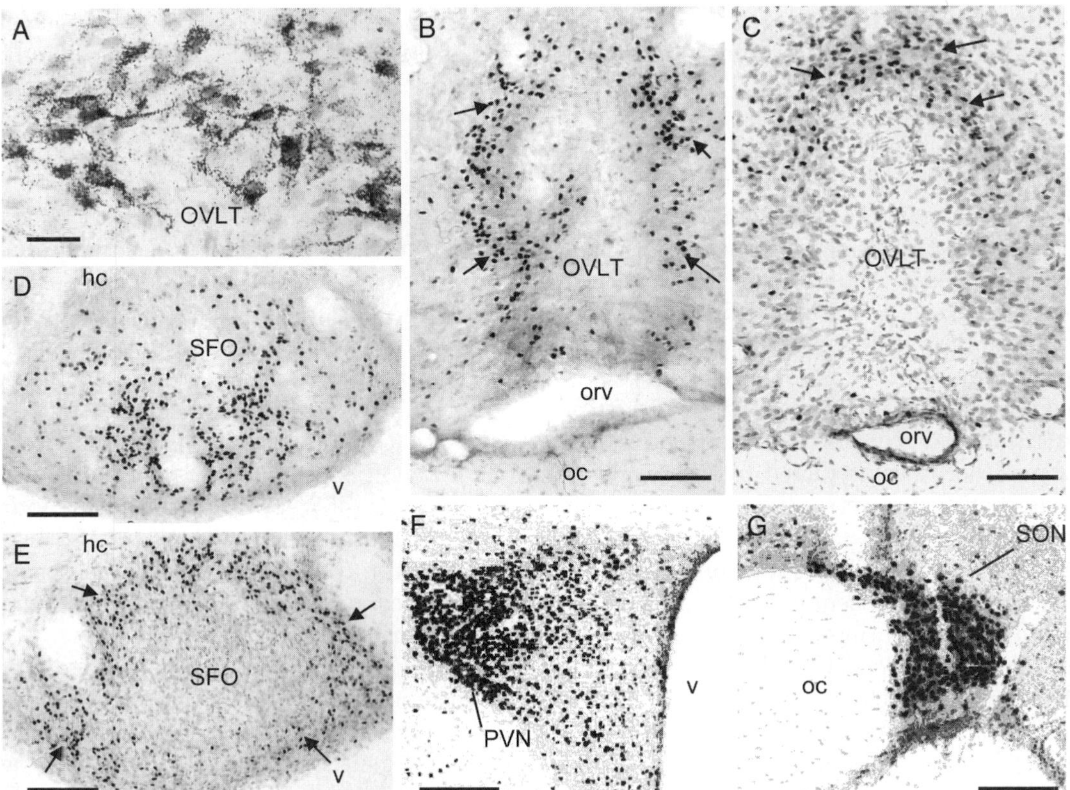

FIG. 29. Fos immunoreactivity in cell nuclei within the organum vasculosum of the lamina terminalis (OVLT) (**A–C**), the subfornical organ (SFO) (**D, E**), and the supraoptic (SON) and paraventricular (PVN) nuclei of rats that had been infused intravenously with hypertonic saline (**A**), angiotensin II (**B, D**), or H2 relaxin (**C–G**). In (**A**), neurons in the dorsal cap of the OVLT projecting to the SON were labeled with cholera toxin B gold conjugate (which can be seen as black particles filling the cytoplasm) that has been retrogradely transported from the SON. Fos immunoreactivity is seen in the dorsal cap of the OVLT (**C,** *arrow*), periphery of the SFO (**E**), hypothalamic PVN (**F**), and SON (**G**) in response to intravenous infusion of relaxin. [See Sunn et al. (596) for further details.] Scale bar = 20 μm in **A** and 100 μm in **B–F.** hc, hippocampal commissure; oc, optic chiasm; orn, optic recess of the third ventricle; v, third ventricle. (From McKinley, M. J., Mathai, M. L., McAllen, R. M., McClear, R. C., Miselis, R. R., Pennington, G. L., Vivas, L., Wade, J. D., and Oldfield, B. J. [2004]. Vasopressin secretion: osmotic and hormonal regulation by the lamina terminalis. *J. Neuroendocrinol.* 16, 340–347, with permission from Blackwell Publishing.)

resistance (606,607). Similar cardiovascular changes occur relatively later in rat pregnancy (590,608,609). The actions of relaxin on the kidney, vasculature, and heart potentially involved in these adaptations are outlined in the following sections.

Kidney. The work of Conrad and colleagues (38) studying the renal circulation during pregnancy led to the discovery of relaxin's renal vasodilatory actions. They postulated that the circulations of nonreproductive organs such as the kidney serve as arteriovascular shunts that bring about a fall in ventricular afterload during pregnancy (38). Hence, the decrease in ventricular afterload initiates the increase in cardiac output and expansion of plasma volume that occurs during rat and human pregnancy. This increase in cardiac output is essential for the tremendous increase in uteroplacental blood flow and the requisite oxygen and nutrient demands of the fetus. In pregnant rats, glomerular filtration rate and effective renal plasma flow increase while effective renal vascular resistance declines (586), and these renal adaptations are maximal during the second half of pregnancy (610). The high serum levels of relaxin at this time of pregnancy implicated a role for relaxin in this process, and relaxin was tested for its ability to vasodilate the renal vasculature. Prolonged administration of porcine or recombinant H2 relaxin to conscious female rats resulted in increases in both effective renal plasma flow and glomerular filtration rate to levels observed during mid-pregnancy (245). The effect was independent of the presence of the ovaries and was also seen in male rats (588). Furthermore, plasma osmolality was reduced in these animals, a well-documented action of relaxin outlined previously (see section on Plasma Osmolality Regulation). Relaxin promoted both a reduction in myogenic reactivity of renal small arteries and attenuation of the vasoconstrictive response of the renal vasculature to angiotensin II (245,586,611,612), both comparable with changes seen in pregnant rats. Importantly, the renal adaptations of pregnancy failed to occur in relaxin-deficient rats, where circulating relaxin was neutralized or the animals were ovariectomized (586). Therefore, in rats, relaxin is essential for these renal adaptations during pregnancy. It is not known whether relaxin contributes to the adaptations in the renal circulation in pregnant women, although results from recent clinical trials demonstrate that relaxin treatment increases predicted creatinine clearance (181), suggesting an action of relaxin on the human kidney.

The potential mechanism of action of relaxin on small renal arteries and subsequent effects on renal vasodilation and hyperfiltration are shown in Fig. 30. An essential role for nitric oxide (245,611) and the endothelin B receptor (588,611) has been established for the actions of relaxin on renal vasodilation, hyperfiltration, and reduced myogenic reactivity in relaxin-treated nonpregnant rats. Furthermore, Conrad and coworkers demonstrated that relaxin's effects on the renal vasculature depend on increased activity of vascular MMP-2 (612): relaxin acting on its receptor induces this vascular gelatinase, which cleaves big endothelin to yield endothelin$_{1-32}$, which in turn act on the endothelin B receptor to induce nitric oxide production, causing renal vasodilation (38).

Vasculature. The aforementioned studies indicate that relaxin can act on small renal arteries to induce vasodilation. It would appear that this is a common action of relaxin because it has been reported to dilate not only microvessels such as arterioles that are surrounded by a smooth muscle coat, but capillaries and postcapillary venules in numerous sites throughout the body. In rodents, relaxin promotes dilation of microvessels not only in reproductive organs (467,561,613,614) but in nonreproductive sites, including the heart (615–617), liver (618), and mesocecum (619). Relaxin has also been reported to blunt responses to vasoconstrictors in perfused mesenteric artery of spontaneously hypertensive rats (620,621), primary cultures of bovine aorta smooth muscle cells (242), rat coronary endothelial cells (622), and segments of rat uterine artery (623). In these primary cell cultures, pretreatment with porcine relaxin markedly reduced the intracellular rise in Ca^{2+} induced by the vasoactive agonists α-thrombin

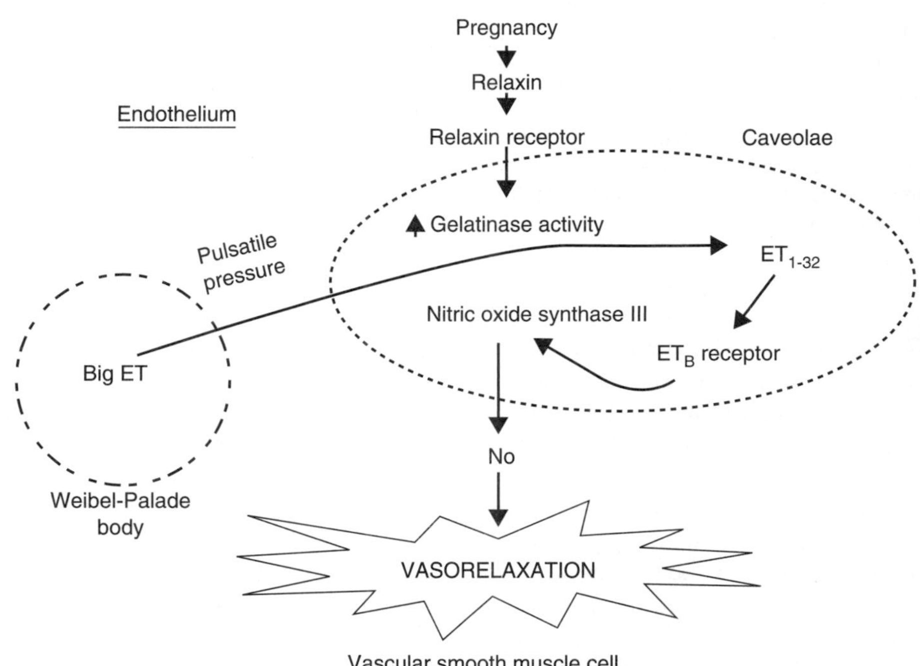

FIG. 30. Working hypothesis of the cellular mechanisms underlying relaxin-induced renal vasodilation and hyperfiltration and reduced myogenic reactivity of small renal arteries. Big endothelin (ET) is released from Weibel-Palade bodies in response to pulsatile pressure in vivo or increases of intraluminal pressure in vitro (809–811). Matrix metalloproteinase-2 (MMP-2) (812), endothelial nitric oxide synthase (813), and possibly the endothelin B (ETB) receptor (814) are localized in caveolae of endothelial cells. [See Conrad (815) for further details.] (From Conrad, K. P. [2004]. Mechanisms of renal vasodilation and hyperfiltration during pregnancy. *J. Soc. Gynecol. Investig.* 11, 438–448, with permission.)

and angiotensin II (242,622). As in the kidney, relaxin appears to mediate its effects on vasodilation in mesenteric arteries and other sites, at least in part, by acting on endothelial cells to upregulate ET_B receptors and also to stimulate the activity of inducible nitric oxide synthase II, thereby increasing nitric oxide production (35,245,611,615,616,622–624). However, it would seem that relaxin does not promote vasodilation in all blood vessels. Whereas in vitro treatment with recombinant H2 relaxin was reported to cause a rapid relaxation of human preconstricted gluteal arteries in an endothelium-dependent manner, it did not influence small pulmonary resistance arteries, uterine myometrial arteries, or placental stem villus arteries (531,624).

Because of the actions of relaxin on the kidney and other vasculature described previously and the plasma profile of relaxin during pregnancy, it is conceivable that endogenous circulating relaxin may contribute to general cardiovascular changes during pregnancy. Conrad and colleagues have tested whether relaxin modifies systemic cardiovascular function when chronically administered to nonpregnant female rats (625). The authors tested the actions of relaxin on cardiac output, global arterial compliance, and steady arterial load (systemic vascular resistance), all critical cardiovascular changes during pregnancy (38). They demonstrated that relaxin is a general vasodilator, reducing systemic vascular resistance and increasing cardiac output. The changes are similar to those seen during pregnancy in humans (607) and rats (626), and although some of the effects are clearly due to the actions of relaxin on the renal vasculature (see earlier), other vasculature is likely to be involved (625). The authors further suggested that relaxin was acting not only by reducing vascular smooth muscle tone through a vasodilatory action, but by modifying the extracellular matrix of the vessel walls. This dual action may be critical for the maintenance of cardiovascular homeostasis during pregnancy (38). A more recent study from this group has demonstrated that the effects of relaxin on systemic hemodynamics and arterial mechanical properties are gender independent and, furthermore, that the dose response is biphasic (627). The authors postulate that relaxin may be a vasculature-derived, locally acting compliance and relaxing factor.

Heart. The heart is clearly a target organ for relaxin in rodents. High-affinity binding sites for relaxin are localized to the atria of male and female rats as early as 1 day after birth (206,207,210). LGR7 mRNA has also been detected in the rat (237,241,250), mouse (256,258), and human (21) heart. Rat atrial cardiomyocytes express high-affinity relaxin binding sites (208), and both atrial and ventricular rat cardiomyocytes express LGR7 mRNA (237). Relaxin binding sites in the female rat atria are at a higher density than in the uterine myometrium and, unlike the uterine sites, the concentrations are not influenced by estrogen (206). Numerous in vitro and in vivo studies have demonstrated that relaxin has potent, direct, and concentration-dependent chronotropic effects on the rat heart (Fig. 31). In addition, relaxin demonstrates inotropic effects on the heart (Fig. 31); relaxin not only increased the rate of spontaneous contractions in perfused intact hearts (240,616,628,629) and isolated right atria (133,170,219,630,631), but increased the force of electrically stimulated contractions in isolated left atria (133,170,219,630,631). Relaxin may act on both atrial and ventricular pacemakers because relaxin increased heart rate in heart preparations from which the atria had been removed (628).

Although relaxin demonstrates clear actions on the heart, little is known about the mechanism of action or which cell types are responsible for the actions. Relaxin's chronotropic effects on isolated perfused rat hearts were accompanied by the secretion of atrial natriuretic peptide (240). Both effects of relaxin appeared to involve cellular signal transduction pathways involving PKA and protein kinase C, whereas the chronotropic actions alone were associated with calcium/calmodulin–dependent protein kinases. Studies in Brattleboro rats demonstrated significantly reduced chronotropic responses to relaxin injections, suggesting that AVP may mediate relaxin's effects (632). When individual cells were examined with whole-cell patch clamp, relaxin was found to inhibit outward potassium currents, increase action potential duration, and enhance calcium entrance into rat atrial myocytes (633,634). In similar experiments, relaxin increased the rate of action potentials and L-type calcium current in rabbit sinoatrial node cells (239). These effects involved the activation of PKA (239,633,634).

Acute intravenous or intracerebroventricular administration of porcine relaxin increased heart rate in urethane-anesthetized nonpregnant rats (632,635), and infusion of recombinant H2 increased heart rate in conscious nonpregnant rats (631). However, acute arterial administration of recombinant H2 relaxin on day 19 did not influence heart rate in conscious pregnant rats (636). Hence, the role of endogenous circulating relaxin on heart rate in pregnant rats is not known. Similarly, the effect of relaxin on the heart of humans or other species is unknown. Neither recombinant H2 relaxin nor synthetic sheep INSL3 was able to induce positive inotropic responses in isolated sheep atrial preparations, whereas isoprenaline had a clear stimulatory action (417). Furthermore, recombinant H2 relaxin at concentrations producing positive chronotropic

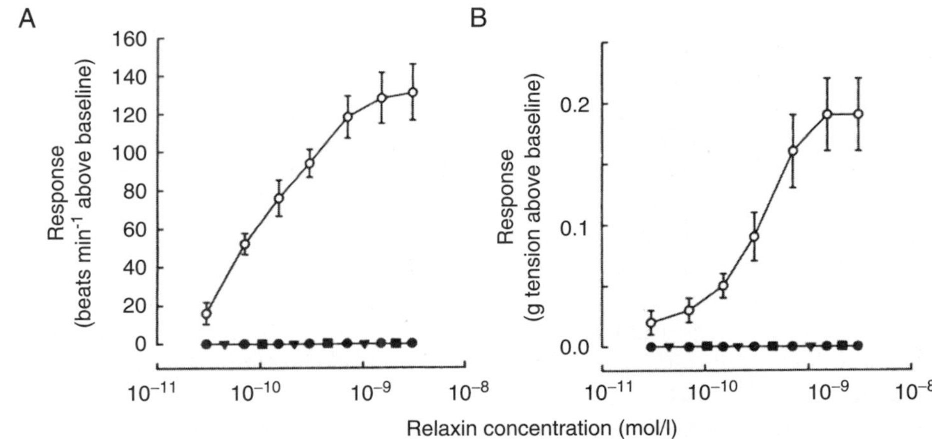

FIG. 31. Chronotropic (**A**) and inotropic (**B**) responses to H2 relaxin (*open circle*), H2 relaxin A (*closed circle*), and B (*closed triangle*) chains and trypsin-digested H2 relaxin (*closed square*) in rat isolated right and left atria. (Adapted from Kakouris, H., Eddie, L. W., and Summers, R. J. [1992]. Cardiac effects of relaxin in rats. *Lancet* 339, 1076–1078, with permission from Elsevier.)

and inotropic responses in rat atria had no effect in human atria (637). [^{33}P]-labeled H2 relaxin does not bind to the sheep or human atria (Y. Y. Tan and R. J. Summers, personal communication). Therefore, as for its actions in the uterus, the actions of relaxin on the atria may be species specific.

Interestingly, the multiple central and cardiovascular effects of relaxin outlined in the foregoing appear to be as profound in male as in female rodents. In males, relaxin was reported to decrease plasma osmolality (588) and increase drinking (592,594,596,599), glomerular filtration rate, effective renal plasma flow (588), vasodilation (617–619), and heart rate (628,630,631). Thus, unlike relaxin's effects on the female rodent reproductive tract, its effects on the brain, kidney, general vasculature, and heart do not require elevated circulating estrogen.

Lactation

The second biological effect of relaxin to be reported was its putative growth-promoting effects on the mammary gland. Hamolsky and Sparrow (638) used crude preparations of porcine relaxin to demonstrate that relaxin, when administered together with estrogen and progesterone, promoted growth and lobulation of the mammary gland in ovariectomized immature rats. Since then, numerous studies have demonstrated that relaxin has effects on the development of the mammary apparatus in pregnant pigs, rats, and mice. However, there are important species differences in the effects. Relaxin clearly has an important role in the development of the mammary gland in pigs, but in rodents the major effect is on the development of the mammary nipple.

Nipple Development

The initial surprising discovery that relaxin is essential for the development of the nipple during pregnancy came from studies in relaxin-deficient rats. Immunoneutralization of circulating rat relaxin throughout the second half of pregnancy with monoclonal antibody MCA1 resulted in the total lack of nipple development (639). The nipples in relaxin-deficient rats are so small at term that the pups cannot grasp them to obtain milk (639,640). An identical phenotype is obtained in relaxin KO mice (255); pups from the KO mothers are unable to suckle milk and die within 24 hours unless cross-fostered to wild-type mothers. The inability of the pups to obtain milk is due to the defect in nipple development because the mammary gland appeared normal in the relaxin KO mothers, milk was produced, and there was no abnormality in the myoepithelial cells (255). Furthermore, one KO female that had slightly larger nipples than the other KO females was able to express milk to her pups (255). Recent studies in LGR7 KO mice (256,262) confirm that this action of relaxin is through LGR7 because the LGR7 KO females display an identical phenotype (Fig. 32). Furthermore, data from LGR7 KO mice indicate that nonpregnant mice also have significantly smaller nipples (262). Overexpression of INSL3 in LGR7 KO mice did not rescue the phenotype, highlighting that there was no role for the INSL3/LGR8 system in this physiology (262). Studies using the soluble ectodomain of LGR7 to block relaxin's action in late pregnancy also demonstrated abnormal nipple development (21).

Biochemical analysis of relaxin KO nipples demonstrated a reduced wet and dry weight of the tissue associated with a marked increase in collagen

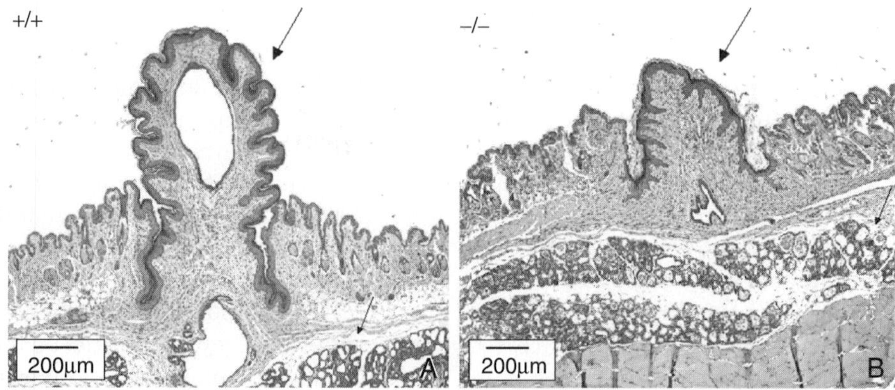

FIG. 32. Histology of nipples and mammary glands from LGR7 knockout (KO;–/–) and wild-type (+/+) mothers using hematoxylin and eosin staining. **A:** Results from the wild-type female; **B:** results from the LGR7 KO female. The lactating nipple (*large arrow*) of the wild-type mother (a) was much longer than the nipple (*large arrow*) of the LGR7 KO mother (b), which was not able to support suckling. Lactating mammary glands (lobular hyperplasia with prominent secretion, *small arrows*) in both wild-type and LGR7 KO mice are also shown. (From Krajnc-Franken, M. A., van Disseldorp, A. J., Koenders, J. E., Mosselman, S., van Duin, M., and Gossen, J. A. [2004]. Impaired nipple development and parturition in LGR7 knockout mice. *Mol. Cell. Biol.* 24, 687–696, with permission.)

content (454). All parameters were similar to those seen in virgin mice. In addition, histological analysis of rat and mouse nipples demonstrated that relaxin promoted a reduction in the density of collagen fiber bundles that was similar to that which occurs in the cervix, vagina, and mammary glands (454,640). Morphometric analysis of rat nipples indicated that relaxin reduced the length of elastin fibers and enlarged the cross-sectional area of arteries (640). Hence, the mechanism of relaxin action on the rodent nipple likely involves roles in both growth and collagen turnover. In contrast to studies in rodents, the nipples of relaxin-deficient primiparous pigs are moderately well developed at term (641,642), and piglets can grasp them and obtain milk (642). There is evidence that relaxin acts directly on the nipples to bring about its effects. Specific relaxin binding sites were reported in nipples from rats (214,643), pigs (215), and humans (217). LGR7 mRNA expression was reported in the connective tissue at the base of the nipple and just beneath the epithelium in mice (256). In addition, subcutaneous injection of porcine relaxin at the base of a left abdominal rat nipple promoted significant growth of that nipple relative to its right counterpart (643). Little is known concerning the roles of ovarian steroids on nipple development. A single study indicated that, unlike the rat cervix, relaxin-induced growth of the rat nipple occurs in the absence of estrogen (643).

Mammary Gland

Although many early studies using crude preparations of porcine relaxin indicated that relaxin influences growth and differentiation of the mammary gland, they were neither consistent nor definitive (25). Later studies with highly purified porcine relaxin indicated that relaxin was able to induce the growth and differentiation of the mammary gland in mice (614,644–647) and rats (648,649). However, relaxin has no apparent effect on the weight of the mammary glands during the second half of pregnancy in relaxin-deficient rats (639,650). Similarly, despite a trend toward lower wet weight, there was no significant decrease in mammary gland weight in relaxin KO mice on the day of birth (454). Despite the lack of nipple development in relaxin and LGR7 KO mice, some animals can still express milk to their pups (255,262), suggesting that mammary gland function is normal. Indeed, it was reported that there were no abnormalities in glandular tissue or fat composition in LGR7 KO mice (256) (Fig. 32). In contrast, it was reported that there was less glandular tissue, alveoli, and fat in the relaxin KO mouse at term and that the ducts were grossly dilated (255). In rats, relaxin reduced the density and organization of collagen fiber bundles, reduced the length of elastin fibers, and increased the cross-sectional area of arterioles (639,650) as well as increasing alveolar development (639). Hence, although relaxin is not likely essential for mammary development in rodents, it probably has an influence on differentiation of the tissue, as in the vagina and cervix.

In contrast, relaxin is likely to have an important role in mammary development in pigs. In primiparous pigs, mammary lobuloalveolar development begins on approximately day 80, and it continues until term (641), a period that coincides not only with rising levels of estrogen and relaxin but relaxin-dependent growth and softening of the cervix (466,651). Studies that used ovariectomized primiparous pigs and hormone replacement demonstrated that endogenous relaxin plays a major role in promoting the development of the mammary gland parenchyma (641,642). A subsequent study in ovariectomized nonpregnant gilts provided evidence that relaxin's effects on mammary parenchymal development are estrogen dependent in pigs, and that they are accompanied by a reduction in the organization of collagen fiber bundles in the stroma (652).

The actions of relaxin are likely mediated by relaxin receptors present in the mammary glands. Relaxin binding sites have been reported in epithelial cells associated with lactiferous ducts and lobuloalveolar structures in rats (214), pigs (215), and humans (217). One study has demonstrated LGR7 immunoreactivity in the human breast that was prominent in the stromal connective tissue between the glandular lobules (653). Staining was not observed in the epithelial cells of the normal breast, but sporadic staining was observed in breast cancer tissue. Numerous studies have investigated the potential association of relaxin with breast cancer [reviewed in Silvertown et al. (37) and Bani (654)]. Breast cancer cell lines express relaxin receptors and respond to exogenous relaxin (655,656). Furthermore, treatment of breast cancer cells with relaxin increased their invasive potential (179,657). Studies also have demonstrated an association between higher relaxin plasma concentrations and the progression of breast cancer (658). The authors suggest that their data support a role for relaxin in tissue remodeling during breast cancer progression.

There is limited evidence that relaxin suppresses the milk ejection reflex by acting either on the rat brain to inhibit oxytocin secretion (659) or directly on the myoepithelial cells of the goat mammary glands to inhibit milk expression (660). Because serum levels of relaxin are extremely low to undetectable during lactation, it was postulated that locally produced relaxin in the brain of the rat (659) and in the mammary glands of the goat (660) brings about the effects on milk release. However, relaxin does not appear to influence milk ejection in the pig (297). Therefore, the physiological significance, if any, of relaxin for the milk ejection reflex remains to be demonstrated.

Implantation

Early investigations examined the influence of prolonged administration of partially purified porcine relaxin, progesterone, and estrogen on the histological characteristics of the endometrium in juvenile or ovariectomized rhesus monkeys (605,661). These original studies suggested that relaxin promotes growth of the endometrium that included proliferation of the endothelial cells located in the distal portions of the spiral arteries, and led these workers to postulate that one of the functions of relaxin may be to assist in the preparation of the endometrium for implantation. More recent studies using recombinant H2 relaxin have demonstrated clear effects of relaxin on the endometrium in humans and monkeys; recombinant H2 relaxin promoted growth of the uterus in a model of early pregnancy in the rhesus monkey (559) and in ovariectomized marmoset monkeys (560). These studies in the rhesus (559) (Fig. 33) and marmoset monkey (560) uterus demonstrated increased endometrial angiogenesis in response to relaxin. Relaxin's effects in the rhesus monkey were further associated with increases in cytokine-containing lymphocyte number (Fig. 33) and maintenance of endometrial connective tissue integrity, all actions consistent with a significant role for relaxin in the establishment or maintenance of pregnancy (559). A study in the cynomolgus macaque (*M. fascicularis*) examining the effects of recombinant H2 relaxin treatment in animals after in vitro fertilization and embryo transfer further highlights a potential role for relaxin in implantation (662). Administration of relaxin to cycling females during the peri-implantation period resulted in a transient increase in endometrial thickness and increased implantation-related bleeding, results consistent

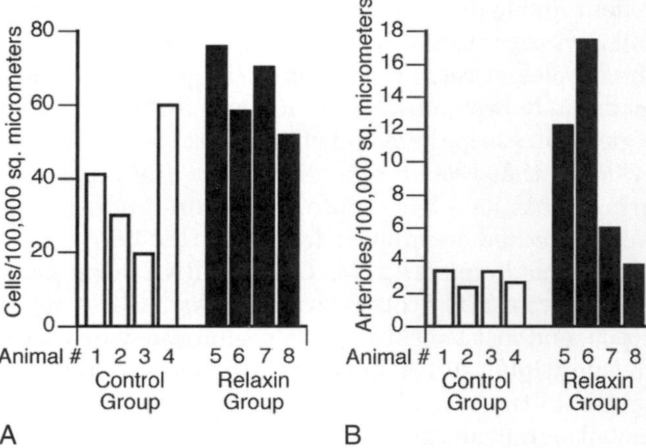

FIG. 33. The effects of relaxin on rhesus monkey endometrial lymphocyte (**A**) and arteriole (**B**) number. Morphological assessments were performed as described (559). Each *bar* shows the mean value for one animal in either the control (*open bar*) or relaxin-treated (*filled bar*) group; at least five fields for each of at least three uterine sections per animal were assessed. Relaxin treatment increases both endometrial lymphocyte and arteriole number. (From Goldsmith, L. T., Weiss, G., Palejwala, S., Plant, T. M., Wojtczuk, A., Lambert, W. C., Ammur, N., Heller, D., Skurnick, J. H., Edwards, D., and Cole, D. M. [2004]. Relaxin regulation of endometrial structure and function in the rhesus monkey. *Proc. Natl. Acad. Sci. U S A* 101, 4685–4689, with permission. Copyright 2004, National Academy of Sciences, USA.)

with a role for relaxin in modulating endometrial phenotype during implantation. Furthermore, rates of implantation and multiple pregnancies were higher in the relaxin-treated groups. Results from the clinical trial in women with diffuse scleroderma provided additional evidence that relaxin may promote angiogenesis in the human endometrium. During the course of 24 weeks of continuous subcutaneous infusion, women receiving recombinant H2 relaxin reported heavy, irregular, or prolonged menstrual bleeding more often than women receiving a placebo (663).

The pattern of relaxin secretion in humans and monkeys, with higher concentrations in the first trimester (see section on Human; and section on Primates) coinciding with embryo implantation, has also led to the hypothesis that relaxin is involved in implantation in these species. Furthermore, relaxin levels rise in close association with chorionic gonadotropin in conceptive cycles (334,391) and are significantly altered in women with early pregnancy loss (664), and relaxin secretion by human granulosa cell cultures was predictive of in vitro fertilization success (665). In the marmoset, a New World monkey, serum relaxin levels rise a few days before implantation, and they are higher than in humans and Old World monkeys (393). However, it is likely that circulating relaxin is not essential for implantation in humans and other primates. Women who experience premature ovarian failure can become pregnant through ovum donation; hence, implantation occurs in these women in spite of the fact that they have neither a corpus luteum nor detectable serum relaxin levels (345). Similarly, in macaque monkeys subjected to bilateral ovariectomy at embryo transfer, implantation and normal pregnancy occur (666). It is still possible that locally produced relaxin can compensate in the absence of circulating relaxin in humans and other primates. Indeed, locally produced relaxin may be the endogenous regulator of the effects of relaxin on the endometrium, especially in species who do not show increases in plasma relaxin during implantation. Relaxin mRNA and immunoactivity have been localized in the endometrium during the luteal phase of the cycle or early pregnancy in humans (337,338,667), marmosets (96), pigs (296), rabbits (422), and guinea pigs (440). Detailed studies of relaxin transcripts in human endometrium in pregnant and nonpregnant cycles clearly indicated that relaxin expression preceded the expression of IGFBP-1 and prolactin (667), two classic markers of decidualization (see later).

The actions of relaxin on the uterus during implantation are mediated by specific receptors in the endometrium. LGR7 mRNA expression colocalizing with relaxin binding sites was demonstrated in the human endometrium during the secretory phase of the menstrual cycle (339). LGR7 immunoreactivity has been shown in the endometrial stroma as well as the glandular and luminal epithelium of the human uterus (544). However, an independent study found LGR7 immunoreactivity only in the endometrial stroma using a variety of antisera raised against different LGR7 epitopes (653). Furthermore, using the same antisera, LGR7 immunoreactivity was demonstrated in the endometrial stroma in the macaque and marmoset monkey uterus. Staining was strongest in stromal cells adjacent to the endometrial glands and was not present in the glandular or luminal epithelium (653). In the marmoset, LGR7 immunostaining increased from the proliferative to the luteal phase of the estrous cycle and increased further in early pregnancy, all consistent with a potential role for relaxin in decidualization of endometrial cells and hence implantation.

Also consistent with a role for relaxin in implantation is the observation that relaxin induced the decidualization of human endometrial stromal cells in culture (668). Relaxin was equally effective as progesterone in inducing the cell differentiation associated with decidualization, and induced the process in the absence of progesterone (669). These primary cultures of human endometrial stromal cells express high-affinity relaxin binding sites (208) as well LGR7 mRNA (235,670). Tseng and coworkers (671) demonstrated that relaxin exerts a synergistic effect with progestin on the expression of prolactin, aromatase activity, estrone sulfate sulfatase, and IGFBP-1, which is a major protein synthesized and secreted by decidualized cells. One study suggests that progestin is necessary to upregulate LGR7 in decidualized cells, whereas the actions of relaxin are independent of progestin (670). Relaxin also induced the expression of VEGF in cultures of human endometrium stromal and epithelial cells (338,663,672), an action that may be associated with the new vessel growth and vasodilation that occur in the endometrium at implantation (673). Glycodelin, a glycoprotein produced and secreted by the secretory endometrium, has been postulated to contribute locally to immunosuppression at implantation (674). Stewart and coworkers demonstrated that there was a close temporal and quantitative relationship between circulating relaxin and glycodelin profiles during the luteal phase of the menstrual cycle and early pregnancy, and that subcutaneous administration of recombinant H2 relaxin for 28 days increases the secretion of glycodelin in women demonstrating ovarian cyclicity (675). It was also reported that porcine relaxin promotes the expression of glycodelin mRNA and protein in primary cultures of human endometrial epithelial cells (676). There is also limited evidence that relaxin increases immunostaining for cyclooxygenase-2 (672) and inhibits the expression of collagenase (338) in primary cultures of human endometrial cells. Relaxin appeared to mediate its effects on human endometrial stromal cells, at least in part, through the generation of cAMP (227,228,668). In fact, the sustained generation of cAMP is an essential part of the decidualization process, and relaxin may be an essential mediator of these sustained increases (235). There is limited evidence that relaxin acts in concert with estrogen and progesterone to enhance the expression of HOXA-10, an endometrial transcription factor that is required at implantation in mice (677).

In contrast to the human and primate studies outlined in this section that demonstrate a critical role for relaxin in implantation, studies in rodents do not support an important role for relaxin. Porcine relaxin was reported to promote increased endometrial thickness, loosening of collagen framework, and dilation of blood vessels in rats and mice (561,613). However, Finn et al. (678) reported that the subcutaneous administration of porcine relaxin to estrogen-plus-progesterone–treated mice did not promote decidualization of the endometrium. Both relaxin and LGR7 KO mice demonstrated normal fertility, and the average litter size did not differ from that in wild-type controls (255,256,262). Hence, relaxin may not be important for implantation in mice.

Intraovarian Effects of Relaxin

Although relaxin is an endocrine hormone produced by the corpus luteum, it could also exert local paracrine actions in ovarian follicles. Although relaxin immunoactivity has not been detected in the peripheral blood during the follicular phase of the estrous cycle in any species, relaxin mRNA or immunoactivity has been detected in preovulatory follicles in pigs (679), rats (270,680), horses (408), tammar wallabies (103), camels (108), marmoset monkeys (96), and humans (681). Bagnell and coworkers reported that the theca interna in prepubertal pigs produced relaxin, and the levels of relaxin in follicular fluid increased with follicular size (679). Moreover, they found that low doses of porcine relaxin promoted growth of both granulosa and theca cells in vitro (682–684). Subsequently, specific relaxin binding sites were localized using biotinylated porcine relaxin in both the theca and granulosa cells of developing follicles (216). These findings led to the hypothesis that relaxin may promote follicular growth through intraovarian autocrine or paracrine mechanisms (679,685).

There is also limited evidence that relaxin may act locally to promote follicular development and ovulation in rats. Proteinase enzymes, including plasminogen activator and collagenase, play an essential role in bringing about the extracellular matrix remodeling required for follicular development and ovulation, and it was demonstrated that porcine relaxin promoted the secretion of these and possibly other proteinases from primary cultures of rat granulosa cells and theca-interstitial cells (686,687). In addition, it was reported that recombinant H2 relaxin induced ovulation in the in vitro perfused rat ovary (688), and that passive immunization of circulating rat relaxin with a monoclonal antibody for rat relaxin reduced the number of ovulated oocytes when immature rats are induced to superovulate with gonadotropins (687).

A more recent study has also demonstrated LGR7 mRNA and immunostaining in the granulosa cells of primordial, primary, and secondary follicles of the human ovary (681). The authors also demonstrated that recombinant H2 relaxin treatment of cultured slices of human ovarian cortical tissue resulted in a higher proportion of secondary follicles and a consequent decrease in primordial follicles. Interestingly, an earlier study demonstrated no effect of H2 relaxin on human granulosa cells in culture (689).

However, studies in mice do not support a role for relaxin in either follicular development or ovulation. Treatment with relaxin did not increase follicular growth or antrum formation in cultures of mouse preantral follicles (690). Moreover, studies in relaxin and LGR7 KO mice indicated that the relaxin–relaxin receptor system was not required for either follicular development or ovulation in mice because these animals were fertile, and the average litter size did not differ from that in controls (255,256,262). Now that the relaxin receptor has been identified, future studies on ovarian expression of the relaxin receptor could provide additional evidence on the potential role of relaxin in ovarian folliculogenesis and ovulation. Furthermore, because INSL3 is important for follicular development in rodents (see section on Gonadal Functions), it is possible that there are species differences in the actions of relaxin family peptides in folliculogenesis.

Male

In 1959, Steinetz et al. (399) reported that a bioactive equivalent of relaxin was present in rooster testicular extracts and subsequently reported that its growth-promoting effect on the mouse pubic symphysis was neutralized with an antiserum to impure porcine relaxin (691). There is now good evidence that small amounts of relaxin are produced in the reproductive tract in male mammals, birds (25), and elasmobranchs (692). Relaxin mRNA has been detected in the rat testis (299), and both relaxin and relaxin-3 mRNA are expressed in the human (23) and mouse testis (18,693) (Fig. 24). Both H1 and H2 relaxin gene expression were reported in the human prostate (110,132,694). Relaxin mRNA has also been detected in the rat (299) and mouse prostate (18,693) (Fig. 24), and relaxin immunostaining was reported in the pig seminal vesicles (695). Because circulating relaxin levels in males are low or nondetectable (78,696), it is unlikely that relaxin from the male reproductive tract is an endocrine hormone. However, relaxin immunoactivity was reported in human seminal plasma by many workers (77,697–703). The male human definitively produces relaxin because Winslow et al. (81) isolated the relaxin peptide from human seminal plasma and found that its structure is the same as that of H2 relaxin produced by the human corpus luteum.

There is considerable evidence that relaxin increases sperm motility and sperm penetration into oocytes, and it has been postulated to be a potential therapeutic agent in male infertility (704). Treatment of human sperm with rabbit anti-porcine relaxin antibody caused a rapid and marked decline in the percentage of motile washed human sperm (705). Also, a monoclonal antibody against H2 relaxin decreased the percentage of motile washed human sperm (706), whereas anti-insulin antiserum had no effect (707). Both sperm motility and fertility have been correlated with levels of relaxin immunoactivity in porcine seminal plasma (708). Similar correlations were not obtained with human semen samples (697,703). There have been conflicting results on the effect of relaxin treatment on sperm motility in vitro (25). LGR7 mRNA has also been localized on spermatids in mice (256). Furthermore, one study demonstrated that [^{125}I]-H2 relaxin binds with high affinity to human sperm, and treatment with H2 relaxin increased sperm motility (709).

Both the testis and the prostate in humans (21), rats (250), and mice (693) express LGR7 mRNA, so it is possible that relaxin has local actions in these tissues. The phenotypes of the relaxin and LGR7 KO mice provide strong evidence that endogenous relaxin plays a role in growth and development of the male reproductive system in mice. By the age of sexual maturity, male relaxin and LGR7 KO mice show retarded growth of the reproductive tract (256,693). Histological examination indicated that sperm maturation was markedly decreased in the testis of relaxin and LGR7 KO mice, and that this may be attributable, at least in part, to an increase in the rate of apoptosis in early stages of spermatogenesis (256,693). The epididymis, seminal vesicles, and prostate in relaxin KO mice (693), and the epididymis in LGR7 KO mice (256), were reported to be smaller than in wild-type controls. Histological analysis revealed developmental anomalies in the epididymis and prostate of relaxin KO mice (693) and epididymis of LGR7 KO mice (256). In the epididymis tubule, compactness, collagen staining, and apoptosis of epithelial cells was greater than in wild-type controls (256,693). The reduced growth of the prostate in relaxin KO mice was accompanied by increased collagen, increased apoptosis of epithelial cells, and a decrease in glandular epithelium relative to wild-type controls (693). These developmental deficiencies are of considerable importance. Whereas both relaxin and LGR7 KO mice are fertile, fertility is markedly lower than in wild-type controls (256,693) (Fig. 34). Importantly, the reproductive phenotype of the LGR7 KO mouse was lost in later generations of animals (256), and LGR7 KO mice developed independently (262) have no deleterious male reproductive tract phenotype. Therefore, the actions of relaxin on the male reproductive tract may be redundant or be dependent on the genetic background of the animals.

PHYSIOLOGICAL ROLES OF RELAXIN IN NONREPRODUCTIVE PROCESSES

It has become increasingly clear that relaxin has important roles outside of reproduction. These roles have been determined from physiological observations in studies where animals have been treated with relaxin; from the discovery of unique sites of relaxin peptide and mRNA expression; and from novel phenotypes in the relaxin and LGR7 KO mice. Furthermore, relaxin shows enormous potential as a therapeutic agent in many conditions unrelated to reproduction. Some of the main nonreproductive actions of relaxin are outlined briefly in the following sections, and reference to relevant detailed reviews on the topics is made where possible.

Fibrosis

The actions of relaxin on the connective tissue of the reproductive tract, particularly its role in inhibiting collagen biosynthesis and promoting collagen breakdown, are well established and have been outlined in detail in the previous sections. In addition, it has been clearly established that relaxin also acts to break down collagen accumulation in nonreproductive tissues, especially tissues associated with tissue damage and excessive accumulation of extracellular matrix components, including collagen (fibrosis) (31,34). These actions have led researchers to postulate that relaxin would be an effective treatment for fibrotic diseases. Relaxin has been demonstrated to act directly

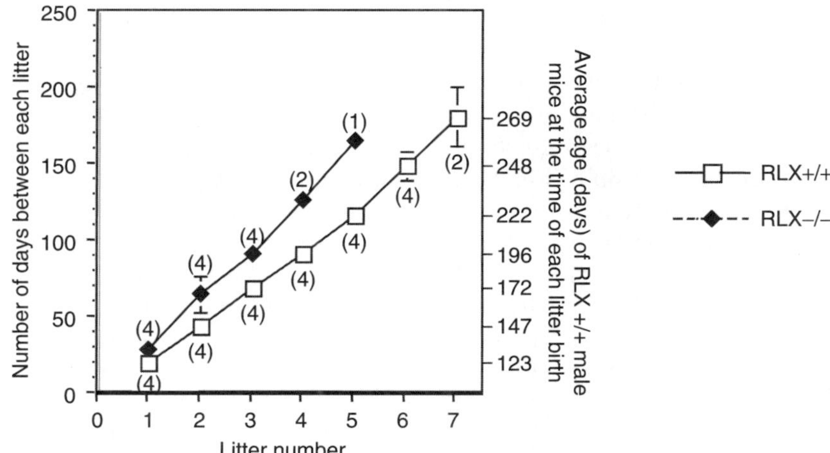

FIG. 34. The mean (±SE) number of days taken by relaxin wild-type (RLX+/+) male and female pairs and relaxin knockout (KO; RLX+/+) male/RLX+/+ female pairs to produce pups over a 5-month period. The corresponding age of the relaxin wild-type mice (RLX+/+) at each point is plotted on the right. As the age of the male RLX+/+ mice increased (over 4 months of age), the number of days taken by their female RLX+/+ breeding partners to give birth to pups increased with each litter compared with that produced by RLX+/+ pairs. (From Samuel, C. S., Tian, H., Zhao, L., and Amento, E. P. [2003]. Relaxin is a key mediator of prostate growth and male reproductive tract development. *Lab. Invest.* 83, 1055–1067, with permission. Copyright 2003, Nature Publishing Group.)

on transforming growth factor-β–stimulated human dermal fibroblasts (710), lung fibroblasts (711), and cardiac fibroblasts (237) to promote both a decrease in type I and type III collagen synthesis and an increase in MMP expression and activation (Fig. 35). Furthermore, it has been used successfully in a number of animal models of fibrosis in vivo, to modify the extracellular matrix in the dermis (712,713), lung (711), liver (714,715), and kidney (716,717). In all these experimental paradigms, relaxin induced its actions on the extracellular matrix only during increased collagen deposition that was artificially induced by profibrotic stimuli, surgery, or chemical means. Hence, relaxin could rapidly prevent excessive collagen deposition in diseased states characterized by fibrosis.

Relaxin was first used as an antifibrotic agent in 1958 (718), when partially purified relaxin was used to treat patients with scleroderma. Scleroderma (systemic sclerosis) is a connective tissue disease of unknown etiology in which tissue fibrosis is the predominant clinical feature and largely determines morbidity and mortality (719). The disease is characterized by fibrotic intimal hyperplasia of small arteries and arterioles (Raynaud's phenomenon, renal crisis, pulmonary hypertension), as well as extravascular fibrosis of the skin and internal organs, including the lung, kidneys, and heart (719). The generally positive results of this and other studies (25), as well as the in vivo and in vitro findings outlined previously, led to the initiation of clinical trials using recombinant H2 relaxin in scleroderma. A phase II trial that used approximately 20 patients per group examined the continuous subcutaneous infusion of 25 μg/kg or 100 μg/kg of body weight recombinant H2 relaxin per day for a 24-week period (720). Portions of the results of this study were favorable and the administration of recombinant H2 relaxin was found to be safe and well tolerated. Moreover, 24 weeks of treatment with the 25 μg/kg dose of recombinant H2 relaxin, but not the 100 μg/kg dose, had significant beneficial effects on skin thickness and mobility. In response to these encouraging results, a phase II/III study that used more than 40 patients per group was conducted at doses of 25 μg/kg and 10 μg/kg recombinant H2 relaxin (181). Unfortunately, skin elasticity, hand extension, oral aperture, cutaneous ulcers, and pulmonary function in recombinant H2–treated patients did not differ from those in the placebo control subjects (181). Although the trial failed to meet the primary efficacy end points, it highlighted the fact that we still know very little about the exact physiological role and actions of relaxin in the human. Furthermore, relaxin did have beneficial actions in some of the patients (181). The results of the phase II clinical trials, and positive results in numerous in vivo and in vitro studies where relaxin has been able to reverse fibrosis, suggest that relaxin still has enormous potential as an antifibrotic agent.

Studies using the relaxin KO mouse have demonstrated that relaxin is an essential endogenous mediator of collagen turnover in nonreproductive tissues.

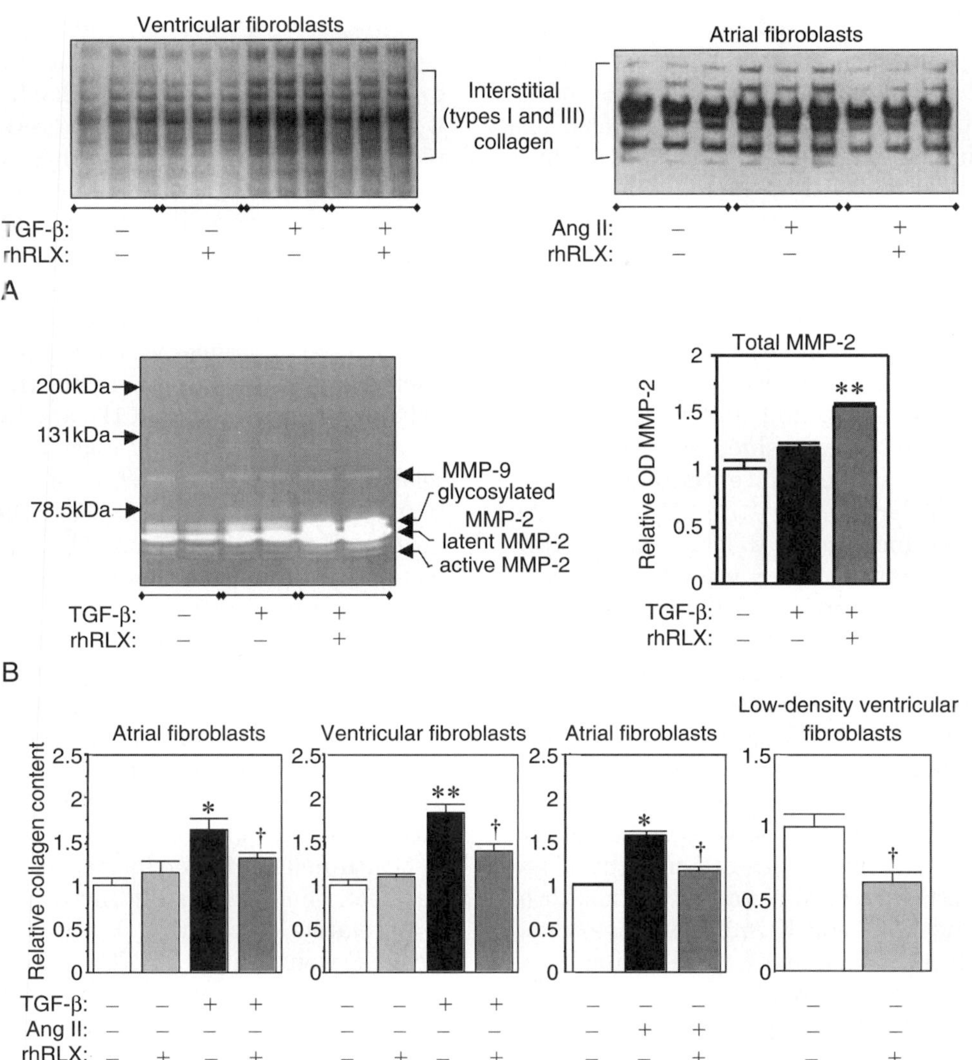

FIG. 35. Modulation of collagen synthesis, degradation, and deposition by recombinant H2 relaxin (rhRLX). **A:** Biosynthetically labeled interstitial collagen from untreated cardiac fibroblasts (2×10^5/cm^2) and cells treated with either rhRLX (100 ng/mL) alone, transforming growth factor (TGF)-β (1 ng/mL) alone or TGF-β (1 ng/mL) and rhRLX (100 ng/mL), or with angiotensin II (Ang II; 5×10^{-7} M) alone or Ang II (5×10^{-7} M) and rhRLX (100 ng/mL), after 72 hours of culture. Shown are representative figures of triplicate samples from three separate experiments. **B:** Matrix metalloproteinase (MMP)-2 and -9 expression and activity were determined by gelatin zymography of media from untreated cultures and cells treated with either TGF-β (2 ng/mL) or TGF-β (2 ng/mL) and rhRLX (100 ng/mL) over 72 hours. Shown is a representative zymograph of duplicate samples from each group, from four sets of samples per group. Also shown are the mean ± SE "relative OD (optical density) MMP-2" of the total MMP-2 (derived from the latent and active forms of MMP-2), as determined by densitometry scanning. **C:** Collagen content of cell layers from untreated fibroblasts and cells treated with rhRLX (100 ng/mL) alone, TGF-β (2 ng/mL) alone, or TGF-β (2 ng/mL) and rhRLX (100 ng/mL), after 72 hours of culture. Also shown is the collagen content of cell layers from untreated atrial fibroblasts and cells treated with Ang II (10^{-7} M) alone or Ang II (10^{-7} M) and rhRLX (100 ng/mL), after 72 hours of culture. Results are presented as the mean ± SE "relative collagen content" from three or four separate experiments. *$p < 0.05$ and **$p < 0.01$ compared with values from untreated cells; †$p < 0.05$ compared with values from TGF-β– or Ang II–treated cells. In addition, rhRLX (100 ng/mL) treatment of low-density cells (5/mm^2) over 7 days caused an inhibition of collagen deposition. Results are presented as the mean ± SE "relative collagen content" from three separate experiments (six assays per group from each experiment). †$p < 0.05$ compared with values from untreated cells. (From Samuel, C. S., Unemori, E. N., Mookerjee, I., Bathgate, R. A., Layfield, S. L., Mak, J., Tregear, G. W., and Du, X. J. [2004]. Relaxin modulates cardiac fibroblast proliferation, differentiation and collagen production and reverses cardiac fibrosis in vivo. *Endocrinology* 145, 4125–4133, with permission. Copyright 2004, The Endocrine Society.)

Hence, similar to the increased collagen accumulation seen in reproductive organs of the relaxin KO mouse outlined in previous sections, in nonreproductive organs, increased interstitial collagen also has been detected in the heart (721), lung (722), kidney (723), and skin (724). Studies in the LGR7 KO mouse have demonstrated a similar increase in interstitial collagen in the lung, whereas other organs have not yet been examined (262). This effect was particularly prevalent in male relaxin KO mice and was associated with abnormal function of these organs. In the heart, an increase in atrial hypertrophy and impeded left ventricular diastolic filling and venous return were observed in older mice, which were attributed to a significant increase in left ventricular collagen content and ventricular chamber stiffness (721). In the lung, a progressive increase in weight, collagen content, and collagen concentration was observed with age, resulting in the distortion of alveolar structure and thickening of the bronchioles. These structural changes in relaxin KO mice were associated with a significantly altered peak expiratory flow and lung recoil (lung function) (722). The increased collagen in the kidney of male relaxin KO mice was associated with increased cortical thickening, focal increases in interstitial fibrosis, and a more general diffuse increase in glomerulosclerosis, from 6 months of age onward (723). Furthermore, the increased renal fibrosis measured in aging relaxin KO mice was associated with impaired renal function of these animals, as demonstrated by significantly increased serum creatinine and urinary protein, compared with that measured in wild-type control mice (723). Similarly, a progressive increase in collagen was observed to invade the subcutaneous layer of the ageing dermis in the relaxin KO mice (724). Therefore, in the absence of relaxin in these mice there is a progressive buildup of collagen deposition (fibrosis) in tissues, resulting in a partial loss of structure and function. Importantly, treatment of relaxin KO mice with recombinant H2 relaxin in early and developed stages of fibrosis resulted in the reversal of collagen deposition in the lung (722), heart (237), and kidney (723), consistent with its actions reported previously and its potential as an antifibrotic agent.

Wound Healing

Wound healing is a well-organized, complex series of events involving inflammation, granular tissue formation and re-epithelialization, and tissue remodeling that is governed by a common underlying necessity for adequate blood supply (725). Clinical studies in the late 1950s in which partially purified porcine relaxin brought about the healing of ischemic ulcers on fingers and toes provided the initial encouragement that relaxin might enhance wound healing (718,726). More recent studies using purified porcine relaxin or recombinant H2 relaxin in rodents have also provided evidence of actions of relaxin on the vasculature and platelet aggregation that supports a possible role of relaxin in wound healing. Relaxin potentially acts at wound sites by increasing blood supply through multiple mechanisms. Hence, relaxin induces vasodilation not only in reproductive organs (561,613,614), but the heart (615,616), kidney (245), mesocecum (619), and liver (727), through a nitric oxide–mediated mechanism. Furthermore, relaxin induces the production of VEGF in numerous tissues (38,338,565,663,728,729), thereby promoting new blood vessel formation. Also consistent with the possibility that relaxin enhances the availability of blood at wound sites is the limited evidence that relaxin counteracts hypercoagulation. Porcine relaxin not only inhibited collagen- or thrombin-induced aggregation of human and rabbit platelets in vitro (730), but reduced circulating platelets after 4 days of intraperitoneal administration to male rats (731), and these effects were mediated, at least in part, by nitric oxide.

Continuous infusion of recombinant H2 relaxin has been demonstrated to increase blood vessel content relative to controls in three rodent models of ischemic wound sites (725,728). Moreover, the authors provided evidence that relaxin increased the expression of VEGF and basic fibroblast growth factor, two angiogenic factors, in both cells removed from wound chambers in a rat model of subcutaneous wound healing and cultures of human monocyte/macrophages (THP-1 cells) (728). Further evidence for a role for relaxin in wound healing is the observation that full-thickness wound size was smaller in diabetic mice that received relaxin than in controls after 14 days of hormone treatment (725).

Cardiac Protection

There is considerable evidence that relaxin has actions on the heart and is an important mediator of the cardiovascular changes during pregnancy (see section on Cardiovascular Function). Moreover, there is now considerable evidence that relaxin may be a locally acting cardioprotective agent. This topic has been reviewed extensively (29,35,36,38,39) and therefore is not covered in detail here.

Porcine relaxin has been demonstrated to protect against myocardial injury caused by ischemia and reperfusion in the hearts of rats (617) and guinea pigs (615). Relaxin's cardioprotective effects are potentially mediated by increased coronary flow and improved cardiac contractility (615). Indeed, relaxin increases dilation of small vessels in the rat heart (617) and increases coronary blood flow in isolated and perfused male rat and guinea pig hearts (616). Available evidence indicates that relaxin brings about its vasodilatory effects, at least in part, by acting directly on coronary endothelial cells and neutrophils to upregulate inducible nitric oxide synthase II, thereby stimulating production of the potent vasodilator nitric oxide (615,616,622,732,733). Studies in a rat model of myocardial infarction suggest that relaxin may influence the perfusion of ischemic sites by inducing VEGF and basic fibroblast growth factor, thus augmenting collateral vessel formation (729). Relaxin may also mediate its cardioprotective effects through actions on mast cells to inhibit degranulation and hence reduce cardiac damage (615). In addition, relaxin may act on neutrophils and endothelial cells to exerts its cardioprotective effects (732,733). The physiological significance of relaxin's apparent cardioprotective effects is not known. It was postulated that circulating relaxin emanating from the corpus luteum could prevent ischemic heart disease in women during nonconceptive cycles and pregnancy (615).

There is evidence that relaxin produced locally may provide cardioprotective actions or predict the severity of chronic heart failure. Relaxin, relaxin-3, and LGR7 are expressed in the rat heart (241,250,734), and both relaxin and relaxin-3 mRNA are expressed in the mouse heart (18). In the relaxin KO mouse heart there is increased ventricular collagen buildup as the animals age, which leads to impaired cardiac function (see section on Fibrosis). Hence, relaxin is a locally acting agent to reduce collagen accumulation in the mouse heart. Furthermore, the human heart has been reported to secrete relaxin. Both H1 and H2 relaxin mRNA expression increased in failing atrial and ventricular tissue, and plasma concentrations of H2 relaxin (Fig. 36) and myocardial expression of the H1 and H2 relaxin genes correlated positively with the severity of congestive heart failure (390). A further study demonstrated that H2 relaxin was secreted by the heart in increased amounts in patients with chronic heart failure, but plasma H2 relaxin concentrations are not a predictor of clinical outcome (735). In vitro studies have demonstrated that relaxin can suppress the hemodynamically induced increase in endothelin-1 secretion, a potent vasoconstrictor in the pathophysiology of heart failure (390). Correspondingly, in patients with severe heart failure there was an inverse correlation between circulating endothelin-1 and H2 relaxin (35). Hence, relaxin potentially acts as a cardioprotective agent through multiple mechanisms.

Other Actions

Relaxin also has a potential role in allergic responses. Relaxin inhibits granule exocytosis and histamine release by both mast cells and basophils which are mediators of allergic reactions and inflammation (733,736,737). Furthermore, there is evidence that relaxin protects against the pathogenic events underlying allergic reactions and may do so through effects on resident mast cells (738), neutrophils (733), and endothelial cells (732). The protective effects of relaxin in these models of allergy were associated with prevention of mast cell degranulation (738), decreased histamine release (739), and reduced interstitial neutrophil content (738). In addition, relaxin may have a protective effect in a cardiac anaphylaxis model by mediating nitric oxide release and subsequently not only reducing the release of histamine from cardiac mast cells but decreasing vascular tone of coronary vessels (739).

There is limited evidence that relaxin may have action on gut motility and may be involved in the changes in gut motility seen during pregnancy (740). Highly purified rat relaxin, but not insulin, was reported to inhibit spontaneous contractions of the rat ileum both in vivo and in vitro (741). Furthermore, relaxin markedly inhibited ileal motility in mice by exerting a direct action on smooth muscle through the activation of intrinsic nitric oxide biosynthesis (244). One study demonstrated that relaxin influenced relaxant responses of the mouse stomach by a nitric oxide–mediated pathway (740). Relaxin may also have direct effects on the microcirculation in the gastrointestinal tract. Topical administration of porcine relaxin and human decidual extract produced rapid dilation of the veins of the rat mesocecum (619).

PHYSIOLOGICAL ROLES OF INSL3

Expression of the *INSL3* Gene

INSL3 was originally called Leydig insulin-like peptide (Ley-I-L) based on its high level of expression in the Leydig cells of fetal and adult testes (12). Subsequently, INSL3 has been shown to be expressed in the Leydig cells of every species in which the *INSL3* gene has been characterized (see section on The *INSL3* Gene). However, INSL3 is also expressed in numerous other tissues, although the expression levels tend to be much lower than that in the testis.

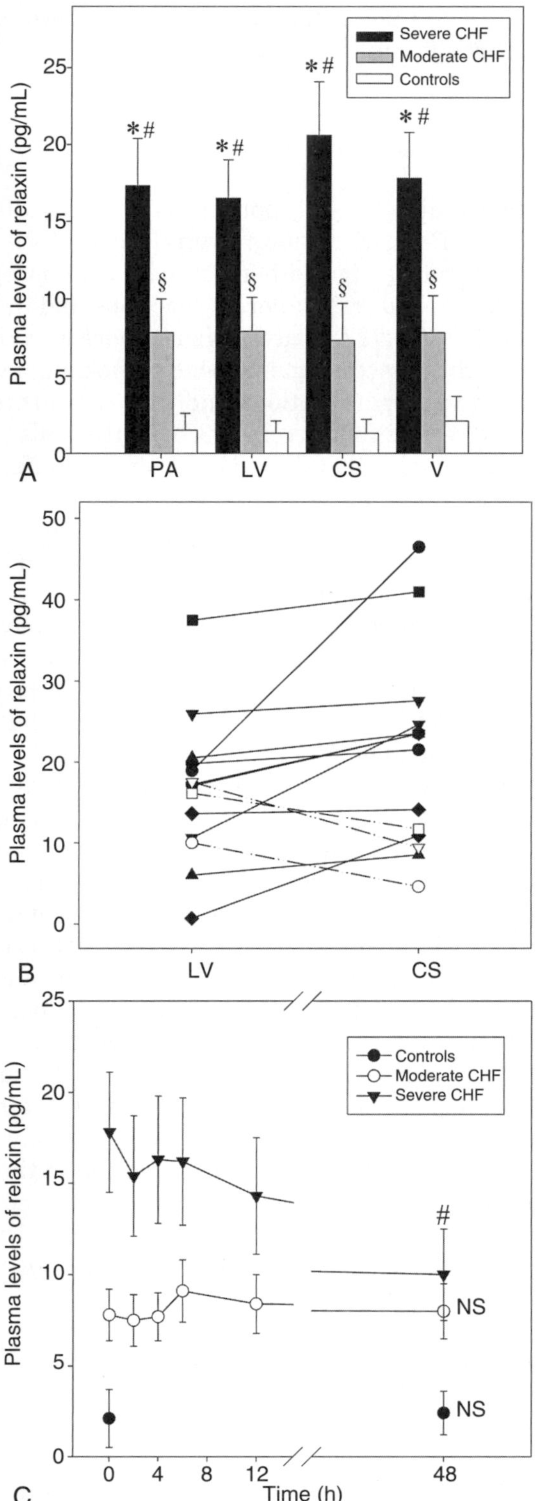

FIG. 36. In patients, plasma levels of relaxin increase with the severity of congestive heart failure (CHF) and respond to acute hemodynamic improvement. Levels of relaxin were detected in left ventricle (LV), pulmonary artery (PA), coronary sinus (CS), and antecubital vein (V). Data (mean ± SE) are given in picograms per milliliter. Detection limit of the relaxin enzyme-linked immunosorbent assay was 0.40 pg/mL. **A:** Baseline levels. $p < 0.05$; #, severe versus moderate CHF; ∗, severe CHF versus control subjects; §, moderate CHF versus control subjects. **B:** Eleven of 14 patients with severe CHF (i.e., 79%) revealed coronary net release of relaxin (i.e., plasma levels are higher in CS than in LV). **C:** Course of circulating venous relaxin during and after 12 hours of vasodilator therapy with sodium nitroprusside. $p < 0.05$; #, versus baseline. (From Dschietzig, T., Richter, C., Bartsch, C., Laule, M., Armbruster, F. P., Baumann, G., and Stangl, K. [2001]. The pregnancy hormone relaxin is a player in human heart failure. *FASEB J.* 15, 2187–2195, with permission.)

The only exceptions are the sheep (416) and the cow (117), where levels of INSL3 mRNA in the corpus luteum and thecal cells are just as high if not higher than in the testis (Fig. 37). In fact, levels of INSL3 mRNA in the corpus luteum of the cow show a pattern of expression similar to that of relaxin in other species (742). Because it is presumed that cows, like sheep, lack a relaxin gene (see section on Evolution of the Relaxin Genes), it has been postulated that INSL3 might be substituting for relaxin in these species (417).

INSL3 mRNA and peptide have been detected in the ovary of other species at lower levels than in the ruminant. Hence, INSL3 mRNA has been detected in human corpus luteum of the cycle by Northern blotting (743), and INSL3 peptide in the corpus luteum and theca cells by immunocytochemistry (744). In the mouse, INSL3 has been detected in the ovary by RT-PCR (114), and the INSL3 peptide has been immunolocalized in the corpus luteum (745). INSL3 mRNA is localized to the theca cells in rats (246), where Northern blotting has previously demonstrated ovarian INSL3 mRNA expression (115). The marmoset monkey also expresses INSL3 mRNA in the corpus luteum and immunoreactive INSL3 peptide in the theca (116). Similarly, immunoreactive INSL3 peptide has been localized in the dog corpus luteum and theca cells (120). INSL3 mRNA is also expressed in the theca cells and corpus luteum of the pregnant fallow deer (121). The pattern of expression of INSL3 is usually distinct from that of relaxin, except in the case of the cow and sheep. The potential function of ovarian INSL3 is discussed later.

The human placenta expresses INSL3 mRNA (743,746) and peptide immunoreactivity (746) that has been localized to trophoblast cells. The dog placenta is also a source of immunoreactive INSL3 peptide (120), and placental trophoblast cells in the deer express INSL3 mRNA (121). In humans, INSL3 mRNA and protein have been localized in the uterus in luminal and glandular epithelium of the cycling endometrium (746). The luminal and glandular epithelium of the pregnant deer uterus also expresses INSL3 mRNA (121). The dog uterus also expresses immunoreactive INSL3 peptide (120), and the cow uterus expresses INSL3 mRNA (117). The potential function of INSL3 in the uterus and placenta is unknown; however, the mouse uterus has INSL3

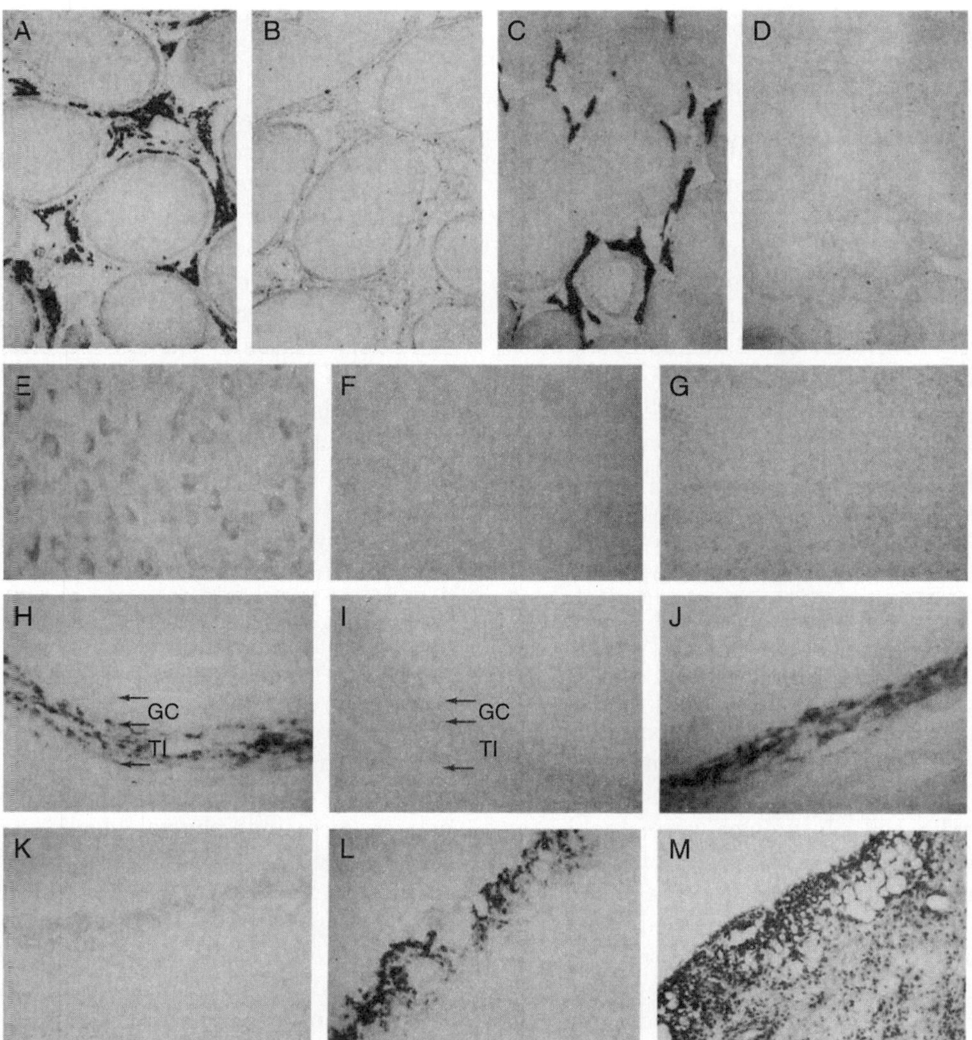

FIG. 37. Nonradioactive in situ hybridization histochemistry for specific *INSL3* gene transcripts in tissues fixed in Bouin's solution and embedded in paraffin. **A:** Adult bovine testis (antisense probe, original magnification × 200). **B:** Adult bovine testis (sense probe as control, 200). **C:** Adult ovine testis (antisense probe, original magnification × 200). **D:** Adult ovine testis (sense probe as control, original magnification × 200). **E:** Bovine corpus luteum (mid- to late cycle; antisense probe, original magnification × 400). **F:** Bovine corpus luteum (mid- to late cycle; sense probe as control, × 400). **G:** Bovine corpus luteum (late cycle; antisense probe, original magnification × 200). **H:** Bovine healthy antral follicle (early cycle; antisense probe, original magnification × 200; GC, granulosa cell layer; TI, theca interna). **I:** As in (**H**) but with sense probe used as control. **J:** Bovine healthy preovulatory follicle (late cycle; antisense probe, original magnification × 200). **K:** Bovine atretic follicle (mid-cycle; sense probe as control, original magnification × 200). **L:** As in (**K**), but an antisense probe was used. **M:** As in (**K**), but histologically stained only with hematoxylin-eosin. Sections of corpus luteum shown in **E–G** required substantially higher-stringency washing to remove nonspecific background, resulting in weaker specific signals with the antisense probe, compared with the other sections. (From Bathgate, R., Balvers, M., Hunt, N., and Ivell, R. [1996]. Relaxin-like factor gene is highly expressed in the bovine ovary of the cycle and pregnancy: sequence and messenger ribonucleic acid analysis. *Biol. Reprod.* 55, 1452–1457, with permission.)

binding sites (122) and the human uterus expresses LGR8 mRNA (21).

Further sites of expression of INSL3 are the human breast (747), human thyroid gland, and thyroid carcinoma cells lines (748), demonstrated via RT-PCR, in situ hybridization, and immunocytochemistry. Northern blot analysis has shown INSL3 mRNA expression in the bovine hypothalamus (742). In the male, INSL3 mRNA has been shown in the mouse (745) and marmoset monkey (116) epididymis, and INSL3 immunoreactivity in the dog epididymis (120).

Testis INSL3 Expression and Secretion

INSL3 gene transcripts represent one of the most abundant mRNAs in the adult Leydig cell (42) (Fig. 37). Studies in rodents indicate that the *INSL3* gene is poorly expressed in mesenchymal prepubertal Leydig cell precursors and is progressively upregulated on pubertal differentiation (111,114,745). In addition, INSL3 mRNA is upregulated in the fetal Leydig cell population, achieving maximum mRNA expression shortly before birth (12,114,115,168) at a time that correlates with gubernacular development (see later). This expression is lost once the fetal Leydig cell population is lost. In hypogonadal mice, there is poor expression of INSL3 mRNA in the adult testis, but normal expression levels are recovered with hCG treatment (745). This treatment induces the differentiation of the adult Leydig cells and, hence, INSL3 mRNA expression. In the fetal testis, where differentiation of the Leydig cells is independent of gonadotropins, INSL3 mRNA expression is normal.

Many studies have indicated that the expression of INSL3 mRNA in the adult testis is constitutive once the adult Leydig cells differentiate (12,114,115,745). Effectors of Leydig cell physiology were unable to influence the levels of INSL3 mRNA in primary cell cultures (745), and no significant differences in INSL3 mRNA or protein expression were observed in human testicular biopsies from patients with various severe testicular lesions (749). In Leydig cell tumors or hyperplasias, where proliferation and dedifferentiation of the Leydig cells occurs (750), INSL3 mRNA and protein are downregulated. A similar pattern of downregulation of INSL3 mRNA and protein expression is seen in the aging rat testis, where Leydig cell function is compromised (751). In the testis of the seasonal breeder, the roe deer, levels of INSL3 mRNA and protein increase during the mating season, mirroring the cyclic seasonal dedifferentiation and redifferentiation of Leydig cells (752). Hence, the downregulation of INSL3 expression is correlated to changes in Leydig cell function and adult Leydig cell INSL3 expression is postulated to be independent of direct regulation by gonadotropins. However, one study demonstrated a direct effect of LH on Leydig cell INSL3 expression (246), correlating with the proposed action of INSL3 in mediating LH action on germ cells (outlined later).

INSL3 has been demonstrated to be a circulating hormone in both rats (753) and humans (754,755). Levels of INSL3 in the rat plasma are highest shortly before parturition, after which they decrease and then begin to increase again as the males mature, to reach constant high levels in adults (753). Therefore, plasma levels of INSL3 mirror the expression of INSL3 in fetal and adult Leydig cells, indicating that this is the source of circulating INSL3. In humans, plasma INSL3 levels are highest in adult men (754,755) and are almost undetectable in orchidectomized men, indicating that the INSL3 is of testicular origin (755). Furthermore, plasma INSL3 levels in men with severe testicular damage, including men with severe infertility, are very low (755). There is also a positive correlation between the levels of INSL3 and LH in the plasma. Hence, INSL3 plasma levels reflect the functional status of the Leydig cells in an identical fashion to INSL3 mRNA and protein levels in the Leydig cells, as discussed previously. Because INSL3 levels were decreased in patients with severe hypospermatogenesis, in whom LH and testosterone levels were normal, the authors postulated that INSL3 may be a more sensitive marker of Leydig cell function than testosterone (755).

Role of INSL3 in Testis Descent

Although the *INSL3* gene was discovered in 1993 and was known to be highly expressed in the testis, suggesting an important role in reproductive physiology, its principal physiological function was determined only by analysis of the phenotype of the KO mouse. In 1999, two independent groups reported the development and characterization of INSL3 KO mice, Nef and Parada in the United States (251), and Adham and colleagues in Germany (252). INSL3 KO mice developed normally and had no obvious deleterious phenotypes. However, male KO mice were infertile despite normal sexual behavior toward female mice and production of copulation plugs. Although the German group reported that female INSL3 KO mice were fertile, Nef and Parada reported an ovarian phenotype that was subsequently confirmed by the German group (see later). Anatomical examination of the male KO mice indicated that animals were all bilaterally cryptorchid with the testis located high in the abdomen (Fig. 38). The vas deferens, epididymis, and other accessory glands were all normally developed. The size of the intra-abdominal INSL3 KO testis was equivalent to the wild-type testis at birth and was identical histologically. However, because they were retained in the abdomen, the adult testes were atrophied and the mice were infertile. Surgical reversal of cryptorchidism was reported to result in normal fertility (252,756), suggesting that testicular function was normal. The testis and genital tracts in the INSL3 KO mice were freely mobile in the abdominal cavity, indicating a lack of cranial and caudal attachment of the testis to the inguinal region by the gubernaculum and complete regression of the cranial suspensory ligament (CSL). Detailed analysis of the gubernaculum

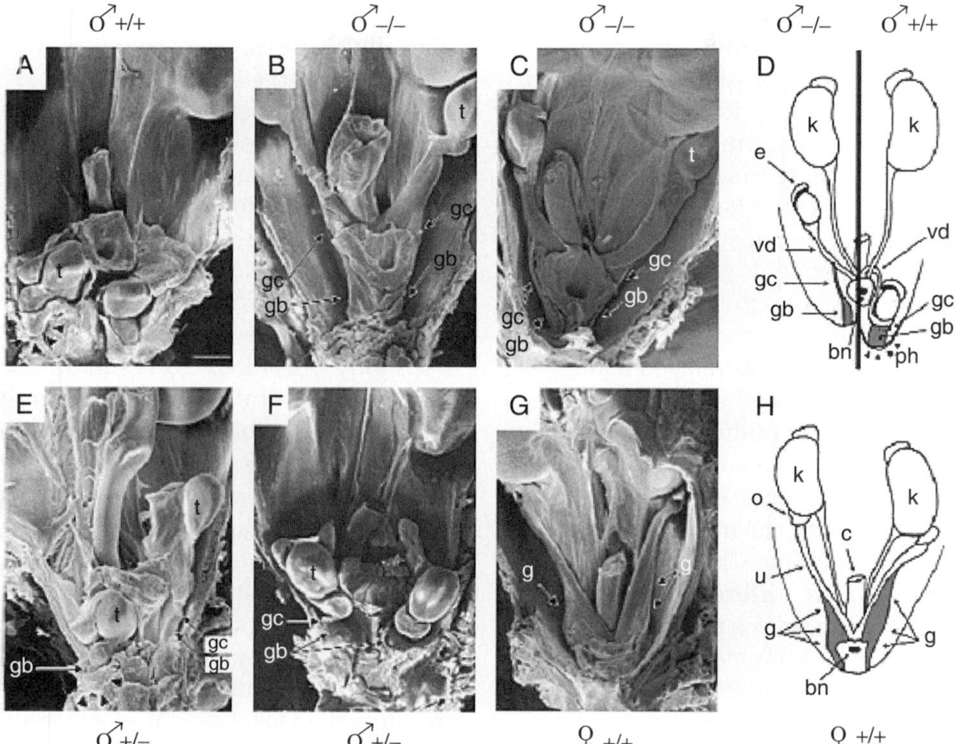

FIG. 38. Scanning electron microscopy of the urogenital tract of insulin-like peptide 3 (INSL3) knockout (KO) mice (–/–) compared with wild-type mice (+/+). **A:** Male wild-type pup. The testes have moved beyond the bladder neck and the gubernaculum is embedded in the peritoneal hollow (*arrowheads*). **B, C:** Male mutant pups have normal-sized testes, but they are located caudal to the kidneys in a position similar to the ovaries in females (compare with **G**). The gubernaculum is a thin, flat bulb with a thin, elongated cord similar to the one found in wild-type female pups. No peritoneal hollows are present in mutant mice. **E, F:** Examples of partial unilateral (**E**) and bilateral (**F**) cryptorchidism in male heterozygous pups. Note the swollen gubernacular bulbs and the contracted gubernacular cords but the absence of peritoneal hollow. **D:** Schematic drawing of mutant (*left*) and wild-type gubernacula and testes. **H:** Schematic drawing of female wild type. The gubernacula are in *solid black*. bn, bladder neck; c, colon; e, epididymis; g, gubernaculum; gb, gubernacular bulb; gc, gubernacular cord; o, ovary; k, kidney; ph, peritoneal hollow; t, testes; u, uterus; vd, vas deferens. (From Nef, S., and Parada, L. F. [1999]. Cryptorchidism in mice mutant for Insl3. *Nat. Genet.* 22, 295–299, with permission. Copyright 1999, Nature Publishing Group.)

indicated that it was poorly developed, being thin and elongated with some muscle development but no central core of mesenchyme (757). Studies in LGR8 KO mice demonstrated an identical phenotype (253,259,758). Therefore, INSL3, acting through LGR8, is essential for the development of the gubernaculum during embryonic development.

Testicular descent is proposed to involve two main stages. In the first, or "transabdominal" phase, the gubernaculum enlarges caudally, holding the testis near the inguinal region during abdominal growth (759). During this time, the CSL regresses under the influence of androgens. The second, or "inguinoscrotal" phase involves the testis descending into the scrotum and is also under androgen control (759). It has long been postulated that an unknown factor was involved in the first phase of testicular descent. A low–molecular-weight protein was previously isolated and was suggested to be the factor necessary for testicular descent. It was named *descendin* (760), but was never fully characterized. It would seem that INSL3 is this elusive "descendin" and is responsible for the development of the gubernaculum during the transabdominal phase of testicular descent.

Further studies have demonstrated that INSL3 has a direct action on the gubernaculum. Cocultures of gubernacular explants with testis from wild-type or INSL3 KO mice demonstrate that growth of the gubernaculum is achieved only with wild-type testis (761). In addition, studies using rat gubernacular explants from day 17 rat fetuses have demonstrated that a synthetic INSL3 peptide can induce growth of the gubernaculum. These cultured explants express LGR8, but not LGR7, and this was the first demonstration that INSL3 can act directly on the gubernaculum through LGR8 to induce growth (260). This was confirmed in dispersed rat gubernacular cell cultures, where the synthetic INSL3 peptide was able to induce cAMP production and proliferation of gubernacular cells (22). The explant cultures indicated that androgens, and also potentially müllerian inhibiting substance (MIS), may also be necessary for the actions of INSL3 on the gubernaculum. In light of these findings, two independent groups generated INSL3-overexpressing mice to establish further the essential actions of INSL3 on the gubernaculum. INSL3-overexpressing mice were developed using two different transgenes, one produced specifically in the pancreas (762) and the other produced ubiquitously in all tissues (763). Both transgenic mice have similar phenotypes, with descended ovaries in a position close to the bladder and attached to the abdominal well by a developed gubernaculum and CSL. The authors report that the gubernacular differentiation seen in these transgenic females excludes a role for androgen or MIS in gubernacular development. In addition, expression of the transgene in male INSL3 KO mice was able to rescue the cryptorchid phenotype (762), and transgene expression in LGR8 KO mice did not result in descended ovaries (261). Therefore, the actions of INSL3 on the gubernaculum are exclusively through LGR8. A further phenotype of the INSL3-overexpressing mice in females was the development of bilateral inguinal hernias, suggesting that a combination of estrogen and INSL3 action disrupts proper development of the muscular and connective tissue structures of the abdomen (763).

INSL3 heterozygous KO mice demonstrate partial cryptorchidism (251), suggesting dosage sensitivity to INSL3 in testis descent. The authors indicated that this heterozygous phenotype seems to resemble most cases of human cryptorchidism, where partial testicular descent at birth often self-corrects (764). Cryptorchidism is one of the most common congenital birth defects of the urogenital tract in humans, occurring in 1% to 2% of full-term boys. It is also a major problem in domestic animal species (765). Clearly, mutations in INSL3 or LGR8 are potential causes of cryptorchidism in humans and domestic species. Therefore, numerous groups have been looking for mutations in the *INSL3* and *LGR8* genes in patient populations from various countries (259,766–776). Although several allelic variations of the *INSL3* gene were found, the frequency of mutations was very low. Most of the mutations are unlikely to confer a functional change in the INSL3 peptide, although two mutations do affect the activity of the INSL3 peptide. A nonsense mutation discovered in the C-peptide region of pro-INSL3 (R49X) clearly results in a nonfunctional peptide (768). A further mutation in the B chain of INSL3 (P69L of the pro-INSL3 sequence) results in reduced activity of the INSL3 peptide (261). This highly conserved proline residue has been previously demonstrated to be essential for the activity of the INSL3 peptide (192) because it is postulated to direct the correct orientation of the receptor binding motif (see section on INSL3). These mutations, like all the INSL3 and LGR8 mutations discovered, are found in the heterozygous state only. However, these nonfunctional INSL3 mutations represent a very low frequency of *INSL3* gene mutations (1.4%). Four mutations have so far been found in the *LGR8* gene, with only one likely to result in a change in LGR8 function. A mutation of proline to threonine at position 222 in the fourth LRR results in a receptor that does not respond to ligand in vitro (259). It is likely that this mutation results in a conformational change that disrupts ligand binding or receptor trafficking. Patients with this mutation demonstrated different phenotypes, including bilateral and unilateral cryptorchidism, retractile testis, normozoospermia, and complete azoospermia (259,775). Although INSL3 and LGR8 KO mice show bilateral abdominal cryptorchidism, this type of cryptorchidism is rare in humans (776). Most human defects in testicular descent are associated with the inguinoscrotal phase (759), which does not seem to involve INSL3.

Another potential association between cryptorchidism and INSL3/LGR8 is a link with the effects of environmental estrogens. The link between environmental estrogens and testicular descent abnormalities is well established (44,777). Hence, treatment of pregnant women with the nonsteroidal estrogenic substance diethylstilbestrol (DES) is associated with undescended testes in male offspring (778). Treatment of mice with 17α-estradiol, 17β-estradiol, estriol, or DES at day 13.5 of pregnancy resulted in cryptorchidism in male fetuses due to undeveloped gubernacula and is associated with downregulation of INSL3 mRNA in the testis (777). An independent study in mice using DES obtained identical results (779). Therefore, environmentally derived estrogenic compounds could be a cause of cryptorchidism through actions on INSL3 expression in the fetal testis. In addition, phthalate esters, a broad class of synthetic, high–production-volume chemicals, have been associated with cryptorchidism (780). One study in pregnant rats has demonstrated that treatment of animals with phthalate esters from days 13 to 18 of pregnancy results in cryptorchidism in male fetuses associated with decreases in INSL3 mRNA in the testis (781). Because these compounds are not estrogenic, the authors propose that the chemicals act to delay fetal Leydig cell maturation. Clearly, there are other mechanisms whereby alterations in INSL3/LGR8 function may be associated with cryptorchidism in humans and especially domestic animal species.

Gonadal Functions

Although the exact physiological role of INSL3 in the testis could not be analyzed in male INSL3 null mice because these animals exhibited bilateral cryptorchidism (251,252), a potential role of INSL3 in ovarian physiology was suggested by the impaired fertility found in female INSL3 KO mice (251). These mice demonstrated lengthening of the normal estrous cycle from an average of 6 to 7 days to 14 to 20 days, and the litter sizes were significantly smaller. Although initial studies on an INSL3 KO mouse developed independently by the German group (252) demonstrated no female phenotype, a subsequent, more detailed study showed a higher apparent rate of apoptosis in both follicles and corpora lutea, resulting in fewer follicles, corpora lutea, and offspring (782). These data imply that INSL3 protects follicular and luteal cells from entering the standard default pathway of apoptosis and atresia. In the mouse ovary, INSL3 mRNA expression is higher in the follicular phase (proestrous and estrous) than in the luteal phase (metestrous and diestrous), coincident with follicular growth. Furthermore, in pregnant mice INSL3 mRNA expression is also correlated with follicular growth, which is also indicative of a role in follicular development (114). Further evidence for an antiapoptotic role for INSL3 comes from studies investigating INSL3 expression in bovine follicles in vivo (783). The expression of INSL3 in the thecal cell layer is lost in cells undergoing atresia during the process of follicular selection. Therefore, thecal cell INSL3 expression in secondary follicles during the selection process is a clear indicator that these follicles will not enter the default atretic pathway, suggesting that INSL3 may be at least protective, if not a determinant in follicle selection.

Another potential role for INSL3 in the ovary, as well as a potential role in the adult testis, has been determined. Gonadotropins secreted by the anterior pituitary are essential for follicle development and spermatogenesis (784–787). In the ovary, oocytes show prolonged arrest at the prophase of meiosis I (G_2/M transition) and undergo meiotic maturation only when exposed to the preovulatory LH surge. The dissolution of the nuclear membrane (germinal vesicle breakdown) is followed by the extrusion of the first polar body. Although the preovulatory increases in LH stimulate cAMP production in follicular somatic cells, a decrease in intraoocyte cAMP levels is required for the resumption of meiosis in oocytes (788,789). In the testis, up to 75% of male germ cells undergo apoptosis, perhaps as a mechanism to delete superfluous or defective germ cells (790). Capable of suppressing apoptosis, LH is a survival factor for male germ cells (791).

LH binds specific receptors in ovarian theca and mature granulosa cells as well as testicular Leydig cells. Because LH acts exclusively on somatic cells in both female and male gonads, local paracrine factors are likely to be involved in the regulation of oocyte meiosis arrest (788,792,793) and optimal spermatogenesis (794). Based on the expression of the INSL3 receptor LGR8 in the oocyte of the ovary and the male germ cells of the testis, potential paracrine roles of INSL3 in the gonads were recently elucidated (246). In contrast to its stimulation of cAMP production by gubernaculum cells (22), binding of INSL3 to LGR8 expressed in the oocyte of the ovary and male germ cells of the testis led to activation of the inhibitory G protein and decreases in cAMP production. In vitro and in vivo studies further indicated that INSL3 treatment initiates meiotic progression of the arrested oocytes in cultured preovulatory follicles obtained from gonadotropin-treated female rats. Furthermore, INSL3 treatment suppresses male germ cell apoptosis induced by gonadotropin withdrawal in male rats treated with a gonadotropin-releasing hormone antagonist. Of interest, the induction of ovulation by hCG, an agonist of LH, in females resulted in increases in ovarian INSL3 expression, and treatment with hCG in males led to increases in testicular INSL3 expression. Thus, LH stimulates ovarian thecal and testicular Leydig cells to produce INSL3. In both female and male gonads, INSL3 and its receptor represent a paracrine system important for meiosis induction in the ovary and male germ cell survival in the testis (Fig. 39).

PHYSIOLOGICAL ROLES OF OTHER RELAXIN FAMILY PEPTIDES

Relaxin-3, a Putative Neuropeptide?

The cloning of an additional relaxin gene over 70 years after the original identification of relaxin was a great surprise to the field of relaxin investigators. Relaxin-3 binds and activates the relaxin receptor LGR7, but also activates two other GPCRs that have been postulated to be specific relaxin-3 receptors, GPCR135 and GPCR142 (see section on Relaxin Family Peptide Receptors). The *RLN3* gene alone is present in lower vertebrate genomes and hence appears to be the ancestral gene of the entire relaxin peptide family. The high sequence homology of the relaxin-3 peptide between species and its specific expression in the brain indicate that it is likely to have highly conserved functions in the central nervous system.

Although RT-PCR demonstrates that the mouse *RLN3* gene is expressed in numerous mouse tissues, Northern blotting indicates that the primary site of

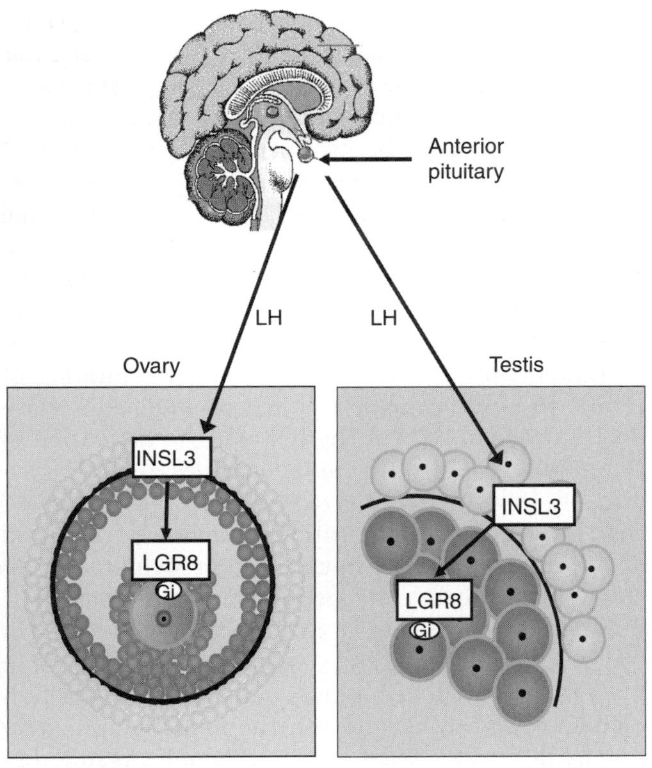

FIG. 39. Gonadotropin stimulation of oocyte maturation and male germ cell survival involves an intragonadal paracrine system mediated by insulin-like peptide 3 (INSL3) and its receptor LGR8. Pituitary luteinizing hormone (LH) stimulates theca or Leydig cells to produce INSL3, which, in turn, activates LGR8 receptors and the inhibitory G protein in oocytes to induce meiotic maturation or in male germ cells to suppress apoptosis.

expression is the brain (18). Levels of mouse relaxin-3 mRNA in the ovary are 500 times lower than relaxin mRNA levels, and do not change throughout pregnancy (795). Furthermore, H3 relaxin mRNA is not expressed in the human ovary (18), and therefore it seems unlikely that relaxin-3 is a luteal cell product similar to relaxin (796). There is no change in the expression of mouse relaxin-3 mRNA in the lung of relaxin KO mice (722), suggesting there is no compensatory relaxin-3 expression in these mice. Relaxin-3 mRNA expression is highest in the human brain, as indicated by RT-PCR, although there is some expression in the testis, the significance of which is unknown (23). A zebrafish relaxin-3 ortholog identified from the EST databases is from a brain cDNA library (129).

Relaxin peptides have clear effects in the brain distinct from actions of the circulating hormone, and there are relaxin receptors in numerous regions of the brain (see section on Relaxin Binding Sites; and section on Plasma Osmolality Regulation). Anatomical studies have successfully identified enriched expression of relaxin-3 mRNA in a population of neurons located in a restricted region of the pons in mouse (129) and rat (23,127). Thus, in both mouse and rat brain, relaxin-3 mRNA is abundantly expressed in a small nucleus located in the caudoventral regions of the pontine periventricular gray, adjacent to the ventromedial border of the dorsal tegmental nucleus (23,127,129). The nucleus is variously known as the ventromedial dorsal tegmental nucleus (797), nucleus O (798), or the nucleus incertus (799). Cells in the midline nucleus incertus region in the rat also show strong immunostaining with a polyclonal anti-porcine relaxin antiserum shown to cross-react with synthetic H3 relaxin (129).

Relaxin is also localized in the brain, as is LGR7, GPCR135, and GPCR142 (see section on Relaxin Family Peptide Receptors), as well as INSL3/LGR8 (R. A. D. Bathgate and A. L. Gundlach, unpublished data). Hence, there is likely a complex interaction of relaxin family peptide receptor–ligand systems in the brain, potentially mediating important central nervous system functions.

INSLs 4 to 6

The potential functions of INSL4, INSL5, and INSL6 are unknown, as are the native structures of the peptides. Our current knowledge of these novel relaxin-like peptides is outlined in the following sections.

INSL4

INSL4 was initially cloned by independent groups because of its high expression in the human placenta (13,124). The highest expression of INSL4 mRNA is seen in the first trimester of pregnancy (124) and is localized to syncytiotrophoblast cells (800). The authors postulated that INSL4 may have a role in trophoblast development. A human endogenous retrovirus element has been identified in the human *INSL4* promoter, and the authors presented evidence that it was responsible for the placental specific expression of the *INSL4* gene (130). The *INSL4* gene has also been demonstrated to be expressed in fetal tissues, specifically in the perichondrium of all four limbs, vertebrae, and ribs, as well as in interbone ligaments (800). It was proposed that INSL4 may play a role in bone development.

Immunoreactive INSL4 protein has been demonstrated in human plasma using a sandwich ELISA with antibodies directed against the C and A chains (801). Although this would suggest that the immunoreactive INSL4 protein is unprocessed, the identity of the immunoreactive protein has not been confirmed.

Synthetic INSL4 peptides have been produced based on the structure of relaxin and predicted C-chain cleavage sites (172,176). One of these peptides has been tested for its ability to interact with LGR7, LGR8, and an LGR7 splice variant (172) (Fig. 40). The peptide did not bind to the receptors or modulate their ability to signal through cAMP accumulation. However, these peptides were demonstrated to possess very little structure in solution, suggesting that the native peptide has a different primary structure potentially as a prohormone rather than a processed peptide. In fact, stimulation of placental cells with conditioned media from cells transfected with INSL4 resulted in tyrosine phosphorylation of a protein of a similar size to the insulin receptor β (124). Although the structure of the recombinant INSL4 protein was not investigated, immunoblotting with an anti-INSL4 antiserum suggested that both pro-forms and processed forms of INSL4 were produced from the cells (124). It has therefore been postulated that the receptor for INSL4 could be a receptor kinase like the receptors for insulin and IGF-2 and 2. However, phylogenetic analysis indicates that INSL4 is likely to interact with LGR7, LGR8, or a splice variant of these receptors (19,20).

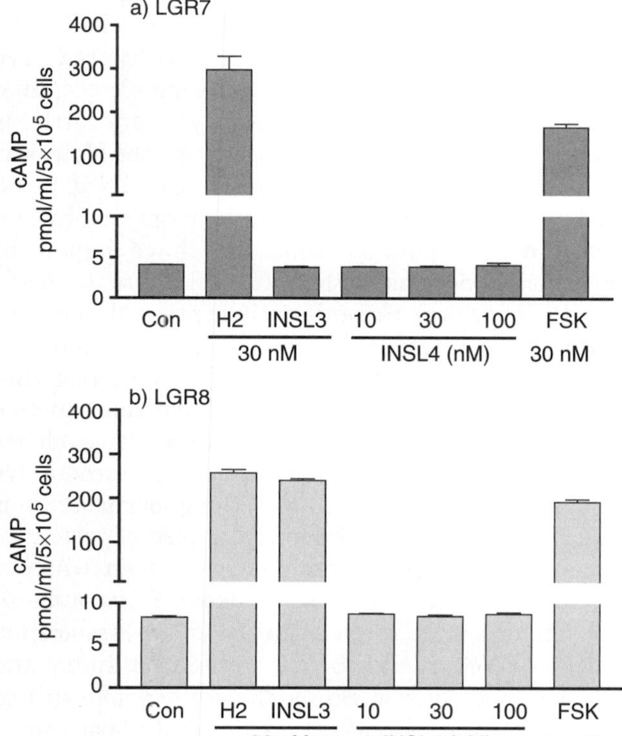

FIG. 40. Cyclic adenosine monophosphate (cAMP) accumulation (pmol/mL/5 × 10^5 cells) in LGR7 (**a**) and LGR8 (**b**) transfected cells in response to various ligands. Data are mean ± SEM of at least three independent experiments performed in triplicate. Con, control; H2, human gene 2 relaxin; INSL3, human insulin-like peptide 3; INSL4, human insulin-like peptide 4; FSK, forskolin. (From Lin, F., Otvos, L., Jr., Kumagai, J., Tregear, G. W., Bathgate, R. A., and Wade, J. D. [2004]. Synthetic human insulin 4 does not activate the G-protein–coupled receptors LGR7 or LGR8. *J. Pept. Sci.* 10, 257–264, with permission. Copyright 2004 John Wiley & Sons.)

Physiological Roles of INSL5?

The *INSL5* gene was identified from an EST library of the sigmoid colon tissue of a 40-year-old man with Crohn's disease and a mouse EST from a (Barstead) mouse (irradiated) colon library (14). The highest expression of human INSL5 mRNA was subsequently demonstrated to be in the colon, with expression also found in the rectum and testis (14). Although there is limited expression of INSL5 mRNA in the human uterus, there was no expression in the mouse uterus. Highest expression of mouse INSL5 mRNA was also demonstrated to be in the colon, with lower expression in the testis and thymus. In contrast, Hsu (15) demonstrated highest expression of mouse INSL5 mRNA in the kidney, with lower expression in the brain, testis, and heart. In addition, the mouse *INSL5* transcript was cloned from ovarian cDNA, suggesting it is also expressed in the ovary (15). The specific cell types associated with the expression of INSL5 in these tissues is not known, although in the kidney, INSL5 protein seems to be expressed in the loops of Henley, the significance of which is unknown (15).

Although the native INSL5 peptide has not been isolated, a synthetic INSL5 peptide has been produced based on the predicted cleavage of the C peptide, and INSL5 prohormone has been produced recombinantly (125). However, these proteins have not been tested for their potential bioactivity or their potential interaction with the relaxin (LGR7) or INSL3 (LGR8) receptors. An INSL5 A-B heterodimeric peptide was produced recombinantly and demonstrated to bind to and activate GPCR142, but not GPCR135 or LGR7 and LGR8 (24). The similar expression profiles of INSL5 and GPCR142 indicate that INSL5 is likely the endogenous ligand for GPCR142 and that INSL5 is likely cleaved to an A-B heterodimer in vivo.

Transgenic mice overexpressing the *INSL5* gene under the influence of the metallothionein or insulin promoter have not demonstrated an obvious effect of this gene on murine physiology (125). An INSL5 KO mouse has also been developed that included the replacement of the deleted gene with lacZ reporter gene (125). This insertion allows the staining of cells that would normally express INSL5 mRNA. The phenotype of these animals was not reported. However, studies on lacZ expression in the INSL5 KO mouse, consistent with INSL5 mRNA studies, have demonstrated that the *INSL5* gene is expressed in a discrete population of cells in the colon. The specific identity of these cells is unknown, as is the potential function of INSL5 in the colon (125). The potential functions of INSL5 await further characterization of the KO mouse.

Physiological Roles of INSL6?

Partial EST clones for mouse INSL6 were found in cDNA libraries from mammary gland, thymus, colon, and pooled organs, whereas human INSL6 was found only in testicular cDNA libraries (15). However, Northern blot analysis indicated that the predominant site of expression of INSL6 mRNA in both the mouse and human is the testis (15). Lok et al. (17) also demonstrated that INSL6 mRNA is highly expressed in the rat and human testis, although they also demonstrated that INSL6 mRNA is present in the rat prostate. Furthermore, they demonstrated using in situ hybridization that INSL6 mRNA is localized on the germ cells in the seminiferous tubules. Subsequent studies have demonstrated that in the rat testis, INSL6 mRNA expression is highest in the adult testis, although low expression is seen in the neonatal testis (802). INSL6 mRNA expression was subsequently investigated in purified populations of rat testicular cells (802) and demonstrated to be present in both pachytene spermatocytes and round spermatids. The potential role of INSL6 in the testis is unknown.

The structure of the native INSL6 peptide is unknown. A synthetic INSL6 peptide and a recombinant INSL6 protein, both based on the structure of relaxin, have been produced and were shown not to activate LGR7 and LGR8 (261), or GPCR135 and GPCR142 (23,247), respectively. Hence, it is unlikely that INSL6 interacts with receptors for other relaxin peptide members, although this has yet to be proven definitively.

CONCLUSION

Since the mid-1990s, there have been major advances in relaxin research. It is now widely accepted that relaxin is a member of a larger relaxin peptide family that now includes seven members in humans. There are three relaxin genes in humans, with the peptide products of these genes, H1, H2, and H3 relaxin, all able to interact with the relaxin receptor LGR7. Nonprimate species only have *RLN1* and *RLN3* genes, and again the peptide products of these genes interact with LGR7. The product of the *RLN1* gene in nonprimates is the well-characterized peptide that was first identified by Hisaw in 1926 and should be referred to as "relaxin." The human form of this reproductive hormone is the product of the *RLN2* gene and should be referred to as "H2 relaxin." The physiological role of the human *RLN1* gene product H1 relaxin is still unknown; however, it activates human LGR7 and LGR8 in vitro. Importantly, although relaxin was the first peptide of this family to be discovered, it is relaxin-3 that has highest sequence similarity with the ancestral peptide. Relaxin-3 is predominantly expressed in distinct nuclei in the brain and it also has its own receptor, GPCR135, unrelated to LGR7. Both GPCR135 and LGR7 (and also GPCR142) are expressed in the brain, so the determination of the physiological roles of relaxin-3 in the brain will be complicated by actions on multiple receptors. The remarkable conservation of the relaxin-3 peptide sequence across species, as well as its sites of expression, suggest that it has important functions in the brain. Although relaxin-3 is also expressed in the testis, there is no evidence that it is a circulating hormone in any species, and certainly no evidence that it has a role in reproductive biology.

Although the other relaxin family peptides have been designated INSLs 3 to 6, they have clearly evolved from the ancestral relaxin-3 gene and show higher homology to relaxin. Furthermore, the INSL3 and INSL5 receptors, LGR8 and GPCR142, are closely related to the relaxin and relaxin-3 receptors, LGR7 and GPCR135, respectively. Very little is known about INSL4, INSL5, and INSL6. However, the predominant sites of expression of INSL4 and INSL6 are the placenta and testis, respectively, and therefore it is possible that they have important reproductive actions. Although INSL5 and its receptor GPCR142 are predominantly expressed in nonreproductive tissues, they are also expressed in the ovary, testis, placenta, and prostate, suggesting that this ligand–receptor system may have reproductive functions. In contrast, there is considerable evidence that INSL3 plays an important role in reproductive biology. The action of INSL3 on the gubernaculum is essential for the transabdominal phase of testicular descent in rodents. It remains to be seen whether INSL3 is essential for testis descent in humans, although there is some evidence of an association between *INSL3* and *LGR8* gene mutations and human cryptorchidism. Studies have also demonstrated that INSL3 is an important mediator of LH action on germ cell maturation. It may also have an important role in follicular growth. INSL3 analogs therefore show great potential as therapeutic agents for regulating fertility in humans and domestic animals.

Research on relaxin has advanced considerably since its discovery in 1926, and it is now firmly established as a crucial hormone of pregnancy and parturition. It is also clear that circulating relaxin is essential during pregnancy in at least three species—rats, mice, and pigs. Although, as outlined in this review, relaxin induces its effects on multiple tissues, two vital roles have been identified. Relaxin promotes growth and softening of the cervix and vagina and thereby enables rapid and safe delivery of fetuses. It also promotes growth and development of the mammary apparatus in these three species. Relaxin's effects on the nipples are required for normal lactational performance in rats and mice. Relaxin has far more striking effects on mammary parenchymal tissue development during late pregnancy in pigs than in rats and mice. These essential actions of relaxin in rats, mice, and pigs encourage the view that relaxin may have similar effects in other species. However, there is great diversity in the physiology of relaxin among species, which is reflected in the varied blood relaxin profiles during pregnancy in different species. For example, relaxin does not seem to be important for implantation in rodents. However, there is considerable evidence that relaxin is an essential hormone in the implantation process in humans and primates. Although clinical trials for the use of relaxin in cervical ripening in humans were unsuccessful, there is considerable evidence from these and earlier trials that relaxin will be an effective therapeutic agent for cervical ripening if delivered successfully.

Studies in rodents have demonstrated that endogenous circulating relaxin has additional crucial actions during pregnancy. Relaxin is essential for the normal changes in plasma osmolality as well as other cardiovascular adaptations of pregnancy, including increases in both glomerular filtration rate and effective renal plasma flow. It also has important actions on the uterus to inhibit myometrial contractile activity during pregnancy. However, unlike the actions outlined previously, the physiological significance of these actions during pregnancy remains to be demonstrated. In both relaxin and LGR7 KO mice, and in relaxin-deficient pigs and rats, pregnancy is maintained and both litter size and viability do not differ from those in controls. Studies suggesting that relaxin may be important in follicular development and ovulation also are not supported by the phenotype of the relaxin and LGR7 KO mice.

There is little evidence that relaxin has important actions in male reproduction, and circulating relaxin levels are very low in males. However, reports of underdeveloped testes, underdeveloped reproductive tract, and reduced fertility in both relaxin and LGR7 KO male mice provide compelling evidence that low levels of circulating or locally produced relaxin may play a role in the development of the male reproductive system. What has also become clear from studies in relaxin and LGR7 KO mice is that relaxin is an essential mediator of collagen turnover in nonreproductive tissues in both males and females. It seems that the actions of relaxin on collagen turnover, which are evident in relaxin's actions on the reproductive tract during pregnancy, are well conserved in rodents and have been adapted to whole-animal collagen regulation. Relaxin KO mice display increased fibrosis in their hearts, lungs, kidney, and skin as they age, and this fibrosis can be reversed by relaxin treatment. Therefore, although clinical trials on the use of relaxin for the treatment of scleroderma failed, relaxin still has great potential as a therapeutic agent to treat fibrotic conditions.

It is clear that both relaxin and INSL3 are key regulators of reproductive processes in both males and females. Future studies should elucidate whether other members of this fascinating peptide hormone family also have important reproductive functions. It is evident that relaxin has enormous potential for therapeutic actions in both reproduction and disease.

ACKNOWLEDGMENTS

The authors thank the following people for their assistance with the manuscript: Professor Geoffrey Tregear for reviewing the manuscript and for advice and support; Dr. Chrishan Samuel for reviewing the manuscript; Tracey Wilkinson and Daniel Scott for preparation of some of the figures; and those researchers who provided figures from their published work for this chapter. Finally, and most important, we thank all relaxin researchers for their excellent work that fills these pages.

REFERENCES

1. Hisaw, F. (1926). Experimental relaxation of the pubic ligament of the guinea pig. *Proc. Soc. Exp. Biol. Med.* 23, 661–663.
2. Hisaw, F. L. (1927). Experimental relaxation of the symphysis pubis of the guinea-pig. *Anat. Rec.* 37.
3. Fevold, H. L., Hisaw, F. L., and Meyer, R. K. (1930). The relaxative hormone of the corpus luteum: its purification and concentration. *J. Am. Chem. Soc.* 52, 3340–3348.
4. Hall, K. (1947). The effects of pregnancy and relaxin on the histology of the pubic symphysis of the mouse. *J. Endocrinol.* 5, 174–185.
5. Hall, K. (1948). Further notes on the action of oestrone and relaxin on the pelvis of the spayed mouse, including a single-dose test of potency of relaxin. *J. Endocrinol.* 5, 314–321.
6. Krantz, J. C., Bryant, H. H., and Carr, C. J. (1950). The action of aqueous corpus luteum extract upon uterine activity. *Surg. Gynecol. Obstet.* 90, 372–375.
7. Graham, E. F., and Dracy, A. E. (1953). The effect of relaxin and mechanical dilatation of the bovine cervix. *J. Dairy Sci.* 36, 772–777.
8. Hudson, P., Haley, J., Cronk, M., Shine, J., and Niall, H. (1981). Molecular cloning and characterization of cDNA sequences coding for rat relaxin. *Nature* 291, 127–131.

9. Haley, J., Hudson, P., Scanlon, D., John, M., Cronk, M., Shine, J., Tregear, G., and Niall, H. (1982). Porcine relaxin: molecular cloning and cDNA structure. *DNA* 1, 155–162.
10. Hudson, P., Haley, J., John, M., Cronk, M., Crawford, R., Haralambidis, J., Tregear, G., Shine, J., and Niall, H. (1983). Structure of a genomic clone encoding biologically active human relaxin. *Nature* 301, 628–631.
11. Hudson, P., John, M., Crawford, R., Haralambidis, J., Scanlon, D., Gorman, J., Tregear, G., Shine, J., and Niall, H. (1984). Relaxin gene expression in human ovaries and the predicted structure of a human preprorelaxin by analysis of cDNA clones. *EMBO J.* 3, 2333–2339.
12. Adham, I. M., Burkhardt, E., Benahmed, M., and Engel, W. (1993). Cloning of a cDNA for a novel insulin-like peptide of the testicular Leydig cells. *J. Biol. Chem.* 268, 26668–26672.
13. Chassin, D., Laurent, A., Janneau, J. L., Berger, R., and Bellet, D. (1995). Cloning of a new member of the insulin gene superfamily (INSL4) expressed in human placenta. *Genomics* 29, 465–470.
14. Conklin, D., Lofton-Day, C. E., Haldeman, B. A., Ching, A., Whitmore, T. E., Lok, S., and Jaspers, S. (1999). Identification of INSL5, a new member of the insulin superfamily. *Genomics* 60, 50–56.
15. Hsu, S. Y. (1999). Cloning of two novel mammalian paralogs of relaxin/insulin family proteins and their expression in testis and kidney. *Mol. Endocrinol.* 13, 2163–2174.
16. Kasik, J., Muglia, L., Stephan, D. A., and Menon, R. K. (2000). Identification, chromosomal mapping, and partial characterization of mouse Insl6: a new member of the insulin family. *Endocrinology* 141, 458–461.
17. Lok, S., Johnston, D. S., Conklin, D., Lofton-Day, C. E., Adams, R. L., Jelmberg, A. C., Whitmore, T. E., Schrader, S., Griswold, M. D., and Jaspers, S. R. (2000). Identification of INSL6, a new member of the insulin family that is expressed in the testis of the human and rat. *Biol. Reprod.* 62, 1593–1599.
18. Bathgate, R. A., Samuel, C. S., Burazin, T. C., Layfield, S., Claasz, A. A., Reytomas, I. G., Dawson, N. F., Zhao, C., Bond, C., Summers, R. J., Parry, L. J., Wade, J. D., and Tregear, G. W. (2002). Human relaxin gene 3 (H3) and the equivalent mouse relaxin (M3) gene: novel members of the relaxin peptide family. *J. Biol. Chem.* 277, 1148–1157.
19. Hsu, S. Y. (2003). New insights into the evolution of the relaxin-LGR signaling system. *Trends Endocrinol. Metab.* 14, 303–309.
20. Wilkinson, T. N., Speed, T. P., Tregear, G. W., and Bathgate, R. A. D. (2005). Evolution of the relaxin-like peptide family. *BMC Evol. Biol.* 5, 14.
21. Hsu, S. Y., Nakabayashi, K., Nishi, S., Kumagai, J., Kudo, M., Sherwood, O. D., and Hsueh, A. J. (2002). Activation of orphan receptors by the hormone relaxin. *Science* 295, 671–674.
22. Kumagai, J., Hsu, S. Y., Matsumi, H., Roh, J. S., Fu, P., Wade, J. D., Bathgate, R. A., and Hsueh, A. J. (2002). INSL3/Leydig insulin-like peptide activates the LGR8 receptor important in testis descent. *J. Biol. Chem.* 277, 31283–31286.
23. Liu, C., Eriste, E., Sutton, S., Chen, J., Roland, B., Kuei, C., Farmer, N., Jornvall, H., Sillard, R., and Lovenberg, T. W. (2003). Identification of relaxin-3/INSL7 as an endogenous ligand for the orphan G-protein-coupled receptor GPCR135. *J. Biol. Chem.* 278, 50754–50764.
24. Liu, C., Kuei, C., Sutton, S., Chen, J., Bonaventure, P., Wu, J., Nepomuceno, D., Wilkinson, T., Bathgate, R., Eriste, E., Sillard, R., and Lovenberg, T. W. (2005). INSL5 is a high affinity specific agonist for GPCR142 (GPR100). *J. Biol. Chem.* 280, 292–300.
25. Sherwood, O. D. (1994). Relaxin. In *The Physiology of Reproduction* (E. Knobil and J. D. Neill, Eds.), pp. 861–1009. Raven Press, New York.
26. Schwabe, C., and Büllesbach, E. E. (1994). Relaxin: structures, functions, promises, and nonevolution. *FASEB J.* 8, 1152–1160.
27. Weiss, G. (1995). Relaxin used to produce the cervical ripening of labor. *Clin. Obstet. Gynecol.* 38, 293–300.
28. Goldsmith, L. T., Weiss, G., and Steinetz, B. G. (1995). Relaxin and its role in pregnancy. *Endocrinol. Metab. Clin. North Am.* 24, 171–186.
29. Bani, D. (1997). Relaxin: a pleiotropic hormone. *Gen. Pharmacol.* 28, 13–22.
30. Schwabe, C., and Büllesbach, E. E. (1998). *Relaxin and the Fine Structure of Proteins*. Springer-Verlag, New York.
31. Gavino, E. S., and Furst, D. E. (2001). Recombinant relaxin: a review of pharmacology and potential therapeutic use. *BioDrugs* 15, 609–614.
32. Weiss, G., and Goldsmith, L. T. (2001). Relaxin and the cervix. *Front Horm. Res.* 27, 105–112.
33. Ivell, R., and Einspanier, A. (2002). Relaxin peptides are new global players. *Trends Endocrinol. Metab.* 13, 343–348.
34. Bathgate, R. A., Samuel, C. S., Burazin, T. C., Gundlach, A. L., and Tregear, G. W. (2003). Relaxin: new peptides, receptors and novel actions. *Trends Endocrinol. Metab.* 14, 207–213.
35. Dschietzig, T., and Stangl, K. (2003). Relaxin: a pregnancy hormone as central player of body fluid and circulation homeostasis. *Cell Mol. Life. Sci.* 60, 688–700.
36. Samuel, C. S., Parry, L. J., and Summers, R. J. (2003). Physiological or pathological—a role for relaxin in the cardiovascular system? *Curr. Opin. Pharmacol.* 3, 152–158.
37. Silvertown, J. D., Summerlee, A. J., and Klonisch, T. (2003). Relaxin-like peptides in cancer. *Int. J. Cancer* 107, 513–519.
38. Conrad, K. P., and Novak, J. (2004). Emerging role of relaxin in renal and cardiovascular function. *Am. J. Physiol. Regul. Integr. Comp. Physiol.* 287, R250–R261.
39. Sherwood, O. D. (2004). Relaxin's physiological roles and other diverse actions. *Endocr. Rev.* 25, 205–234.
40. Ivell, R. (1997). Biology of the relaxin-like factor (RLF). *Rev. Reprod.* 2, 133–138.
41. Nef, S., and Parada, L. F. (2000). Hormones in male sexual development. *Genes Dev.* 14, 3075–3086.
42. Ivell, R., and Bathgate, R. A. (2002). Reproductive biology of the relaxin-like factor (RLF/INSL3). *Biol. Reprod.* 67, 699–705.
43. Ivell, R., and Hartung, S. (2003). The molecular basis of cryptorchidism. *Mol. Hum. Reprod.* 9, 175–181.
44. Adham, I. M., and Agoulnik, A. I. (2004). Insulin-like 3 signaling in testicular descent. *Int. J. Androl.* 27, 257–265.
45. Foresta, C., and Ferlin, A. (2004). Role of INSL3 and LGR8 in cryptorchidism and testicular functions. *Reprod. Biomed. Online* 9, 294–298.
46. McLennan, A. H., Tregear, G., and Bryant-Greenwood, G. D. (1995). *Progress in Relaxin Research*. World Scientific Publishing, Singapore.
47. Tregear, G. W., Ivell, R., Bathgate, R. A. D., and Wade, J. D. (2001). *Relaxin 2000: Proceedings of the Third International Conference on Relaxin and Related Peptides*. Kluwer, Dordrecht.
48. Sherwood, O. D., Steinetz, B., and Fields, P. A. (2004). *Relaxin 2004: Proceedings of the Fourth International Conference on Relaxin and Related Peptides*. New York Academy of Sciences, New York.
49. Fevold, H. L., Hisaw, F. L., and Leonard, S. L. (1932). The hormones of the corpus luteum: the separation and purification of three active substances. *J. Am. Chem. Soc.* 54, 254–263.
50. Albert, A., Money, W. L., and Zarrow, M. X. (1947). An improved method of extraction and purification of relaxin from fresh whole ovaries of the sow. *Endocrinology* 40, 370–374.
51. Abramowitz, A. A., Money, W. L., Zarrow, M. X., Talmage, R. V. N., Kleinholz, L. H., and Hisaw, F. L. (1944). Preparation, biological assay and properties of relaxin. *Endocrinology* 34, 103–114.
52. Frieden, E. H., and Hisaw, F. L. (1950). The purification of relaxin. *Arch. Biochem.* 29, 166–178.

53. Frieden, E. H., and Layman, N. W. (1957). Non-sterid ovarian hormones: II. An improved method for the preparation of relaxin. *J. Biol. Chem.* 229, 569–573.
54. Kroc, R. L., Steinetz, B. G., and Beach, V. L. (1959). The effects of estrogens, progestogens, and relaxin in pregnant and non-pregnant laboratory rodents. *Ann. N Y Acad. Sci.* 75, 942–980.
55. Steinetz, B. G., Beach, V. L., Kroc, R. L., Stasilli, N. R., Nussbaum, R. E., Nemith, P. J., and Dun, R. K. (1960). Bioassay of relaxin using a reference standard: a simple and reliable method utilizing direct measurement of interpubic ligament formation in mice. *Endocrinology* 67, 102–115.
56. Frieden, E. H., Stone, N. R., and Layman, N. W. (1960). Nonsteroid ovarian hormones: 3. The properties of relaxin preparations purified by counter-current distribution. *J. Biol. Chem.* 235, 2267–2271.
57. Frieden, E. H. (1963). Purification and electrophoretic properties of relaxin preparations. *Trans. N Y Acad. Sci.* 25, 331–336.
58. Cohen, H. (1963). Relaxin: studies dealing with isolation, purification, and characterization. *Trans. N Y Acad. Sci.* 25, 313–330.
59. Griss, G., Keck, J., Engelhorn, R., and Tuppy, H. (1967). The isolation and purification of an ovarian polypeptide with uterine-relaxing activity. *Biochim. Biophys. Acta.* 140, 45–54.
60. Frieden, E. H., and Hisaw, F. L. (1953). The biochemistry of relaxin. *Rec. Prog. Horm. Res.* 8, 333–378.
61. Sherwood, C. D., and O'Byrne, E. M. (1974). Purification and characterization of porcine relaxin. *Arch. Biochem. Biophys.* 160, 185–196.
62. Niall, H. D., James, R., John, M., Walsh, J., Kwok, S., Bryant-Greenwood, G. D., Tregear, G. W., and Bradshaw, R. A. (1982). Chemical studies on relaxin. *Adv. Exp. Med. Biol.* 143, 163–169.
63. Schwabe, C., Steinetz, B., Weiss, G., Segaloff, A., McDonald, J. K., O'Byrne, E., Hochman, J., Carriere, B., and Goldsmith, L. (1978). Relaxin. *Recent Prog. Horm. Res.* 34, 123–211.
64. Frieden, E. H., Rawitch, A. B., Wu, L. H., and Chen, S. W. (1980). The isolation of two proline-containing relaxin species from a porcine relaxin concentrate. *Proc. Soc. Exp. Biol. Med.* 163, 521–527.
65. Walsh, J. R., and Niall, H. D. (1980). Use of an octadecylsilica purification method minimizes proteolysis during isolation of porcine and rat relaxins. *Endocrinology* 107, 1258–1260.
66. Büllesbach, E. E., and Schwabe, C. (1985). Naturally occurring porcine relaxins and large-scale preparation of the B29 hormone. *Biochemistry* 24, 7717–7722.
67. Sherwood, O. D. (1979). Purification and characterization of rat relaxin. *Endocrinology* 104, 886–892.
68. Bradshaw, J. M., Downing, S. J., Moffatt, A., Hinton, J. C., and Porter, D. G. (1981). Demonstration of some of the physiological properties of rat relaxin. *J. Reprod. Fertil.* 63, 145–153.
69. Stewart, D. R., and Papkoff, H. (1986). Purification and characterization of equine relaxin. *Endocrinology* 119, 1093–1099.
70. Stewart, D. R., Nevins, B., Hadas, E., and Vandlen, R. (1991). Affinity purification and sequence determination of equine relaxin. *Endocrinology* 129, 375–383.
71. O'Byrne, E. M., Weiss, G., and Steinetz, B. G. (1983). The isolation of human relaxin from the corpus luteum. In *Biology of Relaxin and Its Role in the Human* (M. Bigazzi, F. C. Greenwood, and F. Gasparri, Eds.), pp. 370–376. Excerpta Medica, Amsterdam.
72. Eldridge, R. K., and Fields, P. A. (1983). Isolation of human placental relaxin with octadecylsilica. In *Biology of Relaxin and Its Role in the Human* (M. Bigazzi, F. Greenwood, and F. Gasparri, Eds.), pp. 389–391. Excerpta Medica, Amsterdam.
73. Entenmann, A. H., Lippert, T. H., and Voelter, W. (1983). Isolation of relaxin from human placenta and its biological activity. In *Biology of Relaxin and Its Role in the Human* (M. Bigazzi, F. C. Greenwood, and F. Gasparri, Eds.), pp. 392–394. Excerpta Medica, Amsterdam.
74. Fields, P. A., and Larkin, L. H. (1981). Purification and immunohistochemical localization of relaxin in the human term placenta. *J. Clin. Endocrinol. Metab.* 52, 79–85.
75. Yamamoto, S., Kwok, S. C., Greenwood, F. C., and Bryant-Greenwood, G. D. (1981). Relaxin purification from human placental basal plates. *J. Clin. Endocrinol. Metab.* 52, 601–604.
76. Bigazzi, M., Bruni, P., Nardi, E., Petrucci, F., Pollicino, G., Franchini, M., Scarselli, G., and Farnararo, M. (1982). Human decidual relaxin. *Ann. N Y Acad. Sci.* 380, 87–99.
77. Weiss, G., Goldsmith, L. T., Schoenfeld, C., and D'Eletto, R. (1986). Partial purification of relaxin from human seminal plasma. *Am. J. Obstet. Gynecol.* 154, 749–755.
78. O'Byrne, E. M., Carriere, B. T., Sorensen, L., Segaloff, A., Schwabe, C., and Steinetz, B. G. (1978). Plasma immunoreactive relaxin levels in pregnant and nonpregnant women. *J. Clin. Endocrinol. Metab.* 47, 1106–1110.
79. Drolet, D. W., Henzel, W. J., and Johnston, P. D. (1987). Purification, amino-terminal sequencing and demonstration of biological activity of human relaxin from corpora lutea. In *Program of the 69th Annual Meeting of the Endocrine Society,* p. 196 (abstract).
80. Winslow, J., Shih, A., Laramee, G., Bourell, J., Stults, J., and Johnston, P. (1989). Purification and structure of human pregnancy relaxin from corpora lutea, serum and plasma. In *Program of the 71st Annual Meeting of the Endocrine Society,* p. 245 (abstract)
81. Winslow, J. W., Shih, A., Bourell, J. H., Weiss, G., Reed, B., Stults, J. T., and Goldsmith, L. T. (1992). Human seminal relaxin is a product of the same gene as human luteal relaxin. *Endocrinology* 130, 2660–2668.
82. Büllesbach, E. E., Schwabe, C., and Callard, I. P. (1987). Relaxin from an oviparous species, the skate *(Raja erinacea). Biochem. Biophys. Res. Commun.* 143, 273–280.
83. Stewart, D. R., Henzel, W. J., and Vandlen, R. (1992). Purification and sequence determination of canine relaxin. *J. Protein Chem.* 11, 247–253.
84. Büllesbach, E. E., Gowan, L. K., Schwabe, C., Steinetz, B. G., O'Byrne, E., and Callard, I. P. (1986). Isolation, purification, and the sequence of relaxin from spiny dogfish *(Squalus acanthias). Eur. J. Biochem.* 161, 335–341.
85. Reinig, J. W., Daniel, L. N., Schwabe, C., Gowan, L. K., Steinetz, B. G., and O'Byrne, E. M. (1981). Isolation and characterization of relaxin from the sand tiger shark *(Odontaspis taurus). Endocrinology* 109, 537–543.
86. Renegar, R. H., Owens, C. R., and Chalovich, J. M. (1993). Purification and partial characterization of relaxin and relaxin precursors from the hamster placenta. *Biol. Reprod.* 49, 154–161.
87. Renegar, R. H., and Owens, C. R. III. (2002). Measurement of plasma and tissue relaxin concentrations in the pregnant hamster and fetus using a homologous radioimmunoassay. *Biol. Reprod.* 67, 500–505.
88. Schwabe, C., Büllesbach, E. E., Heyn, H., and Yoshioka, M. (1989). Cetacean relaxin: isolation and sequence of relaxins from *Balaenoptera acutorostrata* and *Balaenoptera edeni. J. Biol. Chem.* 264, 940–943.
89. Woods, A. S., Cotter, R. J., Yoshioka, M., Büllesbach, E., and Schwabe, C. (1991). Enzymatic digestion on the sample foil as a method for sequence determination by plasma desorption mass spectrometry: the primary structure of porpoise relaxin. *Int. J. Mass Spectrom Ion Processes* 111, 77–88.
90. Kemp, B. E., and Niall, H. D. (1984). Relaxin. *Vitam. Horm.* 41, 79–115.
91. Soloff, M. S., Gal, S., Hoare, S., Peters, C. A., Hunzicker-Dunn, M., Anderson, G. D., and Wood, T. G. (2003). Cloning, characterization, and expression of the rat relaxin gene. *Gene* 323, 149–155.

92. Haley, J., Crawford, R., Hudson, P., Scanlon, D., Tregear, G., Shine, J., and Niall, H. (1987). Porcine relaxin. Gene structure and expression. *J. Biol. Chem.* 262, 11940–11946.
93. Crawford, R. J., Hammond, V. E., Roche, P. J., Johnston, P. D., and Tregear, G. W. (1989). Structure of rhesus monkey relaxin predicted by analysis of the single-copy rhesus monkey relaxin gene. *J. Mol. Endocrinol.* 3, 169–174.
94. Evans, B. A., Fu, P., and Tregear, G. W. (1994). Characterization of two relaxin genes in the chimpanzee. *J. Endocrinol.* 140, 385–392.
95. Evans, B. A., Fu, P., and Tregear, G. W. (1994). Characterization of primate relaxin genes. *Endocr. J.* 2, 81–86.
96. Einspanier, A., Zarreh-Hoshyari-Khah, M. R., Balvers, M., Kerr, L., Fuhrmann, K., and Ivell, R. (1997). Local relaxin biosynthesis in the ovary and uterus through the oestrous cycle and early pregnancy in the female marmoset monkey (*Callithrix jacchus*). *Hum. Reprod.* 12, 1325–1337.
97. Klonisch, T., Froehlich, C., Tetens, F., Fischer, B., and Hombach-Klonisch, S. (2001). Molecular remodeling of members of the relaxin family during primate evolution. *Mol. Biol. Evol.* 18, 393–403.
98. Lee, Y. A., Bryant-Greenwood, G. D., Mandel, M., and Greenwood, F. C. (1992). The complementary deoxyribonucleic acid sequence of guinea pig endometrial prorelaxin. *Endocrinology* 130, 1165–1172.
99. Jetten, A. M., Bernacki, S. H., Floyd, E. E., Saunders, N. A., Pieniazek, J., and Lotan, R. (1992). Expression of a preprorelaxin-like gene during squamous differentiation of rabbit tracheobronchial epithelial cells and its suppression by retinoic acid. *Cell. Growth Differ.* 3, 549–556.
100. Fields, P. A., Lee, V. H., Jetten, A., Chang, S. M., and Fields, M. J. (1999). B-chain sequence and in situ hybridization of the rabbit placental relaxin-like gene product. *Biol. Reprod.* 61, 527–532.
101. Evans, B. A., John, M., Fowler, K. J., Summers, R. J., Cronk, M., Shine, J., and Tregear, G. W. (1993). The mouse relaxin gene: nucleotide sequence and expression. *J. Mol. Endocrinol.* 10, 15–23.
102. McCaslin, R. B., and Renegar, R. H. (1995). Determination of the prorelaxin nucleotide sequence and expression of prorelaxin messenger ribonucleic acid in the golden hamster. *Biol. Reprod.* 53, 454–461.
103. Parry, L. J., Rust, W., and Ivell, R. (1997). Marsupial relaxin: complementary deoxyribonucleic acid sequence and gene expression in the female and male tammar wallaby, *Macropus eugenii*. *Biol. Reprod.* 57, 119–127.
104. Bathgate, R. A., Siebel, A. L., Tovote, P., Claasz, A., Macris, M., Tregear, G. W., and Parry, L. J. (2002). Purification and characterization of relaxin from the tammar wallaby (*Macropus eugenii*): bioactivity and expression in the corpus luteum. *Biol. Reprod.* 67, 293–300.
105. Klonisch, T., Ryan, P. L., Yamashiro, S., and Porter, D. G. (1995). Partial complementary deoxyribonucleic acid cloning of equine relaxin messenger ribonucleic acid, and its localization within the equine placenta. *Biol. Reprod.* 52, 1307–1315.
106. Klonisch, T., Hombach-Klonisch, S., Froehlich, C., Kauffold, J., Steger, K., Steinetz, B. G., and Fischer, B. (1999). Canine preprorelaxin: nucleic acid sequence and localization within the canine placenta. *Biol. Reprod.* 60, 551–557.
107. Klonisch, T., Hombach-Klonisch, S., Froehlich, C., Kauffold, J., Steger, K., Huppertz, B., and Fischer, B. (1999). Nucleic acid sequence of feline preprorelaxin and its localization within the feline placenta. *Biol. Reprod.* 60, 305–311.
108. Hombach-Klonisch, S., Abd-Elnaeim, M., Skidmore, J. A., Leiser, R., Fischer, B., and Klonisch, T. (2000). Ruminant relaxin in the pregnant one-humped camel (*Camelus dromedarius*). *Biol. Reprod.* 62, 839–846.
109. Bell, G. I., Pictet, R. L., Rutter, W. J., Cordell, B., Tischer, E., and Goodman, H. M. (1980). Sequence of the human insulin gene. *Nature* 284, 26–32.
110. Gunnersen, J. M., Fu, P., Roche, P. J., and Tregear, G. W. (1996). Expression of human relaxin genes: characterization of a novel alternatively-spliced human relaxin mRNA species. *Mol. Cell. Endocrinol.* 118, 85–94.
111. Pusch, W., Balvers, M., and Ivell, R. (1996). Molecular cloning and expression of the relaxin-like factor from the mouse testis. *Endocrinology* 137, 3009–3013.
112. Burkhardt, E., Adham, I. M., Brosig, B., Gastmann, A., Mattei, M. G., and Engel, W. (1994). Structural organization of the porcine and human genes coding for a Leydig cell-specific insulin-like peptide (LEY I-L) and chromosomal localization of the human gene (INSL3). *Genomics* 20, 13–19.
113. Koskimies, P., Spiess, A. N., Lahti, P., Huhtaniemi, I., and Ivell, R. (1997). The mouse relaxin-like factor gene and its promoter are located within the 3′ region of the JAK3 genomic sequence. *FEBS Lett.* 419, 186–190.
114. Zimmermann, S., Schottler, P., Engel, W., and Adham, I. M. (1997). Mouse Leydig insulin-like (Ley I-L) gene: structure and expression during testis and ovary development. *Mol. Reprod. Dev.* 47, 30–38.
115. Spiess, A. N., Balvers, M., Tena-Sempere, M., Huhtaniemi, I., Parry, L., and Ivell, R. (1999). Structure and expression of the rat relaxin-like factor (RLF) gene. *Mol. Reprod. Dev.* 54, 319–325.
116. Zarreh-Hoshyari-Khah, M. R., Einspanier, A., and Ivell, R. (1999). Differential splicing and expression of the relaxin-like factor gene in reproductive tissues of the marmoset monkey (*Callithrix jacchus*). *Biol. Reprod.* 60, 445–453.
117. Bathgate, R., Balvers, M., Hunt, N., and Ivell, R. (1996). Relaxin-like factor gene is highly expressed in the bovine ovary of the cycle and pregnancy: sequence and messenger ribonucleic acid analysis. *Biol. Reprod.* 55, 1452–1457.
118. Roche, P. J., Crawford, R. J., and Tregear, G. W. (1993). A single-copy relaxin-like gene sequence is present in sheep. *Mol. Cell. Endocrinol.* 91, 21–28.
119. Hombach-Klonisch, S., Tetens, F., Kauffold, J., Steger, K., Fischer, B., and Klonisch, T. (1999). Molecular cloning and localization of caprine relaxin-like factor (RLF) mRNA within the goat testis. *Mol. Reprod. Dev.* 53, 135–141.
120. Klonisch, T., Kauffold, J., Steger, K., Bergmann, M., Leiser, R., Fischer, B., and Hombach-Klonisch, S. (2001). Canine relaxin-like factor: unique molecular structure and differential expression within reproductive tissues of the dog. *Biol. Reprod.* 64, 442–450.
121. Hombach-Klonisch, S., Kauffold, J., Rautenberg, T., Steger, K., Tetens, F., Fischer, B., and Klonisch, T. (2000). Relaxin-like factor (RLF) mRNA expression in the fallow deer. *Mol. Cell. Endocrinol.* 159, 147–158.
122. Büllesbach, E. E., and Schwabe, C. (1995). A novel Leydig cell cDNA-derived protein is a relaxin-like factor. *J. Biol. Chem.* 270, 16011–16015.
123. Büllesbach, E. E., and Schwabe, C. (2002). The primary structure and the disulfide links of the bovine relaxin-like factor (RLF). *Biochemistry* 41, 274–281.
124. Koman, A., Cazaubon, S., Couraud, P. O., Ullrich, A., and Strosberg, A. D. (1996). Molecular characterization and in vitro biological activity of placentin, a new member of the insulin gene family. *J. Biol. Chem.* 271, 20238–20241.
125. Jaspers, S., Lok, S., Lofton-Day, C. E., Haldeman, B. A., Whitmore, T. E., Foley, K. P., and Conklin, D. (2001). The genomics of INSL5. In *Relaxin 2000: Proceedings of the Third International Conference on Relaxin and Related Peptides* (G. W. Tregear, R. Ivell, R. A. Bathgate, and J. D. Wade, Eds.). Kluwer, Dordrecht.

126. Büllesbach, E. E., and Schwabe, C. (2000). The relaxin receptor-binding site geometry suggests a novel gripping mode of interaction. *J. Biol. Chem.* 275, 35276–35280.
127. Burazin, T. C., Bathgate, R. A., Macris, M., Layfield, S., Gundlach, A. L., and Tregear, G. W. (2002). Restricted, but abundant, expression of the novel rat gene-3 (R3) relaxin in the dorsal tegmental region of brain. *J. Neurochem.* 82, 1553–1557.
128. Kizawa, H., Nishi, K., Ishibashi, Y., Harada, M., Asano, T., Ito, Y., Suzuki, N., Hinuma, S., Fujisawa, Y., Onda, H., Nishimura, O., and Fujino, M. (2003). Production of recombinant human relaxin 3 in AtT20 cells. *Regul. Pept.* 113, 79–84.
129. Bathgate, R. A. D., Scott, D., Chung, S., Ellyard, D., Garreffa, A., and Tregear, G. W. (2002). Searching the human genome database for novel relaxin-like peptides. *Lett. Pept. Sci.* 8, 129–132.
130. Bieche, I., Laurent, A., Laurendeau, I., Duret, L., Giovangrandi, Y., Frendo, J. L., Olivi, M., Fausser, J. L., Evain-Brion, D., and Vidaud, M. (2003). Placenta-specific INSL4 expression is mediated by a human endogenous retrovirus element. *Biol. Reprod.* 68, 1422–1429.
131. Veitia, R., Laurent, A., Quintana-Murci, L., Ottolenghi, C., Fellous, M., Vidaud, M., and McElreavey, K. (1998). The INSL4 gene maps close to WI-5527 at 9p24.1→p23.3 clustered with two relaxin genes and outside the critical region for the monosomy 9p syndrome. *Cytogenet. Cell. Genet.* 81, 275–277.
132. Hansell, D. J., Bryant-Greenwood, G. D., and Greenwood, F. C. (1991). Expression of the human relaxin H1 gene in the decidua, trophoblast, and prostate. *J. Clin. Endocrinol. Metab.* 72, 899–904.
133. Tan, Y. Y., Wade, J. D., Tregear, G. W., and Summers, R. J. (1998). Comparison of relaxin receptors in rat isolated atria and uterus by use of synthetic and native relaxin analogues. *Br. J. Pharmacol.* 123, 762–770.
134. Schwabe, C., LeRoith, D., Thompson, R. P., Shiloach, J., and Roth, J. (1983). Relaxin extracted from protozoa (*Tetrahymena pyriformis*): molecular and immunologic properties. *J. Biol. Chem.* 258, 2778–2781.
135. Georges, D., Tashima, L., Yamamoto, S., and Bryant-Greenwood, G. D. (1990). Relaxin-like peptide in ascidians. I. Identification of the peptide and its mRNA in ovary of *Herdmania momus*. *Gen. Comp. Endocrinol.* 79, 423–428.
136. Georges, D., and Schwabe, C. (1999). Porcine relaxin, a 500 million-year-old hormone? The tunicate *Ciona intestinalis* has porcine relaxin. *FASEB J.* 13, 1269–1275.
137. Hartung, S., Kondo, S., Abend, N., Rust, W., Balvers, M., Bryant-Greenwood, G., and Ivell, R. (1995). The search for ruminant relaxin. In *Progress in Relaxin Research* (A. H. McLennan, G. Tregear, and G. D. Bryant-Greenwood, Eds.), pp. 439–456. World Scientific Publishing, Singapore.
138. Musah, A. I., Schwabe, C., Willham, R. L., and Anderson, L. L. (1986). Pelvic development as affected by relaxin in three genetically selected frame sizes of beef heifers. *Biol. Reprod.* 34, 363–369.
139. Porter, D. G., Lye, S. J., Bradshaw, J. M., and Kendall, J. Z. (1981). Relaxin inhibits myometrial activity in the ovariectomized non-pregnant ewe. *J. Reprod. Fertil.* 61, 409–414.
140. Fields, M. J., Fields, P. A., Castro-Hernandez, A., and Larkin, L. H. (1980). Evidence for relaxin in corpora lutea of late pregnant cows. *Endocrinology* 107, 869–876.
141. Steinetz, B. (2001). Hooked on relaxin. In *Relaxin 2000: Proceedings of the Third International Conference on Relaxin and Related Peptides* (G. W. Tregear, R. Ivell, R. A., Bathgate, and J. D. Wade, Eds.). Kluwer, Dordrecht.
142. Wathes, D. C., Rees, J. M., and Porter, D. G. (1988). Identification of relaxin in the placenta of the ewe. *J. Reprod. Fertil.* 84, 247–257.
143. Bravo, P. W., Stewart, D. R., Lasley, B. L., and Fowler, M. E. (1996). Hormonal indicators of pregnancy in llamas and alpacas. *J. Am. Vet. Med. Assoc.* 208, 2027–2030.
144. Riehle, M. A., Garczynski, S. F., Crim, J. W., Hill, C. A., and Brown, M. R. (2002). Neuropeptides and peptide hormones in *Anopheles gambiae*. *Science* 298, 172–175.
145. Rulifson, E. J., Kim, S. K., and Nusse, R. (2002). Ablation of insulin-producing neurons in flies: growth and diabetic phenotypes. *Science* 296, 1118–1120.
146. Frieden, E. H., and Yeh, L. (1977). Evidence for a "pro-relaxin" in porcine relaxin concentrates. *Proc. Soc. Exp. Biol. Med.* 154, 407–411.
147. Kwok, S. C., Chamley, W. A., and Bryant-Greenwood, G. D. (1978). High molecular weight forms of relaxin in pregnant sow ovaries. *Biochem. Biophys. Res. Commun.* 82, 997–1005.
148. Gast, M. J. (1982). Studies of luteal generation and processing of the high molecular weight relaxin precursor. *Ann. N Y Acad. Sci.* 380, 111–125.
149. Gast, M. J. (1983). Characterization of preprorelaxin by tryptic digestion and inhibition of its conversion to prorelaxin by amino acid analogs. *J. Biol. Chem.* 258, 9001–9004.
150. Tager, H. S., and Steiner, D. F. (1972). Primary structures of the proinsulin connecting peptides of the rat and the horse. *J. Biol. Chem.* 247, 7936–7940.
151. Yang, S., Heyn, H., Zhang, Y. Z., Büllesbach, E. E., and Schwabe, C. (1993). The expression of human relaxin in yeast. *Arch. Biochem. Biophys.* 300, 734–737.
152. Vandlen, R., Winslow, J., Moffat, B., and Rinderknecht, E. (1995). Human relaxin: purification, characterization and production of recombinant relaxins for structure function studies. In *Progress in Relaxin Research: The Proceedings of the Second International Congress on the Hormone Relaxin* (A. H. MacLennan, G. W. Tregear, and G. D. Bryant-Greenwood, Eds.), pp. 59–72. Global Publications, Adelaide, Australia.
153. Reddy, G. K., Gunwar, S., Green, C. B., Fei, D. T., Chen, A. B., and Kwok, S. C. (1992). Purification and characterization of recombinant porcine prorelaxin expressed in *Escherichia coli*. *Arch. Biochem. Biophys.* 294, 579–585.
154. Vu, A. L., Green, C. B., Roby, K. F., Soares, M. J., Fei, D. T., Chen, A. B., and Kwok, S. C. (1993). Recombinant porcine prorelaxin produced in Chinese hamster ovary cells is biologically active. *Life Sci.* 52, 1055–1061.
155. Layden, S. S., and Tregear, G. W. (1996). Purification and characterization of porcine prorelaxin. *J. Biochem. Biophys. Methods* 31, 69–80.
156. Zarreh-Hoshyari-Khah, R., Bartsch, O., Einspanier, A., Pohnke, Y., and Ivell, R. (2001). Bioactivity of recombinant prorelaxin from the marmoset monkey. *Regul. Pept.* 97, 139–146.
157. Nakayama, K. (1997). Furin: a mammalian subtilisin/Kex2p-like endoprotease involved in processing of a wide variety of precursor proteins. *Biochem. J.* 327, 625–635.
158. Stults, J. T., Bourell, J. H., Canova-Davis, E., Ling, V. T., Laramee, G. R., Winslow, J. W., Griffin, P. R., Rinderknecht, E., and Vandlen, R. L. (1990). Structural characterization by mass spectrometry of native and recombinant human relaxin. *Biomed. Environ. Mass Spectrom* 19, 655–664.
159. Marriott, D., Gillece-Castro, B., and Gorman, C. M. (1992). Prohormone convertase-1 will process prorelaxin, a member of the insulin family of hormones. *Mol. Endocrinol.* 6, 1441–1450.
160. Schwabe, C., and McDonald, J. K. (1977). Relaxin: a disulfide homolog of insulin. *Science* 197, 914–915.
161. Canova-Davis, E., Kessler, T. J., Lee, P. J., Fei, D. T., Griffin, P., Stults, J. T., Wade, J. D., and Rinderknecht, E. (1991). Use of recombinant DNA derived human relaxin to probe the structure of the native protein. *Biochemistry* 30, 6006–6013.

162. Sherwood, O. D., and Crnekovic, V. E. (1979). Development of a homologous radioimmunoassay for rat relaxin. *Endocrinology* 104, 893–897.
163. Eigenbrot, C., Randal, M., Quan, C., Burnier, J., O'Connell, L., Rinderknecht, E., and Kossiakoff, A. A. (1991). X-ray structure of human relaxin at 1.5 A. Comparison to insulin and implications for receptor binding determinants. *J. Mol. Biol.* 221, 15–21.
164. Bedarkar, S., Turnell, W. G., Blundell, T. L., and Schwabe, C. (1977). Relaxin has conformational homology with insulin. *Nature* 270, 449–451.
165. Isaacs, N., James, R., Niall, H., Bryant-Greenwood, G., Dodson, G., Evans, A., and North, A. C. (1978). Relaxin and its structural relationship to insulin. *Nature* 271, 278–281.
166. Bedarkar, S., Blundell, T., Gowan, L. K., McDonald, J. K., and Schwabe, C. (1982). On the three-dimensional structure of relaxin. *Ann. N Y Acad. Sci.* 380, 22–33.
167. Dawson, N. F., Tan, Y. Y., Macris, M., Otvos, L. Jr., Summers, R. J., Tregear, G. W., and Wade, J. D. (1999). Solid-phase synthesis of ovine Leydig cell insulin-like peptide: a putative ovine relaxin? *J. Pept. Res.* 53, 542–547.
168. Smith, K. J., Wade, J. D., Claasz, A. A., Otvos, L. Jr., Temelcos, C., Kubota, Y., Hutson, J. M., Tregear, G. W., and Bathgate, R. A. (2001). Chemical synthesis and biological activity of rat INSL3. *J. Pept. Sci.* 7, 495–501.
169. Tregear, G. W., Mathieu, M. N., Dawson, N., Smith, K. J., Claasz, A., Clippingdale, A. B., Fu, P., and Wade, J. D. (2001). Chemical synthesis of relaxin analogues: current status and future developments. In *Relaxin 2000: Proceedings of the Third International Conference on Relaxin and Related Peptides* (G. W. Tregear, R. Ivell, R. A. Bathgate, and J. D. Wade, Eds.). Kluwer, Dordrecht.
170. Wade, J. D., Layden, S. S., Lambert, P. F., Kakouris, H., and Tregear, G. W. (1994). Primate relaxin: synthesis of gorilla and rhesus monkey relaxins. *J. Protein Chem.* 13, 315–321.
171. Wade, J. D., Lin, F., Salvatore, D., Otvos, L. Jr., and Tregear, G. W. (1996). Synthesis and characterization of human gene 1 relaxin peptides. *Biomed. Pept. Proteins Nucleic Acids* 2, 27–32.
172. Lin, F., Otvos, L. Jr., Kumagai, J., Tregear, G. W., Bathgate, R. A., and Wade, J. D. (2004). Synthetic human insulin 4 does not activate the G-protein-coupled receptors LGR7 or LGR8. *J. Pept. Sci.* 10, 257–264.
173. Büllesbach, E. E., and Schwabe, C. (1991). Total synthesis of human relaxin and human relaxin derivatives by solid-phase peptide synthesis and site-directed chain combination. *J. Biol. Chem.* 266, 10754–10761.
174. Büllesbach, E. E., and Schwabe, C. (1993). Mouse relaxin: synthesis and biological activity of the first relaxin with an unusual crosslinking pattern. *Biochem. Biophys. Res. Commun.* 196, 311–319.
175. Büllesbach, E. E., and Schwabe, C. (1996). The chemical synthesis of rat relaxin and the unexpectedly high potency of the synthetic hormone in the mouse. *Eur. J. Biochem.* 241, 533–537.
176. Büllesbach, E. E., and Schwabe, C. (2001). Synthesis and conformational analysis of the insulin-like 4 gene product. *J. Pept. Res.* 57, 77–83.
177. Stewart, A. G., Richards, H., Roberts, S., Warwick, J., Edwards, K., Bell, L., Smith, J., and Derbyshire, R. (1983). Cloning and expression of a porcine prorelaxin gene in *E. coli. Nucleic Acids Res.* 11, 6597–6609.
178. Shire, S. J., Holladay, L. A., and Rinderknecht, E. (1991). Self-association of human and porcine relaxin as assessed by analytical ultracentrifugation and circular dichroism. *Biochemistry* 30, 7703–7711.
179. Silvertown, J. D., Geddes, B. J., and Summerlee, A. J. (2003). Adenovirus-mediated expression of human prorelaxin promotes the invasive potential of canine mammary cancer cells. *Endocrinology* 144, 3683–3691.
180. Bell, R. J., Permezel, M., MacLennan, A., Hughes, C., Healy, D., and Brennecke, S. (1993). A randomized, double-blind, placebo-controlled trial of the safety of vaginal recombinant human relaxin for cervical ripening. *Obstet. Gynecol.* 82, 328–333.
181. Erikson, M. S., and Unemori, E.N. (2001). Relaxin clinical trials in systemic sclerosis. In *Relaxin 2000: Proceedings of the Third International Conference on Relaxin and Related Peptides* (G. W. Tregear, R. Ivell, R. A. Bathgate, and J. D. Wade, Eds.), pp. 373–382. Kluwer, Amsterdam.
182. Tregear, G. W., Fagan, C., Reynolds, H., Scanlon, D., Jones, P., Kemp, B., and Niall, H. D. (1981). Porcine relaxin: synthesis and structure activity relationships. In *Peptides: Synthesis, Structure and Function* (D. H. Rich and E. Gross, Eds.), pp. 249–252. Pierce Chemical, Rockford, IL.
183. Büllesbach, E. E., and Schwabe, C. (1986). Preparation and properties of porcine relaxin derivatives shortened at the amino terminus of the A chain. *Biochemistry* 25, 5998–6004.
184. Büllesbach, E. E., and Schwabe, C. (1987). Relaxin structure: quasi allosteric effect of the NH_2-terminal A-chain helix. *J. Biol. Chem.* 262, 12496–12501.
185. Büllesbach, E. E., and Schwabe, C. (1994). Functional importance of the A chain loop in relaxin and insulin. *J. Biol. Chem.* 269, 13124–13128.
186. Büllesbach, E. E., Yang, S., and Schwabe, C. (1992). The receptor-binding site of human relaxin: II. A dual prong-binding mechanism. *J. Biol. Chem.* 267, 22957–22960.
187. Claasz, A. A., Bond, C. P., Bathgate, R. A., Otvos, L., Dawson, N. F., Summers, R. J., Tregear, G. W., and Wade, J. D. (2002). Relaxin-like bioactivity of ovine insulin 3 (INSL3) analogues. *Eur. J. Biochem.* 269, 6287–6293.
188. Tan, Y. Y., Dawson, N. F., Kompa, A. R., Bond, C. P., Claasz, A., Wade, J. D., Tregear, G. W., and Summers, R. J. (2002). Structural requirements for the interaction of sheep insulin-like factor 3 with relaxin receptors in rat atria. *Eur. J. Pharmacol.* 457, 153–160.
189. Büllesbach, E. E., Steinetz, B. G., and Schwabe, C. (1996). Chemical synthesis of a Zwitterhormon, insulaxin, and of a relaxin-like bombyxin derivative. *Biochemistry* 35, 9754–9760.
190. Dodson, G. G., Eliopoulos, E. E., Isaacs, N. W., McCall, M. J., Niall, H. D., and North, A. C. T. (1982). Rat relaxin: insulin-like fold predicts a likely receptor binding region. *Int. J. Biol. Macromol.* 4, 399–405.
191. Sudo, S., Kumagai, J., Nishi, S., Layfield, S., Ferraro, T., Bathgate, R. A., and Hsueh, A. J. (2003). H3 relaxin is a specific ligand for LGR7 and activates the receptor by interacting with both the ectodomain and the exoloop 2. *J. Biol. Chem.* 278, 7855–7862.
192. Büllesbach, E. E., and Schwabe, C. (1999). Tryptophan B27 in the relaxin-like factor (RLF) is crucial for RLF receptor-binding. *Biochemistry* 38, 3073–3078.
193. Sherwood, O. D., Rosentreter, K. R., and Birkhimer, M. L. (1975). Development of a radioimmunoassay for porcine relaxin using ^{125}I-labeled polytyrosyl-relaxin. *Endocrinology* 96, 1106–1113.
194. Bolton, A. E., and Hunter, W. M. (1973). The labelling of proteins to high specific radioactivities by conjugation to a ^{125}I-containing acylating agent. *Biochem. J.* 133, 529–539.
195. McMurtry, J. P., Kwok, S. C., and Bryant-Greenwood, G. D. (1978). Target tissues for relaxin identified in vitro with ^{125}I-labelled porcine relaxin. *J. Reprod. Fertil.* 53, 209–216.
196. Cheah, S. H., and Sherwood, O. D. (1980). Target tissues for relaxin in the rat: tissue distribution of injected ^{125}I-labeled relaxin and tissue changes in adenosine 3′,5′-monophosphate levels after in vitro relaxin incubation. *Endocrinology* 106, 1203–1209.

197. Mercado-Simmen, R. C., Bryant-Greenwood, G. D., and Greenwood, F. C. (1980). Characterization of the binding of ^{125}I-relaxin to rat uterus. *J. Biol. Chem.* 255, 3617–3623.
198. Yang, S., Rembiesa, B., Büllesbach, E. E., and Schwabe, C. (1992). Relaxin receptors in mice: demonstration of ligand binding in symphyseal tissues and uterine membrane fragments. *Endocrinology* 130, 179–185.
199. Mercado-Simmen, R. C., Goodwin, B., Ueno, M. S., Yamamoto, S. Y., and Bryant-Greenwood, G. D. (1982). Relaxin receptors in the myometrium and cervix of the pig. *Biol. Reprod.* 26, 120–128.
200. Weiss, T. J., and Bryant-Greenwood, G. D. (1982). Localization of relaxin binding sites in the rat uterus and cervix by autoradiography. *Biol. Reprod.* 27, 673–679.
201. Koay, E. S., Bryant-Greenwood, G. D., Yamamoto, S. Y., and Greenwood, F. C. (1986). The human fetal membranes: a target tissue for relaxin. *J. Clin. Endocrinol. Metab.* 62, 513–521.
202. McMurtry, J. P., Floersheim, G. L., and Bryant-Greenwood, G. D. (1980). Characterization of the binding of ^{125}I-labelled succinylated porcine relaxin to human and mouse fibroblasts. *J. Reprod. Fertil.* 58, 43–49.
203. Tang, X. M., and Chegini, N. (1995). Human fallopian tube as an extraovarian source of relaxin: messenger ribonucleic acid expression and cellular localization of immunoreactive protein and ^{125}I-relaxin binding sites. *Biol. Reprod.* 52, 1343–1349.
204. Palejwala, S., Stein, D., Wojtczuk, A., Weiss, G., and Goldsmith, L. T. (1998). Demonstration of a relaxin receptor and relaxin-stimulated tyrosine phosphorylation in human lower uterine segment fibroblasts. *Endocrinology* 139, 1208–1212.
205. Osheroff, P. L., Ling, V. T., Vandlen, R. L., Cronin, M. J., and Lofgren, J. A. (1990). Preparation of biologically active ^{32}P-labeled human relaxin: displaceable binding to rat uterus, cervix, and brain. *J. Biol. Chem.* 265, 9396–9401.
206. Tan, Y. Y., Wade, J. D., Tregear, G. W., and Summers, R. J. (1999). Quantitative autoradiographic studies of relaxin binding in rat atria, uterus and cerebral cortex: characterization and effects of oestrogen treatment. *Br. J. Pharmacol.* 127, 91–98.
207. Osheroff, P. L., Cronin, M. J., and Lofgren, J. A. (1992). Relaxin binding in the rat heart atrium. *Proc. Natl. Acad. Sci. U S A* 89, 2384–2388.
208. Osheroff, P. L., and King, K. L. (1995). Binding and cross-linking of ^{32}P-labeled human relaxin to human uterine cells and primary rat atrial cardiomyocytes. *Endocrinology* 136, 4377–4381.
209. Osheroff, P. L., and Phillips, H. S. (1991). Autoradiographic localization of relaxin binding sites in rat brain. *Proc. Natl. Acad. Sci. U S A* 88, 6413–6417.
210. Osheroff, P. L., and Ho, W. H. (1993). Expression of relaxin mRNA and relaxin receptors in postnatal and adult rat brains and hearts: localization and developmental patterns. *J. Biol. Chem.* 268, 15193–15199.
211. Garibay-Tupas, J. L., Maaskant, R. A., Greenwood, F. C., and Bryant-Greenwood, G. D. (1995). Characteristics of the binding of ^{32}P-labelled human relaxins to the human fetal membranes. *J. Endocrinol.* 145, 441–448.
212. Parsell, D. A., Mak, J. Y., Amento, E. P., and Unemori, E. N. (1996). Relaxin binds to and elicits a response from cells of the human monocytic cell line, THP-1. *J. Biol. Chem.* 271, 27936–27941.
213. Büllesbach, E. E., and Schwabe, C. (1990). Monobiotinylated relaxins: preparation and chemical properties of the mono(biotinyl-epsilon-aminohexanoyl) porcine relaxin. *Int. J. Pept. Protein Res.* 35, 416–423.
214. Kuenzi, M. J., and Sherwood, O. D. (1995). Immunohistochemical localization of specific relaxin-binding cells in the cervix, mammary glands, and nipples of pregnant rats. *Endocrinology* 136, 1367–1373.
215. Min, G., and Sherwood, O. D. (1996). Identification of specific relaxin-binding cells in the cervix, mammary glands, nipples, small intestine, and skin of pregnant pigs. *Biol. Reprod.* 55, 1243–1252.
216. Min, G., and Sherwood, O. D. (1998). Localization of specific relaxin-binding cells in the ovary and testis of pigs. *Biol. Reprod.* 59, 401–408.
217. Kohsaka, T., Min, G., Lukas, G., Trupin, S., Campbell, E. T., and Sherwood, O. D. (1998). Identification of specific relaxin-binding cells in the human female. *Biol. Reprod.* 59, 991–999.
218. Einspanier, A., Muller, D., Lubberstedt, J., Bartsch, O., Jurdzinski, A., Fuhrmann, K., and Ivell, R. (2001). Characterization of relaxin binding in the uterus of the marmoset monkey. *Mol. Hum. Reprod.* 7, 963–970.
219. Mathieu, M. N., Wade, J. D., Catimel, B., Bond, C. P., Nice, E. C., Summers, R. J., Otvos, L. Jr., and Tregear, G. W. (2001). Synthesis, conformational studies and biological activity of N(alpha)-mono-biotinylated rat relaxin. *J. Pept. Res.* 57, 374–382.
220. Nice, E., Catimel, B., Rothacker, J., Bathgate, R. A., Mathieu, M. N., Claasz, A. A., Tregear, G. W., and Wade, J. D. (2001). The use of biosensor technology to search for orphan biomolecules: towards the relaxin receptor. In *Relaxin 2000: Proceedings of the Third International Conference on Relaxin and Related Peptides* (G. W. Tregear, R. Ivell, R. A. Bathgate, and J. D. Wade, Eds.), pp. 301–308. Kluwer, Dordrecht.
221. Büllesbach, E. E., and Schwabe, C. (1999). Specific, high affinity relaxin-like factor receptors. *J. Biol. Chem.* 274, 22354–22358.
222. Braddon, S. A. (1978). Stimulation of ornithine decarboxylase by relaxin. *Biochem. Biophys. Res. Commun.* 80, 75–80.
223. Sanborn, B. M., Kuo, H. S., Weisbrodt, N. W., and Sherwood, O. D. (1980). The interaction of relaxin with the rat uterus: I. Effect on cyclic nucleotide levels and spontaneous contractile activity. *Endocrinology* 106, 1210–1215.
224. Judson, D. G., Pay, S., and Bhoola, K. D. (1980). Modulation of cyclic AMP in isolated rat uterine tissue slices by porcine relaxin. *J. Endocrinol.* 87, 153–159.
225. Hsu, C. J., McCormack, S. M., and Sanborn, B. M. (1985). The effect of relaxin on cyclic adenosine 3′,5′-monophosphate concentrations in rat myometrial cells in culture. *Endocrinology* 116, 2029–2035.
226. Kramer, S. M., Gibson, U. E., Fendly, B. M., Mohler, M. A., Drolet, D. W., and Johnston, P. D. (1990). Increase in cyclic AMP levels by relaxin in newborn rhesus monkey uterus cell culture. *In Vitro Cell. Dev. Biol.* 26, 647–656.
227. Chen, G. A., Huang, J. R., and Tseng, L. (1988). The effect of relaxin on cyclic adenosine 3′,5′-monophosphate concentrations in human endometrial glandular epithelial cells. *Biol. Reprod.* 39, 519–525.
228. Fei, D. T., Gross, M. C., Lofgren, J. L., Mora-Worms, M., and Chen, A. B. (1990). Cyclic AMP response to recombinant human relaxin by cultured human endometrial cells: a specific and high throughput in vitro bioassay. *Biochem. Biophys. Res. Commun.* 170, 214–222.
229. Cronin, M. J., Malaska, T., and Bakhit, C. (1987). Human relaxin increases cyclic AMP levels in cultured anterior pituitary cells. *Biochem. Biophys. Res. Commun.* 148, 1246–1251.
230. Dodge, K. L., Carr, D. W., and Sanborn, B. M. (1999). Protein kinase A anchoring to the myometrial plasma membrane is required for cyclic adenosine 3′,5′-monophosphate regulation of phosphatidylinositide turnover. *Endocrinology* 140, 5165–5170.

231. Huang, J. R., Tseng, L., Bischof, P., and Janne, O. A. (1987). Regulation of prolactin production by progestin, estrogen, and relaxin in human endometrial stromal cells. *Endocrinology* 121, 2011–2017.
232. Tabanelli, S., Tang, B., and Gurpide, E. (1992). In vitro decidualization of human endometrial stromal cells. *J. Steroid Biochem. Mol. Biol.* 42, 337–344.
233. Bartsch, O., Bartlick, B., and Ivell, R. (2001). Relaxin signal transduction couples tyrosine phosphorylation to cAMP upregulation. In *Relaxin 2000: Proceedings of the Third International Conference on Relaxin and Related Peptides* (G. W. Tregear, R. Ivell, R. A. Bathgate, and J. D. Wade, Eds.), pp. 309–315. Kluwer, Dordrecht.
234. Zhang, Q., Liu, S. H., Erikson, M., Lewis, M., and Unemori, E. (2002). Relaxin activates the MAP kinase pathway in human endometrial stromal cells. *J. Cell. Biochem.* 85, 536–544.
235. Bartsch, O., Bartlick, B., and Ivell, R. (2004). Phosphodiesterase 4 inhibition synergizes with relaxin signaling to promote decidualization of human endometrial stromal cells. *J. Clin. Endocrinol. Metab.* 89, 324–334.
236. Nguyen, B. T., Yang, L., Sanborn, B. M., and Dessauer, C. W. (2003). Phosphoinositide 3-kinase activity is required for biphasic stimulation of cyclic adenosine 3′,5′-monophosphate by relaxin. *Mol. Endocrinol.* 17, 1075–1084.
237. Samuel, C. S., Unemori, E. N., Mookerjee, I., Bathgate, R. A., Layfield, S. L., Mak, J., Tregear, G. W., and Du, X. J. (2004). Relaxin modulates cardiac fibroblast proliferation, differentiation and collagen production and reverses cardiac fibrosis in vivo. *Endocrinology* 145, 4125–4133.
238. Palejwala, S., Stein, D. E., Weiss, G., Monia, B. P., Tortoriello, D., and Goldsmith, L. T. (2001). Relaxin positively regulates matrix metalloproteinase expression in human lower uterine segment fibroblasts using a tyrosine kinase signaling pathway. *Endocrinology* 142, 3405–3413.
239. Han, X., Habuchi, Y., and Giles, W. R. (1994). Relaxin increases heart rate by modulating calcium current in cardiac pacemaker cells. *Circ. Res.* 74, 537–541.
240. Toth, M., Taskinen, P., and Ruskoaho, H. (1996). Relaxin stimulates atrial natriuretic peptide secretion in perfused rat heart. *J. Endocrinol.* 150, 487–495.
241. Kompa, A. R., Samuel, C. S., and Summers, R. J. (2002). Inotropic responses to human gene 2 (B29) relaxin in a rat model of myocardial infarction (MI): effect of pertussis toxin. *Br. J. Pharmacol.* 137, 710–718.
242. Bani, D., Failli, P., Bello, M. G., Thiemermann, C., Bani Sacchi, T., Bigazzi, M., and Masini, E. (1998). Relaxin activates the L-arginine-nitric oxide pathway in vascular smooth muscle cells in culture. *Hypertension* 31, 1240–1247.
243. Bani, D., Baccari, M. C., Nistri, S., Calamai, F., Bigazzi, M., and Sacchi, T. B. (1999). Relaxin up-regulates the nitric oxide biosynthetic pathway in the mouse uterus: involvement in the inhibition of myometrial contractility. *Endocrinology* 140, 4434–4441.
244. Bani, D., Baccari, M. C., Quattrone, S., Nistri, S., Calamai, F., Bigazzi, M., and Bani Sacchi, T. (2002). Relaxin depresses small bowel motility through a nitric oxide-mediated mechanism: studies in mice. *Biol. Reprod.* 66, 778–784.
245. Danielson, L. A., Sherwood, O. D., and Conrad, K. P. (1999). Relaxin is a potent renal vasodilator in conscious rats. *J. Clin. Invest.* 103, 525–533.
246. Kawamura, K., Kumagai, J., Sudo, S., Chun, S. Y., Pisarska, M., Morita, H., Toppari, J., Fu, P., Wade, J. D., Bathgate, R. A., and Hsueh, A. J. (2004). Paracrine regulation of mammalian oocyte maturation and male germ cell survival. *Proc. Natl. Acad. Sci. U S A* 101, 7323–7328.
247. Liu, C., Chen, J., Sutton, S., Roland, B., Kuei, C., Farmer, N., Sillard, R., and Lovenberg, T. W. (2003). Identification of relaxin-3/INSL7 as a ligand for GPCR142. *J. Biol. Chem.* 278, 50765–50770.
248. Hsu, S. Y., and Hsueh, A. J. (2000). Discovering new hormones, receptors, and signaling mediators in the genomic era. *Mol. Endocrinol.* 14, 594–604.
249. Hsu, S. Y., Liang, S. G., and Hsueh, A. J. (1998). Characterization of two LGR genes homologous to gonadotropin and thyrotropin receptors with extracellular leucine-rich repeats and a G protein-coupled, seven-transmembrane region. *Mol. Endocrinol.* 12, 1830–1845.
250. Hsu, S. Y., Kudo, M., Chen, T., Nakabayashi, K., Bhalla, A., van der Spek, P. J., van Duin, M., and Hsueh, A. J. (2000). The three subfamilies of leucine-rich repeat-containing G protein-coupled receptors (LGR): identification of LGR6 and LGR7 and the signaling mechanism for LGR7. *Mol. Endocrinol.* 14, 1257–1271.
251. Nef, S., and Parada, L. F. (1999). Cryptorchidism in mice mutant for Insl3. *Nat. Genet.* 22, 295–299.
252. Zimmermann, S., Steding, G., Emmen, J. M., Brinkmann, A. O., Nayernia, K., Holstein, A. F., Engel, W., and Adham, I. M. (1999). Targeted disruption of the Insl3 gene causes bilateral cryptorchidism. *Mol. Endocrinol.* 13, 681–691.
253. Overbeek, P. A., Gorlov, I. P., Sutherland, R. W., Houston, J. B., Harrison, W. R., Boettger-Tong, H. L., Bishop, C. E., and Agoulnik, A. I. (2001). A transgenic insertion causing cryptorchidism in mice. *Genesis* 30, 26–35.
254. Osuga, Y., Kudo, M., Kaipia, A., Kobilka, B., and Hsueh, A. J. (1997). Derivation of functional antagonists using N-terminal extracellular domain of gonadotropin and thyrotropin receptors. *Mol. Endocrinol.* 11, 1659–1668.
255. Zhao, L., Roche, P. J., Gunnersen, J. M., Hammond, V. E., Tregear, G. W., Wintour, E. M., and Beck, F. (1999). Mice without a functional relaxin gene are unable to deliver milk to their pups. *Endocrinology* 140, 445–453.
256. Krajnc-Franken, M. A., van Disseldorp, A. J., Koenders, J. E., Mosselman, S., van Duin, M., and Gossen, J. A. (2004). Impaired nipple development and parturition in LGR7 knockout mice. *Mol. Cell. Biol.* 24, 687–696.
257. Guico-Lamm, M. L., and Sherwood, O. D. (1988). Monoclonal antibodies specific for rat relaxin: II. Passive immunization with monoclonal antibodies throughout the second half of pregnancy disrupts birth in intact rats. *Endocrinology* 123, 2479–2485.
258. Scott, D. S., Layfield, S., Riesewijk, A., Morita, H., Tregear, G., and Bathgate, R. (2004). The identification and characterisation of the mouse and rat relaxin receptors as the novel orthologs of human LGR7. *Clin. Exp. Pharmacol. Physiol.* 31, 828–832.
259. Gorlov, I. P., Kamat, A., Bogatcheva, N. V., Jones, E., Lamb, D. J., Truong, A., Bishop, C. E., McElreavey, K., and Agoulnik, A. I. (2002). Mutations of the GREAT gene cause cryptorchidism. *Hum. Mol. Genet.* 11, 2309–2318.
260. Kubota, Y., Temelcos, C., Bathgate, R. A., Smith, K. J., Scott, D., Zhao, C., and Hutson, J. M. (2002). The role of insulin 3, testosterone, mullerian inhibiting substance and relaxin in rat gubernacular growth. *Mol. Hum. Reprod.* 8, 900–905.
261. Bogatcheva, N. V., Truong, A., Feng, S., Engel, W., Adham, I. M., and Agoulnik, A. I. (2003). GREAT/LGR8 is the only receptor for insulin-like 3 peptide. *Mol. Endocrinol.* 17, 2639–2646.
262. Kamat, A. A., Feng, S., Bogatcheva, N. V., Truong, A., Bishop, C. E., and Agoulnik, A. I. (2004). Genetic targeting of relaxin and insulin-like factor 3 receptors in mice. *Endocrinology* 145, 4712–4720.
263. Evans, G., Wathes, D. C., King, G. J., Armstrong, D. T., and Porter, D. G. (1983). Changes in relaxin production by the

theca during the preovulatory period of the pig. *J. Reprod. Fertil.* 69, 677–683.
264. Bagnell, C. A., Frando, L. B., Downey, B. R., Tsang, B. K., and Ainsworth, L. (1987). Localization of relaxin in the pig follicle during preovulatory development. *Biol. Reprod.* 37, 235–240.
265. Bagnell, C. A., Ayau, E., Downey, B. R., Tsang, B. K., and Ainsworth, L. (1989). Localization of relaxin during formation of the porcine corpus luteum. *Biol. Reprod.* 40, 835–841.
266. Denning-Kendall, P. A., Guldenaar, S. E., and Wathes, D. C. (1989). Evidence for a switch in the site of relaxin production from small theca-derived cells to large luteal cells during early pregnancy in the pig. *J. Reprod. Fertil.* 85, 261–271.
267. Bagnell, C. A., Tsark, W., Tashima, L., Downey, B. R., Tsang, B. K., and Ainsworth, L. (1990). Relaxin gene expression in the porcine follicle during preovulatory development induced by gonadotrophins. *J. Mol. Endocrinol.* 5, 211–219.
268. Ohleth, K. M., and Bagnell, C. A. (1999). Relaxin secretion and gene expression in porcine granulosa and theca cells are stimulated during in vitro luteinization. *Biol. Reprod.* 60, 499–507.
269. Hisaw, F. L., Zarrow, M. X. (1948). Relaxin in the ovary of the domestic sow. *Proc. Soc. Exp. Biol. Med.* 69, 395–398.
270. Sherwood, O. D., and Rutherford, J. E. (1981). Relaxin immunoactivity levels in ovarian extracts obtained from rats during various reproductive states and from adult cycling pigs. *Endocrinology* 108, 1171–1177.
271. Ali, S. M., McMurtry, J. P., Bagnell, C. A., and Bryant-Greenwood, G. D. (1986). Immunocytochemical localization of relaxin in corpora lutea of sows throughout the estrous cycle. *Biol. Reprod.* 34, 139–143.
272. Bagnell, C. A., Tashima, L., Tsark, W., Ali, S. M., and McMurtry, J. P. (1990). Relaxin gene expression in the sow corpus luteum during the cycle, pregnancy, and lactation. *Endocrinology* 126, 2514–2520.
273. Anderson, L. L., Ford, J. J., Melampy, R. M., and Cox, D. F. (1973). Relaxin in porcine corpora lutea during pregnancy and after hysterectomy. *Am. J. Physiol.* 225, 1215–1219.
274. Kendall, J. Z., Richards, G. E., and Shih, L. N. (1983). Effect of haloperidol, suckling, oxytocin and hand milking on plasma relaxin and prolactin concentrations in cyclic and lactating pigs. *J. Reprod. Fertil.* 69, 271–277.
275. Arakaki, R. F., Kleinfeld, R. G., and Bryant-Greenwood, G. D. (1980). Immunofluorescence studies using antisera to crude and to purified porcine relaxin. *Biol. Reprod.* 23, 153–159.
276. Fields, P. A., and Fields, M. J. (1985). Ultrastructural localization of relaxin in the corpus luteum of the nonpregnant, pseudopregnant, and pregnant pig. *Biol. Reprod.* 32, 1169–1179.
277. Larkin, L. H., Fields, P. A., and Oliver, R. M. (1977). Production of antisera against electrophoretically separated relaxin and immunofluorescent localization of relaxin in the porcine corpus luteum. *Endocrinology* 101, 679–685.
278. Belt, W. D., Anderson, L. L., Cavazos, L. F., and Melampy, R. M. (1971). Cytoplasmic granules and relaxin levels in porcine corpora lutea. *Endocrinology* 89, 1–10.
279. Kendall, J. Z., Plopper, C. G., and Bryant-Greenwood, G. D. (1978). Ultrastructural immunoperoxidase demonstration of relaxin in corpora lutea from a pregnant sow. *Biol. Reprod.* 18, 94–98.
280. Belt, W. D., Cavazos, L. F., Anderson, L. L., and Kraeling, R. R. (1970). Fine structure and progesterone levels in the corpus luteum of the pig during pregnancy and after hysterectomy. *Biol. Reprod.* 2, 98–113.
281. Anderson, L. L., Adair, V., Stromer, M. H., and McDonald, W. G. (1983). Relaxin production and release after hysterectomy in the pig. *Endocrinology* 113, 677–686.
282. Felder, K. J., Klindt, J., Bolt, D. J., and Anderson, L. L. (1988). Relaxin and progesterone secretion as affected by luteinizing hormone and prolactin after hysterectomy in the pig. *Endocrinology* 122, 1751–1760.
283. Felder, K. J., Molina, J. R., Benoit, A. M., and Anderson, L. L. (1986). Precise timing for peak relaxin and decreased progesterone secretion after hysterectomy in the pig. *Endocrinology* 119, 1502–1509.
284. Gooneratne, A. D., Bryant-Greenwood, G., Walker, F. M., Nottage, H. M., and Hartmann, P. E. (1983). Pre-partum changes in the plasma concentrations of progesterone, relaxin, prostaglandin F-2 alpha and 13,14-dihydro-15-keto prostaglandin F-2 alpha in meclofenamic acid-treated sows. *J. Reprod. Fertil.* 68, 33–40.
285. Li, Y. F., Huang, C. J., Klindt, J., and Anderson, L. L. (1991). Divergent effects of antiprogesterone, RU 486, on progesterone, relaxin, and prolactin secretion in pregnant and hysterectomized pigs with aging corpora lutea. *Endocrinology* 129, 2907–2914.
286. Sherwood, O. D., Chang, C. C., Bevier, G. W., and Dziuk, P. J. (1975). Radioimmunoassay of plasma relaxin levels throughout pregnancy and at parturition in the pig. *Endocrinology* 97, 834–837.
287. Sherwood, O. D., Nara, B. S., Welk, F. A., First, N. L., and Rutherford, J. E. (1981). Relaxin levels in the maternal plasma of pigs before, during, and after parturition and before, during, and after suckling. *Biol. Reprod.* 25, 65–71.
288. Taverne, M., Bevers, M., Bradshaw, J. M., Dieleman, S. J., Willemse, A. H., and Porter, D. G. (1982). Plasma concentrations of prolactin, progesterone, relaxin and oestradiol-17 beta in sows treated with progesterone, bromocriptine or indomethacin during late pregnancy. *J. Reprod. Fertil.* 65, 85–96.
289. Thomford, P. J., Sander, H. K., Kendall, J. Z., Sherwood, O. D., and Dziuk, P. J. (1984). Maintenance of pregnancy and levels of progesterone and relaxin in the serum of gilts following a stepwise reduction in the number of corpora lutea. *Biol. Reprod.* 31, 494–498.
290. Wathes, D. C., King, G. J., Porter, D. G., and Wathes, C. M. (1989). Relationship between pre-partum relaxin concentrations and farrowing intervals in the pig. *J. Reprod. Fertil.* 87, 383–390.
291. Watts, A. D., Flint, A. P., Foxcroft, G. R., and Porter, D. G. (1988). Plasma steroid, relaxin and dihydro-keto-prostaglandin F-2 alpha changes in the minipig in relation to myometrial electrical and mechanical activity in the pre-partum period. *J. Reprod. Fertil.* 83, 553–564.
292. Whitely, J. L., Hartmann, P. E., Willcox, D. L., Bryant-Greenwood, G. D., and Greenwood, F. C. (1990). Initiation of parturition and lactation in the sow: effects of delaying parturition with medroxyprogesterone acetate. *J. Endocrinol.* 124, 475–484.
293. Sherwood, O. D., Chang, C. C., BeVier, G. W., Diehl, J. R., and Dziuk, P. J. (1976). Relaxin concentrations in pig plasma following the administration of prostaglandin F2alpha during late pregnancy. *Endocrinology* 98, 875–879.
294. Sherwood, O. D., Nara, B. S., Crnekovic, V. E., and First, N. L. (1979). Relaxin concentrations in pig plasma after the administration of indomethacin and prostaglandin F2 alpha during late pregnancy. *Endocrinology* 104, 1716–1721.
295. Sherwood, O. D., Wilson, M. E., Edgerton, L. A., and Chang, C. C. (1978). Serum relaxin concentrations in pigs with parturition delayed by progesterone administration. *Endocrinology* 102, 471–475.
296. Knox, R. V., Zhang, Z., Day, B. N., and Anthony, R. V. (1994). Identification of relaxin gene expression and protein

localization in the uterine endometrium during early pregnancy in the pig. *Endocrinology* 135, 2517–2525.

297. Porter, D. G., Ryan, P. L., and Norman, L. (1992). Lack of effect of relaxin on oxytocin output from the porcine neural lobe in vitro or in lactating sows in vivo. *J. Reprod. Fertil.* 96, 251–260.
298. Crish, J. F., Soloff, M. S., and Shaw, A. R. (1986). Changes in relaxin precursor messenger ribonucleic acid levels in ovaries of rats after hysterectomy and removal of conceptuses, and during the estrous cycle. *Endocrinology* 119, 1222–1228.
299. Gunnersen, J. M., Crawford, R. J., and Tregear, G. W. (1995). Expression of the relaxin gene in rat tissues. *Mol. Cell. Endocrinol.* 110, 55–64.
300. Anderson, L. L., Bast, J. D., and Melampy, R. M. (1973). Relaxin in ovarian tissue during different reproductive stages in the rat. *J. Endocrinol.* 59, 371–372.
301. Sherwood, O. D., Crnekovic, V. E., Gordon, W. L., and Rutherford, J. E. (1980). Radioimmunoassay of relaxin throughout pregnancy and during parturition in the rat. *Endocrinology* 107, 691–698.
302. Anderson, M. L., and Long, J. A. (1978). Localization of relaxin in the pregnant rat. Bioassay of tissue extracts and cell fractionation studies. *Biol. Reprod.* 18, 110–117.
303. Anderson, M. L., Long, J. A., and Hayashida, T. (1975). Immunofluorescence studies on the localization of relaxin in the corpus luteum of the pregnant rat. *Biol. Reprod.* 13, 499–504.
304. Anderson, M. B., and Sherwood, O. D. (1984). Ultrastructural localization of relaxin immunoreactivity in corpora lutea of pregnant rats. *Endocrinology* 114, 1124–1127.
305. Fields, P. A. (1984). Intracellular localization of relaxin in membrane-bound granules in the pregnant rat luteal cell. *Biol. Reprod.* 30, 753–762.
306. Fields, P. A., Lee, A. B., Haab, L. M., Hwang, J. J., and Sherwood, O. D. (1992). Evidence for a dual source of relaxin in the pregnant rat: immunolocalization in the corpora lutea and endometrium. *Endocrinology* 130, 2985–2990.
307. Golos, T. G., Weyhenmeyer, J. A., and Sherwood, O. D. (1984). Immunocytochemical localization of relaxin in the ovaries of pregnant rats. *Biol. Reprod.* 30, 257–261.
308. Hilliard, J. (1973). Corpus luteum function in guinea pigs, hamsters, rats, mice and rabbits. *Biol. Reprod.* 8, 203–221.
309. Uchida, K., Kadowaki, M., Nomura, Y., Miyata, K., and Miyake, T. (1970). Relationship between ovarian progestin secretion and corpora lutea function in pregnant rats. *Endocrinol. Jpn.* 17, 499–507.
310. Soloff, M. S., Shaw, A. R., Gentry, L. E., Marquardt, H., and Vasilenko, P. (1992). Demonstration of relaxin precursors in pregnant rat ovaries with antisera against bacterially expressed rat prorelaxin. *Endocrinology* 130, 1844–1851.
311. Long, J. A. (1973). Corpus luteum of pregnancy in the rat: ultrastructural and cytochemical observations. *Biol. Reprod.* 8, 87–99.
312. Golos, T. G., and Sherwood, O. D. (1982). Control of corpus luteum function during the second half of pregnancy in the rat: a direct relationship between conceptus number and both serum and ovarian relaxin levels. *Endocrinology* 111, 872–878.
313. Gordon, W. L., and Sherwood, O. D. (1982). Evidence that luteinizing hormone from the maternal pituitary gland may promote antepartum release of relaxin, luteolysis, and birth in rats. *Endocrinology* 111, 1299–1310.
314. Pepe, G. J., and Rothchild, I. (1974). A comparative study of serum progesterone levels in pregnancy and in various types of pseudo-pregnancy in the rat. *Endocrinology* 95, 275–279.
315. Sherwood, O. D., Downing, S. J., Golos, T. G., Gordon, W. L., and Tarbell, M. K. (1983). Influence of light-dark cycle on antepartum serum relaxin and progesterone immunoactivity levels and on birth in the rat. *Endocrinology* 113, 997–1003.
316. Sherwood, O. D., Downing, S. J., Rieber, A. J., Fraley, S. W., Bohrer, R. E., Richardson, B. C., and Shanks, R. D. (1985). Influence of litter size on antepartum serum relaxin and progesterone immunoactivity levels and on birth in the rat. *Endocrinology* 116, 2554–2562.
317. Kubota, M., and Mizuhira, V. (1988). Fine structure and functions of the rat metrial gland: I. Effect of prostaglandin F2a on the granulated metrial gland cells. *Acta. Chem. Cytochem.* 21, 1–13.
318. Burazin, T. C. D., Davern, P. J., McKinley, M. J., and Tregear, G. W. (2001). Identification of relaxin and relaxin responsive cells in the rat brain. In *Relaxin 2000: Proceedings of the Third International Conference on Relaxin and Related Peptides* (G. W. Tregear, R. Ivell, R., R. A. Bathgate, and J. D. Wade, Eds.). Kluwer, Dordrecht.
319. Noci, I., Nardi, E., Scarselli, G., Petrucci, F., Chelo, E., La Malfa, A., Tantini, C., Bruni, V., and Bigazzi, M. (1983). Immuno- and bio-active relaxin in human ovarian follicles. In *Biology of Relaxin and Its Role in the Human* (M. Bigazzi, F. C. Greenwood, and F. Gasparri, Eds.), pp. 270–272. Excerpta Medica, Amsterdam.
320. Wathes, D. C., Wardle, P. G., Rees, J. M., Mitchell, J. D., McLaughlin, E. A., Hull, M. G., and Porter, D. G. (1986). Identification of relaxin immunoreactivity in human follicular fluid. *Hum. Reprod.* 1, 515–517.
321. Balboni, G. C., and Vannelli, G. B. (1983). Relaxin in ovarian follicles. *Bull. Assoc. Anat. (Nancy)* 67, 149–162.
322. Schmidt, C. L., Kendall, J. Z., Dandekar, P. V., Quigley, M. M., and Schmidt, K. L. (1984). Characterization of long-term monolayer cultures of human granulosa cells from follicles of different size and exposed in vivo to clomiphene citrate and hCG. *J. Reprod. Fertil.* 71, 279–287.
323. Gagliardi, C. L., Goldsmith, L. T., Saketos, M., Weiss, G., and Schmidt, C. L. (1992). Human chorionic gonadotropin stimulation of relaxin secretion by luteinized human granulosa cells. *Fertil. Steril.* 58, 314–320.
324. Blankenship, T., Stewart, D. R., Benirschke, K., King, B., and Lasley, B. L. (1994). Immunocytochemical localization of nonluteal ovarian relaxin. *J. Reprod. Med.* 39, 235–240.
325. O'Byrne, E. M., Flitcraft, J. F., Sawyer, W. K., Hochman, J., Weiss, G., and Steinetz, B. G. (1978). Relaxin bioactivity and immunoactivity in human corpora lutea. *Endocrinology* 102, 1641–1644.
326. Szlachter, N., O'Byrne, E., Goldsmith, L., Steinetz, B. G., and Weiss, G. (1980). Myometrial inhibiting activity of relaxin-containing extracts of human corpora lutea of pregnancy. *Am. J. Obstet. Gynecol.* 136, 584–586.
327. Loumaye, E., Teuwissen, B., and Thomas, K. (1978). Characterization of relaxin radioimmunoassay using Bolton-Hunter reagent: first results in plasma during pregnancy, and in placenta, corpora lutea and ovarian cysts in woman. *Gynecol. Obstet. Invest.* 9, 262–267.
328. Schmidt, C. L., Black, V. H., Sarosi, P., and Weiss, G. (1986). Progesterone and relaxin secretion in relation to the ultrastructure of human luteal cells in culture: effects of human chorionic gonadotropin. *Am. J. Obstet. Gynecol.* 155, 1209–1219.
329. Yki-Jarvinen, H., Wahlstrom, T., and Seppala, M. (1983). Immunohistochemical demonstration of relaxin in gynecologic tumors. *Cancer* 52, 2077–2080.
330. Yki-Jarvinen, H., Wahlstrom, T., and Seppala, M. (1983). Immunohistochemical demonstration of relaxin in the genital tract of pregnant and nonpregnant women. *J. Clin. Endocrinol. Metab.* 57, 451–454.
331. Yki-Jarvinen, H., Wahlstrom, T., Tenhunen, A., Koskimies, A. I., and Seppala, M. (1984). The occurrence of relaxin in

hyperstimulated human preovulatory follicles collected in an in vitro fertilization program. *J. In Vitro Fertil. Embryo Transf.* 1, 180–182.

332. Stoelk, E., Chegini, N., Lei, Z. M., Rao, C. V., Bryant-Greenwood, G., and Sanfilippo, J. (1991). Immunocytochemical localization of relaxin in human corpora lutea: cellular and subcellular distribution and dependence on reproductive state. *Biol. Reprod.* 44, 1140–1147.
333. Ivell, R., Hunt, N., Khan-Dawood, F., and Dawood, M. Y. (1989). Expression of the human relaxin gene in the corpus luteum of the menstrual cycle and in the prostate. *Mol. Cell. Endocrinol.* 66, 251–255.
334. Stewart, D. R., Celniker, A. C., Taylor, C. A. Jr., Cragun, J. R., Overstreet, J. W., and Lasley, B. L. (1990). Relaxin in the peri-implantation period. *J. Clin. Endocrinol. Metab.* 70, 1771–1773.
335. Eddie, L. W., Martinez, F., Healy, D. L., Sutton, B., Bell, R. J., and Tregear, G. W. (1990). Relaxin in sera during the luteal phase of in-vitro fertilization cycles. *Br. J. Obstet. Gynaecol.* 97, 215–220.
336. Chen, J., Qiu, Q., Lohstroh, P. N., Overstreet, J. W., and Lasley, B. L. (2003). Hormonal characteristics in the early luteal phase of conceptive and nonconceptive menstrual cycles. *J. Soc. Gynecol. Investig.* 10, 27–31.
337. Yki-Jarvinen, H., Wahlstrom, T., and Seppala, M. (1985). Human endometrium contains relaxin that is progesterone-dependent. *Acta. Obstet. Gynecol. Scand.* 64, 663–665.
338. Palejwala, S., Tseng, L., Wojtczuk, A., Weiss, G., and Goldsmith, L. T. (2002). Relaxin gene and protein expression and its regulation of procollagenase and vascular endothelial growth factor in human endometrial cells. *Biol. Reprod.* 66, 1743–1748.
339. Bond, C. P., Parry, L. J., Samuel, C. S., Gehring, H. M., Lederman, F. L., Rogers, P. A., and Summers, R. J. (2004). Increased expression of the relaxin receptor (LGR7) in human endometrium during the secretory phase of the menstrual cycle. *J. Clin. Endocrinol. Metab.* 89, 3477–3485.
340. MacLennan, A. H., Grant, P., and Borthwick, A. C. (1991). Relaxin and relaxin C-peptide levels in human reproductive tissues. *Reprod. Fertil. Dev.* 3, 577–583.
341. Weiss, G., O'Byrne, E. M., Hochman, J., Steinetz, B. G., Goldsmith, L., and Flitcraft, J. G. (1978). Distribution of relaxin in women during pregnancy. *Obstet. Gynecol.* 52, 569–570.
342. Khan-Dawood, F. S., Goldsmith, L. T., Weiss, G., and Dawood, M. Y. (1989). Human corpus luteum secretion of relaxin, oxytocin, and progesterone. *J. Clin. Endocrinol. Metab.* 68, 627–631.
343. Goldsmith, L. T., Essig, M., Sarosi, P., Beck, P., and Weiss, G. (1981). Hormone secretion by monolayer cultures of human luteal cells. *J. Clin. Endocrinol. Metab.* 53, 8980–8982.
344. Emmi, A. M., Skurnick, J., Goldsmith, L. T., Gagliardi, C. L., Schmidt, C. L., Kleinberg, D., and Weiss, G. (1991). Ovarian control of pituitary hormone secretion in early human pregnancy. *J. Clin. Endocrinol. Metab.* 72, 1359–1363.
345. Johnson, M. R., Abdalla, H., Allman, A. C., Wren, M. E., Kirkland, A., and Lightman, S. L. (1991). Relaxin levels in ovum donation pregnancies. *Fertil. Steril.* 56, 59–61.
346. Seki, K., and Kato, K. (1987). Evidence for the luteal source of circulating relaxin in molar pregnancy. *Acta. Obstet. Gynecol. Scand.* 66, 319–320.
347. Anderson, M. B., Vaupel, M. R., and Sherwood, O. D. (1984). Pregnant mouse corpora lutea: immunocytochemical localization of relaxin and ultrastructure. *Biol. Reprod.* 31, 391–397.
348. Abramson, D., Hurwitt, E., and Lesnick, G. (1937). Relaxin in human serum as a test of pregnancy. *Surg. Gynecol. Obstet.* 65, 335–339.
349. Eddie, L. W., Bell, R. J., Lester, A., Geier, M., Bennett, G., Johnston, P. D., and Niall, H. D. (1986). Radioimmunoassay of relaxin in pregnancy with an analogue of human relaxin. *Lancet* 1, 1344–1346.
350. Bell, R. J., Eddie, L. W., Lester, A. R., Wood, E. C., Johnston, P. D., and Niall, H. D. (1987). Relaxin in human pregnancy serum measured with an homologous radioimmunoassay. *Obstet. Gynecol.* 69, 585–589.
351. Bell, R. J., Sutton, B., Eddie, L. W., Healy, D. L., Johnston, P. D., and Tregear, G. W. (1989). Relaxin levels in antenatal patients following in vitro fertilization. *Fertil. Steril.* 52, 85–87.
352. Johnson, M. R., Abbas, A., Nicolaides, K. H., and Lightman, S. L. (1992). Distribution of relaxin between human maternal and fetal circulations and amniotic fluid. *J. Endocrinol.* 134, 313–317.
353. MacLennan, A. H., Nicolson, R., and Green, R. C. (1986). Serum relaxin in pregnancy. *Lancet* 2, 241–243.
354. Pommerenke, W. T. (1934). Experimental ligamentous relaxation in the guinea pig pelvis. *Am. J. Obstet. Gynecol.* 27, 708–713.
355. Quagliarello, J., Nachtigall, R., Goldsmith, L. T., Hochman, J., Steinetz, B. G., O'Bryen, E. M., and Weiss, G. (1979). Serum immunoreactive relaxin concentrations in human pregnancy, labor and the puerperium. *Adv. Exp. Med. Biol.* 112, 743–748.
356. Schmidt, C. L., Goldsmith, L. T., Carr, B. R., Weiss, G., Parker, C. R. Jr., and Illingworth, D. R. (1988). Peripheral relaxin levels during pregnancy in a woman with homozygous familial hypobetalipoproteinemia. *Fertil. Steril.* 50, 815–817.
357. Seki, K., Uesato, T., Tabei, T., and Kato, K. (1985). The secretory patterns of relaxin and human chorionic gonadotropin in human pregnancy. *Endocrinol. Jpn.* 32, 741–744.
358. Szlachter, B. N., Quagliarello, J., Jewelewicz, R., Osathanondh, R., Spellacy, W. N., and Weiss, G. (1982). Relaxin in normal and pathogenic pregnancies. *Obstet. Gynecol.* 59, 167–170.
359. Zarrow, M. X., Holmstrom, E. G., and Salhanick, H. A. (1955). The concentration of relaxin in the blood serum and other tissues of women during pregnancy. *J. Clin. Endocrinol. Metab.* 15, 22–27.
360. Quagliarello, J., Steinetz, B. G., and Weiss, G. (1979). Relaxin secretion in early pregnancy. *Obstet. Gynecol.* 53, 62–63.
361. Quagliarello, J., Lustig, D. S., Steinetz, B. G., and Weiss, G. (1980). Absence of a prelabor relaxin surge in women. *Biol. Reprod.* 22, 202–204.
362. Seki, K., Uesato, T., Kato, K., and Tabei, T. (1987). Decline in serum relaxin levels before labor in women. *Nippon Sanka Fujinka Gakkai Zasshi* 39, 2073–2074.
363. Stewart, D. R., Cragun, J. R., Boyers, S. P., Oi, R., Overstreet, J. W., and Lasley, B. L. (1992). Serum relaxin concentrations in patients with out-of-phase endometrial biopsies. *Fertil. Steril.* 57, 453–455.
364. Seki, K., Uesato, T., Tabei, T., and Kato, K. (1988). Serum relaxin and steroid hormones in spontaneous abortions. *Acta. Obstet. Gynecol. Scand.* 67, 483–486.
365. Garcia, A., Skurnick, J. H., Goldsmith, L. T., Emmi, A., and Weiss, G. (1990). Human chorionic gonadotropin and relaxin concentrations in early ectopic and normal pregnancies. *Obstet. Gynecol.* 75, 779–783.
366. Witt, B. R., Wolf, G. C., Wainwright, C. J., Johnston, P. D., and Thorneycroft, I. H. (1990). Relaxin, CA-125, progesterone, estradiol, Schwangerschaft protein, and human chorionic gonadotropin as predictors of outcome in threatened and nonthreatened pregnancies. *Fertil. Steril.* 53, 1029–1036.
367. Petersen, L. K., Skajaa, K., and Uldbjerg, N. (1992). Serum relaxin as a potential marker for preterm labour. *Br. J. Obstet. Gynaecol.* 99, 292–295.
368. Weiss, G., Goldsmith, L. T., Sachdev, R., Von Hagen, S., and Lederer, K. (1993). Elevated first-trimester serum relaxin

concentrations in pregnant women following ovarian stimulation predict prematurity risk and preterm delivery. *Obstet. Gynecol.* 82, 821–828.
369. Vogel, I., Salvig, J. D., Secher, N. J., and Uldbjerg, N. (2001). Association between raised serum relaxin levels during the eighteenth gestational week and very preterm delivery. *Am. J. Obstet. Gynecol.* 184, 390–393.
370. MacLennan, A. H., Nicolson, R., Green, R. C., and Bath, M. (1986). Serum relaxin and pelvic pain of pregnancy. *Lancet* 2, 243–245.
371. Kristiansson, P., Nilsson-Wikmar, L., von Schoultz, B., Svardsudd, K., and Wramsby, H. (2001). Back pain in IVF-induced and spontaneous pregnancies. In *Relaxin 2000: Proceedings of the Third International Conference on Relaxin and Related Peptides* (G. W. Tregear, R. Ivell, R. A. Bathgate, and J. D. Wade, Eds.), pp. 415–420. Kluwer, Dordrecht.
372. Steinetz, B. G., Whitaker, P. G., and Edwards, J. R. (1992). Maternal relaxin concentrations in diabetic pregnancy. *Lancet* 340, 752–755.
373. Whittaker, P. G., Edwards, J. R., Randolph, C., Büllesbach, E. E., Schwabe, C., and Steinetz, B. G. (2003). Abnormal relaxin secretion during pregnancy in women with type 1 diabetes. *Exp. Biol. Med.* 228, 33–40.
374. Bigazzi, M., Nardi, E., Bruni, P., and Petrucci, F. (1980). Relaxin in human decidua. *J. Clin. Endocrinol. Metab.* 51, 939–941.
375. Bryant-Greenwood, G. D., Rees, M. C., and Turnbull, A. C. (1987). Immunohistochemical localization of relaxin, prolactin and prostaglandin synthase in human amnion, chorion and decidua. *J. Endocrinol.* 114, 491–496.
376. Koay, E. S., Bagnell, C. A., Bryant-Greenwood, G. D., Lord, S. B., Cruz, A. C., and Larkin, L. H. (1985). Immunocytochemical localization of relaxin in human decidua and placenta. *J. Clin. Endocrinol. Metab.* 60, 859–863.
377. Sakbun, V., Ali, S. M., Greenwood, F. C., and Bryant-Greenwood, G. D. (1990). Human relaxin in the amnion, chorion, decidua parietalis, basal plate, and placental trophoblast by immunocytochemistry and northern analysis. *J. Clin. Endocrinol. Metab.* 70, 508–514.
378. Yki-Jarvinen, H., and Wahlstrom, T. (1984). Immunohistochemical demonstration of relaxin in the placenta after removal of the corpus luteum. *Acta. Endocrinol. (Copenh)* 106, 544–547.
379. Bryant-Greenwood, G. D. (1991). The human relaxins: consensus and dissent. *Mol. Cell. Endocrinol.* 79, C125–C132.
380. Bryant-Greenwood, G. D. (1991). Human decidual and placental relaxins. *Reprod. Fertil. Dev.* 3, 385–389.
381. Sakbun, V., Koay, E. S., and Bryant-Greenwood, G. D. (1987). Immunocytochemical localization of prolactin and relaxin C-peptide in human decidua and placenta. *J. Clin. Endocrinol. Metab.* 65, 339–343.
382. Gunnersen, J. M., Roche, P. J., Tregear, G. W., and Crawford, R. J. (1995). Characterization of human relaxin gene regulation in the relaxin-expressing human prostate adenocarcinoma cell line LNCaP.FGC. *J. Mol. Endocrinol.* 15, 153–166.
383. Nardi, E., Bigazzi, M., Agrimonti, F., Dogliotti, L., Massi, G. B., Ferrari, A. M., Ciardetti, P., De Luca, V., and Angeli, A. (1983). Relaxin and fibrocystic disease of the mammary gland. In *Biology of Relaxin and Its Role in the Human* (M. Bigazzi, F. C. Greenwood, and F. Gasparri, Eds.), pp. 417–419. Excerpta Medica, Amsterdam.
384. Eddie, L. W., Sutton, B., Fitzgerald, S., Bell, R. J., Johnston, P. D., and Tregear, G. W. (1989). Relaxin in paired samples of serum and milk from women after term and preterm delivery. *Am. J. Obstet. Gynecol.* 161, 970–973.
385. Bongers-Binder, S., Burgardt, A., Seeger, H., Voelter, W., and Lippert, T. H. (1991). Distribution of immunoreactive relaxin in the genital tract and in the mammary gland of non-pregnant women. *Clin. Exp. Obstet. Gynecol.* 18, 161–164.
386. Mazoujian, G., and Bryant-Greenwood, G. D. (1990). Relaxin in breast tissue. *Lancet* 335, 298–299.
387. Tashima, L. S., Mazoujian, G., and Bryant-Greenwood, G. D. (1994). Human relaxins in normal, benign and neoplastic breast tissue. *J. Mol. Endocrinol.* 12, 351–364.
388. Fuchs, U., Seeger, H., Volter, W., and Lippert, T. H. (1989). The hormone profile in serum and cervix secretions in the course of a menstrual cycle. *Geburtshilfe Frauenheilkd.* 49 Suppl 1, 125–126.
389. Fuchs, U., Seeger, H., Voelter, W., and Lippert, T. H. (1988). Immunoreactive relaxin in human cervico-vaginal secretion. *Arch. Gynecol. Obstet.* 243, 37–39.
390. Dschietzig, T., Richter, C., Bartsch, C., Laule, M., Armbruster, F. P., Baumann, G., and Stangl, K. (2001). The pregnancy hormone relaxin is a player in human heart failure. *FASEB J.* 15, 2187–2195.
391. Stewart, D. R., Stouffer, R., Overstreet, J. W., Hendrickx, A., and Lasley, B. L. (1993). Measurement of periimplantational relaxin concentrations in the macaque using a homologous assay. *Endocrinology* 132, 6–12.
392. Steinetz, B. G., Randolph, C., and Mahoney, C. J. (1992). Serum concentrations of relaxin, chorionic gonadotropin, estradiol-17 beta, and progesterone during the reproductive cycle of the chimpanzee (*Pan troglodytes*). *Endocrinology* 130, 3601–3607.
393. Einspanier, A., Nubbemeyer, R., Schlote, S., Schumacher, M., Ivell, R., Fuhrmann, K., and Marten, A. (1999). Relaxin in the marmoset monkey: secretion pattern in the ovarian cycle and early pregnancy. *Biol. Reprod.* 61, 512–520.
394. Nixon, W. E., Reid, R., Abou-Hozaifa, B. M., Williams, R. F., Steinetz, B. G., and Hodgen, G. D. (1983). Origin and regulation of relaxin secretion in monkeys: effects of chorionic gonadotropin, luteectomy, fetectomy, and placentectomy. In *Factors Regulating Ovarian Function* (G. S. Greenwald and P. F. Terranova, Eds.), pp. 427–431. Raven Press, New York.
395. Ottobre, J. S., Nixon, W. E., and Stouffer, R. L. (1984). Induction of relaxin secretion in rhesus monkeys by human chorionic gonadotropin: dependence on the age of the corpus luteum of the menstrual cycle. *Biol. Reprod.* 31, 1000–1006.
396. Weiss, G., Steinetz, B. G., Dierschke, D. J., and Fritz, G. (1981). Relaxin secretion in the rhesus monkey. *Biol. Reprod.* 24, 565–567.
397. Castracane, V. D., Lessing, J., Brenner, S., and Weiss, G. (1985). Relaxin in the pregnant baboon: evidence for local production in reproductive tissues. *J. Clin. Endocrinol. Metab.* 60, 133–136.
398. Steinetz, B. G., Randolph, C., and Mahoney, C. J. (1995). Patterns of relaxin and steroids in the reproductive cycle of the common marmoset (*Callithrix jacchus*): effects of prostaglandin F2 alpha on relaxin and progesterone secretion during pregnancy. *Biol. Reprod.* 53, 834–839.
399. Steinetz, B. G., Beach, V. L., and Kroc, R. L. (1959). The physiology of relaxin in laboratory animals. In *Recent Progress in the Endocrinology of Reproduction* (C. W. Lloyd, Ed.), pp. 389–423. Academic Press, New York.
400. Vaupel, M. R., Sherwood, O. D., and Anderson, M. B. (1985). Immunocytochemical studies of relaxin in ovaries of pregnant and cycling mice. *J. Histochem. Cytochem.* 33, 303–308.
401. Psalti, I., Rahier, J., Loumaye, E., Haumont, S., and Thomas, K. (1990). Changes of relaxin concentrations determined by immunodensitometry in ovaries of NMRI mice during pregnancy. *Gynecol. Obstet. Invest.* 30, 133–138.

402. O'Byrne, E. M., and Steinetz, B. G. (1976). Radioimmunoassay (RIA) of relaxin in sera of various species using an antiserum to porcine relaxin. *Proc. Soc. Exp. Biol. Med.* 152, 272–276.
403. Stewart, D. R., Stabenfeldt, G. H., Hughes, J. P., and Meagher, D. M. (1982). Determination of the source of equine relaxin. *Biol. Reprod.* 27, 17–24.
404. Stewart, D. R. (1986). Development of a homologous equine relaxin radioimmunoassay. *Endocrinology* 119, 1100–1104.
405. Madej, A., Kindahl, H., Nydahl, C., Edqvist, L. E., and Stewart, D. R. (1987). Hormonal changes associated with induced late abortions in the mare. *J. Reprod. Fertil. Suppl.* 35, 479–484.
406. Stewart, D. R., Addiego, L. A., Pascoe, D. R., Haluska, G. J., and Pashen, R. (1992). Breed differences in circulating equine relaxin. *Biol. Reprod.* 46, 648–652.
407. Ryan, P. L. Bennett-Wimbush, K., Vaala, W. E., and Bagnell, C. A. (2001). Systemic relaxin in pregnant pony mares grazed on endophyte-infected fescue: effects of fluphenazine treatment. *Theriogenology* 56, 471–483.
408. Ryan, P. L., Klonisch, T., Yamashiro, S., Renaud, R. L., Wasnidge, C., and Porter, D. G. (1997). Expression and localization of relaxin in the ovary of the mare. *J. Reprod. Fertil.* 110, 329–338.
409. Anderson, L. L., Perezgrovas, R., O'Byrne, E. M., and Steinetz, B. G. (1982). Biological actions of relaxin in pigs and beef cattle. *Ann. N Y Acad. Sci.* 380, 131–150.
410. Fields, M. J., Fields, P. A., and Larkin, L. H. (1982). Chemistry of bovine relaxin. *Adv. Exp. Med. Biol.* 143, 191–207.
411. Fields, M. J., Roberts, R., and Fields, P. A. (1982). Octadecylsilica and carboxymethyl cellulose isolation of bovine and porcine relaxin. *Ann. N Y Acad. Sci.* 380, 36–46.
412. Musah, A. I., Schwabe, C., and Anderson, L. L. (1987). Acute decrease in progesterone and increase in estrogen secretion caused by relaxin during late pregnancy in beef heifers. *Endocrinology* 120, 317–324.
413. Dubois, M. P., and Dacheux, J. L. (1978). Relaxin, a male hormone? Immunocytological localization of a related antigen in the boar testis. *Cell. Tissue Res.* 187, 201–214.
414. Renegar, R. H., and Larkin, L. H. (1985). Relaxin concentrations in endometrial, placental, and ovarian tissues and in sera from ewes during middle and late pregnancy. *Biol. Reprod.* 32, 840–847.
415. Bryant-Greenwood, G. D., and Greenwood, F. C. (1979). Specificity of radioimmunoassays for relaxin. *J. Endocrinol.* 81, 239–247.
416. Roche, P. J., Butkus, A., Wintour, E. M., and Tregear, G. (1996). Structure and expression of Leydig insulin-like peptide mRNA in the sheep. *Mol. Cell. Endocrinol.* 121, 171–177.
417. Bathgate, R. A. D., Moniac, N., Bartlick, B., Claasz, A., Dawson, N., Tan, Y. Y., Wade, J. D., Tregear, G. W., and Ivell, R. (2001). The relaxin-like factor (insulin 3) is highly expressed in the ruminant ovary: a putative ruminant relaxin? In *Relaxin 2000: Proceedings of the Third International Conference on Relaxin and Related Peptides* (G. W. Tregear, R. Ivell, R. A. Bathgate, and J. D. Wade, Eds.), pp. 349–356. Kluwer, Dordrecht.
418. Eldridge, R. K., and Fields, P. A. (1985). Rabbit placental relaxin: purification and immunohistochemical localization. *Endocrinology* 117, 2512–2519.
419. Fields, P. A., and Lee, V. H. (1991). Conceptus-mediated integrity of endometrial epithelial cells and maintenance of relaxin synthesis in pregnant rabbits: effects of unilateral oviduct ligation. *Biol. Reprod.* 44, 364–374.
420. Fields, P., Kondo, S., Tashima, L., Bryant-Greenwood, G., and Greenwood, F. (1995). Expression of SQ10 (a preprorelaxin-like gene) in the pregnant rabbit placenta and uterus. *Biol. Reprod.* 53, 1139–1145.
421. Hisaw, F. L., Zarrow, M. X., Money, W. L., Talmage, R.V. N., and Abramowitz, A. A. (1944). Importance of the female reproductive tract in the formation of relaxin. *Endocrinology* 34, 122–134.
422. Lee, V. H., and Fields, P. A. (1990). Rabbit endometrial relaxin: immunohistochemical localization during preimplantation, pregnancy, and lactation. *Biol. Reprod.* 42, 737–745.
423. Lee, V. H., and Fields, P. A. (1991). Rabbit relaxin: the influence of pregnancy and ovariectomy during pregnancy on the plasma profile. *Biol. Reprod.* 45, 209–214.
424. Steinetz, B. G., O'Byrne, E. M., Goldsmith, L. T., and Anderson, M. B. (1988). The source of relaxin in pregnant Syrian hamsters. *Endocrinology* 122, 795–798.
425. Johns, T. C., and Renegar, R. H. (1990). Ultrastructural morphology and relaxin immunolocalization in giant trophoblast cells of the golden hamster placenta. *Am. J. Anat.* 189, 167–178.
426. Renegar, R. H., Cobb, A. D., and Leavitt, W. W. (1987). Immunocytochemical localization of relaxin in the golden hamster (*Mesocricetus auratus*) during the last half of gestation. *Biol. Reprod.* 37, 925–934.
427. Renegar, R. H., Southard, J. N., and Talamantes, F. (1990). Immunohistochemical co-localization of placental lactogen II and relaxin in the golden hamster (*Mesocricetus auratus*). *J. Histochem. Cytochem.* 38, 935–940.
428. Stewart, D. R., and Stabenfeldt, G. H. (1985). Relaxin activity in the pregnant cat. *Biol. Reprod.* 32, 848–854.
429. Addiego, L. A., Tsutsui, T., Stewart, D. R., and Stabenfeldt, G. H. (1987). Determination of the source of immunoreactive relaxin in the cat. *Biol. Reprod.* 37, 1165–1169.
430. Steinetz, B. G., Goldsmith, L. T., Hasan, S. H., and Lust, G. (1990). Diurnal variation of serum progesterone, but not relaxin, prolactin, or estradiol-17 beta in the pregnant bitch. *Endocrinology* 127, 1057–1063.
431. Tsutsui, T., and Stewart, D. R. (1991). Determination of the source of relaxin immunoreactivity during pregnancy in the dog. *J. Vet. Med. Sci.* 53, 1025–1029.
432. Steinetz, B. G., Goldsmith, L. T., and Lust, G. (1987). Plasma relaxin levels in pregnant and lactating dogs. *Biol. Reprod.* 37, 719–725.
433. Steinetz, B. G., Goldsmith, L. T., Harvey, H. J., and Lust, G. (1989). Serum relaxin and progesterone concentrations in pregnant, pseudopregnant, and ovariectomized, progestin-treated pregnant bitches: detection of relaxin as a marker of pregnancy. *Am. J. Vet. Res.* 50, 68–71.
434. Hisaw, F. L., Talmage, R. V. N., Money, W. L., and Abramowitz, A. A. (1942). Relation of progesterone to the formation of relaxin. *Anat. Rec.* 84, 457.
435. Zarrow, M. X. (1948). The role of steroid hormones in the relaxation of the symphysis pubis of the guinea pig. *Endocrinology* 42, 129–140.
436. Larkin, L. H., Ogilvie, S., Wubbel, L., and Welch, D. E. (1987). Effects of estradiol and progesterone on accumulation of relaxin- and carbohydrate-containing granules in endometrial gland cells of the guinea pig. *Am. J. Anat.* 179, 333–341.
437. Larkin, L. H., Welch, D. E., Ogilvie, S., and Wubbel, L. (1987). Cytochemical detection of carbohydrate and immunocytochemical detection of relaxin in the same secretory granule. *J. Histochem. Cytochem.* 35, 693–697.
438. Pardo, R., Larkin, L. H., and Fields, P. A. (1980). Immunocytochemical localization of relaxin in endometrial glands of the pregnant guinea pig. *Endocrinology* 107, 2110–2112.
439. Pardo, R. J., and Larkin, L. H. (1982). Localization of relaxin in endometrial gland cells of pregnant, lactating, and ovariectomized hormone-treated guinea pigs. *Am. J. Anat.* 164, 79–90.

440. Bryant-Greenwood, G. D., Tashima, L., Greenwood, F. C., Taylor, E., and Peaker, M. (1991). Endometrial relaxin: effects of mastectomy in the cyclic and pregnant guinea pig. *Endocrinology* 129, 2119–2125.
441. Larkin, L. H., and Renegar, R. H. (1986). Immunochemical and cytochemical studies of relaxin-containing cells in the guinea pig uterus. *Am. J. Anat.* 176, 353–365.
442. Pardo, R. J., Larkin, L. H., and Renegar, R. H. (1984). Immunoelectron microscopic localization of relaxin in endometrial gland cells of the pregnant guinea pig. *Anat. Rec.* 209, 373–379.
443. Tashima, L., Greenwood, F. C., Bryant-Greenwood, G. D., and Peaker, M. (1988). Identification of mRNA for relaxin in the endometrium of the pregnant guinea-pig. *J. Endocrinol.* 118, R9–R11.
444. Boyd, S., Kendall, J. Z., Mento, N., and Bryant-Greenwood, G. D. (1981). Relaxin immunoactivity in plasma during the reproductive cycle of the female guinea pig. *Biol. Reprod.* 24, 405–414.
445. Peaker, M., Taylor, E., Tashima, L., Redman, T. L., Greenwood, F. C., and Bryant-Greenwood, G. D. (1989). Relaxin detected by immunocytochemistry and northern analysis in the mammary gland of the guinea pig. *Endocrinology* 125, 693–698.
446. Steinetz, B. G., Randolph, C., Weldele, M., Frank, L. G., Licht, P., and Glickman, S. E. (1997). Pattern and source of secretion of relaxin in the reproductive cycle of the spotted hyena (*Crocuta crocuta*). *Biol. Reprod.* 56, 1301–1306.
447. Exner, C., Wehrend, A., Hospes, R., Einspanier, A., Hoffmann, B., and Heldmaier, G. (2003). Hormonal and behavioural changes during the mating season and pregnancy in Alpine marmots (*Marmota marmota*). *Reproduction* 126, 775–782.
448. Niemuller, C. A., Gray, C., Cummings, E., and Liptrap, R. M. (1998). Plasma concentrations of immunoreactive relaxin activity and progesterone in the pregnant Asian elephant (*Elephas maximus*). *Anim. Reprod. Sci.* 53, 119–131.
449. O'Connor, W. B., Cain, G. D., and Zarrow, M. X. (1966). Elongation of the interpubic ligament in the little brown bat (*Myotis lucifugus*). *Proc. Soc. Exp. Biol. Med.* 123, 935–937.
450. Steinetz, B. G., O'Byrne, E. M., Butler, M. C., and Hickman, L. B. (1983). Hormonal regulation of the connective tissue of the symphysis pubis. In *Biology of Relaxin and Its Role in the Human* (M. Bigazzi, F. C. Greenwood, and F. Gasparri, Eds.), pp. 71–92. Excerpta Medica, Amsterdam.
451. Crelin, E. S., and Brightman, M. W. (1957). The pelvis of the rat: its response to estrogen and relaxin. *Anat. Rec.* 128, 467–483.
452. Bassett, E. G., and Phillips, D. S. (1954). Pelvic relaxation in sheep. *Nature* 174, 1020–1021.
453. Samuel, C. S., Coghlan, J. P., and Bateman, J. F. (1998). Effects of relaxin, pregnancy and parturition on collagen metabolism in the rat pubic symphysis. *J. Endocrinol.* 159, 117–125.
454. Zhao, L., Samuel, C. S., Tregear, G. W., Beck, F., and Wintour, E. M. (2000). Collagen studies in late pregnant relaxin null mice. *Biol. Reprod.* 63, 697–703.
455. Crelin, E. S. (1954). Prevention of innominate bone separation during pregnancy in the mouse. *Proc. Soc. Exp. Biol. Med.* 86, 22–24.
456. MacLennan, A. H. (2000). Pelvic girdle relaxation, developmental dysplasia of the hip and the hormone relaxin—are they loosely connected? In *Relaxin 2000: Proceedings of the Third International Conference on Relaxin and Related Peptides* (G. W. Tregear, R. Ivell, R. A. Bathgate, and J. D. Wade, Eds.), pp. 407–413. Kluwer, Dordrecht.
457. Harkness, M. L., and Harkness, R. D. (1959). Changes in the physical properties of the uterine cervix of the rat during pregnancy. *J. Physiol. (Lond)* 148, 524–547.
458. Hollingsworth, M., Gallimore, S., and Isherwood, C. N. (1980). Effects of prostaglandins F-2 alpha and E-2 on cervical extensibility in the late pregnant rat. *J. Reprod. Fertil.* 58, 95–99.
459. Parry, D. M., and Ellwood, D. A. (1981). Ultrastructural aspects of cervical softening in the sheep. In *The Cervix in Pregnancy and Labour* (D. A. Ellwood and A. B. M. Anderson, Eds.), pp. 74–84. Churchill Livingstone, London.
460. Ellwood, D. A., and Anderson, A. B. M. (1981). *The Cervix in Pregnancy and Labour*. Churchill Livingstone, London.
461. Burger, L. L., and Sherwood, O. D. (1995). Evidence that cellular proliferation contributes to relaxin-induced growth of both the vagina and the cervix in the pregnant rat. *Endocrinology* 136, 4820–4826.
462. Burger, L. L., and Sherwood, O. D. (1998). Relaxin increases the accumulation of new epithelial and stromal cells in the rat cervix during the second half of pregnancy. *Endocrinology* 139, 3984–3995.
463. Hwang, J. J., and Sherwood, O. D. (1988). Monoclonal antibodies specific for rat relaxin: III. Passive immunization with monoclonal antibodies throughout the second half of pregnancy reduces cervical growth and extensibility in intact rats. *Endocrinology* 123, 2486–2490.
464. Hwang, J. J., Shanks, R. D., and Sherwood, O. D. (1989). Monoclonal antibodies specific for rat relaxin: IV. Passive immunization with monoclonal antibodies during the antepartum period reduces cervical growth and extensibility, disrupts birth, and reduces pup survival in intact rats. *Endocrinology* 125, 260–266.
465. Min, G., Hartzog, M. G., Jennings, R. L., Winn, R. J., and Sherwood, O. D. (1997). Evidence that endogenous relaxin promotes growth of the vagina and uterus during pregnancy in gilts. *Endocrinology* 138, 560–565.
466. O'Day, M. B., Winn, R. J., Easter, R. A., Dziuk, P. J., and Sherwood, O. D. (1989). Hormonal control of the cervix in pregnant gilts: II. Relaxin promotes changes in the physical properties of the cervix in ovariectomized hormone-treated pregnant gilts. *Endocrinology* 125, 3004–3010.
467. Lee, A. B., Hwang, J. J., Haab, L. M., Fields, P. A., and Sherwood, O. D. (1992). Monoclonal antibodies specific for rat relaxin: VI. Passive immunization with monoclonal antibodies throughout the second half of pregnancy disrupts histological changes associated with cervical softening at parturition in rats. *Endocrinology* 130, 2386–2391.
468. Lee, H.-L., Zhao, S., Fields, P. A., and Sherwood, O. D. (2005). The extent to which relaxin promotes proliferation and inhibits apoptosis of cervical epithelial and stromal cells is greatest during late pregnancy in rats. *Endocrinology* 146, 1511–1518.
469. Zhao, S., Fields, P. A., and Sherwood, O. D. (2001). Evidence that relaxin inhibits apoptosis in the cervix and the vagina during the second half of pregnancy in the rat. *Endocrinology* 142, 2221–2229.
470. Downing, S. J., and Sherwood, O. D. (1986). The physiological role of relaxin in the pregnant rat: IV. The influence of relaxin on cervical collagen and glycosaminoglycans. *Endocrinology* 118, 471–479.
471. Downing, S. J., and Sherwood, O. D. (1985). The physiological role of relaxin in the pregnant rat: III. The influence of relaxin on cervical extensibility. *Endocrinology* 116, 1215–1220.
472. Luque, E. H., Munoz de Toro, M. M., Ramos, J. G., Rodriguez, H. A., and Sherwood, O. D. (1998). Role of relaxin and estrogen in the control of eosinophilic invasion and collagen remodeling in rat cervical tissue at term. *Biol. Reprod.* 59, 795–800.

473. Winn, R. J., O'Day-Bowman, M. B., and Sherwood, O. D. (1993). Hormonal control of the cervix in pregnant gilts: IV. Relaxin promotes changes in the histological characteristics of the cervix that are associated with cervical softening during late pregnancy in gilts. *Endocrinology* 133, 121–128.
474. O'Day-Bowman, M. B., Winn, R. J., Dziuk, P. J., Lindley, E. R., and Sherwood, O. D. (1991). Hormonal control of the cervix in pregnant gilts: III. Relaxin's influence on cervical biochemical properties in ovariectomized hormone-treated pregnant gilts. *Endocrinology* 129, 1967–1976.
475. Mushayandebvu, T. I., and Rajabi, M. R. (1995). Relaxin stimulates interstitial collagenase activity in cultured uterine cervical cells from nonpregnant and pregnant but not immature guinea pigs: estradiol-17 beta restores relaxin's effect in immature cervical cells. *Biol. Reprod.* 53, 1030–1037.
476. Hwang, J. J., Macinga, D., and Rorke, E. A. (1996). Relaxin modulates human cervical stromal cell activity. *J. Clin. Endocrinol. Metab.* 81, 3379–3384.
477. Shi, L., Shi, S. Q., Saade, G. R., Chwalisz, K., and Garfield, R. E. (2000). Studies of cervical ripening in pregnant rats: effects of various treatments. *Mol. Hum. Reprod.* 6, 382–389.
478. Wing, S. S., and Jain, P. (1995). Molecular cloning, expression and characterization of a ubiquitin conjugation enzyme (E2(17)kB) highly expressed in rat testis. *Biochem. J.* 305 (Pt 1), 125–132.
479. Xenakis, E. M., and Piper, J. M. (1997). Chemotherapeutic induction of labour: a rational approach. *Drugs* 54, 61–68.
480. Sherwood, O. D., Jungheim, E. S., Masferrer, J. L., and Cramer, J. M. (1998). Evidence that relaxin's effects on growth and softening of the cervix are not mediated through prostaglandins in the rat. *Endocrinology* 139, 867–873.
481. Buhimschi, I., Ali, M., Jain, V., Chwalisz, K., and Garfield, R. E. (1996). Differential regulation of nitric oxide in the rat uterus and cervix during pregnancy and labour. *Hum. Reprod.* 11, 1755–1766.
482. Sherwood, O. D., Olson, L. M., Zhao, S., and Little, H. R. (2000). Inhibition of nitric oxide synthase activity diminishes the acute effects of relaxin on growth, but not softening, of the cervix in the rat. *Endocrinology* 141, 2458–2464.
483. Cullen, B. M., and Harkness, R. D. (1960). The effect of hormones on the physical properties and collagen content of the rat's uterine cervix. *J. Physiol.* 152, 419–436.
484. Winn, R. J., Baker, M. D., and Sherwood, O. D. (1994). Individual and combined effects of relaxin, estrogen, and progesterone in ovariectomized gilts: I. Effects on the growth, softening, and histological properties of the cervix. *Endocrinology* 135, 1241–1249.
485. Hall, J. A., and Anthony, R. V. (1993). Influence of ovarian steroids on relaxin-induced distensibility and compositional changes in the porcine cervix. *Biol. Reprod.* 48, 1348–1353.
486. Eddie, L. W., Cameron, I. T., Leeton, J. F., Healy, D. L., and Renou, P. (1990). Ovarian relaxin is not essential for dilatation of cervix. *Lancet* 336, 243.
487. Eichner, E., Waltner, C., Goodman, M., and Post, S. (1956). Relaxin, the third ovarian hormone: its experimental use in women. *Am. J. Obstet. Gynecol.* 71, 1035–1048.
488. Stone, M. L., Sedlis, A., and Zuckerman, M. B. (1959). Effects of relaxin on term and premature labor. *Ann. N Y Acad. Sci.* 75, 1011–1015.
489. MacLennan, A. H., Green, R. C., Bryant-Greenwood, G. D., Greenwood, F. C., and Seamark, R. F. (1980). Ripening of the human cervix and induction of labour with purified porcine relaxin. *Lancet* 1, 220–223.
490. MacLennan, A. H., Green, R. C., Bryant-Greenwood, G. D., Greenwood, F. C., and Seamark, R. F. (1981). Cervical ripening with combinations of vaginal prostaglandin F2-alpha estradiol, and relaxin. *Obstet. Gynecol.* 58, 601–604.
491. Evans, M. I., Dougan, M. B., Moawad, A. H., Evans, W. J., Bryant-Greenwood, G. D., and Greenwood, F. C. (1983). Ripening of the human cervix with porcine ovarian relaxin. *Am. J. Obstet. Gynecol.* 147, 410–414.
492. MacLennan, A. H., Green, R. C., Grant, P., and Nicolson, R. (1986). Ripening of the human cervix and induction of labor with intracervical purified porcine relaxin. *Obstet. Gynecol.* 68, 598–601.
493. Brennand, J. E., Calder, A. A., Leitch, C. R., Greer, I. A., Chou, M. M., and MacKenzie, I. Z. (1997). Recombinant human relaxin as a cervical ripening agent. *Br. J. Obstet. Gynaecol.* 104, 775–780.
494. Steinetz, B. G., Beach, V. L., Blye, R. P., and Kroc, R. L. (1957). Changes in the composition of the rat uterus following a single injection of relaxin. *Endocrinology* 61, 287–292.
495. Brenner, S. H., Lessing, J. B., and Weiss, G. (1984). The effect of in vivo progesterone administration on relaxin-inhibited rat uterine contractions. *Am. J. Obstet. Gynecol.* 148, 946–950.
496. Chamley, W. A., Bagoyo, M. M., and Bryant-Greenwood, G. D. (1977). In vitro response of relaxin-treated rat uterus to prostaglandins and oxytocin. *Prostaglandins* 14, 763–769.
497. Downing, S. J., Bradshaw, J. M., and Porter, D. G. (1980). Relaxin improves the coordination of rat myometrial activity in vivo. *Biol. Reprod.* 23, 899–903.
498. Lessing, J. B., Brenner, S. H., and Weiss, G. (1984). Effect of prolactin and relaxin on in vitro rat uterine contractions and prolactin interaction with relaxin. *Obstet. Gynecol.* 64, 97–100.
499. Miller, J. W., Kisley, A., and Murray, W. J. (1957). The effects of relaxin-containing ovarian extracts on various types of smooth muscle. *J. Pharmacol. Exp. Ther.* 120, 426–437.
500. Porter, D. G., Downing, S. J., and Bradshaw, J. M. (1979). Relaxin inhibits spontaneous and prostaglandin-driven myometrial activity in anaesthetized rats. *J. Endocrinol.* 83, 183–192.
501. Porter, D. G., Downing, S. J., and Bradshaw, J. M. (1981). Inhibition of oxytocin- or prostaglandin F2alpha-driven myometrial activity by relaxin in the rat is oestrogen-dependent. *J. Endocrinol.* 89, 399–404.
502. Sarosi, P., Schmidt, C. L., Essig, M., Steinetz, B. G., and Weiss, G. (1983). The effect of relaxin and progesterone on rat uterine contractions. *Am. J. Obstet. Gynecol.* 145, 402–405.
503. Sawyer, W. H., Frieden, E. H., and Martin, A. C. (1953). In vitro inhibition of spontaneous contractions of the rat uterus by relaxin-containing extracts of sow ovaries. *Am. J. Physiol.* 172, 547–552.
504. St-Louis, J. (1981). Relaxin inhibition of KCl-induced uterine contractions in vitro: an alternative bioassay. *Can. J. Physiol. Pharmacol.* 59, 507–512.
505. St-Louis, J. (1982). Pharmacological studies on the action of relaxin upon KCl-contracted rat uterus. *Pharmacology* 25, 327–337.
506. Wiqvist, N. (1959). Desensitizing effect of exo- and endogenous relaxin on the immediate uterine response to relaxin. *Acta. Endocrinol. (Copenh)* 32 (Suppl 46), 3–14.
507. Wiqvist, N. (1959). The effect of prolonged administration of relaxin on some functional properties of the non-pregnant mouse and rat uterus. *Acta. Endocrinol. (Copenh)* 32 (Suppl 46), 15–32.
508. Wiqvist, N., and Paul, K. G. (1958). Inhibition of the spontaneous uterine motility in vitro as a bioassay of relaxin. *Acta. Endocrinol. (Copenh)* 29, 135–146.
509. Porter, D. G. (1971). The action of relaxin on myometrial activity in the guinea-pig in vivo. *J. Reprod. Fertil.* 26, 251–253.
510. Porter, D. G. (1972). Myometrium of the pregnant guinea pig: the probable importance of relaxin. *Biol. Reprod.* 7, 458–464.
511. Khaligh, H. S. (1968). Inhibition by relaxin of spontaneous contractions of the uterus of the hamster in vitro. *J. Endocrinol.* 40, 125–126.

512. Bigazzi, M., and Nardi, E. (1981). Prolactin and relaxin: antagonism on the spontaneous motility of the uterus. *J. Clin. Endocrinol. Metab.* 53, 665–667.
513. Chamley, W. A., Bagoyo, M. M., and Bryant-Greenwood, G. D. (1981). Potencies of porcine relaxins using two bioassays. *J. Endocrinol.* 88, 89–96.
514. Chamley, W. A., and Parkington, H. C. (1984). Relaxin inhibits the plateau component of the action potential in the circular myometrium of the rat. *J. Physiol.* 353, 51–65.
515. Cheah, S. H., and Sherwood, O. D. (1981). Effects of relaxin on in vivo uterine contractions in conscious and unrestrained estrogen-treated and steroid-untreated ovariectomized rats. *Endocrinology* 109, 2076–2083.
516. Downing, S. J., and Hollingsworth, M. (1991). Antagonism of relaxin by glibenclamide in the uterus of the rat in vivo. *Br. J. Pharmacol.* 104, 71–76.
517. Downing, S. J., and Hollingsworth, M. (1992). Interaction between myometrial relaxants and oxytocin: a comparison between relaxin, cromakalim and salbutamol. *J. Endocrinol.* 135, 29–36.
518. Downing, S. J., and Hollingsworth, M. (1992). Influence of ovarian steroids on myometrial sensitivity and tolerance to relaxin in the rat in vivo: lack of cross-tolerance between relaxin, salbutamol and cromakalim. *J. Endocrinol.* 135, 17–28.
519. Downing, S. J., McIlwrath, A., and Hollingsworth, M. (1992). Cyclic adenosine 3′5′-monophosphate and the relaxant action of relaxin in the rat uterus in vivo. *J. Reprod. Fertil.* 96, 857–863.
520. Fox, D., Handberg, G. M., Hartley, M. L., Monagle, J., and Pennefather, J. N. (1989). Actions of some autacoids and peptides, including relaxin, on costo-uterine muscle from rats. *Clin. Exp. Pharmacol. Physiol.* 16, 561–569.
521. Ginsburg, F. W., Rosenberg, C. R., Schwartz, M., Colon, J. M., and Goldsmith, L. T. (1988). The effect of relaxin on calcium fluxes in the rat uterus. *Am. J. Obstet. Gynecol.* 159, 1395–1401.
522. Goldsmith, L. T., Skurnick, J. H., Wojtczuk, A. S., Linden, M., Kuhar, M. J., and Weiss, G. (1989). The antagonistic effect of oxytocin and relaxin on rat uterine segment contractility. *Am. J. Obstet. Gynecol.* 161, 1644–1649.
523. Grazi, R. V., Goldsmith, L. T., Schmidt, C. L., Von Hagen, S., and Weiss, G. (1988). Synergistic effect of relaxin and progesterone on cyclic adenosine 3′,5′-monophosphate levels in the rat uterus. *Am. J. Obstet. Gynecol.* 159, 1402–1406.
524. Inoue, H., Osa, T., and Okabe, K. (1992). Effects of porcine relaxin on contraction, membrane response and cyclic AMP content in rat myometrium. *Jpn. J. Pharmacol.* 58 Suppl 2, 353P.
525. Lippert, T. H., Schneider-Zeh, S., and Voelter, W. (1987). The action of relaxin, fenoterol and etilefrine on uterine motility of the rat. *Int. J. Clin. Pharmacol. Ther. Toxicol.* 25, 565–566.
526. MacLennan, A. H., Grant, P., Ness, D., and Down, A. (1986). Effect of porcine relaxin and progesterone on rat, pig and human myometrial activity in vitro. *J. Reprod. Med.* 31, 43–49.
527. McGovern, P. G., Goldsmith, L. T., Schmidt, C. L., Von Hagen, S., Linden, M., and Weiss, G. (1992). Effects of endothelin and relaxin on rat uterine segment contractility. *Biol. Reprod.* 46, 680–685.
528. Nishikori, K., Weisbrodt, N. W., Sherwood, O. D., and Sanborn, B. M. (1982). Relaxin alters rat uterine myosin light chain phosphorylation and related enzymatic activities. *Endocrinology* 111, 1743–1745.
529. Nishikori, K., Weisbrodt, N. W., Sherwood, O. D., and Sanborn, B. M. (1983). Effects of relaxin on rat uterine myosin light chain kinase activity and myosin light chain phosphorylation. *J. Biol. Chem.* 258, 2468–2474.
530. Osa, T., Inoue, H., and Okabe, K. (1991). Effects of porcine relaxin on contraction, membrane response and cyclic AMP content in rat myometrium in comparison with the effects of isoprenaline and forskolin. *Br. J. Pharmacol.* 104, 950–960.
531. Petersen, L. K., Svane, D., Uldbjerg, N., and Forman, A. (1991). Effects of human relaxin on isolated rat and human myometrium and uteroplacental arteries. *Obstet. Gynecol.* 78, 757–762.
532. Sanborn, B. M., Kuo, H. S., Weisbrodt, N. W., and Sherwood, O. D. (1982). Effect of porcine relaxin on cyclic nucleotide levels and spontaneous contractions of the rat uterus. *Adv. Exp. Med. Biol.* 143, 273–287.
533. Sanborn, B. M., and Sherwood, O. D. (1981). Effect of relaxin on bound cAMP in rat uterus. *Endocr. Res. Commun.* 8, 179–192.
534. Porter, D. G., and Watts, A. D. (1986). Relaxin and progesterone are myometrial inhibitors in the ovariectomized non-pregnant mini-pig. *J. Reprod. Fertil.* 76, 205–213.
535. Pupula, M., and MacLennan, A. H. (1989). Effect of porcine relaxin on spontaneous, oxytocin-driven and prostaglandin-driven pig myometrial activity in vitro. *J. Reprod. Med.* 34, 819–823.
536. Schramm, W., Einer-Jensen, N., Brown, M. B., and McCracken, J. A. (1984). Effect of four primary prostaglandins and relaxin on blood flow in the ovine endometrium and myometrium. *Biol. Reprod.* 30, 523–531.
537. MacLennan, A. H., Grant, P., and Bryant-Greenwood, G. (1995). hRLX-1: in vitro response of human and pig myometrium. *J. Reprod. Med.* 40, 703–706.
538. Downing, S. J., and Sherwood, O. D. (1985). The physiological role of relaxin in the pregnant rat: II. The influence of relaxin on uterine contractile activity. *Endocrinology* 116, 1206–1214.
539. Taverne, M. A., Naaktgeboren, C., Elsaesser, F., Forsling, M. L., van der Weyden, G. C., Ellendorff, F., and Smidt, D. (1979). Myometrial electrical activity and plasma concentrations of progesterone, estrogens and oxytocin during late pregnancy and parturition in the miniature pig. *Biol. Reprod.* 21, 1125–1134.
540. Downing, S. J., and Hollingsworth, M. (1993). Action of relaxin on uterine contractions: a review. *J. Reprod. Fertil.* 99, 275–282.
541. Mercado-Simmen, R. C., Bryant-Greenwood, G. D., and Greenwood, F. C. (1982). Relaxin receptor in the rat myometrium: regulation by estrogen and relaxin. *Endocrinology* 110, 220–226.
542. Hsu, C. J., and Sanborn, B. M. (1986). Relaxin treatment alters the kinetic properties of myosin light chain kinase activity in rat myometrial cells in culture. *Endocrinology* 118, 499–505.
543. Gates, G. S., Flynn, J. J., Ryan, R. J., and Sherwood, O. D. (1981). In vivo uptake of ^{125}I-relaxin in the guinea pig. *Biol. Reprod.* 25, 549–554.
544. Luna, J. J., Riesewijk, A., Horcajadas, J. A., Van Os Rd, R., Dominguez, F., Mosselman, S., Pellicer, A., and Simon, C. (2004). Gene expression pattern and immunoreactive protein localization of LGR7 receptor in human endometrium throughout the menstrual cycle. *Mol. Hum. Reprod.* 10, 85–90.
545. Sanborn, B. M., Anwer, K., Monga, M., Wen, Y., Singh, S. P., Meera, P., Oberti, C., Toro, L., and Stefani, E. (1995). Mechanisms controlling the acute effects of relaxin on the myometrium. In *Progress in Relaxin Research* (A. H. MacLennan, G. W. Tregear, and G. D. Bryant-Greenwood, Eds.), pp. 289–297. Kluwer, Dordrecht.
546. Sanborn, B. M. (2001). Hormones and calcium: mechanisms controlling uterine smooth muscle contractile activity. *Exp. Physiol.* 86, 223–237.
547. Hollingsworth, M., Rudkin, S., and Downing, S. (2001). Myometrial relaxant action of relaxin. In *Relaxin 2000:*

Proceedings of the Third International Conference on Relaxin and Related Peptides (G. W. Tregear, R. Ivell, R. A. Bathgate, and J. D. Wade, Eds.), pp. 291–299. Kluwer, Dordrecht.

548. Meera, P., Anwer, K., Monga, M., Oberti, C., Stefani, E., Toro, L., and Sanborn, B. M. (1995). Relaxin stimulates myometrial calcium-activated potassium channel activity via protein kinase A. *Am. J. Physiol.* 269, C312–C317.
549. Yue, C., Dodge, K. L., Weber, G., and Sanborn, B. M. (1998). Phosphorylation of serine 1105 by protein kinase A inhibits phospholipase Cβ3 stimulation by Gαq. *J. Biol. Chem.* 273, 18023–18027.
550. Breenan, D. M., and Zarrow, M. X. (1959). Water and electrolyte content of the uterus of the intact and adrenalectomized rat treated with relaxin and various steroid hormones. *Endocrinology* 64, 907–913.
551. Hall, K. (1960). Modification by relaxin of the response of the reproductive tract of mice to oestradiol and progesterone. *J. Endocrinol.* 20, 355–364.
552. Jablonski, W. J., and Velardo, J. T. (1957). Effects of relaxin on uterine weight of immature rats. *Endocrinology* 61, 474–475.
553. Schmidt, J. E., and Leonard, S. L. (1960). The effect of relaxin on uterine phosphorylase in the rat. *Endocrinology* 67, 663–667.
554. Wada, H., and Turner, C. W. (1961). Interaction of relaxin and ovarian steroid hormones of uterus of rat. *Endocrinology* 68, 1059–1063.
555. Hall, J. A., Cantley, T. C., Galvin, J. M., Day, B. N., and Anthony, R. V. (1992). Influence of ovarian steroids on relaxin-induced uterine growth in ovariectomized gilts. *Endocrinology* 130, 3159–3166.
556. Huang, C. J., Li, Y., and Anderson, L. L. (1997). Relaxin and estrogen synergistically accelerate growth and development in the uterine cervix of prepubertal pigs. *Anim. Reprod. Sci.* 46, 149–158.
557. Zaleski, H. M., Winn, R. J., Jennings, R. L., Dziuk, P. J., and Sherwood, O. D. (1995). Effects of relaxin administration in early gestation or prior to mating on uterine length and fetal survival in gilts. *Biol. Reprod.* 52, 1389–1394.
558. Ohleth, K. M., Lenhart, J. A., Ryan, P. L., Radecki, S. V., and Bagnell, C. A. (1997). Relaxin increases insulin-like growth factors (IGFs) and IGF-binding proteins of the pig uterus in vivo. *Endocrinology* 138, 3652–3658.
559. Goldsmith, L. T., Weiss, G., Palejwala, S., Plant, T. M., Wojtczuk, A., Lambert, W. C., Ammur, N., Heller, D., Skurnick, J. H., Edwards, D., and Cole, D. M. (2004). Relaxin regulation of endometrial structure and function in the rhesus monkey. *Proc. Natl. Acad. Sci. U S A* 101, 4685–4689.
560. Einspanier, A. (2001). Relaxin is an important factor for uterine differentiation and implantation in the marmoset monkey. In *Relaxin 2000: Proceedings of the Third International Conference on Relaxin and Related Peptides* (G. W. Tregear, R. Ivell, R. A. Bathgate, and J. D. Wade, Eds.), pp. 73–82. Kluwer, Dordrecht.
561. Vasilenko, P., Mead, J. P., and Weidmann, J. E. (1986). Uterine growth-promoting effects of relaxin: a morphometric and histological analysis. *Biol. Reprod.* 35, 987–995.
562. Lenhart, J. A., Ryan, P. L., Ohleth, K. M., and Bagnell, C. A. (1999). Expression of connexin-26, -32, and -43 gap junction proteins in the porcine cervix and uterus during pregnancy and relaxin-induced growth. *Biol. Reprod.* 61, 1452–1459.
563. Ryan, P. L., Baum, D. L., Lenhart, J. A., Ohleth, K. M., and Bagnell, C. A. (2001). Expression of uterine and cervical epithelial cadherin during relaxin-induced growth in pigs. *Reproduction* 122, 929–937.
564. Lenhart, J. A., Ohleth, K. M., Ryan, P. L., Palmer, S. S., and Bagnell, C. A. (1999). Effect of relaxin on tissue inhibitor of metalloproteinase-1 and -2 in the porcine uterus and cervix. *Ann. N Y Acad. Sci.* 878, 565–566.
565. Pillai, S. B., Rockwell, L. C., Sherwood, O. D., and Koos, R. D. (1999). Relaxin stimulates uterine edema via activation of estrogen receptors: blockade of its effects using ICI 182,780, a specific estrogen receptor antagonist. *Endocrinology* 140, 2426–2429.
566. Adams, W. C., Hanousek, C. A., and Frieden, E. H. (1989). Progesterone inhibits the uterotrophic effect of relaxin in immature rats. *Proc. Soc. Exp. Biol. Med.* 191, 159–162.
567. Downing, S. J., and Hollingsworth, M. (1993). Uptake of relaxin in the uterus and cervix of rats in vivo: influence of ovarian steroids and tolerance. *J. Reprod. Fertil.* 99, 121–129.
568. Pillai, S. B., Jones, J. M., and Koos, R. D. (2002). Treatment of rats with 17beta-estradiol or relaxin rapidly inhibits uterine estrogen receptor beta1 and beta2 messenger ribonucleic acid levels. *Biol. Reprod.* 67, 1919–1926.
569. Schink, W., and Struck, H. (1968). Relaxin in the Allen-Doisy test. *Zentralbl. Gynakol.* 90, 675–678.
570. Jagiello, G. (1967). The effect of several relaxin preparations on the hysterectomized guinea-pig. *J. Reprod. Fertil.* 13, 175–177.
571. Zhao, S., Kuenzi, M. J., and Sherwood, O. D. (1996). Monoclonal antibodies specific for rat relaxin: IX. Evidence that endogenous relaxin promotes growth of the vagina during the second half of pregnancy in rats. *Endocrinology* 137, 425–430.
572. Zhao, S., and Sherwood, O. D. (1998). Monoclonal antibodies specific for rat relaxin: X. Endogenous relaxin induces changes in the histological characteristics of the rat vagina during the second half of pregnancy. *Endocrinology* 139, 4726–4734.
573. Summerlee, A. J., and Wilson, B. C. (1994). Role of the subfornical organ in the relaxin-induced prolongation of gestation in the rat. *Endocrinology* 134, 2115–2120.
574. Summerlee, A. J., Ramsey, D. G., and Poterski, R. S. (1998). Neutralization of relaxin within the brain affects the timing of birth in rats. *Endocrinology* 139, 479–484.
575. Summerlee, A. J., Hornsby, D. J., and Ramsey, D. G. (1998). The dipsogenic effects of rat relaxin: the effect of photoperiod and the potential role of relaxin on drinking in pregnancy. *Endocrinology* 139, 2322–2328.
576. Way, S. A., and Leng, G. (1992). Relaxin increases the firing rate of supraoptic neurones and increases oxytocin secretion in the rat. *J. Endocrinol.* 132, 149–158.
577. Way, S. A., Douglas, A. J., Dye, S., Bicknell, R. J., Leng, G., and Russell, J. A. (1993). Endogenous opioid regulation of oxytocin release during parturition is reduced in ovariectomized rats. *J. Endocrinol.* 138, 13–22.
578. Nishimori, K., Young, L. J., Guo, Q., Wang, Z., Insel, T. R., and Matzuk, M. M. (1996). Oxytocin is required for nursing but is not essential for parturition or reproductive behavior. *Proc. Natl. Acad. Sci. U S A* 93, 11699–11704.
579. Downing, S. J., and Sherwood, O. D. (1985). The physiological role of relaxin in the pregnant rat: I. The influence of relaxin on parturition. *Endocrinology* 116, 1200–1205.
580. Nara, B. S., Welk, F. A., Rutherford, J. E., Sherwood, O. D., and First, N. L. (1982). Effect of relaxin on parturition and frequency of live births in pigs. *J. Reprod. Fertil.* 66, 359–365.
581. Zhao, S., and Sherwood, O. D. (2004). Induction of labor with RU 486 (mifepristone) in relaxin-deficient rats: antepartum administration of relaxin facilitates delivery and increases pup survival. *Am. J. Obstet. Gynecol.* 190, 229–238.
582. Cho, S. J., Dlamini, B. J., Klindt, J., Schwabe, C., Jacobson, C. D., and Anderson, L. L. (1998). Antiporcine relaxin (antipRLX540) treatment decreases relaxin plasma concentration and disrupts delivery in late pregnant pigs. *Anim. Reprod. Sci.* 52, 303–316.
583. Bryant-Greenwood, G. D., and Schwabe, C. (1994). Human relaxins: chemistry and biology. *Endocr. Rev.* 15, 5–26.

584. Bryant-Greenwood, G. D., and Millar, L. K. (2000). Human fetal membranes: their preterm premature rupture. *Biol. Reprod.* 63, 1575–1579.
585. Lindheimer, M. D., Barron, W. M., and Davison, J. M. (1989). Osmoregulation of thirst and vasopressin release in pregnancy. *Am. J. Physiol.* 257, F159–F169.
586. Novak, J., Danielson, L. A., Kerchner, L. J., Sherwood, O. D., Ramirez, R. J., Moalli, P. A., and Conrad, K. P. (2001). Relaxin is essential for renal vasodilation during pregnancy in conscious rats. *J. Clin. Invest.* 107, 1469–1475.
587. Weisinger, R. S., Burns, P., Eddie, L. W., and Wintour, E. M. (1993). Relaxin alters the plasma osmolality-arginine vasopressin relationship in the rat. *J. Endocrinol.* 137, 505–510.
588. Danielson, L. A., Kercher, L. J., and Conrad, K. P. (2000). Impact of gender and endothelin on renal vasodilation and hyperfiltration induced by relaxin in conscious rats. *Am. J. Physiol. Regul. Integr. Comp. Physiol.* 279, R1298–R1304.
589. Johnson, M. R., Brooks, A. A., and Steer, P. J. (1996). The role of relaxin in the pregnancy associated reduction in plasma osmolality. *Hum. Reprod.* 11, 1105–1108.
590. Atherton, J. C., Dark, J. M., Garland, H. O., Morgan, M. R., Pidgeon, J., and Soni, S. (1982). Changes in water and electrolyte balance, plasma volume and composition during pregnancy in the rat. *J. Physiol. (Lond)* 330, 81–93.
591. Omi, E. C., Zhao, S., Shanks, R. D., and Sherwood, O. D. (1997). Evidence that systemic relaxin promotes moderate water consumption during late pregnancy in rats. *J. Endocrinol.* 153, 33–40.
592. Thornton, S. M., and Fitzsimons, J. T. (1995). The effects of centrally administered porcine relaxin on drinking behaviour in male and female rats. *J. Neuroendocrinol.* 7, 165–169.
593. Summerlee, A. J. S., and Robertson, G. F. (1995). Central administration of porcine relaxin stimulates drinking behaviour in rats: an effect mediated by central angiotensin II. *Endocr. J.* 3, 377–381.
594. Sinnayah, P., Burns, P., Wade, J. D., Weisinger, R. S., and McKinley, M. J. (1999). Water drinking in rats resulting from intravenous relaxin and its modification by other dipsogenic factors. *Endocrinology* 140, 5082–5086.
595. Sunn, N., McKinley, M. J., and Oldfield, B. J. (2001). Identification of efferent neural pathways from the lamina terminalis activated by blood-borne relaxin. *J. Neuroendocrinol.* 13, 432–437.
596. Sunn, N., Egli, M., Burazin, T. C., Burns, P., Colvill, L., Davern, P., Denton, D. A., Oldfield, B. J., Weisinger, R. S., Rauch, M., Schmid, H. A., and McKinley, M. J. (2002). Circulating relaxin acts on subfornical organ neurons to stimulate water drinking in the rat. *Proc. Natl. Acad. Sci. U S A* 99, 1701–1706.
597. Zhao, S., Malmgren, C. H., Shanks, R. D., and Sherwood, O. D. (1995). Monoclonal antibodies specific for rat relaxin: VIII. Passive immunization with monoclonal antibodies throughout the second half of pregnancy reduces water consumption in rats. *Endocrinology* 136, 1892–1897.
598. Ma, S., Burazin, T. C., Bathgate, R. A. D., Tregear, G. W., and Gundlach, A. L. (2003). Relaxin peptides and receptors in rat brain: distribution, regulation and function. *Proc. Aust. Neurosci. Soc.* 14, 152.
599. McKinley, M. J., Burns, P., Colvill, L. M., Oldfield, B. J., Wade, J. D., Weisinger, R. S., and Tregear, G. W. (1997). Distribution of Fos immunoreactivity in the lamina terminalis and hypothalamus induced by centrally administered relaxin in conscious rats. *J. Neuroendocrinol.* 9, 431–437.
600. McKinley, M. J., Allen, A. M., Burns, P., Colvill, L. M., and Oldfield, B. J. (1998). Interaction of circulating hormones with the brain: the roles of the subfornical organ and the organum vasculosum of the lamina terminalis. *Clin. Exp. Pharmacol. Physiol. Suppl.* 25, S61–S67.
601. Oldfield, B. J., Badoer, E., Hards, D. K., and McKinley, M. J. (1994). Fos production in retrogradely labelled neurons of the lamina terminalis following intravenous infusion of either hypertonic saline or angiotensin II. *Neuroscience* 60, 255–262.
602. Parry, L. J., and Summerlee, A. J. (1991). Central angiotensin partially mediates the pressor action of relaxin in anesthetized rats. *Endocrinology* 129, 47–52.
603. Geddes, B. J., Parry, L. J., and Summerlee, A. J. (1994). Brain angiotensin-II partially mediates the effects of relaxin on vasopressin and oxytocin release in anesthetized rats. *Endocrinology* 134, 1188–1192.
604. McKinley, M. J., Allen, A. M., Mathai, M. L., May, C., McAllen, R. M., Oldfield, B. J., and Weisinger, R. S. (2001). Brain angiotensin and body fluid homeostasis. *Jpn. J. Physiol.* 51, 281–289.
605. Hisaw, F. L., Hisaw, F. L. Jr., and Dawson, A. B. (1967). Effects of relaxin on the endothelium of endometrial blood vessels in monkeys (*Macaca mulatta*). *Endocrinology* 81, 375–385.
606. Clapp, J. F. III, Seaward, B. L., Sleamaker, R. H., and Hiser, J. (1988). Maternal physiologic adaptations to early human pregnancy. *Am. J. Obstet. Gynecol.* 159, 1456–1460.
607. Robson, S. C., Hunter, S., Boys, R. J., and Dunlop, W. (1989). Serial study of factors influencing changes in cardiac output during human pregnancy. *Am. J. Physiol.* 256, H1060–H1065.
608. Gilson, G. J., Mosher, M. D., and Conrad, K. P. (1992). Systemic hemodynamics and oxygen transport during pregnancy in chronically instrumented, conscious rats. *Am. J. Physiol.* 263, H1911–H1918.
609. Conrad, K. P., and Russ, R. D. (1992). Augmentation of baroreflex-mediated bradycardia in conscious pregnant rats. *Am. J. Physiol.* 262, R472–R477.
610. Conrad, K. P. (1984). Renal hemodynamics during pregnancy in chronically catheterized, conscious rats. *Kidney Int.* 26, 24–29.
611. Novak, J., Ramirez, R. J., Gandley, R. E., Sherwood, O. D., and Conrad, K. P. (2002). Myogenic reactivity is reduced in small renal arteries isolated from relaxin-treated rats. *Am. J. Physiol. Regul. Integr. Comp. Physiol.* 283, R349–R355.
612. Jeyabalan, A., Novak, J., Danielson, L. A., Kerchner, L. J., Opett, S. L., and Conrad, K. P. (2003). Essential role for vascular gelatinase activity in relaxin-induced renal vasodilation, hyperfiltration, and reduced myogenic reactivity of small arteries. *Circ. Res.* 93, 1249–1257.
613. Bani, G., Maurizi, M., Bigazzi, M., and Bani Sacchi, T. (1995). Effects of relaxin on the endometrial stroma. Studies in mice. *Biol. Reprod.* 53, 253–262.
614. Bani, G., Bani Sacchi, T., Bigazzi, M., and Bianchi, S. (1988). Effects of relaxin on the microvasculature of mouse mammary gland. *Histol. Histopathol.* 3, 337–343.
615. Masini, E., Bani, D., Bello, M. G., Bigazzi, M., Mannaioni, P. F., and Sacchi, T. B. (1997). Relaxin counteracts myocardial damage induced by ischemia-reperfusion in isolated guinea pig hearts: evidence for an involvement of nitric oxide. *Endocrinology* 138, 4713–4720.
616. Bani-Sacchi, T., Bigazzi, M., Bani, D., Mannaioni, P. F., and Masini, E. (1995). Relaxin-induced increased coronary flow through stimulation of nitric oxide production. *Br. J. Pharmacol.* 116, 1589–1594.
617. Bani, D., Masini, E., Bello, M. G., Bigazzi, M., and Sacchi, T. B. (1998). Relaxin protects against myocardial injury caused by ischemia and reperfusion in rat heart. *Am. J. Pathol.* 152, 1367–1376.
618. Bani, D., Nistri, S., Quattrone, S., Bigazzi, M., and Bani Sacchi, T. (2001). The vasorelaxant hormone relaxin induces changes in liver sinusoid microcirculation: a morphologic study in the rat. *J. Endocrinol.* 171, 541–549.

619. Bigazzi, M., Del Mese, A., Petrucci, F., Casali, R., and Novelli, G. P. (1986). The local administration of relaxin induces changes in the microcirculation of the rat mesocaecum. *Acta. Endocrinol. (Copenh)* 112, 296–299.
620. St-Louis, J., and Massicotte, G. (1985). Chronic decrease of blood pressure by rat relaxin in spontaneously hypertensive rats. *Life Sci.* 37, 1351–1357.
621. Massicotte, G., Parent, A., and St-Louis, J. (1989). Blunted responses to vasoconstrictors in mesenteric vasculature but not in portal vein of spontaneously hypertensive rats treated with relaxin. *Proc. Soc. Exp. Biol. Med.* 190, 254–259.
622. Failli, P., Nistri, S., Quattrone, S., Mazzetti, L., Bigazzi, M., Sacchi, T. B., and Bani, D. (2002). Relaxin up-regulates inducible nitric oxide synthase expression and nitric oxide generation in rat coronary endothelial cells. *FASEB J.* 16, 252–254.
623. Longo, M., Jain, V., Vedernikov, Y. P., Garfield, R. E., and Saade, G. R. (2003). Effects of recombinant human relaxin on pregnant rat uterine artery and myometrium in vitro. *Am. J. Obstet. Gynecol.* 188, 1468–1474; discussion 1474–1466.
624. Fisher, C., MacLean, M., Morecroft, I., Seed, A., Johnston, F., Hillier, C., and McMurray, J. (2002). Is the pregnancy hormone relaxin also a vasodilator peptide secreted by the heart? *Circulation* 106, 292–295.
625. Conrad, K. P., Debrah, D. O., Novak, J., Danielson, L. A., and Shroff, S. G. (2004). Relaxin modifies systemic arterial resistance and compliance in conscious, nonpregnant rats. *Endocrinology* 145, 3289–3296.
626. Slangen, B. F., Out, I. C., Verkeste, C. M., and Peeters, L. L. (1996). Hemodynamic changes in early pregnancy in chronically instrumented, conscious rats. *Am. J. Physiol.* 270, H1779–H1784.
627. Debrah, D. O., Conrad, K. P., Danielson, L. A., and Shroff, S. G. (2005). Effects of relaxin on systemic arterial hemodynamics and mechanical properties in conscious rats: sex dependency and dose response. *J Appl Physiol*, 98, 1013–1020.
628. Thomas, G. R., and Vandlen, R. (1993). The purely chronotropic effects of relaxin in the rat isolated heart. *J. Pharm. Pharmacol.* 45, 927–928.
629. Coulson, C. C., Thorp, J. M. Jr., Mayer, D. C., and Cefalo, R. C. (1996). Central hemodynamic effects of recombinant human relaxin in the isolated, perfused rat heart model. *Obstet. Gynecol.* 87, 610–612.
630. Kakouris, H., Eddie, L. W., and Summers, R. J. (1992). Cardiac effects of relaxin in rats. *Lancet* 339, 1076–1078.
631. Ward, D. G., Thomas, G. R., and Cronin, M. J. (1992). Relaxin increases rat heart rate by a direct action on the cardiac atrium. *Biochem. Biophys. Res. Commun.* 186, 999–1005.
632. Parry, L. J., Wilson, B. C., Poterski, R. S., and Summerlee, A. J. (1998). The cardiovascular effects of porcine relaxin in Brattleboro rats. *Endocrine* 8, 317–322.
633. Piedras-Renteria, E. S., Sherwood, O. D., and Best, P. M. (1997). Effects of relaxin on rat atrial myocytes: I. Inhibition of I(to) via PKA-dependent phosphorylation. *Am. J. Physiol.* 272, H1791–H1797.
634. Piedras-Renteria, E. S., Sherwood, O. D., and Best, P. M. (1997). Effects of relaxin on rat atrial myocytes: II. Increased calcium influx derived from action potential prolongation. *Am. J. Physiol.* 272, H1798–H1803.
635. Mumford, A. D., Parry, L. J., and Summerlee, A. J. (1989). Lesion of the subfornical organ affects the haemotensive response to centrally administered relaxin in anaesthetized rats. *J. Endocrinol.* 122, 747–755.
636. Ward, D. G., Cronin, M. J., and Baertschi, A. J. (1991). Lack of cardiovascular and vasopressin responses to human relaxin in conscious, late-pregnant rats. *Am. J. Physiol.* 261, H206–H211.
637. Castro, J. M., Eddie, L. W., and Summers, R. J. (1992). Effects of relaxin on human isolated atrial trabeculae and rat isolated atria. *Clin. Exp. Pharmacol. Physiol. Suppl.* 21, 12.
638. Hamolsky, M., and Sparrow, R. C. (1945). Influence of relaxin on mammary development in sexually immature female rats. *Proc. Soc. Exp. Biol. Med.* 60, 8–9.
639. Hwang, J. J., Lee, A. B., Fields, P. A., Haab, L. M., Mojonnier, L. E., and Sherwood, O. D. (1991). Monoclonal antibodies specific for rat relaxin: V. Passive immunization with monoclonal antibodies throughout the second half of pregnancy disrupts development of the mammary apparatus and, hence, lactational performance in rats. *Endocrinology* 129, 3034–3042.
640. Kuenzi, M. J., and Sherwood, O. D. (1992). Monoclonal antibodies specific for rat relaxin: VII. Passive immunization with monoclonal antibodies throughout the second half of pregnancy prevents development of normal mammary nipple morphology and function in rats. *Endocrinology* 131, 1841–1847.
641. Hurley, W. L., Doane, R. M., O'Day-Bowman, M. B., Winn, R. J., Mojonnier, L. E., and Sherwood, O. D. (1991). Effect of relaxin on mammary development in ovariectomized pregnant gilts. *Endocrinology* 128, 1285–1290.
642. Zaleski, H. M., Winn, R. J., Jennings, R. L., and Sherwood, O. D. (1996). Effects of relaxin on lactational performance in ovariectomized gilts. *Biol. Reprod.* 55, 671–675.
643. Kuenzi, M. J., Connolly, B. A., and Sherwood, O. D. (1995). Relaxin acts directly on rat mammary nipples to stimulate their growth. *Endocrinology* 136, 2943–2947.
644. Bani, G., and Bigazzi, M. (1984). Morphological changes induced in mouse mammary gland by porcine and human relaxin. *Acta. Anat. (Basel)* 119, 149–154.
645. Bani, G., Bigazzi, M., and Bani, D. (1985). Effects of relaxin on the mouse mammary gland: I. The myoepithelial cells. *J. Endocrinol. Invest.* 8, 207–215.
646. Bani, G., Bigazzi, M., and Bani, D. (1986). The effects of relaxin on the mouse mammary gland: II. The epithelium. *J. Endocrinol. Invest.* 9, 145–152.
647. Bianchi, S., Bani, G., and Bigazzi, M. (1986). Effects of relaxin on the mouse mammary gland: III. The fat pad. *J. Endocrinol. Invest.* 9, 153–160.
648. Wright, L. C., and Anderson, R. R. (1982). Effect of relaxin on mammary growth in the hypophysectomized rat. *Adv. Exp. Med. Biol.* 143, 341–355.
649. Wahab, I. M., and Anderson, R. R. (1989). Physiologic role of relaxin on mammary gland growth in rats. *Proc. Soc. Exp. Biol. Med.* 192, 285–289.
650. Kass, L., Ramos, J. G., Ortega, H. H., Montes, G. S., Bussmann, L. E., Luque, E. H., and Munoz de Toro, M. (2001). Relaxin has a minor role in rat mammary gland growth and differentiation during pregnancy. *Endocrine* 15, 263–269.
651. Eldridge-White, R., Easter, R. A., Heaton, D. M., O'Day, M. B., Petersen, G. C., Shanks, R. D., Tarbell, M. K., and Sherwood, O. D. (1989). Hormonal control of the cervix in pregnant gilts: I. Changes in the physical properties of the cervix correlate temporally with elevated serum levels of estrogen and relaxin. *Endocrinology* 125, 2996–3003.
652. Winn, R. J., Baker, M. D., Merle, C. A., and Sherwood, O. D. (1994). Individual and combined effects of relaxin, estrogen, and progesterone in ovariectomized gilts: II. Effects on mammary development. *Endocrinology* 135, 1250–1255.
653. Ivell, R., Balvers, M., Pohnke, Y., Telgmann, R., Bartsch, O., Milde-Langosch, K., Bamberger, A. M., and Einspanier, A. (2003). Immunoexpression of the relaxin receptor LGR7 in breast and uterine tissues of humans and primates. *Reprod. Biol. Endocrinol.* 1, 114.
654. Bani, D. (1997). Relaxin and breast cancer. *Bull. Cancer* 84, 179–182.
655. Sacchi, T. B., Bani, D., Brandi, M. L., Falchetti, A., and Bigazzi, M. (1994). Relaxin influences growth, differentiation and cell-cell adhesion of human breast-cancer cells in culture. *Int. J. Cancer* 57, 129–134.

656. Bani, D., Masini, E., Bello, M. G., Bigazzi, M., and Sacchi, T. B. (1995). Relaxin activates the L-arginine-nitric oxide pathway in human breast cancer cells. *Cancer Res.* 55, 5272–5275.
657. Binder, C., Hagemann, T., Husen, B., Schulz, M., and Einspanier, A. (2002). Relaxin enhances in-vitro invasiveness of breast cancer cell lines by up-regulation of matrix metalloproteases. *Mol. Hum. Reprod.* 8, 789–796.
658. Binder, C., Simon, A., Binder, L., Hagemann, T., Schulz, M., Emons, G., Trumper, L., and Einspanier, A. A. (2004). Elevated concentrations of serum relaxin are associated with metastatic disease in breast cancer patients. *Breast Cancer Res. Treat.* 87, 157–166.
659. Summerlee, A. J., O'Byrne, K. T., and Poterski, R. S. (1998). Relaxin inhibits the pulsatile release of oxytocin but increases basal concentrations of hormone in lactating rats. *Biol. Reprod.* 58, 977–981.
660. Peaker, M., Fleet, I. R., Davis, A. J., and Taylor, E. (1995). The effects of relaxin on the response of intramammary pressure and mammary blood flow to exogenous oxytocin in the goat. *Exp. Physiol.* 80, 1047–1052.
661. Hisaw, F. L. (1964). Effects of relaxin on the uterus of monkeys (*Macaca mulatta*) with observations on the cervix and symphysis pubis. *Am. J. Obstet. Gynecol.* 89, 141–155.
662. Hayes, E. S., Curnow, E. C., Trounson, A. O., Danielson, L. A., and Unemori, E. N. (2004). Implantation and pregnancy following in vitro fertilization and the effect of recombinant human relaxin administration in *Macaca fascicularis*. *Biol. Reprod.* 71, 1591–1597.
663. Unemori, E. N., Erikson, M. E., Rocco, S. E., Sutherland, K. M., Parsell, D. A., Mak, J., and Grove, B. H. (1999). Relaxin stimulates expression of vascular endothelial growth factor in normal human endometrial cells in vitro and is associated with menometrorrhagia in women. *Hum. Reprod.* 14, 800–806.
664. Stewart, D. R., Overstreet, J. W., Celniker, A. C., Hess, D. L., Cragun, J. R., Boyers, S. P., and Lasley, B. L. (1993). The relationship between hCG and relaxin secretion in normal pregnancies vs peri-implantation spontaneous abortions. *Clin. Endocrinol. (Oxf)* 38, 379–385.
665. Stewart, D. R., and VandeVoort, C. A. (1999). Relaxin secretion by human granulosa cell culture is predictive of in-vitro fertilization-embryo transfer success. *Hum. Reprod.*14, 338–344.
666. Ghosh, D., De, P., and Sengupta, J. (1994). Luteal phase ovarian oestrogen is not essential for implantation and maintenance of pregnancy from surrogate embryo transfer in the rhesus monkey. *Hum. Reprod.* 9, 629–637.
667. Bryant-Greenwood, G. D., Rutanen, E. M., Partanen, S., Coelho, T. K., and Yamamoto, S. Y. (1993). Sequential appearance of relaxin, prolactin and IGFBP-1 during growth and differentiation of the human endometrium. *Mol. Cell. Endocrinol.* 95, 23–29.
668. Telgmann, R., Maronde, E., Tasken, K., and Gellersen, B. (1997). Activated protein kinase A is required for differentiation-dependent transcription of the decidual prolactin gene in human endometrial stromal cells. *Endocrinology* 138, 929–937.
669. Telgmann, R., and Gellersen, B. (1998). Marker genes of decidualization: activation of the decidual prolactin gene. *Hum. Reprod. Update* 4, 472–479.
670. Mazella, J., Tang, M., and Tseng, L. (2004). Disparate effects of relaxin and TGFβ1: relaxin increases, but TGFβ1 inhibits, the relaxin receptor and the production of IGFBP-1 in human endometrial stromal/decidual cells. *Hum. Reprod.* 19, 1513–1518.
671. Lane, B., Oxberry, W., Mazella, J., and Tseng, L. (1994). Decidualization of human endometrial stromal cells in vitro: effects of progestin and relaxin on the ultrastructure and production of decidual secretory proteins. *Hum. Reprod.* 9, 259–266.
672. Unemori, E. N., Lewis, M., Grove, B. H., and Deshpande, U. (2001). Relaxin induces specific alterations in gene expression in the human endometrium. In *Relaxin 2000: Proceedings of the Third International Conference on Relaxin and Related Peptides* (G. W. Tregear, R. Ivell, R. A. Bathgate, and J. D. Wade, Eds.). Kluwer, Dordrecht.
673. Sunder, S., and Lenton, E. A. (2000). Endocrinology of the peri-implantation period. *Baillieres Best. Pract. Res. Clin. Obstet. Gynaecol.* 14, 789–800.
674. Seppala, M., Koistinen, H., and Koistinen, R. (2001). Glycodelins. *Trends Endocrinol. Metab.* 12, 111–117.
675. Stewart, D. R., Erikson, M. S., Erikson, M. E., Nakajima, S. T., Overstreet, J. W., Lasley, B. L., Amento, E. P., and Seppala, M. (1997). The role of relaxin in glycodelin secretion. *J. Clin. Endocrinol. Metab.* 82, 839–846.
676. Tseng, L., Zhu, H. H., Mazella, J., Koistinen, H., and Seppala, M. (1999). Relaxin stimulates glycodelin mRNA and protein concentrations in human endometrial glandular epithelial cells. *Mol. Hum. Reprod.* 5, 372–375.
677. Gui, Y., Zhang, J., Yuan, L., and Lessey, B. A. (1999). Regulation of HOXA-10 and its expression in normal and abnormal endometrium. *Mol. Hum. Reprod.* 5, 866–873.
678. Finn, C. A., Pope, M. D., and Milligan, S. R. (1996). Relaxin and decidualization in mice: a reappraisal. *Biol. Reprod.* 55, 1415–1418.
679. Bagnell, C. A., Zhang, Q., Downey, B., and Ainsworth, L. (1993). Sources and biological actions of relaxin in pigs. *J. Reprod. Fertil. Suppl.* 48, 127–138.
680. Crish, J. F., Soloff, M. S., and Shaw, A. R. (1986). Changes in relaxin precursor mRNA levels in the rat ovary during pregnancy. *J. Biol. Chem.* 261, 1909–1913.
681. Shirota, K., Tateishi, K., Koji, T., Hishikawa, Y., Hachisuga, T., Kuroki, M., and Kawarabayashi, T. (2005). Early human preantral follicles have relaxin and relaxin receptor (LGR7) and relaxin promotes their development. *J. Clin. Endocrinol. Metab.* 90, 516–521.
682. Zhang, Q., and Bagnell, C. A. (1993). Relaxin stimulation of porcine granulosa cell deoxyribonucleic acid synthesis in vitro: interactions with insulin and insulin-like growth factor I. *Endocrinology* 132, 1643–1650.
683. Zhang, Q., and Bagnell, C. A. (1994). Trophic actions of relaxin on porcine theca cells: interactions with insulin and insulin-like growth factor I in vitro. *Endocr. J.* 2, 349–355.
684. Ohleth, K. M., and Bagnell, C. A. (1995). Relaxin-induced deoxyribonucleic acid synthesis in porcine granulosa cells is mediated by insulin-like growth factor-I. *Biol. Reprod.* 53, 1286–1292.
685. Ohleth, K. M., Zhang, Q., and Bagnell, C. A. (1998). Relaxin protein and gene expression in ovarian follicles of immature pigs. *J. Mol. Endocrinol.* 21, 179–187.
686. Too, C. K., Bryant-Greenwood, G. D., and Greenwood, F. C. (1984). Relaxin increases the release of plasminogen activator, collagenase, and proteoglycanase from rat granulosa cells in vitro. *Endocrinology* 115, 1043–1050.
687. Hwang, J. J., Lin, S. W., Teng, C. H., Ke, F. C., and Lee, M. T. (1996). Relaxin modulates the ovulatory process and increases secretion of different gelatinases from granulosa and theca-interstitial cells in rats. *Biol. Reprod.* 55, 1276–1283.
688. Brannstrom, M., and MacLennan, A. H. (1993). Relaxin induces ovulations in the in-vitro perfused rat ovary. *Hum. Reprod.* 8, 1011–1014.
689. Greenberg, L. H., Stouffer, R. L., Brenner, R. M., Molskness, T. A., Hild-Petito, S. A., and Yu, Q. (1990). Are human luteinizing granulosa cells a site of action for progesterone and relaxin? *Fertil. Steril.* 53, 446–453.

690. Hartshorne, G. M., Sargent, I. L., and Barlow, D. H. (1994). Growth rates and antrum formation of mouse ovarian follicles in vitro in response to follicle-stimulating hormone, relaxin, cyclic AMP and hypoxanthine. *Hum. Reprod.* 9, 1003–1012.
691. Steinetz, B. G., Beach, V. L., Tripp, L. V., and Defalco, R. J. (1964). Reactions of antisera to porcine relaxin with relaxin-containing tissues of other species in vivo and in vitro. *Acta. Endocrinol. (Copenh)* 47, 371–384.
692. Steinetz, B. G., Schwabe, C., Callard, I. P., and Goldsmith, L. T. (1998). Dogfish shark (*Squalus acanthias*) testes contain a relaxin. *J. Androl.* 19, 110–115.
693. Samuel, C. S., Tian, H., Zhao, L., and Amento, E. P. (2003). Relaxin is a key mediator of prostate growth and male reproductive tract development. *Lab. Invest.* 83, 1055–1067.
694. Garibay-Tupas, J. L., Bao, S., Kim, M. T., Tashima, L. S., and Bryant-Greenwood, G. D. (2000). Isolation and analysis of the 3′-untranslated regions of the human relaxin H1 and H2 genes. *J. Mol. Endocrinol.* 24, 241–252.
695. Kohsaka, T., Takahara, H., Sasada, H., Kawarasaki, T., Bamba, K., Masaki, J., and Tagami, S. (1992). Evidence for immunoreactive relaxin in boar seminal vesicles using combined light and electron microscope immunocytochemistry. *J. Reprod. Fertil.* 95, 397–408.
696. Lucas, C., Bald, L. N., Martin, M. C., Jaffe, R. B., Drolet, D. W., Mora-Worms, M., Bennett, G., Chen, A. B., and Johnston, P. D. (1989). An enzyme-linked immunosorbent assay to study human relaxin in human pregnancy and in pregnant rhesus monkeys. *J. Endocrinol.* 120, 449–457.
697. Brenner, S. H., Lesing, J. B., Schoenfeld, C., Goldsmith, L. T., Amelar, R., Dubin, L., and Weiss, G. (1987). Human semen relaxin and its correlation with the parameters of semen analysis. *Fertil. Steril.* 47, 714–716.
698. Colon, J. M., Ginsburg, F., Lessing, J. B., Schoenfeld, C., Goldsmith, L. T., Amelar, R. D., Dubin, L., and Weiss, G. (1986). The effect of relaxin and prostaglandin E2 on the motility of human spermatozoa. *Fertil. Steril.* 46, 1133–1139.
699. De Cooman, S., Gilliaux, P., and Thomas, K. (1983). Immunoreactive relaxin-like substance in human split ejaculates. *Fertil. Steril.* 39, 111–113.
700. Essig, M., Schoenfeld, C., D'Eletto, R. T., Amelar, R., Steinetz, B. G., O'Byrne, E. M., and Weiss, G. (1982). Relaxin in human seminal plasma. *Ann. N Y Acad. Sci.* 380, 224–230.
701. Harris, M. A., Rees, J. M., McLaughlin, E. A., Ford, W. C., Wardle, P. G., Hull, M. G., and Wathes, D. C. (1988). An evaluation of the role of relaxin in the penetration of cervical mucus by spermatozoa. *Hum. Reprod.* 3, 856–860.
702. Loumaye, E., De Cooman, S., and Thomas, K. (1980). Immunoreactive relaxin-like substance in human seminal plasma. *J. Clin. Endocrinol. Metab.* 50, 1142–1143.
703. Schieferstein, G., Voelter, W., Seeger, H., and Lippert, T. H. (1989). Immunoreactive relaxin in seminal plasma of man. *Int. J. Fertil.* 34, 215–218.
704. Weiss, G. (1989). Relaxin in the male. *Biol. Reprod.* 40, 197–200.
705. Sarosi, P., Schoenfeld, C., Berman, J., Basch, R., Randolph, G., Amelar, R., Dubin, L., Steinetz, B. G., and Weiss, G. (1983). Effect of anti-relaxin antiserum on sperm motility in vitro. *Endocrinology* 112, 1860–1861.
706. Sokol, R. Z., Okuda, H., Johnston, P. D., and Swerdloff, R. S. (1988). Videomicrographic analysis of the effects of antihuman relaxin antibody on human sperm motility. *Fertil. Steril.* 49, 729–731.
707. Lessing, J. B., Brenner, S. H., Schoenfeld, C., Sarosi, P., Amelar, R., Dubin, L., and Weiss, G. (1984). The effect of an anti-insulin antiserum on human sperm motility. *Fertil. Steril.* 42, 309–311.
708. Sasaki, Y., Kohsaka, T., Kawarasaki, T., Sasada, H., Ogine, T., Bamba, K., and Takahara, H. (2001). Immunoreactive relaxin in seminal plasma of fertile boars and its correlation with sperm motility characteristics determined by computer-assisted digital image analysis. *Int. J. Androl.* 24, 24–30.
709. Carrell, D. T., Peterson, C. M., and Urry, R. L. (1995). The binding of recombinant human relaxin to human spermatozoa. *Endocr. Res.* 21, 697–707.
710. Unemori, E. N., and Amento, E. P. (1990). Relaxin modulates synthesis and secretion of procollagenase and collagen by human dermal fibroblasts. *J. Biol. Chem.* 265, 10681–10685.
711. Unemori, E. N., Pickford, L. B., Salles, A. L., Piercy, C. E., Grove, B. H., Erikson, M. E., and Amento, E. P. (1996). Relaxin induces an extracellular matrix-degrading phenotype in human lung fibroblasts in vitro and inhibits lung fibrosis in a murine model in vivo. *J. Clin. Invest.* 98, 2739–2745.
712. Kibblewhite, D., Larrabee, W. F. Jr., and Sutton, D. (1992). The effect of relaxin on tissue expansion. *Arch. Otolaryngol. Head Neck Surg.* 118, 153–156.
713. Unemori, E. N., Beck, L. S., Lee, W. P., Xu, Y., Siegel, M., Keller, G., Liggitt, H. D., Bauer, E. A., and Amento, E. P. (1993). Human relaxin decreases collagen accumulation in vivo in two rodent models of fibrosis. *J. Invest. Dermatol.* 101, 280–285.
714. Williams, E. J., Benyon, R. C., Trim, N., Hadwin, R., Grove, B. H., Arthur, M. J., Unemori, E. N., and Iredale, J. P. (2001). Relaxin inhibits effective collagen deposition by cultured hepatic stellate cells and decreases rat liver fibrosis in vivo. *Gut* 49, 577–583.
715. Bennett, R. G., Kharbanda, K. K., and Tuma, D. J. (2003). Inhibition of markers of hepatic stellate cell activation by the hormone relaxin. *Biochem. Pharmacol.* 66, 867–874.
716. Garber, S. L., Mirochnik, Y., Brecklin, C. S., Unemori, E. N., Singh, A. K., Slobodskoy, L., Grove, B. H., Arruda, J. A., and Dunea, G. (2001). Relaxin decreases renal interstitial fibrosis and slows progression of renal disease. *Kidney Int.* 59, 876–882.
717. Garber, S. L., Mirochnik, Y., Brecklin, C., Slobodskoy, L., Arruda, J. A., and Dunea, G. (2003). Effect of relaxin in two models of renal mass reduction. *Am. J. Nephrol.* 23, 8–12.
718. Casten, G. G., and Boucek, R. J. (1958). Use of relaxin in the treatment of scleroderma. *JAMA* 166, 319–324.
719. Seibold, J. R. (1997). Scleroderma. In *Textbook of Rheumatology* (W. M. Kelley, E. D. Harris, S. Ruddy, and C. B. Sledge, Eds.), pp. 1133–1162. WB Saunders, Philadelphia.
720. Seibold, J. R., Korn, J. H., Simms, R., Clements, P. J., Moreland, L. W., Mayes, M. D., Furst, D. E., Rothfield, N., Steen, V., Weisman, M., Collier, D., Wigley, F. M., Merkel, P. A., Csuka, M. E., Hsu, V., Rocco, S., Erikson, M., Hannigan, J., Harkonen, W. S., and Sanders, M. E. (2000). Recombinant human relaxin in the treatment of scleroderma: a randomized, double-blind, placebo-controlled trial. *Ann. Intern. Med.* 132, 871–879.
721. Du, X. J., Samuel, C. S., Gao, X. M., Zhao, L., Parry, L. J., and Tregear, G. W. (2003). Increased myocardial collagen and ventricular diastolic dysfunction in relaxin deficient mice: a gender-specific phenotype. *Cardiovasc. Res.* 57, 395–404.
722. Samuel, C. S., Zhao, C., Bathgate, R. A., Bond, C. P., Burton, M. D., Parry, L. J., Summers, R. J., Tang, M. L., Amento, E. P., and Tregear, G. W. (2003). Relaxin deficiency in mice is associated with an age-related progression of pulmonary fibrosis. *FASEB J.* 17, 121–123.
723. Samuel, C. S., Zhao, C., Bond, C. P., Hewitson, T. D., Amento, E. P., and Summers, R. J. (2004). Relaxin-1-deficient mice develop an age-related progression of renal fibrosis. *Kidney Int.* 65, 2054–2064.

724. Amento, E. P. e. a. (2001). Deletion of the relaxin gene causes age-related progressive dermal fibrosis. *Arthritis Rheum.* 44, A847.
725. Huang, X., Arnold, G., Lewis, M., Guzman, L., Grove, B. H., Unemori, E. N., and Zsebo, K. (2001). Effects of relaxin on normal and impaired wound healing in rodents. In *Relaxin 2000: Proceedings of the Third International Conference on Relaxin and Related Peptides* (G. W. Tregear, R. Ivell, R. A. Bathgate, and J. D. Wade, Eds.). Kluwer, Dordrecht.
726. Casten, G. G., Gilmore, H. R., Houghton, F. E., and Samuels, S. S. (1960). A new approach to the management of obliterative peripheral arterial disease. *Angiology* 11, 408–414.
727. Bani, D., Nistri, S., Quattrone, S., Bigazzi, M., and Sacchi, T. B. (2001). Relaxin causes changes of the liver: in vivo studies in rats. *Horm. Metab. Res.* 33, 175–180.
728. Unemori, E. N., Lewis, M., Constant, J., Arnold, G., Grove, B. H., Normand, J., Deshpande, U., Salles, A., Pickford, L. B., Erikson, M. E., Hunt, T. K., and Huang, X. (2000). Relaxin induces vascular endothelial growth factor expression and angiogenesis selectively at wound sites. *Wound Repair Regen.* 8, 361–370.
729. Lewis, M., Deshpande, U., Guzman, L., Grove, B. H., Huang, X., Erikson, M. E., Pickford, L. B., and Unemori, E. N. (2001). Systemic relaxin administration stimulates angiogenic cytokine expression and vessel formation in a rat myocardial infarct model. In *Relaxin 2000: Proceedings of the Third International Conference on Relaxin and Related Peptides* (G. W. Tregear, R. Ivell, R. A. Bathgate, and J. D. Wade, Eds.), pp. 159–167. Kluwer, Dordrecht.
730. Bani, D., Bigazzi, M., Masini, E., Bani, G., and Sacchi, T. B. (1995). Relaxin depresses platelet aggregation: in vitro studies on isolated human and rabbit platelets. *Lab Invest.* 73, 709–716.
731. Bani, D., Maurizi, M., and Bigazzi, M. (1995). Relaxin reduces the number of circulating platelets and depressed platelet release from megakaryocytes: studies in rats. *Platelets* 6, 330–335.
732. Nistri, S., Chiappini, L., Sassoli, C., and Bani, D. (2003). Relaxin inhibits lipopolysaccharide-induced adhesion of neutrophils to coronary endothelial cells by a nitric oxide-mediated mechanism. *FASEB J.* 17, 2109–2111.
733. Masini, E., Nistri, S., Vannacci, A., Bani Sacchi, T., Novelli, A., and Bani, D. (2004). Relaxin inhibits the activation of human neutrophils: involvement of the nitric oxide pathway. *Endocrinology* 145, 1106–1112.
734. Taylor, M. J., and Clark, C. L. (1994). Evidence for a novel source of relaxin: atrial cardiocytes. *J. Endocrinol.* 143, R5–8.
735. Fisher, C., Berry, C., Blue, L., Morton, J. J., and McMurray, J. (2003). N-terminal pro B type natriuretic peptide, but not the new putative cardiac hormone relaxin, predicts prognosis in patients with chronic heart failure. *Heart* 89, 879–881.
736. Bani, D., Baronti, R., Vannacci, A., Bigazzi, M., Sacchi, T. B., Mannaioni, P. F., and Masini, E. (2002). Inhibitory effects of relaxin on human basophils activated by stimulation of the Fc epsilon receptor: the role of nitric oxide. *Int. Immunopharmacol.* 2, 1195–1204.
737. Masini, E., Bani, D., Bigazzi, M., Mannaioni, P. F., and Bani-Sacchi, T. (1994). Effects of relaxin on mast cells. In vitro and in vivo studies in rats and guinea pigs. *J. Clin. Invest.* 94, 1974–1980.
738. Bani, D., Ballati, L., Masini, E., Bigazzi, M., and Sacchi, T. B. (1997). Relaxin counteracts asthma-like reaction induced by inhaled antigen in sensitized guinea pigs. *Endocrinology* 138, 1909–1915.
739. Masini, E., Zagli, G., Ndisang, J. F., Solazzo, M., Mannaioni, P. F., and Bani, D. (2002). Protective effect of relaxin in cardiac anaphylaxis: involvement of the nitric oxide pathway. *Br. J. Pharmacol.* 137, 337–344.
740. Baccari, M. C., Nistri, S., Quattrone, S., Bigazzi, M., Bani Sacchi, T., Calamai, F., and Bani, D. (2004). Depression by relaxin of neurally induced contractile responses in the mouse gastric fundus. *Biol. Reprod.* 70, 222–228.
741. del Angel Meza, A. R., Beas-Zarate, C., Alfaro, F. L., and Morales-Villagran, A. (1991). A simple biological assay for relaxin measurement. *Comp. Biochem. Physiol. C.* 99, 35–39.
742. Bathgate, R., Moniac, N., Bartlick, B., Schumacher, M., Fields, M., and Ivell, R. (1999). Expression and regulation of relaxin-like factor gene transcripts in the bovine ovary: differentiation-dependent expression in theca cell cultures. *Biol. Reprod.* 61, 1090–1098.
743. Tashima, L. S., Hieber, A. D., Greenwood, F. C., and Bryant-Greenwood, G. D. (1995). The human Leydig insulin-like (hLEY I-L) gene is expressed in the corpus luteum and trophoblast. *J. Clin. Endocrinol. Metab.* 80, 707–710.
744. Bamberger, A. M., Ivell, R., Balvers, M., Kelp, B., Bamberger, C. M., Riethdorf, L., and Loning, T. (1999). Relaxin-like factor (RLF): a new specific marker for Leydig cells in the ovary. *Int. J. Gynecol. Pathol.* 18, 163–168.
745. Balvers, M., Spiess, A. N., Domagalski, R., Hunt, N., Kilic, E., Mukhopadhyay, A. K., Hanks, E., Charlton, H. M., and Ivell, R. (1998). Relaxin-like factor expression as a marker of differentiation in the mouse testis and ovary. *Endocrinology* 139, 2960–2970.
746. Hombach-Klonisch, S., Seeger, S., Tscheudschilsuren, G., Buchmann, J., Huppertz, B., Seliger, G., Fischer, B., and Klonisch, T. (2001). Cellular localization of human relaxin-like factor in the cyclic endometrium and placenta. *Mol. Hum. Reprod.* 7, 349–356.
747. Hombach-Klonisch, S., Buchmann, J., Sarun, S., Fischer, B., and Klonisch, T. (2000). Relaxin-like factor (RLF) is differentially expressed in the normal and neoplastic human mammary gland. *Cancer* 89, 2161–2168.
748. Hombach-Klonisch, S., Hoang-Vu, C., Kehlen, A., Hinze, R., Holzhausen, H. J., Weber, E., Fischer, B., Dralle, H., and Klonisch, T. (2003). INSL-3 is expressed in human hyperplastic and neoplastic thyrocytes. *Int. J. Oncol.* 22, 993–1001.
749. Ivell, R., Balvers, M., Domagalski, R., Ungefroren, H., Hunt, N., and Schulze, W. (1997). Relaxin-like factor: a highly specific and constitutive new marker for Leydig cells in the human testis. *Mol. Hum. Reprod.* 3, 459–466.
750. Klonisch, T., Ivell, R., Balvers, M., Kliesch, S., Fischer, B., Bergmann, M., and Steger, K. (1999). Expression of relaxin-like factor is down-regulated in human testicular Leydig cell neoplasia. *Mol. Hum. Reprod.* 5, 104–108.
751. Paust, H. J., Wessels, J., Ivell, R., and Mukhopadhyay, A. K. (2002). The expression of the RLF/INSL3 gene is reduced in Leydig cells of the aging rat testis. *Exp. Gerontol.* 37, 1461–1467.
752. Hombach-Klonisch, S., Schon, J., Kehlen, A., Blottner, S., and Klonisch, T. (2004). Seasonal expression of INSL3 and Lgr8/Insl3 receptor transcripts indicates variable differentiation of Leydig cells in the roe deer testis. *Biol. Reprod.* 71, 1079–1087.
753. Boockfor, F. R., Fullbright, G., Büllesbach, E. E., and Schwabe, C. (2001). Relaxin-like factor (RLF) serum concentrations and gubernaculum RLF receptor display in relation to pre- and neonatal development of rats. *Reproduction* 122, 899–906.
754. Büllesbach, E. E., Rhodes, R., Rembiesa, B., and Schwabe, C. (1999). The relaxin-like factor is a hormone. *Endocrine* 10, 167–169.
755. Foresta, C., Bettella, A., Vinanzi, C., Dabrilli, P., Meriggiola, M.-C., Garolla, A., and Ferlin, A. (2004). Insulin-like factor:

a novel circulating hormone of testis origin in humans. *J. Clin. Endocrinol. Metab.* 89, 5952–5958.

756. Nguyen, M. T., Showalter, P. R., Timmons, C. F., Nef, S., Parada, L. F., and Baker, L. A. (2002). Effects of orchiopexy on congenitally cryptorchid insulin-3 knockout mice. *J. Urol.* 168, 1779–1783; discussion 1783.
757. Kubota, Y., Nef, S., Farmer, P. J., Temelcos, C., Parada, L. F., and Hutson, J. M. (2001). Leydig insulin-like hormone, gubernacular development and testicular descent. *J. Urol.* 165, 1673–1675.
758. Tomiyama, H., Hutson, J. M., Truong, A., and Agoulnik, A. I. (2003). Transabdominal testicular descent is disrupted in mice with deletion of insulinlike factor 3 receptor. *J. Pediatr. Surg.* 38, 1793–1798.
759. Hutson, J. M., Hasthorpe, S., and Heyns, C. F. (1997). Anatomical and functional aspects of testicular descent and cryptorchidism. *Endocr. Rev.* 18, 259–280.
760. Fentener van Vlissingen, J. M., van Zoelen, E. J., Ursem, P. J., and Wensing, C. J. (1988). In vitro model of the first phase of testicular descent: identification of a low molecular weight factor from fetal testis involved in proliferation of gubernaculum testis cells and distinct from specified polypeptide growth factors and fetal gonadal hormones. *Endocrinology* 123, 2868–2877.
761. Emmen, J. M., McLuskey, A., Adham, I. M., Engel, W., Grootegoed, J. A., and Brinkmann, A. O. (2000). Hormonal control of gubernaculum development during testis descent: gubernaculum outgrowth in vitro requires both insulin-like factor and androgen. *Endocrinology* 141, 4720–4727.
762. Adham, I. M., Steding, G., Thamm, T., Büllesbach, E. E., Schwabe, C., Paprotta, I., and Engel, W. (2002). The overexpression of the Insl3 in female mice causes descent of the ovaries. *Mol. Endocrinol.* 16, 244–252.
763. Koskimies, P., Suvanto, M., Nokkala, E., Huhtaniemi, I. T., McLuskey, A., Themmen, A. P., and Poutanen, M. (2003). Female mice carrying a ubiquitin promoter-Insl3 transgene have descended ovaries and inguinal hernias but normal fertility. *Mol. Cell. Endocrinol.* 206, 159–166.
764. Berkowitz, G. S., Lapinski, R. H., Dolgin, S. E., Gazella, J. G., Bodian, C. A., and Holzman, I. R. (1993). Prevalence and natural history of cryptorchidism. *Pediatrics* 92, 44–49.
765. Klonisch, T., Fowler, P. A., and Hombach-Klonisch, S. (2004). Molecular and genetic regulation of testis descent and external genitalia development. *Dev. Biol.* 270, 1–18.
766. Koskimies, P., Virtanen, H., Lindstrom, M., Kaleva, M., Poutanen, M., Huhtaniemi, I., and Toppari, J. (2000). A common polymorphism in the human relaxin-like factor (RLF) gene: no relationship with cryptorchidism. *Pediatr. Res.* 47, 538–541.
767. Krausz, C., Quintana-Murci, L., Fellous, M., Siffroi, J. P., and McElreavey, K. (2000). Absence of mutations involving the INSL3 gene in human idiopathic cryptorchidism. *Mol. Hum. Reprod.* 6, 298–302.
768. Tomboc, M., Lee, P. A., Mitwally, M. F., Schneck, F. X., Bellinger, M., and Witchel, S. F. (2000). Insulin-like 3/relaxin-like factor gene mutations are associated with cryptorchidism. *J. Clin. Endocrinol. Metab.* 85, 4013–4018.
769. Lim, H. N., Raipert-de Meyts, E., Skakkebaek, N. E., Hawkins, J. R., and Hughes, I. A. (2001). Genetic analysis of the INSL3 gene in patients with maldescent of the testis. *Eur. J. Endocrinol.* 144, 129–137.
770. Marin, P., Ferlin, A., Moro, E., Garolla, A., and Foresta, C. (2001). Different insulin-like 3 (INSL3) gene mutations not associated with human cryptorchidism. *J. Endocrinol. Invest.* 24, RC13–15.
771. Marin, P., Ferlin, A., Moro, E., Rossi, A., Bartoloni, L., Rossato, M., and Foresta, C. (2001). Novel insulin-like 3 (INSL3) gene mutation associated with human cryptorchidism. *Am. J. Med. Genet.* 103, 348–349.
772. Takahashi, I., Takahashi, T., Komatsu, M., Matsuda, J., and Takada, G. (2001). Ala/Thr60 variant of the Leydig insulin-like hormone is not associated with cryptorchidism in the Japanese population. *Pediatr. Int.* 43, 256–258.
773. Baker, L. A., Nef, S., Nguyen, M. T., Stapleton, R., Nordenskjold, A., Pohl, H., and Parada, L. F. (2002). The insulin-3 gene: lack of a genetic basis for human cryptorchidism. *J. Urol.* 167, 2534–2537.
774. Canto, P., Escudero, I., Soderlund, D., Nishimura, E., Carranza-Lira, S., Gutierrez, J., Nava, A., and Mendez, J. P. (2003). A novel mutation of the insulin-like 3 gene in patients with cryptorchidism. *J. Hum. Genet.* 48, 86–90.
775. Ferlin, A., Simonato, M., Bartoloni, L., Rizzo, G., Bettella, A., Dottorini, T., Dallapiccola, B., and Foresta, C. (2003). The INSL3-LGR8/GREAT ligand-receptor pair in human cryptorchidism. *J. Clin. Endocrinol. Metab.* 88, 4273–4279.
776. Roh, J., Virtanen, H., Kumagai, J., Sudo, S., Kaleva, M., Toppari, J., and Hsueh, A. J. (2003). Lack of LGR8 gene mutation in Finnish patients with a family history of cryptorchidism. *Reprod. Biomed. Online* 7, 400–406.
777. Nef, S., Shipman, T., and Parada, L. F. (2000). A molecular basis for estrogen-induced cryptorchidism. *Dev. Biol.* 224, 354–361.
778. Gill, W. B., Schumacher, G. F., Bibbo, M., Straus, F. H. 2nd, and Schoenberg, H. W. (1979). Association of diethylstilbestrol exposure in utero with cryptorchidism, testicular hypoplasia and semen abnormalities. *J. Urol.* 122, 36–39.
779. Emmen, J. M., McLuskey, A., Adham, I. M., Engel, W., Verhoef-Post, M., Themmen, A. P., Grootegoed, J. A., and Brinkmann, A. O. (2000). Involvement of insulin-like factor 3 (Insl3) in diethylstilbestrol-induced cryptorchidism. *Endocrinology* 141, 846–849.
780. Gray, L. E. Jr., Ostby, J., Furr, J., Price, M., Veeramachaneni, D. N., and Parks, L. (2000). Perinatal exposure to the phthalates DEHP, BBP, and DINP, but not DEP, DMP, or DOTP, alters sexual differentiation of the male rat. *Toxicol. Sci.* 58, 350–365.
781. Wilson, V. S., Lambright, C., Furr, J., Ostby, J., Wood, C., Held, G., and Gray, L. E. Jr. (2004). Phthalate ester-induced gubernacular lesions are associated with reduced insl3 gene expression in the fetal rat testis. *Toxicol. Lett.* 146, 207–215.
782. Spanel-Borowski, K., Schafer, I., Zimmermann, S., Engel, W., and Adham, I. M. (2001). Increase in final stages of follicular atresia and premature decay of corpora lutea in Insl3-deficient mice. *Mol. Reprod. Dev.* 58, 281–286.
783. Irving-Rodgers, H. F., Bathgate, R. A., Ivell, R., Domagalski, R., and Rodgers, R. J. (2002). Dynamic changes in the expression of relaxin-like factor (INSL3), cholesterol side-chain cleavage cytochrome p450, and 3beta-hydroxysteroid dehydrogenase in bovine ovarian follicles during growth and atresia. *Biol. Reprod.* 66, 934–943.
784. Richards, J. S., Russell, D. L., Ochsner, S., and Espey, L. L. (2002). Ovulation: new dimensions and new regulators of the inflammatory-like response. *Annu. Rev. Physiol.* 64, 69–92.
785. McGee, E. A., and Hsueh, A. J. (2000). Initial and cyclic recruitment of ovarian follicles. *Endocr. Rev.* 21, 200–214.
786. Huhtaniemi, I. T. (1983). Gonadotrophin receptors: correlates with normal and pathological functions of the human ovary and testis. *Clin. Endocrinol. Metab.* 12, 117–132.
787. Heckert, L. L., and Griswold, M. D. (2002). The expression of the follicle-stimulating hormone receptor in spermatogenesis. *Recent Prog. Horm. Res.* 57, 129–148.
788. Tsafriri, A., and Pomerantz, S. H. (1986). Oocyte maturation inhibitor. *Clin. Endocrinol. Metab.* 15, 157–170.

789. Wassarman, P. M., Schultz, R. M., Letourneau, G. E., LaMarca, M. J., Josefowicz, W. J., and Bleil, J. D. (1979). Meiotic maturation of mouse oocytes in vitro. *Adv. Exp. Med. Biol.* 112, 251–268.
790. Huckins, C., and Oakberg, E. F. (1978). Morphological and quantitative analysis of spermatogonia in mouse testes using whole mounted seminiferous tubules: I. The normal testes. *Anat. Rec.* 192, 519–528.
791. Tapanainen, J. S., Tilly, J. L., Vihko, K. K., and Hsueh, A. J. (1993). Hormonal control of apoptotic cell death in the testis: gonadotropins and androgens as testicular cell survival factors. *Mol. Endocrinol.* 7, 643–650.
792. Eppig, J. J., and Downs, S. M. (1987). The effect of hypoxanthine on mouse oocyte growth and development in vitro: maintenance of meiotic arrest and gonadotropin-induced oocyte maturation. *Dev. Biol.* 119, 313–321.
793. Byskov, A. G., Andersen, C. Y., Nordholm, L., Thogersen, H., Xia, G., Wassmann, O., Andersen, J. V., Guddal, E., and Roed, T. (1995). Chemical structure of sterols that activate oocyte meiosis. *Nature* 374, 559–562.
794. Griswold, M. D. (1995). Interactions between germ cells and Sertoli cells in the testis. *Biol. Reprod.* 52, 211–216.
795. Reytomas, I. G. T., Burazin, T.C.D., Tregear, G.W., and Parry, L.J. (2002). Differential expression of relaxin-1 and relaxin-3 messenger RNA and localization of relaxin receptors in the pregnant mouse. *Biology of Reproduction* 66, 505.
796. Evans, B. A., John, M., Fowler, K. J., Summers, R. J., Cronk, M., Shine, J., and Tregear, G. W. (1993). The mouse relaxin gene: nucleotide sequence and expression. *J. Mol. Endocrinol.* 10, 15–23.
797. Sutin, E. L., and Jacobowitz, D. M. (1988). Immunocytochemical localization of peptides and other neurochemicals in the rat laterodorsal tegmental nucleus and adjacent area. *J. Comp. Neurol.* 270, 243–270.
798. Paxinos, G., and Watson, C. (1986). *The Rat Brain in Stereotaxic Coordinates*. Academic Press, Sydney.
799. Goto, M., Swanson, L. W., and Canteras, N. S. (2001). Connections of the nucleus incertus. *J. Comp. Neurol.* 438, 86–122.
800. Laurent, A., Rouillac, C., Delezoide, A. L., Giovangrandi, Y., Vekemans, M., Bellet, D., Abitbol, M., and Vidaud, M. (1998). Insulin-like 4 (INSL4) gene expression in human embryonic and trophoblastic tissues. *Mol. Reprod. Dev.* 51, 123–129.
801. Mock, P., Frydman, R., Bellet, D., Diawara, D. A., Lavaissiere, L., Troalen, F., and Bidart, J. M. (1999). Pro-EPIL forms are present in amniotic fluid and maternal serum during normal pregnancy. *J. Clin. Endocrinol. Metab.* 84, 2253–2256.
802. Johnson, D. S., Friel, P. J., Wright, W. W., Jaspers, S. R., Lok, S., and Griswold, M. D. (2000). Expression of insulin-like gene 6 (INSL6) in the rat testis. *Biol. Reprod.* 62, 339.
803. Eldridge, R. K., and Fields, P. A. (1986). Rabbit placental relaxin: ultrastructural localization in secretory granules of the syncytiotrophoblast using rabbit placental relaxin antiserum. *Endocrinology* 119, 606–615.
804. Bryant-Greenwood, G., Ali, S., Mandel, M., and Greenwood, F. (1987). Ovarian and decidual relaxins in human pregnancy. *Adv. Exp. Med. Biol.* 219, 709–713.
805. Okada, T., and Palczewski, K. (2001). Crystal structure of rhodopsin: implications for vision and beyond. *Curr. Opin. Struct. Biol.* 11, 420–426.
806. Kobe, B., and Deisenhofer, J. (1993). Crystal structure of porcine ribonuclease inhibitor, a protein with leucine-rich repeats. *Nature* 366, 751–756.
807. Huang, W., Dolmer, K., and Gettins, P. G. (1999). NMR solution structure of complement-like repeat CR8 from the low density lipoprotein receptor-related protein. *J. Biol. Chem.* 274, 14130–14136.
808. Nakabayashi, K., Matsumi, H., Bhalla, A., Bae, J., Mosselman, S., Hsu, S. Y., and Hsueh, A. J. (2002). Thyrostimulin, a heterodimer of two new human glycoprotein hormone subunits, activates the thyroid-stimulating hormone receptor. *J. Clin. Invest.* 109, 1445–1452.
809. Hishikawa, K., Nakaki, T., Marumo, T., Suzuki, H., Kato, R., and Saruta, T. (1995). Pressure enhances endothelin-1 release from cultured human endothelial cells. *Hypertension* 25, 449–452.
810. Macarthur, H., Warner, T. D., Wood, E. G., Corder, R., and Vane, J. R. (1994). Endothelin-1 release from endothelial cells in culture is elevated both acutely and chronically by short periods of mechanical stretch. *Biochem. Biophys. Res. Commun.* 200, 395–400.
811. Russell, F. D., Skepper, J. N., and Davenport, A. P. (1998). Human endothelial cell storage granules: a novel intracellular site for isoforms of the endothelin-converting enzyme. *Circ. Res.* 83, 314–321.
812. Puyraimond, A., Fridman, R., Lemesle, M., Arbeille, B., and Menashi, S. (2001). MMP-2 colocalizes with caveolae on the surface of endothelial cells. *Exp. Cell. Res.* 262, 28–36.
813. Shaul, P. W. (2002). Regulation of endothelial nitric oxide synthase: location, location, location. *Annu. Rev. Physiol.* 64, 749–774.
814. Teixeira, A., Chaverot, N., Schroder, C., Strosberg, A. D., Couraud, P. O., and Cazaubon, S. (1999). Requirement of caveolae microdomains in extracellular signal-regulated kinase and focal adhesion kinase activation induced by endothelin-1 in primary astrocytes. *J. Neurochem.* 72, 120–128.
815. Conrad, K. P. (2004). Mechanisms of renal vasodilation and hyperfiltration during pregnancy. *J. Soc. Gynecol. Investig.* 11, 438–448.

Male Reproductive System

Knobil and Neill's Physiology of Reproduction, Third Edition
edited by Jimmy D. Neill,
Elsevier © 2006

CHAPTER 17

Anatomy, Vasculature, and Innervation of the Male Reproductive Tract

B. P. Setchell and W. G. Breed

Anatomy, 771
Testis, 771
Excurrent Ducts, 781
Accessory Sex Glands, 785
Penis, 789
Scrotum, 790
Vasculature, 791
Spermatic Cord, 791
Testis, 794
Excurrent Ducts, 800
Accessory Sex Glands, 801
Penis, 802
Scrotum, 803
Innervation, 803
Testis, 803
Epididymis and Ductus Deferens, 805
Accessory Sex Glands, 805
Penis, 807
Scrotum, 807
Conclusion, 807
Acknowledgments, 808
References, 808

The male reproductive tract in mammals consists of testes, epididymides, ductus deferentes, accessory sex glands, and penis. The testes differentiate in the fetus from indifferent gonads after expression of the SRY gene on the short arm of the Y chromosome (see Chapter 6). Fetal Leydig cells then secrete androgens that induce differentiation of the mesonephric (or wolffian) duct into the epididymis and ductus deferens and some of the accessory sex glands and of the indifferent external genitalia into a penis and in most eutherians, a scrotum. In the first part of this chapter, we present a comparative account of the anatomy of this system in the adult mammal. This information is presented using the four major groupings of eutherian mammals (409): Afrotheria (elephants, hyraxes, dugongs and elephant shrews), Xenarthra (armadillos, sloths, and anteaters), Laurasiatheria (insectivores, bats, ungulates, cetaceans, and carnivores), and Euarchontoglires (primates, rodents, and lagomorphs). A brief summary for marsupials and monotremes is also given. These data are presented as a background to the subsequent chapters, where more in-depth coverage is given to form and function of the various regions of the male reproductive tract using laboratory rodents, domestic mammals, and humans. In the second half of the chapter, a discussion of the vasculature and innervation is based largely on studies using common experimental and domestic species.

ANATOMY

Testis

Position

Eutherians. Apart from members of the Afrotheria, in which the testes remain close to the kidneys (see below), in other mammals caudal migration of testes takes place late in fetal life. In some species the testes remain near the bladder close to the dorsal abdominal body wall (e.g., whales, Order Cetacea; nine-banded armadillo and giant

Department of Anatomical Sciences, University of Adelaide, Adelaide, SA, Australia.

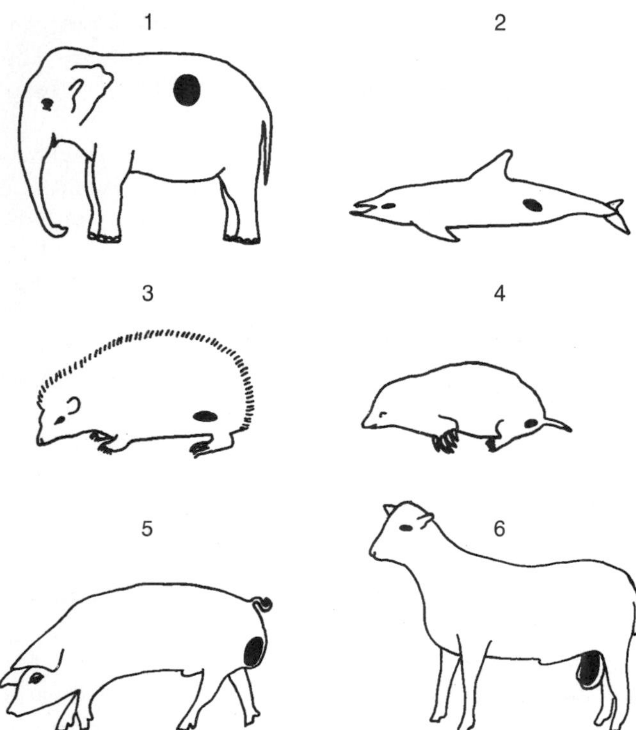

FIG. 1. A diagram illustrating the position of the testes in different mammals, with the testes near the kidneys in elephants (1), at the caudal end of the abdominal cavity in Cetacea (2), in the inguinal region in hedgehogs (3), in a cremasteric sac in moles (4), in a subanal scrotum in pigs (5), and in a pendulous scrotum in sheep (6). [Reproduced with permission from (76).]

anteater, Superorder Xenartha), in others they move close to the ventral abdominal body wall (e.g., most Laurasiatheria such as hedgehogs, Family Erinaceidae; insectivorous bats, Sub-Order Microchiroptera; and the true seals, Family Phocidae), although in most insectivores (e.g., shrews, Family Soricidae; and moles, Family Talpidae) the testes are located in shallow coelomic evaginations, the cremaster sacs, in the posterior region of the abdominal cavity (28,33–36,76,136, 191,384,513,514, 524,654). In other eutherian groups, the testes pass caudally into a scrotum, which may become pendulous. This occurs in both artiodactyls and primates (Fig. 1).

The biological significance of the evolution of the scrotum is an area of debate. It has been argued that migration of the testis into the scrotum helps to externalize the cauda epididymidis, which needs a cool environment for sperm storage [see reviews in (28,30,37,174,188,646)]. However, if the temperature of scrotal testes is raised to body temperature for any length of time, spermatogenesis is disrupted (521), and the requirement for a lower than abdominal temperature may be a secondary adaptation of metabolic needs for meiosis and spermiogenesis (31). It has been proposed (547) that testicular descent into a cool environment may have evolved to minimize the mutation rate in male germ cells. One of the more eccentric suggestions, first made early in the 20th century (667) and subsequently by Chase (80), was that scrotal testes were a result of the animals jumping or leaping. Another eccentric suggestion is that scrotal testes are involved in stimulation of the clitoris during mating (660).

In elephants, the testes remain next to the kidneys (286,549). Early work suggested that elephants might have a lower body temperature than species with scrotal testes, thus precluding the need for testicular descent into a scrotum for spermatogenesis to occur (659). However, subsequent measurements have shown elephants to have a body temperature of 36–37°C (549), like that of scrotal mammals. The testes are also at the same temperature (628), with normal germ cell associations being present (279,524), indicating that germ cell development occurs in these species at typical mammalian body temperature. In shrews, thermistor measurements have shown that the testicular temperature is only about 1.6°C lower than that of core body temperature (222), whereas in hedgehogs, which hibernate during winter, a temperature differential of 1.4°C between the testes and the abdominal region was found to occur late in the hibernating period such that as body temperature rose toward the end of hibernation, the testicular temperature remained low, perhaps to facilitate the initiation of spermatogenesis at this time (170). A slightly smaller differential of 0.7°C was found in anesthetized hedgehogs in the breeding season (517).

Marsupials. The testes of nearly all marsupials also descend into a scrotal sac. An exception occurs in marsupial moles, in which the testes are found between the skin and abdominal body wall (278,574). In Tammar wallabies, in which sexual differentiation has been investigated in some detail, it appears that in contrast to eutherians, formation of the scrotum may be independent of gonadal androgen secretion (472,657). The marsupial scrotum lies in a prepenile position, and its development appears, at least in part, to be dependent on the expression of genes on the X chromosome. As exemplified by Tammar wallabies, marsupial testes are appreciably cooler than body temperature (529).

Monotremes. In all three species of monotremes the testes are intraabdominal, where they are suspended from the abdominal body wall by folds of peritoneum (205,293).

The fact that the testes are abdominal or inguinal in the eutherian Afrotheria, cetaceans, insectivores, and edentates as well as in marsupial moles and monotremes but yet scrotally placed in two eutherian super orders and marsupial mammals invites speculation as to whether the scrotal or nonscrotal condition is the ancestral state. It has been suggested that the most parsimonious view is that scrotal testes evolved in a common ancestor of the eutherian and marsupial lineages with the occasional nonscrotal condition being a secondarily derived state (646). The alternative scenario would necessitate the scrotal condition at least evolving separately in each of the marsupial and eutherian lineages.

Mechanism of Testicular Descent

The descent of the testis from its origin near the kidney to the scrotum in those mammals in which it occurs has been the subject of many investigations (236,472,655). Initially, in both sexes the gonad is displaced caudally by the developing metanephros, a process referred to as nephric displacement. Subsequent descent of the testis can be divided into two stages, transabdominal migration and an inguinoscrotal phase. The transabdominal stage is completed early in fetal development, but the inguinoscrotal stage occurs late in fetal life in humans, pigs, sheep, and cattle and after birth in mice, rats, and dogs (25,145,262,643,645).

Two structures other than the testis itself are involved in its descent, the cranial suspensory ligament and the gubernaculum. The cranial suspensory ligament attaches the indifferent gonad to the cranial part of the dorsal wall of the abdominal cavity or the caudal pole of the kidney and normally regresses in males but not in females (609,610). In females, the ovaries ascend as the fetus grows, so that the transabdominal phase of testicular descent may in fact be a failure of the testis to ascend (542,605), because the testis is immobilized in the inguinal region by the gubernaculum, whereas the kidney and its associated organs move dorsocranially. The gubernaculum (608) runs between the testis and the inguinal area and consists of a cranial part, consisting of the gubernacular cord and a caudal bulb, which can be subdivided into intraabdominal and extraabdominal segments. The intraabdominal part is often referred to as the gubernacular cone. The whole complex is initially present in both male and female fetuses, but in females the gubernacular cord lengthens and the bulb does not enlarge; in males, however, the cord shortens and thickens and the bulb enlarges.

In rodents, during the second phase of testicular descent, the gubernaculum regresses and the bulb starts to invert or prolapse, forming a sac with the muscular layer of the cone as its wall. This sac forms the vaginal process, and the muscular wall becomes the cremaster muscle. However, the inguinal canal does not constrict so that, even in adults, the testis can readily pass from the scrotum into the abdominal cavity. In cows, sheep, and humans, which have a strip-like cremaster, the extraabdominal gubernaculum develops partly inside the vaginal process, which has already formed, and is partly in the tissue outside, beneath, and alongside the vaginal process. The vaginal process constricts, and in humans the inguinal canal is normally closed off after the testis has descended (147,251,262,643,644,645).

In Afrotheria, such as elephants and dugongs (Sirenia), there is no gubernaculum or inguinal canal (177), and hyraxes have no gubernaculum but do have an inguinal canal (607). In whales and dolphins, there is a gubernacular primordium but this does not develop (606). In cattle, when a female twin calf fetus is masculinized to become a free-martin through a vascular anastomosis with her male co-twin, a male-like gubernaculum develops and the bottom of the vaginal process becomes inverted, although the ovotestis does not descend because of the persistence of the cranial suspensory ligament in these animals (611). In the sheath-tail bat straps of muscle connect both the cranial and caudal poles of the testes, so that the testes return to the abdomen outside the breeding season, whereas the cauda epididymidis remains in the scrotum (282).

In marsupials, young are born at an early stage of development before full differentiation of the gonads has occurred, and both stages of testis descent occur after birth. Testicular descent starts early in pouch life, but in species such as possums and macropods, descent is not completed until around week 10 after birth (394,471,597). The vaginal process, like the scrotum, begins to develop independently of androgenic stimulation, before the embryonic gonad undergoes sexual differentiation. As the testis descends, the scrotum enlarges and so does the gubernaculum, especially its extraabdominal portion and the vaginal canal closes as in humans (103,471).

The control of testicular descent is still controversial. It appears to involve at least four factors; (a) insulin-like factor 3 (INSL3, also known as Leydig cell insulin-like factor or relaxin-like factor); (b) calcitonin gene-related peptide (CGRP); (c) müllerian-inhibiting substance (MIS), also known as anti-müllerian hormone; and (d) androgens (3,221,261,268, 269,316,324). Transgenic mice lacking INSL3 were cryptorchid (412,687), and INSL3 had a direct effect on cell proliferation of rat gubernacular cells in vitro, especially when combined with androgen (324). In female mice overexpressing INSL3, the ovaries "descend" because they are anchored in the inguinal region (4).

Testis descent was prevented by sectioning of the genitofemoral nerve, the nucleus of which is sexually dimorphic. The nerve contains CGRP, which can stimulate contractions of mouse gubernacula in vitro (262). Furthermore, treatment of cryptorchid neonatal pigs with CGRP stimulated migration of inguinal, but not abdominal, testes (263). However, treatment of mice with CGRP had no effect on postnatal testis descent (249). The involvement of the genitofemoral nerve is supported by evidence that cutting the gubernaculum in newborn rats near the abdominal wall prevented subsequent descent of the testis, whereas cutting it near the epididymis was without effect (43). The former operation would have severed the nerve to the gubernaculum, whereas the latter would not.

MIS appears in the fetal testis at about the same time as descent commences, and cryptorchidism is common in humans with persistent müllerian ducts, caused by absence of MIS. However, animal models of this condition do not always have cryptorchid testes unless they are also androgen resistant. Testis descent was normal in mice deficient in MIS itself (38), and development of the gubernaculum appeared normal in mice deficient in the receptor for MIS (23). Furthermore, rabbits treated with antibody to MIS had normal testis descent, although the müllerian ducts persisted (591).

It is generally agreed that androgens have an important effect on the inguinal–scrotal phase of testicular descent, and there is evidence for this from animals lacking gonadotropins or androgen receptors and those treated with antiandrogens. However, the site of action is still controversial, although it is now clear that androgens suppress development of the cranial suspensory ligament (148,337). Androgen action probably also involves indirect effects via the spinal nuclei of the genitofemoral nerve. Prenatal treatment of rats with the antiandrogen flutamide had no effect in fetuses up to 20 days postcoitum but had inhibited downward growth of the vaginal process in 5-day-old animals (543). Derangement of the gubernacular migration and delayed regression occurred in only about 50% of animals, although the dose was sufficient to block development of the prostate completely, suggesting that the effect on the gubernaculum was not direct but probably via the genitofemoral nerve (195,262). The androgen disruptor vinclozolin, administered prenatally, also caused aberrant migration of the gubernaculum and cryptorchidism in rats (545). In pigs, prenatal treatment with flutamide resulted in cryptorchidism, again in only about half the male offspring, largely due to failure in gubernacular regression (370). Flutamide given to pouch young Tammar wallabies did not affect testicular descent but prevented closure of the inguinal canal (105). Cryptorchidism was produced in mice by treatment with estrogens, either estradiol benzoate (544) or diethylstilbestrol, probably through an action on INSL3 (149,413). Estrogens also blocked testis descent in marsupials (104).

It is clear from the above that control of testis descent is multifactorial. There is still no complete explanation for the large proportion of cases of cryptorchidism that are unilateral, not bilateral.

Variation in Size of the Testis

The size and structural organization of the testes may vary with season, especially in those species that reproduce at a restricted time of year. This is particularly evident in species that occur at high latitudes where testicular development and activity may be regulated photoperiodically or where there is an endogenous circannual rhythm of testis recrudescence, activity, regression, and inactivity. Variation in testis size between species can also relate to differences in the ratio of spermatogenetic-to-interstitial tissue.

Comparative analyses of differences in testis weight in sexually mature males of a wide range of species have revealed an allometric relationship of about 0.7 between testis and body weight, that is, testis weight as a percentage of body weight becomes less with an increase in body weight (309,403). Nevertheless, species exhibiting a breeding system where two or more males are likely to mate with a female at estrus tend to evolve larger relative testis size than those with monogamous or single-male mating systems (178,215,309), which is probably due to higher levels of intermale sperm competition. It has also been suggested that species with internal testes have relatively larger testes and those with sessile scrota have larger testes than species with pendulous scrota (174).

In this chapter, seasonal differences in testis size and activity are summarized for the major groups of mammals, together with variation in testis weight and cellular organization in sexually mature males in relation to mating systems. Common evolutionary trends for the various groups are briefly indicated, but no attempt is made to cover all the original observations. Others have given more comprehensive coverage of the early literature on relative testis size (174,214,309), and on seasonal differences (Glover et al., 192). In this chapter, testis weight refers to the combined weight of the two testes.

Eutherians

Afrotheria. This group of mammals, as its name implies, occurs largely in Africa and is characterized by species having intraabdominal testes. In African elephants, the testes make up about 0.1% of body weight, grow throughout life, and do not show marked seasonal variation (279,290), whereas in Indian elephants, although there are temporal changes in behavior and androgen levels (350), there seems to be no information on whether there is seasonal variation in testis size. In two genera of rock hyraxes, *Heterohyrax* and *Procavia*, the males show a very restricted breeding season and marked seasonal changes in testis weight, with maximal values between 0.6% and 0.8% of body weight (191,399,400). In elephant shrews (*Elephantulus*), although spermatogenesis occurs in all months of the year, there are seasonal changes in testis activity with maximal size of the testes being relatively small (0.12% of body weight) (665).

Laurasiatheria

Order Eulipotyphla. Many species of "core" insectivores undergo a marked seasonal cycle of testicular size and activity [e.g., various species of hedgehog (6), shrew (58), and mole (466)] with maximum testis weights ranging from about 2% of body weight in the moles (genus *Scalopus* and *Talpa*) to 4% of body weight in the genus *Condylura* (35,37,466).

Order Chiroptera. Most species of Microchiroptera occurring in temperate regions have a distinct breeding season with hibernation taking place during winter. Associated with this, there is an annual cycle of changes in testicular weight and activity and, in some species, testicular migration to and from the abdominal cavity (282,315). Typically, spermatogenesis takes place in spring and summer during which time there is marked increase in testis weight with a maximum occurring in mid to late summer, for example, the Noctule bat (465) and the Pipistrelle bat (467). In most of these species spermatozoa pass from the testis to the epididymis in late summer, with sperm storage in the cauda epididymidis taking place during the autumn and winter while the animals hibernate (210). Even though testis size has markedly decreased at this time and no spermatogenesis is occurring, there appears to be some Leydig cell secretion, which is necessary to maintain the environment of the cauda for sperm storage. One exception to this is the bent-wing bat, *Miniopterus*, where both sperm production and mating occur in autumn (102,478).

Most species of Megachiroptera, which occur mainly in the tropics, also breed seasonally and undergo a circannual cycle of testicular activity, with maximal testicular size occurring either in autumn (e.g., *Pteropus poliocephalus* and *P. alecto*) or in late spring (e.g., *P. scapulatus*) (369,425,426). In Africa, it was found that bats near the equator, where there is an abundance of food throughout the year, have prolonged or continuous spermatogenesis, whereas those in the temperate zones have seasonal sperm production (44). Thus, in tropical species, mating and insemination generally appear to coincide with spermatogenesis, although at least in some species this may cease in the cooler months of the year (281). Many species of bats hibernate for a period between sperm production in autumn and embryonic development and birth in spring and summer. An exception to this is seen in molossid bats, and it is suggested that this may be due to their different foraging behavior that gives them access to abundant food throughout the year (44).

Interspecific differences are evident in size of the testes of both microchiropteran and megachiropteran bats during breeding period. Hosken (246,247) showed that testis size appeared to correlate with "social group" or roost size and suggested that a greater relative testis size may have been selected when a female mates with two or more males due to a higher level of intermale sperm competition. Wilkinson and McCracken (653), in an extension of these studies to 84 microchiropteran and 20 megachiropteran species, found an allometric relationship of 0.95 between testis weight and body weight, a relationship much nearer to isometric than in most other groups of mammals (309). Nevertheless, they found that relative testis weight (0.1%–8.4% of body weight) correlated with the mating system, with promiscuous species generally having the largest relative testis size. Such results support the view that sperm competition influences relative testis size.

Orders Artiodactyla and Perissodactyla. Species in these groups occurring in temperate regions have a circannual cycle of reproductive activity, with mating typically taking place in the autumn. However, in some species only a small percentage of males are successful in this regard. Clearly, whereas social status, nutrition, health, and age can influence testicular activity and hence male fertility, many species also have an annual cycle of testicular growth and regression. In northern Europe and the United Kingdom, rut occurs in roe, red, and fallow deer between mid-July and the end of August. In roe deer spermatogenesis is complete by May, testis weight is maximal (0.2% of body weight) in August/September, and then Leydig cells decrease in size followed by disorganization of spermatogenesis in late autumn (548). Similar trends occur in Fallow deer (maximum testis weight 0.4% of body weight) (196) and Red deer (0.14%) (346), with the latter species probably having a circannual endogenous rhythm of testicular activity that is entrained by the environment (192). The ram, too, particularly in the more primitive breeds, displays seasonality of reproductive activity with the change in testicular activity clearly being controlled photoperiodically (192,348,349,351,444,515). Similarly, in all four species of antelopes in southern Africa, The Impala, Kudu, Blesbok, and Springbok, an annual cycle of testicular activity is evident with greater activity taking place in autumn, although in these last-mentioned species, some spermatogenesis occurs throughout the whole year (553). In Asian tropical species, such as *Axis axis* (363) and *Muntiacus reevesi* (79), males nevertheless appear to be fertile throughout the year, even though antlers are cast annually and changes in testis volume occur. Testes weights in these species were 0.04% and 0.12% of body weight, respectively.

An analysis of data for body and testes weight (presumably taken at the height of the breeding season) from 35 ungulate species revealed an allometric relationship of 0.81, but with testis weight above the regression line in those species that exhibit polyandry (187). However, this study did not include data from domesticated species. Although selective factors operate in these animals, it is interesting that the value for domestic cattle (testis weight 0.10% of body weight) falls right on the line, whereas the values for rams of three domesticated breeds of sheep [0.84% for Australian Merinos (15), 0.60% for British Clun Forest (515), and 0.66% for Ile de France (535)], and even the semi-domesticated Soay breed (0.64%) (349) fall well above the regression line, supporting the idea that large testes are associated with multiple male mating systems. However, this does not extend to some other ungulate species that also normally occur in herds, such as the South African species Impala (0.33%), Kudu (0.04%), Blesbok (0.22%), Springbok (0.19%), Wildebeest (0.11%), or Hartebeest (0.08%) (553,554).

Among perissodactyls, the polyandrous zebra species, the Grevy's zebra, has significantly larger testes than either the Plains or Mountain zebra where apparently a female generally mates with only one male at any one estrus. Because the Grevy's and Plains zebras are closely related, rapid evolution of interspecific differences in testis weight may have taken place (187). In wild members of the Suidae, the testes are relatively small, around 0.1% to 0.2% of body weight in Tayassu (309), whereas some domestic breeds of pigs have somewhat larger testes, 0.25% to 0.49% of body weight, (365,533).

Order Cetacea. Several species of odontocetes have very large testes, with those of the porpoise, *Phocoena phocoena*, being about 4% of body weight. Although the massive Baleen whales have the largest recorded testes of any mammal, these are only about 0.01% of body weight (309).

Order Carnivora. In contrast to the domestic dog and cat, which are often fertile throughout most of the year, most wild carnivores have a well-defined breeding season. In the fox (362) and mink (385), spermatogenesis is initiated in late autumn and attains maximal activity around mid to late winter, whereas in the weasel, polecat (634), ferret, stoat (208), wolverine (396), and skunks (301,302), spermatogenesis occurs from late winter to early summer. Because male ferrets kept under either continuous light or constant dark still exhibit a circannual rhythm of changes in testis weight, there appears, at least in this species, to be an endogenous rhythm of testicular activity that is entrained photoperiodically (52). In the mink, by contrast, even though body weight and pelage changes exhibit an endogenous circannual rhythm, the changes in weight of the testes appear to be almost totally photoperiodically controlled (385). Among marine carnivores, testis size in the Southern Elephant Seal is maximal (0.034%) from September to November and smallest in winter (201). The testis weight for six terrestrial carnivores ranged from 0.05% to 0.8% body weight (309). Domestic dogs have testes that are about 0.13% of body weight, and testes weight showed an isometric relationship with body weight over a more than tenfold range, whereas epididymal weight and sperm reserves showed significant positive allometry (664). Testis weight is comparable in domestic dogs and dingos, although the latter have a discrete breeding season and are less promiscuous (100,666).

Euarchontoglires

Order Primates. Apes display marked differences in testes weight with the orangutans and gorillas having small testes of around 0.02% to 0.05% body weight, whereas chimpanzees have much larger testes of around 0.27% body weight. It has been suggested that such interspecific differences in testes weights were due to differences in breeding systems (546). Extending these observations to many more primate species, Harcourt et al. (214) found an allometric relationship between testis weight and body weight of 0.66, with genera that include species with multimale breeding systems consistently having a larger relative testis weight (of around 0.3%–0.7% body weight) than genera in which single male mating systems or monogamy occurs (129,221).

In addition, testis histology showed that a higher volume of seminiferous tubules to interstitial tissue ratio was present in chimpanzees, baboons, and macaques than in langurs, gibbons, or humans (221). This suggests that variation in mating system, through variation in intermale sperm competition, may influence not only relative testis size, but also the cellular organization of the testis.

Order Rodentia. Although the Brown or Norway rat, *R. norvegicus*, the Black rat, *Rattus rattus*, and the house mouse, *Mus musculus*, have the potential to reproduce at any time of year, reproductive activity of most temperate zone species of native rodents is generally restricted to the spring, summer, and autumn with temperature, plane of nutrition, and/or photoperiod determining the time of onset and duration of the breeding season. Environmental control of testicular activity has been studied in a number of rodent species particularly voles (206), White-footed mice (280,448), Deer mice (118,648), hamsters (22,596,688), and the Australian Bush rat (266,647). In most of these species, exposure to short days and/or long nights induced gonadal regression and cessation of spermatogenesis. However, in two hamster species and in White-footed mice, *Peromyscus leucopus*, testis recrudescence occurred even when animals were kept in short days, thus suggesting an endogenous testicular emergence from the involuted state (22,280,448,688).

In Grey squirrels in the United Kingdom, a circannual cycle of testicular activity is evident (5,138,579), but with some asynchrony such that at least some males have sperm at all times of year (458). Some authors have found a period of testicular regression in autumn. This was also the case for the Indian Palm squirrel (470).

Testis size may vary not only with season, but in some groups of rodents, dramatic variation is seen between sexually active males of related species. Although most sciurids have a testis weight of 0.8% to 2.2% of body weight, marmots have markedly smaller testes of only around 0.2% of body weight. In muvoid rodents, relative testis weight of sexually mature males ranges from around 0.8% in *Sigmodon hispidis* to about 8% of body weight in the species of *Tatera* in the Gerbillinae. Among murine rodents, there is a near-isometric relationship of 0.92 between testes weight and body weight (63), although marked interspecific differences in relative testis weight nevertheless occur. For instance, in Australian murine rodents, differences in testis size in sexually mature males range from about 2% to 4% of body weight in most *Pseudomys* species and various other genera, with a maximum of 5.0% in the New Guinea Chestnut Tree mouse, *Pogonomys macrourus*, to around only 0.15% in four of the five extant *Notomys* species as well as in *Pseudomys apodemoides*, *P. delicatulus*, and *P. shortridgei* (62).

In contrast to other orders, the association between relative testis size and breeding system is less clear-cut in rodents. There are both avuicoline (233,661) and murine species (59,62,63) with small testes, yet the data available suggest that multimale mating can sometimes take place at least within a laboratory environment (64). Such variation in relative testis weight may be due, at least in part, to differences in the physiology of sperm transport in the female reproductive tract as well as differences in intermale sperm competition.

Order Lagomorpha. At least in the United Kingdom, the testes of wild rabbits regress in late summer, probably due to the reduction in day length with recrudesce in mid to late winter (54,55). The Brown hare also has a breeding season from January to August, but males can be fertile in all months of the year except in October and November when the testes involute. As testis development reaches a maximum in spring and early summer, the males become very aggressive toward each other (347).

Marsupials. In marsupials, as in eutherian mammals, there is an allometric relationship of 0.75 between adult testis weight and body weight, but considerable differences in relative testis weight occur between species of similar body weight (493,577). Many marsupial species show seasonal variation in the testis size. In small dasyurids like *Antechinus*, testis weight reaches a maximum when the animals are about 9 months old, about 2 months before their one and only breeding season, by which time spermatogenesis ceases and after which all the males die (312,656). Complete mortality of the males after mating also occurs in some large dasyurids, but without spermatogenic failure before mating (422,424). In some longer lived marsupials, such as Ring-tail possum (257), Eastern quoll (70), and Greater glider (558), there is an annual cycle of growth and regression of the testes, with complete arrest of spermatogenesis occurring outside the breeding season. In marsupials, such as Brush-tail possum (186), bandicoot (590), and Tammar (265) and Bennett's wallabies (77), there is little seasonal variation in testis weight or spermatogenesis, despite large changes in secretion of androgens and possibly other hormones (107).

Small species that have relatively large testes include *Tarsipes* (4.1% of body weight), *Acrobates* (1.4%), and various dasyurid marsupials (e.g., *Antechinus agilis*, formerly *stuartii*), whereas relatively small testes are found in the pigmy possums *Cercatetus* (0.4%) and *Burramys* (0.1%). Among the species of medium body weight, species with relatively large testes include the Ringtail possum, *Pseudocheirus* (0.8%), with small testes being present in the genus *Petaurus* (0.1%–0.2%). Among species with large body weight, several macropod species, in particular the Tammar wallaby, *Macropus eugenii* (0.56%), *M. agilis* (0.2%), and *M. rufogriseus* (0.3%), have relatively large testes, whereas the Swamp wallaby, *Wallabia bicolor*, as well as wombats, *Lasiorhinus latifrons* and *Vombatus*, and koalas, *Phascolarctos cinereus*, have a much smaller relative testis weight (about 0.05%).

The fragmentary data available suggest that this interspecific variation in relative testis size may also relate to mating system and differences in intermale sperm competition (493,577,578). For instance, *Antechinus agilis*, with relatively large testes, displays a promiscuous mating system, whereas *Cercatatus* and *Petaurus* appear to pair monogamously. Large species with relatively large testes generally live gregariously (e.g., *M. eugenii* and *M. rufogriseus*) and thus probably have a promiscuous mating system, whereas *Wallabia bicolor* is generally solitary and may be monogamous. In koalas, one dominant male monopolizes all females within his home range, reducing the chance of mating of a female in estrus with more than one male.

Monotremes. Both the platypus and Short-beaked Echidna show seasonality of breeding, with mating taking place in echidnas between June and early

October (mid-winter to mid-spring in Australia) (202), and the testes of echidnas show marked seasonal changes in weight. In sexually mature animals in the breeding season, testes are relatively large (1%–1.5% of body weight) (203,204,577). It has been suggested that such testis size may relate to a multimale mating system, although observations of Short-beaked Echidnas have indicated that even though "trains" of males follow a female at estrus, only one male mates with her in any one receptive period (482,483). Little is known of the mating behavior of platypus.

Structure of Testis

Capsule of Tunica Albuginea. Testes are encased by the tunica albuginea, a tough fibrous capsule. This comprises an outer layer of visceral peritoneum made up of mesothelial cells, and inside this there is a layer of fibroblasts, collagen fibers, and bundles interspersed by smooth muscle cells. The thickness of the muscle layer varies between species, with a single layer in the laboratory rat and two distinct layers and many more cells in rabbits (120). In humans, myofibroblasts predominate in the outer layer and smooth muscle cells in the inner layer. There are also myelinated and unmyelinated nerve fibers and some Leydig cells. Inside the muscle layers, in the tunica vasculosa, there are relatively large arteries, veins, and lymphatic vessels (397). The muscle cells in the capsule become obvious at about 30 days of age in laboratory rats, although they can be recognized by electron microscopy much earlier than this (338).

The testicular capsule in several species has been shown to contract spontaneously in vitro. These contractions were found to be much more obvious in rabbits (119) and pigs (427) than they were in rodents. Capsular contractions in rabbits were maximal at 32°C (119) and were stimulated by oxytocin, acetylcholine, catecholamines, histamine, and prostaglandin $F_{2\alpha}$, as well as by stimulation of sympathetic nerves, and were inhibited by isopropyl-noradrenaline and PGE_1 [(503), see (120,537) for early references]. Strips of capsule from human testes, if taken from near the rete testis, also contracted spontaneously in vitro; strips taken from parts of the testis remote from the rete did not contract spontaneously but responded to noradrenaline. This response was inhibited by nitric oxide donors, atrial natriuretic peptide, or cyclic GMP (397). Contractions of the capsule may be important in expelling the sperm from the testis into the epididymis, but the flow of fluid from the rete testis was not affected by removal of the capsule (173). However, incising the outer layers of the capsule without interfering with the tunica vasculosa led to accumulation of sperm in the transitional zone of seminiferous tubules (460), reduction in testosterone concentrations in testicular venous blood (461), reductions in size of the Leydig cells and binding of luteinizing hormone (LH) (463,464), and infertility of the rats (462). Capsular contractions may also play a role in maintaining interstitial tissue pressure in the testis (see section on Vascular Permeability and Interstitial Extracellular Fluid).

Interstitial Tissue. The interstitial spaces between the seminiferous tubules contain the blood and lymph vessels, macrophages, and Leydig cells. Macrophages may influence the function of the Leydig cells (260,401), and testicular macrophages secrete a different range of cytokines than do peritoneal macrophages (311). Histological observations have shown that whereas rams, bulls, cats, and moles have an interstitial organization that includes small clusters of Leydig cells in an extensive intercellular matrix (161,171,570), in pigs there is an abundance of Leydig cells throughout the interstitial tissue that make up about 30% of the total testicular volume (Fig. 2), with males of the Meishan breed having a value as high as 50% (365). A similar abundance of Leydig cells occurs in the collared peccary of the New World as well as in the warthog and forest hog of Africa (438), indicating that this is probably characteristic of all suids. The functional significance of the abundant Leydig cells is not clear, but Fawcett et al. (161) pointed out that pig testes secrete high levels of estrogens [see also (533)] and that this may, perhaps, relate to the unusual arrangement of the interstitial tissue.

Among murine rodents, in addition to marked differences in relative testis weight, the organization of the testicular interstitial tissue also varies dramatically. In those species with relatively large testes, there is invariably sparse interstitial tissue made up of small groups of Leydig cells clustered around blood vessels. By contrast, in some species with relatively small testes (e.g., *Pseudomys delicatulus*, *P. apodemoides*, and *P. albocinereus*) the interstitial tissue is composed of abundant Leydig cells (Fig. 3) and represents at least 30% of the total testicular volume (59,62,587). A similar abundance of Leydig cells in testicular interstitial tissue also occurs in the naked mole-rat, *Heterocephalus nanus* (161).

Among marsupials the microstructure of the testes also differs. In phalangerids and macropods, interstitial tissue makes up no more than 5% of the total testicular volume, whereas in didelphids and dasyurids, and to a lesser extent peramelids and wombats, there is a much greater proportion of interstitial tissue of up to 20% or more of the total volume in dasyurids. In these species, most of the interstitial tissue is composed of abundant Leydig cells with blood vessels and lymphatic spaces being relatively

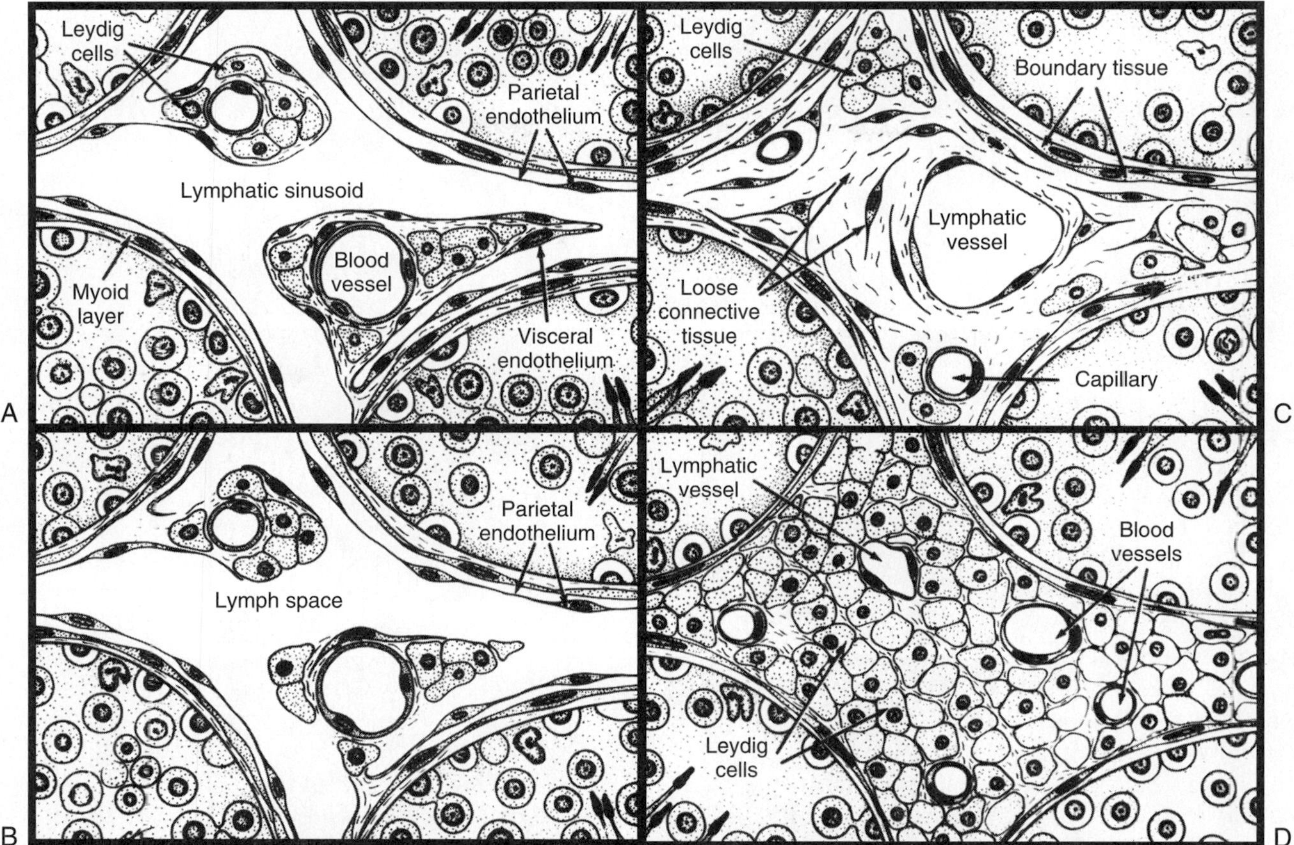

FIG. 2. A diagram illustrating the anatomy of the interstitial tissue of the testes of various mammals. **A:** Guinea pig and chinchilla. Perivascular Leydig cells and seminiferous tubules are invested by endothelium, which bounds the extensive lymphatic sinusoids. **B:** Rat and mouse. Essentially the same as A, but the visceral endothelium is discontinuous, so that the Leydig cells are directly exposed to lymph. **C:** Human, monkey, bull, and ram. Clusters of Leydig cells are scattered in an abundant loose connective tissue, with one or two lymphatic vessels in each intertubular space. **D:** Pigs and zebra. A large volume of closely packed Leydig cells, little connective tissue and occasional small lymphatic vessels. [Reproduced with permission from (158).]

sparse (Fig. 2) (514,668), an arrangement that shows convergence to suids and the few rodent species described above.

Seminiferous Tubules. Each testis of a laboratory rat contains about 30 convoluted seminiferous tubules, with an external diameter of about 250 μm and a total length of about 20 m per testis (12 m/g) and a surface area of 340 cm^2 (658). In contrast, in the Hopping mouse, *Notomys alexis*, the very small testis contains only seven to nine tubules, each about 50 mm long (62). In rams, the total length of the tubules in a testis is about 3,000 m (11 m/g) with a surface area of 85 cm^2/g (238), whereas in cats, the total length of the tubules is 23 m/g testis (171).

As well as the various germ cells, the seminiferous tubules also contain the somatic Sertoli cells, which have essential nutritive and physiological functions essential for the developing germ cells (see Chapter 6 for details). The tubules are enclosed in peritubular tissue comprising a basement membrane and one (in rats) or several layers (in humans) of myoepithelial cells or myofibroblasts, interspersed with layers of collagen fibers and glycosaminoglycan and proteoglycan matrices (244,603); in rodents, there is an outer layer of lymphatic endothelial cells (161). The basement membrane generally consists of an innermost finely filamentous material and an outer network of reticular fibers (124,244) and in rams and boars a number of lamellae (404,449).

In most marsupials, the seminiferous tubules resemble those of eutherians (514). However, each testis in dasyurids, bandicoots, and honey possums contains only one to four seminiferous tubules (Fig. 4), with much larger cross-sectional diameters (668) than in any eutherian mammal (Fig. 5). The significance of these large tubules is obscure, but it may relate to the very large spermatozoa these species produce, one consequence being that these animals produce fewer sperm per unit testis weight than species with smaller tubules.

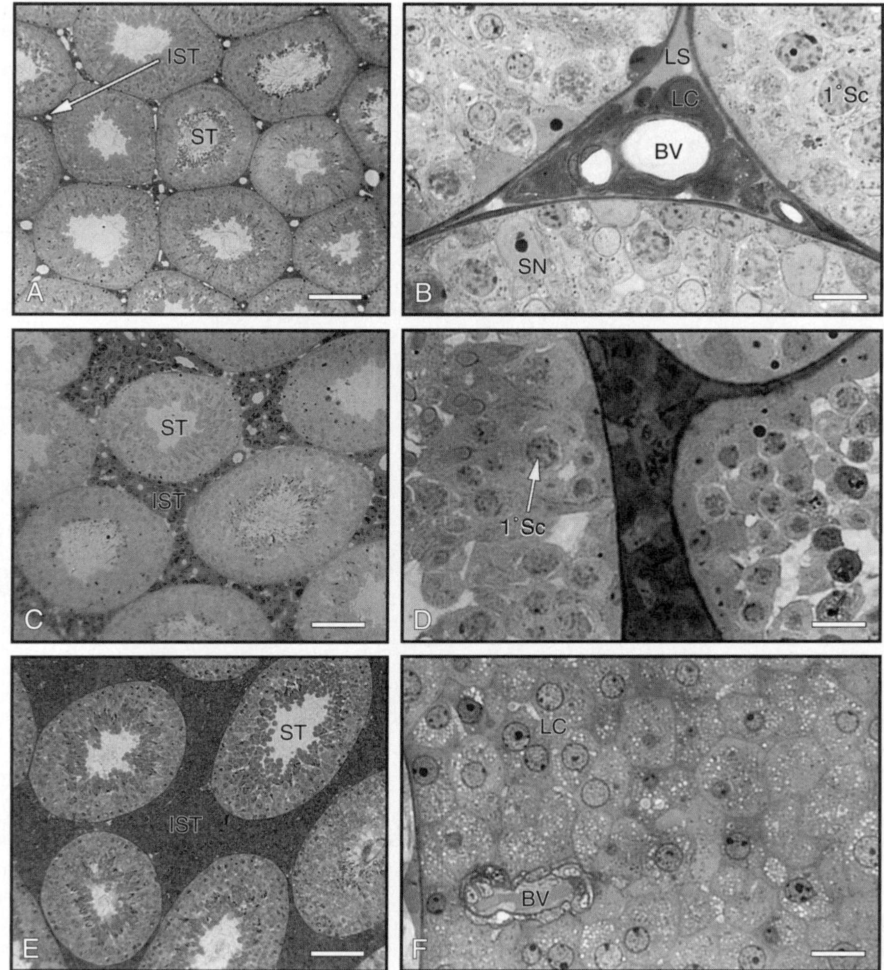

FIG. 3. Light micrograph transverse sections of testes from four species of Australian rodents. **A, B:** *Pseudomys nanus*. **C, D:** *P. hermannsburgensis*. **E:** *P. apodemoides*. **F:** *P. delicatulus*. Note major difference in relative volume of seminiferous tubules (ST) and interstitial tissue (IST) between the species. BV, blood vessels; LS, lymphatic space; LC, Leydig cells; 1°Sc, primary spermatocyte; SN, Sertoli cells nucleus. Scale bars: A, C, E = 50 μm; B, D, F = 10 μm. [Mostly from (59).]

Contractions of the peritubular myoid cells are responsible for moving the luminal fluid and spermatozoa out of the seminiferous tubules through the rete and the efferent ducts into the epididymis. These cells contract in response to oxytocin (416), vasopressin (250), prostaglandin $F_{2\alpha}$ (594), and endothelin (165, 592,593). As the name implies, endothelin is formed in other tissues by the endothelial cells, but in the testis the principal source appears to be Sertoli cells (157). The response to endothelin is enhanced if the cells are treated with platelet-derived growth factor BB (436).

Tubuli Recti and Rete Testis. Both ends of each seminiferous tubule open into tubuli recti and then to the rete testis through short transitional zones lined by cells resembling Sertoli cells, which appear to form a valve or plug. However, there always seems to be a narrow open channel (141,420). Moreover, because cell suspensions injected into the rete through the efferent ducts readily enter the seminiferous tubules to reestablish spermatogenesis (66), these valves do not appear to be particularly effective. The tubuli recti or straight tubules are lined with cells similar to those lining the rete, and in humans up to six seminiferous tubules can join a single tubulus rectus (492).

In most rodents, elephant-shrews, and cetaceans, the rete testis (Figs. 6 and 7) is located on one side of the testis; in humans, gorillas, insectivores, and bats it lies along one edge of the testis in a fibrous mediastinum; and in monkeys, ungulates, carnivores, rabbits, guinea pig, elephants, and edentates it is centrally located and surrounded by a fibrous mediastinum (45,142,159,227,228,286,492,513,675).

In humans and cats, the rete can be divided into three parts, septal, tunical, and extratesticular; the septal rete is made up largely of structures like tubuli recti, the tunical rete is a network of channels embedded in the tunica albuginea, and the extratesticular

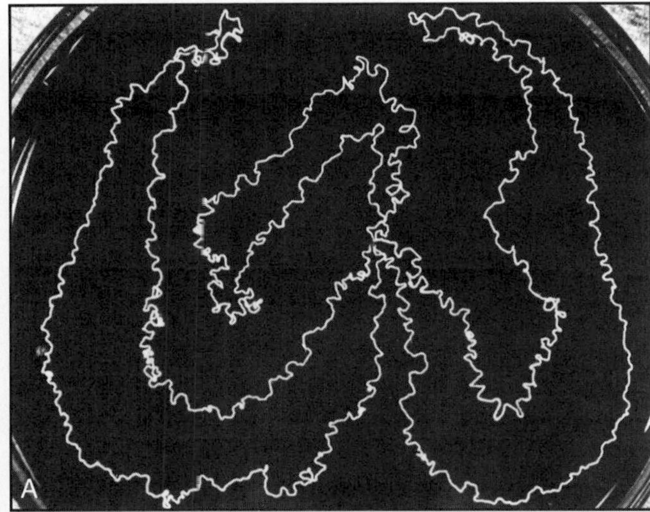

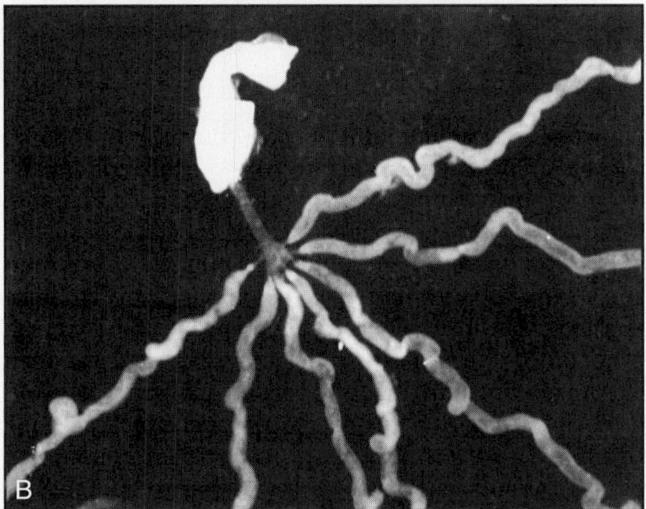

FIG. 4. A: The three seminiferous tubules and rete testis of a testis of *Dasyuroides byrnei*. **B:** An enlarged view of the connection of the four seminiferous tubules of a testis of another *D. byrnei* with the rete testis and the single efferent duct. Note the narrow tubuli recti. One end of one tubule is missing because it was broken during dissection. [Reproduced with permission from (668).]

rete comprises large well-defined dilatations called bullae retis (492,618). In cattle fetuses, the rete testis is derived from a blastema distinct from that giving rise to the testis and can first be recognized at about 40 days postcoitum. Initially, the rete lies outside the testis, but from 55 days postcoitum onward, the rete grows into the center of the testis to finally reach a point about three-fourths along its length (676). The rete is generally lined by cells forming a cuboidal to columnar epithelium (142), but in humans the lining cells are squamous and prismatic (492). In guinea pigs, the cells lining the rete testis contain large amounts of glycogen (159), and in rats, they may absorb proteins (237). Rete testis cells from rams have been isolated and cultured in vitro (595), and they appear to secrete proteins such as clusterin (494).

Among marsupials, the rete is horseshoe shaped in macropods and phalangerids, extending on either surface of the testis under the arteries on one side and the veins on the other to about halfway along from the efferent ducts and the point where the blood vessels enter the testis (Fig. 8). In dasyurids, peramelids, Honey possum, and the American marsupials (Fig. 9), a short simple rete is situated at the hilus (485,514,670,671,673,674).

Excurrent Ducts

Efferent Ducts

The rete testis is linked to the epididymal duct by the efferent ducts (ductuli efferentes). These ducts form a cord that is anatomically differentiated into a proximal (initial) zone, where the individual ducts run roughly parallel to each other, and a distal zone (coni vasculosi), where the course of the ducts is more sinuous. The ducts anastomose with one another, particularly in the distal zone. There are between 1 and more than 30 according to species, arranged in different patterns. These ducts either join to form a common duct or open directly into the epididymal duct (Fig. 10). Humans have between 6 and 15 ducts and rats between 5 and 7, with a total length of 20 cm (289). Tammar wallabies have 12 to 16 ducts with a total length of 6.7 m (81). There are multiple efferent ducts in those marsupial species that have a horseshoe-shaped rete, but there is only a single efferent duct where a simple rete occurs (485,514,668,670,671,673). Short-beaked echidnas have seven efferent ducts with a total length of 39 cm (133).

The efferent ducts develop from mesonephric tubules and consequently are homologous to the proximal tubules of the kidney, with which they share similar functions of fluid and protein resorption (88). The lining epithelium in rodents is made up of ciliated and nonciliated cells, with ciliated cells in rats occurring in groups of up to three, making up about 15% of the epithelial cells in the initial and distal zones and 33% in the common duct (289). The cilia do not appear to be involved in moving the spermatozoa and the fluid in which they are suspended from the rete toward the epididymis, but they presumably stir the fluid in the duct lumen (9,234,264, 498,562,563,565,675). The nonciliated cells have microvilli on their apical surface that increase the surface area, as in the proximal tubules of the kidney, but the basolateral membrane is not folded in the efferent duct cells, as it is in the kidney (639).

The efferent duct epithelium has androgen and estrogen receptors, although the estrogen receptors appear to be more important (235,686).

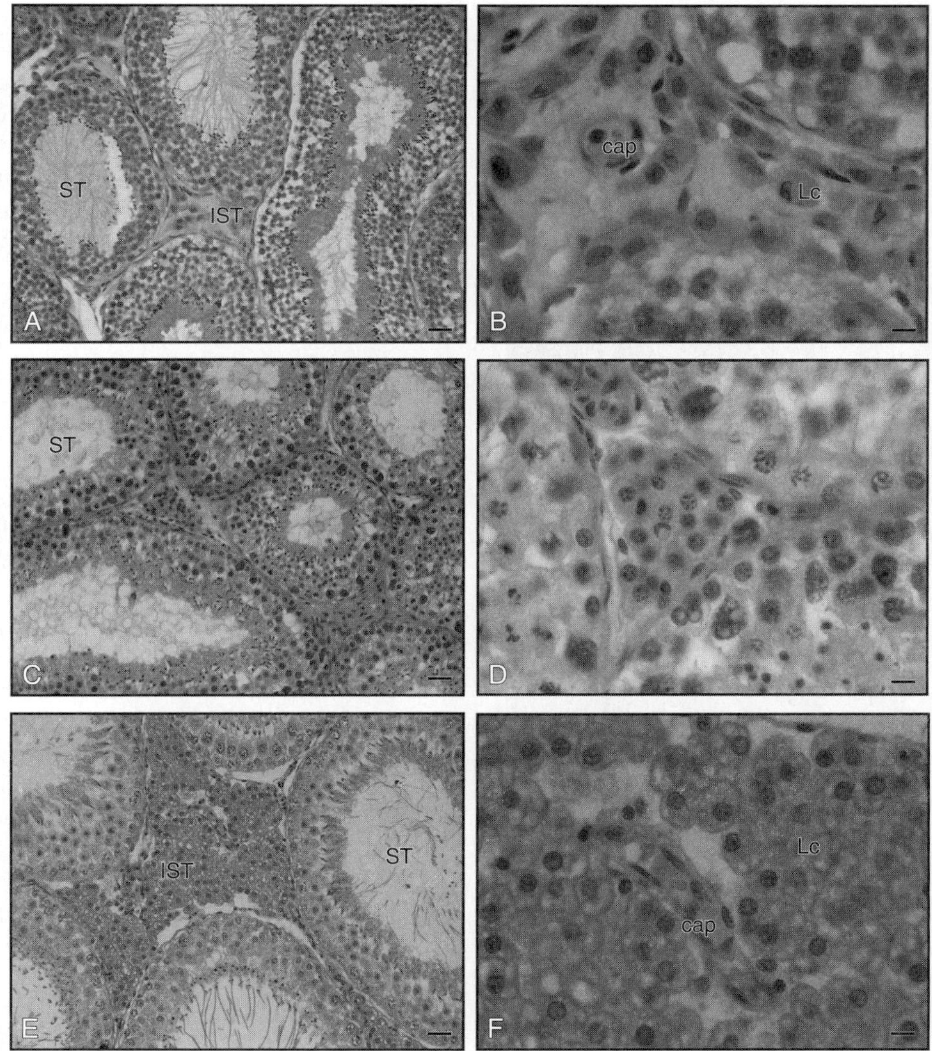

FIG. 5. Marsupial testis transverse sections from brush-tailed possum (**A, B**), bandicoot (**C, D**), and the dasyurid fat-tailed dunnart (**E, F**). Note larger seminiferous tubules (ST) and greater abundance of interstitial tissue (IST) in the dunnart compared with the other species. LC, Leydig cells; cap, capillary. Scale bars: A, C, E = 50 μm; B, D, F = 10 μm.

Estrogen receptor α is abundant, and mice that have the gene for this receptor disrupted exhibit much reduced fluid resorption in the efferent ducts, with the consequence that spermatogenesis is deranged and the animals become infertile soon after puberty; mice with no estrogen receptor β appear to be normal (235).

Epididymis and Ductus Deferens

The epididymis consists of a single highly convoluted duct that develops from the mesonephric (or wolffian) duct. It is arranged in lobules and terminates in the vas (more correctly ductus) deferens. Sperm transport along the epididymal duct is affected by its tunic of smooth muscle (26). In many mammals, the epididymis has been classified into distinct regions based on the structure of its epithelium, as well as on gross morphology, into caput, corpus, and cauda (head, body, and tail), with the corpus being the thinner segment joining the two other wider segments (39). An alternative division has been proposed, based on functional and anatomical grounds, into initial, middle, and terminal segments, which do not necessarily correspond with head, body, and tail (190). The epididymal duct is lined by principal, apical, narrow (mitochondria rich), light, and basal cells and by intraepithelial leucocytes (231). On the basis of the structure of the epithelium, the duct has been subdivided into 12 segments in some species, with gradual changes between segments (99). There is a characteristic initial segment of the epididymis in all mammals that has been studied (39,285), with a wide duct containing few sperm and lined by a tall actively secretory epithelium containing narrow and principal cells, with a characteristic cytology of the apical

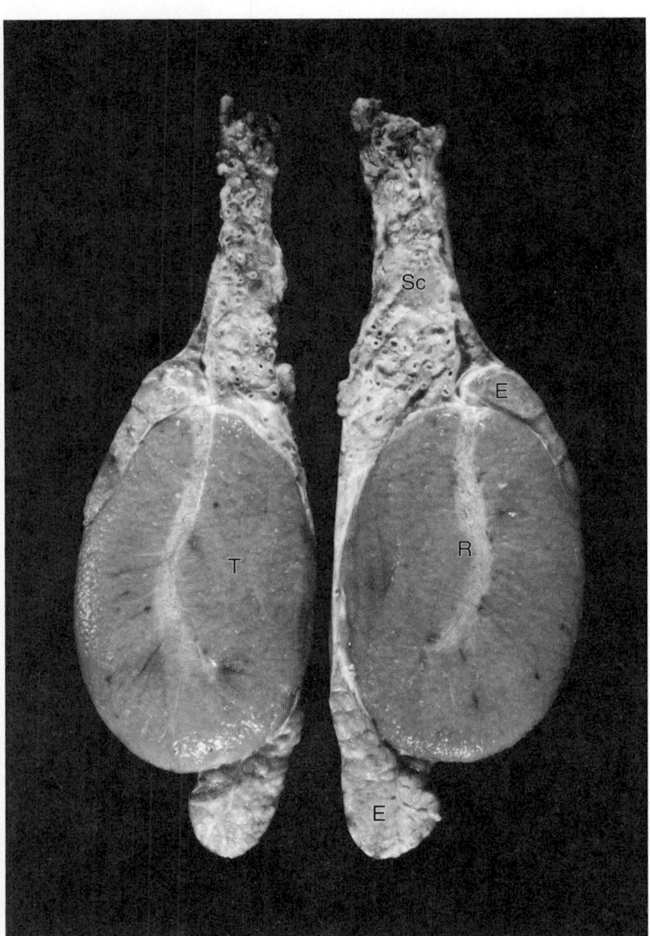

FIG. 6. A longitudinal section through the testis (T), epididymis (E), and spermatic cord (Sc) of a ram, showing the central rete (R) extending about three-fourths of the length of the testis. In the spermatic cord, the single testicular artery is cut at many points along its length because it is so tightly coiled.

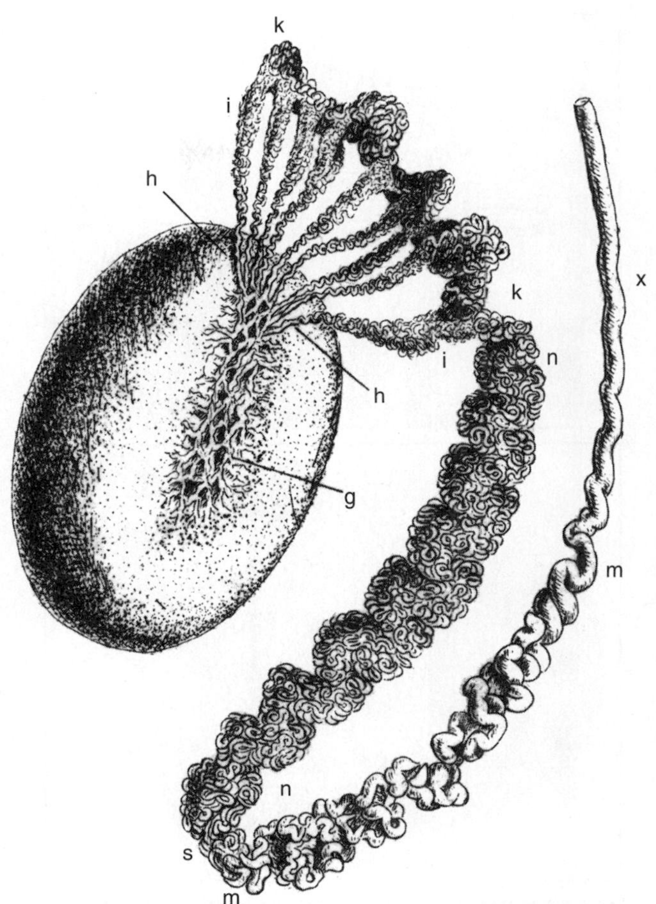

FIG. 7. A drawing of a human testis, epididymis, and ductus deferens filled by injection with mercury, showing the rete testis (g), ductuli efferentes (h), parts of which are inflected to form the vascular cones (i), the head (k), body (n), and tail (s) of the epididymis, and the inflected (m) and straight (x) parts of the ductus deferens. [Reproduced from (331).]

cytoplasm and coarse branched stereocilia (133). In most species, there is a reduction in the height of the epithelial cells lining the duct from the initial segment to the cauda, a coincident widening of the lumen, and an increase in the sperm concentration and in the amount of smooth muscle around the duct. The muscle is well developed in the cauda, from where emission of stored sperm occurs at ejaculation. The cauda is often separated from the testis, lying beneath pigmented and hairless skin (28). The coiled duct of the cauda opens into the ductus deferens, which is an unconvoluted duct leading to the urethra. In most species it is a simple thick-walled duct, with a pseudostratified epithelium and a thick smooth muscle coat. The cell types present are similar to principal, narrow, and basal cells of the epididymis (232).

Differences in the structure of the epididymis occur between mammalian groups. These are described below.

Eutherians

Afrotheria. In African elephants, the anatomical division between the epididymis and ductus deferens, which is highly coiled throughout its length, is not clear, and Short et al. (549) in their study referred to the whole excurrent duct as the wolffian duct. Later studies (245,284,286,288) showed regional differences in structure, sperm maturity, and composition of the luminal fluid similar to those found in the epididymis of scrotal mammals, although more sperm were found in the proximal half, than in the distal half, of the duct. The epididymal duct is more than 50 m long, and its epithelium is folded at both ends, forming villi that protrude into the lumen (286). In the hyrax, Glover and Sale (191) did not find a ductus deferens that was distinct from the sperm storage region of the epididymis, whereas in the dugong, the main storage area for sperm, unlike in elephants and hyraxes, lies close to the distal pole of the testes, although there were some sperm in the straight ductus deferens (382). In the elephant-shrew, *Elephantulus myurus*, the caput epididymidis lies on the dorsal surface of the testis, near the kidney, and a thin elongated corpus extends back to the cauda, a wide coiled duct lying between the rectum

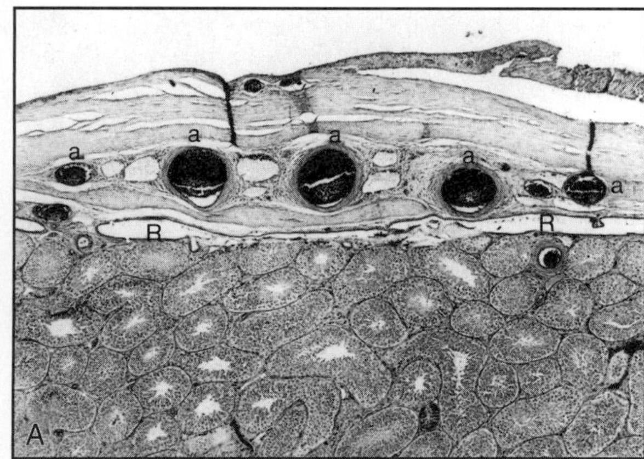

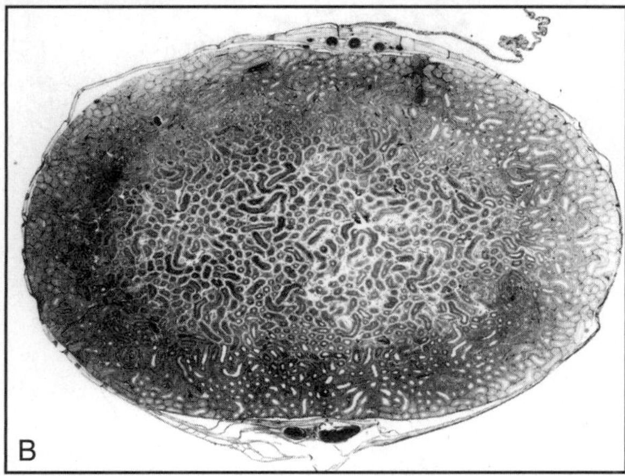

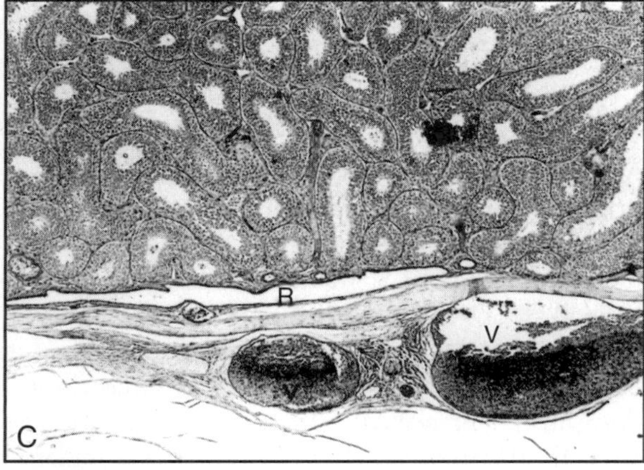

FIG. 8. A cross-section through the same testis of a Tammar wallaby, as shown in Fig. 18. **A** and **C** are enlargements of the upper and lower parts of **B**, showing the branches of the testicular artery (a) and testicular veins (v) overlying part of the horseshoe-shaped rete (R).

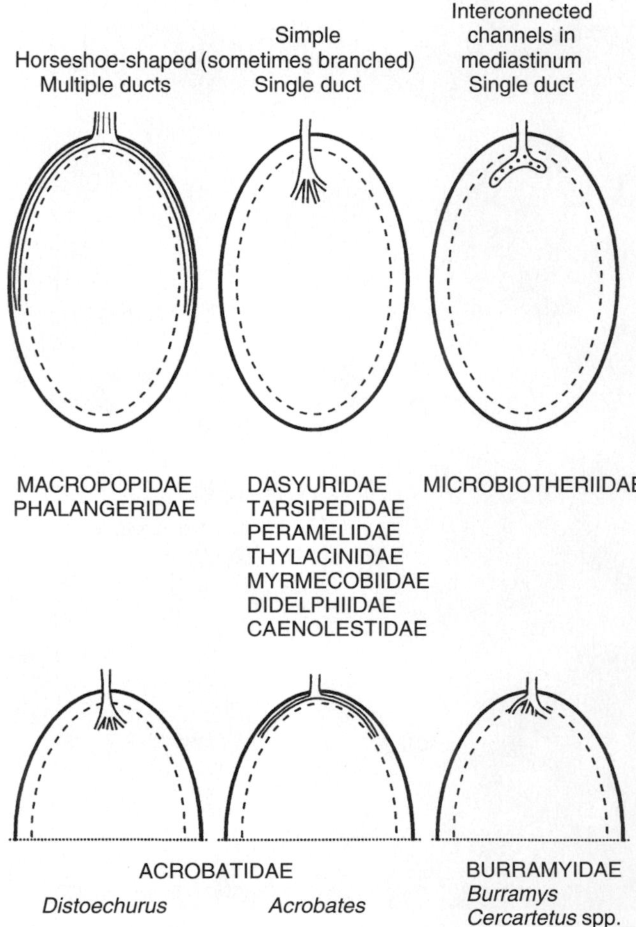

FIG. 9. A diagram showing the arrangement of the rete in the testis of different marsupials species. [Reproduced with permission from (674).]

and urethra. The cauda is the main site for sperm storage in this species. A short muscular ductus deferens joins the sperm storage region of the epididymis to the urethra (663).

Laurasiatheria. In pigs, the epididymal duct is about 60 m long, compressed into about 20 cm (185). The cauda epididymidis is relatively large in bulls, rams, and pigs and normally contains large numbers of sperm (285). In rams, up to 90% of extragonadal sperm are stored in the cauda (87,285,290). In some species of shrews, much of the ductus deferens acts as an accessory sperm storage region, and it appears that these sperm may make up most of the ejaculate population at least for the first mating (32–34,571), although this does not appear to be the case in the African Pygmy Hedgehog, *Atelerix albiventris* (36).

Euarchontoglires. In some primates including humans, the cauda is relatively small, and a comparatively small number of sperm are stored in it (285). Changes in the height of the epithelium and in the thickness of the muscle layer occur along the length of the human epididymis (Fig. 11) (26).

In the laboratory rat, the epididymal duct is about 34 cm long (133), and there is a prominent initial segment that is dependent on constituents of the luminal fluid for normal function (289). Laboratory mice without an initial segment, generated by gene knockout of the c-ros tyrosine kinase receptor, are infertile (559), suggesting that factors critical for sperm maturation are produced in this region. In all

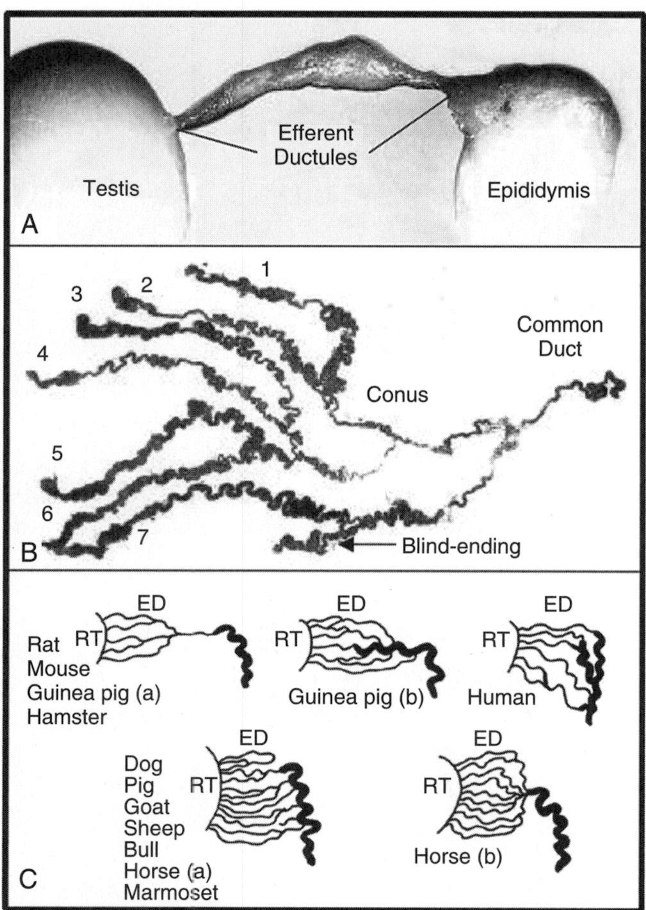

FIG. 10. A diagram of the efferent ducts (ED) and the first part of the epididymal duct of rats (**A,B**) and various mammals (**C**). [Reproduced with permission from (264).]

laboratory rodents, the cauda is comparatively large and is the principal site of sperm storage, with about 90% of the extragonadal sperm in rats (87,285,287). In the Australian hopping mouse the epididymal duct is only about 14 cm long (441), and in this species only about 60% of the extragonadal sperm reserves are stored in the cauda (61). In laboratory rats, the cauda is 3 to 4°C cooler than the testis (67).

The muscle around the ductus deferens normally collapses the duct, almost obliterating the lumen. In the Australian hopping mouse, however, there is a highly unusual ductus, with an enlarged lumen that acts as an accessory sperm storage, particularly at its urethral end (61,441,442).

Marsupials. The epididymis in marsupials, as in eutherians, is anatomically differentiated, except that the corpus is relatively larger. In the Tammar wallaby and Brush-tail possum, the cauda protrudes beyond the posterior end of the testis and lies in a separate scrotal pouch. The duct mucosa in the Tammar is composed of principal, apical, narrow, and basal cells and intraepithelial leucocytes, as in eutherians (291). The proportions of the cell types and the structure of the principal cells vary along the length of the epididymis. The initial segment makes up about half of the length of the epididymal duct and extends into the caput, and its epithelium is lower than in most eutherian mammals (81,285,292). The diameter of the duct lumen, and the thickness of surrounding smooth muscle, increases along its length, whereas the height of the epithelium decreases (291,589). In contrast, the diameter of the epididymal duct in the dasyurid, *Antechinus agilis*, does not increase and, by the time the cauda is reached, the lumen has become slit-like and the height of the epithelium increased. This unusual morphology is associated with the highest concentration of sperm occurring in the distal corpus and proximal cauda region with only small numbers being stored in the distal cauda (575,576). In *Antechinus*, as spermatogenesis ceases a month or two before the onset of the breeding season (423,576), their fertility depends on stored epididymal sperm (312).

Monotremes. In monotremes, the epididymis is differentiated into two regions: the caput, which runs over the testis, and the cauda, which passes caudally. The caput, all of which can be classed as an initial segment, makes up more than 90% of the length of the epididymal duct and the epithelium is cytologically similar throughout its length. The cauda is lined by a low epithelium that is folded to form villi that project into the lumen. Sperm are concentrated as they pass through the initial segment, but the cauda contains only about 25% of the extragonadal sperm. The cauda lies close to the penile urethra, and there is no clear anatomical distinction between it and the ductus deferens (132,133).

Accessory Sex Glands

The accessory sex glands arise in part from the mesonephric or wolffian duct and in part from the prostatic and penile urethra. Vesicular glands (= seminal vesicles) and ampullary glands develop as diverticula from the mesonephric duct, the area of whose secretory epithelium is increased by the formation of villous infoldings. The prostate and bulbourethral (= Cowper's) glands arise from the proximal and distal urethra, respectively, as compound tubular alveolar secretory glands. Depending on species, the prostate may be disseminate or lobular (144,213). In addition, in some species there are also mucous glands, the Littre glands, that develop from the urethra and preputial glands, which are modified sebaceous glands. The secretion of the accessory sex glands contains various substances (e.g., fructose, citric acid, zinc, acid phosphatase) that are added to semen at ejaculation. In many species, the size and

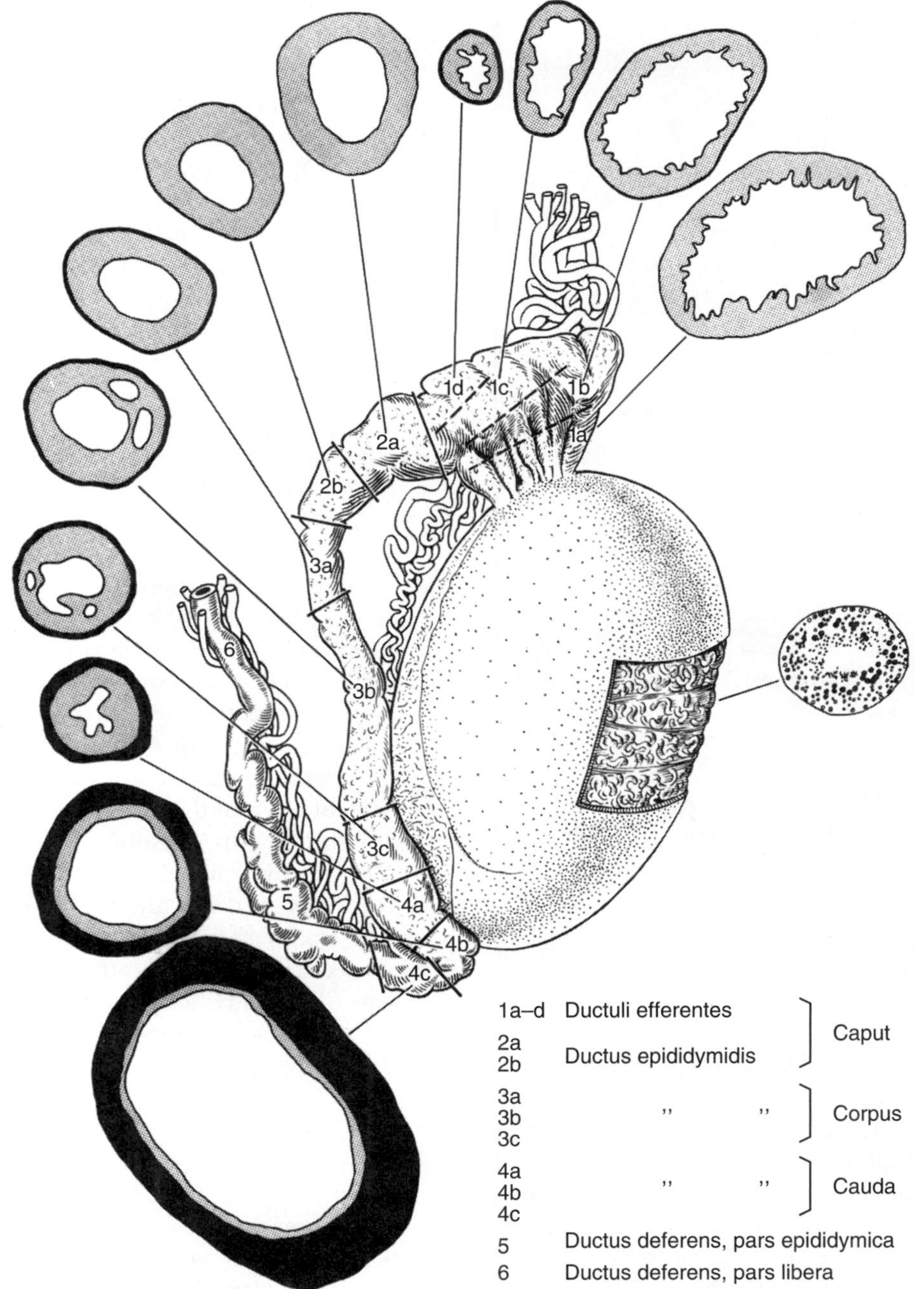

FIG. 11. A diagram of a human testis and epididymis, showing the variation in height of the epididymal epithelium (*shaded*) and the thickness of the muscle coat (*black*). [Reproduced with permission from (26).]

functioning of these glands are androgen dependent; thus, they change in size during the year, depending on levels of circulating androgens (380). The importance of accessory sex gland secretions to male reproductive tract function is somewhat enigmatic because cauda epididymal sperm are capable of fertilizing oocytes in vitro. However, removal of some of the accessory sex glands, in rodents at least, has been found to reduce fertility in vivo (421,437,443,459), and the formation of vaginal plugs from their secretions is consistent with facilitating sperm transport through the cervix. Considerable differences occur in the morphology of the various accessory sex glands between, and even within, some orders of eutherian mammals (144,213,380,454). These are briefly summarized.

Eutherians

Afrotheria. In the African elephant, the urethral end of the ductus deferens forms an ampulla, a large muscular organ lying ventral to the vesicular glands, with which it forms a common ejaculatory duct that opens into the prostatic urethra. A prostate, covered by a thick layer of smooth muscle, is present just posterior to the vesicular glands, and distally there is a pair of large bulbourethral glands (549). In the hyrax, *Procavia capensis*, a similar arrangement of accessory sex glands is present although, like the testes, marked seasonal differences in size are evident (399). In the dugong, the vesicular glands are large, but apparently there is no discrete prostate, and the bulbourethral glands are "diffuse" (382). In the elephant-shrew, *Elephantulus myurus*, Woodall (663) identified five pairs of prostatic glands and a pair of bulbourethral glands.

Laurasiatheria

Order Eulipotyphla. All species of shrews examined have been found to have an unusual modification of the mid-region of the ductus deferens that is enlarged to form an ampullary gland composed of alveoli that open into a centrally located excurrent duct [e.g., in the Common shrew (321,571) and Asian Musk shrew (32)]. In *Suncus*, the secretion of this gland is essential for the ejaculate to form a plug in the vagina (32). In moles, the prostate appears to be the only accessory sex gland in the abdominal cavity, and no enlargement of the urethral end of the ductus deferens is evident (466).

Order Chiroptera. In some families of bats (e.g., Phyllostomatidae) there is consistency in the arrangement of the accessory sex glands, with ampullary glands, vesicular glands, prostates, and bulbourethral glands being present. This also appears to be the case for mollosid and fruit bats, except that species in these groups appear to lack ampullary glands (323). However, in vespertilionid and rhinolophid bats, interspecific differences in arrangement of accessory sex glands are evident at least in relation to the presence or absence of vesicular, ampullary, and urethral glands (323,387,402). Pipistrelle and noctule bats of northern Europe have paired ampullary glands, anterior and median prostates, and paired bulbourethrals but no vesicular glands (465,467), an arrangement that is evident in a number of other species in this family.

Order Artiodactyla. All cervids examined have been found to have ampullae, vesicular glands, prostates, and bulbourethral glands. In species that have a circannual cycle of reproductive activity, such as the Red deer, these glands change in size and activity during the year coincident with fluctuating levels of circulating testosterone, although there is a lag period between the change in androgen level and in the response of the gland, at least in the case of the vesicular glands (548). In bulls, a cylindrical ampulla extends along the last 10 to 12 cm of the ductus deferens, which joins the duct of the vesicular glands to form a short ejaculatory duct opening into the urethra. The vesicular glands, which are lobulated and irregularly shaped and lie lateral to the ductus deferens, are large and produce most of the ejaculate volume. The prostate has a compact part that lies in the proximal region of the urethra and a disseminate part, with glandular elements in the wall of the pelvic urethra. The bulbourethral glands are small and lie almost level with the ischial arch, where they are largely covered by the bulbospongiosus muscle. The accessory glands in the sheep are similar to those in the bull, except that the vesicular glands are smaller and the prostate is entirely disseminated.

In boars, there is no obvious ampulla at the urethral end of the ductus deferens, but both the vesicular glands and the bulbourethral glands are very large and contribute much of the enormous volume of the ejaculate, which can be up to half a liter. The vesicular glands protrude into the abdominal cavity and conceal the small irregular body of the prostate. There is also a disseminate prostate in the wall of the urethra. The large bulbourethral glands lie dorsolateral to the pelvic urethra, into which they discharge through single ducts near the ischial arch (140,144).

Order Carnivora. In carnivores the prostate is closely associated with the bladder musculature and is the only accessory sex gland within the abdominal cavity, although in some species, such as the dog, which lack vesicular glands, there is a small ampulla at the end of the ductus deferens. All prostate tubules are present within the same capsule and, unlike in many other eutherian species, no distinct lobes are apparent. In dogs, the compact prostate is large and completely surrounds the urethra, although there is also a disseminate part in the wall of the urethra. In cats, the ventral aspect of the urethra is not covered by the prostate and there is no ampulla (140,144).

Euarchontoglires

Order Primates. In humans, there is an ampulla at the end of the ductus deferens, but the largest accessory gland is the prostate, which is derived from five independent groups of tubules, whereas the vesicular glands are comparatively small. The prostate surrounds the urethra at the neck of the bladder and, like that of the carnivores, does not generally have distinct lobes except for a shallow midline groove on its dorsal surface. The secretory units form zones defined by connective tissue septae. In some other primate species, the prostate can be divided into cranial and caudal lobes (144,213).

Order Scandentia. In Tupaids, there is considerable variation in the arrangement of accessory sex glands. In the Tree shrew, *Tupaia glis*, the vesicular glands give rise to small globules that pass out in the ejaculate to form a vaginal plug (29).

Order Rodentia. The arrangement of the accessory sex glands varies greatly between groups of rodents. For instance, the Grey squirrel has a single-lobed prostate, whereas the guinea pig has dorsal and lateral lobes. In muroid rodents there are generally three or four prostatic lobes, namely anterior (= coagulating gland), ventral, lateral, and dorsal, but there are significant differences between and within the various lobes in the arrangement and development of their ducts (223,334,567) and in their response to androgens (135,414). In muroids, marked interspecific differences are evident in the arrangement of the male accessory sex glands. In the sigmodontine rodents of South America, most species have a pair of large vesicular glands, small ampullary glands, two pairs of ventral prostates, and a single pair of anterior prostates, dorsal prostates, and bulbourethral glands and one or two pairs of preputial glands (14,620). The secretions from the vesicular glands and anterior prostates, perhaps with those from the bulbourethral glands (219), combine to form a large, hard, copulatory plug in the vagina (619). In a few species, there are no anterior prostates (e.g., *Nyctomys sumichrasti*) or vesicular glands and hence no plug is formed. Five functions have been proposed for the copulatory plug: (a) to facilitate gradual release of sperm within the female tract, (b) to prevent sperm leakage, (c) to induce pseudopregnancy, (d) to facilitate sperm transport, and (e) to prevent, or at least reduce, subsequent fertilizations from subsequent matings by other males. Of these various possible functions, there is good evidence that the plug facilitates sperm transport through the cervix in laboratory rats and *Suncus* (29,51,75,388) and acts as a "chastity enforcing device" in guinea pigs (277,383,619).

Among the North American neotomine rodents, much greater interspecific variation in the size and arrangement of the various accessory sex glands is evident. In species where the vesicular glands are small, this is generally also the case for both anterior and dorsal prostates, whereas, by contrast, either the ampullary glands or ventral prostates are usually large (e.g., *Podomys floridanus*, *Osgoodomys banderamus*, and *Habromys lepturus*) (354). Such a reduced complement of accessory sex glands in neotomines appears not only to be associated with absence of a copulatory plug, but also the occurrence of locking during copulation (27,220).

A similar variation in arrangement of accessory sex glands is also apparent in species of Australian hydromyine rodents. Within this assemblage, most species have a full complement of large glands that includes large vesicular glands, anterior prostates, together with ventral and dorsal prostates, and bulbourethral glands. However, in the genus *Notomys*, four of five extant species have minute vesicular glands, anterior and dorsal prostates, but very large ventral prostates (Fig. 12) (61). The full significance of this divergent arrangement of the accessory sex glands is not clear, but the absence of a large vesicular gland and anterior prostate precludes the formation of a large vaginal plug after mating. Perhaps related to this, the cervical luminal diameter in the female, unlike that in most murines, is similar to that of the vagina, presumably precluding the need for large copulatory plug to facilitate sperm transport through this region (60).

Marsupials

In marsupials, the accessory sex glands are composed of two groups of glands: the prostate and the bulbourethral (= Cowper's) glands. The prostate in most species is carrot shaped, but in peramelids (486), *Acrobates*, and the South American *Caenolestes* it is heart shaped (485). The distinction between these two prostate types is determined by the proportion of membranous urethra surrounded by prostatic tissue relative to the total length of the urethra. The carrot-shaped prostate surrounds most of the membranous urethra, whereas the heart-shaped prostate surrounds only half or less of its length (252,486).

The marsupial prostate is "disseminate" because all the glands occur within the surrounding capsule of smooth muscle. It is composed of three anatomically distinct regions that, in most species, run transversely. Branched tubular glands, which end blindly beneath the muscle layer, drain into the urethra, which runs down the center of the prostate. In the anterior region the glands have a low cuboidal epithelium, wide lumen, and probably exhibit apocrine secretion. The central region has branched glands that have a cuboidal, or low, columnar epithelium in which there are two cell types, one of which secretes *N*-acetyl glucosamine (486,487). The third, posterior, region has low cuboidal epithelium and produces glycogen. The heart-shaped prostate of peramelids is divided into dorsal and ventral components; whether these regions are homologous to the three regions of prostate in the other species does not appear to have been determined.

The bulbourethral glands lie beneath the pelvis and are enclosed in striated muscle. In most marsupial species there are three pairs of these glands, which are bulb-shaped branched tubular glands with wide lumina. The microstructure of the different glands appears to be similar. In a few species only one or two pairs of glands are present (514).

Monotremes

Whether monotremes possess a prostate appears to be unresolved at the present time, but if it does exist, it produces only modest amounts of secretion (203,588). A pair of bulbourethral glands is clearly present and secretes a viscous fluid.

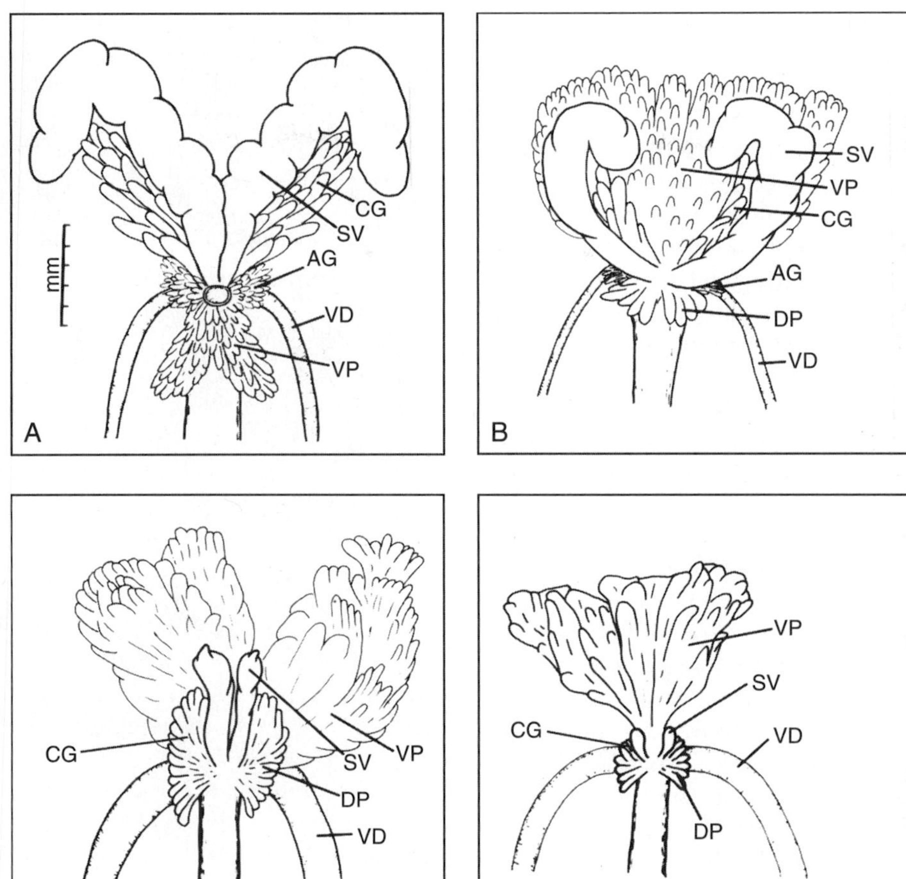

FIG. 12. A diagram of the male accessory sex glands of Australian hydromyine rodents. **A:** Ventral view from *Leggadina forresti*, another Australian hydromyine rodent, which exhibits the ancestral muroid rodent condition. **B:** Dorsal view from *Notomys cervinus*. **C:** Dorsal view from *N. mitchelli*. **D:** Dorsal view from *N. alexis*. Note the very large difference in size of the vesicular glands (SV) and anterior prostate (CG) between the three species of *Notomys*. Ventral prostate (VP) is large in all species of this genus. Note also that the diameter of the ductus deferens (VD) is much greater in the latter two species. An ampullary gland (AG) is absent from *N. mitchelli* and *N. alexis*, but is present in *N. cervinus*. The dorsal prostate (DP) is present but cannot be seen in A. [Reproduced with permission from (61).]

Penis

In all mammals, semen is deposited in the female reproductive tract by an intromittent organ, the penis. There are two basic types of penis, a vascular penis, as in most groups including primates, and a fibroelastic penis, as in artiodactyls (633). In many eutherians, an os penis, or baculum, is also present (440), and in many marsupials, the tip of the penis is bifid (47,669,672).

The vascular penis contains two corpora cavernosa that are united by a septum and, inferiorly, a single corpus spongiosum, or cavernous urethra, that surrounds the urethra. The corpora cavernosa are covered by a tunica albuginea with trabeculae of elastic fibers and smooth muscle passing inward to subdivide the cavernous bodies into endothelial-lined cavities that are continuous with blood vessels. Upon erection, these cavities become filled with blood and the blood vessels extend. The distal ends of the corpora cavernosa are covered by the glans, which is an extension of the corpus spongiosum. Caudally, the corpus spongiosum is enlarged to form the urethral bulb, which is surrounded by a bulbocavernous muscle, whereas the root of the corpora cavernosa is surrounded by ischiocavernosus muscles.

The fibroelastic penis of artiodactyls has a thick tunica of dense collagen fibers, with erectile tissue lined by fibroelastic tissue, only occurring at the root of the penis. During erection, the penis becomes more rigid, but unlike the vascular penis, it does not increase in size. Instead, protrusion is brought about by relaxation of the retractor penis muscles and straightening of the sigmoid flexure.

Although the internal structural organization of the penis falls into these two categories, it has long been known that the external morphology of the penis, and the morphology of the baculum, is highly variable between species (36,143,248, 662,669). A baculum occurs in five eutherian orders—Insectivora, Chiroptera, Carnivora, Rodentia, and Primates. Its form, when present, generally correlates with that of the glans penis. It has been suggested that the interspecific variation in bacular shape is due either to pleiotropic effects (i.e., a gene having two or more apparently independent phenotypic effects) or that it acts as a reproductive isolating mechanism. After consideration of the available data, Patterson and Thaeler (440) concluded that bacular structure, as well as that of the glans penis, correlates, at least in some murid rodents, with vaginal length and that the morphology of both these structures relates to the pattern of copulation. These authors thus rejected the pleiotropic theory for the interspecific variation in bacular form and suggested that its variation in size and shape was one element of a battery of reproductive isolating mechanisms.

Several hypotheses have been proposed to account for the interspecific variation in morphology of the penis. The lock-and-key hypothesis proposes that a female will only allow a male to mate with her if his external genitalia has the "correct" form and that this therefore prevents, or at least minimizes, interspecific hybridization. A second hypothesis states that genital divergence is due to sexual selection as result of "female choice" (128,143,248). Much of the evidence for interspecific variation in morphology of external genitalia being due to sexual selection comes from studies on insects. Within this group, several investigations have shown an association between genital morphology and fertilization success [for summary see (248)]. In mammals, an effect of sexual selection on the morphology of external genitalia has been proposed for primates and carnivores by Dixson (130), who found in primate species that occur in multimale groups, where intermale sperm competition is likely to occur, the external genitalia tended to be more elaborate and variable in form, with a longer baculum, than in males of monogamous species. Furthermore, there seemed to be longer periods of intromission during mating and/or multiple intromissions in both primates and carnivores in which males have a long baculum and/or larger penile spines (129,130,131,561).

An association between the morphology of external genitalia and copulatory behavior has also been found in both North American neotomine and Australian hydromyine rodents where there appears to be an association between the occurrence of locking during mating (a fairly rare event in rodents), the presence of large penile spines (Fig. 13), and a vagina with a relatively thick muscle coat and a small lumen (60,61,125,156). Such coevolution of reproductive morphology of the penis in the male and the vagina in the female has also been found in the African Spring-haas (90) and Warthog (89) as well as in the Stump-tailed macaque (167). Clearly, further studies on how sexual selection influences the morphology of male external genitalia and its coevolution with that of the vagina are needed, but it is evident that, in at least several lineages of eutherian mammals, divergent morphology of both male and female reproductive anatomy have evolved and that these anatomical features relate to a particular type of copulatory behavior.

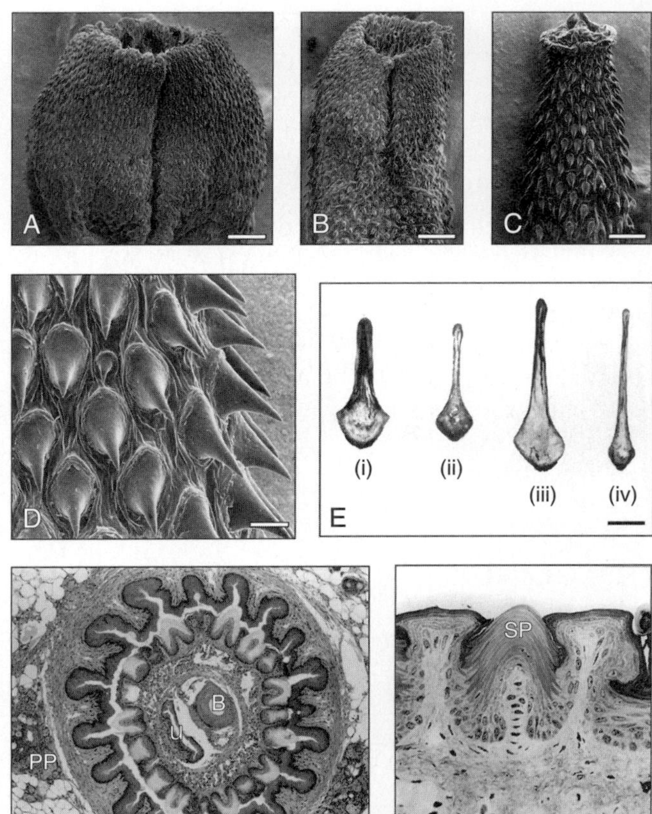

FIG. 13. The structure of the penis of some Australian rodents. **A–D:** Scanning electron micrographs of the penis of *Pseudomys nanus* (A), *P. hermannsburgensis* (B), and *Notomys alexis* (C, D). **E:** Bacula from *Melomys burtoni* (i), *Pseudomys apodemoides* (ii), *Notomys cervinus* (iii), and *N. alexis* (iv), showing variation in shape and size between the species. **F, G:** Light micrographs of cross-section of the penis of *N. alexis*, showing baculum (B), urethra (u), preputial gland (PP), and large penile spine (SP). Scale bars: A–C = 400 μm; D = 150 μm; E = 1.2 mm; F = 250 μm; G = 25 μm. [From (61).]

Scrotum

The scrotum is essentially an outpouching of skin, but this is often quite different from skin elsewhere in the body. It is thinner, often with less hair or wool, lacks subcutaneous fat, and has abundant sweat glands, even in species such as sheep in which sweat glands are not plentiful or large elsewhere. The sweat glands in rams discharge synchronously when the temperature of the scrotal skin is raised (629,630) and a marked polypnea ensues, continuing even when hypothalamic temperature falls below normal (621,622) or during fever (378,379). Beneath the skin there is a layer of smooth muscle, the tunica dartos, that contracts in response to cold and draws the testes up toward the body wall, whereas in a warm environment it relaxes to increase the surface area of the scrotum by up to 20% (378,626). The cremaster muscle can also raise the testes toward the body wall if the individual is alarmed, but because it is a striated muscle, it is less capable than the dartos of sustained contraction.

VASCULATURE

Spermatic Cord

Blood Vessels

In species with a well-developed scrotum, the long and convoluted testicular artery is surrounded by the multiple veins of the pampiniform plexus. In rodents, these vessels are surrounded by the cremaster muscle, but in domestic mammals, the cremaster is present on only one side of the cord (140). The testicular artery originates from the abdominal aorta, and it elongates more than is necessary to compensate for the migration of the testis into the scrotum, with the consequence that it becomes extensively coiled before birth (513,682). In the large domestic ruminants, as much as 5 m of artery is coiled into a spermatic cord about 20 cm in length and is surrounded by many venous branches (Figs. 6 and 14), which are largely free of valves (8,53,216,240,314,432). In laboratory rodents, the degree of coiling (Fig. 15) is less extensive (86,429,473,474,642) and varies considerably between species. In insectivores, the testicular artery is only slightly convoluted and is accompanied by one or two veins, which do not form an obvious pampiniform plexus (477,517). In primates, the degree of coiling of the artery is variable (216), but in Tree-shrews, there is a convoluted artery with a well-developed pampiniform plexus (475). In humans, the artery is only moderately coiled (258), and two classes of veins are present in the cord, one of which forms a venous plexus around the artery, whereas the other does not show a close topographical relationship, although there are venovenous anastomoses between veins of the two types. Valves are well separated (153).

From measurements of blood flow (419) and testosterone concentrations (374,375), there is evidence in a number of species for arteriovenous anastomoses

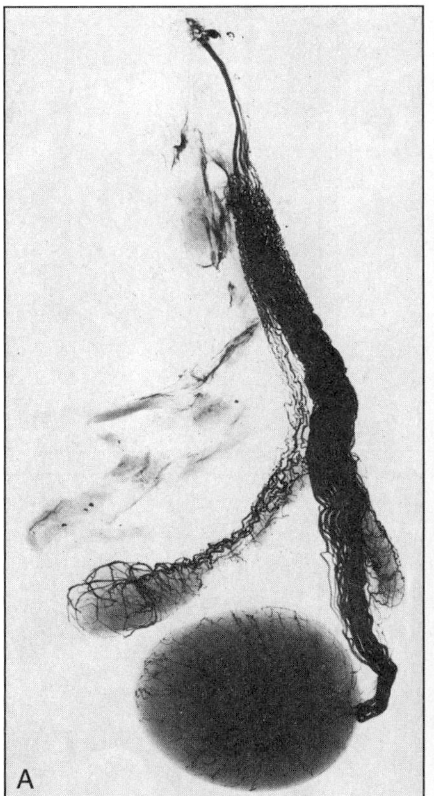

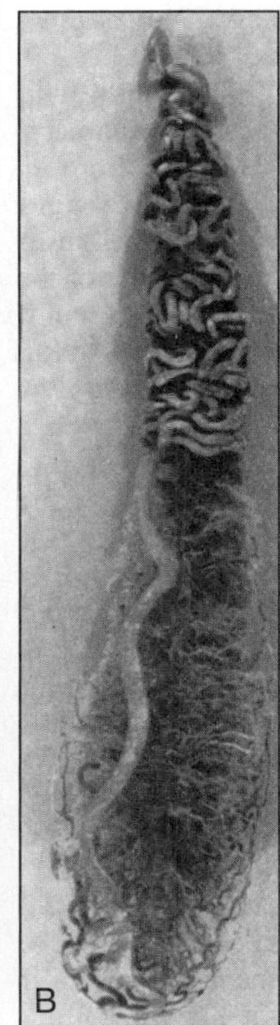

FIG. 14. The anatomy of the testicular artery of a Tammar wallaby (**A**) and a bull (**B**). [A reproduced with permission from (217). B prepared by Professor H. P. Godinho, University of Minas Gerais, Belo Horizonte, Brazil, and reproduced with permission from (537).]

in the cord. Although these were not apparent anatomically in boars (476) or hamsters (474), they are present in bulls [(226), cf. (8)]. Arteriovenous anastomoses have also been found in the spermatic cord of Tree-shrews (475) and humans (153).

In most marsupials with scrotal testes, the spermatic cord is a pendulous stalk, containing numerous parallel arterial branches as a rete (Fig. 14) interspersed among a similar number of veins (194,335,513, 514,601). The cord also includes a cremaster muscle, which is also present in females attached to the mammary glands (612). In monotremes, however, there is a single unconvoluted testicular artery that enters the abdominal testis at its caudal pole (513).

As noted earlier, African and Asian elephants and hyraxes have intraabdominal testes that lie close to the posterior pole of the kidney. In these species the testicular artery is relatively straight (188,189,515), and there is no venous pampiniform plexus, no cremaster muscle, and no inguinal canal (549). One marine representative of the Afrotheria also have testes that lie close to the caudal region of the kidney with a straight testicular artery and no pampiniform plexus (382), although in manatees there is an inguinal venous plexus overlying the hypogastric fossa (Fig. 16B) where the cauda

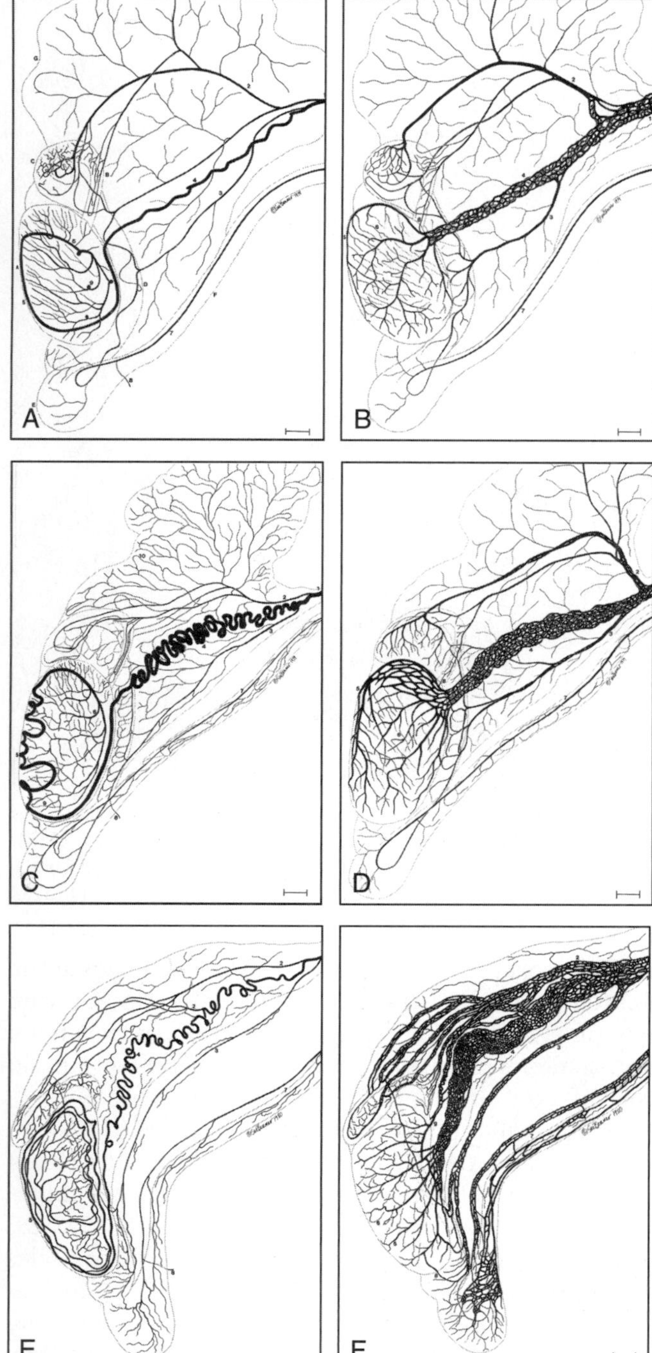

FIG. 15. Diagrams of the arterial (*left*) and venous (*right*) supply to the testes of mouse (*top*), rat (*middle*), and rabbit (*bottom*). Scale bars: 1.3, 2.7, and 3.8 mm, respectively. [Reproduced with permission from (85).]

epididymidis lies. This plexus carries cooled venous blood from the periphery via the superficial thoraco-caudal plexus (491).

In phocid seals, the testicular artery is straight and the testicular veins are single or double throughout and lie on either side of the artery (218). However, cooled venous blood from the hind flippers drains into an inguinal venous plexuses (Fig. 16C), which cup the medial aspect of the testis and epididymis (490).

Dolphins are unusual in that approximately 40 separate arteries leave the aorta and run parallel to each other in a single layer to the testis (Fig. 16A). A lumbosacral venous plexus, supplied by blood from the superficial veins of the dorsal fin and flukes, is juxtaposed, dorsolateral to the arterial layer (435,488). From measurements of temperature in the colon near the testis, this arrangement seems to cool the testicular arterial blood (434,489).

The vascular arrangement of the spermatic cord in scrotal mammals facilitates a countercurrent heat exchange (Fig. 17) that cools the descending arterial blood by 2–6°C before its entry into the testis (109,624). The analogous arrangement in marsupials also facilitates countercurrent heat exchange between the arterial and venous blood (194,529) so that the testes, and epididymides, are kept cool, as in eutherians.

As the blood passes along the testicular artery through the spermatic cord in sheep and rodents, there is also some drop in mean pressure and almost complete elimination of the pulse pressure (Fig. 17), the difference between systolic and diastolic pressures (513,572,623). A similar effect is seen in marsupials (529). The difference between the systemic arterial pressure and that in the artery on the surface of the testis is much greater in hamsters [80 vs. 40 mm Hg (572)] than in rams (99 vs. 90 mm Hg), rats [130 vs. 110 mm Hg (513)], or Tammar wallabies [115 vs. 80 mm Hg (529)]. Doppler sonography of blood flow in the testicular artery above the cord and on the surface of the testis in dogs showed that at both sites there is only about a 30% reduction in blood flow velocity between systole and diastole, compared with more than 90% in prostate arteries (209)

Much attention has been directed to the possibility that substances may pass from arterial to venous blood or vice versa in the spermatic cord. Although this may be significant for highly permeable substances, such as inert gases like krypton and xenon and probably also tritiated water, the extent of transfer of substances, such as testosterone, for which there would normally be an appreciable gradient between venous and arterial blood appears to be minimal. Transfer of steroids from vein to artery does lead to a small increase in the testosterone concentration in arterial blood as it passes through the cord (Fig. 17), but this is trivial compared with the increase in concentration as the blood passes through the testis itself (270).

Varicocele

In approximately 15% of humans, the veins of the pampiniform plexus dilate to form a varicocele, and

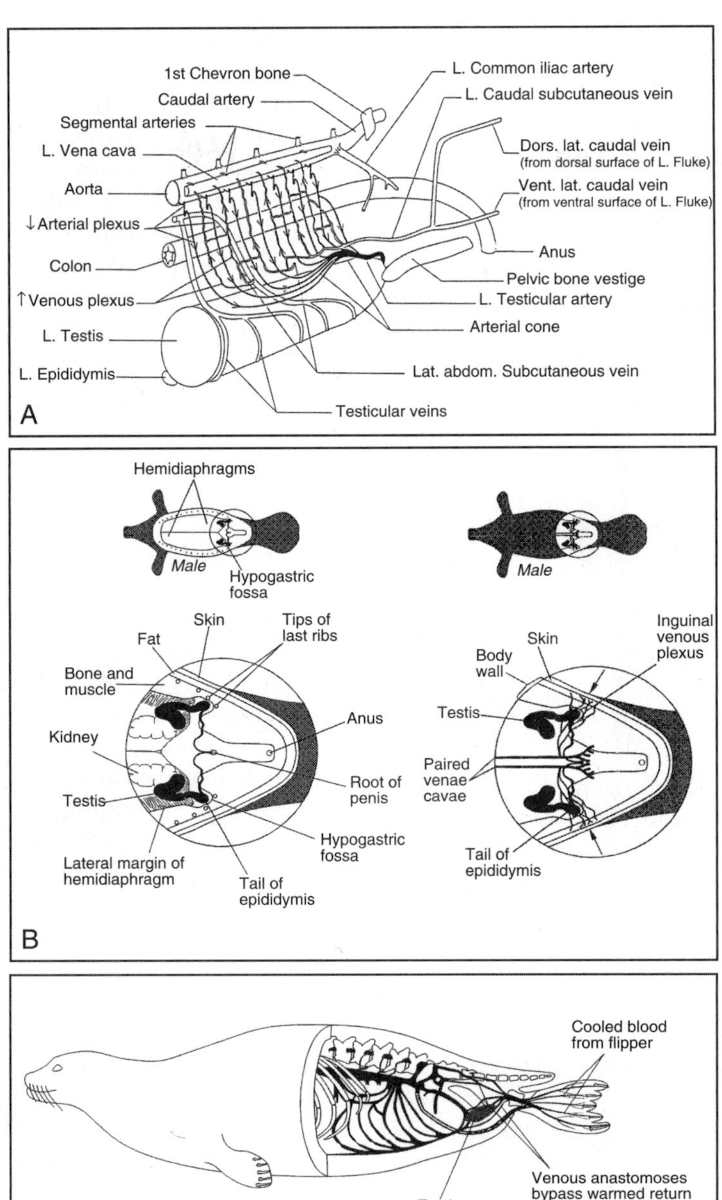

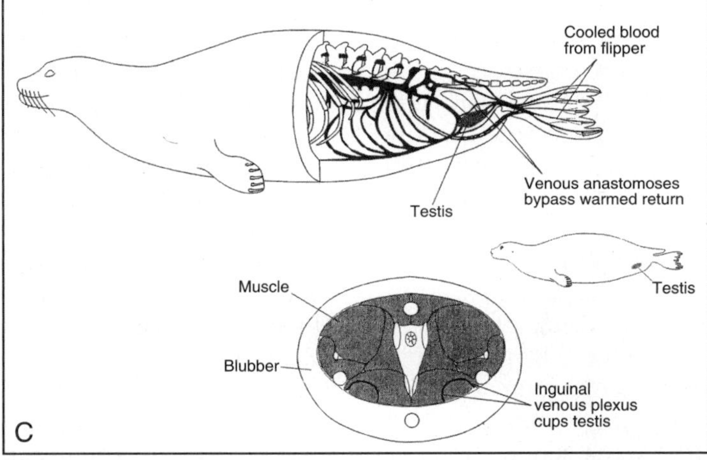

FIG. 16. A: The anatomy of the arteries supplying the left testis of a bottle-nosed dolphin and the veins from the tail that run close to the testicular arteries. **B:** Ventral view of the testes, epididymides, and kidneys of a Florida manatee and the relationship of the tail of the epididymis in the hypogastric fossa to the inguinal venous plexus. **C:** The venous drainage of the flipper of a seal forms an inguinal venous plexus cupping the medial surface of the testis. This plexus thermally isolates the testis and epididymis from the adjacent thermogenic muscles and provides local cooling. [A and B reproduced with permission from (488,491); C kindly supplied by Dr. S. A. Rommel.]

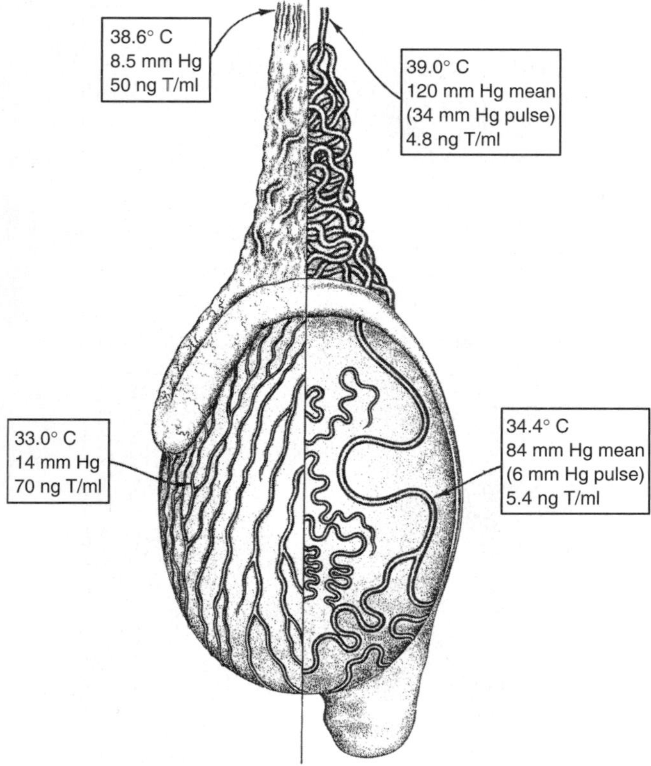

FIG. 17. A diagram of the blood vessels of a ram testis, with arteries on the right side and veins on the left. The figures in the boxes give the temperature, mean blood pressure (and pulse pressure), and the concentration of testosterone at the point shown. [Reproduced with permission from (520).]

the incidence is appreciably higher in men attending infertility clinics. Varicocele is more common on the left side and can range from a palpable varicosity less than 1 cm in diameter to a venous mass that fills the hemiscrotum (98,505,616,617). Varicoceles are often ligated in an attempt to treat male infertility, but results have been inconclusive (19). There is no general agreement as to why a varicocele causes reduced fertility (if indeed it does), although increases in venous pressure, reflux of vasoactive agents, and/or temperature elevation of the testes have been suggested (582). A number of experimental animal models have been developed in an attempt to replicate this condition, involving ligation of veins, usually the left renal vein (176,506), but their relevance to the human condition is uncertain.

Lymphatic Vessels

In sheep, cattle, and pigs, on the outside of the pampiniform plexus there are usually several prominent lymphatic vessels, which can be cannulated for the collection of lymph (20,106,516,525,533). Lymphatic vessels in the spermatic cord of the guinea pig and hamster run in the connective tissue of the walls of the arteries and veins (199,474).

At the top of the spermatic cord, the lymphatic vessels empty into abdominal lymph nodes, principally renal and iliac nodes in the rat (225), although it has also been reported (303,304) that on the right side, in some rats, the testicular lymphatics emptied directly into the cisterna chyli, without passing through a lymph node. This is not the case in mice or guinea pigs, where the testicular lymphatics on both sides invariably drain into the lower paraaortic or renal nodes (267,304).

Testis

Blood Vessels

Arteries. One of the first differences between XX and XY gonads in mouse fetuses is the formation in males of a celomic vessel from endothelial cells that have migrated into the gonad from the mesonephros. Cells in this vessel between days 12.5 and 14.5 postcoitum express the marker ephrin B2, which is specific for developing arteries (65). Development of the vasculature in the male fetal gonad is probably required for the export of steroids from the testis to the rest of the body to ensure masculinization. In females, it is repressed by the signaling molecule WNT4 (273,294).

In adults of most species with scrotal testes, as the testicular artery emerges from the spermatic cord near the proximal pole of the testis, it straightens and extends down the epididymal margin of the testis just under the capsule, usually in a reasonably straight course without branching. The artery on the surface of the testis is appreciably thinner walled and has a larger internal diameter than above the spermatic cord (632,642). In humans, the artery commonly branches into two, or occasionally three, before it reaches the testis and often anastomoses with the deferential or cremasteric arteries (271,319). As it rounds the distal pole of the testis, the testicular artery forms, in domestic ruminants and pigs, several branches that run within the capsule up the free margin of the testis, with appreciable convolutions before entering the parenchyma (Figs. 13 and 16). These branches then run inward, without further subdivision, as centripetal arteries to near the mediastinum where they form coils, turn back on themselves, and run as centrifugal arteries into the parenchyma to branch and supply the interstitial tissue of the testis (10,53,126,127,229,513,515). In donkeys, branches supplying the parenchyma arise from both centripetal and centrifugal arteries (163). In rats and hamsters, after rounding the distal pole of the testis, the artery runs up the free margin of the testis still without branching (Fig. 14), in serpentine loops to the proximal pole near the rete testis, where the artery enters the parenchyma and begins to branch (85,216,317,429,513). The muscle layer in the walls of arteries within the rat testis contains androgen receptors (41), and the arterial endothelial cells modulate vascular tone through nitric oxide and prostaglandins (101). The anatomy of the testicular artery in the guinea pig (199) and mouse is similar (Fig. 14), except that the artery on the free margin of the testis of the latter is not convoluted (85). In rabbits, the main artery encircles the testis and then forms two branches, which in turn run around the testis again without further branching (85,513). In humans, the artery or arteries usually run around the distal pole of the testis but sometimes run across its lateral face to the free margin, from which centripetal branches pass in toward the mediastinum (122,258,318,319). Most of the subtunical branches are found in the anterior inferior sector of the testis, with fewer in the superior anterior, inferior lateral, and inferior medial sectors and fewest in the superior medial and lateral sectors. These branches are not obvious through the opaque tunica albuginea, and therefore there is a risk that they may be inadvertently obstructed if a ligature is placed through the capsule during orchidopexy (271,272). The centripetal arteries run inside the testicular septa in a relatively straight course toward the rete, sometimes, but not always, dividing into two or more branches. Near the rete, the arteries turn back to form centrifugal or recurrent branches, which run in the center of a lobule to supply blood to the parenchyma (150). Fetal testes in humans are also commonly supplied by deferential and cremasteric arteries as well as by the testicular artery (502). The walls of the arterioles and some venules in the testes of infertile men are sometimes thickened (275).

In dolphins, the multiple arteries running from the aorta unite to form a single trunk (Fig. 16A), which enters the testis at its caudal pole (488). In marsupials, the multiple arterial branches in the spermatic cord reunite as they reach the testis to form a single or double vessel, which runs along the epididymal margin over one arm of the horseshoe shaped rete for about half the length of the testis (Fig. 18) before branching and entering the parenchyma (513,514).

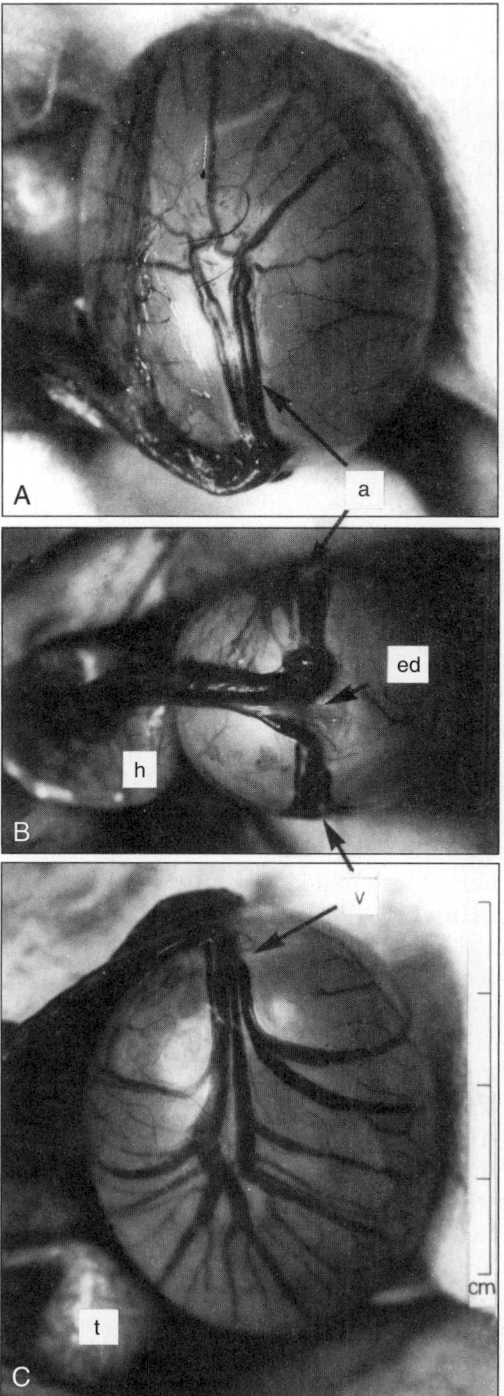

FIG. 18. The blood vessels on the surfaces of the testis of a Tammar wallaby. Note that the arteries (a) and veins (v) at the bottom of the spermatic cord separate to either side of the efferent ducts (ed) and then run on the epididymal and free surfaces of the testis, respectively, over part of the horseshoe-shaped rete before entering the parenchyma. h, t: head and tail of the epididymis. [Reproduced with permission from (513).]

Veins. The veins draining the interstitial tissue of the testis do not accompany the arteries. In humans, the venules empty into intralobular veins, which are found in the periphery of the lobules. These empty into veins in the septa, which run mainly to the rete but in some instances run to the surface of the testis (150). In horses and donkeys, most of the septal veins run to those near the central rete (162,513). In domestic ruminants and pigs, most of the veins run directly toward the surface, to a subalbugineal venous plexus (10). This in turn joins to an intraalbugineal venous plexus within the capsule and to surface veins.

The surface veins run toward the proximal pole of the testis, where they join the distal end of the spermatic cord (513,515). In rats, most veins run to the capsule of the testis where they join two surface veins on the free margin of the testis; these unite just below the efferent ducts and then divide again to form branches that run along either side of the efferent ducts (374). The point at which they join is a convenient point from which testicular venous blood samples can be collected by puncturing the vein with a needle and collecting the blood into a heparinized capillary tube (179). Proximal to the efferent ducts, the veins drain into the branches of the pampiniform plexus. The veins of the mediastinal venous plexus overlying the rete are constricted after treatment of rats with a 5-hydroxytryptamine agonist (450).

Above the pampiniform plexus, the right testicular vein opened directly into the right common iliac vein in most rats. Although, in a few individuals it was found to open directly into the posterior vena cava. The left testicular vein drains into the left common iliac vein in all species investigated, but in most there is also an accessory branch into the left renal vein (439). In wallabies, the veins drain from the parenchyma to the free margin of the testis, where they form several large veins (Fig. 18), which, like the arteries on the other side of the testis, overlie one arm of the horseshoe-shaped rete (513,514).

Capillaries. The capillaries in the testis are confined to the interstitial tissue and in rodent, dog, and pig testes, unlike those in all other endocrine tissues, are unfenestrated (137,161,449,580). They form two networks, one running parallel to the seminiferous tubules and the other around the tubules in a rope ladder-like arrangement (317,406,408,568). In human testes, this pattern is not as clear-cut (274,569,581), and some fenestrated capillaries appear to penetrate the lamina propria of the seminiferous tubules (151,152). The typical pattern of the microvasculature appears in rodents only at puberty (317,389) and reverts to the immature form in hamsters sexually suppressed by being transferred to short days (390).

In the rat testis, the endothelial cells lining the blood vessels show a high rate of proliferation (599), which is reduced if Leydig cells are eliminated but is not restored by testosterone treatment (91,361). An angiogenic mitogen (endocrine gland vascular endothelial growth factor [EG-VEGF]) selective for endocrine gland endothelial cells has recently been identified (332), and receptors for its homologue Bv8 are present on testicular endothelial cells. Bv8, which, like VEGF and EG-VEGF, induces a potent angiogenic response in the testis (333), appears to be principally expressed in primary spermatocytes (641), so it is not clear how it would reach the blood vessels. VEGF itself is secreted by the Leydig and other testicular cells, and its receptors VEGF-R1 and -R2 are expressed on testicular blood vessels (154,495,496). Levels of VEGF in the testes of *Peromyscus* are reduced by short photoperiods (683). Transgenic mice overexpressing VEGF had increased capillary density in the testes but also showed spermatogenic arrest (320). Endothelial cells in the rat testis also have receptors for endothelin (376,499), the principal source of which appears to be the Sertoli cells (157), and glial cell-line derived neurotrophic factor, which appears to be secreted by α-smooth muscle actin positive cells in the vessels and around the seminiferous tubules (300).

Testicular endothelial cells have several other characteristics that are unlike those of other endothelial cells, except in the brain. They have a high content of γ-glutamyl transpeptidase (241,417) and an endothelial barrier antigen, which was thought to be specific for vessels in the central nervous system (184). They also contain a glucose transporter isoform GLUT-1, usually associated with brain and retina, and also the membrane efflux transporter P-glycoprotein (241), the inhibition of which enhances the distribution of human immunodeficiency virus type 1 protease inhibitors in the testis as well as in the brain (83). In mice in which the mdr1a P-glycoprotein gene has been knocked out, the tissue concentrations of ivermectin, digoxin, cyclosporin A, ondansetron, and loperamide were much higher than in wild-type mice, the ratio (–/–)/(+/+) being higher for the testis than all other tissues, except brain (and gallbladder for ivermectin) (509–511). In mice with both mdr1a and mdr1b P-glycoprotein genes knocked out, the penetration of the steroid hormones corticosterone, cortisol, aldosterone, and progesterone into the testis was enhanced (602), raising the possibility that the endothelial cells may be involved in transporting testosterone out of the testis (523). The endothelial cells in the rat testis are the only testicular cells to express claudin-5, a protein constituent of tight-junction strands, whereas occludin, another such protein, is expressed in myoid and Sertoli cells as well as the endothelial cells (300).

The vesicular density in endothelial cells of capillaries in the rat testis is slightly lower than those in brain and much lower than those of all other tissues examined except ovary (560). The endothelial cells of the larger vessels in rat testes transport amino acids, in particular leucine in a saturable manner, with kinetics similar to those for similar systems in brain and different from those in other tissues in the body (73). The large amino acid transporter LAT1 is also present in rat testis as well as brain and heart, but not other tissues (139).

Endothelial cells can be isolated from rat testes and kept in culture for several days while maintaining their characteristics. When cocultured with interstitial cells, they enhance the production of testosterone by several-fold (523,527). The role of endothelial cells from the mesonephros in establishing male-specific vasculature in the testis and steroidogenesis is discussed above.

Testicular Blood Flow

Blood flow through the testis has been measured by a number of techniques, some of which estimate nutrient flow through the capillaries, where exchange of substances with the tissue itself takes place, and some of which estimate total flow, that is, capillary flow plus any flow through arteriovenous anastomoses in the tissue itself and/or in the spermatic cord. For measurements of capillary flow, the preferred method involves injection of microspheres, either radioactively or dye-labeled into the left ventricle, while an arterial blood sample is removed at a known rate from another site (112,305). Earlier results obtained by measuring the distribution of soluble markers, such as iodoantipyrine injected intravenously (531,625), probably gave results that may not be quantitatively accurate but are still useful for indicating changes. More recently, a laser-Doppler probe has been used to monitor blood flow continuously in rats (113,114) and several other species (97), and, although this method does not give absolute values for flow, it does indicate even very rapid changes. This method has revealed that testicular blood flow (TBF) shows vasomotion, that is, rhythmic fluctuations by about 20% of mean flow, with a frequency of about 10/min (114). These fluctuations are linked with changes in oxygen tension in the interstitial tissue (366), but their significance is not known.

Total blood flow is best measured by indicator dilution. This involves infusing an appropriate indicator at a constant rate, either into the arterial supply above or below the cord or into a vein on the surface of the testis, while sampling blood from a vein above the cord, relying on the veins of the pampiniform plexus to ensure complete mixing (15,16,398, 535). Alternatively, flowmeters of various types have been implanted around the testicular artery (419), and more recently, a pulsatility index of the transmediastinal artery in human testes has been assessed by color Doppler scanning, with higher values reported in a nonobstructive than in an obstructive azoospermic group (24,46).

TBF is lower per unit tissue weight than in most active tissues in the body and varies inversely with the body weight of the species, from about 10 ml/100 g/min in rams and boars to about 30 ml/100 g/min in rats (537,627), and total blood flow was about 30% higher than nutrient flow, when both were measured simultaneously (166,193). Measurements of total blood flow with electromagnetic flowmeters through the testicular artery above and below the cord, which would indicate transfer of blood through anastomoses in the cord, suggested that about half the blood entering the cord did not pass through the testis, a conclusion supported by measurements of oxygen saturation of the hemoglobin and testosterone concentrations (374,375,419). These anastomoses appeared to be closed after treatment of rats with a 5-hydroxytryptamine agonist (450).

The trophic hormones LH or human chorionic gonadotropin (hCG) had variable effects on TBF, although there were decreases after surgical hypophysectomy (50,108,115,116,535). No changes in TBF were seen during spontaneous peaks of LH, and pharmacological doses of LH or hCG caused either no effect or first a decrease and then an increase (7,16,330,519,651). Destruction of the Leydig cells with ethane dimethane sulfonate reduced TBF and abolished vasomotion, effects that could be reversed with testosterone (91,117,600). Estradiol was without effects on TBF or vasomotion (650).

Vasodilators had no effect on TBF (418,532) except for adenosine, which produced an increase (166). Vasoconstrictors such as epinephrine, norepinephrine, 5-hydroxytryptamine, or prostaglandins reduced TBF and inhibited vasomotion (42,94,113, 366,532,600).

Stimulation of nerves in the spermatic cord of rams caused vasoconstriction (353), as did stimulation of the superior spermatic nerve in cats; stimulation of the inferior nerve was followed by an increase in TBF (82). Blockade of α-receptors (532) or sympathetic blockade (111) had no effect on TBF, suggesting that there is normally little resting sympathetic tone, although left sympathectomy caused a bilateral increase (198). Inhibition of nitric oxide synthase had no effect on TBF or vasomotion, suggesting that nitric oxide has only a limited role in regulation of TBF (356).

Various peptides, some of which have been shown to be present in the testis, had variable effects on TBF and vasomotion in the rat testis (Table 1). However, the significance of these observations must await information of the concentrations of these factors normally found within the testis (523).

Heating the testis had no effect on nutrient blood flow, except in marsupials, until the temperature of the testis rose well above body temperature [see (521) for references]. Total blood flow did increase in rams when testis temperature was raised (398). Vasomotion disappeared as the testis temperature in rats reached between 36 and 41°C, but there was considerable variation between animals. Vasomotion returned if the testis was cooled and became extreme in amplitude and with a very low frequency as the temperature

TABLE 1. *Effects of various peptides on TBF and vasomotion in rat testes*

Peptide	Local L or parenteral P	TBF	Vasomotion	References
LH-RH	P	+	–	(115)
	L	+	–	(115, 649)
AVP	L	– then +	NC	(652)
Endothelin-1	L	–	–	(93,305)
ANP, BNP, or CNP	L	+	NC	(95)
CGRP				
Low dose	L	+	NC	(357)
High dose	L	–	–	(357)
VIP	L	+	–	(355)
Neuropeptide Y	L	–	–	(96)
PACAP	L	NC	–	(358)

+, increase; –, decrease; ANP, atrial natriuretic peptide; AVP, arginine vasopressin; BNP, brain natriuretic peptide; CNP, C-type natriuretic peptide; NC, no change; PACAP, pituitary adenylate cyclase-activating peptide; RH, releasing hormone.

fell below physiological levels (538,600). At testis temperatures below about 30°C, periodic stoppages of red cell movement in surface vessels were observed by intravital epi-illumination (410).

If the blood supply to a rat testis was clamped for 2 minutes and then released, TBF returned to normal over the next 8 to 10 minutes, but vasomotion did not resume until TBF has been at normal levels for some time. Even after 24 hours, many of these testes did not show vasomotion (366,600).

Spontaneous varicocele in humans was associated with reduced TBF (586). In rats and dogs with experimental varicocele, testis blood flow was increased not only in the operated but also in the contralateral testes with some methods (197,198,259,506,598), but was less (342) or not different using other techniques (600).

Injection of cadmium salts causes a catastrophic decrease in TBF, probably secondary to increases in vascular permeability (530,625). When damage to spermatogenesis is produced in other ways, TBF per testis fell in proportion to the drop in testis weight, even when there was no effect on the Leydig cells or testosterone levels (179,585,635,636). It would therefore appear that when spermatogenesis is disrupted, the seminiferous tubules, because of their greater weight, are a more important determinant of TBF than the interstitial tissue, although the mechanism for this control is not known. Removal of one testis leads to an increase in the volume density and surface density of capillaries and postcapillary venules in the remaining testis without any change in TBF or vasomotion (40).

Vascular Permeability and Interstitial Extracellular Fluid

Lymph from a catheter in a vessel in the spermatic cord is believed to be derived largely from extracellular fluid in the interstitial tissue (IEF), but a number of techniques have been used to obtain IEF directly from the testis for analysis. Both lymph and IEF have a high concentration of total protein, almost as much as in blood plasma. However, in fluid collected by the most commonly used technique, there were major differences in testosterone concentrations between IEF and blood plasma from the testicular veins, which are not apparent using more refined procedures. There is now good evidence that stored or newly formed testosterone continues to leak from Leydig cells into fluid collected from testes after removal from rats (373,518,519,539). In IEF collected from testes in situ in living rats, testosterone concentration is only slightly above that in blood from the testicular vein; the concentration of LH in this IEF did not change significantly before the Leydig cells increased their production of testosterone, suggesting that either they respond to much smaller changes in the concentration of LH in their vicinity or that another cell population, namely the endothelial cells, are mediators in the response to LH (539). Evidence for the latter suggestion has recently been obtained using cultures of endothelial and interstitial cells from rat testes (527).

The volume of the IEF can be estimated either from the amount of fluid that can be collected from testes after removal from rats or from the volume of distribution in vivo of markers known not to enter interstitial cells or the seminiferous tubules, and similar results of about 80 μl/g for normal rat testes been obtained with both techniques (373,518,519,539). However, there has been an unfortunate tendency to use the volume of IEF as an estimate of vascular permeability. In fact, the volume of IEF in a testis is the result of the balance between a number of factors regulating not only fluid formation but also removal, namely capillary and tissue pressures, differences in protein concentrations between blood plasma and IEF (which are admittedly small), and vascular permeability.

The factors affecting fluid removal from the testis in lymph are not known, but interstitial hydrostatic pressure is probably important. Fluid formation or reabsorption by the capillaries is determined according to the formula $J_v = L_pS\ (P_c - P_i - \sigma[\pi_p - \pi_i])$, where J_v is the amount of fluid filtered or resorbed across the capillary wall in unit time, L_p is the hydraulic conductivity of the vessel wall, S is the surface area of the vessel wall, P_c and P_i are the hydrostatic pressures in the capillary and the interstitial tissues, respectively, π_c and π_i are colloid osmotic pressures in the capillary and the interstitial tissues, respectively, and σ is the reflection coefficient of the vessel wall for protein. The hydrostatic pressure in testicular capillaries is comparatively low (572), although 90% of a venous pressure elevation is transmitted to the capillaries (573). This, together with the small difference between colloid osmotic pressure between the capillary blood plasma and IEF, suggests that hydrostatic pressure in the interstitial tissue is an important factor in determining fluid filtration and volume. If spermatogenesis is disrupted and the tubules shrink, it is likely that hydrostatic pressure in the interstitial tissue would be reduced, leading to an increase in interstitial fluid volume without any change in vascular permeability.

Vascular permeability can be measured from the percentage of a marker taken up by the tissue from the blood during a single passage of the marker. This is most easily achieved in the isolated perfused testis, although an estimate can be made if the arterial supply to the testis is cannulated and the testis is removed before recirculation has become significant. Using the isolated perfused testis of the rat, about 1.3% of a dose of radioactive albumin was removed in comparison with chromium-labeled red cells (72), and in rams, 3.1% of a dose injected into the testicular artery on the surface of the testis lodged in the tissue during a single passage (524). Permeability-surface area product (PS) in μl/g/min can be calculated using the formula $\text{PS} = -Q/\ln(1 - E)$, where Q is perfusate or blood flow and E is extraction determined from the area under the curves of the concentrations of the two markers in the effluent or from the proportion of the dose lodging in the tissue, giving a value of about 3 μl/g/min for both rats and rams. This value was not increased in rats around puberty or after treatment with hCG but was dramatically increased after cadmium (536). With smaller markers, the formula is modified to use E_0, which is the extraction at time 0 obtained by extrapolation as the extraction falls with time due to return of the marker to the circulation, instead of E. Using this procedure, values for the permeability to sodium, Cr-EDTA, and vitamin B_{12} were obtained (2230, 1482, and 850 μl/g/min, respectively), which, when allowance was made for the much smaller vascular surface area in the testis, compared with other tissues in which this technique was applied, indicates that the microvasculature in the testis is much more permeable to these markers than in pancreas, muscle, or heart (72).

PS to albumin can also be calculated from the rate at which the volume of distribution (in μl/g) of the marker approaches its equilibrium value. Because this relies on the distribution of the marker throughout the interstitial tissue, it also involves the permeability of the tissue itself, not just the permeability of the microvasculature, and therefore it is not surprising that values obtained with this method (0.38 μl/g/min for albumin, 0.31 for γ-globulin) (519,537) are considerably lower than those given above. PS measured by the volume of distribution method to both markers is considerably higher in rats during puberty (452,534); the values quoted in the original studies are slightly different from the above, because calculation were made with the final concentration of marker in the blood, not the mean concentration during the measurement (519). PS to albumin measured in this way is also increased after treatment with hCG (526), again in contrast to measurements made of microvascular permeability (see above). The PS to γ-globulin was much higher for the capsule than for the parenchyma of the testis (452), although if PS was calculated for just the interstitial tissue, not the whole parenchyma, the reverse was true (451). The permeability to the Fab fragment of IgG was higher than that for IgG for both the testis parenchyma and capsule (453).

The permeability to γ-globulin and to the Fab fragment was also measured in the epididymis. Both values were comparable in the caput and the testicular capsule and were lower in the corpus and cauda. Permeability in the caput increased sharply between 30 and 44 days of age, with smaller changes in the other two segments (451,453).

Lymphatic Vessels

In the testes of rodents, there are large lymphatic spaces in the interstitial tissue. In the testes of domestic ruminants, monkeys, elephant, and hyraxes, there are discrete lymphatic vessels, which were more obvious in these animals than in species such as pig, warthog, zebra, and opossum, which have abundant Leydig cells (86,160,161,212). However, in human testes, the lymphatic vessels appear to be confined to the septula (242,243). In bulls, there are lymph sinuses between the seminiferous tubules that drain into larger lymph capillaries in the septa and tunica albuginea. Numerous elastic fibers were observed between the endothelial wall of the lymphatic vessels and the adjoining connective tissue (684).

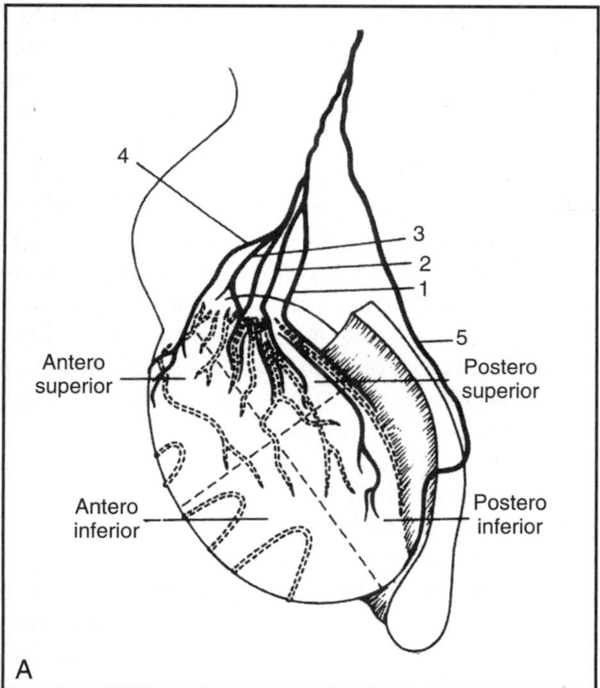

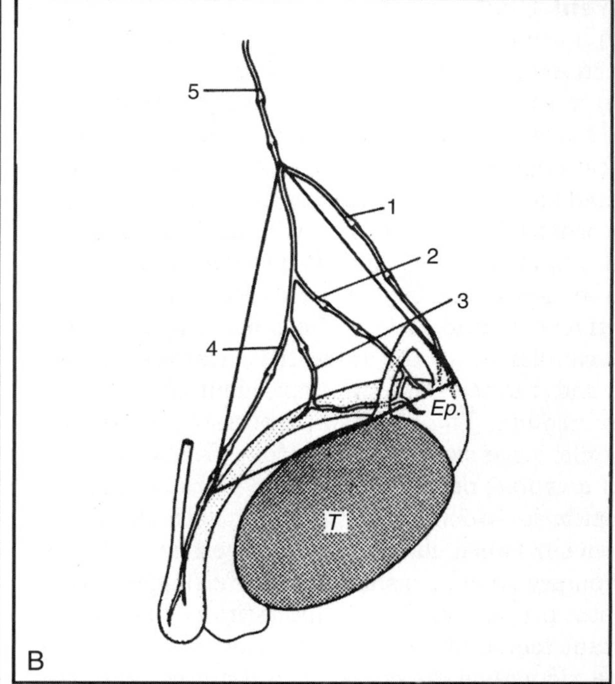

FIG. 19. The lymphatic vessels draining the testis (**A**) and epididymis (**B**) of the rat. The numbers in A refer to the order in which colloidal carbon injected into the lower pole of the testis appeared in the vessels. The labeling in B is as follows: 1, ductular lymphatic trunk; 2, superior epididymal trunk; 3, corporeal epididymal trunk; 4, inferior epididymal trunk; 5, main testicular lymphatic trunk; T, testis; Ep, epididymis. [Reproduced with permission from (446,447).]

The 11 to 15 lymphatic vessels leaving the testis in the rat run in three groups: a superior group surrounding the efferent ducts at the level of the upper pole and then running through the epididymal fat pad, a middle group from the epididymal surface of the testis running around the testicular artery, and an inferior group from the lower pole of the testis and running along the medial face of the corpus epididymidis (Fig. 19) (446). A similar pattern is found in mice, except that the inferior group was missing in some testes, and in others the caudal vessel joined a trunk running along the ductus deferens (267). In rams, there is large lymphatic draining the lower pole of the testis that runs across the cauda of the epididymis and then alongside the ductus deferens (518).

Lymph Flow

Lymph from the testis and epididymis can be collected in rams, bulls, stallions, and boars by cannulating one or more of the lymphatic vessels in the spermatic cord, and if all the other lymphatic vessels there are ligated, an estimate of flow can be obtained. In rams, bulls, and stallions, lymph flow is about 10 ml/h (20,106,405,525,528), and in boars, it is about 40 ml/h (20,516,533). Lymph flow was increased if venous pressure was increased (516) or during exercise or warming the testis, whereas cooling the testis reduced lymph flow (106). Colloidal carbon injected into the anterior–inferior quadrant of the rat testis was cleared more rapidly from the subtunical lymphatics than from those deep in the parenchyma, suggesting that lymph flow was not uniform throughout the tissue (212)

If radioactive albumin is injected into the interstitial space of the testis, almost all leaves in the lymph. Therefore, lymph flow (Q_L) can also be estimated from the half-time of the clearance of the marker from the testis and its volume of distribution, using the formula $Q_L = 0.693 \times V_{alb}/T_{half}$, where V_{alb} is the volume of distribution of albumin in the testis and T_{half} is the half-time of clearance of albumin from the testis. When compared with direct collection, the resultant values are slightly higher (533) but the differences between species remain. This technique has also been applied to animals like rats and ferrets, in which the lymphatic vessels are too small to cannulate. Lymph flow in the rat is the slowest of all, about 0.5 μl/g/min, whereas in ferrets lymph flow is intermediate between boars and rams. Lymph flow through the pig testis is large enough to cause a significant increase in the hematocrit of about 5%, which is what would be predicted from the known values for blood and lymph flow (180).

Excurrent Ducts

Blood Vessels

The blood supply to the epididymis is derived from two sources, with anastomoses in the corpus. The caput is supplied by branches of an epididymal artery, which in most species of scrotal eutherians branches from the testicular artery just above or in the spermatic cord and, like the testicular artery, is surrounded by a venous plexus, in some species separate from the testicular pampiniform plexus. Near the epididymis in the mouse, the epididymal artery divides into four branches, the superior epididymal going to the caput, the inferior epididymal to the corpus, and two branches to the efferent ducts (1). The cauda is supplied by the deferential artery, a branch of the internal iliac or hypogastric artery, that runs alongside and also supplies the ductus deferens.

Within the epididymis, branches of the arteries enter the connective tissue septa and then give rise to capillary networks. In the efferent ducts and the middle segment of the mouse epididymis, the capillaries run around the ducts as in the testis, but in the initial segment, the capillaries form a dense cylindrical network with frequent intercommunications (568). The capillaries in the initial segment are fenestrated, whereas in other regions they are unfenestrated. The characteristic appearance of the capillaries in the initial segment was lost after efferent duct ligation, suggesting that it depends on some constituent of the luminal fluid entering from the testis (1). In the human epididymis, the microvascular arrangement is similar throughout its length. The larger vessels pass through the intertubular connective tissue and capillaries penetrate into the muscular or subepithelial layer of each tubule. The arteries are coiled to a greater extent than the veins (569). In the efferent ducts in the bull, the capillaries are of the continuous type (9,78).

In rats, there is a decrease in capillary size from the initial segment to the cauda, a pattern that was not affected by age (381). In the boar, there are two superimposed vascular networks. Capillaries surrounding the epididymal duct form polygonal meshes around the efferent ducts, whereas circularly arranged capillaries predominate in the caput. This feature is progressively lost between the caput and cauda, where the capillary network again becomes polygonal. Outside the capillary network there is a second layer of larger vessels (562,564). In the buffalo, the microvascularization is more dense in the corpus and cauda than in the caput, and none of the capillaries is fenestrated, although in the postcapillary venules there are large fenestrations connecting the blood and lymphatic systems (507). Receptors for VEGF are present on endothelial cells in the epididymis, and VEGF is found in peritubular and ciliated cells in the efferent ducts and in peritubular and basal cells in the epididymis. VEGF could induce fenestrations and transendothelial gaps in epididymal tissue in vitro (155). In the human ductus deferens, there is a venous network in the connective tissue around the muscle coat, a microvascular plexus with the muscle, and a peritubular capillary plexus (569). In the rat ductus deferens, there are four layers of blood vessels, an adventitial layer, a vascular network in the muscle, a sinusoidal layer in the lamina propria, and a subepithelial capillary network. The sinusoidal layer is present only in the urethral part of the ductus (428).

Blood Flow

Blood flow through the epididymis has been measured by soluble indicator dilution and by microspheres. Values for the whole epididymis for the rat range from 14 to 30 ml/100 g/min, and there appears to be no effect of hemicastration (40), LH–releasing hormone agonist (115), removal of the Leydig cells with ethane dimethanesulfonate or subsequent treatment with testosterone (117), vasopressin (652), or the application of various vasoactive drugs to the surface of the testis (94). Blood flow through the caput epididymidis showed vasomotion but with a slower frequency (3–5/min) and a smaller amplitude than the testis; it was increased by local injection of pituitary adenylate cyclase-activating peptide (358).

Blood flow is higher in the caput and lower in the cauda than in the rest of the rat epididymis, and in rams and rabbits the highest flow is found at the flexure overlying the anterior face of the testis (69,531). The relatively high flow in the caput in rats and rabbits is reduced by unilateral castration or efferent duct ligation (69), showing that luminal factors can influence blood flow, either directly or indirectly.

Lymphatic Vessels

Lymphatic vessels drain the whole length of the epididymis and the ductus deferens. In laboratory rats, several vessels leave the caput, one or two from the corpus and one from the cauda (Fig. 19B). These unite to form a single trunk that joins the testicular lymphatic trunk in the spermatic cord (368,447). These authors disagree on whether the epididymal lymphatics join with those of the ductus in the rat, but in the mouse a lymphatic vessel from the cauda runs along the ductus to join the main testicular lymphatic trunk in about half the cases and in the rest empties separately into the afferent lymph duct of a lumbar lymph node (267). In rams, there is a large lymph vessel running along the ductus, which receives lymph from the cauda epididymidis as well as the caudal pole of the testis (518,522). Lymph from vessels in the spermatic cord of vasectomized animals contains significant numbers of spermatozoa for up to 3 months after the operation (20).

Accessory Sex Glands

The blood supply to the accessory glands is derived from the internal iliac (= hypogastric) artery. The prostatic (= superior vesical) artery runs dorsal to the vesicular gland, where it branches to supply the ventral and dorsolateral prostate, the vesicular gland, together with the anterior prostate (= coagulating gland), with some branches from the inferior vesical artery (276,473). Each prostatic duct in rats appears to have its own vascular supply, originating as a branch of the main artery. The arterial branches continue to the intermediate region, where further branching occurs, and secondary branches continue to the distal portions of the ducts and also turn back to supply the proximal duct region. There is an exceptionally dense capillary plexus on the anterior–lateral surface of the prostate, consisting of two intermingled layers of continuous and fenestrated capillaries (541). In rats, androgen receptors are present on the smooth muscle layer of the blood vessels of the ventral prostate, and VEGF expression has been noted in epithelial and stromal cells of human and rat prostates (68,360). Testosterone stimulated regrowth of the vasculature and endothelial proliferation in the prostate of castrated rats (359).

The veins from the ventral and dorsolateral prostate drain into a single, large, circular anastomosis around the neck of the bladder (Fig. 20). On either side, the circle drains into the superior vesical (= hypogastric) vein, which also receives the deferential vein and the veins draining the vesicular gland and anterior prostate (341). The capsule of the prostate in rat, dog, and human has a rich lymphatic vessel network, but there is doubt whether lymphatics are present inside the gland in dog or humans (367,557).

Blood flow through the rat prostate measured by soluble indicator fractionation (625) or microspheres (115,117,339,340,431,540,652) is about 50 ml/100 g/min for the ventral and slightly less for the dorsolateral prostate and the vesicular glands. Blood flow was reduced within 18 hours after castration (339,431,540) or by elimination of the Leydig cells with ethane dimethane sulfonate and was restored by treatment with testosterone (117). The reduction in blood flow was associated with reduction in the area of smooth muscle–coated blood vessels and nitric oxide synthase activity in the gland (224) and preceded involution of the gland. Blood flow through the ventral and dorsal

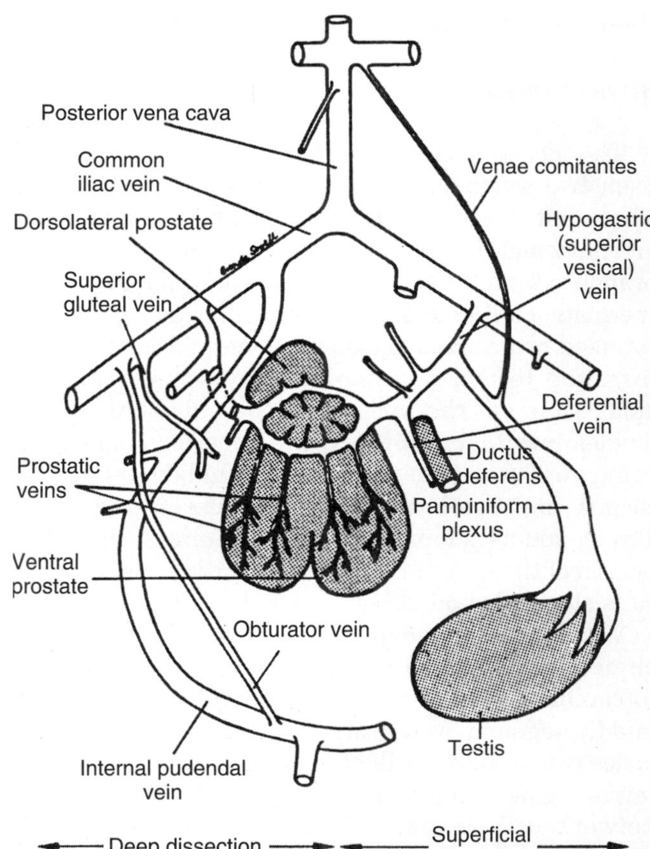

FIG. 20. Venous drainage of the male accessory sex gland of the rat showing a superficial dissection on the right and a deep dissection of the left. [Reproduced with permission from (341).]

prostates was also reduced by treatment with finasteride, a 5α-reductase inhibitor, or bicalutamide, an androgen receptor blocker (340). Prostate and vesicular gland blood flow was less in hypophysectomized rats and in these animals was unaffected by treatment of the animals with LH–releasing hormone agonist 2 or 4 hours earlier in a dose sufficient to raise testosterone levels in blood (115).

Penis

The arterial supply to the penis is derived from a branch of the hypogastric artery, the internal pudendal artery, that divides to form three arteries of the penis: the bulbourethral, dorsal, and cavernosal arteries. The bulbourethral artery supplies blood to the corpus spongiosum, the urethra, and the glans as well as to the bulbourethral muscle, whereas the cavernosal arteries lie in the center of the corpora cavernosa where they give rise to the helicine arteries, which feed the individual trabeculae. These arteries also supply a capillary network that drains into a venular plexus just below the tunica albuginea. The paired dorsal penile arteries supply blood to superficial penile structures, together with some branches of the external pudendal artery, and also the corpora cavernosa via the circumflex arteries. In some men, as well as in bulls and stallions, the dorsal arteries also supply helicine arteries in the cavernous tissue. Three sets of veins, superficial, intermediate, and deep, drain the blood from the penis. The deep veins drain both the corpora cavernosa and the corpus spongiosum, through circumflex veins into the deep dorsal vein, which empties into the internal pudendal vein. There is also a cavernous vein that opens directly into the internal pudendal vein. The intermediate veins lie superficial to the tunica albuginea and also open into the deep dorsal vein. The superficial veins drain the skin and subcutaneous tissue of the penis and open into the external pudendal vein (12,140,551). In rats, the corpora cavernosa and the corpus spongiosum have independent vascular supplies. Each penile crus is penetrated near their union by a deep penile artery. The dorsal artery of the penis, from which arise the deep penile artery and an artery to the penile bulb, continues along near the surface of the penis adjacent to the deep dorsal vein in a large groove on the shaft of the penis. The deep penile artery divides into anterior and posterior branches, and helicine arteries arise from the latter (164). In rabbits, the penile artery arises from the internal pudendal artery and ends by dividing into the deep penile and dorsal penile arteries. The former penetrates the tunica albuginea and forms the arterial network of the corpus cavernosum (433). Blood flow through the penis and its role in erection is covered in Chapter 24.

Scrotum

The scrotum is supplied by the external pudic (pudendal) vessels, and there are arteriovenous anastomoses present with diameters up to 200 µm (626). In rats, blood flow through the scrotal skin at normal scrotal temperatures of 33°C was about 15 ml/100 g/min and approximately doubled when the temperature was raised to 37°C, with further increases at even higher temperatures (631). In rams, blood flow was higher in the skin on the back of the scrotum and particularly the lower half and, as in the rat, showed a marked increase when the temperature was raised from 33 to 37°C or 40°C (169). By cannulating a branch of the artery supplying the scrotal skin in rams, radioactive krypton could be introduced into the tissue and blood flow estimated by measuring its rate of clearance. At normal temperatures there were two components of the flow: 90% of the tissue was supplied by capillaries with a flow of about 2.3 ml/100 g/min, whereas the remaining 10% had a higher flow of about 20 ml/100 g/min. This latter compartment doubled in size if the temperature was raised (168).

INNERVATION

The organs of the male reproductive tract receive a visceral afferent and efferent nerve supply, which is derived from a group of ganglia near the spinal cord, the celiac, aortic, caudal mesenteric, hypogastric, and pelvic ganglia. The scrotum and external cremaster muscle also receive a somatic innervation (239). The caudal mesenteric and hypogastric ganglia are combined in cats, dogs, rats, and rabbits but are separate in humans, apes, and monkeys (504). The cells in the spinal cord and brain linked to the testis, epididymis, or ductus deferens can be revealed using a transsynaptic viral tracing technique (181–183).

Testis

The testis is supplied by the superior and inferior spermatic nerves (Fig. 21). The superior nerve is derived mainly from the caudal mesenteric ganglion in most species except the human, with some contribution from the celiac and aortic plexuses, and directly

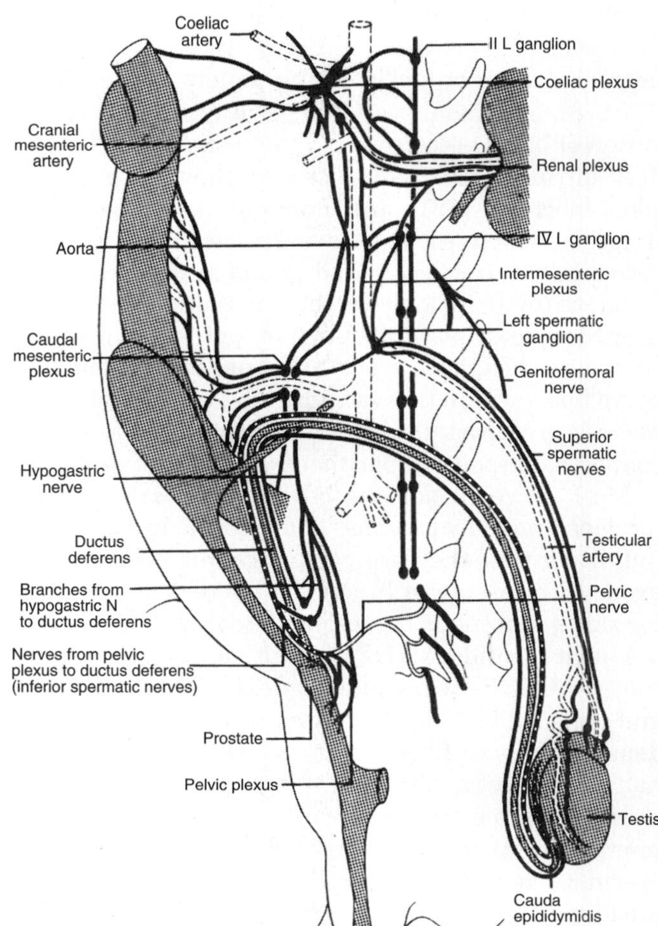

FIG. 21. The nerve supply to the testis and epididymis of the cat. The middle spermatic nerve is labeled "Branches from hypogastric N to ductus deferens." [Reproduced with permission from (239).]

from lumbar splanchnic nerves; in humans, the superior spermatic nerve is derived only from the celiac and aortic plexuses and lumbar splanchnics. The superior spermatic nerve forms a plexus around the testicular artery, or in some species (e.g., humans, sheep, and horses) is a separate nerve (239). The fibers of the superior spermatic nerve enter the testis either with the blood vessels or across the mesorchium (500,677). The inferior spermatic nerve derives from the pelvic plexus and may contain both sympathetic and parasympathetic fibers. It runs along the ductus deferens to the cauda epididymidis (239), from where nerves reach the testis via the ligamentous connection between the tail of the epididymis and the caudal extremity of the testis (500,677,685). It was generally believed that the superior spermatic nerve was the principal source of nerve fibers to the testis (583,681), but studies using retrograde axonal tracing suggest that the nerves from the pelvic ganglion may be more important (469).

Within the testis, the nerves are concentrated in the stromal compartments of the gonad, the tunica albuginea, mediastinum, and perivascular tissue. In adult bulls and pigs, the seminiferous tubules, Leydig cells, and parenchymal microcirculation have no obvious innervation, although there is a higher degree of innervation in young piglets, calves (327, 677,678), and monkeys (392). The degree of innervation of blood vessels within the testis is also less in deer during the breeding season than during seasonal inactivity (681), and innervation was increased in pigs in which testis size was drastically reduced by active immunization against gonadotropin-releasing hormone (550). In donkeys, the innervation is greatest near the epididymis (680). Cats, in contrast to all other species studied, have septula that are innervated and nerve fibers within the testicular lobules; most of these nerves are perivascular, but some are found within the connective tissue compartments of the testis (679).

Most nerves inside the bovine testis are positive for dopamine β-hydroxylase and tyrosine hydroxylase, indicating that they are postganglionic sympathetic axons with vasomotor functions. There is no evidence for parasympathetic nerves, but about half the nerves are neuropeptide Y (NPY) positive, and there are solitary CGRP fibers independent of blood vessels and some solitary fibers containing substance P. The density of nerve fibers is much lower in the caudal half of the testis, although in this part of the testis there are some vasoactive intestinal peptide (VIP) immunoreactive fibers derived from the inferior spermatic nerve (677). Similar types of nerve fibers are found in pig (678), donkey (680), camel (500), deer (681), and buffalo testis (121). In dogs, CGRP and substance P fibers coexist in peripheral spermatic nerve fibers in the spermatic cord and superficially in the tunica albuginea (584). In the cat testis, there are some parasympathetic fibers and many more CGRP positive fibers (566). In rats, the testis is much less well innervated than the rest of the male tract (457), but NPY and its receptors are present in the tunica and around the intracapsular blood vessels, and catecholamines and neuropeptides coexist in the same nerve fibers (469,685).

In monkeys, there are catecholaminergic/peptidergic extrinsic nerves and intrinsic neuron-like cells in the testis, both of which are more prominent in young animals (175,392). The neuronal elements are in close proximity to mast cells, the numbers of which increase after puberty (175). Neuron-like cells are also present in human testes and, together with extrinsic nerve fibers, appear to be more prominent in testes with deranged spermatogenesis (393). In the testes of children, there are nerve fibers and varicosities in proximity to Leydig cells (455) and the lamina propria of the seminiferous tubules (456).

In the dog and rat testes, there are nociceptors concentrated just under the tunica albuginea that react to mechanical, thermal, and chemical stimuli (322,325,326). The afferent fibers carrying pain sensation after injury to the testis run with the efferent sympathetic fibers in the spermatic cord, through the abdominal ganglia to segments T10-11 of the spinal cord (239).

Nerves are important in the control of TBF (see section on Testicular Blood Flow), and testicular innervation is necessary for the testosterone response to stress (172). Also, stimulation of the superior spermatic nerve in rats increased testosterone secretion, whereas stimulation of the inferior spermatic nerve was without effect (82). Nerves may also be involved in the different response to left- or right-sided hemicastration (21).

A number of effects of denervation of the male reproductive tract have also been described. Male infertility results from spinal cord injury, with decreases in sperm production and motility but normal morphology of the sperm (468), and contrary to earlier suggestions, these effects do not appear to be related to changes in scrotal temperature or serum gonadotropin levels (56). In experimental rats, transection of the spinal cord at T-9 or L-1 led to a fall within 3 days in testosterone in the testes and blood, with abnormalities in spermatogenesis appearing between 2 and 4 weeks after surgery, although testis weight was not reduced until later (253–256,352). In contrast, cutting just the superior spermatic nerve led to a decrease in testis weight after 4 weeks, and although cutting the inferior spermatic nerve did not initially have the same effect, it amplified the effect of cutting the superior spermatic nerve. At first, spermatids appeared to be the main cell type affected, but eventually by 12 weeks total regression of spermatogenesis had occurred. Testicular testosterone content (per testis) was similarly reduced, but the concentration per gram of testis fell only after spermatogenesis was affected. Gonadotropin levels in blood were not changed (84).

Section of the superior spermatic nerve in young rats reduced norepinephrine content of the testes, LH receptor number, and in vitro production by whole decapsulated testes of testosterone and 3α-androstanediol before any fall in testis weight (74). Testicular denervation in hamsters by intratesticular injection of 6-hydroxydopamine decreased binding of hCG and reduced testicular testosterone and spermatogenesis (391). Lesions of the paraventricular nucleus in the hypothalamus blocked the inhibitory effect of intraventricular injections of interleukin-1β, corticotropin-releasing factor, or the β-adrenergic agonist isoproterenol and reduced the effects of ethanol in blunting the testosterone response to hCG, suggesting a hypothalamic–testicular neural pathway independent of the pituitary (336,512).

Epididymis and Ductus Deferens

The nerve supply to the epididymis and ductus deferens is derived largely from the inferior (caudal) mesenteric ganglion and pelvic plexus via the hypogastric and then the inferior and middle spermatic nerves, although some fibers to the caput come with the superior spermatic nerve (Fig. 21). The nerve fibers in the epididymis and ductus are of the adrenergic, cholinergic, purinergic, and peptidergic types and are associated with vascular, muscular, and, to a lesser extent, epithelial elements (522). The density of fibers, particularly adrenergic fibers, and the amount of catecholamines were much greater in the cauda than in the caput (552). The enzymes involved in the synthesis of catecholamines are also present in the ductus (295,296,297,299). There is also a cholinergic innervation in the epididymis and ductus, although it may involve cholinergic postganglionic sympathetic fibers, and these are concentrated in the lamina propria rather than in the muscular layer, where most of the noradrenergic fibers are found. There is also abundant evidence for many neuropeptides (200,297,299), including VIP, NPY substance P, and CGRP (Fig. 22), and there may also be purinergic terminals using ATP as a cotransmitter with noradrenaline. In the cauda and ductus, there is a high level of nitric oxide synthase, an enzyme forming nitric oxide, a neurotransmitter in the autonomic nervous system (522).

The noradrenergic nerves in the epididymis and ductus deferens are excitatory for the smooth muscle, particularly during ejaculation (295,313). The cholinergic nerves seem to be more closely associated with epithelial cells and may influence their secretory activity or fluid resorption (295,313,552). The initial contraction of the ductus deferens during ejaculation is probably due to purines released from the noradrenergic nerves, whereas the slower second phase of the contraction is probably under the control of noradrenaline (71,614). In mice lacking the receptor for an ATP-gated cation channel, there was reduced fertility despite normal copulatory behavior (407). Of the neuropeptides, VIP inhibited electrically induced as well as noradrenaline-induced contractions, whereas NPY inhibited only nerve-induced contractions (146). The CGRP and substance P fibers probably have a sensory function (297). The nitrergic nerves concentrated at the urethral end of the ductus are probably involved in relaxing the duct to allow unidirectional sperm transport (613).

In chemically sympathectomized rats, the cauda and ductus deferens were distended with sperm, many of which had attached cytoplasmic droplets (307,328,329). The treated males did not impregnate any of the females to which they were mated and did not produce any copulatory plugs despite the presence of sperm in the vagina. However, intrauterine insemination with sperm from these rats produced normal litters (308). Ablation of the caudal mesenteric plexus caused epididymal distention and reduced sperm motility (48,49,308) and sperm transport through the epididymis (480). Most females mated to these males were not pregnant, and those that did become pregnant had smaller litters. After intrauterine insemination with equivalent numbers of sperm, fertilization rate was unaffected, but there were fewer implantation sites in two experiments (479,481) but not in a subsequent one (308).

Accessory Sex Glands

The sympathetic and parasympathetic nerve supply to the accessory glands is derived from the hypogastric and pelvic nerves, respectively, through the pelvic plexus (310,615). Sensory nerves accompany both types of fibers. The sympathetic supply to the glands is characterized by short postganglionic neurons, located close to the target organs (110, 372,445,552). The noradrenergic innervation in the prostate is concentrated in the thick stromal smooth muscle cell layer around the prostatic ducts and acini, but the glandular epithelium appears to lack noradrenergic innervation. Administration of catecholamines or stimulation of the noradrenergic nerves causes contraction of the muscle layer and expulsion of preformed secretion into the urethra,

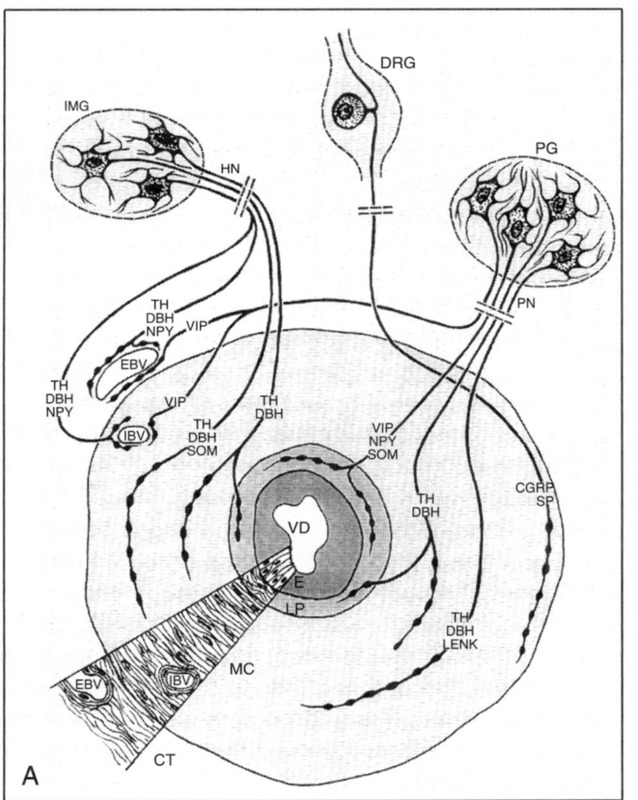

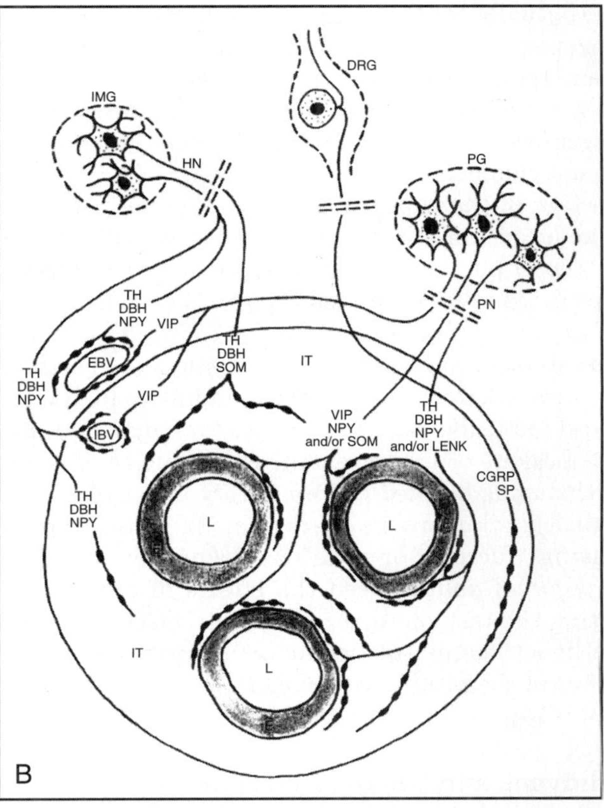

FIG. 22. The nerve supply to the ductus deferens (*left*) and vesicular gland (*right*) of the pig showing the various neurotransmitters involved. IMG, inferior mesenteric ganglion; DRG, dorsal root ganglion; PG, pelvic ganglion; HN, hypogastric nerve; PN, pelvic nerve; CT, connective tissue; MC, muscle coat; EBV, extrinsic blood vessel; IBV, intrinsic blood vessel; IT, interstitial tissue; E, epithelium; L, lumen; TH, tyrosine hydrolase; DBH, dopamine β-hydrolase (catecholamine-synthesizing enzymes); VIP, vasoactive intestinal peptide; NPY, neuropeptide Y; SOM, somatostatin; LENK, Leu-enkephalin; CGRP, calcitonin gene-related peptide; SP, substance P. [Reproduced with permission from (297,298).]

and these effects can be blocked by appropriate antagonists (445). There is also a parasympathetic supply to the epithelium and stroma, acting via muscarinic receptors, and this is probably involved in the control of secretory activity and possibly contraction of the stromal tissue. There is also evidence for purinergic and nitrergic innervation as well as the involvement of many neuropeptides, NPY, VIP, CGRP, and enkephalins (207,296,298,445). The prostate of the opossum, *Didelphis albvientris*, receives an abundant sympathetic and parasympathetic innervation through the pelvic plexus, as in the rat (386).

In the human prostate, there are appreciable numbers of neuroendocrine cells (17,18). Some are also present in the guinea pig (2), but they are absent in rats and cats (13). Their significance is not understood.

Although receiving much less attention, innervation of the vesicular glands appears to be essentially similar. In pigs, where the vesicular and bulbourethral glands are relatively large, the innervation of the vesicular glands and the body of the prostate is comparable and better developed than that for the disseminate prostate and the bulbourethral glands. All tissues have three types of nerve fibers: noradrenergic fibers, which also contain enkephalin and NPY; cholinergic fibers, most of which also contain VIP, NPY, and/or somatostatin (Fig. 22); and nonnoradrenergic sensory fibers, which colocalize CGRP and substance P (298).

Copious secretion by the dog prostate can be induced by stimulation of the hypogastric nerve or by the injection of pilocarpine, a cholinomimetic drug (555,556,640). The use of various blocking agents indicated that the secretion is stimulated by postganglionic sympathetic fibers, not parasympathetic neurons (555). In rats, pilocarpine was ineffective, but secretion from the prostate could be stimulated by adrenergic agonists and muscarinic agonists, the former by causing contraction of smooth muscle surrounding the ducts and the latter by a mechanism other than contraction of the muscles (637).

Sympathetic denervation by guanethidine in immature and mature rats was followed by a reduction in weight of the ventral prostate and vesicular glands and in the fructose content of the ventral prostate, but not of the vesicular gland. This may have been secondary to a reduction in testosterone secretion (306), whereas temporary removal of the sympathetic nerves with 6-hydroxydopamine led to distension of the prostatic alveoli, presumably due to reduced emptying of the gland (604).

Unilateral or bilateral surgical removal of the major pelvic ganglion in rats led to atrophy of the ipsilateral prostate (364,371,638), as did section of the sympathetic hypogastric nerve. However, curiously and inexplicably, section of the parasympathetic pelvic nerve led to an increase in size, DNA, and protein of the contralateral ventral prostate, with no effect on the ipsilateral gland (371).

Penis

The penis is supplied by sympathetic, parasympathetic, and somatic fibers, which are carried in the dorsal and cavernous nerves. The dorsal nerve receives parasympathetic input from the pudendal nerve but also sympathetic fibers from the paravertebral chain. The cavernous nerve originates in the pelvic plexus, which in turn is linked through the pelvic nerve to the sacral spinal cord and the paravertebral sympathetic chain and through the hypogastric nerve to the superior hypogastric plexus (or in some species to the inferior mesenteric ganglion) and the lumbar spinal cord. Because this nerve runs along the posterolateral aspect of the prostate, it is often damaged during prostatectomy, leading to erectile problems. The pudendal nerve carries efferent fibers innervating the ischiocavernosus and bulbocavernosus muscles but also sensory fibers from the many specialized sensory receptors, which are particularly abundant in the glans penis (12,110,123,497,551).

The nerve populations in the penis have been categorized as adrenergic, cholinergic and nonadrenergic, noncholinergic. The adrenergic innervation may also contain endothelins and angiotensins, whereas cholinergic nerves may also contain VIP, NPY, and nitric oxide synthase. Nitric oxide synthase and heme oxygenases are also present in the nonadrenergic, noncholinergic innervation (11,377). The significance of the nerves in the penis in erection is covered in Chapter 25.

Scrotum

The scrotum is supplied by branches from the genitofemoral (cranial and caudal inguinal) nerves, the superficial (superior) perineal nerve, and the caudal scrotal nerve (Fig. 23) (239). In the scrotal skin, there are thermal receptors that transmit information on scrotal temperature to neurons in the thalamus, hypothalamus, and cortex. These receptors respond to temperature increases of as little as 2°C (230,343–345,411,415,508) and provoke respiratory and metabolic reactions, which can appreciably lower body temperature (211,378,379,621,622,627), and stimulate the release of sweat from the scrotal sweat glands. Local anesthesia of the superior perineal nerve in rams blocks the afferent input from these receptors (630).

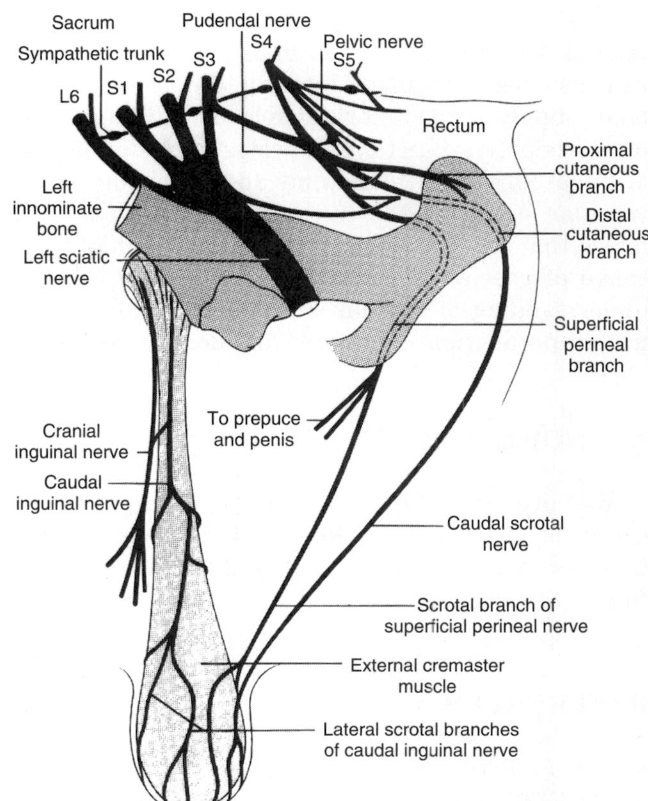

FIG. 23. The nerve supply to the scrotum of the bull. [Reproduced with permission from (239).]

CONCLUSION

The variation between different mammalian species in the anatomy of the various parts of the male reproductive tract has been emphasized. Particular reference to position and size of the testes is alluded to. The reason for descent of the testes into a scrotum that occurs in many, but not all, mammals remains one of the great mysteries of male reproductive biology. In species in which the testes are normally scrotal, failure of one or both testes to descend or exposure of the testes to body temperature for any length of time generally leads to gross abnormalities in function.

The importance of the vascular and nervous systems in the function of the male reproductive tract has also been detailed. Interruption of either the blood supply or the innervation leads to disturbances of reproductive function. However, the significance of the peculiarities of anatomy and physiology of the vascular supply to the testis is still not clear. It is hoped that studies with a comparative emphasis on a range of species of mammals will lead to a better understanding of normal and abnormal function of male reproduction in humans and domestic species.

ACKNOWLEDGMENTS

We are grateful to Michael Bedford, John Curlewis, Richard Ivell, Russell Jones, and the late Geoffrey Waites for their comments during preparation of this chapter.

REFERENCES

1. Abe, K., Takano, H., and Ito, T. (1984). Microvasculature of the mouse epididymis, with special reference to fenestrated capillaries localized in the initial segment. *Anat. Rec.* 209, 209–218.
2. Acosta, S., Dizeyi, N., Pierzynowski, S., Alm, P., and Abrahamsson, P. A. (2001). Neuroendocrine cells and nerves in the prostate of the guinea pig: effects of peripheral denervation and castration. *Prostate* 46, 191–199.
3. Adham, I. M., Emmen, J. M. A., and Engel, W. (2000). The role of the testicular factor INSL3 in establishing the gonadal position. *Mol. Cell. Endocrinol.* 160, 11–16.
4. Adham, I. M., Steding, G., Thamm, T., Bullesbach, E. E., Schwabe, C., Paprotta, I., and Engel, W. (2002). The overexpression of the INSL3 in female mice causes descent of the ovaries. *Mol. Endocrinol.* 16, 244–252.
5. Allanson, M. (1933). The reproductive processes of certain mammals. Part V. Changes in the reproductive organs of the male grey squirrel (*Sciurus carolinensis*). *Phil. Trans. R. Soc. B* 222, 76–96.
6. Allanson, M. (1934). The reproductive processes of certain mammals. VII. Seasonal variation in the reproductive organs of the male hedgehog. *Phil. Trans. R. Soc. B* 222, 288–303.
7. Amann, R. P., Nett, T. M., and Niswender, G. D. (1978). Effect of LH, FSH, prolactin and PGF2α on testicular blood flow and testosterone secretion in the ram. *J. Anim. Sci.* 47, 1307–1313
8. Amselgruber, W., and Sinowatz, F. (1986). Zur Beziehing zwischen der *Arteria testicularis* und den Venen des *Plexus pampiniformis* beim Bullen. *Anat. Histol. Embryol.* 16, 363–370.
9. Amselgruber, W., and Sinowatz, F. (1991). Raumliche Anordnung und Mikrovaskularisation der *Ductuli efferentes testis* vom Bullen (*Bos Taurus*). *Anat. Histol. Embryol.* 20, 139–151.
10. Amselgruber, W., Sinowatz, F., and Spanel-Borowski, K. (1986). Rasterelektronenmikroskopische Untersuchungen zur Mikrovaskularisation der Tubuli seminiferi contorti des Bullenhodens. *Berl. Munch. Tierarztl. Wschr.* 99, 166–170.
11. Andersson, K. E. (2001). Pharmacology of penile erection. *Pharmacol. Rev.* 53, 417–450.
12. Andersson, K. E., and Wagner, G. (1995). Physiology of penile erection. *Physiol. Rev.* 75, 191–236.
13. Angelsen, A., Mecsei, R., Sanvik, A. K., and Waldum, H. L. (1997). Neuroendocrine cells in the prostate of the rat, guinea pig, cat, and dog. *Prostate* 33, 18–25.
14. Arata, A. A. (1964). The anatomy and taxonomic significance of the male accessory reproductive glands of muroid rodents. *Bull. Florida State Mus.* 9, 1–42.
15. Auclair, D., Sowerbutts, S. F., and Setchell, B. P. (1995). Effect of active immunization against oestradiol in developing ram lambs on plasma gonadotrophin and testosterone concentrations, time of onset of puberty and testicular blood flow. *J. Reprod. Fertil.* 104, 7–16.
16. Auclair, D., Sowerbutts, S. F., and Setchell, B. P. (1995). Effect of active immunization against oestradiol on plasma gonadotrophin concentrations, spermatogenic function, testicular blood flow, epididymal mass and mating behaviour in adult rams. *J. Reprod. Fertil.* 104, 17–26.
17. Aumuller, G., Leonhardt, M., Janssen, M., Konrad, L., Bjartell, A., and Abrahamsson, P. A. (1999). Neurogenic origin of human prostate endocrine cells. *Urology* 53, 1041–1048.
18. Aumuller, G., Leonhardt, M., Renneberg, H., von Rahden, B., Bjartell, A., and Abrahamsson, P. A. (2001). Semiquantitative morphology of human prostatic development and regional distribution of prostatic neuroendocrine cells. *Prostate* 46, 108–115.
19. Baker, H. W. G., Burger, H. G., de Kretser, D. M., Hudson, B., Renire, G. C., and Straffon, W. G. E. (1985). Testicular vein ligation and fertility in men with varicoceles. *Br. Med. J.* 21, 1678–1680.
20. Ball, R. Y., and Setchell, B. P. (1983). The passage of spermatozoa to regional lymph nodes in testicular lymph following vasectomy in rams and boars. *J. Reprod. Fertil.* 68, 145–153.
21. Banczerowski, P. Csaba, Z., Csernus, V., and Gerendai, I. (2000). The effect of callostomy on testicular steroidogenesis in hemiorchidectomized rats: a pituitary-independent regulatory mechanism. *Brain Res. Bull.* 53, 227–232.
22. Bartke, A., Klemcke, H. G., Amador, A., Goldman, B. D., and Siler-Khodr, T. M. (1983). Relationship of the length of exposure to short photoperiod to the effect of prolactin on pituitary and testicular function in the golden hamster. *J. Reprod. Fertil.* 69, 587–595.
23. Bartlett, J. E., Lee, S. M. Y., Mishina, Y., Behringer, R. R., Yang, N., Wolf, J., Temelcos, C., and Hutson, J. M. (2002). Gubernacular development in Müllerian inhibiting substance receptor-deficient mice. *BJU Int.* 89, 113–118.
24. Battaglia, C., Giulini, S., Regnani, G., Madgar, I., Facchinetti, F., and Volpe, A. (2001). Intratesticular Doppler flow, seminal plasma nitrite/nitrates, and non-obstructive sperm extraction from patients with obstructive and nonobstructive azoospermia. *Fertil. Steril.* 75, 1088–1094.
25. Baumans, V., Dijkstra, G., and Wensing, C. J. G. (1982). The effect of orchidectomy on gubernacular outgrowth and regression in the dog. *Int. J. Androl.* 5, 387–400.
26. Baumgarten, H. G., Holstein, A. F., and Rosengren, E. (1971). Arrangement, ultrastructure and adrenergic innervation of smooth musculature of ductuli efferentes, ductus epididymis and ductus deferens of man. *Z. Zellforsch.* 120, 37–79.
27. Baumgartner, D. J., Hartung, T. G., Sawrey, D. K., Webster, D. G., and Dewsbury, D. A. (1982). Muroid copulatory plugs and female reproductive tracts: a comparative investigation. *J. Mamm.* 63, 110–117.
28. Bedford, J. M. (1977). Evolution of the scrotum: the epididymis as the prime mover? In *Reproduction and Evolution* (J. Calaby and C. H. Tyndale-Biscoe, Eds.), pp. 171–182. Australian Academy of Science.

29. Bedford, J. M. (1997). Unusual nature and possible evolutionary implications of the male vesicular gland secretion in the tree shrew, *Tupaia glis*. *Anat. Rec.* 247, 199–205.
30. Bedford, J. M. (2004). Enigmas of mammalian gamete form and function. *Biol. Rev.* 79, 429–460.
31. Bedford, J. M., Berrios, M., and Dryden, G. L. (1982). Biology of the scrotum. IV. Testis location and temperature sensitivity. *J. Exp. Zool.* 224, 379–388.
32. Bedford, J. M., Cooper, G. W., Phillips, D. M., and Dryden, G. L. (1994). Distinctive features of the gametes and reproductive tracts of the Asian musk shrew, *Suncus murinus*. *Biol. Reprod.* 50, 820–834.
33. Bedford, J. M., Mock, O. B., and Phillips, D. M. (1997). Unusual ampullary crypts, and behavior and role of the cumulus oophorus, in the oviduct of the Least shrew, *Cryptotis parva*. *Biol. Reprod.* 56, 1255–1267.
34. Bedford, J. M., Rernard R. T. F., and Baxter, R. M. (1998). The "hybrid" character of the gametes and reproductive tracts of the African shrew, *Myosorex varius*, supports its classification in the Crocidosoricinae. *J. Reprod. Fertil.* 112, 165–173.
35. Bedford, J. M., Mock, O. B., Nagdas, S. K., Winfrey, V. P., and Olson, G. E. (1999). Reproductive features of the eastern mole (*Scalopus aquaticus*) and star-nosed mole (*Condylura cristata*). *J. Reprod. Fertil.* 117, 345–353.
36. Bedford, J. M., Mock, O. B., Nagdas, S. K., Winfrey, V. P., and Olson, G. E. (2000). Reproductive characteristics of the African pygmy hedgehog, *Atelerix albiventris*. *J. Reprod. Fertil.* 120, 143–150.
37. Bedford, J. M., Mock, O. B., and Goodman, S. M. (2004). Novelties of conception in insectivorous mammals (Lipotyphla), particularly shrews. *Biol. Rev.* (in press).
38. Behringer, R. R., Fingold, M. J., and Cate, R. L. (1994). Müllerian-inhibiting substance function during mammalian sexual development. *Cell* 79, 415–425.
39. Benoit, M. J. (1926). Recherches anatomiques, cytologiques et histophysiologiques sur les voies excretices du testicule, chez les mammiferes. *Arch. Anat. Histol. Embryol.* 5, 173–412.
40. Bergh, A., and Damber, J. E. (1991). Does unilateral orchidectomy influence blood flow, microcirculation and vascular morphology in the remaining testis? *Int. J. Androl.* 14, 453–460.
41. Bergh, A., and Damber, J. E. (1992). Immunohistochemical demonstration of androgen receptors on testicular blood vessels. *Int. J. Androl.* 15, 425–434.
42. Bergh, A., and Damber, J. E. (1993). Vascular controls in testicular physiology. In *Molecular Biology of the Male Reproductive System* (D. de Kretser, Ed.), pp. 439–468. Academic Press, San Diego.
43. Bergh, A., Helander, H. F., and Wahlqvist, L. (1978). Studies on factors governing testicular descent in the rat—particularly the role of the gubernaculum testis. *Int. J. Androl.* 1, 342–356.
44. Bernard, R. F. F., and Cumming, G. S. (1997). African bats: evolution of reproductive patterns and delays. *Q. Rev. Biol.* 72, 253–274.
45. Beu, C. C. L., Orsi, A. M., Gregorio, E. A., Matheus, S. M. M., and Bassa, N. A. (1998). A scanning electron microscopic study of the rete testis of the guinea pig. *Acta Anat.* 162, 194–198.
46. Biagotti, G., Cavallini, G., Modenini, F., Vitali, G., and Gianaroli, L. (2002). Spermatogenesis and spectral echo-colour Doppler traces from the main testicular artery. *BJU Int.* 90, 903–908.
47. Biggers, J. D. (1966). Reproduction in male marsupials. *Symp. Zool. Soc. Lond.* 15, 251–280.
48. Billups, K. L., Tillman, S., and Chang, T. S. K. (1991). Ablation of the inferior mesenteric plexus in the rat: alteration of sperm storage in the epididymis and vas deferens. *J. Urol.* 143, 625–629.
49. Billups, K. L., Tillman, S., and Chang, T. S. K. (1991). Reduction of epididymal sperm motility after ablation of the inferior mesenteric plexus in the rat. *Fertil. Steril.* 53, 1076–1082.
50. Bindon, B. M., and Waites, G. M. H. (1968). Discrepancy in weight and blood flow of the testis and epididymis of the mouse before and after hypophysectomy. *J. Endocrinol.* 40, 385–386.
51. Blandau, R. J. (1945). On the factors involved in sperm transport through the cervix uteri of the albino rat. *Am. J. Anat.* 77, 253–272.
52. Boissin-Agasse, L., Richard, P., and Boissin, J. (1985). Activité testiculaire du furet maintenu pendant plus de 4 ans en lumière permanente et température constante. *Comp. R. Acad. Sci. Paris* 300, 685–690.
53. Bottger, M., and Lange, W. (1987). Untersuchungen am arteriellen Gefasssystem des Eberhodens. *Arch. Exp. Vet. Med.* 41, 58–64.
54. Boyd, I. L. (1985). Effect of photoperiod and melatonin on testis development and regression in wild European rabbits (*Oryctolagus cuniculus*). *Biol. Reprod.* 33, 21–29.
55. Boyd, I. L. (1986). Photoperiodic regulation of seasonal testicular regression in the wild European rabbit (*Oryctolagus cuniculus*). *J. Reprod. Fertil.* 77, 463–470.
56. Brackett, N. L., Lynne, C. M., Weizman, M. S., Bloch, W. E., and Padron, O. F. (1994). Scrotal and oral temperatures are not related to semen quality or serum gonadotropin levels in spinal cord-injured men. *J. Androl.* 15, 614–619.
58. Brambell, F. W. R. (1935). Reproduction in the common shrew (*Sorex araneus* Linnaeus). II. Seasonal changes in the reproductive organs of the male. *Phil. Trans. R. Soc. Lond. B* 225, 51–62.
59. Breed, W. G. (1982). Morphological variation in the testes and accessory sex organs of Australian rodents in the genera *Pseudomys* and *Notomys*. *J. Reprod. Fertil.* 66, 607–613.
60. Breed, W. G. (1985). Morphological variation in the female reproductive tract of Australian rodents in the genera *Pseudomys* and *Notomys*. *J. Reprod. Fertil.* 73, 379–384.
61. Breed, W. G. (1986). Comparative morphology and evolution of the male reproductive in the Australian hydromyine rodents (Muridae). *J. Zool. Lond. (A)* 209, 607–629.
62. Breed, W. G. (1997). Interspecific variation of testis size and epididymal sperm numbers in Australasian rodents with special reference to the genus *Notomys*. *Aust. J. Zool.* 45, 651–669.
63. Breed, W. G., and Taylor, J. (2000). Body mass, testes mass, and sperm size in murine rodents. *J. Mamm.* 81, 758–768.
64. Breed, W. G., and Washington, J. M. (1991). Mating behaviour and insemination in the hopping mouse (*Notomys alexis*). *J. Reprod. Fertil.* 93, 187–194.
65. Brennan, J., Karl, J., and Capel, B. (2002). Divergent vascular mechanisms downstreatm of *Sry* establish the arterial system in the XY gonad. *Dev. Biol.* 244, 418–428.
66. Brinster, R. L., and Zimmermann, J. W. (1994). Spermatogenesis following male germ-cell transplantation. *Proc. Natl. Acad. Sci. U S A* 91, 11298–11302.
67. Brooks, D. E. (1973). Epididymal and testicular temperature in the unrestrained conscious rat. *J. Reprod. Fertil.* 35, 157–160.
68. Brown, L. F., Yeo, K. T., Berse, B., Morgentaler, A., Dvorak, H. F., and Rosen, S. (1995). Vascular permeability factor (vascular endothelial growth factor) is strongly expressed in the normal male genital tract and is present in substantial quantities in semen. *J. Urol.* 154, 576–579.
69. Brown, P. D. C., and Waites, G. M. H. (1972). Regional blood flow in the epididymis of the rat and rabbit: effect of efferent duct ligation and orchidectomy. *J. Reprod. Fertil.* 28, 221–233.
70. Bryant, S. L. (1986). Seasonal variation of plasma testosterone in a wild population of male Eastern Quoll, *Dasyurus viverrinus* (Masupialia: Dasyuridae), from Tasmania. *Gen. Comp. Endocrinol.* 64, 75–79.

71. Burnstock, G., and Sneddon, P. (1985). Evidence for ATP and noradrenaline as cotransmitters in sympathetic nerves. *Clin. Sci.* Suppl 10, 89S–92S.
72. Bustamante, J. C., and Setchell, B. P. (2000). The permeability of the microvasculature of the perfused rat testis to small hydrophilic substances. *J. Androl.* 21, 444–451.
73. Bustamante, J. C., and Setchell, B. P. (2000). The uptake of amino acids, in particular leucine, by isolated perfused testes of rats. *J. Androl.* 21, 452–463.
74. Campos, M. B., Chioccio, S. R., Calandra, R. S., and Ritta, M. N. (1993). Effect of bilateral denervation of the immature rat testis on testicular gonadotropin receptors and in vitro androgen production. *Neuroendocrinology* 57, 189–194.
75. Carballada, R., and Esponda, P. (1992). Role of fluid from seminal vesicles and coagulating glands in sperm transport into the uterus and fertility in rats. *J. Reprod. Fertil.* 95, 639–648.
76. Carrick, F. N., and Setchell, B. P. (1977). The evolution of the scrotum. In *Reproduction and Evolution* (J. Calaby and C. H. Tyndale-Biscoe, Eds.), pp. 165–170. Australian Academy of Science.
77. Catt, D. C. (1977). The breeding biology of Bennett's wallaby (*Macropus rufogriseus fructicus*) in South Canterbury, New Zealand. *N. Z. J. Zool.* 4, 401–411.
78. Cendrowska, I., Jedrzejewski, K. S., and Okraszewska, E. (1996). Angioarchitecture of the epididymis in men bull and ram. *Folia Morph. (Warsz.)* 55, 227–228.
79. Chapman, N. G., and Harris, S. (1991). Evidence that the seasonal antler cycle of adult Reeves' muntjac (*Muntiacus reevesi*) is not associated with reproductive quiescence. *J. Reprod. Fertil.* 92, 361–369.
80. Chase, M. R. A. (1996). Reason for externalisation of the testis of mammals. *J. Zool.* 239, 691–695.
81. Chaturapanich, G., and Jones, R. C. (1991). Morphometry of the epididymis of the Tammar wallaby, *Macropus eugenii*, and estimation of some physiological parameters. *Reprod. Fertil. Dev.* 3, 651–658.
82. Chiocchio, S. R., Suburo, A. M., Vladucic, E., Zhu, B. C., Charreau, E., Decima, E. E., and Tramezzi, J. H. (1999). Differential effects of superior and inferior spermatic nerves on testosterone secretion and spermatic blood flow in cats. *Endocrinology* 140, 1036–1043.
83. Choo, E. F., Leake, B., Wandel, C., Imamura, H., Wood, A. J. J., Wilkinson, G. R., and Kim, R. B. (2000). Pharmacological inhibition of P-glycoprotein enhances the distribution of HIV-1 protease inhibitors into brain and testis. *Drug Metab. Dispos.* 28, 655–660.
84. Chow, S. H., Giglio, W., Anesetti, R., Ottenweiler, J. E., Pogach, L. M., and Huang, H. F. S. (2000). The effects of testicular denervation on spermatogenesis in the Sprague-Dawley rat. *Neuroendocrinology* 72, 37–45.
85. Chubb, C., and Desjardins, C. (1982). Vasculature of the mouse, rat, and rabbit testis-epididymis. *Am. J. Anat.* 165, 357–372.
86. Clark, R. V. (1976). Three-dimensional organization of testicular interstitial tissue and lymphatic space in the rat. *Anat. Rec.* 184, 203–226.
87. Clulow, J., and Jones, R. C. (1982). Production, transport, maturation, storage and survival of spermatozoa in the male Japanese quail, *Coturnix coturnix*. *J. Reprod. Fertil.* 64, 259–266.
88. Clulow, J., Jones, R. C., Hansen, L. A., and Man, S. Y. (1998). Fluid and electrolyte reabsorption in the ductuli efferentes testis. *J. Reprod. Fertil.* Suppl 53, 1–14.
89. Clough, G. (1969). Some preliminary observations on reproduction in the warthog, *Phacochoerus aethiopicus* Pallas. *J. Reprod. Fertil.* Suppl 6, 323–337.
90. Coe, M. J. (1969). The anatomy of the reproductive tract and breeding in the spring haas, *Pedetes surdaster larvalis* Hollister. *J. Reprod. Fertil.* Suppl 6, 159–174.
91. Collin, O., and Bergh, A. (1996). Leydig cells secrete factors which increase vascular permeability and endothelial proliferation. *Int. J. Androl.* 19, 221–228.
92. Collin, O., Bergh, A., Damber, J. E., and Widmark, A. (1993). Control of testicular vasomotion by testosterone and tubular factors in rats. *J. Reprod. Fertil.* 97, 115–121.
93. Collin, O., Damber, J. E., and Bergh, A. (1996). Effects of endothelin-1 on rat testicular vasculature. *J. Androl.* 17, 360–366.
94. Collin, O, Damber, J. E., and Bergh, A. (1996). 5-Hydroxytryptamine—a local regulator of testicular blood flow and vasomotion in rats. *J. Reprod. Fertil.* 106, 17–22.
95. Collin, O., Lissbrant, E., and Bergh, A. (1997). Atrial natriuretic peptide, brain natriuretic peptide and C-type natriuretic peptide: localisation in the testis and effects on testicular blood flow and vasomotion. *J. Androl.* 20, 55–60.
96. Collin, O., Enfalt, E., Astrom, M., Lissbrant, E., and Bergh, A. (1998). Unilateral injection of neuropeptide Y decreases blood flow in the injected testis but may also increase blood flow in the contralateral testis. *J. Androl.* 19, 580–584.
97. Collin, O., Zupp, J. L., and Setchell, B. P. (2000). Testicular vasomotion in different mammals. *Asian J. Androl.* 2, 290–300.
98. Comhaire, F. H. (1998). Varicocele. In *Textbook of Genitourinary Surgery* (H. N. Whitfield, W. F. Hendry, R. S. Kirby, and J. W. Duckett, Eds.), pp. 1431–1440. Blackwell Science, Oxford.
99. Cooper, T. G. (1986). *The Epididymis, Sperm Maturation and Fertilisation*. Springer-Verlag, Berlin.
100. Corbett, L. (1995). *The Dingo in Australia and Asia*. UNSW Press, Sydney.
101. Costa, G., Jimenez, E., Labadia, A., and Garcia-Pascual, A. (1996). Endothelial modulation of resting and stimulated vascular tone in the pig capsular testicular artery. *Eur. J. Physiol.* 433, 65–70.
102. Courrier, R. (1927). Etude sur le déterminisme des caractères sexuels secondaries chez quelques mammifères à activité testiculaire périodique. *Arch. Biol. Paris* 137, 173–334.
103. Coveney, D., Shaw, G., Hutson, J. M., and Renfree, M. B. (2002). The development of the gubernaculum and inguinal closure in the marsupial *Macropus eugenii*. *J. Anat.* 201, 239–256.
104. Coveney, D., Shaw, G., and Renfree, M. B. (2002). Effects of oestrogen treatment on testicular descent, inguinal closure and prostatic development in a male marsupial, *Macropus eugenii*. *Reproduction* 124, 73–83
105. Coveney, D., Shaw, G., Hutson, J. M., and Renfree, M. B. (2002). Effect of an anti-androgen on testicular descent and inguinal closure in a marsupial, the tammar wallaby (*Macropus eugenii*). *Reproduction* 124, 865–874.
106. Cowie, A. T., Lascelles, A. K., and Wallace, J. C. (1964). Flow and protein content of testicular lymph in conscious rams. *J. Physiol.* 171, 176–187.
107. Curlewis, J. D. (1991). Seasonal changes in the reproductive organs and plasma and pituitary hormone contents of the male Bennett's wallaby (*Macropus rufogriseus rufogriseus*). *J. Zool.* 223, 223–231.
108. Daehlin, L., Damber, J. E., Selstam, G., and Bergman, B. (1985). Effect of human chorionic gonadotrophin, oestradiol and estromustine on testicular blood flow in hypophysectomized rats. *Int. J. Androl.* 8, 58–68.
109. Dahl, E. V., and Herrick, J. F. (1959). A vascular mechanism for maintaining testicular temperature by counter-current exchange. *Surg. Gynecol. Obstet.* 108, 697–705.
110. Dail, W. G. (1993). Autonomic innervation of male reproductive genitalia. In *Neural Control of the Urogenital System* (C. A. Maggi, Ed.), pp. 69–101. Harwood Academic Publishers, Chur.

111. Damber, J. E. (1990). The effect of guanethidine treatment on testicular blood flow and testosterone production in rats. *Experientia* 46, 486–487.
112. Damber, J. E., and Janson, P. O. (1977). Methodological aspects of testicular blood flow measurements in rats. *Acta Physiol. Scand.* 101, 278–285.
113. Damber, J. E., Lindahl, O., Selstam, G., and Tenland, T. (1982). Testicular blood flow measured with a laser-Doppler flowmeter: acute effects of catecholamines. *Acta Physiol. Scand.* 115, 209–215.
114. Damber, J. E., Lindahl, O., Selstam, G., and Tenland, T. (1983). Rhythmical oscillations in rat testicular microcirculation as recorded by laser Doppler flowmetry. *Acta Physiol. Scand.* 118, 117–123.
115. Damber, J. E., Bergh, A., and Daehlin, L. (1984). Stimulatory effect of an LHRH-agonist on testicular blood flow in hypophysectomized rats. *Int. J. Androl.* 7, 236–243.
116. Damber, J. E., Bergh, A., and Widmark, A. (1987). Effect of an LHRH-agonist on testicular microcirculation in hypophysectomized rats. *Int. J. Androl.* 10, 785–791.
117. Damber, J. E., Maddocks, S., Widmark, A., and Bergh, A. (1992). Testicular blood flow and vasomotion can be maintained by testosterone in Leydig cell-depleted rats. *Int. J. Androl.* 15, 385–393.
118. Dark, J., Johnston, P. G., Healy, M., and Zucker, I. (1983). Latitude of origin influences photoperiodic control of reproduction of deer mice (*Peromyscus maniculatus*). *Biol. Reprod.* 28, 213–220.
119. Davis, J. R., and Horowitz, A. M. (1979). Effect of various exposure times to hyperthermia and hypothermia on spontaneous contractions of the adult rabbit isolated testicular capsule. *Biol. Reprod.* 21, 413–417.
120. Davis, J. R., Jangford, G. A., and Kirby, P. J. (1970). The testicular capsule. In *The Testis* (A. D. Johnson, W. R. Gomes, and N. L. Vandemark, Eds.), Vol. I, pp. 281–337. Academic Press, New York.
121. de Girolamo, P., Costagliola, C., Lucini, C., Gargiulo, G., and Castaldo, L. (2003). The autonomous innervation of the buffalo (*Bubalus bubalis*). testis. An immunohistochemical study. *Eur. J. Histochem.* 47, 159–164.
122. de Graaf, R. (1668). *Tractus de virorum organis generationi inservientibus*, translated by Jocelyn, H. D., and Setchell, B. P. (1973). Treatise concerning the generative organs of men. *J. Reprod. Fertil.* Suppl 17, 9–76.
123. de Groat, W. C., and Booth, A. M. (1993). In *Neural Control of the Urogenital System* (C. A. Maggi, Ed.), pp. 467–524. Harwood Academic Publishers, Chur.
124. de Kretser, D. M., Kerr, J. B., and Paulsen, C. A. (1975). The peritubular tissue in the normal and pathological human testis. An ultrastructural study. *Biol. Reprod.* 12, 317–324.
125. Dewsbury, D. A. (1975). Diversity and adaptation in rodent copulatory behavior. *Science* 190, 947–954.
126. Dhingra, L. D. (1979). Angioarchitecture of the arteries of the testis of goat (*Capra aegagrus*). *Anat. Histol. Embryol.* 8, 193–199.
127. Dhingra, L. D. (1979). Angioarchitecture of the buffalo testis. *Anat. Anz.* 146, 60–68.
128. Diamond, M. (1970). Intromission pattern and species vaginal code in relation to induction of pseudopregnancy. *Science* 169, 995–997.
129. Dixson, A. F. (1987). Observations on the evolution of the genitalia and copulatory behaviour in male primates. *J. Zool. Lond.* 213, 423–443.
130. Dixson, A. F. (1987). Baculum length and copulatory behaviour in primates. *Am. J. Primatol.* 13, 51–60.
131. Dixson, A. F. (1995). Baculum length and copulatory behaviour in carnivores and pinnipeds (Grand Order Ferae). *J. Zool. Lond.* 235, 67–76.
132. Djakiew, D., and Jones, R. C. (1981). Structural differentiation of the male genital ducts of the echidna, *Tachyglossus aculeatus*. *J. Anat.* 132, 187–202.
133. Djakiew, D., and Jones, R. C. (1982). Stereological analysis of the epididymis of the echidna, *Tachyglossus aculeatus*, and Wistar rat. *Aust. J. Zool.* 30, 865–875.
134. Djakiew, D., and Jones, R. C. (1983). Sperm maturation, fluid transport, and secretion and absorption of protein in the epididymis of the echidna, *Tachyglossus aculeatus*. *J. Reprod. Fertil.* 68, 445–456.
135. Donjacour, A. A., and Cunha, G. R. (1988). The effect of androgen deprivation on branching morphogenesis in the mouse prostate. *Dev. Biol.* 128, 1–14.
136. Dryden, G. L. (1969). Reproduction in *Suncus murinus*. *J. Reprod. Fertil.* Suppl 6, 377–396.
137. Duarte, H. E., de Oliviera, C., Orsi, A. M., and Vicentini, C. A. (1995). Ultrastuctural characteristics of the testicular capillaries in the dog (*Canis familiaris,* L.) *Anat. Histol. Embryol.* 24, 73–76.
138. Dubock, A. C. (1979). Male grey squirrel (*Sciurus carolinensis*) reproductive cycles in Britain. *J. Zool. Lond.* 188, 41–51.
139. Duelli, R., Enerson, B. E., Gerhardt, D. Z., and Drewes, L. R. (2000). Expression of large amino acid transporter LAT1 in rat brain endothelium. *J. Cereb. Blood Flow Metab.* 20, 1557–1562.
140. Dyce, K. M., Sack, W. O., and Wensing, C. J. G. (2002). *Textbook of Veterinary Anatomy*. WB Saunders, Philadelphia.
141. Dym, M. (1974). The fine structure of monkey Sertoli cells in the transitional zone at the junction of the seminiferous tubules with the tubuli recti. *Am. J. Anat.* 140, 1–26.
142. Dym, M. (1976). The mammalian rete testis—a morphological examination. *Anat. Rec.* 186, 493–524.
143. Eberhard, W. G. (1985). *Sexual Selection and Animal Genitalia*. Harvard University Press, Cambridge.
144. Eckstein, P., and Zuckerman, S. (1956). Morphology of the reproductive tract. In *Marshall's Physiology of Reproduction* (A. S. Parkes, Ed.), Vol. 1, Part 1, pp. 43–155. Longmans, Green & Co, London, New York.
145. Edwards, M. J., Smith, M. S. R., and Freeman, B. (2003). Measurement of the linear dynamics of the descent of the bovine fetal testis. *J. Anat.* 203, 133–142.
146. Ellis, J. L., and Burnstock, G. (1988). Neuropeptide Y neuromodulation of sympathetic co-transmission in the guinea-pig vas deferens. *Br. J. Pharmacol.* 100, 457–462.
147. Emmen, J. M. A. (2000). *Testicular Factors Involved in Testis Descent*. Doctoral Thesis, Erasmus University, Rotterdam.
148. Emmen, J. M. A., McCluskey, A., Grooetgoed, J. A., and Brinkmann, A. O. (1998). Androgen action during male sex differentiation includes suppression of cranial suspensory ligament development. *Hum. Reprod.* 13, 1272–1280.
149. Emmen, J. M. A., McCluskey, A., Adham, I. M., Engel, W., Verhoef-Post, M., Themmen, A. P. N., Grooetgoed, J. A., and Brinkmann, A. O. (2000). Involvement of insulin-like factor 3 (Insl3) in diethylstilbestrol-induced cryptorchidism. *Endocrinology* 141, 846–849.
150. Ergun, S, Stingl, J., and Holstein, A. F. (1994). Segmental angioarchitecture of the testicular lobule in man. *Andrologia* 26, 143–150.
151. Ergun, S., Stingl, J., and Holstein, A. F. (1994). Microvasculature of the human testis in correlation to Leydig cells and seminiferous tubules. *Andrologia* 26, 255–262.
152. Ergun, S., Davidoff, M., and Holstein, A. F. (1996). Capillaries in the lamina propria of human seminiferous tubules are partly fenestrated. *Cell. Tissue Res.* 286, 93–102.
153. Ergun, S., Bruns, T., Soyka, A., and Tauber, R. (1997). Angioarchitecture of the human spermatic cord. *Cell. Tissue Res.* 288, 391–398.

154. Ergun, S., Kilic, N., Fiedler, W., and Mukhopadhyay, A. K. (1997). Vascular endothelial growth factor and its receptors in normal human testicular tissue. *Mol. Cell. Endocrinol.* 131, 9–20.
155. Ergun, S, Luttmer, W., Fiedler, W., and Holstein, A. F. (1998). Functional expression and localization of vascular endothelial growth factor and its receptors in the human epididymis. *Biol. Reprod.* 58, 160–168.
156. Estep, D. Q., and Dewsbury, D. A. (1976). Copulatory behavior of *Neotoma lepida* and *Baromys taylori*: relationships between penile morphology and behavior. *J. Mamm.* 57, 570–573.
157. Fantoni, G., Morris, P. L., Forti, G., Vannelli, G. B., Orlando, C., Barni, T., Sestini, R., Danza, G., and Maggi, M. (1993). Endothelin-1: a new autocrine/paracrine factor in rat testis. *Am. J. Physiol.* 265, E267–E274.
158. Fawcett, D. W. (1973). Observations on the organization of the interstitial tissue of the testis and on the occluding cell junctions in the seminiferous epithelium. *Adv. Biosci.* 10, 83–99.
159. Fawcett, D. W., and Dym, M. (1974). A glycogen-rich segment of the tubuli recti and proximal portion of the rete testis in the guinea pig. *J. Reprod. Fertil.* 38, 401–409.
160. Fawcett, D. W., Heidger, P. M., and Leak, L. V. (1969). Lymph vascular system of the interstitial tissue of the testis as revealed by electron microscopy. *J. Reprod. Fertil.* 19, 109–119.
161. Fawcett, D. W., Neaves, W. B., and Flores, M. N. (1973). Comparative observations on intertubular lymphatics and the organization of the interstitial tissue of the mammalian testis. *Biol. Reprod.* 9, 500–532.
162. Fehlings, K. (1976). *Korrosions- und rontgenanatomische Untersuchungen der Arteria testicularis von Katze, Hund, Schwein, Schaf, Rind und Pferd.* Diss. Hannover Tierarztl. Hochschule.
163. Fehlings, K., and Pohlmeyer, K. (1978). Die Arteria testicularis und ihr Aufzweiging im Hoden und Nebenhoden des Esels (Equus africanus f. asinus). Korrosionanatomische und rontgenologische Untersuchungen. *Anat. Histol. Embryol.* 7, 74–78.
164. Fernandez, E., Dail, W. G., Walton, G., and Martinez, G. (1991). The vasculature of the rat penis: a scanning electron microscopic study. *Am. J. Anat.* 192, 307–318.
165. Filippini, A., Tripiciano, A., Palombi, F., Teti, A., Paniccia, R., Stefanini, M., and Ziparo, E. (1993). Rat testicular myoid cells respond to endothelin: characterization of binding and signal transduction pathway. *Endocrinology* 133, 1789–1796.
166. Fleet, I. R., Laurie, M. S., Noordhuizen-Stassse, E., Setchell, B. P., and Wensing, C. J. G. (1982). The flow of blood from artery to vein in the spermatic cord of the ram with some observations on reactive hyperaemia in the testis and the effects of adenosine and noradrenaline. *J. Physiol.* 332, 44P–45P.
167. Fooden, J. (1970). Complementary specialization of male and female reproductive structures in the Bear Macaque, *Macaca arctoides. Nature* 214, 939–941.
168. Fowler, D. G., quoted by Waites, G. M. H., and Setchell, B. P. (1969). Physiology of the testis, epididymis and scrotum. In *Recent Advances in Reproductive Physiology* (A. McLaren, Ed.), pp. 1–63. Logos Press, London.
169. Fowler, D. G., and Setchell, B. P. (1971). Effect of local heating on blood flow through the scrotal skin and testes of rams selected for different degrees of skin fold. *Aust. J. Exp. Agric.* 11, 143–147.
170. Fowler, P. A., and Racey, P. A. (1987). Relationship between body and testis temperatures in the European hedgehog, *Erinaceus europaeus*, during hibernation and sexual reactivation. *J. Reprod. Fertil.* 81, 567–573.
171. França, L.R., and Godinho, C. L. (2003). Testis morphometry, seminiferous epithelium cyclelength, and daily sperm production in domestic cats (*Felis catus*). *Biol. Reprod.* 68, 1554–1561.
172. Frankel, A. I., and Ryan, E. L. (1981). Testicular innervation is necessary for the response of plasma testosterone levels to acute stress. *Biol. Reprod.* 24, 491–495.
173. Free, M. J., Jaffe, R. A., and Morford, D. E. (1980). Sperm transport through the rete testis of anesthetized rats: role of the testicular capsule and effect of gonadotropins and prostaglandins. *Biol. Reprod.* 22, 1073–1078.
174. Freeman, S. (1990). The evolution of the scrotum: a new hypothesis. *J. Theor. Biol.* 145, 429–445.
175. Frungieri, M. B., Urbanski, H. F., Hohne-Zell, B., and Mayerhofer, A. (2000). Neuronal elements in the testis of the rhesus monkey: ontogeny, characterization and relationship to testicular cells. *Neuroendocrinology* 71, 43–50.
176. Fussell, E. N., Lewis, R. W., Roberts, J. A., and Harrison, R. M. (1981). Early ultrastructural findings in experimentally produced varicocele in the monkey testis. *J. Androl.* 2, 111–119.
177. Gaeth, A. P., Short, R. V., and Renfree, M. B. (1999). The developing renal, reproductive and respiratory systems of the African elephant suggest an aquatic ancestry. *Proc Natl. Acad. Sci. U S A* 96, 5555–5558.
178. Gage, M. J. G., and Freckleton, R. P. (2003). Relative testis size and sperm morphometry across mammals: no evidence for an association between sperm competition and sperm length. *Proc. R. Soc. Lond. B* 270, 625–632.
179. Galil, K. A. A., and Setchell, B. P. (1988). Effects of local heating of the testis on testicular blood flow and testosterone secretion in the rat. *Int. J. Androl.* 11, 73–85.
180. Galil, K. A. A., Laurie, M. S., Main, S. J., and Setchell, B. P. (1981). The measurement of the flow of lymph from the testis. *J. Physiol.* 319, 17P.
181. Gerendai, I., Toth, I. E., Boldogkoi, Z., Medveczky, I., and Halasz, B. (2000). Central nervous system structures labeled from the testis using the transsynaptic viral tracing technique. *J. Neuroendocrinol.* 12, 1087–1095.
182. Gerendai, I., Toth, I. E., Kocsis, K., Boldogkoi, Z., Rusvai, M., and Halasz, B. (2001). Identification of CNS neurons involved in the innervation of the epididymis: a viral transneuronal tracing study. *Auton. Neurosci.* 92, 1–10.
183. Gerendai, I., Wiesel, O., Toth, I. E., Boldogkoi, Z., Rusvai, M., and Halasz, B. (2003). Identification of neurones in the brain and spinal cord involved in the innervation of the ductus deferens using the viral tracing method. *Int. J. Androl.* 26, 91–100.
184. Ghabriel, M. N., Lu, J. J., Hermanis, G., Zhu, C., and Setchell, B. P. (2002). Expression of a blood-brain barrier specific antigen in the male reproductive tract. *Reproduction* 123, 389–397.
185. Ghetie, V. (1939). Praparation und Lange des Ductus epididymidis beim Pferd und Schwein. *Anat. Anz.* 87, 369–374.
186. Gilmore, D. P. (1969). Seasonal reproductive periodicity in the male Australian Brush-tailed possum (*Trichosurus vulpecula*). *J. Zool.* 157, 75–98.
187. Ginsberg, J. R., and Rubenstein, D. I. (1990). Sperm competition and variation in zebra mating behaviour. *Behav. Ecol. Sociob.* 26, 427–434.
188. Glover, T. D. (1973). Aspects of sperm production in some East African mammals. *J. Reprod. Fertil.* 35, 45–53.
189. Glover, T. D., and Hopkirk, W., quoted by Setchell, B. P. (1970). Testicular blood supply, lymphatic drainage and secretion of fluid. In *The Testis* (A. D. Johnson, W. R. Gomes, and N. L. Vandemark, Eds.), Vol. I, pp. 101–239. Academic Press, New York.
190. Glover, T. D., and Nicander, L. (1971). Some aspects of structure and function in the mammalian epididymis. *J. Reprod. Fertil.* Suppl 13, 13–51.
191. Glover, T. D., and Sale, J. B. (1968). The reproductive system of the male rock hyrax (*Procavia* and *Heterohyrax*). *J. Zool. Lond.* 156, 351–362.

192. Glover, T. D., D'Occhio, M. J., and Millar, R. P. (1990). Male life cycle and seasonality. In *Marshall's Physiology of Reproduction* (G. E. Lamming, Ed.), Vol. 2, pp. 213–378. Churchill Livingstone, Edinburgh.
193. Godinho, H. P., and Setchell, B. P. (1975). Total and capillary blood flow through the testes of anaesthetized rams. *J. Physiol.* 251, 19P–20P.
194. Godinho, H. P., Cardoso, F. M., and Nogueira, J. C. (1977). Blood supply to the testis of a Brazilian marsupial (*Didelphis azarae*) and its abdominotesticular temperature gradient. *Acta Anat.* 99, 204–208.
195. Goh, D. W., Middlesworth, W., Farmer, P. J., and Hutson, J. M. (1994). Prenatal androgen blockade with flutamide inhibits masculinization of the genitofemoral nerve and testicular descent. *J. Pediat. Surg.* 29, 836–838.
196. Gosch, B., and Fischer, K. (1989). Seasonal changes of testis volume and sperm quality in adult fallow deer (*Dama dama*) and their relationship to the antler cycle. *J. Reprod. Fertil.* 85, 7–17.
197. Green, K. F., Turner, T. T., and Howards, S. S. (1984). Varicocele: reversal of the testicular blood flow and temperature effects by varicocele repair. *J. Urol.* 131, 1208–1211.
198. Green, K. F., Turner, T. T., and Howards, S. S. (1985). Effects of varicocele after unilateral orchiectomy and sympathectomy. *J. Urol.* 134, 378–383.
199. Greenberg, J., Forssmann, W. G., and Gorgas, K. (1985). Morphology and innervation of a testicular "rete mirabile" in the guinea-pig. *Anat. Embryol.* 173, 225–235.
200. Greenberg, J., Schubert, W., Metz, J., Yanaihara, N., and Forssmann, W. G. (1985). Studies of the guinea-pig epididymis. III. Innervation of epididymal segments. *Cell. Tissue Res.* 239, 395–404.
201. Griffiths, D. J. (1984). The annual cycle of the testis of the elephant seal (*Mirounga leonina*) at Macquarie Island. *J. Zool. Lond* 203, 193–204.
202. Griffiths, M. (1968). *Echidnas*. Pergamon Press, Oxford.
203. Griffiths, M. (1978). *The Biology of the Monotremes*. Academic Press, New York.
204. Griffiths, M. (1984). Mammals: monotremes. In *Marshall's Physiology of Reproduction* (G. E. Lamming, Ed.), Vol. 1. pp. 351–385. Churchill Livingstone, Edinburgh.
205. Griffiths, M. (1989). Tachyglossidae. In *Fauna of Australia* (B. J. Walton and B. J. Richardson, Eds.), pp. 407–445. Australian Government and Printing Service.
206. Grocock, C. A., and Clarke, J. R. (1981). Photoperiodic control of testis activity in the vole, *Microtus agrestis*. *J. Reprod. Fertil.* 39, 337–347.
207. Gu, J., Polak, J. M., Probert, L., Islam, K. N., Marangos, P. J., Mina, S., Adrian, T. E., McGregor, G. P., O'Shaughnessy, D. J., and Bloom, S. R. (1983). Peptidergic innervation of the human male genital tract. *J. Urol.* 130, 386–391.
208. Gulamhusein, A. P., and Tam, W. H. (1974). Reproduction in the male stoat, *Mustela erminea*. *J. Reprod. Fertil.* 41, 303–312.
209. Gunzel-Apel, A. R., Mohrke, C., and Poulsen Nautrup, C. (2001). Colour-coded and pulsed Doppler sonography of the canine testis, epididymis and prostate gland: physiological and pathological findings. *Reprod. Dom. Anim.* 36, 236–240.
210. Gustafson, A. W. (1979). Male reproductive patterns in hibernating bats. *J. Reprod. Fertil.* 56, 317–331.
211. Hales, J. R. S., and Hutchinson, J. C. D. (1971). Metabolic, respiratory and vasomotor resonses to heating the scrotum of the ram. *J. Physiol.* 212, 353–375.
212. Hamasaki, M., and Kumabe, T. (1994). Three-dimensional structure of two different lymphatic spaces in rat testis, and the route of flow fluxes of their lymphatic fluids. *Acta Anat. Nippon* 69, 699–683.
213. Hamilton, D. W. (1990). Anatomy of mammalian male accessory reproductive organs. In *Marshall's Physiology of Reproduction* (G. E. Lamming, Ed.), Vol. 2, pp. 691–746. Churchill Livingstone, Edinburgh.
214. Harcourt, A. H., Harvey, P. H., Larson, S. G., and Short, R. V. (1981). Testis weight, body weight and breeding system in primates. *Nature* 293, 55–56.
215. Harcourt, A. H., Purvis, A., and Liles, L. (1995). Sperm competition: mating system, not breeding season, affects testes size of primates. *Funct. Ecol.* 9, 468–476.
216. Harrison, R. G. (1949). The comparative anatomy of the blood-supply of the mammalian testis. *Proc. Zool. Soc. Lond.* 119, 325–344.
217. Harrison, R. G. (1951). Applications of microradioagraphy: the testis. In *Microarteriography* (A. E. Barclay, Ed.), pp. 89–90. Blackwell, Oxford.
218. Harrison, R. J., Matthews, L. H., and Roberts, J. M. (1952). Reproduction in some Pinnepedia. *Trans. Zool. Soc. Lond.* 27, 447–531.
219. Hart, R. G., and Greenstein, J. S. (1968). A newly discovered role for Cowper's gland secretion in rodent semen coagulation. *J. Reprod. Fertil.* 17, 87–94.
220. Hartung, T.G., and Dewsbury, D. A. (1978). A comparative analysis of copulatory plugs in muroid rodents and their relationship to copulatory behavior. *J. Mamm.* 59, 717–723.
221. Harvey, P. H., and Harcourt, A. H. (1984). Sperm competition, testes size, and breeding systems in primates. In *Sperm Competition and the Evolution of Animal Mating Systems* (R. L. Smith, Ed.), pp. 589–600. Academic Press, Orlando, FL.
222. Hasler, M. J., and Nalbandov, A. V. (1974). Body and peritesticular temperatures of musk shrews (*Suncus murinus*). *J. Reprod. Fertil.* 36, 397–399.
223. Hayashi, N., Sugimura, Y., Kawamura, J., Donjacour, A. A., and Cunha, G. R. (1991). Morphological and functional heterogeneity in the rat prostate gland. *Biol. Reprod.* 45, 308–321.
224. Hayek, O. R., Shabsigh, A., Kaplan, S. A., Kiss, A. J., Chen, M. W., Burchardt, T., Burchardt, M., Olsson, C. A., and Buttyan, R. (1999). Castration induces acute vasoconstriction of blood vessels in the rat prostate concominant with a reduction of prostatic nitric oxide synthase activity. *J. Urol.* 162, 1527–1531.
225. Head, J. R., Neaves, W. B., and Billingham, R. E. (1983). Reconsideration of the lymphatic drainage of the rat testis. *Transplantation* 35, 91–95.
226. Hees, H., Leiser, R., Kohler, T., and Wrobel, K. H. (1984). Vascular morphology of the bovine spermatic cord and testis. I. Light- and scanning electron-microscopic studies on the testicular artery and pampiniform plexus. *Cell. Tissue Res.* 237, 31–38.
227. Hees, H., Wrobel, K. H., Kohler, T., Leiser, R., and Rothbacher, I. (1987). Spatial topography of the excurrent duct system in the bovine testis. *Cell. Tissue Res.* 248, 143–151.
228. Hees, H., Wrobel, K. H., Kohler, T., Elmagd, A. A., and Hees, I. (1989). The mediastium of the bovine testis. *Cell. Tissue Res.* 255, 29–39.
229. Hees, H., Kohler, T., Leiser, R., Hees, I., and Lips, T. (1990). Vascular morphology of the bovine testis Light- and scanning electron microscopic studies. *Anat. Anz.* 170, 119–132.
230. Hellon, R. F., Hensel, H., and Schafer, K. (1975). Thermal receptors in the scrotum of the rat. *J. Physiol.* 248, 349–357.
231. Hermo, L., and Robaire, B. (2002). Epididymal cell types and their functions. In *The Epididymis from Molecules to Clinical Practice* (B. Robaire and B. T. Hinton, Eds.), pp. 81–102. Kluwer Academic/Plenum Publishers, New York.

232. Hermo, L., Badran, H., and Robaire, B. (2002). The structural organization and the function of the epithelium of the vas deferens. In *The Epididymis from Molecules to Clinical Practice* (B. Robaire and B. T. Hinton, Eds.), pp. 233–250. Kluwer Academic/Plenum Publishers, New York.
233. Heske, E. J., and Ostfeld, R. S. (1990). Sexual dimorphism in size, relative size of testes, and mating systems in North American voles. *J. Mamm.* 71, 510–519.
234. Hess, R. A. (2002). Efferent ductules: structure and functions. In *The Epididymis from Molecules to Clinical Practice* (B. Robaire and B. T. Hinton, Eds.), pp. 49–80. Kluwer Academic/Plenum Publishers, New York.
235. Hess, R. A., Zhou, Q., and Nie, R. (2002). The role of estrogens in the endocrine and paracrine regulation of the efferent ductules, epididymis and vas deferens. In *The Epididymis from Molecules to Clinical Practice* (B. Robaire and B. T. Hinton, Eds.), pp. 317–337. Kluwer Academic/Plenum Publishers, New York.
236. Heyns, C. F., and Hutson, J. M. (1995). Historical review of theories on testicular descent. *J. Urol.* 153, 754–767.
237. Hinton, B. T., and Keefer, D. A. (1983). Evidence for protein absorption from the lumen of the seminiferous tubule and rete of the rat testis. *Cell. Tissue Res.* 230, 367–375.
238. Hochereau-de Reviers, M. T., Locatelli, A., Perreau, C., Pisselet, C., and Setchell, B. P. (1993). Effects of a single brief period of moderate heating of the testes on seminiferous tubules in hypophysectomized rams treated with pituitary extract. *J. Reprod. Fertil.* 97, 381–387.
239. Hodson, N. (1970). The nerves of the testis, epididymis and scrotum. In *The Testis* (A. D. Johnson, W. R. Gomes, and N. L. Vandemark, Eds.), Vol. I, pp. 47–99. Academic Press, New York.
240. Hofmann, R. (1960). Die Gefassarchitektur des Bullenhodens, zugleich ein Versuch ihrer funktionellen Deutung. *Zentralblatt. Veterinaermed.* 7, 59–63.
241. Holash, J. A., Harik, S. I., Perry, G., and Stewart, P. A. (1993). Barrier properties of testis microvessels. *Proc. Natl. Acad. Sci. U S A* 90, 11069–11073.
242. Holstein, A. F., and Davidoff, M. (1997). Compartmentalization of the intertubular space of the human testis. *Adv. Exp. Med. Biol.* 424, 161–162.
243. Holstein, A. F., Orlandini, G. E., and Moller, R. (1979). Distribution and fine stucture of the lymphatic system in the human testis. *Cell. Tissue Res.* 200, 15–27.
244. Holstein, A. F., Maekawa, M., Nagano, T., and Davidoff, M. S. (1996). Myofibroblasts in the lamina propria of human seminiferous tubules are dynamic structures of heterogeneous phenotype. *Arch. Histol. Cytol.* 59, 109–125.
245. Holt, W. V., Jones, R. C., and Skinner, J. D. (1980). Studies of the deferent ducts from the testis of the African elephant, *Loxodonta africana*. II. Histochemistry of the epididymis. *J. Anat.* 139, 367–379.
246. Hosken, D. J. (1997). Sperm competition in bats. *Proc. R. Soc. Lond. B* 264, 385–392.
247. Hosken, D. J. (1998). Testes mass in megachiropteran bats varies in accordance with sperm competition theory. *Beh. Ecol. Sociob.* 44, 169–177.
248. Hosken, D. J., and Stockley, P. (2004). Sexual selection and genital evolution. *Trends Ecol. Evol.* 19, 87–93.
249. Houle, A. M., and Gagne, D. (1995). Human chorionic gonadotropin but not calcitonin gene related peptide induces postnatal testicular descent in mice. *J. Androl.* 16, 143–147.
250. Howl, J., Rudge, S. A., Lavis, R. A., Davies, R. L., Parslow, R. A., Hughes, P. J., Kirk, C. J., Michell, R. H., and Wheatly, M. (1995). Rat testicular myoid cells express vasopressin receptors: receptor structure, signal transduction, and developmental regulation. *Endocrinology* 136, 2206–2213.
251. Hrabovszky, Z., Di Pilla, N., Yap, T., Farmer, P. J., Hutson, J. M., and Carlin, J. B. (2002). Role of the gubernacular bulb in cremaster muscle development of the rat. *Anat. Rec.* 267, 159–165.
252. Hruban, Z., Martan, J., Slesers, A., Steiner, D. F., Lubran, M., and Rechcigl, M. Jr. (1965). Fine structure of the prostatic epithelium of the opossum (*Didelphis virginiana* Kerr). *J. Exp. Zool.* 160, 81–106.
253. Huang, H. S. F., Linsenmeyer, T. A., Li, M. T., Giglio, W., Anesetti, R., von Hagen, J., Ottenweiler, J. E., Serenas, C., and Pogach, L. (1995). Acute effects of spinal cord injury on the pituitary-testicular hormone axis and Sertoli cell functions: a time course study. *J. Androl.* 16, 148–157.
254. Huang, H. F. S., Linsenmeyer, T. A., Anesetti, R., Giglio, W., Ottenweiler, J. E., and Pogach, L. (1998). Suppression and recovery of spermatogenesis following spinal cord injury in the rat. *J. Androl.* 19, 72–80.
255. Huang, H. F. S., Li, M. T., Anesetti, R., Giglio, W., Ottenweiler, J. E., and Pogach, L. M. (1999). Effects of spinal cord injury on spermatogenesis and the expression of messenger ribonucleic acid for Sertoli cell proteins in rat Sertoli cell-enriched testes. *Biol. Reprod.* 60, 635–641.
256. Huang, H. F. S., Li, M. T., Giglio, W., Anesetti, R., Ottenweiler, J. E., and Pogach, L. M. (1999). The detrimental effects of spinal cord injury on spermatogenesis in the rat is partially reversed by testosterone, but enhanced by follicle-stimulating hormone. *Endocrinology* 140, 1349–1355.
257. Hughes, R. L., Thomson, J. A., and Owen, W. H. (1965). Reproduction in natural populations of the Australian Ringtail Possum, *Pseudocheirus peregrinus* (Marsupialia: Phalangeridae) in Victoria. *Aust. J. Zool.* 13, 383–406.
258. Hundeiker, M., and Keller, L. (1963). Die Gefassarchitektur des menschlichen Hodens. *Morphol. Jb.* 105, 26–73.
259. Hurt, G. S., Howards, S. S., and Turner, T. T. (1986). Repair of experimental varicoceles in the rat long-term effects on testicular blood flow and temperature and cauda epididymal sperm concentration and motility. *J. Androl.* 7, 261–276.
260. Hutson, J. C. (1992). Development of cytoplasmic digitations between Leydig cells and testicular macrophages of the rat. *Cell. Tissue Res.* 267, 385–389.
261. Hutson, J. M., Terada, M., Zhou, B., and Williams, M. P. L. (1995). Normal testicular descent and the aetiology of cryptorchidism. *Adv. Anat. Embryol. Cell Biol.* 132, 1–56.
262. Hutson, J. M., Hastorpe, S., and Heyns, C. F. (1997). Anatomical and functional aspects of testicular descent and cryptorchidism. *Endocr. Rev.* 18, 259–280.
263. Hutson, J. M., Watts, L. M., and Farmer, P. J. (1998). Congenital undescended testes in neonatal pigs and the effect of exogenous calcitonin gene-related peptide. *J. Urol.* 159, 1025–1030.
264. Ilio, K. Y., and Hess, R. A. (1994). Structure and function of the ductuli efferentes: a review. *Microsc. Res. Tech.* 29, 432–467.
265. Inns, R. W. (1982). Seasonal changes in the accessory reproductive system and plasma testosterone levels of the male tammar wallaby, *Macropus eugenii*, in the wild. *J. Reprod. Fertil.* 66, 675–680.
266. Irby, D. C., Kerr, J. B., Risbridger, G. P., and de Kretser, D. M. (1984). Seasonally and experimentally induced changes in testicular function of the Australian bush rat (*Rattus fuscipes*). *J. Reprod. Fertil.* 70, 657–666.
267. Itoh, M., Li, X. Q., Yano, A., Xie, Q., and Takeuchi, Y. (1998). Patterns of efferent lymphatics of the mouse testis. *J. Androl.* 19, 466–472.
268. Ivell, R., and Bathgate, R. A. D. (2002). Reproductive biology of the relaxin-like factor (RLF/INSL3). *Biol. Reprod.* 67, 699–705.

269. Ivell, R., and Hartung, S. (2003). The molecular basis of cryptorchidism. *Mol. Hum. Reprod.* 9, 175–181.
270. Jacks, F., and Setchell, B. P. (1973). A technique for studying the transfer of substances from venous to arterial blood in the spermatic cord of wallabies and rams. *J. Physiol.* 233, 17P–18P.
271. Jarow, J. P. (1990). Intratesticular arterial anatomy. *J. Androl.* 11, 255–259.
272. Jarow, J. P. (1991). Clinical significance of intratesticular arterial anatomy. *J. Urol.* 145, 777–779.
273. Jeays-Ward, K., Hoyle, C., Brennan, J., Dandonneau, M., Alldus, G. Capel, B., and Swain, A. (2003). Endothelial and sterodogenic cell migration are regulated by WNT4 in the developing mammalian gonad. *Development* 130, 3663–3670.
274. Jedrzejewski, K. S., Cendrowska, I., Okraszewska, E., and Bienias, R. (1996). Comparative study of the intratesticular vascular rete in men and certain animals. *Fol. Morph. (Warsz.)* 55, 303–305.
275. Jesek, D., Schulze, W., Rogatsch, H., and Hittmair, A. (1996). Structure of small blood vessels in the testes of infertile men. *Int. J. Androl.* 19, 299–306.
276. Jesik, C. J., Holland, J. M., and Lee, C. (1982). An anatomic and histologic study of the rat prostate. *Prostate* 3, 81–97.
277. Jia, Z., Duan, E., Jiang, Z., and Wang, Z. (2002). Copulatory plugs in masked palm civets: prevention of semen leakage, sperm storage, or chastity enhancement? *J. Mamm.* 83, 1035–1038.
278. Johnson, K. A., and Walton, D. W. (1989). Notoryctidae. In *Fauna of Australia* (B. J. Walton and B. J. Richardson, Eds.), pp. 591–602. Australian Government and Printing Service.
279. Johnson, O. W., and Buss, I. O. (1967). The testis of the African Elephant (*Loxodonta africana*). II. Development, puberty and weight. *J. Reprod. Fertil.* 13, 23–30.
280. Johnston, P. G., and Zucker, I. (1980). Photoperiodic regulation of the testis of adult white-footed mice (*Peromyscus leucopus*). *Biol. Reprod.* 23, 859–866.
281. Jolly, S. E., and Blackshaw, A. W. (1987). Prolonged epididymal sperm storage, and the temporal dissociation of testicular and accessory gland activity in the common sheath-tail bat, *Taphozous georgianus*, of tropical Australia. *J. Reprod. Fertil.* 81, 205–211.
282. Jolly, S. E., and Blackshaw, A. W. (1988). Testicular migration, spermatogenesis, temperature regulation and environment of the sheath-tail bat, *Taphozous georgianus*. *J. Reprod. Fertil.* 84, 447–455.
283. Jones, R. C. (1980). Luminal compostion and maturation of spermatozoa in the genital ducts of the African elephant (*Loxodonta africana*). *J. Reprod. Fertil.* 60, 87–93.
284. Jones, R. C. (1998). Evolution of the vertebrate epididymis. *J. Reprod. Fertil.* Suppl 53, 163–182.
285. Jones, R. C. (1999). To store or mature spermatozoa? The primary role of the epididymis. *Int. J. Androl.* 22, 57–67.
286. Jones, R. C., and Brosnan, M. F. (1981). Studies of the deferent ducts from the testis of the African elephant, *Loxodonta africana*. I. Structural differentiation. *J. Anat.* 132, 371–386.
287. Jones, R. C., and Djakiew, D. (1978). The role of the excurrent ducts from the testis of the testicond mammals. *Aust. Zool.* 20, 201–210.
288. Jones, R. C., and Holt, W. V. (1981). Studies of the deferent ducts from the testis of the African elephant, *Loxodonta africana*. III. Ultrastructure and cytochemistry of the ductuli efferentes. *J. Anat.* 133, 247–255.
289. Jones, R. C., and Jurd, K. M. (1987). Structural differentiation and fluid resorption in the ductuli efferentes of the rat. *Aust. J. Biol. Sci.* 40, 79–90.
290. Jones, R. C., Rowlands, I. W., and Skinner, J. D. (1974). Spermatozoa in the genital ducts of the African elephant, *Loxodonta africana*. *J. Reprod. Fertil.* 41, 189–192.
291. Jones, R.C., Hinds, L. A., and Tyndale-Biscoe, C. H. (1984). Ultrastructure of the epididymis of the tammar, *Macropus eugenii*, and its relationship to sperm maturation. *Cell. Tissue Res.* 237, 525–535.
292. Jones, R. C., Stone, G. M., Hinds, L. A., and Setchell, B. P. (1988). Distribution of 5α–reductase in the epididymis of the tammar wallaby (*Macropus eugenii*) and dependence of the epididymis on systemic testosterone and luminal fluids from the testis. *J. Reprod. Fertil.* 83, 779–783.
293. Jones, R. C., Stone, G. M., and Zupp, J. (1992). Reproduction in the male echidna. In *Platypus and Echidnas* (M. L. Augee, Ed.), pp. 115–125. Royal Zoological Society of NSW, Sydney.
294. Jordan, B. K., Shen, J. H. C., Olaso, R., Ingraham, H. A., and Vilain, E. (2003). Wnt4 overexpression disrupts normal testicular vasculature and inhibits testosterone synthesis by repressing steroidogenic factor 1/β-catenin synergy. *Proc. Nal. Acad. Sci. U S A* 100, 10866–10871.
295. Kaleczyc, J. (1998). Origin and neurochemical characteristics of nerve fibers supplying the mammalian vas deferens. *Microsc. Res. Tech.* 42, 409–422.
296. Kaleczyc, J., Timmermans, J. P., Majewski, M., Lakomy, M., and Scheuermann, D. W. (1995). Distribution and immunohistochemical characteristics of neurons in the porcine caudal mesenteric ganglion projecting to the vas deferens and seminal vesicle. *Cell. Tissue Res.* 282, 59–68.
297. Kaleczyc, J., Timmermans, J. P., Majewski, M., Lakomy, M., and Scheuermann, D. W. (1997). Immunohistochemical characteristics of nerve fibers supplying the porcine vas deferens. *Histochem. Cell Biol.* 107, 229–241.
298. Kaleczyc, J., Timmermans, J. P., Majewski, M., Lakomy, M., and Scheuermann, D. W. (1999). Immunohistochemical properties of nerve fibers supplying accessory male genital glands in the pig, a colocalization study. *Histochem. Cell Biol.* 111, 217–228.
299. Kaleczyc, J., Scheuermann, D.W., Pilsudko, Z., Majewski, M., Lakomy, M., and Timmerman, J. P. (2002). Distribution, immunohistochemical characteristics and nerve pathways of primary sensory neurones supplying the porcine vas deferens. *Cell. Tissue Res.* 310, 9–17.
300. Kamimura, Y., Chiba, H., Utsumi, H., Gotoh, T., Tobioka, H., and Sawada, N. (2002). Barrier function of microvessels and roles of glial cell-line derived neurotrophic factor in the rat testis. *Med. Electron. Microsc.* 35, 139–145.
301. Kaplan, J. B., and Mead, R. A. (1993). Influence of season on seminal characteristics, testis size and serum testosterone in the Western Spotted Skunk (*Spilogale gracilis*). *J. Reprod. Fertil.* 98, 321–326.
302. Kaplan, J. B., and Mead, R. A. (1994). Seasonal changes in testicular function and seminal characteristics of the male Eastern Spotted Skunk (*Apilogale putorius ambarvilus*). *J. Mamm.* 75, 1013–1020.
303. Kazeem, A. (1986). Reexamination of testicular lymphatic drainage in the rat. *Lymphology* 19, 172–174.
304. Kazeem, A. A. (1991). Species variation in the extrinsic lymphatic draining of the rodent testis: its role within the context of an immunologically privileged site. *Lymphology* 24, 140–144.
305. Kemp, P. A., Gardiner, S. M., March, J. E., Rubin, P. C., and Bennet, T. (1999). Assessment of the effects of endothelin-1 and magnesium sulfate on regional blood flows in conscious rats, by the coloured microsphere reference technique. *Br. J. Pharmacol.* 126, 621–626.
306. Kempinas, W. G., Petenusci, S. O., Rosa e Silva, A. A. M., Favaretto, A. L. V., and Lamano Carvalho, T. L. (1995). The hypophyseal-testicular axis and sex accessory gland following chemical sympathectomy with guanethidine of pre-pubertal to mature rats. *Andrologia* 27, 121–125.

307. Kempinas, W. G., Suarez, J. D., Roberts, N. L., Strader, L., Ferrell, J., Goldman, J. M., and Klinefelter, G. R. (1998). Rat epididymal sperm quantitiy, quality, and transit time after guanethidine-induced sympathectomy. *Biol. Reprod.* 59, 890–896.
308. Kempinas, W. G., Suarez, J. D., Roberts, N. L., Strader, L., Ferrell, J., Goldman, J. M., Narotsky, M. G., Perrault, S. D., Evenson, D. P., Ricker, D. D., and Klinefelter, G. R. (1998). Fertility of rat epididymal sperm after chemically and surgically induced sympathectomy. *Biol. Reprod.* 59, 897–904.
309. Kenagy, G. J., and Trombulak, S. C. (1986). Size and function of mammalian testes in relation to body size. *J. Mamm.* 67, 1–22.
310. Kepper, M., and Keast, J. (1995). Immunohistochemical properties and spinal connection of pelvic autonomic neurons that innervate the rat prostate gland. *Cell. Tissue Res.* 281, 533–542.
311. Kern, S., Robertson, S. A., Mau, V. J., and Maddocks, S. (1995). Cytokine secretion by macrophages in the rat testis. *Biol. Reprod.* 53, 1407–1416.
312. Kerr, J. B., and Hedger, M. P. (1983). Spontaneous spermatogenic failure in the marsupial mouse *Antechinus stuartii* Macleay (Dasyuridae: Marsupialia). *Aust. J. Zool.* 31, 445–466.
313. Kihara, K., Sato, K., and Oshima, H. (1998). Sympathetic efferent pathways projecting to the vas deferens. *Microsc. Res. Tech.* 42, 398–408.
314. Kirby, A. (1953). Observations on the blood supply of the bull testis. *Br. Vet. J.* 109, 464–472.
315. Kitchener, D. J. (1973). Reproduction in the Common Sheath-tailed Bat, *Taphozous georgianus* (Thomas) (Microchiroptera: Emballonuridae), in western Australia. *Aust. J. Zool.* 21, 375–389.
316. Klonisch, T., Fowler, P. A., and Hombach-Klonisch, S. (2004). Molecular and genetic regulation of testis descent and external genitalia development. *Dev. Biol.* 270, 1–18.
317. Kormano, M. (1967). An angiographic study of the testicular vasculature in the postnatal rat. *Z. Anat. Entwickl.* 126, 138–153.
318. Kormano, M., and Nordmark, L. (1977). Angiography of the testicular artery. III. Testis and epididymis analysed with a magnification technique. *Acta Radiol. Diag.* 18, 625–633.
319. Kormano, M., and Suoranto, H. (1971). An angiographic study of the arterial pattern of the human testis. *Anat. Anz.* 128, 69–76.
320. Korpelainen, E. I., Karkkainen, M. J., Tenhunen, A., Lakso, M., Rauvala, H., Vierula, M., Parvinen, M., and Alitalo, K. (1998). Overexpression of VEGF in testis and epididymis causes infertility in transgenic mice: evidence for nonendothelial targets for VEGF. *J. Cell Biol.* 143, 1705–1712.
321. Kowalska-Dyrcz, A. (1966). A comparative study of the male genital tract in some species of shrews. *Acta Theriol.* 11, 71–77.
322. Kruger, L., Kavookjian, A. M., Kumazawa, T., Light, A. R., and Mizumura, K. (2003). Nociceptor structural specialization in canine and rodent testicular "free" nerve endings. *J. Comp. Neurol.* 463, 197–211.
323. Krutzsch, P. H. (1979). Male reproductive patterns in nonhibernating bats. *J. Reprod. Fertil.* 56, 333–344.
324. Kubota, Y., Temelcos, C., Bathgate, R. A. D., Smith, K. J., Scott, D., Zhao, C., and Hutson, J. M. (2002). The role on insulin 3, testosterone, Müllerian inhibiting substance and relaxin in rat gubernacular growth. *Mol. Hum. Reprod.* 8, 900–905.
325. Kumazawa, T., and Mizumura, K. (1980). Mechanical and thermal reponses of polymodal receptors recorded from the superior spermatic nerve of dogs. *J. Physiol.* 299, 233–245.
326. Kumazawa, T., Mizumura, K., and Sato, J. (1987). Response properties of polymodal receptors studies using in vitro superior spermatic nerve preparations of dogs. *J. Neurophysiol.* 57, 702–711.
327. Lakomy, M., Kaleczyc, J., and Majewski, M. (1997). Noradrenergic and peptidergic innervation of the testis and epididymis in the male pig. *Fol. Histochem. Cytobiol.* 35, 19–27.
328. Lamano Carvalho, T. L., Hodson, N., Blank, M. A., Watson, P. F., Mulderry, P. K., Bishop, A. E., Gu, J., Bloom, S. R., and Polak, J. M. (1986). Occurrence, distribution and origin of peptide-containing nerves of guinea pig and rat male genitalia and the effects of denervation on sperm characteristics. *J. Anat.* 149, 121–141.
329. Lamano Carvalho, T. L., Kempinas, W. G., and Favoretto, A. L. V. (1993). Morphometric evaluation of the rat testis, epididymis and vas deferens follwing chemical sympathectomy with guanethidine. *Ann. Anat.* 175, 453–457.
330. Laurie, M. S., and Setchell, B. P. (1979). The continuous measurement of testicular blood flow in the ram, in relation to the pulsatile secretion of testosterone. *J. Physiol.* 287, 10P.
331. Lauth, E. A. (1830). Memoire sur le testicule humain. *Mem. Soc. Hist. Nat. Strasbourg* 1, 1–42.
332. LeCouter, J., Kowalski, J., Foster, J., Hass, P., Zhang, Z., Dillar-Telm, L., Frantz, G., Rangell, L., DeGuzman, L., Keller, G. A., Peale, F., Gurney, A., Hillan K. J., and Ferrara, N. (2001). Identification of an angiogenic mitogen selective for endocrine gland endothelium. *Nature* 412, 877–884.
333. LeCouter, J., Lin, R., Tejada, M., Frantz, G., Peale, F., Hillan, K. J., and Ferrara, N. (2003). The endocrine-gland-derived VEGF homologue Bv8 promotes angiogenesis in the testis: localization of Bv8 receptors to endothelial cells. *Proc. Natl. Acad. Sci. U S A* 100, 2865–2690.
334. Lee, C., Sensibar, J. A., Dudek, S. M., Hiipakka, R. A., and Liao, S. (1990). Prostatic ductal system in rats: regional variation in morphological and functional activities. *Biol. Reprod.* 43, 1079–1086.
335. Lee, C. S., and O'Shea, J. D. (1977). Observations on the vasculature of the reproductive tract in some Australian marsupials. *J. Morph.* 154, 95–114.
336. Lee, S., Miselis, R., and Rivier, C. (2002). Anatomical and functional evidence for a neural hypothalamic-testicular pathway that is independent of the pituitary. *Endocrinology* 143, 4447–4454.
337. Lee, S. M. Y., and Hutson, J. M. (1999). Effect of androgen on the cranial suspensory ligament and ovarian position. *Anat. Rec.* 255, 306–315.
338. Leeson, T. S. (1975). Smooth muscle cells in the rat testicular capsule: a developmental study. *J. Morph.* 147, 171–186.
339. Lekas, E., Johansson, M., Widmark, A., Bergh, A., and Damber, J. E. (1997). Decrement of blood flow precedes the involution of the ventral prostate in the rat after castration. *Urol. Res.* 25, 309–314.
340. Lekas, E., Bergh, A., and Damber, J. E. (2000). Effects of finasteride and bicalutamide on prostatic blood flow in the rat. *BJU Int.* 85, 962–965.
341. Lewis, M. H., and Moffat, D. B. (1975). The venous drainage of the accessory reproductive organs of the rat with special reference to prostatic metabolism. *J. Reprod. Fertil.* 42, 497–502.
342. Li, H., Dubocq, F., Jiang, Y., Tiguert, R., Gheiler, E. L., and Dhubwala, C. B. (1999). Effect of surgically induced varicocele on testicular blood flow and Sertoli cell function. *Urology* 53, 1258–1262.
343. Li, Q., and Thornhill, J. (1996). Specific thermal responsiveness of ventromedial hypothalamic neorons to localized scrotal heating and cooling in rats. *J. Physiol.* 492, 851–865.

344. Li, Q., and Thornhill, J. (1997). Differences in brown adipose tissue effector response and assocaiated thermoresponsiveness of ventromedial hypothalamic (VMH) neurons of 21°C vs 4°C acclimitized rats to scrotal thermal stimulation. *Brain Res.* 770, 18–25.
345. Li, Q., and Thiornhill, J. (1998). Thermoresponsiveness of posterio hypothalamic (PH) neurons of rats to scrotal and abdominal thermal stimulation. *Brain Res.* 794, 80–87.
346. Lincoln, G. A. (1971). The seasonal reproductive changes in the Red Deer stag (*Cervus elaphus*). *J. Zool. Lond.* 163, 105–123.
347. Lincoln, G. A. (1974). Reproduction and "March madness" in the Brown Hare, *Lepus europaeus*. *J. Zool. Lond.* 174, 1–14.
348. Lincoln, G. A. (1978). Hypothalamic control of the testis in the ram. *Int. J. Androl.* 1, 331–341.
349. Lincoln, G. A. (1981). Seasonal aspects of testicular function. In *The Testis* (H. Burger and D. de Kretser, Eds.), pp. 255–302. Raven Press, New York.
350. Lincoln, G. A., and Ratnasooriya, W. D. (1996). Testosterone secretion, musth behaviour and social dominance in captive male Asian elephants living near the equator. *J. Reprod. Fertil.* 108, 107–113.
351. Lincoln, G. A., and Short, R. V. (1980). Seasonal breeding: nature's contraceptive. *Rec. Prog. Horm. Res.* 36, 1–52.
352. Linsenmeyer, T. A., Pogach, L. M., Ottenweiler, J. E., and Huang, H. S. F. (1994). Spermatogenesis and the pituitary-testicular hormone axis in rats during the acute phase of spinal cord injury. *J. Urol.* 152, 1302–1307.
353. Linzell, J. L., and Setchell, B. P. (1969). Metabolism, sperm and fluid production of the isolated perfused testis of the sheep and goat. *J. Physiol.* 201, 129–143.
354. Linzey, A. V., and Layne, J. N. (1969). Comparative morphology of the male reproductive tract in the rodent genus *Peromyscus* (Muridae). *Am. Mus. Nov.* 2355, 1–47.
355. Lissbrant, E., and Bergh, A. (1997). Effects of vasoactive intestinal peptide (VIP) on the testicular vasculature of the rat. *Int. J. Androl.* 20, 356–360.
356. Lissbrant, E., Lofmark, U., Collin, O., and Bergh, A. (1997). Is nitric oxide involved in the regulation of the rat testicular vasculature? *Biol. Reprod.* 56, 1221–1227.
357. Lissbrant, E., Collin, O., and Bergh, A. (1997). Localization and effects of calcitonin gene-related peptide in the testicular vasculature of the rat. *J. Androl.* 18, 385–392.
358. Lissbrant, E., Collin, O., and Bergh, A. (1999). Pituitary adenylate cyclase-activating peptide (PACAP): effects on blood flow in the testis and caput epididymidis of the rat. *J. Androl.* 20, 366–374.
359. Lissbrant, I. F., Haggstrom, S., Damber, J. E., and Bergh, A. (1998). Testosterone stimulates angiogenesis and vascular regrowth in the ventral prostate in castrated adult rats. *Endocrinology* 139, 451–456.
360. Lissbrant, I. F., Lissbrant, E., Damber, J. E., and Bergh, A. (2001). Blood vessels are regulators of growth, diagnostic markers and therapeutic targets in prostate cancer. *Scand. J. Urol. Nephrol.* 35, 437–452.
361. Lissbrant, I. F., Lissbrant E., Persson, A., Damber, J. E., and Bergh, A. (2003). Endothelial cell proliferation in male reproductive organs of adult rat is high and regulated by testicular factors. *Biol. Reprod.* 68, 1107–1111.
362. Lloyd, H. G., and Englund, J. (1973). Reproductive cycle of the Red Fox in Europe. *J. Reprod. Fertil.* Suppl 19, 119–130.
363. Loudon, A. S. I., and Curlewis, J. D. (1988). Cycles of antler and testicular growth in an aseasonal tropical deer (*Axis axis*). *J. Reprod. Fertil.* 83, 729–738.
364. Lujan, M., Paez, A., Llanes, L., Angulo, J., and Berenguer, A. (1998). Role of autonomic innervation in rat prostatic structure maintenance: a morphometric analysis. *J. Urol.* 160, 1919–1923.
365. Lunstra, D. D., Ford, J. J., Klindt, J., and Wise, T. H. (1997). Physiology of the Meishan boar. *J. Reprod. Fertil.* Suppl 52, 181–193.
366. Lysiak, J. L., Nguyen, Q. A. T., and Turner, T. T. (2000). Fluctuation in rat testicular interstitial oxygen tension are linked to testicular vasomotion: persistence after repair of torsion. *Biol. Reprod.* 63, 1383–1389.
367. McCullough, D. L. (1975). Experimental lymphangiography. Experience with direct medium injection into the parenchyma of the rat testis and prostate. *Invest. Urol.* 13, 211–219.
368. McDonald, S. W., and Scothorne, R. J. (1988). The lymphatic drainage of the epididymis and of the ductus deferens of the rat, with reference to the immune response to vasectomy. *J. Anat.* 158, 57–64.
369. McGuckin, M. A., and Blackshaw, A. W. (1991). Seasonal changes in testicular size, plasma testosterone concentration and body weight in captive flying foxes (*Pteropus poliocephalus* and *P. scapulatus*). *J. Reprod. Fertil.* 92, 339–346.
370. McMahon, D. R., Kramer, S. A., and Husmann, D. A. (1995). Antiandrogen induced cryptorchidism in the pig is associated with failed gubernacular regression and epididymal malformations. *J. Urol.* 154, 553–557.
371. McVary, K. T., Razzaq, A., Lee, C., Venegas, M. F., Rademaker, A., and McKenna, K. E. (1994). Growth of the rat prostate gland is facilitated by the autonomic nervous system. *Biol. Reprod.* 51, 99–107.
372. McVary, K. T., McKenna, K. E., and Lee, C. (1998). Prostate innervation. *Prostate* Suppl 8, 2–13.
373. Maddocks, S., and Setchell, B. P. (1988). The physiology of the endocrine testis. *Oxf. Rev. Reprod. Biol.* 10, 53–123.
374. Maddocks, S., and Sharpe, R. M. (1989). Dynamics of testosterone secretion by the rat testis: implications for measurement of the intratesticular levels of testosterone. *J. Endocrinol.* 122, 323–329.
375. Maddocks, S. Hargreave, T. B., Reddie, K., Fraser, H. M., Kerr, J. B., and Sharpe, R. M. (1993). Intratesticular hormone levels and the route of secretion of hormones from the testis of the rat, guinea pig, monkey and human. *Int. J. Androl.* 16, 272–278.
376. Maggi, M., Barni, T., Orlando, C., Fantoni, G., Finetti, G., Vannelli, G. B., Mancina, R., Gloria, L., Bonaccorsi, L., Yanagisawa, M., and Forti, G. (1995). Endothelin-1 and its receptors in human testis. *J. Androl.* 16, 213–224.
377. Majewski, M., Kaleczyc, J., Mayer, B., Schemann, M., Weihe, E., and Lakomy, M. (1999). Innervation of the fibro-elastic type of penis: an immunohistochemical study in the male pig. *Acta Histochem.* 101, 71–101.
378. Maloney, S. K., and Mitchell, D. (1996). Regulation of ram scrotal temperature during heat exposure, cold exposure, fever and exercise. *J. Physiol.* 496, 421–430.
379. Maloney, S. K., Bonomelli, J. M., and DeSouza, J. (2003). Scrotal heating stimulates panting and reduces body temperature similarly in febrile and non-febrile rams (*Ovis aries*). *Comp. Biochem. Physiol.* A 135, 565–573.
380. Mann, T., and Lutwak-Mann, C. (1981). *Male Reproductive Function and Semen*. Springer-Verlag, Berlin.
381. Markey, C. M., and Meyer, G. T. (1992). A quantitative description of the epididymis and its microvasculature: an age-related study in the rat. *J. Anat.* 180, 255–262.
382. Marsh, H., Heinsohn, G. E., and Glover, T. D. (1984). Changes in the male reproductive organs of the dugong, *Dugong dugon* (Sirenia: Dugondidae) with age and reproductive activity. *Aust. J. Zool.* 32, 721–742.
383. Martan, J., and Shepherd, B. A. (1976). The role of the copulatory plug in reproduction in the guinea pig. *J. Exp. Zool.* 196, 79–83.
384. Martin, R. D. (1969). The evolution of reproductive mechanisms in primates. *J. Reprod. Fertil.* Suppl 6, 49–66.

385. Martinet, L., Mondain-Monval, M., and Monnerie, R. (1992). Endogenous circannual rhythms and photorefractoriness of testis activity, moult and prolactin concentrations in mink (*Mustela vison*). *J. Reprod. Fertil.* 95, 325–338.
386. Maruch, S. M. G., Alves, H. J., and Machado, C. R. S. (1989). Sympathetic innervation of the reproductive organs of the male opossum, *Didelphis albiventris* (Lund, 1841). *Acta Anat.* 134, 257–262.
387. Matthews, L. H. (1941). Notes of the genitalia and reproduction of some African bats. *Proc. Zool. Soc. Lond.* 111, 289–346.
388. Matthews, M. K., and Adler, N. T. (1978). Systematic interrelationship of mating, vaginal plug position, and sperm transport in the rat. *Physiol. Behav.* 20, 302–309.
389. Mayerhofer, A., and Bartke, A. (1990). Developing testicular microvasculature in the Golden Hamster, *Mesocricetus auratus*: a model for angiogenesis under physiological conditions. *Acta Anat.* 139, 78–85.
390. Mayerhofer, A., Sinha Hikim, A. P., Bartke, A., and Russell, L. D. (1989). Changes in the testicular microvasculature during photoperiod-related seasonal transition from reproductive quiescene to reproductive activity in the adult Golden Hamster. *Anat. Rec.* 224, 495–507.
391. Mayerhofer, A., Amador, A. G., Steger, R. S., and Bartke, A. (1990). Testicular function after local injection of 6-hydroxydopamine or norepinephrine in the Golden Hamster (*Mesocricetus auratus*). *J. Androl.* 11, 301–311.
392. Mayerhofer, A., Danilchik, M., Pau, K. Y. F., Lara, H. E., Russell, L. D., and Ojeda, S. R. (1996). Testis of prepubertal rhesus monkeys receives a dual catecholaminergic input provided by the extrinsic innervation and an intragonadal source of catecholamines. *Biol. Reprod.* 55, 509–518.
393. Mayerhofer, A., Frungieri, M. B., Fritz, S., Bulling, A., Jessberger, B., and Vogt, H. J. (1999). Evidence for catecholaminergic, neuronlike cells in the adult human testis: changes associated with testicular pathologies. *J. Androl.* 20, 341–347.
394. Maynes, G. M. (1973). Reproduction in the Parma Wallaby *Macropus parma* Waterhouse. *Aust. J. Zool.* 21, 331–351.
395. Mead, R. A. (1968). Reproduction in western forms of the Spotted Skunk (Genus: *Spilojale*). *J. Mamm.* 49, 373–390.
396. Mead, R. A., Rector, M., Starypan, G., Neirinckx, S., Jones, M., and DonCarlos, M. N. (1991). Reproductive biology of captive wolverines. *J. Mamm.* 72, 807–814.
397. Middendorff, R., Muller, D., Mewe, M., Mukhopadhyay, A. K., Holstein, A. F., and Davidoff, M. S. (2002). The tunica albuginea of the human testis is characterized by complex contraction and relaxation activities regulated by cyclic GMP. *J. Clin. Endocrinol. Metab.* 87, 3486–3499.
398. Mieusset, R., Sowerbutts, S. F., Zupp, J. L., and Setchell, B. P. (1992). Increased testicular blood flow during local heating of the testes of rams. *J. Reprod. Fertil.* 94, 345–352.
399. Millar, R. P., and Glover, T. D. (1970). Seasonal changes in the reproductive tract of the male Rock Hyrax, *Procavia capensis*. *J. Reprod. Fertil.* 23, 497–499.
400. Millar, R. P., and Glover, T. D. (1973). Regulation of seasonal activity in an ascrotal mammal, the Rock Hyrax, *Procavia capensis*. *J. Reprod. Fertil.* Suppl 19, 203–220.
401. Miller, S. C., Rowman, B. M., and Rowland, H. G. (1983). Structure, cytochemistry, endocytic activity and immunoglobulin (Fc) receptors of the rat interstitial tissue macrophages. *Am. J. Anat.* 168, 1–13.
402. Mokkapati, S., and Dominic, C. J. (1977). Morphology of the accessory reproductive glands of some male Indian chiropterans. *Anat. Anz.* 141, 391–397.
403. Moller, A. P. (1989). Ejaculate quality, testes size and sperm production in mammals. *Func. Ecol.* 3, 91–96.
404. Morris, B., quoted by Setchell, B. P. (1970). Testicular blood supply, lymphatic drainage and secretion of fluid. In *The Testis* (A. D. Johnson, W. R. Gomes, and N. L. Vandemark, Eds.), Vol. I, pp. 101–239. Academic Press, New York.
405. Morris, B., and McIntosh, G. H. (1971). Techniques for the collection of lymph with special reference to the testis and ovary. *Acta Endocrinol.* Suppl 158, 145–168.
406. Muller, I. (1957). Kanalchen-und Capillararchitektonik des Rattenhodens. *Z. Zellforsch.* 45, 522–537.
407. Mulryan, K., Gitterman, D. P., Lewis, C. J., Vial, C., Leckie, B. J., Cobb, A. L., Brown, J. E., Conley, E. C., Buell, G., Pritchard, C. A., and Evans, J. (2000). Reduced vas deferens contraction and male infertility in mice lacking P2X1 receptors. *Nature* 403, 86–89.
408. Murakami, T., Uno, Y., Ohtsuka, A., and Taguchi, T. (1989). The blood vascular architecture of the rat testis: a scanning electron microscopic study of corrosion casts followed by light microscopy of tissue sections. *Arch. Histol. Cytol.* 52, 151–172.
409. Murphy, W. J., Elzirik, E., O'Brien, S. J., Madesen, O., Scally, M., Douady, C. J., Teeling, E., Ryder, O. A., Stanhope, M. J., de Jong, W. W., and Springer, M. S. (2001). Resolution of the early placental mammal radiation using Bayesian phylogenetics. *Science* 294, 2349–2351.
410. Nagler, H. M., Lizza, E. F., House, S. D., Tomashefsky, P., and Lipowsky, H. H. (1987). Testicular hemodynamic changes after the surgical creation of a varicocele in the rat. Intravital microscopic observations. *J. Androl.* 8, 292–298.
411. Nakayama, T., Ishikawa, Y., and Tsurutani, T., (1979). Projection of scrotal thermal afferents to the preoptic and hypothalamic neurons in rats. *Pflugers Arch.* 380, 59–64.
412. Nef, S., and Parada, L. F. (1999). Cryptorchidism in mice mutant for *INSL3*. *Nat. Genet.* 22, 295–299.
413. Nef, S., Shipman, T., and Parada, L. F. (2000). A molecular basis for estrogen-induced cryptorchidism. *Dev. Biol.* 224, 354–361.
414. Nemeth, J. A., and Lee, C. (1996). Prostatic ductal system in rats: regional variation in stromal organization. *Prostate* 28, 124–128.
415. Neya, T., and Pierau, F. K. (1980). Activity patterns of temperature-reactive dorsal horn neurons and their reactions to peripheral reactions to peripheral receptor stimulation by Ca. *Jpn. J. Physiol.* 30, 921–934.
416. Niemi, M., and Kormano, M. (1965). Contractility of the seminiferous tubules of the postnatal rat testis and its response to oxytocin. *Ann. Med Exp. Fenn.* 43, 40–42.
417. Niemi, M., and Setchell, B. P. (1986). Gamma-glutamyl transpeptidase in the vasculature of the rat testis. *Biol. Reprod.* 35, 385–391.
418. Noordhuizen-Stassen, E. N., Beijer, H. J. M., Charbon, G. A., and Wensing, C. J. G. (1983). The effect of norepinephrine, isoprenaline and acetylcholine on the testicular and epididymal circulation in the pig. *Int. J. Androl.* 6, 44–56.
419. Noordhuizen-Stassen, E. N., Charbon, G. A., de Jong, F. H., and Wensing, C. J. G. (1985). Functional arterio-venous anastomoses between the testicular artery and the pampiniform plexus in the spermatic cord of rams. *J. Reprod. Fertil.* 75, 193–201.
420. Nykanen, M. (1979). Fine structure of the transitional zone of the rat seminiferous tubule. *Cell. Tissue Res.* 198, 441–454.
421. O, W. S., Chen, H. Q., and Chow, P. H. (1988). Effects of male accessory sex gland secretions on early embryonic development in the Golden Hamster. *J. Reprod. Fertil.* 84, 341–344.
422. Oakwood, M. (2000). Reproduction and demography of the Northern Quoll, *Casyurus hallucatus*, in the lowland savanna or northern Australia. *Aust. J. Zool.* 48, 519–510.
423. Oakwood, M. (2004). Death after sex. *Biologist* 51, 5–8.
424. Oakwood, M., Bradley, A. J., and Cockburn, A. (2001). Semelparity in a large marsupial. *Proc. R. Soc. Lond. B* 268, 407–411.

425. O'Brien, G. M. (1993). Seasonal reproduction in flying foxes, reviewed in the context of other tropical mammals. *Reprod. Fertil. Dev.* 5, 499–521.
426. O'Brien, G. M., Curlewis, J. D., and Martin, L. (1993). Effect of photoperiod on the annual cycle of testis growth in a tropical mammal, the Little Red Flying Fox, *Pteropus scapulatus*. *J. Reprod. Fertil.* 98, 121–127.
427. Ohanian, C., Rodriguez, H., Piriz, H., Martino, I., Rieppi, G., Garafalo, E. G., and Roca, R. A. (1979). Studies on the contractile activity and ultrastructure of the boar testicular capsule. *J. Reprod. Fertil.* 57, 79–85.
428. Ohtani, O., and Gannon, B. J. (1984). The microvasculature of the rat vas deferens: a scanning electron and light microscopic study. *J. Anat.* 135, 521–529.
429. Ohtsuka, A. (1984). Microvascular architecture of the pampiniform plexus-testicular artery system in the rat: a scanning electron microscope study of corrosion casts. *Am. J. Anat.* 169, 285–293.
430. Okraszewska, E., Cendrowska, I., and Jedrzejewski, K. S. (1996). The intratesticular lymphatic network in men, bulls and rams. *Fol. Morph. (Warsz.)* 55, 401–402.
431. Ono, Y., Suzuki, K., Kashiwagi, B., Shibata, Y., Ito, K., Fukabori, Y., and Yamanaka, H. (2004). Androgen-dependent blood flow control and morphological changes in the capillaries in rat prostate. *Int. J. Androl.* 27, 50–56.
432. Osman, D. I., Tingari, M. D., and Moniem, K. A. (1979). Vascular supply of the testis of the camel (*Camelus dromedaries*). *Acta Anat.* 104, 16–22.
433. Ozgel, O., Dursun, N., Cengelci, A., and Ates, S. (2003). Arterial supply of the penis in the New Zealand rabbit (*Oryctolagus cuniculus* L.). *Anat. Histol. Embryol.* 32, 6–8.
434. Pabst, D. A., Rommel, S. A., McLellan, W. A., Williams, T. M., and Rowles, T. K. (1995). Thermoregulation of the intra-abdominal testes of the Bottlenose Dolphin (*Tursiops truncates*) during exercise. *J. Exp. Biol.* 198, 221–226.
435. Pabst, D. A., Rommel, S. A., and McLellan, W. A. (1998). Evolution of thermoregulatory function in Cetacean reproductive systems. In *The Emergence of Whales* (Thewissen, Ed.), pp. 379–396. Plenum Press, New York.
436. Palombi, F., Filippini, A., and Chiarenza, C. (2002). Cell-cell interactions in the local control of seminiferous tubule contractility. *Contraception* 65, 289–291.
437. Pang, S. F., Chow, P. H., and Wong, T. M. (1979). The role of the seminal vesicles, coagulating glands and prostate glands on the fertility and fecundity of mice. *J. Reprod. Fertil.* 56, 129–132.
438. Parkes, A. S. (1966). The testes of certain choeromorpha. In *Comparative Biology of Reproduction in Mammals* (I. W. Rowlands, Ed.), pp. 141–154. Academic Press, Zoological Society of London.
439. Pascual, J. A., Lemmi, C., and Rajfer, J. (1992). Variability of venous anatomy of rat testis: application to experimental testicular surgery. *Microsurgery* 13, 335–337.
440. Patterson, B. D., and Thaeler, C. S. Jr. (1982). The mammalian baculum: hypotheses on the nature of bacular variability. *J. Mamm.* 63, 1–15.
441. Peirce, E. J., and Breed, W. G. (1989). Light microscopical structure of the excurrent ducts and distribution of spermatozoa in the Australian rodents *Pseudomys australis* and *Notomys alexis*. *J. Anat.* 162, 195–213.
442. Peirce, E. J., Moore, H. D. M., Leigh, C. M., and Breed, W. G. (2003). Studies on sperm storage in the vas deferens of the Spinifex Hopping Mouse (*Notomys alexis*). *Reproduction* 125, 233–240.
443. Peitz, B., and Olds-Clarke, P. (1986). Effects of seminal vesicle removal on fertility and uterine sperm motility in the House Mouse. *Biol. Reprod.* 35, 608–617.
444. Pelletier, J. (1986). Contribution of increasing and decreasing daylength to the photoperiodic control of LH secretion in the Ile-de-France ram. *J. Reprod. Fertil.* 77, 505–512.
445. Pennefather, J. N., Lau, W. A. K., Mitchelson, F., and Ventura, S. (2000). The autonomic and sensory innervation of the smooth muscle of the prostate gland: a review of pharmacological and histological studies. *J. Autonom. Pharmacol.* 20, 193–206.
446. Perez-Clavier, R., and Harrison, R. G. (1978). The pattern of lymphatic drainage of the rat testis. *J. Anat.* 127, 93–100.
447. Perez-Clavier, R., Harrison, R. G., and MacMillan, E. W. (1982). The pattern of the lymphatic drainage of the rat epididymis. *J. Anat.* 134, 667–675.
448. Petterborg, L. F., and Reiter, R. J. (1980). Effect of photoperiod and melatonin on testicular development in the white-footed mouse, *Peromyscus leucopus*. *J. Reprod. Fertil.* 60, 209–212.
449. Pinart, E., Bonet, S., Briz, M. D., Pastor, L. M., Sancho, S., Garcia, N., and Badia, E. (2001). Morphologic and histochemical study of blood capillaries in boar testes: effects of abdominal cryptorchidism. *Teratology* 63, 42–51.
450. Piner, J., Sutherland, M., Millar, M., Turner, K., Newall, D., and Sharpe, R. M. (2002). Changes in vascular dynamics of the adult rat testis leading to transient accumulation of seminiferous tubule fluid after administration of a novel 5-hydroxytryptamine (5-HT) agonist. *Reprod. Toxicol.* 16, 141–150.
451. Pollanen, P., and Cooper, T. G. (1995). Vascular permeability to effectors of the immune system in the male reproductive tract at puberty. *J. Reprod. Immunol.* 28, 85–109.
452. Pollanen, P., and Setchell, B. P. (1989). Microvascular permeability to IgG in the rat testis at puberty. *Int. J. Androl.* 12, 206–218.
453. Pollanen, P., Cooper, T. G., Kokk, K., Saari, T., and Setchell, B. P. (1997). Microvascular permeability to the $F(ab')_2$ fragment of IgG in the male reproductive tract at puberty. *J. Reprod. Immunol.* 32, 221–240.
454. Price, D., and Williams-Ashman, H. G. (1961). The accessory reproductive glands of mammals. In *Sex and Internal Secretions* (W. C. Young, Ed.), Vol. I, pp. 366–448. Williams & Wilkins, Baltimore.
455. Prince, F. P. (1992). Ultrastructural evidence of indirect and direct autonomic innervation of human Leydig cells: comparison of neonatal, childhood and pubertal ages. *Cell. Tissue Res.* 269, 383–390.
456. Prince, F. P. (1996). Ultrastructural evidence of adrenergic, as well as cholinergic, nerve varicosities in relation to the lamina propria of the human seminiferous tubules during childhood. *Tissue Cell* 28, 507–513.
457. Properzi, G., Cordeschi, G., and Francavilla, S. (1992). Postnatal development and distribution of peptide-containing nerves in the genital system of the male rat. *Histochemistry* 97, 61–68.
458. Pudney, J. (1976). Seasonal changes in the testis and epididymis of the American Grey Squirrel, *Sciurus carolinensis*. *J. Zool. Lond.* 179, 107–120.
459. Queen, K., Dhabuwala, C. B., and Pierrepoint, C. G. (1981). The effect of the removal of the various accessory sex glands on the fertility of male rats. *J. Reprod. Fertil.* 62, 423–426.
460. Qin, D. N., and Lung, M. A. (2000). Studies on relationship between testicular capsule and sperm transport in rat testis. *Asian J. Androl.* 2, 191–198.
461. Qin, D. N., and Lung, M. A. (2000). Effect of testicular capsulotomy on secretion of testosterone and gonadotrophins in rats. *Asian J. Androl.* 2, 257–261.
462. Qin, D. N., and Lung, M. A. (2001). Effect of testicular capsulotomy on fertility of rats. *Asian J. Androl.* 3, 21–25.
463. Qin, D. N., and Lung, M. A. (2001). Immunohistochemical observation on luteinizing hormone in rat testes before and after testicular capsulotomy. *Asian J. Androl.* 3, 227–230.

464. Qin, D. N., and Lung, M. A. (2002). Morphometric study on Leydig cells in capsulotomized testis of rats. *Asian J. Androl.* 4, 49–53.
465. Racey, P. A. (1974). The reproductive cycle in male Noctule Bats, *Nyctalus noctula*. *J. Reprod. Fertil.* 41, 169–182.
466. Racey, P. A. (1978). Seasonal changes in testosterone levels and androgen-dependent organs in male moles (*Talpa europaea*). *J. Reprod. Fertil.* 52, 195–200.
467. Racey, P. A., and Tam, W. H. (1974). Reproduction in male *Pipistrellus pipistrellus* (Mammalia: Chiroptera). *J. Zool. Lond.* 172, 101–122.
468. Rajasekaran, M., and Monga, M. (1999). Cellular and molecular causes of male infertility in spinal cord injury. *J. Androl.* 20, 326–330.
469. Rauchenwald, M., Steers, W. D., and Desjardins, C. (1995). Efferent innervation of the rat testis. *Biol. Reprod.* 52, 1136–1143.
470. Reddi, A. H., and Prasad, M. R. N. (1968). The reproductive cycle of the male Indian Palm Squirrel, *Funambulus pennanti* Wroughton. *J. Reprod. Fertil.* 17, 235–245.
471. Renfree, M. B., O, W. S., Short, R. V., and Shaw, G. (1996). Sexual differentiation of the urogenital system of the fetal and neonatal Tammar wallaby, *Macropus eugenii*. *Anat. Embryol.* 194, 111–134.
472. Renfree, M. B., Pask, A. J., and Shaw, G. (2001). Sex down under: the differentiation of sexual dimorphisms during marsupial development. *Reprod. Fertil. Dev.* 13, 679–690.
473. Rerkamnuaychoke, W., Kurohmaru, M., and Nishida, T. (1988). The arterial supply of the male reproductive system in the hamster. *Anat. Histol. Embryol.* 17, 301–311.
474. Rerkamnuaychoke, W., Nishida, T., Kurohmaru, M., and Hayashi, Y. (1989). Vascular morphology of the Golden Hamster spermatic cord. *Arch. Histol. Cytol.* 52, 183–190.
475. Rerkamnuaychoke, W., Nishida, T., Kuromahru, M., and Hayashi, Y. (1991). Evidence for a direct arteriovenous connection (A-V shunt) between the testicular artery and pampiniform plexus in the spermatic cord of the Tree Shrew (*Tupaia glis*). *J. Anat.* 178, 1–9.
476. Rerkamnuaychoke, W., Nishida, T., Kurohmaru, M., and Hayashi, Y. (1991). Morphological studies on the vascular architecture in the boar spermatic cord. *J. Vet. Med. Sci.* 53, 233–239.
477. Rerkamnuaychoke, W., Yokota, K., Kurohmaru, M., Hayashi, Y., Watanabe, G., Taya, K., Isomura, G., and Nishida, T. (1999). Morphological features of the spermatic cord in the Musk Shrew (*Suncus murinus*) with special reference to extratesticular Leydig cells. *J. Vet Med. Sci.* 61, 1209–1214.
478. Richardson, E. G. (1977). The biology and evolution of the reproductive cycle of *Miniopterus schreibersii* and *M. australis* (Chiroptera: Vespertilionidae). *J. Zool. Lond.* 183, 353–375.
479. Ricker, D. D. (1998). The autonomic innervation of epididymis: its effects on epididymal function and fertility. *J. Androl.* 19, 1–4.
480. Ricker, D. D., and Chang, T. S. K. (1996). Neuronal input from the inferior mesenteric ganglion (IMG) affects sperm transport within the rat cauda epididymis. *Int. J. Androl.* 19, 371–376.
481. Ricker, D. D., Crone, J. K., Chamness, S. L., Klinefelter, G. R., and Chang, T. S. K. (1997). Partial sympathetic denervation of the rat epididymis permits fertilization but inhibits embryo development. *J. Androl.* 18, 131–138.
482. Rismiller, P. D. (1992). Field observations on Kangaroo Island echidnas (*Tachyglossus aculeatus multiaculeatus)* during the breeding season. In *Platypus and Echidnas* (M. H. Augee, Ed.), pp. 101–105. Royal Zoological Society of NSW, Sydney.
483. Rismiller, P. D. (1993). Overcoming a prickly problem. *Aust. Nat. Hist.* 24, 22–29.
484. Rodger, J. C. (1976). Comparative aspects of the accessory sex glands and seminal biochemistry of mammals. *Comp. Biochem. Physiol.* 55B, 1–8.
485. Rodger, J. C. (1982). The testis and its excurrent ducts in American Caenolestid and Didelphid marsupials. *Am. J. Anat.* 163, 269–282.
486. Rodger, J. C., and Hughes, R. L. (1973). Studies of the accessory glands of male marsupials. *Aust. J. Zool.* 21, 303–320.
487. Rodger, J. C., and White, I. G. (1976). Source of seminal N-acetylglucosamine in Australian marsupials and further studies of free sugars of the marsupial prostate gland. *J. Reprod. Fertil.* 46, 467–469.
488. Rommel, S. A., Pabst, D. A., McLellan, W. A., Mead, J. G., and Potter, C. W. (1992). Anatomical evidence for a countercurrent heat exchanger associated with dolphin testes. *Anat. Rec.* 232, 150–156.
489. Rommel, S. A., Pabst, D. A., McLellan, W. A., Williams, T. M., and Friedl, W. A. (1994). Temperature regulation of the testes of the Bottlenose Dolphin (*Tursiops truncates*): evidence from colonic temperatures. *J. Comp. Physiol. B* 164, 130–134.
490. Rommel, S. A., Early, G. A., Matassa, K. A., Pabst, D. A., and McLellan, W. A. (1995). Venous structures associated with thermoregulation of phocid seal reproductive organs. *Anat. Rec.* 243, 390–402.
491. Rommel, S. A., Pabst, D. A., and McLellan, W. A. (2001). Functional morphology of venous structures associated with male and female reproductive systems in Florida Manatees (*Trichechus manatus latirostris*). *Anat. Rec.* 264, 339–347.
492. Roosen-Runge, E. C., and Holstein, A. F. (1978). The human rete testis. *Cell. Tissue Res.* 189, 409–433.
493. Rose, R. W., Nevison, C. M., and Dixson, A. F. (1997). Testes weight, body weight and mating systems in marsupials and monotremes. *J. Zool. Lond.* 243, 523–531.
494. Rosenior, J., Tung, P. S., and Fritz, I. B. (1987). Biosynthesis and secretion of clusterin by ram rete testis cell-enriched preparations in culture. *Biol. Reprod.* 36, 1313–1320.
495. Rudolfsson, S. H., Johasson, A., Lisbrant, I. F., Wikstrom, P., and Bergh, A. (2003). Localized expression of angiopoietin 1 and 2 may explain unique characteristics of the rat testicular vasculature. *Biol. Reprod.* 69, 1232–1237.
496. Rudolfsson, S. H., Wikstrom, P., Jonsson, A., Collin, O., and Bergh, A. (2004). Hormonal regulation and functional role of vascular endothelial growth factor A in the rat testis. *Biol. Reprod.* 70, 340–347.
497. Russell, S., and Nehra, A. (2003). The physiology of erectile dysfunction. *Herz* 28, 277–283.
498. Saitoh, K., Terada, T., and Hatakeyama, S. (1990). A morphological study of the efferent ducts of the human epididymis. *Int. J. Androl.* 13, 369–376.
499. Sakaguchi, H., Kozuka, M., Hirose, S., Ito, T., and Hagiwara, H. (1992). Properties and localization of endothelin-1-specific receptors in rat testicles. *Am. J. Physiol.* 263, R15–R18.
500. Saleh, A. M. M., Alameldin, M. A., Abdelmoniem, M. E., Hassouna, E. M., and Wrobel, K. H. (2002). Immunohistochemical investigation of the autonomous nerve distribution in the testis of the camel (*Camelus dromedaries*). *Ann. Anat.* 184, 209–220.
501. Saleh, A. M. M., Alameldin, M. A., Abdelmoniem, M. E., Hassouna, E. M., and Wrobel, K. H. (2002). On the intrinsic innervation of the epididymis of the camel (*Camelus dromedaries*). *Ann. Anat.* 184, 305–315.
502. Sampaio, F. J. B., Favorito, L. A., Freitas, M. A., Damaio, R., and Gouveia, E. (1999). Arterial supply of the human fetal testis during its migration. *J. Urol.* 161, 1603–1605.

503. Sanchez, M., Menendez, L., Garcia de Boto, M. J., and Hidalgo, A. (1996). Role of cyclic nucleoides in contraction induced by oxytocin in the testicular capsule of the rat in vitro. *Pharmacology* 53, 296–301.
504. Sato, K., and Kihara, K. (1998). Spinal cord segments controlling the canine vas deferens and differentiation of the primate sympathetic pathways to the vas deferens. *Microsc. Res. Tech.* 42, 390–397.
505. Saypol, D. C. (1981). Varicocele. *J. Androl.* 2, 61–71.
506. Saypol, D. C., Howards, S. S., Turner, T. T., and Miller, E. D. (1981). Influence of surgically induced varicocele on testicular blood flow, temperature and histology in adult rats and dogs. *J. Clin. Invest.* 68, 39–45.
507. Scala, G., de Girolamo, P., Corona, M., and Pelagalli, G. V. (2002). Microvasculature of the buffalo epididymis. *Anat. Rec.* 266, 58–68.
508. Schingnitz, G., and Werner, J. (1986). Significance of scrotal afferents within the general thermoafferent system. *J. Therm. Biol.* 11, 181–189.
509. Schinkel, A. H., Smit, J. J. M., van Tellingene, O., Beijnen, J. H., Wagenaary, E., van Deemter, L., Mol, C. A. A. M., van der Valk, L., Robanus-Maandag, E. C., te Riele, H. P. J., Berns, A. J. M., and Borst, P. (1994). Disruption of the mouse *mdr1a* P-glycoprotein gene leads to a deficiency in the blood-brain barrier and to increased sensitivity to drugs. *Cell* 77, 491–502.
510. Schinkel, A. H., Wagenaar, E., van Deemter, L., Mol, C. A. A. M., and Borst, P. (1995). Absence of the mdr1a P-glycoprotein in mice affects tissue distribution and pharmokinetics of dexamethasone, digoxin, and cyclosporin A. *J. Clin. Invest.* 96, 1698–1705.
511. Schinkel, A. H., Wagenaar, E., Mol, C. A. A. M., and van Deemter, L. (1996). P-glycoprotein in the blood-brain barrier of mice influences the brain penetration and pharmacological activity of many drugs. *J. Clin. Invest.* 97, 2517–2524.
512. Selvage, D. J., Lee, S. Y., Parsons, L. H., Seo, D. O., and Rivier, C. L. (2004). A hypothalamic-testicular neural pathway is influenced by brain catecholamines, but not testicular blood flow. *Endocrinology* 145, 1750–1759.
513. Setchell, B. P. (1970). Testicular blood supply, lymphatic drainage and secretion of fluid. In *The Testis* (A. D. Johnson, W. R. Gomes, and N. L. Vandemark, Eds.), Vol. I, pp. 101–239. Academic Press, New York.
514. Setchell, B. P. (1977). Reproduction in male marsupials. In *The Biology of Mammals* (B. Stonehouse and D. Gilmore, Eds.), pp. 411–457. Macmillan Press, London.
515. Setchell, B. P. (1978). *The Mammalian Testis*. Elek Books, London, and Cornell University Press, Ithaca.
516. Setchell, B. P. (1982). The flow and compostion of lymph from the testes of pigs, with some observations on the effect of raised venous pressure. Comp. Biochem. Physiol. 73A, 201–205.
517. Setchell, B. P. (1985). Spermatogenesis in some insectivores. In *Current Trends in Comparative Endocrinology* (B. Lofts and W. N. Holmes, Eds.), pp. 277–280. Hong Kong University Press, Hong Kong.
518. Setchell, B. P. (1986). The movement of fluids and substances in the testis. *Aust. J. Biol. Sci.* 39, 193–207.
519. Setchell, B. P. (1990). Local control of testicular fluids. *Reprod. Fertil. Dev.* 2, 291–309.
520. Setchell, B. P. (1991). Male reproductive organs and semen. In *Reproduction in Domestic Animals* (P. T. Cupps, Ed.), pp. 221–249. Academic Press, San Diego.
521. Setchell B. P. (1998). Heat and the testis. *J. Reprod. Fertil.* 114, 179–194.
522. Setchell, B. P. (2002). Innervation and vasculature of the excurrent duct system. In *The Epididymis from Molecules to Clinical Practice* (B. Robaire and B. T. Hinton, Eds.), pp. 35–48. Kluwer Academic/Plenum Publishers, New York.
523. Setchell, B. P. (2004). Hormones: what the testis really sees. *Reprod. Fertil. Dev.* 16, 535–545.
524. Setchell, B.P., unpublished observations.
525. Setchell, B. P., and Cox, J. C. (1982). Secretion of free and conjugated steroids by the horse testis into lymph and venous blood. *J. Reprod. Fertil.* Suppl 32, 123–127.
526. Setchell, B. P., and Sharpe, R. M. (1981). Effect of injected human chorionic gonadotrophin on capillary permeability, extracellular fluid volume and the flow of lymph in the testes of rats. *J. Endocrinol.* 91, 245–254.
527. Setchell, B. P., and Palombi, F. (2004). Isolation of endothelial cells from the rat testis, and their effect on testosterone secretion by interstitial cells. Miniposter, Presented at the 13th European Workshop on Molecular & Cellular Endocrinology of the Testis, C6.
528. Setchell, B. P., and Ploen, L. Unpublished results.
529. Setchell, B. P., and Waites, G. M. H. (1969). Pulse attenuation and countercurrent heat exchange in the internal spermatic artery of some Australian marsupials. *J. Reprod Fertil.* 20, 165–169.
530. Setchell, B. P., and Waites, G. M. H. (1970). Changes in the permability of the testicular capillaries and of the "blood-testis barrier" after injection of cadmium chloride in the rat. *J. Endocrinol.* 47, 81–86.
531. Setchell, B. P., Waites, G. M. H., and Till, A. R. (1964). Variations in flow of blood with the epididymis and testis of the sheep and rat. *Nature* 203, 317–318.
532. Setchell, B. P., Waites, G. M. H., and Thorburn, G. D. (1966). Blood flow in the testis of the conscious ram, measured with krypton-85: effects of heat, catecholamines and acetylcholine. *Circ. Res.* 18, 755–765.
533. Setchell, B. P., Laurie, M. S., Flint, A. P. F., and Heap, R. B. (1983). Transport of free and conjugated steroids from the boar testis in lymph, venous blood and rete testis fluid. *J. Endocrinol.* 96, 127–136.
534. Setchell, B. P., Pollanen, P., and Zupp, J. L. (1988). The development of the blood-testis barrier and changes in vascular permeability at puberty in rats. *Int. J. Androl.* 11, 225–233.
535. Setchell, B. P., Locatelli, A., Perreau, C., Pisselet, C., Fontaine, I., Kuntz, D., Saumonde, J., Fontaine, J., and Hochereau-de Reviers, M. T. (1991). Form and function of the Leydig cells in hypophysectomized rams treated with pituitary extract when spermatogenesis is disrupted by heating the testis. *J. Endocrinol.* 131, 101–112.
536. Setchell, B. P., Maddocks, S., Tao, L., and Zupp, J. L. (1993). Vascular functions in the testis and infertility. In *Understanding Male Infertility: Basic and Clinical Approaches* (R. W. Whitcomb and B. R. Zirkin, Eds.), pp. 81–93. Raven Press, New York.
537. Setchell, B. P., Maddocks, S., and Brooks, D. E. (1994). Anatomy, vasculature, innervation, and fluids of the male reproductive tract. In *The Physiology of Reproduction* (E. Knobil and J.D. Neill, Eds.), pp. 1063–1175. Raven Press, New York.
538. Setchell, B. P., Bergh, A., Widmark, A., and Damber, J. E. (1995). Effect of temperature of the testis on vasomotion and blood flow. *Int. J. Androl.* 18, 120–126.
539. Setchell, B. P., Pakarinen, P., and Huhtaniemi, I. (2002). How much LH do the Leydig cells see? *J. Endocrinol.* 175, 375–382.
540. Shabsigh, A., Chang, D. T., Heitjan, D. F., Kiss, A., Olsson, C. A. Puchner, P. J., and Buttyan, R. (1998). Rapid reduction in blood flow to the rat ventral prostate gland after castration: preliminary evidence tht androgens influence prostate size

by regulating blood flow to the prostate gland and prostatic endothelial cell survival. *Prostate* 36, 201–206.

541. Shabsigh, A., Tanji, N., D'Agati, V., Burchardt, T., Burchardt, M., Hayek, O., Shabsigh, R., and Buttyan, R. (1999). Vascular anatomy of the rat ventral prostate. *Anat. Rec.* 256, 403–411.
542. Shono, T., Ramm-Anderson, S., and Hutson, J. M. (1994). Transabdominal testicular descent is really ovarian ascent. *J. Urol.* 152, 781–784.
543. Shono, T., Ramm-Anderson, S., Goh, D. W., and Hutson, J. M. (1994). The effect of flutamide on testicular descent in rats examined by scanning electron microscopy. *J. Pediatr. Surg.* 29, 839–844.
544. Shono, T., Hutson, J. M., Watts, L., Goh, D. W., Momose, Y., Middlesworth, B., Zhou, B., and Ramm-Anderson, S. (1996). Scanning electron microscopy shows inhibited gubernacular development in relation to undescended testes in oestrogen-treated mice. *Int. J. Androl.* 19, 263–270.
545. Shono, T., Suita, S., Kai, H., and Yamaguchi, Y. (2004). The effect of a prenatal androgen disruptor, vinclozolin, on gubernacular migration and testicular descent in rats. *J. Pediatr. Surg.* 39, 213–216.
546. Short, R. V. (1980). The origins of human sexuality. In *Reproduction in Mammals. Book 8. Human Sexuality* (C. R. Austin and R. V. Short, Eds.), pp. 1–33. Cambridge University Press, Cambridge.
547. Short, R. V. (1997). The testis: the witness of the mating system, the site of mutation and the engine of desire. *Acta Paediatr.* Suppl 422, 3–7.
548. Short, R. V., and Mann, T. (1966). The sexual cycle of a seasonally breeding mammal, the roebuck (*Capreolus capreolus*). *J. Reprod. Fertil.* 12, 337–351.
549. Short, R. V., Mann, T., and Hay, M. F. (1967). Male reproductive organs of the African Elephant, *Loxodonta africana. J. Reprod. Fertil.* 13, 517–536.
550. Sienkiewicz, W., Molenaar, G. J., Kaleczyc, J., Falkowski, J., and Lakomy, M. (2000). Has active immunization against gonadotrophin-releasing hormone any effect on testis innervation in the pig? An immunohistochemical study. *Anat. Histol. Embryol.* 29, 247–254.
551. Simonsen, U., Garcia-Sacristan, A., and Prieto, D. (2002). Penile arteries and erection. *J. Vasc. Res.* 39, 283–303.
552. Sjostrand, N. O. (1965). The adrenergic innervation of the vas deferens and the accessory male genital glands. *Acta Physiol. Scand.* Suppl 257, 1–82.
553. Skinner, J. D. (1971). The effect of season on spermatogenesis in some ungulates. *J. Reprod. Fertil.* Suppl 13, 29–37.
554. Skinner, J. D., Van Zyl, J. H. M., and Van Heerden, J. A. H. (1973). The effect of season on reproduction in the Black Wildeseest and Red Hartebeest in South Africa. *J. Reprod. Fertil.* Suppl 19, 101–110.
555. Smith, E. R. (1975). The canine prostate and its secretion. In *Molecular Mechanisms of Gonadal Hormone Action* (J. A. Thomas and R. L. Singhal, Eds.), Vol. I, pp. 167–204. University Park Press, Baltimore.
556. Smith, E. R., and Lebeaux, M. I. (1970). The mediation of the canine prostatic secretion provoked by hypogastric nerve stimulation. *Invest. Urol.* 7, 313–318.
557. Smith, M. J. V. (1966). The lymphatics of the prostate. *Invest. Urol.* 3, 439–444.
558. Smith, R. F. C. (1969). Studies on the marsupial glider, *Schinobates volans* (Kerr). I. Reproduction. *Aust. J. Zool.* 17, 625–636.
559. Sonnenberg-Riethmacher, E., Walter, B., Riethmacher, D., Godecke, S., and Birchmeier, C. (1996). The c-*ros* tyrosine linase receptor controls regionalization and differentiation of epithelial cells in the epididymis. *Genes Dev.* 10, 1184–1193.
560. Stewart, P. A. (2000). Endothelial vesicles in the blood-brain barrier: are they related to permeability? *Cell. Mol. Neurobiol.* 20, 149–163.
561. Stockley, P. (2002). Sperm competition risk and male genital anatomy: comparative evidence for reduced duration of female sexual receptivity in primates with penile spines. *Evol. Ecol.* 16, 123–137.
562. Stoffel, M. H., and Friess, A. E. (1994). Morphological characteristics of boar efferent ductules and epididymal duct. *Microsc. Res. Tech.* 29, 411–431.
563. Stoffel, M.H., and Friess, A.E. (1997). The junctions between the efferent ductules and epididymal duct in the boar. *Andrologia* 29, 283–285.
564. Stoffel, M. H., Kohler, T., Friess, A. E., and Zimmermann, W. (1990). Microvasculature of the epididymis of the boar. *Cell. Tissue Res.* 259, 495–501.
565. Stoffel, M. H., Friess, A. E., and Kohler, T. (1991). Efferent ductules of the boar—a morphological study. *Acta Anat.* 142, 272–280.
566. Suburo, A. M., Chiocchio, S. R., Canto Soler, M. V., Nieponice, A., and Tramezzani, J. H. (2002). Peptidergic innervation of blood vessels and interstitial cells in the testis of the cat. *J. Androl.* 23, 121–134.
567. Sugimura, Y., Cunha, G. R., and Donjacour, A. A. (1986). Morphogenesis of ductal networks in the mouse prostate. *Biol. Reprod.* 34, 961–971.
568. Suzuki, F. (1982). Microvasculature of the mouse testis and excurrent duct system. *Am. J. Anat.* 163, 309–325.
569. Suzuki, F., and Nagano, T. (1986). Microvasculature of the human testis and excurrent duct system. Resin-casting and electron-microscopic studies. *Cell. Tissue Res.* 243, 79–89.
570. Suzuki, F., and Racey, P. A. (1978). The organization of testicular interstitial tissue and changes in the fine structure of the Leydig cells of European moles (*Talpa europaea*) throughout the year. *J. Reprod. Fertil.* 52, 189–194.
571. Suzuki, F., and Racey, P. A. (1984). Light and electron microscopical observations on the male excurrent duct system of the Common Shrew (*Sorex araneus*). *J. Reprod. Fertil.* 70, 419–428.
572. Sweeney, T. E., Rozum, J. S., Desjardins, C., and Gore, R. W. (1991). Microvascular pressure distribution in the hamster testis. *Am. J. Physiol.* 260, H1581–H1589.
573. Sweeney, T. E., Rozum, J. S., and Gore, R. W. (1995). Alteration of testicular microvascular pressures during venous pressure elevation. *Am. J. Physiol.* 269, H37–H45.
574. Sweet, G. (1907). Contributions to our knowledge of the anatomy of *Notoryctes typhlops*. Parts IV and V. The skin, hair, and reproductive organs of *Notoryctes*. *Q. J. Micros. Sci.* 51, 325–344.
575. Taggart, D. A., and Temple-Smith, P. D. (1989). Structural features of the epididymis in a dasyurid marsupial (*Antechinus stuartii*). *Cell. Tissue Res.* 258, 203–210.
576. Taggart, D. A., and Temple-Smith, P. D. (1994). Comparative studies of epididymal morphology and sperm distribution in dasyurid marsupials during the breeding season. *J. Zool. Lond.* 232, 365–381.
577. Taggart, D. A., Breed, W. G., Temple-Smith, P. D., Purvis, A., and Shimmin, G. (1998). Sperm competition and mating strategies in marsupials and monotremes. In *Sperm Competition and Sexual Selection* (T.R. Birkhead and A.P. Moller, Eds.), pp. 667–752. Academic Press, London.
578. Taggart, D. A., Shimmin, G. A., Dickman, C. R., and Breed, W. G. (2003). Reproductive biology of carnivorous marsupials: clues to the likelihood of sperm competition. In *Predators with Pouches* (M. Jones, C. Dickman, and M. Archer, Eds.), pp. 358–375.

579. Tait, A. J., and Johnson, E. (1982). Spermatogenesis in the Grey Squirrel (*Sciurus carolinensis*) and changes during sexual regression. *J. Reprod. Fertil.* 65, 53–58.

580. Takayama, H. (1986). Ultrastructure of testicular capillaries as a permeability barrier (in Japanese). *Nippon Hinyokika Gakkai Zasshi* 77, 1840–1850.

581. Takayama, H., and Tomoyoshi, T. (1981). Microvascular architecture of rat and human testes. *Invest. Urol.* 18, 341–344.

582. Takihara, H., Sakatoku, J., and Cockett, A. T. (1991). The pathophysiology of varicocele in male infertility. *Fertil. Steril.* 55, 861–868.

583. Tamura, R., Mizumura, K., Sato, J., Kitoh, J., and Kumazawa, T. (1996). Segmental distribution of afferent neurons innervating the canine testis. *J. Auton. Nerv. Syst.* 58, 101–107.

584. Tamura, R., Hanesch, U., Schmidt, R. F., Kumazawa, T., and Mizumura, K. (1997). Calcitonin gene-related peptide- and substance P-like immunoreactive fibers in the spermatic nerve and testis of the dog. *Neurosci. Lett.* 235, 113–116.

585. Tao, L., Zupp, J. L., and Setchell, B. P. (2000). Effect of efferent duct ligation on the function of the blood-testis barrier in rats. *J. Reprod. Fertil.* 120, 13–18.

586. Tarhan, S., Gumus, B., Gunduz, I., Ayyildiz, V., and Goktan, C. (2003). Effect of varicocele on testicular artery blood flow in men. *Scand J. Urol. Nephrol.* 37, 38–42

587. Taylor, J. M., and Horner, B. E. (1970). Observations on reproduction in *Leggadina* (Rodentia: Muridae). *J. Mamm.* 51, 10–17.

588. Temple-Smith, P., and Grant, T. (2001). Uncertain breeding: a short history of reproduction in monotremes. *Reprod. Fertil. Dev.* 13, 487–497.

589. Temple-Smith, P. D. (1984). Reproductive structures and strategies in male possums and gliders. In *Possums and Gliders* (A. P. Smith and I. D. Hume, Eds.), pp. 89–106. Australian Mammal Society, Sydney.

590. Todhunter, R., and Gemmell, R. T. (1987). Seasonal changes in the reproductive tract of the male marsupial bandicoot, *Isoodon macrourus*. *J. Anat.* 154, 173–186.

591. Tran, D., Picard, J. Y., Vigier, B., Berger, R., and Josso, N. (1986). Persistence of Müllerian ducts in male rabbits passively immunized against bovine anti-Müllerian hormone during fetal life. *Dev. Biol.* 116, 160–167.

592. Tripiciano, A., Filippini, A., Giustiniani, Q., and Palombi, F. (1996). Direct visualization of rat peritubular myoid cell contraction in response to endothelin. *Biol. Reprod.* 55, 25–31.

593. Tripiciano, A., Palombi, F., Ziparo, E., and Filippini, A. (1997). Dual control of seminiferous tubule contractility mediated by ET_A and ET_B endothelin receptor subtypes. *FASEB J.* 11, 276–286.

594. Tripiciano, A., Filippini, A., Ballarini, F., and Palombi, F. (1998). Contractile response of peritubular myoid cells to prostaglandin F2α. *Mol. Cell Endocrinol.* 138, 143–150.

595. Tung, P. S., Rosenior, J., and Fritz, I. B. (1987). Isolation and culture of ram rete testis epithelial cells: structural and biochemical characteristics. *Biol. Reprod.* 36, 1297–1312.

596. Turek, F. W., Elliott, J. A., Alvis, J. D., and Menaker M. (1975). Effect of prolonged exposure to nonstimulatory photoperiods on the activity of the neuroendocrine-testicular axis of Golden Hamsters. *Biol. Reprod.* 13, 475–481.

597. Turnbull, K. E., Mattner, P. E., and Hughes, R. L. (1981). Testicular descent in the marsupial *Trichosurus vulpecula* (Kerr). *Aust. J. Zool.* 29, 189–198.

598. Turner, T. T., and Lopez, T. J. (1990). Testicular blood flow in peripubertal and older rats with unilateral experimental varicocele and investigation into the mechanism of the bilateral response to the unilateral lesion. *J. Urol.* 144, 1018–1021.

599. Turner, T. T., Brown, K. J., and Spann, C. L. (1993). Testicular intravascular volume and microvessel mitotic activity: effect of experimental varicocele. *J. Androl.* 14, 180–186.

600. Turner, T. T., Caplis, L., and Miller, D. W. (1996). Testicular microvascular blood flow: alteration after Leydig cell eradication and ischemia but not experimental varicocele. *J. Androl.* 17, 239–248.

601. Tyndale-Biscoe, C. H., and Renfree, M. B. (1987). *Reproductive Physiology of Marsupials*. Cambridge University Press, Cambridge.

602. Uhr, M., Holsboer, F., and Muller, M. B. (2002). Penetration of endogenous steroid hormones corticosterone, cortisol, aldosterone and progesterone into the brain is enhanced in mice deficient for both *mdr1a* and *mdr1b* P-glycoproteins. *J. Neuroendocrinol.*14, 753–759.

603. Ungefroren, H., Ergun, S., Krull, N. B., and Holstein, A. F. (1995). Expression of the small proteoglycans biglycan and decorin in the adult human testis. *Biol. Reprod.* 52, 1095–1105.

604. Vaalasti, A., Alho, A. M., Tainio, H., and Hervonen, A. (1986). The effect of sympathetic denervation with 6-hydroxydopamine on the ventral prostate of the rat. *Acta Histochem.* 79, 49–54.

605. van der Schoot, P. (1993). Doubts about the "first phase of testis descent" in the rat as a valid concept. *Anat. Embryol.* 187, 203–208.

606. van der Schoot, P. (1995). Studies on the fetal development of the gubernaculum in Cetacea. *Anat. Rec.* 243, 449–460.

607. van der Schoot, P. (1996). Foetal genital development in *Hyrax capensis*, a species with primary testicondia: proposal for the evolution of Hunter's gubernaculum. *Anat. Rec.* 244, 386–401.

608. van der Schoot, P. (1996). Towards a rational terminology in the study of the gubernaculum testis: arguments in support of the notion that the cremasteric sac should be considered the gubernaculum in postnatal rats and other mammals. *J. Anat.* 189, 97–108.

609. van der Schoot, P., and Elger, W. (1992). Androgen-induced prevention of the outgrowth of cranial gonadal suspensory ligaments in fetal rats. *J. Androl.* 13, 534–542.

610. van der Schoot, P., and Emmen, J. M. A. (1996). Development, structure and function of the cranial suspensory ligaments of the mammalian gonads in a cross-species perspective; their possible role in effecting disturbed testicular descent. *Hum. Reprod. Update* 2, 399–418.

611. van der Schoot, P., Vigier, B., Prepin, J., Perchellet, J. P., and Gittenberger-de Groot, A. (1995). Development of the gubernaculum and processus vaginalis in freemartinism: further evidence in support of a specific fetal testis hormone governing male-specific gubernacular development. *Anat. Rec.* 241, 211–224.

612. van der Schoot, P., Payne, A. P., and Kersten, W. (1999). Sex difference in target seeking behavior of developing cremaster muscles and the resulting first visible sign of somatic sexual differentiation in marsupial mammals. *Anat. Rec.* 255, 130–141.

613. Ventura, S., and Burnstock, G. (1996). Variation in nitric oxide synthase-immunoreactive nerve fibers with age and along the length of the vas deferens in the rat. *Cell. Tissue Res.* 285, 427–434.

614. Ventura, S., and Pennefather, J. N. (1994). α2-Adrenoreceptor binding sites vary along the length of the male reproductive tract: a possible basis for the regional variation in response to field stimulation. *Eur. J. Pharmacol.* 254, 167–173.

615. Ventura, S., Pennefather, J. N., and Mitchelson, F. (2002). Cholinergic innervation and function in the prostate gland. *Pharmacol. Ther.* 94, 93–112.

616. Verstoppen, G. R., and Steeno, O. P. (1977). Varicocele and the pathogenesis of the associated subfertility: a review of the various theories I. Varicocelogenesis. *Andrologia* 9, 133–140.
617. Verstoppen, G. R., and Steeno, O. P. (1977). Varicocele and the pathogenesis of the associated subfertility: a review of the various theories II. Results of surgery. *Andrologia* 9, 293–305.
618. Viotto, M. J. S., Orsi, A. M., Vicentini, C. A., Mello Dias, S., and Gregorio, E. A. (1993). Ultrastructure of the rete testis in the cat (*Felis domestica,* L.). *Anat. Histol. Embryol.* 22, 114–122.
619. Voss, R. (1979). Male accessory glands and the evolution of copulatory plugs in rodents. *Occas. Pap. Mus. Zool. Univ. Mich.* 689, 1–27.
620. Voss, R. S., and Linzey, A. V. (1981). Comparative gross morphology of male accessory glands among neotropical Muridae (Mammalia: Rodentia) with comments on systematic implications. *Misc. Pub. Mus. Zool. Univ. Mich.* 159, 1–41.
621. Waites, G. M. H. (1962). The effect of heating the scrotum of the ram on respiration and body temperature. *Q. J. Exp. Physiol.* 47, 314–323.
622. Waites, G. M. H. (1991). Thermoregulation of the scrotum and testis: studies in animals and significance for man. *Adv. Exp. Med. Biol.* 286, 9–17.
623. Waites, G. M. H., and Moule, G. R. (1960). Blood pressure in the internal spermatic artery of the ram. *J. Reprod. Fertil.* 1, 223–229.
624. Waites, G. M. H., and Moule, G. R. (1961). Relation of vascular heat exchange to temperature regulation in the testis of the ram. *J. Reprod. Fertil.* 2, 213–224.
625. Waites, G. M. H., and Setchell, B. P. (1966). Changes in blood flow and vascular permeability of the testis, epididymis and accessory reproductive organs of the rat after administraion of cadmium chloride. *J. Endocrinol.* 34, 329–342.
626. Waites, G. M. H., and Setchell, B. P. (1969). Physiology of the testis, epididymis and scrotum. In *Recent Advances in Reproductive Physiology* (A. McLaren, Ed.), pp. 1–63. Logos Press, London.
627. Waites, G. M. H., and Setchell, B. P. (1990). Physiology of the mammalian testis. In *Marshall's Physiology of Reproduction* (G. E. Lamming, Ed.), Vol. 2, pp. 1–105. Churchill Livingstone, Edinburgh.
628. Waites, G. M. H., and Setchell, B. P., quoted by Waites, G. M. H. (1970). Temperature regulation and the testis. In *The Testis* (A. D. Johnson, W. R. Gomes, and N. L. Vandemark, Eds.), Vol. I, pp. 241–279. Academic Press, New York.
629. Waites, G. M. H., and Voglmayr, J. K. (1962). Apocrine sweat glands of the scrotum of the ram. *Nature* 196, 965–967.
630. Waites, G. M. H., and Voglmayr, J. K. (1963). The functional activity and control of the apocrine sweat glands of the scrotum of the ram. *Aust. J. Agric. Res.* 14, 839–851.
631. Waites, G. M. H., Setchell, B. P., and Quinlan, D. (1973). The effect of local heating of the scrotum, testes and epididymides on cardiac output and regional blood flow. *J. Reprod. Fertil.* 34, 41–49.
632. Waites, G. M. H., Archer, V., and Langford, G. A. (1975). Regional sensitivity of the testicular artery to noradrenaline in ram, rabbit, rat and boar. *J. Reprod. Fertil.* 45, 159–163.
633. Walton, A. (1960). Copulation and natural insemination. In *Marshall's Physiology of Reproduction* (A. S. Parkes, Ed.), Vol. 1, Part 2, pp. 130–160. Longmans, Green & Co, London.
634. Walton, K. C. (1976). The reproductive cycle in the male polecat *Putorius putorius* in Britain. *Notes from the Mammal Society* No. 33, 498–503.
635. Wang, J., Galil, K. A. A., and Setchell, B. P. (1983). Changes in testicular blood flow and testosterone production during aspermatogenesis after irradiation. *J. Endocrinol.* 98, 35–46.
636. Wang, J., Gu, C. H., Tao, L., Wu, X. L., and Qiu, J. P. (1986). Electrolyte composition of rete testis fluid and cauda epididymal plasma and spermatozoa from rats following gossypol treatment. *Andrologia* 18, 43–49.
637. Wang, J. M., McKenna, K. E., and Lee, C. (1991). Determination of prostatic secretion in rats: effect of neurotransmitters and testosterone. *Prostate* 18, 289–301.
638. Wang, J. M., McKenna, K. E., McVary, K. T., and Lee, C. (1991). Requirement of innervation for maintenance of structural and functional integrity in the rat prostate. *Biol. Reprod.* 44, 1171–1176.
639. Wang, S., Jones, R. C., and Clulow, J. (1994). Surface area of apical and basolateral plasmalemma of epithelial cells of ductuli efferentes testis of the rat. *Cell. Tissue Res.* 276, 581–586.
640. Watanabe, H., Shima, M., Kojima, M., and Ohe, H. (1988). Dynamic study of nervous control on prostatic contraction and fluid excretion in the dog. *J. Urol.* 140, 1567–1570.
641. Wechselberger, C., Puglisi, R., Engel, E., Leperdinger, G., Boitani, C., and Kreil, G. (1999). The mammalian homologues of frog Bv8 are mainly expressed in spermatocytes. *FEBS Lett.* 462, 177–181.
642. Weerasooriya, T. R., and Yamamoto, T. (1985). Three-dimensional organization of the vasculature of the rat spermatic cord and testis. *Cell. Tissue Res.* 241, 317–323.
643. Wensing, C. J. G. (1986). Testicular descent in the rat and a comparison of this process in the rat with that in the pig. *Anat. Rec.* 214, 154–160.
644. Wensing, C. J. G. (1988). The embryology of testicular descent. *Horm. Res.* 30, 144–152.
645. Wensing, G. J. G., and Colenbrander, B. (1986). Normal and abnormal testicular descent. *Oxf. Rev. Reprod. Biol.* 8, 130–164.
646. Werdelin, L., and Nilsonne, A. (1999). The evolution of the scrotum and testicular descent in mammals: a phylogenetic view. *J. Theor. Biol.* 196, 61–72.
647. White, R. M., Kennaway, D. J., and Seamark, R. F. (1996). Reproductive seasonality of the Bush Rat (*Rattus fuscipes greyi*) in South Australia. *Wildlife Res.* 23, 317–336.
648. Whitsett, J. M., Noden, F., Cherry, J., and Lawton, A. D. (1984). Effect of transitional photoperiods on testicular development and puberty in male Deer Mice. *J. Reprod. Fertil.* 72, 277–286.
649. Widmark, A., Damber, J. E., and Bergh, A. (1986). Testicular vascular resistance in the rat after intratesticular injection of an LRH-agonist. *Int. J. Androl.* 9, 416–423.
650. Widmark, A., Damber, J. E., and Bergh, A. (1987). Effects of oestradiol on testicular microcirculation in rats. *J. Endocrinol.* 115, 489–495.
651. Widmark, A., Damber, J. E., and Bergh, A. (1989). High and low doses of luteinizing hormone induce different changes in testicular microcirculation. *Acta Endocrinol.* 121, 621–627.
652. Widmark, A., Damber, J. E., and Bergh, A., (1991). Arginine-vasopressin induced changes in testicular blood flow. *Int. J. Androl.* 14, 58–65.
653. Wilkinson, G. S., and McCracken, G. F. (2003). Bats and balls: sexual selection and sperm competition in the Chiroptera. In *Bat Ecology* (T. H. Kunz and M. B. Fenton, Eds.), pp. 128–155. The University of Chicago Press, Chicago.
654. Williams, M. P. L., and Hutson, J. M. (1991). The phylogeny of testicular descent. *Pediatr. Surg. Int.* 6, 162–166.
655. Williams, M. P. L., and Hutson, J. M. (1991). The history of ideas about testicular descent. *Pediatr. Surg. Int.* 6, 180–184.

656. Wilson, B. A., and Bourne, A. R. (1984). Reproduction in the male dasyurid *Antechinus minimus maritimus* (Marsupialia: Dasyuridae). *Aust. J. Zool.* 32, 311–318.
657. Wilson, J. D., George, F. W., and Renfree, M. B. (1995). The endocrine role in mammalian sexual differentiation. *Rec. Prog. Horm. Res.* 50, 349–364.
658. Wing, T. Y., and Christensen, A. K. (1982). Morphometric studies on rat seminiferous tubules. *Am. J. Anat.* 165, 13–25.
659. Wislocki, G. B. (1933). Location of the testes and body temperature in mammals. *Q. Rev. Biol.* 8, 385–396.
660. Wojciechowski, A. P. (1992). Evolutionary aspects of mammalian secondary sexual characteristics. *J. Theor. Biol.* 155, 271–272.
661. Wolff, J. O., Edge, W. D., and Bentley, R. (1994). Reproductive and behavioural biology of the Gray-tailed Vole. *J. Mamm.* 75, 873–879
662. Woodall, P. F. (1995). The penis of elephant-shrews (Mammalia: Macroscelididae). *J. Zool. Lond.* 237, 399–410.
663. Woodall, P. F. (1997). The male reproductive system and the phylogeny of elephant-shrews (Macroscelidea). *Mammal Rev.* 25, 87–93.
664. Woodall, P. F., and Johnstone, I. P. (1988). Dimensions and allometry of testes, epididymides and spermatozoa in the domestic dog (*Canis familiaris*). *J. Reprod. Fertil.* 82, 603–609.
665. Woodall, P. F., and Skinner, J. D. (1989). Seasonality of reproduction in male Rock Elephant Shrews, *Elephantulus myurus*. *J. Zool. Lond.* 217, 203–212.
666. Woodall, P. F., Pavlov, P., and Tolley, L. K. (1993). Comparative dimensions of testes, epididymides and spermatozoa of Australian dingoes (*Canis familiaris dingo*) and domestic dogs (*Canis familiaris familiaris*): some effects of domestication. *Aust. J. Zool.* 41, 133–140.
667. Woodland, W. (1903). On the phylogenetic cause of the transposition of the testes in mammalia: with remarks on the evolution of the diaphragm and the metanephric kidney. *Proc. Zool Soc.* 1, 319–340.
668. Woolley, P. (1975). The seminiferous tubules in dasyurid marsupials. *J. Reprod. Fertil.* 45, 255–261.
669. Woolley, P. A. (1982). Phallic morphology of the Australian species of *Antechinus* (Dasyuridae, Marsupialia): a new taxonomic tool? In *Carnivorous Marsupials* (M. Archer, Ed.), pp. 767–781. Royal Zoological Society NSW, Sydney.
670. Woolley, P. A. (1987). The seminiferous tubules, rete testis and efferent ducts in Didelphid, Caenolestid and Microbiotheriid marsupials. In *Possums and Opossums* (M. Archer, Ed.), pp. 217–227. Royal Zoological Society of NSW, Sydney.
671. Woolley, P. A. (1987). The seminiferous tubules, rete testis and efferent ducts in Peramelid marsupials. In *Bandicoots and Bilbies* (J. H. Seebeck, P. R. Brown, R. L. Wallis, and C. M. Kemper, Eds.), pp. 229–234. Royal Zoological Society of NSW, Sydney.
672. Woolley, P. A., and Webb, S. J. (1977). The penis of dasyurid marsupials. In *The Biology of Mammals* (B. Stonehouse and D. Gilmore, Eds.), pp. 307–323. Macmillan Press, London.
673. Woolley, P. A., and Scarlett, G. (1984). Observations on the reproductive anatomy of male *Tarsipes rostratus* (Marsupialia: Tarsipedidae). In *Possums and Gliders* (A. P. Smith and I. D. Hume, Eds.), pp. 445–450. Australian Mammal Society, Sydney.
674. Woolley, P. A., and Vanderveen, A. E. (2002). The rete testis of Acrobatid and Burramyid marsupials. *Aust. J. Zool.* 50, 237–247.
675. Wrobel, K. H. (1972). Zur Morphologie der Ductuli efferentes des Bullen. *Z. Zellforsch.* 135, 129–148.
676. Wrobel, K. H. (2000). Morphogenesis of the bovine rete testis: the intratesticular rete and its connection to the seminiferous tubules. *Anat. Embryol.* 202, 475–490.
677. Wrobel, K. H., and Abu-Ghali, N. (1997). Autonomic innervation of the bovine testis. *Acta Anat.* 160, 1–14.
678. Wrobel, K. H., and Brandl, B. (1998). The autonomous innervation of the porcine testis in the period from birth to adulthood. *Ann. Anat.* 180, 145–156.
679. Wrobel, K. H., and Gurtler, A. (2001). The nerve distribution in the testis of the cat. *Ann. Anat.* 183, 297–308.
680. Wrobel, K. H., and Moustafa, M. N. K. (2000). On the innervation of the donkey testis. *Ann. Anat.* 182, 13–22.
681. Wrobel, K. H., and Schenk, E. (2003). Immunohistochemical investigations of the autonomous innervation of the cervine testis. *Ann. Anat.* 185, 493–506.
682. Wyrost, P., Radek, J., and Radek, T. (1990). Morphology and development of bovine testicular vein during the fetal and neonatal period. *Pol. Arch. Weter.* 30, 17–38.
683. Young, K. A., and Nelson, R. J. (2000). Short photoperiods reduce vascular endothelial growth factor in the testes of *Peromyscus leucopus*. *Am. J. Physiol.* 279, R1132–R1137.
684. Zhang, M., Mendoza, A. S., Schramm, U., and Kuhnel, W. (1996). Electron microscopic observation of the lymphatics of the mammalian testis. *Ann. Anat.* 178, 461–465.
685. Zhou, B. C., Chiocchio, S. R., Suburo, A. M., and Tramezzani, J. H. (1995). Monoaminergic and peptidergic contributions of the superior and the inferior spermatic nerves to the innervation of the testis in the rat. *J. Androl.* 16, 248–258.
686. Zhou, Q., Nie, R., Prins, G. S. Saunders, P. T. K., Katzelellenbogen, B. S., and Hess, R. A. (2002). Localization of androgen and estrogen receptors in adult male mouse reproductive tract. *J. Androl.* 23, 870–881.
687. Zimmermann, S., Steding, G., Emmen, J. M. A., Brinkmann, A. O., Nayernia, K., Holstein, A. F., Engel, W., and Adham, I. M. (1999). Targeted disruption of the *Insl3* gene causes bilateral cryptorchidism. *Mol. Endocrinol.* 13, 681–691.
688. Zucker, I., and Morin, L. P. (1977). Photoperiodic influences on testicular regression, recrudescence and the induction of scotorefractoriness in male Golden Hamsters. *Biol. Reprod.* 17, 493–498.

Knobil and Neill's Physiology of Reproduction, Third Edition
edited by Jimmy D. Neill,
Elsevier © 2006

CHAPTER 18

Cytology of the Testis and Intrinsic Control Mechanisms

J. B. Kerr, K. L. Loveland, M. K. O'Bryan, and D. M. de Kretser

Historical Aspects, 827
General Structure of the Testis, 828
Intratesticular Ducts, 828
Transitional Distal Segment of Seminiferous Tubule, 828
Rete Testis, 828
Spermatogenesis, 829
Developmental Considerations, 831
Spermatogonia, 836
Spermatocytes, 839
Spermiogenesis, 843
Sertoli Cell, 864
Sertoli Cell Shape, 864
Ultrastructural Features, 866
Surface Specializations, 876
Changes Associated with Seasonal Breeding and Age, 886
Response to Injury, 887
Peritubular Tissue, 889
Cycle of the Seminiferous Epithelium, 889
Duration of Spermatogenesis, 892
Coordination within the Seminiferous Epithelium, 894
Spermatogonial Stem Cell Renewal, 895
Spermatogonial Wave, 898
Germ Cell Degeneration, 899
Leydig Cells, 900
Historical Background, 901
Organization of Intertubular Tissue, 901
Ultrastructure, 905
Nucleus, 905
Cytoplasm, 907
Life History, 914
Response to Testicular Damage, 918
References, 920

HISTORICAL ASPECTS

Although the effects of castration were recognized in antiquity, probably dating back to the Neolithic Age (ca. 7000 B.C.) (1), the association of the testis with fertility did not emerge until the 17th century. Reasonably accurate diagrammatic representations of testicular anatomy can be attributed to Aristotle in 400 B.C. (2), but the link between the testes and fertility did not emerge at that time. de Graaf (3) recorded accurately the general structure and functions of the testis, and his observations were soon followed by the observation of spermatozoa in seminal fluid by van Leeuwenhoek in 1667 (4). The link between the seminiferous tubules and the production of spermatozoa did not fully emerge until von Koelliker (5) concluded that spermatozoa were formed by cellular development within the tubules. The cellular changes resulting in the production of spermatozoa thus constitute spermatogenesis. Although descriptions of the hormonal effects caused by castration were known in Aristotle's time, experimental proof emerged when Berthold (6) showed that the loss of comb size and crowing that occurred in roosters after castration could be reversed by testicular transplantation. The site of production of the masculinizing factor was attributed to the Leydig cells by Bouin and Ancel (7), but the isolation of testosterone did not occur until 1935 by David et al. (8). The improvements in microscopy in the late 19th century expanded our knowledge of the light

Monash Institute of Reproduction and Development, Monash University, Victoria, Australia.

microscopic features of spermatogenesis and the identification of chromosomes, and the processes of mitosis and meiosis greatly improved our understanding of gamete production in the male. The results of those studies provide the foundation of our knowledge.

GENERAL STRUCTURE OF THE TESTIS

The testis is surrounded by a dense connective tissue capsule, the tunica albuginea, which is covered on its anterior and lateral aspects with the remnants of the process vaginalis, forming the visceral and parietal layers of the tunica vaginalis (9). From the internal surface of the tunica albuginea, connective tissue septa extend posteriorly toward a region of the testis termed the mediastinum. This area consists of connective tissue within which an anastomotic network of ducts can be identified, the rete testis.

The tunica albuginea is formed by dense connective tissue within which smooth muscle fibers can be found (10), the latter being responsible for the capacity of the capsule to contract in response to pharmacological stimuli (11). The inner surface of the tunica albuginea is apposed to loose highly vascular connective tissue sometimes termed the tunica vasculosa. The degree of lobulation of the testis varies between species, and within these lobules lie the seminiferous tubules within which spermatogenesis occurs. The tubules extend as loops from the mediastinum testis, both ends of each loop communicating via single straight tubules, the tubuli recti (12). In most mammals this simple arrangement is obscured in the adult testis, because the tubule forming each loop becomes extensively folded, thereby extending its surface area (13–15). Each of the 200–300 lobules in the normal adult human testis contain one to three loops of seminiferous tubules each about 70–80 cm in length. The total length of seminiferous tubules ranges between 450 and 650 m, or 22–32 m/cm^3 of testicular tissue. Comparable data for other species are as follows: mouse, 16–26 m/cm^3 (16–18); rat, 11 m/cm^3 (19); rabbit, 20–30 m/cm^3 (20); rhesus monkey, 30–55 m/cm^3 (21,22); and cynomolgus monkey, 18–24 m/cm^3 (23,24). The extensive lengthening of these loops that occurs during development results principally from the mitotic activity of immature Sertoli cells. Relative testis size among primates, expressed as a percentage of body weight, reveals some surprising differences due to relative spermatogenic efficiency, daily sperm production, phylogenetic lineages, and patterns of mating behavior. In general, body mass is not a predictor of testis size, as can be seen from the following examples: for orangutans and gorillas, 0.05% to 0.02%; langurs and humans, 0.06% to 0.1%; gibbons, 0.1%; chimpanzee, 0.27%; slender loris, 0.7%; macaques, 0.5% to 0.8%; and the grey mouse lemur, 1.2% to 5.6% (25–29).

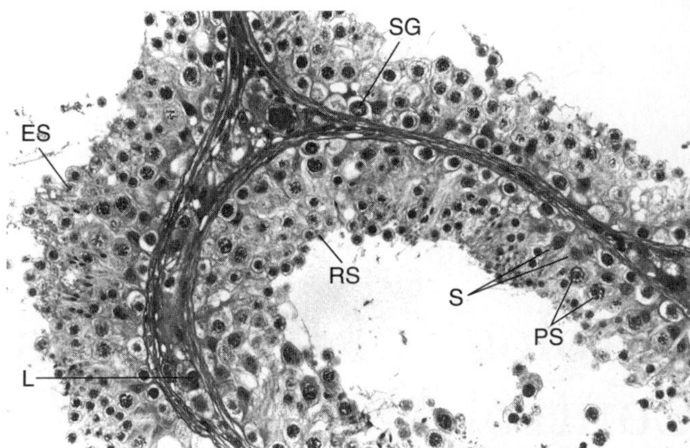

FIG. 1. This light micrograph of the normal human testis illustrates spermatogonia (SG), primary spermatocytes (PS), round spermatids (RS), and elongating spermatids (ES). Note Sertoli cells (S), and Leydig cells (L).

The organization of the intertubular tissue varies dramatically between species (30), but it contains the blood vessels, lymphatics, and nerve fibers (see later discussion). The Leydig cells are scattered in groups in the intertubular tissue in relation to the vasculature and the lamina propria of the seminiferous tubules, the outer layers of which consist of modified smooth muscle cells termed myoid cells (Fig. 1).

INTRATESTICULAR DUCTS

Transitional Distal Segment of Seminiferous Tubule

The segment of the seminiferous tubule that establishes continuity with the rete testis is lined by an epithelium devoid of germ cells. The segment, termed the transitional distal seminiferous segment (Fig. 2), narrows, forming an “epithelial plug” that projects slightly into the tubuli recti, which has a wider lumen. The epithelium consists of Sertoli cells, which contain more extensive rough endoplasmic reticulum and lipid inclusions but less smooth endoplasmic reticulum. Prominent bundles of filaments are present in the cells and may confer some structural rigidity to this region, which may function as a plug to regulate the movement of cells and fluid into the rete testis (14). Dym (31) suggested that the slope of the apices of the cells toward the rete testis made it unlikely that reflux could occur into the seminiferous tubules.

Rete Testis

This anastomotic series of ducts, into which the transitional distal segment of the seminiferous tubules

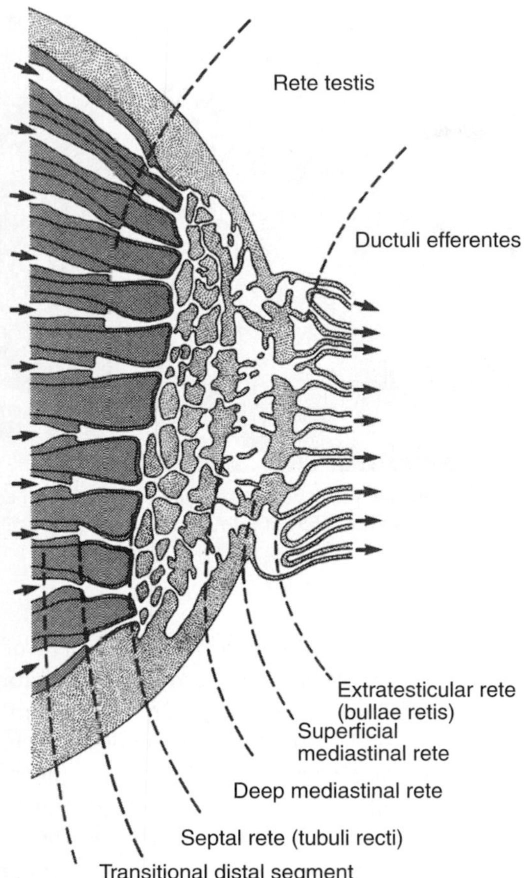

FIG. 2. The method by which the seminiferous tubules terminate in the rete testis is shown. The zones of the rete testis are denoted. [From (1112), with permission.]

opens, drains at their cranial pole via the ductuli efferentes to form the duct of the epididymis. It lies in the mediastinum of the testis parallel to the axis of the epididymis and has been divided into several zones (Fig. 2) by Roosen-Runge and Holstein (32): (a) septal rete, (b) mediastinal rete, and (c) extratesticular rete. The septal rete consists of the zone of the straight ducts, the tubule recti, which drain the seminiferous tubules. The mediastinal rete can be subdivided further into a deep zone consisting of an anastomotic maze of tubules draining the septal rete and opening into a superficial zone, a series of relatively wide longitudinal channels emptying into the extratesticular rete or bullae retia. The latter is characterized by macroscopically visible dilated spaces drained by the ductuli efferentes.

The region of the mediastinal rete is characterized by an irregular network of cylindrical strands, the chordae retia, which range in diameter from 5 to 40 μm (32). The thinner cords are avascular, but each contains a fibroelastic matrix containing myofibroblastic cellular elements, the myoid cells showing similar features to those surrounding the seminiferous tubules. It is postulated that their strut-like features may prevent overdistension of the thin-walled rete channels and, together with the contractile properties of the myoid elements, may provide a mechanism for raising intrarete pressure, thereby forcing fluid into the extratesticular ducts. Two distinct cell types are present, the squamous cells, which line most of the rete, and the prismatic cells, which occur in clusters at bends and corners in the channels (33). The squamous cells are studded with short microvilli and contain a long single cilium, cytoplasm-containing lipid inclusions, glycogen, and a few areas of smooth endoplasmic reticulum. The prismatic cells stain lightly, contain an irregular nucleus with fine granular chromatin, and also contain a single cilium. Their cytoplasmic organelles are polarized, with prominent basal collections of lipid and glycogen and a supranuclear Golgi complex, the surface showing a few thick microvilli that are shorter than the squamous cells. External to the epithelium is a connective tissue layer containing fibroblasts and smooth muscle cells and bundles of collagen and elastin fibers. The extratesticular rete is characterized by vesicular dilatations that are visible macroscopically and act as the vestibules for the excurrent duct system, which extends as the ductuli efferentes.

SPERMATOGENESIS

The sequence of cytological events that result in the formation of mature spermatozoa from precursor cells is known as spermatogenesis (Figs. 3 and 4). In many mammals this process takes place within the seminiferous tubule throughout the reproductive life span of the male. In some species it is interrupted or subdivided into a series of distinct phases based on environmental cues that are transduced into hormonal signals stimulating or inhibiting spermatogenesis

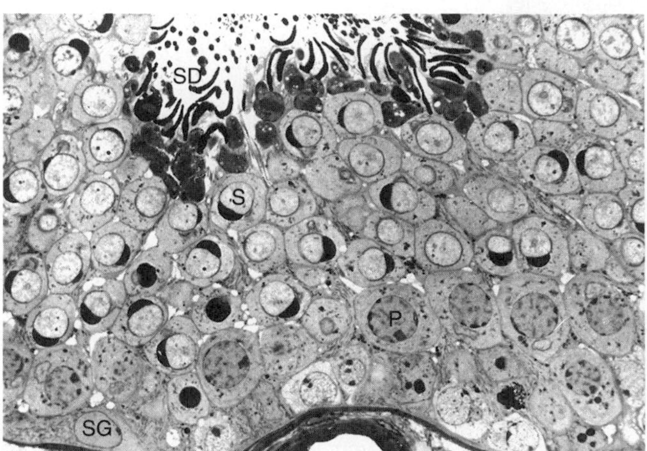

FIG. 3. Guinea pig seminiferous epithelium showing spermatogonia (SG), primary spermatocytes (P), round spermatids (S), and mature spermatids (SD).

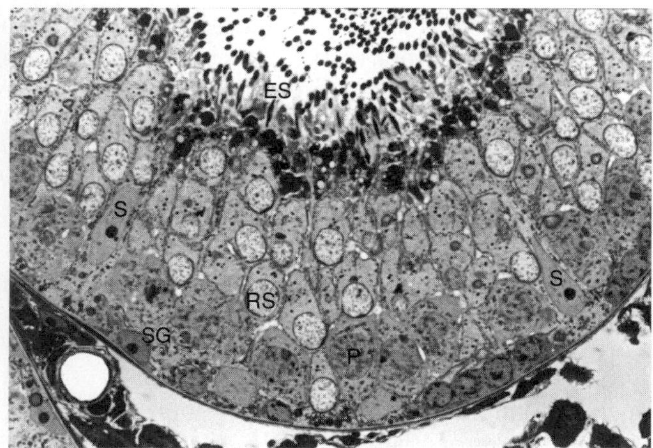

FIG. 4. Rat seminiferous epithelium showing Sertoli cells (S), spermatogonia (SG), primary spermatocytes (P), and round (RS) and elongated spermatids (ES).

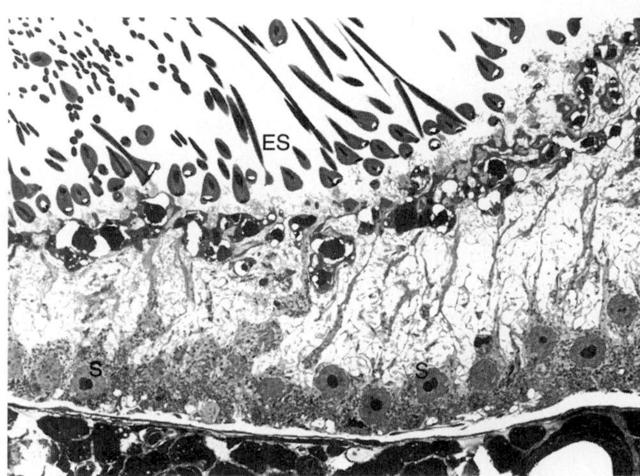

FIG. 6. *Antechinus* seminiferous epithelium in July. Sertoli cells (S) and elongated spermatids (ES) are noted, but other germ cell types are not present.

(e.g., seasonal breeders) (34). In others, unique processes occur such that a single wave of spermatogenic development is seen within the testis, subsequent to which the animal is sterile; such a process occurs in marsupial mice, *Antechinus strudii* and *swainsonii*, wherein the stem cells, after a burst of mitotic activity, differentiate and proceed through the spermatogenic process, resulting in a single wave of reproductive activity (35). Thus, it is possible to see within a seminiferous tubule spermatids of varying maturity without any other basally placed germ cells (Figs. 5, 6, and 7).

Spermatozoa can be viewed as the "secretory product" of the spermatogenic process. To enable a continuous production of "cells" as a product, the process of spermatogenesis must involve a continuous replication of stem cells to produce cohorts of cells that can proceed through the subsequent changes. Furthermore, because the nucleus of the sperm fuses with that of the oocyte to form the zygote, a reduction of the number of chromosomes to the haploid state must occur in gametogenesis such that the diploid state is restored on syngamy. There are four major elements that together constitute spermatogenesis: (a) stem cell renewal by the process of mitosis, (b) stem cell amplification by mitosis and differentiation, (c) reduction of chromosomal number by meiosis, and (d) the transformation of a conventional cell into the complex structure of the spermatozoon by a series of changes involving no further cell division; this represent a metamorphic process termed spermiogenesis. The stem cells for the spermatogenic process are

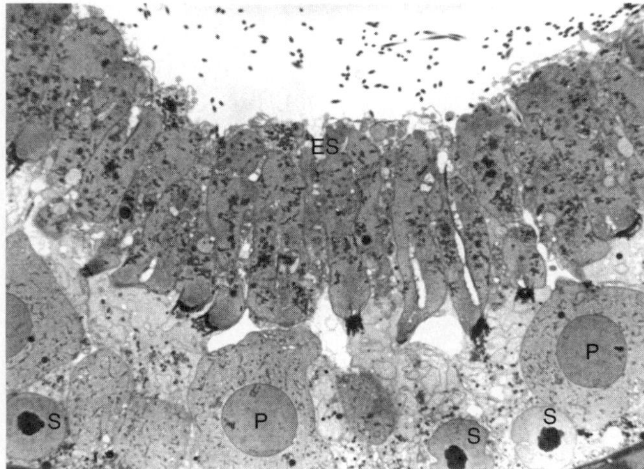

FIG. 5. *Antechinus* seminiferous epithelium in late June showing Sertoli cells (S), pachytene primary spermatocytes (P), and elongated spermatids (ES). Spermatogonia and early spermatocytes not visible.

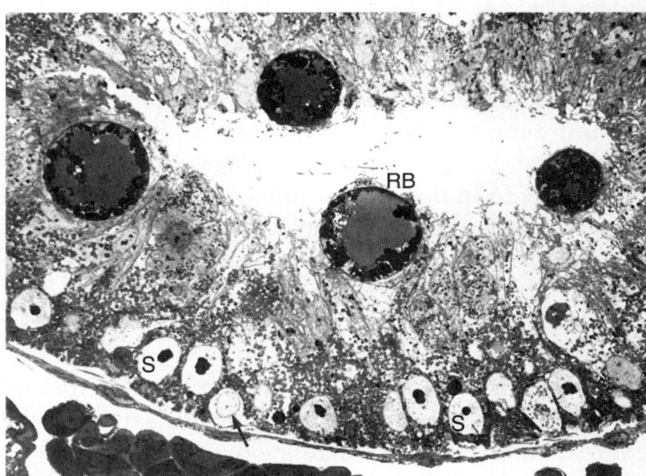

FIG. 7. *Antechinus* seminiferous epithelium in August at the time of breeding. Sertoli cells (S) and spermatid residual bodies (RB) are noted, together with a single basally located spermatogonium (*arrow*).

termed spermatogonial stem cells, and these undergo infrequent mitotic division (Figs. 1 and 3). They are conventionally recognized as single cells randomly arranged along the basement membrane. In response to differentiation cues that are just being defined (see section on Spermatogonial Stem Cell Renewal), they can undergo several rounds of mitosis without completing cytokinesis, beginning the pathway by which groups of spermatogonia then proceed to enter meiosis and first are termed primary spermatocytes and then secondary spermatocytes. The latter divides to form spermatids, haploid cells that are transformed during spermatogenesis into spermatozoa.

The conditions under which spermatogenesis is successfully completed are relatively specialized and necessitate the creation of a unique environment within the seminiferous tubule. This is achieved by the organization of the non–germ cell or somatic elements within the tubules, namely the Sertoli cells (Fig. 8), which form the blood–testis barrier [see below (36)].

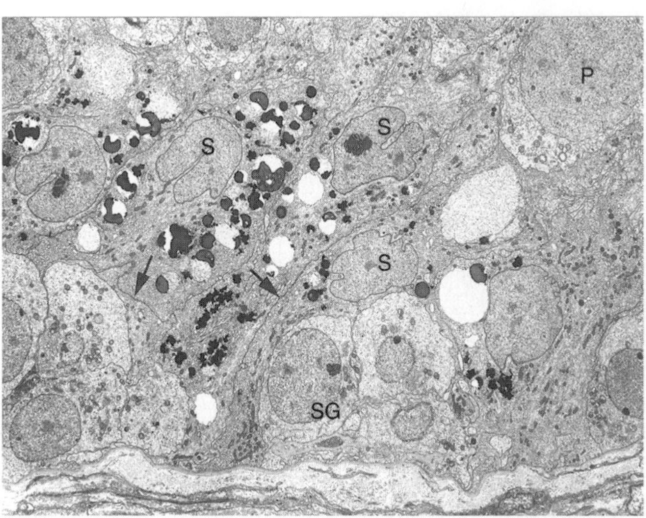

FIG. 9. Ultrastructure of human seminiferous epithelium showing Sertoli cell nuclei (S), spermatogonia (SG), pachytene primary spermatocyte (P), and inter-Sertoli cell junctional complexes (*arrows*).

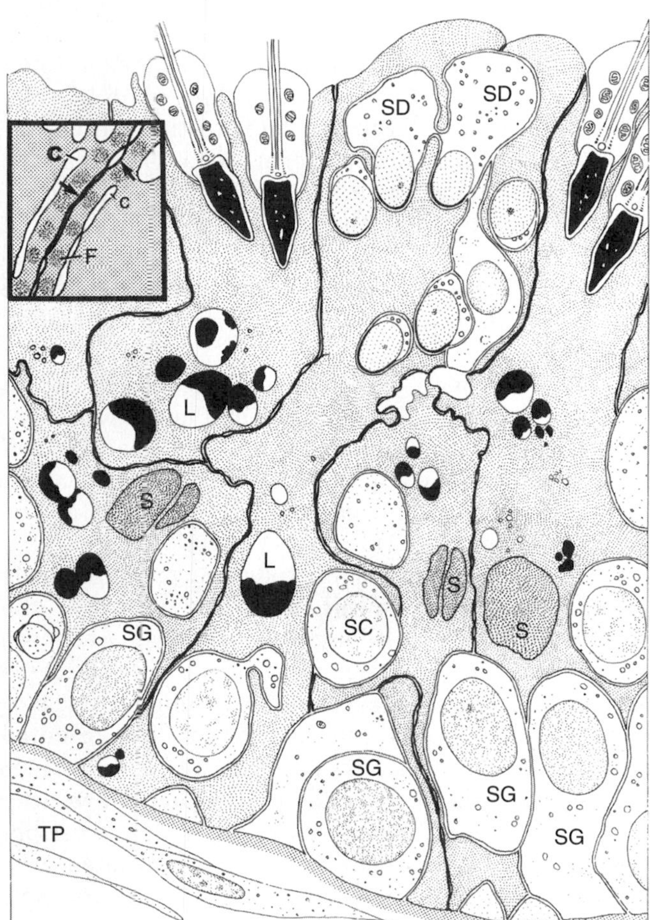

FIG. 8. The relationships between germ cells and Sertoli cells are shown for the human seminiferous epithelium. Note spermatogonia (SG), primary spermatocytes (SC), spermatids (SD), Sertoli cells (S), and their cytoplasm containing lipids (L). The specialized inter-Sertoli cell junctions are shown by a *heavy line* and diagrammatically illustrated at higher magnification in the *inset*. C, cisternae; F, fibrils.

Furthermore, the unusual cellular nature of the product of spermatogenesis, namely spermatozoa, creates a specific requirement within the epithelium that supports the migration of a population of cells that proliferate at the base of the tubule and move progressively toward the lumen of the tubule as the cells differentiate (Figs. 9 and 10). Again, this is achieved by the nature of the nondividing stable population of Sertoli cells found in the adult testis.

Developmental Considerations

Although a detailed development of the testis is the subject of Chapter 6, a brief outline helps to understand the relationships of the two populations of cells within the seminiferous epithelium, the germ cells and the Sertoli cells. The development of the testis is associated with the formation of a series of sex cords, arranged as a series of C-shaped arches running at right angles to the long axis of the testis, each end of the arch connecting with developing rete testis (12,13). The cords are composed of two types of cells, the supporting cells and the primordial germ cells (Fig. 11). With further development, the orientation of the plane of the cords perpendicular to the longitudinal axis of the testis changes, and elongation is achieved by infolding of the cords to form a series of convolutions, mostly in a craniocaudal orientation (12).

The mechanisms by which these cords arise from their mesenchymal precursors are complex and are known to rely on signals from the somatic cells which rely on the expression of *Sry*, the male-determining factor encoded on the Y chromosome (37–40). A cascade

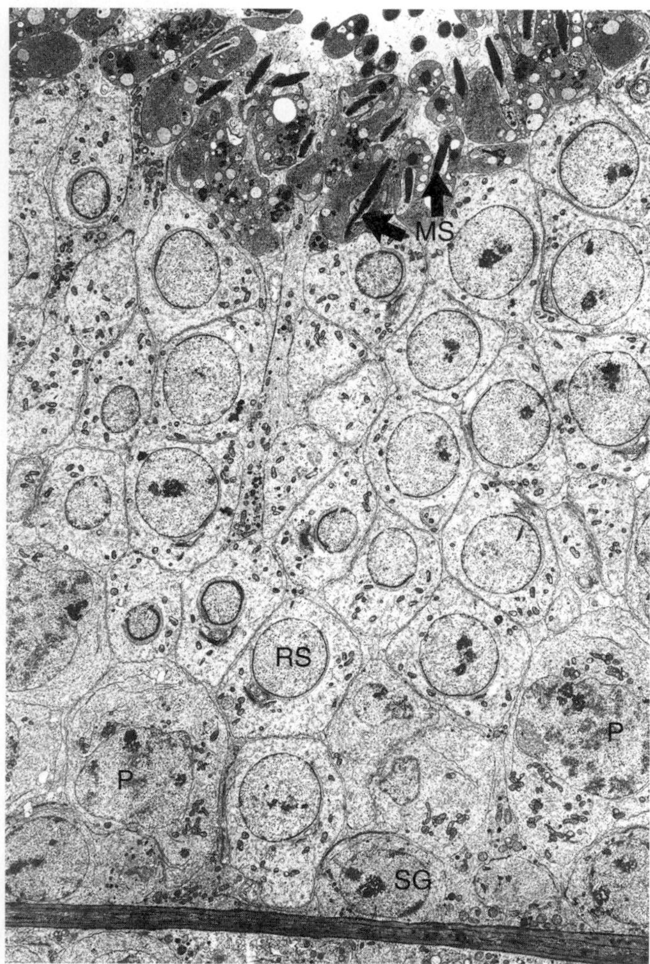

FIG. 10. Mouse seminiferous epithelium showing spermatogonia (SG), pachytene primary spermatocytes (P), and round (RS) and mature (MS) spermatids.

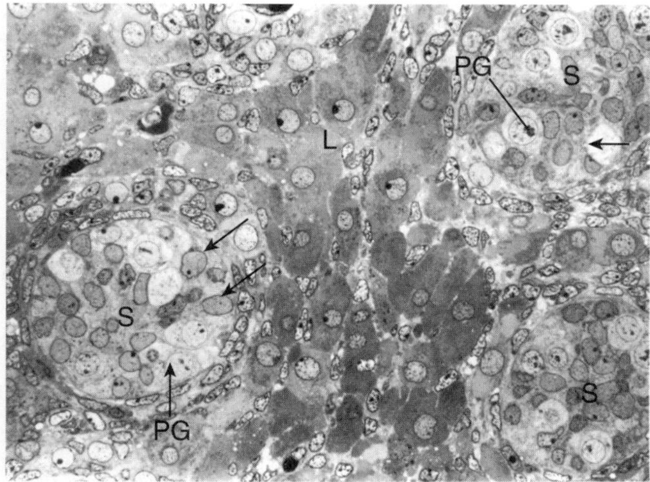

FIG. 11. Leydig cells (L) seminiferous cords (S) from a 16-week human fetus. Note primordial germ cells (PG) and immature Sertoli cells (*arrows*).

of transcription factor activity drives expression of growth factors and proteins, including Müllerian Inhibiting Substance (also known as anti-Müllerian hormone), fibroblast growth factor-9, prostaglandin D_2, and platelet-derived growth factor, which direct development of the somatic and germ cells into the testicular pathway (41–44).

The Sertoli cells and the germ cells arise from distinct cell populations in the embryo (45–49). The original studies of Clermont and Perey (50) provided strong evidence that the Sertoli cells are formed by differentiation of the supporting cells, and it is now understood that the Sertoli cells are the site of expression of Sry between embryonic day 10.5 (E10.5) and E12.5 (51). Reduced levels or altered timing of *Sry* expression can lead to male-to-female sex reversal, highlighting the dominant role of Sertoli cell function in establishment of the testis. Although Sertoli cells appear to be present at the site of the gonadal ridge at the time of sex determination, some of these cells are recruited into the developing murine testis from the coelomic epithelium during a brief (approximately 2-hour) interval (52), acquiring the phenotype of first *Sry* and then *Sox9* expression once in the gonad (53) as they continue to proliferate. The supporting cells in the fetal testis greatly outnumber the primordial germ cells. Though the proportion changes as a result of mitotic multiplication of the primordial germ cells up to about E13.5 in the mouse and E15.5 in the rat, the number of supporting cells remains considerably greater than the number of germ cells at birth in most mammalian species (50).

Concordant with gonadal masculinization by Sertoli cell *Sry* expression, proliferation, and migration, there is a period of peritubular myoid and vascular cell migration from the adjacent mesonephros that is also required for cord and vascular development as the testis develops (54,55).

Before birth, the Sertoli cells are predominantly arranged adjacent to the boundary tissue of the seminiferous cords, although some are displaced more centrally by the close packing of cells within the cords (45,56–64). Primitive Sertoli cells exhibit a conical or polygonal shape, with their cytoplasm often oriented radially within the seminiferous cord (Fig. 11). Their nuclei are variable in shape and seldom show deep indentations. A nucleolus is often present, and peripheral clumps of heterochromatin are associated with the nuclear membrane. The cytoplasm of Sertoli cells in the fetus is unremarkable, but tubular membranes of endoplasmic reticulum are well represented, often bearing variable amounts of ribosomes (56,65). For this reason, these organelles are referred to as a transitional form between conventional smooth and rough endoplasmic reticulum.

A gradual expansion in the diameter of seminiferous cords during fetal life is a reflection of the changing internal organization caused by proliferation of the Sertoli cells and the gonocytes (65). This pattern continues after birth. For instance, the number of Sertoli cells increases from 1.5 ± 0.4 million/testis at birth to 30 ± 2.5 million/testis at 20 days (66). As the gonocytes proliferate and differentiate into spermatogonia, they often form groups of cells that are accommodated in recesses provided by the Sertoli cells. This allows the immature germ cells to gain attachment to the basement membrane of the seminiferous cords, a location that, in postnatal life, becomes their exclusive domain. Morphometric studies and experiments evaluating tritiated thymidine incorporation have provided data on the proliferative behavior of Sertoli cells in the fetus (57,58,65) showing that the number of fetal Sertoli cells increases steadily until shortly before birth, and thereafter their proliferative capacity declines to produce a stable population of nondividing Sertoli cells. Sertoli cells cease proliferating on postnatal days 10–12 in the mouse (50,67–72) and on postnatal days 14–16 in the rat (73–76). In contrast to rodents, higher primate species show two periods of Sertoli cell expansion between birth and puberty. The first expansion in rhesus monkeys occurs up to 5 months of age where Sertoli cells increase fourfold and then slowly expand up to puberty (77). During puberty a second phase of Sertoli cell expansion is initiated as spermatogenesis is activated and cell numbers are further elevated sixfold before proliferation ceases (76–78). In humans a similar increase in Sertoli cell numbers occurs during infancy followed by another doubling during puberty (76,78,79).

A major factor controlling Sertoli cell proliferation is follicle-stimulating hormone (FSH) in the fetus, given the known trophic actions of FSH on the seminiferous epithelium in postnatal life (80). The pituitary–testicular axis appears to be activated by day 17 of fetal life in the rat, because at or before this time the hypothalamus contains gonadotropin-releasing hormone (81) and FSH and luteinizing hormone (LH) are detectable in the pituitary (81–83). Further, the maximum FSH binding to fetal testes occurs at precisely the time of maximum proliferation of Sertoli cells (84). Moreover, addition of FSH to cultures of Sertoli cells derived from postnatal testes results in increased mitotic activity (52,85).

The role of FSH, acting via cAMP, in the control of fetal Sertoli cell proliferation was confirmed because removal of the fetal pituitary or treatment with an antiserum to FSH produced a dramatic reduction in Sertoli cell mitosis (58). Local interactions within the fetal testis may be involved in Sertoli cell proliferation before birth because maximal proliferation shortly before parturition is coincident with the attainment of maximum fetal Leydig cells numbers (86,87). Levels of circulating testosterone (88), the activity of Leydig cell steroidogenic enzymes (89), the production of testicular testosterone, and total steroid content per Leydig cell are greatest during this period (87,90,91). These data raise the possibility that a paracrine relationship exists between fetal Sertoli and Leydig cells and emphasize that, in addition to the demonstrated role of FSH, other factors may be of importance in influencing the expansion of the Sertoli cell population.

Many in vivo and in vitro studies have illustrated the trophic action of FSH on Sertoli cells in postnatal life (75,79,89,92–97). In situations of chronic FSH deprivation, such as mice lacking the FSH-β subunit or FSH receptor genes, adult Sertoli cell numbers are lower than normal. Although spermatogenesis does proceed to completion in these animals, their lowered sperm output illustrates the vital role of FSH in promoting Sertoli cell development and function (98–100). In addition to FSH, however, local production of growth factors appears to influence a second phase of Sertoli cell proliferation. Studies of rat testis fragments and of isolated Sertoli cells in culture have identified an interval from about days 7–11 postpartum when Sertoli cell proliferation is uniquely enhanced by the combination of the transforming growth factor superfamily ligand, activin, with FSH but not by either factor alone (92,96,97). Sertoli cells themselves appear to undergo distinct changes in their synthesis of ligands and receptors during the first wave of spermatogenesis, the functional impact of which appears to be growth in both somatic and germ cell populations (95,101–103).

The attainment of adequate numbers of Sertoli cells achieved during their proliferative phase in the fetal and neonatal growth periods influences the subsequent spermatogenic potential of the mature testis. Administration of the antimitotic agent cytosine arabinoside to neonatal rats partly arrests Sertoli cell proliferation, and the reduced number of Sertoli cells is accompanied by a significant decline in spermatid production by the developing testis (104). Conversely, experimental induction of transient hypothyroidism in neonatal rats extends Sertoli cell proliferation (up to 30 days postnatally) but suppresses spermatogenesis up to 60 days of age (77,105). However subsequently in these rats, the mature testes contain approximately twice the normal number of Sertoli cells (106), and at 160 days of age, testis weight is increased by 80% and daily sperm production by 140% (105,107,108). Paradoxically, circulating gonadotropin and testosterone levels were similar or lowered compared with age-matched control animals, yet thyroid-stimulating hormone levels were markedly elevated coincident with the expansion of Sertoli cell numbers. These findings highlight the importance of nongonadotropic factors in establishing the developmental capacity of the Sertoli cell population, which ultimately affects the spermatogenic potential of the fully mature testis. This concept has been extended by recent studies using recombinant inbred mouse strains that suggest that at least two autosomal genes regulate the number of Sertoli cells and associated testis size (109).

During postnatal testicular growth, the immature Sertoli cells continue to proliferate, albeit at a steadily declining rate, until the adult population is established. The appearance of pachytene primary spermatocytes in the rat testis is associated with this restriction of Sertoli cell division, which, according to numerous studies, marks the establishment of the stable Sertoli cell population present throughout adult life (110–112). With the initiation of spermatogenesis soon after birth in rodents and at various later times in domestic, ruminant, and primate species, the immature Sertoli cells undergo morphological maturation to attain the adult-type structural features seen in the mature testis (112–116). These changes include a large increase in cell size, the movement of the nucleus to a basal position within the cell, the development of a complex nuclear shape and a tripartite nucleolus, and the proliferation of organelles, particularly the smooth endoplasmic reticulum.

Several studies of Sertoli cell maturation have focused on the morphological differentiation of the blood–testis barrier. Light microscopic analysis indicated that the barrier, detected by the admittance or exclusion of acridine dyes, appears at the onset of spermatogenesis (117). Electron microscopy (113,116–121) confirmed that electron-opaque tracers freely penetrate into the seminiferous cords of newborn animals but are subsequently restricted from entry at the time inter-Sertoli cell junctional complexes appear. In the immature rat, the barrier forms during the 16th to 19th postnatal day, which is after the cessation of Sertoli cell mitoses and the commencement of the meiotic maturation of germ cells. Because pachytene primary spermatocytes are formed before the effective establishment of the blood–testis barrier (116), it seems that the onset of meiotic maturation process during the first wave of spermatogenesis is independent of the formation of an adluminal epithelial compartment.

The factors controlling the formation of the blood–testis barrier are not fully known, although the possible involvement of gonadotropins has been proposed. Certainly the appearance of inter-Sertoli cell junctions is not dependent on spermatogenesis, because the elimination of germ cells from the testis (80,122–124) cannot prevent development of the junctions. A similar conclusion can be drawn from studies of the permeability of the Sertoli cell junctional complexes in seasonally breeding species such as the mink (125), in which the barrier undergoes structural decay and reformation in association with respective phases of testicular regression and recrudescence. The fact that germ cells survive and are capable of development in the absence of a functional permeability barrier suggests that the tightness or leakiness of the junctional complexes is related more to the fluid secretory properties of the Sertoli cell than to the extent of germ cell development. However, the appearance of the blood–testis barrier in immature rats was delayed when rats were treated daily from birth with clomiphene citrate or estradiol benzoate, agents that are thought to suppress gonadotropin secretion (118). This finding has remained difficult to evaluate, because measurements of gonadotropin levels were not available and the possible direct effects of clomiphene or estradiol on the testis were not considered. It was concluded that the formation of inter-Sertoli cell junctions was not directly dependent on gonadotropins.

Similar views were expressed when it was found that long-term hypophysectomy failed to disrupt the blood–testis barrier in rats (126–129). In the human testis, however, it has been shown (130) that the formation of inter-Sertoli cell junctions in men with hypogonadotropic hypogonadism is related to the administration of exogenous gonadotropins. The morphological changes that appear in the Sertoli cells during sexual maturation require further study with particular emphasis on the role of the gonadotropic and steroid hormones.

The primordial germ cells originate outside the embryo proper in response to cues from neighboring somatic cells; in the mouse, this occurs at about E6–E6.5. Clark and Eddy (131) noted that their fine-structural features resembled underlying mesodermal cells rather than the surrounding endoderm. Local production of bone morphogenetic protein 4 (BMP4) and BMP8b is required for germ cell allocation (132–134), and their action appears to be restricted to a discrete number of cells by the secretion of BMP antagonists from adjacent parts of the embryo (132). The primordial germ cells migrate from endoderm and dorsal yolk sac epithelium via the dorsal mesentery into the gonadal ridge (135,136). They are morphologically recognizable at E7.0 as large round cells with a large round nucleus. The presence of tissue nonspecific alkaline phosphatase on the plasma membrane (137) provides a marker that has been used to track these cells as they move from their extragonadal site through the embryo hindgut to the gonadal ridge, arriving at the morphologically indifferent gonad by E11.0 [reviewed in (138)]. Their migration and ongoing proliferation is mediated by interactions of the tyrosine kinase receptor, c-*kit*, and its ligand, variably called stem cell factor or mast cell growth factor (139,140); mutations in genes encoding either one of these result in reduced numbers of germ cells arriving at the gonadal ridge and an increased proportion of them migrating to other parts of the embryo (141). Governance of primordial germ cell migration by cues from their extracellular matrix has also been demonstrated, with the cells undergoing progressive changes in adherence to discrete matrix molecules before, during, and after their migratory period (142). A "highway" of laminin has been visualized along their migratory pathway, and upon arrival at the genital ridge, these germ cells become associated with the precursors of the Sertoli cells to form the primitive testicular cords that are in turn surrounded by laminin and other components of the extracellular matrix (143). By E11.5, signals from the somatic cell compartment of the male gonad trigger primordial germ cell commitment to form gonocytes (and not oogonia) at this time (43). Evidence exists that if the structure of the cords is disrupted or if male germ cells lodge in aberrant sites such as the adrenal gland (144), they commence meiosis during fetal life—a characteristic of germ cells in the developing ovary (145).

Subsequently, the primordial germ cells undergo a defined period of mitotic cell division, during which daughter cells remain connected by intercellular bridges (146,147). In the rat, the period of cell division occurs for approximately 48 hours from days 14 to 16 of fetal life, and because the daughter cells differ slightly in structure from the primordial germ cells, they have been called M-prospermatogonia (148). Hilscher and colleagues (148) identified yet another period of fetal mitotic activity that gave rise to T-prospermatogonia, in turn giving rise to the adult A-spermatogonial stage. In the mouse testis, a simple classification scheme is commonly applied in which the proliferating and migrating primordial germ cells arriving in the testis are termed gonocytes, as they become committed to the male pathway of differentiation (43). The gonocytes proliferate until approximately E13.5 and then enter mitotic arrest within the center of the developing seminiferous cord, surrounded by the Sertoli cells. Within a day after birth, the gonocyte reenter the cell cycle and migrate to contact the basement membrane, transforming into the spermatogonial stem cell and differentiating spermatogonia populations. In other species there is general agreement that a period of mitotic activity occurs within the prenatal testis that increases germ cell numbers, but they still remain the minority cell population within the seminiferous cords, usually lying in a central position. The numbers of gonocytes and spermatogonia decrease in fetal life as a result of a degenerative process that has been estimated to reduce numbers between 30% and 40% (149). The degenerating cells are phagocytosed by the immature Sertoli cells.

Gondos (150) suggested a simpler terminology, dividing the germ cells in the fetal testis into (a) primitive germ cells, which are part of the undifferentiated gonad; (b) gonocytes, when the germ cells are located within the seminiferous cords in a central position; and (c) spermatogonia, when they move to the periphery of the tubule. The primordial germ cells are relatively large rounded cells with an irregular horseshoe-shaped nucleus and filamentous centrally placed nucleolus (131). The mitochondria are large, rounded, and contain few cristae. There is sparse endoplasmic reticulum, plentiful free ribosomes, and characteristic membrane-bounded granules with a central dense core separated from the membrane by a flocculent zone (131). In the human, the gonocytes have similar features with globular mitochondria whose cristae are dilated (151). Apart from perinuclear smooth-membraned vesicles, there is paucity of other organelles. Gondos and colleagues (146,151) noted that processes extend from the gonocytes as they migrate toward the basement membrane of the cords to become spermatogonia. The processes contain microtubules, presumably aiding the migratory process. In the testes of newborn rats, gonocyte proliferation occurs before their movement toward the basement membrane of the seminiferous cord. Because both events can occur in vitro in the absence of serum- and hormone-supplemented media, it seems likely that local intratesticular factors regulate the initial maturation of these cells (152). More recently, this migration was shown to involve the c-kit and N-CAM receptor proteins on the germ cell surface (153), and the formation of pseudopods to enable migration has been positively correlated with acquisition of stem cell properties at the onset of the first wave of spermatogenesis (154).

The pattern of development in the sex cords after birth varies considerably in mammals depending on the time span between birth and the acquisition of sexual maturity. In species such as the rat, spermatogenesis effectively commences at birth because the duration of spermatogenesis in this species is 49 days (155) and spermatozoa are present in the testis at about 50 days (156). Hence, the seminiferous cords undergo rapid development after birth, and this first wave of spermatogenesis results in the sequential production of each germ cell stage present in spermatogenesis. In other species such as humans, there is an extensive prepubertal period during which the testes show little change from their appearance at birth. Recent detailed morphometric studies of testes from children dying suddenly have shown that little change occurs until 7 to 9 years, after which mitotic activity of the gonocytes occurs, populating the base of the seminiferous tubules with spermatogonia in numbers equal to those of the Sertoli cells (157). The spermatogonia subsequently undergo the spermatogenic process, with spermatozoa first being released into the lumen between 11 and 13 years.

Spermatogonia

The cells that divide by mitosis and constitute the pool of cells from which meiosis and spermatogenesis proceed are termed spermatogonia. Although we now have a good understanding of their differentiation pathway in the scheme of spermatogenesis [reviewed by de Rooij and Russell (158)], the mechanisms by which their cell fate decisions are driven are not yet fully understood. Spermatogonia were first identified as separate entities from the Sertoli cells by von Ebner in 1871 (159), but the term spermatogonia was first applied to this class of cells by von La Valette St. George in 1876 (160). Regaud (49) defined two types of spermatogonia in the rat, "dusty" cells and "crusty" cells, on the basis of differences in the chromatin patterns of their nuclei. The dusty cells showed a nucleus with fine palely stained chromatin granulation, whereas the crusty cells had nuclei with coarse granules of heavily stained chromatin close to the nuclear membrane.

Allen (47) used a different terminology that persists today, calling the equivalent of the dusty cells spermatogonia type A and the crusty cells type B spermatogonia. In the rat, spermatogonia with nuclear characteristics intermediate between type A and type B could be identified principally on the presence of fine plaques of chromatin close to the nuclear membrane; these were termed intermediate spermatogonia (68). Similar subtypes can also be identified in other species such as the mouse (161), ram (69), bull (162), and guinea pig (163). More recent studies in the rat identified further spermatogonial types on the basis of their nuclear morphology, separating four classes of type A spermatogonia as well as the intermediate and type B (164). Similar types of observations are now available for other species, the number of generations of spermatogonia varying significantly [see reviews (155,158)].

The ability to differentiate different classes of spermatogonia is critically dependent on the fixation used, because identification is based on the morphological characteristics of the nuclei. The type of fixation precipitates chromatin to varying degrees. Zenker-formol provides optimal chromatin patterns for identification, because the chromatin remains widely dispersed (155); on the other hand, Bouin's fixative creates larger chromatin clumps close to the nuclear membrane, rendering spermatogonial classification more difficult. Additionally, the characteristic features of the nuclei that are used to identify each type are only present at certain phases of the cell cycle, usually acquiring the typical nuclear morphology at the S and G_2 phases (68,165). In humans and in some primate species, somewhat different nuclear characteristics were noted. Branca (166) described palely staining areas in some nuclei, termed nuclear vacuoles. Subsequently, types A and B spermatogonia were identified in humans, with type A being subdivided into the dark and pale types (167–169). The A dark spermatogonia are characterized by a densely staining chromatin usually containing a central pale-stained area termed the nuclear vacuole. Close to the nuclear membrane, one or more nucleoli are found. In contrast, the A pale spermatogonia contain an ovoid nucleus with palely staining granular chromatin and exhibit one or two nucleoli lying close to the nuclear membrane. The type B spermatogonia exhibit the characteristics described for other species, although the human cells are somewhat smaller. Similar spermatogonial types have also been identified in monkeys (*Macacus rhesus*; *Ceropithecus aethiops*) but do not show the nuclear vacuole (170,171).

It is self-evident that to enable spermatogenesis to proceed as a continuous process, the spermatogonia must not only provide the precursors for meiosis but must also renew themselves. The mechanisms by which this is achieved are discussed later in this review, together with the cycle of spermatogenesis.

The cytoplasmic features of spermatogonia visible by light microscopy are relatively unremarkable. They have a poorly staining cytoplasm, and studies of whole mounts of seminiferous tubules demonstrate that they remain connected by intercellular bridges such that large numbers are effectively linked together (172–174). By the periodic acid–Schiff reaction, glycogen is found in the A dark spermatogonia. A useful summary of the cytological classifications of spermatogonia in stained paraffin sections, whole mounts, and with electron microscopy (158) also reviews the interrelationships between spermatogonial types across a range of species. These observations have been extended in detail to the mouse testis with the recognition of the potential for spermatogonial stem cells to be studied using germ cell transplantation techniques. In the neonatal mouse (6 days old), type A spermatogonia, the only germ cell present at this time, shows three subtypes that on morphological features resemble the subtypes described for primates. Subtype I is smallest with a nuclear vacuole, subtype II is larger and more basophilic with heterochromatin clumps, and subtype III is larger again but pale with large heterochromatin clumps (175). Intercellular bridges occur between subtypes II and III, but subtype I spermatogonia are single cells. Subtypes II and III are found in approximately equal proportions, about 45% each of the total measured, with subtype I representing the remaining 10%. The respective classification for these germ cells in the mouse lineage, as described later in the section on Spermatogonial Stem Cell Renewal, is As (stem cells), Apr (paired), and Aal (aligned). In the adult mouse, very detailed descriptions of 10 different types of spermatogonia are available for each of the 12 stages of the spermatogenic cycle (176). These studies are based on high-resolution light microscopy of 1-μm-thick plastic sections stained with toluidine blue. A yes or no decision tree is presented that is used to determine spermatogonial types according to cell shape and size, overall density, and chiefly the pattern and extent of heterochromatin within the nucleus.

In their role as the stem cell for spermatogenesis, the spermatogonia must be physically located at appropriate sites along the basal lamina to allow formation of germ cell clones that, in the adult testis, are histologically recognized as stages of the spermatogenic cycle. Such segments become apparent during testicular maturation and after spermatogonial transplantation where spermatogenesis in a tubule is at first asymmetrical but then becomes uniformly established around the circumference of the entire tubule (177). Therefore, spermatogonia are mobile, assisting early spermatogenesis at the rate of 60 μm per day along the length of the recipient tubule (178). This has suggested that spermatogonia find preferential sites within tubules, a notion that is influenced by the ability of Sertoli cells to accommodate a finite number of spermatogonia along the basal lamina. Such an arrangement has given rise to the concept of niches for the settlement and occupancy of stem cells in association with Sertoli cells (179–182). A niche provides factors that maintain the survival of stem cells and prevents their differentiation by exclusion or inhibition of relevant growth factors, an example of which are the proposed niches that play a role in the properties of transplanted hemopoietic stem cells (183) and other organs. Because the space available within a seminiferous tubule is limited both longitudinally and, especially, in circumference, by definition the number of stem cells that may be accommodated in niches is also limited. Analysis of the location of mouse and rat spermatogonia has confirmed their nonrandom distribution within seminiferous tubules (176,184). The more primitive spermatogonia in both species are nonrandomly distributed around the periphery of seminiferous tubules for most of the cycle of the seminiferous epithelium. They preferentially occur adjacent to areas of intertubular tissue where three tubules come into closest proximity. Spread of their progeny by cell division results in lateral displacement, as indicated in Fig. 12. The nature of the factors that attract spermatogonia to niches remains unknown, but it is speculated that they may originate from the interstitium and/or the properties of the local Sertoli cells and basal lamina.

Fine Structure

In the few detailed ultrastructural studies of spermatogonia, investigators have often had difficulty in identifying the ultrastructural counterparts of the

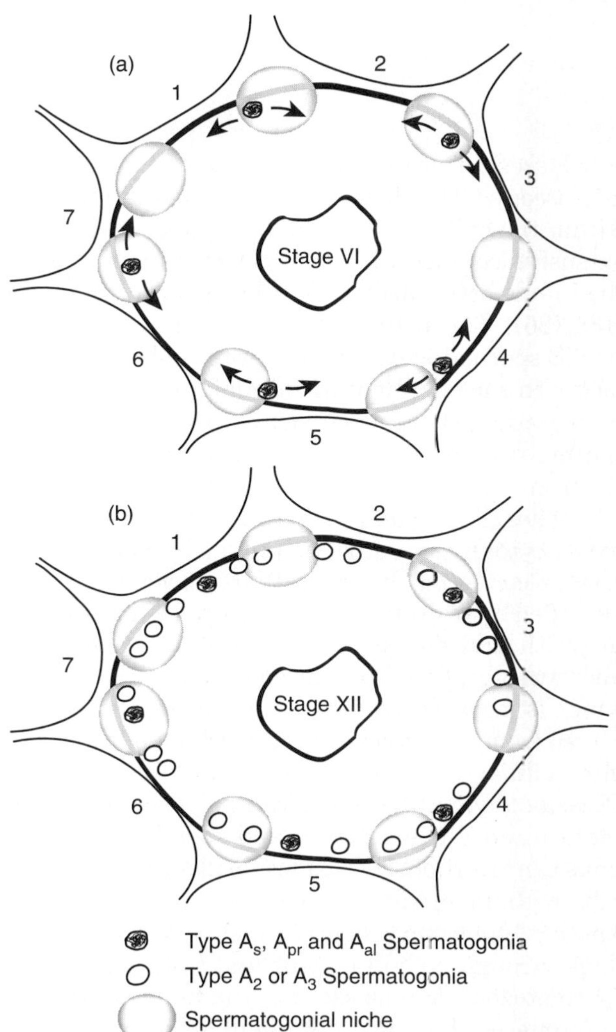

FIG. 12. Drawings of rat seminiferous tubule cross-sections illustrating the location of niches in the interstitial regions of the basal compartment (*large circles*). **A:** In stage VI, the most primitive type A spermatogonia (A_{single}-$A_{aligned}$) are located preferentially in or close to the interstitial areas. After successive mitoses, the daughter cells spread out laterally from this region (*arrows*). **B:** In the basal compartment, resulting in the even distribution of spermatogenesis throughout the circumference of the seminiferous tubule. *Arabic numbers*, adjacent seminiferous tubule. [From (184).]

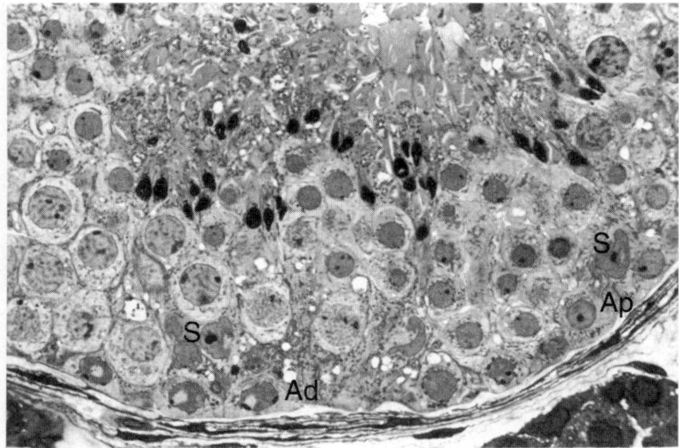

FIG. 13. Human seminiferous epithelium showing type A dark (Ad) and type A pale (Ap) spermatogonia. Note Sertoli cell nuclei (S).

subclasses of spermatogonia identified by light microscopy. The basal position of spermatogonia within the epithelium (Figs. 9, 10, and 13) and their extensive contact with the basement membrane of the tubule are clearly evident by electron microscopy (185,186). The extent of this contact decreases in type B spermatogonia, which eventually lose all contact with the basement membrane to become preleptotene primary spermatocytes (187). All types of spermatogonia are characterized by a relatively electron-lucent cytoplasm and a paucity of cytoplasmic organelles (186–189). Furthermore, because of incomplete cytokinesis during mitosis, spermatogonia remain connected by intercellular bridges originally identified by Watson (190). These bridges are 2 to 3 μm in width and do not usually contain organelles or microtubules; they are limited by the cell membrane, which is more electron-dense in this region, and separated from the adjacent Sertoli cell membrane by an intercellular space of approximately 200 Å (174,191). Occasionally, microtubules can be observed crossing these intercellular bridges and membranous partitions can transiently cross the bridge, separating the cells without breaking down the connections (192). These cytoplasmic bridges persist between clonal siblings through all subsequent maturation steps until spermatozoa are released into the lumen of the seminiferous epithelium. Movement of specific macromolecules through these, including messenger RNAs (mRNAs) and proteins, is known to occur, rendering cells that would otherwise be haploid (i.e., the spermatids) functionally diploid (193). It is a mechanism that therefore enables transport of Y chromosome–encoded gene products into X-bearing sperm cells and vice versa; also, this mechanism is of significance to scientists making knockout models as protein from "wild-type" germ cells can enter "knockout" germ cells.

The classification of spermatogonia in the human by electron microscopy enabled the identification of type A dark, type A pale, and type B spermatogonia, though intermediate forms were noted to occur (187,194–196). Similar findings are also present in the mouse (197). This classification is based on (a) the nuclear and nucleolar features, (b) the presence of aggregations of mitochondria, (c) the presence or absence of glycogen granules, and (d) the presence of crystalloids known as the Lubarsch crystal found in human spermatogonia (187,198). The A dark spermatogonia have an oval-shaped nucleus and demonstrate a relatively electron-translucent region consistent with the appearance of the nuclear vacuole by light microscopy. The nucleoli are small and peripherally placed adjacent to the nuclear membrane and consist of a nucleolonema; occasionally, they are surrounded by the nuclear vacuole. The decreased electron density of these regions results from the absence of chromatin fibrils from the area. The mitochondria lie close to the nucleus, and between the mitochondria finely granular moderately electron-dense material is found. The cristae extend transversely across the matrix, although some areas of the mitochondria are devoid of cristae; in some mitochondria the intracristal space is dilated. Profiles of rough endoplasmic reticulum are largely seen, although some smooth-membraned vesicles are present. The Golgi complex is poorly developed. Glycogen granules are present, often forming aggregations (187). The crystalloids of Lubarsch are present, represented by collections of fibrils and tubules aligned along the long axis of the structure, which may be up to 3 μm in length. Linear arrays of small electron-dense granules separate the fibrillar elements, and small collections of similar granular material are sometimes present in the adjacent cytoplasm (189,199,200). Their function is unknown, though they show some similarities to the structure of the crystals of Charcot-Böttcher found in Sertoli cells.

The A pale spermatogonia have less contact with the basement membrane of the tubule, and their nuclei do not show nuclear vacuoles (Fig. 14). Their nucleoli are peripherally placed and consist of a nucleolonema and pars amorpha. The mitochondria rarely form perinuclear collections, often lying together as pairs separated by the granular intermitochondrial material found in A dark spermatogonia. Glycogen granules are infrequently found, although Lubarsch crystals can be present.

Type B spermatogonia have the least contact with the basement membrane and contain peripherally placed nuclear chromatin aggregations that, because of the thin sections, are not as prominent as visualized by light microscopy. The nucleolus is centrally placed

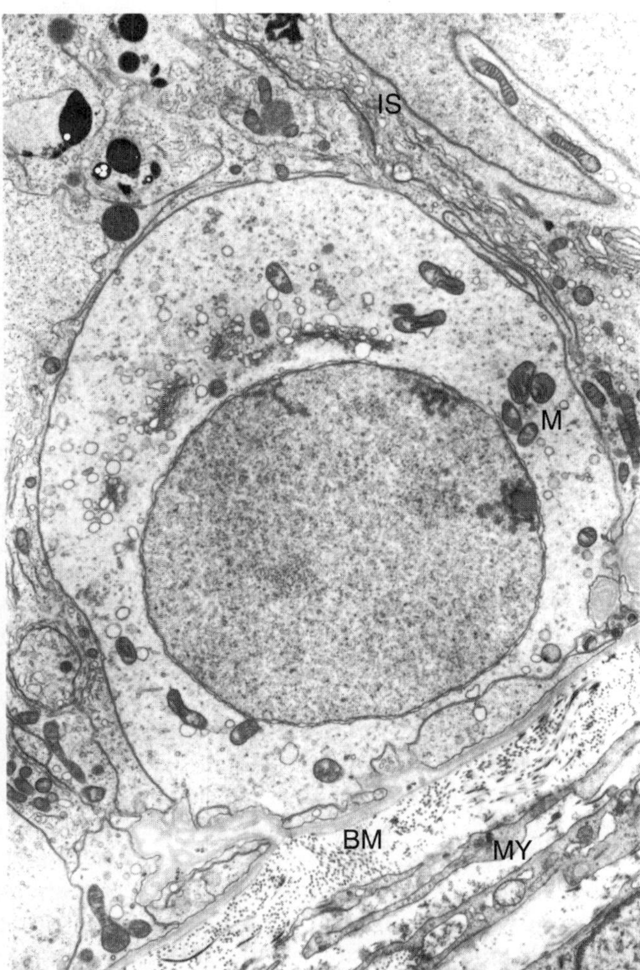

FIG. 14. This electron micrograph illustrates a type A pale spermatogonium situated on the basement membrane (BM) of a seminiferous tubule from the human testis. Note the paucity of organelles and the mitochondria (M) sometimes clustered around an intermitochondrial electron-dense matrix. The adjacent Sertoli cells show specialized inter-Sertoli-cell junctions (IS). Note myoid cell processes (MY).

and consists of nucleolonema and pars amorpha. The mitochondria are scattered singly in the cytoplasm with an apparent disappearance of the intermitochondrial material seen in type A cells. This is of interest because it reappears in primary spermatocytes and has been linked to the perinuclear dense bodies or nuage of invertebrate and amphibian oogenesis [see review (201)]. Scanty profiles of smooth and rough endoplasmic reticulum are present, and the Golgi complex, although not dramatic, is better developed than in the type A spermatogonia.

Rowley et al. (187) identified yet another spermatogonial cell type characterized by its shape—long, very flat, with contact to the basal lamina extending up to 30 μm in length. The nucleus was irregular in shape with a peripherally placed nucleolus. The mitochondria occurred in large collections (up to 10) in the perinuclear region and were joined together by the granular intermitochondrial material. Glycogen and Lubarsch crystalloids were noted. They commented that this type may correspond to the A_0 spermatogonial type of other species (165). A detailed study of mouse spermatogonia using electron microscopy enabled classification of various types A, intermediate (In), and B spermatogonia using ultrastructural characteristics that are supplemented with a yes or no decision tree to aid in their identification (197). In the group of most primitive spermatogonia (As, Apr, Aal), the nuclear chromatin is mottled, a feature that disappears with the transition to type A_1 in which the chromatin becomes homogeneous and finely granular. For A_2, A_3, A_4, and In types there are progressively more and larger chromatin flecks associated with the nuclear membrane, these becoming fewer but denser in the nuclei of type B spermatogonia. Contact with the basal lamina is generally reduced with advancing spermatogonial maturation and is consistent with the fact that the more mature types of spermatogonia, A_4 to B types, are gradually displaced away from the basal lamina in preparation for their development into preleptotene primary spermatocytes.

Spermatocytes

The germ cells undergoing meiosis are the primary and secondary spermatocytes. The term spermatocyte was first used by von La Valette St. George (163) to designate the cells previously termed growing cells by other investigators [cited in (163)]. Although the drawings of Brown (202) are remarkably accurate in depicting the nuclear patterns in these cells, true appreciation of their significance did not become apparent until the studies of Winiwater (203) and Montgomery (204).

Meiosis actually involves two cell divisions (Fig. 15). In the first, the chromosomes of primary spermatocytes appear as pairs of chromatids, subsequent to which heterologous chromosomes pair by synapsis to form bivalents. Each member of the bivalent pair subsequently moves to the daughter cells, termed secondary spermatocytes, which contain half the number of chromosomes (haploid number). However, because each chromosome is composed of a pair of daughter chromatids, the actual total DNA content is equivalent to that of somatic cells. After a relatively short duration, the second division results in the chromatids of each chromosome separating to daughter cells by mechanisms similar to those of mitotic division. The daughter cells, termed spermatids, contain the haploid number of chromosomes and half the DNA content of somatic cells.

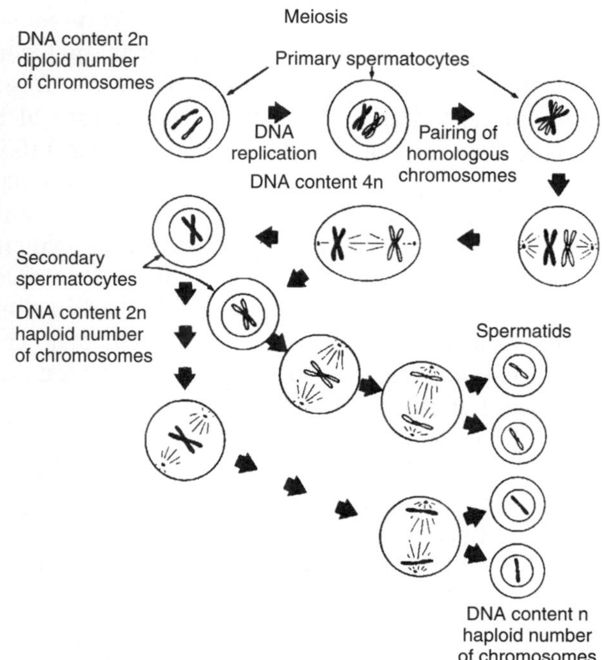

FIG. 15. This diagram illustrates the changes occurring during the meiotic divisions involving primary and secondary spermatocytes.

Primary Spermatocytes

In these cells during the first meiotic division, early studies by Montgomery (204) showed that distinct pairs of chromosomes were visible in primary spermatocytes and suggested that each pair was composed of a chromosome of maternal or paternal origin. Their pairing was originally suggested to be end to end (204,205) but was subsequently shown to be side to side (203). The primary spermatocytes arise from type B spermatogonia that lose contact with the basement membrane of the seminiferous tubule. They are characterized by a spherical nucleus with features similar to the type B spermatogonia lacking any details of chromosomal structure. During this phase they are termed preleptotene spermatocytes and are actively engaged in DNA synthesis (206). This DNA synthesis is the mechanism by which each chromosome, when it condenses, is composed of a pair of chromatids, the total DNA content representing twice the diploid content (Fig. 15). The period of DNA synthesis represents the last in the process of spermatogenesis, and these cells represent the most mature cells capable of being labeled with markers such as [^{3}H]-thymidine, which are incorporated into the DNA.

During the prophase of the first meiotic division, the primary spermatocytes show nuclear features that are based on the morphology of the chromosomes and their process of pairing. The prophase of the first division is characteristically of long duration, and condensation of chromosomes results in their appearance in the leptotene stage (Fig. 16) as single filamentous strands shown by electron microscopic studies to be attached at each end to the nuclear membrane by attachment plaques (207). These attachments probably represent the reasons for the bouquet arrangement of chromosomes seen in leptotene. Though at this stage the chromosomes are already composed of two chromatids, these do not become evident until later in prophase. The zygotene stage is characterized by thickening of the chromosomal elements, which commences the process of pairing known as synapsis (Fig. 16). The mechanisms whereby homologous chromosomes recognize each other to pair as bivalents depends on the chromatin organization that facilitates recognition of homologous sequences on each chromosome.

The long pachytene stage commences with the completion of synapsis and is associated with further thickening and shortening of the chromosomes, which by careful study can be shown to be paired (208). During this phase, exchange of chromosomal material between maternal and paternal homologous chromosomes occurs by crossing over, with the chromosomes linked at such sites by chiasmata. The pachytene phase is characterized by nuclear and cytoplasmic growth, resulting in these cells becoming the largest

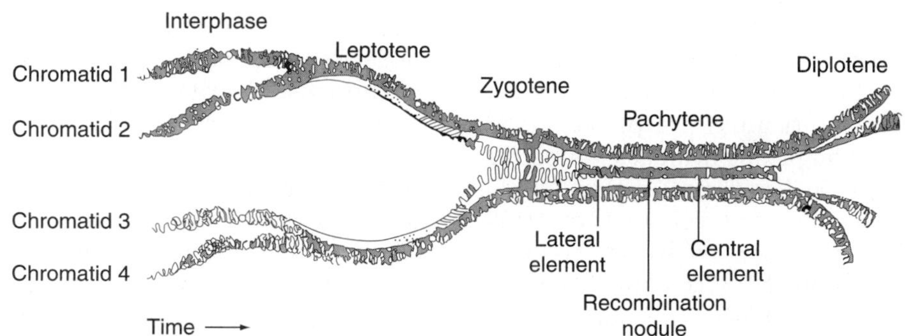

FIG. 16. The manner in which pairs of homologous chromosomes undergo synapsis during meiosis is shown. Note how the lateral elements of the synaptonemal complex come together at pachytene, separating at the diplotene stage.

of the germ cell line. As desynapsis occurs during the next phase, known as diplotene, the paired chromosomes partially separate but remain joined at their chiasmata. Subsequently, in the diakinetic phase, further shortening of chromosomes occurs, and they detach from the nuclear membrane. It is at this stage that each chromosome can be seen to be composed of two chromatids.

Diakinesis is rapidly followed by the dissolution of the nuclear membrane, appearance of the spindle, and the attachment of the bivalents to equator of the spindle during metaphase. Anaphase subsequently results in the movement of the members of each bivalent to opposite poles of the spindle, resulting in daughter cells, termed secondary spermatocytes, that contain the haploid number of chromosomes, each composed of two chromatids. As with other cell divisions within the seminiferous epithelium, cytokinesis is incomplete, and the secondary spermatocytes remain joined by intercellular bridges.

Fine Structure. Electron microscopy has added greatly to our knowledge of events in these cells. Moses (209) and Fawcett (210) independently described the existence of elements, termed synaptonemal complexes, in the nuclei of primary spermatocytes from a number of species. This complex consisted of two lateral elements that appeared as electron-dense fibrils equidistant from a central element consisting of a delicate linear region of increased electron density (Figs. 16 and 17). Woollam and Ford (211) suggested that the lateral elements represented cores of the paired chromosomes and that the central element was composed of microfibrillar processes that extend inward from each lateral element. These views were supported by the observations, from serial electron microscopic reconstructions of the spermatocyte nucleus, that the number of synaptonemal complexes was equal to the number of bivalents (207). Each lateral element first appears in the leptotene stage, representing elements of the single unpaired chromosomes that attach to the nuclear membrane at each end (212,213). Pairing of the elements commences in zygotene, and fully formed synaptonemal complexes over the entire length of the chromosomes are seen in pachytene. Study of these events has been greatly facilitated by the observation that the protein that forms the lateral elements of the complex stains with silver (214). The DNA of each chromosome forms a series of loops extending away from the lateral element, and the process of synapsis is associated with the assembly of the central component, probably by a recognition process involving proteins of the microfibrillar elements projecting centrally from the lateral elements (215,216).

Heyting and colleagues (217,218) attempted biochemical characterization of the synaptonemal complexes after their initial isolation (216) and the development of several monoclonal antibodies.

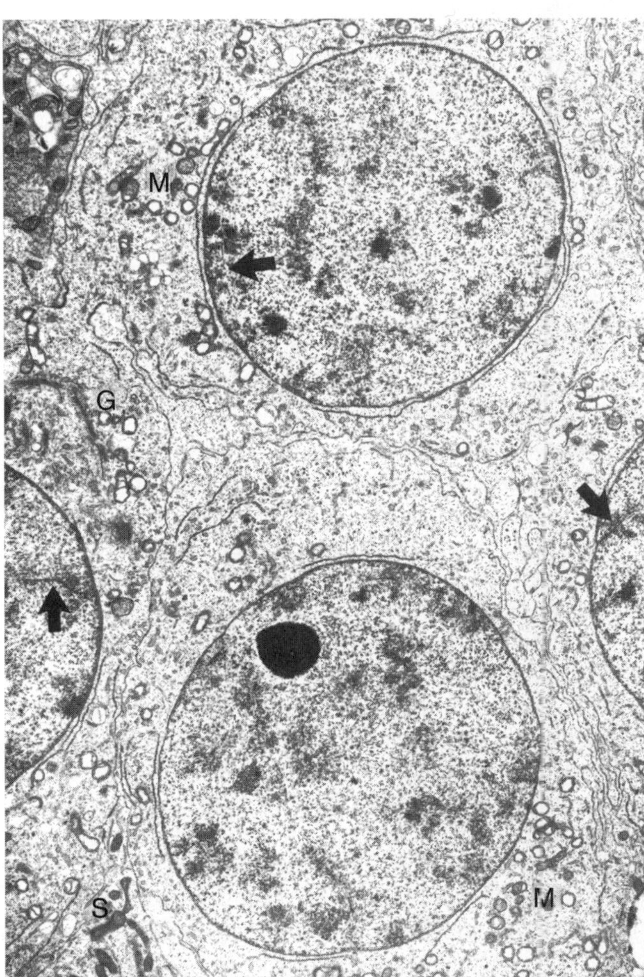

FIG. 17. Two human primary spermatocytes are shown. Note mitochondria (M), Golgi complex (G), leptotene synaptonemal threads (*arrows*), and Sertoli cell cytoplasm (S).

These studies demonstrated that the complexes are not assembled from preexisting components but are synthesized during zygotene. Using one of these monoclonal antibodies, these investigators isolated a cDNA encoding a protein involved in the lateral elements of the synaptonemal complex (219). This protein, SCP1, shares several features with nuclear laminins and some nuclear matrix proteins. It contains long stretches capable of forming amphipathic α helices, and these regions show amino acid sequence similarities to the coiled-coil region of myosin heavy chain. Although the synaptonemal complex provides the framework necessary for synapsis, it is thought that the recombination nodules are the vital component in crossing over of genetic material. These represent ellipsoidal to spherical protein globules approximately 90 nm in diameter that sit along the central portion of the synaptonemal complexes at a number of sites consistent with the number of chiasmata (Fig. 16). Events in diplotene result in disruption of the synaptonemal complexes and separation of the bivalents, which still remain connected at chiasmata.

During meiosis, the sex chromosomes were noted by Painter (208) to be associated with a "chromatin nucleolus," a structure subsequently termed the sex vesicle because it was shown not to represent a true nucleolus (220). It is formed at the end of leptotene and disappears after diakinesis. It is ovoid in shape with a diameter of 2–3 μm. It consists of an area of increased electron density composed of chromatin fibrils through which linear elements similar in appearance to the lateral elements of synaptonemal complexes are scattered (221–224). These take the form of linear arrays or circular profiles, and only occasional short full-formed synaptonemal complexes are seen. This is consistent with the behavior of the sex chromosomes, which only pair over relatively short portions of their length (225).

Although morphologically well described, the precise molecular dynamics remain relatively poorly understood. The use of mouse models, achieved by targeted disruption of genes, has demonstrated significant differences between intrinsic regulators of meiosis and mitosis and significant differences between meiosis in the male and female germline (226,227).

Nucleoli are frequently seen in primary spermatocytes and consist of peripherally placed collections of extremely electron-dense granules. These are sometimes associated with chromatin collections close to synaptonemal complexes. A structure termed the "round body" has been found in spermatocyte nuclei (222,228), occurring as a spherical electron-dense structure adjacent to the nuclear membrane. It is composed of nonhistone protein, appears in leptotene, and increases in size to 1 μm in diameter in diplotene. It persists in secondary spermatocytes and spermatids, decreasing in size progressively and disappearing at step 8 spermatids (228). Schultz and colleagues (228) noted that nucleoli actively incorporated [^{3}H]-thymidine while they were associated with a round body that was increasing in size but not during its decline, thereby suggesting that the round body may be a controller of nucleolar activity in meiotic cells.

The cytoplasm of primary spermatocytes is more electron dense than that of spermatogonia and contains evenly scattered polysomes and ribosomes (229). Profiles of rough or smooth endoplasmic reticulum are sparse, although as prophase progresses, the Golgi complex progressively enlarges. It is perinuclear in position and consists of concentric arrays of membranous lamellae and vesicles, some of which contain a prominent electron-dense core. A detailed analysis of the Golgi complex of rat pachytene spermatocytes demonstrated a marked increase in size of this organelle late in pachytene (230) predominantly from the accumulation of *trans*-Golgi. It is unclear whether this increase is associated with secretory activity of late pachytene spermatocytes or whether it represents an unusual mechanism to ensure that partitioning of this organelle with secondary spermatocytes and subsequently spermatids occurs without disassembly. The mitochondria are ovoid (Fig. 17), and in the leptotene and zygotene stages are frequently aggregated into groups of two or three with electron-dense intermitochondrial material similar to that seen in spermatogonia (231,232). The intracristal spaces are dilated, and later in meiosis the cristae are often displaced toward the periphery as concentric membranous layers, leaving a central electron-lucid area. Other types of collections of finely granular electron-dense material have also been seen in spermatocytes and are termed nuage; Russell and Frank (232) identified six types. Their exact role remains unclear, though one type is similar in appearance to the chromatoid body (233); Russell and Frank suggest that this structure disappears in late diplotene, reforming by coalescence in secondary spermatocytes.

Unusual aggregations of parallel membranous lamellae or cisternae are seen, the adjacent lamellae sometimes joined to each other at regular intervals by annuli. These resemble the structures termed annulate lamellae that are found in oocytes of many species (234). Similar aggregations have been observed, adjacent to or often attached to the nuclear membrane, supporting views that they originate from that site (235,236).

Primary spermatocytes are joined to each other by intercellular bridges similar to those found between spermatogonia. They are separated from adjacent Sertoli cells by a distinct intercellular space that is modified in some regions by desmosome-like structures (237,238). As preleptotene spermatocytes lose their contact with the basement membrane, the processes of Sertoli cell cytoplasm that intervene develop specialized inter-Sertoli cell junctions, discussed later in this review (239).

Secondary Spermatocytes

These cells undertake the second meiotic division, and the first description of their characteristics is credited to von Ebner (240). The relatively infrequent appearance of these cells in sections of the testis (241) indicates a short life span before they complete meiosis to form spermatids. Montgomery (242) correctly noted that the secondary spermatocytes have the haploid number of chromosomes, although their DNA content is still diploid (Fig. 15), and when they complete meiosis the resultant spermatids have both the haploid DNA and chromosomal content. The cells are spherical, intermediate in size between primary spermatocytes and spermatids, with a diameter of 10–12 μm. They are situated close to the lumen of the seminiferous tubules, and their spherical nuclei contain a homogeneous chromatin network throughout which large globular chromatin masses are dispersed. Centrally placed nucleoli are often seen. Light and electron microscopic studies have shown that these cells are joined by intercellular bridges identical to those found between other germ cells (174,243,244).

Fine Structure. There have been few detailed studies of the ultrastructure of secondary spermatocytes, partly because of the difficulty of identifying sections of tubules within which they are present (231,244,245). Their cytoplasm contains scattered cisternal profiles of endoplasmic reticulum arranged concentrically around the nucleus. The Golgi complex is prominent and often contains vesicles demonstrating electron-dense granules. Collections of membranes with the features of annulate lamellae are found sometimes embedded in electron-dense granular material (235,244). The mitochondria are dispersed within the cytoplasm and show dilated intracristal spaces, which result in the cristal membranes being pushed to the periphery of the organelle.

Spermiogenesis

The transformation of spermatids to spermatozoa involves a fascinating but complex sequence of events that constitute the process of spermiogenesis. No cell division is involved, but the process is in essence a metamorphosis in which a conventional cell is converted into a highly organized motile structure. The major features of spermiogenesis are common to all species, but the details differ in each species because there are distinguishing morphological features between spermatozoa determined by genetic factors. Many aspects are visible by light microscopy, but the finer details require the magnification of the electron microscope. Hence, in this section of the review, the light and electron microscopic features are described together (Figs. 18 and 19). For convenience, it is possible to divide the cytological changes during spermiogenesis into a series of developmental steps involving different cellular organelles, but it is

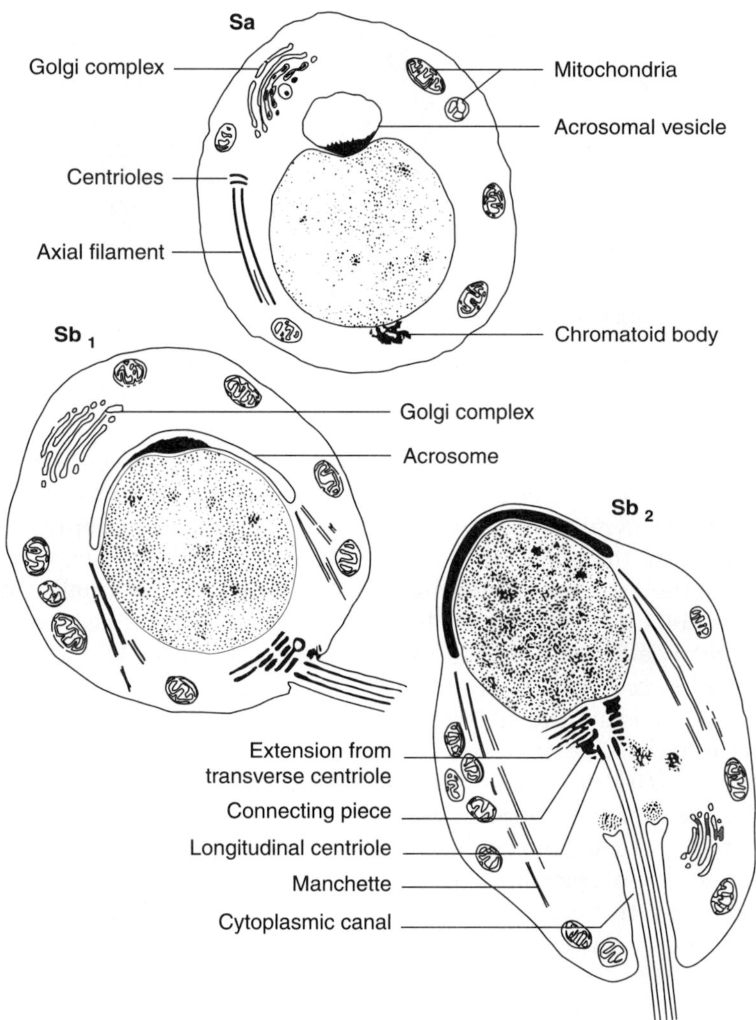

FIG. 18. The electron microscopic features of the stages of human spermatogenesis (Sa to Sb_2) are shown. [From (1113), with permission.]

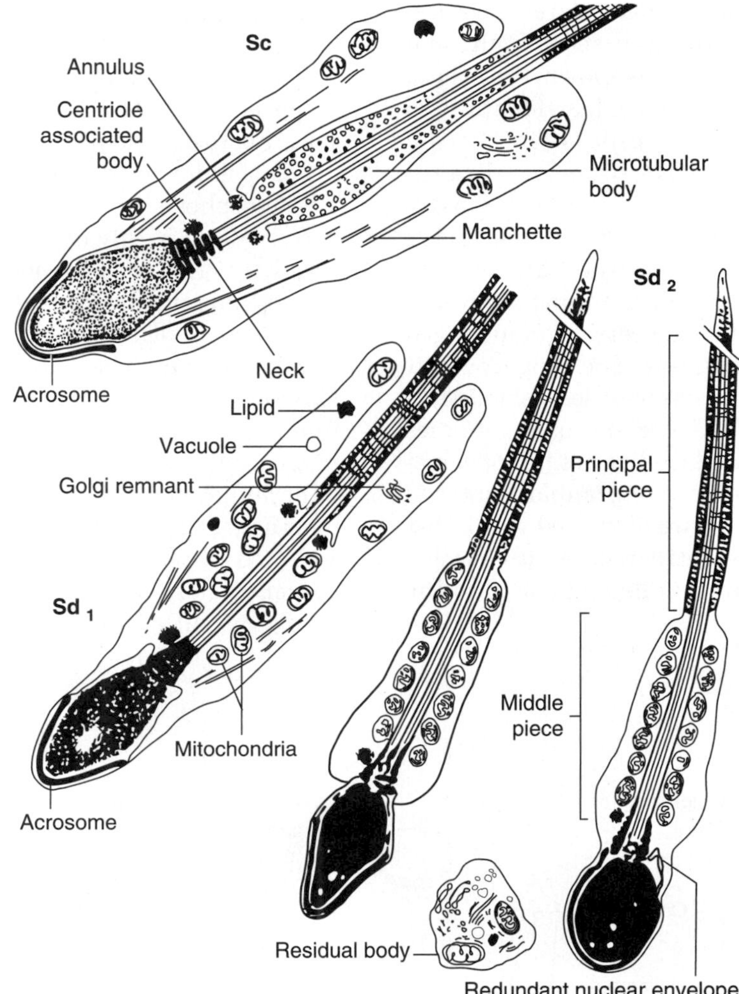

FIG. 19. The electron microscopic features of the stages of human spermatogenesis (Sc to Sd_2) are shown. [From (1113), with permission.]

important to recognize that some steps occur contemporaneously. The changes can be grouped into (a) formation of the acrosome, (b) nuclear changes, (c) development of the flagellum, (d) reorganization of the cytoplasm and cell organelles, and (e) spermiation relationships of the Sertoli cell and spermatids.

Spermatids, like other germ cells, were named by von La Valette St. George (246), and some investigators use the term spermateliosis as an alternative to spermiogenesis. A number of early studies described the general changes characterizing spermiogenesis (166,204,247–249), and these were expanded and used by Clermont and colleagues (250–252) in developing a classification for the stages of spermiogenesis. This classification was aided by the demonstration that the acrosome stained clearly with the periodic acid–Schiff reaction. Recently formed spermatids are spherical cells that are smaller in size than secondary spermatocytes and are found at the luminal aspect of the seminiferous epithelium. They have a centrally placed spherical nucleus, a well-developed Golgi complex, and adjacent centrioles. The mitochondria are dispersed and lie peripherally close to the plasma membrane. Additionally, a chromophilic electron-dense mass, the chromatoid body, can be observed adjacent to the Golgi complex in a perinuclear location (250,253).

Formation of the Acrosome

This structure arises from the Golgi complex, a fact established early by Bowen (254–256) from studies in mammals and other classes. Gatenby and Beams (249) identified proacrosomic granules in the Golgi regions of primary spermatocytes, though it is still unclear today whether the granules in these cells are transmitted to daughter cells; more likely they represent some form of secretory protein that is packaged in the Golgi complex (Figs. 20 and 21). These investigators noted that similar granules were elaborated in the Golgi apparatus of newly formed spermatids, and the vacuole and granule were deposited at one pole of the nucleus, where they spread out to form the acrosome cap. Once the acrosomal cap had been formed, the remainder of the Golgi complex migrated to the opposite pole of the spermatid (249).

Early electron microscopic studies demonstrated that several proacrosomic granules are often formed in the Golgi complex, coalescing to form a single large granule that comes into contact with the nuclear membrane (256–258) (Fig. 20). It is closely applied as a cap-like structure spreading over approximately 25% to 60% of the nuclear surface. During this phase, additional material appears to be transferred from Golgi to the acrosome by vesicles, a process that has been analyzed in considerable detail (259–261). The Golgi has been divided into cortical and medullary zones, the cortex being limited externally by cisternae of rough endoplasmic reticulum. The transition from cortex to medulla is marked by a change from cisternal profiles to vesicles, and it is from this aspect that vesicles are transported to the acrosomal cap. Clermont and Tang (261) showed that glycoproteins are transferred from the Golgi complex to the acrosome, and in contrast to other cells, this accumulation occurs slowly (1 hour) in comparison with other cells (2–10 minutes) (262,263).

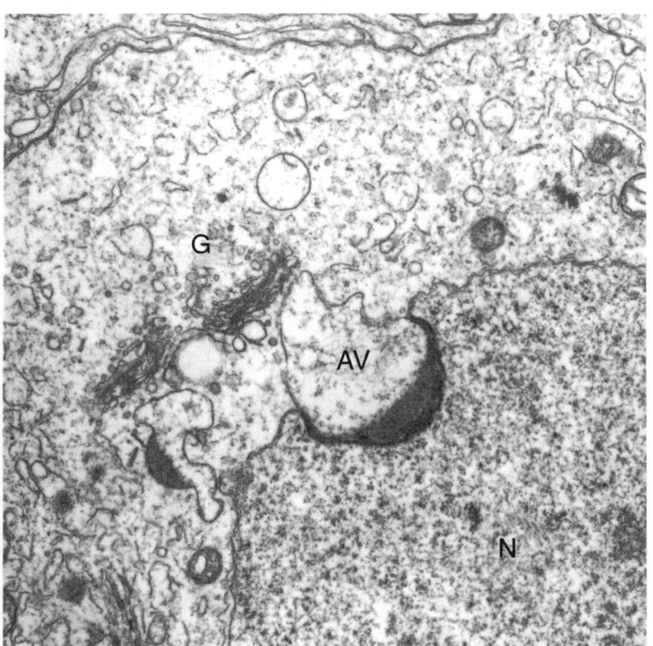

FIG. 20. The developing acrosomal cap of a human spermatid is illustrated. Note nucleus (N), Golgi complex (G), and acrosomal vesicle (AV).

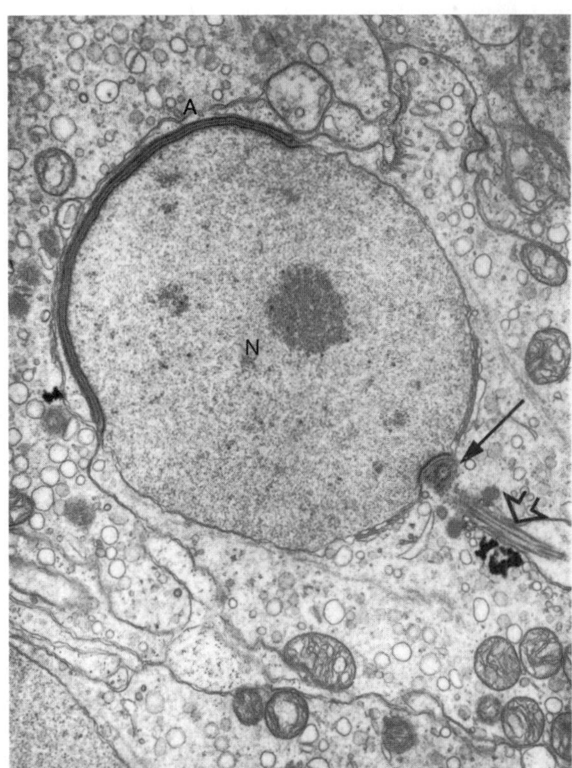

FIG. 21. A human spermatid (stage $Sb_{1,}$) shows acrosome (A), nucleus (N), developing connecting piece (*arrow*), and axial filament (AF, *open arrow*).

For a time the acrosome contains a centralized electron-dense granule and a less electron-dense periphery, but in the human this difference is progressively lost as the cap spreads over the nuclear surface (Fig. 22). In other species this zonal arrangement persists for a longer period of time. In many species, such as the guinea pig, chinchilla, and ground squirrel, a conspicuous thickening of the acrosomal cap extends beyond the nucleus (Fig. 23) and is termed the apical segment (264), but the reasons for this specialization remain unknown. Similarly, the caudal region of the acrosome in many species is partly attenuated and is termed the equatorial segment, and in the human some lamination has been noted in this area (265,266). Although the reason for this specialization is unknown, this region of the acrosome persists after the rest of the acrosomal contents are lost after the acrosomal reaction (267). Furthermore, it represents the region in which binding occurs to the cell membrane of the oocyte during fertilization (268). The size of the acrosome varies significantly between species, being closely applied to the nucleus in humans but being very large and elaborate in species such as the guinea pig (269,270). Furthermore, modifications to this structure may occur after spermatozoa leave the epithelium and

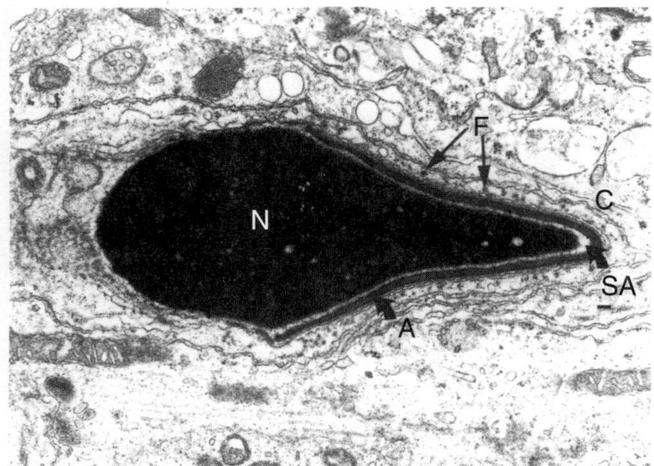

FIG. 22. This view illustrates the human spermatid head composed of the nucleus (N), acrosome (A), and subacrosomal space (SA). Note the cisternae (C) delineating a zone of Sertoli cell cytoplasm containing fibrils (F).

pass through the epididymis (269). The glycoprotein nature of the acrosome is consistent with the observation that it contains a variety of lysosomal enzymes important for penetration of the zona pellucida of the ovum and other specific proteins such as acrosin (271,272).

With subsequent changes in nuclear position, the acrosome is closely applied to the cell membrane of the spermatid and lies between this and the nuclear membrane. Between the inner acrosomal membrane and nuclear membrane, the human spermatid contains a thin layer of moderately electron-dense material (Fig. 22), which Bedford (273) suggested represents the perforatorium. The subacrosomal space contains a variable amount of electron-dense material whose organization differs markedly between species. In rodents, this space is extensive, particularly at the cranial aspect, and the material is organized to form a rod-like perforatorium that appears to be prolonged backward over the nucleus as three prongs, possibly thickenings of the nuclear membrane (274,275) (Figs. 24 and 25). In the toad, the perforatorium consists of strands of electron-dense material in the subacrosomal space (276), and a similar structure has since been found in many species (277,278). The protein composition of the perforatorium in the rat has been partially characterized by Oko and Clermont (279), who identified a number of components. Using antibodies to these components, they demonstrated regional differences in the protein composition of the perforatorium (280). Furthermore, although they could not identify actin in the perforatorium, they noted actin and seven perforatorium proteins in the subacrosomal space before the condensation of the perforatorium.

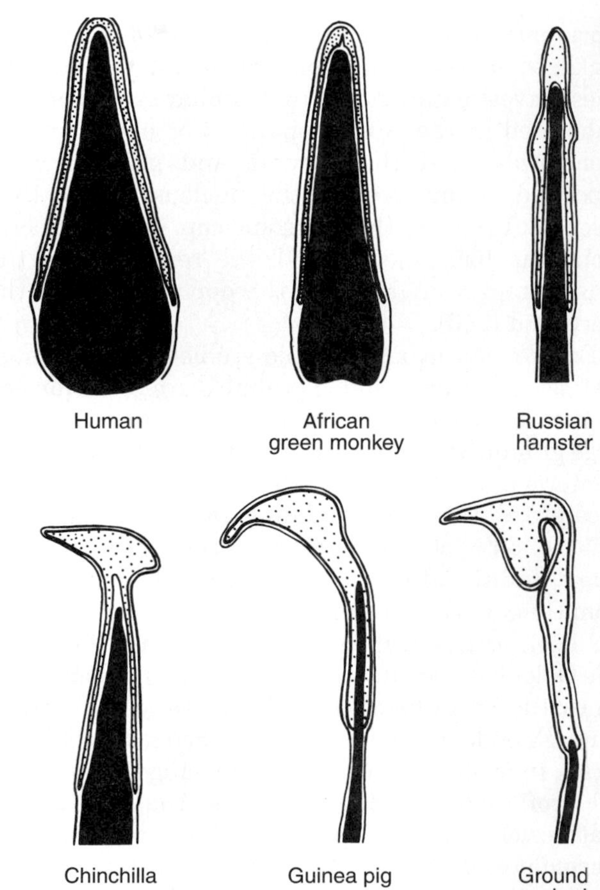

FIG. 23. The variation in the shape of the components of the head of spermatozoa from a number of species is illustrated. [Some views are drawn from micrographs in (273,303).]

Some investigators postulated a mechanical role for the perforatorium, and others suggested it carries a lysin distinct from that found in the acrosome (281). Actin has been identified as a component of the subacrosomal space in a number of species (282–284). The number of actin filaments increases during steps 9 to 13 in the rat and decreases in the later stages of spermiogenesis (283).

In men in whom the acrosome is absent, the head of the sperm takes on a globular appearance, referred to as "globozoospermia" (285). It is likely that this defect arises from a mutation in a gene(s) controlling acrosomal development, but as yet the specific mutation has not been identified. Several studies in mice have shown that interruption of acrosome formation through targeted disruption of the *Hrb* gene or the casein kinase IIα catalytic subunit resulted in globozoospermia and, in the former, the absence of mitochondria from the mid-piece of sperm. (286,287). In the latter, there was also evidence of motility disturbances resulting from bent flagella and uncoordinated flagellar movement. The mechanism by which

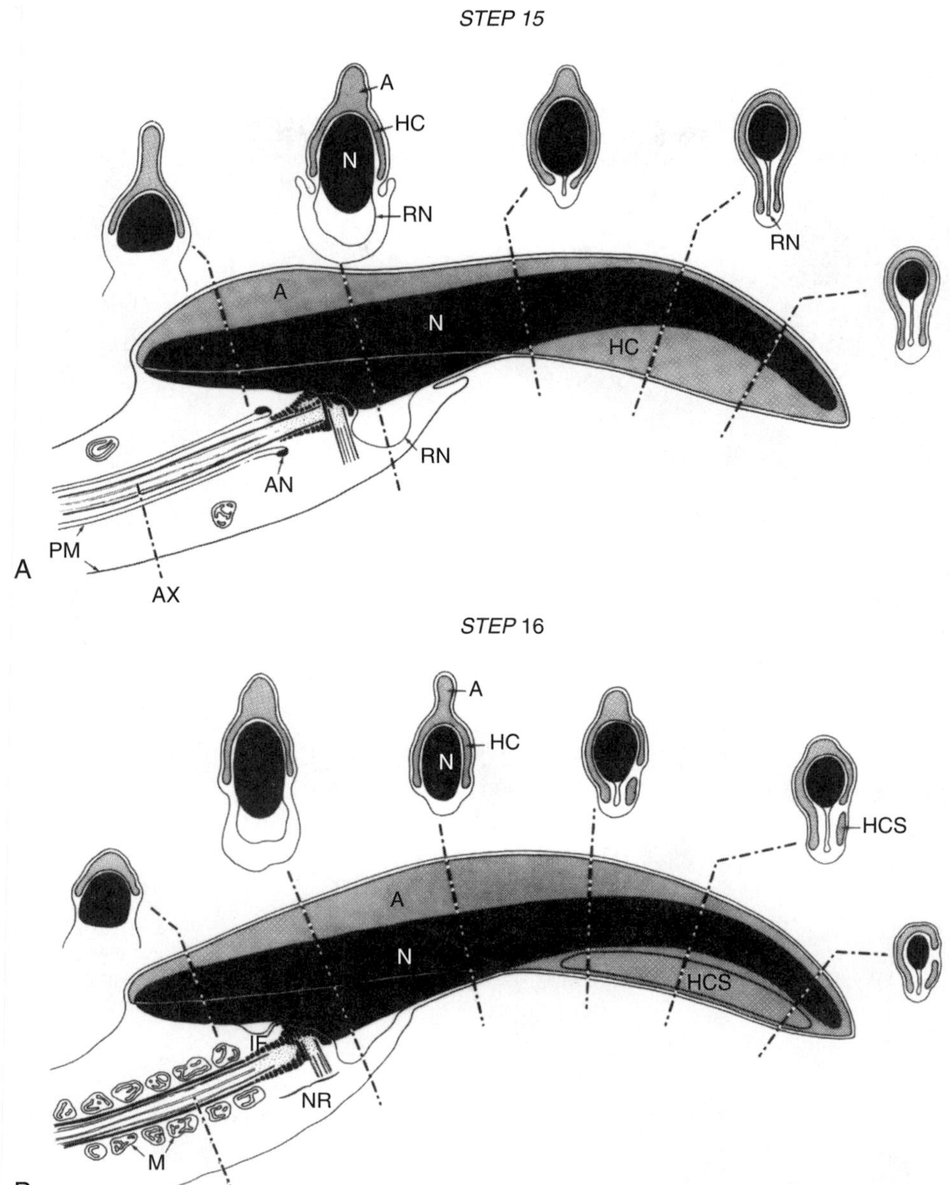

FIG. 24. This diagram demonstrates the changes occurring in the relationships and shape of the structures composing the head of the rat spermatid during spermatogenesis at stages 15 and 16. A, acrosome; AN, annulus; AX, axoneme; HC, headcap; HCS, separated head cap segment; IF, implantation fossa; M, mitochondria; N, nucleus; NR, neck region; PM, plasma membrane; RN, redundant nuclear envelope. [From (275), with permission.]

the successful attainment of motility and normal tail structure are linked to acrosome development needs to be explored.

Nuclear Changes

In most species, the nucleus changes its position during spermiogenesis from a central to an eccentric position. The region of the nucleus that first comes into close apposition with the cell membrane is that segment covered with the developing acrosomal cap. Subsequently, during the rearrangement of spermatid cytoplasm, larger segments of the nucleus come into close association with the plasma membrane.

Associated with the change in nuclear position is a progressive condensation of chromatin to form larger and more electron-dense granules. These increase in number and condense eventually to form an osmiophilic electron-dense homogeneous mass. The degree

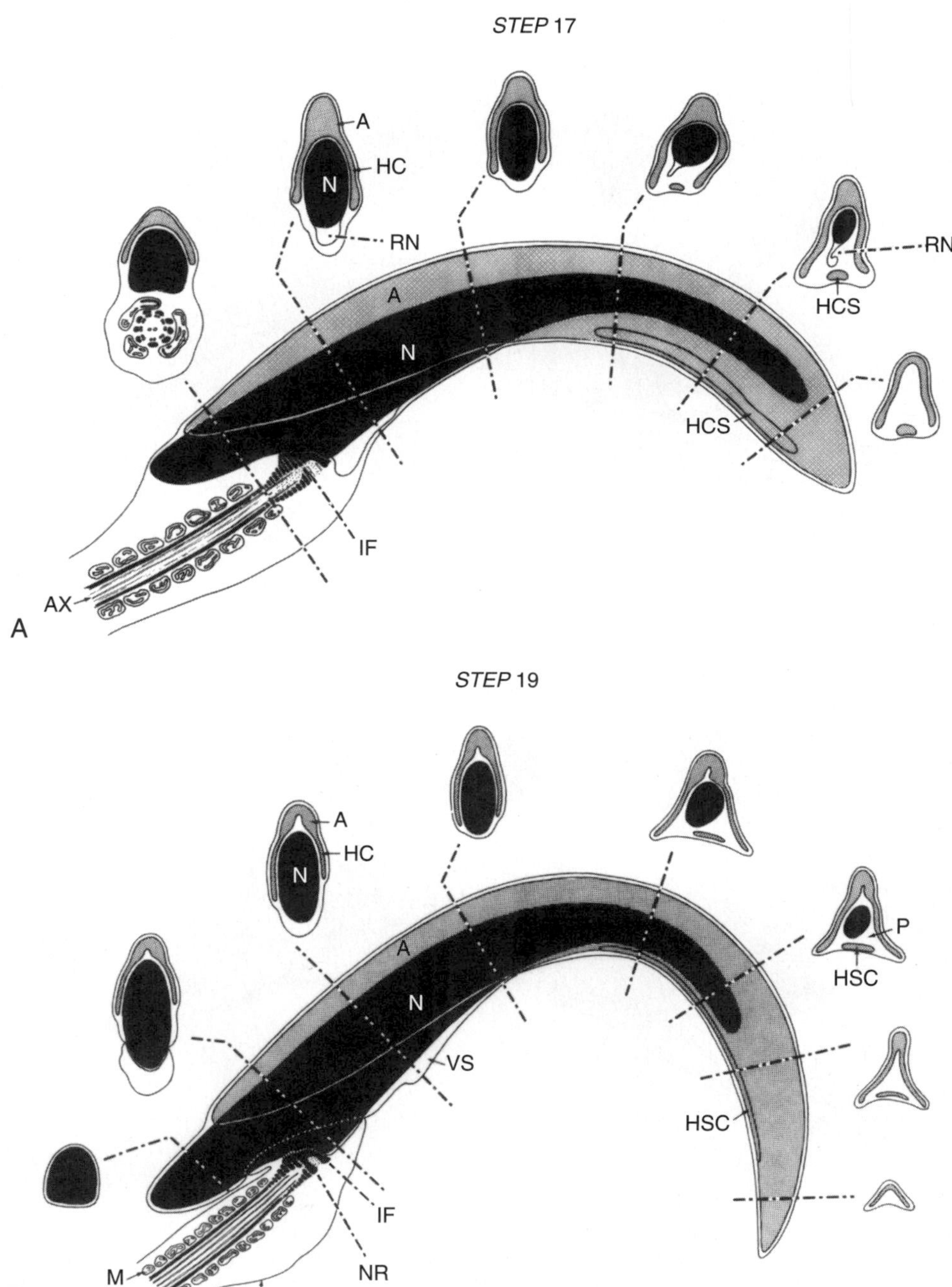

FIG. 25. This diagram demonstrates the changes occurring in the relationships and shape of the structures comprising the head of the rat gamete during spermiogenesis at stages 17 and 19. A, acrosome; AX, axoneme; HC, head cap; HCS, separated head cap segment; IF, implantation fossa; M, mitochondria; N, nucleus; NR, neck region; P, perforatorium; PM, plasma membrane; RN, redundant nuclear envelope; VS, ventral spur. [From (275), with permission.]

of condensation can vary with species and is particularly variable in human spermiogenesis (259,288,289). In different species the temporal linkage between the changes in nuclear shape and chromatin condensation vary. In the rat and the mouse the phase of nuclear elongation precedes the principal condensation events (290). In the rat there is a progressive pattern of development that is not complete until late in spermiogenesis (Figs. 24 and 25). Together with the acrosomal modifications, the final structure of the head of the rat spermatid takes some time to evolve (275). Additionally, in humans, electron-lucid spaces occur with the nucleus and, although they are not limited by a membrane, are termed nuclear vacuoles (276). The coalescence of chromatin granules is associated with ill-defined chemical changes in the

DNA, which is stabilized and is resistant to digestion by the enzyme DNase (253,291). These changes involve the nuclear proteins, termed histones, that in somatic cells bind with DNA to form nucleosomes and consist of several types namely, H1, H2A, H2B, H3, and H4. They are replaced in germ cells before and during meiosis with testis-specific variants that are arginine-rich (H1t, TH2A, TH2B, TH3) (292–294). These testis-specific histones are replaced subsequently by several basic transition proteins, TP_1 and TP_2 (295–301). Initially, the loss of H1t results in a change in the chromatin from the nucleosomal packaging in spermatid steps 1–8 to a filamentous form. As the transition proteins enter the nucleus, the filamentous chromatin packaging changes to a thicker fibrillar form (302)

Several studies of nuclear proteins have shown dramatic changes during spermiogenesis with the appearance of species-specific protamines rich in arginine and cysteine that replace the transition proteins at a time in spermiogenesis when the chromatin is highly condensed (295–301). In all eutherian mammals, sperm protamines are encoded by two genes, termed P_1 and P_2, with most species containing P_1 and the hamster, mouse, stallion, and human containing principally P_2 (298). It is possible that the posttranslational modification of the histones such as the acetylation of H4 in the rat may, through decompaction of nucleosomes and chromatin, facilitate displacement of histones (290,301). These changes and the loss of virtually all nonhistone nuclear proteins late in spermiogenesis occur at a time of complete repression of gene transcription and shaping of the sperm nucleus (303,304). The precise mechanisms of the histone substitution remain unknown, although it is known to be directly regulated by Ca^{2+}/calmodulin-dependent protein kinase IV (305).

The recent development of heterozygous mice carrying protamine 1 or protamine 2 nullizygous alleles confirmed the importance of obtaining the correct balance in protamine and nonprotamine nuclear content. Despite the existence of intercellular bridges between sister spermatids that allow for the sharing of mRNA and proteins between cells in heterozygous knockout germ cells and thus usually mask a phenotype in genetically nullizygous spermatids, these lines of mice both exhibited haploinsufficiency resulting in complete male infertility (306), that is, an inability to pass on either the nullizygous or the wild-type genome. The importance of sperm protamine content is further supported by association studies in men (307) showing an inverse correlation between sperm protamine content and infertility. Surprisingly, however, transition protein 1 and 2 knockout mice showed only reduced fertility, indicating the transition proteins are not crucially important for head condensation and the histone–protamine exchange (308–310). These mice, however, did show some degree of spermatogenic abnormalities and significantly impaired motility and/or morphology, suggesting that nuclear events are intricately linked to the subsequent development and function of sperm tail function (311).

During the chromatin condensation, there is a progressive reduction in nuclear volume and, in some species, dramatic changes in shape that result in sperm heads with shapes characteristic for each species (303,312). These striking differences in head shape have prompted investigators to seek the mechanisms responsible. The appearance of a microtubular sheath, termed the manchette (see below), close to the nucleus at a time when significant elongation and shaping occur (Fig. 26) led a number of investigators to suggest a causal relationship (313,314). However, from their detailed studies, Fawcett and colleagues (303) could not support this view. They noted that significant variations in the form of the manchette did not occur, a fact that would be necessary to explain the nuclear changes if they were causally related. They noted the observations of Beatty (315) that head shape could vary between strains of the same species and could be varied experimentally by selective breeding. In view of this and their own observations, they proposed that the remarkable diversity of head shape is the result of genetically determined patterns of the molecular aggregation that takes place during chromatin condensation. In contrast, Meistrich and colleagues (316), from their studies of abnormal spermiogenesis in *azh* mutant mice, concluded that the abnormal head shape may in part relate to the abnormal shape and size of the manchette. Russell and colleagues (317) used observations on normal and abnormal rats and mice in which altered head shape was induced by treatments such as Cytoxan, 5-fluorouracil, and Taxol. They noted the presence of rod-like elements extending between the innermost microtubules of the manchette and the outer nuclear envelope and raised the possibility that these structures may represent a mechanism by which these microtubules could exert a deforming force on the nucleus. They also suggested that the chromatin condensation reinforces the changes in shape already brought about by the manchette.

More recent studies demonstrating that the manchette contains microtubule motor proteins such as kinesin and dynein as well as nucleocytoplasmic transport agents such as RanGTPase have been linked to the concept of intramanchette transport for both proteins and organelles (318–322). The importance of this concept has been supported by studies

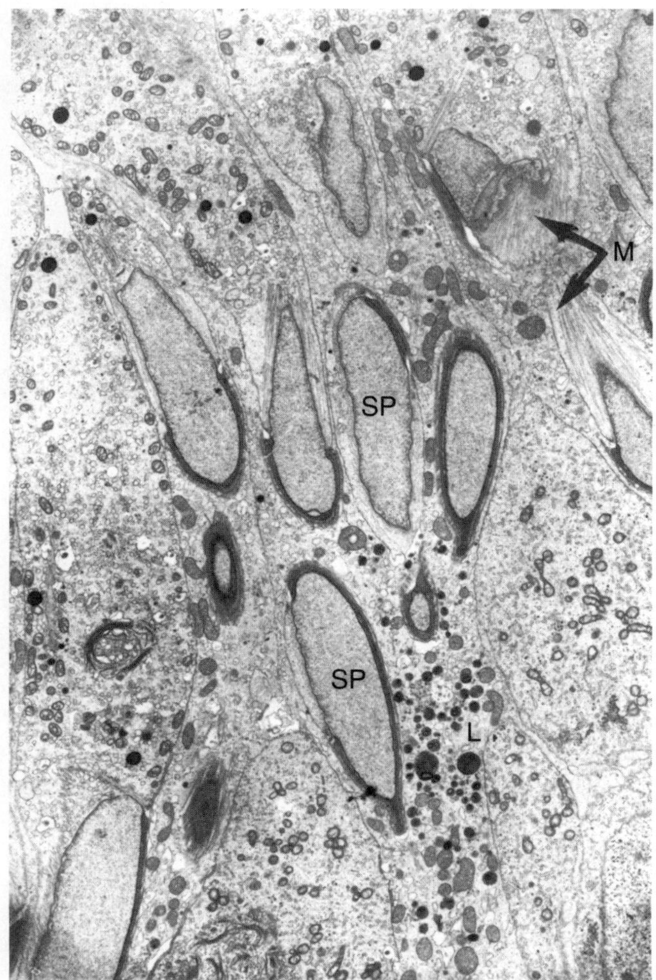

FIG. 26. Rat seminiferous epithelium, illustrating elongated spermatids (SP) and the manchette (M). The Sertoli cell cytoplasm contains numerous lysosomes (L).

evaluating the localization of non-nuclear somatic histones in bovine spermiogenesis (323). They propose that the manchette contributes to the formation of perinuclear theca, a structure that encases the spermatid nucleus except at the base (324). By showing that these histones were found at sites that suggest they are synthesized in the distal cytoplasm of the spermatid and transported via the manchette to the postacrosomal segment of the perinuclear theca, the concept of a transport role for the manchette is strengthened.

A further link between the development of the sperm head and the tail and a role for motor proteins emerges from the targeted disruption of the *act* (activator of cAMP-responsive element modulator in the testis) gene in mice (325). This gene, as its name implies, enhances CREM transcription activity, which is required for the expression of many genes during spermiogenesis (326). It was shown that these mice demonstrated aberrant head shape abnormalities and folded tails. These studies link the disruption of this gene with alterations in the pattern of a kinesin motor protein termed Kif 17b that is localized in the fibrous sheath of the principal piece.

The reduction in nuclear volume is associated with the probable loss of materials from the nucleus via nuclear pores (288) and also results in the formation of redundant folds of nuclear membrane (289,327). This is most evident at the abacrosomal pole of the nucleus, where these folds are separated from the condensed chromatin by an electron-lucid zone containing flocculent material. In this region the nuclear membrane is studded with nuclear pores and frequently exhibits the formation of lamellae and vesicles, the latter containing flocculent material (Fig. 24).

Formation of the Tail

The concept that the sperm tail arose from the centrioles of spermatids was recognized by Meves (247,248), who noted that the two centrioles moved to the periphery of the cell, where the axial filament arose from the more peripherally placed centriole. These observations were substantiated by other investigators early this century (171,242,249). However, Gatenby and Beams (249) implicated both centrioles in the formation of the axial filament and noted that the flagellum moved inward toward the nucleus, where the complex lodged at its caudal pole opposite the developing acrosome. These observations were confirmed by early electron microscopic studies (268). Both Anberg (328) and Fawcett (329) noted the presence of a centriole within the connecting piece of the neck of human spermatozoa and termed it the proximal or transverse centriole. However, they failed to find evidence of the distal or longitudinal centriole. Fawcett (329) also showed that the structure termed the ring centriole by light microscopists appears in ultrastructural studies as the annulus that marks the caudal limit of the middle piece of the tail. It is clear now that the axial filament arises from the distal or longitudinal centriole and that this basic structure is modified by specialization in the different regions of the sperm tail. The subsequent description of cytological events is most easily subdivided into (a) development of the axial filament, (b) formation of the neck or connecting piece, (c) formation of the dense fiber system, (d) development of the principal piece, and (e) formation of the middle piece.

Development of the Axial Filament. Several studies have shown that the axial filament develops from the centriole that is aligned to the axis of the flagellum and is termed the longitudinal or distal centriole (289,330,331). The other centriole is oriented perpendicular to the axial filament and is termed the proximal or transverse centriole. The basic structure of the axial filament is common to flagella and cilia and consists of nine peripheral doublet microtubules arranged equidistant from each other around a circle at whose center two single microtubules are found (Figs. 20 and 27).

The doublets consist of two subfibers: one (subfiber A) is a complete microtubule and is circular in cross-section, whereas the other (subfiber B) is C-shaped with its concavity attached to subfiber A (312). The walls of both subfibers are composed of protofilaments of tubulin (332–334). The axoneme demonstrates a highly organized substructure, which is considered in more detail elsewhere (see Chapter 1). Briefly, subfiber A extends a pair of hooklike arms that project toward subfiber B of the adjacent doublet; these extensions, termed dynein arms, are composed of a protein, dynein, with ATPase activity (335).

Additional links between the doublets are provided by nexin links (336), and the doublets are connected to the helical sheath surrounding the two central microtubules by radial spokes (337). The nature of the substructure of the flagellum has taken on additional significance because of genetically determined abnormalities in its structure that result in immotility of sperm and consequent infertility (338,339). Cilial dysfunction in such circumstances may be limited to sperm or may extend to all, or some, cilia containing tissues in the body, a condition called primary ciliary dyskinesia (340,341). Fawcett (312) proposed a numbering system for the doublets, designating the doublet bisected by a plane passing between the central two microtubules as 1 and the others sequentially in a clockwise direction.

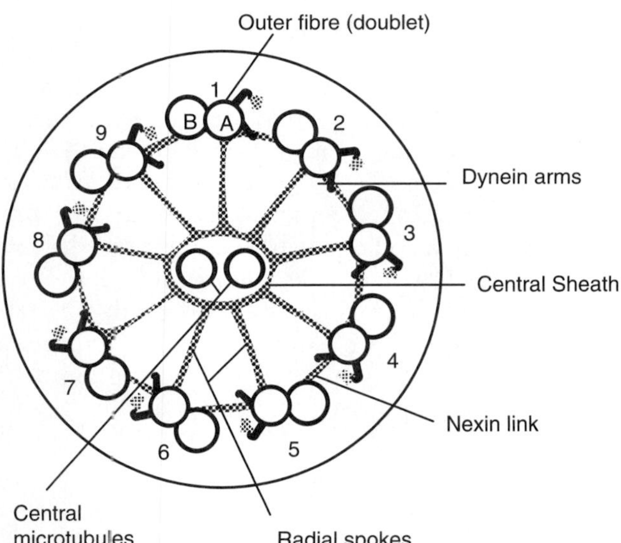

FIG. 27. This diagram illustrates the components of the axial filament when cut in perfect cross-section.

Development of the axial filament commences early in spermiogenesis and projects from the surface of round spermatids, the microtubular core being separated from a finger-like protrusion of the cell membrane by a thin layer of cytoplasm. An invagination of the cell membrane forms a cleft, the cytoplasmic canal, that surrounds the proximal portion of the axial filament and whose proximal limit is formed by the attachment of the annulus (330). Later, the centriolar complex and axial filament become lodged at the abacrosomal pole of the nucleus to form the neck or connecting piece, the complex articulation of the future head and tail of the spermatozoon (Fig. 21). Flagellum formation was successfully initiated by in vitro culture of spermatids (342).

The final form of the connecting piece varies significantly between species, but in all mammalian forms a basic structural organization can be discerned. The connecting piece can be regarded as a truncated cone, modified to contain the proximal and distal centrioles (Fig. 28). The truncated apex points distally and from it emerges the tail of the sperm. The base forms an arched sheet, termed the capitulum, composed of electron-dense material, which is lodged in a shallow depression, called the implantation fossa, in the caudal pole of the nucleus. The truncated apex of the cone surrounds the distal or longitudinal centriole, from which the axial filament of the tail is derived. Superior or rostral to the distal centriole, the conical structure surrounds the proximal centriole, which lies at an angle

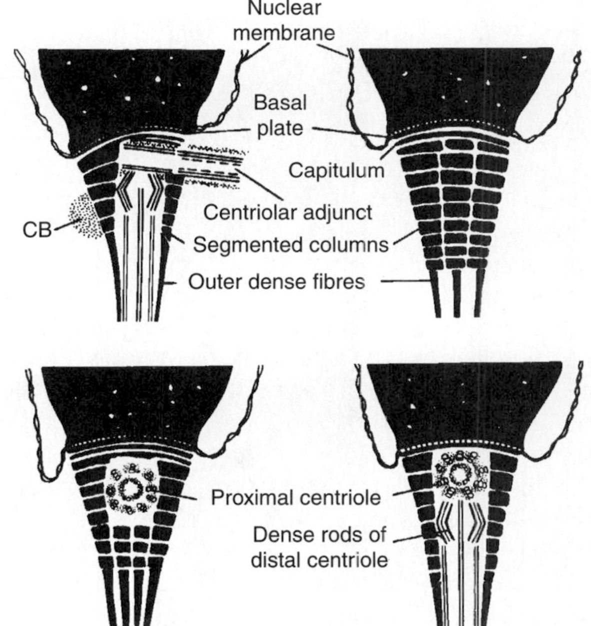

FIG. 28. This diagram illustrates the relationships of the components of the neck of the human spermatid during spermiogenesis. The different appearances represent views seen from different planes of section.

of 75 to 90 degrees to the longitudinal axis of the sperm tail, except in one region opposite the distal end of the proximal centriole from which an extension, termed the centriolar adjunct, emerges (Figs. 28 and 29). The actual structure of the wall of the truncated cone is dependent on the level at which it is examined. The truncated apex at the level of the distal centriole is composed of nine longitudinal cross-striated columns that cranially fuse to varying degrees. Those columns opposite the centriolar extension coalesce and lose their individuality and, together with the others (except those interrupted by the centriolar adjunct or extension), extend cranially to fuse with the capitulum.

The development of the connecting piece commences in round or Golgi-phase spermatids before the association of the developing axial filament with the nucleus. Although the ultimate structure of the connecting piece differs considerably between species,

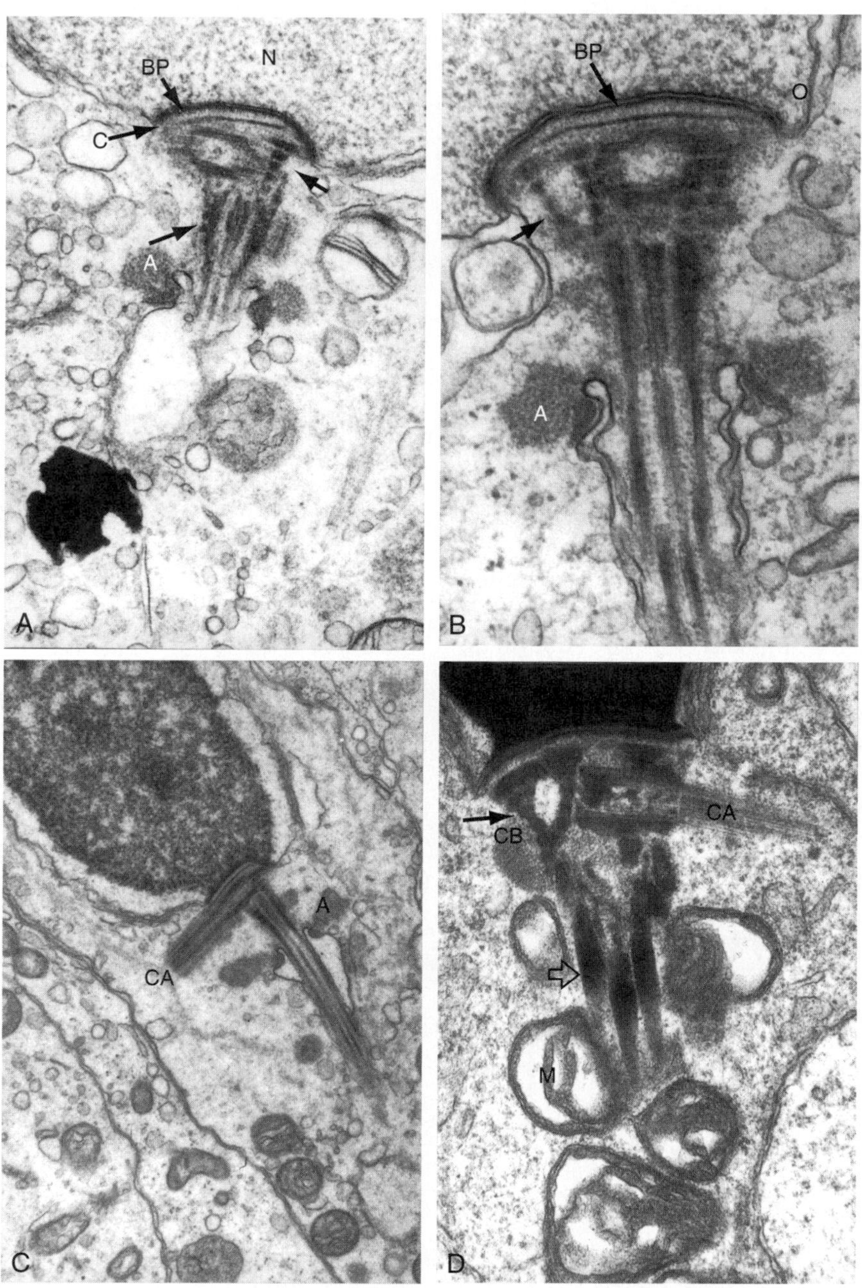

FIG. 29. The developing connecting piece of the human spermatid is shown at the Sb_2 (**A, B**), Sc (**C**), and Sd_2 stages (**D**). Note progressive condensation of the nucleus (N); also, basal plate (BP), capitulum (C), segmented columns of connecting piece (*arrows*), annulus (A), centriolar adjunct (CA), centriole-associated body (CB), mitochondrial sheath (M), and outer dense fibers of mid-piece (*open arrow*).

there is general agreement as to the roles played by the proximal and distal centrioles during its formation. This has arisen from the detailed studies of this region during spermatogenesis in a number of different species (244,289,331). During formation of the connecting piece, the nine cross-striated columns arise from electron-dense material that accumulates adjacent to the triplets of the distal centriole (Fig. 30). This material is augmented by material of similar appearance that develops adjacent to the triplets of the proximal centriole and also extends to contribute to the capitulum. The cross-banding of these fibers is relatively imprecise early in spermatogenesis but becomes very evident later in the process (331).

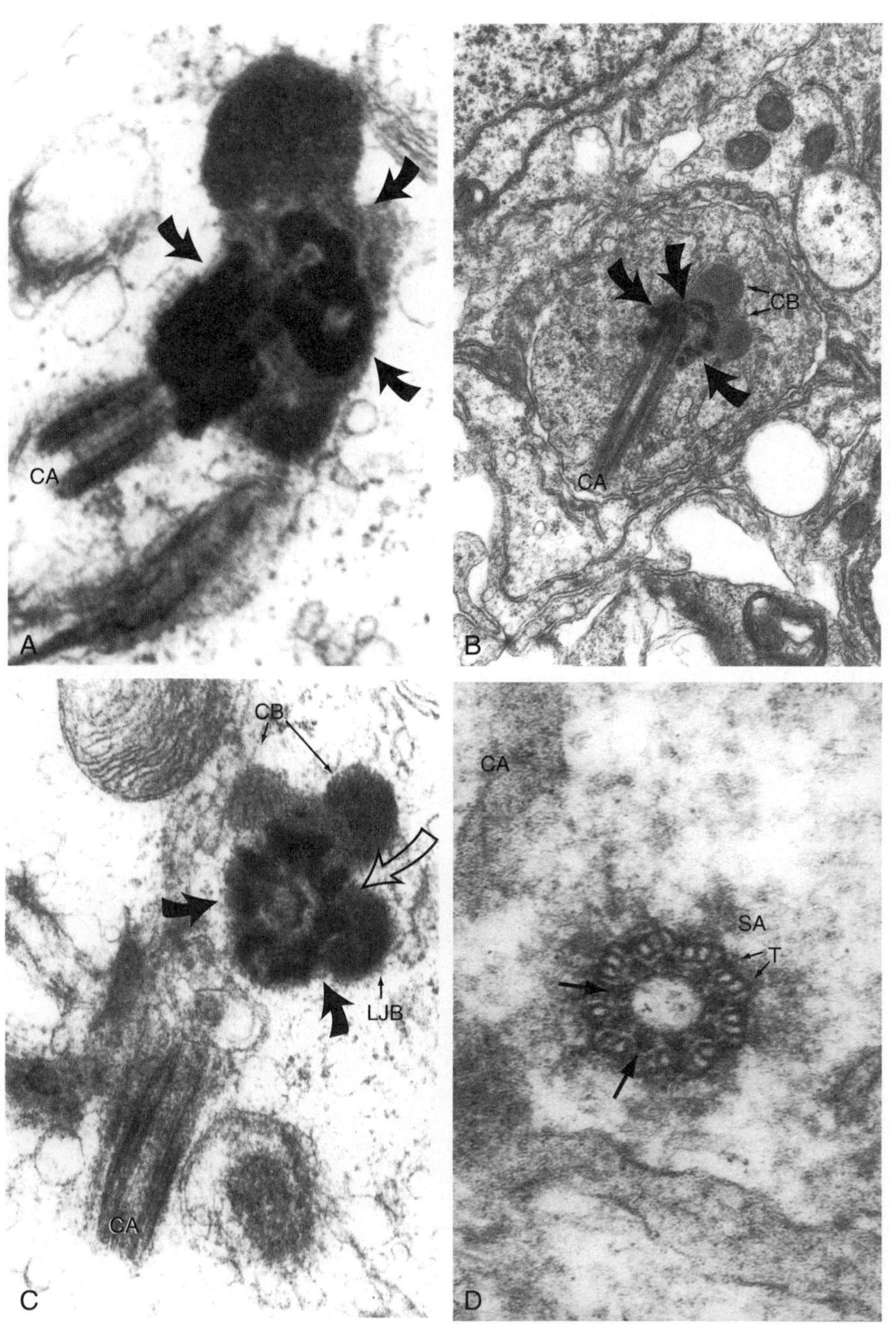

FIG. 30. These micrographs (**A–C**) illustrate sections taken at approximately right angles to the axial filament at progressively descending levels of the connecting piece of the spermatid. Note the columns of the connecting piece (*large arrows*), centriolar adjunct (CA), triplets of distal centriole (*open arrow*, **C**), centriole-associated body (CB), and lateral junctional body (LJB). **C:** Section at right angles through the distal centriole, triplets of longitudinal centriole (*open arrow*), centriole-associated body (CB), and lateral junctional body (LJB). **D** represents a section at right angles through the centriolar adjunct showing triplets (T) of proximal centriole satellite arms (SA) and the inner ring (*arrows*).

The dimensions of the cross-striated columns are usually smaller in the region of the distal centriole, expanding to varying degrees depending on the species as they pass rostrally to insert into the basal plate of the implantation fossa. The latter occurs except for those columns that are interrupted by the extension from the proximal centriole, termed the centriolar adjunct, which takes the form of a "mini-flagellum" extending into the cytoplasm adjacent to the neck of the spermatid (Fig. 31). The cross-striated columns adjacent to the proximal centriole opposite the centriolar adjunct tend to lose their individuality, fusing together to variable degrees.

The centriolar adjunct is preferred as the term to denote the structure that emerges from the distal end of the proximal centriole, because it is not a direct extension of the triplet structure of the centriole (331). The centriolar adjunct extends into the cytoplasm adjacent to the neck of the spermatid (289). This extension retains the basic organization of the centriole with the nine triplets, forming a major part of the adjunct (Figs. 30 and 31). However, the outer pair of subunits, termed B and C, of the triplets is incomplete, presenting a J-shaped profile (331). Within this ring of nine modified triplets, a second ring of nine collections of minute tubular units is found (244,289). These collections are not continuous, as longitudinal views of the centriolar adjunct demonstrate distinct breaks, suggesting a series of 10 to 14 rings in human spermatids (244). Internal to these collections, a circular rim of tubular profiles surrounds an empty cylindrical space.

External to the modified triplets, a moderately electron-dense material is organized to form satellite arms (244). Despite its structural similarity to a centriole, the centriolar adjunct appears to lack the capacity to organize cross-striated material (331). The purpose of forming a centriolar adjunct remains obscure, but it may represent a response of the proximal centriole to influences that cause the distal centriole to form the axial filament. Later in spermatogenesis, the adjunct disappears, because it is only occasionally observed in ejaculated spermatozoa.

The structure of the distal centriole is significantly modified during spermatogenesis, and in some species no evidence of its structure is discernible in mature spermatozoa (289,343). Early in spermatogenesis, periodic densities can be observed in the walls of the distal centriole and adjacent cytoplasm that are the precursors of the distal ends of the longitudinal cross-striated columns of the connecting piece (Fig. 21). The electron-dense material accumulates around the triplets of the centriole and appears to expand the diameter of the centriole (331). The dense material extends outward between the triplets, forming the longitudinal columns of the connecting piece (Fig. 30). Additionally, a pair of rod-like masses develops between the centriole wall and the central pair of microtubules of the axial filament and traverses this region to make contact with the proximal centriole (331). In humans, these dense rods appear to consist of two electron-dense laminae separated by an electron-lucent region and to bow symmetrically (244).

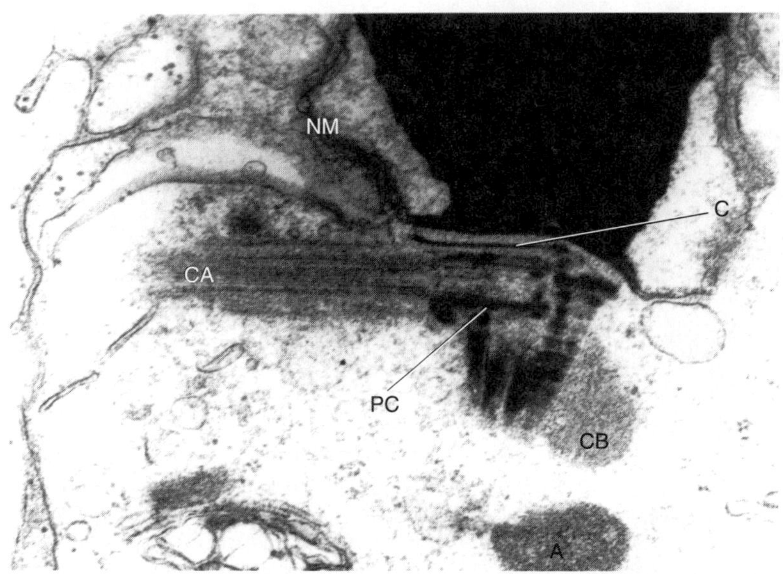

FIG. 31. The neck of the developing human spermatid at the Sd, stage shows the proximal centriole (PC), centriolar adjunct (CA), capitulum (C), centriole-associated body (CB), annulus (A), and nuclear membrane (NM).

There is general agreement that the distal centriole does not give rise to the nine outer dense fibers that characterize the middle piece of spermatozoa (331,344). They arise in continuity with the doublet microtubules of the axial filament and diverge cranially to appear as separate entities (331,345). They develop out of synchrony with the longitudinal striated columns of the connecting piece, but the two systems are closely apposed at an oblique junction at the level of the distal centriole.

Although the method of formation of the connecting piece has been elucidated, the actual mechanisms remain unknown. Fawcett and Phillips (331) proposed that the centrioles act as sites of assembly of precursor molecules formed by normal mechanisms of ribosomal protein synthesis and that the nine striated columns originally present in the developing connecting piece are an expression of the nine-part symmetry of the centriole established by the nine triplets. Fawcett and Phillips (331) suggested that the cross-striated appearance of the columns arises from the alignment of successive segments of cross-striated fibers composed of a fibrous protein with a repeating period of 665 Å measured from the middle of one light band to the corresponding next light band. The dark segments demonstrate less obvious banding, each exhibiting 10 minor bands similar to the cross-striated elements of goshopod spermatozoa (346). The source of precursor material remains conjectural, but several electron-dense aggregations of fibrillar granular material are seen in the vicinity of the developing connecting piece. In human spermatozoa one of these consists of electron-dense granular material found to be associated with one side of the distal centriole and often exhibits a less dense central zone (289). It is also present in bandicoot spermatids (*Perameles nasuta*) and was termed the lateral junctional body by Sapsford and coworkers (330).

The lateral junctional body appears early in spermatogenesis, before the articulation of the developing tail with the nucleus (Figs. 28, 29, and 30). A second roughly spherical body, slightly less electron dense than the lateral junctional body, is seen at the same level, closely associated with the distal centriole; this has been termed the centriole-associated body (289). In some views, the finely granular material demonstrates periodicity, giving the appearance of linear arrays (244). Both the centriole-associated body and lateral junctional body disappear very late in spermatogenesis.

A third structure of moderate electron density is the annulus, which also appears early in spermatogenesis at the stage in which the intracytoplasmic portion of the axial filament is largely exteriorized. This occurs by the formation of a cleft, termed the cytoplasmic canal, which results in an invagination of the cell membrane around the flagellum. The annulus forms a ring-shaped structure that limits the proximal boundary of the cytoplasmic canal. Early in spermatogenesis it is relatively small, but it is augmented later by yet another structure, the chromatoid body, which has migrated from the vicinity of the Golgi complex to the region of the annulus (233). Here it forms a relatively large, ring-shaped, electron-dense, granular structure that is in contact with the annulus and in some views appears to contribute material to that structure. The combined annulus and ring-shaped chromatoid body correspond to what was termed by early investigators as the ring centriole. Fawcett et al. (233) note a progressive decrease in size of the chromatoid body with its eventual disappearance, whereas others suggested it was lost in the residual body (347). The actual nature of the chromatoid body is under debate. Early work suggested that it contained basic proteins and may be a source of RNA (253,348,349); however, Eddy (349) was not able to confirm this conclusion. More recently, Walt and Armbruster (350) again proposed that RNA is found in the chromatoid body and also demonstrated that it contained actin, perhaps reflecting its remarkable motility in spermatids.

The form of the annulus varies between species (344). In some such as humans, guinea pig, chinchilla, mouse, and ram, the ring-shaped electron density has a convex outer border in contact with the cell membrane in a fold at the proximal limit of the cytoplasmic canal (Fig. 32). Its free edge is in contact rostrally with the lower edge of the mitochondrial helix of the mid-piece. In other species (doormouse, Chinese hamster, suni antelope), the annulus is wedge shaped, is located at the proximal limit of the cytoplasmic canal, and its apex extends toward the axial filament.

The articulation of the connecting piece with the nucleus occurs at the implantation fossa. In this region the nuclear membranes are closely apposed and separated by a space of 60 to 80 Å that in favorable views is crossed by uniformly spaced densities, possibly indicating structures that bind the membranes together (345). The outer nuclear membrane appears to be thickened by the deposition of electron-dense material to form the basal plate (Figs. 28 and 29). The anlage of the connecting piece lodges in the implantation fossa in such a manner that the capitulum is aligned parallel to the basal plate but separated from it by an electron-lucent space approximately 40 Å wide. Fawcett and Phillips (331) observed many fine filaments traversing this zone, oriented perpendicular to the capitulum and basal plate, and suggested that they were responsible for attaching the flagellum to the head.

Formation of the Outer Dense Fiber System. Over a large portion, the axial filament is modified by

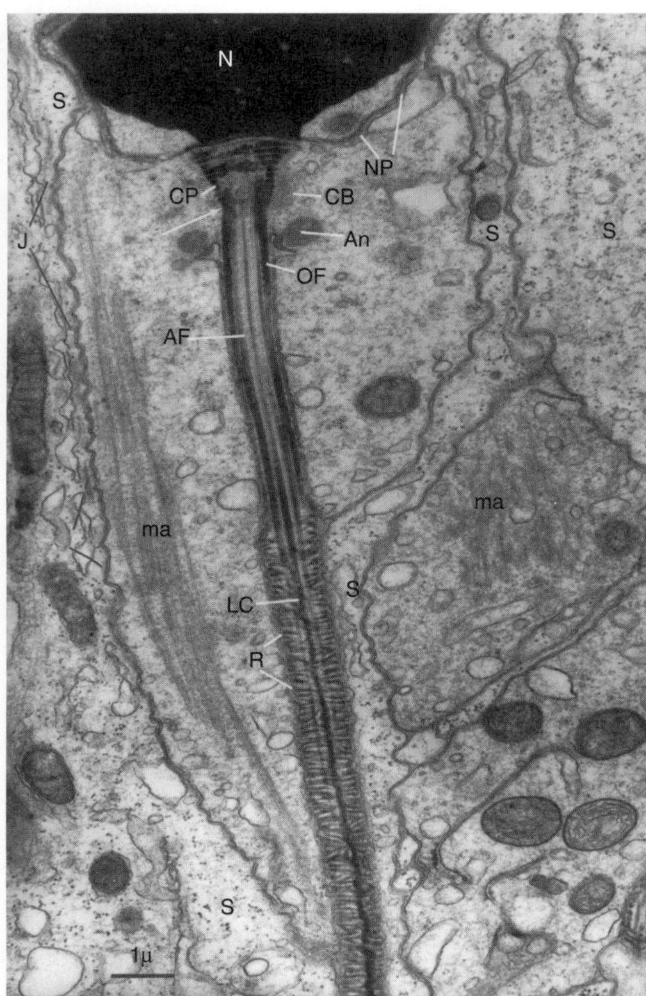

FIG. 32. The structure of the developing human spermatid at the Sd1 stage illustrates the longitudinal segmented columns (LC), annulus (A), nucleus (N), nuclear pores (NP), connecting piece (CP), cytoplasmic canal (CC), axial filament (AF), outer dense fibers (OF), ribs (R) of principal piece, and manchette (M). Note the cisternae (C) delineating a mantle of Sertoli cell (S) cytoplasm containing fibrils.

the development of a set of nine electron-dense fibers termed the outer dense fiber system (Figs. 20 and 32). They appear to develop as very thin fibers that are attached to the outer wall of the doublet microtubules of the axial filament (331) and seem to separate proximally to develop further as independent fibers by further accretion of material. This relationship to the doublets of the axoneme persists, the outer dense fibers being separated by a greater distance from the doublets cranially, whereas distally they retain their close association with the doublets and taper to eventually disappear. Irons and Clermont (351) showed that [^{3}H]-proline and [^{3}H]-cystine are incorporated into the outer dense fiber system over an extensive period of rat spermatid development from steps 8 to 19, indicating the presence of a peak of protein synthetic activity in mid-spermiogenesis. This view is supported by the studies of O'Brien and Bellve (352), who approached the problem in the mouse by different techniques. Oko (353) undertook a comparative analysis of the proteins forming the outer dense fibers and produced polyclonal antibodies to some of the major fractions. Using these antibodies raised against components of 32, 26, and 14.4 kDa, they noted that granular cytoplasmic localization was present over spermatids from steps 9 to 19 with peak activity at step 16 (354). At the electron microscopic level, the antibodies bound to components of the granulated bodies that appear in relationship to the cisternae of endoplasmic reticulum from stages 10 to 14, increasing to a peak at stages 15 to 17 and decreasing thereafter. These results support the view that the granulated bodies represent transitory stores of proteins en route to the outer dense fibers (355). This conclusion has been supported by studies using an antiserum directed against other outer dense fiber components (356,357).

There are significant differences in the size and length of the outer dense fiber system between species (312,328). In human sperm the outer fibers end in the proximal part of the principal piece, fibers 3 and 8 being the first to terminate. In some they are prominent and extend throughout the mid-piece and principal piece, whereas in others they terminate more proximally. Unlike the doublets of the axial filament, which are similar in size and appearance, the outer dense fiber system, numbered according to the adjacent doublet, shows significant differences in size (312). Fibers 1, 5, and 6 are usually larger in many species.

The outer dense fibers possess cortical and central zones that differ in electron density. The portion of the cortical zone immediately adjacent to axial filament consists of electron-dense punctate granules in guinea pig spermatozoa, leading Fawcett (312) to call them satellite fibrils. The outer dense fiber system must be considered as a set of structures that is distinct from the striated fibers of the connecting piece, because they develop independently and asynchronously (331). However, they do join each other at the lower level of the connecting piece at an oblique junction.

Advances in molecular biology and proteomics have circumvented some of the difficulties of relative insolubility and N-terminal blockages originally encountered when trying to determine the composition of the outer dense fibers (358). A significant proportion of the outer dense fiber is composed of a network of leucine zipper and/or coiled-coil proteins. Specifically, a major component of the outer dense fibers, Odf2 (also named Odf27, rt7, rts) (359), associates not only with itself (360), but also with the most abundant outer dense fiber component, Odf1 (alias Odf84), via a leucine zipper motif (361). Further, Odf1 interacts with Spag4 via leucine zipper motifs at the interface of the outer dense fibers and the axoneme during spermatogenesis but not in mature sperm (362). Odf1 also interacts with Spag5 via a leucine zipper motif during spermatogenesis at an undefined location (363).

In addition to serving as a molecular network for outer dense fiber formation, the major components of the outer dense fiber and the fibers themselves may also serve to anchor different components in the tail. van der Hoorn and colleagues (362) predicted that proteins in the outer dense fibers interact with the axoneme, the fibrous sheath in the principal piece and/or the mitochondria of the mid-piece. In support of this hypothesis, Odf1 is found on the surface of the outer dense fibers facing the axoneme, thus allowing an interaction with Spag4 and the developing axoneme. Odf2, on the other hand, is present in the cortex and consequently faces the mitochondria in the mid-piece and the fibrous sheath in the principal piece (357).

It should be noted, however, that not all outer dense fiber components are linked to Odf1 or to each other via leucine zipper motifs. Indeed, the purification of molecules such as cdk5 (364), which is believed to be involved in the phosphorylation of outer dense fiber component proteins during spermatogenesis, voltage-dependent, anion-selective channels 2 and 3 (365), and several proteins of unknown function, suggests that the outer dense fibers may act as anchoring structures for proteins with more dynamic roles in sperm motility. This hypothesis remains to be proven, however. Several outer dense fiber associated proteins, including Spag4 and Sak57 (362,366), are transiently found in the manchette during spermatogenesis, suggesting that the manchette is a scaffold structure involved in the transport of proteins along the length of the developing sperm tail (318).

Development of the Principal Piece. The region of the tail between the annulus and the termination of the fibrous sheath or tail helix is termed the principal piece (329). The presence of a fibrous sheath is peculiar to mammalian spermatozoa, and early ultrastructural studies demonstrated that the sheath was not helical but consisted of rib-like structures joined to two longitudinal columns (329,367,368) (Fig. 33). The closely spaced ribs sometimes branch, attaching to adjacent ribs. Proximally the longitudinal columns consist of the outer dense fibers 3 and 8, but these terminate distally, and the ribs are attached to a thin electron-dense ridge that projects from doublets 3 and 8. The longitudinal columns vary in prominence with species, being well developed in rodents but insignificant in humans (328,329,367).

In the human, the anlage of the fibrous sheath appears to be a system of transversely oriented microtubules (289). The hollow cores of these tubules are progressively obliterated by an electron-dense material, a process that gives rise to the transversely oriented ribs (Fig. 34). Progressive addition of electron-dense material thickens the ribs. More extensive studies of the development of the principal piece in human spermatids have confirmed the microtubular origin of the ribs (368). They showed that a large accumulation of microtubules occurs in the region of the developing principal piece (Figs. 34, 35) to form a microtubular body at the Sc stage of spermatid development (252). Subsequently, with elongation of the axial filament and the distal migration of the annulus, this collection of microtubules is diminished and eventually disappears (244). A microtubular body similar to human spermatids was described by Nicander (369) in cat and rabbit spermatids; it occurred at the anterior extremity of the principal piece and eventually disappeared.

An alternative method of formation of the ribs of the principal piece was proposed by Sapsford and colleagues (370) from their studies of the bandicoot (*Perameles nasuta*). They could find no evidence of microtubules forming the analage of the ribs; instead, they observed that a series of fine filaments joined the longitudinal columns that had appeared as electron-dense thickenings adjacent to doublets 5 and 8. These filaments aggregate and converge to form the definitive ribs. Similar fine filaments were also noted in the lizard (371) and in the mouse (372). The two methods of rib formation are strikingly dissimilar, and further studies in other species are required to determine which process is more common. Irons and Clermont (373) demonstrated by the use of [^{3}H]-proline incorporation into proteins that the longitudinal columns of the principal piece take 15 days to be formed. They also showed that in the rat, the ribs develop independently and asychronously from the longitudinal columns over a period of 4 to 5 days from proteinaceous filaments. Recent studies using immunocytochemical techniques localized fibrous sheath proteins to the cytoplasm of spermatids at steps 9 to 19 (354,355). This diffuse cytoplasmic material did not appear in granulated bodies and did not localize at the electron microscopic level over the dense material that accumulates in the regions where the ribs of the fibrous sheath form (355). However, positive immunocytochemical localization was noted over the ribs of the fibrous sheath, suggesting a direct transfer of the cytoplasmic protein to the ribs.

Recently, there have been huge advances in our understanding of the composition and function of the fibrous sheath. This is described in detail in Chapter 1, but a brief description is warranted here. Data from biochemical analyses and animal models suggest that in addition to providing a structural role in sperm tail directionality, the fibrous sheath is a scaffolding structure for molecules involved in the regulation of

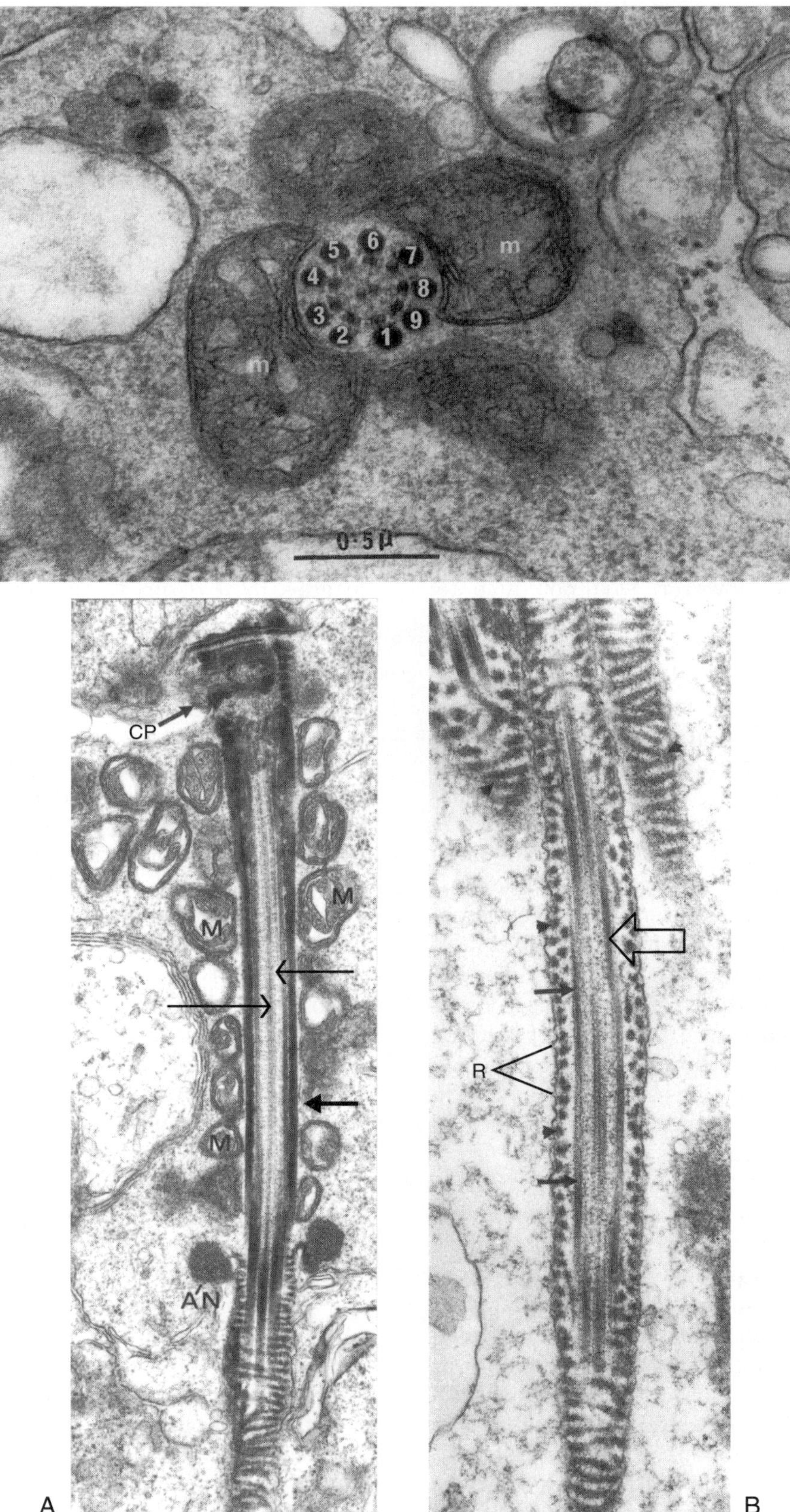

FIG. 33. Top: Cross-section of the mid-piece of a human spermatid illustrating mitochondria (M), outer dense fibers (numbered), and the axial filament. **Bottom:** Two longitudinal sections through developing spermatid tails in the region of the mid-piece (**A**) and principal piece (**B**). Note the ribs (R), microtubules of axial filament (*long arrows* in A, *open arrow* in B), annulus (AN), connecting piece (CP), mitochondria (M), and outer dense fibers (*short arrows* in B).

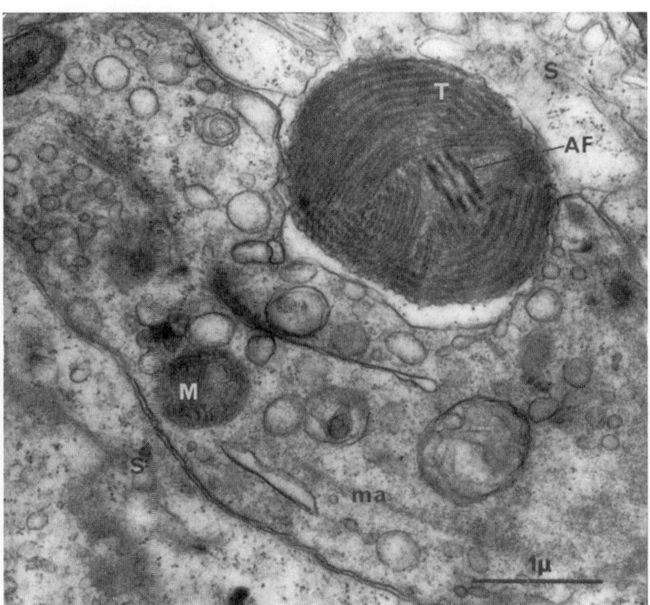

FIG. 34. A human spermatid at the Sc stage shows the axial filament (AF) surrounded by the microtubular body (T) and mitochondria (M).

axoneme function and consequently motility. The major component of the fibrous sheath, A-kinase anchoring protein (AKAP) 4 (374–376), along with AKAP3 (377,378) and TAKAP-90 (379), has anchoring sites for cAMP-dependent protein kinase. Further, AKAP3 binds to ropporin, a sperm-specific protein that in turn binds to rhophilin and then the GTPase Rho (380,381). In addition, the fibrous sheath contains sperm cell–specific components of the glycolytic pathway, including glyceraldehyde 3-phosphate dehydrogenase-S (382) and hexokinase type 1-s (383), suggesting a further role for the fibrous sheath in the energy production required for hyperactivated motility (384,385).

Mitochondrial Sheath. The aggregation of mitochondria was described in spermatids around the axial filament in the region now called the middle piece (386). Similar findings were noted for the human testis. Early ultrastructural studies demonstrated that the mitochondria actually form a spiral sheath that in mammalian spermatozoa was of variable length (329,387). The number of spirals forming the sheath varies from 5 to 14 for humans (388,389) and up to 40 in the guinea pig (312). The mouse (up to 90), bat (up to 15), and rat (up to 350) represent the largest number of spirals encountered (264). Andre (231) called attention to the fact that the helical configuration provides the least resistance to bending.

The actual formation of the middle piece occurs late in spermiogenesis (330). Earlier, the mitochondria are distributed evenly in the cytoplasm in some species, and in others, such as humans, they are peripherally placed close to the cell membrane (256,288). Formation of the mitochondrial sheath is preceded by caudal migration of the annulus, and the mitochondria associate with the axial filament between the connecting piece and the annulus (Figs. 32 and 33). The mechanisms involved in the migration and aggregation of mitochondria remain unknown. However, the organization of the manchette, a cylindrical collection of microtubules extending from the nuclear region to the caudal region of the spermatid, excludes the mitochondria from the region of the flagellum. When this collection of microtubules disperses late in spermatogenesis, the mitochondria are free to aggregate to form helical arrays (303). The end-to-end arrangement of mitochondria is usually random, but Fawcett (264) called attention to the observation that in some species there is a remarkable regularity in spacing (303).

In many species the organization of the cristae of mitochrondria in germ cells takes on certain unique features. As described earlier, the intracristal space of mitochondria in primary and secondary spermatocytes is dilated, often giving a vacuolated appearance to these organelles (231). This process persists in spermatids and results in peripheral margination of the cristae such that in some species a central clear zone appears in the mitochondria (209). In other species the entire mitochondrion appears to be filled by membranes that are concentrically orientated (345,389,390).

Reorganization of Cytoplasm and Organelles and Spermiation

The process of spermatogenesis is characterized by dramatic changes in the relationship of the nucleus and cytoplasm and the remarkable movements of organelles within spermatids. Many of these changes were described by the classical cytologists and were used by Le Blond and Clermont (250) to stage the process of spermatogenesis. The changes in nuclear shape and position were described earlier, but the movement of cytoplasm toward the caudal end of the spermatid requires some discussion.

It was noted that when the caudal movement of cytoplasm occurred, a system of cytoplasmic filaments appeared, extending from the nuclear membrane and terminating freely in the caudal cytoplasm (250). This system was given the name manchette by von Lenhossek (391).

McIntosh and Porter (313) demonstrated that the manchette actually was composed of a cylindrical array of microtubules that was noted to arise from a ring-like structure surrounding the postacrosomal

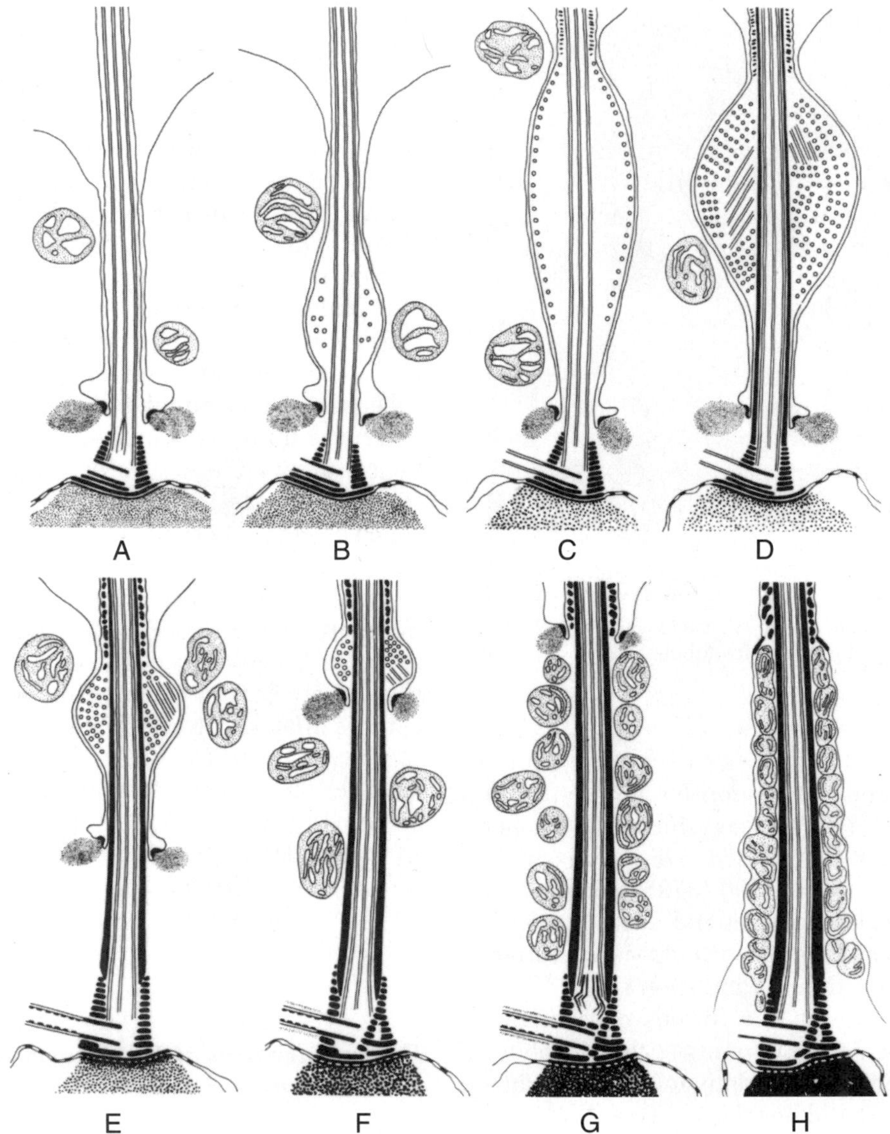

FIG. 35. This diagram illustrates the contribution of the "microtubular body" to the formation of the principal piece in human spermiogenesis. [From (244), with permission.]

region of spermatids in the cat (188). This nuclear ring, originally described by Gresson and Zlotnik (392), consists of electron-dense material that appears to thicken the cell membrane adjacent to the postacrosomal region of the nuclear membrane (303).

The proximal ends of the microtubules of the manchette are embedded in this dense fibrillar material. However, this electron-dense deposit has not been observed in all species, being absent in humans and the bandicoot (289,330). The microtubular nature of the manchette (Figs. 22 and 26) led to suggestions that it is responsible for shaping of the nucleus and reorganization of the cytoplasm (313). However, from a detailed study of spermatogenesis in a number of species, Fawcett and coworkers (303) produced evidence to indicate that nuclear shape was dependent on genetic factors rather than on the manchette. They also suggested that rather than being involved in the physical movement of cytoplasm, the microtubules of the manchette may act as a framework or conveyor for the transport of cytoplasmic vesicles. These vesicles were closely associated with the manchette and sometimes were physically linked by slender linear densities (303). These linear densities are similar in appearance to those linking adjacent microtubules composing the manchette (Fig. 22). The fate of the manchette was unclear, but de Kretser (289) suggested that the microtubules were incorporated in the residual body, a view confirmed by studies of equine spermatogenesis (393).

Associated with the caudal movement of the cytoplasm, the Golgi complex and chromatoid body migrate to the abacrosomal pole. The former can be identified as a component of the residual body, that portion of the spermatid cytoplasm that is shed when the spermatid leaves the seminiferous epithelium (Fig. 3) in a process called spermiation. Early studies demonstrated that the greater part of the spermatid cytoplasm is shed as the residual body, which, in most instances, is phagocytosed by the adjacent Sertoli cells (394,395). Some cytoplasm remains to form a droplet that surrounds the middle piece and contains a few vesicular profiles. Lacy (395) noted that the residual body contained remnants of the Golgi complex and endoplasmic reticulum, and a number of studies have confirmed their phagocytosis by the Sertoli cell (396,397).

The method by which the spermatid sheds the residual body and leaves the epithelium has emerged from the results of a number of ultrastructural studies (289,398,399). de Kretser (289) noted that late in spermatogenesis, the caudal spermatid cytoplasm is invaginated by processes of Sertoli cell cytoplasm and postulated that these processes actually were responsible for pulling off the residual body (Figs. 36 and 37). Similar processes of Sertoli cell cytoplasm within spermatids have been observed in the studies of rat spermatogenesis (399). The residual cytoplasm, in addition to containing remnants of the Golgi complex, also contains ribosomes, lipid inclusions, mitochondria, microtubular remnants of the manchette, and electron-dense remnants of the chromatoid body. The residual bodies from human spermatids were also noted to contain flower-like structures noted in the cytoplasm of spermatids earlier in spermatogenesis (400). These consist of a core of densely packed osmiophilic granules surrounded by a translucent vesicle and originally appear near the nucleus in association with the chromatoid body but their function is unknown. Associated with the invagination of spermatid cytoplasm by Sertoli cell processes, there is progressive movement of the spermatid toward the lumen; the cytoplasm attached to the Sertoli cell is linked to the spermatid by progressively attenuating connections (398). All connections are lost eventually, most of the residual bodies being retained with the Sertoli cell. Breucker et al. (401), however, suggested that in the human many residual bodies are shed into the tubule lumen.

The mechanisms involved in spermiation remain unclear. In mammals, the late spermatids appear to progressively lose their contact with the Sertoli cell, being anchored by the tubulobulbar complex in the region of the head (402). The structure of this device is described later, but it represents the final attachment device by which the spermatid retains contact with the Sertoli cell; although originally described in the rat, it has also been found in humans (244). Loss of this final contact represents the completion of spermatogenesis, and by definition, the cell is now termed a spermatozoon (398,403). In amphibians, Burgos and Vitale-Calpe (404,405) suggested that spermiation resulted in a swelling of the terminal cytoplasmic processes of the Sertoli cells, resulting in the evagination of the lacunae that housed the spermatids, thereby shedding them into the lumen of the tubule. This process was suggested to be a response to human chorionic gonadotropin (hCG) or LH, but similar actions have not been confirmed in mammals.

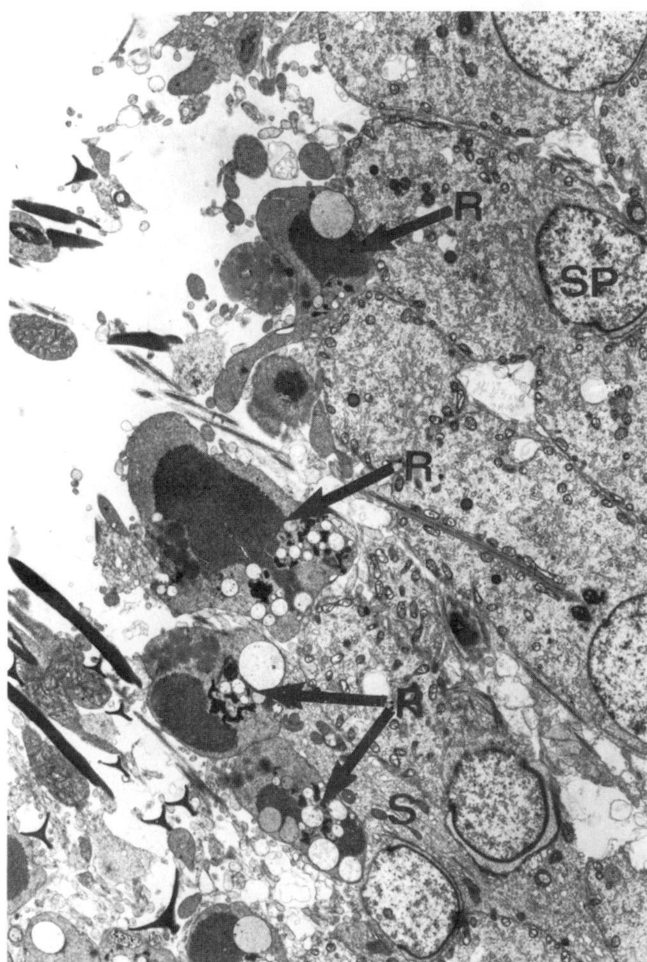

FIG. 36. Adluminal aspect of the seminiferous epithelium of the rat testis, illustrating Sertoli cell cytoplasm (S), spermatids (SP), and lobes of residual cytoplasm (R).

Relationships between the Sertoli Cell and Spermatids

During the early stages of spermatogenesis, the spermatids lie centrally and immediately adjacent to the lumen of the seminiferous tubules, surrounded

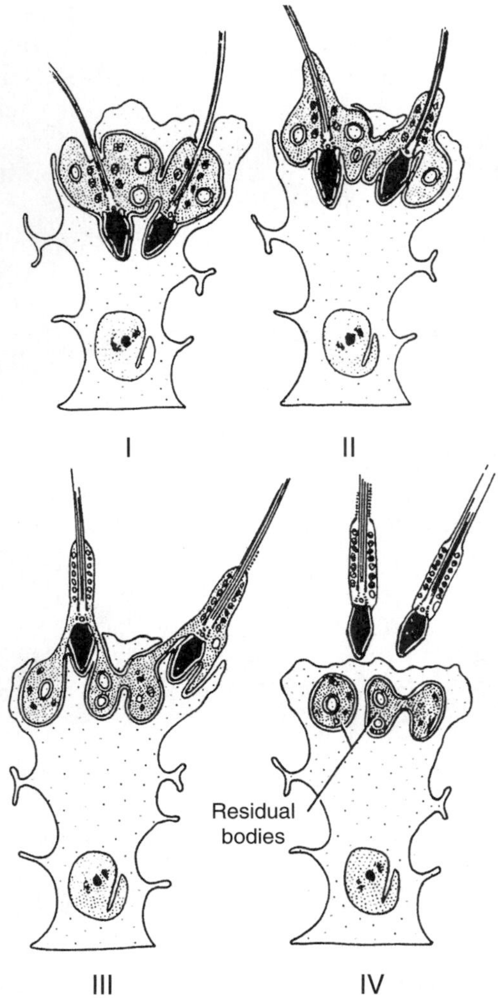

FIG. 37. The final stages of human spermiogenesis are illustrated diagrammatically. Note the manner in which processes of Sertoli cell cytoplasm essentially "pull off" the residual cytoplasm (residual body) and retain it within their cytoplasm.

by processes of Sertoli cell cytoplasm. At this time small punctate desmosome-like cell junctions occur with the adjacent Sertoli cell (237). The spermatids do not demonstrate any specific orientation to the Sertoli cell, but in later stages they are deeply embedded within the epithelium, oriented such that their heads, covered by the acrosomal cap, point basally. Subsequently, they progressively move toward the lumen of the tubule, eventually to lose all contact after spermiation. Studies in procarbazine-treated rats have shown that acrosome formation is disrupted but that this process did not alter the movement of the nucleus into its eccentric position (406). However, the altered acrosome formation disrupted alignment of the heads of the spermatid within the epithelium, possibly by altering the distribution of the Sertoli cell–spermatid junctional specializations described below.

The formation of certain specialized junctions with the adjacent Sertoli is probably related to the specific orientation taken by the spermatids in mid-spermatogenesis. Early studies noted that the cytoplasm of the Sertoli cell adjacent to the heads of spermatids late in spermatogenesis (Fig. 22) was separated into a mantle layer by the presence of an array of vesicles (288,407). A number of studies in different species have demonstrated the existence of these specialized cell attachments in which there is no reduction in the intercellular space but the demarcation of a thin zone of Sertoli cell cytoplasm by the presence of a series of cisternae, the cytoplasmic layer so defined demonstrating the presence of numerous fibrils (243,408–410). The cell junctions have some similarity to the inter-Sertoli cell junctions, but there is no modification of the intercellular space. In many species these specialized Sertoli cell–spermatid cell junctions are located only in the region of the head, but in the human they can be demonstrated over larger areas of the spermatid cell surface (Fig. 32). The cell junctions first appear when the spermatid nucleus takes up an eccentric position with the cell coming into close association with the cell membrane.

A further junctional specialization occurs between late spermatids and the Sertoli cells wherein the cell membrane in the region of the head of the spermatid projects into the surrounding Sertoli cell cytoplasm (410). These tubulobulbar processes provide a mechanism of anchoring spermatids immediately before spermiation (Fig. 38). In the rat, they appear to be limited to the cell membrane in relation to the concave portion of the nucleus (410,411), but in humans they are more irregularly distributed (244).

Reference was made earlier to the processes of Sertoli cell cytoplasm that appears to invaginate the caudal spermatid cytoplasm immediately preceding spermiation. Morales and Clermont (399) described two types of Sertoli cell processes in the rat, one that is essentially devoid of organelles and one that contains vesicles. From these observations, they proposed that the latter may represent a mechanism of transferring materials between these two cell types.

Quantitative Studies of Spermatogenesis

Quantitative assessments of the cytology of the seminiferous epithelium have in the past encountered considerable difficulties with regard to finding unbiased yet reliable methods for cell counting due to fixation artifacts, accurate cell identification, and changes in size and shape of the germ cells. In the past decade new techniques have been developed for quantitation of tissue composition, based on their distribution in the three-dimensional space of cells

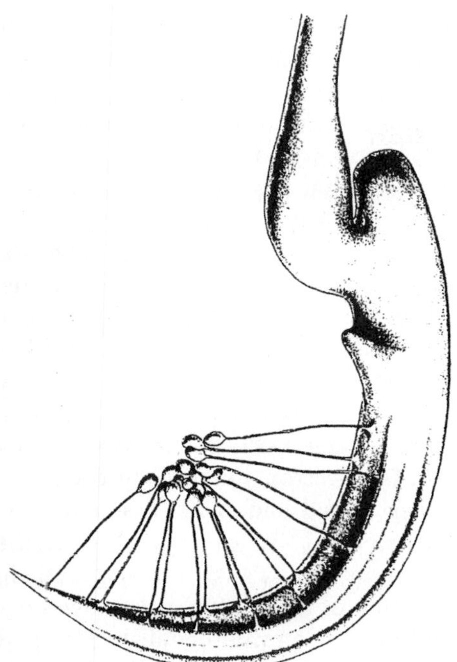

FIG. 38. A diagrammatic representation of the three-dimensional features of a step 19 spermatid of the rat. The concave aspect of the sperm head extends elaborate tubulobulbar complexes into the surrounding Sertoli cell cytoplasm, the latter not shown here. The complexes are congregated together and form bulbous terminal dilations. They are said to act as an anchoring device to temporarily retain the spermatid head within apical extensions of Sertoli cell cytoplasm. [From (402), with permission.]

that together make up a tissue or an organ. The presence of recognizable cells within fixed tissues is determined using thick (20–25 μm) acrylic resin sections in which cell nuclei are visible at various focal planes when viewed by microscopy. By counting all nuclei within computer-generated counting frames throughout the depth of sets of defined tissue sections, referred to as the "optical dissector" approach, an unbiased estimate of cell numbers per organ may be determined. The theoretical basis and practical applications of morphometric techniques such as the optical dissector are classified as stereology (412–414). Methods such as these have revolutionized the quantitative evaluation of the testis (415) because they offer efficiency and importantly unbiased methods that are assumption free in relation to cell numbers and shape. The latter is particularly relevant to germ cells that change from spherical entities to irregular elongated pyriform shapes in the form of the late spermatids.

Table 1 gives examples of the composition of the seminiferous epithelium in a range of species based on these new methods of stereology. To highlight the similarities and/or differences in the cytology between species, the data for cell types are expressed per cm^3 of testis to allow comparisons between extremes of testis size, for example, mouse testes (less than 100 mg) and primate testes (20 g or more). Where necessary, the original data sets were converted to averages between studies and in most cases were recalculated from the usual reporting method of millions of cells per testis. Values in Table 1 are in part agreement with earlier estimates in 12 mammalian species of the number of Sertoli cells/cm^3 of testis (416). Where the values differ is likely a result of those studies being based on quantitative methods that yield biased data that are dependent on measuring the size and shape of individual Sertoli cells.

The optical dissector–derived data show a nearly fivefold difference in the number of Sertoli cells/cm^3 (numerical density) between the rat and the rhesus monkey, but most other species show reasonably similar densities of Sertoli cells. Not surprisingly, based on previous knowledge of sperm concentrations and volume within the ejaculate, there are considerable differences in spermatid numerical densities within testicular tissue. Total germ cells/cm^3 also show much variation, the data for humans tending to be low but comparable with rat, with some strains of mice and the lemur exhibiting high overall germ cell densities in the testis. The average number of germ cells associated with a single Sertoli cell, a measure

TABLE 1. *Composition of seminiferous epithelium assessed by quantitative histology*

	Mouse	Rat	Rabbit	Lemur	Marmoset	Rhesus	Cynomolgus	Human
Sertoli cells	27	19	22	38	40	99	30	32
Spermatogonia A	17	9	17	36	8	24	17	30
Spermatogonia B	45	20	16	83	18	17	16	5
Spermatocytes Leptotene Zygotene	57	16	37	128	60	33	55	15
Spermatocytes Pachytene	230	65	160	260	106	97	98	55
Round spermatids	500	135	360	787	226	330	190	102
Elongate spermatids	490	117	160	761	255	148	385	115
Total germ cells	870–1050	240–410	640–870	2050	670	650	720–800	280–360
Testis volume cm^3	0.062–0.077	1.7	2.0–4.0	0.5–1.5	0.3–0.6	9–25	13–25	20–22
Germ cells/Sertoli cell	35	20	27–39	54	17	7	26	11

Cell data in millions per cm^3 of testis representing average values.

of the "cell load" carried by the Sertoli cell, is somewhat less variable, although the ratio is low in humans and higher in species such as the mouse, rabbit, and lemur. Interestingly, the data show that although the rhesus monkey testis has an extraordinary density of Sertoli cells, each of them on average carries only seven germ cells. Contrast this with the grey mouse lemur, where the load exceeds 50 germ cells per Sertoli cell. The disassociation between the total germ cell population per testis and the numbers supported by the Sertoli cells is reported for a group of New World monkeys of the *Platyrrhini* species again using the optical dissector method (417). For the cotton top tamarind, total germ cells per testis is 940 million with a germ cell-to-Sertoli cell ratio (germ cell load) of 33; the saddle-back tamarin with 1,350 million germ cells and Sertoli cell load of 26; and squirrel monkey with 2,750 million germ cells and Sertoli cell load of 21. Comparisons of quantitative estimates such as these data emphasize the interspecies variability in the composition of the seminiferous epithelium and underline the need for caution in extrapolating the cytological characteristics of one species to another.

SERTOLI CELL

The Sertoli cells were originally described in 1865 by the Italian physiologist Enrico Sertoli (418). He identified the cells as being individual elements extending from the basement membrane to the lumen of the seminiferous tubule and, in doing so, enveloping the many clusters of associated germ cells. The Sertoli cell is a tall columnar cell extending perpendicularly through the seminiferous epithelium (Fig. 8), and von Ebner (159,240) first proposed that a physiological relationship existed between the Sertoli cells and the germ cells.

Without the advantage of modern optical equipment and methods of tissue preservation, Sertoli (419) and Brown (202) constructed diagrams of the shape of Sertoli cells indicating the pronounced length of the centripetal axis compared with the circumferential axis. Sertoli proposed that these cells were independent cellular units and did not form a syncytial relationship with each other, but von Ebner (159) suggested that a symbiotic relationship existed between the Sertoli cells and the more mature generations of germ cells, thereby forming a functional unit, which he termed the spermatoblast (240).

The precise morphological organization of the Sertoli cell remained poorly understood for many years (190,420–424). Electron microscopy clarified the fact that each Sertoli cell had distinct cellular boundaries, and ultrastructural investigations contributed by Fawcett [see reviews (425,426]) have greatly increased our appreciation of the central role played by this cell in the regulation of spermatogenesis. Detailed studies of the topography of the Sertoli cell, its relationship to germ cells, and the changes in Sertoli cell morphology in health and disease (427–437) have each added substantially to a deeper understanding of Sertoli cell histology (Table 2). The book, *The Sertoli Cell*, edited by Russell and Griswold (438), is a superb source of information covering many aspects of Sertoli cell biology.

Sertoli Cell Shape

The early histologists recognized that the unusual columnar shape of the Sertoli cell reflected a three-dimensional configuration of great complexity (202,419). In their studies defining the seminiferous cycle, Le Blond and Clermont (251) noted that the Sertoli cells alter their shape in relation to the 14 stages of the spermatogenic cycle. With a silver

TABLE 2. *Quantitative morphology of Sertoli cells*

Species	Volume (μl/m³)	Surface area (μm²)	Numerical density (10⁶/cm³ testis)	References
Opossum	7,000			(1090)
Mouse	3,300		36	(1090)
Rat	5,000–8,000	9,000–20,000	26–28	(430–432)
Guinea pig	3,300		55	(1090)
Hamster	4,200–4,700	13,000	35–44	(1090,682)
Woodchuck	3,000		47	(1090)
Degu	2,000		80	(1090)
Rabbit	4,500		25	(1090)
Dog	5,000		43	(1090)
Water buffalo	7,000–9,000	11,000–14,000		(468)
Stallion	3,300		24–60	(670,1090)
Monkey	2,600–4,100	2,400	102	(612,1090)
Human	4,800		33–49	(458,1090)

staining method applied to paraffin sections, Elftman (435) concluded that in the rat testis, the Sertoli cells were tall, columnar in shape, and their distribution within the seminiferous epithelium resembled the pattern of trees planted in an orchard where the basal aspect of the Sertoli cells, most readily visible by light microscopy, was likened to a tree trunk (Fig. 39) and the cytoplasmic ramifications between the surrounding germ cells being analogous to the branches. Many subsequent ultrastructural studies have confirmed that although this appealing topographical description may not be entirely applicable to all species so far studied, a better portrayal has not emerged.

The strategic position of the Sertoli cell within the seminiferous epithelium and its intrinsic relationships to neighboring Sertoli cells and germ cells (Figs. 39 to 43) has repeatedly been emphasized in fine structural analyses of a great variety of vertebrate and invertebrate species (426,436–442). The cytoplasmic extensions of the Sertoli cell characterize the topography of the cell as possibly the most complex yet described in any epithelium. The Sertoli cells must continually alter their shape to accommodate the structural transformations and mobilization of germ cells from the base to the free surface of the seminiferous epithelium. Wong and Russell (430) serially sectioned a single Sertoli cell from a stage V rat seminiferous tubule and showed that the cell was best characterized as a short body resting on the basal lamina with many upward-projecting sheet-like cytoplasmic extensions forming cone, cup, and cylindrical configurations. When the seminiferous epithelium was viewed perpendicular to the basal lamina, the Sertoli cells form a hexagonal array (i.e., they presented six membrane surfaces near their base) (419,423,443), although other studies suggest a pentagonal arrangement (431,444).

Ultrastructural studies of rat Sertoli cells at different stages of the spermatogenic cycle depicted variations in shape (Fig. 40) that seem at first to indicate highly irregular and disordered configurations (445). However, the margins of the Sertoli cell are obliged to undergo transformations to remain in

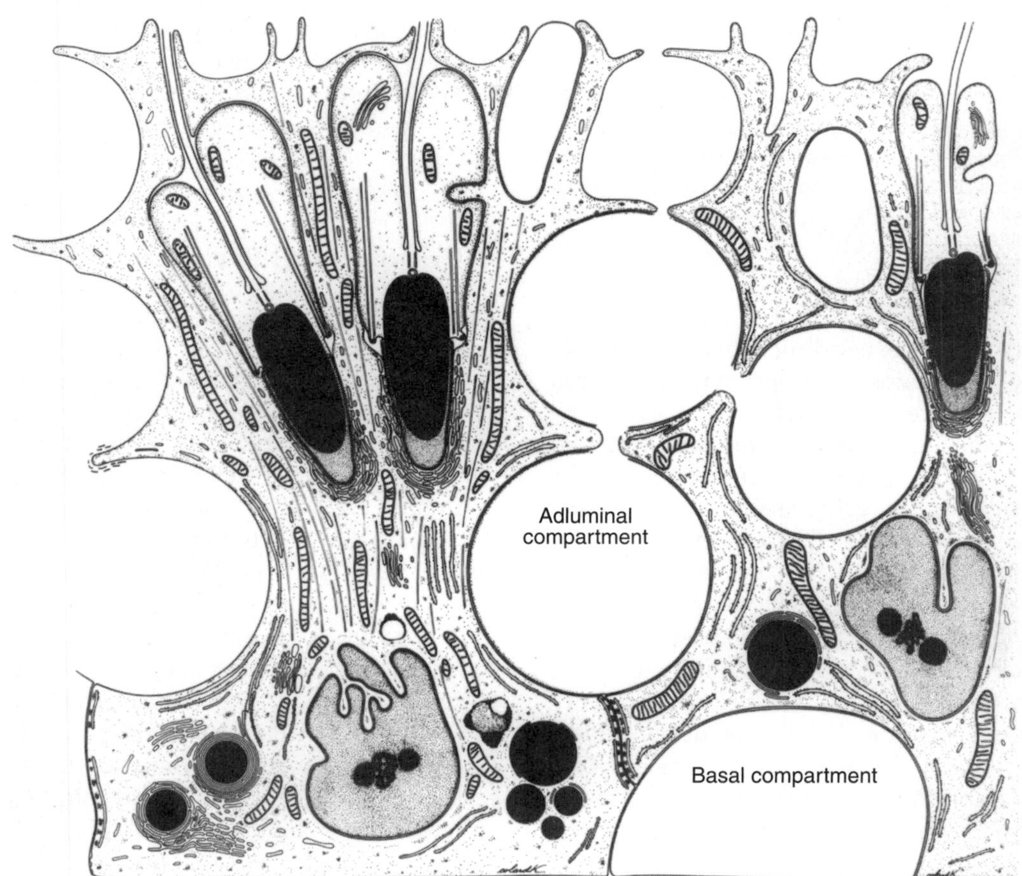

FIG. 39. A diagram illustrating how the occluding inter-Sertoli cell junctions divide the seminiferous epithelium into a basal compartment (containing spermatogonia and early primary spermatocytes) and an adluminal compartment (containing more advanced germ cells). [From (426), with permission.]

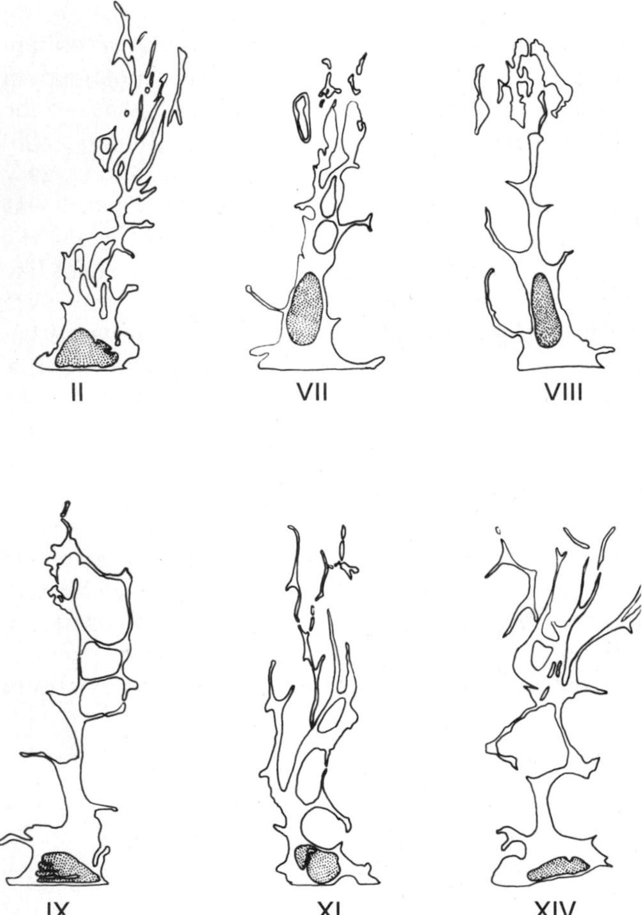

FIG. 40. India ink tracings of the plasma membrane of individual Sertoli cells at various stages of the rat spermatogenic cycle. The surface of each Sertoli cell was identified from assembled electron micrograph montages and emphasizes the complex and varying shape of the Sertoli cell.

association with (a) the expanding spherical volumes exemplified by spermatogonia, spermatocytes, and spermatids; (b) mature spermatids; (c) spermatogonial mitoses and the meiotic maturation divisions; and (d) the formation and subsequent resorption of the excess spermatid residual cytoplasm. Reconstruction of entire profiles of Sertoli cells from many smaller micrographs of each cell reveal that for the major proportion of the rat spermatogenic cycle (Figs. 41, 42, and 43), the overall shape of the cell is best described as tall, irregularly columnar, and possessing numerous very thin lateral processes and cylindrical recesses to accommodate the penetration of elongated spermatids. Similar findings noted by Wong and Russell (430) emphasized that for the remainder of the spermatogenic cycle, these extremely attenuated cytoplasmic processes are absent because the elongated spermatids adopt positions at the extreme apical margins of the Sertoli cells and because of their subsequent release from the seminiferous epithelium (stages VII–IX). Cyclic variations in Sertoli cell shape (446) logically lead to studies of changes in Sertoli cell volume that may also reveal morphological alterations in the internal composition of the Sertoli cell, a subject dealt with in detail later in this section.

Morphometric techniques have been used to estimate the numerical density and the total number of Sertoli cells in the rat testis (446). Earlier quantitative studies on rat seminiferous tubules using light microscopy reported a 10% to 13% occupancy of the seminiferous epithelium by the Sertoli cells (447). In the human and rat testis, Sertoli cells were reported to occupy 36% and 11%, respectively, of the seminiferous epithelium (448). Ultrastructural examination of monkey Sertoli cells revealed that the relative volume of the Sertoli cell within the seminiferous epithelium ranged from 24% at stage I to 32% at stage VII of the monkey spermatogenic cycle (449,450). Morphometric analysis of the rat testis (451,452) has shown that the proportion occupied by Sertoli cells during the spermatogenic cycle ranges from 20% to 29%, representing a cyclical volume change, lowest during stages VI to VIII (5,300–5,500 μm^3) and increasing to maximum volume during stages XII to XIV (7,000–7,700/μm^3). A single stage V rat Sertoli cell reconstructed as a plexiglas model had a volume of about 6,000 μm^3 (430). Thus, the changing shape and volume of the Sertoli cell suggests that the cell engages in continual motor activity and exhibits a high degree of plasticity synchronized with the ever-mobile population of germ cells.

Ultrastructural Features

Nucleus

For many animals, the Sertoli cell nucleus exhibits a characteristic morphology readily visible within the basal aspect of the seminiferous epithelium, for example, for rodents (426,453), monkeys (31,453,454), humans (189,427,428,455–459), and other species (426,433). The Sertoli cell nucleus is large, irregular, and usually at the basal aspect of the cell, which rests on the basement membrane of the seminiferous tubule (Figs. 9, 10, and 44). Usually they form a single row, although at times they are displaced toward the luminal regions of the seminiferous tubule (Figs. 41, 42, and 43). Often, the nuclear membrane is highly infolded, a feature recognized by light microscopy of plastic-embedded sections and confirmed by electron microscopy (426,433,454,457,459).

In rodents, but not for the monkey or human testis (427,454–459), the shape and position of the Sertoli cell nucleus vary according to the stage of the spermatogenic cycle. The nucleus may often be triangular

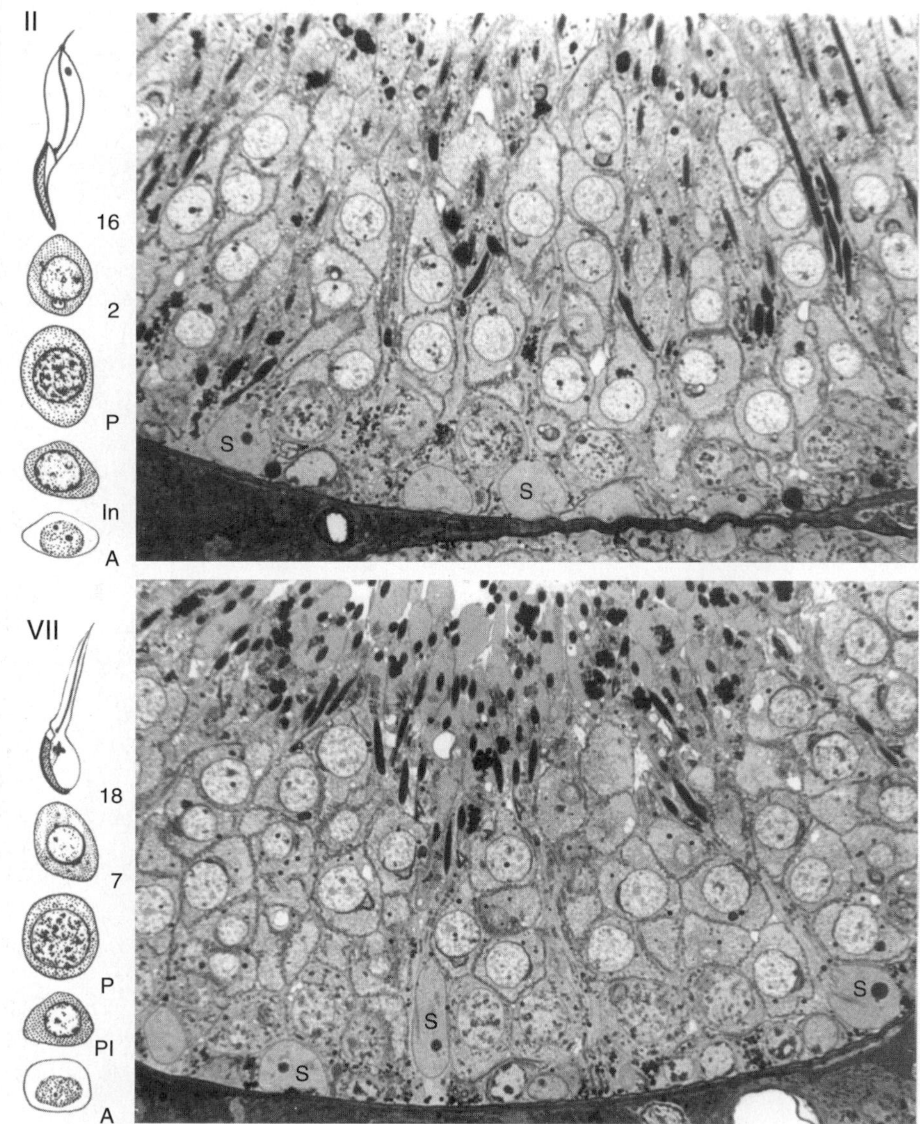

FIG. 41. The cell associations forming stages II and VII of the rat seminiferous cycle are shown. Note Sertoli cells (S). The numbers beside drawings of spermatids indicate stage.

or polygonal in appearance after release of mature spermatids, or alternatively they may appear elongated as the developing elongated spermatids penetrate deeply into the basal aspects of the epithelium (Figs. 34 to 45). Generally, most of the Sertoli cell nuclei remain flattened along the basal lamina of the seminiferous tubule throughout the spermatogenic cycle (235). Within the terminal portions of the seminiferous tubules near their junctions with the tubuli recti, the Sertoli cells exclusively occupy these terminal segments, where their nuclei become closely packed, elongated, and highly irregular in profile (31,460).

The fine structure of the adult Sertoli cell nucleoplasm reveals a homogeneous distribution of euchromatin with a fine fibrillogranular texture. Densely stained and compact masses of heterochromatin are principally confined to the region of the nucleolar complex, although on rare occasions small heterochromatic masses may occur elsewhere within the nucleoplasm or associated with the nuclear membrane (113,430,445,454). Sertoli cells of the human testis contain a relatively large nucleolus, and its intense staining with basophilic dyes facilitates the identification of Sertoli cell nuclei within the seminiferous epithelium (Fig. 9). Its usual configuration (428,461,462) resembles a tripartite structure with a central dense compact nucleolar mass, the nucleolonemma proper. Variable projections often extend outward to gain association with two patches of heterochromatin material that flank the central nucleolonemma in a symmetrical relationship. These two laterally associated bodies are at times referred to as

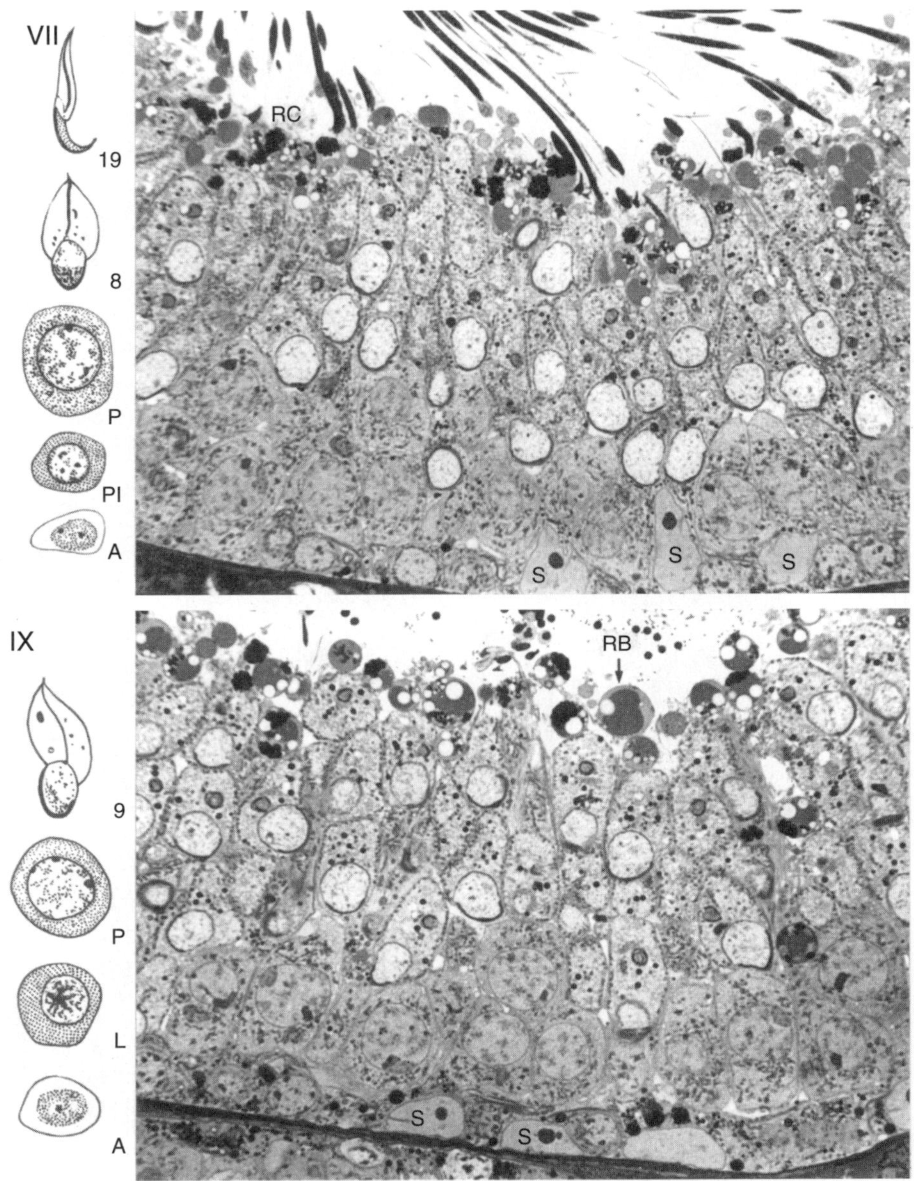

FIG. 42. Cell associations forming stages VIII and early IX of the rat seminiferous cycle. Note residual cytoplasm (RC) and residual bodies (RB). For symbols, see Fig. 41.

the pars amorpha, perinucleolar spheres, satellite karyosomes, or heteropyknotic bodies, which stain positively for DNA with the Feulgen reaction, whereas the nucleolonemma material remains negative (112,426).

Although the tripartite arrangement of the Sertoli cell nucleolus is maintained in rodents and other laboratory species, no close association or contact is established between the central nucleolonemma and the satellite heterochromatin. Nevertheless, the nucleolus is readily identified in rat, mouse, guinea pig, and hamster Sertoli cells (426) (Fig. 39). In mouse Sertoli cell nucleoli, the nucleolonemma contains serial ring-shaped fibrillar components (463) that rapidly incorporate [^{3}H]-uridine, indicating sites of RNA synthesis. With the passage of time, the incorporation of labeled uridine occurred within clusters of interchromatin granules at sites distant from the nucleolus, and at later intervals evidence of labeling disappears from the nucleus. Other studies (464) showed that hybridization with [^{3}H]-rRNA has localized to the central nucleolar mass, within which were many fibrillar centers (26%–41% cell). The latter, together with an interconnecting nucleolar fibrillar network, constitute the site of nucleolus-organizing regions. Autoradiographic studies after [^{3}H]-uridine incorporation indicated that rDNA transcription occurred only in the fibrillar nucleolar network, implying that the above-mentioned fibrillar centers are not the site of rRNA gene transcription. Bustos-Obregon and Esponda (456) described unusual granular nuclear

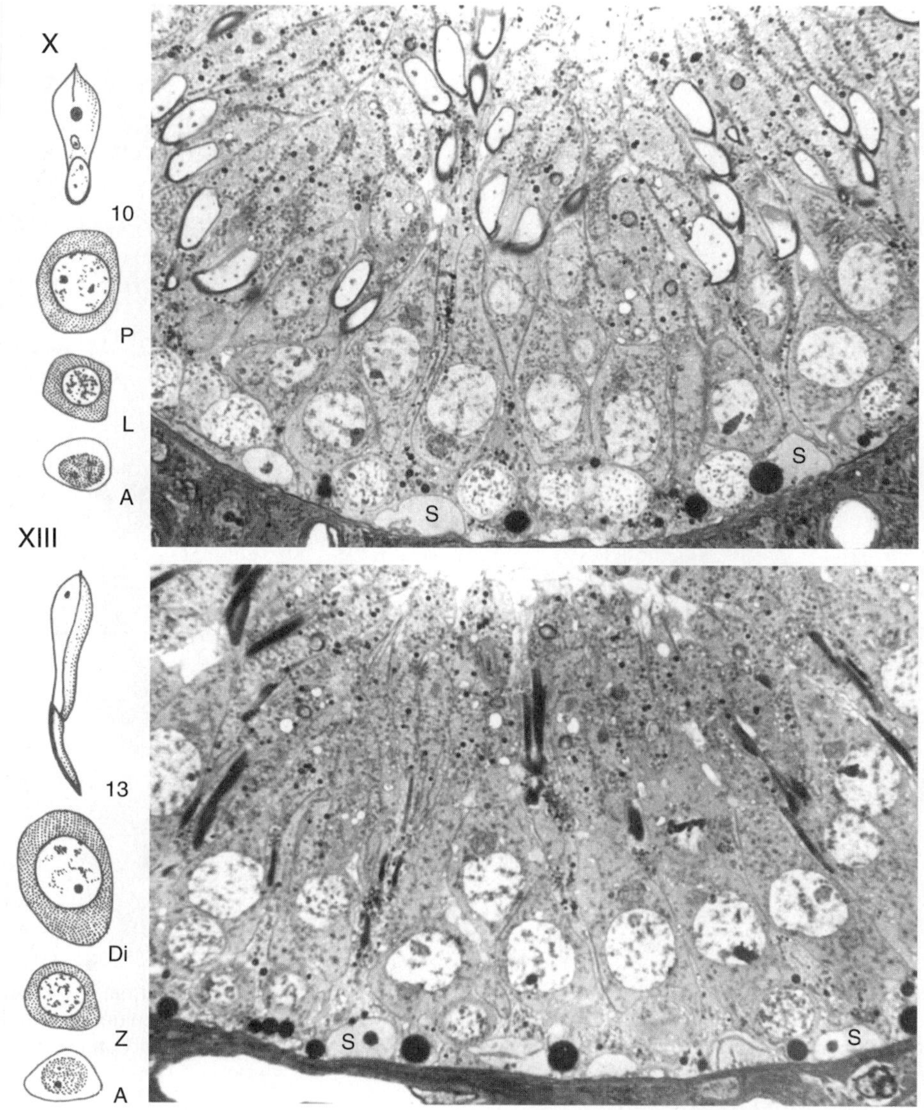

FIG. 43. Cell associations forming stages X and XIII of the rat seminiferous cycle. For symbols, see Fig. 41.

bodies, termed sphaeridia, in human Sertoli cells, 0.4–1.9 μm in diameter and surrounded by a clear halo 0.08–0.1 μm in width (462). Their functional significance remains unknown.

A highly unusual nucleolar morphology has been described for the Sertoli cells of the bull (465–468), which display an aggregation of many membrane-limited vesicles averaging 0.2 to 0.35 μm in diameter and bearing 150-A granules on their outer surface. Similar structures have been noted in the Sertoli cells of the ram, African buffalo, and gerenuk (a rare antelope of East Africa), and it has been suggested (426) that other ruminant species may display these multivesicular nucleolar bodies. Although these membranous forms of the nucleolus in Sertoli cells resemble those observed in the human endometrium (469) and dark type A spermatogonia (123), their functional duties remain unclarified. Sertoli cell nuclei of the seasonally breeding mallard duck contain bundles of intermediate-size filaments, which have also been observed in mouse and pig Sertoli cells when spermatogenesis is quiescent (470). Many nuclear pores traverse the nuclear membrane (471), which itself is invested with a thin zone of 70-Å cytoplasmic filaments, forming a sheath 150 to 250 nm in thickness, thought to confer structural rigidity to the nucleus and prevent close incursion of cytoplasmic organelles.

Cytoplasm

In general, the cytoplasmic components of the Sertoli cell show a polarized distribution: The basal

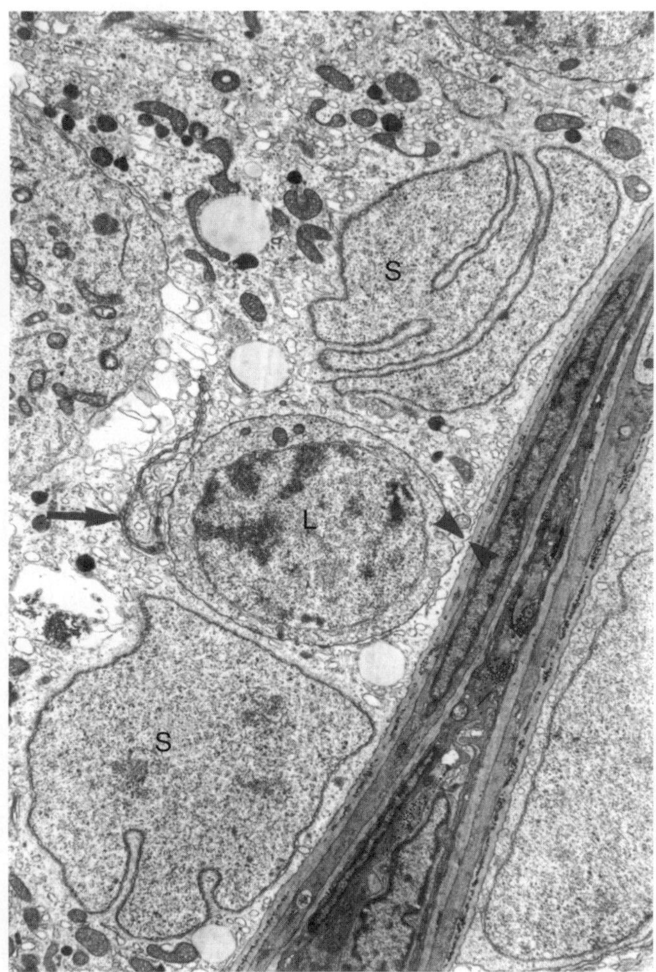

FIG. 44. Basal aspects of two Sertoli cells (S) showing an inter-Sertoli cell tight junction (*arrow*) above a leptotene primary spermatocyte (L). Note that below the germ cell the opposing surfaces of the Sertoli cells are not associated with specialized junctional complexes (*arrowheads*).

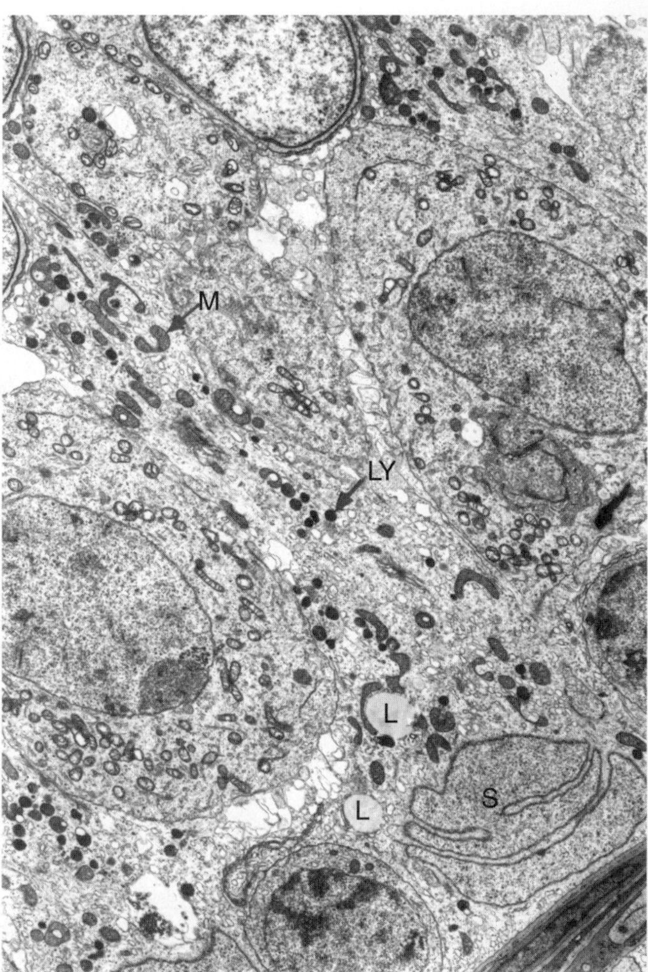

FIG. 45. Stage IX of the rat spermatogenic cycle illustrating the Sertoli cell nucleus (S) and the columnar trunk of the Sertoli cell containing lysosomes (LY), mitochondria (M), and lipid inclusions (L).

and lower trunk regions of the cytoplasm contain an abundance of organelles and inclusions, whereas the apical extensions usually exhibit a paucity of such structures. Exceptions to this rule can be illustrated by the preferential distribution of mitochondria, smooth endoplasmic reticulum, and glycogen to the uppermost apical extensions of Sertoli cell cytoplasm in the mouse, squirrel, and human (427,472,473). Sertoli cell mitochondria vary in shape from slender or spherical and cup shaped (426,427). In rat Sertoli cells, mitochondria exhibit S shapes and dumb-bell or doughnut-type profiles in addition to the usual elongated and round forms (Fig. 44). In all species so far examined, the mitochondria have transversely oriented foliate cristae, but the tubular form is often encountered.

Detailed three-dimensional morphology studies of the Golgi apparatus using high-voltage electron microscopic analysis of sections stained with a variety of heavy metal salts (474,475) showed that in rat and mouse testes, the Sertoli cell Golgi apparatus consists of a primary network of perforated membrane sheets interconnected by narrow bridges. When considered in its entirely, the secondary network, detectable by light microscopy, forms a large three-dimensional structure adopting the overall shape of a cylinder that extends in the main body of Sertoli cell cytoplasm from the juxtanuclear region toward the lumen of the seminiferous epithelium. These findings were chiefly confined to stages V–VIII of the spermatogenic cycle, so it is possible that the general architecture of the Golgi apparatus is modified in other stages.

However, the Golgi membranes of the Sertoli cell are devoid of any appreciable numbers of vesicles or condensing vacuoles usually associated with cells actively engaged in the synthesis of proteins destined for transport or secretion (426). Similarly, the Sertoli cells contain only limited amounts of rough or granular endoplasmic reticulum, which occurs as several short lengths of parallel cisternae or alternatively takes the form of small individual tubules principally in the base or trunk of the Sertoli cell cytoplasm (427,428,454,461). This is surprising in view of the evidence that the Sertoli cells produce numerous proteins (263,476–478), some of which, such as androgen binding protein, are secreted into the lumen of the tubules. Others, such as the glycoprotein hormone inhibin, are secreted in both basal and luminal directions to influence FSH secretion (479–482).

The occurrence of smooth or agranular endoplasmic reticulum has been exhaustively described in studies of the fine structure of the Sertoli cell in vertebrate and invertebrate animals [see reviews (427,428,433,436,441,454,457)]. There is no general morphological pattern that can be applied to a description of the Sertoli cell smooth endoplasmic reticulum, because it has been referred to as vesicular, tubular, cisternal, fenestrated, or lamellar (Figs. 44, 45, and 46). Species differences may account for such variations in fine structure, but equally, the differential effects of tissue fixation probably contribute substantially to the observed changes in vesicular structure (483–485).

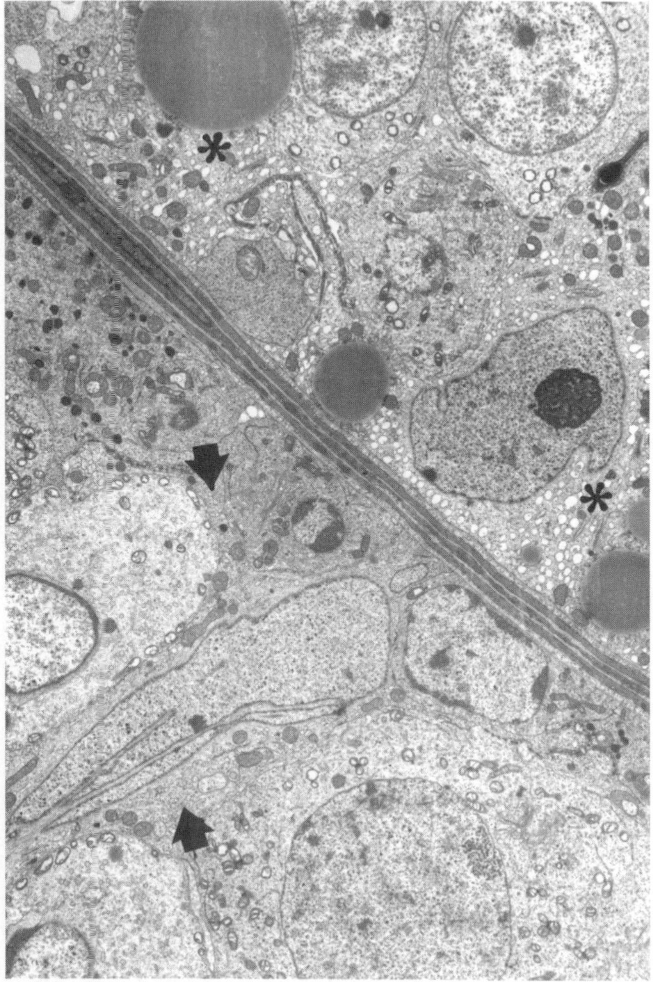

FIG. 46. Rat seminiferous epithelium showing stage 1 (*upper*) with dilated Sertoli cell smooth endoplasmic reticulum (*asterisks*) and stage VII (*lower*) exhibiting tubulovesicular profiles (*arrows*) of Sertoli cell smooth endoplasmic reticulum.

The most striking arrangements of Sertoli cell smooth endoplasmic reticulum occur in several artiodactyl species (even-toed quadrupeds such as the bull, boar, ram, antelope, and gazelle), where large compacted masses of smooth membranes invest and surround the developing heads of elongating spermatids (428,468,486). In the squirrel, just before sperm release, the spermatid head is retained by the Sertoli cell via its association with very large bulbous projections of Sertoli cell cytoplasm filled with smooth endoplasmic reticulum. After spermiation these membranous masses are transported toward the base of each Sertoli cell (473,487). Multiple concentric layers of smooth endoplasmic reticulum are found in the basal aspects of ruminant Sertoli cells, where they surround lipid inclusions (427) or form whorls adjacent to the basement membrane (468,486). In rat Sertoli cells, smooth endoplasmic reticulum undergoes cyclic change in morphology, from tubular to vesicular, in association with the spermatogenic cycle (488,489). To date, however nothing is known about the functional significance of the often rich supplies of smooth endoplasmic reticulum within the Sertoli cells. However, the ubiquitous occurrence of many or large lipid inclusions within the Sertoli cell cytoplasm has reinforced a long-standing view that in some way the metabolism or synthesis of certain steroid compounds known to occur in isolated Sertoli cells (490,491) is mediated via their distinctive cytoplasmic components, characteristically found in accredited steroidogenic cells.

Some confusion exists as to the formation and fate of Sertoli cell lipid inclusions. The size and abundance of lipid inclusions often show cyclic changes in different stages of the spermatogenic cycle (Figs. 41, 42, and 43). von Ebner (240) originally suggested that the seminiferous epithelium exhibited a cyclic variation in lipid content, and there have been a number of studies of Sertoli cell lipids (395,397,426,427,492–495). These observations have been confirmed by morphometric analysis of Sertoli cells during the spermatogenic cycle in the rat, where a cyclic accumulation and decline in Sertoli cell lipid inclusions has been described (445). Dramatic increases in Sertoli cell lipid content are evident in situations of spermatogenic arrest (e.g., seasonal breeders) or conditions that cause germ cell damage (34,436,496–498), but with the reinitiation of spermatogenesis the lipids gradually disappear, and the inclusions return to their normal size and number.

Degenerating germ cells and the end products of degenerate lobes of spermatid residual cytoplasm are the likely source of these lipid inclusions (Fig. 47)

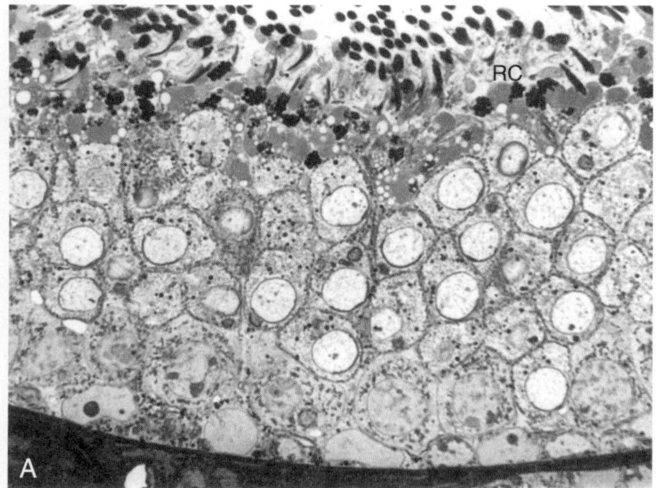

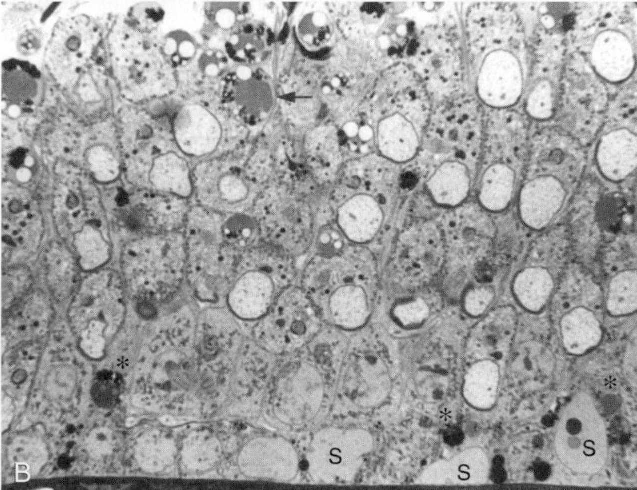

FIG. 47. A: Late stage VII of the rat spermatogenic cycle showing extensive lobes of excess residual cytoplasm (RC) retained by the epithelium as the elongated spermatids are preparing for release into the tubule lumen. **B:** Early stage IX of the rat spermatogenic cycle showing intact (*arrow*) and disintegrating residual bodies (*asterisks*) at various levels within the epithelium. Note Sertoli cell nuclei (S).

that accumulate within the Sertoli cell (395,496,499). However, no definite link between the phagocytosis of residual cytoplasm and the content of Sertoli cell lipid inclusions has been demonstrated. In the mouse (396), rat (500), human (427), and other species (426), the lipid components of degenerating residual bodies do not appear to be released into the Sertoli cell but are probably degraded entirely. After the destruction of residual bodies, Sertoli cells in the rat testis actually begin to accumulate lipid inclusions (445).

Supporting the concept that cyclic variations in Sertoli cell lipid inclusions represent a balance between lipolysis and synthesis, Bergh (123) found that an increase in Sertoli cell lipid inclusions in the cryptorchid rat testis preceded the phase of germ cell degeneration. Accumulation of Sertoli cell lipid also occurred after induction of cryptorchidism in testes lacking germ cells as a result of prior fetal irradiation (123). Because the Sertoli cells can synthesize lipid in the absence of a contribution of substrates from degenerating germ cells, it is likely that the smooth endoplasmic reticulum enzymes are perhaps responsible for lipid synthesis from glycerol and fatty acids, which, on esterification within the Sertoli cell, become visible as lipid inclusions. This hypothesis is supported by the finding that in mammals and a variety of vertebrate species, Sertoli cell lipid inclusions contain a higher ratio of esterified to unesterified cholesterol than germ cells (411,501,502). However, a role for lipid inclusions within the Sertoli cell is still obscure.

In the human Sertoli cell, filamentous inclusions of two main types are found. Lubarsch (198) discovered large crystals up to 25 μm in length in Sertoli cell cytoplasm, naming these the crystals of Charcot-Böttcher. Later, Spangaro (503) observed much smaller crystals within the Sertoli cell, and all of these inclusions are now collectively referred to as Charcot-Böttcher crystals (504,505). Readily visible by light microscopy, their ultrastructural features can be summarized as follows: (a) perinuclear, often obliquely oriented in relation to the basal lamina; (b) elongated, fusiform shape, up to 5 μm in width and 10–25 μm long; (c) often form simple bifurcations, between which are found glycogen or 10-nm filaments; (d) consist of dense parallel filaments, approximately 150 Å in diameter that in cross-section exhibit zig-zag or meandering profiles; and (e) their terminal spike-like ends may be continuous with 9- to 12-nm cytoplasmic filaments (401,403). Crystalloids are also found in Sertoli cells of the pig (506), although they are smaller than those seen in the human and consist of parallel filaments, 50 Å in diameter. These filaments contain actin (507) and thus may be derivatives of the cytoskeleton of the Sertoli cell.

Rod-like crystalloid inclusions have also been described in Sertoli cells of some marsupials, notably the koala (508–510) and the American opossum (511). Ultrastructural analysis of koala Sertoli cells (512) confirmed these earlier observations and showed that koala Sertoli cells exhibit extraordinary cytoplasmic crystals, often aligned perpendicular to the basal lamina and in the vicinity of the Sertoli cell nucleus. Visible by light microscopy, they occur in prepubertal and adult specimens (Fig. 48), and although their ultrastructure resembles human Charcot-Böttcher crystals, they exhibit a more highly ordered substructure in which the filaments are arranged in tubules, thereby forming a regular latticework (Figs. 48 and 49). The function of these crystalloids is not clear.

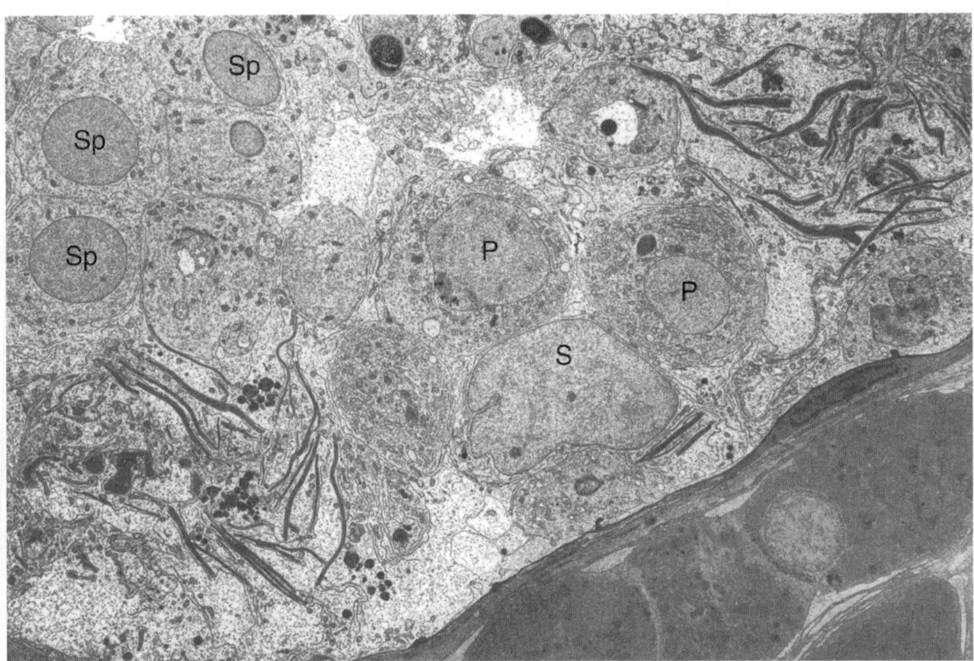

FIG. 48. Ultrastructure of koala seminiferous epithelium showing numerous crystalloid inclusions within the basal Sertoli cell cytoplasm. Sertoli cell nucleus (S), pachytene primary spermatocytes (P), and round spermatids (Sp) are shown. [From (1111), with permission.]

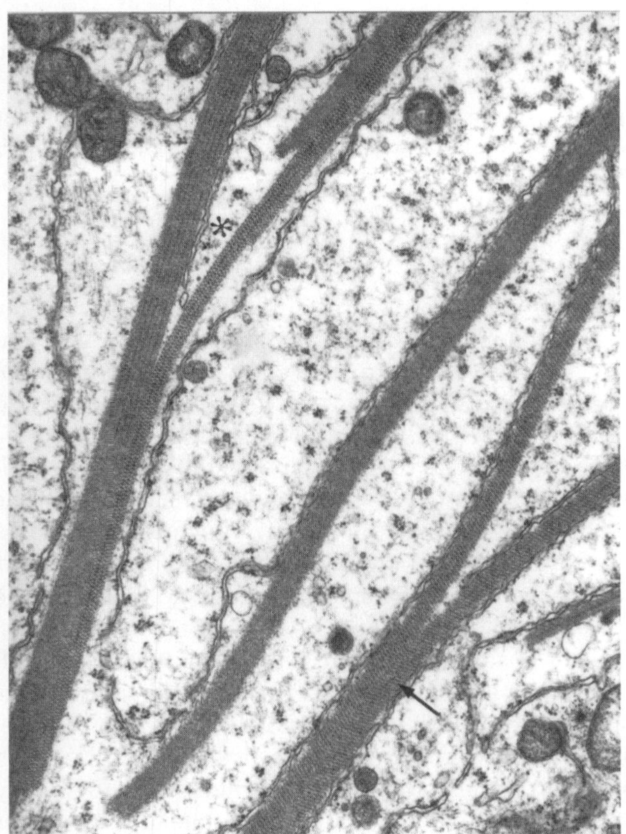

FIG. 49. Bifurcation of koala Sertoli cell crystalloids showing that their individual arms may present differing substructures (*asterisk*). A bipennate crystalloid is shown (*arrow*). [From (1111), with permission.]

Sertoli cells contain variable amounts of dense bodies usually referred to as collections of lysosomes, multivesicular bodies, and heterophagic vacuoles (426,427,454,457). Often, these components of the lysosomal system are sequestered in the deepest regions of the Sertoli cell, where they flank the nucleus, but, alternatively, they are seen in large numbers within the upward columnar trunk of the Sertoli cell, where they lie close to developing spermatids (Figs. 44 and 45).

Historically, Maximow (513) originally suggested that the Sertoli cell is active in the phagocytosis of germ cells, and more recent histochemical analysis revealed the presence of strong hydrolytic enzyme activity in the Sertoli cell (500,514–519), which is principally localized within membrane-limited dense bodies (lysosomes) (500). Sertoli cells can ingest injected dyes and certain foreign particulate matter (520–522) and participate in the removal of degenerating germ cells (161,370). In keeping with the general mechanism by which phagocytic cells such as macrophages recognize, engulf, and destroy degenerating cells, the Sertoli cells recognize germ cells that are undergoing apoptosis. They achieve this through the expression of a surface scavenger receptor, not found in germ cells, that is specific for phosphatidylserine. Apoptotic germ cells externalize phosphatidylserine to their plasma membrane, which acts as a signal for phagocytosis by the Sertoli cell (523).

The disposal of the excess spermatid cytoplasm left behind by the mature sperm as they are released from the luminal surface of the Sertoli cells (Fig. 47) was reported by Regaud in 1901 (49). He proposed that the Sertoli cells phagocytosed and digested the excess residual cytoplasm and described these events in four steps: (a) release into the lumen, (b) resorption within the Sertoli cells, (c) peripheral migration deep into the epithelium, and (d) transformation into "Sertoli hyaline" spheres that blacken in the presence of osmic acid. Since then, the formation (Figs. 50 and 51) and fate of residual bodies has been studied by light and electron microscopy (395,396,426,445,454,488,524), but the mechanisms by which they are eliminated by the Sertoli cell has not been resolved.

Examination of the lysosomal apparatus of the Sertoli cells has provided some insights. When electron-dense tracers were introduced into the lumen of seminiferous tubules, they were actively incorporated into the apical regions of the Sertoli cells by fluid-phase endocytosis (pinocytosis) (525,526). The tracers internalized by this process are eliminated by the lysosomes in the columnar and basal regions of Sertoli cell cytoplasm (527). Other studies (528) investigated the endocytotic and phagocytic properties of the Sertoli cell using native ferritin and protein–gold complexes to show fluid-phase endocytosis and cationic ferritin and concanavalin A ferritin to identify absorptive endocytosis. In the latter, molecules initially bind to the cell surface before internalization by small vesicles. The results indicate that fluid-phase endocytosis by rat Sertoli cells occurs in all stages of the spermatogenic cycle. At stages VIII and IX, the lysosomes formed as a consequence of this process fuse with the newly formed residual bodies and transform them into phagolysosomes, whereupon they disintegrate at the base of the Sertoli cell (Fig. 52). The formation of lysosomes and their participation in the dissolution of residual bodies thus provide a link between the endocytic and phagocytic activities of the Sertoli cell. Adsorptive endocytosis occurred principally during stage VII, when various phagocytic vacuoles form close to the heads of late spermatids, where they probably play a role in resorption of the specialized tubulobulbar complexes that anchor the spermatid head to the apical cytoplasmic processes of the Sertoli cells.

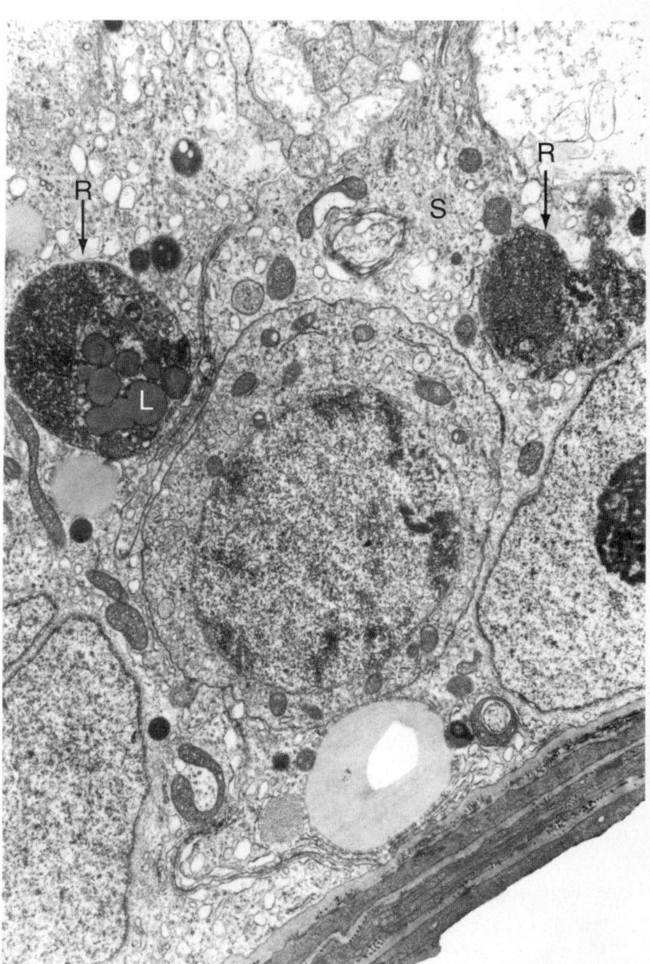

FIG. 51. Degenerated residual bodies (R) deep within the Sertoli cell cytoplasm (S). The lipid component (L) of the residual body is resistant to the phagocytic action of the Sertoli cell.

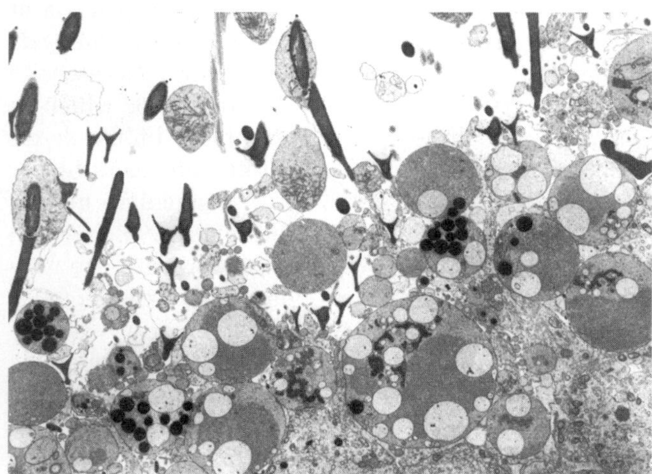
FIG. 50. At stage VIII of the rat spermatogenic cycle, lobes of residual cytoplasm adopt spherical shapes, which characterize the formation of residual bodies.

The basal aspects of the human Sertoli cell may show concentric layers of smooth membranes (224,225,458,505), and due to the presence of pore-like complexes that form a bridge-work between parallel profiles of these membranes (427,457), they have been likened to annulate lamellae found in other tissues (529–532). Although their function in the Sertoli cell remains unknown, the annulate lamellae have been implicated in RNA transport from the

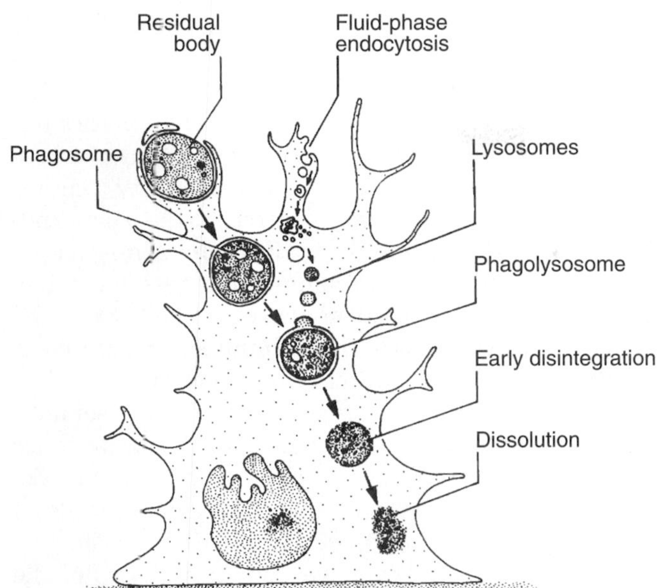

FIG. 52. Representation of phagocytosis of residual bodies and fluid-phase endocytosis by the Sertoli cell. These processes may be integrated to allow disintegration of the residual bodies as they proceed through the body of the Sertoli cell. [Modified from (528).]

nucleus (533), protein synthesis (531,532,534), and the site of tubulin synthesis or polymerization of microtubules (535).

As expected for a cell that is obliged to alter its shape radically in conforming to the ever-changing events within the seminiferous epithelium, all Sertoli cells are endowed with an elaborate cytoskeleton together with contractile elements occupying most parts of the cytoplasmic matrix (395,396). The structure and function of the Sertoli cell cytoskeleton is discussed in several excellent reviews (536–538). The filamentous component, responsible for maintenance of cell shape and the redistribution of the cytostolic gel matrix, is attributed to an often extensive and intricate system of microfilaments (60–70 nm) and intermediate filaments (10 nm). Together these dense filamentous networks are thought to play a major role in structural support of the Sertoli cell when rigidity is necessary, and at other times it is thought that they engineer changes in cell matrix viscosity, allowing variable degrees of plasticity that are essential to accommodate the constant mobility of the germ cells. These filaments are rich in actin (539–541) and vimentin (540–543). Concentrations of filaments occur at the very base of the Sertoli cell adjacent to the basal lamina, around the Sertoli cell nucleus in the columnar cytoplasm, where they course parallel to the cell axis, and also in association with numerous ectoplasmic specializations (ESn). The latter attain a close proximity to developing spermatids and associate with junctional complexes between neighboring Sertoli cells in the basal aspects of the epithelium.

Rich supplies of filaments are seen to fill the cytoplasm of many monkey Sertoli cells in the terminal segments of the seminiferous tubules (31), although the reason for their abundance is not clear. Study of the structure–function relationships of the Sertoli cell cytoskeleton in the ground squirrel has added much to an understanding of the mechanisms underlying shape change (444). The Sertoli cells of the squirrel were chosen for investigation because of their relatively small numbers of germ cells compared with other species, and the Sertoli cells undergo dramatic shape changes. Based on the results obtained from squirrel Sertoli cells (444,487), actin-rich filaments in ectoplasmic specializations are devoid of myosin and, rather than fulfilling a contractile role, probably stabilize the cortical cytoplasm of Sertoli cells at the sites where they occur. Perhaps the filaments maintain the shape of Sertoli cell crypts embracing the penetrating clones of spermatids (404,544) and may add reinforcement to the zone of Sertoli cell ectoplasm at the level of the basal junctional complexes between adjacent Sertoli cells.

After hypophysectomy and loss of germ cells, the actin cytoskeleton is disorganized and diffuse throughout the Sertoli cell cytoplasm, and the content of F-actin declines, although β-actin protein and mRNA expression are unchanged (545). In contrast, expression of profilin, an actin-promoting protein, is significantly decreased, suggesting that hormone withdrawal and/or loss of germ cells causes depolymerization of the actin cytoskeleton. The functions of vimentin filaments are unclear, but they vary dynamically in length during the rat spermatogenic cycle (546) becoming shortest during stages VI–VII, coincident with sperm release and smallest volume of the Sertoli cell (451,452). When intratesticular testosterone levels are experimentally reduced below that which is required to maintain spermatogenesis, the vimentin cytoskeleton collapses around the Sertoli cell nucleus and biochemical analysis showed cleavage of vimentin into smaller subunits (547). Whether this degradation causes or is a result of germ cell apoptosis or exfoliation is unknown, but a similar loss of vimentin structure in Sertoli cells has been reported in cryptorchid testes (548) and after chemical insult associated with disruption of spermatogenesis (549–552).

Studies of microtubule function in Sertoli cells have suggested several roles, including alteration in the shape of cytoplasmic extensions and crypt-type regions and translocation of germ cells, basally and apically, within the seminiferous epithelium (537,553,554). Exposure of Sertoli cells to colchicine destroys most of the microtubular cytoskeleton, causing severe disruption of germ cell movements and blockade of intracellular transport of membranous organelles (473). Additionally, these studies offered a new interpretation of the mechanism by which Sertoli cells participate in sperm release with some studies favoring a role for (555–557) or against (558) the involvement of microtubules. When the effects of colchicine were reinvestigated (473), late spermatids were not shed from the Sertoli cells and the special ectoplasmic specialization adjacent to spermatid heads and to basal tight junctions between Sertoli cells remained intact.

The role of microtubules in generating shape changes within the Sertoli cells is based on structural, molecular, and in vitro motility assays. Microtubules show regional variations in their distribution being abundant in crypts accommodating elongating spermatids, in the long axis of the Sertoli cell trunk, and around the developing spermatid heads (538,559). Electrophoretic studies of isolated preparations of microtubule-associated proteins have shown the presence of dynein-like proteins similar to those with axons (560). Sertoli cell microtubules are nucleated apically (plus end directed basally and minus end oriented toward the cell apex), an arrangement that facilitates motor protein-directed movement of organelles or, in the case of whole germ cells, via transport of the ectoplasmic specialization associated with all spermatids (537,554). The results suggest that bidirectional germ cell translocation through the seminiferous epithelium is dependent on microtubule orientation, coupled with associated motor proteins dynein and kinesin known to be concentrated in the Sertoli cell (554,561).

Surface Specializations

A well-recognized property of the testis is the maintenance, within the seminiferous tubules, of a highly specialized microenvironment created by the Sertoli cells that partitions young and more mature germ cells into two compartments within the seminiferous epithelium. Numerous observations led to the discovery of a blood–testis barrier that exists in the testes of phylogenetically distant species and thus reinforces the fundamental importance of the Sertoli cell in the regulation of spermatogenesis.

Early in the 20th century, physiological studies by Ribbert (561) and Bouffard (562) indicated that the testis was one of the few organs into which intravenously injected dyes did not gain entry. No significance was assigned to these initial observations or to more recent work reporting similar results (563,564). The vascular network of the testicular interstitial tissue is highly permeable to large molecules, as demonstrated by Everett and Simmons (565) when they noted intravenously injected serum albumin had a rapid rate of extravascular transfer and readily permeated into the interstitium. Many other substrates were found to diffuse readily from the testicular blood vessels into the interstitial lymphatics but did not appear in the fluid collected from the rete testis. Thus, a blood–testis permeability barrier seemed to be anatomically located either surrounding or actually within the wall of the seminiferous tubules (115,566–569). In common laboratory rodents, the penetration of electron-dense markers from the blood vasculature into the seminiferous tubules is partly retarded by peritubular myoid cells.

The blood–testis barrier resides within the seminiferous epithelium and is due principally to the presence of various types of tight junctions or desmosome-like structures associated with the plasma membrane of the Sertoli cell (245,465,505), originally described by Brokelmann (257). Whenever the boundaries of the Sertoli cell were apposed, they formed complex cytoplasmic membranes immediately adjacent to and parallel with the limiting plasma membrane. These profiles represent cisternae of endoplasmic reticulum that increased in numbers with the onset of meiosis and the formation of a tubule lumen (114).

These morphological studies were complemented by physiological work indicating that the barrier was absent at birth, but with sexual maturation, the previously unrestricted transport of acriflavine dyes was prevented concomitant with meiotic maturation of the primary spermatocytes (115). Since then, the ultrastructural features of inter-Sertoli cell junctions have been documented in many excellent descriptions (118,409,425,426,454,570–574) that have been extended to detailed analysis of the constituent proteins and molecular anatomy [reviewed in (537,575–583)]. The morphology of junctions are summarized in Figs. 53 and 54 as follows:

1. Junctional specializations between Sertoli cells are particularly prominent in the basal regions, the seminiferous epithelium, and usually occur when adjacent Sertoli cell cytoplasmic processes meet.
2. Sometimes, at the very base of the Sertoli cell adjacent to the basal lamina, the meeting of two apposed Sertoli cell cytoplasmic processes may be devoid of junctional specialization.
3. The basal location of junctional specializations circumscribing the lateral margins of Sertoli cells is anatomically reversed compared with tight junctional specializations present in many other epithelia.
4. Normally, a space of 150–200 Å separates the outer leaflets of adjacent Sertoli cell membranes.
5. Occasionally, this space is narrowed to 20 Å, similar to the traditional gap junction or nexus of other cells.

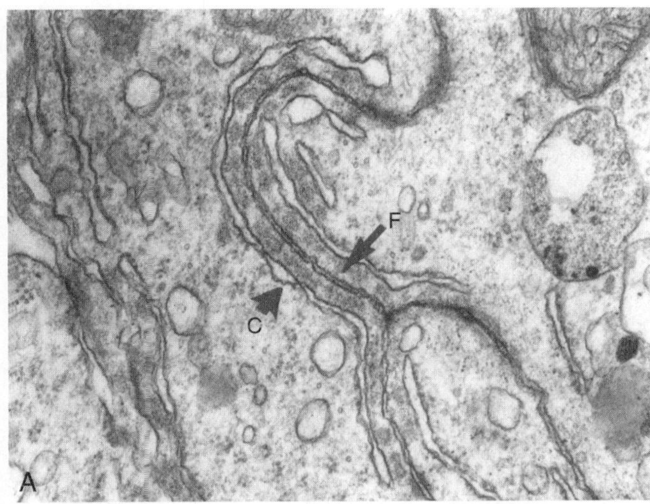

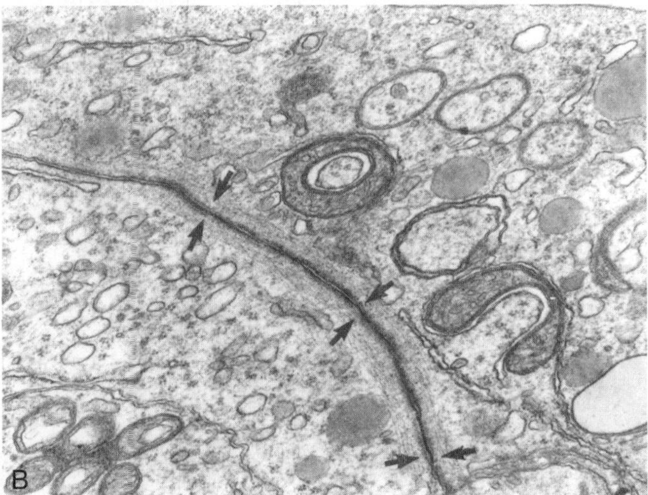

FIG. 53. A: Inter-Sertoli cell junction illustrating bundles of filaments (F) adjacent to the Sertoli cell plasma membrane. Smooth-surfaced cisternae (C) run parallel to the opposed surfaces of the Sertoli cells. **B:** Longitudinal profile of an inter-Sertoli cell junction showing parallel arrays of filaments (*arrows*).

6. In thin sections studied by transmission electron microscopy, multiple sites of fusion of the outer membrane leaflets are seen at regular intervals along the cell-to-cell interfaces.
7. Sertoli cell junctions are often flanked by parallel cisternae of endoplasmic reticulum, irregularly fenestrated and exhibiting ribosomes toward the cell body, but are agranular on the ectoplasmic face.
8. Sandwiched between the cisternae and cell surface, bundles of fine filaments are oriented parallel to the cell surface and, in transverse views, are packed in hexagonal arrays.

Use of freeze-cleaving methods showed that multiple focal sites of fusion between Sertoli cell membranes extend entirely around the circumference of the cell and consisted of long parallel rows of intramembranous particles intercalated with matching grooves (118,425,426,471,584,585). This collar of junctional specialization occupies approximately 4% of the surface area of the plasma membrane in the rat Sertoli cell (431). Up to 50 parallel lines of membrane fusion between adjacent surfaces of Sertoli cells together constitute a highly effective barrier against the intraepithelial penetration of the spaces between cells residing in the base of the seminiferous epithelium (454,570,572). The capacity of inter-Sertoli cell junctions to maintain cell adhesion was convincingly demonstrated after ligation of the efferent ductules, a procedure causing marked distention of the seminiferous tubules. Junctions between Sertoli cells fail to separate regardless of the degree of tubular distension (571). Studies of the permeability properties of inter-Sertoli cell tight junctions have shown a progressive loss of impermeability from the base to the lumen of the seminiferous tubule, which is correlated with increasing incidence of disintegration of the junctional complex (119).

The strategic and unusual location of Sertoli cell tight junctions in relation to the germ cells has given rise to the concept of an anatomic and functional subdivision of the seminiferous epithelium into (a) a basal compartment containing spermatogonia, preleptotene, and leptotene primary spermatocytes and (b) an adluminal component beyond the level of the tight junctions that sequesters the more differentiated germ cells into a unique physiological environment (425,454,572). Some species differ in the precise manner in which germ cells ascend from the basal to the adluminal compartment, thereby breaching the barrier maintained by the Sertoli cell tight junctions. In the macaque testis, leptotene primary spermatocytes reside in the basal subdivision and, with their upward mobilization, mature into zygotene primary spermatocytes on reaching the adluminal compartment (586). In the rat, the transition from basal to adluminal regions has been suggested to occur through the agency of a short-lived intermediate compartment embodying some leptotene primary spermatocytes (587). This transit chamber is flanked above and below by tight junctional complexes, that is, neither truly basal (permeable to bloodborne substances) nor adluminal (impermeable to bloodborne substances). Further research is necessary to establish whether the so-called intermediate compartment provides a special physiological milieu different from either the basal or the adluminal compartment.

Whatever the histological organization of the testis, some form of intraepithelial junctional specialization is always localized in the somatic cells surrounding the germ cells, suggesting that the blood–testis barrier is an ancient evolutionary trait of central importance for the successful development of viable gametes. In broad terms, the ultrastructural organization of

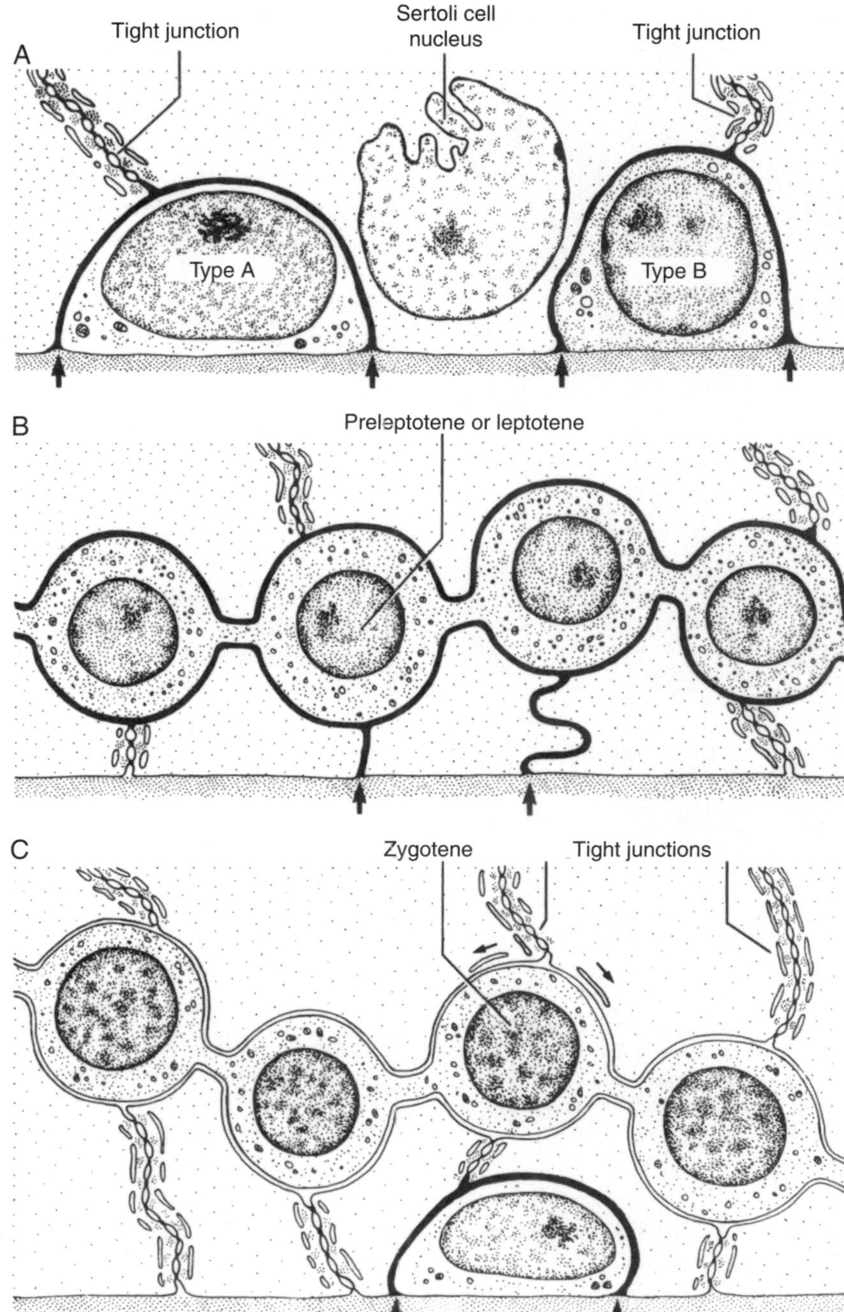

FIG. 54. Diagram illustrating the upward migration of germ cells from the basal aspect of the seminiferous epithelium toward the lumen. In each diagram, tracer entry into the intercellular spaces is shown by *large arrows*. [Modified from Dym and Cavicchia (585).] **A:** The entry of tracers into the seminiferous epithelium is prevented by the occluding inter-Sertoli-cell tight junctions. **B:** Preleptotene and lepototene primary spermatocytes are displaced above the basal lamina. Tight junctions are seen below, above, or both below and above these cells, and consequently tracers are able to surround the interconnected cells via cytoplasmic bridges. Tight junctions above the germ cells prevent further passage of tracers into the seminiferous epithelium. **C:** When leptotene primary spermatocytes enter the zygotene phase of meiotic prophase, tracers do not surround zygotene spermatocytes, since tight junctions are formed beneath these cells. Dissociation of inter-Sertoli cell junctions into hemijunctions has been suggested and is indicated by *small arrows*.

the inter-Sertoli cell junctions can be classified into three main types based principally on the type of membrane specialization. The testes of mammals show a complex organization of the blood–testis barrier, and in species described to date, the arrangement described above is virtually unchanged. Examples can be found in studies of rodents (mouse, rat, guinea pig, squirrels), ruminants (bull, ram, goat), domesticated carnivores (dog and cat), and primates (macaque and human). A group exhibiting somewhat simpler organization of the junctional complexes in that they often lack the bundles of actin-rich filaments and subsurface membranous cisternae is illustrated by birds and reptiles (423,441,587). A simple arrangement serving the function of a blood–testis barrier is peculiar to amphibians and fish, where only desmosome-like and short tight junctions are observed between Sertoli cells (404,437,588–593). In other groups, however, the form of barrier is less well defined, and in nematodes and insects, the occurrence of small septate, desmosome-like, and tight junctions is seen as the structural basis of the blood–testis barrier (590,594–596).

It is clear from the above discussion that the junctional specializations must have the capacity to be regulated to allow transit of germ cells between the basal and adluminal compartments. The plasticity of these junctions has been demonstrated in a study that showed that a single dose of a synthetic-occluding peptide, administered by intratesticular injection, caused a temporary disruption of the blood–testis barrier for a period of about 4 weeks (597). This was followed by profound losses of germ cells but a subsequent return of spermatogenesis after the integrity of the barrier was restored.

A number of studies evaluating the composition of the junctional specializations have been recently reviewed (598). They proposed a "biochemical junction disassembly/assembly theory" in which the integrity of these cell junctions is controlled by a series of cytokines, proteases, and protease inhibitors. They proposed that cytokines such as interleukin-1α and -1β, tumor necrosis factor (TNF)-α, transforming growth factor-α and -β, and basic fibroblast growth factor produced by Sertoli cells and germ cells can stimulate proteases to cleave the cell junctions and, in turn through a similar process, initiate protease inhibitors to synthesize protein production resulting in reassembly of the junctional complexes [see review (598)].

Despite all the above-mentioned studies, very little is known about the actual physiological role of the blood–testis barrier (126). The formation of inter-Sertoli cell junctions coincides with the cessation of Sertoli cell proliferation in the immature testis (71), and junctions make their appearance as germ cells proceed through the zygotene to pachytene steps of meiotic maturation (116,118,120,585). However, the junctions also appear in the absence of germ cells (80,122,123) or in response to gonadotropic stimulation of the Sertoli cells (130), but their formation is dependent on the transformation of Sertoli cells from an immature to an adult-type morphology. In early postnatal growth of mouse and rat testes, junction formation and its sealing properties are retarded by exogenous diethylstilbestrol treatment (599,600), but whether this effect was exerted via suppression of gonadotropins or directly through intratesticular estrogen receptors is not known. It should be noted that the presence of structurally intact Sertoli cell tight junctions is not necessarily indicative of their barrier properties. In cases of germ cell arrest or hypospermatogenesis in the human, the junctions may appear structurally normal yet are leaky to electron-dense tracers (601). Quite opposite findings also have been reported (602) indicating there is no simple correlation between spermatogenic activity and the permeability of Sertoli cell tight junctions. Because early primary spermatocytes (preleptotene or leptotene) reside in the basal compartment of the mammalian seminiferous epithelium and similar observations are reported in lower orders of animals (437), evidently the initiation of meiotic maturation does not require the specialized intratubular milieu provided within the confines of the adluminal epithelial compartment.

In support of this concept, recovery of spermatogenesis by treatment with vitamin A after long-term vitamin A deficiency is accompanied by restoration of the blood–testis barrier in which basal preleptotene and leptotene spermatocytes remain viable, whereas zygotene and later germ cells show marked apoptosis, suggesting a link between barrier permeability and meiotic maturation (603). In the rat testis as the primary spermatocytes enter prophase, cell surface antigens specific to germ cells appear on pachytene primary spermatocytes and all subsequent stages of germ cell differentiation (604–607). Specific antigens also appear on the surface of Sertoli cells, and it has been suggested that the processes of meiotic maturation and differentiation of spermatids occurs in an immunologically privileged adluminal microenvironment via the inter-Sertoli cell tight junctions (126). Isolation of these germ cells by the blood–testis barrier either restricts leakage of antigen or prevents entry of antibodies or immune cells from the vascular system. That the germ cells are not absolutely dependent on an intact blood–testis barrier was demonstrated in the seasonally breeding mink, where the blood–testis barrier appears to undergo cyclic formation and decay related to tubule lumen formation in association with respective periods of activity and inactivity of the seminiferous epithelium (608).

The relationship between the formation of Sertoli cell junctional complexes and the establishment of spermatogenesis during postnatal growth of the testis has been studied using recent physiological and histological techniques. Permeability of the blood–testis barrier is related to the progressive development of a lumen within the growing seminiferous tubules in which the capacity of the junctional complexes to restrict entry of exogenous hypertonic fluids or water-soluble markers is attained gradually rather than abruptly (609,610). The tightness of the barrier is maximally efficient coincident with the complete canalization of the seminiferous tubules with a recognizable lumen, which in the rat testis occurs at about day 30 of postnatal development.

The upward movement of early spermatocytes into the adluminal compartment requires that some mechanism be available to allow the migrating germ cells to traverse the specialized inter-Sertoli cell junctions. These junctions have been observed both above and below young spermatocytes (243,585), and an orderly breakdown and formation of tight junctions has been proposed to permit cell transfer from the basal to the adluminal compartment (243,571). Further studies have indicated that new tight junctions between Sertoli cells form below young spermatocytes, whereas those previously above these cells are thought to dissociate, thus ensuring the patency of the permeability barrier (544,585,586). Thus, the barrier is flexible and deformable and compatible with the movements of migrating germ cells. Early stages of formation of tight junctions beneath migrating germ cells are characterized by increasing numbers of intramembranous junctional strands, which increase in length and begin to assume parallel orientations, thus collectively contributing to an increasing degree of continuity, culminating in a typical junctional complex (119).

The complex yet highly ordered arrangement of multiple clones of germ cells, each embraced by the highly branched Sertoli cell, demands not only that the Sertoli cell should be capable of conforming to the remarkable shape changes of the germ cells but also that it play a role in conferring stability throughout the seminiferous epithelium. In reviewing this topic, a wealth of morphological information is now available from comparative studies of Sertoli cell–germ cell relationships that together indicate the central role played by the Sertoli cell in maintaining an attachment to germ cells in addition to providing potential avenues of intercellular communication. The literature concerned with this aspect of the biology of Sertoli cells is voluminous, but the reader is directed to a number of excellent articles that provide the basis for our present understanding (426,431,444,472,473, 487,572,611,612).

Prevention of premature disengagement and sloughing of germ cells into the lumen of the seminiferous tubule is thought to rely, at least in part, on cell adhesion proteins and on regions of ectoplasmic specializations of the Sertoli cell that face the surface of certain germ cells. The latter consist of a dense band of actin-rich filaments sandwiched between the Sertoli cell plasma membrane and a cistern of endoplasmic reticulum and thus resemble one-half of the paired ectoplasmic specializations that constitute the inter-Sertoli cell tight junctions at the base of the Sertoli cells. Originally described between adjacent Sertoli cells and spermatids in the lumen testis (257,407), their widespread occurrence in other species has subsequently been confirmed (426,442,472,540, 544,614).

Visualization of the sites of apposition of germ cells to ectoplasmic specializations at the lateral and apical surfaces of the Sertoli cell is dependent on the orientation and plane of section when the seminiferous epithelium is examined with the electron microscope. This limitation not only restricts an objective appraisal of what type and how many germ cells are associated with ectoplasmic specialization, but it also seems likely that this relationship is variable between species. In the hamster, monkey, and human testis, regions of ectoplasmic specialization have been reported facing zygotene primary spermatocytes (146,428,454,613), but other studies of a number of species have suggested that this association occurs very rarely. Where most reports seem to be in agreement is in the occurrence of ectoplasmic specializations facing mid-pachytene primary spermatocytes and round spermatids, but as the spermatids begin to elongate (Figs. 20 and 55) and undergo their final phase of maturation, a mantle of ectoplasmic specialization is always positioned around the spermatid head (35,544). Whenever elongated spermatids are seen to penetrate the Sertoli cell deeply, the resultant recess that contains the spermatid head is lined by ectoplasmic specialization that is preferentially associated with the acrosome.

The function of Sertoli cell ectoplasmic specializations adjacent to germ cells remains speculative. Differences in the adhesive properties between Sertoli cells and various germ cells have provided circumstantial evidence that ectoplasmic specializations may actually bind to adjacent germ cells. Round germ cells (spermatocytes and early round spermatids) are easily separated from Sertoli cells (540,544,614), but elongated germ cells (maturing spermatids) are more resistant to dislodgement, and enzymatic digestion with trypsin is required to separate them from the Sertoli cell (112,614). Because desmosomes are not observed between elongated spermatids and Sertoli cells, the adhesive property that exists between them

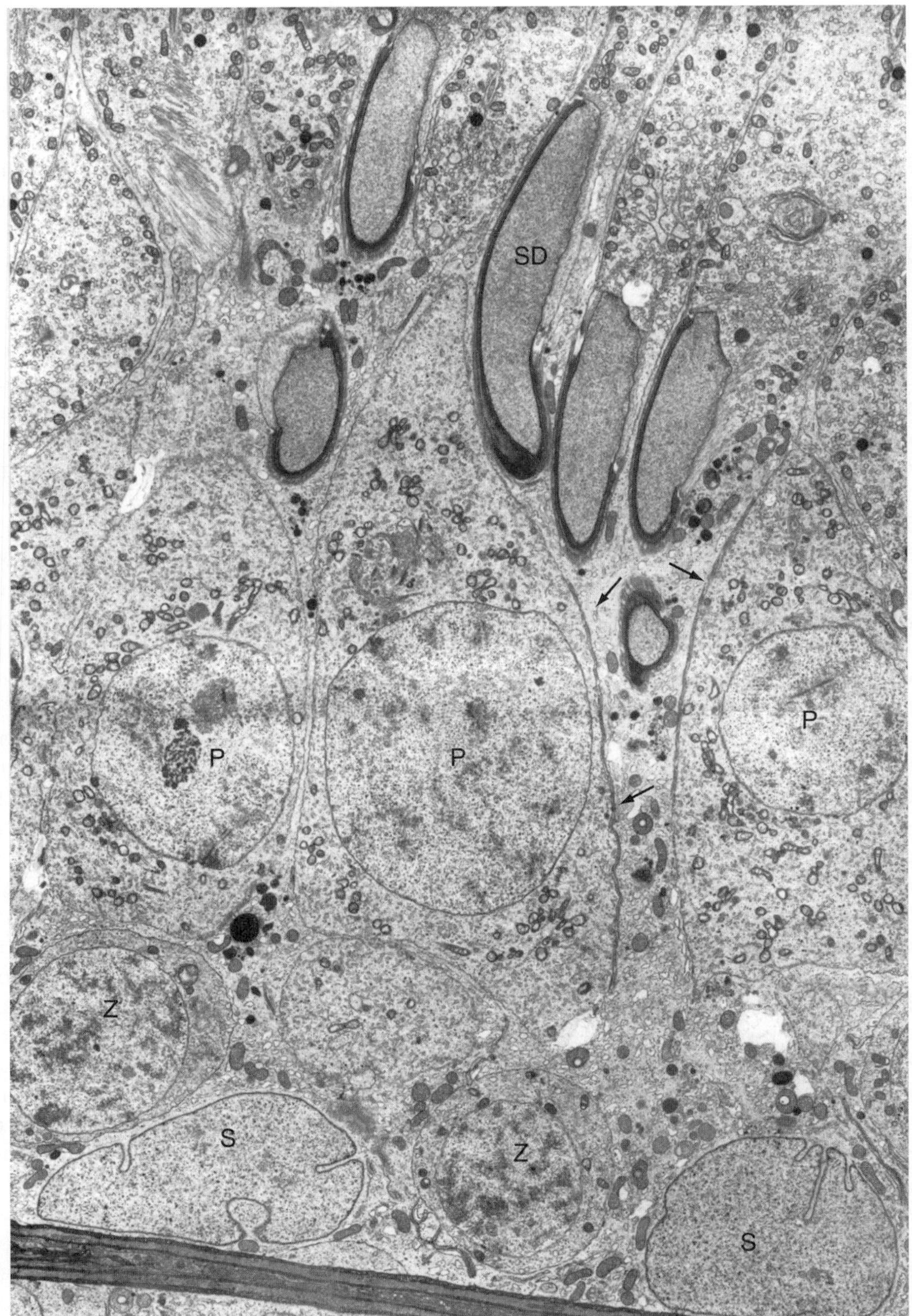

FIG. 55. Stage XI of the rat spermatogenic cycle illustrating Sertoli cells (S), zygotene primary spermatocytes (Z), pachytene primary spermatocytes (P), and elongated spermatids (SD). Note hemijunctions (*arrows*) associated with the plasma membrane of the pachytene primary spermatocytes.

has implicated the ectoplasmic specialization in partly fulfilling this role. In experiments designed to encourage premature sloughing of round or elongated spermatids from the Sertoli cells, the ectoplasmic specializations remain intact at the time of spermatid detachment (615), and ectoplasmic specializations were disassociated from elongated spermatids some 30 hours before final disengagement (616). These findings suggest that disruption of spermiation is due not to ectoplasmic specialization removal but to some other loss of adhesion possibly mediated by integrin protein complexes, discussed again below. Additional theories of the activities of the ectoplasmic specializations include (a) structural support for the Sertoli cell during germ cell mobilization (429,544), (b) participation in sperm release and acrosome shaping (472,614), (c) a contractile role (539,617), and (d) sites of intercellular attachment (472,540,614,618). Actin and espin (an actin-bundling protein) are especially abundant within ectoplasmic specializations associated with junctional complexes and flanks the spermatids that form recesses within the plasma membrane of the Sertoli cell (619–623). Vinculin is codistributed with bundles of actin (624,625), adding support to the concept that ectoplasmic specializations serve as sites for intercellular adhesion between adjacent Sertoli cells and between Sertoli cells and spermatids.

A general concept of the origin of ectoplasmic specializations and their role in sperm release was put forward by Ross and Dobler (442) and Ross (472). Based on their observations on the mouse Sertoli cell, they suggested that as primary spermatocytes begin to ascend through the area of the blood–testis barrier, the inter-Sertoli cell tight junctions disengage and each intact free ectoplasmic specialization so formed would gain attachment to the spermatocyte on entering the adluminal compartment of the seminiferous epithelium. This hypothesis has been questioned (429) on the grounds that (a) in the rat testis, no ectoplasmic specializations were observed in association with the ascending young spermatocytes, and (b) such round germ cells are not actually attached to ectoplasmic specializations as discussed above. Further studies are necessary to clarify the origin and early location of ectoplasmic specializations.

There is no doubt that ectoplasmic specializations are seen facing mid-pachytene primary spermatocytes and round spermatids (428,429,472,585), where they occupy approximately 1% of the surface area of the Sertoli cell in the rat testis at stage V of the spermatogenic cycle (431). All elongated and maturing spermatids are associated with ectoplasmic specializations, which comprise approximately 3% of Sertoli cell membrane surface area. The participation in sperm release and ultimate fate of ectoplasmic specializations are the subject of some discussion. After elongation of mature spermatids and their ascent toward the lumen of the seminiferous tubule, the spermatid head disengages from the ectoplasmic specializations and allows the release of sperm from the apical surface of the Sertoli cells. Profiles of ectoplasmic specialization are retained by the Sertoli cell, and in studies of the mouse Sertoli cell (472) they were seen facing the engulfed lobes of excess spermatid residual cytoplasm; they were also found free at the surface of the Sertoli cell or as isolated nonsurface elements within the cytoplasm. This suggested that the occurrence of ectoplasmic specializations free within the cytoplasm signaled an early stage in their disintegration and subsequent disappearance. These findings have formed the basis for additional study of the ectoplasmic specializations (544,586,626), and three theories have been advanced to account for the origin, behavior, and fate of ectoplasmic specializations. These events are depicted in Fig. 56: (a) they may degenerate and, after internalization by the Sertoli cell, then disappear (472,617,627,628); (b) they become adherent to the cytoplasmic lobe of the sperm and thus selectively retain the lobes of residual cytoplasm (613); and (c) they are removed intact from the Sertoli cell surface and, when free in the cytoplasm, became available for reutilization by the next generation of germ cells. Recycling of ectoplasmic specializations could occur by their distribution from a common pool or by intercellular transfer, for example, from round spermatids to pachytene spermatocytes (544,629).

As yet, the factors regulating the formation and degradation of ectoplasmic specializations and their suggested role in germ cell movement, sperm release, and structural support of the Sertoli cell remain to be determined. As the mature spermatids are released from their attachments to Sertoli cell ectoplasmic specializations, they are retained at the surface of the seminiferous epithelium via very slender upward projections of the apical Sertoli cell cytoplasm. In this position they await their release into the lumen of the seminiferous tubule (Figs. 38 and 57). In favorable semithin sections examined by light microscopy, spermatid heads lay at the very tip of the seminiferous epithelium, and their flagellum projects into the lumen. Immunocytochemical and molecular biology studies have shown that β_1 integrins may be involved in the formation of these ectoplasmic specializations in both inter-Sertoli cell junctions and Sertoli cell–spermatid contacts (630–632) and the stage-specific identification of integrins and their cellular localization fit with the above description. The sperm appear to retain contact with the seminiferous epithelium via two structures. First, and well illustrated in the rat, the apical stalk of Sertoli cell cytoplasm balloons out within the concave aspect of the spermatid head, and second, the cytoplasm of the elongated spermatid becomes progressively attenuated along the long axis of the sperm and thus forms a slender cytoplasmic lobe (future residual cytoplasm) extending from the neck of the spermatid and coursing toward the apical regions of the Sertoli cell. Returning to the delicate Sertoli cell cytoplasmic stalk associated only with the head and neck of the spermatid, faint striations are observed within the concave recess of Sertoli cell cytoplasm. In a series of excellent studies of this relationship between the two cells (402,410,411, 429, 611), these radially oriented spokes (Figs. 38 and 58) form a so-called tubulobulbar complex. This region of

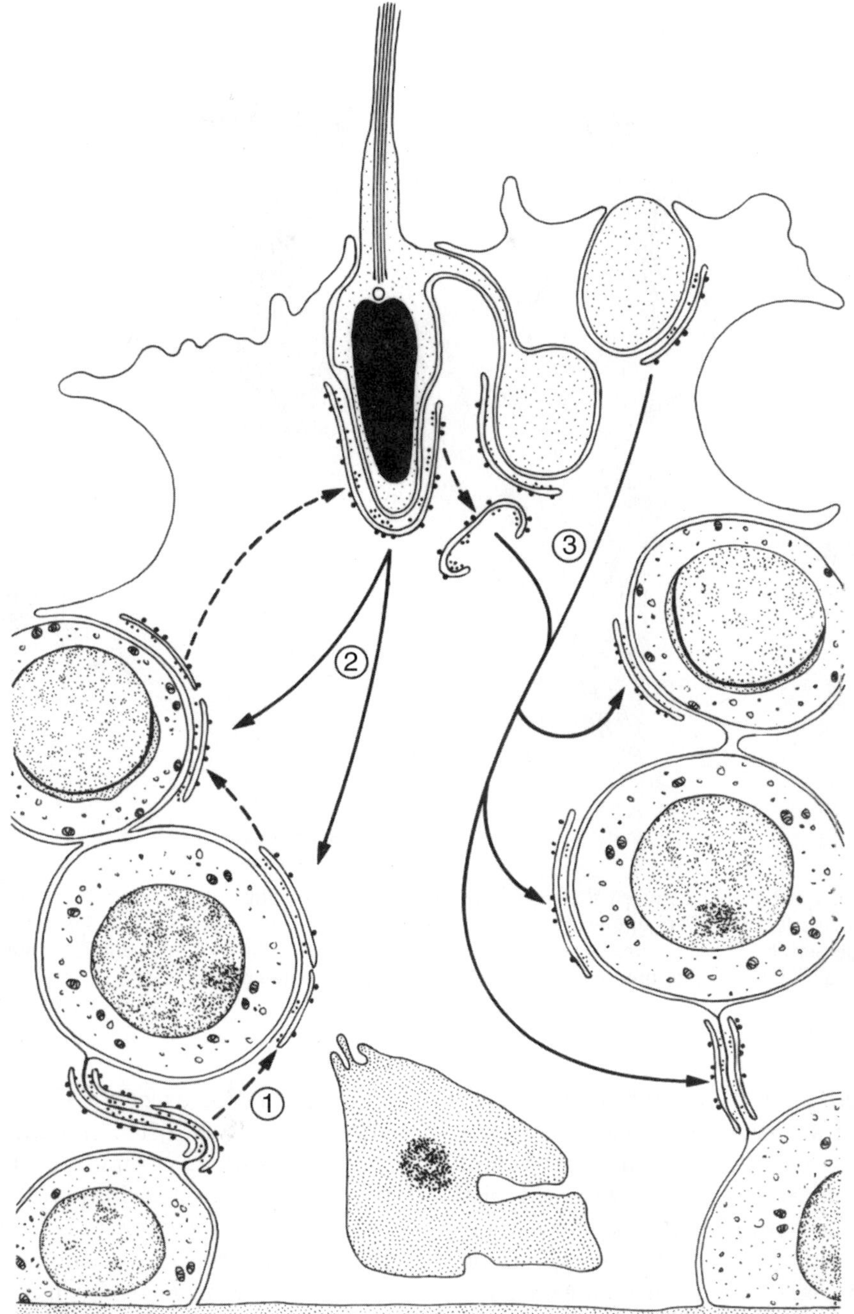

FIG. 56. Diagram demonstrating possible pathways for redistribution of inter-Sertoli cell junctions. In route 1, basally situated junctions may separate to form hemijunctions associated with the surface of primary spermatocytes and early and late spermatids. Route 2 indicates proposed recycling of crypt-shaped hemijunctions from the late spermatids back to earlier generations of germ cells. Route 3 illustrates possible formation of pools of free hemijunctions, which recycle to earlier germ cells and may be involved in the formation of basal inter-Sertoli cell junctions. [Based on (472).]

specialization consists of a series of long narrow cytoplasmic stalks that protrude, in the case of rat spermatids, from the spermatid head and invaginate the recess of Sertoli cell cytoplasm that fills out the concave environs of the spermatid head (Fig. 58). As the spermatid tubular evaginations penetrate into the Sertoli cell, the Sertoli cell plasma membrane closely conforms to that of the spermatid except at the tips of the projections, where small bulbous knobs are formed. Tubulobulbar complexes are described in numerous mammals (611); their number ranges from 4 to 24 per spermatid head and they attain lengths

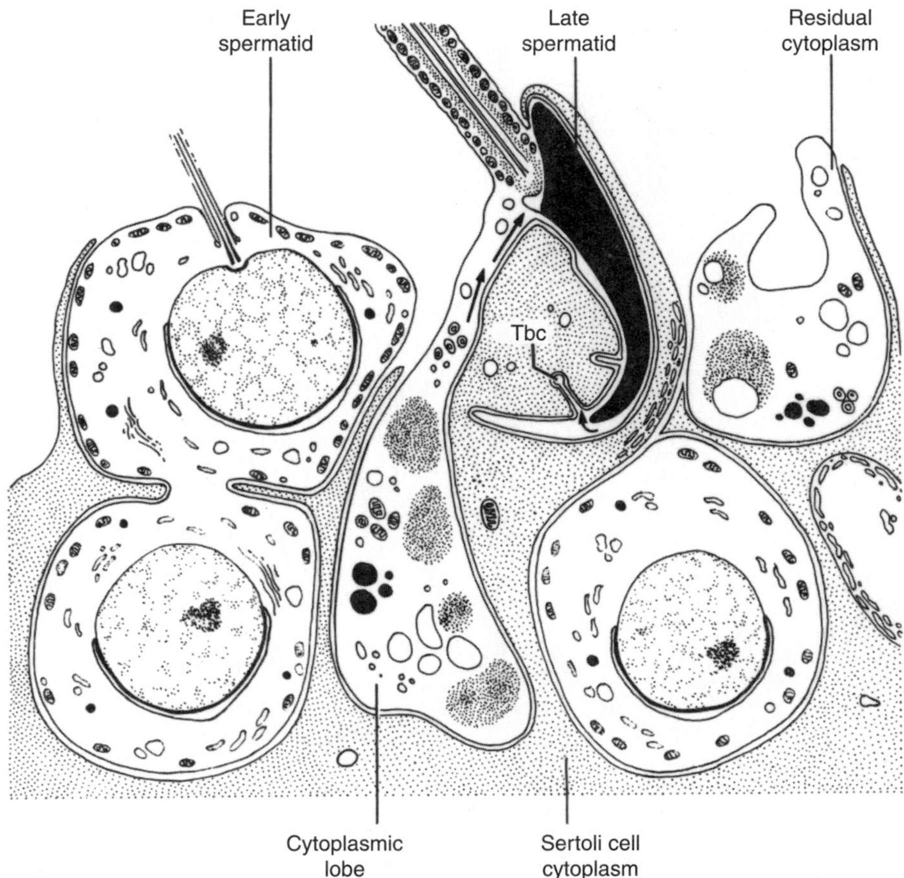

FIG. 57. Relationship between the mature generation of spermatids and the apical surface of the Sertoli cell. Tubulobulbar complexes (Tbc) extend from the concave surface of the spermatid head into the Sertoli cell cytoplasm. Cytoplasm within the cytoplasmic lobe of the spermatid is thought to flow into tubulobulbar complexes as indicated by *arrows*. [Modified from (402).]

of 1 to 8 μm. The complexes are not formed from round spermatids, nor are they seen emanating from elongated spermatids located deep within recesses of Sertoli cell cytoplasm. However, when spermatids rise to the very tips of the Sertoli cell, they lose their relationship to ectoplasmic specializations, whereupon tubulobulbar complexes make their initial appearance. As the Sertoli cell gradually withdraws from the spermatid head, these complexes are maintained until only the extreme tip of the late spermatid head is related to the slender processes of the Sertoli cell. This finding has led to the suggestion that the complexes anchor the spermatid head and participate in its stabilization.

Because several generations of tubulobulbar complexes are formed and then degraded in succession by the Sertoli cells (411), it seems plausible that a proportion of spermatid cytoplasm is resorbed through the degradation of numerous tubulobulbar complexes. The flow of spermatid cytoplasm into these complexes has also been suggested to trigger the sperm release mechanism (629). Although the mechanism by which the Sertoli cell is able to selectively retain, sculpture, and engulf the excess residual cytoplasm remains unclear, Russell (411) advanced the concept that the tubulobulbar complexes indirectly engineer the formation of the residual bodies. This suggestion is based on his observation that in the period immediately preceding sperm release, the volume of the spermatid cytoplasm is reduced as much as 70%, and this disposal of cytoplasm coincides with the successive formation and degradation of tubulobulbar complexes. The cytoplasm eliminated with the resorption of the tubulobulbar complexes by the Sertoli cell is usually watery and lacks organelles, suggesting that the remaining spermatid cytoplasm could become progressively more condensed and filled with spermatid organelles and inclusion bodies. In all species so far studied, the excess cytoplasmic lobes always exhibit these features, and after the sperm are shed, they form deeply staining residual bodies (Figs. 52 and 57). It is not yet known how the Sertoli cell releases the sperm but selectively retains the excess residual cytoplasm, although a report (473) showed that the Sertoli cell cytoplasmic projections that surround and invaginate the residual

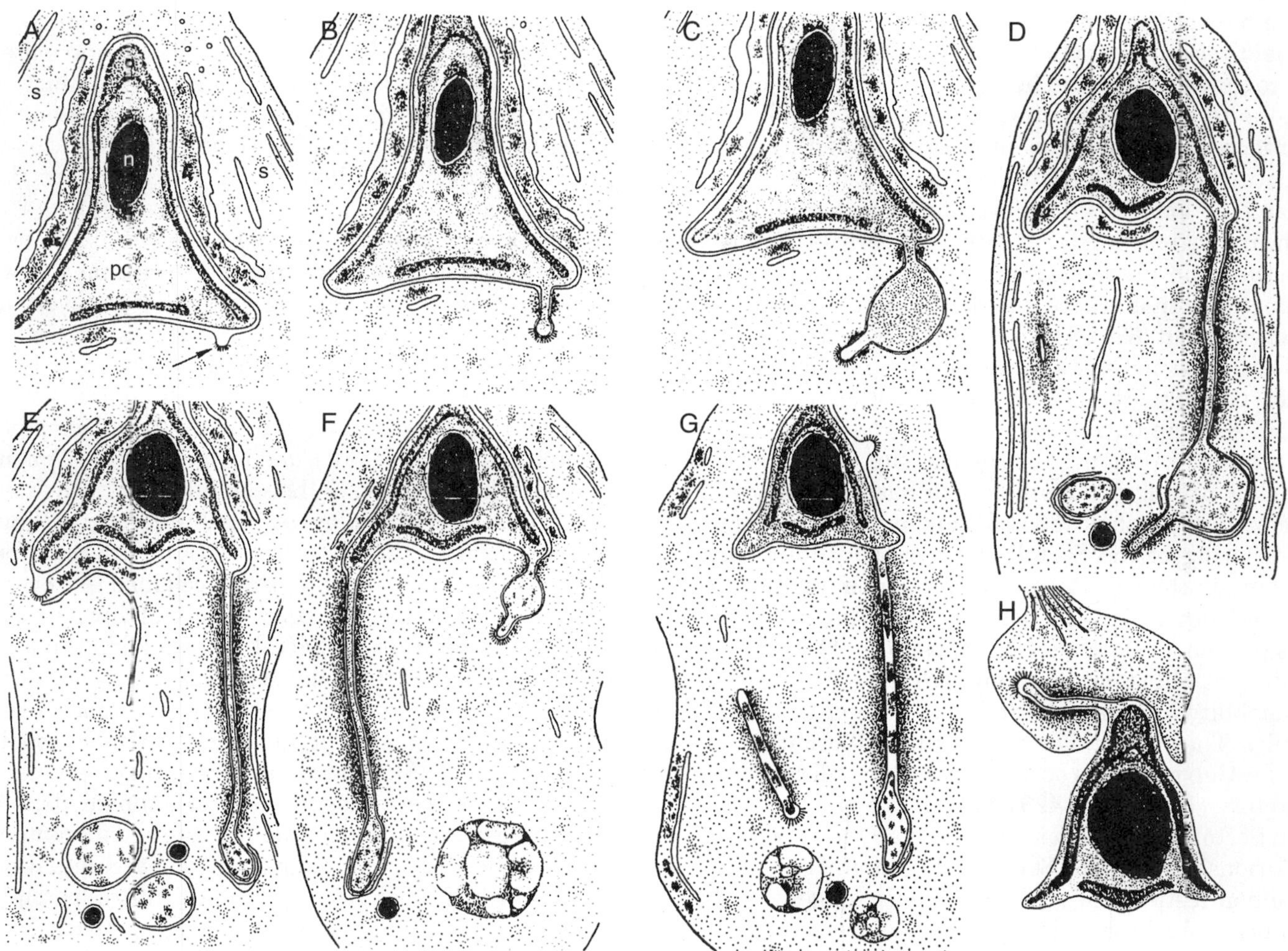

FIG. 58. Illustrations demonstrating the temporal sequence in the formation and disappearance of tubulobulbar complexes in the rat testis. Sertoli cell cytoplasm is *stippled*; spermatid head drawn in transverse section. **A:** A coated pit (*arrow*) of the Sertoli cell occurs in areas devoid of junctional specialization. Perinuclear cytoplasm (pc), Sertoli cell (S), and spermatid nucleus (N) are indicated. **B:** A tube forms by evagination of spermatid into the Sertoli cell. **C:** Mid-portion of the tube develops a bulbous dilatation. **D:** Proximal elongation of the tube, with the distal portion associated with lysosomes. **E:** Separation of bulbs from the tube with an associated increase in their watery consistency. **F:** Individual bulbs sequestered into Sertoli cell phagocytic vacuoles. New tubulobular complexes form. **G:** Dissolution of tubular portions. **H:** Before sperm release, the Sertoli cell cytoplasm withdraws from the spermatid head, and new tubulobulbar complexes develop at the convex margin of the spermatid head. These are subsequently resorbed, and the sperm is released (429).

cytoplasm are richly supplied with filaments and microtubules, which probably pull the residual cytoplasm from the lumen to the apical trunk of the Sertoli cell.

Tubulobulbar complexes also occur between adjacent Sertoli cells at the level of the blood–testis barrier and occasionally give rise to tight and gap junctions in the region of interdigitation (402,626). The tubulobulbar complex is a protrusion of cytoplasm of one Sertoli cell into a neighboring Sertoli cell and resembles a narrow tube, 2–4 μm in length terminating in a bulbous dilation. The complexes undergo a cyclic variation in number during the spermatogenic cycle in the rat, and their degradation and disappearance at stages VI to VII may represent a mechanism for the development or turnover of junctional contacts between the Sertoli cell.

Other types of specialized junctional regions found within the seminiferous epithelium include desmosomes, hemidesmosomes, and gap junctions. Desmosomes (or macula adherens) are traditionally recognized as small apposed plaques on adjacent cells flanked by a thin dense layer just beneath the plaque and represent strong adhesion sites between cells. They have been described between Sertoli cells and all round germ cells but are absent at the interface with elongated spermatids (119,237,238, 243,257,428,633). In the rat they are oriented toward the body of the Sertoli cell and are often seen toward the base of the seminiferous tubule, suggesting they could resist any force tending to dislodge germ cells into the lumen (432). Their general function is believed to be to assist the surface of the Sertoli cell in the orderly upward movement of germ cells during their maturation. A quantitative analysis of desmosome gap junctions between Sertoli cells and germ cells of the rat testis showed that germ cells early in the meiotic maturation process exhibit larger numbers of these junctions than those in more advanced phases of development (634). The finding suggests that whatever their functional significance, desmosome gap junctions are possibly more important for the initiation of meiosis than for its continuance.

Hemidesmosomes occupy the site where the base of the Sertoli cell rests on the basal lamina of the seminiferous tubule (586,635) and probably confer firm anchoring of the Sertoli cell necessary for them to maintain their pentagonal or hexagonal relationships to each other. Gap junctions (or nexus) are represented as extremely close sites of apposition (20 Å) between adjacent cell membranes and are thought to offer the capability of ionic flux and exchange of small molecules. They also confer sites of firm adhesion between cells. Gap junctions occur between Sertoli cells in laboratory species (118,426,517,636) but are not present on Sertoli cells of the normal human testis (428,471). Curiously, the inter-Sertoli cell gap junctions in the human testis occur in cases of spermatogenic disturbances and in certain testicular cancers (637–640). Their presence has also been noted between Sertoli cells and round germ cells (304,641), although they are often seen in association with desmosomes and are occasionally referred to as desmosome gap junctional complexes (432). Gap junctions are not observed between Sertoli cells and elongating spermatids. A major connexin protein of gap junctions, Cx43, is strongly expressed in the mouse and rat testis and expression of the mRNA and protein vary with the spermatogenic cycle (642–644). Immunostaining for Cx43 and cell–cell communication properties using Lucifer yellow dye coupling showed high activity in the rat spermatogenic cycle in all stages, except for stages IX–X when activity is almost abolished (645). The physiological function of this finding may reflect alterations in the Sertoli cell plasma membrane as germ cells translocate from the basal to the adluminal compartment in the seminiferous epithelium or the temporary loss of gap junctions, thereby allowing spermatogonial mitosis at these stages.

Changes Associated with Seasonal Breeding and Age

Most mammals are subjected to wide variations in the environmental conditions in which they live, and thus seasonal variations in photoperiod, ambient temperatures, and the availability of food and water together contribute to seasonal cycles of reproductive activity of many wild and domestic animals. Seasonally breeding animals, unlike their laboratory-bred counterparts, provide an opportunity to investigate the structure and function of the Sertoli cell during the cyclic waxing and waning of spermatogenesis. A great deal of information has accumulated over many years that attests to the annual regression and recovery of spermatogenic activity within the testes of a wide range of vertebrates (34,436,496,646–649).

Histological investigations of the changes in testicular activity among seasonally breeding fish, amphibians, reptiles, birds, and mammals have demonstrated that regression of the seminiferous epithelium is often accompanied by an accumulation of lipid inclusions within the Sertoli cells (436). With reactivation of spermatogenesis, the Sertoli cell lipid inclusions disappear, although the precise role of lipid in the presumed alterations of Sertoli cell function remains unknown. Examples of the changing structure of the Sertoli cell in a seasonally breeding wild rat have been documented at the light and electron microscopic level (497,648,649). In seasonally breeding mammals, the arrest of spermatogenesis is accompanied by a marked reduction in Sertoli cell volume and its content of organelles (650–652). Simultaneous accumulation of lipid inclusions is also a common feature, possibly relating to deposition and storage of lipid-rich substances contributed in part from the degeneration of many germ cells (653–655). These structural changes in Sertoli cells during the nonbreeding season are thought to occur in response to a decline in the levels of FSH, LH, and testosterone (656–658) and resemble the morphological changes in Sertoli cells seen after hypophysectomy (659,660).

Little is known concerning the integrity of the blood–testis barrier in seasonally breeding mammals, and although the inter-Sertoli cell junctions remain during the regression of the seminiferous epithelium (381,648,661), their effectiveness as a permeability barrier requires additional study. However, in lower orders of vertebrates such as reptiles and amphibians (437,441), tight junctions between Sertoli cells disappear during the nonbreeding season. In testes exhibiting regional differences in spermatogenic activity, germ cell maturation is accompanied by Sertoli cell junctions, whereas in regressive zones of the testis not associated with germ cell activity, tight junctions disappear. Thus, at least in some species, the formation of the barrier is correlated with spermatogenic activity.

The basic tenet that Sertoli cell numbers remain stable in the adult testis (50,71) has been brought into question in recent years. Because sperm production in many animals may vary with season and because daily sperm production is subject to an age-related decline, the effect of variation in Sertoli cell numbers offers a theoretical mechanism to allow these changes to occur. Variation in the Sertoli cell population has been documented, although this phenomenon has in the past attracted little attention, possibly for lack of statistical significance or because of large changes in the dimensions of the seminiferous tubules. Nevertheless, when mouse testes were X-irradiated to partially destroy the germ cell population, the restoration of testicular function was associated with an increased number of Sertoli cells per seminiferous tubule cross-section (662).

When rat testes were made artificially cryptorchid (663,664), the Sertoli cells were reported to transiently increase in number, and with prolonged periods of cryptorchidism, Sertoli cells disappear from the seminiferous tubules (665). Sertoli cells of the immature rat testis apparently continue their division processes for longer than is otherwise indicated by their mitotic index (666), and in the absence of germ cells, Sertoli cells in prepubertal rats continue to proliferate beyond day 20, when their numbers normally stabilize (123). Increases in the Sertoli cell population have been reported to occur during FSH treatment of rats after hypophysectomy (667). The lack of rigid quantitative assessment of Sertoli cell numbers in some of these and other studies has been voiced as a criticism of their conclusions. However, other studies of seasonally breeding red deer (668,669) and stallions (664,670) have presented clear evidence that the Sertoli cell population is capable of increasing in the adult by fluctuating with season. Seasonal differences in Sertoli cell numbers of 60 adult stallions were characterized by a 36% increase in the breeding versus the nonbreeding season. Because the horse Sertoli cells only have a limited capacity to alter the ratio of germ cells to single Sertoli cells, a seasonal variation in total numbers of Sertoli cells provides an additional mechanism for increased daily sperm production per testis, which is known to occur in the breeding season.

The source of additional Sertoli cells found in the breeding season and the factors promoting their appearance and disappearance remain unclear. Mitotic figures at the base of the seminiferous epithelium in the stallion were seen at stages II–III of the spermatogenic cycle, whereas spermatogonia are thought to divide only between stages V and VIII. Concentrations of FSH, LH, and testosterone measured in the same study (670) were altered not only by season but also with age, and no clear relationship has emerged to link hormone changes with Sertoli cell numbers. The numbers of type A spermatogonia also exhibit changes with season in adult stallions (671), and it seems possible that the Sertoli cells may influence these cells through hormonal action (672) or perhaps directly via mitogenic peptides secreted by the Sertoli cells (673). Although the adult stallion appears to be an exception to the rule that Sertoli cell numbers remain stable after puberty, additional research may well indicate that Sertoli cells are a dynamic rather than sessile population.

A very different account of the biology of Sertoli cells has been revealed when their numbers are related to increasing age. With morphometric techniques to evaluate cell numbers (674), it was shown that the number of Sertoli cells in the human testis exhibits an age-related decline. In a study of men aged 20 to 85 years who were in apparent good health before sudden death, young adult men (20–48 years, $n = 37$) had approximately 500 million Sertoli cells per testis, and this number declined significantly to a mean of 300 million per testis for older men (50–85 years, $n = 34$). When the relationship between age and numbers of Sertoli cells per testis was examined, there was a significant age-related decline in the Sertoli cell population (458). In addition, the same study reported a significant correlation between Sertoli cell numbers and daily sperm production, although the numbers of germ cells accommodated by the Sertoli cells at any age remain unchanged. Taken together, the analysis of changes in spermatogenesis and Sertoli cell numbers in the special examples of old age and in seasonal breeders indicate that variation in daily sperm production can be attributable to alterations in the total numbers of Sertoli cells and/or to changes in the numbers of germ cells that are associated with an individual Sertoli cell.

Response to Injury

In the past, the persistence of Sertoli cells after damage or elimination of the germ cells was taken to indicate that the Sertoli cells were resistant to a wide variety of treatments that otherwise caused spermatogenic disruption. This belief was based on light microscopic observations of Sertoli cells after testicular damage in which their morphology appeared little changed from normal. However, as pointed out by Fawcett (426), the availability of semithin sections of testicular tissues embedded in plastic has revealed far more morphological details of the Sertoli cell than previously appreciated from studies on paraffin sections. The sensitivity of the Sertoli cell to damage has been emphasized by the early ultrastructural and functional changes of this cell in response to a great range of unrelated treatments that exert adverse effects on the testis (498).

The literature relating to morphological alterations of Sertoli cells is voluminous and encompasses a wide range of treatments too numerous to describe in this chapter. We chose, therefore, to discuss and illustrate changes in Sertoli cell structure in a general manner, because it seems likely that the reaction of this cell to injury is often reflected by very similar morphological alterations. An early indication of altered Sertoli cell cytology is the appearance of many clear watery vacuoles or vesicles within the basal aspects of the cell. These are readily visible by light microscopy (Fig. 59), and when examined by electron microscopy, they consist of membrane-limited vacuoles arising from three locations. First, the basal cytoplasm exhibits intracellular vacuoles of variable configuration (Fig. 59), but whether these form by dilation of endoplasmic reticulum or endocytosis of fluids has not been clarified. Second, vacuoles occasionally appear in the intercellular spaces between Sertoli cells and neighboring germ cells, and third, vacuoles clearly visible by light microscopy are associated with the regions of junctional specializations between adjacent Sertoli cells (Fig. 60). These vacuoles are of specific interest because they involve radical disorganization of the inter-Sertoli cell junctions in which progressive expansion of the intercellular spaces at the site of junctional complexes gives rise to multiple extracellular vacuoles along the pathway of each junctional complex. The formation of these vacuoles in relation to seminiferous tubule damage has been closely studied after the induction of experimental cryptorchidism (675), in which it was suggested that they contribute to complex membranous bodies within the Sertoli cells. The sequence of morphological changes to the inter-Sertoli cell junctions during cryptorchidism is illustrated by light and electron micrographs in Fig. 61 and summarized diagrammatically in Fig. 62. Vacuolization is probably a nonspecific response to the Sertoli cells to injury, because it occurs in unrelated conditions of experimental damage to the seminiferous epithelium (123,127,558,676–684).

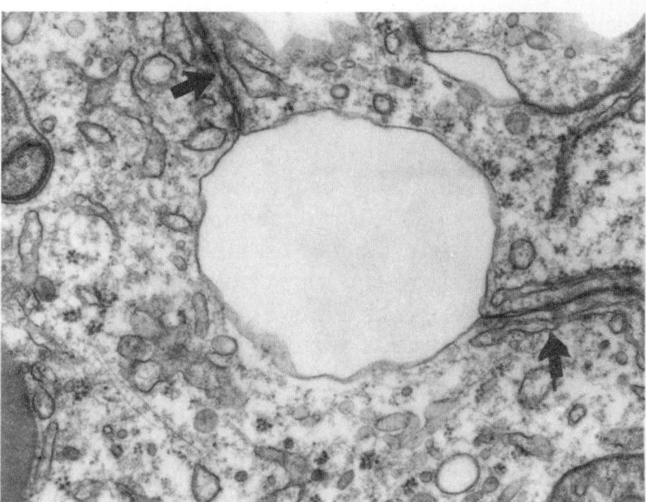

FIG. 60. Five-day experimentally cryptorchid testis showing a dilation of the intercellular space between two Sertoli cells. At opposite sides of the extracellular space, the pathway of inter-Sertoli cell junctional complexes can be noted (*arrows*).

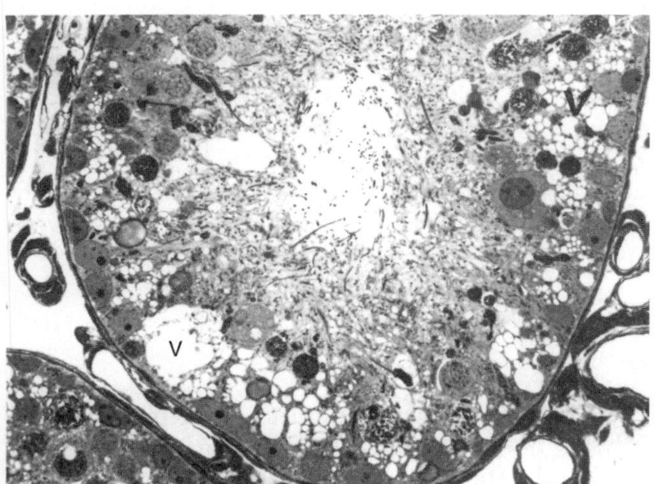

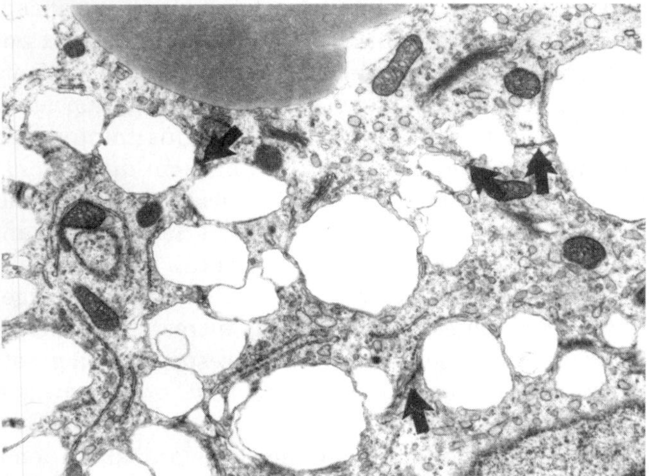

FIG. 59. A: Five-day experimentally cryptorchid rat testis showing many vacuoles (V) within the seminiferous epithelium. **B:** Many of these vacuoles in the 5-day cryptorchid testis are interconnected (*arrows*) by junctional specializations between adjacent Sertoli cells.

An additional sign of morphological disturbance to the Sertoli cell is the rapid accumulation of cytoplasmic lipid droplets, which very often accompany impairment of spermatogenesis. Although lipid inclusions are a common feature of Sertoli cells in the normal testis, they appear to increase in size and number coincident with morphological evidence of germ cell degeneration, suggesting that the degradation products of effete germ cells contribute to ever-increasing numbers of Sertoli cell lipid inclusions. However, the autolysis and phagocytosis of germ cells by Sertoli cells may not be the exclusive source of these lipid droplets, because they are known to accumulate within the Sertoli cell in the absence of germ cells (123). In conditions of prolonged atrophy of the seminiferous tubules, some Sertoli cells retain their content of lipid, whereas others exhibit few if any lipid inclusions. The latter situation of complete degeneration and disappearance of germ cells from the seminiferous epithelium results in further structural modifications to the Sertoli cells,

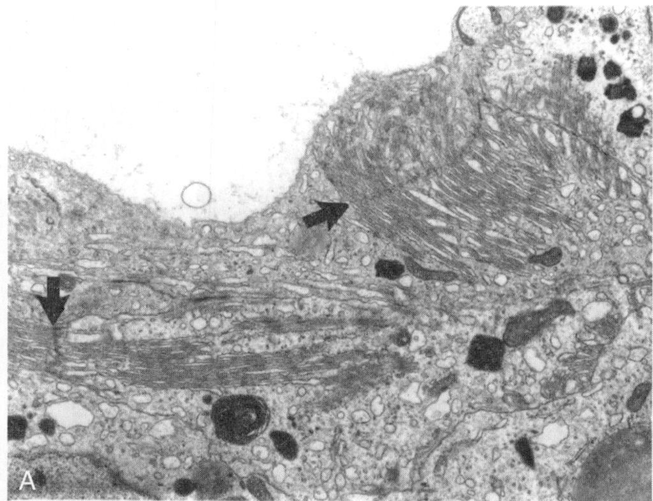

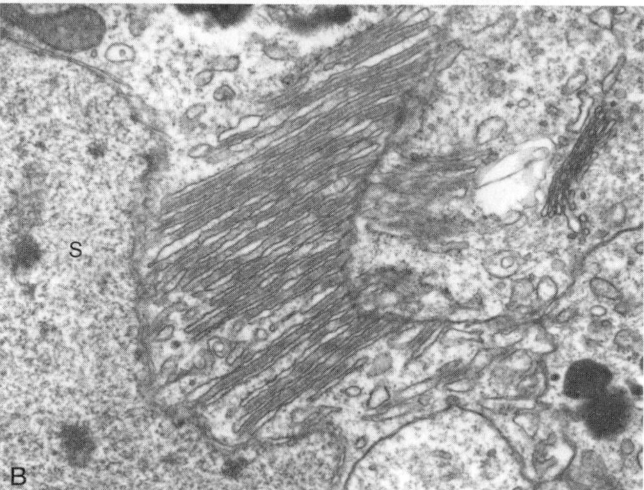

FIG. 61. A: Four-week experimentally cryptorchid testis showing complex membranous bodies within the Sertoli cell cytoplasm. The membranes are arranged perpendicular to a central axis (*arrows*). **B:** Sertoli cell membranous complex illustrating parallel cisternae of smooth membranes, occasionally bearing ribosomes, and electron-dense materials representing bundles of filaments. Sertoli cell nucleus (S).

which involve retraction and convolution of their apical cytoplasm. Such seminiferous tubules exhibit a thin rim of flattened Sertoli cells. Alternatively, removal of the germ cells allows for shrinkage of the seminiferous tubule, whereupon the Sertoli cell cytoplasm becomes interdigitated as the cells are closely packed into the reduced volume of the atrophic tubule. Destruction of Sertoli cells has not often been reported, although this was clearly demonstrated some years ago (685) when it was shown that cadmium salts were acutely toxic to the seminiferous epithelium, causing the destruction of germ cells and Sertoli cells. Other agents that bring about Sertoli cell necrosis include LH, LH plus LH–releasing hormone agonists, and hCG (686,687), although the mechanism by which these hormones destroy the Sertoli cells has not been studied. All the above-mentioned alterations to the Sertoli cell have been described in disorders of the human testis (428,459,461,688–690) and in testes of men of advanced age (674,691).

PERITUBULAR TISSUE

The seminiferous tubules are supported by a circumferential layer of attenuated cells (fibroblasts and myoid cells) and extracellular materials (collagen and glycoproteins), collectively termed the lamina propria, whose ultrastructure features have been extensively documented (Fig. 44) (570,571,675, 692–697). The functions of the lamina propria can be summarized as follows: (a) provision of mechanical support and contractile function for the seminiferous tubule, (b) passive paracrine influence on spermatogenesis via secretory products stimulated by testosterone, (c) partial and species-specific restriction of macromolecules entering the seminiferous tubules via the intertubular tissues, and (d) a source of precursor cells capable of differentiating into Leydig cells (698–704). The contractile properties of peritubular tissues are illustrated in the rat by induced or spontaneous contractions commencing between days 4 and 7 of postnatal life (705). These myoid-type cells contain actin, desmin, and vimentin filaments (610, 619,620,699,700,706), and the outermost peritubular cells in the human testis are probably derived from the intertubular fibroblastic tissue, because they contain vimentin but lack desmin (701). In the human testis, the biochemical properties of myoid cells suggests local mechanisms controlling their contractile function due to the presence of nitric oxide–producing enzymes (nitric oxide synthase), nitric oxide, and soluble heme-containing guanylyl cyclase, the latter an intracellular target for nitric oxide (707,708). Nitric oxide production and activity may cause relaxation of the myoid cells, assisting with peristalsis of the seminiferous tubules. Evidence for innervation of the peritubular tissue of the human testis is available (690,709,710), and autonomic nerve fibers together with cholinergic and adrenergic nerve varicosities have been described at the ultrastructural level (711). However, the precise role of these neural elements remains unknown. More information on the receptor, synthetic, and secretory properties of the peritubular tissue is required to gain a better understanding of its role in the physiology of the testis.

CYCLE OF THE SEMINIFEROUS EPITHELIUM

The organization of the cell types that comprise the process of spermatogenesis within the seminiferous

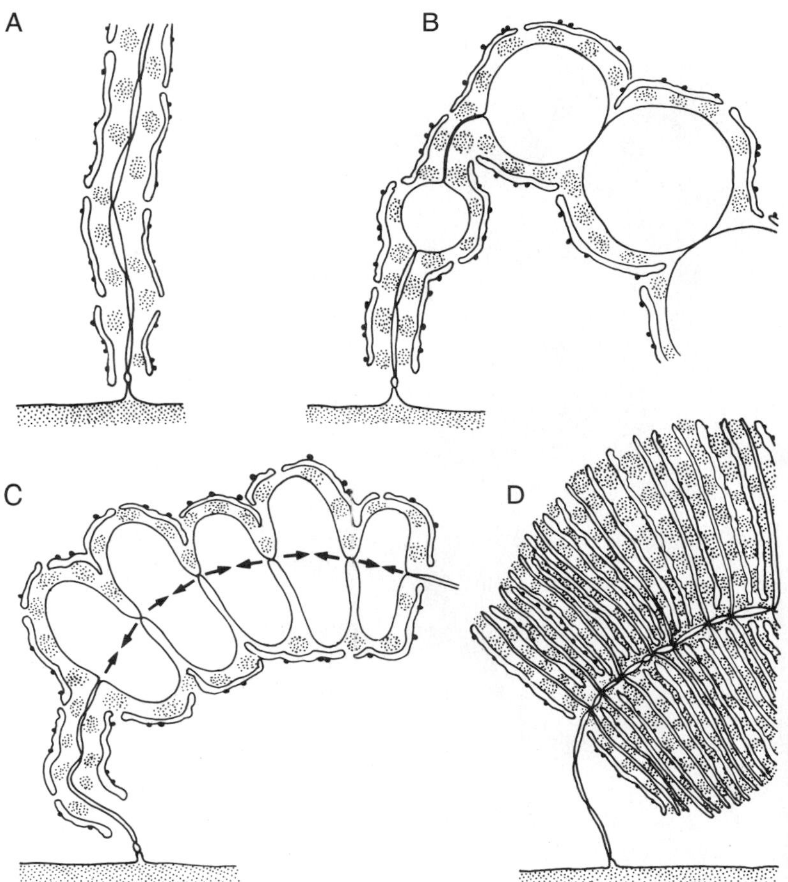

FIG. 62. Suggested sequence of formation of extracellular vacuoles and complex membranous cisternae in the seminiferous epithelium of the experimentally cryptorchid rat testis. **A:** Inter-Sertoli cell junction in normal testis. **B:** In short-term cryptorchidism, vacuoles develop in the intercellular regions between sites of membrane occlusion. **C:** At later time intervals, the degeneration of germ cells allows shrinkage of the seminiferous tubule and progressive collapse of the vacuoles, as indicated by *arrows*. **D:** With continued compression of the Sertoli cells, it is suggested that the rows of collapsing vacuoles undergo a "connective-type" compression, forming elaborate parallel stacks of flattened vacuoles interposed between the cisternae associated with the tight junctions.

epithelium is not random but in fact highly organized. This fact was elucidated toward the end of the 19th century by a number of investigators (49,159,202, 240,712). They noted that the germ cells at different developmental phases formed easily identifiable collections, termed cell associations. Subsequent studies (713,714) were extended by the detailed investigations of Le Blond and Clermont (250,251), who defined the cycle of the seminiferous epithelium as the series of changes in a given area of the seminiferous epithelium between two appearances of the same developmental stage. Using the periodic acid–Schiff reaction to stain the acrosome of spermatids together with nuclear morphology, they divided spermatogenesis in the rat into 14 stages or cell associations (Fig. 63). The complete sequence of 14 stages or cell associations constitutes one cycle of the seminiferous epithelium. If we could observe a segment of the seminiferous epithelium at stage 1 of the cycle by time-lapse photography over a period of days, the epithelium at this site would pass through the 14 stages with the eventual reappearance of stage 1. A remarkable presentation of this process has been produced by creating a movie using images from adult rat testis sections that enable the viewer to visualize the dynamic association of maturing germ cells with a single Sertoli cell (715). Le Blond and Clermont also established that the transformation of a type A spermatogonium into a mature spermatid, released at stage 7 of the cycle, occupied four such cycles. In the rat, the entire tubule cross-section is at the same stage of the seminiferous cycle, which, in fact, extends over several millimeters of adjacent seminiferous tubule.

Investigations in other mammals subsequently established a seminiferous cycle in the mouse (161,716), the ram (717), and the monkey (157).

different colonies establish different stage associations even within the same seminiferous tubule (colony a, stages X–XII; colony b, stages VI–VII; colony c, stages VIII–IX), suggesting a local mechanism in operation. With the passage of time after transplantation, long colonies develop additional germ cell stages by fusion of separate segments. For a complete wave of spermatogenesis to occur requires fusion of these individual colonies and probable adjustment of their individual germ cell stages. Fusion of colonies that are not in synchrony with the normally orderly progression of stages that constitute a spermatogenic wave may explain why, in the normal testis, the wave exhibits "site reversals" as originally described in 1961 by Perey et al. (752).

The orderly distribution of stages of the seminiferous cycle along the tubule is the basis for the innovative studies of Parvinen (819), which involved dissection of the tubule into lengths according to the stage of the cycle present. This was achieved by transillumination of the tubule, the stages providing particular appearances based principally on the light scattering achieved by the condensing spermatid nucleus. The studies of Parvinen demonstrated that the nature of the germ cells surrounding the Sertoli cell can significantly influence certain metabolic activities of the Sertoli cells (820). Additional evidence has been provided by the studies of Jegou et al. (821), which showed that modification of Sertoli cell function after exposure of the rat testis to a single episode of heat did not occur until a loss of spermatids occurred. During recovery, Sertoli cell function remained abnormal until the return of spermatids. Furthermore, the studies of Kerr and colleagues (397,445) showed significant changes in the morphology of the Sertoli cell according to the stage of the cycle of the rat. These studies emphasize the fact that the seminiferous tubule compartment cannot be viewed as homogeneous tissue in terms of either structure or function.

New information about stem cell multiplication indicates that mouse spermatogonial stem cells, the As and Aal cells, are found randomly distributed around the tubule perimeter at stage V but redistribute by stage VI to positions opposed to the interstitium, where they are found through all subsequent stages (822). These authors provide evidence indicating that the progeny of these most undifferentiated spermatogonia spread laterally along the base of the seminiferous tubules, and in both rats and mice, the arrangement of stem cells is predominantly in areas of tubule–interstitium, rather than tubule–tubule, contact (822,823). Visualization of the pattern of donor germ cell colonization after transfer to germ cell–depleted recipient mouse testes reinforces the concept that germ cells migrate along the basement membrane to an appropriate niche before differentiating (178).

Germ Cell Degeneration

Several quantitative studies in a number of species have shown that a significant degree of germ cell degeneration occurs during normal spermatogenesis (824–826), with estimates that up to 75% of the full potential output of sperm is lost during maturation of germ cells from their stem cell precursors (827,828). This process is now believed to be essential to maintain the balance between germ cell and Sertoli cell numbers that is required for normal spermatogenesis, in addition to providing a mechanism for the removal of damaged germ cells. Information about the mechanisms that underpin germ cell loss is relevant to understanding both normal development and the response of the testis to injury. Most germ cell loss occurs through apoptosis, a process triggered by damage or through the absence of survival cues, such as might occur with overcrowding of germ cells [reviewed in (829)]. The degree of degeneration influences the efficiency of spermatogenesis, which can be measured as daily sperm production per gram testis, and this efficiency has been shown in humans to be 25% to 35% of that in other species (830). Johnson et al. (831) suggested this degeneration is evidenced by the paucity of cells within each generation and may contribute to the relatively disorganized appearance of the human seminiferous epithelium.

The cell types involved in the degeneration include spermatogonia, spermatocytes, and spermatids. Spermatogonia and spermatocytes display the classical features of apoptosis, including chromatin condensation around the nuclear envelope and labeling of nicked DNA by the terminal deoxynucleotidal transferase enzyme in the Terminal deoxynucleotidyl transferase Biotin-dUTP Nick End Ligation (TUNEL) assay (832). Spermatids undergoing apoptosis are not readily detected by the TUNEL method, and their unique compacted DNA and nuclear structure makes analysis of the incidence or pathway of their degeneration difficult, though the classic caspase-mediate pathway does not appear to be involved (833). Apoptotic cells are often visible as multinucleated symplasts in the lumen of the testis, as they lose their attachment to the seminiferous epithelium or are phagocytosed by the Sertoli cell.

One pathway for triggering apoptosis involves a signal from an extracellular tumor necrosis superfamily ligand, such as FasL or TNF-α, binding to a receptor, such as Fas or TNFR1. In response to receptor activation, an adaptor protein, FADD, is recruited and activated to trigger a pathway of proteolytic cleavage events mediated by caspases, the various targets of which include lamins of the nuclear envelope and cytoskeletal elements. FasL on Sertoli cells and TNF-α on spermatocytes could therefore act on

Leydig cells, primordial germ cells of the fetus, and spermatocytes in the adult testis through the Fas receptor protein or on Sertoli cells, which express the TNFR1 (834,835). Injuries that result specifically in germ cell apoptosis, including radiation, hyperthermia, and ischemia-reperfusion, have been shown to cause up-regulation of Fas alone, whereas certain toxicants that elevate Sertoli cell apoptosis increase both Fas and FasL expression (836).

In a separate pathway, triggered by environmental assaults including cytotoxic drugs and irradiation, the intracellular balance in expression of prosurvival and prodeath members of the Bcl-2 protein family governs cellular fate. The relative increase in proapoptotic proteins drives release of cytochrome c from the mitochondrion and formation of a complex called the apoptosome that triggers a distinct downstream cascade of caspase activation. The importance of the balanced expression of prosurvival and proapoptosis family members for testicular homeostasis has been illustrated in several transgenic and knockout mouse models. Over-expression of prosurvival Bcl-2 in spermatogonia disrupted the seminiferous epithelium after the onset of spermatogenesis (837), because the elevated numbers of undifferentiated germ cells could not be supported by the normal complement of Sertoli cells. Deletion of prosurvival Bcl-w or proapoptotic Bax causes similar disruption to the integrity of the seminiferous epithelium, leading to loss of germ cells initially between the third and fourth postpartum weeks in the mouse (838–840). The timing of this loss most probably relates to the shifting pattern of expression of growth factors, ligands, and apoptotic regulators, including Bcl-2 family members, that occurs during the first wave of spermatogenesis. For example, differentiating germ cells that express the c-kit tyrosine kinase cell surface receptor rely on a cue from the Sertoli cell–encoded stem cell factor for survival (841–843). Expression of both the ligand and receptor are understood to be up-regulated during the onset of the first wave of spermatogenesis (844–846). In addition, the expression of mRNAs encoding several Bcl-2 family members changes through the first spermatogenic wave, so that Bcl-2 is present in the immature but not adult testis and Bcl-XL and Bax shift from spermatogonia to spermatocytes during the second postpartum week in the developing mouse (847).

LEYDIG CELLS

In reviewing the cytology of the Leydig cells and the intertubular tissues in which they reside, we recognized that for the purposes of this chapter it was not possible to acknowledge the work supplied by every contributor to this field of study since a review in 1921 by Stieve (848) was based on approximately 1,000 earlier studies. Thus, the literature on Leydig cell morphology and its functional significance is vast. Our objectives are subsequently directed toward two major topics: (a) to review the functional morphology of Leydig cells based on authoritative accounts that together form an accurate survey of the light and electron microscopic features of the Leydig cells and the interstitial tissue and (b) to relate these features to illustrations on the morphology of Leydig cells. Particularly useful dissertations are available in the literature (30,570,848–855).

Historical Background

In 1850, Leydig (856) reported that the spaces between the seminiferous tubules were occupied by conspicuous masses of cells containing fatty vacuoles and pigment inclusions, supporting observations by von Koellicker (5), who some years earlier had provided a similar description of the cellular composition of the intertubular tissue. Both investigators considered the cells to be a specialized form of connective tissue.

Although Berthold (6) demonstrated that the testis secreted substances into the bloodstream to influence the growth of anatomically distant tissues, the question of which tissue in the testis contributed to this secretion remained unanswered for many years. The prevailing view of Leydig cells as connective tissue was extended by von Ebner (159) and Hofmeister (857), but thereafter a new concept gained favor in which Leydig cells were thought to produce fats, nutrients, or other substances for consumption by the seminiferous tubules (49,858,859). Several investigators proposed that Leydig cell lipid droplets provided the substrate from which testicular hormones were synthesized. This idea received support from the studies of Bouin and Ancel (7,860,861), who coined the term “la glande interstitielle” and offered convincing evidence that the Leydig cells of both normal and cryptorchid testes provided the hormonal stimulus for the production of sperm and for maintenance of secondary sexual characteristics.

Leydig cell lipid inclusions were shown to contain cholesterol esters (862–864), and McGee (865) showed that lipid extracts from bovine testes could stimulate masculine development when administered to other animals. Testosterone was finally identified in these extracts (8,866) and soon was synthesized using cholesterol as the starting material (867,868). Pathways for steroidogenesis and their control by various hormones are reviewed elsewhere in this volume, but many years passed by until it became clear that the

Leydig cells represented the predominant site for steroidogenic enzymes (869–871) and the chief source of testicular androgens (872,873).

Organization of Intertubular Tissue

In routine histological sections of the testes of common laboratory rodents (rat, mouse, guinea pig, hamster), a clear space of variable extent is commonly observed separating individual seminiferous tubules. Within this space are found blood vessels, often forming a central core around which numerous Leydig cells and other cellular constituents. This histological organization is consistent with an endocrine function, namely, a rich vascular supply and a close association between Leydig cells and blood vessels, the latter facilitating the passage of steroids into the circulation. Empty spaces surrounding the seminiferous tubules have often been interpreted as preparation artifacts brought about by shrinkage during tissue fixation and dehydration before paraffin embedding. Although this is no doubt correct, these clear spaces, at least in the above-mentioned species, are not entirely exaggerated in their extent but represent species variations in the size and architecture of intertubular lymphatic vessels. For many years the intertubular tissue was regarded simply as loose connective tissue harboring Leydig cells, fibroblasts, and extracellular elements. The presence of a system of lymphatic channels penetrating beyond the regions of the septula, rete, and tunica albuginea and into the interior of the testis is supported by improvements in tissue preservation that showed specific patterns for species are present characterized by a lymph vascular system and its relative volume, architecture, and relationship to Leydig cells (570,852,874,875).

These patterns have been categorized into four more-or-less distinct patterns of organization (16) (Figs. 69 to 72). In rodents such as the guinea pig and chinchilla, only a small fraction of the testicular volume is taken up by the Leydig cells, which are commonly clustered together around blood vessels. The remaining and greater part of the intertubular area is occupied by extensive peritubular lymphatic sinusoids, whose limiting walls consist of a delicate layer of endothelium. Lymphatic endothelium investing the perivascular Leydig cells is called visceral (30) or interstitial (874) endothelium. Closely applied and coursing parallel to the boundary tissues of the seminiferous tubules, the lymphatic endothelial wall here is termed parietal (30) or peritubular (874) endothelium. In general, the endothelial walls of the lymphatic sinusoids are continuous, although on some occasional small gaps or discontinuities have been noted in the visceral wall flanking the Leydig cells.

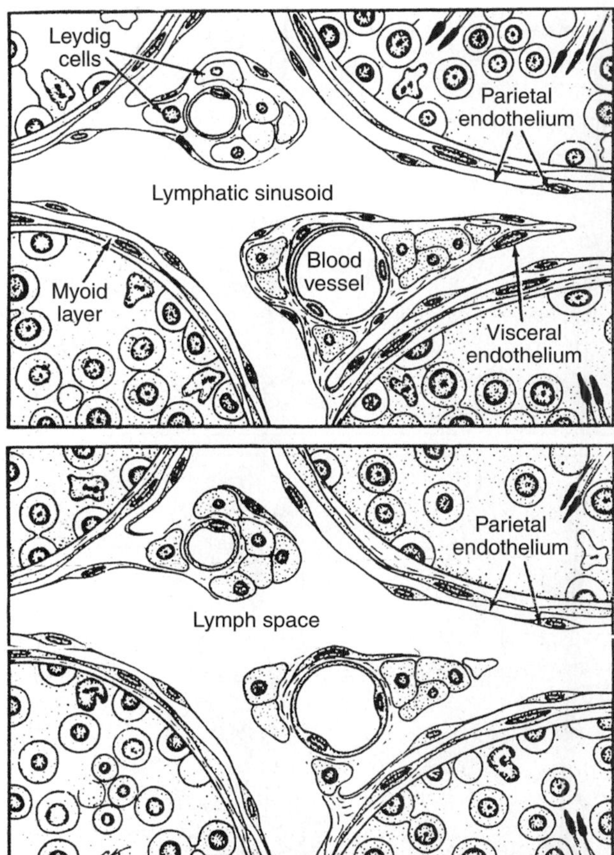

FIG. 69. Architecture of the intertubular tissue from a variety of species. **Top:** Chincilla and guinea pig. Leydig cells associated with blood vessels and enclosed by lymphatic endothelium. **Bottom:** Rat and mouse, showing discontinuities of the lymphatic endothelium partly surrounding perivacular Leydig cells (425).

Structural support in the intertubular tissue is provided by bundles of collagen fibers and occasional fibroblast-like cells forming a bridge linking the clusters of Leydig cells with the peritubular tissues.

In the rat and mouse, the absence of the visceral layer of lymphatic endothelium from the Leydig cell clusters permits a direct communication between loose connective tissues containing collagen and the lymph space (Fig. 73). In these species the architecture of the intertubular lymphatic channels has been aptly classified as lymphatic sinusoids (Fig. 73), implying a continuous admixture of lymph fluids and the ground substance of perivascular connective tissues. Additionally, to the organization described (30), Leydig cells of the rat and mouse are not exclusively confined to perivascular locations but are often seen in peritubular positions or, alternatively, in more central regions of the intertubular tissue, where they appear to lack any association with connective tissue and are in fact almost entirely bathed by lymph. This observation emphasizes the role played by the lymphatic sinusoids in providing the medium by which Leydig cells receive and secrete substances carried by

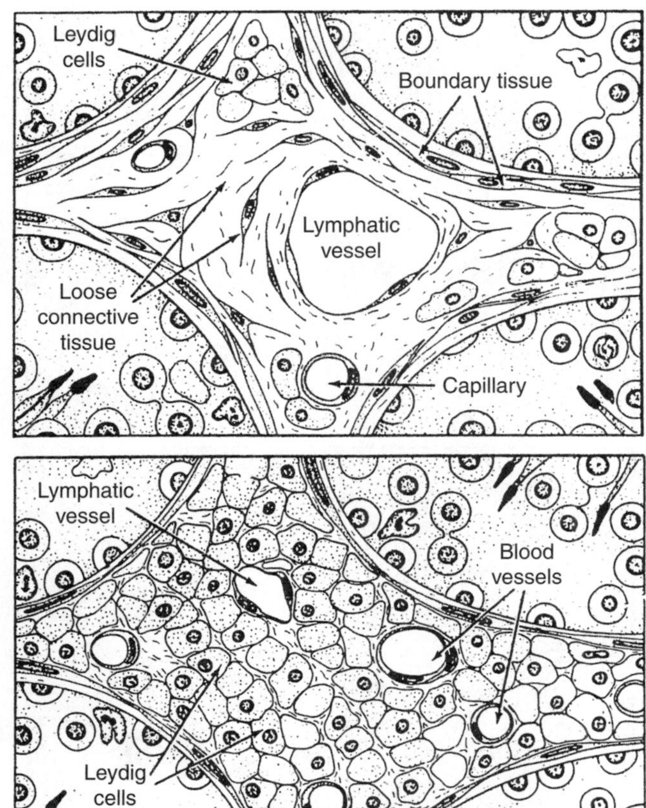

FIG. 70. Architecture of the intertubular tissue from a variety of species. **Top:** Larger animals including ram, bull, monkey, and humans. Randomly scattered clusters of Leydig cells appear within loose connective tissue rich in interstitial fluid. Lymphatic vessels are also demonstrated. **Bottom:** Boar, wart hog, zebra, naked mole, rat, and numerous marsupial species. Many large Leydig cells occupy the intertubular space. Connective tissue and lymphatic vessels are not prominent. [From Fawcett (425), with permission.]

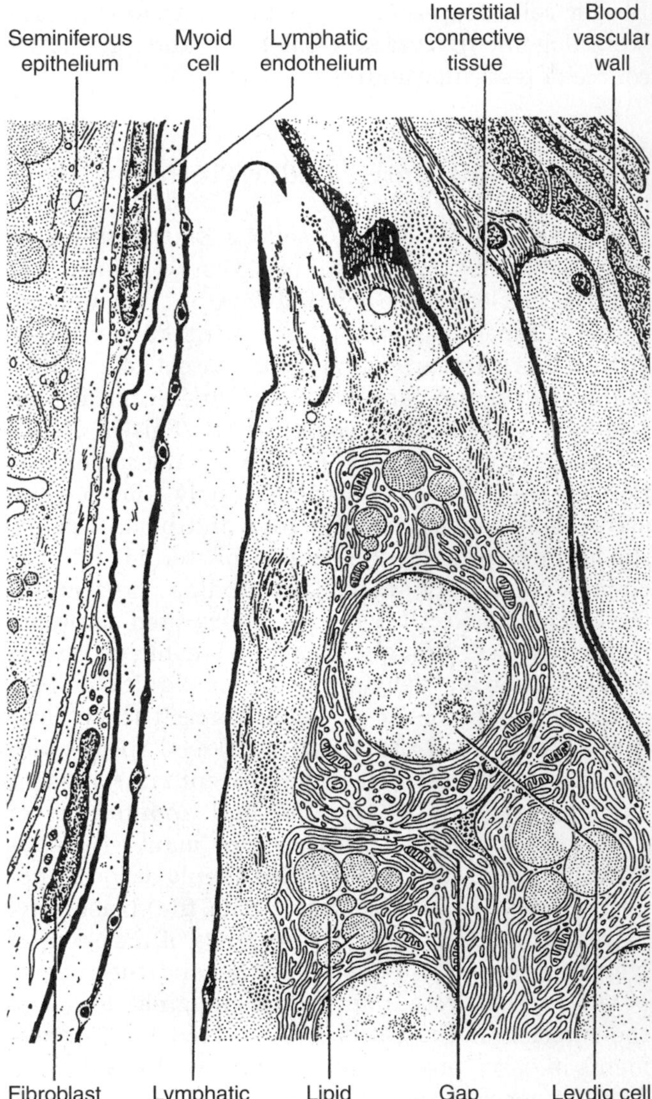

FIG. 71. Schematic representation of the intertubular tissue of the testis. Note the discontinuity in the endothelial wall of a lymphatic vessel, allowing its contents of interstitial fluid to gain access to the loose connective tissues containing the Leydig cells. [Based on micrographs from (570,852).]

the bloodstream. However, these special features of the histology of intertubular tissues in rat and mouse are quite different from those of many other mammals and raise a note of caution in extrapolating their associated physiological properties to cover other species in general. The presence of these extensive lymphatic sinusoids explains the ease that Christensen and Mason (872) found in manually separating seminiferous tubules from Leydig cells, whereas they were unsuccessful in similar attempts with material obtained from the cat and the human in which the lack of extensive peritubular sinusoids prevented a clean separation of the two testicular compartments.

In the second category, the intertubular tissue contains large areas of very loose connective tissues containing small aggregations of Leydig cells often associated with blood vessels or, less often, occupying positions rather distant from the vascular supply. Lymphatic drainage is achieved through prominent lymphatic vessels placed centrally or eccentrically within each intertubular area. These lymphatic vessels are bounded by continuous unbroken endothelial cells and, together with the Leydig cells, are supported by collagen and fibroblasts. To this broad but distinct category belong species such as the ram, bull, hyrax, elephant, monkey, and humans (Fig. 70). The relative paucity of Leydig cells in this group together with their often wide separation from blood vessels suggest that steroids secreted from the Leydig cells must gain access to the seminiferous tubules and venous system via diffusion through the edematous loose connective tissues.

In the third variation, some animals display an extraordinary abundance of Leydig cells but very little intertubular connective tissue. Examples falling

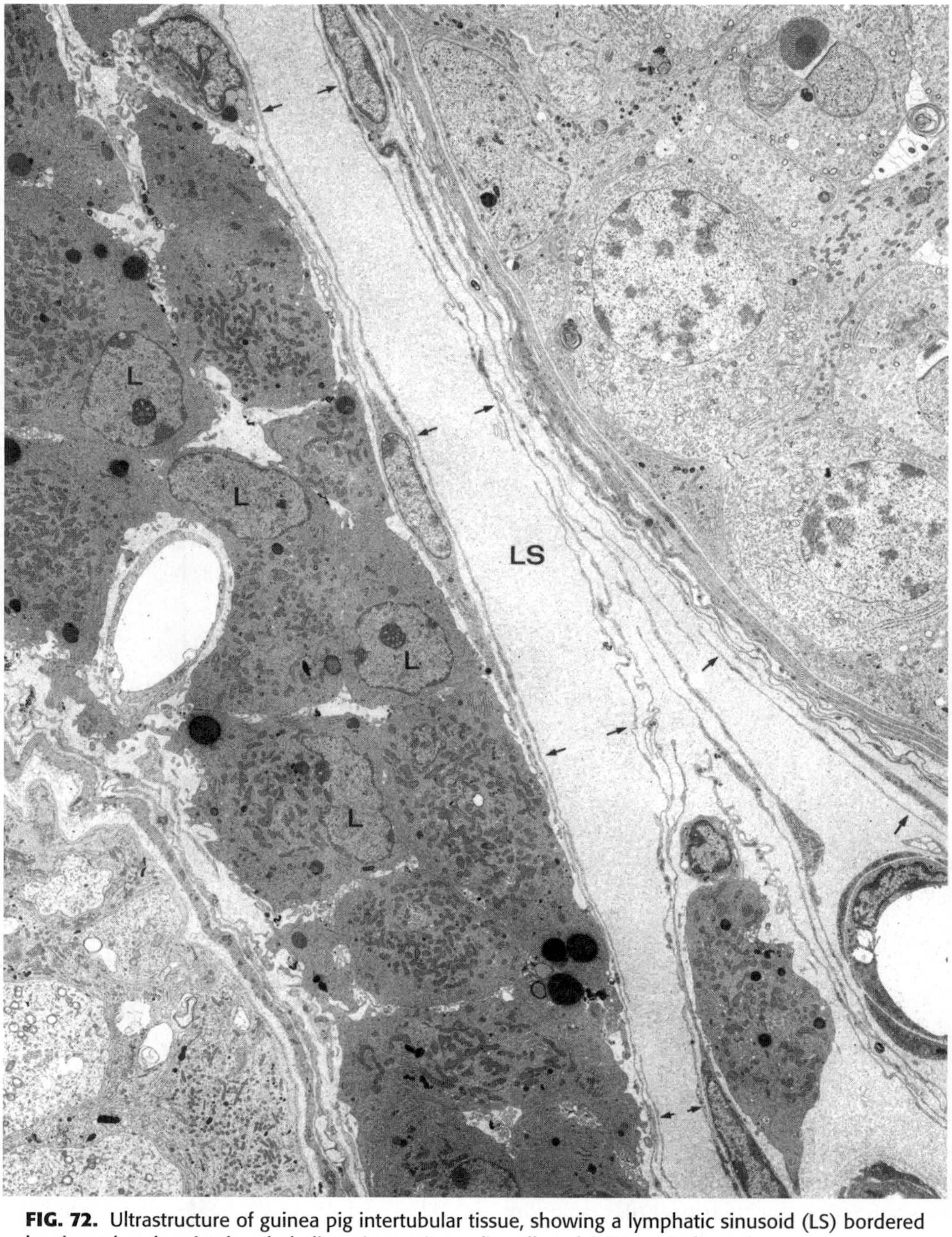

FIG. 72. Ultrastructure of guinea pig intertubular tissue, showing a lymphatic sinusoid (LS) bordered by visceral and parietal endothelium (*arrows*). Leydig cell nuclei (L) are indicated.

into this category include the zebra, domestic boar, wart hog, dog, opossum, and naked mole rat, where Leydig cells occupy 20% to 60% of the testis volume (Fig. 70) and include some Australian marsupials, such as the brush and ring-tailed possum and two marsupial mice (*Antechinus* spp.), and the koala. Small lymphatic vessels are infrequent, and many Leydig cells, by virtue of their large numbers and close packing, often lie at considerable distances from the nearest blood vessel. Why these animals develop large masses of Leydig cells is not understood, although some interesting speculations have been put forward by Fawcett (30,425) and others (35).

The intertubular tissue also contains macrophages, fibroblasts, lymphocytes, plasma cells, and more rarely mast cells, the latter being more commonly seen beneath the tunica albuginea. Because of their capacity for phagocytosis, macrophages have been readily identified in the interstitial tissue after application of various dyes such as pyrrol blue, chlorazol fast pink, and acid fuchsin and thionin (876–878). Subcutaneous injection of trypan blue had been a favored method

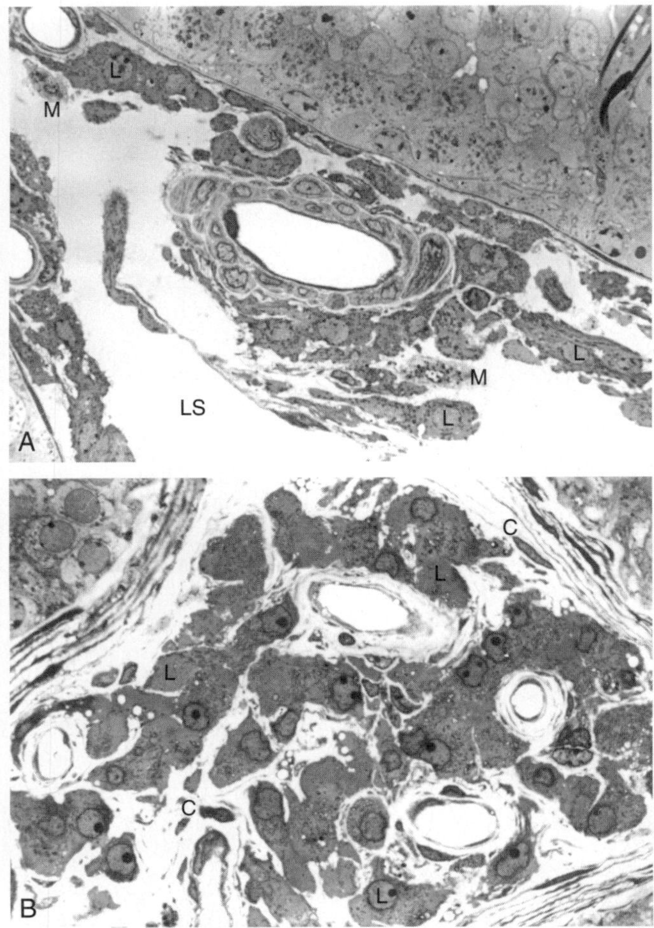

FIG. 73. A: Rat intertubular tissue showing Leydig cells (L), macrophages (M), and extensive lymphatic sinusoid (LS). **B:** Human intertubular tissue containing Leydig cells (L) and various connective tissue cells (C).

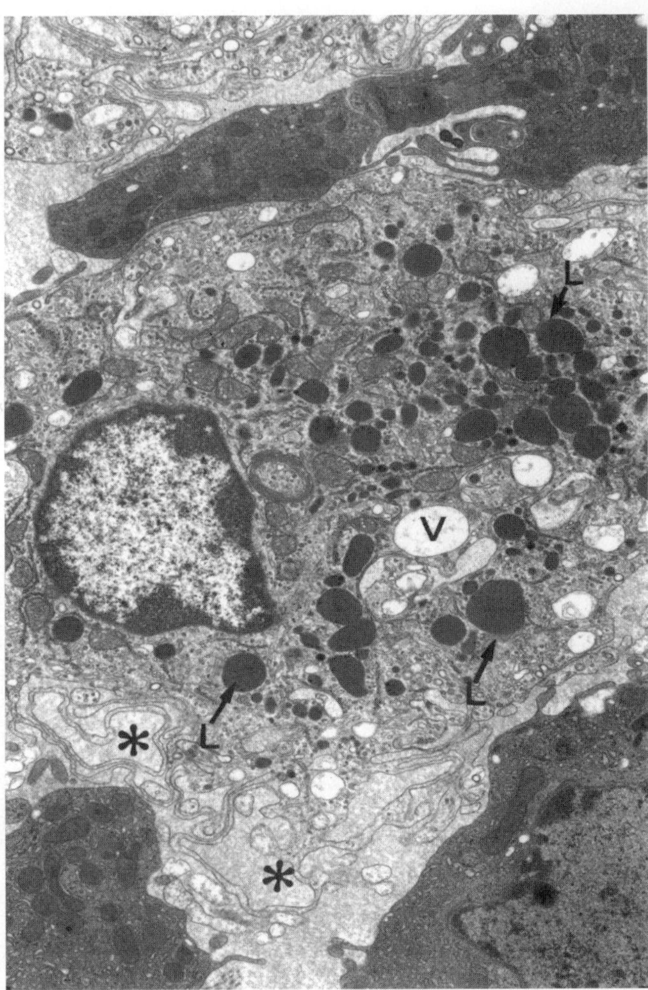

FIG. 74. Intertubular macrophage from the rat testis, showing cytoplasmic lysosomes and dense bodies (L), vacuoles (V), and long surface filopodia and lamellipodia (*asterisks*).

with which to identify macrophages (853,787,879). Additional methods have become available to demonstrate macrophages using nonspecific esterase histochemistry (880) and the uptake of latex beads (881). Numerous ultrastructural descriptions of testicular macrophages [for review see (882)] have stimulated renewed interest in their function, and it is known that testicular macrophages are endocytotically active, avidly incorporating a variety of exogenous dyes, radiolabeled plutonium, and FSH (883–885).

Macrophages in rat intertubular tissue were shown to take up albumin (886). The presence of coated vesicles beneath the surface membrane of macrophages (Fig. 74) has been linked to a receptor-mediated transport of specific proteins, because plutonium is known to bind to transferrin. Alternatively, fluid-phase endocytosis (pinocytosis) is thought to be the mechanism by which macrophages incorporate albumin from the interstitial lymphatics (886). Specialized contacts have been noted between Leydig cells and macrophages (882,887), and Bergh (888) showed that an approximate ratio of 4:1 between these cells occurs in the normal rat testis and after destruction of the seminiferous epithelium induced by artificial cryptorchidism. These findings suggest a functional coupling of these cell types. Rat testicular macrophages express class II major histocompatibility complex androgens (889) and are increased in numbers in response to hCG stimulation (683,684,890); in the developing rat testis they are abundant in fetal life and again at 3 weeks postpartum (891). Activated macrophages secrete inhibitory cytokines that significantly lower testosterone synthesis [reviewed by (892)]. Macrophages have been difficult to detect in ram testes (893) because of their paucity and small slender morphology, which has likened them to histiocytes or dendritic-type cells.

Under special conditions, the interstitial macrophages become activated and they engulf and phagocytose Leydig cells (703,894). This destruction of the

Leydig cells has been verified by ultrastructural studies (895) showing that once this activity has commenced, all Leydig cells disappear from the testis, leaving the interstitial volume practically empty except for the presence of the macrophages and connective tissue cells. In the rat, evidence is available for a role of resident macrophages in the differentiation and proliferation of new Leydig cells (896,897); these findings are discussed further below in Response to Testicular Damage. The role of macrophages in the immunological response of the testis is covered in Chapter 25.

The supporting connective tissues of the interstitial space have received little attention in the past, principally because of their unremarkable morphology and relative scarcity compared with the often much larger size and numbers of Leydig cells (853). However, their importance is highlighted when consideration is given to the mechanisms by which Leydig cells develop within the testis and the question of dedifferentiation of Leydig cells into fibroblast-like cells (429). A fuller account of the morphology of fibroblastic tissue and its proposed relationship to Leydig cell differentiation is given below in section on Life History.

Mononucleated wandering cells such as monocytes together with infrequently encountered lymphocytes and plasma cells have been accorded scant attention in studies of the morphology of the intertubular tissue. At present we can add little to this topic beyond mentioning that severe insults to the testis are unmistakably accompanied by increased numbers of these cells (894,898). Mast cells are occasionally observed in human intertubular tissues (899,900) but are rarely encountered in species such as the rat except in the peripheral margins of the intertubular tissue, where it joins within the subtunical areolar tissues. Mast cells may be functionally linked to Leydig cells, because steroidogenesis by hamster Leydig cells is enhanced in the presence of mast cells (901). In the normal testis their functional significance remains obscure, but their plentiful occurrence in rat interstitial tissue after recovery of testicular function after earlier impairment indicates a hitherto unsuspected role in the restoration of normal activity.

Although this resume of the organizational patterns of intertubular tissues has been confined to mammalian species, a substantial body of information on this topic is available for nonmammalian vertebrates and many invertebrates (647).

TABLE 4. *Selected examples of Leydig cell ultrastructure in adult mammalian testis since 1975*

Species	References
Mouse	(913,923,938)
Rat	(908,937,955,1099)
Bush rat	(649,944)
Guinea pig	(910,1100,1101)
Hamster	(910,1041,1102,1103)
Squirrel	(654)
Hare	(901)
Hyrax	(1038)
Mole	(1042)
Armadillo	(915)
Iranian vole	(1104)
Bat	(949)
Seal	(1105)
Boar	(1106)
Ox	(938)
Dog	(914)
Monkey	(1014)
Human	(428,457,912,1107)

Ultrastructure

Excellent detailed descriptions of Leydig cell structure in relation to function have been published in the excellent reviews by Christensen (853) and in a text on Leydig cells (902). There are marked variations in the ultrastructure of Leydig cells in different species emphasizing the diversity of their internal composition (Tables 4 and 5). Therefore, this review emphasizes the subcellular structure of Leydig cells in relation to known or suggested physiological activities. A summary of selected species indicating quantitative morphological features of their Leydig cells is presented in Table 5.

Nucleus

Leydig cell nuclei are often eccentrically placed within the cell and usually display a round or irregularly oval cells shape but adopt elliptical profiles when close to blood vessels or the tubule. In the human testis, Leydig cells with two or more nuclei have been noted (428). Heterochromatin associated with the inner nuclear envelope is a universal feature of Leydig cells (Fig. 75); the nucleolus is usually conspicuous, comprising the nucleolonemma and dense granular and amorphous areas. Leydig cells of the human testis frequently show duplicate nucleoli, and annular nucleoli are commonly found in the nuclei of adult mouse Leydig cells (903). This unusual configuration is correlated with Leydig cell development being absent in immature animals. Extrapolating from their occurrence in other cell types (904), it has been suggested that the central core of annular nucleoli is associated with partial cessation of RNA synthesis coupled with a maturational shift in the conversion of rough to smooth endoplasmic reticulum in the cytoplasm.

TABLE 5. *Quantitative morphology of Leydig cells*

Species	Volume (μl/cm³ testis)	Numerical density (10⁶/cm³ testis)	Volume/cell (μm³)	Volume smooth ER (% cytoplasm)	Volume smooth ER (μm³/cell)	Surface area of plasma membrane (μm²/cell)	Surface area smooth ER (μm²/cell)	Surface area smooth ER (% cell membrane)	References
Mouse	38–52	25	1,500–4,000	6	1240	1150	10,000–12,000	60	(913,1108,1109)
Rat	22–34	14–22	1,200–2,400	13–44	160–750	1500	30,000–70,000	54–64	(923,961,1014)
Guinea pig	19–30	22	1,600	62	900		135,000	68	(923,926) (650,652,660)
Hamster	10–23	15	970–1,100	18–20	160–180	960	22,000	19–65	(926)
Rock hyrax			3,000–5,000						(1037)
Woodchuck	480	89	5,300						(1044)
Rabbit	22			42					(923)
Dog	32			37					(923)
Stallion	120–170	21–25	5,800–7,000						(669)
Monkey	8	6	1,300	14	165				(1004)
Human	32	8	3,000–4,500	15	550	2400	32,000	71	(910,912,1056) (1108,1110)

ER, endoplasmic reticulum.

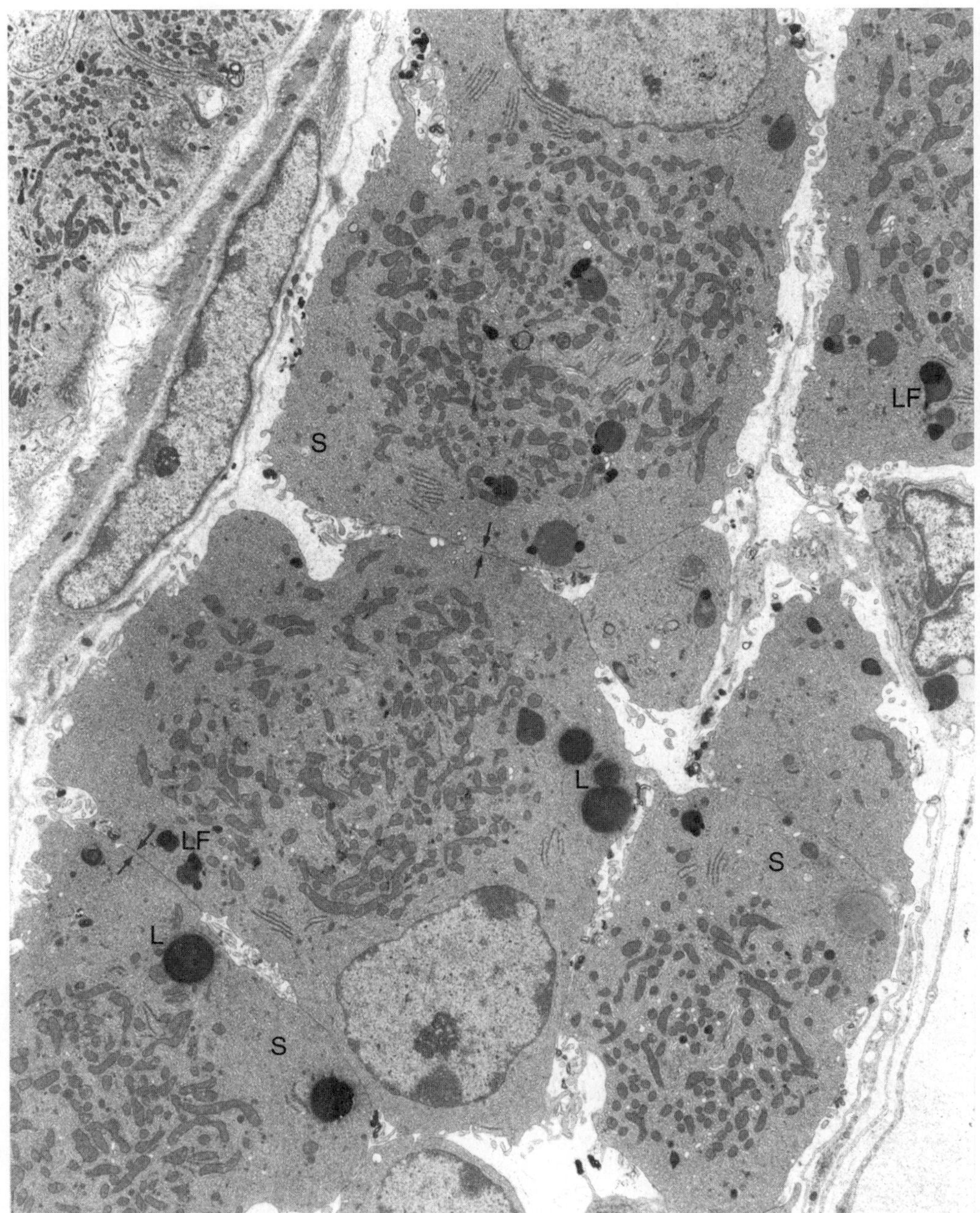

FIG. 75. Guinea pig Leydig cells illustrating central aggregations of mitochondria and peripheral concentrations of smooth endoplasmic reticulum (S). Lipid inclusions (L) and lipofuscin bodies (LF) are shown. Note close apposition of adjacent Leydig cells (*arrows*).

Cytoplasm

The dominant cytoplasmic organelle of the Leydig cells is the smooth endoplasmic reticulum, which provides binding sites on its surface for numerous enzymes necessary for steroidogenesis. Great diversity in the architecture of the smooth endoplasmic reticulum is well documented (853). The qualitative and quantitative differences in the morphology of the smooth endoplasmic reticulum has been described in the guinea pig and mouse, in which it is most abundant (853,905), and the rat and human, where it may vary considerably in amount and concentration (906,907). Examples of these between-species ultrastructural differences are illustrated in Figs. 75, 76, and 77.

Our own studies of the internal organization of the Leydig cell ultrastructure were the first to examine quantitatively the proportion of cytoplasm that was occupied by smooth endoplasmic reticulum (908), and for the rat, this approximated 39% of Leydig cell volume. Since then, the volume occupancy and surface area of this and all other cytoplasmic components in

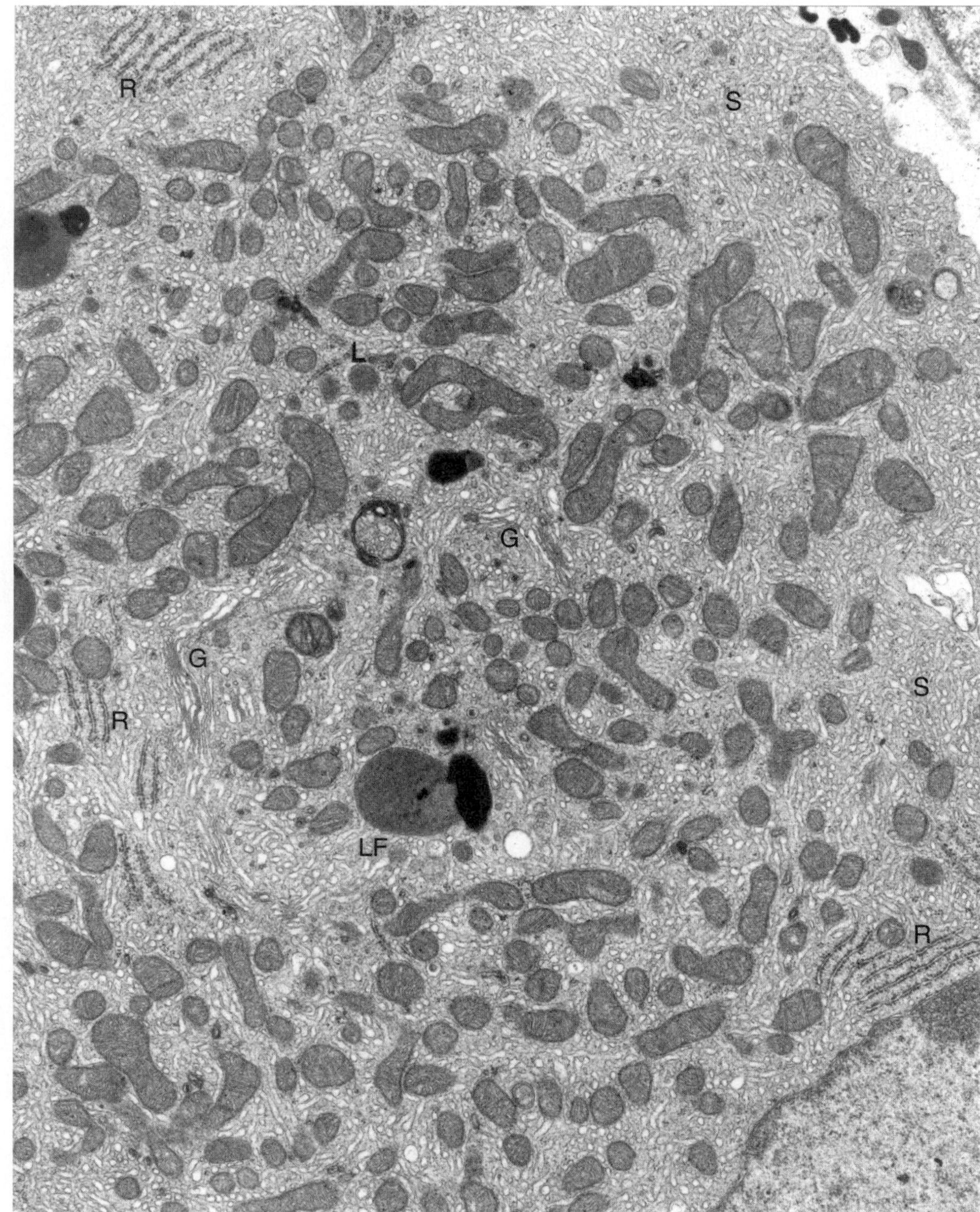

FIG. 76. Detail of cytoplasmic components within the guinea pig Leydig cell, showing numerous mitochondria, tubules of smooth endoplasmic reticulum (S), Golgi membranes (G), rough endoplasmic reticulum (R), lysosomal bodies (L), and lipofuscin inclusions (LF).

Leydig cells of the rat, mouse, hamster, dog, rabbit, guinea pig, and human have been documented (909–913). The surface area of smooth endoplasmic reticulum is vast. For example, if a single rat Leydig cell were represented as a golf ball, then an unraveling of its membranes of smooth endoplasmic reticulum would generate a square with sides of about 30 m and about half this size for an equivalent human Leydig cell. In situ, therefore, the membranes of smooth endoplasmic reticulum appear as a vast concentration of tubules interconnected in various patterns to form a meshwork extending throughout the entire cytoplasm of the Leydig cell. The membranes thus enclose a cavity constituting an intracellular compartment separated from the ground cytoplasm.

A scheme suggesting possible interconversions of the different ultrastructural configurations of smooth endoplasmic reticulum was based on analysis of guinea pig Leydig cells (483). The most common form of smooth endoplasmic reticulum consists of a random network of interconnected tubules (Fig. 76), which may transform into loosely packed sheets of tubules

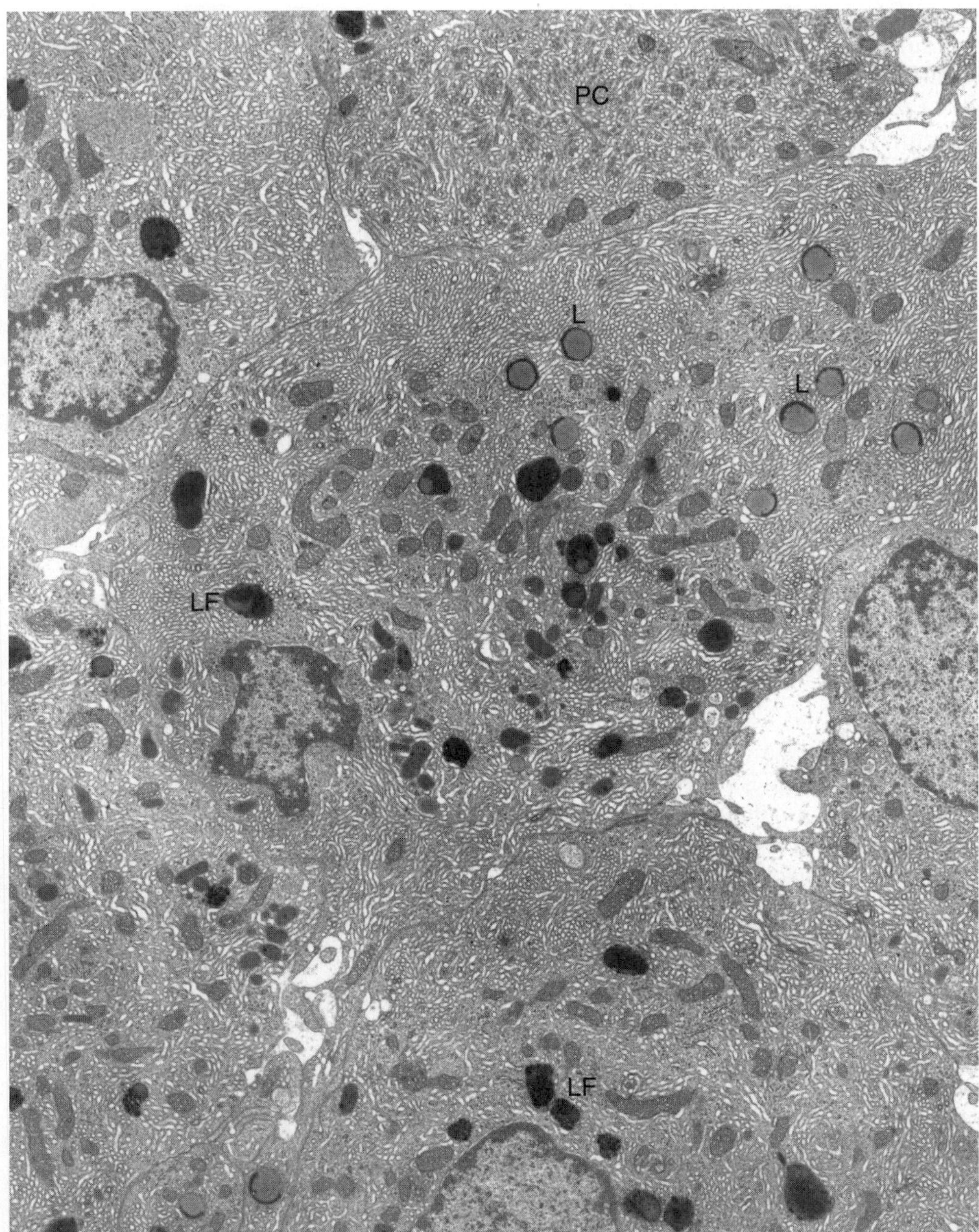

FIG. 77. In human Leydig cells fixed via perfusion, the smooth endoplasmic reticulum is the dominant organelle and presents as tightly packed anastomotic tubules. Lipofuscin pigments (LF) are common, and small lipid inclusions (L) are noted. Paracrystalline inclusions (PC) are shown.

or form more regular arrays of fenestrated cisternae. Leydig cells of the mouse exhibit all these various forms and in addition show a marked segregation of their cytoplasmic membranes, a spectacular example being seen in the appearance of extensive concentric whorls of smooth membranes. These may surround lipid inclusions, mitochondria, or dense membrane-bound bodies, and similar whorls are observed in the guinea pig, rat, dog, armadillo, and monkey (675,914–916). The absence of these whorled membranes in many other species raises the possibility that they provide a specific function to meet the biochemical requirements of these particular cells. In cultures of kidney cells, the conversion of smooth endoplasmic reticulum from branching tubules to stacked membrane arrays can be induced with cytochrome b5, although the binding interactions on apposing stacked membranes are of surprisingly low affinity (917).

A similar concept could be used to explain the infrequent occurrence of other unusual membranous bodies such as single- or double-walled tubules, which occur in the mouse, opossum, and rat (903,905,918,919), and structures resembling annulate lamellae, seen in adult mouse and rabbit Leydig cells (918–922). The functional significance of these organelles remains unknown. A major objective of stereological analysis of the Leydig cell has been to correlate the relative abundance of cellular components with the amount of steroid compounds synthesized and secreted by the same cells (909).

One approach adopted by Ewing and coworkers was to link changes in Leydig cell ultrastructure and various pathways of steroidogenesis. They examined changes in Leydig cell morphology and the capacity to synthesize various steroids by manipulating the degree to which LH is available to act on the Leydig cell (923–926). When different species were examined, the production of testosterone after LH stimulation was significantly different between species but was not related to the total mass of Leydig cells within the testis. Instead, they found a strong correlation between testosterone secretion and the amount of smooth endoplasmic reticulum and Golgi membranes within the Leydig cell cytoplasm, indicating the central role played by these organelles in determining the secretory capacity of a single Leydig cell. Furthermore, they showed a subdivision of enzyme function in the steroidogenic pathway that exists within the membranes of smooth endoplasmic reticulum, raising the interesting possibility that reactions such as pregnenolone to progesterone may be sequestered to specialized regions of the Leydig cell. These findings implicate the concentric whorls and other varieties of smooth cytoplasmic membranes in subserving this function. The levels of testosterone in plasma during the development of the monkey testis are also correlated with the volume of smooth endoplasmic reticulum within the Leydig cell or whole testis rather than the total volume or numbers of these cells (927). A similar study of the Leydig cells of the human testis has shown that the volume of smooth endoplasmic reticulum within the Leydig cells is positively correlated with daily sperm production (928).

Mitochondria vary in size and form both in a given Leydig cell and between different species, and their ultrastructure has been well described in earlier studies (853,929). Cholesterol is known to undergo side-chain cleavage within the mitochondrion, possibly on the surface of the cristae, which exhibit various profiles ranging from foliate, tubular, or intermediate forms (930,931). The ability of Leydig cells to elevate steroidogenesis up to 100-fold within several minutes after stimulation is dependent on cholesterol import from the outer to the inner mitochondrial membranes (932), a process regulated by steroid acute regulatory protein (933,934). Details of the reactions and enzyme systems with Leydig cell mitochondria appear elsewhere in this volume. Electron-dense mitochondrial granules often occur within the matrix of the mitochondrion, but their function is obscure (853,929).

The number of lipid inclusions in Leydig cells varies with species and even across a single section from one animal (Fig. 78) (907), suggesting variable degrees of functional activity. Each mouse Leydig cell contained an average of about 147 lipid inclusions, diameter approximately 1 μm, and collectively occupying about 6% of cytoplasmic volume (912). However, when mouse Leydig cells are centrifuged in Percoll density gradients (935), they separate according to buoyant density, and cells of low and high specific gravity show, respectively, a great many or very few cytoplasmic lipid inclusions (Fig. 78). Similar observations may be made that rat Leydig cells show a similar pattern and most are devoid of lipid inclusions.

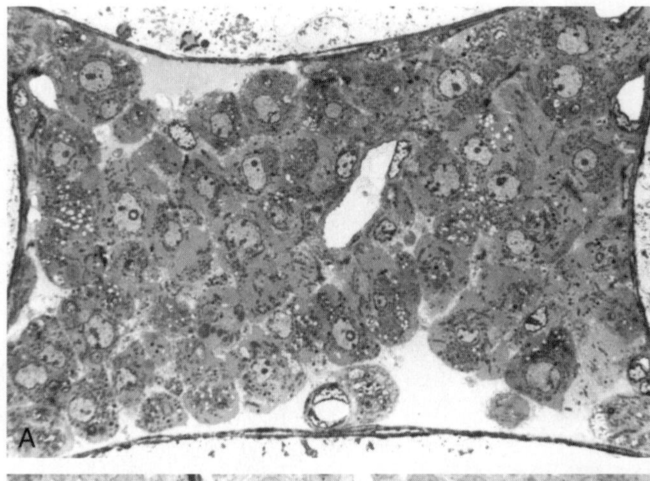

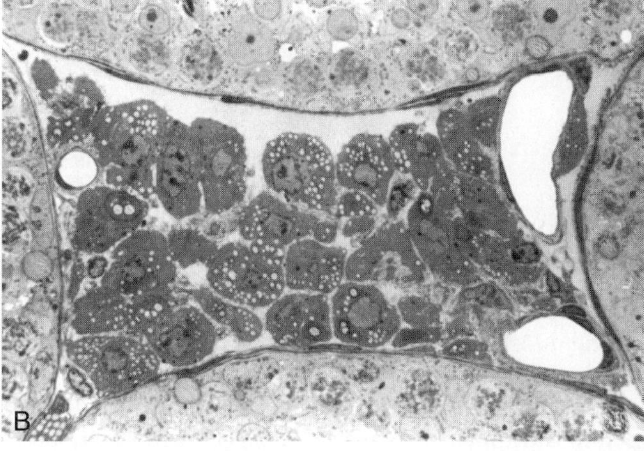

FIG. 78. A: Adult mouse testis showing a large cluster of Leydig cells containing numerous dense inclusions representing mitochondria together with homogeneously stained cytoplasmic regions filled with smooth endoplasmic reticulum. **B:** Adult mouse testis showing Leydig cells with prominent aggregations of pale-staining cytoplasmic lipid inclusions.

Some show plentiful lipid inclusions and may represent a few surviving fetal-type Leydig cells, known to persist within the adult rat testis (936).

The Leydig cells of the dog, cat, mole, rat, mouse, elephant, and rhesus monkey often contain a high proportion of lipid (30,777,914); guinea pig, rabbit, and boar exhibit somewhat less (30,483,910), and Leydig cells of the human, rat, hamster, opossum, ox, African green monkey, squirrel monkey, and several marsupial species found in Australia contain very few lipid inclusions (909,910,912,919,937–939). A discussion of the likely functional role of Leydig cell lipid inclusions (853) has reviewed the evidence that they represent sites of cholesterol storage and/or synthesis. The belief that lipid inclusions are intimately involved in fatty metabolism and, by inference, certain steroidogenic reactions has been supported by morphological data. In rat fetuses, for example, a major feature of the fetal Leydig cells is their rich supply of lipid inclusions, but in postnatal life, adult Leydig cells rarely contain lipid. That these changes in lipid content reflect alterations in Leydig cell function is further emphasized, because trophic stimulation (with LH/hCG) of Leydig cells rapidly depletes their stores of morphologically recognizable lipid droplets (940–944) and heralds an increase in testosterone secretion. Animals fed a cholesterol-rich diet showed a significant depletion of lipid inclusions in the Leydig cells, indicating that if available, cholesterol is taken up by the Leydig cell, which obviates the need for lipid storage in the cytoplasm (911). Opposite changes occur when Leydig cells were examined after withdrawal of gonadotropins, where lipid inclusions became plentiful (937,945), but replacement therapy with exogenous LH resulted in a marked depletion of the accumulated lipid. The Golgi apparatus of Leydig cells is moderately well developed (Fig. 76), and although its peripheral vesicles are usually devoid of any internal material, under conditions of Leydig cell hyperactivity, the saccules of the Golgi complex often appear swollen and contain distinct electron-dense flocculent material (Fig. 75). Nothing is at present known about the role played by the Golgi apparatus, except its participation in the intracellular passage of radiolabeled fucose, traced by electron microscope autoradiography (946,947). It was tentatively concluded from a brief summary of this study that Leydig cells secrete glycoproteins, with the Golgi complex featuring prominently in their early synthesis.

It is uncertain how testosterone or any other secretory product of the Leydig cell is transported through the cytoplasm and released into the extracellular space. This intriguing topic has been discussed many times (853) but at present remains only speculative. However, in other steroid-secreting tissues such as the luteal cells of the ovary, evidence has been put forward to suggest that progesterone is secreted in granule form (947). Secretion of corticosteroids from the adrenal tissue is thought to occur in a similar fashion (948). For most species, Leydig cells do not exhibit typical secretory granules or vacuoles in their cytoplasm. However, in a species of seasonally breeding bat, maximum testosterone secretion in the breeding season is accompanied by a marked increase in the abundance of small, dense, membrane-bound granules, whereas in periods of Leydig cell quiescence, the granules largely disappear from the cytoplasm (949). Because most of these granules do not share the enzymatic or morphological properties attributable to lysosomes or peroxisomes, they were thought to be involved in the transport of testosterone through the cytoplasm before its secretion. Other studies (950) on cytoplasmic granules in luteal cells of guinea pigs failed to detect granule exocytosis and could show no relationship between the abundance of small granules and maximum progesterone secretion. The functional significance of Leydig cell granules in particular species thus requires further clarification. Furthermore, there is increasing evidence that the ovary secretes relaxin and oxytocin, which, being peptides, are more likely to be stored as secretory granules (951,952).

Lysosomes are commonly observed in Leydig cells. These organelles are about 0.5 μm in diameter, bounded by a single membrane, show circular or irregular profiles by electron microscopy, and are also referred to as dense bodies because of their notable staining properties. The lysosomes present in Leydig cells conform to their usual pleomorphic occurrence in other cells, exhibiting various states of fusion with cytoplasmic vacuoles, thus categorizing them as secondary lysosomes as distinct from the individual primary lysosome (853). Endosomes are regarded as structural variations within the lysosomal system and probably represent packaging of recently internalized vesicles destined for lysosomal disposal. The function of Leydig cell lysosomes has been investigated by studying the response of Leydig cells to exogenous tracers known to be incorporated into lysosomes of other cell types (953). Leydig cells are actively endocytic cells using both fluid-phase and adsorptive endocytic mechanisms to take up extracellular macromolecules. The destruction of the internalized tracers followed different routes of transport within the Leydig cell cytoplasm, depending on the ionic charge of the individual tracer. As seen diagrammatically in Fig. 79, fluid-phase endocytosis involved the lysosomal tracers following this same route or, alternatively, was carried to the Golgi region whereupon they ultimately disappeared. The existence of these separate pathways for internalized macromolecules

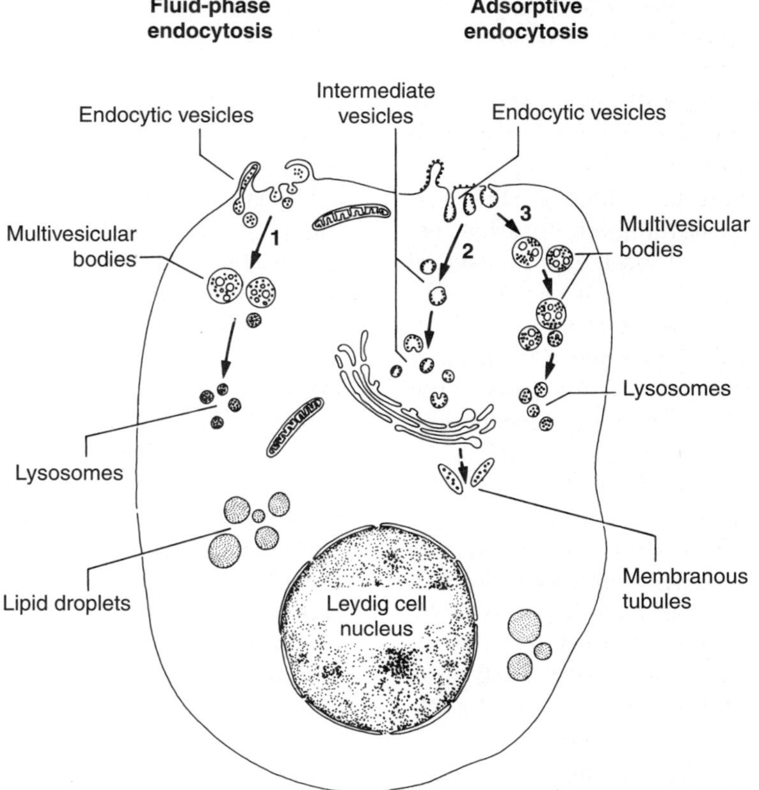

FIG. 79. Diagram illustrating intracellular pathways of endocytosed tracers in the Leydig cell. Pathway 1 shows tracers destined for lysosomal incorporation. Pathway 2 illustrates absorptive endocytosis of molecules initially bound to the cell surface and ultimately transported to the Golgi region via intermediate vesicles. In pathway 3, these molecules follow an intracellular route involving only the lysosomes. Some tracer molecules also appear in membranous tubules associated with the Golgi membranes, although their subsequent fate is not known. [Data from (953).]

suggests that the disposal of unwanted material can be dealt with in a variety of ways. Leydig cells contain autophagolysosomes (954), which arise independently within the cytoplasm and acquire lysosomal enzymes from secondary lysosomes by fusion with them. They are eliminated from the cell by exocytosis.

Poorly digested substances often remain within the Leydig cell lysosomal system and are recognized as dense membrane-bound granules referred to as lipofuscin pigment granules. They are commonly seen in human Leydig cells as heterogeneous conglomerates of myelin figures and particulate matter (853,906,955–957). Stimulation of Leydig cells with hCG tended to increase their abundance (907,945). Although the biochemistry of lipofuscin granules is unclear, they contain acid phosphatases (907,945), suggesting a derivation from lysosomes. Their appearance in other species was reviewed by Ohata (922). Several studies (958,959) identified the peroxisome within Leydig cells as a homogeneous pale-staining granule, approximately 0.2 μm in diameter. These organelles contain catalase and oxidases, and although they increase in number after hCG stimulation (861), their functional role has remained unclear until morphometric and immunolabeling studies suggested their involvement in steroidogenesis. The relationship between testosterone secretory capacity and cellular organelles has identified a positive correlation between steroid secretion and the volume and surface area of peroxisomes (926). Stimulation of Leydig cells with LH is followed rapidly by a three- to fivefold increase in total volume of peroxisomes, whereas other organelles are unchanged (960,961). Peroxisomes contain a sterol carrier protein thought to be involved with the transport of cholesterol to the inner mitochondrial membrane for subsequent side-chain cleavage reactions in the steroid pathway. A role for peroxisomes in intracellular trafficking of cholesterol has been suggested. After LH stimulation of rat Leydig cells, peroxisomes fuse with lipid droplets to acquire free cholesterol and subsequently deliver the cholesterol to mitochondria again by a fusion mechanism (962).

Centrioles at times associated with the Golgi complex have been identified, and cilia may be noted developing from the paired centrioles (914,922,963).

Because the fibril pattern of the cilium has a 9 + 0 pattern, it could reflect a sensory or chemoreceptor property (964). Microtubules and filaments are present within the Leydig cell cytoplasm; the latter occur throughout the cytoplasm and at times form a network subjacent to the plasma membrane (428). When microtubular systems of steroid-secreting cells are disrupted with inhibitors such as colchicine or vinblastine (965,966), an increase in steroidogenesis is observed. In isolated rat Leydig tumor cell lines, the resting cells exhibit tubulin in discrete granular units distributed throughout the cytoplasm, but when steroidogenesis was induced by treatment with cAMP, vast microtubular networks appear, forming radial patterns around the central cell nucleus (967). The small cytoplasmic granules in unstimulated cells were rich in tubulin and cholesterol, indicating that one role of microtubular organization is to make intracellular cholesterol freely available for conversion into pregnenolone. An additional role of cytoskeletal systems in Leydig cells is in the process of LH/hCG receptor down-regulation, which does not occur after treatment with microtubule inhibitors (968). When vinblastine was administered, the Leydig cells showed remarkable increases in Golgi-associated vesicles and cytoplasmic filamentous material, suggesting alterations of the transport mechanisms within the cytoplasm. Filaments with diameters of 5–7 nm or 10–12 nm are always seen in Leydig cells, and their ultrastructure has been reviewed earlier (853,914).

A remarkable specialization of human Leydig cells is the crystal of Reinke, originally described by Reinke (969). Readily visible by light microscopy, they frequently present as rod-shaped structures up to 20 μm in length, although they may exhibit triangular, rhombic, or polygonal shapes (Fig. 80). Not every Leydig cell contains these crystals. Their ultrastructure and crystalline nature have been repeatedly documented (906,907,955,956,970–973). Although they are known to contain protein (972) and to increase in size and number in aging testes (973), their functional role is unclear. A variety of other crystalline structures, termed paracrystalline inclusions, has also been noted in human Leydig cells (428). Because the subunits of Reinke crystals are similar to the paracrystalline inclusions, the latter have been considered precursors of the much larger Reinke crystals (906). Occasionally, either type of crystalline material may occur in the Leydig cell nucleus (905,971,974).

Reinke crystals were commonly believed to be unique to human Leydig cells until similar cytoplasmic crystalloids were described in Leydig cells of the sexually regressed wild bush rat, *Rattus fuscipes* (497). These rats are seasonal breeders, and in the nonbreeding season, serum gonadotropins and testosterone are decreased. Using modified laboratory conditions mimicking the natural environment of the nonbreeding season, Irby et al. (649) induced Leydig cell atrophy in the same species, whereupon many large crystalloid inclusions made their appearance in the Leydig cell cytoplasm. Again, serum hormone levels declined significantly compared with breeding rats that do not display these crystalloid inclusions. When gonadotropin secretion was experimentally suppressed by hypophysectomy or testosterone implants (975), the Leydig cells atrophied, and many crystalloids appeared in the cytoplasm. Formation of nuclear vesicles, their subsequent transfer to the Leydig cell cytoplasm, and the concurrent development of paracrystalline material within the vesicles have been suggested as a mechanism to explain the growth of crystalloids by assembly and fusion of smaller subunits. Because these crystals occur only in regressive forms of Leydig cells that produce very little testosterone, they may represent a storage facility of steroid precursors and enzymes, although further work is necessary to elucidate their precise function.

The surface area of Leydig cell plasma membrane is very large because of extensive surface projections of filopodia and microvilli (853,922,937). In a morphometric analysis of rat Leydig cells (909), their volume was estimated to be about 1,200 μm^3, and if considered to be spherical their surface area would have been approximately 500 μm^2. However, the actual figure is 1,500 μm^2, reflecting the complex topography of its surface. Leydig cells found in clusters form junctions with neighboring Leydig cells and occasionally form junctions with themselves. Gap junctions are frequently observed where the intercellular space is narrowed to 20 Å (853,914). These junctions are normally 0.2 to 2 μm in greater diameter (471) and in thin sections appear as ring-shaped, circular, or double-circle structures. Septate-like junctions have occasionally been described (853), and these are partially permeable to electron-dense tracers such as lanthanum. Tight junctions or desmosomes are generally not considered to form between Leydig cells, although rudimentary desmosomes have been observed in canine Leydig cells (914). Special sites of contact between Leydig cells and macrophages occur frequently (882,887) and consist of short projections of Leydig cell cytoplasm that invaginate nearby macrophages, suggesting the possibility that small amounts of Leydig cell cytoplasm may be endocytosed by the macrophage. A thin discontinuous layer of basal lamina–type material may partly surround the surface of the Leydig cell (853,914,922,956) and may provide structural support to the cell in conjunction with scattered bundles of collagen fibers. Autonomic nerves have been observed in close association with Leydig cells of the immature human testis (976), but their functional significance is not clear.

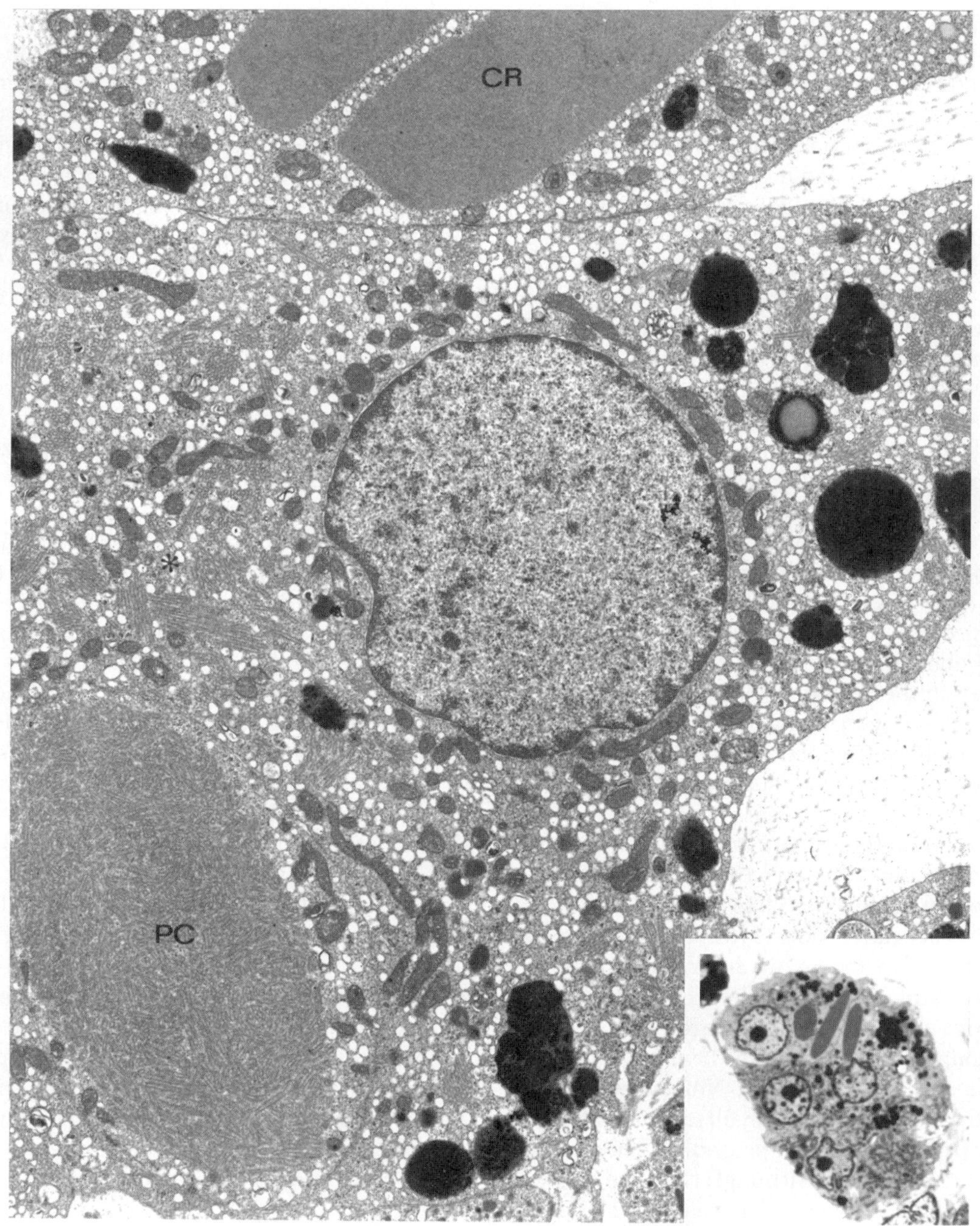

FIG. 80. Leydig cells of the human testis prepared by immersion fixation of biopsy material. The smooth endoplasmic reticulum shows individual vesicles. Crystals of Reinke (CR) are seen, together with paracrystalline inclusions (PC), forming large aggregations or scattered within the cytoplasm (*asterisk*). **Inset:** Crystals of Reinke observed by light microscopy.

Life History

The origin, differentiation, proliferation, and attrition of Leydig cells have been the subject of hundreds of investigations, and yet many unsolved problems remain. We emphasize some of the fundamental aspects of Leydig cell biology available from selected earlier studies and more recent investigations. The history of Leydig cell life in the normal mammalian testis can be temporally subdivided into six categories: (a) the development of fetal Leydig cells, (b) changes in Leydig cells at birth, (c) pubertal development, (d) the adult population, (e) seasonal alterations in Leydig cell biology, and (f) the status of Leydig cells in old age.

In those species so far studied, the fetal testis develops a spectacularly high proportion of active Leydig cells whose morphology and secretory function are unique to this phase of their growth. The debate concerning the precise embryological derivation of fetal Leydig cells has been amply discussed in the past (45,977–979), with two concepts being put forward to explain their origin. The most generally accepted theory favors a mesenchymal origin from the gonadal ridge, although alternative evidence is available (980) favoring their derivation from mesonephric cells. A good deal of uncertainty surrounds the validity of both theories because the process of gonadal differentiation is extremely complex, and repeated attempts to identify the origin and cell type giving rise to histologically recognizable Leydig cells have fallen short of providing clear answers. Useful reviews of this problem are available (45,981–984).

The notion of a mesenchymal origin of Leydig cells has gained acceptance on a number of grounds. Ultrastructural analysis of the early fetal testis reveals numerous mesenchymal-like interstitial cells between the seminiferous cords that are believed to differentiate into fetal Leydig cells (Fig. 11) (56,62,985,986). Similar observations have been presented many times for the postnatal development of Leydig cells, where they are thought to arise chiefly by differentiation from simple bipolar interstitial cells within the interstitial tissue and to a lesser extent by mitotic proliferation of fully formed Leydig cells (851,853,978,987–990). Some confusion still exists as to the precise histological classification of these Leydig stem cells (990,991). They have been designated as pure mesenchymal cells (especially in the fetus) or fibroblasts, myofibroblasts (having affinity with peritubular smooth muscle contractile cells), fibroblastic mesenchymal cells, and many other classifications. A further complication pertaining to Leydig cell differentiation concerns the dual site of their origin, some arising in peritubular positions and others developing in association with capillaries and in more central regions of the interstitial tissue, a feature well illustrated in the pig testis (992–995).

Which cell type(s) represents the Leydig cell precursor? The answer to this question is not yet available, but here we return to the alternative theory of Leydig cell differentiation, namely their development from mesonephrogenic tissues. During early gonadal differentiation, mesenchyme-like cells of mesonephric origin invade the entire testis and occupy the presumptive interstitial tissue (45). These cells, together with others migrating in from the gonadal blastema, could thus form the stem cells for later development of Leydig cells. Because mesonephric cells exhibit the capacity to "transdifferentiate" into completely different cells during fetal life, in theory this behavior may be retained by the intertubular tissue beyond intrauterine life. Examples of transdifferentiation of testicular interstitial cells are available. At birth, the seminiferous cords are surrounded by a homogeneous population of fibroblasts, but as the tubules expand during testicular enlargement, these cells transform into an inner peritubular myoid cell, acquiring epitheloid characteristics, and an outer fibroblastic layer of similar ultrastructure to the newborn situation (996). Alternatively, cells lining the seminiferous tubules are known to differentiate into Leydig cells (997,998), and intermediate steps in the route of differentiation have been suggested (188,914,955,998–1000).

This intriguing topic of Leydig cell differentiation is intimately related to another well-known feature of the Leydig cell population in late fetal life and shortly after birth, namely their disappearance from the testis. A biphasic pattern of Leydig cell growth in rodent species has been recognized for many years, a feature not shared by primate Leydig cells. Among the first to describe peaks of Leydig cell development in the fetus and in postnatal life was Whitehead (1001,1002), who studied the pig testis. In fact, the pig testis is a special case, because it exhibits three consecutive stages of Leydig cell development, fetal, perinatal, and postpubertal, whereas common laboratory rodents and other species show a spurt of Leydig cell proliferation in late fetal life and postnatally coincident with initiation of spermatogenesis (86,990,991,1003,1004). The stimulus for rapid proliferation of fetal Leydig cells in the human testis is chorionic gonadotropin and, possibly to a lesser extent, pituitary LH, which result in large elevations of intratesticular testosterone and other steroids, reflected a short time later by raised serum levels of testosterone (87,1005,1006). A sharp decline in the amount of steroids secreted by each fetal Leydig cell occurs just before birth in the rat, and thereafter a further fall is a reflection of a considerable loss of Leydig cells from the testis (87).

Two suggestions have been offered to account for the dramatic perinatal decline in Leydig cell numbers: a process of cell death (62,990,998,999,1007,1008) or dedifferentiation back to fibroblastic-type cells (1009,1010). Because the morphology of the fetal Leydig cells is well recognized, it has been possible to trace their fate within the developing testis using morphometric analysis (936). Contrary to the notion of cell differentiation or degeneration, fetal Leydig cells persist in the postnatal testis and survive in small numbers within the adult testis, suggesting they are a distinct line of cell development not related to the development of immature and adult-type Leydig cells. At this time it is also not clear if changes in the pituitary–testicular hormonal axis and/or alterations in the cellular composition of the fetal testis may influence this pattern of Leydig cell survival, at least in the rat testis.

There are many descriptions of the morphological features of the intertubular tissue between birth and puberty. The point of interest here has been to characterize, in morphological terms, the process of Leydig cell differentiation and proliferation that parallels the establishment of spermatogenesis and to ascertain the identity and mechanisms of action of hormones that support renewed growth of Leydig cells at puberty [see reviews (1011–1013)]. In the developing testis, there is general agreement that the resident population of interstitial fibroblasts or mesenchymal cells is stimulated by LH to differentiate into the adult-type generation of Leydig cells (853,906,987,988,990,995). Differentiation of adult-type Leydig cells in postnatal life is thought to occur as the numbers of mesenchymal cells decline (978,987,988), with the latter signaling their transformation into Leydig cells by synthesizing increasing amounts of smooth endoplasmic reticulum (990,991,993). Information on the dynamics of Leydig cell development is available (1014–1016). Leydig cells become increasingly apparent within the intertubular tissue soon after birth, and studies using tritiated thymidine methods to trace the pattern of cell proliferation among intertubular cells of the rat show that mesenchymal cells provide the source of new Leydig cells up to 4 weeks of postnatal development. Thereafter, the greatly increased numbers of Leydig cells seen in more mature testes arise via one or two rounds of mitoses of existing Leydig cells. The previously high proportion of mesenchymal cells in fetal and neonatal testes (greater than 50%) is subject to a significant decline to around 3% within the mature testis. Approximately one-third of the normal adult Leydig cell population arises by development from mesenchymal cells. Morphological maturation of mesenchymal cells into accredited Leydig cells is accompanied by their acquisition of receptors for LH and androgen (1017). Thus, the postnatal development of Leydig cells in rodent species (855,1018) begins with primitive LH-independent mesenchymal cells that transform into progenitor cells by acquisition of LH receptors and expression of steroidogenic enzymes. These spindle-shaped cells differentiate into newly formed but small adult-type Leydig cells with polygonal shapes and with gradual increase in LH receptors, cytoplasmic organelles, and steroidogenic capacity expand in volume, finally attaining maximum size and secretory characteristics of mature Leydig cells.

The complex hormonal control of sexual maturation has been the subject of intensive clinical and experimental investigations, which have been continually updated as new information has become available (1019,1020). However, some data relating to the physiology of testicular maturation have been difficult to interpret. First, mature or partly differentiated Leydig cells in humans were thought to be absent from the testis from about 1 year after birth until puberty (45,62,998,999,1021–1024). However, the concentration of testosterone in spermatic venous blood of boys is five times higher than in peripheral blood (1024). Second, acute hCG treatment of prepubertal boys induces a rapid increase in serum testosterone levels (1025), which can probably be explained by the recent observation that differentiated Leydig cells do exist in the neonatal human testis (1026). Third, elevation of serum FSH levels during sexual maturation is known to increase testis weight via stimulation of spermatogenesis. Some data favoring a stimulatory effect of FSH on the interstitial tissue have demonstrated FSH-induced enhancement of LH receptors (1027–1029) and increased testosterone secretion in response to LH in vivo and in vitro (1027,1030–1032). Furthermore, these findings have been difficult to evaluate because Leydig cells and the primitive mesenchymal cells lack receptors for FSH.

The results of other investigations have provided additional information on the structure–function relationships of Leydig cells in the immature testis. Ultrastructural analysis of testicular biopsies collected from 30 boys aged 3 to 8 years has revealed the presence of immature Leydig cells comprising about 9% of the total interstitial cell population (990). Furthermore, acute hCG stimulation of the testes of cryptorchid boys caused a precocious secretion of testosterone within the testis and induced rapid differentiation of immature Leydig cells from primitive fibroblastic cells (991).

The role of FSH in induction of Leydig cell function during testicular maturation has been studied in immature rats, where FSH treatment induces rapid hypertrophy and hyperplasia of interstitial cells that transform into mature Leydig cells (1033,1034). These effects could not be duplicated with LH given in an amount thought to contaminate the FSH preparation. Similar findings in the immature rat have been repeated using highly purified FSH (1035). With FSH treatment, numerous Leydig cells appeared to arise from the peritubular tissues, again implicating a reserve stock of mesenchymal or fibroblastic cells capable of differentiation into Leydig cells. Because the target tissue for FSH is the seminiferous epithelium, the effects on Leydig cells of FSH treatment suggest that secretion of factors by the seminiferous tubules may mediate the maturation of Leydig cells.

Studies with a single dose of the substance ethane dimethane sulfonate (EDS) have demonstrated that the Leydig cells in the adult rat testis can be totally destroyed, but they subsequently regenerate 2 to 3 weeks later (703,894,1036). This regeneration appears to occur from a multifocal differentiation of connective tissue cells consisting of pericytes, fibroblasts, and lymphatic endothelial cells. These data indicate that with appropriate stimulation, the connective tissue of the intertubular region and the testis can be induced to develop into Leydig cells (894). During adult life, very little is known about the dynamics of the total Leydig cell population, because in the common laboratory species (mouse, rat, guinea pig) the histology of the interstitial tissue appears unchanged during the reproductive lifetime of the animal. However, most mammalian species do not remain reproductively active throughout the year and show varying degrees of seasonal regression of spermatogenesis and Leydig cell function (34,649). Studies of the morphological and functional changes in Leydig cells of pronounced seasonal breeders have not been prolific, and additional work is warranted. Examples of a combined ultrastructural–functional investigation of changes in the Leydig cells of seasonal breeders include the amphibians, the rock hyrax (1037–1039), wild rats (497,649,975), bats (949,1040), hamsters (1041), squirrels (654), moles (1042), and monkeys (939).

Leydig cells usually undergo some degree of atrophy during periods of diminished sperm production; in the hamster, Leydig cell volume may vary two- to threefold (651,652,1041). Morphometric studies also show significant changes in Leydig cell numbers in the hamster (651,1043) and woodchuck (1044), although this does not seem to occur in the ram testis (1045). The source of new Leydig cells accompanying the restoration of spermatogenic activity is the mesenchymal cell (1043). Involution of the Leydig cells is often marked by accumulation of cytoplasmic lipid inclusions, suggestive of Leydig cell inactivity (939,1038), whereas in seasonally breeding white-tailed deer and European moles, Leydig cell lipid content is maximal in the sexually active phase (1046,1047). Clearly, the morphological features of Leydig cells in seasonal breeding species represent a spectrum of changes that emphasize caution in extrapolating observations from a given species to another. The effects of season on numerous Leydig cell parameters were reported in adult stallions (669), 48 of which were selected at the onset and during the mid-phase of the breeding season. The latter stallions exhibited a 30% increase in testis weight and 50% increase in daily sperm production per testis, compared with the values obtained in the nonbreeding period. When the interstitial compartment was analyzed by quantitative light microscopy and electron microscopy, the total volume of Leydig cells per testis also showed a corresponding 50% increase, yet the numbers of Leydig cells remained unchanged, indicating, at least for this species, that hypertrophy rather than hyperplasia characterizes the Leydig cell response to seasonal changes in testis function. However, in a further reexamination of the stallion testis, evidence is available that in the nonbreeding season, Leydig cell numbers per testis decline by 36% compared with the fully active testis (1048).

The fate of Leydig cells with increasing age has been discussed for many years (428). A considerable degree of controversy over the morphological and functional status of the Leydig cell population, particularly in the human, has been evident from conflicting reports claiming an age-related increase (1049,1050) or decline in Leydig cell numbers or mass (674,1051,1052). Other claims have been put forward favoring no change in Leydig cell abundance with increasing age (998,1053) or a progressive atrophy in elderly men (1054) commonly characterized by increasing degrees of vacuolization, lipid accumulation, and pigmentation [see review (428)]. In a reinvestigation of this subject in 25 men ranging from ages 18 to 87 years in whom spermatogenesis was normal, it was shown that a significant negative correlation existed between age and total Leydig cell number (1055). Young adult men, 20 years of age on average, contained over 700 million Leydig cells in paired testes; by age 60 this figure had declined to about 200 million per paired testes, representing a loss of 8 million Leydig cells per annum beyond the age of 20. These findings were confirmed and extended (1056) in a subsequent study of 30 men aged between 20 and 76 years in which the average 60-year-old man had fewer than half as many Leydig cells as an average 20-year-old man. However, the question remained as to the mechanism of attrition, either through Leydig cell death or via dedifferentiation back to a primitive interstitial cell. If the latter process occurred, then the abundance of primitive (or fibroblastic-type) cells might be expected to increase as the Leydig cells declined. When this idea was tested by morphometric examination of cell types in the interstitial tissue of testes from the above-mentioned group (1057), the population of interstitial cells (neither Leydig cells nor macrophages) decreased significantly with increasing age. This finding thus makes it unlikely that the loss of Leydig cells from aging results from Leydig cell regression into other interstitial cells. Still unresolved, however, is the process by which Leydig cells undergo a slow yet continual loss from the testis.

Although Leydig cells in the aging human testis may develop abundant lipid inclusions, lysosomes, and lipofuscin bodies (428), there is no evidence to suggest that these morphological features reflect an exhausted functional state, nor have there been any data to show that the disappearance of Leydig cells might occur by autolysis, culminating in self-destruction. It seems more likely that dysfunctional Leydig cells would be disposed of by cells equipped to fulfill this role, namely the interstitial macrophages. At present, little is known about the functional duties of testicular macrophages, but from evidence reviewed in the next section, it is clear that their phagocytic capacity can be directed toward the Leydig cells. These observations may help to explain the age-related disappearance of the Leydig cell.

No such attrition of the Leydig cell population has been recorded in aging rats (1058), but in distinct contrast to this species and the human, the aging stallion testes show a striking increase in Leydig cell numbers (669). A two- to threefold increase in Leydig cell number per testis occurred in stallions between 2–3 years and 20 years of age, yet in the examination of 48 testes in this study, no mitotic figures were observed. A similar failure to observe mitotic figures in the interstitial tissue of rats was reported in a study in which Leydig cell number per testis was tripled during 5 weeks of chronic hCG stimulation (879). However, as mentioned in the study of Leydig cells in stallions (669), if mitosis was responsible for Leydig cell proliferation at the measured rate, no more than 1 Leydig cell of every 20,000 observed should exhibit a mitotic figure. Hence, failure to detect mitosis of Leydig cells was not considered as supporting the notion that cell division does not occur. An alternative mechanism of Leydig cell proliferation from other interstitial cells has been reviewed earlier in this section and remains a viable possibility to account for these age and treatment-related changes in Leydig cell populations.

Response to Testicular Damage

Impairment of testicular function can be manifested by changes in either seminiferous tubules or the interstitial tissue, and for many years it was thought that treatments compromising one particular compartment of the testis were virtually without harmful effect on the adjacent compartment. This conclusion seemed valid at the time because microscopists had to rely heavily on the relatively imprecise technique of histological evaluation of the testis in paraffin section and, on the other hand, biochemists and physiologists interested in pathological changes to the testis had very few methods available with which to monitor alterations in testicular function. Significant advances in the morphological sciences together with an ever-increasing number of sensitive and specific assays of cell and tissue function in the testis have together greatly improved our understanding of testicular pathology (498,1059–1062). In this section we present an overview of the morphological changes of Leydig cells in three broad areas where testicular damage is sustained. First, we discuss nonspecific and unrelated treatments that precipitate impairment of spermatogenesis in varying degrees of severity; second, the deleterious effects on spermatogenesis of endogenous hormones or their synthetic analogues are contrasted with alterations of the Leydig cell population; and third, the extreme sensitivity of Leydig cells to selected toxic compounds is reviewed.

A considerable body of evidence is now available that clearly indicates that disruption of spermatogenesis induced by unrelated agents is accompanied by morphological and function changes of the Leydig cells indicative of a state of stimulation. Those treatments that result in this peculiar response of the Leydig cells share a common feature, namely the induction of varying degrees of seminiferous tubule damage. Some examples of these experimental models of spermatogenic damage include surgical cryptorchidism, ligation of the efferent ducts, temporary heat treatment, a vitamin A–deficient diet, treatment with the cytostatic agent hydroxyurea, and irradiation in utero; the details of these experiments have been reviewed elsewhere (498,819,820,1012,1013,1063). The stimulated condition of Leydig cells is reversible (820) and can occur independently of any involvement by their known endogenous trophic hormone (1064–1066); in addition, this same response is confined only to the immediate vicinity of damaged seminiferous tubules (1067). Thus, the concept has emerged of a paracrine relationship between the seminiferous tubules and the Leydig cells, a subject that is attracting increasing attention (1013,1068,1069). Judging from the morphological response of the Leydig cells to spermatogenic damage, it seems likely that their enlargement and proliferation of organelles are a reflection of the actions of some potent intratesticular factor(s) that, for reasons not yet understood, constitute a pathway of a short-loop feedback mechanism within the testis.

It is a little known fact that the naturally occurring gonadotropic hormones LH and hCG can exert a deleterious effect on the testis (683,684,1070) in direct contrast to their usual mode of action, namely the stimulation of intertubular tissue with subsequent support or enhancement of spermatogenesis. Data in support of these findings are new, and examples of the toxic effects of hCG on spermatogenesis in immature rats and guinea pigs are available (944,1071). However, the impairment of spermatogenesis in these experimental situations is again accompanied by hypertrophy and/or hyperplasia of the Leydig cells, an effect similar to that described above for nonspecific induction of spermatogenic damage. This phenomenon is not confined to hCG treatment, because hCG and pregnant mare's serum gonadotropin given to adult rats for 1 to 2 weeks caused degeneration of the seminiferous epithelium (1071,1072). Also, in investigations of the antispermatogenic effects of LH and LH–releasing hormone agonists in adult hypophysectomized rats (686), it was shown that LH alone or in combination with LH–releasing hormone agonist can focally inhibit spermatogenesis and, in certain areas of the testis, destroy all the cells of the seminiferous epithelium. When the intertubular tissue was examined using morphometric techniques, the adjacent Leydig cells exhibited significant hyperplasia compared with saline-treated controls not experiencing seminiferous tubular damage. The observations emphasize that agents that disrupt the spermatogenic process also interfere with the Leydig cells.

The number of substances known to be specifically toxic to the intertubular tissue is large, and relevant reviews are available (898,1062,1073–1075). Of particular interest is the influence of cadmium salts and the group of alkylating agents, diesters of methane sulfonic acid. For many years cadmium administration, even in minute doses, has been recognized to cause acute testicular necrosis by rapid disturbance of the testicular circulation (685,1073). Within several hours after cadmium exposure, morphological signs of intertubular disruption are noted, including edema, hemorrhage, and infiltration of mononucleated cells. Finally, complete testicular necrosis occurs in which the seminiferous tubules degenerate totally and the intertubular tissue resembles loose connective tissue mainly containing extracellular materials. A surprising long-term consequence of cadmium-induced testicular degeneration is the slow but unmistakable regeneration of Leydig cells within the intertubular tissue. The new generation of Leydig cells is functionally active, as indicated by restoration of previously reduced weights of seminal vesicles and prostate (685,898). It is of interest that regeneration of Leydig cells occurs initially just beneath the tunica albuginea and is accompanied by regeneration of blood vessels and intertubular macrophages. Continued proliferation of Leydig cells gives rise to interstitial cell tumors, but the seminiferous tubules remain permanently sclerosed. This observation raises the possibility that Leydig cell differentiation and hyperplasia may be dependent on vascular proliferation. The blood vessels that regenerate within those testes are evidently quite different from those in the nontreated testis, because they do not react to a second cadmium treatment (1076–1078), thus conferring resistance on the regenerated Leydig cells to further cadmium insult.

In many aspects, the toxic effects of EDS on the intertubular tissue of the rat testis bear some resemblance to those described above for cadmium: EDS specifically destroys Leydig cells within 3 days after a single treatment (703,893,895,1079–1082). Macrophages phagocytose the degenerating Leydig cells and bring about the complete elimination of Leydig cells from the testis. However, beginning several weeks after initial treatment with EDS, new Leydig cells regenerate within the intertubular tissue and bear a striking morphological resemblance to Leydig cells seen in the fetal rat testis. The fetal-type Leydig cells arise from perivascular and peritubular positions within the intertubular tissue and, over 8–10 weeks post-EDS treatment, transform into adult-type Leydig cells occupying these above-mentioned positions and more centrally placed locations.

In a series of experiments aimed at identifying the hormonal control and cellular dynamics of Leydig cell regeneration in the rat testis, it has been shown that hCG causes rapid proliferation and differentiation of precursor mesenchymal cells into Leydig cells. New Leydig cells arise from perivascular regions in response to hCG treatment, whereas in EDS-treated testes not stimulated with hCG, the preferred origin of Leydig cells seems to be a peritubular location (1083,1084). The same laboratory also showed that proliferation of putative Leydig cell precursor cells can occur in association with suppressed endogenous LH (1084), suggesting that local intratesticular or paracrine factors are involved in the early phases of Leydig cell development. After the establishment of a pool of precursors destined to provide an initial wave of Leydig cell development, further expansion of the Leydig cell population occurs via Leydig cell mitosis, which is dependent on LH stimulation (1085). Clearly, there must be some mechanism to regulate this proliferative activity to attain, but not to exceed, the normal number of 28 to 30 million Leydig cells per testis. Studies have shown that administration of 17β-estradiol plus hCG blocks Leydig cell regeneration in the EDS-treated rat if given between 5 and 16 days post-EDS (1086), implicating the Leydig precursor cell as the estrogen target.

Because Leydig cells produce estrogen themselves, perhaps estrogen acts locally within the testes as a paracrine regulator of precursor cell differentiation. However, it has also been shown (1082) that the fetal-type Leydig cells are resistant to additional EDS exposure when it is given at weekly intervals up to 6 weeks from initial treatment.

The inability of EDS to destroy these Leydig cells has also been shown for the Leydig cells of the immature rat testis (1079), suggesting that only the adult-type Leydig cell is sensitive to the toxic effects of EDS. Data available from the response of Leydig cells to indirect or direct assaults on them therefore indicate that although Leydig cells are actually sensitive to many agents causing testicular damage, the intertubular tissue retains the capacity to ensure their persistence in the testis (702).

Interesting results supporting the presence of a paracrine regulation of Leydig cells are also available from studies using EDS. If given to cryptorchid rats, EDS causes a destruction of Leydig cells as seen in normal animals, but the recovery of the Leydig cells is more rapid in the cryptorchid testes (1087–1089). These results support the view that factors are present in testes with damaged seminiferous tubules that not only cause Leydig cell hypertrophy (908,1065,1066) but also stimulate a more rapid regeneration of Leydig cells after their destruction.

REFERENCES

1. Steinach, E. (1940). *Sex and Life, Forty Years of Biological and Medical Experiments*. Viking Press, New York.
2. Aristotle (1910). Historia animalium. In *The Works of Aristotle*, Vol. IV. D. W. Thompson, translator. Clarendon Press, Oxford.
3. de Graaf, R. (1972). On the human reproductive organs. N. D. Jocelyn and B. P. Setchell, translators. *J. Reprod. Fertil.* Suppl 17.
4. van Leeuwenhoek, A. (1960). Observations de natis e semine genitali animaliulis. *Phil. Trans.* 1667). 12, 1040. Cited in *Marshall's Physiology of Reproduction,* pp. 1–129. Longmans & Green, London.
5. von Koelliker, R. Z. (1841). *Bertrage zur Kenntniss der Geschlechtver-hallsmisse und der Samen-flussigkeit wirbelloser Thiere und die Bedeutung de sogenannten Samenthiere*. Berlin.
6. Berthold, A. A. (1849). Transplantation der Hoden. *Arch. Anat. Physiol.* Wiss. Med. 16, 42–46.
7. Bouin, P., and Ancel, P. (1903). Recherches sur les cellules interstitielles de testicule des mammiferes. *Arch. Zool. (Stockh.)* 1, 437–523.
8. David, K., Dingemanse, E., Freud, J., and Laquer, F. (1935). Über krystallinisches mannliches Hormon aus Hoden (Testosterone), wirk-samer als aus Harn oder aus Cholesterin hereites Androsteron. *Z. Physiol. Chem.* 233, 281–282.
9. Leeson, C. R., and Adamson, L. (1962). The mammalian tunica vaginalis testis and its fine structure and function. *Acta. Anat.* 51, 226–240.
10. Holstein, A. F., and Weiss, C. (1967). Über die Wirkung der glatten Muskulatur in der Tunica Albuginea in Hoden des Kaninchens Messungen des interstitiellen Druckes. *Z. Ges. Exp. Med.* 142, 334–337.
11. Davis, J. R., and Langford, G. A. (1970). Pharmacological studies on the testicular capsule in relation to sperm transport. In *The Human Testis* (E. Rosenberg and C. A. Paulsen, Eds.), pp. 495–514. Plenum Press, New York.
12. Clermont, Y., and Huckins, C. (1961). Microscopic anatomy of the sex cords and seminiferous tubules in growing and adult male albino rats. *Am. J. Anat.* 108, 79–97.
13. Bremer, J. T. J. (1911) Morphology of the tubules of the human testis and epididymis. *Am. J. Anat.* 11, 393–417.
14. Roosen-Runge, E. C. (1961). The rete testis in the albino rat, its structure, development and morphological significance. *Acta. Anat.* 45, 1–30.
15. Scott, U. G., and Scott, P. P. (1957). Postnatal development of the testis and epididymis in the cat. *J. Physiol. (Lond.)* 136, 40–45.
16. Ebling, F. J., Brooks, A. N., Cronin, A. S., Ford, H., and Kerr, J. B. (2000). Estrogenic induction of spermatogenesis in the hypogonadal mouse. *Endocrinology* 141, 2861–2869.
17. Robertson, K. M, O'Donnell, L., Simpson, E. R., and Jones, M. E. (2002). The phenotype of the aromatase knockout mouse reveals dietary phytoestrogens impact significantly on testis function. *Endocrinology* 143, 2913–2921.
18. Myers, M., Ebling, F. J. P., Stewart, J., and Kerr, J. B. (2005). Atypical development of Sertoli cells inhibits spermatogenesis in the hypogonadal (hpg) mouse. *J. Anat.* (in press).
19. McLachlan, R. I., Wreford, N. G., deKretser, D. M., and Robertson, D. M. (1995). The effects of recombinant follicle-stimulating hormone on the restoration of spermatogenesis in the gonadotropin-releasing hormone-immunized adult rat. *Endocrinology* 136, 4035–4043.
20. Kong, L. S., Huang, A. P, Deng, X. Z., and Yang, Z. W. (2004). Quantitative (stereological) study of the effects of vasectomy on spermatogenesis in rabbits. *J. Anat.* 205, 147–156.
21. Peng, B., Zhang, R. D., Dai, X. S., Deng, X. Z., Wan, Y., and Yang, Z. W. (2002). Quantitative (stereological) study of the effects of vasectomy on spermatogenesis in rhesus monkeys (*Macaca mulatta*). *Reproduction* 124, 847–856.
22. Zhang, R. D., Peng, B., Deng, X. Z., Wan, Y., and Yang, Z. W. (2003). A stereological study of the effects of experimental inguinal cryptorchidism and subsequent orchiopexy on spermatogenesis in adult monkeys. *Int. J. Androl.* 26, 180–186.
23. Zhengwei, Y., McLachlan, R. I., Bremner, W. J., and Wreford, N. G. (1997). Quantitative (stereological) study of normal spermatogenesis in the adult monkey (*Macaca fascicularis*). *J. Androl.* 18, 681–687.
24. Zhengwei, Y., Wreford, N. G., Schlatt, S., Weinbauer, G. F., Nieschlag, E., and McLachlan, R. I. (1998). Acute and specific impairment of spermatogonial development by GnRH antagonist-induced gonadotrophin withdrawal in the adult macaque (Macaca fascicularis). *J. Reprod. Fertil.* 112, 139–147.
25. Aslam, H., Schneiders, A., Perret, M., Weinbauer, G. F, and Hodges, J. K. (2002). Quantitative assessment of testicular germ cell production and kinematic and morphometric parameters of ejaculated spermatozoa in the grey mouse lemur, *Microcebus murinus. Reproduction* 123, 323–332.
26. Schultz, A. H. (1938). The relative weight of testes in primates. *Anat. Rec.* 72, 387–394.
27. Harvey, P. H., and Harcourt, A. H. (1984). Sperm competition, testes size and breeding systems in primates. *Sperm Competition and the Evolution of Animal Mating Systems* 18, 589–601.
28. Harcourt, A. H., Harvey, P. H., Larson, S. G., and Short, R. V. (1981). Testis weight, body weight and breeding system in primates. *Nature* 293, 55–57.
29. Glander, K. E., Wright, P. C., Daniels, P. S., and Merenlender, A. M. (1992). Morphometrics and testicle size of rain forest lemur species from southeastern Madagascar. *J. Human Evol.* 22, 1–17.
30. Fawcett, D. W., Neaves, W. R., and Flores, M. N. (1973). Comparative observations on intertubular lymphatics and the organisation of the interstitial tissue of the mammalian testis. *Biol. Reprod.* 9, 500–532.
31. Dym, M. (1974).The fine structure of monkey Sertoli cells in the transitional zone at the junction of the seminiferous tubules with the tubuli recti. *Am. J. Anat.* 140, 1–6.
32. Roosen-Runge, E. C., and Holstein, A. F. (1978). The human rete testis. *Cell. Tissue Res.* 189, 409–433.
33. Bustos Obregon, E., and Holstein, A. F. (1976). The rete testis in man, ultrastructural aspects. *Cell. Tissue Res.* 175, 1–15.
34. Lincoln, G. A. (1989). Seasonal aspects of testicular function. In *The Testis* (H. G. Burger and D. M. de Kretser, Eds.), pp. 329–385. Raven Press, New York.
35. Kerr, J. B., and Hedger, M. P. (1983). Spontaneous spermatogenic failure in the marsupial mouse *Antechinus stuartii.* Macleay (*dasyuride, Marsupialia*). *Aust. J. Zool.* 31, 445–466.
36. Setchell, B. P., and Waites, G. M. H. (1975). The blood–testis barrier. In *Handbook of Physiology* (D. W. Hamilton and R. O. Greep, Eds.), Section 7, pp. 143–172. Williams & Wilkins, Baltimore.
37. Sinclair, A. H., Berta, P., Palmer, M. S., Hawkins, J. R., Griffiths, B. L., Smith, M. J., Foster, J. W., Frischauf, A. M., Lovell-Badge, R., and Goodfellow, P. N. (1990). A gene from the human sex-determining region encodes a protein with homology to a conserved DNA-binding motif. *Nature* 346, 240–244.
38. Kent, J., Wheatley, S. C., Andrews, J. E., Sinclair, A. H., and Koopman, P. (1996). A male-specific role for SOX9 in vertebrate sex determination. *Development* 122, 2813–2822.

39. Morrish, B. C., and Sinclair, A. H. (2002). Vertebrate sex determination, many means to an end. *Reproduction* 124, 447–457.
40. Schmahl, J., Eicher, E. M., Washburn, L. L., and Capel, B. (2000). Sry induces cell proliferation in the mouse gonad. *Development* 127, 65–73.
41. Behringer, R. R., Finegold, M. J., and Cate, R. L. (1994). Müllerian inhibiting substance is required for normal male development and fertility. *Cell* 79, 415–425.
42. Colvin, J. S., Green, R. P., Schmahl, J., Capel, B., and Ornitz, D. M. (2001). Male-to-female sex reversal in mice lacking fibroblast growth factor 9. *Cell* 104, 875–889.
43. Adams, I. R., and McLaren, A. (2002). Sexually dimorphic development of mouse primordial germ cells, switching from oogenesis to spermatogenesis. *Development* 129, 1155–1164.
44. Brennan, J., Tilmann, C., and Capel, B. (2003). Pdgfr-alpha mediates testis cord organization and fetal Leydig cell development in the XY gonad. *Genes Dev.* 17, 800–810.
45. Wartenberg, H. (1989). Differentiation and development of the testis. In *The Testis* (H. G. Burger and D. M. de Kretser, Eds.), pp. 67–118. Raven Press, New York.
46. Kirkham, W. B. (1915). The germ cell cycle in the mouse. *Anat. Rec.* 10, 217–219.
47. Allen, E. (1918). Studies on cell division in the albino rat. *J. Morphol.* 31, 133–185.
48. Hargitt, G. T. (1926). The formation of the sex gland and germ cells of mammals. 1. The origin of the germ cells in the albino rat. *J. Morphol. Physiol.* 40, 517–558.
49. Regaud, C. (1901). Etudes sur la structure des tubes seminiferes et sur la spermatogenese chez les mammiferes. *Arch. Anat. Microsc.* 4, 101–156.
50. Clermont, Y., and Perey, B. (1957). Quantitative study of the cell population of the seminiferous tubules in immature rats. *Am. J. Anat.* 100, 241–267.
51. Hacker, A., Capel, B., Goodfellow, P., and Lovell-Badge, R. (1995). Expression of Sry, the mouse sex determining gene. *Development* 121, 1603–1614.
52. Karl, J., and Capel, B. (1998). Sertoli cells of the mouse testis originate from the coelomic epithelium. *Dev. Biol.* 203, 323–333.
53. Sekido, R., Bar, I., Narvaez, V., Penny, G., and Lovell-Badge, R. (2004). SOX9 is up-regulated by the transient expression of SRY specifically in Sertoli cell precursors. *Dev. Biol.* 274, 271–279.
54. Martineau, J., Nordqvist, K., Tilmann, C., Lovell-Badge, R., and Capel, B. (1997). Male-specific cell migration into the developing gonad. *Curr. Biol.* 7, 958–968.
55. Tilmann, C., and Capel, B. (1999). Mesonephric cell migration induces testis cord formation and Sertoli cell differentiation in the mammalian gonad. *Development* 126, 2883–2890.
56. Black, V., and Christensen, A. K. (1969). Differentiation of interstitial cells and Sertoli cells in fetal guinea pig testes. *Am. J. Anat.* 124, 211–238.
57. Orth, J. (1982). Proliferation of Sertoli cells in fetal and postnatal rats, a quantitative autoradiographic study. *Anat. Rec.* 203, 485–492.
58. Orth, J. M (1984). The role of follicle stimulating hormone in controlling Sertoli cell proliferation in testes of fetal rats. *Endocrinology* 115, 1248–1255.
59. Magre, S., and Jost, A. (1980). The initial phases of testicular organogenesis in the rat. *Arch. Anat. Microsc. Morphol. Exp.* 69, 297–317.
60. Pelliniemi, L. J. (1975). Ultrastructure of the indifferent gonad in male and female pig embryos. *Tissue Cell.* 8, 162–174.
61. Pelliniemi, L. J. (1975). Ultrastructure of the gonadal ridge in male and female pig embryos. *Anat. Embryol.* 147, 19–43.
62. Pelliniemi, L. J., and Niemi, M. (1969). Fine structure of the human foetal testis. *Z. Zellforsch.* 99, 507–522.
63. Jost, A., Vigier, B., Prepin, J., and Perchelet, J. P. (1973). Studies on sex differentiation in mammals. *Rec. Prog. Horm. Res.* 29, 1–35.
64. Almond, G. D., and Singh, R. P. (1980). Development of the Sertoli cell in the fetal mouse. *Acta. Anat.* 106, 276–280.
65. van Vorstenbosch, C. J. A. H. V., Spek, E., Colenbrander, R., and Wensin, C. J. G. (1984). Sertoli cell development of pig testis in the fetal and neonatal period. *Biol. Reprod.* 31, 565–577.
66. Zhengwei, Y., Wreford, N. G., and de Kretser, D. M. (1990). A quantitative study of spermatogenesis in the developing rat testis. *Biol. Reprod.* 43, 629–625.
67. Kluin, P. M., Kramer, M. F., and de Rooij, D. G. (1984). Proliferation of spermatogonia and Sertoli cells in maturing mice. *Anat. Embryol.* 169, 73–78.
68. Clermont, Y., and Le Blond, C. P. (1953). Renewal of spermatogonia in the rat. *Am. J. Anat.* 93, 475–502.
69. Ortavant, R. (1959). Spermatogenesis and morphology of the spermatozoon. In *Reproduction in Domestic Animals* (H. H. Cole and P. T. Cupps, Eds.), Vol. 2. Academic Press, New York.
70. Hilscher, W., and Makoski, H. B. (1968). Histollogisch und Autoradiographische Untersuchungen zur "Praspermatogenese" und "Spermatogenese" der Ratte. *Z. Zellforsch.* 86, 327–350.
71. Steinberger, A., and Steinberger, E. (1971). Replication pattern of Sertoli cells in maturing rat testis *in vivo* and in organ culture. *Biol. Reprod.* 4, 84–87.
72. Steinberger, A., and Steinberger, E. (1977). The Sertoli cell. In *The Testis* (A. D. Johnson and W. R. Gomes, Eds.), Vol. 4, pp. 371–399. Academic Press, New York.
73. Wang, Z. X., Wreford, N. G., and de Kretser, D. M. (1989). Determination of Sertoli cell numbers in the developing rat testis by stereological methods. *Int. J. Androl.* 12, 58–64.
74. Vergouwen, R., Jacobs, S., Huiskamp, R., Davids, J., and de Rooij, D. (1991). Proliferative activity of gonocytes, Sertoli cells and interstitial cells during testicular development in mice. *J. Reprod. Fertil.* 93, 233–243.
75. Griswold, M., Solari, A., Tung, P., and Fritz, I. (1977). Stimulation by FSH of DNA synthesis and of mitosis in cultured Sertoli cells prepared from testes of immature rats. *Mol. Cell. Endocrinol.* 7, 151–165.
76. Cortes, D., Muller, J., and Skakkebaek, N. E. (1987). Proliferation of Sertoli cells during development of the human testis assessed by stereological methods. *Int. J. Androl.* 10, 589–596.
77. Simorangkir, D. R., Marshall, G. R., and Plant, T. M. (2003). Sertoli cell proliferation during prepubertal development in the rhesus monkey (*Macaca mulatta*) is maximal during infancy when gonadotropin secretion is robust. *J. Clin. Endocrinol. Metab.* 88, 4984–4989.
78. Marshall, G. R., and Plant, T. M. (1996). Puberty occurring either spontaneously or induced precociously in rhesus monkey (*Macaca mulatta*) is associated with a marked proliferation of Sertoli cells. *Biol. Reprod.* 54, 1192–1199.
79. Sharpe, R. M., McKinnell, C., Kivlin, C., and Fisher, J. S. (2003). Proliferation and functional maturation of Sertoli cells, and their relevance to disorders of testis function in adulthood. *Reproduction* 125, 769–784.
80. Means, A. R., Fakunding, J. L., Huckins, C., Tindall, D. J., and Vitale, R. (1976). Follicle-stimulating hormone, the Sertoli cell, and spermatogenesis. *Rec. Prog. Horm. Res.* 32, 477–522.
81. Chiappa, S., and Fink, G. (1977). Releasing factor and hormonal changes in the hypothalamic-pituitary-gonadotrophin and adrencorticotrophin systems before and after birth and puberty in male, female and androgenized female rats. *J. Endocrinol.* 72, 211–224.

82. Chowdhury, J., and Steinberger, E. (1976).Pituitary and plasma levels of gonadotropin in fetal and newborn male and female rats. *J. Endocrinol.* 69, 381–384.
83. Begeot, M., Dupouy, J. P., Dubois, M. P., and Dubois, P. M. (1981). Immunocytological determination of gonadotrophic and thyrotrophic cells in fetal rat anterior pituitary during normal development and under experimental conditions. *Neuroendocrinology* 32, 285–294.
84. Warren, D., Huhtaniemi, I., Tapanainen, J., Dufau, M., and Catt, K. (1984). Ontogeny of gonadotrophin receptors in the fetal and neonatal rat testis. *Endocrinology* 114, 470–476.
85. Solari, A. J., and Fritz, I. B. (1978). The ultrastructure of immature Sertoli cells. Maturation-like changes during culture and the maintenance of mitotic potentiality. *Biol. Reprod.* 18, 329–345.
86. Lording, D. W., and de Kretser, D. M. (1972). Comparative ultrastructural and histochemical studies of the interstitial cells of the rat testis during fetal and postnatal development. *J. Reprod. Fertil.* 29, 261–269.
87. Tapanainen, J., Kuopio, T., Pelliniemi, L. J., and Huhtaniemi, I. (1984). Rat testicular endogenous steroids and number of Leydig cells between the fetal period and sexual maturity. *Biol. Reprod.* 31, 1027–1035.
88. Weisz, J., and Ward, I. (1980). Plasma testosterone and progesterone titres of pregnant rats, their male and female fetuses and neonatal offspring. *Endocrinology* 106, 306–316.
89. Orth, J., and Weisz, J. (1980). Development of Δ_5–3β hydroxysteroid dehydrogenase and glucose-6-phosphate dehydrogenase activity in Leydig cells of the fetal rat testis, a quantitative cytochemical study. *Biol. Reprod.* 22, 1201–1209.
90. Nomura, T., Weisz, J., and Lloyd, C. W. (1966). *In vitro* conversion of 7-3H progesterone to androgens by the rat testis during the second half of fetal life. *Endocrinology* 78, 245–253.
91. Warren, C., Haltmeyer, G. C., and EikNes, K. R. (1973). Testosterone in the fetal testis. *Biol. Reprod.* 8, 560–565.
92. Boitani, C., Stefanini, M., Fragale, A., and Morena, A. R. (1995). Activin stimulates Sertoli cell proliferation in a defined period of rat testis development. *Endocrinology* 136, 5438–5444.
93. Singh, J., and Handelsman, D. J. (1996). Neonatal administration of FSH increases Sertoli cell numbers and spermatogenesis in gonadotropin-deficient (*hpg*) mice. *Endocrinology* 136, 5311–5321.
94. Meachem, S. J., Mclachlan, R. I., Stanton, P. G., Robertson, D. M., and Wreford, N. G. (1999). FSH immunoneutralization acutely impairs spermatogonial development in normal adult rats. *J. Androl.* 20, 756–762.
95. Meehan, T., Schlatt, S., O'Bryan, M. K., de Kretser, D. M., and Loveland, K. L. (2000). Regulation of germ cell and Sertoli cell development by activin, follistatin, and FSH. *Dev. Biol.* 220, 225–237.
96. Fragale, A., Puglisi, R., Morena, A. R., Stefanini, M., and Boitani, C. (2001). Age-dependent activin receptor expression pinpoints activin A as a physiological regulator of rat Sertoli cell proliferation. *Mol. Hum. Reprod.* 7, 1107–1114.
97. Buzzard, J. J., Wreford, N. G., and Morrison, J. R. (2002). Marked extension of proliferation of rat Sertoli cells in culture using recombinant human FSH. *Reproduction* 124, 633–641.
98. Kumar, T. R., Wang, Y., Lu, N., and Matzuk, M. (1997). Follicle stimulating hormone is required for ovarian follicle maturation but not male fertility. *Nat. Genet.* 15, 201–204.
99. Dierich, A., Sairam, M. R., Monaco, L., Fimia, G. M., Fansmuller, A., LeMeur, M., and Sassone-Corsi, P. (1998). Impairing follicle-stimulating hormone (FSH) signalling *in vivo,* targeted disruption of the FSH receptor leads to aberrant gametogenesis and hormonal imbalance. *Proc. Natl. Acad. Sci. U S A* 95, 13612–13617.
100. Abel, M. H., Wootton, A. N., Wilkins, V., Huhtaniemi, I., Knight, P. G., and Charlton, H. M. (2000). The effect of a null mutation in the follicle-stimulating hormone receptor gene on mouse reproduction. *Endocrinology* 141, 1795–1803.
101. Pellegrini, M., Grimaldi, P., Rossi, P., Geremia, R., and Dolci, S. (2003). Developmental expression of BMP4/ ALK3/SMAD5 signaling pathway in the mouse testis, a potential role of BMP4 in spermatogonia differentiation. *J. Cell. Sci.* 116, 3363–3372.
102. Buzzard, J. J., Farnworth, P. G., de Kretser, D. M., O'Connor, A. E., Wreford, N. G., and Morrison, J. R. (2003). Proliferative phase Sertoli cells display a developmentally regulated response to activin in vitro. *Endocrinology* 144, 474–483.
103. Puglisi, R., Montanari, M., Chiarella, P., Stefanini, M., and Boitani, C. (2004). Regulatory role of BMP2 and BMP7 in spermatogonia and Sertoli cell proliferation in the immature mouse. *Eur. J. Endocrinol.* 151, 511–20.
104. Orth, J. M., Gunsalus, G. L., and Lamperti, A. A. (1988). Evidence from Sertoli cell-depleted rats indicates that spermatid number in adults depends on numbers of Sertoli cells produced during perinatal development. *Endocrinology* 122, 787–794.
105. Kirby, J. D., Jetton, A. E., and Cooke, P. S., Hess, R. A., Bunick, D., Ackland, J. F., Turek, F. W., and Schwartz, N. B. (1992). Developmental hormonal profiles accompanying the neonatal hypothyroidism-induced increase in adult-testicular size and sperm production in the rat. *Endocrinology* 131, 559–565.
106. van Haaster, L. H., de Jong, F. H., Docter, R., and de Rooij, D. G. (1992). The effect of hypothyroidism on Sertoli cell proliferation and differentiation and hormone levels during testicular development in the rat. *Endocrinology* 131, 1574–1576.
107. Cooke, P. S., and Meisami, E. (1991). Early hypothyroidism in rats causes increased adult testis and reproductive organ size but does not change testosterone levels. *Endocrinology* 129, 237–243.
108. Cooke, P. S., Hess, R. A., Porcelli, J., and Meisami, E. (1991). Increased sperm production in adult rats after transient neonatal hypothyroidism. *Endocrinology* 129, 244–248.
109. Chubb, C. (1992). Genes regulating testis size. *Biol. Reprod.* 47, 29–36.
110. Bishop, M. W. H., and Walton, A. (1960). Spermatogenesis and the structure of mammalian spermatozoa. In *Marshall's Physiology of Reproduction*, Vol 1, pp. 1–29. Longmans Green, London.
111. Attal, J., and Courot, M. (1963). Development testiculaire et establissement de la spermatogenese chez le taureau. *Ann. Biol. Anim. Biochem. Biophys.* 3, 219–241.
112. Sapsford, C. S. (1963). The development of the Sertoli cell of the rat and mouse, its existence as a mononucleate unit. *J. Anat.* 97, 225–238.
113. Ramos, A. S., and Dym, M. (1979). Ultrastructural differentiation of rat Sertoli cells. *Biol. Reprod.* 21, 909–922.
114. Flickinger, C. J. (1967). The postnatal development of the junctional complexes of the mouse Sertoli cells of the mouse. *Z. Zellforsch.* 78, 92–113.
115. Nagano, T., and Suzuki, F. (1976). The postnatal development of the junctional complexes of the mouse Sertoli cells as revealed by freeze-fracture. *Anat. Rec.* 185, 403–418.
116. Vitale, R., Fawcett, D. W., and Dym, M. (1973).The normal development of the blood-testis barrier and the effects of clomiphene and estrogen treatment. *Anat. Rec.* 176, 333–344.
117. Kormano, M. (1967). Dye permeability and alkaline phosphatase activity of testicular capillaries in the postnatal rat. *Histochemie* 9, 327–338.

118. Gilula, W. B., Fawcett, D. W., and Aoki, A. (1976). The Sertoli cell occluding junctions and gap junctions in mature and developing mammalian testis. *Dev. Biol.* 50, 142–168.
119. Pelletier, R. M., and Friend, D. S. (1983). The Sertoli cell junctional complex, structure and permeability to filipin in the neonatal and adult guinea pig. *Am. J. Anat.* 168, 213–228.
120. Connell, C. J. (1980). Blood-testis barrier formation and the initiation of meiosis in the dog. In *Testicular Development, Structure and Function* (A. Steinberger and E. Steinberger, Eds.), pp. 71–78. Raven Press, New York.
121. Hagenas, L., Ploen, L., Ritzen, E. M., and Ekwall, H. (1977). Blood-testis barrier, maintained function of inter-Sertoli cell junctions in experimental cryptorchidism in the rat, as judged by a simple lanthanum-immersion technique. *Andrologia* 9, 250–254.
122. Vitale, R. (1975). The development of the blood-testis barrier in Sertoli cell only rats. *Anat. Rec.* 181, 501–508.
123. Bergh, A. (1981). Morphological signs of a direct effect of experimental cryptorchidism on the Sertoli cells in rats irradiated as fetuses. *Biol. Reprod.* 24, 145–152.
124. Rich, K. A., Kerr, J. B., and de Kretser, D. M. (1979). Evidence for Leydig cell dysfunction in rats with seminiferous tubule damage. *Mol. Cell. Endocrinol.* 13, 123–135.
125. Pelletier, R. M. (1986). Cyclic formation and decay of the blood-testis barrier in the mink (*Mustela isori*), a seasonal breeder. *Am. J. Anat.* 175, 91–117.
126. Waites, G. M. H. (1982). Gladwell RT. Physiological significance of fluid secretion in the testis and blood-testis barrier. *Physiol. Rev.* 62, 624–671.
127. Setchell, B. P., Voglmayr, J. K., and Waites, G. M. H. (1969). A blood-testis barrier restricting passage from blood into rete testis fluid but not into lymph. *J. Physiol. (Lond.)* 200, 73–85.
128. Hagenas, L., Ploen, L., and Ekwall, H. (1978). Blood-testis barrier, evidence for intact inter-Sertoli cell junctions after hypophysectomy in the adult rat. *J. Endocrinol.* 76, 87–91.
129. Johnson, M. H. (1973). The role of the pituitary in the development of the blood-testis barrier in mice. *J. Reprod. Fertil.* 32, 509–511.
130. de Kretser, D. M., and Burger, H. G. (1972). Ultrastructural studies of human Sertoli cells in normal men and males with hypogonadotrophic hypogonadism before and after gonadotrophic treatment. In *Gonadotropins* (B. B. Saxena, C. G. Beling, and H. M. Gandy, Eds.), pp. 640–656. Wiley Interscience, New York.
131. Clark, J. M., and Eddy, E. M. (1975). Fine structural observations on the origin and associations of primordial germ cells of the mouse. *Dev. Biol.* 47, 136–155.
132. Lawson, K. A., Dunn, N. R., Roelen, B. A., Zeinstra, L. M., David, A. M., Wright, C. V., Korving, J. P., and Hogan, B. L. (1999). *Bmp4* is required for the generation of primordial germ cells in the mouse embryo *Genes Dev.* 13, 424–436.
133. Ying, Y., Liu, X. M., Marble, A., Lawson, K. A., and Zhao, G. Q. (2000). Requirement of Bmp8b for the generation of primordial germ cells in the mouse. *Mol. Endocrinol.* 14, 1053–1063.
134. Ying, Y., Qi, X., and Zhao, G. Q. (2001). Induction of primordial germ cells from murine epiblasts by synergistic action of BMP4 and BMP8B signaling pathways. *Proc. Natl. Acad. Sci. U S A* 98, 7858–7862.
135. Fuss, A. (1912). Über die Geschlechtzellen des Menschen und der Sauge-tiere. *Arch. Mikrosc. Anat.* 81, 1–23.
136. Witschi, E. (1948). Migration of the germ cells of human embryos from the yolk sac to the primitive gonadal fold. *Carnegie Inst. Wash. Contrib. Embryol.* 209, 67–80.
137. McKay, D. G., Hertig, A. T., Adams, E. C., and Danziger, S. (1953). Histochemical observations on the germ cells of human embryos. *Anat. Rec.* 117, 201–291.
138. McCarrey, J. R. (1993). Development of the germ cell. In *Cell and Molecular Biology of the Testis* (C. Desjardins and L. L. Ewing, Eds.), pp. 58–89. Oxford University Press, Oxford.
139. Yoshinaga, K., Nishikawa, S., Ogawa, M., et al. (1991). Role of *c-kit* in mouse spermatogenesis, identification of spermatogonia as a specific site of *c-kit* expression and function. *Development* 113, 689–699.
140. Brannan, C. I., Bedell, M. A., Resnick, J. L., et al. (1992). Developmental abnormalities in Steel^{17H} mice result from a splicing defect in the steel factor cytoplasmic tail. *Genes Dev.* 6, 1832–1842.
141. Mintz, B., and Russell, E. S. (1957). Gene-induced embryological modifications of primordial germ cells in the mouse. *J. Exp. Zool.* 134, 207–237.
142. Garcia-Castro, M. I., Anderson, R., Heasman, J., and Wylie, C. (1997). Interactions between germ cells and extracellular matrix glycoproteins during migration and gonad assembly in the mouse embryo. *J. Cell. Biol.* 138, 471–480.
143. Pelliniemi, L. J., and Frojdman, K. (2001). Structural and regulatory macromolecules in sex differentiation of gonads. *J. Exp. Zool.* 290, 523–588.
144. Zamboni, L., and Upadhyay, S. (1983). Germ cell differentiation in mouse adrenal glands. *J. Exp. Zool.* 228, 173–193.
145. McLaren, A. (1983). Studies on mouse germ cells inside and outside the gonad. *J. Exp. Zool.* 228, 167–171.
146. Gondos, B., and Conner, L. A. (1973). Ultrastructure of developing germ cells in the fetal rabbit testis. *Am. J. Anat.* 136, 23–42.
147. Wartenburg, H. (1976). Comparative cytomorphologic aspects of the male germ cells especially of the "gonia." *Andrologia* 8, 117–130.
148. Hilscher, B., Hilscher, W., Bulthoff-Obnolz, B., Krāmer, U., Birke, A., Pelzer, H., and Gauss, G. (1974). Kinetics of gametogenesis. I. Comparative histological and autoradiographic studies of oocytes and transitional prospermatogonia during oogenesis and prespermatogenesis. *Cell. Tissue Res.* 154, 443–470.
149. Roosen-Runge, E. C., and Leik, J. (1975). Gonocyte degeneration in the postnatal male rat. *Am. J. Anat.* 122, 275–300.
150. Gondos, B. (1977). Testicular development. In *The Testis* (A. D. Johnson and W. R. Gomes, Eds.), Vol. IV, pp. 1–25. Academic Press, New York.
151. Gondos, B, and Hobel, C. J. (1971). Ultrastructure of germ cell development in the human fetal testis. *Z. Zellforsch.* 119, 1–20.
152. McGuiness, M. P., and Orth, J. M. (1992). Reinitiation of gonocyte mitosis and movement of gonocytes to the basement membrane in testes of newborn rats *in vivo* and *in vitro*. *Anat. Rec.* 233, 527–537.
153. McLean, D. J., Friel, P. J., Johnston, D. S., and Griswold, M. D. (2003). Characterisation of spermatogonial stem cell maturation and differentiation in neonatal mice. *Biol. Reprod.* 69, 2085–2091.
154. Orwig, K. E., Avarbock, M. R., and Brinster, R. L. (2002). Retrovirus-mediated modification of male germline stem cells in rats. *Biol. Reprod.* 67, 874–879.
155. Clermont, Y. (1972). Kinetics of spermatogenesis in mammals, seminiferous epithelium cycle and spermatogonial renewal. *Physiol. Rev.* 52, (198–236.
156. Lee, V. W. K., de Kretser, D. M., Hudson, B. H., and Wang, C. (1975). Variations in serum FSH, LH and testosterone levels in male rats from birth to sexual maturity. *J. Reprod. Fertil.* 42, 121–126.
157. Muller, J., and Skakkebaek, N. E. (1983). Quantification of germ cells and seminiferous tubules by stereological examination of testicles from 50 boys who suffered from sudden death. *Int. J. Androl.* 6, 143–156

158. de Rooij, D. G., and Russell, L. D. (2000). All you wanted to know about spermatogonia but were afraid to ask. *J. Androl.* 21, 776–798.
159. von Ebner, V. (1871). Untersuchungen iiber den Bau der Samenkanalchen und die Entwicklung der Spermtozoiden bei den Sangentieren und beim Menschen. *Rollet Untersuch Inst. Physiol. Histol.* 2, 200–236.
160. von La Valette, and St. George, A. J. H. (1876). Ueber die Genese der Samenkorper. *Arch. Mikrosk. Anat.* 12, 797–825.
161. Oakberg, E. F. (1956). A description of spermiogenesis in the mouse and its use in analysis of the cycle of the seminiferous epithelium and germ cell renewal. *Am. J. Anat.* 99, 391–413.
162. Hochereau, M. T. (1968). Etude des divisions spermatogiales et du renou-vellement de la spermatogonie souche chez le taureau. *Int. Cong. Anim. Reprod.* 1, 149–152.
163. Clermont, Y. (1960). Cycle of the seminiferous epithelium of the guinea pig. A method for identification of stages. *Fertil. Steril.* 11, 563–573.
164. Clermont, Y. (1962). Quantitative analysis of spermatogenesis of the rat, a revised model for the renewal of spermatogonia. *Am. J. Anat.* 111, 111–129.
165. Clermont Y., and Bustos Obregon, E. (1968). Re-examination of spermatogonial renewal in the rat by means of seminiferous tubules mounted "in toto." *Am. J. Anat.* 122, 237–247.
166. Branca, A. (1924). Les canilicules testiculaires et spermatogenese de l'homme. *Arch. Zool. Exp. Genet.* 62, 53–252.
167. Roosen-Runge, E. C., and Barlow, F. D. (1953). Quantitative studies on human spermatogenesis. I. Spermatogonia. *Am. J. Anat.* 93, 143–169.
168. Clermont, Y. (1966). Renewal of spermatogonia in man. *Am. J. Anat.* 118, 509–524.
169. Clermont, Y. (1966). Spermatogenesis in man. A study of the spermatogonial population. *Fertil. Steril.* 17, 705–721.
170. Clermont, Y., and Le Blond, C. P. (1959). Differentiation and renewal of spermatogonia in the monkey *Macacus rhesus. Am. J. Anat.* 104, 237–272.
171. Clermont, Y. (1969). Two classes of spermatogonial stem cells in the monkey (*Ceropithecus aethiops*). *Am. J. Anat.* 126, 57–72.
172. Fawcett, D. W., Ito, S., and Slautterback, D. B. (1959). The occurrence of intercellular bridges in groups of cells exhibiting synchronous differentiation. *J. Biophys. Biochem. Cytol.* 5, 453–458.
173. Huckins, C. (1971). The spermatogonial stem cell population in adult rats. I. Their morphology, proliferation and maturation. *Anat. Rec.* 169, 533–558.
174. Dym, M., and Fawcett, D. W. (1971). Further observations on the numbers of spermatogonia, spermatocytes and spermatids connected by intercellular bridges in the mammalian testis. *Biol. Reprod.* 4, 195–215.
175. Dettin, L., Ravindranath, N., Hofmann, M. C., and Dym, M. (2003). Morphological characterization of the spermatogonial subtypes in the neonatal mouse testis. *Biol. Reprod.* 69, 1565–1571.
176. Chiarini-Garcia, H., and Russell, L. D. (2001). High-resolution light microscopic characterization of mouse spermatogonia. *Biol. Reprod.* 65, 1170–1178.
177. Parreira, G. G., Ogawa, T., Avarbock, M. R., Franca, L. R., Brinster, R. L., and Russell, L. D. (1998). Development of germ cell transplants. *Biol. Reprod.* 59, 1360–1370.
178. Nagano, M., Avarbock, M. R., and Brinster, R. L. (1999). Patterns and kinetics of mouse donor spermatogonial stem cell colonization in recipient testes. *Biol. Reprod.* 60, 1429–1436.
179. Meachem, S., von Schonfeldt, V., and Schlatt, S. (2001). Spermatogonia, stem cells with great potential. *Reproduction* 121, 825–834.
180. Shinohara, T., Orwig, K. E., Avarbock, M. R., and Brinster, R. L. (2001). Remodeling of the postnatal testis is accompanied by dramatic changes in stem cell number and niche accessibility. *Proc. Natl. Acad. Sci. U S A* 98, 6186–6191.
181. Shinohara, T., Orwig, K. E., Avarbock, M. R., and Brinster, R. L. (2002). Germ line stem cell competition in postnatal mouse testes. *Biol. Reprod.* 66, 1491–1497.
182. Ryu, B. Y., Orwig, K. E., Avarbock, M. R., and Brinster, R. L. (2003). Stem cell and niche development in the postnatal rat testis. *Dev. Biol.* 263, 253–263.
183. Schofield, R. (1978). The relationship between the spleen colony-forming cell and the hemopoietic stem cell. *Blood Cells* 4, 7–25.
184. Chiarini-Garcia, H., Raymer, A. M., and Russell, L. D. (2003). Non-random distribution of spermatogonia in rats, evidence of niches in the seminiferous tubules. *Reproduction* 126, 669–680.
185. Vilar, O., Perez del Cerro, and Mancini, R. E. (1962). The Sertoli cell as a "bridge cell" between the basal membrane and the germinal cells. *Exp. Cell. Res.* 27, 158–161.
186. Nicander, L., Abdel-Raouf, M., and Crabo, B. (1961). On the ultrastructure of the seminiferous tubules in bull calves. *Acta. Morphol. Neerl. Scand.* 4, 127–135.
187. Rowley, M. J., Berlin, J. D., and Heller, C. G. (1971). The Uutrastructure of four types of human spermatogonia. *Z. Zellforsch.* 112, 139–157.
188. Leeson, C. R. (1966). An electron microscopic study of cryptorchid and scrotal human testes, with special reference to pubertal maturation. *Invest. Urol.* 3, 498–511.
189. de Kretser, D. M. (1968). The fine structure of the immature human testis in hypogonadotrophic hypogonadism. *Virchows Arch. Zellpathol.* 1, 283–296.
190. Watson, M. L. (1952). Spermatogenesis in the albino rat as revealed by electron microscopy. *Biochim. Biophys. Acta.* 8, 369–374.
191. Moens, P. B., and Go, V. L. W. (1972). Intercellular bridges and division patterns of rat spermatogonia. *Z. Zellforsch.* 127, 201–208.
192. Fawcett, D. W. (1979). The cell biology of gametogenesis in the male. *Perspect. Biol. Med.* 2, S56–S73.
193. Caldwell, K. A., and Handel, M. A. (1991). Protamine sharing among postmeiotic spermatids. *Proc. Natl. Acad. Sci. U S A* 88, 2407–2411.
194. Tres, L. L., and Solari, A. J. (1968). The Ultrastructure of the nuclei and the behaviour of the sex chromosomes of human spermatogonia. *Z. Zellforsch.* 91, 75–89.
195. Schulze, W. (1978). Licht und elektronen-mikroskopische Studien an den A-Spermatogonien von Mannern mit intakter Spermatogenese und bei Patienten nach Behandlung mit Antiandrogenen. *Andrologia* 10, 307–320.
196. Schulze, W. (1978). Zum Problem der morphologischen Characterisierung von Spermatogonien typen beim Erwachsenen. *Verh. Anat. Ges.* 72, 539–540.
197. Chiarini-Garcia, H., and Russell, L. D. (2002). Characterization of mouse spermatogonia by transmission electron microscopy. *Reproduction* 123, 567–577.
198. Lubarsch, O. (1896). Ueber da Vorkommen krystallinischer und krystalloider Bildungen in den Zellen des menschilchen Hodens. *Virchows Arch. Pathol. Anat.* 145, 316–338.
199. Sohval, A. R., Suzuki, Y., Gabrilove, J. L., and Churg, J. (1971). Ultrastructure of crystalloids in spermatogonia and Sertoli cells of normal human testis. *J. Ultrastruct. Res.* 34, 83–102.
200. Nagano, T. (1969). The crystalloid of Lubarsch in the human spermatogonium. *Z. Zellforsch.* 97, 491–501.
201. Fawcett, D. W. (1972). Observations on cell differentiation and organelle continuity in spermatogenesis. In *Proceedings*

International Symposium on the Genetics of Spermatozoon (R. A. Beatty and S. Gluecksohn-Waelsch, Eds.), pp. 37–67. Bogtrykkeriet Forum, Copenhagen.
202. Brown, H. H. (1885). On spermatogenesis in the rat. *Q. J. Microsc. Sci.* 25, 343–370.
203. Winiwater, H. (1901). Recherche sur l'ovogenese et Forganogenese de l'ovaire des mamiferes. *Arch. Biol.* 17, 33–199.
204. Montgomery, T. H. (1901). A study of the chromosomes of the germ cells of the metazoa. *Trans. Am. Phil. Soc.* 20, 154–236.
205. Sutton, W. S. (1903). The chromosomes in heredity. *Biol. Bull.* 4, 231–251.
206. Swift, H. H. (1950). The deoxyribose nucleic acid content of animal nuclei. *Physiol. Zool.* 23, 169–200.
207. Wettstein, R., and Sotelo, J. R. (1967). Electron microscope serial reconstruction of the spermatocyte 1 nuclei at pachytene. *J. Microsc.* 6, 557–576.
208. Painter, T. S. (1923). Studies in mammalian spermatogenesis. II. The spermatogenesis of man. *J. Exp. Zool.* 37, 291–338.
209. Moses, M. J. (1956). Chromosomal structures in crayfish spermatocytes. *J. Biophys. Biochem. Cytol.* 12, 215–218.
210. Fawcett, D. W. (1956). The fine structure of chromosomes in the meiotic prophase of vertebrate spermatocytes. *J. Biophys. Biochem. Cytol.* 2, 403–406.
211. Woollam, D. H. M., and Ford, E. H. R. (1956). The fine structure of the mammalian chromosome, in meiotic prophase with special reference to the synaptinemal complex. *J. Anat. (Lond.)* 98, 163–173.
212. Moses, M. J. (1968). Synaptinemal complex. *Annu. Rev. Genet.* 2, 363–412.
213. Comings, D. E., and Okada, T. A. (1971). Fine structure of the synaptonemal complex. *Exp. Cell. Res.* 65, 104–116.
214. Dresser, M. E., and Moses, M. J. (1979). Silver staining of synaptonemal complexes in surface spreads for light and electron microscopy. *Exp. Cell. Res.* 121, 416–419.
215. Moses, M. J, and Solari, A. J. (1976). Positive contrast staining and protected drying of surface spreads, electron microscopy of the synaptinemal complex by a new method. *J. Ultrastruct. Res.* 54, 109–114.
216. Heyting, C., Dietrich, A. J., Redeker, E. J., and Vink, A. C. (1985). Structure and composition of synaptonemal complexes isolated from rat spermatocytes. *Eur. J. Cell. Biol.* 36, 307–314.
217. Heyting, C., Dettmers, R. J., Dietrich, A. J. J., Redeker, E. J. W., and Vink, A. C. G. (1988). Two major components of synaptonemal complexes are specific for meiotic prophase nuclei. *Chromosoma* 96, 325–332.
218. Offenberg, H. H., Dietrich, A. J. J., and Heyting, C. (1991). Tissue distribution of two major components of synaptonemal complexes of the rat. *Chromosoma* 101, 83–91.
219. Meuwissen, R. L. J., Offenberg, H. H., Dietrich, A. J. J., Riesewijk, A., van Lersel, M., and Heyting, C. A. (1992). coiled-coil related protein specific for synapsed regions of meiotic prophase chromosomes. *EMBO J.* 11, 5091–5100.
220. Sachs, L. (1954). Sex linkage and sex chromosomes in man. *Ann. Eugen. (Lond.)* 18, 255–261.
221. Solari, A. J. (1964). The morphology and ultrastructure of the sex vesicle in the mouse. *Exp. Cell. Res.* 36, 160–168.
222. Solari, A. J., and Tres, L. (1967). The ultrastructure of the human sex vesicle. *Chromosoma* 22, 16–31.
223. Solari, A. J. (1974). The behaviour of the XY pair in mammals. *Int. Rev. Cytol.* 38, 273–317.
224. Moses, M. J., Counce, S. J., and Paulson, D. F. (1975). Synaptonemal complex complement of man in spreads of spermatocytes with details of the sex chromosome pair. *Science* 187, 363–365.
225. Chandley, A. C., Gaetz, P., Hargreave, T. B., Joseph, A. M., and Speed, R. M. (1984). On the nature and extent of XY pairing at meiotic prophase in man. *Cytogenet. Cell. Genet.* 38, 241–247.
226. Dickins, R. A., Frew, I. J., House, C. M., O'Bryan, M. K., Holloway, A. J., Haviv, I., Traficante, N., de Kretser, D. M., and Bowtell, D. D. (2002). The ubiquitin ligase component Siah 1a is required for completion of meiosis I in male mice. *Mol. Cell. Biol.* 22, 2294–2303.
227. Libby, B. J., De La Fuente, R., O'Brien, M. J., Wigglesworth, K., Cobb, J., Inselman, A., Eaker, S., Handel, M. A., Eppig, J. J., and Schimenti, J. C. (2002). The mouse meiotic mutation disrupts chromosome synapsis with sexually dimorphic consequences for meiotic progression. *Dev. Biol.* 242, 174–187.
228. Schultz, M. C., Hermo, L., and Le Blond, C. P. (1984). Structure, development and cytochemical properties of the nucleolus-associated "round body" in rat spermatocytes and early spermatids. *Am. J. Anat.* 171, 41–57.
229. Nicander, L., and Ploen, L. (1969). Fine structure of spermatogonia and primary spermatocytes in rabbits. *Z. Zellforsch.* 99, 221–234.
230. Suarez-Quian, C. A., Qu, A., Jelesoff, N., and Dym, M. (1991). The Golgi apparatus of rat pachytene spermatocytes during spermatogenesis. *Anat. Rec.* 229, 16–26.
231. Andre, J. (1962). Contribution a la connaissance du chondriome. Etude de ses modifications la spermatogenese. *Ultrastruct. Res.* Suppl 3, 1–185.
232. Russell, L. D., and Frank, B. (1978). Ultrastructural characterization of nuage in spermatocytes in the rat testis. *Anat. Rec.* 190, 79–98.
233. Fawcett, D. W., Eddy, E., and Phillips, D. M. (1970). Observations on the fine structure and relationships of the chromatoid body in mammalian spermatogenesis. *Biol. Reprod.* 2, 129–153.
234. Rehbun, L. I. (1961). Some electron microscope observations on membranous basophilic elements of invertebrate eggs. *J. Ultrastruct. Res.* 5, 208–225.
235. Smith, P. E., and Berlin, J. D. (1977). Cytoplasmic annulate lamellae in human spermatogenesis. *Cell. Tissue Res.* 176, 235–242.
236. Chemes, H. E., Fawcett, D. W., and Dym, M. (1978).Unusual features of the nuclear envelope in human spermatogenic cells. *Anat. Rec.* 192, 495–512.
237. Russell, L. D. (1977). Desmosome-like junctions between Sertoli and germ cells in the rat testis. *Am. J. Anat.* 148, 301–312.
238. Kaya, M., and Harrison, R. G. (1976). The ultrastructural relationships between Sertoli cells and spermatogenic cells in the rat. *J. Anat.* 121, 279–290.
239. Russell, L. D. (1978). The blood-testis barrier and its formation relative to spermatocyte maturation in the adult rat, a lanthanum study. *Anat. Rec.* 190, 99–112.
240. von Ebner, V. (1888).Zur Spermatogenese bei den Saugthieren. *Arch. Mikr. Anat.* 31, 236–292.
241. Heller, C. G., and Clermont, Y. (1964). Kinetics of the germinal epithelium in man. *Rec. Prog. Horm. Res.* 20, 545–575.
242. Montgomery, T. H. (1912). Human spermatogenesis, spermatocytes and spermiogenesis. A study in inheritance. *J. Acad. Natl. Sci. Phil.* 15, 1–22.
243. Nicander, L. (1967). An electron microscopical study of cell contacts in the seminiferous tubules of some mammals. *Z. Zellforsch.* 83, 375–397.
244. Holstein, A. F, and Roosen-Runge, E. C. (1981). *Atlas of Human Spermatogenesis*, pp. 1–224. Grosse Verlag, Berlin.
245. Gardner, P. H., and Holyoke, E. A. (1964). Fine structure of the seminiferous tubule of Swiss mouse. I. The limiting

membrane, Sertoli cell, spermatogonia and spermatocytes. *Anat. Rec.* 150, 391–404.
246. von La Valette, and St. George, A. J. H. (1885). Spermatologische Beitrage. *Arch. Mikr. Anat.* 25, 581–593.
247. Meves, F. (1898). Über das Verhalten der Zentralkorper bei der histogenese der Samenfaden von Mensch und Ratte. *Arch. Mikr. Anat.* 54, 329–402.
248. Meves, F. (1898). Zur Entstehung der Achsenfaden menschlicher Sper-matozoen. *Anat. Anz.* 14, 168–170.
249. Gatenby, J. B., and Beams, H. W. (1936). The cytoplasmic inclusions in the spermatogenesis of man. *Q. J. Microsc. Sci.* 78, 1–33.
250. Le Blond, C. P., and Clermont, Y. (1952). Spermiogenesis of rat, mouse, hamster and guinea pig as revealed by the "periodic acid-fuschin sulphurous acid" technique. *Am. J. Anat.* 90, 167–206.
251. Le Blond, C. P., and Clermont, Y. (1952). Definition of the stages of the cycle of the seminiferous epithelium in the rat. *Am. N Y Acad. Sci.* 55, 548–573.
252. Clermont, Y. (1963). The cycle of the seminiferous epithelium in man. *Am. J. Anat.* 112, 35–51.
253. Daoust, R., and Clermont, Y. (1955). Distribution of nucleic acids in germ cells during the cycle of the seminiferous epithelium in the rat. *Am. J. Anat.* 96, 255–283.
254. Bowen, R. H. (1922). On the idiosome, Golgi apparatus and acrosome in the male germ cells. *Anat. Rec.* 24, 158–180.
255. Bowen, R. H. (1924). On the acrosome of the animal sperm. *Anat. Rec.* 28, 1–13.
256. Burgos, M. H., and Fawcett, D. W. (1955). Studies of the fine structure of the mammalian testis. I. Differentiation of the spermatids in the cat (*Felis domestica*). *J. Biophys. Biochem. Cytol.* 2, 223–240.
257. Brokelmann, J. (1963). Fine structure of germ cells and Sertoli cells during the cycle of the seminiferous epithelium in the rat. *Z. Zellforsch.* 59, 820–850.
258. Gardner, P. (1966). Fine structure of the seminiferous epithelium of the Swiss mouse. The spermatid. *Anat. Rec.* 155, 235–250.
259. Holstein, A. F. (1976). Ultrastructural observations on the differentiation of spermatids in man. *Andrologia* 8, 157–165.
260. Hermo, L., Rambourg, A., and Clermont, Y. (1980). Three-dimensional architecture of the cortical region of the Golgi apparatus in the rat spermatids. *Am. J. Anat.* 157, 357–373.
261. Clermont, Y., and Tang, X. M. (1985). Glycoprotein synthesis in the Golgi apparatus of spermatids during spermiogenesis of the rat. *Anat. Rec.* 213, 33–43.
262. Sandoz, D. (1972). Etude autoradiographique de Incorporation *in vitro* de galactose-3H dans les spermatides de souris. *J. Microsc.* 15, 403–408.
263. Lalli, M. F., Tang, X. M., and Clermont, Y. (1984). Glycoprotein synthesis in Sertoli cells during the cycle of the seminiferous epithelium of adult rats. A radioautographic study. *Biol. Reprod.* 30, 493–505.
264. Fawcett, D. W. (1975). The mammalian spermatozoon. *Dev. Biol.* 44, 394–436.
265. Pederson, H. (1972). The postacrosomal region of man and *Macaco artoides*. *J. Ultrastruct. Res.* 40, 366–377.
266. Pederson, H. (1972). Further observations of the fine structure of the human spermatozoon. *Z. Zellforsch.* 123, 305–315.
267. Bedford, J. M. (1968). Ultrastructural changes in the sperm head during fertilization in the rabbit. *Am. J. Anat.* 123, 329–358.
268. Stefanini, M., Oura, C., and Zamboni, L. (1969). Ultrastructure of fertilization in the mouse. 2. Penetration of sperm into the ovum. *J. Submicrosc. Cytol.* 1, 1–23.
269. Fawcett, D. W., and Hollenberg, R. D. (1963). Changes in the acrosome of guinea pig spermatozoa during passage through the epididymis. *Z. Zellforsch.* 60, 276–292.
270. Harding, H. R., Carrick, F. N., and Shorey, C. D. (1976). Ultrastructural changes in spermatozoa of the brush-tailed possum *Trichosurus vulpecula* (Marsupialia) during epididymal transit. Part II. The acrosome. *Cell. Tissue Res.* 171, 61–73.
271. Allison, A. C., and Hartree, E. F. (1970). Lysosomal enzymes in the acrosome and their possible role in fertilization. *J. Reprod. Fertil.* 21, 501–515.
272. Chang, M. C., and Hunter, R. H. F. (1975). Capacitation of mammalian sperm, biological and experimental aspects. In *Handbook of Physiology* (D. W. Hamilton and R. O. Greep, Eds.), Section 7, Vol. 5, pp. 339–352. Williams & Wilkins, Baltimore.
273. Bedford, J. M. (1967). Observations on the fine structure of spermatozoa of the bush baby (*Galago senegalensis*), the African green monkey (*Cercopithecus aethiops*) and man. *Am. J. Anat.* 121, 443–460.
274. Clermont, Y., Einberg, E., Le Blond, C. P., and Wagner, S. (1955). The perforatorium, an extension of the nuclear membrane of the rat spermatozoon. *Anat. Rec.* 121, 1–12.
275. Lalli, M. F., and Clermont, Y. (1981). Structural changes of the head components of the rat spermatid during spermatogenesis. *Am. J. Anat.* 160, 419–434.
276. Burgos, M. H., and Fawcett, D. W. (1956). An electron microscope study of spermatid differentiation in the toad, *Bufo arenarium hensel*. *J. Biophys. Biochem. Cytol.* 2, 223–240.
277. Bedford, J. M. (1964). Fine structure of the sperm head in ejaculate and uterine spermatozoa of the rabbit. *Reprod. Fertil.* 7, 221–228.
278. Bane, A., and Nicander, L. (1963). The structure and formation of the perforatorium in mammalian spermatozoa. *Int. J. Fertil.* 8, 865–869.
279. Oko, R., and Clermont, Y. (1988). Isolation, structure and protein composition of the perforatorium of rat. *Biol. Reprod.* 39, 673–687.
280. Oko, R., Moussakova, L., and Clermont, Y. (1990). Regional differences in composition of the perforatorium and outer periacrosomal layer of the rat spermatozoon as revealed by immunocytochemistry. *Am. J. Anat.* 188, 64–73.
281. Austin, C. R., and Bishop, M. W. H. (1958). Role of the rodent acrosome and perforatorium in fertilization. *Proc. R. Soc. Lond. Biol.* 149, 234–240.
282. Welch, J. E., and O'Rand, M. G. (1985). Identification and distribution of actin in spermatogenic cells and spermatozoa of the rabbit. *Dev. Biol.* 109, 411–417.
283. Russell, L. D., Weber, J. E., and Vogl, A. W. (1986). Characterisation of filaments within the subacrosomal space of rat spermatids during spermiogenesis. *Tissue Cell.* 18, 887–898.
284. Fouquet, J. P., and Kahn, M. L. (1992). Species specific isolation of actin in mammalian spermatozoa, fact or artifact. *Micro. Res. Tech.* 20, 251–258.
285. Holstein, A. F., Roosen-Runge, E. C., and Schirren, C. (1988). *Illustrated Pathology of Human Spermatogenesis*, pp. 1–278. Grosse Verlag, Berlin.
286. Kang-Decker, N., Mantchev, G. T., Juneja, S. C., Mc Niven, M. A., and van Deursen, J. M. A. (2001). Lack of acrosome formation in Hrb-deficient mice. *Science* 294, 1531–1533.
287. Xu, X., Toselli, P. A., Russel, L. D., and Seldin, D. C. (1999). Globozoospermia in mice lacking the casein kinase II a catalytic subunit. *Nat. Genet.* 23, 118–121.
288. Horstmann, E. (1961). Electronemikroskopische Untersuchungen zur Spermiohistogenese beim menschen. *Z. Zellforsch.* 54, 68–89.
289. de Kretser, D. M. (1969). Ultrastructural features of human spermiogenesis. *Z. Zellforsch.* 98, 477–505.
290. Meistrich, M. (1993). Nuclear morphogenesis during spermiogenesis. In *Molecular Biology of Male Reproduction* (D. M. de Kretser, Ed.). Academic Press, New York, pp. 67–97.

291. Gledhill, B. L., Gledhill, M. P., Rigler, R., and Ringertz, N. R. (1966). Changes in deoxyribonucleoprotein during spermatogenesis in the bull. *Exp. Cell. Res.* 41, 652–665.
292. Vaughn, J. C. (1966). The relationship of the sphere chromatophile to the fate of displaced histones following histone transition in rat spermiogenesis. *J. Cell. Biol.* 31, 257–278.
293. Marushinge, Y., and Marusinghe, K. (1975). Transformation of sperm histone during formation and maturation of rat spermatozoa. *J. Biol. Chem.* 250, 39–45.
294. Meistrich, M. L., Brock, W. A., Grimes, S. R., Platz, R. D., and Hnilica, L. S. (1978). Nuclear protein transitions during spermatogenesis. *Fed. Proc.* 37, 2522–2525.
295. Bellve, A. R., Anderson, E., and Hanley-Bowdoin, I. (1975). Synthesis and amino acid composition of basic proteins in mammalian sperm nuclei. *Dev. Biol.* 47, 349–365.
296. Kistler W. S., Nayes, C., Hsu, R., and Heinrikson, R. L. (1975). The amino acid sequence of a testis-specific basic protein that is associated with spermatogenesis. *Biol. Chem.* 250, 1847–1853.
297. Loir, M., and Lanneau, M. (1978). Partial characterization of ram spermatidal basic nuclear proteins. *Biochem. Biophys. Res. Commun.* 80, 975–982.
298. Balhorn, R. (1989). Mammalian protamines. In *Molecular Biology of Chromosome Function* (K. W. Adolph, Ed.), pp. 366–395. Springer Verlag, New York.
299. Meistrich, M. (1989). Histone and basic nuclear protein transections in mammalian spermatogenesis, In *Histones and Other Basic Nuclear Proteins* (L. S. Hnilica, G. S. Stein, and J. L. Stein, Eds.), pp. 165–182. CRC Press, Boca Raton, FL.
300. Alfonso, P. J., and Kistler, W. S. (1993). Immunohistochemical localization of spermatid nuclear transition protein 2 (TP2) in the testes of rats and mice. *Biol. Reprod.* 48, 522–529.
301. Meistrich, M. L., Trostle-Weige, P. K., Lin, R., Bhatnager, Y. M., and Allis, C. D. (1992). Highly acetylated H_4 is associated with histone displacement in rat spermatids. *Mol. Reprod. Dev.* 31, 170–181.
302. Oko, R. J., Jando, V., Wagner, C. L., Kistler, W. S., and Hermo, L. S. (1996). Chromatin reorganization in rat spermatids during the disappearance of testis-specific histone, H1t, and the appearance of transition proteins TP1 and TP2. *Biol. Reprod.* 54, 1141–1157.
303. Fawcett, D. W., Anderson, W. A., and Phillips, D. M. (1971). Morphogenetic factors influencing the shape of the sperm head. *Dev. Biol.* 26, 220–251.
304. Bellve, A. R. (1979). The molecular biology of mammalian spermatogenesis. In *Oxford Reviews in Reproductive Biology* (C. A. Finn, Ed.), Vol. 1, pp. 159–261. Clarendon Press, Oxford.
305. Wu, J. Y., Ribar, T. J., Cummings, D. E., Burton, K. A., McKnight, G. S., and Means, A. R. (2000). Spermiogenesis and exchange of basic nuclear proteins are impaired in male germ cells lacking Camk4. *Nat. Genet.* 25, 448–452.
306. Cho, C., Willis, W. D., Goulding, E. H., Jung-Ha, H., Choi, Y. C., Hecht, N. B., and Eddy, E. M. (2001).Haploinsufficiency of protamine-1 or -2 causes infertility in mice. *Nat. Genet.* 28, 82–86.
307. Balhorn, R., Reed, S., and Tanphaichitr, N. (1988). Aberrant protamine1/protamine 2 ratios in sperm of infertile males. *Experentia* 44, 52–55.
308. Yu, Y. E., Zhang, Y., Unni, E., Shirley, C. R., Deng, J. M., Russell, L. D., Weil, M. M., Behringer, R. R., and Meistrich, M. L. (2000). Abnormal spermatogenesis and reduced fertility in transition nuclear protein 1-deficient mice. *Proc. Natl. Acad. Sci. U S A* 97, 4683–4688.
309. Zhao, M., Shirley, C. R., Yu, Y. E., Mohapatra, B., Zhang, Y., Unni, E, Deng, J. M., Arango, N. A., Terry, N. H., Weil, M. M., Russell, L. D., Behringer, R. R., and Meistrich, M. L. (2001). Targeted disruption of the transition protein 2 gene affects sperm chromatin structure and reduces fertility in mice. *Mol. Cell. Biol.* 21, 7243–7255.
310. Shirley, C. R., Hayashi, S., Mounsey, S., Yanagimachi, R., and Meistrich, M. L. (2004). Abnormalities and reduced reproductive potential of sperm from tnp1 and tnp2-null double mutant mice. *Biol. Reprod.* 71, 1220–1229.
311. Meistrich, M. L., Mohapatra, B., Shirley, C. R., and Zhao, M. (2003). Roles of transition nuclear proteins in spermiogenesis. *Chromosoma* 48, 3–8.
312. Fawcett, D. W. (1965). The anatomy of the mammalian spermatozoon with particular reference to the guinea pig. *Z. Zellforsch.* 67, 279–296.
313. McIntosh, J. R., and Porter, K. R. (1967). Microtubules in the spermatids of the domestic fowl. *J. Cell. Biol.* 35, 153–173.
314. Clark, A. W. (1967). Some aspects of spermiogenesis in a lizard. *Am. J. Anat.* 121, 369–400.
315. Beatty, R. A. (1970). The genetics of the mammalian gamete. *Biol. Rev.* 45, 73–119.
316. Meistrich, M. L., Trostle-Weige, P. K., and Russell, L. D. (1990). Abnormal manchette development in spermatids of *azh/azh* mutant mice. *Am. J. Anat.* 188, 74–86.
317. Russell, L. D., Russell, J. A., MacGregor, G. R., and Meistrich, M. L. (1991). Linkage of manchette microtubules to the nuclear envelope and observations of the role of the manchette in nuclear shaping during spermiogenesis in rodents. *Am. J. Anat.* 192, 97–120.
318. Kierszenbaum, A. L., Gil, M., Rivkin, E., and Tres, L. L. (2002). Ran, a GTP-binding protein involved in nucleocytoplasmic transport and microtubule nucleation, relocates from the manchette to the centrosome region during rat spermiogenesis. *Mol. Reprod. Dev.* 63, 131–140.
319. Yang, W. X., and Sperry, A. O. (2003). C-terminal kinesin motor KIGC1 participates in acrosome biogenesis and vesicle transport. *Biol. Reprod.* 69, 1719–1729.
320. Navolanic, P. M., and Sperry, A. O. (2000). Identification of isoforms of a mitotic motor in mammalian spermatogenesis. *Biol. Reprod.* 62, 1360–1369.
321. Yoshida, T., Ioshii, S. O., Imanaka-Yoshida, K., and Izutsu, K. (1994). Association of cytoplasmic dynein with manchette microtubules and spermatid nuclear envelope during spermatogenesis in rats. *J. Cell. Sci.* 107, 625–633.
322. Kierszenbaum, A. L. (2002). Intramanchette transport (IMT), managing the making of the spermatid head, centrosome, and tail. *Mol. Reprod. Dev.* 63, 1–4.
323. Tovich, P. R., Sutovsky, P., and Oko, R. J. (2004). Novel aspect of perinuclear theca assembly revealed by immunolocalization of non-nuclear somatic histones during bovine spermatogenesis. *Biol. Reprod.* 71, 1182–1194.
324. Mujica, A., Navarro-Garcia, F., Hernandez-Gonzalez, E. O., and Lourdes Juarez-Mosqueda, M. (2003). Perinuclear theca during spermatozoa maturation leading to fertilization. *Microsc. Res. Tech.* 61, 76–87.
325. Kotaja, N., De Cesare, D., Macho, B., Monace, L., Brancorsini, S., Goossens, E., Tournaye, H., Gansmuller, A., and Sassone-Corsi, P. (2004). Abnormal sperm in mice with targeted deletion of the act (activator of cAMP-responsive element modulator in testis) gene. *Proc. Natl. Acad. Sci. U S A* 101, 10620–10625.
326. Foulkes, N. S., Mellstrom, B., Benusiglio, E., and Sassone-Corsi, P. (1996). Developmental switch of CREM function during spermatogenesis from antagonist to activator. *Nature* 355, 80–84.
327. Franklin, L. E. (1968). Formation of the redundant nuclear envelope in monkey spermatids. *Anat. Rec.* 161, 149–162.
328. Anberg, A. (1957). The ultrastructure of the human spermatozoon. *Acta. Obstet. Gynaecol.* XXXVI Suppl 2, 1–133.
329. Fawcett, D. W. (1958). The structure of the mammalian spermatozoon. *Int. Rev. Cytol.* 7, 195–234.
330. Sapsford, C. S., Rae, C. A., and Cleland, K. W. (1967). Ultrastructural studies on spermatids and Sertoli cells

during early spermiogenesis in the Bandicoot *Perameles nasuta geoffroy* (Marsupialia). *Aust. J. Zool.* 15, 881–909.
331. Fawcett, D. W., and Phillips, D. M. (1969). The fine structure and development of the neck region of the mammalian spermatozoon. *Anat. Rec.* 165, 153–184.
332. Grimstone, A. V., and Klug, A. (1966). Observations on the substructure of flagella fibres. *J. Cell. Sci.* 1, 351 –362.
333. Warner, F. D. (1970). New observations on flagella fine structure, the relationship between matrix structure and the microtubule component of the axoneme. *J. Cell. Biol.* 47, 159–182.
334. Tilney, L., Bryan, J., and Bush, D. J., Fujiwara, K., Mooseker, M. S., Murphy, D. B. and Snyder, D. H. (1973). Microtubules, evidence for 13 protofilaments. *J. Cell. Biol.* 59, 267–275.
335. Gibbons, I. R. (1966). Studies on the ATP-base activity of 14S and 30S dynein from cilia of tetrahymena. *J. Biol. Chem.* 241, 5590–5596.
336. Stephens, R. E. (1970). Isolation of nexin-the linkage protein responsible for the maintenance of the 9-fold configuration of flagellar axonemes. *Biol. Bull.* 139, 438–442.
337. Hopkins, J. M. (1970). Subsidiary components of the flagellar of *Chlamydomonas reinhardtii*. *J. Cell. Sci.* 7, 823–839.
338. Afzelius, B. A. (1976). A human syndrome caused by immotile cilia. *Science* 193, 317–319.
339. Sturgess, J. M., Chao, J., Wong, J., Aspin, N., and Turner, J. A. P. (1979). Cilia with defective radial spokes -a cause of human respiratory disease. *N. Engl. J. Med.* 300, 53–56.
340. Chodhari, R., Mitchison, H. M., and Meeks, M. (2004). Cilia, primary ciliary dyskinesia and molecular genetics. *Paediatr. Respir. Rev.* 5, 69–76.
341. Geremek, M., and Witt, M. (2004). Primary ciliary dyskinesia, genes, candidate genes and chromosomal regions. *J. Appl. Genet.* 45, 347–361.
342. Gerton, G. L., and Millette, C. G. (1984). Generation of flagella by cultured mouse spermatids. *Cell. Biol.* 98, 619–628.
343. Woolley, D. M., and Fawcett, D. W. (1973). The degeneration and disappearance of the centrioles during the development of the rat spermatozoon. *Anat. Rec.* 177, 289–301.
344. Fawcett, D. W. (1970). A comparative view of sperm ultrastructure. *Biol. Reprod.* 2, 90–127.
345. Fawcett D. W., and Ito, S. (1965). The fine structure of bat spermatozoa. *Am. J. Anat.* 116, 567–610.
346. Anderson, W. A., and Personne, P. (1967). The fine structure of the neck region of spermatozoa of *Helix aspersa*. *J. Microsc.* 6, 1033–1042.
347. Sud, B. N. (1961). Morphological and histochemical studies of the chromatoid body and related elements in spermatogenesis of the rat. *Q. J. Microsc. Sci.* 102, 495–505.
348. Maillet, P. L., and Gouranton, J. (1965). Sur Fexpulsion de 1'acide ribonucleique nucleaire par les spermatides de *Philaenus spumarius* L (*Hamopttera cucopidae*). *Comp. Rendue.* 261, 1417–1419.
349. Eddy, E. M. (1970). Cytochemical observations on the chromatoid body of the male germ cell. *Biol. Reprod.* 2, 114–119.
350. Walt, H., and Armbruster, B. L. (1984). Actin and RNA are components of the chromatoid body in spermatids of the rat. *Cell. Tissue Res.* 236, 487–490.
351. Irons, M. J., and Clermont, Y. (1982). Formation of the outer dense fibres during spermiogenesis in the rat. *Anat. Rec.* 202, 463–471.
352. O'Brien, D. A., and Bellve, A. R. (1980). Protein constituents of the mouse spermatozoon. II. Temporal synthesis during spermatogenesis. *Dev. Biol.* 75, 405–418.
353. Oko, R. (1988). Comparative analysis of proteins from the fibrous sheath and outer dense fibres of rat spermatozoa. *Biol. Reprod.* 39, 169–182.
354. Oko, R., and Clermont, Y. (1989). Light microscopic immunocytochemical study of fibrous sheath and other dense fibre formation in the rat spermatid. *Anat. Rec.* 225, 46–55.
355. Clermont, Y., Oko, R., and Hermo, L. (1990). Immunocytochemical localization of proteins utilized in the formation of outer dense fibres and fibrous sheath in rat spermatids, an electron microscopic study. *Anat. Rec.* 227, 447–457.
356. Schalles, U., Shao, X., van der Hoorn, F. A., and Oko, R. (1998). Developmental expression of the 84-kDa ODF sperm protein, localization to both the cortex and medulla of outer dense fibers and to the connecting piece. *Dev. Biol.* 199, 250–260.
357. O'Bryan, M. K., Loveland, K. L., Herzfeld, G., McFarlane, J. R., Hearn, M. T., and de Kretser, D. M. (1998). Identification of a rat testis-specific gene encoding a potential oter dense fibre protein. *Mol. Reprod. Dev.* 50, 313–322.
358. Kim, Y. H., McFarlane, J. R., O'Bryan, M. K., Almahbobi, G., Temple-Smith, P. D., and de Kretser, D. M. (1999). Isolation and partial characterization of rat outer dense fibres and comparison with rabbit and human spermatozoa using a polyclonal antiserum. *J. Reprod. Fertil.* 116, 345–353.
359. Morales, C. R., Oko, R., and Clermont, Y. (1994). Molecular cloning and developmental expression of an mRNA encoding the 27 kDa outer dense fiber protein of rat spermatozoa. *Mol. Reprod. Dev.* 37, 229–240.
360. Shao, X., and van der Hoorn, F. A. (1996). Self-interaction of the major 27-kilodalton outer dense fiber protein is in part mediated by a leucine zipper domain in the rat. *Biol. Reprod.* 55, 1343–1350.
361. Shao, X., Tarnasky, H. A., Schalles, U., Oko, R., and van der Hoorn, F. A. (1997). Interactional cloning of the 84-kDa major outer dense fiber protein Odf84. Leucine zippers mediate associations of Odf84 and Odf27. *J. Biol. Chem.* 272, 6105–6113.
362. Shao, X., Tarnasky, H. A., Lee, J. P., Oko, R., and van der Hoorn, F. A. (1999). Spag4, a novel sperm protein, binds outer dense-fiber protein Odf1 and localizes to microtubules of manchette and axoneme. *Dev. Biol.* 211, 109–123.
363. Shao, X., Xue, J., and van der Hoorn, F. A. (2001). Testicular protein Spag5 has similarity to mitotic spindle protein Deepest and binds outer dense fiber protein Odf1. *Mol. Reprod. Dev.* 59, 410–416.
364. Rosales, J. L., Lee, B. C., Modarressi, M., Sarker, K. P., Lee, K. Y., Jeong, Y. G, Oko, R., and Lee, K. Y. (2004). Outer dense fibers serve as a functional target for Cdk5.p35 in the developing sperm tail. Outer dense fibers serve as a functional target for Cdk5.p35 in the developing sperm tail. *J. Biol. Chem.* 279, 1224–1232.
365. Hinsch, K. D., De Pinto, V., Aires, V. A., Schneider, X., Messina, A., and Hinsch, E. (2004). Voltage-dependent anion-selective channels VDAC2 and VDAC3 are abundant proteins in bovine outer dense fibers, a cytoskeletal component of the sperm flagellum. *J. Biol. Chem.* 279, 15281–15288
366. Tres, L. L., and Kierszenbaum, A. L. (1996). Sak57, an acid keratin initially present in the spermatid manchette before becoming a component of para-axonemal structures of the developing tail. *Mol. Reprod. Dev.* 44, 395–407.
367. Bradfield, J. R. G. (1955). Fibre patterns in animal flagella and cilia. *Symp. Soc. Exp. Biol.* 9, 306–322.
368. Wartenberg, H., and Holstein, A. F. (1975). Morphology of the "spindle-shaped body" in the developing tail of human spermatids. *Cell. Tissue Res.* 159, 435–443.
369. Nicander, L. (1962). Development of the fibrous sheath of the mammalian sperm tail. In *Proceedings 5th International Conference for Electron Microscopy*, M4. Academic Press, New York.

370. Sapsford, C. S., Rae, C. A., and Cleland, K. W. (1970). Ultrastructural studies on the development and form of the principal piece sheath of the bandicoot spermatozoon. *Aust. J. Zool.* 18, 21–48.

371. Sotelo, J. R., and Trujillo-Cenoz, O. (1958). Electron microscope study of the kinetic apparatus in animal sperm cells. *Z. Zellforsch.* 565–601.

372. Illison, L. *Studies on the Genetics and Development of Spermatozoon Head Shape in the Mouse.* PhD thesis, University of Sydney, Australia.

373. Irons, M. J., and Clermont, Y. (1982). Kinetics of fibrous sheath formation in the rat spermatid. *Am. J. Anat.* 165, 121–130.

374. Carrera, A., Gerton, G. L., and Moss, S. B. (1994). The major fibrous sheath polypeptide of mouse sperm, structural and functional similarities to the A-kinase anchoring proteins. *Dev. Biol.* 165, 272–284.

375. Fulcher, K. D., Mori, C., Welch, J. E., O'Brien, D. A., Klapper, D. G., and Eddy, E. M. (1995). Characterization of Fsc1 cDNA for a mouse sperm fibrous sheath component. *Biol. Reprod.* 52, 41–49.

376. Miki, K., and Eddy, E. M. (1998). Identification of tethering domains for protein kinase A type Ialpha regulatory subunits on sperm fibrous sheath FSC1. *J. Biol. Chem.* 273, 34384–34390.

377. Vijayaraghavan, S., Liberty, G. A., Mohan, J., Winfrey, V. P., Olson, G. E., and Carr, D. W. (1999). Isolation and molecular characterization of AKAP110, a novel, sperm-specific protein kinase A-anchoring protein. *Mol. Endocrinol.* 13, 705–717.

378. Mandal, A., Naaby-Hansen, S., Wolkowicz, M. J., Klotz, K., Shetty, J., Retief, J. D., Coonrod, S. A., Kinter, M., Sherman, N., Cesar, F., Flickinger, C. J., and Herr, J. C. (1999). FSP95, a testis-specific 95-kilodalton fibrous sheath antigen that undergoes tyrosine phosphorylation in capacitated human spermatozoa. *Biol. Reprod.* 61, 1184–1197.

379. Mei, X., Singh, I. S., Erlichman, J., and Orr, G. A. (1997). Cloning and characterization of a testis-specific, developmentally regulated A-kinase-anchoring protein (TAKAP-80) present on the fibrous sheath of rat sperm. *Eur. J. Biochem.* 246, 425–432.

380 Carr, D. W., Fujita, A., Stentz, C. L., Liberty, G. A., Olson, G. E., and Narumiya, S. (2001). Identification of sperm-specific proteins that interact with A-kinase anchoring proteins in a manner similar to the type II regulatory subunit of PKA. *J. Biol. Chem.* 276, 17332–17338

381. Fujita, A., Nakamura, K., Kato, T., Watanabe, N., Ishizaki, T., Kimura, K., Mizoguchi, A., and Narumiya, S. (2000). Ropporin, a sperm-specific binding protein of rhophilin, that is localized in the fibrous sheath of sperm flagella that is localized in the fibrous sheath of sperm flagella. *J. Cell. Sci.* 113, 103–112.

382. Fenderson, B. A., Toshimori, K., Muller, C. H., Lane, T. F., and Eddy, E. M. (1988). Identification of a protein in the fibrous sheath of the sperm flagellum. *Biol. Reprod.* 38, 345–357.

383. Mori, C., Welch, J. E., Fulcher, K. D., O'Brien, D. A., and Eddy, E. M. (1993). Unique hexokinase messenger ribonucleic acids lacking the porin-binding domain are developmentally expressed in mouse spermatogenic cells. *Biol. Reprod.* 49, 191–203.

384. Hoshi, K., Tsukikawa, S., and Sato, A. (1991). Importance of Ca2+, K+ and glucose in the medium for sperm penetration through the human zona pellucida. *Tohoku. J. Exp. Med.* 165, 99–104.

385. Jensen, O. S. (1887). Untersuchungen über die Samenkorper du Sauge-thiere, Vogel und Amphibien. I. Saugethiere. *Arch. Mikr. Anat.* 30, 379–425.

386. Umer, F., and Sakkas, D. (1996). Glucose participates in sperm-oocyte fusion in the mouse. *Biol. Reprod.* 55, 917–922.

387. Yasuzumi, G. (1956). Spermatogenesis in animals as revealed by electron microscopy. I. Formation and submicroscopic structure of the middle-piece of the albino rat. *Biophys. Biochem. Cytol.* 445–449.

388. Reed, R. I., and Reed, B. P. (1948). Comparative study of human and bovine sperm by electron microscopy. *Anat. Rec.* 100, 1–7.

389. Schultz-Larsen, J. (1958). The morphology of the human sperm. *Acta. Pathol. Microbiol. Scand.* Suppl 128, 1–121.

390. Phillips, D. M. (1970). Development of spermatozoa in the wooly opossum with special reference to the shaping of the sperm head. *J. Ultrastruct. Res.* 33, 369–380.

391. Lenhossek, M. von (1898). Untersuchungen über Spermatogenese. *Arch. Mikr. Anat.* 51, 215–318.

392. Gresson, R. A. R., and Zlotnik, I. (1945). A comparative study of the cytoplasmic components of the male germ cells of certain mammals. *Proc. R. Soc. Edinb. Biol.* 62, 137–170.

393. Goodrowe, K. L., and Heath, E. (1984). Disposition of the manchette in the normal equine spermatid. *Anat. Rec.* 290, 177–183.

394. Kingsley-Smith, B. V., and Lacy, D. (1959). Residual bodies of seminiferous tubules of the rat. *Nature* 184, 249–251.

395. Lacy, D. (1960). Light and electron microscopy and its use in the study of factors influencing spermatogenesis in the rat. *J. Microsc. Soc.* 79, 290–225.

396. Diertert, S. E. (1966). Fine structure of the formation and fate of the residual bodies and degenerating germ cells and the lipid cycle in Sertoli cells in the bandicoot *Perameles nasuta Geoffrey* (Marsupialia). *Aust. J. Zool.* 120, 317–346.

397. Kerr, J. B., and de Kretser, D. M. (1975). Cyclic variations in Sertoli cell lipid content throughout the spermatogenic cycle in the rat. *J. Reprod. Fert.* 43, 1–8.

398. Fawcett, D. W., and Phillips, D. M. (1969). Observations on the release of spermatozoa and on changes in the head during passage through the epididymis. *J. Reprod. Fertil.* Suppl 6, 405–418.

399. Morales, C., and Clermont, Y. (1982). Evolution of Sertoli cell processes invading the cytoplasm of rat spermatids. *Anat. Rec.* 203, 233–244.

400. Holstein, A. F., and Schafer, E. (1978). A further type of transient cytoplasmic organelle in human spermatids. *Cell. Tissue Res.* 192, 359–361.

401. Breucker, H., Schafer, E., and Holstein, A. F. (1985). Morphogenesis and fate of the residual body in human spermiogenesis. *Cell. Tissue Res.* 240, 303–309.

402. Russell, L. D., and Clermont, Y. (1976). Anchoring device between Sertoli cells and late spermatids in rat seminiferous tubules. *Anat. Rec.* 185, 259–278.

403. Vitale-Calpe, R. (1970). Ultrastructural studies of spontaneous spermiation in the guinea pig. *Z. Zellforsch.* 105, 222–223.

404. Burgos M. H. and Vitale-Calpe R. (1967). The fine structure of the Sertoli cell-spermatozoon relationship in the toad. *J. Ultrastruct. Res.* 19, 221–237.

405. Burgos, M. H., and Vitale-Calpe, R. (1967). The mechanism of spermiation in the toad. *Am. J. Anat.* 120, 227–252.

406. Russell, L. D., Lee, I. P., Ettlin, R., and Peterson, R. N. (1983). Development of the acrosome and alignment, elongation and entrenchment of spermatids in procarbazine-treated rats. *Tissue Cell* 15, 615–626.

407. Brokelmann, J. (1961). Surface modifications of Sertoli cells at various stages of spermatogenesis in the rat. *Anat. Rec.* 139, 211.

408. Nagano, T. (1968). Fine structure relation between the Sertoli cell and the differentiating spermatid in the human testis. *Z. Zellforsch.* 89, 39–43.

409. Flickinger, C. J., and Fawcett, D. W. (1967). The junctional specialization of Sertoli cells in the seminiferous epithelium. *Anat. Rec.* 158, 207–222.

410. Russell, L. D. (1979). Spermatid-Sertoli tubulobulbar complexes as devices for elimination of cytoplasm from the head region of late spermatids of the rat. *Anat. Rec.* 194, 233–246.
411. Russell, L. D. (1979). Further observations on tubulobular complexes formed by later spermatids and Sertoli cells in the rat testis. *Anat. Rec.* 194, 213–232.
412. Sterio, D. C. (1984). The unbiased estimation of number and sizes of arbitrary particles using the disector. *J. Microsc.* 134, 127–136.
413. Gunderson, H. J. G. (1977). Notes on the estimation of the numerical density of arbitrary profiles, the edge effect. *J. Microsc.* 111, 219–223.
414. Gunderson, H. J. G, Bagger, P., Bendsten, T. F., Evans, S. M., Korbo, L., Marcussen, N., Nielsen, K., Nyengaard, J. R., Pakkenberg, B., Sorensen, F. B., Vesterby, A., and West, M. J. (1988). The new stereological tools, disector, fractionator, nucleator and point sampled intercepts and their use in pathological research and diagnosis. *APMIS* 96, 857–881.
415. Wreford, N. G. (1995). Theory and practice of stereological techniques applied to the estimation of cell number and nuclear volume in the testis. *Microsc. Res. Tech.* 32, 423–436.
416. Russell, L. D., Ren, H. P., Sinha Hikim, I., Schulze, W., and Sinha Hikim, A. P. (1990). A comparative study in twelve mammalian species of volume densities, volumes, and numerical densities of selected testis components, emphasizing those related to the Sertoli cell. *Am. J. Anat.* 188, 21–30.
417. Wistuba, J., Schrod, A., Greve, B., Hodges, J. K., Aslam, H., Weinbauer, G. F., and Luetjens, C. M. (2003).Organization of seminiferous epithelium in primates, relationship to spermatogenic efficiency, phylogeny, and mating system. *Biol. Reprod.* 69, 582–591.
418. Sertoli, E. (1865). Dell'esistenza di particular cellule ramificate nei canalicoli seminiferi dell'testicolo umano. *Morgagni.* 7, 31–39.
419. Sertoli, E. (1878). Sulla struttura dei canalicoli seminiferi dei testiculo. *Arch. Sci. Med.* 2, 267–295.
420. von La Valette, and St. George, A. J. H. (1878). Über die Genese der Samenkorper. *Arch. Mikrosk. Anat. Entwickl.* 15, 261–314.
421. Hoven, H. (1914). Histogenese du testicule des mammiferes. *Anat. Anz.* 47, 90–109.
422. Stieve, H. (1927). Die Entwicklung der Keimzellen und der Zwischenzellen in der Hodenaniage des Menschen. Ein Beitrag zur Keimba-hufrage. *Z. Mikrosk. Anat. Forsch.* 10, 225–285.
423. Rolshoven, E. (1940). Die funktionelle polymorphic des Sertoli-syncytioms und ihr Zumsammenhang mit der Spermatogenese. *Z. Zellforsch.* 31, 156–164.
424. Elftman, H. (1950). The Sertoli cell cycle in the mouse. *Anat. Rec.* 106, 381–393.
425. Fawcett, D. W. (1973). Observations on the organization of the interstitial tissue of the testis and on the occluding cell junctions in the seminiferous epithelium. *Adv. Biosci.* 10, 83–99.
426. Fawcett, D. W. (1975). Ultrastructure and function of the Sertoli cell. In *Handbook of Physiology* (D. W. Hamilton and R. O. Greep, Eds.), Section 7, Vol. 5, pp. 21–55. American Physiological Society, Washington, DC.
427. Schulze, C. (1974). On the morphology of the human Sertoli cell. *Cell. Tissue Res.* 153, 339–355.
428. Schulze, C. (1984). Sertoli cells and Leydig cells in man. *Adv. Anat. Embryol. Cell. Biol.* 88, 1–104.
429. Russell, L. D. (1980). Sertoli-germ cell interrelations, a review. *Gamete Res.* 3, 179–202.
430. Wong, V., and Russell, L. D. (1983). Three-dimensional reconstruction of a rat stage V Sertoli cell. I. Methods, basic configuration and dimensions. *Am. J. Anat.* 167, 143–161.
431. Weber, J. E., Russell, L. D., Wong, V., and Peterson, R. N. (1983). Three-dimensional reconstruction of a rat stage V Sertoli cell. II. Morphometry of Sertoli-Sertoli and Sertoli-germ cell relationships. *Am. J. Anat.* 167, 163–179.
432. Russell, L. D., Tallon-Doran, M., Weber, J. E., Wong, V., and Peterson, R. N. (1983). Three-dimensional reconstruction of a rat stage V Sertoli cell. III. A study of specific cellular relationships. *Am. J. Anat.* 167, 181–192.
433. Ploen, L., and Ritzen, E. M. (1984). Fine structural features of Sertoli cells. In *Ultrastructure of Reproduction* (J. Van Berkom and P. M. Motta, Eds.), pp. 67–74. Martinus Nijhoff, Boston.
434. Tindall, D. J., Rowley, D. R., Murthy, L., Lipshultz, L. I., and Chang, C. H. (1985). Structure and biochemistry of the Sertoli cell. *Int. Rev. Cytol.* 94, 127–149.
435. Elftman, H. (1963). Sertoli cells and testis structure. *Am. J. Anat.* 113, 25–33.
436. Lofts, B. (1972). The Sertoli cell. *Gen. Comp. Endocrinol.* Suppl 3, 636–648.
437. Bergmann, M., Greven, R., and Schindelmeiser, J. (1983). Observations on the blood-testis barrier in a frog and a salamander. *Cell. Tissue Res.* 232, 189–200.
438. Russell, L. D., and Griswold, M. D., eds. (1993). *The Sertoli Cell.* Cache River Press, Clearwater, FL.
439. Cooksey, E. J., and Rothwell, B. (1973). The ultrastructure of the Sertoli cell and its differentiation in the domestic fowl (*Gallus domesticus*). *J. Anat.* 114, 329–345.
440. Dufaure, J. P. (1971). L'ultrastructure du testicule de lezard vivipare (Reptile, Lacertilien). II. Les cellules de Sertoli. Etude du glycogene. *Z. Zellforsch.* 115, 565–578.
441. Baccetti, B., Bigliardi, E., Vegni Talluri, M., and Burrini, A. G. (1983). The Sertoli cells in lizards. *J. Ultrastruct. Res.* 85, 11–23.
442. Ross, M. R., and Dobler, J. (1975). The Sertoli cell junctional specializations and their relationship to the germinal epithelium as observed after efferent ductule ligation. *Anat. Rec.* 183, 267–292.
443. Rolshoven, E. (1975). Spermatogenese und Sertoli Syncitium. *Z. Zellforsch.* 33, 439–460.
444. Vogl, A. W., and Soucy, L. J. (1985). Arrangement and possible function of actin filament bundles in ectoplasmic specializations of gonad squirrel Sertoli cells. *J. Cell. Biol.* 100, 814–825.
445. Kerr, J. B., Mayberry, R. A., and Irby, D. C. (1984). Morphometric studies on lipid inclusions in Sertoli cells during the spermatogenic cycle in the rat. *Cell. Tissue Res.* 236, 699–709.
446. Wing, T. Y., and Christensen, A. K. (1982). Morphometric studies on rat seminiferous tubules. *Am. J. Anat.* 165, 13–25.
447. Roosen-Runge, E. C. (1955). Quantitative studies on spermatogenesis in the albino rat. III. Volume changes in the cells of the seminiferous tubules. *Anat. Rec.* 123, 385–398.
448. Johnson, L., Petty, C. S., and Neaves, W. B. (1980). A comparative study of daily sperm production and testicular composition in humans and rats. *Biol. Reprod.* 22, 1233–1243.
449. Cavicchia, J. C., and Dym, M. (1977). Relative volume of Sertoli cells in monkey seminiferous epithelium. *Am. J. Anat.* 150, 501–503.
450. Dym, M., and Cavicchia, J. C. (1978). Functional morphology of the testis. *Biol. Reprod.* 18, 1–15.
451. Kerr, J. B. (1988). A light microscopic and morphometric analysis of the Sertoli cell during the spermatogenic cycle of the rat. *Anat. Embryol.* 177, 341–348.
452. Kerr, J. B. (1988). An ultrastructural and morphometric analysis of the Sertoli cell during the spermatogenic cycle of the rat. *Anat. Embryol.* 179, 191–203.

453. Chung, K. W. (1974). Fine structure of Sertoli cells and myoid cells in mice with testicular feminization. *Fertil. Steril.* 25, 325–335.
454. Dym, M. (1973). The fine structure of the monkey (*Macaco*) Sertoli cell and its role in maintaining the blood-testis barrier. *Anat. Rec.* 175, 639–656.
455. Nagano, T. (1966). Some observations on the fine structure of the Sertoli cell in the human testis. *Z. Zellforsch.* 73, 89–106.
456. Bustos-Obregon, E., and Esponda, P. (1974). Ultrastructure of the nucleus of human Sertoli cells in normal and pathological testes. *Cell. Tissue Res.* 152, 467–475.
457. Kerr, J. B., and de Kretser, D. M. (1981). The cytology of the human testis. In *The Testis* (H. G. Burger and D. M. de Kretser, Eds.), pp. 141–170. Raven Press, New York.
458. Johnson, L., Zane, R. S., Petty, G. S., and Neaves, W. B. (1984). Quantification of human Sertoli cell population, Its distribution, relation to germ cell numbers, and age-related decline. *Biol. Reprod.* 31, 785–795.
459. de Kretser, D. M., Kerr, J. B., and Paulsen, C. A. (1981). Evaluation of the ultrastructural changes in the human Sertoli cells in testicular disorders and the relationship of the changes to the levels of serum FSH. *Int. J. Androl.* 4, 124–144.
460. Dym, M , and Romrell, L. J. (1975). Intraepithelial lymphocytes in the male reproductive tract of rats and rhesus monkeys. *J. Reprod. Fertil.* 42, 1–7.
461. Chemes, H. E., Dym, M., Fawcett, D. W., Javadpour, N., and Sherins, R. J. (1977). Patho-physiological observations of Sertoli cells in patients with germinal aplasia or severe germ cell depletion. Ultrastructural findings and hormone levels. *Biol. Reprod.* 17, 108–123.
462. Kerr, J. B. (1991). Ultrastructure of the seminiferous epithelium and inter-tubular tissue of the human testis. *Electron Microsc. Tech.* 19, 215–240.
463. Kierszenbaum, A. L. (1974). RNA synthetic activities of Sertoli cells in the mouse testis. *Biol. Reprod.* 11, 365–376.
464. Mirre, C., and Knibiehler, B. (1982). A re-evaluation of the relationships between the fibrillar centers and the nucleolus-organising regions in reticulated nucleoli, ultrastructural organisation, number and distribution of the fibrillar centres in the nucleolus of the mouse Sertoli cells. *J. Cell. Sci.* 55, 247–259.
465. Nicander, L. (1963). Some Ultrastructural features of mammalian Sertoli cells. *J. Ultrastruct. Res.* 8, 190–191.
466. Zibrin, M. (1971). Some ultrastructural aspects of nuclear morphology in developing spermatids and mature spermatozoa of the bull. *Mikroskopie* 27, 10–16.
467. Zibrin, M. (1972). Multivesicular nuclear body with nucleolar activity in Sertoli cells of bulls. An ultrastructural study. *Z. Zellforsch.* 135, 155–164.
468. Pawar, H. S., and Wrobel, K. H. (1991). The Sertoli cell of the water buffalo (*Bubalus bubalis*) during the spermatogenic cycle. *Cell. Tissue Res.* 265, 45–30.
469. Tersakis, J. A. (1965). The nucleolar channel system in the human endometrium. *J. Cell. Biol.* 166, 37–48.
470. Pelletier, R. M. (1993). A novel perspective: the occluding zonule encircles the apex of the Sertoli cell as observed in birds. *Am. J. Anat.* 235, 191–205.
471. Nagano, T., and Suzuki, F. (1976). Freeze-fracture observations on the intercellular junctions of Sertoli cells and of Leydig cells in the human testis. *Cell. Tissue Res.* 166, 37–48.
472. Ross, M. H. (1976). The Sertoli cell junctional specialization during spermiogenesis and at spermiation. *Anat. Rec.* 186, 79–104.
473. Vogl, A. W., Linck, R. W., and Dym, M. (1983). Colchicine-induced changes in the cytoskeleton of the golden mantled ground squirrel (*Spermophilus lateralis*) Sertoli cells. *Am. J. Anat.* 168, 99–108.
474. Rambourg, A., Clermont, Y., and Marrand, A. (1974). Three-dimensional structure of the osmium-impregnated Golgi apparatus as seen in the high voltage electron microscope. *Am. J. Anat.* 140, 27–46.
475. Rambourg, A., Clermont, Y., and Hermo, L. (1979). Three-dimensional architecture of the Golgi apparatus in Sertoli cells of the rat. *Am. J. Anat.* 154, 455–476.
476. Gunsalus, G. L., Musto, N. A., and Bardin, C. W. (1980). Bidirectional release of a Sertoli cell product, androgen binding protein, into the blood and seminiferous tubule. In *Testicular Development, Structure and Function* (A. Steinberger and E. Steinberger, Eds.), pp. 291–298. Raven Press, New York.
477. Wright, W. W., Parvinen, M., and Musto, N. A., Gunsalus, G. L., Philips, D. M., Mather, J. P., and Bardin, C. W. (1983). Identification of stage-specific proteins synthesized by rat seminiferous tubules. *Biol. Reprod.* 29, 257–270.
478. Ritzen, E. M., Boitani, C., and Parvinen, M. (1981). Cyclic secretion of proteins by the rat seminiferous tubule, depending on the stage of spermatogenesis. *Int. J. Androl.* Suppl 3, 57–58.
479. Robertson, D. M., Foulds, L. M., and Leversha, L., et al. (1985). Isolation of inhibin from bovine follicular fluid. *Biochem. Biophys. Res. Commun.* 126, 220–226.
480. Forage, R. G., Ring, J. M., and Brown, R. W., McInerney, B. V., Cobon G. S., Gregson, R. P., Robertson, D. M., Morgan, F. S., Hearn, M. T. N., Findlay, J. K., Wettenhall, R. E. H., Burger, H. G., and de Krester, D. M. (1986). Cloning and sequence analysis of cDNA species coding for the two subunits of inhibin from bovine follicular fluid. *Proc. Natl. Acad. Sci. U S A* 83, 3091–3095.
481. Mason, A. J., Hayflick, J. S., and Ling, N., Esch, F., Ueno, N., Ying, S. Y., Guillemin, R., Niall, H., and Seeburg, P. H. (1985). Complementary DNA sequences of ovarian follicular fluid inhibin show precursor structure and homology with transforming growth factor B. *Nature* 318, 659–663.
482. McLachlan, R. I., Robertson, D. M., Burger, H. G., and de Kretser, D. M. (1986). The radioimmunoassay of bovine follicular fluid inhibin. *Mol. Cell. Endocrinol.* 46, 175–185.
483. Christensen, A. K. (1965). The fine structure of testicular interstitial cells in guinea pigs. *Cell. Biol.* 26, 911–935.
484. Blanchette, E. J. (1966). Ovarian steroid cells. II. The lutein cell. *J. Cell. Biol.* 31, 517–542.
485. Abrunhosa, R. (1972). Microperfusion fixation of embryos for ultrastructural studies. *J. Ultrastruct. Res.* 41, 176–184.
486. Wrobel, K. H., and Schimmel, M. (1989). Morphology of the bovine Sertoli cell during the spermatogenic cycle. *Cell. Tissue Res.* 257, 93–103.
487. Vogl, A. W., Lin, Y. C., Dym, M., and Fawcett, D. W. (1983). Sertoli cells of the golden-mantled ground squirrel (*Spermophilus lateralis*), a model system for the study of shape change. *J. Reprod. Fertil.* 168, 83–98.
488. Kerr, J. B., and de Kretser, D. M. (1974). The role of the Sertoli cell in phagocytosis of the residual bodies of spermatids. *J. Reprod. Fertil.* 36, 439–440.
489. Olvik, N. M., and Dahl, E. (1981). Stage-dependent variation in volume density and size of Sertoli cell vesicles in the rat testis. *Cell. Tissue Res.* 221, 311–320.
490. Welsh, M. J., and Wiebe, J. P. (1978). Sertoli cell capacity to metabolize C19 steroids. Variation with age and the effect of FSH. *Endocrinology* 103, 838–844.
491. Tcholakian, R. K., and Steinberger, A. (1978). Progesterone metabolism by cultured Sertoli cells. *Endocrinology* 103, 1335–1343.
492. Lacy, D. (1962). Certain aspects of testis structure and function. *Br. Med. Bull.* 18, 205–208.
493. Lacy, D. (1967). The seminiferous tubules in mammals. *Endeavour* 26, 101–108.

494. Lacy, D., and Pettitt, A. J. (1969). Transmission electron microscopy and the production of steroids by the Leydig and Sertoli cells of the human testis. *Micron.* 1, 15–53.
495. Lacy, D., and Pettitt, A. J. (1970). Sites of hormone production in the mammalian testis and their significance in the control of male fertility. *Br. Med. Bull.* 26, 87–911.
496. Johnson, A. D. (1970). Testicular lipids. In *The Testis* (A. D. Johnson, W. R. Gomes, and N. L. van Demark, Eds.), Vol. II, pp. 193–258. Academic Press, New York.
497. Kerr, J. B., Keogh, E. J., Hudson, B., Whipp, G. T., and de Kretser, D. M. (1980). Alterations in spermatogenic activity and hormonal status in a seasonally breeding rat, *Rattus fuscipes. Gen. Comp. Endocrinol.* 40, 78–88.
498. de Kretser, D. M., and Kerr, J. B. (1983). The effect of testicular damage on Sertoli and Leydig cell function. In *The Pituitary and Testis, Clinical and Experimental Studies* (D. M. de Kretser, H. G. Burger, and B. Hudson, Eds.), pp. 133–154. Springer Verlag, Berlin.
499. Collins, P. M., and Lacy, D. (1969). Studies on the structure and function of the mammalian testis. II. Cytological and histochemical observations on the testis of the rat after a single exposure of heat applied for different lengths of time. *Proc. R. Soc. Lond. Biol.* 172, 17–38.
500. Posalaki, Z., Szabo, D., Bacsi, E., and Okros, I. (1968). Hydrolytic enzymes during spermatogenesis in rat. An electron microscopic and histochemical study. *J. Histochem. Cytochem.* 16, 249–262.
501. Beckman, J. K., and Conigliio, J. G. (1979). A comparative study of the lipid composition of isolated rat Sertoli and germinal cells. *Lipids* 14, 262–267.
502. Fleeger, J. L., Bishop, J. P., Gomes, W. R., and van Demark, N. L. (1968). Testicular lipids. I. Effect of unilateral cryptorchidism on lipid classes. *J. Reprod. Fertil.* 15, 1–7.
503. Spangaro, S. (1902). Über die histologischen Veranderungen des Hodens, Nebehodens, und Samenleiters von Geburt an biz zum Grier-salter, mit besonderer Berucksichtigung der Hodenatrophie, des elastischen Gewebes und des Vorkommens von Krystallen in Hoden. *Anat. Rec.* 18, 593–771.
504. Fawcett, D. W., and Burgos, M. H. (1956). The fine structure of Sertoli cells in the human testis. *Anat. Rec.* 124, 401.
505. Bawa, S. R. (1963). The fine structure of the Sertoli cell of the human testes. *J Ultrastruct. Res.* 9, 459–474.
506. Toyama, Y. (1975). Ultrastructure study of crystalloids in Sertoli cells of the normal, intersex and experimental cryptorchid swine. *Cell. Tissue Res.* 158, 205–213.
507. Toyama, Y., Obinata, T., and Holtzen, H. (1979). Crystalloids of actin-like filaments in the Sertoli cell of the swine testis. *Anat. Rec.* 195, 47–62.
508. von Bardeleben, K. (1897). Die Zwischenzellen des Saugetierhodens. *Anat. Anz.* 13, 529–536.
509. Benda, C. (1906). Die spermiogenese der marsupialier. *Denkschr. Medna. Ges. Jena.* 6, 441–458.
510. Greenwood, A. W. (1923). Marsupial spermatogenesis. *Q. J. Microsc. Sci.* 64, 203–218.
511. Duesberg, J. (1918). On the interstitial cells of the testicle in *Didelphys. Biol. Bull.* 35, 175–198.
512. Harding, H. R., Carrick, F. N., and Shorey, C. D. (1982). Crystalloid inclusions in the Sertoli cell of the koala *Phascolarctos cinereus* (Marsupialia). *Cell. Tissue Res.* 22(1), 633–642.
513. Maximow, A. (1899). Die histologische Vorgange bei der Heilung von Hodenvertetzungen und die Regenerationsfahigkeit des Hoden-gwebes. *Beitr. Pathol. Anat.* 26, 230–319.
514. Niemi, M., Harkonen, H., and Kokko, A. (1962). Localisation and identification of testicular esterases in the rat. *J. Histochem. Cytochem.* 10, 186–193.
515. Posalaki, Z. (1964). Histochemische untersuchungen der spermiogenese. *Symp. Biol. Hung.* 4, 83–87.
516. Posalaki, Z. (1965). Activity of different dehydrogenases and diaphorases in the spermatogenesis of the rat and its relation to motility. *Acta. Histochem.* 20, 86–90.
517. Niemi, M., and Kormano, M. (1965). Cyclical changes in and significance of lipids and acid phosphatase activity in the seminiferous tubules of the rat testis. *Anat. Rec.* 157, 159–170.
518. Parvinen, M., and Vanha-Pertulla, T. (1972). Identification and enzyme quantification of the stages of the seminiferous epithelial wave in the rat. *Anat. Rec.* 174, 435–450.
519. Barham, S. S., and Berlin, J. D. (1974). Fine structure and cytochemistry of testicular cells in men treated with testosterone propionate. *Cell. Tissue Res.* 148, 159–182.
520. Clegg, E. J., and MacMillan, E. W. (1965). Uptake of vital dyes and paniculate matter by the Sertoli cells of the rat testis. *J. Anat.* 99, 219–229.
521. Carr, I., Clegg, E. J., and Meek, G. A. (1968). Sertoli cells as phagocytes, an electron microscopic study. *J. Anat.* 102, 501–509.
522. Soares Pessoa, J. F., and David-Ferreira, J. F. (1980). Bidirectional transport of horseradish peroxidase by rat Sertoli cells. An *in vitro* study. *Biol. Cell.* 39, 301–304.
523. Nakanishi, Y., and Shiratsuchi, A. (2004). Phagocytic removal of apoptotic spermatogenic cells by Sertoli cells, mechanisms and consequences. *Biol. Pharm. Bull.* 27, 13–16.
524. Reddy, J. K., and Svoboda, D. J. (1967). Lysosomal activity in Sertoli cells of normal and degenerating seminiferous epithelium of rat testis. *Am. J. Pathol.* 51, 1–17.
525. Morales, C., and Hermo, L. (1983). Demonstration of fluid phase endocytosis in epithelial cells of the male reproductive system by means of horseradish peroxidase colloidal gold complex. *Cell. Tissue Res.* 230, 503–510.
526. Hinton, B. T., and Keefer, D. A. (1983). Evidence for protein absorption from the lumen of the seminiferous tubuli and rete testis of the rat testis. *Cell. Tissue Res.* 230, 367–375.
527. Hermo, L., Morales, C. R., and Clermont, Y. (1982). Endocytic activity of Sertoli cells in the rat. *J. Cell. Biol.* 95, 434a.
528. Morales, C., Clermont, Y., and Hermo, L. (1985). Nature and function of endocytosis in Sertoli cells of the rat. *Am. J. Anat.* 173, 203–217.
529. Golyas, B. J. (1971). The rabbit zygote, formation of annulate lamellae. *J. Ultrastruct. Res.* 35, 112–126.
530. Maul, G. G. (1970). On the relationship between the Golgi apparatus and annulate lamellae. *J. Ultrastruct. Res.* 30, 368–384.
531. Wischmitzer, S. (1970). The annulate lamellae. *Int. Rev. Cytol.* 27, 65–100.
532. Sun, C. N, and White, H. J. (1979). Annulate lamellae in human tumour cells. *Tissue Cell* 11, 139–146.
533. Kessel, R. G. (1968). Annulate lamellae. *J. Ultrastruct. Res.* Suppl 10, 1–82.
534. Merkow, L., and Leighton, J. (1966). Increased numbers of annulate lamellae in myocardium of chick embryo incubated at abnormal temperatures. *J. Cell. Biol.* 28, 127–137.
535. de Brabander, M., and Borgers, M. (1975). The formation of annulated lamellae induced by the disintegration of microtubules. *J. Cell. Sci.* 19, 331–340.
536. Vogl, A. W., Pfeiffer, D. C., Redenbach, D. M., and Grove, B. D. (1993). Sertoli cell cytoskeleton. In *The Sertoli Cell* (L. D. Russell and M. D. Griswold, Eds.), pp. 39–86. Cache River Press, Clearwater, FL.
537. Vogl, A. W., Pfeiffer, D. C., Mulholland, D., Kimel, G., and Guttman, J. (2000). Unique and multifunctional adhesion junctions in the testis, ectoplasmic junctions. *Arch. Histol. Cytol.* 63, 1–15.

538. Amlani, S., and Vogl, A. W. (1988). Changes in the distribution of microtubules and intermediate filaments in mammalian Sertoli cells during spermatogenesis. *Anat. Rec.* 220, 143–160.
539. Toyama, Y. (1976). Actin-like filaments in the Sertoli cell junctional specializations in the swine and mouse testis. *Anat. Rec.* 186, 477–492.
540. Franke, W. W., Grund, C., and Fink, A., Weber, K., Jokusch, B. M., Zentgraf, H., and Osborn, M. (1978). Location of actin in the microfilament bundles associated with the junctional specializations between Sertoli cells and spermatids. *Biol. Cell.* 31, 7–14.
541. Oko, R., Hermo, L., and Hecht, N. B. (1991). Distribution of actin isoforms within cells of the seminiferous epithelium of the rat testis, evidence for a muscle form of actin in spermatids. *Anat. Rec.* 231, 63–81.
542. Franke, W. W., Grund, C., and Schmid, E. (1979). Intermediate-sized filaments present in Sertoli cells are of the vimentin type. *Eur. J. Cell. Biol.* 19, 269–275.
543. Paranko, J., Kallajoki, M., Pelliniemi, L. J., Lehto, V. P., and Virtanen, I. (1986). Transient co-expression of cytokeratin and vimentin in differentiating rat Sertoli cells. *Dev. Biol.* 117, 35–44.
544. Russell, L. D. (1977). Observations on rat Sertoli ectoplasmic ("junctional") specializations in their association with germ cells of the rat testis. *Tissue Cell* 9, 475–498.
545. Show, M. D., Anway, M. D., Folmer, J. S., and Zirkin, B. R. (2003). Reduced intratesticular testosterone concentration alters the polymerization state of the Sertoli cell intermediate filament cytoskeleton by degradation of vimentin. *Endocrinology* 144, 5530–5536.
546. Zhu, L. J., Zong, S. D., Phillips, D. M., MooYoung, A. J., and Bardin, C. W. (1997). Changes in the distribution of intermediate filaments in rat Sertoli cells during the seminiferous epithelium cycle and postnatal development. *Anat. Rec.* 248, 391–405.
547. Show, M. D., Anway, M. D., and Zirkin, B. R. (2004). An ex vivo analysis of Sertoli cell actin dynamics following gonadotropic hormone withdrawal. *J. Androl.* 25, 1013–1021.
548. Wang, Z. Q., Watanabe, Y., Toki, A., and Itano, T. (2002). Altered distribution of Sertoli cell vimentin and increased apoptosis in cryptorchid rats. *J. Pediatr. Surg.* 37, 648–652.
549. Richburg, J. H., and Boekelheide, K. (1996). Mono-(2-ethylhexyl) phthalate rapidly alters both Sertoli cell vimentin filaments and germ cell apoptosis in young rat testes. *Toxicol. Appl. Pharmacol.* 137, 42–50.
550. Dalgaard, M., Nellemann, C., Lam, H. R., Sorensen, I. K., and Ladefoged, O. (2001). The acute effects of mono(2-ethyl)phthalate (MEHP) on testes of prepubertal Wistar rats. *Toxicol. Lett.* 122, 69–79.
551. Allard, E. K., Johnson, K. J., and Boekelheide, K. (1993). Colchicine disrupts the cytoskeleton of the rat testis seminiferous epithelium in a stage-dependent manner. *Biol. Reprod.* 48, 143–153.
552. Hess, R. A., and Nakai, M. (2000). Histopathology of the male reproductive system induced by the fungicide benomyl. *Histol. Histopathol.* 15, 207–224.
553. Beach, S. F., and Vogl, A. W. (1999). Spermatid translocation in the rat seminiferous epithelium, coupling membrane trafficking machinery to a junction plaque. *Biol. Reprod.* 60, 1036–1046.
554. Guttman, J. A., Kimel, G. H., and Vogl, A. W. (2000). Dynein and plus-end microtubule-dependent motors are associated with specialized Sertoli cell junction plaques (ectoplasmic specializations). *J. Cell. Sci.* 113, 2167–2176.
555. Aoki, A. (1980). Induction of sperm release by microtubule inhibitors in rat testis. *Eur. J. Cell. Biol.* 22, 467.
556. Roosen-Runge, E. C. (1951). Quantitative studies on spermatogenesis in the albino rat. II. The duration of spermatogenesis and some effects of colchicine. *Am. J. Anat.* 88, 163–176.
557. Parvinen, L. M., Soderstrom, K. O., and Parvinen, M. (1978). Early effects of vinblastine and vincristine on the rat spermatogenesis, analysis by a new transillumination-phase contrast microscopic method. *Exp. Pathol.* 15, 85–96.
558. Russell, L. D., Malone, J. P., and MacCurdy, D. S. (1981). Effect of the microtubule disrupting agents, colchicine and vinblastine, on seminiferous tubule structure in the rat. *Tissue Cell* 13, 349–367.
559. Vogl, A. W. (1988). Changes in the distribution of microtubules in rat Sertoli cells during spermatogenesis. *Anat. Rec.* 222, 34–41.
560. Neely, M. D., and Boekelheide, K. (1988). Sertoli cell processes have axoplasmic features, an ordered microtubule distribution and an abundant high molecular weight microtubule-associated protein (cytoplasmic dynein). *J. Cell. Biol.* 107, 1767–1776.
561. Ribbert, H. (1904). Die abscheidung intravenos injizierten gelosten Karmins in den Geweben. *Z. Allg. Physiol.* 4, 201–214.
562. Bouffard, G. (1906). Injection des couters de benzidine aux animaux normaux. *Ann Inst Pasteur* 20, 539–546.
563. Goldacre, R. J., and Sylven, B. (1959). A rapid method for studying tumour blood supply using systemic dyes. *Nature* 184, 63–64.
564. Goldacre, R. J., and Sylven, B. (1962). On the access of blood-borne dyes to various tumor regions. *Br. J. Cancer* 16, 306–322.
565. Everett, N. B., and Simmons, B. (1958). Measurement and radioautographic localization of albumin in rat tissues after intravenous injection. *Circ. Res.* 6, 307–313.
566. Kormano, M. (1968). Penetration of intravenous trypan blue into the rat testis and epididymis. *Acta. Histochem.* 30, 133–136.
567. Setchell, B. P. (1967). The blood-testicular fluid barrier in sheep. *J. Physiol. (Lond.)* 189, 63P–65P.
568. Cowie, A. T., Lascelles, A. K., and Wallace, J. C. (1964). Flows and protein content of testicular lymph in conscious rams. *J. Physiol. (Lond.)* 171, 176–187.
569. Waites, G. M. H., and Setchell, B. P. (1966). Changes in blood flow and vascular permeability of the testis, epididymis and accessory reproductive organs of the rat after administration of cadmium chloride. *J. Endocrinol.* 34, 329–342.
570. Fawcett, D. W., Leak, L. V., and Heidger, P. M. (1970). Electron microscopic observations on the structural components of the blood-testis barrier. *J. Reprod. Fertil.* Suppl 10, 105–122.
571. Dym, M., and Fawcett, D. W. (1970). The blood-testis barrier in the rat and the physiological compartmentation of the seminiferous epithelium. *Biol. Reprod.* 3, 308–326.
572. Russell, L. D., and Peterson, R. N. (1985). Sertoli cell junctions, morphological and functional considerations. *Int. Rev. Cytol.* 94, 177–211.
573. Nagano, T. (1980). Freeze-fracture observations on the rat Sertoli cell junctions by metal contact freezing. *J. Electron. Microsc.* 29, 250–255.
574. Parreira, G. G., Melo, R. C. N., and Russell, L. D. (2002). Relationship of Sertoli-Sertoli tight junctions to ectoplasmic specializations in conventional and en face views. *Biol. Reprod.* 67, 1232–1241.
575. Cheng, C. Y., and Mruk, D. D. (2002). Cell junction dynamics in the testis, Sertoli-germ cell interactions and male contraception development. *Physiol. Rev.* 82, 825–874.
576. Lui, W. Y., Mruk, D. D., Lee, W. M., and Cheng, C. Y. (2003). Adherens junction dynamics in the testis and spermatogenesis. *J. Androl.* 24, 14.

577. Lui, W. Y., Lee, W. M., and Cheng, C. Y. (2003). TGF-βs, their role in testicular function and Sertoli cell tight junction dynamics. *Int. J. Androl.* 26, 147–160.
578. Lui, W. Y., Mruk, D., Lee, W. M., and Cheng, C. Y. (2003). Sertoli cell tight junction dynamics, their regulation during spermatogenesis. *Biol. Reprod.* 68, 1087–1097.
579. Toyama, Y., Maekawa, M., and Yuasa, S. (2003). Ectoplasmic specializations in the Sertoli cell, new vistas based on genetic defects and testicular toxicology. *Anat. Sci. Int.* 78, 1–16.
580. Siu, M. K. Y., and Cheng, C. Y. (2004). Dynamic cross-talk between cells and the extracellular matrix in the testis. *BioEssays* 26, 978–992.
581. Siu, M. K. Y., and Cheng, C. Y. (2004). Extracellular matrix, recent advances on its role in junction dynamics in the seminiferous epithelium during spermatogenesis. *Biol. Reprod.* 71, 375–391.
582. Lee, N. P. Y., and Cheng, C. Y. (2004). Adaptors, junction dynamics, and spermatogenesis. *Biol. Reprod.* 71, 392–404.
583. Lee, N. P. Y., and Cheng, C. Y. (2004). Nitric oxide/nitric oxide synthase, spermatogenesis, and tight junction dynamics. *Biol. Reprod.* 267–276.
584. Connell, C. J. (1978). A freeze-fracture and lanthanum tracer study of the complex junction between the Sertoli cells of the canine testis. *J. Cell. Biol.* 76, 57–75.
585. Dym, M., and Caviccia, J. C. (1977). Further observations on the blood-testis barrier in monkeys. *Biol. Reprod.* 17, 390–403.
586. Russell, L. D. (1977). Movement of spermatocytes from the basal to the adluminal compartment of the rat testis. *Am. J. Anat.* 148, 313–328.
587. Osman, D., Ekwall, H., and Ploen, L. (1980). Specialized cell contacts and the blood-testis barrier in the seminiferous tubules of the domestic fowl (*Gallus domesticus*). *Int. J. Androl.* 3, 553–562.
588. Franchi, E., Camatini, M., and de Curtis, I. (1982). Morphological evidence of a permeability barrier in urodele testis. *J. Ultrastruct. Res.* 80, 253–263.
589. Abraham, M., Rahamim, E., Tibika, H., Golenser, E., and Kieselstein, M. (1980). The blood-testis barrier in *Aphanius dispar* (Teleostei). *Cell. Tissue Res.* 211, 207–214.
590. Marcaillou, C., and Szollosi, A. (1980). The blood-testis barrier in a nematode and a fish, a generalizable concept. *J. Ultrastruct. Res.* 70, 128–136.
591. Mattei, X., Mattei, C., Marcand, B., and Leung, T. K. D. (1982). Ultrastructure des cellules de Sertoli d'un teleosteen, *Abudefduf marginatus*. *J. Ultrastruct. Res.* 81, 333–340.
592. Billard, R. (1980). La spermatogenese de *Poecilia reticulata*. HI. Ultrastructure des cellules de Sertoli. *Ann. Biol. Anim. Biochem. Biophys.* 10, 37–50.
593. Parmentier, H. K., van den Boogaart, J. G. M., and Timmermans, L. P. M. (1985). Physiological compartmentation in gonadal tissue of the common carp (*Cyprinus carpio* L). A study with horseradish peroxidase and monoclonal antibodies. *Cell. Tissue Res.* 242, 75–81.
594. Szollosi, A., Reimann, J., and Marcaillou, C. (1980). Localization of the blood-testis barrier in the testis of the moth, *Anagasta kuehnilla*. *J. Ultrastruct. Res.* 72, 189–199.
595. Toshimori, K., Iwashata, T., and Oura, C. (1979). Cell junctions in the cyst envelope in the silkworm testis, *Bombyx mori Linne*. *Cell. Tissue Res.* 202, 63–73.
596. Lane, N. J., and Skaer, H. L. (1980). Intercellular junctions in insect tissues. *Adv. Insect Physiol.* 15, 35–213.
597. Chung, N. P., Mruk, D., Mo, M. Y., Lee, W. M., and Cheng, C. Y. (2001). A 22-amino acid peptide corresponding to the second extracellular loop of rat occluding perturbs the blood-testis barrier and disrupts spermatogenesis reversibly in vivo. *Biol. Reprod.* 65, 1340–1351.
598. Mruk, D. D., and Cheng, Y. C. (2004). Sertoli-Sertoli and Sertoli-germ cell interations and their significance in germ cell movement in the seminiferous epithelium during spermatogenesis. *Endocrine Rev.* 25, 747–806.
599. Toyama, Y., Ohkawa, M., Oku, R., Maekawa, M., and Yuasa, S. (2001). Neonatally administered diethylstilbestrol retards the development of the blood-testis barrier in the rat. *J. Androl.* 22, 413–423.
600. Hosoi, I., Toyama, Y., Maekawa, M., Ito, H., and Yuasa, S. (2002). Development of the blood-testis barrier in the mouse is delayed by neonatally administered diethylstilbestrol but not ß-estradiol 3-benzoate. *Andrologia* 34, 255–262.
601. Meyer, J. M., Mezrahid, P., Vignon, F., Chabrier, G., Reiss, D., and Rumpler, Y. (1996). Sertoli cell barrier dysfunction and spermatogenic cycle breakdown in the human testis, a lanthanum tracer investigation. *Int. J. Androl.* 19, 190–198.
602. Cavicchia, J. C., Sacerdote, F. L., and Ortiz, L. (1996). The human blood-testis barrier in impaired spermatogenesis. *Ultrastruct. Pathol.* 20, 211–218.
603. Morales, A., and Cavicchia, J. C., (2002). Spermatogenesis and blood-testis barrier in rats after long-term vitamin A deprivation. *Tissue Cell* 34, 349–355.
604. O'Rand, M. G., and Romrell, L. H. (1977). Appearance of cell surface auto- and isoantigens during spermatogenesis in the rabbit. *Dev. Biol.* 55, 347–358.
605. Millette, C. F., and Bellve, A. R. (1977). Temporal expression of membrane antigens during mouse spermatogenesis. *J. Cell. Biol.* 74, 86–97.
606. Romrell, L. H., and O'Rand, M. G. (1978). Capping and ultrastructural localization of sperm surface isoantigens during spermatogenesis. *Dev. Biol.* 63, 76–93.
607. Tung, P. S., and Fritz, I. B. (1978). Specific surface antigens on rat pachytene spermatocytes and successive classes of germ cells. *Dev. Biol.* 64, 297–315.
608. Pelletier, R. M. (1986). Cyclic formation and decay of the blood-testis barrier in the mink (*Mustela vison*), a seasonal breeder. *Am. J. Anat.* 175, 91–117.
609. Setchell, B. P., Pollanen, P., and Zupp, J. L. (1988). Development of the blood-testis barrier and changes in vascular permeability at puberty in rats. *Int. J. Androl.* 11, 225–233.
610. Russell, L. D., Bartke, A., and Goh, J. C. (1989). Postnatal development of the Sertoli cell barrier, tubular lumen, and cytoskeleton of Sertoli and myoid cells in the rat, and their relationship to tubular fluid secretion and flow. *Am. J. Anat.* 184, 179–189.
611. Russell, L. D., and Malone, J. P. (1980). A study of Sertoli-spermatid tubulobular complexes in selected mammals. *Tissue Cell* 12, 263–285.
612. Russell, L. D., Gardner, R. J., and Weber, J. E. (1986). Reconstruction of a type-B configuration monkey Sertoli cell, size, shapes and configurational and specialized cell-to-cell relationships. *Am. J. Anat.* 175, 73–90.
613. Gravis, C. J. (1979). Interrelationships between Sertoli cells and germ cells in the Syrian hamster. *Z. Mikrosk. Anat. Forsch. (Leipzig)* 93, 321–342.
614. Romrell, L. H., and Ross, M. H. (1979). Characterization of Sertoli cell-germ cell junctional specialization in dissociated testicular cells. *Anat. Rec.* 193, 23–42.
615. O'Donnell, L., Stanton, P. G., Bartles, J. R., and Robertson, D. M. (2000). Sertoli cell ectoplasmic specializations in the seminiferous epithelium of the testosterone-suppressed adult rat. *Biol. Reprod.* 63, 99–108.
616. Beardsley, A., and O'Donnell, L. (2003). Characterization of normal spermiation and spermiation failure induced

by hormone suppression in adult rats. *Biol. Reprod.* 68, 1299–1307.

617. Gravis, C. J. (1978). Inhibition of spermiation in the Syrian hamster using dibutyryl cyclic AMP. *Cell. Tissue Res.* 192, 241–243.
618. Ross, M. H. (1977). Sertoli-Sertoli junctions and Sertoli-spermatid junctions after efferent ductule ligation and lanthanum treatment. *Am. J. Anat.* 148, 49–56.
619. Oko, R., Hermo, L., and Hecht, N. B. (1991). Distribution of actin isoforms within cells of the seminiferous epithelium of the rat testis, evidence for a muscle form of actin in spermatids. *Anat. Rec.* 231, 63–81.
620. Russell, L. D., Saxena, N. K., and Turner, T. T. (1989). Cytoskeletal involvement in spermiation and sperm transport. *Tissue Cell* 21, 361–379.
621. Weber, J. E., Turner, T. T., Tung, K., and Russell, L. D. (1988). Effect of cytochalasin D on the integrity of the Sertoli cell barrier. *Am. J. Anat.* 182, 130–147.
622. Bartles, J. R., Wierda, A., and Zheng, L. (1996). Identification and characterization of espin, an actin-binding protein localized to the F-actin-rich junctional complexes of Sertoli cell ectoplasmic specializations. *J. Cell. Sci.* 109, 1229–1239.
623. Chen, B., Li, A., Wang, D., Wang, M., Zheng, L., and Bartles, J. R. (1999). Espin contains an additional actin-binding site inits N terminus and is a major actin-bundling protein of the Sertoli cell-spermatid ectoplasmic specialization junctional plaque. *Mol. Biol. Cell.* 10, 4327–4339.
624. Pfeiffer, D. C., and Vogl, A. W. (1991). Evidence that vinculin is co-distributed with actin bundles in ectoplasmic (junctional) specializations of mammalian Sertoli cells. *Anat. Rec.* 231, 89–100.
625. Grove, B. D., Pfeiffer, D. C., Alien, S., and Vogl, A. W. (1990). Immunofluorescence localization of vinculin in ectoplasmic (junctional) specializations of rat Sertoli cells. *Am. J. Anat.* 188, 44–56.
626. Russell, L. D. (1979). Observations on the interrelationships of Sertoli cells at the level of the blood-testis barrier, Evidence for formation and resorption of Sertoli-Sertoli tubulobulbar complexes during the spermatogenic cycle of the rat. *Am. J. Anat.* 155, 259–280.
627. Clermont, Y., McCoshen, J., and Hermo, L. (1980). Evolution of the endoplasmic reticulum in the Sertoli cell cytoplasm encapsulating the heads of late spermatids in the rat. *Anat. Rec.* 196, 83–99.
628. Cooper, C. W., and Bedford, J. M. (1976). Asymmetry of spermiation and sperm surface charge patterns over the giant acrosome in the musk shrew *Suncus murinus*. *J. Cell. Biol.* 69, 415–428.
629. Gravis, C. J. (1980). Ultrastructural observations on spermatozoa retained within the seminiferous epithelium after treatment with dibutyryl cyclic AMP. *Tissue Cell* 12, 309–322.
630. Palombi, F., Salanova, M., Tarone, G., Farini, D., and Stefanini, M. (1992). Distribution of β_1 integrin subunit in rat seminiferous epithelium. *Biol. Reprod.* 47, 1173–1182.
631. Mulholland, D. J., Dedhar, S., and Vogl, A. W. (2001). Rat seminiferous epithelium contains a unique junction (ectoplasmic specialization) with signaling properties both of cell/cell and cell/matrix junctions. *Biol. Reprod.* 64, 396–407.
632. Salanova, M., Ricci, G., Boitani, C., Stefanini, M., DeGrossi, S., and Palombi, F. (1998). Junctional contacts between Sertoli cells in normal and aspermatogenic rat seminiferous epithelium contain a6ß1 integrins, and their formation is controlled by follicle-stimulating hormone. *Biol. Reprod.* 58, 371–378.
633. Altorfer, J., Fukuda, T., and Hedinger, C. (1974). Desmosomes in human seminiferous epithelium. *Virchows Arch. (Cell. Pathol.)* 16, 181–194.
634. Ren, H. P., and Russell, L. D. (1992). Quantitation of Sertoli cell-germ cell desmosome-gap junctions in relation to meiotic divisions in the male rat. *Tissue Cell* 24, 565–573.
635. Connell, C. J. (1974). The Sertoli cell of the sexually mature dog. *Anat. Rec.* 178, 333.
636. McGinley, D., Pozalaky, Z., and Porvanznik, M. (1977). Intercellular junctional complexes of the rat seminiferous tubules, a freeze-fracture study. *Anat. Rec.* 189, 211–232.
637. Bigliardi, E., and Vegni, T. M. (1977). Gap junctions between Sertoli cells in the infertile human testis. *Fertil. Steril.* 28, 755–758.
638. Camatini, M., Franchi, E., deCurtis, I., Anelli, G., and Masera, G. (1982). Chemotherapy does not affect the development of inter-Sertoli cell junctions in childhood leukemia. *Anat. Rec.* 203, 353–363.
639. Brehm, R., Marks, A., Rey, R., Kleisch, S., Bergmann, M., and Steger, K. (2002). Altered expression of connexins 26 and 43 in Sertoli cells in seminiferous tubules infiltrated with carcinoma-in situ or seminoma. *J. Pathol.* 197, 647–653.
640. Roger, C., Mograbi, B., Chevallier, D., Michiels, J. F., Tanaka, H., and Segretain, D. (2004). Disrupted traffic of connexin 43 in human testicular seminoma cells, overexpression induces membrane location and cell proliferation decrease. *J. Pathol.* 202, 241–246.
641. McGinley, D. M., Pozalaky, Z., Porvaznik, M., and Russell, L. D. (1979). Gap junctions between Sertoli and germ cells of rat seminiferous tubules. *Tissue Res.* 11, 741–754.
642. Risley, M. S., Tan, I. P., Roy, C., and Saez, J. C. (1992). Cell, age-, and stage-dependent distribution of connexin43 gap junctions in testes. *J. Cell. Sci.* 103, 81–96.
643. Tan, I. P., Roy, C., Saez, C. G., Paul, D. L., and Risely, M. S. (1996). Regulated assembly of connexin33 and connexin43 into rat Sertoli cell gap junctions. *Biol. Reprod.* 54, 1300–1310.
644. Batias, C., Defamie, N., Lablack, A., Thepot, D., Fenichel, P., Segretain, D., and Pointis, G. (1999). Modified expression of testicular gap-junction connexin 43 during normal spermatogenic cycle and in altered spermatogenesis. *Cell. Tiss. Res.* 298, 113–121.
645. Decrouy, X., Gasc, J. M., Pointis, G., and Segretain, D. (2004). Functional characterization of Cx43 based gap junctions during spermatogenesis. *J. Cell. Physiol.* 200, 146–154.
646. Lodge, J. R., and Salisbury, G. W. (1970). Seasonal variation and male reproductive efficiency. In *The Testis* (A. D. Johnson, A. D. Gomes, and N. L. van Denmark, Eds.), Vol. III, pp. 139–167. Academic Press, New York.
647. Lofts, B., and Bern, H. A. (1972). The functional morphology of steroidogenic tissues. In *Steroids in Non-Mammalian Vertebrates* (D. R. Idler, Ed.), pp. 37–125. Academic Press, New York.
648. Hodgson, Y. M., Irby, D. C., Kerr, J. B., and de Kretser, D. M. (1979). Studies of the structure and function of the Sertoli cell in a seasonally breeding rodent. *Biol. Reprod.* 21, 1091–1098.
649. Irby, D. C., Kerr, J. B., Risbridger, G. P., and de Kretser, D. M. (1984). Seasonally and experimentally induced changes in testicular function of the Australian bush rat, *Rattus fuscipes*. *J. Reprod. Fertil.* 70, 657–666.
650. Sinha Hikim, A. P., Bartke, A., and Russell, L. D. (1988). The seasonal breeding hamster as a model to study structure-function relationships in the testis. *Tissue Cell* 20, 63–78.
651. Sinha Hikim, A. P., Bartke, A., and Russell, L. D. (1988). Morphometric studies on hamster testes in gonadally active

and inactive states, light microscopic findings. *Biol. Reprod.* 39, 1225–1237.
652. Sinha Hikim, A. P., Amador, A. G., Klemcke, H. G., Bartke, A., and Russell, L. D. (1989). Correlative morphology and endocrinology of Sertoli cells in hamster testes in active and inactive states of spermatogenesis. *Endocrinology* 125, 1829–1843.
653. Zamboni, L., Conaway, C. H., and van Pelt, L. (1974). Seasonal changes in production of semen in free-ranging rhesus monkeys. *Biol. Reprod.* 11, 251–267.
654. Pudney, J., and Lacy, D. (1977). Correlation between ultrastructure and biochemical changes in the testes of the American grey squirrel, *Sciurus carolinesis,* during the reproductive cycle. *J. Reprod. Fertil.* 49, 5–16.
655. Vendreley, E., Guerillot, C., and Da Lage, C. (1972). Variations saisonnieres des cellules de Sertoli et de Leydig dans le testicule du hamster dore. Etude caryometrique. *C. R. Acad. Sci. (Paris)* 275, 1143–1146.
656. Berndston, W. E., and Desjardins, C. (1974). Circulating LH and FSH in hamsters during light deprivation and subsequent photoperiodic stimulation. *Endocrinology* 95, 195–205.
657. Lincoln, G. A. (1976). Seasonal variation in the episodic secretion of luteinizing hormone and testosterone in the ram. *Endocrinology* 69, 213–226.
658. Michael, R. P., and Bonsall, R. W. (1977). A 3-year study of an annual rhythm in plasma androgen levels in male rhesus monkey (*Macaco mulatto*) in a constant laboratory environment. *J. Reprod. Fertil.* 49, 129–131.
659. Vilar, O. (1968). Ultrastructural changes observed after hypophysectomy in rat testis. In *Gonadotropins* (E. Rosemberg, Ed.), pp. 205–211. Geron-X, Los Angeles.
660. Ghosn, S., Bartke, A., Grasso, P., Reichert, L. E., and Russell, L. D. (1992). The structural response of the hamster Sertoli cell to hypophysectomy, a correlative morphometric and endocrine study. *Anat. Rec.* 234, 513–529.
661. Gravis, C. J., and Weaker, F. J. (1977). Testicular involution following optic enucleation. *Cell. Tissue Res.* 184, 67–77.
662. Nebel, B. R., and Murphy, C. J. (1960). Damage and recovery of mouse testis after 1000r acute localized X-irradiation, with reference to restitution cells, Sertoli cell increase, and type A spermatogonial recovery. *Radiat. Res.* 12, 626–641.
663. Clegg, E. J. (1963). Studies on artificial cryptorchidism, degenerative and regenerative changes in germinal epithelium of rat testis. *J. Endocrinol.* 27, 241–251.
664. Clegg, E. J. (1963). Studies on artificial cryptorchidism, degenerative and regenerative changes in germinal epithelium of rat testis. *J. Endocrinol.* 26, 567–574.
665. Felizet, G., and Branca, A. (1898). Histologie du testicule ectopique. *J. Anat. Physiol.* 34, 589–641.
666. Nagy, F. (1972). Cell division kinetics and DNA synthesis in the immature Sertoli cells of the rat testis. *J. Reprod. Fertil.* 28, 389–395.
667. Murphy, H. D. (1965). Sertoli cell stimulation following intratesticular injections of FSH in the hypophysectomized rat. *Proc. Soc. Exp. Biol. Med.* 118, 1202–1205.
668. Hochereau-de Reviers, M. T., and Lincoln, G. A. (1978). Seasonal variation in the histology of the testis of the red deer, *Cervus elaphus. J. Reprod. Fertil.* 54, 209–213.
669. Johnson, L., and Neaves, W. B. (1981). Age-related changes in the Leydig cell population, seminiferous tubules and sperm production in stallions. *Biol. Reprod.* 24, 703–712.
670. Johnson, L., and Thompson, D. L. (1983). Age-related and seasonal variation in the Sertoli cell population, daily sperm production and serum concentrations of follicle-stimulating hormone, luteinizing hormone and testosterone in stallions. *Biol. Reprod.* 29, 777–789.
671. Johnson, L. (1985). Increased daily sperm production in the breeding season of stallions is explained by an elevated population of spermatogonia. *Biol. Reprod.* 32, 1181–1190.
672. Hochereau-de Reviers, M. T. (1981). Control of spermatogonial multiplication. In *Reproductive Processes and Contraception* (K. W. McKerns, Ed.), pp. 307–331. Plenum Press, New York.
673. Feig, L. A., Bellve, A. R., Erickson, N. H., and Klagsbrun, M. (1980). Sertoli cells contain a mitogenic polypeptide. *Proc. Natl. Acad. Sci. U S A* 77, 4774–4778.
674. Harbitz, T. B. (1973). Morphometric studies of the Sertoli cells in elderly men with special reference to the histology of the prostate. *Acta. Pathol. Microbiol. Scand.* 81A, 703.
675. Kerr, J. B., Rich, K. A., and de Kretser, D. M. (1979). Effects of experimental cryptorchidism on the ultrastructure and function of the Sertoli cell and peritubular tissue of the rat testis. *Biol. Reprod.* 21, 823–838.
676. Hoffer, A. P. (1983). Effects of gossypol on the seminiferous epithelium in the rat, a light and electron microscopy study. *Biol. Reprod.* 28, 1007–1020.
677. Flickinger, C. J. (1981). Effects of clomiphene on the structure of the testis, epididymis and sex accessory glands of the rat. *Am. J. Anat.* 149, 533–562.
678. Kreuger, P. M., Hodgen, G. D., and Sherins, R. J. (1974). New evidence for the role of the Sertoli cell and spermatogonia in feedback control of FSH secretion in male rats. *Endocrinology* 95, 955–962.
679. Chapin, R. E., Dutton, S. L., Ross, M. D., Sumrell, B. M., and Lamb, J. C. (1984). The effects of ethylene glycol monomethyl ether on testicular histology in rats. *J. Androl.* 5, 369–380.
680. Hildebrandt-Stark, H. E., and Fawcett, D. W. (1978). Effects of deficiency of essential fatty acids and treatment with prostaglandin E_2 on the ultrastructure of the rat testis. *Biol. Reprod.* 19, 736–746.
681. Kierszenbaum, A. L. (1970). Effect of trenimon on the ultrastructure of Sertoli cells in the mouse. *Vichows Arch. (Cell Pathol.)* 5, 1–12.
682. Zhuang, L. Z., Phillips, D. M., Gunsalus, G. L., Bardin, C. W., and Mather, J. P. (1983). Effects of gossypol on rat Sertoli and Leydig cells in primary culture and established cell lines. *J. Androl.* 4, 336–344.
683. Kerr, J. B., and Sharpe, R. M. (1989). Macrophage activation enhances the human chorionic gonadotrophin-induced disruption of spermatogenesis in the rat. *J. Endocrinol.* 121, 285–292.
684. Kerr, J. B., and Sharpe, R. M. (1989). Focal disruption of spermatogenesis in the testis of adult rats after a single administration of human chorionic gonadotrophin. *Cell. Tissue Res.* 257, 163–169.
685. Parizek, J. (1960). Sterilization of the male by cadmium salts. *J. Reprod. Fertil.* 1, 294–309.
686. Kerr, J. B., and Sharpe, R. M. (1986). Effects and interactions of LH and LHRH agonist on testicular morphology and function in hypophysectomized rats. *J. Reprod. Fertil.* 76, 175–192.
687. Vickery, B., McRae, G. I., Bergstrom, K., Briones, W., Worden, A., and Seidenberg, R. (1983). Inability of long-term administration of D-Nal(2)^6-LHRH to abolish fertility in male rats. *J. Androl.* 4, 283–291.
688. Schulze, W., and Schulze, C. (1981). Multinucleate Sertoli cells in aged human testis. *Cell. Tissue Res.* 217, 259–266.
689. Schulze, C., Holstein, A. F., Schirren, C., and Korner, F. (1976). On the morphology of the human Sertoli cells under normal conditions and in patients with impaired fertility. *Andrologia* 8, 167–178.

690. Nistal, M., Paniagua, R., Abaurrea, M. A., and Santamaria, L. (1982). Hyperplasia and the immature appearance of Sertoli cells in primary testicular disorders. *Hum. Pathol.* 13, 1.

691. Paniagua, R., Amat, R., Nistal, M., and Martin, A. (1985). Ultrastructural changes in Sertoli cells in ageing humans. *Int. J. Androl.* 8, 295–312.

692. Bustos-Obregon, E. (1976). Ultrastructure and function of the lamina propria of mammalian seminiferous tubules. *Andrologia* 8, 179–185.

693. Bustos-Obregon, E. (1976). Ultrastructure and function of the lamina propria of human seminiferous tubules. *Z. Zellforsch.* 141, 413–425.

694. de Kretser, D. M., Kerr, J. B., and Paulsen, C. A. (1975). The peritubular tissue in the normal and pathological human testis. An ultrastructural study. *Biol. Reprod.* 12, 317–324.

695. Christl, H. W. (1990). The lamina propria of vertebrate seminiferous tubules, a comparative light and electron microscopic investigation. *Andrologia* 22, 85–94.

696. Hermo, L., Lalli, M., and Clermont, Y. (1977). Arrangement of connective tissue components in the walls of seminiferous tubules of man and monkey. *Am. J. Anat.* 148, 433–436.

697. Chan, F. L., Inoue, S., and Le Blond, C. P. (1993). Cryofixation of basement membranes followed by freeze substitution or freeze drying demonstrates that they are composed of a tridimensional network of irregular cords. *Anat. Rec.* 235, 199–205.

698. Kerr, J. B., and de Kretser, D. M. (2005). Functional morphology of the testis. In *The Basic Endocrinology of the Testis* (L. J. de Groot and L. Jameson, Eds.). Saunders, Philadelphia (in press)

699. Virtanen, I., Kallajoki, M., and Narvanen, O., Paranko, J., Thornell, L. E., Miettinen, M., and Lehto, V. P. (1986). Peritubular myoid cells of human and rat testis are smooth muscle cells that contain desmin-type intermediate filaments. *Anat. Rec.* 215, 10–20.

700. Palombi, F., Farini, D., Salanova, M., De Grossi, S., and Stefanini, M. (1992). Development and cytodifferentiation of peritubular myoid cells in the rat testis. *Anat. Rec.* 233, 32–40.

701. Davidoff, M. S., Breucker, H., Holstein, A. F., and Seidl, K. (1990). Cellular architecture of the lamina propria of human seminiferous tubules. *Cell. Tissue Res.* 262, 253–261.

702. Kerr, J. B., and Knell, C. M. (1987). The regenerative capacity of testicular interstitial tissue. *Wilhelm Roux Arch. Dev. Biol.* 196, 467–471.

703. Jackson, A. E., O'Leary, P., Ayers, M. M., and de Kretser, D. M. (1986). The effects of ethylene dimethane sulphonate (EDS) on rat Leydig cells, evidence to support a connective tissue origin of Leydig cells. *Biol. Reprod.* 35, 425–437.

704. Russell, L. D., de Franca, L. R., Hess, R., and Cooke, P. (1995). Characteristics of mitotic cells in developing and adult testes with observations on cell lineages. *Tissue Cell* 27, 105–128.

705. Worley, R. T. S., Nicholson, H. D., and Pickejing, B. T. (1984). Testicular oxytocin, an initiator of seminiferous tubule movement? In *Recent Progress in Cellular Endocrinology of the Testis* (J. M. Saez, M. G. Forest, A. Dazoid, and J. Bertrand, Eds.), pp. 205–212. INSERM, Paris.

706. Paranko, J., Kallajoki, M., Pelliniemi, L. J., Lehto, V. P., and Virtanen, I. (1986). Transient co-expression of cytokeratin and vimentin in differentiating rat Sertoli cells. *Dev. Biol.* 117, 35–44.

707. Holstein, A. F., Maekawa, M., Nagano, T., and Davidoff, M. (1996). Myofibroblasts in the lamina propria of human seminiferous tubules are dynamic structures of herterogeneous phenotype. *Arch. Histol. Cytol.* 59, 109–125.

708. Middendorff, R., Muller, D., Wichers, S., Holstein, A. F., and Davidoff, M. S. (1997). Evidence for production and functional activity of nitric oxide in seminiferous tubules and blood vessels of the human testis. *J. Clin. Endocrinol. Metab.* 82, 4154–4161.

709. Baumgartner, H. G., and Holstein, A. F. (1967). Catchecolamine-containing nerve fibres in the testis of man. *Z. Zellforsch.* 79, 389–395.

710. Yamamoto, M., Takaba, H., Hashimoto, J., and Miyake, K. (1987). Evidence for innervation of the myoid cell in the human seminiferous tubule. *Urol. Int.* 42, 137–139.

711. Prince, F. P. (1996). Ultrastructural evidence of adrenergic, as well as cholinergic, nerve varicosities in relation to the lamina propria of the human seminiferous tubules during childhood. *Tissue Cell* 28, 507–513.

712. Benda, C. (1887). Untersuchunger iiber den Bau des funktionierenden Samenkanalchens einiger Saugetiere und Folgerungen fur die Spermatogenese dieser Wirbeltierklasse. *Arch. Mikr. Anat.* 30, 49–110.

713. Curtis, G. M. (1918). The morphology of the mammalian seminiferous tubule. *Am. J. Anat.* 24, 339–394.

714. Roosen-Runge, E. C., and Giesel, L. O. (1950). Quantitative studies on spermatogenesis in the albino rat. *Am. J. Anat.* 87, 1–30.

715. Hess, R. A., and Vogl, W. (2004). Symphony of Stages 2003. CD Produced by Vanguard Productions Inc.

716. Oakberg, E. F. (1956). Duration of spermatogenesis in the mouse and timing of stages of the cycle of the seminiferous epithelium. *Am. J. Anat.* 99, 507–516.

717. Ortavant, R. (1958).*La cycle spermatogenetique chez le Belier.* Thesis, University of Paris.

718. Millar, M. R., Sharpe, R. M., Weibauer, G. F., Fraser, H. M., and Saunders, P. T. K. (2000). Marmoset spermatogenesis, organizational similarities to the human. *Int. J. Androl.* 23, 266–277.

719. Weinbauer, G. F., Aslam, H., Krishnamurthy, H., Brinkworth, M. H., Einspanier, E., and Hodges, J. K. (2001). Quantitative analysis of spermatogenesis and apoptosis in the common marmoset (*Callithrix jacchus*) reveals high rates of spermatogonial turnover and high spermatogenic efficiency. *Biol. Reprod.* 64, 120–126.

720. Kerr, J. B. (1995). Macro, micro, and molecular research on spermatogenesis, the quest to understand its control. *Micros. Res. Tech.* 32, 364–384.

721. Smithwick, E. B., and Young, L. G. (1996). Germ cell maturation and cellular associations in the seminiferous epithelial cycle of the chimpanzee. *Tissue Cell* 28, 137–148.

722. Smithwick, E. B., Young, L. G., and Gould, K. G. (1996). Duration of spermatogenesis and relative frequency of each stage in the seminiferous epithelial cycle of the chimpanzee. *Tissue Cell* 28, 357–366.

723. Chowdhury, A. K., and Marshall, G. (1980). Irregular pattern of spermatogenesis in the baboon (*Papio anubis*) and its possible mechanism. In *Testicular Development, Structure and Function* (A. Steinberger and E. Steinberger, Eds.), pp. 129–137. Raven Press, New York.

724. Schulze, W., and Rehder, U. (1984). Organization and morphogenesis of the human seminiferous epithelium. *Cell. Tissue Res.* 237, 395–407.

725. Schulze, W., Reimer, M., Rehder, U., and Hohne, K. H. (1986). Computer-aided three-dimensional reconstructions of the arrangement of primary spermatocytes in human seminiferous tubules. *Cell. Tissue Res.* 244, 1–8.

726. Johnson, L. (1994). A new approach to study the architectural arrangement of spermatogenic stages revealed little evidence of a partial wave along the length of human seminiferous tubules. *J. Androl.* 15, 435–441.

727. Johnson, L. (1995). Efficiency of spermatogenesis. *Micros. Res. Tech.* 32, 385–422.

728. Johnson, L., McKenzie, K. S., and Snell, J. R. (1996). Partial wave in human seminiferous tubules appears to be a random occurrence. *Tissue Cell* 28, 127–136.

729. Chaturvedi, P. K., and Johnson, L. (1993). Architectural arrangement of stages of the spermatogenic cycle within human seminiferous tubules is related to efficiency of spermatogenesis. *Cell. Tissue Res.* 273, 65–70.

730. Sharpe, R. M. (1994). Regulation of spermatogenesis. In *The Physiology of Reproduction* (E. Knobil and J. D. Neill, Eds.), pp. 1363–1394. Raven Press, New York.

731. Zhengwei, Y., Wreford, N. G., Royce, P., de Kretser, D. M., and McLachlan, R. I. (1998). Stereological evaluation of human spermatogenesis after suppression by testosterone treatment, heterogeneous pattern of spermatogenic impairment. *J. Clin. Endocrinol. Metab.* 83, 1284–1291.

732. McLachlan, R. I., O'Donnell, L., Stanton, P. G., Balourdos, G., Frydenberg, M., de Kretser, D. M., and Robertson, D. M. (2002). Effects of testosterone plus medroxytestosterone acetate on semen quality, reproductive hormones, and germ cell populations in normal young men. *J. Clin. Endocrinol. Metab.* 87, 546–556.

733. Ortavant, R. (1954). Etude des generations spermatogoniales chez le Belier. *C. R. Soc. Biol.* 148, 1958–1961.

734. Ortavant, R. (1956). Autoradiographie des cellules germinales du testicule de Belier. Duree des phenomenes spermatogenetiques. *Arch. Anat. Microsc. Morphol. Exp.* 45, 1–10.

735. Clermont, Y., and Harvey, S. G. (1965). Duration of the cycle of the seminiferous epithelium of normal hypophysectomized and hypophysectomized-hormone treated albino rats. *Endocrinology* 76, 80–89.

736. Desclin, J., and Ortavant, R. (1963). Influences des hormones gonadotropes sur la duree des processus spermatogenetiques chez le rat. *Ann. Biol. Anim. Biochem. Biophys.* 3, 329–342.

737. Clouthier, D. E., Avarbock, M. R., Maika, S. D., Hammer, R. E., and Brinster, R. L. (1996). Rat spermatogenesis in mouse testis. *Nature* 381, 418–421.

738. Russell, L. D., and Brinster, R. L., (1996). Ultrastructural observations of spermatogenesis following transplantation of rat testis cells into mouse seminiferous tubules. *J. Androl.* 17, 615–627.

739. Franca, L. R., Ogawa, T., Avarbock, M. R., Brinster, R. L., and Russell, L. D. (1998). Germ cell genotype controls cell cycle during spermatogenesis in the rat. *Biol. Reprod.* 59, 1371–1377.

740. Ogawa, T., Dobrinski, I., and Brinster, R. L. (1999). Recipient preparation is critical for spermatogonial transplantation in the rat. *Tissue Cell* 31, 461–472.

741. Honaramooz, A., Snedaker, A., Boiani, M., Scholer, H., Dobrinski, I., and Schlatt, S. (2002). Sperm from neonatal mammalian testes grafted in mice. *Nature* 418, 778–781.

742. Honaramooz, A., Li, M. W., Penedo, C. T., Meyers, S., and Dobrinski, I. (2004). Accelerated maturation of primate testis xenografting into mice. *Biol. Reprod.* 70, 1500–1503.

743. Dobrinski, I., Megee, S., and Honaramooz, A. (2003). Xenografting of testis tissue from neonatal ram lambs into mouse hosts accelerates testicular maturation and sperm production. *Biol. Reprod.* 68 (Suppl 1), 190.

744. Russell, L. D., Lee, I. P., Ettlin, R., and Malone, J. P. (1983). Morphological pattern of response after administration of procarbazine, alteration of specific cell associations during the cycle of the seminiferous epithelium of the rat. *Tissue Cell* 15, 391–404.

745. Cleland, K. W. (1951). The spermatogenic cycle of the guinea pig. *Aust. J. Biol. Sci.* 4, 344–369.

746. Gnessi, L., Fabbri, A., and Spera, G. (1997). Gonadal peptides as mediators of development and functional control of the testis, an integrated system with hormones and local environment. *Endocr. Rev.* 18, 541–609.

747. Loveland, K. L., and Robertson, D. M. (2005). The transforming growth factor superfamily in Sertoli cell biology. In *The Sertoli Cell* (M. Griswold, M. Skinner, Eds.), pp. 1303–1307. Elsevier Academic Press, Oxford.

748. Van Dissel-Emiliani, F. M., Grootenhuis, A. J., de Jong, F. H., and de Rooij, D. G. (1989). Inhibin reduces spermatogonial numbers in testes of adult mice and Chinese hamsters. *Endocrinology* 125, 1898–1903.

749. Mather, J. P., Attie, K., Woodruff, T., Rice, G., and Phillips, D. (1990). Activin stimulates spermatogonial proliferation in germ-Sertoli cell cocultures from immature rat testis. *Endocrinology* 127, 3206–3214.

750. Rossi, P., Dolci, S., Albanesi, C., Grimaldi, P., Ricca, R., and Geremia, R. (1993). Follicle-stimulating hormone induction of Steel factor (SLF) mRNA in mouse Sertoli cells and stimulation of DNA synthesis in spermatogonia by soluble SLF. *Dev. Biol.* 155, 68–74.

751. Johnsen, S. G. (1964). Studies on the testicular-hypophyseal feedback mechanism in man. *Acta. Endocrinol. (Kbh)* Suppl 90, 99–124.

752. Perey, B., Clermont, Y., and Le Blond, G. P. (1961). The wave of the seminiferous epithelium in the rat. *Am. J. Anat.* 108, 47–77.

753. Dym, M., and Clermont, Y. (1970). Role of spermatogonia in the repair of the seminiferous epithelium following X-irradiation of the rat testis. *Am. J. Anat.* 128, 265–282.

754. Moens, P. B., and Hugenholtz, A. D. (1975). The arrangement of germ cells in the rat seminiferous tubule, an electron microscopic study. *J. Cell. Sci.* 19, 487–507.

755. Ren, H. P., and Russell, L. D. (1991). Clonal development of interconnected germ cells in the rat and its relationship to the segmental and subsegmental organization of spermatogenesis. *Am. J. Anat.* 192, 121–128.

756. Russell, L. D., Vogl, A. W., and Weber, J. E. (1987). Actin localization in male germ cells intercellular bridges in the rat and ground squirrel and disruption of bridges by cytochalasin D. *Am. J. Anat.* 180, 25–40.

757. Weber, J. E., and Russell, L. D. (1987). A study of intercellular bridges during spermatogenesis in the rat. *Am. J. Anat.* 180, 1–24.

758. Jiang, F. X., and Short, R. V. (1995). Male germ cell transplantation in rats, apparent synchronization of spermatogenesis between host and donor seminiferous epithelia. *Int. J. Androl.* 18, 326–330.

759. Brinster, R. L., and Avarbock, M. R. (1994). Germ line transmission of donor haplotype following spermatogonial transplantation. *Proc. Natl. Acad. Sci. U S A* 91, 11303–11307.

760. Brinster, R. L. (2002). Germline stem cell transplantation and transgenesis. *Science* 296, 2174–2176.

761. Dobrinski, I., Avarbock, M. R., and Brinster, R. L. (1999). Transplantation of germ cells from rabbits and dogs into mouse testes. *Biol. Reprod.* 61, 1331–1339.

762. Izadyar, F., Spierenberg, G. T., Creemers, L. B., den Ouden, K., and deRooij, D. G. (2002). Isolation and purification of type A spermatogonia from the bovine testis. *Reproduction* 124, 85–94.

763. Oatley, J. M., de Avila, D. M., McLean, D. J., Griswold, M. D., and Reeves, J. J. (2002). Transplantation of bovine germinal cells into mouse testes. *J. Anim. Sci.* 80, 1925–1931.

764. Dobrinski, I., Avarbock, M. R., and Brinster, R. L. (2000). Germ cell transplantation from large domestic animals into mouse testes. *Mol. Reprod. Dev.* 57, 270–279.

765. Schlatt, S., Foppiani, L., Rolf, C., Weinbauer, G. F., and Nieschlag, E. (2002). Germ cell transplantation into X-irradiated monkey testes. *Hum. Reprod.* 17, 55–62.

766. Nagano, M., McCarry, J. R., and Brinster, R. L. (2001). Primate spermatogonial stem cells colonize mouse testes. *Biol. Reprod.* 64, 1409–1416.

767. Nagano, M., Patrizio, P., and Brinster, R. L. (2002). Long-term survival of human spermatogonial stem cells in mouse testes. *Fert. Steril.* 78, 1225–1233.
768. Alvarez, J. D., Chen, D., Storer, E., and Sehgal, A. (2003). Non-cyclic and developmental stage-specific expression of circadian clock proteins during murine spermatogenesis. *Biol. Reprod.* 69, 81–91.
769. Morse, D., Cermakian, N., Brancorsini, S., Parvinen, M., and Sassone-Corsi, P. (2003). No circadian rhythms in testis, *Period 1* expression is *Clock* independent and developmentally regulated in the mouse. *Mol. Endocrinol.* 17, 141–151.
770. Rolshoven, E. (1941). Zur Frage des "Alterns" der generativen Elemente in den Hodenkanalchen. *Anat. Anz.* 91, 1–8.
771. Huckins, C. (1971). Cell cycle properties of differentiating spermatogonia in adult Sprague Dawley rats. *Cell. Tissue Kinet.* 4, 139–154.
772. Huckins, C. (1971). The spermatogonial stem cell population in adult rats. III. Evidence for a long-cycling population. *Cell. Tissue Kinet.* 4, 335–349.
773. Huckins, C. (1971). The spermatogonial stem cell population in adult rats. II. A radioautographic analysis of their cell cycle properties. *Cell. Tissue Kinet.* 4, 313–334.
774. Tegelenbosch, R. A., and deRooij, D. G. (1993). A quantitative study of spermatogonial multiplication and stem cell renewal in the C3H/101 F1 hybrid mouse. *Mutat. Res.* 290, 193–200.
775. Oakberg, E. F. (1971). Spermatogonial stem-cell renewal in the mouse. *Anat. Rec.* 169, 515–531.
776. Huckins, C., and Oakberg, E. F. (1978). Morphological and quantitative analysis of spermatogonia in mouse testes using whole mounted seminiferous tubules. I. The normal testes. *Anat. Rec.* 192, 519–528.
777. Oud, J. L., and de Rooij, D. G. (1977). Spermatogenesis in the Chinese hamster. *Anat. Rec.* 187, 113–124.
778. de Rooij, D. G., Lok, D., and Weenk, D. (1985). Feedback regulation of the proliferation of the undifferentiated spermatogonia in the Chinese hamster by the differentiating spermatogonia. *Cell. Tissue Kinet.* 18, 71–81.
779. Lok, D., Weenk, D., and de Rooij, D. G. (1982). Morphology, proliferation and differentiation of undifferentiated spermatogonia in the Chinese hamster and the ram. *Anat. Rec.* 203, 83–99.
780. deRooij, D. G., and Grootegoed, J. A. (1998). Spermatogonial stem cells. *Curr. Opin. Cell. Biol.* 10, 694–701.
781. Lin, H. (1998). The self-renewing mechanisms of stem cells in the germline. *Curr. Opin. Cell. Biol.* 10, 687–693.
782. de Rooij, D. G. (1998). Stem cells in the testis. *Int. J. Exp. Pathol.* 79, 67–80.
783. de Rooij, D. G. (2001). Proliferation and differentiation of spermatogonial stem cells. *Reproduction* 121, 347–354.
784. Zhao, G. Q., and Garbers, D. L. (2002). Male germ cell specification and differentiation. *Dev. Cell.* 2, 537–547.
785. Hamra, F. K., Schultz, N., Chapman, K. M., Grellhesl, D. M., Cronkite, J. T., Hammer, R. E., and Garbers, D. L. (2004). Defining the spermatogonial stem cell. *Dev. Biol.* 269, 393–410.
786. Olive, V., and Cuzin, F. (2005). The spermatogonial stem cell, from basic knowledge to trans technology. *Int. J. Biochem. Cell. Biol.* 37, 246–250.
787. Ogawa, T., Dobrinski, I., Avarbock, M. R., and Brinster, R. L. (2000). Transplantation of male germ line stem cells restores fertility in infertile mice. *Nat. Med.* 6, 29–34.
788. Shinohara, T., Avarbock, M. R., and Brinster, R. L. (2000). Functional analysis of spermatogonial stem cells in Steel and cryptorchid infertile mouse models. *Dev. Biol.* 220, 410–411.
789. Kanatsu-Shinohara, M., Ogonuki, N., Inoue, K., Mili, H., Ogura, A., Toyokuni, S., and Shinohara, T. (2003). Long-term proliferation in culture and germline transmission of mouse male germline stem cells. *Biol. Reprod.* 69, 612–616.
790. Shinohara, T., Avarbock, M. R., and Brinster, R. L. (1999). ß1- and a6-Integrin are surface markers on mouse spermaogonial stem cells. *Proc. Natl. Acad. Sci. U S A* 96, 5504–5509.
791. Shinohara, T., Orwig, K. E., Avarbock, M. R., and Brinster, R. L. (2000). Spermatogonial stem cell enrichment by multiparameter selection of mouse testis cells. *Proc. Natl. Acad. Sci. U S A* 97, 8346–8351.
792. Giuili, G., Tomljenovic, A., Labrecque, N., Oulad-Abdelghani, M., Rassoulzadegan, M., and Cuzin, F. (2002). Murine spermatogonial stem cells, targeted transgene expression and purification in an active state. *EMBO Rep.* 3, 753–759.
793. Meng, X., Lindahl, M., Hyvonon, M. E., Parvinen, M., de Rooij, D. G., Hess, M. W., Raatikainen-Ahokas, A., Sainio, K., Rauvala, H., Lakso, M., Pichel, J. G., Wstphal, H., Saarma, K., Sariola, H. (2000). Regulation of cell fate decision of undifferentiated spermatogonia by GDNF. *Science* 287, 1489–1493.
794. Yomogida, K., Yagura, Y., Tadokoro, Y., and Nishimune, Y. (2003). Dramatic expansion of germinal stem cells by ectopically expressed human glial cell line-derived neurotrophic factor in mouse Sertoli cells. *Biol. Reprod.* 69, 1303–1307.
795. Nagano, M., Ryu, B. Y., Brinster, C. J., Avarbock, M. R., and Brinster, R. L. (2003). Maintenance of mouse male germ line stem cells in vitro. *Biol. Reprod.* 68, 2207–2214.
796. Schlatt, S., Honaramooz, A., Boiani, M., Scholer, H. R., and Dobrinski, I. (2003). Progeny from sperm obtained after ectopic grafting of neonatal mouse testes. *Biol. Reprod.* 68, 2331–2335.
797. Snedaker, A. K., Honaramooz, A., and Dobrinski, I. (2004). A game of cat and mouse, xenografting of testis tissue from domestic kittens results in complete cat spermatogenesis in a mouse host. *J. Androl.* 25, 926–30.
798. Lawson, K. A., Dunn, N. R., Roelen, B. A., Zeinstra, L. M., David, A. M., Wright, C. V., Korving, J. P. and Hogan, B. L. (1999). *Bmp4* is required for the generation of primordial germ cells in the mouse embryo *Genes Dev.* 13, 424–436.
799. Pesce, M., Klinger, F. G., and De Felici, M. (2002). Derivation of primordial germ cells from cells of the mouse epiblast, phenotypic induction and growth control by Bmp4 signalling. *Mech. Dev.* 112, 15–24.
800. Toyooka, Y., Tsunekawa, N., Akasu, R., and Noce, T. (2003). Embryonic stem cells can form germ cell *in vitro*. *Proc. Natl. Acad. Sci. U S A* 100, 11457–11462.
801. Koshimizu, U., Watanabe, M., and Nakatsuji, N. (1995). Retinoic acid is a potent growth activator of mouse primordial germ cells *in vitro*. *Dev. Biol.* 168, 683–685.
802. Geijsen, N., Horoshak, M., Kim, K., Gribnau, J., Eggan, K., and Daley, G. Q. (2004). Derivation of embryonic germ cells and male gametes from embryonic stem cells. *Nature* 427, 148–154.
803. Celebi, C., Guillaudeux, T., Auvray, P., Vallet-Erdtmann, V., and Jegou, B. (2003). The making of "transgenic spermatozoa." *Biol. Reprod.* 68, 1477–1483.
804. Yamazaki, Y., Yagi, T., Ozaki, T., and Imoto, K. (2000). In vivo gene transfer to mouse spermatogenic cells using green fluorescent protein as a marker. *J. Exp. Zool.* 286, 212–218.
805. Widlak, W., Scieglinska, D., Vydra, N., Malusecka, E., and Krawczyk, Z. (2003). In vivo electroporation of the testis versus transgenic mice model in functional studies of spermatocyte-specific hst 70 gene promoter, a comparative study. *Mol. Reprod. Dev.* 65, 382–388.
806. Kanatsu-Shinohara, M., Toykuni, S., and Shinohara, T. (2004). Transgenic mice produced by retroviral transduction of male germ line stem cells in vivo. *Biol. Reprod.* 71, 1202–1207.

807. Oately, J. M., de Avila, D. M., Reeves, J. J., and McLean, D. J. (2004). Spermatogenesis and germ cell transgene expression in xenografted bovine tesicular tissue. *Biol. Reprod.* 71, 494–501.
808. Hamra, F. K., Gatlin, J., Chapman, K. M., Grelhesl, D. M., Garcia, J. V., Hammer, R. E., and Garbers, D. L. (2002). Production of transgenic rats by lentiviral transduction of male germ-line stem cells. *Proc. Natl. Acad. Sci. U S A* 99, 14931–14936.
809. Chowdhury, A. K., and Steinberger, E. (1971). A radioautography technique for human and rat seminiferous tubules mounted *in toto*. *Exp. Cell. Res.* 64, 450–456.
810. Fuerst, C. (1887). Über die Entwicklung der Samenkoerperchen bei den Beteltieren. *Arch. Mikrosk. Anat.* 81, 1–23.
811. Hochereau, M. T. (1963). Constance des frequences relatives des etudes du cycle de l'epithelium seminifere chez lez taureau et chez le rat. *Ann. Biol. Anim. Biochem. Biophys.* 3, 93–102.
812. Tiba, T., Ishikawa, T., and Murakami, A. (1968). Histologische Untersuchung der Kinetik des Spermatogenese beim Mink. II. Samenepithel-welle in der Paarungszeit. *Jpn. J. Vet. Res.* 16, 159–187.
813. Brinster, R. L., and Zimmerman, J. W. (1994). Spermatogenesis following male germ-cell transplantation. *Proc. Natl. Acad. Sci. U S A* 91, 11298–11302.
814. Brinster, R. L., Ryu, B. Y., Avarbock, M. R., Karagenc, L., Brinster, R. L., and Orwig, K. E. (2003). Restoration of fertility by germ cell transplantation requires effective recipient preparation. *Biol. Reprod.* 69, 412–420.
815. Ventela, S., Ohta, H., Parvinen, M., and Nishimune, Y. (2002). Development of the stages of the cycle in mouse seminiferous epithelium after transplantation of green fluorescent protein-labeled spermatogonial stem cells. *Biol. Reprod.* 66, 1422–1429.
816. Ogawa, T., Ohmura, M., Yumura, Y., Sawada, H., and Kubota, Y. (2003). Expansion of murine spermatogonial stem cells through serial transplantation. *Biol. Reprod.* 68, 316–322.
817. Ohta, H., Yomogida, K., Yamada, S., Okabe, M., and Nishimune, Y. (2000). Real-time observation of transplanted "green germ cells", proliferation and differentiation of stem cells. *Dev. Growth. Differ.* 42, 105–112.
818. Ohta, H., Wakayama, T., and Nishimune, Y. (2004). Commitment of fetal male germ cells to spermatogonial stem cells during mouse embryonic development. *Biol. Reprod.* 70, 1286–1291.
819. Parvinen, M. (1982). Regulation of the seminiferous epithelium. *Endocr. Rev.* 3, 404–417.
820. Parvinen, M., Vihko, K. K., and Toppari, J. (1987). Cell interactions during the seminiferous epithelial cycle. *Int. Rev. Cytol.* 104, 115–151.
821. Jegou, B., Laws, A. O., and de Kretser, D. M. (1984). Changes in testicular function in the rat testis to heat, further evidence for interaction of germ cells, Sertoli cells and Leydig cells. *Int. J. Androl.* 7, 244–257.
822. Chiarini-Garcia, H., Hornick, J. R., Griswold, M. D., and Russell, L. D. (2001). Distribution of type A spermatogonia in the mouse is not random. *Biol. Reprod.* 65, 1179–1185.
823. Chiarini-Garcia, H., Raymer, A. M., and Russell, L. D. (2003). Non-random distribution of spermatogonia in rats, evidence of niches in the seminiferous tubules. *Reproduction* 126, 669–680.
824. Roosen-Runge, E. C. (1955). Untersuchungen über die degeneration samenbilender zellen in der normalen Spermatogenese der ratte. *Z. Zellforsch.* 41, 221–225.
825. Amann, R. P. (1962). Reproductive capacity of dairy bulls. IV Spermatogenesis and testicular germ cell degeneration. *Am. J. Anat.* 110, 69–78.
826. Barr, A. B., Moore, D. J., and Paulsen, C. A. (1971). Germinal cell loss during human spermatogenesis. *J. Reprod. Fertil.* 25, 75–80.
827. Huckins, C. (1978). The morphology and kinetics of spermatogonial degeneration in normal adult rats, an analysis using a simplified classification of the germinal epithelium. *Anat. Rec.* 190, 905–926.
828. Allan, D. J., Harmon, B. V., and Kerr, J. F. R. (1987). Cell death in spermatogenesis. In *Perspectives on Mammalian Cell Death* (C. S. Potten, Ed.), pp. 229–258. Oxford University Press, London.
829. Print, C. G., and Loveland, K. L. (2000). Germ cell suicide, new insights into apoptosis during spermatogenesis. *BioEssays* 22, 423–430.
830. Johnson, L. (1986). Spermatogenesis and aging in the human. *J. Androl.* 7, 331–354.
831. Johnson, L., Chaturvedi, P. K., and Williams, J. D. (1992). Missing generations of spermatocytes and spermatids in seminiferous epithelium contribute to low efficiency of spermatogenesis in humans. *Biol. Reprod.* 47, 1091–1098.
832. Henriksen, K. and Parvinen, M. (1998). Stage-specific apoptosis of male germ cells in the rat, mechanisms of cell death studied by supravital squash preparations. *Tissue Cell* 30, 692–701.
833. Weil, M., Jacobson, M. D., and Raff, M. C. (1998). Are caspases involved in the death of cells with a transcriptionally inactive nucleus? Sperm and chicken erythrocytes. *J. Cell. Sci.* 111, 2707–2715.
834. Boekelheide, K., Fleming, S. L., Johnson, K. J., Patel, S. R., and Schoenfeld, H. A. (2000). Role of Sertoli cells in injury-associated testicular germ cell apoptosis. *Proc. Soc. Exp. Biol. Med.* 225, 105–115.
835. Pentikainen, V., Erkkila, K., Suomalainen, L., Otala, M., Pentikainen, M. O., Parvinen, M., and Dunkel, L. (2001). TNFalpha down-regulates the Fas ligand and inhibits germ cell apoptosis in the human testis. *J. Clin. Endocrinol. Metab.* 86, 4480–4488.
836. Lee, J., Richburg, J., Shipp, E. B., Meistrich, M. L., and Boekelheide, K. (1999). The Fas system, a regulator of testicular germ cell apoptosis, is differentially up-regulated in Sertoli cell versus germ cell injury of the testis. *Endocrinology* 140, 852–858.
837. Furuchi, T., Masuko, K., Nishimune, Y., Obinata, M., and Matsui, Y. (1996). Inhibition of testicular germ cell apoptosis and differentiation in mice misexpressing Bcl-2 in spermatogonia. *Development* 122, 1703–1709.
838. Knudson, C. M., Tung, K. S., Tourtellotte, W. G., Brown, G. A., and Korsmeyer, S. J. (1995). Bax-deficient mice with lymphoid hyperplasia and male germ cell death. *Science* 270, 96–99.
839. Print, C. G., Loveland, K. L.,Gibson, L., Meehan, T., Stylianou, A., Wreford, N., de Kretser, D. M., Metcalfe, D., Kontgen, F., Adams, J. M., and Cory, S. (1998). Apoptosis regulator Bcl-w is essential for spermatogenesis but appears otherwise redundant. *Proc. Natl. Acad. Sci. U S A* 95, 12424–12431.
840. Ross, A. J., Waymire, K. G., Moss, J. E., Parlow, A. F., Skinner, N. K., Russell, L. D., and MacGregor, G. R. (1998). Testicular degeneration in Bclw-deficient mice. *Nat. Genet.* 18, 251–256.
841. Yoshinaga, K., Nishikawa, S., Ogawa, M., Hayashi, S., Kunisada, T., and Fujimoto, T. (1991). Role of c-kit in mouse spermatogenesis, identification of spermatogonia as a specific site of c-kit expression and function. *Development* 113, 689–699.

842. Packer, A. I., Besmer, P., and Bachvarova, R. F. (1995). Kit ligand mediates survival of type A spermatogonia and dividing spermatocytes in postnatal mouse testes. *Mol. Reprod. Dev.* 42, 303–310.
843. Allard, E. K., Blanchard, K. T., and Boekelheide K. (1996). Exogenous stem cell factor (SCF) compensates for altered endogenous SCF expression in 2.5-hexanedione-induced tesicular atrophy in rats. *Biol. Reprod.* 55, 185–193.
844. Manova, K., Huang, E. J., Angeles, M., De Leon, V., Snchez, S., Pronovost, S. M., Besmer, P., and Bacvarova, R. F. (1993). The expression pattern of the c-kit ligand in gonads of mice supports a role for the c-kit receptor in oocyte growth and in proliferation of spermatogonia. *Dev. Biol.* 157, 85–99.
845. Munsie, M., Schlatt, S., de Kretser, D. M., and Loveland, K. L. (1997). Stem cell factor in the postnatal rat testis. *Mol. Reprod. Dev.* 47, 19–25.
846. Yan, W., Suiminen, J., and Toppari, J. (2000). Stem cell factor protects germ cells from apoptosis in vitro. *J. Cell. Sci.* 113, 161–168.
847. Meehan, T., Loveland, K. L., de Kretser, D., Cory, S., and Print, C. G. (2001). Developmental regulation of the bcl-2 family during spermatogenesis, insights into the sterility of bcl-w-/- male mice. *Cell. Death Differ.* 8, 225–233.
848. Stieve, H. (1921). *Entwickelung Bau und Bedeutung der Keimdrusenzwis-chenzellen.* JF Bergmann, Munich.
849. Hanes, F. M. (1911). The relations of the interstitial cells of Leydig to the production of an internal secretion by the mammalian testis. *J. Exp. Med.* 13, 338–354.
850. Hooker, C. W. (1944). The postnatal history and function of the interstitial cells of the testis of the bull. *Am. J. Anat.* 74, 1–37.
851. Hooker, C. W. (1948). The biology of the interstitial cells of the testis. *Rec. Prog. Horm. Res.* 3, 173–195.
852. Fawcett, D. W., Heidger, D. M., and Leak, L. V. (1969). Lymph vascular system of the interstitial tissue of the testis as revealed by electron microscopy. *J. Reprod. Fertil.* 19, 109–119.
853. Christensen, A. K. (1975). Leydig cells. In *Handbook of Physiology* (D. W. Hamilton and R. O. Greep, Eds.), Section 7, Vol. 5, pp. 57–94. American Physiological Society, Washington, DC.
854. Payne, A. P., Hardy, M. P., and Russell, L. D., eds. (1996). *The Leydig Cell.* Cache River Press, Vienna, IL.
855. Haider, S. G. (2004). Cell biology of Leydig cells in the testis. *Int. Rev. Cytol.* 233, 181–241.
856. Leydig, F. (1850). Zur anatomic der mannlichen Geschlechtsorgane und Analdrusen der Saugetiere. *Z. Wiss. Zool.* 2, 1–57.
857. Hofmeister, H. (1872). Untersuchungen über die Zwischensubstanz im Hoden der Saugethiere. *Sitzungsber. Acad. Wiss. Mathnatur. Classe. Wien.* 5, 77.
858. Plato, J. (1896). Die interstitiellen Zellen des Hodens und ihre physiologische Bedeutung. *Arch. Mikrosk. Anat.* 48, 280.
859. Lenhossek, M. (1897). Beitrage zur Kenntniss der Zwischenzellen des Hodens. *Arch. Anat. Physiol. Anat. Abt.* 51, 65.
860. Bouin, P., and Ancel, P. (1904). La glande interstitielle a seule dans le testicule une action generale sur 1'organisme. *Demonstr. Exp. C. R. Acad. Sci. (Paris)* 138, 110.
861. Ancel, P., and Bouin, P. (1903). L'apparition des caracteres sexuels secondaires est sous la dependance de la gland interstitielle. *C. R. Acad. Sci. (Paris)* 138, 168.
862. Ciaccio, C. (1910). Contribute alia distribuzione ed alia fisiopathologia cellulare dei lipoidi. *Arch. Zellforsch.* 5, 235.
863. Whitehead, R. H. (1912). On the chemical nature of certain granules in the interstitial cells of the testis. *Am. J. Anat.* 14, 63.
864. Lotz, A., and Jaffe, R. (1924). Die Hoden bei Allgemeinerkrankungen. *Z. Konstit.* 10, 99.
865. McGee, L. C. (1972). Effect of injections of lipid fraction of bull testicle in capons. *Proc. Inst. Med. Chicago* 6, 242.
866. Gallagher, T. F., and Koch, F. C. (1929). The testicular hormone. *J. Biol. Chem.* 84, 495–500.
867. Butenandt, A., and Hanisch, G. (1935). Über Testosteron. Umwandlung des Dehydroandrosterons in Androstenediol und Testosteron, ein Weg zur Darstellung des Testosterons ans Cholesterin. *Z. Physiol. Chem.* 237, 89–97.
868. Ruzicka, L., and Wettstein, A. (1935). Kunstliche Herstellung des mannlichen sexual Hormones. Transdehydro-androsten-3,17-dions. *Helv. Chim. Acta.* 18, 986–994.
869. Wattenberg, L. W. (1958). Microscopic histochemical demonstration of steroid-3ß-ol dehydrogenase in tissue sections. *J. Histochem. Cytochem.* 6, 225–232.
870. Levy, H., Dean, H. W., and Rubin, B. L. (1959). Visualization of steroid-3ß-ol dehydrogenase activity in tissues of intact and hypophysectomized rats. *Endocrinology* 65, 932–943.
871. Baillie, A. H. (1964). Further observations on the growth and histochemistry of the Leydig tissue in the postnatal prepubertal mouse testis. *J. Anat.* 98, 403–419.
872. Christensen, A. K., and Mason, N. R. (1965). Comparative ability of seminiferous tubules and interstitial tissue of rat testes to synthesize androgens from progesterone-4-^{14}C *in vitro. Endocrinology* 76, 646–656.
873. Hall, P. F., Irby, D. C., and de Kretser, D. M. (1969). Conversion of cholesterol to androgens by rat testes, comparison of interstitial cells and seminiferous tubules. *Endocrinology* 84, 488–496.
874. Clark, R. V. (1976). Three-dimensional organization of testicular interstitial tissue and lymphatic space in the rat. *Anat. Rec.* 184, 203–226.
875. Holstein, A. F., Orlandini, G. E., and Moller, R. (1979). Distribution and fine structure of the lymphatic system in the human testis. *Cell. Tissue Res.* 200, 15–27.
876. Wagner, K. (1925). Sind die Zwischenzellen des Saugetierhodens Drusenzellen? Ein Beitrag zur Zytologie und Zygotogenese. *Biol. Gen.* 1, 22.
877. Hooker, C. W., Pfieffer, C. A., and de Vita, J. (1946). The significance of the diameter of the interstitial cells of the testis in the aged dog. *Anat. Rec.* 94, 471–472.
878. Evans, H. M., and Schulemann, W. (1914). The action of vital stains belonging to the benzidine group. *Science* 39, 443.
879. Christensen, A. K., and Peacock, K. C. (1980). Increase in Leydig cell number in testis of adult rats treated chronically with an excess of human chorionic gonadotrophin. *Biol. Reprod.* 22, 383–392.
880. Ennist, D. L., and Jones, K. H. (1983). Rapid method for identification of macrophages in suspension by acid alpha-naphthyl acetate esterase activity. *J. Histochem. Cytochem.* 31, 960–963.
881. Molenaar, R., Rommerts, F. F. G., and van der Molen, H. J. (1986). Non-specific esterase, a specific and useful marker enzyme for Leydig cells from mature rats. *J. Endocrinol.* 108, 229–334.
882. Miller, S. C., Bowman, B. M., and Rosland, H. G. (1983). Structure, cytochemistry, endocytic activity and immunoglobulin (Fc) receptors of rat testicular interstitial tissue macrophages. *Am. J. Anat.* 168, 1–13.
883. Miller, S. C. (1982). Localization of plutonium-241 in the testis. An inter-species comparison using light and electron microscope autoradiography. *Int. J. Radiat. Biol.* 41, 633–643.
884. Miller, S. C., and Bowman, B. M. (1983). Tissue, cellular and subcellular distribution of ^{241}Pu in the rat testis. *Radiat Res* 94, 416–426.

885. Orth. J., and Christensen, A. K. (1977). Localization of 125 I-labelled FSH in the testes of hypophysectomized rats by autoradiography at the light and electron microscope levels. *Endocrinology* 101, 262–278.

886. Christensen, A. K., Komorowski, T. E., Wilson, B., Ma, S. F., and Stevens, R. W. (1985). The distribution of serum albumin in the rat testis, studied by electron microscope immunocytochemistry on ultrathin frozen sections. *Endocrinology* 116, 1983–1996.

887. Hutson, J. C. (1992). Development of cytoplasmic digitations between Leydig cells and testicular macrophages of the rat. *Cell. Tissue Res.* 267, 385–389.

888. Berg, A. (1985). Effect of cryptorchidism on the morphology of testicular macrophages, evidence for a Leydig cell-macrophage interaction in the rat testis. *Int. J. Androl.* 8, 86–96.

889. Niemi, M., Sharpe, R. M., Brown, W. R. A. (1986). Macrophages in the interstitial tissue of the rat testis. *Cell. Tissue Res.* 243, 337–344.

890. Raburn, D. J., Coquelin, A., and Hutson, J. C. (1991). Human chorionic gonadotrophin increases the concentration of macrophages in neonatal rat testis. *Biol. Reprod.* 44, 172–177.

891. Hutson, J. C. (1990). Changes in the concentration and size of testicular macrophages during development. *Biol. Reprod.* 43, 885–890.

892. Hales, D. B. (2002). Testicular macrophage modulation of Leydig cell steroidogenesis. *J. Reprod. Immunol.* 57, 3–18.

893. Pollanen, P., and Maddocks, S. (1988). Macrophages, lymphocytes and MHC II antigen in the ram and rat testis. *J. Reprod. Fertil.* 82, 437–445.

894. Kerr, J. B., Donachie, K. I., and Rommerts, F. F. G. (1985). Selective destruction and regeneration of rat Leydig cells *in vivo.* A new method for the study of seminiferous tubular-interstitial tissue interaction. *Cell. Tissue Res.* 242, 145–156.

895. Kerr, J. B., Bartlett, J. M. S., and Donachie, K. (1986). Acute response of testicular interstitial tissue in rats to the cytotoxic drug ethane dimethane sulfonate. *Cell. Tissue Res.* 243, 405–414.

896. Gaytan, F., Bellido, C., Morales, C., Reymundo, C., Aguilar, E., and van Rooijen, N. (1994). Selective depletion of testicular macrophages and prevention of Leydig cell repopulation after treatment with ethylene dimethane sulphonate (EDS) in rat. *J. Reprod. Fertil.* 101, 175–182.

897. Gaytan, F., Bellido, C., Morales, C., Reymundo, C., Aguilar, E., and van Rooijen, N. (1994). Effects of macrophage depletion at different times after treatment with ethylene dimethane sulphonate (EDS) on the regeneration of Leydig cells in the adult rat. *J. Androl.* 15, 558–564.

898. Gunn, S. A., and Gould, T. C. (1975). Vasculature of the testes and adnexa. In *Handbook of Physiology* (D. W. Hamilton and R. O. Greep, Eds.), Section 7, Part 5, pp. 117–142. American Physiological Society, Washington, DC.

899. Maseki, Y., Miyake, K., Mitsuya, H., Kitamura, H., and Yamada, K. (1981). Mastocytosis occurring in the testes from patients with idiopathic male infertility. *Fertil. Steril.* 36, 814–817.

900. Nistal, M., Santamaria, L., and Paniagua, R. (1984). Mast cells in the human testis and epididymis from birth to adulthood. *Acta. Anat.* 119, 155–160.

901. Mayerhofer, A., Bartke, A., Amador, A. G., and Began, T. (1989). Histamine affects testicular steroid production in the golden hamster. *Endocrinology* 125, 560–562.

902. Payne, A. P., Hardy, M. P., and Russell, L. D., eds. (1996). *The Leydig Cell.* Cache River Press, Vienna, IL.

903. Terao, K. (1973). Annular nucleolus in Leydig cells of mouse. *Z. Zellforsch.* 137, 167–175.

904. Terao, L., Sakaibara, Y., Yamazaki, M., and Miyaki, K. (1971). Annular nucleolus in chicken embryonal hepatocyte induced by aflatoxin B. *Exp. Cell. Res.* 66, 81–89.

905. Christensen, A. K., and Fawcett, D. W. (1966). The fine structure of testicular interstitial cells in mice. *Am. J. Anat.* 118, 551–572.

906. de Kretser, D. M. (1967). The fine structure of the testicular interstitial cells in men of normal androgenic status. *Z. Zellforsch.* 80, 594–609.

907. de Kretser, D. M. (1967). Changes in the fine structure of human testicular interstitial cells after treatment with human gonadotrophins. *Z. Zellforsch.* 83, 344–358.

908. Kerr, J. B., Rich, K. A., and de Kretser, D. M. (1979). Alteration of the fine structure and androgen secretion of the interstitial cells in the experimentally cryptorchid rat testis. *Biol. Reprod.* 20, 409–422.

909. Mori, H., and Christensen, A. K. (1980). Morphometric analysis of Leydig cells in the normal rat testis. *J. Cell. Biol.* 84, 340–354.

910. Zirkin, B. R., Ewing, L. L., Kromann, N., and Cochran, R. C. (1980). Testosterone secretion by rat, rabbit, guinea pig, dog and hamster testes perfused *in vitro*, correlation with Leydig cell ultrastructure. *Endocrinology* 107, 1867–1874.

911. Mori, H., Kadota, A., Fukunishi, R., Kukita, H., Takeuchi, N., and Matsumoto, K. (1980). Effects of a cholesterol-rich diet and a hypolipidemic drug (Clofibrate, CP1 B) on Leydig cells in rats. Stereological and biochemical analysis. *Andrologia* 12, 271–291.

912. Mori, H., Hiromoto, N., Nakahara, M., and Shiraishi, T. (1982). Stereological analysis of Leydig cell ultrastructure in aged humans. *J. Clin. Endocrinol. Metab.* 55, 634–641.

913. Mori, H., Shimizu, D., Fukunishi, Y., and Christensen, A. K. (1982). Morphometric analysis of testicular Leydig cells in normal adult mice. *Anat. Rec.* 204, 333–339.

914. Connell, C. J., and Christensen, A. K. (1975). The ultrastructure of the canine testicular interstitial tissue. *Biol. Reprod.* 12, 368–382.

915. Weaker, F. J. (1977). The fine structure of the interstitial tissue of the testis of the nine-banded armadillo. *Anat. Rec.* 187, 11–28.

916. Camatini, M., Franchi, E., and de Curtis, I. (1981). Ultrastructure of Leydig cells in the African green monkey. *J. Ultrastruct. Res.* 76, 224–234.

917. Snapp, E. L., Hegde, R. S., Francolini, M., Lombardo, S., Pedrazzini, E., Borgese, N., and Lippincott-Schwartz, J. (2003). Formation of stacked ER cisternae by low affinity protein interactions. *J. Cell. Biol.* 163, 257–269.

918. Murakami, M., and Kitahara, Y. (1971). Cylindrical bodies derived from endoplasmic reticulum in Leydig cells of the rat testis. *J. Electronmicrosc.* 20, 318–323.

919. Christensen, A. K., and Fawcett, D. W. (1961). The normal fine structure of opossum testicular interstitial cells. *J. Biophys. Biochem. Cytol.* 9, 653–670.

920. Nakai, Y. (1974). Adult interstitial cell. In *An Atlas of Electron Micrographs, Functional Morphology of Endocrine Glands* (K. Kurosumi and H. Fujita, Eds.), pp. 178–179. Igaku Shoin, Tokyo.

921. Emura, S. (1978). Electron microscopic studies on annulate lamellae in Leydig cells of rabbit. *Cell* 10, 752–756.

922. Ohata, M. (1979). Electron microscopic study on the testicular interstitial cells in the mouse. *Arch. Histol. Jpn.* 42, 51–79.

923. Ewing, L. L., Zirkin, B. R., Cochran, R. C., Kromann, N., Peters, C., and Ruiz-Bravo, N. (1979). Testosterone secretion by rat, rabbit, guinea pig, dog, and hamster testes perfused *in vitro*, correlation with Leydig cell mass. *Endocrinology* 105, 1135–1142.

924. Wing, T. Y., Ewing, L. L., and Zirkin, B. R. (1984). Effects of luteinizing hormone withdrawal on Leydig cell smooth endoplasmic reticulum and steroidogenic reactions which convert pregnenolone to testosterone. *Endocrinology* 115, 2290–2296.
925. Wing, T. Y., Ewing, L. L., Zegeye, B., and Zirkin, B. R. (1985). Restoration effects of exogenous luteinizing hormone on the testicular steroidogenesis and Leydig cell ultrastructure. *Endocrinology* 117, 1779–1787.
926. Mendis-Handagama, S. M. L. C., Zirkin, B. R., and Ewing, L. L. (1988). Comparison of components of the testis interstitium with testosterone secretion in hamster, rat, and guinea pig testes perfused *in vitro*. *Am. J. Anat.* 181, 12–22.
927. Fouquet, J. P., Meusy-Dessolle, N., and Dang, D. C. (1984). Relationships between Leydig cell morphometry and plasma testosterone during postnatal development of the monkey, *Macaca fascicularis*. *Reprod. Nutr. Dev.* 24, 281–296.
928. Johnson, L., Grumbles, J. S., Chastain, S., Goss, H. F., and Petty, C. S. (1990). Leydig cell cytoplasmic content is related to daily sperm production in men. *J. Androl.* 11, 155–160.
929. Russell, L. D., and Burguet, S. (1977). Ultrastructure of Leydig cells as revealed by secondary tissue treatment with a ferrocyanide-osmium mixture. *Tissue Cell* 9, 751–766.
930. Prince, F. P. (1999). Mitochondrial cristae diversity in human Leydig cells, a revised look at cristae morphology in these steroid-producing cells. *Anat. Rec.* 254, 534–541.
931. Prince, F. P. (2002). Lamellar and tubular associations of the mitochondrial cristae, unique forms of the cristae present in steroid-producing cells. *Mitochondrion* 1, 381–389.
932. Stocco, D. M. (1996). Acute regulation of Leydig cell steroidogenesis. In *The Leydig Cell* (A. P. Payne, M. P. Hardy, and L. D. Russell, Eds.), 241–258. Cache River Press, Vienna, IL.
933. Stocco, D. M. (1999). An update on the mechanism of action of the steroidogenic acute regulatory (StAR) protein. *Exp. Clin. Endocrinol. Diabetes* 107, 229–235.
934. Bose, H. S., Lingappa, V. R., and Miller, W. L. (2002). Rapid regulation of steroidogenesis by mitochondrial protein import. *Nature* 417, 87–88.
935. Kerr, J. B., Robertson, D. M., and de Kretser, D. M. (1985). Morphological and functional characterization of interstitial cells from mouse testes fractionated on Percoll density gradients. *Endocrinology* 116, 1030–1043.
936. Kerr, J. B., and Knell, C. M. (1988). The fate of fetal Leydig cells during the development of the fetal and postnatal rat testis. *Development* 103, 535–544.
937. Aoki, A., and Massa, E. M. (1972). Early responses of testicular interstitial cells to stimulation by interstitial cell stimulating hormone. *Am. J. Anat.* 134, 239–262.
938. Wrobel, K. H., Sinowatz, F., and Mademann, R. (1981). Intertubular topography in the bovine testis. *Cell. Tissue Res.* 217, 289–310.
939. Belt, W. D., and Cavazos, L. F. (1971). Fine structure of the interstitial cells of Leydig in the squirrel monkey during seasonal regression. *Anat. Rec.* 169, 115–128.
940. Russo, J., and Sacerdote, F. C. (1971). Ultrastructural changes induced by hCG in the Leydig cell of the adult mouse testis. *Z. Zellforsch.* 112, 363–370.
941. Nussdorfer, G. G., Robba, C., Mazzocchi, G., and Rebuffat, P. (1980). Effects of chorionic gonadotrophins on the interstitial cells of the rat testis, a morphometric and radioimmunological study. *Int. J. Androl.* 3, 319–332.
942. Neaves, W. B. (1978). The pattern of gonadotropin-induced changes in plasma testosterone, testicular esterified cholesterol and Leydig cell lipid droplets in immature mice. *Biol. Reprod.* 19, 864–871.
943. Aoki, A. (1970). Hormonal control of Leydig cell differentiation. *Protoplasma* 71, 209–225.
944. Merkow, L., Acevedo, H., Slifkia, M., and Caito, M. (1968). Studies on the interstitial cells of the testis. I. The ultrastructure of the immature guinea pig and the effect of stimulation with human chorionic gonadotropin. *Am. J. Pathol.* 33, 47–62.
945. Aoki, A., and Massa, E. M. (1975). Subcellular compartmentation of free and esterified cholesterol in the interstitial cells in the mouse testis. *Cell. Tissue Res.* 165, 49–62.
946. Lalli, M. F., and Clermont, Y. (1975). Leydig cells and their role in the synthesis and secretion of glycoproteins. *Anat. Rec.* 181, 403–404.
947. Gemmell, R. T., and Stacy, B. D. (1979). Ultrastructural study of granules in the corpora lutea of several mammalian species. *Am. J. Anat.* 155, 1–14.
948. Gemmell, R. T., Laychock, S. G., and Rubin, R. P. (1977). Ultrastructural and biochemical evidence for a steroid-containing secretory organelle in the perfused cat adrenal gland. *J. Cell. Biol.* 72, 209–215.
949. Loh, H. S., and Gemmell, R. T. (1980). Changes in the fine structure of the Leydig cells of the seasonally breeding bat, *Myotis adversus*. *Cell. Tissue Res.* 210, 339–347.
950. Paavola, L. G., and Boyd, C. O. (1981). Cytoplasmic granules in luteal cells of pregnant and nonpregnant guinea pigs. A cytochemical study. *Anat. Rec.* 201, 127–140.
951. Porter, D. G. (1979). Relaxin, old hormone, new prospect. *Oxf. Rev. Reprod. Endocrinol* 1, 1–45.
952. Fields, P. A, Eldrige, R. K., Fuchs, A. R., Roberts, R. F., and Fields, M. J. (1983). Human placenta and bovine corpora luteal oxytocin. *Endocrinology* 112, 1544–1546.
953. Hermo, L., Clermont, Y., and Lalli, M. (1985). Intracellular pathways of endocytosed tracers in Leydig cells of the rat. *J. Androl.* 6, 213–224.
954. Tang, X. M., Clermont, Y., and Hermo, L. (1988). Origin and fate of autophagosomes in Leydig cells of normal adult rats. *J. Androl.* 9, 284–293.
955. Fawcett, D. W., and Burgos, M. H. (1960). Studies on the fine structure of the mammalian testis. II. The human interstitial tissue. *Am. J. Anat.* 107, 245–269.
956. Yamada, E. (1965). Some observations on the fine structure of the interstitial cells and Sertoli cells of the human testis. *Gunma. Symp. Endocrinol.* 2, 1–17.
957. Nagano, T. (1965). Some observations on the structure of interstitial cells and Sertoli cells of the human testis. *Gunma. Symp. Endocrinol.* 2, 19–28.
958. Reddy, J., and Svoboda, D. (1972). Microbodies (peroxisomes) in the interstitial cells of rodent testes. *Lab. Invest.* 26, 657–665.
959. Hruban, Z., Vigil, E. L., Steasers, A., and Hopkins, E. (1972). Microbodies, constituent organelles of animals. *Lab. Invest.* 27, 184–191.
960. Mendis-Handagama, S. M. L. C., Zirkin, B. R., Scallen, T. J., and Ewing, L. L. (1990). Studies on peroxisomes of the adult rat Leydig cells. *J. Androl.* 11, 270–278.
961. Mendis-Handagama, S. M. L. C., Watkins, P. A., Gelber, S. J., Scallen, T. J., Zirkin, B. R., and Ewing, L. L. (1990). Luteinizing hormone causes rapid and transient changes in rat Leydig cell peroxisome volume and intra-peroxisomal sterol carrier protein-2 content. *Endocrinology* 127, 2947–2954.
962. Mendis-Handagama, S. M. (2000). Peroxisomes and intracellular cholesterol trafficking in adult rat Leydig cells following luteinizing hormone stimulation. *Tissue Cell* 32, 102–106.
963. Usui, N. (1976). Fine structure of interstitial cells and macrophages in immature rat testes. *J. Tokyo Womens Med. Coll.* 46, 809–825.
964. Tanuma, Y., and Ohata, M. (1978). Transmission electron microscopic observation of epithelial cells with single cilia in intrahepatic biliary ductules of bats. *Arch. Histol. Jpn.* 41, 367–376.

965. Temple, R., and Wolff, J. (1973). Stimulation of steroid secretion by antimicro-tubular agents. *J. Biol. Chem.* 248, 2691.
966. Ray, P., and Strott, C. A. (1978). Stimulation of steroid synthesis by normal rat adrenocortical cells in response to antimicrotubular agents. *Endocrinology* 103, 1281–1288.
967. Clark, M. A., and Shay, J. W. (1981). The role of tubulin in the steroidogenic response of murine adrenal and rat Leydig cell. *Endocrinology* 109, 2261–2263.
968. Laws, A. O., Kerr, J. B., and de Kretser, D. M. (1984). The role of the microtubular system in LH/hCG receptor down regulation in rat Leydig cells. *Mol. Cell. Endocrinol.* 38, 39–51.
969. Reinke, F. (1896). Beitrage zur Histologie des Menschen. *Arch. Mikrosk. Anat.* 47, 34–44.
970. Nagano, T., and Ohtsuki, I. (1971). Reinvestigation of the fine structure of Reinke's crystals in the human testicular interstitial cell. *J. Cell. Biol.* 51, 148–161.
971. Yasuzumi, G., Nakai, Y., Tsubo, I., Yasuda, M., and Sugioka, T. (1967). The fine structure of nuclei as revealed by electronmicroscopy. IV. The intranuclear inclusion formation of Leydig cells of ageing human testes. *Exp. Cell. Res.* 45, 261–276.
972. Janko, A. B., and Sandberg, E. C. (1970). Histochemical evidence for the protein nature of the Reinke crystalloid. *Obstet. Gynaecol.* 35, 493–503.
973. Mori, H., Fukunishi, R., Fujii, M., Hataji, K., Shiraishi, T., and Matsumoto, K. (1978). Stereological analysis of Reinke crystals in human Leydig cells. *Vichows Arch. (Pathol. Anat.)* 380, 1–10.
974. Sohval, A. R., Gabrilove, J. L., and Chung, L. (1973). Ultrastructure of Leydig cell paracrystalline inclusions, possibly related to Reinke crystals, in the normal human testis. *Z. Zellforsch.* 142, 13–26.
975. Kerr, J. B., Abbenhuys, D. C., and Irby, D. C. (1986). Crystalloid formation in rat Leydig cells. An ultrastructural and hormonal study. *Cell. Tissue Res.* 245, 91–100.
976. Prince, F. P. (1992). Ultrastructural evidence of indirect and direct autonomic innervation of human Leydig cells, comparison of neonatal, childhood and pubertal ages. *Cell. Tissue Res.* 269, 383–390.
977. Satoh, M. (1985). The histogenesis of the gonads in the rat embryos. *J. Anat.* 143, 17–37.
978. Mancini, R. E., Vilar, O., Lavieri, J. C., Andrada, J. A., and Heinrich, J. J. (1963). Development of Leydig cells in the normal human testis. *Am. J. Anat.* 112, 203–214.
979. Gondos, B. (1981). Cellular interrelationships in the human foetal ovary and testis. *Proc. Clin. Biol. Res.* 59B, 373–381.
980. Witschi, E. (1951). Embryogenesis of the adrenal and the reproductive gland. *Rec. Prog. Horm. Res.* 6, 1–27.
981. Merchant-Larios, H., and Moreno-Mendoza, N. (1998). Mesonephric stromal cells differentiate into Leydig cells in the mouse fetal testis. *Exp. Cell. Res.* 244, 230–238.
982. Nishino, K., Yamanouchi, K., Naito, K., and Tojo, H. (2001). Characterization of mesonephric cells that migrate into the XY gonad during testis differentiation. *Exp. Cell. Res.* 267, 225–232.
983. Lejeune, H., Habert, R., and Saez, J. M. (1998). Origin, proliferation and differentiation of Leydig cells. *J. Mol. Endocrinol.* 20, 1–25.
984. Byskov, A. G. (1986). Differentiation of mammalian embryonic gonad. *Physiol. Rev.* 66, 77–112.
985. Holstein, A. F., Wartenburg, H., and Vossmeyer, J. (1971). Zur Cytologie de pranatalen Gonadenentwicklung beim Menschen. III. Die Entwicklung der Leydig-zellen im Hoden von embrynon und Feten. *Z. Anat. Entwickl. Gesch.* 135, 43–66.
986. Gondos, B., Paup, D., Ross, J., and Gorski, R. (1974). Ultrastructural differentiation of Leydig cells in the fetal and postnatal hamster testis. *Anat. Rec.* 178, 551–566.
987. Gondos, B., Renston, R., and Goldstein, D. (1976). Postnatal differentiation of Leydig cells in the rabbit testis. *Am. J. Anat.* 145, 167–182.
988. Gondos, B., Morrison, K., and Renston, R. (1977). Leydig cell differentiation in the prepubertal rabbit testis. *Biol. Reprod.* 17, 745–748.
989. Niemi, M., and Kormano, M. (1964). Cell renewal in the interstitial tissue of postnatal prepubertal rat testis. *Endocrinology* 74, 996–998.
990. Prince, F. P. (1984). Ultrastructure of immature Leydig cells in the human prepubertal testis. *Anat. Rec.* 209, 165–176.
991. Chemes, H. E., Gottlieb, S. E., Pasqualini, T., Domenichini, E., Rivarola, M. A., and Bergada, C. (1985). Response to acute hCG stimulation and steroidogenic potential of Leydig cell fibroblastic precursors in humans. *J. Androl.* 6, 102–112.
992. Dierichs, R., Wrobel, K. H., and Schilling, E. (1973). Licht und elektronen-mikroskopische Untersuchungen an den Leydigzellen des Schweines wahrend der postnatalen Entwicklung. *Z. Zellforsch.* 143, 203–277.
993. Dierichs, R., and Wrobel, K. H. (1973). Licht und electronenmikroskopische Untersuchnugen an den peritubularen Zellen des Schweinhoden wahrend der postnatalen Entwicklung. *Z. Anat. Entwickl. Gesch.* 143, 49–64.
994. van Straaten, H. W. M., and Wensing, C. J. G. (1977). Histomorphometric aspect of testicular morphogenesis in the pig. *Biol. Reprod.* 17, 467–472.
995. van Straaten, H. W. M., and Wensing, C. J. G. (1978). Leydig cell development in the testis of the pig. *Biol. Reprod.* 18, 86–93.
996. Bressler, R. S., and Ross, M. H. (1972). Differentiation of peritubular myoid cells of the testis, effects of intratesticular implantation of newborn mouse testes into normal and hypophysectomized adult. *Biol. Reprod.* 6, 148–159.
997. Mori, H., Shiraishi, T., and Matsumoto, K. (1978). Leydig cells within the lamina propria of seminiferous tubules in four patients with azoospermia. *Andrologia* 10, 444–452.
998. Sniffen, R. C. (1950). The testis. I. The normal testis. *Arch. Pathol.* 50, 259–284.
999. Vilar, O. (1970). Histology of the human testis from neonatal period to adolescence. In *The Human Testis* (E. Rosenburg and C. A. Paulsen, Eds.), pp. 95–108. Plenum Press, New York.
1000. Hadziselimovic, F. (1977). Cryptorchidism, ultrastructure of normal and cryptorchid testis development. *Adv. Anat. Embryol. Cell. Biol.* 53, 1–17.
1001. Whitehead, R. H. (1904). The embryonic development of the interstitial cells of Leydig. *Am. J. Anat.* 3, 167.
1002. Whitehead, R. H. (1905). Studies on the interstitial cells of Leydig. No. 2. Their post-embryonic development in the pig. *Am. J. Anat.* 4, (193.
1003. Roosen-Runge, E. C., and Anderson, D. (1959). The development of the interstitial cells in the testis of the albino rat. *Acta. Anat.* 37, 125–137.
1004. Fouquet, J. P., Meusy-Dessolle, N., and Dang, D. C. (1984). Relationships between Leydig cell morphometry and plasma testosterone during postnatal development of the monkey, *Macaco fascicularis*. *Reprod. Nutr. Dev.* 24, 281–286.
1005. Winter, J. S. D., Faiman, C., and Reyes, F. I. (1977). Sex steroid production by the human fetus, its role in morphogenesis and control by gonadotrophins. *Birth Defects* 13, 41–52.

1006. Faiman, C., Winter, J. S. D., and Reyes, F. I. (1989). Endocrinology of the fetal testis. In *The Testis* (H. G. Burger and D. M. de Kretser, Eds.), pp. 119–142. Raven Press, New York.

1007. Niemi M., Ikonen, M., and Hervonen, A. (1967). Histochemistry and fine structure of the interstitial tissue in the human foetal testis. In *Endocrinology of the Testis* (G. E. W. Wolstenholme and M. O'Connor, Eds.), pp. 31–55. Churchill, London.

1008. Helal, M. A., Mehmet, H., Thomas, N. S. B., Cox, P. M., Ralph D. J., Bajora, R., and Chaterjee, R. (2002). Ontogeny of human fetal testicular apoptosis during first, second, and third trimesters of pregnancy. *J. Clin. Endocrinol. Metab.* 87, 1189–1193.

1009. Gruenwald, P. (1946). Structure of the testis in infancy and in childhood. *Arch. Pathol.* 42, 35–48.

1010. Ottowicz, J. (1963). The stadial development of Leydig cells. *Acta. Med. Pol.* 4, 1–13.

1011. Sharpe, R. M. (1982). The hormonal regulation of the Leydig cell. In *Oxford Reviews of Reproductive Biology* (C. A. Finn, Ed.), pp. 241–317. Oxford University Press, Oxford.

1012. Sharpe, R. M. (1983). Local control of testicular function. *Q. J. Exp. Physiol.* 68, 265–287.

1013. Sharpe, R. M. (1984). Intratesticular factors controlling testicular function. *Biol. Reprod.* 30, 24–49.

1014. Zirkin, B. R., and Ewing, L. L. (1987). Leydig cell differentiation during maturation of the rat testis, a stereological study of cell number and ultrastructure. *Anat. Rec.* 2(19, 157–163.

1015. Hardy, M. P., Zirkin, B. R., and Ewing, L. L. (1989). Kinetic studies on the development of the adult population of Leydig cells of the pubertal rat. *Endocrinology* 124, 762–770.

1016. Vergouwen, R., Jacobs, S., Huiskamp, R., Davids, J., and de Rooij, D. (1991). Proliferative activity of gonocytes, Sertoli cells and interstitial cells during testicular development in mice. *J. Reprod. Fertil.* 93, 233–243.

1017. Shan, L. X., and Hardy, M. P. (1992). Developmental changes in levels of luteinizing hormone receptor and androgen receptor in rat Leydig cells *Endocrinology* 131, 1107–1114.

1018. Faiman, C., and Winter, J. S. D. (1974). Gonadotropins and sex hormone patterns in puberty, clinical data. In *The Control of the Onset of Puberty* (M. M. Grumbach, G. D. Grave, and F. E. Mayer, Eds.), p. 32. Wiley, New York.

1019. Ariyaratne, H. B., Mendis-Handagama, S. M., Hales, D. B., and Mason, J. I. (2000). Studies of the onset of Leydig precursor cell differentiation in the prepubertal rat testis. *Biol. Reprod.* 63, 165–171.

1020. Odell, W. D., and Swerdloff, R. S. (1976). Etiologies of sexual maturation, a model system based on the sexually maturing rat. *Rec. Prog. Horm. Res.* 32, 245.

1021. Charny, C., Conston, A., and Meranze, D. (1952). Development of the testis. *Fertil. Steril.* 3, 461–479.

1022. de la Baize, F., Mancini, R., Arrillaga, F., Andrada, J., and Vilar, O. (1960). Pubertal maturation of the human testis. A histologic study. *J. Clin. Endocrinol. Metab.* 20, 266–285.

1023. Hayashi, H., and Harrison, R. (1971). The development of the interstitial tissue of the human testis. *Fertil. Steril.* 22, 351–355.

1024. Forti, G., Santoro, S., and Grisolla, G. A., Bassi, F., Boninsegni, R., Fiorelli, G., and Serio, M. (1981). Spermatic and peripheral plasma concentrations of testosterone and androstenedione in prepubertal boys. *J. Clin. Endocrinol. Metab.* 53, 883–886.

1025. Rivarola, M. A., Bergada, C., and Cullen, M. (1970). hCG stimulation test in prepubertal boys with cryptorchidism, in bilateral anorchia and in male pseudohermaphroditism. *J. Clin. Endocrinol. Metab.* 31, 526–530.

1026. Prince, F. P. (1990). Ultrastructural evidence of mature Leydig cells and Leydig cell regression in the neonatal human testis. *Anat. Rec.* 228, 405–417.

1027. Chen, Y. D. I., Shaw, M. J., and Payne, A. P. (1977). Steroid and FSH action on LH receptors and LH-sensitive testicular responsiveness during sexual maturation of the rat. *Mol. Cell. Endocrinol.* 8, 291–299.

1028. Odell, W. D., Swerdloff, R. S., Jacobs, H. S., and Hescox, M. A. (1973). FSH induction of sensitivity to LH, one cause of sexual maturation in the male rat. *Endocrinology* 92, 160–165.

1029. van Beurden, W. M. O., Roodnat, B., de Jong, F. H., Mulder, E., van der Molen, and H. J. (1976). Hormonal regulation of LH stimulation of testosterone production in isolated Leydig cells of immature rats, the effects of hypophysectomy, FSH and estradiol-17ß. *Steroids* 28, 847–866.

1030. Chen, Y. D. I., Payne, A. P., and Kelch, R. P. (1976). FSH stimulation of Leydig cell function in the hypophysectomized immature rat. *Proc. Soc. Exp. Biol. Med.* 153, 473–475.

1031. Odell, W. D., and Swerdloff, R. S. (1975). The role of testicular sensitivity to gonadotrophins in sexual maturation of the male rat. *J. Steroid Biochem.* 6, 853–857.

1032. Selin, L. K., and Moger, W. H. (1977). The effect of FSH on LH induced testosterone secretion in the immature hypophysectomized male rat. *Endocrinol. Res. Commun.* 4, 171–182.

1033. Kerr, J. B., and Sharpe, R. M. (1985). Follicle-stimulating hormone induction of Leydig cell maturation. *Endocrinology* 116, 2592–2604.

1034. Kerr, J. B., and Sharpe, R. M. (1985). Stimulatory effect of follicle-stimulating hormone on rat Leydig cells. A morphometric and ultrastructural study. *Cell. Tissue Res.* 239, 405–415.

1035. Teerds, K. J., Closset, J., and Rommerts, F. F. G., et al. (1989). Effects of pure FSH and LH preparations on the number and function of Leydig cells in immature hypophysectomized rats. *J. Endocrinol.* 120, 97–106.

1036. Kerr, J. B., Bartlett, J. M. S., Donachie, K., and Sharpe, R. M. (1987). Origin of regenerating Leydig cells in the testis of the adult rat. An ultrastructural, morphometric and hormonal assay study. *Cell. Tissue Res.* 249, 367–377.

1037. Neaves, W. B. (1973). Changes in testicular Leydig cells and in plasma testosterone levels among seasonally breeding rock hyrax. *Biol. Reprod.* 8, 451–466.

1038. Neaves, W. B. (1979). The annual testicular cycle in an equatorial colony of lesser rock hyrax, *Heterohyrax brucei*. *Proc. R. Soc. Lond. Biol.* 206, 183–189.

1039. Neaves, W. B. (1980). Asynchronous testicular cycles among equatorial colonies of rock hyrax (*Procavia habessinicd*). In *Testicular Development, Structure and Function* (E. Steinberger and A. Steinberger, Eds.), pp. 411–418. Raven Press, New York.

1040. Gustafson, A. W. (1975). Observations on the hydroxysteroid dehydrogenase and lipid histochemistry and ultrastructure of the Leydig cells in adult *Myotis lucifugus* during the annual reproductive cycle. *Anat. Rec.* 181, 366–367,

1041. Wing, T. Y., and Lin, H. S. (1977). The fine structure of testicular interstitial cells in the adult golden hamster with special reference to seasonal changes. *Cell. Tissue Res.* 183, 385–393.

1042. Suzuki, F., and Racey, P. A. (1978). The organization of testicular interstitial tissue and changes in the fine structure of the Leydig cells of European moles (*Tulpa europaed*) throughout the year. *J. Reprod. Fertil.* 52, 189–194.
1043. Hardy, M. P., Mendis-Handagama, S. M. L. C., Zirkin, B. R., and Ewing, L. L. (1987). Photoperiodic variation of Leydig cell numbers in the testis of the golden hamster, a possible mechanism for their renewal during recrudescence. *J. Exp. Zool.* 244, 269–276.
1044. Sinha Hikim, A. P., Sinha Hikim, I., Amador, A. G., Bartke, A., Woolf, A., and Russell, L. D. (1991). Reinitiation of spermatogenesis by exogenous gonadotrophins in a seasonal breeder, the woodchuck (*Marmota monax*), during gonadal activity. *Am. J. Anat.* 192, 194–213.
1045. Hochereau-de Reviers, M. T., Perreau, C., and Lincoln, G. A. (1985). Photoperiodic variations of somatic and germ cell populations in the Soay ram testis. *J. Reprod. Fertil.* 74, 329–334.
1046. Wislocki, G. B. (1949). Seasonal changes in the testes, epididymides and seminal vesicles of deer investigated by histochemical methods. *Endocrinology* 44, 167–189.
1047. Lofts, B. (1960). Cyclical changes in the distribution of the testis lipids of a seasonal mammal (*Talpa europaed*). *J. Microsc. Sci.* 101, 199–205.
1048. Johnson, L., and Thompson, D. L. (1986). Seasonal variation in the total volume of Leydig cells in stallions is explained by variation in cell number rather than cell size. *Biol. Reprod.* 35, 971–979.
1049. Kothari, L. K., and Gupta, A. S. (1974). Effect of ageing on the volume, structure and total Leydig cell content of the human testis. *Int. J. Fertil.* 19, 140–146.
1050. Honore, L. H. (1978). Ageing changes in the human testis, a light microscopic study. *Gerentology* 24, 58–65.
1051. Teem, M. V. B. (1935). The relation of the interstitial cells of the testis to prostatic hypertrophy. *J. Urol.* 34, 692–713.
1052. Sargent, J. W., and McDonald, J. R. (1948). A method for the quantitative estimate of Leydig cells in the human testis. *Proc. Staff Meet. Mayo. Clin.* 23, 249–254.
1053. Sokal, Z. (1964). Morphology of the human testis in various periods of life. *Folia. Morphol.* 23, 102–111 .
1054. Vermeulen, A. (1976). Leydig cell function in old age. In *Hypothalamus, Pituitary and Aging* (J. A. Burgess, Ed.), pp. 458–463. Charles C Thomas, Springfield, IL.
1055. Kaler, L. W., and Neaves, W. B. (1978). Attrition of the human Leydig cell population with advancing age. *Anat. Rec.* 192, 513–518.
1056. Neaves, W. B., Johnson, L., Porter, J. C., Parker, C. R., and Petty, C. S. (1984). Leydig cell numbers, daily sperm production and serum gonadotrophin levels in ageing men. *J. Clin. Endocrinol. Metab.* 59, 756–763.
1057. Neaves, W. B., Johnson, L., and Petty, C. S. (1985). Age-related change in numbers of other interstitial cells in testes of adult men, evidence bearing on the fate of Leydig cells lost with increasing age. *Biol. Reprod.* 33, 259–269.
1058. Kaler, L. W., and Neaves, W. B. (1981). The androgen status of ageing male rats. *Endocrinology* 108, 712–719.
1059. Kerr, J. B., and de Kretser, D. M. (1983). Techniques for detecting and evaluating abnormalities in testicular function. In *Methods for Assessing the Effects of Chemicals on Reproductive Functions* (V. B. Vouk, P. J. Sheehan, Scientific Committee on Problems of the Environment, Eds.), pp. 247–262. Wiley, New York.
1060. Russell, L. D. (1983). Normal testicular structure and methods of evaluation under experimental and disruptive conditions. In *Reproductive and Developmental Toxicity of Metals* (T. W. Clarkson, G. F. Nordberg, and P. R. Sager, Eds.), pp 227–252. Plenum Press, New York.
1061. Mann, T., and Lutwak-Mann, C. (1981). *Male Reproductive Function and Semen. Themes and Trends in Physiology, Biochemistry and Investigative Andrology,* Suppl 5. Springer-Verlag, Berlin.
1062. Vermeulen, A. (1982). Effects of drugs on Leydig cell function. *Int. J. Androl.* Suppl 5, 163–182.
1063. de Kretser, D. M. (1982). Sertoli cell-Leydig cell interaction in the regulation of testicular function. *Int. J. Androl.* Suppl 5, 11–17.
1064. Wilton, L. J., and de Kretser, D. M. (1984). The influence of luteinizing hormone on the Leydig cells of cryptorchid rat testes. *Acta. Endocrinol. (Kbh)* 107, 110–116.
1065. Risbridger, G. P., Kerr, J. B., and de Kretser, D. M. (1981). Evaluation of Leydig cell function and gonadotrophin binding in unilateral and bilateral cryptorchidism, evidence for local control of Leydig cell function by the seminiferous tubule. *Biol. Reprod.* 24, 534–540.
1066. Risbridger, G. P., Kerr, J. B., Peake, R. A., and de Kretser, D. M. (1981). An assessment of Leydig cell function after bilateral or unilateral efferent duct ligation, further evidence for local control of Leydig cell function. *Endocrinology* 109, 1234–1241.
1067. Aoki, A., and Fawcett, D. W. (1978). Is there a local feedback from the seminiferous tubules affecting activity of the Leydig cell? *Biol. Reprod.* 19, 144–158.
1068. Sharpe, R. M. (1985). Intragonadal hormones. *Biol. Reprod.* 133, C1–C5.
1069. Sharpe, R. M. (1986). Paracrine control of the testis. *Clin. Endocrinol. Metab.* 15, 185–207.
1070. van Vlliet, J., Rommerts, F. F. G., de Rooij, D. G., Buwalda, G., and Wensing, C. J. G. (1988). Reduction of testicular blood flow and focal degeneration of tissue in the rat following administration of human chorionic gonadotrophin. *J. Endocrinol.* 117, 51–57.
1071. Chemes, H. E., Rivarola, M. A., and Bergada, C. (1976). Effect of gonadotrophins and testosterone on the seminiferous tubules of the immature rat. *J. Reprod. Fertil.* 46, 283–288.
1072. Rivier, C., Rivier, J., and Vale, W. (1979). Chronic effects of [DTrp^6, Pro^9-NET]-luteinizing hormone-releasing factor on reproductive processes in the male rat. *Endocrinology* 105, 1191–1201.
1073. Johnson, A. D. (1977). The influence of cadmium on the testis. In *The Testis* (A. D. Johnson and W. R. Gomes, Eds.), Vol. IV, pp. 1191–11201. Academic Press, New York.
1074. Patanelli, D. J. (1975). Suppression of fertility in the male. In *Handbook of Physiology* (D. W. Hamilton and R. O. Creep, Eds.), Section 7, pp. 245–258. American Physiological Society, Washington, DC.
1075. Jackson, H., and Ericsson, R. J. (1970). Effect of chemical agents and hormones on spermatogenesis and the epididymis. *Biol. Reprod.* 14, 453–600.
1076. Gunn, S. A., Gould, T. C, and Anderson, W. A. D. (1966). Loss of selective injurious vascular response to cadmium in regenerated blood vessels of testis. *Am. J. Pathol.* 48, 959–969.
1077. Gunn, S. A., and Gould, T. C. (1970). Specificity of the vascular system of the male reproductive tract. *J. Reprod. Fertil.* Suppl 10, 75–95.
1078. Gunn, S. A., and Gould, T. C. (1970). Cadmium and other mineral elements. In The Testis (A. D. Johnson, W. R. Gomes, and N. L. van Demark, Eds.), Vol. III, pp. 377–481. Academic Press, New York.
1079. Rommerts, F. F. G., Grootenhuis, A. J., Hoogerbrugge, J. W., and van der Molen, H. J. (1985). Ethane dimethane sulphonate specifically inhibits LH stimulated steroidogenesis in Leydig cells isolated from mature rats but not in cells from immature rats. *Mol. Cell. Endocrinol.* 42, 105–111.

1080. Molenaar, R., de Rooji, D. G., Rommerts, F. F. G., Reuvers, P. J., and van der Molen, H. J. (1985). Specific destruction of Leydig cells in mature rats after *in vivo* administration of ethane dimethane sulphonate. *Biol. Reprod.* 33, 1213–1222.

1081. Bartlett, J. M. S., Kerr, J. B., and Sharpe, R. M. (1986). The effect of selective destruction and regeneration of rat Leydig cells on seminiferous tubule morphology and the intratesticular distribution of testosterone. *J. Androl.* 7, 240–253.

1082. Morris, I. D. (1985). Leydig cell resistance to the cytotoxic effect of ethylene dimethane sulphonate in the adult rat testis. *J. Endocrinol.* 105, 311–316.

1083. Teerds, K. J., de Rooij, D. G., Rommerts, F. F. G., and Wensing, C. J. G. (1988). The regulation of the proliferation and differentiation of rat Leydig cell precursor cells after EDS administration or daily hCG treatment. *J. Androl.* 9, 343–351.

1084. Teerds, K. J., de Rooij, D. G., Rommerts, F. F. G., van den Hurk R., and Wensing, C. J. G. (1989). Proliferation and differentiation of possible Leydig cell precursors after destruction of the existing Leydig cells with ethane dimethylsulphonate, the role of LH/human chorionic gonadotrophin. *J. Endocrinol.* 122, 689–696.

1085. Teerds, K. J., de Rooij, D. G., Rommerts, F. F. G, and Wensing, C. J. G. (1990). Development of a new Leydig cell population after the destruction of existing Leydig cells by ethane dimethane sulphonate in rats, an autoradiographic study. *J. Endocrinol.* 126, 229–236.

1086. Abney, T. O., and Myers, R. B. (1991). 17ß-Estradiol inhibition of Leydig cell regeneration in the ethane dimethylsulphonate-treated mature rat. *J. Androl.* 12, 295–304.

1087. O'Leary, P. O., Jackson, A. E., Averill, S., and de Kretser, D. M. (1986). The effects of ethane dimethanesulphonate (EDS) on bilaterally cryptorchid rat testes. *Mol. Cell. Endocrinol.* 45, 183–190.

1088. Kerr, J. B., and Donachie, K. (1986). Regeneration of Leydig cells in unilaterally cryptorchid rats, evidence for stimulation by local testicular factors. *Cell. Tissue Res.* 243, 405–414.

1089. Sharpe, R. M., Maddocks, S., and Kerr, J. B. (1990). Cell-cell interactions in the control of spermatogenesis as studied using Leydig cell destruction and testosterone replacement. Am. J. Anat. 188, 3–20.

1090. Russell, L. D., Ren, H. P., Sinha Hikim, I., Schulze, W., and Sinha Hikim, A. P. (1990). A comparative study in twelve mammalian species of volume densities, volumes, and numerical densities of selected testis components, emphasizing those related to the Sertoli cell. *Am. J. Anat.* 188, 21–30.

1091. Clermont, Y., and Troh, M. (1969). Duration of the cycle of the seminiferous epithelium in the mouse and hamster determined by means of ^{3}H-thymidine and radioautography. *Fertil. Steril.* 20, 805–817.

1092. de Rooij, D. G. (1968). Stem cell renewal and duration of spermatogonial cycle in the golden hamster. *Z. Zellforsch.* 89, 133–136.

1093. Swierstra, E. E. (1968). Cytology and duration of the seminiferous epithelium of the boar. Duration of spermatozoon transit through the epididymis. *Anat. Rec.* 161, 171–186.

1094. Clermont, Y., Le Blond, C. P., and Messier, B. (1959). Duree du cycle de l'epithelium seminal du rat. *Arch. Anat. Microsc. Morphol. Exp.* 48, 37–56.

1095. Hochereau, M. T., Courot, M., and Ortavant, R. (1964). Marquage des cellules germinales du belier et du taureau par injection de thymidine tritee dans l'artere spermatique. *Am. Biol. Anim. Biochem. Biophys.* 2, 157–161.

1096. Orgebin-Crist, M. C. (1965). Passage of spermatozoa labelled with ^{3}H-thymidine through the ductus epididymis of the rabbit. *J. Reprod. Fertil.* 10, 241–251.

1097. Swierstra, E. E., and Foote, R. H. (1963). Cytology and kinetics of spermatogenesis in the rabbit. *J. Reprod. Fertil.* 5, 309–322.

1098. Antar, M. (1971). Duration of the cycle of the seminiferous epithelium and of spermatogenesis in the monkey (*Macaca speciosa*). *Anat. Rec.* 169, 268–269.

1099. Arsenieva, N. A., Dubinin, N. P., Orlova, N. N., and Bakulina, E. D. (1961). A radiation analysis of the duration of meiotic phases in the monkey (*Macaca mulatto*). *Dokl. Akad. Nauk. SSSR* 141, 1486–1489.

1100. Mori, H., Shimizu, D., Takeda, A., Takioka, Y., and Fukuhishi, R. (1980). Stereological analysis of Leydig cells in normal guinea pig testes. *J. Electron Microsc.* 29, 8–21.

1101. Ewing, L. L., and Zirkin, B. R. (1983). Leydig cell structure and steroidogenic function. *Rec. Prog. Horm. Res.* 39, 599.

1102. Lin, H. S., and Wing, T. Y. (1979). A dense-cored filamentous body in Leydig cells of the golden hamster. *Cell. Tissue Res.* 201, 369–376.

1103. Payer, A. F., and Parkening, T. A. (1983). Membrane-bound intranuclear inclusions in the Leydig cell of the Chinese hamster (*Cricetulus griseus*). *J. Ultrastruct. Res.* 84, 317–325.

1104. Stefan, Y., and Steimer, T. (1978). The Leydig cell of a hypogonadic rodent (*Ellobius lutescens* Th), correlation between ultrastructure and biosynthetic activity. *Biol. Reprod.* 19, 913–921.

1105. Sinha, A. A., Erickson, A. W., and Seal, U. S. (1977). Fine structure of Leydig cells in crabeater, leopard and Ross seals. *J. Reprod. Fertil.* 49, 51–54.

1106. Osman, D. J., and Ploen, L. (1978). The Ultrastructure of Sertoli cells in the boar *Int. J. Androl.* 1, 162–179.

1107. Gotoh, M., Miyake, K., Mitsuya, H., Hoshino, T., and Yamada, K. (1983). Cytoplasmic inclusion bodies in Leydig cells from the testes of postpubertal cryptorchid patients. *Int. J. Androl.* 6, 221–228.

1108. Mori, H. (1984). Ultrastructure and stereological analysis of Leydig cells. In *Ultrastructure of Endocrine Cells and Tissues* (P. Motta, Ed.), pp. 225–237. Martinus Nijhoff, The Hague.

1109. Mendis-Handagama, S. M. L. C., Kerr, J. B., and de Kretser, D. M. (1990). Experimental cryptorchidism in the adult mouse. I. Quantitative and qualitative light microscopic morphology. *J. Androl.* 11, 539–547.

1110. Mendis-Handagama, S. M. L. C., Kerr, and J. B., de Kretser, D. M. (1991). Experimental cryptorchidism in the adult mouse. III. Quantitative and qualitative electron microscopic morphology of Leydig cells. *J. Androl.* 12, 335–343.

1111. Sinha Hikim, A. P., Chakraborty, J., and Jhunjhunwala, J. S. (1987). Unilateral torsion of spermatic cord in men, effect on Leydig cell. *Urology* 29, 40–44.

1112. Kerr, J. B., Knell, C. M., and Irby, D. C. (1987). Ultrastructure and possible function of giant crystalloids in the Sertoli cell of the juvenile and adult koala (*Phascolarctos cinereus*). *Anat. Embryol.* 176, 213–244.

1113. de Kretser, D. M., Temple-Smith, P. D., and Kerr, J. B. (1982). Anatomical and functional aspects of the male reproductive organs. In *Disturbances in Male Fertility* (K. Band-hauer and J. Frick, Eds.), pp. 1–131. Springer Verlag, Berlin.

Knobil and Neill's Physiology of Reproduction, Third Edition
edited by Jimmy D. Neill,
Elsevier © 2006

CHAPTER 19

The Sertoli Cell

Michael D. Griswold[1] and Derek McLean[2]

Introduction, 949
Testis Formation and Development, 950
Cell Fates and Embryonic Sertoli Cells, 950
Late Embryonic and Early Postnatal Maturation, 951
Functional Maturation of Sertoli Cells at Puberty, 951
Gene Expression in Sertoli Cells, 951
Gene Expression in Sertoli Cells during Testis Formation, 952
Gene Expression in Functionally Mature Sertoli Cells, 953
Transgenic Mice and Sertoli Cell-Specific Expression, 953
Genearray Analysis of Sertoli Cell Gene Expression, 953
FSH, Testosterone, and Sertoli Cells, 956
FSH and Spermatogenesis, 956
Testosterone and Spermatogenesis, 957
Advanced Technologies, 958
Germ Cell Culture and Sertoli Cells, 958
Established Sertoli Cell Lines, 959
Germ Cell Transplantation and Sertoli Cells, 960
Comparative Sertoli Cell Biology, 961
Vertebrate and Invertebrate, 961
Seasonal Breeders, Testicular Regression, and Recrudescence, 963
Conclusion, 965
References, 966

INTRODUCTION

The Italian scientist, Enrico Sertoli, first described the cells that bear his name in 1865 and with primitive morphological evidence alone he proclaimed that these large branched cells were "linked to the production of spermatozoa." Since 1865, the "cells of Sertoli" have provoked the curiosity and interest of reproductive scientists and have stimulated efforts to elaborate on the original proclamation by Sertoli. The National Library of Medicine (Medline) database reveals that in 1970 there were about 60 scientific reports that included the term "Sertoli" in the subject. Since 1970 the annual number has doubled every decade to where the first few years of the new millennium saw 500 to 700 Sertoli-related publications per year. The early studies on Sertoli cells focused on increasingly sophisticated structural observations but in the 1970s the techniques for culture of primary Sertoli cells from rats were developed and perfected in the laboratories of Anna and Emil Steinberger (1) and Jennifer Dorrington and Irving Fritz (2). The ability to maintain relatively pure preparations of primary Sertoli cells in culture was the advancement that led to subsequent molecular and genetic studies on endocrine response and gene expression. Most of these earlier functional studies on cultured Sertoli cells were dedicated to the mechanisms by which Sertoli cells support germ cells and spermatogenesis. These studies revealed a number of the gene products made and secreted by Sertoli cells and speculated on the implications. In addition, many studies focused on the response of Sertoli cells to follicle-stimulating hormone (FSH), testosterone or other growth factors. [For reviews see (3–6).]

The molecular and genetic glimpse we have received into the function of Sertoli cells has made it very clear that they have a dual role in male reproduction. The original assertion of Enrico Sertoli that Sertoli cells are "linked to the production of spermatozoa" has been confirmed. Coincident with the interest in Sertoli cells in spermatogenesis was an increasing

[1]College of Sciences/School of Molecular Biosciences;
[2]Department of Animal Sciences, Washington State University, Pullman, Washington.

awareness of the importance of Sertoli cells in the formation and development of the testis and suppression of the female reproductive tract. Sertoli cells are the first cells to differentiate in the indifferent fetal gonad and this differentiation results in seminiferous cord formation, prevention of germ-cell entry into meiosis, and differentiation and function of the other somatic cells of the testis (7). The initial expression of Sry and other genes involved in testis formation has been shown to be an important function of Sertoli cells and a prerequisite for the formation of a testis (8). The Sertoli cells also ensure regression of the Müllerian ducts via secretion of Müllerian inhibiting substance (MIS) (9). During puberty the Sertoli cell undergoes a "functional maturation" and takes on a functionally separate role of supporting spermatogenesis. Sperm production and reproduction of the species require the physical and metabolic support that Sertoli cells provide to the germ cells.

With all of this scientific interest in Sertoli cells there have been a plethora of important review articles and two books dedicated to the structure and function of Sertoli cells (10,11). It is the goal of this review to highlight the important advances that have provided insights into the biological and molecular mechanisms by which Sertoli cells act in both testis formation and spermatogenesis. The reader is referred to Chapter 19 for details of the complete complex cytological features of these cells.

TESTIS FORMATION AND DEVELOPMENT

Cell Fates and Embryonic Sertoli Cells

The formation of the testis in mammals involves the migration of germ cells to the genital ridge and the active expression of genes from the Y chromosome in developing Sertoli cells (12,13). The most detailed information is known from studies of the mouse, yet it can be assumed that the principles remain the same for other mammals with the timing of each event differing for each species. At about 9.5 dpc in the mouse, the intermediate mesoderm gives rise to the genital ridge composed of somatic cells derived from the mesonephros and primordial germ cells (PGCs) that have migrated through the gut mesentery from extraembryonic mesoderm at the base of the allantois (14). A short time later (10.0 dpc), the gonads start to form on the mesonephros in the ventromedial region. The somatic cells of the gonad originate from the mesonephros and the coelomic epithelium, a single layer of epithelial cells that lines the coelomic cavity (15,16). Sertoli cells are the first somatic cell type that can be identified in the primordial testis and they express Sry, the male sex-determining gene, that is crucial in testis formation. In 1993 Pelliniemi et al. (17) stated that, "The embryonic Sertoli cells are according to present knowledge the primary actors in the differentiation of the indifferent gonad into a testis in genetically male individuals." In 1998 Karl and Capel showed using a dye injected into cells that at 11.2 to 11.4 dpc, the coelomic epithelial cells of both XX and XY individuals migrated into the gonad. In XY gonads, the migrating coelomic epithelial cells initially gave rise to Sertoli cells that formed testis cords. At 11.5 to 11.7 dpc, the coelomic epithelial cells that migrated into the gonad stayed in the interstitium to form primordial Leydig cells. Cellular migration from the coelomic epithelium gradually decreased and was over at 12.5 dpc. By this time Sertoli cells had enclosed the PGCs, interstitial cells were present and the basement membrane layer had thickened to form the tunica albuginea. Until 11.5 dpc, both XX and XY gonads are similar and are capable of forming either testes or ovaries. In XX gonads, coelomic epithelial cells also migrated into the gonad, but there was no enclosure of the PGCs into cords (18). After cord formation the testis then undergoes an impressive increase in growth. Primordial germ cells proliferate until around day 14, after which mitosis is arrested until the postnatal period. When the seminiferous cords are formed the gonocytes are found well towards the center of these structures. Approximately 1 to 3 days after birth the gonocytes begin migrating towards the periphery of the cords and resume proliferation [for review see (19)]. As the gonocytes contact the basement membrane the proliferation continues and final relationship of spermatogonia to the basement membrane is established. This movement of the gonocytes is apparently obligatory as the gonocytes remaining in the center of the tubule undergo apoptosis. The Sertoli cells are key players in the formation of the testis and very likely have a role in the control of the mitosis of primordial germ cells, the prevention of the entrance of primordial germ cells into meiosis, and the distribution of spermatogonia along the basement membrane.

Testis development begins with the expression of Sry in the undifferentiated, bipotential genital ridges of mammalian XY fetuses. This gene expression directs supporting cell precursors to develop as Sertoli cells (20). In the absence of Sry, the somatic supporting cell precursors become follicle cells destined to be part of the ovary. The development of primary germ cells into either pre-sperm or pre-oocytes is dependent on whether their environment is that of a testis or that of an ovary [for review see (21)]. In the absence of any influence from Sertoli cells, both XX and XY PGCs enter meiosis by 13.5 dpc. In the embryonic ovary the PGCs are arrested in the first prophase of meiosis I (21–23). While the formation of follicles is dependent on the presence of PGCs, the differentiation of Sertoli cells and formation of testis cords occurs normally in their absence. The differentiation of Sertoli cells in the gonad leads to the enclosure of PGCs inside testis cords and their arrest in mitosis before 13.5 dpc (21–23). Experiments with XX/XY mosaic mice by Palmer and Burgoyne in 1991 showed that the vast majority of Sertoli cells were derived from the XY genotype (24). However, all the other somatic cell types had an even distribution of cells from the XX and the XY genotypes. From these results, the authors concluded that Sry-expressing Sertoli cells promoted the differentiation of all other cell types within the testis. In another study using transgenic mice expressing a marker gene under control of the Sry promoter it was shown that Sry was expressed exclusively in pre-Sertoli cells in the urogenital ridge, and that Sertoli and granulosa cells develop from a common precursor (20).

As a result of all of the studies described above, the role that embryonic Sertoli cells play in the formation of the testis and sex determination has become less of a mystery. Harley et al. (2003) in their review state that, "Sex determination can be defined as the proliferation, migration, and differentiation of supporting cells to become Sertoli cells, committing the fate of the gonad to the testis pathway" (25).

Late Embryonic and Early Postnatal Maturation

In mammals, the formation of the testis and the enclosure of the gonocytes by the Sertoli cells to form sex cords are followed by a period of testicular growth. During this time the numbers of Sertoli cells expand via cell division (26). The number of Sertoli cells in the adult testis determines the ultimate spermatogenic output for a particular species or individual (27,28). Sertoli cells appear to proliferate first, during fetal or neonatal development and then again, later, in the prepubertal period (7). In rodents, the neonatal period and the prepubertal period are hard to distinguish and thus two periods of proliferation are not obvious. The data is much clearer in other species such as the human where the neonatal development is 10 to 12 years separated from puberty (7).

Since biological factors that regulate the proliferation of Sertoli cells will determine the reproductive capacity of a species or individual there is considerable interest in the identification of these factors. While few of them have been identified there is good evidence for the roles of FSH and thyroid hormone in this process. FSH receptor knockout mice are fertile but have a testis of reduced size and reduced sperm output due to the lack of FSH action (29–31). In addition, the administration of FSH to neonatal or gonadotropin deficient mice restores or increases the number of Sertoli cells (32,33). Hypothyroidism during the critical prepubertal time period leads to an increased number of Sertoli cells while administration of excess thyroid hormone leads to a decrease in the number of Sertoli cells and subsequently in testicular size (34–37). While the mechanisms involved in the action of thyroid hormones on Sertoli cells is not well understood there is some evidence that thyroid hormones alter the timing of the maturation of Sertoli cells (36). Because the final number of Sertoli cells plays such an important role in the final reproductive capacity of an individual or species multiple mechanisms and levels of control are likely.

Functional Maturation of Sertoli Cells at Puberty

Sertoli cells undergo a maturational change in both morphology and function at the time of puberty. They cease the functions of the embryonic or immature Sertoli cells and stop proliferating. The adult morphology consists of a larger irregularly shaped nucleus and, a tripartite nucleolus, abundant smooth and rough endoplasmic reticulum and tight junctional complexes with adjacent Sertoli cells (7,38,39). The end result of this maturation process is the creation of a unique adluminal compartment where the meiotic and postmeiotic germ cells can be sequestered in a specialized environment. The junctional complexes exclude necessary macromolecules and many nutrients and, as a result, the selective secretion of these nutritive factors becomes one of the functions of the mature Sertoli cells. Maturation also involves the conversion of the seminiferous cord into a true fluid filled tubule allowing the release and transport of spermatozoa (40–42).

GENE EXPRESSION IN SERTOLI CELLS

Each type of cell is functionally defined by the expression of a unique subset of their genes (the transcriptome). Information about when and why a subset of genes is expressed in Sertoli cells has provided insight into the role the cells play in testis formation and spermatogenesis. Many of the important gene products are secreted or are external components of the cell membrane and mediate cellular interactions with other Sertoli cells or with germ cells (43). Sertoli cells have this dual role in testis formation and support of spermatogenesis. Many of the genes expressed in Sertoli cells during the period of testis formation are not expressed later during spermatogenesis and vice versa (7).

Gene Expression in Sertoli Cells during Testis Formation

As discussed above, the Sry gene is the key in a cascade of events that lead to male sexual differentiation [for review see (44)]. The Sry protein interacts with DNA through a high mobility group (HMG) box-type of DNA binding domain and presumably activates a cascade of downstream genes that are currently unidentified. The expression of Sry is a form of molecular switch and expression of Sry in the mouse embryonic Sertoli cells is detectable for only a short time and therefore its action must be to initiate but not maintain a pathway of gene activity leading to the differentiation of the testis. Sry action appears to be required between 11 and 11.5 dpc in the mouse and experiments using knockout models are mostly sex reversed but the results depend on the gene dosage or level of expression (45).

A number of genes not located on the Y chromosome are expressed downstream of Sry only in the XY gonad and many of them have been considered to be a part of the sex determination cascade. Sox9 mRNA was found in male (XY) but not female (XX) genital ridges, and was localized to the sex cords of the developing testis. Sox9 expression was shown to be specific to the Sertoli cell lineage and the timing and cell-type specificity of Sox9 expression suggested that the expression of this gene could be directly regulated by Sry (46). From transgenic studies in mice Sox9 was shown to induce testis formation. The homozygous deletion of Sox9 in XY gonads of mice was accomplished using conditional gene targeting (47). This study showed that the homozygous deletion of Sox9 in XY gonads prevented sex cord development and that Sox9 is involved in Sertoli cell differentiation, the stimulation of synthesis of MIS and Sox8, and the inactivation of Sry. MIS, a product of immature Sertoli cells, leads to the regression of the Müllerian ducts from which the female reproductive tract develops (48). The activation of Sox9 can apparently mimic all of the effects of Sry (45).

Gene knockout studies in mice have been important tools for the identification of genes that direct the development of the gonad. Mice lacking gonads were obtained when the following genes were targeted for knockouts: *Lim1* (also known as *Lhx1*), *Sf1* (steroidogenic factor 1; also known as *Zfp162*), *Wt1* (Wilms tumor gene 1), *Emx2*, and *Lhx9*. These results are consistent with a role for these genes in Sertoli cells downstream from Sry and Sox9 but still relating to the initial stages of gonad development (25,49). Another gene whose elimination leads to mice with extremely malformed testes includes Desert hedgehog (Dhh), expressed in Sertoli cells at 12.5 dpc (50), The knockout of fibroblast growth factor 9 (fgf9) genes in male mice leads to sex reversal and the speculation that fgf9 plays an early role in regulating male-specific pathways downstream of Sry (51).

Gene Expression in Functionally Mature Sertoli Cells

The known gene products of Sertoli cells have primarily been secreted proteins and have previously been placed into four major categories including transport or bioprotective proteins, proteases and protease inhibitors, membrane glycoproteins, and growth factors and paracrine factors (43,52). In recent years there has been substantial progress on identification of proteins in each of these categories.

Most of the initial studies on gene expression involved cultured rat Sertoli cells from peripubertal animals. These studies focused on the protein products of Sertoli cells and the endocrine regulation of the synthesis and secretion of those proteins. Many of the first proteins characterized as Sertoli cell products were transport proteins such as androgen binding protein (ABP) or transferrin (52). The transportation of important nutrients and regulatory factors around or across tight junction barriers to the sequestered germ cells has been presumed to be a major function of Sertoli cells. The transport of ferric ions by transferrin was the first such function of Sertoli cells to be described at the molecular level (40).

Many different proteases and protease inhibitors are synthesized by Sertoli cells and one rationale for their function is it involves the continuous tissue remodeling that occurs when tight junctions must be traversed or when spermatozoa are released from the seminiferous epithelium (53). The large number of different proteases and inhibitors made by Sertoli cells appear to be regulated by hormones and growth factors or are present at some stages of the cycle of the seminiferous epithelium but not at others.

Hormone receptors such as the FSH receptor are among membrane glycoproteins that have been characterized as Sertoli cell products (31,54–56). In addition, the components of the junctional complexes between adjacent Sertoli cells or between germ cells and Sertoli cells must be considered as potential membrane glycoproteins (57–60). Inhibin and activin (61–63) are included in the secreted hormones and growth factor category.

Additional categories of gene products found in Sertoli cells that have been of recent interest to investigators includes transcription factors and signaling molecules. Transcription factors or putative transcription factors that are unique products in the testis to Sertoli cells include the GATA family of transcription factors (64–67), and several helix-loop-helix transcription factors (63,68,69). Both the desert hedgehog and tyro-3 families of signaling molecules have been found to be produced by Sertoli cells (70,71).

Table 1 lists reports published since 1993 that characterize one or more products of Sertoli cells. Note that they are listed in categories of interest discussed above.

Transgenic Mice and Sertoli Cell-Specific Expression

One way to test the role of Sertoli cells in spermatogenesis is to utilize the transgenic technology and over-express or eliminate the expression of genes that are unique to Sertoli cells. As a result there has been an interest among investigators for promoters that will direct transcription specifically in Sertoli cells. The first use of the transgenic technology applied to Sertoli cells was reported in 1992 (72). Transgene expression specific to Sertoli cells was accomplished in mice by using the transcriptional regulatory sequences of the human Müllerian inhibiting substance (hMIS) to drive transcription of the SV40 T-antigen gene. This gene is expressed initially in embryonic Sertoli cells but not in functionally mature Sertoli cells. These transgenic mice developed testicular tumors and a cell line derived from one of the tumors expressed several genes characteristic of Sertoli cells such as transferrin, sulfated glycoprotein-2, and inhibin-beta B. It was later shown that the cell line designated MSC-1 did not express detectable levels of inhibin-alpha, MIS, or FSH receptor (73).

Androgen-binding protein is made by the both the liver and the Sertoli cells in the rat (74,75). When a 5.5-kilobase (kb) rat DNA fragment containing the ABP gene with all 8 exon sequences and 1.5 kb upstream of the transcription start site was used to make transgenic mice, the expression of ABP from the transgene was confined to the testis and brain (76). The transgenic testicular rat ABP was shown by immunohistochemistry to be primarily in the cytoplasm of Sertoli cells and lumen of the seminiferous tubules of the transgenic mice at concentration up to 25 times normal. The elevated ABP levels in the transgenic mice were associated with a variable degree of abnormal spermatogenesis leading to reduced numbers of sperm in the epididymis.

The promoter of the FSH receptor gene has been extensively characterized because of the high level of cell specificity for its expression. Since the FSH receptor is expressed exclusively in the Sertoli cells in the male, an understanding of the FSH receptor gene regulatory sequences could lead to an additional way to direct transgene expression to Sertoli cells. However, in both cell transfections and in transgenic mice, the promoter directed the expression of transgenes promiscuously. It was proposed that repression and activation of local chromatin structure are likely to play a major role in the regulation of the FSH receptor gene (54).

The Pem gene promoter and the cathepsin L gene promoter have been used to successfully direct the expression of transgenes in Sertoli cells (77–79). The Pem gene encodes a homeodomain related to those in the Prd/Pax gene family and Pem transcripts were shown to be androgen regulated and preferentially expressed in Sertoli cells in stages VII–VIII of the cycle of the seminiferous epithelium (80). A 0.6-kb 5′-flanking sequence of the Pem gene directed transgene expression specifically in the testis and the epididymis. The transgene expression was androgen dependent and was confined primarily to Sertoli cells during stages IV–VIII of the cycle of the seminiferous epithelium similar to the expression pattern of the endogenous Pem gene. When a 0.3 kb fragment was deleted from the 5′-end of the transgene there was no effect on androgen-dependent Sertoli-specific expression but the stage-specific expression and the timing of the postnatal expression was changed. These results demonstrated that there are separate regions in the Pem promoter to direct expression specifically to Sertoli cells and to impart stage-specific expression (79).

Cathepsin L (also known as cyclic protein 2 in the Sertoli cell studies) is another gene shown to be expressed at high levels by the Sertoli cells (81). The 3-kilobase genomic fragment immediately upstream of the rat cathepsin L translation start site was shown to direct expression of the reporter gene, beta-galactosidase, only in Sertoli cells of transgenic mice (77). The expression pattern of the reporter gene was identical to that of the endogenous gene in Sertoli cells at different developmental stages and during different stages of the cycle of the seminiferous epithelium.

Genearray Analysis of Sertoli Cell Gene Expression

Attempts to characterize Sertoli cells by examination of the genes they express has historically involved techniques which have confined these approaches to a very limited number of relatively abundant expressed sequences. The complete sequencing of the genomes

TABLE 1. *mRNAs and proteins made in Sertoli cells: a summary of the reports concerning the expression of genes by Sertoli cells secreted proteins and secreted binding proteins*

Protein	Ref.
Clusterin	(220)
Transferrin	(3)
Prosaposin	(221)
IGFBP-3	(222)
Riboflavin carrier protein	(223)
Thiamin carrier protein	(224)
IGF-binding proteins	(225)
Pregnancy-specific beta1 glycoprotein-like substance	(226)
Intracellular binding proteins	
Cellular retinol binding protein	(227)
Mannose 6-phosphate receptor	(228)
Hsp27	(229)
stAR	(230)
Neurofilament H, M, and L	(231)
Calgizzarin (Calcium binding)	(232)
FAS ligand	(233)
Rbm3 cold shock protein	(234)
Musashi 1 RNA binding	(235)
Membrane and junctional proteins	
Short-type PB cadherin	(236)
Calponin	(237)
Connexin43	(238)
Connexin26	(239)
Gelsolin	(240)
Iron transporter Nramp2/DMT-1	(241)
Class B scavenger receptor type I	(242)
Cell surface proteoglycans	(243)
Membrane anchored Ceruloplasmin	(244)
Omega-Conotoxin-sensitive Ca2+ voltage-gated channels	(245)
Cell adhesion molecules	(246)
Laminin	(247)
Collagen IV	(247)
Entactin	(248)
Placental cadherin	(249)
N-cadherin	(250)
Vimentin	(251)
Desmin	(251)
Cytokeratin	(251)
Low-density lipoprotein receptor-related protein-1	(252)
Plectin	(253)
Class B scavenger receptor type I	(254,255)
Hormone receptors and growth factors	
Androgen receptor	(256)
Activin	(257)
Inhibin	(258,259)
Adrenoreceptor	(260)
ckit ligand	(261)
Transforming growth factor beta 1	(262)
P2-purinogenic receptors	(263)
FGF receptor I	(264)
IGF-I receptors	(265)
Thyroid hormone	(266)
TNF alpha receptor p55	(267)
Estrogen receptor	(268)
Müllerian-inhibiting substance	(269)
FSH receptor	(270)
LIF and LIF receptor	(271)
REBalpha	(272)
Prolactin receptors	(273)
Serotonin receptor	(274)
Kappa opioid receptor	(275)
Muscarinic acetylcholine receptor	(276)
Anandamide	(277)
Transcription factors	
GATA-1	(67)
GATA-4	(278)
CREB	(279)
cfos	(280)
Pem homeobox gene	(101)
cjun	(281)
cmyb	(281)
DAX-1	(282)
Sry	(283)
Sox9	(284)
Dmrt 1	(285)
Signaling related	
Opioids	(286)
Interleukin 6	(246)
Neuropeptide Y	(287)
Interleukin 1	(288)
Cyclin dependent kinase 5	(289)
14-3-3 theta novel isoform	(290)
Myotubularin (tyr phosphatase)	(291)
Interleukin 1 receptor antagonist	(292)
Interleukin 1 alpha	(293)
Notch receptors and ligands	(294)
Bclw/Bax/Bak	(295)
Testicular protein kinase 2 (TESK2)	(296)
AKT/protein kinase B	(297)
Fyn tyrosine kinase	(298)
Cyclin dependent kinase inhibitor p27(kip1)	(299)
Mer tyrosine kinase	(300)
Proteases and inhibitors	
Plasminogen activator	(301,302)
Plasminogen activator inhibitor	(303)
TIMPs	(304)
Cathepsin L	(81)
Alph2-macroglobulin	(305)
Neuronal ubiquitin C-terminal hydrolase	(306)
Caspase3	(307)
Cathepsin K	(308)
Other enzyme activities	
Surface associated glycosyltransferase	(309)
Dolichyl phosphomannose synthase	(310)
Hormone sensitive lipase	(311)
Asparagines synthetase	(312)
Guanidinoacetate methyltransferase	(313)
Aromatase	(314)
Gamma glutamyl transpeptidase	(315)
UDP-glucuronosyltransferase 1 isoforms	(316)
Nitric oxide synthase	(317)
Ecto-ATPase	(318)
Glucose transporter 1	(319)
Lactate dehydrogenase A	(319)
ATP-binding cassette transporter 1	(320)
Proteins of unknown function	
LRPR1 (leucine-rich protein)	(321)
Testin	(322)
MAGE D	(323)
PASS1	(324)
Erythropoietin	(325)
Amphiphysin 1	(326)
SERZ	(327)
GP130	(328)
Prion like protein Doppel	(329)
Tsx	(330)
Beta-synuclein (Phosphneuroprotein)	(331)

of many species and the subsequent development of oligonucleotide based expression arrays have made it possible to obtain a more global view of gene expression by a tissue or a cell. Affymetrix GeneChips have been used to analyze the gene expression of rat and mouse Sertoli cells from prepubertal animals in vitro (82). The rat Sertoli cells were isolated from day 20 postpartum rats and cultured using standard techniques for 2 days prior to RNA purification. The goal of this study was to identify genes expressed by Sertoli cells in culture and genes whose expression was induced or repressed by FSH. An analysis of the relative levels rat Sertoli cell expressed genes in control samples provided an ordered list of genes expressed in cultured rat Sertoli cells. A comparison analysis was used to determine genes induced or repressed by FSH at each time point in comparison to untreated controls. Subsequent analysis of the lists of genes by clustering or grouping genes based on ontology provided a useful way to determine what genes are involved in Sertoli cell functions (83). In the gene ontology analysis gene products are described in terms of their associated biological processes, cellular components and molecular functions in a species-independent manner. Due to the number of ESTs present in the GeneChip dataset and the incomplete nature of the ontology datasets, a total of 1,014 rat Sertoli genes were included in the ontology analysis. The largest proportion of expressed genes were associated with metabolism with relatively large numbers of genes expressed in the category of "transport." These data were expected due to the proposed "nurse" cell function of Sertoli cells.

Additional results utilizing oligonucleotide arrays on cultured mouse Sertoli cells are available from studies that characterized the entire testicular transcriptome from birth to the adult (84). Data from this study included results from isolated cell types as well as from whole testis. The results from Sertoli cells isolated from 16- to 18-day-old mice and cultured for 2 days can be examined in detail. Using in silico subtraction techniques the authors concluded approximately 150 genes had expression levels that were greatly enriched in the Sertoli cells and approximately 50 of those appeared to be unique to Sertoli cells. The data from this study have been further filtered to remove any expressed sequences called "present" but lacking a signal above 50. Sequences were also eliminated that were not twofold enriched in Sertoli cell versus other testicular cell types. A signal of 50 represents a few molecules of RNA per cell. The resulting lists are shown in Table 2. These lists identify sequences that are enriched in the Sertoli cells and they begin to tell a story about the specialized functions of these cells. The Sertoli cells are enriched in a number of enzymes involved in general metabolism, lipid metabolism, nucleotide metabolism, transcription factors and other protein classes. It has been noted that many of the expressed genes whose abundance is enriched in Sertoli cells are also enriched in brain or neural tissue and in liver. This could be explained by the proposed functions of Sertoli cells. As in liver, Sertoli cells produce some proteins similar to serum transport proteins that provide mechanisms for moving nutrients across junctional barriers. As in brain, Sertoli cells create a barrier to the free movement of macromolecules. While the structure of the barrier is different, it is functionally similar. Perhaps the similar gene products thus reflect partial biological roles that

TABLE 2. *Stages in the life of a Sertoli cell*

Sertoli cells in mammals have the separate functions of forming the testis and supporting the progression of germ cells into sperm. The following description of the stages was derived from a review by Sharpe et al., 2003 (7). The timing of these events can vary greatly between species. In rodents the quiescent period is nonexistent as the neonatal period and the onset of puberty are hard to distinguish while in humans the quiescent period may last a decade or more.

Testis determination

Expression of Sry followed by recruitment of other cells to somatic testicular lineage, surrounding of PGCs to form cords and suppression of entry into meiosis. Expression of Sox9, Wt1, SF-1, Dhh, Dax-1, GATA-4, Fgf-9, suppression of Müllerian duct derivatives via MIH.

Neonatal period

Cell division and growth, germ cells move to basement membrane, expression of cytokeratin 18 until puberty.

Quiescent period

Homeostasis

Puberty

Tubule lumen forms, tight junctions between adjacent cells form, germ cell progression through meiosis begins, secretion of luminal fluid, junctional complexes with advanced germ cells form.

Adulthood

Stages of the cycle of the seminiferous epithelium are apparent, maximal asynchronous sperm production, sperm production determined by number of Sertoli cells. Adult genes expressed include GATA-1, clusterin, transferrin, cathepsin L, genes associated with junctional complexes.

require a unique environment in the testis and the difficulty in transporting macromolecules through that environment.

Genearray analysis provides a unique opportunity to evaluate global gene expression in cells or tissues. Genearray and genomic approaches, along with proteomic and other high throughput techniques, will significantly enhance our understanding of cellular function within the testis. The Sertoli cell transcriptome will provide a key piece to the puzzle of genes necessary for sperm production.

FSH, TESTOSTERONE, AND SERTOLI CELLS

The primary hormonal controls on spermatogenesis involve the action of FSH and testosterone on Sertoli cells. Despite decades of research, questions have remained about the relative roles of both of these hormones on testis development and spermatogenesis. In particular, the functional significance of FSH in the adult mammal and the cellular and gene targets for the action of testosterone have been subjects for discussion and debate for many years (85,86). Recent progress in these areas has been a result of the use of transgenic mice and both targeted and naturally occurring gene knockout mouse models.

FSH and Spermatogenesis

The requirement for FSH in spermatogenesis and the result of the stimulatory action of FSH in Sertoli cells has received a great deal of attention. A number of studies have utilized the gonadotropin (GnRH)-deficient mouse or hypogonadal mouse designated *hpg*. These mice are gonadotropin deficient due to a natural mutation in the GnRH gene, which results in noncirculating LH and FSH levels and hypogonadism. When testosterone alone is given to the GnRH-deficient mouse testicular maturation and fertility was seen (87). GnRH-deficient mutant mice treated with testosterone implants had normal spermatogenesis but reduced testis size and germ cell numbers. If the *hpg* mice were treated with both FSH and testosterone the result was quantitatively normal spermatogenesis and testes of normal size (88). The exogenous FSH treatment increased testis size by 43%. Treatment with FSH during the first two weeks after birth increased both the number of Sertoli cells and total sperm production by the *hpg* mouse testis. In related results, the treatment of prepubertal normal rats with exogenous FSH resulted in larger than normal testes and higher total germ cell numbers (89). Additional information has come from a number of studies that utilized targeted gene knockouts. When the FSH β gene was eliminated the male mice were fertile but had smaller than normal testes and reduced germ cell numbers (31). The FSH null mutants had testes that were about 1/2 normal size and at 6 to 7 weeks of age the number of epididymal sperm was reduced by 75%. The elimination of the FSH receptor in mice resulted in fertile males that had a reduced testis size and sperm production. Two groups have produced FSH receptor null mice (FORKO) and have shown that while the males were fertile, the size of the testis and the total number of sperm produced were decreased significantly (30,90). In humans a FSH receptor null mutation (566C→T) has been described in five male individuals (91). The men showed variable degrees of spermatogenic failure but were fertile. None of the five men had normal sperm parameters and their testicular size was reduced but two of the men had fathered children.

Studies on mice with the null mutations in the GnRH and FSH genes and the description of fertile men with inactivating mutations in the FSH receptor gene lead to the conclusion that FSH is not required for fertility in mice or men. It is clear that FSH does play an important role in the size of the testis and ultimately in the sperm output of an individual. Recent more detailed studies have noted structural and functional problems with both the sperm and the Sertoli cells in the FORKO mice (92–94). The relationship between Sertoli cell proliferation, testis size and spermatogenic capacity has been well established. The proliferation of Sertoli cells is maximal in 20- and 21-day-old rat fetuses and declines steadily until the second week after birth when further cell division is rare (95). The role of FSH in the prenatal and newborn rat is to act as a mitogen to the Sertoli cells, which will ultimately set the adult size of the testis. When the number of Sertoli cells was experimentally restricted there is a corresponding decrease in the size of the testis in the adult rat but the ratio of round spermatids to Sertoli cells was constant (28). One of the basic tenets of testis biology is that the size of the Sertoli cell population limits sperm production. The interaction of FSH and the Sertoli cells is crucial in setting the size of that population.

The key to understanding the action of FSH on Sertoli cells is to understand the changes that occur within the cells when FSH is present. Recent attempts have been made to characterize the response to FSH on a global scale. The mRNA from FSH-treated rat Sertoli cells was used to screen Affymetrix U34A rat GeneChip oligonucleotide microarrays (82). Sertoli cells from 20-day-old rats were cultured in the presence of 25 ng/mL ovine FSH. At 0, 2, 4, 8, and 24 hours after the addition of FSH, total RNA was purified and hybridized to the U34A rat microarray containing oligonucleotides representing approximately 9,000 rat genes. Approximately 100 to 300 transcripts at each time point were up-regulated or down-regulated by twofold or greater. The number of transcripts up-regulated were generally four- to sixfold more than the number of those down-regulated at the same time period. The regulation of genes previously reported to be influenced by FSH or cAMP in rat Sertoli cells were confirmed and numerous additional regulated genes were found. In similar studies, testicular genes that were regulated in vivo in the testes of *hpg* mice following FSH treatment for 4, 8, 12, or 24 hours were identified using the Affymetrix GeneChip U74A arrays (96). An overall activation of gene expression was observed after FSH treatment, which was consistent with the results obtained on FSH treatment of cultured rat Sertoli cells. These studies represented initial attempts to define the transcriptome induced and repressed by FSH in rat Sertoli cells. Large datasets of genes were generated and are available for further analysis of spermatogenesis and Sertoli cell signaling.

Testosterone and Spermatogenesis

The androgenic steroid produced and active in the testis is testosterone synthesized in Leydig cells under the regulation of luteinizing hormone. Testosterone is considered essential for complete spermatogenesis and, as described above, testosterone in the absence of FSH can maintain spermatogenesis. In the testis the androgen receptor AR has been localized to Leydig cells, peritubular cells and Sertoli cells (5,6) so the target cells by which androgens control spermatogenesis are still debated. Despite evidence to the contrary there was also a continuing debate about the direct action of androgen on and the presence of AR in germ cells. However, recent studies in which germ cells from TfmX/Y mice were transplanted into seminiferous tubules of azoospermic mice expressing a functional AR resulted in complete and qualitatively normal donor-derived spermatogenesis (97). This experiment that built on previous studies using chimeric mice clearly showed that germ cells do not require a functional AR. Therefore, androgens must affect spermatogenesis indirectly via actions on somatic testicular cells and very likely, on the Sertoli cells which express the androgen receptor. The Sertoli cells make direct contact with developing germ cells and exert control over the intratubular environment and structure and thus could reflect the actions of androgens on these processes. Several animal models have been examined in an effort to clarify the sites of action and the molecular mechanisms of action of testosterone in the testis. However, all of these animal models lack either testosterone or the androgen receptor in all testicular cells and thus the exact role of the Sertoli cells remained ambiguous. Recently, the Cre/loxP technology has been used to generate mice with a selective knockout of the androgen receptor in Sertoli cells (98). Mice in which exon 2 of the androgen receptor was floxed were crossed with mice expressing Cre recombinase under control of the Müllerian Inhibiting Substance (MIS) promoter specifically in Sertoli cells. This strategy resulted in a frame shift and premature termination of AR transcription The Sertoli cell selective knockout mice had normal urogenital tracts and descended testes showed spermatogenic arrest at the late spermatocyte/ spermatid stage. Two similar studies showed qualitatively similar results (99,100). In one of these studies, the analyses showed the Sertoli-specific androgen receptor knockout mice were indistinguishable from wild-type mice (B6 except for the smaller testes (99). The knockout mice were infertile, with most spermatogenic arrest occurring at the diplotene premeiotic stage. The knockout mice had very few sperm in the epididymis, lower serum testosterone concentrations and higher serum leuteinizing hormone concentrations than wild-type mice. It was also demonstrated that the knockout mice had defects in the expression of the Sertoli cell products MIS and androgen-binding protein. The data from all of these studies utilizing Sertoli cell-specific androgen receptor knockout mice showed that the action of the androgen receptor in Sertoli cells is an absolute requirement for complete spermatogenesis. The development of spermatocytes into spermatids is a major site of Sertoli cell-mediated androgen action.

Despite the known biological actions and target cells of testosterone in the testis there have been very few clues with regard to the actual genes that might be under androgen regulation. One gene that has been shown to respond to testosterone in Sertoli cells is the Pem gene which is an orphan homeobox gene (101). In adult mice, Pem transcripts were preferentially expressed in Sertoli cells in stages VII–VIII of the seminiferous epithelium. These stages correspond to the androgen-dependent stages during which germ cells undergo the first step of meiosis. In addition, it was demonstrated that Pem gene expression was reduced or absent in hypophysectomized mice, gonadotrophin-deficient hypogonadal (hpg) mutant mice, and androgen receptor-deficient (tfm) mutant mice. Injection of either testosterone or luteinizing hormone (LH) into hypophysectomized and hpg/hpg mice restored Pem expression in the testes to normal levels. Of interest was the finding that the response of the Pem gene to testosterone took 24 to 48 hours. The question remains as to whether androgens act directly or indirectly on the PEM gene expression.

While the response of the Pem gene to testosterone could be demonstrated in vivo in several animal models, when Sertoli cells were placed in culture the response to androgens disappeared (80).

The hypogonadal mice (*hpg*) were utilized in a global examination of the action of androgens on the testis (102). Since, this mouse strain lacks significant secretion of FSH and LH, the circulating levels of androgens are very low or nonexistent. These mice exhibit infantile testes with spermatogenesis halted at the first meiosis rendering the mice infertile. Nonetheless, the *hpg* mice possess a hormonally responsive reproductive tract, and it has been shown that androgen replacement is capable of initiating qualitatively complete spermatogenesis and fertility in the GnRH deficient mouse strain. This study was designed to identify genes that were regulated by testosterone in the hypogonadal mouse testis and to monitor their expression over twenty-four hours. Affymetrix Murine oligonucleotide arrays were used to study the expression pattern of 36,700 transcripts in testes of these mice after treatment with testosterone. At all experimental time points earlier than 24 hours, there were significantly more mRNAs with reduced than increased abundance in testes of *hpg* mice after treatment suggesting that in the murine testis, the primary action of T might be to repress gene expression. Many of the genes whose expression was upregulated were ESTs. As a control for the overall response the expression of the Pem gene was monitored on the arrays. Pem was not detected in testes of control *hpg* mice, however, its signal increased in testes of *hpg* mice following treatment with testosterone as predicted.

ADVANCED TECHNOLOGIES

Germ Cell Culture and Sertoli Cells

The contribution of Sertoli cells to support and regulate the differentiation of germ cells in the testis is well established. In the testis, the small population of spermatogonial stem cells (103) produce spermatogonia that undergo several rounds of mitosis before entering meiosis to become spermatocytes. On completion of meiosis, spermatocytes become haploid spermatids that differentiate into sperm that are released by the Sertoli cells into the lumen of the seminiferous tubule. Differentiating germ cells require intimate contact with Sertoli cells to complete this process. Investigation of the interactions between germ cells and Sertoli cells is an active area and new techniques such as spermatogonial stem cell transplantation are being used to further understand this process (82,103). Cultures of germ cells and Sertoli cells are also being utilized, both independently and together, to determine the necessary components that lead to the production of sperm. The development of techniques to culture highly pure populations of primary Sertoli cells has been extensively covered previously (104). In this section the focus will be on the technologies that have enhanced our understanding of how germ cells can influence Sertoli cell function. In addition, we will discuss how culture of spermatogonial stem cells with Sertoli cell lines has increased our understanding of these specialized cells.

Culture of Sertoli cells and germ cells together has been used to investigate many aspects of spermatogenesis in multiple species. The most challenging goal has been to recapitulate the differentiation of diploid spermatogonia into haploid spermatids in vitro (105). Successful production of sperm in vitro has been achieved in the Japanese eel using testicular organ explants treated with hormones to stimulate germ cell differentiation (106). In mammals, the production of haploid germ cells from neonatal testicular cells has been reported in bulls. The testicular germ cells from 3-day-old bull calves were enzymatically dispersed and then reaggregated by phytohemagglutinin and encapsulated by calcium alginate (107). Fourteen weeks after initiation of culture, transcripts for spermatid-specific gene were detected by RT-PCR (107).

The cells and architecture of the testis can be maintained for several weeks with the use of organ or tissue explant culture systems (108,109). This approach maintains a normal testicular structure since the seminiferous tubules remain distinct from the interstitial compartment of the testis. Sertoli-germ cell interactions are intact. Molecular events important for development of gonocytes including migratory ability, resumption of mitosis and interactions with Sertoli cells in neonatal rats have been investigated using neonatal rat testicular organ cultures (110). Testicular explant cultures have been used with bovine testis tissue to investigate the biological activity of spermatogonial stem cells. Culture of small pieces of neonatal bovine testis tissue on floating filters supported the survival and proliferation of spermatogonial stem cells for 2 weeks (109). Maintaining the spermatogonial stem cell-Sertoli cell interactions in vitro appears to support stem cell survival and proliferation. A similar approach, the culture of small pieces of seminiferous tubules has been used to investigate the conditions needed to support the completion of meiosis in rodents (111,112). This system provides a means to investigate differential analysis of factors produced by Sertoli cells that regulate meiosis. In addition, development of in vitro techniques that support the differentiation of germ cells to haploid spermatids is potentially beneficial for human assisted reproductive techniques (ART) such as intracytoplasmic spermatid injection. Perrard et al. (111) have shown that the transition from early to late pachytene spermatocytes is a limiting factor for the completion of meiosis in vitro.

Co-culture of dispersed Sertoli and germ cells has also been used to generate haploid germ cells in vitro for ART for both rodents and humans (113,114). In mice, round spermatids were generated in vitro from pachytene spermatocytes co-cultured with Sertoli cells. Normal and fertile offspring were generated by nuclear injection of the spermatid donors (113). This approach can also be used to investigate factors produced by Sertoli cells that regulate germ cell differentiation. Similarly, the influence of germ cells on Sertoli cell activity and communication can be evaluated. The effect of dispersed Sertoli-germ cell co-culture on cellular activity of both immature and mature Sertoli cells has lead to a better understanding of how these cells communicate (113,115–121). Lastly, identification of genes influenced by Sertoli or germ cells during co-culture has accelerated the characterization of signaling pathways regulating function and differentiation in these cells (122–126).

As discussed above, the complex interactions involved in the regulation of Sertoli cell function have lead to the development of techniques to investigate these cells in vitro. Primary Sertoli cell culture provides the opportunity to evaluate hormonal response and protein secretion (2,82,127,128). Dispersed primary Sertoli cells are usually cultured in flasks as a monolayer. However, this approach has a disadvantage if investigation of the polarity and specialized nature of Sertoli cells in the seminiferous epithelium is the goal. Sertoli cells form highly specialized tight junctions during development (129,130). The specialized tight junctions restrict the passage of molecules from the fluid the surrounds the seminiferous tubules into the adluminal compartment where spermatocytes completing meiosis and haploid spermatids are located. Sertoli cells must translocate the differentiating germ cells from the basal compartment to the adluminal compartment during spermatogenesis. The regulation of this process and the identification of the specialized proteins that make up the tight junctions has been investigated in vitro with the use of multichamber bicameral units (131–133).

Multichamber culture methods were developed because Sertoli cells cultured as monolayers using conventional methods do not form tight junctions so the compartmentalized nature of the seminiferous epithelium is not recapitulated (133). Likewise, the polarity of Sertoli cell secretion toward the lumen and Sertoli cell-germ cell interactions were not maintained using standard culture techniques. Primary Sertoli cells can be cultured on Matrigel (Collaborative Research, Inc., Bedford, MA) coated filters with medium in this chamber and below the filter [see (133) for details]. Transepithelial electrical resistance is used to monitor the integrity of the tight junctions formed by adjacent Sertoli cells (132). Germ cells cultured with Sertoli cells using multichamber units has provided a method to determine how germ cells influence Sertoli cell secretions and function. Germ cells co-cultured with Sertoli cells demonstrated that different populations of germ cells (e.g., spermatocytes or spermatids) influence Sertoli cell secretion (134–136). Germ cells can influence the amount and direction of the secretion of Sertoli cell proteins. Sertoli cell factors regulated by germ cells can influence other testicular cells such as Leydig cells (135). These findings indicate the interactions between germ cells and Sertoli cells can influence the local microenvironment of the seminiferous tubule and the interstitial space of the testis. Continued investigation of the dynamics of Sertoli cell tight junctions will lead to a better understanding of the precise physiological relationship between junction dynamics and spermatogenesis and how germ cells may influence this important structural component of the seminiferous epithelium.

Established Sertoli Cell Lines

Primary Sertoli cells, when cultured as dispersed cells, begin to lose their ability to respond to hormones after approximately 7 days in culture (104). Primary cultures of Sertoli cells are often contaminated with germ or peritubular myoid cells that sometimes complicate the interpretation of results. Co-culture of dispersed primary Sertoli cells and germ cells in order to investigate interaction between these cell types is also limited by time and the necessity to culture germ cells with serum. Primary Sertoli cells cultured in a serum free media provide the most reproducible response to hormones (104). Inclusion of serum to a Sertoli-germ cell co-culture often leads to rapid expansion of contaminating peritubular myoid and fibroblast cells while slowly dividing germ cells and nondividing Sertoli cells are out-competed for resources. These factors have lead to the establishment of multiple immortal Sertoli cell lines that have been selected to mimic the characteristics of Sertoli cells in vivo. Multiple Sertoli cell lines have been established for the mouse (137–145), rat (146,147), zebrafish (148), and sheep (149). The degree of characterization of each of these cell lines varies depending upon how well the line maintains Sertoli cell characteristics and supports germ cell differentiation (73,144). In zebrafish, Sertoli cell lines have been developed that support different aspects of germ cell development when co-cultured. For example, the ZtA6-2 cell line stimulates the proliferation of spermatogonia while the ZtA6-12 stimulates the differentiation to sperm (148).

The development of the spermatogonial stem cell transplantation technique as a functional assay for the presence of spermatogonial stem cells has resulted in an increased interest in the culture of these cells (103). A major goal of this area of research is to define the factors critical in maintaining and supporting the proliferation of spermatogonial stem cells. Co-culture of spermatogonial stem cells with a fibroblast feeder layer (STO cells) has been used to successfully maintain spermatogonial stem cells in culture for extended periods (150,151). Several cell lines derived from Sertoli cells, fibroblasts and bone marrow stroma were used to serve as feeder layers for spermatogonial stem cells to elucidate regulatory mechanisms of the stem cells and to determine what feeder cells provide the best microenvironment for stem cell survival and proliferation (152). Using the embryonic fibroblast STO feeder cell line as a control for spermatogonial stem cell maintenance, the ability of each cell line to support spermatogonial stem cells was evaluated with the use of the stem cell transplantation technique. Interestingly, the Sertoli cell lines SF7 (138) and TM4 (145) significantly reduced the stem cell number in vitro. The Sertoli cell lines MSC-1 (137) and 15P-1 (144) had no effect on stem cell number when compared to STO cells. The fibroblast L cell line and the bone marrow stroma cell line OP9 (153) were the only feeder cells that improved spermatogonial stem cell survival when compared to STO cells (152). These results are in direct contrast to results reported by Kent-Hamra et al. (154). These researchers co-cultured rat spermatogonial stem cells on STO and MSC-1 feeder cells. Rat germ cells cultured on STO cells lost stem cell activity over time while those cultured on MSC-1 cells maintained stem cell activity as evaluated by spermatogonial stem cell transplantation. The nature of this difference could be due to the use of mouse spermatogonial stem cells in Nagano et al. (152) and rat spermatogonial stem cells in Kent-Hamra et al. (154). Differences in the morphology of spermatogonial stem cells and the size of the stem cell populations in mice and rats have been reported (155,156).

Germ Cell Transplantation and Sertoli Cells

Sertoli cells contribute to the formation of the spermatogonial stem cell niche found along the basement membrane of the seminiferous tubule (103). Culture of spermatogonial stem cells with Sertoli cell lines or other feeder cells have begun to characterize the factors potentially produced by Sertoli cells that contribute to the formation and maintenance of the niche. Spermatogonial stem cell transplantation has also been used to investigate other ways Sertoli cells contribute to stem cell biological activity and maintenance of spermatogenesis. Natural mutations and targeted disruption of genes essential for the production of sperm present multiple phenotypes ranging from seminiferous tubules devoid of a lumen and differentiating germ cells to tubules containing haploid cells that undergo apoptosis and degenerate prior to spermiation. It is often difficult to determine if the mutation leads to arrest of spermatogenesis due to a block at some point of germ cell differentiation or because Sertoli cells are unable to provide support needed for germ cells to develop. Spermatogonial stem cell transplantation provides a means to determine which cell of the seminiferous epithelium is directly affected. The strategy employed requires the transplantation of germ cells from the mutant into a permissive host and then a means to detect the transplanted cells. Likewise, mutant animals lacking differentiating germ cells can serve as recipients of germ cells from wild type donors. The extent of germ cell differentiation in recipients may allow the researchers to conclude if the mutation affects the germ or somatic cells of the seminiferous epithelium.

Male mice homozygous for a mutation in the gene encoding estrogen receptor alpha (ERα) are infertile. To determine whether germ cells or somatic cells require ERα, germ cells were transplanted from donor males homozygous for the mutation into testes of wild-type recipients (157). Wild-type recipients sired offspring heterozygous for the mutation. These studies show that male germ cells do not require ERα for development or to function in fertilization. Mahato et al. (157) concluded that ERα mice are infertile due to disruption of estrogen action within somatic cells of the male reproductive system.

Johnston et al. (97) used a similar approach to determine if mice lacking androgen receptor are infertile because germ cells or Sertoli cells require a functional receptor. The natural mutant *tfm* or testicular feminization has a mutation in the androgen receptor that completely eliminates function (158,159). Germ cells from tfm mice were transplanted into testes of wild-type recipients. Donor-derived spermatogenesis was observed in the recipient mice. Thus, these researchers concluded that germ cells do not require androgen receptor to complete spermatogenesis (160). However, androgen receptor is essential for the formation of the testis during development and Sertoli cell function during spermatogenesis.

The interaction between the receptor c-kit and kit ligand (kitl), also called stem cell factor, is one of the best-characterized ligand receptor interactions for spermatogenesis (161). In the seminiferous epithelium Sertoli cells produce a membrane bound and secreted form of kitl while spermatogonia express the c-kit receptor on their surface. Ligand binding to c-kit is critical for spermatogonial mitosis. The natural mouse mutant W/W^v has a mutation in the c-kit gene that prevents spermatogonial division making these mice infertile. Similarly, the Steel (Sl/Sl) mutant has a mutation in kitl, also leading to infertile mice. Ogawa et al. (162) showed that germ cells from the infertile Sl/Sl mutant can colonize the testes of W/W^v mice and generate donor-derived spermatogenesis. This was the first demonstration that transplantation of spermatogonial stem cells from an infertile donor to a permissive testicular environment can restore fertility and result in progeny with the donor genotype. It also demonstrated that W/W^v mice, despite being infertile, are suitable recipients for these type of experiments. W/W^v mice have subsequently used as recipients for other projects to determine if a particular mutation affects the ability of germ cells to differentiate in a permissive testicular environment (163).

The complex process of spermatogenesis requires the coordinated differentiation of germ cells that are intimately associated with Sertoli cells. As a germ cell completes one phase of differentiation all of the other germ cell cohorts associated with the same Sertoli cell also complete a phase of differentiation. This is true even for germ cells at all points of differentiation from spermatogonia to spermatids. Thus, germ cells associated with a particular Sertoli cell pass through unique stages of differentiation. The number of stages of spermatogenesis and the timing of these stages is unique to each animal. The spermatogonial stem cell transplantation technique has shown that the seminiferous epithelium is permissive, to a point, in supporting the differentiation of germ cells from closely related species (164,165). Franca et al. (166) transplanted rat germ cells into the seminiferous tubules of mice to determine if germ cells or Sertoli cells regulated the time required for rat germ cell differentiation. Rat germ cells that were supported by mouse Sertoli cells in recipients differentiated with cell cycle timing characteristic of the rat. These researchers conclude that that the cell differentiation process of spermatogenesis is regulated by germ cells alone (166). Therefore differentiating germ cells appear to regulate, at least in part, the differential gene expression and protein secretion by Sertoli cells during specific stages of spermatogenesis.

The spermatogonial stem cell transplantation technique was utilized to transplant Sertoli cells into infertile recipients (167). Recipient animals were treated with cadmium sulfate to eliminate endogenous Sertoli cells (168). Sertoli cells from the testes of mice at embryonic day 19- to 2-days-old had a greater capacity to generate seminiferous tubules capable of supporting germ cell differentiation than Sertoli cells from adult donors (167). These researchers also demonstrated that Sertoli cells injected into the testes of infertile *Sl/Sl* mutants supported germ cell differentiation. Shinohara et al. (167) concluded from these experiments that the Sertoli cell contribution to the spermatogonial stem cell niche is transferable between animals and can restore fertility in infertile mice.

Spermatogonial stem cell and Sertoli cell transplantation results challenge our conventional thinking about spermatogenesis. This novel tool provides a mechanism to investigate many aspects of spermatogenesis that were previously unobtainable. Translation of this technique and insights from these experiments might provide a means to correct fertility in males suffering from germ cell or Sertoli cell defects.

COMPARATIVE SERTOLI CELL BIOLOGY

Vertebrate and Invertebrate

The testes of vertebrates is essentially two separate compartments—the seminiferous tubules containing the germ cells and somatic Sertoli cells and the interstitial compartment that contains interstitial Leydig cells, blood vessels, connective tissue, lymphatic elements, and immune cells. The organization of the testis varies between vertebrate classes that affect how Sertoli cells conduct their main function, supporting germ cell differentiation.

The testis and the interactions between Sertoli cells and germ cells in fish and amphibians are different on several levels when compared with other vertebrates including mammals. Spermatogenesis occurs in a structure called the spermatocyst or cyst (169,170). The process is initiated when a spermatogonium becomes associated with a somatic Sertoli cell and begins to divide by mitosis and the size of the cyst increases. The Sertoli cell engulfs the dividing germ cells and contributes to a basement membrane that surrounds the cyst. The germ cells continue to differentiate, usually in a synchronous fashion, and the cyst continues to increase in size. Evidence supports the contention that the Sertoli cell associated with the cyst continues to divide while the germ cells divide and the cyst increases in size. Tritium-labeled cells, presumably Sertoli cells, on the periphery of the cyst have been observed in multiple species (169,171). Spermiation occurs with the rupture of the cyst and a degeneration of the Sertoli cell that supported the development of the cohort of sperm. Thus, new Sertoli cells must be recruited to support the formation of cyst and eventually sperm. It is believed that each year Sertoli cells arise from stem cells that are present in the interstitial space of the gonad. This is a major contrast with mammalian Sertoli cells that after mitosis and differentiation during prepubertal development do not divide and remain a stable population throughout the adult lifetime of the animal (172,173). More detailed reviews of teleost Sertoli cell cytology have been published (174,175).

The Sertoli cell in amniotes is a columnar cell characterized by cytoplasm that extends from the basement membrane of the seminiferous tubule to the lumen. The developing germ cells are surrounded by the Sertoli cell cytoplasm such that haploid spermatids are actually embedded within the cytoplasmic processes of the Sertoli cell. In contrast, Sertoli cells in anamniotes such as fish and amphibians appear less differentiated and flat while possessing a ovoid nucleus with a large nucleolus. Polarity of Sertoli cells is established with a distinct basement membrane and more distal adlumina surface that interacts directly with maturing sperm (176,177). Adjacent Sertoli cells interdigitate with each other in a complex manner and in certain species tight junctions or desmosomes have been observed and a blood testis barrier has been described for multiple fish species (178). The blood–testis barrier forms after germ cells enter meiosis.

The morphology of the amphibian testis is similar to teleosts such that spermatogenesis occurs in cysts located in a seminiferous compartment (179). Comprehensive investigation on the fine structure of amphibian Sertoli cells is limited. A blood–testis barrier does exist in amphibians (180). In reptiles there is a permanent population of Sertoli cells (179). There are some differences in reptile Sertoli cells when compared to other vertebrates. In seasonal breeding lizards, there are two types of Sertoli cells. The first type has ovoid nuclei and is lost into the lumen after spermiation while the second type have basally located pyramidal shaped nuclei and these Sertoli cells are permanent throughout the life of the animal (181). Sertoli cells that appear to form syncytia have also been described in lizards and snakes (182). A blood–testis barrier has been described in the testes of lizards (183).

Many songbirds and waterfowl are seasonal breeders whereas domestic fowl are generally continuous breeders. The Sertoli cells in both types of breeders maintain a permanent population and form a blood–testis barrier (184). The nucleus in avian Sertoli cells is basally located, irregularly shaped with a prominent nucleolus (185). Birds and reptile Sertoli cells produce cytoplasmic processes that penetrate spermatid cytoplasm (186). The functional role for these cytoplasmic processes is unknown.

The structure and function of Sertoli cells in mammals is generally consistent between species. The timing of Sertoli cell mitosis and proliferation varies between species based on age at puberty and postnatal development (187). In primates, Sertoli cells undergo two phases of replication due to the extended timing of puberty while rodents undergo a single phase of Sertoli cell proliferation (7). Sertoli cell proliferation in stallions appears to occur over 4 to 5 years, which also includes several periods of seasonal regression of the testis (discussed in detail in following section) since stallions are long-day breeders (188).

Boars contain two populations of Sertoli cells (189). Type A Sertoli cells have fine structural characteristics similar to Sertoli cells observed in other mammals including typical junctions with adjacent Sertoli cells and columnar morphology extending from the basement membrane to the lumen of the seminiferous tubule. Type B Sertoli cells are less numerous and are found only in close apposition to the basement membrane of the seminiferous epithelium. They are small cells with dark nuclei and limited cytoplasm that look similar to spermatogonia using light microscopy (190). The functional role of type B Sertoli cells in boars is unknown.

Investigation of Sertoli cells in invertebrates has been complicated by the lack of a consensus if the somatic cells involved in germ cell differentiation in this wide ranging group of organisms share enough characteristics with mammalian Sertoli cells to indeed be considered the equivalent cell (191). The limited amount of information on the somatic cells in many invertebrates increases the challenge of determining the homology of these cells with Sertoli cells in other species. In the nematode testis, somatic cells form a discontinuous layer at the apical end of the tubular testis. As the germ cells differentiate and are found at the caudal end of the testis, the somatic cells form a barrier likened to the blood–testis barrier (192). The role of the somatic cells in regulating and supporting spermatogenesis is not known. In certain marine worms and earthworms, a germ cell derived cytoplasmic mass called the cytophore is associated with differentiating germ cells. The cytophore may support and aid in synchronization of germ cell development, however, due to its germ cell origin it is not homologous to a Sertoli cell.

Invertebrates are often hermaphroditic leading to the development of sperm and oocytes in the same organ. Mollusks have been used to investigate the morphological and physical characteristics of an ovotestis (193). Sertoli cells derive from the germinal ring of the ovotestis and engulf differentiating spermatogonia and remain associated with germ cells undergoing spermatogenesis. In addition, Sertoli cells form a barrier between the testicular and ovarian regions of the gonad. This barrier also is important in maintaining a specialized environment for germ cell development. Sertoli cells contribute to and provide a barrier to maintain gonadal fluid that has a different osmolarity and protein concentration than the hemolymph (194).

The insect testis consists of follicles, which are divided into an apical and basal compartment. The somatic cells associated with germ cell development in the insect testis are not called Sertoli cells; instead, the spermatogonia undergo mitosis in a structure called a cyst that contains germ cells surrounded by perifollicular cells. As the germ cells differentiate the cyst moves to the basal compartment of the testis and the entire cyst is surrounded by the parietal cell layer (195). The parietal cell layer provides a blood–testis barrier and meiosis begins when the cyst has migrated to the basal compartment and is surrounded by this layer. As with mollusks, the testicular fluid surrounding the germ cells is different than the circulating body fluid suggesting the parietal cells contribute to maintaining a microenvironment suitable for germ cell development (191). Thus, somatic cells in mollusks and insects display homology to Sertoli cells in mammals. These cells maintain a microenvironment suitable for germ cell development with a blood–testis barrier, participate in spermiation and phagocytose residual germ cell cytoplasm.

Seasonal Breeders, Testicular Regression, and Recrudescence

Short- or long-day lengths can inhibit reproductive function depending on the species. Male reproductive quiescence is accomplished via testicular regression and reduction of both steroidogenesis and gametogenesis after 8 to 14 weeks of short photoperiods in several species of rodents (196–198). Likewise, stallions and several bird species are considered long-day breeders (199). In contrast, sheep and goat reproductive activity is stimulated by decreasing day length (200). It is generally believed seasonal breeding strategies were developed to time the production of offspring to the time of year when food and other resources are optimal for survival. Most research in this area has focused on changes in the endocrine parameters and gamete population in relation to reproductive inactivity. Some interesting aspects of Sertoli cell biology have been discovered through these efforts and are discussed in the following section.

A biological shift from growth and differentiation of germ cells to cell death occurs during seasonal testicular regression. The regulation of this process begins in the brain through neuroendocrine pathways regulating hormonal signals that influence gonadal activity. It is generally believed that the directional change of the day length is the cue that leads to seasonal changes in reproductive activity. To some extent the seasonal changes in reproductive activity can be manipulated by, for example, maintaining an animal on a constant light regime. However, animals maintained on inhibitory photoperiods will eventually show seasonal changes in reproductive activity. For example, the long-day breeding Djungarian hamster will undergo spontaneous recrudescence even when exposed to continuous inhibitory short photoperiods (201–203). The neuroendocrine regulation of seasonal reproductive behavior in mammals has been extensively reviewed (196).

Investigation of Sertoli cells during testicular regression and recrudescence has mainly focused on morphological and biochemical changes (204). In adult testes, Sertoli cells are considered to exist as a stable population and are notable for their resistance to stimuli that induce death in other testis cell types (172). Changes in Sertoli cell populations associated with breeding condition have been reported in several seasonal breeding species (e.g., stallions, red deer, Corriedale rams); however, in other seasonal species changes in Sertoli cell numbers have not been reported. When testicular morphology is compared between breeding and nonbreeding seasons in photoperiodic species, the most commonly described changes in Sertoli cells are cell volume and shape.

The Djungarian hamster (*Phodopus sungorus*) has been used as a research model to investigate the impact of daily light period on gonadal function (202). In male hamsters, short photoperiods induce testicular involution and testis weight decreases to 3% of adult active weight. The seminiferous tubule lumen disappears and germ cell differentiation stops at the beginning of meiosis with the degeneration of primary spermatocytes. In the hamster, during testicular regression the Sertoli cells were reduced in size by 72.4%. This was primarily due to a decrease in their cytoplasmic volume. There are major quantitative differences in the structure of the Sertoli cells during active and inactive states of spermatogenesis in Djungarian hamsters (204). Overall, it appears that short photoperiod-induced involution resulted in decreases in volume and surface area of structural components including the plasma membrane. In addition, the absolute volumes of most major organelles including mitochondria, smooth endoplasmic reticulum, rough endoplasmic reticulum and golgi complex decreased in Sertoli cells during photoperiod-induced testicular involution (204).

Morphological changes in Sertoli cells during testicular regression are likely due to reduced gonadotropin secretion and the consequent decrease in testosterone secretion. In addition, the total content of FSH receptors in the testis decreases during photoperiod-induce involution. Indeed, there is a net reduction of 58.5% in FSH receptors on the basal surface of Sertoli cells during photoperiod-induced testicular regression (204). However, due to the decrease in the surface membrane of the Sertoli cell during testicular regression, the receptor number per μm^2 is the same as the active testis.

Sertoli cell structural and surface characteristics increase in size and activity during early phases of photoperiod-related recrudescence (205). Structural features that are enhanced in volume at this time include cytoplasm, outer and inner mitochondrial membranes, smooth and rough endoplasmic reticulum and the basal compartment plasma membrane. As blood levels of FSH increase during early recrudescence there appears to be an increase in the synthetic processes in Sertoli cells that support the reestablishment of spermatogenesis (205).

Both serum FSH levels and testis weight increase more rapidly than serum LH levels after photostimulation of male Djungarian hamsters. Significant changes in serum FSH levels were observed within 3 days, and significant testis growth occurred within 5 days after transfer of male hamsters from a short to long photoperiod, while significant increases in serum LH levels in these same animals did not occur until 21 days after exposure to long days (206). FSH appears to be the primary hormone responsible for initial testis growth during recrudescence (207). Inhibition of the biological activity of FSH during the first 7 days of photostimulation by inhibin prevented testicular growth. Male Djungarian hamsters maintained on a short photoperiod and administered ovine FSH had increases in testis weight. In contrast, administration of ovine LH did not induce testicular growth. Ovine FSH induced testis growth was comparable to testis growth observed when hamsters were exposed to stimulatory photoperiod. These results suggest that the rapid increase in serum FSH after photostimulation is the primary signal for initiating testicular development in the seasonally breeding Djungarian hamster (207). Thus, the Sertoli cell is the first cell to respond to stimulatory photoperiods in the Djungarian hamster.

Testicular recrudescence has been suggested as a model to investigate key aspects of testicular development including the regulation of Sertoli cell activity by FSH. Sertoli cell responsiveness to FSH during development has been measured by the ability to accumulated cAMP in vitro. The ability of cultured Sertoli cells to respond to FSH by accumulating cAMP declines during sexual maturation in both the rat and hamster (208,209). During development the Sertoli cell response to FSH, based on accumulation of cAMP, stops prior to the peak of gonadotropins that occurs during puberty. Following this peak, gonadotropin levels stabilized to an "adult" level. Gonadotropin levels during recrudescence follow a similar trend as during development with low levels during complete regression followed by a gradual increase to a peak then decreasing to a stable "adult" level. In contrast to Sertoli cell response during development, the return of the Sertoli cell FSH response during regression does not occur until after the gonadotropin peak. The gonadotropin peak during recrudescence is higher than during puberty suggesting that this process is similar to sexual development but a different mechanism is regulating this process. Although Sertoli cells from the testes of seasonally regressed animals respond to FSH in culture with the accumulation of cAMP, no studies have indicated Sertoli cells resume mitosis during this process in rodents. It appears FSH initially and later testosterone regulates the dramatic changes in Sertoli cell morphology during testicular regression and recrudescence in the Djungarian hamster that support the resumption of germ cell differentiation and spermatogenesis.

In stallions, seasonal changes in the Sertoli cell population are also affected by the age of the animal (188,210). The number of Sertoli cells increases during each breeding season until 4 to 5 years of age after which the number of Sertoli cells stabilizes. The size of Sertoli cells decreases during photoperiod-induced regression and, similar to hamsters, increases when photoperiod stimulated reproductive activity resumes. These changes are also associated with increases in testicular weights, hormone levels, and number of elongated spermatids per Sertoli cell as stallions mature during the first 4 to 5 years of life (188). In contrast to rodents, Sertoli cells in stallions during the first 4 to 5 years of life can continue to proliferate during recrudescence. However, after an "adult" population of Sertoli cells has populated the testis, proliferation appears to cease.

Adult rams are short day breeders and the influence of Sertoli cells and FSH on testicular regression and recrudescence has been investigated (211). As adult testes become fully active after seasonal regression, Sertoli cells do not increase in number, but their capacity for protein secretion (for example, androgen binding protein and inhibin) and for supporting increased numbers of germ cells does increase (212,213) and that FSH induces gene expression for its own functional, membrane-bound receptor late in recrudescence (211). Sanford (211) reported the number of membrane-bound FSH receptors in the testes (number per Sertoli cell) of adult rams increased by an estimated 70% during recrudescence. The FSH signals provided to Sertoli cells, first by increases in FSH concentration and then by increased FSH receptors, are critical regulators of inhibin expression and production by Sertoli cells during recrudescence. In rams, FSH increases a few weeks before scrotal circumference and inhibin levels increase following seasonal regression. Thus, in rams as in other seasonal breeders, Sertoli cells are important in regulating the initial response and resumption of spermatogenesis following seasonal regression.

Seasonal changes in the testes of several other species have been investigated. The male white-footed mouse (*Peromyscus leucopus*) undergoes testicular regression when exposed to less than 12.5 hours of light per day (214). This animal model was used to investigate the role apoptosis, or programmed cell death, plays in photoperiod induced testicular regression. Spermatocytes and spermatids in testes undergoing regression were positive for Fas receptor staining. Fas receptor mediates apoptotic cell death when bound by the Fas ligand (FasL) in a rapid fashion (215). Sertoli cells appear to regulate this process in a regressing testis by expressing FasL (216). The FasL will bind to germ cells expressing Fas receptor leading to apoptosis of these cells and testicular regression. In mammalian testes undergoing induced regression, Sertoli cells secrete the soluble ligand for the Fas receptor, FasL. Germ cells expressing the death receptor, Fas, receive the local signal for apoptosis initiation from Sertoli cells. Exposure of Sertoli cells to toxins has been shown to up-regulate the expression of both Fas ligand on Sertoli cells as well as Fas in germ cells (216–218). Apoptotic Sertoli death could therefore enhance subsequent induction of apoptosis for local clusters of germ cells. This model of comprehensive programmed cell death would result in the rapid atrophy of the seminiferous epithelium observed in starlings compare with species in which germ cells are exclusively targeted.

The only report in higher vertebrates of Sertoli cell death during seasonal testis regression was for European starlings (*Sturnus vulgaris*) (219). During seasonal regression, starling Sertoli cells were positive for TUNEL labeling. Degenerating cells were identified by irregularly shaped nuclei, marginated chromatin, attenuating cell membranes and gene structural distortion. In addition, dying cells were characterized by darkly staining condensed cytoplasm. These characteristics of cell death were observed in cells identified by light and electron microscopy as Sertoli cells based on their position within the seminiferous tubule and characteristic elongated cytoplasmic branches (219).

Sertoli cells obviously play a role in seasonal testis regression and recrudescence. The details of this process are beginning to be described in a variety of species. Regression appears to be due to cell death because Sertoli cells no longer receive the essential signals from the brain to maintain function. The withdrawal of gonadotropins may induce the Fas system to initiate germ cell death and testicular regression through Sertoli cell-specific signaling pathways. Subsequently, low levels of testosterone resulting from decease LH secretion and the inability of Sertoli cells to support differentiating germ cells. Sertoli cells appear to initially respond to increasing gonadotropins during recrudescence. This process, also poorly understood, leads to an endocrine and cellular cascade reactivating the hypothalamic-pituitary-testicular axis resulting in spermatogenesis. Continued investigation of the testicular aspects of seasonal breeding may lead to a better understanding of the factors involved in development and premature spermatogenic failure in multiple species.

CONCLUSION

The original proclamation of Enrico Sertoli was that the cells he discovered were "linked to the production of spermatozoa." After nearly 150 years of interest and decades of intense investigations it is fitting that we examine how much more we know and how much more we have to learn about the details of and mechanisms involved in this process. The following is a conservative and succinct summary of what we currently understand about Sertoli cells as gleaned from the literature reviewed above: (a) Somatic cells in association with germ cells are recognized across vertebrates and invertebrates. In some species the presumptive Sertoli cells are actively involved in the "production of spermatozoa" while in others their role has not been investigated. (b) In mammals, Sertoli cells have dual roles in the embryology and development of the testis. (c) The appearance of Sertoli cells is the focal point for the formation of the testis and ultimately for the continuance of the male germ line. (d) The maturation of and number of Sertoli cells ultimately determines the sperm output of an adult mammalian testis. (e) The development of germ cells from stem cells to spermatozoa appears to be possible at some level in the absence of Sertoli cells. At least in mammals, it appears that Sertoli cells enhance the efficiency of this process dramatically. (f) In mammals, the actions of hormones such as FSH and testosterone on spermatogenesis occur via actions on Sertoli cells. (g) Also, in mammals, one of the ways that Sertoli cells enhance the efficiency of spermatogenesis is by the creation of a unique environment for meiotic and postmeiotic germ cells.

We still have much to learn about the molecular details and mechanisms involved in each of the principles described above. While we have made great strides in understanding the role of the Sertoli cell in testis formation we know only a handful of the genes that may be involved. Similarly, we know of only a few Sertoli cell gene products that have been shown to directly affect the efficiency of spermatogenesis.

These details will have implications with regard to understanding male infertility that is of somatic cell origin and in aiding research efforts on in vitro spermatogenesis.

REFERENCES

1. Steinberger, A., Heindel, J. J., Lindsey, J. N., Elkington, J. S., Sanforn, B. M., and Steinberger, E. (1975). Isolation and culture of FSH responsive Sertoli cells. *Endocr. Res. Commun.* 2(3), 261–272.
2. Dorrington, J. H., Roller, N. F., and Fritz, I. B. (1975). Effects of follicle-stimulating hormone on cultures of Sertoli cell preparations. *Mol. Cell. Endocrinol.* 3(1), 57–70.
3. Sylvester, S. R., and Griswold, M.D. (1994). The testicular iron shuttle: a "nurse" function of the Sertoli cells. *J. Androl.* 15(5), 381–385.
4. Kierszenbaum, A.L. (1994). Mammalian spermatogenesis *in vivo* and *in vitro*: a partnership of spermatogenic and somatic cell lineages. *Endocr. Rev.* 15(1), 116–134.
5. Grootegoed, J. A., Baarends, W. M., Hendriksen, P. J., Hoogerbrugge, J. W., Slegtenhorst-Eegdeman, K. E., and Themmen, A. P. (1995). Molecular and cellular events in spermatogenesis. *Hum. Reprod.* 10, Suppl 1, 10–14.
6. de Kretser, D. M., Loveland, K. L., Meinhardt, A., Simorangkir, D., and Wreford, M. (1998). Spermatogenesis. *Hum. Reprod.* 13, Suppl 1, 1–8.
7. Sharpe, R. M., McKinnell, C., Kivlin, C., and Fisher, J. S. (2003). Proliferation and functional maturation of Sertoli cells, and their relevance to disorders of testis function in adulthood. *Reproduction* 125(6), 769–784.
8. Lovell-Badge, R. (1992). The role of Sry in mammalian sex determination. *Ciba Found Symp.* 165, 162–179, discussion 179–182.
9. Josso, N., Racine, C., di Clemente, N., Rey, R., and Xavier, F. (1998). The role of anti-Müllerian hormone in gonadal development. *Mol. Cell. Endocrinol.* 145(1–2), 3–7.
10. de Franca, L. R., Ghosh, S., Ye, S. J., and Russell, L. D. (1993). Surface and surface-to-volume relationships of the Sertoli cell during the cycle of the seminiferous epithelium in the rat. *Biol. Reprod.* 49(6), 1215–1228.
11. Skinner, M. K., and Griswold, M. D. (in press). "Sertoli Cell Biology." Elsevier/Academic Press, San Diego.
12. Swain, A., and Lovell-Badge, R. (1999). Mammalian sex determination: a molecular drama. *Genes Dev.* 13(7), 755–767.
13. Tilmann, C., and Capel, B. (2002). Cellular and molecular pathways regulating mammalian sex determination. *Recent Prog. Horm. Res.* 57, 1–18.
14. Ginsburg, M., Snow, M. H., and McLaren, A. (1990). Primordial germ cells in the mouse embryo during gastrulation. *Development* 110(2), 521–528.
15. Byskov, A. G. (1986). Differentiation of mammalian embryonic gonad. *Physiol. Rev.* 66(1), 71–117.
16. Karl, A. F., and Griswold, M. D. (1990). Sertoli cells of the testis: preparation of cell cultures and effects of retinoids. *Methods Enzymol.* 190, 71–75.
17. Pelliniemi, L., Frojdman, K., and Paranko, J. (1993). Embryological and prenatal development and function of Sertoli cells. In *The Sertoli Cell* (L. D. Russell and M. D. Griswold, Eds.), pp. 87–113. Cache River Press, Clearwater, FL.
18. Karl, J., and Capel, B. (1998). Sertoli cells of the mouse testis originate from the coelomic epithelium. *Dev. Biol.* 203(2), 323–333.
19. Orth, J. M., Jester, W. F., Li, L. H., and Laslett, A. L. (2000). Gonocyte-Sertoli cell interactions during development of the neonatal rodent testis. *Curr. Top. Dev. Biol.* 50, 103–124.
20. Albrecht, K. H., and Eicher, E. M. (2001). Evidence that Sry is expressed in pre-Sertoli cells and Sertoli and granulosa cells have a common precursor. *Dev. Biol.* 240(1), 92–107.
21. McLaren, A. (1995). Germ cells and germ cell sex. *Philos. Trans. R. Soc. Lond. B Biol. Sci.* 350(1333), 229–233.
22. McLaren, A., and Southee, D. (1997). Entry of mouse embryonic germ cells into meiosis. *Dev. Biol.* 187(1), 107–113.
23. McLaren, A. (2000). Germ and somatic cell lineages in the developing gonad. *Mol. Cell. Endocrinol.* 163(1–2), 3–9.
24. Palmer, S. J., and Burgoyne, P. S. (1991). In situ analysis of fetal, prepuberal and adult XX-XY chimaeric mouse testes: Sertoli cells are predominantly, but not exclusively, XY. *Development* 112(1), 265–268.
25. Harley, V. R., Clarkson, M. J., and Argentaro, A. (2003). The molecular action and regulation of the testis-determining factors, SRY (sex-determining region on the Y chromosome) and SOX9 (SRY-related high-mobility group (HMG) box 9). *Endocr. Rev.* 24(4), 466–487.
26. Orth, J. M. (1982). Proliferation of Sertoli cells in fetal and postnatal rats: a quantitative autoradiographic study. *Anat. Rec.* 203(4), 485–492.
27. Amann, R. P. (1970). Sperm producion rates. In *The Testis* (A. D. Johnson, W. R. Gomes, and N. L. Van Denmark, Eds.), pp. 433–482. Academic Press, New York.
28. Orth, J. M., Gunsalus, G. L., and Lamperti, A. A. (1988). Evidence from Sertoli cell-depleted rats indicates that spermatid number in adults depends on numbers of Sertoli cells produced during perinatal development. *Endocrinology* 122(3), 787–794.
29. Gromoll, J., Simoni, M., Nordhoff, V., Behre, H. M., De Geyer, C., and Nieschlag, E. (1996). Functional and clinical consequences of mutations in the FSH receptor. *Mol. Cell. Endocrinol.* 125(1–2), 177–182.
30. Dierich, A., Sairam, M. R., Monaco, L., Fimia, G. M., Gansmuller, A., LeMeur, M., and Sassone-Corst, P. (1998). Impairing follicle-stimulating hormone (FSH) signaling *in vivo*: targeted disruption of the FSH receptor leads to aberrant gametogenesis and hormonal imbalance. *Proc. Natl. Acad. Sci. U S A* 95(23), 13612–13617.
31. Kumar, T. R., Wang, Y., Lu, N., and Matzuk, M. M. (1997). Follicle stimulating hormone is required for ovarian follicle maturation but not male fertility. *Nat. Genet.* 15(2), 201–204.
32. Allan, C. M., Garcia, A., Spaliviero, J., Zhang, F. P., Jimenez, M., Huhtaniemi, I., and Handelsman, D. J. (2004). Complete Sertoli cell proliferation induced by follicle-stimulating hormone (FSH) independently of luteinizing hormone activity: evidence from genetic models of isolated FSH action. *Endocrinology* 145(4), 1587–1593.
33. Singh, J., and Handelsman, D. J. (1996). Neonatal administration of FSH increases Sertoli cell numbers and spermatogenesis in gonadotropin-deficient (hpg) mice. *J. Endocrinol.* 151(1), 37–48.
34. Maran, R. R., Sivakumar, R., Arunakaran, J., Ravisankar, B., Ravichandran, K., Sidharthan, V., Jeyaraj, D. A., and Aruldhas, M. M. (1999). Duration-dependent effect of transient neonatal hypothyroidism on sertoli and germ cell number, and plasma and testicular interstitial fluid androgen binding protein concentration. *Endocr. Res.* 25(3–4), 323–340.
35. Majdic, G., Snoj, T., Horvat, A., Mrkun, J., Kosec, M., and Cestnik, V. (1998). Higher thyroid hormone levels in neonatal life result in reduced testis volume in postpubertal bulls. *Int. J. Androl.* 21(6), 352–357.
36. De Franca, L. R., Hess, R. A., Cooke, P. S., and Russell, L. D. (1995). Neonatal hypothyroidism causes delayed Sertoli cell

maturation in rats treated with propylthiouracil: evidence that the Sertoli cell controls testis growth. *Anat. Rec.* 242(1), 57–69.
37. Simorangkir, D. R., de Kretser, D. M., and Wreford, N. G. (1995). Increased numbers of Sertoli and germ cells in adult rat testes induced by synergistic action of transient neonatal hypothyroidism and neonatal hemicastration. *J. Reprod. Fertil.* 104(2), 207–213.
38. Waites, G. M., Speight, A. C., and Jenkins, N. (1985). The functional maturation of the Sertoli cell and Leydig cell in the mammalian testis. *J. Reprod. Fertil.* 75(1), 317–326.
39. Solari, A. J., and Fritz, I. B. (1978). The ultrastructure of immature Sertoli cells. Maturation-like changes during culture and the maintenance of mitotic potentiality. *Biol. Reprod.* 18(3), 329–345.
40. Griswold, M. D. (1998). The central role of Sertoli cells in spermatogenesis. *Semin. Cell. Dev. Biol.* 9(4), 411–416.
41. Griswold, M. D. (1995). Interactions between germ cells and Sertoli cells in the testis. *Biol. Reprod.* 52(2), 211–216.
42. Grootegoed, J. A., Siep, M., and Baarends, W. M. (2000). Molecular and cellular mechanisms in spermatogenesis. *Baillieres Best Pract. Res. Clin. Endocrinol. Metab.* 14(3), 331–343.
43. Griswold, M. D. (1993). Protein secretion by Sertoli cells: general considerations. In *The Sertoli Cell* (L. D. Russell and M. D. Griswold, Eds.), pp. 195–200. Cache River Press, Clearwater, FL.
44. Lovell-Badge, R., and Hacker, A. (1995). The molecular genetics of Sry and its role in mammalian sex determination. *Philos. Trans. R. Soc. Lond. B Biol. Sci.* 350(1333), 205–214.
45. Capel, B. (1998). Sex in the 90s: SRY and the switch to the male pathway. *Annu. Rev. Physiol.* 60, 497–523.
46. Kent, J., Wheatley, S. C., Andrews, J. E., Sinclair, A. H., and Koopman, P. (1996). A male-specific role for SOX9 in vertebrate sex determination. *Development* 122(9), 2813–2822.
47. Chaboissier, M. C., Kobayashi, A., Vidal, V. I., Lutzkendorf, S., van de Kant, H. J., Wenger, M., de Rooij, D. G., Behringer, R. R., and Schedl, A. (2004). Functional analysis of Sox8 and Sox9 during sex determination in the mouse. *Development* 131(9), 1891–1901.
48. Josso, N., di Clemente, N., and Gouedard, L. (2001). Anti-Müllerian hormone and its receptors. *Mol. Cell. Endocrinol.* 179(1–2), 25–32.
49. Clarkson, M. J., and Harley, V. R. (2002). Sex with two SOX on: SRY and SOX9 in testis development. *Trends Endocrinol. Metab.* 13(3), 106–111.
50. Clark, A. M., Garland, K. K., and Russell, L. D. (2000). Desert hedgehog (Dhh) gene is required in the mouse testis for formation of adult-type Leydig cells and normal development of peritubular cells and seminiferous tubules. *Biol. Reprod.* 63(6), 1825–1838.
51. Colvin, J. S., Green, R. P., Schmahl, J., Capel, B., and Ornitz, D. M. (2001). Male-to-female sex reversal in mice lacking fibroblast growth factor 9. *Cell* 104(6), 875–889.
52. Griswold, M. D. (1988). Protein secretions of Sertoli cells. *Int. Rev. Cytol.* 110, 133–156.
53. Fritz, I. B. (1993). Secretion of proteases and inhibitors. In *The Sertoli Cell* (L. D. Russell and M. D. Griswold, Eds.). Cache River Press, Clearwater, FL.
54. Heckert, L. L., and Griswold, M. D. (2002). The expression of the follicle-stimulating hormone receptor in spermatogenesis. *Recent Prog. Horm. Res.* 57, 129–148.
55. Lampa, J., Hoogerbrugge, J. W., Baarends, W. M., Stanton, P. G., Perryman, K. J., Grootegoes, J. A., and Robertson, D. M. (1999). Follicle-stimulating hormone and testosterone stimulation of immature and mature Sertoli cells *in vitro*: inhibin and N-cadherin levels and round spermatid binding. *J. Androl.* 20(3), 399–406.
56. O'Shaughnessy, P. J., Dudley, K., and Rajapaksha, W. R. (1996). Expression of follicle stimulating hormone-receptor mRNA during gonadal development. *Mol. Cell. Endocrinol.* 125(1–2), 169–175.
57. Risley, M. S. (2000). Connexin gene expression in seminiferous tubules of the Sprague-Dawley rat. *Biol. Reprod.* 62(3), 748–754.
58. Gow, A., Southwood, C. M., Li, J. S., Pariali, M., Riordan, G. P., Brodie, S. E., Danias, J., Bronstein, J. M., Kachar, B., and Lazzarini, R. A. (1999). CNS myelin and sertoli cell tight junction strands are absent in Osp/claudin-11 null mice. *Cell* 99(6), 649–659.
59. Batias, C., Siffroi, J. P., Fenichel, P., Pointis, G., and Segretain, D. (2000). Connexin43 gene expression and regulation in the rodent seminiferous epithelium. *J. Histochem. Cytochem.* 48(6), 793–805.
60. Batias, C., Defamie, N., Lablack, A., Thepot, D., Fenichel, P., Segretain, D., and Pointis, G. (1999). Modified expression of testicular gap-junction connexin 43 during normal spermatogenic cycle and in altered spermatogenesis. *Cell. Tissue Res.* 298(1), 113–121.
61. Depuydt, C. E., Mahmoud, A. M., Dhooge, W. S., Schoonjans, F. A., Comhaire, F. H. (1999). Hormonal regulation of inhibin B secretion by immature rat sertoli cells *in vitro*: possible use as a bioassay for estrogen detection. *J. Androl.* 20(1), 54–62.
62. Le Magueresse-Battistoni, B., Morera, A. M., and Benahmed, M. (1995). *In vitro* regulation of rat Sertoli cell inhibin messenger RNA levels by transforming growth factor-beta 1 and tumour necrosis factor alpha. *J. Endocrinol.* 146(3), 501–508.
63. Moore, A., Krummen, L. A., and Mather, J. P. (1994). Inhibins, activins, their binding proteins and receptors: interactions underlying paracrine activity in the testis. *Mol. Cell. Endocrinol.* 100(1–2), 81–86.
64. Ketola, I., Anttonen, M., Vaskivuo, T., Tapanainen, J. S., Toppari, J., and Heikinheimo, M. (2002). Developmental expression and spermatogenic stage specificity of transcription factors GATA-1 and GATA-4 and their cofactors FOG-1 and FOG-2 in the mouse testis. *Eur. J. Endocrinol.* 147(3), 397–406.
65. Lei, N., and Heckert, L. L. (2004). Gata4 regulates testis expression of Dmrt1. *Mol. Cell. Biol.* 24(1), 377–388.
66. Viger, R. S., Mertineit, C., Trasler, J. M., and Nemer, M. (1998). Transcription factor GATA-4 is expressed in a sexually dimorphic pattern during mouse gonadal development and is a potent activator of the Müllerian inhibiting substance promoter. *Development* 125(14), 2665–2675.
67. Yomogida, K., Ohtani, H., Harigae, H., Ito, E., Nichimune, Y., Engel, J. D., and Yamamoto, M. (1994). Developmental stage- and spermatogenic cycle-specific expression of transcription factor GATA-1 in mouse Sertoli cells. *Development* 120(7), 1759–1766.
68. Chaudhary, J., Sadler-Riggleman, I., and Skinner, M. K. (2004). Identification of a novel Sertoli cell gene product SERT that influences follicle stimulating hormone actions. *Gene* 324, 79–88.
69. Chaudhary, J., and Skinner, M. K. (1999). Basic helix-loop-helix proteins can act at the E-box within the serum response element of the c-fos promoter to influence hormone-induced promoter activation in Sertoli cells. *Mol. Endocrinol.* 13(5), 774–786.
70. Bitgood, M. J., Shen, and McMahon, A. P. (1996). Sertoli cell signaling by Desert hedgehog regulates the male germline. *Curr. Biol.* 6(3), 298–304.
71. Lu, Q., Gore, M., Zhang, Q., Camenisch, T., Boast, S., Casagranda, F., Lai, C., Skinner, M. K., Klein, R., Matsushima, G. K., Earp, H. S., Goff, S. P., and Lemke, G. (1999). Tyro-3 family receptors are essential regulators of mammalian spermatogenesis. *Nature* 398(6729), 723–728.

72. Peschon, J. J., Behringer, R. R. Cate, R. L., Harwood, K. A., Idzerda, R. L., Brinster, R. L., and Palmiter, R. D. (1992). Directed expression of an oncogene to Sertoli cells in transgenic mice. *Mol. Endo.* 6(9), 1403–1411.
73. McGuinness, M. P., Linder, C. C., Morales, C. R., Heckert, L. L., Pikus, J., and Griswold, M. D. (1994). Relationship of a mouse Sertoli cell line (MSC-1) to normal Sertoli cells. *Biol. Reprod.* 51(1), 116–124.
74. Fritz, I. B., Kopec, B., Lam, K., and Vernon, R. G. (1974). Effects of FSH on levels of androgen binding protein on the testis. *Curr. Top. Mol. Endocrinol.* 1, 311–327.
75. Tindall, D. J., and Means, A. R. (1976). Concerning the hormonal regulation of androgen binding protein in rat testis. *Endocrinology* 99(3), 809–818.
76. Joseph, D. R., O'Brien, D. A., Sullivan, P. M., and Becchis, M., Tsuruta, J. K., and Petrusz, P. (1997). Overexpression of androgen-binding protein/sex hormone-binding globulin in male transgenic mice: tissue distribution and phenotypic disorders. *Biol. Reprod.* 56(1), 21–32.
77. Charron, M., Folmer, J. S., and Wright, W. W. (2003). A 3-kilobase region derived from the rat cathepsin L gene directs *in vivo* expression of a reporter gene in sertoli cells in a manner comparable to that of the endogenous gene. *Biol. Reprod.* 68(5), 1641–1648.
78. Charron, M., DeCerbo, J. N., and Wright, W. W. (2003). A GC-box within the proximal promoter region of the rat cathepsin L gene activates transcription in Sertoli cells of sexually mature rats. *Biol. Reprod.* 68(5), 1649–1656.
79. Rao, M. K., Wayne, C. M., Meistrich, M. L., and Winkinson, M. F. (2003). Pem homeobox gene promoter sequences that direct transcription in a Sertoli cell-specific, stage-specific, and androgen-dependent manner in the testis *in vivo*. *Mol. Endocrinol.* 17(2), 223–233.
80. Sutton, K. A., Maiti, S., Tribley, W. A., Lindsey, J. S., Meistrich, M. L., Bucana, C. D., Sanborn, B. M., Joseph, D. R., Griswold, M. D., Cornwall, G. A., and Wilkinson, M. F. (1998). Androgen regulation of the Pem homeodomain gene in mice and rat Sertoli and epididymal cells. *J. Androl.* 19(1), 21–30.
81. Penttila, T. L., Hakovirta, H., Mali, P., Wright, W. W., and Parvinen, M. (1995). Follicle-stimulating hormone regulates the expression of cyclic protein-2/cathepsin L messenger ribonucleic acid in rat Sertoli cells in a stage-specific manner. *Mol. Cell. Endocrinol.* 113(2), 175–181.
82. McLean, D. J., Friel, P. J., Pouchnik, D., and Griswold, M. D. (2002). Oligonucleotide microarray analysis of gene expression in follicle-stimulating hormone-treated rat Sertoli cells. *Mol. Endocrinol.* 16(12), 2780–2792.
83. McLean, D., and Griswold, M. (2004). Sertoli cell function and gne expression. In *Sertoli Cell Biology* (M. D. Griswold, Ed.). Elsevier, St. Louis.
84. Shima, J. E., McLean, D. J., McCarrey, J. R., and Griswold, M. D. (2004). The murine testicular transcriptome: characterizing gene expression in the testis during the progression of spermatogenesis. *Biol. Reprod.* 71(1), 319–330.
85. Zirkin, B. R., Awoniyi, C., Griswold, M. D., Russell, L. D., and Sharpe, R. (1994). Is FSH required for adult spermatogenesis? *J. Androl.* 15(4), 273–276.
86. Fritz, I. (1978). Sites of actions of androgens and follicle stimulating hormone on cells of the seminiferous tubule. In *Biochemical Actions of Hormones* (G. Litwack, Ed.), pp. 249–278. Academic Press, New York.
87. Singh, J., O'Neill, C., and Handelsman, D. J. (1995). Induction of spermatogenesis by androgens in gonadotropin-deficient (hpg) mice. *Endocrinology* 136(12), 5311–5321.
88. Singh, J., and Handelsman, D. J. (1996). The effects of recombinant FSH on testosterone-induced spermatogenesis in gonadotrophin-deficient (hpg) mice. *J. Androl.* 17(4), 382–393.
89. Meachem, S. J., McLachlan, R. I., de Kretser, D. M., Robertson, D. M., and Wreford, N. G. (1996). Neonatal exposure of rats to recombinant follicle stimulating hormone increases adult Sertoli and spermatogenic cell numbers. *Biol. Reprod.* 54(1), 36–44.
90. Abel, M. H., Wootton, A. N., Wilkins, V., Huhtaniemi, I., Knight, P. G., and Charlton, H. M. (2000). The effect of a null mutation in the follicle-stimulating hormone receptor gene on mouse reproduction. *Endocrinology* 141(5), 1795–1803.
91. Tapanainen, J. S., Aittomaki, K., Min, J., Vaskivuo, T., and Huhtaniemi, I. T. (1997). Men homozygous for an inactivating mutation of the follicle-stimulating hormone (FSH) receptor gene present variable suppression of spermatogenesis and fertility. *Nat. Genet.* 15(2), 205–206.
92. Krishnamurthy, H., Babu, P. S., Morales, C. R., Sairam, M. R. (2001). Delay in sexual maturity of the follicle-stimulating hormone receptor knockout male mouse. *Biol. Reprod.* 65(2), 522–531.
93. Krishnamurthy, H., Danilovich, N., Morales, C. R., and Sairam, M. R. (2000). Qualitative and quantitative decline in spermatogenesis of the follicle-stimulating hormone receptor knockout (FORKO) mouse. *Biol. Reprod.* 62(5), 1146–1159.
94. Grover, A., Sairam, M. R., Smith, C. E., and Hermo, L. (2004). Structural and functional modifications of sertoli cells in the testis of adult follicle-stimulating hormone receptor knockout mice. *Biol. Reprod.* 71(1), 117–129.
95. Orth, J. M. (1984). The role of follicle-stimulating hormone in controlling Sertoli cell proliferation in testes of fetal rats. *Endocrinology* 115(4), 1248–1255.
96. Sadate-Ngatchou, P. I., Pouchnik, D. J., and Griswold, M. D. (2004). Follicle-stimulating hormone induced changes in gene expression of murine testis. *Mol. Endocrinol.* 18(11), 2805–2816.
97. Johnston, D. S., Russell, L. D., Friel, P. J., and Griswold, M. D. (2001). Murine germ cells do not require functional androgen receptors to complete spermatogenesis following spermatogonial stem cell transplantation. *Endocrinology* 142(6), 2405–2408.
98. De Gendt, K., Swinnen, J. V., Saunders, P. T., Schoonjans, L., Dewerchin, M., Devos, A., Tan, K., Atanassova, N., Claessens, F., Lecureuil, C., Heyns, W., Carmeliet, P., Guillou, F., Sharpe, R. M., and Verhoeven, G. (2004). A Sertoli cell-selective knockout of the androgen receptor causes spermatogenic arrest in meiosis. *Proc. Natl. Acad. Sci. U S A* 101(5), 1327–1332.
99. Chang, C., Chen, Y. T., Yeh, S. D., Xu, Q., Wang, R. S., Guillou, F., Lardy, H., and Yeh, S. (2004). Infertility with defective spermatogenesis and hypotestosteronemia in male mice lacking the androgen receptor in Sertoli cells. *Proc. Natl. Acad. Sci. U S A* 101(18), 6876–6881.
100. Holdcraft, R. W., and Braun, R. E. (2004). Androgen receptor function is required in Sertoli cells for the terminal differentiation of haploid spermatids. *Development* 131(2), 459–467.
101. Lindsey, J. S., and Wilkinson, M. F. (1996). Pem: a testosterone- and LH-regulated homeobox gene expressed in mouse Sertoli cells and epididymis. *Dev. Biol.* 179(2), 471–484.
102. Sadate-Ngatchou, P. I., Pouchnik, D. J., and Griswold, M. D. (2004). Identification of testosterone-regulated genes in testes of hypogonadal mice using oligonucleotide microarray. *Mol. Endocrinol.* 18(2), 422–433.
103. Brinster, R. L. (2002). Germline stem cell transplantation and transgenesis. *Science* 296(5576), 2174–2176.
104. Steinberger, A., and Jakubowiak, A. (1993). Sertoli cell culture: historical perspective and review of methods. In *The Sertoli Cell* (M.D. Griswold, Ed.), pp. 155–179. Cache River Press, Clearwater, FL.

105. Parks, J. E., Lee, D. R., Huang, S., and Kaproth, M. T. (2003). Prospects for spermatogenesis *in vitro*. *Theriogenology* 59(1), 73–86.
106. Miura, T., Ando, N., Miura, C., and Yamauchi, K. (2002). Comparative studies between *in vivo* and *in vitro* spermatogenesis of Japanese eel (Anguilla japonica). *Zoolog Sci.* 19(3), 321–329.
107. Lee, D. R., Kaproth, M. T., and Parks, J. E. (2001). *In vitro* production of haploid germ cells from fresh or frozen-thawed testicular cells of neonatal bulls. *Biol. Reprod.* 65(3), 873–878.
108. Orth, J. M., McGuinness, M. P., Qiu, J., Jester, W. F. Jr., and Li, L. H. (1998). Use of *in vitro* systems to study male germ cell development in neonatal rats. *Theriogenology* 49(2), 431–439.
109. Oatley, J. M., de Avila, D. M., Reeves, J. J., and McLean, D. J. (2004). Testis tissue explant culture supports survival and proliferation of bovine spermatogonial stem cells. *Biol. Reprod.* 70(3), 625–631.
110. McGuinness, M. P., and Orth, J. M. (1992). Reinitiation of gonocyte mitosis and movement of gonocytes to the basement membrane in testes of newborn rats *in vivo* and *in vitro*. *Anat. Rec.* 233(4), 527–537.
111. Perrard, M. H., Hue, D., Staub, C., Le Vern, Y., Kerboeuf, D., and Durand P. (2003). Development of the meiotic step in testes of pubertal rats: comparison between the *in vivo* situation and under *in vitro* conditions. *Mol. Reprod. Dev.* 65(1), 86–95.
112. Hue, D., Staub, C., Perrard-Sapori, M. H., Weiss, M., Nicolle, J. C., Vigier, M., and Durand, P. (1998). Meiotic differentiation of germinal cells in three-week cultures of whole cell population from rat seminiferous tubules. *Biol. Reprod.* 59(2), 379–387.
113. Marh, J., Tres, L. L., Yamazaki, Y., Yanagimachi, R., and Kierszenbaum, A. L. (2003). Mouse round spermatids developed *in vitro* from preexisting spermatocytes can produce normal offspring by nuclear injection into *in vivo*-developed mature oocytes. *Biol. Reprod.* 69(1), 169–176.
114. Sousa, M., Cremades, N., Alves, C., Silva, J., and Barros, A. (2002). Developmental potential of human spermatogenic cells co-cultured with Sertoli cells. *Hum. Reprod.* 17(1), 161–172.
115. Van Dissel-Emiliani, F. M., De Boer-Brouwer, M., and De Rooij, D. G. (1996). Effect of fibroblast growth factor-2 on Sertoli cells and gonocytes in coculture during the perinatal period. *Endocrinology* 137(2), 647–654.
116. Clifton, R. J., O'Donnell, L., and Robertson, D. M. (2002). Pachytene spermatocytes in co-culture inhibit rat Sertoli cell synthesis of inhibin beta B-subunit and inhibin B but not the inhibin alpha-subunit. *J. Endocrinol.* 172(3), 565–574.
117. Le Magueresse, B., and Jegou, B. (1988). *In vitro* effects of germ cells on the secretory activity of Sertoli cells recovered from rats of different ages. *Endocrinology* 122(4), 1672–1680.
118. Le Magueresse, B., and Jegou, B. (1988). Paracrine control of immature Sertoli cells by adult germ cells, in the rat (an *in vitro* study). Cell-cell interactions within the testis. *Mol. Cell. Endocrinol.* 58(1), 65–72.
119. Le Magueresse, B., Pineau, C., Guillou, F., and Jegou, B. (1988). Influence of germ cells upon transferrin secretion by rat Sertoli cells *in vitro*. *J. Endocrinol.* 118(3), R13–16.
120. Boulogne, B., Habert, R., and Levacher, C. (2003). Regulation of the proliferation of cocultured gonocytes and Sertoli cells by retinoids, triiodothyronine, and intracellular signaling factors: differences between fetal and neonatal cells. *Mol. Reprod. Dev.* 65(2), 194–203.
121. Chung, S. S., Zhu, L. J., Mo, M. Y., Silvestrini, B., Lee, W. M., and Cheng, C. Y. (1998). Evidence for cross-talk between Sertoli and germ cells using selected cathepsins as markers. *J. Androl.* 19(6), 686–703.
122. Weiss, M., Vigier, M., Hue, D., Perrard-Sapori, M. H., Marret, C., Avallet, O., and Durand, P. (1997). Pre- and postmeiotic expression of male germ cell-specific genes throughout 2-week cocultures of rat germinal and Sertoli cells. *Biol. Reprod.* 57(1), 68–76.
123. Syed, V., and Hecht, N. B. (1997). Up-regulation and down-regulation of genes expressed in cocultures of rat Sertoli cells and germ cells. *Mol. Reprod. Dev.* 47(4), 380–389.
124. Syed, V., Gomez, E., and Hecht, N. B. (1999). mRNAs encoding a von Ebner's-like protein and the Huntington disease protein are induced in rat male germ cells by Sertoli cells. *J. Biol. Chem.* 274(16), 10737–10742.
125. van der Wee, K. S., Johnson, E. W., Dirami, G., Dym, T. M., and Hofmann, M. C. (2001). Immunomagnetic isolation and long-term culture of mouse type A spermatogonia. *J. Androl.* 22(4), 696–704.
126. Fujioka, H., Fujisawa, M., Tatsumi, N., Kanzaki, M., Okuda, Y., Okada, H., Arakawa, S., and Kamidono, S. (2001). Sertoli cells inhibited apoptosis of pachytene spermatocytes and round spermatids. *Endocr. Res.* 27(1–2), 75–90.
127. Kissinger, C., Skinner, M. K., and Griswold, M. D. (1982). Analysis of Sertoli cell-secreted proteins by two-dimensional gel electrophoresis. *Biol. Reprod.* 27(1), 233–240.
128. Steinberger, A., and Steinberger, E. (1976). Secretion of an FSH-inhibiting factor by cultured Sertoli cells. *Endocrinology* 99(3), 918–921.
129. Dym, M., and Fawcett, D. W. (1970). The blood–testis barrier in the rat and the physiological compartmentation of the seminiferous epithelium. *Biol. Reprod.* 3(3), 308–326.
130. Russell, L. D., and Peterson, R. N. (1985). Sertoli cell junctions: morphological and functional correlates. *Int. Rev. Cytol.* 94, 177–211.
131. Lui, W. Y., Mruk, D., Lee, W. M., and Cheng, C. Y. (2003). Sertoli cell tight junction dynamics: their regulation during spermatogenesis. *Biol. Reprod.* 68(4), 1087–1097.
132. Cheng, C. Y., and Mruk, D. D. (2002). Cell junction dynamics in the testis: Sertoli-germ cell interactions and male contraceptive development. *Physiol. Rev.* 82(4), 825–874.
133. Djakiew, D., and Onoda, M. (1993). Multichamber cell culture and directional secretion. In *The Sertoli Cell* (M.D. Griswold, Ed.), pp. 181–194. Cache River Press, Clearwater, FL.
134. Onoda, M., and Djakiew, D. (1990). Modulation of Sertoli cell secretory function by rat round spermatid protein(s). *Mol. Cell. Endocrinol.* 73(1), 35–44.
135. Onoda, M., and Djakiew, D. (1991). Pachytene spermatocyte protein(s) stimulate Sertoli cells grown in bicameral chambers: dose-dependent secretion of ceruloplasmin, sulfated glycoprotein-1, sulfated glycoprotein-2, and transferrin. *In Vitro Cell. Dev. Biol.* 27A(3 Pt 1), 215–222.
136. Onoda, M., Djakiew, D., and Papadopoulos, V. (1991). Pachytene spermatocytes regulate the secretion of Sertoli cell protein(s) which stimulate Leydig cell steroidogenesis. *Mol. Cell. Endocrinol.* 77(1–3), 207–216.
137. Peschon, J. J., Behringer, R. R., Cate, R. L., Harwood, K. A., Idzerda, R. L., Brinster, R. L., and Palmiter, R. D. (1992). Directed expression of an oncogene to Sertoli cells in transgenic mice using müllerian inhibiting substance regulatory sequences. *Mol. Endocrinol.* 6(9), 1403–1411.
138. Hofmann, M. C., Narisawa, S., Hess, R. A., and Millan, J. L. (1992). Immortalization of germ cells and somatic testicular cells using the SV40 large T antigen. *Exp. Cell. Res.* 201(2), 417–435.
139. Walther, N., Jansen, M., Ergun, S., Kascheike, B., and Ivell, R. (1996). Sertoli cell lines established from H-2Kb-tsA58 transgenic mice differentially regulate the expression of cell-specific genes. *Exp. Cell. Res.* 225(2), 411–421.
140. Dutertre, M., Rey, R., Porteu, A., Josso, N., and Picard, J. Y. (1997). A mouse Sertoli cell line expressing anti-Müllerian hormone and its type II receptor. *Mol. Cell. Endocrinol.* 136(1), 57–65.

141. Bourdon, V., Lablack, A., Abbe, P., Segretain, D., Pointis, G. (1998). Characterization of a clonal Sertoli cell line using adult PyLT transgenic mice. *Biol. Reprod.* 58(2), 591–599.
142. Hofmann, M. C., Van Der Wee, K. S., Dargart, J. L., Dirami, G., Dettin, L., and Dym, M. (2003). Establishment and characterization of neonatal mouse sertoli cell lines. *J. Androl.* 24(1), 120–130.
143. Paquis-Flucklinger, V., Michiels, J. F., Vidal, F., Alquier, C., Pointis, G., Bourdon, V., Cuzin, F., and Rassoulzadegan, M. (1993) Expression in transgenic mice of the large T antigen of polyomavirus induces Sertoli cell tumours and allows the establishment of differentiated cell lines. *Oncogene* 8(8), 2087–2094.
144. Rassoulzadegan, M., Paquis-Flucklinger, V., Bertino, B., Sage, J., Jasin, M., Miyagawa, K., can Heyningen, V., Besmer, P., and Cuzin, F. (1993). Transmeiotic differentiation of male germ cells in culture. *Cell* 75(5), 997–1006.
145. Mather, J. P. (1980). Establishment and characterization of two distinct mouse testicular epithelial cell lines. *Biol. Reprod.* 23(1), 243–252.
146. Roberts, K. P., Banerjee, P. P., Tindall, J. W., and Zirkin, B. R. (1995). Immortalization and characterization of a Sertoli cell line from the adult rat. *Biol. Reprod.* 53(6), 1446–1453.
147. Pognan, F., Masson, M. T., Lagelle, F., and Charuel, C. (1997). Establishment of a rat Sertoli cell line that displays the morphological and some of the functional characteristics of the native cell. *Cell. Biol. Toxicol.* 13(6), 453–463.
148. Kurita, K., and Sakai, N. (2004). Functionally distinctive testicular cell lines of zebrafish to support male germ cell development. *Mol. Reprod. Dev.* 67(4), 430–438.
149. Merhi, R. A., Guillaud, L., Delouis, C., and Cotinot, C. (2001). Establishment and characterization of immortalized ovine Sertoli cell lines. *In Vitro Cell. Dev. Biol. Anim.* 37(9), 581–588.
150. Brinster, R. L., and Nagano, M. (1998). Spermatogonial stem cell transplantation, cryopreservation and culture. *Semin. Cell. Dev. Biol.* 9(4), 401–409.
151. Nagano, M., Avarbock, M. R., Leonida, E. B., Brinster, C. J., and Brinster, R. L. (1998). Culture of mouse spermatogonial stem cells. *Tissue Cell* 30(4),389–397.
152. Nagano, M., Ryu, B. Y., Brinster, C. J., Avarbock, M. R., and Brinster, R. L. (2003). Maintenance of mouse male germ line stem cells *in vitro*. *Biol. Reprod.* 68(6), 2207–2214.
153. Nakano, T., Kodama, H., and Honjo, T. (1994). Generation of lymphohematopoietic cells from embryonic stem cells in culture. *Science* 265(5175), 1098–1101.
154. Kent Hamra, F., Schultz, N., Chapman, K. M., Grellhesl, D. M., Cronkhite, J. T., Hammer, R. E., and Garbers, D. L. (2004). Defining the spermatogonial stem cell. *Dev. Biol.* 269(2), 393–410.
155. Orwig, K. E., Ryu, B. Y., Avarbock, M. R., and Brinster, R. L. (2002). Male germ-line stem cell potential is predicted by morphology of cells in neonatal rat testes. *Proc. Natl. Acad. Sci. U S A* 99(18), 11706–11711.
156. Orwig, K. E., Shinohara, T., Averbock, M. R., and Brinster, R. L. (2002). Functional analysis of stem cells in the adult rat testis. *Biol. Reprod.* 66(4), 944–949.
157. Mahato, D., Goulding, E. H., Korach, K. S., and Eddy, E. M. (2000). Spermatogenic cells do not require estrogen receptor-alpha for development or function. *Endocrinology* 141(3), 1273–1276.
158. Charest, N. J., Zhou, Z. X., Lubahn, D. B., Olsen, K. L., Wilson, E. M., French, F. S. (1991). A frameshift mutation destabilizes androgen receptor messenger RNA in the Tfm mouse. *Mol. Endocrinol.* 5(4), 573–581.
159. Lyon, M. F., and Hawkes, S. G. (1970). X-linked gene for testicular feminization in the mouse. *Nature* 227(5264), 1217–1219.
160. Lee, J., Richburg, J. H., Shipp, E. B., Meistrich, M. L., and Boekelheide, K. (1999). The Fas system, a regulator of testicular germ cell apoptosis, is differentially up-regulated in Sertoli cell versus germ cell injury of the testis. *Endocrinology* 140(2), 852–858.
161. Yoshinaga, K., Nishikawa, S., Ogawa, M., Hayashi, S., Kunisada, T., Fujimoto, T., and Nishikawa, S. (1991). Role of c-kit in mouse spermatogenesis: identification of spermatogonia as a specific site of c-kit expression and function. *Development* 113(2), 689–699.
162. Ogawa, T., Dobrinski, I., Avarbock, M. R., and Brinster, R. L. (2000). Transplantation of male germ line stem cells restores fertility in infertile mice. *Nat. Med.* 6(1),29–34.
163. Buaas, F. W., Kirsh, A. L., Sharma, M., McLean, D. J., Morris, J. L., Griswold, M. D., de Rooij, D. G., and Braun, R. E. (2004). Plzf is required in adult male germ cells for stem cell self-renewal. *Nat. Genet.* 36(6), 647–652.
164. Clouthier, D. E., Avarbock, M. R., Maika, S. D., Hammer, R. E., and Brinster, R. L. (1996). Rat spermatogenesis in mouse testis. *Nature* 381(6581), 418–421.
165. Ogawa, T., Dobrinski, I., Avarbock, M. R., and Brinster, R. L. (1999). Xenogeneic spermatogenesis following transplantation of hamster germ cells to mouse testes. *Biol. Reprod.* 60(2), 515–521.
166. Franca, L. R., Ogawa, T., Avarbock, M. R., Brinster, R. L., and Russell, L. D. (1998). Germ cell genotype controls cell cycle during spermatogenesis in the rat. *Biol. Reprod.* 59(6), 1371–1377.
167. Shinohara, T., Orwig, K. E., Avarbock, M. R., and Brinster, R. L. (2003). Restoration of spermatogenesis in infertile mice by Sertoli cell transplantation. *Biol. Reprod.* 68(3), 1064–1071.
168. Parizek, J. (1957). The destructive effect of cadmium ion on testicular tissue and its prevention by zinc. *J. Endocrinol.* 15(1), 56–63.
169. Koulish, S., Kramer, C. R., and Grier, H. J. (2002). Organization of the male gonad in a protogynous fish, Thalassoma bifasciatum (Teleostei: Labridae). *J. Morphol.* 254(3), 292–311.
170. Grier, H. J., and Linton, J. R. (1977). Ultrastructural identification of the Sertoli cell in the testis of the northern pike, Esox lucius. *Am. J. Anat.* 149(2), 283–288.
171. DeFelice, D. A., and Rasch, E. M. (1969). Chronology of spermatogenesis and spermiogenesis in Poeciliid fishes. *J. Exp. Zool.* 171, 191–208.
172. Clermont, Y., and Perey, B. (1957). Quantitative study of the cell population of the seminiferous tubules in immature rats. *Am. J. Anat.* 100(2), 241–267.
173. Steinberger, A., and Steinberger, E. (1971). Replication pattern of Sertoli cells in maturing rat testis *in vivo* and in organ culture. *Biol. Reprod.* 4(1), 84–87.
174. Grier, H. J., Fitzsimons, J. M., and Linton, J. R. (1978). Structure and ultrastructure of the testis and sperm formation in goodeid teleosts. *J. Morphol.* 156(3), 419–437.
175. Guraya, S. S. (1976). Recent advances in the morphology, histochemistry, and biochemistry of steroid-synthesizing cellular sites in the testes of nonmammalian vertebrates. *Int. Rev. Cytol.* 47, 99–136.
176. Gresik, E. W., Quirk, J. G., and Hamilton, J. B. (1973). Fine structure of the Sertoli cell of the testis of the teleost Oryzias latipes. *Gen. Comp. Endocrinol.* 21(2), 341–352.
177. Russell, L. D., and Griswold, M. D. (1993). *The Sertoli Cell.* Cache River Press, Clearwater, FL.
178. Parmentier, H. K., van den Boogaart, J. G. M., and Timmermans, L. P. (1985). Physiological compartmentation in gonadal tissu of the common carp. *Cell. Tissue Res.* 242, 75–81.

179. Pudney, J. (1993). Comparative cytology of the nonmammalian vertebrate Sertoli cell. In *The Sertoli Cell* (M.D. Griswold, Ed.), pp. 611–657. Cache River Press, Clearwater, FL.
180. Bergmann, M., Greven, H., and Schindelmeiser, J. (1983). Observations on the blood–testis barrier in a frog and a salamander. *Cell. Tissue Res.* 232(1), 189–200.
181. Wilhoft, D. C., and Quay, W. B. (1961). Testicular histology and seasonal changes in the lizard. Sceloporus occidentalis. *J. Morphol.* 108, 95–106.
182. Reynolds, A. E. (1943). The normal seasonal reproductive cycle in the male Eumeces fasciatus together with some observations on the effects of castration and hormone administration. *J. Morphol.* 72, 331–337.
183. Baccetti, B., Bigliardi, E., Vegni Taluri, M., and Burrini, A. G. (1983). The Sertoli cell in lizards. *J. Ultrastruct. Res.* 85(1), 11–23.
184. Pelletier, R. M. (1990). A novel perspective: the occluding zonule encircles the apex of the Sertoli cell as observed in birds. *Am. J. Anat.* 188(1), 87–108.
185. Breuker, H. (1982). Seasonal spermatogenesis in the mute suam (Cygnus olor). *Adv. Anat. Embryol. Cell. Biol.* 72, 1–90.
186. Sprando, R. L., and Russell, L. D. (1988). Spermiogenesis in the red-ear turtle (Pseudemys scripta) and the domestic fowl (Gallus domesticus): a study of cytoplasmic events including cell volume changes and cytoplasmic elimination. *J. Morphol.* 198(1), 95–118.
187. Gondos, B., and Berndston, W. E. (1993). Postnatal and pubertal development. In *The Sertoli Cell* (M.D. Griswold, Ed.), pp. 115–154. Cache River Press, Clearwater, FL.
188. Johnson, L., and Thompson, D. L. Jr. (1983). Age-related and seasonal variation in the Sertoli cell population, daily sperm production and serum concentrations of follicle-stimulating hormone, luteinizing hormone and testosterone in stallions. *Biol. Reprod.* 29(3), 777–789.
189. Chevalier, M. (1978). Sertoli cell ultrastructure. I. A comparative study in immature, pubescent adult and cryptorchid pigs. *Ann. Biol. Anim. Biochem. Biophys.* 18, 1279–1292.
190. McCoard, S. A., Lunstra, D. D., Wise, T. H., and Ford, J. J. (2001). Specific staining of Sertoli cell nuclei and evaluation of Sertoli cell number and proliferative activity in Meishan and White Composite boars during the neonatal period. *Biol. Reprod.* 64(2), 689–695.
191. Hinsch, G. W. (1993). Comparative organization and cytology of Sertoli cells in invertebrates. In *The Sertoli Cell* (M.D. Griswold, Ed.), pp. 659–683. Cache River Press, Clearwater, FL.
192. Marcaillou, C., and Szollosi, A. (1980). The "blood–testis" barrier in a nematode and a fish: a generalizable concept. *J. Ultrastruct. Res.* 70(1), 128–136.
193. de Jong-Brink, M., Boer, H. H., Hommer, T. G., and Kodde, A. (1977). Spermatogenesis and the role of Sertoli cells in the freshwater snail Biomphalaria glabrata. *Cell. Tissue Res.* 181, 37–58.
194. de Jong-Brink, M., de With, N. D., Hurkmans, P. J., and Bergamin Sassen, M. J. (1984). A morphological, enzyme-cytochemical, and physiological study of the blood–gonad barrier in the hermaphroditic snail Lymnaea stagnalis. *Cell. Tissue Res.* 235, 593–600.
195. Szollosi, A., and Marcaillou, C. (1977). Electron microscope study of the blood–testis barrier in an insect: Locusta migratoria. *J. Ultrastruct. Res.* 59(2), 158–172.
196. Gerlach, T., and Aurich, J. E. (2000). Regulation of seasonal reproductive activity in the stallion, ram and hamster. *Anim. Reprod. Sci.* 58(3–4), 197–213.
197. Gorman, M. R., and Zucker, I. (1995). Seasonal adaptations of Siberian hamsters. II. Pattern of change in daylength controls annual testicular and body weight rhythms. *Biol. Reprod.* 53(1), 116–125.
198. Munoz, E. M., Fogal, T., Dominguez, S., Scardapane, L., and Piezzi, R. S. (2001). Ultrastructural and morphometric study of the Sertoli cell of the viscacha (Lagostomus maximus maximus) during the annual reproductive cycle. *Anat. Rec.* 262(2), 176–185.
199. Young, K. A., Ball, G. F., and Nelson, R. J. (2001). Photoperiod-induced testicular apoptosis in European starlings (Sturnus vulgaris). *Biol. Reprod.* 64(2), 706–713.
200. Sanford, L. M., Palmer, W. M., and Howland, B. E. (1977). Changes in the profiles of serum LH, FSH and testosterone, and in mating performance and ejaculate volume in the ram during the ovine breeding season. *J. Anim. Sci.* 45(6), 1382–1391.
201. Turek, F. W., Elliot, J. A., Alvis, J. D., and Menaker, M. (1975). Effect of prolonged exposure to nonstimulatory photoperiods on the activity of the neuroendocrine-testicular axis of golden hamsters. *Biol. Reprod.* 13(4), 475–481.
202. Schlatt, S., De Geyter, M., Kilesch, S., Mieschlag, E., and Bergmann, M. (1995). Spontaneous recrudescence of spermatogenesis in the photoinhibited male Djungarian hamster, Phodopus sungorus. *Biol. Reprod.* 53(5), 1169–1177.
203. Lerchl, A., and Nieschlag, E. (1992). Interruption of nocturnal pineal melatonin synthesis in spontaneous recrudescent Djungarian hamsters (Phodopus sungorus). *J. Pineal Res.* 13(1), 36–41.
204. Hikim, A. P., Amador, A. G., Bartke, A., and Russell, L. D. (1989). Structure/function relationships in active and inactive hamster Leydig cells: a correlative morphometric and endocrine study. *Endocrinology* 125(4), 1844–1856.
205. Russell, L. D., Chandrashekar, V., Bartke, A., and Hikim, A. P. (1994). The hamster Sertoli cell in early testicular regression and early recrudescence: a stereological and endocrine study. *Int. J. Androl.* 17(2), 93–106.
206. Simpson, S. M., Follett, B. K., and Ellis, D. H. (1982). Modulation by photoperiod of gonadotrophin secretion in intact and castrated Djungarian hamsters. *J. Reprod. Fertil.* 66(1), 243–250.
207. Milette, J. J., Schwartz, N. B., and Turek, F. W. (1988). The importance of follicle-stimulating hormone in the initiation of testicular growth in photostimulated Djungarian hamsters. *Endocrinology* 122(3), 1060–1066.
208. Steinberger, A., Hintz, M., and Heindel, J. J. (1978). Changes in cyclic AMP responses to FSH in isolated rat Sertoli cells during sexual maturation. *Biol. Reprod.* 19(3), 566–572.
209. Heindel, J. J. (1981) FSH stimulation of cAMP in hamster Sertoli cells: effect of age and optic enucleation. *J. Androl.* 2, 217.
210. Johnson, L., Varner, D. D., Tatum, M. E., and Scrutchfield, W. L. (1991). Season but not age affects Sertoli cell number in adult stallions. *Biol. Reprod.* 45(3), 404–410.
211. Sanford, L. M. (2002). Role of FSH, numbers of FSH receptors and testosterone in the regulation of inhibin secretion during the seasonal testicular cycle of adult rams. *Reproduction* 123(2), 269–280.
212. Hochereau-de Reviers, M. T., Monet-Kuntz, C., and Courot, M. (1987). Spermatogenesis and Sertoli cell numbers and function in rams and bulls. *J. Reprod. Fertil. Suppl.* 34, 101–114.
213. Lincoln, G. A., Lincoln, C. E., and McNeilly, A. S. (1990). Seasonal cycles in the blood plasma concentration of FSH, inhibin and testosterone, and testicular size in rams of wild, feral and domesticated breeds of sheep. *J. Reprod. Fertil.* 88(2), 623–633.
214. Young, K. A., Zirkin, B. R., and Nelson, R. J. (2000). Testicular regression in response to food restriction and short photoperiod in white-footed mice (Peromyscus leucopus) is mediated by apoptosis. *Biol. Reprod.* 62(2), 347–354.

215. Nagata, S. (1996). Apoptosis mediated by the Fas system. *Prog. Mol. Subcell. Biol.* 16, 87–103.
216. Lee, J., Richburg, J. H., Younkin, S. C., and Boekelheide, K. (1997). The Fas system is a key regulator of germ cell apoptosis in the testis. *Endocrinology* 138(5), 2081–2088.
217. Boekelheide, K., Lee, J., Shipp, E. B., Richburg, J. H., and Li, G. (1998). Expression of Fas system-related genes in the testis during development and after toxicant exposure. *Toxicol. Lett.* 102–103, 503–508.
218. Richburg, J. H., Nanez, A., Williams, L. R., Embree, M. E., and Boekelheide, K. (2000). Sensitivity of testicular germ cells to toxicant-induced apoptosis in gld mice that express a nonfunctional form of Fas ligand. *Endocrinology* 141(2), 787–793.
219. Young, K. A., and Nelson, R. J. (2001). Mediation of seasonal testicular regression by apoptosis. *Reproduction* 122(5), 677–685.
220. Law, G. L., and Griswold, M. D. (1994). Activity and form of sulfated glycoprotein 2 (clusterin) from cultured Sertoli cells, testis, and epididymis of the rat. *Biol. Reprod.* 50(3), 669–679.
221. Morales, C. R., el-Alfy, M., Zhao, Q., and Igdoura, S. (1995). Molecular role of sulfated glycoprotein-1 (SGP-1/prosaposin) in Sertoli cells. *Histol. Histopathol.* 10(4), 1023–1034.
222. Rappaport, M. S., and Smith, E. P. (1995). Insulin-like growth factor (IGF) binding protein 3 in the rat testis: follicle-stimulating hormone dependence of mRNA expression and inhibition of IGF-I action on cultured Sertoli cells. *Biol. Reprod.* 52(2), 419–425.
223. Subramanian, S., and Adiga, P. R. (1996). Hormonal modulation of riboflavin carrier protein secretion by immature rat Sertoli cells in culture. *Mol. Cell. Endocrinol.* 120(1), 41–50.
224. Subramanian, S., and Adiga, P. R. (1999). Characterization and hormonal modulation of immunoreactive thiamin carrier protein in immature rat Sertoli cells in culture. *J. Steroid Biochem. Mol. Biol.* 68(1–2), 23–30.
225. Bardi, G., Bottazzi, C., Demori, I., and Palmero, S. (1999). Thyroid hormone and retinoic acid induce the synthesis of insulin-like growth factor-binding protein-4 in prepubertal pig sertoli cells. *Eur. J. Endocrinol.* 141(6), 637–643.
226. Boockfor, F. R. (1997). Sertoli cells in culture release a pregnancy specific beta1 glycoprotein-like substance. *Proc. Soc. Exp. Biol. Med.* 214(2), 139–145.
227. Kroepelien, C. F., Knutsen, H. K., Haugen, T. B., Hansson, V., and Eskild, W. (1993). Serum factors induce messenger ribonucleic acid levels for cellular retinol-binding protein in rat Sertoli cells. *Endocrinology* 132(3), 968–974.
228. O'Brien, D. A., Childress, A. R., Ehrman, R., and Robbins, S. J. (1994). Expression of mannose 6-phosphate receptor messenger ribonucleic acids in mouse spermatogenic and Sertoli cells. *Biol. Reprod.* 50(2), 429–435.
229. Welsh, M. J., Wu, W., Parvinen, M., and Gilmont, R. R. (1996). Variation in expression of hsp27 messenger ribonucleic acid during the cycle of the seminiferous epithelium and co-localization of hsp27 and microfilaments in Sertoli cells of the rat. *Biol. Reprod.* 55(1), 141–151.
230. Gregory, C. W., and DePhilip, R. M. (1998). Detection of steroidogenic acute regulatory protein (stAR) in mitochondria of cultured rat Sertoli cells incubated with follicle-stimulating hormone. *Biol. Reprod.* 58(2), 470–474.
231. Davidoff, M. S., Middendorff, R., Pusch, W., Muller, D., Wichers, S., and Holstein, A. F. (1999). Sertoli and Leydig cells of the human testis express neurofilament triplet proteins. *Histochem. Cell. Biol.* 111(3), 173–187.
232. Kraszucka, K., Burfeind, P., Nayermia, K., Kohler, M., Schmid, M., Yaylaoglu, M., and Engel, W. (1999). Developmental stage- and germ cell-regulated expression of a calcium-binding protein mRNA in mouse Sertoli cells. *Mol. Reprod. Dev.* 54(3), 232–243.
233. McClure, R. F., Heppelmann, C. J., and Paya, C. V. (1999). Constitutive Fas ligand gene transcription in Sertoli cells is regulated by Sp1. *J. Biol. Chem.* 274(12), 7756–7762.
234. Danno, S., Itoh, K., Matsuda, T., and Fujita, J. (2000). Decreased expression of mouse Rbm3, a cold-shock protein, in Sertoli cells of cryptorchid testis. *Am. J. Pathol.* 156(5), 1685–1692.
235. Saunders, P. T., Maguire, S. M., Macpherson, S., Fenelon, M. C., Sakakibara, S., and Okano, H. (2002). RNA binding protein Musashi1 is expressed in sertoli cells in the rat testis from fetal life to adulthood. *Biol. Reprod.* 66(2), 500–507.
236. Wu, J., Jester, W. F. Jr., Laslett, A. L., Meinhardt, A., and Orth, J. M. (2003). Expression of a novel factor, short-type PB-cadherin, in Sertoli cells and spermatogenic stem cells of the neonatal rat testis. *J. Endocrinol.* 176(3), 381–391.
237. Zhu, Q., Emanuele, N. V., and Van Thiel, D. H. (2004). Calponin is expressed by Sertoli cells within rat testes and is associated with actin-enriched cytoskeleton. *Cell. Tissue Res.* 316(2), 243–253.
238. Bravo-Moreno, J. F., Diaz-Sanchez, V., Montoya-Flores, J. G., Lamoyi, E., Saez, J. C., and Perez-Armendariz, E. M. (2001). Expression of connexin43 in mouse Leydig, Sertoli, and germinal cells at different stages of postnatal development. *Anat. Rec.* 264(1), 13–24.
239. Brehm, R., Marks, A., Rey, R., Kliesch, S., Bergmann, M., and Steger, K. (2002). Altered expression of connexins 26 and 43 in Sertoli cells in seminiferous tubules infiltrated with carcinoma-in-situ or seminoma. *J. Pathol.* 197(5), 647–653.
240. Guttman, J. A., Janmey, P., and Vogl, A. W. (2002). Gelsolin: evidence for a role in turnover of junction-related actin filaments in Sertoli cells. *J. Cell. Sci.* 115(Pt 3), 499–505.
241. Jabado, N., Canonne-Hergaux, F., Gruenheid, S., Picard, V., and Gros, P. (2002). Iron transporter Nramp2/DMT-1 is associated with the membrane of phagosomes in macrophages and Sertoli cells. *Blood* 100(7), 2617–2622.
242. Kawasaki, Y., Nakagawa, A., Nagaosa, K., Shiratsuchi, A., and Nakanishi, Y. (2002). Phosphatidylserine binding of class B scavenger receptor type I, a phagocytosis receptor of testicular sertoli cells. *J. Biol. Chem.* 277(30), 27559–27566.
243. Brucato, S., Harduin-Lepers, A., Godard, F., Bocquest, J., and Villers, C. (2000). Expression of glypican-1, syndecan-1 and syndecan-4 mRNAs protein kinase C-regulated in rat immature Sertoli cells by semi-quantitative RT-PCR analysis. *Biochim. Biophys. Acta* 1474(1), 31–40.
244. Fortna, R. R., Watson, H. A., and Nyquist, S. E. (1999). Glycosyl phosphatidylinositol-anchored ceruloplasmin is expressed by rat Sertoli cells and is concentrated in detergent-insoluble membrane fractions. *Biol. Reprod.* 61(4), 1042–1049.
245. Taranta, A., Morena, A. R., Barbacci, E., and D'Agostino, A. (1997). omega-Conotoxin-sensitive Ca2+ voltage-gated channels modulate protein secretion in cultured rat Sertoli cells. *Mol. Cell. Endocrinol.* 126(2), 117–123.
246. Riccioli, A., Filippini, A., De Cesaris, P., Barbacci, E., Stefanini, M., Starace, G., Ziparo, E. (1995). Inflammatory mediators increase surface expression of integrin ligands, adhesion to lymphocytes, and secretion of interleukin 6 in mouse Sertoli cells. *Proc. Natl. Acad. Sci. U S A* 92(13), 5808–5812.
247. Richardson, L. L., Kleinman, H. K., and Dym, M. (1995). Basement membrane gene expression by Sertoli and peritubular myoid cells *in vitro* in the rat. *Biol. Reprod.* 52(2), 320–330.
248. Ulisse, S., Rucci, N., Piersanti, D., Carosa, E., Graziano, F.M., Pavan, A., Ceddia, P., Arizzi, M., Muzi, P., Cironi, L., Gnessi, L.,

D'Armiento, M., and Jannini, E. A. (1998). Regulation by thyroid hormone of the expression of basement membrane components in rat prepubertal Sertoli cells. *Endocrinology* 139(2), 741–747.

249. Lin, L. H., and DePhilip, R. M. (1996). Differential expression of placental (P)–cadherin in sertoli cells and peritubular myoid cells during postnatal development of the mouse testis. *Anat. Rec.* 244(2), 155–164.
250. MacCalman, C. D., Getsios, S., Farookhi, R., and Blaschuk, O. W. (1997). Estrogens potentiate the stimulatory effects of follicle-stimulating hormone on N-cadherin messenger ribonucleic acid levels in cultured mouse Sertoli cells. *Endocrinology* 138(1), 41–48.
251. Rogatsch, H., Jezek, D., Hittmair, A., Mikuz, G., and Feichtinger, H. (1996). Expression of vimentin, cytokeratin, and desmin in Sertoli cells of human fetal, cryptorchid, and tumour-adjacent testicular tissue. *Virchows Arch.* 427(5), 497–502.
252. Igdoura, S. A., Argraves, W. S., and Morales, C. R. (1997). Low density lipoprotein receptor-related protein-1 expression in the testis: regulated expression in Sertoli cells. *J. Androl.* 18(4), 400–410.
253. Guttman, J. A., Mulholland, D. J., and Vogl, A. W. (1999). Plectin is concentrated at intercellular junctions and at the nuclear surface in morphologically differentiated rat Sertoli cells. *Anat. Rec.* 254(3), 418–428.
254. Steger, K., Rey, R., Louis, F., Kliesch, S., Behre, H. M., Nieschlag, E., Hoepffner, W., Bailey, D., Marks, A., and Bergmann, M. (1999). Reversion of the differentiated phenotype and maturation block in Sertoli cells in pathological human testis. *Hum. Reprod.* 14(1), 136–143.
255. Shiratsuchi, A., Kawasaki, Y., Ikemoto, M., Arai, H., and Nakanishi, Y. (1999). Role of class B scavenger receptor type I in phagocytosis of apoptotic rat spermatogenic cells by Sertoli cells. *J. Biol. Chem.* 274(9), 5901–5908.
256. Hill, C. M., Anway, M. D., Zirkin, B. R., and Brown, T. R. (2004). Intratesticular androgen levels, androgen receptor localization, and androgen receptor expression in adult rat Sertoli cells. *Biol. Reprod.* 71(4), 1348–58.
257. de Winter, J. P., Vanderstichele, H. M., Timmerman, M. A., Blok, L. J., Themmen, A. P., and de Jong, F. H. (1993). Activin is produced by rat Sertoli cells *in vitro* and can act as an autocrine regulator of Sertoli cell function. *Endocrinology* 132(3), 975–982.
258. Najmabadi, H., Rosenberg, L. A., Yuan, Q. X., Reyaz, G., and Bhasin, S. (1993). Transcriptional regulation of inhibin beta B messenger ribonucleic acid levels in TM.4 or primary rat Sertoli cells by 8–bromo-cyclic adenosine monophosphate. *Mol. Endocrinol.* 7(4), 561–569.
259. Carreau, S. (1995) Human Sertoli cells produce inhibin *in vitro*: an additional marker to assess the seminiferous epithelium development. *Hum. Reprod.* 10(8), 1947–1949.
260. Eikvar, L., Bjornerheim, R., Attramadal, H., and Hansson, V. (1993). Beta-adrenoceptor mediated responses and subtypes of beta-adrenoceptors in cultured rat Sertoli cells. *J. Steroid Biochem. Mol. Biol.* 44(1), 85–91.
261. Marziali, G., Lazzaro, D., and Sorrentino, V. (1993). Binding of germ cells to mutant Sld Sertoli cells is defective and is rescued by expression of the transmembrane form of the c-kit ligand. *Dev. Biol.* 157(1), 182–190.
262. Avallet, O., Vigier, M., Leduque, P., Dubois, P. M., and Saez, J. M. (1994). Expression and regulation of transforming growth factor-beta 1 messenger ribonucleic acid and protein in cultured porcine Leydig and Sertoli cells. *Endocrinology* 134(5), 2079–2087.
263. Filippini, A., Riccioli, A., De Cesaris, P., Paniccia, R., Teti, A., Stefanini, M., Conti, M., and Ziparo, E. (1994). Activation of inositol phospholipid turnover and calcium signaling in rat Sertoli cells by P2–purinergic receptors: modulation of follicle-stimulating hormone responses. *Endocrinology* 134(3), 1537–1545.
264. Le Magueresse-Battistoni, B., Wolff, J., Morera, A. M., and Benahmed, M. (1994). Fibroblast growth factor receptor type 1 expression during rat testicular development and its regulation in cultured Sertoli cells. *Endocrinology* 135(6), 2404–2411.
265. Antich, M., Fabian, E., Sarquella, J., Bassas, L. (1995). Effect of testicular damage induced by cryptorchidism on insulin-like growth factor I receptors in rat Sertoli cells. *J. Reprod. Fertil.* 104(2), 267–275.
266. Palmero, S., De Marco, P., and Fugassa, E. (1995). Thyroid hormone receptor beta mRNA expression in Sertoli cells isolated from prepubertal testis. *J. Mol. Endocrinol.* 14(1), 131–134.
267. Mauduit, C., Besset, V., Caussanel, V., and Benahmed, M. (1996). Tumor necrosis factor alpha receptor p55 is under hormonal (follicle-stimulating hormone) control in testicular Sertoli cells. *Biochem. Biophys. Res. Commun.* 224(3), 631–637.
268. Panno, M. L., Sisci, D., Salerno, M., Lanzino, M., Mauro, L., Morrone, E. G., Pezzi, V., Palmero, S., Fugassa, E., and Ando, S. (1996). Effect of triiodothyronine administration on estrogen receptor contents in peripuberal Sertoli cells. *Eur. J. Endocrinol.* 134(5), 633–638.
269. Rey, R., al-Attar, L., Louis, F., Jaubert, F., Barbet, P., Nihoul-Fekete, C. C., Haussain, J. L., and Josso, N. (1996). Testicular dysgenesis does not affect expression of anti-müllerian hormone by Sertoli cells in premeiotic seminiferous tubules. *Am. J. Pathol.* 148(5), 1689–1698.
270. Maguire, S. M., Tribley, W. A., and Griswold, M. D. (1997). Follicle-stimulating hormone (FSH) regulates the expression of FSH receptor messenger ribonucleic acid in cultured Sertoli cells and in hypophysectomized rat testis. *Biol. Reprod.* 56(5), 1106–1111.
271. Jenab, S., and Morris, P. L. (1998). Testicular leukemia inhibitory factor (LIF) and LIF receptor mediate phosphorylation of signal transducers and activators of transcription (STAT)–3 and STAT-1 and induce c-fos transcription and activator protein-1 activation in rat Sertoli but not germ cells. *Endocrinology* 139(4), 1883–1890.
272. Chaudhary, J., Kim, G., and Skinner, M. K. (1999). Expression of the basic helix-loop-helix protein REBalpha in rat testicular Sertoli cells. *Biol. Reprod.* 60(5), 1244–1250.
273. Guillaumot, P., and Benahmed, M. (1999). Prolactin receptors are expressed and hormonally regulated in rat Sertoli cells. *Mol. Cell. Endocrinol.* 149(1–2), 163–168.
274. Syed, V., Gomez, E., and Hecht, N. B. (1999). Messenger ribonucleic acids encoding a serotonin receptor and a novel gene are induced in Sertoli cells by a secreted factor(s) from male rat meiotic germ cells. *Endocrinology* 140(12), 5754–5760.
275. Jenab, S., and Morris, P. L. (2000). Interleukin-6 regulation of kappa opioid receptor gene expression in primary sertoli cells. *Endocrine* 13(1), 11–15.
276. Borges, M. O., Abreu, M. L., Porto, C. S., and Avellar, M. C. (2001). Characterization of muscarinic acetylcholine receptor in rat Sertoli cells. *Endocrinology* 142(11), 4701–4710.
277. Maccarrone, M., Cecconi, S., Rossi, G., Battista, N., Pauselli, R., and Finazzi-Agro, A. (2003). Anandamide activity and degradation are regulated by early postnatal aging and follicle-stimulating hormone in mouse Sertoli cells. *Endocrinology* 144(1), 20–28.
278. Imai, T., Kawai, Y., Tadokoro, Y., Yamamoto, M., Nishimune, Y., and Yomogida, K. (2004). *In vivo* and *in vitro* constant expression of GATA-4 in mouse postnatal Sertoli cells. *Mol. Cell. Endocrinol.* 214(1–2), 107–115.

279. Walker, W. H., Fucci, L., and Habener, J. F. (1995). Expression of the gene encoding transcription factor cyclic adenosine 3′,5′-monophosphate (cAMP) response element-binding protein (CREB): regulation by follicle-stimulating hormone-induced cAMP signaling in primary rat Sertoli cells. *Endocrinology* 136(8), 3534–3545.
280. Jia, M. C., Ravindranath, N., Papadopoulos, V., and Dym, M. (1996). Regulation of c-fos mRNA expression in Sertoli cells by cyclic AMP, calcium, and protein kinase C mediated pathways. *Mol. Cell. Biochem.* 156(1), 43–49.
281. Page, K. C., Heitzman, D. A., and Chernin, M. I. (1996). Stimulation of c-jun and c-myb in rat Sertoli cells following exposure to retinoids. *Biochem. Biophys. Res. Commun.* 222(2), 595–600.
282. Tamai, K. T., Monaco, L., Alastalo, T. P., Lalli, E., Parvinen, M., and Sassone-Corsi, P. (1996). Hormonal and developmental regulation of DAX-1 expression in Sertoli cells. *Mol. Endocrinol.* 10(12), 1561–1569.
283. Salas-Cortes, L., Jaubert, F., Barbaux, S., Nessmann, C., Bono, M. R., Fellous, M., McElreavey, K., and Rosemblatt, M. (1999). The human SRY protein is present in fetal and adult Sertoli cells and germ cells. *Int. J. Dev. Biol.* 43(2), 135–140.
284. Frojdman, K., Harley, V. R., and Pelliniemi, L. J. (2000). Sox9 protein in rat sertoli cells is age and stage dependent. *Histochem. Cell. Biol.* 113(1), 31–36.
285. Chen, J. K., and Heckert, L. L. (2001). Dmrt1 expression is regulated by follicle-stimulating hormone and phorbol esters in postnatal Sertoli cells. *Endocrinology* 142(3), 1167–1178.
286. Fujisawa, M., Bardin, C. W., and Morris, P. L. (1993). Germ cell factor(s) regulates opioid gene expression in Sertoli cells. *Recent Prog. Horm. Res.* 48, 497–503.
287. Kanzaki, M., Fujisawa, M., Okuda, Y., Okada, H., Arakawa, S., and Kamidono, S. (1996). Expression and regulation of neuropeptide Y messenger ribonucleic acid in cultured immature rat Leydig and Sertoli cells. *Endocrinology* 137(4), 1249–1257.
288. Cudicini, C., Lejeune, H., Gomez, E., Bosmans, E., Ballet, D., Saez, J., and Jegou, B. (1997). Human Leydig cells and Sertoli cells are producers of interleukins-1 and -6. *J. Clin. Endocrinol. Metab.* 82(5), 1426–1433.
289. Musa, F. R., Tokuda, M., Kuwata, Y., Ogawa, T., Tomizawa, K., Konishi, R., Takenaka, I., and Hatase, O. (1998). Expression of cyclin-dependent kinase 5 and associated cyclins in Leydig and Sertoli cells of the testis. *J. Androl.* 19(6), 657–666.
290. Chaudhary, J., and Skinner, M. K. (2000). Characterization of a novel transcript of 14–3–3 theta in Sertoli cells. *J. Androl.* 21(5), 730–738.
291. Li, J. C., Samy, E. T., Grima, J., Chung, S. S., Mruk, D., Lee, W. M., Silvestrini, B., and Cheng, C. Y. (2000). Rat testicular myotubularin, a protein tyrosine phosphatase expressed by Sertoli and germ cells, is a potential marker for studying cell-cell interactions in the rat testis. *J. Cell. Physiol.* 185(3), 366–385.
292. Zeyse, D., Lunenfeld, E., Beck, M., Prinsloo, I., and Huleihel, M. (2000). Interleukin-1 receptor antagonist is produced by sertoli cells *in vitro*. *Endocrinology* 141(4), 1521–1527.
293. Zeyse, D., Lunenfeld, E., Beck, M., Prinsloo, I., and Huleihel, M. (2000). Induction of interleukin-1alpha production in murine Sertoli cells by interleukin-1. *Biol. Reprod.* 62(5), 1291–1296.
294. Dirami, G., Ravindranath, N., Achi, M. V., and Dym, M. (2001). Expression of Notch pathway components in spermatogonia and Sertoli cells of neonatal mice. *J. Androl.* 22(6), 944–952.
295. Ross, A. J., Amy, S. P., Mahar, P. L., Lindsten, T., Knudson, C. M., Thompson, C. B., Korsmeyer, S. J., and MacGregor, G. R. (2001). BCLW mediates survival of postmitotic Sertoli cells by regulating BAX activity. *Dev. Biol.* 239(2), 295–308.
296. Toshima, J., Toshima, J. Y., Takeuchi, K., Mori, R., and Mizuno, K. (2001). Cofilin phosphorylation and actin reorganization activities of testicular protein kinase 2 and its predominant expression in testicular Sertoli cells. *J. Biol. Chem.* 276(33), 31449–31458.
297. Khan, S. A., Ndjountche, L., Pratchard, L., Spicer, L. J., and Davis, J. S. (2002). Follicle-stimulating hormone amplifies insulin-like growth factor I-mediated activation of AKT/protein kinase B signaling in immature rat Sertoli cells. *Endocrinology* 143(6), 2259–2267.
298. Maekawa, M., Toyama, Y., Yasuda, M., Yagi, T., and Yuasa, S. (2002). Fyn tyrosine kinase in Sertoli cells is involved in mouse spermatogenesis. *Biol. Reprod.* 66(1), 211–221.
299. Holsberger, D. R., Jirawatnotai, S., Kiyokawa, H., and Cooke, P. S. (2003). Thyroid hormone regulates the cell cycle inhibitor p27Kip1 in postnatal murine Sertoli cells. *Endocrinology* 144(9), 3732–3738.
300. Wong, C. C., and Lee, W. M. (2002). The proximal cis-acting elements Sp1, Sp3 and E2F regulate mouse mer gene transcription in Sertoli cells. *Eur. J. Biochem.* 269(15), 3789–3800.
301. Laurent-Cadoret, V., Guillou, F., and Combarnous, Y. (1993). Involvement of cyclic adenosine monophosphate-dependent protein kinase isozymes in tissue plasminogen activator secretion by rat Sertoli cells stimulated with follicle-stimulating hormone *in vitro*. *Acta Endocrinol. (Copenh.)* 128(6), 555–562.
302. Liu, Y. X., Peng, X. R., Liu, K., and Ny, T. (1993). Hormonal regulation of tissue-type plasminogen activator and plasminogen activator inhibitor type-1 gene expression in cultured mouse Sertoli cells. *Sci. China B* 36(3), 319–328.
303. Liu, Y., Du, Q., Zhou, H., Liu, K., and Hu, Z. (1996). Regulation of tissue-type plasminogen activator and plasminogen activator inhibitor type-1 in cultured rat Sertoli and Leydig cells. *Sci. China C Life Sci.* 39(1), 37–44.
304. Ulisse, S., Farina, A. R., Piersanti, D., Tiberio, A., Cappabianca, L., D'Orazi, G., Jannini, E. A., Malykh, O., Stetler-Stevenson, W. G., and D'Armiento, M., et al. (1994). Follicle-stimulating hormone increases the expression of tissue inhibitors of metalloproteinases TIMP-1 and TIMP-2 and induces TIMP-1 AP-1 site binding complex(es) in prepubertal rat Sertoli cells. *Endocrinology* 135(6), 2479–2487.
305. Braghiroli, L., Silvestrini, B., Sorrentino, C., Grima, J., Mruk, D., and Cheng, C. Y. (1998). Regulation of alpha2–macroglobulin expression in rat Sertoli cells and hepatocytes by germ cells *in vitro*. *Biol. Reprod.* 59(1), 111–123.
306. Kon, Y., Endoh, D., and Iwanaga, T. (1999). Expression of protein gene product 9.5, a neuronal ubiquitin C-terminal hydrolase, and its developing change in sertoli cells of mouse testis. *Mol. Reprod. Dev.* 54(4), 333–341.
307. Giannattasio, A., Angeletti, G., De Rosa, M., Zarrilli, S., Ambrosiino, M., Cimmino, A., Coppola, C., Panza, G., Calafiore, R., Colao, A., Abete, O., and Lombardi, G. (2002). RNA expression bcl-w, a new related protein Bcl-2 family, and caspase-3 in isolated sertoli cells from pre-pubertal rat testes. *J. Endocrinol. Invest.* 25(7), RC23–25.
308. Anway, M. D., Wright, W. W., Zirkin, B. R., Korah, N., Mort, J. S., and Hermo, L. (2004). Expression and localization of cathepsin k in adult rat sertoli cells. *Biol. Reprod.* 70(3), 562–569.
309. Raychoudhury, S. S., and Millette, C. F. (1993). Surface-associated glycosyltransferase activities in rat Sertoli cells *in vitro*. *Mol. Reprod. Dev.* 36(2), 195–202.
310. Guma, F. C., and Bernard, E. A. (1994). Effects of retinol on glycoprotein synthesis by Sertoli cells in culture: dolichyl phosphomannose synthase activation. *Int. J. Androl.* 17(1), 50–55.

311. Holst, L. S., Hoffmann, A. M., Mulder, H., Sundler, F., Holm, C., Bergh, A., and Fredrikcon, G. (1994). Localization of hormone-sensitive lipase to rat Sertoli cells and its expression in developing and degenerating testes. *FEBS Lett.* 355(2), 125–130.
312. Hongo, S., Chiyo, T., and Sato, T. (1994). Induction of asparagine synthetase by follicle-stimulating hormone in primary cultures of rat Sertoli cells. *Arch. Biochem. Biophys.* 313(2), 222–228.
313. Lee, H., Ogawa, H., Fujioka, M., and Gerton, G. L. (1994). Guanidinoacetate methyltransferase in the mouse: extensive expression in Sertoli cells of testis and in microvilli of caput epididymis. *Biol. Reprod.* 50(1), 152–162.
314. Ulisse, S., Jannini, E. A., Carosa, E., Piersanti, D., Graziano, F. M., and D'Armiento, M. (1994). Inhibition of aromatase activity in rat Sertoli cells by thyroid hormone. *J. Endocrinol.* 140(3), 431–436.
315. Meroni, S., Canepa, D., Pellizzari, E., Schteingart, H., and Cigorraga, S. (1997). Regulation of gamma-glutamyl transpeptidase activity by Ca(2+)– and protein kinase C-dependent pathways in Sertoli cells. *Int. J. Androl.* 20(4), 189–194.
316. Magnanti, M., Giuliani, L., Gandini, P, Gazzaniga, P., Santiemma, V., Ciotti, M., Saccani, G., Frati, L., and Agliano, A. M. (2000). Follicle-stimulating hormone, testosterone, and hypoxia differentially regulate UDP-glucuronosyltransferase 1 isoforms expression in rat sertoli and peritubular myoid cells. *J. Steroid Biochem. Mol. Biol.* 74(3), 149–155.
317. Fujisawa, M., Yamanaka, K., Tanaka, H., Tanaka, H., Okada, H., Arakawa, S., and Kamidono, S. (2001). Expression of endothelial nitric oxide synthase in the Sertoli cells of men with infertility of various causes. *BJU Int.* 87(1), 85–88.
318. Lu, Q., Porter, L. D., Cui, X., and Sanborn, B. M. (2001). Ecto-ATPase mRNA is regulated by FSH in Sertoli cells. *J. Androl.* 22(2), 289–301.
319. Riera, M. F., Meroni, S. B., Schteingart, H. F., Pellizzari, E.H., and Cigorraga, S. B. (2002). Regulation of lactate production and glucose transport as well as of glucose transporter 1 and lactate dehydrogenase A mRNA levels by basic fibroblast growth factor in rat Sertoli cells. *J. Endocrinol.* 173(2), 335–343.
320. Selva D. M., Hirsh-Reinshagen, V., Burgess, B., Zhou, S., Chan J., McIssac, S., Hayden, M. R., Hammond, G. L., Vogl, A. W., and Wellington, C. L. (2004). The ATP-binding cassette transporter 1 mediates lipid efflux from Sertoli cells and influences male fertility. *J. Lipid. Res.* 45(6), 1040–1050.
321. Slegtenhorst-Eegdeman, K. E., Post, M., Baarends, W. M., Themmen, A. P., and Grootegoed, J. A. (1995). Regulation of gene expression in Sertoli cells by follicle-stimulating hormone (FSH): cloning and characterization of LRPR1, a primary response gene encoding a leucine-rich protein. *Mol. Cell. Endocrinol.* 108(1–2), 115–124.
322. Grima, J., Zhu, L., and Cheng, C. Y. (1997). Testin is tightly associated with testicular cell membrane upon its secretion by sertoli cells whose steady-state mRNA level in the testis correlates with the turnover and integrity of inter-testicular cell junctions. *J. Biol. Chem.* 272(10), 6499–6509.
323. Hennuy, B., Reiter, E., Cornet, A., Bruyninx, M., Daukandt, M., Houssa, P., N'Guyen, V. H., Closset, J., and Hennen, G. (2000) A novel messenger ribonucleic acid homologous to human MAGE-D is strongly expressed in rat Sertoli cells and weakly in Leydig cells and is regulated by follitropin, lutropin, and prolactin. *Endocrinology* 141(10), 3821–3831.
324. Liu, C., Gilmont, R. R., Benndorf, R., and Welsh, M. J. (2000). Identification and characterization of a novel protein from Sertoli cells, PASS1, that associates with mammalian small stress protein hsp27. *J. Biol. Chem.* 275(25), 18724–18731.
325. Magnanti, M., Gandini, P, Giuliani, L, Gazzaniga, P., Marti, H. H., Gradilone, A., Frati, L., Agliano, A. M., and Gassmann, M. (2001). Erythropoietin expression in primary rat Sertoli and peritubular myoid cells. *Blood* 98(9), 2872–2874.
326. Watanabe, M., Tsutsui, K., Hosoya, O., Tsutsui, K., Kumon, H., and Tokunaga, A. (2001). Expression of amphiphysin I in Sertoli cells and its implication in spermatogenesis. *Biochem. Biophys. Res. Commun.* 287(3), 739–745.
327. Chaudhary, J., and Skinner, M. K. (2002). Identification of a novel gene product, Sertoli cell gene with a zinc finger domain, that is important for FSH activation of testicular Sertoli cells. *Endocrinology* 143(2), 426–435.
328. Fujisawa, M., Okuda, Y., Fujioka, H., and Kamidono, S. (2002). Expression and regulation of gp130 messenger ribonucleic acid in cultured immature rat Sertoli cells. *Endocr. Res.* 28(1–2), 1–8.
329. Peoc'h, K., Serres, C., Frobert, Y., Martin, C., Lehmann, S., Chasseigneaux, S., Sazdovitch, V., Grassi, J., Jouannet, P., Launay, J. M., and Laplanche, J. L. (2002). The human "prion-like" protein Doppel is expressed in both Sertoli cells and spermatozoa. *J. Biol. Chem.* 277(45), 43071–43078.
330. Cunningham, D. B., Segretain, D., Arnaud, D., Rogner, U. C., and Avner, P. (1998). The mouse Tsx gene is expressed in Sertoli cells of the adult testis and transiently in premeiotic germ cells during puberty. *Dev. Biol.* 204(2), 345–360.
331. Shibayama-Imazu, T., Ogane, K., Hasegawa, Y., Nakaajo, S., Shioda, S., Ochiaia, H., Nakai, Y., and Nakaya, K. (1998). Distribution of PNP 14 (beta-synuclein) in neuroendocrine tissues: localization in Sertoli cells. *Mol. Reprod. Dev.* 50(2), 163–169.

Knobil and Neill's Physiology of Reproduction, Third Edition
edited by Jimmy D. Neill,
Elsevier © 2006

CHAPTER 20

Physiology of Testicular Steroidogenesis

Douglas M. Stocco[1] and Michael J. McPhaul[2]

Introduction, 977
Sources of Cholesterol for Steroidogenesis in Leydig Cells, 978
Enzymes Involved in Androgen Synthesis in Leydig Cells, 979
Cytochrome P450 Side Chain Cleavage (P450scc; CYP11A1), 980
3β-Hydroxysteroid Dehydrogenase (3β-HSD), 982
17α-Hydroxylase/17-20 Lyase (P450c17; CYP17), 982
17β-Hydroxysteroid Dehydrogenase (17β-HSD), 983
5α-Reductase, 983
Cytochrome P450 Aromatase (P450arom; CYP19), 984
Regulation of Leydig Cell Steroidogenesis, 984
History, 984
Steroidogenic Acute Regulatory Protein (StAR), 986
Other Proteins Involved in Acute Regulation of Steroidogenesis, 987
Role of the StAR Protein in Leydig Cells, 988
Transcription Factors Involved in Regulation of StAR Gene Expression, 992
Androgen Action, 995
Androgen Receptor and the Nuclear Receptor Family, 995
Androgen Receptor and the Regulation of Gene Expression, 997
AR and Alternate Signaling Pathways, 999
Responses to Androgen that Require the Participation of More than One Cell Type, 999
Studies of Androgen Action in the Testis, 1000
Selective Ablation of the AR in Specific Cell Types of the Testes, 1000
Acknowledgments, 1001
References, 1001

INTRODUCTION

Testicular steroidogenesis consists almost entirely of the production of androgens within the interstitial compartment of the testis (1,2). More specifically, androgens are produced by the approximately 500 million interstitial Leydig cells under the control of the pulsatile release of the pituitary gonadotroph, luteinizing hormone (LH) acting through the LH receptor on the surface of the Leydig cells (3–5). The most biologically important androgen synthesized in the testis is testosterone. However, testosterone itself has little direct biological action, first requiring conversion to its 5α-reduced metabolite, dihydrotestosterone or aromatization to estrogen prior to achieving biological activity. The episodic release of LH from the anterior pituitary is in turn controlled by the episodic release of the decapeptide gonadotropin releasing hormone (GnRH) from the hypothalamus into the hypophyseal-portal system where it is transported to the gonadotrophs of the anterior pituitary and binds to specific receptors and elicits the release of LH (3,5,6). LH released from anterior pituitary gonadotrophs reaches the testis via the bloodstream and interacts with specific LH receptors on the surface of the testicular Leydig cells. The transduction signaling pathway utilized following LH/LH receptor interaction is the G-protein–coupled pathway that results in a rise in intracellular cyclic AMP and the activation of protein kinase A (PKA) (7). The role of the LH receptor in Leydig cell proliferation, differentiation and steroid production is graphically demonstrated in

[1]Department of Cell Biology and Biochemistry, Texas Tech University Health Sciences Center, Lubbock;
[2]Department of Internal Medicine, University of Texas Southwestern Medical Center, Dallas, Texas.

LH receptor null mice (8–11). However, it has also been shown that during the course of fetal development in mice there is no requirement for LH action to maintain Leydig cell numbers or androgen production in the testis (8,10,11). Interestingly, it has recently been demonstrated that adrenocorticotropin hormone (ACTH) is also capable of regulating some Leydig cell functions such as steroidogenesis during fetal development, and indeed this may be the reason that LH is not required during this developmental period (12). In addition to the LH/cAMP signaling pathway, there is considerable evidence that other signaling pathways are also induced in LH stimulated Leydig cells and the specific pathways and the evidence for their existence has been reviewed (13).

While virtually all steroid production occurs in the interstitial compartment of the testis, small amounts of pregnenolone can be synthesized by the Sertoli cells within the seminiferous tubules, but it is not clear that this production has a physiological relevance (14). Also, Sertoli cells contain significant levels of aromatase activity that can convert androgen substrates into estrogens (15–18). As mentioned above, testosterone synthesized by the Leydig cells can be secreted into the circulatory system or it can be metabolized further in the Leydig cells to its 5α-reduced and highly active metabolite, dihydrotestosterone or aromatized to 17β-estradiol (19). Androgens produced by the Leydig cells are responsible for the development and maintenance of male sexual characteristics in the body such as the internal sex organs and external genitalia of the male reproductive system, male secondary sexual characteristics, the muscular and skeletal systems and for the maintenance of spermatogenesis (20).

Testosterone represents the major circulating androgen in the male and is synthesized by a sequence of enzymatic reactions that converts the substrate for all steroid biosynthesis, cholesterol, initially to the first steroid produced, pregnenolone, and then through a number of steroid intermediates to testosterone (19,21). These specific enzymatic reactions and the products so formed will be described later in this chapter.

SOURCES OF CHOLESTEROL FOR STEROIDOGENESIS IN LEYDIG CELLS

In order to assure adequate levels of steroid hormone synthesis in steroidogenic tissues, a constant supply of substrate cholesterol is required. While the sources of this cholesterol can vary between different species and even between different steroidogenic tissues within the same species, in general, steroidogenic cells have access to substrate cholesterol from one of three origins, namely: (a) from de novo cholesterol biosynthesis within the cell, (b) from the serum in the form of circulating lipoproteins that are then processed to yield either esterified cholesterol for storage in lipid droplets or free cholesterol for immediate steroid biosynthesis, or (c) from the hydrolysis of cholesterol esters stored within the lipid droplets of steroidogenic cells. The de novo synthesis of cholesterol involves the multi step conversion of acetate to cholesterol through a complex route that requires both cytosolic and membrane bound enzymes (22,23). Cholesterol synthesis involves the formation of mevalonic acid, squalene, lanosterol and finally cholesterol. The uptake of cholesterol from the serum requires the interaction of circulating lipoproteins with specific cell surface receptors designed to interact with these lipoproteins and internalize them either in whole or in part. Cholesterol esters can be delivered into the cells by two separate processes. One process is an endocytic uptake in which low density lipoprotein (LDL) or other apolipoproteins are bound to the LDL receptor and are rapidly internalized by endocytosis (24,25). In this process LDL binds to receptors located within clathrin-coated vesicles on the surface of the cell and the entire LDL-receptor complex is internalized by endocytosis. The LDL-receptor complex then dissociates and the receptor recycles back to the plasma membrane. The endosome containing the LDL is delivered to a lysosome where it fuses with this organelle and awaits degradation by proteases and lipases. Following degradation of the protein and lipid components of the LDL particle, the free cholesterol is now available either directly for steroid biosynthesis or it can be esterified and transferred to intracellular lipid droplets for later use as a substrate for steroid synthesis. This type of cholesterol ester uptake is a high affinity but low capacity uptake and is referred to as the "endocytic" uptake pathway (26–28). A second pathway utilized in the uptake of circulating cholesterol esters is known as the "selective" uptake pathway (29). Some steroidogenic cells use this pathway almost exclusively as a source of cholesterol ester for steroid synthesis. This pathway utilizes HDL-derived cholesterol esters for uptake through the HDL receptor, but its mechanism differs from that seen in the endocytic pathway. In steroidogenic cells, the HDL receptor is known as the scavenger receptor class B, type I receptor (SR-BI) and it is this receptor that functions in "selective" uptake. The SR-BI has been discovered within the past decade and is being characterized in several laboratories (30–35). This receptor has a wide tissue distribution with steroidogenic cells being only one type of cell in which it is expressed (36). Unlike the endocytic pathway in which the entire receptor-lipoprotein complex is internalized, the selective uptake pathway is quite different. The SR-BI receptor is a high capacity, bulk delivery receptor through which the steroidogenic cell is able to internalize far more of the cholesterol ester component than the lipoprotein component of the complex. This process occurs without the endocytic internalization and degradation of the receptor-lipoprotein moiety as occurs in endocytic uptake. The selective uptake pathway is prominently active in the steroidogenic tissues of species such as rodents, rabbits, cows, and humans (37). Unlike the endocytic pathway involved in cholesterol ester uptake from LDL, the selective uptake pathway is regulated by trophic hormone stimulation in adrenal cells (38–44), ovarian granulosa cells (31,34,45,46), and under certain experimental conditions, isolated Leydig cells (47). Thus, the source of cholesterol for steroid hormone biosynthesis is highly variable in steroidogenic cells. While the preferred route for the majority of tissues and species appears to be the selective uptake pathway, exceptions occur and the Leydig cell appears to be the leading exception.

The sources, the receptors, and the mechanisms involved in the uptake and utilization of substrate cholesterol is a large and complex topic and will not be covered in great detail in this chapter. Rather, only an overview of the sources of cholesterol that are utilized for testicular Leydig cell steroidogenesis will be given. Those wishing to see a more detailed description of cholesterol uptake and processing in steroidogenic tissues such as adrenal and ovarian tissues are directed to several recent excellent reviews on these subjects (36,37,48–50).

From the early work of Srere and colleagues, it was known that cholesterol derived from the de novo synthesis of this sterol from acetate could be found in steroid molecules in adrenal cells (51,52), an observation that was later confirmed in the testis (53). Thus, it was clear from early studies that steroidogenic cells could utilize cholesterol that was synthesized de novo within the cell as a substrate for steroids. Following these studies it was demonstrated that in the rat the primary source of cholesterol for androgen synthesis was endogenous (54). Thus, even in the presence of excess feeding with cholesterol or lipoprotein, rat testis preferentially utilized de novo cholesterol for steroid synthesis. This observation was confirmed in isolated rat Leydig cells treated with human LDL in which their rate of basal or stimulated testosterone synthesis was not increased by the addition of LDL (55). Similar results were found using mouse Leydig cells (56), and thus, it appeared that in rodent Leydig cells, the preferred source of cholesterol for steroid synthesis was derived endogenously from de novo synthesis. These observations are supported by the finding that Leydig cells of normal rats contain only very low levels of SR-BI, a finding that correlates well with the low selective uptake and utilization of HDL cholesterol by these cells (47), and further indicates that they are not dependent on exogenous lipoproteins as a source of cholesterol. However, the machinery to utilize other forms of cholesterol for steroid biosynthesis do exist in rodent Leydig cells. If Leydig cells are isolated from rats pretreated with desensitizing doses of hCG or depleted of intracellular cholesterol, addition of human LDL resulted in a significant increase in testosterone synthesis while the control cells showed no such increase (57,58). Also, even in rodents, the dependence on de novo cholesterol synthesis may be subject to additional qualifications. During the acute phase of stimulated steroid biosynthesis, de novo synthesized cholesterol appeared to be used exclusively, but on chronic stimulation of mouse and rat Leydig cells, LDL-derived cholesterol could be utilized for steroidogenesis (59–63). This observation is supported by the demonstration that chronic stimulation of rat Leydig cells resulted in an increase in SR-BI expression, HDL binding, cholesteryl ester uptake and testosterone synthesis (64). The situation observed in rodent Leydig cells is not seen in all species and, for example, in porcine Leydig cells it was observed that the majority of the cholesterol for steroid synthesis is derived from circulating lipoproteins and that the addition of either human or porcine LDL could increase both basal and hCG stimulated androgen production in these cells (65). Human fetal Leydig cells also appear to be capable of utilizing LDL for steroid biosynthesis in a manner similar to pig Leydig cells (66).

In summary it appears that most species do not utilize cholesterol present in circulating lipoproteins for Leydig cell steroid biosynthesis very efficiently, especially in the acute phase. Conditions such as gonadotropin-induced desensitization and chronic hormone stimulation can result in the expression of lipoprotein receptors and the uptake from the serum and utilization of lipoprotein cholesterol in Leydig cells, but under normal conditions these sources do not appear to be the major source of cholesterol for steroid production.

ENZYMES INVOLVED IN ANDROGEN SYNTHESIS IN LEYDIG CELLS

Leydig cells utilize two separate pathways in the synthesis of androgens, the type of pathway usually being species specific. The Δ4 pathway (cholesterol → pregnenolone → progesterone → androstenedione → testosterone) is found in most rodents including the rat and mouse (67–70). The Δ5 pathway (cholesterol → pregnenolone → 17α-hydroxypregnenolone → dehydroepiandrosterone → androstenedione → testosterone) is found in dogs (71), rabbits (72), pigs (73,74), humans and higher primates (75–79). Both the Δ4 pathway and the Δ5 pathway leading to testosterone production are shown in Fig. 1. While testosterone is the major steroid produced and secreted by the Leydig cell, significant levels of other steroids may also be produced through the further metabolism of testosterone. These include 17β-dehydrogenation to form androstenedione, 5α-reduction to form dihydrotestosterone, aromatization to form estrogen and 7α-hydroxylation to form 7α-hydroxytestosterone. Exceptions to these pathways have been noted in the tammar wallaby pouch young testes and in the immature mouse in which the synthesis of 5α-androstane-3α,17β-diol (androstanediol; 5α-adiol) can occur through two different pathways (80,81). In these studies it was found that one pathway for the formation of androstanediol involves the formation of testosterone and dihydrotestosterone as intermediates while the other involves the formation of 5α-pregnane-3α, 17α-diol-20-one and androsterone as intermediates. It was further found that the latter pathway utilizes 5α-reductase isoenzyme 1 in this pathway. The enzyme for the conversion of cholesterol to pregnenolone is found in the mitochondria while all other steroid biosynthetic and metabolic enzymes in the Leydig cell are found in the smooth endoplasmic reticulum compartment.

Cytochrome P450 Side Chain Cleavage (P450scc; CYP11A1)

The initial enzymatic reaction in the biosynthesis of all steroids in all steroidogenic tissues is the conversion of cholesterol to pregnenolone (82–86). This reaction occurs through the activity of the cytochrome P450 side-chain cleavage enzyme (P450scc; CYP11A1). This enzyme is a heme-containing protein that is located on the matrix-facing side of the inner mitochondrial membrane (87–90) and converts C27 cholesterol to the first steroid formed, C21 pregnenolone. In the testis, this enzyme is expressed almost exclusively in the Leydig cells (91), although under unusual conditions, low levels of expression can be seen in Sertoli cells (14). During the course of pregnenolone production, the C6 compound, isocaproaldehyde, is released and is subsequently oxidized to isocaproic acid (92). Conversion of cholesterol to pregnenolone occurs in three separate, successive and very rapid monooxygenation reactions (since only one atom of molecular oxygen appears in the product), beginning with the hydroxylation of C22 followed by the hydroxylation of C20 to yield 20,22R-hydroxycholesterol (93,94). The final step in the reaction is the cleavage of this hydroxylated intermediate to form pregnenolone and isocaproic acid. Thus, the three reactions consist of a 22-hydroxylase activity followed by a 20-hydroxylase activity and finally a C20-22-lyase activity to yield pregnenolone. While early studies indicated that the two hydroxylations and subsequent cleavage of substrate cholesterol were performed by separate enzymes, it is now clear that all these reactions are performed by one protein (95–98). This was demonstrated when it was determined that biochemically purified P450scc protein (95,96,98), as well as the recombinant protein expressed directly from P450scc cDNA (98), could convert cholesterol to pregnenolone. This observation was further strengthened when it was illustrated that the hydroxylated intermediates of cholesterol bind very tightly to the active site of the P450scc enzyme and it is only following the cleavage of the intermediate

FIG. 1. Steroid biosynthesis in Leydig cells. Testosterone biosynthetic pathway. This figure illustrates the steps present in the enzymatic conversion of cholesterol to testosterone. Once delivered to the cholesterol side chain cleavage cytochrome P450 enzyme, cholesterol is quickly converted into pregnenolone. Then, if the 5-ene pathway is employed, pregnenolone is converted to 17α-hydroxypregnenolone by the enzyme cytochrome P450 17α-hydroxylase/17,20 lyase and eventually to dehydroepiandrosterone by the same enzyme. If the 4-ene pathway is used, pregnenolone is converted to progesterone through the action of 3-β-hydroxysteroid dehydrogenase and then to 17α-hydroxyprogesterone and androstenedione by cytochrome P450 17α-hydroxylase/17,20 lyase. Androstenedione is then converted to testosterone by the action of 17-ketosteroid reductase. Testosterone can also be converted to dihydrotestosterone in target tissues by the enzyme 5α-reductase. In general, cholesterol side chain cleavage cytochrome P450 is found in the mitochondria and the remainder of the enzymes are found in the microsomal compartment of steroidogenic cells. However, there is now good evidence that 3β-hydroxysteroid dehydrogenase can also be found in the mitochondria of some species and some tissues. The enzymes shown are: P450scc: cytochrome P450 cholesterol side chain cleavage. 3βHSD, 3β-hydroxysteroid dehydrogenase/Δ5-Δ4Δisomerase; P450c17, cytochrome P450 17α-hydroxylase/17,20 lyase; 17KSR, 17-ketosteroid reductase/17β-hydroxysteroid dehydrogenase; P450arom, cytochrome P450 aromatase; 5αRED, 5α reductase.

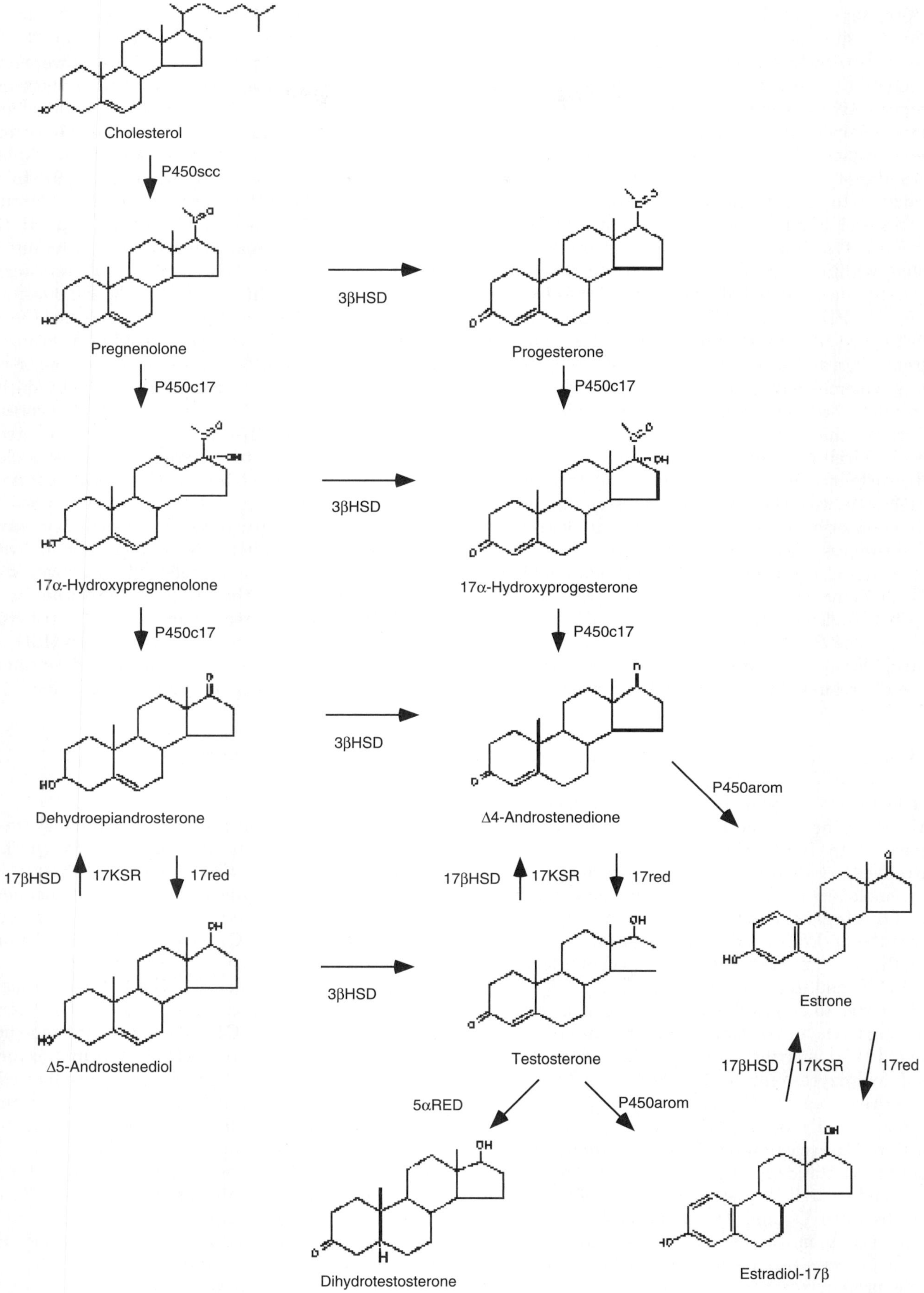

Cholesterol

P450scc

Pregnenolone

3βHSD

Progesterone

P450c17

P450c17

17α-Hydroxypregnenolone

3βHSD

17α-Hydroxyprogesterone

P450c17

P450c17

Dehydroepiandrosterone

3βHSD

Δ4-Androstenedione

P450arom

17βHSD 17KSR

17red

17βHSD 17KSR

17red

Δ5-Androstenediol

3βHSD

Testosterone

Estrone

17βHSD 17KSR

17red

5αRED

P450arom

Dihydrotestosterone

Estradiol-17β

to pregnenolone that the product is liberated from this site (84). The existence of a hydroxylated intermediate, while predicted, was most difficult to demonstrate because of its tight binding to the enzyme active site and it was only through the use of very large scale preparations that it was isolated (94). The formation of pregnenolone requires three molecules of oxygen and three molecules of NADPH in addition to the proteins adrenodoxin, an iron-sulfur protein and adrenodoxin reductase, a flavoprotein reductase, that function as a mini-electron transport system within the mitochondria (99). In these reactions, the adrenodoxin reductase accepts the electrons from NADPH and transfers them to adrenodoxin that in turn transfers them to the P450scc enzyme. Adrenodoxin and adrenodoxin reductase are proteins that are members of a more generic group of electron transport proteins termed ferrodoxins and ferrodoxin reductases that can donate electrons to a number of mitochondrial P450 proteins (100–103).

In early studies it was thought that the activity of P450scc was the rate-limiting step in steroid hormone biosynthesis (104,105). While under certain circumstances, such as in the presence of high levels of cholesterol substrate, it can indeed be rate-limiting, this is no longer thought to be the true rate-limiting step in steroidogenesis as will be discussed later in this chapter. The P450scc enzyme itself can be regulated by trophic hormone in steroidogenic cells, with LH being the regulating hormone in the Leydig cell.

3β-Hydroxysteroid Dehydrogenase (3β-HSD)

The second enzyme in the steroid biosynthetic pathway is the enzyme 3β-hydroxysteroid dehydrogenase/Δ4-Δ5 isomerase (3β-HSD) that catalyzes the conversion of pregnenolone to progesterone. Once pregnenolone is formed in the mitochondria, it diffuses out of this organelle to the microsomal compartment where the 3β-HSD resides. It should be noted, however, that in some cases, there are mitochondrial forms of 3β-HSD and in these instances, pregnenolone can be converted to progesterone in this organelle prior to diffusing to the microsomal compartment (106–111). The 3β-HSD enzyme catalyzes the conversion of Δ5-3β-hydroxysteroids to Δ4-3-ketosteroids. This is a required step in the biosynthesis of all steroid hormones and requires NAD^+ as a cofactor in the reaction. The reaction occurs in two steps, initially resulting in the dehydrogenation of the 3β-hydroxy group and secondly in the isomerization of the double bond between C5 and C6 to the position between C4 and C5. As in the case of P450scc, these separate enzymatic activities were found to reside within a single protein when using either purified protein preparations (112–116), or proteins obtained by expressing the products of the cDNAs for 3β-HSD (117–122). Since the discovery of the conversion of pregnenolone to progesterone by 3β-HSD, there have been a number of isoforms of this enzyme characterized. The cDNAs for these isoforms have been studied in the rodent and human and to date number as high as six isoforms in the mouse (19). In the human, two 3β-HSD isoforms have been identified (117–119,123–125) and it is 3β-HSD type II that is expressed in steroidogenic tissues including the Leydig cell and is thus involved in testosterone biosynthesis (124). In the rat, four 3β-HSD isoforms have been found and it appears that isoforms I and II (now called Ia and Ib) (126) are found in steroidogenic tissues with isoform Ia believed to be the major contributor to testosterone biosynthesis in the Leydig cell. In the mouse, 3β-HSD isoforms I and VI are expressed in the Leydig cell (127–129), and while the contribution of isoform I to testosterone synthesis is clear, much less is understood about the role of isoform VI in this process. It is of interest to note that the 3β-HSD isoforms can catalyze different activities with most being involved in the NAD^+-dependent dehydrogenase-isomerase activity referred to above, while others have shown the ability to function in the NADPH-dependent reduction of 5α-3-ketosteroids (129–131). Thus, some isoforms of 3β-HSD are involved in the biosynthesis of steroid hormones while others are involved in their inactivation.

17α-Hydroxylase/17-20 Lyase (P450c17; CYP17)

The next enzyme in the testosterone biosynthetic pathway is the cytochrome P450 17α-hydroxylase/17-20 lyase enzyme (P450c17; CYP17). This enzyme catalyzes two key reactions in steroidogenesis, namely, the 17α-hydroxylation of C21 steroids followed by the cleavage of the C17–C20 bond of C21 steroids. The latter reaction results in the conversion of the C21 steroid, pregnenolone, to the C19 steroid dehydroepiandrosterone (Δ5 pathway) or the C21 steroid progesterone to the C19 steroid androstenedione (Δ4 pathway). In these reactions 17α-hydroxypregnenolone or 17α-hydroxyprogesterone are formed as intermediates, respectively. The Leydig cell utilizes both these pathways for conversion depending on the species in question with rodents utilizing the Δ4 pathway while humans, primates, dogs, rabbits and pigs utilize the Δ5 pathway for testosterone production. The specificity for either the Δ4 or Δ5 pathway arises, at least in part, from the fact that the P450c17 enzyme appears to have different catalytic properties that are highly dependent on the species of origin. For example, using bovine P450c17, it was shown that while 17α-hydroxylation activity was similar with both pregnenolone and progesterone, the C17–C20 lyase was activity was much higher when 17α-hydroxypregnenolone was the substrate. Similarly, while the human enzyme can convert 17α-hydroxypregnenolone to dehydroepiandrosterone, it cannot convert 17α-hydroxyprogesterone to androstenedione, thus showing a highly specific C17–C20 lyase activity (132). This specificity is further shown in the rat and guinea pig in which the rat can readily convert both 17α-hydroxypregnenolone and 17α-hydroxyprogesterone to C19 steroids while the guinea pig has a much higher affinity for 17α-hydroxyprogesterone in this reaction (133). Thus, it is clear that while the dual activities of this enzyme have been well characterized, the degree of these activities will vary from species to species and will give rise to different products in those species. The reasons for this differing activity have been carefully studied and it has been determined that the presence of additional accessory proteins are highly involved in the process. These accessory proteins, a flavoprotein reductase named P450 oxidoreductase (OR), and a second protein cytochrome *b5* can determine which P450c17 activities will be enhanced. These studies showed that OR was required for both the 17α-hydroxylase and C17–C20 lyase activities, with the lyase activity being much more dependent on adequate levels of OR than the hydroxylase reaction (134). On the other hand, cytochrome *b5* can significantly and selectively increase lyase activity but does so only if the levels of OR are adequate and cannot support lyase activity by itself. Thus, in addition to the presence or the absence of the enzyme itself in steroidogenic tissues, it is the presence and ratio of the accessory proteins OR and cytochrome *b5* which determine the proportion of 17α-hydroxylase and C17–C20 lyase activities and the substrates these reactions prefer. Again, in a manner similar to previously described steroidogenic enzymes, while it was initially believed that two separate enzymes were responsible for the two reactions, it was subsequently determined that both the 17α-hydroxylation and C17–C20 lyase activities were housed in one protein (135–142).

Like other steroidogenic enzymes, the P450c17 enzyme is found in the microsomal compartment of steroidogenic cells. This enzyme requires a flavoprotein reductase to transfer electrons from NADPH to the enzyme (143). The hydroxylation reaction at C17 and the lyase reaction at C17–C20 are both monooxygenation reactions as seen previously with the P450scc enzyme (79,144). Other studies have suggested that both reactions occur within a single active site (137,145). It is of interest to note that the P450c17 enzyme can be considered as a qualitative regulator of steroidogenesis simply through the presence or absence of one or both of its activities in steroidogenic tissues (146). For example, in tissues such as adrenal glomerulosa, P450c17 is absent and thus, these cells cannot produce C19 steroids, instead producing C21, 17-deoxysteroids such as aldosterone. However, if the 17α-hydroxylase activity of the enzyme is present the products will be C21 hydroxysteroids such as cortisol. When both activities are present the result is both the hydroxylation and the cleavage of the C21 steroids to produce the C19 precursors of the sex steroids, which, of course, occurs in the Leydig cell.

17β-Hydroxysteroid Dehydrogenase (17β-HSD)

The final step in the biosynthesis of testosterone occurs via the activity of the enzyme 17β-hydroxysteroid dehydrogenase (17β-HSD), an enzyme also known as 17-keto-steroid reductase. In the testis, this enzyme catalyzes the readily reversible conversion of androstenedione to testosterone. The direction of the reaction is determined by the concentration of the substrates and products, with testosterone favoring the formation of more testosterone and androstenedione favoring the formation of more androstenedione (147,148). In the rat testis, 17β-HSD is found mostly in the Leydig cells, but significant levels of activity are also present in the seminiferous tubules (91). In addition to testicular Leydig cells, this enzyme has also been shown to be present in the theca interna cells of the ovary, further indicating its role in androgen synthesis (149). Three different isoforms of 17β-HSD have been found with the type 1 isoform demonstrating relatively wide distribution (150). The type 2 isoform is found in the placenta and prostate and perhaps other tissues as well (151), and the type 3 isoform is specific for the testis (152). The type 2 isoform is capable of performing both oxidation and reduction reactions in the interconversion of androstenedione and testosterone using NADH/NAD^+ as cofactors (151). Mutations in the type 3 enzyme in humans results in a complete loss only in testicular 17β-HSD activity further indicating that this isoform is the only form found in the testis (152,153).

5α-Reductase

In the testis, testosterone is the major androgen produced by the Leydig cells and is responsible for the normal maintenance of spermatogenesis (154). However, testosterone can also be converted to the more potent androgen, dihydrotestosterone (DHT), through the action of the enzyme 5α-reductase (155).

While the conversion of testosterone to DHT usually occurs in androgenic target tissues, it is well known that the Leydig cells are also able to produce 5α-reduced androgens from testosterone. Indeed, the 5α-reduced metabolites DHT and 5α-androstane-3α, 17β-diol (Adiol) appear to be the dominant androgens in the pubertal testis (156–158). The activity of 5α-reductase peaks at the time of puberty and declines thereafter reaching low levels in adulthood (159–164). It is thought that due to the increased potency of the 5α-reduced metabolites, spermatogenesis can occur normally in times of decreased testosterone levels in the testis, as occurs at the time of puberty (165–167). The irreversible conversion of testosterone to DHT can be accomplished by two isoforms of 5α-reductase, namely 5α-reductases types 1 and 2 (5αR1 and 5αR2) (155). These isoforms are encoded by different genes located on different chromosomes and, while highly homologous, they have distinctly different biochemical characteristics with regards to substrate affinities and pH activity profiles (168,169), and in the regulation of their expression (155,169–173). Expression of the two isoenzymes is differentially regulated by testosterone and follicle-stimulating hormone (FSH) in the rat testis (173). With regards to the identity of the 5α-reductase isoform responsible for the testicular conversion of testosterone to DHT during puberty, there has been some controversy on this issue (169,174–176). A more recent study (177) has concluded that both 5αR1 and 5αR2 are expressed in the testis during the peak of DHT biosynthesis and are both likely involved in the production of 5α-reduced androgens that support spermatogenesis during this crucial time. Interestingly, this study showed that while 5αR1 activity levels were tenfold higher than those of 5αR2, because the 5αR1 isozyme has a Km value that is in the micromolar range while the 5αR2 isozyme is in the nanomolar range, the contribution of the two isoforms would be approximately the same during puberty.

Thus, while the conversion of testosterone to the more potent DHT occurs in many target tissues and is responsible for the expression of androgen regulated genes in these tissues, in the testis it appears that this conversion occurs in the Leydig cells predominantly at the time of puberty and is responsible for the initiation and maintenance of spermatogenesis at a time of low testosterone levels in the testis.

Cytochrome P450 Aromatase (P450arom; CYP19)

The testis is also a source of estrogen. This steroid can be synthesized from androgens present in the testis through the action of the microsomal enzyme cytochrome P450 aromatase (P450arom; CYP19), which is present in a wide variety of tissues in addition to the testis (178-181). The aromatase enzyme is the product of the aromatase gene (*CYP19* gene), and together with the flavoprotein, NADPH-cytochrome P450 reductase, forms the aromatase enzyme complex. P450arom is approximately 58 kDa in size, is a member of the very large cytochrome P450 gene family and has been shown to contain a highly conserved heme binding region as well as a highly conserved substrate binding region (182). Unlike other steroidogenic enzymes, there appears to be only one gene coding for *CYP19* (183,184). The aromatase gene consists of nine coding exons and, in the case of the human, nine additional noncoding exons (185). This highly complex gene spans up to 120 kb and is regulated by tissue specific promoters and alternate splicing mechanisms in different tissues (184,186,187–190). Within the testis, aromatase expression has been shown to occur in Leydig cells, Sertoli cells, and germ cells (18,191–196). The testicular expression of aromatase has been demonstrated to vary with the age of the animals and the species in question (197), however, in the adult in many mammalian species, the bulk of aromatase expression appears to occur in the Leydig cells (193,194,198–203). Considerable evidence exists to illustrate that aromatase expression is significant in male germ cells (204). These studies indicated that aromatase expression was highest in pachytene spermatocytes and somewhat lower in round spermatids and spermatozoa (204–206). While the role of estrogen in the testis is not completely understood, it is informative that sterility results in humans with aromatase deficiencies (207–209). Also, the aromatase null mouse, while fertile for the first few months of age, begins to suffer from irregularities in spermatogenesis and becomes sterile after approximately 1 year, further indicating the role of estrogen for normal male reproduction (210). Similar problems with fertility are also observed in estrogen receptor alpha (ERα) null mice, an observation that provides further corroboration of the need for estrogen in male fertility (211). Thus, estrogen can be produced locally within several different cell types in the testis and, in an intracrine or paracrine manner, can function in the maintenance of spermatogenesis in the male.

REGULATION OF LEYDIG CELL STEROIDOGENESIS

History

Perhaps the most significant advances in our knowledge of the biosynthesis of testicular steroids that have occurred since publication of the previous volume of this series have been in the area of the regulation of steroid hormone biosynthesis. New information concerning the function and regulation of the expression of the steroidogenic enzymes involved in the steroid pathway has certainly been uncovered, but there has been little added that is significantly different from our previous knowledge of these reactions. At the time of the writing of the previous edition (21), one of the main mysteries that still existed in steroidogenesis was the manner in which this important pathway was acutely regulated by trophic hormone stimulation. The synthesis of all steroid hormones shares a common characteristic in that, regardless of the tissue of origin, they are all synthesized from a common precursor substrate, cholesterol. Thus, the biosynthesis of all hormonal steroids in response to steroidogenic stimuli begins with the cleavage of cholesterol to form the first steroid synthesized, the 21 carbon-containing molecule, pregnenolone. This reaction is catalyzed by the P450scc enzyme, that is part of the cholesterol side chain cleavage enzyme system (CSCC) located on the matrix side of the inner mitochondrial membrane (87,88,212,213). Once pregnenolone is formed, it may be metabolized within the mitochondria to progesterone by a mitochondrial form of the 3β-HSD enzyme (106–111), or it may exit the mitochondria and undergo further metabolism, by the microsomal steroid dehydrogenases and cytochrome P450 steroid hydroxylases described earlier in this chapter, with the final steroid hormone product being dependent upon the nature of the tissue in which the subsequent steps take place.

For many years it was the action of the P450scc enzyme in converting cholesterol to pregnenolone that was considered as the rate-limiting step in steroidogenesis. However, it became clear that the activity of the P450scc enzyme was not the rate-limiting step in this process (214), and that in order to initiate and sustain steroidogenesis, first, a constant supply of the substrate cholesterol for steroid biosynthesis must be available within the cell and second, a mechanism must exist for the delivery of this cholesterol to the site of cleavage in the inner mitochondrial membrane. Given adequate cholesterol supplies, two separate but equally important processes must occur in steroidogenic cells. The first is the mobilization of cholesterol from cellular stores such as lipid droplets or other cellular membranes to the outer mitochondrial membrane and the second is the transfer of this cholesterol from the outer to the inner mitochondrial membrane where the P450scc enzyme is located (82,215,216). The factors and mechanisms responsible for the mobilization of cholesterol to the outer mitochondrial membrane are thought to involve changes in cellular architecture and putative transport proteins but their mechanisms of action are not well understood. A number of reviews on the subject of intracellular cholesterol trafficking have been written over the last 20 years (82,217–220). More recently the role of steroidogenic acute regulatory protein (StAR)-related lipid transfer START domain proteins in this process have been hypothesized (221) but the generation of a START domain null mouse that functions normally in cholesterol homeostasis and steroidogenesis has cast some doubt on their involvement in this process (222).

While the action of the P450scc may indeed be considered as the rate limiting *enzymatic* step in steroidogenesis (104,105,223), there were a number of observations that indicated the true rate limiting step in this process was the delivery of cholesterol to the cholesterol deficient inner mitochondrial membrane and the P450scc enzyme (105,224–230). This became apparent when it was shown that hydroxylated analogs of cholesterol such as 22R-hydroxycholesterol, 20α-hydroxycholesterol, or 25-hydroxycholesterol, all of which can readily diffuse across the mitochondrial membranes to the P450scc, resulted in high levels of steroids in the absence of hormone stimulation when placed on steroidogenic cells (231–234). This indicated that the P450scc was fully active and that it was the lack of availability of cholesterol that prevented the biosynthesis of steroids. This led to the understanding that the barrier in the translocation of the hydrophobic cholesterol to the P450scc was the aqueous space between the outer and inner mitochondrial membranes since the diffusion of cholesterol through water is very slow (220,235,236). Thus, steroid synthesis is controlled by events that facilitate the transport of cholesterol from the mitochondrial outer membrane to the inner membrane and it is this process that is accepted as the rate-limiting step in hormone-regulated steroidogenesis. Regardless of the type of steroidogenic cell studied, the acute responses to trophic hormone stimulation share many of the same characteristics. The steroidogenic responses are usually dose-responsive to trophic hormone, increase with time and are sensitive to protein synthesis inhibitors. This last characteristic was one of the first and most fundamental observations concerning steroidogenesis and indicated that acute steroid production had an absolute requirement for the de novo synthesis of proteins. The first of such studies were performed by Ferguson (237,238) who made the prophetic conclusion "of the several possible explanations for the observed effects of puromycin, the most provocative but most difficult to prove is the idea that a specific protein must be synthesized in order for the adrenal to increase steroid output." Also, Garren and co-workers conducted a series of studies that demonstrated steroidogenesis was highly dependent on the synthesis of new proteins (239–242). They also observed that while steroid synthesis was dependent upon de novo protein synthesis, the conversion of cholesterol esters to free cholesterol was not (243), indicating that the hormonally controlled step was distal to cholesterol ester hydrolysis but proximal to its side chain cleavage, i.e., at the delivery of cholesterol to the P450scc enzyme. Following these observations many similar studies were confirmatory of the need for de novo protein synthesis in the hormone regulated, acute production of steroids (105,229,244–251). Studies on the testis by van der Molen and colleagues also confirmed the need for de novo protein synthesis in the stimulation of steroid production in rat Leydig cells and identified two possible protein candidates (252,253). The observation that new protein synthesis was indispensable for the acute production of steroids in response to hormone stimulation was also made more recently in several different steroidogenic tissues (254–259).

Simpson and Boyd (212), determined that the cycloheximide sensitive step in this process was located in the mitochondria and Arthur and Boyd (260), demonstrated that this inhibitor had no effect on the activity of the P450scc enzyme itself. Also, inhibition of protein synthesis had no effect on the delivery of cholesterol to the outer mitochondrial membrane, rather, it was the delivery of this substrate from the outer membrane to the inner mitochondrial membrane that was completely inhibited (229,261). As a result of such studies, the precise cycloheximide sensitive site had been pinpointed to the transfer of cholesterol to the P450scc enzyme. Thus, by compiling many observations a list of the characteristics to describe the acute regulation of steroidogenesis was generated and concluded that the acute regulation of steroid synthesis was dependent on a hormone stimulated, rapidly synthesized, cycloheximide sensitive and highly labile protein whose function appeared to be the mediation of cholesterol transfer from the outer to the inner mitochondrial membrane. The effort to identify and characterize this acute regulatory protein(s) was ongoing since the early observations of Ferguson and Garren and their colleagues. Several candidates emerged from these efforts, and a listing of these proteins and the data supporting their candidacies has been collectively reviewed (216), as has the characteristics for individual candidates (262,263). A summary of the events and the factors involved in the acute production of steroids in response to LH in the Leydig cell is shown in Fig. 2.

Steroidogenic Acute Regulatory Protein (StAR)

One protein proposed as the acute regulator of steroid biosynthesis was initially described by Orme-Johnson and colleagues as an ACTH-induced 30 kDa phosphoprotein in hormone-treated rat and mouse adrenocortical cells, and as an LH-induced protein in rat corpus luteum cells and mouse Leydig cells (254–257,264–267). They observed a close relationship between the appearance of the 30 kDa proteins and steroid hormone biosynthesis and determined that the synthesis of these proteins, as was steroidogenesis, was sensitive to cycloheximide. Identical proteins were later characterized in hormone-stimulated MA-10 mouse Leydig tumor cells (258,259,268–272). These proteins were found to be localized in the mitochondria and consisted of several forms of newly synthesized 37 kDa precursor and 30 kDa processed mature proteins (256,258). Since the initial observation of these proteins, there have been a number of studies in which correlations between the synthesis of steroids and the synthesis of the 37/30 kDa proteins were made (254–257,259,264–267,270–272). Shortly after this time, the cDNA for the 37 kDa protein was cloned from hormone-stimulated MA-10 mouse Leydig tumor cells and when compared with other sequences in the database both the nucleic acid sequence and protein sequence were found to be novel (273). Transient transfection experiments demonstrated that expression of the cDNA-derived protein in MA-10 cells resulted in a significant increase in steroid production in the absence of hormone stimulation. These results substantiated and extended the previous correlative studies and indicated a direct role for the 37/30 kDa proteins in hormone-regulated steroid production. The protein was named the steroidogenic acute regulatory (StAR) protein (273) and it appears to be the best candidate protein for the rapidly synthesized, acute regulator of steroidogenesis as proposed by Ferguson, Garren, and others.

The importance of StAR in the regulation of steroidogenesis has been dramatically demonstrated in studies on the human disease congenital lipoid adrenal hyperplasia (lipoid CAH). Lipoid CAH is a lethal condition that results from an almost complete inability of the newborn infant to synthesize steroids. This condition is manifested by the presence of large adrenals containing very high levels of cholesterol and cholesterol esters and also by an increased amount of lipid accumulation in testicular Leydig cells. Several proteins thought to be involved in cholesterol transport to the P450scc, the P450scc enzyme itself and auxiliary proteins required for its activity were studied and found to be normal in individuals with lipoid CAH (274,275). Therefore, it seemed possible that lipoid CAH may be due to a defect in StAR expression or function. Lin et al. (276) identified nonsense mutations in StAR cDNA prepared from the testicular tissue of two patients with lipoid CAH, representing for the first time an observation of a protein which was altered in this disease.

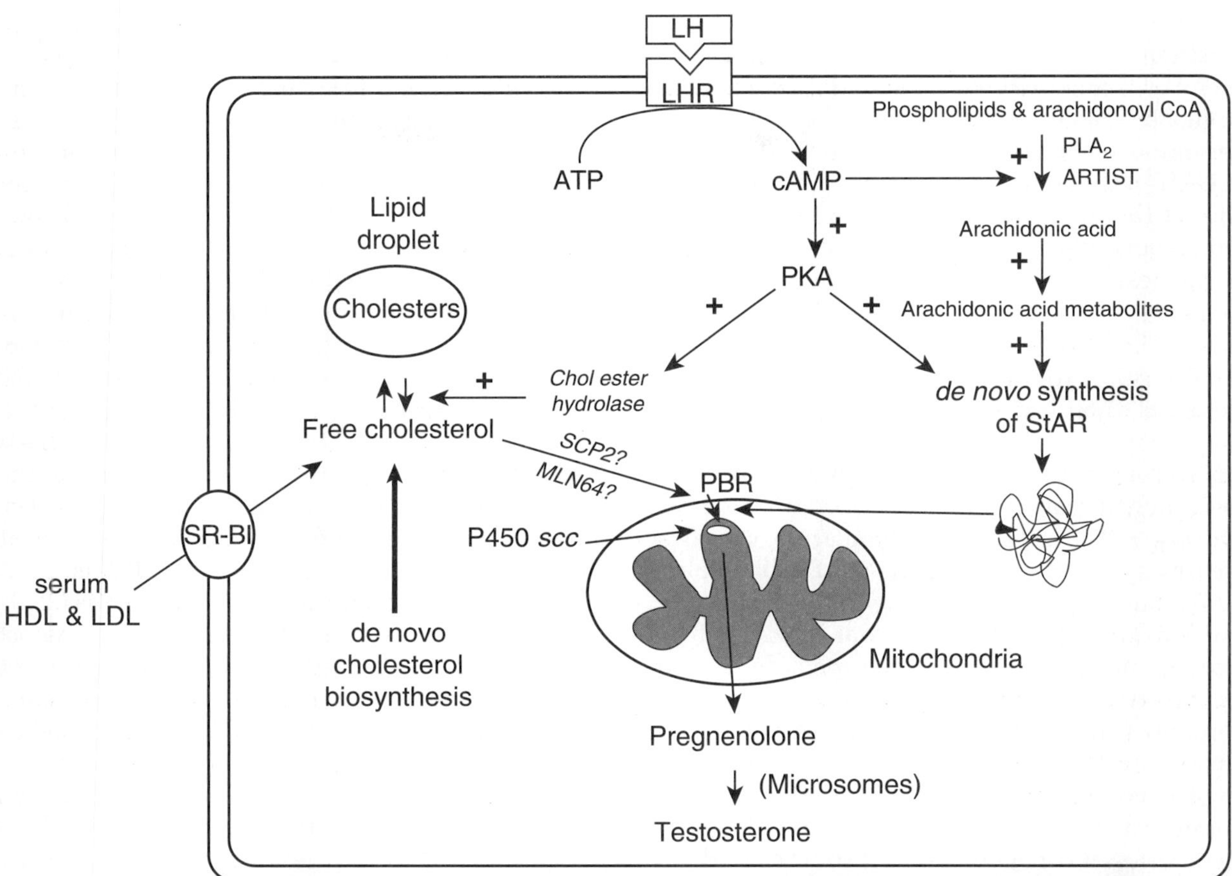

FIG. 2. Regulation of testosterone biosynthesis. This figure illustrates the factors involved in the biosynthesis and regulation of testosterone by the testicular Leydig cell. The sources of substrate cholesterol are mainly from one or more of three sources depending on the species and tissue in question. Serum LDL and HDL, de novo cholesterol biosynthesis and the utilization of cholesterol esters stored in lipid droplets all constitute sources of this steroid hormone precursor. In the Leydig cell, the predominant source of cholesterol for steroid synthesis is from de novo synthesis (*heavy arrow*). LH interaction with its receptor on the cell surface elicits an increase in intracellular cAMP. This in turn activates cholesterol ester hydrolase that results in the formation of free cholesterol. Additional, but not well understood, events result in the delivery of cellular cholesterol to the outer mitochondrial membrane. Various factors such as SCP2, MLN64, and cellular architectural components have been reported to be involved in the delivery of cholesterol to the mitochondria, but the exact mechanisms involved in this delivery have not yet been established. Further, cAMP action results in the liberation of arachidonic acid from both intracellular phospholipids and arachidonyl-CoA through the action of phospholipase A2 and an acyl-CoA thioesterase known as ARTISt. Arachidonic acid release and conversion to one or more of its metabolites are absolutely required in de novo steroid biosynthesis. The combination of PKA activity and arachidonic acid metabolites results in the de novo synthesis of the StAR protein that quickly interacts with the outer mitochondrial membrane and regulates the transfer of cholesterol from the outer mitochondrial membrane to the inner mitochondrial membrane and the P450scc enzyme where it is converted into pregnenolone. It is thought that PKA activity functions in the activation of transcription factors required for the transcription of the StAR gene, while the role of arachidonic acid metabolites is not yet known. The peripheral benzodiazepine receptor (PBR) is also required for cholesterol transfer, but again, the mechanism whereby it acts in cholesterol transfer is unknown. In addition, the conversion of pregnenolone to progesterone may also occur in the mitochondria of some species or tissues. Progesterone then freely diffuses out of the mitochondria to the microsomes where it is further converted to testosterone. In both the Leydig cell and in androgen responsive target tissues, testosterone can be converted to dihydrotestosterone, a biologically more potent form of this steroid.

This disease seems to affect people of Japanese and Palestinian ancestry disproportionately and one estimate places the carrier rate for the most common mutation in the Japanese population to be 1 in 200 (277). A synopsis of the data on lipoid CAH and mutations in the StAR gene has been compiled (277,278).

As further proof of the indispensable role of StAR in steroid biosynthesis, Caron et al. (279) used targeted disruption of the StAR gene in mice to successfully produce StAR null mice. Following birth, all animals failed to grow normally and death within a short period of time occurred, presumably as a result of adrenocortical insufficiency. This was confirmed by the observation that serum levels of corticosterone and aldosterone were depressed while levels of ACTH and CRH were elevated. These animals also demonstrated the buildup of lipid in the adrenals and testes as seen in the human condition. Thus, both lipoid CAH in the human and the phenotypic characteristics of the StAR null mouse clearly indicate the absolute requirement for the StAR protein in regulated steroidogenesis.

Other Proteins Involved in Acute Regulation of Steroidogenesis

The steroidogenesis activator polypeptide (SAP) was originally discovered and described by Pederson and Brownie (280). This small molecule was first purified from rat adrenal cells as a 2.2 kDa peptide (280,281), but was later determined to be a 30 amino acid (3.2 kDa) peptide when it was purified from rat Leydig tumor cells (282). This peptide was an extremely attractive candidate for the acute regulatory protein since it was found to be present only in steroidogenic cells, its levels could be acutely increased by trophic hormone stimulation and this increase was prevented by cycloheximide (283–285). All of these characteristics were consistent with those of the putative regulatory protein. Importantly, addition of SAP to isolated mitochondria was able to increase steroid production by four- to fivefold in a dose-dependent manner, thus indicating that SAP may play a role in cholesterol transfer within this organelle (280,281). SAP was found to be nearly identical to the carboxy terminus of a minor heat shock protein known as glucose related protein 78 (GRP 78) (286,287). To date, the role of SAP in the acute regulation of steroid biosynthesis has not been completely elucidated and there have been no recent studies on this interesting peptide.

Other proteins that have received considerable attention as being involved in acute steroidogenesis are the peripheral benzodiazepine receptor (PBR) and its intracellular ligand the diazepam binding inhibitor (DBI). Studies on the role of PBR in steroid hormone biosynthesis were initiated due to observations that the most pronounced "peripheral" expression of PBR was in steroidogenic tissues and that the highest levels of PBRs were localized to the mitochondrial outer membrane, although other cellular locations for this receptor have also been reported (288–290). Studies that lead to the proposed role for PBR in the intramitochondrial transport of cholesterol in steroidogenesis have been reviewed in detail (262,263,291,292). The endogenous ligand for PBR, DBI, was originally purified from bovine adrenal cells as an 8.2 kDa protein (293), and had an amino acid sequence that was identical to endozepine except for truncation of the carboxy terminal two amino acids (Gly-Ile) (294). The endozepines function to inhibit diazepam binding to the CNS receptor. However, DBI and des-(Gly-Ile)-DBI have also been shown to have stimulatory effects on steroid production in adrenal and gonadal cells, suggesting an involvement in steroidogenesis (295–302). The most convincing data which supported a role for PBR/DBI in steroidogenesis came from the ablation of DBI in MA-10 mouse Leydig tumor cells and in the R2C rat Leydig tumor cell using cholesterol-linked phosphorothioate antisense oligodeoxynucleotides to DBI. This ablation resulted in inhibition of the hormone-stimulated (MA-10), or constitutive (R2C), progesterone production in these cell lines (303,304). An early observation of Yanagibashi indicated the stimulatory effect of DBI at the mitochondrial membrane PBR receptor may be to induce the flux of cholesterol across the mitochondrial membranes (297). The authors report that the addition of DBI increased cholesterol levels in the inner membrane of isolated bovine adrenal mitochondria. These data were confirmed in studies that demonstrated that the cholesterol content of the inner mitochondrial membranes increased in Y1 adrenocortical cells stimulated with PBR agonists (305). Thus, DBI appeared to function at the acutely regulated step in steroidogenesis.

To investigate the regulation of DBI by trophic hormone, DBI levels were measured in rat adrenal and Leydig cells following hypophysectomy. DBI was shown to be decreased in both tissues but could be restored to near normal levels by trophic hormone treatment indicating its expression is dependent on pituitary hormones (306–308). However, DBI synthesis was not acutely up-regulated by hormones, was not sensitive to cycloheximide within in a time frame required to down-regulate steroidogenesis, and the protein had a half-life of greater than 3 hours, all characteristics that are inconsistent with the acute regulatory protein (309). In addition, it has been demonstrated that PK11195 (a PBR ligand)-stimulated pregnenolone production in mouse Y1 adrenocortical cells occurs in the presence of cycloheximide (298). Therefore, the authors concluded that while DBI/PBR function in steroidogenesis, they are not directly under acute hormonal control (309) and thus it would appear that these proteins are not the long sought labile protein factor(s) described by Ferguson (237,238) and Garren (239–242).

Ensuing studies focused on PBR and the effect of hormone on ligand binding to the receptor as the target for hormone action using the drug flunitrazepam (310). Flunitrazepam was shown to bind to PBR with high affinity and inhibit hormone-stimulated pregnenolone production in Y1 mouse adrenal cells, MA-10 mouse Leydig tumor cells, and in isolated mitochondria from these cell lines suggesting that DBI/PBR function is necessary for hormone-induced steroidogenesis. An insight into the possible mechanism of action was reported in studies that demonstrated hormone stimulation of MA-10 mouse Leydig tumor cells causes a very rapid (15 sec) cAMP-dependent increase in binding of ligand to PBR due to the exposure of a second higher affinity benzodiazepine-binding site (311). Flunitrazepam blocked this rapid response, presumably by blocking binding of DBI to this hormone-dependent, higher affinity, binding site. Interestingly, in R2C rat Leydig tumor cells which constitutively express steroids, it was reported that only the higher affinity PBR benzodiazepine-binding sites are present (304). Thus, it was suggested that hormone stimulation mediates a structural change in PBR that "exposes" the higher affinity benzodiazepine-binding site (312), and a model was proposed in which hormone stimulation mediates its effect via this alteration in PBR binding affinity (262,312,313). This structural change in PBR permitted increased binding of DBI and the formation of contact sites between the outer and inner mitochondrial membranes, events that could result in an increased flux of cholesterol to the inner mitochondrial membrane. Targeted disruption of the PBR by homologous recombination in R2C cells (314) resulted in a 90% decrease in constitutive steroid production that could be rescued by transfection of these cells with a PBR cDNA. Separate studies showed that hormone stimulation of steroidogenic cells resulted in a rapid reorganization of PBR molecules on the surface of the mitochondria to form pore-like structures through which cholesterol might pass (313,315). Computer modeling of the three dimensional structure of PBR with other mitochondrial proteins supports, in theory, the transport of cholesterol through such a pore structure (262). A more recent review has modeled the interaction of StAR and PBR in the outer mitochondrial membrane of luteal cells in which StAR binds cholesterol in the cytoplasm or in the outer mitochondrial membrane and then transfers it to the PBR for its subsequent delivery to the inner mitochondrial membrane (316).

In an in vivo study, it was shown that rats treated acutely with the benzodiazepine, diazepam, had higher levels of circulating corticosterone (317,318). It was also demonstrated that the effects on corticosterone levels were more pronounced if the animals were hypophysectomized (300,319). However, another study showed that treatment of female rats with differing PBR ligands resulted in either no change or a decrease in circulating estradiol levels, although this result may have been due to the concentration of ligand used and the length of treatment (317,318,320).

In summary, it appears that PBR is a required component in the transfer of cholesterol to the inner mitochondrial membrane. The exact manner in which it acts to aid in this transfer remains unknown, as does the entire mechanism of cholesterol transfer to the inner mitochondrial membrane.

Role of the StAR Protein in Leydig Cells

As indicated above, the single best candidate to fulfill the role of the putative acute regulator of steroidogenesis based on the earlier published descriptions of its characteristics is the StAR protein. Since the initial characterization of the StAR protein in MA-10 cells, and with the availability of new reagents such as antipeptide antisera and cDNAs to StAR, subsequent studies have also demonstrated close correlations between its presence and steroid hormone biosynthesis in other experimental systems. The details of the acute regulation of steroidogenesis and the role that StAR plays in this regulation as well as speculation on its mechanism of action have been the subject of a number of review articles (216,321–330) and much of what has been written in those reviews will not be repeated here. Rather, this chapter will attempt to confine its scope to what is known about StAR in the Leydig cell as much as possible, though descriptions in other steroidogenic cells, such as the section describing the transcription factors involved in the regulation of the StAR gene, are informative and have thus been utilized.

In early studies using MA-10 Leydig tumor cells, it was clearly demonstrated that both StAR mRNA and protein could be rapidly induced by treatment with cAMP analog, an indication that this induction occurred through the LH signal transduction pathway (331,332). These studies lead to the speculation that the mechanism of hormone action resulting in StAR expression was due to a cAMP-induced change in StAR transcription and/or mRNA stability (331,333). In another study it was determined that while ongoing translation of the StAR protein did not require new transcription, the continued translation of StAR and steroid hormone production was dependent on continued transcription of the StAR gene (334). This study also found that stability of the StAR mRNA played very little role in the mechanism of the acute regulation of steroids in MA-10 cells. Similar results were obtained using primary cultures of rat Leydig cells and demonstrated that expression of StAR increased in response to trophic hormone stimulation in the same manner as it did in Leydig tumor cells (335,336). In another study using the estrogen receptor-α null mouse as a model, it was demonstrated that serum LH levels were approximately threefold higher than in wild type animals and both serum testosterone and Leydig cell StAR levels were significantly elevated as were other enzymes in the steroidogenic pathway (337). Price et al. (338) followed LH levels, testosterone secretion and testicular StAR expression in adult rams during the breeding and nonbreeding seasons. They found that during the breeding season, serum LH was significantly higher than during the nonbreeding season and this increase was concomitant with an increase in testosterone secretion and StAR mRNA expression, again indicating that StAR expression is regulated by the Leydig cell trophic hormone, LH. The precise spatial and temporal relationship between the presence of StAR transcripts and the capacity to produce steroid hormones was also confirmed during mouse embryonic development (331). In situ hybridization of StAR mRNA in embryonic mice clearly demonstrated that the timing and localization of StAR expression was similar to that previously shown for P450scc and the orphan nuclear receptor steroidogenic factor 1 (SF-1). In these studies, StAR was detected in the developing adrenal and testis and was specifically confined to the Leydig cells in the developing testis (331).

Analysis of the StAR mRNA reveals three specific transcripts of 3.4, 2.7, and 1.6 kb in mouse Leydig cells (331). While the functional significance of the different sizes of the transcripts is not yet known, Kim et al. (339) provided some initial discussion on this subject. The different sizes of the transcripts were the results of extended 3′ ends of the StAR mRNA and not alternate splicing as was thought to be a possibility. It was also determined that maintenance of optimal steroid synthesis requires not only ongoing translation of the StAR mRNA, but active transcription of the StAR gene as well (334).

In addition to LH, many factors have been shown to increase steroid production by testicular Leydig cells. In some cases, these factors can act independently while in other cases they act in concert with trophic hormone. These factors include growth factors, cytokines, ions and a variety of other compounds. Thus, it was demonstrated that insulin-like growth factor was able to further enhance chorionic gonadotrophin-induced testosterone synthesis and both StAR mRNA and protein expression in rat Leydig cells, indicating that the regulation of the StAR gene is complex (340). Ramnath et al. (341) demonstrated that stimulation of MA-10 mouse Leydig tumor cells in the absence of chloride resulted in a significant increase in steroid production as well as a significant increase in StAR protein expression. Importantly, this experimental paradigm had no effect on the expression of either P450scc or 3β-HSD indicating that StAR was predominantly responsible for the increase in steroid observed. In another study with MA-10 cells an increase in steroid production and StAR protein expression was found to occur in a dose-dependent manner in response to treatment with corticotropin-releasing hormone (CRH) (342), indicating that Leydig cells contained receptors for this releasing hormone and these receptors were connected to the steroidogenic pathway and StAR expression. The supporting role of extracellular Ca^{2+} in trophic hormone stimulated steroid production in cultured mouse Leydig tumor cells was shown in experiments demonstrating that while extracellular Ca^{2+} had no effect by itself, in the presence of hCG it resulted in significant increases in both steroid production and StAR mRNA expression (343). The role of yet another hormone in regulating steroid biosynthesis was shown in a series of interesting experiments by Manna et al. (344–346) who described the presence of a thyroid-gonadal axis by demonstrating that treatment of both mouse primary Leydig cells and mouse mLTC-1 Leydig tumor cells with thyroxine (T3) resulted in a dose-dependent increase in steroid production and StAR expression. They also found that prolonged treatment with T3 caused a subsequent inhibition of both steroid production and StAR expression within a similar time frame. These studies showed that StAR expression could be mediated by the action of T3 on the StAR promoter most likely acting in concert with the transcription factor steroidogenic factor 1 (SF-1).

Retinoic acids (RA) are known to induce steroid biosynthesis in mouse Leydig cells. Lee et al. (347) showed that treatment of K28 mouse Leydig tumor cells with RA resulted in a dose- and time-dependent increase in steroid synthesis and StAR mRNA expression. They further showed that RA acted at the level of the StAR promoter to control StAR gene transcription. In another study in rat Leydig cells it was interestingly demonstrated that hCG-induced steroidogenesis and StAR gene transcription and translation could be further enhanced by treatment of the cells with the amino acid D-aspartate (348). The mechanism for this observation remains unknown but represents the first observation of an amino acid having an effect on gene expression in mammalian cells. It has been demonstrated that gonadal development and differentiation is partly dependent on the action of growth hormone (GH). Kanzaki and Morris (349) showed that GH treatment of progenitor Leydig cells from 21-day-old rats resulted in an increased production of androgens and the concomitant increased expression of both StAR and 3β-HSD mRNA and protein, indicating these genes are regulated by GH during early Leydig cell development. In a similar study in mLTC-1 mouse Leydig tumor cells it was shown that treatment with epidermal growth factor (EGF) caused an increase in both steroid production and StAR mRNA and protein in a dose- and time-dependent manner (350). Once again it is clear that regulation of the expression of StAR is complex and appears to be able to respond to a large number of factors that can increase both its expression and steroid production. Demonstrating yet another link between a steroidogenic agent and StAR expression, it was illustrated that the steroid end product in MA-10 cells, progesterone, was able to increase StAR expression in these Leydig tumor cells (351). Of further interest in this study was the observation that progesterone appeared to exert this effect on StAR expression acting through a non-classical steroid receptor, thus presenting the exciting possibility that a membrane-bound steroid receptor is directly involved in the regulation of steroid biosynthesis and StAR expression. Continuing to add to the complexity of the signal transduction systems involved in the production of steroids Wang and colleagues have demonstrated that in addition to the cAMP second messenger pathway, another signal transduction pathway requiring the release of intracellular arachidonic acid is an absolute requirement for regulated steroid hormone biosynthesis (352–356). In these studies they demonstrated that in MA-10 Leydig cells release of arachidonic acid and its further metabolism through the lipoxygenase pathway is required to obtain full steroid biosynthesis in response to hormone stimulation. Importantly they illustrated that activation of this pathway was also required for the transcription of the StAR gene and its translation into StAR protein. Further studies indicted that the 5-lipoxygenase metabolites 5-hydroperoxyeicosatetraenoic acid (5-HPETE) and 5-hydroeicosatetraenoic acid (5-HETE) were the compounds responsible for the observed effects on steroid induction and StAR expression. A potential mechanism responsible for the increase in lipoxygenase metabolites was uncovered when it was demonstrated that inhibition of the cyclooxygenase-2 (COX2) pathway resulted in a marked stimulation of steroid biosynthesis and StAR expression at low levels of intracellular cAMP (355).

In summary, while steroid production and StAR are most assuredly up-regulated in the testis by the trophic hormone LH, it is now clear that many additional factors can also regulate and modulate these activities in Leydig cells. The physiological relevance of many of these other factors remains to be determined, but the potential for fine tuning of steroid production and StAR expression is readily observed in the Leydig cell.

In contrast to stimulation of steroid production, a number of agents have been shown to result in a decrease in testosterone synthesis and secretion in the testis. Because of the key role of StAR in regulated steroidogenesis, many studies have focused on the effects of these agents on StAR expression. One of the first of such studies demonstrated that both steroidogenesis and StAR expression were significantly inhibited in mouse MA-10 Leydig tumor cells by dimethylsulfoxide (272). Interestingly, this inhibition did not occur in rat R2C Leydig tumor cells and the reasons for this difference are as yet unknown. Diethylumbelliferyl phosphate is a cholesterol ester hydrolase inhibitor that blocks steroidogenesis by preventing cholesterol transport into the mitochondria. Using MA-10 Leydig cells it was illustrated that this compound acted by inhibiting the expression of the StAR protein (357). In an interesting series of experiments injection of the bacterial endotoxin lipopolysaccharide (LPS) into mice resulted in a decline in serum testosterone levels within 2 hours (358). This rapid decline was caused by a parallel decline in StAR expression in these animals. In a subsequent study this same group demonstrated that while testosterone levels were inhibited by LPS, adrenal steroid levels were increased and that this increase was probably a result of the concomitant increase in adrenal StAR protein expression while the levels of other steroidogenic enzymes remained unchanged (359). Similar findings were reported in a model system in which sepsis was induced in male rats and resulted in a decrease in serum testosterone levels to almost undetectable levels and a parallel decrease in StAR protein levels (360).

It was known that heat shock treatment resulted in a decrease in steroid biosynthesis in steroidogenic cells and appeared to act at the level of cholesterol transfer to the P450scc enzyme (361,362). In MA-10 Leydig tumor cells it was subsequently demonstrated that heat shock resulted in an essentially complete inhibition of StAR protein expression (363). In a later study, Murphy et al. (364) clearly demonstrated that the inhibition of StAR protein expression by heat shock occurred at the level of transcription through the disruption of the binding of nuclear proteins (presumably transcription factors) to the proximal region of the StAR promoter. It has also been reported that the major cytokine, interleukin-1 (IL-1) can inhibit hormone stimulated steroid biosynthesis in testicular Leydig cells. To determine if StAR played a role in this inhibition, Lin et al. (365) treated primary cultures of rat Leydig cells with IL-1 and observed that this treatment had no affect on the expression of StAR mRNA or protein, but that the expression of the P450scc enzyme was severely inhibited. In a contrasting study, Ogilvie et al. (366) showed that the intracerebroventricular injection of IL-1 resulted in a significant time- and dose-dependent decrease in testosterone production in testes isolated from male rats and also resulted in a time-dependent decrease in StAR protein expression.

The reasons for the discrepancies in the two studies are not known, but the route of IL-1 treatment could certainly be a factor. Tumor necrosis factor alpha (TNF-α) has also been shown to inhibit testicular steroid biosynthesis. Studies in porcine, mouse tumor and rat Leydig cells all demonstrated that treatment with TNF-α resulted in an inhibition of both steroid production and StAR expression, indicating that this cytokine has a direct affect on the Leydig cell itself (367–369). Importantly, in two of these studies (367,368) it was shown that addition of hydroxylated cholesterol analogs bypassed the inhibitory effects of TNF-α indicating that inhibition of StAR was primarily responsible for the effects on steroidogenesis. Other cytokines have also been shown to inhibit steroidogenesis in Leydig cells. Lin et al. (370) showed that interferon-γ blocked steroid production and StAR mRNA and protein levels in cultured rat Leydig cells and Mauduit et al. (371) obtained essentially identical results in cultured porcine Leydig cells in response to the cytokine leukemia inhibitory factor (LIF). The compound ethane dimethyl sulfonate (EDS) can induce apoptosis, inhibit steroidogenesis in cultured Leydig cells and selectively kill Leydig cells when injected into mature male rats. It was illustrated that EDS could concomitantly inhibit both steroid production and StAR protein expression when added to cultures of MA-10 Leydig tumor cells (372). In a somewhat puzzling study, King et al. (373) demonstrated that the disruption of one component of the mitochondrial electrochemical gradient (the ΔpH) resulted in a decrease in StAR protein expression, but curiously in no decrease in progesterone production in MA-10 cells. The reasons for this discrepancy are unclear but may be related to the possibility that while disruption of the ΔpH has no effect on StAR synthesis, the mitochondria cannot import and process StAR in the absence of an electrochemical gradient. Without import and processing of StAR, cholesterol transfer cannot occur and would result in the inhibition of steroid synthesis observed.

The indole derivative melatonin is secreted from the pineal gland and has been shown to have inhibitory effects on the reproductive capacity of some mammals. It was demonstrated that treatment of MA-10 Leydig cells with melatonin inhibited hormone-stimulated steroid synthesis and did so by decreasing StAR expression (374). Leptin is an adipocyte-derived hormone that plays a key role in body weight control. It has also been shown to be a factor in regulating the reproductive axis. In adult male rats it was shown that treatment of testicular tissue slices with leptin resulted in an inhibition of testosterone secretion (375). In addition to the inhibition in steroid synthesis leptin resulted in a decrease in the expression of several genes implicated in the steroid pathway, namely, SF-1, P450scc, and StAR. The relative contributions of these factors are difficult to assess but it is clear that without StAR expression steroid synthesis is severely compromised. Propylthiouracil (PTU) is a thioamide drug used to inhibit thyroid hormone production. However, it has also been observed that PTU can interfere with the reproductive axis and inhibit testosterone production in males. Chiao et al. (376) showed that treatment of rat Leydig cells with PTU resulted in a decrease in steroid synthesis and a corresponding decrease in P450scc activity and StAR mRNA expression. The imidazole antifungal drugs econazole and miconazole are commonly used to treat clinical fungal infections and their use has been shown to be associated with decreased steroid production in patients treated with them. Walsh et al. (377) showed that both of these antifungal drugs caused a decrease in steroid production and StAR protein expression in MA-10 cells. Interestingly, these agents had no effects on StAR transcription indicating that the inhibition occurred at the level of translation. Lastly, it has also been demonstrated that exposure of adult male rats to ethanol resulted in a significant decrease in serum levels of testosterone (378). This decrease was paralleled by a decrease in the expression of StAR mRNA in the testes of these animals leading to the conclusion that it was this decrease that was the cause of the decreased serum testosterone. From the studies outlined above, it is clear that testicular steroidogenesis is affected by many factors in addition to LH, and that many of these factors also all serve to regulate StAR expression in the Leydig cell.

Transcription Factors Involved in Regulation of StAR Gene Expression

This section will focus on what has so far been determined with regards to the transcription factors that both positively and negatively regulate the StAR gene. Many putative regulatory elements previously identified and characterized by the presence of consensus binding sequences in various gene promoters have been found in the StAR promoter. In addition, several transcription factors have been shown to be involved in the activity of the StAR promoter in spite of the lack of demonstrated consensus binding sequences. It is clear that proteins that can both bind directly to the StAR promoter and those that may have an indirect role serve to regulate expression of this important gene.

Steroidogenic factor-1 (SF-1) is a member of the steroid hormone receptor family of nuclear transcription factors and was first identified in adrenal cortical cells (379). SF-1 knockout mice revealed an essential role for SF-1 in the development of steroidogenic and other cells types (380–382). In addition, SF-1 knockout mice do not express StAR mRNA in the urogenital ridge during embryonic development in the mouse, suggesting that SF-1 is required for appropriate StAR gene expression (383). Several binding sites for SF-1 have been found in the StAR promoter (383–385), and SF-1 can efficiently transactivate the StAR promoter in transient transfection assays in numerous cell types (383–388). However, although SF-1 plays an important role in StAR gene expression, it is certainly not the sole factor responsible for cAMP-dependent regulation of StAR gene transcription, since mutations in any of the SF-1 binding sites decreased, but did not abolish, basal- and cAMP-mediated StAR reporter activity.

CCAAT/enhancer binding proteins (C/EBP) are a family of basic region/leucine zipper transcription factors involved in the differentiation and function of multiple cell types (389,390). Two family members, C/EBPα and C/EBPβ, are expressed in steroidogenic cells, including Leydig cells and ovarian granulosa cells (335,391,392). Both C/EBPα and C/EBPβ are phosphoproteins and are targeted by PKA (393). In addition, C/EBPβ is a component of LH signaling pathways in the ovary and is required for efficient ovulation (394,395). Based on homology to consensus C/EBP elements, two putative C/EBP binding sites were found in the *StAR* promoter. One, and perhaps both, may be bound by C/EBPβ in MA-10 nuclear extracts. Additionally, the StAR promoter is transactivated by C/EBPβ in transient transfection assays and SF-1 transactivation of the StAR promoter is dependent on the presence of functional C/EBP binding sites, suggesting that SF-1 and C/EBPβ may form a complex on this promoter. However, cAMP-inducibility of the StAR promoter was relatively unaffected by mutations in either or both C/EBP sites, indicating that C/EBPβ may be involved in developmental and basal level expression rather than acute, cAMP-dependent activation of the StAR gene, at least in Leydig cells.

Sp1 is a member of the zinc finger family of transcription factors (396). An Sp1 site centered at –155 and another at –1,156 have been identified in the human StAR promoter (397). Sp1 is involved in the expression of multiple steroidogenic genes (398–401). Sp1 has been shown to interact with a number of transcription factors, that may participate in the regulation of the StAR gene, including the estrogen receptor (402), and C/EBPβ (403,404). Also, Sp1 has been shown to potentiate the binding of sterol regulatory element binding protein 1a (SREBP-1a) and subsequently increase mouse StAR promoter activity in SL2 insect cells (405). Mutation of the Sp1 site resulted in an attenuation of SREBP-1a-mediated activation of rat StAR promoter responsiveness in HTB-9 cells (405). Taken together, these findings indicate that Sp1 plays an active role in StAR gene expression. Another transcription factor involved in StAR expression is GATA-4. Silverman and colleagues (406) conducted StAR promoter assays using FSH-induced primary granulosa cell cultures from prepubertal rat ovaries. This led to the identification of two trans-acting proteins, C/EBPβ and GATA-4, which were required for the transcriptional activation of the StAR promoter. In addition to the consensus binding sequence for C/EBPβ, a nonconsensus binding site for this transcription factor was also located in the StAR promoter 10 nucleotides upstream. Site-directed mutagenesis of both binding sequences essentially ablated both basal and hormone driven promoter activities, reinforcing the notion that these two binding elements are required for the FSH induced transactivation of the StAR promoter in ovarian cells. Western analyses showed that GATA-4 was constitutively expressed in granulosa cells and that three isoforms of C/EBPβ were rapidly induced by FSH. This suggests that GATA-4 may play a permissive role and C/EBPβ a more active and regulated role in the acute rate of StAR transcription (384,406).

SREBP is yet another transcription factor shown to be involved in the regulation of the StAR gene. The 5′-flanking region of the rat StAR gene contains recognition motifs for five SREBP binding sites. Studies have shown that SREBP-1a is capable of binding to and trans-activating both the rat and human StAR promoters (405,407,408). Mutation of the SREBP binding sites has been demonstrated to decrease mouse, rat, and human StAR reporter activity (405,408,409). Over-expression of SREBP-1a, SREBP-1c, or SREBP-2 in the presence of a rat StAR promoter increases reporter responsiveness in HTB-9 human bladder cells (405). SREBP-1a has also been demonstrated to increase human StAR promoter activity in COS-1 monkey kidney and granulosa-lutein cells (408). It has been suggested that the human StAR promoter is conditionally responsive to SREBP-1a. However, SREBPs are known to be weak transcriptional activators and may require co-activators such as Sp1 and/or nuclear factor-Y (NF-Y) for maximal response. Nonetheless, studies performed to date indicate that SREBPs are involved in StAR gene expression.

Sugawara et al. (410) demonstrated that the arylhydrocarbon receptor (AhR) may be involved in the expression of the StAR gene. They found that transfection of Y-I adrenal tumor cells with AhR and its heterodimer binding partner the arylhydrocarbon nuclear translocator (ARNT), resulted in an increase in the activity of the StAR promoter when ligand for the AhR was added to the cells. Since the AhR can act as a receptor for toxins such as dioxin, these results indicate that certain environmental endocrine disruptors may have an effect on the expression of the StAR gene acting through the AhR receptor.

Transcriptional activation by cAMP is mediated through the interaction of the cyclic AMP response element binding protein (CREB) with a consensus CRE found in the promoter of target genes (411–413). Members of the CREB/CREM family, which includes activating transcription factor 1 (ATF-1), are distinguished by their DNA-binding bZip domains and mediate part of the transcriptional response to cAMP signaling (411,414,415). While the StAR gene lacks a consensus CRE, we recently identified three perfect 5′-CRE half-sites within the cAMP responsive region of the mouse StAR promoter (409). These three CRE elements were analyzed to determine their potential involvement in StAR expression and steroidogenesis, and it was found that CREB/CREM was indeed involved in StAR regulation and that the CRE2 site appeared to be of greater importance than the CRE1 or CRE3 sites (409,416). These studies demonstrated that the CRE2 sequence could bind recombinant CREB as well as nuclear extract from MA-10 and Y-1 cells, and that the binding was decreased by consensus CRE and AP-1 sequences (416). These results demonstrated the importance of the CRE2 site in StAR gene expression, but did not rule out the possibility that each CRE element may mediate part of the response.

The AP-1 transcription factor family (Fos and Jun) are members of the bZip family, share many properties, and have been demonstrated to regulate several biological functions (417,418). Three putative AP-1 elements have been identified in the rat StAR promoter (419). Fos and Jun were shown to bind to these sites and it was further demonstrated that c-Fos was able to decrease basal, cAMP and c-Jun induced rat StAR promoter responses in Y-1 cells (419). The CRE2 half-site that is predominantly involved in mouse StAR gene expression is analogous to an AP-1 sequence motif (409,416), and on closer analysis of CRE and AP-1 consensus sequences, it was observed that this AP-1 element was more highly conserved than was the CRE2 element. Moreover, Fos and Jun were found to influence basal StAR gene transcription to varying degrees, but both resulted in repression of cAMP mediated responses (420). Oligonucleotide probes containing AP-1 and/or CRE binding motifs readily bind to MA-10 and Y-1 nuclear extract, and the resulting protein-DNA complexes were significantly affected by addition of consensus AP-1 and CRE sequences and by mutation of the target sites. Antibodies specific to Fos and Jun members demonstrated the involvement of several AP-1 family member proteins, especially c-Fos and Fra-2, in StAR gene expression. In addition, alteration of the AP-1 binding site in combination with other mutated cis-elements that reside within the –151/-1 region of the mouse StAR promoter, resulted in a further decrease in StAR reporter responsiveness, indicating a functional cooperation of these factors in StAR gene expression (420). The involvement of Fos and Jun in epidermal growth factor mediated attenuation of mouse StAR promoter responsiveness has also been demonstrated in mouse Leydig cells (350).

While most of the studies on the StAR promoter have been involved with its positive regulation, several studies have shown that this promoter also contains elements involved in its negative regulation. The unusual member of the nuclear family of transcription factors, DAX-1, plays key roles in adrenal and gonadal differentiation, development, and function (421–423). Mutations in the DAX-1 gene cause impaired adrenal function and hypogonadotropic hypogonadism (424). A DAX-1 binding site has been identified in the human StAR promoter, and it was demonstrated that DAX-1 can bind to a hairpin structure in the human StAR promoter and repress StAR gene expression (425). Over-expression of DAX-1 results in an inhibition of StAR expression and steroid synthesis in Y-1 cells (426). Separate studies on the mechanism of DAX-1 inhibitory action have shown that DAX-1 interacts with SF-1 via LXXLL motifs present in the DAX-1 N-terminus and represses SF-1 trans-activation (427). However, it has also been reported that DAX-1 exerts its negative effects by binding directly to SF-1, resulting in an inhibition of SF-1–mediated StAR gene transcription (343,344,428). Regardless of the mechanism involved, these findings clearly demonstrate that DAX-1 negatively regulates StAR expression.

The multifunctional transcription factor, YY1, has been demonstrated to act both as an activator or a repressor of transcription of sterol responsive genes (429,430). YY1 has also been demonstrated to inhibit SREBP mediated gene transcription (431). Transcriptional repression of the StAR gene by YY1 has also been demonstrated. In the human StAR promoter, a YY1 binding site has been found within one of the SREBP recognition motifs (408). This motif has also been shown to bind with recom SREBPs. Elimination of the YY1 binding site enhanced human StAR promoter responsiveness to SREBP-1a, and, in further support of this observation, YY1 itself was shown to inhibit SREBP-1a–mediated StAR gene expression (408). Identification of three YY1 binding sites and one SREBP/YY1 binding site has been demonstrated in the rat StAR promoter (432). Each of these YY1 binding motifs can bind recom YY1 with high affinity. This study also demonstrated that YY1 was able to repress both basal and SREBP-1a–mediated activation of the rat StAR gene. Mutation of the YY1 binding sites resulted in no significant effects on either basal or SREBP induced StAR reporter responses, suggesting that the binding of YY1 to these sites was not a requirement for repression (432). Indeed, the inhibitory effect of YY1 on StAR gene expression has been demonstrated to occur through the direct inhibition of SREBP-1a and NF-Y–mediated responsiveness at the level of the promoter (408,432). Therefore, YY1 appears to act, at least in part, through the inhibition of SREBP-mediated StAR gene expression.

Taken together, one could envision that it is the balance between the positive and negative effects of many different transcription factors that control the events associated with StAR gene transcription.

ANDROGEN ACTION

In mammals, the principal androgens are testosterone and its 5α–reduced metabolite, 5α-dihydrotestosterone. Although testosterone itself is sufficient to virilize the Wolffian ducts, 5α-dihydrotestosterone formation is required for the normal development of the male external genitalia. A variety of genetic and physiological models have reinforced that both hormones are needed to effect the normal actions of androgen. Genetic males deficient for one of the two 5α reductase enzymes, 5α reductase II, display marked abnormalities of virilization of their external genitalia (155,433,434). In like fashion, the administration of 5α reductase inhibitors to rats results in an impairment of virilization and defects of the development of organs, such as the prostate, which are the dependent on the formation of 5α-dihydrotestosterone (435,436).

The mechanisms that account for this dichotomy have not yet been completely elucidated. In cultured fibroblasts, it has been possible to demonstrate that the androgen receptor possesses three- to fourfold increased affinity for 5α-dihydrotestosterone compared to testosterone. Furthermore, androgen receptor complexes that contain 5α-dihydrotestosterone are more stable than those complexed with testosterone. This combination of the increased affinity of the androgen receptor for 5α-dihydrotestosterone and the increased stability of the complexes formed, lead to preferential accumulation of 5α-dihydrotestosterone-AR complexes in cell nuclei (437). Such observations suggest a model in which the formation of 5α-dihydrotestosterone is required in target tissues where the concentration of the androgen is limiting. Although this model has been formally tested only in a small number of systems, in those instances where it has been examined experimentally, this model has held true. In cells transfected with a model androgen-regulated reporter gene, the difference in activity following stimulation with testosterone and 5α-dihydrotestosterone paralleled the activities of these hormones in binding to the receptor itself in a cell type in which the steroid hormones were not significantly metabolized (438). Similar conclusions were reached using an in vivo rat model to examine the potencies of testosterone and 5α-dihydrotestosterone in restoring spermiogenesis. In these experiments, the investigators demonstrated that the effects of 5α-reductase inhibitors were most notable in the context of limiting testosterone concentrations (166). Both paradigms are consistent with a model in which the regulation of genes by T and DHT are similar at high concentrations, but that DHT is capable of regulating genes more efficiently at lower, limiting concentrations of androgen. It remains to be determined whether this model is valid in all instances and for all genes that are regulated by androgen.

Androgen Receptor and the Nuclear Receptor Family

The androgen receptor is a prototypic member of the nuclear receptor family (439,440). This family, which contains 48 different members in humans (441), includes members that are regulated in response to ligands, such as the steroid receptors. This family also includes other members, "orphans," for which ligands have not yet been identified.

The organization of the androgen receptor is similar to the organization of the other members of the nuclear receptor family and is most closely related to that of the other steroid receptors (Fig. 3). The androgen receptor contains a central DNA-binding domain (DBD) that is highly conserved when compared to other members of the nuclear receptor family. This element mediates the binding of target DNA sequences within the genome. Distinctive portions of the DBD are responsible for the dimerization of this domain and the recognition of its recognition of specific DNA sequences. The carboxyl terminal ligand-binding domain (LBD) is approximately 250 amino acids in length and is responsible for the high-affinity binding of ligand by the receptor molecule. The amino terminus of the AR is large and accounts for nearly half of the primary sequence of the receptor molecule.

The structural insights that have followed the cloning of the androgen receptor and other members of the nuclear receptor family have been extended progressively in recent years. As noted below, detailed structural information is now available relating to the 3-dimensional structure of specific domains of

A

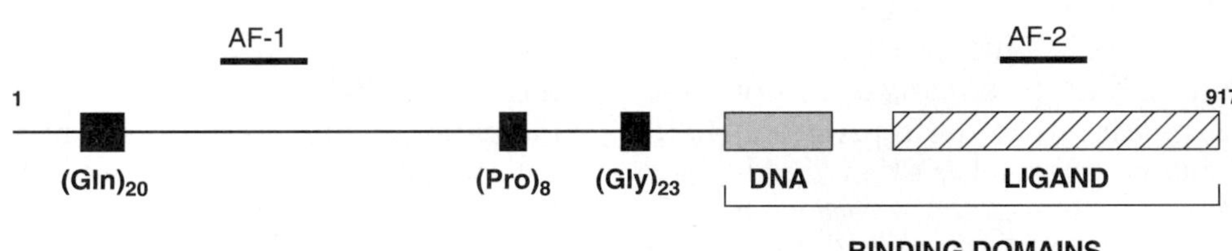

B

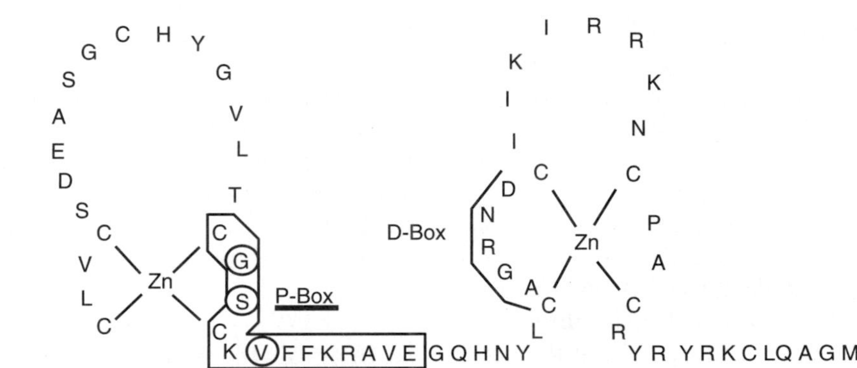

C

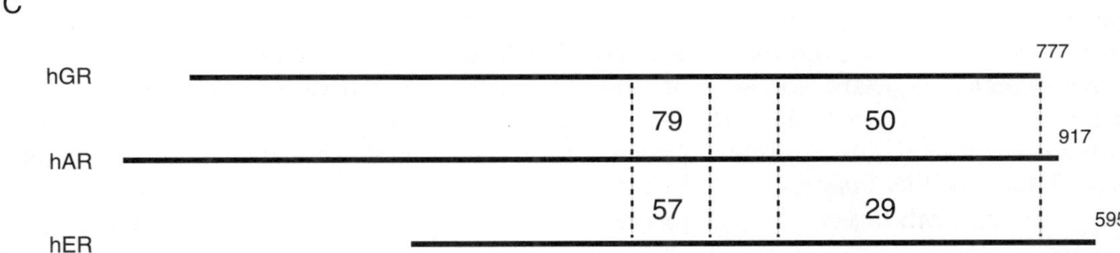

D

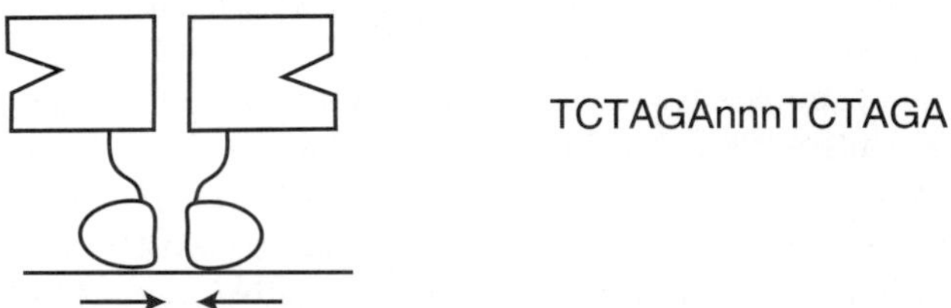

FIG. 3. Modular structure of the nuclear receptor family. Members of the nuclear receptor family contain segments that are conserved to varying degrees among the family members. **A:** A schematic structure of the androgen receptor, a prototypic nuclear receptor, is shown. The conserved ligand binding and DNA binding domains are indicated (*hatched* and *filled rectangles*, respectively). The relative positions of the activating functions (AF)-1 and -2 within the amino terminus and carboxyl terminus of the receptor proteins are shown. Sequences responsible for the nuclear localization (NLS) of these proteins have been localized to the carboxyl terminal end of the DNA-binding domain. **B:** While the relative positions of the individual domains are in most instances maintained the degree of sequence conservation varies widely. Members of the nuclear receptor family exhibit the highest degree of conservation when the amino acid sequences of the DNA-binding domains of the receptor proteins are compared. Lesser degrees of homology are evident between the sequences that comprise the ligand-binding domains of the receptors. In the example shown, the predicted amino acid sequences of three different members of the nuclear receptor family are compared. The degree of relatedness is shown for each of the two receptors when aligned with the predicted amino acid

Continued

sequence of the human glucocorticoid receptor. The amino terminal segments of the receptors differ considerably in size and sequence. The extent of homology is less than 15% when the amino termini of the receptor proteins are compared. **C:** Schematic organization of the human glucocorticoid receptor DNA binding domain. The zinc ions serve as nucleation centers for the two "zinc finger" modules of the DNA binding domain. The regions identified by mutagenesis studies as being important for receptor dimerization (D-box) and for target gene specificity (P-box) are indicated. In some members of the nuclear receptor family, carboxyl terminal extensions (CTE) of the DNA binding domain are important determinants of the high affinity binding of the receptor proteins to its DNA targets (the T- and A-boxes). **D:** Schematic representations of the mechanism by which the AR binds to target DNA sequences. The mechanisms by which members of the steroid receptor family recognize DNA fall into four different classes [reviewed in Mangelsdorf, 1995 (439); Zoppi, 2002 (440)]. The AR bind to sequences similar to the inverted palindromic sequence (shown at right). In this type of recognition sequence, a three-nucleotide "spacer" (nnn) separates the individual half sites. (Modified with permission from Figure 1 in Zoppi, S., Young, M., and McPhaul, M. J. [2002]. Regulation of gene expression by the nuclear receptor family. In *Genetics of Steroid Biosynthesis and Function* [J. I. Mason, Ed.], pp. 376–403. Harwood Academic Publishers.)

many of members of the nuclear receptor family, including the androgen receptor.

The target elements that are recognized by the AR that have been most carefully studied are palindromic sequences that are recognized by the DBD by binding in a dimeric fashion, oriented "head-to-head" (Fig. 4). This mode of binding is similar to that which has been studied in detail for other related steroid receptors, such as the GR and ER. The structural studies that have examined the mechanism of DNA recognition by these domains have employed NMR spectroscopy and x-ray crystallography and have been performed most thoroughly for the DBDs of the ER and GR (442). Although the androgen receptor DNA-binding domain has not been crystallized in association with a palindromic response element, it is likely that the overall structural features of these interactions are also true for the AR. This inference has been reinforced with the finding that a similar mode of DNA recognition is observed even in a context that seem designed to force an alternate orientation (443).

Crystal structures have now been determined for the rat and human AR LBDs complexed to agonist ligands (444–446). These structures have demonstrated that the androgen receptor is organized in a fashion that is consistent with the organizations of the ligand-binding domains of other nuclear receptors. These models have demonstrated that each of these domains is comprised of a highly conserved "3-layered" collection of alpha helices that surround the ligand-binding pocket. It is apparent from mutagenesis mapping and co-crystallization experiments, particularly of the ERα, that the binding of agonist ligands permits the formation of surfaces that are required for the binding of co-activators such as SRC-1 (447–450). By contrast, the binding of ligands with antagonist activity does not permit the movement of the carboxyl terminal helix (helix 12) into a position that exposes these surfaces and does not permit coactivator recruitment. Although crystal structures of the AR LBD bound to antiandrogens have not been reported, it is believed that the position of the carboxyl terminal helix (helix 12) is different when complexed to androgen agonist in androgen receptor antagonist. As detailed below, current models suggest that such changes in the position of the terminal helix are central to the capacity of the receptor molecule to recruit co-activators needed for function of the agonist-bound receptor.

Androgen Receptor and the Regulation of Gene Expression

The mechanisms controlling gene expression have been the subject of intense interest for decades. These investigations have employed a variety of cellular and in vitro models to identify the factors that control gene expression. Insights first made using in vitro transcription systems have permitted the recognition that steroid receptors, including the androgen receptor, recognize specific elements within genomic DNA to permit the nucleation and stabilization of multiprotein complexes required for RNA initiation and transcription. Specific classes of proteins have been identified that serve to activate (co-activators) by stabilizing the preinitiation complex and by recruiting enzymatic activities that modify the configuration of the surrounding chromatin to permit the entrance of the transcription machinery (451–454). Distinct classes of proteins have been identified which serve to repress gene expression (corepressors) by recruiting enzymatic activities that condense the surrounding chromatin. In addition, multi-protein complexes have been identified that co-purify with different nuclear receptors or nuclear receptor domains. These protein complexes are similar to complexes that have been identified in association with a range of transcriptional activators in yeast and in mammalian cells (455). In aggregate, these types of studies have demonstrated that an exceedingly diverse group of proteins participate in regulating the assembly and activity of the transcription machinery at the sites of regulated genes. These findings initially presented a confusing diversity of possibilities by which nuclear receptors, including the AR, acted to regulate the transcription of responsive genes. It was not clear whether these proteins acted in parallel or in series to regulate genes or acted to modify the activities of distinct sets of genes. These questions have been addressed, at least in part, by the use of chromatin precipitation assays to examine the complement of proteins that are recruited to response elements within and adjacent to the promoters of hormonally regulated genes. Such experiments have demonstrated that in response to the addition of hormone, activated nuclear receptors recruit specific groups to specific DNA elements in a precise temporal sequence. These models suggest that a concerted, cyclic exchange takes place at regulated genes in response to agonist ligands. Further, it appears that the exchange and turnover of specific components is required for the normal responsiveness of regulated genes to their cognate hormones (Fig. 4) (456–458).

In addition to such models, it has been recognized for some time that nuclear receptors participate in the regulation of responsive genes by distinct mechanisms. For example, a number of paradigms have been defined in which a nuclear receptor binds to another transcription factor that has bound to its own DNA

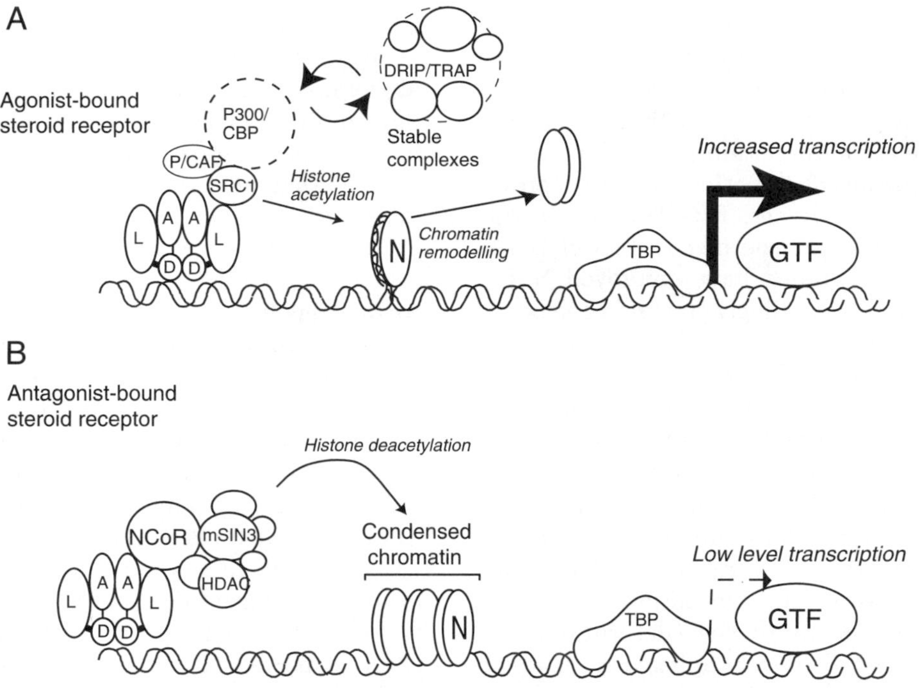

FIG. 4. Co-activators and co-repressors in the action of nuclear receptor. A general model for the activities of co-activators and co-repressors in the modulation of the transcription of responsive genes by nuclear receptors, including steroid receptors, such as the AR. **A:** The binding of an agonist-bound steroid receptor to specific sequences adjacent to the site of transcription initiation of a responsive gene recruits co-activator complexes containing proteins such as SRC-1, p300/CBP, and DRIPs/TRAPs (the order or competition among these different proteins for NR binding is unclear at this time). Enzymatic activities contained within these complexes (e.g., histone acetyl transferase) modify the local chromatin structure. In some instances, these modifications may result in large-scale alterations of chromatin organization. These changes make the transcription unit more accessible to the assembly and stability of transcription initiation complexes and results in an increase in the rate of transcription. Additional complexes, including Mediator, are recruited and dismissed from the specific sites within and adjacent to regulated genes in an ordered and cyclic fashion. **B:** The binding of an antagonist to nuclear receptors results in the recruitment of protein complexes containing corepressors such as NCoR and SMRT. The enzymatic activities (e.g., deacetylases) associated with these co-repressor complexes, which include SIN3 and HDACs, lead to a condensation of chromatin structure and a decreased level of gene transcription. (A, amino terminus of NR with activation functions; D, DNA binding domain of NR; L, ligand binding domain of NR; N, nucleosome; TBP, TATA binding protein; GTF, general transcription factors.) (Modified with permission from Figure 7 in Zoppi, S., Young, M., and McPhaul, M. J. [2002]. Regulation of gene expression by the nuclear receptor family. In *Genetics of Steroid Biosynthesis and Function* [J. I. Mason, Ed.], pp. 376–403. Harwood Academic Publishers.)

response element. In such circumstances, the nuclear receptor itself behaves in a fashion analogous to a coactivator or corepressor (459,460). Although such examples have been well studied for some nuclear receptors (such as the glucocorticoid or estrogen receptors), this type of gene regulation by the androgen receptor has not been explored extensively (461). Nonetheless, additional studies may ultimately reveal genes that are regulated by androgens in analogous fashion.

AR and Alternate Signaling Pathways

Much of the preceding has focused on the direct regulation of responsive genes by members of the nuclear receptor family. In contrast to these types of genetic regulation, which usually require hours for effects to become manifest, other responses have been identified that occur within seconds to minutes of hormone addition. These effects, which have been called "nongenomic," are believed to be mediated by nuclear receptor family members by distinctive mechanisms that involve the direct participation in cellular signaling cascades. Among the best characterized of these are regulation of genes such as nitric oxide synthase by estrogen (462–464) and the regulation of MAP kinase by estrogen and progesterone (465–467). More recently, it has been recognized in the case of estrogen that regulation of this type involves the formation of complexes within the cell with other proteins in order to activate protein kinase activities. In particular, the studies of Wong and colleagues have identified a scaffold protein, MNAR, that associates with the estrogen receptor and which is required for its action in nongenomic responses (468). More recent work has suggested this same protein (MNAR) also associates with the androgen receptor and is important in mediating nongenomic responses following the addition of androgen (469,470). To date, these observations have come from studies using cultured cell models. It remains to be determined what relevance these signaling pathways may have for the regulation of processes in vivo by androgens.

Responses to Androgen that Require the Participation of More than One Cell Type

Much of the preceding has touched on models in which individual genes are regulated by androgens. As described above, these models have permitted important insights into how androgens regulate gene expression at the level of transcription and intracellular signaling. In addition to such models, a number of different systems imply a level of complexity that cannot be explained on the basis of the participation of a single cell type. In fact, these models have suggested that more than one cell type is required in order to see the observed biological responses to androgen (471–474). For example, experiments examining the development of organs, such as the prostate and seminal vesicle, have employed distinctive cell populations (stroma, epithelium) to demonstrate the role that AR signaling plays in each cell type in the growth of these tissues. The use of cell populations from normal mice and from mice deficient for AR function (^{Tfm}X/Y) to prepare tissue chimeras has demonstrated that the normal differentiation of the prostatic and seminal vesicle epithelium requires that a functional AR be expressed in the stroma, even though the cell type that is most responsive in vivo in the adult to androgen is the epithelium. Such experiments have led to the proposal of models in which the binding of androgen to the functional AR expressed in the stromal compartment exerts an influence on the function of the adjacent epithelium by the effect of a secreted intermediary factor.

While such inferences regarding the indirect effects of androgens on cell function have been most clearly delineated in models of androgen action in the prostate and seminal vesicle, experiments in other systems have also suggested the effects of androgens in the control of spermatogenesis may also be indirect. One activity, PModS, has been reported to be produced by the peritubular cells of the testes to modulate Sertoli cell function and has been suggested to indirectly exert the effects of androgen on Sertoli cells. The identity of this activity has not been defined and its nature is somewhat controversial (475).

In addition to insights derived from such in vitro models such as those that defined the activity of PModS, it is clear from in vivo studies that the regulation by androgen of the process of spermiogenesis itself involves mechanisms that are themselves indirect. This conclusion is based on two observations. First, localization studies suggest that the germ cells and spermatocytes themselves do no express the AR. Second, is that a functional AR is not required for the process of spermatogenesis. Studies by Lyon and colleagues demonstrated that chimeric mice (^{Tfm}X/Y/X/Y) were able to produce functional sperm from the ^{Tfm}X/Y contribution (476). More recent studies by Johnston et al. confirmed that this was true by demonstrating that the transplantation of germ cells devoid of a functional AR (derived from a ^{Tfm}X/Y host anima) into a normal seminiferous epithelium led to the production of functional spermatozoa (477). Conclusions based on the results of these transplantation experiments have been amply supported by the elegant studies of Verhoeven (discussed below).

Studies of Androgen Action in the Testis

A number of distinctive cell types are present within the testis and are required for the normal production of testicular hormones and for normal spermatogenesis. Following the cloning of the androgen receptor, specific antibodies have been produced and used to analyze the cell types and patterns of AR expression in order to better determine in which cells the effects of androgen are exerted. Published reports have examined the patterns of AR expression most carefully in the rat, mouse, human, and goat. Although these studies are in general agreement, some inconsistencies are evident. There is agreement that the AR is expressed in the Sertoli cells, peritubular cells, and Leydig cells (478–488). In most analyses, the investigators have noted that AR expression is not present in the germ cells or developing spermatocytes (478,479,482,483,487), although in a few reports in human (484), mouse (485), and rat (480), AR was detected in spermatogonia (484,485) and in developing spermatocytes or spermatids (480,484). Interestingly, in the rat the levels of AR expression in the Sertoli cells have been reported to change in parallel with the stage of spermatogenesis in the tubule. The stages at which androgen action had been inferred to be most active (VI–VIII) were those at which AR expression was highest (479). This pattern has been noted to be dependent on the presence of androgen and to be unaffected by germ cell depletion (479). Conflicting data is present in the literature as to whether these stage-specific alterations of AR expression are present in the human and other species (481,489). When detected, these changes are noted to be specific for the Sertoli cell population and are not observed in the other testicular cell types.

Until recently, the methods used to study the actions of androgens in the testes were often indirect. A number of rodent models have employed the use of hormone inhibitors or states of hormone deprivation and "add back" to study the requirements for androgen in the process of spermatogenesis. These studies have uniformly demonstrated the critical role of androgens in controlling this process and indicated that androgens are required at specific points in the spermatogenic cycle of the rodent testes. Analyses of cell morphology and rates of protein production led to the concept that the effects of androgen are required at specific stages in the spermatogenic cycle (490–496).

Although the importance of androgens in regulating normal spermatogenesis has been clear for quite some time, the mechanisms by which androgens regulate spermatogenesis have not. Specifically, although androgen deprivation leads to defects in the later stages of spermatogenesis, how those effects are exerted is less clear. Studies of androgen receptor immunoreactivity using specific antibodies to the androgen receptor have demonstrated the androgen receptor to be expressed in Sertoli cells, Leydig cells, and myoid peritubular cells (478–488). Although studies using mutant mice and rats carrying defects of the X-chromosomally encoded androgen receptor have demonstrated the importance of the androgen receptor in regulating spermatogenesis, the complex phenotypes associated with these genetic models (defects of testicular descent, abnormalities of Leydig cell differentiation) have led to considerable uncertainty as to the cell type(s) in which the effects of androgen are required to regulate the process of normal spermatogenesis. Furthermore, the recognition that the androgen receptor is not expressed in the cells in which androgens effects are most obvious (germ cells and their derivatives) have supported models in which androgen receptor function is required in one cell type but the effects are observed in distinctive cells types (germ cells and developing spermatocytes). In most models, authors have postulated that androgens exert their effects in binding to ARs in the peritubular and/or Sertoli cells. Transplantation experiments using germs cell prepared from Tfm/Y mice transplanted into normal testis have supported these models and demonstrated that spermatogenesis can occur in seminiferous epithelium reconstructed with germ cells devoid of a functional AR (476,477).

Selective Ablation of the AR in Specific Cell Types of the Testes

These indirect experiments have more recently been supported by experiments in which the expression of androgen receptor has been ablated specifically in Sertoli cells (497). In these Sertoli cell-selective androgen receptor knockout (SCARKO) mice, male mice demonstrated normal external sexual development and were represented in litters at the expected ratio. Growth patterns of SCARKO male mice were similar to non-SCARKO male littermates. Analysis of the anatomy of SCARKO mice demonstrated that the testis were located in the normal position, compared to wild-type littermates, indicating normal testicular descent. The SCARKO males also displayed normal development of classic androgen target tissues, including the epididymis, ductus deferens, coagulating gland, seminal vesicles, and prostate. No significant differences were observed between wild-type and SCARKO animals in levels of T and LH at 50 days, but FSH levels were increased by 1/3 in the SCARKO animals. Testicular weight in the SCARKO animals was reduced approximately threefold compared to wild-type littermates at 50 days of age.

TABLE 1. *Cellular composition of the seminiferous tubules of WT and SCARKO male mice at 50 days of age*

	Sertoli cells	Spermatogonia	Spermatocytes	Round spermatids	Elongated spermatids
WT mm^3 ($n = 5$)	1.29 ± 0.08	0.81 ± 0.07	8.66 ± 0.37	5.75 ± 0.54	4.65 ± 0.40
SCARKO, mm^3 ($n = 3$)	1.13 ± 0.18	0.74 ± 0.13	5.51 ± 0.13*	0.19 ± 0.05*	0.00 ± 0.00*
SCARKO/WT, %	88	91	64	3	0

Data are expressed as nuclear volume per testis. Values are mean ± SEM for the number of animals indicated (n). SCARKO nuclear volume as percentage of the WT is also indicated. *$P < 0.05$ vs. WT. Although the numbers of Sertoli cells and spermatogonia are similar between SCARKO and wild-type animals, a progressive decrease in the number of spermatocytes, round spermatids, and elongating spermatids is evident in the SCARKO animals. (Reprinted with permission from De Gendt, K., et al. [2004]. A Sertoli cell-selective knockout of the androgen receptor causes spermatogenic arrest in meiosis. *PNAS* 101, 1331. Copyright 2004, National Academy of Sciences, USA.)

Examination of testicular histology in the SCARKO animals demonstrated a number of important changes compared to wild-type animals. Specifically, the complement of germ cells was observed to be decreased compared to wild-type littermates. Furthermore, spermatogenesis was clearly abnormal. Careful analysis indicated that although germ cell entry into meiosis was normal, a progressive lost of primary spermatocytes was observed, particularly between stages VI and XII, findings consistent with the stages most dependent on androgen action in other analyses. The examination of the cellular composition of the seminiferous tubules of the SCARKO male mice demonstrated relatively normal numbers of Sertoli cells and spermatogonia (88% and 91%, respectively, compared to wild-type). By contrast, these investigators noted a progressive decrease in the number of spermatocytes, round spermatids, and elongating spermatids, compared to wild-type littermates (Table 1). These findings demonstrate the central requirement for the expression of a functional androgen receptor in Sertoli cells to support the process of normal spermatogenesis.

ACKNOWLEDGMENTS

The authors acknowledge the support of funds from NIH Grant HD17481 and Grant Number B1-0028 from the Robert A. Welch Foundation to D.M.S. and NIH grant DK03892 and Grant Number I-1090 from the Robert A. Welch foundation to M.J.M.

REFERENCES

1. Eik-Nes, K. B., and Hall, P. F. (1965). Secretion of steroid hormones in vivo. *Vitam. Horm.* 23, 153–208.
2. Christensen, A. K., and Mason, N. R. (1965). Comparative ability of seminiferous tubules and interstitial tissue of rat testes to synthesize androgens from progesterone-4-^{14}C in vitro. *Endocrinology* 76, 646–656.
3. Dufau, M. L., Veldhuis, J. D., Fraioli, F., Johnson, M. L., and Beitins, I. Z. (1983). Mode of secretion of bioactive luteinizing hormone in man. *J. Clin. Endocrinol. Metab.* 57, 993–1003.
4. Dufau, M. (1996). The luteinizing hormone receptor. In *The Leydig Cell* (A. H. Payne, M. P. Hardy, and L. D. Russell, Eds.), pp. 333–350. Cache River Press, Vienna, IL.
5. Ascoli, M., Fanelli, F., and Segaloff, D. L. (2002). The lutropin/choriogonadotropin receptor, a 2002 perspective. *Endocr. Rev.* 23, 141–174.
6. Neill, J. D., Patton, J. M., Dailey, R. A., Tsou, R. C., and Tindall, G. T. (1977). Luteinizing hormone releasing hormone (LHRH) in pituitary stalk blood of rhesus monkeys: relationship to level of LH release. *Endocrinology* 101, 430–434.
7. Dufau, M. L., Tsuruhara, T., Horner, K. A., Podesta, E., and Catt, K. J. (1977). Intermediate role of adenosine 3′:5′-cyclic monophosphate and protein kinase during gonadotropin-induced steroidogenesis in testicular interstitial cells. *Proc Natl Acad Sci U S A* 74, 3419–3423.
8. Zhang, F. P., Poutanen, M., Wilbertz, J., and Huhtaniemi, I. (2001). Normal prenatal but arrested postnatal sexual development of luteinizing hormone receptor knockout (LuRKO) mice. *Mol. Endocrinol.* 15, 172–183.
9. Lei, Z. M., Mishra, S., Zou, W., Xu, B., Foltz, M., Li, X., and Rao, C. V. (2001). Targeted disruption of luteinizing hormone/human chorionic gonadotropin receptor gene. *Mol. Endocrinol.* 15, 184–200.
10. O'Shaughnessy, P. J., Baker, P., Sohnius, U., Haavisto, A. M., Charlton, H. M., and Huhtaniemi, I. (1998). Fetal development of Leydig cell activity in the mouse is independent of pituitary gonadotroph function. *Endocrinology* 139, 1141–1146.
11. Baker, P. J., and O'Shaughnessy, P. J. (2001). Role of gonadotrophins in regulating numbers of Leydig and Sertoli cells during fetal and postnatal development in mice. *Reproduction* 122, 227–234.
12. O'Shaughnessy, P. J., Fleming, L. M., Jackson, G., Hochgeschwender, U., Reed, P., and Baker, P. J. (2003). Adrenocorticotropic hormone directly stimulates testosterone production by the fetal and neonatal mouse testis. *Endocrinology* 144, 3279–3284.
13. Cooke, B. A. (1996). Transduction of the luteinizing hormone signal within the Leydig cell. In *The Leydig Cell* (A. H. Payne, M. P. Hardy, and L. D. Russell, Eds.), pp. 351–364. Cache River Press, Vienna, IL.
14. Ford, S. L., Reinhart, A. J., Lukyanenko, Y., Hutson, J. C., and Stocco, D. M. (1999). Pregnenolone synthesis in immature rat Sertoli cells. *Mol. Cell. Endocrinol.* 157, 87–94.
15. Dorrington, J. H., and Armstrong, D. T. (1975). Follicle-stimulating hormone stimulates estradiol-17beta synthesis in cultured Sertoli cells. *Proc. Natl. Acad. Sci. U S A* 72, 2677–2681.
16. Nyman, M. A., Geiger, J., and Goldzieher, J. W. (1959). Biosynthesis of estrogen by the perfused stallion testis. *J. Biol. Chem.* 234, 16–18.
17. Ritzen, E. M., Van Damme, M. P., Froysa, B., Reuter, C., De La Torre, B., and Diczfalusy, E. (1981). Identification of

estradiol produced by Sertoli cell enriched cultures during incubation with testosterone. *J. Steroid Biochem.* 14, 533–535.
18. Canick, J. A., Makris, A., Gunsalus, G. L., and Ryan, K. J. (1979). Testicular aromatization in immature rats: localization and stimulation after gonadotropin administration in vivo. *Endocrinology* 104, 285–288.
19. Payne, A. H., and O'Shaughnessy, P. J. (1996). Structure, function and regulation of steroidogenic enzymes in the Leydig cell. In *The Leydig Cell* (A. H. Payne, M. P. Hardy, and L. D. Russell, Eds.), pp. 259–286. Cache River Press, Vienna, IL.
20. Swerdloff, R. S., and Wang, C. (1996). Clinical evaluation of Leydig cell function. In *The Leydig Cell* (A. H. Payne, M. P. Hardy, and L. D. Russell, Eds.), pp. 663–694. Cache River Press, Vienna, IL.
21. Hall, P. F. (1994). Testicular steroid synthesis: organization and regulation. In *The Physiology of Reproduction* (E. Knobil and J. D. Neill, Eds.), pp. 1335–1362. Raven Press, New York.
22. Rudney, H., and Sexton, R. C. (1986). Regulation of cholesterol biosynthesis. *Annu. Rev. Nutr.* 6, 245–272.
23. Rudney, H., and Pannini, R. (1993). Cholesterol biosynthesis. *Curr. Opin. Lipidol.* 4, 230–237.
24. Goldstein, J. L., Brown, M. S., Anderson, R. G., Russell, D. W., and Schneider, W. J. (1985). Receptor-mediated endocytosis: concepts emerging from the LDL receptor system. *Annu. Rev. Cell. Biol.* 1, 1–39.
25. Brown, M. S., and Goldstein, J. L. (1986). A receptor-mediated pathway for cholesterol homeostasis. *Science* 232, 34–47.
26. Andersen, J. M., and Dietschy, J. M. (1978). Relative importance of high and low density lipoproteins in the regulation of cholesterol synthesis in the adrenal gland, ovary, and testis of the rat. *J. Biol. Chem.* 253, 9024–9032.
27. Kovanen, P. T., Schneider, W. J., Hillman, G. M., Goldstein, J. L., and Brown, M. S. (1979). Separate mechanisms for the uptake of high and low density lipoproteins by mouse adrenal gland in vivo. *J. Biol. Chem.* 254, 5498–5505.
28. Spady, D. K., Meddings, J. B., and Dietschy, J. M. (1986). Kinetic constants for receptor-dependent and receptor-independent low density lipoprotein transport in the tissues of the rat and hamster. *J. Clin. Invest.* 77, 1474–1481.
29. Glass, C., Pittman, R. C., Weinstein, D. B., and Steinberg, D. (1983). Dissociation of tissue uptake of cholesterol ester from that of apoprotein A-I of rat plasma high density lipoprotein: selective delivery of cholesterol ester to liver, adrenal, and gonad. *Proc. Natl. Acad. Sci. U S A* 80, 5435–5439.
30. Acton, S., Rigotti, A., Landschulz, K. T., Xu, S., Hobbs, H. H., and Krieger, M. (1996). Identification of scavenger receptor SR-BI as a high density lipoprotein receptor. *Science* 271, 518–520.
31. Reaven, E., Tsai, L., and Azhar, S. (1996). Intracellular events in the "selective" transport of lipoprotein-derived cholesteryl esters. *J. Biol. Chem.* 271, 16208–16217.
32. Graf, G. A., Connell, P. M., van der Westhuyzen, D. R., and Smart, E. J. (1999). The class B, type I scavenger receptor promotes the selective uptake of high density lipoprotein cholesterol ethers into caveolae. *J. Biol. Chem.* 274, 12043–12048.
33. Azhar, S., Nomoto, A., Leers-Sucheta, S., and Reaven, E. (1998). Simultaneous induction of an HDL receptor protein (SR-BI) and the selective uptake of HDL-cholesteryl esters in a physiologically relevant steroidogenic cell model. *J. Lipid Res.* 39, 1616–1628.
34. Reaven, E., Nomoto, A., Leers-Sucheta, S., Temel, R., Williams, D. L., and Azhar, S. (1998). Expression and microvillar localization of scavenger receptor, class B, type I (a high density lipoprotein receptor) in luteinized and hormone-desensitized rat ovarian models. *Endocrinology* 139, 2847–2856.
35. Ikemoto, M., Arai, H., Feng, D., Tanaka, K., Aoki, J., Dohmae, N., Takio, K., Adachi, H., Tsujimoto, M., and Inoue, K. (2000). Identification of a PDZ-domain-containing protein that interacts with the scavenger receptor class B type I. *Proc. Natl. Acad. Sci. U S A* 97, 6538–6543.
36. Azhar, S., Leers-Sucheta, S., and Reaven, E. (2003). Cholesterol uptake in adrenal and gonadal tissues: the SR-BI and 'selective' pathway connection. *Front Biosci.* 8, s998–s1029.
37. Azhar, S., and Reaven, E. (2002). Scavenger receptor class BI and selective cholesteryl ester uptake: partners in the regulation of steroidogenesis. *Mol. Cell. Endocrinol.* 195, 1–26.
38. Leitersdorf, E., Stein, O., Eisenberg, S., and Stein, Y. (1984). Uptake of rat plasma HDL subfractions labeled with [^{3}H]cholesteryl linoleyl ether or with :I by cultured rat hepatocytes and adrenal cells. *Biochim. Biophys. Acta.* 796, 72–82.
39. Leitersdorf, E., Israeli, A., Stein, O., Eisenberg, S., and Stein, Y. (1986). The role of apolipoproteins of HDL in the selective uptake of cholesteryl linoleyl ether by cultured rat and bovine adrenal cells. *Biochim. Biophys. Acta.* 878, 320–329.
40. Pittman, R. C., Glass, C. K., Atkinson, D., and Small, D. M. (1987). Synthetic high density lipoprotein particles. Application to studies of the apoprotein specificity for selective uptake of cholesterol esters. *J. Biol. Chem.* 262, 2435–2442.
41. Pittman, R. C., Knecht, T. P., Rosenbaum, M. S., and Taylor, C. A. Jr. (1987). A nonendocytotic mechanism for the selective uptake of high density lipoprotein-associated cholesterol esters. *J. Biol. Chem.* 262, 2443–2450.
42. Gwynne, J. T., and Mahaffee, D. D. (1989). Rat adrenal uptake and metabolism of high density lipoprotein cholesteryl ester. *J. Biol. Chem.* 264, 8141–8150.
43. Green, S. R., and Pittman, R. C. (1991). Selective uptake of cholesteryl esters from low density lipoproteins in vitro and in vivo. *J. Lipid Res.* 32, 667–678.
44. Cherradi, N., Bideau, M., Arnaudeau, S., Demaurex, N., James, R. W., Azhar, S., and Capponi, A. M. (2001). Angiotensin II promotes selective uptake of high density lipoprotein cholesterol esters in bovine adrenal glomerulosa and human adrenocortical carcinoma cells through induction of scavenger receptor class B type I. *Endocrinology* 142, 4540–4549.
45. Azhar, S., Tsai, L., and Reaven, E. (1990). Uptake and utilization of lipoprotein cholesteryl esters by rat granulosa cells. *Biochim. Biophys. Acta.* 1047, 148–160.
46. Reaven, E., Tsai, L., and Azhar, S. (1995). Cholesterol uptake by the 'selective' pathway of ovarian granulosa cells: early intracellular events. *J. Lipid Res.* 36, 1602–1617.
47. Reaven, E., Zhan, L., Nomoto, A., Leers-Sucheta, S., and Azhar, S. (2000). Expression and microvillar localization of scavenger receptor class B, type I (SR-BI) and selective cholesteryl ester uptake in Leydig cells from rat testis. *J. Lipid Res.* 41, 343–356.
48. Rigotti, A., Miettinen, H. E., and Krieger, M. (2003). The role of the high-density lipoprotein receptor SR-BI in the lipid metabolism of endocrine and other tissues. *Endocr. Rev.* 24, 357–387.
49. Christenson, L. K., and Devoto, L. (2003). Cholesterol transport and steroidogenesis by the corpus luteum. *Reprod. Biol. Endocrinol.* 1, 90.
50. Maxfield, F. R., and Wustner, D. (2002). Intracellular cholesterol transport. *J. Clin. Invest.* 110, 891–898.
51. Srere, P., Chaikoff, I., and Dauben, W. (1948). The in vitro synthesis of cholesterol from acetate by surviving adrenal cortical tissue. *J. Biol. Chem.* 176, 829–833.
52. Srere, P. (1950). The extrahepatic synthesis of cholesterol. *J. Biol. Chem.* 182, 629–645.
53. Savard, K. (1952). Biogenesis of androgens in the human testis. *J. Clin. Endocrinol.* 12, 935.

54. Morris, M. D., and Chaikoff, I. L. (1959). The origin of cholesterol in liver, small intestine, adrenal gland, and testis of the rat: dietary versus endogenous contributions. *J. Biol. Chem.* 234, 1095–1097.
55. Azhar, S., and Menon, K. M. (1982). Receptor mediated gonadotropin action in gonadal tissues: relationship between blood cholesterol levels and gonadotropin stimulated steroidogenesis in isolated rat Leydig and luteal cells. *J. Steroid Biochem.* 16, 175–184.
56. Hou, J. W., Collins, D. C., and Schleicher, R. L. (1990). Sources of cholesterol for testosterone biosynthesis in murine Leydig cells. *Endocrinology* 127, 2047–2055.
57. Quinn, P. G., Dombrausky, L. J., Chen, Y. D., and Payne, A. H. (1981). Serum lipoproteins increase testosterone production in hCG-desensitized Leydig cells. *Endocrinology* 109, 1790–1792.
58. Charreau, E. H., Calvo, J. C., Nozu, K., Pignataro, O., Catt, K. J., and Dufau, M. L. (1981). Hormonal modulation of 3-hydroxy-3-methylglutaryl coenzyme A reductase activity in gonadotropin-stimulated and -desensitized testicular Leydig cells. *J. Biol. Chem.* 256, 12719–12724.
59. Freeman, D. A., and Ascoli, M. (1982). Studies on the source of cholesterol used for steroid biosynthesis in cultured Leydig tumor cells. *J. Biol. Chem.* 257, 14231–14238.
60. Schreiber, J. R., Weinstein, D. B., and Hsueh, A. J. (1982). Lipoproteins stimulate androgen production by cultured rat testis cells. *J. Steroid Biochem.* 16, 39–43.
61. Klinefelter, G. R., and Ewing, L. L. (1988). Optimizing testosterone production by purified adult rat Leydig cells in vitro. *In Vitro Cell. Dev. Biol.* 24, 545–549.
62. Risbridger, G. P., and Hedger, M. P. (1992). Adult rat Leydig cell cultures: minimum requirements for maintenance of luteinizing hormone responsiveness and testosterone production. *Mol. Cell. Endocrinol.* 83, 125–132.
63. Hedger, M. P., and Risbridger, G. P. (1992). Effect of serum and serum lipoproteins on testosterone production by adult rat Leydig cells in vitro. *J. Steroid. Biochem. Mol. Biol.* 43, 581–589.
64. Landschulz, K. T., Pathak, R. K., Rigotti, A., Krieger, M., and Hobbs, H. H. (1996). Regulation of scavenger receptor, class B, type I, a high density lipoprotein receptor, in liver and steroidogenic tissues of the rat. *J. Clin. Invest.* 98, 984–995.
65. Benahmed, M., Reventos, J., and Saez, J. M. (1983). Steroidogenesis of cultured purified pig Leydig cells: effects of lipoproteins and human chorionic gonadotropin. *Endocrinology* 112, 1952–1957.
66. Carr, B. R., Parker, C. R. Jr., Ohashi, M., MacDonald, P. C., and Simpson, E. R. (1983). Regulation of human fetal testicular secretion of testosterone: low-density lipoprotein-cholesterol and cholesterol synthesized de novo as steroid precursor. *Am. J. Obstet. Gynecol.* 146, 241–247.
67. Slaunwhite, W. R. Jr., and Samuels, L. T. (1956). Progesterone as a precursor of testicular androgens. *J. Biol. Chem.* 220, 341–352.
68. Bell, J. B., Vinson, G. P., Hopkin, D. J., and Lacy, D. (1968). Pathways for androgen biosynthesis from [7 alpha-^{3}H] pregnenolone and [4-^{14}C]progesterone by rat testis interstitium in vitro. *Biochim. Biophys. Acta.* 164, 412–420.
69. Tamaoki, B. (1969). In vitro synthesis and conversion of androgen in testicular tissue. In *The Gonads* (K. W. McKerns, Ed.), pp. 547–613. Appleton, New York.
70. Kwan, T. K., Pertiwi, A. K., Taylor, N. F., and Gower, D. B. (1988). Steroid profiling in the study of rat testicular steroidogenesis. *Biochim. Biophys. Acta.* 962, 214–219.
71. Eik-Nes, K. B. (1970). Synthesis and secretion of androstenedione and testosterone. In *The Androgens of the Testis* (K. B. Eik-Nes, Ed.), pp. 10–14. Dekker, New York.
72. Hall, P. F., Sozer, C. C., and Eik-Nes, K. B. (1964). Formation of dehydroepiandrosterone during in vivo and in vitro biosynthesis of testosterone by testicular tissue. *Endocrinology* 74, 35–43.
73. Booth, W. D. (1975). Changes with age in the occurrence of C19 steroids in the testis and submaxillary gland of the boar. *J. Reprod. Fertil.* 42, 459–472.
74. Vihko, R. (1974). Regulation of steroidogenesis in testis. *J. Steroid Biochem.* 5, 843–847.
75. Preslock, J. P., and Steinberger, E. (1977). Testicular steroidogenesis in the common marmoset, *Callithrix jacchus*. *Biol. Reprod.* 17, 289–293.
76. Preslock, J. P. (1980). In vitro steroidogenesis in subhuman primates. In *Testicular Development, Structure and Function* (E. Steinberger and A. Steinberger, Eds.), pp. 249–268. Raven, New York.
77. Hammar, M., and Petersson, F. (1986). Testosterone production in vitro in human testicular tissue. *Andrologia* 18, 196–200.
78. Rajfer, J., Sikka, S. C., and Swerdloff, R. S. (1987). Lack of a direct effect of gonadotropin hormone-releasing hormone agonist on human testicular steroidogenesis. *J. Clin. Endocrinol. Metab.* 64, 62–67.
79. Rey, R., Campo, S., Ayuso, S., Nagle, C., and Chemes, H. (1995). Testicular steroidogenesis in the Cebus monkey throughout postnatal development. *Biol Reprod.* 52, 997–1002.
80. Wilson, J. D., Auchus, R. J., Leihy, M. W., Guryev, O. L., Estabrook, R. W., Osborn, S. M., Shaw, G., and Renfree, M. B. (2003). 5alpha-androstane-3alpha,17beta-diol is formed in tammar wallaby pouch young testes by a pathway involving 5alpha-pregnane-3alpha,17alpha-diol-20-one as a key intermediate. *Endocrinology* 144, 575–580.
81. Mahendroo, M., Wilson, J. D., Richardson, J. A., and Auchus, R. J. (2004). Steroid 5alpha-reductase 1 promotes 5alpha-androstane-3alpha,17beta-diol synthesis in immature mouse testes by two pathways. *Mol. Cell. Endocrinol.* 222, 113–120.
82. Jefcoate, C. R. (1992). Regulation of cholesterol movement to mitochondrial cytochrome P450scc in steroid hormone synthesis. *J. Steroid Biochem. Mol. Biol.* 43, 751–767.
83. Lambeth, J. D. (1985). Cytochrome P-450scc: Enzymology, and the regulation of intramitochondrial delivery to the enzyme. *Endocr. Res.* 10, 283–309.
84. Orme-Johnson, N. R. (1990). Distinctive properties of adrenal cortex mitochondria. *Biochim Biophys. Acta.* 1020, 213–231.
85. Hanukoglu, I., and Jefcoate, C. R. (1980). Mitochondrial cytochrome P-450scc. Mechanism of electron transport by adrenodoxin. *J. Biol. Chem.* 255, 3057–3061.
86. Hanukoglu, I., Spitsberg, V., Bumpus, J. A., Dus, K. M., and Jefcoate, C. R. (1981). Adrenal mitochondrial cytochrome P-450scc. Cholesterol and adrenodoxin interactions at equilibrium and during turnover. *J. Biol. Chem.* 256, 4321–4328.
87. Yago, N., and Ichii, S. (1969). Submitochondrial distribution of components of the steroid 11 beta-hydroxylase and cholesterol sidechain-cleaving enzyme systems in hog adrenal cortex. *J. Biochem. (Tokyo).* 65, 215–224.
88. Churchill, P. F., and Kimura, T. (1979). Topological studies of cytochromes P-450scc and P-45011 beta in bovine adrenocortical inner mitochondrial membranes. Effects of controlled tryptic digestion. *J. Biol. Chem.* 254, 10443–10448.
89. Farkash, Y., Timberg, R., and Orly, J. (1986). Preparation of antiserum to rat cytochrome P-450 cholesterol side chain cleavage, and its use for ultrastructural localization of the immunoreactive enzyme by protein A-gold technique. *Endocrinology* 118, 1353–1365.
90. Geuze, H. J., Slot, J. W., Yanagibashi, K., McCracken, J. A., Schwartz, A. L., and Hall, P. F. (1987). Immunogold cytochemistry of cytochromes P-450 in porcine adrenal cortex. Two enzymes (side-chain cleavage and 11 beta-hydroxylase) are co-localized in the same mitochondria. *Histochemistry* 86, 551–557.

91. O'Shaughnessy, P. J., and Murphy, L. (1991). Steroidogenic enzyme activity in the rat testis following Leydig cell destruction by ethylene-1,2-dimethanesulphonate and during subsequent Leydig cell regeneration. *J. Endocrinol.* 131, 451–457.
92. Schulster, D. (1976). Biosynthesis of steroid hormones. In *Molecular Endocrinology of the Steroid Hormones* (D. Schulster, S. Burstein, and B. A. Cooke, Eds.), pp. 44–74. John Wiley & Sons, New York.
93. Boyd, G. S., and Simpson, E. R. (1968). Studies on the conversion of cholesterol to pregnenolone in bovine adrenal mitochondria. In *Functions of the Adrenal Cortex* (K. W. McKerns, Ed.), Vol. 1, pp. 49–76. Appleton-Century-Crofts, New York.
94. Burstein, S., and Gut, M. (1976). Intermediates in the conversion of cholesterol to pregnenolone: kinetics and mechanism. *Steroids* 28, 115–131.
95. Shikita, M., and Hall, P. F. (1973). Cytochrome P-450 from bovine adrenocortical mitochondria: an enzyme for the side chain cleavage of cholesterol. I. Purification and properties. *J. Biol. Chem.* 248, 5598–5604.
96. Watanuki, M., Granger, G. A., and Hall, P. F. (1978). Cytochrome P-450 from bovine adrenocortical mitochondria. Immunochemical properties and purity. *J. Biol. Chem.* 253, 2927–2931.
97. Black, S. M., Harikrishna, J. A., Szklarz, G. D., and Miller, W. L. (1994). The mitochondrial environment is required for activity of the cholesterol side-chain cleavage enzyme, cytochrome P450scc. *Proc. Natl. Acad. Sci. U S A* 91, 7247–7251.
98. Lambeth, J. D., and Pember, S. O. (1983). Cytochrome P-450scc-adrenodoxin complex. Reduction properties of the substrate-associated cytochrome and relation of the reduction states of heme and iron-sulfur centers to association of the proteins. *J. Biol. Chem.* 258, 5596–5602.
99. Porter, T. D., and Coon, M. J. (1991). Cytochrome P-450. Multiplicity of isoforms, substrates, and catalytic and regulatory mechanisms. *J. Biol. Chem.* 266, 13469–13472.
100. Picado-Leonard, J., Voutilainen, R., Kao, L. C., Chung, B. C., Strauss, J. F. III, and Miller, W. L. (1988). Human adrenodoxin: cloning of three cDNAs and cycloheximide enhancement in JEG-3 cells. *J. Biol. Chem.* 263, 3240–3244.
101. Solish, S. B., Picado-Leonard, J., Morel, Y., Kuhn, R. W., Mohandas, T. K., Hanukoglu, I., and Miller, W. L. (1988). Human adrenodoxin reductase: two mRNAs encoded by a single gene on chromosome 17cen-q25 are expressed in steroidogenic tissues. *Proc. Natl. Acad. Sci. U S A* 85, 7104–7108.
102. Chang, C. Y., Wu, D. A., Lai, C. C., Miller, W. L., and Chung, B. C. (1988). Cloning and structure of the human adrenodoxin gene. *DNA* 7, 609–615.
103. Lin, D., Shi, Y. F., and Miller, W. L. (1990). Cloning and sequence of the human adrenodoxin reductase gene. *Proc. Natl. Acad. Sci. U S A* 87, 8516–8520.
104. Stone, D., and Hechter, O. (1954). Studies on ACTH action in perfused bovine adrenals: the site of action of ACTH in corticosteroidogenesis. *Arch. Biochem.* 51, 457–469.
105. Karaboyas, G. C., and Koritz, S. B. (1965). The transformation of delta-5-pregnenolone and progesterone to cortisol by rat adrenal slices and the effect of ACTH and adenosine 3′-,5′-monophosphate upon it. *Biochim. Biophys. Acta.* 100, 600–602.
106. Sulimovici, S., Bartoov, B., and Lunenfeld, B. (1973). Localization of 3-beta hydroxysteroid dehydrogenase in the inner membrane subfraction of rat testis mitochondria. *Biochim. Biophys. Acta.* 321, 27–40.
107. Chapman, J. C., Waterhouse, T. B., and Michael, S. D. (1992). Changes in mitochondrial and microsomal 3 beta-hydroxysteroid dehydrogenase activity in mouse ovary over the course of the estrous cycle. *Biol. Reprod.* 47, 992–997.
108. Cherradi, N., Defaye, G., and Chambaz, E. M. (1994). Characterization of the 3 beta-hydroxysteroid dehydrogenase activity associated with bovine adrenocortical mitochondria. *Endocrinology* 134, 1358–1364.
109. Cherradi, N., Chambaz, E. M., and Defaye, G. (1995). Organization of 3 beta-hydroxysteroid dehydrogenase/isomerase and cytochrome P450scc into a catalytically active molecular complex in bovine adrenocortical mitochondria. *J. Steroid Biochem. Mol. Biol.* 55, 507–514.
110. Sauer, L. A., Chapman, J. C., and Dauchy, R. T. (1994). Topology of 3 beta-hydroxy-5-ene-steroid dehydrogenase/delta 5-delta 4-isomerase in adrenal cortex mitochondria and microsomes. *Endocrinology* 134, 751–759.
111. Cherradi, N., Rossier, M. F., Vallotton, M. B., Timberg, R., Friedberg, I., Orly, J., Wang, X. J., Stocco, D. M., and Capponi, A. M. (1997). Submitochondrial distribution of three key steroidogenic proteins (steroidogenic acute regulatory protein and cytochrome p450scc and 3beta-hydroxysteroid dehydrogenase isomerase enzymes) upon stimulation by intracellular calcium in adrenal glomerulosa cells. *J. Biol. Chem.* 272, 7899–7907.
112. Thomas, J. L., Berko, E. A., Faustino, A., Myers, R. P., and Strickler, R. C. (1988). Human placental 3 beta-hydroxy-5-ene-steroid dehydrogenase and steroid 5–4-ene-isomerase: purification from microsomes, substrate kinetics, and inhibition by product steroids. *J. Steroid Biochem.* 31, 785–793.
113. Ishii-Ohba, H., Inano, H., and Tamaoki, B. (1987). Testicular and adrenal 3 beta-hydroxy-5-ene-steroid dehydrogenase and 5-ene-4-ene isomerase. *J. Steroid Biochem.* 27, 775–779.
114. Ishii-Ohba, H., Saiki, N., Inano, H., and Tamaoki, B. I. (1986). Purification and characterization of rat adrenal 3 beta-hydroxysteroid dehydrogenase with steroid 5-ene-4-ene-isomerase. *J. Steroid Biochem.* 24, 753–760.
115. Ford, H. C., and Engel, L. L. (1974). Purification and properties of the delta 5-3 beta-hydroxysteroid dehydrogenase-isomerase system of sheep adrenal cortical microsomes. *J. Biol. Chem.* 249, 1363–1368.
116. Rutherfurd, K. J., Chen, S. A., and Shively, J. E. (1991). Isolation and amino acid sequence analysis of bovine adrenal 3 beta-hydroxysteroid dehydrogenase/steroid isomerase. *Biochemistry* 30, 8108–8116.
117. Rheaume, E., Lachance, Y., Zhao, H. F., Breton, N., Dumont, M., de Launoit, Y., Trudel, C., Luu-The, V., Simard, J., and Labrie, F. (1991). Structure and expression of a new complementary DNA encoding the almost exclusive 3 beta-hydroxysteroid dehydrogenase/delta 5-delta 4-isomerase in human adrenals and gonads. *Mol. Endocrinol.* 5, 1147–1157.
118. Lorence, M. C., Murry, B. A., Trant, J. M., and Mason, J. I. (1990). Human 3 beta-hydroxysteroid dehydrogenase/delta 5-4 isomerase from placenta: expression in nonsteroidogenic cells of a protein that catalyzes the dehydrogenation/isomerization of C21 and C19 steroids. *Endocrinology* 126, 2493–2498.
119. Lachance, Y., Luu-The, V., Labrie, C., Simard, J., Dumont, M., de Launoit, Y., Guerin, S., Leblanc, G., and Labrie, F. (1990). Characterization of human 3 beta-hydroxysteroid dehydrogenase/delta 5-delta 4-isomerase gene and its expression in mammalian cells. *J. Biol. Chem.* 265, 20469–20475.
120. Simard, J., de Launoit, Y., and Labrie, F. (1991). Characterization of the structure-activity relationships of rat types I and II 3 beta-hydroxysteroid dehydrogenase/delta 5-delta 4 isomerase by site-directed mutagenesis and expression in HeLa cells. *J. Biol. Chem.* 266, 14842–14845.
121. Lorence, M. C., Naville, D., Graham-Lorence, S. E., Mack, S. O., Murry, B. A., Trant, J. M., and Mason, J. I. (1991).

3 beta-hydroxysteroid dehydrogenase/delta 5-4-isomerase expression in rat and characterization of the testis isoform. *Mol. Cell. Endocrinol.* 80, 21–31.
122. Clarke, T. R., Bain, P. A., Sha, L., and Payne, A. H. (1993). Enzyme characteristics of two distinct forms of mouse 3 beta-hydroxysteroid dehydrogenase/delta 5-delta 4-isomerase complementary deoxyribonucleic acids expressed in COS-1 cells. *Endocrinology* 132, 1971–1976.
123. Luu-The, V., Lachance, Y., Labrie, C., Leblanc, G., Thomas, J.L., Strickler, R. C., and Labrie, F. (1989). Full length cDNA structure and deduced amino acid sequence of human 3 beta-hydroxy-5-ene steroid dehydrogenase. *Mol. Endocrinol.* 3, 1310–1312.
124. Lachance, Y., Luu-The, V., Verreault, H., Dumont, M., Rheaume, E., Leblanc, G., and Labrie, F. (1991). Structure of the human type II 3 beta-hydroxysteroid dehydrogenase/delta 5-delta 4 isomerase (3 beta-HSD) gene: adrenal and gonadal specificity. *DNA Cell. Biol.* 10, 701–711.
125. Lorence, M. C., Corbin, C. J., Kamimura, N., Mahendroo, M. S., and Mason, J. I. (1990). Structural analysis of the gene encoding human 3 beta-hydroxysteroid dehydrogenase/delta 5-4-isomerase. *Mol. Endocrinol.* 4, 1850–1855.
126. Zhao, H. F., Labrie, C., Simard, J., de Launoit, Y., Trudel, C., Martel, C., Rheaume, E., Dupont, E., Luu-The, V., Pelletier, G., et al. (1991). Characterization of rat 3 beta-hydroxysteroid dehydrogenase/delta 5-delta 4 isomerase cDNAs and differential tissue-specific expression of the corresponding mRNAs in steroidogenic and peripheral tissues. *J. Biol. Chem.* 266, 583–593.
127. Bain, P. A., Yoo, M., Clarke, T., Hammond, S. H., and Payne, A. H. (1991). Multiple forms of mouse 3 beta-hydroxysteroid dehydrogenase/delta 5-delta 4 isomerase and differential expression in gonads, adrenal glands, liver, and kidneys of both sexes. *Proc. Natl. Acad. Sci. U S A* 88, 8870–8874.
128. Abbaszade, I. G., Arensburg, J., Park, C. H., Clarke, T., Kasa-Vubu, J. Z., Orly, J., and Payne, A. H. (1995). Isolation and tissue-specific expression of a mouse 3B-hydroxysteroid dehydrogenase isoform (3βHSD VI) that may be essential for implantation and/or maintenance of early pregnancy. 77th Annual Meeting of the Endocrine Society; 619, Abstract #P613–P601.
129. Clarke, T. R., Bain, P. A., Greco, T. L., and Payne, A. H. (1993). A novel mouse kidney 3 beta-hydroxysteroid dehydrogenase complementary DNA encodes a 3-ketosteroid reductase instead of a 3 beta-hydroxysteroid dehydrogenase/delta 5-delta 4-isomerase. *Mol. Endocrinol.* 7, 1569–1578.
130. Abbaszade, I. G., Clarke, T. R., Park, C. H., and Payne, A. H. (1995). The mouse 3 beta-hydroxysteroid dehydrogenase multigene family includes two functionally distinct groups of proteins. *Mol. Endocrinol.* 9, 1214–1222.
131. de Launoit, Y., Zhao, H. F., Belanger, A., Labrie, F., and Simard, J. (1992). Expression of liver-specific member of the 3 beta-hydroxysteroid dehydrogenase family, an isoform possessing an almost exclusive 3-ketosteroid reductase activity. *J. Biol. Chem.* 267, 4513–4517.
132. Swart, P., Swart, A. C., Waterman, M. R., Estabrook, R. W., and Mason, J. I. (1993). Progesterone 16 alpha-hydroxylase activity is catalyzed by human cytochrome P450 17 alpha-hydroxylase. *J. Clin. Endocrinol. Metab.* 77, 98–102.
133. Tremblay, Y., Fleury, A., Beaudoin, C., Vallee, M., and Belanger, A. (1994). Molecular cloning and expression of guinea pig cytochrome P450c17 cDNA (steroid 17 alpha-hydroxylase/17,20 lyase): tissue distribution, regulation, and substrate specificity of the expressed enzyme. *DNA Cell. Biol.* 13, 1199–1212.
134. Auchus, R. J., Lee, T. C., and Miller, W. L. (1998). Cytochrome b5 augments the 17,20-lyase activity of human P450c17 without direct electron transfer. *J. Biol. Chem.* 273, 3158–3165.
135. Nakajin, S., Shively, J. E., Yuan, P. M., and Hall, P. F. (1981). Microsomal cytochrome P-450 from neonatal pig testis: two enzymatic activities (17 alpha-hydroxylase and c17,20-lyase) associated with one protein. *Biochemistry* 20, 4037–4042.
136. Nakajin, S., and Hall, P. F. (1981). Microsomal cytochrome P-450 from neonatal pig testis. Purification and properties of a C21 steroid side-chain cleavage system (17 alpha-hydroxylase-C17,20 lyase). *J. Biol. Chem.* 256, 3871–3876.
137. Nakajin, S., Hall, P. F., and Onoda, M. (1981). Testicular microsomal cytochrome P-450 for C21 steroid side chain cleavage. Spectral and binding studies. *J. Biol. Chem.* 256, 6134–6139.
138. Nakajin, S., Shinoda, M., Haniu, M., Shively, J. E., and Hall, P. F. (1984). C21 steroid side chain cleavage enzyme from porcine adrenal microsomes. Purification and characterization of the 17 alpha-hydroxylase/C17,20-lyase cytochrome P-450. *J. Biol. Chem.* 259, 3971–3976.
139. Kominami, S., Shinzawa, K., and Takemori, S. (1982). Purification and some properties of cytochrome P-450 specific for steroid 17 alpha-hydroxylation and C17-C20 bond cleavage from guinea pig adrenal microsomes. *Biochem. Biophys. Res. Commun.* 109, 916–921.
140. Suhara, K., Fujimura, Y., Shiroo, M., and Katagiri, M. (1984). Multiple catalytic properties of the purified and reconstituted cytochrome P-450 (P-450sccII) system of pig testis microsomes. *J. Biol. Chem.* 259, 8729–8736.
141. Zuber, M. X., Simpson, E. R., and Waterman, M. R. (1986). Expression of bovine 17 alpha-hydroxylase cytochrome P-450 cDNA in nonsteroidogenic (COS 1) cells. *Science* 234, 1258–1261.
142. Lin, D., Harikrishna, J. A., Moore, C. C., Jones, K. L., and Miller, W. L. (1991). Missense mutation serine106-proline causes 17 alpha-hydroxylase deficiency. *J. Biol Chem.* 266. 15992–15998.
143. Omura, T., Sanders, E., Estabrook, R. W., Cooper, D. Y., and Rosenthal, O. (1966). Isolation from adrenal cortex of a non-heme iron protein and a flavoprotein as a TPNH-cytochrome P-450 reductase. *Arch. Biochem. Biophys.* 117, 660–673.
144. Hochberg, R. B., Ladany, S., and Lieberman, S. (1976). Conversion of a C-20-deoxy-C21, steroid, 5-pregnen-3beta-ol, into testosterone by rat testicular microsomes. *J. Biol. Chem.* 251, 3320–3325.
145. Onoda, M., Haniu, M., Yanagibashi, K., Sweet, F., Shively, J. E., and Hall, P. F. (1987). Affinity alkylation of the active site of C21 steroid side-chain cleavage cytochrome P-450 from neonatal porcine testis: a unique cysteine residue alkylated by 17-(bromoacetoxy)progesterone. *Biochemistry* 26, 657–662.
146. Miller, W. L. (2002). Androgen biosynthesis from cholesterol to DHEA. *Mol. Cell. Endocrinol.* 198, 7–14.
147. Oshima, H., and Ochiai, K. (1973). On testicular 17-hydroxysteroid oxidoreductase product activation of testosterone formation from androstenedione in vitro. *Biochim. Biophys. Acta.* 306, 227–236.
148. Hall, P. F. (1988). Testicular steroid synthesis: organization and regulation. In *The Physiology of Reproduction* (E. Knobil and J. D. Neill, Eds.), Vol. 1, pp. 975–998. Raven Press, New York.
149. Yoshinaga-Hirabayashi, T., Ishimura, K., Fujita, H., Inano, H., Ishii-Ohba, H., and Tamaoki, B. (1987). Immunocytochemical localization of 17 beta-hydroxysteroid dehydrogenase (17 beta-HSD), and its relation to the ultrastructure of steroidogenic cells in immature and mature rat ovaries. *Arch. Histol. Jpn.* 50, 545–556.

150. Martel, C., Rheaume, E., Takahashi, M., Trudel, C., Couet, J., Luu-The, V., Simard, J., and Labrie, F. (1992). Distribution of 17 beta-hydroxysteroid dehydrogenase gene expression and activity in rat and human tissues. *J. Steroid Biochem. Mol. Biol.* 41, 597–603.
151. Wu, L., Einstein, M., Geissler, W. M., Chan, H. K., Elliston, K. O., and Andersson, S. (1993). Expression cloning and characterization of human 17 beta-hydroxysteroid dehydrogenase type 2, a microsomal enzyme possessing 20 alpha-hydroxysteroid dehydrogenase activity. *J. Biol. Chem.* 268, 12964–12969.
152. Geissler, W. M., Davis, D. L., Wu, L., Bradshaw, K. D., Patel, S., Mendonca, B. B., Elliston, K. O., Wilson, J. D., Russell, D. W., and Andersson, S. (1994). Male pseudohermaphroditism caused by mutations of testicular 17 beta-hydroxysteroid dehydrogenase 3. *Nat. Genet.* 7, 34–39.
153. Ademola Akesode, F., Meyer, W. J. III, and Migeon, C. J. (1977). Male pseudohermaphroditism with gynaecomastia due to testicular 17-ketosteroid reductase deficiency. *Clin. Endocrinol. (Oxf).* 7, 443–452.
154. Wright, W. W., and Frankel, A. I. (1979). Endogenous androgen concentrations in nuclei isolated from seminiferous tubules of mature rat testes. *J. Steroid Biochem.* 10, 633–640.
155. Russell, D. W., and Wilson, J. D. (1994). Steroid 5 alpha-reductase: two genes/two enzymes. *Annu. Rev. Biochem.* 63, 25–61.
156. Matsumoto, K., and Yamada, M. (1973). 5-Reduction of testosterone in vitro by rat seminiferous tubules and whole testes at different stages of development. *Endocrinology* 93, 253–255.
157. Corpechot, C., Baulieu, E. E., and Robel, P. (1981). Testosterone, dihydrotestosterone and androstanediols in plasma, testes and prostates of rats during development. *Acta. Endocrinol. (Copenh).* 96, 127–135.
158. Van der Molen, H. J., Grootegoed, J., de Greef-Bijeveld, M., Rommerts, F. F., and van der Vuuse, G. (1975). Distribution of steroids, steroid production and steroid metabolizing enzymes in rat testis. In *Hormonal Regulation of Spermatogenesis* (F. French, V. Hansson, E. M. Ritzen, and S. Nayfey, Eds.), pp. 3–23. Plenum Press, New York.
159. Sheffield, J. W., and O'Shaughnessy, P. J. (1999). Testicular steroid metabolism during development in the normal and hypogonadal mouse. *J. Endocrinol.* 119, 257–264.
160. Ficher, M., and Steinberger, E. (1971). In vitro progesterone metabolism by rat testicular tissue at different stages of development. *Acta. Endocrinol. (Copenh).* 68, 285–292.
161. Chase, D. J., and Payne, A. H. (1983). Changes in distribution and androgen production of Leydig cells of two populations during sexual maturation in the rat. *Endocrinology* 112, 29–34.
162. Tsujimura, T., and Matsumoto, K. (1974). Progesterone metabolism in vitro by mouse testes at different stages of development. *Endocrinology* 94, 288–290.
163. Chase, D. J., and Payne, A. H. (1983). Changes in Leydig cell function during sexual maturation in the mouse. *Biol. Reprod.* 29, 1194–1200.
164. Watkins, W. J., Goldring, C. E., and Gower, D. B. (1988). Properties of 4-ene-5 alpha-reductase and studies on its solubilization from porcine testicular microsomes. *J. Steroid Biochem.* 29, 325–331.
165. Anderson, R. A., Wallace, A. M., and Wu, F. C. (1996). Comparison between testosterone enanthate-induced azoospermia and oligozoospermia in a male contraceptive study. III. Higher 5 alpha-reductase activity in oligozoospermic men administered supraphysiological doses of testosterone. *J. Clin. Endocrinol. Metab.* 81, 902–908.
166. O'Donnell, L., Stanton, P. G., Wreford, N. G., Robertson, D. M., and McLachlan, R. I. (1996). Inhibition of 5 alpha-reductase activity impairs the testosterone-dependent restoration of spermiogenesis in adult rats. *Endocrinology* 137, 2703–2710.
167. O'Donnell, L., Pratis, K., Stanton, P. G., Robertson, D. M., and McLachlan, R. I. (1999). Testosterone-dependent restoration of spermatogenesis in adult rats is impaired by a 5alpha-reductase inhibitor. *J. Androl.* 20, 109–117.
168. Thigpen, A. E., Cala, K. M., and Russell, D. W. (1993). Characterization of Chinese hamster ovary cell lines expressing human steroid 5 alpha-reductase isozymes. *J. Biol. Chem.* 268, 17404–17412.
169. Normington, K., and Russell, D. W. (1992). Tissue distribution and kinetic characteristics of rat steroid 5 alpha-reductase isozymes. Evidence for distinct physiological functions. *J. Biol. Chem.* 267, 19548–19554.
170. Lopez-Solache, I., Luu-The, V., Seralini, G. E., and Labrie, F. (1996). Heterogeneity of rat type I 5 alpha-reductase cDNA: cloning, expression and regulation by pituitary implants and dihydrotestosterone. *Biochim. Biophys. Acta.* 1305, 139–144.
171. Horton, R., Pasupuletti, V., and Antonipillai, I. (1993). Androgen induction of steroid 5 alpha-reductase may be mediated via insulin-like growth factor-I. *Endocrinology* 133, 447–451.
172. Antonipillai, I., Wahe, M., Yamamoto, J., and Horton, R. (1995). Activin and inhibin have opposite effects on steroid 5 alpha-reductase activity in genital skin fibroblasts. *Mol. Cell. Endocrinol.* 107, 99–104.
173. Pratis, K., O'Donnell, L., Ooi, G. T., Stanton, P. G., McLachlan, R. I., and Robertson, D. M. (2003). Differential regulation of rat testicular 5alpha-reductase type 1 and 2 isoforms by testosterone and FSH. *J. Endocrinol.* 176, 393–403.
174. Viger, R. S., and Robaire, B. (1995). Steady state steroid 5 alpha-reductase messenger ribonucleic acid levels and immunocytochemical localization of the type 1 protein in the rat testis during postnatal development. *Endocrinology* 136, 5409–5415.
175. Thigpen, A. E., Silver, R. I., Guileyardo, J. M., Casey, M. L., McConnell, J. D., and Russell, D. W. (1993). Tissue distribution and ontogeny of steroid 5 alpha-reductase isozyme expression. *J. Clin. Invest.* 92, 903–910.
176. Reyes, E. M., Camacho-Arroyo, I., Nava, G., and Cerbon, M. A. (1997). Differential methylation in steroid 5 alpha-reductase isozyme genes in epididymis, testis, and liver of the adult rat. *J. Androl.* 18, 372–377.
177. Killian, J., Pratis, K., Clifton, R. J., Stanton, P. G., Robertson, D. M., and O'Donnell, L. (2003). 5 alpha-reductase isoenzymes 1 and 2 in the rat testis during postnatal development. *Biol. Reprod.* 68, 1711–1718.
178. Frost, P. G., Reed, M. J., and James, V. H. (1980). The aromatization of androstenedione by human adipose and liver tissue. *J. Steroid Biochem.* 13, 1427–1431.
179. Frieden, E. H., Patkin, J. K., and Mills, M. (1968). Effects of follicle stimulating hormone (FSH) upon steroid aromatization in vitro. *Proc. Soc. Exp. Biol. Med.* 129, 606–609.
180. Ryan, K. J., Naftolin, F., Reddy, V., Flores, F., and Petro, Z. (1972). Estrogen formation in the brain. *Am. J. Obstet. Gynecol.* 114, 454–460.
181. Gray, S. A., Mannan, M. A., and O'Shaughnessy, P. J. (1995). Development of cytochrome P450 aromatase mRNA levels and enzyme activity in ovaries of normal and hypogonadal (hpg) mice. *J. Mol. Endocrinol.* 14, 295–301.
182. Hinshelwood, M. M., Corbin, C. J., Tsang, P. C., and Simpson, E. R. (1993). Isolation and characterization of a complementary deoxyribonucleic acid insert encoding bovine aromatase cytochrome P450. *Endocrinology* 133, 1971–1977.
183. Terashima, M., Toda, K., Kawamoto, T., Kuribayashi, I., Ogawa, Y., Maeda, T., and Shizuta, Y. (1991). Isolation of a full-length cDNA encoding mouse aromatase P450. *Arch. Biochem. Biophys.* 285, 231–237.

184. Means, G. D., Mahendroo, M. S., Corbin, C. J., Mathis, J. M., Powell, F. E., Mendelson, C. R., and Simpson, E. R. (1989). Structural analysis of the gene encoding human aromatase cytochrome P-450, the enzyme responsible for estrogen biosynthesis. *J. Biol. Chem.* 264, 19385–19391.
185. Sebastian, S., and Bulun, S. E. (2001). A highly complex organization of the regulatory region of the human CYP19 (aromatase) gene revealed by the Human Genome Project. *J. Clin. Endocrinol. Metab.* 86, 4600–4602.
186. Toda, K., Terashima, M., Kawamoto, T., Sumimoto, H., Yokoyama, Y., Kuribayashi, I., Mitsuuchi, Y., Maeda, T., Yamamoto, Y., Sagara, Y., et al. (1990). Structural and functional characterization of human aromatase P-450 gene. *Eur. J. Biochem.* 193, 559–565.
187. Means, G. D., Kilgore, M. W., Mahendroo, M. S., Mendelson, C. R., and Simpson, E. R. (1991). Tissue-specific promoters regulate aromatase cytochrome P450 gene expression in human ovary and fetal tissues. *Mol. Endocrinol.* 5, 2005–2013.
188. Mahendroo, M. S., Means, G. D., Mendelson, C. R., and Simpson, E. R. (1991). Tissue-specific expression of human P-450AROM. The promoter responsible for expression in adipose tissue is different from that utilized in placenta. *J. Biol. Chem.* 266, 11276–11281.
189. Toda, K., and Shizuta, Y. (1993). Molecular cloning of a cDNA showing alternative splicing of the 5′-untranslated sequence of mRNA for human aromatase P-450. *Eur. J. Biochem.* 213, 383–389.
190. Harada, N., Yamada, K., Saito, K., Kibe, N., Dohmae, S., and Takagi, Y. (1990). Structural characterization of the human estrogen synthetase (aromatase) gene. *Biochem. Biophys. Res. Commun.* 166, 365–372.
191. Pierrepoint, C. G., Griffiths, K., Grant, J. K., and Stewart, J. S. (1966). Neutral steroid sulphation and oestrogen biosynthesis in vitro by a feminizing Leydig cell tumour of the testis *J. Endocrinol.* 35, 409–417.
192. Valladares, L. E., and Payne, A. H. (1979). Acute stimulation of aromatization in Leydig cells by human chorionic gonadotropin in vitro. *Proc. Natl. Acad. Sci. U S A* 76, 4460–4463.
193. Raeside, J. I., and Lobb, D. K. (1984). Metabolism of androstenedione by Sertoli cell enriched preparations and purified Leydig cells from boar testes in relation to estrogen formation. *J. Steroid Biochem.* 20, 1267–1272.
194. Raeside, J. I., and Renaud, R. L. (1983). Estrogen and androgen production by purified Leydig cells of mature boars. *Biol. Reprod.* 28, 727–733.
195. Dorrington, J. H., Fritz, I. B., and Armstrong, D. T. (1978). Control of testicular estrogen synthesis. *Biol. Reprod.* 18, 55–64.
196. Nitta, H., Bunick, D., Hess, R. A., Janulis, L., Newton, S. C., Millette, C. F., Osawa, Y., Shizuta, Y., Toda, K., and Bahr, J. M. (1993). Germ cells of the mouse testis express P450 aromatase. *Endocrinology* 132, 1396–1401.
197. Rommerts, F. F., de Jong, F. H., Brinkmann, A. O., and van der Molen, H. J. (1982). Development and cellular localization of rat testicular aromatase activity. *J. Reprod. Fertil.* 65, 281–288.
198. Eisenhauer, K. M., McCue, P. M., Nayden, D. K., Osawa, Y., and Roser J. F. (1994). Localization of aromatase in equine Leydig cells. *Domest. Anim. Endocrinol.* 11, 291–298.
199. Almadhidi, J., Seralini, G. E., Fresnel, J., Silberzahn, P., and Gaillard, J. L. (1995). Immunohistochemical localization of cytochrome P450 aromatase in equine gonads. *J. Histochem. Cytochem.* 43, 571–577.
200. Payne, A. H., Kelch, R. P., Musich, S. S., and Halpern, M. E. (1976). Intratesticular site of aromatization in the human. *J. Clin. Endocrinol. Metab.* 42, 1081–1087.
201. Raeside, J. I. (1978). Seasonal changes in the concentration of estrogens and testosterone in the plasma of the stallion. *Anim. Reprod. Sci.* 1, 205–212.
202. Carreau, S., Genissel, C., Bilinska, B., and Levallet, J. (1999). Sources of oestrogen in the testis and reproductive tract of the male. *Int. J. Androl.* 22, 211–223.
203. Carreau, S., Lambard, S., Delalande, C., Denis-Galeraud, I., Bilinska, B., and Bourguiba, S. (2003). Aromatase expression and role of estrogens in male gonad: a review. *Reprod. Biol. Endocrinol.* 1, 35.
204. Carreau, S., Bourguiba, S., Lambard, S., Galeraud-Denis, I., Genissel, C., and Levallet, J. (2002). Reproductive system: aromatase and estrogens. *Mol. Cell. Endocrinol.* 193, 137–143.
205. Levallet, J., Bilinska, B., Mittre, H., Genissel, C., Fresnel, J., and Carreau, S. (1998). Expression and immunolocalization of functional cytochrome P450 aromatase in mature rat testicular cells. *Biol. Reprod.* 58, 919–926.
206. Janulis, L., Bahr, J. M., Hess, R. A., Janssen, S., Osawa, Y., and Bunick, D. (1998). Rat testicular germ cells and epididymal sperm contain active P450 aromatase. *J. Androl.* 19, 65–71.
207. Carani, C., Qin, K., Simoni, M., Faustini-Fustini, M., Serpente, S., Boyd, J., Korach, K. S., and Simpson, E. R. (1997). Effect of testosterone and estradiol in a man with aromatase deficiency. *N. Engl. J. Med.* 337, 91–95.
208. Morishima, A., Grumbach, M. M., Simpson, E. R., Fisher, C., and Qin, K. (1995). Aromatase deficiency in male and female siblings caused by a novel mutation and the physiological role of estrogens. *J. Clin. Endocrinol. Metab.* 80, 3689–3698.
209. Smith, E. P., Boyd, J., Frank, G. R., Takahashi, H., Cohen, R. M., Specker, B., Williams, T. C., Lubahn, D. B., and Korach, K. S. (1994). Estrogen resistance caused by a mutation in the estrogen-receptor gene in a man. *N. Engl. J. Med.* 331, 1056–1061.
210. Robertson, K. M., O'Donnell, L., Jones, M. E., Meachem, S. J., Boon, W. C., Fisher, C. R., Graves, K. H., McLachlan, R. I., and Simpson, E. R. (1999). Impairment of spermatogenesis in mice lacking a functional aromatase (cyp 19) gene. *Proc. Natl. Acad. Sci. U S A* 96, 7986–7991.
211. Couse, J. F., and Korach, K. S. (1999). Estrogen receptor null mice: what have we learned and where will they lead us? *Endocr. Rev.* 20, 358–417.
212. Simpson, E. R., and Boyd, G. S. (1966). The cholesterol side-chain cleavage system of the adrenal cortex: a mixed-function oxidase. *Biochem. Biophys. Res. Commun.* 24, 10–17.
213. Simpson, E. R., and Boyd, G. S. (1967). The cholesterol side-chain cleavage system of bovine adrenal cortex. *Eur. J. Biochem.* 2, 275–285.
214. Hanukoglu, I., and Hanukoglu, Z. (1986). Stoichiometry of mitochondrial cytochromes P-450, adrenodoxin and adrenodoxin reductase in adrenal cortex and corpus luteum. Implications for membrane organization and gene regulation. *Eur. J. Biochem.* 157, 27–31.
215. Liscum, L., and Dahl, N. K. (1992). Intracellular cholesterol transport. *J. Lipid Res.* 33, 1239–1254.
216. Stocco, D. M., and Clark, B. J. (1996). Regulation of the acute production of steroids in steroidogenic cells. *Endocr. Rev.* 17, 221–244.
217. Scallen, T. J., Pastuszyn, A., Noland, B. J., Chanderbhan, R., Kharroubi, A., and Vahouny, G. V. (1985). Sterol carrier and lipid transfer proteins. *Chem. Phys. Lipids* 38, 239–261.
218. Pedersen, R. C., and Brownie, A. C. (1986). The mechanism of action of adrenocorticotropic hormone on cholesterol metabolism in the adrenal cortex. In *Biochemical Actions of Hormones* (G. Litwack, Ed.), Vol. 13, pp. 130–166. Academic Press, Orlando, FL.
219. van Meer, G. (1989). Lipid traffic in animal cells. *Annu. Rev. Cell. Biol.* 5, 247–275.
220. Schroeder, F., Jefferson, J. R., Kier, A. B., Knittel, J., Scallen, T. J., Wood, W. G., and Hapala, I. (1991). Membrane cholesterol dynamics: cholesterol domains and kinetic pools. *Proc. Soc. Exp. Biol. Med.* 196, 235–252.

221. Zhang, M., Liu, P., Dwyer, N. K., Christenson, L. K., Fujimoto, T., Martinez, F., Comly, M., Hanover, J. A., Blanchette-Mackie, E. J., and Strauss, J. F. III. (2002). MLN64 mediates mobilization of lysosomal cholesterol to steroidogenic mitochondria. *J. Biol. Chem.* 277, 33300–33310.
222. Kishida, T., Kostetskii, I., Zhang, Z., Martinez, F., Liu, P., Walkley, S. U., Dwyer, N. K., Blanchette-Mackie, E. J., Radice, G. L., and Strauss, J. F. III. (2004). Targeted mutation of the MLN64 START domain causes only modest alterations in cellular sterol metabolism. *J. Biol. Chem.* 279, 19276–19285.
223. Garren, L. D., Gill, G. N., Masui, H., and Walton, G. M. (1971). On the mechanism of action of ACTH. *Recent Prog. Horm. Res.* 27, 433–478.
224. Brownie, A. C., Simpson, E. R., Jefcoate, C. R., Boyd, G. S., Orme-Johnson, W. H., and Beinert, H. (1972). Effect of ACTH on cholesterol side-chain cleavage in rat adrenal mitochondria. *Biochem. Biophys. Res. Commun.* 46, 483–490.
225. Brownie, A. C., Alfano, J., Jefcoate, C. R., Orme-Johnson, W., Beinert, H., and Simpson, E. R. (1973). Effect of ACTH on adrenal mitochondrial cytochrome P-450 in the rat. *Ann. N Y Acad. Sci.* 212, 344–360.
226. Simpson, E. R., McCarthy, J. L., and Peterson, J. A. (1978). Evidence that the cycloheximide-sensitive site of adrenocorticotropic hormone action is in the mitochondrion. Changes in pregnenolone formation, cholesterol content, and the electron paramagnetic resonance spectra of cytochrome P-450. *J. Biol. Chem.* 253, 3135–3139.
227. Crivello, J. F., and Jefcoate, C. R. (1980). Intracellular movement of cholesterol in rat adrenal cells. Kinetics and effects of inhibitors. *J. Biol. Chem.* 255, 8144–8151.
228. Mori, M., and Marsh, J. M. (1982). The site of luteinizing hormone stimulation of steroidogenesis in mitochondria of the rat corpus luteum. *J. Biol. Chem.* 257, 6178–6183.
229. Privalle, C. T., Crivello, J. F., and Jefcoate, C. R. (1983). Regulation of intramitochondrial cholesterol transfer to side-chain cleavage cytochrome P-450 in rat adrenal gland. *Proc. Natl. Acad. Sci. U S A* 80, 702–706.
230. Jefcoate, C. R., DiBartolomeis, M. J., Williams, C. A., and McNamara, B. C. (1987). ACTH regulation of cholesterol movement in isolated adrenal cells. *J. Steroid Biochem.* 27, 721–729.
231. Lambeth, J. D., Kitchen, S. E., Farooqui, A. A., Tuckey, R., and Kamin, H. (1982). Cytochrome P-450scc-substrate interactions. Studies of binding and catalytic activity using hydroxycholesterols. *J. Biol. Chem.* 257, 1876–1884.
232. Tuckey, R. C., and Stevenson, P. M. (1984). Properties of bovine luteal cytochrome P-450scc incorporated into artificial phospholipid vesicles. *Int. J. Biochem.* 16, 497–503.
233. Tuckey, R. C., and Atkinson, H. C. (1989). Pregnenolone synthesis from cholesterol and hydroxycholesterols by mitochondria from ovaries following the stimulation of immature rats with pregnant mare's serum gonadotropin and human choriogonadotropin. *Eur. J. Biochem.* 186, 255–259.
234. Tuckey, R. C. (1992). Cholesterol side-chain cleavage by mitochondria from the human placenta. Studies using hydroxycholesterols as substrates. *J. Steroid Biochem. Mol. Biol.* 42, 883–890.
235. Phillips, M. C., Johnson, W. J., and Rothblat, G. H. (1987). Mechanisms and consequences of cellular cholesterol exchange and transfer. *Biochim. Biophys. Acta.* 906, 223–276.
236. Rennert, H., Y. J. C., and Strauss, J. F. (1993). Intracellular cholesterol dynamics in steroidogenic cells. A contemporary view. In *The Ovary* (E. Y. Adashi and P. C. Leung, Eds.), pp. 147–164. Raven Press, New York.
237. Ferguson, J. J. (1962). Puromycin and adrenal responsiveness to adrenocorticotropic hormone. *Biochim. Biophys. Acta.* 57, 616–617.
238. Ferguson, J. J. (1963). Protein synthesis and adrenocorticotropin responsiveness. *J. Biol. Chem.* 238, 2754–2759.
239. Garren, L. D., Ney, R. L., and Davis, W. W. (1965). Studies on the role of protein synthesis in the regulation of corticosterone production by adrenocorticotropic hormone in vivo. *Proc. Natl. Acad. Sci. U S A* 53, 1443–1450.
240. Garren, L. D., Davis, W. W., Crocco, R. M., and Ney, R. L. (1966). Puromycin analogs: action of adrenocorticotropic hormone and the role of glycogen. *Science* 152, 1386–1388.
241. Davis, W. W., and Garren, L. D. (1968). On the mechanism of action of adrenocorticotropic hormone. The inhibitory site of cycloheximide in the pathway of steroid biosynthesis. *J. Biol. Chem.* 243, 5153–5157.
242. Garren, L. D. (1968). The mechanism of action of adrenocorticotropic hormone. *Vitam. Horm.* 26, 119–145.
243. Davis, W. W., and Garren, L. D. (1966). Evidence for the stimulation by adenocorticotropic hormone of the conversion of cholesterol esters to cholesterol in the adrenal, in vivo. *Biochem. Biophys. Res. Commun.* 22, 805–810.
244. Farese, R. V. (1967). Adrenocorticotrophin-induced changes in the steroidogenic activity of adrenal cell-free preparations. *Biochemistry* 6, 2052–2065.
245. Cooke, B. A., Janszen, F. H., Clotscher, W. F., and van der Molen, H. J. (1975). Effect of protein-synthesis inhibitors on testosterone production in rat testis interstitial tissue and Leydig-cell preparations. *Biochem. J.* 150, 413–418.
246. Paul, D. P., Gallant, S., Orme-Johnson, N. R., Orme-Johnson, W. H., and Brownie, A. C. (1976). Temperature dependence of cholesterol binding to cytochrome P-450scc of the rat adrenal. Effect of adrenocorticotropic hormone and cycloheximide. *J. Biol. Chem.* 251, 7120–7126.
247. Farese, R. V., and Prudente, W. J. (1977). On the requirement for protein synthesis during corticotropin-induced stimulation of cholesterol side chain cleavage in rat adrenal mitochondrial and solubilized desmolase preparations. *Biochim. Biophys. Acta.* 496, 567–570.
248. Crivello, J. F., and Jefcoate, C. R. (1978). Mechanisms of corticotropin action in rat adrenal cells. I. The effects of inhibitors of protein synthesis and of microfilament formation on corticosterone synthesis. *Biochim. Biophys. Acta.* 542, 315–329.
249. Toaff, M. E., Strauss, J. F. III, Flickinger, G. L., and Shattil, S. J. (1979). Relationship of cholesterol supply to luteal mitochondrial steroid synthesis. *J. Biol. Chem.* 254, 3977–3982.
250. Solano, A. R., Neher, R., and Podesta, E. J. (1984). Rat adrenal cycloheximide-sensitive factors and phospholipids in the control of acute steroidogenesis. *J. Steroid Biochem.* 21, 111–116.
251. Stevens, V. L., Xu, T., and Lambeth, J. D. (1993). Cholesterol trafficking in steroidogenic cells. Reversible cycloheximide-dependent accumulation of cholesterol in a pre-steroidogenic pool. *Eur. J. Biochem.* 216, 557–563.
252. Janszen, F. H., Cooke, B. A., van Driel, M. J., and van der Molen, H. J. (1976). LH induction of a specific protein (LH-IP) in rat testis Leydig cells. *FEBS Lett.* 71, 269–272.
253. Janszen, F. H., Cooke, B. A., and van der Molen, H. J. (1977). Specific protein synthesis in isolated rat testis Leydig cells. Influence of luteinizing hormone and cycloheximide. *Biochem. J.* 162, 341–346.
254. Krueger, R. J., and Orme-Johnson, N. R. (1983). Acute adrenocorticotropic hormone stimulation of adrenal corticosteroidogenesis. Discovery of a rapidly induced protein. *J. Biol. Chem.* 258, 10159–10167.
255. Epstein, L. F., and Orme-Johnson, N. R. (1991). Acute action of luteinizing hormone on mouse Leydig cells: accumulation of mitochondrial phosphoproteins and stimulation of testosterone synthesis. *Mol. Cell. Endocrinol.* 81, 113–126.

256. Epstein, L. F., and Orme-Johnson, N. R. (1991). Regulation of steroid hormone biosynthesis. Identification of precursors of a phosphoprotein targeted to the mitochondrion in stimulated rat adrenal cortex cells. *J. Biol. Chem.* 266, 19739–19745.
257. Pon, L. A., and Orme-Johnson, N. R. (1988). Acute stimulation of corpus luteum cells by gonadotrophin or adenosine 3′,5′-monophosphate causes accumulation of a phosphoprotein concurrent with acceleration of steroid synthesis. *Endocrinology* 123, 1942–1948.
258. Stocco, D. M., and Sodeman, T. C. (1991). The 30-kDa mitochondrial proteins induced by hormone stimulation in MA-10 mouse Leydig tumor cells are processed from larger precursors. *J. Biol. Chem.* 266, 19731–19738.
259. Stocco, D. M., and Chen, W. (1991). Presence of identical mitochondrial proteins in unstimulated constitutive steroid-producing R2C rat Leydig tumor and stimulated nonconstitutive steroid-producing MA-10 mouse Leydig tumor cells. *Endocrinology* 128, 1918–1926.
260. Arthur, J. R., and Boyd, G. S. (1974). The effect of inhibitors of protein synthesis on cholesterol side-chain cleavage in the mitochondria of luteinized rat ovaries. *Eur. J. Biochem.* 49, 117–127.
261. Ohno, Y., Yanagibashi, K., Yonezawa, Y., Ishiwatari, S., and Matsuba M. (1983). A possible role of "steroidogenic factor" in the corticoidogenic response to ACTH; effect of ACTH, cycloheximide and aminoglutethimide on the content of cholesterol in the outer and inner mitochondrial membrane of rat adrenal cortex. *Endocrinol. Jpn.* 30, 335–338.
262. Papadopoulos, V. (1993). Peripheral-type benzodiazepine/diazepam binding inhibitor receptor: biological role in steroidogenic cell function. *Endocr. Rev.* 14, 222–240.
263. Lacapere J. J., and Papadopoulos, V. (2003). Peripheral-type benzodiazepine receptor: structure and function of a cholesterol-binding protein in steroid and bile acid biosynthesis. *Steroids* 68, 569–585.
264. Pon, L. A., and Orme-Johnson, N. R. (1986). Acute stimulation of steroidogenesis in corpus luteum and adrenal cortex by peptide hormones. Rapid induction of a similar protein in both tissues. *J. Biol. Chem.* 261, 6594–6599.
265. Pon, L. A., Hartigan, J. A., and Orme-Johnson, N. R. (1986). Acute ACTH regulation of adrenal corticosteroid biosynthesis. Rapid accumulation of a phosphoprotein. *J. Biol. Chem.* 261, 13309–13316.
266. Pon, L. A., Epstein, L. F., and Orme-Johnson, N. R. (1986). Acute cAMP stimulation in Leydig cells: rapid accumulation of a protein similar to that detected in adrenal cortex and corpus luteum. *Endocr. Res.* 12, 429–446.
267. Alberta, J. A., Epstein, L. F., Pon, L. A., and Orme-Johnson, N. R. (1989). Mitochondrial localization of a phosphoprotein that rapidly accumulates in adrenal cortex cells exposed to adrenocorticotropic hormone or to cAMP. *J. Biol. Chem.* 264, 2368–2372.
268. Stocco, D. M., and Kilgore, M. W. (1988). Induction of mitochondrial proteins in MA-10 Leydig tumour cells with human choriogonadotropin. *Biochem. J.* 249, 95–103.
269. Stocco, D. M., and Chaudhary, L. R. (1990). Evidence for the functional coupling of cyclic AMP in MA-10 mouse Leydig tumour cells. *Cell. Signal.* 2, 161–170.
270. Stocco, D. M. (1992). Further evidence that the mitochondrial proteins induced by hormone stimulation in MA-10 mouse Leydig tumor cells are involved in the acute regulation of steroidogenesis. *J. Steroid Biochem. Mol. Biol.* 43, 319–333.
271. Stocco, D. M., and Ascoli, M. (1993). The use of genetic manipulation of MA-10 Leydig tumor cells to demonstrate the role of mitochondrial proteins in the acute regulation of steroidogenesis. *Endocrinology* 132, 959–967.
272. Stocco, D. M., King, S., and Clark, B. J. (1995). Differential effects of dimethylsulfoxide on steroidogenesis in mouse MA-10 and rat R2C Leydig tumor cells. *Endocrinology* 136, 2993–2999.
273. Clark, B. J., Wells, J., King, S. R., and Stocco, D. M. (1994). The purification, cloning, and expression of a novel luteinizing hormone-induced mitochondrial protein in MA-10 mouse Leydig tumor cells. Characterization of the steroidogenic acute regulatory protein (StAR). *J. Biol. Chem.* 269, 28314–28322.
274. Lin, D., Gitelman, S. E., Saenger, P., and Miller, W. L. (1991). Normal genes for the cholesterol side chain cleavage enzyme, P450scc, in congenital lipoid adrenal hyperplasia. *J. Clin. Invest.* 88, 1955–1962.
275. Lin, D., Chang, Y. J., Strauss, J. F. III, and Miller, W. L. (1993). The human peripheral benzodiazepine receptor gene: cloning and characterization of alternative splicing in normal tissues and in a patient with congenital lipoid adrenal hyperplasia. *Genomics* 18, 643–650.
276. Lin, D., Sugawara, T., Strauss, J. F. III, Clark, B. J., Stocco, D. M., Saenger, P., Rogol, A., and Miller, W. L. (1995). Role of steroidogenic acute regulatory protein in adrenal and gonadal steroidogenesis. *Science* 267, 1828–1831.
277. Bose, H. S., Sugawara, T., Strauss, J. F. III, and Miller, W. L. (1996). The pathophysiology and genetics of congenital lipoid adrenal hyperplasia. International Congenital Lipoid Adrenal Hyperplasia Consortium. *N. Engl. J. Med.* 335, 1870–1878.
278. Stocco, D. M. (2002). Clinical disorders associated with abnormal cholesterol transport: mutations in the steroidogenic acute regulatory protein. *Mol. Cell. Endocrinol.* 191, 19–25.
279. Caron, K. M., Soo, S. C., Wetsel, W. C., Stocco, D. M., Clark, B. J., and Parker, K. L. (1997). Targeted disruption of the mouse gene encoding steroidogenic acute regulatory protein provides insights into congenital lipoid adrenal hyperplasia. *Proc. Natl. Acad. Sci. U S A* 94, 11540–11545.
280. Pedersen, R. C., and Brownie, A. C. (1983). Cholesterol side-chain cleavage in the rat adrenal cortex: isolation of a cycloheximide-sensitive activator peptide. *Proc. Natl. Acad. Sci. U S A* 80, 1882–1886.
281. Pedersen, R. C. (1984). Polypeptide activators of cholesterol side-chain cleavage. *Endocr. Res.* 10, 533–561.
282. Pedersen, R. C., and Brownie, A. C. (1987). Steroidogenesis-activator polypeptide isolated from a rat Leydig cell tumor. *Science* 236, 188–190.
283. Pedersen, R. C. (1987). Steroidogenesis activator polypeptide (SAP) in the rat ovary and testis. *J. Steroid Biochem.* 27, 731–735.
284. Mertz, L. M., and Pedersen, R. C. (1989). The kinetics of steroidogenesis activator polypeptide in the rat adrenal cortex. Effects of adrenocorticotropin, cyclic adenosine 3′:5′-monophosphate, cycloheximide, and circadian rhythm. *J. Biol. Chem.* 264, 15274–15279.
285. Frustaci, J., Mertz, L. M., and Pedersen, R. C. (1989). Steroidogenesis activator polypeptide (SAP) in the guinea pig adrenal cortex. *Mol. Cell. Endocrinol.* 64, 137–143.
286. Mertz, L. M., and Pedersen, R. C. (1989). Steroidogenesis activator polypeptide may be a product of glucose regulated protein 78 (GRP78). *Endocr. Res.* 15, 101–115.
287. Li, X. A., Warren, D. W., Gregoire, J., Pedersen, R. C., and Lee, A. S. (1989). The rat 78,000 dalton glucose-regulated protein (GRP78) as a precursor for the rat steroidogenesis-activator polypeptide (SAP): the SAP coding sequence is homologous with the terminal end of GRP78. *Mol. Endocrinol.* 3, 1944–1952.
288. Woods, M. J., Zisterer, D. M., and Williams, D. C. (1996). Two cellular and subcellular locations for the peripheral-type benzodiazepine receptor in rat liver. *Biochem. Pharmacol.* 51, 1283–1292.

289. Garnier, M., Boujrad, N., Oke, B. O., Brown, A. S., Riond, J., Ferrara, P., Shoyab, M., Suarez-Quian, C. A., and Papadopoulos, V. (1993). Diazepam binding inhibitor is a paracrine/autocrine regulator of Leydig cell proliferation and steroidogenesis: action via peripheral-type benzodiazepine receptor and independent mechanisms. *Endocrinology* 132, 444–458.
290. Hardwick, M., Fertikh, D., Culty, M., Li, H., Vidic, B., and Papadopoulos, V. (1999). Peripheral-type benzodiazepine receptor (PBR) in human breast cancer: correlation of breast cancer cell aggressive phenotype with PBR expression, nuclear localization, and PBR-mediated cell proliferation and nuclear transport of cholesterol. *Cancer Res.* 59, 831–842.
291. Verma, A., and Snyder, S. H. (1989). Peripheral type benzodiazepine receptors. *Annu. Rev. Pharmacol. Toxicol.* 29, 307–322.
292. Parola, A. L., Yamamura, H. I., and Laird, H. E. III. (1993). Peripheral-type benzodiazepine receptors. *Life Sci.* 52, 1329–1342.
293. Yanagibashi, K., Ohno, Y., Kawamura, M., and Hall, P. F. (1988). The regulation of intracellular transport of cholesterol in bovine adrenal cells: purification of a novel protein. *Endocrinology* 123, 2075–2082.
294. Besman, M. J., Yanagibashi, K., Lee, T. D., Kawamura, M., Hall, P. F., and Shively, J. E. (1989). Identification of des-(Gly-Ile)-endozepine as an effector of corticotropin-dependent adrenal steroidogenesis: stimulation of cholesterol delivery is mediated by the peripheral benzodiazepine receptor. *Proc. Natl. Acad. Sci. U S A* 86, 4897–4901.
295. Holloway, C. D., Kenyon, C. J., Dowie, L. J., Corrie, J. E., Gray, C. E., and Fraser, R. (1989). Effect of the benzodiazepines diazepam, des-N-methyldiazepam and midazolam on corticosteroid biosynthesis in bovine adrenocortical cells in vitro; location of site of action. *J. Steroid Biochem.* 33, 219–225.
296. Amsterdam, A., and Suh, B. S. (1991). An inducible functional peripheral benzodiazepine receptor in mitochondria of steroidogenic granulosa cells. *Endocrinology* 129, 503–510.
297. Yanagibashi, K., Ohno, Y., Nakamichi, N., Matsui, T., Hayashida, K., Takamura, M., Yamada, K., Tou, S., and Kawamura, M. (1989). Peripheral-type benzodiazepine receptors are involved in the regulation of cholesterol side chain cleavage in adrenocortical mitochondria. *J. Biochem. (Tokyo)* 106, 1026–1029.
298. Papadopoulos, V., Mukhin, A. G., Costa, E., and Krueger, K. E. (1990). The peripheral-type benzodiazepine receptor is functionally linked to Leydig cell steroidogenesis. *J. Biol. Chem.* 265, 3772–3779.
299. Papadopoulos, V., Berkovich, A., Krueger, K. E., Costa, E., and Guidotti, A. (1991). Diazepam binding inhibitor and its processing products stimulate mitochondrial steroid biosynthesis via an interaction with mitochondrial benzodiazepine receptors. *Endocrinology* 129, 1481–1488.
300. Cavallaro, S., Korneyev, A., Guidotti, A., and Costa, E. (1992). Diazepam-binding inhibitor (DBI)-processing products, acting at the mitochondrial DBI receptor, mediate adrenocorticotropic hormone-induced steroidogenesis in rat adrenal gland. *Proc. Natl. Acad. Sci. U S A* 89, 10598–10602.
301. Barnea, E. R., Fares, F., and Gavish, M. (1989). Modulatory action of benzodiazepines on human term placental steroidogenesis in vitro. *Mol. Cell. Endocrinol.* 64, 155–159.
302. Mukhin, A. G., Papadopoulos, V., Costa, E., and Krueger, K. E. (1989). Mitochondrial benzodiazepine receptors regulate steroid biosynthesis. *Proc. Natl. Acad. Sci. U S A* 86, 9813–9816.
303. Boujrad, N., Hudson, J. R. Jr., and Papadopoulos, V. (1993). Inhibition of hormone-stimulated steroidogenesis in cultured Leydig tumor cells by a cholesterol-linked phosphorothioate oligodeoxynucleotide antisense to diazepam-binding inhibitor. *Proc. Natl. Acad. Sci. U S A* 90, 5728–5731.
304. Garnier, M., Boujrad, N., Ogwuegbu, S. O., Hudson, J. R. Jr., and Papadopoulos, V. (1994). The polypeptide diazepam-binding inhibitor and a higher affinity mitochondrial peripheral-type benzodiazepine receptor sustain constitutive steroidogenesis in the R2C Leydig tumor cell line. *J. Biol. Chem.* 269, 22105–22112.
305. Krueger, K. E., and Papadopoulos, V. (1990). Peripheral-type benzodiazepine receptors mediate translocation of cholesterol from outer to inner mitochondrial membranes in adrenocortical cells. *J. Biol. Chem.* 265, 15015–15022.
306. Anholt, R. R., De Souza, E. B., Kuhar, M. J., and Snyder, S. H. (1985). Depletion of peripheral-type benzodiazepine receptors after hypophysectomy in rat adrenal gland and testis. *Eur. J. Pharmacol.* 110, 41–46.
307. Massotti, M., Slobodyansky, E., Konkel, D., Costa, E., and Guidotti, A. (1991). Regulation of diazepam binding inhibitor in rat adrenal gland by adrenocorticotropin. *Endocrinology* 129, 591–596.
308. Cavallaro, S., Pani, L., Guidotti, A., and Costa, E. (1993). ACTH-induced mitochondrial DBI receptor (MDR) and diazepam binding inhibitor (DBI) expression in adrenals of hypophysectomized rats is not cause-effect related to its immediate steroidogenic action. *Life Sci.* 53, 1137–1147.
309. Brown, A. S., Hall, P. F., Shoyab, M., and Papadopoulos, V. (1992). Endozepine/diazepam binding inhibitor in adrenocortical and Leydig cell lines: absence of hormonal regulation. *Mol. Cell. Endocrinol.* 83, 1–9.
310. Papadopoulos, V., Nowzari, F. B., and Krueger, K. E. (1991). Hormone-stimulated steroidogenesis is coupled to mitochondrial benzodiazepine receptors. Tropic hormone action on steroid biosynthesis is inhibited by flunitrazepam. *J. Biol. Chem.* 266, 3682–3687.
311. Boujrad, N., Gaillard, J. L., Garnier, M., and Papadopoulos, V. (1994). Acute action of choriogonadotropin on Leydig tumor cells: induction of a higher affinity benzodiazepine-binding site related to steroid biosynthesis. *Endocrinology* 135, 1576–1583.
312. Papadopoulos, V., and Brown, A. S. (1995). Role of the peripheral-type benzodiazepine receptor and the polypeptide diazepam binding inhibitor in steroidogenesis. *J. Steroid Biochem. Mol. Biol.* 53, 103–110.
313. Papadopoulos, V., Boujrad, N., Ikonomovic, M. D., Ferrara, P., and Vidic, B. (1994). Topography of the Leydig cell mitochondrial peripheral-type benzodiazepine receptor. *Mol. Cell. Endocrinol.* 104, R5–R9.
314. Papadopoulos, V., Amri, H., Li, H., Boujrad, N., Vidic, B., and Garnier, M. (1997). Targeted disruption of the peripheral-type benzodiazepine receptor gene inhibits steroidogenesis in the R2C Leydig tumor cell line. *J. Biol. Chem.* 272, 32129–32135.
315. Boujrad, N., Vidic, B., and Papadopoulos, V. (1996). Acute action of choriogonadotropin on Leydig tumor cells: changes in the topography of the mitochondrial peripheral-type benzodiazepine receptor. *Endocrinology* 137, 5727–5730.
316. Niswender, G. D. (2002). Molecular control of luteal secretion of progesterone. *Reproduction* 123, 333–339.
317. Lazzarini, R., Malucelli, B. E., and Palermo-Neto, J. (2001). Reduction of acute inflammation in rats by diazepam: role of peripheral benzodiazepine receptors and corticosterone. *Immunopharmacol. Immunotoxicol.* 23, 253–265.
318. Lazzarini, R., Malucelli, B. E., Muscara, M. N., de Nucci, G., and Palermo-Neto, J. (2003). Reduction of inflammation in rats by diazepam: tolerance development. *Life Sci.* 72, 2361–2368.

319. Marc, V., and Morselli, P. L. (1969). Effect of diazepam on plasma corticosterone levels in the rat. *J. Pharm. Pharmacol.* 21, 784–786.
320. Weizman, R., Leschiner, S., Schlegel, W., and Gavish, M. (1997). Peripheral-type benzodiazepine receptor ligands and serum steroid hormones. *Brain Res.* 772, 203–208.
321. Stocco, D. M. (1997). A StAR search: implications in controlling steroidogenesis. *Biol. Reprod.* 56, 328–336.
322. Miller, W. L. (1997). Congenital lipoid adrenal hyperplasia: the human gene knockout for the steroidogenic acute regulatory protein. *J. Mol. Endocrinol.* 19, 227–240.
323. Stocco, D. M. (1999). Steroidogenic acute regulatory protein. *Vitam. Horm.* 55, 399–441.
324. Miller, W. L., and Strauss, J. F. III. (1999). Molecular pathology and mechanism of action of the steroidogenic acute regulatory protein, StAR. *J. Steroid Biochem. Mol. Biol.* 69, 131–141.
325. Orly, J., and Stocco, D. M. (1999). The role of the steroidogenic acute regulatory (StAR) protein in female reproductive tissues. *Horm. Metab. Res.* 31, 389–398.
326. Christenson, L. K., and Strauss, J. F. III. (2000). Steroidogenic acute regulatory protein (StAR) and the intramitochondrial translocation of cholesterol. *Biochim. Biophys. Acta.* 1529, 175–187.
327. Stocco, D. (2001). StAR protein and the regulation of steroid hormone biosynthesis. *Annu. Rev. Physiol.* 63, 193–213.
328. Stocco, D. M. (2001). Tracking the role of a StAR in the sky of the new millennium. *Mol. Endocrinol.* 15, 1245–1254.
329. Christenson, L. K., and Strauss, J. F. III. (2001). Steroidogenic acute regulatory protein: an update on its regulation and mechanism of action. *Arch. Med. Res.* 32, 576–586.
330. Strauss, J. F. III, Kishida, T., Christenson, L. K., Fujimoto, T., and Hiroi, H. (2003). START domain proteins and the intracellular trafficking of cholesterol in steroidogenic cells. *Mol. Cell. Endocrinol.* 202, 59–65.
331. Clark, B. J., Soo, S. C., Caron, K. M., Ikeda, Y., Parker, K. L., and Stocco, D. M. (1995). Hormonal and developmental regulation of the steroidogenic acute regulatory protein. *Mol. Endocrinol.* 9, 1346–1355.
332. Clark, B. J., and Stocco, D. M. (1995). Expression of the steroidogenic acute regulatory (StAR) protein: a novel LH-induced mitochondrial protein required for the acute regulation of steroidogenesis in mouse Leydig tumor cells. *Endocr. Res.* 21, 243–257.
333. Sugawara, T., Holt, J. A., Driscoll, D., Strauss, J. F. III, Lin, D., Miller, W. L., Patterson, D., Clancy, K. P., Hart, I. M., Clark, B. J., and Stocco, D. M. (1995). Human steroidogenic acute regulatory protein: functional activity in COS-1 cells, tissue-specific expression, and mapping of the structural gene to 8p11.2 and a pseudogene to chromosome 13. *Proc. Natl. Acad. Sci. U S A* 92, 4778–4782.
334. Clark, B. J., Combs, R., Hales, K. H., Hales, D. B., and Stocco, D. M. (1997). Inhibition of transcription affects synthesis of steroidogenic acute regulatory protein and steroidogenesis in MA-10 mouse Leydig tumor cells. *Endocrinology* 138, 4893–4901.
335. Nalbant, D., Williams, S. C., Stocco, D. M., and Khan, S. A. (1998). Luteinizing hormone-dependent gene regulation in Leydig cells may be mediated by CCAAT/enhancer-binding protein-beta. *Endocrinology* 139, 272–279.
336. Luo, L., Chen, H., Stocco, D. M., and Zirkin, B. R. (1998). Leydig cell protein synthesis and steroidogenesis in response to acute stimulation by luteinizing hormone in rats. *Biol. Reprod.* 59, 263–270.
337. Akingbemi, B. T., Ge, R., Rosenfeld, C. S., Newton, L. G., Hardy, D. O., Catterall, J. F., Lubahn, D. B., Korach, K. S., and Hardy, M. P. (2003). Estrogen receptor-alpha gene deficiency enhances androgen biosynthesis in the mouse Leydig cell. *Endocrinology* 144, 84–93.
338. Price, C. A., Cooke, G. M., and Sanford, L. M. (2000). Influence of season and low-level oestradiol immunoneutralization on episodic LH and testosterone secretion and testicular steroidogenic enzymes and steroidogenic acute regulatory protein in the adult ram. *J. Reprod. Fertil.* 118, 251–262.
339. Kim, Y. C., Ariyoshi, N., Artemenko, I., Elliott, M. E., Bhattacharyya, K. K., and Jefcoate, C. R. (1997). Control of cholesterol access to cytochrome P450scc in rat adrenal cells mediated by regulation of the steroidogenic acute regulatory protein. *Steroids* 62, 10–20.
340. Lin, T., Wang, D., Hu, J., and Stocco, D. M. (1998). Upregulation of human chorionic gonadotrophin-induced steroidogenic acute regulatory protein by insulin-like growth factor-I in rat Leydig cells. *Endocrine* 8, 73–78.
341. Ramnath, H. I., Peterson, S., Michael, A. E., Stocco, D. M., and Cooke, B. A. (1997). Modulation of steroidogenesis by chloride ions in MA-10 mouse tumor Leydig cells: roles of calcium, protein synthesis, and the steroidogenic acute regulatory protein. *Endocrinology* 138, 2308–2314.
342. Huang, B. M., Stocco, D. M., Li, P. H., Yang, H. Y., Wu, C. M., and Norman, R. L. (1997). Corticotropin-releasing hormone stimulates the expression of the steroidogenic acute regulatory protein in MA-10 mouse cells. *Biol. Reprod.* 57, 547–551.
343. Manna, P. R., Pakarinen, P., El-Hefnawy, T., and Huhtaniemi, I. T. (1999). Functional assessment of the calcium messenger system in cultured mouse Leydig tumor cells: regulation of human chorionic gonadotropin-induced expression of the steroidogenic acute regulatory protein. *Endocrinology* 140, 1739–1751.
344. Manna, P. R., Tena-Sempere, M., and Huhtaniemi, I. T. (1999). Molecular mechanisms of thyroid hormone-stimulated steroidogenesis in mouse Leydig tumor cells. Involvement of the steroidogenic acute regulatory (StAR) protein. *J. Biol. Chem.* 274, 5909–5918.
345. Manna, P. R., Kero, J., Tena-Sempere, M., Pakarinen, P., Stocco, D. M., and Huhtaniemi, I. T. (2001). Assessment of mechanisms of thyroid hormone action in mouse Leydig cells: regulation of the steroidogenic acute regulatory protein, steroidogenesis, and luteinizing hormone receptor function. *Endocrinology* 142, 319–331.
346. Manna, P. R., Roy, P., Clark, B. J., Stocco, D. M., and Huhtaniemi, I. T. (2001). Interaction of thyroid hormone and steroidogenic acute regulatory (StAR) protein in the regulation of murine Leydig cell steroidogenesis. *J. Steroid Biochem. Mol. Biol.* 76, 167–177.
347. Lee, H. K., Yoo, M. S., Choi, H. S., Kwon, H. B., and Soh, J. (1999). Retinoic acids up-regulate steroidogenic acute regulatory protein gene. *Mol. Cell. Endocrinol.* 148, 1–10.
348. Nagata, Y., Homma, H., Matsumoto, M., and Imai, K. (1999). Stimulation of steroidogenic acute regulatory protein (StAR) gene expression by D-aspartate in rat Leydig cells. *FEBS Lett.* 454, 317–320.
349. Kanzaki, M., and Morris, P. L. (1999). Growth hormone regulates steroidogenic acute regulatory protein expression and steroidogenesis in Leydig cell progenitors. *Endocrinology* 140, 1681–1686.
350. Manna, P. R., Huhtaniemi, I. T., Wang, X. J., Eubank, D. W., and Stocco, D. M. (2002). Mechanisms of epidermal growth factor signaling: regulation of steroid biosynthesis and the steroidogenic acute regulatory protein in mouse Leydig tumor cells. *Biol. Reprod.* 67, 1393–1404.
351. Schwarzenbach, H., Manna, P. R., Stocco, D. M., Chakrabarti, G., and Mukhopadhyay, A. K. (2003). Stimulatory effect of progesterone on the expression of steroidogenic acute regulatory protein in MA-10 Leydig cells. *Biol. Reprod.* 68, 1054–1063.

352. Wang, X., Walsh, L. P., and Stocco, D. M. (1999). The role of arachidonic acid on LH-stimulated steroidogenesis and steroidogenic acute regulatory protein accumulation in MA-10 mouse Leydig tumor cells. *Endocrine* 10, 7–12.
353. Wang, X., Walsh, L. P., Reinhart, A. J., and Stocco, D. M. (2000). The role of arachidonic acid in steroidogenesis and steroidogenic acute regulatory (StAR) gene and protein expression. *J. Biol. Chem.* 275, 20204–20209.
354. Wang, X. J., Dyson, M. T., Mondillo, C., Patrignani, Z., Pignataro, O., and Stocco, D. M. (2002). Interaction between arachidonic acid and cAMP signaling pathways enhances steroidogenesis and StAR gene expression in MA-10 Leydig tumor cells. *Mol. Cell. Endocrinol.* 188, 55–63.
355. Wang, X., Dyson, M. T., Jo, Y., and Stocco, D. M. (2003). Inhibition of cyclooxygenase-2 activity enhances steroidogenesis and steroidogenic acute regulatory gene expression in MA-10 mouse Leydig cells. *Endocrinology* 144, 3368–3375.
356. Wang, X. J., Dyson, M. T., Jo, Y., Eubank, D. W., and Stocco, D. M. (2003). Involvement of 5-lipoxygenase metabolites of arachidonic acid in cyclic AMP-stimulated steroidogenesis and steroidogenic acute regulatory protein gene expression. *J. Steroid Biochem. Mol. Biol.* 85, 159–166.
357. Choi, Y. S., Stocco, D. M., and Freeman, D. A. (1995). Diethylumbelliferyl phosphate inhibits steroidogenesis by interfering with a long-lived factor acting between protein kinase A activation and induction of the steroidogenic acute regulatory protein (StAR). *Eur. J. Biochem.* 234, 680–685.
358. Bosmann, H. B., Hales, K. H., Li, X., Liu, Z., Stocco, D. M., and Hales, D.B. (1996). Acute in vivo inhibition of testosterone by endotoxin parallels loss of steroidogenic acute regulatory (StAR) protein in Leydig cells. *Endocrinology* 137, 4522–4525.
359. Hales, K. H., Diemer, T., Ginde, S., Shankar, B. K., Roberts, M., Bosmann, H. B., and Hales, D. B. (2000). Diametric effects of bacterial endotoxin lipopolysaccharide on adrenal and Leydig cell steroidogenic acute regulatory protein. *Endocrinology* 141, 4000–4012.
360. Sam, A. D. II, Sharma, A. C., Lee, L. Y., Hales, D. B., Law, W. R., Ferguson, J. L, and Bosmann, H. B. (1999). Sepsis produces depression of testosterone and steroidogenic acute regulatory (StAR) protein. *Shock* 11, 298–301.
361. Khanna, A., Aten, R. F., and Behrman, H. R. (1994). Heat shock protein induction blocks hormone-sensitive steroidogenesis in rat luteal cells. *Steroids* 59, 4–9.
362. Khanna, A., Aten, R. F., and Behrman, H. R. (1995). Physiological and pharmacological inhibitors of luteinizing hormone-dependent steroidogenesis induce heat shock protein-70 in rat luteal cells. *Endocrinology* 136, 1775–1781.
363. Liu, Z., and Stocco, D. M. (1997). Heat shock-induced inhibition of acute steroidogenesis in MA-10 cells is associated with inhibition of the synthesis of the steroidogenic acute regulatory protein. *Endocrinology* 138, 2722–2728.
364. Murphy, B. D., Lalli, E., Walsh, L. P., Liu, Z., Soh, J., Stocco, D. M., and Sassone-Corsi, P. (2001). Heat shock interferes with steroidogenesis by reducing transcription of the steroidogenic acute regulatory protein gene. *Mol. Endocrinol.* 15, 1255–1263.
365. Lin, T., Wang, D., and Stocco, D. M. (1998). Interleukin-1 inhibits Leydig cell steroidogenesis without affecting steroidogenic acute regulatory protein messenger ribonucleic acid or protein levels. *J. Endocrinol.* 156, 461–467.
366. Ogilvie, K. M., Held Hales, K., Roberts, M. E., Hales, D. B., and Rivier, C. (1999). The inhibitory effect of intracerebroventricularly injected interleukin 1beta on testosterone secretion in the rat: role of steroidogenic acute regulatory protein. *Biol. Reprod.* 60, 527–533.
367. Mauduit, C., Gasnier, F., Rey, C., Chauvin, M. A., Stocco, D. M., Louisot, P., and Benahmed, M. (1998). Tumor necrosis factor-alpha inhibits Leydig cell steroidogenesis through a decrease in steroidogenic acute regulatory protein expression. *Endocrinology* 139, 2863–2868.
368. Budnik, L. T., Jahner, D., and Mukhopadhyay, A. K. (1999). Inhibitory effects of TNF alpha on mouse tumor Leydig cells: possible role of ceramide in the mechanism of action. *Mol. Cell. Endocrinol.* 150, 39–46.
369. Morales, V., Santana, P., Diaz, R., Tabraue, C., Gallardo, G., Lopez Blanco, F., Hernandez, I., Fanjul, L. F., and Ruiz de Galarreta, C. M. (2003). Intratesticular delivery of tumor necrosis factor-alpha and ceramide directly abrogates steroidogenic acute regulatory protein expression and Leydig cell steroidogenesis in adult rats. *Endocrinology* 144, 4763–4772.
370. Lin, T., Hu, J., Wang, D., and Stocco, D. M. (1998). Interferon-gamma inhibits the steroidogenic acute regulatory protein messenger ribonucleic acid expression and protein levels in primary cultures of rat Leydig cells. *Endocrinology* 139, 2217–2222.
371. Mauduit, C., Goddard, I., Besset, V., Tabone, E., Rey, C., Gasnier, F., Dacheux, F., and Benahmed, M. (2001). Leukemia inhibitory factor antagonizes gonadotropin induced-testosterone synthesis in cultured porcine leydig cells: sites of action. *Endocrinology* 142, 2509–2520.
372. King, S. R., Rommerts, F. F., Ford, S. L., Hutson, J. C., Orly, J., and Stocco, D. M. (1998). Ethane dimethane sulfonate and NNN'N'-tetrakis-(2-pyridylmethyl)ethylenediamine inhibit steroidogenic acute regulatory (StAR) protein expression in MA-10 Leydig cells and rat Sertoli cells. *Endocr. Res.* 24, 469–478.
373. King, S. R., Walsh, L. P., and Stocco, D. M. (2000). Nigericin inhibits accumulation of the steroidogenic acute regulatory protein but not steroidogenesis. *Mol. Cell. Endocrinol.* 166, 147–153.
374. Wu, C. S., Leu, S. F., Yang, H. Y., and Huang, B. M. (2001). Melatonin inhibits the expression of steroidogenic acute regulatory protein and steroidogenesis in MA-10 cells. *J. Androl.* 22, 245–254.
375. Tena-Sempere, M., Manna, P. R., Zhang, F. P., Pinilla, L., Gonzalez, L. C., Dieguez, C., Huhtaniemi, I., and Aguilar, E. (2001). Molecular mechanisms of leptin action in adult rat testis: potential targets for leptin-induced inhibition of steroidogenesis and pattern of leptin receptor messenger ribonucleic acid expression. *J. Endocrinol.* 170, 413–423.
376. Chiao, Y. C., Cho, W. L., and Wang, P. S. (2002). Inhibition of testosterone production by propylthiouracil in rat Leydig cells. *Biol. Reprod.* 67, 416–422.
377. Walsh, L. P., Kuratko, C. N., and Stocco, D. M. (2000). Econazole and miconazole inhibit steroidogenesis and disrupt steroidogenic acute regulatory (StAR) protein expression post-transcriptionally. *J. Steroid Biochem. Mol. Biol.* 75, 229–236.
378. Kim, J. H., Kim, H. J., Noh, H. S., Roh, G. S., Kang, S. S., Cho, G. J., Park, S. K., Lee, B. J., and Choi, W. S. (2003). Suppression by ethanol of male reproductive activity. *Brain Res.* 989, 91–98.
379. Ikeda, Y., Lala, D. S., Luo, X., Kim, E., Moisan, M. P., and Parker, K. L. (1993). Characterization of the mouse FTZ-F1 gene, which encodes a key regulator of steroid hydroxylase gene expression. *Mol. Endocrinol.* 7, 852–860.
380. Luo, X., Ikeda, Y., and Parker, K. L. (1994). A cell-specific nuclear receptor is essential for adrenal and gonadal development and sexual differentiation. *Cell* 77, 481–490.
381. Shinoda, K., Lei, H., Yoshii, H., Nomura, M., Nagano, M., Shiba, H., Sasaki, H., Osawa, Y., Ninomiya, Y., Niwa, O., et al. (1995). Developmental defects of the ventromedial

hypothalamic nucleus and pituitary gonadotroph in the Ftz-F1 disrupted mice. *Dev. Dyn.* 204, 22–29.

382. Sadovsky, Y., Crawford, P. A., Woodson, K. G., Polish, J. A., Clements, M. A., Tourtellotte, L. M., Simburger, K., and Milbrandt, J. (1995). Mice deficient in the orphan receptor steroidogenic factor 1 lack adrenal glands and gonads but express P450 side-chain-cleavage enzyme in the placenta and have normal embryonic serum levels of corticosteroids. *Proc. Natl. Acad. Sci. U S A* 92, 10939–10943.
383. Caron, K. M., Ikeda, Y., Soo, S. C., Stocco, D. M., Parker, K. L., and Clark, B. J. (1997). Characterization of the promoter region of the mouse gene encoding the steroidogenic acute regulatory protein. *Mol. Endocrinol.* 11, 138–147.
384. Sugawara, T., Kiriakidou, M., McAllister, J. M., Kallen, C. B., and Strauss, J. F. III. (1997). Multiple steroidogenic factor 1 binding elements in the human steroidogenic acute regulatory protein gene 5´-flanking region are required for maximal promoter activity and cyclic AMP responsiveness. *Biochemistry* 36, 7249–7255.
385. Sandhoff, T. W., Hales, D. B., Hales, K. H., and McLean, M. P. (1998). Transcriptional regulation of the rat steroidogenic acute regulatory protein gene by steroidogenic factor 1. *Endocrinology* 139, 4820–4831.
386. Sugawara, T., Holt, J. A., Kiriakidou, M., and Strauss, J. F. III. (1996). Steroidogenic factor 1-dependent promoter activity of the human steroidogenic acute regulatory protein (StAR) gene. *Biochemistry* 35, 9052–9059.
387. Clark, B. J., and Combs, R. (1999). Angiotensin II and cyclic adenosine 3´,5´-monophosphate induce human steroidogenic acute regulatory protein transcription through a common steroidogenic factor-1 element. *Endocrinology* 140, 4390–4398.
388. Wooton-Kee, C. R., and Clark, B. J. (2000). Steroidogenic factor-1 influences protein-deoxyribonucleic acid interactions within the cyclic adenosine 3,5-monophosphate-responsive regions of the murine steroidogenic acute regulatory protein gene. *Endocrinology* 141, 1345–1355.
389. Johnson, P. F., and Williams, S. C. (1994). CCAAT/enhancer binding (C/EBP) proteins. In *Liver Gene Expression* (F. Tronche and M. Yaniv, Eds.), pp. 231–258. R. G. Landes Co., Austin, TX.
390. Lekstrom-Himes, J., and Xanthopoulos, K. G. (1998). Biological role of the CCAAT/enhancer-binding protein family of transcription factors. *J. Biol. Chem.* 273, 28545–28548.
391. Sirois, J., and Richards, J. S. (1993). Transcriptional regulation of the rat prostaglandin endoperoxide synthase 2 gene in granulosa cells. Evidence for the role of a cis-acting C/EBP beta promoter element. *J. Biol. Chem.* 268, 21931–21938.
392. Piontkewitz, Y., Enerback, S., and Hedin, L. (1996). Expression of CCAAT enhancer binding protein-alpha (C/EBP alpha) in the rat ovary: implications for follicular development and ovulation. *Dev. Biol.* 179, 288–296.
393. Roesler, W. J., Park, E. A., and McFie, P. J. (1998). Characterization of CCAAT/enhancer-binding protein alpha as a cyclic AMP-responsive nuclear regulator. *J. Biol. Chem.* 273, 14950–14957.
394. Sterneck, E., Tessarollo, L., and Johnson, P. F. (1997). An essential role for C/EBPbeta in female reproduction. *Genes Dev.* 11, 2153–2162.
395. Pall, M., Hellberg, P., Brannstrom, M., Mikuni, M., Peterson, C. M., Sundfeldt, K., Norden, B., Hedin, L., and Enerback, S. (1997). The transcription factor C/EBP-beta and its role in ovarian function; evidence for direct involvement in the ovulatory process. *EMBO J.* 16, 5273–5279.
396. Berg, J. M. (1992). Sp1 and the subfamily of zinc finger proteins with guanine-rich binding sites. *Proc. Natl. Acad. Sci. U S A* 89, 11109–11110.
397. Sugawara, T., Lin, D., Holt, J. A., Martin, K. O., Javitt, N. B., Miller, W. L., and Strauss, J. F. III. (1995). Structure of the human steroidogenic acute regulatory protein (StAR) gene: StAR stimulates mitochondrial cholesterol 27-hydroxylase activity. *Biochemistry* 34, 12506–12512.
398. Venepally, P., and Waterman, M. R. (1995). Two Sp1-binding sites mediate cAMP-induced transcription of the bovine CYP11A gene through the protein kinase A signaling pathway. *J. Biol. Chem.* 270, 25402–25410.
399. Ahlgren, R., Simpson, E. R., Waterman, M. R., and Lund, J. (1990). Characterization of the promoter/regulatory region of the bovine CYP11A (P-450scc) gene. Basal and cAMP-dependent expression. *J. Biol. Chem.* 265, 3313–3319.
400. Kagawa, N., and Waterman, M. R. (1991). Evidence that an adrenal-specific nuclear protein regulates the cAMP responsiveness of the human CYP21B (P450C21) gene. *J. Biol. Chem.* 266, 11199–11204.
401. Borroni, R., Liu, Z., Simpson, E. R., and Hinshelwood, M. M. (1997). A putative binding site for Sp1 is involved in transcriptional regulation of CYP17 gene expression in bovine ovary. *Endocrinology* 138, 2011–2020.
402. Sun, G., Porter, W., and Safe, S. (1998). Estrogen-induced retinoic acid receptor alpha 1 gene expression: role of estrogen receptor-Sp1 complex. *Mol. Endocrinol.* 12, 882–890.
403. Lee, Y. H., Williams, S. C., Baer, M., Sterneck, E., Gonzalez, F. J., and Johnson, P. F. (1997). The ability of C/EBP beta but not C/EBP alpha to synergize with an Sp1 protein is specified by the leucine zipper and activation domain. *Mol. Cell. Biol.* 17, 2038–2047.
404. Rodenburg, R. J., Holthuizen, P. E., and Sussenbach, J. S. (1997). A functional Sp1 binding site is essential for the activity of the adult liver-specific human insulin-like growth factor II promoter. *Mol. Endocrinol.* 11, 237–250.
405. Shea-Eaton, W. K., Trinidad, M. J., Lopez, D., Nackley, A., and McLean, M. P. (2001). Sterol regulatory element binding protein-1a regulation of the steroidogenic acute regulatory protein gene. *Endocrinology* 142, 1525–1533.
406. Silverman, E., Eimerl, S., and Orly, J. (1999). CCAAT enhancer-binding protein beta and GATA-4 binding regions within the promoter of the steroidogenic acute regulatory protein (StAR) gene are required for transcription in rat ovarian cells. *J. Biol. Chem.* 274, 17987–17996.
407. Christenson, L. K., McAllister, J. M., Martin, K. O., Javitt, N. B., Osborne, T. F., and Strauss, J. F. III. (1998). Oxysterol regulation of steroidogenic acute regulatory protein gene expression. Structural specificity and transcriptional and posttranscriptional actions. *J. Biol. Chem.* 273, 30729–30735.
408. Christenson, L. K., Osborne, T. F., McAllister, J. M., Strauss, J. F. III. (2001). Conditional response of the human steroidogenic acute regulatory protein gene promoter to sterol regulatory element binding protein-1a. *Endocrinology* 142, 28–36.
409. Manna, P. R., Dyson, M. T., Eubank, D. W., Clark, B. J., Lalli, E., Sassone-Corsi, P., Zeleznik, A. J., and Stocco, D. M. (2002). Regulation of steroidogenesis and the steroidogenic acute regulatory protein by a member of the cAMP response-element binding protein family. *Mol. Endocrinol.* 16, 184–199.
410. Sugawara, T., Nomura, E., Sakuragi, N., and Fujimoto, S. (2001). The effect of the arylhydrocarbon receptor on the human steroidogenic acute regulatory gene promoter activity. *J. Ster. Biochem. Mol. Biol.* 78, 253–260.
411. Meyer, T. E., and Habener, J. F. (1993). Cyclic adenosine 3´,5´-monophosphate response element binding protein (CREB) and related transcription-activating deoxyribonucleic acid-binding proteins. *Endocr. Rev.* 14, 269–290.

412. Sassone-Corsi, P. (1995). Transcription factors responsive to cAMP. *Annu. Rev. Cell. Dev. Biol.* 11, 355–377.
413. Montminy, M. (1997). Transcriptional regulation by cyclic AMP. *Annu. Rev. Biochem.* 66, 807–822.
414. Gonzalez, G. A., Menzel, P., Leonard, J., Fischer, W. H., and Montminy, M. R. (1991). Characterization of motifs which are critical for activity of the cyclic AMP-responsive transcription factor CREB. *Mol. Cell. Biol.* 11, 1306–1312.
415. Della Fazia, M. A., Servillo, G., and Sassone-Corsi, P. (1997). Cyclic AMP signalling and cellular proliferation: regulation of CREB and CREM. *FEBS Lett.* 410, 22–24.
416. Manna, P. R., Eubank, D. W., Lalli, E., Sassone-Corsi, P., and Stocco, D. M. (2003). Transcriptional regulation of the mouse steroidogenic acute regulatory protein gene by the cAMP response-element binding protein and steroidogenic factor 1. *J. Mol. Endocrinol.* 30, 381–397.
417. Hai, T., and Curran, T. (1991). Cross-family dimerization of transcription factors Fos/Jun and ATF/CREB alters DNA binding specificity. *Proc. Natl. Acad. Sci. U S A* 88, 3720–3724.
418. Masquilier, D., and Sassone-Corsi, P. (1992). Transcriptional cross-talk: nuclear factors CREM and CREB bind to AP-1 sites and inhibit activation by Jun. *J. Biol. Chem.* 267, 22460–22466.
419. Shea-Eaton, W., Sandhoff, T. W., Lopez, D., Hales, D. B., and McLean, M. P. (2002). Transcriptional repression of the rat steroidogenic acute regulatory (StAR) protein gene by the AP-1 family member c-Fos. *Mol. Cell. Endocrinol.* 188, 161–170.
420. Manna, P. R., Eubank, D. W., and Stocco, D. M. (2004). Assessment of the role of activator protein-1 on transcription of the mouse steroidogenic acute regulatory protein gene. *Mol. Endocrinol.* 18, 558–573.
421. Ikeda, Y., Swain, A., Weber, T. J., Hentges, K. E., Zanaria, E., Lalli, E., Tamai, K. T., Sassone-Corsi, P., Lovell-Badge, R., Camerino, G., and Parker, K. L. (1996). Steroidogenic factor 1 and Dax-1 colocalize in multiple cell lineages: potential links in endocrine development. *Mol. Endocrinol.* 10, 1261–1272.
422. Swain, A., Zanaria, E., Hacker, A., Lovell-Badge, R., and Camerino, G. (1996). Mouse Dax1 expression is consistent with a role in sex determination as well as in adrenal and hypothalamus function. *Nat. Genet.* 12, 404–409.
423. Lalli, E., and Sassone-Corsi, P. (2003). DAX-1, an unusual orphan receptor at the crossroads of steroidogenic function and sexual differentiation. *Mol. Endocrinol.* 17, 1445–1453.
424. Muscatelli, F., Strom, T. M., Walker, A. P., Zanaria, E., Recan, D., Meindl, A., Bardoni, B., Guioli, S., Zehetner, G., Rabl, W., et al. (1994). Mutations in the DAX-1 gene give rise to both X-linked adrenal hypoplasia congenita and hypogonadotropic hypogonadism. *Nature* 372, 672–676.
425. Zazopoulos, E., Lalli, E., Stocco, D. M., and Sassone-Corsi, P. (1997). DNA binding and transcriptional repression by DAX-1 blocks steroidogenesis. *Nature* 390, 311–315.
426. Lalli, E., Melner, M. H., Stocco, D. M., and Sassone-Corsi, P. (1998). DAX-1 blocks steroid production at multiple levels. *Endocrinology* 139, 4237–4243.
427. Suzuki, T., Kasahara, M., Yoshioka, H., Morohashi, K., and Umesono, K. (2003). LXXLL-related motifs in Dax-1 have target specificity for the orphan nuclear receptors Ad4BP/SF-1 and LRH-1. *Mol. Cell. Biol.* 23, 238–249.
428. Ito, M., Yu, R., and Jameson, J. L. (1997). DAX-1 inhibits SF-1-mediated transactivation via a carboxy-terminal domain that is deleted in adrenal hypoplasia congenita. *Mol. Cell. Biol.* 17, 1476–1483.
429. Ericsson, J., Usheva, A., and Edwards, P. A. (1999). YY1 is a negative regulator of transcription of three sterol regulatory element-binding protein-responsive genes. *J. Biol. Chem.* 274, 14508–14513.
430. Galvin, K. M., and Shi, Y. (1997). Multiple mechanisms of transcriptional repression by YY1. *Mol. Cell. Biol.* 17, 3723–3732.
431. Bennett, M. K., Ngo, T. T., Athanikar, J. N., Rosenfeld, J. M., and Osborne, T. F. (1999). Co-stimulation of promoter for low density lipoprotein receptor gene by sterol regulatory element-binding protein and Sp1 is specifically disrupted by the yin yang 1 protein. *J. Biol. Chem.* 274, 13025–13032.
432. Nackley, A. C., Shea-Eaton, W., Lopez, D., and McLean, M. P. (2002). Repression of the steroidogenic acute regulatory gene by the multifunctional transcription factor Yin Yang 1. *Endocrinology* 143, 1085–1096.
433. Wilson, J. D., Griffin, J. E., and Russell, D. W. (1993). Steroid 5 alpha-reductase 2 deficiency. *Endocr. Rev.* 14, 577–593.
434. Russell, D. W., Berman, D. M., Bryant, J. T., Cala, K. M., Davis, D. L., Landrum, C. P., Prihoda, J. S., Silver, R. I., Thigpen, A. E., and Wigley, W. C. (1994). The molecular genetics of steroid 5 alpha-reductases. *Recent Prog. Horm. Res.* 49, 275–284.
435. Imperato-McGinley, J., Binienda, Z., Arthur, A., Mininberg, D. T., Vaughan, E. D. Jr., and Quimby, F. W. (1985). The development of a male pseudohermaphroditic rat using an inhibitor of the enzyme 5 alpha-reductase. *Endocrinology* 116, 807–812.
436. George, F. W., and Peterson, K. G. (1988). 5 alpha-dihydrotestosterone formation is necessary for embryogenesis of the rat prostate. *Endocrinology* 122, 1159–1164.
437. Grino, P. B., Griffin, J. E., and Wilson, J. D. (1990). Testosterone at high concentrations interacts with the human androgen receptor similarly to dihydrotestosterone. *Endocrinology* 126, 1165–1172.
438. Deslypere, J. P., Young, M., Wilson, J. D., and McPhaul, M. J. (1992). Testosterone and 5 alpha-dihydrotestosterone interact differently with the androgen receptor to enhance transcription of the MMTV-CAT reporter gene. *Mol. Cell. Endocrinol.* 88, 15–22.
439. Mangelsdorf, D. J., Thummel, C., Beato, M., Herrlich, P., Schutz, G., Umesono, K., Blumberg, B., Kastner, P., Mark, M., Chambon, P., et al. (1995). The nuclear receptor superfamily: the second decade. *Cell* 83, 835–839.
440. Zoppi, S., Young, M., and McPhaul, M. (2002). Regulation of gene expression by the nuclear receptor family. In *Genetics of Steroid Biosynthesis and Function* (J. Mason, Ed.), pp. 376–403. Harwood Academic Publisher.
441. Robinson-Rechavi, M., Escriva Garcia, H., and Laudet, V. (2003). The nuclear receptor superfamily. *J. Cell. Sci.* 116, 585–586.
442. Freedman, L. P., and Luisi, B. F. (1993). On the mechanism of DNA binding by nuclear hormone receptors: a structural and functional perspective. *J. Cell. Biochem.* 51, 140–150.
443. Shaffer, P. L., Jivan, A., Dollins, D. E., Claessens, F., and Gewirth, D. T. (2004). Structural basis of androgen receptor binding to selective androgen response elements. *Proc. Natl. Acad. Sci. U S A* 101, 4758–4763.
444. Matias, P. M., Donner, P., Coelho, R., Thomaz, M., Peixoto, C., Macedo, S., Otto, N., Joschko, S., Scholz, P., Wegg, A., Basler, S., Schafer, M., Egner, U., and Carrondo, M. A. (2000). Structural evidence for ligand specificity in the binding domain of the human androgen receptor. Implications for pathogenic gene mutations. *J. Biol. Chem.* 275, 26164–26171.
445. Sack, J. S., Kish, K. F., Wang, C., Attar, R. M., Kiefer, S. E., An, Y., Wu, G. Y., Scheffler, J. E., Salvati, M. E., Krystek, S. R. Jr., Weinmann, R., and Einspahr, H. M. (2001). Crystallographic structures of the ligand-binding domains of the androgen receptor and its T877A mutant complexed with the natural agonist dihydrotestosterone. *Proc. Natl. Acad. Sci. U S A* 98, 4904–4909.

446. Matias, P. M., Carrondo, M. A., Coelho, R., Thomaz, M., Zhao, X. Y., Wegg, A., Crusius, K., Egner, U., and Donner, P. (2002). Structural basis for the glucocorticoid response in a mutant human androgen receptor (AR[ccr]) derived from an androgen-independent prostate cancer. *J. Med. Chem.* 45, 1439–1446.
447. Brzozowski, A. M., Pike, A. C., Dauter, Z., Hubbard, R. E., Bonn, T., Engstrom, O., Ohman, L., Greene, G. L., Gustafsson, J. A., and Carlquist, M. (1997). Molecular basis of agonism and antagonism in the oestrogen receptor. *Nature* 389, 753–758.
448. Darimont, B. D., Wagner, R. L., Apriletti, J. W., Stallcup, M. R., Kushner, P. J., Baxter, J. D., Fletterick, R. J., and Yamamoto, K. R. (1998). Structure and specificity of nuclear receptor-coactivator interactions. *Genes Dev.* 12, 3343–3356.
449. Shiau, A. K., Barstad, D., Loria, P. M., Cheng, L., Kushner, P. J., Agard, D. A., and Greene, G. L. (1998). The structural basis of estrogen receptor/coactivator recognition and the antagonism of this interaction by tamoxifen. *Cell* 95, 927–937.
450. He, B., and Wilson, E. M. (2003). Electrostatic modulation in steroid receptor recruitment of LXXLL and FXXLF motifs. *Mol. Cell. Biol.* 23, 2135–2150.
451. Hampsey, M., and Reinberg, D. (1999). RNA polymerase II as a control panel for multiple coactivator complexes. *Curr. Opin. Genet. Dev.* 9, 132–139.
452. Jenster, G. (1998). Coactivators and corepressors as mediators of nuclear receptor function: an update. *Mol. Cell. Endocrinol.* 143, 1–7.
453. McKenna, N. J., Lanz, R. B., and O'Malley, B. W. (1999). Nuclear receptor coregulators: cellular and molecular biology. *Endocr. Rev.* 20, 321–344.
454. Heinlein, C. A., and Chang, C. (2002). Androgen receptor (AR) coregulators: an overview. *Endocr. Rev.* 23, 175–200.
455. Taatjes, D. J., Marr, M. T., and Tjian, R. (2004). Regulatory diversity among metazoan co-activator complexes. *Nat. Rev. Mol. Cell. Biol.* 5, 403–410.
456. Shang, Y., Hu, X., DiRenzo, J., Lazar, M. A., and Brown, M. (2000). Cofactor dynamics and sufficiency in estrogen receptor-regulated transcription. *Cell* 103, 843–852.
457. Shang, Y., Myers, M., and Brown, M. (2002). Formation of the androgen receptor transcription complex. *Mol. Cell.* 9, 601–610.
458. Metivier, R., Penot, G., Hubner, M. R., Reid, G., Brand, H., Kos, M., and Gannon, F. (2003). Estrogen receptor-alpha directs ordered, cyclical, and combinatorial recruitment of cofactors on a natural target promoter. *Cell* 115, 751–763.
459. Webb, P., Nguyen, P., Valentine, C., Lopez, G. N., Kwok, G. R., McInerney, E., Katzenellenbogen, B. S., Enmark, E., Gustafsson, J. A., Nilsson, S., and Kushner, P. J. (1999). The estrogen receptor enhances AP-1 activity by two distinct mechanisms with different requirements for receptor transactivation functions. *Mol. Endocrinol.* 13, 1672–1685.
460. De Bosscher, K., Vanden Berghe, W., and Haegeman, G. (2001). Glucocorticoid repression of AP-1 is not mediated by competition for nuclear coactivators. *Mol. Endocrinol.* 15, 219–227.
461. Lobaccaro, J. M., Poujol, N., Terouanne, B., Georget, V., Fabre, S., Lumbroso, S., and Sultan, C. (1999). Transcriptional interferences between normal or mutant androgen receptors and the activator protein 1–dissection of the androgen receptor functional domains. *Endocrinology* 140, 350–357.
462. Chen, Z., Yuhanna, I. S., Galcheva-Gargova, Z., Karas, R. H., Mendelsohn, M. E., and Shaul, P. W. (1999). Estrogen receptor alpha mediates the nongenomic activation of endothelial nitric oxide synthase by estrogen. *J. Clin. Invest.* 103, 401–406.
463. Mendelsohn, M. E. (2002). Genomic and nongenomic effects of estrogen in the vasculature. *Am. J. Cardiol.* 90, 3F–6F.
464. Chambliss, K. L., and Shaul, P. W. (2002). Estrogen modulation of endothelial nitric oxide synthase. *Endocr. Rev.* 23, 665–686.
465. Coleman, K. M., and Smith, C. L. (2001). Intracellular signaling pathways: nongenomic actions of estrogens and ligand-independent activation of estrogen receptors. *Front Biosci.* 6, D1379–D1391.
466. Castoria, G., Barone, M. V., Di Domenico, M., Bilancio, A., Ametrano, D., Migliaccio, A., and Auricchio, F. (1999). Non-transcriptional action of oestradiol and progestin triggers DNA synthesis. *EMBO J.* 18, 2500–2510.
467. Boonyaratanakornkit, V., Scott, M. P., Ribon, V., Sherman, L., Anderson, S. M., Maller, J. L., Miller, W. T., and Edwards, D. P. (2001). Progesterone receptor contains a proline-rich motif that directly interacts with SH3 domains and activates c-Src family tyrosine kinases. *Mol. Cell.* 8, 269–280.
468. Wong, C. W., McNally, C., Nickbarg, E., Komm, B. S., and Cheskis, B. J. (2002). Estrogen receptor-interacting protein that modulates its nongenomic activity-crosstalk with Src/Erk phosphorylation cascade. *Proc. Natl. Acad. Sci. U S A* 99, 14783–14788.
469. Lutz, L. B., Jamnongjit, M., Yang, W. H., Jahani, D., Gill, A., and Hammes, S. R. (2003). Selective modulation of genomic and nongenomic androgen responses by androgen receptor ligands. *Mol. Endocrinol.* 17, 1106–1116.
470. Unni, E., Sun, S., Nan, B., McPhaul, M., Cheskis, B. J., Mancini, M., and Marcelli, M. (2004). Changes in androgen receptor nongenotropic signaling correlate with transition of LNCaP cells to androgen independence. *Cancer Res.* 64(19), 7156–7168.
471. Donjacour, A. A., Thomson, A. A., and Cunha, G. R. (2003). FGF-10 plays an essential role in the growth of the fetal prostate. *Dev. Biol.* 261, 39–54.
472. Gerdes, M. J., Larsen, M., Dang, .T D., Ressler, S. J., Tuxhorn, J. A., and Rowley, D. R. (2004). Regulation of rat prostate stromal cell myodifferentiation by androgen and TGF-beta1. *Prostate* 58, 299–307.
473. Cunha, G. R. (1996). Growth factors as mediators of androgen action during male urogenital development. *Prostate Suppl.* 6, 22–25.
474. Thomson, A. A. (2001). Role of androgens and fibroblast growth factors in prostatic development. *Reproduction* 121, 187–195.
475. Verhoeven, G., Hoeben, E., and De Gendt, K. (2000). Peritubular cell-Sertoli cell interactions: factors involved in PmodS activity. *Andrologia* 32, 42–45.
476. Lyon, M. F., Glenister, P. H., and Lamoreux, M. L. (1975). Normal spermatozoa from androgen-resistant germ cells of chimaeric mice and the role of androgen in spermatogenesis. *Nature* 258, 620–622.
477. Johnston, D. S., Russell, L. D., Friel, P. J., and Griswold, M. D. (2001). Murine germ cells do not require functional androgen receptors to complete spermatogenesis following spermatogonial stem cell transplantation. *Endocrinology* 142, 2405–2408.
478. Sar, M., Lubahn, D. B., French, F. S., and Wilson, E. M. (1990). Immunohistochemical localization of the androgen receptor in rat and human tissues. *Endocrinology* 127, 3180–3186.
479. Bremner, W. J., Millar, M. R., Sharpe, R. M., and Saunders, P. T. (1994). Immunohistochemical localization of androgen receptors in the rat testis: evidence for stage-dependent expression and regulation by androgens. *Endocrinology* 135, 1227–1234.

480. Vornberger, W., Prins, G., Musto, N. A., and Suarez-Quian, C. A. (1994). Androgen receptor distribution in rat testis: new implications for androgen regulation of spermatogenesis. *Endocrinology* 134, 2307–2316.
481. Suarez-Quian, C. A., Martinez-Garcia, F., Nistal, M., and Regadera, J. (1999). Androgen receptor distribution in adult human testis. *J. Clin. Endocrinol. Metab.* 84, 350–358.
482. Van Roijen, J. H., Van Assen, S., Van Der Kwast, T. H., De Rooij, D. G., Boersma, W. J., Vreeburg, J. T., and Weber, R. F. (1995). Androgen receptor immunoexpression in the testes of subfertile men. *J. Androl.* 16, 510–516.
483. Goyal, H. O., Bartol, F. F., Wiley, A. A., Khalil, M. K., Chiu, J., and Vig, M. M. (1997). Immunolocalization of androgen receptor and estrogen receptor in the developing testis and excurrent ducts of goats. *Anat. Rec.* 249, 54–62.
484. Kimura, N., Mizokami, A., Oonuma, T., Sasano, H., and Nagura, H. (1993). Immunocytochemical localization of androgen receptor with polyclonal antibody in paraffin-embedded human tissues. *J. Histochem. Cytochem.* 41, 671–678.
485. Zhou, X., Kudo, A., Kawakami, H., and Hirano, H. (1996). Immunohistochemical localization of androgen receptor in mouse testicular germ cells during fetal and postnatal development. *Anat. Rec.* 245, 509–518.
486. Pelletier, G., Labrie, C., and Labrie, F. (2000). Localization of oestrogen receptor alpha, oestrogen receptor beta and androgen receptors in the rat reproductive organs. *J. Endocrinol.* 165, 359–370.
487. Zhu, L. J., Hardy, M. P., Inigo, I. V., Huhtaniemi, I., Bardin, C. W., and Moo-Young, A. J. (2000). Effects of androgen on androgen receptor expression in rat testicular and epididymal cells: a quantitative immunohistochemical study. *Biol. Reprod.* 63, 368–376.
488. Zhou, Q., Nie, R., Prins, G. S., Saunders, P. T., Katzenellenbogen, B. S., and Hess, R. A. (2002). Localization of androgen and estrogen receptors in adult male mouse reproductive tract. *J. Androl.* 23, 870–881.
489. Saunders, P., Millar, M., Majdic, G., Bremner, W., McLaren, T., Grigor, K., and Sharpe, R. (1996). Testicular androgen receptor protein: distribution and control of expression. In *Cellular and Molecular Regulation of Testicular Cells* (C. Desjardins, Ed.), pp. 213–229. Springer Verlag, New York.
490. Sharpe, R. M., Maddocks, S., Millar, M., Kerr, J. B., Saunders, P. T., and McKinnell, C. (1992). Testosterone and spermatogenesis. Identification of stage-specific, androgen-regulated proteins secreted by adult rat seminiferous tubules. *J. Androl.* 13, 172–184.
491. Russell, L. D., and Clermont, Y. (1977). Degeneration of germ cells in normal, hypophysectomized and hormone treated hypophysectomized rats. *Anat. Rec.* 187, 347–366.
492. Bartlett, J. M., Kerr, J. B., and Sharpe, R. M. (1986). The effect of selective destruction and regeneration of rat Leydig cells on the intratesticular distribution of testosterone and morphology of the seminiferous epithelium. *J. Androl.* 7, 240–253.
493. Dym, M., and Raj, H. G. (1977). Response of adult rat Sertoli cells and Leydig cells to depletion of luteinizing hormone and testosterone. *Biol. Reprod.* 17, 676–696.
494. McKinnell, C., and Sharpe, R. M. (1995). Testosterone and spermatogenesis: evidence that androgens regulate cellular secretory mechanisms in stage VI–VIII seminiferous tubules from adult rats. *J. Androl.* 16, 499–509.
496. McLachlan, R. I., O'Donnell, L., Meachem, S. J., Stanton, P. G., de Kretser, D. M., Pratis, K., and Robertson, D. M. (2002). Identification of specific sites of hormonal regulation in spermatogenesis in rats, monkeys, and man. *Recent Prog. Horm. Res.* 57, 149–179.
497. De Gendt, K., Swinnen, J. V., Saunders, P. T., Schoonjans, L., Dewerchin, M., Devos, A., Tan, K., Atanassova, N., Claessens, F., Lecureuil, C., Heyns, W., Carmeliet, P., Guillou, F., Sharpe, R. M., and Verhoeven, G. (2004). A Sertoli cell-selective knockout of the androgen receptor causes spermatogenic arrest in meiosis. *Proc. Natl. Acad. Sci. U S A* 101, 1327–1332.

Knobil and Neill's Physiology of Reproduction, Third Edition
edited by Jimmy D. Neill,
Elsevier © 2006

CHAPTER 21

Endocrine Regulation of Spermatogenesis

Liza O'Donnell, Sarah J. Meachem, Peter G. Stanton, and Robert I. McLachlan

Introduction and Scope, 1017
Organization of Spermatogenesis, 1018
Overview, 1018
Germ Cell Development, 1019
Organization of the Spermatogenic Cycle, 1021
Cell–Cell Adhesion Systems Involved in Spermatogenesis, 1022
Ability of the Testis to Respond to Endocrine Signals; Hormone Receptor Expression, 1022
Steroid Receptors, 1023
Pituitary Hormone Receptors, 1025
Androgen Regulation of Spermatogenesis, 1026
Role during Initiation of Spermatogenesis, 1027
Role in the Adult, 1029
FSH Regulation of Spermatogenesis, 1032
Role during Initiation of Spermatogenesis, 1033
Role in the Adult, 1035
FSH and T Cooperativity, 1037
Thyroid Hormone and Spermatogenesis, 1039
Estrogen Regulation of Spermatogenesis, 1039
Estrogen Production in the Testis, 1040
Estrogen and Spermatogenesis during the Neonatal Period, 1041
Estrogen and Spermatogenesis in the Adult, 1043
Importance of Estrogen in Efferent Ductule Development and Function; Impact on Spermatogenesis, 1044
Endocrine Regulation of Human Spermatogenesis, 1045
Hormonal Dependency of Human Spermatogenesis, 1045
Hypogonadotropic Hypogonadism (HH), 1049
Male Hormonal Contraceptive (MHC) Development, 1051
Conclusion and Future Directions, 1053
References, 1054

INTRODUCTION AND SCOPE

Male fertility depends on the continuous daily production of millions of spermatozoa. The process of spermatogenesis is one of exquisite complexity, requiring 6 to 9 weeks for completion, and involves a coordinated series of mitotic and meiotic divisions, elaborate cyto-differentiative steps, and constantly changing intercellular interactions, all overseen by an extraordinary interplay of autocrine, paracrine, and endocrine factors.

The histology of spermatogenesis has been described in detail [e.g., (1–4) and Chapter 18]; and the endocrine factors required for its completion and for optimal male fertility have been extensively explored [see (5–10) for reviews]. At this time it is reasonable to say that we have a good understanding of these endocrine factors; at least we understand which hormones are required for spermatogenesis and how we can impair or restore spermatogenesis by their manipulation. But beneath this "first layer" of control remain many questions about the local control, cell–cell interactions, paracrine and genetic factors essential for normal spermatogenesis, and perhaps of more clinical importance, at which points they fail leading to male infertility, a condition affecting about 5% of the population.

Our knowledge of the endocrine regulation of spermatogenesis arises from numerous studies suppressing and replacing endocrine factors and then

Prince Henry's Institute of Medical Research, Monash Medical Centre, Victoria, Australia.

examining the ensuing effects on testicular function and spermatogenesis in rodents and primates (11–19); such studies have been particularly valuable in elucidating the role of endocrine hormones in adult spermatogenesis. Recent gene knockout studies and the creation of mice congenitally deficient in gonadotropin subunits, steroid hormone biosynthetic enzymes or hormone receptors have been invaluable in revealing which are crucial for the initiation of spermatogenesis, and have provided important clues as to their cellular sites of action [see (20,21) for reviews]. Despite this clear progress, we are still a long way from understanding the molecular targets of these endocrine factors. For example, we know that testosterone (T) is absolutely required for spermatogenesis (22–24), yet few androgen-regulated genes with a known function in spermatogenesis have been identified within the seminiferous tubules of the testis (see section on Role in the Adult, page 1029).

This chapter aims to provide an overview of the current knowledge of the ways in which spermatogenesis is regulated by gonadotropins and steroid hormones. We have endeavored to cite the latest reviews and key original publications rather than give an exhaustive bibliography; and we urge the interested reader to refer to this literature for more in-depth consideration of particular topics. A brief overview of the organization of spermatogenesis is given to orient the reader for the latter discussion on its endocrine regulation and note that a comprehensive account can be found in Chapter 18. This chapter will focus on endocrine hormones and their effects on spermatogenesis; however, it is well known that other factors, such as genetic and paracrine factors, play key roles in sperm production (the reader is referred to Chapter 18). In most cases this chapter will discuss the effects of endocrine factors on the seminiferous epithelium; while the physiology of testicular steroidogenesis and the generation of steroid hormones will be explored elsewhere in this edition. Also, this chapter will not consider the maturation of sperm once they leave the testis; this important posttesticular maturation of sperm in the epididymis is key for male fertility and will be explored in Chapter 22.

We will review the localization of endocrine hormone receptors in the testis in an effort to educate the reader on the likely sites at which hormones act within spermatogenesis. Much of the chapter will provide a comprehensive overview of the two major endocrine factors that regulate spermatogenesis: androgens and follicle-stimulating hormone (FSH). A discussion of the involvement of estrogen in spermatogenesis will also be included, since many recent studies have focused on its potential role in male fertility. In each case, the endocrine regulation will be discussed in terms of the initiation of spermatogenesis (that is, the onset of this process during the neonatal and pubertal periods) and the maintenance of spermatogenesis in the adult, since it is clear that the roles of such hormones change during maturation of the testis. By necessity, much of the information presented will be taken from rodent and primate models, however, our knowledge of human spermatogenesis and the relevance to the application of endocrine hormones to managing spermatogenesis in a clinical setting will be dealt with in a separate section.

ORGANIZATION OF SPERMATOGENESIS

Overview

Spermatogenesis is the process by which immature germ cells undergo division, differentiation and meiosis to give rise to haploid, highly specialized elongated spermatids (2). Spermatogenesis occurs within the seminiferous tubules of the testis, in close association with the somatic cells of the seminiferous epithelium, the Sertoli cells. At the completion of spermatogenesis, mature spermatids are released from the Sertoli cells into the seminiferous tubule lumen, and proceed through the excurrent duct system, known as the rete testis, until they enter the epididymis via the efferent ducts.

The testis containing the seminiferous tubules and interstitial tissue is enclosed by a capsule called the tunica. The interstitial tissue contains the blood and lymphatic vessels that are essential for the movement of hormones and nutrients into, and out of, the testis. Importantly, the interstitium contains the Leydig cell (25) which secretes androgens, notably T, as well as other steroids including estrogen.

The seminiferous tubules are surrounded by the lymphatic endothelium and the peritubular myoid cells (26). Within the tubules, developing germ cells form intimate associations with Sertoli cells, which have numerous cup-shaped processes that encompass the various germ cell types (4). The seminiferous epithelium is a unique and exquisitely complex epithelium (Fig. 1), with both germ cells and Sertoli cells showing major structural and morphological changes during the spermatogenic process.

Sertoli cells divide in the fetal and early neonatal period in rodents. At the end of the proliferative period, which is around day 15 in the rat (27), the Sertoli cells undergo morphological changes to assume a terminally differentiated state capable of supporting germ cell development [see (28) for review]. The pubertal alteration of factors that potentiate Sertoli cell proliferation (29,30) or interfere with their maturation (31,32) produce changes in the testicular size and spermatogenic potential of the adult. Thus, the

FIG. 1. Cross-section of the adult rat seminiferous epithelium. A columnar Sertoli cell nucleus (SC) is visible towards the basement membrane (b) of the epithelium. In this particular cross section the Sertoli cell can be seen to support five different germ cells types: type A spermatogonia (A), preleptotene spermatocytes (pl), pachytene spermatocytes (ps), round spermatids (rs) and elongated spermatids (es).

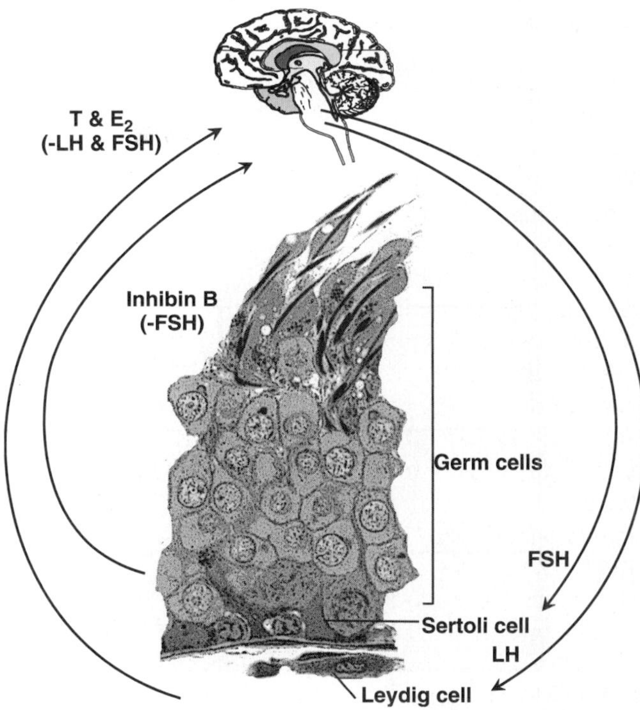

FIG. 2. Overview of the endocrine feedback loop controlling spermatogenesis. Luteinizing hormone (LH) and follicle-stimulating hormone (FSH) are secreted into the blood stream from the anterior pituitary in response to hypothalamic GnRH. LH targets Leydig cells in the testis to produce steroids, primarily testosterone (T) which can be aromatized to estradiol (E_2). While T and E_2 have local effects on spermatogenesis, these steroids also have a negative feedback effect at the level of the hypothalamus (via controlling GnRH secretion) and the pituitary to limit LH and FSH secretion (the steroid negative feedback effect on FSH varies between species). FSH stimulates Sertoli cells to support spermatogenesis. In response to FSH, the seminiferous epithelium produces inhibin B, which feeds back on the pituitary to suppress FSH production.

duration of the proliferative period and the number of Sertoli cells produced, together with the subsequent maturation period, determines the spermatogenic potential of the testis, with each Sertoli cell capable of supporting a finite number of germ cells.

The endocrine regulation of spermatogenesis is accomplished via a classic negative feedback loop (Fig. 2) involving interactions between the hypothalamus, pituitary, and testis (the hypothalamic–pituitary–testis, or HPT, axis). The production of spermatozoa is dependent on stimulation by the pituitary gonadotropins, luteinizing hormone (LH) and FSH, which are secreted in response to hypothalamic gonadotropin-releasing hormone (GnRH). LH stimulates androgen synthesis by the Leydig cells of the testis (see section on Androgen Regulation of Spermatogenesis), which acts locally to regulate sperm production and feeds back on the hypothalamus and pituitary to affect GnRH and LH production in a self-limiting loop. FSH stimulates Sertoli cells (see section on FSH Regulation of Spermatogenesis) to secrete inhibin B that has a negative feed back effect on the pituitary to limit FSH synthesis.

Germ Cell Development

The various germ cell types, from the immature diploid spermatogonia to mature haploid elongated spermatids were originally described in rats based on the morphological changes that occur during maturation (33) (Fig. 3). The most immature germ cells, the spermatogonia, include type A (denoted $A_{1\text{-}4}$), intermediate (found only in rodents), and type B forms, the latter being considered to be committed to

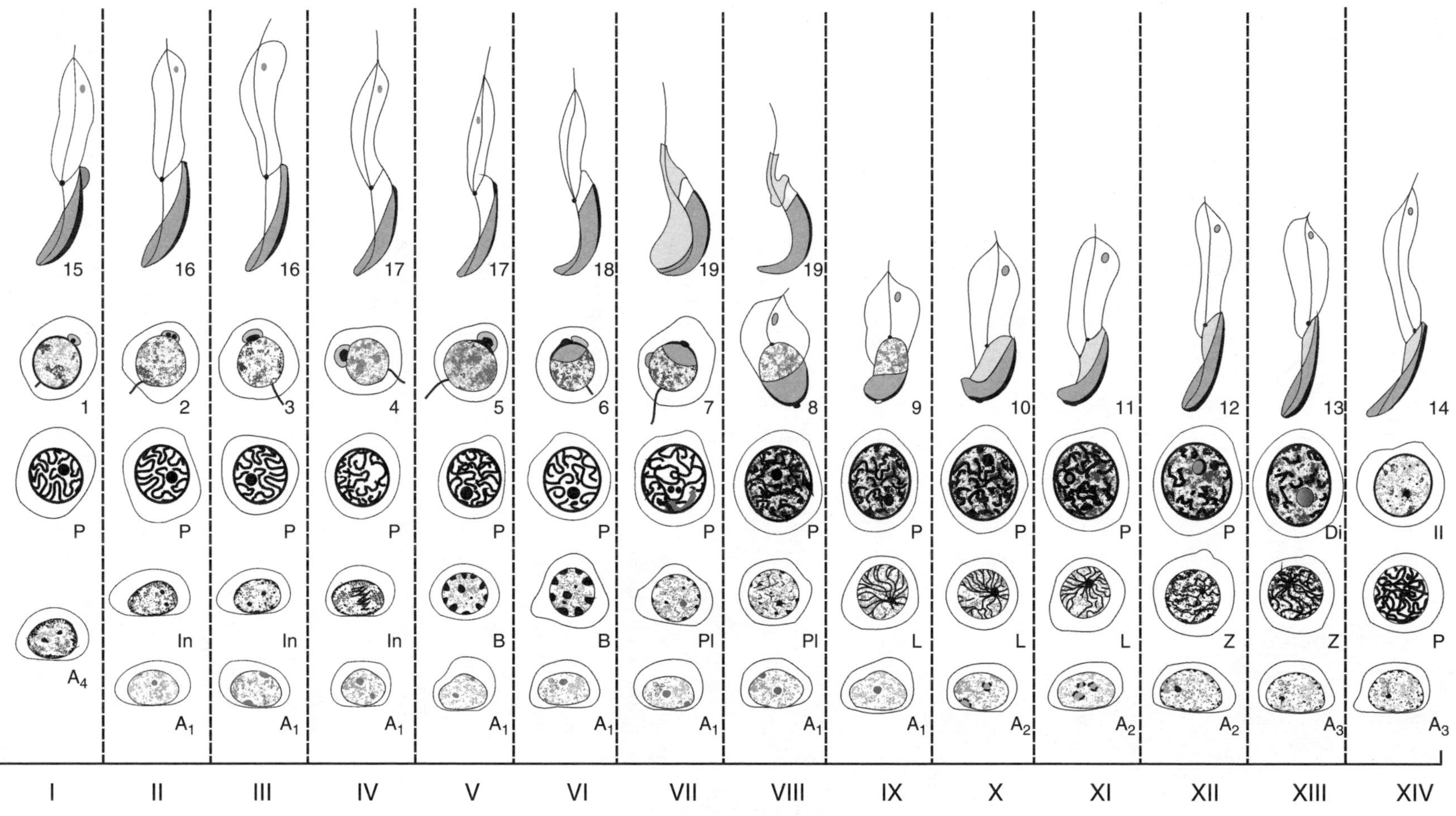

FIG. 3. Diagram of the cycle of rat spermatogenesis. The 14 stages of the rat spermatogenic cycle, denoted I–XIV, are shown in the vertical columns. Germ cell development is shown horizontally. $A_{(1-4)}$, type A_{1-4} spermatogonia; In, intermediate spermatogonia; B, type B spermatogonia; Pl, preleptotene spermatocyte; L, leptotene spermatocyte; Z, zygotene spermatocyte; PS, pachytene spermatocyte; Di, diploid pachytene spermatocyte; II, secondary spermatocyte. The steps of spermiogenesis (1–19) are indicated beside each spermatid. Diagram modified from (3).

differentiation into meiosis (1,34). The actual stem cell pool for the germ cell line is considered to be a subset of the type A spermatogonia population, yet the identity of the "true" stem cell cannot be discerned on the basis of morphology (1,34) nor by biochemical means. However, various technological advances such as the transplantation of germ cells from one animal to another (of the same or different species) are now beginning to yield important new information on the biology of spermatogonial stem cells [see (35,36) for review].

Three models to explain spermatogonial renewal kinetics in rodents have been proposed [see (37)]. In the monkey and human, there are two morphologically distinct type A spermatogonia; Adark (Ad) and Apale (Ap), as well as type B spermatogonia (1). Ap spermatogonia are proposed to divide to give rise to type B spermatogonia as well as to renew their own population, whereas Ad spermatogonia are considered to be the nonproliferative reserve spermatogonial population (38–41). Indeed, Ad spermatogonia have been suggested to be able to undergo transition to Ap following testicular insult, allowing repopulation of the testis in times of injury or insult (40,41). Ap spermatogonia (or a subset of) have been suggested to be the "true" stem cell of the testis, because Ap (not Ad) spermatogonia are seen in humans after radiation therapy (42) and in the postpubertal cryptorchid testes (43). Various studies have suggested that Ap spermatogonia can undergo transition without division into Ad (40,41,44,45).

In all mammals, both type A and B spermatogonia undergo a series of mitotic divisions to produce a large number of germ cells available for entry into meiosis, and thus the size of the spermatogonial population is a key determinant of the number of mature sperm eventually produced. The size of this population is likely controlled by a balance of proliferation and anti-apoptotic events (34) and is influenced by endocrine factors (see section on FSH Regulation of Spermatogenesis).

Meiosis is initiated by the production of preleptotene spermatocytes from type B spermatogonia. During the prophase of the first meiotic division, germ cells undergo morphological transitions that can be classified on the basis of nuclear size and morphology (1–3). DNA is replicated in the preleptotene phase and pairing of homologous chromosomes occurs in the zygotene phase. Cells with completely paired chromosomes are termed pachytene spermatocytes. The pachytene spermatocyte phase takes some weeks and as such these cells are seen frequently within the seminiferous epithelium (1,3). A brief diplotene phase follows in which the chromosome pairs partially separate. The cells then undergo the first meiotic division to yield secondary spermatocytes, which quickly progress through a second meiotic division to yield haploid round spermatids.

Spermiogenesis is the process by which the round spermatid transforms, without further division, into the specialized elongated spermatid via a series of complex cytodifferentiative steps. Nineteen defined steps of spermiogenesis have been categorized in the rat (Fig. 3) (46). These steps involve the formation and development of the acrosome and flagellum, condensation of the chromatin, reshaping and elongation of the nucleus and removal of the cytoplasm prior to release of the spermatid during spermiation (3,46). An important physiological consideration during spermiogenesis is that the spermatid nucleus becomes incapable of transcription as it condenses, and thus round spermatids transcribe high levels of messenger ribonucleic acids (mRNAs) that are subject to translational delay until translation of protein is required during elongation (47).

Spermiation is the final step of spermatogenesis, and involves removal of spermatid cytoplasm to yield the streamlined spermatozoon capable of motility, retraction of the Sertoli cell away from the spermatid and, finally, the release of the mature spermatid into the tubule lumen (48).

Organization of the Spermatogenic Cycle

Germ cell development is not randomly distributed within the seminiferous epithelium, but is arranged in strictly defined cellular associations (1,3,33,46). A particular association of germ cells is referred to as a *stage*, and the number of stages of spermatogenesis in a particular species is defined by the number of morphologically recognizable germ cell associations within the testis; there are 12 stages in the mouse, 14 in the rat, and 6 in the human (3). Roman numerals are always used to denote the different spermatogenic stages. The stages of spermatogenesis are routinely used to describe changes in gene expression and protein production/secretion; for example, expression of the androgen receptor (AR) protein is highest in stage VII in the rat (49), lending support to the contention that the mid-spermatogenic stages are androgen responsive [see (9) for review; and sections on Androgen Receptor (AR) and Androgen Regulation of Spermatogenesis].

These stages occur in a consecutive wave along the length of the seminiferous tubule; in the rat germ cells will progress from stage I through to stage XIV, then again into stage I (Fig. 3). The completion of one series of stages is known as one *cycle* of the seminiferous epithelium (3). Since the progression of stages through the spermatogenic cycle is precisely ordered in a wave-like fashion along the seminiferous tubule in most species, a single stage can be seen in cross sections of the seminiferous epithelium. In the marmoset and in humans, spermatogenesis is arranged in stages that are intertwined in a helical pattern [see (2) for review] such that a single tubule cross section may have all 6 identified stages present (50–52), meaning that, for practical purposes, specific stages cannot be identified.

The duration of the spermatogenic cycle is species-specific (3). Based on interspecies germ cell transplantation experiments, the timing of this cycle appears to be driven by germ cells in the adult (53), although other studies demonstrate that somatic cells of the embryonic and fetal testis show cyclic patterns of gene expression similar to the stage-dependent changes seen in the adult (54), suggesting that embryonic Sertoli cells establish the seminiferous epithelial cycle. Given that there are species-dependent differences in the duration of the spermatogenic cycle, it follows then that the timing to produce spermatozoa from spermatogonia also differs with species; for example in the mouse it takes approximately 35 days, whereas in the human it takes approximately 64 days (3,55). Monkey and human data suggest that the duration of spermatogenesis cannot be altered by modulation of the gonadotropin environment, e.g., hypophysectomy (56), or by GnRH-antagonist treatment of cynomolgus monkeys (57). Limited studies in man also suggest that the length of the spermatogenic cycle is not affected by hormone treatment (58), but clearly this is a very difficult issue to study directly.

Of note is the fact that each stage of the spermatogenic cycle also differs in duration; for example, in the rat, stage VII takes 58 hours to complete whereas stage IX takes only 7 hours (3). Consequently, stage VII tubules will be seen far more frequently in cross sections of the testis than stage IX tubules.

Cell–Cell Adhesion Systems Involved in Spermatogenesis

Cell–cell interactions are key to spermatogenesis, since germ cell development is dependent on their association and communication with the supporting Sertoli cells. One Sertoli cell is in constant contact with the basement membrane, approximately six other Sertoli cells, and five to six different types of germ cells in varying phases of development (Fig. 1). It can thus be appreciated that intercellular adhesion systems play an important role in spermatogenesis and that Sertoli cells participate in numerous intercellular adhesion junctions at any one time. Given that adhesion junctions are likely targets of hormone action (see section on Androgen Regulation of Spermatogenesis and section on FSH Regulation of Spermatogenesis), the types of junctions present in the epithelium are very briefly outlined here. The reader is referred to several other excellent reviews on this topic (2,59–66).

The base of the Sertoli cell is in direct contact with the extracellular matrix of the basal lamina surrounding the tubule via hemidesmosomes, which are intermediate filament-containing (rather than actin-containing) junctions (62,63). Inter-Sertoli cell junctions form the basis of the blood–testis barrier, which physiologically separates germ cells above the inter-Sertoli cell junctions from the blood and lymphatic systems situated in the interstitium, and allows the Sertoli cell to regulate the composition of the adluminal compartment in which the germ cells develop [see (2,26,59,61) for reviews]. The junctions between Sertoli cells are morphologically complex (26,59) and include testis-specific actin-based ectoplasmic specializations [see (64) for review], tight or occluding junctions (65), intermediate filament-based desmosome junctions containing cadherin and catenins (67–69), as well as communicating gap junctions (70).

The Sertoli cell also forms complex junctions with germ cells, and the type of junction formed depends on the stage of germ cell development. Desmosome-like junctions are present between Sertoli cells and spermatogonia, spermatocytes, and round spermatids (62,71); these junctions appear to contain cadherins and catenins yet are linked to a vimentin intermediate-filament system, and as such have features characteristic of adherens junctions and traditional desmosome junctions (67,68). Elongating and elongated spermatids interact with Sertoli cells via the apical ectoplasmic specialization (64,66,72), which appears to be important for germ cell adhesion as well as translocation throughout the seminiferous epithelium (64, 73,74). Prior to spermiation the ectoplasmic specialization is removed and an integrin-containing junction appears to mediate spermatid release from the Sertoli cell (75).

ABILITY OF THE TESTIS TO RESPOND TO ENDOCRINE SIGNALS; HORMONE RECEPTOR EXPRESSION

An important function of the somatic cells of the testis, such as the Sertoli and Leydig cells, is the ability to "receive" endocrine signals from the bloodstream and to transmit these signals in an appropriate manner to the developing germ cells. Expression of the appropriate hormone receptor is the first step to determining whether a cell is capable of responding to an endocrine signal. For the reader to begin to appreciate how spermatogenesis responds to hormones, the following section will give an overview of hormone receptor expression in the testis (Table 1), in particular the cells of the seminiferous epithelium.

TABLE 1. *Expression of hormone receptors in adult testicular cells (rodent and primate)*

	Androgen receptors	Estrogen receptors[b]	LH receptors	FSH receptors	TSH receptors
Leydig cells	+	+[c]	+	–	–
Peritubular cells	+	+ (ER-α,-β)	–	–	–
Sertoli cells	+	+ (ER-β)	–	+	+
Germ cells	–[a]	+ (ER-β)[d]	–	–	+

[a]Immunoexpression of AR has been reported in these cells, however, studies suggest that AR in germ cells is not essential for their development (see section on Androgen Receptor [AR]).
[b]There is some controversy as to the localization of ER subtypes within the testis, however, this table represents our view on the likely sites of ER subtype localization (see section on Estrogen Receptor [ER]).
[c]ER-α is the likely subtype present in rodent Leydig cells, ER-β is the likely subtype present in primates (see section on Estrogen Receptor [ER]).
[d]ER-β is reported in most germ cells types except for elongating and elongated spermatids.

Steroid Receptors

Androgen Receptor (AR)

Androgens are produced by the Leydig cells (see Chapter 20) and regulate spermatogenesis by binding to intracellular ARs that are members of the steroid hormone receptor family and, more broadly, members of the ligand-activated nuclear receptor superfamily. Testosterone, or its 5α-reduced metabolite dihydrotestosterone (DHT), bind to cytosolic AR which subsequently dimerise, translocate to the nucleus, and bind to androgen response elements (AREs) in the promoters of androgen-responsive genes [see (76,77) for review] to modulate gene transcription in a so-called "genomic" pathway of action. It is also worth noting that rapid "nongenomic" signal transduction pathways initiated by androgens in Sertoli cells also appear to be involved in modulating Sertoli cell function [see (78,79) for review].

It is clear that ARs are expressed in Sertoli cells, Leydig cells, and peritubular myoid cells of the testis (49,80–84). The general consensus at present is that germ cells lack AR, despite the reported immunoexpression of AR in germ cells in some studies (82,85,86) that is not demonstrated in others (49,81, 83,84,87). Recent studies suggested that androgens stimulate aromatase mRNA in highly purified germ cell preparations which were shown to be devoid of somatic cell contamination, and that this effect was blocked by treatment with an AR antagonist (88), suggesting that germ cells may be capable of respond directly to androgens in terms of aromatase expression. However, germ cell transplantation experiments showed that germ cells lacking ARs can develop normally in testes which contain functional ARs (89), indicating that if germ cells do express ARs, they are not essential for germ cell development. In the absence of proof that germ cells contain functional ARs, androgen action on spermatogenesis (see section on Androgen Regulation of Spermatogenesis) must be regarded as occurring via the Sertoli cell (Table 1).

During postnatal testicular development in the rat, androgen binding sites in the testis increase after day 15 (80), which can be attributed to the proliferation of AR-expressing Leydig cells as well as increases in the number of androgen binding sites per Sertoli cell (80,90). Sertoli cells in the neonatal period show weak AR immunoreactivity which progressively increases after day 14 (i.e., after the Sertoli cell proliferative period; see Organization of Spermatogenesis) to reach adult levels at day 45 (49). Increases in the intensity of Sertoli cell AR immunostaining was also seen in the testes of the marmoset, with only a subset of Sertoli cells showing weak immunoreactivity in neonatal animals, yet strong staining in all Sertoli cells was progressively attained during late infancy and the pubertal period (87). Thus, although the developing testis can clearly respond to androgens (see section on Androgen Regulation of Spermatogenesis), it seems that AR expression in general is lower at this time as compared to the adult.

In adult animals and humans, ARs are expressed in Sertoli cells in a stage-specific manner. In rats, AR mRNA (91) and immuno-expression is maximal in the mid-spermatogenic stages VII–VIII (49,81,84,85), with an apparent down-regulation during stage VIII (49). Peak expression of AR in the mid-spermatogenic stages is also noted in marmosets (87) and humans (92). These observations fit well with experimental data in rodents demonstrating that stages VII–VIII are particularly androgen-dependent [see (93–97); and section on Androgen Regulation of Spermatogenesis], supporting the view that the modulation of ARs in Sertoli cells is likely to be an important mechanism by which the response of the seminiferous epithelium to androgens is controlled.

FSH stimulates AR expression in pubertal Sertoli cells in vitro (98–101); however, the response to FSH appears to decrease with age (99), probably reflecting the age-related decline in the FSH-responsiveness of these cells (see below). Testosterone appears to positively regulate its own receptor as shown by the stimulation of Sertoli cell AR expression in vitro (98) and the reduction of intratesticular T levels being associated with reduced Sertoli cell AR protein in vivo (49,81). More recent studies suggest that nuclear factor-κB (NF-κB) is a potent transcriptional activator of the AR gene in Sertoli cells (102,103). Interestingly, tumor necrosis factor α (TNFα) stimulated NF-κB binding to the AR promoter and increased AR expression in Sertoli cells, suggesting that endogenous secretion of TNFα by round spermatids in stages VII and VIII may regulate the peak in Sertoli cell AR in these stages (102).

It is clear that testicular expression of AR is essential for spermatogenesis, since mice lacking a functional AR have arrested spermatogenesis (104). While germ cells lacking AR can develop normally when transplanted into mouse testes containing AR (89), the selective removal of AR from Sertoli cells in mice results in impaired spermatogenesis (23,24,105), highlighting the crucial role of AR in Sertoli cells for mediating the androgen-dependent effects on spermatogenesis (see section on Androgen Regulation of Spermatogenesis).

Estrogen Receptor (ER)

The metabolism of T to estrogen is catalyzed by the aromatase cytochrome P450 enzyme (P450arom, the product of the *CYP19* gene). Estrogen receptors (ERs), like ARs, are members of the steroid hormone family of receptors, and modulate transcription by binding to estrogen response elements (EREs) in the promoter region of estrogen-responsive genes; reviewed in (106). The level of estrogen in fluid leaving the testis is higher than that in circulation, supporting the concept that the testis is a major site of estrogen production in the male; reviewed in (107,108) and see section on Estrogen Regulation of Spermatogenesis. Two ER subtypes, named ER-α and ER-β, have been cloned from humans (109,110) and other species (111,112); see (21,106,113) for review. In vitro studies show that homo- and heterodimers of ER-α and ER-β can form (114,115), and it has been proposed that in cells expressing both ER-α and ER-β, ER-β may act to modulate ER-α transcriptional activity and that the relative expression of both subtypes is central to determining cellular responses to estrogenic ligands (116). It should also be noted that there are complex differences in the transcriptional activation of the ER subtypes by agonists and antagonists [e.g., see (117) for review]. Various isoforms of ER-β exist [see (106,118) for review]. Of particular interest is the ER-βcx/ER-β2 variant which is expressed in the testis [see (119) and references therein] and may act as a negative regulator of ER activity (120).

Disruptions of spermatogenesis are noted in mice lacking a functional ER-α (121–124) or both ER-α and ER-β (125,126), highlighting a central role for ER in sperm production; see section on Estrogen Regulation of Spermatogenesis and (8,21,113) for review. There is considerable conflict in the literature concerning the localization of ERs in the male reproductive tract [see (8) for review]; however, the following section will broadly summarize the likely sites of ER localization in the testis.

The fetal rodent testis contains ER-α protein, which is first expressed in the undifferentiated gonad (127) and persists in fetal Leydig cells [e.g., (128), see (8) for review] but does not appear to be present within the fetal seminiferous epithelium, reviewed in (8). In contrast the human fetal testis (mid-gestation) does not appear to contain ER-α mRNA or protein (129). ER-β protein is present in the fetal testis and has been immunolocalized to Sertoli cells, Leydig cells, and gonocytes in rodents [e.g., (130), see (8) for review] and humans (129). Studies on the two human ER-β isoforms (i.e., wild-type ER-β also known as ER-β1, and the ER-β2 [or ER-βcx] splice variant which may act as a dominant negative receptor) in the human fetal testis show that while gonocytes contain mRNA for both splice variants, only ER-β2 protein is present in relatively high levels in these cells, suggesting that the expression of this ER subtype in gonocytes may prevent estrogen action in these cells (129). In contrast, human fetal Sertoli cells, Leydig cells, and peritubular cells appeared to contain both ER-β isoforms, suggesting that these cells are capable of responding to estrogens (129).

In the postnatal developing testis of the rodent, ER-α continues to be expressed in Leydig cells [e.g., (128)], but is generally not expressed in the seminiferous epithelium [see (8) for review]. ER-β is also found in the immature Leydig cells [e.g., (130)] and appears to be the predominant ER in the seminiferous epithelium, being found in immature Sertoli cells, spermatogonia, and spermatocytes of the rodent (130,131).

In the adult testis, ER-α is found in the Leydig cells of rodents [e.g., (84,128)], however, ER-β is likely to be the subtype present in primate (including human) Leydig cells (119,132,133). It is worth noting, however, that some studies suggest ER-α is present in human testis (134,135) whereas others demonstrate that it is absent (119,132,133). ER-β expression is variable in the peritubular cells of adult rodents, primates, and humans (84,119,133).

It seems clear that ER-β is the ER subtype found in the adult seminiferous epithelium of the rodent, and is localized to Sertoli cells, spermatogonia, pachytene spermatocytes and round, but not elongating, spermatids [e.g., (128,130), reviewed in (8)]. This pattern of ER-β expression is also apparent in primates and humans, with ER-β present in Sertoli cells and most germ cells except for elongating and elongated spermatids (132,133). Analysis of the human ER-β splice variants in the adult human testis shows that Sertoli cells show strong immunostaining for the apparent dominant negative receptor ER-β2, but less intense immunostaining for the wild-type ER-β form (119). The ER-β2 variant also appeared more predominant, in comparison to wild-type ER-β, in type B spermatogonia, perhaps suggesting these cells are likely to be less responsive to estrogens, along with adult Sertoli cells (119). Both variants were present in type A spermatogonia and pachytene spermatocytes whereas round spermatids appeared to have only wild-type ER-β protein (119). This latter observation could be interpreted to mean that round spermatids may be the germ cell types best able to respond to estrogens in the adult human testis.

In summary, the somatic and germ cells of the seminiferous epithelium possess ERs throughout development, suggesting that these cells may be responsive to estrogens (see section on Estrogen Regulation of Spermatogenesis). The demonstration of splice variants of human ER-β, which may have inhibitory receptor activity, adds a further level of complexity to the modulation of estrogen action in the human seminiferous epithelium; the ability of this variant to inhibit estrogen activity in testicular cells in vivo requires further investigation.

Pituitary Hormone Receptors

FSH

FSH exerts its biological effect on the testis via FSH receptors (FSH-R) present on the plasma membrane of the Sertoli cell, which is likely to be the only cell in the male that expresses the FSH-R gene; see (136) for review. Despite one old and unconfirmed report suggesting that FSH-R may be present on spermatogonia (137), the general consensus is that FSH actions on spermatogenesis are mediated via the Sertoli cell (Table 1).

The FSH-R is a member of the G-protein–coupled receptor superfamily and has a well-characterized promoter in which various regulatory elements have been identified; see (138) for review. FSH binding to its receptor initiates various signal transduction events including cyclic adenosine 3′,5′-monophosphate (cAMP) stimulation (139) and Ca^{2+} release (140), which leads to various downstream signal transduction cascades within the Sertoli cell, such as protein kinase-A (PKA) and protein kinase-C (PKC)–dependent pathways (141). Sertoli cell responsiveness to FSH is likely regulated at several levels, including FSH-R number and the magnitude of downstream signal transduction cascades (138,141). For example, homologous down-regulation of the FSH-R by FSH or agents that stimulate cAMP decreases Sertoli cell FSH-responsiveness (142,143). Sertoli cell responsiveness to FSH, as assessed by the production of cAMP, decreases markedly as Sertoli cells mature [see (141,144) for review] despite increasing FSH-R number per testis across postnatal development (145,146). The decline in the ability of the Sertoli cell to produce cAMP in response to FSH is likely mediated at least in part by increases in phosphodiesterase activity (147,148) and by the appearance of germ cells (149). Nevertheless adult Sertoli cells in vitro are able to produce inhibin in response to FSH (150) indicating that adult Sertoli cells do retain some degree of FSH-responsiveness.

The FSH-R is found on the Sertoli cell membrane in the basal compartment of the seminiferous epithelium (151), i.e., below the blood–testis barrier, presumably to allow access of FSH from the circulation. FSH-R protein (152) and full-length mRNA (153) has been found in the Sertoli cell as early as day 17 of rat fetal life, with the level of FSH-R rising sharply before birth (152). Sertoli cells express FSH-R throughout postnatal life, with particularly high levels apparent at postnatal day 15 (145,154). Like AR expression, FSH-R in the adult rodent is expressed in a stage-dependent manner. Parvinen (148) used microdissected and cultured staged rat seminiferous tubule segments to show that FSH binding and biochemical responses to exogenous FSH differed across the spermatogenic cycle, with maximal responses noted in the early spermatogenic stages. In agreement, FSH-stimulated cAMP production is highest in seminiferous tubules in stages XIII through to IV, yet lowest in the mid-stages (VII and IX) (155), and FSH-R mRNA (149,156) and protein (157) is maximal in stages XIII–I and lowest in the mid-spermatogenic stages VII–VIII. These results, together with the demonstration of maximal AR expression in stages VII–VIII (see section on Androgen Receptor) suggest that androgens act at the mid-spermatogenic stages VII and VIII, whereas FSH acts at stages XIII–I (see section on Androgen Regulation of Spermatogenesis and section on FSH Regulation of Spermatogenesis).

LH

The LH receptor (LH-R) is a member of the G-protein–coupled receptor family and is encoded by a single gene, however, multiple mRNA transcripts are found in gonadal cells [see (158)]. The expression of LH-R is generally accepted to be restricted to the Leydig cells [e.g., (159)], where it mediates LH actions on Leydig cell number, function, LH-responsiveness, and, importantly, steroidogenesis [see (158,160) for review]. There is an absolute requirement for LH and androgen for the initiation and maintenance of spermatogenesis (9,22,161) (see section on Androgen Regulation of Spermatogenesis). Studies on mice lacking functional LH-R demonstrate that the LH-R is not essential for differentiation and function of fetal Leydig cells, but that postnatal steroidogenesis relies on LH signaling (162). Thus, the ability of LH to act on LH-R on Leydig cells in the postnatal testis to stimulate androgen production is key to successful spermatogenesis.

Binding of LH to LH-R in Leydig cells initiates several intracellular signaling pathways. Most LH-R–mediated effects on the function of differentiated Leydig cells, such as steroidogenesis, are mediated by the Gs/adenylyl cyclase/cAMP/PKA pathway, however, additional signaling pathways such as the phospholipase C and mitogen-activated protein kinase (MAPK) pathways may be involved in other LH-R–dependent events including proliferation and differentiation; reviewed in (158,160).

The appearance of full-length LH-R mRNA in the fetal testis coincides with the developmental onset of LH binding, with fetal Leydig cells expressing full length LH-R mRNA and able to bind LH from around 15.5 days of gestation in the rat (152,159). In the postnatal testis, immature Leydig cells arise from progenitor Leydig cells (163), and then transform into adult-type Leydig cells [see (160)]. Progenitor Leydig cells express truncated forms of the LH-R mRNA, yet more differentiated Leydig cells express the full-length mRNA (164). Consistent with this observation, LH-R numbers are lowest in progenitor Leydig cells but increase as Leydig cells mature into the adult phenotype (165).

In summary, LH-R expression by Leydig cells in the postnatal testis is essential for testicular steroidogenesis. The importance of androgens produced by Leydig cells in response to LH is discussed in section on Androgen Regulation of Spermatogenesis.

Thyroid Hormone

Thyroid hormone effects are mediated by thyroid hormone receptors (TR), which are members of the steroid receptor superfamily (77). TRs are encoded by two different genes: TR-α and TR-β. Alternate splicing leads to the production of several peptide isoforms, five which have been described: TR-α1, TR-α2, TR-α3, TR-β1, and TR-β2. TR-α2 and TR-α3, which lack a hormone binding domain, are thought to function as inhibitors of thyroid hormone action by competition for the binding of thyroid response elements, resulting in suppression of transcription (166–171).

Binding studies have demonstrated specific thyroid hormone binding sites (presumed to be TRs) in developing testes and Sertoli cell-enriched extracts. These binding sites were detected at very high levels in the fetus and early neonate, decreasing toward negligible expression at day 20 postpartum (172,173), a pattern coincident with the proliferative phase of Sertoli cell development. Several molecular techniques have been used to demonstrate the presence of mRNA encoding TR-α isoforms in fetal and neonatal testes (172). Similar techniques indicate that TR-β expression is present but is significantly lower than TR-α (172,174). These studies also suggest that TR gene expression is absent in the adult testis. In contrast, immunohistochemical data show the existence of TR in the adult testis, although the exact cellular localization remains unclear (175–177). More recently immunohistochemistry, immunoblot and RT-PCR were used to demonstrate the expression pattern of TRs in the juvenile and adult rat testis (178). TR-α1 was expressed in proliferating Sertoli cells, however, its expression decreased coincident with the cessation of Sertoli cell proliferation. TR-α2, TR-α3, and TR-β1 mRNA transcripts were expressed in low levels during development, yet protein was not detectable by immunoblot analysis. TR-α1 immuno-expression is also evident in germ cells, from intermediate spermatogonia through to mid-cycle pachytene spermatocytes (178).

ANDROGEN REGULATION OF SPERMATOGENESIS

The role of testicular androgens, mainly T and its 5α-reduced product DHT, in the initiation, maintenance, and restoration of spermatogenesis has been the subject of decades of extensive research [for reviews (5,7,179)], yet the details of the cellular and molecular mechanisms of androgen action in the testis remain ill-defined. The biological actions of androgens within the seminiferous epithelium are principally thought to be mediated via ARs that are likely to be localized exclusively in the Sertoli cell (see section on Androgen Receptor), although there is evidence for nongenomic activation of Sertoli cell function as well [reviewed in (79); see section on Androgen Receptor]. A third androgen metabolite, 5α-androstane-3α,17β-diol (3α-adiol) is formed reversibly from DHT by the action of the enzyme 3α hydroxy-steroid dehydrogenase, however, it interacts weakly with the AR (180) and hence its role in the mammalian testis has been largely ignored. Back-conversion of 3α-adiol to DHT restores its potent androgenic activity in the wallaby testis (180,181), establishing that this metabolite is an important androgenic precursor in this species. 3α-adiol levels are also relatively higher in the immature rat testis (182), mirroring increased DHT levels and 5α-reductase (the enzyme that metabolizes T to DHT) activity (182,183) (see below). Whether or not 3α-adiol has an androgenic role in the rat or any other mammalian species has yet to be established.

Testosterone is essential for the initiation and maintenance of mammalian spermatogenesis [reviewed in (5,179)] and due to its local synthesis by Leydig cells has an exceedingly high testicular concentration [100–200 ng/gm, testis tissue ($3.5–7.0\times10^{-7}$ M)] in rodents, and two- to threefold greater in humans (Table 2), being 50- to 100-fold greater than in the circulation. However, the dissociation constant of the testicular AR is 3×10^{-9} M (184), which suggests that the AR is fully saturated in the normal testis. Studies in rats (7,11,185) and mice (186) have shown that quantitative spermatogenesis, as indicated by the numbers of germ cells compared to Sertoli cells, can occur when intratesticular T concentrations are reduced by 80% to 20 ng/ml. An important distinction needs to be made here, in that the dose of T required to *initiate* spermatogenesis in the mouse is about an order of magnitude greater than that needed for the *maintenance* of spermatogenesis, suggesting that two different mechanisms of androgen action may exist in the testis (186) depending on maturational status. Other studies suggest that binding of T to androgen-binding protein (ABP) may sequester part of the intratesticular free T pool (187,188).

An understanding of the role of androgens in spermatogenesis requires firstly an appreciation of the various models used to study the effects of individual hormones (FSH, LH/T), together with their strengths and caveats (9). Typically, these models have included ablation of gonadotropic support by hypophysectomy [e.g., (189,190)], active immunization against GnRH [e.g., (191,192)] inhibition of GnRH release by administration of GnRH analogs (193), inhibition of gonadotropin release by sex steroids (e.g., [11,97]), or the use of transgenic animal models [e.g., (22,194)]. Hormones are then selectively replaced via implants (steroids) or injection (gonadotropins). While the desired effects of these models have been to selectively remove and replace testicular endocrine support, in practice each of these models have limitations that need to be recognized. These can include ablation of other pituitary hormones by hypophysectomy, incomplete suppression of both gonadotropins by GnRH antagonists or immunization regimes, or re-initiation of pituitary FSH release by exogenous androgen following GnRH immunization in rats, meaning that the effects of T cannot be studied in isolation (191). These limitations must be kept in mind when evaluating data on the role of androgens in the initiation and maintenance of spermatogenesis.

As will be apparent in the coming sections, T and FSH have overlapping as well as distinct roles in the regulation of spermatogenesis. This section and the section on FSH Regulation of Spermatogenesis will discuss the separate roles of T and FSH in the initiation and maintenance of spermatogenesis; however, the commonalities between the two hormones in terms of their sites and mechanism actions will also be considered (see section on FSH and T Cooperativity).

Role during Initiation of Spermatogenesis

The role of androgens in the initiation of spermatogenesis has been elegantly studied using a mouse model congenitally deficient (>95%) in GnRH, but with an otherwise intact reproductive tract,

TABLE 2. *Serum and intratesticular testosterone (T), dihydrotestosterone (DHT), and 5α-androstane-3α,17β-diol (3α-diol) concentrations in various species*

		T		DHT		3α-diol	
Species	**Age**	**Testis (ng/gm)**	**Serum (ng/ml)**	**Testis (ng/gm)**	**Serum (ng/ml)**	**Testis (ng/gm)**	**Serum (ng/ml)**
Mouse	Adult	150–364 (22,477)	3.2 (22)	1–8 (22,477)	0.05 (22)	NA	NA
Rat	Immature (28–35d)	12.1–24.5 (182,183)	0.21–0.65 (182,224)	13.0–19.1 (182,183)	0.64–0.69 (182,224)	22.5–41.0 (182,183)	1.2–1.4 (182,478)
	Adult	100–160 (182,206,212,479)	2.5–4.2 (224,480,481)	1.6–6.9 (182,183,206,212,479)	0.2–0.5 (224,480,481)	4.2–27 (182,183,206,212,479)	0.4–1.6 (182,212,479)
Monkey	Adult	103–247 (482,483)	2.9 (482) 9.3–15.9 (483,484)	1.0 (483)	NA	2.5 (483)	NA
Human	Adult	430–643 (324,421)	5.0–8.0 (324,421)	2.9–6.4 (324,422)	1.2 (324)	6.9 (324)	1.7 (324)

NA, not available.
References to the literature are shown in parentheses.

termed the hypogonadal (*hpg*) mouse (22,186,195). Spermatogenesis in the *hpg* mouse is characterized by reduced numbers of Sertoli cells, which have immature nuclear morphology, and some spermatogonia (~60% to 70% of wild-type when expressed as number per Sertoli cell) and spermatocytes, with pachytene spermatocytes being the most mature germ cell type seen (195). Studies in this mouse model have shown that qualitatively complete spermatogenesis can be initiated by exogenous T alone (22,195). In addition, spermatogenesis can be similarly initiated by DHT which cannot be metabolized to estrogen, establishing that the effects of T are mediated via the AR not the ER (22). As FSH is nondetectable by radioimmunoassay (measuring down to 20% of wild-type controls), it appears that spermatogenesis in the *hpg* mouse can be initiated by T alone. Despite the presence of all germ cell types in the androgen-treated *hpg* mouse, testis size and absolute germ cell numbers are only about 40% of non-*hpg* (wild-type) values, (22), due to the limited proliferation of Sertoli cells in the absence of neonatal and prepubertal pituitary FSH secretion in these animals (see section on FSH Regulation of Spermatogenesis).

The importance of the Sertoli cell as the modulator of the androgenic effects on germ cells during the initiation of mouse spermatogenesis has recently been demonstrated in mice lacking Sertoli cell AR, created using a Cre-Lox conditional knockout strategy (23,24). In mice lacking AR, testicular descent does not occur and germ cell development is severely disrupted (82,104). In contrast, mice lacking AR only in Sertoli cells (SC AR) showed normal testicular descent and urogenital tract development (23,24). In these models, expression of the SC AR was placed under the control of the anti-Müllerian hormone promoter, meaning that the expression of functional AR protein should be switched off from 15 days postcoitum (24). After puberty, SC AR knockouts displayed spermatogenic arrest at the late spermatocyte/spermatid stage, with apparently normal entry of germ cells into meiosis, but a block at the diplotene spermatocyte stage (23,24). A four- to fivefold increase in the number of apoptotic germ cells (pachytene and metaphase spermatocytes) was also observed (23,24). Hence, the Sertoli cell AR is an absolute requirement for the completion of meiosis and spermiogenesis.

Sertoli Cell Proliferation and Maturation

While FSH is well known to control neonatal Sertoli cell proliferation (see section on FSH Regulation of Spermatogenesis), some models suggest T may be involved in modulating Sertoli cell differentiation during the early postnatal period. In vitro data (196) suggests that T, in conjunction with thyroid hormone and retinoic acid, may be involved in suppressing proliferation and inducing terminal differentiation of Sertoli cells via effects on cell cycle regulators [see (197)]. Sertoli cells with a differentiated nuclear morphology were observed in adult *hpg* mice given exogenous T from 21 days of age (195) and in FSH-R deficient mice in which androgens are present (198), suggesting that T in the absence of FSH can induce Sertoli cell differentiation. Surprisingly, *hpg* mice given T alone from 21 days of age showed a significant increase in Sertoli cell numbers (which were still lower than in wild-type mice), suggesting that T can be a mitogen for Sertoli cell proliferation in this model (195). However, normal Sertoli cell numbers were observed in mice lacking AR within Sertoli cells (SC AR mice) (24) and in LH-R null mice which have very low intratesticular T levels (199), suggesting that T is not likely to be a major factor controlling Sertoli cell proliferation.

Germ Cell Development

Various reports suggest that T, like FSH (see section on FSH Regulation of Spermatogenesis), acts as a germ cell survival factor during the initiation of spermatogenesis (200,201). In the immature rat testis, selective immunoneutralization of LH disrupts spermatogenesis (202), and leads to apoptotic cell death, particularly of pachytene spermatocytes, but not of spermatogonia (203). Similarly, the antiapoptotic effects of FSH are potentiated by LH (200) and Leydig cells (204), further indicating that androgens are important for the maintenance of germ cell viability following gonadotropin withdrawal in the immature rat. Other studies have demonstrated that T is required for the completion of spermiogenesis in the pubertal rat (205); administration of the Leydig cell toxicant ethane dimethanesulphonate (EDS) caused a transient decrease in interstitial fluid T concentration and a detachment of step 8–9 round spermatids from the epithelium. This mid-spermiogenic action of T has also been well documented in the adult rat (see below). These studies collectively demonstrate that androgens are strongly implicated in the initiation of spermatogenesis in rodent models, via actions as a germ cell survival factor, probably in synergy with FSH, as well as in the promotion of spermiogenesis.

Role of 5α-Reductase (5αR)

In the adult testis, T is the major androgen as it is present at a ~100-fold greater amount than DHT (Table 2), and the activity of the 5α-reductase (5αR) enzyme (which converts T to DHT) is low (206). However, in the neonatal and pubertal rat testis DHT and 3α-adiol are the predominant androgens present (182,207) (Table 2). Both of these metabolites are produced by the action of 5αR, indicating that the activity of this enzyme is increased in the immature animal (208). DHT has a higher affinity and receptor-bound half-life than T (209) and is therefore a more potent androgen than T by a factor of five- to tenfold (210); thus, the 5α-reduction of T to DHT is thought to amplify androgen action when T concentrations are equivalent to, or lower than, DHT concentrations (209,210). This amplification mechanism may well be important for the androgen-dependent initiation of spermatogenesis when testicular T concentrations are comparatively low (183) (see below). Similarly, this mechanism is involved in maintaining or restoring sperm production in adult rats in a setting of low intratesticular T concentration (211,212) (see section on 5α-Reductase, page 1031). Accordingly, the activity and regulation of testicular 5αR during puberty and in adult spermatogenesis has been extensively studied.

To date, two isoenzymes of 5αR (5αR1 and 5αR2) have been characterized, which are encoded by different genes [for review, see (213)]. Both forms are found in the mammalian testis, but have different biochemical properties. 5αR1 has a micromolar affinity for steroid substrates (1.02 μM) and a broad optimal pH range at neutrality (214), while 5αR2 has a nanomolar substrate affinity (75 nM) and an optimal pH at pH 5.0 in cell lysates (214,215). 5αR1 is present predominantly in Leydig cells (216), but there is also evidence for 5αR activity (probably 5αR1) within the seminiferous epithelium, in both Sertoli cells and germ cells (217). The localization of 5αR2 within the testis is unknown. Various studies have quantitated testicular 5αR1 and 5αR2 mRNA expression, and produced conflicting data regarding their relative abundance during testicular maturation (208,216,218,219). More recently, assays specific for each isoenzyme have demonstrated that 5αR1 activity predominates over 5αR2 in the rat testis at all postnatal ages (183,219). Despite this, during the onset of spermatogenesis (between days 20–40), 5αR2 probably has a significant contribution to the increase in 5α-reduced metabolites, as the testicular T concentration (30–120 nM) is similar to the K_m of this isoenzyme (183,215). After day 40, however, testicular T concentrations increase to μM concentrations by about day 60 (183) and 5αR1 is then likely to be the major isoenzyme for the production of 5α-reduced metabolites (Fig. 4).

Testicular 5αR1 and 5αR2 mRNA and activity are highest between days 20 and 40 in the rat (183), coinciding with the first wave of meiotic and postmeiotic germ cell development, which begins after Sertoli cell division has been completed at day 15 and ends with the release of the first mature spermatids at about day 44 (200) (Fig. 4). This rise in 5αR activity corresponds with the pubertal increase in circulatory FSH (183) and is consistent with reports that 5αR1 and 5αR2 may be regulated by FSH (206,220,221). 5αR1 isoenzyme (both mRNA and activity) is apparently negatively regulated by T (206,222). Thus, both 5αR1 and 5αR2 isoenzymes, which may be regulated by FSH and T, are developmentally regulated during the pubertal initiation of spermatogenesis. Pubertal rats given 5αR inhibitors and 5αR null transgenic mice have high levels of T which are likely to be able to act directly with the AR without conversion to DHT, making it difficult to study the role of 5α-reduced metabolites in the initiation of spermatogenesis (183,223,224).

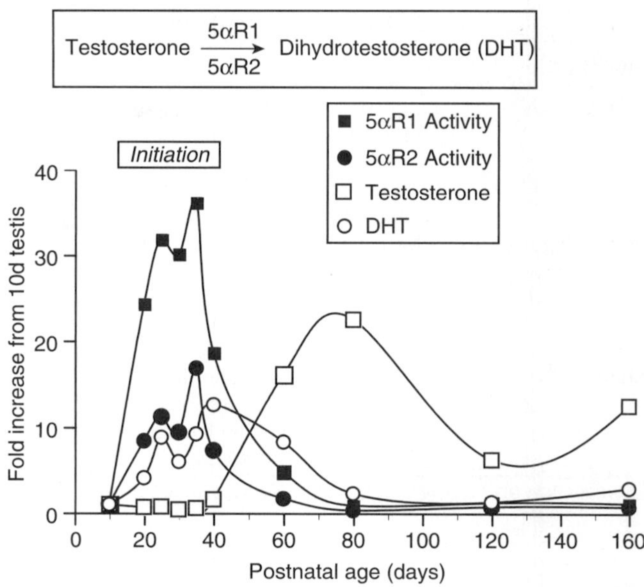

FIG. 4. Developmental pattern of 5α-reductase isoenzyme type 1 (5αR1) and 2 (5αR2) activity and testicular testosterone (T) and dihydrotestosterone (DHT) concentrations in the postnatal rat. Testosterone is 5α-reduced to DHT in an irreversible reaction that is catalyzed by either of the 5α-reductase isoenzymes. The timing of the initiation of the first wave of spermatogenesis in the rat is indicated (*initiation*); this begins with the entry of germ cells into meiosis at the end of the Sertoli cell proliferative period (approximately day 15 in the rat), and ends with the completion of one full spermatogenic cycle at approximately day 44 (200). Data was adapted from (183) and shows the fold increase of each parameter above the levels present in 10-day-old rat testis.

Role in the Adult

Testosterone has profound effects on adult spermatogenesis (summarized in Table 3). Numerous studies indicate that T alone can at least *qualitatively* maintain and restore this process in the rodent

TABLE 3. *Summary of the proposed sites of androgen action in rodent spermatogenesis**

Site of action	Effect
Sertoli cells	Can affect Sertoli cell proliferation and maturation during the neonatal period, but likely not to have a major role
	Modulates androgen receptor signaling to germ cells
	Regulates the localization of androgen receptor protein to the nucleus
	Down-regulates 5α-reductase type I isoenzyme
	Stimulates fluid production
Spermatogonia	No involvement in spermatogonial development
	Can prevent differentiation at high local concentration
Spermatocytes	Can partially support their maturation
	Can promote their survival/prevent apoptosis, particularly in stages VII and VIII
	Supports the final meiotic division into haploid germ cells
Spermatids	Essential for spermatid development
	Can promote their survival/prevent apoptosis, particularly in stages VII and VIII
	Promotes elongation by mediating the adhesion of step 8 round spermatids to Sertoli cells
	Can affect sperm release

*See section on Androgen Regulation of Spermatogenesis.

(i.e., maintain or restore all phases of germ cell development, but at levels less than normal). The sites of action of T include maintenance of earlier germ cell populations, an essential function at meiosis, maintenance of germ cell viability particularly at stage VII of the spermatogenic cycle, and major effects on spermiogenesis.

Testosterone alone appears to be able to maintain qualitatively normal spermatogenesis in mature rats following hypophysectomy (189,225–230) or GnRH-antagonist treatment (231). These animals have undergone normal testicular development that included exposure to FSH and have a full complement of Sertoli cells, as distinct from the *hpg* mouse model. The ability of T alone to *quantitatively* maintain spermatogenesis (i.e., normal numbers of all phases of germ cell development) in the absence of FSH remains more controversial, particularly as it is difficult to design an in vivo model in which FSH action is completely ablated. With this caveat in mind, quantitative maintenance by exogenous androgens has been reported in studies on GnRH-immunized (192), GnRH antagonist-suppressed (232), or hypophysectomized (233) adult rats, whereas other studies have argued that a failure of androgens to quantitatively restore spermatogenesis indicates a requirement for FSH in restoration of spermatogonial numbers (189,234).

The maintenance or restoration of spermatogenesis by T occurs over a narrow dose range which is considerably lower than that normally present within the testis (7) (Table 2). Hence, very small changes in testicular T levels result in large changes in sperm output. For example, studies in intact rats showed that spermatogenesis was suppressed when testicular T was reduced to 4.2% to 4.6% of normal, but that spermatogenesis could be maintained if the T level was held at 7.45% to 8.8% of normal (94). In this model, FSH levels remained at 60% to 80% of normal. Hence, an intratesticular T level of approximately 20 ng/ml is sufficient for the maintenance of quantitatively complete spermatogenesis (7,11), which is fivefold less than in the normal testis. However, this threshold concentration of T is still approximately four- to eightfold greater than serum T levels (Table 2).

Spermatogonia

There is little evidence to suggest that T or Leydig cell factors are required for spermatogonial development (35). In the *hpg* mouse, T alone did not stimulate spermatogonia numbers above those seen in untreated *hpg* mice (195), and in the rat, selective administration of T to long-term gonadotropin-depleted rats did not alter spermatogonial development (15,235). In fact, high testicular T concentrations may be detrimental to spermatogonial development (236), which is supported by data from other models (juvenile spermatogonial depletion, or *jsd*, mutant mouse, and irradiated rat) that demonstrate that suppression of testicular T levels is needed to promote spermatogonial development (237). Similarly, Ogawa and colleagues (238) reported that a transient reduction in testicular T via GnRH-antagonist treatment improved the ability of donor spermatogonia to colonize a recipient testis in transplantation studies.

Meiosis and Spermiogenesis

Two key sites of androgen action are evident specifically in meiosis (22,179,186,195) and spermiogenesis (94,96,97,239). Of these, the completion of meiosis during the onset of spermatogenesis appears to be remarkably sensitive to T as recently observed in the *hpg* mouse (22,186) wherein exogenous T administration, resulting in intratesticular T concentrations 40% of normal, markedly promoted meiotic progression from preleptotene through to pachytene spermatocytes (195). T alone facilitated the completion of meiosis in *hpg* mice, as evidenced by the production of haploid round spermatids from pachytene spermatocytes (195).

While an initiation of meiosis by exogenous T is clear in the *hpg* mouse, a suppression of meiosis is not evident in the various rat models employing T implants to suppress spermatogenesis (94,239), as intratesticular T levels are likely not suppressed sufficiently to see this action (186). However, progression of meiotic spermatocytes through stages VII and VIII is decreased when intratesticular T levels are severely suppressed using a combination of gonadotropin suppression plus the anti-androgen flutamide to eliminate residual androgen action (12,211,235). Thus, it seems likely that T facilitates meiotic progression at least in part by promoting the survival of meiotic spermatocytes through the hormone sensitive stages VII and VIII.

Following T withdrawal, preferential germ cell death in stages VII–VIII, particularly meiotic and postmeiotic cells is reported (95,240,241). After administration of EDS to acutely withdraw intratesticular T, degenerating pachytene spermatocytes and round spermatids were observed in stage VII (95) whereas T replacement suppressed the appearance of apoptotic cells in this stage (241). Similar studies in EDS-treated rats have demonstrated that T withdrawal induced apoptotic cell death in most stages of the cycle (242), but that T appeared to promote apoptosis of pre-meiotic cells in stage XII (242), perhaps a reflection of inhibitory effects of T on spermatogonia as described above. Hence, as observed during the initiation of spermatogenesis, T acts to support spermatogenesis in the adult through mechanisms that regulate germ cell viability, particularly but not exclusively at stage VII and maybe stage VIII of the spermatogenic cycle.

Testosterone has a specific site of action in terms of the initiation (195), maintenance (230) and restoration (96,239) of all phases of spermiogenesis. Studies in *hpg* mice suggest that T alone can initiate and maintain the production of haploid round spermatids from pachytene spermatocytes and their subsequent elongation, albeit at reduced levels compared to wild-type mice (195). Studies in adult rats have pinpointed a particular site in spermiogenesis at which T action is required. Administration of 3 cm Silastic T implants, in combination with a low dose of estradiol (0.1 cm) to intact rats (known as the TE model) results in the suppression of the conversion of round to elongated spermatids, which can be dose-dependently restored by the administration of higher doses of T (11,96,233,239). This treatment does not suppress FSH, hence the model is one of selective LH and androgen withdrawal; reviewed in (9). While a general suppression of most germ cell types (spermatogonia, early spermatocytes, and pachytene spermatocytes) occurs in this model, the most marked changes are in round spermatids at stage VIII (i.e., step 8 round spermatids) that virtually disappear from the epithelium. Subsequent studies have demonstrated that suppression of intratesticular T impairs the conversion of step 7 to step 8 round spermatids due to the premature detachment of step 8 round spermatids from the epithelium (97). Recent studies in mice lacking LH-R, which have testicular T levels only 2% of normal, showed that the conversion of round spermatids to elongated spermatids was abolished by the administration of the AR antagonist flutamide, suggesting that mid-spermiogenesis requires only very low T levels in this model (243).

It has been hypothesized that androgens regulate adhesion between Sertoli cells and step 8 round spermatids (97,205,244), however, definitive proof for this proposition in vivo has not been found; reviewed in (9). Androgen suppression does not appear to disturb the structure of the ectoplasmic specialization (244), which is a Sertoli cell-specific junctional complex (64). Cell adhesion molecules which could be involved in this round spermatid–Sertoli cell junction include $\alpha6\beta1$ integrin (69,245,246), and members of the cadherin (67,247,248) and protocadherin families (249). In vitro studies suggest N-cadherin may be important, as an N-cadherin antibody will block the androgen-stimulated adhesion between rat Sertoli cells and isolated round spermatids, and N-Cad production by Sertoli cells in vitro is stimulated by androgens (250).

5α-Reductase

Given the low threshold of T required for the maintenance of spermatogenesis (see above), several studies in rats have examined the importance of 5αR in the re-initiation of spermatogenesis in situations where the intratesticular concentration of T is low, as observed following T + low dose estradiol (TE) treatment (211,212). The ability of T to restore spermatogenesis, as evidenced by round and elongated spermatid populations and an assessment of the androgen-dependent conversion of step 7 to 8 round spermatids (see above), was compromised by administration of a 5αR inhibitor (a preferential inhibitor of the 5αR-1 isoenzyme). This effect of the 5αR inhibitor was only noted when Silastic T implants of 6 cm or less were used to restore spermatogenesis; at higher T doses the 5αR inhibitor had no effect (211,212), presumably due to the restoration of intratesticular T levels that can then interact directly with the AR (209). These studies thus show that, while T is likely to act directly on the AR to support spermatogenesis in the normal adult rat, the 5α-reduction of T to DHT is important for androgen action on the seminiferous epithelium when intratesticular T levels are reduced. This finding has potential implications for androgen-based male contraceptive development, as other studies have suggested that men who fail to achieve azoospermia in response to T-based contraception have a greater 5αR activity (251,252), suggesting that the maintenance of low levels of spermatogenesis in these men is due to increased androgenic support via higher levels of DHT (see section on Role of 5α-Reductase).

Regulation of Testicular Proteins

Despite the enormous amount of research into the regulation of spermatogenesis by androgens, surprisingly few genes and proteins in the seminiferous epithelium have been shown to be directly regulated by T. Since T has been implicated in maintaining germ cell viability particularly in stages VII and VIII, as well as cell adhesion during these stages, it is likely that T regulates gene expression and protein secretion at these stages. In addition, T also positively regulates seminiferous tubule fluid secretion at stages VII and VIII (93,234,253) and modulates the protein composition of this fluid (93).Various strategies have been employed to identify androgen-regulated proteins, including two-dimensional gel electrophoresis (93) and, more recently, oligonucleotide microarray analysis (254).

One of the few T-regulated proteins to be conclusively identified has been *Pem*, or placentae and embryos oncofetal gene, that codes for a homeobox protein (255). This protein is expressed in both Sertoli cells and in the epididymis, and is up-regulated following androgen administration in the *hpg* model (255). In the adult mouse testis, *pem* mRNA and protein is expressed in Sertoli cells during stages IV–VII of spermatogenesis (255), and is significantly up-regulated in the immature testis when adjacent germ cells are initiating meiosis between days 8 and 9 postpartum (255,256). *Pem* expression is androgen-dependent (254,256,257) and may also act as a stage-specific marker of Sertoli cell function (258).

Testosterone may also regulate AR protein within Sertoli cells. Suppression of intratesticular T concentrations caused a loss of AR protein in Sertoli nuclei, but T replacement restored AR immunostaining within 24 hours (81). Interestingly, steady-state levels of AR mRNA were not regulated by changes in testicular T concentrations. Hence, these authors proposed that T has a primary role in the regulation of translation or posttranslational stability of the AR protein, and also acts to promote nuclear translocation of the AR prior to activation of androgen-dependent gene transcription (81).

A comparison of gene expression in the *hpg* mouse following treatment with T propionate for periods of 4, 8, 12, and 24 hours has demonstrated that ~twofold more mRNAs were down-regulated than up-regulated at the earlier time points (4, 8, 12 hr) compared with control nontreated mice (254). In contrast, *pem* expression in the same model was increased at all time points by up to 500-fold. After 24 hours of treatment, most changes involved an increase in transcript levels, which could reflect an activation of germ cell gene transcription (254).

Clearly, the numbers of testicular genes now able to be identified as regulated by this type of experimental approach is large, necessitating extensive post-analysis validation to sort out those that are directly regulated versus those that are not. Nevertheless, some interesting proteins were identified by this study, including two nuclear proteins (nuclear protein 220, and matrin 3) that have RNA-binding motifs that may be important in translational control of mRNA expression (254).

In summary, while T has marked effects on the seminiferous epithelium and cellular changes can be readily identified, only a few seminiferous epithelial proteins have been confirmed as being directly regulated by androgens. With the recent advent of gene-microarray screening approaches, it is expected that this list will rapidly grow, allowing the mechanism(s) by which T exerts its effects on the seminiferous epithelium to be better understood.

FSH REGULATION OF SPERMATOGENESIS

In the past decade, the development of transgenic mouse models has been instrumental in elucidating the roles of FSH in spermatogenesis. It is clear that FSH plays a key role in the development of the immature testis, particularly by controlling the size of the Sertoli cell population, which is set early in postnatal life. This is of particular importance as Sertoli cell number dictates sperm output in adulthood. We will review the apparently conflicting data about the need for FSH derived from various animal models suggesting that while spermatogenesis can be initiated and proceed to completion in the absence of FSH, quantitatively normal spermatogenesis depends on FSH. We will review the multiple sites of FSH action in the spermatogenic pathway including the promotion of spermatogonial development, the maintenance of key Sertoli cell functions and the viability

TABLE 4. *Summary of the proposed sites of action of FSH in rodent spermatogenesis**

Site of action	Effect
Sertoli cells	Supports proliferation during the neonatal period which determines final testis size
	May play a role in terminating proliferation prior to the onset of spermatogenesis
	Maintains ultrastructural characteristics
	Modulates expression of many gene products
	Facilitates the ability to support the maximal number of germ cells
	Can support the organization of key cytoskeletal Sertoli cell molecules required for the completion of spermiogenesis
Leydig cells	Influences the size of the population and steroidogenic capacity
Spermatogonia	May influence gonocyte differentiation and survival
	Supports their development, proliferation and enhances their survival
Spermatocytes	Can partially support their maturation
	Can promote their survival/prevent apoptosis
Spermatids	Provides some support for development
	Can promote their survival/prevent apoptosis
	Can affect sperm release
	Can affect sperm quality

*See section on FSH Regulation of Spermatogenesis.

of meiotic and postmeiotic germ cells (summarized in Table 4).

Role during Initiation of Spermatogenesis

FSH has key roles in the regulation of Sertoli and germ cell development in the immature testis. The mechanisms by which FSH exerts its effects during the first wave of spermatogenesis appear to be influenced by a complex network of interacting signals, operating via Sertoli cells as only they contain receptors for FSH (see section on FSH on page 1025) (Table 1). A specific role for FSH is apparent early in life (see below) in promoting Sertoli cell proliferation and in setting their eventual population. It is notable that Sertoli cell responsiveness to FSH declines as Sertoli cells cease proliferation and commence differentiation (259) (see section on FSH on page 1025). FSH-regulated genes products, including the retinoblastoma family proteins, activin and activin receptors, and bone morphogenetic protein 4 (BMP4), that potentially regulate the Sertoli cell response to FSH have recently been identified (260–263). It should be noted here that the regulation of adult spermatogenesis may well be mediated differently to the first wave of spermatogenesis, given that shifts in the expression patterns of key regulatory genes (i.e., Bcl-2 family members and stem cell factor) have been observed as Sertoli and germ cells mature (264).

Sertoli Cell Proliferation and Function

FSH is the primary endocrine hormone regulating Sertoli cell function (139) and proliferation. It regulates the expression of many Sertoli cell genes including FSH-R, ABP, transferrin, plasminogen activator, and aromatase (139,265,266). Sertoli cells at the onset of puberty in the rat undergo a radical change in morphology and function, heralding a switch from an immature, proliferative state to a mature, nonproliferative state. The maturation of Sertoli cells can be envisioned as a multistep process, with the precise timing being species-specific; reviewed in (259).

In both the rat and human, Sertoli cell proliferation is under the control of FSH (29,267–269). Withdrawal of FSH in fetal rats by in utero decapitation or immuno-neutralization of FSH both led to similar decreases in DNA synthesis in Sertoli cells (268). Conversely, elevated levels of circulating FSH in neonatal rats, that were induced by hemi-castration, markedly increased Sertoli cell proliferation (268) and exogenous FSH during this period increased Sertoli cell populations and, subsequently, elongated spermatid numbers in adult animals (29). In vitro studies show that FSH and activin A act in a synergistic manner to enhance Sertoli cell proliferation in day 3 testis fragment cultures (270,271). The expression of transgenic FSH in *hpg* mice increases Sertoli cell numbers, whereas mice lacking a functional LH-R (and which have twofold higher levels of FSH compared to wild-type yet very low intratesticular T levels) have normal Sertoli cells numbers, indicating that FSH is the primary hormone required for Sertoli cell proliferation (199).

Whether FSH plays a role in terminating Sertoli cell proliferation (and hence facilitating the switch to the nonproliferative, differentiating form) around 15 days after birth in the rat is unclear, however, in vitro and in vivo studies imply a role (272,273); reviewed in (259). Sertoli cells may not be governed by FSH during the entire proliferative phase as mice deficient in FSH action show normal Sertoli cell number at postnatal day 5 (274) and consistent with this is the observation that Sertoli cells are not responsive to FSH at postnatal day 6 in vitro (275). It has been hypothesized that the first proliferative phase in monkeys and humans (during infancy) may be gonadotropin-independent, while the second proliferative phase (during puberty) may be gonadotropin-dependent (276) [reviewed in (277)], although direct evidence for this is lacking. Deficient Sertoli cell proliferation may underlie the limited spermatogenic potential of congenitally GnRH-deficient men who lack both fetal and pre-pubertal gonadotropin exposure (see section on Hormonal Dependency of Human Spermatogenesis). Thus, although it is clear that FSH can regulate Sertoli cell proliferation, the precise timing of this regulation is unclear.

The pubertal rise in FSH has been shown to be associated with a redistribution of actin to the periphery of Sertoli cells in the rat (278). This peripheral organization of actin, which is a cytoskeletal protein found in ectoplasmic specializations, has been shown to be related to the ability of Sertoli cells to bind round spermatids in vitro and in vivo (228,279,280). The fact that key changes in actin organization, which are thought to be required for normal spermiogenesis (228,279,280) occur during the postnatal rise in FSH but not T suggest that FSH may facilitate the Sertoli cell's ability to support the initiation of spermiogenesis in the rat (278). A similar role for FSH in the adult Sertoli cell has also been proposed (see section on Role in the Adult, below).

FSH may regulate the expression of Sertoli cell genes that are involved in controlling androgen responsiveness and steroid production during the pubertal period. In cultures of Sertoli cells from immature rats, exposure to FSH can increase AR mRNA (98) and protein (100), although the response to FSH appears to decrease with age (99). Microarray analysis of gene expression in cultured Sertoli cells (from 20-day-old rats) with and without FSH (281) identified both the steroidogenic acute regulatory (StAR) gene, which is known to play a central role in steroid hormone synthesis (282), and the ABP gene, which is important for sex steroid movement in circulation (283,284), as potential FSH-responsive genes.

A large number of FSH-responsive genes have been identified in cultured 20-day-old rat Sertoli cells, however, in most cases, their specific contribution to Sertoli and germ cell viability and function is undefined. Recent work using microarray analysis has defined 100 to 300 known transcripts that are regulated by FSH in cultures of Sertoli cells from 20-day-old rats (281). Similar approaches have identified developmentally regulated genes in mouse testicular somatic cells (285), as well as genes with stage-specific expression patterns (286); however, further studies are required to identify which of these genes are regulated by FSH.

Germ Cell Development

FSH plays an important role in germ cell development in the juvenile testis by enhancing germ cell survival and proliferation. FSH has been hypothesized to play a role in determining the number of stem cell spermatogonial niches (the special microenvironment wherein stem cells reside), as their number is determined by Sertoli cell number, that in turn is dependent on FSH (see above) (287,288) [reviewed in (34,35)]. Stem cell renewal is positively regulated by glial cell line-derived neurotrophic factor (GDNF) (289) and the absence of FSH results in lower GDNF levels in the testes (290), which may lead to fewer stem cells. Thus, an effect of FSH on stem cell renewal is possible via the GDNF pathway.

There is some evidence that FSH may contribute to the establishment of the spermatogonial population from gonocytes (precursor spermatogonia present in the neonatal testis) that mature into type A spermatogonia in the early postnatal period of the rodent. New born *hpg* mice lacking both FSH and LH show normal numbers of gonocytes (272), but these fall 5 days after birth, suggesting that gonocyte survival may be in part regulated by gonadotropins but whether this is an effect specific to FSH is unknown (272). It is worth noting that gonocyte apoptosis and mitosis were not affected by FSH treatment in organ cultures of fetal and neonatal rat testes (291). However, gonocyte maturation into type A spermatogonia in short-term culture of postnatal rat testis fragments was promoted by combined FSH plus follistatin treatment, although FSH or follistatin alone were ineffective (271). Gonocyte migration is important for the establishment of the spermatogonial population in the neonatal testis and relies at least in part on the interaction of c-kit and its FSH-regulated ligand, stem cell factor (292); therefore, this might be one way in which FSH may affect the establishment of the spermatogonial population.

In terms of the survival and differentiation of spermatogonia and meiotic cells in the juvenile testis, the administration of FSH to gonadotropin-deplete immature rats reduced the number of degenerating cells (200) and prevented germ cell apoptosis (201). Consistent with this, selective FSH withdrawal in immature (day 21 postnatal) rats enhanced germ cell apoptosis (293). In vitro data also suggest that FSH enhances the proliferation of differentiated spermatogonia in mice (294,295). It has been postulated that there are age-dependent changes in germ cell gene expression that may be regulated by FSH, for example, members of the Bcl-2 family during the first wave of mouse spermatogenesis. Both Bcl-w (296) and Bok (297) mRNAs have been shown to be up- and down-regulated, respectively, by FSH in vitro, rendering them likely mediators of apoptosis responses.

Role in the Adult

FSH acts at multiple sites in the spermatogenic process, and although the profile of action may vary somewhat between species, overall a consensus is developing that FSH is required for quantitatively normal spermatogenesis. Using well-characterized in vivo models of FSH manipulation and unbiased methods for the estimation of testicular cell populations, sites in the spermatogenic process that are sensitive to FSH action have been identified. Mouse models of congenital FSH deficiency have been invaluable in determining the role of FSH in rodent spermatogenic development, but their complete lack of FSH action in fetal and early postnatal life may alter the adult phenotype to a point where the information they provide may not be directly relevant to normal spermatogenesis. Alternative data derived from adult onset rodents models of FSH withdrawal and replacement must also be considered.

In monkeys and humans it is clear that quantitatively normal spermatogenesis is highly dependent on gonadotropins (9,10) (see section on Endocrine Regulation of Human Spermatogenesis), notably spermatogonial development and the final phase of spermiogenesis (spermiation); however, the exclusive role of FSH in these processes are largely undefined. Basic research in rodent models will assist in identifying key FSH-regulated processes that can then be explored in primates.

Intriguingly in one species of rodent, the seasonally breeding Djungarian hamster, spermatogenesis (298) and fertility (299) appears to be exclusively regulated by FSH (298), with T only being reported to be necessary for mounting behavior (299). Other seasonal effects on the ultrastructural changes of the hamster testis have been reported and reviewed elsewhere (300), although specific FSH effects are not described.

Our knowledge of the role of FSH in spermatogenesis has been greatly advanced by studies involving the disruption of the FSH-β subunit (274,301,302) and the FSH-R (198,274,301,303–305) [reviewed in (306)]. Targeted disruption of either the FSH-R gene or the FSH-β subunit gene have clearly demonstrated that FSH signaling is required for maintaining normal testicular size, seminiferous tubule diameter, sperm number and quality. FSH-β subunit gene transgenic males are reportedly fully fertile (307), despite the number of mature spermatids per testis being reduced to 36% of wild-type (302). FSH-R transgenics males exhibit reduced fertility as evidenced by decreased litter sizes, as well as reductions in sperm number and motility and aberrant sperm morphology (198,305). Genetic rescue of the FSH-β subunit-deficient mouse increased testis weight, sperm number, and motility comparable to those seen in control mice (308). FSH-R transgenics show a more marked phenotype compared to FSH-β subunit gene transgenics (274,301), probably because the constitutive activity within the FSH-R may be sufficient to stimulate the spermatogenic process. Other transgenic mouse models, including *hpg* mice expressing varying levels of human transgenic FSH and *hpg* mice expressing an activated form of FSH-R have also been invaluable in revealing specific effects of FSH in the mouse (194,195,199,309).

Leydig Cells

Studies in FSH-R transgenic mice revealed marked decreases in Leydig cell populations and steroidogenesis (301), with the reduction in Leydig cells being more profound than that of Sertoli cells (274), implicating FSH in the regulation of Leydig cells as has been postulated (234). The FSH-R transgenic Leydig cell population failed to increase normally after puberty, leading to a reduced adult population with increased LH and decreased intratesticular T levels (301). Conversely in the FSH-β transgenic, Leydig cell populations, LH and T levels remained at wild-type levels suggesting that constitutive FSH-R activity in Sertoli cells regulates the development of Leydig cells (301). There is evidence in both rats and humans that FSH-stimulated Sertoli cell products act on as yet unidentified paracrine factors that may regulate Leydig cell function. For example, in the rat, ABP from the Sertoli cell can act on the Leydig cells to enhance T production in response to FSH (98). FSH may also be involved in regulating Leydig cell number (310,311), steroidogenic capacity, (190,311) and in mediating Leydig cell maturation (312).

Sertoli Cell Function

The general view of adult mammalian Sertoli cells is that the population is fixed and not modified by hormones. This view largely derives from reports that Sertoli cells neither divide in normal adults or in hypophysectomized or hypohysectomized-hormone treated rats, nor do they degenerate after hypophysectomy (16), a view supported by data obtained from hormonally manipulated rats using modern stereological methods to determine Sertoli cell numbers (14). However, fluctuations in Sertoli cell numbers have been reported in various species of adult seasonal breeders [reviewed in (300)], and in particular the stallion which shows significant seasonal changes in Sertoli cell numbers as well as levels of FSH, LH, and T (313–315). We have also noted significant seasonal changes in Sertoli cell numbers, which appear to be driven by FSH, in the adult Djungarian hamster (316). The mechanism by which the adult Sertoli cell population is altered between seasons remains unknown.

A role for FSH in regulating Sertoli cell ultrastructure arises from adult FSH-R–deficient transgenic mice wherein 40% of seminiferous tubules showed evidence of disruption, which was not stage-specific (304). Fluid-filled focal dilations within the Sertoli cell cytoplasm were seen, indicative of disturbed water balance, and abnormal morphology of Sertoli cell ectoplasmic specializations was noted (304). Thus, FSH is likely important for mediating adult Sertoli cell structure and function. FSH action on Sertoli cells impacts on the ability of adult Sertoli cells to support a full germ cell complement, as suggested by germ cell to Sertoli cell ratios in transgenic mice. Adult FSH-β null males showed a reduction in the ratio of germ cells to Sertoli cells, which showed a gradual attrition across germ cell development (302). In addition, expression of transgenic human FSH in *hpg* mice stimulated germ cell to Sertoli cell ratios (195), indicating that FSH action on Sertoli cells facilitates germ cell development in the adult (see below).

Spermatogonial Development

FSH plays a major role in regulating spermatogonia in the adult animal, as shown by restoration and maintenance studies in rat (12,14,15,17,317,318) and primate models (13,18). The exact sites in which FSH regulates spermatogonial development are largely undefined. In the rat, selective FSH withdrawal by passive immuno-neutralization showed a time-dependent decline in the number of early germ cell types, indicating that spermatogonia of types A_3-A_4 (associated with stages XIV and I) are the principal sites of FSH action in this species (318), a proposal that is supported by the demonstration that rat stages XIV–I are particularly FSH-responsive (see section on FSH, page 1025). Whether type B spermatogonia are also FSH sensitive in the rat is controversial (12,318). Expression of transgenic FSH in *hpg* mice revealed that FSH supported postnatal spermatogonial (type A, intermediate and B) populations when compared to *hpg* mice given T, highlighting a specific action of FSH (195). The mechanism by which FSH exerts its effects on spermatogonia is probably by acting as a survival factor (29,293) rather than a proliferative factor (14), at least in the rat. Studies simultaneously examining proliferative and apoptotic effects in the testis in response to FSH are lacking so that the relative importance of these two processes remains ill defined.

Primate spermatogonia are highly gonadotropin-dependent as evidenced by a 90% reduction in type B spermatogonial number compared to control following GnRH-antagonist treatment for 2 to 3 weeks in monkeys (319), and 20 weeks of gonadotropin suppression induced by an androgen-based contraceptive regimen in men (320). It is unclear whether this reduction in spermatogonia is attributed to the loss of FSH or T, however, data in adult macaque monkeys suggest that the inhibition of type B spermatogonia was associated with serum FSH suppression (45). The specific site of gonadotropin (presumably FSH) action in monkeys and humans is probably in the transition of A pale to B spermatogonia, based on studies of the withdrawal of FSH (321–323) or both gonadotropins (38,45,319,320,324). This may be due to an inhibition of type A spermatogonial proliferation following withdrawal of gonadotropin support (38), although type B spermatogonia rather than A pale spermatogonia have been suggested to be FSH responsive (13).

Spermiogenesis

Testosterone is essential for the completion of spermiogenesis (see section on Androgen Regulation of Spermatogenesis), however, the role of FSH in this process has been difficult to define. Homologous FSH preparations are not available for the long-term study of spermatogenic restoration following gonadotropin suppression, while heterologous FSH rapidly generates neutralizing antibodies. Short-term studies have provided some clues to an involvement of FSH in spermiogenesis. For example, immuno-neutralization of circulating FSH in adult rats for 8.5 days caused a 30% reduction in round spermatid populations, yet did not change serum or testicular androgen concentrations (318). Others have provided data implying that FSH may have a role in supporting early round spermatid maturation (14,189,325). Recently, *hpg* mice expressing a mutated activated human FSH-R or transgenic human FSH have been used to investigate FSH-specific effects (195,199,309). Moderate serum FSH levels in these transgenic mice partially restored Sertoli cell numbers and led to a slight but significant restoration of the ratio of spermatids to

Sertoli cells (195), whereas mice that expressed higher levels of transgenic FSH produced small numbers of elongated spermatids, whose number did not correlate to intratesticular T levels (199). These studies thus show a small but measurable effect of FSH on spermiogenesis.

There are several lines of evidence for the proposition that FSH exerts its permissive effect on spermiogenesis by regulating the ability of Sertoli cells to support spermatid development. Firstly, a reduction of spermatid to Sertoli cell ratios was observed in FSH-β subunit gene-deficient mice, indicating a diminished ability of Sertoli cells to support spermiogenic cell development (302). The acute withdrawal of FSH from gonadotropin-deficient adult rats prevented the restoration of round spermatids by T (15), implicating FSH in the T-mediated restoration of spermiogenesis. However, round spermatids could be restored by higher intratesticular T levels despite an absence of FSH (15), suggesting that the effects of FSH are particularly important when T levels are low. These observations support the proposition that FSH may potentiate the ability of the Sertoli cell to support the T-dependent restoration of spermiogenesis (see section on FSH and T Cooperativity, below).

The final event of spermiogenesis, spermiation (the release of mature sperm from the Sertoli cell), is impaired by FSH suppression, although this process is also affected by T suppression, suggesting that FSH and T have synergistic roles in this process (326) (see section on FSH and T Cooperativity, below). It is worth noting that disruption of FSH action leads to a reduction in sperm count and fertility in rats (327) and bonnet monkeys (323). In addition, human studies show that selective FSH suppression reduces sperm counts and sperm quality (323). Whether these effects result from aberrant spermatid maturation or are due to spermiation failure remain unknown; however, it seems that FSH is needed for normal sperm maturation and release.

FSH and T Cooperativity

It is apparent that FSH and T have distinct roles in spermatogenesis, but at many points they act cooperatively to promote maximal spermatogenic output (summarized in Fig. 5). Such cooperativity may stem from (a) the need for each hormone, acting independently, to support a particular process; (b) the potential for FSH and T to regulate the same metabolic pathway, in which setting either hormone may support a process to a qualitative degree, but both hormones combined being required for a maximal (normal) response. An example of the latter is spermiation (326) [reviewed in (9)]; other pathways are discussed below.

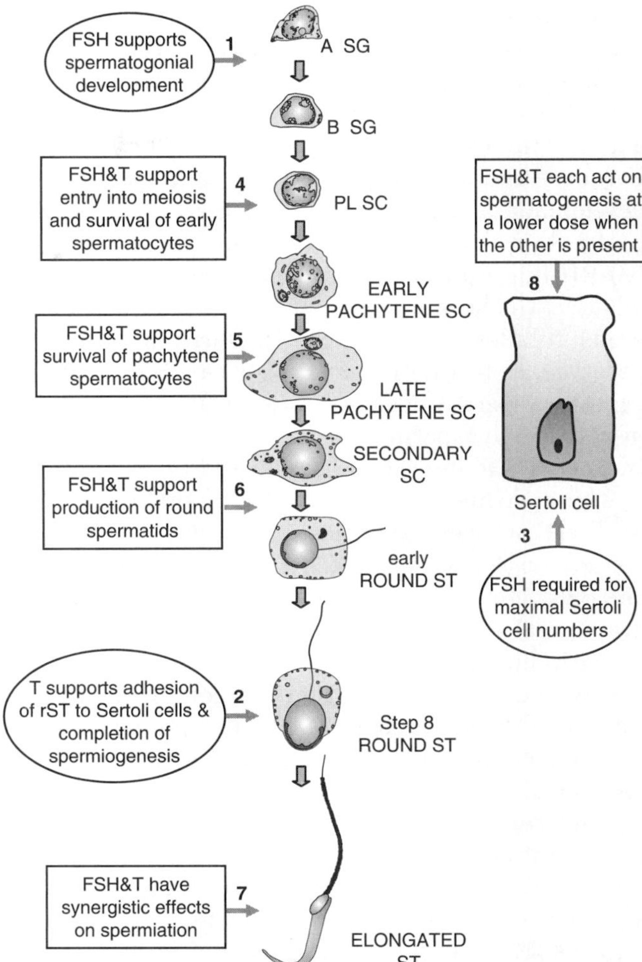

FIG. 5. Diagram illustrating the sites of FSH and testosterone (T) cooperativity in spermatogenesis. FSH and T specific effects are indicated by the *circles* and common and/or synergistic actions of FSH&T are indicated by the *rectangles*. **1.** FSH can support spermatogonial development and survival [e.g., (15,195,318)], whereas T has little role in this process (see section on Spermatogonia, page 1030). Conversely, it seems that **2.** T alone is required for the completion of spermatid elongation [e.g., (195,240,325)]. Therefore, one level of FSH and T cooperativity is likely to involve specific effects at the beginning and end of germ cell development, with FSH being important for increasing the number of spermatogonia available for entry into meiosis, yet T being required for the completion of spermatid elongation. Thus, both hormones act in concert to promote full spermatogenic output. **3.** An FSH-specific action on Sertoli cell division underlies the capacity of the testis to produce elongated spermatids in response to androgens [e.g., (22,29,195,199)]. FSH and T also appear to have similar actions with either additive or synergistic effects on **4.** the entry of germ cells into meiosis and the survival of preleptotene spermatocytes (Pl SC) [e.g., (195,331)], **5.** the development and survival of pachytene spermatocytes (pachytene sc) [e.g., (12,15,325)], **6.** the final meiotic division of spermatocytes into haploid round spermatids and their subsequent development, although T seems slightly more effective in this process [e.g., (14,195,239)], and **7.** the release of elongated spermatids during spermiation (16,326). **8.** Finally numerous studies show that FSH and T potentiate the action of the other (see section on FSH and T Cooperativity) with the androgen requirements of spermatogenesis being lower when FSH is present [e.g., (230,233, 325)] and a stimulatory effect of FSH on spermatogenesis being potentiated by the presence of T [e.g., (317,330)]. SG, spermatogonia; SC, spermatocytes; rST, round spermatids; ST, spermatids.

When considering whether FSH or T can fully maintain or restore any particular event in spermatogenesis, it is essential to define the experimental model and its fidelity in regard to normal spermatogenesis. The choice of experimental regimen, the dose of hormone and the length of treatment, the animals developmental status (e.g., congenital deficiency vs post-pubertal onset), and other hormonal changes (GnRH-immunized versus hypophysectomized) may all affect the hormonal responsiveness of the seminiferous tubules. Therefore, differences in outcomes observed, as to whether FSH or T can fully maintain or restore spermatogenesis, may be attributed at least in part to the experimental model rather than reflect their ability to impact on normal spermatogenesis.

There are many reports suggesting that either FSH or T act on spermatogenesis at a lower dose when the other is present (94,230,317,325); reviewed in (7,9,328). More recent studies in transgenic models also reveal FSH and T cooperativity (194,195,199). In particular there appears to be a strong case for a potentiating effect of FSH on T action. In many studies the androgen requirements of spermatogenesis were much lower in the presence of circulating FSH (94,185,230,233,240). FSH treatment of hypophysectomized rats promoted the effect of T on spermiogenesis and was associated with an increased testicular content of ABP (an extracellular proteins that binds androgens with high affinity) in the testis and epididymis (329). Through this mechanism, it was suggested that FSH has a role in facilitating the transport and localization of T within Sertoli cells. Alternatively, the stimulatory effect of FSH on spermatogenesis may be potentiated by the presence of T. The effect of FSH on the maintenance of germ cell populations in GnRH antagonist-treated rats was diminished when Leydig cells were destroyed by EDS treatment (330) or when the antiandrogen flutamide was added (317), suggesting that the presence of even very low levels of T may somehow facilitate FSH action on spermatogenesis.

FSH and T cooperativity is also apparent in that they act at different sites in spermatogenesis in order to optimize germ cell development (Fig. 5). For example, it is known that FSH is involved in increasing the number of spermatogonia available for entry into meiosis (see section on Role in the Adult, page 1035), which ultimately impacts on the number of spermatids produced, whereas an essential role for T in spermatid maturation (spermiogenesis) is apparent (see section on Androgen Regulation of Spermatogenesis). While research suggests that FSH can support spermatogenesis in the adult rodent up until the round spermatid phase [e.g., (14,280,325)], a general consensus exists that T is necessary for the completion of full spermiogenesis [e.g., (195,228, 240,280)]; see also section on Androgen Regulation of Spermatogenesis.

Testosterone and FSH may also act cooperatively by exerting effects at different stages of spermatogenesis, and therefore acting in collaboration to allow complete germ cell development. As previously discussed, preferential action of FSH on stage XIII–V on the cycle of the seminiferous epithelium has been suggested, whereas T preferentially acts on stages VII–VIII. It is possible that, for example, FSH action on a stage XIV tubule is necessary to initiate transcription of a particular gene product, which will ultimately be required several days later for an androgen-initiated molecular cascade in stage VII tubules. Such hypotheses will only be confirmed by a better understanding of the molecular targets of FSH and T.

There is evidence to suggest that both FSH and T can exert the same biological effect. For example, concomitant FSH and T treatment increased germ cell number (325) and decreased germ cell degeneration more so than either hormone alone (12,331) and either FSH or T was able to support the development of early spermatocytes in *hpg* mice (195). It has also been found that FSH and T can stimulate Sertoli cell products such as ABP (332) and transferrin (333). This view has led to the speculation that they may have common post receptor pathways of action (12,331), although evidence for this is still lacking.

Spermatocyte development appears to be reliant on the co-joint actions of FSH and T. FSH only partially maintains spermatocyte number in both GnRH antagonist-treated rats (17,317) and cynomologous monkeys (18), partially prevents spermatocyte degeneration in rats (17,331) and partially restores spermatocytes in GnRH immunized rats (14) suggesting that T is also required (see section on Androgen Regulation of Spermatogenesis). Conversely the ability of exogenous T to maximally stimulate pachytene spermatocyte number at all stages of spermatogenesis is dependent on concomitant FSH action in rats (15). Recent studies in *hpg* mice expressing transgenic FSH were compared to *hpg* mice given exogenous T treatment. These studies showed that early spermatocyte populations could be supported by either FSH or T, suggesting that the initiation of meiosis can be supported by either hormone, but that FSH and T had additive effects on pachytene spermatocyte populations (195). Thus, FSH and T cooperativity is apparent at the commencement of meiosis as well as in the survival of pachytene spermatocytes during meiosis.

Another process that seems to be supported by both FSH and T is spermiation. Studies in rats showed that 11% and 14% of elongated spermatids fail to spermiate from Sertoli cells after 1 week of

FSH or T suppression, respectively, compared to only 2% in normal rats (326). However, combined suppression of FSH and T for 1 week resulted in 50% of the spermatids failing to be released (326). Thus, the failure of spermiation occurs on suppression of FSH or T alone, but removal of both hormones has a synergistic effect. Spermiation failure induced by gonadotropin suppression is caused by defects in the disengagement of spermatids from the Sertoli cell in a process likely mediated by β1-integrin (75). Studies in the monkey (45) and human (324) also showed that spermiation failure occurs after gonadotropin suppression induced by androgen-based contraceptive administration. Importantly the latter study demonstrated that the induction of spermiation failure in humans is central to the acute and chronic suppression of sperm counts during contraceptive administration and is a key determinant of contraceptive effectiveness (see section on Spermiation, page 1048). Despite the clinical ramifications of spermiation failure, relatively little is known about the molecular events involved in spermiation (75,248,334) and the exclusive roles of FSH and T in this process are unknown.

Another point at which FSH and T cooperativitiy may exist is the Sertoli cell cytoskeleton and associated Sertoli-germ cell junctional interactions. As previously discussed in Androgen Regulation of Spermatogenesis, in-vitro studies of round spermatids and Sertoli cells showed that binding of round spermatids to Sertoli cells was dependent on T only in the presence of FSH (250,279). Similarly, in vivo experiments showed that following the prolonged absence of gonadotropins, T was only able to restore spermiogenesis when FSH was administered for at least 3 weeks (228,280). Both FSH and T are important for maximal Sertoli cell N-Cad production, which is particularly important for round spermatid adhesion to Sertoli cells in vitro (250).

The demonstration that both FSH and T can act similarly to maintain germ cell viability in both immature (201,335) and mature animals (12,17,240,331,336), has led to the suggestion that T and FSH may have a common endpoint, i.e., the survival of germ cells, despite the fact that the hormones act via different intracellular pathways; reviewed in (328). The molecular mechanisms of T and FSH synergy remain largely unknown, but there are several possibilities for T and FSH interaction at the cellular level. These include the potential for FSH to affect AR expression (see section on Androgen Receptor). The regulation of cAMP-responsive element modulator (CREM) and cAMP-responsive element binding (CREB) transcription factors by FSH and T is also likely (337). However, in the absence of identified FSH- and T-dependent signal transduction cascades, one can only speculate as to the role of FSH and T (individual or synergistic) in the production of Sertoli cell factors involved in spermatogenesis.

THYROID HORMONE AND SPERMATOGENESIS

There is a great deal of data demonstrating the role of thyroid hormone in modulating the proliferative phase of the Sertoli cell. Experimentally induced neonatal hypothyroidism (induced by treatment of rats with the goitrogens propylthiouracil or methimazole in the mothers' drinking water) delays the differentiation of Sertoli cells (338,339) and causes significant increases in adult testis weight (up to 80%) and daily sperm production (up to 140%) (340–342). Neonatal hypothyroidism maintained until day 26 after birth prolongs the period of Sertoli cell proliferation and results in an 84% increase in the adult Sertoli cell population (342,343). Conversely, hyperthyroidism (induced by a daily injection of triiodothyronine) leads to the premature cessation of Sertoli cell proliferation and an associated 50% decrease in the Sertoli cell population (344). The demonstration of thyroid hormone receptor expression in proliferative phase, but not nonproliferative phase, Sertoli cells (178) further suggests that thyroid hormone directly affects Sertoli cell proliferation and differentiation.

Recent studies provide evidence for thyroid hormone being an initial signal required to limit Sertoli cell proliferation (197). Studies by Holsberger and colleagues (345) link thyroid hormone stimulation to the induction of cell cycle inhibitors, such as Sertoli cell $p27^{kip1}$ protein. Buzzard et al. (196) used cultured Sertoli cells isolated from 6-day-old rats to demonstrate that thyroid hormone stimulation slowed Sertoli cell proliferation.

ESTROGEN REGULATION OF SPERMATOGENESIS

The involvement of estrogen in testicular physiology, including spermatogenesis, has received considerable attention of late; see (8,21,107,346) for recent reviews. Estrogen is produced locally by several cell types within the testis, suggesting a paracrine action on spermatogenesis. However, circulating estrogen can also impact on testicular function through endocrine feedback mechanisms, which in turn regulate spermatogenesis.

Until the 1990s, research into estrogen and spermatogenesis was limited to investigating the effects of exogenous estrogens on spermatogenesis. Speculation then arose that environmental estrogens may have deleterious effects on male fertility wherein inappropriate intra-uterine and/or postnatal exposure to estrogen adversely affects reproductive tract development, leading to impaired reproductive function in adulthood (347,348). Exposure to exogenous estrogenic substances during the neonatal period permanently alters reproductive tract gene expression (349) and male reproductive function [e.g., (350–354), reviewed in (8)]. Exposure to environmental estrogens, such as xeno- and phytoestrogens, has been hypothesized to be involved in a proposed decline in sperm counts in the latter half of the twentieth century (347,355,356), but whether environmental estrogens can negatively influence human health as well as male fertility has been hotly debated (357,358). Serious questions have been raised as to whether sperm counts have actually fallen on a global scale (359,360), calling into doubt a link between a deleterious effect of environmental estrogens and male fertility. However, it is clear that exogenous estrogens administered in doses ranging from physiological to supraphysiological influence testicular function (see below), and thus the effect of exogenous estrogens on spermatogenesis will be reviewed in this section.

During the 1990s, mice lacking a functional ER-α were produced and the males proved to be infertile, indicating that normal estrogen action is essential for spermatogenesis (121,123–125). Subsequent studies demonstrated that male mice unable to produce estrogen due to a congenital absence of the aromatase enzyme showed an age-related decline in spermatogenesis (361,362) and varying degrees of infertility (363,364). Thus, a physiological role for estrogens in testicular development and function can be examined by assessing the phenotype of transgenic animals, such as mice lacking functional ER-α (ERαKO), ER-β (ERβKO), both ER-α and ER-β (ERαβKO), and aromatase (ArKO) (Table 5).

The following sections will discuss the concept that estrogen can impact on spermatogenesis in the immature and adult testis (summarized in Table 6), as well as the estrogenic regulation of efferent ductule development and function that directly impacts on the ability of the testis to produce sperm. The well-known role for estrogen in the development and maintenance of normal male sexual behavior, which is key to successful fertility, has been reviewed elsewhere (8,365,366) and will not be considered here.

Estrogen Production in the Testis

Estrogen biosynthesis is catalyzed by a microsomal member of the cytochrome P450 superfamily, namely aromatase cytochrome P450 (P450arom, the product of the *CYP19* gene) (367). Aromatase is expressed in a variety of tissues including gonadal, adipose, bone, and brain via the use of tissue-specific promoters; reviewed in (368). The expression of the *CYP19* gene in gonadal tissue utilizes a proximal promoter, promoter II (369), and is regulated by cAMP and gonadotropins via interaction of the transcription factors CREB and SF1 with promoter II (370).

The synthesis of estrogens in the testis has been reviewed extensively elsewhere (107,108,371,372). Of note is the fact that there is a high concentration of estrogen in rete testis fluid of the rat (373) and that the concentrations of estrogen in the testis, rete testis fluid, and testicular venous blood far exceeds the concentration in male serum in various species [see (107,108) for review] indicating a testicular site of estrogen synthesis.

Aromatase is expressed in the testis at all stages of development [reviewed in (8)], and estrogen production has been demonstrated in the fetal testis [e.g., (374)]. In the immature postnatal testis, both Leydig cells and Sertoli cells contribute to estrogen biosynthesis [e.g., (154), reviewed in (371)]. Immature Sertoli cells show very high levels of basal and FSH-stimulated aromatase activity, which is highest during the proliferative period but declines precipitously as Sertoli cells cease division and commence maturation into the terminally differentiated form; reviewed in (8,371). Conversely, Leydig cells produce more estrogen as they mature (154,375) with aromatase being regulated by LH and T in adult cells (376). It is now becoming evident that germ cells may be a major site of estrogen biosynthesis in the adult testis, since the first demonstration of the presence of aromatase mRNA, protein and activity in adult germ cell preparations (377). There is a growing body of evidence that germ cells of numerous species synthesize and secrete estrogen and that germ cells possibly serve as a major site of estrogen production in the male reproductive tract; reviewed in (107,372). Aromatase is first detected in pachytene spermatocytes and remains in germ cells throughout spermiogenesis, becoming restricted to the elongated spermatid cytoplasm and flagella (377); reviewed in (8,107,372). Growth factors and cytokines have been reported to influence germ cell aromatase expression which is also stimulated by cAMP (378). Further studies have suggested that androgens stimulate germ cell aromatase expression whereas estrogen inhibits it (88), pointing to a complex regulatory mechanism in these cells. The production of estrogen by germ cells and secretion into the seminiferous tubule fluid may be important for the function of the efferent ductules and epididymis; see below and (107,108).

TABLE 5. *Comparison of spermatogenic phenotypes in estrogen receptor subtype and aromatase (cyp19) null male mice*

	ArKO[a]	ERαKO[b]	ERβKO[c]	ERαβKO[d]
FSH	Normal, but increased in older animals raised on phytoestrogen-free diet (361,362)	Normal (123,398)	Not described	Not described
LH	Elevated (361,362,399)	Elevated (396,398)	Not described	Not described
Serum T	Elevated (361,399)	Elevated (123,396,398)	Not described	Not described
Leydig cells	Hyperplasia/hypertrophy after 18 wk of age (361)	Increased cell size, T production, steroidogenic enzyme expression and activity (396)	Normal number at 2 days postpartum (383)	Not described
Sertoli cells	Number is normal, seminiferous tubule lumen volume greatly reduced in older mice raised on a phytoestrogen-free diet (361,362)	Seminiferous tubule fluid secretion decreased (124)	Normal number at 2 days postpartum (383)	Not described
Germ cells	Spermiogenic defect after ~14 wks of age, reductions in spermatocytes and spermatids in older mice raised on phytoestrogen-free diet (361,362,364)	Germ cells develop normally when transplanted into wild-type mice (411)	Gonocytes increased at 2 days postpartum due to decreased apoptosis (383)	Reduced numbers of sperm in the epididymis (126)
Efferent ductules	Morphology not described, but lack of enlarged seminiferous tubule lumen at any age suggests function is not markedly impaired (361,362)	Numerous abnormalities and morphological changes; disturbance in reabsorption function, decreases in ion exchange protein expression (107,108,124,407,408)	Not described	Not described, but similar phenotype to ERαKO mice suggests dysfunction (126)
General observations	Testicular histology grossly normal until approximately 14 wk of age. There is an age-related disruption to spermatogenesis which is exacerbated by removing phytoestrogens from the mouse diet (361,362,364)	Dilation of seminiferous tubule lumen precedes germ cell loss and defects to spermatogenesis. Fluid build up due to efferent ductule dysfunction appears to be the primary cause of spermatogenic failure (21,107,108,123–125)	Histology of the testis appears normal (405)	Preliminary observations suggest that the testicular phenotype is similar to that of ERαKO, i.e., disrupted fertility due to fluid build up (125,126)
Overall fertility	Initially fertile (399) but mice show reduced fertility at 14 wks (364) then progressively become infertile (361). Other lines show variable degrees of infertility (363,485)	Infertile (123)	Fertile (125,405)	Infertile (125,126)

[a]Aromatase defic ent mice.
[b]ER-α–deficient mice.
[c]ER-β–deficient mice.
[d]ER-α– and β–deficient mice.

Estrogen and Spermatogenesis during the Neonatal Period

HPT Axis

It is well known that exposure to estrogen during the fetal and neonatal period can impact on the development of the HPT axis; reviewed in (8). For example, exposure to estrogenic substances during the neonatal period can permanently alter the organization of the HPT axis (351,352). Neonatal exposure to even weak estrogenic substances can actually stimulate FSH during puberty in rats and cause long-term changes in the testis (353).

TABLE 6. *Summary of the effects of estrogen exposure and insufficiency on rodent spermatogenesis and efferent ductule function**

	Effect of estrogen exposure[a]	Effect of estrogen insufficiency[b]
Hypothalamic–pituitary–testis axis	• Suppression of LH and T, also FSH in neonatal period	• Increased LH and T in ERαKO and ArKO mice, aromatase inhibitor and ER antagonist-treated mammals
Leydig cells	• Suppression of T production and steroidogenic enzyme expression	• Increased steroidogenic enzyme expression and enhanced androgen biosynthesis in ERαKO, hyperplasia in the ArKO
Sertoli cells	• Neonatal administration affects Sertoli cell maturation and gene expression	• Decreased seminiferous tubule fluid production in ERαKO
	• High dose DES suppresses AR immunoexpression	• ArKO mice show reduced seminiferous lumen diameter suggestive of reduced fluid production
Germ cells	• Stimulates gonocyte proliferation in vitro and gene expression in vivo	• Increased gonocytes in early postnatal ERβKO
	• Prevented human germ cell apoptosis in vitro	• Significant reductions in spermatocytes and spermatids in ArKO mice raised on a phytoestrogen-free diet • Sperm function compromised in ERαKO and ArKO mice
Efferent ductules	• Changes in gene expression, impaired reabsorption function	• ERαKO mice and ER antagonist-treated rodents show numerous changes in morphology, reduced expression of key ion exchangers and impaired reabsorption function

*Described in sections on Estrogen and Spermatogenesis during the Neonatal Period; Estrogen and Spermatogenesis in the Adult; and Importance of Estrogen in Efferent Ductule Development and Function; Impact on Spermatogenesis.
[a]Includes administration of estradiol, diethylstilbestrol (DES), and other estrogenic substances such as phytoestrogens and xenoestrogens.
[b]Includes mice deficient in ERα, ERβ (or both) or aromatase, and normal rodents administered aromatase inhibitors.

Leydig Cells

Exogenous estrogen can inhibit the proliferation of precursor Leydig cells during the neonatal period and thereby modify the steroidogenic capacity of the adult testis, reviewed in (8,379). It can interfere with Leydig cell development and proliferation during puberty and in models of Leydig cell regeneration [see (379) for review] as well as in fetal testis culture (380). Neonatal exposure to diethylstilbestrol (DES) impairs testicular steroidogenesis in adulthood via a mechanism which cannot be explained by effects on the HPT axis (352). The mechanism by which neonatal exposure impacts on Leydig cell development and function is unclear, but has been suggested to involve inhibiting the development of precursor Leydig cells as well as steroidogenic enzymes involved in androgen biosynthesis (349,379,381).

Sertoli Cells

Proliferating Sertoli cells produce high levels of estrogen during the early postnatal period of division (371) suggesting that estrogen is involved in this process. Aromatase activity, under FSH control, is highest in Sertoli cells from prepubertal rats and declines as Sertoli cells cease division and begin maturation [see above and (371) for review]. Exposure to DES during the neonatal Sertoli cell proliferative period decreases Sertoli cell numbers (32), probably by reducing FSH which is a mitogen for Sertoli cell proliferation (see section on FSH Regulation of Spermatogenesis). Estrogen does not appear to suppress Sertoli cell proliferation in the fetal testis, since in utero exposure to estrogenic substances does not change (351) or slightly increases Sertoli cell numbers in rats (382). Consistent with this, ERαKO and ERβKO mice have normal Sertoli cell numbers immediately after birth (383).

Estrogen may inhibit Sertoli cell maturation and thereby prevent premature Sertoli cell differentiation during the proliferative period. For example, neonatal estrogen administration causes a maturational delay in Sertoli cells in rats (32); reviewed in (8). The proposition that estrogen may play an inhibitory role in Sertoli cell differentiation is supported by the observation that cells with the morphological features of a differentiated Sertoli cell are found in the ovaries of ERαβKO (125,126) and ArKO mice (384). Thus, the removal of estrogen action results in an environment in which Sertoli cells are able to differentiate, presumably in this case from granulosa cells.

Finally, estrogen may influence the establishment of germ cell–Sertoli cell adhesion as it regulates expression of the cell adhesion molecule neural cadherin (NCad) in the immature mouse testis (385) and in cultured mouse Sertoli cells (386).

Germ Cell Development

In terms of germ cell development in the early postnatal period, the proliferation of rat gonocytes, which express ER-β (130), is stimulated by estradiol in vitro (387). In addition, exposure of male rat pups to xeno- or phytoestrogens in utero up-regulates the expression of platelet-derived growth factor (PDGF) receptors in gonocytes (388); PDGF is a known stimulator of gonocyte proliferation in vitro (387). But other studies suggest that estrogen may have a negative effect on gonocytes in the perinatal period. Estradiol and DES treatment inhibited gonocyte numbers in cultured fetal rat testes (380), whereas ERβKO mice have increased numbers of gonocytes after birth probably due to a decreased rate of apoptosis (383). Finally the human ER-β2 receptor variant, which may act as a dominant negative receptor, is preferentially expressed in human fetal gonocytes (129) and may protect them from unwanted estrogen action (see section on Estrogen Receptor). There is thus evidence for stimulatory and inhibitory effects of estrogen on gonocyte proliferation; further studies are needed to investigate whether there are subsets or developmental phases of gonocytes and spermatogonia which are differentially affected by estrogen.

Estrogen and Spermatogenesis in the Adult

HPT Axis

Given that the maintenance of spermatogenesis in the adult relies on normal levels of FSH, LH, and T (see section on Androgen Regulation of Spermatogenesis and section on FSH Regulation of Spermatogenesis), it is important to consider that estrogen is a component of the negative feedback action on gonadotropin secretion at the level of the hypothalamus and pituitary [e.g., (389,390), reviewed in (8,21)] (Fig. 2). High-dose estrogen administration to adult rats decreases serum FSH, LH ,and T [see (8)], yet low doses can stimulate FSH (391) as was observed in neonates (see above). Suppression of circulating estrogen in males in various species by administering aromatase inhibitors increases serum FSH, LH, and T (392–395), whereas administration of the ER-antagonist ICI 182,780 increased serum LH and T in adult mice (396) but not in rats (397). Like models of aromatase inhibition and ER antagonism, higher LH and T levels are seen in ERαKO (396, 398) and ArKO mice (361,399), yet both transgenic models show normal FSH. Thus, an absence of ERα or aromatase produces higher circulating levels of LH and T.

Leydig Cells

The higher circulating T levels in adult ERαKO mice result from increased LH drive due to a lack of ER-α–mediated negative feedback on the pituitary (396, 398). Treatment of cultured wild-type Leydig cells with the ER-antagonist ICI 182,780 suppressed in vitro T production; however, this effect was not seen in ERαKO Leydig cells, implicating a role for ERα in Leydig cells in mediating steroidogenesis (396). Estrogen administration is known to impact on adult Leydig cell function, to decrease the activity of steroidogenic enzymes (400) and reduce T biosynthesis; reviewed in (8,379).

Seminiferous Epithelium

Estrogen may have a physiological role in the adult seminiferous epithelium. The induction of testicular atrophy and abnormal spermatogenesis due to dysfunction of efferent ductules in models of ER-α deficiency is covered in a separate section below. However, there are data suggesting effects of estrogen administration or deprivation on seminiferous epithelial function that are independent of effects on efferent ductule function.

First, ArKO mice do not show evidence of efferent ductule dysfunction yet still show abnormal spermatogenesis from about 14 weeks of age (361,362,364); reviewed in (8). The initial ArKO spermatogenic phenotype is primarily spermiogenic arrest (361,362,364); however, older ArKO mice raised on a phytoestrogen (soy protein)-free diet showed wider spermatogenic impairment, extending from spermatocytes through to elongated spermatids (362), suggesting that a prolonged absence of endogenous and dietary estrogen produces more marked effects on spermatogenesis than a prolonged absence of endogenous estrogen alone (361,362). Older ArKO mice raised on a phytoestrogen-free diet also displayed features of compromised Sertoli cell function as evidenced by markedly decreased tubule lumen diameters (suggesting reduced seminiferous tubule fluid production), increased circulating FSH levels (suggesting reduced inhibin B production by Sertoli cells), and a reduction in the number of germ cells per Sertoli cell. This more profound suppression was not associated with a reduced gonadotropic stimulus, supporting a direct action of phytoestrogens on the testis.

Other evidence for a role for estrogen in the seminiferous epithelium comes from reports that reduced spermiogenesis and sperm concentrations were apparent in adult male monkeys administered an aromatase inhibitor for 100 to 150 days (394,401), although administration of an aromatase inhibitor to rats for 133 days produced variable effects on the seminiferous epithelium (395). Apparently physiological doses of estradiol given to adult *hpg* mice stimulates spermatogenesis, although it is unclear whether this is a direct effect of estrogen on the testis, or via the stimulation of subnormal levels of FSH (402). Estradiol has also been shown to prevent germ cell apoptosis in human adult seminiferous tubules in vitro (134). Decreased seminiferous tubule fluid production in male ERαKO mice compared to wild-type males (124) suggests compromised seminiferous epithelial function in the absence of ER-α.

Recent studies showed that long term administration of the ER-α and ER-β–antagonist ICI 182,780 to mice resulted in a decreased testis weight and a significant increase in the number of morphologically abnormal seminiferous tubules in a manner that was independent of effects on efferent ductule function (403). Such effects were not seen when similar studies were performed in rats (397,404), suggesting that long-term ER antagonist treatment in these mice affected spermatogenesis directly, perhaps via effects on ER-β within the seminiferous epithelium (403).

Considering these data and the localization of ER-β in Sertoli and germ cells, it is most surprising that male ERβKO do not show any apparent reduction in spermatogenesis, nor have defects to the seminiferous epithelium been reported (Table 5) (125,405). Mice lacking both ER subtypes (ERαβKO) have been reported to have a phenotype that is grossly similar to ERαKO mice (125,126), which are infertile primarily due to defects in the efferent ductules; see below and Table 5. The lack of an obvious spermatogenic phenotype related to the ablation of ER-β is not readily explicable, but potential explanations include compensatory expression of ER-α, non-ER-β–dependent estrogen actions on germ cells and Sertoli cells [reviewed in (8)], the presence of undocumented ER-encoding genes and compensatory developmental mechanisms within the testis.

Finally it is worth noting that dietary phytoestrogens (soy proteins, particularly genistein) can produce subtle effects on testicular structure and function in normal rodents. Adult mice raised on a phytoestrogen-free diet had an increased seminiferous epithelial volume, longer seminiferous tubules and a decreased volume of interstitium, and some groups showed a small increase in Sertoli cell numbers (362). Adult rats raised on a diet free of phytoestrogens had larger testes and lower circulating FSH levels (353). An effect of dietary phytoestrogens is not surprising, since phytoestrogens such as genistein and coumestrol have a significantly higher affinity for ER-β than ER-α (406), and ER-β is the predominant ER subtype expressed within the seminiferous epithelium (see section on Estrogen Receptor).

Importance of Estrogen in Efferent Ductule Development and Function; Impact on Spermatogenesis

An elegant series of studies on the phenotype of the male ERαKO mouse (123,124,407,408), reviewed extensively in (107,108), revealed a physiological requirement for ERα in efferent ductule function. Since normal efferent ductule function has a profound impact on the ability of the testis to produce sperm, and on the fertility of the animal, the ability of estrogens to regulate efferent ductule development and maintain function will be briefly considered.

The first demonstration of a physiological role for estrogen in male fertility arose from the generation of mice which lacked a functional ER-α (121). Testis and spermatogenic development of these mice was normal but pubertal testes showed dilated seminiferous tubule lumens, which was eventually followed by progressive disruptions to spermatogenesis and a reduction in epididymal sperm concentrations (123).

Subsequent studies pinpointed impaired efferent ductule function as the underlying cause of infertility in ERαKO mice (124,407); reviewed in (107,108,409). Efferent ductules are a series of tubules that connect the rete testis and the epididymis and function to reabsorb over 90% of fluid exiting the testis; thus, the efferent ductules effectively concentrate the sperm prior to entering the epididymis; reviewed in (107, 108,409). Hess and colleagues hypothesized that the enlarged seminiferous tubule lumens and swollen rete testis observed in ERαKO mice was a consequence of impaired efferent ductule function (124). In vitro ligation experiments revealed that the ability of ERαKO efferent ductules to reabsorb fluid was significantly impaired, as was the ability of wild-type efferent ductules treated with the ER-antagonist ICI 182,780 (124). However, while ERαKO efferent ductules swelled in vitro, ICI-treated wild-type ductules did not (124), suggesting developmental defects in ERαKO tissues or ERβ-dependent mechanisms.

ERαKO efferent ductules show numerous structural differences compared to wild-type ductules, all of which are consistent with a decrease in fluid reabsorption (107,108,407). A molecular target for estrogen action in the efferent ductules appears to be Na^+/H^+ exchanger-3 (NHE-3); the ER-dependent expression of this ion exchanger is a primary defect in ERαKO efferent ductules and has downstream effects on the expression of other proteins central to fluid reabsorption (410). Studies in adult rats and mice treated with the ER-antagonist ICI 182,780 also revealed impaired efferent ductule function and decreased expression of NHE-3 (404,410), indicating that the impairment of efferent ductule function in ERαKO mice is not due to developmental abnormalities but can be attributed to a central role for ER-α in adult efferent ductule function.

The compromised function of ERαKO efferent ductules explains the observed infertility in these mice: as the testis develops and spermatogenesis is initiated, the testis swells due to an inability of the efferent ductules to reabsorb the fluid produced by the seminiferous epithelium, resulting in a transient increase in testis weight and enlarged seminiferous tubule lumens. As fluid build-up in the testis continues, the resulting back pressure compromises the function of the seminiferous epithelium and spermatogenesis is disrupted. The proposition that the infertility in ERαKO mice arises from efferent tubule dysfunction rather than a defect in spermatogenesis per se is supported by the demonstration that germ cells from ERαKO mice develop normally when transplanted into wild-type testes (in which germ cells had been removed by busulphan treatment), indicating that ERαKO mice are not infertile as a consequence of ER-α ablation from the germ cells (411). This is not surprising given that there are few reports of ER-α localization in germ cells (see section on Estrogen Receptor).

It is worth noting that exposure to exogenous estrogen during the neonatal period or adulthood can also impair efferent ductule function; reviewed in (8). Neonatal exposure to highly potent estrogenic substances causes dilation of the rete testis and fluid accumulation, permanent reductions in the water channel protein aquaporin-1, changes in the cell- and region-specific expression of ER-α protein in the male reproductive tracts and compromised spermatogenesis in adulthood (412–415). Such studies highlight the fact that the development of the efferent ductules is a target for exogenous estrogen action. Short-term treatment of adult rats with estradiol benzoate also disturbs efferent ductule function (416). The mechanisms by which estrogen exposure and deficiency impair efferent ductule development and function are not clear, but may relate to estrogen exposure changing patterns of AR and ER expression in the male reproductive tract, thereby interfering with the androgen: estrogen balance, as well as the balance of ER subtype expression, that is likely important in the function of such tissues [e.g., (350,412), reviewed in (8)].

ArKO mice did not show obvious fluid retention in the testis and disturbance of fluid resorption (361,362) suggesting that efferent ductule dysfunction was not a major consequence of aromatase deficiency. The reason aromatase deficiency does not produce such defects (as might be predicted by the ERαKO phenotype) is unclear but possible explanations include: (a) that ArKO mice are subject to maternal estrogens in utero, which may facilitate efferent ductule development, (b) that seminiferous tubule fluid secretion is compromised in ArKOs (361,362), and (c) that efferent ductule function is maintained by ligand-independent means. It should also be noted that administration of the ER-antagonist ICI 182,780 to rats and mice in similar treatment regimes produced nearly identical effects on efferent ductule morphology, yet fluid accumulation in the testis was noted only in rats (397), not mice (403). These observations perhaps point to species and even strain-dependent variations in the response of the testis to disturbed fluid dynamics.

ENDOCRINE REGULATION OF HUMAN SPERMATOGENESIS

Hormonal Dependency of Human Spermatogenesis

Overall data supports that both FSH and T have roles in the initiation of spermatogenesis at puberty, and are essential for quantitatively normal spermatogenesis in adulthood. Their shared importance underscores modern approaches to male hormonal contraception (MHC) and to the restoration of fertility in gonadotropin-deficient states. The term hypogonadotropic hypogonadism (HH) is applied to a range of clinical disorders due to deficiency of hypothalamic GnRH drive or intrinsic pituitary FSH and LH secretory ability, that feature poor sexual development (if pre-pubertal in onset), or infertility and androgen deficiency in adulthood.

In terms of spermatogenesis in males congenitally deficient in GnRH, germ cells do not proceed beyond the immature state in which only a few spermatogonia are seen (Fig. 6) (417). Acquired gonadotropin deficiency after puberty leads to testicular regression and marked oligospermia (i.e., reduced but detectable numbers of sperm in the ejaculate) or azoospermia (i.e., an absence of sperm from the ejaculate) and may result from disorders of the HPT axis or from treatment with gonadotropin suppressing agents (Table 7), such as when GnRH analogs are used to achieve therapeutic androgen withdrawal in prostate cancer. When gonadotropin depletion is partial, qualitatively normal spermatogenesis may occur with a reduced total sperm output but yet maintained fertility. When the hormonal depletion is more marked, spermatogenesis fails to progress due to inhibition of several key steps and infertility results.

Treatment aimed at restoring spermatogenesis (Fig. 6) and fertility involves the restoration of endogenous gonadotropin secretion or the use of exogenous gonadotropin replacement therapy. In congenital gonadotropin deficiency, this may take many months or years as long periods are needed for the normal puberty maturational process to occur (418,419). When fertility is not sought, the associated T deficiency requires exogenous T replacement to correct the clinical features of hypoandrogenism that vary depending on the timing of onset (fetal, puberty, adulthood), severity, and duration of deficiency.

Our understanding of gonadotropin effects on spermatogenesis in man comes from "experiments of

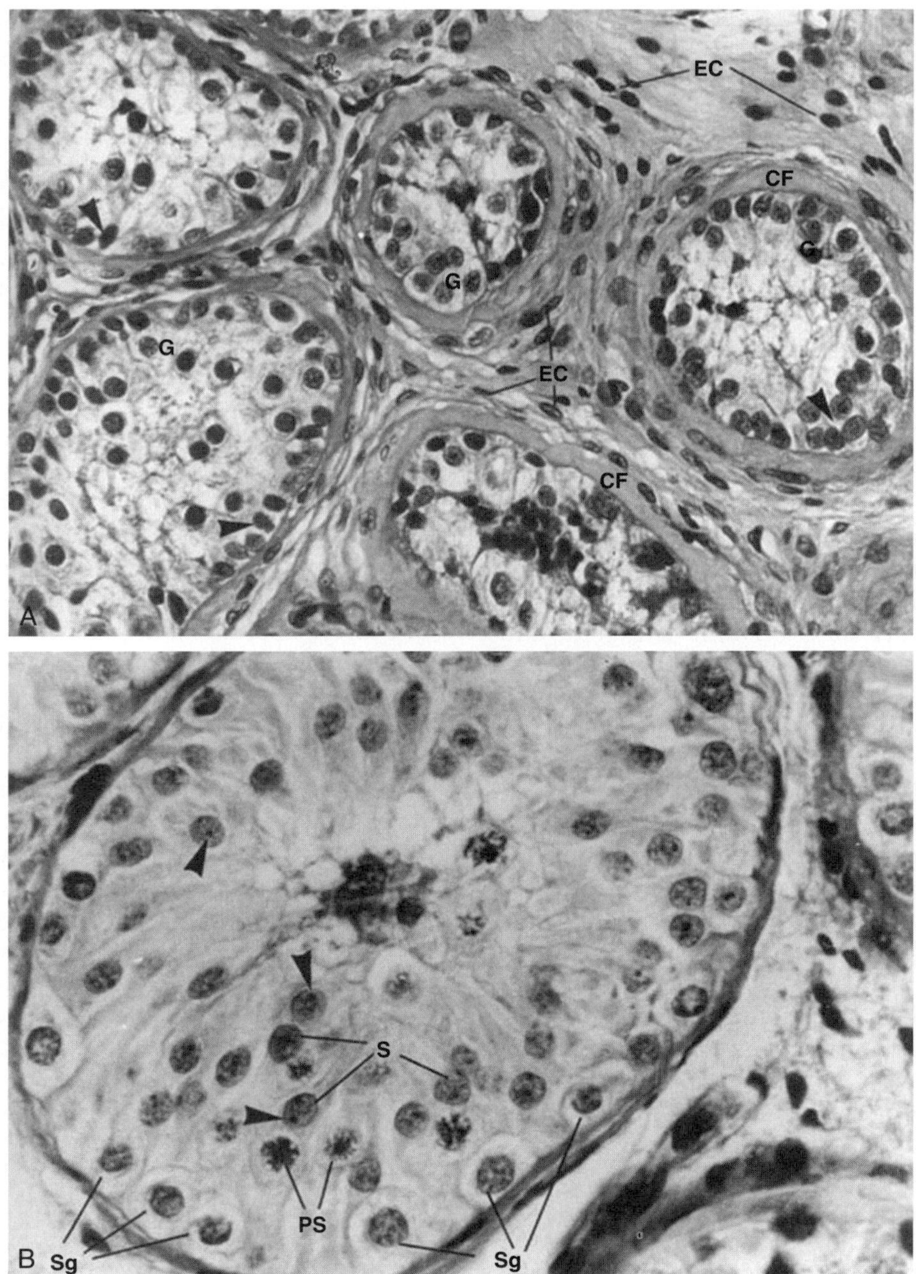

FIG. 6. Panel A shows the features of a testis from a man with congenital hypogonadotropic hypogonadism which displays an immature spermatogenic phenotype. Note the immature Sertoli cells (*arrows*) and a few spermatogonia (G). Some cords show peri-tubular fibrosis (CF) and the inter-tubule areas contain no interstitial cells, but do show some elongated cells (EC) which may act as their precursors. **Panel B** shows a testicular biopsy from the same man given gonadotropin treatment. The seminiferous tubules show an increase in diameter and now possess a lumen. The peri-tubular fibrosis has dispersed and spermatogonia (Sg) form a basal layer on the epithelium with some primary spermatocytes (PS) being present. The Sertoli cells (S) now show well developed nuclei. (Figure courtesy of David de Kretser, PhD.)

nature" (naturally occurring mutations/deficiency of key components) and from the study of spontaneous or experimentally induced HH (e.g., by hormonal contraceptive treatment, see section on Male Hormonal Contraceptive [MHC] Development) and gonadotropin replacement. Earlier descriptive approaches that studied testicular changes have been superseded by stereological approaches to the estimation of germ cell number and structural changes that allow identification of sites in germ cell development that are hormonally regulated and represent targets for fertility regulation.

TABLE 7. *Causes of hypogonadotropic hypogonadism (HH)*

	Acquired HH		
Congenital HH	**Direct damage to elements of the HPT axis**	**Endocrine or drug effects**	**Systemic disorders affecting the HPT axis**
• Idiopathic HH including Kallmann's syndrome *–Partial/complete* • Genetic defects in key proteins: GnRH receptor FGFR1 GPR54 *DAX1* • Mutations of gonadotropin subunit or/receptors • Associated with CNS disorders and obesity, e.g., Prader-Willi • Multiple pituitary hormone deficiency, e.g., Prop 1 complete or partial HH (fertile eunuch syndrome)	• Tumors– *craniopharyngioma, pituitary adenoma, meningioma* • Pituitary hemorrhage (apoplexy) • Infiltrative disorders • Head trauma • Radiation therapy	• Androgen excess *– Leydig cell tumors* *– Choriocarcinoma producing hCG* *– Androgen abuse* • Glucocorticoid excess *– Cushing's syndrome* *– Pharmacological* • Opiates drug-induced hyperprolactinemia, e.g., psychotropics	• Critical illness including burns • Extreme exercise • Malnutrition (anorexia nervosa) • Morbid obesity

Gonadotropin and Androgen Effects on Spermatogenesis

Certain aspects of androgen and FSH physiology in humans require review as background for understanding the effects of gonadotropins in health and disease.

Testicular Sex Hormone Physiology. There is an absolute need for androgens in spermatogenesis (see section on Androgen Regulation of Spermatogenesis). By virtue of its local production, intratesticular T levels in man are ~50-fold higher than those in serum with estimates for intratesticular T of ~400 to 600 ng/g compared to 6 ng/ml in serum (324,420–422) (Table 2). Following exogenous androgen administration and resulting LH withdrawal, intratesticular T levels fall by over 98% to approach those in serum suggesting the virtual abolition of its local production [e.g., (324)].

The dose-response relationship between intratesticular T and human spermatogenesis has not been defined as it has been in rats (see section on Androgen Regulation of Spermatogenesis), but it still appears that, as in other species, spermatogenesis requires an intratesticular T level higher than that in peripheral tissues. Testis-specific interactions among T, AR, and associated co-factors may exist but evidence is lacking. Despite DHT's higher affinity for AR and higher AR-dependent transcriptional activity compared with T (209,210), the vast excess of T over DHT [about 100:1 (324)] (Table 2) in the normal testis suggests that DHT is not important, although it may be when intratesticular T levels are markedly reduced, as in MHC regimens (see section on 5α-Reductase, page 1031). Estrogen is also produced from T in the human testis and is present in high concentrations in testicular venous blood (see section on Estrogen Production in the Testis). Given the presence of ER-β on many testicular cell types (see section on Estrogen Receptor), this also represents a potential pathway of hormonal action. However, no direct effects on human spermatogenesis in vivo have been shown although in vitro studies do show that estradiol can prevent germ cell apoptosis (134). Few patients with congenital aromatase or ER deficiencies have been described, and their phenotypes reveal little information as to the role of estrogen in human spermatogenesis; reviewed in (8).

Role of FSH in Human Spermatogenesis. As in rodent studies suggesting FSH is required for maximal spermatogenic output (see section on FSH Regulation of Spermatogenesis), data in men also supports the need for FSH for a normal sperm output. In a classic series of experiments involving the experimental manipulation of FSH and LH levels in normal men, Matsumoto and colleagues (423–425) demonstrated that sperm production was restored to 50% of baseline values by FSH or human chorionic gonadotropin (hCG, as an LH substitute) treatment, whereas only combined hCG plus FSH treatment lead to quantitative restoration. The restoration of fertility by hCG treatment in hypogonadotropic patients (see below) does not argue against a role for FSH in normal spermatogenesis as fertility often occurs at submaximal sperm production rates.

Naturally occurring mutations of FSH and its receptors are exceedingly rare and reveal discrepant spermatogenic phenotypes that remain unexplained (426).

FSH receptor mutations have been associated with moderately elevated FSH levels, variably severe reductions in sperm counts (0.1–42 million/ml) but retained fertility (427).Yet a man with an inactivating FSH-β mutation was azoospermic (428), unlike the FSH-β knockout mouse (307), arguing that FSH is essential in testicular development and/or in the establishment and maintenance of spermatogenesis in humans. The frequent requirement for FSH in the establishment of spermatogenesis in congenitally and completely hypogonadotropic men (see below) points to the need for FSH to induce permanent maturational effects on the Sertoli cell/seminiferous epithelium, as during normal puberty.

Effects of Gonadotropins on Specific Germ Cell Types in Man

Using stereological approaches, it is possible to quantitate the effects of gonadotropin withdrawal and replacement and to deduce the site(s) of hormone action. While there are similar overall effects on germ cell maturation, relative to rodents, in man there is a striking inhibition in type Ap to B spermatogonial maturation, a maintenance of progression until late spermiogenesis, and a major loss of sperm output in the ejaculate despite the persistence of a reasonable number of elongated spermatids, suggestive of a defect in spermiation (320,324).

Another feature of human spermatogenesis following gonadotropin withdrawal is the heterogeneity both between apparently similar men given identical treatments, and between adjacent tubules in the same individual (Fig. 7). These differences require the application of stereological "whole testis" estimates of germ cell number (429) and appropriate statistical analyses.

Some general comments about changes in individual germ cell types are described below.

Spermatogonia. Impairment of human spermatogonial development appears to be a major rate-limiting step in spermatogenesis when gonadotropic support is withdrawn. Following exogenous T treatment resulting in profound FSH/LH withdrawal, type B spermatogonia are the first cells to decrease, followed by decreases in type Ap (324). The fall in type B spermatogonia may result from an inhibition of Ap spermatogonial mitosis, such as was demonstrated in GnRH antagonist-treated monkeys (38), due to their accelerated apoptosis, or by a direct effect on B spermatogonial mitosis. Stereological assessment reveals a far greater reduction in spermatogonia in man as compared to rodents with type B spermatogonia being suppressed to ~10% of that in control men (320).

It is likely, but not certain, that these changes in spermatogonial subtypes relate to FSH. As discussed in FSH Regulation of Spermatogenesis, there is evidence from studies in monkeys that FSH alone increased B spermatogonia (13,323,430,431). In fact there has been little support for the notion that sex steroids or Leydig cell factors stimulate human spermatogonial development.

Spermatocyte-Round Spermatid Development. Although gonadotropin suppression inhibits spermatogonial development, the subsequent progression through meiosis and spermiogenesis appears relatively unimpeded, albeit at reduced levels, suggesting that an interruption of the development of these cells is not a major feature of experimental gonadotropin suppression (319,320,324) unlike in rodents (see section on Androgen Regulation of Spermatogenesis and section on FSH Regulation of Spermatogenesis). Although some round spermatids are seen in the ejaculates of men following gonadotropin withdrawal, their number is small and the time course of their appearance does not correlate with the fall in sperm count (432) suggesting that the detachment of round spermatids seen in the androgen-deficient rat (see section on Role in the Adult, page 1030) is not observed in men. However, it must be noted that in man, both FSH and intratesticular T levels fall with exogenous T treatment, unlike the similarly treated rat model of spermatid detachment in which FSH levels are maintained (see section on Meiosis and Spermiogenesis, page 1030). Thus, a similar pattern might be induced in man if such a paradigm could be explored.

Studies of human testicular tissue in vitro demonstrate the ability of FSH, androgens, and estrogens to prevent germ cell apoptosis in vitro (134,433,434). Differences in the susceptibility of human germ cells to apoptosis in young vs aged men (435), and between ethnic populations (436), mean that apoptosis of spermatocytes and spermatids could contribute to different responses of spermatogenesis to gonadotropin suppression.

Spermiation. Studies in primates (45) and humans (320) suggest that spermiation failure is a feature of both acute and chronic gonadotropin suppression; see also section on FSH and T Cooperativity. The presence of several stages of spermatogenesis in a single cross-section of the human testis makes this process difficult to study in man. However, features of spermiation failure can be appreciated in men treated with hormone regimes to induce contraception (Fig. 8). In chronically gonadotropin-suppressed monkeys, a negative correlation between the numbers of elongated spermatids retained in the seminiferous epithelium (i.e., failing to spermiate) and the sperm count has been shown (45). Suppression of spermiation probably represents the second most significant spermatogenic lesion leading to azoospermia in response to acute gonadotropin withdrawal.

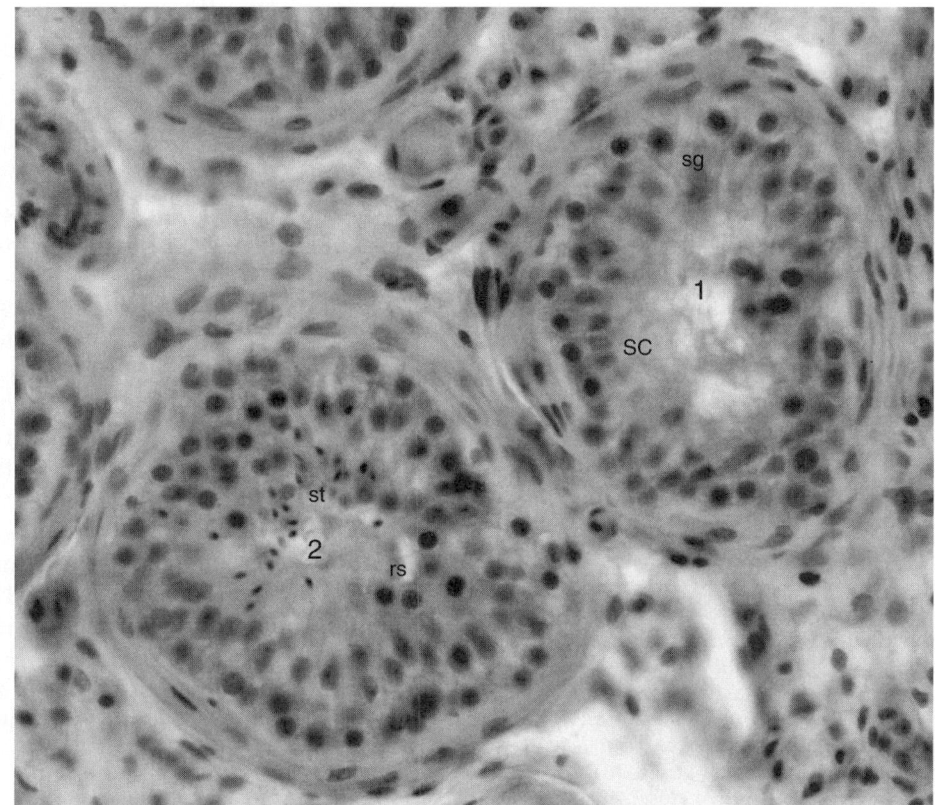

FIG. 7. Example of the heterogeneity observed between adjacent seminiferous tubules in a man undergoing male hormonal contraceptive treatment. While one tubule (1) shows primarily Sertoli cells (SC) and spermatogonia (sg) yet few meiotic and postmeiotic cells, the adjacent tubule (2) shows postmeiotic germ cell production including round spermatids (rs) and elongated spermatids (st).

Hypogonadotropic Hypogonadism (HH)

Etiology and Effects on Spermatogenesis

Congenital idiopathic hypogonadotropic hypogonadism (IHH) results from isolated GnRH deficiency that, when associated with reduced (hyposmia) or absent (anosmia) sense of smell is termed Kallmann's syndrome. IHH is most often sporadic, but also displays X-linked and autosomal patterns of inheritance (419). Other genetic causes for HH are being increasingly recognized, some are associated with other hormonal deficiencies and others with central nervous system disorders (Table 7). Genetic mutations affecting the gonadotropin subunits or their receptors result in functional gonadotropin deficiency.

In complete IHH, severe T deficiency is associated with hypospadias, micropenis, testicular maldescent, and failed puberty. The testes have a pre-pubertal histological appearance and feature small seminiferous tubules containing immature Sertoli cells that are round and may be pseudo-stratified (Fig. 6). A few large spermatogonia with clear cytoplasm and a central nucleolus are present in the center of the seminiferous tubule. The interstitium is composed of loose connective tissue and mature Leydig cells are absent, but fibroblast-like precursor cells are found. It is worthwhile noting that in the congenital absence of gonadotropic stimulus, human spermatogenesis does not proceed beyond the gonocyte stage (Fig. 6), whereas in the *hpg* mouse early meiotic cells are present within the testis (195).

Impaired Sertoli cell maturation due to an absent or attenuated postnatal gonadotropin surge, or to the effect of prolonged gonadotropin deficiency before beginning therapy, limits the maximum testicular response to gonadotropin treatment in adulthood. Despite often achieving fertility with treatment, most patients do not develop normal testicular volumes nor sperm output underlining their reduced potential. Patients with IHH and cryptorchidism tend to respond less well to therapy, although unilateral cryptorchidism does not preclude fertility (437).

In partial IHH, clinical features of poor virilization are less apparent and there is variable testicular growth. Basal LH, FSH, and T levels, and the gonadotropin response to GnRH stimulation, are greater than in men with complete gonadotropin deficiency (438). Spontaneous LH pulses of reduced amplitude and/or frequency can often be detected

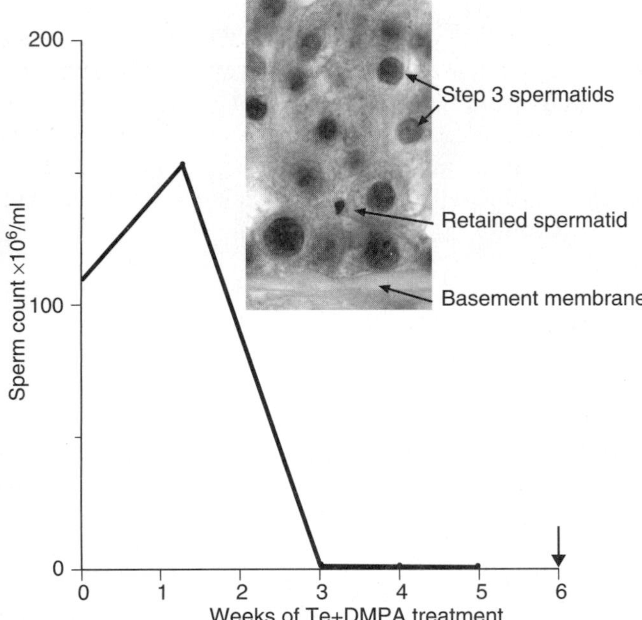

FIG. 8. Spermiation failure during male hormonal contraceptive treatment. Sperm counts (millions/ml) in a man given testosterone enanthate (Te, 200 mg i.m., weekly) in combination with depot medroxyprogesterone acetate (DMPA, 300 mg single injection at commencement of treatment) for 6 weeks prior to testicular biopsy (*arrow*; no sperm count was taken at the time of biopsy). This subject's sperm count drops rapidly to <0.1 million/ml within 3 weeks of treatment, indicative of spermiation failure. *Inset* shows a retained elongated spermatid located toward the basement membrane of the seminiferous epithelium after 6 weeks of treatment. Stage III of spermatogenesis, in which step 3 round spermatids are present, is shown. Spermiation occurs in the preceding stage II, and thus spermatids retained at this stage have failed to spermiate.

in the circulation, sometimes with augmentation during sleep (438,439). Because of evidence of testis growth and spermatogenesis, but with incomplete androgenization, the term "fertile eunuch" has been applied in the past.

Acquired HH (Table 7) arising before puberty results in delayed or absent puberty while its onset after the completion of a normal puberty results in T deficiency and infertility. Many structural disorders are also associated with deficiencies of other pituitary hormones. Systemic disorders may be associated with profound suppression of the HPT axis, e.g., severe wasting disorders or chronic infection. Endocrine causes include the suppression of gonadotropin by high levels of glucocorticoid due to endogenous excess (Cushing's disease) or their pharmacological use. Androgen excess may rarely result from tumorous secretion of androgens (Leydig cell tumors) or hCH (providing excess LH-like bioactivity and thereby T secretion). Most androgen excess results from androgenic steroid abuse for performance enhancement that leads to profound and prolonged HH with infertility and symptomatic androgen deficiency after cessation of abuse until spontaneous recovery of the HPT axis (440).

Restoration of Spermatogenesis in Patients with Hypogonadotropic Hypogonadism

The establishment of high intratesticular T levels is essential for the induction or restoration of spermatogenesis in patients with HH. hCG (1,000–1,500 IU, weekly), used as an LH substitute, is given for a 4 to 6 month period to stimulate Leydig cell T production and testicular growth. Most patients with congenital IHH also require later co-treatment with FSH (100–150 IU subcutaneously, three times weekly) to induce spermatogenesis. This is especially important in those patients with complete IHH and in those men who remain azoospermic in response to hCG alone (418). This likely reflects a requirement for FSH to induce terminal differentiation of Sertoli cells as has been indicated in rodent models (see section on FSH Regulation of Spermatogenesis). In response to hormone treatment, testis volume increases to 12 to 15 ml (normal male range 15–35 ml) and spermatozoa usually appear in the ejaculate within 12 months of starting treatment. Most IHH subjects will produce sperm (>1 million/ml) if hCG/FSH treatment is continued for 18 months (441).

hCG alone is generally successful in restoring fertility in men who develop HH postpubertally (442). hCG treatment is also often successful in men with IHH who previously required both hCG and FSH to initiate spermatogenesis (443), suggesting that the initial passage through full spermatogenesis induces some permanent changes in the testis in response to gonadotropin.

Pulsatile GnRH therapy, delivered by programmable portable infusion pumps, can also be used to stimulate spermatogenesis and testicular growth in men with GnRH deficiency (444), and this treatment may be more effective than hCG/FSH treatment (445) as shown by the earlier appearance of sperm in the ejaculate and greater testicular growth (437) but patient compliance and cost means that hCG/FSH continues to be recommended as initial therapy.

It must be noted that fertility is the end point in clinical management of HH and that the female partners of gonadotropin-treated HH men often conceive with sperm densities well below the normal range of 20 million/ml, with one study reporting a median sperm density of only 5 million/ml (446). These observations do not conflict with the notion that both hCG (LH) and FSH are both required for quantitatively normal spermatogenesis in experimental HH (423–425). It should also be noted that even with prolonged gonadotropin therapy, some men never produce sperm while others only produce very few sperm and may use assisted reproduction (specifically intracytoplasmic sperm injection) as an avenue for conception. The reason spermatogenesis cannot be restored by gonadotropin treatment in some HH patients is unclear. A failure to restore spermatogenesis is more likely in those with profound gonadotropin deficiency as reflected in their very small testicular size and cryptorchidism. In this setting, failure to achieve spermatogenesis despite gonadotropin therapy may reflect irremediable intrinsic defects to spermatogenesis.

Male Hormonal Contraceptive (MHC) Development

The goal of MHC treatment is the uniform and reversible induction of azoospermia in men of all racial groups using safe and acceptable drug formulations. Following on work using androgen and progestin administration in the 1970s, much progress has been made in the past 15 years, particularly the landmark World Health Organization (WHO) studies using weekly T enanthate (Te) injections involving 671 couples in 10 countries (447,448). Testosterone administration to normal men, either alone or in combination with a progestin, reduces pituitary LH and FSH secretion, and thereby sperm production. In Caucasian men, weekly injections of Te at a dose of 100 to 200 mg are effective with the 200 mg dose inducing azoospermia in ~70% of subjects, while rates of 95% are seen in Asian men (447,448).

The relationship between the suppression of sperm density by MHC treatment and fertility outcomes must be assessed in contraceptive efficacy trials in which other methods are avoided. Logically, suppression to azoospermia ought to provide excellent contraception and data supports this contention (447), as rebound of sperm production during continued treatment is uncommon. But it appears that acceptable efficacy rates can be achieved with suppression to severe oligospermia. For example in the WHO trial (448) suppression of sperm density to 0.1 to 3 million/ml (about 25% of the subjects) had a failure rate of 8%, similar to that of the condom in community use. Thus, while azoospermia remains the goal for MHC, sperm densities below 1 million/ml may provide acceptable contraception (449), particularly when compared to other available and reversible methods.

Recent emphasis has been on combination androgen plus progestin regimens, the inclusion of the latter promoting the suppression of gonadotropins and spermatogenesis, and allowing a reduction in the T dosage toward more physiological replacement levels (450). GnRH antagonists also profoundly suppress gonadotropin levels and have been used in conjunction with T where they show promise (451). The effect of various MHC regimens on sperm quality, and on clinical and safety parameters, has been summarized in recent reviews (452–455).

Mechanism of MHC Action on Spermatogenesis

Both clinical parameters (sperm density) and morphological assessment of testicular tissue can be used to understand the regression of spermatogenesis. Impairment of spermatogonial development appears to be a major rate-limiting step when gonadotropic support is withdrawn. MHC results in a rapid decline in type B spermatogonia, followed by decreases in type Ap (324). Type B spermatogonia are reduced to 10% to 20% of control levels by long-term MHC treatment. The maintenance of qualitatively normal meiotic and postmeiotic germ cell development, albeit at reduced levels, suggests that large numbers of germ cells are not lost by apoptosis during meiosis and spermiogenesis (324).

A failure of spermiation appears a feature of both acute and chronic MHC-induced gonadotropin withdrawal (320,324). The rapid fall in sperm density (within 4 weeks) is not consistent with a lesion in spermatogonial development but rather in spermiogenesis or spermiation (324,456,457) (Fig. 8). Indeed allowing for the onset of gonadotropin suppression and epididymal transit time, spermiation failure is likely to occur almost immediately upon gonadotropin suppression. Despite normal numbers of mature elongated spermatids in the epithelium at 6 weeks of treatment, the suppression of sperm counts to less than 1 million/ml suggests nearly complete spermiation failure (324).

Spermiation failure is also important for suppression of sperm counts in the longer term, since mature elongated spermatids persist in the epithelium at 10% to 20% of control despite sperm densities falling to less than 1 million/ml in the majority of cases (320,324). A better understanding of the molecular events involved in sperm release would facilitate the discovery of specific pharmaceutical agents that could be used to perturb this process. There are no data in man regarding the relative roles of FSH and T in the regulation of spermiation, nor of the molecular mechanisms involved.

Mitosis of type B spermatogonia (38) and spermiation occur in stages I–II of the human spermatogenic cycle. The fact that spermatogonial development and spermiation are the two major defects in human spermatogenesis induced by contraceptive treatment suggests these stages may be more hormone sensitive than others, in a similar way to stages VII–VIII in the rat, which are known to be sensitive to hormone suppression (see section on Ability of the Testis to Respond to Endocrine Signals; Hormone Receptor Expression and section on Androgen Regulation of Spermatogenesis). Immunohistochemical analysis of AR showed that staining was more intense in stages I–III than at later stages, suggesting that stage-specific variation in hormone sensitivity may also apply to humans (92).

Determinants of Sperm Suppression with MHC

The uniform induction of azoospermia has not yet been consistently achieved with any MHC regimen. About 5% of men continue to have sperm densities greater than 3 million/ml despite undetectable gonadotropin levels, such sperm densities are unlikely to provide adequate contraceptive effectiveness (448). At this time there is no explanation for their different response. For example, there are no apparent differences in terms of T pharmacokinetics (458) or anthropometric measures (459). Differences between men in the speed of induction and recovery of spermatogenesis also remain unexplained. There are several potential determinants of MHC effects on spermatogenesis that may provide reasons for inadequate suppression and future avenues for MHC development.

Gonadotropin Suppression and Coadministration of Progestins. The rapid and maximal suppression of serum gonadotropins is a logical premise for MHC. Inadequate suppression of FSH is a potential reason for contraceptive failure (460) and complete suppression of LH (and thereby intratesticular T) should also be sought. Predictably a relationship between the degree of gonadotropin suppression and spermatogenic suppression has been reported in many human androgen-based MHC trials (450,461–463). Markedly suppressed gonadotropin levels (to around the level of assay detection, or ~5% baseline levels) are needed to consistently suppress sperm density. The perplexing issue is that despite such suppression, sperm density may not be adequately suppressed in some individuals such that no relationship was evident between gonadotropin levels and sperm densities in the WHO studies (459).

Coadministration of progestin with T appears to accelerate and augment gonadotropin suppression (450,452,453,464). As a result, the inclusion of progestin improves sperm suppression and yields a greater proportion of subjects reaching a likely threshold for adequate contraceptive efficacy (450,456,465). The addition of depot medroxyprogesterone acetate (DMPA) to weekly Te injections, led to a more rapid fall in serum FSH/LH levels, although the maximum extent of their suppression (to ~1.5% and 0.3% baseline, respectively) did not differ (324). Type B spermatogonial and early spermatocytes number also fell more rapidly with DMPA, presumably reflecting the lower gonadotropin levels (324). Efficacy data is limited with as yet only one T plus progestin study demonstrating that T implants and DMPA adequately suppressed sperm density in 95% of men, and prevented pregnancy in 55 couples for up to 1 year (457).

The thresholds of the biological effect of FSH and intratesticular T on spermatogenesis are not known and may differ both between men and between ethnic groups. In other words, there are no clear target levels for their suppression that ensure the abolition of sperm output. Unlike in animal models, this issue is exceedingly difficult to pursue in clinical studies. Alternatively there may be nongonadotropin-dependent processes that may account for different individual or racial responses.

The relationship between low gonadotropin levels and the persistence of sperm production was studied across three T plus progestin MHC regimens [using cyproterone acetate, levonogestrel, and desogestrel (463)]. While sperm densities below 0.1 million/ml were associated with lower gonadotropin levels, multivariate logistic regression was used to identify individual predictors of response. The suppression of serum LH (but not FSH) to less than 5% of baseline values (LH < 0.15 IU/L) was a consistent predictor of sperm suppression. Higher T dose was also a negative predictor of suppression that, as suggested by others, may indicate a direct sustaining effect of exogenous T by diffusion from serum. Together these data point to the need to maximally reduce intratesticular T levels. Finally, the use of progestin was a significant independent predictor of sperm suppression suggesting T plus progestin MHC regimens suppress sperm production by both gonadotropin-dependent and -independent mechanisms, but as yet there is no direct evidence for the latter. In combination with T, the anti-androgenic progestin, cyproterone acetate, markedly suppressed sperm production and has been proposed to act both by gonadotropin suppression and locally by inhibition of AR activity within the seminiferous epithelium (456).

Ethnic and Genetic Factors. Ethnic differences in response to identical MHC regimens suggest the involvement of genetic factors, with Asian populations showing higher rates of azoospermia (447,448,466,467). Chinese men were reported to have reduced testis size, total Sertoli cell number, and daily pachytene spermatocyte production as compared to Hispanic and Caucasian men, suggesting a lower spermatogenic potential that may expose them to greater reductions in sperm output during MHC treatments (468).

Higher CAG repeat number in the AR has been suggested to impair receptor trans-activation and androgen action. The impact of CAG repeat polymorphism in the AR, and of polymorphisms in ER and FSH-R genes, were examined in 85 Caucasian men treated with different MHC regimens (469). Suppression of sperm density was related to the degree of gonadotropin suppression and independently to the coadministration of progestin, but not to any receptor polymorphisms. When gonadotropin levels were profoundly suppressed, none of the polymorphisms explained why some men failed to achieve azoospermia but when only partial, the prospects of attaining azoospermia was 2.5 times higher in men having more than 22 CAG repeats. Another study found no difference in CYP3A4 gene (the major hepatic inactivating enzyme) or in CAG and GGC triplet repeats in the AR in men rendered azoospermic or nonazoospermic by MHC (470). Thus, a clear genetic basis for poor sperm suppression with MHC remains elusive.

Role of 5α-Reductase (5αR). A reduction of intratesticular DHT support for spermatogenesis may be a useful MHC strategy (see section on Role in the Adult, page 1029), but evidence is limited. The production of 5α-reduced androgens was increased in oligospermic as compared to azoospermic men, suggesting that up-regulation of 5αR maintains intratesticular DHT levels and thereby some spermatogenesis (251,252). Asian men are more reliably rendered azoospermic by MHC treatment (~98% vs 70%) (447,448), and have significantly lower levels of serum 5α-reduced androgens (471). The type 1 isoenzyme is predominate in the rat testis (see section on Androgen Regulation of Spermatogenesis), but there are no data in the human. Small pilot studies using the type 2 5αR isoenzyme inhibitor, finasteride, failed to show a better suppression of sperm density (461,472) but combined type 1 and 2 5αR inhibitors, although available, have not been used in MHC regimens. Should further evidence support a role for intratesticular DHT, the use of a non-5α-reducible androgen, such as 7α-methyl-19-nortestosterone (MENT) (473), or coadministration of 5αR inhibitors may be useful strategies.

GnRH Antagonists. GnRH receptor antagonists (GnRH-A), but not GnRH agonists, appear promising MHC agents when used with parenteral T formulations (451,474,475). Initial GnRH-A use was limited by local histamine release, but new agents such as acyline (476) show greater promise in terms of patient acceptability, rapid and profound gonadotropin suppression, and long duration of action. Such agents are expensive but it has been suggested they may find a place as initial induction agents to achieve rapid sperm suppression, while in the longer term, T, with or without progestin, might maintain spermatogenic suppression (451).

CONCLUSION AND FUTURE DIRECTIONS

In this chapter we have summarized current knowledge about the endocrine factors required for spermatogenesis, together with their likely sites of action. Key phases in the ontogeny of testicular development have been identified that set up the essential "infrastructure" for spermatogenesis in adulthood. Certain phases of germ cell development are particularly hormone sensitive such as spermatogonial proliferation, germ cell survival, progression through spermiogenesis and spermiation; all of these processes are fundamental to providing maximal spermatogenic output in adulthood.

Manipulation of hormone action and/or spermatogenic processes in rodent models, particularly transgenic mice, has changed our fundamental understanding of the mechanisms of hormone action on spermatogenesis. Such manipulations often lead to appreciable effects on the seminiferous epithelium yet have no impact on fertility, since rodents have a high fecundity due to efficient spermatogenesis and large numbers of sperm in the ejaculate. This has lead to debate over which factors are "essential" for spermatogenesis and male fertility versus which factors are dispensable and perhaps only required for "optimal" spermatogenesis. However, the distinction between essential and dispensable becomes blurred when considering the potential impact of similar manipulations in humans, with spermatogenesis being much less efficient than in rodents, and there is a much wider variation of sperm output in the general population. A case in point is the observation that mice congenitally deficient in FSH action (FSH-β gene knockouts) are fully fertile (307) yet they exhibit small testes and sperm production is less than 40% of normal (302). One could speculate that such a phenotype in humans would have far more serious consequences for fertility than is revealed in this mouse model. We must now begin to appreciate how human fertility may be affected by intrinsic and extrinsic endocrine factors involved in the control of spermatogenesis.

The establishment of the seminiferous epithelium during the early postnatal period is fundamental to the spermatogenic potential of the adult testis. A major advance in the field of the endocrine regulation of spermatogenesis in the past decade has been the demonstration that the establishment of the Sertoli cell population is controlled by FSH and thyroid hormone, and that the resulting size of this population confers a life-long ability of the testis to produce sperm. Further developments in the past decade have been in the elucidation of distinct roles of FSH and T in spermatogenesis (Tables 3 and 4; Fig. 5), as well as an appreciation of the ability of these hormones to cooperate in terms of the maintenance of germ cell survival and optimal Sertoli cell function. Another key development in the past decade has been the emerging role of estrogen in male fertility, as evidenced by the ability of exogenous and endogenous estrogens to impact on spermatogenesis, and the knowledge that situations of estrogen insufficiency can impair spermatogenesis.

Knowledge of the endocrine regulation of spermatogenesis has contributed to the development of hormone-based contraceptives, with various MHC formulations now being tested in clinical trials of contraceptive effectiveness. Studies investigating the response of the human testis to contraceptive formulations have revealed that key spermatogenic processes known to be regulated by hormones in rodents (such as spermatogonial proliferation and spermiation) are targets for current MHC formulations, leading to the possibility that they may underlie the variable response between individuals to such treatments.

Where to from here? While we have gained a better understanding of which hormones are important for spermatogenesis and their cellular sites of action, the application of comparative genomics to the testis has revealed an array of genes that are potentially hormonally regulated (254,281). The challenge now is to marry these two streams of knowledge to map the molecular mechanisms which control key events in Sertoli cell and germ cell development and function. Newer technologies such as proteomics and specific cell isolation techniques, such as laser-assisted microdissection, will allow us to elucidate the molecular pathways, from gene transcription through to protein–protein interactions, that regulate the key hormone-regulated processes in spermatogenesis. Identification of testis-specific factors in such pathways may well lead to discoveries of next generation nonhormone-based contraceptives that target the testis directly, circumventing the endocrine feedback loop, and hence offering a better potential for maximizing efficacy while minimizing side effects. Lastly, by applying our knowledge of specific genes and gene products involved in spermatogenesis to different causes of male infertility, we may begin to map the genetic basis of idiopathic male infertility and lead to the development of new therapies for conditions of infertility and subfertility in men.

REFERENCES

1. Clermont, Y. (1972). Kinetics of spermatogenesis in mammals: Seminiferous epithelium cycle and spermatogonial renewal. *Physiol. Rev.* 52, 198–236.
2. de Kretser, D. M., and Kerr, J. B. (1988). The cytology of the testis. In *The Physiology of Reproduction* (E. Knobil, and J. D. Neill, Eds.), Vol. 1, pp. 837–932. Raven Press, New York.
3. Russell, L. D., Ettlin, R. A., Sinha Hikim, A. P., and Clegg, E. D. (1990). *Histological and Histopathological Evaluation of the Testis.* Cache River Press, Clearwater, FL.
4. Russell, L. D. (1993). Form, dimensions and cytology of mammalian Sertoli cells. In *The Sertoli Cell* (L. D. Russell, M. D. Griswold, Eds.), pp. 1–38. Cache River Press, Clearwater, FL.
5. Weinbauer, G. F., and Nieschlag, E. (1993). Hormonal control of spermatogenesis. In *Molecular Biology of the Male Reproductive System* (D. M. de Kretser, Ed.), pp. 99–142. Academic Press, San Diego.
6. Weinbauer, G. F., and Nieschlag, E. (1997). Endocrine control of germ cell proliferation in the primate testis. What do we really know? *Adv. Exp. Med. Biol.* 424, 51–58.
7. Zirkin, B. R. (1998). Spermatogenesis: its regulation by testosterone and FSH. *Semin. Cell. Dev. Biol.* 9, 417–421.
8. O'Donnell, L., Robertson, K. M., Jones, M. E., and Simpson, E. R. (2001). Estrogen and spermatogenesis. *Endocr. Rev.* 22, 289–318.
9. McLachlan, R. I., O'Donnell, L., Meachem, S. J., Stanton, P. G., de Kretser, D. M., Pratis, K., and Robertson, D. M. (2002). Identification of specific sites of hormonal regulation in spermatogenesis in rats, monkeys, and man. *Recent Prog. Horm. Res.* 57, 149–179.
10. McLachlan, R. I., O'Donnell, L., Meachem, S. J., Stanton, P. G., de, K., Pratis, K., and Robertson, D. M. (2002). Hormonal regulation of spermatogenesis in primates and man: insights for development of the male hormonal contraceptive. *J. Androl.* 23, 149–162.
11. Awoniyi, C. A., Santulli, R., Sprando, R. L., Ewing, L. L., and Zirkin, B. R. (1989). Restoration of advanced spermatogenic cells in the experimentally regressed rat testis: quantitative relationship to testosterone concentration within the testis. *Endocrinology* 124, 1217–1223.
12. El Shennawy, A., Gates, R. J., and Russell, L. D. (1998). Hormonal regulation of spermatogenesis in the hypophysectomized rat: cell viability after hormonal replacement in adults after intermediate periods of hypophysectomy. *J. Androl.* 19, 320–334; discussion 341–322.
13. Marshall, G. R., Zorub, D. S., and Plant, T. M. (1995). Follicle-stimulating hormone amplifies the population of differentiated spermatogonia in the hypophysectomized testosterone-replaced adult rhesus monkey (Macaca mulatta). *Endocrinology* 136, 3504–3511.
14. McLachlan, R. I., Wreford, N. G., de Kretser, D. M., and Robertson, D. M. (1995). The effects of recombinant follicle-stimulating hormone on the restoration of spermatogenesis in the gonadotropin-releasing hormone- immunized adult rat. *Endocrinology* 136, 4035–4043.
15. Meachem, S. J., Wreford, N. G., Stanton, P. G., Robertson, D. M., and McLachlan, R. I. (1998). Follicle-stimulating hormone is required for the initial phase of spermatogenic restoration in adult rats following gonadotropin suppression. *J. Androl.* 19, 725–735.
16. Russell, L. D., and Clermont, Y. (1977). Degeneration of germ cells in normal, hypophysectomized and hormone treated hypophysectomized rats. *Anat. Rec.* 187, 347–366.
17. Sinha Hikim, A. P., and Swerdloff, R. S. (1995). Temporal and stage-specific effects of recombinant human follicle- stimulating hormone on the maintenance of spermatogenesis in gonadotropin-releasing hormone antagonist-treated rat. *Endocrinology* 136, 253–261.
18. Weinbauer, G. F., Behre, H. M., Fingscheidt, U., and Nieschlag, E. (1991). Human follicle-stimulating hormone exerts a stimulatory effect on spermatogenesis, testicular size, and serum inhibin levels in the gonadotropin-releasing

hormone antagonist-treated nonhuman primate (Macaca fascicularis). *Endocrinology* 129, 1831–1839.

19. Weinbauer, G. F., Gockeler, E., and Nieschlag, E. (1988). Testosterone prevents complete suppression of spermatogenesis in the gonadotropin-releasing hormone antagonist-treated nonhuman primate (Macaca fascicularis). *J. Clin. Endocrinol. Metab.* 67, 284–290.
20. Cooke, H. J., and Saunders, P. T. (2002). Mouse models of male infertility. *Nat. Rev. Genet.* 3, 790–801.
21. Couse, J. F., and Korach, K. S. (1999). Estrogen receptor null mice: what have we learned and where will they lead us? *Endocr. Rev.* 20, 358–417.
22. Singh, J., O'Neill, C., and Handelsman, D. J. (1995). Induction of spermatogenesis by androgens in gonadotropin-deficient (hpg) mice. *Endocrinology* 136, 5311–5321.
23. Chang, C., Chen, Y. T., Yeh, S. D., Xu, Q., Wang, R. S., Guillou, F., Lardy, H., and Yeh, S. (2004). Infertility with defective spermatogenesis and hypotestosteronemia in male mice lacking the androgen receptor in Sertoli cells. *Proc. Natl. Acad. Sci. U S A* 101, 6876–6881.
24. De Gendt, K., Swinnen, J. V., Saunders, P. T., Schoonjans, L., Dewerchin, M., Devos, A., Tan, K., Atanassova, N., Claessens, F., Lecureuil, C., Heyns, W., Carmeliet, P., Guillou, F., Sharpe, R. M., and Verhoeven, G. (2004). A Sertoli cell-selective knockout of the androgen receptor causes spermatogenic arrest in meiosis. *Proc. Natl. Acad. Sci. U S A* 101, 1327–1332.
25. Christensen, A. K. (1975). Leydig cells. In *Handbook of Physiology* (R. O. Greep and E. B. Astwood, Eds.), Vol. 5, Section 7, pp. 57–94. American Physiological Society, Washington, DC.
26. Dym, M., and Fawcett, D. W. (1970). The blood–testis barrier in the rat and the physiological compartmentation of the seminiferous epithelium. *Biol. Reprod.* 3, 308–326.
27. Orth, J. M. (1982). Proliferation of Sertoli cells in fetal and postnatal rats: a quantitative autoradiographic study. *Anat. Rec.* 203, 485–492.
28. Gondos, B., and Berndston, W. E. (1994). Postnatal and pubertal development. In *The Sertoli Cell* (L. D. Russell and M. D. Griswold, Eds.), pp. 116–153. Cache River Press, Clearwater, FL.
29. Meachem, S. J., McLachlan, R. I., de Kretser, D. M., Robertson, D. M., and Wreford, N. G. (1996). Neonatal exposure of rats to recombinant follicle stimulating hormone increases adult Sertoli and spermatogenic cell numbers. *Biol. Reprod.* 54, 36–44.
30. Cooke, P. S., Zhao, Y. D., and Bunick, D. (1994). Triiodothyronine inhibits proliferation and stimulates differentiation of cultured neonatal Sertoli cells: possible mechanism for increased adult testis weight and sperm production induced by neonatal goitrogen treatment. *Biol. Reprod.* 51, 1000–1005.
31. Huhtaniemi, I. T., Nevo, N., Amsterdam, A., and Naor, Z. (1986). Effect of postnatal treatment with a gonadotropin-releasing hormone antagonist on sexual maturation of male rats. *Biol. Reprod.* 35, 501–507.
32. Sharpe, R. M., Atanassova, N., McKinnell, C., Parte, P., Turner, K. J., Fisher, J. S., Kerr, J. B., Groome, N. P., Macpherson, S., Millar, M. R., and Saunders, P. T. (1998). Abnormalities in functional development of the Sertoli cells in rats treated neonatally with diethylstilbestrol: a possible role for estrogens in Sertoli cell development. *Biol. Reprod.* 59, 1084–1094.
33. Leblond, C. P., and Clermont, Y. (1952). Definition of the stages of the cycle of the seminiferous epithelium in the rat. *Ann. N Y Acad. Sci.* 548–573.
34. de Rooij, D. G., and Russell, L. D. (2000). All you wanted to know about spermatogonia but were afraid to ask. *J. Androl.* 21, 776–798.
35. Meachem, S., von Schonfeldt, V., and Schlatt, S. (2001). Spermatogonia: stem cells with a great perspective. *Reproduction* 121, 825–834.
36. McLean, D. J., Johnston, D. S., Russell, L. D., and Griswold, M. D. (2001). Germ cell transplantation and the study of testicular function. *Trends Endocrinol. Metab.* 12, 16–21.
37. Meistrich, M. L., and van Beek, M. E. (1993). Spermatogonial stem cells. In *Cell and Molecular Biology of the Testis* (C. Desjardins and L. L. Ewing, Eds.), pp. 266–295. Oxford University Press, New York.
38. Schlatt, S., and Weinbauer, G. F. (1994). Immunohistochemical localization of proliferating cell nuclear antigen as a tool to study cell proliferation in rodent and primate testes. *Int. J. Androl.* 17, 214–222.
39. Clermont, Y. (1969). Two classes of spermatogonial stem cells in the monkey (Cercopithecus aethiops). *Am. J. Anat.* 126, 57–71.
40. van Alphen, M. M., van de Kant, H. J., and de Rooij, D. G. (1988). Depletion of the spermatogonia from the seminiferous epithelium of the rhesus monkey after X irradiation. *Radiat. Res.* 113, 473–486.
41. van Alphen, M. M., van de Kant, H. J., and de Rooij, D. G. (1988). Repopulation of the seminiferous epithelium of the rhesus monkey after X irradiation. *Radiat. Res.* 113, 487–500.
42. Schulze, C. (1979). Morphological characteristics of the spermatogonial stem cells in man. *Cell. Tissue Res.* 198, 191–199.
43. Schulze, C. (1981). Survival of human spermatogonial stem cells in various clinical conditions. *Fortschritte Der Andrologie.* 7, 58–68.
44. Fouquet, J. P., and Dadoune, J. P. (1986). Renewal of spermatogonia in the monkey (Macaca fascicularis). *Biol. Reprod.* 35, 199–207.
45. O'Donnell, L., Narula, A., Balourdos, G., Gu, Y. Q., Wreford, N. G., Robertson, D. M., Bremner, W. J., and McLachlan, R. I. (2001). Impairment of spermatogonial development and spermiation after testosterone-induced gonadotropin suppression in adult monkeys (Macaca fascicularis). *J. Clin. Endocrinol. Metab.* 86, 1814–1822.
46. Leblond, C. P., and Clermont, Y. (1952). Spermiogenesis of rat, mouse and guinea pig as revealed by the 'periodic acid-fuchsin sulfurus acid' technique. *Am. J. Anat.* 90, 167–206.
47. Braun, R. E. (1998). Post-transcriptional control of gene expression during spermatogenesis. *Semin. Cell. Dev. Biol.* 9, 483–489.
48. Russell, L. (1993). Role in spermiation. In *The Sertoli Cell* (L. D. Russell and M. D. Griswold, Eds.), pp. 269–302. Cache River Press, Clearwater, FL.
49. Bremner, W. J., Millar, M. R., Sharpe, R. M., and Saunders, P. T. (1994). Immunohistochemical localization of androgen receptors in the rat testis: evidence for stage-dependent expression and regulation by androgens. *Endocrinology* 135, 1227–1234.
50. Schulze, W., and Rehder, U. (1984). Organization and morphogenesis of the human seminiferous epithelium. *Cell. Tissue Res.* 237, 395–407.
51. Schulze, W., Riemer, M., Rehder, U., and Hohne, K. H. (1986). Computer-aided three-dimensional reconstructions of the arrangement of primary spermatocytes in human seminiferous tubules. *Cell. Tissue Res.* 244, 1–7.
52. Millar, M. R., Sharpe, R. M., Weinbauer, G. F., Fraser, H. M., and Saunders, P. T. (2000). Marmoset spermatogenesis: organizational similarities to the human. *Int. J. Androl.* 23, 266–277.
53. Franca, L. R., Ogawa, T., Avarbock, M. R., Brinster, R. L., and Russell, L. D. (1998). Germ cell genotype controls cell cycle during spermatogenesis in the rat. *Biol. Reprod.* 59, 1371–1377.
54. Timmons, P. M., Rigby, P. W., and Poirier, F. (2002). The murine seminiferous epithelial cycle is pre-figured in the Sertoli cells of the embryonic testis. *Development* 129, 635–647.

55. Heller, C. G., and Clermont, Y. (1963). Spermatogenesis in man: an estimation of its duration. *Science* 140, 184–186.
56. Clermont, Y., and Harvey, S. C. (1965). Duration of the cycle of the seminiferous epithelium of normal, hypophysectomized and hypophysectomized-hormone treated albino rats. *Endocrinology* 76, 80–89.
57. Aslam, H., Rosiepen, G., Krishnamurthy, H., Arslan, M., Clemen, G., Nieschlag, E., and Weinbauer, G. F. (1999). The cycle duration of the seminiferous epithelium remains unaltered during GnRH antagonist-induced testicular involution in rats and monkeys. *J. Endocrinol.* 161, 281–288.
58. Heller, C. G., and Clermont, Y. (1964). Kinetics of the germinal epithelium in man. *Recent Prog. Horm. Res.* 20, 545–575.
59. Pelletier, R. M., and Byers, S. W. (1992). The blood–testis barrier and Sertoli cell junctions: structural considerations. *Microsc. Res. Tech.* 20, 3–33.
60. Byers, S., Jegou, B., MacCalman, C., and Blaschuk, O. (1993). Sertoli cell adhesion molecules and the collective organization of the testis. In *The Sertoli Cell* (L. D. Russell and M. D. Griswold, Eds.), pp. 461–476. Cache River Press, Clearwater, FL.
61. Cheng, C. Y., and Mruk, D. D. (2002). Cell junction dynamics in the testis: Sertoli-germ cell interactions and male contraceptive development. *Physiol. Rev.* 82, 825–874.
62. Russell, L. D. (1993). Morphological and functional evidence for Sertoli-germ cell relationships. In *The Sertoli Cell* (L. D. Russell and M. D. Griswold, Eds.), pp. 365–390. Cache River Press, Clearwater, FL.
63. Vogl, A. W., Pfeiffer, D. C., Redenbach, D. M., and Grove, B. D. (1993). Sertoli cell cytoskeleton. In *The Sertoli Cell* (L. D. Russell and M. D. Griswold, Eds.), pp. 40–86. Cache River Press, Clearwater, FL.
64. Vogl, A. W., Pfeiffer, D. C., Mulholland, D., Kimel, G., and Guttman, J. (2000). Unique and multifunctional adhesion junctions in the testis: ectoplasmic specializations. *Arch Histol. Cytol.* 63, 1–15.
65. Lui, W. Y., Mruk, D., Lee, W. M., and Cheng, C. Y. (2003). Sertoli cell tight junction dynamics: their regulation during spermatogenesis. *Biol. Reprod.* 68, 1087–1097.
66. Lui, W. Y., Mruk, D. D., Lee, W. M., and Cheng, C. Y. (2003). Adherens junction dynamics in the testis and spermatogenesis. *J. Androl.* 24, 1–14.
67. Johnson, K. J., and Boekelheide, K. (2002). Dynamic testicular adhesion junctions are immunologically unique. II. Localization of classic cadherins in rat testis. *Biol. Reprod.* 66, 992–1000.
68. Johnson, K. J., and Boekelheide, K. (2002). Dynamic testicular adhesion junctions are immunologically unique. I. Localization of p120 catenin in rat testis. *Biol. Reprod.* 66, 983–991.
69. Mulholland, D. J., Dedhar, S., and Vogl, A. W. (2001). Rat seminiferous epithelium contains a unique junction (Ectoplasmic specialization) with signaling properties both of cell/cell and cell/matrix junctions. *Biol. Reprod.* 64, 396–407.
70. Munari-Silem, Y., and Rousset, B. (1996). Gap junction-mediated cell-to-cell communication in endocrine glands—molecular and functional aspects: a review. *Eur. J. Endocrinol.* 135, 251–264.
71. Russell, L. D. (1977). Desmosome-like junctions between Sertoli and germ cells in the rat testis. *Am. J. Anat.* 148, 301–312.
72. Toyama, Y., Maekawa, M., and Yuasa, S. (2003). Ectoplasmic specializations in the Sertoli cell: new vistas based on genetic defects and testicular toxicology. *Anat. Sci. Intl.* 78, 1–16.
73. Beach, S. F., and Vogl, A. W. (1999). Spermatid translocation in the rat seminiferous epithelium: coupling membrane trafficking machinery to a junction plaque. *Biol. Reprod.* 60, 1036–1046.
74. Guttman, J., Kimel, G., and Vogl, A. W. (2000). Dyein and plus-end microtubule-dependent motors are associated with specialized Sertoli cell junction plaques (ectoplasmic specializations). *J. Cell Sci.* 113, 2167–2176.
75. Beardsley, A., and O'Donnell, L. (2003). Characterization of normal spermiation and spermiation failure induced by hormone suppression in adult rats. *Biol. Reprod.* 68, 1299–1307.
76. Lindzey, J., Kumar, M. V., Grossman, M., Young, C., and Tindall, D. J. (1994). Molecular mechanisms of androgen action. *Vitam. Horm.* 49, 383–432.
77. Tsai, M. J., and O'Malley, B. W. (1994). Molecular mechanisms of action of steroid/thyroid receptor superfamily members. *Annu. Rev. Biochem.* 63, 451–486.
78. Silva, F. R., Leite, L. D., and Wassermann, G. F. (2002). Rapid signal transduction in Sertoli cells. *Eur. J. Endocrinol.* 147, 425–433.
79. Walker, W. H. (2003). Nongenomic actions of androgen in Sertoli cells. *Curr. Top. Dev. Biol.* 56, 25–53.
80. Buzek, S. W., and Sanborn, B. M. (1988). Increase in testicular androgen receptor during sexual maturation in the rat. *Biol. Reprod.* 39, 39–49.
81. Hill, C. M., Anway, M. D., Zirkin, B. R., and Brown, T. R. (2004). Intratesticular androgen levels, androgen receptor localization, and androgen receptor expression in adult rat Sertoli cells. *Biol. Reprod.* 71, 1348–1358.
82. Zhou, X., Kudo, A., Kawakami, H., and Hirano, H. (1996). Immunohistochemical localization of androgen receptor in mouse testicular germ cells during fetal and postnatal development. *Anat. Rec.* 245, 509–518.
83. Pelletier, G., Labrie, C., and Labrie, F. (2000). Localization of oestrogen receptor alpha, oestrogen receptor beta and androgen receptors in the rat reproductive organs. *J. Endocrinol.* 165, 359–370.
84. Zhou, Q., Nie, R., Prins, G. S., Saunders, P. T., Katzenellenbogen, B. S., and Hess, R. A. (2002). Localization of androgen and estrogen receptors in adult male mouse reproductive tract. *J. Androl.* 23, 870–881.
85. Vornberger, W., Prins, G., Musto, N. A., and Suarez-Quian, C. A. (1994). Androgen receptor distribution in rat testis: new implications for androgen regulation of spermatogenesis. *Endocrinology* 134, 2307–2316.
86. Kimura, N., Mizokami, A., Oonuma, T., Sasano, H., and Nagura, H. (1993). Immunocytochemical localization of androgen receptor with polyclonal antibody in paraffin-embedded human tissues. *J. Histochem. Cytochem.* 41, 671–678.
87. McKinnell, C., Saunders, P. T., Fraser, H. M., Kelnar, C. J., Kivlin, C., Morris, K. D., and Sharpe, R. M. (2001). Comparison of androgen receptor and oestrogen receptor beta immunoexpression in the testes of the common marmoset (Callithrix jacchus) from birth to adulthood: low androgen receptor immunoexpression in Sertoli cells during the neonatal increase in testosterone concentrations. *Reproduction* 122, 419–429.
88. Bourguiba, S., Lambard, S., and Carreau, S. (2003). Steroids control the aromatase gene expression in purified germ cells from the adult male rat. *J. Mol. Endocrinol.* 31, 83–94.
89. Johnston, D. S., Russell, L. D., Friel, P. J., and Griswold, M. D. (2001). Murine germ cells do not require functional androgen receptors to complete spermatogenesis following spermatogonial stem cell transplantation. *Endocrinology* 142, 2405–2408.
90. Buzek, S. W., Caston, L. A., and Sanborn, B. M. (1987). Evidence for age-dependent changes in Sertoli cell androgen receptor concentration. *J. Androl.* 8, 83–90.
91. Shan, L. X., Zhu, L. J., Bardin, C. W., and Hardy, M. P. (1995). Quantitative analysis of androgen receptor messenger ribonucleic acid in developing Leydig cells and Sertoli cells by in situ hybridization. *Endocrinology* 136, 3856–3862.
92. Suarez-Quian, C. A., Martinez-Garcia, F., Nistal, M., and Regadera, J. (1999). Androgen receptor distribution in adult human testis. *J. Clin. Endocrinol. Metab.* 84, 350–358.

93. Sharpe, R. M., Maddocks, S., Millar, M., Kerr, J. B., Saunders, P. T., and McKinnell, C. (1992). Testosterone and spermatogenesis. Identification of stage-specific, androgen-regulated proteins secreted by adult rat seminiferous tubules. *J. Androl.* 13, 172–184.
94. Sun, Y. T., Wreford, N. G., Robertson, D. M., and de Kretser, D. M. (1990). Quantitative cytological studies of spermatogenesis in intact and hypophysectomized rats: identification of androgen-dependent stages. *Endocrinology* 127, 1215–1223.
95. Bartlett, J. M., Kerr, J. B., and Sharpe, R. M. (1986). The effect of selective destruction and regeneration of rat Leydig cells on the intratesticular distribution of testosterone and morphology of the seminiferous epithelium. *J. Androl.* 7, 240–253.
96. O'Donnell, L., McLachlan, R. I., Wreford, N. G., and Robertson, D. M. (1994). Testosterone promotes the conversion of round spermatids between stages VII and VIII of the rat spermatogenic cycle. *Endocrinology* 135, 2608–2614.
97. O'Donnell, L., McLachlan, R. I., Wreford, N. G., de Kretser, D. M., and Robertson, D. M. (1996). Testosterone withdrawal promotes stage-specific detachment of round spermatids from the rat seminiferous epithelium. *Biol. Reprod.* 55, 895–901.
98. Verhoeven, G., and Cailleau, J. (1988). Follicle-stimulating hormone and androgens increase the concentration of the androgen receptor in Sertoli cells. *Endocrinology* 122, 1541–1550.
99. Blok, L. J., Hoogerbrugge, J. W., Themmen, A. P., Baarends, W. M., Post, M., and Grootegoed, J. A. (1992). Transient down-regulation of androgen receptor messenger ribonucleic acid (mRNA) expression in Sertoli cells by follicle-stimulating hormone is followed by up-regulation of androgen receptor mRNA and protein. *Endocrinology* 131, 1343–1349.
100. Sanborn, B. M., Caston, L. A., Chang, C., Liao, S., Speller, R., Porter, L. D., and Ku, C. Y. (1991). Regulation of androgen receptor mRNA in rat Sertoli and peritubular cells. *Biol. Reprod.* 45, 634–641.
101. Blok, L. J., Mackenbach, P., Trapman, J., Themmen, A. P., Brinkmann, A. O., and Grootegoed, J. A. (1989). Follicle-stimulating hormone regulates androgen receptor mRNA in Sertoli cells. *Mol. Cell. Endocrinol.* 63, 267–271.
102. Delfino, F. J., Boustead, J. N., Fix, C., and Walker, W. H. (2003). NF-kappaB and TNF-alpha stimulate androgen receptor expression in Sertoli cells. *Mol. Cell. Endocrinol.* 201, 1–12
103. Zhang, L., Charron, M., Wright, W. W., Chatterjee, B., Song, C. S., Roy, A. K., and Brown, T. R. (2004). Nuclear factor-kappaB activates transcription of the androgen receptor gene in Sertoli cells isolated from testes of adult rats. *Endocrinology* 145, 781–789.
104. Yeh, S., Tsai, M. Y., Xu, Q., Mu, X. M., Lardy, H., Huang, K. E., Lin, H., Yeh, S. D., Altuwaijri, S., Zhou, X., Xing, L., Boyce, B. F., Hung, M. C., Zhang, S., Gan, L., and Chang, C. (2002). Generation and characterization of androgen receptor knockout (ARKO) mice: an in vivo model for the study of androgen functions in selective tissues. *Proc. Natl. Acad. Sci. U S A* 99, 13498–13503.
105. Holdcraft, R. W., and Braun, R. E. (2004). Androgen receptor function is required in Sertoli cells for the terminal differentiation of haploid spermatids. *Development* 131, 459–467.
106. Pettersson, K., and Gustafsson, J. A. (2001). Role of estrogen receptor beta in estrogen action. *Annu. Rev. Physiol.* 63, 165–192.
107. Hess, R. A. (2003). Estrogen in the adult male reproductive tract: A review. *Reprod. Biol. Endocrinol.* 1, 52.
108. Hess, R. A. (2000). Oestrogen in fluid transport in efferent ducts of the male reproductive tract. *Rev. Reprod.* 5, 84–92.
109. Mosselman, S., Polman, J., and Dijkema, R. (1996). ER beta: identification and characterization of a novel human estrogen receptor. *FEBS Lett.* 392, 49–53.
110. Walter, P., Green, S., Greene, G., Krust, A., Bornert, J. M., Jeltsch, J. M., Staub, A., Jensen, E., Scrace, G., Waterfield, M., and Chambon, P. (1985). Cloning of the human estrogen receptor cDNA. *Proc. Natl. Acad. Sci. U S A* 82, 7889–7893.
111. Kuiper, G. G., Enmark, E., Pelto-Huikko, M., Nilsson, S., and Gustafsson, J. A. (1996). Cloning of a novel receptor expressed in rat prostate and ovary. *Proc. Natl. Acad. Sci. U S A* 93, 5925–5930.
112. White, R., Lees, J. A., Needham, M., Ham, J., and Parker, M. (1987). Structural organization and expression of the mouse estrogen receptor. *Mol. Endocrinol.* 1, 735–744.
113. Nilsson, S., Makela, S., Treuter, E., Tujague, M., Thomsen, J., Andersson, G., Enmark, E., Pettersson, K., Warner, M., and Gustafsson, J. A. (2001). Mechanisms of estrogen action. *Physiol. Rev.* 81, 1535–1565.
114. Pettersson, K., Grandien, K., Kuiper, G. G., and Gustafsson, J. A. (1997). Mouse estrogen receptor beta forms estrogen response element-binding heterodimers with estrogen receptor alpha. *Mol. Endocrinol.* 11, 1486–1496.
115. Cowley, S. M., Hoare, S., Mosselman, S., and Parker, M. G. (1997). Estrogen receptors alpha and beta form heterodimers on DNA. *J. Biol. Chem.* 272, 19858–19862.
116. Hall, J. M., and McDonnell, D. P. (1999). The estrogen receptor beta-isoform (ERbeta) of the human estrogen receptor modulates ERalpha transcriptional activity and is a key regulator of the cellular response to estrogens and antiestrogens. *Endocrinology* 140, 5566–5578.
117. Dutertre, M., and Smith, C. L. (2000). Molecular mechanisms of selective estrogen receptor modulator (SERM) action. *J. Pharmacol. Exp. Ther.* 295, 431–437.
118. Toran-Allerand, C. D. (2004). Minireview: a plethora of estrogen receptors in the brain: where will it end? *Endocrinology* 145, 1069–1074.
119. Saunders, P. T., Millar, M. R., Macpherson, S., Irvine, D. S., Groome, N. P., Evans, L. R., Sharpe, R. M., and Scobie, G. A. (2002). ERbeta1 and the ERbeta2 splice variant (ERbetacx/beta2) are expressed in distinct cell populations in the adult human testis. *J. Clin. Endocrinol. Metab.* 87, 2706–2715.
120. Ogawa, S., Inoue, S., Watanabe, T., Orimo, A., Hosoi, T., Ouchi, Y., and Muramatsu, M. (1998). Molecular cloning and characterization of human estrogen receptor betacx: a potential inhibitor of estrogen action in human. *Nucleic Acids Res.* 26, 3505–3512.
121. Lubahn, D. B., Moyer, J. S., Golding, T. S., Couse, J. F., Korach, K. S., and Smithies, O. (1993). Alteration of reproductive function but not prenatal sexual development after insertional disruption of the mouse estrogen receptor gene. *Proc. Natl. Acad. Sci. U S A* 90, 11162–11166.
122. Korach, K. S. (1994). Insights from the study of animals lacking functional estrogen receptor. *Science* 266, 1524–1527.
123. Eddy, E. M., Washburn, T. F., Bunch, D. O., Goulding, E. H., Gladen, B. C., Lubahn, D. B., and Korach, K. S. (1996). Targeted disruption of the estrogen receptor gene in male mice causes alteration of spermatogenesis and infertility. *Endocrinology* 137, 4796–4805.
124. Hess, R. A., Bunick, D., Lee, K. H., Bahr, J., Taylor, J. A., Korach, K. S., and Lubahn, D. B. (1997). A role for oestrogens in the male reproductive system. *Nature* 390, 509–512.
125. Dupont, S., Krust, A., Gansmuller, A., Dierich, A., Chambon, P., and Mark, M. (2000). Effect of single and compound knockouts of estrogen receptors alpha (ERalpha) and beta

(ERbeta) on mouse reproductive phenotypes. *Development* 127, 4277–4291.

126. Couse, J. F., Hewitt, S. C., Bunch, D. O., Sar, M., Walker, V. R., Davis, B. J., and Korach, K. S. (1999). Postnatal sex reversal of the ovaries in mice lacking estrogen receptors a and b. *Science* 286, 2328–2331.
127. Greco, T. L., Furlow, J. D., Duello, T. M., and Gorski, J. (1992). Immunodetection of estrogen receptors in fetal and neonatal male mouse reproductive tracts. *Endocrinology* 130, 421–429.
128. Fisher, J. S., Millar, M. R., Majdic, G., Saunders, P. T., Fraser, H. M., and Sharpe, R. M. (1997). Immunolocalisation of oestrogen receptor-alpha within the testis and excurrent ducts of the rat and marmoset monkey from perinatal life to adulthood. *J. Endocrinol.* 153, 485–495.
129. Gaskell, T. L., Robinson, L. L., Groome, N. P., Anderson, R. A., and Saunders, P. T. (2003). Differential expression of two estrogen receptor beta isoforms in the human fetal testis during the second trimester of pregnancy. *J. Clin. Endocrinol. Metab.* 88, 424–432.
130. Saunders, P. T., Fisher, J. S., Sharpe, R. M., and Millar, M. R. (1998). Expression of oestrogen receptor beta (ER beta) occurs in multiple cell types, including some germ cells, in the rat testis. *J. Endocrinol.* 156, R13–R17.
131. van Pelt, M. M., de Rooij, D. G., van der Burg, B., van der Saag, P. T., Gustafsson, J. A., and Kuiper, G. J. M. (1999). Ontogeny of estrogen receptor-beta expression in rat testis. *Endocrinology*. 140, 478–483.
132. Pais, V., Leav, I., Lau, K. M., Jiang, Z., and Ho, S. M. (2003). Estrogen receptor-beta expression in human testicular germ cell tumors. *Clin. Cancer Res.* 9, 4475–4482.
133. Saunders, P. T., Sharpe, R. M., Williams, K., Macpherson, S., Urquart, H., Irvine, D. S., and Millar, M. R. (2001). Differential expression of oestrogen receptor alpha and beta proteins in the testes and male reproductive system of human and non-human primates. *Mol. Hum. Reprod.* 7, 227–236.
134. Pentikainen, V., Erkkila, K., Suomalainen, L., Parvinen, M., and Dunkel, L. (2000). Estradiol acts as a germ cell survival factor in the human testis in vitro. *J. Clin. Endocrinol. Metab.* 85, 2057–2067.
135. Pelletier, G., and El-Alfy, M. (2000). Immunocytochemical localization of estrogen receptors a and b in the human reproductive organs. *J. Clin. Endocrinol. Metab.* 85, 4835–4840.
136. Heckert, L., and Griswold, M. D. (1993). Expression of the FSH receptor in the testis. *Recent Prog. Horm. Res.* 48, 61–77.
137. Orth, J., and Christensen, A. K. (1978). Autoradiographic localization of specifically bound 125I-labeled follicle-stimulating hormone on spermatogonia of the rat testis. *Endocrinology* 103, 1944–1951.
138. Heckert, L. L., and Griswold, M. D. (2002). The expression of the follicle-stimulating hormone receptor in spermatogenesis. *Recent Prog. Horm. Res.* 57, 129–148.
139. Means, A. R., Dedman, J. R., Tash, J. S., Tindall, D. J., van Sickle, M., and Welsh, M. J. (1980). Regulation of the testis Sertoli cell by follicle stimulating hormone. *Annu. Rev. Physiol.* 42, 59–70.
140. Lalevee, N., Rogier, C., Becq, F., and Joffre, M. (1999). Acute effects of adenosine triphosphates, cyclic 3′,5′-adenosine monophosphates, and follicle-stimulating hormone on cytosolic calcium level in cultured immature rat Sertoli cells. *Biol. Reprod.* 61, 343–352.
141. Griswold, M. D. (1993). Actions of FSH on mammalian Sertoli cells. In *The Sertoli Cell* (L. D. Russell and M. D. Griswold, Eds.), pp. 493–508. Cache River Press, Clearwater, FL.
142. Griswold, M. D., Kim, J. S., and Tribley, W. A. (2001). Mechanisms involved in the homologous down-regulation of transcription of the follicle-stimulating hormone receptor gene in Sertoli cells. *Mol. Cell. Endocrinol.* 173, 95–107.
143. Themmen, A. P., Blok, L. J., Post, M., Baarends, W. M., Hoogerbrugge, J. W., Parmentier, M., Vassart, G., and Grootegoed, J. A. (1991). Follitropin receptor down-regulation involves a cAMP-dependent post-transcriptional decrease of receptor mRNA expression. *Mol. Cell. Endocrinol.* 78, R7–R13.
144. Zirkin, B. R., Awoniyi, C., Griswold, M. D., Russell, L. D., and Sharpe, R. (1994). Is FSH required for adult spermatogenesis? *J. Androl.* 15, 273–276.
145. Ketelslegers, J. M., Hetzel, W. D., Sherins, R. J., and Catt, K. J. (1978). Developmental changes in testicular gonadotropin receptors: plasma gonadotropins and plasma testosterone in the rat. *Endocrinology* 103, 212–222.
146. Bortolussi, M., Zanchetta, R., Belvedere, P., and Colombo, L. (1990). Sertoli and Leydig cell numbers and gonadotropin receptors in rat testis from birth to puberty. *Cell. Tissue Res.* 260, 185–191.
147. Conti, M., Jin, S. L., Monaco, L., Repaske, D. R., and Swinnen, J. V. (1991). Hormonal regulation of cyclic nucleotide phosphodiesterases. *Endocr. Rev.* 12, 218–234.
148. Parvinen, M. (1982). Regulation of the seminiferous epithelium. *Endocr. Rev.* 3, 404–417.
149. Rannikko, A., Penttila, T. L., Zhang, F. P., Toppari, J., Parvinen, M., and Huhtaniemi, I. (1996). Stage-specific expression of the FSH receptor gene in the prepubertal and adult rat seminiferous epithelium. *J. Endocrinol.* 151, 29–35.
150. Lampa, J., Hoogerbrugge, J. W., Baarends, W. M., Stanton, P. G., Perryman, K. J., Grootegoed, J. A., and Robertson, D. M. (1999). Follicle-stimulating hormone and testosterone stimulation of immature and mature Sertoli cells in vitro: inhibin and N-cadherin levels and round spermatid binding. *J. Androl.* 20, 399–406.
151. Misrahi, M., Beau, I., Ghinea, N., Vannier, B., Loosfelt, H., Meduri, G., Vu Hai, M. T., and Milgrom, E. (1996). The LH/CG and FSH receptors: different molecular forms and intracellular traffic. *Mol. Cell. Endocrinol.* 125, 161–167.
152. Warren, D. W., Huhtaniemi, I. T., Tapanainen, J., Dufau, M. L., and Catt, K. J. (1984). Ontogeny of gonadotropin receptors in the fetal and neonatal rat testis. *Endocrinology* 114, 470–476.
153. Rannikki, A. S., Zhang, F. P., and Huhtaniemi, I. T. (1995). Ontogeny of follicle-stimulating hormone receptor gene expression in the rat testis and ovary. *Mol. Cell. Endocrinol.* 107, 199–208.
154. Tsai-Morris, C. H., Aquilano, D. R., and Dufau, M. L. (1985). Cellular localization of rat testicular aromatase activity during development. *Endocrinology* 116, 38–46.
155. Kangasniemi, M., Kaipia, A., Mali, P., Toppari, J., Huhtaniemi, I., and Parvinen, M. (1990). Modulation of basal and FSH-dependent cyclic AMP production in rat seminiferous tubules staged by an improved transillumination technique. *Anat. Rec.* 227, 62–76.
156. Heckert, L. L., and Griswold, M. D. (1991). Expression of follicle-stimulating hormone receptor mRNA in rat testes and Sertoli cells. *Mol. Endocrinol.* 5, 670–677.
157. Kangasniemi, M., Kaipia, A., Toppari, J., Perheentupa, A., Huhtaniemi, I., and Parvinen, M. (1990). Cellular regulation of follicle-stimulating hormone (FSH) binding in rat seminiferous tubules. *J. Androl.* 11, 336–343.
158. Ascoli, M., Fanelli, F., and Segaloff, D. L. (2002). The lutropin/choriogonadotropin receptor, a 2002 perspective. *Endocr. Rev.* 23, 141–174.
159. Zhang, F. P., Hamalainen, T., Kaipia, A., Pakarinen, P., and Huhtaniemi, I. (1994). Ontogeny of luteinizing hormone receptor gene expression in the rat testis. *Endocrinology* 134, 2206–2213.

160. Saez, J. M. (1994). Leydig cells: endocrine, paracrine, and autocrine regulation. *Endocr. Rev.* 15, 574–626.
161. Huhtaniemi, I., Zhang, F. P., Kero, J., Hamalainen, T., and Poutanen, M. (2002). Transgenic and knockout mouse models for the study of luteinizing hormone and luteinizing hormone receptor function. *Mol. Cell. Endocrinol.* 187, 49–56.
162. Zhang, F. P., Pakarainen, T., Zhu, F., Poutanen, M., and Huhtaniemi, I. (2004). Molecular characterization of postnatal development of testicular steroidogenesis in luteinizing hormone receptor knockout mice. *Endocrinology* 145, 1453–1463.
163. Hardy, M. P., Zirkin, B. R., and Ewing, L. L. (1989). Kinetic studies on the development of the adult population of Leydig cells in testes of the pubertal rat. *Endocrinology* 124, 762–770.
164. Tena-Sempere, M., Zhang, F. P., and Huhtaniemi, I. (1994). Persistent expression of a truncated form of the luteinizing hormone receptor messenger ribonucleic acid in the rat testis after selective Leydig cell destruction by ethylene dimethane sulfonate. *Endocrinology* 135, 1018–1024.
165. Shan, L. X., and Hardy, M. P. (1992). Developmental changes in levels of luteinizing hormone receptor and androgen receptor in rat Leydig cells. *Endocrinology* 131, 1107–1114.
166. Fondell, J. D., Brunel, F., Hisatake, K., and Roeder, R. G. (1996). Unliganded thyroid hormone receptor alpha can target TATA-binding protein for transcriptional repression. *Mol. Cell. Biol.* 16, 281–287.
167. Brent, G. A., Dunn, M. K., Harney, J. W., Gulick, T., Larsen, P. R., and Moore, D. D. (1989). Thyroid hormone aporeceptor represses T3-inducible promoters and blocks activity of the retinoic acid receptor. *New Biol.* 1, 329–336.
168. Damm, K., Thompson, C. C., and Evans, R. M. (1989). Protein encoded by v-erbA functions as a thyroid-hormone receptor antagonist. *Nature* 339, 593–597.
169. Graupner, G., Wills, K. N., Tzukerman, M., Zhang, X. K., and Pfahl, M. (1989). Dual regulatory role for thyroid-hormone receptors allows control of retinoic-acid receptor activity. *Nature* 340, 653–656.
170. Sap, J., Munoz, A., Schmitt, J., Stunnenberg, H., and Vennstrom, B. (1989). Repression of transcription mediated at a thyroid hormone response element by the v-erb-A oncogene product. *Nature* 340, 242–244.
171. Baniahmad, A., Kohne, A. C., and Renkawitz, R. (1992). A transferable silencing domain is present in the thyroid hormone receptor, in the v-erbA oncogene product and in the retinoic acid receptor. *EMBO J.* 11, 1015–1023.
172. Jannini, E. A., Dolci, S., Ulisse, S., and Nikodem, V. M. (1994). Developmental regulation of the thyroid hormone receptor alpha 1 mRNA expression in the rat testis. *Mol. Endocrinol.* 8, 89–96.
173. Palmero, S., Maggiani, S., and Fugassa, E. (1988). Nuclear triiodothyronine receptors in rat Sertoli cells. *Mol. Cell. Endocrinol.* 58, 253–256.
174. Palmero, S., De Marco, P., and Fugassa, E. (1995). Thyroid hormone receptor beta mRNA expression in Sertoli cells isolated from prepubertal testis. *J. Mol. Endocrinol.* 14, 131–134.
175. Tagami, T., Nakamura, H., Sasaki, S., Mori, T., Yoshioka, H., Yoshida, H., and Imura, H. (1990). Immunohistochemical localization of nuclear 3,5,3′-triiodothyronine receptor proteins in rat tissues studied with antiserum against C-ERB A/T3 receptor. *Endocrinology* 127, 1727–1734.
176. Macchia, E., Nakai, A., Janiga, A., Sakurai, A., Fisfalen, M. E., Gardner, P., Soltani, K., and DeGroot, L. J. (1990). Characterization of site-specific polyclonal antibodies to c-erbA peptides recognizing human thyroid hormone receptors alpha 1, alpha 2, and beta and native 3,5,3′-triiodothyronine receptor, and study of tissue distribution of the antigen. *Endocrinology* 126, 3232–3239.
177. Falcone, M., Miyamoto, T., Fierro-Renoy, F., Macchia, E., and DeGroot, L. J. (1992). Antipeptide polyclonal antibodies specifically recognize each human thyroid hormone receptor isoform. *Endocrinology* 131, 2419–2429.
178. Buzzard, J. J., Morrison, J. R., O'Bryan, M. K., Song, Q., and Wreford, N. G. (2000). Developmental expression of thyroid hormone receptors in the rat testis. *Biol. Reprod.* 62, 664–669.
179. Steinberger, E. (1971). Hormonal control of mammalian spermatogenesis. *Physiol. Rev.* 51, 1–22.
180. Wilson, J. D., Leihy, M. W., Shaw, G., and Renfree, M. B. (2003). Unsolved problems in male physiology: studies in a marsupial. *Mol. Cell. Endocrinol.* 211, 33–36.
181. Wilson, J. D., Auchus, R. J., Leihy, M. W., Guryev, O. L., Estabrook, R. W., Osborn, S. M., Shaw, G., and Renfree, M. B. (2003). 5alpha-androstane-3alpha,17beta-diol is formed in tammar wallaby pouch young testes by a pathway involving 5alpha-pregnane-3alpha,17alpha-diol-20-one as a key intermediate. *Endocrinology* 144, 575–580.
182. Corpechot, C., Baulieu, E. E., and Robel, P. (1981). Testosterone, dihydrotestosterone and androstanediols in plasma, testes and prostates of rats during development. *Acta Endocrinol. (Copenh).* 96, 127–135.
183. Killian, J., Pratis, K., Clifton, R. J., Stanton, P. G., Robertson, D. M., and O'Donnell, L. (2003). 5alpha-reductase isoenzymes 1 and 2 in the rat testis during postnatal development. *Biol. Reprod.* 68, 1711–1718.
184. Isomaa, V., Parvinen, M., Janne, O. A., and Bardin, C. W. (1985). Nuclear androgen receptors in different stages of the seminiferous epithelial cycle and the interstitial tissue of rat testis. *Endocrinology* 116, 132–137.
185. Zirkin, B. R., Santulli, R., Awoniyi, C. A., and Ewing, L. L. (1989). Maintenance of advanced spermatogenic cells in the adult rat testis: quantitative relationship to testosterone concentration within the testis. *Endocrinology* 124, 3043–3049.
186. Handelsman, D. J., Spaliviero, J. A., Simpson, J. M., Allan, C. M., and Singh, J. (1999). Spermatogenesis without gonadotropins: maintenance has a lower testosterone threshold than initiation. *Endocrinology* 140, 3938–3946.
187. Hansson, V., Weddington, S. C., French, F. S., McLean, W., Smith, A., Nayfeh, S. N., Ritzen, E. M., and Hagenas, L. (1976). Secretion and role of androgen-binding proteins in the testis and epididymis. *J. Reprod. Fertil. Suppl.* 17–33.
188. Selva, D. M., Tirado, O. M., Toran, N., Suarez-Quian, C. A., Reventos, J., and Munell, F. (2000). Meiotic arrest and germ cell apoptosis in androgen-binding protein transgenic mice. *Endocrinology* 141, 1168–1177.
189. Huang, H. F., Marshall, G. R., Rosenberg, R., and Nieschlag, E. (1987). Restoration of spermatogenesis by high levels of testosterone in hypophysectomized rats after long-term regression. *Acta Endocrinol. (Copenh).* 116, 433–444.
190. Vihko, K. K., LaPolt, P. S., Nishimori, K., and Hsueh, A. J. (1991). Stimulatory effects of recombinant follicle-stimulating hormone on Leydig cell function and spermatogenesis in immature hypophysectomized rats. *Endocrinology* 129, 1926–1932.
191. McLachlan, R. I., Wreford, N. G., Tsonis, C., De Kretser, D. M., and Robertson, D. M. (1994). Testosterone effects on spermatogenesis in the gonadotropin-releasing hormone-immunized rat. *Biol. Reprod.* 50, 271–280.
192. Awoniyi, C. A., Zirkin, B. R., Chandrashekar, V., and Schlaff, W. D. (1992). Exogenously administered testosterone maintains spermatogenesis quantitatively in adult rats actively immunized against gonadotropin-releasing hormone. *Endocrinology* 130, 3283–3288.

193. Bhasin, S., Fielder, T. J., and Swerdloff, R. S. (1987). Testosterone selectively increases serum follicle-stimulating hormonal (FSH) but not luteinizing hormone (LH) in gonadotropin-releasing hormone antagonist-treated male rats: evidence for differential regulation of LH and FSH secretion. *Biol. Reprod.* 37, 55–59.
194. Allan, C. M., Haywood, M., Swaraj, S., Spaliviero, J., Koch, A., Jimenez, M., Poutanen, M., Levallet, J., Huhtaniemi, I., Illingworth, P., and Handelsman, D. J. (2001). A novel transgenic model to characterize the specific effects of follicle-stimulating hormone on gonadal physiology in the absence of luteinizing hormone actions. *Endocrinology* 142, 2213–2220.
195. Haywood, M., Spaliviero, J., Jimemez, M., King, N. J., Handelsman, D. J., and Allan, C. M. (2003). Sertoli and germ cell development in hypogonadal (hpg) mice expressing transgenic follicle-stimulating hormone alone or in combination with testosterone. *Endocrinology* 144, 509–517.
196. Buzzard, J. J., Wreford, N. G., and Morrison, J. R. (2003). Thyroid hormone, retinoic acid, and testosterone suppress proliferation and induce markers of differentiation in cultured rat sertoli cells. *Endocrinology* 144, 3722–3731.
197. Walker, W. H. (2003). Molecular mechanisms controlling Sertoli cell proliferation and differentiation. *Endocrinology* 144, 3719–3721.
198. Dierich, A., Sairam, M. R., Monaco, L., Fimia, G. M., Gansmuller, A., LeMeur, M., and Sassone-Corsi, P. (1998). Impairing follicle-stimulating hormone (FSH) signaling in vivo: targeted disruption of the FSH receptor leads to aberrant gametogenesis and hormonal imbalance. *Proc. Natl. Acad. Sci. U S A* 95, 13612–13617.
199. Allan, C. M., Garcia, A., Spaliviero, J., Zhang, F. P., Jimenez, M., Huhtaniemi, I., and Handelsman, D. J. (2004). Complete Sertoli cell proliferation induced by follicle-stimulating hormone (FSH) independently of luteinizing hormone activity: evidence from genetic models of isolated FSH action. *Endocrinology* 145, 1587–1593.
200. Russell, L. D., Alger, L. E., and Nequin, L. G. (1987). Hormonal control of pubertal spermatogenesis. *Endocrinology* 120, 1615–1632.
201. Tapanainen, J. S., Tilly, J. L., Vihko, K. K., and Hsueh, A. J. (1993). Hormonal control of apoptotic cell death in the testis: gonadotropins and androgens as testicular cell survival factors. *Mol. Endocrinol.* 7, 643–650.
202. Madhwa Raj, H. G., and Dym, M. (1976). The effects of selective withdrawal of FSH or LH on spermatogenesis in the immature rat. *Biol. Reprod.* 14, 489–494.
203. Marathe, G. K., Shetty, J., and Dighe, R. R. (1995). Selective immunoneutralisation of luteinizing hormone results in the apoptotic cell death of pachytene spermatocytes and spermatids in the rat testis. *Endocr. J.* 3, 705–709.
204. Matikainen, T., Toppari, J., Vihko, K. K., and Huhtaniemi, I. (1994). Effects of recombinant human FSH in immature hypophysectomized male rats: evidence for Leydig cell-mediated action on spermatogenesis. *J. Endocrinol.* 141, 449–457.
205. Cameron, D. F., Muffly, K. E., and Nazian, S. J. (1993). Reduced testosterone during puberty results in a midspermiogenic lesion. *Proc. Soc. Exp. Biol. Med.* 202, 457–464.
206. Pratis, K., O'Donnell, L., Ooi, G. T., Stanton, P. G., McLachlan, R. I., and Robertson, D. M. (2003). Differential regulation of rat testicular 5alpha-reductase type 1 and 2 isoforms by testosterone and FSH. *J. Endocrinol.* 176, 393–403.
207. Sheffield, J. W., and O'Shaughnessy, P. J. (1988). Testicular steroid metabolism during development in the normal and hypogonadal mouse. *J. Endocrinol.* 119, 257–264.
208. Matsumoto, K., and Yamada, M. (1973). 5alpha-Reduction of testosterone in vitro by rat seminiferous tubules and whole testes at different stages of development. *Endocrinology* 93, 253–255.
209. Grino, P. B., Griffin, J. E., and Wilson, J. D. (1990). Testosterone at high concentrations interacts with the human androgen receptor similarly to dihydrotestosterone. *Endocrinology* 126, 1165–1172.
210. Deslypere, J. P., Young, M., Wilson, J. D., and McPhaul, M. J. (1992). Testosterone and 5 alpha-dihydrotestosterone interact differently with the androgen receptor to enhance transcription of the MMTV-CAT reporter gene. *Mol. Cell. Endocrinol.* 88, 15–22.
211. O'Donnell, L., Pratis, K., Stanton, P. G., Robertson, D. M., and McLachlan, R. I. (1999). Testosterone-dependent restoration of spermatogenesis in adult rats is impaired by a 5alpha-reductase inhibitor. *J. Androl.* 20, 109–117.
212. O'Donnell, L., Stanton, P. G., Wreford, N. G., Robertson, D. M., and McLachlan, R. I. (1996). Inhibition of 5 alpha-reductase activity impairs the testosterone- dependent restoration of spermiogenesis in adult rats. *Endocrinology* 137, 2703–2710.
213. Russell, D. W., and Wilson, J. D. (1994). Steroid 5 alpha-reductase: two genes/two enzymes. *Annu. Rev. Biochem.* 63, 25–61.
214. Normington, K., and Russell, D. W. (1992). Tissue distribution and kinetic characteristics of rat steroid 5 alpha- reductase isozymes. Evidence for distinct physiological functions. *J. Biol. Chem.* 267, 19548–19554.
215. Thigpen, A. E., Cala, K. M., and Russell, D. W. (1993). Characterization of Chinese hamster ovary cell lines expressing human steroid 5 alpha-reductase isozymes. *J. Biol. Chem.* 268, 17404–17412.
216. Viger, R. S., and Robaire, B. (1995). Steady state steroid 5 alpha-reductase messenger ribonucleic acid levels and immunocytochemical localization of the type 1 protein in the rat testis during postnatal development. *Endocrinology* 136, 5409–5415.
217. Dorrington, J. H., and Fritz, I. B. (1975). Cellular localization of 5alpha-reductase and 3alpha-hydroxysteroid dehydrogenase in the seminiferous tubule of the rat testis. *Endocrinology* 96, 897–889.
218. Folman, Y., Ahmad, N., Sowell, J. G., and Eik-Nes, K. B. (1973). Formation in vitro of 5alpha-dihydrotestosterone and other 5alpha-reduced metabolites of 3H-testosterone by the seminiferous tubules and interstitial tissue from immature and mature rat testes. *Endocrinology* 92, 41-47.
219. Pratis, K., O'Donnell, L., Ooi, G. T., McLachlan, R. I., and Robertson, D. M. (2000). Enzyme assay for 5alpha-reductase type 2 activity in the presence of 5alpha-reductase type 1 activity in rat testis. *J. Steroid Biochem. Mol. Biol.* 75, 75–82.
220. Nayfeh, S. N., Coffey, J. C., Hansson, V., and French, F. S. (1975). Maturational changes in testicular steroidogenesis: hormonal regulation of 5alpha-reductase. *J. Steroid Biochem.* 6, 329–335.
221. Welsh, M. J., and Wiebe, J. P. (1976). Sertoli cells from immature rats: in vitro stimulation of steroid metabolism by FSH. *Biochem. Biophys. Res. Commun.* 69, 936–941.
222. Lopez-Solache, I., Luu-The, V., Seralini, G. E., and Labrie, F. (1996). Heterogeneity of rat type I 5 alpha-reductase cDNA: cloning, expression and regulation by pituitary implants and dihydrotestosterone. *Biochim. Biophys. Acta.* 1305, 139–144.
223. Mahendroo, M. S., Cala, K. M., Hess, D. L., and Russell, D. W. (2001). Unexpected virilization in male mice lacking steroid 5 alpha-reductase enzymes. *Endocrinology* 142, 4652–4662.
224. George, F. W., Johnson, L., and Wilson, J. D. (1989). The effect of a 5 alpha-reductase inhibitor on androgen

physiology in the immature male rat. *Endocrinology*. 125, 2434–2438.

225. Ahmad, N., Haltmeyer, G. C., and Eik-nes, K. B. (1975). Maintenance of spermatogenesis with testosterone or dihydrotestosterone in hypophysectomized rats. *J. Reprod. Fertil*. 44, 103–107.
226. Buhl, A. E., Cornette, J. C., Kirton, K. T., and Yuan, Y. D. (1982). Hypophysectomized male rats treated with polydimethylsiloxane capsules containing testosterone: effects on spermatogenesis, fertility, and reproductive tract concentrations of androgens. *Biol. Reprod*. 27, 183–188.
227. Chowdhury, A. K. (1979). Dependence of testicular germ cells on hormones: a quantitative study in hypophysectomized testosterone-treated rats. *J. Endocrinol*. 82, 331–340.
228. Muffly, K. E., Nazian, S. J., and Cameron, D. F. (1993). Junction-related Sertoli cell cytoskeleton in testosterone-treated hypophysectomized rats. *Biol. Reprod*. 49, 1122–1132.
229. Santulli, R., Sprando, R. L., Awoniyi, C. A., Ewing, L. L., and Zirkin, B. R. (1990). To what extent can spermatogenesis be maintained in the hypophysectomized adult rat testis with exogenously administered testosterone? *Endocrinology* 126, 95–101.
230. Sun, Y. T., Irby, D. C., Robertson, D. M., and de Kretser, D. M. (1989). The effects of exogenously administered testosterone on spermatogenesis in intact and hypophysectomized rats. *Endocrinology* 125, 1000–1010.
231. Pogach, L., Giglio, W., Nathan, E., and Huang, H. F. (1993). Maintenance of spermiogenesis by exogenous testosterone in rats treated with a GnRH antagonist: relationship with androgen-binding protein status. *J. Reprod. Fertil*. 98, 415–422.
232. Rea, M. A. Marshall, G. R., Weinbauer, G. F., and Nieschlag, E. (1986). Testosterone maintains pituitary and serum FSH and spermatogenesis in gonadotrophin-releasing hormone antagonist-suppressed rats. *J. Endocrinol*. 108, 101–107.
233. Awoniyi, C. A., Sprando, R. L., Santulli, R., Chandrashekar, V., Ewing, L. L., and Zirkin, B. R. (1990). Restoration of spermatogenesis by exogenously administered testosterone in rats made azoospermic by hypophysectomy or withdrawal of luteinizing hormone alone. *Endocrinology* 127, 177–184.
234. Sharpe, R. M. (1994). Regulation of spermatogenesis. In *The Physiology of Reproduction* (E. Knobil and J. D. Neill, Eds.), Vol. 1, pp. 1363–1434. Raven Press, New York.
235. Meachem, S. J., Wreford, N. G., Robertson, D. M., and McLachlan, R. I. (1997). Androgen action on the restoration of spermatogenesis in adult rats: effects of human chorionic gonadotrophin, testosterone and flutamide administration on germ cell number. *Int. J. Androl*. 20, 70–79.
236. Meistrich, M. L., and Kangasniemi, M. (1997). Hormone treatment after irradiation stimulates recovery of rat spermatogenesis from surviving spermatogonia. *J. Androl*. 18, 80–87.
237. Matsumiya, K., Meistrich, M. L., Shetty, G., Dohmae, K., Tohda, A., Okuyama, A., and Nishimune, Y. (1999). Stimulation of spermatogonial differentiation in juvenile spermatogonial depletion (jsd) mutant mice by gonadotropin-releasing hormone antagonist treatment. *Endocrinology* 140, 4912–4915.
238. Ogawa, T., Dobrinski, I., and Brinster, R. L. (1999). Recipient preparation is critical for spermatogonial transplantation in the rat. *Tissue Cell*. 31, 461–472.
239. McLachlan, R. I., Wreford, N. G., Meachem, S. J., De Kretser, D. M., and Robertson, D. M. (1994). Effects of testosterone on spermatogenic cell populations in the adult rat. *Biol. Reprod*. 51, 945–955.
240. Kerr, J. B., Maddocks, S., and Sharpe, R. M. (1992). Testosterone and FSH have independent, synergistic and stage-dependent effects upon spermatogenesis in the rat testis. *Cell. Tissue Res*. 268, 179–189.
241. Kerr, J. B., Millar, M., Maddocks, S., and Sharpe, R. M. (1993). Stage-dependent changes in spermatogenesis and Sertoli cells in relation to the onset of spermatogenic failure following withdrawal of testosterone. *Anat. Rec*. 235, 547–559.
242. Henriksen, K., Hakovirta, H., and Parvinen, M. (1995). Testosterone inhibits and induces apoptosis in rat seminiferous tubules in a stage-specific manner: in situ quantification in squash preparations after administration of ethane dimethane sulfonate. *Endocrinology* 136, 3285–3291.
243. Zhang, F. P., Pakarainen, T., Poutanen, M., Toppari, J., and Huhtaniemi, I. (2003). The low gonadotropin-independent constitutive production of testicular testosterone is sufficient to maintain spermatogenesis. *Proc. Natl. Acad. Sci. U S A* 100, 13692–13697.
244. O'Donnell, L., Stanton, P. G., Bartles, J. R., and Robertson, D. M. (2000). Sertoli cell ectoplasmic specializations in the seminiferous epithelium of the testosterone-suppressed adult rat. *Biol. Reprod*. 63, 99–108.
245. Palombi, F., Salanova, M., Tarone, G., Farini, D., and Stefanini, M. (1992). Distribution of beta 1 integrin subunit in rat seminiferous epithelium. *Biol. Reprod*. 47, 1173–1182.
246. Salanova, M., Stefanini, M., De Curtis, I., and Palombi, F. (1995). Integrin receptor alpha 6 beta 1 is localized at specific sites of cell- to-cell contact in rat seminiferous epithelium. *Biol. Reprod*. 52, 79–87.
247. Byers, S. W., Sujarit, S., Jegou, B., Butz, S., Hoschutzky, H., Herrenknecht, K., MacCalman, C., and Blaschuk, O. W. (1994). Cadherins and cadherin-associated molecules in the developing and maturing rat testis. *Endocrinology* 134, 630–639.
248. Wine, R. N., and Chapin, R. E. (1999). Adhesion and signaling proteins spatiotemporally associated with spermiation in the rat. *J. Androl*. 20, 198–213.
249. Johnson, K. J., Patel, S. R., and Boekelheide, K. (2000). Multiple cadherin superfamily members with unique expression profiles are produced in rat testis. *Endocrinology* 141, 675–683.
250. Perryman, K. J., Stanton, P. G., Loveland, K. L., McLachlan, R. I., and Robertson, D. M. (1996). Hormonal dependency of neural cadherin in the binding of round spermatids to Sertoli cells in vitro. *Endocrinology* 137, 3877–3883.
251. Anderson, R. A., Kelly, R. W., and Wu, F. C. (1997). Comparison between testosterone enanthate-induced azoospermia and oligozoospermia in a male contraceptive study. V. Localization of higher 5 alpha-reductase activity to the reproductive tract in oligozoospermic men administered supraphysiological doses of testosterone. *J. Androl*. 18, 366–371.
252. Anderson, R. A., Wallace, A. M., and Wu, F. C. (1996). Comparison between testosterone enanthate-induced azoospermia and oligozoospermia in a male contraceptive study. III. Higher 5 alpha-reductase activity in oligozoospermic men administered supraphysiological doses of testosterone. *J. Clin. Endocrinol. Metab*. 81, 902–908.
253. O'Leary, P. C., Jackson, A. E., Irby, D. C., and de Kretser, D. M. (1987). Effects of ethane dimethane sulphonate (EDS) on seminiferous tubule function in rats. *Int. J. Androl*. 10, 625–634.
254. Sadate-Ngatchou, P. I., Pouchnik, D. J., and Griswold, M. D. (2004). Identification of testosterone-regulated genes in testes of hypogonadal mice using oligonucleotide microarray. *Mol. Endocrinol*. 18, 422–433.
255. Lindsey, J. S., and Wilkinson, M. F. (1996). Pem: a testosterone- and LH-regulated homeobox gene expressed in mouse Sertoli cells and epididymis. *Dev. Biol*. 179, 471–484.

256. Maiti, S., Doskow, J., Li, S., Nhim, R. P., Lindsey, J. S., and Wilkinson, M. F. (1996). The Pem homeobox gene. Androgen-dependent and -independent promoters and tissue-specific alternative RNA splicing. *J. Biol. Chem.* 271, 17536–17546.
257. Lindsey, J. S., and Wilkinson, M. F. (1996). An androgen-regulated homeobox gene expressed in rat testis and epididymis. *Biol. Reprod.* 55, 975–983.
258. Rao, M. K., Wayne, C. M., Meistrich, M. L., and Wilkinson, M. F. (2003). Pem homeobox gene promoter sequences that direct transcription in a Sertoli cell-specific, stage-specific, and androgen-dependent manner in the testis in vivo. *Mol. Endocrinol.* 17, 223–233.
259. Sharpe, R. M., McKinnell, C., Kivlin, C., and Fisher, J. S. (2003). Proliferation and functional maturation of Sertoli cells, and their relevance to disorders of testis function in adulthood. *Reproduction* 125, 769–784.
260. Yan, W., Kero, J., Suominen, J., and Toppari, J. (2001). Differential expression and regulation of the retinoblastoma family of proteins during testicular development and spermatogenesis: roles in the control of germ cell proliferation, differentiation and apoptosis. *Oncogene* 20, 1343–1356.
261. Fragale, A., Puglisi, R., Morena, A. R., Stefanini, M., and Boitani, C. (2001). Age-dependent activin receptor expression pinpoints activin A as a physiological regulator of rat Sertoli cell proliferation. *Mol. Hum. Reprod.* 7, 1107–1114.
262. Migrenne, S., Racine, C., Guillou, F., and Habert, R. (2003). Pituitary hormones inhibit the function and differentiation of fetal Sertoli cells. *Endocrinology* 144, 2617–2622.
263. Pellegrini, M., Grimaldi, P., Rossi, P., Geremia, R., and Dolci, S. (2003). Developmental expression of BMP4/ALK3/SMAD5 signaling pathway in the mouse testis: a potential role of BMP4 in spermatogonia differentiation. *J. Cell Sci.* 116, 3363–3372.
264. Meehan, T., Loveland, K. L., de Kretser, D., Cory, S., and Print, C. G. (2001). Developmental regulation of the bcl-2 family during spermatogenesis: Insights into the sterility of bcl-w-/- male mice. *Cell Death Differ.* 8, 225–233.
265. Skinner, M. K. (1991). Cell-cell interactions in the testis. *Endocr. Rev.* 12, 45–77.
266. Griswold, M. D. (1998). The central role of Sertoli cells in spermatogenesis. *Semin. Cell Dev. Biol.* 9, 411–416.
267. Griswold, M. D., Solari, A., Tung, P. S., and Fritz, I. B. (1977). Stimulation by follicle-stimulating hormone of DNA synthesis and of mitosis in cultured Sertoli cells prepared from testes of immature rats. *Mol. Cell. Endocrinol.* 7, 151–165.
268. Orth, J. M. (1984). The role of follicle-stimulating hormone in controlling Sertoli cell proliferation in testes of fetal rats. *Endocrinology* 115, 1248–1255.
269. Griswold, M. D., Mably, E. R., and Fritz, I. B. (1976). FSH stimulation of DNA synthesis in Sertoli cells in culture. *Mol. Cell. Endocrinol.* 4, 139–149.
270. Boitani, C., Stefanini, M., Fragale, A., and Morena, A. R. (1995). Activin stimulates Sertoli cell proliferation in a defined period of rat testis development. *Endocrinology* 136, 5438–5444.
271. Meehan, T., Schlatt, S., O'Bryan, M. K., de Kretser, D. M., and Loveland, K. L. (2000). Regulation of germ cell and Sertoli cell development by activin, follistatin, and FSH. *Dev. Biol.* 220, 225–237.
272. Baker, P. J., and O'Shaughnessy, P. J. (2001). Role of gonadotrophins in regulating numbers of Leydig and Sertoli cells during fetal and postnatal development in mice. *Reproduction.* 122, 227–234.
273. Buzzard, J. J., Farnworth, P. G., De Kretser, D. M., O'Connor, A. E., Wreford, N. G., and Morrison, J. R. (2003). Proliferative phase Sertoli cells display a developmentally regulated response to activin in vitro. *Endocrinology* 144, 474–483.
274. Johnston, H., Baker, P. J., Abel, M., Charlton, H. M., Jackson, G., Fleming, L., Kumar, T. R., and O'Shaughnessy, P. J. (2004). Regulation of Sertoli cell number and activity by follicle-stimulating hormone and androgen during postnatal development in the mouse. *Endocrinology* 145, 318–329.
275. Buzzard, J. J., Wreford, N. G., and Morrison, J. R. (2002). Marked extension of proliferation of rat Sertoli cells in culture using recombinant human FSH. *Reproduction* 124, 633–641.
276. Simorangkir, D. R., Marshall, G. R., and Plant, T. M. (2003). Sertoli cell proliferation during prepubertal development in the rhesus monkey (Macaca mulatta) is maximal during infancy when gonadotropin secretion is robust. *J. Clin. Endocrinol. Metab.* 88, 4984–4989.
277. Plant, T. M., and Marshall, G. R. (2001). The functional significance of FSH in spermatogenesis and the control of its secretion in male primates. *Endocr. Rev.* 22, 764–786.
278. Cameron, D. F., Muffly, K. E., and Nazian, S. J. (1998). Development of Sertoli cell binding competency in the peripubertal rat. *J. Androl.* 19, 573–579.
279. Cameron, D. F., and Muffly, K. E. (1991). Hormonal regulation of spermatid binding. *J. Cell Sci.* 100, 623–633.
280. Muffly, K. E., Nazian, S. J., and Cameron, D. F. (1994). Effects of follicle-stimulating hormone on the junction-related Sertoli cell cytoskeleton and daily sperm production in testosterone-treated hypophysectomized rats. *Biol. Reprod.* 51, 158–166.
281. McLean, D. J., Friel, P. J., Pouchnik, D., and Griswold, M. D. (2002). Oligonucleotide microarray analysis of gene expression in follicle-stimulating hormone-treated rat Sertoli cells. *Mol. Endocrinol.* 16, 2780–2792.
282. Manna, P. R., Wang, X. J., and Stocco, D. M. (2003). Involvement of multiple transcription factors in the regulation of steroidogenic acute regulatory protein gene expression. *Steroids* 68, 1125–1134.
283. Larriba, S., Esteban, C., Toran, N., Gerard, A., Audi, L., Gerard, H., and Reventos, J. (1995). Androgen binding protein is tissue-specifically expressed and biologically active in transgenic mice. *J. Steroid Biochem. Mol. Biol.* 53, 573–578.
284. Joseph, I. B., Nelson, J. B., Denmeade, S. R., and Isaacs, J. T. (1997). Androgens regulate vascular endothelial growth factor content in normal and malignant prostatic tissue. *Clin. Cancer Res.* 3, 2507–2511.
285. O'Shaughnessy, P. J., Fleming, L., Baker, P. J., Jackson, G., and Johnston, H. (2003). Identification of developmentally regulated genes in the somatic cells of the mouse testis using serial analysis of gene expression. *Biol. Reprod.* 69, 797–808.
286. Yu, Z., Guo, R., Ge, Y., Ma, J., Guan, J., Li, S., Sun, X., Xue, S., and Han, D. (2003). Gene expression profiles in different stages of mouse spermatogenic cells during spermatogenesis. *Biol. Reprod.* 69, 37–47.
287. Kanatsu-Shinohara, M., Morimoto, T., Toyokuni, S., and Shinohara, T. (2004). Regulation of mouse spermatogonial stem cell self-renewing division by the pituitary gland. *Biol. Reprod.* 70, 1731–1737.
288. Zhang, Z., Renfree, M. B., and Short, R. V. (2003). Successful intra- and interspecific male germ cell transplantation in the rat. *Biol. Reprod.* 68, 961–967.
289. Meng, X., Lindahl, M., Hyvonen, M. E., Parvinen, M., de Rooij, D. G., Hess, M. W., Raatikainen-Ahokas, A., Sainio, K., Rauvala, H., Lakso, M., Pichel, J. G., Westphal, H., Saarma, M., and Sariola, H. (2000). Regulation of cell fate decision of undifferentiated spermatogonia by GDNF. *Science* 287, 1489–1493.

290. Tadokoro, Y., Yomogida, K., Ohta, H., Tohda, A., and Nishimune, Y. (2002). Homeostatic regulation of germinal stem cell proliferation by the GDNF/FSH pathway. *Mech. Dev.* 113, 29–39.
291. Boulogne, B., Olaso, R., Levacher, C., Durand, P., and Habert, R. (1999). Apoptosis and mitosis in gonocytes of the rat testis during foetal and neonatal development. *Int. J. Androl.* 22, 356–365.
292. Orth, J. M., Jester, W. F., Li, L. H., and Laslett, A. L. (2000). Gonocyte-Sertoli cell interactions during development of the neonatal rodent testis. *Curr. Top. Dev. Biol.* 50, 103–124.
293. Shetty, J., Marathe, G. K., and Dighe, R. R. (1996). Specific immunoneutralization of FSH leads to apoptotic cell death of the pachytene spermatocytes and spermatogonial cells in the rat. *Endocrinology* 137, 2179–2182.
294. Boitani, C., Politi, M. G., and Menna, T. (1993). Spermatogonial cell proliferation in organ culture of immature rat testis. *Biol. Reprod.* 48, 761–767.
295. Haneji, T., Maekawa, M., and Nishimune, Y. (1984). Vitamin A and follicle-stimulating hormone synergistically induce differentiation of type A spermatogonia in adult mouse cryptorchid testes in vitro. *Endocrinology* 114, 801–805.
296. Yan, W., Kero, J., Huhtaniemi, I., and Toppari, J. (2000). Stem cell factor functions as a survival factor for mature Leydig cells and a growth factor for precursor Leydig cells after ethylene dimethane sulfonate treatment: implication of a role of the stem cell factor/c-Kit system in Leydig cell development. *Dev. Biol.* 227, 169–182.
297. Suominen, J. S., Yan, W., Toppari, J., and Kaipia, A. (2001). The expression and regulation of Bcl-2-related ovarian killer (Bok) mRNA in the developing and adult rat testis. *Eur. J. Endocrinol.* 145, 771–778.
298. Lerchl, A., Sotiriadou, S., Behre, H. M., Pierce, J., Weinbauer, G. F., Kliesch, S., and Nieschlag, E. (1993). Restoration of spermatogenesis by follicle-stimulating hormone despite low intratesticular testosterone in photoinhibited hypogonadotropic Djungarian hamsters (Phodopus sungorus). *Biol. Reprod.* 49, 1108–1116.
299. Niklowitz, P., Lerchl, A., and Nieschlag, E. (1997). In vitro fertilizing capacity of sperm from FSH-treated photoinhibited Djungarian hamsters (Phodopus sungorus). *J. Endocrinol.* 154, 475–481.
300. Russell, L. D. (1993). Sertoli cell structure and function in seasonally breeding mammals. In *The Sertoli Cell* (L. D. Russell and M. D. Griswold, Eds.), pp. 350–364. Cache River Press, Clearwater, FL.
301. Baker, P. J., Pakarinen, P., Huhtaniemi, I. T., Abel, M. H., Charlton, H. M., Kumar, T. R., and O'Shaughnessy, P. J. (2003). Failure of normal Leydig cell development in follicle-stimulating hormone (FSH) receptor-deficient mice, but not FSHbeta-deficient mice: role for constitutive FSH receptor activity. *Endocrinology* 144, 138–145.
302. Wreford, N. G., Rajendra Kumar, T., Matzuk, M. M., and de Kretser, D. M. (2001). Analysis of the testicular phenotype of the Follicle-Stimulating Hormone b-subunit knockout and the activin type II receptor knockout mice by stereological analysis. *Endocrinology* 142, 2916–2920.
303. Abel, M. H., Wootton, A. N., Wilkins, V., Huhtaniemi, I., Knight, P. G., and Charlton, H. M. (2000). The effect of a null mutation in the follicle-stimulating hormone receptor gene on mouse reproduction. *Endocrinology* 141, 1795–1803.
304. Grover, A., Sairam, M. R., Smith, C. E., and Hermo, L. (2004). Structural and functional modifications of Sertoli cells in the testis of adult follicle-stimulating hormone receptor knockout mice. *Biol. Reprod.* 71, 117–129.
305. Krishnamurthy, H., Danilovich, N., Morales, C. R., and Sairam, M. R. (2000). Qualitative and quantitative decline in spermatogenesis of the follicle- stimulating hormone receptor knockout (FORKO) mouse. *Biol. Reprod.* 62, 1146–1159.
306. Sairam, M. R., and Krishnamurthy, H. (2001). The role of follicle-stimulating hormone in spermatogenesis: lessons from knockout animal models. *Arch. Med. Res.* 32, 601–608.
307. Kumar, T. R., Wang, Y., Lu, N., and Matzuk, M. M. (1997). Follicle stimulating hormone is required for ovarian follicle maturation but not male fertility. *Nat. Genet.* 15, 201–204.
308. Kumar, T. R., Low, M. J., and Matzuk, M. M. (1998). Genetic rescue of follicle-stimulating hormone beta-deficient mice. *Endocrinology* 139, 3289–3295.
309. Haywood, M., Tymchenko, N., Spaliviero, J., Koch, A., Jimenez, M., Gromoll, J., Simoni, M., Nordhoff, V., Handelsman, D. J., and Allan, C. M. (2002). An activated human follicle-stimulating hormone (FSH) receptor stimulates FSH-like activity in gonadotropin-deficient transgenic mice. *Mol. Endocrinol.* 16, 2582–2591.
310. Kerr, J. B., and Sharpe, R. M. (1985). Follicle-stimulating hormone induction of Leydig cell maturation. *Endocrinology* 116, 2592–2604.
311. Teerds, K. J., Closset, J., Rommerts, F. F., de Rooij, D. G., Stocco, D. M., Colenbrander, B., Wensing, C. J., and Hennen, G. (1989). Effects of pure FSH and LH preparations on the number and function of Leydig cells in immature hypophysectomized rats. *J. Endocrinol.* 120, 97–106.
312. Kerr, J. B., and Sharpe, R. M. (1985). Stimulatory effect of follicle-stimulating hormone on rat Leydig cells. A morphometric and ultrastructural study. *Cell. Tissue Res.* 239, 405–415.
313. Johnson, L., and Thompson, D. L., Jr. (1983). Age-related and seasonal variation in the Sertoli cell population, daily sperm production and serum concentrations of follicle-stimulating hormone, luteinizing hormone and testosterone in stallions. *Biol. Reprod.* 29, 777–789.
314. Johnson, L., and Nguyen, H. B. (1986). Annual cycle of the Sertoli cell population in adult stallions. *J. Reprod. Fertil.* 76, 311–316.
315. Johnson, L., Varner, D. D., Tatum, M. E., and Scrutchfield, W. L. (1991). Season but not age affects Sertoli cell number in adult stallions. *Biol. Reprod.* 45, 404–410.
316. Meachem, S. J., Stanton, P. G., and Schlatt, S. (2005). Follicle-stimulating hormone regulates both Sertoli cell and spermatogonial populations in the adult photoinhibited Djungarian hamster testis. *Biol. Reprod.* 72, 1187–1193.
317. Chandolia, R. K., Weinbauer, G. F., Fingscheidt, U., Bartlett, J. M., and Nieschlag, E. (1991). Effects of flutamide on testicular involution induced by an antagonist of gonadotrophin-releasing hormone and on stimulation of spermatogenesis by follicle-stimulating hormone in rats. *J. Reprod. Fertil.* 93, 313–323.
318. Meachem, S. J., McLachlan, R. I., Stanton, P. G., Robertson, D. M., and Wreford, N. G. (1999). FSH immunoneutralization acutely impairs spermatogonial development in normal adult rats. *J. Androl.* 20, 756–762; discussion 755.
319. Zhengwei, Y., Wreford, N. G., Schlatt, S., Weinbauer, G. F., Nieschlag, E., and McLachlan, R. I. (1998). Acute and specific impairment of spermatogonial development by GnRH antagonist-induced gonadotrophin withdrawal in the adult macaque (Macaca fascicularis). *J. Reprod. Fertil.* 112, 139–147.
320. Zhengwei, Y., Wreford, N. G., Royce, P., de Kretser, D. M., and McLachlan, R. I. (1998). Stereological evaluation of human spermatogenesis after suppression by testosterone treatment: heterogeneous pattern of spermatogenic impairment. *J. Clin. Endocrinol. Metab.* 83, 1284–1291.

321. Aravindan, G. R., Gopalakrishnan, K., Ravindranath, N., and Moudgal, N. R. (1993). Effect of altering endogenous gonadotrophin concentrations on the kinetics of testicular germ cell turnover in the bonnet monkey (Macaca radiata). *J. Endocrinol.* 137, 485–495.
322. Schlatt, S., Arslan, M., Weinbauer, G. F., Behre, H. M., and Nieschlag, E. (1995). Endocrine control of testicular somatic and premeiotic germ cell development in the immature testis of the primate Macaca mulatta. *Eur. J. Endocrinol.* 133, 235–247.
323. Moudgal, N. R., Sairam, M. R., Krishnamurthy, H. N., Sridhar, S., Krishnamurthy, H., and Khan, H. (1997). Immunization of male bonnet monkeys (M. radiata) with a recombinant FSH receptor preparation affects testicular function and fertility. *Endocrinology* 138, 3065–3068.
324. McLachlan, R. I., O'Donnell, L., Stanton, P. G., Balourdos, G., Frydenberg, M., de Kretser, D. M., and Robertson, D. M. (2002). Effects of testosterone plus medroxyprogesterone acetate on semen quality, reproductive hormones, and germ cell populations in normal young men. *J. Clin. Endocrinol. Metab.* 87, 546–556.
325. Bartlett, J. M., Weinbauer, G. F., and Nieschlag, E. (1989). Differential effects of FSH and testosterone on the maintenance of spermatogenesis in the adult hypophysectomized rat. *J. Endocrinol.* 121, 49–58.
326. Saito, K., O'Donnell, L., McLachlan, R. I., and Robertson, D. M. (2000). Spermiation failure is a major contributor to early spermatogenic suppression caused by hormone withdrawal in adult rats. *Endocrinology* 141, 2779–2785.
327. Davies, R. V., Main, S. J., Laurie, M. S., and Setchell, B. P. (1979). The effects of long-term administration of either a crude inhibin preparation or an antiserum to FSH on serum hormone levels, testicular function and fertility of adult male rats. *J. Reprod. Fertil. Suppl.* 183–191.
328. McLachlan, R. I., Wreford, N. G., O'Donnell, L., de Kretser, D. M., and Robertson, D. M. (1996). The endocrine regulation of spermatogenesis: independent roles for testosterone and FSH. *J. Endocrinol.* 148, 1–9.
329. Huang, H. F., Pogach, L. M., Nathan, E., Giglio, W., and Seebode, J. J. (1991). Synergistic effects of follicle-stimulating hormone and testosterone on the maintenance of spermiogenesis in hypophysectomized rats: relationship with the androgen-binding protein status. *Endocrinology* 128, 3152–3161.
330. Spiteri-Grech, J., Weinbauer, G. F., Bolze, P., Chandolia, R. K., Bartlett, J. M., and Nieschlag, E. (1993). Effects of FSH and testosterone on intratesticular insulin-like growth factor-I and specific germ cell populations in rats treated with gonadotrophin-releasing hormone antagonist. *J. Endocrinol.* 137, 81–89.
331. Russell, L. D., Corbin, T. J., Borg, K. E., De Franca, L. R., Grasso, P., and Bartke, A. (1993). Recombinant human follicle-stimulating hormone is capable of exerting a biological effect in the adult hypophysectomized rat by reducing the numbers of degenerating germ cells. *Endocrinology* 133, 2062–2070.
332. Hansson, V., Weddington, S. C., Naess, O., Attramadal, A., French, F. S., Kotite, N., and Nayfeh, S. N. (1975). Testicular androgen binding protein (ABP)—a parameter of Sertoli cell secretory function. *Curr. Top. Mol. Endocrinol.* 2, 323–336.
333. Skinner, M. K., and Griswold, M. D. (1982). Secretion of testicular transferrin by cultured Sertoli cells is regulated by hormones and retinoids. *Biol. Reprod.* 27, 211–221.
334. Chapin, R. E., Wine, R. N., Harris, M. W., Borchers, C. H., and Haseman, J. K. (2001). Structure and control of a cell-cell adhesion complex associated with spermiation in rat seminiferous epithelium. *J. Androl.* 22, 1030–1052.
335. Billig, H., Furuta, I., Rivier, C., Tapanainen, J., Parvinen, M., and Hsueh, A. J. (1995). Apoptosis in testis germ cells: developmental changes in gonadotropin dependence and localization to selective tubule stages. *Endocrinology* 136, 5–12.
336. Kerr, J. B., Savage, G. N., Millar, M., and Sharpe, R. M. (1993). Response of the seminiferous epithelium of the rat testis to withdrawal of androgen: evidence for direct effect upon intercellular spaces associated with Sertoli cell junctional complexes. *Cell. Tissue Res.* 274, 153–161.
337. West, A. P., Sharpe, R. M., and Saunders, P. T. (1994). Differential regulation of cyclic adenosine 3′,5′-monophosphate (cAMP) response element-binding protein and cAMP response element modulator messenger ribonucleic acid transcripts by follicle-stimulating hormone and androgen in the adult rat testis. *Biol. Reprod.* 50, 869–881.
338. Francavilla, S., Cordeschi, G., Properzi, G., Di Cicco, L., Jannini, E. A., Palmero, S., Fugassa, E., Loras, B., and D'Armiento, M. (1991). Effect of thyroid hormone on the pre- and post-natal development of the rat testis. *J. Endocrinol.* 129, 35–42.
339. Simorangkir, D. R., Wreford, N. G., and De Kretser, D. M. (1997). Impaired germ cell development in the testes of immature rats with neonatal hypothyroidism. *J. Androl.* 18, 186–193.
340. Cooke, P. S., Hess, R. A., Porcelli, J., and Meisami, E. (1991). Increased sperm production in adult rats after transient neonatal hypothyroidism. *Endocrinology* 129, 244–248.
341. Cooke, P. S., and Meisami, E. (1991). Early hypothyroidism in rats causes increased adult testis and reproductive organ size but does not change testosterone levels. *Endocrinology* 129, 237–243.
342. Simorangkir, D. R., de Kretser, D. M., and Wreford, N. G. (1995). Increased numbers of Sertoli and germ cells in adult rat testes induced by synergistic action of transient neonatal hypothyroidism and neonatal hemicastration. *J. Reprod. Fertil.* 104, 207–213.
343. Van Haaster, L. H., De Jong, F. H., Docter, R., and De Rooij, D. G. (1992). The effect of hypothyroidism on Sertoli cell proliferation and differentiation and hormone levels during testicular development in the rat. *Endocrinology* 131, 1574–1576.
344. van Haaster, L. H., de Jong, F. H., Docter, R., and de Rooij, D. G. (1993). High neonatal triiodothyronine levels reduce the period of Sertoli cell proliferation and accelerate tubular lumen formation in the rat testis, and increase serum inhibin levels. *Endocrinology* 133, 755–760.
345. Holsberger, D. R., Jirawatnotai, S., Kiyokawa, H., and Cooke, P. S. (2003). Thyroid hormone regulates the cell cycle inhibitor p27Kip1 in postnatal murine Sertoli cells. *Endocrinology* 144, 3732–3738.
346. Sharpe, R. M. (1998). The roles of oestrogen in the male. *Trends Endocrinol. Metab.* 9, 371–377.
347. Sharpe, R. M., and Skakkebaek, N. E. (1993). Are oestrogens involved in falling sperm counts and disorders of the male reproductive tract? *Lancet* 341, 1392–1395.
348. Sharpe, R. M. (1993). Declining sperm counts in men—is there an endocrine cause? *J. Endocrinol.* 136, 357–360.
349. Saunders, P. T., Majdic, G., Parte, P., Millar, M. R., Fisher, J. S., Turner, K. J., and Sharpe, R. M. (1997). Fetal and perinatal influence of xenoestrogens on testis gene expression. *Adv. Exp. Med. Biol.* 424, 99–110.
350. McKinnell, C., Atanassova, N., Williams, K., Fisher, J. S., Walker, M., Turner, K. J., Saunders, P. T. K., and Sharpe, R. M. (2001). Suppression of androgen action and the induction of gross abnormalities of the reproductive tract in male rats treated neonatally with diethylstilbestrol. *J. Androl.* 22, 323–338.

351. Cook, J. C., Johnson, L., O'Connor, J. C., Biegel, L. B., Krams, C. H., Frame, S. R., and Hurtt, M. E. (1998). Effects of dietary 17 beta-estradiol exposure on serum hormone concentrations and testicular parameters in male Crl:CD BR rats. *Toxicol. Sci.* 44, 155–168.
352. Atanassova, N., McKinnell, C., Walker, M., Turner, K. J., Fisher, J. S., Morley, M., Millar, M. R., Groome, N. P., and Sharpe, R. M. (1999). Permanent effects of neonatal estrogen exposure in rats on reproductive hormone levels, Sertoli cell number, and the efficiency of spermatogenesis in adulthood. *Endocrinology* 140, 5364–5373.
353. Atanassova, N., McKinnell, C., Turner, K. J., Walker, M., Fisher, J. S., Morley, M., Millar, M. R., Groome, N. P., and Sharpe, R. M. (2000). Comparative effects of neonatal exposure of male rats to potent and weak (environmental) estrogens on spermatogenesis at puberty and the relationship to adult testis size and fertility: evidence for stimulatory effects of low estrogen levels. *Endocrinology* 141, 3898–3907.
354. Khan, S. A., Ball, R. B., and Hendry, W. J. III. (1998). Effects of neonatal administration of diethylstilbestrol in male hamsters: disruption of reproductive function in adults after apparently normal pubertal development. *Biol. Reprod.* 58, 137–142.
355. Toppari, J., Larsen, J. C., Christiansen, P., Giwercman, A., Grandjean, P., Guillette, L. J., Jr., Jegou, B., Jensen, T. K., Jouannet, P., Keiding, N., Leffers, H., McLachlan, J. A., Meyer, O., Muller, J., Rajpert-De Meyts, E., Scheike, T., Sharpe, R., Sumpter, J., and Skakkebaek, N. E. (1996). Male reproductive health and environmental xenoestrogens. *Environ. Health Perspect.* 104, Suppl 4, 741–803.
356. Auger, J., Kunstmann, J. M., Czyglik, F., and Jouannet, P. (1995). Decline in semen quality among fertile men in Paris during the past 20 years. *N. Engl. J. Med.* 332, 281–285.
357. Sonnenschein, C., and Soto, A. M. (1998). An updated review of environmental estrogen and androgen mimics and antagonists. *J. Steroid Biochem. Mol. Biol.* 65, 143–150.
358. Daston, G. P., Gooch, J. W., Breslin, W. J., Shuey, D. L., Nikiforov, A. I., Fico, T. A., and Gorsuch, J. W. (1997). Environmental estrogens and reproductive health: a discussion of the human and environmental data. *Reprod. Toxicol.* 11, 465–481.
359. Handelsman, D. J. (2000). Myth and methodology in the evaluation of human sperm output. *Int. J. Androl.* 23, Suppl 2, 50–53.
360. Jouannet, P., Wang, C., Eustache, F., Kold-Jensen, T., and Auger, J. (2001). Semen quality and male reproductive health: the controversy about human sperm concentration decline. *Apmis.* 109, 333–344.
361. Robertson, K. M., O'Donnell, L., Jones, M. E., Meachem, S. J., Boon, W. C., Fisher, C. R., Graves, K. H., McLachlan, R. I., and Simpson, E. R. (1999). Impairment of spermatogenesis in mice lacking a functional aromatase (cyp 19) gene. *Proc. Natl. Acad. Sci. U S A* 96, 7986–7991.
362. Robertson, K. M., O'Donnell, L., Simpson, E. R., and Jones, M. E. (2002). The phenotype of the aromatase knockout mouse reveals dietary phytoestrogens impact significantly on testis function. *Endocrinology* 143, 2913–2921.
363. Honda, S., Harada, N., Ito, S., Takagi, Y., and Maeda, S. (1998). Disruption of sexual behavior in male aromatase-deficient mice lacking exons 1 and 2 of the cyp19 gene. *Biochem. Biophys. Res. Commun.* 252, 445–449.
364. Robertson, K. M., Simpson, E. R., Lacham-Kaplan, O., and Jones, M. E. (2001). Characterization of the fertility of male aromatase knockout mice. *J. Androl.* 22, 825–830.
365. Scordalakes, E. M., Imwalle, D. B., and Rissman, E. F. (2002). Oestrogen's masculine side: mediation of mating in male mice. *Reproduction* 124, 331–338.
366. Murata, Y., Robertson, K. M., Jones, M. E., and Simpson, E. R. (2002). Effect of estrogen deficiency in the male: the ArKO mouse model. *Mol. Cell. Endocrinol.* 193, 7–12.
367. Simpson, E. R., Mahendroo, M. S., Means, G. D., Kilgore, M. W., Hinshelwood, M. M., Graham-Lorence, S., Amarneh, B., Ito, Y., Fisher, C. R., Michael, M. D., et al. (1994). Aromatase cytochrome P450, the enzyme responsible for estrogen biosynthesis. *Endocr. Rev.* 15, 342–355.
368. Simpson, E. R., Zhao, Y., Agarwal, V. R., Michael, M. D., Bulun, S. E., Hinshelwood, M. M., Graham-Lorence, S., Sun, T., Fisher, C. R., Qin, K., and Mendelson, C. R. (1997). Aromatase expression in health and disease. *Recent Prog. Horm. Res.* 52, 185–213.
369. Bulun, S. E., Rosenthal, I. M., Brodie, A. M., Inkster, S. E., Zeller, W. P., DiGeorge, A. M., Frasier, S. D., Kilgore, M. W., and Simpson, E. R. (1994). Use of tissue-specific promoters in the regulation of aromatase cytochrome P450 gene expression in human testicular and ovarian sex cord tumors, as well as in normal fetal and adult gonads. *J. Clin. Endocrinol. Metab.* 78, 1616–1621.
370. Carlone, D. L., and Richards, J. S. (1997). Functional interactions, phosphorylation, and levels of 3′,5′-cyclic adenosine monophosphate-regulatory element binding protein and steroidogenic factor-1 mediate hormone-regulated and constitutive expression of aromatase in gonadal cells. *Mol. Endocrinol.* 11, 292–304.
371. Dorrington, J. H., and Khan, S. A. (1993). Steroid production, metabolism and release by Sertoli cells. In *The Sertoli Cell* (L. D. Russell and M. D. Griswold, Eds.), pp. 538–549. Cache River Press, Clearwater, FL.
372. Carreau, S., Genissel, C., Bilinska, B., and Levallet, J. (1999). Sources of oestrogen in the testis and reproductive tract of the male. *Int. J. Androl.* 22, 211–223.
373. Free, M. J., and Jaffe, R. A. (1979). Collection of rete testis fluid from rats without previous efferent duct ligation. *Biol. Reprod.* 20, 269–278.
374. Weniger, J. P., Chouraqui, J., and Zeis, A. (1993). Production of estradiol by the fetal rat testis. *Reprod. Nutr. Dev.* 33, 121–127.
375. Papadopoulos, V., Carreau, S., Szerman-Joly, E., Drosdowsky, M. A., Dehennin, L., and Scholler, R. (1986). Rat testis 17 beta-estradiol: identification by gas chromatography-mass spectrometry and age related cellular distribution. *J. Steroid Biochem.* 24, 1211–1216.
376. Genissel, C., Levallet, J., and Carreau, S. (2001). Regulation of cytochrome P450 aromatase gene expression in adult rat Leydig cells: comparison with estradiol production. *J. Endocrinol.* 168, 95–105.
377. Nitta, H., Bunick, D., Hess, R. A., Janulis, L., Newton, S. C., Millette, C. F., Osawa, Y., Shizuta, Y., Toda, K., and Bahr, J. M. (1993). Germ cells of the mouse testis express P450 aromatase. *Endocrinology* 132, 1396–13401.
378. Bourguiba, S., Chater, S., Delalande, C., Benahmed, M., and Carreau, S. (2003). Regulation of aromatase gene expression in purified germ cells of adult male rats: effects of transforming growth factor beta, tumor necrosis factor alpha, and cyclic adenosine 3′,5′-monosphosphate. *Biol. Reprod.* 69, 592–601.
379. Abney, T. O. (1999). The potential roles of estrogens in regulating Leydig cell development and function: a review. *Steroids* 64, 610–617.
380. Lassurguere, J., Livera, G., Habert, R., and Jegou, B. (2003). Time- and dose-related effects of estradiol and diethylstilbestrol on the morphology and function of the fetal rat testis in culture. *Toxicol. Sci.* 73, 160–169.
381. Abney, T. O., and Myers, R. B. (1991). 17 beta-estradiol inhibition of Leydig cell regeneration in the ethane dimethylsulfonate-treated mature rat. *J. Androl.* 12, 295–304.

382. Wistuba, J., Brinkworth, M. H., Schlatt, S., Chahoud, I., and Nieschlag, E. (2003). Intrauterine bisphenol A exposure leads to stimulatory effects on Sertoli cell number in rats. *Environ. Res.* 91, 95–103.
383. Delbes, G., Levacher, C., Pairault, C., Racine, C., Duquenne, C., Krust, A., and Habert, R. (2004). Estrogen receptor beta-mediated inhibition of male germ cell line development in mice by endogenous estrogens during perinatal life. *Endocrinology* 145, 3395–3403.
384. Britt, K. L., Kerr, J., O'Donnell, L., Jones, M. E., Drummond, A. E., Davis, S. R., Simpson, E. R., and Findlay, J. K. (2002). Estrogen regulates development of the somatic cell phenotype in the eutherian ovary. *FASEB J.* 16, 1389–1397.
385. MacCalman, C. D., and Blaschuk, O. W. (1994). Gonadal steroids regulate N-cadherin mRNA levels in the mouse testis. *Endocr. J.* 2, 157–163.
386. MacCalman, C. D., Getsios, S., Farookhi, R., and Blaschuk, O. W. (1997). Estrogens potentiate the stimulatory effects of follicle-stimulating hormone on N-cadherin messenger ribonucleic acid levels in cultured mouse Sertoli cells. *Endocrinology* 138, 41–48.
387. Li, H., Papadopoulos, V., Vidic, B., Dym, M., and Culty, M. (1997). Regulation of rat testis gonocyte proliferation by platelet-derived growth factor and estradiol: identification of signaling mechanisms involved. *Endocrinology* 138, 1289–1298.
388. Thuillier, R., Wang, Y., and Culty, M. (2003). Prenatal exposure to estrogenic compounds alters the expression pattern of platelet-derived growth factor receptors alpha and beta in neonatal rat testis: identification of gonocytes as targets of estrogen exposure. *Biol. Reprod.* 68, 867–880.
389. Hayes, F. J., Seminara, S. B., Decruz, S., Boepple, P. A., and Crowley, W. F., Jr. (2000). Aromatase inhibition in the human male reveals a hypothalamic site of estrogen feedback. *J. Clin. Endocrinol. Metab.* 85, 3027–3035.
390. Bagatell, C. J., Dahl, K. D., and Bremner, W. J. (1994). The direct pituitary effect of testosterone to inhibit gonadotropin secretion in men is partially mediated by aromatization to estradiol. *J. Androl.* 15, 15–21.
391. De Jong, F. H., Uilenbroek, T. J., and Van der Molen, H. J. (1975). Oestradiol-17beta, testosterone and gonadotrophins in oestradiol-17beta-treated intact adult male rats. *J. Endocrinol.* 65, 281–282.
392. Hayes, F. J., DeCruz, S., Seminara, S. B., Boepple, P. A., and Crowley, W. F. Jr. (2001). Differential regulation of gonadotropin secretion by testosterone in the human male: absence of a negative feedback effect of testosterone on follicle-stimulating hormone secretion. *J. Clin. Endocrinol. Metab.* 86, 53–58.
393. Mauras, N., O'Brien, K. O., Klein, K. O., and Hayes, V. (2000). Estrogen suppression in males: metabolic effects. *J. Clin. Endocrinol. Metab.* 85, 2370–2377.
394. Shetty, G., Krishnamurthy, H., Krishnamurthy, H. N., Bhatnagar, A. S., and Moudgal, N. R. (1998). Effect of long-term treatment with aromatase inhibitor on testicular function of adult male bonnet monkeys (M. radiata). *Steroids* 63, 414–420.
395. Turner, K. J., Morley, M., Atanassova, N., Swanston, I. D., and Sharpe, R. M. (2000). Effect of chronic administration of an aromatase inhibitor to adult male rats on pituitary and testicular function and fertility. *J. Endocrinol.* 164, 225–238.
396. Akingbemi, B. T., Ge, R., Rosenfeld, C. S., Newton, L. G., Hardy, D. O., Catterall, J. F., Lubahn, D. B., Korach, K. S., and Hardy, M. P. (2003). Estrogen receptor-alpha gene deficiency enhances androgen biosynthesis in the mouse Leydig cell. *Endocrinology* 144, 84–93.
397. Oliveira, C. A., Zhou, Q., Carnes, K., Nie, R., Kuehl, D. E., Jackson, G. L., Franca, L. R., Nakai, M., and Hess, R. A. (2002). ER function in the adult male rat: short- and long-term effects of the antiestrogen ICI 182,780 on the testis and efferent ductules, without changes in testosterone. *Endocrinology* 143, 2399–2409.
398. Lindzey, J., William, W. C., Couse, J. F., Stoker, T., Cooper, R., and Korach, K. S. (1998). Effects of castration and chronic steroid treatments on hypothalamic gonadotropin-releasing hormone content and pituitary gonadotropins in male wild-type and estrogen receptor a- knockout mice. *Endocrinology* 139, 4092–4101.
399. Fisher, C. R., Graves, K. H., Parlow, A. F., and Simpson, E. R. (1998). Characterization of mice deficient in aromatase (ArKO) because of targeted disruption of the cyp19 gene. *Proc. Natl. Acad. Sci. U S A* 95, 6965–6970.
400. Brinkmann, A. O., Leemborg, F. G., Roodnat, E. M., De Jong, F. H., and Van der Molen, H. J. (1980). A specific action of estradiol on enzymes involved in testicular steroidogenesis. *Biol. Reprod.* 23, 801–809.
401. Shetty, G., Krishnamurthy, H., Krishnamurthy, H. N., Bhatnagar, S., and Moudgal, R. N. (1997). Effect of estrogen deprivation on the reproductive physiology of male and female primates. *J. Steroid Biochem. Mol. Biol.* 61, 157–166.
402. Ebling, F. J., Brooks, A. N., Cronin, A. S., Ford, H., and Kerr, J. B. (2000). Estrogenic induction of spermatogenesis in the hypogonadal mouse. *Endocrinology* 141, 2861–2869.
403. Cho, H. W., Nie, R., Carnes, K., Zhou, Q., Sharief, N. A., and Hess, R. A. (2003). The antiestrogen ICI 182,780 induces early effects on the adult male mouse reproductive tract and long-term decreased fertility without testicular atrophy. *Reprod. Biol. Endocrinol.* 1, 57.
404. Oliveira, C. A., Carnes, K., Franca, L. R., and Hess, R. A. (2001). Infertility and testicular atrophy in the antiestrogen-treated adult male rat. *Biol. Reprod.* 65, 913–920.
405. Krege, J. H., Hodgin, J. B., Couse, J. F., Enmark, E., Warner, M., Mahler, J. F., Sar, M., Korach, K. S., Gustafsson, J. A., and Smithies, O. (1998). Generation and reproductive phenotypes of mice lacking estrogen receptor beta. *Proc. Natl. Acad. Sci. U S A* 95, 15677–15682.
406. Kuiper, G. G., Lemmen, J. G., Carlsson, B., Corton, J. C., Safe, S. H., van der Saag, P. T., van der Burg, B., and Gustafsson, J. A. (1998). Interaction of estrogenic chemicals and phytoestrogens with estrogen receptor beta. *Endocrinology* 139, 4252–4263.
407. Hess, R. A., Bunick, D., Lubahn, D. B., Zhou, Q., and Bouma, J. (2000). Morphologic changes in efferent ductules and epididymis in estrogen receptor-alpha knockout mice. *J. Androl.* 21, 107–121.
408. Lee, K. H., Hess, R. A., Bahr, J., Lubahn, D. B., Taylor, J., and Bunick, D. (2000). Estrogen receptor a has a functional role in the mouse rete testis and efferent ductules. *Biol. Reprod.* 63, 1873–1880.
409. Hess, R. A., Zhou, Q., Nie, R., Oliveira, C., Cho, H., Nakaia, M., and Carnes, K. (2001). Estrogens and epididymal function. *Reprod. Fertil. Dev.* 13, 273–283.
410. Zhou, Q., Clarke, L., Nie, R., Carnes, K., Lai, L. W., Lien, Y. H., Verkman, A., Lubahn, D., Fisher, J. S., Katzenellenbogen, B. S., and Hess, R. A. (2001). Estrogen action and male fertility: roles of the sodium/hydrogen exchanger-3 and fluid reabsorption in reproductive tract function. *Proc. Natl. Acad. Sci. U S A* 98, 14132–14137.
411. Mahato, D., Goulding, E. H., Korach, K. S., and Eddy, E. M. (2000). Spermatogenic cells do not require estrogen receptor-alpha for development or function. *Endocrinology* 141, 1273–1276.

412. Atanassova, N., McKinnell, C., Williams, K., Turner, K. J., Fisher, J. S., Saunders, P. T. K., Millar, M. R., and Sharpe, R. M. (2001). Age-, cell- and region-specific immunoexpression of estrogen receptor alpha (but not estrogen receptor beta) during postnatal development of the epididymis and vas deferens of the rat and disruption of this pattern by neonatal treatment with diethylstilbestrol. *Endocrinology* 142, 874–886.

413. Aceitero, J., Llanero, M., Parrado, R., Pena, E., and Lopez-Baltran, A. (1998). Neonatal exposure of male rats to estradiol benzoate causes rete testis dilation and backflow impairment of spermatogenesis. *Anat. Rec.* 252, 17–33.

414. Fisher, J. S., Turner, K. J., Brown, D., and Sharpe, R. M. (1999). Effect of neonatal exposure to estrogenic compounds on development of the excurrent ducts of the rat testis through puberty to adulthood. *Environ. Health Perspect.* 107, 397–405.

415. Fisher, J. S., Turner, K. J., Fraser, H. M., Saunders, P. T., Brown, D., and Sharpe, R. M. (1998). Immunoexpression of aquaporin-1 in the efferent ducts of the rat and marmoset monkey during development, its modulation by estrogens, and its possible role in fluid resorption. *Endocrinology* 139, 3935–3945.

416. Hansen, L. A., Clulow, J., and Jones, R. C. (1997). Perturbation of fluid reabsorption in the efferent ducts of the rat by testosterone propionate, 17beta-oestradiol 3-benzoate, flutamide and tamoxifen. *Int. J. Androl.* 20, 264–273.

417. de Kretser, D. M., Taft, H. P., Brown, J. B., Evans, J. H., and Hudson, B. (1968). Endocrine and histological studies on oligospermic men treated with human pituitary and chorionic gonadotrophin. *J. Endocrinol.* 40, 107–115.

418. Burris, A. S., Rodbard, H. W., Winters, S. J., and Sherins, R. J. (1988). Gonadotropin therapy in men with isolated hypogonadotropic hypogonadism: the response to human chorionic gonadotropin is predicted by initial testicular size. *J. Clin. Endocrinol. Metab.* 66, 1144–1151.

419. Hayes, F., and Pitteloud, N. (2004). Hypogonadotropic hypogonadism and gonadotropin therapy. In *Endocrinology of Male Reproduction* (R. I. McLachlan, Ed.). Available at: www.endotext.org (MDText.com, Inc.).

420. Huhtaniemi, I., Nikula, H., and Rannikko, S. (1987). Pituitary-testicular function of prostatic cancer patients during treatment with a gonadotropin-releasing hormone agonist analog I. Circulating hormone levels. *J. Androl.* 8, 355–362.

421. Morse, H. C., Horike, N., Rowley, M. J., and Heller, C. G. (1973). Testosterone concentrations in testes of normal men: effects of testosterone propionate administration. *J. Clin. Endocrinol. Metab.* 37, 882–886.

422. Hammond, G. L., Ahonen, V., and Vikho, R. (1978). The radioimmunoassay of testosterone, 5a-dihydrotestosterone and their precursors in the human testis. *Int. J. Androl.* Suppl, 391–399.

423. Matsumoto, A. M., Karpas, A. E., and Bremner, W. J. (1986). Chronic human chorionic gonadotropin administration in normal men: evidence that follicle-stimulating hormone is necessary for the maintenance of quantitatively normal spermatogenesis in man. *J. Clin. Endocrinol. Metab.* 62, 1184–1192.

424. Matsumoto, A. M., and Bremner, W. J. (1985). Stimulation of sperm production by human chorionic gonadotropin after prolonged gonadotropin suppression in normal men. *J. Androl.* 6, 137–143.

425. Matsumoto, A. M., Karpas, A. E., Paulsen, C. A., and Bremner, W. J. (1983). Reinitiation of sperm production in gonadotropin-suppressed normal men by administration of follicle-stimulating hormone. *J. Clin. Invest.* 72, 1005–1015.

426. Huhtaniemi, I. (2003). Mutations affecting gonadotropin secretion and action. *Horm. Res.* 60 Suppl 3, 21–30.

427. Tapanainen, J. S., Aittomaki, K., Min, J., Vaskivuo, T., and Huhtaniemi, I. T. (1997). Men homozygous for an inactivating mutation of the follicle-stimulating hormone (FSH) receptor gene present variable suppression of spermatogenesis and fertility. *Nat. Genet.* 15, 205–206.

428. Phillip, M., Arbelle, J. E., Segev, Y., and Parvari, R. (1998). Male hypogonadism due to a mutation in the gene for the beta-subunit of follicle-stimulating hormone. *N. Engl. J. Med.* 338, 1729–1732.

429. Wreford, N. G. (1995). Theory and practice of stereological techniques applied to the estimation of cell number and nuclear volume in the testis. *Microsc. Res. Tech.* 32, 423–436.

430. van Alphen, M. M., van de Kant, H. J., and de Rooij, D. G. (1988). Follicle-stimulating hormone stimulates spermatogenesis in the adult monkey. *Endocrinology* 123, 1449–1455.

431. Ramaswamy, S., Plant, T. M., and Marshall, G. R. (2000). Pulsatile stimulation with recombinant single chain human luteinizing hormone elicits precocious Sertoli cell proliferation in the juvenile male rhesus monkey (Macaca mulatta). *Biol. Reprod.* 63, 82–88.

432. Zhengwei, Y., Wreford, N. G., Bremner, W. J., Matsumoto, A. M., Anawalt, B. A., and McLachlan, R. I. (1998). Immature spermatids are not prevalent in semen from men who are receiving androgen-based contraceptive regimens. *Fertil. Steril.* 69, 89–95.

433. Tesarik, J., Mendoza, C., and Greco, E. (2000). The effect of FSH on male germ cell survival and differentiation in vitro is mimicked by pentoxifylline but not insulin. *Mol. Hum. Reprod.* 6, 877–881.

434. Erkkila, K., Henriksen, K., Hirvonen, V., Rannikko, S., Salo, J., Parvinen, M., and Dunkel, L. (1997). Testosterone regulates apoptosis in adult human seminiferous tubules in vitro. *J. Clin. Endocrinol. Metab.* 82, 2314–2321.

435. Johnson, L., Grumbles, J. S., Bagheri, A., and Petty, C. S. (1990). Increased germ cell degeneration during postprophase of meiosis is related to increased serum follicle-stimulating hormone concentrations and reduced daily sperm production in aged men. *Biol. Reprod.* 42, 281–287.

436. Sinha Hikim, A. P., Wang, C., Lue, Y., Johnson, L., Wang, X. H., and Swerdloff, R. S. (1998). Spontaneous germ cell apoptosis in humans: evidence for ethnic differences in the susceptibility of germ cells to programmed cell death. *J. Clin. Endocrinol. Metab.* 83, 152–156.

437. Buchter, D., Behre, H. M., Kliesch, S., and Nieschlag, E. (1998). Pulsatile GnRH or human chorionic gonadotropin/human menopausal gonadotropin as effective treatment for men with hypogonadotropic hypogonadism: a review of 42 cases. *Eur. J. Endocrinol.* 139, 298–303.

438. Boyar, R. M., Wu, R. H., Kapen, S., Hellman, L., Weitzman, E. D., and Finkelstein, J. W. (1976). Clinical and laboratory heterogeneity in idiopathic hypogonadotropic hypogonadism. *J. Clin. Endocrinol. Metab.* 43, 1268–1275.

439. Spratt, D. I., Carr, D. B., Merriam, G. R., Scully, R. E., Rao, P. N., and Crowley, W. F., Jr. (1987). The spectrum of abnormal patterns of gonadotropin-releasing hormone secretion in men with idiopathic hypogonadotropic hypogonadism: clinical and laboratory correlations. *J. Clin. Endocrinol. Metab.* 64, 283–291.

440. Gazvani, M. R., Buckett, W., Luckas, M. J., Aird, I. A., Hipkin, L. J., and Lewis-Jones, D. I. (1997). Conservative management of azoospermia following steroid abuse. *Hum. Reprod.* 12, 1706–1708.

441. Burris, A. S., Clark, R. V., Vantman, D. J., and Sherins, R. J. (1988). A low sperm concentration does not preclude fertility

in men with isolated hypogonadotropic hypogonadism after gonadotropin therapy. *Fertil. Steril.* 50, 343–347.
442. Finkel, D. M., Phillips, J. L., and Snyder, P. J. (1985). Stimulation of spermatogenesis by gonadotropins in men with hypogonadotropic hypogonadism. *N. Engl. J. Med.* 313, 651–655.
443. Johnsen, S. G. (1978). Maintenance of spermatogenesis induced by HMG treatment by means of continuous HCG treatment in hypogonadotrophic men. *Acta Endocrinol. (Copenh).* 89, 763–769.
444. Whitcomb, R. W., and Crowley, W. F., Jr. (1990). Clinical review 4: Diagnosis and treatment of isolated gonadotropin-releasing hormone deficiency in men. *J. Clin. Endocrinol. Metab.* 70, 3–7.
445. Schopohl, J., Mehltretter, G., von Zumbusch, R., Eversmann, T., and von Werder, K. (1991). Comparison of gonadotropin-releasing hormone and gonadotropin therapy in male patients with idiopathic hypothalamic hypogonadism. *Fertil. Steril.* 56, 1143–1150.
446. Liu, P. Y., Gebski, V. J., Turner, L., Conway, A. J., Wishart, S. M., and Handelsman, D. J. (2002). Predicting pregnancy and spermatogenesis by survival analysis during gonadotrophin treatment of gonadotrophin-deficient infertile men. *Hum. Reprod.* 17, 625–633.
447. World Health Organization Task Force on methods for the regulation of male fertility. (1990). Contraceptive efficacy of testosterone-induced azoospermia in normal men. *Lancet* 336, 955–959.
448. World Health Organization Task Force on methods for the regulation of male fertility. (1996). Contraceptive efficacy of testosterone-induced azoospermia and oligozoospermia in normal men. *Fertil. Steril.* 65, 821–829.
449. Nieschlag, E. (2002). Sixth Summit Meeting Consensus: Recommendations for Regulatory Approval for Hormonal Male Contraception. *Int. J. Androl.* 25, 375.
450. Handelsman, D. J., Conway, A. J., Howe, C. J., Turner, L., and Mackey, M. A. (1996). Establishing the minimum effective dose and additive effects of depot progestin in suppression of human spermatogenesis by a testosterone depot. *J. Clin. Endocrinol. Metab.* 81, 4113–4121.
451. Swerdloff, R. S., Bagatell, C. J., Wang, C., Anawalt, B. D., Berman, N., Steiner, B., and Bremner, W. J. (1998). Suppression of spermatogenesis in man induced by Nal-Glu gonadotropin releasing hormone antagonist and testosterone enanthate (TE) in maintained by TE alone. *J. Clin. Endocrinol. Metab.* 83, 3527–3533.
452. Anderson, R. A., and Baird, D. T. (2002). Male contraception. *Endocr. Rev.* 23, 735–762.
453. Kamischke, A., and Nieschlag, E. (2004). Progress towards hormonal male contraception. *Trends Pharmacol. Sci.* 25, 49–57.
454. Amory, J. K., and Bremner, W. J. (2003). Regulation of testicular function in men: implications for male hormonal contraceptive development. *J. Steroid Biochem. Mol. Biol.* 85, 357–361.
455. Handelsman, D. J. (2004). Male contraception. In *Endocrinology of Male Reproduction* (R. I. McLachlan, Ed.). Available at: www.endotext.org (MDText.com, Inc.).
456. Meriggiola, M. C., Bremner, W. J., Paulsen, C. A., Valdiserri, A., Incorvaia, L., Motta, R., Pavani, A., Capelli, M., and Flamigni, C. (1996). A combined regimen of cyproterone acetate and testosterone enanthate as a potentially highly effective male contraceptive. *J. Clin. Endocrinol. Metab.* 81, 3018–3023.
457. Turner, L., Conway, A. J., Jimenez, M., Liu, P. Y., Forbes, E., McLachlan, R. I., and Handelsman, D. J. (2003). Contraceptive efficacy of a depot progestin and androgen combination in men. *J. Clin. Endocrinol. Metab.* 88, 4659–4667.
458. Anderson, R. A., and Wu, F. C. (1996). Comparison between testosterone enanthate-induced azoospermia and oligozoospermia in a male contraceptive study. II. Pharmacokinetics and pharmacodynamics of once weekly administration of testosterone enanthate. *J. Clin. Endocrinol. Metab.* 81, 896–901.
459. Handelsman, D. J., Farley, T. M., Peregoudov, A., and Waites, G. M. (1995). Factors in nonuniform induction of azoospermia by testosterone enanthate in normal men. World Health Organization Task Force on Methods for the Regulation of Male Fertility. *Fertil. Steril.* 63, 125–133.
460. Nieschlag, E., Simoni, M., Gromoll, J., and Weinbauer, G. F. (1999). Role of FSH in the regulation of spermatogenesis: clinical aspects. *Clin. Endocrinol. (Oxf).* 51, 139–146.
461. McLachlan, R. I., McDonald, J., Rushford, D., Robertson, D. M., Garrett, C., and Baker, H. W. (2000). Efficacy and acceptability of testosterone implants, alone or in combination with a 5alpha-reductase inhibitor, for male hormonal contraception. *Contraception* 62, 73–78.
462. Matsumoto, A. M. (1990). Effects of chronic testosterone administration in normal men: safety and efficacy of high dosage testosterone and parallel dose-dependent suppression of luteinizing hormone, follicle-stimulating hormone, and sperm production. *J. Clin. Endocrinol. Metab.* 70, 282–287.
463. McLachlan, R. I., Robertson, D. M., Pruysers, E., Ugoni, A., Matsumoto, A. M., Anawalt, B. D., Bremner, W. J., and Meriggiola, C. (2004). Relationship between serum gonadotropins and spermatogenic suppression in men undergoing steroidal contraceptive treatment. *J. Clin. Endocrinol. Metab.* 89, 142–149.
464. Meriggiola, M. C., and Bremner, W. J. (1997). Progestin-androgen combination regimens for male contraception. *J Androl.* 18, 240–244.
465. Bebb, R. A., Anawalt, B. D., Christensen, R. B., Paulsen, C. A., Bremner, W. J., and Matsumoto, A. M. (1996). Combined administration of levonorgestrel and testosterone induces more rapid and effective suppression of spermatogenesis than testosterone alone: a promising male contraceptive approach. *J. Clin. Endocrinol. Metab.* 81, 757–762.
466. Pangkahila, W. (1991). Reversible azoospermia induced by an androgen-progestagen combination regimen in Indonesian men. *Int. J. Androl.* 44, 248–256.
467. World Health Organization task force on methods for the regulation of male fertility. (1993). Comparison of two androgens plus depot-medroxyprogesterone acetate for suppression to azoospermia in indonesian men. *Fertil. Steril.* 60, 1062–1068.
468. Johnson, L., Barnard, J. J., Rodriguez, L., Smith, E. C., Swerdloff, R. S., Wang, X. H., and Wang, C. (1998). Ethnic differences in testicular structure and spermatogenic potential may predispose testes of Asian men to a heightened sensitivity to steroidal contraceptives. *J. Androl.* 19, 348–357.
469. Eckardstein, S. V., Schmidt, A., Kamischke, A., Simoni, M., Gromoll, J., and Nieschlag, E. (2002). CAG repeat length in the androgen receptor gene and gonadotrophin suppression influence the effectiveness of hormonal male contraception. *Clin. Endocrinol. (Oxf).* 57, 647–655.
470. Yu, B., and Handelsman, D. J. (2001). Pharmacogenetic polymorphisms of the AR and metabolism and susceptibility to hormone- induced azoospermia. *J. Clin. Endocrinol. Metab.* 86, 4406–4411.
471. Lookingbill, D. P., Demers, L. M., Wang, C., Leung, A., Rittmaster, R. S., and Santen, R. J. (1991). Clinical and biochemical parameters of androgen action in normal healthy Caucasian versus Chinese subjects. *J. Clin. Endocrinol. Metab.* 72, 1242–1248.
472. Kinniburgh, D., Anderson, R. A., and Baird, D. T. (2001). Suppression of spermatogenesis with desogestrel and testosterone pellets is not enhanced by addition of finasteride. *J. Androl.* 22, 88–95.

473. Sundaram, K., and Kumar, N. (2000). 7alpha-methyl-19-nortestosterone (MENT): the optimal androgen for male contraception and replacement therapy. *Int. J. Androl.* 23, Suppl 2, 13–15.
474. Tom, L., Bhasin, S., Salameh, W., Steiner, B., Peterson, M., Sokol, R., Rivier, J., Vale, W. W., and Swerdloff, R. S. (1992). Induction of azoospermia in normal men with combined Nal-Glu GnRH antagonist and testosterone enanthate. *J. Clin. Endocrinol. Metab.* 75, 476–483.
475. Bremner, W. J., Bagatell, C. J., and Steiner, R. A. (1991). Gonadotropin-releasing hormone antagonist plus testosterone: a potential male contraceptive. *J. Clin. Endocrinol. Metab.* 73, 465–469.
476. Herbst, K. L., Anawalt, B. D., Amory, J. K., and Bremner, W. J. (2002). Acyline: the first study in humans of a potent, new gonadotropin-releasing hormone antagonist. *J. Clin. Endocrinol. Metab.* 87, 3215–3220.
477. Jean-Faucher, C., Berger, M., Gallon, C., de Turckheim, M., Veyssiere, G., and Jean, C. (1986). Regional differences in the testosterone to dihydrotestosterone ratio in the epididymis and vas deferens of adult mice. *J. Reprod. Fertil.* 76, 537–543.
478. Moger, W. H. (1977). Serum 5alpha-androstane-3alpha, 17beta-diol, androsterone, and testosterone concentrations in the male rat. Influence of age and gonadotropin stimulation. *Endocrinology* 100, 1027–1032.
479. Punjabi, U., Deslypere, J. P., Verdonck, L., and Vermeulen, A. (1983). Androgen and precursor levels in serum and testes of adult rats under basal conditions and after hCG stimulation. *J. Steroid Biochem.* 19, 1481–1490.
480. Chen, H., Chandrashekar, V., and Zirkin, B. R. (1994). Can spermatogenesis be maintained quantitatively in intact adult rats with exogenously administered dihydrotestosterone? *J. Androl.* 15, 132–138.
481. Podesta, E. J., and Rivarola, M. A. (1974). Concentration of androgens in whole testis, seminiferous tubules and interstitial tissue of rats at different stages of development. *Endocrinology* 95, 455–461.
482. Ramaswamy, S., Marshall, G. R., McNeilly, A. S., and Plant, T. M. (2000). Dynamics of the follicle-stimulating hormone (FSH)-inhibin B feedback loop and its role in regulating spermatogenesis in the adult male rhesus monkey (Macaca mulatta) as revealed by unilateral orchidectomy. *Endocrinology* 141, 18–27.
483. Narula, A., Yi-Qun, G., O'Donnell, L., Stanton, P., Robertson, D., McLachlan, R., and Bremner, W. (2002). Variability in sperm suppression during testosterone administration to adult monkeys is related to FSH suppression and not to intra-testicular androgens. *J. Clin. Endocrinol. Metab.* 87, 3399–3406.
484. Weinbauer, G. F., Schlatt, S., Walter, V., and Nieschlag, E. (2001). Testosterone-induced inhibition of spermatogenesis is more closely related to suppression of FSH than to testicular androgen levels in the cynomolgus monkey model (Macaca fascicularis). *J. Endocrinol.* 168, 25–38.
485. Toda, K., Okada, T., Takeda, K., Akira, S., Saibara, T., Shiraishi, M., Onishi, S., and Shizuta, Y. (2001). Oestrogen at the neonatal stage is critical for the reproductive ability of male mice as revealed by supplementation with 17beta-oestradiol to aromatase gene (Cyp19) knockout mice. *J. Endocrinol.* 168, 455–463.

Knobil and Neill's Physiology of Reproduction, Third Edition
edited by Jimmy D. Neill,

CHAPTER 22

The Epididymis

Bernard Robaire,[1] Barry T. Hinton,[2] and Marie-Claire Orgebin-Crist[3]

Introduction, 1072
Historical Perspective, 1072
Development of the Epididymis, 1073
Formation of the Mesonephric/Wolffian/Nephric Duct and Tubules, 1073
Postnatal Development, 1074
Structural Organization of the Epididymis, 1076
Anatomy, 1076
Epididymal Cell Types and Specific Markers, 1076
The Blood–Epididymis Barrier, 1080
Functions Taking Place in the Luminal Compartment, 1081
Transport of Spermatozoa, 1081
Maturation of Spermatozoa, 1083
Storage of Spermatozoa, 1087
Protection of Spermatozoa, 1088
Microenvironment for Maturation, Protection, and Storage, 1089
Regulation of Epididymal Functions, 1092
Hormones, 1092
Testicular Factors, 1093
Major Protein Families in the Epididymis and Their Regulation, 1095
General Principles, 1095
Functional Families, 1098
Aging, 1114
Effects of Aging on the Appearance of the Epididymal Epithelium, 1114
Molecular and Functional Effects of Aging in the Epididymis, 1115
Mechanisms Underlying Aging of the Epididymis, 1115
Changes in Spermatozoa during Aging, 1116
The Epididymis as Target for Xenobiotics, 1117
Chemicals for Which the Epididymis is an Explicit Target, 1117
Chemicals Targeted Principally at Other Organs that Also Act on the Epididymis, 1119
Perspective and Future Directions, 1120
Acknowledgments, 1120
References, 1120

ABSTRACT

Throughout embryonic and early postnatal development, the mammalian epididymis changes from a straight tube to a highly coiled, complex duct that links the efferent ducts to the vas deferens. Overwhelming evidence points to the importance of this tissue in transforming spermatozoa leaving the testis as immotile cells, unable to fertilize oocytes, into fully mature cells that have the ability both to swim and to recognize and fertilize eggs. Under normal conditions, the acquisition of these functions is essentially completed by the time sperm enter the proximal cauda epididymidis. In addition to sperm maturation, the epididymis also plays an important role in sperm transport, concentration, protection, and storage. A highly specialized and region-specific microenvironment is created along the epididymal lumen by active secretion and absorption of water, ions, organic solutes, and proteins as well as by the blood–epididymis barrier. The primary factor regulating epididymal function is androgens, but there is mounting evidence that estrogens, retinoids, and other factors coming directly into the epididymis from the testis through the efferent ducts, such as growth factors, also play specific regulatory roles. Several epithelial cell types,

[1]Department of Pharmacology and Therapeutics, McGill University, Montreal, Quebec, Canada;
[2]Department of Cell Biology, University of Virginia Health System, School of Medicine, Charlottesville, Virginia;
[3]Department of Obstetrics and Gynecology, Vanderbilt University School of Medicine, Center for Reproductive Biology Research, Nashville, Tennessee.

each showing selective expression of genes and proteins, are differentially distributed along the duct; each cell type shows highly regionalized expression of a wide array of markers. Both epididymal epithelial cells and spermatozoa in the lumen are targets for xenobiotics; such exposures can result in undesirable toxic effects or may provide the basis for the development of novel male contraceptive agents. During aging, both the epididymal epithelium and the germ cells in the lumen undergo a series of dramatic changes. The explosion of knowledge we are witnessing regarding all aspects of epididymal structure and function is likely to lay the basis for a new fundamental understanding of epididymal cell biology and novel therapeutic approaches targeted at this organ.

INTRODUCTION

Since the publication of the comprehensive review on the male excurrent duct system (efferent ducts, epididymis, and vas deferens) in the first edition of *Physiology of Reproduction* in 1988 (1), a remarkable series of events has marked the growth in our knowledge and understanding of this duct system. At that time, the epididymis, a long, complex, convoluted duct connecting the efferent ducts to the vas deferens, was viewed as having moved from being "an abandoned child" of the male reproductive system (2), to a point of maturity, where its basic structures, functions, and regulation were beginning to be understood.

Since 1988, three international conferences dedicated to this tissue have been held, with proceedings ensuing for two of these (3,4), and the first comprehensive, multiauthored volume dedicated to the epididymis has been published (5). The yearly number of research publications on this tissue has increased by more than one order of magnitude (from less than 500 to more than 6,000) since the review appeared in the first edition of *Physiology of Reproduction*. Therefore, in preparing the current review, we have chosen first to narrow the scope by placing most of the emphasis on the epididymis itself, as opposed to the other components of the excurrent duct system; second, to exclude some topics that have been extensively reviewed recently, such as changes in spermatozoa during epididymal transit, innervation and vasculature of the duct system, or pathology of the epididymis; and third, to focus on some of the more exciting recent developments in this field and provide supplemental information in an alternative format. Several of the subjects covered in the first edition are updated, but, because of space limitations, we have chosen not to include most of the plates that depict the histological organization and structure of the various epididymal epithelial cells from the first edition, but rather to place them on a web site (www.medicine.mcgill.ca/PhysiolReprodThirdEd/Epididymis). Along with these plates, this web site contains color versions of plates from the current chapter and a number of tables, including some containing detailed information about specific epididymal proteins and genes as well as their regulation. We have intentionally chosen not to include such tables in the chapter not only because of space limitations but because of the need to update this information regularly. We also felt that the review should focus on our ideas, philosophy, and biased judgment on what we believe is important and where the field should go.

HISTORICAL PERSPECTIVE

As early as the 4th century B.C., the epididymis was described by Aristotle in his *Historia Animalium*, whereas the first recorded description of a dissected epididymis was made by de Graaf in 1668 in his monograph *Tractatus de Virorum Organis Generationi Inservientibus* [reviewed in Orgebin-Crist (6)]. De Graaf noted that "the semen ... was watery and ash-like in the testis and becomes milky and thick in the epididymis" (7). Between 1888 and 1928, a number of scientists described the histological features of the epididymal epithelium, and surmised, from the images of secretion they observed, that epididymal secretions "nourished" the spermatozoa in the epididymal lumen. In 1913, Tournade (8) showed that spermatozoa released from the proximal epididymis were not motile when diluted in saline, but spermatozoa released from the distal epididymis were fully motile. In the bat, spermatozoa were observed to survive for several months in the cauda epididymidis during hibernation, presumably protected by epididymal secretions (9), whereas rabbit spermatozoa survived in the epididymis for 30 to 60 days (10), and bull spermatozoa for 2 months (11), after ligation of the efferent ducts. The implication of these observations was that the epididymis conditioned the development of sperm motility and was important for sperm survival.

The consensus of these early investigations can be summarized by the last sentence of Benoit's classic 1926 (12) monograph: "The role of epididymal secretions is to maintain sperm vitality, to permit the development of sperm motility, and possibly to protect them against noxious agents." In a series of four papers between 1929 and 1931, Young (13–16) showed that during epididymal transit spermatozoa not only acquire a mature motility pattern but become fertile; based on some poorly designed and interpreted studies, he concluded, erroneously, that the changes spermatozoa undergo during their transit through the epididymis represent a continuation of changes that start while spermatozoa are still attached to the germinal epithelium, and are not conditioned by some specific action of the epididymal secretion (6). Nevertheless, by 1931 most of the problems and questions relating to epididymal physiology had been recognized, and most of the work done since then has been to provide the experimental evidence to flesh out the insights of earlier investigators.

Few studies were published on epididymal physiology until the 1960s. Those that appeared focused primarily on establishing the length of time necessary for spermatozoa to transit through the epididymis (17,18). The apparent lack of interest in the epididymis during this 20- to 30-year span is puzzling. After the work of Benoit in 1926 (12) and Young in 1931 (13–16), there was a clear controversy that needed to be resolved. In 1965, only 43 papers on the epididymis were published. Whatever the reason for this disaffection, in 1964 Thaddeus Mann, in his book *The Biochemistry of Semen* (2), refers to the epididymis as the "abandoned child" of the reproductive system.

In the mid- and late 1960s, a resurgence of interest in the epididymis was spearheaded independently by Orgebin-Crist and Bedford; they demonstrated that the key event in sperm maturation was not the passage of time, as proposed by Young, but exposure to the luminal environment of the epididymis (19,20). Thanks to these and other studies, by the end of the 1960s it was established that the potential for sperm motility and fertilizing ability is acquired as spermatozoa pass from the proximal to the distal epididymis; that the maturation process does not end with the acquisition of fertilizing ability, because spermatozoa that have just become fertile induce a higher rate of embryonic mortality when inseminated in vivo; that the maturation process depends on an androgen-stimulated epididymis; and that the maturation process includes changes in sperm organelles [reviewed in Orgebin-Crist (21)].

Since the 1980s, we have seen an explosion of studies on the presence, characterization, immunolocalization, and regulation of a large number of proteins and their RNAs known to be specific to the epididymis, to be expressed at particularly high levels in some segments of the tissue, or to control key molecules that regulate epididymal function. The objective of many of these studies has been to develop ways of modifying epididymal function with respect to rendering spermatozoa fully mature and motile, thus providing potential leads for male contraception or the management of male infertility. In the course of these studies, it has become clear that this tissue presents a novel model for studying cell-, segment-, and region-specific gene expression, for understanding mechanisms of aging, and for identifying processes conferring selective protection from infections and cancer.

DEVELOPMENT OF THE EPIDIDYMIS

Formation of the Mesonephric/Wolffian/Nephric Duct and Tubules

Many of the studies focusing on the specification and regulation of mesonephric duct and tubule formation have been conducted on chick, frog, and zebrafish embryos, with some more recent studies using the mouse embryo. Space limitations prevent a comprehensive review of the development of the urogenital system, but there are several excellent resources with detailed descriptions of these events (22,23). What follows is a basic summary of the embryonic and postnatal development of the epididymis.

The urogenital system is derived from the intermediate mesoderm, which, in the chick and mouse, is the result of a complex interaction between the intermediate and paraxial mesoderm. The expression of two key transcription factors, Pax 2 and Pax 8, plays a vital role in this process because mice that lack these two genes fail to form a mesonephros and, therefore, later structures of the urogenital system; Bouchard et al. (2002) (24) suggest that Pax2 and Pax8 are critical regulators that specify the nephric lineage. At approximately gestational days 8 to 9 in the mouse and stage 16 in the developing chick, the mesonephric (Wolffian/nephric) duct develops as a cord of epithelial cells that express c-Sim-1 (chick) and Pax2 (25,26), undergoes the formation of a lumen, and rapidly extends throughout the length of the embryo in a cranial-to-caudal direction. There is considerable evidence to suggest that, in some species, elongation of the nephric duct is the result of cell rearrangements rather than proliferation or changes in cell shape (27). Interestingly, for proper nephric duct formation but not elongation, bone morphogenetic protein-4 expression in the surface ectoderm appears to be critical (28). Retinoic acid is also crucial for nephric duct formation because the duct does not form in RALDH2, a retinoic acid synthetic enzyme, knockout mice (29). As the mesonephric duct elongates, it induces the nearby mesenchyme to form the mesonephric tubules (30). The tubules have a characteristic J- or S-shape and resemble developing nephrons. Without the Wolffian duct, the mesonephric tubules do not form (31).

It is clear that those cranial mesonephric tubules that lie close to the testis survive, and the distal end grows toward the gonad, whereas the proximal end contacts the wolffian/mesonephric duct. However, several fundamental questions remain. What are mechanisms by which the mesonephric tubules are induced by the mesonephric duct? This process involves a complex mesenchyme-to-epithelial transition, and there is evidence to suggest that leukemia inhibitory factor, members of the Wnt family (Wnt6 and Wnt4), and fibroblast growth factor (FGF)-2 are responsible for mesenchymal cell aggregation and the formation of the renal nephron [reviewed in Gilbert (22)]. Whether this is recapitulated during the formation of the mesonephric tubules is unclear. Alarid et al. (1991) (32) have shown that FGF-2 is important for epididymal development because the epididymis fails to develop when embryonic urogenital ridges and sinuses are cultured under the kidney capsule in the presence of anti–FGF-2 antibodies. Studies by Sainio et al. (1997) (33) showed that the Wilms tumor-1 (*WT1*) gene is important in the formation of the most caudal mesonephric tubules.

What are the mechanisms by which only those tubules close to the gonad survive, yet others cranial and caudal to the gonad degenerate? Evidence suggests that the caudal tubules normally undergo apoptosis, so presumably those tubules close to the gonad express the antiapoptotic genes required to survive, but this has not been established.

What controls the migration and differentiation of cells of the distal tubule into the rete testis? Earlier studies by Upadhyay et al. (34) provided evidence that in the mouse, the cells in the distal end of the mesonephric tubules contribute directly to the rete testis and undergo differentiation, and that fusion between the mesonephric tubules and the testis did not occur. This finding is perhaps not too surprising in light of the extensive contribution of the mesonephros to the developing testis.

Which portions of the mesonephric tubule and duct form the discrete regions of the initial segment and the first part of the caput epididymidis? There is some evidence to suggest that cells of the wolffian duct contribute to a small part of the proximal portion of the mesonephric tubule but, again, it is not clear whether this represents the junction between the caput and the initial segment. For a more complete overview of these processes, the reader should refer to the reviews by Sainio (2003) (31), Vazquez et al. (2003) (35), and Jones (2003) (36).

It is intriguing that the development of the mesonephric tubules may recapitulate the development of the renal vesicle into the renal proximal and distal tubules. It would not be too surprising to find that many genes that regulate this process also play a role in the formation of the efferent ducts or initial segment. For example, members of the Wnt family [Wnt 4 (37)] and cadherins (38) play a role in the proximal/distal patterning of the renal proximal and distal tubules. Cadherin 6 is expressed in the proximal tubule progenitors, E-cadherin is expressed in the distal tubule, and P-cadherin is expressed in the glomerulus (39). The initial segment fails to develop in c-Ros mutants (40) and in Sxr (XXSxr) mice (41), suggesting that the defect may lie in the origin or development of the mesonephric tubules. Although it is well recognized that androgens play an important role in epididymal development, the initial segment fails to develop in a normal androgen environment in the Sxr mouse (41). This would suggest that additional factors, such as growth factors, may be important for mesonephric tubule/initial segment development. In both of these mutants the prominent vascular system that supplies the initial segment also fails to develop (40,42), indicating that there is an intimate relationship between epithelial and endothelial development. Again, this is perhaps not surprising in view of the formation of the renal glomerulus at the tip of the renal proximal tubule where endothelial cells are recruited to this site.

The embryonic origins of the different regions of the epididymis have received some interest; it appears that the efferent ducts are derived from the mesonephric tubules, whereas the caput epididymidis to the vas deferens regions are derived from the mesonephric duct [reviewed in Jones (36)]. However, the origins of the initial segment warrant further investigation because it is not entirely clear if the cells are of mesonephric tubule or duct origin. The cells of the initial segment are quite distinct from either the efferent ducts or the distal epididymal regions, but function similarly to the efferent ducts in that they are actively involved in water reabsorption. Hence, it is possible that the initial segment may be of mesonephric tubule origin.

Postnatal Development

A more complete picture of the postnatal development of the epididymis has emerged over the years, with the majority of studies focused on the mouse, rat, and human, although other species have been studied, including the bull, dog, rabbit, marmoset, and boar [reviewed in Rodriguez et al. (43)]. Perhaps the most detailed description of postnatal epididymal development is that of the rat, and the studies by Sun and Flickinger (1979) (44) and Hermo et al. (1992) (45) provide the basis for the following brief overview. There appear to be three major stages of epididymal development: the undifferentiated period, the period of differentiation, and the period of expansion.

By birth, the epididymis has undergone considerable coiling in the proximal regions (initial segment to corpus) and the cauda has yet to complete coiling (Fig. 1). In the next 1 to 2 days, individual

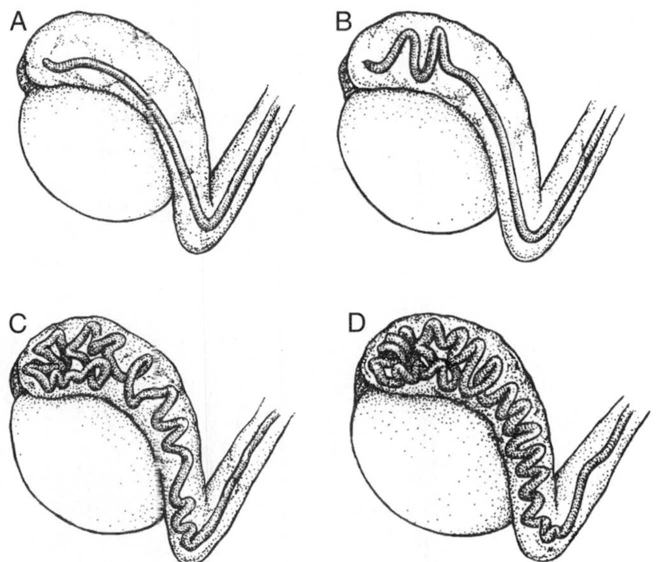

FIG. 1. Coiling of the epididymal duct from mouse E14 to P1. Note that coiling proceeds in a caput-to-cauda direction and is completed in the early postnatal period. The efferent duct–initial segment coiling is not shown here, but proceeds independently of the coiling of the main epididymal duct. To help visualize the changes, the size of the testes and epididymides at different ages are not drawn to scale. (From D. Bomgardner and B. T. Hinton, unpublished observations.)

septa are seen delineating the tubule into segments (D. Bomgardner and B. T. Hinton, unpublished observations). The epithelium is undifferentiated and is characterized by columnar cells containing numerous mitotic figures. There is considerable growth of the rat epididymal duct from its embryological origins to postnatal day 15, when it reaches almost 2 m in length (45,46). The period of differentiation has been examined extensively. Undifferentiated cells differentiate into the classically described cells of the adult epididymis: principal, halo, narrow (pencil), basal, and clear (light) cells; these cells are described later. Key changes in development are following: at postnatal day 14 the halo cells appear, on day 15 one observes narrow and columnar cells, by day 28 the columnar cells differentiate into basal and principal cells, from day 36 onward the narrow and clear cells appear, and by approximately day 49 all epididymal cells are fully differentiated. Figure 2 outlines the differentiation of the rat epididymal epithelium from birth to adulthood. The period of expansion describes the continued growth of the duct and the appearance of spermatozoa in the lumen.

The mechanisms that regulate the growth and differentiation of the epididymal duct are unknown, although it is clear that the expression of many epididymal genes is developmentally regulated. Certainly, luminal and circulating androgens play a critical role, but luminal fluid factors other than androgens also may be important. Early studies by Alexander (1972) (47) and Abe et al. (1984) (48) provided evidence that luminal contents produced by the testis or proximal regions of the epididymal duct may regulate epididymal epithelial differentiation, although the candidate regulators were not identified. Because there is some evidence to suggest that testicular luminal fluid growth

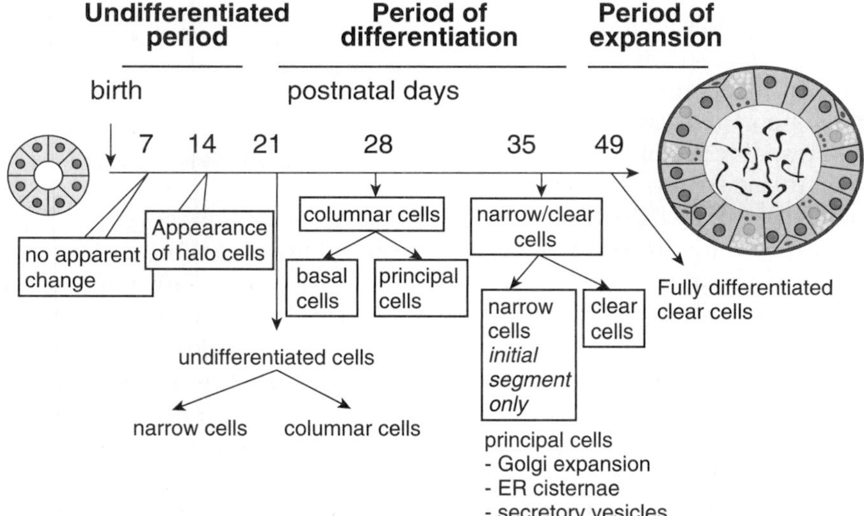

FIG. 2. Diagram outlining the differentiation of the rat epididymal epithelium from birth until adulthood. Epididymal epithelial cells are undifferentiated until approximately day 21, when both narrow and columnar cells are first observed. At approximately day 28, the columnar cells differentiate into principal cells and basal cells. Narrow cells are seen in the initial segment only, but clear cells are observed throughout the epithelium from approximately day 36 onward. By day 49, all epididymal epithelial cells are fully differentiated. (Adapted from Rodriguez, C. M., Kirby, J. L., and Hinton, B. T. [2002]. The development of the epididymis. In *The Epididymis: From Molecules to Clinical Practice* [B. Robaire and B. T. Hinton, Eds.], pp. 251–267. Kluwer Academic/Plenum, New York.)

factors (e.g., FGFs) may regulate initial segment function and that the same growth factors regulate embryonic wolffian duct development, it is tempting to speculate that luminal growth factors (e.g., bone morphogenetic proteins, FGFs, nerve growth factors) may also regulate epididymal epithelial differentiation. One might also envisage a more complex cell–cell interaction in which multiple luminal factors are secreted in the proximal regions of the duct (and testis), which in turn cause the cells in the mid-distal regions (e.g., corpus) to secrete factors, which in turn regulate cell function in the more distal epididymal region (e.g., cauda). Such multiple cascades may act to coordinate proper differentiation along the duct.

STRUCTURAL ORGANIZATION OF THE EPIDIDYMIS

Anatomy

Seminiferous tubules converge to form the rete testis, which in turn gives rise to the efferent ducts (ductuli efferentes), a series of 4 to 20 tubules, the number depending on the species (49,50). These tubules converge to form a single highly coiled duct, the epididymis (from the Greek meaning "on or adjacent to the testis"), which is extremely long and varies in length from 1 m in mice, (51), 3 m in rats, (52), 3 to 6 m in humans, (53), and up to 80 m in horses, (54).

The epididymis is usually divided into four gross anatomical regions: the initial segment, head (caput), body (corpus), and tail (cauda), as first described by Benoit (1926) (12). Subsequently, a number of other schemes have been proposed for dividing the epididymis into different regions or segments, including a number of zones and the demonstration of the presence of an intermediate zone between the initial segment and caput epididymidis that has characteristic cells (55–59). However, in this review, we retain the most commonly used nomenclature for the four regions described previously. In all mammalian species examined to date, each region of the epididymis is further organized into lobules separated by connective tissue septa. These septa not only as serve internal support for the organ but have been proposed to provide a functional separation between lobules that allows selective expression of genes and proteins within individual lobules (60). The extension of the epididymis is a straight tube, the vas deferens, which is surrounded by a very thick muscular layer. The vas deferens connects with the urethra, which empties to the outside of the body. A schematic representation of the testis, efferent ducts, epididymis, and vas deferens is shown in Fig. 3.

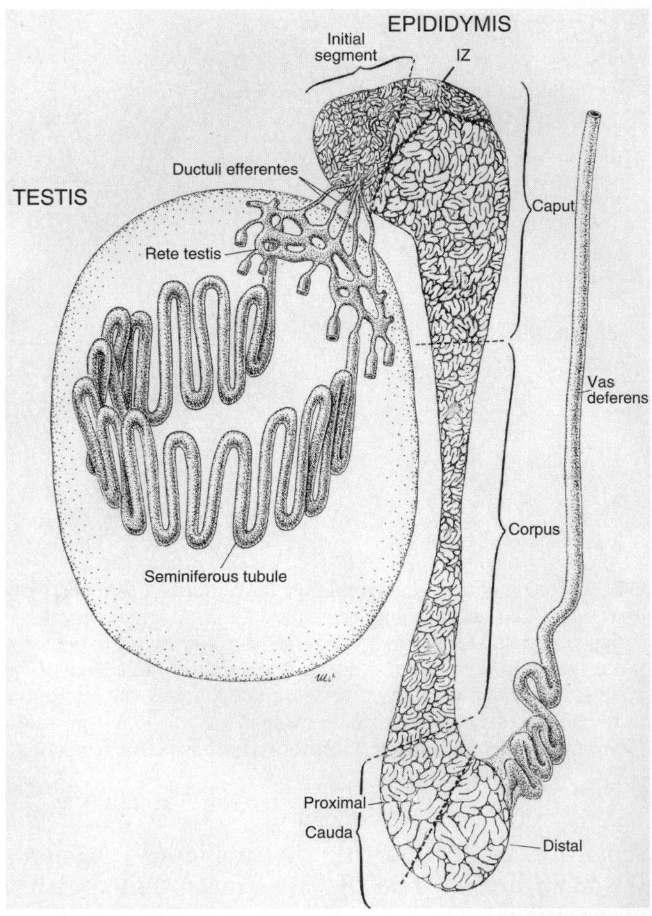

FIG. 3. Diagrammatic representation of the testis showing a seminiferous tubule and the rete testis, the ductuli efferentes, the epididymis, and vas deferens. The major regions of the epididymis (i.e., the initial segment, intermediate zone, caput, corpus, and proximal and distal cauda) are indicated. (Adapted from Robaire, B., and Hermo, L. [1988]. Efferent ducts, epididymis and vas deferens: structure, functions and their regulation. In *The Physiology of Reproduction* [E. Knobil and J. D. Neill, Eds.], pp. 999–1080. Raven Press, New York.)

Epididymal Cell Types and Specific Markers

There are several types of epithelial cells that line the epididymis; some are located throughout the duct (e.g., principal cells), whereas others are found either exclusively or primarily in specific regions (e.g., narrow cells). We provide a brief description of the major cell types along the epididymis, discuss the potential functions of these cells, and demonstrate the presence of specific markers for each cell type. Several detailed reviews of the histology of the epididymis for a number of species ranging from mouse to human, and including dog, camel, elephant, opossum, bull, ram, hamster, mouse, and monkey, have been published (1,12,48,56,57,61–73). Together, these publications provide increasing evidence that similar regions and cell types are present in most mammals, including humans.

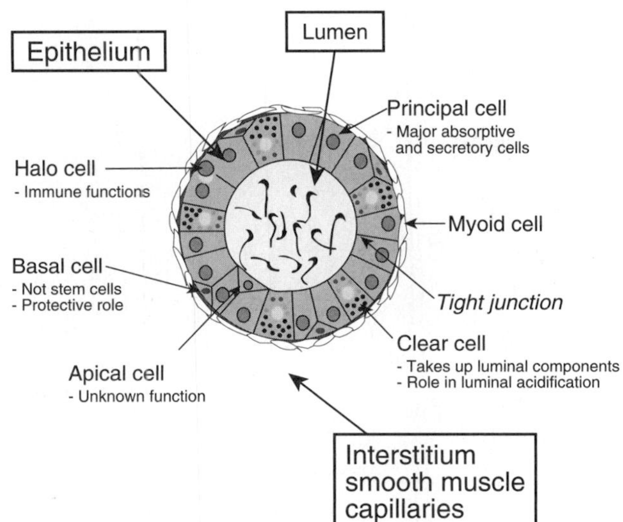

FIG. 4. Schematic organization of the major cell types in the epididymis as observed at the light microscope. The three epididymal compartments as well as the relative position and distribution of each of the main cell types are illustrated. The major functions associated with each cell type are also identified.

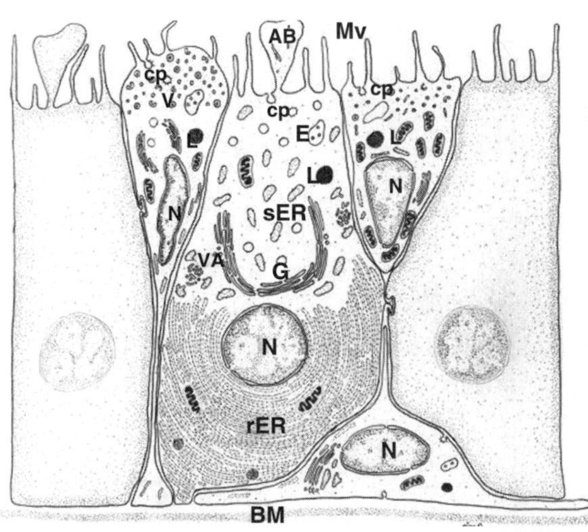

FIG. 5. Schematic representation of cells found in the epithelium of the initial segment of the epididymis, as visualized by the electron microscope. Narrow cells, represented on the left, are elongated cells tapering toward the basement membrane (BM); they show numerous, small, apical cup-shaped vesicles (v), coated pits (cp), and occasional endosomes (E) and lysosomes (L). Principal cells are columnar in appearance and show coated pits (cp), endosomes (E), and lysosomes (L), components of the endocytic apparatus. They also contain numerous parallel cisternae of rough endoplasmic reticulum (rER) located basally and dilated, irregularly shaped cisternae of ER sporadically dispersed in the apical and supranuclear regions of the cell (sER). The Golgi apparatus (G) is elaborate and located supranuclearly. Principal cells show blebs of cytoplasm emanating from their apical cell surface, referred to as apical blebs (AB), and small vesicular tubular aggregates (VA) occupying regions adjacent to the Golgi apparatus. Also illustrated on the left is an apical cell that, unlike the narrow cell, does not extend to the basement membrane and shows few coated pits (cp), apical vesicles, or endosomes and lysosomes (L). A basal cell stretches along the basement membrane and sends a thin process up toward the lumen. N, nucleus; Mv, microvilli. (Reproduced with permission from Hermo, L., and Robaire, B. [2002]. Epididymis cell types and their function. In *The Epididymis: From Molecules to Clinical Practice* [B. Robaire and B. T. Hinton, Eds.], pp. 81–102. Kluwer Academic/Plenum, New York.)

These cell types appear in the appropriate regions, share similar structural features and functions, including regional differences, and show very similar patterns of expression of some secretory proteins (74–79). The comprehensive set of light and electron microscopy plates prepared for this volume are available at www.medicine.mcgill.ca/PhysiolReprodThirdEd/-Epididymis. Schematic representations of the organization of all the cell types and their known functions are shown in Fig. 4. Key components of epithelial cells in the initial segment and in the remainder of the epididymis as observed at the electron microscope are depicted in Figs. 5 and 6, respectively.

Principal Cells

The main cell type in the epididymis of all mammals is referred to as the *principal cell*. These cells appear along the entire duct but show structural differences in each region (1,63). The most striking feature of these cells is their highly developed secretory and endocytic machinery and their basally aligned nuclei. Depending on the segment examined, principal cells comprise approximately 65% to 80% of the total epithelial cell population of the epididymis (80). Both their structure and functions vary dramatically between the different segments (1,63,81). These differences are reflected in the appearance and organization of their secretory apparatus (endoplasmic reticulum, Golgi apparatus, and secretory granules) and endocytic apparatus (coated pits, endosomes, multivesicular bodies, and lysosomes). The precise characterization of the secretory granules and Golgi apparatus of these cells, as well as the identification of organelles and mechanism of uptake of luminal substances, have been well described (82). One other major difference is the abundance of lipid droplets found only in the corpus epididymidis (Fig. 6), the exact significance of which is still poorly understood.

Principal cells synthesize a large number of proteins that are then either retained in the cells or actively secreted into the luminal compartment (1,81,83–89) [reviewed in Robaire et al. (90), Cornwall et al. (92), and Kirchhoff (93)]. They also play an active role in endocytosing proteins found in the luminal compartment of the epididymis [reviewed in Hermo and

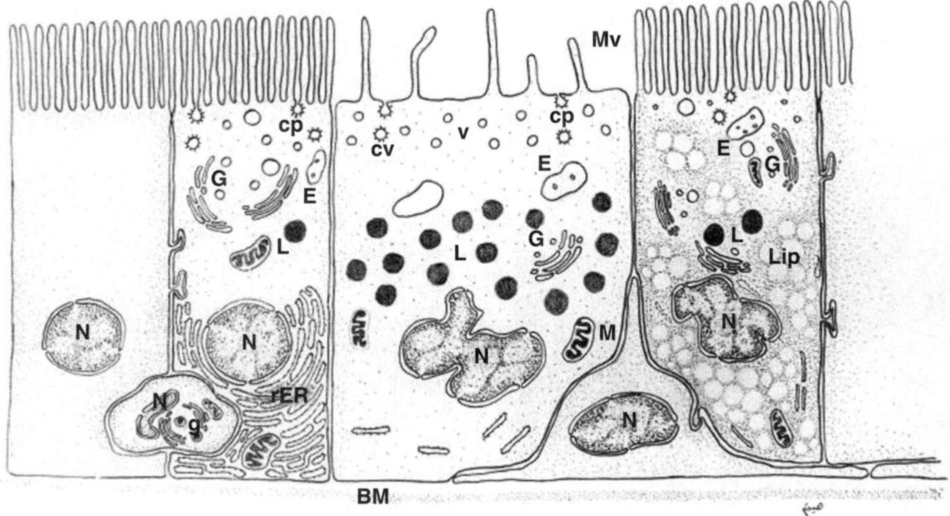

FIG. 6. Schematic representation of a principal cell of the caput epididymidis on the left and a principal cell of the corpus epididymidis on the right, with a clear cell in between, as visualized by the electron microscope. Also represented is a halo cell and a basal cell. Principal cells of both regions contain coated pits (cp), endosomes (E) and lysosomes (L), and an elaborate Golgi apparatus (G). Rough endoplasmic reticulum (rER) occupies the basal region of the principal cell of the caput, whereas numerous lipid droplets (lip) occupy the cytoplasm of the principal cells of the corpus region. The clear cell shows few microvilli (Mv), but numerous coated pits (cp), small apical vesicles (v), endosomes (E), and lysosomes (L), all involved in endocytosis. The halo cell is inserted between adjacent principal cells, is located basally, and contains small dense core granules (g), whereas the basal cell stretches itself along the basement membrane (BM). N, nucleus. (Reproduced with permission from Hermo, L., and Robaire, B. [2002]. Epididymis cell types and their function. In *The Epididymis: From Molecules to Clinical Practice* [B. Robaire and B. T. Hinton, Eds.], pp. 81–102. Kluwer Academic/Plenum, New York.)

Robaire (2002) (82)]. A schematic representation of these processes is depicted in Fig. 7.

Apical Cells

Apical cells are found primarily in the epithelium of the initial segment and intermediate zone (94,95), although they have been seen occasionally in other segments in aging rats (96). These cells have a characteristic apically located spherical nucleus and do not contact the basement membrane (Fig. 5). They differ clearly from adjacent narrow and principal cells in terms of their protein expression profile (95). However, little is known about the specific functions of these cells, aside from their ability to endocytose substances from the lumen, as revealed by the examination of β-hexosaminidase A knockout mice (69), and the observation that they contain many proteolytic enzymes (95).

Narrow Cells

In the rat and mouse, narrow (pencil) cells of the adult epididymis appear only within the epithelium of the initial segment and intermediate zone (94,95). These cells are narrower than the adjacent principal cells, attenuated, and send a thin process of cytoplasm to reach the basement membrane (Fig. 5). They are characterized by numerous apically located cup-shaped vesicles that are involved in endocytosis and function in secreting H^+ ions into the lumen by recycling to and from the apical plasma membrane (97). Similar cells have also been reported in the same regions in numerous other species, including bovine, hamster, echidna, and human (61,65,66,76,95,98). Narrow cells are distinct from apical cells in their morphological appearance, relative distribution, and expression of different proteins. They also differ dramatically from neighboring principal cells and display region-specific expression of proteins such as the glutathione S-transferases and lysosomal enzymes (95).

Clear Cells

Clear cells are large, active endocytic cells present only in the caput, corpus, and cauda regions of the epididymis and are found in many species, including humans (1,63,99). These cells are characterized by an apical region containing numerous coated pits, vesicles, endosomes, multivesicular bodies, and lysosomes and a basal region containing the nucleus and

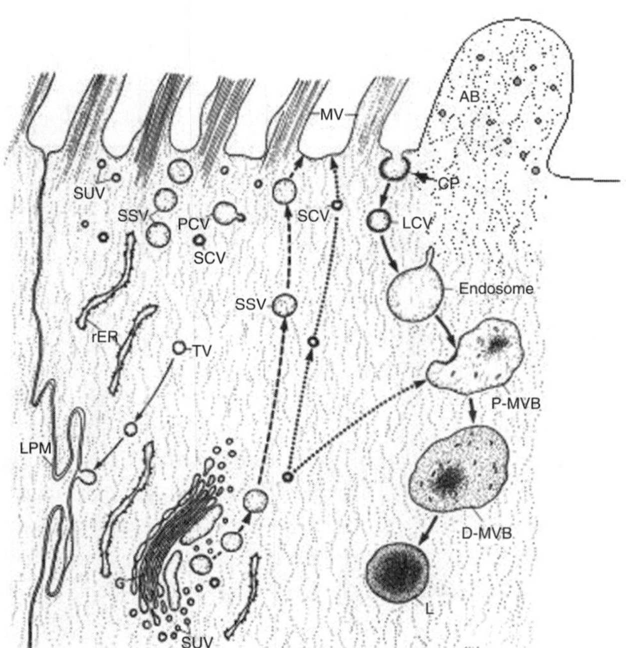

FIG. 7. Schematic representation of the apical and supranuclear cytoplasm of a principal cell of the epididymis. The Golgi apparatus (G) consists of several saccules packed on top of each other, from which the smooth-surfaced secretory vesicles (SSV; 150–300 nm) and small coated vesicles (SCV; 60–70 nm) are derived. The SSV are destined for the apical cell surface with which they will fuse, delivering their content into the epididymal lumen in a regulated manner, by a process termed *merocrine secretion*. The SCV are targeted from the Golgi apparatus to multivesicular bodies (MVBs), delivering their lysosomal contents therein, although some may be destined for the cell surface. The endocytic apparatus consists of coated pits (CP), large coated vesicles (LCV), endosomes, pale (P) and dense (D) MVBs, and lysosomes (L). Tubules emanating from endosomes serve to recycle receptors back to the cell surface. Small uncoated vesicles (SUV; 60–70 nm) are also indicated, some of which may be Golgi derived and involved in constitutive merocrine secretion by fusion with the apical cell surface. An apical bleb (AB) is also indicated and contains mainly polysomes and small vesicles, the origin of which may be Golgi derived. ABs appear to detach from the apical cell surface and upon fragmentation in the epididymal lumen liberate their contents therein for functions involved in sperm maturation. LPM, lateral plasma membrane; Mv, microvilli; PCV, partially coated vesicle; rER, rough endoplasmic reticulum; TV, small vesicles involved in the transcytosis or passage of substances from the lumen to the lateral intercellular space. (Reproduced with permission from Hermo, L., and Robaire, B. [2002]. Epididymis cell types and their function. In *The Epididymis: From Molecules to Clinical Practice* [B. Robaire and B. T. Hinton, Eds.], pp. 81–102. Kluwer Academic/Plenum, New York.).

a variable amount of lipid droplets (Fig. 6) (1,98,100). After injection of tracers into the lumen of the cauda epididymidis, the apical vesicles, endosomes, multivesicular bodies, and lysosomes of clear cells are labeled. This indicates an endocytic role for these cells; this endocytic activity is much greater in clear cells than in the adjacent principal cells, particularly in the cauda epididymidis (100,101). Clear cells normally take up the contents of cytoplasmic droplets released by spermatozoa as they traverse the duct (1,100). Cytoplasmic droplets are formed at the time of the release of spermatozoa and contain Golgi saccular elements that may be involved in modification of the plasma membrane of spermatozoa (102). Clear cells also endocytose a number of different proteins, but often in a region-specific manner (83,87,103,104). These cells become abnormally large and filled with lysosomes after various experimental conditions that disrupt the normal functioning of the testis and epididymis (80).

The acidification of the luminal fluid (see later) is thought to be mediated by clear (and narrow) cells. Indeed, key proteins for this process, H^+-adenosine triphosphatase ([ATPase] or vacuolated [V]-ATPase), carbonic anhydrase II, and soluble adenylate cyclase, are selectively localized to these cells. Further, ClC-5, a member of the voltage-gated ClC chloride channel family, is also expressed exclusively in clear cells and partially colocalizes with the H^+-ATPase in their apical region. Breton's group has proposed that, in clear cells, apical membrane accumulation of V-ATPase is triggered by a soluble adenylate cyclase–dependent rise in cyclic adenosine monophosphate (cAMP) in response to alkaline luminal pH (105–107).

Basal Cells

Basal cells appear in all species studied to date, including humans (1,63,108). Hemispherical in appearance, they adhere to the basement membrane and do not have direct access to the lumen of the duct, although processes of these cells extend at times toward the lumen (Figs. 5 and 6) (109). Like principal cells, basal cells are not thought to divide in adults, nor are they thought to act as stem cells to replenish principal cells (110).

Basal cells possess thin, attenuated processes that extend along the basement membrane from their main hemispherical cell body collectively to cover a large proportion of the circumference of the epididymal tubule (109). However, in rats, after ligation and castration, these cells transform into large, bulbous, dome-shaped cells that are closely packed together and show few short lateral processes (111). Because of the dramatic decrease in size of the epididymal tubules after these treatments, it is likely that the shape and arrangement of basal cells in normal untreated animals is governed in part by the volume and pressure exerted on the tubular epithelium by luminal fluids and sperm derived from the testis. Consistent with this hypothesis is the observation that during postnatal development in the rat, basal cells are transformed from dome-shaped cells to flattened cells that exhibit processes as spermatozoa and fluids arrive in the corpus and cauda epididymidis by days 49 and 56, respectively (108).

Basal cells possess coated pits on the plasma membrane face opposing the basement membrane and overlying principal cells, suggesting the receptor-mediated endocytosis of factors derived from the blood or principal cells. Basal cells also show an accumulation of a secretory material in Golgi saccules, and distinct secretory granules appear next to the Golgi apparatus (L. Hermo and B. Robaire, unpublished results), as seen in other typical secretory cells (112). The destiny of the secretory material may be to regulate principal cell function or enter the circulation for functions as yet to be determined. Basal cells have been shown also to express apolipoprotein E and alcohol dehydrogenases (113).

It has also been proposed that basal cells may have a role as immune cells because of their ability to respond in number and macrophage antigen expression to the presence of sperm autoantigens in the lumen (114), and it has been postulated that these cells may have an extratubular origin (115).

Studies from Wong's group [reviewed in Leung et al. (116)] have proposed an additional function for basal cells; they suggest that these cells may have a role in regulating electrolyte and water transport by principal cells. This process is proposed to be mediated by the local formation of prostaglandins (PGs) and require the participation of the transient receptor potential (Trp) proteins. The latter serve as transmembrane pathways for Ca^{2+} influx, whereas cyclooxygenase-1 (COX-1) is a key enzyme in the formation of PGs. Both of these proteins are exclusively expressed in basal cells.

Halo Cells

Halo cells are small cells with a narrow rim of clear cytoplasm that are present throughout the epididymal epithelium (Fig. 6) (1). These cells are usually located at the base of the epithelium and contain variable numbers of dense core granules. Halo cells have been described either as lymphocytes (1) or monocytes (117); these two cell types are difficult to distinguish by light microscopy because of their similarity in size and nuclear morphology. Although the exact nature of halo cells has been controversial since they were first described by Reid and Cleland (1957) (57), studies by Flickinger et al. (1997) (118) and Serre and Robaire (1999) (119) have resolved this issue by immunolabeling the main types of immunocompetent cells. It is now clear that, in young adult animals, halo cells consist of helper T lymphocytes, cytotoxic T lymphocytes, and monocytes, but not B lymphocytes. With age, there is a region-specific increase in the number of each of these immune cell types, as well as the occasional appearance of eosinophils (96) and B lymphocytes. In the epididymal epithelium of young rats, the number of cells that stain for antibodies against monocytes–macrophages (ED1+), helper T lymphocytes (CD4+), and cytotoxic T lymphocytes (CD8+) is equivalent to the number of halo cells (119), suggesting that halo cells are, under normal conditions, the primary immune cell in the epididymis.

The Blood–Epididymis Barrier

Given the presence in spermatozoa of proteins that are recognized by the body as foreign, it stands to reason that there should be a continuation beyond the testis of a functional barrier. The probable existence of a blood–epididymis barrier was discussed as early as 1976 (120), and several reviews describing different aspects of the barrier have appeared (121–123).

Structure

The junctional complex between adjacent epididymal principal cells is composed of apically located gap, adherens, and tight junctions. Tight junctions between adjacent principal epithelial cells at their luminal surface form the blood–epididymis barrier (122,124), whereas gap junctions allow communication between adjacent principal cells. These tight junctions form a continuous zonule around the cell, sealing the spaces between the epithelial cells, so that the luminal compartment and the intercellular spaces become separate physiological compartments (125). The tight junctions begin to form at the time of differentiation of the Wolffian duct (126). Using lanthanum nitrate as an electron-opaque tracer that is blocked at tight junctions, the postnatal development of the blood–epididymis barrier was shown to be gradual; its formation is virtually complete by postnatal day 21 in rats (125).

Electron microscopic changes in the structure of the junctional complex of the initial segment have been observed compared with the other segments of the epididymis. In the initial segment, the tight junctions span a considerable length of the apical plasma membrane and have few desmosomes (122). With progress toward the caudal end of the epididymis, a general decrease in the number of tight junctional strands is noted; the span of merging plasma membranes is considerably reduced, but numerous desmosomes are found in the apical region (122,126).

Junctional Proteins

Adherens junctions form a continuous belt and hold neighboring cells together through a family of calcium-dependent cell–cell adhesion molecules called *cadherins*, which mediate calcium-dependent homotypic interactions (127). The cadherins have also been implicated in the formation and maintenance of tight junctions (127–133). The cytoplasmic domain of cadherins forms a tight complex with several proteins, which either link cadherins to the cytoskeleton or are involved in signal transduction pathways. These include catenins, actinin, vinculin, and zonula occludens-1. The cadherin–catenin complex is essential for cadherin-mediated cell adhesion. Cyr et al. (1992) (134) reported the presence of E-cadherin and P-cadherin messenger RNA (mRNA) in the rat epididymis. Electron microscopy with immunogold labeling indicates that E-cadherin is localized in the extracellular space between the lateral plasma membranes of adjacent principal cells at the level of apical junctional complexes in the adult rat epididymis (122,135,136), as well as in the deeper underlying regions of the extracellular space between the lateral plasma membranes. Similar observations were noted in the human and mouse epididymis (137–139). The composition of the catenin-adhering junctional family of proteins and their relationship with cadherins remain to be established in the epididymis; however, in one study it was shown that, in the normal adult rat epididymis, there was immunostaining for three anti-catenin antibodies (alpha-, beta-, and p120ctn) along the lateral plasma membranes between adjacent epithelial cells (140).

In addition to adherens and tight junctions, the epididymal junctional complex also contains gap junctions (124,141). Gap junctions, made up of proteins termed *connexins*, mediate communication between cells by allowing small molecules to pass from cytoplasm to cytoplasm of neighboring cells, thereby metabolically and electrically coupling them (142). Connexin subunits oligomerize in the trans-Golgi network to form hemichannels or connexons. In the epididymis, gap junctions containing connexin 43 were first localized between principal and basal cells (143). Using a reverse transcriptase-polymerase chain reaction (RT-PCR) strategy, Finnson and Cyr (unpublished observations) have identified at least seven different connexin transcripts in the rat epididymis. Although the presence of multiple connexins in specific cell types is not unique to the epididymis, the large number of different connexins is suggestive of complex communication between epididymal cells.

Functions

The composition of epididymal luminal fluid is distinctly different from that of blood plasma. The blood–epididymis barrier keeps the two fluids in separate compartments (144). The blood–epididymis barrier also maintains a specialized luminal microenvironment for the maturing spermatozoa by restricting the passage of a number of ions, solutes, and macromolecules across the epididymal epithelium (121,144). For instance, molecules such as inositol and carnitine can be concentrated ten- to 100-fold in the lumen of the caput epididymidis, whereas others, such as inulin, L-glucose, and bovine serum albumin, are effectively excluded [reviewed in Robaire and Hermo (1) and Turner (85)]. The blood–epididymis barrier carefully controls the microenvironment so that the spermatozoa are bathed in an appropriate fluid milieu at each stage of maturation as they travel through each segment of the epididymis (85).

This barrier also serves as an extension of the blood–testis barrier. Spermatozoa are immunogenic; they contain proteins on their surfaces that would be recognized as foreign if they were to leave the epididymis (1). The exact function of the blood–epididymis barrier in protecting spermatozoa from the immune system is unclear at this time. The barrier prevents the passage of spermatozoa between epithelial cells, but cellular elements of spermatozoa can be taken up by epithelial cells. However, additional studies are needed to clarify whether, in the adult, any epididymal epithelial cell is capable of acting as an antigen-presenting cell.

Although it would appear that the blood–epididymis barrier is resistant to some foreign substances [e.g., gossypol (144a), estradiol, (144b)], administration of cyclophosphamide to efferent duct–ligated rats resulted in the production of damaged spermatozoa, suggesting that this drug could enter the epididymal lumen and modify spermatozoa. However, very little is known about the role played by this barrier in protecting spermatozoa from toxic substances and immunoglobulins (145). The inability of this barrier to maintain its tightness under conditions of stress, such as aging (136), may play help explain some of the deleterious effects of stressors on sperm function and fertility.

FUNCTIONS TAKING PLACE IN THE LUMINAL COMPARTMENT

The four main functions of the epididymis are transport of spermatozoa, development of sperm motility, development of sperm fertilizing ability, and the creation of a specialized luminal environment conducive of the maturation process through the absorptive and secretory activities of the epididymal epithelium.

Transport of Spermatozoa

Once released in the lumen of the seminiferous tubule, spermatozoa are transported through the efferent ducts and begin their journey down the epididymis. Several methods have been used to measure the duration of this transit. The most direct method is to incorporate a labeled isotope into the DNA of germ cells at the spermatogonia and preleptotene spermatocyte stages and follow the progression of the first wave of labeled spermatozoa down the epididymis. This approach gives the minimal time required for sperm passage through the epididymis. Total transit time, or transit through each epididymal segment, may also be estimated from the ratio of epididymal sperm reserves and daily testicular sperm production (146), assuming that there is no sperm resorption and no difference in transit speed between segments. Despite some minor discrepancies between estimates obtained with these two techniques, it appears that, regardless of the size of the animal, its sperm production, or its epididymal sperm reserves, the minimal time required for spermatozoa to transit through the epididymis is approximately 10 days (Fig. 8; Table 1), supplemental material on web site). In most species, the average transit time of the majority of labeled spermatozoa is longer than that of the sperm vanguard [e.g., 14 days versus 8 days in bulls (146)]. As a result, there is some mixing of spermatozoa of different ages in the distal part of the epididymis (147). There are two notable exceptions to the 10-day minimum transit time: in the human and chimpanzee, the first labeled spermatozoa pass through the epididymis in 1 and 2 ± 1 days, respectively (Table 1, supplemental material on web site). These values give the most rapid transit time, but the average transit time is 12 days in humans (148), a value comparable with that found in other species.

When one follows the progression of labeled spermatozoa in the different regions of the epididymis, the transit time through the caput and the corpus is quite similar in all species studied, including humans. Most of the difference among species comes from transit through the cauda epididymidis. Surprisingly, in species where this was tested, frequency of semen collection did not influence transit time through the epididymis markedly (146,149,150). In bulls, daily sperm collection must exceed twice the daily testicular sperm production to lead to an acceleration of sperm transit of only 3 days for the first wave of labeled spermatozoa (146), whereas in rabbits different frequencies of ejaculation ranging from once per week to four times in 1 day had no effect on sperm reserves in the caput and corpus epididymidis, but dramatically reduced sperm reserves in the cauda epididymis and vas deferens (151).

Spermatozoa enter the epididymis propelled by testicular fluid and possibly the beat of the ciliated cells of the efferent ducts. However, in the epididymis, the epithelium is lined by immotile stereocilia and the massive fluid uptake taking place in the ductuli efferentes and the initial segment of the epididymis drastically reduces this fluid flow (152). Transport takes place against an increasing hydrostatic pressure gradient from testis to cauda epididymidis (153), and proceeds even when fluid flow from the testis is prevented by ligation of the ductuli efferentes (16,154). It is unlikely that these mechanisms alone are responsible for sperm transport.

The epididymis is surrounded by a smooth muscle layer of increasing thickness and adrenergic innervation from proximal to more distal regions (155).

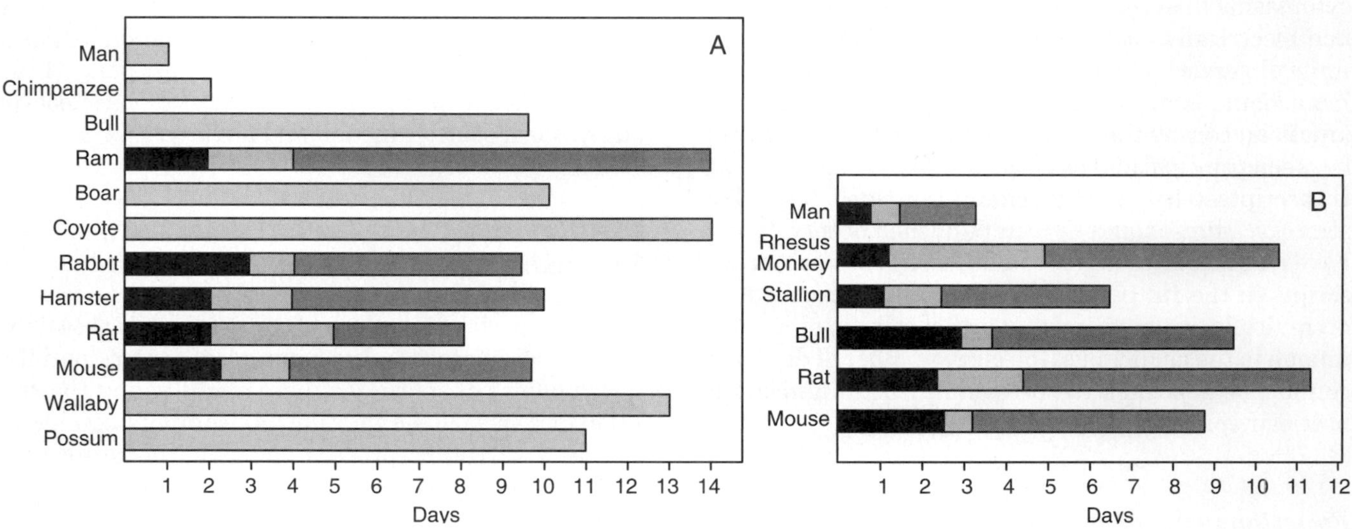

FIG. 8. A: Minimal time for epididymal transit determined by the progression of the first wave of labeled spermatozoa after isotope injection [human (148); chimpanzee (927); bull (146); ram (150); boar (928); coyote (929); rabbit (147); hamster (930); rat (165); mouse (166); tammar wallaby and brushtailed possum (931)]. **B:** Duration of sperm transit through the whole epididymis and each segment determined by the ratio of epididymal sperm reserves to daily testicular sperm production [human (932), (933); rhesus monkey (934); stallion (935); bull (18), (936); rat (937); mouse (166)].

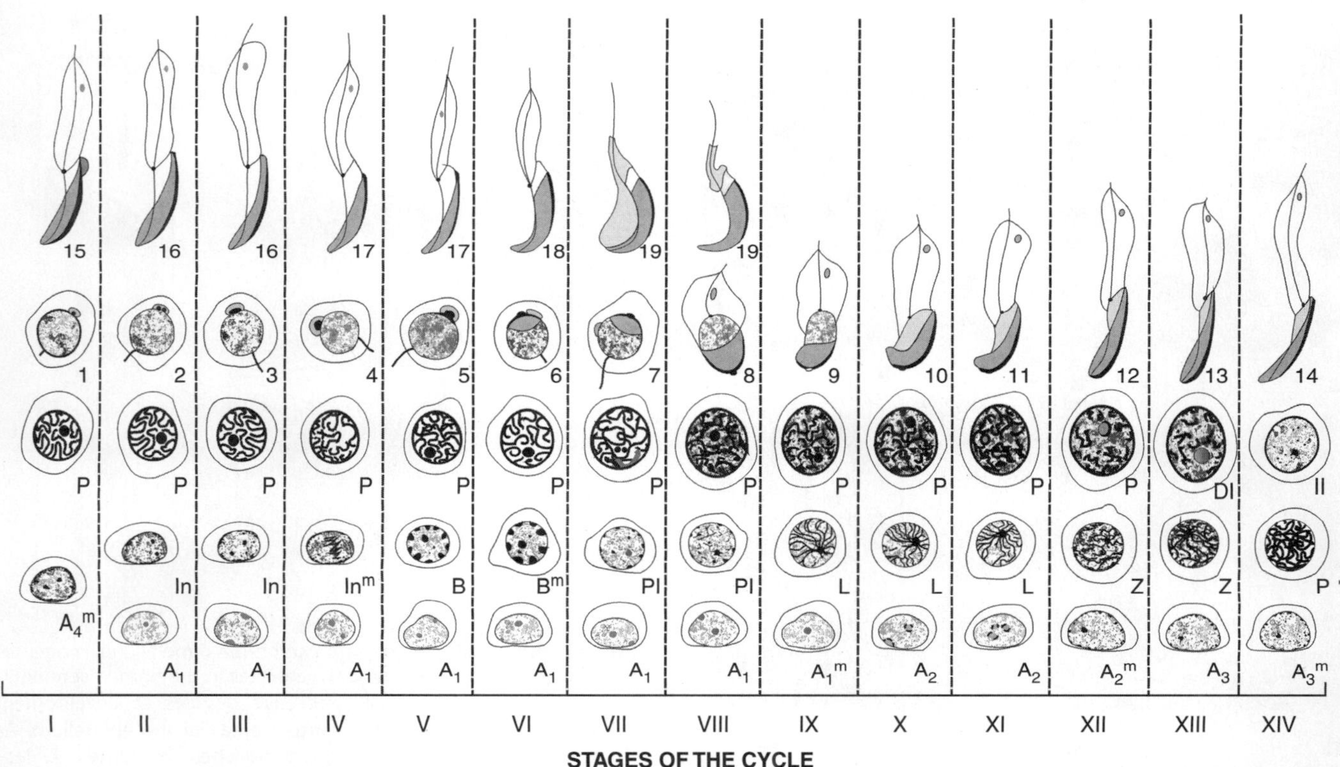

FIG. 63. This diagram illustrates the stages of the rat seminiferous cycle and the cell associations comprising them. [From (753), with permission.]

Because spermatogenesis is a continuous sequence of changes, the subdivision into stages is artificial, and the number of stages depends on the criteria used. The more extensive the criteria used, the more specific and detailed become the stages that can be identified. For light microscopic studies, the conventional staging of Clermont (163) is usually applied to kinetic studies of the seminiferous epithelium. In each species studied, the cycle has a characteristic pattern, duration, and number of stages (Table 3).

In the human, considerable difficulty was experienced in identifying the cell associations or stages because multiple stages could be seen in a single cross-section of a tubule and because of inadequate care in the fixation and handling of tissue. However, by detailed studies, Clermont (252) identified six cell associations as stages that formed the cycle of the human seminiferous epithelium (Fig. 64). Multistaged seminiferous tubules have been described in the marmoset (417,718,719), macaque (417), chimpanzee (417,720–722), and baboon (417,723). The significance of this unique arrangement, which differs from the more simple serial/segmental arrangement of stages common to rodents, was not appreciated until details of the spatial and temporal organization of the human testis was studied in greater depth. Schulze and colleagues (724,725) analyzed the human seminiferous epithelium in a different manner and proposed that germ cells at the same stage of development are distributed in a helical pattern. As they progress through spermatogenesis, the diameter of the helix decreases such that they are overlapped on the external surface by a gyration from another helical sequence (Fig. 65). They proposed that this concept of an Archimedian spiral provides the best explanation of the human data. They also suggested that the greater spermatogonal mitotic rate in species other than humans provides a larger spermatocyte population and hence the conventional appearance of stages of a seminiferous cycle.

The attractive concept of a spiral/helical arrangement of spermatogenic stages along the seminiferous

TABLE 3. *Duration of seminiferous cycle*

Species	Days	References
Mouse	34–35	(1091)
Hamster *(Cricetus auratus)*	35–36	(1091,1092)
Boar	34–35	(1093)
Rat	48–53	(1093,1094)
Ram	49	(734,1095)
Bull	54	(811)
Rabbit	48–51	(1096,1097)
Monkey		
Macaca speciosa	45	(1098)
Macaca mulatta	70	(1099)
Human	64	(241)

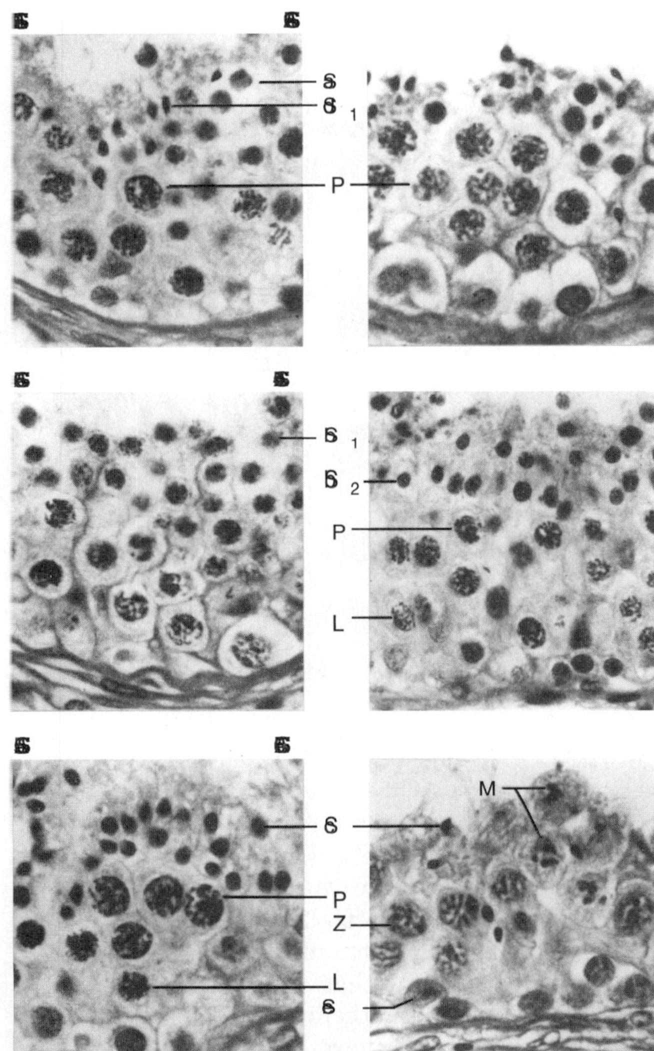

FIG. 64. This series of light micrographs illustrates the stages of the human seminiferous cycle as defined by Clermont (188). [From (1112), with permission.]

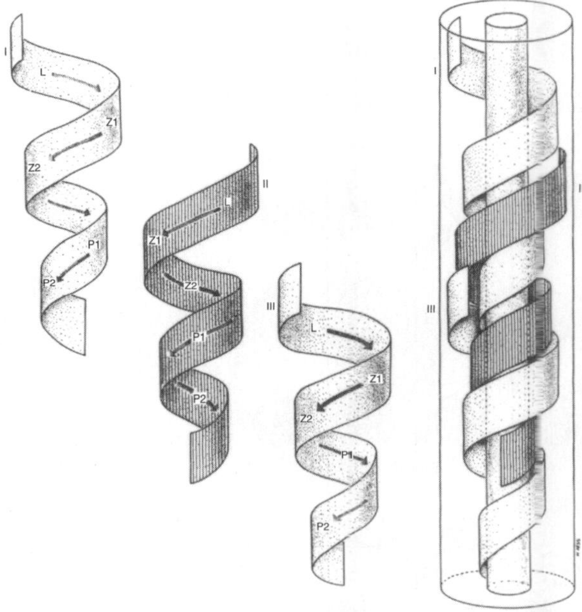

FIG. 65. This illustrates the proposed three-dimensional model for the arrangement of primary spermatocytes in the human seminiferous tubule. Populations of successive degrees of development occupy helically running strip-shaped areas of the epithelium. As development proceeds the strip approaches the lumen. L, leptotene; Z1, early zygotene; Z2, late zygotene; P1, early/mid-pachytene; P2, late pachytene. [From (724), with permission.]

tubule has been questioned by other studies of the human testis, showing that the arrangement is more irregular than previously thought (726) and more likely to be a random occurrence (727,728). The biological significance of irregular/random/helical disposition of stages was then linked to the efficiency of spermatogenesis where fewer stages per tubule cross-section were linked with "low efficiency" of spermatogenesis (729). In men with high daily sperm production per gram of testis, tubules tended to have more stages and showed more atypical cell types in these tubules. Not unreasonably, it was assumed that multistage arrangements in the human were associated with comparatively low sperm production efficiency in contrast to the single-stage tubules of many rodents where the opposite is the case (727,728,730). However, with the introduction of more reliable methods for cell quantitation, such as the optical dissector technique, primate and human spermatogenesis may be more efficient than previously thought (23,719,731,732). Comparative data are presented in Table 1. Further studies in 17 species of primates including humans by Wistuba and colleagues (417) clearly showed that spermatogenic efficiency (i.e., the capacity of germ cells to survive the complete process of spermatogenesis) and sperm output (millions per gram of testis) showed no correlations with single- or multistaged tubule architecture. The organization of the tubules in the testes of primates is not closely related to their phylogenetic classification. When the mean number of stages per tubule is expressed in relation to the observed percentage of stages (Fig. 66), the prosimian species (lemur, bushbaby) tend to show the ancestral type of tubular cytology (i.e., closer to rodent stage organization), whereas the great apes and humans exhibit more complex architecture of the seminiferous epithelium. The data point to an evolutionary trend toward more complex multistaged tubules such as in the chimpanzee and human.

Duration of Spermatogenesis

Studies to determine the duration of spermatogenesis centered initially on attempts to destroy sensitive germ cell stages and analyze the rate of depletion of

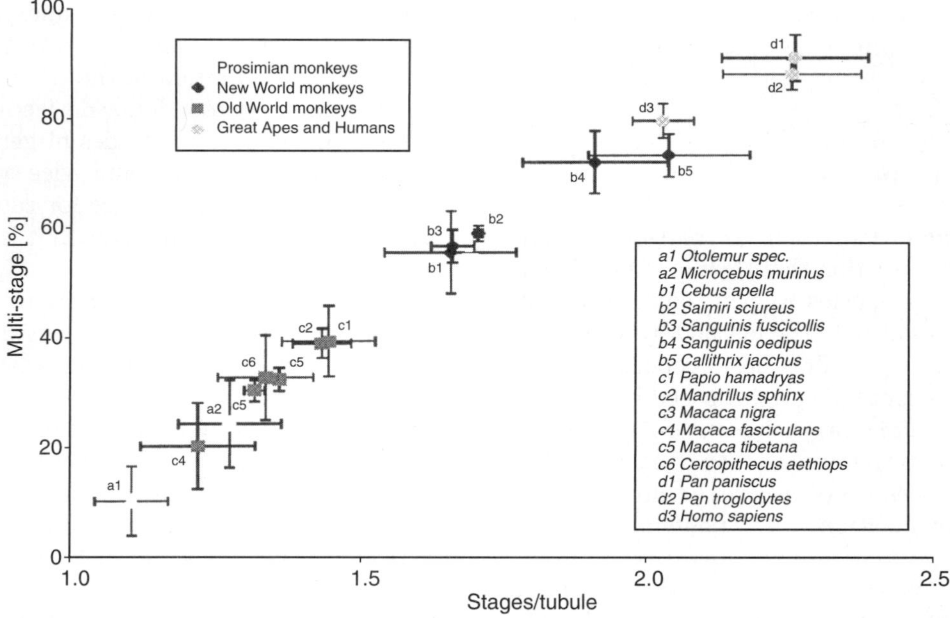

FIG. 66. Correlation between the percentage of multistage tubules and the average number of stages per tubule cross-section. The linear correlation elucidates that the species belonging to a certain systematic entity are grouped together. Overlapping is rare. The *error bars* give SD. [From (417).]

the seminiferous epithelium. Such approaches defined the duration of the cycle in the mouse as 8.6 days, with the entire process taking 34.6 days (716). Alternative approaches emerged with the availability of radioactive tracers such as ^{32}P- and ^{3}H-labeled thymidine (733,734). Using these techniques, studies of spermatogenesis in a number of species demonstrated a duration (Table 3) that appears to be specific for each species, and attempts to influence this biological constant by numerous approaches have been unsuccessful. Ortavant (733) showed that although photoperiod could affect the yield of spermatozoa, it could not affect the rate of spermatogenesis. Removal of the pituitary to assess the influence of the hypophyseal hormones was shown to be without effect (735,736). Strong support for the species-specific constancy of the duration of spermatogenesis has come from studies on transplantation of spermatogonia from rat to mouse (737–739) and from mouse to rat (740). Rat germ cells that colonize recipient mouse testes establish spermatogenesis with a duration identical to the rat spermatogenic cycle, and this occurs in association with the recipient mouse Sertoli cells. The reverse experiment (i.e., mouse germ cells into rat testes) showed that mouse spermatogenesis occurred within rat seminiferous tubules. These observations suggest that the germ cell plays a dominant role in the timing of the differentiation processes in spermatogenesis independent of the genotype of the Sertoli cells that provide structural and functional associations with all germ cells. It is important to emphasize that the duration of spermatogenesis, or the time required for a spermatogonium to produce a spermatozoon, is not confused with reports of accelerated maturation of spermatogenesis (741,742). In these studies, fragments of testicular tissue from neonatal mice, pigs, or goats or from prepubertal monkeys (aged 13 months) were transplanted into murine hosts. Spermatogenesis developed 2–3 weeks earlier in grafted pig tissue compared with age-matched normal testis development, and in grafted monkey fragments, full spermatogenesis occurred 7 months after transplantation, whereas in normal monkeys this occurs at 4 years of age. A similar acceleration of maturation of neonatal testicular tissue from ram lambs occurs when xenografted into host mice (743).

In the normal adult-type seminiferous epithelium, it appears that the germ cells progress through the spermatogenic process at predetermined rates or subsequently degenerate. This view has been challenged by Russell et al. (744), who noted that spermatids in procarbazine-treated rats were found out of phase with the normal cycle of the seminiferous epithelium. The existence of a fixed rate of development of the germ cells results in an important corollary, namely, the volume occupancy (or absolute volume if referring to the testis) of a germ cell in sections of the epithelium is a reflection of the length that the cell occupies in the spermatogenic process. Thus, secondary spermatocytes have a short life span, and the stage of the cycle at which they are seen (stage 14 in

the rat, stage 6 in human) are represented infrequently in sections of the seminiferous epithelium.

Coordination within the Seminiferous Epithelium

The remarkable organization of the seminiferous cycle and the demonstration that the duration of spermatogenesis in each species is a biological constant have led to speculation as to the method by which this coordination is achieved. Two basic concepts of the method of synchronization were proposed by Roosen-Runge and Giesel (714): (a) that the regulated development depended on the precise rate of each stage of cell development because of inherent timing devices or (b) that synchronization was dependent on external factors.

Of external factors, the Sertoli cell has emerged as a potential candidate for coordination (745). The demonstration of the existence of specialized inter-Sertoli cell junctions (409) forming the blood–testis barrier (571) emphasizes the close relationship of this cell to the germ cells, particularly those located on the luminal side of the barrier. Though favorably placed to exert a coordinating influence based in a radial direction in the epithelium, the Sertoli cell is less able to extend its coordinating influence along the length of the tubule unless the specialized inter-Sertoli cell junctions facilitate communication between adjacent Sertoli cells. The extensive arborization of the Sertoli cell extends a potential influence over a number of germ cells in various phases of development, raising the potential of it influencing these cells by the secretion of local regulators, a hypothesis put forward many years ago but still lacking evidence (395). Many Sertoli cell products have been shown to influence distinct steps in germ cell maturation, including stem cell factor, inhibin, activin, and BMPs [reviewed in (92,101,746–750)]. The phagocytosis of the residual bodies of spermatids has been proposed to constitute a signal to the Sertoli cell that may be important in controlling spermatogenesis (395,751). Clermont (155), however, discounted the Sertoli cell as a factor based on (a) the fact that cyclic activity is established within the epithelium before the Sertoli cells attaining their mature cytological characteristics (752), and (b) the observation that in hypophysectomized rats, the Sertoli cells are regressed yet the germ cell stages that are present in the epithelium maintain their characteristic cell associations (736). Furthermore, coordination within the epithelium exists in the absence of residual bodies either before their formation (752) or subsequent to irradiation (753). He concluded that the cycle of the epithelium is the direct consequence of the entrance, at fixed intervals, of spermatogonial stem cells in spermatogenesis and of the fixed duration of the various steps of spermatogenesis. The demonstrated retention of the normal rate and cytological organization of donor-derived spermatogenesis in the transplantation studies of germ cells between rat and mouse testes, and vice versa (737–740), provides convincing evidence for coordination of cell proliferation and maturation that is intrinsic to the germ cells.

The nature of such inherent timing devices has not been resolved, but more recent studies have emphasized the extensive links between synchronously developing germ cells through the formation of cytoplasmic bridges associated with incomplete cytokinesis (174). These bridges, which link adjacent cells at all stages of germ cell development, effectively create a syncytium, and, based on calculations, up to 512 spermatids may be linked in such a manner (174). The existence of large numbers of cells joined in this manner was amply demonstrated by Moens and Hugenholtz (754), who observed 80 spermatocytes joined by cytoplasmic bridges. By incomplete cytokinesis of spermatogonial stem cells, cohorts of cells remain linked in a syncytium, enabling coordinated subsequent cell division.

In an extensive reexamination of this hypothesis, Ren and Russell (755) examined the number of germ cells linked by intercellular bridges by treating rat testis in vitro with cytochalasin D, leading to the formation of symplasts. They concluded that because the clone size so determined was greater than the smallest known unit of synchrony within the epithelium, the intercellular bridges were the probable mechanism to allow coordination. This coordination is likely to result from the relative ease of cytoplasmic transport of coordinating molecules, and in fact, "cytoplasmic flow" was found to occur in some of the changes noted after cytochalasin treatment (756). In other studies, Weber and Russell (757) demonstrated that the intercellular bridges are not static structures but vary considerably in form during spermatogenesis. The germ cell transplantation techniques have provided new insights into the relative roles of germ cell factors versus external influences in the coordination of spermatogenesis, and the available evidence indicates that both parameters are responsible for synchrony and completion of spermatogenesis. When testicular tissue obtained from fetal or neonatal Sprague-Dawley rats was transplanted into the testes of recipient Long Evans rats, the transplants colonized the host seminiferous tubules by establishing "tubules-within-tubules." The minitubules showed synchronization of spermatogenesis with that of the surrounding recipient tubules, suggesting that the microenvironment within the tubule is the governing factor in determining the coordination of the

spermatogenic cycle (758). Germ cell transplantation experiments from Brinster's group have shown that rat spermatogenesis can proceed in association with mouse Sertoli cells, and vice versa, but it is in these two species only that xenogeneic germ cell transfer results in the production of donor-derived progeny to date (737,740,759,760). Transplants of donor germ cells into mainly immunodeficient nude mice from rabbit, cat and dog (761), farm animals (741,762–764), and monkeys and humans (765–767) result in long-term establishment of colonies within recipient seminiferous tubules, but none shows evidence of spermatogenesis, the most advanced cell type being one or more types of spermatogonia. An exception is noted for hamster donor germ cells that colonize recipient mouse tubules and establish full spermatogenesis where it is believed that the hamster germ cells develop in association with mouse Sertoli cells. However, the hamster germ cell associations are less well organized compared with the normal hamster seminiferous epithelium and show abnormal spermiogenesis and elongate spermatids that lack fertilizing ability (740). Together these observations indicate that with increasing phylogenetic separation, donor germ cells from one species lose the capacity to establish autonomous spermatogenesis in unrelated recipient testes because the local intratubular conditions, most likely regulated by the particular type of Sertoli cell, are incompatible with xenogeneic germ cells. It seems likely that coordination of spermatogenesis is dependent on the support of species-specific Sertoli cells that interact with germ cells, the success of which is determined by the nature of the molecular relationships between these cell populations.

The possibility that the intrinsic regulation and therefore coordination of germ cell development is due to a circadian clock mechanism based on cyclic expression of local clock genes has been investigated in the mouse testis (768,769). In relation to known clock gene transcript and protein levels, none of the core circadian genes shows cycles within the seminiferous epithelium. The absence or suspension of clock genes in the testis is similar to the absence of their cycling in the thymus (768), suggesting that organs with stepwise highly differentiated cell types rely on other as yet unknown developmental clock genes for coordination of cell development.

A final aspect of coordination of the seminiferous epithelium to consider is the difference in the testicular environment during the first wave of spermatogenesis in comparison with that present in adulthood. As noted previously (see sections on Developmental Considerations), during the period of testicular growth, somatic cells are dividing and differentiating in response to hormonal and locally derived factors, and the production, and hence bioavailability, of these factors changes as the testis matures. For example, mouse Sertoli cells produce relatively high levels of BMP4 during the first postnatal week of life, but the level is substantially reduced at 14 dpp and relatively negligible thereafter (101). This factor appears to drive entrance of germ cells into the first wave of differentiation by enhancing production of cell surface c-kit receptor protein, and the dramatic reduction in BMP4 synthesis as Sertoli cells mature suggests that other cues govern subsequent germ cell development. In the rat testis, the related members of the transforming growth factor-β family of ligands, activin A and inhibin B, decline sharply on a per Sertoli cell basis during the first 2 weeks after birth (102). On the basis of these data, it becomes plausible to suggest that the influence of the Sertoli cell on germ cell maturation is greater during the first wave of spermatogenesis, as the niche for germ cells is undergoing establishment, and that intrinsic control mechanisms are established with the precise cellular arrangements into stages which characterize the adult testis.

Spermatogonial Stem Cell Renewal

Because it is likely that one of the major factors responsible for the coordination seen in spermatogenesis is the regular division of spermatogonia, it seems pertinent to consider their pattern of division in more detail. Spermatogonial renewal is also of importance for the continuation of spermatogenesis throughout adult life, because there must be a mechanism by which they are renewed and yet provide a pool of cells to enter the later stages of the process. Early studies proposed that this was achieved by the products of spermatogonial division being different (i.e., bivalent), with one daughter cell remaining a spermatogonium and the other a primary spermatocyte (770). This concept ignored the fact of different types of spermatogonia as described by Regaud (49). Further studies by Roosen-Runge and Giesel (714) recognized that multiple spermatogonial divisions occur throughout the seminiferous cycle. Subsequently, Clermont and Le Blond (68) proposed a scheme to account for the five peaks of mitoses observed in the epithelium during the seminiferous cycle. In their view, some type A spermatogonia divided in "bivalent" fashion to form stem cells, which remained dormant throughout the cycle, and type A spermatogonia, which entered the cycle.

Subsequent development of concepts in this field was the direct result of improved methods of identifying spermatogonial cell types and the study of spermatogonia in whole mounts of seminiferous tubules (164,165,173). In the rat, five different types

of spermatogonia were identified, A_0 and A_1 through A_4, with A_0 viewed as a reserve stem cell that did not divide unless the epithelium was damaged by agents such as irradiation (164,165). The renewing stem cells were proposed to arise by one of the A_4 spermatogonia dividing to form two A_1 spermatogonia. This proposal seemed unlikely in view of the observation that groups of dividing spermatogonia remained linked by cytoplasmic bridges, making it difficult to accept that one member of such a cohort could act independently of the rest (174).

Subsequent studies of Huckins (173,771–773) using whole mounts of rat seminiferous tubules in combination with autoradiography with [^{3}H]-thymidine identified three spermatogonial compartments (Fig. 67), the stem cells (As), the proliferating cells (Apr and Aal), and the differentiating cells (A_{1-4}, In, B). The difficulty in identifying and studying the behavior of the spermatogonial stem cells is emphasized by their rarity: They comprise only 1 in about 3,500 cells in the adult mouse testis (774) and 1 in about 500 cells of the adult rat testis (154,173). The stem cells divide sporadically to replicate themselves as isolated entities and to provide pairs of spermatogonia (Apr). The latter engage in a series of synchronous divisions leading to the formation of chains of spermatogonia joined by intercellular bridges, the aligned spermatogonia (A_{al}). The proliferating compartment, on approaching its final size, ceases mitotic division, and the Aal cells differentiate synchronously into A_1 spermatogonia, which then mature synchronously into the more differentiated spermatogonial types (A_2, A_3, A_4, In, B). The type B spermatogonia subsequently differentiate to form preleptotene spermatocytes. This view has been supported by studies in the mouse (775,776), the Chinese hamster (777,778), and the ram (779). The concept proposed by Huckins (173) fits well with the observed data of increased numbers of proliferating cells joined by intercellular bridges (174).

Numerous studies were performed from 1960 to 1970 on different species, quantifying the number of spermatogonial divisions and proposing schemes of spermatogonial multiplication (Table 3). These have been extensively reviewed by Clermont (155) and de Rooij and Russell (158), and it is generally accepted that mouse or rat spermatogonia require a series of 9–11 divisions to produce a spermatocyte. The number of spermatogonial divisions in primates and humans is less well characterized. With the recent advances in syngeneic, allogeneic, and xenogeneic transplantation techniques for studying the identity and behavior of spermatogonial stem cells and the germ cells derived from them, there has been intense interest in defining more clearly the morphology and kinetic properties of the various subsets of spermatogonia. A number of reviews are available with emphasis on the biology of mouse spermatogonia (158,179,780–783). An important outcome of recent studies of spermatogonia in culture is the emerging data that define the signaling pathways that maintain germline stem cells (784) and the identification of gene transcripts that correlate with stem cell quiescence, proliferative activity, or cell death (785).

The past 10 years have produced a revolution in our understanding of male germline stem cells as the direct result of an assay developed to identify these cells [reviewed in (760,786)]. The pioneering studies of Brinster and colleagues used transplantation of germ cells (donor cells) into the testes of recipient animals to examine the nature of the stem cells and the properties of their environment that influence their development (759). Spermatogonial stem cells

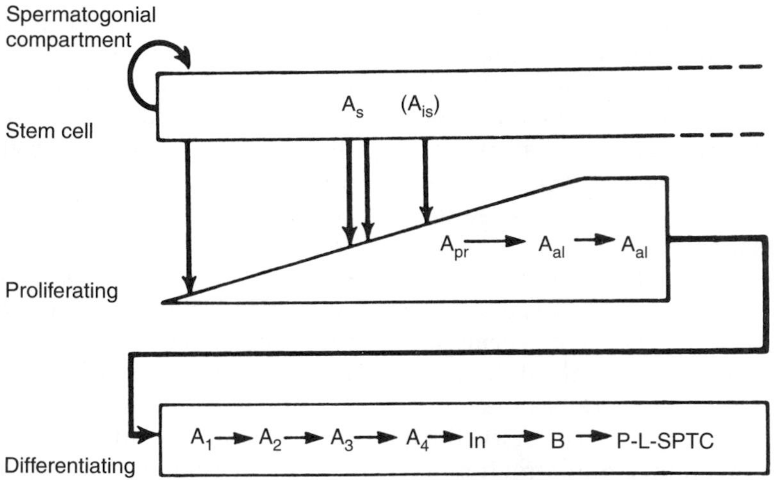

FIG. 67. This figure diagrammatically illustrates the proposed model for renewal and differentiation of spermatogonia. [From Huckins (173), with permission.]

can be maintained in culture for months, they can be frozen in a simple solution and recovered, they can establish their numbers in a different species, and they can restore spermatogenesis in infertile recipients (178,740,741,761,762,787–789). The latter point highlights the species-specific differences in germ cell cues, and this provides an opportunity to uncover more information about the nature of the mechanisms that underpin spermatogenesis. It must be noted that the property to colonize a recipient testis is the only definitive test available to identify spermatogonial stem cells, and it is both labor intensive and protracted.

Tremendous progress has been made toward identification of the surface properties of spermatogonial stem cells through two distinct experimental paths, and this provides the opportunity to develop methods to isolate them in sufficient numbers for study and manipulation. In one approach, beginning with differential adhesion to extracellular matrix components and subsequent analysis by fluorescence activated cell sorting, murine spermatogonial stem cells were shown to express integrin $\alpha6$ and integrin $\beta1$ on their surface, while at the same time they lacked cell surface c-kit protein, which is a marker of differentiating type A spermatogonia (790). Parameters for fluorescence activated cell sorting of germ cells enables significant enrichment (791), and there are now mice that express transgenic reporter constructs in the spermatogonial stem cell population that can be used to isolate and study these cells (792).

A second valuable route to identification of male germline stem cells and their biology came from the observation that glial-derived neurotropic factor (GDNF) regulated the number of stem cells in the developing mouse testis (793). Normally a Sertoli cell–derived product, over-expression of GDNF in germ cells or Sertoli cells led to production of excess numbers of spermatogonia, which bear the receptors for this protein, GFR1α and c-ret, on their surface (793,794). Reduction of GDNF synthesis in a heterozygous GDNF knockout mouse produced mice with abnormally low numbers of spermatogonia. (793). In both of these mouse models, spermatogenesis became severely impaired, demonstrating the importance of both an adequate number of germ cells and the correct ratio of germ cells to Sertoli cells for the maintenance of spermatogenesis. The application of GDNF to support spermatogonial stem cells in culture has been demonstrated using 7-day cultures that were subsequently transplanted into recipient testes (795). In contrast, the addition of BMP4 or activin A to spermatogonial cultures decreased spermatogonial colonization potential, and this may result from the enhancement of spermatogonial differentiation by these factors (101).

Growth of gonadal tissue in ectopic sites has been used to demonstrate that the integrity of testicular function with regard to spermatogenesis can be maintained in tissue fragments (741,796). This experimental approach has recently been exploited to grow testicular xenografts from species as diverse as humans, monkeys, pigs, cats, and goats in the backs of nude mice (742,797). Fragments of about 1 mm^3 taken from an immature (i.e., prepubertal) testis have the capacity to develop full spermatogenesis, and the rate of this development is notably accelerated for species that normally would have a prolonged period of prepubertal stasis (742).

Arising from the highly dynamic area of stem cell research, with novel insights and applications constantly emerging, additional opportunities for producing and studying male germline stem cells have emerged. Based on the principle that the embryonic stem (ES) cell has the property to develop into any cell lineage (i.e., it is pluripotent), several research teams have endeavored to produce male germline cells from murine ES cells. Two approaches have successfully yielded sperm with the capacity to fertilize a mouse oocyte. One strategy is based on the requirement for BMP4 signaling to drive establishment of the germline in the embryo (798–800). Growth of ES cells containing a germline promoter (from the mouse *vasa* homolog) linked to a reporter (enhanced green fluorescent protein or β-galactosidase) in aggregates with BMP4-expressing cell lines resulted in enhanced frequency of ES cell entrance into the germline lineage within 24 hours, as assessed by measuring and selecting cells expressing the reporter gene. The selected cells were next left to aggregate overnight with a dispersed population of cells from an E13.5 gonad, and the formed aggregates were transplanted under the testis capsule of recipient mice. Within 3 months, sperm could be retrieved from these grafts that contained the transgene, indicating that this protocol enabled derivation of male germline stem cells from ES cells. A second approach used the production of embryoid bodies from ES cells that were subsequently cultured with retinoic acid for 5 days, the latter treatment having been previously shown to enhance primordial germ cell proliferation (801,802). In experiments confined entirely to the culture dish, haploid cells were isolated and shown to be capable of driving oocyte development and contributing to blastocyst formation after their intracytoplasmic injection into recipient oocytes. These studies have illustrated new avenues for generating genetically modified male germ cells for biochemical and cell biological analyses.

A serious impediment to the study of male germ cells, which is in contrast to most other cellular systems, has been the difficulty of introducing ectopic

DNA for expression. Reports of chemical and electrical methods of DNA transfection into male germ cells are uncommon, and the efficiency of this approach is low (803–807). However, successes have been reported using retroviruses with both rat and mouse germline cells (154,766,808). Clearly, this is an area where much effort is currently directed, with the expectation that we shall have much more powerful tools with which to study the intrinsic control of spermatogenesis than ever before.

Spermatogonial Wave

von Ebner (159) recognized that the stages of the seminiferous cycle were distributed along the length of the seminiferous tubule in an orderly sequence, thus introducing the concept of the spermatogenic wave. Regaud (49) correctly interpreted the significance of this wave by the statement that the wave is in space what the cycle is in time. Detailed histological studies have analyzed the distribution of stages along the length of the seminiferous tubule (Fig. 68) and found that the subdivisions between stages are irregular but distinct (752). Each rat seminiferous tubule contains approximately 12 complete spermatogenic waves, each approximately 2–6 cm in length (752). Studies using autoradiography of whole mounts of seminiferous tubules demonstrated orderly mitotic activity of spermatogonia according to the stage of the cycle, providing further evidence to support the concept that the cycle of the seminiferous epithelium is coordinated by the mitotic activity of the spermatogonia (809). The wave has been identified in a number of mammalian species such as the mouse, bull, mink, guinea pig, rabbit, boar, dog, cat, and marsupials (712,752,810–812). A significant step forward in the study of the establishment of the spermatogenic wave has come from the pioneering work of Brinster and coworkers, who successfully transplanted germ cells isolated from a donor testis into an appropriately prepared (germ cell–deficient) recipient testis (759,813). Identification, cell kinetics, and lineage tracing of transplanted donor germ cells was achieved by the addition of a transgene consisting of metallothionein DNA sequences together with the *Escherichia coli* lacZ structural gene (*lacZ*). This fusion gene is expressed in both Sertoli cells and germ cells and allows them to be identified by blue staining after incubation with a β-galactoside enzyme (X-gal). Thus, any blue cell in a nontransgenic recipient testis is of donor origin. Transplanted mouse germ cells colonize recipient seminiferous tubules as seen by variable lengths of proliferating and maturing germ cells that stain blue (181,182,737,740,814).

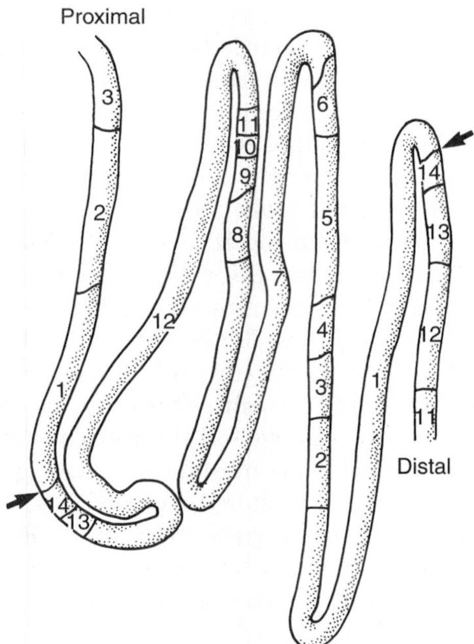

FIG. 68. The distribution of the stages of the rat seminiferous cycle that constitute the seminiferous wave along the tubule. [Redrawn from (752).]

With increasing time after transplantation, individual colonizing segments spread laterally along the seminiferous tubules and establish individual germ cell colonies capable of full qualitative spermatogenesis. The ability of transplanted germ cells to establish small and then increasingly large colonies of differentiating germ cells is very likely a consequence of spermatogonia settling upon the basal lamina of the recipient tubule where in association with one or more Sertoli cells, the so-called niche is founded as mentioned earlier in the section on spermatogonial morphology. In some cases the expansion and synchronization of colony activity may occupy more than 40% of recipient seminiferous tubules (814). Labeling of mouse donor germ cells has been further explored with the use spermatogonia carrying green fluorescent protein gene (GFP) as a marker (814–818). These studies have shown that transplant-derived germ cell colonies keep expanding along seminiferous tubules for more than a year, and such colonies may be serially transplanted up to four times from one recipient to the next (816).

Obviously, the proliferative capacity of the stem cell spermatogonia, the only donor cell type able to establish renewed proliferation of spermatogenesis, is particularly robust and within a suitable microenvironment within the seminiferous tubule, possibly inexhaustible with regard to proliferative potential. Such colonies do establish new waves of spermatogenesis (815). Early colonies usually show the same germ cell association or stage throughout their lengths, but

Therefore, the mechanism responsible for driving the contents through the lumen of the resting epididymis has been attributed primarily to the rhythmic muscular contractions of the smooth muscle lining the epididymal tubule (156–159). These elegant studies have shown a relationship between the electrical and contractile activities of the epididymal tubule and the progression of oil droplets injected in the lumen. The droplets do not move in a linear fashion, but back and forth. The movement forward starts when the spread of electrical activity approaches the droplet and stops or reverses when the electrical activity wanes or changes direction. Because the electrical pacemakers are randomly distributed and the electrical pulse spreads in both directions, droplets injected at the same time may spread in the lumen and follow different courses. The net distance covered during each pendular movement is small. Nevertheless, the droplets progress downward towards the vas deferens, where the frequency of electrical activity is lower than in the caput epididymidis (156,158). The rate of luminal flow is not uniform in the different segments of the epididymis (158–160). The progression of droplets decreases from 420 mm/2 hours in the initial segment to 64 mm/2 hours in the distal caput and 25 mm/2 hours in the cauda epididymidis and vas deferens (158). It is likely that the progression of oil droplets mimics that of spermatozoa in the epididymis. In bulls, labeled gold-coated beads injected in the rete testis are grouped in the distal caput epididymidis 2 days after injection and in the cauda 5 days after injection, a transit time comparable with that of labeled spermatozoa, but 6 days after injection, beads are spread throughout the cauda (146). Therefore, the progression of the epididymal luminal content is a dynamic process, resulting in a mixing of luminal content, and is controlled by the electrical and contractile activity of the epididymal tubule.

The smooth muscle contractions of the epididymal tubule and transit of spermatozoa therein are influenced by several factors, both hormonal and neuronal. Castration depletes epididymal sperm reserves (161–163) and increases intraluminal pressure, contractility of the epididymis (164), and sperm transport (165). Testosterone treatment reverses the effect of castration, indicating that androgens control the contractility of the epididymal tubule to ensure an optimal rate of sperm transport. Estrogen, on the other hand, speeds up murine sperm transport drastically from 9.7 to 2.1 days (166).

Contractility of the epididymal tubule is also influenced by PGs (167,168). $PGF_{2\alpha}$ increases the frequency and amplitude of contractions in proximal epididymis tubules in vitro, whereas PGE_2 decreases these contractions (168). The endogenous levels of PGs are consistent with their regulation of basal contractility of the proximal epididymis (168).

Neurohypophysial peptides, such as oxytocin or vasopressin, mediated by receptors present in the epididymis (169–173), also increase epididymal contractility both in vitro (174–176) and in vivo (177–180). In several species, including humans (169), this results in an increase in the number of ejaculated spermatozoa or spermatozoa transported through the epididymis (179,181–187). Although the relative effect of oxytocin and vasopressin may vary among species, depending on the dose of the peptide or on the epididymal segment used in the in vitro studies (180,187), it is clear that neurohypophysial peptides regulate both basal contractility of the epididymis and, on release in the peripheral circulation around the time of ejaculation (188–192), transport of spermatozoa through the vas deferens. Interestingly, the effect of oxytocin on epididymal motility is regulated by estrogens in part by an upregulation of the oxytocin receptor gene and protein (175). This may account for the estrogen-induced accelerated sperm transport (166), and suggests an interplay between steroids and neurohypophysial hormones in the regulation of epididymal contractility.

Neuronal regulation is also involved in epididymal contractility and sperm transport. This was first demonstrated by Simeone in 1933 (193) using surgical sympathectomy and confirmed in later studies using either surgical (168) or guanethidine-induced chemical sympathectomy (194–197). Surgical removal of a single neuronal ganglion, the inferior mesenteric ganglion that provides sympathetic innervation to the cauda epididymidis, is sufficient to slow sperm transport through the epididymis (198). Indeed, adrenergic and cholinergic drugs affect contractility of the epididymis both in vitro (199) and in vivo (179,200–203). Temperature also affects epididymal contractility: a switch from scrotal to body temperature increases the frequency and spread of electrical activity of the smooth muscle of the epididymis (204), and significantly speeds up sperm transport through the epididymis (205), thus having potential deleterious effects on sperm maturation and fertility.

Although a neuromuscular mechanism may not be the only one responsible for sperm transport in the various segments of the epididymis and in all species, it appears to be the main mechanism responsible for sperm transport through the epididymis in mammals.

Maturation of Spermatozoa

Fertilizing Ability

In lower vertebrates, such as cyclostomes, fish, and amphibians, spermatozoa released from the testis are fully motile and competent to fertilize. In contrast, in higher vertebrates, spermatozoa become functionally mature as they pass through the epididymis. They acquire the ability to move forward when released from the epididymis, ascend the female genital tract, undergo the acrosome reaction, bind to and penetrate the egg vestments, and achieve syngamy with the female gamete.

In all species examined thus far, a gradient of fertilizing potential is observed as spermatozoa traverse the epididymis (Fig. 9). Although there is some species variation with respect to the exact site at which spermatozoa first gain their fertilizing potential, it is clear that to be competent to fertilize in vivo, spermatozoa leaving the testis have to pass through some part of the proximal epididymis.

This fertility profile can be altered somewhat using in vitro fertilization (IVF) techniques that bypass some or all steps in the normal fertilization process. For example, rabbit spermatozoa from the caput epididymis, which do not fertilize more than 2% of eggs after in vivo insemination (20), fertilize 8% of eggs after in vitro insemination (206). A similar shift in fertility profile is observed when mouse spermatozoa are inseminated in vivo, in vitro, or after subzona injection (Fig. 10). Testicular spermatozoa injected directly into the oocytes can even achieve a fertilization rate of 94% (207). Fertilization also can be achieved when round spermatids are injected or electrofused with oocytes (208), or when secondary or

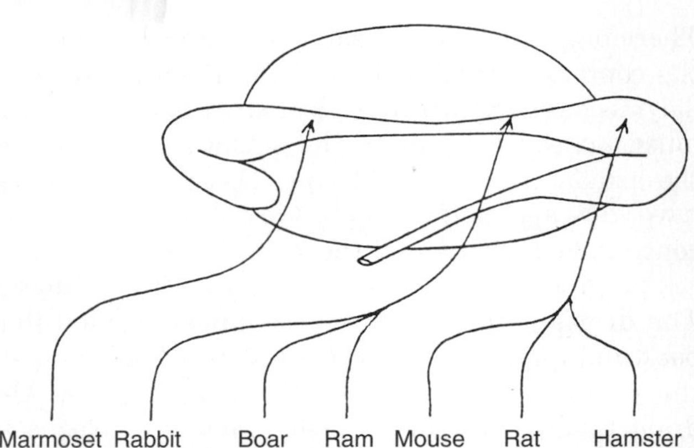

FIG. 9. Approximate site in the epididymis where spermatozoa acquire their fertilizing ability. [Information for each species, as brought together by Orgebin-Crist (938), was obtained from the following sources: marmoset (939), rabbit (20,210,940) boar (941), ram (942), mouse (943,944), rat (161,211,244), and hamster (945–948).]

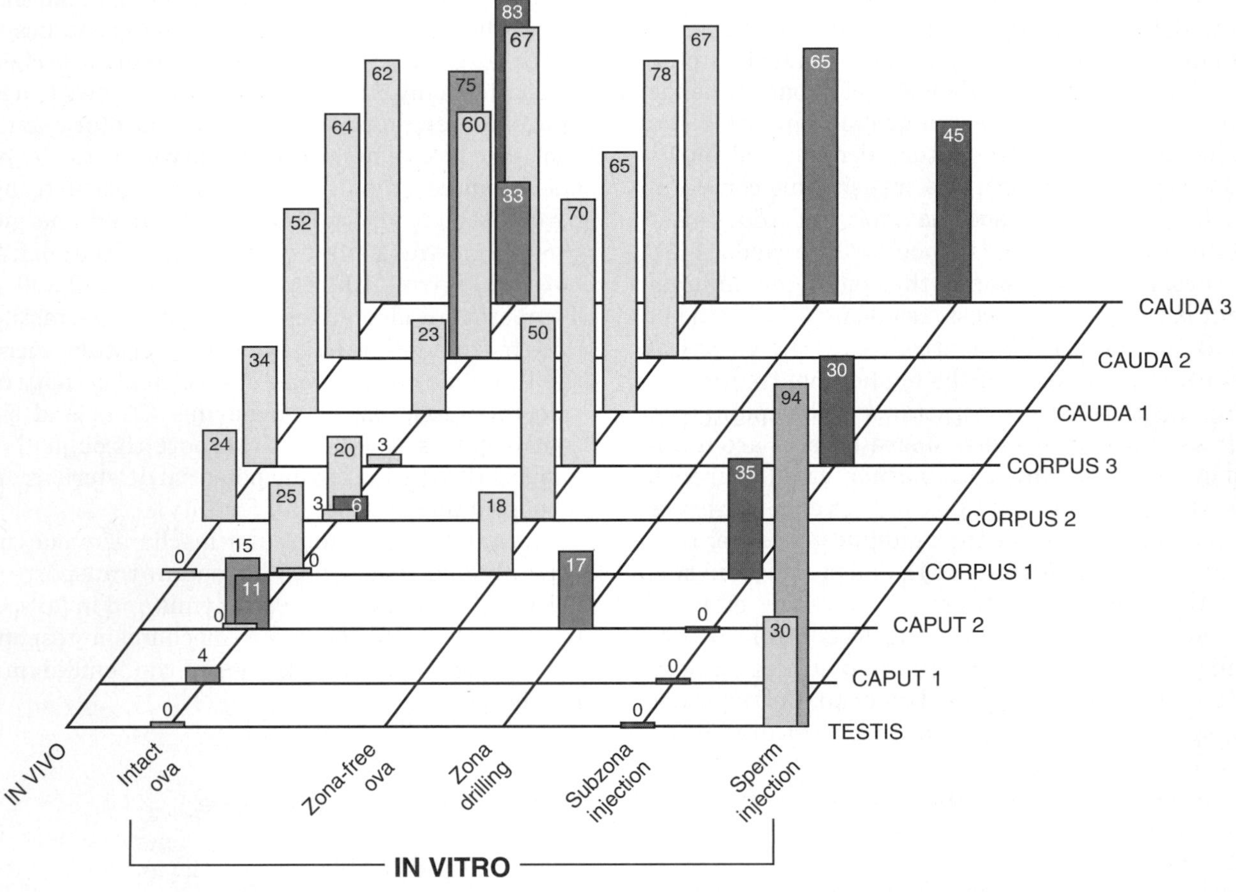

FIG. 10. Fertilizing capacity of mouse spermatozoa from successive segments of the epididymis inseminated either in vivo (943) or in vitro with cumulus-encased or cumulus-free ova (218), with intact ova (943,944), with zona-free ova (944), after zona drilling (217), or by subzonal sperm injection (949).

even primary spermatocytes are injected into the ooplasm (207,209). However, the fertilization rates are lower than after injection of testicular spermatozoa.

Moreover, it appears that the acquisition of fertilizing potential is not a simple "on" or "off" state. Initial studies with rabbits (20,210), rats (211), and rams (212,213) showed that spermatozoa first gain the ability to fertilize eggs, and only after further transit through the epididymis acquire the ability to produce complete litters of viable offspring. Orgebin-Crist noted an increase in both preimplantation and postimplantation loss when rabbits were inseminated with corpus versus ejaculated spermatozoa (20); this is apparently due to a delay in the fertilization of eggs (20,214) and in the first zygotic division (215). Although Overstreet and Bedford (216) were unable to confirm these observations, more recent studies in mice (217) reported that only 8% of oocytes fertilized by caput epididymal spermatozoa were capable of developing into blastocysts in vitro, compared with 48% of oocytes fertilized by cauda spermatozoa; Lacham-Kaplan and Trounson (218) confirmed that a high rate of embryonic arrest and retarded development occurs in mouse oocytes fertilized by epididymal spermatozoa that have just gained their fertilizing capacity. Although testicular spermatozoa and spermatids fertilize 94% and 37% of eggs, respectively, after injection into the oocytes, only 54% and 28%, respectively, of the fertilized eggs develop into live offspring when transferred into foster mothers (207). Collectively, these experiments show that passage through the epididymis endows spermatozoa with the ability to ascend the female genital tract and interact with the egg. This maturation process can be circumvented because spermatid-injected oocytes develop into live offspring, but the chances of normal development are higher after injection of more mature spermatozoa.

The experimental protocol of insemination of equal numbers of spermatozoa from succeeding segments of the epididymis from one individual is obviously not possible in humans. However, in cases of obstructive azoospermia, epididymovasostomies have been done to recanalize the epididymis. The cumulative results show a low pregnancy rate when the anastomosis is done proximally and an increased incidence of pregnancies with more distal connections (219,220). These results, although not directly comparable with the animal studies, imply that in human, as in other species, there is a progressive maturation of sperm fertilizing ability in the epididymis. They also suggest that the fertility profile may be shifted more proximally compared with other species. Pregnancies have been achieved even after reanastomosis of the efferent ducts to the vas, but the postoperative interval before pregnancy occurred was longer (2 years) (221) than after corpus–vas anastomosis (6 to 13 months) (220). If spermatozoa from the efferent ducts had the same level of maturity as spermatozoa from the corpus, one would expect the same postoperative interval, but the small number of cases precludes any generalization.

In cases of obstructive azoospermia or congenital vas agenesis, spermatozoa can also be aspirated from the epididymis and inseminated in vitro (IVF). There is a statistical difference in fertilization rates with spermatozoa retrieved from different levels of the epididymis (222), confirming the progressive maturation of human spermatozoa in the epididymis observed after reanastomosis (219). After IVF and transfer, pregnancies have been reported with spermatozoa recovered from the corpus epididymidis (223–225), and even the proximal caput epididymidis (226,227). However, the consequences of either obstructive azoospermia or congenital vas agenesis on the structure and functions of the epididymal epithelium have not been investigated.

If immature human spermatozoa bypass normal sperm ascent in the female genital tract and sperm–egg interaction, and are injected directly into the ooplasm (intracytoplasmic sperm injection [ICSI]), they can form zygotes. As in other species, the overall fertilization rate is higher than after conventional IVF (45% versus 6.9%) (228). Even testicular spermatozoa achieve fertilization rates of approximately 50%. Nevertheless, in all studies, there is a small but consistent difference in fertilization rates after injection of testicular and epididymal ejaculated spermatozoa (229–236). In all studies but one (234) that reported statistical analyses, the differences in fertilization rates were significant.

Delayed fertilization and cleavage arrest were reported in one case of IVF with spermatozoa from the corpus epididymidis (224). In a large series (236) comparing outcome after ICSI with testicular, epididymal, and ejaculated spermatozoa, testicular spermatozoa had not only statistically significantly lower fertilization rates (53.4%) than epididymal (58.5%) or ejaculated spermatozoa (64.8%), but statistically lower yields of good-quality embryos (47%, 40%, and 59%, respectively). The embryos chosen for transfer yielded a similar pregnancy rate per transfer (23.1%, 31.3%, and 27.4%, respectively). The congenital malformation rate and the developmental outcome of children born after ICSI with testicular, epididymal, or ejaculated spermatozoa are similar to that of children born after conventional IVF (237–239) or in the general population. Collectively, these clinical studies indicate that, as in other mammalian species, human sperm capacity for optimal fertility in vivo increases during epididymal transit, and procedures bypassing critical steps of the natural process of fertilization permit fertilization with immature spermatozoa. However, unlike other species, the fertility profile of human epididymal spermatozoa appears to be shifted more proximally because pregnancies have been reported with spermatozoa from the proximal epididymis.

Motility

Concomitant with the acquisition of fertilizing ability, a number of sperm characteristics undergo maturational changes. The first one to be recognized last century, by Tournade (8), was the epididymal maturation of the sperm potential for motility. This was subsequently confirmed in numerous species of higher vertebrates, from lizard (240,241), to mouse (242), rat (211,243–251), hamster (252,253,253a), guinea pig (254,255), rabbit (20,251,256,257), boar (258,259), goat (260), ram (261–263), bull (264,265), monkey (266–268), and human (75,269–271). The acquisition of motility is observed whether the testis and epididymis are located in the scrotum or in the abdomen, as in the elephant (272), the hyrax (272,273), and the armadillo (273). The maturation of sperm motility potential involves both a quantitative increase in the percentage of motile spermatozoa and a qualitative difference in motility pattern. Testicular spermatozoa are either immotile or display only a faint twitch of the flagellum. Spermatozoa released from the caput epididymidis swim in a circular pattern, whereas spermatozoa released from the cauda move progressively and vigorously forward.

Other Maturational Changes

In addition to the capacity for progressive motility, epididymal spermatozoa develop the capacity to undergo the acrosome reaction [mouse (274), ram (275), pig (276), dog (277), monkey (278), and human (279)], recognize and bind to the zona pellucida [mouse (280), pig (276)], and fuse with the vitelline membrane as tested with zona-free hamster eggs [pig (281), human (281a,282)]. Concomitant with these functional changes, spermatozoa undergo structural changes during epididymal transit: migration of the cytoplasmic droplet along the sperm flagellum, acrosomal reshaping, changes in the sperm nuclear chromatin and some tail organelles, and changes in the sperm plasma membrane [reviewed in Bedford (283,284)]. Collectively, these changes underpin the functional maturation and subsequent storage of spermatozoa in the epididymis, but none of these changes has been shown to be the only determinant of the acquisition of fertilizing ability.

Regulation of Sperm Maturation

Benoit (12) demonstrated that the function of the epididymis, and the maturation and survival of spermatozoa within it, depended on hormones secreted by the testis. After bilateral castration, not only did the epididymal epithelium dedifferentiate, but epididymal spermatozoa died quickly, whereas after unilateral castration spermatozoa maintained their potential for motility on the castrated side 2 months after the operation. Experiments conducted by Bedford (19) and Orgebin-Crist (21,285) showed that although immature rabbit spermatozoa develop the capacity for motility and fertility when retained by ligatures in the proximal corpus, they do not become fertile when retained in the caput, although they develop a mature pattern of forward motility. This indicates that sperm maturation is a multistep process and shows that although the development of a mature pattern of motility is necessary, it is not sufficient to render a spermatozoon capable of fertilizing. It also shows that whereas some of the maturational changes, like motility, may be intrinsic to the sperm cell and develop with time, others, such as the ability to interact with the egg, depend on the epididymal environment. The latter is conditioned by testicular androgens because sperm maturation occurs after castration, hypophysectomy, or in organ culture only when androgen is administered in vivo or included in the culture medium (161,286–288).

Studies of coincubation of epididymal spermatozoa with epididymal epithelial cell cultures have confirmed the role of the epididymis in promoting sperm maturation. In such cultures, brushtail possum and tammar wallaby spermatozoa from the proximal caput underwent the morphological maturational change in head orientation accompanied by the development of progressive motility, normally only observed in vivo (289,290). In hamsters, mice, and humans, coculturing of immature spermatozoa from the caput or corpus epididymis with cauda cells not only increased their motility (291–293), but increased their capacity to bind to salt-stored zona pellucida (293), to fertilize (291,292), and to support the development of embryos (sired by corpus spermatozoa) (292). These maturational changes are promoted by androgen-dependent factors from epididymal principal cells because cocultures maintained in the absence of androgens fail to induce maturation (292,293). The overwhelming evidence from all these experimental studies, both in vivo and in vitro, is that the final stages of the sperm maturation process in all mammalian species studied to date, including humans, depend on the epididymis.

However, reports of human pregnancies in two cases where the vasa efferentia was anastomosed to the vas deferens, thereby bypassing the entire epididymis, led to a challenge to the concept that the epididymis is necessary for the acquisition of fertilizing ability (221). As discussed previously, subsequent reports noted that the more distal the anastomotic site, the greater the chances of fertility, confirming that the epididymis provides the milieu necessary for optimal fertility. Nevertheless, Temple-Smith (294) revisited the issue in elegant experiments using epididymovasostomies at different levels of the epididymis in rabbits and rats to assess the fertility of these animals. In both of these species, anastomosis of the distal initial segment to the vas, bypassing the distal caput, corpus, and cauda epididymis, resulted in a significant reduction in sperm viability, motility, and fertility, confirming that spermatozoa require exposure to the epididymal environment for normal development of fertilizing ability. Interestingly, in rats, bypass of the distal caput and proximal corpus, either by epididymo-epididymostomy or graft of a vas bridge, had no effect on sperm motility, viability, or fertility. These elegant studies show clearly that the epididymis plays a major role in the post-testicular maturation of spermatozoa, but they also show that exposure to the initial segment may be sufficient or that the lack of exposure to the epididymal regions where spermatozoa normally become fertile may be compensated by secretions from other epididymal regions. As discussed later, epididymal spermatozoa are exposed to a highly complex environment, but the nature of the changes induced by epididymal transit that are required to promote optimal fertilization remains to be established.

Storage of Spermatozoa

The major site for storage of spermatozoa in the excurrent duct system of mammals is the cauda epididymidis. Although normal transit time in mammals through the cauda epididymidis is in the range of 3 to 10 days (Fig. 8; Table 1, supplemental material on web site), spermatozoa can be stored in this tissue for periods extending beyond 30 days (295); in bats, spermatozoa may be stored in this tissue for many months and retain their function (296). With storage in the cauda epididymidis, a loss in fertilizing ability was found to occur before a loss in motility (297). Interestingly, spermatozoa that were aged in the male reproductive tract of rabbits, presumably in the cauda epididymidis, induced a tenfold higher incidence of chromosomal abnormalities in the resulting blastocysts than did their fresh counterparts (298).

Based on studies in a number of mammalian species [reviewed in Amann (299)], it was noted that 50% to 80% of spermatozoa present in the excurrent ducts were found in the cauda epididymidis and that approximately 50% of these spermatozoa were available for ejaculation (299). When animals (rabbits, stallions, and bulls) were at sexual rest (i.e., 7 or more days without ejaculation), the number of stored spermatozoa that were available for ejaculation was three- to fivefold greater than the daily sperm production rate and two- to threefold greater than that found in a "typical" ejaculate; frequent ejaculation, that is, one ejaculate every 1 to 2 days, did not result in a change in sperm production rate but did markedly decrease caudal epididymal sperm reserves and the number of spermatozoa in the ejaculate (299). In contrast, the human, whose sperm production rate is well below that of most other mammals, has a sperm reserve that is only approximately equal to the number of spermatozoa found in an ejaculate, whether at sexual rest or not (299).

In a series of studies on the evolution of the scrotum, Bedford proposed that the prime moving force behind the formation of the scrotum was the need to store spermatozoa at a low temperature (205,300,301); he suggested "that migration to a scrotal site has been ordained primarily by the sperm storage region—the cauda epididymidis—and that the function of the testis has only been an incidental factor in this evolutionary development" (301). This proposal was based on anatomical observations and on some physiological studies where the epididymis was placed abdominally while the testis remained in the scrotum; the rate of sperm passage through the cauda epididymidis was found to increase approximately twofold in the rabbit after such a procedure.

One would expect that the luminal environment required for the storage of spermatozoa in the cauda epididymidis versus that required for the acquisition of fertilizing ability would be different. Although many differences have been described in the luminal makeup of ions, small organic molecules, proteins, and glycoproteins between the cauda and the rest of the epididymis (see later), those special conditions that allow for the storage of spermatozoa in a quiescent state for long periods in this tissue have not yet been elucidated, although changes in pH (302) in several species and the presence of immobilin in the rat and hamster (302,303) have been proposed to be major factors.

Although it is clear that the major function of the distal segments of the epididymis is to store mature, live spermatozoa, other functions have been ascribed to this tissue. Abnormal-appearing or dead spermatozoa have often been seen in the cauda epididymidis (304,305), and the ability to recognize such spermatozoa and then develop a mechanism to neutralize or destroy them has been proposed recently by several groups.

Sutovsky et al. (2001) (306) proposed that the cell-surface ubiquitination of defective spermatozoa supplies the necessary signal for these cells to be phagocytosed by epididymal principal cells, thus providing a mechanism for sperm quality control. However, based on extensive histological observations of the normal epididymal epithelium and the rarity with which sperm phagocytosis is observed, it is apparent that phagocytosis of spermatozoa is an unlikely mechanism for the disposal of the large number of spermatozoa produced daily (307). If a quality control system is present in the epididymis, then it is likely to be more subtle than the phagocytosis of marked spermatozoa.

Studies from Olson's group (308,309), have identified a fibrinogen-like protein (fgl2) that is secreted selectively in the proximal cauda epididymidis of the hamster and that binds to and coats nonviable, but not viable, luminal spermatozoa. Unlike the hypothesis of Sutovsky's group, these investigators propose that the epididymis possesses a specific mechanism to identify and envelop defective spermatozoa with a protein complex containing fgl2 that remains within the epididymal lumen. Both of these novel mechanisms require that epididymal epithelial cells have the ability somehow to recognize and tag defective spermatozoa. Although conceptually appealing, more extensive studies using different species and means of creating "abnormal" spermatozoa, such as drug treatments, are required to establish this as a function of the normal epididymis.

In a review, Jones (2004) (310) has hypothesized that after androgen withdrawal, either by orchidectomy or as a result of seasonal variation in seasonal breeders, a death pathway is activated that leads to the dissolution of spermatozoa. Such a mechanism would ensure that the epididymis is cleansed of defective and dead spermatozoa in preparation for the next breeding period. How androgen withdrawal would activate such a pathway remains to be determined, but it is interesting to note that many genes are activated in the epididymis after orchidectomy (311).

Protection of Spermatozoa

It is clear that the blood–epididymis barrier allows for the production of a specialized luminal fluid microenvironment that is important for sperm maturation. However, another critical role played by this barrier is protection of the maturing spermatozoa. In addition, a series of elaborate defense mechanisms that help to protect spermatozoa from the immune system, harmful xenobiotics, and reactive oxygen species (ROS) have been developed in the epididymis. The defense mechanisms include restricting the types of compounds that can enter the epididymal lumen, the synthesis and secretion of specific proteins such as the defensins and defensin-like molecules, rapid elimination of potential harmful agents through the synthesis and secretion of antioxidant and conjugating enzymes, and the synthesis and secretion of antioxidant compounds such as glutathione and taurine. Because spermatozoa mature in a hyperosmotic environment, the epididymis also ensures that the spermatozoa are protected from potential rapid changes in osmolality and can regulate their cell volume.

It was shown some 50 years ago that spermatozoa are highly susceptible to oxidative damage and that hydrogen peroxide was responsible for loss of motility and cell death (312,313). Later, through a series of well-designed studies, Jones and Mann clearly showed that when spermatozoa were incubated under aerobic conditions, they produced an organic peroxide and released a substance that was believed to be a lipid (314–317). When incubated in the presence of lipid peroxides, spermatozoa became irreversibly immotile and released numerous intracellular enzymes. From this series of studies, it was suggested that incubation of spermatozoa under aerobic conditions resulted in lipid peroxidation by peroxides or lipid radicals that were responsible for structural damage, loss of motility, decline in metabolic activity, and release of intracellular enzymes. Because mammalian spermatozoa have a high content of polyunsaturated fatty acids in their membranes, they are highly susceptible to lipid peroxidation by ROS (317–320). It is crucial that spermatozoa be protected from the deleterious effects of ROS as they progress along the epididymal duct; lipid peroxidation of membranes has been correlated with midpiece defects (321), decreased motility due to axonemal defects, and reduced intracellular ATP levels (322–324), as well as impaired capacity for fertilization [reviewed in Vernet et al. (2004) (320)]. Spermatozoa are protected from ROS by superoxide dismutase (325–328). This enzyme protects spermatozoa from lipid peroxidation by ROS through dismutation of the reactive oxygen to hydrogen peroxide and water. Hydrogen peroxide is then rapidly converted to water by the enzymes catalase or glutathione peroxidase (328,329,330).

In a continuing effort to prevent the oxidative damage of spermatozoa, the epididymis has developed an elaborate system to ensure that spermatozoa are protected as they mature along the epididymal duct. Each region or segment of the epididymis has developed its own sperm protective mechanisms primarily because (a) the metabolic activity differs from one region to the next, thereby producing different ROS species that need to be eliminated appropriately; and (b) spermatozoa are in a different state of maturity in each region, and therefore differ with respect to susceptibility to oxidative damage. For example, spermatozoa need to be especially protected from ROS as they enter the initial segment. The luminal fluid in this region is highly oxygenated (331), the epithelial cells are surrounded by a dense capillary network (332), with blood flow exceeding that of the distal epididymal regions (333,334), and the initial segment cells are highly metabolically active [reviewed in Hinton et al. (335)]. This unique combination results in the generation of ROS from several sources, including the endothelial and initial segment epithelial cells [reviewed in Vernet et al. (320)]. Therefore, it is not surprising that the initial segment expresses antioxidant enzymes. In the distal epididymal regions, the epithelial cells are still metabolically active, and the spermatozoa are continually exposed to an oxygen-rich environment; however, there is less vascularization compared with the initial segment. Hence, spermatozoa are still subjected to ROS production, but the type of ROS may well be different from the types produced in the initial segment. Hence, the distal epididymal regions have alternative strategies to protect spermatozoa.

The major antioxidant enzymes present in the epididymis (320) include superoxide dismutase (336,337), γ-glutamyl transpeptidase (338–344), glutathione peroxidases (345–349), glutathione transferases (109,350–354), and indolamine dioxygenase (355). In addition, the lumen of the epididymis contains antioxidant molecules such as glutathione, taurine, and tryptophan as the substrate for indolamine dioxygenase. Each of these antioxidant enzymes and molecules is found to varying degrees throughout the length of the epididymis.

Hence, the epididymis plays a critical role in the protection of spermatozoa from oxidative stress and harmful xenobiotics. The processes by which the epididymis protects spermatozoa have also been considered to be a prime target for the development of a male contraceptive (320).

Microenvironment for Maturation, Protection, and Storage

From the previous discussion it can be seen that the epididymis plays an important role in sperm maturation. Although the actual time spent in the epididymis may be important for the maturation process, spermatozoa must still undergo considerable remodeling and be protected and stored in a specialized luminal fluid milieu. Crabo (1965) (152) first recognized the importance of the epididymal microenvironment: "The consensus today is doubtless that the epididymis actively promotes maturation of spermatozoa. The natural way for the epididymis to exert this influence would be via regulation of the sperm cells' environment, i.e., epididymal plasma: the environment in the caudal portion should favor maximal survival time of the spermatozoa which may be stored there for relatively long periods, without losing their fertilizing power and motility." In view of several published comprehensive reviews on the formation of the epididymal luminal microenvironment (144,356–361), this section summarizes the contributions of ions, organic solutes, and proteins to the specialized luminal fluid milieu. Although much is known concerning the composition of the epididymal luminal fluid, more studies are needed to understand the precise contribution of each component to sperm maturation.

Perhaps the most underappreciated role of the epididymis is the manner by which it forms the microenvironment. To accomplish this formidable task, the epithelium must be structured in a way such that (a) it prevents and regulates the entry of blood-borne substances into the lumen; (b) it has the ability to synthesize, secrete, and absorb components; (c) it is arranged so that spermatozoa come into contact with the appropriate environment at the appropriate time—this is the most challenging aspect to understand because it implies that each cell along the duct knows precisely what its neighbors are doing; and (d) it can respond to spermatozoa, which in turn regulate the microenvironment. Hence, there is a series of complex and ever-changing interactions between the epididymal epithelium, the microenvironment, and spermatozoa from the beginning to the end of this tube that is several meters in length.

Water Movement and Its Influence on the Formation of the Luminal Microenvironment

Before discussing the composition of the luminal fluid microenvironment, it is important to consider water movement. The efferent ducts and the epididymis are water-transporting epithelia. Hence, the concentrations of ions, organic solutes, and proteins depend on the movement of water into and out of the lumen. Thus, the measured concentrations of any ion, solute, or protein in epididymal luminal fluid are not only due to direct secretion into the lumen but also to the amount of water that is being reabsorbed. Water movement out of the duct is quite extensive.

For example, the water-transporting ability of the rat efferent ducts is 5 to 10 times less than that of the proximal kidney tubule, but the water-transporting rate of the efferent ducts of the quail almost matches that of the rat nephron (356). Remarkably, 70% to 96% of the fluid leaving the testis is reabsorbed in the efferent ducts in several species; for example, 96% is reabsorbed in rat (362), 90% in boar (152), 87% in the tammar wallaby (363), and 96% in the elephant (364), with concomitant declines in sodium and chloride concentrations but increases in potassium and hydrogen ion concentrations from the rete testis to the caput region. Within 24 hours of efferent duct ligation in the rat, there is a remarkable increase of nearly 50% in testicular weight due to accumulated fluid, thus demonstrating how effective the efferent ducts and epididymis are at removing large volumes of water (365). From the caput region onward, 95% of the remaining 5% to 10% water is reabsorbed, ultimately increasing the concentration of spermatozoa from approximately 10^4/mL in rete testis to 10^9/mL in the cauda epididymidis.

Hence, when analyzing the concentration of various ions, solutes, and proteins, it is important to take into account the reabsorption of water. For example, the increase in the concentration of some organic solutes seen from the caput to cauda epididymidis may not necessarily be due to an increase in the secretion of that solute, but entirely to water movement. Examples of this include carnitine and inositol (366,367). However, if water reabsorption is taken into account when calculating what the total protein concentration should be, a much higher value is calculated compared with the actual value measured (368). This would indicate that considerable protein is being reabsorbed, a well-known role of the epididymis. A simple analysis of this kind may help determine the extent, if any, of movement of ions, solutes, and proteins across the epididymal epithelia from proximal to distal regions. It should be emphasized that the measured concentration of any ion, solute, or protein in the luminal fluid of any epididymal region is a product not only of secretion, absorption, and water reabsorption but of sperm uptake, degradation, and metabolism.

Ionic Microenvironment

Crabo (1965) (152) and Levine and Marsh (1971) (369) were perhaps the first to show that the ionic luminal fluid composition of the epididymal duct is distinctly different from blood plasma. However, through a more thorough analysis of epididymal luminal fluid collected by micropuncture, we now know the exact ionic composition to which spermatozoa are exposed as they progress along the duct (369,370,371). Briefly, spermatozoa move from an environment in the seminiferous tubule that has a pH of approximately 7.3 and lower sodium concentration but higher potassium concentration than blood plasma [reviewed in Hinton and Setchell (372)], to an environment in the initial segment/caput region with pH 6.5 and sodium and potassium concentrations closer to those in blood plasma. Spermatozoa then move into a progressively lower ionic environment from the caput to the cauda but a progressively higher organic solute and protein environment [reviewed in Hinton and Palladino (144), Rodriguez and Hinton (361), and Turner (359,373)]. Sodium, chloride, calcium, and magnesium concentrations decline from the caput to cauda epididymidis, but phosphorus (as phosphate) and potassium increase in concentration (Fig. 11). The physiological importance of the ionic microenvironment is unclear, but it may be involved in the development of sperm motility and in keeping spermatozoa in a quiescent state as they mature (374–377). More studies are needed to test the

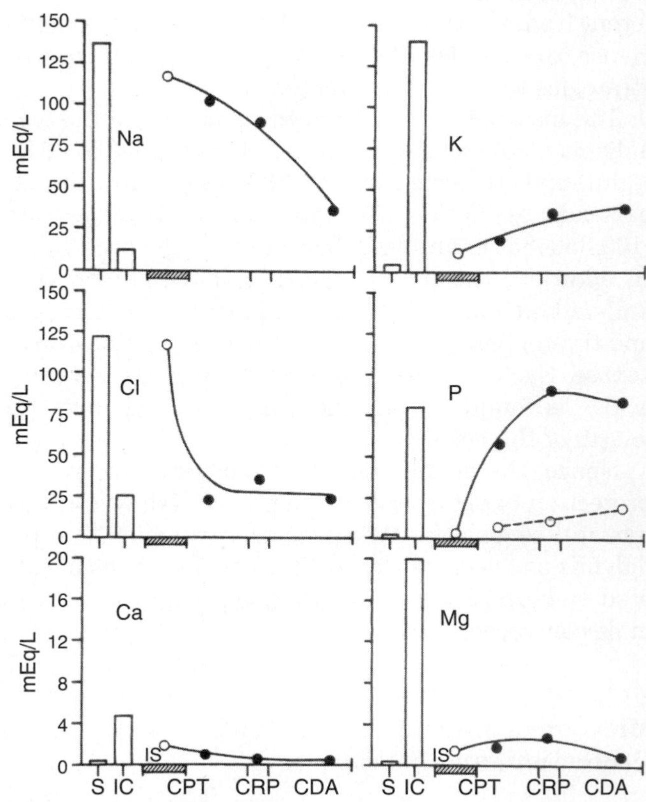

FIG. 11. Serum, intracellular, and intraluminal concentrations of sodium, potassium, chloride, phosphorus, calcium, and magnesium in the rat epididymis. Serum (S) and intraluminal values are from Jenkins et al. (371). Intracellular (IC) values are those expected in typical somatic cells (950) and represent anticipated values in the epididymal epithelial cells (initial segment [IS], caput [CPT], corpus [CRP], and cauda [CDA] epididymidis). Data connected by *hatched lines* in the P panel are P_i data from Hinton and Setchell (951).

hypothesis that ions play a critical role in sperm maturation.

Organic Solute Microenvironment

As well as analyzing the luminal fluid ionic microenvironment, Levine and Marsh (1971) (369) also measured the osmolality of fluid collected from each epididymal region and then compared these values with the sum of osmotically active ions at each region. They noted a difference. This osmotic difference increased from the caput to cauda epididymidis to reach over 250 mOsm/kg water; the investigators suggested that this difference was due to the presence of organic solutes. Later studies by several groups showed that this was indeed the case, with the identification of solutes such as glutamate, taurine, L-carnitine, myoinositol, glycerylphosphorylcholine, phosphorylcholine, sialic acids, and many other amino acids present in different concentrations through the epididymal duct of many species, including the human [reviewed in Turner (85,359), Hinton and Palladino (144), and Rodriguez and Hinton (361)].

The total concentration of organic solutes in the luminal fluid of the rat testis and epididymis increases from approximately 10 to 20 mM in the seminiferous tubule to 100 to 150 mM in the caput and to over 200 mM in the cauda epididymidis. Although virtually nothing is known about the roles of each solute, evidence suggest that they may play a critical role in the protection of spermatozoa and the epididymal epithelium from osmotic stress, similar to the role played by these solutes in the kidney (377a). The manner by which the epididymal epithelium forms an organic solute–rich environment is not completely known, but there is considerable evidence to suggest that at least several transporters are involved in moving the solute from the blood into the lumen; in addition, the epididymis synthesizes and secretes solutes into the lumen. For example, L-carnitine is transported into the epididymal cells and lumen, reaching concentrations as high as 50 to 60 mM in some species (378–382). The high intraluminal concentration of L-carnitine translates into a remarkable 2,000-fold gradient of L-carnitine across the epididymal epithelium. Studies suggest that the distal caput, corpus, and proximal cauda epididymidis are the regions involved in the active transport of L-carnitine (383–388). The mechanism of L-carnitine transport in the epididymis is unknown, but has been suggested to involve an active transport system (383,384,386,389) that is androgen dependent (378,380,390,391).

More recent studies have suggested that L-carnitine is transported from blood into cells by the organic cation/carnitine transporter OCTN2 (392), and then presumably from cells into the lumen by the carnitine transporter CT2, which has been identified in the human epididymis (393). Several transporters, such as OCTN2 and SMIT (a sodium–myoinositol transporter), have TonE binding motifs on their promoters that bind a transcription factor, TonE binding protein (TonEBP; Rodriguez and Hinton, unpublished observations), that is expressed during periods of hypertonicity; these data suggest that the organic solutes play a role in the regulation of water movement across the epididymal cell membranes. This is particularly important because in the epididymis of most species, spermatozoa are exposed to a hyperosmotic environment. Disruption of the ability of spermatozoa to regulate their cell volume has been suggested to cause male infertility [reviewed in Cooper et al. (394)].

The high levels of organic solutes measured in the cauda epididymal fluid of several species suggest that these solutes play a similar role in the long-term storage of spermatozoa in this region. Likewise, in bats that store spermatozoa in the cauda epididymidis for up to 6 months, the spermatozoa are stored in a supraphysiologically high osmotic environment with a luminal fluid osmolality measured at greater than 1,000 mOsm/kg water (395); the mechanisms for generating and maintaining such a high osmolality are unknown. Taurine and glutathione are both found in epididymal luminal fluid (396,397), and because both of these compounds have been implicated in protecting cells from oxidative stress, it would not be surprising if they have a similar role in the epididymis.

The transport of sugars and amino acids, including inositol, 3-O-methyl-D-glucose and 2-deoxy-D-glucose (nonmetabolizable forms of glucose), and α-aminoisobutyric acid (AIB; a nonmetabolizable neutral amino acid), has been studied in the rat epididymis (398–403), yielding several observations. First, glucose entered by a facilitated diffusion mechanism, and this uptake appeared to be similar across all epididymal regions. Second, glucose transport was inhibited by the male infertility agents α-chlorhydrin, 5-thio-D-glucose, and 6-chloro-6-deoxy-glucose. Third, transport of inositol and AIB was highest in the initial segment compared with the rest of the epididymis. Finally, transport of inositol and AIB was saturable and, presumably, transport of inositol was through an active mechanism, whereas that of AIB was through a facilitated diffusion mechanism. Although the GLUT family of glucose transporters has been well studied in numerous tissues, only GLUT8 has been identified in epididymal spermatozoa (404) and tissue (43). Presumably, the glucose and amino acid needs of the epididymis reflect the metabolic needs of the epididymis and maturing spermatozoa. However, more studies are needed to examine the roles played by each solute. The role of known specific ion and organic solute transporters that have been identified in the epididymis is discussed in greater detail later in this chapter.

The androgen luminal fluid microenvironment has received little attention recently, but past studies revealed that the spermatozoa are exposed to a rich and complex androgen environment as they progress along the duct [reviewed in Turner (359)]. In several species, including rat, human, hamster, stallion, and bull, the epididymal luminal fluid concentrations of testosterone and dihydrotestosterone (DHT) exceed the blood concentrations (405–407), with DHT concentration being very high in the caput region. The high luminal fluid androgen concentrations are maintained by androgen-binding protein (ABP) because studies by Turner et al. (1984) (407) showed that intraluminal ABP concentrations are equimolar with total androgen levels. It has been proposed that the ABP–androgen complex plays a role in epididymal function (408), but confirmatory experiments have yet to be designed to test this hypothesis.

Protein Microenvironment

The protein composition of epididymal luminal fluid has been well studied for a number of species [reviewed in Dacheux et al. (93,409)]. Spermatozoa are exposed to varying protein concentrations as they move along the seminiferous tubule (approximately 6 to 10 μg/μL), to the rete testis (approximately 1 μg/μL), to the caput and cauda epididymidis (approximately 25 to 30 μg/μL). Because water is being absorbed along the epididymal duct, a higher protein concentration would have been expected in the cauda compared with the caput epididymidis. To account for this difference, proteins are presumably being absorbed or degraded, or are being utilized by the spermatozoa. However, even though the net movement of protein is by absorption, many proteins are still being synthesized and secreted into the duct.

The proteins in epididymal luminal fluid have been analyzed by one- and two-dimensional gel electrophoresis (410), generating characteristic patterns of protein profiles for several species; many of the proteins have now been identified using mass spectrophotometric methods. Using an in vivo split-drop, stopped-flow microperfusion technique together with two-dimensional gel electrophoresis and both silver staining and Western blot analyses, Hinton and colleagues (411,412) showed in vivo protein secretion into the lumen of discrete epididymal segments and regions. Later studies by Turner and colleagues (413,414) obtained similar results when the interstitium was perfused with radiolabeled methionine and luminal fluid collected by micropuncture and analyzed for radiolabeled proteins.

Epididymal luminal fluid contains a large repertoire of proteins, including enzymes, growth factors, lipid-binding proteins, iron-binding proteins, proteins that may be involved in protection or in sperm–egg binding, and other proteins such as the lipocalins, clusterin, lactoferrin, and the cholesterol transport protein CTP/HE1 [reviewed in Dacheux and Dacheux (93)]. Many of these families of proteins are discussed in later sections.

REGULATION OF EPIDIDYMAL FUNCTIONS

Hormones

In 1926, pioneering studies by Benoit demonstrated that the epididymis depends on an unknown testicular substance for maintenance of its structure and functions; he showed that the ratio of nucleus to cytoplasm in the mouse epididymis decreased after birth as the epididymis differentiated, and increased dramatically after orchidectomy, as the epididymis dedifferentiated (12). This regulatory substance was identified 5 years later as testosterone (415). Since then, over a thousand papers have appeared on the response of the epididymis in numerous mammals to androgen withdrawal and replacement on a wide variety of end points, ranging from morphological to biochemical and molecular.

The main approach to understanding the effects of androgen withdrawal on the epididymis has been removal of the testis. It is clear that this approach causes loss not only of androgens but of estrogens and any other testicular factor that may affect the epididymis.

Orchidectomy causes a decrease in epididymal weight that is less marked than that of sex accessory tissues such as the prostate or seminal vesicles (416,417). Unlike the case with other androgen-dependent male reproductive tissues, testosterone replacement, even at supraphysiological levels, only partially restores epididymal weight; this is presumably due to the large proportion (nearly half) of epididymal weight that is attributable to spermatozoa and the luminal fluid bathing them (417,418). In the androgen-deprived state, spermatozoa become immotile, lose the ability to fertilize, and die (12,161,419). After orchidectomy, the luminal diameter and epithelial cell height decrease and the intertubular stroma increases (420). The smooth endoplasmic reticulum content is dramatically reduced, whereas the extent of decline in the Golgi apparatus is less pronounced (421–423). Morphological changes in principal cells suggest that these cells are particularly sensitive to androgen levels, in contrast to the other epithelial cell types, which appear to be less affected by orchidectomy (101). The secretory function of principal cells becomes compromised in the androgen-deprived state. In addition to the virtual disappearance of endoplasmic reticulum from their apical cytoplasm, principal cells undergo a striking loss of apical microvilli from their surface, as well as lysosome accumulation, vacuolization, disappearance of vesicles from the cell apex, and increased endocytosis (101,287). Epididymal androgen receptors and 5α-reductase activity are both decreased in the androgen-deficient state (417,424–426), suggesting that the mechanisms of androgen action are compromised by androgen withdrawal. Total epididymal protein, RNA, and DNA content are reduced after orchidectomy, but the DNA concentration is apparently increased (427). The increase in DNA concentration is attributed to the concomitant decline in cell volume; this is thought to be the principal mechanism by which the epididymal epithelium regresses after orchidectomy.

Restoration of circulating testosterone levels appears sufficient to reverse regressive changes in the caput, corpus, and cauda epididymidis after orchidectomy but not in the initial segment, even when supraphysiological doses of testosterone are administered (417,423,428–432). At 3 days after orchidectomy with testosterone or DHT replacement to circulating levels, a time at which the prostate shows maximum rates of DNA synthesis, the epididymal labeling index, measured by [^{3}H]-thymidine incorporation and mitotic index, is low except in the corpus region (433). The lack of effect of androgens on mitotic rate in the epididymis of the adult animal distinguishes this tissue from other androgen-dependent tissues such as the prostate and seminal vesicles. This characteristic of the epididymal epithelium suggests that it may contain antiproliferative signals that inhibit cellular proliferative capacity in response to androgen stimulation or other stimuli. Interestingly, B-myc, a transcription factor known to inhibit cellular proliferation, is highly expressed in the epididymal epithelium (434).

Withdrawal of androgen stimulation by orchidectomy induces a wave of apoptotic cell death in the epididymis, beginning in the initial segment and moving over several days to the cauda epididymidis (435,436). Apoptosis in the initial segment seems to be caused by withdrawal of androgens and luminal components from the testis, and appears to be p53 independent. Using the entire epididymis, Bcl-2, an antiapoptotic factor (437), was found to be suppressed by 36 hours after orchidectomy; this was followed by the appearance of Fas and DNA fragmentation in the epididymal epithelium at 48 hours postorchidectomy (438). Mutant mice null for Fas (*lpr*) showed no epididymal regression or DNA fragmentation after orchidectomy. These data suggest that the regression of the epididymal epithelium after orchidectomy may be regulated through the Fas pathway (438).

In addition to androgens, many other hormonal factors have been postulated to play a role in regulating epididymal function; these are listed in Table 4 (supplemental material on web site). Of special note, however, is estradiol. It has been known for many years that the administration of estrogens could affect the male reproductive system in general (439–448) and the epididymis in particular (166,449). With the advent of aromatase-specific inhibitors and null mutant mice for the two isoforms of the estrogen receptor (discussed later), it is likely that estrogens will be shown to act as regulators of some specific epididymal functions; however, much remains to be done to resolve exactly what role this steroid family plays.

Testicular Factors

In addition to depending on the presence of circulating androgens, the epididymis also depends on the presence of luminal fluid factors originating from the testis or the epididymis itself. Without testicular luminal fluid factors, many cells in the initial segment undergo apoptosis within 24 hours (435,436,450). Therefore, it is not surprising that ligating the efferent ducts results in changes in morphology and gene expression in the initial segment [reviewed in Robaire and Hermo (1), Cornwall et al. (91), and Hinton et al. (451,452)]. From these studies, it has been postulated that a factor or factors originating from the testis is responsible for maintaining the integrity and survival of cells in the initial segment. This type of paracrine secretion has been termed *lumicrine* because this mode of regulation occurs in a duct/tubal system (452). A general scheme depicting how factors secreted directly into the epididymis through the efferent ducts can affect activities of cells in the initial segment and how such effects can cascade down the duct is shown in Fig. 12. Perhaps the first study demonstrating lumicrine regulation in the male excurrent duct system was that by Skinner and Rowson (1967; 1968) (453,454). These investigators showed that the ampulla of the vas deferens weighed less, contained less fructose, and was smaller in diameter after unilateral vasectomy compared with the ampulla on the intact control side. Skinner and Rowson (454) suggested that vasectomy

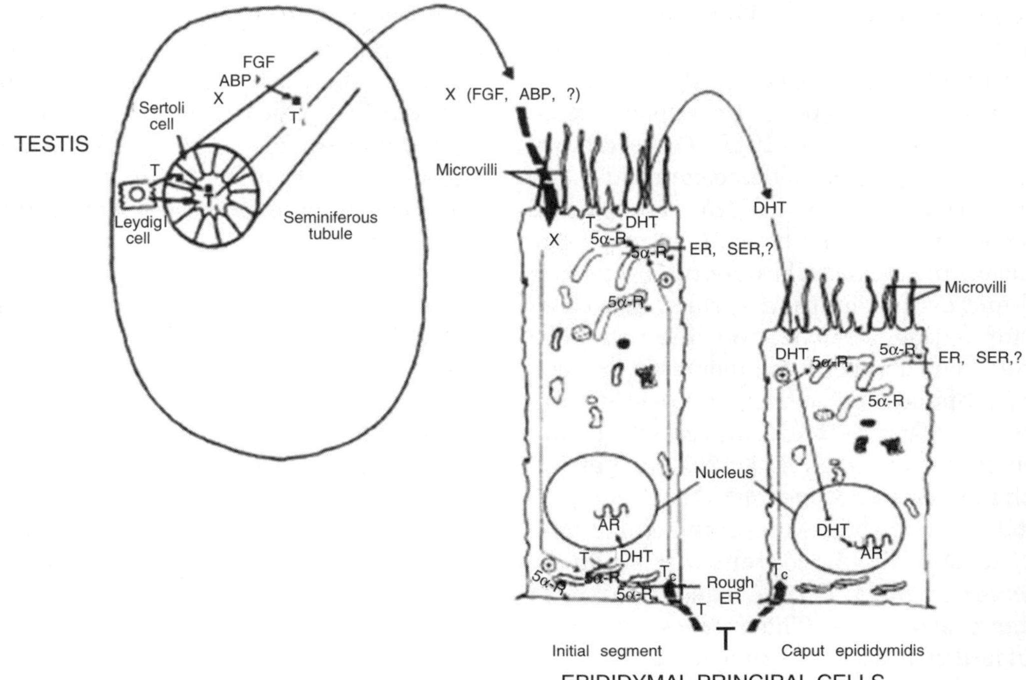

FIG. 12. Proposed mechanism for the regulation of 4-ene steroid-5α-reductase and androgen action in the rat epididymis. Leydig cells synthesize testosterone (T), which stimulates the Sertoli cell to secrete androgen-binding protein (ABP) and other proteins into the lumen of the seminiferous tubule. In the initial segment of the epididymis, the paracrine/lumicrine regulator (e.g., fibroblast growth factor [FGF]), ABP is proposed to be the paracrine factor regulating the infranuclear localized 5α-reductase enzyme. In contrast, the apically localized 5α-reductase enzyme (microsomal 5α-reductase) found throughout the epididymis is under the control of circulating androgens. The synthesis of locally high concentrations of dihydrotestosterone (DHT) in the initial segment of the epididymis provides an excellent mechanism for stimulating critically important androgen-dependent genes in this region of the epididymis and for providing the most active androgen to mediate androgen action in the rest of the tissue. 5ur-RN, nuclear 5α-reductase; 5-RM, microsomal 5α-reductase; ER, endoplasmic reticulum; SER, smooth endoplasmic reticulum. (Modified from Robaire, B., and Viger, R. S. [1995]. Regulation of epididymal epithelial functions. *Biol. Reprod.* 52, 226–236.)

prevented luminal fluid testosterone from reaching the ampulla, and perfused the vas deferens of an orchidectomized ram with testosterone to test this hypothesis. Their data show that the ampulla of the perfused side weighed more and contained more fructose and citric acid compared with the nonperfused control side. Similarly, studies by Abe et al. (1984) (48) show that ligation of the mouse corpus epididymidis results in abnormal differentiation of the principal cells downstream from the ligature. Hence, lumicrine regulation may occur not only between the testis and epididymis but between regions of the epididymis, including the vas deferens.

The cells in zones 1a, 1b, and 1c [see Reid and Cleland (57) for zones] of the rat initial segment have been shown to be the cells most highly regulated by testicular luminal fluid factors. Whether there are more distal effects along the epididymis and the extent to which this regulation occurs in other species need further investigation. Although testicular luminal fluid steroids, androgens and estrogens, are the most obvious candidates for regulating initial segment function, several lines of evidence suggest that additional factors are important. Using a histological approach, Fawcett and Hoffer (1979) (423), supported the idea that factors other than androgens are responsible for the maintenance of initial segment function because the morphology of the initial segment cells failed to return to normal after efferent duct ligation and testosterone treatment. Likewise, Nicander and colleagues (1983) (450), suggested that because the cells in the initial segment are highly mitotic, they are more dependent on the presence of a mitogen in testicular luminal fluid.

The first nonsteroidal molecule proposed to be a testicular factor acting on the epididymis was ABP (408,455). This hypothesis stemmed from a series of studies demonstrating that the steroid 5α-reductase activity in the initial segment of the epididymis fell after efferent duct ligation or unilateral orchidectomy;

it could not be maintained by administration of testosterone, even at very high doses. Luminal fluid ABP was also proposed as the factor that enhances the function of ovine epididymal principal cells in culture (456), and appears to play a role in protein synthesis in the caput region (457). However, although γ-glutamyl transpeptidase IV mRNA is also controlled in a lumicrine manner in the initial segment of the epididymis, ABP does not play a role in its expression (458). Instead, FGFs may be partly responsible for regulation of the expression of this gene (458,459). It has been postulated that luminal fluid growth factors interact with their cognate receptors on the apical cell surface of initial segment cells, whereupon a signal transduction pathway is initiated (e.g., mitogen-activated protein kinase [MAPK], phosphatidylinositol-3 kinase); transcription factors are then activated, presumably by phosphorylation, leading to transcription of target genes (452,459). In support of this hypothesis, soluble FGFs (2, 4, and 8) have been identified in testicular luminal fluid (458,459), and FGF receptor (FGFR)-1 IIIc (both long and short forms) has been identified in the principal cells in the initial segment. Further, the activity of the MAPK pathway and the expression of members of the polyomavirus enhancer activator 3 (PEA3) transcription factor family (PEA3, ERM, ER81) are decreased after loss of testicular factors [(460); J. L. Kirby, B. V. Troan, and B. T. Hinton, unpublished observations]. Interestingly, several genes, including steroid 5α-reductase (461), expressed in the rat initial segment contain multiple PEA3 family member–binding motifs, and most of these genes are regulated by testicular luminal factors. In addition, neurotrophins (nerve growth factor, NT-3) and their cognate receptors, Trks and p75, may also play a role in the regulation of initial segment function (S. Crenshaw and B. T. Hinton, unpublished observations). Crucial to the coordinated functions of the FGFR and Trk receptor pathways may be adaptor proteins such as FRS2 and Shc (S. Crenshaw and B. T. Hinton, unpublished observation).

Another "factor" that has been postulated to regulate initial segment function is the presence of spermatozoa (462). At first, cells seem to be an unusual "ligand," yet, on closer inspection, there may be a sound basis for this hypothesis. Spermatozoa themselves may not be the ligand, but it is possible that the ligands they carry and transport to the different epididymal regions are important; many molecules are known to bind to spermatozoa and potentially shed in the epididymis. Interestingly, testicular spermatozoa stain positively for growth factor receptors (463); it is tempting to speculate that on entering the initial segment, growth factors dissociate from the sperm surface and become available to stimulate their cognate receptors on the epididymal apical cell surface.

If it is assumed that the lumicrine mode of regulation occurs throughout the epididymis, there are several potential key points that emerge regarding the regulation of the initial segment, and possibly the entire epididymis. First, the cells in the adult initial segment are being constantly stimulated by FGFs, yet the cells are not proliferating; likewise, the MAPK pathway is constitutively switched on in these cells. However, not all cells in the initial segment show MAPK pathway activity (J. L. Kirby et al., unpublished observation). Hence, this raises a second point that cells may be in communication with each other such that they cycle between "on" and "off" states with respect to the response to growth factors. This may explain the characteristic checkerboard pattern seen when the tissue is stained for many different proteins; it would enable the initial segment to function normally as a whole, yet with each cell autonomous with respect to the regulation of its repertoire of target genes but susceptible to influence by the state of its neighboring cells. This theory may also explain why cells do not proliferate in the initial segment after constant stimulation with growth factors. At any time, there is sufficient stimulation of growth factors to upregulate target genes, yet, because of complex negative feedback systems (e.g., involving the brain-derived neurotrophic factor/NT-3–Trk receptors and adaptor proteins), those cells do not enter the cell cycle pathway. However, the cells need to be responsive to growth factors fairly rapidly after the "off" state because lack of stimulation by testicular factors leads to apoptosis (S. Crenshaw and B. T. Hinton, unpublished observations). Future experiments will be needed to test this hypothesis, but it will also be interesting to know whether the recycling hypothesis for many of the genes applies in all the cells throughout the epididymis. This hypothesis may also have implications with respect to the low rate of primary or secondary cancers observed in the epididymis.

MAJOR PROTEIN FAMILIES IN THE EPIDIDYMIS AND THEIR REGULATION

General Principles

The division of the epididymis into segments has more than anatomical implications. For many years, the differential expression of proteins and mRNA along the epididymis has become a hallmark of this tissue. Indeed, the large number of gene expression profiling studies (311,353,464–468a), proteomic analyses [reviewed in Dacheux and Dacheux (93), Gatti et al. (469), Chaurand et al. (470), and Umar et al. (471)], and immunohistochemical studies [e.g., (107,109, 472–474)] have all revealed that nearly all molecules have characteristic longitudinal expression profiles in the epididymis of all species examined. Data on the gene expression profiles of the different segments of rat and mouse epididymis are available at the website for this article (www.ttuhsc.edu/cbb/faculty/cornwall/default.asp), and at www.wsu.edu/%7Egriswold/microarray/epididymis_dht/. In addition, tables of genes encoding distinct classes of proteins (enzymes, signaling, DNA binding) that exhibit regionalized expression in the epididymis are provided in the supplemental material on this chapter's web site. Table 2 lists genes encoding secretory proteins that may play direct or indirect roles in the sperm maturation process, whereas Table 3 provides a list of genes that encode cellular proteins that likely perform regulatory functions.

Several comprehensive reviews describing the longitudinal profiles of both proteins and mRNA have been published recently. Although all of these studies clearly demonstrate that there is a highly regionalized activation and deactivation of gene expression, the underlying mechanisms that allow for this regulation are unclear. Regionalized expression of transcription factors (see later) is beginning to provide some insight into this question, but the relationship between the luminal, endocrine, and cellular components that allow for such highly regionalized expression of genes and proteins will require many more years to resolve.

Perhaps as challenging to understand is the checkerboard pattern of expression observed in cross sections of individual tubules. This characteristic pattern is usually seen when a given protein or mRNA is first expressed along the duct or when the expression terminates. It must be assumed that adjacent principal cells either use different sensing mechanisms for the triggering signal, or more likely function as a syncytium to optimize coordinately the degree of expression of a given gene or protein. Examples of this checkerboard pattern of expression are shown in Fig. 13.

As might be expected based on their appearance and proposed functions, the different cells of the epididymal epithelium express different genes and proteins.

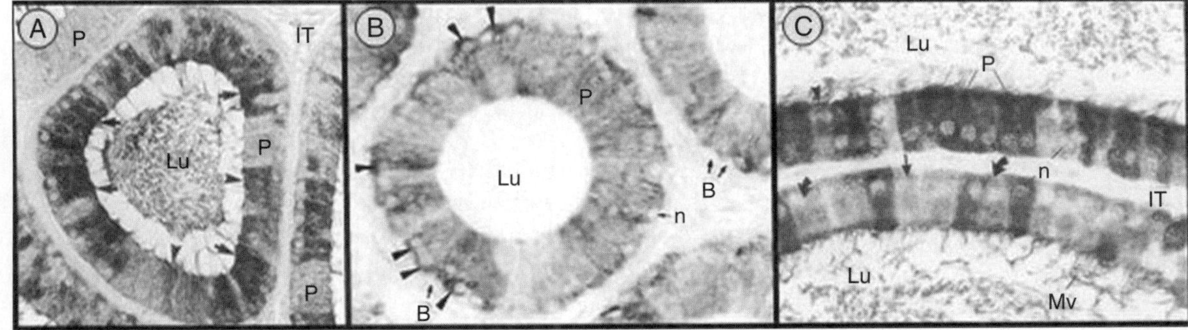

FIG. 13. Representative examples of the checkerboard pattern of expression seen immunohistochemically along the epididymis. **A:** High-power light micrograph of tubules of the distal area of the initial segment immunostained with anti-immobilin. Although many principal cells are reactive, they show a variable staining pattern. Several principal cells are intensely stained throughout their cytoplasm (*arrows*), whereas others are moderately or weakly stained (*arrowheads*); a few are unreactive (P). The luminal content (Lu) is well stained. IT, intertubular space. (Magnification ×350; modified from Hermo, L., Oko, R., and Robaire, B. [1992]. Epithelial cells of the epididymis show regional variations with respect to the secretion or endocytosis of immobilin as revealed by light and electron microscope immunocytochemistry. *Anat. Rec.* 232, 202–220.) **B:** High-power light micrographs of the proximal initial segment of the epididymis. Epithelial principal cells (P), nuclei of principal cells (n), basal cells (B), and lumen of the duct (Lu) are shown. Intense staining in the infranuclear region of the principal cells is designated by the *arrowheads*. Immunostaining reaction is confined to an oval region present above the nuclei of the principal cells (*large arrows*). (Magnification ×400; modified from Robaire, B., and Viger, R. S. [1995]. Regulation of epididymal epithelial functions. *Biol. Reprod.* 52, 226–236.) **C:** Higher-power micrograph of portions of adjacent tubules of the caput epididymis immunostained with anti–SGP-2. Note intense immunoperoxidase staining reaction product over some epithelial principal cells (P). Other principal cells are moderately (*arrowheads*) or weakly (*curved arrows*) immunoreactive. A few principal cells are virtually unreactive (*arrow*). Note that the reaction product extends throughout the entire cell cytoplasm (i.e., basal, supranuclear, and apical regions), leaving only the nucleus (n) unstained. The microvillar (Mv) border of the principal cells and sperm in the lumen (Lu) of the tubules are reactive. IT, intertubular space. (Magnification ×600; modified from Hermo, L., Wright, J., Oko, R., and Morales, C. R. [1991]. Role of epithelial cells of the male excurrent duct system of the rat in the endocytosis or secretion of sulfated glycoprotein-2 (clusterin). *Biol. Reprod.* 44, 1113–1131.)

Isolating each of the different cell types has been a challenge, but with the development of tools such as laser capture microscopy, it should become possible to resolve the profiles of genes and proteins expressed in each type of epithelial cell. Using immunohistochemistry, however, it has been possible to demonstrate the highly specific expression of certain proteins in principal, basal, clear, narrow, and halo cells. Figure 14 presents examples of the selective marking of each of these cell types. To provide potential tools to help identify cell types in vivo and in vitro, representative proteins expressed selectively in individual cell types of different regions of the epididymis are listed in Table 4 (supplemental material on web site); some of these proteins are secreted, whereas others are retained in the cell.

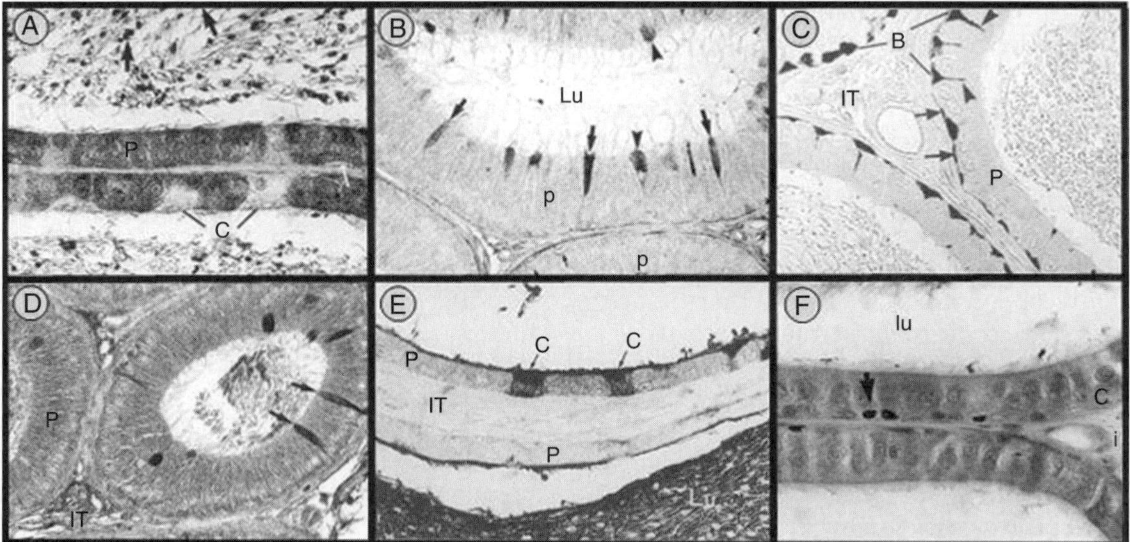

FIG. 14. Cell-specific expression of proteins along the rat epididymis. Examples of immunolocalization of proteins in each of the major cell types are presented. **A:** Principal cells (junction of the caput–corpus of the epididymis) are immunostained for the glutathione S-transferase (GST) Yo subunit. The epithelial principal cells (P) are well stained, but basal (*white arrow*) and clear (C) cells are unreactive. Cytoplasmic droplets of spermatozoa (*arrows*) in the lumen are intensely reactive. (Magnification ×365; modified from Veri, J. P., Hermo, L., and Robaire, B. [1994]. Immunocytochemical localization of glutathione S-transferase Yo subunit in the rat testis and epididymis. *J. Androl.* 15, 415–434.) **B:** Apical cells (*arrowheads*) and narrow cells (*large arrows*) of the proximal initial segment of the epididymis are intensely reactive when immunostained with anti-Yo GST antibody, whereas principal cells (P) and basal cells (*small arrows*) are unreactive. Lu, lumen. (Magnification ×320; modified from Adamali, H. I., and Hermo, L. [1996]. Apical and narrow cells are distinct cell types differing in their structure, distribution, and functions in the adult rat epididymis. *J. Androl.* 17, 208–222.) **C:** Basal cells. The only cell type of the epithelium of the corpus epididymidis that is intensely reactive is the basal cell (B) when immunostained with antibodies to the Yf (Yp) subunit of GST. These cells often show slender processes (*arrows*) extending from the expanded main cell body along the basement membrane and, occasionally, toward the lumen (*arrowheads*). Principal cells (P) appear unreactive. Spermatozoa in the lumen are unreactive. IT, intertubular space. (Magnification ×440; modified from Veri, J. P., Hermo, L., and Robaire, B. [1993]. Immunocytochemical localization of the Yf subunit of glutathione S-transferase P shows regional variation in the staining of epithelial cells of the testis, efferent ducts, and epididymis of the male rat. *J. Androl.* 14, 23–44.) **D:** Narrow cells. Initial segment of epididymides immunostained with anti–carbonic anhydrase II antibody. Intense reaction is localized over narrow cells. P, principal cell; IT, intertubular space; *arrows*, narrow cells. (Magnification ×350; modified from Hermo, L., Adamali, H. I., and Andonian, S. [2000]. Immunolocalization of carbonic anhydrase (CA) II and H^+ V-ATPase in epithelial cells of the mouse and rat epididymis. *J. Androl.* 21, 376–391.) **E:** Clear cells. Tubules of the distal area of the cauda epididymidis immunostained with anti-immobilin. Note that several epithelial clear cells (C) are intensely reactive throughout their cytoplasm, whereas other cells remain unreactive. IT, intertubular space; P, principal cells. (Magnification ×375; modified from Hermo, L., Oko, R., and Robaire, B. [1992]. Epithelial cells of the epididymis show regional variations with respect to the secretion or endocytosis of immobilin as revealed by light and electron microscope immunocytochemistry. *Anat. Rec.* 232, 202–220.) **F:** Halo cells. Immunostaining with an antibody for helper T lymphocytes (CD4) in the cauda epididymidis. *Arrows* indicate CD4+ cells. lu, lumen. (Magnification ×450; modified from Serre, V., and Robaire, B. [1999]. The distribution of immune cells in the epithelium of the epididymis of the aging brown Norway rat is segment-specific and related to the luminal content. *Biol. Reprod.* 61, 705–714.)

Functional Families

Receptors and Their Ligands

Androgen Receptor, Androgen-binding Protein, and Steroid 5α-Reductases. The limiting factor in determining the action of testosterone on cells is not clear. The presence of both the androgen receptor and an agonist are obviously necessary, but there are several lines of evidence that indicate that it is not primarily the absolute number of androgen receptors present but rather the amount of the ligand that is limiting. In addition, testosterone, a highly lipophilic molecule, is carried to sites distant from where it is made by binding proteins such as ABP. Testosterone can act directly on the androgen receptor in some tissues, such as muscle; in other tissues, such as some brain nuclei, testosterone needs to be aromatized to estradiol to mediate its action through the estrogen receptor; whereas in still other tissues, it is converted to a 5α-reduced metabolite, DHT. DHT binds to the androgen receptor with higher affinity than testosterone [reviewed in Blanchard and Robaire (475)]. The latter mode operates in many androgen-dependent tissues such as the epididymis, prostate, seminal vesicles, and skin. The rate-limiting enzyme in the pathway leading from testosterone to its 5α-reduced metabolites is steroid 5α-reductase (5α-R, EC 1.3.1.22).

Androgen Receptor. The first report that exogenously administered testosterone was converted to DHT and bound to a cytosolic protein in the epididymis (the androgen receptor) (476) was nearly coincident with the discovery of a second androgen-binding component (ABP) in testis and epididymal cytoplasmic fractions (476a). Although both of these proteins bind androgen with high affinity, they are distinct molecules. For example, although both of these proteins have high affinity for DHT, the androgen receptor has a slow dissociation rate for DHT ($t_{1/2}$ 0°C > 4 days), whereas ABP dissociates at a much faster rate ($t_{1/2}$ 0°C = 6 minutes). Androgen receptors have been found in the epididymis of all species examined to date (rat, rabbit, dog, ram, monkey, and human) (476,476b–476h). In other mammalian tissues, a single gene for the androgen receptor has been identified. In humans, the gene spans approximately 90 kilobases, and it consists of 8 exons on chromosome Xq11-12. The coded protein is approximately 110 kDa, and multiple transcripts are produced with variable lengths of a polymorphic region of CAG repeats in exon 1 (477,478). Recently, a variant of the human androgen receptor (AR45), lacking the entire region encoded by exon 1 of the gene but retaining the DNA-binding and steroid-binding domains, has been identified in heart and muscle tissues (479). The potential presence of further variants in other tissues may help explain the tissue specificity of androgen action.

There are relatively small changes in the concentration of mRNA for the androgen receptor along the epididymis (480,481). When changes are noted, they often reflect a decline in the concentration of androgen receptors in the progression toward the distal end of the epididymis. For example, in humans, both the mRNA protein for the androgen receptor are found throughout the tissue, with high levels in the distal caput and proximal corpus and declining in the caudal region, including the vas deferens. Interestingly the efferent ducts do not show any immunostaining, whereas the vas deferens is only weakly positive (476d). An exception is the ram, where androgen receptor concentrations are similar in the caput and cauda epididymidis but are less than 25% of these levels in the corpus epididymidis (482). In both the goat and the rat, epithelial cells stain fairly evenly for androgen receptors throughout the duct (483,484).

It is clear that principal cells stain heavily for androgen receptors, but the relative concentrations of this receptor in basal, clear, narrow, and apical cells are not known. However, the immunolocalization of androgen receptors in all epididymal epithelial cells has been reported for monkey, mouse, and goat (476h,485,486), whereas in the rat, it is apparent that clear cells do not stain for the androgen receptor (426). Because of their origin, and as one might anticipate, halo cells do not stain for the androgen receptor in any species examined. Peritubular cells occasionally display weak signals, whereas interstitial cells or blood vessels are consistently negative. When noted, the intracellular localization of the androgen receptor is distinctly nuclear.

Very few studies have focused directly on the mechanism of action of the androgen receptor in the epididymis; however, with the advent of several stable cell lines from this tissue (487–489), such studies are likely to be undertaken in the near future. The characteristics and mode of action of the androgen receptor as a transcription factor (490,491) in the epididymis seem to resemble those in other tissues. However, androgen signaling, although essential, is not sufficient for epididymis-specific gene expression because the androgen receptor is expressed in many tissues. One mechanism proposed to explain the specificity of transcriptional responses to androgen is the tissue-specific combination of transcription factors binding to the promoter region of androgen-regulated genes (492). For example, the epididymal and prostatic lipocalin genes for epididymal retinoic acid–binding protein (E-RABP) and probasin are regulated by the androgen receptor. In the prostate, the transcription factor Foxa1 (forkhead box A1) is required for probasin gene expression (493). In the epididymis, Foxa2, but not Foxa1, is expressed, interacts with the androgen receptor, and binds to the promoter of the E-RABP gene. Interestingly, overexpression of Foxa2 suppresses androgen activation of the E-RABP promoter, but increases activation of the prostate-specific promoter in an androgen-independent manner (494). Because Foxa1 and Foxa2 expression is restricted to the prostate and epididymis, respectively, it suggests that these transcription factors have distinct and different action on androgen receptor–regulated genes in different male reproductive organs. It is tempting to speculate that these differential responses play a role in the different responses of these two organs to oncogenic transformation.

Little is known about the regulation of expression of the androgen receptor in the epididymis. In the goat, circulating androgens seem to be sufficient to regulate the expression of both androgen and estrogen receptors (486). In the rat, androgen withdrawal by the administration of a luteinizing hormone–releasing hormone antagonist results in a more extensive and rapid decline in androgen receptors in stromal than in epithelial cells; androgen replacement results in a complete restoration of androgen receptors, indicating that stromal cells are more sensitive than epithelial cells to the regulation of androgen receptor concentrations by androgen (426).

Androgen-binding Protein and Its Receptor. ABP is one of the first proteins shown to be synthesized by Sertoli cells of the testis (495) and has been found in several species, including the rat (476a), rabbit (496), guinea pig (497), ram (476b), monkey (498), and human (499). It is secreted in both the luminal and basal compartments of the seminiferous tubules, but the overwhelming majority, more than 80%, is directed toward the epididymis (500). Plasma sex hormone–binding globulin (SHBG) shares the same amino acid sequence with ABP; these molecules differ only in the types of oligosaccharides associated with them (501). The biological significance of these differences is not clear in light of the fact that SHBG and a glycosylated mutant have similar binding characteristics to sex steroids (502). ABP concentrations are highest in the caput and lowest in the cauda epididymal segments of most species examined, including primates; this distribution correlates well with DHT content (85,503,504). ABP is synthesized and secreted by epididymal principal cells all along the epididymis (505), but in several species it is also taken up into the epididymal principal cells of the initial segment and caput epididymidis (495,506,507) through a receptor-mediated process (505,508–511). Testicular ABP production is regulated by androgens (512), whereas that of the epididymis seems to be under the control of a testicular factor (505).

Several functions have been ascribed to ABP. It has been proposed to act as a carrier of androgens to the epididymis (513). In the principal cells, testosterone is released from ABP and converted to the more potent androgen, DHT, by the enzyme 5α-reductase. ABP has high affinity for 5α-DHT ($K_a = 1.25 \times 10^9$ M^{-1}) and testosterone ($K_a = 0.5 \times 10^9$ M^{-1}); rapid dissociation of testosterone from ABP (6 minutes) is consistent with a role for ABP as a carrier molecule for testosterone. In addition to its role as a carrier molecule, ABP has been suggested to be an androgen sink in the seminiferous tubules of the testis (513) because the concentration of androgens in the seminiferous tubules needs to be higher than that in the peripheral circulation (514–516) to maintain spermatogenesis. ABP has also been proposed to regulate steroid 5α-reductase type 1 (408,456,517).

To demonstrate unequivocally a function for ABP in spermatogenesis or in the epididymis would require the development of a null mutation model; this has not yet been accomplished. However, mice overexpressing ABP (518) show fertility defects stemming from spermatogenic dysfunction in the testis, which results in reduced numbers of spermatozoa in the cauda epididymidis. Although in these transgenic lines, ABP levels are increased in the epididymis, the consequence of this for epididymal function is not evident.

Although it is clear that testicular ABP enters the principal cells of the proximal segments of the epididymis, the mechanism of internalization is still not fully understood. The presence of high-affinity binding sites (K_{ass} = 3.7 nM^{-1}; 4.5 × 10^{11} sites/mg protein) on the apical surface of epididymal principal cells and the demonstration that the number of these receptors is fivefold lower in the cauda than in the caput of the epididymis have led to the proposal that ABP internalization is a receptor-mediated process in both rats and monkeys (504,509,510).

Steroid 5α-Reductases. The selective presence of radioactive DHT in epididymal cell nuclei after injection of radiolabeled testosterone (519), and the demonstration using micropuncture studies that, beyond the efferent ducts, the predominant androgens in epididymal fluid are 5α-reduced metabolites of testosterone [reviewed in Turner (85)], are two of the key findings establishing the central role of DHT, and hence steroid 5α-reductase, in mediating androgen action in the epididymis. The discovery that testosterone is rapidly converted by the epididymis to 5α-reduced products was first made almost simultaneously by two different groups (520,521). The epididymis is one of the tissues with the highest level of activity anywhere in the body (521).

In the mouse, rat, dog, monkey, and human, two 5α-reductases (isozymes), with different tissue distributions, have been identified. Although several theories have been put forward regarding the relative role of these two isozymes (522,523), there are few solid data to allow resolution of their relative roles. The consequences on the female of a null mutation in 5α-reductase 1 have been described (524). Null mutations for these two genes are compatible with male fertility; however, no detailed studies on spermatozoa or the epididymis of mice carrying null mutations for these genes have yet appeared. To understand the regulation of this enzyme activity in the epididymis, it is necessary to determine where and how the proteins are expressed, as well as the translation and, ultimately, the factors regulating transcription of the genes for this enzymatic activity.

Enzyme activity is expressed in a striking positional gradient as well as in different subcellular fractions in the adult rat epididymis (455,525). The activity associated with the nuclear fraction is highest in the initial segment, where it is higher than in any other male reproductive tissue, and declines dramatically in more distal segments of the tissue. The hormonal regulation of epididymal 5α-reductase activity is complex [reviewed in Robaire and Viger (517) and Ezer and Robaire (526)]. This was the first enzyme shown to be regulated by testicular factors as well as by circulating testosterone, depending on which segment and subcellular fractions were examined. Whereas nuclear fraction epididymal 5α-reductase activity is regulated in a paracrine or lumicrine manner by a substance directly entering the epididymis through the efferent ducts, microsomal 5α-reductase activity is found throughout the epididymis but is present at a lower level and is regulated by circulating androgens.

The two mRNAs for 5α-reductase types 1 and 2 are the products of separate genes (527,528). The type 1 mRNA isozyme is most abundantly expressed in the initial segment of the epididymis, with a positional gradient that is similar to the enzyme activity. Studies of the endocrine and developmental regulation of the 5α-reductase type 1 mRNA show that (a) orchidectomy results in a decrease in type 1 mRNA levels in all epididymal segments; (b) after orchidectomy, high-dose exogenous testosterone maintains 5α-reductase type 1 mRNA levels at control levels in all regions of the epididymis except the initial segment; (c) unilateral orchidectomy and efferent duct ligation cause a dramatic decrease in type 1 5α-reductase, selectively in the initial segment of the epididymis; and (d) only the type 1 transcript is developmentally regulated. Therefore, with regard to enzyme activity, the primary regulator of 5α-reductase type 1 mRNA expression in the initial segment is a paracrine/lumicrine factor of testicular origin [reviewed in Robaire and Viger (517) and Ezer and Robaire (526)].

The 5α-reductase type 2 mRNA is found in high concentrations throughout the epididymis, with a somewhat elevated concentration in the caput epididymal segment (523). Although 5α-reductase type 2 mRNA is present in abundance in the epididymis relative to the type 1 mRNA, its activity as a functional enzyme appears to be relatively poor. Unlike the developmental pattern of 5α-reductase type 1 mRNA expression, characterized by dramatic increases occurring just before the first appearance of spermatozoa in the epididymis, 5α-reductase type 2 mRNA expression does not show any significant developmental changes in any epididymal segment (523). Efferent duct ligation results in a dramatic decrease in 5α-reductase type 1 mRNA levels in the initial segment, but causes a near doubling in type 2 mRNA levels. This differential regulation of 5α-reductase type 1 and type 2 transcripts in one tissue has not been observed in other rat tissues such as the prostate (523).

In comparing the regulation of epididymal 5α-reductase enzyme activity and the expression of its mRNAs, it is apparent that a potential regulatory step is at the level of the protein itself. Posttranscriptional regulation has been proposed for several other epididymal proteins (79,135,345,466,529,530). In the adult rat, 5α-reductase type 1 protein expression is intensely immunolocalized in discrete lobules of the proximal initial segment of the epididymis; a sharp decline in staining intensity occurs between the proximal initial segment and its adjacent region, followed by a progressive decrease in intensity beyond that point [reviewed in Robaire and Viger (517)]. In all epididymal regions, 5α-reductase immunoreactive protein is localized specifically in the epithelial principal cells and is uniquely associated with membranous cytoplasmic elements; the intracellular localization of the 5α-reductase protein changes with movement down the epididymis. Savory et al. (1995) (531) have localized type 1 5α-reductase to the nuclear envelope of both transfected cells and rat tissues. The infranuclear localized form of the 5α-reductase type 1 protein is regulated in a similar fashion to the type 1 mRNA and enzyme activity, that is, by a paracrine factor of Sertoli cell origin (517,532).

The genomic regulation of the expression of both 5α-reductase isozyme genes is still poorly understood. Cloning and characterization of the proximal 2.2 kilobases of the 5′ upstream region of rat 5α-reductase type 2 have revealed that there are regions with strong repressor and enhancer activity; potential transcription factors (Sp1 and Sp3) regulating gene expression have been identified and immunolocalized to principal cells of the epididymis. The promoter region of rat 5α-reductase type 2 is GC-rich with a noncanonical TATA box located upstream of the transcriptional start site (461). The potential role of this GC-rich region is intriguing in light of the observation that the methylation pattern of the two genes appears to differ not only between each other but between reproductive and nonreproductive tissues (533). The genes for 5α-reductase types 1 and 2 are also differentially methylated in lymphocytes from normal and 5α-reductase–deficient patients (534).

Estrogen Receptors and Cytochrome P450 Aromatase

Estrogen Receptors. Starting in the 1970s, substantial evidence accumulated for the direct action of estrogens on androgen target tissues. Using binding studies, specific, high-affinity cytosolic and nuclear estrogen-binding proteins, presumably receptors, were identified by a number of investigators in mice (535), rats (536–538), guinea pigs (539), rabbits (540–543), dogs (472,544), rams (545), monkeys (546,547), and men (548). Through a comprehensive series of studies, Danzo's group established the presence of estrogen receptors in the rabbit epididymis and demonstrated both its developmental (549) and regional distribution (539); however, enthusiasm for further studies on the estrogen receptor in the epididymis diminished when this same group demonstrated the presence of a protease, in the adult, that cleaves the DNA binding region of the estrogen receptor (550).

The histolocalization of tritiated DHT and estradiol in the adult mouse epididymis revealed differential distribution of grains over nuclei in different segments of the excurrent duct system, as well as over different cell types (535,551). These studies confirmed the earlier binding experiments and provided a more detailed localization of the putative receptors. Very strong labeling of the efferent ducts and initial segment of the epididymis was noted, with lesser binding in the more distal segments of the epididymis; nuclei of clear cells were more densely labeled with estradiol than with DHT, in sharp contrast to what was seen over nuclei of principal cells. Based on such data, and before the studies with null mutant mice, specific functions for estradiol in the epididymis were proposed (1,552).

With the discovery of the existence of two estrogen receptors (ERα and ERβ), and the availability of specific antibodies for these receptors as well as mice with null mutations for either or both receptors, our understanding of the localization of estrogen receptors and their potential role(s) has grown rapidly [reviewed in Hess (553)]. ERα and ERβ are remarkably similar to one another in some respects (e.g., 95% identity in the DNA binding domain), but they have individual characteristics (e.g., <55% homology in the ligand binding domain) that clearly indicate functional differences (554,555). The two isoreceptors have different binding affinity for estrogenic and antiestrogenic compounds (537,556), different tissue distributions (557), and different response elements (558,559).

Although there are some conflicting data on the immunolocalization of ERα and ERβ along the efferent ducts and the epididymis, a general pattern is emerging. In the mouse, the epithelium of the efferent ducts show strong ERα and weaker ERβ immunoreaction (553,560,561). The initial segment of the epididymis shows some reactivity for ERα, primarily in narrow, apical, and some basal cells (485); in the caput epididymidis, principal cells and clear cells stain positively, whereas in the distal regions only clear cells are immunoreactive. In contrast, ERβ immunoreactivity is clearly observed along the entire epididymis, with more intense staining in the more distal regions (corpus and cauda) (553,562). In the rat, in spite of some earlier controversies relating to the method of section preparation and antibody used for immunolocalization studies (551,563), it appears that ERα and ERβ immunolocalize in patterns similar to those of the mouse. The presence of ERα and the strong immunoreactivity of ERβ all along the epididymis have been confirmed using RT-PCR data (551,564) [reviewed in Hess (553)]. In monkeys, baboons, and humans, both forms of the estrogen receptor are present along the epididymis (565–568).

The first clear insight into the functional role of ERα in the efferent ducts and epididymis came through null mutation studies. The ERα knockout mouse was developed by Lubahn et al. (1993) (569) and is infertile in spite of the presence until puberty of apparently normal seminiferous tubules; in the adult, testicular atrophy becomes evident (570). Through an elegant series of studies, Hess's group (553,571,572) has demonstrated that the cause of infertility is back-pressure atrophy of the seminiferous tubules owing to the inability of the efferent ducts and initial segment of the epididymis to reabsorb the large volume of fluid secreted by the testis. This reabsorption process is under the control of estradiol, acting as a paracrine regulator of the efferent ducts. Further, it has been proposed that the action of estrogen on fluid reabsorption is mediated by the cystic fibrosis transmembrane conductance regulator (CFTR) receptor (573). In contrast to the clear role of the ERα receptor, function of ERβ in the male reproductive tract awaits further investigation because the ERβ knockout mouse is fertile and appears to have a normal testis and epididymis (574).

Cytochrome P450 Aromatase. An ongoing source of ligand is necessary to activate estrogen receptors. Cytochrome P450 aromatase, localized to the smooth endoplasmic reticulum, is a terminal enzyme that irreversibly catalyzes the conversion of testosterone to estradiol. Both the absence of aromatase and an excess of estrogen result in male infertility [reviewed in Carreau (575)]. There are several sources of estradiol in the adult male reproductive tract. In the testis of adults, both Leydig cells (576,577) and germ cells (578) have been reported actively to synthesize estradiol. In germ cells, the enzyme is localized in the Golgi of round spermatids and throughout the cytoplasm of elongating and late spermatids [reviewed in Hess (553), Carreau (575), and O'Donnell et al. (579)].

In the epididymis, aromatization of the very high concentrations of testosterone to produce estradiol, particularly in the proximal segments, could occur either in the luminal or the epithelial compartments. The presence of cytochrome P450 aromatase in epididymal spermatozoa has been demonstrated in several species, including mouse, rat, bear, and rooster (578,580–582). The enzyme is localized to cytoplasmic droplets, and the staining becomes less intense as spermatozoa traverse the epididymis (583).

Early investigations of the presence of cytochrome P450 aromatase in epididymal epithelial cells indicated that this activity was absent (584); however, no direct assessment of the enzyme was done in these studies. In several subsequent investigations, direct detection of the P450 aromatase mRNA, protein, and enzyme activity (formation of estradiol from testosterone), as well as the ability of specific aromatase inhibitors to inhibit estradiol formation, all led to the conclusion that epididymal principal cells, but not the smooth muscle cells surrounding the duct, have the ability to synthesize estradiol from testosterone in a number of species, including the rat (585), monkey (586), and human (587).

Retinoic Acid Receptors. Retinoids are highly potent molecules with pleiotropic action in a wide range of biological processes during development and in the adult (588). Vitamin A deficiency results in both keratinizing and nonkeratinizing squamous metaplasia of numerous epithelia, including the epididymal epithelium (589,590).

Given the importance of retinoids, it is not surprising that a complex system has evolved to control retinoic acid biosynthesis and metabolism (591–593). Because retinoids are hydrophobic molecules insoluble in water, they require chaperones to ensure proper storage, transport, and uptake in the tissues. These chaperones, the retinoid-binding proteins, are members of either a large superfamily of small intracellular proteins that includes numerous fatty acid–binding proteins or another superfamily of secreted proteins, the lipocalins, which bind hydrophobic molecules. There are four intracellular retinoid-binding proteins that are members of the fatty acid–binding protein family (cellular retinol-binding protein [CRBP] I and II, and cellular retinoic acid–binding protein [CRABP] I and II), and two extracellular retinoid-binding proteins that are members of the lipocalin family (retinol-binding proteins and epididymal retinoic acid–binding protein [E-RABP]).

Within the nucleus, two classes of nuclear retinoic acid receptors, RAR and RXR, have been identified and cloned. Each receptor class consists of three receptor subtypes, α, β, and γ, and their isoforms. RXRs form heterodimers with RARs that are thought to be the functional units transducing the retinoid signal. The binding of RARs/RXRs to retinoic acid response elements modulates gene expression and is likely to be responsible for the observed biological effects of retinoids (594).

Most of the components of the retinoid signaling pathway have been identified in the epididymis. Retinoids were quantified in the rat epididymis by high-performance liquid chromatography (HPLC) followed by derivation of the HPLC-purified retinoids to confirm their identities (595). In the epididymal tissue, the major retinoids identified included retinol, retinyl ester, all-*trans* retinoic acid, and 9-*cis*-retinoic acid (9-*cis*-RA). The concentration of retinol and retinyl esters decreased from the caput to the cauda epididymidis from 2.7 to 1.7 nmole/g of tissue and from 6.2 to 1.4 nmole/g tissue, respectively. The active retinoids, all-*trans* retinoic acid and 9-*cis*-RA, showed an inverse gradient, with lowest concentrations in the caput (all-*trans* retinoic acid: 13 pmole/g tissue; 9-*cis*-RA: 2.9 pmole/g tissue) and highest concentrations in the cauda (all-*trans* retinoic acid: 35 pmole/g tissue; 9-*cis*-RA: 6.8 pmole/g tissue). Using a bioassay, levels of 7 pmole/g of tissue have been reported in the murine epididymis (595a).

The CRBPs and CRABPs are also present in the epididymis, but their localization is regionalized. CRBP is most abundant in the columnar epithelial cells of the proximal caput epididymidis of the rat; this tissue has one of the highest CRBP expressions in the body (596–598). The mouse has the highest concentrations of CRBP in the proximal caput epithelium, but also has significant levels in the epithelium throughout the epididymis. CRABP II is found in the epithelium of the cauda epididymidis coexpressed with CRBP in both the mouse and the rat, whereas CRABP I is present in the smooth muscle of the distal cauda epididymidis (599).

The retinoid signaling pathway in the epididymis includes the retinoic acid receptors and the retinoic acid carrier protein E-RABP, synthesized and secreted into the lumen in the distal caput and found in the lumen from the caput to the cauda epididymidis. In the rat epididymis, RARα, β, and γ transcripts have been detected by RT-PCR (600). In situ hybridization and immunolocalization revealed region-specific patterns of RARα mRNA and protein expression (601). The highest levels of RARα transcripts and RARα immunostaining were seen in the proximal caput epididymidis. The levels of RARα transcripts and RARα protein staining were higher in the initial segment, proximal caput, and distal cauda. The lowest levels were seen in distal caput, corpus, and proximal cauda epididymidis.

It is clear that a remarkable retinoid homeostasis system has evolved to permit the transport of these hydrophobic molecules and their metabolism, extracellular and intracellular sequestration, and delivery to the nuclear chromatin for optimum retinoid signaling in the epididymis. However, knowledge of the pattern of expression of the various components of the retinoid biosynthesis and metabolism pathway, although suggestive, is not sufficient to deduce the precise role of these regulatory factors or the site of action of retinoic acid in the epididymis. Vitamin A deficiency has pervasive effects, affecting many organs during prenatal and postnatal development and adult life. Reproduction is affected both at the level of the testis, where spermatogenesis is arrested, and at the level of the epididymis, where the epithelium develops squamous metaplasia. Because of this dual effect, it is difficult to ascertain from vitamin A deficiency studies the physiological role of retinoids in epididymal function.

Gene inactivation has produced surprising results. The transthyretin null mutant has a virtual loss of serum retinol-binding protein, yet the mice seem normal (602). CRABP I and CRABP II null mutant mice also appear normal (603,604). The double knockouts of CRABP I and CRABP II have a higher death rate at 6 weeks than the wild-type mice, but the survivors appear normal (604). Although each retinoid-binding protein separately appears to be dispensable, the lack of phenotype may simply reflect that redundant mechanisms are built into the organism.

The functional redundancy is well illustrated in the case of the retinoic acid nuclear receptors, of which six RXR isoforms and six RAR isoforms have been identified (605). Null mutations of RARβ, $RAR\alpha_1$, $RAR\beta_2$, or $RAR\gamma_2$ appear normal. Null mutations of all RARα isoforms or all RARγ isoforms display some of the defects of the postnatal vitamin A deficiency syndrome. Only double-null mutants of the RAR subtypes (RARα/γ, $RAR\alpha/\beta_2$, $RAR\alpha_1/\beta_2$, $RAR\beta_2/\gamma$) reproduce most of the features of vitamin A deficiency syndrome. Null mutation of RXRα induces some abnormalities, but double inactivation of RXRα and RARβ or RARγ increases the severity of the RXR phenotype. It appears that even though the different receptor isoforms are not truly functionally equivalent, they can substitute for one another under certain conditions. This functional redundancy may not be surprising in any gene family that has been conserved throughout vertebrate evolution and has evolved by duplications from an ancestral gene.

Even with these caveats, an epididymal phenotype has been observed in transgenic mice after disruption of the RAR gene. Expression of a dominant negative mutant of RARα induced loss of organization of the columnar epithelium lining the cauda and its transformation by squamous metaplasia. The aberrant transformation resulted in blockage or rupture of the duct with concomitant inflammatory response and, ultimately, infertility (606). The RARα null mutant mice display not only aspermatogenesis, but vacuolization of the epididymal epithelium (607), and the RARα/γ double-null mutants display either severe dysplasia or complete agenesis of the epididymis and the vas deferens (608).

In summary, given the pathological changes seen in the epididymal epithelium in vitamin A deficiency, the epididymal phenotypes seen in RAR null mutant mice, and the presence of most of the components of the retinoid signaling pathway, it is highly likely that retinoids play a regulatory role in epididymal function.

Oxytocin Receptors. The first demonstration of the presence of oxytocin receptors in the epididymis was by Maggi et al. (1987) (170). Since then, several reports on the localization, distribution, regulation, and potential functions of these receptors in a variety of species have appeared (169,171–173).

Although in the marmoset monkey, the oxytocin receptor seems to be primarily expressed in peritubular cells of the epididymis, in the human, macaque monkey, and ram, its localization is in the epididymal epithelium and varies along the duct (169,172). The expression in principal cells is seen only in the initial segment of the epididymis of primates and has a checkerboard appearance; throughout the rest of the duct, it is confined to peritubular cells of the epididymis (172). In the ram, however, oxytocin receptors are found in both principal and basal cells throughout the epididymal duct, whereas peritubular staining is seen only in the more distal segments (173).

Two functions have been ascribed to the oxytocin/-oxytocin receptor system in the epididymis. As discussed previously, the first is that of promoting contractility of the duct, as demonstrated by both in vitro (174) and in vivo (179) studies, thus promoting sperm transport through the epididymis (187). The second function of this system involves the promotion of the formation of DHT by stimulation of 5α-reductase activity in the initial segment of the epididymis (609). This function is of particular interest, given the very high 5α-reductase enzymatic activity in this segment of the epididymis and the presence of oxytocin in the luminal fluid of the ram and the rat (610,611). The mechanism for this type of stimulation is not clear but has been postulated to be mediated by phosphorylation and activation of existing 5α-reductase due to tyrosine kinase activation through a coupling to the oxytocin receptor by G_i signal transduction (612).

Growth Factor Receptors. FGF receptors (FGFR), platelet-derived growth factor (PDGF) receptors, and vascular endothelial growth factor (VEGF) receptors have all been identified in the epididymis. For FGFR, two splice variants, FGFR-1 IIIc (long and short forms), have been identified in the principal cells of the initial segment (459). Although all four major splice variants of FGFR (1 through 4) are present in initial segment tissue, only the FGFR-I IIIc variants are present in principal cells. The remaining splice variants are presumably localized to the endothelial and interstitial cells. FGFs 2, 4 and 8, present in rete testis and epididymal luminal fluid (458,459), are potential ligands for the apically placed receptor (S. Crenshaw and B. T. Hinton, unpublished observations).

PDGF α and β receptors and their ligands (PDGF-A through D) have been localized in the rat and mouse epididymis (613). In view of the specific localization of PDGF-A in the epithelium and the PDGF α receptor in the mesenchyme, the authors suggested that the PDGF-A/α receptor system may be involved in epithelial–mesenchymal interactions during epididymal development. In the adult epididymis, PDGF-A and B and α and β receptors were identified in the epithelium only. Interestingly, in PDGF-A null mutants, an epididymal phenotype was not observed during development, but epididymal epithelial disruption was observed at postnatal day 25 (613).

VEGF and its receptors are highly localized in both the testis and epididymis, with fms-like tyrosine kinase (flt-1) and fetal liver kinase (designated as kinase insert domain–containing receptor [KDR]) VEGF receptors being identified by immunohistochemistry in the ciliated cells of the efferent ducts and vascular endothelial cells, respectively, of the human epididymis (614). Flt-1 was also observed in the lymphatics. Furthermore, addition of VEGF28 to epididymal tissue in culture resulted in endothelial fenestrations and opening of interendothelial junctions. Because the ligand, VEGF, was localized to the peritubular and ciliated cells of the efferent ducts and to the peritubular and basal cells of the epididymis, it was suggested that it may play a paracrine role in the permeability of blood vessels through KDR and lymphatics through flt-1 (614). Overexpression of VEGF in the testis and epididymis results in male infertility due to spermatogenic arrest and increased capillary density in the testis and epididymis (615).

Although not specifically expressed in the epididymal epithelium, the hepatocyte growth factor receptor, c-met, is localized to epididymal spermatozoa, with the ligand, hepatocyte growth factor, being secreted into the lumen by the epididymal epithelium. The putative action of hepatocyte growth factor is maintenance of sperm motility (616). Interestingly, the hepatocyte growth factor–like protein receptor, Ron mRNA, which is a receptor tyrosine kinase, is expressed in the epididymal epithelium (617). Additional growth factor–binding proteins and growth factor receptors that are expressed in epididymal epithelial cells include insulin-like growth factor–binding protein-rP1/mac25 (618), latent transforming growth factor-β–binding protein 2 (619), epidermal growth factor receptor (620), and growth hormone receptor–binding protein (621).

c-Ros. c-Ros is an orphan tyrosine kinase receptor that has received considerable attention since it was shown that c-Ros null male mice were infertile (40). The kinase and extracellular domains of the *c-Ros* gene show similarities to the *Drosophila sevenless* (*sev*) tyrosine kinase receptor (622). The *sev* gene was originally discovered in a *Drosophila* mutant that failed to express the R7 photoreceptor cell in the ommatidium (623,624). Further studies demonstrated that the adjacent R8 photoreceptor cell expressed *boss*, the ligand for sev. Interaction between these two cell types was crucial for the development of these photoreceptor cells. After this cell interaction, a signal transduction cascade involving Ras and activation of downstream target genes was initiated in the R7 photoreceptor cell. Interestingly, ectopic expression of *boss* or of constitutively active forms of *sev* causes the formation of ectopic R7 cells (625,626). The overall sequence homology between *c-Ros* and *sev* is quite low, but there are some very interesting structural similarities between the two. For example, (a) both genes encode a large extracellular domain and a kinase domain; (b) sequence homology extends into the putative extracellular ligand-binding domain; (c) there is a short insert of five to seven amino acids in the same position in the kinase domain of both genes that is not found in other receptor tyrosine kinases; (d) there are several common exon–intron junction points between the two genes; (e) the kinase domains of both genes are 70% homologous (622,627–630); and (f) particularly striking is the analysis of Springer (1998) (631), who showed a colinearity and spacing of β-propeller and epidermal growth factor-repeat structures in c-Ros and *sev*. Interaction of ligand with β-propellers has been shown for other proteins (632). The ligand of sev is bride of sevenless (boss) (626,633), a seven-transmembrane G-coupled protein kinase, which has signal transduction properties of its own. Presumably, a *boss* homolog exists in mammals if *c-Ros* has a similar function to *sev*.

Although c-Ros is expressed in several tissues throughout development, including the lung, kidney, and epididymis, it remains highly expressed in the initial segment of the adult mouse epididymis (40,634). A clue to the function of c-Ros came from c-Ros knockout mice: the animals displayed an underdeveloped initial segment and male infertility. The infertility phenotype was due to spermatozoa having an angulated flagellum at the end of the midpiece (635); it was suggested that this defect was caused by the failure of spermatozoa to regulate their cell volume (635). This infertility phenotype was also shown to be cell autonomous; the defect in the sperm phenotype was due to the passage of normal spermatozoa through the underdeveloped initial segment, and not to some inherent defect in the spermatozoa themselves. Further, the infertility phenotype was observed if either the kinase domain or the extracellular domain of c-Ros was lacking (40).

Although the functional role of c-Ros is not known, it has been suggested that it plays a role in the regulation of cell volume (636,637), and although c-Ros itself may not have a direct effect on cell volume regulation, gene expression profiling suggests that several osmolyte transporters are compromised in the initial segment of the c-Ros null animals (467). It also appears that at least inositol and glutamate sperm concentrations are lower in the c-Ros knockout mice compared with controls, even though there are no differences in luminal fluid concentrations of the two osmolytes between the two groups (377). Whether the lower intracellular levels of these osmolytes in spermatozoa is the cause of the infertility phenotype remains to be determined. Adding to the complexity of the problem is that the infertility phenotype may be related to an underdeveloped initial segment rather than directly to c-Ros. Because the adult mouse initial segment still expresses c-Ros, the challenge will be to silence this gene in the adult and then examine the fertility phenotype. This experiment needs to be done, especially if c-Ros and its downstream signaling pathways are to be considered as potential targets for the development of a male contraceptive. An additional important finding is that the adult human epididymis also expresses c-Ros along its length (638).

In an attempt to identify downstream targets of c-Ros, Keilhack and colleagues (2001) (639) used a yeast two-hybrid assay and identified several proteins that associated with the kinase domain of c-Ros. These included the phosphatase SHP-1 (Src homology region 2–containing protein tyrosine phosphatase-1), phospholipase Cγ, the SH2 domain of c-Abl, Grb2, the phosphoinositide-3 kinase subunit p85α and the SH2 domains of SHP-2. To demonstrate the potential importance of the SHP-1 phosphatase in male fertility, *viable motheaten* (*me^v*) mice were tested for fertility because these mice contain a mutation in the SH2 domain of SHP-1. The male mice were infertile. Hence, downstream targets of c-Ros are clearly excellent candidates for male contraceptive development. Although Keilhack et al. (2001) (639) claim that the infertility phenotype is similar to that seen in the c-Ros knockout animals, the authors do not present convincing data that there is a reduction in epithelial cell height in the initial segment of the SHP-1 knockout animals, similar to that seen in the initial segment of c-Ros knockout mice. Their data suggest that the initial segment has developed normally in this knockout. Hence, although male infertility is observed in both the c-Ros and SHP-1 knockouts, the mechanisms by which infertility is achieved in each case are probably different.

HE6. The HE6 gene encodes a novel human epididymal protein that comprises a seven transmembrane component with a large extracellular domain and shows high homology to the secretin/pituitary adenylate cyclase–activating polypeptide (PACAP) superfamily of G-protein–coupled receptors (GPCRs) (640). In view of the large ectodomain, HE6 has also been considered to be a member of the LNB-TM7 proteins (family-B GPCR–related TM7 receptors with long N-terminal extensions) (641). This large domain also contains a GPCR proteolysis site, and therefore HE6 may exist as a two-subunit receptor. HE6 is located on the human X chromosome, and both rat and mouse homologs of HE6 have been cloned (640,641). Immunolocalization studies show human and mouse HE6 localized to the apical surface of human epididymal cells, with staining more pronounced in the efferent ducts–caput region compared with the distal epididymal regions. A mosaic pattern of staining was also observed, as is often seen with localization of other epididymal proteins. Mouse HE6 was localized to the apical cell surface but highly restricted to the initial segment region (641). The function of HE6 is yet to be determined, although a knockout study suggests that the protein may be involved in fluid reabsorption in the efferent ducts (642).

Other Receptors. In addition to those receptors discussed in the preceding sections, the epididymis also expresses orexin receptors (643–645), the urokinase-type plasminogen activator receptor (646), apolipoprotein E receptor-2 (647), adrenomedullin receptors (648), purinergic receptors (649), and guanylyl cyclase C receptor (650).

Transcription Factors

An understanding of transcriptional regulation of epididymal genes has been limited to a few genes [reviewed in Robaire et al. (90), Cornwall et al. (91), Kirchhoff (92), Rodriguez et al. (651), Rao and Wilkinson (652), and Suzuki et al. (653)]. However, in these studies, several key transcription factors have been identified that may play an important role in the tissue-specific, region-specific, or cell-specific expression of epididymal genes. For the sake of simplicity, steroid hormone receptor coregulators are not discussed here, but are discussed in the section on Receptors and Their Ligands.

PEA3 Family. One of the first major epididymal transcription factors to be examined was polyomavirus enhancer activator 3 (PEA3), which is a member of the Ets transcription factor family. This family is characterized by an 85–amino acid Ets domain (654) that encodes for a helix-turn-helix DNA binding motif. It is also part of a subfamily, referred to as the PEA3 family, that comprises PEA3, ERM, and ER81 (655–660). The consensus site for PEA3 is 5′-AGGAAG-3′, with GGAA being the core binding motif for the Ets family. Xin et al. (1992) (655) were the first to show expression of PEA3 in the epididymis, and later studies by Lan et al. (1997) (460) and Drevet et al. (1998) (661) clearly showed a role for members of the PEA3 family in regulating GGTIV and GPX5 gene expression in the epididymis. Standard in vitro promoter analyses of both GGTIV and GPX5 revealed that PEA3 acts as both a repressor and an activator of transcription. This was confirmed more recently by Kirby et al. (2004) (662) using a novel in vivo electroporation technique. There were differences in promoter activity between the in vivo and in vitro studies when the same constructs were used, highlighting the importance of performing such an analysis under as near-physiological conditions as possible. The arrangement of PEA3 binding sites along the GGTIV promoter may be unique to this gene because of the absence of binding motifs (e.g., AP-1) that are known to regulate the activity of PEA3 in many other gene promoters (e.g., urokinase plasminogen activator and matrix metalloproteinase) (663,664).

There appears to be redundancy among the PEA3 family members because SouthWestern analysis (460) showed that multiple proteins identified as PEA3, and possibly ERM and ER81, bound to GGTIV mRNA PEA3 binding motifs. Further, PEA3 knockouts had a normal epididymal phenotype, supporting the idea that ERM and ER81 may take over the functions of PEA3. PEA3 knockout male mice are infertile, but their infertility is due to impotence or failure to ejaculate spermatozoa (665). Interestingly, the promoters of several genes that are highly expressed in the initial segment, including 5α-reductase type 2 (461), have PEA3 family binding motifs, and each gene, including PEA3 family members, is regulated specifically by testicular factors. It will be important to generate initial segment-specific PEA3 family member knockouts to determine whether PEA3 family members play an important role in the function of the initial segment and hence male fertility.

CCAAT/Enhancer-binding Protein β. CCAAT/enhancer-binding protein β (C/EBPβ) is a member of the CCAAT/enhancer-binding proteins that regulate genes that have various functions, including cell cycle control, fatty acid metabolism, cellular differentiation, growth, tumorigenesis, and apoptosis [reviewed in Ranji and Foka (666)]. This six-member transcription factor family is characterized by a highly conserved basic-leucine zipper domain at the carboxyl terminus that is involved in dimerization and DNA binding. Interestingly, although heterodimerization can occur, the heterodimer C/EBPβ-C/EBPζ does not bind to the standard C/EBP consensus sequence, but binds to a subset of promoters in genes involved in cellular stress. Hence, this family of transcription factors can function as either repressors or enhancers depending on the physiological state of the cell (667).

C/EBPβ is expressed in the proximal regions of the mouse epididymis and plays a role in the regulation of expression of the CRES (cystatin-related epididymal spermatogenic) gene (668). Two putative C/EBP DNA binding motifs were identified in the first 135 base pairs of the CRES promoter. Data from gel shift assays, in vitro and in vivo promoter analyses, and analysis of C/EBPβ knockout mice suggest that expression and activation of C/EBP is necessary for complete activation of CRES in the proximal caput region (662,668). Further analysis showed that although C/EBPβ was the predominant form expressed in the epididymis, C/EBPα and δ were also expressed, but to a lesser degree. C/EBPε was not identified (668). Because the expression of members of the CRES gene family depends on the presence of testicular luminal fluid factors (669), it is assumed that C/EBP family members are also regulated in a similar manner. Because members of C/EBP family play a role during cellular stress, it would be interesting to examine their potential role during apoptosis of the initial segment after loss of testicular factors.

Homeodomain-containing Transcription Factors. Several classes of homeodomain-containing proteins are expressed in the mammalian epididymis; these include hox (homeobox), PEPP-family (PEM), Prd-class, and Pou-class genes [reviewed in Rao and Wilkinson (652) and Bomgardner et al. (670)]. Homeodomain-containing proteins are transcription factors that contain a highly conserved 60–amino acid

DNA binding domain composed of three α-helices, with the third helix interacting closely with DNA. The major role of hox genes is in organ development and the establishment of segment identity along the anteroposterior body axis (671,672). The epididymis is a highly segmented organ, and therefore it is not surprising that many hox genes are expressed in the developing epididymis (670,673–675). Because the epididymis forms later in development, it is the 5′ hox genes that appear to play a major role in epididymal development compared to the earlier expressed 3′ hox genes. Interestingly, hox genes are also expressed in the adult epididymis (675); their precise function in the adult warrants further study.

A first indication that hox genes were expressed in the epididymis and played a role in epididymal development came from an analysis of reproductive tissues from hoxa10 and hoxa11 knockout animals (673,676). One of the most intriguing phenotypes to be produced from these knockouts was the homeotic transformation of the distal epididymal regions. Hoxa11 was expressed in the initial segment and in the more distal epididymal regions (675), so the observation of a distal epididymal phenotype in the hoxa11 knockout was not surprising. However, it is not clear if the epididymis developed appropriately in this mutant because the initial segments of these animals were not analyzed. The regulation of hox genes in the epididymis is poorly understood; however, it is known that in many tissues cofactors must be involved in the transcriptional activation of genes by hox transcription factors. Two examples include Meis family members and PBX (677), both of which contain a homeodomain; Meis 1a and 1b are expressed in the mammalian epididymis (675).

More recently, a new homeobox cluster termed *Rhox* has been identified on the X chromosome. Although most of the members of this family are expressed in the testis, several are also expressed in the epididymis (678). A putative homeodomain transcription factor called Phtf1 has been identified in the principal cells of the initial segment and caput region (679). Although this protein contains domains that suggest it is a transcription factor, immunolocalization studies show it to be localized to the endoplasmic reticulum saccules applied to the *trans* face of the Golgi system.

One the best-studied homeodomain-containing transcription factors is PEM, a member of the PEPP homeobox subfamily. Rao and Wilkinson (2002) (652) suggested that the PEPP genes were derived from the Drosophila *aristaless* gene. Of the PEPP gene family members, only PEM was expressed in the epididymis as well as in the testis (652). PEM, like hoxa11, was expressed throughout development and into adulthood (680,681) from a "male-specific" proximal promoter (Pp). The expression of PEM in the epididymis was species dependent. In the mouse, PEM was found mainly in the caput, whereas in the rat it is expressed throughout the duct, increasing from caput to cauda. However, PEM was not expressed in the initial segment of either the rat or mouse (652). Further, expression of PEM was androgen dependent (680,682). The function of PEM is not known because PEM knockout animals did not display an epididymal phenotype (683), although Rao and Wilkinson (2002) (652) suggested that the function of PEM may be context dependent (e.g., under different environmental conditions).

Other transcription factors that have been identified in the epididymis include sp1 and sp3 (461,684), c-Fos (91), Tst-1 (685), Emx-2 (686), Lbx-2 (687), pax-2 (688), zfy (689), and nor-1 (690).

Transporters and Channels

Water and Ion Transport. Although the efferent ducts serve as a conduit for spermatozoa from the testis to the epididymis, a major function of the efferent ducts is the reabsorption of water. The major driving force for water movement across the efferent ducts and the cauda epididymidis is the Na^+/K^+-ATPase located on the basolateral membrane (691,692). However, because the activity of the Na^+/K^+-ATPase in the efferent ducts is less than that in the cauda of the epididymis, the paracellular pathway also plays a critical role in sodium and water movement (356) across the epithelium of the efferent ducts.

The involvement of an apically placed Na^+-H^+ exchanger that drives sodium out of the lumen with water following, was proposed by Wong and colleagues (2002) (358) because fluid absorption is amiloride inhibitable, isosmotic, and flow dependent. So far, four Na^+-H^+ exchangers (NHE1 through 4) have been identified in several tissues, and three of them have been found in the epididymis (693–696); such transporters appear to play a role in sodium and bicarbonate reabsorption and proton secretion. NHE1 and NHE2 are expressed on the basolateral surface and apical surface, respectively, throughout the epididymal duct, except that NHE2 is not expressed in the initial segment. Cheng-Chew et al. (2000) (694) suggested that NHE1 is involved in bicarbonate secretion, whereas NHE2 may be involved in sodium reabsorption. Studies by Breton et al. (695) and others (693,696) have shown that the expression of NHE3 varies throughout the length of the epididymis and is localized to the apical surface of principal cells only. Important to bicarbonate reabsorption is the presence of apical and cytoplasmic carbonic anhydrases. NHE3 appears to participate in bicarbonate reabsorption, similar to its role in the renal proximal tubule. Through the activity of apically placed NHE3, protons are secreted and sodium is reabsorbed. Carbonic anhydrase IV, which is also apically positioned, then converts the luminal protons into water and carbon dioxide, allowing the carbon dioxide to enter the cells. Cytosolic carbonic anhydrase II catalyzes the production of new intracellular bicarbonate and generates a proton. The proton is recycled back into the lumen by NHE3. Intracellular bicarbonate is then transported out of the cell across the basolateral membrane by the bicarbonate exchanger (AE2) (473), or the electrogenic sodium-bicarbonate cotransporter, NBC1 (473). The driving force for this whole process is provided by the low intracellular sodium concentration achieved by the basolateral Na^+/K^+-ATPase. In addition, maintenance of a high potassium concentration by the Na^+/K^+-ATPase induces an intracellular negative potential of approximately −70 mV. This drives bicarbonate across the basolateral membrane through the NBC1 transporter, which transports three molecules of bicarbonate for every sodium molecule, resulting in the net transfer of two negative charges.

A major player in hydrogen ion secretion is vacuolar-type H^+-ATPase, which is highly expressed in apical/-narrow cells in the proximal regions of the epididymis and in the clear cells throughout the epididymis (97,697–700). Using a proton-selective electrode, Breton and colleagues (1996) (699) showed that the epithelium of the vas deferens secreted protons and that this was inhibited by bafilomycin, an inhibitor of H^+-ATPase. This study clearly demonstrated that acidification of the luminal fluid in the vas deferens was due to this transporter. To maintain an acidic luminal fluid microenvironment, clear cells probably respond to a luminal increase in bicarbonate by increasing their rate of proton secretion (106). Further, the rapid recycling of the H^+-ATPase in clear cells suggests that the regulation of proton secretion may involve unique clathrin- and caveolin-independent mechanisms, as well as members of the SNARE protein family; using colchicine, a microtubule network has been found to be important [reviewed in Breton (360)].

The aquaporins also play a role in water reabsorption and, of the 10 isoforms, AQP1, 2, 3, 8, and 9 have been identified throughout the epididymis, including the efferent ducts (701–708), but not necessarily in every region and epithelial cells. Although very few functional studies have been done on the aquaporins, studies by Wong and colleagues (707) have shown that the cystic fibrosis transmembrane conductance regulator (CFTR) potentiates water permeability of AQP9 in oocytes and in the rat epididymis. CFTR is expressed in the rat and human epididymis. Besides CFTR's known role as a cAMP-activated chloride channel, its role as a regulator of AQP9 may have ramifications in men with cystic fibrosis (707). Other chloride channels have received little attention. However, a study by Isnard-Bagnis et al. (2003) (107) showed that the epididymis expresses members of the CIC gene family of chloride channels. Studies from this group suggest that CIC-3 may be involved in chloride transport by principal cells, and CIC-5 might play a role in the acidification of H^+-ATPase–containing vesicles in narrow and clear cells.

Although net water movement is from lumen to cell, water is still secreted and plays an important role in ion and water homeostasis. Chloride and bicarbonate are again responsible for driving water from the cell into the lumen, and Wong and colleagues (2002) (358) have undertaken extensive investigations recognizing the importance of peptide hormones in this process. These investigators have presented a novel model to explain anion secretion by multiple peptide hormones. For example, angiotensin II, bradykinin, endothelin, and arginine vasopressin interact with their cognate receptors on the basolateral membrane, which in turn activates phospholipase A2 on membrane phospholipids, resulting in the release of arachidonic acid. PGs are subsequently produced through the actions of COX-1 and isomerases. This process occurs in the basal cells. PGs then diffuse out of the basal cells and interact with their receptors on the basolateral membranes of principal cells, which in turn causes an increase in cAMP resulting in activation of the apically placed CFTR channel and chloride secretion (Wong et al., 2002) (358). With the identification of the family of Trp channels in the epididymis, Leung et al. (2004) (116) suggest that Trp3 provides the increase in intracellular calcium levels, which translocates phospholipase A2 from the cytosol to membrane phospholipids, where COX-1 has been identified (Fig. 15).

In summary, water secretion is driven by chloride transport, whereas fluid reabsorption is principally driven by sodium transport. As suggested by Wong and colleagues (2002) (358), secretion of water may play a role in fine-tuning water movement across the epididymal epithelium. This may be particularly important because for now very few hormones appear to play a role in regulating water reabsorption. However, estrogen has been implicated as a major regulator of water reabsorption in the efferent ducts (551,571,709). As discussed previously, the efferent ducts in an ERα knockout mouse failed to reabsorb water sufficiently, resulting in the accumulation of fluid and spermatozoa, testicular blockage, and atrophy. Later studies by Hess and others suggest that androgens may also play a role in the regulation of water reabsorption by the efferent ducts (553). If hormones such as estrogen and androgen play a

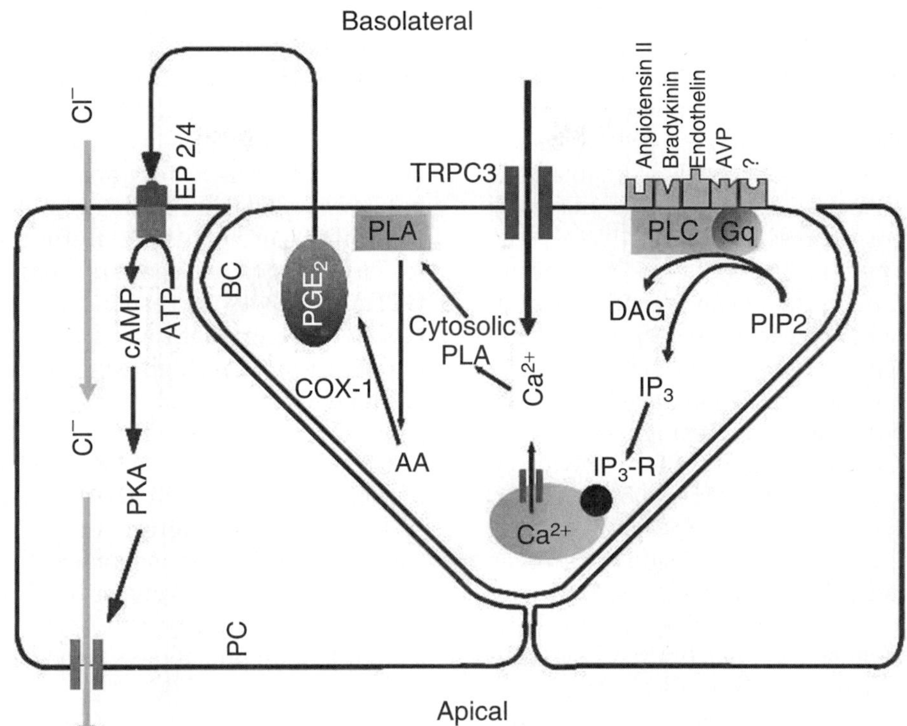

FIG. 15. Schematic diagram illustrating the roles of cyclooxygenase (COX)-1 and transient receptor potential 3 (Trp3) in the regulation of principal cell functions by basal cells in the rat epididymis. (Modified from Leung, G. P. H., Cheung, K. H., Leung, T., Tsabg, M. W., and Wong, P. Y. D. [2004]. Regulation of epididymal principal cell functions by basal cells: role of transient receptor potential [Trp] proteins and cyclooxygenase-1 [COX-1]. *Mol. Cell. Endocrinol.* 216, 5–13; kindly supplied by Dr. Patrick Wong.)

role in water reabsorption in the efferent ducts, it will be of interest to identify their downstream targets to uncover the mechanisms by which these hormones regulate water movement.

Organic Solute Transporters. Few studies have focused on the physiology of transport, although some organic transporters, such as ion transporters, have been identified in gene arrays (353,467–468a). This aspect of the review focuses only on those transporters that have been identified as functioning in the epididymis.

L-Carnitine Transporters. In 1998, two independent laboratories reported the cloning of a high-affinity L-carnitine transporter, OCTN2, isolated from human and rat small intestine complementary DNA (cDNA) libraries (710,711). OCTN2 is a member of the organic cation transporter family. Thus far, only three members of this family, OCTN1, OCTN2, and OCTN3, have been shown to transport L-carnitine. OCTN2 transports L-carnitine against a concentration gradient in a sodium-dependent manner. In contrast, OCTN1 is a low-affinity, sodium-independent carnitine transporter (712,713), and OCTN3 is a high-affinity, sodium-independent carnitine transporter (714,715).

The OCTN2 cDNA encodes a 557–amino acid protein with a predicted molecular mass of 63 kDa. Mutations in OCTN2 are responsible for primary carnitine deficiency, an autosomal recessive disorder characterized by cardiomyopathy and muscle weakness (716–718). Hence, OCTN2 is a key protein involved in L-carnitine transport. OCTN2 is localized to the basolateral membranes of the rat epididymis in a region-specific manner, with highest mRNA and protein expression observed in epididymal regions involved in L-carnitine transport (392). Therefore, OCTN2 is the prime candidate for transporting L-carnitine into epididymal cells.

The function of L-carnitine in the epididymis is unknown, but the hypothesis that OCTN2 plays a role in protection against osmotic stress in the epididymis is supported by an intriguing observation. Toshimori et al. (1999) (719) described an animal model for primary carnitine deficiency, the juvenile visceral steatosis (*jvs*) mouse. These mice exhibit symptoms similar to those of patients with primary carnitine deficiency. Studies by Lu et al. (1998) (720) demonstrated that mutations in OCTN2 were responsible for the *jvs* phenotype, and complementation with human OCTN2, but not OCTN1, rescues the phenotype in the *jvs* mice (721). Loss of OCTN2 results in the loss of integrity of the epididymal epithelium in *jvs* mice, especially in the corpus region, a site known to express high levels of OCTN2 and that readily transports L-carnitine. Therefore, it would seem plausible that loss of epididymal epithelial integrity in *jvs* mice may be the result of osmotic stress.

CT2, the putative protein that transports L-carnitine from cells into the lumen, has been identified (393). Although initially thought to be a testis-specific transporter, immunohistochemical studies clearly demonstrate its presence on the apical surface of human epididymal cells (393). The transporting characteristics and kinetics of CT2 are similar to those previously reported for an apical L-carnitine transporter in the rat (722); we propose that they are, in fact, the same transporter. CT2 is a somewhat unusual in that it can transport L-carnitine in either direction; transport partially depends on the presence of sodium (in both directions), and the transporter has limited substrate specificity (393). CT2 may have multiple functions despite its narrow substrate specificity. Certainly, from the distal caput to proximal cauda epididymidis, it functions to transport L-carnitine from cells into the lumen. However, CT2 can also transport solutes from the lumen into cells (722), suggesting that it ensures that any L-carnitine that does enter the lumen proximal to the distal caput is rapidly removed or that it can transport other solutes yet to be identified. Studies have revealed that flipt2 is the mouse homolog of human CT2, and that the mRNA and protein are expressed throughout the length of the mouse epididymis (B. T. Hinton, unpublished observations).

Taurine Transporters. Although not shown specifically to transport taurine into the epididymis, the taurine transporter TauT and the putative taurine channel phospholemman are expressed in the epididymis, with the more distal epididymal regions showing higher expression (723). Although these studies were performed using the mouse epididymis, these data coincide well with the measurement of intraluminal concentrations of taurine in the rat epididymis; the values measured were corpus ($\sim$6 mM) > cauda ($\sim$3mM) > caput ($\sim$2 mM) (396).

Glutamate Transporters. Expression of glutamate transporter EAAC1 and the glutamate transporter–associated protein (GTRAP) has been identified in the epididymis (724). Although the expression of both of these genes is high in the caput region, where the highest concentration of glutamate is observed [$\sim$50 mM (396)], expression is also high in the distal epididymal regions. This suggests that the protein is transporting glutamate into the lumen in the proximal epididymal regions, that the protein is transporting glutamate out of the lumen in the distal epididymal regions because the intraluminal concentration of glutamate rapidly declines from the caput to cauda epididymidis (396), or that the protein may be transporting other amino acids. Because many amino acid transporters are not specific for individual amino acids and some are bidirectional, all three possibilities are plausible. Definitive functional studies are needed to confirm that either EAAC1 or GTRAP (or both) is responsible for the high intraluminal concentrations of glutamate. An alternative possibility for the high concentrations of glutamate is the metabolism of intraluminal glutathione by γ-glutamyl transpeptidase (335).

CE11. More recently, a transporter called CE11 has been identified in the canine epididymis. This 12 transmembrane cotransporter has homology to the thymic stromal cotransporter (TSCOT), with some similarity to a sugar transporter (725). The transporter has been localized to the apical membranes of the canine epididymis and its function is not known.

Antimicrobial Peptides

Groups of epididymal proteins belonging to the defensin and defensin-like family and other proteins having antimicrobial properties have been identified in the epididymis. The presence of antimicrobial peptides in both reproductive and nonreproductive tissues has long been acknowledged, but more recently several antimicrobial peptides have been shown to be epididymis specific. Some of these are members of the β-defensin family, β-defensin–like peptides (e.g., bin1b, EP2, and HE2 variants), lactoferrin, cystatins, Eppin, and the cathelicidins (human cathelicidin antimicrobial protein-18 [hCAP18], cathelin-related antimicrobial peptide [CRAMP]) (726–745).

The vertebrate defensin family of genes comprises two major groups, the α- and β-defensins. The classic defensin motif consists of six cysteines linked by three intramolecular cysteine disulfide bonds, which differ in topology in the α and β groups. α-Defensins are produced mainly in neutrophils and Paneth cells, whereas the β-defensins are widely expressed in epithelial tissues, including those of the respiratory, gastrointestinal, and reproductive systems and the skin. Defensins are synthesized as preproproteins that are subsequently cleaved to generate the active cationic peptides of approximately 3 to 5 kDa [reviewed in Lehrer and Ganz (732)]. It has been suggested that, in the epididymis, a furin-like proprotein convertase is responsible for the cleavage (733). The defensin genes are also clustered on certain chromosomes—chromosome 8 in humans (746,747). A recent study by Zaballos et al. (2004) (741) using the Celera mouse genome database found exons encoding 23 different β-defensins. Using a combination of RT-PCR and in situ hybridization, these workers found nine of these genes to be highly specific for the epididymis and expressed in distinct epididymal regions.

Members of the β-defensin and β-defensin–like family are expressed in the epididymal epithelium in a region-specific manner and appear to be secreted into the epididymal fluid as demonstrated by variety of techniques, including RT-PCR, in situ hybridization, Western blot analysis, and immunolocalization (726,730,733–739,741–744,748). The androgen regulation of epididymal β-defensins is still unclear. The expression of bin1b in the rat epididymis (730) and HE2 expression in the nonhuman primate epididymis are androgen dependent (726). Although rat β-defensin-1 is androgen regulated in the initial segment and caput regions, this may not be the case in the more distal epididymal regions. To add to the complexity, rat β-defensin-1 expression in the initial segment/caput regions appears to respond slowly to androgen deprivation, suggesting that the action of androgens may be indirect (736). HE2 expression in the human epididymis does not appear to be androgen regulated, but the subcellular distribution of the HE2 variants is androgen regulated (733). Expression of human β-defensin-118 (previously called ESC42) in the nonhuman primate is primarily in the caput region and is androgen regulated. Hence, more studies are warranted to define clearly the androgen regulation of these genes. Although one would suspect that members of the β-defensin family are upregulated during inflammation, it turns out that, at least for rat β-defensin-1, this gene was not upregulated during inflammation of the epididymis induced by lipopolysaccharide from *Pseudomonas aeruginosa* or *Escherichia coli* (736).

The structure–function relationship of the defensins is complex. Outside of the defensin motif, their sequences vary greatly (748a). The antimicrobial activities of human β-defensin-3 (749) and HE2 (733,743) depend very much on both charge and structure. In the case of HE2 expressed in the chimpanzee and the human (726,733), two major isoforms are found in the epididymis, HE2α1 and HE2β1, each related to the β-defensin gene family. HE2β1 clearly has the β-defensin motif and antibacterial activity, yet the processed form of HE2α1 does not have the β-defensin motif but does have antibacterial activity. This suggests that the human epididymis secretes a novel class of antimicrobial peptides (733).

Cathelicidins are another group of antimicrobial peptides that has received attention recently. The human cathelicidin hCAP18 is expressed in the epididymis and associates with prostasomes (exosomes) in seminal fluid and with spermatozoa (727,728,750). Although hCAP18 is processed in neutrophils by proteinase 3 (751), it is processed in seminal fluid by a prostate-derived protease, gastricsin, to generate a 38–amino acid antimicrobial peptide ALL-38. Presumably, this antimicrobial peptide plays an important role in preventing vaginal bacterial infections after intercourse. Cathelicidins have also been identified in other species, for example, rat and mouse CRAMP, and have been shown to have antimicrobial properties (752–754). Members of the family of lipocalins are actively synthesized and secreted by the epididymis of most mammals and have been proposed to play an active role in immune and inflammatory responses (755).

Several of these peptides have functions in addition to their antimicrobial activity. For example, bin1b is potentially involved in the initiation of sperm motility in the epididymis (745), and Eppin (756) may be involved in the ability of spermatozoa to fertilize an egg. It is intriguing that these peptides have evolved to play multiple roles in the epididymis. Future studies should enhance our understanding of these bifunctional peptides and the mechanisms by which they promote sperm maturation and maintain antimicrobial activity.

Lipocalins

Several proteins binding and transporting small hydrophobic ligands and belonging to the lipocalin family have been identified in the epididymis.

Epididymal Retinoic Acid–binding Protein. E-RABP is one of the major epididymal-secreted proteins. It was first identified as one of the four prealbumin proteins migrating after polyacrylamide gel electrophoresis of rat and mouse epididymal cytosol (757) and luminal fluid (88,758), with an apparent molecular weight of 18,500 Da. The mouse and rat proteins were subsequently purified (529,759) and the cDNA isolated (529,760,761). In other studies, two retinoic acid–binding activities were found after separation by ion exchange chromatography of rat and mouse caudal epididymal fluid (762–764). After purification and N-termini sequencing (763,765), these proteins were found to be identical to proteins identified by Brooks et al. (529) and Rankin et al. (764). The predicted amino acid sequence, the ligand-binding specificity, and the x-ray crystal structure showing that the protein has a β-barrel structure and a hydrophobic ligand pocket, indicated that E-RABP belonged to the lipocalin family of small extracellular hydrophobic binding proteins (766,767).

E-RABP is present as a single-copy gene in both the rat and mouse. The gene structures are similar but not identical. The rat E-RABP gene is unusual compared with other lipocalins in that it has eight exons instead of seven, the more usual number. The cDNA encodes a protein of 188 residues with alternative sites for signal sequence cleavage to give secreted proteins composed of 166 and 169 amino acid residues. In contrast, the murine gene has seven exons, as seen in other lipocalins (767). The cDNA encodes a protein of 185 residues, also with alternative sites for cleavage of the signal sequence to give secreted proteins of 163 and 166 residues. The exon–intron boundaries of the first seven exons are identical for rat and mouse, and the first five exons are quite similar in size to the exons of other members of the lipocalin family. The murine gene is located in the (A3) region of chromosome 2.

A computer analysis of the nucleotide sequence of the flanking regions of the mouse E-RABP gene (*Lcn 5*) showed the presence of 6 new genes, *Lcn 8*, *Lcn 9*, *Lcn 10*, *Lcn 11*, *Lcn 12*, and *Lcn 13*, that evolved by gene duplication. The new genes encode proteins of 175, 178, 182, 170, 193, and 176 amino acids, respectively, with a signal peptide of 16 to 21 amino acids, consistent with these proteins being secreted proteins (768,769). Within this cluster of genes, *Lcn 5*, *Lcn 8*, *Lcn 9*, *Lcn 10*, *Lcn 12*, and *Lcn 13* are specifically expressed in the mouse epididymis. However, each gene has a distinct spatial expression in the epididymis and a different regulation. Lcn 8 and Lcn 9 are expressed in the initial segment only, Lcn 5 is expressed in the distal caput only, whereas Lcn 10 is expressed in the initial segment but also in the upper margin of the distal caput. The *Lcn 8*, *Lcn 9*, and *Lcn 13* genes depend strictly on testicular factors circulating in the luminal fluid. The *Lcn 5* and *Lcn 12* genes depend on androgens circulating in the serum, whereas *Lcn 10* gene expression is regulated both by circulating androgens and by testicular factors supplied from the luminal fluid.

A computational analysis of the human genome identified five putative genes (*E-RABP* [*LCN5*], *LCN8*, *LCN9*, *LCN10*, and *LCN12*) encoding homologous lipocalins. These genes were localized on human chromosome 9q34 (769,770). The genomic organization, chromosomal arrangement, and orientation of these human genes were similar to those of the murine genes.

The cluster of lipocalin genes expressed specifically in the epididymis is localized in an area rich in lipocalin genes on mouse chromosome 2 and human chromosome 9 (771). Interestingly, this epididymis-specific lipocalin cluster is flanked by other members of the lipocalin family, such as prostaglandin D synthase (PGDS) and Lcn 2 (769,771), that are expressed in other tissues besides the epididymis (772,773). This suggests that the epididymis-specific lipocalin cluster may contain a locus control region that enhances lipocalin gene expression only in the epididymis, whereas genes flanking this epididymal gene cluster show expression in other tissues besides the epididymis.

Lipocalin-type Prostaglandin D Synthase. PGDS, formerly identified as β-trace, was first shown to be localized in the central nervous system and secreted in cerebrospinal fluid (774–779). PGDS is also localized in the male genital tract, where it constitutes one of the major secreted epididymal proteins in rodents, domestic animals, and humans [mouse (780–782), rat (772,782,783), hamster (781), ram, stallion (784,785), bull (786), monkey (787), and man (788)]. The cDNA for PGDS was first isolated from a rat brain library (789), and subsequently from many species, including humans and zebrafish (784,786,790–794). The cDNA for PGDS encodes a protein of 180 to 190 amino acid residues with a signal peptide of approximately 20 amino acids removed in the mature protein and 2 N-glycosylation sites. The tertiary structure of the rat and mouse recombinant proteins identified PGDS as a lipocalin (787,795). The gene organization is similar to that of other lipocalins in terms of number and size of exons and phase of splicing introns (796–798).

In the epididymis, PGDS distribution varies among species. In the mouse, PGDS mRNA is not detected in the initial segment, but is expressed in all other regions of the epididymis (781), whereas in the rat (772,799), ram, stallion (784), and bull (800), there is a higher expression of PGDS mRNA in the caput than in the cauda epididymidis. In rodents, the secreted protein accumulates in the caudal fluid of mouse, rat, and hamster (781,786), whereas in ram and stallion the highest concentration of the protein is found in the fluid of the proximal epididymis (784).

PGDS expression is regulated by androgens; it increases in parallel with endogenous androgen levels during sexual maturation, decreases after castration, and is restored with exogenous androgens (772,782, 784,799).

Lipocalin 2. Lipocalin 2 (Lcn 2), also known as NGAL (neutrophil gelatinase-associated lipocalin precursor), uterocalin, *neu*-related lipocalin, and 24p3, was originally identified as a component of neutrophil granules, but is also expressed in epithelia in response to inflammatory signals (801). It is present in epididymal fluid, although it is not one of its major components. This protein is expressed in various organs but, in the male reproductive tract, is restricted to the epididymis (773). Lcn 2 is a 25-kDa protein originally purified from human neutrophils (802). It is a 178–amino acid protein with glycosylation sites and a 20–amino acid signal peptide (803–806). X-ray crystallography revealed that the protein has the typical lipocalin structure, but the calyx is shallower and broader than in other lipocalins (807).

In the epididymis, Lcn 2 is expressed in the caput (773) and is under androgen control (769); the protein accumulates in the caudal fluid (773). Unlike E-RABP, sequential washing of spermatozoa with low salt and high salt left a substantial amount of Lcn2 immunoreactivity in the Triton X-100 and the sodium dodecyl sulfate (SDS) sperm extract, suggesting that Lcn 2 is bound to spermatozoa (773).

Lipocalin Function. Lipocalins comprise an ancient family of proteins widely distributed among species, with the phylogenetic tree rooted in bacteria (808). L-PGDS gene is found in fish, amphibians, and many mammalian species (787), and is believed to be the ancestral gene of the epididymal lipocalin cluster (769), whereas the E-RABP gene is found in reptiles (809) and mammals, including humans (769). Genes preserved in different lineages are likely to possess important biological functions that are conserved from reptiles to primates. In addition, gene duplication provides genetic redundancy and improves the fitness of the organism (810). However, gene duplication, through neofunctionalization or subfunctionalization, also allows adaptation and functional innovation (811,812). Species comparison, even over short evolutionary distances, reveals dramatic expansions and contractions of gene families (813–817). It is all the more remarkable that the organization of the lipocalin gene cluster is conserved in mouse and human, with similar gene number, order, and orientation (769). This implies that functional constraints have maintained the integrity of the cluster since the last common ancestor of the two species, 75 million years ago.

The basic function of lipocalins, determined by x-ray crystallography, is to bind small hydrophobic ligands. The folding of the protein into a barrel structure with a deep calyx is well suited for this function. As extracellular secreted proteins, lipocalins function as carriers for these ligands, and their biological properties depend on their respective ligand. However, endogenous ligands are known for few lipocalins because the methods of isolation used disrupt the complexes. The endogenous ligands of LCN 1 (human tear lipocalin) belong to a class of lipophilic molecules such as fatty acids, phospholipids, glycolipids, and cholesterols (818). By scavenging these molecules, LCN 1 may act as a general protection factor for epithelial surfaces. In vitro, LCN 1 also binds retinol (819). In addition, LCN 1 binds to microbial siderophores with high affinities and effectively inhibits bacterial and fungal growth under iron-limiting conditions. Therefore, as discussed previously, LCN 1 appears to be part of the complex defense system involved in protection against harmful molecules and microbial and fungal infections (820). LCN 2, the only lipocalin with a known endogenous ligand, is part of the lipocalin cluster on human chromosome 9. Its ligand is a breakdown product of enterobactin, a siderophore that is used by bacteria for uptake of the essential nutrient iron (821). This is consistent with the potent antimicrobial action of LCN 2 (822). LCN 2 is also upregulated in response to inflammatory stimuli (823,824). This suggests that LCN 2 may have a role in the innate immune response to bacterial challenges in the epididymis. Mouse Lcn 2 has also been reported to deliver ferric iron to spermatozoa through internalization (825), and to enhance sperm motility through an elevation in intracellular pH and increased intracellular cAMP accumulation (826). Lcn 2 may have a dual function in the epididymis.

Ligand affinity of the other lipocalins has been determined in vitro. L-PGDS binds all-*trans* or 9-*cis* retinoic acid and all-*trans* or 13-*cis* retinaldehyde, but not all-*trans* retinol (827). It also binds thyroid hormone, biliverdin, and bilirubin (828). However, similar to bin1b and Eppin, discussed earlier, PGDS is a dual-function protein. In the cell, it acts as an enzyme producing PGD_2, a potent endogenous somnogen and pain modulator as well as anti-inflammatory agent; after secretion, it functions as a lipophilic ligand carrier. This dual function may reflect subfunctionalization in the lipocalin cluster. After inactivation of the PGDS gene, the main abnormalities that occur are in the regulation of sleep and pain responses, but no obvious phenotype was reported in the male genital tract, reflecting perhaps a built-in redundancy in the lipocalin cluster (829). Nevertheless, PGDS has been reported to be a biochemical marker of sperm quality in human semen (830) and to be correlated with fertility in bulls (786). Another study failed to find the correlation statistically significant, although the lowest PGDS contents were found in rams and bulls with the lowest fertility and the highest content in animals with normal or high fertility (781). This suggests that PGDS affects male fertility, but its role is either not essential or can be compensated by other lipocalins in the cluster.

It is likely that retinoic acid is the endogenous ligand for E-RABP, which binds in vitro all-*trans* and 9-*cis* retinoic acid, but not retinol. E-RABP inactivation produced a phenotype similar to that seen in transgenic mice expressing a dominant negative mutation of RARα, albeit with a lower severity and penetrance [K. Suzuki, unpublished observations, cited in Costa et al. (606)]. The transgenic mice expressing the dominant negative RARα were either infertile or had reduced fertility. The epithelium of the cauda epididymidis had undergone squamous metaplasia with tubule rupture, sperm leakage into the connective tissue, and inflammatory reaction. This indicates that interaction of retinoic acid with its receptor is essential for the maintenance of the epithelium of the cauda epididymidis. These observations are consistent with E-RABP functioning as an intraluminal carrier protein delivering retinoic acid to the epithelium of the cauda, and the fact that the epithelium of the cauda epididymidis is most affected in vitamin A deficiency (831).

In summary, gene duplication of a PGDS-like lipocalin gave rise to a new epididymal lipocalin family as amniotes appeared during evolution. Internal fertilization is a common feature of amniotes and this process leads to complex and changing environmental conditions for male gametes. Gene duplication, which led to an epididymal lipocalin gene cluster, may have allowed an early adaptive response to a new environment; its conservation through 250 million years strongly suggests that the epididymal lipocalins have important functions.

Lipocalins on mouse chromosome 2 and human chromosome 9 are involved in immunoregulatory, anti-inflammatory, and antibacterial responses (755). As discussed previously, the epididymis has a complex host defense system involving, in particular, the defensins. It is likely that the lipocalins of the epididymal cluster are part of this network, either through their antimicrobial action, as with Lcn 2, or by preserving the integrity of the luminal compartment of the epididymis, as with E-RABP (Lcn 5).

AGING

Effects of Aging on the Appearance of the Epididymal Epithelium

As discussed previously, numerous studies have focused on the structure, function, and regulation of the epididymis during prenatal and postnatal development and in the adult, yet very few studies have investigated the effects of senescence on epididymal structure and function. Less than 20 publications have either directly or indirectly addressed this question. In the aging rabbit, Cran and Jones (1980) (832) showed that there was an absence of spermatozoa in the epididymal lumen, an accumulation of lipofuscin pigmented bodies in principal and basal cells, cytoplasmic vacuoles in the distal segments, and intranuclear inclusions. In the aging cat epididymis, Elcock and Scoeming (833) observed occasional intraepithelial cysts and hyperplasia of the epithelium, as well as eosinophilic to amphophilic round cytoplasmic bodies in interstitial cells, and proteinaceous luminal debris. In the hamster, Calvo et al. (833a) found that the cauda epididymidis was particularly affected by aging. Aged hamsters presented involutive changes in the epididymis, including a decrease in tubular diameter in the cauda epididymidis. Principal cell ultrastructure showed the appearance of damaged mitochondria and bundles of filaments. The presence of large, electron-dense vacuoles in some clear cells was unusual. Decreases in sperm concentration and an apparent decrease in motility were noted also. Decreased epididymal sperm counts, brown patches or lipofuscin, arteriosclerosis, lymphocyte infiltration, and the emergence of sperm autoantibodies have been reported in other studies [rat (834,835), human (836,837), monkey (838)].

Results from many of these studies are difficult to interpret owing to the difficulty in establishing whether the age-dependent effects observed are due to aging of the tissue per se or to other conditions, such as arthroscleroses or carcinoma in other tissues impinging on the epididymis. The brown Norway rat has been established as a valuable model for the study of the aging male reproductive system (90,839–842). This strain of rat has a long life span, does not exhibit many of the age-related pathologies found in other rat strains, such as pituitary and Leydig cell tumors, does not become obese, and is the first rat strain to have had its genome sequenced (839,840,843); similarly, in humans, aging in men leads to dramatic changes in the seminiferous epithelium and to decreases in spermatogenesis and steroidogenesis [reviewed in Neaves et al. (844), Johnson et al. (845), Zirkin and Chen (846), and Viger and Robaire (847)], with no decrease in the concentrations of gonadotropins.

Using the brown Norway rat as a model to study aging of the epididymis, several clear observations have emerged (96,119,136). Although the luminal diameter and epithelial height were not affected by age, there was a marked progressive increase in the thickness of the basement membrane during aging. In young rats, the relative contribution of clear cells increased in the distal segments, resulting in a relative decrease in the numbers of principal cells, whereas halo cells were rare. With increasing age, there was a decrease in the proportion of principal cells and basal cells in all segments of the epididymis, of narrow cells in the initial segment, and of clear cells in the corpus epididymidis. This decrease was accompanied by a proportional striking increase in the number of halo cells in each segment (96).

Several dramatic changes occurred along the epididymis in specific cellular structures during aging (96). In the initial segment, basal cells emitted an extensive network of pseudopods extending downward toward the basement membrane, and the number and size of halo cells increased progressively during aging. In the caput epididymidis, principal and basal cells did not appear to change, whereas clear cells changed dramatically: they appeared swollen and bulged into the lumen; their nuclei were irregularly shaped and displaced to the upper half of the cell; lysosomes, enlarged and dense, were often fused with lipid droplets; and the presence of lipofuscin became evident. In the corpus epididymidis, a remarkable increase in the size and number of lysosomes was seen in principal cells. The emergence of a localized region with large vacuoles reflected the major effect of age in the proximal distal corpus/proximal cauda epididymidis. Although some principal cells in old animals had a normal morphology, many others contained extremely large vacuoles. Endosomes and lysosomes often appeared to be emptying their contents into the large vacuoles, and debris from spermatozoa was found in giant vacuoles. Serial sections of these vacuoles revealed that entire spermatozoa were found in the vacuoles, thus leading to the hypothesis that mechanisms controlling intracellular trafficking are impaired as animals age (96).

There was a dramatic increase with increasing age in the number of halo cells (composed of monocytes, helper T lymphocytes, and cytotoxic T lymphocytes) throughout the epididymis in a segment-specific manner (119). There was a segment-specific recruitment of cytotoxic T lymphocytes and monocytes–macrophages in the epididymal epithelium of aged rats whose epididymal lumen contained few spermatozoa. Thus, accumulation of damaged epithelial cells and antigens of germ cell origin, leaking through a dysfunctional blood–epididymis barrier, may contribute to the active recruitment of immune cells with age (136). This age-dependent increase in the number of halo cells in the epididymis, in conjunction with the fact that the blood–epididymis barrier should protect from immunological attack, led to investigation of whether there are changes in the structure and function of this barrier with age in the brown Norway rat model (136). Based on changes in the distribution of several immunocytochemical markers of this barrier (occludin, zonula occludens-1, and E-cadherin), as well as in lanthanum nitrate permeability, a clear breakdown of the barrier during aging was shown to occur selectively in the corpus epididymidis.

Molecular and Functional Effects of Aging in the Epididymis

Using both gene expression profiling (cDNA microarrays) and Northern blot analysis, segment-specific changes have been found to occur along the epididymis during aging (464,847). The overwhelming effect was a decrease in gene expression during aging. In the initial segment, corpus, and cauda epididymidis, more genes decreased in expression with age than did not change. Interestingly, the magnitude of the decreases in expression was considerably larger in the corpus and cauda epididymidis, where expression of 83% (211 of 254) and 62% (157 of 254), respectively, of the genes decreased by greater than 50% with age. This is in contrast to the initial segment, in which only 31% of the genes (78 of 254) expressed decreased by at least 50%; the caput epididymidis was the only segment in which the expression of a large proportion of the genes did not change with age (less than 33% changed between young and old). No genes had increased expression with age in any of the four segments of the tissue; however, the expression of four transcripts was increased in a segment-specific manner (464).

Changes in the expression of specific genes/gene families are of particular interest. Expression of the mRNAs for 5α-reductases (particularly the type 1 isozyme) was among the first to decline during aging, whereas the message for the androgen receptor was not affected (847). Changes in the expression of components of intracellular degradation pathways (i.e., proteasome components) were most dramatic in the distal epididymis. Although in the initial segment and caput epididymidis, either no change or small decreases (<33%) were seen with age, in the corpus and cauda epididymidis essentially all proteasome components showed remarkable decreases in intensity of expression (>67%). Cathepsins, a family of cysteine proteases implicated in proteolytic processes, are abundantly expressed in the epididymis (353,474,847). Of the six cathepsin transcripts detected in the epididymis, the relative intensity of cathepsin K was affected by age to the greatest extent in the initial segment, whereas that of cathepsin H and B decreased most in the more distal segments of the epididymis; cathepsin E transcripts decreased to below the level of detection in all regions of aged epididymides. Oxidative stress–related genes, such as copper-zinc superoxide dismutase, showed a small decrease with age in the initial segment, but a very large decrease in the corpus epididymidis. The relative intensity of expression of glutathione synthetase decreased dramatically with age in the initial segment, corpus and cauda epididymides, whereas epididymal secretory glutathione peroxidase (GPX5) showed the largest decrease in relative intensity in the initial segment and GPX4 showed the largest decrease in relative intensity in the cauda epididymidis (464). The changes seen with age for many of these gene transcripts were consistent with what has been observed by others on the effects of aging in various tissues in a wide spectrum of animals, from flies (848) to mice [muscle (849); brain (850)].

Mechanisms Underlying Aging of the Epididymis

Numerous hypotheses have been put forward regarding the underlying cause or causes of aging. These include the specific gene theory (851), such as the *klotho* gene (852), the telomere theory (853), altered gene expression due either to mutations (854) or altered DNA methylation status (855), the network or immune theory of aging (856), and the free radical (oxidative stress) theory (857). However, there is no consensus regarding any one hypothesis, and too few studies have been done on the epididymis to demonstrate the primary role of any one or combination of these mechanisms. There is a significant recruitment of immune cells into the epididymis as animals age, and this recruitment is greatest in parts of the tissue exhibiting greatest damage (119,136). Although it is possible that this immune response is a cause of the epithelial damage, it is perhaps more likely that the damaged tissue and ruptured blood– epididymis barrier are the attractants of immune cells. It was also shown that the amount of apoptosis seen in epididymides was higher in aged than in young mice, and that this increased rate of apoptosis could be prevented by the administration of testosterone (858).

There are several lines of suggestive evidence that point to the association of oxidative stress with changes induced by aging in the epididymis. These include region-specific accumulation of lipofuscin (859), changes in the distribution of glutathione S-transferases (860), formation of intracellular vacuoles (96), and altered expression of genes associated with oxidative stress (464). Therefore, it is reasonable to speculate that oxidative stress is a key factor in mediating the dramatic response of the epididymis to age. However, little is known about potential changes with age in the ability of the epididymis to respond to stress or altered endocrine environments.

In addition, two issues have been addressed with respect to the effects of advancing age on epididymal functions: the first relates to tissue weight, sperm number, and the effect of sexual experience on sperm number in this tissue (861), and the second to the effects of testosterone administration on sperm number and sperm maturation status ("sperm profile") (40). Sexual experience in rats was shown to lead to an increase in epididymal sperm counts, which peaked at 9 months but returned to within 5% of that at 3 months when animals were 24 months old. Only the weight of the epididymides increased between 3 and 24 months, reaching a value at 24 months that was 28% greater than that of the 3-month-old animals. However, sexually inexperienced rats had a 31% decrease in epididymal spermatozoa between the ages of 3 and 24 months, although epididymal weight still increased by 25% (861). The lack of correlation between epididymal weight and epididymal sperm content is completely unexpected and provides an important clue to potential histological changes taking place in the tissue. The administration of exogenous testosterone could not prevent the loss in sperm number, but did prevent the changes seen in sperm profile, suggesting that the epididymis of aged animals retains its ability to respond to endocrine manipulations (834).

Changes in Spermatozoa during Aging

Several clinical studies, as well as studies using models of animal aging, indicate that significant changes occur in spermatozoa as males enter advanced age and that such changes have consequences for progeny (862–866). Determining whether these changes are due to alterations taking place in the testis or the epididymis is challenging. The existence of a paternal age effect on semen parameters, fertility, and anomalies in children demonstrates that the quality of spermatozoa and the genetic integrity of otherwise apparently healthy spermatozoa are not immune to the effects of time.

Although genetic changes in spermatozoa during aging, such as nondysjunction and autosomal dominant disorders that exhibit paternal age effects [e.g., Down's (867) and Klinefelter's syndromes (868) and achondroplasia (869)] are clearly of testicular origin, others, such as changes in the DNA methylation pattern (epigenetic changes) (870), altered sperm motility, and retention of cytoplasmic droplets, are more likely to have an origin in the epididymis. Using the brown Norway rat as an model for male reproductive aging, it was shown that there is a marked increase with age in the number of abnormal flagellar midpieces of spermatozoa, that the percentage of motile spermatozoa was significantly decreased in the cauda epididymidis of old rats, and that the proportion of spermatozoa that retained their cytoplasmic droplet was markedly elevated (871). The origin of these changes may still be ascribable to altered proteins made during spermatogenesis, but the possibility that altered epididymal epithelial functions contribute to such changes must be considered.

Another mechanism that may allow for sperm quality control has been observed to act at the level of the epididymis and involves the cell surface ubiquitination and subsequent phagocytosis of abnormal spermatozoa (306). However, large-scale phagocytosis of spermatozoa by the epididymal epithelium is not observed in young or old animals (307). To date, it has not been clearly demonstrated that genetic damage–specific apoptosis or phagocytosis does occur or that there is an age-dependent increase in this process.

The multiple effects of aging on progeny outcome, sperm morphology, the acquisition of motility, and the shedding of the cytoplasmic droplet all indicate that the quality of spermatozoa is affected by aging.

THE EPIDIDYMIS AS TARGET FOR XENOBIOTICS

Xenobiotics that have been shown to act on the epididymis can be broadly divided into two major, somewhat overlapping, categories. The first consists of those chemicals for which the epididymis is the explicit target; a number of compounds that have the potential to act as part of a male contraceptive formulation fall into this group. The second is made up of therapeutic agents or known toxicants that are targeted principally at other organs but also act, as a side effect, on the epididymis. For some of these compounds, their effects on the epididymis are at the level of the epididymal epithelium; for others, effects have been demonstrated on sperm traversing the epididymis; and for others still the site of action is mixed or not resolved. Various aspects of the action of drugs on the epididymis have been thoroughly reviewed (872–877).

Demonstrating the direct action of chemicals on the epididymis is challenging because if, subsequent to administration or exposure to a chemical, there is an effect on the epididymal epithelium or on epididymal spermatozoa, it becomes essential to establish that such an effect is not mediated at the level of the hypothalamus, pituitary, or testis. Furthermore, because epididymal histopathological study and sperm function tests have not been standard in drug development and the assessment of toxicants, the action of xenobiotics on this tissue may often have been overlooked.

Chemicals for Which the Epididymis is an Explicit Target

Selective modification of functions carried out either in or by the epididymis should be of major value not only in the development of male contraceptives, but for the treatment of male fertility problems with an etiology that arises in the epididymis. Because spermatozoa reside a relatively short time in the epididymis and because no genetic transformation is known to occur as spermatozoa undergo epididymal maturation, the epididymis can be viewed as an ideal target for the development of male contraceptives or drugs that can enhance the maturation process. Although no therapeutic agent has yet been developed that meets these objectives, a large number of approaches have been pursued to attain these goals. Most of these agents can be grouped into drugs that act by modulating hormonal dependence, neuronal activity, or metabolism.

Hormonal Regulation

Although, as described previously, many hormones and hormone receptors have been found in the epididymis in various species, we focus on androgens and estrogens, hormones for which the consequences of manipulating the endocrine signal have been reported.

Blocking Androgen Action. Near the beginning of the 20th century it was established that the epididymis is highly dependent on androgens (12). Several laboratories have determined the effects on the epididymis of agents that act as androgen receptor antagonists, such as cyproterone acetate or hydroxyflutamide (878,879). These compounds successfully block epididymal functions and cause a reduction in tissue weight. Interestingly, unlike the case with other androgen-dependent tissues, such as the prostate or seminal vesicles, exposure to androgen receptor antagonists does not result in changes in intracellular androgen receptor localization in the epididymis (880). The therapeutic potential of these compounds as male contraceptives targeting epididymal functions is limited, however, because, as expected, the action of androgens in other target tissues was also blocked, producing side effects equivalent to a chemical orchidectomy.

Blocking the Formation of Dihydrotestosterone. Testosterone is normally not synthesized de novo in the epididymis but is converted to the more biologically active androgen, DHT, by steroid 5α-reductase (517). There are two isozymes of 5α-reductase (types 1 and 2), both of which are abundantly expressed in the epididymis (523,881). It has been established that DHT is essential for the maturation of spermatozoa and that the androgen found in the nuclei of epididymal cells is DHT (476,882). Thus, one approach to block androgen action in the epididymis is to inhibit DHT formation. Several compounds have been reported to inhibit either one or both forms of 5α-reductase in many tissues, including the epididymis (883–889). The first report was of a family of 5,10-secosteroids that are suicide substrates (irreversible inhibitors) of the enzyme in vitro and have been shown to inhibit the epididymal enzyme (884). In a series of studies, Rasmusson's group (885,890) found that 4-aza steroids were effective inhibitors of steroid-5α-reductase in the prostate and a variety of other tissues. The lead compound (diethyl-4-methyl-3-oxo-4-aza-5α-androstane-17 β-carboxamide [4-MA]), a competitive inhibitor of the epididymal enzyme (886), ushered in a new era in potent inhibitors of this enzyme activity; this inhibitor was shown in one study to alter the fertility of male rats (891). The first commercially available inhibitor of 5α-reductase, finasteride, also a member of the azasteroid family, has a predominant effect on the type 2 isozyme (892), whereas more recently developed agents, such as dutasteride or PNU157706, act as dual inhibitors (887,893). Using PNU157706 as a model inhibitor, it was shown that treatment of adult rats with this agent results in pronounced effects on the expression of genes involved in signal transduction, fatty acid and lipid metabolism, regulation of ion and fluid transport, luminal acidification, oxidative defense, and protein processing and degradation, all processes essential to the formation of the optimal luminal microenvironment required for proper sperm maturation (889). Treatment of adult rats with this dual inhibitor had effects on both sperm quality and pregnancy outcome but did not result in an arrest of fertility (894).

Blocking Estrogen Action. Although the presence and selective localization of both ERα and ERβ along the epididymis has now been established in several species (485,567,568), few studies have focused on elaborating the consequences of blocking estrogen action in the epididymis. It is clear that the removal of estrogen action (null mutations) in the efferent ducts resulted in a dramatic reduction in the fluid uptake capacity of that tissue and consequent infertility (571), but less is known about the consequences of withdrawing estrogen from the epididymis. Some tantalizing lines of evidence are that selective blockade of estrogen action, whether by treating with an aromatase inhibitor or an ER antagonist, resulted in blocking the oxytocin receptor in the rabbit epididymis (175), and that by using an ERα blocker, ICI182,780, the apical cytoplasm of narrow and clear cells was affected (895).

Neuronal Regulation

An alternate means by which xenobiotics can affect epididymal functions is by increasing or decreasing sperm transit time. Either administration of guanethidine, a drug that causes chemical sympathetectomy, or surgical sympathetectomy resulted in a marked increase in sperm transit time as well as in cauda epididymal weight without other apparent effects (896,896a,897). In contrast, several drugs accelerate transport of spermatozoa through the epididymis. In 1975, Meistrich et al. (166) demonstrated that treatment of mice with estradiol resulted in an increased rate of transport of sperm through the epididymis. Administration of chloroethylmethanesulfonate, a chemical that causes a reduction in serum testosterone, or hydroxyflutamide also led to accelerated sperm transit (897). Sulfapyridine, a metabolite of the antibiotic sulfasalazine, caused a marked reduction in fertility in both rat and human; although its mechanism of action has yet to be elucidated, it was also found to accelerate sperm transit through the epididymis (898).

Metabolism

Several compounds that act on enzymes in the cellular metabolic pathways have been tested as potential male contraceptives. The ones that hold the greatest potential are reviewed in the following.

α-Chlorohydrin. In the late 1960s, a series of studies appeared on the ability of α-chlorohydrin, a monochloro derivative of glycerol, to inhibit fertility reversibly in an array of species (899,900). This drug is converted by glycerol dehydrogenase to 3-phosphoglyceraldehyde, an inhibitory analog of the substrate for glyceraldehyde-3-phosphate dehydrogenase (GADPH), and acts, therefore, at low doses to inhibit this key metabolic enzyme in spermatozoa residing in the cauda epididymidis (901) and reduces sperm motility (902). At higher doses, however, α-chlorohydrin caused the formation of spermatoceles in the caput epididymidis of rodents, resulting in the blockade of sperm transport through the tissue and the consequent atrophy of the seminiferous tubules due to increases in back pressure (903). Wong and colleagues (1977) (904) demonstrated that α-chlorohydrin blocked the resorption of sodium and water from the cauda epididymidis. Although α-chlorohydrin inhibits a ubiquitous metabolic enzyme, its selectivity of action on spermatozoa is based on the fact that spermatozoa have a GADPH isoenzyme with a higher molecular weight that is not cytosolic (905). Nevertheless, severe toxicity (bone marrow depression, kidney, liver) was associated with this compound when it was tested in primates (900,906), and therefore it will not be developed as a contraceptive in humans.

6-Chloro-6-Deoxy-Glucose. In 1978, Ford and Waites (907) described another antifertility compound, 6-chloro-6-deoxy-glucose, with the ability to inactivate spermatozoa from the cauda epididymides; it also acts by blocking GADPH (903). This compound was also found to affect different functions of the epididymis, but because of its neurotoxicity in primates (associated with its effects on glucose transport and inhibition of GADPH), it is not likely to be accepted as a male contraceptive in humans.

Gossypol. Although gossypol, a chemical found in cotton seed oil, has been studied extensively as a potential male contraceptive agent (908,909), its mechanism of action remains poorly understood. Although it has been shown to have a high contraceptive efficacy, its toxicity, probably mediated through a hypokalemic action, is likely to preclude the clinical use of gossypol as a "stand-alone" contraceptive (910); it may have a role at low does in combination with steroidal contraception (911). Gossypol caused shedding of spermatozoa from the seminiferous tubules, separation of sperm heads from tails, and a decrease in sperm number in the cauda epididymides (908). Studies by Romualdo et al. (2002) (912) have also shown that treatment of rats during the pubertal time window resulted in an increase in the appearance of round cells in the epididymal lumen; these cells were most likely of epididymal epithelial origin because they stained for cysteine-rich secretory protein (CRISP)-1, also known as protein E, a marker of epididymal principal cells. CRISP-1 staining was not found in testicular cells, providing further evidence that the epididymis is indeed a target organ for gossypol (912).

Chemicals Targeted Principally at Other Organs that Also Act on the Epididymis

The epididymis has not usually been studied as one of the target tissues for the toxicological effects of drugs or toxicants, and consequently little is known about the importance of this tissue as a target for the toxic effects of xenobiotics. However, there are a few instances where the epididymis has been demonstrated to be a target of drug action.

Cyclophosphamide

Cyclophosphamide, a widely used anticancer and immunosuppressive drug, has been shown to act as a male-mediated developmental toxicant [reviewed in Robaire and Hales (913)]. Administration of this drug resulted in an increase in the number of clear cells and in the number of spermatozoa with abnormal tails in the epididymis of rats (80). By using several complementary approaches, Qiu et al. (1992) (145) demonstrated that the increase in postimplantation embryo loss observed among progeny sired by male rats subjected to short-term (4 or 7 days) treatment with cyclophosphamide was mediated by the action of this drug on spermatozoa while transiting through the epididymis. Whether the drug action was mediated by an effect on the epididymal epithelium or acted directly on spermatozoa is unresolved.

Fungicides

Several fungicides, including mancozeb, ornidazole, carbendazim, and benzimidazole, have effects on male fertility (914–917). For some of these compounds, sufficiently detailed studies have been undertaken to indicate that there is a selective action at the level of the epididymis. For example, benzimidazole causes occlusion of the epididymis and the efferent ducts, and ornidazole causes a decrease in sperm motility. Whether the members of this family of drugs have a similar mechanism of action is not clear, but treatment with ornidazole caused an inhibition of GADPH, a similar action to that of 6-chloro,6-deoxy fructose and α-chlorohydrin (918).

Methyl Chloride

Another mechanism by which a xenobiotic can affect the epididymis is by increasing the inflammatory response. Exposure of male rats to methyl chloride, an industrial gas, resulted in infertility (a high rate of dominant lethal mutations) that was associated with testicular degeneration, epididymal inflammation, and sperm granuloma formation. Chellman et al. (1986) (919) demonstrated that the infertility was associated with epididymal inflammation and that treatment with an anti-inflammatory drug resulted in a reversal of these effects.

Sulfonates

Exposure to either ethyldimethylsulfonate or chloroethylmethanesulfonate resulted in what is commonly viewed as a selective destruction of Leydig cells and commensurate reduction in testosterone production (920). These chemicals also caused a reduction in fertility. Although the effect on fertility may be ascribed to the reduction in testosterone, Klinefelter et al. (1994) (921) clearly demonstrated that these chemicals have, at similar doses, a direct effect on the epididymis, independent of their action on Leydig cells. The height of epithelial cells along the corpus and proximal cauda epididymidis was reduced, there was a reduction in the fertilizing ability of cauda epididymal spermatozoa, and these spermatozoa had a reduction in protein content. Although the specific mechanism of action of this family of chemicals on the epididymis is still unresolved, a selective reduction in SP22, a protein that is specifically acquired by spermatozoa as they traverse the epididymis, has been reported (922).

In addition to the effects of xenobiotics seen after the exposure of adults, a number of studies have demonstrated that exposure during fetal, early postnatal, or pubertal time windows can result in abnormalities in the epididymis (weight, histology) or in the spermatozoa (number, motility) present in this tissue. Some of these chemicals include dibutyl phthalate, a plasticizer (923,924), 2,3,7,8-tetrachlorodibenzo-rho-dioxin (925), and pesticides such as PCB 169 and methoxychlor (926).

PERSPECTIVE AND FUTURE DIRECTIONS

Since the early 1990s, we have witnessed an explosion of information on a wide range of facets of epididymal gene and protein expression and their regulation, effects of null mutation and targeted transgenic expression of specific molecules, secretion of proteins by epididymal epithelial cells and their interactions with spermatozoa, and mechanisms for creating an ever-changing luminal microenvironment in the fluid surrounding spermatozoa in the duct lumen. Many of these advances have been attained thanks to the application of novel technologies that range from laser capture microscopy to gene expression profiling by microarrays, from in situ electroporation to tissue-targeted gene deletions and the development of immortalized cell lines.

We are beginning to grasp the remarkable complexity of the epithelium lining the epididymal duct and the milieu it creates in the lumen that allows spermatozoa to mature and be stored and protected. However, in spite of these great strides, some of the major questions relating to the epididymis remain unanswered. What mechanisms regulate the lobe-/segment-/region-specific expression of genes and proteins along the epididymis? How can adjacent principal cells have such dramatically different levels of expression of various proteins? What are the actual functions of each of the highly differentiated cell types in the duct? Why have we not yet succeeded in isolating and immortalizing each of these cell types? Is there a specific protein secreted by the epididymis that allows spermatozoa to attain a mature state, or is there a series of multiple, inter-related steps involving changes in proteins, sugars, and lipids? What aspects of the cauda epididymal fluid allow sperm to remain dormant and functional for protracted periods? What is the role of the assortment of antimicrobials being discovered in this tissue? Why is primary cancer of the epididymis so rare? These and other questions have been challenging reproductive biologists for many decades. With the plethora of new tools that have been developed and that will certainly continue to emerge, we are now poised to elucidate the working of this remarkable tissue.

ACKNOWLEDGMENTS

We acknowledge the support of the Canadian Institutes for Health Research, the National Institutes of Health (NICHD HD 32979, HD 042089, and HD 045890), The Ernst Schering Research Foundation, and CONRAD for support of our ongoing research. We also express our thanks to Charles Flickinger, Sylvie Breton, Louis Hermo, and Barbara Hales for helpful comments and suggestions, and to Elise Boivin-Ford for excellent assistance in preparation of the final manuscript.

REFERENCES

1. Robaire, B., and Hermo, L. (1988). Efferent ducts, epididymis and vas deferens: structure, functions and their regulation. In *The Physiology of Reproduction* (E. Knobil and J. Neill, Eds.), pp. 999–1080. Raven Press, New York.
2. Mann, T. (1964). *The Biochemistry of Semen and of the Male Reproductive Tract*. Methuen, London.
3. Jones, R. C., Holland, M. K., and Doberska, C. (1998). *The Epididymis: Cellular and Molecular Aspects. J. Reprod. Fertil. Suppl.* 53, 1–292.
4. Hinton, B. T., and Turner, T. T. (2003). *The Third International Conference on the Epididymis*. Van Doren, Charlottesville, VA.
5. Robaire, B., and Hinton, B. T. (Eds.) (2002). *The Epididymis: From Molecules to Clinical Practice: A Comprehensive Survey of the Efferent Ducts, the Epididymis and Vas Deferens*. Kluwer Academic/Plenum, New York.
6. Orgebin-Crist, M.-C. (1998). The epididymis across 24 centuries. *J. Reprod. Fertil. Suppl.* 53, 285–292.
7. Jocelyn, H. D., and Setchell, B. P. (1972). A treatise concerning the generative organs of men. An annotated translation of "Tractatus de Virorum Organis Generationi Inservientibus" (1968). *J. Reprod. Fertil. Suppl.* 17, 1–76.
8. Tournade, A. (1913). Différence de motilité des spermatozoïdes prélevés dans les divers segments de l'épididyme. *C. R. Soc. Biol.* 74, 738–739.
9. Courrier, R. (1920). Sur l'existence d'une sécrétion épididymaire chez la Chauve-Souris hibernante et sa signification. *C. R. Soc. Biol.* 83, 67–69.
10. Hammond, J., and Asdell, S. A. (1926). The vitality of the spermatozoa in the male and female reproductive tracts. *Br. J. Exp. Biol.* 4, 155–185.
11. Kirillov, V. S., and Morozov, V. A. (1936). Durée de conservation de la vitalité des spermatozoides de Taureau dans les épididymes isolés du testicule. *Uspekhi zootekh, Nauk* 2, 19–22.
12. Benoit, J. (1926). Recherches anatomiques, cytologiques et histophysiologiques sur les voies excrétrices du testicule chez les mammifères. *Arch. Anat. Histol. Embryol. (Strasb).* 5, 173–412.
13. Young, W. C. (1929). A study of the function of the epididymis: I. Is the attainment of full spermatozoon maturity attributable to some specific action of the epididymal secretion? *J. Morphol. Physiol.* 47, 479–495.
14. Young, W. C. (1929). A study of the function of the epididymis: II. The importance of an aging process in sperm for the length of the period during which fertilizing capacity is retained by sperm isolated in the epididymis of the guinea pig. *J. Morphol. Physiol.* 48, 475–491.
15. Young, W. C. (1931). A study of the function of the epididymis: III. Functional changes undergone by spermatozoa during their passage through the epididymis and vas deferens in the guinea pig. *J. Exp. Biol.* 8, 151–160.
16. Toothill, M. C., and Young, W. C. (1931). The time consumed by spermatozoa in passing through the epididymis of the guinea-pig determined by India-ink injections. *Anat. Rec.* 50, 95–107.
17. Ortavant, R. (1954). Détermination de la vitesse de transfert des spermatozoïdes dans l'épididyme de Bélier à l'aide de ^{32}P. *C. R. Soc. Biol.* 143, 866–871.
18. Orgebin-Crist, M.-C. (1961). Étude du transit épididymaire des spermatozoïdes de Taureau marqués à l'aide du ^{32}P. *Ann. Biol. Anim. Biochem. Biophys.* 1, 117–120.

19. Bedford, J. M. (1967). Effect of duct ligation on the fertilizing ability of spermatozoa from different regions of the rabbit epididymis. *J. Exp. Zool.* 166, 271–281.
20. Orgebin-Crist, M.-C. (1967). Maturation of spermatozoa in the rabbit epididymis: fertilizing ability and embryonic mortality in does inseminated with epididymal spermatozoa. *Ann. Biol. Anim. Biochem. Biophys.* 7, 373–389.
21. Orgebin-Crist, M.-C. (1969) Studies on the function of the epididymis. *Biol. Reprod.* 1, 155–175.
22. Gilbert, S. F. (2003). *Developmental Biology*. Sinauer, Sunderland, MA.
23. Vize, P. D., Woolf, A. S., and Bard, J. B. L. (Eds.) (2003). *The Kidney: From Normal Development to Congenital Disease*. Academic Press, New York.
24. Bouchard, M., Souabni, A., Mandler, M., Neubuser, A., and Busslinger, M. (2002). Nephric lineage specification by Pax2 and Pax8. *Genes Dev.* 16, 2958–2970.
25. Dressler, G. R., Deutsch, U., Chowdhury, K., Nornes, H. O., and Gruss, P. (1990). Pax2: a new murine paired-box-containing gene and its expression in the developing excretory system. *Development* 109, 787–795.
26. Pourquie, O., Fan, C. M., Coltey, M., Hirsinger, E., Watanabe, Y., Breant, C., Francis-West, P., Brickell, P., Tessier-Lavigne, M., and Le Douarin, N. M., (1996). Lateral and axial signals involved in avian somite patterning: a role for BMP4. *Cell* 84, 461–471.
27. Schultheiss, T. M., James, R. G., Listopadova, A., and Herzlinger, D. (2003). Formation of the nephric duct. In *The Kidney: From Normal Development to Congenital Disease* (P. D. Vize, A. S. Woolfe, and J. B. L. Bard, Eds.), pp. 51–60. Academic Press, New York.
28. Obara-Ishihara, T., Kuhlman, J., Niswander, L., and Herzlinger, D. (1999). The surface ectoderm is essential for nephric duct formation in intermediate mesoderm. *Development* 126, 1103–1108.
29. Niederreither, K., Subbarayan, V., Dolle, P., and Chambon, P. (1999). Embryonic retinoic acid synthesis is essential for early mouse post-implantation development. *Nat. Genet.* 21, 444–448.
30. Saxen, L. (1987). *Organogenesis of the Kidney*. Cambridge University Press, Cambridge.
31. Sainio, K. (2003). Development of the mesonephric kidney. In *The Kidney: From Normal Development to Congenital Disease* (P. D. Vize, A. S. Woolfe, and J. B. L. Bard, Eds.), pp. 75–86. Academic Press, New York.
32. Alarid, E. T., Cunha, G. R., Young, P., and Nicoll, C. S. (1991). Evidence for an organ- and sex-specific role of basic fibroblast growth factor in the development of the fetal mammalian reproductive tract. *Endocrinology* 129, 2148–2154.
33. Sainio, K., Hellstedt, P., Kreidberg, J., Saxen, L., and Sariola, H. (1997). Differential regulation of two sets of mesonephric tubules by WT-1. *Development* 124, 1293–1299.
34. Upadhyay, S., Luciani, J.-M., and Zamboni, L. (1981) The role of the mesonephros in the development of the mouse testis and its excurrent pathways. In *Development and Function of Reproductive Organs* (A. G. Byskov and H. Peters, Eds.), pp. 18–27. Excerpta Medica, Amsterdam.
35. Vazquez, M. D., Bouchet, P., and Vize, P. D. (2003). Three-dimensional anatomy of mammalian mesonephroi. In *The Kidney: From Normal Development to Congenital Disease* (P. D. Vize, A. S. Woolfe, and J. B. L. Bard, Eds.), pp. 87–92. Academic Press, New York.
36. Jones, E. A. (2003). Molecular control of pronephric development: an overview. In *The Kidney: From Normal Development to Congenital Disease* (P. D. Vize, A. S. Woolfe, and J. B. L. Bard, Eds.), pp. 93–118. Academic Press, New York.
37. Stark, K., Vainio, S., Vassileva, G., and McMahon, A. P. (1994). Epithelial transformation of metanephric mesenchyme in the developing kidney regulated by Wnt-4. *Nature* 372, 679–683.
38. Dahl, U., Sjodin, A., Larue, L., Radice, G. L., Cajander, S., Takeichi, M., Kemler, R., and Semb, H. (2002). Genetic dissection of cadherin function during nephrogenesis. *Mol. Cell Biol.* 22, 1474–1487.
39. Cho, E. A., Patterson, L. T., Brookhiser, W. T., Mah, S., Kintner, C., and Dressler, G. R. (1998). Differential expression and function of cadherin-6 during renal epithelium development. *Development* 125, 4806–4815.
40. Sonnenberg-Riethmacher, E., Walter, B., Riethmacher, D., Godecke, S., and Birchmeier, C. (1996). The c-ros tyrosine kinase receptor controls regionalization and differentiation of epithelial cells in the epididymis. *Genes Dev.* 10, 1184–1193.
41. Le Barr, D. K., Blecher, S. R., and Moger, W. H. (1986). Androgen levels and androgenization in sex-reversed (XXSxr pseudomale) mouse: absence of initial segment of epididymis is independent of androgens. *Arch. Androl.* 17, 195–205.
42. Le Barr, D. K., and Blecher, S. R. (1987). Decreased arterial vasculature of the epididymal head in XXSxr pseudomale ("sex-reversed") mice. *Acta Anat.* 129, 123–126.
43. Rodriguez, C. M., Kirby, J. L., and Hinton, B. T. (2002). The development of the epididymis. In *The Epididymis: From Molecules to Clinical Practice* (B. Robaire and B. T. Hinton, Eds.), pp. 251–267. Kluwer Academic/Plenum, New York.
44. Sun, E. L., and Flickinger, C. J. (1979). Development of cell types and of regional differences in the postnatal rat epididymis. *Am. J. Anat.* 154, 27–55.
45. Hermo, L., Barin, K., and Robaire, B. (1992). Structural differentiation of the epithelial cells of the testicular excurrent duct system of rats during postnatal development. *Anat. Rec.* 233, 205–228.
46. Jiang, F. X., Temple-Smith, P., and Wreford, N. G. (1994). Postnatal differentiation and development of the rat epididymis: a stereological study. *Anat. Rec.* 238, 191–198.
47. Alexander, N. J. (1972). Prenatal development of the ductus epididymis in the rhesus monkey: the effects of fetal castration. *Am. J. Anat.* 135, 119–134.
48. Abe, K., Takano, H., and Ito, T. (1984). Interruption of the luminal flow in the epididymal duct of the corpus epididymidis in the mouse, with special reference to differentiation of the epididymal epithelium. *Arch. Histol. Jpn.* 47, 137–147.
49. Hemeida, N. A., Sack, W., and McEntee, K. (1978). Ductuli efferentes in the epididymis of boar, goat, ram, bull, and stallion. *Am. J. Vet. Res.* 39, 1892–1900.
50. Nistal, M., and Paniagua, R. (1984). Development of the male genital tract. In *Testicular and Epididymal Pathology* (M. Nistal and R. Paniagua, Eds.), pp. 1–13. Thieme-Stratton, New York.
51. Takano, H., Abe, K., and Ito, T. (1981). Changes in the mouse epididymis after ligation of the ductuli efferentes or proximal epididymal duct: qualitative and quantitative histological studies [author's translation]. *Kaibogaku Zasshi* 56, 79–90.
52. Turner, T. T., Gleavy, J. L., and Harris, J. M. (1990). Fluid movement in the lumen of the rat epididymis: effect of vasectomy and subsequent vasovasostomy. *J. Androl.* 11, 422–428.
53. von Lanz, T., and Neuhauser, G. (1964). Morphometrische Analyse des menschlichen Nebenhodens. *Z. Anat. Entwickl.* 124, 126–152.
54. Maneely, R. B. (1959). Epididymal structure and function: a historical and critical review. *Acta Zool.* 40, 1–21.
55. Nicander, L. (1956–58). Studies on the regional histology and cytochemistry of the ductus epididymis in stallions, rams, and bulls. *Acta Morphol. Neerl. Scand.* 1, 337–362.
56. Hoffer, A. P., and Karnovsky, M. L. (1981). Studies on zonation in the epididymis of the guinea pig: I. Ultrastructural and biochemical analysis of the zone rich in large lipid droplets (zone II). *Anat. Rec.* 201, 623–633.

57. Reid, B. L., and Cleland, K. W. (1957). The structure and function of the epididymis: histology of the rat epididymis. *Aust. J. Zool.* 5, 223–246.
58. Hermo, L., (1995). Structural features and functions of principal cells of the intermediate zone of the epididymis of adult rats. *Anat. Rec.* 242, 515–530.
59. Holstein, A. F. (1969). Morphologische studien am Nebenhoden des Menschen. In *Zwanglose Abhandlungen aus dem Gebiet der normalen un pathologischen Anatomie* (W. Bargmann and W. Doerr, Eds.), Heft 20. Georg Thieme, Stuttgart.
60. Turner, T. T., Bomgardner, D., Jacobs, J. P., and Nguyen, Q. A. (2003). Association of segmentation of the epididymal interstitium with segmented tubule function in rats and mice. *Reproduction* 125, 871–878.
61. Flickinger, C. J., Howards, S. S., and English, H. F. (1978). Ultrastructural differences in efferent ducts and several regions of the epididymis of the hamster. *Am. J. Anat.* 152, 557–586.
62. Ramos, A. S. Jr. (1979). Morphologic variations along the length of the monkey vas deferens. *Arch. Androl.* 3, 187–196.
63. Hamilton, D. W. (1975). Structure and function of the epithelium lining the ductuli efferentes, ductus epididymidis and ductus deferens in the rat. In *Handbook of Physiology* (R. O. Greep and E. B. Astwood, Eds.), Sec. 7, Vol. 5, pp. 259–301. American Physiological Society, Washington, DC.
64. Nilnophakoon, N. (1978). Histological studies on regional postnatal differentiation of the epididymis in the ram. *Zbl. Vet. Med. C. Anat. Histol. Embryol.* 7, 253–272.
65. Djakiew, D., and Jones, R. C. (1982). Ultrastructure of the ductus epididymidis of the echidna, *Tachyglossus aculeatus*. *J. Anat.* 135, 625–634.
66. Goyal, H. O. (1985). Morphology of the bovine epididymis. *Am. J. Anat.* 172, 155–172.
67. Tingari, M. D., and Moniem, K. A. (1979). On the regional histology and histochemistry of the epididymis of the camel (*Camelus dromedarius*). *J. Reprod. Fertil.* 57, 11–20.
68. Orsi, A. M., DeMelo, V. R., Ferreira, A. L., and Campos, V. J. M. (1980). Morphology of the epithelial cells of the epididymal duct of the South American opossum (*Didelphis azarae*). *Anat. Anz.* 148, 7–13.
69. Adamali, H. I., Somani, I. H., Huang, J. Q., Mahuran, D., Gravel, R. A., Trasler, J. M., and Hermo, L. (1999). I: Abnormalities in cells of the testis, efferent ducts, and epididymis in juvenile and adult mice with beta-hexosaminidase A and B deficiency. *J. Androl.* 20, 779–802.
70. Orsi, A. M. (1983). Regional histology of the epididymis of the dog: a light microscope study. *Anat. Anz.* 153, 441–445.
71. Jones, R. C., and Brosman, M. F. (1981). Studies of the deferent ducts from the testis of the African elephant, *Loxodonta africana:* I. Structural differentiation. *J. Anat.* 132, 376–386.
72. Vendrely, E. (1981). Histology of the epididymis in the human adult. In *Epididymis and Fertility: Biology and Pathology* (C. Bollack and A. Clavert, Eds.), pp. 21–33. S. Karger, Basel.
73. Smithwick, E. B., and Young, L. G. (1997). Sequential histology of the adult chimpanzee epididymis. *Tissue Cell* 29, 383–412.
74. Yeung, C. H., Cooper, T. G., Bergmann, M., and Schulze, H. (1991). Organization of tubules in the human caput epididymidis and the ultrastructure of their epithelia. *Am. J. Anat.* 191, 261–279.
75. Yeung, C. H., Cooper, T. G., Oberpenning, F., Schulze, H., and Nieschlag, E. (1993). Changes in movement characteristics of human spermatozoa along the length of the epididymis. *Biol. Reprod.* 49, 274–280.
76. Palacios, J., Regadera, J., Nistal, M., and Paniagua, R. (1991). Apical mitochondria-rich cells in the human epididymis: an ultrastructural, enzymohistochemical, and immunohistochemical study. *Anat. Rec.* 231, 82–88.
77. Krull, N., Ivell, R., Osterhoff, C., and Kirchhoff, C. (1993). Region-specific variation of gene expression in the human epididymis as revealed by in situ hybridization with tissue-specific cDNAs. *Mol. Reprod. Dev.* 34, 16–24.
78. O'Bryan, M. K., Mallidis, C., Murphy, B. F., and Baker, H. W. (1994). Immunohistological localization of clusterin in the male genital tract in humans and marmosets. *Biol. Reprod.* 50, 502–509.
79. Kirchhoff, C. (1999). Gene expression in the epididymis. *Int. Rev. Cytol.* 188, 133–202.
80. Trasler, J. M., Hermo, L., and Robaire, B. (1988). Morphological changes in the testis and epididymis of rats treated with cyclophosphamide: a quantitative approach. *Biol. Reprod.* 38, 463–479.
81. Hermo, L., Oko, R., and Morales, C. R. (1994). Secretion and endocytosis in the male reproductive tract: a role in sperm maturation. *Int. Rev. Cytol.* 154, 106–189.
82. Hermo, L., and Robaire, B. (2002). Epididymis cell types and their function. In *The Epididymis: From Molecules to Clinical Practice* (B. Robaire and B. T. Hinton, Eds.), pp. 81–102. Kluwer Academic/Plenum, New York.
83. Lea, O. A., Petrusz, P., and French, F. S. (1978). Purification and localization of acidic epididymal glycoprotein (AEG). A sperm coating protein secreted by the rat epididymis. *Int. J. Androl.* (Suppl. 2), 592–607.
84. Holland, M. K., and Orgebin-Crist, M.-C. (1988). Characterization and hormonal regulation of protein synthesis by the murine epididymis. *Biol. Reprod.* 38, 487–496.
85. Turner, T. T. (1991). Spermatozoa are exposed to a complex microenvironment as they traverse the epididymis. *Ann. N. Y. Acad. Sci.* 637, 364–383.
86. Garrett, S. H., Garrett, J. E., and Douglass, J. (1991). In situ histochemical analysis of region-specific gene expression in the adult rat epididymis. *Mol. Reprod. Dev.* 30, 1–17.
87. Hermo, L., Oko, R., and Robaire, B. (1992). Epithelial cells of the epididymis show regional variations with respect to the secretion or endocytosis of immobilin as revealed by light and electron microscope immunocytochemistry. *Anat. Rec.* 232, 202–220.
88. Rankin, T. L., Tsuruta, K. J., Holland, M. K., Griswold, M. D., and Orgebin-Crist, M.-C. (1992). Isolation, immunolocalization, and sperm-association of three proteins of 18, 25, and 29 kilodaltons secreted by the mouse epididymis. *Biol. Reprod.* 46, 747–766.
89. Vierula, M. E., Araki, Y., Rankin, T. L., Tulsiani, D. R., and Orgebin-Crist, M.-C. (1992). Immunolocalization of a 25-kilodalton protein in mouse testis and epididymis. *Biol. Reprod.* 47, 844–856.
90. Robaire, B., Syntin, P., and Jervis, K. (2000). The coming of age of the epididymis. In *Testis, Epididymis and Technologies in the Year 2000* (B. Jégou, C. Pineau, and J. Saez, Eds.), pp. 229–262. Springer-Verlag, New York.
91. Cornwall, G. A., Lareyre, J.-J., Matusik, R. J., Hinton, B. T., and Orgebin-Crist, M.-C. (2002). Gene expression and epididymal function. In *The Epididymis: From Molecules to Clinical Practice* (B. Robaire and B. T. Hinton, Eds.), pp. 169–199. Kluwer Academic/Plenum, New York.
92. Kirchhoff, C. (2002). Specific gene expression in the human and non-human primate epididymis. In *The Epididymis: From Molecules to Clinical Practice* (B. Robaire and B. T. Hinton, Eds.), pp. 201–218. Kluwer Academic/Plenum, New York.
93. Dacheux, J.-L., and Dacheux, F. (2002). Protein secretion in the epididymis. In *The Epididymis: From Molecules to Clinical Practice* (B. Robaire and B. T. Hinton, Eds.), pp. 151–168. Kluwer Academic/Plenum, New York.
94. Sun, E. L, and Flickinger, C. J. (1980). Morphological characteristics of cells with apical nuclei in the initial segment of the adult rat epididymis. *Anat. Rec.* 196, 285–293.

95. Adamali, H. I., and Hermo, L. (1996). Apical and narrow cells are distinct cell types differing in their structure, distribution, and functions in the adult rat epididymis. *J. Androl.* 17, 208–222.
96. Serre, V., and Robaire, B. (1998). Segment specific morphological changes in the aging brown Norway rat epididymis. *Biol. Reprod.* 58, 497–513.
97. Hermo, L., Adamali, H. I., and Andonian, S. (2000). Immunolocalization of CA II and H^+ V-ATPase in epithelial cells of the mouse and rat epididymis. *J. Androl.* 21, 376–391.
98. Abou-Haila, A., and Fain-Maurel, M. A. (1984). Regional differences of the proximal part of mouse epididymis: morphological and histochemical characterization. *Anat. Rec.* 209, 197–208.
99. Cooper, T. G. (1986). *The Epididymis, Sperm Maturation and Fertilisation.* Springer-Verlag, New York.
100. Hermo, L., Dworkin, J., and Oko, R. (1988). Role of epithelial clear cells of the rat epididymis in the disposal of the contents of cytoplasmic droplets detached from spermatozoa. *Am. J. Anat.* 183, 107–124.
101. Moore, H. D. M., and Bedford, J. M. (1979). Short-term effects of androgen withdrawal on the structure of different epithelial cells in the rat epididymis. *Anat. Rec.* 193, 293–311.
102. Oko, R., Hermo, L., Chan, P. T., Fazel, A., and Bergeron, J. J. (1993). The cytoplasmic droplet of rat epididymal spermatozoa contains saccular elements with Golgi characteristics. *J. Cell. Biol.* 123, 809–821.
103. Flickinger, C. J., Herr, J. C., and Klotz, K. L. (1988). Immunocytochemical localization of the major glycoprotein of epididymal fluid from the cauda in the epithelium of the mouse epididymis. *Cell. Tissue Res.* 251, 603–610.
104. Vierula, M. E., Rankin, T. L., and Orgebin-Crist, M.-C. (1995). Electron microscopic immunolocalization of the 18 and 29 kilodalton secretory proteins in the mouse epididymis: evidence for differential uptake by clear cells. *Microsc. Res. Tech.* 30, 24–36.
105. Jensen, L. J., Stuart-Tilley, A. K., Peters, L. L., Lux, S. E., Alper, S. L., and Breton, S. (1999). Immunolocalization of AE2 anion exchanger in rat and mouse epididymis. *Biol. Reprod.* 61, 973–980.
106. Pastor-Soler, N., Beaulieu, V., Litvin, T. N., Da Silva, N., Chen, Y., Brown, D., Buck, J., Levin, L. R., and Breton, S. (2003). Bicarbonate-regulated adenylyl cyclase (sAC) is a sensor that regulates pH-dependent V-ATPase recycling. *J. Biol. Chem.* 278, 49523–49529.
107. Isnard-Bagnis, C., Da Silva, N., Beaulieu, V., Yu, A. S., Brown, D., and Breton, S. (2003). Detection of ClC-3 and ClC-5 in epididymal epithelium: immunofluorescence and RT-PCR after LCM. *Am. J. Physiol. Cell. Physiol.* 284, C220–C232.
108. Hermo, L., Papp, S., and Robaire, B. (1994). Developmental expression of the Yf subunit of glutathione S-transferase P in epithelial cells of the testis, efferent ducts, and epididymis of the rat. *Anat. Rec.* 239, 421–440.
109. Veri, J. P., Hermo, L., and Robaire, B. (1993). Immunocytochemical localization of the Yf subunit of glutathione S-transferase P shows regional variation in the staining of epithelial cells of the testis, efferent ducts, and epididymis of the male rat. *J. Androl.* 14, 23–44.
110. Clermont, Y., and Flannery, J. (1970). Mitotic activity in the epithelium of the epididymis in young and old adult rats. *Biol. Reprod.* 3, 283–292.
111. Hermo, L., and Papp, S. (1996). Effects of ligation, orchidectomy, and hypophysectomy on expression of the Yf subunit of GST-P in principal and basal cells of the adult rat epididymis and on basal cell shape and overall arrangement. *Anat. Rec.* 244, 59–69.
112. Clermont, Y., Rambourg, A., and Hermo, L. (1995). Trans-Golgi network (TGN) of different cell types: three-dimensional structural characteristics and variability. *Anat. Rec.* 242, 289–301.
113. Olson, L. M., Zhou, X., and Schreiber, J. R. (1995). Cell-specific localization of apolipoprotein E messenger ribonucleic acid in the testis and epididymis of the rat. *Biol. Reprod.* 52, 1003–1011.
114. Seiler, P., Cooper, T. G., and Nieschlag, E. (2000). Sperm number and condition affect the number of basal cells and their expression of macrophage antigen in the murine epididymis. *Int. J. Androl.* 23, 65–76.
115. Holschbach, C., and Cooper, T. G. (2002). A possible extratubular origin of epididymal basal cells in mice. *Reproduction* 123, 517–525.
116. Leung, G. P. H., Cheung, K. H., Leung, T., Tsabg, M. W., and Wong, P. Y. D. (2004). Regulation of epididymal principal cell functions by basal cells: role of transient receptor potential (Trp) proteins and cyclooxygenase-1 (COX-1). *Mol. Cell. Endocrinol.* 216, 5–13.
117. Hamilton, D. W. (1972). The mammalian epididymis. In *Reproductive Biology* (H. Balin and S. Glassner, Eds.), pp. 268–337. Excerpta Medica, Amsterdam.
118. Flickinger, C. J., Bush, L. A., Howards, S. S., and Herr, J. C. (1997). Distribution of leucocytes in the epithelium and interstitium of four regions of the Lewis rat epididymis. *Anat. Rec.* 248, 380–390.
119. Serre, V., and Robaire, B. (1999). The distribution of immune cells in the epithelium of the epididymis of the aging brown Norway rat is segment-specific and related to the luminal content. *Biol. Reprod.* 61, 705–714.
120. Howards, S. S., Jessee, S. J., and Johnson, A. L. (1976). Micropuncture studies of the blood–seminiferous tubule barrier. *Biol. Reprod.* 14, 264–269.
121. Hinton, B. T. (1985). Physiological aspects of the blood–epididymis barrier. In *Male Fertility and its Regulation* (T. J. Lobl and E. S. E. Hafez, Eds.), pp. 371–382. MTP Press, Boston.
122. Cyr, D. G., Robaire, B., and Hermo, L. (1995). Structure and turnover of junctional complexes between principal cells of the rat epididymis. *Microsc. Res. Tech.* 30, 54–66.
123. Cyr, D. G., Finnson, K., Dufresne, J., and Gregory, M. (2002). Cellular interactions and the blood–epididymal barrier. In *The Epididymis: From Molecules to Clinical Practice* (B. Robaire and B. T. Hinton, Eds.), pp. 103–118. Kluwer Academic/Plenum, New York.
124. Friend, D. S., and Gilula, N. B. (1972). Variations in tight and gap junctions in mammalian tissues. *J. Cell. Biol.* 53, 758–776.
125. Agarwal, A., and Hoffer, A. P. (1989). Ultrastructural studies on the development of the blood–epididymis barrier in immature rats. *J. Androl.* 10, 425–431.
126. Suzuki, F., and Nagano, T. (1978). Development of tight junctions in the caput epididymal epithelium of the mouse. *Dev. Biol.* 63, 321–334.
127. Gumbiner, B. M. (1996). Cell adhesion: the molecular basis of tissue architecture and morphogenesis. *Cell* 84, 345–357.
128. Takeichi, M. (1988). The cadherins: cell–cell adhesion molecules controlling animal morphogenesis. *Development* 102, 639–655.
129. Takeichi, M. (1990). Cadherins: a molecular family important in selective cell–cell adhesion. *Annu. Rev. Biochem.* 59, 237–252.
130. Vestweber, D., Kemler, R., and Ekblom, P. (1985). Cell-adhesion molecule uvomorulin during kidney development. *Dev. Biol.* 112, 213–221.
131. Bollier, K., Vestweber, D., and Kemler, R. (1985). Cell-adhesion molecule uvomorulin is localized in the intermediate junctions of adult intestinal epithelial cells. *J. Cell. Biol.* 100, 327–332.

132. Ogou, S. I., Yoshida-Noro, C., and Takeichi, M. (1983). Calcium-dependent cell–cell adhesion molecules common to hepatocytes and teratocarcinoma stem cells. *J. Cell. Biol.* 97, 944–948.
133. Cyr, D. G., Blaschuk, O. W., and Robaire, B. (1992). Identification and developmental regulation of cadherin messenger ribonucleic acids in the rat testis. *Endocrinology* 131, 139.
134. Cyr, D. G., and Robaire, B. (1991). Developmental regulation of epithelial and placental-cadherin mRNA in the rat epididymis. *Ann. N. Y. Acad. Sci.* 637, 399–408.
135. Cyr, D. G., Hermo, L., and Robaire, B. (1993). Developmental changes in epithelial cadherin messenger ribonucleic acid and immunocytochemical localization of epithelial cadherin during postnatal epididymal development in the rat. *Endocrinology* 132, 1115–1124.
136. Levy, S., and Robaire, B. (1999). Segment-specific changes in the expression of junctional proteins and the permeability of the blood–epididymis barrier with age. *Biol. Reprod.* 60, 1392–1401.
137. Andersson, A. M., Edvardsen, K., and Skakkebaek, N. E. (1994). Expression and localization of N- and E-cadherin in the human testis and epididymis. *Int. J. Androl.* 17, 174–180.
138. Byers, S. W., Citi, S., Anderson, J. M., and Hoxter, B. (1992). Polarized functions and permeability properties of rat epididymal epithelial cells in vitro. *J. Reprod. Fertil.* 95, 385–396.
139. Byers, S., Jegou, B., MacCalman, C., and Blaschuk, O. (1993). Sertoli cell adhesion molecules and the collective organization of the testis. In *The Sertoli Cell* (L. D. Russell and M. D. Griswold, Eds.), p. 461. Cache River Press, Clearwater, FL.
140. DeBellefeuille, S., Hermo, L., Gregory, M., Dufresne, J., and Cyr, D. G. (2003). Catenins in the rat epididymis: their expression and regulation in adulthood and during postnatal development. *Endocrinology* 144, 5040–5049.
141. Pelletier, R.-M. (1995). Freeze-fracture study of cell junctions in the epididymis and vas deferens of a seasonal breeder: the mink (*Mustela vison*). *Microsc. Res. Tech.* 30, 37–53.
142. Soranzo, L., Dadoune, J.-P., and Fain-Maurel, M.-A. (1982). Segmentation of the epididymal duct in mouse: an ultrastructural study. *Reprod. Nutr. Dev.* 22, 999–1012.
143. Cyr, D. G., Hermo, L., and Laird, D. W. (1996). Immunocytochemical localization and regulation of connexin43 in the adult rat epididymis. *Endocrinology* 137, 1474–1484.
144. Hinton, B. T., and Palladino, M. A. (1995). Epididymal epithelium: its contribution to the formation of a luminal fluid microenvironment. *Microsc. Res. Tech.* 30, 67–81.
144a. Hoffer, A. P., and Hinton, B. T. (1984). Morphological evidence for a blood–epididymis barrier and the effects of gossypol on its integrity. *Biol. Reprod.* 30, 991–1004.
144b. Turner, T. T., Giles, R. D., and Howards, S. S. (1981). Effect of oestradiol valerate on the rat blood–testis and blood–epididymal barriers to [^{3}H]inulin. *J. Reprod. Fertil.* 63, 355–358.
145. Qiu, J., Hales, B. F., and Robaire, B. (1992). Adverse effects of cyclophosphamide on progeny outcome can be mediated through post-testicular mechanisms in the rat. *Biol. Reprod.* 46, 926–931.
146. Orgebin-Crist, M.-C. (1962). Recherches expérimentales sur la durée de passage des spermatozoïdes dans l'épididyme du taureau. *Ann. Biol. Anim. Biochem. Biophys.* 2, 51–108.
147. Orgebin-Crist, M.-C. (1965). Passage of spermatozoa labelled with thymidine-^{3}H through the ductus epididymidis of the rabbit. *J. Reprod. Fertil.* 10, 241–251.
148. Rowley, M. J., Teshima, F., and Heller, C. G. (1970). Duration of transit of spermatozoa through the human male ductular system. *Fertil. Steril.* 21, 390–396.
149. Koefoed-Johnsen, H. H. (1960). Influence of ejaculation frequency on the time required for sperm formation and epididymal passage in the full. *Nature* 185, 49–50.
150. Amir, D., and Ortavant, R. (1968). Influence de la fréquence des collectes sur la durée du transit des spermatozoïdes dans le canal épididymaire du bélier. *Ann. Biol. Anim. Biochem. Biophys.* 8, 195–207.
151. Kirton, K. T., Desjardins, C., and Hafs, H. D. (1967). Distribution of sperm in male rabbits after various ejaculation frequencies. *Anat. Rec.* 158, 287–292.
152. Crabo, B. (1965). Studies on the composition of epididymal content in bulls and boars. *Acta Vet. Scand. Suppl.* 5, 1–94.
153. Johnson, A. L., and Howards, S. S. (1975). Intratubular hydrostatic pressure in testis and epididymis before and after vasectomy. *Am. J. Physiol.* 228, 556–564.
154. Macmillan, E. W., and Aukland, J. (1960). The transport of radiopaque medium through the initial segment of the rat epididymis. *J. Reprod. Fertil.* 1, 139–145.
155. Baumgarten, H. G., Holstein, A. F., and Rosengren, E. (1971). Arrangement, ultrastructure and adrenergic innervation of smooth musculature of the ductuli efferentes, ductus epididymis and ductus deferens of man. *Z. Zellforsch. Mikrosk. Anat.* 120, 37–79.
156. Talo, A., Jaakkola, U.-M., and Markkula-Viitanen, M. (1979). Spontaneous electrical activity of the rat epididymis in vitro. *J. Reprod. Fertil.* 57, 423–429.
157. Markkula-Viitanen, M., Nikkanen, V., and Talo, A. (1979). Electrical activity and intraluminal pressure of the cauda epididymidis of the rat. *J. Reprod. Fertil.* 57, 431–435.
158. Jaakkola, U.-M., and Talo, A. (1982). Relation of electrical activity to luminal transport in the cauda epididymidis of the rat in vitro. *J. Reprod. Fertil.* 64, 121–126.
159. Jaakkola, U.-M. (1983). Regional variations in transport of the luminal contents of the rat epididymis in vivo. *J. Reprod. Fertil.* 68, 465–470.
160. Jaakkola, U.-M., and Talo, A. (1983). Movements of the luminal contents in two different regions of the caput epididymidis of the rat in vitro. *J. Physiol. (Lond.)* 336, 453–463.
161. Dyson, A. L. M. B., and Orgebin-Crist, M.-C. (1973). Effect of hypophysectomy, castration and androgen replacement upon the fertilizing ability of rat epididymal spermatozoa. *Endocrinology* 93, 391–402.
162. Lubicz-Nawrocki, C. M. (1974). Effects of castration and testosterone replacement on the number of spermatozoa in the cauda epididymidis of hamsters. *J. Reprod. Fertil.* 39, 97–100.
163. Foldesy, R. G., and Bedford, J. M. (1982). Biology of the scrotum: I. Temperature and androgen as determinants of the sperm storage capacity of the rat cauda epididymidis. *Biol. Reprod.* 26, 673–682.
164. Din-Udom, A., Sujarit, S., and Pholpramool, C. (1985). Short-term effect of androgen deprivation on intraluminal pressure and contractility of the rat epididymis. *J. Reprod. Fertil.* 73, 405–410.
165. Sujarit, S., and Pholpramool, C. (1985). Enhancement of sperm transport through the rat epididymis after castration. *J. Reprod. Fertil.* 74, 497–502.
166. Meistrich, M. L., Hughes, T. H., and Bruce, W. R. (1975). Alteration of epididymal sperm transport and maturation in mice by oestrogen and testosterone. *Nature* 258, 145–147.
167. Hib, J., and Oscar, P. (1978). Effects of prostaglandins and indomethacin on rat epididymal responses to norepinephrine and acetylcholine. *Arch. Androl.* 1, 43–47.
168. Cosentino, M. J., Takihara, H., Burhop, J. W., and Cockett, A. T. K. (1984). Regulation of rat caput epididymidis contractility by prostaglandins. *J. Androl.* 5, 216–222.
169. Filippi, S., Vannelli, G. B., Granchi, S., Luconi, M., Crescioli, C., Mancina, R., Natali, A., Brocchi, S., Vignozzi, L., Bencini, E., Noci, I., Ledda, F., Forti, G., and Maggi, M. (2002). Identification, localization and functional activity of oxytocin receptors in epididymis. *Mol. Cell. Endocrinol.* 193, 89–100.
170. Maggi, M., Malozowski, S., Kassis, S., Guardabasso, V., and Rodbard, D. (1987). Identification and characterization of

two classes of receptors for oxytocin and vasopressin in porcine tunica albuginea, epididymis, and vas deferens. *Endocrinology* 120, 986–994.

171. Einspanier, A., and Ivell, R. (1997). Oxytocin and oxytocin receptor expression in reproductive tissues of the male marmoset monkey. *Biol. Reprod.* 56, 416–422.
172. Frayne, J., and Nicholson, H. D. (1998). Localization of oxytocin receptors in the human and macaque monkey male reproductive tracts: evidence for a physiological role of oxytocin in the male. *Mol. Hum. Reprod.* 4, 527–532.
173. Whittington, K., Assinder, S. J., Parkinson, T., Lapwood, K. R., and Nicholson, H. D. (2001). Function and localization of oxytocin receptors in the reproductive tissue of rams. *Reproduction* 122, 317–325.
174. Hib, J. (1974). The in vitro effects of oxytocin and vasopressin on spontaneous contractility of the mouse cauda epididymidis. *Biol. Reprod.* 11, 436–439.
175. Filippi, S., Luconi, M., Granchi, S., Vignozzi, L., Bettuzzi, S., Tozzi, P., Ledda, F., Forti, G., and Maggi, M. (2002). Estrogens, but not androgens, regulate expression and functional activity of oxytocin receptor in rabbit epididymis. *Endocrinology* 143, 4271–4280.
176. Studdard, P. W., Stein, J. L., and Cosentino, M. J. (2002). The effects of oxytocin and arginine vasopressin in vitro on epididymal contractility in the rat. *Int. J. Androl.* 25, 65–71.
177. Melin, P. (1970). Effects in vivo of neurohypophysial hormones on the contractile activity of accessory sex organs in male rabbits. *J. Reprod. Fertil.* 22, 283–292.
178. Hibb, J. (1977). The in vivo effects of oxytocin and vasopressin on spontaneous contractility of the rat epididymis. *Int. J. Fertil.* 22, 63–64.
179. Knight, T. W. (1974). A qualitative study of factors affecting the contractions of the epididymis and ductus deferens of the ram. *J. Reprod. Fertil.* 40, 19–29.
180. Jaakkola, U.-M., and Talo, A. (1981). Effects of oxytocin and vasopressin on electrical and mechanical activity of the rat epididymis in vitro. *J. Reprod. Fertil.* 63, 47–51.
181. Kihlstrom, J. M., and Melin, P. (1963). The influence of oxytocin upon some seminal characteristics in the rabbit. *Acta Physiol. Scand.* 59, 363–369.
182. Knight, T. W., and Lindsay, D. R. (1970). Short- and long-term effects of oxytocin on quality and quantity of semen from rams. *J. Reprod. Fertil.* 21, 523–529.
183. Voglmayr, J. K. (1975). Output of spermatozoa and fluid by the testis of the ram and its response to oxytocin. *J. Reprod. Fertil.* 43, 119–122.
184. Sharma, O. P., and Hays, R. L. A. (1976). A possible role for oxytocin in sperm transport in the male rabbit. *J. Endocrinol.* 68, 43–47.
185. Agmo, A., Andersson, R., and Johansson, C. (1978). Effect of oxytocin on sperm numbers in spontaneous rat ejaculates. *Biol. Reprod.* 18, 346–349.
186. Berndtson, W. E., and Igboeli, G. (1988). Spermatogenesis, sperm output and seminal quality of Holstein bulls electroejaculated after administration of oxytocin. *J. Reprod. Fertil.* 82, 467–475.
187. Nicholson, H. D., Parkinson, T. J., and Lapwood, K. R. (1999). Effects of oxytocin and vasopressin on sperm transport from the cauda epididymis in sheep. *J. Reprod. Fertil.* 117, 299–305.
188. Murphy, M. R., Seckl, J. R., Burton, S., Checkley, S. A., and Lightman, S. L. (1987). Changes in oxytocin and vasopressin secretion during sexual activity in men. *J. Clin. Endocrinol. Metab.* 65, 738–741.
189. Sharma, S. C., Fitzpatrick, R. J., and Ward, W. R. (1972). Coital-induced release of oxytocin in the ram. *J. Reprod. Fertil.* 31, 488–489.
190. Ogawa, S., Kudo, S., Kitsunai, Y., and Fukuchi, S. (1980). Increase in oxytocin secretion at ejaculation in male. *Clin. Endocrinol.* 13, 95–97.
191. Peeters, G., Legros, J. J., Piron-Bossuyt, C., Reynaert, R., Vanden Driessche, R., and Vannieuwenhuyse, E. (1983). Release of neurophysin I and oxytocin by stimulation of the genital organs in bulls. *J. Endocrinol.* 99, 161–171.
192. Stoneham, M. D., Everitt, B. J., Hansen, S., Lightman, S. L., and Todd, K. (1985). Oxytocin and sexual behavior in the male rat and rabbit. *J. Endocrinol.* 107, 97–106.
193. Simeone, F. A. (1933). A neuromuscular mechanism in the ductus epididymidis and its impairment by sympathetic denervation. *Am. J. Physiol.* 103, 582–591.
194. Zankl, H., and Leidl, W. (1969). Effect of vasoligation and a sympatholytic agent on the number of sperm cells in the epididymis in rabbit. *J. Reprod. Fertil.* 18, 181–182.
195. Evans, B., Gannon, B. J., Heath, J. W., and Burnstock, G. (1972). Long-lasting damage to the internal male genital organs and their adrenergic innervation in rats following chronic treatment with the anthihypertensive drug guanethidine. *Fertil. Steril.* 23, 657–667.
196. Bhathal, P. S., Gerkens, J. F., and Mashford, M. L. (1974). Spermatic granuloma of the epididymis in rats treated with guanethidine. *J. Pathol.* 112, 19–26.
197. Hib, J., Ponzio, R. O., and Vilar, O. (1979). Contractile behaviour of rat epididymis after sympathectomy produced by the administration of guanethidine. *Andrologia* 11, 461–465.
198. Ricker, D. D., and Chang, T. S. K. (1996). Neuronal input from the inferior mesenteric ganglion (IMG) affects sperm transport within the rat cauda epididymis. *Int. J. Androl.* 19, 371–376.
199. Laitinen, L., and Talo, A. (1981). Effects of adrenergic and cholinergic drugs on electrical and mechanical activities of the rat cauda epididymidis in vitro. *J. Reprod. Fertil.* 63, 205–209.
200. Pholpramool, C., and Triphrom, N. (1984). Effects of cholinergic and adrenergic drugs on intraluminal pressures and contractility of the rat testis and epididymis in vivo. *J. Reprod. Fertil.* 71, 181–188.
201. Da Silva e Souza, M. C., Gimeno, M. F., and Gimeno, A. L. (1975). Physiologic and pharmacologic studies on the motility of isolated guinea pig cauda epididymidis. *Fertil. Steril.* 26, 1250–1256.
202. Hib, J. (1976). Effects of autonomic drugs on epididymal contractions. *Fertil. Steril.* 27, 951–956.
203. Yamamoto, M., Hibi, H., and Miyake, K. (1995). Effects of alpha-blocker on daily testicular sperm production and sperm concentration, motility, intraluminal pressure and fluid movement in the rat epididymis. *Tohoku J. Exp. Med.* 177, 25–37.
204. Jaakkola, U.-M., and Talo, A. (1980). Effect of temperature on the electrical activity of the rat epididymis in vitro. *J. Therm. Biol.* 5, 207–210.
205. Bedford, J. M. (1978). Influence of abdominal temperature on epididymal function in the rat and rabbit. *Am. J. Anat.* 152, 509–522.
206. Brackett, B. G., Hall, J. L., and Oh, Y.-K. (1978). In vitro fertilizing ability of testicular, epididymal, and ejaculated rabbit spermatozoa. *Fertil. Steril.* 29, 571–582.
207. Kimura, Y., and Yanagimachi, R. (1995). Development of normal mice from oocytes injected with secondary spermatocyte nuclei. *Biol. Reprod.* 53, 855–862.
208. Ogura, A., Matsuda, J., and Yanagimachi, R. (1994). Birth of normal young after electrofusion of mouse oocytes with round spermatids. *Proc. Natl. Acad. Sci. U S A* 91, 7460–7462.
209. Ogura, A., Suzuki, O., Tanemura, K., Mochida, K., Kobayashi, Y., and Matsuda, J. (1998). Development of normal mice from metaphase I oocytes fertilized with primary spermatocytes. *Proc. Natl. Acad. Sci. U S A* 95, 5611–5615.

210. Nishikawa, Y., and Waide, Y. (1952). Studies on the maturation of spermatozoa. I. Mechanism and speed of transition of spermatozoa in the epididymis and their functional changes. *Bull. Natl. Inst. Agr. Sci. (G)* 3, 69–81.
211. Paz (Frenkel), G., Kaplan, R., Yedwab, G., Homonnai, Z. T., and Kraicer, P. F. (1978). The effect of caffeine on rat epididymal spermatozoa: motility, metabolism and fertilizing capacity. *Int. J. Androl.* 1, 145–152.
212. Fournier-Delpech, S., Colas, G., Courot, M., Ortavant, R., and Brice, G. (1979). Epididymal sperm maturation in the ram: motility, fertilizing ability and embryonic survival after uterine artificial insemination in the ewe. *Ann. Biol. Anim. Biochem. Biophys.* 19, 597–605.
213. Fournier-Delpech, S., Colas, G., and Courot, M. (1981). Observations sur les premiers clivages des œufs intratubaires de brebis après fécondation avec des spermatozoides épididymaires ou éjaculés. *C. R. Séances Acad. Sci. (D) (Paris)* 292: 515–517.
214. Orgebin-Crist, M.-C. (1968). Maturation of spermatozoa in the rabbit epididymis: delayed fertilization in does inseminated with epididymal spermatozoa. *J. Reprod. Fertil.* 16, 29–33.
215. Orgebin-Crist, M.-C., and Jahad, N. (1977). Delayed cleavage of rabbit ova after fertilization by young epididymal spermatozoa. *Biol. Reprod.* 16, 358–362.
216. Overstreet, J. W., and Bedford, J. M. (1976). Embryonic mortality in the rabbit is not increased after fertilization by young epididymal spermatozoa. *Biol. Reprod.* 15, 54–57.
217. Wazzan, W. C., Gwatkin, R. B. L., and Thomas, A. J., Jr. (1990). Zona drilling enhances fertilization by mouse caput epididymal sperm. *Mol. Reprod. Dev.* 27, 332–336.
218. Lacham-Kaplan, O., and Trounson, A. O. (1994). Embryo development capacity of oocytes fertilized by immature sperm and sperm treated with motility stimulants. *Reprod. Fertil. Dev.* 6, 113–116.
219. Schoysman, R. J., and Bedford, J. M. (1986). The role of the human epididymis in sperm maturation and sperm storage as reflected in the consequences of epididymovasostomy. *Fertil. Steril.* 46, 293–299.
220. Silber, S. J. (1989). Results of microsurgical vasoepididymostomy: role of epididymis in sperm maturation. *Hum. Reprod.* 4, 298–303.
221. Silber, S. J. (1988). Pregnancy caused by sperm from vasa efferentia. *Fertil. Steril.* 49, 373–375.
222. Patrizio, P., Ord, T., Silber, S. J., and Asch, R. H. (1994). Correlation between epididymal length and fertilization rate in men with congenital absence of the vas deferens. *Fertil. Steril.* 61, 265–268.
223. Pryor, J. P. (1987). Surgical retrieval of epididymal spermatozoa. *Lancet* 2, 1341.
224. Mahadevan, M. M., and Trounson, A. O. (1985). Removal of the cumulus oophorus from the human oocyte for in vitro fertilization. *Fertil. Steril.* 43, 263–267.
225. Temple-Smith, P. D., Southwick, G. J., Yates, C. A., Trounson, A. O., and De Kretser, D. M. (1985). Human pregnancy by IVF using sperm aspirated from the epididymis. *J.* In *Vitro Fertil. Embryo Transf.* 2, 119–122.
226. Silber, S. J., Balmaceda, J., Borrero, C., Ord, T., and Asch, R. (1988). Pregnancy with sperm aspiration from the proximal head of the epididymis: a new treatment for congenital absence of the vas deferens. *Fertil. Steril.* 50, 525–528.
227. Jequier, A. M., Cummins, J. M., Gearon, C., Apted, S. L., Yovich, J. M., and Yovich, J. L. (1990). A pregnancy achieved using sperm from the epididymal caput in idiopathic obstructive azoospermia. *Fertil. Steril.* 53, 1104–1105.
228. Silber, S. J., Nagy, Z. P., Liu, J., Godoy, H., Devroey, P., and Van Steirteghem, A. C. (1994). Conventional in-vitro fertilization versus intracytoplasmic sperm injection for patients requiring microsurgical sperm aspiration. *Hum. Reprod.* 9, 1705–1709.
229. Devroey, P., Liu, J., Nagy, Z., Tournaye, H., Silber, S. J., and Van Steirteghem, A. C. (1994). Normal fertilization of human oocytes after testicular sperm extraction and intracytoplasmic sperm injection. *Fertil. Steril.* 62, 639–641.
230. Tournaye, H., Devroey, P., Liu, J., Nagy, Z., Lissens, W., and Van Steirteghem, A. (1994). Microsurgical epididymal sperm aspiration and intracytoplasmic sperm injection: a new effective approach to infertility as a result of congenital bilateral absence of the vas deferens. *Fertil. Steril.* 61, 1045–1051.
231. Nagy, Z., Liu, J., Cecile, J., Silber, S., Devroey, P., and Van Steirteghem, A. (1995). Using ejaculated, fresh, and frozen-thawed epididymal and testicular spermatozoa gives rise to comparable results after intracytoplasmic sperm injection. *Fertil. Steril.* 63, 808–815.
232. Tarlatzis, B. C. (1996). Report on the activities of the ESHRE task force on intracytoplasmic sperm injection. *Hum. Reprod.* 11 (Suppl. 4), 160–186.
233. Watkins, W., Nieto, F., Bourne, H., Wutthiphan, B., Speirs, A., and Baker, H. W. G. (1997). Testicular and epididymal sperm in a microinjection program: methods of retrieval and results. *Fertil. Steril.* 67, 527–535.
234. Ghazzawi, I. M., Sarraf, M. G., Taher, M. R., and Khalifa, F. A. (1998). Comparison of the fertilizing capability of spermatozoa from ejaculates, epididymal aspirates and testicular biopsies using intracytoplasmic sperm injection. *Hum. Reprod.* 13, 348–352.
235. Tarlatzis, B. C., and Bili, H. (1998). Survey on intracytoplasmic sperm injection: report from the ESHRE ICSI task force. *Hum. Reprod.* 13 (Suppl. 1), 165–177.
236. Van Steirteghem, A., Nagy, P., Joris, H., Janssenswillen, C., Staessen, C., Verheyen, G., Camus, M., Tournaye, H., and Devroey, P. (1998). Results of intracytoplasmic sperm injection with ejaculated, fresh and frozen-thawed epididymal and testicular spermatozoa. *Hum. Reprod.* 13 (Suppl. 1), 134–142.
237. Bonduelle, M., Wilikens, A., Buysse, A., Van Assche, E., Wisanto, A., Devroey, P., Van Steirteghem, A. C., and Liebaers, I. (1996). Prospective follow-up study of 877 children born after intracytoplasmic sperm injection (ICSI), with ejaculated epididymal and testicular spermatozoa and after replacement of cryopreserved embryos obtained after ICSI. *Hum. Reprod.* 11 (Suppl. 4), 131–155.
238. Bonduelle, M., Wilikens, A., Buysse, A., Van Assche, E., Devroey, P., Van Steirteghem, A. C., and Liebaers, I. (1998). A follow-up study of children born after intracytoplasmic sperm injection (ICSI) with epididymal and testicular spermatozoa and after replacement of cryopreserved embryos obtained after ICSI. *Hum. Reprod.* 13 (Suppl. 1), 196–207.
239. Devroey, P., and Van Steirteghem, A. (2004). A review of ten years experience of ICSI. *Hum. Reprod. Update* 10, 19–28.
240. Depeiges, A., and Dacheux, J. L. (1985). Acquisition of sperm motility and its maintenance during storage in the lizard, *Lacerta vivipara*. *J. Reprod. Fertil.* 74, 23–27.
241. Nirmal, B. K., and Rai, U. (1997). Epididymal influence on acquisition of sperm motility in the gekkonid lizard *Hemidactylus flaviviridis*. *Arch. Androl.* 39, 105–110.
242. Soler, C., Yeung, C. H., and Cooper, T. G. (1994). Development of sperm motility patterns in the murine epididymis. *Int. J. Androl.* 17, 271–278.
243. Yochem, D. E. (1930). A study of the motility and resistance of rat spermatozoa at different levels in the reproductive tract. *Physiol. Zool.* 3, 309–329.
244. Blandau, R. J., and Rumery, R. E. (1964). The relationship of swimming movements of epididymal spermatozoa to their fertilizing capacity. *Fertil. Steril.* 15, 571–579.

245. Fray, C. S., Hoffer, A. P., and Fawcett, D. W. (1972). Reexamination of motility patterns of rat epididymal spermatozoa. *Anat. Rec.* 173, 301–308.
246. Wyker, R., and Howards, S. S. (1977). Micropuncture studies of the motility of rete testis and epididymal spermatozoa. *Fertil. Steril.* 28, 108–112.
247. Hinton, B. T., Dott, H. M., and Setchell, B. P. (1979). Measurement of the motility of rat spermatozoa collected by micropuncture from the testis and from different regions along the epididymis. *J. Reprod. Fertil.* 55, 167–172.
248. Turner, T. T., and Giles, R. D. (1981). The effects of carnitine, glycerylphosphorylcholine, caffeine, and egg yolk on the motility of rat epididymal spermatozoa. *Gamete Res.* 4, 283–295.
249. Pholpramool, C., Lea, O. A., Burrow, P. V., Dott, H. M., and Setchell, B. P. (1983). The effects of acidic epididymal glycoprotein (AEG) and some other proteins on the motility of rat epididymal spermatozoa. *Int. J. Androl.* 6, 240–248.
250. Jeulin, C., Lewin, L. M., Chevrier, C., and Schoevaert-Brossault, D. (1996). Changes in flagellar movement of rat spermatozoa along the length of the epididymis: manual and computer-aided image analysis. *Cell Motil. Cytoskeleton* 35, 147–161.
251. Pholpramool, C., and Chaturapanich, G. (1979). Effect of sodium and potassium concentrations and pH in the maintenance of motility of rabbit and rat epididymal spermatozoa. *J. Reprod. Fertil.* 57, 245–251.
252. Kann, M.-L., and Serres, C. (1980). Development and initiation of sperm motility in the hamster epididymis. *Reprod. Nutr. Dev.* 20, 1739–1749.
253. Cornwall, G. A., Smyth, T. B., Vindivich, D., Harter, C., Robinson, J., and Chang, T. S. K. (1986). Induction and enhancement of progressive motility in hamster caput epididymal spermatozoa. *Biol. Reprod.* 35, 1065–1074.
253a. Yeung, C. H., Oberländer, G., and Cooper, T. G. (1994). Maturation of hamster epididymal sperm motility and influence of the thiol status of hamster and rat spermatozoa on their motility patterns. *Mol. Reprod. Dev.* 38, 347–355.
254. Frenkel, G., Peterson, R. N., and Freund, M. (1973). Changes in the metabolism of guinea pig sperm from different segments of the epididymis (37508). *Proc. Soc. Exp. Biol. Med.* 143, 1231–1236.
255. Shilon, M., Paz (Frenkel), G., Homonnai, Z. T., and Schoenbaum, M. (1978). The effect of caffeine on guinea pig epididymal spermatozoa: motility and fertilizing capacity. *Int. J. Androl.* 1, 416–423.
256. Gaddum, P. (1968). Sperm maturation in the male reproductive tract: development of motility. *Anat. Rec.* 161, 471–482.
257. Pérez-Sánchez, F., Tablado, L., Yeung, C.-H., Cooper, T. G., and Soler, C. (1996). Changes in the motility patterns of spermatozoa from the rabbit epididymis as assessed by computer-aided sperm motion analysis. *Mol. Reprod. Dev.* 45, 364–371.
258. Dacheux, J. L., O'Shea, T., and Paquignon, M. (1979). Effects of osmolality, bicarbonate and buffer on the metabolism and motility of testicular, epididymal and ejaculated spermatozoa of boars. *J. Reprod. Fertil.* 55, 287–296.
259. Bork, K., Chevrier, C. Paquignon, M., Jouannet, P., and Dacheux, J. L. (1988). Analyse de la motilité et du mouvement flagellaire des spermatozoides de verrat au cours du transit epididymaire. *Reprod. Nutr. Dev.* 28, 1307–1315.
260. Jaiswal, B. S., and Majumder, G. C. (1996). Cyclic AMP phosphodiesterase: a regulator of forward motility initiation during epididymal sperm maturation. *Biochem. Cell. Biol.* 74, 669–674.
261. Amann, R. P., Hay, S. R., and Hammerstedt, R. H. (1982). Yield, characteristics, motility and cAMP content of sperm isolated from seven regions of ram epididymis. *Biol. Reprod.* 27, 723–733.
262. Pariset, C. C., Feinberg, J. M. F., Dacheux, J. L., and Weinman, S. J. (1985). Changes in calmodulin level and cAMP-dependent protein kinase activity during epididymal maturation of ram spermatozoa. *J. Reprod. Fertil.* 74, 105–112.
263. Chevrier, C., and Dacheux, J.-L. (1992). Evolution of the flagellar waveform of ram spermatozoa in relation to the degree of epididymal maturation. *Cell. Motil. Cytoskeleton* 23, 8–18.
264. Acott, T. S., Katz, D. F., and Hoskins, D. D. (1983). Movement characteristics of bovine epididymal spermatozoa: effects of forward motility protein and epididymal maturation. *Biol. Reprod.* 29, 389–399.
265. Pholpramool, C., Zupp, J. L., and Setchell, B. P. (1985). Motility of undiluted bull epididymal spermatozoa collected by micropuncture. *J. Reprod. Fertil.* 75, 413–420.
266. Yeung, C. H., Morrell, J. M., Cooper, T. G., Weinbauer, G. F., Hodges, J. K., and Nieschlag, E. (1996). Maturation of sperm motility in the epididymis of the common marmoset (*Callithrix jacchus*) and the cynomolgus monkey (*Macaca fascicularis*). *Int. J. Androl.* 19, 113–121.
267. Van der Horst, G., Seier, J. V., Spinks, A. C., and Hendricks, S. (1999). The maturation of sperm motility in the epididymis and vas deferens of the vervet monkey, *Cercopithecus aethiops*. *Int. J. Androl.* 22, 197–207.
268. Mahony, M. C., Oehninger, S., Doncel, G., Morshedi, M., Acosta, A., and Hodgen, G. D. (1993). Functional and morphological features of spermatozoa microaspirated from the epididymal regions of cynomolgus monkeys (*Macaca fascicularis*). *Biol. Reprod.* 48, 613–620.
269. Mooney, J. K., Horan, A. H., and Lattimer, J. K. (1972). Motility of spermatozoa in the human epididymis. *J. Urol* 108, 443–445.
270. Bedford, J. M., Calvin, H., and Cooper, G. W. (1973). The maturation of spermatozoa in the human epididymis. *J. Reprod. Fertil. Suppl.* 18, 199–213.
271. Dacheux, J. L., Chevrier, C., and Lanson, Y. (1987). Motility and surface transformations of human spermatozoa during epididymal transit. *Proc. Natl. Acad. Sci. U S A* 513, 560–563.
272. Glover, T. D. (1973). Aspects of sperm production in some East African mammals. *J. Reprod. Fertil.* 35, 45–53.
273. Bedford, J. M., and Millar, R. P. (1978). The character of sperm maturation in the epididymis of the ascrotal hyrax, *Procavia capensis* and armadillo, *Dasypus novemcinctus*. *Biol. Reprod.* 19, 396–406.
274. Lakoski, K. A., Carron, C. P., Cabot, C. L., and Saling, P. M. (1988). Epididymal maturation and the acrosome reaction in mouse sperm: response to zona pellucida develops coincident with modification of M42 antigen. *Biol. Reprod.* 38, 221–233.
275. Williams, R. M., Graham, J. K., and Hammerstedt, R. H. (1991). Determination of the capacity of ram epididymal and ejaculated sperm to undergo the acrosome reaction and penetrate ova. *Biol. Reprod.* 44, 1080–1091.
276. Burkin, H., and Miller, D. J. (2000). Zona pellucida protein binding ability of porcine sperm during epididymal maturation and the acrosome reaction. *Dev. Biol.* 222, 99–109.
277. Sirivaidyapong, S., Bevers, M. M., Gadella, B. M., and Colenbrander, B. (2001). Induction of the acrosome reaction in dog sperm cells is dependent on epididymal maturation: the generation of a functional progesterone receptor is involved. *Mol. Reprod. Dev.* 58, 451–459.
278. Yeung, C. H., Cooper, T. G., and Weinbauer, G. F. (1996). Maturation of monkey spermatozoa in the epididymis with respect to their ability to undergo the acrosome reaction. *J. Androl.* 17, 427–432.
279. Yeung, C. H., Perez-Sanchez, F., Soler, C., Poser, D., Kliesch, S., and Cooper, T. G. (1997). Maturation of human spermatozoa (from selected epididymides of prostatic carcinoma patients) with respect to their morphology and ability to undergo the acrosome reaction. *Hum. Reprod. Update* 3, 205–213.

280. Saling, P. M. (1982). Development of the ability to bind to zonae pellucidae during epididymal maturation: reversible immobilization of mouse spermatozoa by lanthanum. *Biol. Reprod.* 26, 429–436.
281. Harayama, H., Kusunoki, H., and Kato, S. (1993). Capacity of rete testicular and cauda epididymal boar spermatozoa to undergo the acrosome reaction and subsequent fusion with egg plasma membrane. *Mol. Reprod. Dev.,* 35, 62–68.
281a. Hinrichsen, M. J., and Blaquier, J. A. (1980). Evidence supporting the existence of sperm maturation in the human epididymis. *J. Reprod. Fertil.* 60, 291–294.
282. Moore, H. D. M., Hartman, T. D., and Pryor, J. P. (1983). Development of the oocyte-penetrating capacity of spermatozoa in the human epididymis. *Int. J. Androl.* 6, 310–318.
283. Bedford, J. M. (1973). Components of sperm maturation in the human epididymis. *Adv. Biosci.* 10, 145–155.
284. Bedford, J. M. (2004). Enigmas of mammalian gamete form and function. *Biol. Rev.* 79, 429–460.
285. Orgebin-Crist, M.-C. (1967). Sperm maturation in rabbit epididymis. *Nature* 216, 816–818.
286. Orgebin-Crist, M.-C. (1973). Maturation of spermatozoa in the rabbit epididymis: effect of castration and testosterone replacement. *J. Exp. Zool.* 185, 301–309.
287. Orgebin-Crist, M.-C., and Davies, J. (1974). Functional and morphological effects of hypophysectomy and androgen replacement in the rabbit epididymis. *Cell. Tissue Res.* 148, 183–201.
288. Orgebin-Crist, M.-C., and Tichenor, P. L. (1973). Effect of testosterone on sperm maturation in vitro. *Nature* 245, 328–329.
289. Lin, M., Zhang, X., Murdoch, R., and Aitken, R. J. (2000). In vitro culture of brushtail possum (*Trichosurus vulpecula*) epididymal epithelium and induction of epididymal sperm maturation in co-culture. *J. Reprod. Fertil.* 119, 1–4.
290. Lin, M., Hess, R., and Aitken, R. J. (2002). Induction of sperm maturation in vitro in epididymal cell cultures of the tammar wallaby (*Macropus eugenii*): disruption of motility initiation and sperm morphogenesis by inhibition of actin polymerization. *Reproduction* 124, 107–117.
291. Moore, H. D. M., and Hartman, T. D. (1986). In-vitro development of the fertilizing ability of hamster epididymal spermatozoa after co-culture with epithelium from the proximal cauda epididymidis. *J. Reprod. Fertil.* 78, 347–352.
292. Bongso, A., and Trounson, A. (1996). Evaluation of motility, fertilizing ability and embryonic development of murine epididymal sperm after coculture with epididymal epithelium. *Hum. Reprod.* 11, 1451–1456.
293. Moore, H. D. M., Curry, M. R., Penfold, L. M., and Pryor, J. P. (1992). The culture of human epididymal epithelium and in vitro maturation of epididymal spermatozoa. *Fertil. Steril.* 58, 776–783.
294. Temple-Smith, P. D., Zheng, S. S., Kadioglu, T., and Southwick, G. J. (1998). Development and use of surgical procedures to bypass selected regions of the mammalian epididymis: effects on sperm maturation. *J. Reprod. Fertil. Suppl.* 53, 183–195.
295. Orgebin-Crist, M.-C., Danzo, B. J., and Davies, J. (1975). Endocrine control of the development and maintenance of sperm fertilizing ability in the epididymis. In *Handbook of Physiology* (R. O. Greep and E. B. Astwood, Eds.), Section 7, Vol. 5, pp. 319–338. American Physiological Society, Washington, DC.
296. Gopalakrishna, A., and Bhatia, D. (1980). Storage of spermatozoa in the epididymis of the bat, *Hipposideros speoris* (Schneider). *Curr. Sci.* 49, 951–953.
297. Depeiges, A., Betail, G., Coulet, M., and Dufaure, J.-P. (1985): Histochemical study of epididymal secretions in the lizard, *Lacerta vivipara*. *Cell. Tissue Res.* 239, 463–466.
298. Martin-DeLeon, P. A., Shaver, E. L., and Gammal, E. B. (1973). Chromosome abnormalities in rabbit blastocysts resulting from spermatozoa aged in the male tract. *Fertil. Steril.* 24, 212–220.
299. Amann, R. P. (1981). A critical review of methods for evaluation of spermatogenesis from seminal characteristics. *J. Androl.* 2, 37–58.
300. Bedford, J. M. (1977). Evolution of the scrotum: the epididymis as the prime mover. In *Reproduction and Evolution* (J. H. Calaby, Ed.), pp. 171–182. Australian Academy of Science, Canberra.
301. Bedford, J. M. (1978). Anatomical evidence for the epididymis as the prime mover in the evolution of the scrotum. *Am. J. Anat.* 152, 483–508.
302. Carr, D. W., Usselman, M. C., and Acott, T. S. (1985). Effects of pH, lactate, and viscoelastic drag on sperm motility: a species comparison. *Biol. Reprod.* 33, 588–595.
303. Usselman, M. C., and Cone, R. A. (1983). Rat sperm are mechanically immobilized in the caudal epididymis by "immobilin," a high molecular weight glycoprotein. *Biol. Reprod.* 29, 1241–1253.
304. Cooper, T. G., and Hamilton, D. W. (1977). Observations on destruction of spermatozoa in the cauda epididymidis and proximal vas deferens of non-seasonal male mammals. *Am. J. Anat.* 149, 93–110.
305. Weissenberg, R., Yossefi, S., Oschry, Y., Madgar, I., and Lewin, L. M. (1995). Investigation of epididymal sperm maturation in the golden hamster. *Int. J. Androl.* 18, 55.
306. Sutovsky, P., Moreno, R., Ramalho-Santos, J., Dominko, T, Thompson, W. E., and Schatten, G. (2001). A putative, ubiquitin-dependent mechanism for the recognition and elimination of defective spermatozoa in the mammalian epididymis. *J. Cell. Sci.* 114, 1665–1675.
307. Cooper, T. G., Yeung, C. H., Jones, R., Orgebin-Crist, M.-C., and Robaire, B. (2002). Rebuttal of a role for the epididymis in sperm quality control by phagocytosis of defective sperm. *J. Cell. Sci.* 115, 5–7.
308. NagDas, S. K., Winfrey, V. P., and Olson, G. E. (2000). Identification of a hamster epididymal region-specific secretory glycoprotein that binds nonviable spermatozoa. *Biol. Reprod.* 63, 1428–1436
309. Olson, G. E., Winfrey, V. P., NagDas, S. K., and Melner, M. H. (2004). Region-specific expression and secretion of the fibrinogen-related protein, fgl2, by epithelial cells of the hamster epididymis and its role in disposal of defective spermatozoa. *J. Biol. Chem.* 279, 51266–51274.
310. Jones, R. (2004). Sperm survival versus degradation in the mammalian epididymis: a hypothesis. *Biol. Reprod.* 71, 1405–1411.
311. Ezer, N., and Robaire, B. (2003). Gene expression is differentially regulated in the epididymis after orchidectomy. *Endocrinology* 144, 975–988.
312. MacLeod, J. (1943). The role of oxygen in the metabolism and motility of human spermatozoa. *Am. J. Physiol.* 138, 512–518.
313. Tosic, J., and Walton, A. (1950). Metabolism of spermatozoa: formation of hydrogen peroxide by spermatozoa and its effects on motility and survival. *J. Biochem. (Tokyo)* 47, 199–212.
314. Jones, R., and Mann, T. (1973). Lipid peroxidation in spermatozoa. *Proc. R. Soc. Lond. B. Biol. Sci.* 184, 103–107.
315. Jones, R., and Mann, T. (1977). Toxicity of exogenous fatty acid peroxides towards spermatozoa. *J. Reprod. Fertil.* 50, 255–260.
316. Jones, R., Mann, T., and Sherins, R. J. (1978). Adverse effects of peroxidized lipid on human spermatozoa. *Proc. R. Soc. Lond. B. Biol. Sci.* 201, 413–417.
317. Jones, R., Mann, T., and Sherins, R. J. (1979). Peroxidative breakdown of phospholipids by human spermatozoa,

spermicidal properties of fatty acid peroxides and protective action of seminal plasma. *Fertil. Steril.* 31, 531–537.
318. Poulos, A., Darin-Bennett, A., and White, I. G. (1973). The phospholipids-bound fatty acids and aldehydes of mammalian spermatozoa. *Comp. Biochem. Physiol. B. Biochem. Mol. Biol.* 46, 541–549.
319. Aitken, R. J., and Clarkson, J. S. (1987). Cellular basis of defective sperm function and its association with the genesis of reactive oxygen species by human spermatozoa. *Biol. Reprod.* 40, 183–197.
320. Vernet, P., Aitken, R. J., and Drevet, J. R. (2004). Antioxidant strategies in the epididymis. *Mol. Cell. Endocrinol.* 216, 31–39.
321. Rao, B., Soufir, J. C., Martin, M., and David, G. (1989). Lipid peroxidation in human spermatozoa as related to midpiece abnormalities and motility. *Gamete Res.* 24, 127–134.
322. Alvarez, J. G., and Storey, B. T. (1982). Spontaneous lipid peroxidation in rabbit epididymal spermatozoa: its effects on sperm motility. *Biol. Reprod.* 27, 1102–1108.
323. De Lamirande, E., and Gagnon, C. (1992). Reactive oxygen species and human spermatozoa: I. Effects on the motility of intact spermatozoa and on sperm axonemes. *J. Androl.* 13, 368–378.
324. De Lamirande, E., and Gagnon, C. (1992). Reactive oxygen species and human spermatozoa: II. Depletion of adenosine triphosphate plays an important role in the inhibition of sperm motility. *J. Androl.* 13, 379–386.
325. Holland, M. K., and Storey, B. T. (1981). Oxygen metabolism of mammalian spermatozoa: generation of hydrogen peroxide by rabbit spermatozoa. *Biochem. J.* 198, 273–280.
326. Holland, M. K., Alvarez, J. G., and Storey, B. T. (1982). Production of superoxide and activity of superoxide dismutase in rabbit epididymal spermatozoa. *Biol. Reprod.* 27, 1109–1118.
327. Alvarez, J. G., and Storey, B. T. (1983). Role of superoxide dismutase in protecting rabbit spermatozoa from O_2 toxicity due to lipid peroxidation. *Biol. Reprod.* 28, 1129–1108.
328. Alvarez, J. G., and Storey, B. T. (1989). Role of glutathione peroxidase in protecting mammalian spermatozoa from loss of motility caused by spontaneous lipid peroxidation. *Gamete Res.* 23, 77–90.
329. Jeulin, C., Soufir, J. C., Weber, P., Laval-Martin, D., and Calvayrac, R. (1989). Catalase activity in human spermatozoa and seminal plasma. *Gamete Res.* 24, 185–196.
330. Alvarez, J. G., Touchstone, J. C., Blasco, L., and Storey, B. T. (1987). Spontaneous lipid peroxidation and production of hydrogen peroxide and superoxide in human spermatozoa: superoxide dismutase as major enzyme protectant against oxygen toxicity. *J. Androl.* 8, 338–348.
331. Free, M. J., Schluntz, G. A., and Jaffe, R. A. (1976). Respiratory gas tensions in tissues and fluids of the male reproductive tract. *Biol. Reprod.* 14, 481–488.
332. Suzuki, F. (1982). Microvasculature of the mouse testis and excurrent duct system. *Am. J. Anat.* 163, 309–325.
333. Setchell, B. P., Waites, G. M. H., and Till, A. R. (1964). Variations in blood flow within the epididymis and testis of the sheep and rat. *Nature* 203, 317–318.
334. Waites, G. M. H., Setchell, B. P., and Quinlan, D. (1973). Effects of local heating of the scrotum, testis and epididymides of rats on cardiac output and regional blood flow. *J. Reprod. Fertil.* 34, 41–49.
335. Hinton, B. T. Palladino, M. A., Rudolph, D., Lan, Z. J., and Labus, J. C. (1996). The role of the epididymis in the protection of spermatozoa. *Curr. Topics Dev. Biol.* 33, 61–102.
336. Nonogaki, T., Noda, Y., Narimoto, K., Shiotani, M., Mori, T., Matsuda, T., and Yoshida, O. (1992). Localization of CuZn-superoxide dismutase in the human male genital organs. *Hum. Reprod.* 7, 81–85.
337. Perry, A. C., Jones, R., and Hall, L. (1993). Isolation and characterization of a rat cDNA clone encoding a secreted superoxide dismutase reveals the epididymis to be a major site of its expression. *Biochem. J.* 293(Pt. 1), 21–25.
338. DeLap, L. W., Tate, S. S., and Meister, A. (1977). γ-Glutamyl transpeptidase and related enzyme activities in the reproductive system of the male rat. *Life Sci.* 20, 673–680.
339. Kozak, E. M., and Tate, S. S. (1982). Glutathione-degrading enzymes of microvillus membranes. *J. Biol. Chem.* 257, 6322–6327.
340. Kohdaira, T., Kinoshita, Y., Konno, M., and Oshima, H. (1986). Distribution of γ-glutamyl transpeptidase in male reproductive system of rats and its age-related changes. *Andrologia* 18, 610–617.
341. Agrawal, Y. P., Peura, T., and Vanha-Perttula, T. (1989). Distribution of γ-glutamyl transpeptidase in the mouse epididymis and its response to acivicin. *J. Reprod. Fertil.* 86, 185–193.
342. Agrawal, Y. P., and Vanha-Perttula, T. (1989). γ-glutamyl transpeptidase in the rat epididymis: effects of castration, hemicastration and efferent duct ligation. *Int. J. Androl.* 12, 321–328.
343. Hinton, B. T., Palladino, M. A., Mattmueller, D. R., Bard, D., and Good, K. (1991). Expression and activity of gamma-glutamyl transpeptidase in the rat epididymis. *Mol. Reprod. Dev.* 28, 40–46.
344. Palladino, M. A., Laperche, Y., and Hinton, B. T. (1994). Multiple forms of gamma-glutamyl transpeptidase messenger ribonucleic acid are expressed in the adult rat testis and epididymis. *Biol. Reprod.* 50, 320–328.
345. Ghyselinck, N. B., Jimenez, C., Lefrancois, A. M., and Dufaure, J. P. (1990). Molecular cloning of a cDNA for androgen-regulated proteins secreted by the mouse epididymis. *J. Mol. Endocrinol.* 4, 5–12.
346. Perry, A. C., Jones, R., Niang, L. S., Jackson, R. M., and Hall, L. (1992). Genetic evidence for an androgen-regulated epididymal secretory glutathione peroxidase whose transcript does not contain a selenocysteine codon. *Biochem. J.* 285, 863–870.
347. Schwaab, V., Baud, E., Ghyselinck, H., Mattei, M. G., Dufaure, J. P. and Drevet, J. R. (1995). Cloning of the mouse gene encoding plasma glutathione peroxidase: organization. *Gene* 167, 25–31.
348. Vernet, P., Faure, J., Dufaure, J. P., and Drevet, J. R. (1997). Tissue and developmental distribution dependence upon testicular factors and attachment to spermatozoa of GPX5 a murine epididymis-specific glutathione peroxidase. *Mol. Reprod. Dev.* 47, 87–98.
349. Rejraji, H., Vernet, P., and Drevet, J. R. (2002). GPX5 is present in the mouse caput and cauda epididymidis lumen at three different locations. *Mol. Reprod. Dev.* 63, 96–103.
350. Robaire, B., and Hales, B. F. (1982). Regulation of epididymal glutathione S-transferases: effects of orchidectomy and androgen replacement. *Biol. Reprod.* 26, 559–565.
351. Papp, S., Robaire, B., and Hermo, L. (1995). Immunocytochemical localization of the Ya, Yc, Yb1, and Yb2 subunits of glutathione S-transferases in the testis and epididymis of adult rats. *Microsc. Res. Tech.* 30, 1–23.
352. Gandy, J., Primiano, T., Novak, R. F., Kelce, W. R., and York, J. L. (1996). Differential expression of glutathione S-transferase isoforms in compartments of the testis and segments of the epididymis of the rat. *Drug Metab. Dispos.* 7, 725–733.
353. Jervis, K. M., and Robaire, B. (2001). Dynamic changes in gene expression along the rat epididymis. *Biol. Reprod.* 65, 696–703.
354. Montiel, E. E., Huidobro, C. C., and Castellon, E. A. (2003). Glutathione-related enzymes in cell cultures from different regions of human epididymis. *Arch. Androl.* 49, 95–105.

355. Yoshida, R., Nukiwa, T., Watanabe, Y., Fujiwara, M., Hirata, F., and Hayaishi, O. (1980). Regulation of indolamine 2,3-dioxygenase activity in the small intestine and the epididymis of mice. *Arch. Biochem. Biophys.* 203, 343–351.
356. Clulow, J., Jones, R. C., Hansen, L. A., and Man, S. Y. (1998). Fluid and electrolyte reabsorption in the ductuli efferentes testis. *J. Reprod. Fertil. Suppl.* 53, 1–14.
357. Leung, P. S., Chan, H. C., Chung, Y. W., Wong, T. P., and Wong, P. Y. D. (1998). The role of local angiotensins and prostaglandins in the control of anion secretion by the rat epididymis. *J. Reprod. Fertil. Suppl.* 53, 15–22.
358. Wong, P. Y. D., Gong, X. D., Leung, G. P. H., and Cheuk, B. L. Y. (2002). Formation of the epididymal fluid microenvironment. In *The Epididymis: From Molecules to Clinical Practice* (B. Robaire and B. T. Hinton, Eds.), pp. 119–130. Kluwer Academic/Plenum, New York.
359. Turner, T. T. (2002). Necessity's potion: inorganic ions and small organic molecules in the epididymal lumen. In *The Epididymis: From Molecules to Clinical Practice* (B. Robaire and B. T. Hinton, Eds.), pp. 131–150. Kluwer Academic/ Plenum, New York.
360. Breton, S. (2003). Luminal acidification in the epididymis and vas deferens. In *Third International Conference on the Epididymis* (B. T. Hinton and T. T. Turner, Eds.), pp. 60–72. Van Doren, Charlottesville, VA.
361. Rodriguez, C. M., and Hinton, B. T. (2003). The testicular and epididymal luminal fluid microenvironment. In *Introduction to Mammalian Reproduction* (D. Tulsiani, Ed.), pp. 61–77. Kluwer Academic, Norwell, MA.
362. Clulow, J., Jones, R. C., and Hansen, L. A. (1994). Micropuncture and cannulation studies of fluid composition and transport in the ductuli efferentes testis of the rat: comparisons with the homologous metanephric proximal tubule. *Exp. Physiol.* 79, 915–928.
363. Jones, R. C., and Clulow, J. (1987). Regulation of the elemental composition of the epididymal fluids in the tammar. *Macropus eugenii*. *J. Reprod. Fertil.* 81, 583–590.
364. Jones, R. C. (1980). Luminal composition and maturation of spermatozoa in the genital ducts of the African elephant. *Loxodonta africana*. *J. Reprod. Fertil.* 60, 87–93.
365. Tao, L., Zupp, J. L., and Setchell, B. P. (2000). Effect of efferent duct ligation on the function of the blood–testis barrier in rats. *J. Reprod. Fertil.* 120, 13–18.
366. Hinton, B. T., and Setchell, B. P. (1978). Fluid movement in the seminiferous tubules and the epididymal duct of the rat [Proceedings]. *J. Physiol. (Lond.)* 284, 16P-17P.
367. Hinton, B. T., White, R. W., and Setchell, B. P. (1980). The concentration of free myo-inositol in the luminal fluid of the mammalian testis and epididymis. *J. Reprod. Fertil.* 58, 395–399.
368. Hinton, B. T., and Turner, T. T. (1988). Is the epididymis a kidney analog? *News Phys. Sci.* 3, 28–31.
369. Levine, N., and Marsh, D. J. (1971). Micropuncture studies of the electrochemical aspects of fluid and electrolyte transport in individual seminiferous tubules, the epididymis and vas deferens in rats. *J. Physiol. (Lond.)* 213, 557–570.
370. Jessee, S. J., and Howards, S. S. (1976). A survey of sperm, potassium, and sodium concentrations in the tubular fluid of the hamster epididymis. *Biol. Reprod.* 15, 626–631.
371. Jenkins, A. D., Lechene, C. P., and Howards, S. S. (1980). Concentrations of seven elements in the intraluminal fluids of the rat seminiferous tubules, rate testis, and epididymis. *Biol. Reprod.* 23, 981–987.
372. Hinton, B. T., and Setchell. B. P. (1993). Fluid secretion and movement. In *The Sertoli Cell* (L. D. Russell and M. D. Groswold, Eds.), pp. 249–267. Cache River Press, Clearwater, FL.
373. Turner, T. T. (1984). Resorption versus secretion in the rat epididymis. *J. Reprod. Fertil.* 72, 509–514.
374. Morton, B., Harrigan-Lum, J., Albagi, L., and Jooss, T. (1974). The activation of motility in quiescent hamster sperm from the epididymis by calcium and cyclic nucleotides. *Biochem. Biophys. Res. Commun.* 56, 372–379.
375. Turner, T. T., and Howards, S. S. (1978). Factors involved in the initiation of sperm motility. *Biol. Reprod.* 18, 571–578.
376. Pholpramool, C., and Chaturapanich, G. (1979). Effect of sodium and potassium concentrations and pH in the maintenance of motility of rabbit and rat epididymal spermatozoa. *J. Reprod. Fertil.* 57, 245–251.
377. Wong, P. Y. D., Lee, W. M., and Tsang, A. Y. F. (1981). The effects of sodium and amiloride on the motility of caudal epididymal spermatozoa of the rat. *Experientia* 37, 69–71.
377a. Yeung, C. H., Anapolski, M., Setiawan, I., and Cooper, T. G. (2004). Effects of putative epididymal osmolytes on sperm volume regulation of fertile and infertile c-ros transgenic mice. *J. Androl.* 25, 216–223.
378. Marquis, N. R., and Fritz, I. B. (1965). Effects of testosterone on the distribution of carnitine, acetylcarnitine, and carnitine acetyltransferase in tissues of the reproductive system of the male rat. *J. Biol. Chem.* 240, 2197–2200.
379. Casillas, E. R. (1972). The distribution of carnitine in male reproductive tissues and its effect on palmitate oxidation by spermatozoal particles. *Biochim. Biophys. Acta* 280, 545–551.
380. Brooks, D. E., Hamilton, D. W., and Mallek, A. H. (1974). Carnitine and glycerylphosphorylcholine in the reproductive tract of the male rat. *J. Reprod. Fertil.* 36, 141–160.
381. Hinton, B. T., Pryor, J. P., Hirsh, A. V., and Setchell, B. P. (1981). The concentration of some inorganic ions and organic compounds in the luminal fluid of the human ductus deferens. *Int. J. Androl.* 4, 457–461.
382. Casillas, E. R., Villalobos, P., and Gonzales, R. (1984). Distribution of carnitine and acetylcarnitine in the hamster epididymis and in epididymal spermatozoa during maturation. *J. Reprod. Fertil.* 72, 197–201.
383. Brooks, D. E. (1980). Carnitine in the male reproductive tract and its relation to the metabolism of the epididymis and spermatozoa. In *Carnitine Biosynthesis, Metabolism and Functions* (R. A. Frenkel and J. D. McGary, Eds.), pp. 219–235. Academic Press, New York.
384. Hinton, B. T., and Setchell, B. P. (1980). Concentration and uptake of carnitine in the rat epididymis: a micropuncture study. In *Carnitine Biosynthesis, Metabolism and Functions* (R. A. Frenkel and J. D. McGary, Eds.), pp. 237–250. Academic Press, New York.
385. Yeung, C. H., Cooper, T. G., and Waites, G. M. H. (1980). Carnitine transport into the perfused epididymis of the rat: regional differences, stereospecificity, stimulation by choline, and the effect of other luminal factors. *Biol. Reprod.* 23, 294–304.
386. James, M. J., Brooks, D. E., and Snoswell, A. M. (1981). Kinetics of carnitine uptake by rat epididymal cells: androgen-dependence and lack of stereospecificity. *FEBS Lett.* 126, 53–56.
387. Cooper, T. G., Gudermann, T. W., and Yeung, C-H. (1986). Characteristics of the transport of carnitine into the cauda epididymidis of the rat as ascertained by luminal perfusion in vitro. *Int. J. Androl.* 9, 348–358.
388. Cooper, T. G., Yeung, C-H., and Weinbauer, G. F. (1986). Transport of carnitine by the epididymis of the cynomolgus macaque (*Macaca fasicularis*). *J. Reprod. Fertil.* 77, 297–301.
389. Brooks, D. E., Hamilton, D. W., and Mallek, A. H. (1975). The uptake of L-[methyl^{3}H] carnitine by the rat epididymis. *Biochem. Biophys. Res. Commun.* 52, 1354–1360.
390. Bohmer, T., and Hansson, V. (1975). Androgen-dependent accumulation of carnitine by rat epididymis after injection

of [^{3}H]butyrobetaine in vivo. *Mol. Cell. Endocrinol.* 3, 103–115.
391. Bohmer, T. (1978). Accumulation of carnitine in rat epididymis after injection of [^{3}H] butyrobetaine in vivo: quantitative aspects and the effects of androgens and antiandrogens. *Mol. Cell. Endocrinol.* 11, 213–223.
392. Rodriguez, C. M., Labus, J. C., and Hinton, B. T. (2002). Organic cation/carnitine transporter, OCTN2, is differentially expressed in the adult rat epididymis. *Biol. Reprod.* 67, 314–319.
393. Enomoto, A., Wempe, M. F., Tsuchida, H., Shin, H. J., Cha, S. H., Anzai, N., Goto, A., Sakamoto, A., Niwa, T., Kanai, Y., Anders, M. W., and Endou, H. (2002). Molecular identification of a novel carnitine transporter specific to human testis. Insights into the mechanism of carnitine recognition. *J. Biol. Chem.* 277, 36262–36271.
394. Cooper, T. G., Yeung, C-H., Wagenfeld, A., Nieschlag, E., Poutanen, M., Huhtaniemi, I., and Sipila, P. (2004). Mouse models of infertility due to swollen spermatozoa. *Mol. Cell. Endocrinol.* 216, 55–63.
395. Crichton, E. G., Hinton, B. T., Pallone, T. L., and Hammerstedt, R. H. (1994). Hyperosmolality and sperm storage in hibernating bast: prolongation of sperm life by dehydration. *Am. J. Physiol.* 267, 1363–1370.
396. Hinton, B. T. (1990). The testicular and epididymal luminal amino acid microenvironment in the rat. *J. Androl.* 11, 498–505.
397. Hinton, B. T., Palladino, M. A., Rudolph, D., and Labus, J. C. (1995). The epididymis as protector of maturing spermatozoa. *Reprod. Fertil. Dev.* 7, 731–745.
398. Brooks, D. E. (1979). Carbohydrate metabolism in the rat epididymis: Evidence that glucose is taken up by tissue slices and isolated cells by a process of facilitated diffusion. *Biol. Reprod.* 21, 19–26.
399. Cooper, T. G., and Waites, G. M. H. (1979). Investigation by luminal perfusion of the transfer of compounds into the epididymis of the anaesthetized rat. *J. Reprod. Fertil.* 56, 159–164.
400. Turner, T. T., D'Addario, D. A., and Howards, S. S. (1980). [^{3}H]-3-O-methyl-D-glucose transport from blood to the lumina of the seminiferous and epididymal tubules in intact and vasectomized hamsters. *J. Reprod. Fertil.* 60, 285–289.
401. Cooper, T. G. (1982). Secretion of inositol and glucose by the perfused rat cauda epididymis. *J. Reprod. Fertil.* 64, 373–379.
402. Hinton, B. T., and Howards, S. S. (1982). Rat testis and epididymis can transport [^{3}H] 3-O-methyl-D-glucose, [^{3}H] inositol and [^{3}H] α-aminoisobutyric acid across its epithelia in vivo. *Biol. Reprod.* 27, 1181–1189.
403. Hinton, B. T., Hernandez, H., and Howards, S. S. (1983). The male antifertility agents alpha chlorhydrin, 5-thio-D-glucose and 6-chloro-6-deoxy-D-glucose interfere with sugar transport across the epithelium of the rat caput epididymidis. *J. Androl.* 4, 216–221.
404. Schurmann, A., Axer, H., Scheepers, A., Doege, H., and Joost, H. G. (2002). The glucose transport facilitator GLUT8 is predominantly associated with the acrosomal region of mature spermatozoa. *Cell. Tissue Res.* 307, 237–242.
405. Ganjam, V. K., and Amann, R. P. (1976). Steroids in fluids and sperm entering and leaving the bovine epididymis, epididymal tissue, and accessory sex gland secretions. *Endocrinology* 99, 1618–1630.
406. Ganjam, V. K., and Amann, R. P. (1973). Testosterone and dihydrotestosterone concentrations in the fluid milieu of spermatozoa in the reproductive tract of the bull. *Acta Endocrinol. (Copenh.)* 74, 186–200.
407. Turner, T. T., Jones, C. E., Howards, S. S., Ewing, L. L., Zegeye, B., and Gunsalus, G. L. (1984). On the androgen microenvironment of maturing spermatozoa. *Endocrinology* 115, 1925–1932.
408. Scheer, H., and Robaire, B. (1980). Steroid Δ^4-5α-reductase and 3α-hydroxysteroid dehydrogenase in the rat epididymis during development. *Endocrinology* 107, 948–953.
409. Dacheux, J-L., Gatti, J-L., Castella, S., Metayer, S., Fouchecourt, S., and Dacheux, F. (2003). The epididymal proteome. In *Third International Conference on the Epididymis* (B. T. Hinton and T. T. Turner, Eds.), pp. 115–122. Van Doren, Charlottesville, VA.
410. Dacheux, J. L., and Voglmayr, J. K. (1983). Sequence of sperm cell surface differentiation and its relationship to exogenous fluid proteins in the ram epididymis. *Biol. Reprod.* 29, 1033–1046.
411. Hinton, B. T., Olson, G. E., and Good, K. (1987). A novel technique for studying the in vivo secretion of epididymal proteins. *Ann. N. Y. Acad. Sci.* 513, 559.
412. Mattmueller, D. R., and Hinton, B. T. (1991). In vivo secretion and association of clusterin (SGP-2) in luminal fluid with spermatozoa in the rat testis and epididymis. *Mol. Reprod. Dev.* 30, 62–69.
413. Turner, T. T., Avery, E. A., and Sawchuk, T. J. (1994). Assessment of protein synthesis and secretion by rat seminiferous and epididymal tubules in vivo. *Int. J. Androl.* 17, 205–213.
414. Turner, T. T., Riley, T. A., Vagnetti, M., Flickinger, C. J., Caldwell, J. A., and Hunt, D. F. (2000). Postvasectomy alterations in protein synthesis and secretion in the rat caput epididymidis are not repaired after vasovasostomy. *J. Androl.* 21, 276–290.
415. Butenandt, A. (1931). Uber die chemisch Untersuchung der Sexualhormone. *Z. Angnew. Chem.* 44, 905–908.
416. Brooks, D. E. (1976). Control of glycolytic enzymes by androgens in the rat epididymis. *J. Endocrinol.* 71, 355–365.
417. Robaire, B., Ewing L. L., Zirkin, B. R., and Irby, D. C. (1977). Steroid Δ^4-5α-reductase and 3α-hydroxysteroid dehydrogenase in the rat epididymis. *Endocrinology* 101, 1379–1390.
418. Brooks, D. E. (1979). Influence of androgens on the weights of the male accessory reproductive organs and on the activities of mitochondrial enzymes in the epididymis of the rat. *J. Endocrinol.* 82, 293–303.
419. White, W. E. (1932). The effect of hypophysectomy on the survival of spermatozoa in the male rat. *Anat. Rec.* 54, 253–273.
420. Delongeas, J., Gelly, J., Leheup, B., and Grignon, G. (1987). Influence of testicular secretions on differentiation of the rat epididymis: ultrastructural studies after castration, efferent duct ligation and cryptorchidism. *Exp. Cell. Biol.* 55, 74–82.
421. Moore, H. D. M., and Bedford, J. M. (1979). The differential absorptive activity of epithelial cells of the rat epididymis before and after castration. *Anat. Rec.* 193, 313–328.
422. Bartsch, G., Oberholzer, M., Hollinger, O., Weber, J., Weber, A., and Rohr, P. R. (1978). Stereology: a new and quantitative morphological method to study epididymal function. *Andrologia* 10, 31–42.
423. Fawcett, D. W., and Hoffer, A. P. (1979). Failure of exogenous androgen to prevent regression of the initial segments of rat epididymis after efferent duct ligation or orchidectomy. *Biol. Reprod.* 20, 162–181.
424. De Larminat, M., Monsalve, A., Charreau, E., Calandra, R., and Blaquier, J. (1978). Hormonal regulation of 5α-reductase activity in rat epididymis. *J. Endocrinol.* 79, 157–165.
425. Pujol, A., and Bayard, F. (1979). Androgen receptors in the rat epididymis and their hormonal control. *J. Reprod. Fertil.* 56, 217–222.
426. Zhu, L. J., Hardy, M. P., Inigo, I. V., Huhtaniemi, I., Bardin, C. W., and Moo-Young, A. J. (2000). Effects of androgen on androgen receptor expression in rat testicular and epididymal cells: a quantitative immunohistochemical study. *Biol. Reprod.* 63, 368–376.

427. Brooks, D. E. (1977). The androgenic control of the composition of the rat epididymis determined by efferent duct ligation or castration. *J. Reprod. Fertil.* 49, 383–385.
428. Moniem, K., Glover, T., and Lubicz-Nawrocki, C. (1978). Effects of duct ligation and orchidectomy on histochemical reactions in the hamster epididymis. *J. Reprod. Fertil.* 54, 173–176.
429. Ruiz-Bravo, N. (1988). Tissue and cell specificity of immobilin biosynthesis. *Biol. Reprod.* 39, 901–911.
430. Holland, M. K., Vreeburg, J. T., and Orgebin-Crist, M.-C. (1992). Testicular regulation of epididymal protein secretion. *J. Androl.* 13, 266–273.
431. Schwaab, V., Lareyre, J. J., Vernet, P., Pons, E., Faure, J., Dufaure, J. P., and Drevet, J. R. (1998). Characterization, regulation of the expression and putative roles of two glutathione peroxidase proteins found in the mouse epididymis. *J. Reprod. Fertil. Suppl.* 53, 157–162.
432. Palladino, M. A., and Hinton, B. T. (1994). Expression of multiple gamma-glutamyl transpeptidase messenger ribonucleic acid transcripts in the adult rat epididymis is differentially regulated by androgens and testicular factors in a region-specific manner. *Endocrinology* 135, 1146–1156.
433. Niemi, M., and Tuohimaa, P. (1971). The mitogenic activity of testosterone in the accessory sex glands of the rat in relation to its conversion to dihydrotestosterone. In *Basic Actions of Sex Steroids on Target Organs* (P. O. Hubinot and F. Leroy, Eds.), pp. 258–264. S. Karger, Basel.
434. Gregory, M., Xiao, Q., Cornwall, G., Lutterbach, B., and Hann, S. (2000). B-myc is preferentially expressed in hormonally-controlled tissues and inhibits cellular proliferation. *Oncogene* 19, 4886–4895.
435. Fan, X. P., and Robaire, B. (1998). Orchiectomy induces a wave of apoptotic cell death in the epididymis. *Endocrinology* 139, 2128–2136.
436. Turner, T. T., and Riley, T. A. (1999) p53 independent, region-specific epithelial apoptosis in induced in the rat epididymis by deprivation of luminal factors. *Mol. Reprod. Dev.* 53, 188–197.
437. Jara, M., Esponda, P., Carballada, R. (2002). Abdominal temperature induces region-specific p53-independent apoptosis in the cauda epididymidis of the mouse. *Biol. Reprod.* 67, 1189–1196.
438. Suzuki, A., Matsuzawa, A., and Iguchi, T. (1996). Down regulation of Bcl-2 is the first step on Fas-mediated apoptosis of male reproductive tract. *Oncogene* 13, 31–37.
439. Grayhack, J. T. (1965). Effect of testosterone-estradiol administration on citric acid and fructose content of rat prostate. *Endocrinology* 76, 1168–1174.
440. Steinberger,E., and Duckett, G. E. (1965). The effect of estrogen or testosterone on the initiation and maintenance of spermatogenesis in the rat. *Endocrinology* 76, 1184–1189.
441. Gay, V. L., and Dever, N. W. (1971). Effects of testosterone propionate and estradiol benzoate—alone or in combination—on serum LH and FSH in orchidectomized rats. *Endocrinology* 89, 161–168.
442. Oshima, H., Wakabayashi, K., and Tamaoki, I. (1967). The effect of synthetic estrogen on the biosynthesis in vitro of androgen and LH in the rat. *Biochim. Biophys. Acta* 137, 356–366.
443. Andersson, M., and Muntzing, J. (1972). The effect of a long-acting estrogen on the activity and distribution of some hydrolases in the ventral prostate of intact, castrated, and androgen treated castrated adult rats. *Invest. Urol.* 9, 401–407.
444. Swerdloff, R. S., and Walsh, P. C. (1973). Testosterone and oestradiol suppression of LH and FSH in adult male rats: duration of castration, duration of treatment and combined treatment. *Acta Endocrinol.* 73, 11–21.
445. Chowdhury, M., Tcholakian, R., and Steinberger, E. (1974). An unexpected effect of oestradiol-17 on LH and testosterone. *J. Endocrinol.* 60, 375–376.
446. Karr, J. P., Kirdani, R. Y., Murphy, G. P., and Sandberg, A. A. (1974). Effects of testosterone and estradiol on ventral prostate and body weights of castrated rats. *Life Sci.* 15, 501–513.
447. Verjans, H. L., DeJong, F. H., Cooke, B. A., Van der Molen, H. J., and Eik-Nes, K. B. (1974). Effect of oestradiol benzoate on pituitary and testis function in the normal and adult male rat. *Acta Endocrinol.* 77, 636–642.
448. Ewing, L. L., Desjardins, C., Irby, D. C., and Robaire, B. (1977). Synergistic interaction of testosterone and estradiol inhibits spermatogenesis in rats. *Nature* 269, 409–411.
449. Orgebin-Crist, M.-C., Eller, B. C., and Danzo, B. J. (1983). The effects of estradiol, tamoxifen, and testosterone on the weights and histology of the epididymis and accessory sex organs of sexually immature rabbits. *Endocrinology* 113, 1703–1715.
450. Nicander, L., Osman, D. I., Ploen, L., Bugge, H. P., and Kvisgaard, K. N. (1983). Early effects of efferent ductile ligation on the proximal segment of the rat epididymis. *Int. J. Androl.* 6, 91–102.
451. Hinton, B. T., Lan, Z.-J., Rudolph, D. B., Labus, J. C., and Lye, R. J. (1998). Testicular regulation of epididymal gene expression. *J. Reprod. Fertil. Suppl.* 53, 47–57.
452. Hinton, B. T., Lan, Z.-J., Lye, R. J., and Labus, J. C. (2000). Regulation of epididymal function by testicular factors: The lumicrine hypothesis. In *The Testis: From Stem Cell to Sperm Function* (E. Goldberg, Ed.), pp. 163–173. Serono Symposia USA. Springer, New York.
453. Skinner, J. D., and Rowson, L. E. A. (1967). Effect of unilateral cryptorchidism on sexual development in the pubescent male animal. *J. Reprod. Fertil.* 14, 349–350.
454. Skinner, J. D., and Rowson, L. E. A. (1968). Some effects of unilateral cryptorchidism and vasectomy on sexual development of the pubescent ram and bull. *J. Reprod. Fertil.* 42, 311–321.
455. Robaire, B., Scheer, H., and Hachey, C. (1981). Regulation of epididymal steroid metabolizing enzymes. In *Bioregulators of Reproduction* (G. Jagiello and H. J. Vogel, Eds.), pp. 487–498. Academic Press, New York.
456. Brown, D. V., Amann, R. P., and Wagley, L. M. (1983). Influence of rete testis fluid on the metabolism of testosterone by cultured principal cells isolated from the proximal or distal caput of the rat epididymis. *Biol. Reprod.* 28, 1257–1268.
457. Turner, T. T., Miller, D. W., and Avery, E. A. (1995). Protein synthesis and secretion by the rat caput epididymidis in vivo: influence of the luminal microenvironment. *Biol. Reprod.* 52, 1012–1019.
458. Lan, Z.-J., Labus, J. C., and Hinton, B. T. (1998). Regulation of gamma-glutamyl transpeptidase catalytic activity and protein level in the initial segment of the rat epididymis by testicular factors: role of basic fibroblast growth factor. *Biol. Reprod.* 58, 197–206.
459. Kirby, J. L., Yang, L., Labus, J. C., and Hinton, B. T. (2003). Characterization of fibroblast growth factor receptors expressed in principal cells in the initial segment of the rat epididymis. *Biol. Reprod.* 68, 2314–2321.
460. Lan, Z.-J., Palladino, M. A., Rudolph, D. B., Labus, J. C., and Hinton, B. T. (1997). Identification, expression and regulation of the transcription factor polyomavirus enhancer activator 3 and its putative role in regulating the expression of gamma-glutamyltranspeptidase mRNA-IV in the rat epididymis. *Biol. Reprod.* 57, 186–193.
461. Seenundun, S., and Robaire, B. (2005). Cloning and characterization of the 5α-reductase type 2 promoter in the rat epididymis. *Biol. Reprod.* 72, 851–861.

462. Garrett, J. E., Garrett, S. H., and Douglass, J. A. (1990). A spermatozoa-associated factor regulates proenkephalin gene expression in the rat epididymis. *Mol. Endocrinol.* 4, 108–118.

463. Cancilla, B., Davies, A., Ford-Perriss, M., and Risbridger, G. P. (2000). Discrete cell- and stage-specific localization of fibroblast growth factors and receptor expression during testis development. *J. Endocrinol.* 164, 149–159.

464. Jervis, K. M., and Robaire B. (2002). Changes in gene expression during aging in the brown Norway rat epididymis. *Exp. Gerontol.* 37, 897–906

465. Jervis, K. M., and Robaire, B. (2003). Effects of caloric restriction on gene expression along the epididymis of the brown Norway rat during aging. *Exp. Gerontol.* 38, 549–560.

466. Cornwall, G. A., and Hann, S. R. (1995). Specialized gene expression in the epididymis. *J. Androl.* 16, 379–383.

467. Cooper, T. G., Wagenfeld, A., Cornwall, G. A., Hsia, N., Chu, S. T., Orgebin-Crist, M.-C., Drevet, J., Vernet, P., Avram, C., Nieschlag, E., and Yeung, C. H. (2003). Gene and protein expression in the epididymis of infertile c-ros receptor tyrosine kinase-deficient mice. *Biol. Reprod.* 69, 1750–1762.

468. Chauvin, T. R., and Griswold, M. D. (2004). Androgen-regulated genes in the murine epididymis. *Biol. Reprod.* 71, 560–569.

468a. Hsia, N., and Cornwall, G. A. (2004) DNA microarray analysis of region-specific gene expression in the mouse epididymis. *Biol. Reprod.* 70, 448–457.

469. Gatti, J. L., Metayer, S., Belghazi, M., Dacheux, F., and Dacheux, J. L. (2005). Identification, proteomic profiling, and origin of ram epididymal fluid exosome-like vesicles. *Biol. Reprod.* 72, 1452–1465.

470. Chaurand, P., Fouchecourt, S., DaGue, B. B., Xu, B. J., Reyzer, M. L., Orgebin-Crist, M.-C., and Caprioli, R. M. (2003). Profiling and imaging proteins in the mouse epididymis by imaging mass spectrometry. *Proteomics* 3, 2221–2239.

471. Umar, A., Ooms, M. P., Luider, T. M., Grootegoed, J. A., and Brinkmann, A. O. (2003). Proteomic profiling of epididymis and vas deferens: identification of proteins regulated during rat genital tract development. *Endocrinology* 144, 4637–4647.

472. Palladino, M. A., Powell, J. D., Korah, N., and Hermo, L. (2004). Expression and localization of hypoxia-inducible factor-1 subunits in the adult rat epididymis. *Biol. Reprod.* 70, 1121–1130.

473. Jensen, L. J., Schmitt, B. M., Berger, U. V., Nsumu, N. N., Boron, W. F., Hediger, M. A., Brown, D., and Breton, S. (1999). Localization of sodium bicarbonate cotransporter (NBC) protein and messenger ribonucleic acid in rat epididymis. *Biol. Reprod.* 60, 573–579.

474. Luedtke, C. C., Andonian, S., Igdoura, S., and Hermo, L. (2000). Cathepsin A is expressed in a cell- and region-specific manner in the testis and epididymis and is not regulated by testicular or pituitary factors. *J. Histochem. Cytochem.* 48, 1131–1146.

475. Blanchard Y., and Robaire B. (1997). Le mode d'action des androgènes et la 5α-réductase. *Med. Sci.* 13, 467–473.

476. Blaquier, J. A. (1971). Selective uptake and metabolism of androgens by rat epididymis. The presence of a cytoplasmic receptor. *Biochem. Biophys. Res. Commun.* 45, 1076–1082.

476a. Ritzen, E., Nayfeh, S., French, F., and Dobbins, M. (1971). Demonstration of androgen-binding components in rat epididymal cytosol and comparison with binding components in prostate and other tissues. *Endocrinology* 89, 143–151.

476b. Carreau, S., Drosdowsky, M. A., and Courot, M. (1984). Androgen-binding proteins in sheep epididymis: characterization of a cytoplasmic androgen receptor in the ram epididymis. *J. Endocrinol.* 103, 273–279.

476c. Zhang, T., Guo, C. X., Hu, Z. Y., and Liu, Y. X. (1997). Localization of plasminogen activator and inhibitor, LH and androgen receptors and inhibin subunits in monkey epididymis. *Mol. Hum. Reprod.* 3, 945–952.

476d. Ungefroren, H., Ivell, R., and Ergun, S. (1997). Region-specific expression of the androgen receptor in the human epididymis. *Mol. Hum. Reprod.* 3, 933–940.

476e. Hansson, V., Djoseland, O., Reusch, E., Attramadal, A., and Torgensen, O. (1973). Intracellular receptor for 5α-dihydrotestosterone in the epididymis of the adult rats: comparison with the androgenic receptor in the ventral prostate and the androgen binding protein (ABP) in the testicular and epididymal fluid. *Steroids* 22, 19–33.

476f. Danzo, B. J., Orgebin-Crist, M.-C., and Toft, D. O. (1973). Characterization of a cytoplasmic receptor for 5α-dihydrotestosterone in the caput epididymidis of intact rabbits. *Endocrinology* 92, 310–317.

476g. Younes, M., Evans, B. A. J., Chaisiri, N., Valotaire, Y., Pierrepoint, C. G. (1979). Steroid receptors in the canine epididymis. *J. Reprod. Fertil.* 56, 45–52.

476h. Roselli, C. E., West, N. B., and Brenner, R. M. (1991). Androgen receptor and 5α-reductase activity in the ductuli efferentes and epididymis of adult rhesus macaques. *Biol. Reprod.* 44, 739–745.

477. Brown, C., Goss, S., Lubahn, D., Joseph, D., Wilson, E., Frech, F., and Willard, H. F. (1989). Androgen receptor locus on the human X chromosome: regional localization to Xq11-12 and description of a DNA polymorphism. *Am. J. Hum. Genet.* 44, 264–269.

478. Zhou, Z. X., Wong, C. I., Sar, M., and Wilson, E. M. (1994). The androgen receptor: an overview. *Recent Prog. Horm. Res.* 49, 249–274.

479. Ahrens-Fath, I., Politz, O., Geserick, C., and Haendler, B. (2005). Androgen receptor function is modulated by the tissue-specific AR45 variant. *FEBS J.* 72, 74–84.

480. Tindall, D. J., Hansson, V., McLean, W. S., Ritzen, E. M., Nayfeh, S. N., and French, F. S. (1975). Androgen-binding proteins in rat epididymis: properties of a cytoplasmic receptor for androgen similar to the androgen receptor in ventral prostate and different from androgen-binding protein (ABP). *Mol. Cell. Endocrinol.* 3, 83–101.

481. Robaire, B., and Viger, R. S. (1993). Regulation of epididymal epithelial functions. In *Understanding Male Infertility: Basic and Clinical Approaches* (B. R. Zirkin and R. Whitcomb, Eds.), pp. 183–210. Raven Press, New York.

482. Carreau, S., Drosdowsky, M. A., and Courot, M. (1984). Androgen binding proteins in sheep epididymis: age-related effects on androgen-binding protein, cytosolic androgen receptor and testosterone concentrations. Correlations with histological studies. *J. Endocrinol.* 103, 281–286.

483. Cooke, P. S., Young, P., and Cunha, G. R. (1991) Androgen receptor expression in developing male reproductive organs. *Endocrinology* 128, 2867–2873.

484. Goyal, H. O., Bartol, F. F., Wiley, A. A., Khalil, M. K., Chiu, J., and Vig, M. M. (1997). Immunolocalization of androgen receptor and estrogen receptor in the developing testis and excurrent ducts of goats. *Anat. Rec.* 249, 54–62.

485. Zhou, Q., Nie, R., Prins, G. S., Saunders, P. T., Katzenellenbogen, B. S., and Hess, R. A. (2002). Localization of androgen and estrogen receptors in adult male mouse reproductive tract. *J. Androl.* 23, 870–881.

486. Goyal, H. O., Bartol, F. F., Wiley, A. A., Khalil, M. K., Williams, C. S., and Vig, M. M. (1998). Regulation of androgen and estrogen receptors in male excurrent ducts of the goat: an immunohistochemical study. *Anat. Rec.* 250, 164–171.

487. Telgmann, R., Brosens, J. J., Kappler-Hanno, K., Ivell, R., and Kirchhoff, C. (2001). Epididymal epithelium immortalized by simian virus 40 large T antigen: a model to study epididymal gene expression. *Mol. Hum. Reprod.* 7, 935–945.

488. Araki, Y., Suzuki, K., Matusik, R. J., Obinata, M., and Orgebin-Crist, M.-C. (2002). Immortalized epididymal cell lines from transgenic mice overexpressing temperature-sensitive simian virus 40 large T-antigen gene. *J. Androl.* 23, 854–869.
489. Sipila, P., Shariatmadari, R., Huhtaniemi, I. T., and Poutanen, M. (2004). Immortalization of epididymal epithelium in transgenic mice expressing simian virus 40 T antigen: characterization of cell lines and regulation of the polyoma enhancer activator 3. *Endocrinology* 145, 437–446.
490. Lamb, D. J., Weigel, N. L., and Marcelli, M. (2001). Androgen receptors and their biology. *Vitam. Horm.* 62, 199–230.
491. Janne, O. A., Moilanen, A. M., Poukka, H., Rouleau, N., Karvonen, U., Kotaja, N., Hakli, M., and Palvimo, J. J. (2000). Androgen-receptor-interacting nuclear proteins. *Biochem. Soc. Trans.* 28, 401–405.
492. Heinlein, C. A., and Chang, C. (2004). Androgen receptor in prostate cancer [review]. *Endocr. Rev.* 25, 276–308.
493. Gao, N., Zhang, J., Rao, M. A., Case, T. C., Mirosevich, J., Wang, Y., Jin, R., Gupta, A., Rennie, P. S., and Matusik, R. J. (2003). The role of hepatocyte nuclear factor-3 alpha (forkhead box A1) and androgen receptor in transcriptional regulation of prostatic genes. *Mol. Endocrinol.* 17, 1484–1507.
494. Yu, X., Gupta, A., Wang, Y., Suzuki, K., Orgebin-Crist, M.-C., and Matusik, R. (2006). Foxa transcription factor differentially regulates epididymal and prostatic genes. Testis Workshop. *Ann. N. Y. Acad. Sci.* In press.
495. French, F., and Ritzen, E. (1973). A high-affinity androgen-binding protein (ABP) in rat testis: evidence for secretion into efferent duct fluid and absorption by epididymis. *Endocrinology* 93, 88–95.
496. Danzo, B., and Eller, B. (1984). Clearance, metabolic fate and tissue distribution of an injected bolus of photoaffinity-labeled rat androgen binding protein. *Biol. Reprod.* 31, 259–270.
497. Danzo, B., Dunn, J., and Davies J. (1982). The presence of androgen-binding protein in the guinea-pig testis, epididymis and epididymal fluid. *Mol. Cell. Endocrinol.* 28, 513–527.
498. Vigersky, R. A., Loriaux, D. L., Howards, S. S., Hodgen, G. B., Lipsett, M. B., and Chrambach, A. (1976). Androgen binding proteins of testis, epididymis, and plasma in man and monkey. *J. Clin. Invest.* 58, 1061–1068.
499. Hsu, A., and Troen, P. (1978). An androgen binding protein in the testicular cytosol of human testis: comparison with human plasma testosterone-estrogen binding globulin. *J. Clin. Invest.* 61, 1611–1619.
500. Bardin, C. W., Cheng, C. Y., and Musto, N. A. (1988). The Sertoli cell. In *The Physiology of Reproduction* (E. Knobil and J. Neill, Eds.), pp. 933–974. Raven Press, New York.
501. Cheng, C. Y., Gunsalus, G. L., Musto, N. A., and Bardin, C. W. (1984). The heterogeneity of rat androgen-binding protein in serum differs from that in testis and epididymis. *Endocrinology* 114, 1386–1394.
502. Hammond, G. L., and Bocchinfuso, W. P. (1995). Sex hormone-binding globulin/androgen-binding protein: steroid-binding and dimerization domains. *J. Steroid Biochem. Mol. Biol.* 53, 543–552.
503. Purvis, K., and Hansson, V. (1978). Androgens and androgen-binding protein in the rat epididymis. *J. Reprod. Fertil.* 52, 59–63.
504. Gerard, A., Egloff, M., Gerard, H., el Harate, A., Domingo, M., Gueant, J. L., Dang, C. D., and Degrelle, H. (1990). Internalization of human sex steroid-binding protein in the monkey epididymis. *J. Mol. Endocrinol.* 5, 239–251.
505. Hermo, L., Barin, K., and Oko, R. (1998). Androgen binding protein secretion and endocytosis by principal cells in the adult rat epididymis and during postnatal development. *J. Androl.* 19, 527–541.
506. Felden, M., Lea, O., Petrusz, P., Tres, L., Kierszenbaum, A., and French, F. (1981). Androgen binding protein. Purification from rat epididymis, characterization, and immunohistochemical localization. *J. Biol. Chem.* 256, 5170–5175.
507. Pelliniemi. L. J., Dym, M., Gunsalus, G. L., Musto, N. A., Bardin, C. W., and Fawcett, D. W. (1981). Immunocytochemical localization of androgen-binding protein in the male rat reproductive tract. *Endocrinology* 108, 925–931.
508. Gerard, A., Khanfri, J., Gueant, J. L., Fremont, S., Nicolas, J. P., Grignon, G., and Gerard, H. (1988). Electron microscope radioautographic evidence of in vivo androgen-binding protein internalization in the rat epididymis principal cells. *Endocrinology* 122, 1297–1307.
509. Gueant, J. L., Fremont, S., Felden, F., Nicolas, J. P., Gerard, A., Leheup, B., Gerard, H., and Grignon, G. (1991). Evidence that androgen-binding protein endocytosis in vitro is receptor mediated in principal cells of the rat epididymis. *J. Mol. Endocrinol.* 7, 113–122.
510. Felden, F., Leheup, B., Fremont, S., Bouguerne, R., Egloff, M., Nicolas, J. P., Grignon, G., and Gueant, J. L. (1992). The plasma membrane of epididymal epithelial cells has a specific receptor which binds to androgen-binding protein and sex steroid-binding protein. *J. Steroid Biochem. Mol. Biol.* 42, 279–285.
511. Krupenko, S. A., Krupenko, N. I., and Danzo, B. J. (1994). Interaction of sex hormone-binding globulin with plasma membranes from the rat epididymis and other tissues. *J. Steroid Biochem. Mol. Biol.* 51, 115–124.
512. Danzo, B. J., Pavlou, S. N., and Anthony, H. L. (1990). Hormonal regulation of androgen-binding protein in the rat. *Endocrinology* 127, 2829–2838.
513. Hansson, V., Trygstad, O., French, F. S., McLean, W. S., Smith, A. A., Tindall, D. J., Weddington, S. C., Petrusz, P., Nayfeh, S. N., and Ritzen, E. M. (1974). Androgen transport and receptor mechanisms in testis and epididymis. *Nature* 250, 387–391.
514. Dykman, D. D., Cochran, R., Wise, P. M., Barraclough, C. A., Dubin, N. H., and Ewing, L. L. (1981). Temporal effects of testosterone-estradiol polydimethylsiloxane subdermal implants on pituitary, Leydig cell, and germinal epithelium function and daily serum testosterone rhythm in male rats. *Biol. Reprod.* 25, 235–243.
515. Roberts, K. P., and Zirkin, B. R. (1991). Androgen regulation of spermatogenesis in the rat. *Ann. N. Y. Acad. Sci.* 637, 90–106.
516. Coviello, A. D., Bremner, W. J., Matsumoto, A. M., Herbst, K. L., Amory, J. K., Anawalt, B. D., Yan, X., Brown, T. R., Wright, W. W., Zirkin, B. R., and Jarow, J. P. (2004). Intratesticular testosterone concentrations comparable with serum levels are not sufficient to maintain normal sperm production in men receiving a hormonal contraceptive regimen. *J. Androl.* 25, 931–938.
517. Robaire, B., and Viger, R. S. (1995). Regulation of epididymal epithelial functions. *Biol. Reprod.* 52, 226–236.
518. Joseph, D. R., O'Brien, D. A., Sullivan, P. M., Becchis, M., Tsuruta, J. K., and Petrusz, P. (1997). Overexpression of androgen-binding protein/sex hormone-binding globulin in male transgenic mice: tissue distribution and phenotypic disorders. *Biol. Reprod.* 56, 21–32.
519. Tindall, D. J., French, F. S., and Nayfeh, S. N. (1972). Androgen uptake and binding in rat epididymal nuclei, in vivo. *Biochem. Biophys. Res. Commun.* 49, 1391–1397.
520. Inano, H., Machino, A., and Tamaoki, B.-I. (1969). In vitro metabolism of steroid hormones by cell-free homogenates of epididymides of adult rats. *Endocrinology* 84, 997–1003.

521. Gloyna, R. E., and Wilson, J. D. (1969). A comparative study of the conversion of testosterone to 17β-hydroxy-5α-androstan-3-one (dihydrotestosterone) by rat prostate and epididymis. *J. Clin. Endocrinol. Metab.* 29, 970–977.
522. Mahendroo, M. S., and Russell, D. W. (1999). Male and female isoenzymes of steroid 5α-reductase. *Rev. Reprod.* 4, 179–183.
523. Viger, R. S., and Robaire, B. (1996). The mRNAs for the steroid 5α-reductase isozymes, type 1 and type 2, are differentially regulated in the rat epididymis. *J. Androl.* 17, 27–34.
524. Mahendroo, M. S., Porter, A., Russell, D. W., and Word, R. A. (1999). The parturition defect in steroid 5α-reductase type 1 knockout mice is due to impaired cervical ripening. *Mol. Endocrinol.* 13, 981–992.
525. Scheer, H., and Robaire, B. (1983). Subcellular distribution of steroid Δ^4-5α-reductase and 3α-hydroxysteroid dehydrogenase in the rat epididymis during sexual maturation. *Biol. Reprod.* 29, 1–10.
526. Ezer, N., and Robaire, B. (2002). Androgenic regulation of the structure and functions of the epididymis. In *The Epididymis: From Molecules to Clinical Practice* (B. Robaire and B. T. Hinton, Eds.), pp. 297–316. Kluwer Academic/Plenum, New York.
527. Jenkins, E., Hsieh, C.-L., Milatovich, A., Normington, K., Berman, D. M., Francke, U., and Russell, D. W. (1991). Characterization and chromosomal mapping of a human steroid 5α-reductase gene and pseudogene and mapping of the mouse homologue. *Genomics* 11, 1102–1112.
528. Normington, K., and Russell D. W. (1992). Tissue distribution and kinetic characteristics of rat steroid 5α-reductase isozymes. *J. Biol. Chem.* 267, 19548–19554.
529. Brooks, D. E., Means, A. R., Wright, E. J., Singh, S. P., and Tiver, K. K. (1986). Molecular cloning of the cDNA for two major androgen-dependent secretory proteins of 18.5 kilodaltons synthesized by the rat epididymis. *J. Biol. Chem.* 261, 4956–4961.
530. Rutllant, J., and Meyers, S. A. (2001). Posttranslational processing of PH-20 during epididymal sperm maturation in the horse. *Biol. Reprod.* 65, 1324–1331.
531. Savory, J. G., May, D., Reich, T., LaCasse, E. C., Lakins, J., Tenniswood, M., Raymond, Y., Hache, R. J., Sikorska, M., and Lefebvre, Y. A. (1995). 5α-Reductase type 1 is localized to the nuclear membrane. *Mol. Cell. Endocrinol.* 110, 137–147.
532. Viger, R. S., and Robaire, B. (1994). Immunocytochemical localization of 4-ene steroid 5α-reductase type 1 along the rat epididymis during postnatal development. *Endocrinology* 134, 2298–2306.
533. Reyes, E. M., Camacho-Arroyo, I., Nava, G., and Cerbon, M. A. (1997). Differential methylation in steroid 5 alpha-reductase isozyme genes in epididymis, testis, and liver of the adult rat. *J. Androl.* 18, 372–377.
534. Rodriguez-Dorantes, M., Lizano-Soberon, M., Camacho-Arroyo, I., Calzada-Leon, R., Morimoto, S., Tellez-Ascencio, N., and Cerbon, M. A. (2002). Evidence that steroid 5α-reductase isozyme genes are differentially methylated in human lymphocytes. *J. Steroid Biochem. Mol. Biol.* 80, 323–330.
535. Schleicher, G., Drews, U., Stumpf, W. E., and Sar, M. (1984). Differential distribution of dihydrotestosterone and estradiol binding sites in the epididymis of the mouse. An autoradiographic study. *Histochemistry* 81, 139–147.
536. Van Beurden-Lamers, W. M. O., Brinkmann, A. L., Mulder, E., and Van der Molen, H. J. (1974). High-affinity binding of oestradiol-17β by cytosols from testis interstitial tissue, pituitary, adrenal, liver and accessory sex glands of the male rat. *Biochem. J.* 140, 495–502.
537. Kuiper, G. G., Carlsson, B., Grandien, K., Enmark, E., Haggblad, J., Nilsson, S., and Gustafsson, J. A. (1997). Comparison of the ligand binding specificity and transcript tissue distribution of estrogen receptors alpha and beta. *Endocrinology* 138, 863–870.
538. Danzo, B. J., Wolfe, M. S., and Curry, J. B. (1977). The presence of an estradiol binding component in cytosol from immature rat epididymides. *Mol. Cell. Endocrinol.* 6, 271–279.
539. Danzo, B. J., St. Raymond, P. A., and Davies, J. (1981). Hormonally responsive areas of the reproductive system of the male guinea pig: III. Presence of cytoplasmic estrogen receptors. *Biol. Reprod.* 25, 1159–1168.
540. Danzo, B. J., Sutton, W., and Eller, B. C. (1978). Analysis of [H] estradiol binding to nuclei prepared from epididymides of sexually immature intact rabbits. *Mol. Cell. Endocrinol.* 9, 291–301.
541. Danzo, B. J., and Eller, B. C. (1979). The presence of a cytoplasmic estrogen receptor in sexually mature rabbit epididymides: comparison with the estrogen receptor in immature rabbit epididymal cytosol. *Endocrinology* 105, 1128–1134.
542. Danzo, B. J., Eller, B. C., and Hendry, W. J. (1983). Identification of cytoplasmic estrogen receptors in the accessory sex organs of the rabbit and their comparison to the cytoplasmic estrogen receptor in the epididymis. *Mol. Cell. Endocrinol.* 33, 197–209.
543. Hendry, W. J., Eller, B. C., Orgebin-Crist, M.-C., and Danzo, B. J. (1985). Hormonal effects on the estrogen receptor system in the epididymis and accessory sex organs of sexually immature rabbits. *J. Steroid Biochem.* 23, 39–49.
544. Younes, M. A., and Pierrepoint, C. G. (1981). Estrogen steroid–receptor binding in the canine epididymis. *Andrologia* 13, 562–572.
545. Tekpetey, F. R., and Amann, R. P. (1988). Regional and seasonal differences in concentrations of androgen and estrogen receptors in ram epididymal tissue. *Biol. Reprod.* 38, 1051–1060.
546. Kamal, N., Agarwal, A. K., Jehan, Q., and Setty, B. S. (1985). Biological action of estrogen on the epididymis of prepubertal rhesus monkey. *Andrologia* 17, 339–345.
547. West, N. B., and Brenner, R. M. (1990). Estrogen receptor in the ductuli efferentes, epididymis, and testis of rhesus and cynomolgus macaques. *Biol. Reprod.* 42, 533–538.
548. Murphy, J. B., Emmott, R. C., Hicks, L. L., and Walsh, P. C. (1980). Estrogen receptors in the human prostate, seminal vesicle, epididymis, testis, and genital skin: a marker for estrogen-responsive tissues. *J. Clin. Endocrinol. Metab.* 50, 938–948.
549. Toney, T. W., and Danzo, B. J. (1988). Developmental changes in and hormonal regulation of estrogen and androgen receptors present in the rabbit epididymis. *Biol. Reprod.* 39, 818–828.
550. Hendry, W. J. 3rd, and Danzo, B. J. (1986). Further characterization of a steroid receptor-active protease from the mature rabbit epididymis. *J. Steroid Biochem.* 25, 433–443.
551. Hess, R. A., Gist, D. H., Bunick, D., Lubagn, D. B., Farrell, A., Bahr, J., Cooke, P. S., and Green, G. L. (1997). Estrogen receptor (α and β) expression in the excurrent ducts of the adult male rat reproductive tract. *J. Androl.* 18, 602–611.
552. Danzo, B. J., Eller, B. C., Judy, L. A., Trautman, J. R., and Orgebin-Crist, M.-C. (1975). Estradiol binding in cytosol from epididymides of immature rabbits. *Mol. Cell. Endocrinol.* 2, 91–105.
553. Hess, R. A. (2002). The efferent ductules: structure and functions. In *The Epididymis: From Molecules to Clinical Practice* (B. Robaire and B. T. Hinton, Eds.), pp. 49–80. Kluwer Academic/Plenum, New York.
554. Kuiper, G. G., Enmark, E., Pelto-Huikko, M., Nilsson, S., and Gustafsson, J. A. (1996). Cloning of a novel receptor expressed in rat prostate and ovary. *Proc. Natl. Acad. Sci. U S A* 93, 5925–5930.

555. Tremblay, G. B., Tremblay, A., Copeland, N. G., Gilbert, D. J., Jenkins, N. A., Labrie, F., and Giguere, V. (1997). Cloning, chromosomal localization, and functional analysis of the murine estrogen receptor beta. *Mol. Endocrinol.* 11, 353–365.
556. Harris, H. A., Bapat, A. R., Gonder, D. S., and Frail, D. E. (2002). The ligand binding profiles of estrogen receptors alpha and beta are species dependent. *Steroids* 67, 379–384.
557. Gustafsson, J. A. (2003). What pharmacologists can learn from recent advances in estrogen signalling. *Trends Pharmacol. Sci.* 24, 479–485.
558. Paech, K., Webb, P., Kuiper G. G., Nilsson, S., Gustafsson, J., Kushner, P. J., and Scanlan, T. S. (1997). Differential ligand activation of estrogen receptors ERα and ERβ at AP1 sites. *Science* 277, 1508–1510.
559. Loven, M. A., Wood, J. R., and Nardulli, A. M. (2001). Interaction of estrogen receptors alpha and beta with estrogen response elements. *Mol. Cell. Endocrinol.* 181, 151–163.
560. Iguchi, T., Uesugi, Y., Sato, T., Ohta, Y., and Takasugi, N. (1991). Developmental pattern of estrogen receptor expression in male mouse genital organs. *Mol. Androl.* 6, 109–119.
561. Yamashita, S. (2004). Localization of estrogen and androgen receptors in male reproductive tissues of mice and rats. *Anat. Rec. A Discov. Mol. Cell. Evol. Biol.* 279, 768–778.
562. Oliveira, C. A., Nie, R., Carnes, K., Franca, L. R., Prins, G. S., Saunders, P. T., and Hess, R. A. (2003). The antiestrogen ICI 182,780 decreases the expression of estrogen receptor-alpha but has no effect on estrogen receptor-beta and androgen receptor in rat efferent ductules. *Reprod. Biol. Endocrinol.* 1, 75.
563. Fisher, J. S., Millar, M. R., Majdic, G., Saunders, P. T., Fraser, H. M., and Sharpe, R. M. (1997). Immunolocalisation of oestrogen receptor-alpha within the testis and excurrent ducts of the rat and marmoset monkey from perinatal life to adulthood. *J. Endocrinol.* 153, 485–495.
564. Heikinheimo, O., Mahony, M. C., Gordon, K., Hsiu, J. G., Hodgen, G. D., and Gibbons, W. E. (1995). Estrogen and progesterone receptor mRNA are expressed in distinct pattern in male primate reproductive organs. *J. Assist. Reprod. Genet.* 12, 198–204.
565. Saunders, P. T., Sharpe, R. M., Williams, K., Macpherson, S., Urquart, H., Irvine, D. S., and Millar, M. R. (2001). Differential expression of oestrogen receptor alpha and beta proteins in the testes and male reproductive system of human and non-human primates. *Mol. Hum. Reprod.* 7, 227–236.
566. Ergun, S., Ungefroren, H., Holstein, A. F., and Davidoff, M. S. (1997). Estrogen and progesterone receptors and estrogen receptor-related antigen (ER-D5) in human epididymis. *Mol. Reprod. Dev.* 47, 448–455.
567. Kolasa, A., Wiszniewska, B., Marchlewicz, M., and Wenda-Rozewicka, L. (2003). Localisation of oestrogen receptors (ERα and ERβ) in the human and rat epididymides. *Folia Morphol. (Warsz.)* 62, 467–469.
568. Albrecht, E. D., Billiar, R. B., Aberdeen, G. W., Babischkin, J. S., and Pepe, G. J. (2004). Expression of estrogen receptors alpha and beta in the fetal baboon testis and epididymis. *Biol. Reprod.* 70, 1106–1113.
569. Lubahn, D. B., Moyer, J. S., Golding, T. S., Couse, J. F., Korach, K. S., and Smithies, O. (1993). Alteration of reproductive function but not prenatal sexual development after insertional disruption of the mouse estrogen receptor gene. *Proc. Natl. Acad. Sci. U S A* 90, 11162–11166.
570. Eddy, E. M., Washburn, T. F., Bunch, D. O., Goulding, E. H., Gladen, B. C., Lubahn, D. B., and Korach, K. S. (1996). Targeted disruption of the estrogen receptor gene in male mice causes alteration of spermatogenesis and infertility. *Endocrinology* 137, 4796–47805.
571. Hess, R. A., Bunick, D., Lee, K. H., Bahr, J., Taylor, J. A., Korach, K. S., and Lubahn, D. B. (1997). A role for oestrogens in the male reproductive system. *Nature* 390, 509–512.
572. Hess, R. A. (2000). Oestrogen in fluid transport and reabsorption in efferent ducts of the male reproductive tract. *Rev. Reprod.* 5, 84–92.
573. Ruz, R., Andonian, S., and Hermo, L. (2004). Immunolocalization and regulation of cystic fibrosis transmembrane conductance regulator in the adult rat epididymis. *J. Androl.* 25, 265–273.
574. Krege, J. H., Hodgin, J. B., Couse, J. F., Enmark, E., Warner, M., Mahler, J. F., Sar, M., Korach, K. S., Gustafsson, J. A., and Smithies, O. (1998). Generation and reproductive phenotypes of mice lacking estrogen receptor beta. *Proc. Natl. Acad. Sci. U S A* 95, 15677–15682.
575. Carreau, S. (2003). Estrogens: male hormones? *Folia Histochem. Cytobiol.* 41, 107–111.
576. Payne, A., Kelch, R., Musich, S., and Halpern, M. (1976). Intratesticular site of aromatization in the human. *J. Clin. Endocrinol. Metab.* 42, 1081–1087.
577. Levallet, J., and Carreau, S. (1997). In vitro gene expression of aromatase in rat testicular cells. *C. R. Acad. Sci. III* 320, 123–129.
578. Nitta, H., Bunick, D., Hess, R. A., Janulis, L., Newton, S. C., Millette, C. F., Osawa, Y., Shizuta, Y., Toda, K., and Bahr, J. M. (1993). Germ cells of the mouse testis express P450 aromatase. *Endocrinology* 132, 1396–1401.
579. O'Donnell, L., Robertson, K. M., Jones, M. E., and Simpson, E. R. (2001). Estrogen and spermatogenesis. *Endocr. Rev.* 22, 289–318.
580. Tsubota, T., Nitta, H., Osawa, Y., Mason, I., Kita, I., Tiba, T., and Bahr, J. (1993). Immunolocalization of steroidogenic enzymes, P450scc, 3β-HSD, P450c17, and P450arom in the Hokkaido brown bear. *Gen. Comp. Endocrinol.* 92, 439–444.
581. Kwon, S., Hess, R. A., Bunick, D., Nitta, H., Janulis, L., Osawa, Y., and Bahr, J. M. (1995). Rooster testicular germ cells and epididymal sperm contain P450 aromatase. *Biol. Reprod.* 53, 1259–1264.
582. Janulis, L., Bahr, J. M., Hess, R. A., Janssen, S., Asawa, Y., and Bahr, J. M. (1998). Rat testicular germ cells and epididymal sperm contain active P450 aromatase. *J. Androl.* 17, 111–
583. Janulis, L., Hess, R. A., Bunick, D., Nitta, H., Janssen, S., Asawa, Y., and Bahr, J. M. (1996). Mouse epididymal sperm contain active P450 aromatase which decreases as sperm traverse the epididymis. *J. Androl.* 17, 111–116.
584. Schleicher, G., Drews, U., and Stumpf, W. E. (1989). No evidence for aromatization of [^{3}H]testosterone in oestrogen receptor containing cells of the epididymis. *J. Steroid Biochem.* 32, 299–302.
585. Wiszniewska, B. (2002). Primary culture of the rat epididymal epithelial cells as a source of oestrogen. *Andrologia* 34, 180–187.
586. Pereyra-Martinez, A. C., Roselli, C. E., Stadelman, H. L., and Resko, J. A. (2001). Cytochrome P450 aromatase in testis and epididymis of male rhesus monkeys. *Endocrine* 16, 15–19.
587. Carpino, A., Romeo, F., and Rago, V. (2004). Aromatase immunolocalization in human ductuli efferentes and proximal ductus epididymis. *J. Anat.* 204, 217–220.
588. Sporn, M. B., Roberts, A. B., and Goodman, D. S. (1994). *The Retinoids.* Raven Press, New York.
589. Wolbach, S. B., and Howe, P. R. (1925). Tissue changes following deprivation of fat-soluble A vitamins. *J. Exp. Med.* 42, 753–777.
590. Mason, K. E. (1939). Relation of the vitamins to the sex glands. In *Sex and Internal Secretions: A Survey of Recent Research* (E. Allen, C. H. Danforth and E. A. Doisy, Eds.). Williams & Wilkins, Baltimore.

591. Blaner, W. S., and Olson, J. A. (1994). Retinol and retinoic acid metabolism. In *The Retinoids: Biology, Chemistry, and Medicine* (M. B. Sporn, A. B. Roberts and D. S. Goodman, Eds.), pp. 229–255. Raven Press, New York.
592. Giguère, V. (1994). Retinoic acid receptors and cellular retinoid binding proteins: complex interplay in retinoid signaling. *Endocr. Rev.* 15, 61–79.
593. Napoli, J. L. (1999). Interactions of retinoid binding proteins and enzymes in retinoid metabolism. *Biochim. Biophys. Acta* 1440, 139–162.
594. Chambon, P. (1996). A decade of molecular biology of retinoic acid receptors. *FASEB J.* 10, 940–954.
595. Pappas, R. S., Newcomer, M. E., and Ong, D. E. (1993). Endogenous retinoids in rat epididymal tissue and rat and human spermatozoa. *Biol. Reprod.* 48, 235–247.
595a. Deltour, L., Haselbeck, R. J., Ang, H. L., and Duester, G. (1997). Localization of class I and class IV alcohol dehydrogenases in mouse testis and epididymis: potential retinol dehydrogenases for endogenous retinoic acid synthesis. *Biol. Reprod.* 56, 102–109.
596. Porter, S. B., Fraker, L. D., Chytil, F., and Ong, D. E. (1983). Localization of cellular retinol-binding protein in several rat tissues. *Proc. Natl. Acad. Sci. U S A* 80, 6586–6590.
597. Porter, S. B., Ong, D. E., Chytil, F., and Orgebin-Crist, M.-C. (1985). Localization of cellular retinol-binding protein and cellular retinoic acid-binding protein in the rat testis and epididymis. *J. Androl.* 6, 197–212.
598. Kato, M., Blaner, W. S., Mertz, J. R., Das, K., Kato, K., and Goodman, D. S. (1985). Influence of retinoid nutritional status on cellular retinol- and cellular retinoic acid-binding protein concentrations in various rat tissues. *J. Biol. Chem.* 260, 4832–4838.
599. Orgebin-Crist, M.-C., Lareyre, J.-J., Suzuki, K., Araki, Y., Fouchécourt, S., Matusik, R. J., and Ong, D. E. (2002). Retinoids and epididymal function. In *The Epididymis: From Molecules to Clinical Practice* (B. Robaire and B. T. Hinton, Eds.), pp. 339–352. Kluwer Academic/Plenum, New York.
600. Wan, Y.-J., Wang, L., and Wu, T.-C. (1992). Detection of retinoic acid receptor mRNA in rat tissues by reverse transcriptase-polymerase chain reaction. *J. Mol. Endocrinol.* 9, 291–294.
601. Akmal, K. M., Dufour, J. M., and Kim, K. H. (1996). Region-specific localization of retinoic acid receptor-alpha expression in the rat epididymis. *Biol. Reprod.* 54, 1111–1119.
602. Wei, S., Episkopou, V., Piantedosi, R., Maeda, S., Shimada, K., Gottesman, M. E., and Blaner, W. S. (1995). Studies on the metabolism of retinol and retinol-binding protein in transthyretin-deficient mice produced by homologous recombination. *J. Biol. Chem.* 270, 866–870.
603. Gorry, P., Lufkin, T., Dierich, A., Rochette-Egly, C., Decimo, D., Dolle, P., Mark, M., Durand, B., and Chambon, P. (1994). The cellular retinoic acid binding protein I is dispensable. *Proc. Natl. Acad. Sci. U S A* 91, 9032–9036.
604. Lampron, C., Rochette-Egly, C., Gorry, P., Dolle, P., Mark, M., Lufkin, T., LeMeur, M., and Chambon, P. (1995). Mice deficient in cellular retinoic acid binding protein II (CRABPII) or in both CRABPI and CRABPII are essentially normal. *Development* 121, 539–548.
605. Kastner, P., Mark, M., and Chambon, P. (1995). Nonsteroid nuclear receptors: what are genetic studies telling us about their role in real life? *Cell* 83, 859–869.
606. Costa, S. L., Boekelheide, K., Vanderhyden, B. C., Seth, R., and McBurney, M. W. (1997). Male infertility caused by epididymal dysfunction in transgenic mice expressing a dominant negative mutation of retinoic acid receptor alpha 1. *Biol. Reprod.* 56, 985–990.
607. Lufkin, T., Lohnes, D., Mark, M., Dierich, A., Gorry, P., Gaub, M.-P., LeMeur, M., and Chambon, P. (1993). High postnatal lethality and testis degeneration in retinoic acid receptor alpha mutant mice. *Proc. Natl. Acad. Sci. U S A* 90, 7225–7229.
608. Mendelsohn, C., Lohnes, D., Decimo, D., Lufkin, T., LeMeur, M., Chambon, P., and Mark, M. (1994). Function of the retinoic acid receptors (RARs) during development (II): multiple abnormalities at various stages of organogenesis in RAR double mutants. *Development* 120, 2749–2771.
609. Nicholson, H. D., and Jenkin, L. (1994). Oxytocin increases 5α-reductase activity in the rat testis. In *Function of Somatic Cells in the Testis* (A. Bartke, Ed.), pp. 278–285. Springer-Verlag, New York.
610. Veeramachaneni, D. N. R., and Amann, R. P. (1990). Oxytocin in the ovine ductuli efferentes and caput epididymis: immunolocalization and endocytosis from the luminal fluid. *Endocrinology* 126, 1156–1164.
611. Harris, G. C., Frayne, J., and Nicholson, H. D. (1996). Epididymal oxytocin in the rat: its origin and regulation. *Int. J. Androl.* 19, 278–286.
612. Assinder, S. J., Johnson, C., King, K., and Nicholson, H. D. (2004). Regulation of 5α-reductase isoforms by oxytocin in the rat ventral prostate. *Endocrinology* 145, 5767–5773.
613. Basciani, S., Mariani, S., Arizzi, M., Brama, M., Ricci, A., Betsholz, C., Bondjers, C., Ricci, G., Catizone, A., Galdieri, A., Spera, G., and Gnessi, L. (2004). Expression of platelet derived growth factors (PDGF) in the epididymis and the analysis of the epididymal development in PPDGF-A, PDGF-B and PDGF receptor β deficient mice. *Biol. Reprod.* 70, 168–177.
614. Ergun, S., Luttmer, W., Fiedler, W., and Holstein, A. F. (1998). Functional expression and localization of vascular endothelial growth factor and its receptors in the human epididymis. *Biol. Reprod.* 58, 160–168.
615. Korpelainen, E. I., Karkkainen, M. J., Tenhunen, A., Lakso, M., Rauvala, H., Vierula, M., Parvinen, M., and Alitalo, K. (1998). Overexpression of VEGF in testis and epididymis causes infertility in transgenic mice: evidence for non-endothelial targets for VEGF. *J. Cell. Biol.* 143, 1705–1712.
616. Catizone, A., Ricci, G., and Galdieri, M. (2002). Functional role of hepatocyte growth factor during sperm maturation. *J. Androl.* 23, 911–918.
617. Hess, K. A., Waltz, S. E., Chan, E. L., and Degen, S. J. (2003). Receptor tyrosine kinase Ron is expressed in mouse reproductive tissues during embryo implantation and is important in trophoblast cell function. *Biol. Reprod.* 68, 1267–1275.
618. Degeorges, A., Wang, F., Frierson, H. F. Jr, Seth, A., and Sikes, R. A. (2000). Distribution of IGFBP-rP1 in normal human tissues. *J. Histochem. Cytochem.* 48, 747–754.
619. Shipley, J. M., Mecham, R. P., Maus, E., Bonadio, J., Rosenbloom, J., McCarthy, R. T., Baumann, M. L., Frankfater, C., Segade, F., and Shapiro, S. D. (2000). Developmental expression of latent transforming growth factor beta binding protein 2 and its requirement early in mouse development. *Mol. Cell. Biol.* 20, 4879–4887.
620. Radhakrishnan, B., and Suarez-Quian, C. A. (1992). Characterization of epidermal growth factor receptor in testis, epididymis and vas deferens of non-human primates. *J. Reprod. Fertil.* 96, 12–23.
621. Lobie, P. E., Breipohl, W., Aragon, J. G., and Waters, M. J. (1990). Cellular localization of the growth hormone receptor/binding protein in the male and female reproductive systems. *Endocrinology* 126, 2214–2221.
622. Birchmeier, C., O'Neill, K., Riggs, M., and Wigler, M. (1990). Characterization of ROS1 gene in human glioblastoma cell line. *Proc. Natl. Acad. Sci. U S A* 87, 4799–4803.
623. Tomlinson, A., and Ready, D. F. (1986). *Sevenless*: a cell-specific homeotic mutation of the *Drosophila* eye. *Science* 231, 400–402.
624. Tomlinson, A., and Ready, D. F. (1987). Cell fate in the *Drosophila* ommatidium. *Dev. Biol.* 123, 264–275.

625. Basler, K., Christen, B., and Hafen, E. (1991). Ligand-independent activation of the *sevenless* receptor tyrosine kinase changes the fate of cells in the developing *Drosophila* eye. *Cell* 64, 1069–1081.
626. Kramer, H., Cagan, R. L., and Zipursky, S. L. (1991). Interaction of *bride of sevenless* membrane-bound ligand and the *sevenless* tyrosine kinase receptor. *Nature* 352, 207–212.
627. Matsushime, H., and Shibuya, M. (1990). Tissue-specific expression of rat *c-ros-1* gene and partial structural similarity of its predicted products with *sev* protein of *Drosophila melanogaster*. *J. Virol.* 64, 2117–2125.
628. Chen, J., Heller, M., Poon, B., Kang, L., and Wang, L-H. (1991). The proto-oncogene *c-ros* codes for a transmembrane tyrosine kinase sharing sequence and structural homology with *sevenless* protein of *Drosophila melanogaster*. *Oncogene* 6, 257–264.
629. Hanks, S. K. (1991). Eukaryotic protein kinases. *Curr. Opin. Struct. Biol.* 1, 369–383.
630. Tessarollo, L., Nagarajan, L., and Parada, L. F. (1992). C-ros: the vertebrate homolog of the sevenless tyrosine kinase receptor is tightly regulated during organogenesis in mouse embryonic development. *Development* 115, 11–20.
631. Springer, T. A. (1998). An extracellular β-propeller module predicted in lipoprotein and scavenger receptors, tyrosine kinases, epidermal growth factor precursor, and extracellular matrix components. *J. Mol. Biol.* 283, 837–862.
632. Orlicky, S., Tang, X., Willems, A., Tyers, M., and Sicheri, F. (2003). Structural basis for phosphodependent substrate selection and orientation by the SCFCdc4 ubiquitin ligase. *Cell* 112, 243–256.
633. Hart, A. C., Kramer, H., Van Vactor, D. L., Jr., Paidhungat, M., and Zipursky, S. L. (1990). Induction of cell fate in the *Drosophila* retina: the *bride of sevenless* protein is predicted to contain a large extracellular domain and seven transmembrane segments. *Genes Dev.* 4, 1835–1847.
634. Tessarollo, L., Nagarajan, L., and Parada, L. F. (1992). C-ros: the vertebrate homolog of the sevenless tyrosine kinase receptor is tightly regulated during organogenesis in mouse embryonic development. *Development* 115, 11–20.
635. Yeung, C-H., Sonnenberg-Riethmacher, E., and Cooper, T. G. (1999). Infertile spermatozoa of c-ros tyrosine kinase receptor knock-out mice show flagellar angulation and maturational defects in cell volume regulatory mechanisms. *Biol. Reprod.* 61, 1062–1069.
636. Cooper, T. G., and Yeung, C.-H. (2003). Approaches to post-testicular contraception: insights from the infertile c-ros knockout mouse. In *Third International Conference on the Epididymis* (B. T. Hinton and T. T. Turner, Eds.), pp. 208–221. Van Doren, Charlottesville, VA.
637. Yeung, C. H., Anapolski, M., Setiawan, I., Lang, F., and Cooper, T. G. (2004). Effects of putative epididymal osmolytes on sperm volume regulation of fertile and infertile c-ros transgenic Mice. *J. Androl.* 25, 216–223.
638. Legare, C., and Sullivan, R. (2004) Expression and localization of c-ros oncogene along the human excurrent duct. *Mol. Hum. Reprod.* 10, 697–703.
639. Keilhack, H., Muller, M., Bohmer, S-A., Frank, C., Weidner, K. M., Birchmeier, W., Ligensa, T., Berndt, A., Kosmehl, H., Gunther, B., Muller, T., Birchmeier, C., and Bohmer, F. D. (2001). Negative regulation of ros receptor tyrosine kinase signaling: an epithelial function of the SH2 domain protein tyrosine phosphatase SHP-1. *J. Cell. Biol.* 152, 325–334.
640. Osterhoff, C., Ivell, R., and Kirchhoff, C. (1997). Cloning of human epididymis-specific mRNA, HE6, encoding a novel member of the seven transmembrane-domain receptor superfamily. *DNA Cell. Biol.* 16, 379–389.
641. Obermann, H., Samalecos, A., Osterhoff, C., Schroder, B., Heller, R., and Kirchhoff, C. (2003). HE6, a two-subunit heptahelical receptor associated with apical membranes of efferent and epididymal duct epithelia. *Mol. Reprod. Dev.* 64, 13–26.
642. Davies, B., Baumann, C., Kirchhoff, C., Ivell, R., Nubbemeyer, R., Habenicht, U-F., Theuring, F., and Gottwald, U. (2004). Targeted deletion of the epididymal receptor HE6 results in fluid dysregulation and male infertility. *Mol. Cell. Biol.* 24, 8642–8648.
643. de Lecea, L., Kilduff, T. S., Peyron, C., Gao, X., Foye, P. E., Danielson, P. E., Fukuhara, C., Battenberg, E. L., Gautvik, V. T., Bartlett, F. S., Jr., Frankel, W. N., van den Pol, A. N., Bloom, F. E., Gautvik, M., and Sutcliffe, J. G. (1998). The hypocretins: hypothalamus-specific peptides with neuroexcitatory activity. *Proc. Natl. Acad. Sci. U S A* 95, 322–327.
644. Sakurai. T., Amemiya, A., Ishii, M., Matsuzaki, I., Chemelli, R. M., Tanaka, H., Williams, S. C., Richardson, J. A., Kozlowski, G. P., Wilson, S., Arch, J. R., Buckingham, R. E., Haynes, A. C., Carr, S. A., Annan, R. S., McNulty, D. E., Liu, W. S., Terrett, J. A., Elshourbagy, N. A., Bergsma, D. J, and Yanagisawa, M. (1998). Orexins and orexin receptors: a family of hypothalamic neuropeptides and G protein-coupled receptors that regulate feeding behavior. *Cell* 92, 573–585.
645. Karteris. E., Chen. J, and Randeva, H. S. (2004). Expression of human prepro-orexin and signaling characteristics of orexin receptors in the male reproductive system. *J. Clin. Endocrinol. Metab.* 89, 1957–1962.
646. Wang, H., Hicks, J., Khanbolooki, P., Kim, S. J., Yan. C., Wang, Y., and Boyd, D. (2003). Transgenic mice demonstrate novel promoter regions for tissue-specific expression of the urokinase receptor gene. *Am. J. Pathol.* 163, 453–464.
647. Andersen, O. M., Yeung, C. H., Vorum, H., Wellner, M., Andreassen, T. K., Erdmann, B., Mueller, E. C., Herz, J., Otto, A., Cooper, T. G., and Willnow, T. E. (2003). Essential role of the apolipoprotein E receptor-2 in sperm development. *J. Biol. Chem.* 278, 23989–23995.
648. Hwang, I. S., Autelitano, D. J., Wong, P. Y., Leung, G. P., and Tang, F. (2003). Co-expression of adrenomedullin and adrenomedullin receptors in rat epididymis: distinct physiological actions on anion transport. *Biol. Reprod.* 68, 2005–2012.
649. Shariatmadari, R., Sipila, P., Vierula, M., Tornquist, K., Huhtaniemi, I., and Poutanen, M. (2003). Adenosine triphosphate induces Ca^{2+} signal in epithelial cells of the mouse caput epididymis through activation of P2X and P2Y purinergic receptors. *Biol. Reprod.* 68, 1185–1192.
650. Jaleel, M., London, R. M., Eber, S. L., Forte, L. R., and Visweswariah, S. S. (2002). Expression of the receptor guanylyl cyclase C and its ligands in reproductive tissues of the rat: a potential role for a novel signaling pathway in the epididymis. *Biol. Reprod.* 67, 1975–1980.
651. Rodriguez, C. M., Kirby, J. L., and Hinton, B. T. (2001). Regulation of gene transcription in the epididymis. *Reproduction* 122, 41–48.
652. Rao, M., and Wilkinson, M. F. (2002). Homeobox genes and the male reproductive system. In *The Epididymis: From Molecules to Clinical Practice* (B. Robaire and B. T. Hinton, Eds.), pp. 269–283. Kluwer Academic/Plenum, New York.
653. Suzuki, K., Drevet, J., Hinton, B. T., Huhtaniemi, I., Lareyre, J-J., Matusik, R. J., Pons, E., Poutanen, M., Sipila, P., and Orgebin-Crist, M-C. (2004). Epididymis-specific promoter-driven gene targeting: a new approach to control epididymal function? *Mol. Cell. Endocrinol.* 216, 15–22.
654. Karim, F. D., Urness, L. D., Thummel, C. S., Klemsz, M. J., McKercher, S. R., Celada, A., Van Beveren, C., Maki, R. A., Gunther, C. V., and Nye, J. A. (1990). The ETS-domain a

new DNA-binding motif that recognizes a purine-rich core DNA sequence. *Genes Dev.* 4, 1451–1453.

655. Xin, J. H., Cowie, A., Lachance, P., and Hassell, J. A. (1992). Molecular cloning and characterization of PEA3, a new member of the Ets oncogene family that is differentially expressed in mouse embryonic cells. *Genes Dev.* 6, 481–496.
656. Brown, T. A., and McKnight, S. L. (1992). Specificities of protein-protein and protein-DNA interaction of GABP alpha and two newly defined ets-related proteins. *Genes Dev.* 6, 2502–2512.
657. Monte, D., Baert, J. L., Defossez, P. A., de Launoit, Y., and Stehelin, D. (1994). Molecular cloning and characterization of human ERM, a new member of the Ets family closely related to mouse PEA3 and ER81 transcription factors. *Oncogene* 9, 1397–1406.
658. Monte, D., Coutte, L., Baert, J. L. Angeli, I., Stehelin, D., and de Launoit, Y. (1995). Molecular characterization of the ets-related human transcription factor ER81. *Oncogene* 11, 771–779.
659. Jeon, I. S., Davis, J. N., Braun, B. S., Sublett, J. E., Roussel, M. F., Denny, C. T., and Shapiro, D. N. (1995). A variant Ewing's sarcoma translocation (7;22) fuses the EWS gene to the ETS gene ETV1. *Oncogene* 10, 1229–1234.
660. Kurpios, N. A., Sabolic, N. A., Shepherd, T. G., Fidalgo, G. M., and Hassell, J. A. (2003). Function of PEA3 Ets transcription factors in mammary gland development and oncogenesis. *J. Mammary Gland Biol. Neoplasia* 8, 177–190.
661. Drevet, J. L., Lareyre, J.-J., Schwaab, V., Vernet, P., and Dufaure, J. P. (1998). The PEA3 protein of the Ets oncogene family is a putative transcriptional modulator of the mouse epididymis-specific glutathione peroxidase gene gpx5. *Mol. Reprod. Dev.* 49, 131–140.
662. Kirby, J. L., Labus, J. C., Ling, Y., Lye, R. J., Hsia, N, Day, R., Cornwall, G., and Hinton, B. T. (2004). Characterization of epididymal epithelial cell-specific gene promoters by in vivo electroporation. *Biol. Reprod.* 71, 613–619.
663. Lengyel, E., Stepp, E., Gum, R., and Boyd, D. (1995). Involvement of a mitogen-activated protein kinase signaling pathway in the regulation of urokinase promoter activity by c-Ha-ras. *J. Biol. Chem.* 270, 23007–23012.
664. Gum, R., Lengyel, E., Juarez, J., Chen, J. H., Sato, H., Seiki, M., and Boyd, D. (1996). Stimulation of 92-kDa gelatinase B promoter activity by ras is mitogen-activated protein kinase kinase 1-independent and requires multiple transcription factor binding sites including closely spaced PEA3/ets and AP-1 sequences. *J. Biol. Chem.* 271, 10672–10680.
665. Laing, M. A., Coonrod, S., Hinton, B. T., Downie, J. W., Tozer, R., Rudnicki, M. A., and Hassell, J. A. (2000). Male sexual dysfunction in mice bearing targeted mutant alleles of the PEA3 ets gene. *Mol. Cell. Biol.* 20, 9337–9345.
666. Ranji, D. P., and Foka, P. (2002). CCAAT/enhancer-binding proteins: structure, function and regulation. *Biochem. J.* 365, 561–575.
667. Ubeda, M., Wang, X. Z., Zinszner, H., Wu, I., Habener, J. F., and Ron, D. (1996) Stress-induced binding of the transcriptional factor CHOP to a novel DNA control element. *Mol. Cell. Biol.* 16, 1479–1489.
668. Hsia, N., and Cornwall, G. A. (2001). CCAAT/enhancer binding protein beta regulates expression of the cystatin-related epididymal spermatogenic (Cres) gene. *Biol. Reprod.* 65, 1452–1461.
669. Hsia. N., and Cornwall, G. A. (2003). Cres2 and Cres3: new members of the cystatin-related epididymal spermatogenic subgroup of family 2 cystatins. *Endocrinology* 144, 909–915.
670. Bomgardner, D., Hinton, B. T., and Turner, T. T. (2001). The role of homeobox genes in the adult epididymis. *J. Androl.* 22, 527–531.
671. Favier, B., and Dolle, P. (1997). Developmental functions of mammalian Hox genes. *Mol. Hum. Reprod.* 3, 115–131.
672. Popperl, H., Rikhof, H., Chang, H., Haffter, P., Kimmel, C. B., and Moens, C. B. (2000). Lazarus is a novel pbx gene that globally mediates hox gene function in zebrafish. *Mol. Cell.* 6, 255–267.
673. Benson, G. V., Lim, H., Paria, B. C., Satokata, I., Dey, S. K., and Maas, R. L. (1996). Mechanisms of reduced fertility in Hoxa-10 mutant mice: uterine homeosis and loss of maternal Hoxa-10 expression. *Development* 122, 2687–2696.
674. Podlasek, C. A., Seo, R. M., Clemens, J. Q., Ma, L., Maas, R. L., and Bushman, W. (1999). Hoxa-10 deficient male mice exhibit abnormal development of the accessory sex organs. *Dev. Dyn.* 214, 1–12.
675. Bomgardner, D., Hinton, B. T., and Turner, T. T. (2003). 5′ Hox genes and Meis 1, a hox-DNA binding cofactor, are expressed in the adult mouse epididymis. *Biol. Reprod.* 68, 644–650.
676. Hsieh-Li, H. M., Witte, D. P., Weinstein, M., Branford, W., Li, H., Small, K., and Potter, S. S. (1995). Hoxa 11 structure, extensive antisense transcription, and function in male and female fertility. *Development* 121, 1373–1385.
677. Shen, W. F., Rozenfeld, S., Kwong, A., Kom ves, L. G., Lawrence, H. J., and Largman, C. (1999). HOXA9 forms triple complexes with PBX2 and MEIS1 in myeloid cells. *Mol. Cell. Biol.* 19, 3051–3061.
678. Maclean, J. A., Chen, M. A., Wayne, C. M., Bruce, S. R., Rao, M., Meistrich, M. L., Macleod, C., and Wilkinson, M. F. (2005). Rhox: a new homeobox gene cluster. *Cell* 120, 369–382.
679. Oyhenart, J., Dacheux, J. L., Dacheux, F., Jegou, B., and Raich, N. (2005). Expression, regulation, and immunolocalization of putative homeodomain transcription factor 1 (PHTF1) in rodent epididymis: evidence for a novel form resulting from proteolytic cleavage. *Biol. Reprod.* 72, 50–57.
680. Lindsey, J. S., and Wilkinson, M. F. (1996). An androgen-regulated homeobox gene expressed in rat testis and epididymis. *Biol. Reprod.* 55, 975–983.
681. Lindsey, J. S., and Wilkinson, M. F. (1996). Homeobox genes and male reproductive development. *J. Assist. Reprod. Genet.* 13,182–192.
682. Sutton, K. A., Maiti, S., Tribley, W. A., Lindsey, J. S., Meistrich, M. L., Bucana, C. D., Sanborn, B. M., Joseph, D. R., Griswold, M. D., Cornwall, G. A., and Wilkinson, M. F. (1998). Androgen regulation of the Pem homeodomain gene in mice and rat Sertoli and epididymal cells. *J. Androl.* 19, 21–30.
683. Pitman, J. L., Lin, T. P., Kleeman, J. E., Erickson, G. F., and MacLeod, C. L. (2002). Normal reproductive and macrophage function in Pem homeobox gene-deficient mice. *Dev. Biol.* 202, 196–214.
684. Saffer, J. D., Jackson, S. P., and Annarella, M. B. (1991). Developmental expression of Sp1 in the mouse. *Mol. Cell. Biol.* 11, 2189–2199.
685. Pearse, R. V., 2nd, Drolet, D. W., Kalla, K. A., Hooshmand, F., Bermingham. J. R., Jr., and Rosenfeld, M. G. (1997). Reduced fertility in mice deficient for the POU protein sperm-1. *Proc. Natl. Acad. Sci. U S A* 94, 7555–7560.
686. Miyamoto, N., Yoshida, M., Kuratani, S., Matsuo, I., and Aizawa, S. (1997). Defects of urogenital development in mice lacking Emx2. *Development* 124, 1653–1664.
687. Chen, M. Y., Carpenter, D., and Zhao, G. Q. (1999). Expression of bone morphogenetic protein 7 in murine epididymis is developmentally regulated. *Biol. Reprod.* 60, 1503–1508.
688. Oefelein, M., Grapey, D., Schaeffer, T., Chin-Chance, C., and Bushman, W. (1996). Pax-2: a developmental gene constitutively expressed in the mouse epididymis and ductus deferens. *J. Urol.* 156, 1204–1207.

689. Mastrangelo, P., Zwingman, T., Erickson, R. P., and Blecher, S. R. (1994). Zfy is transcribed in the normal mouse epididymis and in the XXSxr ("sex reversed") testis. *Dev. Genet.* 15, 129–138.
690. Maruyama, K., Tsukada, T., Bandoh, S., Sasaki, K., Ohkura, N., and Yamaguchi, K. (1997). Expression of the putative transcription factor NOR-1 in the nervous, the endocrine and the immune systems and the developing brain of the rat. *Neuroendocrinology* 65, 2–8.
691. Byers, S., and Graham, R. (1990). Distribution of sodium-potassium ATPase in the rat testis and epididymis. *Am. J. Anat.* 188, 21–43.
692. Ilio, K. Y., and Hess, R. A. (1992). Localization and activity of Na^+-K^+-ATPase in the ductuli efferentes of the rat. *Anat. Rec.* 234, 190–200.
693. Pushkin, A., Clark, I., Kwon, T. H., Nielsen, S., and Kurtz, I. (2000). Immunolocalization of NBC3 and NHE3 in the rat epididymis: colocalization of NBC3 and the vacuolar H^+-ATPase. *J. Androl.* 21, 708–720.
694. Cheng-Chew, S. B., Leung, G. P. H., Leung, P. Y., Tse, C. M., and Wong, P. Y. D. (2000). Polarized distribution of NHE1 and NHE2 in the rat epididymis. *Biol. Reprod.* 62, 755–758.
695. Bagnis, C., Marsolais, M., Biemesderfer, D., Laprade, R., and Breton, S. (2001). Na(+)/H(+)-exchange activity and immunolocalization of NHE3 in rat epididymis. *Am. J. Physiol. Renal Physiol.* 280, F426–F436.
696. Kaunisto, K., Moe, O. W., Pelto-Huikko, M., Traebert, M., and Rajaniemi, H. (2001). An apical membrane Na+/H+ exchanger form, NHE3, is present in the rat epididymis epithelium. *Pflugers Arch.* 442, 230–236.
697. Miller, R. L., Zhang, P., Smith, M., Beaulieu, V., Paunescu, T. G., Brown, D., Breton, S., and Nelson, R. D. (2005). V-ATPase B1 subunit promoter drives expression of EGFP in intercalated cells of kidney, clear cells of epididymis and airway cells of lung in transgenic mice. *Am. J. Physiol. Cell. Physiol.* 288, C1134–C1144.
698. Brown, D., Lui, B., Gluck, S., and Sabolic, I. (1992). A plasma membrane proton ATPase in specialized cells of rat epididymis. *Am. J. Physiol.* 263, C913–C916.
699. Breton, S., Smith, P. J. S., Lui, B., and Brown, D. (1996). Acidification of the male reproductive tract by a proton pumping (H^+)-ATPase. *Nat. Med.* 2, 470–472.
700. Herak-Kramberger, C. M., Breton, S., Brown, D., Kraus, O., and Sabolic, I. (2001). Distribution of the vacuolar H^+ ATPase along the rat and human male reproductive tract. *Biol. Reprod.* 64, 1699–1707.
701. Brown, D., Verbavatz, J. M., Valenti, G., Lui, B., and Sabolic, I. (1993). Localization of the Chip28 water channel in reabsorptive segments of the rat male reproductive tract. *Eur. J. Cell. Biol.* 61, 264–273.
702. Fisher, J. S., Turner, K. J., Fraser, H. M., Saunders, P. T. K., Brown, D., and Sharpe, R. M. (1998). Immunoexpression of aquaporin-1 in the efferent ducts of that rat and marmoset monkey during development, its modulation by estrogens, and its possible role in fluid resorption. *Endocrinology* 139, 3935–3945.
703. Elkjær, M-L., Vajda, Z., Nejsum, L. N., Kwon, T-H., Jensen, U. B., Amiry-Moghaddam, M., Frøkiær, J, and Nielsen, S. (2000). Immunolocalization of AQP9 in liver, epididymis, testis, spleen and brain. *Biochem. Biophys. Res. Commun.* 276, 1118–1128.
704. Pastor-Soler, N., Bagnis, C., Sabolic, I., Tyszkowski, R., McKee, M., Van Hoek, A., Breton, S., and Brown, D. (2001). Aquaporin 9 expression along the male reproductive tract. *Biol. Reprod.* 65, 384–393.
705. Badran, H. H., and Hermo, L. S. (2002). Expression and regulation of aquaporins 1, 8, and 9 in the testis, efferent ducts, and epididymis of adult rats and during postnatal development. *J. Androl.* 23, 358–373.
706. Pastor-Soler, N., Isnard-Bagnis, C., Herak-Kramberger, C., Sabolic, I., Van Hoek, A., Brown, D., and Breton, S. (2002). Expression of aquaporin 9 in the adult rat epididymal epithelium is modulated by androgens. *Biol. Reprod.* 66, 1716–1722.
707. Cheung, K. H., Leung, C. T., Leung, G. P. H., and Wong, P. Y. D. (2003). CFTR regulates water permeability by interacting with AQP-9. In *The Third International Conference on the Epididymis* (B. T. Hinton and T. T. Turner, Eds.), pp. 23–33. Van Doren, Charlottesville, VA.
708. Hermo, L., Krzeczunowicz, D., and Ruz, R. (2004). Cell specificity of aquaporins 0, 3, and 10 expressed in the testis, efferent ducts, and epididymis of adult rats. *J. Androl.* 25, 494–505.
709. Hess, R. A., Bunick, D., Lubahn, D. B., Zhou, Q., and Bouma, J. (2000). Morphological changes in efferent ductules and epididymis in estrogen receptor-alpha knockout mice. *J. Androl.* 21, 107–121.
710. Tamai, I., Ohashi, R., Nezu, J., Yabuuchi, H., Oku, A., Shimane, M., Sai, Y., and Tsuji, A. (1998). Molecular and functional identification of sodium ion-dependent, high affinity human carnitine transporter OCTN2. *J. Biol. Chem.* 273, 20378–20382.
711. Sekine, T., Kusuhara, H., Utsunomiya-tate, N., Tsuda, M., Sugiyama, Y., Kanai, Y., and Endou, H. (1998). Molecular cloning and characterization of high-affinity carnitine transporter from rat intestine. *Biochem. Biophys. Res. Commun.* 251, 586–591.
712. Yabuuchi, H., Tamai, I., Nezu, J., Sakamoto, K., Oku, A., Shimane, M., Sai, Y., and Tsuji, A. (1999). Novel membrane transporter OCTN1 mediates multispecific, bidirectional, and pH-dependent transport of organic cations. *J. Pharmacol. Exp. Ther.* 289, 768–773.
713. Kobayashi, D., Aizawa, S., Maeda, T., Tsuboi, I., Yabuuchi, H., Nezu, J., Tsuji, A., and Tamai, I. (2004). Expression of organic cation transporter OCTN1 in hematopoietic cells during erythroid differentiation. *Exp. Hematol.* 32, 1156–1162.
714. Lamhonwah, A. M., Skaug, J., Scherer, S. W., and Tein, I. (2003). A third human carnitine/organic cation transporter (OCTN3) as a candidate for the 5q31 Crohn's disease locus (IBD5). *Biochem. Biophys. Res. Commun.* 301, 98–101.
715. Duran, J. M., Peral, M. J., Calonge, M. L., and Ilundain, A. A. (2005). OCTN3: A Na(+)-independent L-carnitine transporter in enterocytes basolateral membrane. *J. Cell. Physiol.* 202, 929–935.
716. Nezu, J-I., Tamai, I., Oku, A., Ohashi, R., Yabuuchi, H., Hashimoto, N., Nikaido, H., Sai, Y., Koizumi, A., Shoji, Y., Takada, G., Matsuishi, T., Yoshino, M., Kato, H., Ohura, T., Tsujimoto, G., Hayakawa, J-I., Shimane, M., and Tsuji, A. (1999). Primary systemic carnitine deficiency is caused by mutations in a gene encoding sodium ion-dependent carnitine transporter. *Nat. Genet.* 21, 91–94.
717. Tang, N. L. S., Ganapathy, V., Wu, X., Hui, J., Seth, P., Yuen, P. M. P., Fok, T. F., and Hjelm, N. M. (1999). Mutations of OCTN2, an organic cation/carnitine transporter, lead to deficient cellular carnitine uptake in primary carnitine deficiency. *Hum. Mol. Genet.* 8, 655–660.
718. Wang, Y., Ye, J., Ganapathy, V., and Longo, N. (1999). Mutations in the organic cation/carnitine transporter OCTN2 in primary carnitine deficiency. *Proc. Natl. Acad. Sci. U S A* 96, 2356–2360.
719. Toshimori, K., Kuwajima, M., Yoshinaga, K., Wakayama, T., and Shima, K. (1999). Dysfunctions of the epididymis as a result of primary carnitine deficiency in juvenile visceral steatosis in mice. *FEBS Lett.* 446, 323–326.
720. Lu, K.-M., Nishimori, H., Nakamura, Y., Shima, K., and Kuwajima, M. (1998). A missense mutation of mouse OCTN2,

a sodium-dependent carnitine cotransporter, in the juvenile visceral steatosis mouse. *Biochem. Biophys. Res. Commun.* 252, 590–594.

721. Zhu, Y. W., Jong, M. C., Frazer, K. A., Gong, E., Krauss, R. M., Cheng, J. F., Boffelli, D., and Rubin, E. M. (2000). Genomic interval engineering of mice identifies a novel modulator of triglyceride production. *Proc. Natl. Acad. Sci. U S A* 97, 1137–1142.
722. Hinton, B. T., and Hernandez, H. (1985). Selective luminal absorption of L-carnitine from the proximal regions of the rat epididymis: possible relationships to development of sperm motility. *J. Androl.* 6, 300–305.
723. Xu, Y. X., Wagenfeld, A., Yeung, C. H., Lehnert, W., and Cooper, T. G. (2003). Expression and location of taurine transporters and channels in the epididymis of infertile c-ros receptor tyrosine kinase-deficient and fertile heterozygous mice. *Mol. Reprod. Dev.* 64, 144–151.
724. Wagenfeld, A., Yeung, C. H., Lehnert, W., Nieschlag, E., and Cooper, T. G. (2002). Lack of glutamate transporter EAAC1 in the epididymis of infertile c-ros receptor tyrosine-kinase deficient mice. *J. Androl.* 23, 772–782.
725. Obermann, H., Wingbermuhle, A., Munz, S., and Kirchhoff, C. (2003). A putative 12-transmembrane domain cotransporter associated with apical membranes of the epididymal duct. *J. Androl.* 24, 542–556.
726. Frohlich, O., Po, C., Murphy, T., and Young, L. G. (2000). Multiple promoter and splicing mRNA variants of the epididymis-specific gene EP2. *J. Androl.* 21, 421–430.
727. Malm, J., Sorensen, O., Persson, T., Frohm-Nilsson, M., Johansson, B., Bjartell, A., Lilja, H., Stahle-Backdahl, M., Borregaard, N., and Egesten, A. (2000). The human cationic antimicrobial protein (hCAP-18) is expressed in the epithelium of human epididymis, is present in seminal plasma at high concentrations, and is attached to spermatozoa. *Infect. Immun.* 68, 4297–4302.
728. Hammami-Hamza, S., Doussau, M., Bernard, J., Rogier, E., Duquenne, C., Richard, Y., Lefevre, A., and Finaz, C. (2001). Cloning and sequencing of SOB3, a human gene coding for a sperm protein homologous to an antimicrobial protein and potentially involved in zona pellucida binding. *Mol. Hum. Reprod.* 7, 625–632.
729. Jia, H. P., Schutte, B. C., Schudy, A., Linzmeier, R., Guthmiller, J. M., Johnson, G. K., Tack, B. F., Mitros, J. P., Rosenthal, A., Ganz, T., and McCray, P. B., Jr. (2001). Discovery of new human beta-defensins using a genomics-based approach. *Gene* 263, 211–218.
730. Li, P., Chan, H. C., He, B., So, S. C., Chung, Y. W., Shang, Q., Zhang, Y. D., and Zhang, Y. L. (2001). An antimicrobial peptide gene found in the male reproductive system of rats. *Science* 291, 1783–1785.
731. Hamil, K. G., Liu, Q., Sivashanmugam, P., Yenugu, S., Soundararajan, R., Grossman, G., Richardson, R. T., Zhang, Y. L., O'Rand, M. G., Petrusz, P., French, F. S., and Hall, S. H. (2002). Cystatin 11: a new member of the cystatin type 2 family. *Endocrinology* 143, 2787–2796.
732. Lehrer, R. I., and Ganz, T. (2002). Cathelicidins: a family of endogenous antimicrobial peptides. *Curr. Opin. Hematol.* 9, 18–22.
733. von Horsten, H. H., Derr, P., and Kirchhoff, C. (2002). Novel antimicrobial peptide of human epididymal duct origin. *Biol. Reprod.* 67, 804–813.
734. Yamaguchi, Y., Nagase, T., Makita, R., Fukuhara, S., Tomita, T., Tominaga, T., Kurihara, H., and Ouchi, Y. (2002). Identification of multiple novel epididymis-specific beta-defensin isoforms in humans and mice. *J. Immunol.* 169, 2516–2523.
735. Com, E., Bourgeon, F., Evrard, B., Ganz, T., Colleu, D., Jegou, B., and Pineau, C. (2003). Expression of antimicrobial defensins in the male reproductive tract of rats, mice, and humans. *Biol. Reprod.* 68, 95–104.
736. Palladino, M. A, Mallonga, T. A., and Mishra, M. S. (2003). Messenger RNA (mRNA) expression for the antimicrobial peptides beta-defensin-1 and beta-defensin-2 in the male rat reproductive tract: beta-defensin-1 mRNA in initial segment and caput epididymidis is regulated by androgens and not bacterial lipopolysaccharides. *Biol. Reprod.* 68, 509–515.
737. Rao, J., Herr, J. C., Reddi, P. P., Wolkowicz, M. J., Bush, L. A., Sherman, N. E., Black, M., and Flickinger, C. J. (2003). Cloning and characterization of a novel sperm-associated isoantigen (E-3) with defensin- and lectin-like motifs expressed in rat epididymis. *Biol. Reprod.* 68, 290–301.
738. Rodriguez-Jimenez, F. J., Krause, A., Schulz, S., Forssmann, W. G., Conejo-Garcia, J. R., Schreeb, R., and Motzkus, D. (2003). Distribution of new human beta-defensin genes clustered on chromosome 20 in functionally different segments of epididymis. *Genomics* 81, 175–183.
739. Zanich, A., Pascall, J. C., and Jones, R. (2003). Secreted epididymal glycoprotein 2D6 that binds to the sperm's plasma membrane is a member of the beta-defensin superfamily of pore-forming glycopeptides. *Biol. Reprod.* 69, 1831–1842.
740. Avellar, M. C., Honda, L., Hamil, K. G., Yenugu, S., Grossman, G., Petrusz, P., French, F. S., and Hall, S. H. (2004). Differential expression and antibacterial activity of epididymis protein 2 isoforms in the male reproductive tract of human and rhesus monkey (*Macaca mulatta*). *Biol. Reprod.* 71, 1453–1460.
741. Zaballos, A., Villares, R., Albar, J. P., Martinez-A. C., and Marquez, G. (2004). Identification on mouse chromosome 8 of new beta-defensin genes with regionally specific expression in the male reproductive organ. *J. Biol. Chem.* 279, 12421–12426.
742. Yenugu, S., Hamil, K. G., French, F. S., and Hall, S. H. (2004). Antimicrobial actions of the human epididymis 2 (HE2) protein isoforms, HE2alpha, HE2beta1 and HE2beta2. *Reprod. Biol. Endocrinol.* 2, 61.
743. Yenugu, S., Hamil, K. G., Radhakrishnan, Y., French, F. S., and Hall, S. H. (2004). The androgen-regulated epididymal sperm-binding protein, human beta-defensin 118 (DEFB118) (formerly ESC42), is an antimicrobial beta-defensin. *Endocrinology* 145, 3165–3173.
744. Yenugu, S., Richardson, R. T., Sivashanmugam, P., Wang, Z., O'Rand, M. G., French, F. S., and Hall, S. H. (2004). Antimicrobial activity of human EPPIN, an androgen-regulated, sperm-bound protein with a whey acidic protein motif. *Biol. Reprod.* 71, 1484–1490.
745. Zhou, C. X., Zhang, Y. L., Xiao, L., Zheng, M., Leung, K. M., Chan, M. Y., Lo, P. S., Tsang, L. L., Wong, H. Y., Ho, L. S., Chung, Y. W., and Chan, H. C. (2004). An epididymis-specific beta-defensin is important for the initiation of sperm maturation. *Nat. Cell. Biol.* 6, 458–464.
746. Li, L., Zhao, C., Heng, H. H., and Gantz, T. (1997). The human beta-defensin-1 and alpha-defensins are encoded by adjacent genes: two peptide families with differing disulfide topology share common ancestry. *Genomics* 43, 316–320.
747. Maxwell, A. I., Morrison, G. M., and Dorin, J. R. (2003). Rapid sequence divergence in mammalian beta-defensins by adaptive evolution. *Mol. Immunol.* 40, 413–421.
748. Hamil, K. G., Sivashanmugam, P., Richardson, R. T., Grossman, G., Ruben, S. M., Mohler, J. L., Petrusz, P., O'Rand, M. G., French, F. S., and Hall, S. H. (2000). HE2β and HE2γ, new members of an epididymis-specific family of androgen-regulated proteins in the human. *Endocrinology* 141, 1245–1253.

748a. Bauer, F., Schweimer, K., Kluver, E., Conejo-Garcia, J. R., Forssmann, W. G., Rosch, P., Adermann, K., and Sticht, H. (2001). Structure determination of human and murine

beta-defensins reveals structural conservation in the absence of significant sequence similarity. *Protein Sci.* 10, 2470–2479.
749. Hoover, D. M., Wu, Z., Tucker, K., Lu, W., and Lubkowski, J. (2003). Antimicrobial characterization of human beta-defensin 3 derivatives. *Antimicrob. Agents Chemother.* 47, 2804–2809.
750. Andersson, E., Sorensen, O. E., Frohm, B., Borregaard, N., Egesten, A., and Malm, J. (2002). Isolation of human cationic antimicrobial protein-18 from seminal plasma and its association with prostasomes. *Hum. Reprod.* 17, 2529–2534.
751. Sorensen, O. E., Follin, P., Johnsen, A. H., Calafat, J., Tjabringa, G. S., Hiemstra, P. S., and Borregaard, N. (2000). Human cathelicidin, hCAP-18, is processed to the antimicrobial peptide LL-37 by extracellular cleavage with proteinase 3. *Blood* 97, 3951–3959.
752. Popsueva, A. E., Zinovjeva, M. V., Visser, J. W., Zijlmans, J. M., Fibbe, W. E., and Belyavsky, A. V. (1996). A novel murine cathelin-like protein expressed in bone marrow. *FEBS Lett.* 391, 5–8.
753. Gallo, R. L., Kim, K. J., Bernfield, M., Kozak, C. A., Zanetti, M., Merluzzi, L., and Gennaro, R. (1997). Identification of CRAMP a cathelin-related antimicrobial peptide expressed in the embryonic and adult mouse. *J. Biol. Chem.* 272, 13088–13093.
754. Termen, S., Tollin, M., Olsson, B., Svenberg, T., Agerberth, B., and Gudmundsson, G. H. (2003). Phylogeny, processing and expression of the rat cathelicidin rCRAMP: a model for innate antimicrobial peptides. *Cell. Mol. Life Sci.* 60, 536–549.
755. Lögdberg, L., and Wester, L. (2000). Immunocalins: a lipocalin subfamily that modulates immune and inflammatory responses. *Biochim. Biophys. Acta* 1482, 284–297.
756. O'Rand, M. G., Widgren, E. E., Sivashanmugam, P., Richardson, R. T., Hall, S. H., French, F. S., Vandevoort, C. A., Ramachandra, S. G., Ramesh, V., and Jagannadha Rao, A. (2004). Reversible immunocontraception in male monkeys immunized with Eppin. *Science* 306, 1189–1190.
757. Cameo, M. S., and Blaquier, J. A. (1976). Androgen-controlled specific proteins in rat epididymis. *J. Endocrinol.* 69, 47–55.
758. Brooks, D. E., and Higgins, S. J. (1980). Characterization and androgen-dependence of proteins associated with luminal fluid and spermatozoa in the rat epididymis. *J. Reprod. Fertil.* 59, 363–375.
759. Zwain, I. H., Grima, J., and Cheng, C. Y. (1992). Rat epididymal retinoic acid-binding protein development of a radioimmunoassay, its tissue distribution, and its changes in selected androgen-dependent organs after orchiectomy. *Endocrinology* 131, 1511–1526.
760. Girotti, M., Jones, R., Emery, D. C., Chia, W., and Hall, L. (1992). Structure and expression of the rat epididymal secretory protein I gene: an androgen-regulated member of the lipocalin superfamily with a rare splice donor site. *Biochem. J.* 281, 203–210.
761. Lareyre, J.-J., Zheng, W. L., Zhao, G. Q., Kasper, S., Newcomer, M. E., Matusik, R. J., Ong, D. E., and Orgebin-Crist, M.-C. (1998). Molecular cloning and hormonal regulation of a murine epididymal retinoic acid-binding protein messenger ribonucleic acid. *Endocrinology* 139, 2971–2981.
762. Ong, D. E., and Chytil, F. (1988). Presence of novel retinoic acid-binding proteins in the lumen of rat epididymis. *Arch. Biochem. Biophys.* 267, 474–478.
763. Nishiwaki, S., Kato, M., Okuno, M., Moriwaki, H., Kanai, M., and Muto, Y. (1991). Purification and partial characterization of rat epididymal retinoic acid-binding protein, and its immunohistochemical localization. *J. Nutr. Sci. Vitaminol. (Tokyo)* 37, 461–471.
764. Rankin, T. L., Ong, D. E., and Orgebin-Crist, M.-C. (1992). The 18-kDa mouse epididymal protein (MEP 10) binds retinoic acid. *Biol. Reprod.* 46, 767–771.
765. Newcomer, M. E., and Ong, D. E. (1990). Purification and crystallization of a retinoic acid-binding protein from rat epididymis: identity with the major androgen-dependent epididymal proteins. *J. Biol. Chem.* 265, 12876–12879.
766. Newcomer, M. E. (1993). Structure of the epididymal retinoic acid binding protein at 2.1 A resolution. *Structure* 1, 7–18.
767. Lareyre, J.-J., Mattei, M. G., Kasper, S., Ong, D. E., Matusik, R. J., and Orgebin-Crist, M.-C. (1998). Genomic organization and chromosomal localization of the murine epididymal retinoic acid-binding protein (mE-RABP) gene. *Mol. Reprod.* 50, 387–395.
768. Lareyre, J.-J., Winfrey, V. P., Kasper, S., Ong, D. E., Matusik, R. J., Olson, G. E., and Orgebin-Crist, M.-C. (2001). Gene duplication gives rise to a new 17 kilodalton lipocalin that shows epididymal region-specific expression and testicular factor(s) regulation. *Endocrinology* 142, 1296–1308.
769. Suzuki, K., Lareyre, J.-J., Sánchez, D., Gutierrez, G., Araki, Y., Matusik, R. J., and Orgebin-Crist, M.-C. (2004). Molecular evolution of epididymal lipocalin genes localized on mouse chromosome 2. *Gene* 339, 49–59.
770. Hamil, K. G., Liu, Q., Sivashanmugam, P., Anbalagan, M., Yenugu, S., Soundararajan, R., Grossman, G., Rao, A. J., Birse, C. E., Ruben, S. M., Richardson, R. T., Zhang, Y. L., O'Rand, M. G., Petrusz, P., French, F. S., and Hall, S. H. (2003). LCN6, a novel human epididymal lipocalin. *Reprod. Biol. Endocrinol.* 1, 112.
771. Chan, P., Simon-Chazottes, D., Mattei, M. G., Guenet, J. L., and Salier, J. P. (1994). Comparative mapping of lipocalin genes in human and mouse: the four genes for complement C8 γ chain, prostaglandin-D-synthase, oncogene-24p3 and progestagen-associated endometrial protein map to HSA9 and MMU2. *Genomics* 23, 145–150.
772. Sorrentino, C., Silvestrini, B., Braghiroli, L., Chung, S. S., Giacomelli, S., Leone, M. G., Xie, Y., Sui, Y., Mo, M., and Cheng, C. Y. (1998). Rat prostaglandin D_2 synthetase: its tissue distribution, changes during maturation, and regulation in the testis and epididymis. *Biol. Reprod.* 59, 843–853.
773. Chu, S.-T., Lee, Y.-C., Nein, K.-M., and Chen, Y.-H. (2000). Expression, immunolocalization and sperm-association of a protein derived from 24p3 gene in mouse epididymis. *Mol. Reprod. Dev.* 57, 26–36.
774. Clausen, J. (1961). Proteins in normal cerebrospinal fluid not found in serum. *Proc. Soc. Exp. Biol. Med.* 107, 170–172.
775. Urade, Y., Fujimoto, N., and Hayaishi, O. (1985). Purification and characterization of rat brain prostaglandin D synthetase. *J. Biol. Chem.* 260, 12410–12415.
776. Kuruvilla, A. P., Hochwald, G. M., Ghiso, J., Castano, E. M., Pizzolato, M., and Frangione, B. (1991). Isolation and amino terminal sequence of β-trace, a novel protein from human cerebrospinal fluid. *Brain Res.* 565, 337–340.
777. Zahn, M., Mäder, M., Schmidt, B., Bollensen, E., and Felgenhauer, K. (1993). Purification and N-terminal sequence of β-trace, a protein abundant in human cerebrospinal fluid. *Neurosci. Lett.* 154, 93–95.
778. Hoffmann, A., Conradt, H. S., Gross, G., Nimtz, M., Lottspeich, F., and Wurster, U. (1993). Purification and chemical characterization of β-trace protein from human cerebrospinal fluid: its identification as prostaglandin D synthase. *J. Neurochem.* 61, 451–456.
779. Watanabe, K., Urade, Y., Mäder, M., Murphy, C., and Hayaishi, O. (1994). Identification of β-trace as prostaglandin D synthase. *Biochem. Biophys. Res. Commun.* 203, 1110–1116.
780. Gerena R. L., Eguchi, N., Urade, Y., and Killian, G. J. (2000). Stage and region-specific localization of lipocalin-type prostaglandin D synthase in the adult murine testis and epididymis. *J. Androl.* 21, 848–854.

781. Fouchécourt, S., Chaurand, P., DaGue, B. B., Lareyre, J.-J., Matusik, R. J., Caprioli, R. M., and Orgebin-Crist, M.-C. (2002). Epididymal lipocalin-type prostaglandin D_2 synthase: identification using mass spectrometry, messenger RNA localization, and immunodetection in mouse, rat, hamster, and monkey. *Biol. Reprod.* 66, 524–533.
782. Zhu, H., Ma, H., Ni, H., Ma, X.-H., Mills, N., and Yang, Z.-M. (2004). L-prostaglandin D synthase expression and regulation in mouse testis and epididymis during sexual maturation and testosterone treatment after castration. *Endocrine* 24, 39–45.
783. Ujihara, M., Urade, Y., Eguchi, N., Hayashi, H., Ikai, K., and Hayaishi, O. (1988). Prostaglandin D_2 formation and characterization of its synthetases in various tissues of adult rats. *Arch. Biochem. Biophys.* 260, 521–531.
784. Fouchécourt, S., Dacheux, F., and Dacheux, J.-L. (1999). Glutathione-independent prostaglandin D_2 synthase in ram and stallion epididymal fluid: origin and regulation. *Biol. Reprod.* 60, 558–566.
785. Fouchécourt, S., Metayer, S., Locatelli, A., Dacheux, F., and Dacheux, J. L. (2000). Stallion epididymal fluid proteome: qualitative and quantitative characterization; secretion and dynamic changes of major proteins. *Biol. Reprod.* 62, 1790–1803.
786. Gerena, R. L., Irikura, D., Urade, Y., Eguchi, N., Chapman, D. A., and Killian, G. J. (1998). Identification of a fertility-associated protein in bull seminal plasma as lipocalin-type prostaglandin D synthase. *Biol. Reprod.* 58, 826–833.
787. Urade, Y., and Hayaishi, O. (2000). Biochemical, structural, genetic, physiological, and pathophysiological features of lipocalin-type prostaglandin D synthase. *Biochim. Biophys. Acta* 1482, 259–271.
788. Tokugawa, Y., Kunishige, I., Kubota, Y., Shimoya, K., Nobunaga, T., Kimura, T., Saji, F., Murata, Y., Eguchi, N., Oda, H., Urade, Y., and Hayaishi, O. (1998). Lipocalin-type prostaglandin D synthase in human male reproductive organs and seminal plasma. *Biol. Reprod.* 58, 600–607.
789. Urade, Y., Nagata, A., Suzuki, Y., Fujii, Y., and Hayaishi, O. (1989). Primary structure of rat brain prostaglandin D synthetase deduced from cDNA sequence. *J. Biol. Chem.* 264, 1041–1045.
790. Nagata, A., Suzuki, Y., Igarashi, M., Eguchi, N., Toh, H., Urade, Y., and Hayaishi, O. (1991). Human brain prostaglandin D synthase has been evolutionarily differentiated from lipophilic-ligand carrier proteins. *Proc. Natl. Acad. Sci. U S A* 88, 4020–4024.
791. Hoffmann, A., Bächner, D., Betar, N., Lauber, J., and Gross, G. (1996). Developmental expression of murine β-trace in embryos and adult animals suggests a function in maturation and maintenance of blood-tissue barriers. *Dev. Dynamics* 207, 332–343.
792. Hoffmann, A., Gath, U., Gross, G., Lauber, J., Getzlaff, R., Hellwig, S., Galla, H. J., and Conradt, H. S. (1996). Constitutive secretion of beta-trace protein by cultivated porcine choroid plexus epithelial cells: elucidation of its complete amino acid and cDNA sequences. *J. Cell. Physiol.* 169, 235–241.
793. Lepperdinger, G., Strobl, B., Jilek, A., Weber, A., Thalhamer, J., Flöckner, H., and Mollay, C. (1996). The lipocalin Xlcpll expressed in the neural plate of *Xenopus laevis* embryos is a secreted retinaldehyde binding protein. *Protein Sci.* 5, 1250–1260.
794. Achen, M. G., Harms, P. J., Thomas, T., Richardson, S. J., Wettenhall, R. E. H., and Schreiber, G. (1992). Protein synthesis at the blood–brain barrier: the major protein secreted by amphibian choroid plexus is a lipocalin. *J. Biol. Chem.* 267, 23170–23174.
795. Urade, Y., Tanaka, T., Eguchi, N., Kikuchi, M., Kimura, H., Toh, H., and Hayaishi, O. (1995). Structural and functional significance of cysteine residues of glutathione-independent prostaglandin D synthase: identification of Cys^{65} as an essential thiol. *J. Biol. Chem.* 270, 1422–1428.
796. Igarashi, M., Nagata, A., Toh, H., Urade, Y., and Hayaishi, O. (1992). Structural organization of the gene for prostaglandin D synthase in the rat brain. *Proc. Natl. Acad. Sci. U S A* 89, 5376–5380.
797. White, D. M., Mikol, D. D., Espinosa, R., Weimer, B., Le Beau, M. M., and Stefansson, K. (1992). Structural and chromosomal localization of the human gene for a brain form of prostaglandin D_2 synthase. *J. Biol. Chem.* 267, 23202–23208.
798. Salier, J.-P. (2000). Chromosomal location, exon/intron organization and evolution of lipocalin genes. *Biochim. Biophys. Acta* 1482, 25–34.
799. Zhu, H., Ma, H., Ni, H., Ma, X.-H., Mills, N., and Yang, Z.-M. (2004). Expression and regulation of lipocalin-type prostaglandin D synthase in rat testis and epididymis. *Biol. Reprod.* 70, 1088–1095.
800. Rodríguez, C. M., Day, J. R., and Killian, G. J. (2000). Expression of the lipocalin-type prostaglandin D synthase gene in the reproductive tracts of Holstein bulls. *J. Reprod. Fertil.* 120, 303–309.
801. Kjeldsen, L., Cowland, J. B., and Borregaard, N. (2000). Human neutrophil gelatinase-associated lipocalin and homologous proteins in rat and mouse. *Biochim. Biophys. Acta* 1482, 272–283.
802. Xu, S. Y., Carlson, M., Engström, Å, Garcia, R., Peterson, C. G. B., and Venge, P. (1994). Purification and characterization of a human neutrophil lipocalin (HNL) from the secondary granules of human neutrophils. *Scand. J. Clin. Lab. Invest.* 54, 365–376.
803. Kjeldsen, L., Johnsen, A. H., Sengelov, H., and Borregaard, N. (1993). Isolation and primary structure of NGAL, a novel protein associated with human neutrophil gelatinase. *J. Biol. Chem.* 268, 10425–10432.
804. Chu, S. T., Lin, H.-J., and Chen, Y.-H. (1997). Complex formation between a formyl peptide and 24p3 protein with a blocked N-terminus of pyroglutamate. *J. Pept. Res.* 49, 582–585.
805. Bundgaard, J. R., Sengelov, H., Borregaard, N., and Kjeldsen, L. (1994). Molecular cloning and expression of a cDNA encoding NGAL: a lipocalin expressed in human neutrophils. *Biochem. Biophys. Res. Commun.* 202, 1468–1475.
806. Bartsch, S., and Tschesche, H. (1995). Cloning and expression of human neutrophil lipocalin cDNA derived from bone marrow and ovarian cancer cells. *FEBS Lett.* 357, 255–259.
807. Goetz, D. H., Willie, S. T., Armen, R. S., Bratt, T., Borregaard, N., and Strong, R. K. (2000). Ligand preference inferred from the structure of neutrophil gelatinase associated lipocalin. *Biochemistry* 39, 1935–1941.
808. Gutiérrez, G., Ganfornina, M. D., and Sánchez, D. (2000). Evolution of the lipocalin family as inferred from a protein sequence phylogeny. *Biochim. Biophys. Acta* 1482, 35–45.
809. Morel, L., Dufaure, J. P., and Depeiges, A. (1993). LESP, an androgen-regulated lizard epididymal secretory protein family identified as a new member of the superfamily. *J. Biol. Chem.* 268, 10274–10281.
810. Gue, Z., Stelnmetz, L. M., Gu, X., Scharfe, C., Davis, R. W., and Li, W.-H. (2003). Role of duplicate genes in genetic robustness against null mutations. *Nature* 421, 63–66.
811. Ohno, S. (1970). *Evolution by Gene Duplication.* Berlin, Springer-Verlag.
812. Lynch, M., and Force, A. (2000). The probability of duplicate gene preservation by subfunctionalization. *Genetics* 154, 459–473.

813. Waterston, R. H., Lindblad-Toh, K., Birney, E., Rogers, J., Abril, J. F., Agarwal, P., Agarwala, R., Ainscough, R., Alexandersson, M., An, P., Antonarakis, S. E., Attwood, J., Baertsch, R., Bailey, J., Barlow, K., Beck, S., Berry, E., Birren, B., Bloom, T., Bork, P., Botcherby, M., Bray, N., Brent, M. R., Brown, D. G., Brown, S. D., Bult, C., Burton, J., Butler, J., Campbell, R. D., Carninci, P., Cawley, S., Chiaromonte, F., Chinwalla, A. T., Church, D. M., Clamp, M., Clee, C., Collins, F. S., Cook, L. L., Copley, R. R., Coulson, A., Couronne, O., Cuff, J., Curwen, V., Cutts, T., Daly, M., David, R., Davies, J., Delehaunty, K. D., Deri, J., Dermitzakis, E. T., Dewey, C., Dickens, N. J., Diekhans, M., Dodge, S., Dubchak, I., Dunn, D. M., Eddy, S. R., Elnitski, L., Emes, R. D., Eswara, P., Eyras, E., Felsenfeld, A., Fewell, G. A., Flicek, P., Foley, K., Frankel, W. N., Fulton, L. A., Fulton, R. S., Furey, T. S., Gage, D., Gibbs, R. A., Glusman, G., Gnerre, S., Goldman, N., Goodstadt, L., Grafham, D., Graves, T. A., Green, E. D., Gregory, S., Guigo, R., Guyer, M., Hardison, R. C., Haussler, D., Hayashizaki, Y., Hillier, L. W., Hinrichs, A., Hlavina, W., Holzer, T., Hsu, F., Hua, A., Hubbard, T., Hunt, A., Jackson, I., Jaffe, D. B., Johnson, L. S., Jones, M., Jones, T. A., Joy, A., Kamal, M., Karlsson, E. K., Karolchik, D., Kasprzyk, A., Kawai, J., Keibler, E., Kells, C., Kent, W. J., Kirby, A., Kolbe, D. L., Korf, I., Kucherlapati, R. S., Kulbokas, E. J., Kulp, D., Landers, T., Leger, J. P., Leonard, S., Letunic, I., Levine, R., Li, J., Li, M., Lloyd, C., Lucas, S., Ma, B., Maglott, D. R., Mardis, E. R., Matthews, L., Mauceli, E., Mayer, J. H., McCarthy, M., McCombie, W. R., McLaren, S., McLay, K., McPherson, J. D., Meldrim, J., Meredith, B., Mesirov, J. P., Miller, W., Miner, T. L., Mongin, E., Montgomery, K. T., Morgan, M., Mott, R., Mullikin, J. C., Muzny, D. M., Nash, W. E., Nelson, J. O., Nhan, M. N., Nicol, R., Ning, Z., Nusbaum, C., O'Connor, M. J., Okazaki, Y., Oliver, K., Overton-Larty, E., Pachter, L., Parra, G., Pepin, K. H., Peterson, J., Pevzner, P., Plumb, R., Pohl, C. S., Poliakov, A., Ponce, T. C., Ponting, C. P., Potter, S., Quail, M., Reymond, A., Roe, B. A., Roskin, K. M., Rubin, E. M., Rust, A. G., Santos, R., Sapojnikov, V., Schultz, B., Schultz, J., Schwartz, M. S., Schwartz, S., Scott, C., Seaman, S., Searle, S., Sharpe, T., Sheridan, A., Shownkeen, R., Sims, S., Singer, J. B., Slater, G., Smit, A., Smith, D. R., Spencer, B., Stabenau, A., Stange-Thomann, N., Sugnet, C., Suyama, M., Tesler, G., Thompson, J., Torrents, D., Trevaskis, E., Tromp, J., Ucla, C., Ureta-Vidal, A., Vinson, J. P., Von Niederhausern, A. C., Wade, C. M., Wall, M., Weber, R. J., Weiss, R. B., Wendl, M. C., West, A. P., Wetterstrand, K., Wheeler, R., Whelan, S., Wierzbowski, J., Willey, D., Williams, S., Wilson, R. K., Winter, E., Worley, K. C., Wyman, D., Yang, S., Yang, S. P., Zdobnov, E. M., Zody, M. C., and Lander E. S. (2002). Initial sequencing and comparative analysis of the mouse genome. *Nature* 420, 520–562.

814. Dujon, B., Sherman, D., Fischer, G., Durrens, P., Casaregola, S., Lafontaine, I., De Montigny, J., Marck, C., Neuveglise, C., Talla, E., Goffard, N., Frangeul, L., Aigle, M., Anthouard, V., Babour, A., Barbe, V., Barnay, S., Blanchin, S., Beckerich, J. M., Beyne, E., Bleykasten, C., Boisrame, A., Boyer, J., Cattolico, L., Confanioleri, F., De Daruvar, A., Despons, L., Fabre, E., Fairhead, C., Ferry-Dumazet, H., Groppi, A., Hantraye, F., Hennequin, C., Jauniaux, N., Joyet, P., Kachouri, R., Kerrest, A., Koszul, R., Lemaire, M., Lesur, I., Ma, L., Muller, H., Nicaud, J. M., Nikolski, M., Oztas, S., Ozier-Kalogeropoulos, O., Pellenz, S., Potier, S., Richard, G. F., Straub, M. L., Suleau, A., Swennen, D., Tekaia, F., Wesolowski-Louvel, M., Westhof, E., Wirth, B., Zeniou-Meyer, M., Zivanovic, I., Bolotin-Fukuhara, M., Thierry, A., Bouchier, C., Caudron, B., Scarpelli, C., Gaillardin, C., Weissenbach, J., Wincker, P., and Souciet, J. L. (2004). Genome evolution in yeasts. *Nature* 430, 35–44.

815. Kellis, M., Birren, B. W., and Lander, E. S. (2004). Proof and evolutionary analysis of ancient genome duplication in the yeast *Saccharomyces cerevisiae*. *Nature* 428, 617–624.

816. Dietrich, F. S., Voegeli, S., Brachat, S., Lerch, A., Gates, K., Steiner, S., Mohr, C., Pohlmann, R., Luedi, P., Choi, S., Wing, R. A., Flavier, A., Gaffney, T. D., and Philippsen, P. (2004). The *Ashbya gossypii* genome as a tool for mapping the ancient *Saccharomyces cerevisiae* genome. *Science* 304, 304–307.

817. Fortna, A., Kim, Y., MacLaren, E., Marshall, K., Hahn, G., Meltesen, L., Brenton, M., Hink, R., Burgers, S., Hernandez-Boussard, T., Karimpour-Fard, A., Glueck, D., McGavran, L., Berry, R., Pollack, J., and Sikela, J. M. (2004). Lineage-specific gene duplication and loss in human and great ape evolution. *PLoS Biol.* 2, 0937–0954.

818. Glasgow, B. J., Abduragimov, A. R., Farahbakhsh, Z. T., Faull, K. F., and Hubbell, W. L. (1995). Tear lipocalins bind a broad array of lipid ligands. *Curr. Eye Res.* 14, 363–372.

819. Gasymov, O. K., Abduragimov, A. R., Yusifov, T. N., and Glasgow, B. J. (1999). Binding studies of tear lipocal: the role of the conserved tryptophan in maintaining structure, stability and ligand affinity. *Biochim. Biophys. Acta* 1433, 307–320.

820. Fluckinger, M., Haas, H., Merschak, P., Glasgow, B. J., and Redl, B. (2004). Human tear lipocalin exhibits antimicrobial activity by scavenging microbial siderophores. *Antimicrob. Agents Chemother.* 48, 3367–3372.

821. Doneanu, C. E., Strong, R. K., and Howald, W. N. (2004). Characterization of a noncovalent lipocalin complex by liquid chromatography/electrospray ionization mass spectrometry. *J. Biomol. Tech.* 15, 208–212.

822. Goetz, D. H., Holmes, M. A., Borregaard, N., Bluhm, M. E., Raymond, K. N., and Strong, R. K. (2002). The neutrophil lipocalin NGAL is a bacteriostatic agent that interferes with siderophore-mediated iron acquisition. *Mol. Cell.* 10, 1033–1043.

823. Cowland, J. B., Sorensen, O. E., Sehested, M., and Borregaard, N. (2003). Neutrophil gelatinase-associated lipocalin is up-regulated in human epithelial cells by IL-1β, but not by TNF-α. *J. Immunol.* 171, 6630–6639.

824. Flo, T. H., Smith, K. D., Sato, S., Rodriguez, D. J., Holmes, M. A., Strong, R. K., Akira, S., and Aderem, A. (2004). Lipocalin 2 mediates an innate immune response to bacterial infection by sequestrating iron. *Nature* 432, 917–921.

825. Elangovan, N., Lee, Y.-C., Tzeng, W.-F., and Chu, S.-T. (2004). Delivery of ferric ion to mouse spermatozoa is mediated by lipocalin internalization. *Biochem. Biophys. Res. Commun.* 319, 1096–1104.

826. Lee, Y.-C., Liao, C., Jr., Li, P.-T., Tzeng, W.-F., and Chu, S.-T. (2003). Mouse lipocalin as an enhancer of spermatozoa motility. *Mol. Biol. Rep.* 30, 165–172.

827. Tanaka, T., Urade, Y., Kimura, H., Eguchi, N., Nishikawa, A., and Hayaishi, O. (1997). Lipocalin-type prostaglandin D synthase (β-trace) is a newly recognized type of retinoid transporter. *J. Biol. Chem.* 272, 15789–15795.

828. Beuckmann, C. T., Aoyagi, M., Okazaki, I., Hiroike, T., Toh, H., Hayaishi, O., and Urade, Y. (1999). Binding of biliverdin, bilirubin, and thyroid hormones to lipocalin-type prostaglandin D synthase. *Biochemistry* 38, 8006–8013.

829. Eguchi, N., Minami, T., Shirafuji, N., Kanaoka, Y., Tanaka, T., Nagata, A., Yoshida, N., Urade, Y., Ito, S., and Hayaishi, O. (1999). Lack of tactile pain (allodynia) in lipocalin-type prostaglandin D synthase-deficient mice. *Proc. Natl. Acad. Sci. U S A* 96, 726–730.

830. Diamandis, E. P., Arnett, W. P., Foussias, G., Pappas, H., Ghandi, S., Melegos, D. N., Mullen, B., Yu, H., Srigley, J., and Jarvi, K. (1999). Seminal plasma biochemical markers and their association with semen analysis findings. *Urology* 53, 596–603.

831. Ghannam, S., Shehata, O., Deeb, S., and al-Alily, H. (1969). The effect of vitamin A depletion on the vasa deferentia of young bulls. *Res. Vet. Sci.* 10, 79–82.

832. Cran, D. G., and Jones, R. (1980). Aging of male reproductive system: changes in the epididymis. *Exp. Gerontol.* 15, 93–101.

833. Elcock, L. H., and Schoning, P. (1984). Age-related changes in the cat testis and epididymis. *Am. J. Vet. Res.* 45, 2380–2384.

833a. Calvo, A., Pastor, L. M., Martinez, E., Vazquez, J. M., and Roca, J. (1999). Age-related changes in the hamster epididymis. *Anat. Rec.* 256, 335–346.

834. Taylor, G. T., Weiss, J., and Pitha, J. (1988). Epididymal sperm profiles in young adult, middle-aged, and testosterone-supplemented old rats. *Gamete Res.* 19, 401–409.

835. Markey, C. M., and Meyer, G. T. (1992). A quantitative description of the epididymis and its microvasculature: an age-related study in the rat. *J. Anat.* 180, 255–262.

836. Regadera, J., Nistal, M., and Paniagua, R. (1985). Testis, epididymis, and spermatic cord in elderly men: correlation of angiographic and histologic studies with systemic arteriosclerosis. *Arch. Pathol. Lab. Med.* 109, 663–667.

837. Mitchinson, M. J., Sherman, K. P., and Stainer-Smith, A. M. (1975). Brown patches in the epididymis. *J. Pathol.* 115, 57–62.

838. Baskerville, A., Cook, R. W., Dennis, M. J., Cranage, M. P., and Greenaway, P. J. (1992). Pathological changes in the reproductive tract of male rhesus monkeys associated with age and simian AIDS. *J. Comp. Pathol.* 107, 49–57.

839. Zirkin, B. R., Santulli, R., Strandberg, J. D., Wright, W. W., and Ewing, L. L. (1993). Testicular steroidogenesis in the aging brown Norway rat. *J. Androl.* 14, 118–123.

840. Wang, C., Leung, A., and Sinha-Hikim, A. (1993). Reproductive aging in the male brown-Norway rat: a model for the human. *Endocrinology* 133, 2773–2281.

841. Gruenewald, D. A., Naai, M. A., Hess, D. L., and Matsumoto, A. M. (1994). The brown Norway rat as a model of male reproductive aging: evidence for both primary and secondary testicular failure. *J. Gerontol.* 49, 842–850.

842. Wright, W. W., Fiore, C., and Zirkin, B. R. (1993). The effects of aging in the seminiferous epithelium of the brown Norway rat. *J. Androl.* 14, 110–117.

843. Gibbs, R. A., Weinstock, G. M., Metzker, M. L., Muzny, D. M., Sodergren, E. J., Scherer, S., Scott, G., Steffen, D., Worley, K. C., Burch, P. E., Okwuonu, G., Hines, S., Lewis, L., DeRamo, C., Delgado, O., Dugan-Rocha, S., Miner, G., Morgan, M., Hawes, A., Gill, R., Celera, Holt, R. A., Adams, M. D., Amanatides, P. G., Baden-Tillson, H., Barnstead, M., Chin, S., Evans, C. A., Ferriera, S., Fosler, C., Glodek, A., Gu, Z., Jennings, D., Kraft, C. L., Nguyen, T., Pfannkoch, C. M., Sitter, C., Sutton, G. G., Venter, J. C., Woodage, T., Smith, D., Lee, H. M., Gustafson, E., Cahill, P., Kana, A., Doucette-Stamm, L., Weinstock, K., Fechtel, K., Weiss, R. B., Dunn, D. M., Green, E. D., Blakesley, R. W., Bouffard, G. G., De Jong, P. J., Osoegawa, K., Zhu, B., Marra, M., Schein, J., Bosdet, I., Fjell, C., Jones, S., Krzywinski, M., Mathewson, C., Siddiqui, A., Wye, N., McPherson, J., Zhao, S., Fraser, C. M., Shetty, J., Shatsman, S., Geer, K., Chen, Y., Abramzon, S., Nierman, W. C., Havlak, P. H., Chen, R., Durbin, K. J., Egan, A., Ren, Y., Song, X. Z., Li, B., Liu ,Y., Qin, X., Cawley, S., Worley, K. C., Cooney, A. J., D'Souza, L. M., Martin, K., Wu, J. Q., Gonzalez-Garay, M. L., Jackson, A. R., Kalafus, K. J., McLeod, M. P., Milosavljevic, A., Virk, D., Volkov, A., Wheeler, D. A., Zhang, Z., Bailey, J. A., Eichler, E. E., Tuzun, E., Birney, E., Mongin, E., Ureta-Vidal, A., Woodwark, C., Zdobnov, E., Bork, P., Suyama, M., Torrents, D., Alexandersson, M., Trask, B. J., Young, J. M., Huang, H., Wang, H., Xing, H., Daniels, S., Gietzen, D., Schmidt, J., Stevens, K., Vitt, U., Wingrove, J., Camara, F., Mar Alba, M., Abril, J. F., Guigo, R., Smit, A., Dubchak, I., Rubin, E. M., Couronne, O., Poliakov, A., Hubner, N., Ganten, D., Goesele, C., Hummel, O., Kreitler, T., Lee, Y. A., Monti, J., Schulz, H., Zimdahl, H., Himmelbauer, H., Lehrach, H., Jacob, H. J., Bromberg, S., Gullings-Handley, J., Jensen-Seaman, M. I., Kwitek, A. E., Lazar, J., Pasko, D., Tonellato, P. J., Twigger, S., Ponting, C. P., Duarte, J. M., Rice, S., Goodstadt, L., Beatson, S. A., Emes, R. D., Winter, E. E., Webber, C., Brandt, P., Nyakatura, G., Adetobi, M., Chiaromonte, F., Elnitski, L., Eswara, P., Hardison, R. C., Hou, M., Kolbe, D., Makova, K., Miller, W., Nekrutenko, A., Riemer, C., Schwartz, S., Taylor, J., Yang, S., Zhang, Y., Lindpaintner, K., Andrews, T. D., Caccamo, M., Clamp, M., Clarke, L., Curwen, V., Durbin, R., Eyras, E., Searle, S. M., Cooper, G. M., Batzoglou, S., Brudno, M., Sidow, A., Stone, E. A., Venter, J. C., Payseur, B. A., Bourque, G., Lopez-Otin, C., Puente, X. S., Chakrabarti, K., Chatterji, S., Dewey, C., Pachter, L., Bray, N., Yap, V. B., Caspi, A., Tesler, G., Pevzner, P. A., Haussler, D., Roskin, K. M., Baertsch, R., Clawson, H., Furey, T. S., Hinrichs, A. S., Karolchik, D., Kent, W. J., Rosenbloom, K. R., Trumbower, H., Weirauch, M., Cooper, D. N., Stenson, P. D., Ma, B., Brent, M., Arumugam, M., Shteynberg, D., Copley, R. R., Taylor, M. S., Riethman, H., Mudunuri, U., Peterson, J., Guyer, M., Felsenfeld, A., Old, S., Mockrin, S., and Collins, F. (2004). Genome sequence of the brown Norway rat yields insights into mammalian evolution. *Nature* 428, 493–521.

844. Neaves, W. B., Johnson, L., and Petty, C. S. (1987). Seminiferous tubules and daily sperm production in older adult with varied numbers of Leydig cells. *Biol. Reprod.* 36, 301–308.

845. Johnson, L., Varner, D. D., Roberts, M. E., Smith, T. L., Keillor, G. E., and Scrutchfield, W. L. (2000). Efficiency of spermatogenesis: a comparative approach. *Anim. Reprod. Sci.* 60–61, 471–480.

846. Zirkin, B. R., and Chen, H. (2000). Regulation of Leydig cell steroidogenic function during aging. *Biol. Reprod.* 63, 977–981.

847. Viger, R. S., and Robaire, B. (1995). Gene expression in the aging brown Norway rat epididymis. *J. Androl.* 16, 108–117.

848. Zou, S., Meadows, S., Sharp, L., Jan, L. Y., and Jan, Y. N. (2000). Genome-wide study of aging and oxidative stress response in *Drosophila melanogaster*. *Proc. Natl. Acad. Sci. U S A* 97, 13726–13731.

849. Jiang, C. H., Tsien, J. Z., Schultz, P. G., and Hu, Y. (2001). The effects of aging on gene expression in the hypothalamus and cortex of mice. *Proc. Natl. Acad. Sci. U S A* 98, 1930–1934.

850. Weindruch, R., Kayo, T., Lee, C. K., and Prolla, T. A. (2001). Microarray profiling of gene expression in aging and its alteration by caloric restriction in mice. *J. Nutr.* 131, 918S–923S.

851. Guarente, L., and Kenyon, C. (2000). Genetic pathways that regulate ageing in model organisms. *Nature* 408, 255–262.

852. Ohyama, Y., Kurabayashi, M., Masuda, H., Nakamura, T., Aihara, Y., Kaname, T., Suga, T., Arai, M., Aizawa, H., Matsumura, Y., Kuro-o, M., Nabeshima, Y., and Nagail, R. (1998). Molecular cloning of rat klotho cDNA: markedly decreased expression of klotho by acute inflammatory stress. *Biochem. Biophys. Res. Commun.* 251, 920–925.

853. Klapper, W., Parwaresch, R., and Krupp, G. (2001). Telomere biology in human aging and aging syndromes. *Mech. Ageing Dev.* 122, 695–712.

854. Martin, G. M. (1997). Genetics and the pathobiology of ageing. *Philos. Trans. R. Soc. Lond. B. Biol. Sci.* 352, 1773–1780.

855. Tollefsbol, T. O., and Andrews, L. G. (1993). Mechanisms for methylation-mediated gene silencing and aging. *Med. Hypotheses* 41, 83–92.

856. Franceschi, C., Valensin, S., Bonafe, M., Paolisso, G., Yashin, A. I., Monti, D., and De Benedictis, G. (2000). The network and the remodeling theories of aging: historical background and new perspectives. *Exp. Gerontol.* 35, 879–896.

857. Finkel, T., and Holbrook, N. J. (2000). Oxidants, oxidative stress and the biology of ageing. *Nature* 408, 239–247.
858. Jara, M., Carballada, R., and Esponda, P. (2004). Age-induced apoptosis in the male genital tract of the mouse. *Reproduction* 127, 359–366.
859. Robaire, B., and Serre, V. (2000). Aging causes structural and functional alterations. In *The Testis: From Stem Cell to Sperm Function* (E. Goldberg, Ed.), pp. 174–185. Springer-Verlag, New York.
860. Mueller, A., Hermo, L., and Robaire, B. (1998). The effects of aging on the expression of glutathione S-transferases in the testis and epididymis of the brown Norway rat. *J. Androl.* 19, 450–465.
861. Taylor, G. T., Weiss, J., and Frechmann, T. (1986). Ontogeny of epididymal sperm reserved during the reproductive lifespan of rats after previous sexual experiences. *J. Reprod. Fertil.* 77, 419–423.
862. Sloter, E., Nath, J., Eskenazi, B., and Wyrobek, A. J. (2004). Effects of male age on the frequencies of germinal and heritable chromosomal abnormalities in humans and rodents. *Fertil. Steril.* 81, 925–943.
863. Kidd, S. A., Eskenazi, B., and Wyrobeck, A. J. (2001). Effects of male age on semen quality and fertility: a review of the literature. *Fertil. Steril.* 75, 237–248.
864. Plas, E., Berger, P., Hermann, M., and Pfluger, H. (2000). Effect of aging on male fertility? *Exp. Gerontol.* 35, 543–551.
865. Hassan, M. A. M., and Killick, S. R. (2003). Effect of male age on fertility: evidence for the decline in male fertility with increasing age. *Fertil. Steril.* 79 (Suppl. 3), 1520–1527.
866. Robaire B. (2001). Aging of the epididymis. In *The Epididymis: From Molecules to Clinical Practice* (B. Robaire, B. T. Hinton, Eds.), pp. 285–296. Kluwer Academic/Plenum, New York.
867. Muller, F., Rebiffe, M., Taillandier, A., Oury, J. F., and Mornet, E. (2000). Parental origin of the extra chromosome in prenatally diagnosed fetal trisomy 21. *Hum. Genet.* 106, 340–344.
868. MacDonald, M., Hassold, T., Harvey, J., Wang, L. H., Morton, N. E., and Jacobs, P. (1994). The origin of 47,XXY and 47,XXX aneuploidy: heterogeneous mechanisms and role of aberrant recombination. *Hum. Mol. Genet.* 3, 1365–1371.
869. Crow, J. F. (2000). The origins, patterns and implications of human spontaneous mutation. *Nat. Rev. Genet.* 1, 40–47.
870. Geyer C. B., Kiefer C. M., Yang T. P., and McCarrey J. R. (2004). Ontogeny of a demethylation domain and its relationship to activation of tissue-specific transcription. *Biol. Reprod.* 71, 837–844.
871. Syntin, P., and Robaire, B. (2001). Sperm structural and motility changes during aging in the brown Norway rat. *J. Androl.* 22, 235–244.
872. Wu, F. C. (1988). Male contraception: current status and future prospects. *Clin. Endocrinol. (Oxf.)* 29, 443–465.
873. Hess, R. A. (1998). Effects of environmental toxicants on the efferent ducts, epididymis and fertility. *J. Reprod. Fertil. Suppl.* 53, 247–259.
874. Cooper, T. G., and Yeung, C. H. (1999). Recent biochemical approaches to post-testicular, epididymal contraception. *Hum. Reprod. Update* 5, 141–152.
875. Klinefelter, G. R. (2002). Action of toxicants on the structure and functions of the epididymis. In *The Epididymis: From Molecules to Clinical Practice* (B. Robaire and B. T. Hinton, Eds.), pp. 353–369. Kluwer Academic/Plenum, New York.
876. Cooper, T. G. (2002). The epididymis as a target for male contraception. In *The Epididymis: From Molecules to Clinical Practice* (B. Robaire and B. T. Hinton, Eds.), pp. 483–502. Kluwer Academic/Plenum, New York.
877. Robaire, B. (2003). Advancing towards a male contraceptive: a novel approach from an unexpected direction. *Trends Pharmacol. Sci.* 24, 326–328.
878. Toyoda, K., Shibutani, M., Tamura, T., Koujitani, T., Uneyama, C., and Hirose, M. (2000). Repeated dose (28 days) oral toxicity study of flutamide in rats, based on the draft protocol for the "Enhanced OECD Test Guideline 407" for screening for endocrine-disrupting chemicals. *Arch. Toxicol.* 74, 127–132.
879. Schneider, H. P. (2003). Androgens and antiandrogens. *Ann. N. Y. Acad. Sci.* 997, 292–306.
880. Paris, F., Weinbauer, G. F., Blum, V., and Nieschlag, E. (1994). The effect of androgens and antiandrogens on the immunohistochemical localization of the androgen receptor in accessory reproductive organs of male rats. *J. Steroid Biochem. Mol. Biol.* 48, 129–137.
881. Silver, R. I., Wiley, E. L., Thigpen, A. E., Guileyardo, J. M., McConnell, J. D., and Russell, D. W. (1994). Cell type specific expression of steroid 5 alpha-reductase 2. *J. Urol.* 152, 438–442.
882. Orgebin-Crist, M.-C., Jahad, N., and Hoffman, L. H. (1976). The effects of testosterone, 5α-dihydrotestosterone, 3α-androstanediol, and 3β-androstanediol on the maturation of rabbit epididymal spermatozoa in organ culture. *Cell. Tissue Res.* 167, 515–525.
883. Katashima, M., Irino, T., Shimojo, F., Kawamura, A., Kageyama, H., Higashi, N., Miyao, Y., Tokuma, Y., Hata, T., Yamamoto, K., Sawada, Y., and Iga, T. (1998). Pharmacokinetics and pharmacodynamics of FK143, a nonsteroidal inhibitor of steroid 5 alpha-reductase, in healthy volunteers. *Clin. Pharmacol. Ther.* 63, 354–366.
884. Robaire, B., Covey, D. F., Robinson, C. H., and Ewing, L. L. (1977). Selective inhibition of rat epididymal steroid Δ4-5α-reductase by conjugated allenic 3-oxo-5,10-secosteroids. *J. Steroid. Biochem.* 8, 307–310.
885. Rasmusson, G. H., Reynolds, G. F., Utne, T., Jobson, R. B., Primka, R. L., Berman, C., and Brooks, J. R. (1984). Azasteroids as inhibitors of rat prostatic 5α-reductase. *J. Med. Chem.* 27, 1690–1701.
886. Cooke, G. M., and Robaire, R. (1986). The effects of diethyl-4-methyl-3-oxo-4-aza-5α-androstane-17β-carboxamide (4-MA) and (4R)-5,10-seco-19-norpregna-4,5-diene-3,10,20-trione (SECO) on androgen biosynthesis in the rat testis and epididymis. *J. Steroid Biochem.* 24, 877–886.
887. Zaccheo, T., Giudici, D., and di Salle, E. (1998). Effect of the dual 5α-reductase inhibitor PNU 157706 on the growth of dunning R3327 prostatic carcinoma in the rat. *J. Steroid Biochem. Mol. Biol.* 64, 193–198.
888. Occhiato, E. G., Guarna, A., Danza, G., and Serio, M. (2004). Selective non-steroidal inhibitors of 5 alpha-reductase type 1. *J. Steroid Biochem. Mol. Biol.* 88, 1–16.
889. Henderson, N. A., Cooke, G. M., and Robaire, B. (2004). Effects of PNU157706, a dual 5α-reductase inhibitor on gene expression in the rat epididymis. *J. Endocrinol.* 181, 245–261.
890. Liang, T., Brooks, J. R., Cheung, A., Reynolds, G. F., and Rasmusson, G. H. (1984). 4-Azasteroids as inhibitors of 5α-reductase. In *Hormones and Cancer 2* (F. Bresciani, J. B. K. King, M. E. Lippman, M. Namer, and J. P. Raynaud, Eds.), pp. 497–505. Raven Press, New York.
891. Brooks, J. R., Berman, C., Hichens, M., Primka, R. L., Reynolds, G. F., and Rasmusson, G. H. (1982). Biological activities of a new steroidal inhibitor of Δ^4-5α-reductase (41309). *Proc. Soc. Exp. Biol. Med.* 169, 67–73.
892. Wilde, M. I., and Goa, K. L. (1999). Finasteride: an update of its use in the management of symptomatic benign prostatic hyperplasia. *Drugs* 57, 557–581.
893. Evans, H. C., and Goa, K. L. (2003). Dutasteride. *Drugs Aging* 20, 905–916.
894. Henderson, N. A., and Robaire, B. (2005). Effects of PNU157706, a dual 5α-reductase inhibitor, on rat epididymal sperm maturation and fertility. *Biol. Reprod.* 72, 436–443.

895. Cho, H. W., Nie, R., Carnes, K., Zhou, Q., Sharief, N. A., and Hess, R. A. (2003). The antiestrogen ICI 182,780 induces early effects on the adult male mouse reproductive tract and long-term decreased fertility without testicular atrophy. *Reprod. Biol. Endocrinol.* 1, 57.
896. Kempinas, W. D., Suarez, J. D., Roberts, N. L., Strader, L. F., Ferrell, J., Goldman, J. M., Narotsky, M. G., Perreault, S. D., Evenson, D. P., Ricker, D. D., and Klinefelter, G. R. (1998). Fertility of rat epididymal sperm after chemically and surgically induced sympathectomy. *Biol. Reprod.* 59, 897–904.
896a. Kempinas, W. D., Suarez, J. D., Roberts, N. L., Strader, L., Ferrell, J., Goldman, J. M., and Klinefelter, G. R. (1998). Rat epididymal sperm quantity, quality, and transit time after guanethidine-induced sympathectomy. *Biol. Reprod.* 59, 890–896.
897. Klinefelter, G. R., and Suarez, J. D. (1997). Toxicant-induced acceleration of epididymal sperm transit: androgen-dependent proteins may be involved. *Reprod. Toxicol.* 11, 511–519.
898. Chaturapanich, G., Sujarit, K., and Pholpramool, C. (1999). Effects of sulphapyridine on sperm transport through the rat epididymis and contractility of the epididymal duct. *J. Reprod. Fertil.* 117, 199–205.
899. Lobl, T. J. (1980). Chlorohydrin: review of a model post-testicular antifertility agent. In *Regulation of Male Fertility* (G. R. Cunningham, W. B. Schill, and E. S. E. Hafez, Eds.), pp. 109–122. Martinus Nijhoff, Hague.
900. Jones, A. R. (1983). Antifertility actions of alpha-chlorohydrin in the male. *Aust. J. Biol. Sci.* 36, 333–350.
901. Jones, A. R. (1998). Chemical interference with sperm metabolic pathways. *J. Reprod. Fertil. Suppl.* 53, 227–234.
902. Slott, V. L., Jeffay, S. C., Dyer, C. J., Barbee, R. R., and Perreault, S. D. (1997). Sperm motion predicts fertility in male hamsters treated with alpha-chlorohydrin. *J. Androl.* 18, 708–716.
903. Ford, W. C., and Waites, G. M. (1982). Activities of various 6-chloro-6-deoxysugars and (S) alpha-chlorohydrin in producing spermatocoeles in rats and paralysis in mice and in inhibiting glucose metabolism in bull spermatozoa in vitro. *J. Reprod. Fertil.* 65, 177–183.
904. Wong, P. Y. D., and Yeung, C. H. (1977). Inhibition by α-chlorohydrin of fluid reabsorption in the rat cauda epididymidis. *J. Reprod. Fertil.* 51, 469–471.
905. Welch, J. E., Schatte, E. C., O'Brien, D. A., and Eddy, E. M. (1992). Expression of a glyceraldehyde 3-phosphate dehydrogenase gene specific to mouse spermatogenic cells. *Biol. Reprod.* 46, 869–878.
906. Kirton, K. T., Ericsson, R. J., Ray, J. A., and Forbes, A. D. (1970). Male antifertility compounds: efficacy of U-5897 in primates (*Macaca mulatta*). *J. Reprod. Fertil.* 21, 275–278.
907. Ford, W. C., and Waites, G. M. (1978). A reversible contraceptive action of some 6-chloro-6-deoxy sugars in the male rat. *J. Reprod. Fertil.* 52, 153–157.
908. Segal, S. (1985). *Gossypol: A Potential Contraceptive for Men.* Plenum, New York.
909. Waites, G. M., Wang, C., and Griffin, P. D. (1998). Gossypol: reasons for its failure to be accepted as a safe, reversible male antifertility drug. *Int. J. Androl.* 21, 8–12.
910. Kumar, M., Sharma, S., and Lohiya, N. K. (1997). Gossypol-induced hypokalemia and role of exogenous potassium salt supplementation when used as an antispermatogenic agent in male langur monkey. *Contraception* 56, 251–256.
911. Yang, Z. J., Ye, W. S., Cui, G. H., Guo, Y., and Xue, S. P. (2004). Combined administration of low-dose gossypol acetic acid with desogestrel/mini-dose ethinylestradiol/testosterone undecanoate as an oral contraceptive for men. *Contraception* 70, 203–211.
912. Romualdo, G. S., Klinefelter, G. R., and de, K. (2002). Postweaning exposure to gossypol results in epididymis-specific effects throughout puberty and adulthood in rats. *J. Androl.* 23, 220–228.
913. Robaire, B., and Hales, B. F. (2003). Mechanisms of action of cyclophosphamide as a male-mediated developmental toxicant. *Adv. Exp. Med. Biol.* 518, 169–180.
914. McClain, R. M., and Downing, J. C. (1988). The effect of ornidazole on fertility and epididymal sperm function in rats. *Toxicol. Appl. Pharmacol.* 92, 488–496.
915. Kackar, R., Srivastava, M. K., and Raizada, R. B. (1997). Induction of gonadal toxicity to male rats after chronic exposure to mancozeb. *Ind. Health.* 35, 104–111.
916. Hess, R. A., and Nakai, M. (2000). Histopathology of the male reproductive system induced by the fungicide benomyl. *Histol. Histopathol.* 15, 207–224.
917. Akbarsha, M. A., Kadalmani, B., Girija, R., Faridha, A., and Hamid, K. S. (2001). Spermatotoxic effect of carbendazim. *Indian J. Exp. Biol.* 39, 921–924.
918. Bone, W., and Cooper, T. G. (2000). In vitro inhibition of rat cauda epididymal sperm glycolytic enzymes by ornidazole, alpha-chlorohydrin and 1-chloro-3-hydroxypropanone. *Int. J. Androl.* 23, 284–293.
919. Chellman, G. J., Bus, J. S., and Working, P. K. (1986). Role of epididymal inflammation in the induction of dominant lethal mutations in Fischer 344 rat sperm by methyl chloride. *Proc. Natl. Acad. Sci. U S A* 83, 8087–8091.
920. Risbridger, G., Kerr, J., and de Kretser, D. (1989). Differential effects of the destruction of Leydig cells by administration of ethane dimethane sulphonate to postnatal rats. *Biol. Reprod.* 40, 801–809.
921. Klinefelter, G. R., Laskey, J. W., Kelce, W. R., Ferrell, J., Roberts, N. L., Suarez, J. D., and Slott, V. (1994). Chloroethylmethanesulfonate-induced effects on the epididymis seem unrelated to altered Leydig cell function. *Biol. Reprod.* 51, 82–91.
922. Klinefelter, G. R., Laskey, J. W., Ferrell, J., Suarez, J. D., and Roberts, N. L. (1997). Discriminant analysis indicates a single sperm protein (SP22) is predictive of fertility following exposure to epididymal toxicants. *J. Androl.* 18, 139–150.
923. Foster, P. M., Mylchreest, E., Gaido, K. W., and Sar, M. (2001). Effects of phthalate esters on the developing reproductive tract of male rats. *Hum. Reprod. Update* 7, 231–235.
924. Zhang, Y., Jiang, X., and Chen, B. (2004). Reproductive and developmental toxicity in F1 Sprague-Dawley male rats exposed to di-n-butyl phthalate in utero and during lactation and determination of its NOAEL. *Reprod. Toxicol.* 18, 669–676.
925. Barthold, J. S., Kryger, J. V., Derusha, A. M., Duel, B. P., Jednak, R., and Skafar, D. F. (1999). Effects of an environmental endocrine disruptor on fetal development, estrogen receptor(alpha) and epidermal growth factor receptor expression in the porcine male genital tract. *J. Urol.* 162, 864–871.
926. Gray, L. E. Jr., and Kelce, W. R. (1996). Latent effects of pesticides and toxic substances on sexual differentiation of rodents. *Toxicol. Ind. Health* 12, 515–531.
927. Smithwick, E. B., Gould, K. G., and Young, L. G. (1996). Estimate of epididymal transit time in the chimpanzee. *Tissue Cell* 28, 485–493.
928. Swierstra, E. E. (1968). Cytology and duration of the cycle of the seminiferous epithelium of the boar: duration of spermatozoan transit through the epididymis. *Anat. Rec.* 161, 171–186.
929. Kennelly, J. J. (1972). Coyote reproduction: I. The duration of the spermatogenic cycle and epididymal sperm transport. *J. Reprod. Fertil.* 31, 163–170.

930. Sinha Hikim, A. P., and Hoffer, A. P. (1988). Duration of epididymal sperm transit in hamster: an autoradiographic study. *Gamete Res.* 19, 411–416.
931. Setchell, B. P., and Carrick, F. N. (1973). Spermatogenesis in some Australian marsupials. *Aust. J. Zool.* 21, 491–499.
932. Amann, R. P., and Howards, S. S. (1980). Daily spermatozoal production and epididymal spermatozoal reserves of the human male. *J. Urol.* 124, 211–215.
933. Johnson, L., and Varner, D. D. (1988). Effect of daily spermatozoan production but not age on transit time of spermatozoa through the human epididymis. *Biol. Reprod.* 39, 812–817.
934. Amann, R. P., Johnson, L., Thompson, D. L., Jr., and Pickett, B. W. (1976). Daily spermatozoal production, epididymal spermatozoal reserves and transit time of spermatozoa through the epididymis of the rhesus monkey. *Biol. Reprod.* 15, 586–592.
935. Gebauer, M. R., Pickett, B. W., and Swierstra, E. E. (1974). Reproductive physiology of the stallion: III. Extra-gonadal transit time and sperm reserves. *J. Anim. Sci.* 39, 737–742.
936. Amann, R. P., Kavanaugh, J. F., Griel, L. C. Jr., and Voglmayr, J. K. (1974). Sperm production of Holstein bulls determined from testicular spermatid reserves, after cannulation of rete testis or vas deferens, and by daily ejaculation. *J. Dairy Sci.* 57, 93–99.
937. Sommer, R. J., Ippolito, D. L., and Peterson, R. E. (1996). In utero and lactational exposure of the male Holtzman rat to 2,3,7,8-tetrachlorodibenzo-p-dioxin: decreased epididymal and ejaculated sperm numbers without alterations in sperm transit rate. *Toxicol. Appl. Pharmacol.* 140, 146–153.
938. Orgebin-Crist, M.-C., and Olson, G. E. (1984). Epididymal sperm maturation. In *The Male in Farm Animal Reproduction* (M. Courot, Ed.), pp. 80–102. Martinus Nijhoff, Amsterdam.
939. Moore, H. D. M. (1981). An assessment of the fertilizing ability of spermatozoa in the epididymis of the marmoset monkey (*Callithrix jacchus*). *Int. J. Androl.* 4, 321–330.
940. Bedford, J. M. (1966). Development of the fertilizing ability of spermatozoa in the epididymis of the rabbit. *J. Exp. Zool.*, 163, 319–330.
941. Holtz, W., and Smidt, D. (1976). The fertilizing capacity of epididymal spermatozoa in the pig. *J. Reprod. Fertil.* 46, 227–229.
942. Fournier-Delpech, S., Colas, G., Courot, M., and Ortavant, R. (1977). Observations on the motility and fertilizing ability of ram epididymal spermatozoa. *Ann. Biol. Anim. Biochem. Biophys.* 17, 987–990.
943. Pavlok, A. (1974). Development of the penetration activity of mouse epididymal spermatozoa in vivo and in vitro. *J. Reprod. Fertil.* 36, 203–205.
944. Hoppe, P. C. (1975). Fertilizing ability of mouse sperm from different epididymal regions and after washing and centrifugation. *J. Exp. Zool.* 192, 219–222.
945. Horan, A. H., and Bedford, J. M. (1972). Development of the fertilizing ability of spermatozoa in the epididymis of the Syrian hamster. *J. Reprod. Fertil.* 30, 417–423.
946. Cummins, J. M. (1976). Effects of epididymal occlusion on sperm maturation in the hamster. *J. Exp. Zool.* 197, 187–190.
947. Moore, H. D. M. (1981). Glycoprotein secretions of the epididymis in the rabbit and hamster: Localization on epididymal spermatozoa and the effects of specific antibodies on fertilization in vivo. *J. Exp. Zool.* 215, 77–85.
948. Weissenberg, R., Yossefi, S., Oschry, Y., Madgar, I., and Lewin, L. M. (1994). Investigation of epididymal sperm maturation in the golden hamster. *Int. J. Androl.* 17, 256–261.
949. Lacham, O., and Trounson, A. (1991). Fertilizing capacity of epididymal and testicular spermatozoa microinjected under the zona pellucida of the mouse oocyte. *Mol. Reprod. Dev.* 29, 85–93.
950. Duling, B. R. (1988). Components of renal function. In *Physiology* (R. M. Berne and J. L. Levy, Eds.), pp. 745–756. CV Mosby, St. Louis.
951. Hinton, B. T., and Setchell, B. P. (1980). Concentrations of glycerylphosphorylcholine, phosphocholine and free inorganic phosphate in the luminal fluid of the rat and some other mammals. *J. Reprod. Fertil.* 58, 401–406.

Knobil and Neill's Physiology of Reproduction, Third Edition
edited by Jimmy D. Neill,
Elsevier © 2006

CHAPTER 23

Physiology of the Male Accessory Sex Structures: The Prostate Gland, Seminal Vesicles, and Bulbourethral Glands

Gail P. Risbridger and Renea A. Taylor

Introduction, 1149
Structure of the Human Prostate Gland, 1150
Embryology and Fetal Development of the Human (and Primate) Prostate Gland, 1150
Pubertal Maturation of the Prostate Gland, 1152
Adult Gland: The Mature Gland, 1152
Structure of the Human and Rodent Prostate Gland, 1153
Histology of Mature Prostate Gland, 1158
Function of the Prostate Gland, 1161
Disease of the Prostate Gland, 1161
Breakdown of Reciprocal Signaling in Disease, 1161
Stem Cell Proliferation and Disordered Growth, 1163
Structure and Function of Seminal Vesicles, 1163
Development of the Seminal Vesicles, 1163
Structure of the Mature Seminal Vesicles, 1164
Function of the Seminal Vesicles, 1164
Bulbourethral (Cowper's) Glands and Skene's Gland, 1165
Emerging Issues and Research Priorities, 1165
References, 1166

ABSTRACT

This review of the male accessory sex organs, including the prostate gland, seminal vesicles, and bulbourethral gland, focuses on their development, normal function, and disease from the viewpoint of a physiologist/biologist. An initial overview of the development and structure of these organs is followed by details of the cell–cell interactions involved in normal homeostasis and perturbations thereof that lead to disease. Emerging research priorities include the implications of specific cell–cell interactions and tissue heterogeneity, the identity of stem cells in normal and diseased tissue, and the early fetal or postnatal origins of adult disease.

INTRODUCTION

The physiology of the male accessory sex structures includes the study of the prostate gland, seminal vesicles, and bulbourethral glands (also known as Cowper's glands). They are organs of the male reproductive system that are responsive to androgens whose function is intimately linked to that of the male gonads or testes; they provide the constituents of seminal plasma in the semen.

The importance or essential nature of these organs and their role in producing constituents of semen is equivocal. The absence of the prostate or seminal vesicles does not render the male infertile, so what is the essential nature of these organs? If the sex accessory structures are nonessential but able to increase fertility in situations of fluctuating or marginal fertility, do they confer an evolutionary advantage to a species? In humans, such a role is becoming increasingly unimportant. Tremendous advances in assisted reproductive technologies have resulted in the ability to achieve fertilization with testicular sperm that have never been exposed to seminal fluid.

So, why study the male accessory sex structures at all? Is the answer simply that there is a need to

Centre for Urology Research, Monash Institute of Reproduction and Development, Monash University, Victoria, Australia.

understand the physiology of these organs, especially the prostate gland, during aging, when benign disease occurs as well as malignancies? Our fundamental understanding of how the sex accessory organs develop and function normally underpins our understanding of how benign and malignant disease develops as men age. Over the last decade, humans have increased their life span so that the older population is getting older. The ability to age well and live a healthy life is a goal of many governments and health care organizations in the Western world. To achieve this aim, it is necessary to understand the normal physiology of organs such as the prostate gland, seminal vesicles, and bulbourethral gland; from this knowledge base, the etiology of benign and malignant disease will emerge.

This review of the male accessory sex structures focuses on their development, normal function, and disease from the viewpoint of a physiologist/biologist. An initial overview of the development and structure of these organs is followed by more detailed discussion of cell–cell interactions during development and in the adult male. The chapter concludes with a discussion of the relevance of normal prostate homeostasis and perturbations leading to disease states, to identify future directions for research in this field.

STRUCTURE OF THE HUMAN PROSTATE GLAND

Embryology and Fetal Development of the Human (and Primate) Prostate Gland

Prostate development begins in the human during the 11th to 12th week of gestation as five pairs of epithelial buds emerge from the urethral portion of the urogenital sinus (UGS), above and below the entrance of the mesonephric ducts (Fig. 1). The five pairs of epithelial buds (anterior, posterior, medial, and two lateral ones) undergo branching to form a lobular arrangement of tubuloalveolar glands surrounded by stroma that encircle the developing urethra and ejaculatory ducts. The top pairs of the buds are composed of epithelia that are believed to be mesodermal in origin and form the inner zones of the mature prostate; the lower buds, which form the outer zones

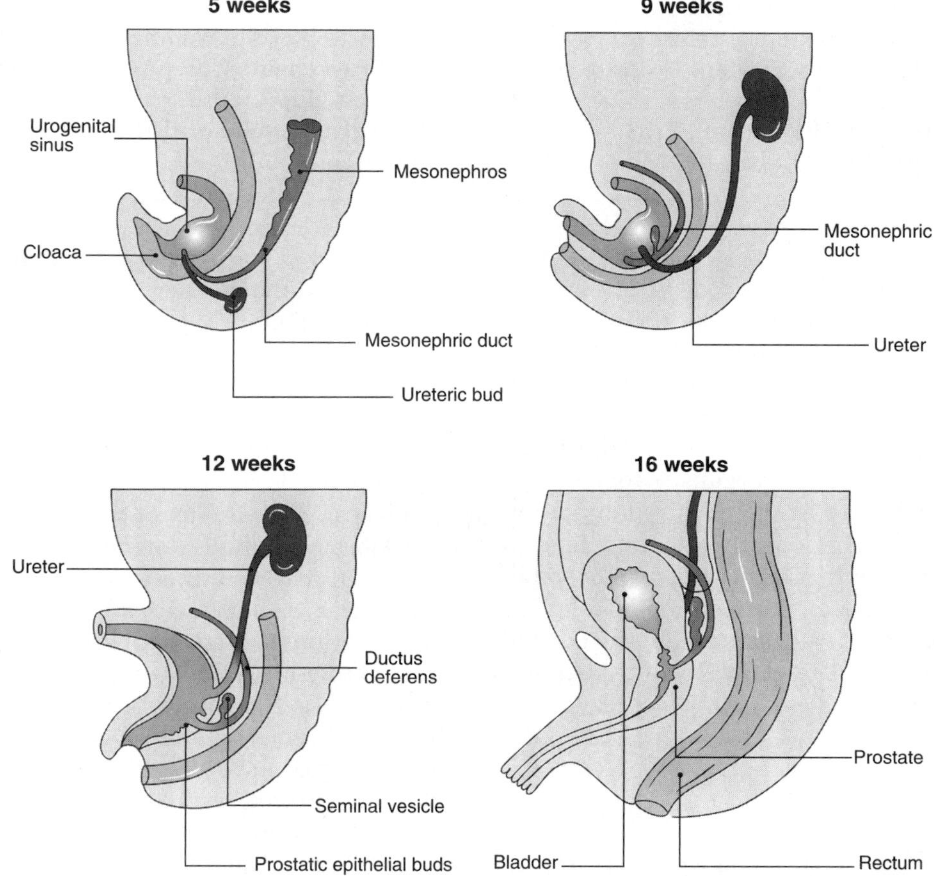

FIG. 1. Embryology of the human prostate. The gland develops at the base of the bladder as epithelial buds emerge from the urethral portion of the urogenital sinus, above and below the entrance of the mesonephric ducts. (Reproduced with permission from Kirby, R. S. [2003]. *An Atlas of Prostatic Diseases*, 3rd ed. The Parthenon Publishing Group, New York.)

of the mature prostate, are endodermal in origin (1). The different origins may be of significance because the zones are susceptible to different disease processes; benign prostatic hyperplasia (BPH) arises mainly in the inner zone and prostate cancer in the outer zone (2). The inner and outer zones of the prostate expand as concentric circles around the urethra and become fused, although the tubules of the individual lobes remain separate, lying adjacent and not intertwining with one another.

Although lobe development is initiated simultaneously, it is not uniform. The posterior tubules are fewer in number but larger, and show evidence of more extensive branching as they extend into the posterior region of the gland. The initial anterior buds develop similarly and form large tubules with many branches. However, this expansion is not maintained and gradually they regress, so that by birth they appear as small, solid epithelial buds. Thus, Lowsley's embryological description of five pairs of buds emerging from the UGS to form a lobular structure was inconsistent with subsequent development of the gland; today, the human prostate is regarded as consisting of inner and outer concentric zones that completely encircle the urethra, rather than as a lobular structure. A smaller central region includes the mucosal and submucosal ducts and the ejaculatory ducts, with the latter extending into the peripheral region as well.

In the primate, the prostate develops into a cranial and caudal prostate; the cranial prostate is believed to be analogous to the central zone of the human prostate and the caudal prostate to the peripheral zone. Unlike the human prostate, the tubules in these regions do not coalesce and lie side by side, but remain separate (Fig. 2).

The initial outgrowth of the epithelial buds is an androgen-driven process that requires androgen receptor expression in the surrounding urogenital mesenchyme to facilitate the reciprocal interactions between the epithelia and mesenchyme (3,4). As budding of the epithelial outgrowths occurs and the tubules branch, the combination of ductal invagination and mesenchymal differentiation results in the formation of a glandular organ composed of tubuloalveolar glands surrounded by mesenchyme containing fibrous tissue and differentiating smooth muscle. At the periphery of the organ, the mesenchyme thickens to encapsulate the gland, although a true capsule is not formed (5).

At mid-gestation (~22 weeks' gestational age), the gland consists of small ducts lined by undifferentiated epithelial cells. Increasing levels of maternal estrogen cause squamous metaplasia of the epithelium (i.e., multilayering of the epithelial cells), and at birth the epithelial cells lining the immature glands vary in both the incidence and extent of squamous metaplasia. Upon birth and removal of maternal estrogen, this histological picture is reversed within approximately 4 weeks (6), and the neonatal gland consists of differentiated pseudostratified epithelia.

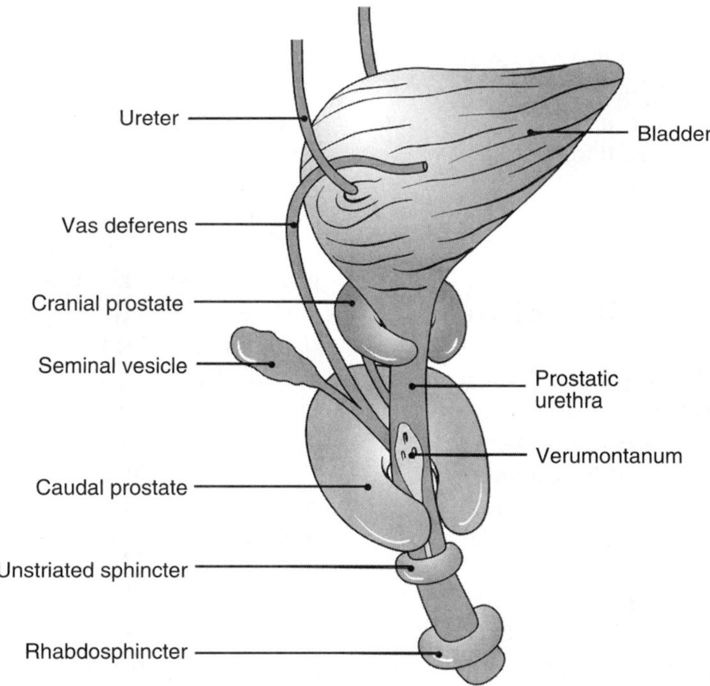

FIG. 2. Schematic representation of the primate prostate, which is divided into caudal prostate and cranial prostate. (Reproduced with permission from Kirby, R. S. [2003]. *An Atlas of Prostatic Diseases*, 3rd ed. The Parthenon Publishing Group, New York.)

In the mid-20th century, the synthetic estrogen diethylstilbestrol (DES) was given to millions of women to prevent spontaneous abortion. The consequences of exposing the male fetuses to high levels of estrogen in utero is apparent from several clinical studies that cited an increase in structural abnormalities of the genitourinary tract, including the prostate gland. Autopsy studies of infants from DES-treated mothers reported more than 90% with abnormalities of the utricles, dilation of the prostatic ductal structures, and squamous metaplasia (7). It remains unknown and controversial as to whether any adverse consequences of DES exposure will emerge with aging of the male cohort of fetuses treated from 1948 to 1971. Experimental studies using human fetal prostate tissues suggest the metaplastic changes in response to DES are deleterious and predict an increase in the risk of prostate disease (8).

In the male human of 2 to 3 months of age, the postnatal surge in testosterone reaches levels that can be within the adult range and are 60-fold higher than normal, prepubertal levels (9,10). In the rodent, many studies demonstrate the importance of postnatal hormone imprinting, especially by estrogens. Hormonal imprinting is an important determinant of the long-term growth regulation of the gland because the effects are permanent and long ranging (11–13) (see section on Hormonal Imprinting of the Developing Rodent Prostate Gland, for more detail). Although there is little evidence in the human, neonatal imprinting may be important in setting the growth response of the mature and aged prostate gland (14). In this context, it is reasonable to predict that inappropriate hormone exposure in the neonatal infant (or the fetus exposed to DES) or disruption of the endocrine system at this critical phase might result in the development of abnormalities of prostate growth upon maturity and aging, such as BPH or adenocarcinoma.

Pubertal Maturation of the Prostate Gland

The immature, prepubertal prostate weighs approximately 2 g. As a result of significant growth at puberty, including more extensive arborization of the glands, the prostate eventually reaches a mature size of approximately 20 g (15). In the prepubertal boy, the glands show little evidence of complex ductal branching and appear as tubular structures surrounded by fibrous stroma (16). At puberty, as maturation and growth proceed, the prostate is composed of increasingly complex tubuloalveolar glands arranged in lobules and surrounded by stroma in which fibroblasts, smooth muscle cells, vasculature, nerves, and lymphatics are located. Although prostate growth is exponential during puberty, the human prostate gland is a relatively slow-growing organ and the doubling time was calculated as approximately 2.76 years during adulthood (15).

The growth of the pubertal prostate gland is regulated by androgens, but more complex regulatory mechanisms must also contribute to growth at this, and other, times of a man's life. The levels of androgens rise during puberty and the prostate gland grows to full size; at maturity androgen levels are maintained, prostate size remains fairly constant, and the organ is considered to be relatively growth quiescent until the fourth decade of life, when growth is reinitiated. Because rising levels of serum androgens stimulate prostate growth in the pubertal boy, how and in what way is the response to the same level of androgens modified in the mature adult man, so that further growth does not occur and prostate size is maintained? Because androgen levels gradually decline with aging, how and in what way is the response to falling androgen levels altered so that growth processes are reactivated and the prostate becomes enlarged? These key questions relating to cell–cell interactions in the prostate are addressed in later sections of this chapter.

Adult Gland: The Mature Gland

The mature human prostate gland is located below the base of the bladder and completely encircles the proximal portion of the urethra. It consists of numerous alveoli lined by pseudostratified epithelium with infolding to the lumen and surrounded by stroma. The epithelium is polarized and a continuous layer of basal cells is positioned at the basal aspect, whereas the secretory products from the tall columnar secretory cells enter the lumen. The secretions of the alveoli drain into the urethra through a system of branched ducts and tubules. At ejaculation, when the secretions are exuded, they mix with secretions from the seminal vesicles and sperm.

The urethra does not pass through the prostate gland in a straight line, as if it were a straight tube. Instead, the midpoint of the urethra is angled so that the proximal urethral segment runs 35 degrees anterior to its distal portion (Fig. 3). Further, the prostate gland is not spherical but is a truncated cone shape (or shaped like a top lying on its side), so that its apex is located at the urogenital diaphragm and its base at the bladder neck. Hence, the prostate gland is most accurately described in three-dimensional rather than two-dimensional configuration.

McNeal's zonal description of the anatomy of the prostate becomes more obvious and accurate with a three-dimensional organization of the gland in mind. Using a combination of sagittal, coronal, and oblique sections, rather than more conventional

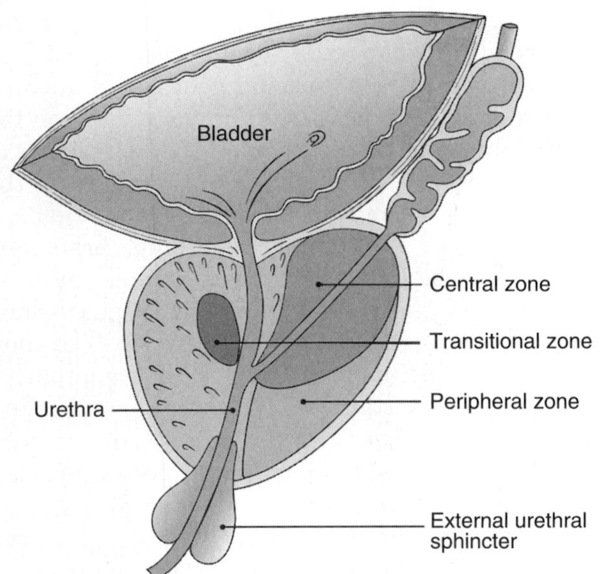

FIG. 3. Anatomy of the human prostate gland. The human prostate is a spinning top–shaped organ that surrounds the urethra, which does not pass through the gland in a straight line, but has a kink in it. The prostate gland is divided into zones, including the central zone, peripheral zone, and transitional zone. (Reproduced with permission from Kirby, R. S. [2003]. *An Atlas of Prostatic Diseases*, 3rd ed. The Parthenon Publishing Group, New York.)

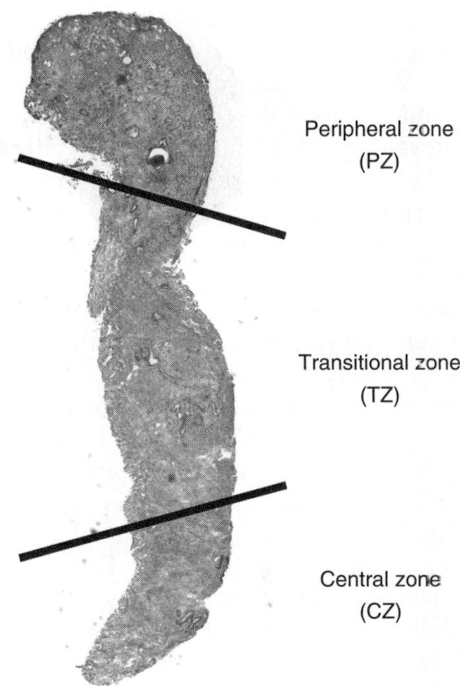

FIG. 4. Human prostate needle biopsy specimen showing peripheral zone (PZ), transitional zone (TZ), and central zone (CZ). A variation in ductal and glandular density can be readily observed histologically, although the distinction between the central and peripheral zones can be difficult. The specimen is approximately 5 mm long.

transverse sections, McNeal divided the gland into five parts. These include the anterior fibromuscular stroma, a peripheral zone, a central zone, and smaller transition and preprostatic zones. The reader is referred to the detailed description of each zone by McNeal (17–21) and summaries such as those provided by Partin and Coffey (1).

Briefly, the peripheral zone is the largest region of the prostate, containing approximately 75% of glandular the tissue; this is the region where adenocarcinoma of the gland most commonly occurs (19). The central zone is the smaller region, comprising approximately 25% of the glandular mass of the prostate, and is an uncommon site of carcinoma. The periurethral area of the gland or the transition zone accounts for less than 5% of the prostatic mass, but this small region of ducts is remarkable because it is the site at which nodular hyperplasia or BPH is usually initiated. Nodules of BPH most prominent in the transition zone may vary in size (22,23), and there is some evidence to suggest that transition zone tumors arise from within BPH nodules (19). The preprostatic tissue surrounds the urethra and consists of glandular and nonglandular structures, including a ring of smooth muscle that functions to prevent retrograde ejaculation. The anterior fibromuscular stroma completely covers the surface of the prostate gland and does not contain any ductal structures.

In ordinary pathology specimens, a variation in ductal and glandular density is readily observed (Fig. 4), although the distinction between the central and peripheral zones can be difficult. Hence, McNeal's schema is frequently reduced to a simple two-zone concept corresponding to the inner (transitional) and outer (peripheral and central) regions of the prostate gland.

Structure of the Human and Rodent Prostate Gland

Embryology and Development of the Rodent Prostate Gland

The embryology of the human and rodent (including mouse and rat) prostates are similar, with both arising from the endodermal UGS, which is derived from the caudal terminus of the hindgut called the cloaca (24). The UGS is first seen at day 14 of gestation in rat, but this structure cannot be visually subdivided into a bladder and definitive UGS until day 17 of gestation (25). By day 19 of gestation, solid cords of urogenital epithelium (UGE; prostatic buds) emanate from the UGS as bilateral ventral buds and invade the surrounding mesenchyme (25,26). Coincidental with the ventral budding, paired dorsocranial outgrowths appear. During the next 24-hour period, extensive development of ventral, lateral, and dorsal prostate buds occurs (26). Anterior buds also grow in close association with the seminal vesicles, which develop within a common mass of mesenchyme (27). The epithelial budding and branching patterns were modeled in mice (Swiss Webster) and rats (Sprague Dawley and Wistar), and it appears that both species share common developmental characteristics (26) (Fig. 5). These events depend on testicular androgens that stimulate androgen receptors (ARs) localized in urogenital mesenchyme from as early as day 16 of gestation in the rodent (28–32).

The process by which these initial solid buds of UGE elongate, bifurcate at branch points, and form branches with terminal branch tips is collectively termed *branching morphogenesis*. Branching morphogenesis is completed by the end of puberty in the rodent (33,34), although significant branching occurs in the first 3 days of life during the period of neonatal imprinting (35). The three-dimensional patterns of branching morphogenesis are precisely regulated and highly conserved to determine the final volume, length, and morphological patterns of the arborized prostatic epithelial ducts; all the rodent prostate lobes have a specific three-dimensional ductal network arising from the branching process (34,36). However, at this age, the small size of the rodent organ restricts the descriptive detail to the period of initial budding from the UGE. Using image analysis and three-dimensional computer-assisted reconstruction of serial histological sections, accurate measurements were made of the initial stages of prostate budding; more recently, a computer-based volume-rendering method of image analysis (similar to that developed to quantify branching morphogenesis in kidney) provided additional parameters to quantitate branching (26,37–41). It is now possible objectively to assess and delineate the process of development in normal as well as genetically manipulated mice from fetal life to adulthood (35). This type of information doubtless will reveal the roles of specific genes in branching morphogenesis, elucidate the differences between the prostate lobes, and ultimately identify the features of branching morphogenesis that are unique to the prostate and common with other branched organs.

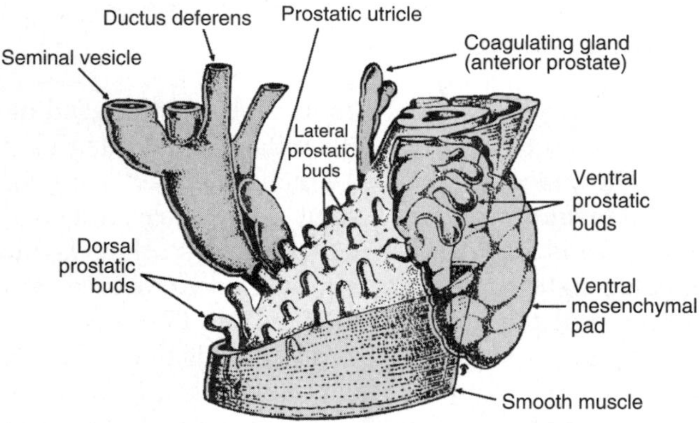

FIG. 5. Schematic illustration of typical ductal budding pattern in early stages of development of fetal rodent prostate. By day 19 of gestation, solid cords of epithelium emanate from the urogenital sinus as bilateral ventral, lateral, and dorsal buds and invade the surrounding mesenchyme. (Reproduced with permission from Timms, B. G., Mohs, T. J., and Didio, L. J. [1994]. Ductal budding and branching patterns in the developing prostate. *J. Urol.* 151, 1427–1432.)

The result of the complex budding, branching morphogenesis and reciprocal mesenchymal–epithelial interactions in the rodent is a multilobular gland arranged around the base of the bladder. The individual lobes are defined according to their anatomical position: ventral prostate (VP), anterior prostate (alternatively named *coagulating gland* or *cranial prostate*), dorsal prostate, and lateral prostate (collectively termed *dorsolateral prostate*) (Fig. 6).

Although the developmental processes between human and rodent are mechanistically similar, the timing is different. Unlike the human prostate, the rodent prostate is rudimentary at birth and branching and differentiation occur postnatally, during the first 15 days of life (36). During this time, solid prostatic ducts begin to canalize, initially at the urethra, and progressing distally toward the ductal tips (42). Concurrently, the epithelium undergoes a complex series of changes, including extensive proliferation and segregation and polarization into basal and luminal secretory epithelial cells (25,36,43,44). Concurrent with epithelial differentiation, the associated mesenchyme differentiates into dense stroma composed of interfascicular fibroblasts and smooth muscle fibers (25,33,45).

Cell–cell communication is critical for prostate branching morphogenesis. The classic work of Franks and colleagues (46), together with the tissue recombination studies of Cunha and associates (47,48), emphasized the important role played by the mesenchymal elements in controlling prostate growth and differentiation during the developmental period. These studies showed prostatic morphogenesis and differentiation are the result of the localized

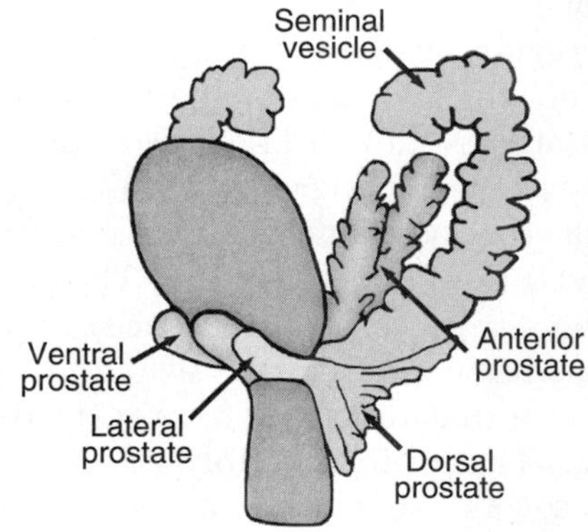

FIG. 6. Lateral view of the rodent prostate. Lobes include ventral prostate (VP), anterior prostate (AP), dorsal prostate (DP), lateral prostate (LP), and seminal vesicles (SV).

instructive potential of the mesenchyme and the receptive epithelia in the UGE [for detailed reviews of mesenchymal–epithelial interactions during development, see Cunha et al. (3,43), Franks et al. (49), and Chung and Cunha (50)].

Briefly, UGM specifies prostatic epithelial identity, induces epithelial bud formation, elicits prostatic bud growth, regulates ductal branching, promotes differentiation of a secretory epithelium, and specifies the types of secretory proteins expressed (42,48,51). In turn, the differentiation of UGM relies on interactions and reciprocal signaling from the adjacent epithelia (48). For example, when human prostate epithelium was recombined with rat mesenchyme from UGM, thick sheaths of smooth muscle were formed, characteristic of human smooth muscle (52). Therefore, in addition to the stroma controlling epithelial differentiation, the epithelium is critical in directing spatial patterning of the stroma, and these data highlight the intimate relationship between the epithelium and surrounding stromal cells. The bidirectional nature of the relationship between epithelium and stroma is not restricted to development and continues, albeit in different ways, during normal and abnormal function of the gland (53).

The process of branching morphogenesis depends on androgens. Androgenic effects are mediated through ARs located solely in the UGM before and during prostatic bud formation. Under the influence of androgens, mesenchymally derived paracrine signals induce epithelial cell differentiation (54); direct androgen binding to epithelial ARs is not required for initial epithelial development (30,43,50,55). The requirement of androgens for prostate development was evident from the observations that the homologous region of the female UGS, which normally forms the vagina, undergoes differentiation into prostate if appropriately exposed to testosterone (56–58). Thus, androgens override the influence of genetic sex in the development of the prostate. As well as testicular androgens, estrogens are also present during the developmental period and influence prostatic growth (see section on Hormonal Imprinting of the Developing Rodent Prostate Gland).

Although the prostate is highly sensitive to hormones during development, most ductal branching morphogenesis occurs before puberty, when circulating androgen levels are low (59). In addition to hormonal influences, it is well accepted that growth factors and developmental genes play a critical role in the local control of prostate development. In neonatal rat organ cultures, several growth factors have shown direct effects on prostate branching morphogenesis (60). Studies have revealed that several factors, such as activin A (61), transforming growth factor (TGF)-β (3,62), bone morphogenetic protein-4 (63), retinoic acid (64), sonic hedgehog (65–67), and *Nkx* (68–71) or *Hox* genes (72–74), inhibit prostate branching morphogenesis, whereas others, including fibroblast growth factor (FGF)-10 (75–77), FGF-7 (77), and keratinocyte growth factor (78,79), stimulate the developmental processes. A more comprehensive listing of the findings is summarized and compiled in Table 1 [for review, see Cunha et al. (49)].

At puberty, there is a growth spurt characterized by an increase in rodent prostatic size, but only a small increase in the number of ductal tips (34). These data indicate the developing prostate is sensitive to low androgen titers for ductal morphogenesis and that the number of branches is determined very early in life. Therefore, if androgen signaling is altered before puberty, this may potentially predispose the adult prostate to abnormal growth and disease. Although branching morphogenesis is mediated by mesenchymal ARs, the induction and maintenance of secretory activity, which first occurs around postnatal day 12 in the mouse (80,81), requires the presence of epithelial ARs (4,30,50).

Hormonal Imprinting of the Developing Rodent Prostate Gland

The prostate is sensitive to estrogen exposure during the neonatal period, and estrogens have both acute and long-term effects (12,14,44,82,83). Exposure to estrogens during the developmental period has been termed *developmental estrogenization* or *neonatal imprinting* (84), and has become a popular model in rodents to investigate the actions of estrogens in the prostate gland. Interestingly, imprinting was defined as "a process by which genomic signals generated in utero can induce greater propensity for cancer in later life" (85), and this appears to be the case when high doses of estrogen are used.

Neonatal imprinting using pharmacological doses of DES in rats leads to permanent, irreversible aberrations in the prostate gland, characterized by a reduction in prostate size and permanent alterations in prostatic growth, morphology, cellular organization, and secretory functions, leading to an increased incidence of prostatic lesions on aging, including hyperplasia, inflammation, and dysplasia similar to prostatic intraepithelial neoplasia (11,12,14,82,86–91). Similar observations have been reported in the mouse prostate (13,92). In the rat, the responses are lobe specific, with differentiation abnormalities occurring with highest frequency and severity in the VP, whereas the dorsal and lateral prostate lobes show greater inhibition of ductal branching complexity (12,93).

Extensive investigation of a rat model of developmental estrogenization allowed Prins and coworkers to

TABLE 1. *Examples of some of the developmental genes known to regulate prostate development*

	Expression and effects on prostate branching morphogenesis	Refs.
TGF-β superfamily		
Activin A	Activin βA and FS are expressed in epithelium Activin βA is also expressed in mesenchymal cap of tip of ducts Activin A inhibits prostate BM in organ culture	(61)
Follistatin	FS stimulates prostate BM in organ culture	(61)
TGF-β	In fetal and neonatal mouse prostate, TGF-β_1, TGF-β_2, and TGF-β_3 are expressed in mesenchyme TGF-β is a growth-inhibitory factor on prostate development—TGF-β_1 and TGF-β_2 inhibited prostate BM in organ culture	(3,62)
Bone morphogenetic proteins		
BMP-4	BMP-4 is expressed primarily in the mesenchyme and exerts an inhibitory influence on epithelial proliferation BMP-4 restricts ductal budding and BM	(63)
Retinoic acid	Retinoic acid inhibits prostatic ductal growth and branching Inhibitory effect of retinoic acid is reversible because prostatic ducts resumed branching and growth after removal from culture Retinoic acid receptors are present from day 0 in the rat prostate	(64)
Sonic hedgehog		(65–67)
Shh	Expressed in the urogenital epithelium and prostatic ducts during BM and in epithelial cells at the distal tips of elongating ducts *Shh* reduces ductal growth and branching in organ culture; decreases FGF-10 levels while increasing BMP-4 in the adjacent mesenchyme *Shh*-induced growth suppression is reversed by exogenous FGF-10, but not noggin, indicating that FGF-10 suppression is the cause of growth inhibition Blockade of function by neutralizing antibody in organ culture abrogates BM Downstream targets are BMP4, FGF-10, and *Hoxd-13*; all involved in prostate development	
NK gene family (homeobox transcription factors)		
Nkx3.1	*Nkx3.1* is expressed in UGS epithelium during development Expressed in newly forming bud sites before bud formation Plays a role in epithelial differentiation and determination Target disruption of *Nkx3.1* results in BM defects	(68–71)
Hox gene family (homeobox transcription factors)		
Hoxa-13, Hoxd-13, 10	*Hoxa-13* and *Hoxd-13* are expressed in UGS mesenchyme during development Loss of *Hoxd-13* function results in a diminished number of main prostatic ducts and decreased prostate size	(72–74)
Fibroblast growth factors		(75–77)
FGF-10	FGF-10 mRNA is present in embryonic and neonatal rat prostate mesenchyme, but not in adult prostate FGF-10 is not androgen regulated FGF-10 protein stimulates the development of ventral prostate in organ culture (not through androgen receptor) FGF-10 is mitogenic on developing prostate epithelium FGF-10 is not the inducer of prostatic bud formation, but instead may be required for bud stabilization FGF-10 knockout mice develop UGS but fail to develop prostate	
FGF-7	FGF-7 is produced by prostate mesenchyme FGF-7's receptor (FGFR2iiib) is expressed in prostate epithelium Neutralization of FGF-7 with a monoclonal antibody or a soluble fragment of FGFR2iiib inhibits androgen-stimulated prostatic epithelial growth and ductal branching.	(77)
Keratinocyte growth factor	Expressed in prostate mesenchyme Stimulates BM in organ culture in the absence of testosterone	(79,80)

BM, branching morphogenesis; BMP, bone morphogenetic protein; FGF, fibroblast growth factor; FS, follistatin; TGF, transforming growth factor; UGS, urogenital sinus.

speculate on the factors that contribute to this process. The temporal regulation of steroid receptors is one of the mechanisms of estrogen action. Developmental estrogenization permanently reduces the levels of AR protein, leading to reduced responsiveness to androgens (44,94–96). Interestingly, the decreased expression of AR protein after neonatal estrogenization was not the result of transcriptional alterations, but of the proteolytic degradation of the AR protein itself (97). Concurrent with downregulation of AR, estrogen receptor α (ERα), progesterone receptor, and retinoic acid receptors (RARα, RARβ, and RARγ) are significantly upregulated in a cell-specific manner (12,66,84,97–100). Studies using ER knockout mice demonstrated that the long-term effects induced by developmental estrogenization are directly mediated through ERα and not ERβ (99).

In addition to altered steroid receptor expression, developmental estrogenization results in increased levels of retinoid-metabolizing enzymes and retinoid levels (100), as well as disrupted expression of critical developmental genes, including TGF-β (101), *Hox-13* (84), and *Nkx3.1* (84). It is believed that the molecular and cellular changes caused by inappropriate estrogen exposure and an imbalance between estrogens and androgens predispose the prostate to the neoplastic state, particularly on aging (84).

The prostate gland is also sensitive to low doses of estrogens during development, but the details of the effects are much more controversial. Vom Saal and coworkers proposed that an increase in reproductive organ disorders was linked to in utero exposure to endocrine-disrupting chemicals in the environment (102). Initially, this group described the phenomenon of intrauterine positioning as important in determining prostate gland size. They showed male mice positioned in utero between two female fetuses (2F males) were exposed to higher local estrogen levels than males positioned between two males (2M males) or those positioned between a male and a female because of transport from adjacent female fetuses (103,104). They demonstrated 2F males had significantly larger prostate lobes in adulthood as well as a threefold increase in prostate ARs (105,106), and associated the late life differences to intrauterine positioning and estrogen effects from the female littermates. Extending the implications of these data, vom Saal and colleagues investigated the effects of fetal exposure to a large range of estrogen (DES) doses on the prostates of male CF1 mice. An inverted-U dose–response relationship between estrogen dose and prostate size was recorded. Adult prostate size was increased after low-dose maternal estrogen administration (days 11 to 17 of gestation), whereas high-dose estrogen exposure decreased adult male prostate weight, consistent with previous observations (38). In addition to increased prostate size, mice exposed to low doses of estrogens in utero had increased AR expression. The pioneering work of vom Saal and colleagues was confirmed by Nagel and coworkers, who showed similar effects, including prostate enlargement in 6-month-old CF1 mice after in utero exposure to low doses of bisphenol A (BPA; an environmental estrogen) (107). The implications of these studies are significant because most toxicological testing does not reach into the lower dose range and will not detect these adverse effects of estrogenic compounds, natural or synthetic. However, these conclusions remain particularly controversial. At least two independent groups of investigators (108,109) were unable to confirm the effects of prenatal exposure to DES (38) and BPA (107) on male CF1 mice. A study conducted in F344 and Sprague Dawley rats, administering low and high doses of EB (β-estradiol-3-benzoate), demonstrated a similar inverted-U–shaped response curve for prostate sizes during puberty, as previously described (38), but this effect was not permanent and was absent from adult animals, suggesting a "transient" effect was induced by low-dose estrogen treatment (110).

The conflicting data regarding low-dose estrogens remain unresolved, but these reports generated a great deal of interest in the prostate biology field as it seeks to evaluate the potential effects of low-dose estrogen exposure, particularly that due to environmental endocrine-disrupting chemicals. Because estrogens play a role in imprinting of the prostate gland during postnatal development, it is important to resolve the controversies to understand the permanent nature of early-life events that may influence the onset of late-life disease.

Anatomy of the Rodent Prostate Gland at Maturity

The rodent prostate is a multilobular organ consisting of separate lobes, distinct from the alobular human prostate. Many investigators commonly use the rodent to study prostate physiology, and it is appropriate to consider the analogy between rodent and human prostate structures. It was believed the VP lobe in the rodent was analogous to the human prostate. However, in humans, the ventral buds regress during development, so there is no direct human analog of the rodent VP lobe (111). Thus, it is important to characterize biological effects in all of the murine prostate lobes rather than focus exclusively on any particular lobe, especially the VP (44,112).

At maturity, each individual lobe of the rodent prostate exhibits its own unique complex ductal anatomy. Sugimura and colleagues manually dissected and enzymatically dispersed the lobes of the mouse prostate to examine the ductal structures (34). Each lobe is a complex system of ducts that arise from the urethra and terminate in many branches at what is known as the distal tips (34,113). Because of lobe-specific differences in the pattern of branching morphogenesis, the final shape of each lobe is distinct (49). Each lobe exhibits a characteristic branching pattern and expresses different secretory proteins (36). Within the ductal system of each lobe, three segments were defined as proximal, intermediate, and distal with respect to the urethra, and along this axis there are regional differences in cell morphology, rates of DNA synthesis, and secretory activity (114). Therefore, any examination of prostatic structures in rodents should take into consideration the lobular and the regional orientation of the ducts. Overall, the prostate gland matures into a complex structure composed of heterogeneous tissues, and cannot be regarded as a uniform organ. The spatial organization of the gland needs to be considered when reporting change to the prostate, and investigators using human tissue appear to be more vigilant in this regard. However, when using rodent specimens, a select region of prostate tissue is more often regarded as representative of the entire gland, or change to one lobe as representative of that in all lobes of the rodent prostate.

Histology of Mature Prostate Gland

Cellular Organization of the Glandular Ducts

Although anatomically distinct, the human and rodent prostate glands are very similar at the cellular level. Ductal glands are lined by a pseudostratified epithelium consisting of histologically and functionally distinct cell types, including the secretory cells and basal cells (21,25,115). Neuroendocrine cells, a third type, are present, but they are rare and scattered throughout acini and ducts (116,117).

The *secretory cells* comprise the exocrine compartment of the prostate epithelium, characterized by columnar luminal cells that synthesize and secrete proteins, including prostate-specific antigen (PSA) and prostatic acid phosphatase (PAP), into the glandular lumen and ductal network. The apical aspect of the secretory cell projects into the lumen, and the basal aspect rests on the basal cells and basement membrane (20,21,118). Secretory cells express high levels of AR and require testosterone for their survival, although the reduced metabolite dihydrotestosterone is considered to be more potent than testosterone in maintaining secretory activity (119). The biomarkers commonly used to identify secretory epithelial cells include PSA, PAP, AR (120–122), and Nkx3.1 (123). Secretory epithelial cells also express cytokeratins (CKs) 8 and 18, but not CKs 5 and 15, thus distinguishing them from basal cells (124–126).

The *basal cells* lie beneath the secretory cells and appear as flattened, nonsecretory epithelial cells. The ultrastructural characteristics of basal cells in human tissues are described in a number of reports (20,127–129). Basal cells are distinguished by their morphology, ranging from small, flattened cells with condensed chromatin and small amounts of cytoplasm, to more cuboidal cells with increased amounts of cytoplasm and more open-appearing chromatin. In the human prostate, basal cells form a continuous layer of cells abutting the basement membrane (130). However, in other species, the basal cells are more scattered in appearance (130,131), as reflected by the ratio of basal to luminal cells, which is approximately 1:1 in human prostate, whereas the average ratio in the other animal species examined (mouse, dog, monkey, rat) is approximately 1:7 (129). Prostatic basal cells express markers in common with basal epithelial cells in other tissues, including CKs 5 and 14 (124,126,132) as well as p63 (133). They are also identified by a lack of expression of the major prostatic secretory proteins, such as PSA and PAP (134).

Unlike secretory epithelial cells, the basal cells are androgen independent, yet they are androgen responsive (135); they are not reliant on androgens for maintenance and survival, but their growth and differentiation is stimulated by androgens. AR expression is low in basal cells compared with secretory epithelial cells, consistent with their ability to be androgen independent, but also androgen responsive (120,121).

The exact role of basal epithelial cells with respect to prostatic function, development, and carcinogenesis is essentially unknown (136). Based on the intimate location of the basal cells between the secretory epithelia and stroma, basal cells are believed to play a role in modulating and mediating endocrine, paracrine, or other regulatory functions (129). The presence of junction-like structures between adjacent basal cells in the human suggests these cells may form a physical "blood–prostate barrier" (129), preventing substances derived from the blood or stroma from coming into direct contact with the luminal cells. Gap junctional proteins (connexins 32 and 43) are present between basal and luminal cells in the human, suggesting these cells communicate directly with each other (137). It is not known if a similar blood–prostate barrier exists in the rodent prostate because the basal cells are more sparse compared with the human prostate; however, gap junctional proteins are located in rat prostate basal cells, specifically connexin 43 in undifferentiated and mature basal cells (138). A last and important function of basal cells (often described as undifferentiated cells without secretory activity) relates to the location of the putative stem cell population that gives rise to all epithelial cell types (see section on Adult Stem Cells in the Prostate) (135,139–142).

In addition to basal and secretory epithelial cells, the human prostate epithelium contains *neuroendocrine cells* [for detailed reviews, see Abrahamsson (143,144), Noordzij et al. (145), and Xue et al. (146)]. Very little is known about the presence of neuroendocrine cells and their functional significance, although they express a broad variety of peptide hormones, biogenic amines, and carrier proteins (147). Studies of human fetal prostates revealed neuroendocrine cells in the UGE at gestational week 9 in close association with urethral prostatic buds (148). Neuroendocrine cells are present in all regions of the prostate at birth, but rapidly disappear from the peripheral regions after birth and then reappear at puberty (147). They are located in the luminal layer of the epithelium, together with secretory epithelial cells, and tend to be more abundant in the major ducts and more sparse in acinar tissue (148).

Neuroendocrine cells can be identified by expression of the neuropeptide chromogranin A (149) and calcitonin (150). Neuroendocrine cells do not express PSA or AR, and are therefore distinguished from secretory epithelial cells in the luminal layer of epithelium. Although the function of neuroendocrine cells is unknown, the secretion of neuropeptides such as chromogranin A and synaptophysin (149) may induce proliferation of adjacent cells (137).

Adult Stem Cells in the Prostate

The lineage of prostate epithelial cells is widely debated, but the identity or immunophenotype of prostatic stem cells is not clearly defined. Based on several functional studies, the basal cell layer is believed to house the stem cells of the epithelium. First, tissue recombinants composed of epithelia from ductal tips of adult rat prostate and embryonic prostate mesenchyme generate new chimeric prostatic ductal structures. The epithelia are obtained from growth-quiescent adult tissues that do not regenerate without the fetal or newborn stroma to induce and instruct differentiation. However with the correct signals from the mesenchyme, the stem cells in the adult epithelium exhibit a high proliferative capacity, pluripotency, and an ability to regenerate tissue—the classic hallmarks of stem cells (151). Second, clonal analysis on adult human prostate epithelium revealed a putative stem cell population (152). Finally, a human prostate primary xenograft model showed that the residual stem cell population in the tissue survived transplantation and androgen deprivation, and maintained pluripotentiality. This capacity to generate progeny that differentiate along multiple lineages in response to microenvironmental signals provides functional evidence of a stem cell population in the human prostate (153).

In rapidly renewing tissues, such as bone marrow, skin, intestinal tract, and squamous epithelium, tissue integrity is maintained by proliferation of stem cells and their differentiation into transiently proliferating cells and mature, postmitotic, terminally differentiated cells; each is found in discrete locations, often forming stratified layers (154,155). Using immunohistochemical techniques, Bonkhoff and coworkers identified a subset of basal cells that expressed basal cell–specific CKs (5 and 14), in addition to PSA (156). This was interpreted as proof of the existence of intermediate stages of differentiation between undifferentiated basal and terminally differentiated secretory epithelial cell types. Hudson and colleagues used cytokeratin antibodies to identify a subpopulation of CK19-immunopositive cells (of the putative intermediate cell type) that were localized between the basal and luminal layers of the epithelium (157). The presence of a transiently proliferating cell population in the prostate epithelium suggests that a typical stem cell–driven hierarchical arrangement, similar to that found in rapidly renewing tissues, also exists in the more slowly renewing adult prostate (158).

There is general agreement that stem cells exist in the prostate epithelium, but there is less congruence on the markers to locate them and the pathway of epithelial lineages. Based on the model described previously, basal cells are widely believed to contain the putative adult stem cells that give rise to all epithelial cell types and can differentiate into secretory, basal, and neuroendocrine cells (127,135,139,140,142, 152,156–170). This theory is based on two fundamental biological properties of basal cells. First, the bulk of the proliferating pool in the normal human prostate is located in the basal cell compartment (158,171, 172). This high proliferative capacity, along with the relatively undifferentiated state and long life span of basal cells, endows them with the classic biological properties of stem cells. Second, the basal cells are resistant to the effects of androgen withdrawal (158, 160,173) and are able to proliferate and regenerate the prostate to normal size and morphology on readministration of androgens. Together, these findings led several groups to propose a stem cell model for the prostate epithelium.

Assuming the basal cell layer housed the stem cells, Collins and coworkers identified and isolated stem cells from nonmalignant prostate epithelia based on their expression of high levels of integrin $\alpha_2\beta_1$ (159) and CD133 (174). Integrin $\alpha_2\beta_1$ immunopositive cells comprise approximately 1% of basal cells and have a fourfold greater ability to form colonies in vitro than the total basal population (159). $CD133^+$ cells are restricted to the $\alpha_2\beta_1^{hi}$ population and are located in the basal layer, often at the base of a budding region or branching point. $\alpha_2\beta_1^{hi}/CD133^+$ cells exhibit two important attributes of epithelial stem cells: they possess a high in vitro proliferative potential and can reconstitute prostate-like acini in immunocompromised male nude mice (174).

An alternative stem cell model was proposed that assumed the cells of an intermediate phenotype (transiently proliferating cells) housed the stem cell population (126,142). This model was developed by Cunha and colleagues (126), who comprehensively mapped the pattern of CK expression in mouse and human tissues during development and in the mature organ as a means of identifying the adult stem cell. Their hypothesis revolved around the idea that embryonic progenitor cells, initially located in the UGS epithelium, are retained in the mature epithelium as adult stem cells. They therefore proposed that the prostate stem cells must be present in the prostate during development. Examination of human, mouse, and rat fetal prostates demonstrated that only cells of the intermediate/transiently proliferating phenotype were present in UGS epithelium (126). The immunophenotype of these cells was consistent in fetal and adult prostates, including expression of CKs 5, 14, 8, 18, and 19. Interestingly, CK5- and CK14-positive basal cells, previously thought to be stem cells, were detected only in the mature organ, suggesting they represent a later stage of differentiation (126). The two commonly proposed models of stem cell differentiation in the prostate are represented in Fig. 7.

Other theories have emerged more recently that challenge both of these models. One group suggested stem cells also reside in the luminal cell compartment as a slow-proliferating/self-reserve population in the proximal part of prostatic ducts (175). Another group suggested luminal and neuroendocrine cells differentiate in the absence of prostatic basal cells.

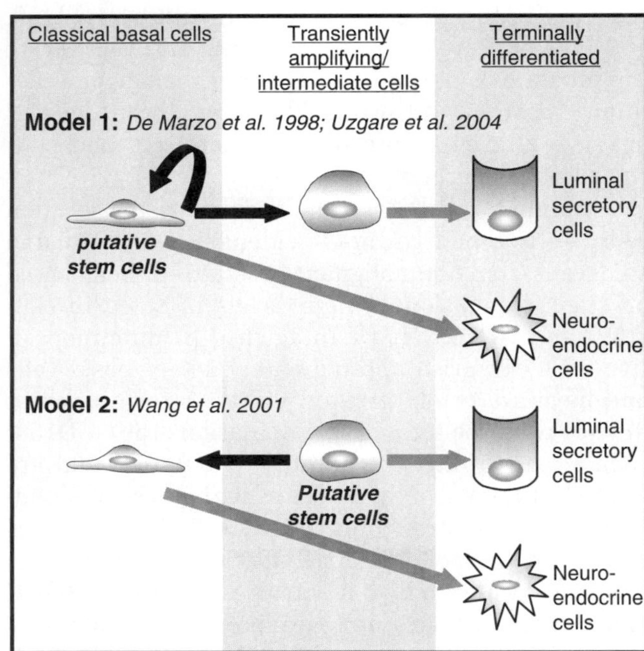

FIG. 7. Two commonly proposed prostate stem cell models. Model 1 depicts the putative stem cells as classic basal cells that give rise to a transiently amplifying or intermediate population and subsequent secretory cells. Model 2 depicts the putative stem cells as an intermediate cell type residing in the basal cell compartment, giving rise to the basal cells and secretory cells.

This concept was based on studies using $p63^{-/-}$ mice that are remarkable because they do not have a basal cell layer; differentiation of terminal cell types in the absence of basal cells suggested the stem cell population was housed outside the basal cell compartment (176). Finally, it was proposed that prostatic neuroendocrine cells arise from a stem cell lineage that is different from the UGS-derived prostate secretory and basal cells, being of neurogenic origin (148,177). This second cell lineage is thought to be important for proliferation regulation of the UGS-derived lineage (148).

Given the importance of identifying stem cells of the prostatic epithelium, additional studies are required to determine unequivocally the correct model(s) of stem cell renewal in the prostate. Once this information is known, new therapies designed to target these cells can be used for the treatment prostate diseases, particularly androgen-independent prostate cancer.

Reciprocal Cell–Cell Interactions in Adulthood to Maintain Homeostasis

After epithelial differentiation, it is important to consider epithelial homeostasis, which results from a steady-state equilibrium of programmed cell death (apoptosis) and cellular proliferation (142). The human prostate is an extremely slow-growing organ; the slow rate of proliferation is balanced by correspondingly low rates of apoptosis in the normal adult human prostate, resulting in a growth-quiescent gland (178,179). In the prostate epithelium, apoptosis and proliferative activity are discordant in the basal and luminal cell compartments.

In the rodent prostate lobes, the incidence of apoptosis and proliferation varies according to regional location. The ducts of the rodent prostate lobe are divided into proximal, intermediate, and distal regions (114), and in each of these regions, there are morphological and functional differences between the epithelial cells. In the most differentiated proximal region, the secretory epithelial cells are of cuboidal or low-columnar shape and appear degenerative, with many cells undergoing apoptosis, whereas cells lining distal and intermediate segments are tall-columnar in shape and undergo extensive proliferation (43,114,180).

The regional differences in the ductal morphology suggest that although the epithelial and stromal cells in each region are exposed to the same level of circulating androgens, the reciprocal cell–cell interactions are variable. The prostatic stroma is composed of smooth muscle and fibroblast cells (32,181), providing both structural and biochemical support to the epithelium. The main role of smooth muscle cells relates to expulsion of the secretions from the secretory glands into the prostate ductal network during ejaculation (32). However, the stromal cells play a more critical role in controlling the growth of adjacent epithelia through the production of soluble growth factors. For example, growth factors show regional variation of expression along prostatic ducts and are associated with differential biological responses in the epithelium [i.e., stromal expression of TGF-β1 is associated with epithelial cell death in the proximal ducts (182)]. The differences in stromal composition give some clues as to the biological actions exerted on the epithelium.

Not surprisingly, the regional variations in epithelial proliferation and apoptosis along the ducts are associated with differing organization of the stroma. Sheaths of stromal smooth muscle are widely distributed throughout the human prostate gland (183), but in the adult rodent prostate, the distribution of smooth muscle and fibroblasts along the prostatic duct is variable (184). It appears that signalling from fibroblasts promotes proliferation (in the distal region), whereas signals from smooth muscle cells alone and the absence of proliferative signals from fibroblasts promote cell death (in the proximal region). Fibroblasts are abundant throughout the stroma, but in the distal regions there is a periductal fibroblast layer between the smooth muscle cells that forms discontinuous layers (one to two cells deep) surrounding the glands. In the intermediate segment, fibroblasts are mainly in the intraductal spaces, with very few dispersed between the continuous layers of smooth muscle (one to four layers deep). Smooth muscle around proximal segments is continuous and consists of many more layers forming a sheath around the ductal structures; only a few fibrous cells are present in the proximal stroma.

In the mature prostate, androgens are believed to act through paracrine signaling on smooth muscle (which expresses AR) to maintain the fully differentiated growth-quiescent epithelium. AR expression varies in stromal cell types; smooth muscle cells consistently express AR, whereas fibroblasts do not (93,184–186). The variation in AR expression, together with varied organization of the stroma along the duct, contributes to the different epithelial activity despite the overall regulation by circulating androgen levels. It appears fibroblastic signals promote proliferation (in the distal region), whereas predominantly smooth muscle signals (or lack of fibroblastic signals) promote cell death (in the proximal region).

FUNCTION OF THE PROSTATE GLAND

The prostate gland is an accessory organ of the male reproductive tract and its main function is to contribute secretions to the ejaculate. This role was recognized many years ago, and the reader is referred to a chapter in the previous edition of *Physiology of Reproduction* that provides a comprehensive description of the secretory products synthesized by the prostate epithelium (187). Briefly, the seminal plasma contains very high concentrations of potassium, zinc, citric acid, fructose, phosphorylcholine, spermine, free amino acids, prostaglandins, and enzymes, and much of this originates from the prostate (187).

The more immediate research priority relates to understanding how normal prostate function becomes altered in aging men when benign and malignant prostate diseases become more prevalent. Nevertheless, the identification of specific and novel secretory proteins will lead to the generation of new biomarkers for diagnosis and prognosis of prostate function and disease.

At present, detection of PSA in the blood is commonly used for the detection of prostate cancer [for review, see Risbridger and Frydenberg (188)]. The introduction of this tool had a tremendous impact on all aspects of the management of prostate carcinoma. Although PSA-based screening has resulted in diagnosis of most commonly organ-confined tumors, and is associated with a decrease in prostate carcinoma mortality, PSA screening has its limitations, and not all countries in the world advocate PSA screening for prostate cancer (189,190). As the search for more accurate and reliable biomarkers of prostate disease continues, new members of the kallikrein family are of interest, such as kallikrein 3 (191). Their potential as new biomarkers of prostate disease eclipses the functional role of tissue kallikreins, which are serine proteases from a highly conserved multigene family in rodent species and humans. It is believed that the biological action of the kallikreins is to participate in an enzymatic cascade similar to that of the coagulation cascade (191).

DISEASE OF THE PROSTATE GLAND

Breakdown of Reciprocal Signaling in Disease

Controlled communication between the stroma and epithelia is critical for prostate development and maintenance. Aberrant interactions or communications play a major role in the initiation and progression of disease, particularly prostate cancer.

Because prostate carcinoma arises from epithelial cells lining the glands, the focus of the majority of research to date is on the epithelial cells and the genetic changes that occur in them as they progress from normality to malignancy. However, it is evident the surrounding microenvironment plays a major role in prostate tumorigenesis (192,193). To reach a point where the tumor microenvironment can be targeted by new therapeutics for prostate cancer, we must understand the role of the tumor stroma in the initiation, promotion, and metastasis of the disease. Specifically, this requires a more complete understanding of the events mediating the interactions between the cancer cell and its immediate microenvironment (192,193).

Prostate carcinogenesis is surmised to involve a sequential disruption of the reciprocal homeostatic interactions between prostate cell types, observed as dedifferentiation of both emerging prostatic carcinoma cells and surrounding smooth muscle cells (53). Thus, the epithelium fails to signal appropriately to the surrounding smooth muscle, leading to dedifferentiation and the formation of a fibroblastic phenotype. The altered stroma is referred to as *tumor stroma* or *tumor microenvironment*, based on its altered morphology and genetic composition. As smooth muscle differentiation begins to deviate, signaling from smooth muscle to epithelium becomes abnormal and there is a progressive loss of control over epithelial differentiation and proliferation. This hypothesis is based partly on the progressive diminution of smooth muscle in human prostatic adenocarcinomas during progression from low- to high-grade cancers (53), and partly on the reverse process (i.e., the ontogeny of prostatic smooth muscle differentiation during development). For decades, pathologists recognized the stroma surrounding tumors was altered, specifically with respect to cellular composition and extracellular matrix. In fact, stromal cells surrounding a prostate carcinoma are more typically fibroblastic or myofibroblastic (194) as a result of smooth muscle dedifferentiation. In addition to morphological differences, gene expression is altered in tumor stroma. It is believed that the altered tumor stroma responds to androgenic stimulation by producing paracrine-acting mitogens, fueling a cycle of cancer cell proliferation and stromal dedifferentiation. Thus, prostate cancer cells are under the tight control of their surrounding tumor stroma or microenvironment.

One of the first experiments to test the idea that stromal cells themselves might induce prostatic carcinogenesis was performed in 1993 by Thompson and coworkers (195). Using transgenic mice transfected with a virus carrying the *myc* and *ras* oncogenes, tissue recombination experiments were performed using combinations of infected and uninfected UGM and UGE. Infected UGE recombined with normal UGM resulted in epithelial hyperplasia, whereas normal UGE recombined with infected UGM resulted in stromal desmoplasia (195). However, it was not until infected UGE was combined with infected UGM that prostatic carcinomas were generated (195). Taken together, these initial data indicate genetic alteration was required in *both* epithelium and stroma for malignant transformation to occur, at least with respect to the activation of oncogenes (*myc* and *ras*).

More recently, the hypothesis was tested using human tissue fibroblasts obtained from the vicinity of the malignant lesion in the prostate, as well as breast (196,197) and bladder (197) tissues. In the prostate, the stromal cell type within and surrounding a tumor was designated *carcinoma-associated fibroblasts* (CAFs) (198). Primary cultures of CAFs from patients with prostate tumors were collected at the time of radical prostatectomy and reported to be indistinguishable from normal prostatic fibroblasts morphologically, immunocytochemically, and by growth characteristics (198). However, the CAFs demonstrated the ability to induce malignant transformation in other cells, whereas normal prostatic fibroblasts did not. Others, notably Rowley and coworkers, described the same phenotype as “reactive stroma” (199,200). As described by Rowley's group, the peritumoral stromal cells have many features of cells present during wound repair, including the expression of a myofibroblastic phenotype and the deposition of extracellular matrix components.

To determine whether CAFs influence the progression of prostate epithelial cells, CAFs were grown with a human prostate epithelial cell line, BPH-1. The BPH-1 cell line was clonally derived and generated by SV40-T immortalization, and is regarded as an “initiated” rather than “normal” epithelial cell type. The BPH-1 cells have an abnormal karyotype with a mean chromosome count of 76 (201). Although this cell line itself is nontumorigenic (201), when combined with CAFs and subrenally grafted for up to 85 days, the BPH-1 cells generated poorly differentiated adenocarcinomas that were almost 500 times the wet weight of any other tissue recombinants (198). In contrast, when normal fibroblasts not associated with carcinoma were used in place of CAFs, minimal growth was observed. These data indicated that CAFs stimulate progression of tumorigenesis in “initiated” prostate cells.

When these studies were extended to normal human prostate epithelium that was not subjected to genetic alteration, CAFs were unable to initiate tumorigenesis or malignant transformation (198). However, the histological appearance of the epithelial cells was different from that seen when they were combined with normal fibroblasts, and epithelial differentiation was found to be abnormal; specifically, normal tall columnar cells were replaced with an abnormal, multilayered, stratified squamous appearance (198). The epithelial cells were benign in appearance, exhibiting a normal nuclear-to-cytoplasmic ratio and ductal morphology, and were not locally invasive. Therefore, CAFs did not induce malignant transformation of normal human prostate epithelial cells, but epithelial histology was altered. This was a surprising finding because it was previously predicted that tumor stroma would be capable of inducing malignant transformation in normal prostate epithelial cells. These data suggest that paracrine signaling from CAFs can affect both normal and genetically altered epithelial cells, but that the outcome of the signaling is dictated by the genetic composition of the epithelial cells, that is, CAFs are capable of facilitating the development of malignancy in cells that had previously suffered an "initiating" event. The involvement of CAFs in the initiation of prostate carcinogenesis remains equivocal.

Stem Cell Proliferation and Disordered Growth

Alternative targets in the treatment of prostate cancer are the tumor cells themselves. To date, there is little consensus about the target in prostate epithelium that is susceptible to malignant transformation. Currently, all cancer cells are treated as though they have an unlimited proliferative potential and can acquire the ability to metastasize. However, as in other types of cancers, it is becoming clear that there is a population of cells in the tumor that comprise a putative cancer stem cell population. Most cancers comprise a heterogeneous population of cells with marked differences in their phenotypic characteristics and proliferative potential, as well as in the ability to reconstitute the tumor on transplantation (202–204). Cancer stem cells are a minor population in the tumor that possess the stem cell property of self-renewal, and dysregulation of the renewal process is an additional requirement for the development of cancer. The identification of the adult prostate stem cell is as critical as the discovery of putative prostate cancer stem cells (142); once this is resolved, new cancer treatments can directly target the cancer stem cell population.

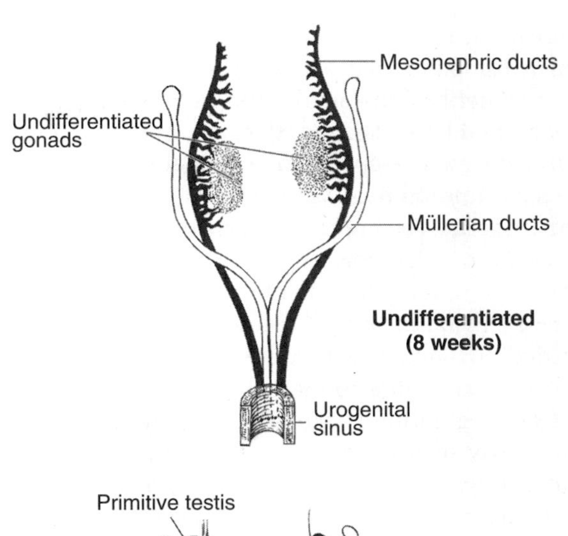

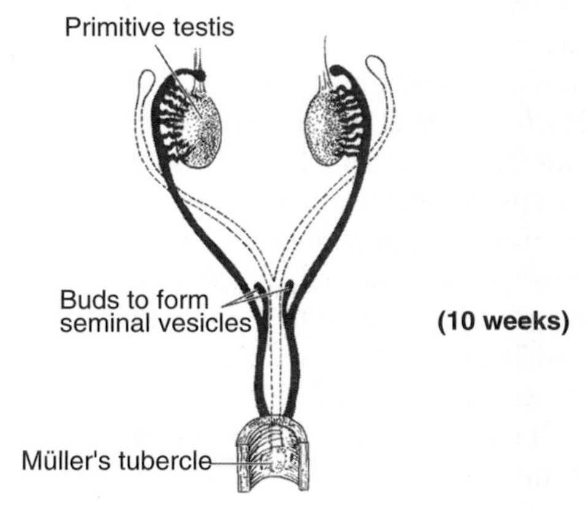

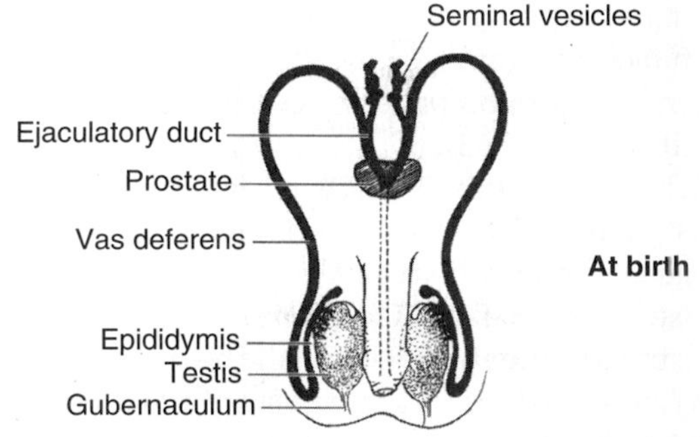

FIG. 8. Embryology of seminal vesicle. In the male human, the development of the seminal vesicles begins at approximately 12 weeks of fetal age from the mesonephric or wolffian duct. Buds that form the seminal vesicles emerge as dorsolateral swellings or dilations of the mesonephric duct. (Reproduced with permission from Tanagho, E. A. [2004]. Embryology of the genitourinary system. In *Smith's General Urology* [E. A. Tanagho and J. W. McAninch, Eds.]. Lange Medical Books/McGraw-Hill, New York.)

STRUCTURE AND FUNCTION OF SEMINAL VESICLES

Development of the Seminal Vesicles

In the male human, the development of the seminal vesicles begins at approximately 12 weeks of fetal age from the mesonephric or wolffian duct, as does the development of the epididymis and the vas deferens. Buds that form the seminal vesicles emerge as dorsolateral swellings or dilations of the mesonephric duct proximal to the region where it joins the urethra (Fig. 8). Morphogenesis of the seminal vesicles depends on fetal testicular androgens (205) and the organ is unique to the male, with no known female homologue.

Studies using the rodent seminal vesicle describe the development process in more detail. Seminal vesicle morphogenesis begins on day 15 of fetal life with the dilation of the lower regions of the wolffian duct, although rudiments are not seen until days 17 to 18 of gestation (25). Just before birth, the seminal vesicle is essentially a hollow tube surrounded by mesenchyme, and a lumen is present at the very early stages of development of the seminal vesicles (34,206). As the organ develops, the complexity of the epithelium increases as the epithelial buds grow laterally into the surrounding mesenchyme, elongate and bifurcate, and undergo secondary branching. Simultaneously, infolding of the epithelium into the ductal lumen increases, and at the end of development the seminal vesicle epithelia are characteristically described as "folded." Hence, the complex topography of the seminal vesicles is due to lateral budding and branching as well as glandular infolding into the ductal lumen.

A study of mice bearing a spontaneous mutation that results in altered seminal vesicle shape [*svs* mice (207)] provided further insights into the mechanisms of ductal development (208). Although the seminal vesicles of *svs* mice are folded, this is largely due to epithelial infolding and not to the growth of lateral buds into the surrounding mesenchyme. Thus, epithelial infolding is independent of lateral budding and branching, and each of these processes is under separate genetic control (208).

The differentiation of seminal vesicle stroma to smooth muscle also proceeds in an ordered process and is controlled by androgens. Castration reduces the smooth muscle layer and the expression of smooth muscle cell markers is lost in the reverse order of their expression during developmental maturation (32). The smooth muscle stroma is separated from the glandular epithelium by a lamina propria, including a periductal layer of fibroblasts (32).

As with the prostate, there is evidence that the seminal vesicles are responsive to estrogens as well as androgens and may be regarded as target tissues for both classes of steroid. This was most clearly demonstrated using another mouse model, the gonadotropin-deficient *hpg* mouse, which has postnatal deficiency in gonadotropins and testosterone but remains hormone sensitive (209). The seminal vesicles of adult *hpg* mice develop into rudimentary structures, and treatment with estrogen provided a means to distinguish between the direct actions of estrogens and indirect effects due to androgen suppression (210). With administration of estrogen, dual (proliferative and dedifferentiating) actions of estrogen on the seminal vesicle epithelia and stroma were observed, including smooth muscle regression, fibroblast proliferation, inflammation, and basal epithelial cell proliferation and metaplasia (210).

It is not known if rising maternal estrogen levels in late gestation or the postnatal hormone surge of androgens and estrogens have any permanent effects on vesicular development. However, exposure to exogenous phthalates (which act as antiandrogens) results in permanent changes in the endocrine system and abnormal reproductive tract development, including that of the seminal vesicles. The mechanism of action of phthalates probably involves aberrant expression of the AR or ERβ (211).

Structure of the Mature Seminal Vesicles

At maturity, when growth and cytodifferentiation are complete, the seminal vesicle consists of a highly folded glandular epithelium with tall columnar luminal secretory cells and a discontinuous layer of basal cells, surrounded by a stromal layer of smooth muscle. In contrast to the prostate, neuroendocrine cells are absent from the seminal vesicle epithelium (212), and the epithelial cell layer consists only of basal and luminal cell types. Segregation of the epithelial cell types occurs in a distal-to-proximal manner early in development and because there is canalization before basal cells differentiate, it must be concluded they are not required for the formation of a duct or lumen. Instead, the basal cells appear as the luminal space becomes enlarged, consistent with a role in establishing and maintaining, rather than forming, the large ductal space of the mature seminal vesicles (25).

The human mature organ is 5 to 10 cm long and has a volume capacity of approximately 13 mL (213). The secretions from the seminal vesicles contribute from 50% to 80% of the volume of the ejaculate, and the contractile response of the smooth muscle layer is necessary for exuding the vesicular secretions at ejaculation. Serotonin is one of the factors known to regulate contractility of the seminal vesicles, at least in the rodent (214).

Function of the Seminal Vesicles

The essential function of the mature vesicles is unclear. Seminal vesicle agenesis requires no treatment, and absence of the seminal vesicles may be asymptomatic. However, unilateral agenesis of the seminal vesicles commonly occurs with the unilateral absence of the vas deferens and ipsilateral renal abnormalities (215,216). Bilateral agenesis of the seminal vesicles frequently occurs in association with congenital absence of the vas deferens and may be associated with cystic fibrosis (217–219).

The secretory proteins and other constituents of the seminal vesicles are well characterized; they are thought to be important for sperm motility and metabolism and include fructose and prostaglandins [for review, see Luke and Coffey (187)]. Human seminal fluid coagulates after ejaculation and semenogelin proteins are a major component of the coagulum. The seminal vesicles of the human and higher primate species are the primary source of semenogelin proteins [for review, see Robert and Gagnon (220)]. After ejaculation, a series of events takes place that includes cleavage of semenogelin protein by PSA, liquefaction of the coagulum, progressive release of motile spermatozoa in the female tract, as well as the generation of a cleavage product of semenogelin that inhibits sperm motility (221).

Although the function of the individual components of the vesicular secretions may be inconsequential in the normal adult male, it is possible they assume greater importance in the subfertile male or aging male. For example, antioxidant secretions from the seminal vesicle may protect sperm in the ejaculate from reactive oxygen species that are known to reduce sperm motility and hence play a role in male infertility. During aging of the brown Norway rat, the seminal vesicles and other reproductive tract tissues show an age-related decline in antioxidant enzyme activities. It is tempting to speculate that vesicular antioxidant enzyme depletion is one of the mechanisms that contribute to a reduction in fertility, especially in older men (222).

BULBOURETHRAL (COWPER'S) GLANDS AND SKENE'S GLAND

The bulbourethral glands are also known as Cowper's glands, and are present in most mammals. This pair of glands produces secretions that empty into the bulbous urethra and contribute to the seminal plasma (223). In rodents, the secretions are thought to be important in plug formation after copulation.

During development, the bulbourethral glands arise from epithelial projections emanating from the UGS that invade the surrounding mesenchyme adjacent to the primitive urethra. They share a common developmental history with the prostate gland and are under the regulatory control of androgens (224,225).

At maturity, the bulbourethral glands are tubuloalveolar structures consisting of acini lined by mucus-secreting epithelial cells surrounded by stroma containing smooth muscle cells (226). Human specimens obtained at surgery showed mucous cells in endpieces of these glands at different stages of the secretory cycle, and cuboidal cells line the ductal network emptying into the urethra (227,228).

Cystic dilations of the ducts of the bulbourethral glands (also known as Cowper's cysts or syringoceles) may be congenital and are usually diagnosed in infants or children; only rarely are they detected in adults. Congenital cysts occur because the paired ducts fuse, and there is some evidence in mice that TGF-β is important in ductal development, together with androgens. Heterozygous TGF-β mutant mice develop cysts as well as epithelial glandular hyperplasia, suggesting that the correct stromal–epithelial cell interactions are as important in the bulbourethral glands as they are in the prostate (229).

Less complex tubular structures are located in the invaginations of mucosa in the penile urethra and are known as the glands of Littre in males or Skene's gland in females. In women, Skene's gland is regarded as either vestigial or as a functional organ involved in female orgasm (230–233). More recently, Skene's (or the paraurethral) gland has been proposed to be the histologic homolog to the male prostate (234). The common protein markers of male prostate secretory function, PSA and PAP (235), are both localized to Skene's gland, and hence it is also termed "the female prostate gland" by some investigators. Skene's gland has aroused scientific interest because it is subject to the same diseases as the male prostate during aging, including prostatitis, hypertrophy, and carcinoma (236–241).

EMERGING ISSUES AND RESEARCH PRIORITIES

Androgens stimulate prostate growth in the male in a temporal manner, with the greatest effects during fetal development and puberty, and upon aging. At each of these times (and even in the intervening periods, especially at maturity when the adult prostate gland is relatively growth quiescent), cell–cell interactions play a significant role in the growth process. Cell–cell interactions between the stroma and epithelia are essential for maintaining the anatomical and functional organization of the prostate gland and result in an organ that is heterogeneous. In studies of prostate physiology, whether normal or abnormal, the organization of the gland is crucial for the interpretation of any observations. Investigators using human specimens are usually vigilant in identifying the zonal region of origin of tissues, but in studies using rodent specimens, a select region of prostate tissue is more often regarded as representative of the entire gland, or change to one lobe as representative of that in all lobes of the rodent prostate. In any species, the prostate gland must be regarded as a heterogeneous organ so that the anatomical differences in the organ reflect the regional organization and varied cellular interactions. Whether these considerations apply to seminal vesicle structure and function is unknown, but the same complexity has not been described.

What then are the implications of organ heterogeneity and specific cell–cell interactions in the prostate gland? The most obvious is the modification of the growth response to androgens; during development, androgens stimulate growth and differentiation but, in the adult, net growth of the prostate does not occur although androgens maintain its function. Conversely, as androgen levels gradually decline with aging, growth processes are reactivated and the prostate becomes enlarged. Aberrant cell–cell communications play a major role in the initiation and progression of disease, particularly prostate cancer. Although changes to the epithelia are well described, less is known about the important role of the stroma in controlling growth and differentiation except during the developmental period, where there are numerous studies demonstrating stromal action in prostatic morphogenesis and differentiation. In the context of disease, the notion of the stroma being a key regulator of normal and abnormal epithelial physiology is a current focus of research activity. The identification of specific stromal components that contribute to the initiation and progression of disease will be a significant breakthrough and provide new approaches to therapies for prostate disease. Revealing the complexity of cell–cell interactions in the seminal vesicles will enhance our understanding of their physiology, but is not regarded as a high research priority because disease of the seminal vesicles is rare.

The identification of the prostate stem cells in normal and diseased tissue and the inability to distinguish them from differentiating progeny is another key area of research in the field. Identification of prostate stem cells, and perhaps prostate cancer stem cells, will have significant consequences for the identification of new targets and the development of new approaches to regulate prostate disease.

Finally, the early origins of adult disease are an emerging research priority. As for other organs and tissues, there is substantial evidence that early changes to the prostate gland lay the foundations for adult disease. The best-characterized change in the prostate is during hormonal imprinting, when the balance between androgens and estrogens is perturbed. Most of the research has involved rodent studies in which neonatal estrogenization predisposes the prostate gland to disease. The implications of these results for human prostate health are unknown, but they are an impetus for continuing research to assess the effects of environmental estrogens or xenoestrogens in our diet.

REFERENCES

1. Partin, A. W., and Coffey, D. S. (1998). The molecular biology, endocrinology, and physiology of the prostate and seminal vesicles. In *Campbell's Urology* (P. C. Walsh, A. B. Retik, E. J. D. Vaughan, and A. J. Wein, Eds.), pp. 1381–1428. WB Saunders, Philadelphia.
2. McNeal, J. E. (1983). Relationship of the origin of benign prostatic hypertrophy to prostatic structure of man and other mammals. In *Benign Prostatic Hypertrophy* (F. Hinman, Ed.), pp. 152–166. Springer-Verlag, New York.
3. Cunha, G. R., Alarid, E. T., Turner, T., Donjacour, A. A., Boutin, E. L., and Foster, B. A. (1992). Normal and abnormal development of the male urogenital tract: role of androgens, mesenchymal-epithelial interactions, and growth factors. *J. Androl.* 13, 465–475.
4. Cunha, G. R. (1994). Role of mesenchymal-epithelial interactions in normal and abnormal development of the mammary gland and prostate. *Cancer* 74, 1030–1044.
5. Ayala, A. G., Ro, J. Y., Babaian, R., Troncoso, P., and Grignon, D. J. (1989). The prostatic capsule: does it exist? Its importance in the staging and treatment of prostatic carcinoma. *Am. J. Surg. Pathol.* 13, 21–27.
6. Andrews, G. (1951). The histology of the human foetal and prepubertal prostates. *J. Anat.* 85, 44–54.
7. Driscoll, S. G., and Taylor, S. H. (1980). Effects of prenatal maternal estrogen on the male urogenital system. *Obstet. Gynecol.* 56, 537–542.
8. Yonemura, C. Y., Cunha, G. R., Sugimura, Y., and Mee, S. L. (1995). Temporal and spatial factors in diethylstilbestrol-induced squamous metaplasia in the developing human prostate: II. Persistent changes after removal of diethylstilbestrol. *Acta Anat.* 153, 1–11.
9. Pang, S. F., Chow, P. H., and Wong, T. M. (1979). The role of the seminal vesicles, coagulating glands and prostate glands on the fertility and fecundity of mice. *J. Reprod. Fertil.* 56, 129–132.
10. Forest, M. G. (1979). Plasma androgens (testosterone and 4-androstenedione) and 17-hydroxyprogesterone in the neonatal, prepubertal and peripubertal periods in the human and the rat: differences between species. *J. Steroid Biochem.* 11, 543–548.
11. Prins, G. S. (1992). Neonatal estrogen exposure induces lobe-specific alterations in adult rat prostate androgen receptor expression. *Endocrinology* 130, 3703–3714.
12. Prins, G. S., and Birch, L. (1997). Neonatal estrogen exposure up-regulates estrogen receptor expression in the developing and adult rat prostate lobes. *Endocrinology* 138, 1801–1809.
13. Singh, J., and Handelsman, D. J. (1999). Imprinting by neonatal sex steroids on the structure and function of the mature mouse prostate. *Biol. Reprod.* 61, 200–208.
14. Naslund, M. J., and Coffey, D. S. (1986). The differential effects of neonatal androgen, estrogen and progesterone on adult rat prostate growth. *J. Urol.* 136, 1136–1140.
15. Berry, S. J., Coffey, D. S., Walsh, P. C., and Ewing, L. L. (1984). The development of human benign prostatic hyperplasia with age. *J. Urol.* 132, 474–479.
16. Glenister, T. W. (1962). The development of the utricle and of the so-called "middle" or "median" lobe of the human prostate. *J. Anat.* 96, 443–455.
17. McNeal, J. E. (1980). The anatomic heterogeneity of the prostate. *Prog. Clin. Biol. Res.* 37, 149–160.
18. McNeal, J. E. (1980). Anatomy of the prostate: an historical survey of divergent views. *Prostate* 1, 3–13.
19. McNeal, J. E., Redwine, E. A., Freiha, F. S., and Stamey, T. A. (1988). Zonal distribution of prostatic adenocarcinoma: correlation with histologic pattern and direction of spread. *Am. J. Surg. Pathol.* 12, 897–906.
20. McNeal, J. E. (1988). Normal histology of the prostate. *Am. J. Surg. Pathol.* 12, 619–633.
21. McNeal, J. E. (1968). Regional morphology and pathology of the prostate. *Am. J. Clin. Pathol.* 49, 347–357.
22. Mostofi, F. K., and Price, E. B. J. (1973). Tumors of the male genital system. In *Atlas of Tumor Pathology, 2nd Series, Fascicle 8*, pp. 182–194. Armed Forces Institute of Pathology, Washington, DC.
23. McNeal, J. E. (1978). Origin and evolution of benign prostatic enlargement. *Invest. Urol.* 15, 340–345.

24. Cunha, G. R., Donjacour, A. A., and Sugimura, Y. (1986). Stromal-epithelial interactions and heterogeneity of proliferative activity within the prostate. *Biochem. Cell Biol.* 64, 608–614.
25. Hayward, S. W., Baskin, L. S., Haughney, P. C., Cunha, A. R., Foster, B. A., Dahiya, R., Prins, G. S., and Cunha, G. R. (1996). Epithelial development in the rat ventral prostate, anterior prostate and seminal vesicle. *Acta Anat.* 155, 81–93.
26. Timms, B. G., Mohs, T. J., and Didio, L. J. (1994). Ductal budding and branching patterns in the developing prostate. *J. Urol.* 151, 1427–1432.
27. Cunha, G. R. (1972). Epithelio-mesenchymal interactions in primordial gland structures which become responsive to androgenic stimulation. *Anat. Rec.* 172, 179–195.
28. Price, D., and Ortiz, E. (1965). The role of fetal androgens in sex differentiation in mammals. In *Organogenesis* (R. L. Ursprung and H. DeHaan H, Eds.), pp. 629–652. Holt, Rinehart and Winston, New York.
29. Price, D. (1963). Comparative aspects of development and structure in the prostate. In *Biology of the Prostate and Related Tissues* (E. P. Vollmer and G. Kauffman, Eds.), pp. 1–28. U.S. Government Printing Office, Washington, DC.
30. Donjacour, A. A., and Cunha, G. R. (1993). Assessment of prostatic protein secretion in tissue recombinants made of urogenital sinus mesenchyme and urothelium from normal or androgen-insensitive mice. *Endocrinology* 132, 2342–2350.
31. Cunha, G. R., and Donjacour A. A. (1989). Mesenchymal-epithelial interactions in the growth and development of the prostate. *Cancer Treat. Res.* 46, 159–175.
32. Hayward, S. W., Baskin, L. S., Haughney, P. C., Foster, B. A., Cunha, A. R., Dahiya, R., Prins, G. S., and Cunha, G. R. (1996). Stromal development in the ventral prostate, anterior prostate and seminal vesicle of the rat. *Acta Anat.* 155, 94–103.
33. Kinbara, H., and Cunha, G. R. (1996). Ductal heterogeneity in rat dorsal-lateral prostate. *Prostate* 28, 58–64.
34. Sugimura, Y., Cunha, G. R., and Donjacour, A. A. (1986). Morphogenesis of ductal networks in the mouse prostate. *Biol. Reprod.* 34, 961–971.
35. Almahbobi, G., Hedwards, S., Fricout, G., Jeulin, D., Bertram, J. F., and Risbridger, G. P. (2005). Computer-based detection of neonatal changes to branching morphogenesis reveals different mechanisms of and predicts prostate enlargement in bone morphogenetic protein 4 haploinsufficient mice. *J. Pathol.* 206, 52–61.
36. Hayashi, N., Sugimura, Y., Kawamura, J., Donjacour, A. A., and Cunha, G. R. (1991). Morphological and functional heterogeneity in the rat prostatic gland. *Biol. Reprod.* 45, 308–321.
37. Marker, P. C., Stephan, J. P., Lee, J., Bald, L., Mather, J. P., and Cunha, G. R. (2001). Fucosyltransferase1 and H-type complex carbohydrates modulate epithelial cell proliferation during prostatic branching morphogenesis. *Dev. Biol.* 233, 95–108.
38. vom Saal, F. S., Timms, B. G., Montano, M. M., Palanza, P., Thayer, K. A., Nagel, S. C., Dhar, M. D., Ganjam, V. K., Parmigiani, S., and Welshons, W. V. (1997). Prostate enlargement in mice due to fetal exposure to low doses of estradiol or diethylstilbestrol and opposite effects at high doses. *Proc. Natl. Acad. Sci. U S A* 94, 2056–2061.
39. Timms, B. G., Petersen, S. L., and vom Saal, F. S. (1999). Prostate gland growth during development is stimulated in both male and female rat fetuses by intrauterine proximity to female fetuses. *J. Urol.* 161, 1694–1701.
40. Fricout, G., Cullen-McEwen, L., Harper, I. S., Jeulin, D., and Bertram, J. F. (2001). A quantitative method for analysing 3D branching in embryonic kidneys: development of a technique and preliminary data. *Image Anal. Stereol.* 20, 36–41.
41. Cullen-McEwen, L. A., Fricout, G., Harper, I. S., Jeulin, D., and Bertram, J. F. (2002). Quantitation of 3D ureteric branching morphogenesis in cultured embryonic mouse kidney. *Int. J. Dev. Biol.* 46, 1049–1055.
42. Marker, P. C., Donjacour, A. A., Dahiya, R., and Cunha, G. R. (2003). Hormonal, cellular, and molecular control of prostatic development. *Dev. Biol.* 253, 165–174.
43. Cunha, G. R., Donjacour, A. A., Cooke, P. S., Mee, S., Bigsby, R. M., Higgins, S. J., and Sugimura, Y. (1987). The endocrinology and developmental biology of the prostate. *Endocr. Rev.* 8, 338–362.
44. Prins, G. S., and Birch, L. (1995). The developmental pattern of androgen receptor expression in rat prostate lobes is altered after neonatal exposure to estrogen. *Endocrinology* 136, 1303–1314.
45. Hayward, S. W., Brody, J. R., and Cunha, G. R. (1996). An edgewise look at basal epithelial cells: three-dimensional views of the rat prostate, mammary gland and salivary gland. *Differentiation* 60, 219–227.
46. Franks, L. M., Riddle, P. N., Carbonell, A. W., and Gey, G. O. (1970). A comparative study of the ultrastructure and lack of growth capacity of adult human prostate epithelium mechanically separated from its stroma. *J. Pathol.* 100, 113–119.
47. Chung L. W., and Cunha, G. R. (1983). Stromal-epithelial interactions: II. Regulation of prostatic growth by embryonic urogenital sinus mesenchyme. *Prostate* 4, 503–511.
48. Cunha, G. R., Fujii, H., Neubauer, B. L., Shannon, J. M., Sawyer, L., and Reese, B. A. (1983). Epithelial-mesenchymal interactions in prostatic development: I. Morphological observations of prostatic induction by urogenital sinus mesenchyme in epithelium of the adult rodent urinary bladder. *J. Cell Biol.* 96, 1662–1670.
49. Cunha, G. R., Ricke, W., Thomson, A., Marker, P. C., Gail, R., Hayward, S. W., and Wang, Y. Z. (2004). Hormonal, cellular, and molecular regulation of normal and neoplastic prostatic development. *J. Steroid Biochem. Mol. Biol.* 92, 221–226.
50. Cunha, G. R., Chung, L. W., Shannon, J. M., Taguchi, O., and Fujii, H. (1983). Hormone-induced morphogenesis and growth: role of mesenchymal-epithelial interactions. *Recent Prog. Horm. Res.* 39, 559–598.
51. Hayashi, N., Cunha, G. R., and Parker, M. (1993). Permissive and instructive induction of adult rodent prostatic epithelium by heterotypic urogenital sinus mesenchyme. *Epithel. Cell Biol.* 2, 66–78.
52. Hayward, S. W., Haughney, P. C., Rosen, M. A., Greulich, K. M., Weier, H. U., Dahiya, R., and Cunha, G. R. (1998). Interactions between adult human prostatic epithelium and rat urogenital sinus mesenchyme in a tissue recombination model. *Differentiation* 63, 131–140.
53. Hayward, S. W., Rosen, M. A., and Cunha, G. R. (1997). Stromal-epithelial interactions in the normal and neoplastic prostate. *Br. J. Urol.* 79 (Suppl. 2), 18–26.
54. Shima, H., Tsuji, M., Elfman, F., and Cunha, G. R. (1995). Development of male urogenital epithelia elicited by soluble mesenchymal factors. *J. Androl.* 16, 233–241.
55. Takeda, H., Chodak, G., Mutchnik, S., Nakamoto, T., and Chang, C. (1990). Immunohistochemical localization of androgen receptors with mono- and polyclonal antibodies to androgen receptor. *J. Endocrinol.* 126, 17–25.
56. Cunha, G. R., Lung, B., and Reese, B. (1980). Glandular epithelial induction by embryonic mesenchyme in adult bladder epithelium of BALB/c mice. *Invest. Urol.* 17, 302–304.
57. Cunha, G. R., Chung, L. W., Shannon, J. M., and Reese, B. A. (1980). Stromal-epithelial interactions in sex differentiation. *Biol. Reprod.* 22, 19–42.
58. Takeda, H., Lasnitzki, I., and Mizuno, T. (1986). Analysis of prostatic bud induction by brief androgen treatment in the fetal rat urogenital sinus. *J. Endocrinol.* 110, 467–470.
59. Donjacour, A. A., and Cunha, G. R. (1988). The effect of androgen deprivation on branching morphogenesis in the mouse prostate. *Dev. Biol.* 128, 1–14.

60. Lipschutz, J. H., Foster, B. A., and Cunha, G. R. (1997). Differentiation of rat neonatal ventral prostates grown in a serum-free organ culture system. *Prostate* 32, 35–42.
61. Cancilla, B., Jarred, R. A., Wang, H., Mellor, S. L., Cunha, G. R., and Risbridger, G. P. (2001). Regulation of prostate branching morphogenesis by activin a and follistatin. *Dev. Biol.* 237, 145–158.
62. Timme, T. L., Truong, L. D., Merz, V. W., Krebs, T., Kadmon, D., Flanders, K. C., Park, S. H., and Thompson, T. C. (1994). Mesenchymal-epithelial interactions and transforming growth factor-beta expression during mouse prostate morphogenesis. *Endocrinology* 134, 1039–1045.
63. Lamm, M. L. G., Podlasek, C. A., Barnett, D. H., Lee, J., Clemens, J. Q., Hebner, C. M., and Bushman, W. (2001). Mesenchymal factor bone morphogenetic protein 4 restricts ductal budding and branching morphogenesis in the developing prostate. *Dev. Biol.* 232, 301–314.
64. Aboseif, S. R., Dahiya, R., Narayan, P., and Cunha, G. R. (1997). Effect of retinoic acid on prostatic development. *Prostate* 31, 161–167.
65. Podlasek, C. A., Barnett, D. H., Clemens, J. Q., Bak, P. M., and Bushman, W. (1999). Prostate development requires Sonic hedgehog expressed by the urogenital sinus epithelium. *Dev. Biol.* 209, 28–39.
66. Pu, Y., Huang, L., and Prins, G. S. (2004). Sonic hedgehog-patched Gli signaling in the developing rat prostate gland: lobe-specific suppression by neonatal estrogens reduces ductal growth and branching. *Dev. Biol.* 273, 257–275.
67. Lamm, M. L., Catbagan, W. S., Laciak, R. J., Barnett, D. H., Hebner, C. M., Gaffield, W., Walterhouse, D., Iannaccone, P., and Bushman, W. (2002). Sonic hedgehog activates mesenchymal Gli1 expression during prostate ductal bud formation. *Dev. Biol.* 249, 349–366.
68. Bieberich, C. J., Fujita, K., He, W. W., and Jay, G. (1996). Prostate-specific and androgen-dependent expression of a novel homeobox gene. *J. Biol. Chem.* 271, 31779–31782.
69. Schneider, A., Brand, T., Zweigerdt, R., and Arnold, H. (2000). Targeted disruption of the Nkx3.1 gene in mice results in morphogenetic defects of minor salivary glands: parallels to glandular duct morphogenesis in prostate. *Mech. Dev.* 95, 163–174.
70. Tanaka, M., Komuro, I., Inagaki, H., Jenkins, N. A., Copeland, N. G., and Izumo, S. (2000). Nkx3.1, a murine homolog of *Drosophila* bagpipe, regulates epithelial ductal branching and proliferation of the prostate and palatine glands. *Dev. Dyn.* 219, 248–260.
71. Sciavolino, P. J., Abrams, E. W., Yang, L., Austenberg, L. P., Shen, M. M., and Abate-Shen, C. (1997). Tissue-specific expression of murine Nkx3.1 in the male urogenital system. *Dev. Dyn.* 209, 127–138.
72. Podlasek, C. A., Duboule, D., and Bushman, W. (1997). Male accessory sex organ morphogenesis is altered by loss of function of Hoxd-13. *Dev. Dyn.* 208, 454–465.
73. Podlasek, C. A., Clemens, J. Q., and Bushman, W. (1999). Hoxa-13 gene mutation results in abnormal seminal vesicle and prostate development. *J. Urol.* 161, 1655–1661.
74. Oefelein, M., Chin-Chance, C., and Bushman, W. (1996). Expression of the homeotic gene Hox-d13 in the developing and adult mouse prostate. *J. Urol.* 155, 342–346.
75. Thomson, A. A., and Cunha, G. R. (1999). Prostatic growth and development are regulated by FGF10. *Development* 126, 3693–3701.
76. Donjacour, A. A., Thomson, A. A., and Cunha, G. R. (2003). FGF-10 plays an essential role in the growth of the fetal prostate. *Dev. Biol.* 261, 39–54.
77. Thomson, A. A. (2001). Role of androgens and fibroblast growth factors in prostatic development. *Reproduction* 121, 187–195.
78. Sugimura, Y., Foster, B. A., Hom, Y. K., Lipschutz, J. H., Rubin, J. S., Finch, P. W., Aaronson, S. A., Hayashi, N., Kawamura, J., and Cunha, G. R. (1996). Keratinocyte growth factor (KGF) can replace testosterone in the ductal branching morphogenesis of the rat ventral prostate. *Int. J. Dev. Biol.* 40, 941–951.
79. Thomson, A. A., Foster, B. A., and Cunha, G. R. (1997). Analysis of growth factor and receptor mRNA levels during development of the rat seminal vesicle and prostate. *Development* 124, 2431–2439.
80. Price, D. (1936). Normal development of the prostate and seminal vesicles of the rat with a study of experimental postnatal modifications. *Am J. Anat.* 60, 79–127.
81. Lopes, E. S., Foster, B. A., and Cunha, G. R. (1994). Expression of prostatic secretory proteins in rat prostates. *Cancer Res.* 35, 282.
82. Rajfer, J., and Coffey, D. S. (1979). Effects of neonatal steroids on male sex tissues. *Invest. Urol.* 17, 3–8.
83. Piacsek, B. E., and Hostetter, M. W. (1984). Neonatal androgenization in the male rat: evidence for central and peripheral defects. *Biol. Reprod.* 30, 344–351.
84. Prins, G. S., Birch, L., Habermann, H., Chang, W. Y., Tebeau, C., Putz, O., and Bieberich, C. (2001). Influence of neonatal estrogens on rat prostate development. *Reprod. Fertil. Dev.* 13, 241–252.
85. Griffiths, K. (2000). Estrogens and prostatic disease. International Prostate Health Council Study Group. *Prostate* 45, 87–100.
86. Pylkkanen, L., Makela, S., Valve, E., Harkonen, P., Toikkanen, S., and Santti, R. (1993). Prostatic dysplasia associated with increased expression of c-myc in neonatally estrogenized mice. *J. Urol.* 149, 1593–1601.
87. Rajfer, J., and Coffey, D. S. (1978). Sex steroid imprinting of the immature prostate: long-term effects. *Invest. Urol.* 16, 186–190.
88. Prins, G. S., Woodham, C., Lepinske, M., and Birch, L. (1993). Effects of neonatal estrogen exposure on prostatic secretory genes and their correlation with androgen receptor expression in the separate prostate lobes of the adult rat. *Endocrinology* 132, 2387–2398.
89. Higgins, S. J., Brooks, D. E., Fuller, F. M., Jackson, P. J., and Smith, S. E. (1981). Functional development of sex accessory organs of the male rat: use of oestradiol benzoate to identify the neonatal period as critical for development of normal protein-synthetic and secretory capabilities. *Biochem. J.* 194, 895–905.
90. Gaytan, F., Bellido, C., Aguilar, R., and Lucena, M. C. (1986). Morphometric analysis of the rat ventral prostate and seminal vesicles during prepubertal development: effects of neonatal treatment with estrogen. *Biol. Reprod.* 35, 219–225.
91. Chung, L. W., and MacFadden, D. K. (1980). Sex steroids imprinting and prostatic growth. *Invest. Urol.* 17, 337–342.
92. Singh, J., Zhu, Q., and Handelsman, D. J. (1999). Stereological evaluation of mouse prostate development. *J. Androl.* 20, 251–258.
93. Prins, G. S., Cooke, P. S., Birch, L., Donjacour, A. A., Yalcinkaya, T. M., Siiteri, P. K., and Cunha, G. R. (1992). Androgen receptor expression and 5 alpha-reductase activity along the proximal-distal axis of the rat prostatic duct. *Endocrinology* 130, 3066–3073.
94. Prins, G. S. (1989). Differential regulation of androgen receptors in the separate rat prostate lobes: androgen independent expression in the lateral lobe. *J. Steroid Biochem.* 33, 319–326.
95. Prins, G. S., and Birch, L. (1993). Immunocytochemical analysis of androgen receptor along the ducts of the separate rat prostate lobes after androgen withdrawal and replacement. *Endocrinology* 132, 169–178.

96. Prins, G. S. (1997). Developmental estrogenization of the prostate gland. In *Prostate: Basic and Clinical Aspects* (R. K. Naz, Ed.), pp. 245–263. CRC Press, New York.
97. Woodham, C., Birch, L., and Prins, G. S. (2003). Neonatal estrogen down-regulates prostatic androgen receptor through a proteosome-mediated protein degradation pathway. *Endocrinology* 144, 4841–4850.
98. Prins, G. S., Marmer, M., Woodham, C., Chang, W., Kuiper, G., Gustafsson, J. A., and Birch, L. (1998). Estrogen receptor-beta messenger ribonucleic acid ontogeny in the prostate of normal and neonatally estrogenized rats. *Endocrinology* 139, 874–883.
99. Prins, G. S., Birch, L., Couse, J. F., Choi, I., Katzenellenbogen, B., and Korach, K. S. (2001). Estrogen imprinting of the developing prostate gland is mediated through stromal estrogen receptor α: studies with αERKO and βERKO mice. *Cancer Res.* 61, 6089–6097.
100. Prins, G. S., Chang, W. Y., Wang, Y., and Van Breemen, R. B. (2002). Retinoic acid receptors and retinoids are up-regulated in the developing and adult rat prostate by neonatal estrogen exposure. *Endocrinology* 143, 3628–3640.
101. Chang, W. Y., Birch, L., Woodham, C., Gold, L. I., and Prins, G. S. (1999). Neonatal estrogen exposure alters the transforming growth factor-beta signaling system in the developing rat prostate and blocks the transient p21(cip1/waf1) expression associated with epithelial differentiation. *Endocrinology* 140, 2801–2813.
102. Sharpe, R. M. (1993). Declining sperm counts in men: is there an endocrine cause? *J. Endocrinol.* 136, 357–360.
103. vom Saal, F. S. (1989). Sexual differentiation in litter-bearing mammals: influence of sex of adjacent fetuses in utero. *J. Anim. Sci.* 67, 1824–1840.
104. Even, M. D., Dhar, M. G., and vom Saal, F. S. (1992). Transport of steroids between fetuses via amniotic fluid in relation to the intrauterine position phenomenon in rats. *J. Reprod. Fertil.* 96, 709–716.
105. vom Saal, F. S., Grant, W. M., McMullen, C. W., and Laves, K. S. (1983). High fetal estrogen concentrations: correlation with increased adult sexual activity and decreased aggression in male mice. *Science* 220, 1306–1309.
106. Nonneman, D. J., Ganjam, V. K., Welshons, W. V., and vom Saal, F. S. (1992). Intrauterine position effects on steroid metabolism and steroid receptors of reproductive organs in male mice. *Biol. Reprod.* 47, 723–729.
107. Nagel, S. C., vom Saal, F. S., Thayer, K. A., Dhar, M. G., Boechler, M., and Welshons, W. V. (1997). Relative binding affinity-serum modified access (RBA-SMA) assay predicts the relative in vivo bioactivity of the xenoestrogens bisphenol A and octylphenol. *Environ. Health Perspect.* 105, 70–76.
108. Ashby, J., Tinwell, H., and Haseman, J. (1999). Lack of effects for low dose levels of bisphenol A and diethylstilbestrol on the prostate gland of CF1 mice exposed in utero. *Regul. Toxicol. Pharmacol.* 30, 156–166.
109. Cagen, S. Z., Waechter, J. M., Jr., Dimond, S. S., Breslin, W. J., Butala, J. H., Jekat, F. W., Joiner, R. L., Shiotsuka, R. N., Veenstra, G. E., and Harris, L. R. (1999). Normal reproductive organ development in CF-1 mice following prenatal exposure to bisphenol A. *Toxicol. Sci.* 50, 36–44.
110. Putz, O., Schwartz, C. B., Kim, S., LeBlanc, G. A., Cooper, R. L., and Prins, G. S. (2001). Neonatal low- and high-dose exposure to estradiol benzoate in the male rat: I. Effects on the prostate gland. *Biol. Reprod.* 65, 1496–1505.
111. vom Saal, F., Montano, M., and Wang, M. (1992). Chemically-induced alterations in sexual and functional development: the wildlife/human connection. In *Sexual Differentiation in Mammals* (T. Colburn and C. Clement, Eds.), pp. 17–83. Princeton Scientific, Princeton, NJ.
112. Risbridger, G. P., Wang, H., Frydenberg, M., and Cunha, G. (2001). The metaplastic effects of estrogen on mouse prostate epithelium: proliferation of cells with basal cell phenotype. *Endocrinology* 142, 2443–2450.
113. Jesik, C. J., Holland, J. M., and Lee, C. (1982). An anatomic and histologic study of the rat prostate. *Prostate* 3, 81–97.
114. Lee, C., Sensibar, J. A., Dudek, S. M., Hiipakka, R. A., and Liao, S. T. (1990). Prostatic ductal system in rats: regional variation in morphological and functional activities. *Biol. Reprod.* 43, 1079–1086.
115. El-Alfy, M., Luu-The, V., Huang, X. F., Berger, L., Labrie, F., and Pelletier, G. (1999). Localization of type 5 17β-hydroxysteroid dehydrogenase, 3β-hydroxysteroid dehydrogenase, and androgen receptor in the human prostate by in situ hybridization and immunocytochemistry. *Endocrinology* 140, 1481–1491.
116. di Sant'Agnese, P. A. (1992). Neuroendocrine differentiation in human prostatic carcinoma. *Hum. Pathol.* 23, 287–296.
117. di Sant'Agnese, P. A., and Cockett, A. T. (1996). Neuroendocrine differentiation in prostatic malignancy. *Cancer* 78, 357–361.
118. McNeal, J. E. (1981). The zonal anatomy of the prostate. *Prostate* 2, 35–49.
119. Wright, A. S., Thomas, L. N., Douglas, R. C., Lazier, C. B., and Rittmaster, R. S. (1996). Relative potency of testosterone and dihydrotestosterone in preventing atrophy and apoptosis in the prostate of the castrated rat. *J. Clin. Invest.* 98, 2558–2563.
120. Bonkhoff, H., and Remberger, K. (1993). Widespread distribution of nuclear androgen receptors in the basal cell layer of the normal and hyperplastic human prostate. *Virchows Arch. A Pathol. Anat. Histopathol.* 422, 35–38.
121. Leav, I., McNeal, J. E., Kwan, P. W., Komminoth, P., and Merk, F. B. (1996). Androgen receptor expression in prostatic dysplasia (prostatic intraepithelial neoplasia) in the human prostate: an immunohistochemical and in situ hybridization study. *Prostate* 29, 137–145.
122. Sar, M., Lubahn, D. B., French, F. S., and Wilson, E. M. (1990). Immunohistochemical localization of the androgen receptor in rat and human tissues. *Endocrinology* 127, 3180–3186.
123. Bhatia-Gaur, R., Donjacour, A. A., Sciavolino, P. J., Kim, M., Desai, N., Young, P., Norton, C. R., Gridley, T., Cardiff, R. D., Cunha, G. R., Abate-Shen, C., and Shen, M. M. (1999). Roles for Nkx3.1 in prostate development and cancer. *Genes Dev.* 13, 966–977.
124. Sherwood, E. R., Berg, L. A., Mitchell, N. J., McNeal, J. E., Kozlowski, J. M., and Lee, C. (1990). Differential cytokeratin expression in normal, hyperplastic and malignant epithelial cells from human prostate. *J. Urol.* 143, 167–171.
125. Yang, Y., Hao, J., Liu, X., Dalkin, B., and Nagle, R. B. Differential expression of cytokeratin mRNA and protein in normal prostate, prostatic intraepithelial neoplasia, and invasive carcinoma. *Am. J. Pathol.* 150, 693–704.
126. Wang, Y., Hayward, S., Cao, M., Thayer, K., and Cunha, G. (2001). Cell differentiation lineage in the prostate. *Differentiation* 68, 270–279.
127. Mao, P., and Angrist, A. (1966). The fine structure of the basal cell of human prostate. *Lab. Invest.* 15, 1768–1782.
128. Brandes, D., Kirchleim, D., and Scott, W. (1964). Ultrastructure of the human prostate: normal and neoplastic. *Lab. Invest.* 13, 1541.
129. El-Alfy, M., Pelletier, G., Hermo, L. S., and Labrie, F. (2000). Unique features of the basal cells of human prostate epithelium. *Microsc. Res. Tech.* 51, 436–446.
130. Fawcett, J. W., and Keynes, R. J. (1986). Muscle basal lamina: a new graft material for peripheral nerve repair. *J. Neurosurg.* 65, 354–363.

131. Gartner, L. P., and Hiatt, J. L. (1997). Male reproductive systems. In *Colour Textbook of Histology*, pp. 403–420. WB Saunders, Philadelphia.
132. Brawer, M. K., Peehl, D. M., Stamey, T. A., and Bostwick, D. G. (1985). Keratin immunoreactivity in the benign and neoplastic human prostate. *Cancer Res.* 45, 3663–3667.
133. Yang, A., Kaghad, M., Wang, Y., Gillett, E., Fleming, M. D., Dotsch, V., Andrews, N. C., Caput, D., and McKeon, F. (1998). p63, a p53 homolog at 3q27-29, encodes multiple products with transactivating, death-inducing, and dominant-negative activities. *Mol. Cell* 2, 305–316.
134. Epstein, J. I. (1993). PSA and PAP as immunohistochemical markers in prostate cancer. *Urol. Clin. North Am.* 20, 757–770.
135. De Marzo, A. M., Meeker, A. K., Epstein, J. I., and Coffey, D. S. (1998). Prostate stem cell compartments: expression of the cell cycle inhibitor p27Kip1 in normal, hyperplastic, and neoplastic cells. *Am. J. Pathol.* 153, 911–919.
136. Danielpour, D. (1999). Transdifferentiation of NRP-152 rat prostatic basal epithelial cells toward a luminal phenotype: regulation by glucocorticoid, insulin-like growth factor-I and transforming growth factor-beta. *J. Cell Sci.* 112, 169–179.
137. Habermann, H., Ray, V., Habermann, W., and Prins, G. S. (2002). Alterations in gap junction protein expression in human benign prostatic hyperplasia and prostate cancer. *J. Urol.* 167, 655–660.
138. Habermann, H., Chang, W. Y., Birch, L., Mehta, P., and Prins, G. S. (2001). Developmental exposure to estrogens alters epithelial cell adhesion and gap junction proteins in the adult rat prostate. *Endocrinology* 142, 359–369.
139. Bonkhoff, H., and Remberger, K. (1996). Differentiation pathways and histogenetic aspects of normal and abnormal prostatic growth: a stem cell model. *Prostate* 28, 98–106.
140. Robinson, E. J., Neal, D. E., and Collins, A. T. (1998). Basal cells are progenitors of luminal cells in primary cultures of differentiating human prostatic epithelium. *Prostate* 37, 149–160.
141. Dermer, G. B. (1978). Basal cell proliferation in benign prostatic hyperplasia. *Cancer* 41, 1857–1862.
142. van Leenders, G. J., and Schalken, J. A. (2001). Stem cell differentiation within the human prostate epithelium: implications for prostate carcinogenesis. *BJU Int.* 88 (Suppl. 2), 35–42; discussion 49–50.
143. Abrahamsson, P. A. (1999). Neuroendocrine differentiation in prostatic carcinoma. *Prostate* 39, 135–148.
144. Abrahamsson, P. A. (1999). Neuroendocrine cells in tumor growth of the prostate. *Endocr. Relat. Cancer* 6, 503–519.
145. Noordzij, M. A., van Steenbrugge, G. J., van der Kwast, T. H., and Schroder, F. H. (1995). Neuroendocrine cells in the normal, hyperplastic and neoplastic prostate. *Urol. Res.* 22, 333–341.
146. Xue, Y., Smedts, F., Verhofstad, A., Debruyne, F., de la Rosette, J., and Schalken, J. (1998). Cell kinetics of prostate exocrine and neuroendocrine epithelium and their differential interrelationship: new perspectives. *Prostate Suppl.* 8, 62–73.
147. Cohen, R. J., Glezerson, G., Taylor, L. F., Grundle, H. A., and Naude, J. H. (1993). The neuroendocrine cell population of the human prostate gland. *J. Urol.* 150, 365–368.
148. Aumuller, G., Leonhardt, M., Renneberg, H., von Rahden, B., Bjartell, A., and Abrahamsson, P. A. (2001). Semiquantitative morphology of human prostatic development and regional distribution of prostatic neuroendocrine cells. *Prostate* 46, 108–115.
149. Bonkhoff, H., Wernert, N., Dhom, G., and Remberger, K. (1991). Relation of endocrine-paracrine cells to cell proliferation in normal, hyperplastic, and neoplastic human prostate. *Prostate* 19, 91–98.
150. Abrahamsson, P. A., Dizeyi, N., Alm, P., di Sant'Agnese, P. A., Deftos, L. J., and Aumuller, G. (2000). Calcitonin and calcitonin gene-related peptide in the human prostate gland. *Prostate* 44, 181–186.
151. Kinbara, H., Cunha, G. R., Boutin, E., Hayashi, N., and Kawamura, J. (1996). Evidence of stem cells in the adult prostatic epithelium based upon responsiveness to mesenchymal inductors. *Prostate* 29, 107–116.
152. Hudson, D. L., O'Hare, M., Watt, F. M., and Masters, J. R. (2000). Proliferative heterogeneity in the human prostate: evidence for epithelial stem cells. *Lab. Invest.* 80, 1243–1250.
153. Huss, W. J., Gray, D. R., Werdin, E. S., Funkhouser, W. K., Jr., and Smith, G. J. (2004). Evidence of pluripotent human prostate stem cells in a human prostate primary xenograft model. *Prostate* 60, 77–90.
154. Potten, C. S., and Morris, R. J. (1988). Epithelial stem cells in vivo. *J. Cell Sci. Suppl.* 10, 45–62.
155. Potten, C. S., and Loeffler, M. (1990). Stem cells: attributes, cycles, spirals, pitfalls and uncertainties: lessons for and from the crypt. *Development* 110, 1001–1020.
156. Bonkhoff, H., Stein, U., and Remberger, K. (1994). Multidirectional differentiation in the normal, hyperplastic, and neoplastic human prostate: simultaneous demonstration of cell-specific epithelial markers. *Hum. Pathol.* 25, 42–46.
157. Hudson, D. L., Guy, A. T., Fry, P., O'Hare, M. J., Watt, F. M., and Masters, J. R. (2001). Epithelial cell differentiation pathways in the human prostate: identification of intermediate phenotypes by keratin expression. *J. Histochem. Cytochem.* 49, 271–278.
158. Uzgare, A. R., Xu, Y., and Isaacs, J. T. (2004). In vitro culturing and characteristics of transit amplifying epithelial cells from human prostate tissue. *J. Cell. Biochem.* 91, 196–205.
159. Collins, A. T., Habib, F. K., Maitland, N. J., and Neal, D. E. (2001). Identification and isolation of human prostate epithelial stem cells based on $\alpha_2\beta_1$-integrin expression. *J. Cell Sci.* 114, 3865–3872.
160. Verhagen, A. P., Aalders, T. W., Ramaekers, F. C., Debruyne, F. M., and Schalken, J. A. (1988). Differential expression of keratins in the basal and luminal compartments of rat prostatic epithelium during degeneration and regeneration. *Prostate* 13, 25–38.
161. Timms, B. G., Chandler, J. A., and Sinowatz, F. (1976). The ultrastructure of basal cells of rat and dog prostate. *Cell. Tissue Res.* 173, 543–554.
162. Foster, C. S., and Ke, Y. (1997). Stem cells in prostatic epithelia. *Int. J. Exp. Pathol.* 78, 311–329.
163. Brandes, D. (1996). The fine structure and histochemistry of prostatic glands in relation to sex hormones. *Int. Rev. Cytol.* 20, 207–276.
164. Heatfield, B. M., Sanefuji, H., and Trump, B. F. (1982). Studies on carcinogenesis of human prostate: III. Long-term explant culture of normal prostate and benign prostatic hyperplasia: transmission and scanning electron microscopy. *J. Natl. Cancer Inst.* 69, 757–766.
165. Merchant, D. J., Clarke, S. M., Ives, K., and Harris, S. (1983). Primary explant culture: an in vitro model of the human prostate. *Prostate* 4, 523–542.
166. Isaacs, J. T., and Coffey, D. S. (1989). Etiology and disease process of benign prostatic hyperplasia. *Prostate Suppl.* 2, 33–50.
167. Bonkhoff, H. (1996). Role of the basal cells in premalignant changes of the human prostate: a stem cell concept for the development of prostate cancer. *Eur. Urol.* 30, 201–205.
168. De Marzo, A. M., Nelson, W. G., Meeker, A. K., and Coffey, D. S. (1998). Stem cell features of benign and malignant prostate epithelial cells. *J. Urol.* 160, 2381–2392.

169. Bui, M., and Reiter, R. E. (1998). Stem cell genes in androgen-independent prostate cancer. *Cancer Metastasis Rev.* 17, 391–399.
170. Foster, C. S., Dodson, A., Karavana, V., Smith, P. H., and Ke, Y. (2002). Prostatic stem cells. *J. Pathol.* 197, 551–565.
171. Bonkhoff, H., Stein, U., and Remberger, K. (1994). The proliferative function of basal cells in the normal and hyperplastic human prostate. *Prostate* 24, 114–118.
172. McNeal, J. E., Haillot, O., and Yemoto, C. (1995). Cell proliferation in dysplasia of the prostate: analysis by PCNA immunostaining. *Prostate* 27, 258–268.
173. Montpetit, M., Abrahams, P., Clark, A. F., and Tenniswood, M. (1988). Androgen-independent epithelial cells of the rat ventral prostate. *Prostate* 12, 13–28.
174. Richardson, G. D., Robson, C. N., Lang, S. H., Neal, D. E., Maitland, N. J., and Collins, A. T. (2004). CD133, a novel marker for human prostatic epithelial stem cells. *J. Cell Sci.* 117, 3539–3545.
175. Tsujimura, A., Koikawa, Y., Salm, S., Takao, T., Coetzee, S., Moscatelli, D., Shapiro, E., Lepor, H., Sun, T. T., and Wilson, E. L. (2002). Proximal location of mouse prostate epithelial stem cells: a model of prostatic homeostasis. *J. Cell Biol.* 157, 1257–1265.
176. Kurita, T., Medina, R. T., Mills, A. A., and Cunha, G. R. (2004). Role of p63 and basal cells in the prostate. *Development* 131, 4955–4964.
177. Aumuller, G., Leonhardt, M., Janssen, M., Konrad, L., Bjartell, A., and Abrahamsson, P. A. (1999). Neurogenic origin of human prostate endocrine cells. *Urology* 53, 1041–1048.
178. Isaacs, J. T. (1985). Control of cell proliferation and death in the normal and neoplastic prostate: a stem cell model. In *Benign Prostatic Hyperplasia* (C. H. Rogers, D. S. Coffey, and G. R. Cunha, Eds.), pp. 85–94. National Institutes of Health, Bethesda, MD.
179. Isaacs, J. T., Furuya, Y., and Berges, R. (1994). The role of androgen in the regulation of programmed cell death/apoptosis in normal and malignant prostatic tissue. *Semin. Cancer Biol.* 5, 391–400.
180. Sugimura, Y., Cunha, G. R., Donjacour, A. A., Bigsby, R. M., and Brody, J. R. (1986). Whole-mount autoradiography study of DNA synthetic activity during postnatal development and androgen-induced regeneration in the mouse prostate. *Biol. Reprod.* 34, 985–995.
181. Ichihara, I., Kallio, M., and Pelliniemi, L. J. (1978). Light and electron microscopy of the ducts and their subepithelial tissue in the rat ventral prostate. *Cell. Tissue Res.* 192, 381–390.
182. Nemeth, J. A., Sensibar, J. A., White, R. R., Zelner, D. J., Kim, I. Y., and Lee, C. (1997). Prostatic ductal system in rats: tissue-specific expression and regional variation in stromal distribution of transforming growth factor-beta 1. *Prostate* 33, 64–71.
183. Shapiro, E., Becich, M. J., Hartanto, V., and Lepor, H. (1992). The relative proportion of stromal and epithelial hyperplasia is related to the development of symptomatic benign prostate hyperplasia. *J. Urol.* 147, 1293–1297.
184. Nemeth, J. A., and Lee, C. (1996). Prostatic ductal system in rats: regional variation in stromal organization. *Prostate* 28, 124–128.
185. Prins, G. S., Birch, L., and Greene, G. L. (1991). Androgen receptor localization in different cell types of the adult rat prostate. *Endocrinology* 129, 3187–3199.
186. Cunha, G. R., Hayward, S. W., Dahiya, R., and Foster, B. A. (1996). Smooth muscle-epithelial interactions in normal and neoplastic prostatic development. *Acta Anat.* 155, 63–72.
187. Luke, M. C., and Coffey, D. S. (1994). The male sex accessory tissues: structure, androgen action, and physiology. In *The Physiology of Reproduction* (E. Knobil and J. D. Neill, Eds.), pp. 1435–1487. Raven Press, New York.
188. Risbridger, G., and Frydenberg, M. (2005). Endocrinology of prostate cancer. In *Endocrinology* (L. J. D. Groot, Ed.). WB Saunders, Philadelphia.
189. Hernandez, J., and Thompson, I. M. (2004). Prostate-specific antigen: a review of the validation of the most commonly used cancer biomarker. *Cancer* 101, 894–904.
190. Stamey, T. A., Caldwell, M., McNeal, J. E., Nolley, R., Hemenez, M., and Downs, J. (2004). The prostate specific antigen era in the United States is over for prostate cancer: what happened in the last 20 years? *J. Urol.* 172, 1297–1301.
191. Clements, J. A., Willemsen, N. M., Myers, S. A., and Dong, Y. (2004). The tissue kallikrein family of serine proteases: functional roles in human disease and potential as clinical biomarkers. *Crit. Rev. Clin. Lab. Sci.* 41, 265–312.
192. Cunha, G. R., Hayward, S. W., and Wang, Y. Z. (2002). Role of stroma in carcinogenesis of the prostate. *Differentiation* 70, 473–485.
193. Cunha, G. R., and Matrisian, L. M. (2002). It's not my fault, blame it on my microenvironment. *Differentiation* 70, 469–472.
194. Arnold, J. T., and Isaacs, J. J. (2002). Mechanisms involved in the progression of androgen-independent prostate cancers: it is not only the cancer cell's fault. *Endocr. Relat. Cancer* 9, 61–73.
195. Thompson, T. C., Timme, T. L., Kadmon, D., Park, S. H., Egawa, S., and Yoshida, K. (1993). Genetic predisposition and mesenchymal-epithelial interactions in ras+myc-induced carcinogenesis in reconstituted mouse prostate. *Mol. Carcinog.* 7, 165–179.
196. Noel, A., Hajitou, A., L'Hoir, C., Maquoi, E., Baramova, E., Lewalle, J. M., Remacle, A., Kebers, F., Brown, P., Calberg-Bacq, C. M., and Foidart, J. M. (1998). Inhibition of stromal matrix metalloproteases: effects on breast-tumor promotion by fibroblasts. *Int. J. Cancer* 76, 267–273.
197. Camps, J. L., Chang, S. M., Hsu, T. C., Freeman, M. R., Hong, S. J., Zhau, H. E., von Eschenbach, A. C., and Chung, L. W. (1990). Fibroblast-mediated acceleration of human epithelial tumor growth in vivo. *Proc. Natl. Acad. Sci. U S A* 87, 75–79.
198. Olumi, A. F., Grossfeld, G. D., Hayward, S. W., Carroll, P. R., Tlsty, T. D., and Cunha, G. R. (1999). Carcinoma-associated fibroblasts direct tumor progression of initiated human prostatic epithelium. *Cancer Res.* 59, 5002–5011.
199. Rowley, D. R. (1998). What might a stromal response mean to prostate cancer progression? *Cancer Metastasis Rev.* 17, 411–419.
200. Tuxhorn, J. A., Ayala, G. E., and Rowley, D. R. (2001). Reactive stroma in prostate cancer progression. *J. Urol.* 166, 2472–2483.
201. Hayward, S. W., Dahiya, R., Cunha, G. R., Bartek, J., Deshpande, N., and Narayan, P. (1995). Establishment and characterization of an immortalized but non-transformed human prostate epithelial cell line: BPH-1. *In Vitro Cell Dev. Biol. Anim.* 31, 14–24.
202. Al-Hajj, M., Becker, M. W., Wicha, M., Weissman, I., and Clarke, M. F. (2004). Therapeutic implications of cancer stem cells. *Curr. Opin. Genet. Dev.* 14, 43–47.
203. Heppner, G. H. (1984). Tumor heterogeneity. *Cancer Res.* 44, 2259–2265.
204. Nowell, P. C. (1986). Mechanisms of tumor progression. *Cancer Res.* 46, 2203–2207.
205. Jost, A., Prepin, J., and Vigier, B. (1977). Hormones in the morphogenesis of the genital system. In *Morphogenesis and Malformation of the Genital System* (R. J. Blandau and D. Bergsma, Eds.), pp. 85–98. Alan R. Liss, New York.
206. Lung, B., and Cunha, G. R. (1981). Development of seminal vesicles and coagulating glands in neonatal mice: I. The morphogenetic effects of various hormonal conditions. *Anat. Rec.* 199, 73–88.

207. Shukri, N. M., Grew, F., and Shire, J. G. (1988). Recessive mutation in a standard recombinant-inbred line of mice affects seminal vesicle shape. *Genet. Res.* 52, 27–32.
208. Marker, P. C., Dahiya, R., and Cunha, G. R. (2003). Spontaneous mutation in mice provides new insight into the genetic mechanisms that pattern the seminal vesicles and prostate gland. *Dev. Dyn.* 226, 643–653.
209. Cattanach, B. M., Iddon, C. A., Charlton, H. M., Chiappa, S. A., and Fink, G. (1977). Gonadotrophin-releasing hormone deficiency in a mutant mouse with hypogonadism. *Nature* 269, 338–340.
210. Bianco, J., Handelsman, D. J., and Pedersen, J. (2002). Direct response of the murine prostate gland and seminal vesicles to estradiol. *Endocrinology* 143, 4922–4933.
211. Kim, H., Kim, T., Shin, J. H., Moon, H., Kang, I., Kim, I., Oh, J., and Han, S. (2004). Neonatal exposure to di(n-butyl) phthalate (dbp) alters male reproductive-tract development. *J. Toxicol. Environ. Health A* 67, 2045–2060.
212. Laczko, I. I., Hudson, D. L., Freeman, A., Feneley, M. R., and Masters, J. R. (2005). Comparison of the zones of the human prostate with the seminal vesicle: morphology, immunohistochemistry, and cell kinetics. *Prostate* 62, 260–266.
213. Williams, R. D., and Sandlow, J. I. (1998). Surgery of the seminal vesicles. In *Campbell's Urology* (P. C. Walsh, A. B. Retik, E. J. D. Vaughan, and A. J. Wein, Eds.), pp. 3299–3315. WB Saunders, Philadelphia.
214. Kim, S. W., and Paick, J. S. (2004). Peripheral effects of serotonin on the contractile responses of rat seminal vesicles and vasa deferentia. *J. Androl.* 25, 893–899.
215. Hall, S., and Oates, R. D. (1993). Unilateral absence of the scrotal vas deferens associated with contralateral mesonephric duct anomalies resulting in infertility: laboratory, physical and radiographic findings, and therapeutic alternatives. *J. Urol.* 150, 1161–1164.
216. Donohue, R. E., and Fauver, H. E. (1989). Unilateral absence of the vas deferens: a useful clinical sign. *J. A. M. A.* 261, 1180–1182.
217. Anguiano, A., Oates, R. D., Amos, J. A., Dean, M., Gerrard, B., Stewart, C., Maher, T. A., White, M. B., and Milunsky, A. (1992). Congenital bilateral absence of the vas deferens: a primarily genital form of cystic fibrosis. *J. A. M. A.* 267, 1794–1797.
218. Chillon, M., Casals, T., Mercier, B., Bassas, L., Lissens, W., Silber, S., Romey, M. C., Ruiz-Romero, J., Verlingue, C., Claustres, M., Nunes, V., Ferec, C., and Estivill, X. (1995). Mutations in the cystic fibrosis gene in patients with congenital absence of the vas deferens. *N. Engl. J. Med.* 332, 1475–1480.
219. Holsclaw, D. S., Perlmutter, A. D., Jockin, H., and Shwachman, H. (1971). Genital abnormalities in male patients with cystic fibrosis. *J. Urol.* 106, 568–574.
220. Robert, M., and Gagnon, C. (1999). Semenogelin I: a coagulum forming, multifunctional seminal vesicle protein. *Cell. Mol. Life Sci.* 55, 944–960.
221. Yoshida, K., Yamasaki, T., Yoshiike, M., Takano, S., Sato, I., and Iwamoto, T. (2003). Quantification of seminal plasma motility inhibitor/semenogelin in human seminal plasma. *J. Androl.* 24, 878–884.
222. Zubkova, E. V., and Robaire, B. (2004). Effect of glutathione depletion on antioxidant enzymes in the epididymis, seminal vesicles, and liver and on spermatozoa motility in the aging brown Norway rat. *Biol. Reprod.* 71, 1002–1008.
223. Epstein, J. I., and Murphy, W. M. (1997). Diseases of the prostate gland and seminal vesicles. In *Urological Pathology* (W. M. Murphy, Ed.), pp. 148–241. WB Saunders, Philadelphia.
224. Cooke, P. S., Young, P. F., and Cunha, G. R. (1987). A new model system for studying androgen-induced growth and morphogenesis in vitro: the bulbourethral gland. *Endocrinology* 121, 2161–2170.
225. Parr, M. B., Ren, H. P., Kepple, L., Parr, E. L., and Russell, L. D. (1993). Ultrastructure and morphometry of the urethral glands in normal, castrated, and testosterone-treated castrated male mice. *Anat. Rec.* 236, 449–458.
226. Sirigu, P., Turno, F., Usai, E., and Perra, M. T. (1993). Histochemical study of the human bulbourethral (Cowper's) glands. *Andrologia* 25, 293–299.
227. Riva, A., Usai, E., Cossu, M., Scarpa, R., and Testa-Riva, F. (1988). The human bulbo-urethral glands: a transmission electron microscopy and scanning electron microscopy study. *J. Androl.* 9, 133–141.
228. Riva, A., Usai, E., Cossu, M., Lantini, M. S., Scarpa, R., and Testa-Riva, F. (1990). Ultrastructure of human bulbourethral glands and of their main excretory ducts. *Arch. Androl.* 24, 177–184.
229. Dunker, N., and Aumuller, G. (2002). Transforming growth factor-beta 2 heterozygous mutant mice exhibit Cowper's gland hyperplasia and cystic dilations of the gland ducts (Cowper's syringoceles). *J. Anat.* 201, 173–183.
230. Zaviacic, M., and Ablin, R. J. (1998). The female prostate. *J. Natl. Cancer Inst.* 90, 713–714.
231. Wernert, N., Albrech, M., Sesterhenn, I., Goebbels, R., Bonkhoff, H., Seitz, G., Inniger, R., and Remberger, K. (1992). The 'female prostate': location, morphology, immunohistochemical characteristics and significance. *Eur. Urol.* 22, 64–69.
232. Folsom, A. I., and O'Brien, H. A. (1943). The female obstructing prostate. *J. A. M. A.* 121, 573–580.
233. Huffman, J. W. (1948). The detailed anatomy of the paraurethral ducts in the adult human female. *Am. J. Obstet. Gynecol.* 55, 86–101.
234. Flamini, M. A., Barbeito, C. G., Gimeno, E. J., and Portiansky, E. L. (2002). Morphological characterization of the female prostate (Skene's gland or paraurethral gland) of *Lagostomus maximus maximus*. *Ann. Anat.* 184, 341–345.
235. Zaviacic, M., and Ablin, R. J. (2000). The female prostate and prostate-specific antigen: immunohistochemical localization, implications of this prostate marker in women and reasons for using the term "prostate" in the human female. *Histol. Histopathol.* 15, 131–142.
236. Gittes, R. F., and Nakamura, R. M. (1996). Female urethral syndrome: a female prostatitis? *West. J. Med.* 164, 435–438.
237. Sloboda, J., Zaviacic, M., Jakubovsky, J., Hammar, E., and Johnsen, J. (1998). Metastasizing adenocarcinoma of the female prostate (Skene's paraurethral glands): histological and immunohistochemical prostate markers studies and first ultrastructural observation. *Pathol. Res. Pract.* 194, 129–136.
238. Dodson, M. K., Cliby, W. A., Pettavel, P. P., Keeney, G. L., and Podratz, K. C. (1995). Female urethral adenocarcinoma: evidence for more than one tissue of origin? *Gynecol. Oncol.* 59, 352–357.
239. Zaviacic, M., Sidlo, J., and Borovsky, M. (1993). Prostate specific antigen and prostate specific acid phosphatase in adenocarcinoma of Skene's paraurethral glands and ducts. *Virchows Arch. A Pathol. Anat. Histopathol.* 423, 503–505.
240. Ali, S. Z., Smilari, T. F., Gal, D., Lovecchio, J. L., and Teichberg, S. (1995). Primary adenoid cystic carcinoma of Skene's glands. *Gynecol. Oncol.* 57, 257–261.
241. Ejnes, L., Toullalan, O., Fayad, S., Bongain, A., Reville, M. D., and Gillet, J. Y. (2004). Very large Skene's duct cyst. *Acta Obstet. Gynecol. Scand.* 83, 598.0

Knobil and Neill's Physiology of Reproduction, Third Edition
edited by Jimmy D. Neill,
Elsevier © 2006

CHAPTER 24

Male Sexual Function

Shalender Bhasin and George S. Benson

Introduction, 1173
Human Sexual Response Cycle, 1174
Clinical Implications of the Sexual Response Cycle, 1174
Mechanisms of Penile Erection, 1175
Penile Anatomy, 1175
Blood Supply to the Penis, 1176
Neural Input into the Penis, 1176
Hemodynamic Changes during Penile Erection, 1177
Biochemical Mechanisms of Penile Erection, 1177
Central Regulation of Sexual Function by Sex Steroids, 1179
Testosterone Effects on Mood, 1181
Clinical Translation of Physiological Principles and Basic Science Advances, 1181
Biochemical Basis of Pharmacological Therapy of ED, 1182
Ejaculation and Emission, 1186
Anatomy, 1186
Neural Control of Emission and Ejaculation, 1186
Clinical Application of Physiological Principles, 1187
References, 1187

INTRODUCTION

Although Vatsyayana, the author of *Kama Sutra*, wrote a treatise on sexology almost 2,000 years ago, the knowledge of human sexuality remained shrouded in myth throughout most of human history and was influenced greatly by the prevalent religious doctrine—most of it appallingly erroneous. In fact, the fervent assertiveness of the Victorian era views on human sexuality was matched only by the glaring lack of evidence to support these views. It wasn't until Michael Kinsey conducted some of the first epidemiological surveys on sexual behavior of ordinary Americans that the complexity and variability of human sexual behavior began to be recognized. The publication of Kinsey's *Sexual Behavior in the Human Male* (1) dispelled many Victorian myths about human sexual behavior and eased the way for future generations of researchers to explore this hitherto forbidden territory. Several additional epidemiological surveys, such as the Hite report (2), the Redbook report (3), and the Janus report (4), provided additional insights into the patterns of sexual activity and behaviors of American men and women. Subsequent research by Masters and Johnson (5) formed the basis of our current understanding of the physiological changes during the human sexual response.

A large body of research in the 1990s led to the recognition that penile erections are the result of cavernosal smooth muscle relaxation and increased blood flow (6–9). It was also recognized that nitric oxide served as an important signaling molecule in inducing cavernosal smooth muscle relaxation (10–12). In 1998, the Nobel Assembly at the Karolinska Institute in Stockholm, Sweden awarded the Nobel Prize in Physiology or Medicine to Robert F. Furchgott, Louis J. Ignarro, and Ferid Murad for their discoveries concerning "the nitric oxide as a signaling molecule in the cardiovascular system" (13). Subsequent research revealed that actions of nitric oxide on the cavernosal smooth muscle are mediated through

Division of Endocrinology, University of the City of Los Angeles, Los Angeles, California.

the activation of guanylyl cyclase and the production of cyclic guanosine monophosphate (cGMP) (7,8, 14–19). Because cGMP acts as an important intracellular second messenger and causes smooth muscle relaxation by lowering intracellular calcium (16), it became apparent that strategies that prevent cGMP degradation might have potential as therapies for erectile dysfunction (ED). The translation of these concepts into the development of selective phosphodiesterase inhibitors as therapeutic agents for the treatment of ED is a remarkable story that ushered in the modern era of effective oral therapy (20–32). Elucidation of the mechanisms that regulate penile erections has helped identify many attractive targets for the discovery of therapeutic agents for ED; not surprisingly, at least half a dozen new drugs for ED are currently in various phases of development. The World Health Organization appropriately recognized the importance of sexual health as a determinant of quality of life when it declared sexual health a fundamental right of all human beings.

Freud postulated that sexual problems in adult men and women have their origin in disturbances in maturation of childhood sexuality and development of the child–parent relationship. With the recognition that ED has a vascular etiology in most men and is amenable to mechanism-specific drug therapy, many psychoanalytical theories, including those put forth by Freud, have lost favor.

HUMAN SEXUAL RESPONSE CYCLE

William Masters and Virginia E. Johnson (5), in a pioneering effort, studied physical changes in response to sexual stimulation in male and female volunteers. These investigators found that both men and women display predictable, sequential, physiological responses after sexual stimulation. These responses can be categorized in four phases: the excitement phase, the plateau phase, the orgasmic phase, and the resolution phase (Fig. 1).

During the excitation phase, the heart and breathing rates increase, blood pressure rises, nipples become erect, penis achieves varying degrees of erection, the testes are drawn up, and the skin may undergo a sex flush (5). The excitation phase can last from few minutes to several hours.

The excitation phase is followed by the plateau phase in which heart rate increases further, sexual pleasure increases, bladder sphincter closes, the muscles at the base of the penis contract rhythmically, and men begin to secrete a small amount of seminal fluid. Tension in the muscles increases, and muscle spasms in the feet, face, and hands may occur.

Achievement of orgasm releases the sexual tension and is associated with contractions of the pelvic muscles and anal sphincter, ejaculation of seminal fluid, and a perception of intense pleasure. In the initial emission phase, the seminal fluid collects in the urethral bulb and is associated with the sensation that orgasm is imminent. The ejaculation of the semen from the penis requires contractions of the periurethral and other pelvic muscles. The skin may also appear to have a "sex flush" over the entire body, although the flush can appear during earlier phases as well.

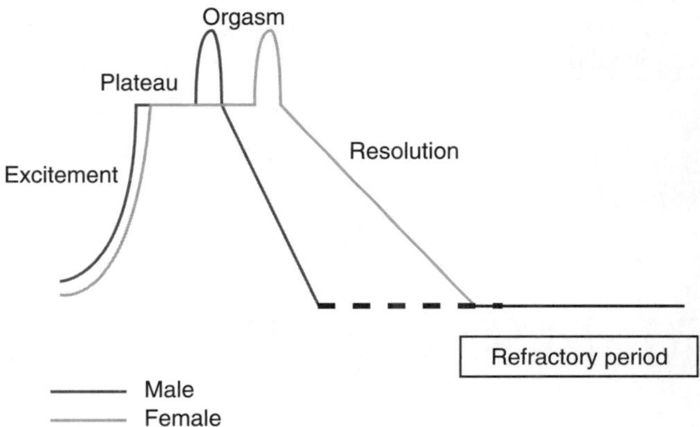

FIG. 1. The human sexual response cycle. Masters and Johnson studied the physiological changes in response to sexual stimulation in male and female volunteers. These physiological changes are predictable and sequential and occur in four phases: the excitation phase, the plateau phase, orgasm, and relaxation phase. The timing of the four phases varies in men and women. The achievement of orgasm is followed by a refractory phase in which men and women cannot achieve another orgasm.

The resolution phase is characterized by a period of muscle relaxation, lowering of heart rate and blood pressure, and the loss of penile erection. After orgasm and ejaculation, men enter into a refractory period in which they are unable to achieve orgasm again (5). The amount of time in the refractory period varies among men.

Clinical Implications of the Sexual Response Cycle

The description of the sexual response cycle greatly facilitated the classification of sexual disorders in men and women. Because abnormalities can exist in each phase of the sexual response cycle, there are four categories of sexual disorders (Table 1) (6,33):

1. Hypoactive sexual desire disorder
2. ED
3. Ejaculatory and orgasmic disorders
4. Disorders of pain

Appropriate classification of the patient's disorder into these categories is important because etiological factors, diagnostic tests, and therapeutic strategies vary for each class of sexual disorder. For sexual

TABLE 1. *Classification of sexual disorders in men and women based on the sexual response cycle*

Type of disorder	Sexual dysfunction	
	Men	**Women**
Desire	Hypoactive sexual desire disorder Sexual aversion disorder	Hypoactive sexual desire disorder Sexual aversion disorder
Arousal	Male erectile disorder	Female sexual arousal disorder
Orgasm	Inhibited male orgasm Premature ejaculation	Inhibited female orgasm
Pain	Dyspareunia	Dyspareunia Vaginismus

An understanding of the human sexual response cycle has helped the development of rational classification system for sexual disorders in men and women. Appropriate classification of the patient's disorder into these categories is important as the etiological factors, diagnostic tests, and the therapeutic strategies vary for each class of sexual disorder.

Adapted from Wincze and Carey. (1995). Sexual and gender identity disorders. In *Abnormal Psychology* (D. H. Barlow and V. M. Durand, Eds.). Brooks/Cole, Pacific Grove, CA.

dysfunction to be classified as a clinical disorder, sexual problems must be persistent or recurrent and cause distress or interpersonal problems.

Hypoactive sexual desire disorder is the persistent or recurrent deficiency (or absence) of sexual fantasies and desire for sexual activity that causes marked distress or interpersonal difficulty and that is not better explained by another disorder, direct physiological effects of a substance (drug, medication), or general medical condition (34). Hypoactive sexual desire disorder is often a multifactorial disorder that can result from androgen deficiency (35–40); as an adverse effect of medications (e.g., selective serotonin reuptake inhibitors, antiandrogens, gonadotropin-releasing hormone analogues, antihypertensives, cancer chemotherapeutic agents, anticonvulsants), systemic illness, depression, and other psychological problems; secondary to other causes of sexual dysfunction, fear of humiliation (41,42), or relationship and differentiation problems (35–40,43,44). Androgen deficiency is an important treatable cause of hypoactive sexual desire disorder and should be excluded by measuring serum total testosterone levels.

ED, previously referred to as impotence, is the inability of the male to attain and/or maintain an erection sufficient for satisfactory sexual intercourse (45). Sexual dysfunction is a more general term that also includes libidinal, orgasmic, and ejaculatory dysfunction in addition to the inability to attain or maintain penile erection (46). The Massachusetts Male Aging Study (47,48) and the National Health and Social Life Survey (49,50) revealed a surprisingly high prevalence of ED in men; for instance, in the Massachusetts Male Aging Study (47,48), a random probability sample of men in Boston, 52% of men between the ages of 40 and 70 were affected by ED of some degree, 17.2% of surveyed men reported minimal ED, 25.2% moderate ED, and 9.6% complete ED.

Ejaculatory disorders are a heterogeneous group of disorders that includes anejaculation, anorgasmia, delayed ejaculation, retrograde ejaculation, premature ejaculation, and painful ejaculation (51). A national survey highlighted the high prevalence and clinical importance of ejaculatory disorders (50,52,53); this survey recognized premature ejaculation as the most prevalent sexual disorder in men 18 to 59 years of age. Retrograde ejaculation due to diabetes-associated autonomic neuropathy is the second most prevalent ejaculatory disorder. Ejaculatory disorders can also lead to infertility among men.

MECHANISMS OF PENILE ERECTION

Penile erection results from a series of biochemical and hemodynamic events that are associated with relaxation of cavernosal smooth muscle, increased blood flow into cavernosal sinuses, and venous occlusion resulting in penile engorgement and rigidity (7,8,14–16). Normal penile erection requires coordinated involvement of intact central and peripheral nervous systems, corpora cavernosa and spongiosa, and a normal arterial blood supply and venous drainage (7,8,14–16).

Penile Anatomy

The erectile tissue of the penis consists of two dorsally positioned corpora cavernosa and a ventrally placed corporum spongiosum (Fig. 2) (54,55). The erectile tissue of both the corpora cavernosa and corpus spongiosum is composed of numerous cavernous spaces separated by trabeculae (56). These trabeculae are composed mainly of smooth muscle cells that are

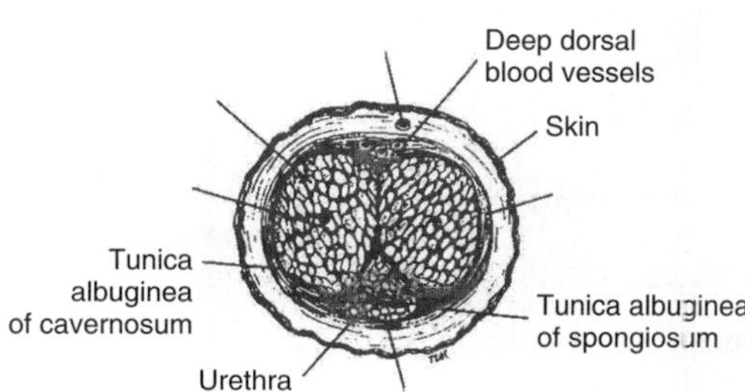

FIG. 2. Cross-sectional anatomy of the human penis. The human penis contains three corporal bodies: two dorsally located corpora cavernosa and a ventrally located corpus spongiosum. Each of the three erectile corpora is surrounded by a thick layer of fibrous tissue, the tunica albuginea. Bucks' fascia surrounds all three corpora. The urethra is located within the corpus spongiosum.

arranged in a syncytium but also contain fibroblasts, collagen, and elastic fibers. Endothelial cells, which resemble those found in blood vessels, cover the surfaces of the trabeculae. The corpus spongiosum differs from the corpus cavernosum in that the spongiosum contains larger cavernous spaces and the trabeculae are smaller and contain fewer smooth muscle cells (45). The trabeculae of corpora cavernosa surround interconnected sinusoidal spaces.

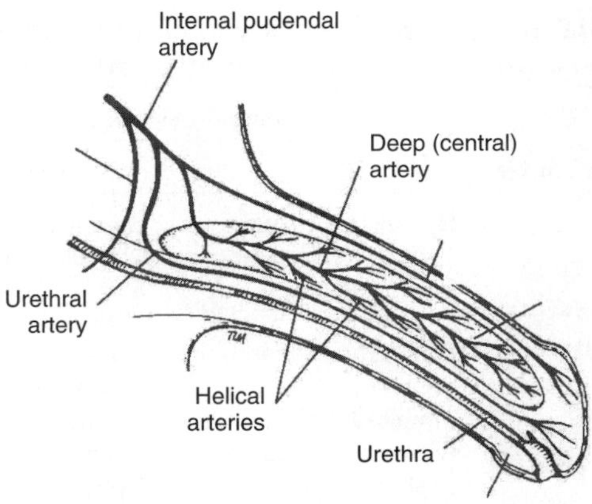

FIG. 3. Blood supply of the human penis. The internal pudendal arteries, which provide the blood supply to the penis, branches off from the bulbar and urethral arteries to the corpus spongiosum and divides into dorsal and deep (central) arteries. The deep (central) penile arteries branch off into a number of helical arteries that supply the corpora cavernosa.

Blood Supply to the Penis

The penile arterial blood supply is derived from the pudendal arteries, which are branches of the internal iliac (57) arteries (Fig. 3). Each of the paired internal pudendal arteries supplies two arteries to the corpus spongiosum (45): a bulbar branch, which supplies the proximal corpus spongiosum, and the urethral artery, which courses from the perineum to the glans penis. The internal pudendal artery divides into two terminal branches, the deep penile and the dorsal penile artery.

The central or deep arteries supply the corpora cavernosa through numerous small vessels, the helical arteries (16). Dilatation of the helicine arteries increases the blood flow and pressure in the cavernosal sinuses (14,15,58). The central artery also sends branches to the contralateral corpus cavernosum; this makes it possible for the two corpora cavernosa to act as one functional unit (58).

Neural Input into the Penis

The neural input to the penis is conveyed through sympathetic (T11-L2), parasympathetic (S2-S4), and somatic nerves (Table 2). Sympathetic and parasympathetic fibers converge in the inferior hypogastric plexus where the autonomic input to the penis is integrated and communicated to the penis through cavernosal nerves. In humans, this ganglionic plexus is located retroperitoneally near the rectum (59). The cavernosal nerves leave the pelvic plexus, course between the rectum and urethra, enter the urogenital diaphragm in proximity to the muscular wall of the urethra, and finally enter the dorsal medial aspect of the corpora cavernosa. These nerves presumably innervate the smooth musculature and vasculature located within the corpora cavernosa (14).

In general, parasympathetic input is proerectile and sympathetic input is inhibitory. The stimuli from the perineum and lower urinary tract are carried to the penis through the sacral reflex arc. Several brain regions, including amygdala, medial preoptic area, paraventricular nucleus of the hypothalamus, and periaqueductal gray matter, act coordinately to affect penile erections. The medial preoptic area of the hypothalamus serves as the integration site for the central nervous system control of erections; it receives sensory input from amygdala and sends impulses to the paraventricular nuclei of the hypothalamus and the periaqueductal gray matter. Neurons in paraventricular nuclei project on to the thoracolumbar and sacral nuclei associated with erections.

Penile parasympathetic innervation, originating in the sacral spinal cord (S2-S4), is conveyed by the pelvic and cavernous nerves. It is widely recognized as the proerectile pathway.

TABLE 2. *Innervation of the penis*

Types of fibers	Location of neurons in the spinal cord	Nerves carrying the fibers	General function
Sympathetic	T10-L2	Prevertebral outflow through the hypogastric and cavernous nerves; additionally, paravertebral outflow through the parasympathetic ganglia and pudendal or pelvic and cavernous nerves	Generally antierectile; sympathetic innervation plays an important role in regulating seminal emission
Parasympathetic	S2-S4	Cavernosal and pelvic nerves	Proerectile
Somatic	S2-S4	Pudendal nerve	Penile sensation, contraction of the striated muscles during ejaculation

Sympathetic nerves originate in the low thoracic and upper lumbar regions (T10-L2) of the spinal cord and course retroperitoneally to condense into the superior hypogastric plexus located just inferior to the aortic bifurcation. Nerve fibers leave the superior hypogastric plexus as the paired hypogastric nerves, which fuse distally and then enter the inferior hypogastric plexus (45). The sympathetic fibers are generally antierectile, and they exert an antierectile role by increasing in the tone of penile smooth muscle fibers surrounding the sinusoidal spaces.

The somatic innervation to the penis is carried via the pudendal nerve, which arises from the sacral region of the spinal cord. Penile sensation is carried in the dorsal nerve of the penis, a branch of the pudendal nerve.

Hemodynamic Changes during Penile Erection

The crucial importance of increased arterial flow to the penis in the genesis of penile erections is now universally recognized (7,8,14,45,60–65). Semans and Langworthy (60) demonstrated almost 60 years ago in classic experiments in cats that aortic occlusion prevents the development of penile erection produced by sacral nerve root stimulation. Furthermore, after an erection had been produced by nerve stimulation, aortic occlusion resulted in prompt detumescence. Not surprisingly, men with occlusive aortoiliac disease suffer from both intermittent claudication and impotence (66).

Relaxation of the cavernosal smooth muscle that surrounds the cavernosal sinuses along with increased blood flow results in pooling of blood in the cavernosal spaces and penile engorgement (7,8,14–16). The expanding corpora cavernosa compress the venules against the rigid tunica albuginea, restricting the venous outflow from the cavernosal spaces. This facilitates entrapment of blood in the cavernosal sinuses and achievement of a rigid erection (14).

The role of venous occlusion in generation and maintenance of penile erection is not as well understood. However, studies in the rat and men have demonstrated an increase in venous resistance during penile erections (8). The increased venous resistance is likely the result of venous compression between an engorged and expanded corpora cavernosa and an inflexible tunica albuginea. The net result of increased arterial blood inflow and increased venous resistance is entrapment of blood in the corpora cavernosa, marked increase in intracavernosal pressure, and achievement of tumescence and rigidity.

Peak intracavernosal pressures are achieved during the contractions of the ischiocavernosus and bulbospongiosus muscles just before ejaculation. It is possible that skeletal muscle contraction produces compression and therefore blockade of venous flow; it is also possible that ischiocavernosus muscle contraction compresses the corpora cavernosa, thereby raising intracavernosal pressure. The ischiocavernosus and bulbospongiosus muscles are highly sexually dimorphic and androgen sensitive. Thus, after induction of androgen deficiency, these muscles atrophy; we do not know whether this contributes to the decrease in peak penile rigidity in hypogonadal men.

Biochemical Mechanisms of Penile Erection

The erectile state of the penis is determined by the tone of the corporal smooth muscle cells (67). When the cavernosal smooth muscle cells are relaxed, the tone is low and the penis is engorged with blood and erect. Conversely, when the cavernosal smooth muscle tone is high, there is predominance of sympathetic neural activity, and the penis is flaccid. The smooth muscle tone in the corpora cavernosa is maintained by agonist-stimulated release of intracellular calcium into the cytoplasm and influx of calcium through membrane channels (68). An increase in intracellular calcium through its binding to calmodulin activates myosin light chain kinase, resulting in phosphorylation of myosin light chain, actin–myosin interactions, and muscle contraction. The transmembrane and intracellular calcium flux in the cavernosal smooth muscle cells is regulated by a number of cellular processes and signaling molecules such as K^+ flux through potassium channels, connexin43-derived gap junctions, norepinephrine, prostaglandin E_1 (PGE_1), and nitric oxide (Figs. 4 to 7).

The adjacent smooth cells are interconnected in a syncytium through connexin43-derived gap junctions (Fig. 4) (69–72). Therefore, changes in K^+ channel activity in one myocyte affect the membrane potential of adjacent cells, resulting in rapid transmission of electrical and biochemical signaling throughout the syncytium (69–72). In addition, the two corpora cavernosa are also connected by anastomotic vascular channels, which allow the two contralateral corpora cavernosa to act as a single functional unit.

The vascular smooth muscle contraction in the corpora cavernosa is regulated by the adrenergic pathway. Norepinephrine binds to adrenergic receptors, resulting in generation of diacyl glycerol and inositol triphosphate. Diacyl glycerol activates protein kinase C, which inhibits K^+ channels, whereas inositol triphosphate increases intracellular calcium and calcium influx through the membrane (Fig. 5). The net increase in intracellular calcium promotes actin–myosin interaction, resulting in smooth muscle contraction.

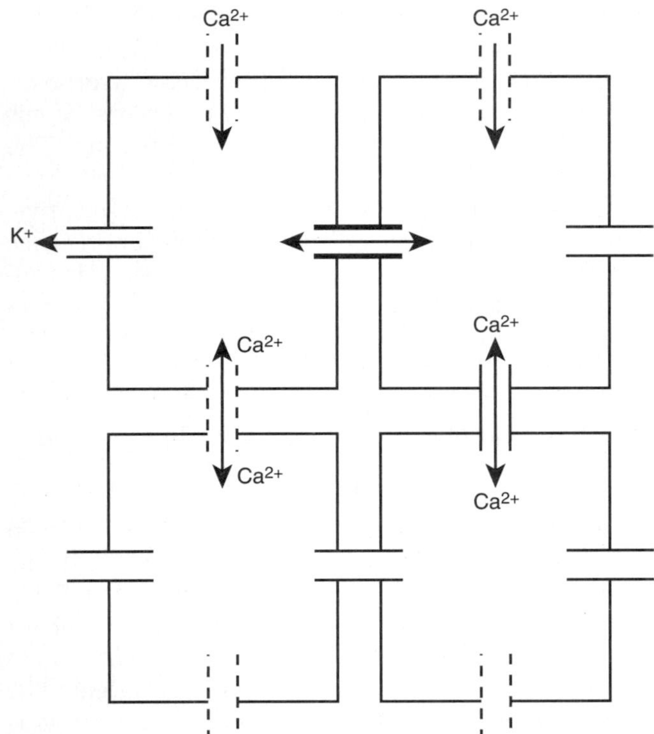

FIG. 4. The smooth muscle cells (myocytes) in the corpora cavernosa are interconnected through connexin43 gap junction channels. These channels allow flow of ions among interconnected myocytes; thus, the corpus cavernosum acts as a functional syncytium. In addition, anastomotic blood vessels connect the contralateral corpora cavernosa. Thus, the two corpora cavernosa operate as a single functional unit.

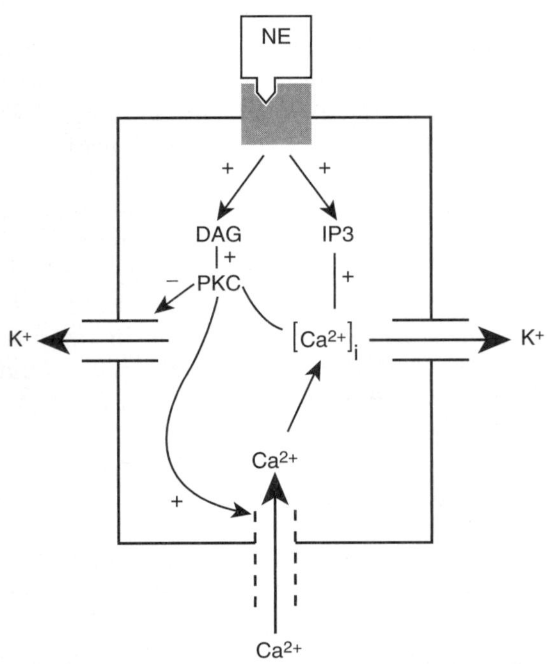

FIG. 5. Adrenergic stimuli are communicated through norepinephrine which after binding to adrenergic receptors stimulates diacyl glycerol (DAG) and inositol-3 phosphate (IP3). DAG stimulates protein kinase C, which along with IP3 causes increase in cytosolic calcium and inhibition of K^+ channels. The net result is cavernosal smooth muscle contraction and loss of penile erection.

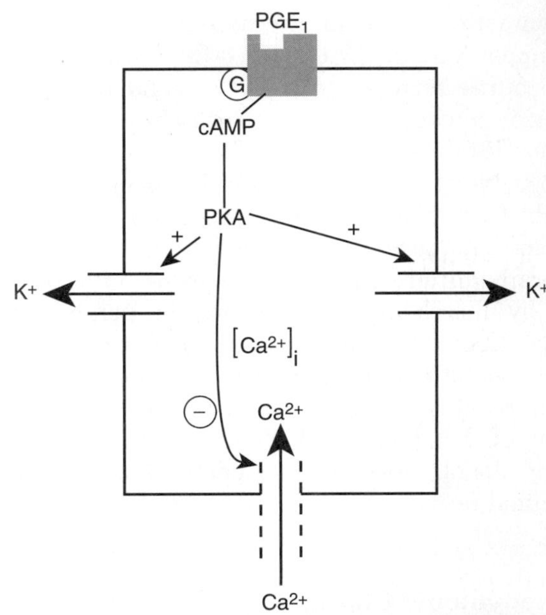

FIG. 6. PGE_1, by binding to PGE_1 receptor, stimulates adenylyl cyclase and increases the intracellular concentrations of cAMP, which activates protein kinase A. This results in sequestration of intracellular calcium, retards calcium influx, and stimulates K^+ channels. The net result is a reduction in cytosolic calcium and cavernosal smooth muscle relaxation.

Prostaglandin E_1 results in generation of cyclic AMP, which activates protein kinase A (73). Activated protein kinase A stimulates K^+ channels, resulting in K^+ efflux from the cell (Fig. 6). In addition, protein kinase A–mediated processes also result in a net decrease in intracellular calcium, favoring smooth muscle cell relaxation (68). Nitric oxide through activation of intracellular cGMP also decreases intracellular calcium and K^+ efflux (74,75).

The relaxation of the cavernosal smooth muscle trabeculae is under the regulation of the autonomic nervous system (7,8,14–18). A number of cholinergic, adrenergic, and noradrenergic noncholinergic mediators regulate cavernosal smooth muscle relaxation. The nonadrenergic noncholinergic mediators include vasoactive intestinal peptide, calcitonin gene-related peptide, and nitric oxide. Nitric oxide is derived from the nerve terminals innervating the corpora cavernosa, endothelial lining of penile arteries, and cavernosal sinuses (7,8,14–18) and is an important biochemical regulator of cavernosal smooth muscle relaxation.

Nitric oxide, released from the nonadrenergic noncholinergic nerve terminals and possibly endothelium, also induces arterial dilatation. The actions of nitric oxide on the cavernosal smooth muscle and the arterial blood flow are mediated through the activation of guanylyl cyclase (Fig. 7), the production of cGMP, and activation of cGMP-dependent protein kinase (7,8,14–19). cGMP acts as an intracellular second messenger and causes smooth muscle relaxation by lowering intracellular calcium (16).

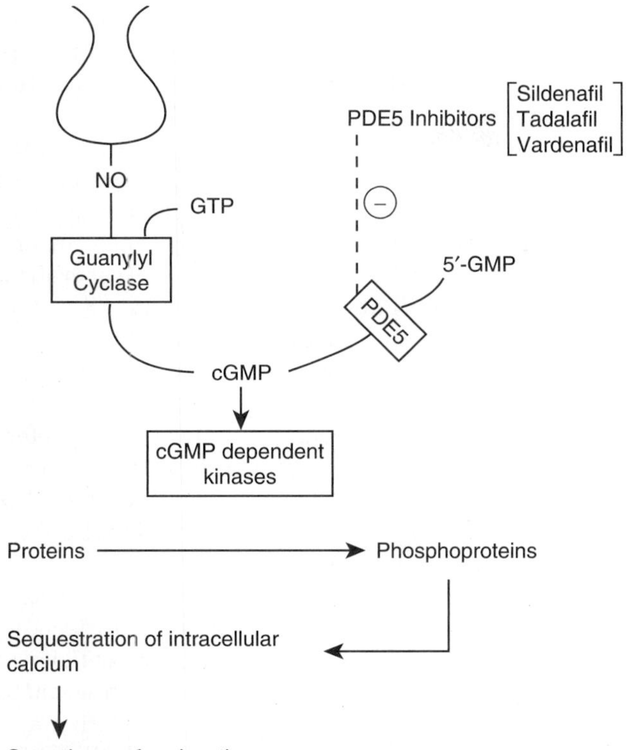

FIG. 7. Cavernosal smooth muscle relaxation and contraction are regulated by intracellular cAMP and cGMP. These intracellular second messengers, by activating specific protein kinases, cause phosphorylation of specific proteins, resulting in sequestration of intracellular calcium. The resulting decrease in intracellular calcium concentrations causes smooth muscle relaxation. Nitric oxide is released from the noradrenergic noncholinergic nerve terminals and possibly from the endothelium and activates guanylyl cyclase, resulting in generation of cGMP. cGMP is degraded by a class of enzymes called cyclic nucleotide phosphodiesterases. cAMP is degraded by phosphodiesterase isoforms 2, 3, and 4, whereas cGMP is degraded by phosphodiesterase isoform 5. Sildenafil, vardenafil, and tadalafil are relatively specific inhibitors of type 5 phosphodiesterase.

cGMP-dependent protein kinase phosphorylates numerous ion channels and pumps, each promoting a reduction in cytosolic calcium. In particular, cGMP-dependent protein kinase activates high-conductance Ca^{2+}-sensitive K^+ channels (45,76), which hyperpolarize the arterial and cavernosal smooth muscle cell membranes, causing relaxation. Movement of potassium ions across the membrane determines the membrane potential of the cavernosal smooth muscle cells; at least four subtypes of potassium channels mediate this K^+ efflux (67). This mechanism involving potassium channels appears to be compromised with age and with vascular disease, contributing to ED (67). Thus, strategies that increase cavernosal Ca^{2+}-sensitive K^+ channel expression might be useful as effective therapies for ED (67). Efforts to augment K^+ channel expression by gene transfer are currently in progress.

Role of Cyclic Nucleotide Phosphodiesterases

Three classes of enzymes—adenylyl cyclase, guanylyl cyclase, and phosphodiesterases—play an important role in regulating the intracavernosal concentrations of cAMP and cGMP. Phosphodiesterases hydrolyze cAMP and cGMP, thus reducing their concentrations within the cavernosal smooth muscle (7,77–81). At least 13 different isoforms of cyclic nucleotide phosphodiesterases have been identified (82–84). These isoforms are widely distributed throughout the body; the predominant isoform of this enzyme in the cavernosal smooth muscle is cyclic nucleotide phosphodiesterase type 5 (PDE5) (7,14,16,85). Although phosphodiesterase isofroms 2, 3, 4, and 5 also are expressed in the penis, only PDE5 (80) is specific to the nitric oxide/cGMP pathway in the corpora cavernosa (86). Hydrolysis of cGMP by this enzyme results in reversal of the smooth muscle relaxation and reversal of penile erection (Fig. 7). Sildenafil, vardenafil, and tadalafil are potent and selective inhibitors of the activity of PDE5 that prevent breakdown of cGMP and thereby enhance penile erection (20–32,87).

CENTRAL REGULATION OF SEXUAL FUNCTION BY SEX STEROIDS

Sexual function is a complex, multicomponent, biological process that comprise central mechanisms for regulation of libido and arousability and local mechanisms for the generation of penile tumescence, rigidity, orgasm, and ejaculation.

The term libido is frequently used interchangeably with sexual desire, but libido is a more complex construct, involving sexual thoughts, fantasies, attentiveness to erotic stimuli, and sex-seeking behavior. Although sexual desire in men and women is affected by a sense of well-being, satisfaction with relationship, sexual identity, and differentiation issues and by the occurrence of comorbidities, testosterone is an important regulator of libido. The idea that testosterone is an important regulator of libido in men is not new; humans have known since antiquity that you can take away the vigor and sexual drive of men by removing their testes. Thus, it was not uncommon for royalty in the middle ages to appoint eunuchs (castrated men) to guard the harem. However, the notion that penile erections can occur in the absence of testosterone is also embedded in ancient texts. For instance, *One Thousand and One Arabian Nights* documents the tales of eunuchs who cohabited with the Queens.

Although androgen-deficient men can achieve penile erections, their overall sexual activity is decreased (88–93). In a series of pioneering experiments,

Davidson and coworkers (91,92) demonstrated that compared with eugonadal men, hypogonadal men had lower self-reported sexual activity, feelings, and thoughts and lesser number of spontaneous erections. However, when hypogonadal men were shown erotic pictures and videos, they were able to achieve normal erections. Testosterone replacement for hypogonadal men increased sexual feelings and thoughts, overall sexual activity scores, and the number of spontaneous erections but did not change erectile response to visual erotic stimulus (92). Davidson's experiments led to the prevalent dogma that spontaneous, but not stimulus-induced, erections are testosterone dependent and that testosterone stimulates sexual thoughts and feelings.

Subsequent studies have provided additional evidence of the important role of testosterone in regulating sexual behavior in men. Men whose serum testosterone levels are lowered pharmacologically for sexual offenses experience decreased sexual desire and activity (94). Men treated with gonadotropin-releasing hormone analogues for prostate cancer or contraception also report reduced sexual desire and activity (35,95–99).

A number of open-label trials have shown that testosterone replacement of androgen-deficient men increases overall sexual activity, sexual desire, and fantasies (35,37,88,90,100–103). Testosterone treatment increases the number of spontaneous sexual thoughts and spontaneous erections in hypogonadal men.

Nocturnal erections are temporally related to peaks of nighttime testosterone secretion (104). Although nocturnal penile erections occur in androgen-deficient men, testosterone replacement therapy in hypogonadal men increases the frequency, size, and duration of nocturnal penile tumescence (88,89).

Alexander et al. (36,106) reported that testosterone increased sexual arousal and enjoyment in response to erotic auditory stimulus. In a series of elegant studies, they also demonstrated that attentiveness to sexual stimulus in dichotic listening/selective attention task is androgen dependent. Thus, in comparison with androgen-deficient men, androgen-replete men were more distractible by auditory erotic stimulus administered during the performance of a computational task (106).

The concept that libido is testosterone dependent and the local mechanisms for penile erection are androgen independent may be simplistic. There is emerging evidence that testosterone is a regulator of nitric oxide synthase activity in the cavernosal smooth muscle (9,107–111). Testosterone also exerts trophic effects on cavernosal smooth muscle and ischiocavernosus and bulbospongiosus muscles. Therefore, it is possible that physiologically normal testosterone concentrations might be required for optimum penile rigidity. Testosterone increases blood flow to the penis and is necessary for optimum veno-occlusive response (111). Orgasm and ejaculation are not androgen dependent and can occur without a full penile erection.

The mechanisms by which testosterone regulates sexual desire and behavior are poorly understood. In male mammals, testosterone administration appears to have rewarding properties. For instance, in one experiment, male rats were given intracerebral injections of graded doses of testosterone or saline. Intracerebral injection of testosterone (0.1 μg) into the medial preoptic area produced a conditioned place preference, whereas saline injection did not (112–115). The effects of testosterone in promoting place preference were blocked by a dopamine receptor antagonist (115). These data suggest that the rewarding effects of testosterone may mediate its motivational effects on sexual behavior.

The conversion of testosterone to dihydrotestosterone (DHT) is necessary for phenotypic differentiation of the male genital tract. Congenital deficiency of steroid 5α-reductase type 2 isoenzyme is associated with male pseudohermaphroditism (2,34). Both major metabolites of testosterone, estradiol and DHT, are necessary for sexual differentiation of the brain. A large body of data supports the view that aromatization of testosterone to estradiol within the brain plays an important role in regulating sexual differentiation of dimorphic regions of the brain (116–122). Similarly, administration of a steroid 5α-reductase inhibitor to hatchling zebra finches demasculinizes the song system, although the data are inconsistent across different studies (123–125). However, the role of steroid 5α reduction in mediating androgen effects on sexual function in adult mammals remains controversial because of inconsistencies in published data.

Important insights into the role of 5α-reduction in regulating male sexual behavior have emerged from the studies of a synthetic steroid, 7α-methyl-19-nortestosterone (MENT). This androgen binds the androgen receptor and exerts biological effects at some peripheral tissues with a greater potency than testosterone. In vivo, MENT does not undergo 5α-reduction. Morali et al. (128) reported that MENT, when administered to castrated rats, maintains full copulatory behavior similar to that observed in testosterone-treated rats. MENT also has been shown to maintain sexual function in hypogonadal men (126–128). These studies of MENT suggest that 5α-reduction of testosterone is not necessary for maintaining normal mating behaviors in the male rat. However, Lugg et al. (107) reported that penile nitric oxide synthase activity in the rat is stimulated by testosterone and that conversion of testosterone to DHT is required for optimal activity of this enzyme.

The pivotal study of finasteride in men with benign prostatic hypertrophy reported a higher incidence of decreased libido, impotence, and ejaculatory disorders in the finasteride-treated men as compared with placebo-treated men (46). However, in this finasteride efficacy trial, the sexual function data were not systematically collected using a validated sexual function instrument. Even the placebo-treated men reported less than 5% incidence of sexual dysfunction, far lower than the baseline rates of sexual dysfunction expected in healthy men of this age range. Other studies in men with benign prostatic hypertrophy have also reported a higher prevalence of sexual dysfunction in men treated with 5α-reductase inhibitors than in placebo-treated men (58,59). Similarly, steroid 5α-reductase activity is necessary, but not sufficient, for maintaining courtship behavior in the green azole lizard (60). Similarly, MENT, an androgen analogue that is not 5α reduced, normalizes sexual function in healthy hypogonadal men. Nonsteroidal selective androgen receptor modulators that do not undergo 5α-reduction have also been reported to normalize mating behavior in castrated rats (129,130). Therefore, it appears that androgen effects on libido do not require its obligatory conversion to DHT. In contrast, optimal induction of penile nitric oxide synthase activity by testosterone requires its 5α-reduction (62).

Testosterone Effects on Mood

In addition to its direct effects on sexual behavior, testosterone may also affect sexual desire indirectly through its effects on mood and sense of well-being. Testosterone replacement improves positive aspects of mood and decreases negative aspects of mood in healthy hypogonadal men (38,131) and in men infected with human immunodeficiency virus (132–136). A higher prevalence of low testosterone levels has been reported in men with clinical depression (137–140). Conversely, in a placebo-controlled, randomized, single-center trial of testosterone replacement in 23 men with refractory depression on antidepressant therapy, with serum testosterone < 350 ng/dl, administration of 10 g testosterone gel daily for 8 weeks was associated with greater improvements in Hamilton Depression score (141) than those associated with placebo administration. Anecdotal clinical experience indicates that hypogonadal men often report considerable improvement in sense of well-being after initiation of testosterone replacement therapy.

CLINICAL TRANSLATION OF PHYSIOLOGICAL PRINCIPLES AND BASIC SCIENCE ADVANCES

Epidemiological studies, such as the Massachusetts Male Aging Study (47,48,142,143) and the National Health and Social Life Survey (49,50), have revealed that ED is a common medical problem that affects 25 to 30 million men in the United States alone. It has been estimated that there are 600,000 to 700,000 new cases of ED each year in the United States alone (142,144). Age, diabetes mellitus, hypertension, medication use, cardiovascular disease, depression, and neurological disorders are important risk factors for ED (47). Advancing age is also an important risk factor for ED: Less than 10% of men below the age of 40 and over 50% of men over the age of 70 will have ED (144). Most patients presenting with ED have an organic cause, although there is often an overlay of complex psychosocial factors.

The diagnostic workup of the patient with ED should start with an evaluation of general health (16,28) to identify etiological factors as well as factors that might affect the selection and response to therapy. The presence of diabetes mellitus, coronary artery disease, peripheral vascular disease, hypertension, stroke, spinal cord or back injury, multiple sclerosis, depression, or dementia should be verified. Information about use of recreational drugs such as alcohol, marijuana, cocaine, and tobacco; prescription medications, particularly antihypertensives; antiandrogens, antidepressants and antipsychotic drugs; and nonprescription over-the-counter supplements is important because almost a quarter of all cases of impotence can be attributed to medications. A detailed sexual history, including the nature of relationships, partner expectations, situational erectile failure, performance anxiety, and marital discord, needs to be elicited (16,28). It is important to distinguish between inability to achieve erection, changes in sexual desire, failure to achieve orgasm and ejaculation, and dissatisfaction with the sexual relationship because the etiological factors vary with the type of sexual disorder.

A directed physical examination should focus on secondary sex characteristics, the presence or absence of breast enlargement and testicular volume, evaluation of femoral and pedal pulses, neurological examination to determine the presence of motor weakness, perineal sensation, anal sphincter tone and bulbocavernosus reflex, and examination of the penis to evaluate any unusual curvature, palpable plaques, or superficial lesions.

Although dynamic infusion cavernosometry, angiography, and penile duplex Doppler ultrasonography can provide information about the blood supply to the penis, over the last decade there has been a general shift in most male sexual dysfunction clinics away from expensive, time-consuming, and invasive techniques and toward the use of simple, noninvasive, self-reporting questionnaires (7,145). Thus, the initial diagnostic workup in most men presenting with ED consists of general health evaluation; evaluation of cardiovascular risk by the measurements of blood glucose, plasma lipids, and blood chemistries in accordance with community standards; and measurement of serum testosterone levels. The brachioponderal index does not have sufficient sensitivity or specificity to be an accurate diagnostic tool in men with ED (146,147). Further evaluation using more invasive diagnostic test is limited to those men who do not respond to an empiric trial of oral PDE5 inhibitors; these patients should be referred to a specialist for detailed urological evaluation.

The self-reporting questionnaires are useful because many men with ED do not voluntarily come forward to their physicians and state their sexual complaints for a variety of reasons. The International Index of Erectile Function is a multidimensional scale consisting of 15 questions that address relevant domains of male sexual function, including sexual desire, intercourse satisfaction, orgasmic function, and overall satisfaction (148–150). It has been validated in several languages and has adequate sensitivity and specificity for detecting treatment-related changes.

Biochemical Basis of Pharmacological Therapy of ED

Advances in our understanding of the biochemical pathways that mediate cavernosal smooth muscle relaxation and increased blood flow have facilitated the development of mechanism-specific therapies for ED. Most important of these advances has been the approval of several selective PDE5 inhibitors for the treatment of ED in men.

Selective PDE5 Inhibitors for the Treatment of ED in Men

The selective PDE5 inhibitors block the hydrolysis of cGMP induced by nitric oxide (86,151,152), thus promoting cavernosal smooth muscle relaxation. The action of these drugs requires an intact nitric oxide response as well as constitutive synthesis of cGMP by the smooth muscle cells of the corpora cavernosa. By selectively inhibiting cGMP catabolism in the cavernosal smooth muscle cells, PDE5 inhibitors restore the natural erectile response to sexual stimulation but do not produce an erection in the absence of sexual stimulation.

Although the three currently available phosphodiesterase inhibitors have some structural similarities, they differ in their selectivity and pharmacokinetics. The common adverse events of the available PDE5 inhibitors—headache, visual problems, and flush—are related to nonselective inhibition of phosphodiesterase isoforms 6 and 11 in other organ systems (32). The selectivity of PDE5 inhibitors is the ratio of its inhibitory potency for phosphodiesterase isoforms other than type 5 relative to its inhibitory potency for PDE5 (32). For PDE6, tadalafil is the most selective and sildenafil is the least selective; for PDE11, vardenafil is the most selective and tadalafil is the least selective (32). The retinal side effects of sildenafil are related to inhibition of PDE6 in the retina, whereas muscle aches experienced by a small fraction of men using tadalafil may be related to inhibition of PDE11 in the skeletal muscle (26,153–155).

After oral administration of sildenafil, peak plasma concentrations are achieved within 30 to 120 minutes, after which plasma concentrations decline with a half-life of 4 hours (156,157). Vardenafil achieves peak concentrations within 0.7 to 0.9 hours and has a half-life of 4–5 hours (157–160). In contrast, peak concentrations of tadalafil are achieved at 2 hours, and its half-life of 16.9 hours in young men is significantly longer than half-lives of sildenafil and vardenafil (157,161–164). The half-life of tadalafil is even longer in older men (21.6 hours) compared with young men (16.9 hours). Because of the relatively short half-lives of vardenafil and sildenafil, these drugs should be taken 2–4 hours before the planned intercourse; in contrast, tadalafil, because of its longer half-life, can be, but does not have to be, taken on demand.

Food, particularly high fat meals, and alcohol can delay and decrease the absorption of sildenafil. However, early pharmacokinetic studies have not reported changes in maximum serum concentrations or absorption rates of vardenafil or tadalafil due to food or moderate alcohol ingestion (165).

Introduced to the U.S. market in March 1998, sildenafil (Viagra, Pfizer, New York) was the first effective oral agent for the treatment of ED (86,87, 151,152,166–169). The efficacy of all three orally active PDE5 inhibitors has been proven in randomized clinical trials.

All three commercially available PDE5 inhibitors—sildenafil, vardenafil, and tadalafil—have been shown to be efficacious in randomized clinical trials in improving erectile function in men with ED. The efficacy of sildenafil was demonstrated in a randomized dose–response study (19) in which men with organic, psychogenic, or mixed ED were randomized to receive placebo or 25, 50, or 100 mg sildenafil for 24 weeks. Patients on sildenafil performed better in terms of increased rigidity, frequency of vaginal penetration, and maintenance of erection (19). Increasing doses of sildenafil were associated with higher mean scores for the questions assessing frequency of penetration and maintenance of erections after sexual penetration. In a follow-up dose escalation study (87), men with ED were randomly assigned to receive placebo or 50 mg of sildenafil for 12 weeks. At each follow-up, the dose of sildenafil was increased or decreased by 50% depending on the therapeutic response or side effects. Sixty-four percent of attempts at intercourse were successful for the men receiving sildenafil, as compared with 22% of men receiving placebo. The mean number of successful attempts per month was 5.9 for men receiving sildenafil and 1.5 for those receiving placebo. The mean scores for orgasms, intercourse satisfaction, and overall satisfaction domains were also significantly higher in the sildenafil group compared with placebo (87).

In a separate randomized clinical trial, 268 men with diabetes mellitus and ED received either placebo or sildenafil for 12 weeks (152). Fifty-six percent of men receiving sildenafil reported improved erections compared with 10% of those receiving placebo ($P < 0.001$). This study (152) demonstrated that sildenafil is an effective treatment for ED in patients with diabetes mellitus.

In the vardenafil efficacy trials, 5-, 10-, and 20-mg doses of vardenafil were all superior to placebo in improving erectile function domain scores; the improvements in erectile function scores were dose related (170–172). Vardenafil improved rates of vaginal penetration, intercourse success, and overall satisfaction with sexual experience in a broad population of men with ED (170–172).

Similarly, in randomized clinical trials, 2.5-, 5-, 10-, and 20-mg doses of tadalafil were each superior to placebo in improving erectile function domain scores (173–175). The beneficial effects of tadalafil were dose related.

PDE5 inhibitors are also effective in men with ED due to a variety of other causes, including spinal cord injury and postradical prostatectomy (167,168). In general, baseline sexual function correlates positively with response to sildenafil, and patients with diabetes mellitus or previous prostate surgery respond less well than patients with psychogenic or vasculogenic ED (168). Because there is no baseline characteristic that predicts the likelihood of failure to respond to sildenafil therapy, a therapeutic trial of PDE5 inhibitors is warranted in all patients except in those in whom it is contraindicated (168).

Adverse Effects Associated with PDE5 Inhibitors. In clinical trials, the adverse effects reported with greater frequency in men treated with PDE5 inhibitors than in those treated with placebo include headaches, flushing, rhinitis, dyspepsia, and visual disturbances (176). The occurrence of headache, flushing, and rhinitis is a direct consequence of nonselective PDE5 inhibition in other organ systems and is related to the administered dose. These drugs do not affect semen characteristics (177). No cases of priapism were noted in any of the pivotal clinical trials.

Cardiovascular and Hemodynamic Effects of PDE5 Inhibitors. In postmarketing surveillance of adverse events associated with sildenafil use, several instances of myocardial infarction and sudden death were reported (178–182) in men using sildenafil; many of these reported deaths occurred in temporal relation to the ingestion of sildenafil (178–182) in individuals who were taking nitrates. Because most men presenting with ED also have high prevalence of cardiovascular risk factors, it is unclear whether these events were causally related to the ingestion of sildenafil, underlying heart disease, or both (181). In a rigorously controlled study (183), oral administration of 100-mg sildenafil to men with severe coronary artery disease produced only small decreases in systemic blood pressure and no significant changes in cardiac output, heart rate, coronary blood flow, and coronary artery diameter. In a separate analysis of five randomized placebo-controlled trials of vardenafil, Kloner (184) pooled the data on cardiovascular safety profile. The overall frequency of cardiovascular events was similar in vardenafil-treated men and placebo-treated men. However, vardenafil treatment was associated with a mild reduction in blood pressure (4.6 mm Hg decrease in systolic blood pressure) and a small increase in heart rate (2 beats/min). This led the American Heart Association to conclude that the preexistence of coronary artery disease by itself does not constitute a contraindication for the use of sildenafil (181).

Before prescribing PDE5 inhibitors, cardiovascular risk factors should be assessed. If the patient has hypertension or symptomatic coronary artery disease, the treatment of those clinical disorders should be addressed first (7,181). The use of nitrates must be ascertained because PDE5 inhibitors are contraindicated in individuals taking any form of nitrates. PDE5 inhibitors should not be used within 24 hours of the use of nitrates (181).

Because the data on the effects of PDE5 inhibitors on exercise-induced ischemia is limited, it is prudent to warn patients that sexual activity can induce coronary ischemia in men with preexisting coronary artery disease (182); these individuals should undergo assessment of their exercise tolerance. If the individual can safely climb one or two flights of stairs without angina or excessive shortness of breath, he can likely engage in sexual intercourse with a stable partner without similar symptoms. Exercise testing before prescribing PDE5 inhibitors may be indicated in some men with significant heart disease to assess the risk of inducing cardiac ischemia during sexual activity (181). In one double-blind crossover study, vardenafil was shown not to impair the ability of patients with stable coronary artery disease to engage in exercise at levels equivalent to that attained during sexual intercourse. Similarly, each of the three PDE5 inhibitors has not been shown to have significant adverse effect on hemodynamics and cardiac events in carefully selected men with ED who did not have any contraindication for the use of PDE5 inhibitors (185). None of the PDE5 inhibitors adversely affects total exercise time or time to ischemia during exercise testing in men with stable angina. Tadalafil and vardenafil are contraindicated in men using alpha-blockers; the sildenafil label also has a precautionary warning for use with alpha-blockers. In men with congestive heart failure, those receiving vasodilator drugs, or those who are using complex regimens of antihypertensive drugs, blood pressure should be monitored after initial administration of PDE5 inhibitors (181,184,186–191).

A number of drugs can affect the metabolism of sildenafil by the P-450 2C9 and the P-450 3A4 pathways (181). Cimetidine and erythromycin, inhibitors of P-450 3A4, increase the plasma concentrations of sildenafil. Protease inhibitors may also alter the activity of the P-450 3A4 pathway and affect the clearance of sildenafil (181). Sildenafil is an inhibitor of the P-450 2C9 metabolic pathway, and its administration could potentially affect the metabolism of drugs metabolized by this system, such as warfarin (181). Combined administration of sildenafil and ritonavir in combination results in significantly increased plasma levels of sildenafil than sildenafil given alone (192). There are similar reactions with other drugs, including saquinavir, erythromycin, and itraconazole. Therefore, the doses of PDE5 inhibitors should be reduced appropriately in men taking protease inhibitors or erythromycin.

Grapefruit juice can alter oral drug pharmacokinetics by different mechanisms. Grapefruit juice given as a single normal amount or by whole fresh fruit segments can inactivate irreversibly intestinal cytochrome P-450 3A4, thus reducing presystemic metabolism and oral drug bioavailability of PDE5 inhibitors (193). Although the magnitude of this problem in clinical practice is unknown, it seems prudent to warn men who are contemplating the use of PDE5 inhibitors not to ingest more than a small amount of grapefruit juice.

The vasodilator effects of nitrates are augmented by PDE5 inhibitors; this also applies to inhaled forms of nitrates such as amyl nitrate or nitrite that are sold under the street name "poppers." Concomitant administration of the two drugs can cause a potentially fatal decrease in blood pressure (181).

Sildenafil and vardenafil are taken "on demand" at least 1 hour before sexual intercourse and not more than once in any 24-hour period; because of its longer half-life, tadalafil does not need to be taken immediately before intercourse.

Intraurethral Therapies

Alprostadil is a stable synthetic form of PGE_1 that causes an increase in cAMP levels and a decrease in intracellular calcium and thereby promotes cavernosal smooth muscle relaxation and penile erection (194–204). An intraurethral system for delivery of alprostadil called MUSE (medicated urethral system for erection; VIVUS, Menlo Park, CA) was released in 1997 (194–203). Alprostadil, when applied into the urethra, is absorbed through the urethral mucosa and ventral side of the tunica albuginea into the corpus cavernosum. Intraurethral alprostadil has several advantages over intracavernosal injection of PGE_1: Intraurethral administration is easier than intracavernosal injection and has a lower frequency of adverse effects, particularly penile fibrosis.

Alprostadil is available in 125-, 250-, 500-, and 1,000-μg strengths. Typically, the initial alprostadil dose of 500 μg is applied in the clinician's office to observe any changes in blood pressure or urethral bleeding secondary to misapplication of the device into the urethra. Common side effects of transurethral alprostadil are penile pain and urethral burning in up to 30% of patients (205–207). Initial randomized placebo-controlled studies reported 40% to 60% success rates, determined as having at least one successful sexual intercourse during a 3-month study period (205–207). In clinical practice, approximately one-third of men using intraurethral alprostadil respond. The use of a constriction device (Actis, VIVUS) at the time of application of transurethral alprostadil has been shown to increase efficacy.

The use of intraurethral alprostadil is associated with penile pain in 32% of men and dizziness, hypotension, and syncope in a small fraction of users. Intraurethral alprostadil can also cause mild burning or itching in the vagina of the sexual partner. Intraurethral alprostadil should not be used by men whose partners are pregnant or planning to get pregnant.

Intracavernosal Injection of Vasoactive Agents

The use of intracavernosal injections of vasoactive agents has been a cornerstone of the medical management of ED since the early 1980s. Patients are taught how to self-inject a vasoactive agent into their corpora cavernosa with a 27- or 30-gauge needle up to three times a week for sexual intercourse.

Erections occur typically 15 minutes after intracorporal injection and last 45 to 90 minutes. When appropriately titrated, the success rate of this therapy in producing a rigid erection is 80% to 90%. Early studies with intracavernosal injection therapy report patient and partner satisfaction rates of 70% and 67%, respectively.

The main adverse effects include penile pain, occurrence of hematoma, formation of corporal nodules, and the possibility of prolonged erections (priapism, if longer than 4 hours). Despite the effectiveness of this approach in producing rigid erections, most patients do not relish injecting a needle into their penis; therefore, it is not surprising that long-term drop-out rates approach 60% to 70%.

Three different agents, PGE_1, papaverine, and phentolamine, are commonly used alone or in combination by clinicians who prescribe injection therapy for the treatment of ED. Several formulations of alprostadil (PGE_1) are commercially available (Caverject, Pharmacia; Prostin VR, Pharmacia; Edex, Schwarz Pharma). PGE_1 binds to PGE_1 receptors on the cavernosal smooth muscle cells, stimulates adenylyl cyclase, increases the concentrations of cAMP, and is a powerful smooth muscle relaxant. In one efficacy trial, intracavernosal alprostadil injection resulted in satisfactory sexual performance in over 80% of users (208). The usual dose is 5 to 20 μg. Common side effects of intracavernosal PGE_1 injections include penile pain, fibrosis, and prolonged erections (208). Priapism occurs less commonly compared with other vasoactive agents.

Papaverine, derived originally from the poppy seed, is a nonspecific phosphodiesterase inhibitor that increases both intracellular cAMP and cGMP. As a single agent it is efficacious, inexpensive, and does not need to be refrigerated; however, intracavernosal injections of papaverine are associated with a greater propensity to induce priapism and fibrosis with long-term use than alprostadil (209). Phentolamine is a competitive α_1- and α_2-adrenergic antagonist that contributes to smooth muscle relaxation. As a single agent it is minimally efficacious, but it is commonly used in combination to potentiate the effects of papaverine and/or PGE_1 (209). In an attempt to maximize efficacy and minimize side effects, many clinicians use a combination of α_1, papaverine, and phentolamine as a triple mix, which allows the use of a lower dose of each agent.

The most serious complication of intracavernosal injection therapy is priapism (210). In the event the patient develops a prolonged or painful erection with PGE_1, either 5 mg brethane or 60 mg pseudoephedrine, self-administered orally, may be of benefit. If priapism persists longer than 4 hours, the patient should be instructed to seek medical care in which aspiration alone or with the injection of an α-adrenergic agent is used to induce detumescence (210). Vital signs should be closely monitored during this procedure. If this fails, surgical therapy may be indicated to reverse a prolonged erection; otherwise, anoxic damage to the cavernosal smooth muscle cells, fibrosis, and permanent ED can occur.

Vacuum Devices for Inducing Erection

Vacuum devices consist of a plastic cylinder, a vacuum pump, and an elastic constriction band (7,211). The negative pressure created by the vacuum within the cylinder draws blood into the penis, producing an erection. An elastic band slipped around the base of the penis traps the blood in the penis, maintaining an erection as long as the rubber band is retained around the base.

These devices are safe, relatively inexpensive, and reasonably effective. They can impair ejaculation, resulting in entrapment of semen, and are difficult and awkward for some patients to use. Some couples dislike the lack of spontaneity engendered by the use of these devices. Partner cooperation is usually important for successful use of these devices (212–216). With the availability of effective oral therapy, the use of vacuum devices has decreased.

Development of Procedures for Modified Nerve-Sparing Prostatectomy

The description of the anatomy of the corporal nerves in men has allowed the technique of several surgical procedures to be modified. Previously, most patients undergoing radical prostatectomy for prostate cancer experienced postoperative impotence. Following the precise anatomical description of the cavernosal nerves, it was recognized (217) that impotence in these patients results from injury to the autonomic nerves during transection of the prostatic apex and the urethra or during division of the lateral pelvic fascia and lateral pedicle. The modification of the surgical technique to preserve these nerves has allowed a much greater proportion of patients who have undergone radical prostatectomy to preserve their sexual function than was possible before (217–222).

Novel Targets for the Development of Therapeutic Agents for Men with ED

Several novel classes of drugs, including soluble guanylate cyclase activators (223–227), Rho kinase inhibitors (228–230), centrally active agents that stimulate hypothalamic dopamine or melanocortin receptors (231), combinations of different types of drugs, and new facilitators of tissue uptake of active agents, are being investigated (232). Agonists of melanocortin receptor, MC4R, which is expressed primarily in the central nervous system, are being explored for the treatment of ED in men. Administration of one such MC4R agonist, PT-141, to rats and nonhuman primates resulted in penile erections. Administration of PT-141 to normal men and to patients with ED has been reported to improve erectile function (231). Thus, MCR4 agonists hold promise as a new treatment for sexual dysfunction.

Rho kinase is a signaling molecule that occurs downstream of the small GTPase Rho, which plays an important role in pathophysiology and progression of various cardiovascular diseases such as hypertension, coronary vasospasm, angina pectoris, restenosis after percutaneous coronary intervention, and ED (228–230). Inhibition of endogenous Rho kinase initiates an erectile response in the rat model, providing rationale for the further exploration of Rho kinase inhibitors for the treatment of ED.

EJACULATION AND EMISSION

Ejaculation involves a series of events triggered by central nervous system activation of the sympathetic nervous system that creates an emission containing the semen and the secretions from seminal vesicles, prostate, and bulbourethral glands (233–237). This emission is ejaculated out of the urethra by the contractions of the bulbocavernosus and levator ani muscles. The sensation associated with the rhythmic contractions of these pelvic floor muscles is referred to as the orgasm (233–237). Emission, the deposition of seminal fluid into the posterior urethra, is dependent on the integrity of the vasa deferentia, seminal vesicles, prostate gland, and bladder neck. Ejaculation is dependent primarily on the function of the striated perineal musculature. Although orgasm and ejaculation often occur contemporaneously, the two processes are distinct and regulated by separate mechanisms. The normal ejaculate includes secretions from the prostate, seminal vesicle, Cowper's glands, and the vas deferens.

Anatomy

Spermatozoa are transported from the testes and epididymides to the posterior urethra by the vas deferens. The tubular vas deferens is approximately 35 cm long in the human adult. It extends from the tail of the epididymis to the region of the verumontanum in the posterior urethra (Fig. 5); it is joined by the ducts of the seminal vesicles. The vas deferens is derived from the wolffian ductal system and is absent in some patients with cystic fibrosis and renal agenesis.

At the time of seminal fluid emission and ejaculation, the bladder neck closes to prevent the retrograde ejaculation of seminal fluid into the bladder (109). The anatomy of the bladder neck is controversial, and the anatomical basis for bladder neck closure in preventing not only retrograde ejaculation but also urinary incontinence is unclear. Most investigators agree that no anatomical sphincter consisting of circular smooth muscle fibers exists in the area of the bladder outlet (110,111). A physiological sphincter, however, does exist. The bladder neck probably remains collapsed because of tension created by the large amount of elastic tissue in this area as well as smooth muscle contraction under α-adrenergic control (177).

After emission, the seminal fluid that has been deposited in the posterior urethra is expelled out the urethral meatus by the process of ejaculation. Clonic contractions of the perineal striated musculature, primarily the bulbocavernosus and ischiocavernosus muscles, are responsible for this event (114). The bulbocavernosus muscle takes its origin from the central tendon, encircles the corpus spongiosum, and inserts into the corpus cavernosum (233–237).

Neural Control of Emission and Ejaculation

The sympathetic innervation is the primary regulator of a number of physiological processes involved in emission (55,238). Sympathetic stimulation causes contraction of the smooth muscle of vas deferens, and the adrenergic neurons that mediate this control use norepinephrine as the neurotransmitter (Fig. 8) (239,240). Preparations of human vas deferens in vitro demonstrate contractile response to norepinephrine; this contractile response to norepinephrine is blocked by phentolamine and phenoxybenzapine, indicating that this is mediated by α-adrenergic receptors (239,240). Increased sympathetic stimulation causes the bladder neck to contract and the smooth muscle in the body of the urinary bladder to relax (241–245). Norepinephrine-mediated relaxation of the bladder wall and contraction of the bladder neck prevents seminal fluid from entering the urinary bladder during ejaculation.

Electrical stimulation of the superior hypogastric ganglion or the hypogastric nerves in humans causes the contraction of the bladder neck, ejaculatory ducts, and the prostatic smooth muscle (238). We do not know the exact origin of the sympathetic nerves in

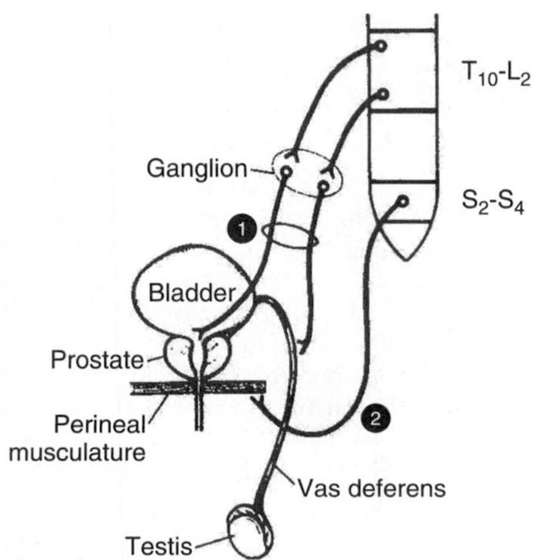

FIG. 8. 1, Sympathetic innervation to the bladder neck and vas deferens; 2, somatic innervation of the striated perineal muscles. Sympathetic activation causes the smooth muscle of the vas deferens to contract during seminal emission, the bladder neck to contract, and the bladder wall smooth muscle to relax. This prevents retrograde passage of the semen into the urinary bladder.

the spinal cord, although it is generally believed that the efferent sympathetic nerves emerge at T10-L2, form the lumbar sympathetic ganglia, encircle the aorta, and combine to form the superior hypogastric plexus (238). Postganglionic fibers from the superior hypogastric plexus play an important role in emission.

The ejaculation is accompanied by the rhythmic contracts of the periurethral and anal sphincter muscles (238). The somatic nerve fibers that originate at S2-4 and are carried in the pudendal nerve are important in ejaculation. In monkeys, stimulation of the preoptic region of the hypothalamus can produce ejaculation (58,246). Intrathecal injection of cholinesterase inhibitor, Prostigmin, can induce ejaculation in humans with neurological disorders (58).

After retroperitoneal lymph node dissection for testicular cancer, which might result in removal of the sympathetic nerves and paravertebral ganglia, some patients report a dry ejaculate even though they are able to achieve penile erection and orgasm (247–249). In most patients with retroperitoneal lymph node dissection, the cause of dry ejaculate is the absence of seminal fluid emission, although some patients might also suffer from retrograde ejaculation. These clinical observations provide additional evidence for the important role of sympathetic innervation in regulating contraction of the vas deferens during seminal fluid emission and bladder neck contraction (58).

Clinical Application of Physiological Principles

Data from a recent national survey have highlighted the high prevalence and clinical importance of ejaculatory disorders (50,52,53); this survey recognized premature ejaculation as the most prevalent sexual disorder in men 18 to 59 years of age. Ejaculatory disorders can also lead to infertility among men.

Ejaculatory disorders are a heterogeneous group of disorders that includes anejaculation, anorgasmia, delayed ejaculation, retrograde ejaculation, premature ejaculation, and painful ejaculation (51). Of these, premature ejaculation due to psychological factors and retrograde ejaculation due to diabetes-associated autonomic neuropathy are the most frequent ejaculatory disorders. An ejaculation is usually due to the lesions involving the nerve supply to the penis as a result of spinal cord injury, pelvic surgery, degenerative diseases, or to complications of drugs, such as antihypertensives, antipsychotics, or antidepressants. Retrograde ejaculation is the result of autonomic dysfunction due to autonomic neuropathy associated with diabetes mellitus, sympathectomy, or therapy with adrenergic antagonists; some types of antihypertensives, antipsychotics, or anti-depressants; bladder neck incompetence; or urethral obstruction.

After transurethral resection of the prostate, the bladder neck closure mechanism is destroyed. Patients remain continent because of a second, more distal, continence mechanism that is present in the region of the membranous urethra (113) (Fig. 6). Most patients who have undergone transurethral resection of the prostate, however, do experience retrograde ejaculation.

REFERENCES

1. Kinsey, A. C., Pomeroy, W. B., and Martin, C. E. (1948). *Sexual Behavior in the Human Male*. WB Saunders, Philadelphia.
2. Hite, S. (1976). *The Hite Report*. Dell, New York.
3. Tavris, C., and Sadd, S. (1975). *The Redbook Report on Female Sexuality*. Delacorte, New York.
4. Janus, S. S., and Janus, C. L. (1993). *The Janus Report on Sexual Behavior*. John Wiley & Sons, New York.
5. Masters, W., and Johnson, V. E. (1966). *Human Sexual Response*. Little, Brown, Boston.
6. Meuleman, E. (2000). Clinical evaluation and the doctor-patient dialogue. In *Erectile Dysfunction* (A. Jardin, G. Wagner, and S. Khoury, Eds.), pp. 115–138. Plymbridge Distributors, Plymouth, UK.
7. Lue, T. F. (2000). Erectile dysfunction. *N. Engl. J. Med.* 342, 1802–1813.
8. Lue, T. F., and Tanagho, E. A. (1988). Hemodynamics of erection. In *Contemporary Management of Impotence and Infertility* (E. A. Tanagho, T. F. Lue, and R. D. McClure, Eds.), pp. 28–38. Williams & Wilkins, Baltimore.
9. Lugg, J. A., and Rajfer, J. (1996). Drug therapy for erectile dysfunction. *AUA Update* 15, 290.
10. Rajfer, J., Aronson, W. J., Bush, P. A., Dorey, F. J., and Ignarro, L. J. (1992). Nitric oxide as a mediator of relaxation of the corpus cavernosum in response to nonadrenergic, noncholinergic neurotransmission. *N. Engl. J. Med.* 326, 90–94.

11. Gonzalez-Cadavid, N. F., Ignarro, L. J., and Rajfer, J. (1999). Nitric oxide and the cyclic GMP system in the penis. *Mol. Urol.* 3, 51–59.
12. Trigo-Rocha, F., Aronson, W. J., Hohenfellner, M., Ignarro, L. J., Rajfer, J., and Lue, T. F. (1993). Nitric oxide and cGMP: mediators of pelvic nerve-stimulated erection in dogs. *Am. J. Physiol.* 264, H419–H422.
13. Smith, O. (1998). Nobel Prize for NO research. *Nat. Med.* 4, 1215.
14. Andersson, K. E., and Wagner, G. (1995). Physiology of penile erection. *Physiol. Rev.* 75, 191–236.
15. Christ, G. J. (1995). The penis as a vascular organ. The importance of corporal smooth muscle tone in the control of erection. *Urol. Clin. North Am.* 22, 727–745.
16. Naylor, A. M. (1998). Endogenous neurotransmitters mediating penile erection. *Br. J. Urol.* 81, 424–431.
17. Rajfer, J., Aronson, W. J., Bush, P. A., Dorey, F. J., and Ignarro, L. J. (1992). Nitric oxide as a mediator of relaxation of the corpus cavernosum in response to nonadrenergic, noncholinergic neurotransmission. *N. Engl. J. Med.* 326, 90–94.
18. McDonald, L. J., and Murad, F. (1996). Nitric oxide and cyclic GMP signaling. *Proc. Soc. Exp. Biol. Med.* 211, 1–6.
19. Nehra, A., Barrett, D. M., and Moreland, R. B. (1999). Pharmacotherapeutic advances in the treatment of erectile dysfunction. *Mayo Clin. Proc.* 74, 709–721.
20. Taher, A., Meyer, M., Stief, C. G., Jonas, U., and Forssmann, W. G. (1997). Cyclic nucleotide phosphodiesterase in human cavernous smooth muscle. *World J. Urol.* 15, 32–35.
21. Jeremy, J. Y., Ballard, S. A., Naylor, A. M., Miller, M. A., and Angelini, G. D. (1997). Effects of sildenafil, a type-5 cGMP phosphodiesterase inhibitor, and papaverine on cyclic GMP and cyclic AMP levels in the rabbit corpus cavernosum in vitro. *Br. J. Urol.* 79, 958–963.
22. Stief, C. G., Uckert, S., Becker, A. J., Truss, M. C., and Jonas, U. (1998). The effect of the specific phosphodiesterase (PDE) inhibitors on human and rabbit cavernous tissue in vitro and in vivo. *J. Urol.* 159, 1390–1393.
23. Moreland, R. B., Goldstein, I., and Traish, A. (1998). Sildenafil, a novel inhibitor of phosphodiesterase type 5 in human corpus cavernosum smooth muscle cells. *Life Sci.* 62, 309–318.
24. Carter, A. J., Ballard, S. A., and Naylor, A. M. (1998). Effect of the selective phosphodiesterase type 5 inhibitor sildenafil on erectile dysfunction in the anesthetized dog. *J. Urol.* 160, 242–246.
25. Chuang, A. T., Strauss, J. D., Murphy, R. A., and Steers, W. D. (1998). Sildenafil, a type-5 CGMP phosphodiesterase inhibitor, specifically amplifies endogenous cGMP-dependent relaxation in rabbit corpus cavernosum smooth muscle in vitro. *J. Urol.* 160, 257–261.
26. Wallis, R. M., Corbin, J. D., Francis, S. H., and Ellis, P. (1999). Tissue distribution of phosphodiesterase families and the effects of sildenafil on tissue cyclic nucleotides, platelet function, and the contractile responses of trabeculae carneae and aortic rings in vitro. *Am. J. Cardiol.* 83, 3C–12C.
27. Wallis, R. M. (1999). The pharmacology of sildenafil, a novel and selective inhibitor of phosphodiesterase (PDE) type 5. *Nippon Yakurigaku Zasshi* 114 Suppl. 1, 22P–26P.
28. Stief, C. G., Uckert, S., Becker, A. J., Harringer, W., Truss, M. C., Forssmann, W. G., and Jonas, U. (2000). Effects of sildenafil on cAMP and cGMP levels in isolated human cavernous and cardiac tissue. *Urology* 55, 146–150.
29. Maggi, M., Filippi, S., Ledda, F., Magini, A., and Forti, G. (2000). Erectile dysfunction: from biochemical pharmacology to advances in medical therapy. *Eur. J. Endocrinol.* 143, 143–154.
30. Uckert, S., Kuthe, A., Stief, C. G., and Jonas, U. (2001). Phosphodiesterase isoenzymes as pharmacological targets in the treatment of male erectile dysfunction. *World J. Urol.* 19, 14–22.
31. Yu, G., Mason, H. J., Wu, X., Wang, J., Chong, S., Dorough, G., Henwood, A., Pongrac, R., Seliger, L., He, B., et al. (2001). Substituted pyrazolopyridines as potent and selective PDE5 inhibitors: potential agents for treatment of erectile dysfunction. *J. Med. Chem.* 44, 1025–1027.
32. Saenz de Tejada, I., Angulo, J., Cuevas, P., Fernandez, A., Moncada, I., Allona, A., Lledo, E., Korschen, H. G., Niewohner, U., Haning, H., et al. (2001). The phosphodiesterase inhibitory selectivity and the in vitro and in vivo potency of the new PDE5 inhibitor vardenafil. *Int. J. Impot. Res.* 13, 282–290.
33. Vroege, J. A., Gijs, L., and Hengeveld, M. W. (1998). Classification of sexual dysfunctions: towards DSM-V and ICD-11. *Compr. Psychiatry* 39, 333–337.
34. Association, A. P. (1994). *Diagnostic and Statistical Manual of Mental Disorders*. American Psychiatric Association, Washington, DC.
35. Bagatell, C. J., Heiman, J. R., Rivier, J. E., and Bremner, W. J. (1994). Effects of endogenous testosterone and estradiol on sexual behavior in normal young men [published erratum appears in *J. Clin. Endocrinol. Metab.* 1994 Jun;78(6):1520]. *J. Clin. Endocrinol. Metab.* 78, 711–716.
36. Alexander, G. M., Swerdloff, R. S., Wang, C., Davidson, T., McDonald, V., Steiner, B., and Hines, M. (1997). Androgen-behavior correlations in hypogonadal men and eugonadal men. I. Mood and response to auditory sexual stimuli. *Horm. Behav.* 31, 110–119.
37. Wang, C., Eyre, D. R., Clark, R., Kleinberg, D., Newman, C., Iranmanesh, A., Veldhuis, J., Dudley, R. E., Berman, N., Davidson, T., et al. (1996). Sublingual testosterone replacement improves muscle mass and strength, decreases bone resorption, and increases bone formation markers in hypogonadal men—a clinical research center study. *J. Clin. Endocrinol. Metab.* 81, 3654–3662.
38. Alexander, G. M., Swerdloff, R. S., Wang, C., Davidson, T., McDonald, V., Steiner, B., and Hines, M. (1998). Androgen-behavior correlations in hypogonadal men and eugonadal men. II. Cognitive abilities. *Horm. Behav.* 33, 85–94.
39. Buena, F., Swerdloff, R. S., Steiner, B. S., Lutchmansingh, P., Peterson, M. A., Pandian, M. R., Galmarini, M., and Bhasin, S. (1993). Sexual function does not change when serum testosterone levels are pharmacologically varied within the normal male range. *Fertil. Steril.* 59, 1118–1123.
40. Wang, C., Berman, N., Longstreth, J. A., Chuapoco, B., Hull, L., Steiner, B., Faulkner, S., Dudley, R. E., and Swerdloff, R. S. (2000). Pharmacokinetics of transdermal testosterone gel in hypogonadal men: application of gel at one site versus four sites: a General Clinical Research Center Study. *J. Clin. Endocrinol. Metab.* 85, 964–969.
41. Schover, L. R., and LoPiccolo, J. (1982). Treatment effectiveness for dysfunctions of sexual desire. *J. Sex. Marital Ther.* 8, 179–197.
42. LoPiccolo, J. (1985). Diagnosis and treatment of male sexual dysfunction. *J. Sex. Marital Ther.* 11, 215–232.
43. Meuleman, E. J., and van Lankveld, J. J. (2005). Hypoactive sexual desire disorder: an underestimated condition in men. *BJU Int.* 95, 291–296.
44. Corona, G., Mannucci, E., Petrone, L., Giommi, R., Mansani, R., Fei, L., Forti, G., and Maggi, M. (2004). Psycho-biological correlates of hypoactive sexual desire in patients with erectile dysfunction. *Int. J. Impot. Res.* 16, 275–281.
45. NIH Consensus Development Panel on Impotence. (1993). NIH Consensus Conference. Impotence. *J. A. M. A.* 270, 83–90.
46. Benet, A. E., and Melman, A. (1995). The epidemiology of erectile dysfunction. *Urol. Clin. North Am.* 22, 699–709.
47. Feldman, H. A., Goldstein, I., Hatzichristou, D. G., Krane, R. J., and McKinlay, J. B. (1994). Impotence and its medical and

psychosocial correlates: results of the Massachusetts Male Aging Study. *J. Urol.* 151, 54–61.

48. Feldman, H. A., Johannes, C. B., Derby, C. A., Kleinman, K. P., Mohr, B. A., Araujo, A. B., and McKinlay, J. B. (2000). Erectile dysfunction and coronary risk factors: prospective results from the Massachusetts male aging study. *Prev. Med.* 30, 328–338.
49. Laumann, E. O., Paik, A., and Rosen, R. C. (1999). The epidemiology of erectile dysfunction: results from the National Health and Social Life Survey. *Int. J. Impot. Res.* 11 Suppl. 1, S60–S64.
50. Laumann, E. O., Paik, A., and Rosen, R. C. (1999). Sexual dysfunction in the United States: prevalence and predictors. *J. A. M. A.* 281, 537–544.
51. Hendry, W. F. (1998). Disorders of ejaculation: congenital, acquired and functional. *Br. J. Urol.* 82, 331–341.
52. Brannigan, R. E. (2004). Ejaculatory disorders and lower urinary tract symptoms. *Curr. Urol. Rep.* 5, 280–286.
53. Rosen, R., Altwein, J., Boyle, P., Kirby, R. S., Lukacs, B., Meuleman, E., O'Leary, M. P., Puppo, P., Robertson, C., and Giuliano, F. (2003). Lower urinary tract symptoms and male sexual dysfunction: the multinational survey of the aging male (MSAM-7). *Eur. Urol.* 44, 637–649.
54. Benson, G. (1981). Mechanisms of penile erection. *Invest. Urol.* 19, 65–69.
55. Benson, G. S., McConnell, J., Lipshultz, L. I., Corriere, J. N. Jr., and Wood, J. (1980). Neuromorphology and neuropharmacology of the human penis: an in vitro study. *J. Clin. Invest.* 65, 506–513.
56. Deysach, L. J. (1939). The comparative morphology of the erectile tissue of the penis with special emphasis on the probable mechanism of erection. *Am. J. Anat.* 64, 111–131.
57. Goldstein, I., and Berman, J. R. (1998). Vasculogenic female sexual dysfunction: vaginal engorgement and clitoral erectile insufficiency syndromes. *Int. J. Impot. Res.* 10 Suppl. 2, S84–S90; discussion, S98–S101.
58. Lipshultz, L. I., McConnell, J., and Benson, G. S. (1981). Current concepts of the mechanisms of ejaculation. Normal and abnormal states. *J. Reprod. Med.* 26, 499–507.
59. Abbasi, A. A., Prasad, A. S., Ortega, J., Congco, E., and Oberleas, D. (1976). Gonadal function abnormalities in sickle cell anemia. Studies in adult male patients. *Ann. Intern. Med.* 85, 601–605.
60. Semans, J., and Langworthy, O. 1939. Observations on the neurophysiology of sexual function in the male cat. *J. Urol.* 40, 836–846.
61. Dorr, L., and Brody, M. (1967). Hemodynamic mechanisms of erection in the canine penis. *Am. J. Physiol.* 213, 1526–1531.
62. Marletta, M. A. (1991). Nitric oxide, nitrovasodilators, and L-arginine—an unusual relationship. *West J. Med.* 154, 107–109.
63. Bush, P. A., Aronson, W. J., Buga, G. M., Rajfer, J., and Ignarro, L. J. (1992). Nitric oxide is a potent relaxant of human and rabbit corpus cavernosum. *J. Urol.* 147, 1650–1655.
64. Burnett, A. L., Lowenstein, C. J., Bredt, D. S., Chang, T. S., and Snyder, S. H. (1992). Nitric oxide: a physiologic mediator of penile erection. *Science* 257, 401–403.
65. Aboseif, S. R., and Lue, T. F. (1988). Hemodynamics of penile erection. *Urol. Clin. North Am.* 15, 1–7.
66. Leriche, R., and Morel, A. (1948). Syndrome of thrombotic obliteration of the aortic bifurcation. *Ann. Surg.* 127, 193–206.
67. Melman, A., and Christ, G. J. (2001). Integrative erectile biology. The effects of age and disease on gap junctions and ion channels and their potential value to the treatment of erectile dysfunction. *Urol. Clin. North Am.* 28, 217–231.
68. Purvis, K., Egdetveit, I., and Christiansen, E. (1999). Intracavernosal therapy for erectile failure—impact of treatment and reasons for drop-out and dissatisfaction. *Int. J. Impot. Res.* 11, 287–299.
69. Christ, G. J., Moreno, A. P., Melman, A., and Spray, D. C. (1992). Gap junction-mediated intercellular diffusion of Ca^{2+} in cultured human corporal smooth muscle cells. *Am. J. Physiol.* 263, C373–C383.
70. Moreno, A. P., Campos de Carvalho, A. C., Christ, G., Melman, A., and Spray, D. C. (1993). Gap junctions between human corpus cavernosum smooth muscle cells: gating properties and unitary conductance. *Am. J. Physiol.* 264, C80–C92.
71. Campos de Carvalho, A. C., Roy, C., Moreno, A. P., Melman, A., Hertzberg, E. L., Christ, G. J., and Spray, D. C. (1993). Gap junctions formed of connexin43 are found between smooth muscle cells of human corpus cavernosum. *J. Urol.* 149, 1568–1575.
72. Tsai, H., Werber, J., Davia, M. O., Edelman, M., Tanaka, K. E., Melman, A., Christ, G. J., and Geliebter, J. (1996). Reduced connexin 43 expression in high grade, human prostatic adenocarcinoma cells. *Biochem. Biophys. Res. Commun.* 227, 64–69.
73. Porst, H. (1996). The rationale for prostaglandin E1 in erectile failure: a survey of worldwide experience. *J. Urol.* 155, 802–815.
74. Seftel, A. D., Viola, K. A., Kasner, S. E., and Ganz, M. B. (1996). Nitric oxide relaxes rabbit corpus cavernosum smooth muscle via a potassium-conductive pathway. *Biochem. Biophys. Res. Commun.* 219, 382–387.
75. Recio, P., Lopez, P. G., Hernandez, M., Prieto, D., Contreras, J., and Garcia-Sacristan, A. (1998). Nitrergic relaxation of the horse corpus cavernosum. Role of cGMP. *Eur. J. Pharmacol.* 351, 85–94.
76. Kun, A., Martinez, A. C., Tanko, L. B., Pataricza, J., Papp, J. G., and Simonsen, U. (2003). Ca^{2+}-activated K^{+} channels in the endothelial cell layer involved in modulation of neurogenic contractions in rat penile arteries. *Eur. J. Pharmacol.* 474, 103–115.
77. Bivalacqua, T. J., Champion, H. C., Mehta, Y. S., Abdel-Mageed, A. B., Sikka, S. C., Ignarro, L. J., Kadowitz, P. J., and Hellstrom, W. J. (2000). Adenoviral gene transfer of endothelial nitric oxide synthase (eNOS) to the penis improves age-related erectile dysfunction in the rat. *Int. J. Impot. Res.* 12 Suppl. 3, S8–S17.
78. Kim, N. N., Huang, Y. H., Goldstein, I., Bischoff, E., and Traish, A. M. (2001). Inhibition of cyclic GMP hydrolysis in human corpus cavernosum smooth muscle cells by vardenafil, a novel, selective phosphodiesterase type 5 inhibitor. *Life Sci.* 69, 2249–2256.
79. Haning, H., Niewohner, U., and Bischoff, E. (2003). Phosphodiesterase type 5 (PDE5) inhibitors. *Prog. Med. Chem.* 41, 249–306.
80. Bischoff, E. (2004). Potency, selectivity, and consequences of nonselectivity of PDE inhibition. *Int. J. Impot. Res.* 16 Suppl. 1, S11–S14.
81. Kuhn, R., Uckert, S., Stief, C. G., Truss, M. C., Lietz, B., Bischoff, E., Schramm, M., and Jonas, U. (2000). Relaxation of human ureteral smooth muscle in vitro by modulation of cyclic nucleotide-dependent pathways. *Urol. Res.* 28, 110–115.
82. Beavo, J. A. (1988). Multiple isozymes of cyclic nucleotide phosphodiesterase. *Adv. Second Messenger Phosphoprotein Res.* 22, 1–38.
83. Beavo, J. A. (1995). Cyclic nucleotide phosphodiesterases: functional implications of multiple isoforms. *Physiol. Rev.* 75, 725–748.
84. Beavo, J. A., Conti, M., and Heaslip, R. J. (1994). Multiple cyclic nucleotide phosphodiesterases. *Mol. Pharmacol.* 46, 399–405.
85. Rybalkin, S. D., Yan, C., Bornfeldt, K. E., and Beavo, J. A. (2003). Cyclic GMP phosphodiesterases and regulation of smooth muscle function. *Circ. Res.* 93, 280–291.

86. Boolell, M., Allen, M. J., Ballard, S. A., Gepi-Attee, S., Muirhead, G. J., Naylor, A. M., Osterloh, I. H., and Gingell, C. (1996). Sildenafil: an orally active type 5 cyclic GMP-specific phosphodiesterase inhibitor for the treatment of penile erectile dysfunction. *Int. J. Impot. Res.* 8, 47–52.
87. Goldstein, I., Lue, T. F., Padma-Nathan, H., Rosen, R. C., Steers, W. D., and Wicker, P. A. (1998). Oral sildenafil in the treatment of erectile dysfunction. Sildenafil Study Group [see comments] [published erratum appears in *N. Engl. J. Med.* 1998 Jul 2;339(1):59]. *N. Engl. J. Med.* 338, 1397–1404.
88. Carani, C., Granata, A. R., Bancroft, J., and Marrama, P. (1995). The effects of testosterone replacement on nocturnal penile tumescence and rigidity and erectile response to visual erotic stimuli in hypogonadal men. *Psychoneuroendocrinology* 20, 743–753.
89. Cunningham, G. R., Hirshkowitz, M., Korenman, S. G., and Karacan, I. (1990). Testosterone replacement therapy and sleep-related erections in hypogonadal men. *J. Clin. Endocrinol. Metab.* 70, 792–797.
90. Carani, C., Zini, D., Baldini, A., Della, C.L., Ghizzani, A., and Marrama, P. (1990). Effects of androgen treatment in impotent men with normal and low levels of free testosterone. *Arch. Sex Behav.* 19, 223–234.
91. Davidson, J. M., Camargo, C. A., and Smith, E. R. (1979). Effects of androgen on sexual behavior in hypogonadal men. *J. Clin. Endocrinol. Metab.* 48, 955–958.
92. Kwan, M., Greenleaf, W. J., Mann, J., Crapo, L., and Davidson, J. M. (1983). The nature of androgen action on male sexuality: a combined laboratory- self-report study on hypogonadal men. *J. Clin. Endocrinol. Metab.* 57, 557–562.
93. Salmimies, P., Kockott, G., Pirke, K. M., Vogt, H. J., and Schill, W. B. (1982). Effects of testosterone replacement on sexual behavior in hypogonadal men. *Arch. Sex Behav.* 11, 345–353.
94. Bradford, J. M. (1983). The hormonal treatment of sexual offenders. *Bull. Am. Acad. Psychiatry Law* 11, 159–169.
95. da Silva, F. C., Fossa, S. D., Aaronson, N. K., Serbouti, S., Denis, L., Casselman, J., Whelan, P., Hetherington, J., Fava, C., Richards, B., et al. (1996). The quality of life of patients with newly diagnosed M1 prostate cancer: experience with EORTC clinical trial 30853. *Eur. J. Cancer* 32A, 72–77.
96. Kasimis, B., Wilding, G., Kreis, W., Feuerman, M., Chang, V., Hwang, S., Steafather, H., Cogswell, J., Rae, C., and Blumenfrucht, M. (2000). Survival of patients who had salvage castration after failure on bicalutamide monotherapy for stage (D2) prostate cancer. *Cancer Invest.* 18, 602–608.
97. Lucas, M. D., Strijdom, S. C., Berk, M., and Hart, G. A. (1995). Quality of life, sexual functioning and sex role identity after surgical orchidectomy in patients with prostatic cancer. *Scand. J. Urol. Nephrol.* 29, 497–500.
98. Potosky, A. L., Knopf, K., Clegg, L. X., Albertsen, P. C., Stanford, J. L., Hamilton, A. S., Gilliland, F. D., Eley, J. W., Stephenson, R. A., and Hoffman, R. M. (2001). Quality-of-life outcomes after primary androgen deprivation therapy: results from the Prostate Cancer Outcomes Study. *J. Clin. Oncol.* 19, 3750–3757.
99. Rousseau, L., Dupont, A., Labrie, F., and Couture, M. (1988). Sexuality changes in prostate cancer patients receiving antihormonal therapy combining the antiandrogen flutamide with medical (LHRH agonist) or surgical castration. *Arch. Sex Behav.* 17, 87–98.
100. Snyder, P. J., Peachey, H., Berlin, J. A., Hannoush, P., Haddad, G., Dlewati, A., Santanna, J., Loh, L., Lenrow, D. A., Holmes, J. H., et al. (2000). Effects of testosterone replacement in hypogonadal men. *J. Clin. Endocrinol. Metab.* 85, 2670–2677.
101. Arver, S., Dobs, A. S., Meikle, A. W., Allen, R. P., Sanders, S. W., and Mazer, N. A. (1996). Improvement of sexual function in testosterone deficient men treated for 1 year with a permeation enhanced testosterone transdermal system. *J. Urol.* 155, 1604–1608.
102. Wang, C., Swedloff, R. S., Iranmanesh, A., Dobs, A., Snyder, P. J., Cunningham, G., Matsumoto, A. M., Weber, T., and Berman, N. (2000). Transdermal testosterone gel improves sexual function, mood, muscle strength, and body composition parameters in hypogonadal men. Testosterone Gel Study Group. *J. Clin. Endocrinol. Metab.* 85, 2839–2853.
103. Wang, C., Alexander, G., Berman, N., Salehian, B., Davidson, T., McDonald, V., Steiner, B., Hull, L., Callegari, C., and Swerdloff, R. S. (1996). Testosterone replacement therapy improves mood in hypogonadal men—a clinical research center study. *J. Clin. Endocrinol. Metab.* 81, 3578–3583.
104. Hirshkowitz, M., and Moore, C. A. (1996). Sleep-related erectile activity. *Neurol. Clin.* 14, 721–737.
105. Alexandersen, P., Haarbo, J., and Christiansen, C. (1996). The relationship of natural androgens to coronary heart disease in males: a review. *Atherosclerosis* 125, 1–13.
106. Alexander, G. M., and Sherwin, B. B. (1991). The association between testosterone, sexual arousal, and selective attention for erotic stimuli in men. *Horm. Behav.* 25, 367–381.
107. Lugg, J. A., Rajfer, J., and Gonzalez-Cadavid, N.F. (1995). Dihydrotestosterone is the active androgen in the maintenance of nitric oxide-mediated penile erection in the rat. *Endocrinology* 136, 1495–1501.
108. Lewis, R. W., and Mills, T. M. (2004). Effect of androgens on penile tissue. *Endocrine* 23, 101–105.
109. Mills, T. M., Dai, Y., Stopper, V. S., and Lewis, R. W. (1999). Androgenic maintenance of the erectile response in the rat. *Steroids* 64, 605–609.
110. Mills, T. M., and Lewis, R. W. (1999). The role of andorgens in the erectile response: a 1999 perspective. *Mol. Urol.* 3, 75–86.
111. Mills, T. M., Lewis, R. W., and Stopper, V. S. (1998). Androgenic maintenance of inflow and veno-occlusion during erection in the rat. *Biol. Reprod.* 59, 1413–1418.
112. Alexander, G. M., Packard, M. G., and Hines, M. (1994). Testosterone has rewarding affective properties in male rats: implications for the biological basis of sexual motivation. *Behav. Neurosci.* 108, 424–428.
113. King, B. E., Packard, M. G., and Alexander, G. M. (1999). Affective properties of intra-medial preoptic area injections of testosterone in male rats. *Neurosci. Lett.* 269, 149–152.
114. Packard, M. G., Cornell, A. H., and Alexander, G. M. (1997). Rewarding affective properties of intra-nucleus accumbens injections of testosterone. *Behav. Neurosci.* 111, 219–224.
115. Schroeder, J. P., and Packard, M. G. (2000). Role of dopamine receptor subtypes in the acquisition of a testosterone conditioned place preference in rats. *Neurosci. Lett.* 282, 17–20.
116. Balthazart, J., and Schumacher, M. (1984). Organization and activation of behavior in quail: role of testosterone metabolism. *J. Exp. Zool.* 232, 595–604.
117. Gorski, R. A. (1985). Sexual dimorphisms of the brain. *J. Anim. Sci.* 61 Suppl. 3, 38–61.
118. Gorski, R. A. (1986). Sexual differentiation of the brain: a model for drug-induced alterations of the reproductive system. *Environ. Health Perspect* 70, 163–175.
119. Lieberburg, I., Wallach, G., and McEwen, B. S. (1977). The effects of an inhibitor of aromatization (1,4,6-androstatriene-3,17-dione) and an anti-estrogen (Cl-628) on in vivo formed testosterone metabolites recovered from neonatal rat brain tissues and purified cell nuclei. Implications for sexual differentiation of the rat brain. *Brain Res.* 128, 176–181.
120. McEwen, B. S., Lieberburg, I., Chaptal, C., and Krey, L. C. (1977). Aromatization: important for sexual differentiation of the neonatal rat brain. *Horm. Behav.* 9, 249–263.

121. McGivern, R. F., Roselli, C. E., and Handa, R. J. (1988). Perinatal aromatase activity in male and female rats: effect of prenatal alcohol exposure. *Alcohol Clin. Exp. Res.* 12, 769–772.
122. Toran-Allerand, C. D. (1980). Sex steroids and the development of the newborn mouse hypothalamus and preoptic area in vitro. II. Morphological correlates and hormonal specificity. *Brain Res.* 189, 413–427.
123. Grisham, W., Tam, A., Greco, C. M., Schlinger, B. A., and Arnold, A. P. (1997). A putative 5 alpha-reductase inhibitor demasculinizes portions of the zebra finch song system. *Brain Res.* 750, 122–128.
124. Gurney, M. E. (1981). Hormonal control of cell form and number in the zebra finch song system. *J. Neurosci.* 1, 658–673.
125. Schlinger, B. A., and Arnold, A. P. (1991). Androgen effects on the development of the zebra finch song system. *Brain Res.* 561, 99–105.
126. Anderson, R. A., Martin, C. W., Kung, A. W., Everington, D., Pun, T. C., Tan, K. C., Bancroft, J., Sundaram, K., Moo-Young, A. J., and Baird, D. T. (1999). 7Alpha-methyl-19-nortestosterone maintains sexual behavior and mood in hypogonadal men. *J. Clin. Endocrinol. Metab.* 84, 3556–3562.
127. Anderson, R. A., Wallace, A. M., Sattar, N., Kumar, N., and Sundaram, K. (2003). Evidence for tissue selectivity of the synthetic androgen 7 alpha-methyl-19-nortestosterone in hypogonadal men. *J. Clin. Endocrinol. Metab.* 88, 2784–2793.
128. Morali, G., Lemus, A. E., Munguia, R., Arteaga, M., Perez-Palacios, G., Sundaram, K., Kumar, N., and Bardin, C. W. (1993). Induction of male sexual behavior in the rat by 7 alpha-methyl-19-nortestosterone, an androgen that does not undergo 5 alpha-reduction. *Biol. Reprod.* 49, 577–581.
129. Negro-Vilar, A. (1999). Selective androgen receptor modulators (SARMs): a novel approach to androgen therapy for the new millennium. *J. Clin. Endocrinol. Metab.* 84, 3459–3462.
130. Rosen, J., and Negro-Vilar, A. (2002). Novel, non-steroidal, selective androgen receptor modulators (SARMs) with anabolic activity in bone and muscle and improved safety profile. *J. Musculoskelet. Neuron. Interact.* 2, 222–224.
131. Wang, C., Eyre, D. R., Clark, R., Kleinberg, D., Newman, C., Iranmanesh, A., Veldhuis, J., Dudley, R. E., Berman, N., Davidson, T., et al. (1996). Sublingual testosterone replacement improves muscle mass and strength, decreases bone resorption, and increases bone formation markers in hypogonadal men—a clinical research center study. *J. Clin. Endocrinol. Metab.* 81, 3654–3662.
132. Grinspoon, S., Corcoran, C., Stanley, T., Baaj, A., Basgoz, N., and Klibanski, A. (2000). Effects of hypogonadism and testosterone administration on depression indices in HIV-infected men. *J. Clin. Endocrinol. Metab.* 85, 60–65.
133. Goggin, K., Engelson, E. S., Rabkin, J. G., and Kotler, D. P. (1998). The relationship of mood, endocrine, and sexual disorders in human immunodeficiency virus positive (HIV+) women: an exploratory study. *Psychosom. Med.* 60, 11–16.
134. Rabkin, J. G., Wagner, G. J., and Rabkin, R. (1999). Testosterone therapy for human immunodeficiency virus-positive men with and without hypogonadism. *J. Clin. Psychopharmacol.* 19, 19–27.
135. Rabkin, J. G., Wagner, G. J., and Rabkin, R. (2000). A double-blind, placebo-controlled trial of testosterone therapy for HIV-positive men with hypogonadal symptoms. *Arch. Gen. Psychiatry* 57, 141–147; discussion, 155–146.
136. Seidman, S. N., and Rabkin, J. G. (1998). Testosterone replacement therapy for hypogonadal men with SSRI-refractory depression. *J. Affect. Disord.* 48, 157–161.
137. Levitt, A. J., and Joffe, R. T. (1988). Total and free testosterone in depressed men. *Acta Psychiatr. Scand.* 77, 346–348.
138. Seidman, S. N. (2001). Testosterone deficiency and depression in aging men: pathogenic and therapeutic implications. *J. Gend. Specif. Med.* 4, 44–48.
139. Seidman, S. N. (2003). Testosterone deficiency and mood in aging men: pathogenic and therapeutic interactions. *World J. Biol. Psychiatry* 4, 14–20.
140. Seidman, S. N., Araujo, A. B., Roose, S. P., Devanand, D. P., Xie, S., Cooper, T. B., and McKinlay, J. B. (2002). Low testosterone levels in elderly men with dysthymic disorder. *Am. J. Psychiatry* 159, 456–459.
141. Kaplan, S. A., Reis, R. B., Kohn, I. J., Shabsigh, R., and Te, A. E. (1998). Combination therapy using oral alpha-blockers and intracavernosal injection in men with erectile dysfunction. *Urology* 52, 739–743.
142. Johannes, C. B., Araujo, A. B., Feldman, H. A., Derby, C. A., Kleinman, K. P., and McKinlay, J. B. (2000). Incidence of erectile dysfunction in men 40 to 69 years old: longitudinal results from the Massachusetts male aging study. *J. Urol.* 163, 460–463.
143. McKinlay, J. B. (2000). The worldwide prevalence and epidemiology of erectile dysfunction. *Int. J. Impot. Res.* 12 Suppl. 4, S6–S11.
144. Lewis, R. W. (2001). Epidemiology of erectile dysfunction. *Urol. Clin. North Am.* 28, 209–216.
145. Ruutu, M. L., Virtanen, J. M., Lindstrom, B. L., and Alfthan, O. S. (1987). The value of basic investigations in the diagnosis of impotence. *Scand. J. Urol. Nephrol.* 21, 261–265.
146. Aitchison, M., Aitchison, J., and Carter, R. (1990). Is the penile brachial index a reproducible and useful measurement? *Br. J. Urol.* 66, 202–204.
147. Takasaki, N., Kotani, T., Miyazaki, S., and Saitou, S. (1989). Measurement of penile brachial index (PBI) in patients with impotence. *Hinyokika Kiyo* 35, 1365–1368.
148. Cappelleri, J. C., Rosen, R. C., Smith, M. D., Mishra, A., and Osterloh, I. H. (1999). Diagnostic evaluation of the erectile function domain of the International Index of Erectile Function. *Urology* 54, 346–351.
149. Rosen, R. C., Riley, A., Wagner, G., Osterloh, I. H., Kirkpatrick, J., and Mishra, A. (1997). The international index of erectile function (IIEF): a multidimensional scale for assessment of erectile dysfunction. *Urology* 49, 822–830.
150. Cappelleri, J. C., and Rosen, R. C. (2003). A comparison of the International Index of Erectile Function and erectile dysfunction studies. *BJU Int.* 92, 654.
151. Goldstein, I. (1999). A 36-week, open label, non-comparative study to assess the long-term safety of sildenafil citrate (VIAGRA) in patients with erectile dysfunction. *Int. J. Clin. Pract.* 102 Suppl. 8–9.
152. Rendell, M. S., Rajfer, J., Wicker, P. A., and Smith, M. D. (1999). Sildenafil for treatment of erectile dysfunction in men with diabetes: a randomized controlled trial. Sildenafil Diabetes Study Group [see comments]. *J. A. M. A.* 281, 421–426.
153. Yu, G., Mason, H., Wu, X., Wang, J., Chong, S., Beyer, B., Henwood, A., Pongrac, R., Seliger, L., He, B., et al. (2003). Substituted pyrazolopyridopyridazines as orally bioavailable potent and selective PDE5 inhibitors: potential agents for treatment of erectile dysfunction. *J. Med. Chem.* 46, 457–460.
154. Seftel, A. D. (2004). Phosphodiesterase type 5 inhibitor differentiation based on selectivity, pharmacokinetic, and efficacy profiles. *Clin. Cardiol.* 27, I14–I19.
155. Crowe, S. M., and Streetman, D. S. (2004). Vardenafil treatment for erectile dysfunction. *Ann. Pharmacother.* 38, 77–85.
156. Grossman, E. B., Swan, S. K., Muirhead, G. J., Gaffney, M., Chung, M., DeRiesthal, H., Chow, D., and Raij, L. (2004). The pharmacokinetics and hemodynamics of sildenafil citrate in male hemodialysis patients. *Kidney Int.* 66, 367–374.

157. Sussman, D. O. (2004). Pharmacokinetics, pharmacodynamics, and efficacy of phosphodiesterase type 5 inhibitors. *J. Am. Osteopath. Assoc.* 104, S11–S15.
158. Bischoff, E. (2004). Vardenafil preclinical trial data: potency, pharmacodynamics, pharmacokinetics, and adverse events. *Int. J. Impot. Res.* 16 Suppl. 1, S34–S37.
159. Keating, G. M., and Scott, L. J. (2003). Vardenafil: a review of its use in erectile dysfunction. *Drugs* 63, 2673–2703.
160. Kendirci, M., Bivalacqua, T. J., and Hellstrom, W. J. (2004). Vardenafil: a novel type 5 phosphodiesterase inhibitor for the treatment of erectile dysfunction. *Expert Opin. Pharmacother.* 5, 923–932.
161. Corbin, J. D., and Francis, S. H. (2002). Pharmacology of phosphodiesterase-5 inhibitors. *Int. J. Clin. Pract.* 56, 453–459.
162. Curran, M., and Keating, G. (2003). Tadalafil. *Drugs* 63, 2203–2212; discussion, 2213–2204.
163. Holmes, S. (2003). Tadalafil: a new treatment for erectile dysfunction. *BJU Int.* 91, 466–468.
164. Meuleman, E. J. (2003). Review of tadalafil in the treatment of erectile dysfunction. *Expert Opin. Pharmacother.* 4, 2049–2056.
165. Rajagopalan, P., Mazzu, A., Xia, C., Dawkins, R., and Sundaresan, P. (2003). Effect of high-fat breakfast and moderate-fat evening meal on the pharmacokinetics of vardenafil, an oral phosphodiesterase-5 inhibitor for the treatment of erectile dysfunction. *J. Clin. Pharmacol.* 43, 260–267.
166. Moreland, R. B., Goldstein, I., and Traish, A. (1998). Sildenafil, a novel inhibitor of phosphodiesterase type 5 in human corpus cavernosum smooth muscle cells. *Life Sci.* 62, PL309–PL318.
167. Giuliano, F., Hultling, C., el Masry, W. S., Smith, M. D., Osterloh, I. H., Orr, M., and Maytom, M. (1999). Randomized trial of sildenafil for the treatment of erectile dysfunction in spinal cord injury. Sildenafil Study Group. *Ann. Neurol.* 46, 15–21.
168. Jarow, J. P., Burnett, A. L., and Geringer, A. M. (1999). Clinical efficacy of sildenafil citrate based on etiology and response to prior treatment [see comments]. *J. Urol.* 162, 722–725.
169. Dinsmore, W. W., Hodges, M., Hargreaves, C., Osterloh, I. H., Smith, M. D., and Rosen, R. C. (1999). Sildenafil citrate (Viagra) in erectile dysfunction: near normalization in men with broad-spectrum erectile dysfunction compared with age-matched healthy control subjects [published erratum appears in *Urology* 1999 May;53(5):1072]. *Urology* 53, 800–805.
170. Giuliano, F., Donatucci, C., Montorsi, F., Auerbach, S., Karlin, G., Norenberg, C., Homering, M., Segerson, T., and Eardley, I. (2005). Vardenafil is effective and well-tolerated for treating erectile dysfunction in a broad population of men, irrespective of age. *BJU Int.* 95, 110–116.
171. Markou, S., Perimenis, P., Gyftopoulos, K., Athanasopoulos, A., and Barbalias, G. (2004). Vardenafil (Levitra) for erectile dysfunction: a systematic review and meta-analysis of clinical trial reports. *Int. J. Impot. Res.* 16, 470–478.
172. Montorsi, F., Hellstrom, W. J., Valiquette, L., Bastuba, M., Collins, O., Taylor, T., Thibonnier, M., Homering, M., and Eardley, I. (2004). Vardenafil provides reliable efficacy over time in men with erectile dysfunction. *Urology* 64, 1187–1195.
173. Brock, G. B., McMahon, C. G., Chen, K. K., Costigan, T., Shen, W., Watkins, V., Anglin, G., and Whitaker, S. (2002). Efficacy and safety of tadalafil for the treatment of erectile dysfunction: results of integrated analyses. *J. Urol.* 168, 1332–1336.
174. Carson, C. C., Rajfer, J., Eardley, I., Carrier, S., Denne, J. S., Walker, D. J., Shen, W., and Cordell, W. H. (2004). The efficacy and safety of tadalafil: an update. *BJU Int.* 93, 1276–1281.
175. Padma-Nathan, H. (2003). Efficacy and tolerability of tadalafil, a novel phosphodiesterase 5 inhibitor, in treatment of erectile dysfunction. *Am. J. Cardiol.* 92, 19M–25M.
176. Morales, A., Gingell, C., Collins, M., Wicker, P. A., and Osterloh, I. H. (1998). Clinical safety of oral sildenafil citrate (ViagrA) in the treatment of erectile dysfunction. *Int. J. Impot. Res.* 10, 69–73.
177. Aversa, A., Mazzilli, F., Rossi, T., Delfino, M., Isidori, A. M., and Fabbri, A. (2000). Effects of sildenafil (Viagra) administration on seminal parameters and post-ejaculatory refractory time in normal males. *Hum. Reprod.* 15, 131–134.
178. Feenstra, J., Drie-Pierik, R. J., Lacle, C. F., and Stricker, B. H. (1998). Acute myocardial infarction associated with sildenafil [letter] [see comments]. *Lancet* 352, 957–958.
179. Zusman, R. M., Morales, A., Glasser, D. B., and Osterloh, I. H. (1999). Overall cardiovascular profile of sildenafil citrate. *Am. J. Cardiol.* 83, 35C–44C.
180. Arora, R. R., Timoney, M., and Melilli, L. (1999). Acute myocardial infarction after the use of sildenafil [letter]. *N. Engl. J. Med.* 341, 700.
181. Cheitlin, M. D., Hutter, A. M. Jr., Brindis, R. G., Ganz, P., Kaul, S., Russell, R. O. Jr., and Zusman, R. M. (1999). Use of sildenafil (Viagra) in patients with cardiovascular disease. Technology and Practice Executive Committee [published erratum appears in *Circulation* 1999 Dec 7;100(23):2389] [see comments]. *Circulation* 99, 168–177.
182. Muller, J. E., Mittleman, A., Maclure, M., Sherwood, J. B., and Tofler, G. H. (1996). Triggering myocardial infarction by sexual activity. Low absolute risk and prevention by regular physical exertion. Determinants of Myocardial Infarction Onset Study Investigators [see comments]. *J. A. M. A.* 275, 1405–1409.
183. Herrmann, H. C., Chang, G., Klugherz, B. D., and Mahoney, P. D. (2000). Hemodynamic effects of sildenafil in men with severe coronary artery disease. *N. Engl. J. Med.* 342, 1622–1626.
184. Kloner, R. A. (2000). Cardiovascular risk and sildenafil. *Am. J. Cardiol.* 86, 57F–61F.
185. Kloner, R. A. (2004). Novel phosphodiesterase type 5 inhibitors: assessing hemodynamic effects and safety parameters. *Clin. Cardiol.* 27, I20–I25.
186. Padma-Nathan, H., Steers, W. D., and Wicker, P. A. (1998). Efficacy and safety of oral sildenafil in the treatment of erectile dysfunction: a double-blind, placebo-controlled study of 329 patients. Sildenafil Study Group [see comments]. *Int. J. Clin. Pract.* 52, 375–379.
187. Goldenberg, M. M. (1998). Safety and efficacy of sildenafil citrate in the treatment of male erectile dysfunction. *Clin. Ther.* 20, 1033–1048.
188. Conti, C. R., Pepine, C. J., and Sweeney, M. (1999). Efficacy and safety of sildenafil citrate in the treatment of erectile dysfunction in patients with ischemic heart disease. *Am. J. Cardiol.* 83, 29C–34C.
189. Osterloh, I. H., Collins, M., Wicker, P., and Wagner, G. (1999). Sildenafil citrate (Viagra): overall safety profile in 18 double-blind, placebo controlled, clinical trials. *Int. J. Clin. Pract.* 102 Suppl. 3–5.
190. Young, J. (1999). Sildenafil citrate (Viagra) in the treatment of erectile dysfunction: a 12-week, flexible-dose study to assess efficacy and safety. *Int. J. Clin. Pract.* 102 Suppl. 6–7.
191. McMahon, C. G., Samali, R., and Johnson, H. (2000). Efficacy, safety and patient acceptance of sildenafil citrate as treatment for erectile dysfunction [In Process Citation]. *J. Urol.* 164, 1192–1196.
192. Highleyman, L. (1999). Protease inhibitors and sildenafil (Viagra) should not be combined. *Beta* 12, 3.
193. Bailey, D. G., and Dresser, G. K. (2004). Interactions between grapefruit juice and cardiovascular drugs. *Am. J. Cardiovasc. Drugs* 4, 281–297.
194. (1997). Intraurethral alprostadil for impotence. *Med. Lett. Drugs Ther.* 39, 32.

195. Benevides, M. D., and Carson, C. C. (2000). Intraurethral application of alprostadil in patients with failed inflatable penile prosthesis. *J. Urol.* 163, 785–787.
196. Bodner, D. R., Haas, C.A., Krueger, B., and Seftel, A. D. (1999). Intraurethral alprostadil for treatment of erectile dysfunction in patients with spinal cord injury. *Urology* 53, 199–202.
197. Engelhardt, P. F., Plas, E., Hubner, W. A., and Pfluger, H. (1998). Comparison of intraurethral liposomal and intracavernosal prostaglandin-E1 in the management of erectile dysfunction. *Br. J. Urol.* 81, 441–444.
198. Guay, A. T., Perez, J. B., Velasquez, E., Newton, R. A., and Jacobson, J. P. (2000). Clinical experience with intraurethral alprostadil (MUSE) in the treatment of men with erectile dysfunction. A retrospective study. Medicated urethral system for erection. *Eur. Urol.* 38, 671–676.
199. Jaffe, J. S., Antell, M. R., Greenstein, M., Ginsberg, P. C., Mydlo, J. H., and Harkaway, R. C. (2004). Use of intraurethral alprostadil in patients not responding to sildenafil citrate. *Urology* 63, 951–954.
200. Kim, S. C., Ahn, T. Y., Choi, H. K., Choi, N. G., Chung, T. G., Chung, W. S., Hwang, T. K., Hyun, J. S., Jung, G. W., and Kim, C. I., et al. (2000). Multicenter study of the treatment of erectile dysfunction with transurethral alprostadil (MUSE) in Korea. *Int. J. Impot. Res.* 12, 97–101.
201. Lewis, R. (2000). Review of intraurethral suppositories and iontophoresis therapy for erectile dysfunction. *Int. J. Impot. Res.* 12 Suppl. 4, S86–S90.
202. Lewis, R. W. (2000). Intraurethral and topical agents. *Drugs Today (Barc.)* 36, 113–119.
203. Shokeir, A. A., Alserafi, M. A., and Mutabagani, H. (1999). Intracavernosal versus intraurethral alprostadil: a prospective randomized study. *BJU Int.* 83, 812–815.
204. Padma-Nathan, H., Hellstrom, W. J., Kaiser, F. E., Labasky, R. F., Lue, T. F., Nolten, W. E., Norwood, P. C., Peterson, C. A., Shabsigh, R., and Tam, P. Y. (1997). Treatment of men with erectile dysfunction with transurethral alprostadil. Medicated Urethral System for Erection (MUSE) Study Group [see comments]. *N. Engl. J. Med.* 336, 1–7.
205. Engelhardt, P. F., Plas, E., Hubner, W. A., and Pfluger, H. (1998). Comparison of intraurethral liposomal and intracavernosal prostaglandin-E1 in the management of erectile dysfunction. *Br. J. Urol.* 81, 441–444.
206. Kim, E. D., and McVary, K. T. (1995). Topical prostaglandin-E1 for the treatment of erectile dysfunction [see comments]. *J. Urol.* 153, 1828–1830.
207. Peterson, C. A., Bennett, A. H., Hellstrom, W. J., Kaiser, F. E., Morley, J. E., Nemo, K. J., Padma-Nathan, H., Place, V. A., Prendergast, J. J., Tam, P. Y., et al. (1998). Erectile response to transurethral alprostadil, prazosin and alprostadil-prazosin combinations. *J. Urol.* 159, 1523–1527.
208. Lakin, M. M., Montague, D. K., VanderBrug Medendorp, S., Tesar, L., and Schover, L. R. (1990). Intracavernous injection therapy: analysis of results and complications. *J. Urol.* 143, 1138–1141.
209. Leungwattanakij, S., Flynn, V. Jr., and Hellstrom, W. J. (2001). Intracavernosal injection and intraurethral therapy for erectile dysfunction. *Urol. Clin. North Am.* 28, 343–354.
210. Lue, T. F., Hellstrom, W. J. G., McAninch, J. W., and Tanagho, E. A. (1986). Priapism: a refined approach to diagnosis and treatment. *J. Urol.* 136, 104–108.
211. Conference, N. C. (1993). Consensus Development Panel on Impotence. *J. A. M. A.* 270, 83–90.
212. Witherington, R. (1991). Vacuum devices for the impotent. *J. Sex. Marital Ther.* 17, 69–80.
213. Lewis, J. H., Sidi, A. A., and Reddy, P. K. (1991). A way to help your patients who use vacuum devices. *Contemp. Urol.* 3, 15–21.
214. Lewis, R. W., and Witherington, R. (1997). External vacuum therapy for erectile dysfunction: use and results. *World J. Urol.* 15, 78–82.
215. Ganem, J. P., Lucey, D. T., Janosko, E. O., and Carson, C. C. (1998). Unusual complications of the vacuum erection device. *Urology* 51, 627–631.
216. Morales, A. (1993). Nonsurgical management options in impotence. *Hosp. Pract.* 28, 15–20, 23.
217. Walsh, P. C., and Mostwin, J. L. (1984). Radical prostatectomy and cystoprostatectomy with preservation of potency. Results utilizing new nerve sparing technique. *Br. J. Urol.* 56, 694–697.
218. Bigg, S. W., Kavoussi, L. R., and Catalona, W. J. (1990). Role of nerve-sparing radical prostatectomy for clinical stage B2 prostate cancer. *J. Urol.* 144, 1420–1424.
219. Catalona, W. J., and Bigg, S. W. (1990). Nerve-sparing radical prostatectomy: evaluation of results after 250 patients. *J. Urol.* 143, 538–543; discussion, 544.
220. Rossignol, G., Leandri, P., Gautier, J. R., Quintens, H., Gabay-Torbiero, L., and Tap, G. (1991). Radical retropubic prostatectomy: complications and quality of life (429 cases, 1983–1989). *Eur. Urol.* 19, 186–191.
221. Walsh, P. C. (1988). Nerve sparing radical prostatectomy for early stage prostate cancer. *Semin. Oncol.* 15, 351–358.
222. Walsh, P. C., and Mostwin, J. L. (1984). Radical prostatectomy and cystoprostatectomy with preservation of potency. Results using a new nerve-sparing technique. *Br. J. Urol.* 56, 694–697.
223. Garthwaite, G., Goodwin, D. A., Neale, S., Riddall, D., and Garthwaite, J. (2002). Soluble guanylyl cyclase activator YC-1 protects white matter axons from nitric oxide toxicity and metabolic stress, probably through Na(+) channel inhibition. *Mol. Pharmacol.* 61, 97–104.
224. Horio, Y., and Murad, F. (1991). Solubilization of guanylyl cyclase from bovine rod outer segments and effects of lowering Ca^{2+} and nitro compounds. *J. Biol. Chem.* 266, 3411–3415.
225. Hwang, T. L., Hung, H. W., Kao, S. H., Teng, C. M., Wu, C. C., and Cheng, S. J. (2003). Soluble guanylyl cyclase activator YC-1 inhibits human neutrophil functions through a cGMP-independent but cAMP-dependent pathway. *Mol. Pharmacol.* 64, 1419–1427.
226. Nakane, M. (2003). Soluble guanylyl cyclase: physiological role as an NO receptor and the potential molecular target for therapeutic application. *Clin. Chem. Lab. Med.* 41, 865–870.
227. Whitworth, J. A. (2003). Emerging drugs in the management of hypertension. *Expert Opin. Emerg. Drugs* 8, 377–388.
228. Chitaley, K., Webb, R. C., and Mills, T. M. (2001). RhoA/Rho-kinase: a novel player in the regulation of penile erection. *Int. J. Impot. Res.* 13, 67–72.
229. Hirooka, Y., and Shimokawa, H. (2005). Therapeutic potential of rho-kinase inhibitors in cardiovascular diseases. *Am. J. Cardiovasc. Drugs* 5, 31–39.
230. Wingard, C. J., Johnson, J. A., Holmes, A., and Prikosh, A. (2003). Improved erectile function after Rho-kinase inhibition in a rat castrate model of erectile dysfunction. *Am. J. Physiol. Regul. Integr. Comp. Physiol.* 284, R1572–R1579.
231. Molinoff, P. B., Shadiack, A. M., Earle, D., Diamond, L. E., and Quon, C. Y. (2003). PT-141: a melanocortin agonist for the treatment of sexual dysfunction. *Ann. N Y Acad. Sci.* 994, 96–102.
232. Gonzalez-Cadavid, N. F., and Rajfer, J. (2004). Therapy of erectile dysfunction: potential future treatments. *Endocrine* 23, 167–176.
233. Marberger, H. (1974). The mechanisms of ejaculation. *Basic Life Sci.* 4, 99–110.
234. deGroat, W. C., and Booth, A. M. (1980). Physiology of male sexual function. *Ann. Intern. Med.* 92, 329–331.
235. Yeates, W. K. (1990). Ejaculation and its disorders. *Arch. Ital. Urol. Nefrol. Androl.* 62, 137–148.

236. Gil-Vernet, J. M. Jr., Alvarez-Vijande, R., Gil-Vernet, A., and Gil-Vernet, J. M. (1994). Ejaculation in men: a dynamic endorectal ultrasonographical study. *Br. J. Urol.* 73, 442–448.
237. Argiolas, A., and Melis, M. R. (2003). The neurophysiology of the sexual cycle. *J. Endocrinol. Invest.* 26, 20–22.
238. Master, V. A., and Turek, P. J. (2001). Ejaculatory physiology and dysfunction. *Urol. Clin. North Am.* 28, 363–375.
239. Owman, C., and Sjoberg, N. O. (1972). The importance of short adrenergic neurons in the seminal emission mechanism of rat, guinea-pig and man. *J. Reprod. Fertil.* 28, 379–387.
240. Stefanick, M. L., Smith, E. R., Szumowski, D. A., and Davidson, J. M. (1985). Reproductive physiology and behavior in the male rat following acute and chronic peripheral adrenergic depletion by guanethidine. *Pharmacol. Biochem. Behav.* 23, 55–63.
241. Alm Pelmer, M. (1975). Adrenergic and cholinergic innervation of the rag urinary bladder. *Acta Physiol. Scand.* 94, 36–45.
242. Benson, G. S., Wein, A. J., Raezer, D. M., and Corriere, J. N. Jr. (1976). Adrenergic and cholinergic stimulation and blockade of the human bladder base. *J. Urol.* 116, 174–175.
243. Ek, A. (1977). Innervation and receptor functions of the human urethra. *Scand. J. Urol. Nephrol.* 45, Suppl. 1–50.
244. Gosling, J. A., Dixon, J. S., and Lendon, R. G. (1977). The autonomic innervation of the human male and female bladder neck and proximal urethra. *J. Urol.* 118, 302–305.
245. Raezer, D. M., Wein, A. J., Jacobowitz, D., and Corriere, J. N. Jr. (1973). Autonomic innervation of canine urinary bladder. Cholinergic and adrenergic contributions and interaction of sympathetic and parasympathetic nervous systems in bladder function. *Urology* 2, 211–221.
246. Robinson, B. W., and Mishkin, M. (1966). Ejaculation evoked by stimulation of the preoptic area in monkey. *Physiol. Behav.* 1, 269–272.
247. Brenner, J., Vugrin, D., and Whitmore, W. F. Jr. (1985). Effect of treatment on fertility and sexual function in males with metastatic nonseminomatous germ cell tumors of testis. *Am. J. Clin. Oncol.* 8, 178–182.
248. Nijman, J. M., Schraffordt Koops, H., Kremer, J., and Sleijfer, D. T. (1987). Gonadal function after surgery and chemotherapy in men with stage II and III nonseminomatous testicular tumors. *J. Clin. Oncol.* 5, 651–656.
249. Stephenson, W. T., Poirier, S. M., Rubin, L., and Einhorn, L. H. (1995). Evaluation of reproductive capacity in germ cell tumor patients following treatment with cisplatin, etoposide, and bleomycin. *J. Clin. Oncol.* 13, 2278–2280.

Knobil and Neill's Physiology of Reproduction, Third Edition
edited by Jimmy D. Neill,
Elsevier © 2006

CHAPTER 25

Immunophysiology of the Male Reproductive Tract

M. P. Hedger[1] and D. B. Hales[2]

General Introduction, 1196
Historical Aspects, 1197
Structural Organization and Functions of the Male Reproductive Tract Relevant to Immunology, 1199
The Testis, 1199
The Excurrent Ducts and Accessory Glands, 1201
Endocrine Regulation of the Male Reproductive Tract, 1201
The Immune System and Its Endocrine Control, 1203
General Principles, 1203
Endocrine Regulation of Immunity, 1208
Immune Cells in the Male Reproductive Tract, 1209
The Testicular Macrophages, 1210
Lymphocytes and Polymorphonuclear Cells in the Testis, 1213
Immune Cells of the Epididymis, 1214
Immune Cells of the Vas Deferens, Accessory Glands, and Urethra, 1215
Effects of Inflammation, Infection, and Immunity on Male Reproduction, 1216
Inflammatory and Immunoregulatory Cytokines and Testis Function, 1216
Effects of the Immune System on Testicular Function: Androgen Production, 1221
Effects of the Immune System on Testicular Function: Spermatogenesis, 1229
Role of Macrophages in Leydig Cell Development and Steroidogenesis, 1231
Immunological Responses in the Male Reproductive Tract, 1232
Sperm Antibodies, 1232
Orchitis, 1233
Immune Privilege of the Testis I: The Blood–Testis Barrier, 1235
Immune Privilege of the Testis II: Transplantation and Immunoregulation, 1236
Germ Cell and Testis Transplantation, 1243
Immune Privilege as a "Testis-Specific" Phenomenon, 1244
Immunology of the Epididymis and Genital Tract, 1245
Response to Infections and Tumors: Innate Immunity in the Male Reproductive Tract, 1247
Implications, Applications, and the Future, 1249
Immunocontraception, 1249
Implications for Transplantation Medicine, 1249
The Future, 1249
References, 1250

ABSTRACT

The necessity for a close functional relationship between the male reproductive tract and the immune system is self-evident. Spermatogenic cells are particularly susceptible to immunological responses, and activation of immunity against sperm or other elements of the reproductive tract can lead to androgen insufficiency, infertility, or chronic inflammatory conditions. Inflammation and immune activation directly inhibit the hypothalamic–pituitary–Leydig cell axis, interfere with essential interactions between the Sertoli cells and spermatogenic cells, and increase the potential for sperm antibody formation, a major cause of infertility in men. It is fortunate, therefore, that spermatogenic cells normally are

[1]Monash Institute of Reproduction and Development, Monash Medical Centre, Victoria, Australia; [2]Department of Physiology and Biophysics, University of Illinois at Chicago, Chicago, Illinois.

ignored by the immune system, as are grafts of foreign tissues placed within the testicular capsule. Traditional explanations for the protection of these cells, based on "immune privilege" of the testis maintained by the blood–testis barrier or by exclusion of immune cells, are not consistent with either the histological organization of the reproductive tract or modern concepts of immunoregulation. A more realistic understanding of the control of immune responses in the male reproductive tract should focus on the activity of immunoregulatory macrophages and lymphocytes, as well as active suppression of antigen-specific immunity by somatic cells involving regulatory cytokines, androgenic steroids, and other antiinflammatory and immunosuppressive factors. Equally important for maintaining fertility, it appears that the restraints on antigen-specific immunity in the male reproductive tract are counterbalanced by enhanced local innate immune mechanisms and conventional mucosal immunity. In the testis, at least, there is convincing evidence that establishment of reproductive function during development is linked to the normal development of the local immune environment.

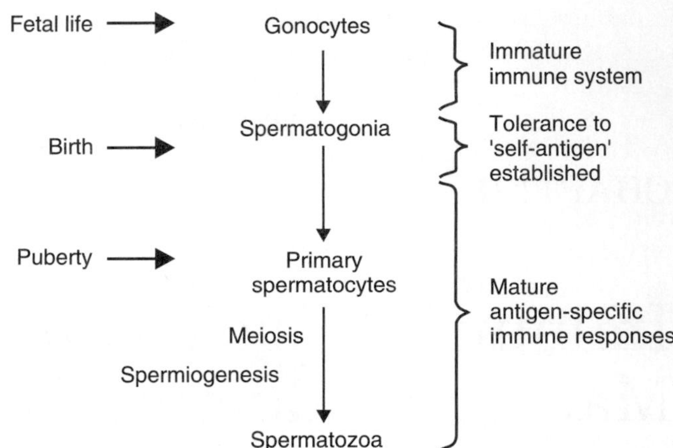

FIG. 1. Developmental time-line of spermatogenesis and the maturation of the immune system. The majority of spermatogenic cells do not appear in the testis until the initiation of meiosis at puberty, whereas tolerance to "self" antigen is largely established at or around the time of birth. Because the development of spermatozoa from the spermatogonial stem cell population involves complex processes of nuclear reorganization (meiosis) and cellular differentiation (spermiogenesis) that are unique to the testis, there exists enormous potential for antigens of spermatogenesis to evade conventional tolerance mechanisms.

GENERAL INTRODUCTION

The male reproductive tract, and the male gamete especially, poses a challenge to the immune system not presented by most other organ systems and tissues. The human testis produces approximately 100 million highly differentiated sperm each day from a pool of no more than a few hundred thousand spermatogonial stem cells (1–3), a level of productivity matched in output and complexity only by the hematopoietic system. In contrast to the hematopoietic tissues, or any other organ in the body for that matter, these differentiated cells first appear very late in developmental life, at the time of sexual maturation, long after the maturation of the immune system and the establishment of systemic immune tolerance (4–7) (Fig. 1). In humans, the period between the perinatal editing of the lymphocyte repertoire and the first appearance of significant numbers of the earliest meiotic germ cells (the spermatocytes) is usually more than 10 years. As a consequence, these cells and their progeny express a broad range of novel structural proteins, surface receptors, signaling molecules, and enzymes that have the potential to be seen as foreign by the mature immune system.

We know that the immune system does view some of these antigens as foreign because of the relatively high incidence of autoimmune infertility among human populations. In developed countries, the presence of sperm autoantibodies accounts for 5% to 12% of all male infertility (8–12). In other forms of autoimmune disease, such as type 1 diabetes mellitus or gastritis, the development of autoimmunity is perceived to be due to a breakdown or dysfunction of the normal regulatory controls of the immune system, leading to attack against antigens that previously were perceived as "self." Usually only a few antigens are involved, and for many of these diseases a specific dominant antigen has been identified (13,14). In the case of the male reproductive tract, however, the autoimmunity is directed against cells expressing numerous antigens that have not been edited out of the lymphocyte repertoire in the first place. Hence, autoimmune infertility involves multiple antigens as well as different antigens from individual to individual (15–25).

Another major urological problem with an immunological basis is chronic prostatitis, which may be accompanied by recurrent and even debilitating perineal or scrotal pain (26–31). Although infections certainly account for the majority of such conditions, it is by no means clear in many circumstances what are the underlying causes and mechanisms. At least some of these cases may involve an autoimmune inflammatory component.

Unraveling the origin of these immunologically based disturbances in reproductive function is an important clinical task, but the equally important scientific question is: What is it about the male reproductive tract that permits the continuous production of huge numbers of immunogenic cells expressing multiple autoantigens without apparent problems in most individuals? In most respects, the immune system in the reproductive tract appears relatively normal, with effective lymphatic drainage of the testis and epididymis and relatively free access of immune cells to all the tissues of the male reproductive tract (32–38). Certainly, there are some unique structural characteristics of the male reproductive tract that may be important. In species with scrotal testes and epididymides, the male gametes mature and are stored at a substantially lower temperature than the remainder of the body. There also exists a very effective blood–testis barrier located in the meiotic cell compartment (39–44). Nonetheless, these physical constraints do not appear to explain all the manifestations of immunological protection in the male reproductive tract across a range of species.

Studies over many years have gradually exposed the fact that the male reproductive tract, and the testis in particular, constitutes a unique immunoregulatory environment. There is now clear evidence that there are communication pathways, cytokine networks, and regulatory mediators common to both the male reproductive tract and the immune system, with many striking overlaps between the control of spermatogenesis and the processes of inflammation and immune activation. Moreover, there is evidence that part of this network includes local immunoregulatory and immunosuppressive mechanisms that provide protection for the endogenous antigens of spermatogenesis, protection that also extends to foreign antigens on grafts inserted into the testicular environment (45–52). It now appears that the immune cells that find their way into the testis, and presumably the rest of the male reproductive tract, are functionally modified to restrict their proinflammatory activity and provide an immunologically "constrained" environment where antigen-specific immune responses are suppressed or limited. Balanced against this, of course, is the question of how the male reproductive tract is able to protect itself from recurrent infections and tumor development under such circumstances.

Further evidence for an intimate interaction between the male reproductive tract and the immune system is the fact that reproductive dysfunction is associated not only with local infection and its accompanying inflammation, but with systemic disease and inflammation (53–58). Consequently, men with a wide range of systemic illnesses generally display a reduction in both serum androgen levels and sperm output. This indicates that male sexual function and general well-being operate in a reciprocal relationship, which may even represent a physiologically important regulatory mechanism. The yin-yang relationship between the male reproductive endocrine system and immune system has been described in terms of "sickness behavior" when the immune system is dominant and "testosterone behavior" under normal reproductive conditions when libido and feelings of well-being dominate (59) (Fig. 2). The possibility that local infection or inflammatory dysregulation of reproductive function can lead to more permanent problems, such as autoimmune infertility or chronic inflammatory disease, in some men also must be considered. Indeed, it may well be the case that preexisting hypogonadism predisposes men to inflammatory disease.

Consideration of decades of research has led to a growing acceptance that immune regulation in the male reproductive tract possesses unique features that may provide important clues to understanding how immune responses can be regulated in complex tissues. For this reason, unraveling the immune system in the male reproductive tract may have important implications for understanding autoimmune disease in other tissues, as well as obvious practical lessons for the field of transplantation biology. It should be self-evident that the ability of the immune system and the male reproductive system to coexist is no less important for reproductive success than the normal operation of the hypothalamic–pituitary axis. Moreover, the physiological mechanisms that allow the immune and male reproductive systems to operate together are likely to be similar in many ways to the mechanisms involved in cell–cell interactions in any other reproductive process.

HISTORICAL ASPECTS

Although often considered a relatively recent field of research, studies concerning the immunology of the male reproductive tract actually go back to the very beginnings of endocrinology and immunology. Inflammation of the testis and its association with mumps parotitis has been known since ancient times, having been described in the writings of Hippocrates (c. 460–377 B.C.). Studies on the transplantation of testes in domestic chickens, generally considered to be the earliest studies in the field of endocrinology, were undertaken in the 18th century by Hunter and by Michaelis (60), although the first systematic experiments actually were reported by Berthold in 1849 (61). "Successful" transplantations of mammalian testes had to wait until some time later (62–64). In the early part of the last century, before the discovery of the male sex hormone, testosterone, transplantation of male gonads, including transplants of animal gonads into humans, was practiced by both charlatans and serious researchers in the search for treatments to increase male health and virility (65,66). Although the function of the interstitial tissue appeared to be preserved in these transplant experiments, seminiferous tubule function was not (67). While there is no doubt that temperature and revascularization issues were important limitations, we now know that many of these experiments also faced a crucial immunological impediment.

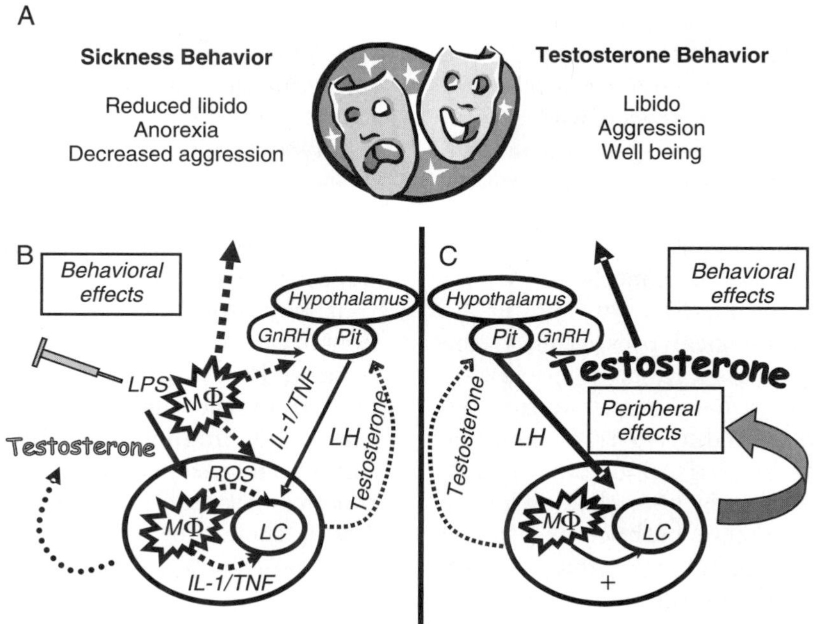

FIG. 2. Macrophage–Leydig cell interactions and effects of testosterone on behavior. **A:** Two behavioral paradigms symbolized by "comedy" and "tragedy." Sickness behavior represents an organized behavioral strategy and is not a specific response to an illness-like condition. An important feature of sickness behavior is depressed intramale aggression without altered defensive behavior. Sickness behavior is mediated in part by repression of testosterone-mediated aggression and specific cytokine pathways in the brain. Testosterone behavior represents the normal physiological state typified by intramale aggression and mating behavior. **B:** Current working model of the effects of lipopolysaccharide (LPS)-induced immune activation on the hypothalamic–pituitary–testicular (HPT) axis. Injection of LPS causes an acute and immediate inhibition of testosterone production, mediated in part by reactive oxygen species (ROS) elaborated by testicular macrophages (MΦ) during an LPS-triggered respiratory burst (685,688,712,761,776). Adjacent Leydig cells (LC) in close proximity to macrophages are subjected to oxidative stress resulting in perturbation of mitochondria and inhibition of steroidogenic acute regulatory protein (StAR) expression. LPS activation of testicular macrophages and peripheral monocyte/macrophages also induces proinflammatory cytokine production: interleukin-1 (IL-1) and tumor necrosis factor (TNF). IL-1 and TNF are potent transcriptional repressors of steroidogenic enzyme gene expression (396,582,712,724,727,731,1280). Collectively, these inflammatory mediators produced by testicular macrophages shut off testosterone production, resulting in decreased plasma levels. Despite decreased circulating levels of testosterone, luteinizing hormone (LH) secretion is either unchanged or decreased, indicating that immune activation results in inhibition of the hypothalamus–pituitary as well (446,685,818,1281). Sickness behavior is mediated in part by an inducible brain cytokine compartment that is activated by peripheral cytokines acting through neural afferent pathways (519-521,700,1281). **C:** The HPT axis under normal physiological conditions when LH primarily drives testosterone production. Testosterone feedback inhibits the hypothalamic–pituitary axis, modulating LH secretion. Testicular macrophages may provide factors that have positive effects on Leydig cell functions (385,841,842,847,850). Testosterone is secreted by the testis into the peripheral circulation to support androgen-dependent tissues, including positive reinforcement of "testosterone behavior" such as libido, aggression, and feelings of well-being. (Reproduced from Hales, D. B. [2002]. Testicular macrophage modulation of Leydig cell steroidogenesis. *J. Reprod. Immunol.* 57, 3–18, with permission.)

Even today, successful grafts of testicular tissue usually require an immunologically compatible or immunocompromised host (68-71). The science of immunoregulation entered this arena with the discovery that allogeneic testicular grafts could be made to survive much more successfully if tolerance had been induced by injection of the donor allogeneic cells at the time of birth in the donor (72), or when transferred into the anterior chamber of the eye (73,74). Later, Billingham and others conducted a series of investigations into so-called immunologically privileged sites that led to recognition of the testis as one of several tissues that were particularly favorable for graft survival (75).

Evidence that the spermatozoon itself was immunogenic to its autologous host dates back to Lansteiner (76) and Metchnikoff (77). In the 1920s, Guyer (78) was able to produce infertility in rabbits and guinea pigs by passive immunization with sperm-specific antisera, whereas Kennedy (79) reported degenerative changes in the testis after active immunization with autologous sperm. By the middle of the century, Voisin and colleagues (80,81) were able to produce aspermatogenesis in guinea pigs by active immunization with testicular extracts. Numerous experiments have since confirmed this process in several other species. Eventually, an association between sperm autoantibody formation and infertility in humans was noted (82,83). Further experimental evidence confirming the possibility of autoimmune reactions against the sperm led to a general acceptance that these are an important cause of infertility in men (8,10). In turn, this acceptance gradually led to the general concept that the testicular environment must provide protection for these developing sperm cells through very specific regulatory mechanisms.

Sex differences in health and the greater prevalence, earlier onset, and greater severity of autoimmune diseases in women, such as systemic lupus erythematosus (lupus) and rheumatoid arthritis, were recognized even before the concept of autoimmunity existed (84–88). A specific effect of the testes on the immune system was reported by Calzolari (89) after he found that castration of rabbits before puberty led to an increase or maintenance in the size of the thymus, an effect that has been repeatedly confirmed by subsequent studies (90–93). Many subsequent studies have shown that various immunological processes have sex-specific differences, or can be affected by castration or sex steroid replacement (92,94–99). These data clearly establish that products of the testis, and the sex steroids in particular, regulate the immune system either directly or indirectly.

In spite of the long history of research and longstanding evidence that the immunological environment of the male reproductive tract may be unique in a number of ways, there is still only a very superficial understanding at best of what is actually going on. However, from the 1960s onward there has been an exponential increase in activity in this area, and the accumulation of knowledge underpinning this research continues to accelerate. Although the present review contains many open questions and speculations, there is little doubt that a similar review on this topic written in 10 years' time will be able to answer most of these uncertainties. No doubt, given the complexity of the two systems involved, new issues and questions will have arisen to take their place.

STRUCTURAL ORGANIZATION AND FUNCTIONS OF THE MALE REPRODUCTIVE TRACT RELEVANT TO IMMUNOLOGY

The Testis

The main functions of the male reproductive system are to produce the male sex steroid hormones (androgens), to manufacture the male gametes (spermatozoa), and to deliver these successfully to the female reproductive tract. The testes are anatomically and functionally compartmentalized to accomplish the first two of these tasks, whereas production of semen and delivery of spermatozoa to the female depend on maintenance of the remainder of the reproductive tract by androgens secreted by the testis.

Functionally and anatomically, the testis is separated into an endocrine compartment and a gametogenic compartment where androgen production and spermatogenesis take place, respectively. The vascularized intertubular or interstitial tissue represents the endocrine compartment and the avascular seminiferous tubules are the site of spermatogenesis. In this regard, the organization of the testis is analogous to the organization of the ovary, where steroids are produced in the ovarian interstitial tissue and oogenesis occurs in the follicle. The testis itself is contained within a tough fibrous capsule, but there is considerable species variation in the amount of connective tissue extending from the capsule into the parenchyma of the testis. In the human, the testis is actually partitioned by connective tissue septa into discrete lobules containing the loops of the seminiferous tubules, which themselves connect at both ends to a reservoir structure called the *rete testis* located along one pole of the testis in the mediastinum (100,101) (see Chapter 18). In contrast, rodent species such as rat and mouse display only very loose intertubular connective tissue with no distinct septa separating the seminiferous tubules (102,103).

The interstitial compartment lies within the connective tissue septa and completely surrounds the seminiferous tubules (Fig. 3). The interstitial tissue contains the vasculature, lymphatic vessels, and nerves of the testis. A considerably more detailed review of this aspect of testis function is provided in Chapter 17. The vascular supply to the testis arises from the abdominal aorta and, in species with scrotal testes, this results in a comparatively long and highly coiled spermatic artery that is particularly susceptible to physical insult and torsion. The arterioles, capillaries, and venules of the testis completely permeate the interstitial tissue surrounding the seminiferous tubules and rete testis. Consequently, in spite of the completely avascular nature of the spermatogenic compartment, these structures are close to an effective blood supply at all times. Unlike the capillaries of other endocrine glands, most testicular capillaries are not fenestrated (38,104), and the mechanisms whereby molecules such as proteins and steroids enter and exit the testis by this route remain surprisingly poorly understood, even a source of some controversy (105–107). Nonetheless, there appears to be very little functional restriction on the exchange of even large molecules across this barrier, and the interstitial fluid is very similar in its overall composition to that of the circulating blood (108–110). The venous drainage of

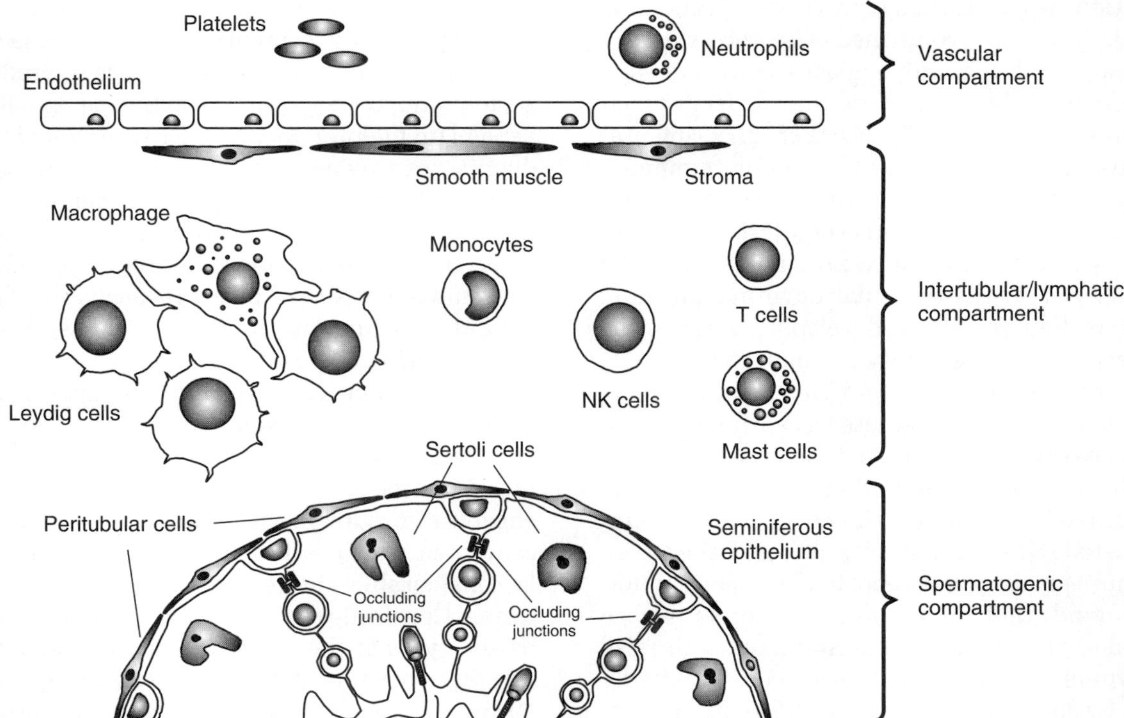

FIG. 3. Immunological compartmentalization of the testis. The mammalian testis comprises three immunologically distinct compartments: the vascular compartment and intertubular (or interstitial) compartment are separated by a layer of nonfenestrated endothelium, whereas the intertubular and spermatogenic compartments are separated by a layer of peritubular myoid cells and by occluding junctions between adjacent Sertoli cells. These junctions constitute the blood–testis barrier, which further divides the seminiferous epithelium into a basal and an adlumenal region. The adlumenal region contains the meiotic germ cells within a highly specialized microenvironment. Under normal conditions, monocytes, macrophages, T cells, natural killer (NK) cells, and, in some species, mast cells or eosinophils have relatively free access to the intertubular compartment, but are entirely excluded from the adlumenal region of the seminiferous epithelium. Neutrophils are confined to the vascular compartment except during specific immunological events. (See text for details.)

the testis by the spermatic veins is very closely associated with the arterial supply and involves a very effective countercurrent heat and solute exchange structure, called the *pampiniform plexus* (111–114).

The lymphatics of the testis are quite variable between species, ranging from irregular channels or sinusoids that are incompletely bounded by endothelial cells in rodents, to large, discrete lymphatic vessels in humans, to very small, rapid-flow lymphatics in porcine species (102,115,116). These lymphatics completely invest the entire interstitial tissue and pass without restriction to local-draining lymph nodes, principally the lumbar or para-aortic lymph nodes (36,117). In the laboratory rat, drainage to the iliac and renal nodes predominates, and there may be some lymphatic drainage directly to the thoracic ducts without passage through regional lymph nodes (33,35,37,118,119).

Among the normal connective tissue cells, the most abundant cell type present in the interstitium is the Leydig cell (120,121). This actively steroidogenic cell produces the vast majority of circulating androgens, but is also a prolific producer of many secreted proteins (122). Macrophages are commonly observed in the interstitium of most, if not all, species and many testes also contain variable numbers of mast cells or eosinophils (123,124). Less numerous, but ubiquitous nonetheless, are the intratesticular lymphocytes (125–131). The presence of these cells clearly demonstrates that immune cells have relatively free access to the testicular parenchyma (Fig. 3).

The seminiferous tubules and their associated draining ducts comprise the other discrete compartment of the testis. The seminiferous tubules are bounded by a circumferential layer of peritubular cells, mainly comprising myoid cells, and the acellular components of the basal lamina, which together form the limiting tissue on which the Sertoli cells and most basally located spermatogenic cells rest. Immune cells are occasionally seen in this boundary layer as well (132,133). The Sertoli cell provides the structural framework for the organization of the seminiferous epithelium, but also plays a crucial role in supporting and directing the development of the spermatogenic cells. These cells remain in intimate contact with their adjacent Sertoli cells at all times during the spermatogenic process, with junctional and membrane specializations providing physical contact and communication (134,135). The spermatogenic cells start out as mitotically dividing precursors called *spermatogonia* sitting on the basal lamina. Commencing at the time of puberty, cohorts of these mitotically dividing cells begin to enter into meiosis at regular intervals, moving away from the periphery of the tubule and becoming spermatocytes in the process. Meiosis is a lengthy procedure of chromosomal rearrangements leading to production of haploid round spermatids, which subsequently undergo considerable structural differentiation to become mature or elongated spermatids (see Chapter 18 for details). Once these cells are released by the Sertoli cell into the tubule lumen, they are called *spermatozoa*, and fluid secreted by the Sertoli cells sweeps the released spermatozoa toward the rete testis.

Occluding junctions between adjacent Sertoli cells and their associated membrane specializations form an intercellular barrier that is completely impermeable even to small molecules (42,134,135). This blood–testis barrier separates the spermatogonia and early meiotic cells in the basal region of the seminiferous epithelium from the adlumenal spermatocytes and spermatids. In this way, a large majority of the developing germ cells are sequestered behind a physical barrier within a highly specialized environment, and effectively isolated from the immune system (Fig. 3). On the other hand, the rete testis epithelium lacks both Sertoli cells and their highly specialized junctional specializations. The epithelial barrier restricting movement from the blood into the rete testis appears to be substantially less effective than that of the seminiferous epithelium, with the result that immunoglobulin and possibly even immune cells are able to cross the epithelium of this structure (43,125,136–139).

The Excurrent Ducts and Accessory Glands

Sperm production occurs in the testis, but sperm maturation and storage take place in the epididymis. Spermatozoa released into the lumen of the seminiferous tubule are neither motile nor capable of fertilizing an egg. The epididymis, which lies adjacent to the testis and is connected to the rete testis by a series of efferent ducts, comprises a long, single, highly convoluted epididymal duct lined by columnar principal cells with extensive apical stereocilia (see Chapter 22 for details). Testicular fluid secreted by the Sertoli cells is largely reabsorbed by the epithelial cells of the efferent ducts and the proximal regions (caput) of the epididymis (140,141). Sperm maturation occurs during transit through the epididymal duct and sperm are stored before ejaculation in the caudal region of the epididymis (142,143). The cauda epididymis in turn is connected to the vas deferens, a highly muscularized duct that drives the epididymal contents toward the urethra at the time of ejaculation. The accessory glands and their associated ductal structures, like the testis, depend on androgens for their normal development and maintenance of adult function. The testicular and epididymal secretions constitute only approximately 10% of the ejaculate, with the remaining 90% of the semen coming from the accessory glands, the seminal vesicles, and prostate in particular (100). It might be assumed that the most sperm spend the majority of their time in the testis and epididymis and appear only briefly in the vas deferens and urethra (i.e., the genital tract) during ejaculation. Critically, however, some sperm may be retained in the genital tract for much longer periods, because spermatozoa continue to appear in the ejaculates of vasectomized men for several months even after a successful procedure (144). In fact, the presence of intact sperm has been noted even in human prostate glands collected after prostatic surgery or postmortem, with the implication that such ectopic sperm may play a role in the etiology of prostatic inflammation and possibly even sperm autoimmunity (145,146).

There is a blood–epididymis barrier restricting movement of molecules across the epididymal epithelium, although the evidence suggests that this barrier is not entirely analogous to, nor as completely effective as, the blood–testis barrier (147–151). In normal adults, circulating immunoglobulin appears to be restricted, although perhaps not entirely excluded, from passage into the epididymal fluid (151). However, the most striking contrast to the seminiferous epithelium is the presence of macrophages and lymphocytes in the epididymal epithelium (126,152–158). The presence of these cells suggests that a very different immunological environment operates in this organ compared with that of the testis. Similar epithelial barriers exist throughout the remainder of the genital tract, although these have been subject to considerably less examination (135,147,159–161).

Endocrine Regulation of the Male Reproductive Tract

The hormonal regulation of the male reproductive tract is described extensively in Chapters 18 and 21, and only a very brief summary is presented here.

Male reproduction is regulated and maintained by pulsatile secretion of gonadotropin-releasing hormone (GnRH) by the hypothalamus under the control of the central nervous system (CNS), which stimulates concordant pulses of luteinizing hormone (LH) and follicle-stimulating hormone (FSH) from the anterior pituitary (162,163) (Fig. 4). These circulating gonadotropins exert direct effects on the two main testicular compartments. LH binds to specific receptors on the surface of the Leydig cell, stimulating Leydig cell development, morphology, and secretion of androgens, chiefly testosterone (164–167). Both testosterone and FSH bind to specific Sertoli cell receptors to regulate spermatogenesis and Sertoli cell functions, such as secretion of the protein hormone, inhibin (168,169). It is generally considered that the spermatogenic cells do not respond directly to these hormones. In turn, androgens and inhibin operate through a negative feedback loop to regulate LH and FSH synthesis and secretion at the pituitary

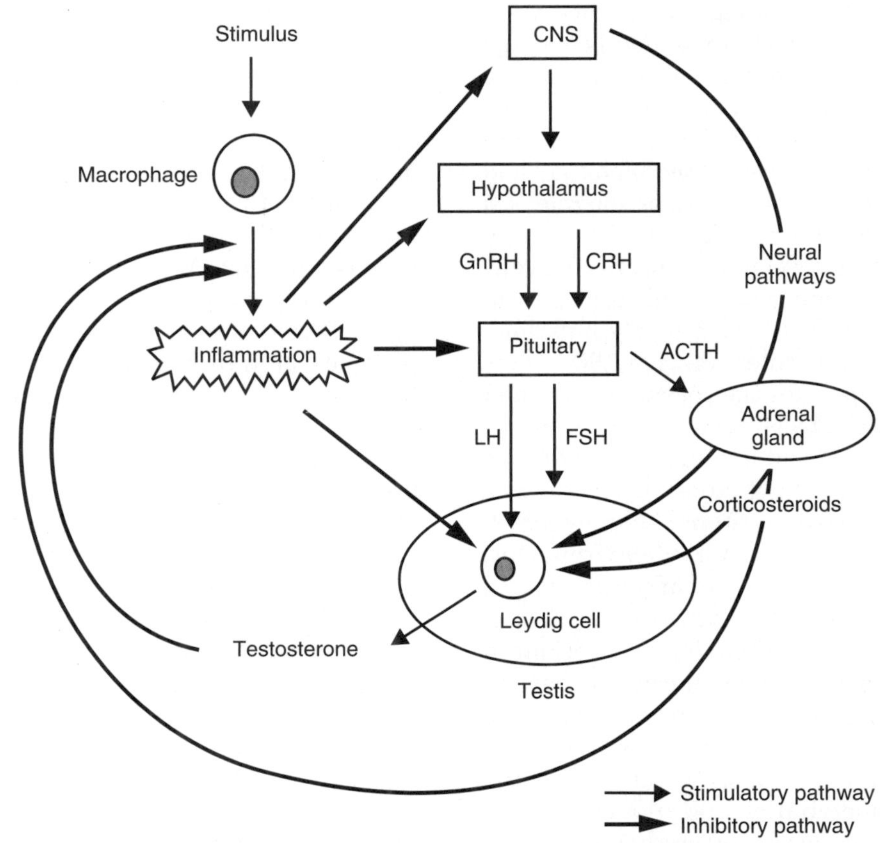

FIG. 4. The interface between the hypothalamic–pituitary–gonadal–adrenal axis and the inflammatory reaction. Regulation of the testes and adrenal glands is under the control of the central nervous system (CNS), which integrates information from the periphery and modulates secretion of the gonadotropin-releasing hormone (GnRH) and corticotropin-releasing hormone (CRH) by the hypothalamus. These hypothalamic peptides stimulate secretion of the gonadotropins (luteinizing hormone [LH] and follicle-stimulating hormone [FSH]) and adrenocorticotropic hormone (ACTH), respectively. LH acts directly on the Leydig cells to stimulate steroidogenesis, whereas FSH controls the development and activity of the Sertoli cells. ACTH stimulates secretion of corticosteroids by the adrenals, which generally exert direct inhibitory effects on Leydig cell steroidogenic activity (180,181,287,792). Activation of macrophage function by an inflammatory stimulus triggers a cascade of events and secretions, which interact with the hypothalamic–pituitary–gonadal axis at all levels, inhibiting gonadotropin secretion and steroidogenesis and stimulating the hypothalamic–pituitary–adrenal axis. There is evidence that inhibitory regulation of steroidogenesis in response to inflammation also involves direct neural pathways from the CNS (519–521,700,1281). As a consequence, inflammation profoundly inhibits the ability of the Leydig cell to produce testosterone. Eventually, steroids produced by the adrenals and testis exert feedback inhibitory effects on the inflammatory process, bringing about the resolution of the inflammation and recovery of testicular testosterone production.

and hypothalamic levels (163,170). Withdrawal of androgens rapidly leads to cessation of spermatogenesis, although it is evident that the levels of intratesticular androgen required to maintain qualitatively normal spermatogenesis are considerably lower than the elevated intratesticular concentration that normally exists (171–173). Consequently, spermatogenesis can tolerate even relatively large declines in testicular androgen production with only minor losses of efficiency. In contrast, peripheral levels of androgens are critical and even small reductions can have profound effects on many androgen-dependent functions, including accessory gland function, secondary sex characteristics, and libido (169). Peripheral androgen levels are a product of both Leydig cell production and testicular vascular function; so that interference with the vasculature of the testis can alter circulating testosterone levels significantly (174).

Aside from the gonadotropins, thyroid hormones have effects on either testicular development or adult function (175–179). The adrenal steroids have important effects on testicular development and adult function (180–183), as well as important immunological significance for the testis because the adrenal steroids are involved in the modulation and resolution of the inflammatory response (Fig. 4).

THE IMMUNE SYSTEM AND ITS ENDOCRINE CONTROL

General Principles

Fundamentally, the immune system provides protection for more complex animals from organisms that seek to exploit opportunities or weaknesses of their host. This protection involves a complex suite of organs, cells, and molecules that allows the host to identify and then eliminate the invading pathogen. In vertebrates, the immune system comprises an innate immune system, which generally recognizes uniquely conserved molecular patterns expressed by various pathogens, and the adaptive (or acquired) immune system, which specifically recognizes molecular patterns that are foreign to the host. Neither system operates in isolation, and many of the cellular and molecular mechanisms involved in each system overlap (Table 1).

TABLE 1. *Characteristics of innate immunity versus adaptive immunity*

Innate immunity	Adaptive immunity
"Immediate" response	Delayed response
Macrophages, granulocytes, NK cells are main effector cells	Antigen-presenting cells (dendritic cells and macrophages), T cells, B cells are involved
Signals mediated by pattern recognition receptors: • Toll-like receptors, NK receptors	Signals mediated by specific molecular receptors: • T-cell receptor, immunoglobulins, major histocompatibility complex
Important in: • Early response to infection and tumors • Development of adaptive responses • Acute inflammation • Acute graft rejection	Important in: • Effective responses to infection and tumors • Autoimmune disease • Chronic inflammatory disease • Extended graft rejection • Immunological memory • Immunization

NK, natural killer.

The cellular components of the immune system are the leukocytes, or white blood cells. These cells and their products circulate continuously through the blood, lymph, and tissues in surveillance and effector modes. The innate immune system principally comprises the mononuclear phagocytes (monocytes and macrophages) and granulocytes or polymorphonuclear cells (neutrophils, eosinophils, basophils, and mast cells), but also involves cells more closely aligned with the adaptive responses (i.e., natural killer [NK] cells and dendritic cells). The cellular components of the adaptive immune system are the lymphocytes (T cells, B cells, and NK cells), and the professional antigen-presenting cells (APCs; dendritic cells and macrophages). In modern immunology, the cells of the immune system and their various functional subsets are primarily identified and even defined by expression of specific antigens, called cluster designation (CD) markers, recognized by well-characterized monoclonal antibodies (184) (Table 2).

Innate Immune Response

Activation of the innate immune response fundamentally involves a family of transmembrane receptors called the Toll-like receptors (TLRs) expressed on wide variety of cell types (Fig. 5). At the time of writing, there are 11 known family members that are able to recognize various viral and bacterial molecular structures (pathogen-associated molecular patterns [PAMPs]) (185). The most relevant to the testis are TLR4, TLR2, and TLR3, which are responsible for recognition of lipopolysaccharide (LPS) and lipoteichoic acid (gram-negative and gram-positive bacteria), peptidoglycan, bacterial lipoproteins, and zymosan (gram-positive bacteria and yeast), and double-stranded RNA (viruses), respectively (186,187). Activation of the TLRs involves a number of accessory

TABLE 2. *Cluster designation (CD) markers relevant to immunology in the male reproductive tract*[a]

Marker	Common or alternative designation(s)	Function
CD1	Ly-38, R3 (CD1d)	Nonclassic MHC, presentation of lipid and glycolipid antigens
CD3	T3, Leu 4	Signaling component of the TCR complex
CD4	L3T4, Leu 3	Coreceptor for MHC class II, component of the TCR complex
CD8	Ly-2, Ly-3, T8, Leu 2	Coreceptor for MHC class I, component of the TCR complex
CD11a	LFA-1, Ly-15, Ly-21	Integrin α chains, adhesion molecules, expressed on leukocytes
CD11b		
CD11c	Mac-1, Ly-40	
CD14	LPS-R	Lipopolysaccharide-binding protein complex coreceptor
CD16	FcγRIII, Ly-17	Immunoglobulin Fc receptor, expressed on subset of NK cells
CD25	Ly-43	Interleukin-2 receptor α chain, marker for activated and regulatory T cells
CD28		Receptor for B7-1 and B7-2, expressed on activated T cells and NK cells
CD40		Costimulatory receptor, tumor necrosis factor receptor family, expressed on APCs
CD45	LCA, Ly-5	Leukocyte common antigen, membrane tyrosine phosphatase, expressed on all leukocytes
CD54	ICAM-1, Ly-47	Intercellular adhesion molecule-1, expressed on activated endothelial cells
CD56	N-CAM	Neural cell adhesion molecule, specific variant expressed on NK cells
CD68		Lysosomal membrane glycoprotein, expressed by dendritic cells, monocytes, and some macrophages
CD80	B7/BB1, B7-1, Ly-53	Costimulatory coreceptor, ligand for CD28, expressed on APCs
CD86	B7-2, Ly-58	Costimulatory coreceptor, ligand for CD28, expressed on APCs
CD95	Fas, APO-1	Receptor for Fas ligand, mediates apoptosis, expressed on activated lymphocytes
CD106	VCAM-1	Vascular cell adhesion molecule-1, expressed on activated endothelial cells
CD154	CD40L	Costimulatory coreceptor, ligand for CD40, expressed on T cells
CD163		Scavenger receptor, expressed on macrophage subset

APC, antigen-presenting cell; MHC, major histocompatibility complex; NK, natural killer; TCR, T-cell receptor.
[a]Note that many designations represent multiple protein members.

proteins, including MD2 and the LPS-binding protein CD14 in the case of TLR4 (188,189). Signaling occurs principally through the interleukin (IL)-1 receptor-associated kinase (IRAK)/tumor necrosis factor receptor-associated factor (TRAF) pathway leading to activation of nuclear factor (NF)-κB, c-*jun* N-terminal kinase (Jnk), and p38 and eventually production of proinflammatory cytokines and mediators, including IL-1, tumor necrosis factor-α (TNF-α), IL-6, IL-8, IL-12, interferon (IFN; Fig. 6), inducible nitric oxide synthase (iNOS), and major histocompatibility complex (MHC) genes (186,187,190,191). More recently, two novel intracellular PAMP receptors belonging to the nucleotide binding site leucine-rich repeat protein family, called Nod1 and Nod2, have been identified (192). These proteins detect bacterial peptidoglycans within the cytosol and signal to the NF-κB and Jnk pathways.

Inflammation is the immune system's earliest response to a pathological challenge and involves major changes in homeostasis: fever, activation of immune cells, increased blood flow at the site of inflammation, and enhanced pain sensitivity (193,194). In addition to specific pathogens, the inflammatory cascade can be triggered by tissue injury and activation of the plasma complement and clotting/fibrinolytic protease pathways. Mononuclear phagocytes (monocytes and macrophages) are by far the most effective promoters of the inflammatory response, but many other immune and somatic cells can be involved. In addition to the proinflammatory cytokines, the main mediators of the inflammatory process are prostaglandins and leukotrienes, neuropeptides such as substance P, vasoactive amines produced by basophils and mast cells (histamine and serotonin), and acute-phase proteins from the liver (e.g., C-reactive protein and serum amyloid A). Local production of chemoattractive cytokines (chemokines), such as IL-8 and monocyte chemoattractant protein-1 (MCP-1), together with upregulation of specific adhesion molecules on the endothelium (E-selectin, P-selectin, intercellular adhesion molecule-1 [ICAM-1], vascular cell adhesion molecule-1 [VCAM-1]) and on the leukocytes (L-selectin, integrins), allow circulating neutrophils, monocytes, and lymphocytes specifically to target and enter the affected tissues (195–198) (Table 2). Resolution of the inflammatory response involves production of antiinflammatory cytokines, such as transforming growth factor-β (TGF-β; Fig. 6), prostaglandins, and late acute-phase proteins, and activation of the hypothalamic–pituitary–adrenal (HPA) axis to produce antiinflammatory glucocorticoids (199–205). Long-term consequences of inflammation include increased tissue fibrosis and persistent alterations in the type and activity of immune cells in the tissue.

The main effector mechanisms of the innate immune system involve production of hydrolytic enzymes (e.g., lysozyme, serprocidins), antimicrobial

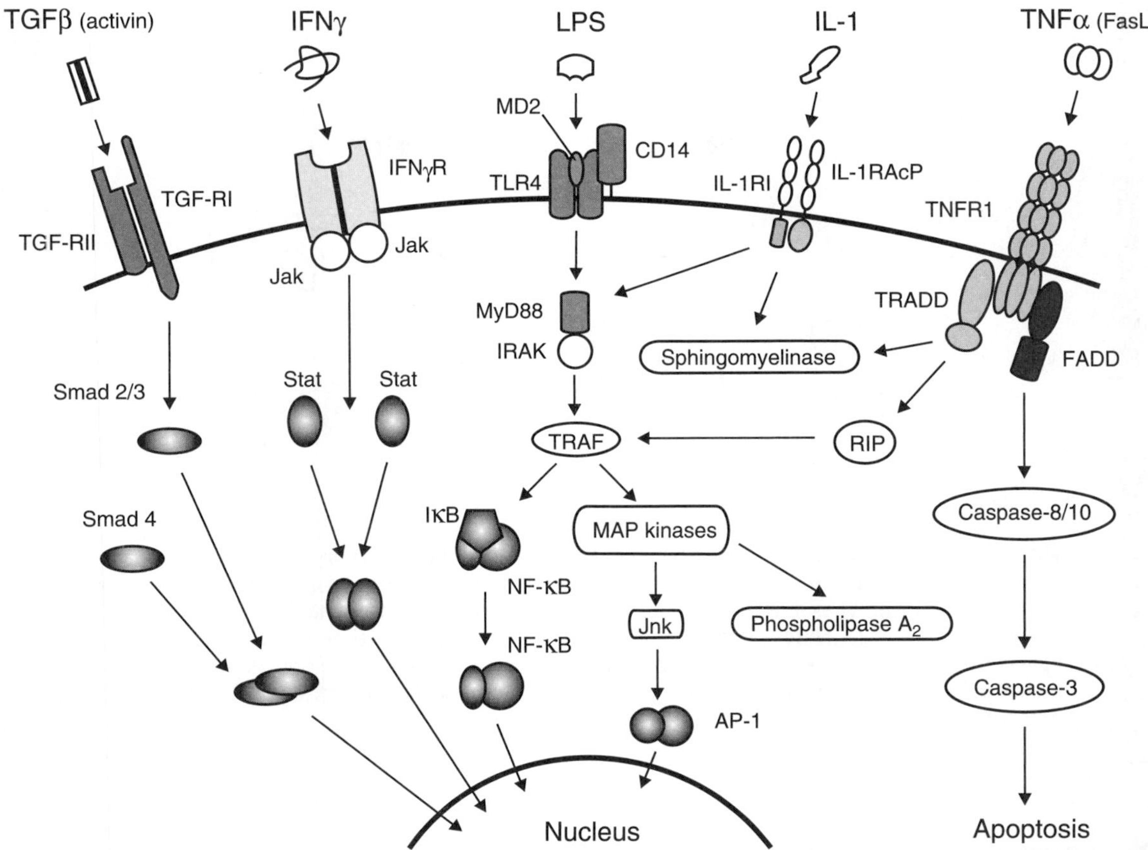

FIG. 5. Cytokine and pathogen signaling in the immune system. Binding of immune ligands to their specific receptors leads to activation of convergent and divergent signaling pathways in responsive cells. Binding of bacterial lipopolysaccharide (LPS) to Toll-like receptor 4 (TLR4), which depends on the presence of the secreted protein MD2 and help from the LPS coreceptor CD14, or binding of interleukin (IL)-1α or IL-1β to the IL-1 receptor (IL-1R) and engagement of the IL-1R acceptor protein (IL-1RAcP), leads to interaction with the adaptor molecule MyD88. Signaling by MyD88 occurs through IL-1R–associated kinases (IRAK) and tumor necrosis factor (TNF) receptor–associated factors (TRAF), leading to degradation of the nuclear factor (NF)-κB repressor protein IκB, and activation of mitogen-activated protein (MAP) kinases. MAP kinases activate multiple downstream events, including production of the transcription factor AP-1 through the c-*jun* N-terminal kinase (Jnk). Depending on which adaptor molecule is engaged, binding of TNF-α to its receptor can lead to activation of TRAF through the TNFR-associated death domain protein (TRADD)/receptor-interacting protein (RIP) pathway, or to the caspase activation cascade and apoptosis through the Fas-associated death domain protein (FADD). Binding of Fas ligand (FasL) to its receptor leads to apoptosis by a similar mechanism. Interferons (INF) act through the Jak-Stat pathway. Binding of transforming growth factor-β (TGF-β) family members causes dimerization of type II and type I signal-transducing receptors (TGF-RII and RI) and activation of the Smad family of transcription factors. The various transcription factors produced translocate to the nucleus to control a number of transcriptionally regulated genes, including IL-1, IL-2, IL-6, IL-8, IL-12, TNF-α, and inducible nitric oxide synthase (iNOS). Other signaling pathways involving bioactive lipid intermediates produced by the action of sphingomyelinase (the ceramide pathway) and phospholipase A_2 (the arachidonic acid metabolite pathways) are also activated. Note that the pathways shown are highly simplified—not all intermediates or potential interactions are depicted here. (See text for details.)

proteins (e.g., complement, defensins), cytotoxic reactive oxygen species (ROS), cytotoxic cytokines such as TNF-α, and the antiviral interferons (194). Moreover, activation of the innate immune system usually leads to the recruitment of the adaptive immune system as well.

Adaptive Immune Response

The adaptive immune system depends on the unique ability of lymphocytes to generate a vast repertoire of cell surface receptors that can bind to almost any conceivable molecular surface (antigen),

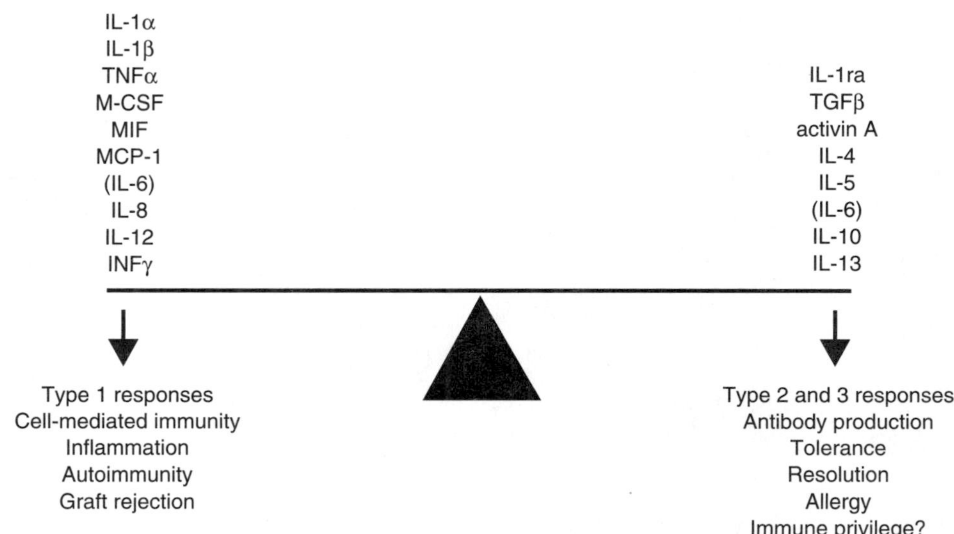

FIG. 6. The cytokine balance. Most cytokines can be designated either proinflammatory or anti-inflammatory/immunoregulatory, depending on their predominant activities. The former group is associated with type 1 responses (cell-mediated immunity, autoimmunity) and the latter with type 2 or type 3 responses (antibody production, allergy, tolerance). Immunological responses involve multiple representatives of one or the other cytokine group, but rarely members of both groups. However, most cytokines, and interleukin-6 in particular, possess both proinflammatory and antiinflammatory properties under different circumstances. (See text for details.)

without having encountered the molecule before. The receptors on the B cells are simply surface-bound immunoglobulins. The core proteins of the T-cell receptor (TCR) are structurally related to the immunoglobulins, but the TCR itself is a complex of a number of interacting surface proteins (206,207). The diversity of these receptors involves extensive genetic rearrangements of the basic immunoglobulin and TCR gene regions, which produce randomly reassembled genes encoding proteins, each with a very specific and unique topography (208–211). As a result of this process, each precursor T cell and B cell expresses a surface receptor that is specific for a single antigenic determinant, and all their clonal daughter cells will express the same receptor and specificity (212,213). B cells interact more or less directly with the antigenic molecule in situ. However, more precise regulation of the immune response involving T cells is determined by proteins of the highly polymorphic MHC, the expression of which is upregulated during inflammation (214,215). APCs process exogenous protein antigens into short antigenic peptides, which are then incorporated into a structural groove on the extracellular surface of the MHC protein complex during its assembly (216,217). The TCR subsequently binds to the antigen–MHC complex on the surface of the APC, leading to the activation and proliferation of the T cell.

In general, circulating T cells express one or other of the coreceptors CD4 and CD8 as part of their TCR, which permits them to recognize antigens associated with MHC class II or MHC class I molecules, respectively (218–220). Antigens are presented to the $CD4^+$ T cells by "professional" APCs that express MHC class II antigens (i.e., dendritic cells, macrophages, and B cells) (221–223). $CD8^+$ T cells are activated by MHC class I antigens, which are expressed by almost all cell types, in contrast to the very restricted expression of MHC class II. Activation of the T cell also requires physical interaction between the APC and T cell involving costimulatory ligand–receptor pairs, particularly CD28:B7 and CD40:CD40L, and production of either type 1 cytokines (IL-2, IL-12, and IFN-γ) or type 2 cytokines (IL-4, IL-5, IL-10, and IL-13) (221,224–226) (Fig. 6). As a result of the complexity of the interactions involved, the activated T cell can have a number of different fates depending on the costimulatory molecules engaged and cytokines produced (Fig. 7). Thus, $CD4^+$ T cells may become type 1 helper (Th1) cells, which direct the development of the cellular immune response involving cytotoxic $CD8^+$ T cells, or type 2 helper (Th2) cells, which promote B-cell and antibody responses. Moreover, absence of appropriate costimulatory molecule interactions or the presence of specific cytokines may lead to T-cell inactivation and deletion or even the generation of regulatory (i.e., suppressor) T cells (4,227–229). The development of B cells into antibody-secreting plasma cells after interaction with antigen requires specific Th2-cell help (230). Once activated, these cells initially secrete multivalent immunoglobulin M (IgM), but

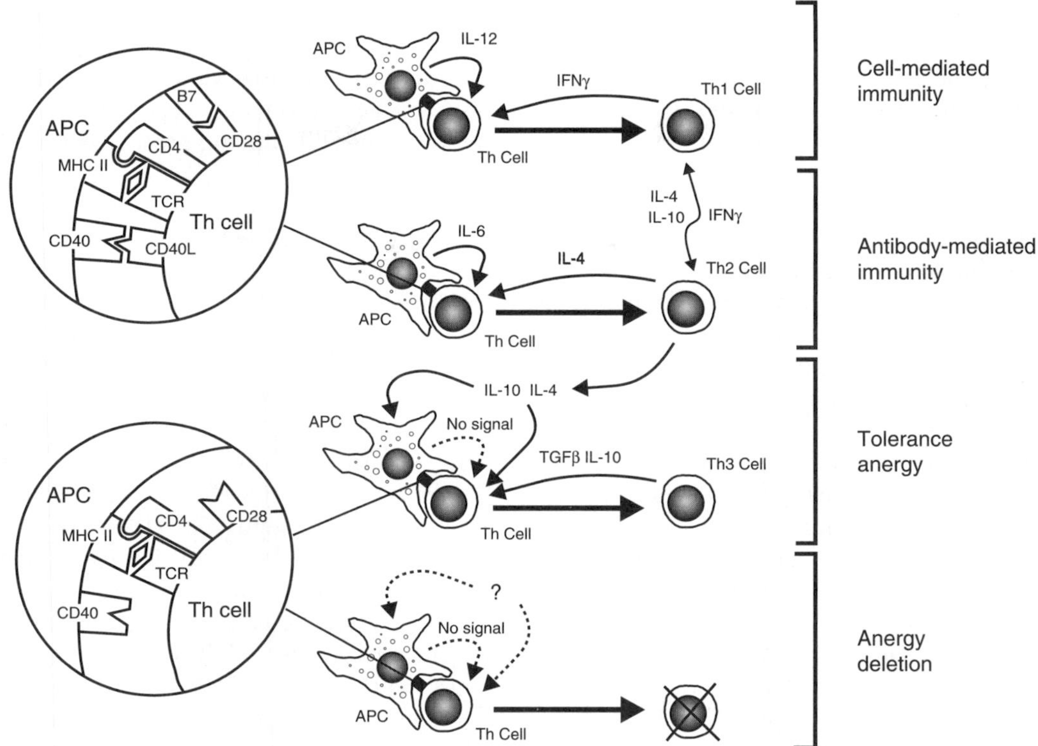

FIG. 7. Activation of the adaptive immune response. Interaction between antigen-presenting cells (APC) and helper T (Th) cells can have different outcomes depending on the cytokine environment and costimulatory surface molecules expressed at the time of interaction. Recognition of the MHC class II–peptide antigen complex by the naive Th cell together with engagement of the B7/CD28 and CD40/CD40L receptor/coreceptor pairs can lead to generation of Th1 cells if type 1 cytokines (interleukin [IL]-12 and interferon [IFN]-γ) are present, or a Th2 response if IL-6 and type 2 cytokines (IL-4 and IL-10) are present (221, 224–226). Presentation of antigen in the absence of costimulation or in the presence of type 2 and "type 3" cytokines such as transforming growth factor (TGF)-β results in a tolerogenic response, involving deletion or inactivation (anergy) of the Th cell, or production of "Th3 cells" producing TGF-β, IL-4, and Il-10 (1161). Similar mechanisms may be responsible for generation of antigen-specific regulatory/suppressor T cells (4,227–229).

the cells gradually mature to produce high-affinity IgG with the same antigenic specificity (209,231). Finally, after resolution of the immune response, at least some activated T- and B-cell clones persist as memory cells, with the result that lymphocyte responses to antigens generally develop much faster on second exposure to the antigen (232–234).

Presentation of nonpeptide antigens by the MHC also can occur. Some nonclassic MHC class I molecules, designated CD1, are able to present lipids and glycolipids to certain T-cell subsets (235,236). NK cells are a third minor subset of cytotoxic lymphocytes that bridge the gap between the innate and adaptive immune systems. These cells are characterized by expression of specific NK receptors that allow them to recognize ligands uniquely expressed by virally transformed cells and tumors (237). In addition, NK cells have the ability to recognize and kill transformed cells lacking MHC class I expression, which is a characteristic of many tumor cells (238,239). Their main role appears to be to provide rapid mobilization against pathogenic challenges in the period before T- and B-cell responses become fully developed. More recently, NK cells have been shown to interact with dendritic cells and regulate the activity of these APCs in either a positive or negative manner (240). Another important minor lymphocyte subset that recently has been identified is the NK T-cell subset (241,242). These cells are generally considered to be T cells with NK activity, rather than NK cells per se, but they also appear to play an important role in immune regulation through production of either IFN-γ or IL-4/IL-10 (243–245). In humans and mice, NK T cells respond to nonpeptide ligands presented by CD1d, and are usually defined by this specific characteristic (242).

The ability of randomly generated T- and B-cell clones to avoid attacking antigens expressed by the host organism depends on an effective cell-editing and inactivating process, known as *tolerance*.

Central tolerance occurs primarily in the thymus, usually with maximal activity around the time of birth, when expression of antigens by the thymic epithelium leads to functional deletion of self-reactive T cells by apoptosis (4–7) (Fig. 7). A similar process appears to operate in the bone marrow for B-cell deletion. There is now compelling evidence that induction of central tolerance involves specific transcription factors that induce promiscuous expression of much of the genome in the thymic epithelium (246). In fact, even sperm-specific antigens have been found to be expressed in thymic epithelial cells under this system (247). However, it is not clear that this mechanism has complete efficacy because autoreactive lymphocytes persist and may expand or become activated later in life, resulting in autoimmune disease (248–251). The mechanisms of peripheral tolerance, which are believed to operate throughout life, are less well understood. Peripheral tolerance appears to involve the deletion or inactivation of autoreactive lymphocytes in the peripheral (secondary) lymphoid tissues, development of regulatory/suppressor lymphocytes, and production of blocking and anti-idiotypic antibodies (252–256).

Autoimmune disease usually represents the failure of existing tolerance. Failure to induce tolerance in the first place lies behind the transplantation rejection response, as the leukocytes of the graft and the recipient react against their respective antigens in a process called *graft-versus-host disease* (257,258). The response is said to be "allogeneic" if it occurs across genetic boundaries within the same species (allograft), whereas "xenogeneic" responses involve a graft and host belonging to different species (xenograft). The extensive polymorphism of the MHC, which is also called the human leukocyte antigen (HLA) complex in the human, is a major contributor to the allogeneic rejection response. However, the nonclassic MHC class I gene, HLA-G, actually appears to be associated with suppression of the adaptive immune response. This molecule is expressed by the human placenta and has been implicated in protecting the fetal alloantigens from the maternal immune system (259–261).

It is generally accepted that the immune response is initiated in the secondary lymphoid tissues, particularly in the follicles and germinal centers of the draining lymph nodes and spleen or lymphoepithelial aggregates of mucosal tissues, where APCs come into contact with large numbers of T and B cells (262–266). Thus, for an immune response to occur against a specific antigen, the antigen normally must first travel through the lymphatics, either by simple diffusion or carried by an APC, to one of these organs. Most immunologists do not consider that primary immune activation can occur within "nonlymphoid" tissues, or at least that any response generated there is likely to be vigorous enough to lead to full participation by the immune system. However, there is some evidence that primary activation can occur outside the lymphoid tissues, especially during graft rejection (267–269).

Mucosal Immunity

Mucosal immunity concerns immunity at the interface between the external and internal environments and is very relevant to understanding the immunology of the male reproductive tract. Many of the mechanisms involved in mucosal immune responses are very similar to those of other tissues and lymphoid organs. Discrete lymphoepithelial aggregates (mucosa-associated lymphoepithelial tissue [MALT]) in mucosal surface tissues allow interaction between APCs and lymphocytes to facilitate local immune responses (270,271). One key difference, however, is the isotype of the antibodies involved at mucosal surfaces. To cross epithelial barriers, a unique mechanism for transport of dimers of IgA through the epithelial cells involving a protein called *secretory component* is necessary (272–274). Thus, IgA is usually the major immunoglobulin in secretions of mucosal surfaces. In mucosa, the epithelial cells also play an important role in host defense by producing many molecules associated with innate and adaptive immunity, including defensins, regulatory cytokines, TLRs, MHC antigens, leukocyte adhesion molecules, and costimulatory molecules (275–280).

There is strong evidentiary support for inclusion of the male reproductive tract as a constituent of the mucosal immune system (151,281,282). As far as the male reproductive tract is concerned, both IgA and serum-derived immunoglobulins (IgG) appear in semen and bound to the sperm of men with autoimmune infertility (8,9,11,283,284). Thus, it would be expected that the epithelial cells of the male genital tract might play a similar active role in regulating local immune responses. Although discrete MALTs have never been described in the male reproductive tract, the existence of similar or analogous structures cannot be entirely excluded.

Endocrine Regulation of Immunity

Glucocorticoid Control of Inflammation and Immunity

Glucocorticoids have a profound inhibitory effect on immunity, and are widely used in the treatment of autoimmune and inflammatory disease. During inflammation, production of proinflammatory cytokines activates the HPA axis, leading to increased secretion of glucocorticoids by the adrenal cortex (202,203). In turn, glucocorticoids acting through the ubiquitously expressed glucocorticoid receptor exert a negative feedback on the inflammatory response, primarily by repression of NF-κB, thereby inhibiting the production and actions of the proinflammatory cytokines, suppressing adhesion molecule expression, and stimulating the production of the antiinflammatory cytokines, IL-4 and IL-10 (204,285). In fact, glucocorticoids suppress innate and acquired immunity at many levels through effects on the function of nearly all immune cell types and are an essential physiological regulator of the immune response (Fig. 4). The particular relevance of this feedback system for male reproductive function is the fact that glucocorticoids also inhibit Leydig cell steroidogenesis through actions at all levels of the hypothalamic-pituitary–testicular (HPT) axis (286-289), in particular by direct inhibitory effects on the Leydig cells (180,181,290), with the result that activation of the HPA axis during inflammation has a negative impact on testicular function.

Effects of Gender on Immune Function

There is absolutely no doubt that the male reproductive system exerts a profound inhibitory effect on the development and function of the immune system. Compared with females, males have lower serum immunoglobulin levels, reduced cellular immunity, and less effective responses to antigenic challenge, whereas autoimmune diseases such as rheumatoid arthritis and lupus are generally far less common and less severe in men than in women (87,291–299). The fact that this reduced immunity is due to products secreted by the testis has been established in numerous ablation and steroid replacement studies, confirming that testosterone inhibits many immune parameters, including the size of the thymus and other immune tissues, graft rejection, and resistance to infection (90–93,96,99,300–304). After much effort on the part of the early researchers, however, it soon became apparent that functional androgen receptors of the classic (i.e., cytoplasmic) type were expressed on the stromal and epithelial cells of the immune tissues rather than on the immune cells themselves (92,305–312). A more recent study using a highly sensitive reverse transcriptase-polymerase chain reaction (RT-PCR) technique did detect androgen receptor messenger RNA (mRNA) in purified mouse spleen macrophages and T cells, but failed to confirm that the androgen receptor protein also was expressed (313). In all, the data suggested that the effects of androgens on immunity were being exerted indirectly through actions on the immune tissues, rather than through direct effects on the circulating leukocytes.

Evidence has accumulated, however, that steroids can also interact with membrane-bound G-protein–coupled receptors to trigger nongenomic responses (314,315). Studies have shown that androgens can alter [Ca] fluxes in lymphocytes and macrophages through such membrane-mediated interactions (310,311,316). This signaling affects gene expression and function in the target cells (317), and although the physiological significance of these discoveries is yet to be fully assessed, the existence of such receptors certainly abrogates the need for classic steroid receptors to be expressed in a responding cell type. These studies resurrect the intriguing possibility that androgens, and for that matter other testicular steroids such as estrogens, might directly modify the functions of immune cells in the testis and adjacent draining lymph nodes.

Apart from the sex steroids, there is evidence that protein hormones and neuropeptides produced by the testis also have direct effects on immunity. The pro-opiomelanocortin gene-derived peptides, β-endorphin and α-melanocyte–stimulating hormone, which are produced by the Leydig cells and macrophages of the testis (318–321), have both stimulatory and inhibitory effects on lymphocyte and macrophage activity (322–328). These and other neuropeptides with known immunoregulatory actions produced by the testis, such as GnRH, could potentially interact with their specific receptors expressed by cells of the immune system in the testis environment (329–332). Moreover, the testis is a source of protein hormones and cytokines with immunoregulatory activity, most importantly members of the immunosuppressive/antiinflammatory TGF-β family, including TGF-β 1 through 3 and the activin-related proteins (333–335). There is no doubt that exposure of circulating lymphocytes to these molecules within the testis, or their secretion into the blood and lymph, plays a significant role in both local immunoregulation and sexual dimorphism of immune function.

IMMUNE CELLS IN THE MALE REPRODUCTIVE TRACT

Far from being a site where immune cells are restricted entry, even casual observation quickly establishes that immune cells are a characteristic feature of the testicular tissue (Table 3). Moreover, macrophages, lymphocytes, and granulocytes (mast cells and eosinophils) are ubiquitous throughout the male tract, although there are variations in number and type within the different tissues and from species to species.

TABLE 3. *Quantification of immune cells in the normal testis of adult rats and humans*

Cell type	Rat testis ($\times 10^6$/g tissue)	Human testis ($\times 10^6$/g tissue)
Macrophages	5–10	10–25[a]
Dendritic cells	0.2–0.3	ND
T cells	1–2	1.4–2.4
CD8+ T cells	0.6–1.8	ND
CD4+ T cells	0.2–0.3	ND
NK cells	0.6–1.0	1.0–2.8

ND, no quantitative data available.

Data were largely obtained from stereological analysis of testes from Sprague-Dawley rats (130,131,397,431) and from adult human testes with normal spermatogenesis (M. P. Hedger, unpublished data). The study of Vergouwen and colleagues (357) indicates that adult CBA/P mouse testes contain approximately 2 to 4×10^6 macrophages/g tissue, but there are as yet no definitive quantitative studies of other leukocyte subsets in the mouse testis.

[a]Upper limit calculated from data obtained by Frungieri and colleagues (360) using a well-characterized and reliable monoclonal antibody against human CD68. The observation that macrophage numbers in the normal human testis are at least as large, if not larger, than those found in either the rat or mouse testis is consistent with observations in several earlier nonquantitative studies using other macrophage markers (128,388,430,433).

The Testicular Macrophages

Macrophages and dendritic cells are found in every tissue of the body. Both cell types are derived from the circulating monocyte pool and are directed along one developmental pathway or the other by the influence of specific cytokines (336–340), although dendritic cells of putative lymphoid origin have been identified more recently (341,342). Macrophages are extremely heterogeneous in appearance and function, and their characteristics are primarily dictated by the tissue site in which they are found (343,344). At first glance, the microglial cells of the brain and the Kupffer cells of the liver have little in common, but both cell types are macrophages and share many common features: they express markers of the mononuclear phagocyte lineage, are actively mobile, phagocytic, and cytotoxic, and are involved in tissue restructuring and antigen presentation to CD4+ T cells. However, in spite of these shared characteristics, macrophages at different sites may produce very different patterns of cytokines, proteases, and other bioactive secretions (343,345). In fact, macrophages are almost certainly the most versatile and diverse single cell type in the entire body.

The testicular macrophages themselves have been extensively studied, initially by Miller and colleagues (346–348) and subsequently by Hutson and colleagues (349–354). These cells were found to share all the classic characteristics of macrophages in other sites, but to possess testis-specific features as well. Testicular macrophages have been most extensively studied in the rat, and to a much lesser extent in the mouse, with relatively limited investigations in other species. Because these cells appear to be a fixture of the testes of all mammalian species, it has generally been assumed that the rat macrophages are representative. In fact, some data suggest that there are significant differences even between rat and mouse testis macrophages—for example, in resting MHC class II expression and the ability to produce proinflammatory cytokines in response to an inflammatory stimulus. Further study on the macrophages of other species, in particular the human, is clearly necessary. The following discussion of this complex cell type in the testis necessarily reflects the rather narrow knowledge base that is available.

Number, Phenotype, and Heterogeneity

There are substantial populations of resident or tissue-fixed macrophages in both the rat and mouse testes and estimates of the number of these cells by different laboratories generally agree (130,355–358). Other species with large numbers of testicular macrophages include the guinea pig, hamster, horse, bull, and human (359–362). In the boar, which has a relatively sparse testicular interstitial connective tissue and very large numbers of Leydig cells, macrophages appear to represent a smaller proportion of total interstitial cells (363). Curiously, in spite of the general similarity of the testicular interstitial tissue and lymphatic organization in this species with that of the bull or human, the ram testis appears to possess only small numbers of recognizable macrophages (102,129).

In the rat and mouse, the ratio of macrophages to Leydig cells appears to be relatively fixed at approximately one macrophage to every four or five Leydig cells (358,364), and macrophages display a very close physical and functional relationship with the Leydig cell clusters. Ultrastructural studies have established the existence of highly specialized cytoplasmic interdigitations linking the two cell types, indicating the potential for direct exchange of information and material (121,286,347,349), whereas macrophages and Leydig cells undergo parallel alterations in morphology and cytoplasmic volume in experimental models of cryptorchidism and vasectomy in the adult rat (365–367). During aging, testicular macrophages retain their close morphological association with Leydig cells, but the cytoplasmic interdigitations are lost (368). Aging testicular macrophages also acquire lipofuscin granules similar to those observed in aged Leydig cells (368).

Early interest in testicular macrophages arose from studies on the accumulation of nondigestible tracers and radionuclides in the testicular interstitial tissue (346,369–371). In spite of early speculation that these cells might have a nonhematopoietic origin, numerous studies since have established that the testicular resident macrophages possess all the structural and functional characteristics of macrophages found in other connective tissues (124). They display the characteristic nuclear and cytoplasmic morphology of the mononuclear phagocyte lineage, are actively phagocytic, bactericidal, and adherent in culture, and they express many macrophage-specific enzymes, cytokine receptors, and surface antigen markers (130,347,348,352,364,372,373). Early studies of these cells in vitro by Hutson and colleagues also indicated that the macrophages possess a number of testis-specific features, most notably the ability to respond directly to stimulation by FSH (350,351,354,374). This apparent action of FSH was later shown by the same group to be due to the relatively impure nature of the FSH preparations used in earlier studies, and subsequent studies with pure recombinant FSH did not support a direct effect of FSH on the testicular macrophage (375). Nonetheless, the concept of a testis-specific phenotype for these macrophages has survived. In addition to presumptive roles in connective tissue remodeling, phagocytosis of cellular debris, and maintenance of innate immunity in the testis, studies by several laboratories suggest that these cells play an important role in local immunoregulation (376–378). More surprising, however, was the discovery that macrophages may play an essential role in normal Leydig cell development and in maintaining Leydig cell steroidogenic function in the adult (307,379–385). Thus, it seems that the testicular macrophages have an integral role in testicular biology that goes beyond the anticipated functions of "normal" tissue-fixed macrophages.

The immunological functions of the testicular macrophage population have been investigated indirectly in a number of studies, but the data are not entirely consistent and many questions still remain. Although several studies have shown that rat macrophages express MHC class II molecules under normal conditions, indicating a capacity for antigen presentation (49,129,372), Tung and colleagues reported that strong MHC class II expression in the normal mouse testis was restricted to cells located around the rete testis and was upregulated in the interstitial tissue only after immunization with testicular antigens in the presence of adjuvant (386,387). This seems to indicate a fundamental difference in immunological activity between the testicular macrophages of the rat and mouse. Significantly, human testicular macrophages appear to be more like those of the rat, with significant MHC class II expression found throughout interstitial tissue samples from normal men (128,388). In the mouse testis, a lack of expression of the essential costimulatory molecules for T-cell activation, B7-1 (CD80) and B7-2 (CD86), also has been reported (389). Moreover, studies of the ability of rat testicular macrophages to provide help for T-cell activation in vitro indicated that these cells are deficient in costimulatory activity (376). With respect to inflammation, selective depletion of testicular macrophages with a macrophage-specific toxin exacerbated the testicular inflammatory response induced by hyperstimulation of the Leydig cells with pathological doses of the LH agonist, human chorionic gonadotropin (hCG), in rats (377). Finally, production of several key inflammatory mediators, including IL-1β, TNF-α, IL-6, iNOS, and macrophage migration inhibitory factor (MIF), by rat testicular macrophages in response to stimulation by LPS is relatively poor or absent (376,390–394). Elsewhere, macrophages displaying such properties have been termed "M2 macrophages" (analogous to "Th2 cells" or "type 2 cytokines"), and are characteristic of tumors and late-stage inflammation responses (345). It remains to be confirmed that this is a feature of testicular macrophages from other species, however, because studies on inflammation in the mouse testis have not been as supportive of this concept. Murine macrophages have been shown to express IL-1β mRNA and produce bioactive TNF-α at fairly significant levels upon LPS stimulation (395,396). Thus, although the balance of data suggests that testicular macrophages possess an antiinflammatory phenotype, more study is required to verify and establish this concept.

Quantitative studies of macrophage numbers using specific antigenic markers in the rat testis suggest that testicular macrophages do not represent a functionally homogeneous population (130,392,397). The majority of testicular macrophages in the rat express a molecule that is specific to tissue-fixed macrophages in most nonlymphoid tissues, recognized by the monoclonal antibody ED2 (398,399). The antigen for the ED2 antibody has been identified as CD163, a member of the group B scavenger receptor cysteine-rich protein superfamily (400). This receptor has been implicated in the stimulation of IL-10 production by monocytes and macrophages after the endocytosis of hemoglobin–haptoglobin complexes (401). However, a significant subset of testicular macrophages (approximately 15% to 20% of the total) do not express this marker. The latter cells can be identified by expression of the lysosomal antigen, CD68, recognized by the antibody ED1, which suggests they may be intratesticular monocytes or newly arrived macrophages that have yet to become resident (130,378,387,392). Moreover, approximately half of the $ED2^+$ subset in the rat testis does not react with ED1.

These data indicate the existence of several populations of macrophages in the rat testis, putatively representing different stages of development or functional states (378) (Fig. 8). There is also evidence to suggest that this heterogeneity may have a functional correlate: compared with the ED1⁻ macrophages, the ED1⁺ subset possesses a distinctly smaller nuclear diameter and displays significant levels of expression of iNOS in both normal and LPS-stimulated rats (392). Similarly, expression of IL-1β protein after LPS treatment appears to be confined to a relatively small subset of rat testicular macrophages (394), although it is yet to be shown that this is the same subset that expresses iNOS (Fig. 8). One publication has provided evidence of heterogeneity in the ability to produce TGF-β in vitro among macrophages isolated from mouse testes (402), whereas in the human testis, Frungieri and colleagues found that CD68⁺ macrophages displayed different levels of expression of CD163 (360). All these observations are consistent with the hypothesis that the testis contains subsets of macrophages with varying proinflammatory capacities. If this is indeed the case, it is also important to recognize that most investigations on testicular macrophages to date have used mixed populations to study their function. Moreover, these studies highlight the importance of using multiple markers to identify the entire macrophage population of the testis.

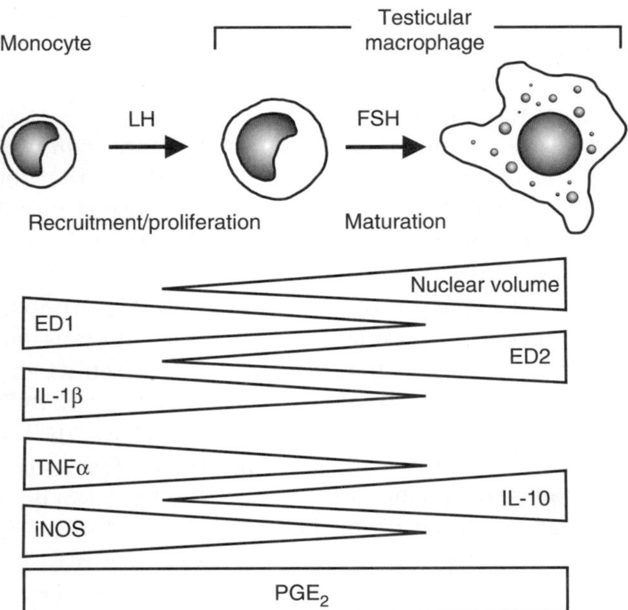

FIG. 8. Maturation of the resident macrophage population of the rat testis. Macrophages in the rat testis are heterogeneous, corresponding to different stages of maturation from circulating monocytes through to a distinct testicular resident macrophage phenotype (378). This testicular phenotype is characterized by an increased nuclear and cytoplasmic volume, loss of the CD68 marker recognized by antibody ED1, upregulation of the resident macrophage surface marker ED2 (CD163), loss of ability to produce several proinflammatory mediators, and upregulation of the ability to produce the immunoregulatory cytokine, interleukin-10. Recruitment of macrophages to the testis is under the control of luteinizing hormone (LH), acting through the Leydig cells (130,397,406,411,412,414), whereas maturation to the mature testicular phenotype appears to be follicle-stimulating hormone (FSH) dependent, indicating regulation by the Sertoli cells (410,413).

In spite of the common lineage, dendritic cells are morphologically distinct from macrophages and lack the efficient phagocytic and cell-killing capabilities of macrophages, although they are much more effective as APCs (221). Cells that express dendritic cell markers or possess dendritic cell morphology have been observed in testes of the mouse (403,404), rat (49), and human (388,405). The functional characteristics of these cells in the testis have yet to be studied in any detail, although it is highly probable that these cells are involved in directing immune responses in the testis and adjacent lymph nodes.

Recruitment and Regulation

In the "prepubertal" testis there are relatively few macrophages, but macrophage numbers expand dramatically during development, coinciding with the proliferation of the adult Leydig cell population and the appearance of the meiotic spermatocytes (355–357, 406–409). Indeed, the development of both the testicular macrophage and the adult Leydig cell populations appears to be interdependent. There is considerable evidence that macrophages influence adult Leydig cell development and function (307,379,381,382,384). Conversely, studies from a number of different groups have shown that the accumulation of macrophages during testicular development and maintenance of their numbers in the adult depend on the action of pituitary LH and, more specifically, the Leydig cells themselves (130,397,406,410–415). Depletion of the Leydig cells or suppression of their activity in the adult testis by various means causes a progressive decline in the number of testicular macrophages (130,397,410,411,413,415,416). There is some evidence that macrophages in the testis and other reproductive tissues may possess LH receptors (417,418), but the balance of the data suggests that this gonadotropin is not directly responsible for regulating testicular macrophage proliferation. Testicular macrophage depletion occurs in models in which serum LH levels are either increased, after treatment with the Leydig cell toxin, ethane dimethane sulfonate (EDS) (130), or decreased (i.e., hypophysectomy, GnRH immunization, suppression of endogenous androgen with subcutaneous testosterone implants) (410,411,413,415,416), clearly indicating that it is the Leydig cell that is responsible, rather than a direct action of LH itself.

The actual mechanisms involved in maintaining the testicular macrophage population are unknown. Complete depletion of spermatogenic cells by cryptorchidism has absolutely no effect on testicular macrophage numbers in the rat (397), suggesting that regulation does not involve the seminiferous epithelium. Moreover, it does not appear that androgens are directly involved (397,406,411). It is likely that direct contact with the Leydig cell membrane (347,349) or nonandrogenic products of the Leydig cells may be responsible. Studies have clearly indicated that normal development of the testicular macrophage population involves the macrophage growth factor, macrophage colony-stimulating factor (M-CSF or CSF-1), as in other tissues (373,384,419,420). Production of the chemoattractant cytokine MIF also may be involved in recruitment of circulating monocytes to the testis (393,397). This chemokine is constitutively expressed by the Leydig cells and is not under acute gonadotropin control (393,397,421).

In most tissues, macrophages are largely derived and sustained from the circulating monocytes, and resident macrophages display minimal proliferation in the tissue itself under normal conditions (344,422). As already described, there appears to be a significant population of "monocyte-like" macrophages in the testis at any one time, suggesting that there is constant recruitment of these cells from the monocyte pool (378). On the other hand, there is clear evidence that the testicular resident macrophages may undergo active proliferation by mitosis, at least under certain circumstances: during the early inflammatory phase after destruction of the Leydig cells by EDS (423) and during testis development in rats (406) and mice (424). However, it is still not clear whether the resident macrophage population of the testis, once established, is maintained by recruitment of new monocytes, or by cell proliferation, or both. Although there have been a number of kinetic studies on the turnover and replacement of macrophages in several other tissues (425–429), no such studies have been carried out in the testis to date. Based on the extended persistence of radionuclides in rodent, canine, and human testicular macrophages, however, it appears that resident macrophages do not leave the testis alive (346,370,371).

Although the reported ability of testicular macrophages to respond directly to FSH has been shown to be due to an experimental artifact (375), a stereological study on macrophage recruitment to the testis in GnRH-immunized rats given recombinant FSH replacement has shown that FSH does stimulate an increase in the mean nuclear diameter of testicular macrophages (410). This response indicates an effect of the gonadotropin on macrophage activity in the testis. This effect is almost certainly mediated by the Sertoli cell, the only testicular cell type able to respond directly to FSH. Thus, it appears that although Leydig cells are responsible for recruiting and maintaining the testicular macrophage population, the Sertoli cell may play a role in directing at least some of the testis-specific functions of these cells. It is not unreasonable to speculate that these two somatic cells act together to recruit and modify the function of the macrophages, thereby bringing about the unique resident macrophage phenotype found in this organ (Fig. 8).

Lymphocytes and Polymorphonuclear Cells in the Testis

Lymphocytes

Although it is frequently stated by some research groups that the normal testis does not contain lymphocytes, this is most definitely incorrect. Lymphocytes are relatively prominent in and adjacent to the epithelium of the rete testis and are sparsely distributed throughout the testicular interstitial tissue in all species so far studied (126–131,430). In the rat and human testis, lymphocytes actually represent approximately 10% to 20% of the total leukocyte population, although the proportion of these cells in the normal mouse testis appears to be somewhat lower (130,431) (M. P. Hedger, unpublished data; Table 3). Considerable differences may exist among different strains of animals and between animals raised under different levels of exposure to environmental pathogens because this cell population is significantly affected by differences in genetic background and past immunological events. In general, there has been a tendency to overlook or underestimate the potential importance of these cells in the normal testis and, consequently, their functional characterization has received relatively little attention. T cells (both $CD4^+$ and $CD8^+$) and NK cells, but not B cells, comprise the specific subsets of lymphocytes that have been described in the normal rat, mouse, and human testis (126,128–131,431). In the rat and human testis, at least, $CD8^+$ T cells and NK cells predominate (126,130,131) (M. P. Hedger, unpublished data).

In fact, it is not surprising that T cells are found in the testis because these cells circulate throughout all tissues as part of their surveillance function. It is widely believed that activated T cells tend to recirculate through tissues where they initially encountered an antigen (263,432), implying that at least some of the T cells in the testis may be specific for testicular autoantigens. Alternatively, some of these cells may represent T cells that have been exposed to exogenous antigens within the testis environs during past infections. There is clear evidence that lymphocyte numbers increase in the testes of men with infertility and sperm autoimmunity (127,433), and our own studies indicate that lymphocyte numbers in the rat testis gradually increase with age while macrophage numbers remain relatively constant (M. P. Hedger, unpublished data). Certainly, the phenotype of the majority of the T cells of the rat testis is consistent with activated or memory T cells (131).

Particularly interesting is the observation that there are significant numbers of cells expressing NK markers in the testis of rats (131), mice, and humans (M. P. Hedger, unpublished data). After isolation of these cells from the rat testis, we have been able to show that these cells have specific NK activity against transformed cells in vitro (M. P. Hedger, unpublished data). In contrast to T cells, which are part of the adaptive immune system, NK cells play an important role in innate immunity and therefore may be important for protection of the testis against viruses and tumors. However, NK cells also are involved in immunoregulation through their ability to direct the activity of dendritic cells (240). Equally significant is the fact that a proportion of the cells expressing NK markers in the rat and mouse are actually NK T cells (M. P. Hedger, unpublished data), because this cell subset has been shown to play an essential role in controlling graft rejection responses (243–245). Further study of these cells is vital, and certain to produce very important insights into the unique immunological environment of the testis.

As opposed to the situation in the normal testis, the number of mononuclear cells (lymphocytes and macrophages) increases dramatically in the testicular interstitial tissue during testicular infection, experimental autoimmune or allergic orchitis (EAO), leukemic relapse, and testicular tumorigenesis (434–439). This infiltration is usually associated with failure of spermatogenesis and the immune cells may eventually breach the junctions of the blood–testis barrier and invade the seminiferous epithelium. In the absence of any overt immune event, lymphocytes and macrophages are rarely observed in the epithelium or lumen of the seminiferous tubules, although lymphocytes appear to be able to cross the epithelium of the rete testis (125–127).

Mast Cells and Eosinophils

The distribution of granulocytes in the testis appears to be quite species specific (123,440). In the testes of the rat, mouse, dog, cat, bull, and deer, both types are largely absent from the testicular parenchyma, but are frequently associated with blood vessels in the testicular capsule. In contrast, mast cells are found throughout the interstitial tissue in equine and human testes (123,133,441,442), and both mast cells and eosinophils are present in porcine species (123). The functional significance of these cells in the testis is not known, but it is reasonable to assume that they play a role in local innate immunity and possibly in the fine control of testicular blood flow. In humans, testicular mast cell numbers increase in various types of male infertility (442–444), but decline with advancing age (441). In contrast to mast cells and eosinophils, neutrophils are found in the testis only in conditions of testicular inflammation or damage (392,437,445-448).

In the adult rat, mast cells are normally confined to the subcapsular region, but these cells proliferate dramatically throughout the testicular parenchyma after ablation of the Leydig cells by EDS (130,449–451). The degree of proliferation appears to be under the control of the Leydig cells, suggesting that these cells produce an inhibitor of mast cell activity (130,451). However, the increased mast cell numbers persist even after the Leydig cells recover and testis function returns to normal after EDS treatment. The specific mechanisms responsible are unknown. The species distribution of mast cells and eosinophils suggests some relationship with the level of aromatase activity of the Leydig cells (452-454), and neonatal estrogen treatment increases mast cell numbers in the rat testis (451,455,456). Stem cell factor, which is a growth factor for both mast cells and Leydig cells that is produced by the Sertoli cell (457,458), also may be involved. Although eosinophils are not a feature of the normal rat or mouse testis, we have observed these cells occasionally in the testes of GnRH-immunized rats (410). It appears that the regulatory mechanisms for the two different granulocytes in the testis may be separate and distinct.

Immune Cells of the Epididymis

In contrast to the seminiferous epithelium, macrophages and lymphocytes are frequently observed in the epithelium as well as the interstitial tissue of the epididymis (125–127,152–157,459–464). The published quantitative leukocyte subset staining data for this tissue is somewhat confused because some studies have not clearly differentiated between lymphocytes and macrophages, which share the CD4 marker in the rat and human (465,466). Studies in the mouse, however, indicate that macrophages are the main leukocyte present in the epididymis, located chiefly in the interstitial and peritubular regions, and that there appears to be a slight preponderance of $CD4^+$ T cells over $CD8^+$ T cells (153,154,157). As in the testis, the development of the epididymal macrophages is M-CSF dependent (467). MHC class II is expressed by the interstitial tissue macrophages, but not by those in the epididymal epithelium (154,157). The intraepithelial lymphocytes in the epididymis are predominantly CD8+ (155,156,463,464). There appear to be no major regional differences in interstitial or epithelial leukocytes along the length of the normal adult epididymis (154–157). The lymphatics of the epididymis have been studied in detail only in the rat, but these appear to be unremarkable (32,34,468).

The very different distribution of leukocytes in the epididymis compared with the testis clearly indicates the existence of a different immunological environment (Fig. 9). Obviously, separation of antigens and immune cells is not the principal mechanism protecting sperm from immunological attack in this organ. In fact, it has been speculated that the macrophages in the genital tract may be responsible for phagocytosis of senescent and excess sperm, although the evidence for this is equivocal (461,469,470). The immunological functions of the macrophages and lymphocytes in the epididymis are almost completely unknown and represent a potentially fertile area for future research. Not surprisingly, all epididymal leukocyte subsets increase during epididymal infection, in orchitis, or after vasectomy, implicating the epididymis in the etiology of the autoimmune response that occurs in these conditions (153,155,460).

Immune Cells of the Vas Deferens, Accessory Glands, and Urethra

Sperm are normally present in the vas deferens only transiently at the time of ejaculation, but this structure plays a special role in our understanding of autoimmunity in the male tract because of the widespread use of vasectomy as a contraceptive method. Sperm antibodies develop in approximately 70% of all vasectomized men, and in many of these men the antibodies are persistent, thereby frustrating attempts to restore fertility in reversal patients (471–477). Sperm antibodies are also associated with obstructive azoospermia and congenital absence of the epididymis or vas, and the incidence of antibody formation is directly related to the distance of the lesion from the testis (478). Similar antibody responses occur in experimental animal models (460,479–482). These observations clearly indicate that obstruction or damage to the epididymis and vas is a potent inducer of sperm autoimmunity, but that proximity to the testis reduces the severity of the response. There is evidence from animal models and patients that vasectomy can have deleterious effects on the testis and epididymis as well, and that at least some of these effects may have an immunological basis (433,460,474,479,480,483–492). However, the severity of these responses to vasectomy appears to be highly specific to both species and genetic background.

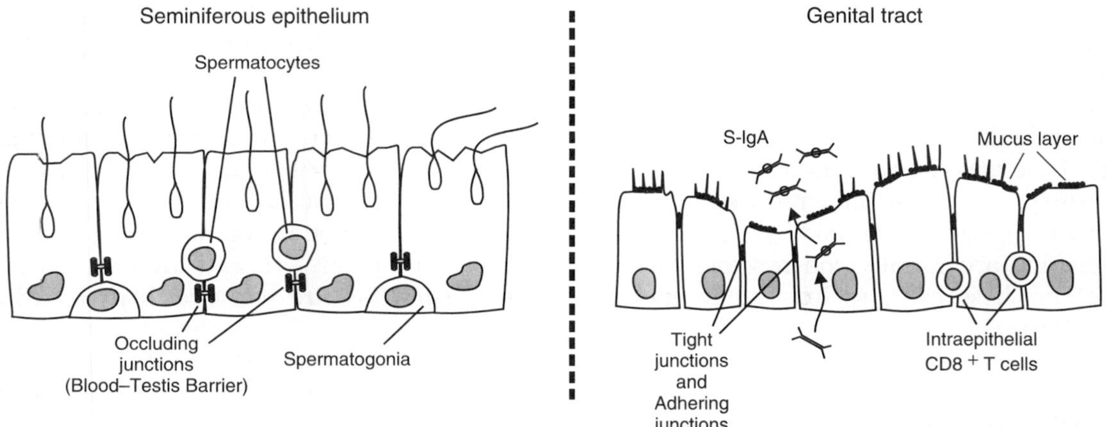

FIG. 9. Comparison of immunoregulation in the testis and male genital tract. Immunoregulation in the testis is primarily associated with the occluding junctional complexes comprising the blood–testis barrier and the presence of immunoregulatory cells in the intertubular tissue, most notably macrophages with an immunosuppressive phenotype. Immunoglobulins, complement components, and immune cells are unable to cross the epithelium in the testis, but protection from infection may be provided by innate immune functions of the testicular macrophages, intertubular lymphocytes, and the Sertoli cells themselves. The remainder of the genital tract, from the rete testis to the urethra, appears to belong to the common mucosal immune system. Immune protection here is probably provided by mucoid secretions, secretory immunoglobulin A (S-IgA) and intraepithelial CD8+ T cells possessing cytotoxic activities. The latter cells, together with immunoregulatory cells (CD4+ T cells, natural killer [NK] cells, and NK T cells) and resident macrophages in the interstitial tissue, may contribute to local inhibition of antigen-specific immune responses and protection of spermatogenic cells throughout the male reproductive tract. (See text for details.)

Intraepithelial lymphocytes, principally CD8+ T cells, are found throughout the vas deferens, seminal vesicles, prostate, and urethra (125,127,282,493–497) (Fig. 9). Macrophages appear to be relatively rare in the epithelia, but they are found in the stroma of these tissues even under noninflammatory conditions, where they depend on M-CSF for their normal development (467,493). As with the epididymis, little is known about the immunological properties of these immune cells. At one time, it was believed that the CD8 phenotype was a marker of "suppressor" T cells, and some of the older literature in particular invokes the CD8+ T cells in the epithelia of the male reproductive tract as being responsible for protection of sperm antigens. It is now known that that CD8 is not a specific marker of any immunoregulatory lymphocyte subset and is most closely associated with cytotoxic T cells (498). In fact, a predominance of CD8+ T cells is a common characteristic of all mucosal epithelia. Although an immunoregulatory role of these lymphocytes in the reproductive epithelium remains to be determined, these cells are almost certainly involved in mediating cellular immunity at these sites.

It is self-evident that the leukocytes found in semen, even from men with no obvious infection or other inflammatory condition, must originate from the epithelium of the ducts and organs of the male reproductive tract (499). Logically, immune cells pass with relative ease across the epithelium in certain areas, thereby coming into physical contact with the sperm at the time of ejaculation. It is nonetheless curious that, although macrophages and lymphocytes are usually present as well, the leukocytes most commonly found in the semen are neutrophils, which are not a major leukocyte subset in either the stroma or epithelia of the male reproductive tract (499–505). Because vasectomy dramatically reduces leukocyte numbers in the ejaculate, the epididymis and proximal vas deferens are usually considered to be the main source of these cells, at least under normal conditions (505–507).

EFFECTS OF INFLAMMATION, INFECTION, AND IMMUNITY ON MALE REPRODUCTION

Inflammation and activation of the immune response have immediate and mostly negative effects on male reproductive function (54,56,57,508–511). There has been a tendency to assume that reproductive failure is related to fever and the negative effects of raised body temperature on spermatogenesis, but there is very little evidence that increased body temperature is the principal cause of testicular failure in febrile patients. In fact, it is much more likely that other inflammatory events are involved. The effects of inflammation on male reproduction largely can be separated into effects on testicular steroidogenesis leading to androgen deficiency problems and effects on spermatogenesis resulting in reduced sperm output, although there is an obvious link between the two outcomes. This may involve immune cells in the male reproductive tract, circulating immune mediators, or even activation of the somatic cells of the testis itself.

Inflammatory and Immunoregulatory Cytokines and Testis Function

Numerous inflammatory mediators and other immunoregulatory molecules are produced by Sertoli cells, Leydig cells, and spermatogenic cells either in response to an inflammatory stimulus or to orchestrate cellular interactions during normal physiological function. Such molecules include the cytokines IL-1, IL-6, TNF-α, and TGF-β family members, but also low–molecular-weight mediators such as nitric oxide (NO) and prostaglandins of the E series (PGE) and F series (PGF) (512–515). Not all inflammatory mediators that have been found in testis extracts or cultures of testis cells must necessarily play a role in normal functions: many early studies in particular may have been compromised by failure to eliminate endotoxin (i.e., endogenous LPS) contamination or by the extreme sensitivity and nonquantitative nature of the detection methods used (i.e., RT-PCR). However, it can be assumed with some confidence that certain mediators that have been shown repeatedly to be constitutively expressed in the normal testis (e.g., IL-1α) play some role in normal function, whereas other mediators may be important only during inflammation, infection, or other immune activation events. Cytokines in particular have been implicated as growth and differentiation factors in both compartments of the testis (516–518). They also play a role in modifying the endocrine control of the testis, through actions at multiple levels in the HPT axis; for example, IL-1β inhibits steroidogenesis by suppressing hypothalamic GnRH production and Leydig cell steroidogenesis directly (519–521).

Interleukin-1 and Interleukin-6

The archetypical proinflammatory cytokine, IL-1 exists as two isoforms (IL-1α and IL-1β) with quite different structures able to bind to the same receptor to exert an almost identical range of effects (522,523). Both isoforms induce the expression of numerous proinflammatory proteins, such as iNOS, cyclooxygenase-2 (COX-2), IL-6, and TNF-α (522), and stimulate their own production as part of an amplification loop (522,523). IL-1 is produced by many cells types, including fibroblasts, endothelial cells, keratinocytes, and smooth muscle cells, but activated monocytes and macrophage are the major source of secreted IL-1 in the body (522).

Both IL-1α and IL-1β are single-chain proteins, with only approximately 25% homology at the mature protein level. They are synthesized as 31- to 33-kDa precursor proteins that are cleaved enzymatically to active 17-kDa forms (Fig. 10). Both the precursor and mature forms of IL-1α are biologically active, but the IL-1β precursor is functionally inactive (524–526). Conversion of IL-1β to the mature protein involves the action of a specific protease, called IL-1–converting enzyme or caspase-1, and processing is linked to its secretion into the extracellular space (527,528). Conversion of IL-1α does not involve caspase-1, but the precursor can be cleaved by the calcium-dependent membrane-associated cysteine proteases, or calpains, and extracellular proteases (524). In contrast to IL-1β, most IL-1α tends to remain within the cell or associated with the cell membrane, although it can also be found in secretions (522). It is generally believed to act as an autocrine growth factor or as a mediator of direct cell–cell communication. IL-1α and IL-1β differ from most cytokines in that they lack a signal sequence, and their mechanism of secretion is poorly characterized (529).

The actions of IL-1 are mediated by the IL-1 type I receptor (IL-1RI), a member of the TLR superfamily (530) (Fig. 5). In addition, there is an IL-1 type II receptor, which is not linked to signal transduction but acts as a decoy receptor (531). Both the type I and type II receptors also belong to the immunoglobulin superfamily of receptors. After ligand binding to the receptor, a protein called IL-1R acceptor protein is recruited to form a complex that initiates signal transduction at the cell surface (532,533). This complex triggers the activation of IRAK, which ultimately leads to gene activation (534). It is thought that IL-1 normally does not act through activation of adenylate cyclase (522). The biochemical changes associated with IL-1 signal transduction are hydrolysis of guanosine triphosphate, phosphatidylcholine, phosphatidylserine, or phosphatidylethanolamine and the release of ceramide by neutral sphingomyelinase (535–537).

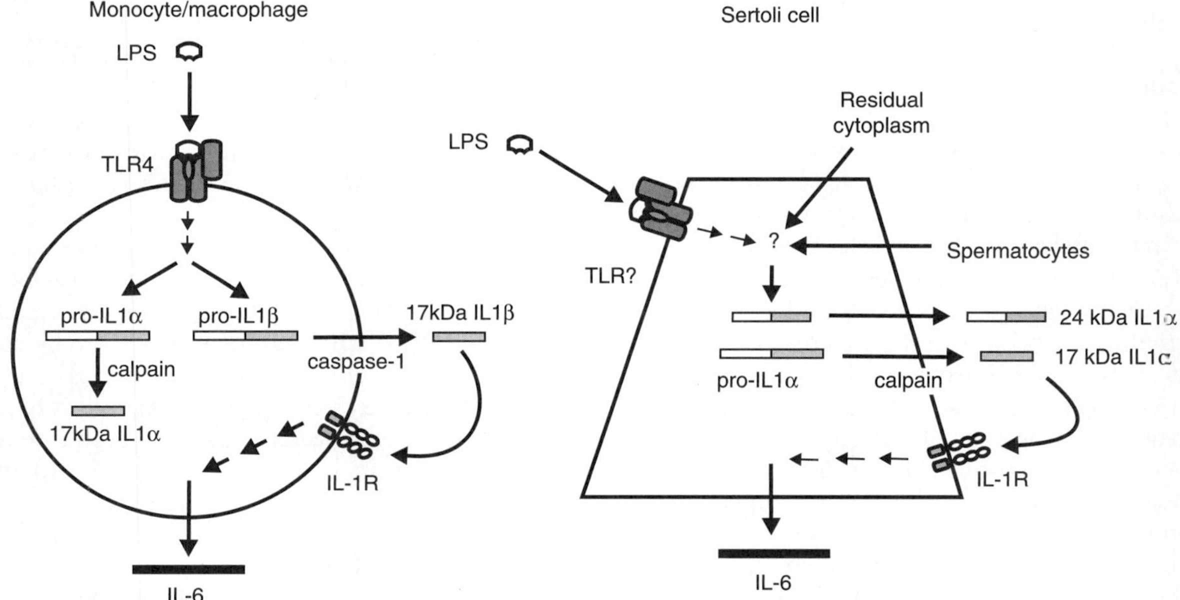

FIG. 10. Differential regulation of interleukin (IL)-1 and IL-6 production and secretion in the monocyte/macrophage and Sertoli cell. Binding of lipopolysaccharide (LPS) to the Toll-like receptor 4 (TLR4) on the surface of the macrophage upregulates expression of both IL-1α and IL-1β, which subsequently are processed to their mature bioactive 17-kDa forms by the action of calpain and caspase-1, respectively. IL-1α tends to remain associated with the cell, but IL-1β is secreted upon cleavage and binds to the IL-1 receptor (IL-1R) to stimulate production of IL-6 (522). The Sertoli cell also responds to LPS, presumably through TLR4. LPS, spermatocytes, and the residual cytoplasm of released spermatozoa are potent stimulators of IL-1α (but not IL-1β) in the Sertoli cell (542,544,545,551,552,556,568). Two alternate transcripts of IL-1α are produced by the Sertoli cell, including a transcript lacking the calpain cleavage site domain, which encodes a 24-kDa form of IL-1α (547,548). Both isoforms appear to be secreted by the Sertoli cell and both possess IL-1 bioactivity. IL-1 subsequently signals through the IL-1R to stimulate the production of IL-6 by the Sertoli cell (544,552,568,569).

Phosphorylation of phospholipase A_2 (PLA_2)–activating protein also occurs in the first few minutes, leading to rapid release of arachidonic acid (538,539).

There is a third member of the IL-1 cytokine group that is homologous with IL-1α and IL-1β and binds to the IL-1 receptors, but lacks the ability to transduce a signal. This cytokine is called IL-1 receptor antagonist (IL-1ra), and appears to act as a competitive antagonist of IL-1 action (540,541). Thus, regulation of IL-1 bioactivity can occur either through competition between IL-1 and the IL-1ra for the IL-1RI, or by competition between the type II and type I receptors for bioactive IL-1. Unlike IL-1α and IL-1β, IL-1ra does have a signal sequence and is secreted by the endoplasmic reticulum–Golgi pathway (541).

In vitro and in vivo studies have confirmed that IL-1 is expressed in the human and rodent testis and has complex effects on testis function. IL-1α appears during development in the rat testis at approximately days 15 to 20 of age, more or less coinciding with the onset of mature spermatogenesis (542–545). This is believed to be almost entirely due to the Sertoli cells, which synthesize and secrete both the mature 17-kDa IL-1α molecule and a slightly larger 24-kDa testis-specific form of IL-1α, the latter being the product of an altered mRNA transcript lacking the calpain cleavage site necessary for processing of the precursor (546–548) (Fig. 10). This testis-specific transcript is biologically active, although it is less potent than mature 17-kDa IL-1α (547,549). There is some evidence that spermatocytes and spermatids also express IL-1α constitutively (550). Sertoli cell production and secretion of IL-1α is induced by inflammatory stimuli, including LPS (545,551,552). Synthesis and secretion of IL-1α by the Sertoli cells also appears to be under control of the spermatogenic cells, and shows a distinct cyclical variation in the seminiferous epithelium (553–555). It is particularly strongly stimulated by phagocytosis of the residual cytoplasm of spermatids when they are released into the tubule lumen (544,545,556).

In contrast to IL-1α, IL-1β does not appear to be produced in substantial amounts in the testis under normal conditions. Studies have shown that Leydig cells can express the mRNA for both IL-1α and IL-1β in vitro in response to LPS or exogenous IL-1 (557–559), and both whole testes and isolated macrophages collected from LPS-treated rats or mice also express IL-1β mRNA (396,560,561) and the immunoreactive protein (562). Thus, although IL-1β expression in the normal testis is relatively low, it is upregulated during inflammation because of production by interstitial cells rather than by the Sertoli cells. The response of testicular IL-1α to inflammation is less clear because production by the Sertoli cell is upregulated by inflammatory stimuli in vitro (545,552,557), but in vivo studies report either no response or downregulation during systemic inflammation (394,562). The regulatory IL-1 family member, IL-1ra, has been shown to be produced by mouse Sertoli cells, and its production is stimulated by FSH, LPS, and IL-1 (564).

The IL-6–type cytokines, IL-6, IL-11, leukemia inhibitory factor (LIF), oncostatin M, ciliary neurotropic factor, cardiotropin-1, and cardiotropin-like cytokine, represent an important family of mediators involved in the regulation of the acute-phase response to injury and infection (205,565). Besides their functions in inflammation and the immune response, these cytokines also play a crucial role in hematopoiesis, liver and neuronal regeneration, embryonic development, and fertility. IL-6 is the main circulating cytokine, and possesses both proinflammatory and anti-inflammatory properties (205). Dysregulation of IL-6–type cytokine signaling contributes to the onset and maintenance of several diseases, such as rheumatoid arthritis, inflammatory bowel disease, osteoporosis, multiple sclerosis, and various types of cancer (e.g., multiple myeloma and prostate cancer) (566). IL-6–type cytokines exert their action by binding to specific receptors that associate with a common membrane signal transducer gp130, leading to the activation of the Jak/Stat and mitogen-activated protein (MAP) kinase cascades (567).

The Sertoli cells produce IL-6 (552,557), although Leydig cells actually may be the major source of this cytokine in the testis (557,568,569). IL-6 is produced by isolated rat Sertoli cells in response to stimulation by FSH, testosterone, neuropeptides, and phagocytosis (544,551,552,557,568,569). In the seminiferous epithelium, IL-6 is produced in a coordinated manner with spermatogenesis, and FSH differentially stimulates IL-6 secretion during the cycle of the seminiferous epithelium (544,570). Production of IL-6 by cultured mouse Sertoli cells is stimulated by IL-1α, IL-1β, TNF-α, and LPS (544,552,568,571), and inhibited by IFN-γ (572). Leydig cells are strong producers of IL-6 after stimulation by LH, LPS, and IL-1β in vitro (557,568,569,571).

The data from many studies indicate that that IL-1 and IL-6 are integrated in a complex network of endocrine and local regulatory mechanisms in the seminiferous epithelium. Phagocytosis of residual bodies shed by late spermatids at the time of spermiation triggers Sertoli cell IL-1α mRNA expression and production, which in turn stimulates IL-6 secretion through activation of leukotriene production (544). This results in an endogenous cyclical pattern of secretion that corresponds with the changes in stages of the spermatogenic cycle, and suggests a physiological role for these cytokines unrelated to normal inflammatory processes.

Studies in vivo and in vitro have indicated that IL-1α stimulates DNA synthesis in intermediate and type B spermatogonia, which suggests that IL-1α may act as a paracrine growth factor for spermatogonia (554,573). There is also evidence that IL-1 acts as an autocrine regulator of Sertoli cell function. For example, IL-1β inhibits FSH-induced aromatase activity in immature rat Sertoli cells (574) and stimulates production of lactate and transferrin by these cells (575–579). More recently, IL-1 has been shown to stimulate the proliferation of developing rat Sertoli cells in culture (580), and IL-1α stimulates IL-6 production by the Sertoli cell (544,552). IL-6 has been found to act as an inhibitor of meiotic DNA synthesis in preleptotene spermatocytes (570) and increases basal and FSH-induced transferrin and cyclic guanosine monophosphate secretion by Sertoli cell (579,581). In the interstitial tissue, IL-1 is a regulator of androgen production (549,560,582). All three forms of IL-1α stimulate testosterone production by immature Leydig cells, suggesting that Sertoli cell–derived IL-1α may be an important paracrine regulator of Leydig cells during testicular development. In contrast, locally injected IL-1β, but not IL-1α, induces acute inflammatory-like changes in the testicular microcirculation of adult rats (583).

The effects of IL-1 and IL-6 on testicular function are almost entirely deduced from studies with cultured cells or tubules; direct evidence for physiological roles for these cytokines in normal testicular function in vivo remains to be confirmed. Of importance is the fact that both IL-1 receptor knockout and IL-6–deficient mice are considered to be "fertile" (584,585), suggesting that these cytokines do not play a unique or essential role in maintaining spermatogenesis or that the reproductive phenotype is very subtle. However, not all cytokine deficiency phenotypes are obvious, and defects that increase susceptibility to inflammatory or immunological dysfunction may not be evident under normal animal housing conditions. Effects may become evident only in older animals or in animals living in the wild exposed to environmental stresses, dietary deficiency, or specific pathogens. Evidence for this caveat comes from studies in male mice infected with *Taenia crassiceps* (584). Wild-type mice with this infection have increased IL-6 expression and become feminized owing to excessive aromatase activity. This steroidogenic deviation is completely prevented in IL-6–deficient mice, which otherwise have similar reproductive functions to those of wild-type mice. This observation supports other studies implicating IL-6 as a regulator of aromatase activity (586,587), but also provides an example of a reproductive "knockout" phenotype that manifests itself only after specific immunological events. There is no doubt that IL-1 and IL-6 are produced in the testis and exert effects in the testis, with actions on both the seminiferous tubule and interstitial tissue compartments. As a result, changes in the production of IL-1 and IL-6 in the testis, such as might occur during inflammation, will almost certainly initiate alterations in testicular function.

LIF is a pleiotropic cytokine known to control the proliferation and survival of stem cells, including primordial germ cells and gonocytes (588,589). The LIF mRNA transcript has been detected in the rat testis from 13.5 days postcoitum until adulthood (590,591). LIF was found to be produced by peritubular cells and, to a much lesser extent, by the other testicular somatic cell types. No LIF was detected in meiotic and postmeiotic germ cell–conditioned medium, and only low levels of LIF protein were detected in spermatogonia-conditioned medium. Large amounts of bioactive LIF are present in testicular lymph, possibly because of peritubular cell production. Although LIF production was greatly enhanced in the presence of serum, LPS and TNF-α further increased production in peritubular and Sertoli cell cultures, and hCG enhanced LIF production by Leydig cells. Given the proliferative effect of LIF on immature germ cells, it is likely that peritubular LIF plays an important role in the regulation of normal spermatogenesis (591). The LIF receptor has been identified in Leydig cells and Sertoli cells and spermatogonia (592). The presence of both LIF and its receptor in adult and neonatal Leydig cells suggests that LIF signaling also may be involved in Leydig cell development.

Transforming Growth Factor-β Family

At least five distinct dimeric proteins comprise the core TGF-β cytokine family, three of which are expressed in mammals (types β_1, β_2, and β_3; Fig. 11). These cytokines, in turn, are the archetypical members of a superfamily of mostly homodimeric and heterodimeric proteins that includes activin, inhibin, and anti-müllerian hormone (593–595). The TGF-βs are multifunctional growth and differentiation factors involved in many aspects of tissue remodeling and repair as well as regulation of the immune system. Nearly all cells synthesize a form of TGF-β and possess functional membrane receptors for this cytokine family. TGF-β superfamily ligands bind to and activate specific transmembrane serine/threonine kinase receptors, and signals are then transduced from the cytoplasm to the nucleus through a network involving the Sma- and Mad-related (Smad) proteins (596–599) (Fig. 5).

The three mammalian TGF-β forms are highly expressed by Sertoli cells, peritubular cells, and

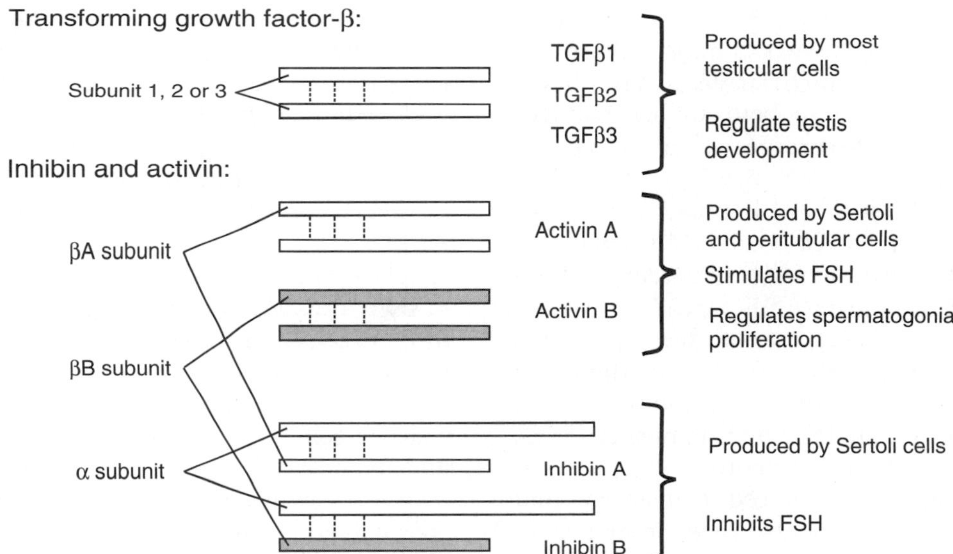

FIG. 11. The transforming growth factor (TGF)-β and activin/inhibin family of cytokines. These are dimers of disulfide-linked subunits with considerable sequence and structural homology. Three distinct subunits lead to production of homodimers (or, less frequently, heterodimers) of TGF-β types 1, 2, or 3, which have similar regulatory activities in the testis and immune system. In the testis, TGF-β appears to be most important during development, but continues to be produced in the adult (334,335,601–603). Dimerization of two homologous subunits lead to activin A, activin B, or, less commonly, activin AB. Activin A has been most thoroughly studied, although data indicate that activin B has a similar range of effects, including the defining ability to stimulate follicle-stimulating hormone (FSH) secretion by the anterior pituitary. Activin A has regulatory effects on spermatogonial proliferation (635,648–651) and possesses antiinflammatory and immunosuppressive activity (622,626,628,631,633). Dimerization of the activin subunits with the closely related α subunit produces the inhibins, which are produced by the Sertoli cells and are secreted into the blood to regulate FSH at the pituitary.

Leydig cells in the fetal and immature testis, although production declines considerably around the time of puberty (600–602). In the postpubertal testis they also have been localized to the spermatogenic cells in a development-specific pattern of expression (335,603). The receptors for TGF-β are found in both somatic and spermatogenic cells (334,335,604). TGF-β_1 and TGF-β_2 have been identified as inducing agents of apoptosis in gonocytes during developmental and proliferative stages, but they do not exert the same effect on Sertoli or Leydig cells (605). It has been suggested that the TGF-βs participate in the morphological differentiation of immature Leydig cells into adult Leydig cells in the rat testis by inducing the expression of extracellular matrix proteins (606). Sertoli cells and Leydig cells apparently regulate the concentration of TGF-β to physiological levels to prevent effects, such as apoptosis of gonocytes, that have been observed with nonphysiological levels in experimental studies (605,607). TGF-β_3 regulates Sertoli cell tight junction dynamics in vitro through the p38 MAP kinase pathway, suggesting that this cytokine plays a crucial role in regulating the opening and closing of the blood–testis barrier (608). This in turn regulates the passage of preleptotene and leptotene spermatocytes across the blood–testis barrier. Moreover, because TGF-β is a potent immunomodulatory cytokine, it may be an important factor contributing to the suppression of immunological events in the testis (131,609,610).

Activin is a dimer of the β subunits of the Sertoli cell protein hormone inhibin and, consequently, activin A ($\beta_A\beta_A$), activin B ($\beta_B\beta_B$), and activin AB ($\beta_A\beta_B$) forms exist, although most studies have concentrated on activin A, which may be the most physiologically relevant form (611). Activins display a high degree of sequence and structural homology with the TGF-βs, but have a different range of biological actions. Activin A circulates in the blood and is produced by a wide variety of tissues beside the testis, particularly the liver and hematopoietic system (612–617). The biological activity of activin A is regulated by a high-affinity inhibitory binding protein, follistatin, that has no homology with the TGF-β superfamily (618–620).

Although first identified as a regulator of reproductive function, activin A has a broad range of actions on the cell growth and differentiation of many cells types, including hepatocytes, red blood cells, and immune cells (613,614,616,617,621–628). Activin A is produced by activated monocytes, macrophages, and bone marrow stromal cells (614,615,629), and has been shown to be an inhibitor of T- and B-cell growth and peripheral mononuclear cell cytokine production (616,621,622,626,628,630–632). The production of activin A is stimulated by IL-1 (615,629), and expression of activin A has been shown to be increased in gut mucosa and submucosa during ulcerative colitis (626) and in the synovial fluid of patients with rheumatoid arthritis (617,633).

Studies from several laboratories have shown that activin A inhibits many proinflammatory actions stimulated by IL-1 and IL-6 in vitro, including T- and B-cell proliferation, monocyte phagocytosis, and production of acute-phase proteins (616,617,625–628,630,631,633). The mode of action involved is yet to be fully clarified. Nonetheless, these actions indicate an antiinflammatory and immunoregulatory role for activin A, consistent with its homology to the TGF-β proteins, which are themselves potent inhibitory regulators of inflammation and immunity (199,593,634).

In situ hybridization and immunohistochemistry have demonstrated that gonocytes, but not spermatogonia, contain the activin A subunit mRNA and protein in the normal rat testis (635). In the immature rat testis, the activin A dimer is produced by the Sertoli cells, peritubular cells, and Leydig cells (636–638). Activin A protein is produced by the testis throughout development, and the Sertoli and peritubular cells appear to be the principal sources in the adult (639,640). Activin receptors are present in Sertoli cells, spermatogonia, primary spermatocytes, and round spermatids (641–646), and in the peritubular cells and Leydig cells (637,641,644,645,647).

In vitro, activin exerts both stimulatory and inhibitory effects on spermatogonial cell proliferation (635,648,649) and on Sertoli cell proliferation (635,650), depending on the culture system used. One study also identified the capacity for activin A specifically to maintain the condensed mitochondrial morphology found in germ cells beyond the leptotene stage of the first meiotic prophase (651). These data indicate a direct or indirect effect of activin on early germ cell development, which may be modulated by changes in the specific environment. There is also evidence from a small number of in vitro studies that activin inhibits steroidogenesis by the Leydig cells, at least in the immature testis (652–655).

Effects of the Immune System on Testicular Function: Androgen Production

Leydig Cell Steroidogenesis

Androgens are produced by Leydig cells in the vascularized interstitial tissue for maintenance of the seminiferous tubule compartment of the testis and the extratesticular androgen-dependent tissues. A lack of androgens ultimately results in failure of reproductive function, not only because spermatogenesis depends on androgens for support, but because maturation, transport, and ejaculation of spermatozoa rely on the activity of the androgen-dependent accessory organs and tissues. Moreover, because the male phenotype itself, including muscle mass, fat distribution, and metabolic activity, is androgen dependent, loss of androgens also has an impact on general male health (Fig. 2).

The biosynthesis of testosterone by the Leydig cells is primarily under the control of LH (see Chapter 20 for details). Circulating LH binds to specific G-coupled receptors on the surface of Leydig cells and stimulates adenylate cyclase to produce cyclic adenosine monophosphate (cAMP), the intracellular second messenger for LH action (656,657). cAMP has two principal actions in the control of Leydig cell steroidogenesis. The first action of cAMP is the acute stimulation of testosterone biosynthesis through mobilization and transport of cholesterol into the steroidogenic pathway, an action that takes place within minutes. The cAMP-dependent protein kinase A activates cholesterol mobilization from intracellular cholesterol pools, extracellular lipoprotein sources, or cholesterol synthesized de novo from acetate. Regardless of its origin, cholesterol transfer to the inner mitochondrial membrane is a cAMP-dependent process requiring the action of the steroidogenic acute regulatory protein (StAR) (658,659). The second action of cAMP in Leydig cell steroidogenesis is a chronic and prolonged stimulation of expression of the steroidogenic enzyme genes and upregulation of their activity (660,661). Once cholesterol is transferred into the mitochondrion, it is converted to pregnenolone through the action of the cholesterol side-chain cleavage P450 (P450scc) enzyme residing on the inside face of the mitochondrial inner matrix membrane. Pregnenolone subsequently diffuses out of the mitochondrion to the smooth endoplasmic reticulum, where it is converted to progesterone through the action of 3β-hydroxysteroid dehydrogenase-Δ^4-Δ^5 isomerase (3β-HSD). Progesterone in turn is converted to 17α-hydroxyprogesterone and then androstenedione by the action of 17α-hydroxylase/C17-20 lyase (P450c17). Androstenedione is finally converted to testosterone through the action of 17β-hydroxysteroid dehydrogenase (17β-HSD) and eventually secreted from the cell (Fig. 12).

Immune–Endocrine Interactions in the Testicular Interstitium

Inhibition of testosterone production is an important mechanism mediating inflammatory disease–associated decreases in male fertility. Men with critical illness, burn trauma, sepsis, and rheumatoid

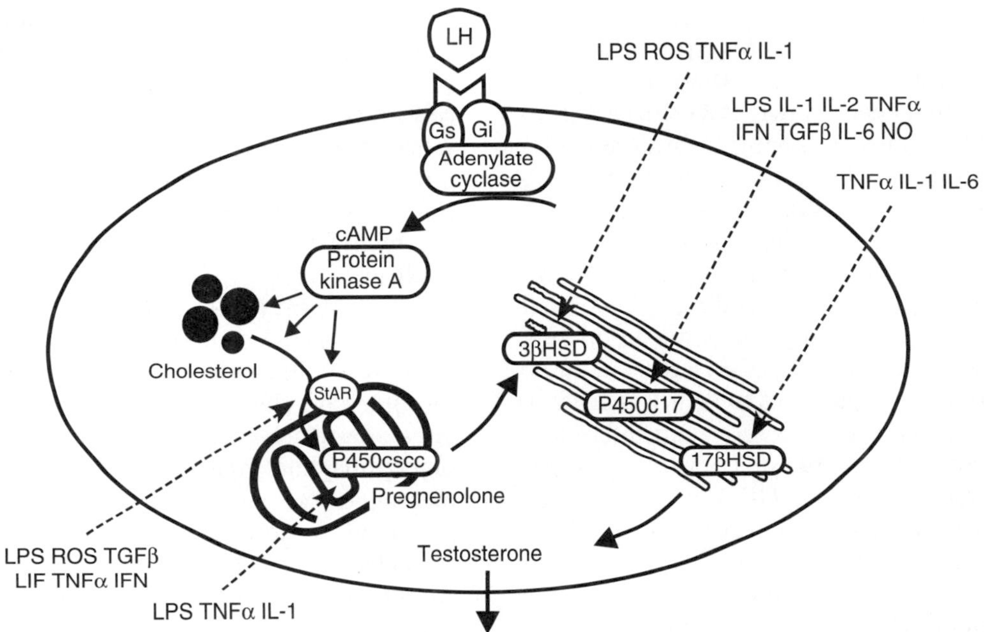

FIG. 12. Regulation of the testosterone biosynthetic pathway in Leydig cells and sites of inhibition by cytokines and inflammatory mediators. Steroidogenic acute regulatory protein (StAR) is inhibited by lipopolysaccharide (LPS) (685), transforming growth factor (TGF)-β (1282), tumor necrosis factor (TNF)-α (727), interferon (IFN) (1283), leukemia inhibitory factor (LIF) (747), and reactive oxygen species (ROS) (685,712). 3β-Hydroxysteroid dehydrogenase (3β-HSD) is inhibited by LPS (560,759), TNF-α (712), interleukin (IL)-1 (712), and ROS (761). P450cscc is inhibited by LPS (560), TNF-α, and IL-1 (396,582). P450c17 is inhibited by LPS (560), TNF-α, IL-1 (396,582,724,731), IL-6 (1280), IFN-γ (737), IL-2 (746), TGF-β (748,749), and nitric oxide (NO) (776). 17β-Hydroxysteroid dehydrogenase (17β-HSD) is inhibited by TNF-α, IL-1, and IL-6 (1280).

arthritis have elevated serum TNF-α and IL-1 levels and reduced testosterone levels (662–681). Similar decreases in gonadal function have been reproduced in experimental animal models of chronic inflammation and systemic immune activation. Experimental adjuvant-induced arthritis (682,683) results in a dramatic reduction in serum testosterone levels in rodents, and conditioned medium from testicular macrophages isolated from adjuvant-induced arthritic rats is inhibitory to Leydig cell testosterone production in vitro (684). Intraperitoneal or intravenous injection of the gram-negative bacterial cell wall endotoxin, LPS, in rats, mice, and rams or induction of sepsis with cecal slurry in male rats is also typified by a significant decrease in intratesticular or serum testosterone levels (446,508,685–688). There appear to be distinct species differences in the dynamics and severity of this inhibition, however, because responses in mice generally are more rapid, larger, and prolonged than those in rats, which show a biphasic response at approximately 6 hours and 24 hours after treatment, with a brief recovery period in between (446,561,685,686,689). Moreover, serum testosterone levels actually appear to rise in the boar after endotoxin treatment as a result of increases in pulsatile LH secretion (509,510,690). In rats, the early inhibition of testosterone is due to direct inhibition of Leydig cell steroidogenic function by LPS or its intermediates, whereas the later inhibitory phase may involve extratesticular effects, such as increased circulating inflammatory mediators or the release of endogenous glucocorticoids (446,561). This is an appropriate place, as well, to insert a word of caution about the use of LPS in such studies. LPS from different bacterial strains can have a significantly variable chemical composition and is usually contaminated to varying degrees by ligands of other TLRs (i.e., bacterial lipoproteins and peptidoglycans). Indeed, even different batches of LPS from the same bacterial strain can show considerable variation in potency and contamination. It is wise not to be too quick to conclude that all differences in experimental results obtained using LPS are simply due to differences between the animal models used.

In another model, testicular torsion resulting in a twisting of the spermatic cord that renders the testis ischemic followed by reperfusion causes an increase in TNF-α and IL-1β expression, which is correlated with activation of the stress-related kinase signaling pathway, leading to neutrophil recruitment and infiltration into the testis, oxidative stress, germ cell apoptosis, and significantly decreased serum testosterone levels (691,692). Similar inhibitory effects on serum testosterone levels can be induced in male rats by infusion of TNF-α alone (693).

To assess directly the effects of TNF-α on the HPT axis in humans, van der Poll and colleagues (668) injected six healthy men with recombinant TNF-α and measured serum concentrations of gonadotropins, testosterone, and sex hormone–binding globulin (SHBG). The TNF-α induced an early and transient increase in serum LH, whereas the concentrations of FSH remained unchanged. The increase in LH concentrations was followed by a transient decrease in serum testosterone levels after 4 hours. LH levels had returned to control values when the testosterone levels reached a nadir. SHBG levels were not affected. The results suggest that TNF-α affects the HPT axis at multiple levels in men and may be involved either directly or indirectly in the decrease in circulating testosterone concentrations in systemic illnesses (668). In another study, male patients treated with high doses of IL-2 as therapy for metastatic cancer were found to have significantly reduced serum testosterone levels (694). IL-2 causes the elevation of numerous serum cytokines, in particular TNF-α, IL-1, and IFN-γ (695,696). Finally, TNF-α levels are elevated in obese men and men with type 2 diabetes (697), conditions known to be associated with hypogonadism and decreased circulating testosterone levels.

Collectively, these experimental and clinical observations predict that in conditions associated with elevated inflammatory mediators there is a concomitant decrease in serum testosterone levels. One important group of inflammatory mediators that are known to inhibit Leydig cell steroidogenic function are the macrophage-secreted proinflammatory cytokines, such as IL-1 and TNF-α, and lymphocyte-secreted cytokines, such as IL-2 and IFN-γ (516). Cytokines are also important for integration of the neural–immune–endocrine network that controls testicular function under normal and pathophysiological conditions (698–700). More recently, there has been an increase in interest in the role of ROS in these processes as well.

Interleukin-1

A review of the literature dealing with the direct effects of IL-1 on testosterone production by Leydig cells in vitro reveals various reports of stimulation, inhibition, or even no effect (396,582,701–706). These apparent discrepancies may come from differences in the experimental conditions or methodologies used. On balance, studies suggest that IL-1 is either stimulatory or has no effect on basal testosterone production by adult Leydig cells (701,707,708), and there is general agreement that IL-1 inhibits LH/hCG- or cAMP-stimulated testosterone production (59,396,582,701,703,704,707,709–711). In the mouse, the major site of inhibition occurs at the level of the P450c17 enzyme (Fig. 12). In primary cultures of macrophage-depleted mouse Leydig cells, IL-1 inhibited the cAMP-stimulated induction of P450c17 expression and testosterone biosynthesis in a dose-dependent manner (582). Only the highest concentrations of IL-1 inhibited the cAMP-induced levels of P450scc and 3β-HSD, but IL-1 did not inhibit the basal or constitutive expression of any of the steroidogenic enzymes (582,712). In the rat, IL-1 appears primarily to affect Leydig cell steroidogenesis at the level of P450ssc, whereas StAR gene expression and protein synthesis are unaffected (705,713).

There is some evidence to suggest that the effects of IL-1 on the Leydig cell may involve, in part, regulation of COX-2 enzyme expression and PGE_2 acting as an intermediate (703,714,715). The effects of testicular IL-1α on Leydig cell steroidogenesis also appear to depend on both the variant of IL-1α involved and on the stage of development of the Leydig cell itself. The 32-kDa IL-1α precursor and the 17-kDa mature IL-1α are both inhibitory to gonadotropin-stimulated or cAMP-stimulated steroidogenesis in mature Leydig cells, whereas the testis-specific 24-kDa IL-1α splice variant has no effect (549). In contrast, all three forms of IL-1α stimulate testosterone production by immature rat Leydig cells.

Tumor Necrosis Factor-α

TNF-α is a 17-kDa glycosylated polypeptide secreted principally by activated monocytes and macrophages (716). It binds as a trimer to either of the two TNF receptors (TNFR1 and TNFR2), which are found on most cells in the body, and plays a central role in the initiation of the inflammatory response (717) (Fig. 5). TNF-α stimulates the release of IL-1 and IL-6 from activated monocytes and macrophages, and its synthesis and release is enhanced by IFN-γ secreted from activated T cells (718).

There are two families of TNFRs. The TNFR type 1 family includes TNFR1, Fas (CD95), and the death receptors 3 through 6 (719). The type 1 TNFRs induce cell death through a motif in their cytoplasmic regions called the death domain (719). When TNF-α binds to TNFR1, there is recruitment to the death domain of the intracellular adaptor proteins TRADD (TNFR-associated death domain protein) and FADD (Fas-associated death domain protein), leading to activation of the cell death caspase pathway (719,720). However, the binding of TRADD can also lead to recruitment of cIAP (cellular inhibitor of apoptosis) or RIP (receptor interacting protein), enabling the binding of TRAF-2. This complex mediates activation of the NF-κB pathway or MAP kinase, Jnk, or p38 (719). Whether signaling through TNFR1 leads to cell death or survival depends on the adaptor proteins involved. The TNFR type 2 receptors, which include TNFR2, CD30, CD40, and the lymphotoxin receptor, do not contain a death domain in their intracellular domains, and instead associate with TRAFs, leading to the activation of cell signaling events (719,720).

Results from in situ hybridization studies in mice have confirmed the presence of TNF-α mRNA in the round spermatids, pachytene spermatocytes, and testicular interstitial macrophages (721). Bioactive TNF-α was produced by the round spermatids in vitro and mRNA for the corresponding receptor was located on Sertoli and Leydig cells. Treatment of isolated rat or mouse testicular macrophages with LPS induced TNF-α secretion (395,722), indicating that TNF-α is produced by testicular macrophages under inflammatory conditions.

There are a number of reports in the literature describing the effects of TNF-α on Leydig cell steroidogenesis (560,693,694,702,723–726). Although one group concluded that TNF-α stimulates steroidogenesis (702), most of these reports describe inhibitory effects and a decrease in the production of testosterone. These studies have been performed in a variety of systems, including intact animals (693,694), isolated primary cultures of Leydig cells (725,726), and in MA-10 tumor Leydig cells transfected with Cyp17 (the P450c17 gene) reporter constructs (724). Inhibition of hCG binding by TNF-α has been reported (725), but most of these studies suggest that TNF-α inhibition occurs downstream of cAMP production, at the level of steroidogenic gene expression (Fig. 12). In mouse Leydig cells in primary culture, TNF-α caused a decrease in P450scc, P450c17, and 3β-HSD expression (726). TNF-α had no effect on the basal expression of P450scc but did inhibit basal expression of 3β-HSD (712). This TNF-α inhibitory affect on the cAMP-stimulated testosterone production in Leydig cells was found to be due to a decrease in mRNA and protein levels of P450scc and P450c17 (726). Furthermore, it was demonstrated that the inhibitory affect of TNF-α on P450c17/Cyp17 gene expression was mediated by protein kinase C (724). The inhibitory affect of TNF-α on LH/hCG-induced testosterone secretion has also been reported in porcine Leydig cells (727). In these studies, the mechanism of the inhibitory affect of TNF-α was reported to be due to a decrease in StAR protein mRNA and protein levels (727).

Results from several independent studies also have shown that the TNF-α inhibitory effects on Leydig cell testosterone secretion may involve a sphingomyelin/ceramide–dependent pathway (728–730). Intratesticular delivery of TNF-α induced a rapid (4 hours) and sustained (up to 24 hours) reduction in StAR protein expression and testosterone biosynthesis in nonstimulated or hCG-treated intact or hypophysectomized rats (730). Bilateral treatment with cell-permeant short-chain ceramides (C2-cer or C6-cer) reproduced the early inhibitory action of TNF-α on testosterone biosynthesis and testicular StAR expression. More recently, the inhibitory action of TNF-α on Leydig cell steroidogenesis was shown to occur through the NF-κB pathway (731), possibly in response to activation through protein kinase C (724). The transactivation of the P450c17/Cyp17 gene by SF-1 and NUR-77 was inhibited by NF-κB activation, suggesting that perturbation of these transcription factors is the most distal event in TNF-α–mediated suppression of Cyp17 promoter activity (731).

Together, these data support the concept that TNF-α triggers different effector mechanisms directly to inhibit Leydig cell steroidogenic enzyme gene expression and steroidogenesis, which ultimately contributes to the global reproductive failure associated with chronic inflammation and sepsis. However, it remains unclear whether Leydig cells are exposed to TNF-α under normal testicular conditions or only during inflammation.

Besides the effects on Leydig cell testosterone production, TNF-α has also been reported to increase plasminogen activator inhibitor-1 expression in rat testicular peritubular cells, indicating that it may be involved in controlling testicular protease activity (732). The authors suggested, however, that the biological effects of TNF-α on plasminogen activator inhibitor-1 may be secondary to epidermal growth factor (EGF) receptor signaling because TNF-α also increased EGF receptor mRNA and EGF binding in the peritubular cells (732). Similar to IL-1, TNF-α stimulates lactate production by cultured Sertoli cells (575,733).

Although a definitive role for TNF-α in normal testicular function, if any, has yet to be established, TNF-α has been implicated as a major causative agent in the development of EAO (734). In rats with EAO, there is a significant increase in the number of TNF-α–positive testicular macrophages and the number of TNFR1-positive germ cells (735). Sixty percent of TNFR1-positive germ cells were apoptotic, supporting the suggestion that, acting together with other local factors such as Fas–Fas ligand (FasL), TNF-α could trigger germ cell apoptosis in this model.

Interferons

The interferons are a group of functionally related protein cytokines that comprise three main groups (α, β, and γ), based on their structural relationships and major cellular sources: IFN-α is produced by monocyte and macrophages, IFN-β is produced by fibroblasts and epithelial cells, and IFN-γ is principally produced by T cells (736). The best-known effects of interferons are their antiviral, antiproliferative, and immunomodulatory actions (Table 4), but interferons also have been shown to have effects on the endocrine system. In particular, both IFN-α and IFN-γ inhibit testosterone production in primary cultures of porcine Leydig cells (737,738). IFN-γ exerts its inhibitory effect on testosterone production at the level of cholesterol transport into the mitochondria. IFN-γ also inhibits the expression of both P450scc and P450c17 (737), similar to the effect of TNF-α on mouse Leydig cells (726) (Fig. 12). Normal healthy men who were treated with human IFN-α had significantly decreased serum testosterone levels (739). The observed decrease in testosterone was most likely due to a direct inhibition of Leydig cell steroidogenesis because serum gonadotropin levels were unaffected by the treatment. In other experimental studies, steroidogenesis in rat ovarian cells and testicular Leydig cells was compromised by human and murine IFN-α, and this effect could be reversed by addition of specific interferon antibodies (740).

These data indicate that interferons may contribute to the overall decline in steroidogenic function of patients with viral infections (741–744). However, Dejucq and colleagues (745) have shown that Sertoli and Leydig cells in the rat testis strongly expressed IFN-α and IFN-γ during infection with Sendai virus, and that this elevation in expression was associated with an increase in testosterone production. This result suggests that there may be indirect or secondary stimulatory effects of interferon on steroidogenesis in the testis, indicating a more complex role for these cytokines, at least as far as testicular steroidogenesis is concerned.

TABLE 4. *Biological properties of interferon-γ*

General property	Specific activities
Antiviral	Inherent cytotoxicity for viruses
Antiproliferative	Induces G_1 arrest
Macrophage activation	Increases cytotoxicity to microbes and neoplastic cells
	Increases inducible nitric oxide synthase expression
	Increases reactive oxygen species and reactive nitrogen production
Antigen presentation	Increases MHC class I and II expression
	Stimulates peptide degradation, processing and presentation
Cell-mediated immunity	Necessary but not sufficient for type 1 helper T-cell development
	Increases IL-12 receptor expression
	Suppresses IL-4 production
Tumor immunity	Inhibits tumor cell proliferation
	Increases MHC class I on tumor cells
B-cell and humoral immunity	Suppresses activity

IL, interleukin; MHC, major histocompatibility complex.

Other Cytokines

In addition to the cytokines already discussed in detail, there is evidence that many other cytokines have direct inhibitory effects on Leydig cell steroidogenesis. The autocrine T-cell growth factor, IL-2, inhibits gonadotropin-stimulated testosterone production by rat Leydig cells at the level of the P450c17 enzyme, which is similar to the actions of TNF-α and IL-1 (746). In patients treated with IL-2, serum testosterone levels are reduced (694), but IL-2 also causes a robust peripheral elevation of several other cytokines, including TNF-α, IL-1, and IFN-γ, which would contribute to the overall inhibition (695,696). Subcutaneous IL-6 administration in men produced prolonged suppression of serum testosterone levels, without apparent changes in gonadotropin levels, suggesting that this effect of IL-6 may be mediated by direct effects on Leydig cells (592). In vitro studies indicate that LIF inhibits Leydig cell steroidogenesis at the level of StAR mRNA expression and may inhibit cholesterol delivery to the mitochondria (747), and TGF-β inhibits Leydig cell steroidogenesis at the level of LH receptor number and signaling as well as distal to cAMP production at the level of P450c17 expression (748,749) (Fig. 12).

Reactive Oxygen and Nitrogen Species

Recognition of an inflammatory or immunogenic signal by both immune and nonimmune cells triggers gene expression for cytokines, adhesion proteins, and enzymes that produce very low–molecular-weight inflammatory mediators. The products of these enzymes, the reactive oxygen and nitrogen species, are referred to collectively as ROS and participate in eliminating the infection (750). Important ROS include superoxide anion (O_2^-) hydrogen peroxide (H_2O_2), hydroxyl radical ($HO^•$), nitric oxide ($NO^•$), and peroxynitrite anion ($ONOO^-$; Fig. 13). ROS can react directly with and modify cellular macromolecules such as protein, lipids, and DNA. Repair systems exist to correct oxidative damage, but excess

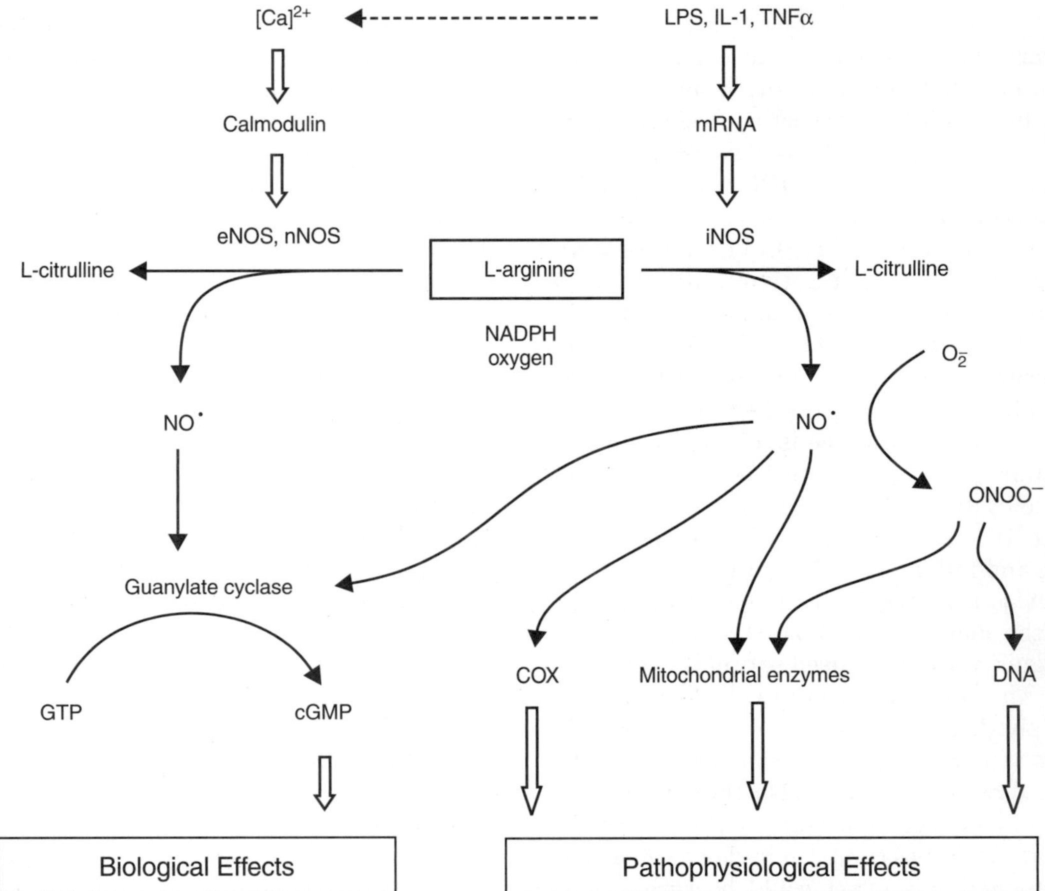

FIG. 13. Regulation and actions of nitric oxide synthase (NOS) and NO production in normal physiology and pathophysiology. NO is produced by enzymic conversion of L-arginine to L-citrulline by NOS, and the freely diffusible NO radical (NO˙) exerts a variety of physiological functions through regulation of cyclic guanosine monophosphate (cGMP)–dependent protein kinases. The activity of the constitutively expressed isoforms, endothelial NOS (eNOS) and neuronal NOS (nNOS), is regulated through calcium–calmodulin. Synthesis of a constitutively active from of NOS, called inducible NOS (iNOS), is induced by inflammatory mediators and is responsible for the large upregulation of NO production during inflammation. High levels of NO˙ can be detrimental due because of direct reaction with the heme group of cyclooxygenase (COX), mitochondrial enzymes, and DNA, effects that can be potentiated by interaction with superoxide to produce the extremely reactive peroxynitrite anion (514,751,1284).

and cumulative oxidative damage results in cellular dysfunction and contributes to the pathological process of many diseases (751). ROS are produced continuously in cells as the byproducts of mitochondrial and microsomal electron transport reactions and other metabolic processes.

Mitochondrial respiration consumes 85% to 90% of the oxygen used by cells, and represents the greatest potential source of ROS in the cell. The steroidogenic cytochrome P450 enzymes also produce ROS as a byproduct of their catalytic reaction mechanism (752–755). Cellular antioxidant systems that normally protect cells from oxidative damage include superoxide dismutase (SOD), catalase, and glutathione peroxidases. Other antioxidant molecules include ascorbic acid, α-tocopherol, β-carotene, retinoic acid, and glutathione. Mitochondrial SOD and redox cycling of glutathione and oxidized glutathione through the action of glutathione peroxidase afford the cell the greatest protection from ROS generated intracellularly by mitochondria (756). Under normal physiological conditions, ROS generation is controlled and oxidative damage is minimized. However, during oxidative stress such as that due to reperfusion after ischemia/hypoxia, inflammation, or exposure to extracellular sources of ROS, the antioxidant protective mechanisms are overwhelmed and cellular oxidative damage results. Oxidative damage often results in initiation of apoptosis or necrosis of affected cells (751).

The elaboration of reactive oxygen and nitrogen species during inflammation subjects Leydig cells to oxidative stress. LPS has been shown to activate ROS production from testicular macrophages in vitro (352). Exposure to LPS causes oxidative damage to Leydig cells in vivo, resulting in a collapse of the mitochondrial electrochemical gradient ($\Delta\Psi m$), lipid peroxidation, decreased StAR protein, and a precipitous decline in serum testosterone (757). Hydrogen peroxide, a potent oxidant, inhibits steroidogenesis in Leydig cells (757–759), MA-10 tumor Leydig cells (760,761), and in corpora luteal cells of the ovary (762–765). In addition, the production of reactive oxygen and nitrogen species during postischemic reperfusion contributes to tissue damage in the testis. LPS stimulates the immediate production of NO from endothelial nitric oxide synthase in endothelial cells (766) and induces the expression of iNOS in macrophages, resulting in the production of significant quantities of NO (767,768) (Fig. 13). These enzymes also are expressed by several cell types in the normal testis, and iNOS in particular is upregulated by LPS treatment in Leydig, Sertoli, peritubular, and spermatogenic cells, as well as a subset of the testicular macrophages (392,769,770). NO has been shown to inhibit Leydig cell steroidogenesis in vivo (771–775) and in vitro (774,776–778). Treatment of immunologically challenged rodents with NOS inhibitors counteracts the decrease in serum testosterone levels (688,772,779).

The mechanism through which NO inhibits Leydig cells is not known. One possibility is through oxidative damage produced by generation of reactive nitrogen species such as peroxynitrite anion (780). Peroxynitrite is the reaction product of NO and superoxide, and is one of the major cytotoxic agents produced during sepsis, inflammation, and ischemia/reperfusion (781). Peroxynitrite decomposes to produce the hydroxyl radicals, which are potent oxidants (782), and NO enhances ROS-mediated toxicity (783). Moreover, reactive oxygen and nitrogen species produced during postischemic reperfusion of the testis in experimental models of testicular torsion have been implicated as causing destruction of the seminiferous epithelium and germ cell apoptosis (784). Together, these studies demonstrate the importance of ROS and NO as mediators of the pathophysiological consequences of aberrant immune–endocrine interactions in the testes (Fig. 12).

The Neural–Immune–Endocrine Axis in the Control of Testicular Steroidogenesis

A hallmark of immune–endocrine interactions is immune activation of the HPA axis, resulting in the activation of the "stress response" (785,786). The interaction between the stress and reproductive axes has been studied extensively, and the reproductive axis can be suppressed by hormones from all levels of stress-related events (699,787). Glucocorticoids have been shown to inhibit LH and FSH secretion (788,789). Such suppression requires days to become evident, suggesting a more chronic role for adrenal corticosteroids in the inhibition of gonadotropin secretion and suppression of other reproductive functions such as spermatogenesis and steroidogenesis. Similarly, the direct inhibitory effects of glucocorticoids on Leydig cell steroidogenesis acting through specific receptors on the Leydig cells (286,790) are manifested over a longer time course and at the level of transcriptional repression of the steroidogenic enzymes P450scc, 3β-HSD, and P450c17 (180,181,791,792). In contrast, the hypothalamic hormone, corticotropin-releasing hormone, appears to repress the reproductive axis more acutely (699).

Rivest and colleagues demonstrated that intracerebroventricular injection of IL-1β inhibited GnRH release into the median eminence of the rat (787). Others studies have shown that injection of the bacterial endotoxin LPS into ovariectomized sheep inhibited pulsatile GnRH secretion into the hypothalamic–pituitary portal blood, further supporting the hypothesis that inflammatory stimuli inhibit the reproductive axis by acting centrally (793,794). Previously, Rivier and colleagues (795,796) and subsequently Kalra and colleagues demonstrated that IL-1 inhibits LH secretion (797,798). There is evidence that other cytokines, including TNF-α and IL-6, also inhibit gonadotropin release at the hypothalamus or pituitary (799), whereas NO appears to have the opposite effect (771,800,801).

Although perturbation of LH secretion by central mechanisms would certainly result in a concomitant inhibition of gonadal steroidogenesis, evidence suggests that, in addition, a neural pathway exists through which intracerebroventricular IL-1 directly inhibits steroidogenesis. During stress conditions that lower LH levels, decreased testosterone levels are undoubtedly due at least in part to a pituitary-mediated event. However, there are many stresses that lead to low testosterone levels in the absence of decreased LH secretion. These observations led Rivier and colleagues to postulate that there may be a direct neuronal connection between the brain and the testis that is activated by cytokines derived from the CNS (519,520,700). Direct support for this hypothesis comes from the demonstration that intracerebroventricular administration of IL-1β decreases testicular responsiveness to hCG in rats pretreated with a GnRH antagonist and therefore lacking LH secretion (521). This inhibitory effect of centrally injected

IL-1β precedes elevation of peripheral cytokine levels or decreases in plasma LH. The possible involvement of prolactin and opioids as mediators of the IL-1β effect has been ruled out; however, it appears that central catecholamine pathways may be involved. Of the pharmacological interventions tested, propranolol effectively reversed the inhibitory effect of intracerebroventricular injection of IL-1β.

Testosterone secretion is under the primary control of pituitary LH, but secretagogues present in the general circulation or manufactured in the testis can also alter Leydig cell activity independently of the pituitary. Indeed, Rivier and colleagues have shown that there is a neural brain–testicular circuit that directly regulates testosterone synthesis independently from LH release (802,803). They demonstrated that injection of the transganglionic retrograde tracer pseudorabies virus into the testes caused viral staining in the spinal cord, the brain stem, and the hypothalamus, thus establishing the presence of a neural pathway between the CNS and the testis. Moreover, spinal cord injury significantly interfered with this staining, thus supporting the hypothesis that the proposed circuit travels through the cord. Spinal cord injury completely abolished the ability of intracerebroventricular injection of IL-1β or corticotropin-releasing hormone to blunt the testosterone response to hCG, indicating that these two secretagogues act in the brain to stimulate a neural pathway that interferes with Leydig cell function independently of the pituitary. The existence of this brain–testicular circuit may play a role in pathological processes, so far unexplained, that are characterized by decreased testosterone levels despite normal LH production (802).

In light of the large body of evidence demonstrating the existence of neuronal peptide signaling pathways that affect Leydig cell function (749,804), it is tempting to speculate that these neuronal peptides are the efferent effectors of the direct brain-to-gonad signal produced in response to immune activation. In support of this hypothesis, surgical or pharmacological denervation of the testes blocks the effects of oxytocin, which has been shown to stimulate steroidogenesis in immature rat testis. Serotonergic elements were destroyed by treatment with 5,6-dihydroxytryptamine. Transsection of the inferior testicular nerve by vasectomy resulted in similar effects. Electrical stimulation of spermatic nerves results in a significant increase in levels of testosterone in venous blood from the spermatic vein in cats (805). These results support the hypothesis that testicular innervation is involved in the control of local peptide effects (806). Acute spinal cord injury in experimental animals is followed by immediate decrease of serum testosterone levels and profound reductions in spermatogenesis, indicating the important role of efferent nerves for testicular function. These events are accompanied by elevation of gonadotropins. Defects in spermatogenesis and steroidogenesis can be rescued by external administration of testosterone (807,808). However, lesions in spermatogenesis after spinal cord injury are not evident throughout the testis and appear to recover in a period of months. Local regulatory mechanisms of the testis appear to be able to regulate spermatogenesis and steroidogenesis without neuronal assistance after a period of adaptation (807). These mechanisms evident in experimental animals are supported by findings among patients with chronic spinal cord injury in whom spermatozoa for assisted reproduction can be retrieved from the testis or the ejaculate after electroejaculation, although disorders in spermatogenesis also are evident in many patients (809).

The testis lacks somatic nerves and is supplied only with autonomic nerves, and the bulk of testicular nerves are sympathetic (698). The neurotransmitter associated with the sympathetic fibers innervating the testis appears to be norepinephrine. Other neuropeptides are coexpressed with norepinephrine in nerve fibers innervating the testis (810). Recently, the GABAergic system (γ-aminobutyric acid and receptors) has been identified in the interstitial compartment of the testis, and is hypothesized to be an important efferent of the peripheral neuroendocrine system that may function in the regulation of testosterone production (811). Considerable evidence demonstrates the importance of catecholaminergic control of Leydig cell function and development; notably, pharmacological denervation by intratesticular injection of 6-hydroxydopamine inhibited LH responsiveness and testosterone production in the hamster (812). Surgical denervation of the rat testis induced a decline in gonadotropin responsiveness and resulted in a decrease in LH receptor numbers on the surface of Leydig cells (813). However, high concentrations of catecholamines are correlated with decreased androgen production (812). These findings suggest that local intratesticular actions of catecholamines acting in concert with other central and peripheral mechanisms may be involved with suppression of testicular functions during times of acute stress and activation of the sympathetic nervous system (698).

Under normal physiological conditions, pituitary gonadotropins, direct neural innervation, and local neuroactive peptides and catecholamines act in concert to control testicular function, especially androgen biosynthesis in testicular Leydig cells. During stress and inflammation, perturbation of pituitary gonadotropin secretion, activation or suppression of direct neural connections, production of local inflammatory mediators, and inhibition or activation of local neuroactive peptide pathways act in concert to suppress testicular androgen production. The highly complex integration of the neural–immune–endocrine signals that control testis function awaits complete elucidation (Fig. 4).

Effects of the Immune System on Testicular Function: Spermatogenesis

There is no doubt that spermatogenesis is affected directly and indirectly by inflammation, although this aspect of male reproductive function has not been studied quite so extensively as the effects of inflammation on testicular steroidogenesis. Specifically, this discussion is concerned with the effects of the innate immune response on Sertoli cell function and spermatogenesis, as distinct from antigen-specific reactions to the cells of the seminiferous epithelium. In general, the experimental models that have helped to elucidate this aspect of immune–testis interactions have involved administration of the inflammatory mediator LPS or studies on the effects of agents and treatments that alter the vasculature of the testis.

Lipopolysaccharide-Induced Inflammation and Spermatogenesis

Acute systemic administration of LPS has been found to affect spermatogenesis in a number of species, resulting in apoptosis and progressive loss of spermatogonia, spermatocytes, and spermatids within days of administration in rats, boars, and rams (446,508–511,690). In the adult rat, treatment with relatively large doses of LPS causing an endotoxic shock–like condition leads to apoptosis of spermatogonia and spermatocytes, and sloughing of spermatocytes and early round spermatids from the epithelium several days later, after the endotoxic condition itself has resolved (446). These effects occur more rapidly and are quite different in specificity from the effects of androgen withdrawal, which inhibits the release of mature spermatids into the tubule lumen and the integrity of the junctions between the Sertoli cells and midphase round spermatids (814). Moreover, even very high doses of LPS do not reduce intratesticular levels of testosterone in the rat much below 30% of control, well above the threshold necessary to sustain spermatogenesis in this species (172,173,814). These observations suggest that alternative explanations to androgen deficiency must be sought to explain spermatogenic failure, at least for the LPS-treated rat model.

Although there are elements of the LPS-mediated response that resemble heat damage, in fact, rats have a very poor fever response to LPS and the doses of LPS that affect spermatogenesis in rats actually cause a fall in body temperature (446,511). Nor is the pattern of spermatogenic degeneration entirely consistent with vascular disruption. Ischemia/reperfusion injury causes rapid apoptosis of the spermatogenic cells entering mitosis—spermatogonia and the very early spermatocytes (815–817). Administration of LPS does appear to alter blood flow through the rat testis, possibly as a result of vasodilation of the testicular arteries due to upregulation of testicular iNOS expression (446,770), but there is no evidence of ischemia (392,446). Paradoxically, although LPS treatment causes an increase in vascular endothelial cell leakage in the testis, the inflammation is not accompanied by an increase in testicular edema; in fact, interstitial fluid volume in the testis falls quite dramatically in this model (446).

The other potential cause of damage in the LPS treatment model is the action of proinflammatory cytokines and mediators on the seminiferous epithelium itself. This could involve increased levels of circulating cytokines as well as local production by leukocytes and somatic cells in the testis. In addition to the resident macrophage population, LPS stimulates an increase in intratesticular monocytes in the rat (392), and neutrophils in the boar testis interstitium (510,690). Moreover, although TLR expression on these cells has yet to be confirmed, there is clear evidence that LPS stimulates IL-1, IL-6, or activin A production by Leydig cells and Sertoli cells in vitro (394,552,557,617) (M. P. Hedger, unpublished data). In vivo, LPS causes the upregulation of testicular IL-1β, TNF-α, IL-6, and iNOS expression (394,561,562,722,770,818). These inflammatory mediators are produced at relatively low levels during the normal cycle of the seminiferous epithelium and have direct and complex effects on both Sertoli cell and spermatogenic cell function. Most significant, however, is the observation that formation of Sertoli cell tight junctions and junctional protein levels are inhibited by the action of TNF-α and NO (819,820). In summary, it is highly likely that LPS-induced inflammation may have its most profound effects on spermatogenesis by (a) altering critical Sertoli cell functions, (b) overriding the signals normally produced by the Sertoli cell to regulate spermatogenesis, and (c) disrupting the junctional complexes responsible for integrity of the blood–testis barrier and Sertoli–germ cell attachment (Fig. 14).

Vascular Inflammation Models: The Role of Neutrophils

High doses of hCG are administered in the treatment of delayed testicular descent in young boys, and

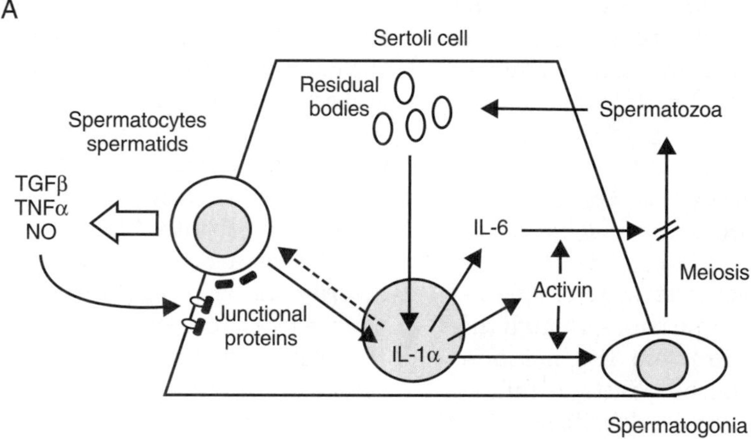

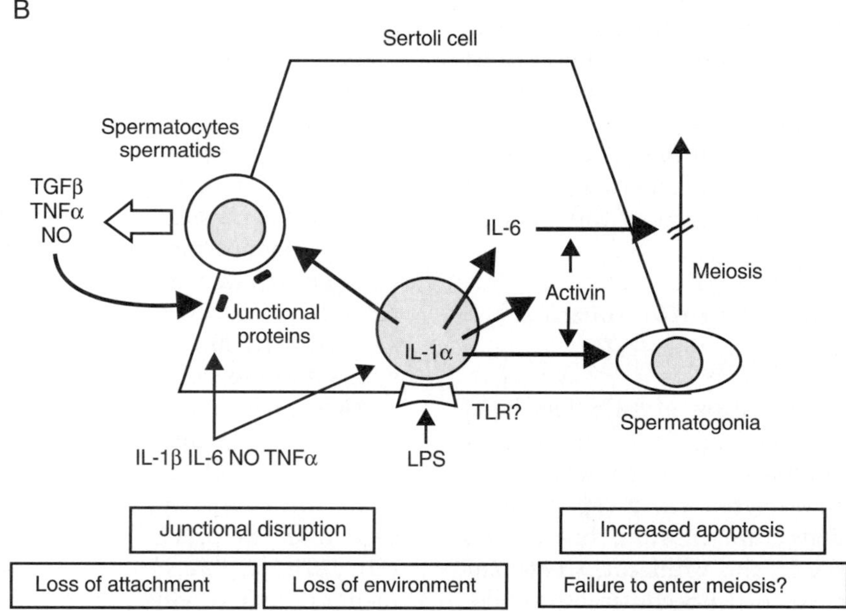

FIG. 14. Cytokine networks in control of Sertoli–germ cell interactions. **A:** Normal interactions in the seminiferous epithelium. Residual bodies produced by released spermatozoa and contact with the spermatocytes and spermatids stimulate production of interleukin (IL)-1α by the Sertoli cell, which in turn stimulates IL-6 and activin A (542,544,545,551,552,556,568,569,640). IL-1 and IL-6 have stimulatory and inhibitory actions, respectively, on spermatogonial proliferation and spermatocyte survival (553,554,570,573). Activin exerts both stimulatory and inhibitory effects on spermatogonia, suggesting a modulatory role (573). Spermatocytes and spermatids produce tumor necrosis factor (TNF)-α, transforming growth factor (TGF)-β, and nitric oxide (NO) (335,603,721,770), which are regulators of Sertoli cell tight junction formation and degradation (608,819,820). It is hypothesized that these interactions coordinate spermatogenesis and junctional reorganization throughout the cycle of the seminiferous epithelium. **B:** Effect of inflammation. The action of lipopolysaccharide (LPS) and inflammation in general induces a large and sustained increase in exogenous and locally produced inflammatory cytokines and other mediators in the seminiferous epithelium. This leads to loss of junctional integrity, sloughing of spermatocytes and spermatids, disruption of the blood–testis barrier, loss of epithelial signaling, and increased germ cell apoptosis (446,508–511,690). Sustained inflammation could potentially lead to complete disruption of the seminiferous epithelium and increased susceptibility to further damage.

have been extensively studied in adult rats (821). This treatment causes a hyperstimulation syndrome comprising a transient decrease in testicular blood flow, which is immediately followed by increased testicular blood flow and pressure, opening of the vascular endothelial cell junctions, and an increase in testicular interstitial fluid volume some 16 to 24 hours later (822–825). The syndrome is accompanied by accumulation of intravascular and interstitial neutrophils in the testis (826,827), failure of mature sperm release, vacuolation of the seminiferous epithelium, and apoptosis and loss of spermatogonia and primary spermatocytes (828,829). The hCG response can be eliminated by depletion of the Leydig cells with EDS (830–832) or depletion of neutrophils with a specific antiserum (826), and IL-1β is able to replicate most of the effects (558,583,833–835). These observations suggest that this is an inflammatory response, possibly mediated by IL-1β secreted by the Leydig cells, but in fact the actual mechanisms are very poorly understood. Significantly, both depletion of the testicular macrophages using liposome-encapsulated dichloromethylene diphosphonate (377) and stimulation of the macrophages with latex beads (829) exacerbate the effects of hCG on the testis, suggesting that the resident macrophages normally play a role in limiting the inflammatory response. Interestingly, this response is quite different from that of LPS-induced inflammation, which causes a massive reduction in interstitial fluid volume in the testis (446), although some of the germ cell damage seen is similar in both models.

Another inflammation model implicating intratesticular neutrophil accumulation as a cause of spermatogenic damage comes from studies on the response to transient testicular ischemia in mice (692,817). Germ cell damage in this model, specifically apoptosis of spermatogonia, was correlated with an increase in neutrophils in the testicular subcapsular venules and an increase in ROS in the interstitium. Ischemia causes a similar accumulation of neutrophils in rats (836), and of both mononuclear cells and polymorphonuclear cells in rams (837). However, the important observation in the mouse model was the fact that E-selectin knockout mice and neutropenic wild-type mice displayed a significant decrease in both neutrophil recruitment to the testis and germ cell–specific apoptosis. Locally produced IL-1β and TNF-α have been implicated as mediators of the response (691,838). Thus, disruption of the seminiferous epithelium and spermatogenic cell apoptosis in both the ischemia/reperfusion mouse model and hCG hyperstimulation rat model are directly linked to recruitment of neutrophils to the vasculature or interstitial tissue of the testis. Exactly how these cells exert their damage on the spermatogenic cells is not yet known.

In summary, it is clear that several different testicular inflammation models have direct effects on one or more stages of spermatogenic cell development. In addition, as in most other tissues, inflammation responses in the testis do not appear to lead automatically to autoimmune complications.

Role of Macrophages in Leydig Cell Development and Steroidogenesis

There seems little doubt that macrophages play an important role in Leydig cell development. There is a close temporal link between the maturation of the adult Leydig cell population and the increase in the number of testicular resident macrophages during puberty (356,357,406). Moreover, specific depletion of the intratesticular macrophages inhibits the development of Leydig cells in immature tats, and the recovery of Leydig cells after EDS treatment in adult rats (379,381,382). The *op/op* mouse, which has an inactivating mutation in the M-CSF gene and consequently has reduced numbers of macrophages throughout the body, including the testis, also displays very poor fertility and reduced testicular testosterone production because of developmental and steroidogenic abnormalities of the Leydig cells (384,420).

The role of the testicular macrophages in Leydig cell growth and differentiation may involve the intercytoplasmic specializations between the two cell types, which appear very early in adult testicular development in the rat (349), or specific macrophage-derived cytokines. In cultures of isolated Leydig cells from rats 10 to 20 days of age, IL-1β caused a dose-dependent increase in DNA synthesis as measured by incorporation of [^{3}H]-thymidine (839). IL-1α also had an effect, although it was much less potent than IL-1β. The effect of IL-1β was not observed in Leydig cells isolated from older animals, suggesting that macrophage IL-1β may play a role in the proliferation of Leydig cells during prepubertal development (839). The involvement of prostaglandins in this developmental regulation is indicated by the observations that IL-1β stimulates the expression of COX-2, both IL-1 isoforms, and IL-6 in rat progenitor Leydig cells, and that prostaglandins act as intermediates in this stimulatory pathway (840).

Although it is fairly consistently observed that activated and inflammatory macrophages inhibit Leydig cell function, largely owing to the production of inflammatory cytokines and ROS (59), and possibly also prostaglandins (714), the effects of noninflammatory macrophages on Leydig cell steroidogenesis are not so consistent. Stimulation of macrophage phagocytosis with an intratesticular injection of latex beads increased the steroidogenic capacity of Leydig cells in adult rats (829). However, in macrophage-depleted adult rat testes, both a fall (377,383) and an increase (380,385) in testosterone production have been reported by different groups using essentially the same model. Moreover, media collected from testicular macrophages in culture have been shown both to stimulate (841–843) and inhibit (706,708,843–845) Leydig cell steroidogenesis in vitro. Similar discrepancies were observed in Leydig–macrophage cocultures. The most likely explanation for these differences is that partial activation of the macrophages during isolation, or possibly endotoxin contamination, may have distorted the results. Neuroendocrine influences also may be important in some species. Studies of the hibernating rodent, the bank vole, have revealed that Leydig cells from long-photoperiod animals produced more testosterone and were more sensitive to stimulation by testicular macrophage-conditioned medium than were Leydig cells from short-photoperiod animals (846). Long-photoperiod cells were more sensitive to IL-1α–mediated inhibition as well. On balance, the published data seem to indicate that resting testicular macrophages have a positive or trophic effect on Leydig cell steroidogenesis.

Testicular macrophages have been shown to express the cholesterol 25-hydroxylase enzyme and produce 25-hydroxycholesterol (842,847). It has been proposed that testicular macrophages, because of their close association with the Leydig cells, provide 25-hydroxycholesterol as a substrate for testosterone biosynthesis, bypassing the need for StAR and supporting basal steroidogenesis. Interestingly, the 25-hydroxylase enzyme is negatively regulated by testosterone, suggesting that there may be a feedback loop between the macrophages and Leydig cells (848,849). Furthermore, it has been suggested that one of the factors that macrophages secrete in support of Leydig cell proliferation and development may be 25-hydroxycholesterol (850).

IMMUNOLOGICAL RESPONSES IN THE MALE REPRODUCTIVE TRACT

Sperm Antibodies

Clinically, the most common immunological "dysfunction" of the male reproductive tract is the presence of sperm antibodies, which may inhibit sperm motility through agglutination reactions, target the sperm for immunological destruction in the male or female tract, or block essential surface receptors and molecules required for fertilization (851–854) (Fig. 15). Multiple antigens are usually involved, and there is as yet no clear evidence for a single dominant antigen associated with sperm autoimmunity (15–25). There is no doubt that physical damage to the male reproductive tract can lead to sperm antibody formation. This is best exemplified by the high incidence of sperm antibodies associated with vasectomy, where antibodies develop in up to 70% of patients after the procedure (471–477), but blunt trauma and testicular biopsy may cause sperm antibodies to form as well (855,856). Sperm antibodies also are associated with obstructive azoospermia and congenital absence of the vas deferens (478,857). Another association of sperm antibodies is with reproductive tract infections involving orchitis or epididymitis, although there is evidence that inflammation alone may not be sufficient to cause antibody formation (853,858–864). There is a definite relationship between sperm antibodies and the existence of other autoantibodies, suggesting an important role for genetic factors in predisposition to sperm antibody formation (865,866). A separate subset of antibodies directed against steroidogenesis-specific antigens expressed by the Leydig cells is associated with the type I polyglandular autoimmune (PGA) syndrome, although sperm antibodies also may be found in patients with these conditions (867,868).

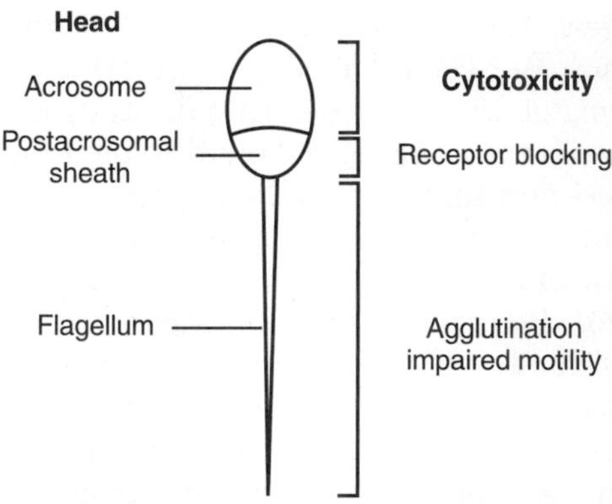

FIG. 15. Sperm antibody binding sites and effects. The principal sequelae of antibody binding to spermatozoa are (a) activation of immune cells against the sperm through complement or through interaction with immunoglobulin Fc receptors on phagocytes leading to cytotoxicity; (b) blocking or interfering with surface recognition molecules, in particular the egg binding receptors in the postacrosomal region; and (c) agglutination caused by cross-linking of sperm by multivalent antibody, which may impede the ability of sperm to swim freely in the female tract.

In animal models, sperm antibodies can be induced by active immunization with spermatogenic cells or whole sperm, or with testicular or epididymal extracts. This procedure is usually performed using adjuvants to boost the immune response, but protocols leading to antibody formation without the use of adjuvants also have been reported in male and female animals (148,869–873). As in humans, antibodies may develop in rats, guinea pigs, mice, rabbits, and monkeys after experimental vasectomy or vasal obstruction (155,460,479,481,482,874–878). The strength of the response and its outcomes are highly strain dependent. In some of these experimental models, the formation of antibodies is associated with the development of orchitis and variable degrees of damage to the seminiferous epithelium, an outcome that is rarely, if ever, observed in humans with sperm antibodies.

The antigens recognized by sperm antibodies have been extensively studied in actively immunized or vasectomized experimental animals and in infertile patients (15,16,18–25,878–887). Some of these antigens have been identified. They include numerous sperm surface and, paradoxically, intracellular proteins, glycoproteins, and glycolipids. In neither studies of human sperm antibodies nor the experimental studies in animals have any universally dominant antigens associated with the induction of sperm autoimmunity been identified.

There has been considerable debate regarding the clinical significance of sperm antibodies or, perhaps more precisely, the tests used to measure them. Although many studies have measured antibody levels in seminal plasma or even blood, it has been suggested that only antibodies that are actually bound to the sperm, and more specifically the anterior portion of the sperm, should be considered to be of real significance, and then only when a majority of the sperm are affected (10,852,888,889). Hence, studies that use tests that detect sperm bound antibodies (e.g., agglutination and immunobead binding tests) are most likely to be reliable, whereas measurement of antibody titers by enzyme-linked immunosorbent assay are considered to be of less utility (851,890,891). Although IgG can cross the epithelium at certain sites such as the rete testis to bind to the sperm in the male reproductive tract, it also has been argued that only IgA antibodies are evidence of a specific mucosal immune response and should be considered of greater prognostic value in semen samples. The most rigorous approach to this issue involves confirmation that the antibody binding has functional consequences using a sperm function test, such as the mucus penetration assay, before accepting the pathological significance of the antibodies (10,852,889,892).

Even allowing for the variation in methods used, it appears that populations vary widely in the incidence of sperm antibodies. This may be due in part to genetic differences in susceptibility to antibody formation, although the prevalence of reproductive tract infections in the population and a lack of suitable treatment options are probably a stronger determinant. In developed nations, the incidence of sperm antibodies is approximately 5% to 12% of all infertility cases (8–12,851). On the other hand, in parts of central Africa, where treatment for reproductive tract infection is limited or there is a reliance on self-medication and alternative therapies, incidences of sperm antibodies among infertile populations as high as 44% or more have been reported (893). In considering these figures, it also should be borne in mind that sperm antibodies may be present in the semen or blood of men of proven normal fertility (894). Moreover, in the absence of evidence for any other underlying cause, the presence of sperm antibodies tends to be assumed to be the cause of the infertility. It is quite possible that the incidence of infertility due primarily to sperm antibody reactions is lower than these values might suggest. Although treatment with high-dose corticosteroids has been used with limited success in the past to treat sperm antibodies, sperm washing coupled with intrauterine insemination, in vitro fertilization, or intracytoplasmic sperm injection are the current methods of choice (851,856,895).

Equally serious for the infertile couple is the development of sperm antibodies in the female tract. Although largely outside the scope of this review, it is important to recognize that histocompatibility differences between partners are a potential problem. In fact, given that sperm deposited in the female tract are essentially an allograft, the lack of an overt immune response to sperm in all but a very small subset of women is somewhat surprising (896–900). Although specific immunological regulation in the female tract plays a role here, the immunoregulatory properties of the sperm and seminal plasma are believed to play the most important role in this protection (901,902).

Orchitis

In contrast to sperm autoantibodies, autoimmune reactions against the germ cells in the testis appear to be comparatively rare, at least among humans. However, there are several experimental animal models of autoimmune orchitis that have been extensively studied by a number of groups (80,81,139,148,251,434, 436–438,459,872,903–914). Most of these models involve immunization with testicular antigens (including spermatogenic cell, somatic cell, and extracellular matrix antigens) using "complete" adjuvants (i.e., adjuvants in combination with active or inactivated microbial agents), or adoptive transfer of lymphocytes from animals with ongoing EAO. A particularly interesting model of orchitis development occurs subsequent to vasectomy in some animals. Whereas vasectomy leads to sperm antibody formation alone in most strains of rats, in the Lewis rat vasectomy results in a rapid progression to full-blown orchitis (479–481). Certain strains of mice are also susceptible to orchitis postvasectomy (915), and similar responses to vasectomy or epididymal obstruction have been observed in the rabbit (486,487,916), guinea pig (917), and rhesus monkey (918). There also are a number of animal models of spontaneous autoimmune orchitis, including naturally arising conditions in strains of mink (919,920), rats (921), and dogs (922). Finally, removal of the thymus gland around postnatal day 3 causes a polyglandular autoimmunity similar to that found in human PGA syndrome, which can include orchitis (251,903,904), indicating that immunoregulation by T cells is involved in the normal protection of at least some testis-specific antigens. These models have been crucial to the development of our concepts of the control of immune responses in the testis.

Although the majority of this research is focused on autoimmune reactions to the germ cells, testicular autoimmune disease also may involve the somatic cells. Autoimmune polyglandular diseases, most notably Addison's disease, are associated with specific immune reactions to steroidogenic cells throughout the body, including Leydig cells, thereby leading to hypogonadism (867,923,924). This autoimmunity primarily involves reactions against the steroidogenic enzymes and hormone receptors (925–928), although germ cell autoantibodies also are commonly observed (868).

The onset of EAO in animals usually involves the appearance of testicular antibodies and immune complexes, followed by infiltration of mononuclear cells (macrophages and lymphocytes), invasion of the seminiferous tubules by these cells, sloughing of germ cells and focal necrosis, accumulation of neutrophils or eosinophils, and, finally, aspermatogenesis (434,437,459,929,930). The epididymis and vas deferens are frequently involved as well. In adoptive transfer studies, it has been shown that $CD4^+$ T cells, but not antibody, can transfer orchitis from one animal to another, clearly implicating a cellular rather than humorally mediated mechanism of disease development (386,387,931,932). In adoptive transfer of orchitis in mice, moreover, the early disease almost invariably occurs in the region of the rete testis (386,387), consistent with the fact that this is the area where access to the antigens of spermatogenesis is likely to be greatest (43,125,136,137,139). However, in the mouse, the interstitial tissue around the rete testis is the region where there is a larger concentration of MHC class II–positive macrophages, implicating these cells in the initiation of the disease as well (386,387,437,459,933). The inflammatory cytokines TNF-α and IFN-γ have been specifically implicated in the development of orchitis in experimental models (734,735,934), whereas administration of IL-6 has been shown to inhibit the progression of orchitis in LPS-resistant mice (910).

There is a small but compelling body of data to suggest that immunological events in one testis, caused by a variety of localized events, lead to changes in the contralateral testis, a condition sometimes called *sympathetic orchiopathia* (436,486,905,935–941). This is not surprising because autoreactive lymphocytes and antibodies generated by events on one side will be able to travel to the other testis to initiate the disease, in a manner comparable with adoptive transfer from one animal to another.

In humans, orchitis is most commonly associated with a number of viral infections, most notably postpubertal "mumps" (942–944), but also coxsackievirus B infection (945,946), varicella (947), human immunodeficiency virus (HIV) infection (445,948), dengue fever (949) and Epstein-Barr virus–associated infectious mononucleosis (950,951). Bacterial orchitis is usually caused by spreading of urethral pathogens such as *Escherichia coli*, *Chlamydia trachomatis*, and *Neisseria gonorrhoeae* up the genital tract, and is almost invariably associated with inflammation of the epididymis (i.e., epididymo-orchitis) (952). However, a number of systemic bacterial and mycobacterial infections, including syphilis, tuberculosis, and leprosy, can cause a granulomatous orchitis (953–960). Sterility after orchitis is relatively rare, but bilateral disease can lead to aspermatogenesis in severe cases. In rabbits, rats, and mice, infectious orchitis has been shown to occur after peripheral or intratesticular injection of several viral and bacterial agents (436,741,931,961–965). In most, if not all of these cases, the precipitating event is believed to be an immunological response to the infectious agent. This in turn may lead to collateral damage to the testicular cells by normal cellular immunity reactions causing obstruction, fibrosis, and displacement or ischemia, although development of a direct autoimmune attack against testicular antigens due to antigen cross-reactivity is a possibility as well (942,959,966).

Evidence that spontaneous autoimmune orchitis occurs in humans is scant. This is not entirely surprising, given that the event and its diagnosis might be separated by many years, and testicular biopsies will not necessarily pick up focal lesions. Nonetheless, there is some evidence for immune complex formation in the basement membrane of the seminiferous epithelium (967–971), and a high prevalence of asymptomatic testicular inflammatory lesions (972), among infertile men. The problem lies in establishing whether these lesions are evidence of a past autoimmune reaction or of a secondary event. In spite of the lack of definitive evidence, the existence of "spontaneous" animal models of testicular autoimmunity clearly indicates that it is a possibility in humans as well. There also is some evidence of an association of orchitis with other inflammatory diseases, particularly vasculitis (i.e., inflammation of the blood vessels), in men (973).

The precise reasons why some animal strains are more susceptible to the development of autoimmune orchitis remain incompletely understood, but there is no doubt that susceptibility to autoimmune reactions to sperm and testis antigens is genetically determined (908,974–977). Clearly, there is a clinical association between sperm antibodies and other autoimmune antibodies (865,866), and the development of sperm antibodies after vasectomy is strongly associated with the MHC (978). In mice, specific orchitis susceptibility genes have been mapped to both MHC and non-MHC regions, and the loci linked to development of orchitis also have been shown to govern susceptibility to other autoimmune diseases, such as encephalomyelitis and diabetes (976,979–981). Finally, orchitis frequently forms part of the type I PGA syndrome, which has been linked to mutations in one of the transcription factors (Aire) responsible for the promiscuous thymic expression of peripheral antigens (982,983). Together, these data clearly indicate that there is a genetic predisposition for development of autoimmune infertility in humans, which might be triggered by precipitating infection, physical trauma to the reproductive tract, or other inflammatory event.

Orchitis is actually much less common than inflammation in the remainder of the genital tract (i.e., epididymitis, vasitis, and prostatitis). If the inflammatory response is severe enough this can lead to damage to the mucosal epithelium or obstructive lesions that may inhibit fertility. As in the case of orchitis, these conditions are usually caused by infections (952), their increased incidence reflecting the fact that the organs are closer to the point of access from the external environment; however, differences in the inflammatory, immunoregulatory, or antimicrobial mechanisms of the testis may contribute to the reduced incidence as well. Sometimes chronic asymptomatic or idiopathic inflammation of the male reproductive tract can occur, usually manifesting as persistent scrotal or perineal pain (27–31). This inflammation may arise after infection or physical trauma, such as postvasectomy in a subpopulation of patients, or it may arise spontaneously. Such chronic inflammatory conditions are particularly poorly understood, and frequently difficult to treat.

Immune Privilege of the Testis I: The Blood–Testis Barrier

It is fairly obvious that the testis is an immunological target, and that autoimmune responses to spermatogenic antigens occur given the right conditions. What then are the normal systems that exist in the testis to prevent or regulate development of autoimmunity there? What is the functional and physiological basis of "immune privilege" in the testis?

In fact, the concept of immunological privilege in the testis is fraught with immense confusion and misconceptions among researchers. The term has generally been used by immunologists to describe tissues that are not readily accessible to APCs and circulating lymphocytes or that have deficient lymphatic drainage to local lymph nodes, most notably the brain, the anterior chamber of the eye, and the placenta (75,973,984). Because neither of these properties applies to the testis, the term also has been used in reference to the fact that antigen-specific immune responses against both endogenous antigens (spermatogenic cells) and exogenous antigens (grafts) appear to be suppressed in this tissue (48,49,984). However, the most common misconception centers on the role of the blood–testis barrier in this process.

The blood–testis barrier was originally identified as a physiological barrier restricting the passage of lipid-insoluble molecules from the blood into the seminiferous tubular fluid (39,41). Restricted agents include the immunoglobulins and, of course, all immune cells. The actual location of the barrier lies in tight occluding junctional specializations (zonulae occludens) between adjacent Sertoli cells in the seminiferous epithelium (40,42) (Fig. 3). The junctions are quite distinct from the normal adhering junctions and gap junctions that exist in most epithelia and are composed of a number of specific adhesion molecule proteins, the occludins and claudins (134). These Sertoli cell tight junctions are assembled at the time of puberty and form a complete seal separating the basal and adlumenal compartments of the epithelium. Their primary role is to create a unique physiological environment for the meiotic and postmeiotic germ cells (168,985). The location of the junctions means that all spermatogenic cells beyond the leptotene or zygotene spermatocyte stage are completely surrounded by the cytoplasm of adjacent Sertoli cells and are entirely dependent on the Sertoli cells for their support and regulation (134,135). The process of movement of the germ cells themselves through the junctions during spermatogenesis remains poorly understood, but does not involve any breach of the junctional integrity (135). Indeed, loss of the integrity of this barrier is a major factor in several forms of testicular failure, including spermatogenic damage during inflammation and infection.

Although there are other cell layers separating the spermatogenic cells from the circulation, specifically the vascular endothelium and the peritubular cell layer, these appear to provide little or no barrier against circulating cytokines, immunoglobulins, complement components, or even immune cells (108–110,986–993). Much has been made of the lack of fenestrations of the testicular capillaries, as well as a number of similarities with the endothelium comprising the blood–brain barrier (38,160,994). In fact, some testicular capillaries are fenestrated (104). In practice, testicular interstitial fluid differs little in background composition from that of the blood, even though transport of many proteins and other large molecules across this endothelium appears to be relatively slow (105–107,987). At least in immunological terms, the endothelium of the testis appears to afford little more protection than the vascular endothelium of any other large organ. Tracer labeling studies also have shown that the peritubular cell layer does not restrict intercellular molecular or cellular passage, and there is no evidence for occluding-type junctions between these cells. They do not comprise functional components of the blood–testis barrier.

More recently, a broader interpretation of the blood–testis barrier that goes beyond the physiological barrier concept has been proposed by a group of researchers working in the field of oncology (995). These researchers postulated that the blood–testis barrier exists to protect the germ cells from harmful influences and consists not only of the structural elements already discussed above, but also the efflux-pump barrier system involving expression of the drug-transport proteins p-glycoprotein and the multidrug resistance associated protein-1 on the capillary endothelium, peritubular myoid cells and the basal aspect of the Sertoli cells (996–998). While these elements may certainly be important for protection of the testis against toxic agents, their contribution to immunoregulation is likely to be marginal. Unfortunately, this appropriation of the term "blood–testis barrier" to include a much wider concept will probably serve to add even more confusion to the field, with the term potentially meaning something entirely different to reproductive biologists, immunologists and oncologists.

There are several lines of evidence that the blood–testis barrier does not account for all the manifestations of immune privilege in the testis. Studies have shown that the spermatogenic cell autoantigens are not confined behind the Sertoli cell tight junctions, and are expressed by spermatogonia and the early spermatocytes (999,1000). Moreover, there is clear evidence that the barrier is incomplete in the rete testis epithelium. Significantly, orchitis can be passively transferred to naive mice by lymphocytes from mice with active autoimmune orchitis, with the initial reaction concentrated in the interstitium surrounding the rete testis (386,387). A similar initial pattern of development of orchitis around the rete testis region has been observed in mice immunized with viable spermatogenic cells (933,1001). In many seasonally breeding species, annual regression of both the later spermatogenic cells and blood-testis barrier occurs without inducing overt inflammation or autoimmunity (919,1002,1003). Finally, the blood-testis barrier cannot explain the enhanced survival of grafts in the interstitial tissue (45–52). On balance, the data support the conclusion that the blood–testis barrier does not prevent exposure of germ cell antigens to the immune system; indeed, antibodies and lymphocytes specific for spermatogenic antigens may be a normal feature of the circulating immune repertoire (894). Although the sequestration of a large proportion of the antigenic burden behind the blood–testis barrier no doubt plays a role in this process, it cannot be the primary explanation for functional immune privilege in the testis.

Immune Privilege of the Testis II: Transplantation and Immunoregulation

Perhaps the most intriguing aspect of immune responses in the male tract is the observation that allografts and even xenografts into the testicular interstitial tissue outside the blood–testis barrier are preserved for extended periods of time, possibly even indefinitely (45–51). This enhanced survival is not simply due to the reduced ambient temperature of the testis because grafts to the equally hypothermic skin of the ear are not preserved (49), whereas grafts continue to survive in testes that have been translocated to the abdominal cavity (49,51,52). Moreover, the efferent lymphatics of the testis do not show any evidence of functional deficiency (35). However, intratesticular parathyroid allografts have failed to survive in rats sensitized against the donor antigens, and long-established intratesticular allografts were rapidly rejected after active immunization of the recipient with donor tissue (48). In addition, humoral and cellular responses to testicular antigens occur normally after they have been initiated by passive transfer of lymphocytes or by active immunization (386,387). These observations have led to the hypothesis that local immunoregulation mechanisms are responsible for graft survival and, more specifically, that the inductive phase of the immune response is suppressed in the testis or its draining lymph nodes. In other words, the immune system may be unable to recognize or respond to foreign antigens in the testicular environment, thereby implicating local APCs or lymphocyte-mediated tolerance.

Although studies on graft survival in the testes of laboratory rodents have been very convincing, the data in other species are less consistent. In fact, similar studies in the ram (1004) and cynomolgus monkey (1005) have been unsuccessful. Moreover, the type of graft also may be a factor. Salawry and colleagues were unable to get pancreatic islet allografts or xenografts to survive in the normal rat testes, but they did survive in abdominally located testes (52,1006). Whether these differences in outcome are due to differences in the effectiveness of local immunoregulatory mechanisms, species differences in overall systemic immunity, differences in testicular architecture, or physical features of the graft tissue itself remains to be answered. In spite of these difficulties, there remains compelling evidence that there is something unique about the local immunological environment of the testis.

Major Histocompatibility Complex Expression and Antigen-Presenting Cells of the Testis

Although at least one study detected MHC class I expression on human Sertoli cells (1007), most studies have reported a characteristic absence of expression of both MHC class I and II proteins on the cells of the seminiferous epithelium under normal conditions (49,128,129,386,388,433,438,1008–1012). This indicates that spermatogenic cells are able to avoid direct recognition by $CD4^+$ and $CD8^+$ T cells, which might be important for reducing the potential for antigen-specific immune responses in the seminiferous epithelium. On the other hand, there are at least some data to indicate that mRNA and protein for both MHC class I and II molecules are expressed in human spermatozoa (1013-1018), suggesting that mRNA present in the spermatogenic cells may be translated into protein at some time after they are released from the testis. This later expression of MHC molecules may play a role in protection of the sperm against infection or immune cells in the reproductive tract.

In contrast to the seminiferous epithelium, both MHC class I and II proteins are expressed throughout the testicular interstitial tissue. As would be expected, MHC class I expression is found on most interstitial cells, including the Leydig cells (388,1008). Studies on MHC class II cell expression, however, indicate that there may be species differences in the number and distribution of these cells in the testis under normal conditions. In the rat and human, testicular macrophages and dendritic cells express MHC class II throughout the interstitium (49,348,372,1008), but studies on the mouse testis indicate that expression of MHC class II is concentrated on cells in regions adjacent to the rete testis (386,387). A greatly reduced number of MHC class II–positive cells in the ram testis is consistent with the relatively low number of resident macrophages in this species (129). In light of all the observations, however, it appears unlikely that a lack of MHC class II–positive APCs is a contributing factor in testicular immune privilege. Differences in the distribution of such cells, however, may be reflected in differences in susceptibility and development of autoimmune reactions between species, or possibly even between different strains.

The MHC class II–positive macrophages and dendritic cells of the testis actually may play a pivotal role in controlling the immune response to intratesticular antigens. Certainly, interaction between an APC and an antigen-specific $CD4^+$ T cell can lead to development of either a Th1 or Th2 cell, largely depending on whether the interaction occurs in the presence of type 1 or type 2 cytokines, respectively (224). The type of helper T cell generated clearly has an important influence on the nature and consequences of the subsequent immune response, which may be either cellular or antibody mediated (Fig. 5). On the other hand, one of the mechanisms by which peripheral tolerance is believed to be induced involves an ongoing process of T-cell downregulation by low-dose exposure of T cells to their antigen, accompanied by either modified costimulatory or specific immunoregulatory signals from the APC or other regulatory lymphocytes in the vicinity (1019–1022). For example, engagement of the TCR by the peptide–MHC class II complex in the absence of linkage of B7 (on the APC) to CD28 (on the T cell) leads to T-cell inactivation, followed by either functional anergy or apoptosis (1020). Furthermore, there is increasing evidence that engagement of the TCR and CD28 in the presence of various factors, such as IL-10 or TGF-β, or in the absence of signaling by CD40L, can lead to induction of regulatory/suppressor T-cell activity (227).

Unfortunately, there has been almost no investigation of the antigen-presenting and costimulatory activity of the macrophages and dendritic cells of the testis, although absence of the classic B7 isoforms (CD80 and CD86) has been reported in the mouse testis (389). The possibility that other members of the B7 family, such as the potentially immunoinhibitory variant B7-H1 that is expressed on the placental trophoblast (1023), may be found in the testis also warrants investigation. In addition, a soluble form of the nonclassic MHC class I molecule HLA-G has been found in Sertoli cells, spermatocytes, spermatids, and small numbers of interstitial cells in the rhesus monkey testis (261). This molecule also is associated with the placenta and its soluble form has been implicated in apoptosis of alloreactive $CD8^+$ cytotoxic T cells (1024), but its significance to the testis is entirely unknown. Indeed, it is fair to say that the immunological role of the MHC- and

APC-mediated regulation of T helper subset and T regulatory/suppressor cell activity in the testis is a vitally important area that has been almost entirely neglected by researchers in this field.

Fas Ligand

In 1995, Bellgrau and colleagues (1025) presented data that suggested that expression of FasL (CD95L), a cell death signal for activated T cells, was expressed by mouse Sertoli cells, and that mice deficient in either FasL or its receptor, Fas, did not show evidence of testicular immune privilege. In this study, Fas–FasL interaction was implicated in the prevention of antigen-specific responses in the testis, and subsequently, in other immune-privileged or immune-deficient sites (1026,1027). This was an attractive hypothesis, which seemed to answer many of the issues regarding protection of antigens in the testis, particularly those lying outside the blood–testis barrier. Unfortunately, it was not long before a number of studies appeared indicating problems with the FasL expression hypothesis as an explanation for immune privilege. Several groups showed that expression of FasL did not confer immunoprotection in their own studies in the testis or in other systems, and in fact caused quite virulent inflammatory reactions in some cases (1028–1030). FasL also appeared to be expressed at relatively high levels in epithelia of human tissues not generally thought of as immunologically privileged, such as the esophagus, prostate, lung, and nonpregnant uterus (1031). Finally, Kimmel and colleagues reported being unable to detect FasL expression at all in normal human testes that had been flushed of all peripheral blood cell contamination before collection of RNA (1032).

Several studies have suggested that although FasL mRNA can be detected in Sertoli cells using very sensitive methods such as RT-PCR, the protein is not significantly expressed by the Sertoli cells in the normal adult testis, and in fact it may be only the spermatogenic cells that express the ligand constitutively (1028,1033–1035). In the immature rat and porcine testis, FasL protein has been detected in Sertoli cells by immunohistochemistry (1036,1037), but the specificity of FasL antisera used for this purpose has been challenged in at least one major review of the issue (1038). Although it appears that modulation of Fas and FasL in the seminiferous epithelium may play an important role in regulating spermatogenic cell apoptosis, particularly in various testicular damage models (1036,1039,1040), a role in maintaining immune privilege in the testis remains questionable on the weight of evidence.

Immunoregulatory Lymphocyte Subsets

There is no doubt that T cells play a vital role in controlling testis autoimmunity. Orchitis is part of the cluster of autoimmune diseases induced by thymectomy at day 3 of age in mice and rats, which has been attributed to the elimination of regulatory T cells (251,903,904,982,1041–1043). It has now been established that the human type 2 PGA syndrome, which incorporates testicular autoimmunity and hypogonadism in a subset of sufferers, is related to a defect in regulatory T-cell function (1044). On the other hand, type 1 PGA syndrome is associated with a mutation in a transcription factor that controls thymic antigen expression and tolerance induction (983). These disease models in humans and experimental rodents, which frequently involve testicular autoimmune responses, are linked to a failure of tolerance and, more precisely, a shift in the balance between autoreactive T cells and specific regulatory T cells.

After a number of years in the wilderness, suppressor or regulatory T cells have experienced a major revival in interest recently, although many details of the biology of these cells remain to be elucidated. Most attention has been directed toward $CD4^+$ regulatory T cells, although there is evidence for $CD8^+$ regulatory T-cell populations as well (1045–1049). Although details of their mechanism of action remain unknown, they act largely through direct cell–cell contact (e.g., through the B7 inhibitory ligand, cytotoxic T-lymphocyte–associated antigen-4, or CTLA-4) rather than by secretion of immunoregulatory cytokines (1048,1050–1052). How specificity of regulation is generated is not clear because the suppression itself is not antigen specific. Both $CD4^+$ and $CD8^+$ T cells circulate through the normal testis and the possibility that at least some of these cells are testis-specific regulatory/suppressor T cells must be considered. In at least one study, isolation of a $CD4^+$ T-cell line that was able to downregulate the development of adoptive transfer of EAO in mice has been reported (1053). The importance of exposure to testicular antigens during immune system maturation for the development of tolerance to the testis also has been confirmed in adult mice with severe combined immunodeficiency that have had their immune cells reconstituted from fetal liver cells (909). Moreover, although there is still much debate regarding the induction of regulatory/suppressor T cells in the periphery, production of cell-mediated tolerance to autoimmune uveoretinitis by injection of a dominant retinal antigen into the rat testis has been observed (1054–1056). More recently, similar studies have been successfully carried out using adjuvant-induced arthritis and autoimmune encephalomyelitis as end points (1057,1058).

Investigations of the rat testis have shown that significant populations of both NK and NK T cells are present and these lymphocytes appear to be a feature of the normal testes of other species (131) (M. P. Hedger, unpublished data). NK cells form part of the innate immune system, but studies have shown that they are able to modulate dendritic cell function and survival to control adaptive immune responses as well (1059–1061). In contrast to "classic" NK cells, NK T cells are T cells with NK activity, which display unique restriction to glycolipid antigens presented by the MHC-like molecule CD1d (241,242,245). These cells play a key role in promoting graft survival (244,258) and have been implicated in the generation of $CD8^+$ regulatory/suppressor T cells in another immune-privileged site, the anterior chamber of the eye (243,1062,1063). Consequently, there is a strong possibility that testicular NK and NK T cells are involved in production of regulatory/suppressor T cells specific to the testis. Finally, a role for γδ T cells, a minor subset of T cells that possess an alternate TCR structure, has been invoked in suppressing autoimmune reactions in a bilateral model of bacteria-induced autoimmune orchitis, in part through production of IL-10 and TGF-β (436,936). This lymphocyte subset also appears to be involved in maintaining immune privilege in the eye (1064).

Local Immunoregulation

The suggestion that suppression of immune responses in the testicular environment involves regulation of immune cell function by the somatic and spermatogenic cells of the testis goes back many years. Historically, germ cells and Leydig cells have been implicated (1065–1068), but more recently most attention has been directed toward the role of the Sertoli cell.

Rat testicular interstitial fluid is a potent inhibitor of T-cell activation responses in vitro, in spite of the presence of substantial levels of IL-1α secreted from the Sertoli cells, clearly indicating that soluble immunosuppressive factors are a principal influence on lymphocytes circulating through the interstitial tissue (1008,1069–1071). Similar inhibitory effects on lymphocytes in vitro have been observed using whole-testis extracts from mice (1072,1073). The molecules responsible for this activity remain to be identified, although a role for TGF-β has been suggested (609). In fact, neither the various TGF-β isoforms alone nor the other member of the TGF-β family produced by the adult testis, activin A, can account for all the immunosuppressive activity present in this fluid (1069). A contribution by locally produced IL-10 also has been proposed (1074), and there is evidence that this immunoregulatory cytokine is a product of normal testicular cells (394). In fact, although a large number of immunoregulatory factors are known to be produced by the testis under various conditions (Table 5), identification of the active immunosuppressive fraction of testicular interstitial fluid and its cellular source continues to be an important and ongoing area of research.

TABLE 5. *Immunoregulatory factors identified in the testis*

Factor	Source	Main immunoregulatory actions
TGF-β	Sertoli cells, peritubular cells, (resident macrophages)	Anti-inflammatory, immunosuppressive Inhibition of T- and B-cell function
Activin A	Sertoli cells, peritubular cells	Anti-inflammatory, immunosuppressive Inhibition of T- and B-cell function
IL-1α	Sertoli cells, (spermatogenic cells)	Stimulates NO and prostaglandin production Stimulates Th2 responses
α-MSH	Resident macrophages, Leydig cells	Stimulates IL-10 production Inhibits IL-2 responses
M-CSF	not known	Stimulates resident macrophage development
IL-6	Sertoli cells, Leydig cells	Regulates dendritic cell and macrophage development Stimulates Th2 responses
IL-10	Resident macrophages	Th2 cytokine Anti-inflammatory, immunosuppressive
iNOS/NO	Leydig cells, Sertoli cells, spermatocytes	Regulates Th1/Th2 balance Inhibits lymphocyte adhesion Stimulates COX-2 expression
MIF	Leydig cells, (Sertoli cells)	Inhibits T cell and NK cell cytotoxicity
Fas ligand	Spermatogenic cells, (Sertoli cells)	Causes apoptosis of activated T cells
PGE_2	Resident macrophages, Leydig cells	Modulates inflammatory functions of macrophages
IL-1ra	Sertoli cells	Blocks actions of IL-1
Clusterin	Sertoli cells	Inhibits T cell activation and function

Refer to text for complete names and details.

In spite of the questions hanging over the FasL hypothesis of testicular immune privilege, there appears to be little doubt that the Sertoli cell possesses unique immunoregulatory properties. Studies from several groups have shown that Sertoli cells from immature rat, murine, or porcine testes display extended survival as allografts or xenografts, and that cotransplantation of Sertoli cells or testis cell mixtures containing these cells confers increased survival on neural cell xenografts, and pancreatic islet allografts and xenografts (610,1075–1079). There is some evidence that Fas–FasL may play a role in this survival, and that the number of Sertoli cells transplanted is a critical determinant (1080,1081). Several groups have shown that Sertoli cells secrete lymphocyte-inhibiting activity in culture (1082–1086), and these cells are major testicular sites of production of both TGF-β and activin A (333,335,601–603, 638,640). Production of TGF-β_1 by cotransplanted Sertoli cells has been implicated in protection of pancreatic islets implanted under the kidney capsule of syngeneic (genetically identical) recipients (610). Furthermore, the proinflammatory cytokine IL-6, which is secreted by the Sertoli cell under hormonal control, also has a number of immunoregulatory properties, stimulating the production or processing of antiinflammatory cytokines by T cells (205) and inhibiting the maturation of dendritic cells from circulating monocytes through stimulation of M-CSF and the resident macrophage phenotype (336).

Other protective mechanisms also have been suggested, including the ability of the Sertoli cell to form a physical barrier through formation of tight junctions in the mixed graft, and the inherent ability of the Sertoli cell to provide a fully supportive environment for cell growth and differentiation (1078,1087). It appears that Sertoli cells express no MHC class II and comparatively low levels of MHC class I, enhancing their potential to avoid detection by T cells and subsequent immune activation (388,1007,1012,1088). On the other hand, expression of ICAM-1 and VCAM-1 by the Sertoli cells suggests that these cells can selectively bind and interact directly with circulating lymphocytes (1089). Finally, Sertoli cells are similar to macrophages in that they possess an enormous capacity for phagocytosis of senescent cells, cell debris, and other potentially antigenic complexes. Some or all of these characteristics no doubt contribute to the unique graft-protecting abilities of the Sertoli cell, and may play a role inhibiting adaptive immune responses in the intact testis as well.

Leydig cells contribute to the inflammation-related responses of the testis because they produce several inflammatory mediators constitutively (MIF, iNOS) or in response to inflammatory stimuli (IL-1, IL-6, MCP-1) (392,393,397,419,557–559,568,569,770,1090), and regulate interstitial fluid formation through control of endothelial cell permeability (563,830,835, 991,992,1091,1092). However, a converse role in the regulation of immunity in the testis also has been suggested. Early studies established that mouse Leydig cells, but not other testicular cells, bound lymphocytes, macrophages, and eosinophils specifically in vitro (1068,1093,1094), and that the presence of rat or mouse Leydig cells could inhibit lymphocyte proliferation responses in vitro (1067). This binding may have been due, at least in part, to a variant of VCAM-1 expressed on the Leydig cell surface (1095). Moreover, tight junctional specializations between Leydig cells and resident macrophages mediate close physical attachment of these cells in the normal testis (347,349). Inhibition of lymphocytes by Leydig cells in vitro appears to involve both direct contact and secretion of soluble inhibitory factors (1096).

Testosterone concentrations are extremely high in the testicular interstitial tissue (171,173). There is some in vitro evidence that androgens exert direct inhibitory effects on lymphocyte activity, including stimulation of IL-10 production by $CD4^+$ T cells, which may be mediated either by the classic cytoplasmic androgen receptor or through a novel membrane-bound G-protein–coupled receptor (310,311,313,316,1097). A more general effect of androgens on the type 1/type 2 cytokine balance was demonstrated by the fact that administration of testosterone to hypogonadal men stimulated baseline IL-10 levels and inhibited IL-1β and TNF-α levels in serum (1098). However, in vivo evidence for local immunoregulation by androgens in the testis is conflicting. Manipulation of Leydig cell function and androgen production by a number of methods had no effect on survival of parathyroid allografts in the normal rat testis (50,51,1099) or the survival of pancreatic cells allografts in the abdominal testis model of Selawry and colleagues (1100,1101). However, inhibition of Leydig cell function by estrogen treatment of the recipient rats several days before grafting did abrogate survival of parathyroid grafts in the normal scrotal testis (49). The problem with the latter model is that it does not exclude a role for direct estrogen-mediated effects on graft rejection in the testis. In the absence of any comprehensive study of this issue, it remains difficult to say whether androgens play an important local role in immune regulation in the testis. However, in addition to androgens, Leydig cells produce several other factors with lymphocyte-regulating activity. These include the antiproliferative proopiomelanocortin peptides (319,320,323,328,1102) and MIF, which notwithstanding its proinflammatory functions, inhibits the cell-killing activity of cytotoxic T cells and NK cells (1103–1105).

The most significant role of the Leydig cells in mediating immune control of the testis may lie in their ability to recruit large numbers of resident macrophages into the testis (130,397,406,413).

The unique immune environment of the testes appears to be due, at least in part, to suppression of the proinflammatory functions and upregulation of antiinflammatory functions of the resident testicular macrophages (378). In macrophages from other tissues, regulation of this phenotype has been shown to be a consequence of exposure to type 2 cytokines (IL-4, IL-10, and IL-13) (345) or PGE_2 (1106–1109). It is significant, therefore, that prostaglandins, including PGE_2, are produced at significant levels in the testis even under normal conditions (1110–1112), and the rate-limiting enzyme of prostaglandin synthesis during inflammation, COX-2, is expressed constitutively in a wide range of testicular cell types (1113) (M. P. Hedger, unpublished data). These data suggest that local production of prostaglandins may be responsible for the antiinflammatory/immunosuppressive phenotype of the testicular macrophages. In support of this hypothesis, studies by Kern and colleagues confirmed that the addition of the COX inhibitor indomethacin to cultures of rat testicular macrophages restored their production of IL-1β, IL-6 and TNF-α in response to LPS stimulation and blocked their inhibitory effect on peripheral blood lymphocyte proliferation in vitro (376,390).

Prostaglandins are synthesized from arachidonic acid, which is itself produced by cleavage from membrane phospholipids by the action of PLA_2, an enzyme that is expressed in both Sertoli cells and Leydig cells and is hormonally regulated (1114,1115). The arachidonic acid produced can be converted through the lipoxygenase pathway to leukotrienes and lipoxins, or through the action of COX to produce the various bioactive prostaglandin series, prostacyclins, or thromboxanes (1116–1118) (Fig. 16). COX, lipoxygenase, and other enzymes of the main synthetic pathways are present in the testis (1110,1113,1119–1125), although their cellular localization, regulation,

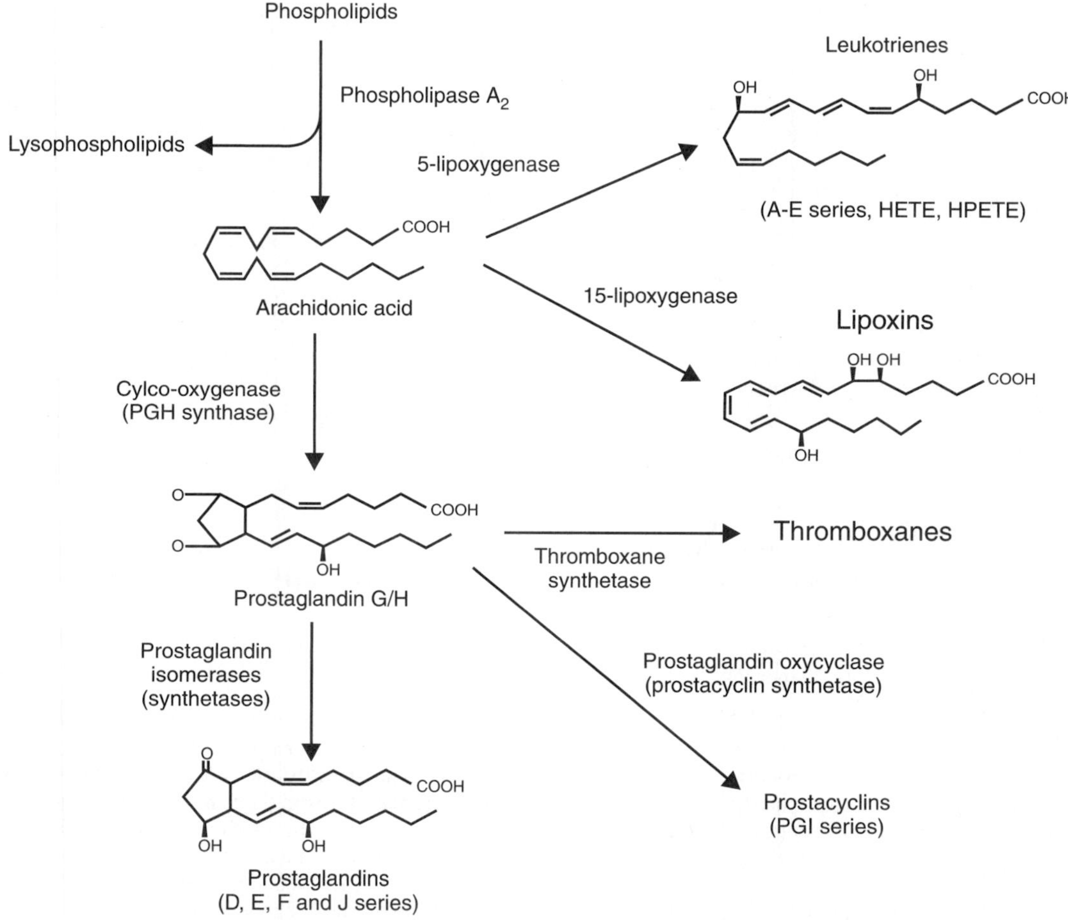

FIG. 16. Synthesis of bioactive lipids. Cleavage of membrane-bound phospholipids by the action of phospholipase A_2 leads to production of arachidonic acid and lysophospholipids. Arachidonic acid may be converted to prostaglandins, prostacyclins, or thromboxanes through the intermediates prostaglandin G and H by the action of the rate-limiting cyclooxygenase (COX) enzymes. Two major forms of COX have been identified: a constitutively expressed form (COX-1) and an inducible form (COX-2), which is upregulated during inflammation (1116,1285). Alternatively, lipoxygenases may catalyze the conversion of arachidonic acid to leukotrienes or lipoxins (1286). Most of the products in these complex pathways have specific regulatory activities, including profound effects on immune responses (512,1129,1131,1132,1135).

and significance to testicular function are still relatively poorly defined. Acting through a number of different receptor subtypes, E series prostaglandins can exert proinflammatory and regulatory effects on T cells, macrophages, and dendritic cells (512,1106–1109,1126–1128), whereas prostaglandins D and J and the lipoxins possess specifically anti-inflammatory actions (512,1128–1132). Moreover, phosphatidylcholine-containing lipids, produced by the cleavage of arachidonic acid from phospholipids by the action of PLA_2, possess antiinflammatory and immunoregulatory functions (1133–1135). Although the role of bioactive lipids in testicular immunoregulation has been largely ignored in favor of locally produced cytokines, this is an area of research that potentially has much to offer in the future.

In summary, there are a number of potential mechanisms for local control of adaptive immune responses in the testis involving specific regulatory functions of the Sertoli and Leydig cells, and recruitment of specific testicular macrophage and lymphocyte subsets with regulatory phenotypes (Fig. 17).

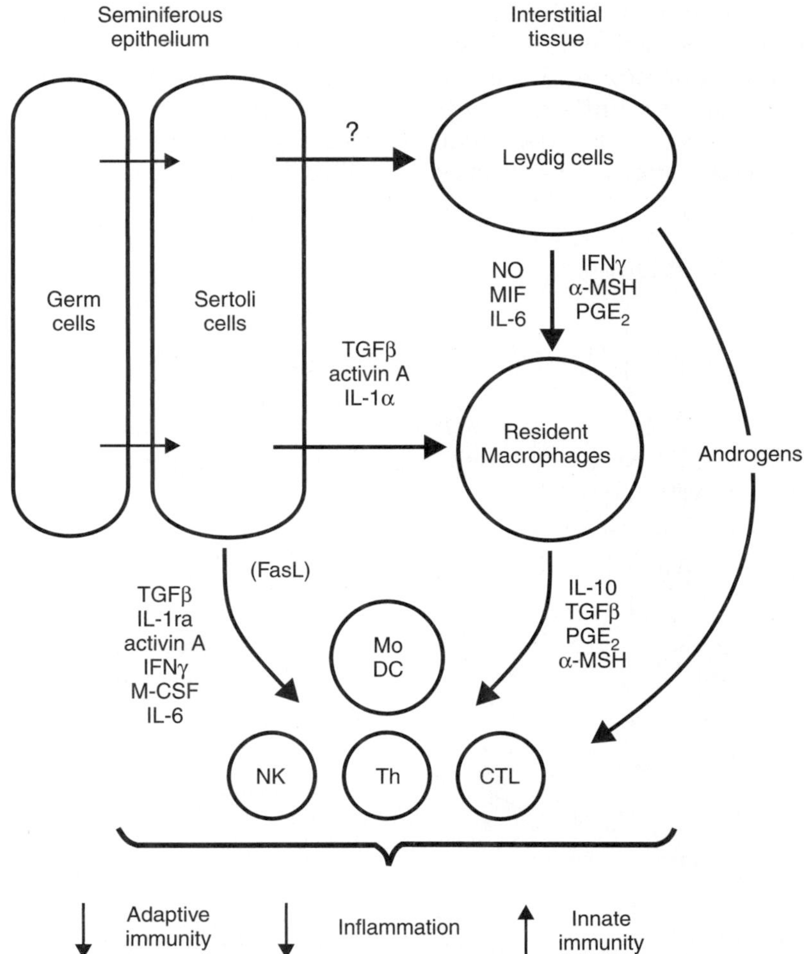

FIG. 17. A paradigm for understanding the intercompartmental interactions regulating immunity in the testis (see text for more details). In the seminiferous epithelium, the Sertoli cell produces a number of cytokines, including interleukin (IL)-1α, IL-6, activin A, transforming growth factor (TGF)-β, interferon (IFN)-γ, and IL-1 receptor antagonist (IL-1ra). Evidence for expression of Fas ligand (FasL) by the Sertoli cells in the normal testis is controversial. Production of these cytokines is stimulated by the presence of the spermatogenic cells. It is hypothesized that these cytokines modulate the activity of the resident macrophages and circulating immune cells in the testis to create an environment where inflammation and antigen-specific immunity are suppressed, while innate immune functions are retained or enhanced. The role of the Leydig cell and its regulation by the seminiferous epithelium is less well defined, but these cells are responsible for recruiting macrophages into the testis and may have other immunomodulatory actions through production of macrophage migration inhibitory factor (MIF), IL-6, nitric oxide (NO), IFN-γ, prostaglandins, pro-opiomelanocortin–derived peptides such as α-melanocyte–stimulating hormone (α-MSH), and androgens. The resident macrophages have a predominantly immunosuppressive phenotype, producing IL-10, TGF-β, prostaglandins and α-MSH, which also exert their effects on the circulating immune cells. Mo, monocytes; DC, dendritic cells; NK, natural killer cells; CTL, cytotoxic lymphocytes; Th, helper T cells.

This regulation may involve physical sequestration of antigen by the blood–testis barrier, specific membrane-bound receptors and adhesion molecules, and locally produced cytokines, steroids, and bioactive lipids. A role for the spermatogenic cells in this process is less certain because translocation of the testis to the abdominal cavity and subsequent disruption of spermatogenesis does not abrogate survival of intratesticular allografts or xenografts (49,52,1100). However, there is evidence that spermatogenic cells can inhibit lymphocyte activity in vitro and in vivo (1065,1066), and it cannot be excluded entirely that these cells participate in local immunoregulation either directly, or through stimulation of the functions of the Sertoli and Leydig cells. Although no doubt important in its own right, local immunoregulation may act to back up the normal tolerance to testicular antigens provided by thymic selection and regulatory lymphocytes. These mechanisms may help to explain why orchitis affects only a minor subset of patients with PGA syndrome, and why orchitis is not universally induced in the day 3 thymectomy model in rats and mice.

Germ Cell and Testis Transplantation

Since the mid-1990s, there has been a virtual explosion in the number of studies dealing with the transplantation of spermatogenic cells. After the pioneering work of Brinster and colleagues (1994) (1136,1137), the restoration of fertility in mice with congenital or chemically induced absence of spermatogenesis through transfer of isolated spermatogonial stem cells from fertile donors is more or less routine (70,1138,1139). This procedure usually involves injection of spermatogonia directly into the rete testis or efferent ducts of the recipient. The only fully successful cross-species transplantation model to date has been that of rat spermatogonia to the mouse testis (1140,1141), although spermatogonial proliferation or partial spermatogenesis has been observed for hamster, dog, rabbit, bovine, equine, porcine, monkey, and human spermatogonia in the murine testis (1142–1146). Spermatogenic development of transplanted stem cells in testis of species other than the mouse also has been achieved, including rat into rat (1147,1148) and mouse into rat, bovine, or monkey testes (1148,1149). Most recently, successful transplantations of Sertoli cells (1150) and Leydig cell stem cells (1151) have been reported in mice, using similar protocols.

There is some evidence that spermatogonial transplantation can occur across allogeneic boundaries in the rat and mouse (1147,1148,1152,1153), but prolonged restoration of spermatogenesis without rejection usually requires a histocompatible or immunodeficient recipient (70). A characteristic feature of transplantation in immunologically intact animals appears to be an increase in peritubular macrophages and macrophage invasion into the seminiferous epithelium and tubule lumen (1152,1154). Whether these cells are serving an immunoprotective function, or are simply responding to an inflammatory stimulus from degenerating spermatogenic cells, is unclear. In a detailed study by Kanatsu-Shinohara and colleagues (70), allogeneic transplantation of mouse spermatogonia was only partially successful, with arrest at the round spermatid stage in highly disorganized tubules, mononuclear cell infiltration in the interstitial tissue, and eventual rejection after several months. However, mature spermatozoa and successful natural mating were achieved by the use of donor stem cells from immature testes and treatment of the recipient animal with anti-CD4 and anti-CD8 antibodies or the immunosuppressive agent, rapamycin, to inhibit T-cell activity.

With respect to the immunology of spermatogonial cell transplantation, the most obvious question to be asked is why do allogeneic spermatogenic cells in the seminiferous epithelium induce a rejection response, when the testis is able to support allogeneic grafts in the interstitial tissue (46–51)? No consistent, properly controlled study has been carried out to address this question. However, the spermatogonial cells normally are injected through the rete testis or efferent ducts, and into testes that lack mature germ cells. Such an injection might seem to be no more likely to cause a sensitization reaction than the insertion of a graft under the testis capsule, but it is clear that the rete testis is the most susceptible region of the testis for initiation of autoimmune reactions (125,127,136,386,387,433,437,933). It is quite possible that the cytotoxic treatment used to deplete the endogenous spermatogenic cells may increase the risk of graft failure and rejection, or that the lack endogenous spermatogenic cells may contribute to an absence or failure of central or peripheral tolerance to these cells. Prolonged absence of spermatogenic cells might alter the local immunoprotective environment. Perhaps the key to understanding lies in the fact that different groups have reported variable success rates with transplantation of spermatogenesis across allogeneic barriers, and that there simply remain some minor technical issues to be overcome. Regardless of the reasons, it is clear that implanting grafts into the interstitial tissue of the intact adult testis and injecting isolated spermatogonial stem cells into a germ cell–depleted seminiferous tubule usually lead to quite different immunological sequelae.

More recently, a method for transplantation of fragments of intact testicular tissue from mice, pigs, bulls, goats, or monkeys under the skin of immunocompromised mice has been developed (68,69,71). Surprisingly, such grafts tend to be very successful, with vascularization, normal steroidogenesis, and even complete spermatogenesis taking place, although the "testes" are nonfunctional because there is no exit for the spermatozoa that are produced. The ability of spermatogenesis to occur in such grafts is believed to depend on the reduced temperature of the skin, and the immunodeficient nude mouse has proved to be the most successful recipient model.

The fact that immunocompromised recipients must be used also indicates that fragments of adult testis tissue are not inherently immunoprotective. This is consistent with early observations that fetal and postnatal testis tissue are viable as grafts under the kidney capsule of outbred rats, but that adult testis tissue is quickly rejected under the same conditions (1155,1156). A much more recent study showed a very similar lack of protection for human testis xenografts in the murine kidney, although survival was observed in recipient mice with an inactivating deletion of MHC class II expression, thereby implicating a $CD4^+$ T-cell–mediated rejection process (1032). Such results are in direct contrast to studies that have successfully used isolated Sertoli cells to extend neural cell and pancreatic islet survival in similar sites (610,1075,1077–1079,1087). Once again, there are insufficient experimental data to explain this apparent discrepancy with confidence. Obviously, the cellular composition and structural integrity of the transplants are quite different in the two models. For example, whole-testis fragments will contain MHC class II–positive macrophages and dendritic cells, whereas isolated Sertoli cells are MHC class II negative (388,1007,1012,1088). The presence or absence of MHC class II expression, and possibly the level of MHC class I expression as well, may be the critical determinant for survival of foreign testicular tissue in an immunocompetent host.

Immune Privilege as a "Testis-Specific" Phenomenon

Based on the evidence presented in the previous sections, it appears that immune privilege is not an absolute property of the testis, but is provisional on certain criteria that remain relatively poorly understood. Why are grafts involving testicular tissue successful in some models, but not in others? There is no doubt that the antigenicity, structural integrity, inherent viability, and "passenger" immune cells of the graft are likely to be important, and that the cellular composition and structural organization of the testis itself plays a role. The Sertoli cells, Leydig cells, resident testicular macrophages, and possibly the spermatogenic cells all contribute to the creation of a unique testicular environment that also includes the suppression of antigen-specific immune responses. There is evidence that grafts of testicular tissue are more likely to avoid rejection in some sites than in others. There are genetic determinants that regulate autoimmunity in the testis, so that experiments involving different species and strains of animals produce different results. Any or all of these parameters may affect the outcome in studies on testicular immune privilege.

The actual mechanisms that might contribute to immune privilege in the testis include (a) the blood–testis barrier, (b) the lack of MHC expression on spermatogenic cells, (c) tolerogenic APC activity, (d) regulatory lymphocyte subsets, (e) antiinflammatory and immunosuppressive functions of the resident macrophages, (f) local production of immunoregulatory/immunosuppressive cytokines and other mediators, and (g) androgens. Considerable work is required to establish which of these mechanisms are truly involved and essential. However, examination of the available data for the testis suggests that tissues traditionally identified as being immunologically privileged, such as the testis and anterior chamber of the eye, may simply lie at one extreme of a range, because the mechanisms implicated in the testis apply to a greater or lesser extent in most tissues. For example, regulatory/suppressor T cells are involved in the control of autoimmune disease in many tissues (255,1048,1050,1051), whereas the enhanced success rate of liver transplants may be attributable to the presence of large numbers of immunoregulatory NK T cells (1157,1158), which are now also implicated in testicular immune privilege. One might properly think of immune privilege as simply a manifestation of a particularly effective immunoregulatory environment. The need for enhanced immunoregulation in tissues such as the testis or the eye may be related to the burden of the autoantigens present, or the need to prevent active inflammation causing damage to particularly susceptible tissues. Although it is debatable whether autoimmunity to spermatogenic cells is more catastrophic to the organism than autoimmune gastritis, diabetes, or thyroiditis, there must have been considerable evolutionary selection pressure applied to ensure that the immune system was restrained from attacking the postmeiotic germ cells the moment they appeared during sexual maturation.

Immunology of the Epididymis and Genital Tract

The Epididymis, Vas Deferens, and Accessory Glands

Compared with the testis, there have been relatively few studies dealing with the immunology of the remainder of the male reproductive tract. It is obvious that the distribution of immune cells in the ducts of the epididymis, vas deferens, accessory glands, and urethra is quite different from that seen in the testis. Most distinctive is the presence of many intraepithelial lymphocytes, which are predominantly CD8+ T cells (126,127,154,156,463,494,496). Second, although there are epithelial cell tight junctions similar to those found in other epithelia (135,150), there is no distinct structural correlate to the occluding junctions of the blood–testis barrier. Nonetheless, the fact that sperm can survive in the epididymis for considerable periods of time without eliciting an autoimmune response indicates that some immunoregulatory mechanisms operate in this tissue and in the remainder of the tract as well. It can be assumed that the patency of the tract is essential to this process, because blockage, trauma, or rupture of the ducts rapidly leads to sperm autoimmunity after leakage of the luminal contents, inflammation, and granuloma formation (474,877).

It is most likely that protection of spermatozoa in the male reproductive tract is roughly identical to immunoregulation in the other elements of the common mucosal system (i.e., the gastrointestinal and respiratory tracts). The mucosal surface represents the immediate interface between the host organism and its external environment, and close regulation of immune responses at this interface is essential. Tight junctions and adherence between the epithelial cells restrict passage of both antigen and antibody, so that antibody responses are dominated by secretory IgA, which lacks the ability to activate complement and possesses strong antiinflammatory properties (1159,1160) (Fig. 9). Reactions to inhaled or digested antigens are controlled through a process called *mucosal tolerance* (277,278,1161,1162), and failure of tolerance in the gastrointestinal and respiratory tracts results in allergies. Presumably, the same applies to sperm antigens in the genital tract and sperm autoimmunity. The mechanisms of mucosal tolerance involve clonal T-cell deletion or anergy and active suppression by regulatory/suppressor T cells (1163–1165). Although cytotoxic CD8+ T cells, comprising both $\alpha\beta$ TCR and $\gamma\delta$ TCR subsets, predominate in the mucosal epithelium, it is the CD4+ T cells that are responsible for mucosal tolerance, producing IL-4 and IL-10 (Th2 cells), and TGF-β (Th3 cells) (277,1161,1164,1166). Mucosal intraepithelial lymphocytes also express CD1d constitutively and can activate immunoregulatory NK T cells in the epithelium (1167,1168).

There is no evidence for distinct mucosal lymphoid structures or MALTs in the male genital tract. However, there is considerable evidence that epithelial cells themselves can act as APCs in the mucosal immune system (280,1169,1170), and in this regard it is relevant that the principal cells of the epididymis are actively pinocytotic cells constantly absorbing large quantities of luminal fluid and its contents, including sperm products (1171–1173). Consequently, presentation of antigen to the intraepithelial lymphocytes by the principal cells of the epididymis might be an important mechanism for controlling sperm autoimmunity. Indeed, it is highly likely that most cases of sperm autoimmunity develop, not through failure of immunoprotection in the testis, but through disruption of mucosal tolerance in the epididymis and remainder of the genital tract.

In addition to the common mucosal regulatory mechanisms, it is conceivable that there are other immunoregulatory mechanisms that are more specific to the genital tract. For example, immunosuppressive factors produced by the testis may diffuse into the epididymal fluid, and it is significant that the risk of antibody formation in obstructive azoospermia and congenital absence of the epididymal duct increases as the distance of the lesion from the testis increases (478). Modifications of the sperm surface membrane by epididymal secretions may act to obscure sperm antigens (142), and expression of immunoregulatory molecules on the surface of the sperm itself, such as both classic and nonclassic MHC antigens (1013,1014,1018), a CD4-like MHC ligand (1174,1175), and FasL (1035), also may play a role. Finally, the possibility that specific immunosuppressive cells and secretions are involved in the remainder of the genital tract, as they appear to be in the testis, should not be dismissed simply for lack of evidence for their involvement.

Immunosuppression by Seminal Plasma

Seminal plasma is profoundly immunosuppressive, as defined by the ability to inhibit various T-cell and NK-cell activities in vitro (1176–1179). This immunosuppressive activity has been proposed by a number of researchers to play a role in preventing lymphocyte responses against sperm autoantigens in the male and female reproductive tracts at the time of ejaculation and immediately after (897,1180). This activity can be attributed to a number of factors, including prostasomes (1181,1182), oxidized polyamines (1183), prostaglandins of the E series (1178,1184–1186), "nonspecific" lymphocyte-suppressing proteins (1187,1188), and immunoregulatory cytokines (1189–1192).

Prostasomes are multilaminar vesicles secreted by the normal prostate, and are a major component of human semen (1182). Pure preparations of prostasomes inhibit mitogen-induced T-cell proliferation and inhibit macrophage phagocytic activity in vitro (1181). Seminal plasma also contains very high concentrations of the polyamines spermine and spermidine (1183,1193,1194). These polyamines are not immunosuppressive themselves, but are converted to oxidized forms that inhibit cell growth by the action of polyamine oxidase, an enzyme found in serum used in culture media (1195). Oxidized polyamines are unstable and rapidly metabolized to the cytotoxic molecules, acrolein and putrescine (1196). Prostasomes and polyamines are responsible for much of the apparent immunosuppressive activity of the ejaculate measured using lymphocyte cultures, but whether these factors have any physiological significance in terms of controlling immune responses in vivo remains speculative (1197,1198). On the other hand, after removal of the prostasomes and inactivation of polyamine activity in human seminal plasma samples from infertility clinic patients, an inverse relationship between T-cell inhibitory activity and the incidence autoimmune infertility associated with sperm antibodies has been observed (1199). This observation suggests that immunosuppression by seminal plasma could be an important determinant of male autoimmune infertility.

In contrast to the prostasomes and seminal polyamines, prostaglandins and cytokines are well-characterized regulators of immunity and inflammation and are likely to play a significant immunological role in the male and female reproductive tracts. Human seminal plasma in particular contains extraordinarily high concentrations of PGE_2, PGE_1, and their 19-hydroxylated forms (1185,1200,1201). Aside from their well-characterized effects on vascular permeability and smooth muscle contractility, these molecules inhibit T-cell proliferation, NK-cell cytotoxicity, and the proinflammatory activities of macrophages and T cells (512,1128). Interacting with specific receptors on the surface of these cells, the PGEs are able to switch cytokine responses from type 1 to type 2 and the functional phenotype of macrophages from proinflammatory (M1) to anti-inflammatory (M2) (345,1106–1109). They are produced by the seminal vesicles and prostate, but the vas deferens also may be a major source because this tissue expresses extremely high levels of COX-2 under normal conditions (1119,1202). After removal of the prostasomes and inhibition of polyamine oxidation, prostaglandins are found to be responsible for most of the immunosuppressive activity of human seminal plasma.

The cytokines with immunosuppressive activity that have been positively identified in human seminal plasma are TGF-β_1 and TGF-β_2 (1189,1192,1203,1204), IL-10 (1190,1191,1205,1206), and activin A (1207). The TGF-β in seminal plasma is derived from the distal genital tract, comprising the seminal vesicles and the prostate gland (1204). IL-10 is a product of monocyte/macrophages and T cells, although other cellular sources in the male tract may be possible because epithelial cells can produce IL-10 constitutively (1208). Although activin A subunit mRNA and protein is expressed in the human prostate (1209), activin A levels in normal seminal plasma are effectively eliminated by vasectomy, indicating a principally testicular or epididymal origin (1207). However, TGF-β_1 is by far the most abundant and effective of the immunosuppressive cytokines in seminal plasma (1210) (M. P. Hedger, unpublished data). Other immunosuppressive molecules that are neither prostaglandins nor regulatory cytokines have been detected in seminal plasma or accessory gland secretions from various species (1187,1188). Some of these molecules have been identified, including the protein clusterin and a prostatic steroid-binding protein in rats (1211,1212). The actual origins of many of these molecules and their possible role in immune regulation must await either their identification or further functional characterization.

Although the relationship between immunosuppressive molecules in seminal plasma and immunological infertility in men remains to be firmly established, there is no doubt that these conditions lead to significant alterations in both inflammatory and immunosuppressive cytokines in seminal plasma. These molecules also have direct effects on the epithelia, stroma, and leukocytes of the female genital tract (1210,1213). Animal studies show an intense but transient inflammatory response in the endometrium at mating stimulated by seminal plasma, and one consequence of this inflammatory response is the induction of a transient state of hyporesponsiveness to paternal MHC class I antigens (1214). Insemination is causally linked to the activation and expansion of populations of lymphocytes mediating active immune tolerance in the implantation site, a process in which seminal plasma TGF-β_1 has been particularly implicated (902,1210,1214). Semen may therefore play a critical role in providing the antigenic and environmental signals necessary to initiate an appropriate maternal immune response to the conceptus during pregnancy.

Apart from immunosuppressive factors, many proinflammatory cytokines are found in samples of seminal plasma, even those from men with apparently normal fertility, including IL-1α, IL-2, IL-6, IL-8, IL-11, IL-12, TNF-α, IFN-γ, and M-CSF (1191,1215–1219). The presence of soluble IL-2 and IL-6 receptors, molecules that modulate the activity of their respective cytokines, also have been found (1191). This is not particularly surprising because these cytokines are present in most biological fluids and are modulated by the general health status of the individual. Elevated cytokine concentrations, both proinflammatory and immunosuppressive, in seminal plasma are associated with inflammation, infection, and increased leukocyte numbers in the male tract (1190,1205,1216–1221). However, specific positive and negative associations with infertility of a noninflammatory nature also have been reported (1189,1191,1215,1216,1222,1223).

Inflammation, Infertility, and Seminal Leukocytes

In humans, seminal plasma is usually the only window available for investigating the immunology of the male reproductive tract. More invasive procedures such as biopsies are rarely performed and highly selective, and have the potential to cause inflammatory reactions themselves. Much is made, therefore, of the relationship between measurements of leukocytes, antibodies, and cytokines in seminal plasma, and fertility. However, although there is little doubt that sperm antibodies contribute to infertility, and that elevated cytokines are an indicator of ongoing infection, inflammation, or other immune event, the significance of leukocytes in the semen is a matter of ongoing debate.

An elevated number of leukocytes in the semen is usually considered to be an indication of infection, and the World Health Organization (WHO) has set an arbitrary level of 1×10^6/mL as the threshold of normality. In fact, leukocytes are present even in semen of fully fertile men, with many studies agreeing that 10^4/mL to 10^5/mL is a relatively normal value, and even in men with urogenital infections leukocyte numbers do not always reach the WHO level (499,1224). In fact, leukocyte numbers alone may be a very unreliable indicator of the presence or absence of infection. The origin of these cells is also somewhat obscure. Evidence suggests that the epididymis or vas may be a major source (505–507), but the major leukocyte subset present in most semen samples is neutrophils (499–505). Neutrophils are not a normal feature of the tissues of the genital tract. Finally, the impact of these cells on fertility is poorly understood. Some studies have shown a relationship between leukocytospermia and impaired sperm function (1225–1228), and data suggest that these cells might be a major source of ROS or other products causing sperm damage (1229–1231). Other studies have failed to confirm a clear link between seminal leukocytes and infertility or sperm antibody formation (503,1231–1233). It even has been suggested that they might play a beneficial role by the removal by phagocytosis of abnormal sperm or assisting in sperm capacitation (469,503). It is fair to say that the role of leukocytes in semen is more complicated than might be expected, and many questions still remain (499,505,1234).

RESPONSE TO INFECTIONS AND TUMORS: INNATE IMMUNITY IN THE MALE REPRODUCTIVE TRACT

If the male reproductive tract, and the testis in particular, is a site of reduced antigen-specific immune responses, then the question must be asked: how does the genital tract avoid recurrent infections or the development of tumors? Interest in the question has been stimulated by the fact that the male reproductive tract is a major source of transmission of HIV (1235,1236), and by the observation that relapsing lymphoblastic leukemia in the testis after treatment is a frequent problem in male patients (1237–1239). Although the progression of HIV infection in the male reproductive tract is very poorly defined, destruction of the spermatogenic cells is a characteristic feature of HIV infection in men (948,1240). Studies by Jahnukainen and colleagues in rats have suggested that testicular relapse of leukemia may be due to the unique immunoregulatory environment of the testis and specifically the ability of Leydig cells to bind lymphocytes (1241–1243). Moreover, it appears that the immunological protection in the testis is extended not just to spermatogenic cell antigens and graft antigens, but to tumor-specific antigens that would induce immune reactions elsewhere in the body (1244).

As a result of these and other observations, it has been suggested that virally or tumorigenically transformed cells might be able to evade both the immune system and cytotoxic drugs by "hiding out" in the testis. The reality, however, is that infection of the testis is relatively rare compared with the remainder of the genital tract (952), and testicular tumors are less frequent than tumors in many other parts of the body (1245). Moreover, when they do occur, testicular tumors are accompanied by the expected mononuclear cell infiltrates, which are related to the size, progression, and type of the tumor (435,439). The apparent deficiency in adaptive immune responses in the testis, therefore, implies that innate (pathogen-specific) immunity might have increased importance for dealing with tumors and infected cells at this site. There is some evidence for this assertion, both from studies of the male reproductive tract and from analogy with the rest of the common mucosal system.

As noted previously, IgA is the main immunoglobulin isotype in the common mucosal system, and possesses specific antiinflammatory properties (1159,1160). This immunoglobulin is found in the fluid of all tissues of the male reproductive tract, but because vasectomy does not substantially alter IgA levels in semen, it appears that most comes from the urethra and accessory glands, in particular the prostate gland (1246–1249). IgA-secreting plasma cells have been found in the urethral gland (494,1250), but most IgA in the reproductive tract is believed to be derived initially from the circulation (281,282). Production of secretory component by the prostatic epithelium, which is required for transport of IgA across the mucosa, is androgen regulated (1249). In men, bacterial infection of the genital tract is associated with a large increase in secretion of IgA in prostatic fluid (1247). This antiinflammatory immunoglobulin is an important first line of defense against infection in the male tract (Fig. 9).

Lymphocytes expressing CD8, a marker expressed by both cytotoxic T cells and NK cells, predominate in the testicular interstitial tissue and in the epithelium of the remainder of the genital tract (126,127,131, 154,156,431,463,494,496). In the rat, mouse, and possibly the human testis, a significant number of these cells are either NK or NK T cells (131) (M. P. Hedger, unpublished data). Whereas T cells specifically mediate antigen-specific immune responses, NK cells are able to recognize and destroy infected or transformed cells without prior immunization and therefore are an important part of innate immunity. However, NK and NK T cells also inhibit antigen-specific immunity, autoimmunity, and graft rejection responses, and secrete the antiinflammatory type 2 cytokines, IL-4 and IL-10, as well as TGF-β (243–245,258, 1059–1061,1063). The Leydig cell–secreted cytokine MIF is a potent inhibitor of cytotoxic T-cell and NK-cell activity (1103–1105), and it may be that a function of this cytokine in the testis is to restrain cell killing in the absence of an overt immunological challenge. In the remainder of the genital tract, the role of NK cells is poorly defined, although mucosal intraepithelial lymphocytes express CD1d constitutively and consequently may activate NK T cells at the epithelium (1167,1168). Together, the data clearly indicate that an active cellular response to viruses and tumors is possible in the male reproductive tract, involving lymphocyte subsets that also can regulate inflammation and adaptive immunity responses.

In addition to the testicular macrophages, both the Sertoli cell and Leydig cell are able to recognize and respond to bacterial pathogens directly (545,551,552,557,559,1089), presumably through expression of specific TLRs on the surface of these somatic cells. However, the expression of TLRs and their roles in the male reproductive tract have yet to be studied in any detail. A novel member of the TLR family called TLR11 was found to be strongly and specifically expressed in kidney and bladder epithelium, although not in the testis (185). The kidneys of mice with deletions of TLR11 were enormously more susceptible to intraurethral infection with *E. coli* than were wild-type mice, suggesting that this TLR plays a specific role in protection of the urogenital tract against infection.

All mucosa protect themselves through a number of nonimmunological mechanisms that support the integrity of the epithelial barrier. These include tight junctions that restrict passage between cells, secretion of mucins and other antigenically inert molecules, control of local pH, recruitment of phagocytes (macrophages and neutrophils) to the epithelium, and production of lysozyme, ROS, proteases, complement components, and antimicrobial cytokines and peptides (271,281,1161,1162,1170). There appears to be no reason to assume that the genital mucosa is any different (281,282).

The role of interferons in the rest of the male reproductive tract remains to be determined, but both type I (α and β) and type II (γ) interferons are produced by the rat testis (745,1251–1253). These molecules exert a vast range of immunoregulatory and antiviral actions, including suppression of proliferation, cytotoxicity, and stimulation of MHC class I and class II expression (190,1254,1255) (Table 4). Testicular macrophages produce type I interferons constitutively, but viral infections upregulate their expression in macrophages, Sertoli cells, and Leydig cells, as well as expression a number of interferon-inducible genes involved in viral protection (745,1252, 1253). The strongest interferon response to infection occurs in the Leydig and Sertoli cells, but only the Leydig cells appear to express IFN-γ under such conditions. Peritubular cells also respond to viral infection by producing interferon-inducible genes. In contrast, spermatogenic cells display little or no interferon secretion or interferon-induced protein expression (745).

Finally, defensins are small (3 to 4 kDa), positively charged antimicrobial peptides that are able to disrupt bacteria, fungi, parasites, and some enveloped viruses by forming multimeric pores in the distinctive cell membranes of these pathogens (1256). The defensins belong to one of two families, the α- and β-defensins (1257). The β-defensins in particular are produced by most mucosal epithelial tissues, including the tissues of the urogenital tract such as the testis and epididymis, and their production is stimulated by TLR and cytokines (1258–1261). A number of epididymis-specific β-defensins have been identified in the mouse and the rat (1262,1263), and a novel β-defensin called Bin-1b was shown to be exclusively produced and secreted by the rat caput epididymis (1263,1264). This peptide was not expressed in the testis, prostate, or seminal vesicles, but was developmentally regulated, with expression first appearing at approximately 30 days of age (i.e., late puberty in the rat). Interestingly, Bin-1b also bound to sperm, and blocking the expression or activity of Bin-1b reduced sperm motility in vivo, suggesting a role for this molecule in sperm maturation (1263).

IMPLICATIONS, APPLICATIONS, AND THE FUTURE

Although this review has concentrated almost exclusively on the study of inflammation and immunity in the male reproductive tract as a means of understanding immune-based infertility and the impact of infection and inflammation on male reproduction, there are a number of other reasons to study these interactions. They involve the application of the knowledge obtained by this research to clinical issues of contraceptive development and transplantation medicine.

Immunocontraception

Immunocontraception through the application of a vaccine that targets sperm antigens or reproductive hormones has been a topic of active research almost since sperm antibodies were first discovered (1265,1266). This approach has the advantages of a potentially high degree of specificity and convincing "proof-of-principle" from the many patients with preexisting autoimmune infertility. A large number of candidate antigens in the male have been proposed, mostly directed toward critical sperm function antigens identified from both human and animal studies (883,886,887). Although vaccines against reproductive hormones actually have been trialed (1267), it is unlikely that such an approach would be readily applicable to men given that androgen is required for many functions besides fertility, whereas ablation of FSH is unlikely to reduce fertility sufficiently. However, a GnRH vaccine approach might be a suitable alternative to the use of GnRH analogs for reducing prostatic hypertrophy in men with hormone-dependent prostate cancer (1268–1270).

It is recognized that the concept of immunocontraception generates serious concerns related to safety, efficacy, and reversibility. The major concern would be the risk of more widespread autoimmune disease because autoantigens are involved and there is an established relationship between sperm autoimmunity and autoimmunity in general (865,866). A much more rigorous standard of safety and efficacy would apply for a contraceptive vaccine than for vaccines that have been developed to deal with life-threatening and catastrophic diseases. Genetic differences in the immune response genes in the human population and the complexity of the immune system itself might make it difficult to develop a single vaccine that works effectively enough in all men to compete successfully with currently available hormone-based approaches to contraception. Finally, it may be difficult to develop a contraceptive vaccine that is readily reversible.

Although such considerations may inhibit development of antireproductive vaccines in humans, they do not limit the potential use of immunocontraception for controlling pest animal species in the wild. Some progress has been made toward development of contraceptive vaccines that are effective in such species, although these studies have yet to deal adequately with the difficulties associated with obtaining effective dispersal of the vaccine vector in the wild, while limiting the impact on domestic or commercial stocks (1271–1273). Nonetheless, the targeting of critical reproductive antigens to control fertility is a very attractive and powerful idea that will continue to engage reproductive immunologists for some time in the future. Likely to be of most benefit for contraceptive development is the use of sperm autoimmunity and antigen identification studies to characterize molecules with crucial function that might be used as molecular targets for other, nonimmunological drug agents.

Implications for Transplantation Medicine

One of the most obvious potential benefits from understanding the principles of immunoregulation that operate in the male reproductive tract is the possibility that the same mechanisms might be used to control immune responses at other sites. Obviously, many of the immunoregulatory mechanisms that operate in the male reproductive tract are not specific to these organs. However, if there are immunoregulatory mechanisms that are unique to the testis, for example, where extended graft survival has already been shown to be feasible, then discovery of these mechanisms has the potential to lead to the development of entirely novel methods of treating graft rejection in general. The current methods used in transplantation medicine involve broad-spectrum immunosuppressives, principally cyclic peptides of fungal origin that interfere with T-cell signaling and prevent proliferation of these immune cells in a nonselective manner (1274–1277). These drugs have considerable harmful side effects, such as generalized immune suppression, nephrotoxicity, and inherent tumorigenicity (1278,1279). The treatment also requires close monitoring and adjustment for the remainder of life. Viable alternatives or adjuncts to these agents would be most desirable.

The Future

In this review, we have attempted to establish a case for the concept that normal male reproductive function and the response to disease represent different facets of the same regulatory environment, involving complex interactions between somatic cells, resident immune cells, and the circulating cellular elements of the immune system. These interactions have been best studied in the testis, where the Sertoli and Leydig cells respond to the presence of the spermatogenic cells and interact with a population of testicular resident macrophages. Under normal conditions, immunoregulatory mediators dominate, protecting the spermatogenic cells by suppressing antigen-specific immunity and maintaining innate immunity, leading to an "immunologically privileged" environment (Fig. 18). The balance of data suggests that the specialized testis environment normally acts to prevent creation of germ cell antigen-specific lymphocytes or to maintain the circulating pool of regulatory cells, or both. Dysregulation of the normal environment caused by infection, local or systemic inflammation, toxic insult, active immunization, or experimental deletion of regulatory T cells may activate the circulating immune cells, leading to a range of effects from temporary disturbance of spermatogenesis and steroidogenesis, all the way through to the creation of testis-reactive T cells and autoimmunity.

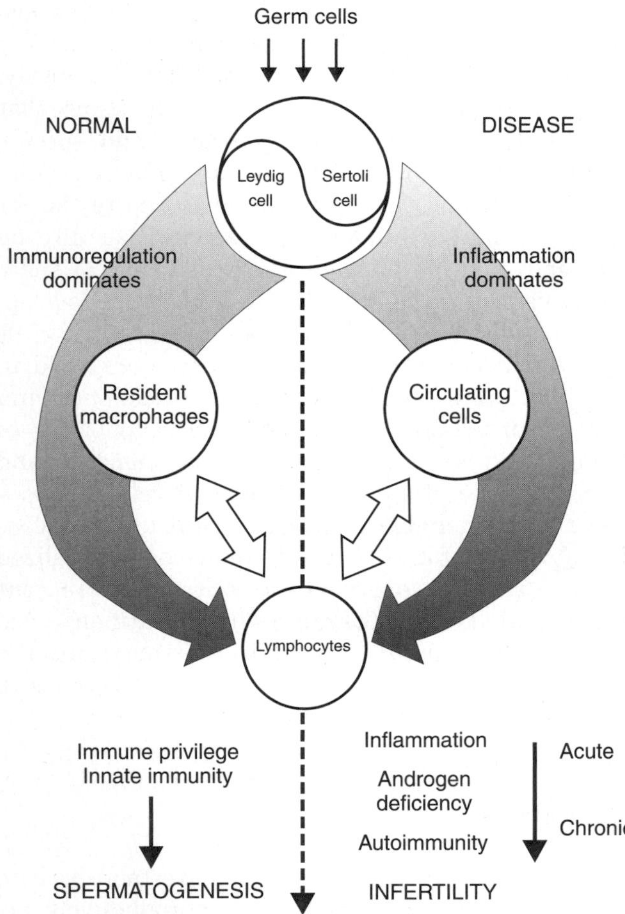

FIG. 18. Final summary. (See text for details.)

There is no doubt that the immune system and male reproductive tract interact at many levels. There is equally no doubt that we still have a long way to go to understand these interactions completely, and how they affect health and physiology. The reader will notice that many questions have been raised in this review, often without a convincing answer being provided. This predominantly reflects the current state of knowledge in the field. However, we can be certain that the dramatic increase in interest and awareness of this erstwhile "niche" area of male reproductive biology will lead to considerable new discoveries in the near future, with exciting and perhaps even totally unexpected implications and benefits.

REFERENCES

1. de Rooij, D. G., Creemers, L. B., de Ouden, K., and Izadyar, F. (2002). Spermatogonial stem cell development. In *Testicular Tangrams* (F. F. G. Rommerts and K. J. Teerds, Eds.), pp. 121–138. Springer, Berlin.
2. de Rooij, D. G. (1998). Stem cells in the testis. *Int. J. Exp. Pathol.* 79, 67–80.
3. Johnson, L., Petty, C. S., and Neaves, W. B. (1984). Influence of age on sperm production and testicular weights in men. *J. Reprod. Fertil.* 70, 211–218.
4. Gershon, R. K., and Kondo, K. (1970). Cell interactions in the induction of tolerance: the role of thymic lymphocytes. *Immunology* 18, 723–737.
5. Kappler, J. W., Roehm, N., and Marrack, P. (1987). T cell tolerance by clonal elimination in the thymus. *Cell* 49, 273–280.
6. Nossal, G. J. (1989). Immunological tolerance then and now: was the Medawar school right? *Immunology Suppl.* 2, 2–5; discussion 6.
7. Nossal, G. J. (1994). Negative selection of lymphocytes. *Cell* 76, 229–239.
8. Stedronska, J., and Hendry, W. F. (1983). The value of the mixed antiglobulin reaction (MAR test) as an addition to routine seminal analysis in the evaluation of the subfertile couple. *Am. J. Reprod. Immunol.* 3, 89–91.
9. Pattinson, H. A., and Mortimer, D. (1987). Prevalence of sperm surface antibodies in the male partners of infertile couples as determined by immunobead screening. *Fertil. Steril.* 48, 466–469.
10. Baker, H. W., Clarke, G. N., Hudson, B., McBain, J. C., McGowan, M. P., and Pepperell, R. J. (1983). Treatment of sperm autoimmunity in men. *Clin. Reprod. Fertil.* 2, 55–71.
11. Lenzi, A., Gandini, L., Lombardo, F., Rago, R., Paoli, D., and Dondero, F. (1997). Antisperm antibody detection: 2. Clinical, biological, and statistical correlation between methods. *Am. J. Reprod. Immunol.* 38, 224–230.
12. Crosignani, P. G., and Rubin, B. L. (2000). Optimal use of infertility diagnostic tests and treatments. The ESHRE Capri Workshop Group. *Hum. Reprod.* 15, 723–732.
13. Toh, B. H., Sentry, J. W., and Alderuccio, F. (2000). The causative H^+/K^+ ATPase antigen in the pathogenesis of autoimmune gastritis. *Immunol. Today* 21, 348–354.
14. Adorini, L., Gregori, S., and Harrison, L. C. (2002). Understanding autoimmune diabetes: insights from mouse models. *Trends Mol. Med.* 8, 31–38.
15. Hjort, T., and Hargreave, T. B. (1994). Immunity to sperm and fertility. In *Male Infertility* (T. B. Hargreave, Ed.), pp. 269–290. Springer-Verlag, London.

16. Auer, J., Senechal, H., Desvaux, F. X., Albert, M., and De Almeida, M. (2000). Isolation and characterisation of two sperm membrane proteins recognised by sperm-associated antibodies in infertile men. *Mol. Reprod. Dev.* 57, 393–405.
17. Menge, A. C., and Fuller, B. (1975). Testis antigens of man and some other primates. *Fertil. Steril.* 26, 473–479.
18. Aitken, R. J., Hulme, M. J., Henderson, C. J., Hargreave, T. B., and Ross, A. (1987). Analysis of the surface labelling characteristics of human spermatozoa and the interaction with antisperm antibodies. *J. Reprod. Fertil.* 80, 473–485.
19. Parslow, J. M., Poulton, T. A., and Hay, F. C. (1987). Characterization of sperm antigens reacting with human antisperm antibodies. *Clin. Exp. Immunol.* 69, 179–187.
20. Lee, C. Y., Lum, V., Wong, E., Menge, A. C., and Huang, Y. S. (1983). Identification of human sperm antigens to antisperm antibodies. *Am. J. Reprod. Immunol.* 3, 183–187.
21. Shetty, J., Naaby-Hansen, S., Shibahara, H., Bronson, R., Flickinger, C. J., and Herr, J. C. (1999). Human sperm proteome: immunodominant sperm surface antigens identified with sera from infertile men and women. *Biol. Reprod.* 61, 61–69.
22. Primakoff, P., Lathrop, W., and Bronson, R. (1990). Identification of human sperm surface glycoproteins recognized by autoantisera from immune infertile men, women, and vasectomized men. *Biol. Reprod.* 42, 929–942.
23. Bohring, C., Krause, E., Habermann, B., and Krause, W. (2001). Isolation and identification of sperm membrane antigens recognized by antisperm antibodies, and their possible role in immunological infertility disease. *Mol. Hum. Reprod.* 7, 113–118.
24. Bohring, C., and Krause, W. (2001). Differences in the antigen pattern recognized by antisperm antibodies in patients with infertility and vasectomy. *J. Urol.* 166, 1178–1180.
25. Koide, S. S., Wang, L., and Kamada, M. (2000). Antisperm antibodies associated with infertility: properties and encoding genes of target antigens. *Proc. Soc. Exp. Biol. Med.* 224, 123–132.
26. Comhaire, F., Verschraegen, G., and Vermeulen, L. (1980). Diagnosis of accessory gland infection and its possible role in male infertility. *Int. J. Androl.* 3, 32–45.
27. Schaeffer, A. J., Knauss, J. S., Landis, J. R., Propert, K. J., Alexander, R. B., Litwin, M. S., Nickel, J. C., O'Leary, M. P., Nadler, R. B., Pontari, M. A., Shoskes, D. A., Zeitlin, S. I., Fowler, J. E., Jr., Mazurick, C. A., Kusek, J. W., Nyberg, L. M., and The Chronic Prostatitis Collaborative Research Network Study Group. (2002). Leukocyte and bacterial counts do not correlate with severity of symptoms in men with chronic prostatitis: the National Institutes of Health Chronic Prostatitis Cohort Study. *J. Urol.* 168, 1048–1053.
28. Chan, P. T., and Schlegel, P. N. (2002). Inflammatory conditions of the male excurrent ductal system. Part II. *J. Androl.* 23, 461–469.
29. Chan, P. T., and Schlegel, P. N. (2002). Inflammatory conditions of the male excurrent ductal system. Part I. *J. Androl.* 23, 453–460.
30. Krieger, J. N., Egan, K. J., Ross, S. O., Jacobs, R., and Berger, R. E. (1996). Chronic pelvic pains represent the most prominent urogenital symptoms of "chronic prostatitis." *Urology* 48, 715–721; discussion 721–712.
31. Nickel, C. J. (2000). Prostatitis syndromes: an update for urologic practice. *Can. J. Urol.* 7, 1091–1098.
32. Perez-Clavier, R., Harrison, R. G., and Macmillan, E. W. (1982). The pattern of the lymphatic drainage of the rat epididymis. *J. Anat.* 134, 667–675.
33. Kazeem, A. (1986). Reexamination of testicular lymphatic drainage in the rat. *Lymphology* 19, 172–174.
34. McDonald, S. W., and Scothorne, R. J. (1988). The lymphatic drainage of the epididymis and of the ductus deferens of the rat, with reference to the immune response to vasectomy. *J. Anat.* 158, 57–64.
35. Head, J. R., Neaves, W. B., and Billingham, R. E. (1983). Reconsideration of the lymphatic drainage of the rat testis. *Transplantation* 35, 91–95.
36. Itoh, M., Li, X. Q., Yano, A., Xie, Q., and Takeuchi, Y. (1998). Patterns of efferent lymphatics of the mouse testis. *J. Androl.* 19, 466–472.
37. Kazeem, A. A. (1991). Species variation in the extrinsic lymphatic drainage of the rodent testis: its role within the context of an immunologically privileged site. *Lymphology* 24, 140–144.
38. Fawcett, D. W., Heidger, P. M., and Leak, L. V. (1969). Lymph vascular system of the interstitial tissue of the testis as revealed by electron microscopy. *J. Reprod. Fertil.* 19, 109–119.
39. Waites, G. M., and Gladwell, R. T. (1982). Physiological significance of fluid secretion in the testis and blood-testis barrier. *Physiol. Rev.* 62, 624–671.
40. Flickinger, C., and Fawcett, D. W. (1967). The junctional specializations of Sertoli cells in the seminiferous epithelium. *Anat. Rec.* 158, 207–221.
41. Setchell, B. P., Voglmayr, J. K., and Waites, G. M. (1969). A blood-testis barrier restricting passage from blood into rete testis fluid but not into lymph. *J. Physiol. (Lond.)* 200, 73–85.
42. Dym, M., and Fawcett, D. W. (1970). The blood-testis barrier in the rat and the physiological compartmentation of the seminiferous epithelium. *Biol. Reprod.* 3, 308–326.
43. Johnson, M. H. (1973). Physiological mechanisms for the immunological isolation of spermatozoa. *Adv. Reprod. Physiol.* 6, 297–324.
44. Setchell, B. P. (1967). The blood-testicular fluid barrier in sheep. *J. Physiol. (Lond.)* 189, 63P-65P.
45. Selawry, H., Fojaco, R., and Whittington, K. (1987). Extended survival of MHC-compatible islet grafts from diabetes-resistant donors in spontaneously diabetic BB/W rat. *Diabetes* 36, 1061–1067.
46. Ferguson, J., and Scothorne, R. J. (1977). Extended survival of pancreatic islet allografts in the testis of guinea-pigs. *J. Anat.* 124, 1–8.
47. Bobzien, B., Yasunami, Y., Majercik, M., Lacy, P. E., and Davie, J. M. (1983). Intratesticular transplants of islet xenografts (rat to mouse). *Diabetes* 32, 213–216.
48. Head, J. R., Neaves, W. B., and Billingham, R. E. (1983). Immune privilege in the testis: I. Basic parameters of allograft survival. *Transplantation* 36, 423–431.
49. Head, J. R., and Billingham, R. E. (1985). Immune privilege in the testis: II. Evaluation of potential local factors. *Transplantation* 40, 269–275.
50. Whitmore III, W. F., and Gittes, R. F. (1978). Intratesticular grafts: the testis as an exceptional immunologically privileged site. *Trans. Am. Assoc. Genito-urinary Surg.* 70, 76–80.
51. Whitmore III, W. F., Karsh, L., and Gittes, R. F. (1985). The role of germinal epithelium and spermatogenesis in the privileged survival of intratesticular grafts. *J. Urol.* 134, 782–786.
52. Selawry, H. P., and Whittington, K. (1984). Extended allograft survival of islets grafted into intra-abdominally placed testis. *Diabetes* 33, 405–406.
53. Baker, H. W. (1998). Reproductive effects of nontesticular illness. *Endocrinol. Metab. Clin. North Am.* 27, 831–850.
54. Buch, J. P., and Havlovec, S. K. (1991). Variation in sperm penetration assay related to viral illness. *Fertil. Steril.* 55, 844–846.
55. Carlsen, E., Andersson, A. M., Petersen, J. H., and Skakkebaek, N. E. (2003). History of febrile illness and variation in semen quality. *Hum. Reprod.* 18, 2089–2092.
56. Cutolo, M., Balleari, E., Giusti, M., Monachesi, M., and Accardo, S. (1988). Sex hormone status of male patients with

rheumatoid arthritis: evidence of low serum concentrations of testosterone at baseline and after human chorionic gonadotropin stimulation. *Arthritis Rheum.* 31, 1314–1317.

57. Adamopoulos, D. A., Lawrence, D. M., Vassilopoulos, P., Contoyiannis, P. A., and Swyer, G. I. (1978). Pituitary-testicular interrelationships in mumps orchitis and other viral infections. *BMJ* 1, 1177–1180.
58. Dong, Q., Hawker, F., McWilliam, D., Bangah, M., Burger, H., and Handelsman, D. J. (1992). Circulating immunoreactive inhibin and testosterone levels in men with critical illness. *Clin. Endocrinol.* 36, 399–404.
59. Hales, D. B. (2002). Testicular macrophage modulation of Leydig cell steroidogenesis. *J. Reprod. Immunol.* 57, 3–18.
60. Setchell, B. P. (1990). The testis and tissue transplantation: historical aspects. *J. Reprod. Immunol.* 18, 1–8.
61. Berthold, A. A. (1849). Transplantation der Hoden. *Arch. Anat. Physiol. Wiss. Med.* 16, 42–46.
62. Sand, K. (1919). Experiments on the internal secretion of the sexual glands, especially on experimental hermaphroditism. *J. Physiol. (Lond.)* 53, 257–263.
63. Steinach, E. (1923). Geschelchtstrieb und echtsekundäre Geschelchtsmerkmale als folgender innersekretorischen Funktion der Keimdrüsen. *Zentralbl. Physiol.* 24, 551–566.
64. Cevelotto, G. (1909). Über Verpfanzungen und Gefuerungen der Hoden, Frank. *Z. Pathol.* 3, 331–337.
65. Barten, E. J., and Newling, D. W. (1996). Transplantation of the testis; from the past to the present. *Int. J. Androl.* 19, 205–211.
66. Sengoopta, C. (2003). 'Dr Steinach coming to make old young!': sex glands, vasectomy and the quest for rejuvenation in the roaring twenties. *Endeavour* 27, 122–126.
67. Moore, C. R. (1926). On the properties of the gonads as controllers of somatic and physical characteristics: IX. Testis graft reactions in different environments (rat). *Am. J. Anat.* 37, 351–416.
68. Schlatt, S., Kim, S. S., and Gosden, R. (2002). Spermatogenesis and steroidogenesis in mouse, hamster and monkey testicular tissue after cryopreservation and heterotopic grafting to castrated hosts. *Reproduction* 124, 339–346.
69. Honaramooz, A., Snedaker, A., Boiani, M., Scholer, H., Dobrinski, I., and Schlatt, S. (2002). Sperm from neonatal mammalian testes grafted in mice. *Nature* 418, 778–781.
70. Kanatsu-Shinohara, M., Ogonuki, N., Inoue, K., Ogura, A., Toyokuni, S., Honjo, T., and Shinohara, T. (2003). Allogeneic offspring produced by male germ line stem cell transplantation into infertile mouse testis. *Biol. Reprod.* 68, 167–173.
71. Oatley, J. M., de Avila, D. M., Reeves, J. J., and McLean, D. J. (2004). Spermatogenesis and germ cell transgene expression in xenografted bovine testicular tissue. *Biol. Reprod.* 71, 494–501.
72. Leckband, E. (1964). Testicular homografts in tolerant mice. *Transplantation* 2, 522–533.
73. Turner, C. D. (1938). Intra-ocular homotransplantation of prepuberal testes in the rat. *Am. J. Anat.* 63, 101–159.
74. Dameron, J. T. (1951). The anterior chamber of the eye for investigative purposes: a site for transplantation of fetal endocrine tissues and cancer, and for the study of tissue reaction. *Surgery* 30, 787–799.
75. Barker, C. F., and Billingham, R. E. (1977). Immunologically privileged sites. *Adv. Immunol.* 25, 1–54.
76. Lansteiner, K. (1899). Zur Kenutuis der spezifisch auf-blut-Korperchen wirkenden Sera. *Z. Bacteriol.* 25, 546–549.
77. Metchinikoff, E. L. (1900). L'influence de l'organisme lex toxines: sur la spermotoxine et l'antispermotoxine. *Ann. Inst. Pasteur* 14, 1–12.
78. Guyer, M. F. (1922). Studies of cytolysins: III. Experiments with spermatotoxins. *J. Exp. Zool.* 35, 207–223.
79. Kennedy, W. P. (1924). The production of spermatoxins. *Q. J. Exp. Physiol.* 14, 279–283.
80. Voisin, G., and Delaunay, A. (1955). Sur les lésions testiculaires observeés chez des animaux soumis à des injections de substances adjuvantes, seules ou mélangées avec des extraits de tissus homologues. *Ann. Inst. Pasteur* 89, 307–317.
81. Voisin, G., Delaunay, A., and Barber, M. (1951). Sur les lésions testiculaires provoqueés chez le cobaye par iso- et autosensibilisation. *Ann. Inst. Pasteur* 81, 48–63.
82. Rümke, P. (1954). The presence of sperm antibodies in the serum of two patients with oligospermia. *Vox Sang.* 4, 135–140.
83. Wilson, L. (1954). Sperm agglutinins in human semen, blood. *Proc. Soc. Exp. Biol. Med.* 85, 652–655.
84. Grossman, C. (1989). Possible underlying mechanisms of sexual dimorphism in the immune response, fact and hypothesis. *J. Steroid Biochem.* 34, 241–251.
85. Grossman, C. J. (1985). Interactions between the gonadal steroids and the immune system. *Science* 227, 257–261.
86. Schuurs, A. H., and Verheul, H. A. (1990). Effects of gender and sex steroids on the immune response. *J. Steroid Biochem.* 35, 157–172.
87. Ansar Ahmed, S., Penhale, W. J., and Talal, N. (1985). Sex hormones, immune responses, and autoimmune diseases: mechanisms of sex hormone action. *Am. J. Pathol.* 121, 531–551.
88. Whitacre, C. C. (2001). Sex differences in autoimmune disease. *Nat. Immunol.* 2, 777–780.
89. Calzolari, A. (1898). Rescherches experimentales sur un rapport probable entre la function du thymus et cells des testicules. *Arch. Ital. Biol.* 30, 71.
90. Castro, J. E. (1974). Orchidectomy and the immune response: I. Effect of orchidectomy on lymphoid tissues of mice. *Proc. R. Soc. Lond. B* 185, 425–436.
91. Chiodi, H. (1940). The relationship between the thymus and the sexual organs. *Endocrinology* 126, 107–116.
92. Sasson, S., and Mayer, M. (1981). Effect of androgenic steroids on rat thymus and thymocytes in suspension. *J. Steroid Biochem.* 14, 509–517.
93. Castro, J. E., and Hamilton, D. N. (1972). Adrenalectomy and orchidectomy as immunopotentiating procedures. *Transplantation* 13, 615–616.
94. Kimura, M., Tomita, Y., Watanabe, H., Sato, S., and Abo, T. (1995). Androgen regulation of intra-and extra-thymic T cells and its effect on sex differences in the immune system. *Int. J. Androl.* 18, 127–136.
95. Araneo, B. A., Dowell, T., Diegel, M., and Daynes, R. A. (1991). Dihydrotestosterone exerts a depressive influence on the production of interleukin-4 (IL-4), IL-5, and γ-interferon, but not IL-2 by activated murine T cells. *Blood* 78, 688–699.
96. Kotani, M., Nawa, Y., and Fujii, H. (1974). Inhibition by testosterone of immune reactivity and of lymphoid regeneration in irradiated and marrow reconstituted mice. *Experientia* 30, 1343–1345.
97. Fujii, H., Nawa, Y., Tsuchiya, H., Matsuno, K., Fukumoto, T., Fukuda, S., and Kotani, M. (1975). Effect of a single administration of testosterone on the immune response and lymphoid tissues in mice. *Cell. Immunol.* 20, 315–326.
98. Weinstein, Y., and Berkovich, Z. (1981). Testosterone effect on bone marrow, thymus, and suppressor T cells in the (NZB X NZW)F1 mice: its relevance to autoimmunity. *J. Immunol.* 126, 998–1002.
99. Berczi, I., Nagy, E., Kovacs, K., and Horvath, E. (1981). Regulation of humoral immunity in rats by pituitary hormones. *Acta Endocrinol.* 98, 506–513.

100. de Kretser, D. M., Temple-Smith, P. D., and Kerr, J. B. (1982). Anatomical and functional aspects of the male reproductive organs. In *Disturbances of Male Fertility* (K. Bandhauer and J. Frick, Eds.), pp. 1–131. Springer-Verlag, Berlin.
101. Kerr, J. B. (1989). The cytology of the human testis. In *The Testis* (2nd ed. H. Burger and D. M. de Kretser, Eds.), pp. 197–229. Raven Press, New York.
102. Fawcett, D. W., Neaves, W. B., and Flores, M. N. (1973). Comparative observations on intertubular lymphatics and the organization of the interstitial tissue of the mammalian testis. *Biol. Reprod.* 9, 500–532.
103. Clermont, Y., and Huckins, C. (1961). Microscopic anatomy of the sex cords and seminiferous tubules in growing and adult male albino rats. *Am. J. Anat.* 108, 79–97.
104. Ergun, S., Davidoff, M., and Holstein, A. F. (1996). Capillaries in the lamina propria of human seminiferous tubules are partly fenestrated. *Cell Tissue Res.* 286, 93–102.
105. Turner, T. T., and Rhoades, C. P. (1995). Testicular capillary permeability: the movement of luteinizing hormone from the vascular to the interstitial compartment. *J. Androl.* 16, 417–423.
106. Ghinea, N., and Milgrom, E. (1995). Transport of protein hormones through the vascular endothelium. *J. Endocrinol.* 145, 1–9.
107. Setchell, B. P., Pakarinen, P., and Huhtaniemi, I. (2002). How much LH do the Leydig cells see? *J. Endocrinol.* 175, 375–382.
108. Setchell, B. P. (1986). The movement of fluids and substances in the testis. *Aust. J. Biol. Sci.* 39, 193–207.
109. Maddocks, S., and Setchell, B. P. (1988). The composition of extracellular interstitial fluid collected with a push-pull cannula from the testes of adult rats. *J. Physiol. (Lond.)* 407, 363–372.
110. Hinton, B. T. (1990). The testicular and epididymal luminal amino acid microenvironment in the rat. *J. Androl.* 11, 498–505.
111. Kormano, M. (1967). An angiographic study of the testicular vasculature in the postnatal rat. *Z. Anat. Entwicklungsgesch.* 126, 138–153.
112. Hees, H., Leiser, R., Kohler, T., and Wrobel, K. H. (1984). Vascular morphology of the bovine spermatic cord and testis: I. Light- and scanning electron-microscopic studies on the testicular artery and pampiniform plexus. *Cell Tissue Res.* 237, 31–38.
113. Ohtsuka, A. (1984). Microvascular architecture of the pampiniform plexus-testicular artery system in the rat: a scanning electron microscope study of corrosion casts. *Am. J. Anat.* 169, 285–293.
114. Waites, G. M. H., and Moule, G. R. (1961). Relation of vascular heat exchange to temperature regulation in the testis of the ram. *J. Reprod. Fertil.* 2, 213–224.
115. Cowie, A. T., Lascelles, A. K., and Wallace, J. C. (1964). Flow and protein content of testicular lymph in conscious rams. *J. Physiol. (Lond.)* 171, 176–187.
116. Setchell, B. P. (1982). The flow and composition of lymph from the testes of pigs with some observations on the effect of raised venous pressure. *Comp. Biochem. Physiol. A* 73, 201–205.
117. Moller, R. (1980). Arrangement and fine structure of lymphatic vessels in the human spermatic cord. *Andrologia* 12, 564–576.
118. Engeset, A. (1959). The route of peripheral lymph to the bloodstream. A X-ray study of the barrier theory. *J. Anat.* 93, 96–100.
119. Tilney, N. L. (1971). Patterns of lymphatic drainage in the adult laboratory rat. *J. Anat.* 109, 369–383.
120. Leydig, F. (1850). Zur antomie der männlichen Geschlechtsorgane und Analdrüsen der Säugatiere [On the anatomy of the male sex organs and anal glands of mammals]. *Z. Wiss. Zool.* 2, 1–57.
121. Christensen, A. K., and Gillim, S. W. (1969). The correlation of fine structure and function in steroid-secreting cells, with emphasis on those of the gonad. In *The Gonads* (K. W. McKearns, Ed.), pp. 415–488. Appleton-Century-Crofts, New York.
122. Hedger, M. P., and de Kretser, D. M. (2000). Leydig cell function and its regulation. *Results Probl. Cell Differ.* 28, 69–110.
123. Anton, F., Morales, C., Aguilar, R., Bellido, C., Aguilar, E., and Gaytan, F. (1998). A comparative study of mast cells and eosinophil leukocytes in the mammalian testis. *Zentralbl. Veterin. A* 45, 209–218.
124. Hutson, J. C. (1994). Testicular macrophages. *Int. Rev. Cytol.* 149, 99–143.
125. Dym, M., and Romrell, L. J. (1975). Intraepithelial lymphocytes in the male reproductive tract of rats and rhesus monkeys. *J. Reprod. Fertil.* 42, 1–7.
126. Ritchie, A. W., Hargreave, T. B., James, K., and Chisholm, G. D. (1984). Intra-epithelial lymphocytes in the normal epididymis: a mechanism for tolerance to sperm autoantigens? *Br. J. Urol.* 56, 79–83.
127. el-Demiry, M. I., Hargreave, T. B., Busuttil, A., James, K., Ritchie, A. W., and Chisholm, G. D. (1985). Lymphocyte subpopulations in the male genital tract. *Br. J. Urol.* 57, 769–774.
128. Pöllänen, P., and Niemi, M. (1987). Immunohistochemical identification of macrophages, lymphoid cells and HLA antigens in the human testis. *Int. J. Androl.* 10, 37–42.
129. Pöllänen, P., and Maddocks, S. (1988). Macrophages, lymphocytes and MHC II antigen in the ram and the rat testis. *J. Reprod. Fertil.* 82, 437–445.
130. Wang, J., Wreford, N. G., Lan, H. Y., Atkins, R., and Hedger, M. P. (1994). Leukocyte populations of the adult rat testis following removal of the Leydig cells by treatment with ethane dimethane sulfonate and subcutaneous testosterone implants. *Biol. Reprod.* 51, 551–561.
131. Tompkins, A. B., Hutchinson, P., de Kretser, D. M., and Hedger, M. P. (1998). Characterization of lymphocytes in the adult rat testis by flow cytometry: effects of activin and transforming growth factor β on lymphocyte subsets in vitro. *Biol. Reprod.* 58, 943–951.
132. Hermo, L., and Clermont, Y. (1976). Light cells within the limiting membrane of rat seminiferous tubules. *Am. J. Anat.* 145, 467–483.
133. Hermo, L., and Lalli, M. (1978). Monocytes and mast cells in the limiting membrane of human seminiferous tubules. *Biol. Reprod.* 19, 92–100.
134. Cheng, C. Y., and Mruk, D. D. (2002). Cell junction dynamics in the testis: Sertoli-germ cell interactions and male contraceptive development. *Physiol. Rev.* 82, 825–874.
135. Pelletier, R. M. (2001). The tight junctions in the testis, epididymis and vas deferens. In *Tight Junctions* (2nd ed. M. Cereijido and J. Anderson, Eds.), pp. 599–628. CRC Press, Boca Raton, FL.
136. Koskimies, A. I., Kormano, M., and Lahti, A. (1971). A difference in the immunoglobulin content of seminiferous tubule fluid and rete testis fluid of the rat. *J. Reprod. Fertil.* 27, 463–465.
137. Kormano, M., Koskimies, A. I., and Hunter, R. L. (1971). The presence of specific proteins, in the absence of many serum proteins in the rat seminiferous tubule fluid. *Experientia* 27, 1461–1463.

138. Koskimies, A. I., and Kormano, M. (1973). The proteins in fluids from the seminiferous tubules and rete testis of the rat. *J. Reprod. Fertil.* 34, 433–434.
139. Tung, K. S., Unanue, E. R., and Dixon, F. J. (1970). The immunopathology of experimental allergic orchitis. *Am. J. Pathol.* 60, 313–328.
140. Jones, R. C., and Jurd, K. M. (1987). Structural differentiation and fluid reabsorption in the ductuli efferentes testis of the rat. *Aust. J. Biol. Sci.* 40, 79–90.
141. Clulow, J., Jones, R. C., Hansen, L. A., and Man, S. Y. (1998). Fluid and electrolyte reabsorption in the ductuli efferentes testis. *J. Reprod. Fertil. Suppl.* 53, 1–14.
142. Eddy, E. M., Vernon, R. B., Muller, C. H., Hahnel, A. C., and Fenderson, B. A. (1985). Immunodissection of sperm surface modifications during epididymal maturation. *Am. J. Anat.* 174, 225–237.
143. Hinton, B. T., and Palladino, M. A. (1995). Epididymal epithelium: its contribution to the formation of a luminal fluid microenvironment. *Microsc. Res. Tech.* 30, 67–81.
144. Marwood, R. P., and Beral, V. (1979). Disappearance of spermatozoa from ejaculate after vasectomy. *BMJ* 1, 87.
145. Nelson, G., Culberson, D. E., and Gardner, W. A., Jr. (1988). Intraprostatic spermatozoa. *Hum. Pathol.* 19, 541–544.
146. McClinton, S., Eremin, O., and Miller, I. D. (1990). Inflammatory infiltrate in prostatic hyperplasia: evidence of a host response to intraprostatic spermatozoa? *Br. J. Urol.* 65, 606–610.
147. Pelletier, R. M. (1994). Blood barriers of the epididymis and vas deferens act asynchronously with the blood barrier of the testis in the mink (*Mustela vison*). *Microsc. Res. Tech.* 27, 333–349.
148. Tung, K. S., Unanue, E. R., and Dixon, F. J. (1971). Pathogenesis of experimental allergic orchitis: II. The role of antibody. *J. Immunol.* 106, 1463–1472.
149. Friend, D. S., and Gilula, N. B. (1972). Variations in tight and gap junctions in mammalian tissues. *J. Cell Biol.* 53, 758–776.
150. Levy, S., and Robaire, B. (1999). Segment-specific changes with age in the expression of junctional proteins and the permeability of the blood-epididymis barrier in rats. *Biol. Reprod.* 60, 1392–1401.
151. Beagley, K. W., Wu, Z. L., Pomering, M., and Jones, R. C. (1998). Immune responses in the epididymis: implications for immunocontraception. *J. Reprod. Fertil. Suppl.* 53, 235–245.
152. Yeung, C. H., Nashan, D., Sorg, C., Oberpenning, F., Schulze, H., Nieschlag, E., and Cooper, T. G. (1994). Basal cells of the human epididymis: antigenic and ultrastructural similarities to tissue-fixed macrophages. *Biol. Reprod.* 50, 917–926.
153. Nashan, D., Jantos, C., Ahlers, D., Bergmann, M., Schiefer, H. G., Sorg, C., and Nieschlag, E. (1993). Immuno-competent cells in the murine epididymis following infection with *Escherichia coli*. *Int. J. Androl.* 16, 47–52.
154. Nashan, D., Malorny, U., Sorg, C., Cooper, T., and Nieschlag, E. (1989). Immuno-competent cells in the murine epididymis. *Int. J. Androl.* 12, 85–94.
155. Hooper, P., Smythe, E., Richards, R. C., Howard, C. V., Lynch, R. V., and Lewis-Jones, D. I. (1995). Total number of immunocompetent cells in the normal rat epididymis and after vasectomy. *J. Reprod. Fertil.* 104, 193–198.
156. Flickinger, C. J., Bush, L. A., Howards, S. S., and Herr, J. C. (1997). Distribution of leukocytes in the epithelium and interstitium of four regions of the Lewis rat epididymis. *Anat. Rec.* 248, 380–390.
157. Nashan, D., Cooper, T. G., Knuth, U. A., Schubeus, P., Sorg, C., and Nieschlag, E. (1990). Presence and distribution of leucocyte subsets in the murine epididymis after vasectomy. *Int. J. Androl.* 13, 39–49.
158. Serre, V., and Robaire, B. (1999). Distribution of immune cells in the epididymis of the aging Brown Norway rat is segment-specific and related to the luminal content. *Biol. Reprod.* 61, 705–714.
159. Fulmer, B. R., and Turner, T. T. (2000). A blood-prostate barrier restricts cell and molecular movement across the rat ventral prostate epithelium. *J. Urol.* 163, 1591–1594.
160. Ghabriel, M. N., Lu, J. J., Hermanis, G., Zhu, C., and Setchell, B. P. (2002). Expression of a blood-brain barrier-specific antigen in the reproductive tract of the male rat. *Reproduction* 123, 389–397.
161. Suzuki, F., and Nagano, T. (1978). Regional differentiation of cell junctions in the excurrent duct epithelium of the rat testis as revealed by freeze-fracture. *Anat. Rec.* 191, 503–519.
162. Huhtaniemi, I. (1995). Molecular aspects of the ontogeny of the pituitary-gonadal axis. *Reprod. Fertil. Dev.* 7, 1025–1035.
163. Leung, P. C. K., and Steele, G. L. (1992). Intracellular signaling in the gonads. *Endocrine Rev.* 13, 476–497.
164. Coquelin, A., and Desjardins, C. (1982). Luteinizing hormone and testosterone secretion in young and old male mice. *Am. J. Physiol.* 243, E257–E263.
165. Ellis, G. B., and Desjardins, C. (1982). Male rats secrete luteinizing hormone and testosterone episodically. *Endocrinology* 110, 1618–1627.
166. Sisk, C. L., and Desjardins, C. (1986). Pulsatile release of luteinizing hormone and testosterone in male ferrets. *Endocrinology* 119, 1195–1203.
167. Mendis-Handagama, S. M. (1997). Luteinizing hormone on Leydig cell structure and function. *Histol. Histopathol.* 12, 869–882.
168. Griswold, M. D. (1988). Protein secretions of Sertoli cells. *Int. Rev. Cytol.* 110, 133–156.
169. Desjardins, C. (1978). Endocrine regulation of reproductive development and function in the male. *J. Anim. Sci.* 47 (Suppl. 2), 56–79.
170. Plant, T. M., Winters, S. J., Attardi, B. J., and Majumdar, S. S. (1993). The follicle stimulating hormone-inhibin feedback loop in male primates. *Hum. Reprod.* 8 (Suppl. 2), 41–44.
171. Zirkin, B. R. (1993). Regulation of spermatogenesis in the adult mammal: gonadotropins and androgens. In *Cell and Molecular Biology of the Testis* (C. Desjardins and L. L. Ewing, Eds.), pp. 166–188. Oxford University Press, New York.
172. Zirkin, B. R., Santulli, R., Awoniyi, C. A., and Ewing, L. L. (1989). Maintenance of advanced spermatogenic cells in the adult rat testis: quantitative relationship to testosterone concentration within the testis. *Endocrinology* 124, 3043–3049.
173. Sharpe, R. M., Donachie, K., and Cooper, I. (1988). Re-evaluation of the intratesticular level of testosterone required for quantitative maintenance of spermatogenesis in the rat. *J. Endocrinol.* 117, 19–26.
174. Turner, T. T., Caplis, L. A., and Rhoades, C. P. (1996). Testicular vascular permeability: effects of experimental lesions associated with impaired testis function. *J. Urol.* 155, 1078–1082.
175. Ariyaratne, H. B., Mason, J., and Mendis-Handagama, S. M. (2000). Effects of thyroid and luteinizing hormones on the onset of precursor cell differentiation into Leydig progenitor cells in the prepubertal rat testis. *Biol. Reprod.* 63, 898–904.
176. Mendis-Handagama, S. M., Ariyaratne, H. B., Teunissen van Manen, K. R., and Haupt, R. L. (1998). Differentiation of adult Leydig cells in the neonatal rat testis is arrested by hypothyroidism. *Biol. Reprod.* 59, 351–357.

177. Mendis-Handagama, C., and Ariyaratne, S. (2004). Prolonged and transient neonatal hypothyroidism on Leydig cell differentiation in the postnatal rat testis. *Arch. Androl.* 50, 347–357.

178. Manna, P. R., Kero, J., Tena-Sempere, M., Pakarinen, P., Stocco, D. M., and Huhtaniemi, I. T. (2001). Assessment of mechanisms of thyroid hormone action in mouse Leydig cells: regulation of the steroidogenic acute regulatory protein, steroidogenesis, and luteinizing hormone receptor function. *Endocrinology* 142, 319–331.

179. Manna, P. R., Tena-Sempere, M., and Huhtaniemi, I. T. (1999). Molecular mechanisms of thyroid hormone-stimulated steroidogenesis in mouse Leydig tumor cells: involvement of the steroidogenic acute regulatory (StAR) protein. *J. Biol. Chem.* 274, 5909–5918.

180. Hales, D. B., and Payne, A. H. (1989). Glucocorticoid-mediated repression of P450scc mRNA and *de novo* synthesis in cultured Leydig cells. *Endocrinology* 124, 2099–2104.

181. Monder, C., Miroff, Y., Marandici, A., and Hardy, M. P. (1994). 11β-Hydroxysteroid dehydrogenase alleviates glucocorticoid-mediated inhibition of steroidogenesis in rat Leydig cells. *Endocrinology* 134, 1199–1204.

182. Hardy, M. P., and Ganjam, V. K. (1997). Stress, 11β-HSD, and Leydig cell function. *J. Androl.* 18, 475–479.

183. Page, K. C., Sottas, C. M., and Hardy, M. P. (2001). Prenatal exposure to dexamethasone alters Leydig cell steroidogenic capacity in immature and adult rats. *J. Androl.* 22, 973–980.

184. Mason, D., Simmons, D., Buckley, C., Schwartz-Albiez, R., Hadam, M., Saalmuller, A., Clark, E., Malavasi, F., Morrissey, J. A., Vivier, E., Horejsi, V., De Haas, M., Van der Schoot, E., Uguccioni, M., Hart, D., Hadam, M., Goyert, S., Zola, H., Civin, C., Meuer, S., Shaw, S., Turni, L., P., A., and Bensussan, A. (2002). *Leucocyte Typing VII.* Oxford University Press, Oxford.

185. Zhang, D., Zhang, G., Hayden, M. S., Greenblatt, M. B., Bussey, C., Flavell, R. A., and Ghosh, S. (2004). A toll-like receptor that prevents infection by uropathogenic bacteria. *Science* 303, 1522–1526.

186. Akira, S., Takeda, K., and Kaisho, T. (2001). Toll-like receptors: critical proteins linking innate and acquired immunity. *Nat. Immunol.* 2, 675–680.

187. Medzhitov, R. (2001). Toll-like receptors and innate immunity. *Nat. Rev. Immunol.* 1, 135–145.

188. Haziot, A., Ferrero, E., Kontgen, F., Hijiya, N., Yamamoto, S., Silver, J., Stewart, C. L., and Goyert, S. M. (1996). Resistance to endotoxin shock and reduced dissemination of gram-negative bacteria in CD14-deficient mice. *Immunity* 4, 407–414.

189. Schromm, A. B., Lien, E., Henneke, P., Chow, J. C., Yoshimura, A., Heine, H., Latz, E., Monks, B. G., Schwartz, D. A., Miyake, K., and Golenbock, D. T. (2001). Molecular genetic analysis of an endotoxin nonresponder mutant cell line: a point mutation in a conserved region of MD-2 abolishes endotoxin-induced signaling. *J. Exp. Med.* 194, 79–88.

190. Hertzog, P. J., O'Neill, L. A., and Hamilton, J. A. (2003). The interferon in TLR signaling: more than just antiviral. *Trends Immunol.* 24, 534–539.

191. Cook, D. N., Pisetsky, D. S., and Schwartz, D. A. (2004). Toll-like receptors in the pathogenesis of human disease. *Nat. Immunol.* 5, 975–979.

192. Athman, R., and Philpott, D. (2004). Innate immunity via Toll-like receptors and Nod proteins. *Curr. Opin. Microbiol.* 7, 25–32.

193. Decker, K. (1991). Basic mechanisms of the inflammatory response. In *Molecular Aspects of Inflammation* (H. Sies, L. Flohé, and G. Zimmer, Eds.), pp. 1–23. Springer-Verlag, Berlin.

194. Rosenberg, H. F., and Gallin, J. I. (2003). Inflammation. In *Fundamental Immunology* (5th ed. W. E. Paul, Ed.), pp. 1151–1169. Lippincott Williams & Wilkins, Philadelphia.

195. Murphy, P. M. (1996). Chemokine receptors: structure, function and role in microbial pathogenesis. *Cytol. Growth Factor Rev.* 7, 47–64.

196. Springer, T. A. (1994). Traffic signals for lymphocyte recirculation and leukocyte emigration: the multistep paradigm. *Cell* 76, 301–314.

197. Moser, B., and Loetscher, P. (2001). Lymphocyte traffic control by chemokines. *Nat. Immunol.* 2, 123–128.

198. Mantovani, A. (1999). The chemokine system: redundancy for robust outputs. *Immunol. Today* 20, 254–257.

199. Wahl, S. M., McCartney-Francis, N., and Mergenhagen, S. E. (1989). Inflammatory and immunomodulatory roles of TGF-β. *Immunol. Today* 10, 258–261.

200. Ashcroft, G. S. (1999). Bidirectional regulation of macrophage function by TGF-β. *Microbes Infect.* 1, 1275–1282.

201. Kulkarni, A. B., Huh, C. G., Becker, D., Geiser, A., Lyght, M., Flanders, K. C., Roberts, A. B., Sporn, M. B., Ward, J. M., and Karlsson, S. (1993). Transforming growth factor β1 null mutation in mice causes excessive inflammatory response and early death. *Proc. Natl. Acad. Sci. U. S. A.* 90, 770–774.

202. Buckingham, J. C., Loxley, H. D., Christian, H. C., and Philip, J. G. (1996). Activation of the HPA axis by immune insults: roles and interactions of cytokines, eicosanoids, glucocorticoids. *Pharmacol. Biochem. Behav.* 54, 285–298.

203. Imura, H., Fukata, J., and Mori, T. (1991). Cytokines and endocrine function: an interaction between the immune and neuroendocrine systems. *Clin. Endocrinol.* 35, 107–115.

204. Kapcala, L. P., Chautard, T., and Eskay, R. L. (1995). The protective role of the hypothalamic-pituitary-adrenal axis against lethality produced by immune, infectious, and inflammatory stress. *Ann. N. Y. Acad. Sci.* 771, 419–437.

205. Tilg, H., Dinarello, C. A., and Mier, J. W. (1997). IL-6 and APPs: anti-inflammatory and immunosuppressive mediators. *Immunol. Today* 18, 428–432.

206. Clevers, H., Alarcon, B., Wileman, T., and Terhorst, C. (1988). The T cell receptor/CD3 complex: a dynamic protein ensemble. *Annu. Rev. Immunol.* 6, 629–662.

207. Weiss, A. (1991). Molecular and genetic insights into T cell antigen receptor structure and function. *Annu. Rev. Genet.* 25, 487–510.

208. Davies, D. R., and Metzger, H. (1983). Structural basis of antibody function. *Annu. Rev. Immunol.* 1, 87–117.

209. Tonegawa, S. (1983). Somatic generation of antibody diversity. *Nature* 302, 575–581.

210. Chien, Y. H., Gascoigne, N. R., Kavaler, J., Lee, N. E., and Davis, M. M. (1984). Somatic recombination in a murine T-cell receptor gene. *Nature* 309, 322–326.

211. Hood, L., Kronenberg, M., and Hunkapiller, T. (1985). T cell antigen receptors and the immunoglobulin supergene family. *Cell* 40, 225–229.

212. Nossal, G. J. V., and Lederberg, J. (1958). Antibody production by single cells. *Nature* 182, 1383–1384.

213. Burnet, F. M. (1957). A modification of Jerne's theory of antibody formation. *Aust. J. Sci.* 20, 67–68.

214. Zinkernagel, R. M., and Doherty, P. C. (1979). MHC-restricted cytotoxic T cells: studies on the biological role of polymorphic major transplantation antigens determining T-cell restriction-specificity, function, and responsiveness. *Adv. Immunol.* 27, 51–177.

215. Ploegh, H. L., Orr, H. T., and Strominger, J. L. (1981). Major histocompatibility antigens: the human (HLA-A, -B, -C) and murine (H-2K, H-2D) class I molecules. *Cell* 24, 287–299.

216. Cresswell, P., Bangia, N., Dick, T., and Diedrich, G. (1999). The nature of the MHC class I peptide loading complex. *Immunol. Rev.* 172, 21–28.

217. Sant, A. J., Beeson, C., McFarland, B., Cao, J., Ceman, S., Bryant, P. W., and Wu, S. (1999). Individual hydrogen bonds

play a critical role in MHC class II: peptide interactions: implications for the dynamic aspects of class II trafficking and DM-mediated peptide exchange. *Immunol. Rev.* 172, 239–253.
218. Janeway, C. A., Jr. (1992). The T cell receptor as a multicomponent signalling machine: CD4/CD8 coreceptors and CD45 in T cell activation. *Annu. Rev. Immunol.* 10, 645–674.
219. Gao, G. F., Tormo, J., Gerth, U. C., Wyer, J. R., McMichael, A. J., Stuart, D. I., Bell, J. I., Jones, E. Y., and Jakobsen, B. K. (1997). Crystal structure of the complex between human CD8αα and HLA-A2. *Nature* 387, 630–634.
220. Wang, J. H., Meijers, R., Xiong, Y., Liu, J. H., Sakihama, T., Zhang, R., Joachimiak, A., and Reinherz, E. L. (2001). Crystal structure of the human CD4 N-terminal two-domain fragment complexed to a class II MHC molecule. *Proc. Natl. Acad. Sci. U. S. A.* 98, 10799–10804.
221. Banchereau, J., and Steinman, R. M. (1998). Dendritic cells and the control of immunity. *Nature* 392, 245–252.
222. Unanue, E. R. (1984). Antigen-presenting function of the macrophage. *Annu. Rev. Immunol.* 2, 395–428.
223. Sprent, J., and Schaefer, M. (1989). Antigen-presenting cells for unprimed T cells. *Immunol. Today* 10, 17–23.
224. Constant, S. L., and Bottomly, K. (1997). Induction of Th1 and Th2 $CD4^+$ T cell responses: the alternative approaches. *Annu. Rev. Immunol.* 15, 297–322.
225. Moser, M., and Murphy, K. M. (2000). Dendritic cell regulation of T_H1-T_H2 development. *Nat. Immunol.* 1, 199–205.
226. Kourilsky, P., and Truffa-Bachi, P. (2001). Cytokine fields and the polarization of the immune response. *Trends Immunol.* 22, 502–509.
227. Piccirillo, C. A., and Thornton, A. M. (2004). Cornerstone of peripheral tolerance: naturally occurring $CD4^+$ $CD25^+$ regulatory T cells. *Trends Immunol.* 25, 374–380.
228. Gilliet, M., and Liu, Y. J. (2002). Generation of human CD8 T regulatory cells by CD40 ligand-activated plasmacytoid dendritic cells. *J. Exp. Med.* 195, 695–704.
229. Sakaguchi, S. (2000). Regulatory T cells: key controllers of immunologic self-tolerance. *Cell* 101, 455–458.
230. DeFranco, A. L. (1987). Molecular aspects of B-lymphocyte activation. *Annu. Rev. Cell Biol.* 3, 143–178.
231. Davis, M. M., Kim, S. K., and Hood, L. (1980). Immunoglobulin class switching: developmentally regulated DNA rearrangements during differentiation. *Cell* 22, 1–2.
232. Schittek, B., and Rajewsky, K. (1990). Maintenance of B-cell memory by long-lived cells generated from proliferating precursors. *Nature* 346, 749–751.
233. L'Age-Stehr, J., and Herzenberg, L. A. (1970). Immunological memory in mice: I. Physical separation and partial characterization of memory cells for different immunoglobulin classes from each other and from antibody-producing cells. *J. Exp. Med.* 131, 1093–1108.
234. Dutton, R. W., Bradley, L. M., and Swain, S. L. (1998). T cell memory. *Annu. Rev. Immunol.* 16, 201–223.
235. Moody, D. B., Besra, G. S., Wilson, I. A., and Porcelli, S. A. (1999). The molecular basis of CD1-mediated presentation of lipid antigens. *Immunol. Rev.* 172, 285–296.
236. Calabi, F., Jarvis, J. M., Martin, L., and Milstein, C. (1989). Two classes of CD1 genes. *Eur. J. Immunol.* 19, 285–292.
237. Trinchieri, G. (1989). Biology of natural killer cells. *Adv. Immunol.* 47, 187–376.
238. Kraus, E., Lambracht, D., Wonigeit, K., and Hünig, T. (1996). Negative regulation of rat natural killer cell activity by major histocompatibility complex class I recognition. *Eur. J. Immunol.* 26, 2582–2586.
239. Kärre, K., Ljunggren, H. G., Piontek, G., and Kiessling, R. (1986). Selective rejection of H-2-deficient lymphoma variants suggests alternative immune defence strategy. *Nature* 319, 675–678.
240. Raulet, D. H. (2004). Interplay of natural killer cells and their receptors with the adaptive immune response. *Nat. Immunol.* 5, 996–1002.
241. Bendelac, A., Rivera, M. N., Park, S. H., and Roark, J. H. (1997). Mouse CD1-specific NK1 T cells: development, specificity, and function. *Annu. Rev. Immunol.* 15, 535–562.
242. Godfrey, D. I., Hammond, K. J., Poulton, L. D., Smyth, M. J., and Baxter, A. G. (2000). NKT cells: facts, functions and fallacies. *Immunol. Today* 21, 573–583.
243. Sonoda, K. H., Faunce, D. E., Taniguchi, M., Exley, M., Balk, S., and Stein-Streilein, J. (2001). NK T cell-derived IL-10 is essential for the differentiation of antigen-specific T regulatory cells in systemic tolerance. *J. Immunol.* 166, 42–50.
244. Baker, J., Verneris, M. R., Ito, M., Shizuru, J. A., and Negrin, R. S. (2001). Expansion of cytolytic $CD8^+$ natural killer T cells with limited capacity for graft-versus-host disease induction due to interferon γ production. *Blood* 97, 2923–2931.
245. Moodycliffe, A. M., Nghiem, D., Clydesdale, G., and Ullrich, S. E. (2000). Immune suppression and skin cancer development: regulation by NKT cells. *Nat. Immunol.* 1, 521–525.
246. Derbinski, J., Schulte, A., Kyewski, B., and Klein, L. (2001). Promiscuous gene expression in medullary thymic epithelial cells mirrors the peripheral self. *Nat. Immunol.* 2, 1032–1039.
247. Gotter, J., Brors, B., Hergenhahn, M., and Kyewski, B. (2004). Medullary epithelial cells of the human thymus express a highly diverse selection of tissue-specific genes colocalized in chromosomal clusters. *J. Exp. Med.* 199, 155–166.
248. Ermann, J., and Fathman, C. G. (2001). Autoimmune diseases: genes, bugs and failed regulation. *Nat. Immunol.* 2, 759–761.
249. Kumar, V., Kono, D. H., Urban, J. L., and Hood, L. (1989). The T-cell receptor repertoire and autoimmune diseases. *Annu. Rev. Immunol.* 7, 657–682.
250. Dziarski, R. (1988). Autoimmunity: polyclonal activation or antigen induction? *Immunol. Today* 9, 340–342.
251. Sakaguchi, S., and Sakaguchi, N. (1990). Thymus and autoimmunity: capacity of the normal thymus to produce pathogenic self-reactive T cells and conditions required for their induction of autoimmune disease. *J. Exp. Med.* 172, 537–545.
252. Miller, J. F., and Morahan, G. (1992). Peripheral T cell tolerance. *Annu. Rev. Immunol.* 10, 51–69.
253. Schwartz, R. H. (1996). Models of T cell anergy: is there a common molecular mechanism? *J. Exp. Med.* 184, 1–8.
254. Taylor, P. A., Noelle, R. J., and Blazar, B. R. (2001). $CD4^+$ $CD25^+$ immune regulatory cells are required for induction of tolerance to alloantigen via costimulatory blockade. *J. Exp. Med.* 193, 1311–1318.
255. Sakaguchi, S., Sakaguchi, N., Shimizu, J., Yamazaki, S., Sakihama, T., Itoh, M., Kuniyasu, Y., Nomura, T., Toda, M., and Takahashi, T. (2001). Immunologic tolerance maintained by $CD25^+$ $CD4^+$ regulatory T cells: their common role in controlling autoimmunity, tumor immunity, and transplantation tolerance. *Immunol. Rev.* 182, 18–32.
256. Coutinho, A. (1995). The network theory: 21 years later. *Scand. J. Immunol.* 42, 3–8.
257. Shlomchik, W. D., Couzens, M. S., Tang, C. B., McNiff, J., Robert, M. E., Liu, J., Shlomchik, M. J., and Emerson, S. G. (1999). Prevention of graft versus host disease by inactivation of host antigen-presenting cells. *Science* 285, 412–415.
258. Baker, M. B., Altman, N. H., Podack, E. R., and Levy, R. B. (1996). The role of cell-mediated cytotoxicity in acute GVHD after MHC-matched allogeneic bone marrow transplantation in mice. *J. Exp. Med.* 183, 2645–2656.
259. Carosella, E. D., Rouas-Freiss, N., Paul, P., and Dausset, J. (1999). HLA-G: a tolerance molecule from the major histocompatibility complex. *Immunol. Today* 20, 60–62.

260. Kovats, S., Main, E. K., Librach, C., Stubblebine, M., Fisher, S. J., and DeMars, R. (1990). A class I antigen, HLA-G, expressed in human trophoblasts. *Science* 248, 220–223.
261. Ryan, A. F., Grendell, R. L., Geraghty, D. E., and Golos, T. G. (2002). A soluble isoform of the rhesus monkey nonclassical MHC class I molecule Mamu-AG is expressed in the placenta and the testis. *J. Immunol.* 169, 673–683.
262. Smith, M. E., and Ford, W. L. (1983). The recirculating lymphocyte pool of the rat: a systematic description of the migratory behaviour of recirculating lymphocytes. *Immunology* 49, 83–94.
263. Picker, L. J., and Butcher, E. C. (1992). Physiological and molecular mechanisms of lymphocyte homing. *Annu. Rev. Immunol.* 10, 561–591.
264. Garside, P., Ingulli, E., Merica, R. R., Johnson, J. G., Noelle, R. J., and Jenkins, M. K. (1998). Visualization of specific B and T lymphocyte interactions in the lymph node. *Science* 281, 96–99.
265. Romani, N., Ratzinger, G., Pfaller, K., Salvenmoser, W., Stossel, H., Koch, F., and Stoitzner, P. (2001). Migration of dendritic cells into lymphatics—the Langerhans cell example: routes, regulation, and relevance. *Int. Rev. Cytol.* 207, 237–270.
266. Gretz, J. E., Anderson, A. O., and Shaw, S. (1997). Cords, channels, corridors and conduits: critical architectural elements facilitating cell interactions in the lymph node cortex. *Immunol. Rev.* 156, 11–24.
267. Pedersen, N. C., and Morris, B. (1970). The role of the lymphatic system in the rejection of homografts: a study of lymph from renal transplants. *J. Exp. Med.* 131, 936–969.
268. Ascher, N. L., Hoffman, R., Hanto, D. W., and Simmons, R. L. (1983). Cellular events within the rejecting allograft. *Transplantation* 35, 193–197.
269. Gruber, S. A. (1992). The case for local immunosuppression. *Transplantation* 54, 1–11.
270. Bienenstock, J., McDermott, M., Befus, D., and O'Neill, M. (1978). A common mucosal immunologic system involving the bronchus, breast and bowel. *Adv. Exp. Biol. Med.* 107, 53–59.
271. McGhee, J. R., Fujihashi, K., Xu-Amano, J., Jackson, R. J., Elson, C. O., Beagley, K. W., and Kiyono, H. (1993). New perspectives in mucosal immunity with emphasis on vaccine development. *Semin. Hematol.* 30, 3–12; discussion 13–15.
272. Brandtzaeg, P. (1985). Role of J chain and secretory component in receptor-mediated glandular and hepatic transport of immunoglobulins in man. *Scand. J. Immunol.* 22, 111–146.
273. Hendrickson, B. A., Rindisbacher, L., Corthesy, B., Kendall, D., Waltz, D. A., Neutra, M. R., and Seidman, J. G. (1996). Lack of association of secretory component with IgA in J chain-deficient mice. *J. Immunol.* 157, 750–754.
274. Mestecky, J. (1987). The common mucosal immune system and current strategies for induction of immune responses in external secretions. *J. Clin. Immunol.* 7, 265–276.
275. Christ, A. D., and Blumberg, R. S. (1997). The intestinal epithelial cell: immunological aspects. *Springer Semin. Immunopathol.* 18, 449–461.
276. Jung, H. C., Eckmann, L., Yang, S. K., Panja, A., Fierer, J., Morzycka-Wroblewska, E., and Kagnoff, M. F. (1995). A distinct array of proinflammatory cytokines is expressed in human colon epithelial cells in response to bacterial invasion. *J. Clin. Invest.* 95, 55–65.
277. Weiner, H. L. (1997). Oral tolerance: immune mechanisms and treatment of autoimmune diseases. *Immunol. Today* 18, 335–343.
278. Brandtzaeg, P., Halstensen, T. S., Huitfeldt, H. S., Krajci, P., Kvale, D., Scott, H., and Thrane, P. S. (1992). Epithelial expression of HLA, secretory component (poly-Ig receptor), and adhesion molecules in the human alimentary tract. *Ann. N. Y. Acad. Sci.* 664, 157–179.
279. Kvale, D., Krajci, P., and Brandtzaeg, P. (1992). Expression and regulation of adhesion molecules ICAM-1 (CD54) and LFA-3 (CD58) in human intestinal epithelial cell lines. *Scand. J. Immunol.* 35, 669–676.
280. Hershberg, R. M., Cho, D. H., Youakim, A., Bradley, M. B., Lee, J. S., Framson, P. E., and Nepom, G. T. (1998). Highly polarized HLA class II antigen processing and presentation by human intestinal epithelial cells. *J. Clin. Invest.* 102, 792–803.
281. Clifton, V. L., Husband, A. J., and Kay, D. J. (1992). Local immunity in the male reproductive tract. *Immunol. Cell Biol.* 70, 301–307.
282. Husband, A. J., and Clifton, V. L. (1989). Role of intestinal immunization in urinary tract defence. *Immunol. Cell Biol.* 67, 371–376.
283. Bronson, R., Cooper, G., Rosenfeld, D., and Witkin, S. S. (1984). Detection of spontaneously occurring sperm-directed antibodies in infertile couples by immunobead binding and enzyme-linked immunosorbent assay. *Ann. N. Y. Acad. Sci.* 438, 504–507.
284. Bronson, R. A., Cooper, G. W., and Rosenfeld, D. L. (1982). Sperm-specific isoantibodies and autoantibodies inhibit the binding of human sperm to the human zona pellucida. *Fertil. Steril.* 38, 724–729.
285. McKay, L. I., and Cidlowski, J. A. (2000). CBP (CREB binding protein) integrates NF-κB (nuclear factor-κB) and glucocorticoid receptor physical interactions and antagonism. *Mol. Endocrinol.* 14, 1222–1234.
286. Bambino, T. H., and Hsueh, A. J. (1981). Direct inhibitory effect of glucocorticoids upon testicular luteinizing hormone receptor and steroidogenesis *in vivo* and *in vitro*. *Endocrinology* 108, 2142–2148.
287. Saez, J. M., Morera, A. M., Haour, F., and Evain, D. (1977). Effects of *in vivo* administration of dexamethasone, corticotropin and human chorionic gonadotropin on steroidogenesis and protein and DNA synthesis of testicular interstitial cells in prepuberal rats. *Endocrinology* 101, 1256–1263.
288. Smith, E. R., Johnson, J., Weick, R. F., Levine, S., and Davidson, J. M. (1971). Inhibition of the reproductive system in immature rats by intracerebral implantation of cortisol. *Neuroendocrinology* 8, 94–106.
289. Chantaraprateep, P., and Thibier, M. (1978). Effects of dexamethasone on the responses of luteinizing hormone and testosterone to two injections of luteinizing hormone releasing hormone in young postpubertal bulls. *J. Endocrinol.* 77, 389–395.
290. Welsh, T. H., Jr., Bambino, T. H., and Hsueh, A. J. (1982). Mechanism of glucocorticoid-induced suppression of testicular androgen biosynthesis in vitro. *Biol. Reprod.* 27, 1138–1146.
291. Butterworth, M., McClellan, B., and Allansmith, M. (1967). Influence of sex in immunoglobulin levels. *Nature* 214, 1224–1225.
292. Terres, G., Morrison, S. L., and Habicht, G. S. (1968). A quantitative difference in the immune response between male and female mice. *Proc. Soc. Exp. Biol. Med.* 127, 664–667.
293. Rhodes, K., Scott, A., Markham, R. L., and Monk-Jones, M. E. (1969). Immunological sex differences: a study of patients with rheumatoid arthritis, their relatives, and controls. *Ann. Rheum. Dis.* 28, 104–120.
294. London, W. T., and Drew, J. S. (1977). Sex differences in response to hepatitis B infection among patients receiving chronic dialysis treatment. *Proc. Natl. Acad. Sci. U. S. A.* 74, 2561–2563.

295. Wormald, P. J. (1977). Age-sex incidence in symptomatic allergies: an excess of females in the child-bearing years. *J. Hygiene* 79, 39–42.
296. Masi, A. T., and Kaslow, R. A. (1978). Sex effects in systemic lupus erythematosus: a clue to pathogenesis. *Arthritis Rheum.* 21, 480–484.
297. Inman, R. D. (1978). Immunologic sex differences and the female predominance in systemic lupus erythematosus. *Arthritis Rheum.* 21, 849–852.
298. Michaels, R. M., and Rogers, K. D. (1971). A sex difference in immunologic responsiveness. *Pediatrics* 47, 120–123.
299. Cutolo, M., Sulli, A., Capellino, S., Villaggio, B., Montagna, P., Seriolo, B., and Straub, R. H. (2004). Sex hormones influence on the immune system: basic and clinical aspects in autoimmunity. *Lupus* 13, 635–638.
300. Lunn, S. F., Cowen, G. M., and Fraser, H. M. (1997). Blockade of the neonatal increase in testosterone by a GnRH antagonist: the free androgen index, reproductive capacity and postmortem findings in the male marmoset monkey. *J. Endocrinol.* 154, 125–131.
301. Steinberg, A. D., Melez, K. A., Raveche, E. S., Reeves, J. P., Boegel, W. A., Smathers, P. A., Taurog, J. D., Weinlein, L., and Duvic, M. (1979). Approach to the study of the role of sex hormones in autoimmunity. *Arthritis Rheum.* 22, 1170–1176.
302. Weinstein, Y., and Isakov, Y. (1983). Effects of testosterone metabolites and of anabolic androgens on the bone marrow and thymus in castrated female mice. *Immunopharmacology* 5, 239–250.
303. Aboudkhil, S., Bureau, J. P., Garrelly, L., and Vago, P. (1991). Effects of castration, Depo-Testosterone and cyproterone acetate on lymphocyte T subsets in mouse thymus and spleen. *Scand. J. Immunol.* 34, 647–653.
304. Kiyota, M., Korenaga, M., Nawa, Y., and Kotani, M. (1984). Effect of androgen on the expression of the sex difference in susceptibility to infection with Strongyloides ratti in C57BL/6 mice. *Aust. J. Exp. Biol. Med. Sci.* 62, 607–618.
305. Grossman, C. J., Nathan, P., Taylor, B. B., and Sholiton, L. J. (1979). Rat thymic dihydrotestosterone receptor: preparation, location and physiochemical properties. *Steroids* 34, 539–553.
306. McCruden, A. B., and Stimson, W. H. (1981). Androgen binding cytosol receptors in the rat thymus: physicochemical properties, specificity and localisation. *Thymus* 3, 105–117.
307. Cohen, J. H., Danel, L., Cordier, G., Saez, S., and Revillard, J. P. (1983). Sex steroid receptors in peripheral T cells: absence of androgen receptors and restriction of estrogen receptors to OKT8-positive cells. *J. Immunol.* 131, 2767–2771.
308. Kovacs, W. J., and Olsen, N. J. (1987). Androgen receptors in human thymocytes. *J. Immunol.* 139, 490–493.
309. Gulshan, S., McCruden, A. B., and Stimson, W. H. (1990). Oestrogen receptors in macrophages. *Scand. J. Immunol.* 31, 691–697.
310. Benten, W. P., Lieberherr, M., Giese, G., Wrehlke, C., Stamm, O., Sekeris, C. E., Mossmann, H., and Wunderlich, F. (1999). Functional testosterone receptors in plasma membranes of T cells. *FASEB J.* 13, 123–133.
311. Benten, W. P., Lieberherr, M., Stamm, O., Wrehlke, C., Guo, Z., and Wunderlich, F. (1999). Testosterone signaling through internalizable surface receptors in androgen receptor-free macrophages. *Mol. Biol. Cell* 10, 3113–3123.
312. Olsen, N. J., Olson, G., Viselli, S. M., Gu, X., and Kovacs, W. J. (2001). Androgen receptors in thymic epithelium modulate thymus size and thymocyte development. *Endocrinology* 142, 1278–1283.
313. Liva, S. M., and Voskuhl, R. R. (2001). Testosterone acts directly on CD4+ T lymphocytes to increase IL-10 production. *J. Immunol.* 167, 2060–2067.
314. Braun, A. M., and Thomas, P. (2004). Biochemical characterization of a membrane androgen receptor in the ovary of the Atlantic croaker (*Micropogonias undulatus*). *Biol. Reprod.* 71, 146–155.
315. Zhu, Y., Bond, J., and Thomas, P. (2003). Identification, classification, and partial characterization of genes in humans and other vertebrates homologous to a fish membrane progestin receptor. *Proc. Natl. Acad. Sci. U. S. A.* 100, 2237–2242.
316. Benten, W. P., Lieberherr, M., Sekeris, C. E., and Wunderlich, F. (1997). Testosterone induces Ca^{2+} influx via non-genomic surface receptors in activated T cells. *FEBS Lett.* 407, 211–214.
317. Benten, W. P., Guo, Z., Krucken, J., and Wunderlich, F. (2004). Rapid effects of androgens in macrophages. *Steroids* 69, 585–590.
318. He, L., Hedger, M. P., Clements, J. A., and Risbridger, G. P. (1991). Localization of immunoreactive β-endorphin and adrenocorticotropic hormone and pro-opiomelanocortin mRNA to rat testicular interstitial tissue macrophages. *Biol. Reprod.* 45, 282–289.
319. Margioris, A. N., Liotta, A. S., Vaudry, H., Bardin, C. W., and Krieger, D. T. (1983). Characterization of immunoreactive proopiomelanocortin-related peptides in rat testes. *Endocrinology* 113, 663–671.
320. Chen, C. L., Mather, J. P., Morris, P. L., and Bardin, C. W. (1984). Expression of pro-opiomelanocortin-like gene in the testis and epididymis. *Proc. Natl. Acad. Sci. U. S. A.* 81, 5672–5675.
321. Boitani, C., Chen, C. L., Margioris, A. N., Gerendai, I., Morris, P. L., and Bardin, C. W. (1986). Pro-opiomelanocortin-derived peptides in the testis: evidence for a possible role in Leydig and Sertoli cell function. *Med. Biol.* 63, 251–258.
322. Grabbe, S., Bhardwaj, R. S., Mahnke, K., Simon, M. M., Schwarz, T., and Luger, T. A. (1996). α-Melanocyte-stimulating hormone induces hapten-specific tolerance in mice. *J. Immunol.* 156, 473–478.
323. Bhardwaj, R. S., Schwarz, A., Becher, E., Mahnke, K., Aragane, Y., Schwarz, T., and Luger, T. A. (1996). Pro-opiomelanocortin-derived peptides induce IL-10 production in human monocytes. *J. Immunol.* 156, 2517–2521.
324. Gilman, S. C., Schwartz, J. M., Milner, R. J., Bloom, F. E., and Feldman, J. D. (1982). β-Endorphin enhances lymphocyte proliferative responses. *Proc. Natl. Acad. Sci. U. S. A.* 79, 4226–4230.
325. Brown, S. L., and Van Epps, D. E. (1985). Suppression of T lymphocyte chemotactic factor production by the opioid peptides β-endorphin and met-enkephalin. *J. Immunol.* 134, 3384–3390.
326. Payan, D. G., McGillis, J. P., Renold, F. K., Mitsuhashi, M., and Goetzl, E. J. (1987). Neuropeptide modulation of leukocyte function. *Ann. N. Y. Acad. Sci.* 496, 182–191.
327. Peck, R. (1987). Neuropeptides modulating macrophage function. *Ann. N. Y. Acad. Sci.* 496, 264–270.
328. van den Bergh, P., Rozing, J., and Nagelkerken, L. (1991). Two opposing modes of action of β-endorphin on lymphocyte function. *Immunology* 72, 537–543.
329. Rao, L. V., Cleveland, R. P., Kimmel, R. J., and Ataya, K. M. (1996). Gonadotropin-releasing hormone agonist influences absolute levels of lymphocyte subsets *in vivo* in male mice. *Immunol. Cell Biol.* 74, 134–143.
330. Batticane, N., Morale, M. C., Gallo, F., Farinella, Z., and Marchetti, B. (1991). Luteinizing hormone-releasing hormone signaling at the lymphocyte involves stimulation of interleukin-2 receptor expression. *Endocrinology* 129, 277–286.

331. Morale, M. C., Batticane, N., Bartoloni, G., Guarcello, V., Farinella, Z., Galasso, M. G., and Marchetti, B. (1991). Blockade of central and peripheral luteinizing hormone-releasing hormone (LHRH) receptors in neonatal rats with a potent LHRH-antagonist inhibits the morphofunctional development of the thymus and maturation of the cell-mediated and humoral immune responses. *Endocrinology* 128, 1073–1085.
332. Standaert, F. E., Chew, B. P., De Avila, D., and Reeves, J. J. (1992). Presence of luteinizing hormone-releasing hormone binding sites in cultured porcine lymphocytes. *Biol. Reprod.* 46, 997–1000.
333. de Kretser, D. M., Buzzard, J. J., Okuma, Y., O'Connor, A. E., Hayashi, T., Lin, S. Y., Morrison, J. R., Loveland, K. L., and Hedger, M. P. (2004). The role of activin, follistatin and inhibin in testicular physiology. *Mol. Cell. Endocrinol.* 225, 57–64.
334. Le Magueresse-Battistoni, B., Morera, A. M., Goddard, I., and Benahmed, M. (1995). Expression of mRNAs for transforming growth factor-β receptors in the rat testis. *Endocrinology* 136, 2788–2791.
335. Caussanel, V., Tabone, E., Hendrick, J. C., Dacheux, F., and Benahmed, M. (1997). Cellular distribution of transforming growth factor betas 1, 2, and 3 and their types I and II receptors during postnatal development and spermatogenesis in the boar testis. *Biol. Reprod.* 56, 357–367.
336. Chomarat, P., Banchereau, J., Davoust, J., and Palucka, A. K. (2000). IL-6 switches the differentiation of monocytes from dendritic cells to macrophages. *Nat. Immunol.* 1, 510–514.
337. Palucka, K. A., Taquet, N., Sanchez-Chapuis, F., and Gluckman, J. C. (1998). Dendritic cells as the terminal stage of monocyte differentiation. *J. Immunol.* 160, 4587–4595.
338. Becker, S., Warren, M. K., and Haskill, S. (1987). Colony-stimulating factor-induced monocyte survival and differentiation into macrophages in serum-free cultures. *J. Immunol.* 139, 3703–3709.
339. Traver, D., Akashi, K., Manz, M., Merad, M., Miyamoto, T., Engleman, E. G., and Weissman, I. L. (2000). Development of CD8α-positive dendritic cells from a common myeloid progenitor. *Science* 290, 2152–2154.
340. Peters, J. H., Gieseler, R., Thiele, B., and Steinbach, F. (1996). Dendritic cells: from ontogenetic orphans to myelomonocytic descendants. *Immunol. Today* 17, 273–278.
341. Grabbe, S., Kampgen, E., and Schuler, G. (2000). Dendritic cells: multi-lineal and multi-functional. *Immunol. Today* 21, 431–433.
342. Briere, F., Bendriss-Vermare, N., Delale, T., Burg, S., Corbet, C., Rissoan, M. C., Chaperot, L., Plumas, J., Jacob, M. C., Trinchieri, G., and Bates, E. E. (2002). Origin and filiation of human plasmacytoid dendritic cells. *Hum. Immunol.* 63, 1081–1093.
343. Rutherford, M. S., Witsell, A., and Schook, L. B. (1993). Mechanisms generating functionally heterogeneous macrophages: chaos revisited. *J. Leukoc. Biol.* 53, 602–618.
344. van Furth, R. (1988). Phagocytic cells: development and distribution of mononuclear phagocytes in normal steady state and inflammation. In *Inflammation: Basic Principles and Clinical Correlates* (J. I. Gallin, I. M. Goldstein, and R. Snyderman, Eds.), pp. 281–295. Raven Press, New York.
345. Mantovani, A., Sozzani, S., Locati, M., Allavena, P., and Sica, A. (2002). Macrophage polarization: tumor-associated macrophages as a paradigm for polarized M2 mononuclear phagocytes. *Trends Immunol.* 23, 549–555.
346. Miller, S. C. (1982). Localization of plutonium-241 in the testis: an interspecies comparison using light and electron microscope autoradiography. *Int. J. Radiat. Biol. Relat. Stud. Phys. Chem. Med.* 41, 633–643.
347. Miller, S. C., Bowman, B. M., and Rowland, H. G. (1983). Structure, cytochemistry, endocytic activity, and immunoglobulin (Fc) receptors of rat testicular interstitial-tissue macrophages. *Am. J. Anat.* 168, 1–13.
348. Miller, S. C., Bowman, B. M., and Roberts, L. K. (1984). Identification and characterization of mononuclear phagocytes isolated from rat testicular interstitial tissues. *J. Leukoc. Biol.* 36, 679–687.
349. Hutson, J. C. (1992). Development of cytoplasmic digitations between Leydig cells and testicular macrophages of the rat. *Cell Tissue Res.* 267, 385–389.
350. Yee, J. B., and Hutson, J. C. (1983). Testicular macrophages: isolation, characterization and hormonal responsiveness. *Biol. Reprod.* 29, 1319–1326.
351. Yee, J. B., and Hutson, J. C. (1985). In vivo effects of follicle-stimulating hormone on testicular macrophages. *Biol. Reprod.* 32, 880–883.
352. Wei, R. Q., Yee, J. B., Straus, D. C., and Hutson, J. C. (1988). Bactericidal activity of testicular macrophages. *Biol. Reprod.* 38, 830–835.
353. Hutson, J. C. (1989). Leydig cells do not have Fc receptors. *J. Androl.* 10, 159–165.
354. Hutson, J. C., and Stocco, D. M. (1989). Comparison of cellular and secreted proteins of macrophages from the testis and peritoneum on two-dimensional polyacrylamide gels: evidence of tissue specific function. *Reg. Immunol.* 2, 249–253.
355. Mendis-Handagama, S. M., Risbridger, G. P., and de Kretser, D. M. (1987). Morphometric analysis of the components of the neonatal and the adult rat testis interstitium. *Int. J. Androl.* 10, 525–534.
356. Hardy, M. P., Zirkin, B. R., and Ewing, L. L. (1989). Kinetic studies on the development of the adult population of Leydig cells in testes of the pubertal rat. *Endocrinology* 124, 762–770.
357. Vergouwen, R. P., Huiskamp, R., Bas, R. J., Roepers-Gajadien, H. L., Davids, J. A., and de Rooij, D. G. (1993). Postnatal development of testicular cell populations in mice. *J. Reprod. Fertil.* 99, 479–485.
358. Hume, D. A., Halpin, D., Charlton, H., and Gordon, S. (1984). The mononuclear phagocyte system of the mouse defined by immunohistochemical localization of antigen F4/80: macrophages of endocrine organs. *Proc. Natl. Acad. Sci. U. S. A.* 81, 4174–4177.
359. Wrobel, K. H., Dostal, S., and Schimmel, M. (1988). Postnatal development of the tubular lamina propria and the intertubular tissue in the bovine testis. *Cell Tissue Res.* 252, 639–653.
360. Frungieri, M. B., Calandra, R. S., Lustig, L., Meineke, V., Kohn, F. M., Vogt, H. J., and Mayerhofer, A. (2002). Number, distribution pattern, and identification of macrophages in the testes of infertile men. *Fertil. Steril.* 78, 298–306.
361. Mendis-Handagama, S. M., Zirkin, B. R., and Ewing, L. L. (1988). Comparison of components of the testis interstitium with testosterone secretion in hamster, rat, and guinea pig testes perfused in vitro. *Am. J. Anat.* 181, 12–22.
362. Clemmons, A. J., Thompson, D. L., Jr., and Johnson, L. (1995). Local initiation of spermatogenesis in the horse. *Biol. Reprod.* 52, 1258–1267.
363. Sur, J. H., Doster, A. R., Christian, J. S., Galeota, J. A., Wills, R. W., Zimmerman, J. J., and Osorio, F. A. (1997). Porcine reproductive and respiratory syndrome virus replicates in testicular germ cells, alters spermatogenesis, and induces germ cell death by apoptosis. *J. Virol.* 71, 9170–9179.
364. Niemi, M., Sharpe, R. M., and Brown, W. R. (1986). Macrophages in the interstitial tissue of the rat testis. *Cell Tissue Res.* 243, 337–344.

365. Bergh, A. (1987). Treatment with hCG increases the size of Leydig cells and testicular macrophages in unilaterally cryptorchid rats. *Int. J. Androl.* 10, 765–772.
366. Bergh, A. (1985). Effect of cryptorchidism on the morphology of testicular macrophages: evidence for a Leydig cell-macrophage interaction in the rat testis. *Int. J. Androl.* 8, 86–96.
367. Geierhaas, B., Bornstein, S. R., Jarry, H., Scherbaum, W. A., Herrmann, M., and Pfeiffer, E. F. (1991). Morphological and hormonal changes following vasectomy in rats, suggesting a functional role for Leydig-cell associated macrophages. *Horm. Metab. Res.* 23, 373–378.
368. Giannessi, F., Giambelluca, M. A., Scavuzzo, M. C., and Ruffoli, R. (2005). Ultrastructure of testicular macrophages in aging mice. *J. Morphol.* 263, 39–46.
369. Clegg, E. J., and Macmillan, E. W. (1965). The phagocytic nature of Schiff-positive interstitial cells in the rat testis. *J. Endocrinol.* 31, 299–300.
370. Brooks, A. L., Diel, J. H., and McClellan, R. O. (1979). The influence of testicular microanatomy on the potential genetic dose from internally deposited ^{239}Pu citrate in Chinese hamster, mouse, and man. *Radiat. Res.* 77, 292–302.
371. Russell, J. J., and Lindenbaum, A. (1979). One-year study of nonuniformly distributed plutonium in mouse testis as related to spermatogonial irradiation. *Health Phys.* 36, 153–157.
372. Hedger, M. P., and Eddy, E. M. (1987). The heterogeneity of isolated adult rat Leydig cells separated on Percoll density gradients: an immunological, cytochemical, and functional analysis. *Endocrinology* 121, 1824–1838.
373. Sasmono, R. T., Oceandy, D., Pollard, J. W., Tong, W., Pavli, P., Wainwright, B. J., Ostrowski, M. C., Himes, S. R., and Hume, D. A. (2003). A macrophage colony-stimulating factor receptor-green fluorescent protein transgene is expressed throughout the mononuclear phagocyte system of the mouse. *Blood* 101, 1155–1163.
374. Yee, J. B., and Hutson, J. C. (1985). Biochemical consequences of follicle-stimulating hormone binding to testicular macrophages in culture. *Biol. Reprod.* 32, 872–879.
375. Carpenter, A. M., Lukyanenko, Y. O., Lee, V. H., and Hutson, J. C. (1998). FSH does not directly influence testicular macrophages. *J. Androl.* 19, 420–427.
376. Kern, S., and Maddocks, S. (1995). Indomethacin blocks the immunosuppressive activity of rat testicular macrophages cultured in vitro. *J. Reprod. Immunol.* 28, 189–201.
377. Bergh, A., Damber, J. E., and van Rooijen, N. (1993). The human chorionic gonadotrophin-induced inflammation-like response is enhanced in macrophage-depleted rat testes. *J. Endocrinol.* 136, 415–420.
378. Hedger, M. P. (2002). Macrophages and the immune responsiveness of the testis. *J. Reprod. Immunol.* 57, 19–34.
379. Gaytan, F., Bellido, C., Aguilar, E., and van Rooijen, N. (1994). Requirement for testicular macrophages in Leydig cell proliferation and differentiation during prepubertal development in rats. *J. Reprod. Fertil.* 102, 393–399.
380. Gaytan, F., Bellido, C., Aguilar, E., and van Rooijen, N. (1995). Pituitary-testicular axis in rats lacking testicular macrophages. *Eur. J. Endocrinol.* 132, 218–222.
381. Gaytan, F., Bellido, C., Morales, C., Reymundo, C., Aguilar, E., and Van Rooijen, N. (1994). Effects of macrophage depletion at different times after treatment with ethylene dimethane sulfonate (EDS) on the regeneration of Leydig cells in the adult rat. *J. Androl.* 15, 558–564.
382. Gaytan, F., Bellido, C., Morales, C., Reymundo, C., Aguilar, E., and van Rooijen, N. (1994). Selective depletion of testicular macrophages and prevention of Leydig cell repopulation after treatment with ethylene dimethane sulfonate in rats. *J. Reprod. Fertil.* 101, 175–182.
383. Bergh, A., Damber, J. E., and van Rooijen, N. (1993). Liposome-mediated macrophage depletion: an experimental approach to study the role of testicular macrophages in the rat. *J. Endocrinol.* 136, 407–413.
384. Cohen, P. E., Chisholm, O., Arceci, R. J., Stanley, E. R., and Pollard, J. W. (1996). Absence of colony-stimulating factor-1 in osteopetrotic (*csfm^op^/csfm^op^*) mice results in male fertility defects. *Biol. Reprod.* 55, 310–317.
385. Gaytan, F., Bellido, C., Morales, C., Garcia, M., van Rooijen, N., and Aguilar, E. (1996). *In vivo* manipulation (depletion versus activation) of testicular macrophages: central and local effects. *J. Endocrinol.* 150, 57–65.
386. Tung, K. S., Yule, T. D., Mahi-Brown, C. A., and Listrom, M. B. (1987). Distribution of histopathology and Ia positive cells in actively induced and passively transferred experimental autoimmune orchitis. *J. Immunol.* 138, 752–759.
387. Mahi-Brown, C. A., Yule, T. D., and Tung, K. S. (1987). Adoptive transfer of murine autoimmune orchitis to naive recipients with immune lymphocytes. *Cell. Immunol.* 106, 408–419.
388. Haas, G. G., Jr., D'Cruz, O. J., and De Bault, L. E. (1988). Distribution of human leukocyte antigen-ABC and -D/DR antigens in the unfixed human testis. *Am. J. Reprod. Immunol. Microbiol.* 18, 47–51.
389. Sainio-Pöllänen, S., Saari, T., Simell, O., and Pöllänen, P. (1996). CD28-CD80/CD86 interactions in testicular immunoregulation. *J. Reprod. Immunol.* 31, 145–163.
390. Kern, S., Robertson, S. A., Mau, V. J., and Maddocks, S. (1995). Cytokine secretion by macrophages in the rat testis. *Biol. Reprod.* 53, 1407–1416.
391. Hayes, R., Chalmers, S. A., Nikolic-Paterson, D. J., Atkins, R. C., and Hedger, M. P. (1996). Secretion of bioactive interleukin 1 by rat testicular macrophages *in vitro*. *J. Androl.* 17, 41–49.
392. Gerdprasert, O., O'Bryan, M. K., Muir, J. A., Caldwell, A. M., Schlatt, S., de Kretser, D. M., and Hedger, M. P. (2002). The response of testicular leukocytes to lipopolysaccharide-induced inflammation: further evidence for heterogeneity of the testicular macrophage population. *Cell Tissue Res.* 308, 277–285.
393. Meinhardt, A., Bacher, M., McFarlane, J. R., Metz, C. N., Seitz, J., Hedger, M. P., de Kretser, D. M., and Bucala, R. (1996). Macrophage migration inhibitory factor production by Leydig cells: evidence for a role in the regulation of testicular function. *Endocrinology* 137, 5090–5095.
394. O'Bryan, M. K., Gerdprasert, O., Nikolic-Paterson, D. P., Meinhardt, A., Muir, J. A., Foulds, L. K., Phillips, D. J., de Kretser, D. M., and Hedger, M. P. (2005). Cytokine profiles in the testes of rats treated with lipopolysaccharide reveal localized suppression of inflammatory responses. *Am. J. Physiol. Reg. Int. Comp. Physiol.* 288, R1774–R1775.
395. Xiong, Y., and Hales, D. B. (1993). Expression, regulation, and production of tumor necrosis factor-α in mouse testicular interstitial macrophages *in vitro*. *Endocrinology* 133, 2568–2573.
396. Hales, D. B., Xiong, Y., and Tur-Kaspa, I. (1992). The role of cytokines in the regulation of Leydig cell *P450c17* gene expression. *J. Steroid Biochem. Mol. Biol.* 43, 907–914.
397. Meinhardt, A., Bacher, M., Metz, C., Bucala, R., Wreford, N., Lan, H., Atkins, R., and Hedger, M. (1998). Local regulation of macrophage subsets in the adult rat testis: examination of the roles of the seminiferous tubules, testosterone, and macrophage-migration inhibitory factor. *Biol. Reprod.* 59, 371–378.
398. Barbé, E., Damoiseaux, J. G., Döpp, E. A., and Dijkstra, C. D. (1990). Characterization and expression of the antigen present on resident rat macrophages recognized by monoclonal antibody ED2. *Immunobiology* 182, 88–99.

399. Barbé, E., Huitinga, I., Döpp, E. A., Bauer, J., and Dijkstra, C. D. (1996). A novel bone marrow frozen section assay for studying hematopoietic interactions in situ: the role of stromal bone marrow macrophages in erythroblast binding. *J. Cell Sci.* 109, 2937–2945.
400. van den Berg, T. K., Döpp, E. A., and Dijkstra, C. D. (2001). Rat macrophages: membrane glycoproteins in differentiation and function. *Immunol. Rev.* 184, 45–57.
401. Philippidis, P., Mason, J. C., Evans, B. J., Nadra, I., Taylor, K. M., Haskard, D. O., and Landis, R. C. (2004). Hemoglobin scavenger receptor CD163 mediates interleukin-10 release and heme oxygenase-1 synthesis: antiinflammatory monocyte-macrophage responses in vitro, in resolving skin blisters in vivo, and after cardiopulmonary bypass surgery. *Circ. Res.* 94, 119–126.
402. Bryniarski, K., Szczepanik, M., Maresz, K., Ptak, M., and Ptak, W. (2004). Subpopulations of mouse testicular macrophages and their immunoregulatory function. *Am. J. Reprod. Immunol.* 52, 27–35.
403. Hoek, A., Allaerts, W., Leenen, P. J., Schoemaker, J., and Drexhage, H. A. (1997). Dendritic cells and macrophages in the pituitary and the gonads: evidence for their role in the fine regulation of the reproductive endocrine response. *Eur. J. Endocrinol.* 136, 8–24.
404. Itoh, M., De Rooij, D. G., Jansen, A., and Drexhage, H. A. (1995). Phenotypical heterogeneity of testicular macrophages/dendritic cells in normal adult mice: an immunohistochemical study. *J. Reprod. Immunol.* 28, 217–232.
405. Derrick, E. K., Barker, J. N., Khan, A., Price, M. L., and Macdonald, D. M. (1993). The tissue distribution of factor XIIIa positive cells. *Histopathology* 22, 157–162.
406. Raburn, D. J., Coquelin, A., Reinhart, A. J., and Hutson, J. C. (1993). Regulation of the macrophage population in postnatal rat testis. *J. Reprod. Immunol.* 24, 139–151.
407. Ariyaratne, H. B., and Mendis-Handagama, S. M. (2000). Changes in the testis interstitium of Sprague Dawley rats from birth to sexual maturity. *Biol. Reprod.* 62, 680–690.
408. Li, X. Q., Itoh, M., Yano, A., Miyamoto, K., and Takeuchi, Y. (1998). Immunohistochemical detection of testicular macrophages during the period of postnatal maturation in the mouse. *Int. J. Androl.* 21, 370–376.
409. Itoh, M., Xie, Q., Miyamoto, K., and Takeuchi, Y. (1999). F4/80-positive cells rapidly accumulate around tubuli recti and rete testis between 3 and 4 weeks of age in the mouse: an immunohistochemical study. *Am. J. Reprod. Immunol.* 42, 321–326.
410. Duckett, R. J., Hedger, M. P., McLachlan, R. I., and Wreford, N. G. (1997). The effects of gonadotropin-releasing hormone immunization and recombinant follicle-stimulating hormone on the Leydig cell and macrophage populations of the adult rat testis. *J. Androl.* 18, 417–423.
411. Duckett, R. J., Wreford, N. G., Meachem, S. J., McLachlan, R. I., and Hedger, M. P. (1997). Effect of chorionic gonadotropin and flutamide on Leydig cell and macrophage populations in the testosterone-estradiol-implanted adult rat. *J. Androl.* 18, 656–662.
412. Raburn, D. J., Coquelin, A., and Hutson, J. C. (1991). Human chorionic gonadotropin increases the concentration of macrophages in neonatal rat testis. *Biol. Reprod.* 45, 172–177.
413. Gaytan, F., Bellido, C., Romero, J. L., Morales, C., Reymundo, C., and Aguilar, E. (1994). Decreased number and size and the defective function of testicular macrophages in long-term hypophysectomized rats are reversed by treatment with human gonadotrophins. *J. Endocrinol.* 140, 399–407.
414. Gaytan, F., Pinilla, L., Romero, J. L., and Aguilar, E. (1994). Differential effects of the administration of human chorionic gonadotropin to postnatal rats. *J. Endocrinol.* 142, 527–534.
415. Gaytan, F., Romero, J. L., Bellido, C., Morales, C., Reymundo, C., and Aguilar, E. (1994). Effects of growth hormone and prolactin on testicular macrophages in long-term hypophysectomized rats. *J. Reprod. Immunol.* 27, 73–84.
416. Dombrowicz, D., Sente, B., Closset, J., and Hennen, G. (1992). Dose-dependent effects of human prolactin on the immature hypophysectomized rat testis. *Endocrinology* 130, 695–700.
417. Jackson, A. E., and de Kretser, D. M. (1991). Ultrastructural immunoperoxidase investigations of hCG binding to isolated testicular intertubular cells. *Histochem. J.* 23, 517–528.
418. Zhang, Y. M., Rao Ch, V., and Lei, Z. M. (2003). Macrophages in human reproductive tissues contain luteinizing hormone/chorionic gonadotropin receptors. *Am. J. Reprod. Immunol.* 49, 93–100.
419. Gerdprasert, O., O'Bryan, M. K., Nikolic-Paterson, D. J., Sebire, K., de Kretser, D. M., and Hedger, M. P. (2002). Expression of monocyte chemoattractant protein-1 and macrophage colony-stimulating factor in normal and inflamed rat testis. *Mol. Hum. Reprod.* 8, 518–524.
420. Cohen, P. E., Hardy, M. P., and Pollard, J. W. (1997). Colony-stimulating factor-1 plays a major role in the development of reproductive function in male mice. *Mol. Endocrinol.* 11, 1636–1650.
421. Meinhardt, A., Bacher, M., O'Bryan, M. K., McFarlane, J. R., Mallidis, C., Lehmann, C., Metz, C. N., de Kretser, D. M., Bucala, R., and Hedger, M. P. (1999). A switch in the cellular localization of macrophage migration inhibitory factor in the rat testis after ethane dimethane sulfonate treatment. *J. Cell Sci.* 112, 1337–1344.
422. Westermann, J., Ronneberg, S., Fritz, F. J., and Pabst, R. (1989). Proliferation of macrophage subpopulations in the adult rat: comparison of various lymphoid organs. *J. Leukoc. Biol.* 46, 263–269.
423. Schlatt, S., de Kretser, D. M., and Hedger, M. P. (1999). Mitosis of resident macrophages in the adult rat testis. *J. Reprod. Fertil.* 116, 223–228.
424. Vergouwen, R. P., Jacobs, S. G., Huiskamp, R., Davids, J. A., and de Rooij, D. G. (1991). Proliferative activity of gonocytes, Sertoli cells and interstitial cells during testicular development in mice. *J. Reprod. Fertil.* 93, 233–243.
425. Crofton, R. W., Diesselhoff-den Dulk, M. M., and van Furth, R. (1978). The origin, kinetics, and characteristics of the Kupffer cells in the normal steady state. *J. Exp. Med.* 148, 1–17.
426. Gassmann, A. E., and van Furth, R. (1975). The effect of azathioprine (Imuran) on the kinetics of monocytes and macrophages during the normal steady state and an acute inflammatory reaction. *Blood* 46, 51–64.
427. Blussé van Oud Alblas, A., Mattie, H., and van Furth, R. (1983). A quantitative evaluation of pulmonary macrophage kinetics. *Cell Tissue Kinet.* 16, 211–219.
428. Blussé van oud Alblas, A. B., and van Furth, R. (1979). Origin, kinetics, and characteristics of pulmonary macrophages in the normal steady state. *J. Exp. Med.* 149, 1504–1518.
429. van Furth, R., and Diesselhoff-den Dulk, M. M. (1984). Dual origin of mouse spleen macrophages. *J. Exp. Med.* 160, 1273–1283.
430. el-Demiry, M., and James, K. (1988). Lymphocyte subsets and macrophages in the male genital tract in health and disease: a monoclonal antibody-based study. *Eur. Urol.* 14, 226–235.

431. Hedger, M. P., and Meinhardt, A. (2000). Local regulation of T cell numbers and lymphocyte-inhibiting activity in the interstitial tissue of the adult rat testis. *J. Reprod. Immunol.* 48, 69–80.
432. Kantele, A., Zivny, J., Hakkinen, M., Elson, C. O., and Mestecky, J. (1999). Differential homing commitments of antigen-specific T cells after oral or parenteral immunization in humans. *J. Immunol.* 162, 5173–5177.
433. el-Demiry, M. I., Hargreave, T. B., Busuttil, A., Elton, R., James, K., and Chisholm, G. D. (1987). Immunocompetent cells in human testis in health and disease. *Fertil. Steril.* 48, 470–479.
434. Doncel, G. F., Di Paola, J. A., and Lustig, L. (1989). Sequential study of the histopathology and cellular and humoral immune response during the development of an autoimmune orchitis in Wistar rats. *Am. J. Reprod. Immunol.* 20, 44–51.
435. Jahnukainen, K., Jørgensen, N., Pöllänen, P., Giwercman, A., and Skakkebæk, N. E. (1995). Incidence of testicular mononuclear cell infiltrates in normal human males and in patients with germ cell neoplasia. *Int. J. Androl.* 18, 313–320.
436. Mukasa, A., Hiromatsu, K., Matsuzaki, G., O'Brien, R., Born, W., and Nomoto, K. (1995). Bacterial infection of the testis leading to autoaggressive immunity triggers apparently opposed responses of $\alpha\beta$ and $\gamma\delta$ T cells. *J. Immunol.* 155, 2047–2056.
437. Kohno, S., Munoz, J. A., Williams, T. M., Teuscher, C., Bernard, C. C., and Tung, K. S. (1983). Immunopathology of murine experimental allergic orchitis. *J. Immunol.* 130, 2675–2682.
438. Lustig, L., Lourtau, L., Perez, R., and Doncel, G. F. (1993). Phenotypic characterization of lymphocytic cell infiltrates into the testes of rats undergoing autoimmune orchitis. *Int. J. Androl.* 16, 279–284.
439. Torres, A., Casanova, J. F., Nistal, M., and Regadera, J. (1997). Quantification of immunocompetent cells in testicular germ cell tumours. *Histopathology* 30, 23–30.
440. Qo, S. (1994). [Mast cell induction to the mouse testicular interstitium]. *Jpn. J. Urol.* 85, 747–752.
441. Nistal, M., Santamaria, L., and Paniagua, R. (1984). Mast cells in the human testis and epididymis from birth to adulthood. *Acta Anat.* 119, 155–160.
442. Yamanaka, K., Fujisawa, M., Tanaka, H., Okada, H., Arakawa, S., and Kamidono, S. (2000). Significance of human testicular mast cells and their subtypes in male infertility. *Hum. Reprod.* 15, 1543–1547.
443. Nagai, T., Takaba, H., Miyake, K., Hirabayashi, Y., and Yamada, K. (1992). Testicular mast cell heterogeneity in idiopathic male infertility. *Fertil. Steril.* 57, 1331–1336.
444. Meineke, V., Frungieri, M. B., Jessberger, B., Vogt, H., and Mayerhofer, A. (2000). Human testicular mast cells contain tryptase: increased mast cell number and altered distribution in the testes of infertile men. *Fertil. Steril.* 74, 239–244.
445. Nistal, M., Santana, A., Paniaqua, R., and Palacios, J. (1986). Testicular toxoplasmosis in two men with the acquired immunodeficiency syndrome (AIDS). *Arch. Pathol. Lab. Med.* 110, 744–746.
446. O'Bryan, M. K., Schlatt, S., Phillips, D. J., de Kretser, D. M., and Hedger, M. P. (2000). Bacterial lipopolysaccharide-induced inflammation compromises testicular function at multiple levels *in vivo*. *Endocrinology* 141, 238–246.
447. Anosa, V. O., and Kaneko, J. J. (1984). Pathogenesis of *Trypanosoma brucei* infection in deer mice (*Peromyscus maniculatus*): light and electron microscopic study of testicular lesions. *Vet. Pathol.* 21, 238–246.
448. Coelho, K. I., Takeo, K., Yamaguchi, M., Sano, A., Kurita, N., Yoshida, S., Nishimura, K., and Miyaji, M. (1994). Experimental paracoccidioidomycosis in hamster: transmission electron microscopy of inoculation site lesion. *Rev. Inst. Med. Trop. Sao Paulo* 36, 217–223.
449. Jackson, A. E., O'Leary, P. C., Ayers, M. M., and de Kretser, D. M. (1986). The effects of ethylene dimethane sulphonate (EDS) on rat Leydig cells: evidence to support a connective tissue origin of Leydig cells. *Biol. Reprod.* 35, 425–437.
450. Gaytan, F., Aceitero, J., Lucena, C., Aguilar, E., Pinilla, L., Garnelo, P., and Bellido, C. (1992). Simultaneous proliferation and differentiation of mast cells and Leydig cells in the rat testis: are common regulatory factors involved? *J. Androl.* 13, 387–397.
451. Gaytan, F., Bellido, C., Aceitero, J., Aguilar, E., and Sanchez-Criado, J. E. (1990). Leydig cell involvement in the paracrine regulation of mast cells in the testicular interstitium of the rat. *Biol. Reprod.* 43, 665–671.
452. Raeside, J. I., and Renaud, R. L. (1983). Estrogen and androgen production by purified Leydig cells of mature boars. *Biol. Reprod.* 28, 727–733.
453. Claus, R., and Hoffmann, B. (1980). Oestrogens, compared to other steroids of testicular origin, in blood plasma of boars. *Acta Endocrinol.* 94, 404–411.
454. Eisenhauer, K. M., McCue, P. M., Nayden, D. K., Osawa, Y., and Roser, J. F. (1994). Localization of aromatase in equine Leydig cells. *Domest. Anim. Endocrinol.* 11, 291–298.
455. Gaytan, F., Carrera, G., Pinilla, L., Aguilar, R., and Bellido, C. (1989). Mast cells in the testis, epididymis and accessory glands of the rat: effects of neonatal steroid treatment. *J. Androl.* 10, 351–358.
456. Gaytan, F., Bellido, C., Carrera, G., and Aguilar, E. (1990). Differentiation of mast cells during postnatal development of neonatally estrogen-treated rats. *Cell Tissue Res.* 259, 25–31.
457. Maurer, M., Echtenacher, B., Hültner, L., Kollias, G., Männel, D. N., Langley, K. E., and Galli, S. J. (1998). The c-kit ligand, stem cell factor, can enhance innate immunity through effects on mast cells. *J. Exp. Med.* 188, 2343–2348.
458. Yan, W., Kero, J., Huhtaniemi, I., and Toppari, J. (2000). Stem cell factor functions as a survival factor for mature Leydig cells and a growth factor for precursor Leydig cells after ethylene dimethane sulfonate treatment: implication of a role of the stem cell factor/c-Kit system in Leydig cell development. *Dev. Biol.* 227, 169–182.
459. Zhou, Z. Z., Zheng, Y., Steenstra, R., Hickey, W. F., and Teuscher, C. (1989). Actively-induced experimental allergic orchitis (EAO) in Lewis/NCR rats: sequential histo- and immunopathologic analysis. *Autoimmunity* 3, 125–134.
460. Flickinger, C. J., Herr, J. C., Caloras, D., Sisak, J. R., and Howards, S. S. (1990). Inflammatory changes in the epididymis after vasectomy in the Lewis rat. *Biol. Reprod.* 43, 34–45.
461. Barratt, C. L., and Cohen, J. (1987). Quantitation of sperm disposal and phagocytic cells in the tract of short- and long-term vasectomized mice. *J. Reprod. Fertil.* 81, 377–384.
462. Mullen, T. E., Jr., Kiessling, R. L., and Kiessling, A. A. (2003). Tissue-specific populations of leukocytes in semen-producing organs of the normal, hemicastrated, and vasectomized mouse. *AIDS Res. Hum. Retrovir.* 19, 235–243.
463. Yakirevich, E., Yanai, O., Sova, Y., Sabo, E., Stein, A., Hiss, J., and Resnick, M. B. (2002). Cytotoxic phenotype of intraepithelial lymphocytes in normal and cryptorchid human testicular excurrent ducts. *Hum. Reprod.* 17, 275–283.
464. Wang, Y. F., and Holstein, A. F. (1983). Intraepithelial lymphocytes and macrophages in the human epididymis. *Cell Tissue Res.* 233, 517–521.

465. Jefferies, W. A., Green, J. R., and Williams, A. F. (1985). Authentic T helper CD4 (W3/25) antigen on rat peritoneal macrophages. *J. Exp. Med.* 162, 117–127.
466. Crocker, P. R., Jefferies, W. A., Clark, S. J., Chung, L. P., and Gordon, S. (1987). Species heterogeneity in macrophage expression of the CD4 antigen. *J. Exp. Med.* 166, 613–618.
467. Pollard, J. W., Dominguez, M. G., Mocci, S., Cohen, P. E., and Stanley, E. R. (1997). Effect of the colony-stimulating factor-1 null mutation, osteopetrotic (*csfm^op*), on the distribution of macrophages in the male mouse reproductive tract. *Biol. Reprod.* 56, 1290–1300.
468. Kazeem, A. A. (1983). The assessment of epididymal lymphatics within the concept of immunologically privileged sites. *Lymphology* 16, 168–171.
469. Tomlinson, M. J., White, A., Barratt, C. L., Bolton, A. E., and Cooke, I. D. (1992). The removal of morphologically abnormal sperm forms by phagocytes: a positive role for seminal leukocytes? *Hum. Reprod.* 7, 517–522.
470. Jones, R. (2004). Sperm survival versus degradation in the Mammalian epididymis: a hypothesis. *Biol. Reprod.* 71, 1405–1411.
471. Ansbacher, R. (1973). Vasectomy: sperm antibodies. *Fertil. Steril.* 24, 788–792.
472. Hellema, H. W., Samuel, T., and Rümke, P. (1979). Sperm autoantibodies as a consequence of vasectomy: II. Long-term follow-up studies. *Clin. Exp. Immunol.* 38, 31–36.
473. Linnet, L. (1983). Clinical immunology of vasectomy and vasovasostomy. *Urology* 22, 101–114.
474. McDonald, S. W. (2000). Cellular responses to vasectomy. *Int. Rev. Cytol.* 199, 295–339.
475. Alexander, N. J., and Anderson, D. J. (1979). Vasectomy: consequences of autoimmunity to sperm antigens. *Fertil. Steril.* 32, 253–260.
476. Witkin, S. S., Brandslund, I., Svehag, S. E., Linnet, L., and Moller, N. P. (1983). Comparison of different assays for circulating immune complexes in age matched vasectomized and non-vasectomized men. *J. Clin. Lab. Immunol.* 10, 193–197.
477. Bigazzi, P. (1981). Immunologic effects of vasectomy in men and experimental animals. *Prog. Clin. Biol. Res.* 70, 461–476.
478. de Kretser, D. M., Huidobro, C., Southwick, G. J., and Temple-Smith, P. D. (1998). The role of the epididymis in human infertility. *J. Reprod. Fertil. Suppl.* 53, 271–275.
479. Herr, J. C., Flickinger, C. J., Howards, S. S., Yarbro, S., Spell, D. R., Caloras, D., and Gallien, T. N. (1987). The relation between antisperm antibodies and testicular alterations after vasectomy and vasovasostomy in Lewis rats. *Biol. Reprod.* 37, 1297–1305.
480. Flickinger, C. J., Herr, J. C., Howards, S. S., Sisak, J. R., Gleavy, J. M., Fusia, T. J., Vailes, L. D., and Handley, H. H. (1990). Early testicular changes after vasectomy and vasovasostomy in Lewis rats. *Anat. Rec.* 227, 37–46.
481. Bigazzi, P. E., Kosuda, L. L., and Harnick, L. L. (1977). Sperm autoantibodies in vasectomized rats of different inbred strains. *Science* 197, 1282–1283.
482. Flickinger, C. J., Herr, J. C., Baran, M. L., and Howards, S. S. (1995). Temporal appearance of antisperm antibodies during sexual maturation of rats after obstruction of the vas deferens. *J. Androl.* 16, 75–79.
483. Dobson, C. C., Reid, O., Bennett, N. K., and McDonald, S. W. (2000). Effect of vasectomy on the seminiferous tubule boundary zone in the albino Swiss rat. *Clin. Anat.* 13, 277–286.
484. Flickinger, C. J. (1985). The effects of vasectomy on the testis. *N. Engl. J. Med.* 313, 1283–1285.
485. Flickinger, C. J. (1975). Fine structure of the rabbit testis after vasectomy. *Biol. Reprod.* 13, 61–67.
486. Alexander, N. J., and Tung, K. S. (1977). Immunological and morphological effects of vasectomy in the rabbit. *Anat. Rec.* 188, 339–350.
487. Tumboh-Oeri, A. G., and Roberts, T. K. (1979). Immunological and morphological consequences of vasectomy in the rabbit. *Experientia* 35, 675–676.
488. Lohiya, N. K., Mathur, N., Tiwari, S. N., and Shipstone, A. C. (1983). Ultrastructural changes in rabbit testis and epididymis following vasectomy: a long term study. *Acta Eur. Fertil.* 14, 141–146.
489. Salerno, R. G., Albert, P. S., and Davis, J. E. (1974). Effect of vasectomy on guinea pig testes. *Urology* 4, 190–192.
490. Hutson, J. C., Gardner, P. J., and Lacy, S. S. (1976). Changes in testis of guinea pig after vasectomy. *Urology* 7, 287–291.
491. Aitken, H., Kumarakuru, S., Orr, R., Reid, O., Bennett, N. K., and McDonald, S. W. (1999). Effect of long-term vasectomy on seminiferous tubules in the guinea pig. *Clin. Anat.* 12, 250–263.
492. Raleigh, D., O'Donnell, L., Southwick, G. J., de Kretser, D. M., and McLachlan, R. I. (2004). Stereological analysis of the human testis after vasectomy indicates impairment of spermatogenic efficiency with increasing obstructive interval. *Fertil. Steril.* 81, 1595–1603.
493. Theyer, G., Kramer, G., Assmann, I., Sherwood, E., Preinfalk, W., Marberger, M., Zechner, O., and Steiner, G. E. (1992). Phenotypic characterization of infiltrating leukocytes in benign prostatic hyperplasia. *Lab. Invest.* 66, 96–107.
494. Pudney, J., and Anderson, D. J. (1995). Immunobiology of the human penile urethra. *Am. J. Pathol.* 147, 155–165.
495. Quayle, A. J., Pudney, J., Munoz, D. E., and Anderson, D. J. (1994). Characterization of T lymphocytes and antigen-presenting cells in the murine male urethra. *Biol. Reprod.* 51, 809–820.
496. Bostwick, D. G., de la Roza, G., Dundore, P., Corica, F. A., and Iczkowski, K. A. (2003). Intraepithelial and stromal lymphocytes in the normal human prostate. *Prostate* 55, 187–193.
497. Lohiya, N. K., Sharma, R. S., Ansari, A. S., and Anand Kumar, T. C. (1988). Structure of rete testis, vas efferens, epididymis and vas deferens of langur monkey (*Presbytis entellus entellus* Dufresne). *Acta Eur. Fertil.* 19, 167–173.
498. O'Rourke, A. M., and Mescher, M. F. (1993). The roles of CD8 in cytotoxic T lymphocyte function. *Immunol. Today* 14, 183–188.
499. Aitken, R. J., and Baker, H. W. (1995). Seminal leukocytes: passengers, terrorists or good Samaritans? *Hum. Reprod.* 10, 1736–1739.
500. el-Demiry, M. I., Young, H., Elton, R. A., Hargreave, T. B., James, K., and Chisholm, G. D. (1986). Leucocytes in the ejaculate from fertile and infertile men. *Br. J. Urol.* 58, 715–720.
501. el-Demiry, M. I., Hargreave, T. B., Busuttil, A., James, K., and Chisholm, G. D. (1986). Identifying leucocytes and leucocyte subpopulations in semen using monoclonal antibody probes. *Urology* 28, 492–496.
502. Wolff, H., and Anderson, D. J. (1988). Immunohistologic characterization and quantitation of leukocyte subpopulations in human semen. *Fertil. Steril.* 49, 497–504.
503. Tomlinson, M. J., Barratt, C. L., and Cooke, I. D. (1993). Prospective study of leukocytes and leukocyte subpopulations in semen suggests they are not a cause of male infertility. *Fertil. Steril.* 60, 1069–1075.
504. Denny, T. N., Scolpino, A., Garcia, A., Polyak, A., Weiss, S. N., Skurnick, J. H., Passannante, M. R., and Colon, J. (1995). Evaluation of T-lymphocyte subsets present in

semen and peripheral blood of healthy donors: a report from the heterosexual transmission study. *Cytometry* 20, 349–355.
505. Wolff, H. (1995). The biologic significance of white blood cells in semen. *Fertil. Steril.* 63, 1143–1157.
506. Anderson, D. J., Politch, J. A., Martinez, A., Van Voorhis, B. J., Padian, N. S., and O'Brien, T. R. (1991). White blood cells and HIV-1 in semen from vasectomised seropositive men. *Lancet* 338, 573–574.
507. Schwartz, G. G. (1990). Can vasectomy reduce the sexual transmission of HIV? *J. Clin. Epidemiol.* 43, 1433.
508. Wallgren, M., Kindahl, H., and Larsson, K. (1989). Clinical, endocrinological and spermatological studies after endotoxin in the ram. *Zentralbl. Veterin. A* 36, 90–103.
509. Wallgren, M. (1989). Clinical, endocrinological and spermatological studies after endotoxin injection in the boar. *Zentralbl. Veterin. A* 36, 664–675.
510. Wallgren, M., Kindahl, H., and Rodriguez-Martinez, H. (1993). Alterations in testicular function after endotoxin injection in the boar. *Int. J. Androl.* 16, 235–243.
511. Tulassay, Z., Viczián, M., Böjthe, L., and Czeizel, A. (1970). Quantitative histological studies on the injury of spermatogenesis induced by endotoxin in rats. *J. Reprod. Fertil.* 22, 161–164.
512. Harris, S. G., Padilla, J., Koumas, L., Ray, D., and Phipps, R. P. (2002). Prostaglandins as modulators of immunity. *Trends Immunol.* 23, 144–150.
513. Langenbach, R., Loftin, C. D., Lee, C., and Tiano, H. (1999). Cyclooxygenase-deficient mice: a summary of their characteristics and susceptibilities to inflammation and carcinogenesis. *Ann. N. Y. Acad. Sci.* 889, 52–61.
514. Schmidt, H. H., and Walter, U. (1994). NO at work. *Cell* 78, 919–925.
515. Bogdan, C. (2001). Nitric oxide and the immune response. *Nat. Immunol.* 2, 907–916.
516. Hales, D. B., Diemer, T., and Hales, K. H. (1999). Role of cytokines in testicular function. *Endocrinology* 10, 201–217.
517. Jégou, B., Cudicini, C., Gomez, E., and Stéphan, J. P. (1995). Interleukin-1, interleukin-6 and the germ cell-Sertoli cell cross-talk. *Reprod. Fertil. Dev.* 7, 723–730.
518. Hedger, M. P., and Meinhardt, A. (2003). Cytokines and the immune-testicular axis. *J. Reprod. Immunol.* 58, 1–26.
519. Ogilvie, K., and Rivier, C. (1998). The intracerebroventricular injection of interleukin-1β blunts the testosterone response to human chorionic gonadotropin: role of prostaglandin- and adrenergic-dependent pathways. *Endocrinology* 139, 3088–3095.
520. Ogilvie, K. M., Held Hales, K., Roberts, M. E., Hales, D. B., and Rivier, C. (1999). The inhibitory effect of intracerebroventricularly injected interleukin 1β on testosterone secretion in the rat: role of steroidogenic acute regulatory protein. *Biol. Reprod.* 60, 527–533.
521. Turnbull, A. V., and Rivier, C. (1997). Inhibition of gonadotropin-induced testosterone secretion by the intracerebroventricular injection of interleukin-1β in the male rat. *Endocrinology* 138, 1008–1013.
522. Dinarello, C. A. (1996). Biologic basis for interleukin-1 in disease. *Blood* 87, 2095–2147.
523. Stylianou, E., and Saklatvala, J. (2000). Interleukin-1. *Int. J. Biochem. Cell Biol.* 30, 1075–1079.
524. Watanabe, N., and Kobayashi, Y. (1994). Selective release of a processed form of interleukin 1α. *Cytokine* 6, 597–601.
525. Hazuda, D. J., Lee, J. C., and Young, P. R. (1988). The kinetics of interleukin 1 secretion from activated monocytes. Differences between interleukin 1α and interleukin 1β. *J. Biol. Chem.* 263, 8473–8479.
526. Black, R. A., Kronheim, S. R., Cantrell, M., Deeley, M. C., March, C. J., Prickett, K. S., Wignall, J., Conlon, P. J., Cosman, D., and Hopp, T. P. (1988). Generation of biologically active interleukin-1β by proteolytic cleavage of the inactive precursor. *J. Biol. Chem.* 263, 9437–9442.
527. Thornberry, N. A., Bull, H. G., Calaycay, J. R., Chapman, K. T., Howard, A. D., Kostura, M. J., Miller, D. K., Molineaux, S. M., Weidner, J. R., Aunins, J., Elliston, K. O., Ayala, J. M., Casano, F. J., Chin, J., Ding, G. J., Egger, L. A., Gaffney, E. P., Limjuco, G., Palyha, O. C., Raju, S. M., Rolando, A. M., Salley, J. P., Yamin, T. T., Lee, T. D., Shively, J. E., MacCross, M., Mumford, R. A., Schmidt, J. A., and Tocci, M. J. (1992). A novel heterodimeric cysteine protease is required for interleukin-1β processing in monocytes. *Nature* 356, 768–774.
528. Schönbeck, U., Mach, F., and Libby, P. (1998). Generation of biologically active IL-1β by matrix metalloproteinases: a novel caspase-1-independent pathway of IL-1β processing. *J. Immunol.* 161, 3340–3346.
529. Rubartelli, A., Cozzolino, F., Talio, M., and Sitia, R. (1990). A novel secretory pathway for interleukin-1 β, a protein lacking a signal sequence. *EMBO J.* 9, 1503–1510.
530. Fitzgerald, K. A., and O'Neill, L. A. (2000). The role of the interleukin-1/Toll-like receptor superfamily in inflammation and host defence. *Microbes Infect.* 2, 933–943.
531. Sims, J. E., Gayle, M. A., Slack, J. L., Alderson, M. R., Bird, T. A., Giri, J. G., Colotta, F., Re, F., Mantovani, A., and Shanebeck, K. (1993). Interleukin 1 signaling occurs exclusively via the type I receptor. *Proc. Natl. Acad. Sci. U. S. A.* 90, 6155–6159.
532. Sims, J. E., March, C. J., Cosman, D., Widmer, M. B., MacDonald, H. R., McMahan, C. J., Grubin, C. E., Wignall, J. M., Jackson, J. L., and Call, S. M. (1988). cDNA expression cloning of the IL-1 receptor, a member of the immunoglobulin superfamily. *Science* 241, 585–589.
533. Stylianou, E., O'Neill, L. A., Rawlinson, L., Edbrooke, M. R., Woo, P., and Saklatvala, J. (1992). Interleukin 1 induces NF-κB through its type I but not its type II receptor in lymphocytes. *J. Biol. Chem.* 267, 15836–15841.
534. Janssens, S., and Beyaert, R. (2003). Functional diversity and regulation of different interleukin-1 receptor-associated kinase (IRAK) family members. *Mol. Cell* 11, 293–302.
535. Rosoff, P. M. (1989). Characterization of the interleukin-1-stimulated phospholipase C activity in human T lymphocytes. *Lymphokine Res.* 8, 407–413.
536. Rosoff, P. M., Savage, N., and Dinarello, C. A. (1988). Interleukin-1 stimulates diacylglycerol production in T lymphocytes by a novel mechanism. *Cell* 54, 73–81.
537. Schutze, S., Machleidt, T., and Kronke, M. (1994). The role of diacylglycerol and ceramide in tumor necrosis factor and interleukin-1 signal transduction. *J. Leukoc. Biol.* 56, 533–541.
538. Bomalaski, J. S., Steiner, M. R., Simon, P. L., and Clark, M. A. (1992). IL-1 increases phospholipase A_2 activity, expression of phospholipase A_2-activating protein, and release of linoleic acid from the murine T helper cell line EL-4. *J. Immunol.* 148, 155–160.
539. Gronich, J., Konieczkowski, M., Gelb, M. H., Nemenoff, R. A., and Sedor, J. R. (1994). Interleukin 1α causes rapid activation of cytosolic phospholipase A_2 by phosphorylation in rat mesangial cells. *J. Clin. Invest.* 93, 1224–1233.
540. Eisenberg, S. P., Evans, R. J., Arend, W. P., Verderber, E., Brewer, M. T., Hannum, C. H., and Thompson, R. C. (1990). Primary structure and functional expression from complementary DNA of a human interleukin-1 receptor antagonist. *Nature* 343, 341–346.

541. Arend, W. P. (1993). Interleukin-1 receptor antagonist. *Adv. Immunol.* 54, 167–227.
542. Jonsson, C. K., Zetterström, R. H., Holst, M., Parvinen, M., and Söder, O. (1999). Constitutive expression of interleukin-1α messenger ribonucleic acid in rat Sertoli cells is dependent upon interaction with germ cells. *Endocrinology* 140, 3755–3761.
543. Syed, V., Söder, O., Arver, S., Lindh, M., Khan, S., and Ritzén, E. M. (1988). Ontogeny and cellular origin of an interleukin-1-like factor in the reproductive tract of the male rat. *Int. J. Androl.* 11, 437–447.
544. Syed, V., Stéphan, J. P., Gérard, N., Legrand, A., Parvinen, M., Bardin, C. W., and Jégou, B. (1995). Residual bodies activate Sertoli cell interleukin-1α (IL-1α) release, which triggers IL-6 production by an autocrine mechanism, through the lipoxygenase pathway. *Endocrinology* 136, 3070–3078.
545. Gérard, N., Syed, V., and Jégou, B. (1992). Lipopolysaccharide, latex beads and residual bodies are potent activators of Sertoli cell interleukin-1α production. *Biochem. Biophys. Res. Commun.* 185, 154–161.
546. Gustafsson, K., Sultana, T., Zetterström, C. K., Setchell, B. P., Siddiqui, A., Weber, G., and Söder, O. (2002). Production and secretion of interleukin-1α proteins by rat testis. *Biochem. Biophys. Res. Commun.* 297, 492–497.
547. Sultana, T., Svechnikov, K., Weber, G., and Söder, O. (2000). Molecular cloning and expression of a functionally different alternative splice variant of prointerleukin-1α from the rat testis. *Endocrinology* 141, 4413–4418.
548. Sultana, T., Wahab-Wahlgren, A., Assmus, M., Parvinen, M., Weber, G., and Söder, O. (2003). Expression and regulation of the prointerleukin-1α processing enzymes calpain I and II in the rat testis. *Int. J. Androl.* 26, 37–45.
549. Svechnikov, K. V., Sultana, T., and Söder, O. (2001). Age-dependent stimulation of Leydig cell steroidogenesis by interleukin-1 isoforms. *Mol. Cell. Endocrinol.* 182, 193–201.
550. Haugen, T. B., Landmark, B. F., Josefsen, G. M., Hansson, V., and Högset, A. (1994). The mature form of interleukin-1 α is constitutively expressed in immature male germ cells from rat. *Mol. Cell. Endocrinol.* 105, R19–R23.
551. Cudicini, C., Kercret, H., Touzalin, A. M., Ballet, F., and Jégou, B. (1997). Vectorial production of interleukin 1 and interleukin 6 by rat Sertoli cells cultured in a dual culture compartment system. *Endocrinology* 138, 2863–2870.
552. Stéphan, J. P., Syed, V., and Jégou, B. (1997). Regulation of Sertoli cell IL-1 and IL-6 production in vitro. *Mol. Cell. Endocrinol.* 134, 109–118.
553. Hakovirta, H., Pentitilä, T. L., Pöllänen, P., Fröysa, B., Söder, O., and Parvinen, M. (1993). Interleukin-1 bioactivity and DNA synthesis in X-irradiated rat testes. *Int. J. Androl.* 16, 159–164.
554. Parvinen, M., Söder, O., Mali, P., Fröysa, B., and Ritzén, E. M. (1991). *In vitro* stimulation of stage-specific deoxyribonucleic acid synthesis in rat seminiferous tubule segments by interleukin-1α. *Endocrinology* 129, 1614–1620.
555. Söder, O., Syed, V., Callard, G. V., Toppari, J., Pöllänen, P., Parvinen, M., Fröysa, B., and Ritzén, E. M. (1991). Production and secretion of an interleukin-1-like factor is stage-dependent and correlates with spermatogonial DNA synthesis in the rat seminiferous epithelium. *Int. J. Androl.* 14, 223–231.
556. Wang, J. E., Josefsen, G. M., Hansson, V., and Haugen, T. B. (1998). Residual bodies and IL-1α stimulate expression of mRNA for IL-1α and IL-1 receptor type I in cultured rat Sertoli cells. *Mol. Cell. Endocrinol.* 137, 139–144.
557. Cudicini, C., Lejeune, H., Gomez, E., Bosmans, E., Ballet, F., Saez, J., and Jégou, B. (1997). Human Leydig cells and Sertoli cells are producers of interleukins-1 and -6. *J. Clin. Endocrinol. Metab.* 82, 1426–1433.
558. Lin, T., Wang, D., and Nagpal, M. L. (1993). Human chorionic gonadotropin induces interleukin-1 gene expression in rat Leydig cells in vivo. *Mol. Cell. Endocrinol.* 95, 139–145.
559. Wang, D. L., Nagpal, M. L., Calkins, J. H., Chang, W. W., Sigel, M. M., and Lin, T. (1991). Interleukin-1β induces interleukin-1α messenger ribonucleic acid expression in primary cultures of Leydig cells. *Endocrinology* 129, 2862–2866.
560. Xiong, Y., and Hales, D. B. (1994). Immune-endocrine interactions in the mouse testis: cytokine-mediated inhibition of Leydig cell steroidogenesis. *Endocr. J.* 2, 223–228.
561. Gow, R. M., O'Bryan, M. K., Canny, B. J., Ooi, G. T., and Hedger, M. P. (2001). Differential effects of dexamethasone treatment on lipopolysaccharide-induced testicular inflammation and reproductive hormone inhibition in adult rats. *J. Endocrinol.* 168, 193–201.
562. Jonsson, C. K., Setchell, B. P., Martinelle, N., Svechnikov, K., and Söder, O. (2001). Endotoxin-induced interleukin 1 expression in testicular macrophages is accompanied by downregulation of the constitutive expression in Sertoli cells. *Cytokine* 14, 283–288.
563. Hedger, M., Klug, J., Fröhlich, S., Müller, R., and Meinhardt, A. (2005). Regulatory cytokine expression and fluid formation in the normal and inflamed rat testis is under Leydig cell control. *J. Androl.* 26, 379–386.
564. Zeyse, D., Lunenfeld, E., Beck, M., Prinsloo, I., and Huleihel, M. (2000). Interleukin-1 receptor antagonist is produced by Sertoli cells *in vitro*. *Endocrinology* 141, 1521–1527.
565. Bravo, J., and Heath, J. K. (2000). Receptor recognition by gp130 cytokines. *EMBO J.* 19, 2399–2411.
566. Akira, S., Taga, T., and Kishimoto, T. (1993). Interleukin-6 in biology and medicine. *Adv. Immunol.* 54, 1–78.
567. Heinrich, P. C., Behrmann, I., Haan, S., Hermanns, H. M., Muller-Newen, G., and Schaper, F. (2003). Principles of interleukin (IL)-6-type cytokine signalling and its regulation. *Biochem. J.* 374, 1–20.
568. Okuda, Y., Sun, X. R., and Morris, P. L. (1994). Interleukin-6 (IL-6) mRNAs expressed in Leydig and Sertoli cells are regulated by cytokines, gonadotropins and neuropeptides. *Endocrine J.* 2, 617–624.
569. Boockfor, F. R., Wang, D., Lin, T., Nagpal, M. L., and Spangelo, B. L. (1994). Interleukin-6 secretion from rat Leydig cells in culture. *Endocrinology* 134, 2150–2155.
570. Hakovirta, H., Syed, V., Jégou, B., and Parvinen, M. (1995). Function of interleukin-6 as an inhibitor of meiotic DNA synthesis in the rat seminiferous epithelium. *Mol. Cell. Endocrinol.* 108, 193–198.
571. Okuda, Y., Bardin, C. W., Hodgskin, L. R., and Morris, P. L. (1995). Interleukins-1α and -1β regulate interleukin-6 expression in Leydig and Sertoli cells. *Recent Prog. Horm. Res.* 50, 367–372.
572. Riccioli, A., Starace, D., D'Alessio, A., Starace, G., Padula, F., de Cesaris, P., Filippini, A., and Ziparo, E. (2000). TNF-α and IFN-γ regulate expression and function of the Fas system in the seminiferous epithelium. *J. Immunol.* 165, 743–749.
573. Pöllänen, P., Söder, O., and Parvinen, M. (1989). Interleukin-1α stimulation of spermatogonial proliferation *in vivo*. *Reprod. Fertil. Devel.* 1, 85–87.
574. Khan, S. A., and Nieschlag, E. (1991). Interleukin-1 inhibits follitropin-induced aromatase activity in immature rat Sertoli cells in vitro. *Mol. Cell. Endocrinol.* 75, 1–7.
575. Riera, M. F., Meroni, S. B., Gómez, G. E., Schteingart, H. F., Pellizzari, E. H., and Cigorraga, S. B. (2001). Regulation of

lactate production by FSH, IL1β, and TNFα in rat Sertoli cells. *Gen. Comp. Endocrinol.* 122, 88–97.

576. Nehar, D., Mauduit, C., Boussouar, F., and Benahmed, M. (1998). Interleukin 1α stimulates lactate dehydrogenase A expression and lactate production in cultured porcine Sertoli cells. *Biol. Reprod.* 59, 1425–1432.
577. Huleihel, M., Zeyse, D., Lunenfeld, E., Zeyse, M., and Mazor, M. (2002). Induction of transferrin secretion in murine Sertoli cells by FSH and IL-1: the possibility of different mechanism(s) of regulation. *Am. J. Reprod. Immunol.* 47, 112–117.
578. Huleihel, M., and Lunenfeld, E. (2002). Involvement of intratesticular IL-1 system in the regulation of Sertoli cell functions. *Mol. Cell. Endocrinol.* 187, 125–132.
579. Hoeben, E., Wuyts, A., Proost, P., Van Damme, J., and Verhoeven, G. (1997). Identification of IL-6 as one of the important cytokines responsible for the ability of mononuclear cells to stimulate Sertoli cell functions. *Mol. Cell. Endocrinol.* 132, 149–160.
580. Petersen, C., Boitani, C., Fröysa, B., and Söder, O. (2002). Interleukin-1 is a potent growth factor for immature rat Sertoli cells. *Mol. Cell. Endocrinol.* 186, 37–47.
581. Boockfor, F. R., and Schwarz, L. K. (1991). Effects of interleukin-6, interleukin-2, and tumor necrosis factor alpha on transferrin release from Sertoli cells in culture. *Endocrinology* 129, 256–262.
582. Hales, D. B. (1992). Interleukin-1 inhibits Leydig cell steroidogenesis primarily by decreasing 17α-hydroxylase/C17-20 lyase cytochrome P450 expression. *Endocrinology* 131, 2165–2172.
583. Bergh, A., and Söder, O. (1990). Interleukin-1β, but not interleukin-1α, induces acute inflammation-like changes in the testicular microcirculation of adult rats. *J. Reprod. Immunol.* 17, 155–165.
584. Morales-Montor, J., Baig, S., Mitchell, R., Deway, K., Hallal-Calleros, C., and Damian, R. T. (2001). Immunoendocrine interactions during chronic cysticercosis determine male mouse feminization: role of IL-6. *J. Immunol.* 167, 4527–4533.
585. Cohen, P. E., and Pollard, J. W. (1998). Normal sexual function in male mice lacking a functional type I interleukin-1 (IL-1) receptor. *Endocrinology* 139, 815–818.
586. Reed, M. J., Coldham, N. G., Patel, S. R., Ghilchik, M. W., and James, V. H. (1992). Interleukin-1 and interleukin-6 in breast cyst fluid: their role in regulating aromatase activity in breast cancer cells. *J. Endocrinol.* 132, R5–R8.
587. Noble, L. S., Simpson, E. R., Johns, A., and Bulun, S. E. (1996). Aromatase expression in endometriosis. *J. Clin. Endocrinol. Metab.* 81, 174–179.
588. de Miguel, M. P., de Boer-Brouwer, M., Paniagua, R., van den Hurk, R., de Rooij, D. G., and van Dissel-Emiliani, F. M. (1996). Leukemia inhibitory factor and ciliary neurotropic factor promote the survival of Sertoli cells and gonocytes in coculture system. *Endocrinology* 137, 1885–1893.
589. Cheng, L., Gearing, D. P., White, L. S., Compton, D. L., Schooley, K., and Donovan, P. J. (1994). Role of leukemia inhibitory factor and its receptor in mouse primordial germ cell growth. *Development* 120, 3145–3153.
590. Jenab, S., and Morris, P. L. (1998). Testicular leukemia inhibitory factor (LIF) and LIF receptor mediate phosphorylation of signal transducers and activators of transcription (STAT)-3 and STAT-1 and induce *c-fos* transcription and activator protein-1 activation in rat Sertoli but not germ cells. *Endocrinology* 139, 1883–1890.
591. Piquet-Pellorce, C., Dorval-Coiffec, I., Pham, M. D., and Jégou, B. (2000). Leukemia inhibitory factor expression and regulation within the testis. *Endocrinology* 141, 1136–1141.
592. Bornstein, S. R., Rutkowski, H., and Vrezas, I. (2004). Cytokines and steroidogenesis. *Mol. Cell. Endocrinol.* 215, 135–141.
593. Letterio, J. J., and Roberts, A. B. (1998). Regulation of immune responses by TGF-β. *Annu. Rev. Immunol.* 16, 137–161.
594. Chang, H., Lau, A. L., and Matzuk, M. M. (2001). Studying TGF-β superfamily signaling by knockouts and knockins. *Mol. Cell. Endocrinol.* 180, 39–46.
595. Josso, N., and di Clemente, N. (1999). TGF-β family members and gonadal development. *Trends Endocrinol. Metab.* 10, 216–222.
596. Chen, X., Weisberg, E., Fridmacher, V., Watanabe, M., Naco, G., and Whitman, M. (1997). Smad4 and FAST-1 in the assembly of activin-responsive factor. *Nature* 389, 85–89.
597. Bilezikjian, L. M., Corrigan, A. Z., Blount, A. L., Chen, Y., and Vale, W. W. (2001). Regulation and actions of Smad7 in the modulation of activin, inhibin, and transforming growth factor-β signaling in anterior pituitary cells. *Endocrinology* 142, 1065–1072.
598. de Caestecker, M. P., Parks, W. T., Frank, C. J., Castagnino, P., Bottaro, D. P., Roberts, A. B., and Lechleider, R. J. (1998). Smad2 transduces common signals from receptor serine-threonine and tyrosine kinases. *Genes Dev.* 12, 1587–1592.
599. Datto, M. B., Frederick, J. P., Pan, L., Borton, A. J., Zhuang, Y., and Wang, X. F. (1999). Targeted disruption of Smad3 reveals an essential role in transforming growth factor β-mediated signal transduction. *Mol. Cell. Biol.* 19, 2495–2504.
600. Avallet, O., Vigier, M., Leduque, P., Dubois, P. M., and Saez, J. M. (1994). Expression and regulation of transforming growth factor-β1 messenger ribonucleic acid and protein in cultured porcine Leydig and Sertoli cells. *Endocrinology* 134, 2079–2087.
601. Mullaney, B. P., and Skinner, M. K. (1993). Transforming growth factor-β (β1, β2, and β3) gene expression and action during pubertal development of the seminiferous tubule: potential role at the onset of spermatogenesis. *Mol. Endocrinol.* 7, 67–76.
602. Konrad, L., Albrecht, M., Renneberg, H., and Aumüller, G. (2000). Transforming growth factor-β2 mediates mesenchymal-epithelial interactions of testicular somatic cells. *Endocrinology* 141, 3679–3686.
603. Teerds, K. J., and Dorrington, J. H. (1993). Localization of transforming growth factor β_1 and β_2 during testicular development in the rat. *Biol. Reprod.* 48, 40–45.
604. Goddard, I., Bouras, M., Keramidas, M., Hendrick, J. C., Feige, J. J., and Benahmed, M. (2000). Transforming growth factor-β receptor types I and II in cultured porcine Leydig cells: expression and hormonal regulation. *Endocrinology* 141, 2068–2074.
605. Olaso, R., Pairault, C., Boulogne, B., Durand, P., and Habert, R. (1998). Transforming growth factor β1 and β2 reduce the number of gonocytes by increasing apoptosis. *Endocrinology* 139, 733–740.
606. Dickson, C., Webster, D. R., Johnson, H., Cecilia Millena, A., and Khan, S. A. (2002). Transforming growth factor-β effects on morphology of immature rat Leydig cells. *Mol. Cell. Endocrinol.* 195, 65–77.
607. Gautier, C., Levacher, C., Saez, J. M., and Habert, R. (1997). Expression and regulation of transforming growth factor β1 mRNA and protein in rat fetal testis in vitro. *Biochem. Biophys. Res. Commun.* 236, 135–139.
608. Lui, W. Y., Lee, W. M., and Cheng, C. Y. (2003). TGF-βs: their role in testicular function and Sertoli cell tight junction dynamics. *Int. J. Androl.* 26, 147–160.
609. Pöllänen, P., von Euler, M., Jahnukainen, K., Saari, T., Parvinen, M., Sainio-Pöllänen, S., and Söder, O. (1993). Role

of transforming growth factor β in testicular immunosuppression. *J. Reprod. Immunol.* 24, 123–137.

610. Suarez-Pinzon, W., Korbutt, G. S., Power, R., Hooton, J., Rajotte, R. V., and Rabinovitch, A. (2000). Testicular Sertoli cells protect islet β-cells from autoimmune destruction in NOD mice by a transforming growth factor-β1-dependent mechanism. *Diabetes* 49, 1810–1818.
611. Phillips, D. J. (2003). The activin/inhibin family. In *The Cytokine Handbook* (4th ed. A. W. Thomson and M. T. Lotze, Eds.), pp. 1153–1177. Academic Press, Amsterdam.
612. Thomsen, G., Woolf, T., Whitman, M., Sokol, S., Vaughan, J., Vale, W., and Melton, D. A. (1990). Activins are expressed early in *Xenopus* embryogenesis and can induce axial mesoderm and anterior structures. *Cell* 63, 485–493.
613. Corrigan, A. Z., Bilezikjian, L. M., Carroll, R. S., Bald, L. N., Schmelzer, C. H., Fendly, B. M., Mason, A. J., Chin, W. W., Schwall, R. H., and Vale, W. (1991). Evidence for an autocrine role of activin B within rat anterior pituitary cultures. *Endocrinology* 128, 1682–1684.
614. Erämaa, M., Hurme, M., Stenman, U. H., and Ritvos, O. (1992). Activin A/erythroid differentiation factor is induced during human monocyte activation. *J. Exp. Med.* 176, 1449–1452.
615. Shao, L., Frigon, N. L., Jr., Sehy, D. W., Yu, A. L., Lofgren, J., Schwall, R., and Yu, J. (1992). Regulation of production of activin A in human marrow stromal cells and monocytes. *Exp. Hematol.* 20, 1235–1242.
616. Yu, J., and Dolter, K. E. (1997). Production of activin A and its roles in inflammation and hematopoiesis. *Cytokines Cell. Mol. Ther.* 3, 169–177.
617. Gribi, R., Tanaka, T., Harper-Summers, R., and Yu, J. (2001). Expression of activin A in inflammatory arthropathies. *Mol. Cell. Endocrinol.* 180, 163–167.
618. Robertson, D. M., Klein, R., de Vos, F. L., McLachlan, R. I., Wettenhall, R. E., Hearn, M. T., Burger, H. G., and de Kretser, D. M. (1987). The isolation of polypeptides with FSH suppressing activity from bovine follicular fluid which are structurally different to inhibin. *Biochem. Biophys. Res. Commun.* 149, 744–749.
619. Robertson, D. M., Farnworth, P. G., Clarke, L., Jacobsen, J., Cahir, N. F., Burger, H. G., and de Kretser, D. M. (1990). Effects of bovine 35 kDa FSH-suppressing protein on FSH and LH in rat pituitary cells *in vitro*: comparison with bovine 31 kDa inhibin. *J. Endocrinol.* 124, 417–423.
620. Phillips, D. J., and de Kretser, D. M. (1998). Follistatin: a multifunctional regulatory protein. *Front. Neuroendocrinol.* 19, 287–322.
621. Broxmeyer, H. E., Lu, L., Cooper, S., Schwall, R. H., Mason, A. J., and Nikolics, K. (1988). Selective and indirect modulation of human multipotential and erythroid hematopoietic progenitor cell proliferation by recombinant human activin and inhibin. *Proc. Natl. Acad. Sci. U. S. A.* 85, 9052–9056.
622. Hedger, M. P., Drummond, A. E., Robertson, D. M., Risbridger, G. P., and de Kretser, D. M. (1989). Inhibin and activin regulate (^{3}H)thymidine uptake by rat thymocytes and 3T3 cells *in vitro*. *Mol. Cell. Endocrinol.* 61, 133–138.
623. Schwall, R. H., and Lai, C. (1991). Erythroid differentiation bioassays for activin. *Methods Enzymol.* 198, 340–346.
624. Schwall, R. H., Robbins, K., Jardieu, P., Chang, L., Lai, C., and Terrell, T. G. (1993). Activin induces cell death in hepatocytes *in vivo* and *in vitro*. *Hepatology* 18, 347–356.
625. Sternberg, D., Honigwachs-sha'anani, J., Brosh, N., Malik, Z., Burstein, Y., and Zipori, D. (1995). Restrictin-P/stromal activin A, kills its target cells via an apoptotic mechanism. *Growth Factors* 12, 277–287.
626. Hübner, G., Brauchle, M., Gregor, M., and Werner, S. (1997). Activin A: a novel player and inflammatory marker in inflammatory bowel disease? *Lab. Invest.* 77, 311–318.
627. Russell, C. E., Hedger, M. P., Brauman, J. N., de Kretser, D. M., and Phillips, D. J. (1999). Activin A regulates growth and acute phase proteins in the human liver cell line, HepG2. *Mol. Cell. Endocrinol.* 148, 129–136.
628. Hedger, M. P., Phillips, D. J., and de Kretser, D. M. (2000). Divergent cell-specific effects of activin-A on thymocyte proliferation stimulated by phytohemagglutinin, and interleukin 1β or interleukin 6 in vitro. *Cytokine* 12, 595–602.
629. Yu, J., Shao, L. E., Frigon, N. L., Jr., Lofgren, J., and Schwall, R. (1996). Induced expression of the new cytokine, activin A, in human monocytes: inhibition by glucocorticoids and retinoic acid. *Immunology* 88, 368–374.
630. Brosh, N., Sternberg, D., Honigwachs-Sha'anani, J., Lee, B. C., Shav-Tal, Y., Tzehoval, E., Shulman, L. M., Toledo, J., Hacham, Y., Carmi, P., Wen, J., Sasse, J., Horn, F., Burstein, Y., and Zipori, D. (1995). The plasmacytoma growth inhibitor restrictin-P is an antagonist of interleukin 6 and interleukin 11: identification as a stroma-derived activin A. *J. Biol. Chem.* 270, 29594–29600.
631. Ohguchi, M., Yamato, K., Ishihara, Y., Koide, M., Ueda, N., Okahashi, N., Noguchi, T., Kizaki, M., Ikeda, Y., Sugino, H., and Nisihara, T. (1998). Activin A regulates the production of mature interleukin-1β and interleukin-1 receptor antagonist in human monocytic cells. *J. Interferon Cytokine Res.* 18, 491–498.
632. Shao, L. E., Frigon, N. L., Jr., Yu, A., Palyash, J., and Yu, J. (1998). Contrasting effects of inflammatory cytokines and glucocorticoids on the production of activin A in human marrow stromal cells and their implications. *Cytokine* 10, 227–235.
633. Yu, E. W., Dolter, K. E., Shao, L. E., and Yu, J. (1998). Suppression of IL-6 biological activities by activin A and implications for inflammatory arthropathies. *Clin. Exp. Immunol.* 112, 126–132.
634. Shull, M. M., Ormsby, I., Kier, A. B., Pawlowski, S., Diebold, R. J., Yin, M., Allen, R., Sidman, C., Proetzel, G., Calvin, D., Annunziata, N., and Doetschman, T. (1992). Targeted disruption of the mouse transforming growth factor-β1 gene results in multifocal inflammatory disease. *Nature* 359, 693–699.
635. Meehan, T., Schlatt, S., O'Bryan, M. K., de Kretser, D. M., and Loveland, K. L. (2000). Regulation of germ cell and Sertoli cell development by activin, follistatin, and FSH. *Dev. Biol.* 220, 225–237.
636. Lee, W., Mason, A. J., Schwall, R., Szonyi, E., and Mather, J. P. (1989). Secretion of activin by interstitial cells in the testis. *Science* 243, 396–398.
637. de Winter, J. P., Vanderstichele, H. M., Verhoeven, G., Timmerman, M. A., Wesseling, J. G., and de Jong, F. H. (1994). Peritubular myoid cells from immature rat testes secrete activin-A and express activin receptor type II *in vitro*. *Endocrinology* 135, 759–767.
638. de Winter, J. P., Vanderstichele, H. M., Timmerman, M. A., Blok, L. J., Themmen, A. P., and de Jong, F. H. (1993). Activin is produced by rat Sertoli cells *in vitro* and can act as an autocrine regulator of Sertoli cell function. *Endocrinology* 132, 975–982.
639. Buzzard, J. J., Loveland, K. L., O'Bryan, M. K., O'Connor, A. E., Bakker, M., Hayashi, T., Wreford, N. G., Morrison, J. R., and de Kretser, D. M. (2004). Changes in circulating and testicular levels of inhibin A and B and activin A during postnatal development in the rat. *Endocrinology* 145, 3532–3541.
640. Okuma, Y., Saito, K., O'Connor, A. E., Phillips, D. J., de Kretser, D. M., and Hedger, M. P. (2005). Reciprocal regulation of activin A and inhibin B by interleukin-1 (IL-1) and follicle-stimulating hormone (FSH) in rat Sertoli cells *in vitro*. *J. Endocrinol.* 185, 99–110.

641. de Winter, J. P., Themmen, A. P., Hoogerbrugge, J. W., Klaij, I. A., Grootegoed, J. A., and de Jong, F. H. (1992). Activin receptor mRNA expression in rat testicular cell types. *Mol. Cell. Endocrinol.* 83, R1–R8.
642. Kaipia, A., Penttila, T. L., Shimasaki, S., Ling, N., Parvinen, M., and Toppari, J. (1992). Expression of inhibin β_A and β_B, follistatin and activin-A receptor messenger ribonucleic acids in the rat seminiferous epithelium. *Endocrinology* 131, 2703–2710.
643. Kaipia, A., Parvinen, M., and Toppari, J. (1993). Localization of activin receptor (ActR-IIB_2) mRNA in the rat seminiferous epithelium. *Endocrinology* 132, 477–479.
644. Feng, Z. M., Madigan, M. B., and Chen, C. L. (1993). Expression of type II activin receptor genes in the male and female reproductive tissues of the rat. *Endocrinology* 132, 2593–2600.
645. Cameron, V. A., Nishimura, E., Mathews, L. S., Lewis, K. A., Sawchenko, P. E., and Vale, W. W. (1994). Hybridization histochemical localization of activin receptor subtypes in rat brain, pituitary, ovary, and testis. *Endocrinology* 134, 799–808.
646. Anderson, R. A., Cambray, N., Hartley, P. S., and McNeilly, A. S. (2002). Expression and localization of inhibin α, inhibin/activin βA and βB and the activin type II and inhibin β-glycan receptors in the developing human testis. *Reproduction* 123, 779–788.
647. Chen, C. L., Pignataro, O. P., and Feng, Z. M. (1993). Inhibin/activin subunits and activin receptor are co-expressed in Leydig tumor cells. *Mol. Cell. Endocrinol.* 94, 137–143.
648. Hakovirta, H., Kaipia, A., Söder, O., and Parvinen, M. (1993). Effects of activin-A, inhibin-A, and transforming growth factor-β1 on stage-specific deoxyribonucleic acid synthesis during rat seminiferous epithelial cycle. *Endocrinology* 133, 1664–1668.
649. Mather, J. P., Attie, K. M., Woodruff, T. K., Rice, G. C., and Phillips, D. M. (1990). Activin stimulates spermatogonial proliferation in germ-Sertoli cell cocultures from immature rat testis. *Endocrinology* 127, 3206–3214.
650. Boitani, C., Stefanini, M., Fragale, A., and Morena, A. R. (1995). Activin stimulates Sertoli cell proliferation in a defined period of rat testis development. *Endocrinology* 136, 5438–5444.
651. Meinhardt, A., McFarlane, J. R., Seitz, J., and de Kretser, D. M. (2000). Activin maintains the condensed type of mitochondria in germ cells. *Mol. Cell. Endocrinol.* 168, 111–117.
652. Lejeune, H., Chuzel, F., Sanchez, P., Durand, P., Mather, J. P., and Saez, J. M. (1997). Stimulating effect of both human recombinant inhibin A and activin A on immature porcine Leydig cell functions *in vitro*. *Endocrinology* 138, 4783–4791.
653. Hsueh, A. J., Dahl, K. D., Vaughan, J., Tucker, E., Rivier, J., Bardin, C. W., and Vale, W. (1987). Heterodimers and homodimers of inhibin subunits have different paracrine action in the modulation of luteinizing hormone-stimulated androgen biosynthesis. *Proc. Natl. Acad. Sci. U. S. A.* 84, 5082–5086.
654. Lin, T., Calkins, J. K., Morris, P. L., Vale, W., and Bardin, C. W. (1989). Regulation of Leydig cell function in primary culture by inhibin and activin. *Endocrinology* 125, 2134–2140.
655. Mauduit, C., Chauvin, M. A., de Peretti, E., Morera, A. M., and Benahmed, M. (1991). Effect of activin A on dehydroepiandrosterone and testosterone secretion by primary immature porcine Leydig cells. *Biol. Reprod.* 45, 101–109.
656. Mendelson, C., Dufau, M., and Catt, K. (1975). Dependence of gonadotropin-induced steroidogenesis upon RNA and protein synthesis in the interstitial cells of the rat testis. *Biochim. Biophys. Acta* 411, 222–230.
657. Mendelson, C., Dufau, M., and Catt, K. (1975). Gonadotropin binding and stimulation of cyclic adenosine 3′:5′-monophosphate and testosterone production in isolated Leydig cells. *J. Biol. Chem.* 250, 8818–8823.
658. Stocco, D. M. (2000). Intramitochondrial cholesterol transfer. *Biochim. Biophys. Acta* 1486, 184–197.
659. Christenson, L. K., and Strauss, J. F. (2000). Steroidogenic acute regulatory protein (StAR) and the intramitochondrial translocation of cholesterol. *Biochim. Biophys. Acta* 1529, 175–187.
660. Payne, A. H., and O'Shaughnessy, P. J. (1996). Structure, function and regulation of steroidogenic enzymes in the Leydig cell. In *The Leydig Cell* (L. D. Russell, Ed.), pp. 259–285. Cache River Press, Vienna, IL.
661. Payne, A. H., and Hales, D. B. (2004). Overview of steroidogenic enzymes in the pathway from cholesterol to active steroid hormones. *Endocr. Rev.* 25, 947–970.
662. Calandra, T., Baumgartner, J. D., Grau, G. E., Wu, M. M., Lambert, P. H., Schellekens, J., Verhoef, J., and Glauser, M. P. (1990). Prognostic values of tumor necrosis factor/cachectin, interleukin-1, interferon-α, and interferon-γ in the serum of patients with septic shock. Swiss-Dutch J5 Immunoglobulin Study Group. *J. Infect. Dis.* 161, 982–987.
663. Calandra, T., Bochud, P. Y., and Heumann, D. (2002). Cytokines in septic shock. *Curr. Clin. Top. Infect. Dis.* 22, 1–23.
664. Dinarello, C. A., and Cannon, J. G. (1993). Cytokine measurements in septic shock. *Ann. Intern. Med.* 119, 853–854.
665. Cannon, J. G., Tompkins, R. G., Gelfand, J. A., Michie, H. R., Stanford, G. G., van der Meer, J. W., Endres, S., Lonnemann, G., Corsetti, J., and Chernow, B. (1990). Circulating interleukin-1 and tumor necrosis factor in septic shock and experimental endotoxin fever. *J. Infect. Dis.* 161, 79–84.
666. Damas, P., Reuter, A., Gysen, P., Demonty, J., Lamy, M., and Franchimont, P. (1989). Tumor necrosis factor and interleukin-1 serum levels during severe sepsis in humans. *Crit. Care Med.* 17, 975–978.
667. Damas, P., Canivet, J. L., de Groote, D., Vrindts, Y., Albert, A., Franchimont, P., and Lamy, M. (1997). Sepsis and serum cytokine concentrations. *Crit. Care Med.* 25, 405–412.
668. van der Poll, T., Romijn, J. A., Endert, E., and Sauerwein, H. P. (1993). Effects of tumor necrosis factor on the hypothalamic-pituitary-testicular axis in healthy men. *Metabolism* 42, 303–307.
669. Vogel, A. V., Peake, G. T., and Rada, R. T. (1985). Pituitary-testicular axis dysfunction in burned men. *J. Clin. Endocrinol. Metab.* 60, 658–665.
670. Woolf, P. D., Hamill, R. W., McDonald, J. V., Lee, L. A., and Kelly, M. (1985). Transient hypogonadotropic hypogonadism caused by critical illness. *J. Clin. Endocrinol. Metab.* 60, 444–450.
671. Christeff, N., Auclair, M. C., Benassayag, C., Carli, A., and Nunez, E. A. (1987). Endotoxin-induced changes in sex steroid hormone levels in male rats. *J. Steroid Biochem.* 26, 67–71.
672. Christeff, N., Benassayag, C., Carli-Vielle, C., Carli, A., and Nunez, E. A. (1988). Elevated oestrogen and reduced testosterone levels in the serum of male septic shock patients. *J. Steroid Biochem.* 29, 435–440.
673. Cutolo, M., Balleari, E., Giusti, M., Intra, E., and Accardo, S. (1991). Androgen replacement therapy in male patients with rheumatoid arthritis. *Arthritis Rheum.* 34, 1–5.
674. Fourrier, F., Jallot, A., Leclerc, L., Jourdain, M., Racadot, A., Chagnon, J. L., Rime, A., and Chopin, C. (1994). Sex steroid hormones in circulatory shock, sepsis syndrome, and septic shock. *Circ. Shock* 43, 171–178.
675. Handelsman, D. J. (1994). Testicular dysfunction in systemic disease. *Clin. Androl.* 23, 839–856.

676. Lephart, E. D., Baxter, C. R., and Parker, C. R. J. (1987). Effect of burn trauma on adrenal and testicular steroid hormone production. *J. Clin. Endocrinol. Metab.* 64, 842–848.
677. Lindh, A., Carlstrom, K., Eklund, J., and Wilking, N. (1992). Serum steroid and prolactin during and after major surgical trauma. *Acta Anesth. Scand.* 36, 119–121.
678. Martens, H. F., Sheets, P. K., Tenover, J. S., Dugowson, C. E., Bremner, W. J., and Starkebaum, G. (1994). Decreased testosterone levels in men with rheumatoid arthritis: effect of low dose prednisone therapy. *J. Rheumatol.* 21, 1427–1431.
679. Spector, T. D., Ollier, W., Perry, L. A., Silman, A. J., Thompson, P. W., and Edwards, A. (1989). Free and serum testosterone levels in 276 males: a comparative study of rheumatoid arthritis, ankylosing spondylitis and healthy control. *Clin. Rheum.* 8, 37–41.
680. Spratt, D. I., Bigos, S. T., Beitins, I., Cox, P., Longcope, C., and Orav, J. (1992). Both hyper- and hypogonadotropic hypogonadism occur transiently in acute illness: bio- and immunoactive gonadotropins. *J. Clin. Endocrinol. Metab.* 75, 1562–1570.
681. Spratt, D. I., Cox, P., Orav, J., Moloney, J., and Bigos, T. (1993). Reproductive axis suppression in acute illness is related to disease severity. *J. Clin. Endocrinol. Metab.* 76, 1548–1554.
682. Bruot, B. C., and Clemens, J. W. (1987). Effect of adjuvant-induced arthritis on serum luteinizing hormone and testosterone concentrations in the male rat. *Life Sci.* 41, 1559–1565.
683. Bruot, B. C., and Clemens, J. W. (1992). Regulation of testosterone production in the adjuvant-induced arthritic rat. *J. Androl.* 13, 87–92.
684. Clemens, J. W., and Bruot, B. C. (1989). Testicular dysfunction in the adjuvant-induced arthritic rat. *J. Androl.* 10, 419–424.
685. Hales, K. H., Diemer, T., Ginde, S., Shankar, B. K., Roberts, M., Bosmann, H. B., and Hales, D. B. (2000). Diametric effects of bacterial endotoxin lipopolysaccharide on adrenal and Leydig cell steroidogenic acute regulatory protein. *Endocrinology* 141, 4000–4012.
686. Bosmann, H. B., Hales, K. H., Li, X., Liu, Z., Stocco, D. M., and Hales, D. B. (1996). Acute in vivo inhibition of testosterone by endotoxin parallels loss of steroidogenic acute regulatory (StAR) protein in Leydig cells. *Endocrinology* 137, 4522–4525.
687. Sharma, A. C., Bosmann, H. B., Motew, S. J., Hales, K. H., Hales, D. B., and Ferguson, J. L. (1996). Steroid hormone alterations following induction of chronic intraperitoneal sepsis in male rats. *Shock* 6, 150–154.
688. Sharma, A. C., Sam, A. D., 2nd, Lee, L. Y., Hales, D. B., Law, W. R., Ferguson, J. L., and Bosmann, H. B. (1998). Effect of N^G-nitro-L-arginine methyl ester on testicular blood flow and serum steroid hormones during sepsis. *Shock* 9, 416–421.
689. Sewer, M. B., and Morgan, E. T. (1998). Down-regulation of the expression of three major rat liver cytochrome P450S by endotoxin in vivo occurs independently of nitric oxide production. *J. Pharmacol. Exp. Ther.* 287, 352–358.
690. Wallgren, M., Kindahl, H., and Rodriguez-Martinez, H. (1995). Modulation of endotoxin-induced andrological alterations by flunixin meglumine in the boar. *Zentralbl. Veterin. A* 42, 357–369.
691. Lysiak, J. J. (2004). The role of tumor necrosis factor-alpha and interleukin-1 in the mammalian testis and their involvement in testicular torsion and autoimmune orchitis. *Reprod. Biol. Endocrinol.* 2, 9.
692. Lysiak, J. J., Nguyen, Q. A., Kirby, J. L., and Turner, T. T. (2003). Ischemia-reperfusion of the murine testis stimulates the expression of proinflammatory cytokines and activation of c-jun N-terminal kinase in a pathway to E-selectin expression. *Biol. Reprod.* 69, 202–210.
693. Mealy, K., Robinson, B., Millette, C. F., Majzoub, J., and Wilmore, D. W. (1990). The testicular effects of tumor necrosis factor. *Ann. Surg.* 211, 470–475.
694. Meikle, A. W., Cardoso de Sousa, J. C., Ward, J. H., Woodward, M., and Samlowski, W. E. (1991). Reduction of testosterone synthesis after high dose interleukin-2 therapy of metastatic cancer. *J. Clin. Endocrinol. Metab.* 73, 931–935.
695. Kasahara, T., Hooks, J. J., Dougherty, S. F., and Oppenheim, J. J. (1983). Interleukin-2–mediated immune interferon (IFN-γ) production by human T cells and T cell subsets. *J. Immunol.* 130, 1784–1789.
696. Nedwin, G. E., Svedersky, L. P., Bringman, T. S., Palladino, M. A. J., and Goeddel, D. V. (1985). Effect of interleukin-2, interferon-γ, and mitogens on the production of tumor necrosis factors α and β. *J. Immunol.* 135, 2492–2497.
697. Tsigos, C., Kyrou, I., Chala, E., Tsapogas, P., Stavridis, J. C., Raptis, S. A., and Katsilambros, N. (1999). Circulating tumor necrosis factor α concentrations are higher in abdominal versus peripheral obesity. *Metabolism* 48, 1332–1335.
698. Mayerhofer, A. (1996). Leydig cell regulation by catecholamines and neuroendocrine messengers. In *The Leydig Cell* (L. D. Russell, Ed.), pp. 407–417. Cache River Press, Vienna, IL.
699. Rivier, C., and Rivest, S. (1991). Effect of stress on the activity of the hypothalamic-pituitary-gonadal axis: peripheral and central mechanisms. *Biol. Reprod.* 45, 523–532.
700. Turnbull, A., and Rivier, C. (1995). Brain-periphery connections: do they play a role in mediating the effect of centrally injected interleukin-1β on gonadal function. *Neuroimmunomodulation* 2, 224–235.
701. Verhoeven, G., Cailleau, J., Van Damme, J., and Billiau, A. (1988). Interleukin-1 stimulates steroidogenesis in cultured rat Leydig cells. *Mol. Cell. Endocrinol.* 57, 51–60.
702. Warren, D. W., Pasupuleti, V., Lu, Y., Platler, B. W., and Horton, R. (1990). Tumor necrosis factor and interleukin-1 stimulate testosterone secretion in adult male rat Leydig cells in vitro. *J. Androl.* 11, 353–360.
703. Calkins, J. H., Sigel, M. M., Nankin, H. R., and Lin, T. (1988). Interleukin-1 inhibits Leydig cell steroidogenesis in primary culture. *Endocrinology* 123, 1605–1610.
704. Mauduit, C., Chauvin, M. A., Hartmann, D. J., Revol, A., Morera, A. M., and Benahmed, M. (1992). Interleukin-1α as a potent inhibitor of gonadotropin action in porcine Leydig cells: site(s) of action. *Biol. Reprod.* 46, 1119–1126.
705. Lin, T., Wang, T. L., Nagpal, M. L., Calkins, J. H., Chang, W. W., and Chi, R. (1991). Interleukin-1 inhibits cholesterol side-chain cleavage cytochrome P450 expression in primary cultures of Leydig cells. *Endocrinology* 129, 1305–1311.
706. Sun, X. R., Hedger, M. P., and Risbridger, G. P. (1993). The effect of testicular macrophages and interleukin-1 on testosterone production by purified adult rat Leydig cells cultured under *in vitro* maintenance conditions. *Endocrinology* 132, 186–192.
707. Moore, C., and Moger, W. H. (1991). Interleukin-1α-induced changes in androgen and cyclic adenosine 3',5'-monophosphate release in adult rat Leydig cells in culture. *J. Endocrinol.* 129, 381–390.
708. Watson, M. E., Newman, R. J., Payne, A. M., Abdelrahim, M., and Francis, G. L. (1994). The effect of macrophage conditioned media on Leydig cell function. *Ann. Clin. Lab. Sci.* 24, 84–95.
709. Calkins, J. H., Guo, H., Sigel, M. M., and Lin, T. (1990). Differential effects of recombinant interleukin-1α and β on

Leydig cell function. *Biochem. Biophys. Res. Commun.* 167, 548–553.

710. Fauser, B. C., Galway, A. B., and Hsueh, A. J. (1989). Inhibitory actions of interleukin-1β on steroidogenesis in primary cultures of neonatal rat testicular cells. *Acta Endocrinol.* 120, 401–408.
711. Diemer, T., Hales, D. B., and Weidner, W. (2003). Immune-endocrine interactions and Leydig cell function: the role of cytokines. *Andrologia* 35, 55–63.
712. Xiong, Y., and Hales, D. B. (1997). Differential effects of tumor necrosis factor-α and interleukin-1 on 3 β-hydroxysteroid dehydrogenase/Δ5→Δ4 isomerase expression in mouse Leydig cells. *Endocrine* 7, 295–301.
713. Lin, T., Wang, D., and Stocco, D. M. (1998). Interleukin-1 inhibits Leydig cell steroidogenesis without affecting steroidogenic acute regulatory protein messenger ribonucleic acid or protein levels. *J. Endocrinol.* 156, 461–467.
714. Wang, X., Dyson, M. T., Jo, Y., and Stocco, D. M. (2003). Inhibition of cyclooxygenase-2 activity enhances steroidogenesis and steroidogenic acute regulatory gene expression in MA-10 mouse Leydig cells. *Endocrinology* 144, 3368–3375.
715. Romanelli, F., Valenca, M., Conte, D., Isidori, A., and Negro-Vilar, A. (1995). Arachidonic acid and its metabolites effects on testosterone production by rat Leydig cells. *J. Endocrinol. Invest.* 18, 186–193.
716. Aggarwal, B. B., and Pocsik, E. (1992). Cytokines: from clone to clinic. *Arch. Biochem. Biophys.* 292, 335–359.
717. Cerami, A. (1992). Inflammatory cytokines. *Clin. Immunol. Immunopathol.* 62, S3–S10.
718. Spooner, C. E., Markowitz, N. P., and Sarvolatz, L. D. (1992). The role of tumor necrosis factor in sepsis. *Clin. Immunol. Immunopathol.* 62, S11–S17.
719. Mak, T., and Yeh, W.-C. (2002). Signaling for survival and apoptosis in the immune system. *Arthritis Res.* 4, S243–S252.
720. Hsu, H., Shu, H. B., Pan, M. G., and Goeddel, D. V. (1996). TRADD-TRAF2 and TRADD-FADD interactions define two distinct TNF receptor 1 signal transduction pathways. *Cell* 84, 299–308.
721. De, S. K., Chen, H. L., Pace, J. L., Hunt, J. S., Terranova, P. F., and Enders, G. C. (1993). Expression of tumor necrosis factor-α in mouse spermatogenic cells. *Endocrinology* 133, 389–396.
722. Moore, C., and Hutson, J. C. (1994). Physiological relevance of tumor necrosis factor in mediating macrophage-Leydig cell interactions. *Endocrinology* 134, 63–69.
723. Calkins, J. H., Guo, H., Sigel, M. M., and Lin, T. (1990). Tumor necrosis factor-α enhances inhibitory effects of interleukin-1β on Leydig cell steroidogenesis. *Biochem. Biophys. Res. Commun.* 166, 1313–1318.
724. Li, X., Youngblood, G. L., Payne, A. H., and Hales, D. B. (1995). Tumor necrosis factor-α inhibition of 17α-hydroxylase/C17-20 lyase gene (*Cyp17*) expression. *Endocrinology* 136, 3519–3526.
725. Mauduit, C., Hartmann, D. J., Chauvin, M. A., Revol, A., Morera, A. M., and Benahmed, M. (1991). Tumor necrosis factor α inhibits gonadotropin action in cultured porcine Leydig cells: site(s) of action. *Endocrinology* 129, 2933–2940.
726. Xiong, Y., and Hales, D. B. (1993). The role of tumor necrosis factor-α in the regulation of mouse Leydig cell steroidogenesis. *Endocrinology* 132, 2438–2444.
727. Mauduit, C., Gasnier, F., Rey, C., Chauvin, M. A., Stocco, D. M., Louisot, P., and Benahmed, M. (1998). Tumor necrosis factor-α inhibits Leydig cell steroidogenesis through a decrease in steroidogenic acute regulatory protein expression. *Endocrinology* 139, 2863–2868.
728. Budnik, L. T., Jahner, D., and Mukhopadhyay, A. K. (1999). Inhibitory effects of TNFα on mouse tumor Leydig cells: possible role of ceramide in the mechanism of action. *Mol. Cell. Endocrinol.* 150, 39–46.
729. Meroni, S. B., Pellizzari, E. H., Cánepa, D. F., and Cigorraga, S. B. (2000). Possible involvement of ceramide in the regulation of rat Leydig cell function. *J. Steroid Biochem. Mol. Biol.* 75, 307–313.
730. Morales, V., Santana, P., Diaz, R., Tabraue, C., Gallardo, G., Lopez Blanco, F., Hernandez, I., Fanjul, L. F., and Ruiz de Galarreta, C. M. (2003). Intratesticular delivery of tumor necrosis factor-α and ceramide directly abrogates steroidogenic acute regulatory protein expression and Leydig cell steroidogenesis in adult rats. *Endocrinology* 144, 4763–4772.
731. Hong, C. Y., Park, J. H., Ahn, R. S., Im, S. Y., Choi, H. S., Soh, J., Mellon, S. H., and Lee, K. (2004). Molecular mechanism of suppression of testicular steroidogenesis by proinflammatory cytokine tumor necrosis factor α. *Mol. Cell. Biol.* 24, 2593–2604.
732. Le Magueresse-Battistoni, B., Pernod, G., Kolodié, L., Morera, A. M., and Benahmed, M. (1997). Tumor necrosis factor-α regulates plasminogen activator inhibitor-1 in rat testicular peritubular cells. *Endocrinology* 138, 1097–1105.
733. Nehar, D., Mauduit, C., Boussouar, F., and Benahmed, M. (1997). Tumor necrosis factor-α-stimulated lactate production is linked to lactate dehydrogenase A expression and activity increase in porcine cultured Sertoli cells. *Endocrinology* 138, 1964–1971.
734. Yule, T. D., and Tung, K. S. (1993). Experimental autoimmune orchitis induced by testis and sperm antigen-specific T cell clones: an important pathogenic cytokine is tumor necrosis factor. *Endocrinology* 133, 1098–1107.
735. Suescun, M. O., Rival, C., Theas, M. S., Calandra, R. S., and Lustig, L. (2003). Involvement of tumor necrosis factor-α in the pathogenesis of autoimmune orchitis in rats. *Biol. Reprod.* 68, 2114–2121.
736. Borden, E. C. (1992). Interferons: pleiotropic cellular modulators. *Clin. Immunol. Immunopathol.* 62, S18–S24.
737. Orava, M., Voutilainen, R., and Vihko, R. (1989). Interferon-γ inhibits steroidogenesis and accumulation of mRNA of the steroidogenic enzymes, $P450_{SCC}$ and $P450_{C}17$ in cultured porcine Leydig cells. *Mol. Endocrinol.* 3, 887–894.
738. Orava, M., Cantell, K., and Vihko, R. (1985). Human leukocyte interferon inhibits human chorionic gonadotropin stimulated testosterone production by porcine Leydig cells in culture. *Biochem. Biophys. Res. Commun.* 127, 809–815.
739. Orava, M., Cantell, K., and Vihko, R. (1986). Treatment with preparations of human leukocyte interferon decreases serum testosterone concentrations in men. *Int. J. Cancer* 38, 295–296.
740. Montor, J. M., Mendoza, M. E., and Romano, M. C. (1998). Effect of human and murine interferon-α on steroid production by rat ovarian cells. *Life Sci.* 62, 1733–1744.
741. Fountain, S., Holland, M. K., Hinds, L. A., Janssens, P. A., and Kerr, P. J. (1997). Interstitial orchitis with impaired steroidogenesis and spermatogenesis in the testes of rabbits infected with an attenuated strain of myxoma virus. *J. Reprod. Fertil.* 110, 161–169.
742. Fray, M. D., Mann, G. E., Clarke, M. C., and Charleston, B. (2000). Bovine viral diarrhoea virus: its effects on ovarian function in the cow. *Vet. Microbiol.* 77, 185–194.
743. Lamba, H., Goldmeier, D., Mackie, N. E., and Scullard, G. (2004). Antiretroviral therapy is associated with sexual dysfunction and with increased serum oestradiol levels in men. *Int. J. STD AIDS* 15, 234–237.
744. Hengge, U. R. (2003). Testosterone replacement for hypogonadism: clinical findings and best practices. *AIDS Reader* 13, S15–S21.
745. Dejucq, N., Lienard, M. O., Guillaume, E., Dorval, I., and Jégou, B. (1998). Expression of interferons-α and -γ in

testicular interstitial tissue and spermatogonia of the rat. *Endocrinology* 139, 3081–3087.

746. Guo, H., Calkins, J. H., Sigel, M. M., and Lin, T. (1990). Interleukin-2 is a potent inhibitor of Leydig cell steroidogenesis. *Endocrinology* 127, 1234–1239.
747. Mauduit, C., Goddard, I., Besset, V., Tabone, E., Rey, C., Gasnier, F., Dacheux, F., and Benahmed, M. (2001). Leukemia inhibitory factor antagonizes gonadotropin induced-testosterone synthesis in cultured porcine Leydig cells: sites of action. *Endocrinology* 142, 2509–2520.
748. Avallet, O., Vigier, M., Perrard-Sapori, M. H., and Saez, J. M. (1987). Transforming growth factor β inhibits Leydig cell functions. *Biochem. Biophys. Res. Commun.* 146, 575–581.
749. Saez, J. M., and Lejeune, H. (1996). Regulation of Leydig cell functions by hormones and growth factors other than LH and IGF-1. In *The Leydig Cell* (L. D. Russell, Ed.), pp. 383–406. Cache River Press, Vienna, IL.
750. Ulevitch, R. J., and Tobias, P. S. (1995). Receptor-dependent mechanisms of cell stimulation by bacterial endotoxin. *Annu. Rev. Immunol.* 13, 437–457.
751. Simonian, N. A., and Coyle, J. T. (1996). Oxidative stress in neurodegenerative diseases. *Annu. Rev. Pharmacol. Toxicol.* 36, 83–106.
752. Quinn, P. G., and Payne, A. H. (1984). Microsomal cytochrome P-450 enzyme damage in cultured Leydig cells: relation to steroidogenic desensitization. *Ann. N. Y. Acad. Sci.* 438, 649–651.
753. Quinn, P. G., and Payne, A. H. (1985). Steroid product-induced, oxygen-mediated damage of microsomal cytochrome P-450 enzymes in Leydig cell cultures: relationship to desensitization. *J. Biol. Chem.* 260, 2092–2099.
754. Quinn, P. G., and Payne, A. H. (1984). Oxygen-mediated damage of microsomal cytochrome P-450 enzymes in cultured Leydig cells: role in steroidogenic desensitization. *J. Biol. Chem.* 259, 4130–4135.
755. Hornsby, P. J. (1987). Physiological and pathological effects of steroids on the function of the adrenal cortex. *J. Steroid Biochem.* 27, 1161–1171.
756. Fernandez-Checa, J. C., Kaplowitz, N., Garcia-Ruiz, C., Colell, A., Miranda, M., Mari, M., Ardite, E., and Morales, A. (1997). GSH transport in mitochondria: defense against TNF-induced oxidative stress and alcohol-induced defect. *Am. J. Physiol.* 273, G7–G17.
757. Allen, J. A., Diemer, T., Janus, P., Held Hales, K., and Hales, D. B. (2005). Bacterial endotoxin lipopolysaccharide and reactive oxygen species inhibit Leydig cell steroidogenesis via perturbation of mitochondria. *Endocrine* 25, 265–275.
758. Allen, J., Ginde, S., Choi, J., Diemer, T., Held Hales, K., and Hales, D. B. (2000). Reactive oxygen disrupts mitochondria in Leydig cells and inhibits steroidogenesis. *Biol. Reprod.* 63, 338.
759. Diemer, T., Held Hales, K., Ginde, S., Choi, J., Nardulli, B., Bosmann, H. B., and Hales, D. B. (2000). Immune activation via injection of bacterial lipopolysaccharide (LPS) in mice results in disruption of Leydig cell steroidogenesis due to oxidative mitochondrial damage. *Biol. Reprod.* 63, 343.
760. Stocco, D. M., Wells, J., and Clark, B. J. (1993). The effects of hydrogen peroxide on steroidogenesis in mouse Leydig tumor cells. *Endocrinology* 133, 2827–2832.
761. Diemer, T., Allen, J. A., Hales, K. H., and Hales, D. B. (2003). Reactive oxygen disrupts mitochondria in MA-10 tumor Leydig cells and inhibits steroidogenic acute regulatory (StAR) protein and steroidogenesis. *Endocrinology* 144, 2882–2891.
762. Behrman, H. R., and Aten, R. F. (1991). Evidence that hydrogen peroxide blocks hormone-sensitive cholesterol transport into mitochondria of rat luteal cells. *Endocrinology* 128, 2958–2966.
763. Endo, T., Aten, R. F., Leykin, L., and Behrman, H. R. (1993). Hydrogen peroxide evokes antisteroidogenic and antigonadotropic actions in human granulosa luteal cells. *J. Clin. Endocrinol. Metab.* 76, 337–342.
764. Gatzuli, E., Aten, R. F., and Behrman, H. R. (1991). Inhibition of gonadotropin action and progesterone synthesis by xanthine oxidase in rat luteal cells. *Endocrinology* 128, 2253–2258.
765. Riley, J. C., and Behrman, H. R. (1991). In vivo generation of hydrogen peroxide in the rat corpus luteum during luteolysis. *Endocrinology* 128, 1749–1753.
766. Salvemini, D., Korbut, R., Anggard, E., and Vane, J. (1990). Immediate release of a nitric oxide-like factor from bovine aortic endothelial cells by Escherichia coli lipopolysaccharide. *Proc. Natl. Acad. Sci. U. S. A.* 87, 2593–2597.
767. Thiemermann, C., Wu, C.-C., Szabo, C., Perretti, M., and Vane, J. R. (1993). Role of tumor necrosis factor in the induction of nitric oxide synthase in a rat model of endotoxin shock. *Br. J. Pharmacol.* 110, 177–182.
768. Tracey, W. R., Tse, J., and Carter, G. (1995). Lipopolysaccharide-induced changes in plasma nitrite and nitrate concentrations in rats and mice: pharmacological evaluation of nitric oxide synthase inhibitors. *J. Pharmacol. Exp. Ther.* 272, 1011–1015.
769. Bauché, F., Stéphan, J. P., Touzalin, A. M., and Jégou, B. (1998). In vitro regulation of an inducible-type NO synthase in the rat seminiferous tubule cells. *Biol. Reprod.* 58, 431–438.
770. O'Bryan, M. K., Schlatt, S., Gerdprasert, O., Phillips, D. J., de Kretser, D. M., and Hedger, M. P. (2000). Inducible nitric oxide synthase in the rat testis: evidence for potential roles in both normal function and inflammation-mediated infertility. *Biol. Reprod.* 63, 1285–1293.
771. Gaytán, F., Bellido, C., Aguilar, R., Morales, C., van Rooijen, N., and Aguilar, E. (1997). Role of the testis in the response of the pituitary-testicular axis to nitric oxide-related agents. *Eur. J. Endocrinol.* 137, 301–308.
772. Kostic, T., Andric, S., Kovacevic, R., and Maric, D. (1998). The involvement of nitric oxide in stress-impaired testicular steroidogenesis. *Eur. J. Pharmacol.* 346, 267–273.
773. Adams, M. L., Nock, B., Truong, R., and Cicero, T. J. (1992). Nitric oxide control of steroidogenesis: endocrine effects of N^G-nitro-L-arginine and comparisons to alcohol. *Life Sci.* 50, L35–40.
774. Adams, M. L., Meyer, E. R., Sewing, B. N., and Cicero, T. J. (1994). Effects of nitric oxide-related agents on rat testicular function. *J. Pharmacol. Exp. Ther.* 269, 230–237.
775. Adams, M. L., Meyer, E. R., and Cicero, T. J. (1996). Effects of nitric oxide-related agents on opioid regulation of rat testicular steroidogenesis. *Biol. Reprod.* 54, 1128–1134.
776. Pomerantz, D. K., and Pitelka, V. (1998). Nitric oxide is a mediator of the inhibitory effect of activated macrophages on production of androgen by the Leydig cell of the mouse. *Endocrinology* 139, 922–931.
777. Del Punta, K., Charreau, E. H., and Pignataro, O. P. (1996). Nitric oxide inhibits Leydig cell steroidogenesis. *Endocrinology* 137, 5337–5343.
778. Welch, C., Watson, M. E., Poth, M., Hong, T., and Francis, G. L. (1995). Evidence to suggest nitric oxide is an interstitial regulator of Leydig cell steroidogenesis. *Metab. Clin. Exp.* 44, 234–238.
779. Sharma, A., Motew, S., Farias, S., Alden, K., Bosmann, B., Law, W., and Ferguson, J. (1997). Sepsis alters myocardial and plasma concentrations of endothelin and nitric oxide in rats. *J. Mol. Cell. Cardiol.* 29, 1469–1477.
780. Koppenol, W. H., Moreno, J. J., Pryor, W. A., Ischiropoulos, H., and Beckman, J. S. (1992). Peroxynitrite, a cloaked

oxidant formed by nitric oxide and superoxide. *Chem. Res. Toxicol.* 5, 834–842.
781. Ischiropoulos, H., Zhu, L., Chen, J., Tsai, M., Martin, J. C., Smith, C. D., and Beckman, J. S. (1992). Peroxynitrite-mediated tyrosine nitration catalyzed by superoxide dismutase. *Arch. Biochem. Biophys.* 298, 431–437.
782. Beckman, J. S., Beckman, T. W., Chen, J., Marshall, P. A., and Freeman, B. A. (1990). Apparent hydroxyl radical production by peroxynitrite: implications for endothelial injury from nitric oxide and superoxide. *Proc. Natl. Acad. Sci. U. S. A.* 87, 1620–1624.
783. Guidarelli, A., Clementi, E., Sciorati, C., and Cantoni, O. (1998). The mechanism of the nitric oxide-mediated enhancement of tert-butylhydroperoxide-induced DNA single strand breakage. *Br. J. Pharmacol.* 125, 1074–1080.
784. Zini, A., Abitbol, J., Girardi, S. K., Schulsinger, D., Goldstein, M., and Schlegel, P. N. (1998). Germ cell apoptosis and endothelial nitric oxide synthase (eNOS) expression following ischemia-reperfusion injury to testis. *Arch. Androl.* 41, 57–65.
785. Bateman, A., Singh, A., Kral, T., and Solomon, S. (1989). The immune-hypothalamic-pituitary-adrenal axis. *Endocr. Rev.* 10, 92–112.
786. Besedovsky, H. O., and Del Rey, A. (1996). Immune-neuro-endocrine interactions: facts and hypotheses. *Endocr. Rev.* 17, 64–102.
787. Rivest, S., and Rivier, C. (1993). Interleukin-1β inhibits the endogenous expression of the early gene c-fos located within the nucleus of LHRH neurons and interferes with hypothalamic LHRH release during proestrus in the rat. *Brain Res.* 613, 132–142.
788. Sowers, J. R., Rice, B. F., and Blanchard, S. (1979). Effect of dexamethasone on luteinizing hormone and follicle stimulating hormone responses to LHRH and to clomiphene in the follicular phase of women with normal menstrual cycles. *Horm. Metab. Res.* 11, 478–480.
789. Tohei, A., and Kogo, H. (1999). Dexamethasone increases follicle-stimulating hormone secretion via suppression of inhibin in rats. *Eur. J. Pharmacol.* 386, 69–74.
790. Stalker, A., Hermo, L., and Antakly, T. (1989). Covalent affinity labeling, radioautography, and immunocytochemistry localize the glucocorticoid receptor in rat testicular Leydig cells. *Am. J. Anat.* 186, 369–377.
791. Sapolsky, R. M. (1985). Stress-induced suppression of testicular function in the wild baboon: role of glucocorticoids. *Endocrinology* 116, 2273–2278.
792. Gao, H. B., Shan, L. X., Monder, C., and Hardy, M. P. (1996). Suppression of endogenous corticosterone levels in vivo increases the steroidogenic capacity of purified rat Leydig cells in vitro. *Endocrinology* 137, 1714–1718.
793. Battaglia, D. F., Bowen, J. M., Krasa, H. B., Thrun, L. A., Viguie, C., and Karsch, F. J. (1997). Endotoxin inhibits the reproductive neuroendocrine axis while stimulating adrenal steroids: a simultaneous view from hypophyseal portal and peripheral blood. *Endocrinology* 138, 4273–4281.
794. Refojo, D., Arias, P., Moguilevsky, J. A., and Feleder, C. (1998). Effect of bacterial endotoxin on in vivo pulsatile gonadotropin secretion in adult male rats. *Neuroendocrinology* 67, 275–281.
795. Rivier, C., and Vale, W. (1989). In the rat, interleukin-1α acts at the level of the brain and the gonads to interfere with gonadotropin and sex steroid secretion. *Endocrinology* 124, 2105–2109.
796. Rivier, C., and Vale, W. (1990). Cytokines act within the brain to inhibit luteinizing hormone secretion and ovulation in the rat. *Endocrinology* 127, 849–856.
797. Bonavera, J. J., Kalra, S. P., and Kalra, P. S. (1993). Mode of action of interleukin-1 suppression of pituitary LH release in castrated male rats. *Brain Res.* 612, 1–8.
798. Kalra, P. S., Fuentes, M., Sahu, A., and Kalra, S. P. (1990). Endogenous opioid peptides mediate the interleukin-1–induced inhibition of the release of luteinizing hormone (LH)-releasing hormone and LH. *Endocrinology* 127, 2381–2386.
799. Russell, S. H., Small, C. J., Stanley, S. A., Franks, S., Ghatei, M. A., and Bloom, S. R. (2001). The *in vitro* role of tumour necrosis factor-alpha and interleukin-6 in the hypothalamic-pituitary gonadal axis. *J. Neuroendocrinol.* 13, 296–301.
800. Yu, W. H., Walczewska, A., Karanth, S., and McCann, S. M. (1997). Nitric oxide mediates leptin-induced luteinizing hormone-releasing hormone (LHRH) and LHRH and leptin-induced LH release from the pituitary gland. *Endocrinology* 138, 5055–5058.
801. McCann, S. M., Kimura, M., Karanth, S., Yu, W. H., and Rettori, V. (1997). Nitric oxide controls the hypothalamic-pituitary response to cytokines. *Neuroimmunomodulation* 4, 98–106.
802. Lee, S., Miselis, R., and Rivier, C. (2002). Anatomical and functional evidence for a neural hypothalamic-testicular pathway that is independent of the pituitary. *Endocrinology* 143, 4447–4454.
803. Selvage, D. J., Lee, S. Y., Parsons, L. H., Seo, D. O., and Rivier, C. L. (2004). A hypothalamic-testicular neural pathway is influenced by brain catecholamines, but not testicular blood flow. *Endocrinology* 145, 1750–1759.
804. Gnessi, L., Fabbri, A., and Spera, G. (1997). Gonadal peptides as mediators of development and functional control of the testis: an integrated system with hormones and local environment. *Endocr. Rev.* 18, 541–609.
805. Chiocchio, S. R., Suburo, A. M., Vladucic, E., Zhu, B. C., Charreau, E., Decima, E. E., and Tramezzani, J. H. (1999). Differential effects of superior and inferior spermatic nerves on testosterone secretion and spermatic blood flow in cats. *Endocrinology* 140, 1036–1043.
806. Gerendai, I., Csaba, Z., and Csernus, V. (1996). Testicular injection of 5,6-dihydroxytryptamine or vasectomy interferes with local stimulatory effect of oxytocin on testicular steroidogenesis in immature rats. *Neuroendocrinology* 63, 284–289.
807. Huang, H. F., Li, M. T., Giglio, W., Anesetti, R., Ottenweller, J. E., and Pogach, L. M. (1999). The detrimental effects of spinal cord injury on spermatogenesis in the rat is partially reversed by testosterone, but enhanced by follicle-stimulating hormone. *Endocrinology* 140, 1349–1355.
808. Huang, H. F., Li, M., Anesetti, R., Giglio, W., Ottenweller, J. E., and Pogach, L. M. (1999). Effects of spinal cord injury on spermatogenesis and the expression of messenger ribonucleic acid for Sertoli cell proteins in rat Sertoli cell-enriched testes. *Biol. Reprod.* 60, 635–641.
809. Chen, D., Hartwig, D. M., and Roth, E. J. (1999). Comparison of sperm quantity and quality in antegrade V retrograde ejaculates obtained by vibratory penile stimulation in males with spinal cord injury. *Am. J. Phys. Med. Rehabil.* 78, 46–51.
810. Mayerhofer, A., Frungieri, M. B., Bulling, A., and Fritz, S. (1999). Sources and function of neuronal signalling molecules in the gonads. *Medicina* 59, 542–545.
811. Geigerseder, C., Doepner, R., Thalhammer, A., Frungieri, M. B., Gamel-Didelon, K., Calandra, R. S., Kohn, F. M., and Mayerhofer, A. (2003). Evidence for a GABAergic system in rodent and human testis: local GABA production and GABA receptors. *Neuroendocrinology* 77, 314–323.

812. Mayerhofer, A., Amador, A. G., Steger, R. S., and Bartke, A. (1990). Testicular function after local injection of 6-hydroxydopamine or norepinephrine in the golden hamster. *J. Androl.* 11, 301–311.
813. Campos, M. B., Vitale, R. S., Calandra, R. S., and Chiocchio, S. R. (1990). Effect of bilateral denervation of the immature testis on testicular gonadotropin receptors and in vitro androgen production. *Neuroendocrinology* 57, 189–194.
814. McLachlan, R. I., Wreford, N. G., O'Donnell, L., de Kretser, D. M., and Robertson, D. M. (1996). The endocrine regulation of spermatogenesis: independent roles for testosterone and FSH. *J. Endocrinol.* 148, 1–9.
815. Tjioe, D. Y., and Steinberger, E. (1970). A quantitative study of the effect of ischaemia on the germinal epithelium of rat testes. *J. Reprod. Fertil.* 21, 489–494.
816. Turner, T. T., Tung, K. S., Tomomasa, H., and Wilson, L. W. (1997). Acute testicular ischemia results in germ cell-specific apoptosis in the rat. *Biol. Reprod.* 57, 1267–1274.
817. Lysiak, J. J., Turner, S. D., Nguyen, Q. A., Singbartl, K., Ley, K., and Turner, T. T. (2001). Essential role of neutrophils in germ cell-specific apoptosis following ischemia/reperfusion injury of the mouse testis. *Biol. Reprod.* 65, 718–725.
818. Takao, T., Culp, S. G., and De Souza, E. B. (1993). Reciprocal modulation of interleukin-1β (IL-1β) and IL-1 receptors by lipopolysaccharide (endotoxin) treatment in the mouse brain-endocrine-immune axis. *Endocrinology* 132, 1497–1504.
819. Siu, M. K., Lee, W. M., and Cheng, C. Y. (2003). The interplay of collagen IV, tumor necrosis factor-α, gelatinase B (matrix metalloprotease-9), and tissue inhibitor of metalloproteases-1 in the basal lamina regulates Sertoli cell-tight junction dynamics in the rat testis. *Endocrinology* 144, 371–387.
820. Lee, N. P., and Cheng, C. Y. (2003). Regulation of Sertoli cell tight junction dynamics in the rat testis via the nitric oxide synthase/soluble guanylate cyclase/3′,5′-cyclic guanosine monophosphate/protein kinase G signaling pathway: an *in vitro* study. *Endocrinology* 144, 3114–3129.
821. Hjertkvist, M., Läckgren, G., Plöen, L., and Bergh, A. (1993). Does HCG treatment induce inflammation-like changes in undescended testes in boys? *J. Pediatr. Surg.* 28, 254–258.
822. van Vliet, J., Rommerts, F. F., de Rooij, D. G., Buwalda, G., and Wensing, C. J. (1988). Reduction of testicular blood flow and focal degeneration of tissue in the rat after administration of human chorionic gonadotrophin. *J. Endocrinol.* 117, 51–57.
823. Sharpe, R. M. (1979). Gonadotrophin-induced accumulation of "interstitial fluid" in the rat testis. *J. Reprod. Fertil.* 55, 365–371.
824. Setchell, B. P., and Sharpe, R. M. (1981). Effect of injected human chorionic gonadotrophin on capillary permeability, extracellular fluid volume and the flow of lymph and blood in the testes of rats. *J. Endocrinol.* 91, 245–254.
825. Widmark, A., Damber, J. E., and Bergh, A. (1986). Relationship between human chorionic gonadotrophin-induced changes in testicular microcirculation and the formation of testicular interstitial fluid. *J. Endocrinol.* 109, 419–425.
826. Widmark, A., Bergh, A., Damber, J. E., and Smedegård, G. (1987). Leucocytes mediate the hCG-induced increase in testicular venular permeability. *Mol. Cell. Endocrinol.* 53, 25–31.
827. Bergh, A., Widmark, A., Damber, J. E., and Cajander, S. (1986). Are leukocytes involved in the human chorionic gonadotropin-induced increase in testicular vascular permeability? *Endocrinology* 119, 586–590.
828. Kerr, J. B., and Sharpe, R. M. (1989). Focal disruption of spermatogenesis in the testis of adult rats after a single administration of human chorionic gonadotrophin. *Cell Tissue Res.* 257, 163–169.
829. Kerr, J. B., and Sharpe, R. M. (1989). Macrophage activation enhances the human chorionic gonadotrophin-induced disruption of spermatogenesis in the rat. *J. Endocrinol.* 121, 285–292.
830. Setchell, B. P., and Rommerts, F. F. (1985). The importance of the Leydig cells in the vascular response to hCG in the rat testis. *Int. J. Androl.* 8, 436–440.
831. Veijola, M., and Rajaniemi, H. (1985). The hCG-induced increase in hormone uptake and interstitial fluid volume in the rat testis is not mediated by steroids, prostaglandins or protein synthesis. *Int. J. Androl.* 8, 69–79.
832. Sowerbutts, S. F., Jarvis, L. G., and Setchell, B. P. (1986). The increase in testicular vascular permeability induced by human chorionic gonadotrophin involves 5-hydroxytryptamine and possibly oestrogens, but not testosterone, prostaglandins, histamine or bradykinin. *Aust. J. Exp. Biol. Med. Sci.* 64, 137-147.
833. Bergh, A., Damber, J. E., and Hjertkvist, M. (1996). Human chorionic gonadotrophin-induced testicular inflammation may be related to increased sensitivity to interleukin-1. *Int. J. Androl.* 19, 229–236.
834. Veijola, M., and Rajaniemi, H. (1986). Luteinising hormones activate a factor(s) in testicular interstitial fluid which increases testicular vascular permeability. *Mol. Cell. Endocrinol.* 45, 113–118.
835. Collin, O., and Bergh, A. (1996). Leydig cells secrete factors which increase vascular permeability and endothelial cell proliferation. *Int. J. Androl.* 19, 221–228.
836. Bergh, A., Collin, O., and Lissbrant, E. (2001). Effects of acute graded reductions in testicular blood flow on testicular morphology in the adult rat. *Biol. Reprod.* 64, 13–20.
837. Markey, C. M., Jequier, A. M., Meyer, G. T., and Martin, G. B. (1994). Testicular morphology and androgen profiles following testicular ischaemia in rams. *J. Reprod. Fertil.* 101, 643–650.
838. Turner, T. T., Bang, H. J., and Lysiak, J. L. (2004). The molecular pathology of experimental testicular torsion suggests adjunct therapy to surgical repair. *J. Urol.* 172, 2574–2578.
839. Khan, S. A., Khan, S. J., and Dorrington, J. H. (1992). Interleukin-1 stimulates deoxyribonucleic acid synthesis in immature rat Leydig cells *in vitro*. *Endocrinology* 131, 1853–1857.
840. Walch, L., and Morris, P. L. (2002). Cyclooxygenase 2 pathway mediates IL-1β regulation of IL-1α, -1β, and IL-6 mRNA levels in Leydig cell progenitors. *Endocrinology* 143, 3276–3283.
841. Yee, J. B., and Hutson, J. C. (1985). Effects of testicular macrophage-conditioned medium on Leydig cells in culture. *Endocrinology* 116, 2682–2684.
842. Nes, W. D., Lukyanenko, Y. O., Jia, Z. H., Quideau, S., Howald, W. N., Pratum, T. K., West, R. R., and Hutson, J. C. (2000). Identification of the lipophilic factor produced by macrophages that stimulates steroidogenesis. *Endocrinology* 141, 953–958.
843. Suescun, M. O., Calandra, R. S., and Lustig, L. (2000). Effect of testicular macrophage conditioned media from rats with autoimmune orchitis on Leydig cell function. *Am. J. Reprod. Immunol.* 43, 116–123.
844. Mayerhofer, D., Mayerhofer, A., and Bartke, A. (1992). Isolation and culture of testicular macrophages from a seasonally breeding species, *Phodopus sungorus*: evidence for

functional differences between macrophages from active and regressed testes. *Int. J. Androl.* 15, 263–281.

845. Afane, M., Dubost, J. J., Sauvezie, B., Issoual, D., Dosgilbert, A., Grizard, G., and Boucher, D. (1998). Modulation of Leydig cell testosterone production by secretory products of macrophages. *Andrologia* 30, 71–78.
846. Kmicikiewicz, I., Wojtusiak, A., and Bilinska, B. (1999). The effect of testicular macrophages, macrophage-conditioned medium and interleukin-1α on bank vole Leydig cell steroidogenesis. *Exp. Clin. Endocrinol. Diabetes* 107, 262–271.
847. Lukyanenko, Y. O., Chen, J. J., and Hutson, J. C. (2001). Production of 25-hydroxycholesterol by testicular macrophages and its effects on Leydig cells. *Biol. Reprod.* 64, 790–796.
848. Lukyanenko, Y., Chen, J. J., and Hutson, J. C. (2002). Testosterone regulates 25-hydroxycholesterol production in testicular macrophages. *Biol. Reprod.* 67, 1435–1438.
849. Chen, J. J., Lukyanenko, Y., and Hutson, J. C. (2002). 25-hydroxycholesterol is produced by testicular macrophages during the early postnatal period and influences differentiation of Leydig cells in vitro. *Biol. Reprod.* 66, 1336–1341.
850. Haider, S. G. (2004). Cell biology of Leydig cells in the testis. *Int. Rev. Cytol.* 233, 181–241.
851. Hendry, W. F. (1989). Detection and treatment of antispermatozoal antibodies in men. *Reprod. Fertil. Dev.* 1, 205–220; discussion 220–202.
852. Kremer, J., and Jager, S. (1992). The significance of antisperm antibodies for sperm-cervical mucus interaction. *Hum. Reprod.* 7, 781–784.
853. Fjallbrant, B., and Obrant, O. (1968). Clinical and seminal findings in men with sperm antibodies. *Acta Obstet. Gynecol. Scand.* 47, 451–468.
854. Lombardo, F., Gandini, L., Dondero, F., and Lenzi, A. (2001). Immunology and immunopathology of the male genital tract: antisperm immunity in natural and assisted reproduction. *Hum. Reprod. Update* 7, 450–456.
855. Lin, W. W., Kim, E. D., Quesada, E. T., Lipshultz, L. I., and Coburn, M. (1998). Unilateral testicular injury from external trauma: evaluation of semen quality and endocrine parameters. *J. Urol.* 159, 841–843.
856. McLachlan, R. I. (2002). Basis, diagnosis and treatment of immunological infertility in men. *J. Reprod. Immunol.* 57, 35–45.
857. Patrizio, P., Silber, S. J., Ord, T., Moretti-Rojas, I., and Asch, R. H. (1992). Relationship of epididymal sperm antibodies to their in vitro fertilization capacity in men with congenital absence of the vas deferens. *Fertil. Steril.* 58, 1006–1010.
858. Jalal, H., Bahadur, G., Knowles, W., Jin, L., and Brink, N. (2004). Mumps epididymo-orchitis with prolonged detection of virus in semen and the development of anti-sperm antibodies. *J. Med. Virol.* 73, 147–150.
859. Quesada, E. M., Dukes, C. D., Deen, G. H., and Franklin, R. R. (1968). Genital infection and sperm agglutinating antibodies in infertile men. *J. Urol.* 99, 106–108.
860. Witkin, S. S., and Toth, A. (1983). Relationship between genital tract infections, sperm antibodies in seminal fluid, and infertility. *Fertil. Steril.* 40, 805–808.
861. Shulman, A., Shohat, B., Gillis, D., Yavetz, H., Homonnai, Z. T., and Paz, G. (1992). Mumps orchitis among soldiers: frequency, effect on sperm quality, and sperm antibodies. *Fertil. Steril.* 57, 1344–1346.
862. Eggert-Kruse, W., Probst, S., Rohr, G., Tilgen, W., and Runnebaum, B. (1996). Induction of immunoresponse by subclinical male genital tract infection? *Fertil. Steril.* 65, 1202–1209.
863. Ness, R. B., Markovic, N., Carlson, C. L., and Coughlin, M. T. (1997). Do men become infertile after having sexually transmitted urethritis? An epidemiologic examination. *Fertil. Steril.* 68, 205–213.
864. Kalaydjiev, S., Dimitrova, D., Tsvetkova, P., and Tsvetkov, D. (2001). Serum sperm antibodies unrelated to mumps orchitis. *Andrologia* 33, 69–70.
865. Paschke, R., Bertelsbeck, D. S., Tsalimalma, K., and Nieschlag, E. (1994). Association of sperm antibodies with other autoantibodies in infertile men. *Am. J. Reprod. Immunol.* 32, 88–94.
866. Baker, H. W., Clarke, G. N., McGowan, M. P., Koh, S. H., and Cauchi, M. N. (1985). Increased frequency of autoantibodies in men with sperm antibodies. *Fertil. Steril.* 43, 438–441.
867. Elder, M., Maclaren, N., and Riley, W. (1981). Gonadal autoantibodies in patients with hypogonadism and/or Addison's disease. *J. Clin. Endocrinol. Metab.* 52, 1137–1142.
868. Tsatsoulis, A., and Shalet, S. M. (1991). Antisperm antibodies in the polyglandular autoimmune (PGA) syndrome type I: response to cyclical steroid therapy. *Clin. Endocrinol.* 35, 299–303.
869. Sakamoto, Y., Himeno, K., Sanui, H., Yoshida, S., and Nomoto, K. (1985). Experimental allergic orchitis in mice: I. A new model induced by immunization without adjuvants. *Clin. Immunol. Immunopathol.* 37, 360–368.
870. Bell, E. B. (1969). Immunological control of fertility in the mouse: a comparison of systemic and intravaginal immunization. *J. Reprod. Fertil.* 18, 183–192.
871. Tung, K. S., Goldberg, E. H., and Goldberg, E. (1979). Immunobiological consequence of immunization of female mice with homologous spermatozoa: induction of infertility. *J. Reprod. Immunol.* 1, 145–158.
872. Itoh, M., Hiramine, C., Tokunaga, Y., Mukasa, A., and Hojo, K. (1991). A new murine model of autoimmune orchitis induced by immunization with viable syngeneic testicular germ cells alone: II. Immunohistochemical findings of fully-developed inflammatory lesion. *Autoimmunity* 10, 89–97.
873. Itoh, M., Miki, T., Takeuchi, Y., Miyake, M., and De Rooij, D. G. (1994). Immunohistological localization of autoantigens detected by serum autoantibodies from mice with experimental autoimmune orchitis without using adjuvants. *Arch. Androl.* 32, 45–52.
874. Kosuda, L. L., and Bigazzi, P. E. (1987). Animal models of testis autoimmunity. In *Immunology of the Male Reproductive System* (P. Bigazzi, Ed.), pp. 253–352. Marcel Dekker, New York.
875. Alexander, N. J., and Fulgham, D. L. (1978). Antibodies to spermatozoa in male monkeys: mode of action. *Fertil. Steril.* 30, 334–342.
876. Pedersen, J., Rubenson, A., and Nilsson, L. A. (1983). Formation of antisperm autoantibodies in rats vasectomized at prefertile age. *Scand. J. Urol. Nephrol.* 17, 277–281.
877. Flickinger, C. J., Bush, L. A., Williams, M. V., Naaby-Hansen, S., Howards, S. S., and Herr, J. C. (1999). Post-obstruction rat sperm autoantigens identified by two-dimensional gel electrophoresis and western blotting. *J. Reprod. Immunol.* 43, 35–53.
878. Handley, H. H., Herr, J. C., and Flickinger, C. J. (1991). Localization of post-vasectomy sperm autoantigens in the Lewis rat. *J. Reprod. Immunol.* 20, 205–220.
879. O'Rand, M. G., and Romrell, L. J. (1977). Appearance of cell surface auto- and isoantigens during spermatogenesis in the rabbit. *Dev. Biol.* 55, 347–358.
880. O'Rand, M. G., and Romrell, L. J. (1980). Appearance of regional surface autoantigens during spermatogenesis: comparison of anti-testis and anti-sperm autoantisera. *Dev. Biol.* 75, 431–441.
881. Kurpisz, M., Dobratz, B., and Alexander, N. J. (1993). Sperm antigens and reactivity of antisperm monoclonal antibodies in ELISA. *Andrologia* 25, 175–179.

882. Menge, A. C., Shoultz, G. K., Kelsey, D. E., Rutherford, P., and Lee, C. Y. (1987). Characterization of monoclonal antibodies against human sperm antigens by immunoassays including sperm function assays and epitope evaluation. *Am. J. Reprod. Immunol.* 13, 108–114.
883. Herr, J. C. (1996). Update on the Center for Recombinant Gamete Contraceptive Vaccinogens. *Am. J. Reprod. Immunol.* 35, 184–189.
884. Goldberg, E. (1975). Effects of immunization with LDH-X on fertility. *Acta Endocrinol. Suppl.* 194, 202–222.
885. Diekman, A. B., Norton, E. J., Westbrook, V. A., Klotz, K. L., Naaby-Hansen, S., and Herr, J. C. (2000). Anti-sperm antibodies from infertile patients and their cognate sperm antigens: a review. Identity between SAGA-1, the H6-3C4 antigen, and CD52. *Am. J. Reprod. Immunol.* 43, 134–143.
886. Diekman, A. B., and Herr, J. C. (1997). Sperm antigens and their use in the development of an immunocontraceptive. *Am. J. Reprod. Immunol.* 37, 111–117.
887. O'Hern, P. A., Bambra, C. S., Isahakia, M., and Goldberg, E. (1995). Reversible contraception in female baboons immunized with a synthetic epitope of sperm-specific lactate dehydrogenase. *Biol. Reprod.* 52, 331–339.
888. Barratt, C. L., Dunphy, B. C., McLeod, I., and Cooke, I. D. (1992). The poor prognostic value of low to moderate levels of sperm surface-bound antibodies. *Hum. Reprod.* 7, 95–98.
889. Eggert-Kruse, W., Hofsäss, A., Haury, E., Tilgen, W., Gerhard, I., and Runnebaum, B. (1991). Relationship between local anti-sperm antibodies and sperm-mucus interaction in vitro and in vivo. *Hum. Reprod.* 6, 267–276.
890. Clarke, G. N. (1988). Lack of correlation between the immunobead test and the enzyme-linked immunosorbent assay for sperm antibody detection. *Am. J. Reprod. Immunol. Microbiol.* 18, 44–46.
891. Mettler, L., Czuppon, A. B., Alexander, N., D'Almeida, M., Haas, G. G., Jr., Hjort, T., Moller Jensen, J., Ing, R., Jones, W. R., and Wang, S. X. (1985). Antibodies to spermatozoa and seminal plasma antigens detected by various enzyme-linked immunosorbent (ELISA) assays. *J. Reprod. Immunol.* 8, 301–312.
892. Eggert-Kruse, W., Leinhos, G., Gerhard, I., Tilgen, W., and Runnebaum, B. (1989). Prognostic value of in vitro sperm penetration into hormonally standardized human cervical mucus. *Fertil. Steril.* 51, 317–323.
893. Ekwere, P. D. (1995). Immunological infertility among Nigerian men: incidence of circulating antisperm auto-antibodies and some clinical observations: a preliminary report. *Br. J. Urol.* 76, 366–370.
894. Turek, P. J., and Lipshultz, L. I. (1994). Immunologic infertility. *Urol. Clin. North Am.* 21, 447–468.
895. Clarke, G. N., Bourne, H., and Baker, H. W. (1997). Intracytoplasmic sperm injection for treating infertility associated with sperm autoimmunity. *Fertil. Steril.* 68, 112–117.
896. Witkin, S. S. (1989). Failure of sperm-induced immunosuppression: association with antisperm antibodies in women. *Am. J. Obstet. Gynecol.* 160, 1166–1168.
897. Alexander, N. J., and Anderson, D. J. (1987). Immunology of semen. *Fertil. Steril.* 47, 192–205.
898. Gurka, G., and Rocklin, R. E. (1987). Reproductive immunology. *JAMA* 258, 2983–2987.
899. Billington, W. D. (1989). Maternal immune response to pregnancy. *Reprod. Fertil. Dev.* 1, 183–190.
900. Sacks, G., Sargent, I., and Redman, C. (1999). An innate view of human pregnancy. *Immunol. Today* 20, 114–118.
901. Robertson, S. A., and Sharkey, D. J. (2001). The role of semen in induction of maternal immune tolerance to pregnancy. *Semin. Immunol.* 13, 243–254.
902. Robertson, S. A., Bromfield, J. J., and Tremellen, K. P. (2003). Seminal "priming" for protection from pre-eclampsia-a unifying hypothesis. *J. Reprod. Immunol.* 59, 253–265.
903. Tung, K. S., Smith, S., Teuscher, C., Cook, C., and Anderson, R. E. (1987). Murine autoimmune oophoritis, epididymoorchitis, and gastritis induced by day 3 thymectomy. Immunopathology. *Am. J. Pathol.* 126, 293–302.
904. Tung, K. S., Smith, S., Matzner, P., Kasai, K., Oliver, J., Feuchter, F., and Anderson, R. E. (1987). Murine autoimmune oophoritis, epididymoorchitis, and gastritis induced by day 3 thymectomy: autoantibodies. *Am. J. Pathol.* 126, 303–314.
905. Boughton, B., and Spector, W. G. (1963). "Auto-immune" testicular lesions induced by injury to the contralateral testis and intradermal injection of adjuvant. *J. Pathol. Bacteriol.* 86, 69–74.
906. Freund, J., Lipton, M. M., and Thompson, G. E. (1953). Aspermatogenesis in the guinea pig induced by testicular tissue and adjuvant. *J. Exp. Med.* 97, 711–725.
907. Bishop, D. W., Narbaitz, R., and Lessof, M. (1961). Induced aspermatogenesis in adult guinea pigs injected with testicular antigen and adjuvant in neonatal stages. *Dev. Biol.* 3, 444–485.
908. Teuscher, C., Blankenhorn, E. P., and Hickey, W. F. (1987). Differential susceptibility to actively induced experimental allergic encephalomyelitis and experimental allergic orchitis among BALB/c substrains. *Cell. Immunol.* 110, 294–304.
909. Wakabayashi, A., Eishi, Y., and Nakamura, K. (1997). Regulation of experimental autoimmune orchitis by the presence or absence of testicular antigens during immunological development in SCID mice reconstituted with fetal liver cells. *Immunology* 92, 84–90.
910. Li, L., Itoh, M., Ablake, M., Macri, B., Bendtzen, K., and Nicoletti, F. (2002). Prevention of murine experimental autoimmune orchitis by recombinant human interleukin-6. *Clin. Immunol.* 102, 135–137.
911. Adekunle, A. O., Hickey, W. F., Smith, S. M., Tung, K. S., and Teuscher, C. (1987). Experimental allergic orchitis in mice: IV. Preliminary characterization of the major murine testis specific aspermatogenic autoantigen(s). *J. Reprod. Immunol.* 12, 49–62.
912. Suescun, M. O., Calandra, R. S., and Lustig, L. (1994). Alterations of testicular function after induced autoimmune orchitis in rats. *J. Androl.* 15, 442–448.
913. Tung, K. S., Primakoff, P., Woolman-Gamer, L., and Myles, D. G. (1997). Mechanism of infertility in male guinea pigs immunized with sperm PH-20. *Biol. Reprod.* 56, 1133–1141.
914. Yokochi, T., Ikeda, H., Inoue, Y., Kimura, Y., Ito, H., Fujii, Y., and Kato, N. (1990). Characterization of autoantigens relevant to experimental autoimmune orchitis (EAO) in mice immunized with a mixture of syngeneic testis homogenate and Klebsiella O3 lipopolysaccharide. *Am. J. Reprod. Immunol.* 22, 42–48.
915. Kojima, A., and Spencer, C. A. (1983). Genetic susceptibility to testicular autoimmunity: comparison between postthymectomy and postvasectomy models in mice. *Biol. Reprod.* 29, 195–205.
916. Bigazzi, P. E., Kosuda, L. L., Hsu, K. C., and Andres, G. A. (1976). Immune complex orchitis in vasectomized rabbits. *J. Exp. Med.* 143, 382–404.
917. Tung, K. S. (1978). Allergic orchitis lesions are adoptively transferred from vasoligated guinea pigs to syngeneic recipients. *Science* 201, 833–835.
918. Tung, K. S., and Alexander, N. J. (1980). Monocytic orchitis and aspermatogenesis in normal and vasectomized rhesus macaques (*Macaca mulatta*). *Am. J. Pathol.* 101, 17–30.

919. Tung, K. S., Ellis, L., Teuscher, C., Meng, A., Blaustein, J. C., Kohno, S., and Howell, R. (1981). The black mink (*Mustela vison*): a natural model of immunologic male infertility. *J. Exp. Med.* 154, 1016–1032.
920. Tung, K. S., Teuscher, C., and Meng, A. L. (1981). Autoimmunity to spermatozoa and the testis. *Immunol. Rev.* 55, 217–255.
921. Furbeth, C., Hubner, G., and Thoenes, G. H. (1989). Spontaneous immune complex orchitis in brown Norway rats. *Virchows Arch. B Cell. Pathol.* 57, 37–45.
922. Fritz, T. E., Lombard, S. A., Tyler, S. A., and Norris, W. P. (1976). Pathology and familial incidence of orchitis and its relation to thyroiditis in a closed beagle colony. *Exp. Mol. Pathol.* 24, 142–158.
923. Neufeld, M., Maclaren, N. K., and Blizzard, R. M. (1981). Two types of autoimmune Addison's disease associated with different polyglandular autoimmune (PGA) syndromes. *Medicine (Baltimore)* 60, 355–362.
924. Ahonen, P., Miettinen, A., and Perheentupa, J. (1987). Adrenal and steroidal cell antibodies in patients with autoimmune polyglandular disease type I and risk of adrenocortical and ovarian failure. *J. Clin. Endocrinol. Metab.* 64, 494–500.
925. Uibo, R., Aavik, E., Peterson, P., Perheentupa, J., Aranko, S., Pelkonen, R., and Krohn, K. J. (1994). Autoantibodies to cytochrome P450 enzymes P450scc, P450c17, and P450c21 in autoimmune polyglandular disease types I and II and in isolated Addison's disease. *J. Clin. Endocrinol. Metab.* 78, 323–328.
926. Soderbergh, A., Winqvist, O., Norheim, I., Rorsman, F., Husebye, E. S., Dolva, O., Karlsson, F. A., and Kampe, O. (1996). Adrenal autoantibodies and organ-specific autoimmunity in patients with Addison's disease. *Clin. Endocrinol.* 45, 453–460.
927. Arif, S., Vallian, S., Farzaneh, F., Zanone, M. M., James, S. L., Pietropaolo, M., Hettiarachchi, S., Vergani, D., Conway, G. S., and Peakman, M. (1996). Identification of 3 β-hydroxysteroid dehydrogenase as a novel target of steroid cell autoantibodies: association of autoantibodies with endocrine autoimmune disease. *J. Clin. Endocrinol. Metab.* 81, 4439–4445.
928. Song, Y. H., Li, Y., and Maclaren, N. K. (1996). The nature of autoantigens targeted in autoimmune endocrine diseases. *Immunol. Today* 17, 232–238.
929. Tung, K. S. K., and Alexander, N. J. (1977). Autoimmune reactions in the testis. In *The Testis: Advances in Physiology, Biochemistry and Function* (A. D. Johnson, W. R. Gomes, and L. L. Vandemark, Eds.), Vol. IV, pp. 491–516. Academic Press, New York.
930. Tung, K. S. K. (1980). Autoimmunity of the testis. In *Immunological Aspects of Infertility and Fertility Regulation* (D. S. Dhindsa and G. F. B. Schumacher, Eds.), pp. 33–91. Elsevier/North-Holland Press, New York.
931. Matsuzaki, G., Sonoda, K. H., Mukasa, A., Yamada, H., Nakamura, T., Ikebe, H., Hamano, S., and Nomoto, K. (1997). The characterization of testicular cell (TC)-specific T-cell clones induced by intratesticular *Listeria monocytogenes* infection: TC-specific T cells with atypical cytokine profile transfer orchitis. *Immunology* 91, 520–528.
932. Mahi-Brown, C. A., and Tung, K. S. (1989). Activation requirements of donor T cells and host T cell recruitment in adoptive transfer of murine experimental autoimmune orchitis (EAO). *Cell. Immunol.* 124, 368–379.
933. Itoh, M., De-Rooij, D., and Takeuchi, Y. (1995). Mode of inflammatory cell infiltration in testes of mice injected with syngeneic testicular germ cells without adjuvant. *J. Anat.* 187, 671–679.
934. Itoh, M., Yano, A., Xie, Q., Iwahashi, K., Takeuchi, Y., Meroni, P. L., and Nicoletti, F. (1998). Essential pathogenic role for endogenous interferon-gamma (IFN-γ) during disease onset phase of murine experimental autoimmune orchitis. I. *In vivo* studies. *Clin. Exp. Immunol.* 111, 513–520.
935. Lewis-Jones, D. I., Lynch, R. V., Kerrigan, D. D., and Davies, I. (1987). Long-term study of the immuno-pathological consequences of sympathetic orchiopathia in the rat. *J. Reprod. Fertil.* 80, 641–647.
936. Mukasa, A., Yoshida, H., Kobayashi, N., Matsuzaki, G., and Nomoto, K. (1998). γδ T cells in infection-induced and autoimmune-induced testicular inflammation. *Immunology* 95, 395–401.
937. Kaya, M., and Harrison, R. G. (1975). An analysis of the effect of ischaemia on testicular ultrastructure. *J. Pathol.* 117, 105–117.
938. Wallace, D. M., Gunter, P. A., Landon, G. V., Pugh, R. C., and Hendry, W. F. (1982). Sympathetic orchiopathia: an experimental and clinical study. *Br. J. Urol.* 54, 765–768.
939. Lewis-Jones, D. I., Moreno de Marval, M., and Harrison, R. G. (1982). Impairment of rat spermatogenesis following unilateral experimental ischemia. *Fertil. Steril.* 38, 482–490.
940. Cerasaro, T. S., Nachtsheim, D. A., Otero, F., and Parsons, C. L. (1984). The effect of testicular torsion on contralateral testis and the production of antisperm antibodies in rabbits. *J. Urol.* 132, 577–579.
941. Suominen, J. J. (1995). Sympathetic auto-immune orchitis. *Andrologia* 27, 213–216.
942. Andrada, J. A., von der Walde, F., Hoschoian, J. C., Comini, E., and Mancini, E. (1977). Immunological studies in patients with mumps orchitis. *Andrologia* 9, 207–215.
943. Casella, R., Leibundgut, B., Lehmann, K., and Gasser, T. C. (1997). Mumps orchitis: report of a mini-epidemic. *J. Urol.* 158, 2158–2161.
944. Manson, A. L. (1990). Mumps orchitis. *Urology* 36, 355–358.
945. Artenstein, M. S., Cadigan, F. C., Jr., and Buescher, E. L. (1965). Clinical and epidemiological features of Coxsackie group B virus infections. *Ann. Intern. Med.* 63, 597–603.
946. Willems, W. R., Hornig, C., Bauer, H., and Klingmuller, V. (1978). A case of Coxsackie A9 virus infection with orchitis. *J. Med. Virol.* 3, 137–140.
947. Liu, H. C., Tsai, T. C., Chang, P. Y., and Shih, B. F. (1994). Varicella orchitis: report of two cases and review of the literature. *Pediatr. Infect. Dis. J.* 13, 748–750.
948. Pudney, J., and Anderson, D. (1991). Orchitis and human immunodeficiency virus type 1 infected cells in reproductive tissues from men with the acquired immune deficiency syndrome. *Am. J. Pathol.* 139, 149–160.
949. Weyrauch, H. M., and Gass, H. (1946). Urogenital complications of dengue fever. *J. Urol.* 55, 90–93.
950. Weiner, R. L. (1997). Orchitis: a rare complication of infectious mononucleosis. *Pediatr. Infect. Dis. J.* 16, 1008–1009.
951. Parnes, L. R. (1968). Infectious mononucleosis with orchitis: report of a case. *J. Am. Coll. Health Assoc.* 17, 90–92.
952. Krieger, J. N. (1984). Epididymitis, orchitis, and related conditions. *Sex. Transm. Dis.* 11, 173–181.
953. Marill, F. G., Marill, R. M., and Sayag, J. (1970). Remarques a propos de cinq cas de syphilis du testicule. *Bull. Soc. Franc. Dermatol. Syphil.* 77, 241–245.
954. Singh, R., Kaur, D., and Parameswaran, M. (1971). Infantile congenital syphilis: presenting with bilateral orchitis. *Br. J. Vener. Dis.* 47, 206–208.
955. Chang, A. R., and Penman, H. G. (1970). Granulomatous orchitis: a report of two cases. *Aust. N. Z. J. Surg.* 40, 79–84.
956. Zeithofer, J., Bibus, B., and Swoboda, G. (1970). Orchitis granulomatosa. *Z. Urol. Nephrol.* 63, 841–846.

957. Akhtar, M., Ali, M. A., and Mackey, D. M. (1980). Lepromatous leprosy presenting as orchitis. *Am. J. Clin. Pathol.* 73, 712–715.
958. Watson, R. A., Gangai, M. P., and Skinsnes, O. K. (1974). Genitourinary leprosy. *Urol. Int.* 29, 312–326.
959. Wall, J. R., and Wright, D. J. (1974). Antibodies against testicular germinal cells in lepromatous leprosy. *Clin. Exp. Immunol.* 17, 51–59.
960. Jenkin, G. A., Choo, M., Hosking, P., and Johnson, P. D. (1998). Candidal epididymo-orchitis: case report and review. *Clin. Infect. Dis.* 26, 942–945.
961. Shamaei-Tousi, A., Collin, O., Bergh, A., and Bergström, S. (2001). Testicular damage by microcirculatory disruption and colonization of an immune-privileged site during *Borrelia crocidurae* infection. *J. Exp. Med.* 193, 995–1004.
962. Blanchard, K. T., and Boekelheide, K. (1997). Adenovirus-mediated gene transfer to rat testis in vivo. *Biol. Reprod.* 56, 495–500.
963. Wicher, K., Wicher, V., Nakeeb, S. M., and Dubiski, S. (1983). Studies of rabbit testes infected with *Treponema pallidum*: I. Immunopathology. *Br. J. Vener. Dis.* 59, 349–358.
964. Audring, H., Klug, H., Bollmann, R., Sokolowska-Köhler, W., and Engel, S. (1989). Ureaplasma urealyticum and male infertility: an animal model: II. Morphologic changes of testicular tissue at light microscopic level and electron microscopic findings. *Andrologia* 21, 66–75.
965. Melaine, N., Ruffault, A., Dejucq-Rainsford, N., and Jégou, B. (2003). Experimental inoculation of the adult rat testis with Sendai virus: effect on testicular morphology and leukocyte population. *Hum. Reprod.* 18, 1574–1579.
966. Ang, C. W., Jacobs, B. C., and Laman, J. D. (2004). The Guillain-Barré syndrome: a true case of molecular mimicry. *Trends Immunol.* 25, 61–66.
967. Rabin, B. S., Nankin, H. R., and Troen, P. (1977). Immunologic studies of patients with idiopathic oligospermia. In *The Testis in Normal and Infertile Men* (P. Troen and H. R. Nankin, Eds.), pp. 435–444. Raven Press, New York,.
968. Salomon, F., and Hedinger, C. E. (1987). Histopathology of immunologic lesions of the human testis. In *Immunology of the Male Reproductive Tract* (P. E. Bigazzi, Ed.), pp. 203–231. Marcel Dekker, New York.
969. Lehmann, D., and Emmons, L. R. (1989). Immunological phenomena observed in the testis and their possible role in infertility. *Am. J. Reprod. Immunol.* 19, 43–52.
970. Salomon, F., Saremaslani, P., Jakob, M., and Hedinger, C. E. (1982). Immune complex orchitis in infertile men. Immunoelectron microscopy of abnormal basement membrane structures. *Lab. Invest.* 47, 555–567.
971. Lehmann, D., and Muller, H. (1987). Analysis of the autoimmune response in an "in situ" carcinoma of the testis. *Int. J. Androl.* 10, 163–168.
972. Suominen, J., and Soderstrom, K. O. (1982). Lymphocyte infiltration in human testicular biopsies. *Int. J. Androl.* 5, 461–466.
973. Pannek, J., and Haupt, G. (1997). Orchitis due to vasculitis in autoimmune diseases. *Scand. J. Rheumatol.* 26, 151–154.
974. Kojima, A., and Prehn, R. T. (1981). Genetic susceptibility to post-thymectomy autoimmune diseases in mice. *Immunogenetics* 14, 15–27.
975. Roper, R. J., Doerge, R. W., Call, S. B., Tung, K. S., Hickey, W. F., and Teuscher, C. (1998). Autoimmune orchitis, epididymitis, and vasitis are immunogenetically distinct lesions. *Am. J. Pathol.* 152, 1337–1345.
976. Person, P. L., Snoek, M., Demant, P., Woodward, S. R., and Teuscher, C. (1992). The immunogenetics of susceptibility and resistance to murine experimental allergic orchitis. *Reg. Immunol.* 4, 284–297.
977. Kasai, K., Teuscher, C., Smith, S., Matzner, P., and Tung, K. S. (1987). Strain variations in anti-sperm antibody responses and anti-fertility effects in inbred mice. *Biol. Reprod.* 36, 1085–1094.
978. Law, H. Y., Bodmer, W. F., Mathews, J. D., and Skegg, D. C. (1979). The immune response to vasectomy and its relation to the HLA system. *Tissue Antigens* 14, 115–139.
979. Butterfield, R. J., Sudweeks, J. D., Blankenhorn, E. P., Korngold, R., Marini, J. C., Todd, J. A., Roper, R. J., and Teuscher, C. (1998). New genetic loci that control susceptibility and symptoms of experimental allergic encephalomyelitis in inbred mice. *J. Immunol.* 161, 1860–1867.
980. Meeker, N. D., Hickey, W. F., Korngold, R., Hansen, W. K., Sudweeks, J. D., Wardell, B. B., Griffith, J. S., and Teuscher, C. (1995). Multiple loci govern the bone marrow-derived immunoregulatory mechanism controlling dominant resistance to autoimmune orchitis. *Proc. Natl. Acad. Sci. U. S. A.* 92, 5684–5688.
981. Sudweeks, J. D., Todd, J. A., Blankenhorn, E. P., Wardell, B. B., Woodward, S. R., Meeker, N. D., Estes, S. S., and Teuscher, C. (1993). Locus controlling *Bordetelia pertussis*-induced histamine sensitization (Bphs), an autoimmune disease-susceptibility gene, maps distal to T-cell receptor beta-chain gene on mouse chromosome 6. *Proc. Natl. Acad. Sci. U. S. A.* 90, 3700–3704.
982. Maclaren, N., Chen, Q. Y., Kukreja, A., Marker, J., Zhang, C. H., and Sun, Z. S. (2001). Autoimmune hypogonadism as part of an autoimmune polyglandular syndrome. *J. Soc. Gynecol. Invest.* 8, S52–S54.
983. Pitkanen, J., and Peterson, P. (2003). Autoimmune regulator: from loss of function to autoimmunity. *Genes Immun.* 4, 12–21.
984. Head, J. R., and Billingham, R. E. (1985). Immunologically privileged sites in transplantation immunology and oncology. *Perspect. Biol. Med.* 29, 115–131.
985. Jégou, B. (1992). The Sertoli cell. In *The Testis* (D. M. de Kretser, Ed.), pp. 273–311. Baillière Tindall, London.
986. McLay, R. N., Banks, W. A., and Kastin, A. J. (1997). Granulocyte macrophage-colony stimulating factor crosses the blood-testis barrier in mice. *Biol. Reprod.* 57, 822–826.
987. Pöllänen, P. P., and Setchell, B. P. (1989). Microvascular permeability to IgG in the rat testis at puberty. *Int. J. Androl.* 12, 206–218.
988. Plotkin, S. R., Banks, W. A., Maness, L. M., and Kastin, A. J. (2000). Differential transport of rat and human interleukin-1α across the blood-brain barrier and blood-testis barrier in rats. *Brain Res.* 881, 57–61.
989. Bustamante, J. C., and Setchell, B. P. (2000). The permeability of the microvasculature of the perfused rat testis to small hydrophilic substances. *J. Androl.* 21, 444–451.
990. Banks, W. A., and Kastin, A. J. (1992). Human interleukin-1α crosses the blood-testis barriers of the mouse. *J. Androl.* 13, 254–259.
991. Hedger, M. P., and Hettiarachchi, S. (1994). Measurement of immunoglobulin G levels in adult rat testicular interstitial fluid and serum. *J. Androl.* 15, 583–590.
992. Hedger, M. P., and Muir, J. A. (2000). Differential actions of gonadotropin-releasing hormone and human chorionic gonadotropin on interstitial fluid volume and immunoglobulin G concentrations in adult rat testis. *J. Androl.* 21, 747–752.
993. Hedger, M. P. (1997). Testicular leukocytes: what are they doing? *Rev. Reprod.* 2, 38–47.
994. Holash, J. A., Harik, S. I., Perry, G., and Stewart, P. A. (1993). Barrier properties of testis microvessels. *Proc. Natl. Acad. Sci. U. S. A.* 90, 11069–11073.

995. Bart, J., Groen, H. J., van der Graaf, W. T., Hollema, H., Hendrikse, N. H., Vaalburg, W., Sleijfer, D. T., and de Vries, E. G. (2002). An oncological view on the blood-testis barrier. *Lancet Oncol.* 3, 357–363.
996. Bart, J., Hollema, H., Groen, H. J., de Vries, E. G., Hendrikse, N. H., Sleijfer, D. T., Wegman, T. D., Vaalburg, W., and van der Graaf, W. T. (2004). The distribution of drug-efflux pumps, P-gp, BCRP, MRP1 and MRP2, in the normal blood-testis barrier and in primary testicular tumours. *Eur. J. Cancer* 40, 2064–2070.
997. Wijnholds, J., Scheffer, G. L., van der Valk, M., van der Valk, P., Beijnen, J. H., Scheper, R. J., and Borst, P. (1998). Multidrug resistance protein 1 protects the oropharyngeal mucosal layer and the testicular tubules against drug-induced damage. *J. Exp. Med.* 188, 797–808.
998. Melaine, N., Lienard, M. O., Dorval, I., Le Goascogne, C., Lejeune, H., and Jégou, B. (2002). Multidrug resistance genes and p-glycoprotein in the testis of the rat, mouse, guinea pig, and human. *Biol. Reprod.* 67, 1699–1707.
999. Saari, T., Jahnukainen, K., and Pöllänen, P. (1996). Autoantigenicity of the basal compartment of seminiferous tubules in the rat. *J. Reprod. Immunol.* 31, 65–79.
1000. Yule, T. D., Montoya, G. D., Russell, L. D., Williams, T. M., and Tung, K. S. (1988). Autoantigenic germ cells exist outside the blood testis barrier. *J. Immunol.* 141, 1161–1167.
1001. Itoh, M., Takeuchi, Y., and De Rooij, D. (1995). Histopathology of the seminiferous tubules in mice injected with syngeneic testicular germ cells alone. *Arch. Androl.* 35, 93–103.
1002. Pelletier, R. M., and Byers, S. W. (1992). The blood-testis barrier and Sertoli cell junctions: structural considerations. *Microsc. Res. Tech.* 20, 3–33.
1003. Pelletier, R. M. (1986). Cyclic formation and decay of the blood-testis barrier in the mink (*Mustela vison*), a seasonal breeder. *Am. J. Anat.* 175, 91–117.
1004. Maddocks, S., and Setchell, B. P. (1988). The rejection of thyroid allografts in the ovine testis. *Immunol. Cell Biol.* 66, 1–8.
1005. Setchell, B. P., Granholm, T., and Ritzén, E. M. (1995). Failure of thyroid allografts to function in the testes of cynomolgus monkeys. *J. Reprod. Immunol.* 28, 75–80.
1006. Selawry, H. P., Whittington, K. B., and Forster, H. G. (1988). Intratesticular islet xenograft survival in relation to tissue cyclosporine levels. *Am. J. Med. Sci.* 295, 497–502.
1007. Turek, P. J., Malkowicz, S. B., Tomaszewski, J. E., Wein, A. J., and Peehl, D. (1996). The role of the Sertoli cell in active immunosuppression in the human testis. *Br. J. Urol.* 77, 891–895.
1008. Pöllänen, P., Jahnukainen, K., Punnonen, J., and Sainio-Pöllänen, S. (1992). Ontogeny of immunosuppressive activity, MHC antigens and leukocytes in the rat testis. *J. Reprod. Immunol.* 21, 257–274.
1009. Anderson, D. J., Narayan, P., and DeWolf, W. C. (1984). Major histocompatibility antigens are not detectable on post-meiotic human testicular germ cells. *J. Immunol.* 133, 1962–1965.
1010. Castilla, J. A., Gil, T., Rodriguez, F., Molina, J., Samaniego, F., Vergara, F., and Herruzo, A. J. (1993). Lack of expression of HLA antigens on immature germ cells from ejaculates with antisperm antibodies. *Am. J. Reprod. Immunol.* 30, 9–14.
1011. Tomita, Y., Kimura, M., Tanikawa, T., Nishiyama, T., Morishita, H., Takeda, M., Fujiwara, M., and Sato, S. (1993). Immunohistochemical detection of intercellular adhesion molecule-1 (ICAM-1) and major histocompatibility complex class I antigens in seminoma. *J. Urol.* 149, 659–663.
1012. Kohno, S., Ziparo, E., Marek, L. F., and Tung, K. S. (1983). Murine Sertoli cells: major histocompatibility antigens and glycoconjugates. *J. Reprod. Immunol.* 5, 339–350.
1013. Martin-Villa, J. M., Longas, J., and Arnaiz-Villena, A. (1999). Cyclic expression of HLA class I and II molecules on the surface of purified human spermatozoa and their control by serum inhibin B levels. *Biol. Reprod.* 61, 1381–1386.
1014. Martín-Villa, J. M., Luque, I., Martínez-Quiles, N., Corell, A., Regueiro, J. R., Timón, M., and Arnaiz-Villena, A. (1996). Diploid expression of human leukocyte antigen class I and class II molecules on spermatozoa and their cyclic inverse correlation with inhibin concentration. *Biol. Reprod.* 55, 620–629.
1015. Fellous, M., and Dausset, J. (1970). Probable haploid expression of HL-A antigens on human spermatozoon. *Nature* 225, 191–193.
1016. Halim, A., Abbasi, K., and Festenstein, H. (1974). The expression of the HL-A antigens of human spermatozoa. *Tissue Antigens* 4, 1–6.
1017. Jassim, A., Ollier, W., Payne, A., Biro, A., Oliver, R. T., and Festenstein, H. (1989). Analysis of HLA antigens on germ cells in human semen. *Eur. J. Immunol.* 19, 1215–1220.
1018. Fiszer, D., Ulbrecht, M., Fernandez, N., Johnson, J. P., Weiss, E. H., and Kurpisz, M. (1997). Analysis of HLA class Ib gene expression in male gametogenic cells. *Eur. J. Immunol.* 27, 1691–1695.
1019. Harding, F. A., and Allison, J. P. (1993). CD28–B7 interactions allow the induction of $CD8^+$ cytotoxic T lymphocytes in the absence of exogenous help. *J. Exp. Med.* 177, 1791–1796.
1020. Schwartz, R. H. (1990). A cell culture model for T lymphocyte clonal anergy. *Science* 248, 1349–1356.
1021. Garza, K. M., Agersborg, S. S., Baker, E., and Tung, K. S. (2000). Persistence of physiological self antigen is required for the regulation of self tolerance. *J. Immunol.* 164, 3982–3989.
1022. Garza, K. M., Chan, S. M., Suri, R., Nguyen, L. T., Odermatt, B., Schoenberger, S. P., and Ohashi, P. S. (2000). Role of antigen-presenting cells in mediating tolerance and autoimmunity. *J. Exp. Med.* 191, 2021–2027.
1023. Petroff, M. G., Chen, L., Phillips, T. A., and Hunt, J. S. (2002). B7 family molecules: novel immunomodulators at the maternal-fetal interface. *Placenta* 23 (Suppl. A), S95–S101.
1024. Fournel, S., Aguerre-Girr, M., Huc, X., Lenfant, F., Alam, A., Toubert, A., Bensussan, A., and Le Bouteiller, P. (2000). Cutting edge: soluble HLA-G1 triggers CD95/CD95 ligand-mediated apoptosis in activated $CD8^+$ cells by interacting with CD8. *J. Immunol.* 164, 6100–6104.
1025. Bellgrau, D., Gold, D., Selawry, H., Moore, J., Franzusoff, A., and Duke, R. C. (1995). A role for CD95 ligand in preventing graft rejection. *Nature* 377, 630–632.
1026. Griffith, T. S., Brunner, T., Fletcher, S. M., Green, D. R., and Ferguson, T. A. (1995). Fas ligand-induced apoptosis as a mechanism of immune privilege. *Science* 270, 1189–1192.
1027. Griffith, T. S., and Ferguson, T. A. (1997). The role of FasL-induced apoptosis in immune privilege. *Immunol. Today* 18, 240–244.
1028. Allison, J., Georgiou, H. M., Strasser, A., and Vaux, D. L. (1997). Transgenic expression of CD95 ligand on islet beta cells induces a granulocytic infiltration but does not confer immune privilege upon islet allografts. *Proc. Natl. Acad. Sci. U. S. A.* 94, 3943–3947.
1029. Kang, S. M., Schneider, D. B., Lin, Z., Hanahan, D., Dichek, D. A., Stock, P. G., and Baekkeskov, S. (1997). Fas ligand expression in islets of Langerhans does not confer immune privilege and instead targets them for rapid destruction. *Nat. Med.* 3, 738–743.

1030. Seino, K., Ogino, T., Fukunaga, K., Taniguchi, H., Takada, Y., Yuzawa, K., Otsuka, M., Yagita, H., Okumura, K., and Fukao, K. (1999). Attempts to reveal the mechanism of CD95-ligand-mediated inflammation. *Transplant. Proc.* 31, 1942–1943.

1031. Xerri, L., Devilard, E., Hassoun, J., Mawas, C., and Birg, F. (1997). Fas ligand is not only expressed in immune privileged human organs but is also coexpressed with Fas in various epithelial tissues. *Mol. Pathol.* 50, 87–91.

1032. Kimmel, S. G., Ohbatake, M., Kushida, M., Merguerian, P., Clarke, I. D., and Kim, P. C. (2000). Murine xenogeneic immune responses to the human testis: a presumed immune-privileged tissue. *Transplantation* 69, 1075–1084.

1033. French, L. E., Hahne, M., Viard, I., Radlgruber, G., Zanone, R., Becker, K., Müller, C., and Tschopp, J. (1996). Fas and Fas ligand in embryos and adult mice: ligand expression in several immune-privileged tissues and coexpression in adult tissues characterized by apoptotic cell turnover. *J. Cell Biol.* 133, 335–343.

1034. Woolveridge, I., de Boer-Brouwer, M., Taylor, M. F., Teerds, K. J., Wu, F. C., and Morris, I. D. (1999). Apoptosis in the rat spermatogenic epithelium following androgen withdrawal: changes in apoptosis-related genes. *Biol. Reprod.* 60, 461–470.

1035. D'Alessio, A., Riccioli, A., Lauretti, P., Padula, F., Muciaccia, B., De Cesaris, P., Filippini, A., Nagata, S., and Ziparo, E. (2001). Testicular FasL is expressed by sperm cells. *Proc. Natl. Acad. Sci. U. S. A.* 98, 3316–3321.

1036. Lee, J., Richburg, J. H., Younkin, S. C., and Boekelheide, K. (1997). The Fas system is a key regulator of germ cell apoptosis in the testis. *Endocrinology* 138, 2081–2088.

1037. Sanberg, P. R., Saporta, S., Borlongan, C. V., Othberg, A. I., Allen, R. C., and Cameron, D. F. (1997). The testis-derived cultured Sertoli cell as a natural Fas-L secreting cell for immunosuppressive cellular therapy. *Cell Transplant.* 6, 191–193.

1038. Restifo, N. P. (2000). Not so Fas: Re-evaluating the mechanisms of immune privilege and tumor escape. *Nat. Med.* 6, 493–495.

1039. Nandi, S., Banerjee, P. P., and Zirkin, B. R. (1999). Germ cell apoptosis in the testes of Sprague Dawley rats following testosterone withdrawal by ethane 1,2-dimethanesulfonate administration: relationship to Fas? *Biol. Reprod.* 61, 70–75.

1040. Ohta, Y., Nishikawa, A., Fukazawa, Y., Urushitani, H., Matsuzawa, A., Nishina, Y., and Iguchi, T. (1996). Apoptosis in adult mouse testis induced by experimental cryptorchidism. *Acta. Anat.* 157, 195–204.

1041. Taguchi, O., and Nishizuka, Y. (1981). Experimental autoimmune orchitis after neonatal thymectomy in the mouse. *Clin. Exp. Immunol.* 46, 425–434.

1042. Lam-Tse, W. K., Lernmark, A., and Drexhage, H. A. (2002). Animal models of endocrine/organ-specific autoimmune diseases: do they really help us to understand human autoimmunity? *Springer Semin. Immunopathol.* 24, 297–321.

1043. Lipscomb, H. L., Gardner, P. J., and Sharp, J. G. (1979). The effect of neonatal thymectomy on the induction of autoimmune orchitis in rats. *J. Reprod. Immunol.* 1, 209–217.

1044. Kriegel, M. A., Lohmann, T., Gabler, C., Blank, N., Kalden, J. R., and Lorenz, H. M. (2004). Defective suppressor function of human $CD4^+$ $CD25^+$ regulatory T cells in autoimmune polyglandular syndrome type II. *J. Exp. Med.* 199, 1285–1291.

1045. Kemeny, D. M., Noble, A., Holmes, B. J., Diaz-Sanchez, D., and Lee, T. H. (1995). The role of $CD8^+$ T cells in immunoglobulin E regulation. *Allergy* 50, 9–14.

1046. Chang, C. C., Ciubotariu, R., Manavalan, J. S., Yuan, J., Colovai, A. I., Piazza, F., Lederman, S., Colonna, M., Cortesini, R., Dalla-Favera, R., and Suciu-Foca, N. (2002). Tolerization of dendritic cells by T(S) cells: the crucial role of inhibitory receptors ILT3 and ILT4. *Nat. Immunol.* 3, 237–243.

1047. Sakaguchi, S., Sakaguchi, N., Asano, M., Itoh, M., and Toda, M. (1995). Immunologic self-tolerance maintained by activated T cells expressing IL-2 receptor α-chains (CD25): breakdown of a single mechanism of self-tolerance causes various autoimmune diseases. *J. Immunol.* 155, 1151–1164.

1048. Takahashi, T., Kuniyasu, Y., Toda, M., Sakaguchi, N., Itoh, M., Iwata, M., Shimizu, J., and Sakaguchi, S. (1998). Immunologic self-tolerance maintained by $CD25^+$ $CD4^+$ naturally anergic and suppressive T cells: induction of autoimmune disease by breaking their anergic/suppressive state. *Int. Immunol.* 10, 1969–1980.

1049. Liu, Z., Tugulea, S., Cortesini, R., and Suciu-Foca, N. (1998). Specific suppression of T helper alloreactivity by allo-MHC class I-restricted $CD8^+$ $CD28^-$ T cells. *Int. Immunol.* 10, 775–783.

1050. Thornton, A. M., and Shevach, E. M. (1998). $CD4^+$ $CD25^+$ immunoregulatory T cells suppress polyclonal T cell activation in vitro by inhibiting interleukin 2 production. *J. Exp. Med.* 188, 287–296.

1051. Piccirillo, C. A., Letterio, J. J., Thornton, A. M., McHugh, R. S., Mamura, M., Mizuhara, H., and Shevach, E. M. (2002). $CD4^+CD25^+$ regulatory T cells can mediate suppressor function in the absence of transforming growth factor $\beta 1$ production and responsiveness. *J. Exp. Med.* 196, 237–246.

1052. Nakamura, K., Kitani, A., and Strober, W. (2001). Cell contact-dependent immunosuppression by $CD4^+CD25^+$ regulatory T cells is mediated by cell surface-bound transforming growth factor β. *J. Exp. Med.* 194, 629–644.

1053. Itoh, M., Mukasa, A., Tokunaga, Y., Hiramine, C., and Hojo, K. (1992). Suppression of efferent limb of testicular autoimmune response by a regulatory $CD4^+$ T cell line in mice. *Clin. Exp. Immunol.* 87, 455–460.

1054. Peng, B., Yoshitoshi, T., and Shichi, H. (1992). Suppression of experimental autoimmune uveoretinitis by intraorchidic administration of S-antigen. *Autoimmunity* 14, 149–153.

1055. Ren, J., Singh, A. K., Gregerson, D. S., and Shichi, H. (1996). Induction of immunotolerance in rats by intratesticular administration of an eicosapeptide of bovine S-antigen. *Autoimmunity* 25, 19–31.

1056. Li, H., Ren, J., Dhabuwala, C. B., and Shichi, H. (1997). Immunotolerance induced by intratesticular antigen priming: expression of TGF-β, Fas and Fas ligand. *Ocular Immunol. Inflamm.* 5, 75–84.

1057. Ditzian-Kadanoff, R. (1999). Testicular-associated immune deviation and prevention of adjuvant-induced arthritis by three tolerization methods. *Scand. J. Immunol.* 50, 150–158.

1058. Veräjänkorva, E., Setälä, N., Teros, T., Salmi, A. A., and Pöllänen, P. (2002). Testicular-associated immune deviation: flushing of the testicular lymph sinusoids induces immunosuppression and inhibits formation of EAE in SJL mice. *Scand. J. Immunol.* 55, 478–483.

1059. Piccioli, D., Sbrana, S., Melandri, E., and Valiante, N. M. (2002). Contact-dependent stimulation and inhibition of dendritic cells by natural killer cells. *J. Exp. Med.* 195, 335–341.

1060. Andrews, D. M., Scalzo, A. A., Yokoyama, W. M., Smyth, M. J., and Degli-Esposti, M. A. (2003). Functional interactions between dendritic cells and NK cells during viral infection. *Nat. Immunol.* 4, 175–181.

1061. Mocikat, R., Braumuller, H., Gumy, A., Egeter, O., Ziegler, H., Reusch, U., Bubeck, A., Louis, J., Mailhammer, R., Riethmuller, G., Koszinowski, U., and Rocken, M. (2003).

Natural killer cells activated by MHC class I^{low} targets prime dendritic cells to induce protective CD8 T cell responses. *Immunity* 19, 561–569.

1062. Nakamura, T., Sonoda, K. H., Faunce, D. E., Gumperz, J., Yamamura, T., Miyake, S., and Stein-Streilein, J. (2003). CD4$^+$ NKT cells, but not conventional CD4$^+$ T cells, are required to generate efferent CD8$^+$ T regulatory cells following antigen inoculation in an immune-privileged site. *J. Immunol.* 171, 1266–1271.
1063. Sonoda, K. H., and Stein-Streilein, J. (2002). Ocular immune privilege and CD1d-reactive natural killer T cells. *Cornea* 21, S33–S38.
1064. Skelsey, M. E., Mellon, J., and Niederkorn, J. Y. (2001). γδ T cells are needed for ocular immune privilege and corneal graft survival. *J. Immunol.* 166, 4327–4333.
1065. Hurtenbach, U., and Shearer, G. M. (1982). Germ cell-induced immune suppression in mice: effect of inoculation of syngeneic spermatozoa on cell-mediated immune responses. *J. Exp. Med.* 155, 1719–1729.
1066. Hurtenbach, U., Morgenstern, F., and Bennett, D. (1980). Induction of tolerance *in vitro* by autologous murine testicular cells. *J. Exp. Med.* 151, 827–838.
1067. Born, W., and Wekerle, H. (1982). Leydig cells nonspecifically suppress lymphoproliferation *in vitro*: implications for the testis as an immunologically privileged site. *Am. J. Reprod. Immunol.* 2, 291–295.
1068. Born, W., and Wekerle, H. (1981). Selective, immunologically nonspecific adherence of lymphoid and myeloid cells to Leydig cells. *Eur. J. Cell Biol.* 25, 76–81.
1069. Hedger, M. P., Nikolic-Paterson, D. J., Hutchinson, P., Atkins, R. C., and de Kretser, D. M. (1998). Immunoregulatory activity in adult rat testicular interstitial fluid: roles of interleukin-1 and transforming growth factor β. *Biol. Reprod.* 58, 927–934.
1070. Pöllänen, P., Söder, O., and Uksila, J. (1988). Testicular immunosuppressive protein. *J. Reprod. Immunol.* 14, 125–138.
1071. Sainio-Pöllänen, S., Pöllänen, P., and Setchell, B. P. (1991). Testicular immunosuppressive activity in experimental hypogonadism and cryptorchidism. *J. Reprod. Immunol.* 20, 59–72.
1072. Emoto, M., Nishikawa, F., Oku, D., Hamuro, A., Kita, E., and Kashiba, S. (1991). Suppressive effect of a mouse testicular extract on lymphocyte activation. *Int. J. Androl.* 14, 291–302.
1073. Emoto, M., Yagyu, Y., Nishikawa, F., Katsui, N., Kita, E., and Kashiba, S. (1989). Effects of mouse testicular extract on immunocompetent cells. *Am. J. Reprod. Immunol.* 21, 61–66.
1074. Veräjänkorva, E., Martikainen, M., and Pöllänen, P. (2001). Cytokines in the BALB/c mouse testis in various conditions. *Asian J. Androl.* 3, 9–19.
1075. Korbutt, G. S., Elliott, J. F., and Rajotte, R. V. (1997). Cotransplantation of allogeneic islets with allogeneic testicular cell aggregates allows long-term graft survival without systemic immunosuppression. *Diabetes* 46, 317–322.
1076. Yang, H., and Wright, J. R., Jr. (1999). Co-encapsulation of Sertoli enriched testicular cell fractions further prolongs fish-to-mouse islet xenograft survival. *Transplantation* 67, 815–820.
1077. Saporta, S., Cameron, D. F., Borlongan, C. V., and Sanberg, P. R. (1997). Survival of rat and porcine Sertoli cell transplants in the rat striatum without cyclosporine-A immunosuppression. *Exp. Neurol.* 146, 299–304.
1078. Sanberg, P. R., Borlongan, C. V., Saporta, S., and Cameron, D. F. (1996). Testis-derived Sertoli cells survive and provide localized immunoprotection for xenografts in rat brain. *Nat. Biotechnol.* 14, 1692–1695.
1079. Selawry, H. P., and Cameron, D. F. (1993). Sertoli cell-enriched fractions in successful islet cell transplantation. *Cell Transplant.* 2, 123–129.
1080. Takeda, Y., Gotoh, M., Dono, K., Nishihara, M., Grochowiecki, T., Kimura, F., Yoshida, T., Ohta, Y., Ota, H., Ohzato, H., Umeshita, K., Takeda, T., Matsuura, N., Sakon, M., Kayagaki, N., Yagita, H., Okumura, K., Miyasaka, M., and Monden, M. (1998). Protection of islet allografts transplanted together with Fas ligand expressing testicular allografts. *Diabetologia* 41, 315–321.
1081. Korbutt, G. S., Suarez-Pinzon, W. L., Power, R. F., Rajotte, R. V., and Rabinovitch, A. (2000). Testicular Sertoli cells exert both protective and destructive effects on syngeneic islet grafts in non-obese diabetic mice. *Diabetologia* 43, 474–480.
1082. De Cesaris, P., Filippini, A., Cervelli, C., Riccioli, A., Muci, S., Starace, G., Stefanini, M., and Ziparo, E. (1992). Immunosuppressive molecules produced by Sertoli cells cultured in vitro: biological effects on lymphocytes. *Biochem. Biophys. Res. Commun.* 186, 1639–1646.
1083. Wyatt, C. R., Law, L., Magnuson, J. A., Griswold, M. D., and Magnuson, N. S. (1988). Suppression of lymphocyte proliferation by proteins secreted by cultured Sertoli cells. *J. Reprod. Immunol.* 14, 27–40.
1084. Pöllänen, P., von Euler, M., Sainio-Pöllänen, S., Jahnukainen, K., Hakovirta, H., Söder, O., and Parvinen, M. (1992). Immunosuppressive activity in the rat seminiferous tubules. *J. Reprod. Immunol.* 22, 117–126.
1085. Selawry, H. P., Kotb, M., Herrod, H. G., and Lu, Z. N. (1991). Production of a factor, or factors, suppressing IL-2 production and T cell proliferation by Sertoli cell-enriched preparations: a potential role for islet transplantation in an immunologically privileged site. *Transplantation* 52, 846–850.
1086. Nikolova, D. B., Kancheva, L. S., Surneva, M. D., and Martinova, Y. S. (1992). Species-specific effect of proteins secreted by cultured pre-pubertal rat Sertoli cells on natural killer cell activity. *Immunopharmacology* 23, 15–20.
1087. Willing, A. E., Cameron, D. F., and Sanberg, P. R. (1998). Sertoli cell transplants: their use in the treatment of neurodegenerative disease. *Mol. Med. Today* 4, 471–477.
1088. Housseau, F., Rouas-Freiss, N., Roy, M., Bidart, J. M., Guillet, J. G., and Bellet, D. (1997). Antigen-presenting function of murine gonadal epithelial cell lines. *Cell. Immunol.* 177, 93–101.
1089. Riccioli, A., Filippini, A., De Cesaris, P., Barbacci, E., Stefanini, M., Starace, G., and Ziparo, E. (1995). Inflammatory mediators increase surface expression of integrin ligands, adhesion to lymphocytes, and secretion of interleukin 6 in mouse Sertoli cells. *Proc. Natl. Acad. Sci. U. S. A.* 92, 5808–5812.
1090. Guazzone, V. A., Rival, C., Denduchis, B., and Lustig, L. (2003). Monocyte chemoattractant protein-1 (MCP-1/CCL2) in experimental autoimmune orchitis. *J. Reprod. Immunol.* 60, 143–157.
1091. Collin, O., Bergh, A., Damber, J. E., and Widmark, A. (1993). Control of testicular vasomotion by testosterone and tubular factors in rats. *J. Reprod. Fertil.* 97, 115–121.
1092. Maddocks, S., and Sharpe, R. M. (1989). Interstitial fluid volume in the rat testis: androgen-dependent regulation by the seminiferous tubules? *J. Endocrinol.* 120, 215–222.
1093. Rivenzon, A., Rivenzon, M., and Madden, R. E. (1974). Spontaneous adherence and rosette formation of lymphocytes to Leydig cells: an *in vitro* technique. *Cell. Immunol.* 14, 411–416.
1094. Rivenson, A., Ohmori, T., Hamazaki, M., and Madden, R. (1981). Cell surface recognition: spontaneous identification

of mouse Leydig cells by lymphocytes, macrophages and eosinophils. *Cell. Mol. Biol.* 27, 49–56.
1095. Sainio-Pöllänen, S., Sundström, J., Erkkilä, S., Hänninen, A., Vainiopää, M., Martikainen, M., Salminen, E., Veräjänkorva, E., Antola, H., Nikula, H., Simell, O., and Pöllänen, P. (1997). CD106 (VCAM-1) in testicular immunoregulation. *J. Reprod. Immunol.* 33, 221–238.
1096. Hedger, M. P., Qin, J. X., Robertson, D. M., and de Kretser, D. M. (1990). Intragonadal regulation of immune system functions. *Reprod. Fertil. Dev.* 2, 263–280.
1097. Bebo, B. F., Jr., Schuster, J. C., Vandenbark, A. A., and Offner, H. (1999). Androgens alter the cytokine profile and reduce encephalitogenicity of myelin-reactive T cells. *J. Immunol.* 162, 35–40.
1098. Malkin, C. J., Pugh, P. J., Jones, R. D., Kapoor, D., Channer, K. S., and Jones, T. H. (2004). The effect of testosterone replacement on endogenous inflammatory cytokines and lipid profiles in hypogonadal men. *J. Clin. Endocrinol. Metab.* 89, 3313–3318.
1099. Whitmore III, W. F., and Gittes, R. F. (1981). The effect of physiologic and supraphysiologic concentrations of progesterone on allograft survival. *Surg. Forum* 32, 593–595.
1100. Selawry, H. P., and Whittington, K. B. (1988). Prolonged intratesticular islet allograft survival is not dependent on local steroidogenesis. *Horm. Metab. Res.* 20, 562–565.
1101. Cameron, D. F., Whittington, K., Schultz, R. E., and Selawry, H. P. (1990). Successful islet/abdominal testis transplantation does not require Leydig cells. *Transplantation* 50, 649–653.
1102. Pintar, J. E., Schachter, B. S., Herman, A. B., Durgerian, S., and Krieger, D. T. (1984). Characterization and localization of proopiomelanocortin messenger RNA in the adult rat testis. *Science* 225, 632–634.
1103. Abe, R., Peng, T., Sailors, J., Bucala, R., and Metz, C. N. (2001). Regulation of the CTL response by macrophage migration inhibitory factor. *J. Immunol.* 166, 747–753.
1104. Apte, R. S., Sinha, D., Mayhew, E., Wistow, G. J., and Niederkorn, J. Y. (1998). Role of macrophage migration inhibitory factor in inhibiting NK cell activity and preserving immune privilege. *J. Immunol.* 160, 5693–5696.
1105. Repp, A. C., Mayhew, E. S., Apte, S., and Niederkorn, J. Y. (2000). Human uveal melanoma cells produce macrophage migration-inhibitory factor to prevent lysis by NK cells. *J. Immunol.* 165, 710–715.
1106. Schnyder, J., Dewald, B., and Baggiolini, M. (1981). Effects of cyclooxygenase inhibitors and prostaglandin E2 on macrophage activation in vitro. *Prostaglandins* 22, 411–421.
1107. Kunkel, S. L., Spengler, M., May, M. A., Spengler, R., Larrick, J., and Remick, D. (1988). Prostaglandin E_2 regulates macrophage-derived tumor necrosis factor gene expression. *J. Biol. Chem.* 263, 5380–5384.
1108. Hart, P. H., Whitty, G. A., Piccoli, D. S., and Hamilton, J. A. (1989). Control by IFN-γ and PGE_2 of TNFα and IL-1 production by human monocytes. *Immunology* 66, 376–383.
1109. Scales, W. E., Chensue, S. W., Otterness, I., and Kunkel, S. L. (1989). Regulation of monokine gene expression: prostaglandin E_2 suppresses tumor necrosis factor but not interleukin-1α or β-mRNA and cell-associated bioactivity. *J. Leukoc. Biol.* 45, 416–421.
1110. Reddy, G. P., Prasad, M., Sailesh, S., Kumar, Y. V., and Reddanna, P. (1992). The production of arachidonic acid metabolites in rat testis. *Prostaglandins* 44, 497–507.
1111. Abayasekara, D. R., Kurlak, L. O., Jeremy, J. Y., Dandona, P., Sharpe, R. M., and Cooke, B. A. (1990). The levels and possible involvement of leukotriene B_4 and prostaglandin $F_{2\alpha}$ in the control of interstitial fluid volume in the rat testis. *Int. J. Androl.* 13, 408–418.
1112. Haour, F., Kouznetzova, B., Dray, F., and Saez, J. M. (1979). hCG-induced prostaglandin E_2 and $F_{2\alpha}$ release in adult rat testis: role in Leydig cell desensitization to hCG. *Life Sci.* 24, 2151–2158.
1113. O'Neill, G. P., and Ford-Hutchinson, A. W. (1993). Expression of mRNA for cyclooxygenase-1 and cyclooxygenase-2 in human tissues. *FEBS Lett.* 330, 156–160.
1114. Mele, P. G., Dada, L. A., Paz, C., Neuman, I., Cymeryng, C. B., Mendez, C. F., Finkielstein, C. V., Cornejo Maciel, F., and Podesta, E. J. (1997). Involvement of arachidonic acid and the lipoxygenase pathway in mediating luteinizing hormone-induced testosterone synthesis in rat Leydig cells. *Endocr. Res.* 23, 15–26.
1115. Jannini, E. A., Ulisse, S., Cecconi, S., Cironi, L., Colonna, R., D'Armiento, M., Santoni, A., and Cifone, M. G. (1994). Follicle-stimulating hormone-induced phospholipase A_2 activity and eicosanoid generation in rat Sertoli cells. *Biol. Reprod.* 51, 140–145.
1116. Funk, C. D. (2001). Prostaglandins and leukotrienes: advances in eicosanoid biology. *Science* 294, 1871–1875.
1117. Samuelsson, B. (2000). The discovery of the leukotrienes. *Am. J. Respir. Crit. Care Med.* 161, S2–S6.
1118. Goetzl, E. J., An, S., and Smith, W. L. (1995). Specificity of expression and effects of eicosanoid mediators in normal physiology and human diseases. *FASEB J.* 9, 1051–1058.
1119. Lazarus, M., Munday, C. J., Eguchi, N., Matsumoto, S., Killian, G. J., Kubata, B. K., and Urade, Y. (2002). Immunohistochemical localization of microsomal PGE synthase-1 and cyclooxygenases in male mouse reproductive organs. *Endocrinology* 143, 2410–2419.
1120. Baker, P. J., and O'Shaughnessy, P. J. (2001). Expression of prostaglandin D synthetase during development in the mouse testis. *Reproduction* 122, 553–559.
1121. Shahin, I., Grossman, S., and Sredni, B. (1978). Lipoxygenase-like enzyme in rat testis microsomes. *Biochim. Biophys. Acta* 529, 300–308.
1122. Sullivan, M. H., and Cooke, B. A. (1985). Control and production of leukotriene B4 in rat tumour and testicular Leydig cells. *Biochem. J.* 230, 821–824.
1123. Schon, I., Sofer, Y., Cojacaru, M., and Grossman, S. (1989). Arachidonic acid metabolism by perfused ram testis. *Int. J. Biochem.* 21, 7–13.
1124. Sorrentino, C., Silvestrini, B., Braghiroli, L., Chung, S. S., Giacomelli, S., Leone, M. G., Xie, Y., Sui, Y., Mo, M., and Cheng, C. Y. (1998). Rat prostaglandin D_2 synthetase: its tissue distribution, changes during maturation, and regulation in the testis and epididymis. *Biol. Reprod.* 59, 843–853.
1125. Samy, E. T., Li, J. C., Grima, J., Lee, W. M., Silvestrini, B., and Cheng, C. Y. (2000). Sertoli cell prostaglandin D_2 synthetase is a multifunctional molecule: its expression and regulation. *Endocrinology* 141, 710–721.
1126. Mackrell, P. J., Daly, J. M., Mestre, J. R., Stapleton, P. P., Howe, L. R., Subbaramaiah, K., and Dannenberg, A. J. (2001). Elevated expression of cyclooxygenase-2 contributes to immune dysfunction in a murine model of trauma. *Surgery* 130, 826–833.
1127. Kojima, M., Morisaki, T., Uchiyama, A., Doi, F., Mibu, R., Katano, M., and Tanaka, M. (2001). Association of enhanced cyclooxygenase-2 expression with possible local immunosuppression in human colorectal carcinomas. *Ann. Surg. Oncol.* 8, 458–465.
1128. Morelli, A. E., and Thomson, A. W. (2003). Dendritic cells under the spell of prostaglandins. *Trends Immunol.* 24, 108–111.
1129. Gilroy, D. W., Colville-Nash, P. R., Willis, D., Chivers, J., Paul-Clark, M. J., and Willoughby, D. A. (1999). Inducible

cyclooxygenase may have anti-inflammatory properties. *Nat. Med.* 5, 698–701.
1130. Jiang, C., Ting, A. T., and Seed, B. (1998). PPAR-γ agonists inhibit production of monocyte inflammatory cytokines. *Nature* 391, 82–86.
1131. Takano, T., Fiore, S., Maddox, J. F., Brady, H. R., Petasis, N. A., and Serhan, C. N. (1997). Aspirin-triggered 15-epi-lipoxin A_4 (LXA_4) and LXA_4 stable analogues are potent inhibitors of acute inflammation: evidence for anti-inflammatory receptors. *J. Exp. Med.* 185, 1693–1704.
1132. Sodin-Semrl, S., Taddeo, B., Tseng, D., Varga, J., and Fiore, S. (2000). Lipoxin A_4 inhibits IL-1β-induced IL-6, IL-8, and matrix metalloproteinase-3 production in human synovial fibroblasts and enhances synthesis of tissue inhibitors of metalloproteinases. *J. Immunol.* 164, 2660–2666.
1133. Takeda, A., Palfree, R. G., and Forsdyke, D. R. (1982). Role of serum in inhibition of cultured lymphocytes by lysophosphatidylcholine. *Biochim. Biophys. Acta* 710, 87–98.
1134. Yan, J. J., Jung, J. S., Lee, J. E., Lee, J., Huh, S. O., Kim, H. S., Jung, K. C., Cho, J. Y., Nam, J. S., Suh, H. W., Kim, Y. H., and Song, D. K. (2004). Therapeutic effects of lysophosphatidylcholine in experimental sepsis. *Nat. Med.* 10, 161–167.
1135. Kabarowski, J. H., Xu, Y., and Witte, O. N. (2002). Lysophosphatidylcholine as a ligand for immunoregulation. *Biochem. Pharmacol.* 64, 161–167.
1136. Brinster, R. L., and Avarbock, M. R. (1994). Germline transmission of donor haplotype following spermatogonial transplantation. *Proc. Natl. Acad. Sci. U. S. A.* 91, 11303–11307.
1137. Brinster, R. L., and Zimmermann, J. W. (1994). Spermatogenesis following male germ-cell transplantation. *Proc. Natl. Acad. Sci. U. S. A.* 91, 11298–11302.
1138. Boettger-Tong, H. L., Johnston, D. S., Russell, L. D., Griswold, M. D., and Bishop, C. E. (2000). Juvenile spermatogonial depletion (*jsd*) mutant seminiferous tubules are capable of supporting transplanted spermatogenesis. *Biol. Reprod.* 63, 1185–1191.
1139. Ogawa, T., Dobrinski, I., Avarbock, M. R., and Brinster, R. L. (2000). Transplantation of male germ line stem cells restores fertility in infertile mice. *Nat. Med.* 6, 29–34.
1140. Clouthier, D. E., Avarbock, M. R., Maika, S. D., Hammer, R. E., and Brinster, R. L. (1996). Rat spermatogenesis in mouse testis. *Nature* 381, 418–421.
1141. França, L. R., Ogawa, T., Avarbock, M. R., Brinster, R. L., and Russell, L. D. (1998). Germ cell genotype controls cell cycle during spermatogenesis in the rat. *Biol. Reprod.* 59, 1371–1377.
1142. Ogawa, T., Dobrinski, I., Avarbock, M. R., and Brinster, R. L. (1999). Xenogeneic spermatogenesis following transplantation of hamster germ cells to mouse testes. *Biol. Reprod.* 60, 515–521.
1143. Dobrinski, I., Avarbock, M. R., and Brinster, R. L. (2000). Germ cell transplantation from large domestic animals into mouse testes. *Mol. Reprod. Dev.* 57, 270–279.
1144. Dobrinski, I., Avarbock, M. R., and Brinster, R. L. (1999). Transplantation of germ cells from rabbits and dogs into mouse testes. *Biol. Reprod.* 61, 1331–1339.
1145. Nagano, M., Patrizio, P., and Brinster, R. L. (2002). Long-term survival of human spermatogonial stem cells in mouse testes. *Fertil. Steril.* 78, 1225–1233.
1146. Nagano, M., McCarrey, J. R., and Brinster, R. L. (2001). Primate spermatogonial stem cells colonize mouse testes. *Biol. Reprod.* 64, 1409–1416.
1147. Jiang, F. X., and Short, R. V. (1995). Male germ cell transplantation in rats: apparent synchronization of spermatogenesis between host and donor seminiferous epithelia. *Int. J. Androl.* 18, 326–330.
1148. Ogawa, T., Dobrinski, I., and Brinster, R. L. (1999). Recipient preparation is critical for spermatogonial transplantation in the rat. *Tissue Cell* 31, 461–472.
1149. Schlatt, S., Rosiepen, G., Weinbauer, G. F., Rolf, C., Brook, P. F., and Nieschlag, E. (1999). Germ cell transfer into rat, bovine, monkey and human testes. *Hum. Reprod.* 14, 144–150.
1150. Shinohara, T., Orwig, K. E., Avarbock, M. R., and Brinster, R. L. (2003). Restoration of spermatogenesis in infertile mice by Sertoli cell transplantation. *Biol. Reprod.* 68, 1064–1071.
1151. Lo, K. C., Lei, Z., Rao Ch, V., Beck, J., and Lamb, D. J. (2004). *De novo* testosterone production in luteinizing hormone receptor knockout mice after transplantation of Leydig stem cells. *Endocrinology* 145, 4011–4015.
1152. Parreira, G. G., Ogawa, T., Avarbock, M. R., França, L. R., Hausler, C. L., Brinster, R. L., and Russell, L. D. (1999). Development of germ cell transplants: morphometric and ultrastructural studies. *Tissue Cell* 31, 242–254.
1153. Nagano, M., Brinster, C. J., Orwig, K. E., Ryu, B. Y., Avarbock, M. R., and Brinster, R. L. (2001). Transgenic mice produced by retroviral transduction of male germ-line stem cells. *Proc. Natl. Acad. Sci. U. S. A.* 98, 13090–13095.
1154. Parreira, G. G., Ogawa, T., Avarbock, M. R., França, L. R., Brinster, R. L., and Russell, L. D. (1998). Development of germ cell transplants in mice. *Biol. Reprod.* 59, 1360–1370.
1155. Barksdale, E. M., Jr., McGenis, T. G., and Donahoe, P. K. (1991). Gonadotropins moderate rejection of trophic-specific congenic testes grafts. *J. Pediatr. Surg.* 26, 886–892.
1156. Statter, M. B., Foglia, R. P., Parks, D. E., and Donahoe, P. K. (1988). Fetal and postnatal testis shows immunoprivilege as donor tissue. *J. Urol.* 139, 204–210.
1157. Kamada, N., and Wight, D. G. (1984). Antigen-specific immunosuppression induced by liver transplantation in the rat. *Transplantation* 38, 217–221.
1158. Emoto, M., and Kaufmann, S. H. (2003). Liver NKT cells: an account of heterogeneity. *Trends Immunol.* 24, 364–369.
1159. Russell, M. W., Sibley, D. A., Nikolova, E. B., Tomana, M., and Mestecky, J. (1997). IgA antibody as a non-inflammatory regulator of immunity. *Biochem. Soc. Trans.* 25, 466–470.
1160. Lamm, M. E. (1997). Interaction of antigens and antibodies at mucosal surfaces. *Annu. Rev. Microbiol.* 51, 311–340.
1161. MacDonald, T. T. (1998). T cell immunity to oral allergens. *Curr. Opin. Immunol.* 10, 620–627.
1162. Czerkinsky, C., Anjuere, F., McGhee, J. R., George-Chandy, A., Holmgren, J., Kieny, M. P., Fujiyashi, K., Mestecky, J. F., Pierrefite-Carle, V., Rask, C., and Sun, J. B. (1999). Mucosal immunity and tolerance: relevance to vaccine development. *Immunol. Rev.* 170, 197–222.
1163. Graham, B. S., Bunton, L. A., Wright, P. F., and Karzon, D. T. (1991). Role of T lymphocyte subsets in the pathogenesis of primary infection and rechallenge with respiratory syncytial virus in mice. *J. Clin. Invest.* 88, 1026–1033.
1164. Thorstenson, K. M., and Khoruts, A. (2001). Generation of anergic and potentially immunoregulatory $CD25^+$ CD4 T cells *in vivo* after induction of peripheral tolerance with intravenous or oral antigen. *J. Immunol.* 167, 188–195.
1165. Husby, S., Mestecky, J., Moldoveanu, Z., Holland, S., and Elson, C. O. (1994). Oral tolerance in humans. T cell but not B cell tolerance after antigen feeding. *J. Immunol.* 152, 4663–4670.
1166. Khoo, U. Y., Proctor, I. E., and Macpherson, A. J. (1997). $CD4^+$ T cell down-regulation in human intestinal mucosa: evidence for intestinal tolerance to luminal bacterial antigens. *J. Immunol.* 158, 3626–3634.
1167. Matsuda, J. L., Naidenko, O. V., Gapin, L., Nakayama, T., Taniguchi, M., Wang, C. R., Koezuka, Y., and Kronenberg, M.

(2000). Tracking the response of natural killer T cells to a glycolipid antigen using CD1d tetramers. *J. Exp. Med.* 192, 741–754.

1168. Blumberg, R. S., Terhorst, C., Bleicher, P., McDermott, F. V., Allan, C. H., Landau, S. B., Trier, J. S., and Balk, S. P. (1991). Expression of a nonpolymorphic MHC class I-like molecule, CD1d, by human intestinal epithelial cells. *J. Immunol.* 147, 2518–2524.
1169. Kalb, T. H., Chuang, M. T., Marom, Z., and Mayer, L. (1991). Evidence for accessory cell function by class II MHC antigen-expressing airway epithelial cells. *Am. J. Respir. Cell Mol. Biol.* 4, 320–329.
1170. Campbell, N., Yio, X. Y., So, L. P., Li, Y., and Mayer, L. (1999). The intestinal epithelial cell: processing and presentation of antigen to the mucosal immune system. *Immunol. Rev.* 172, 315–324.
1171. Roussel, J. D., Stallcup, O. T., and Austin, C. R. (1967). Selective phagocytosis of spermatozoa in the epididymis of bulls, rabbits, and monkeys. *Fertil. Steril.* 18, 509–516.
1172. Alexander, N. J. (1973). Ultrastructural changes in rat epididymis after vasectomy. *Z. Zellforsch. Mikrosk. Anat.* 136, 177–182.
1173. Moore, H. D., and Bedford, J. M. (1979). The differential absorptive activity of epithelial cells of the rat epididymis before and after castration. *Anat. Rec.* 193, 313–327.
1174. Ashida, E. R., and Scofield, V. L. (1987). Lymphocyte major histocompatibility complex-encoded class II structures may act as sperm receptors. *Proc. Natl. Acad. Sci. U. S. A.* 84, 3395–3399.
1175. Scofield, V. L., Clisham, R., Bandyopadhyay, L., Gladstone, P., Zamboni, L., and Raghupathy, R. (1992). Binding of sperm to somatic cells via HLA-DR. Modulation by sulfated carbohydrates. *J. Immunol.* 148, 1718–1724.
1176. Stites, D. P., and Erickson, R. P. (1975). Suppressive effect of seminal plasma on lymphocyte activation. *Nature* 253, 727–729.
1177. Petersen, B. H., Lammel, C. J., Stites, D. P., and Brooks, G. F. (1980). Human seminal plasma inhibition of complement. *J. Lab. Clin. Med.* 96, 582–591.
1178. Vallely, P. J., Sharrard, R. M., and Rees, R. C. (1988). The identification of factors in seminal plasma responsible for suppression of natural killer cell activity. *Immunology* 63, 451–456.
1179. Moulik, S., Ranga, G., Meherji, P. K., and Shahani, S. K. (1989). Detection of immunosuppressive activity in human seminal plasma by an immunobioassay. *Int. J. Androl.* 12, 131–138.
1180. James, K., and Hargreave, T. B. (1984). Immunosuppression by seminal plasma and its possible clinical significance. *Immunol. Today* 5, 353–363.
1181. Kelly, R. W., Holland, P., Skibinski, G., Harrison, C., McMillan, L., Hargreave, T., and James, K. (1991). Extracellular organelles (prostasomes) are immunosuppressive components of human semen. *Clin. Exp. Immunol.* 86, 550–556.
1182. Ronquist, G., and Brody, I. (1985). The prostasome: its secretion and function in man. *Biochim. Biophys. Acta* 822, 203–218.
1183. Allen, R. D., and Roberts, T. K. (1986). The relationship between the immunosuppressive and cytotoxic effects of human seminal plasma. *Am. J. Reprod. Immunol. Microbiol.* 11, 59–64.
1184. Skibinski, G., Kelly, R. W., Harrison, C. M., McMillan, L. A., and James, K. (1992). Relative immunosuppressive activity of human seminal prostaglandins. *J. Reprod. Immunol.* 22, 185–195.
1185. Kelly, R. W., Taylor, P. L., Hearn, J. P., Short, R. V., Martin, D. E., and Marston, J. H. (1976). 19-Hydroxyprostaglandin E_1 as a major component of the semen of primates. *Nature* 260, 544–545.
1186. Quayle, A. J., Kelly, R. W., Hargreave, T. B., and James, K. (1989). Immunosuppression by seminal prostaglandins. *Clin. Exp. Immunol.* 75, 387–391.
1187. Fahmi, H. A., Hunter, A. G., Markham, R. J., and Seguin, B. E. (1985). Identification of an immunosuppressive protein in bovine seminal plasma with activity against bovine lymphocytes. *J. Dairy Sci.* 68, 2322–2328.
1188. Veselský, L., Dostál, J., Kraus, M., Peknicová, J., Holán, V., Zajícová, A., Jonáková, V., and Zelezná, B. (2002). Reverse effect of indomethacin on the immunosuppressive activity of boar seminal immunosuppressive fraction. *Anim. Reprod. Sci.* 71, 111–123.
1189. Loras, B., Vételé, F., El Malki, A., Rollet, J., Soufir, J. C., and Benahmed, M. (1999). Seminal transforming growth factor-β in normal and infertile men. *Hum. Reprod.* 14, 1534–1539.
1190. Miller, L. J., Fischer, K. A., Goralnick, S. J., Litt, M., Burleson, J. A., Albertsen, P., and Kreutzer, D. L. (2002). Interleukin-10 levels in seminal plasma: implications for chronic prostatitis-chronic pelvic pain syndrome. *J. Urol.* 167, 753–756.
1191. Huleihel, M., Lunenfeld, E., Horowitz, S., Levy, A., Potashnik, G., Mazor, M., and Glezerman, M. (1999). Expression of IL-12, IL-10, PGE2, sIL-2R and sIL-6R in seminal plasma of fertile and infertile men. *Andrologia* 31, 283–288.
1192. Nocera, M., and Chu, T. M. (1993). Transforming growth factor β as an immunosuppressive protein in human seminal plasma. *Am. J. Reprod. Immunol.* 30, 1–8.
1193. Allen, R. D., and Roberts, T. K. (1987). Role of spermine in the cytotoxic effects of seminal plasma. *Am. J. Reprod. Immunol. Microbiol.* 13, 4–8.
1194. Curry, M. C., Hussain, J. I., Smith, C. J., and Allen, J. C. (1980). Identification of a macromolecular inhibitor of in vitro lymphocyte proliferation in human seminal plasma as bound spermine. *Eur. J. Cell Biol.* 21, 180–182.
1195. Allen, J. C., Smith, C. J., Hussain, J. I., Thomas, J. M., and Gaugas, J. M. (1979). Inhibition of lymphocyte proliferation by polyamines requires ruminant-plasma polyamine oxidase. *Eur. J. Biochem.* 102, 153–158.
1196. Kimes, B. W., and Morris, D. R. (1971). Preparation and stability of oxidized polyamines. *Biochim. Biophys. Acta* 228, 223–234.
1197. Allen, R. D., and Roberts, T. K. (1986). Seminal plasma immunosuppression: an irrelevant biological phenomenon? *Clin. Reprod. Fertil.* 4, 353–355.
1198. Maayan, R., Zukerman, Z., and Shohat, B. (1995). Oxidation of polyamines in human seminal plasma: a possible role in immunological infertility. *Arch. Androl.* 34, 95–99.
1199. Imade, G. E., Baker, H. W., de Kretser, D. M., and Hedger, M. P. (1997). Immunosuppressive activities in the seminal plasma of infertile men: relationship to sperm antibodies and autoimmunity. *Hum. Reprod.* 12, 256–262.
1200. Kelly, R. W., Quayle, A. J., Wallace, E. M., Wu, F. C., Hargreave, T. B., and James, K. (1991). Immunosuppression by seminal plasma from fertile and infertile men: inhibition of natural killer cell function correlates with seminal PG concentration. *Prostaglandins Leukot. Essent. Fatty Acids* 42, 257–260.
1201. Kelly, R. W., Skibinski, G., and James, K. (1994). The immunosuppressive contribution of prostaglandin components of human semen and their ability to elevate cyclic adenosine monophosphate levels in peripheral blood mononuclear cells. *J. Reprod. Immunol.* 26, 31–40.

1202. McKanna, J. A., Zhang, M. Z., Wang, J. L., Cheng, H., and Harris, R. C. (1998). Constitutive expression of cyclooxygenase-2 in rat vas deferens. *Am. J. Physiol.* 275, R227–R233.
1203. Srivastava, M. D., Lippes, J., and Srivastava, B. I. (1996). Cytokines of the human reproductive tract. *Am. J. Reprod. Immunol.* 36, 157–166.
1204. Nocera, M., and Chu, T. M. (1995). Characterization of latent transforming growth factor-β from human seminal plasma. *Am. J. Reprod. Immunol.* 33, 282–291.
1205. Rajasekaran, M., Hellstrom, W., and Sikka, S. (1996). Quantitative assessment of cytokines (GROα and IL-10) in human seminal plasma during genitourinary inflammation. *Am. J. Reprod. Immunol.* 36, 90–95.
1206. Shoskes, D. A., Albakri, Q., Thomas, K., and Cook, D. (2002). Cytokine polymorphisms in men with chronic prostatitis/chronic pelvic pain syndrome: association with diagnosis and treatment response. *J. Urol.* 168, 331–335.
1207. Anderson, R. A., Evans, L. W., Irvine, D. S., McIntyre, M. A., Groome, N. P., and Riley, S. C. (1998). Follistatin and activin A production by the male reproductive tract. *Hum. Reprod.* 13, 3319–3325.
1208. de Waal Malefyt, R., and Moore, K. W. (1998). Interleukin-10. In *The Cytokine Handbook* (3rd ed. A. W. Thomson, Ed.), pp. 333–364. Academic Press, San Diego.
1209. Thomas, T. Z., Chapman, S. M., Hong, W., Gurusingfhe, C., Mellor, S. L., Fletcher, R., Pedersen, J., and Risbridger, G. P. (1998). Inhibins, activins, and follistatins: expression of mRNAs and cellular localization in tissues from men with benign prostatic hyperplasia. *Prostate* 34, 34–43.
1210. Robertson, S. A., Ingman, W. V., O'Leary, S., Sharkey, D. J., and Tremellen, K. P. (2002). Transforming growth factor β: a mediator of immune deviation in seminal plasma. *J. Reprod. Immunol.* 57, 109–128.
1211. O'Bryan, M. K., Baker, H. W., Saunders, J. R., Kirszbaum, L., Walker, I. D., Hudson, P., Liu, D. Y., Glew, M. D., d'Apice, A. J., and Murphy, B. F. (1990). Human seminal clusterin (SP-40,40): isolation and characterization. *J. Clin. Invest.* 85, 1477–1486.
1212. Maccioni, M., Riera, C. M., and Rivero, V. E. (2001). Identification of rat prostatic steroid binding protein (PSBP) as an immunosuppressive factor. *J. Reprod. Immunol.* 50, 133–149.
1213. Jones, R. L., Salamonsen, L. A., and Findlay, J. K. (2002). Potential roles for endometrial inhibins, activins and follistatin during human embryo implantation and early pregnancy. *Trends Endocrinol. Metab.* 13, 144–150.
1214. Robertson, S. A., Mau, V. J., Hudson, S. N., and Tremellen, K. P. (1997). Cytokine-leukocyte networks and the establishment of pregnancy. *Am. J. Reprod. Immunol.* 37, 438–442.
1215. Naz, R. K., and Evans, L. (1998). Presence and modulation of interleukin-12 in seminal plasma of fertile and infertile men. *J. Androl.* 19, 302–307.
1216. Eggert-Kruse, W., Boit, R., Rohr, G., Aufenanger, J., Hund, M., and Strowitzki, T. (2001). Relationship of seminal plasma interleukin (IL) -8 and IL-6 with semen quality. *Hum. Reprod.* 16, 517–528.
1217. Maegawa, M., Kamada, M., Irahara, M., Yamamoto, S., Yoshikawa, S., Kasai, Y., Ohmoto, Y., Gima, H., Thaler, C. J., and Aono, T. (2002). A repertoire of cytokines in human seminal plasma. *J. Reprod. Immunol.* 54, 33–42.
1218. Koumantakis, E., Matalliotakis, I., Kyriakou, D., Fragouli, Y., and Relakis, K. (1998). Increased levels of interleukin-8 in human seminal plasma. *Andrologia* 30, 339–343.
1219. Matalliotakis, I., Kyriakou, D., Fragouli, Y., Loutradis, D., Goumenou, A., and Koumantakis, E. (1998). Determination of interleukin-11 in seminal plasma and elevated IL-11 in seminal plasma of infertile patients with urogenital infection. *Arch. Androl.* 41, 177–183.
1220. Krause, W., Bohring, C., Gueth, A., Horster, S., Krisp, A., and Skrzypek, J. (2003). Cellular and biochemical markers in semen indicating male accessory gland inflammation. *Andrologia* 35, 279–282.
1221. Matalliotakis, I., Kiriakou, D., Fragouli, I., Sifakis, S., Eliopoulos, G., and Koumantakis, E. (1998). Interleukin-6 in seminal plasma of fertile and infertile men. *Arch. Androl.* 41, 43–50.
1222. Matalliotakis, I., Arici, A., Goumenou, A., Koumantakis, G., Selam, B., Matalliotakis, G., and Koumantakis, E. (2002). Distinct expression pattern of cytokines in semen of men with genital infection and oligo-terato-asthenozoospermia. *Am. J. Reprod. Immunol.* 48, 170–175.
1223. Gruschwitz, M. S., Brezinschek, R., and Brezinschek, H. P. (1996). Cytokine levels in the seminal plasma of infertile males. *J. Androl.* 17, 158–163.
1224. Barratt, C. L., Harrison, P. E., Robinson, A., Kessopoulou, E., and Cooke, I. D. (1991). Seminal white blood cells in men with urethral tract infection: a monoclonal antibody study. *Br. J. Urol.* 68, 531–536.
1225. Berger, R. E., Karp, L. E., Williamson, R. A., Koehler, J., Moore, D. E., and Holmes, K. K. (1982). The relationship of pyospermia and seminal fluid bacteriology to sperm function as reflected in the sperm penetration assay. *Fertil. Steril.* 37, 557–564.
1226. Auroux, M. (1984). Nonspermatozoal cells in human sperm: a study of 1243 subfertile and 253 fertile men. *Arch. Androl.* 12, 197–201.
1227. Wolff, H., Politch, J. A., Martinez, A., Haimovici, F., Hill, J. A., and Anderson, D. J. (1990). Leukocytospermia is associated with poor semen quality. *Fertil. Steril.* 53, 528–536.
1228. Aziz, N., Agarwal, A., Lewis-Jones, I., Sharma, R. K., and Thomas, A. J., Jr. (2004). Novel associations between specific sperm morphological defects and leukocytospermia. *Fertil. Steril.* 82, 621–627.
1229. Tomlinson, M. J., East, S. J., Barratt, C. L., Bolton, A. E., and Cooke, I. D. (1992). Preliminary communication: possible role of reactive nitrogen intermediates in leucocyte-mediated sperm dysfunction. *Am. J. Reprod. Immunol.* 27, 89–92.
1230. Hill, J. A., Haimovici, F., Politch, J. A., and Anderson, D. J. (1987). Effects of soluble products of activated lymphocytes and macrophages (lymphokines and monokines) on human sperm motion parameters. *Fertil. Steril.* 47, 460–465.
1231. Aitken, R. J., Buckingham, D. W., Brindle, J., Gomez, E., Baker, H. W., and Irvine, D. S. (1995). Analysis of sperm movement in relation to the oxidative stress created by leukocytes in washed sperm preparations and seminal plasma. *Hum. Reprod.* 10, 2061–2071.
1232. Barratt, C. L., Harrison, P. E., Robinson, A., and Cooke, I. D. (1990). Antisperm antibodies and lymphocyte subsets in semen: not a simple relationship. *Int. J. Androl.* 13, 50–58.
1233. Tomlinson, M. J., Barratt, C. L., Bolton, A. E., Lenton, E. A., Roberts, H. B., and Cooke, I. D. (1992). Round cells and sperm fertilizing capacity: the presence of immature germ cells but not seminal leukocytes are associated with reduced success of in vitro fertilization. *Fertil. Steril.* 58, 1257–1259.
1234. Barratt, C. L., Bolton, A. E., and Cooke, I. D. (1990). Functional significance of white blood cells in the male and female reproductive tract. *Hum. Reprod.* 5, 639–648.
1235. Shattock, R. J., and Moore, J. P. (2003). Inhibiting sexual transmission of HIV-1 infection. *Nat. Rev. Microbiol.* 1, 25–34.
1236. Byrn, R. A., and Kiessling, A. A. (1998). Analysis of human immunodeficiency virus in semen: indications of a genetically distinct virus reservoir. *J. Reprod. Immunol.* 41, 161–176.

1237. Baum, E., Sather, H., Nachman, J., Seinfeld, J., Krivit, W., Leikin, S., Miller, D., Joo, P., and Hammond, D. (1979). Relapse rates following cessation of chemotherapy during complete remission of acute lymphocytic leukemia. *Med. Pediatr. Oncol.* 7, 25–34.
1238. Sather, H., Coccia, P., Nesbit, M., Level, C., and Hammond, D. (1981). Disappearance of the predictive value of prognostic variables in childhood acute lymphoblastic leukemia: a report from Childrens Cancer Study Group. *Cancer* 48, 370–376.
1239. Brecher, M. L. (1986). Treatment of acute lymphoid leukemia in children: current regimens and future prospects. *N. Y. State J. Med.* 86, 188–196.
1240. Shevchuk, M. M., Pigato, J. B., Khalife, G., Armenakas, N. A., and Fracchia, J. A. (1999). Changing testicular histology in AIDS: its implication for sexual transmission of HIV. *Urology* 53, 203–208.
1241. Jahnukainen, K., Morris, I., Roe, S., Salmi, T. T., Makipernaa, A., and Pöllänen, P. (1993). A rodent model for testicular involvement in acute lymphoblastic leukaemia. *Br. J. Cancer* 67, 885–892.
1242. Jahnukainen, K., Saari, T., Morris, I. D., Salmi, T. T., and Pöllänen, P. (1994). Regulation of testicular infiltration in acute lymphoblastic leukaemia of the rat. *Leukemia* 8, 458–464.
1243. Jahnukainen, K., Saari, T., Salmi, T. T., Pöllänen, P., and Pelliniemi, L. J. (1995). Reactions of Leydig cells and blood vessels to lymphoblastic leukemia in the rat testis. *Leukemia* 9, 908–914.
1244. Uyttenhove, C., Godfraind, C., Lethé, B., Amar-Costesec, A., Renauld, J. C., Gajewski, T. F., Duffour, M. T., Warnier, G., Boon, T., and Van den Eynde, B. J. (1997). The expression of mouse gene P1A in testis does not prevent safe induction of cytolytic T cells against a P1A-encoded tumor antigen. *Int. J. Cancer* 70, 349–356.
1245. Giwercman, A., Carlsen, E., Keiding, N., and Skakkebaek, N. E. (1993). Evidence for increasing incidence of abnormalities of the human testis: a review. *Environ. Health Perspect.* 101 (Suppl. 2), 65–71.
1246. Linnet, L., Fogh-Andersen, P., Hjort, T., and Moller, N. P. (1982). Immunoglobulin classes, secretory component, and sperm agglutinins in semen after vasovasostomy. *Am. J. Reprod. Immunol.* 2, 13–17.
1247. Fowler, J. E., Jr., Kaiser, D. L., and Mariano, M. (1982). Immunologic response of the prostate to bacteriuria and bacterial prostatitis: I. Immunoglobulin concentrations in prostatic fluid. *J. Urol.* 128, 158–164.
1248. Fowler, J. E., Jr., and Mariano, M. (1983). Immunoglobulin in seminal fluid of fertile, infertile, vasectomy and vasectomy reversal patients. *J. Urol.* 129, 869–872.
1249. Stern, J. E., Gardner, S., Quirk, D., and Wira, C. R. (1992). Secretory immune system of the male reproductive tract: effects of dihydrotestosterone and estradiol on IgA and secretory component levels. *J. Reprod. Immunol.* 22, 73–85.
1250. Parr, M. B., and Parr, E. L. (1989). Immunohistochemical localization of secretory component and immunoglobulin A in the urogenital tract of the male rodent. *J. Reprod. Fertil.* 85, 115–124.
1251. Dejucq, N., Dugast, I., Ruffault, A., van der Meide, P. H., and Jegou, B. (1995). Interferon-α and -γ expression in the rat testis. *Endocrinology* 136, 4925–4931.
1252. Dejucq, N., Chousterman, S., and Jégou, B. (1997). The testicular antiviral defense system: localization, expression, and regulation of 2′5′ oligoadenylate synthetase, double-stranded RNA-activated protein kinase, and Mx proteins in the rat seminiferous tubule. *J. Cell Biol.* 139, 865–873.
1253. Dejucq, N., Liénard, M. O., and Jégou, B. (1998). Interferons and interferon-induced antiviral proteins in the testis. *J. Reprod. Immunol.* 41, 291–300.
1254. Billiau, A. (1996). Interferon-γ in autoimmunity. *Cytokine Growth Factor Rev.* 7, 25–34.
1255. Farrar, M. A., and Schreiber, R. D. (1993). The molecular cell biology of interferon-γ and its receptor. *Annu. Rev. Immunol.* 11, 571–611.
1256. Boman, H. G. (1995). Peptide antibiotics and their role in innate immunity. *Annu. Rev. Immunol.* 13, 61–92.
1257. Hancock, R. E., and Diamond, G. (2000). The role of cationic antimicrobial peptides in innate host defences. *Trends Microbiol.* 8, 402–410.
1258. Valore, E. V., Park, C. H., Quayle, A. J., Wiles, K. R., McCray, P. B., Jr., and Ganz, T. (1998). Human β-defensin-1: an antimicrobial peptide of urogenital tissues. *J. Clin. Invest.* 101, 1633–1642.
1259. Bals, R., Goldman, M. J., and Wilson, J. M. (1998). Mouse β-defensin 1 is a salt-sensitive antimicrobial peptide present in epithelia of the lung and urogenital tract. *Infect. Immun.* 66, 1225–1232.
1260. Com, E., Bourgeon, F., Evrard, B., Ganz, T., Colleu, D., Jégou, B., and Pineau, C. (2003). Expression of antimicrobial defensins in the male reproductive tract of rats, mice, and humans. *Biol. Reprod.* 68, 95–104.
1261. Yenugu, S., Hamil, K. G., Radhakrishnan, Y., French, F. S., and Hall, S. H. (2004). The androgen-regulated epididymal sperm-binding protein, human beta-defensin 118 (DEFB118) (formerly ESC42), is an antimicrobial β-defensin. *Endocrinology* 145, 3165–3173.
1262. Yamaguchi, Y., Nagase, T., Makita, R., Fukuhara, S., Tomita, T., Tominaga, T., Kurihara, H., and Ouchi, Y. (2002). Identification of multiple novel epididymis-specific β-defensin isoforms in humans and mice. *J. Immunol.* 169, 2516–2523.
1263. Zhou, C. X., Zhang, Y. L., Xiao, L., Zheng, M., Leung, K. M., Chan, M. Y., Lo, P. S., Tsang, L. L., Wong, H. Y., Ho, L. S., Chung, Y. W., and Chan, H. C. (2004). An epididymis-specific β-defensin is important for the initiation of sperm maturation. *Nat. Cell Biol.* 6, 458–464.
1264. Li, P., Chan, H. C., He, B., So, S. C., Chung, Y. W., Shang, Q., Zhang, Y. D., and Zhang, Y. L. (2001). An antimicrobial peptide gene found in the male reproductive system of rats. *Science* 291, 1783–1785.
1265. Alexander, N. J., and Bialy, G. (1994). Contraceptive vaccine development. *Reprod. Fertil. Dev.* 6, 273–280.
1266. McLaughlin, E. A., Holland, M. K., and Aitken, R. J. (2003). Contraceptive vaccines. *Expert Opin. Biol. Ther.* 3, 829–841.
1267. Gupta, S. K. (2003). Status of immunodiagnosis and immunocontraceptive vaccines in India. *Adv. Biochem. Eng. Biotechnol.* 85, 181–214.
1268. Ladd, A. (1993). Progress in the development of anti-LHRH vaccine. *Am. J. Reprod. Immunol.* 29, 189–194.
1269. Sad, S., Chauhan, V. S., Arunan, K., and Raghupathy, R. (1993). Synthetic gonadotrophin-releasing hormone (GnRH) vaccines incorporating GnRH and synthetic T-helper epitopes. *Vaccine* 11, 1145–1150.
1270. Furst, J., Fiebiger, E., Mack, D., Frick, J., and Rovan, E. (1994). The effect of active immunization against gonadotropin-releasing hormone on the ultrastructure of the rat ventral prostate. *Urol. Res.* 22, 107–113.
1271. Bradley, M. P. (1994). Experimental strategies for the development of an immunocontraceptive vaccine for the European red fox, *Vulpes vulpes*. *Reprod. Fertil. Dev.* 6, 307–317.
1272. de Jersey, J., Bird, P. H., Verma, N. K., and Bradley, M. P. (1999). Antigen-specific systemic and reproductive tract

antibodies in foxes immunized with *Salmonella typhimurium* expressing bacterial and sperm proteins. *Reprod. Fertil. Dev.* 11, 219–228.

1273. Boyle, D. B. (1994). Disease and fertility control in wildlife and feral animal populations: options for vaccine delivery using vectors. *Reprod. Fertil. Dev.* 6, 393–400.
1274. Flechner, S. M. (1983). Cyclosporine: a new and promising immunosuppressive agent. *Urol. Clin. North Am.* 10, 263–275.
1275. Rovira, P., Mascarell, L., and Truffa-Bachi, P. (2000). The impact of immunosuppressive drugs on the analysis of T cell activation. *Curr. Med. Chem.* 7, 673–692.
1276. Schreiber, S. L., and Crabtree, G. R. (1992). The mechanism of action of cyclosporin A and FK506. *Immunol. Today* 13, 136–142.
1277. Liu, J. (1993). FK506 and cyclosporin, molecular probes for studying intracellular signal transduction. *Immunol. Today* 14, 290–295.
1278. Pulla, B., Barri, Y. M., and Anaissie, E. (1998). Acute renal failure following bone marrow transplantation. *Ren. Fail.* 20, 421–435.
1279. Ryffel, B. (1992). The carcinogenicity of ciclosporin. *Toxicology* 73, 1–22.
1280. Hales, D. B., Rivier, C., and Shankar, B. K. (1997). Interleukin-6 (IL-6) inhibits cAMP-stimulated testosterone by blocking P450c17 expression in mouse Leydig cells. Proceedings of the 79th Annual Meeting of the Endocrine Society (M. E. Freeman and R. A. Kreisberg, Eds.), p. 209. The Endocrine Society Press, Bethesda, MD.
1281. Rivest, S., and Rivier, C. (1995). The role of corticotropin-releasing factor and interleukin-1 in the regulation of neurons controlling reproductive functions. *Endocr. Rev.* 16, 177–199.
1282. Brand, C., Cherradi, N., Defaye, G., Chinn, A., Chambaz, E. M., Feige, J. J., and Bailly, S. (1998). Transforming growth factor β1 decreases cholesterol supply to mitochondria via repression of steroidogenic acute regulatory protein expression. *J. Biol. Chem.* 273, 6410–6416.
1283. Lin, T., Hu, J., Wang, D., and Stocco, D. M. (1998). Interferon-γ inhibits the steroidogenic acute regulatory protein messenger ribonucleic acid expression and protein levels in primary cultures of rat Leydig cells. *Endocrinology* 139, 2217–2222.
1284. Nathan, C., and Xie, Q. W. (1994). Nitric oxide synthases: roles, tolls, and controls. *Cell* 78, 915–918.
1285. Smith, W. L., DeWitt, D. L., and Garavito, R. M. (2000). Cyclooxygenases: structural, cellular, and molecular biology. *Annu. Rev. Biochem.* 69, 145–182.
1286. Samuelsson, B., Dahlén, S. E., Lindgren, J. A., Rouzer, C. A., and Serhan, C. N. (1987). Leukotrienes and lipoxins: structures, biosynthesis, and biological effects. *Science* 237, 1171–1176.

Pituitary and Hypothalamus

Knobil and Neill's Physiology of Reproduction, Third Edition
edited by Jimmy D. Neill,
Elsevier © 2006

CHAPTER 26

Pituitary and Hypothalamus: Perspectives and Overview*

John W. Everett

Anterior Pituitary Cytology, 1292
Evidence for Neural Control of Pars Distalis Secretion, 1292
Pituitary Portal Vessels, 1293
Neurohumoral Regulation, 1294
Hypothalamic–Pituitary Control of Ovulation, 1295
Brain Stimulation of Gonadotropin in Rats, 1297
Curvilinear Pattern of Ovulatory Gonadotropin Surges, 1298
Actions of Gonadal Secretions on Pituitary Secretion, 1298
Biphasic Action of Progesterone: Interaction with Estrogen, 1299
Episodic Gonadotropin Release, 1299
Steroids and the Phasic Release of Gonadotropins, 1300
Nonsteroid Gonadal Feedback, 1301
Conclusion, 1301
Acknowledgments, 1301
References, 1301

Nearly all our knowledge on the hypothalamic-pituitary system has been gathered during the 20th century, building on fragmentary notions from earlier times. Galen regarded the pituitary as a sump for waste products (phlegm = pituita) derived in the brain from distillation of "animal spirit." Supposedly the phlegm would then filter through openings in the ethmoid bone into the nasal passages. That notion held without question until 1655, when Conrad Victor Schneider of Wittenburg concluded, on anatomical grounds, that the openings in the cribriform plate of the ethmoid bone are for the olfactory nerves and that fluids cannot pass from the cranial cavity into the nose (1). Richard Lower of Oxford confirmed this in 1670 with a case of hydrocephalus and a series of experiments in cadavers. Although intraventricular fluid was greatly increased after injection of water into the jugular veins, no fluid appeared in the nasal cavities. Lower envisioned substances to be conducted from the ventricles through the infundibulum to the pituitary, there to be "distilled" into the blood stream (1).

Recognition of the dual embryonic origin of the pituitary from the diencephalon and the buccal epithelium awaited disclosure in 1838 by Martin Heinrich Rathke (2) of Königsberg, an embryologist noted for discovering the embryonic gill slits and gill arches. Rathke described a dorsal outpocketing from the roof of the stomodeum extending to meet a ventral process from the diencephalic floor. Knowledge of the histological structure of the respective components was many years in the future, however. In the same year as Rathke's report, the cell theory was proposed for plants by Schleiden and extended the following year by Schwann to include animals. Methods for fixing tissues were in use, but histology did not come into its own until the latter part of the 19th century. Staining with hematoxylin (Waldeyer) and aniline dyes (Beneke) was described in 1863, and the microtome was introduced by His in 1870.

*This chapter is reprinted from the 1988 (first) and 1994 (second) editions of this work (E. Knobil and J. D. Neill, Eds. *The Physiology of Reproduction*. Raven Press, New York. Copyright Elsevier.). This now-deceased pioneer in neuroendocrine research describes the history underlying the information presented in the Pituitary and Hypothalamus section. His involvement in many of these developments and discoveries make this chapter a unique and timeless contribution.

Although the distinction between glands with and without ducts was apparent by older methods, the identification of ductless glands structurally specialized for internal secretion required the microscopic observations made possible by these procedures and by improved lens systems.

Claude Bernard first conceptualized "internal secretion," noting that all tissues and organs influence the body as a whole by discharging substances into the blood (3). The idea was extended by Brown-Séquard and d'Arsonval with the view that internal secretions serve to coordinate body functions (3). The concept of control of specific target tissues by circulating messengers ("hormones") was stated by Bayliss and Starling in their Croonian Lecture (4) that reported their discovery of secretin.

Although histology of the pituitary complex (Table 1) clearly showed the glandular features of the pars distalis, the neural lobe did not appear to be a secretory organ. Silver-staining methods devised by Golgi and modified by Ramón by Cajal disclosed a rich content of nerve fibers in the neural lobe that Cajal (5) judged to be sensory fibers with terminals in the surrounding epithelium of the pars intermedia. Camus and Roussy (6), on the other hand, considered the neural lobe to be merely a "fragment nerveux atrophié." Bailey and Bremer (7), as late as 1921, agreed, after observing that experimental removal of the neural lobe had no evident effect. They interpreted pituitrin as nothing more than a "pharmacologically interesting extract." They also noted, with respect to the pars distalis, that in spite of its obviously glandular structure there was "little actual knowledge of its functional significance." Nonetheless, Evans and Long (8) had already produced gigantism in rats by long-term daily treatment with extracts of beef anterior pituitary (AP).

In retrospect, the functional relationship between the pars distalis and body growth is apparent in clinical reports beginning in 1864. The syndrome of acromegaly described by Pierre Marie was shown by Minkowski to be accompanied consistently by the presence of a pituitary tumor (9). Unfortunately, he wrongly assumed that a tumor must represent loss of function and, hence, that the syndrome must be due to impaired pituitary activity. That notion persisted into the 20th century and led to various attempts at experimental hypophysectomy. Most notable of these attempts were reported by Paulesco (9), Cushing and associated (10–12), and Aschner (13,14). Because Paulesco had been unable to obtain survival of hypophysectomized dogs for more than a few days, the concept at first arose that the pituitary is essential for life. There were also arguments about complications from possible damage to the hypothalamus. However, Aschner, operating on puppies, obtained survival for several months in the total absence of the hypophysis and without damage to the brain. The puppies failed to grow. It was by no means clear, however, that this failure represented specific loss of a particular hormone, for there were metabolic defects in the cachectic animals, and the thyroids and adrenal cortex were obviously impaired. Impairment of the reproductive organs could similarly have been due to the general debility. Questions of specificity also applied to the early AP extracts alleged to contain growth hormone. The sorting out of the specific trophic hormones present in the gland and their eventual purification were to occupy investigators for many years to come.

Several other lines of evidence stand in the background of present awareness of the central position of the hypothalamic-pituitary apparatus in regulating gonadal functions. Seasonal influences of the environment on animal reproduction in temperate zones have been well known from time immemorial (15,16) Especially note-worthy is the fact that in some species ovulation and corpus luteum formation require the stimulus of copulation. This was demonstrated in domestic rabbits by both Haighton (17) and Cruikshank (18) in presentations to the Royal Society of London in 1797, confirmed later by Barry (19), Heape (20), and many others. By mating rabbits after cutting the fallopian tubes, Haighton showed that the effect could not be due to contact of the semen with the ovary. Provoked ovulation was later described in ferrets (21) and domestic cats (22). Numerous other species have been added in recent years (1).

Retroactive influence of the gonads on the AP was recognized by Fichera (23), who reported that in several species gonadectomy was followed by AP enlargement and, in the rat, by the occurrence of many enlarged, vacuolated cells having a signet-ring appearance. Such "castration cells" were identified by Addison (24) as belonging to the basophil cell class. Other phenomena now recognized as reflecting the influence of the gonads on the AP were the suppression of estrous and menstrual cycles during pregnancy and the prompt return of cycles after removal of corpora lutea (25). The tendency for lactation to

TABLE 1. *Terminology of the subdivisions of the hypophysis*

Neurohypophysis	Median eminence	
	Infundibular stem	Infundibulum
	Neural lobe (infundibular process)	Posterior lobe
Adenohypophysis (glandular lobe)	Pars intermedia	
	Pars tuberalis	
	Pars distalis	Anterior lobe (AP)

suppress cycles, likewise a hypothalamic-pituitary effect, was well known from early times.

The period from 1916 to 1926 was marked by a series of discoveries fundamental for modern endocrinology and reproductive biology. Development of the vaginal smear method by Stockard and Papanicolaou (26) to reveal the estrous cycle of the guinea pig was a breakthrough that led eventually to the isolation and synthesis of the estrogens. The method was soon applied to the rat by Long and Evans (27), and to the mouse by Edgar Allen (28). Allen and Doisy (28a) and Allen et al. (29) soon reported that oily extracts of porcine follicular fluid when injected into spayed mice brought about full cornification of the vagina; the Allen-Doisy test became a basic procedure for the biochemical studies that followed.

Procedures for hypophysectomy received continuing attention. Both Allen (30,31) and Smith (32) investigated effects of hypophysectomy (removal of Rathke's pouch) in amphibian larvae, nothing atrophy of the adrenals and thyroids and failure to metamorphose. Smith and Smith (33,34) counteracted these losses by intraperitoneal administration of bovine AP material. Smith (35–37) subsequently devised the parapharyngeal technique for pituitary ablation in rats, a breakthrough that enabled isolation, purification, and chemical characterization of the various AP hormones. Smith and Engle (38) promptly demonstrated maintenance and repair of gonads by repeated intramuscular implants of fresh AP tissue in hypophysectomized rats.

No less important was the classic monograph by Long and Evans (27) detailing many aspects of reproduction in the rat. The usual length of the estrous cycles was shown to be either 4 to 5 days, not 10 days or longer, as most earlier reports had indicated. Stages of the vaginal cycle were defined and correlated with events in the ovary, reproductive tract, and mammary glands. Pseudopregnancy was described and named as the result of infertile copulation or mechanical stimulation of the cervix. The production of deciduomata by uterine trauma was described, and the importance of the critical timing of traumatization was noted. Since pseudopregnancy occurred readily in animals whose ovaries had been transplanted, it did not involve the ovarian innervation. In the correlated investigation by Evans and Long (8) of the effects of prolonged daily intraperitoneal administration of bovine AP substance, in addition to gigantism the females showed marked enlargement of the ovaries with masses of corpora lutea, together with suppression of estrous cycles. Here was the first demonstration of the presence of both growth-promoting and gonad-stimulating substances in the anterior lobe.

Evans et al. (39,40) succeeded in separating fractions of AP extracts having these respective activities. Separation of thyroid-stimulating and gonadotropic fractions was achieved by Greep (41). Eventually, by the late 1930s the existence of six AP hormones had been established: hormones controlling body growth, the gonads, mammary glands, thyroids, and adrenal cortex. The advances toward their eventual isolation and determination of their chemical structures required progressive sophistication of in vitro techniques. This was accompanied by progressive refinement of definitions of the biological actions of the respective hormones.

In the case of the gonadotropins, progress was delayed from time to time by certain faulty assumptions, sometimes engendered by impurities in the substances administered. The existence of two gonadotropic principles was first proposed by Zondek (42,43), who had obtained a follicle-stimulating material (Prolan A) from the urine of ovariectomized or postmenopausal women and a luteinizing material (Prolan B) from human pregnancy urine. He assumed both to be secreted by the AP. Although Prolan B was later shown to originate from the placenta (44–47), the idea of a separate luteinizer led to intensive search for distinct pituitary principles having potencies resembling the two Prolans. The dual hormone concept was championed by Fevold, Hisaw, and associates in Wisconsin. Fevold et al. (48) were first to succeed in separating the respective fractions from pyridine extracts. Details of the efforts leading to eventual isolation and determination of the chemical structure of follicle-stimulating hormone (FSH) and luteinizing hormone (LH) have been admirably reviewed by Greep (49). A major misapprehension not emphasized was the early failure to distinguish formation, per se, of corpora lutea (luteinization) and maintenance of their function (luteotropic action). In rats and mice the principal luteotropin is prolactin, a fact unknown until demonstrated in 1941 by Astwood (50) and Evans et al. (51).

Prolactin research stems from an observation by Stricker and Grueter (52,53) after induction of ovulation in rabbits by administration of AP extract. In rabbits that received the material for several days after ovulation the mammary glands were distended with milk. The effect was confirmed by Corner (54), who also obtained it in ovariectomized estrous rabbits. Evans and Simpson (55) saw mammary glands distended with milk in rats that had received an AP extract for 20 to 30 days. A series of studies by Riddle and associates (56–59) between 1931 and 1935 disclosed that, in the pigeon and ring-dove, secretion of crop milk was governed by the AP and that the same hormone, which they named prolactin, caused milk secretion in mammals. It suppressed gonadal function by inhibiting FSH secretion. In both mice and rats, daily treatment interrupted estrous cycles for approximately 2 weeks (60,61). Although Lahr and Riddle (61) saw large corpora lutea, demonstration of the luteotropic action of prolactin was 5 years in the future, as noted above. The growth-promoting power of prolactin was shown in hypophysectomized pigeons, as for years Riddle, Bates, and others questioned "the concept of a growth hormone as an individual entity." The disagreement was finally resolved by Bates et al. (62) through use of highly purified AP hormones. Optimal metabolic effects were obtained in hypophysectomized pigeons by giving a combination of ovine prolactin, bovine growth hormone, thyroxine, and prednisone.

ANTERIOR PITUITARY CYTOLOGY

From the very beginning of microscopic observation of the pituitary, considerable variation was evident in size, shape, and granule content of pars distalis cells. Application of dyes disclosed coloration differences as well. Schönemann (63) is credited with having made the original distinction among basophils ("cyanophils"), acidophils ("eosinophils"), and "chromophobes" as seen in human pituitaries stained with alum hematoxylin and eosin. Holmes and Ball (64) present a useful summary of the complex story of efforts in this century to subdivide these classes and to relate the subclasses to particular hormones and functional states. Underlying these efforts was the developing knowledge of the number of AP hormones, coupled with the feeling that the number of cell types ought to correspond.

Early confusion arose from the finding that in most animals other than humans the basophil cells cannot be stained by hematoxylin. Staining of ergastoplasm by basic dyes like toluidin blue added further confusion. The use of the multiple acid dyes introduced by Mallory (65) produced an array of tinctorial cell types. Modifications of the Mallory procedure were especially useful in some species and appeared to distinguish cells with granules containing simple proteins from cells having granules of glycoprotein. Application of the McManus (66) periodic acid (PAS) histochemical method provided a generally useful distinction between "serous" cells (67) and "mucoid" cells (68). The latter types are those producing FSH, LH, melanocyte-stimulating hormone (MSH), and thyroid-stimulating hormone (TSH), all of which are glycoproteins.

The eventual availability of highly purified AP hormones and antibodies and the advent of immunocytochemical methods (69) in both light microscopy and electron microscopy have led far in recent years toward the ideal of a functional classification. The older terminologies have given way largely to the naming of cells according to the hormone produced, whenever known. Some problems remain, as when a gonadotropic cell contains granules that react with other FSH and LH antibodies. Special difficulties attend the localization of adrenocorticotropic hormone (ACTH), MSH, and the related peptides, several of which may be contained within a parent molecule in some cells (70).

Pituitary histology and cytology have moved hand-in-hand with progress in pituitary physiology and biochemistry. Regional variations of cell types in the bovine AP made it possible for Smith and Smith (33) to distinguish the selective growth-promoting power of the acidophilic lateral portions from the thyrotropic power of the anteromedial portion containing basophils and chromophobes. Enlargement of AP and castration cell formation after gonadectomy provided means for testing the feedback actions of gonadal hormones. The lack of castration cells in rat AP transplants to sites away from the brain was an early sign of the importance of the brain in regulating AP secretion (71).

EVIDENCE FOR NEURAL CONTROL OF PARS DISTALIS SECRETION

Frölich's (72) original description of the syndrome termed urogenital dystrophy was the start of prolonged controversy about the respective roles of damage to the hypothalamus or the pituitary gland in causing the disease. An intracranial cystic tumor thought to be a craniopharyngioma was surgically drained, after which there was satisfactory recovery. Eventually, following experimental studies by Camus and Roussy (6), Bailey and Bremer (7), Smith (36), and Hetherington and Ranson (73–75), it became clear that both obesity and genital atrophy can result from injury to the hypothalamus without direct involvement of the AP. Among 60 clinical cases of hypothalamic pathology assembled by Bauer (76), 43 manifested either hypogonadism or sexual precocity, five showing obesity. Precocity was commonly associated with basal tuberal lesions, while hypogonadism accompanied lesions located more rostrally.

Smith's observations of the effects of hypothalamic damage were made as he developed his method for parapharyngeal hypophysectomy (36). One attempt to ablate the gland was to inject chromic acid into the pituitary capsule, with the result that some of the acid passed beyond and damaged the brain. He then found that selective damage to the hypothalamus itself produced genital atrophy and obesity. Soon afterward Grafe and Grünthal (77) reported similar results.

During the following 25 years, paralleling the progressive purification of the several AP secretions and determination of their actions, there was growing interest in the possible control of these secretions by the nervous system. Although the Moore and Price hypothesis (78) of pituitary–gonad reciprocity did not seem to take the nervous system into account, the authors recognized the modifying influences of environmental factors.

There was also strong implication of neural participation in the special case of the coitally induced ovulation in rabbits. Fee and Parkes (79,80) demonstrated two important features of that process; since the hypophysis must remain in place for only an hour after copulation, the necessary release of hormone must be relatively acute, and since anesthesia of the vagina prevented ovulation, the nervous system must be directly involved. Participation of the brain was more directly indicated by Marshall and Verney (81), who obtained ovulation in estrous rabbits by passing an electric current through the head. With greater precision, Harris (82) and Haterius and Derbyshire (83) produced the same effect by local stimulation of the hypothalamus. The importance of the pituitary stalk to complete the reflex connection to the pars distalis was shown by Westman and Jacobsohn (84). Ovulation failed when the stalk had been cut and a metal foil barrier placed between the hypothalamus and pituitary gland, supposedly interrupting neural connections. The result also seemed to eliminate participation of autonomic fibers directly innervating the pars distalis.

Several workers had described autonomic fibers within the pars distalis parenchyma. However, a careful study by Rasmussen (85) of the pituitaries of human, rat, guinea pig, rabbit, dog, cat, and monkey led him to conclude that the only nerve fibers present in the AP must be vasomotor. Many parts of the gland were free of nerves. This view has been widely corroborated.

Meanwhile, the concept of a "sexual center" in the hypothalamus arose from the studies in rats by Hohlweg and associates (76,86), confirmed and amplified by Westman and Jacobsohn (87–89). As noted above, Hohlweg and Junkmann (71) saw that gonadectomy failed to produce castration cells in hypophyses that had been transplanted to a site away from the brain. Searching for effects of estrogen on the AP, Hohlweg and Chamorro (86) discovered that a single injection of estradiol benzoate into prepubertal rats induced formation of a set of corpora lutea within 7 days, presumptive evidence of increased LH secretion and ovulation. Luteinization was prevented by hypophysectomy 2 days after injection, but not at 4 days. Westman and Jacobsohn (87–89), after confirming the Hohlweg effect, examined the results of stalk section and insertion of a barrier of metal foil. Stalk section less than 2½ days after estrogen injection prevented luteinization, but not when performed later on. As in the rabbit reflex, the estrogen stimulus was thought to operate through an essential neural link from the hypothalamus. The pituitary portal vessels were not yet generally recognized.

Pituitary Portal Vessels

Prominent vessels on the pituitary stalk were first well described by Pietsch (90) and Popa and Fielding (91,92). The latter authors described them as *portal vessels connecting capillary beds in the median eminence and pars distalis*. Erroneously, they concurred with Pietsch that blood moves "upward" from the gland toward the brain. That conclusion was based partly on the experimental observation in rabbits that when a clamp was placed on the stalk the vessels distended below the clamp. Unknown at the time was the fact that rabbits, unlike most mammals, have a direct arterial supply to the pars distalis.

Evidence for "downward" flow in the portal vessels was first reported by Houssay et al. (93) from direct microscopic observation of the exposed vessels in South American toads. Furthermore, lesioning of the infundibulum caused infarction of the pars distalis. On histologic grounds, Wislocki and King (94) and Wislocki (95) reached the conclusion that the primary capillary bed occupies the median eminence and the infundibular stem, whence blood is transported by the portal vessels to the secondary capillaries in the pars distalis. They further recognized the median eminence as a part of the neurohypophysis, not of the hypothalamus. Direct observation of downward blood flow was later reported in several species [amphibia: Green (96); duck: Benoit and Assenmacher (97); rat: Green and Harris (98,99) and others; mouse: Worthington (100,101); dog and cat: Török (102,103); baboon and monkey: Daniel (104)]. Although the possibility of some recurrent flow was indicated recently (105–107), the present view is that the principal flow is downward, transmitting hypophysiotropic agents to the AP.

Considerable territorial specificity exists in the origins of the various portal vessels from the primary capillaries and distributions of the secondary capillaries within the pars distalis. After stalk sectioning in the rat, sheep, goat, monkey, and human the extent of the infarcted area in the AP depends on the number and location of portal vessels interrupted (108). Daniel (104) proposed that the territorial specificity may explain the regional variation of cell types. An experimental basis for such a view was shown by Pasteels (109). In the amphibian *Pleurodeles waltlii*, he removed the hypophysis and reimplanted it in the original location, but inverted and rotated 180 degrees. The cytology of the gland became regionally reversed in accord with the relationship to the portal circulation. A contrary view has been expressed by Porter et al. (110) after they found in rats that when the portal vessels were exposed and a colored fluid was injected into a single vessel through a microcannula, the distribution of the dye in the AP was greatly varied, sometimes reaching the entire gland. One may suggest, however, that under physiological conditions of balanced pressure territorial specificity would be maintained.

Neurohumoral Regulation

The concept of neurohumoral control of the pars distalis can be traced back to a paper by Hinsey and Markee (111). Failing to prevent postcoital ovulation in rabbits by severing the cervical sympathetic trunks, they suggested that the reflex involves a humoral link, transmitting some substance from the "posterior lobe" to the pars distalis. After Wislocki and King (94) presented their strong evidence for downward movement of blood in the portal vessels, there was speculation that these channels might constitute the humoral pathway (see ref. 112 and accompanying discussions). Resistance to the idea continued for a while, partly because in avian and cetacean forms a heavy connective tissue septum separates the neural lobe from the pars distalis. However, it has been fully demonstrated that in such cases a "portotuberal tract" transmits the portal vessels rostral to that septum (97,113). Green (114) concluded from his comparative study that the pituitary portal vessels or their equivalent are a constant feature of all vertebrates.

Green and Harris (98), reversing Harris' earlier view (82), presented several strong arguments in favor of the humoral transmission via the portal vessels. They proposed that the variable effects others had encountered from cutting the stalk may have resulted from ability of the vessels to regenerate. Harris (115) conclusively demonstrated in stalk-sectioned rats that recovery of reproductive function was directly proportional to the extent of portal vessel regeneration. Regeneration was also demonstrated in monkeys (116) and rabbits (117). Thus, to interrupt fully hypothalamic influence on the AP by stalk section, an impermeable barrier is mandatory.

These findings seemed to explain the restoration of function in the pars distalis grafts that Greep (118) placed in the original site after hypophysectomy. Yet there remained some uncertainty, since it would be difficult to distinguish remnants of the gland left in situ. To avoid this, Harris and Jacobsohn (119) modified the experiment with a dual operation, hypophysectomizing female rats by the parapharyngeal route and implanting pars distalis tissue by the transtemporal approach to a location beneath the median eminence. Donors of the grafts were either several of the subjects' own male or female young, adult females, or adult males. Control subjects received the grafts under the temporal lobe. Estrous cycles returned in all rats bearing infantile grafts under the median eminence, often within a week. Several became pregnant and delivered normal young. Adrenals and thyroids were histologically normal. Significantly, similar results appeared in the few subjects bearing adult male grafts, one having a successful pregnancy. No traces of AP tissue were microscopically detectable in the original sites. These highly significant findings were confirmed and supplemented by Nikitovitch-Winer and Everett (120,121) in a two-stage experiment. The pars distalis of adult female rats was first autografted to the renal capsule where capacity for secretion of FSH, LH, ACTH, and TSH was largely lost. Several weeks later, the graft was retransplanted by the transtemporal route beneath the median eminence. Controls either received the graft under the temporal lobe or were left with the graft on the kidney. Estrous cycles reappeared in most rats having grafts revascularized from the median eminence, and in many cases the animals became pregnant when mated. Secretion of ACTH and TSH appeared to be moderately restored, whereas the control grafts continued to secrete mostly prolactin. Grafts under the median eminence contained many large gonadotrophs and thyrotrophs, which were absent from the controls. Thus, in spite of the double surgical insult with considerable loss of tissue due to infarction at each operation, the gland could rapidly renew its several trophic secretions once it received blood from the median eminence. Smith (122,123) contributed a further confirmatory variation of Greep's experiment. In both male and female rats, months after hypophysectomy he introduced homografts of pars distalis tissue near the median eminence through the reopened pituitary capsules. Fertility was restored along with significantly improved thyrotropic and adrenotropic functions. During the long intervals after hypophysectomy the typical apituitary syndrome had shown the essential completeness of that operation. Since hypophysectomy was performed in young subjects around 40 days of age, the return of somatotropin secretion was demonstrable after the grafts were introduced.

Thus it became fully evident that the hypophysial portal circulation is essential for stimulating the pars distalis to secrete FSH, LH, TSH, ACTH, and somatotropin (GH, STH). Although Green and Harris (98) had suggested that there might be separate transmitter agents, excitatory or inhibitory, controlling these respective secretions, there were alternative possibilities (124). A single agent might act permissively or its different concentrations in the portal blood might selectively influence particular secretions. Nevertheless, the territorial variations in cell types argued strongly for multiple stimulative factors. Saffran et al. (125) coined the term *releasing factor* and the acronym RF.

Although inhibitory neural action on the pars intermedia was known, the first evidence of inhibitory neurohumoral control of a pars distalis secretion emerged in experiments involving autotransplantation of the gland to the renal capsule (126,127). When the operation was performed in female rats on the day after ovulation, the newly forming corpora lutea were activated and continued to secrete progesterone for weeks or months until the experiment concluded. Since the principal luteotropin in rats is prolactin, the results demonstrated enhancement of prolactin secretion by removal of the pars distalis from hypothalamic influence. Although not recognized at the time, this effect had previously been obtained by Westman and Jacobsohn (89) after separating the gland from the hypothalamus by stalk section and placement of an impermeable barrier; they interpreted the result as being due to stimulation of the cervix. Desclin (128) also had evidence of prolactin secretion by transplanted pars distalis, but thought it due to the direct stimulative effect of treatment with estrogen. Nikitovitch-Winer (129) repeated the stalk-sectioning experiment of Westman and Jacobsohn, but omitting the cervical stimulation showed clearly that the enhanced prolactin secretion must be due to removal from hypothalamic inhibition. In vitro organ cultures and tissue cultures of pars distalis from several mammals show autonomous secretion of prolactin, increasing with time, while other secretions diminish (130–134). Addition of hypothalamic tissue or extract to the cultures reduces prolactin output, extracts of cerebral cortex having no such effect.

Hypothalamic Hormones

Several different releasing (and inhibiting) factors are recognized today as hypothalamic hormones transmitted to the pars distalis via the portal vessels, each factor having more or less selective action on pituitary secretions. The agents of primary importance for reproduction include first and foremost the luteinizing hormone-releasing hormone LHRH (or GnRH, since it stimulates both FSH and LH secretion). Thyroid-releasing hormone (TRH), the releasing agent for the thyroid-stimulating hormone, has the added capacity for stimulating prolactin secretion. The chemical structures of both LHRH, a decapeptide, and TRH, a tripeptide, are known, and both are available in synthetic form. The remarkable story of the elucidation of their chemistry and synthesis has been fully told by Wade (135). Prolactin-inhibiting factor (PIF), remains to be fully characterized. Dopamine, a potent inhibitor (136), appears in the arcuate nuclei-median eminence and portal vessels (137), but may not be the only inhibiting agent.

Hypothalamic–Pituitary Control of Ovulation

Pharmacologic Blockade

In rabbits, the sequence of events provoked by the copulatory stimulus and culminating in ovulation begins with a very brief triggering period during which the process can be blocked by the intravenous injection of certain drugs (138–140). The α-adrenergic blockers Dibenamine or a congener SKF-501, when injected intravenously within 1 minute after coitus, prevented ovulation in 80% of the subjects. Atropine sulphate or another cholinergic blocker, Banthine, also blocked, but they had to be injected more rapidly still, indicating that a cholinergic mechanism precedes the adrenergic trigger. This was further indicated when ovulation was induced by injecting a high dose of Adrenaline (Parke-Davis) into estrous rabbits protected from the lethal effects by atropine. The initial postulate that the neurohumoral transmitter carried to the pars distalis by the portal vessels is a catecholamine was later abandoned. Sawyer (141) determined that ovulation could be induced in rabbits by intraventricular injection of either epinephrine or norepinephrine [see also (142)]. Thus, both the cholinergic and adrenergic links in the rabbit ovulatory reflex are central neural processes, completed within a very few minutes post coitum. Although no means has been devised for sampling the portal blood content after the mating stimulus, indirect evidence indicates that LHRH rises rapidly during the first hour, and then remains high for several hours (143). The LHRH rise is thought to slightly precede the LH surge, which peaks at 60 to 90 minutes, remains high for approximately an hour, and then declines (143–146). The prolonged high level of LHRH recalls a report by Westman (147), who placed clamps on the pituitary stalks of estrous rabbits, mated them, and removed the clamps 50 to 60 minutes later; ovulation followed unless the pituitaries were also removed. A related observation by Westman and Jacobsohn (84) was that, while hypophysectomy 30 minutes post coitum prevented ovulation, exsanguinations and replacement with blood from another rabbit at the time allowed ovulation to take place. It failed, however, if exsanguinations and replacement were delayed until 75 to 90 minutes post coitum. In the first instance, the continuing LH surge surely added enough for ovulation, whereas in the latter case any added LH was inadequate.

Reflex Ovulators

During the interval between the LH surge and ovulation, in both reflex ovulators and spontaneous ovulators, certain maturation changes take place in the Graafian follicles and ova: follicular hyperemia and swelling, appearance of "secondary liquor," dispersal of cells in the cumulus oophorus, formation of a prominent corona, production of the first polar body (typically), and formation of the second polar division spindle. In retrospect, the relatively brief preovulatory spurt of follicle enlargement described in the guinea pig by Dempsey (148) was an early indication of a spontaneous abrupt increase of gonadotropin secretion.

Spontaneous Ovulators

Clear demonstration of acute timing of a preovulatory gonadotropin surge in the spontaneously ovulating rat was presented in a series of studies by Everett, Sawyer, and associates (149–155) employing a number of drugs including those that blocked the rabbit reflex. In addition to Dibenamine and atropine, drugs effective in rats included several barbiturates, chlorpromazine, reserpine, morphine, and urethane. Their use defined a "critical period" of 2 to 3 hours on the afternoon of proestrus, beginning predictably at approximately 1400 hours. Analysis by injecting atropine sulphate or pentobarbital or by hypophysectomy at progressively later times during the critical period indicated temporal variation in the beginning of LH release among different individuals (156–158). Hormone release could be interrupted while in progress as readily by the drugs as by hypophysectomy. Hence, although an atropine-sensitive mechanism controls the spontaneous surge in rats, the action is essential throughout the surge, not a brief trigger, as in the rabbit reflex. The analysis also indicated that the time needed to release the minimal ovulation quota of LH in rats is approximately 30 minutes. There were indications that the normal surge continues much longer, however. Modern data from radioimmunoassay confirm this finding and show that the full amount of LH released is far greater than the minimum quota. The surge begins during the critical period, but reaches peak level later, remaining high for 2 to 3 hours and returning to baseline typically by 1800 to 1900 hours (159). It is provoked by a rise of LHRH in the pituitary-portal blood (160).

The occurrence of acute preovulatory surges of LH and FSH is now recognized as universal among spontaneous and reflex ovulators alike. The duration of the surges tends to be considerably longer in larger mammals; in the human female and other large primates it exceeds 24 hours.

The use of barbiturates for blocking spontaneous ovulation in rats led to three discoveries: (a) circadian periodicity of the neural stimulus for release of the LH surge; (b) "delayed pseudopregnancy"; and (c) the fact that rats pharmacologically blocked from ovulating spontaneously can nevertheless be made to release an ovulatory surge of LH by stimulating the brain.

Circadian Periodicity of the Luteinizing Hormone Surge Mechanism

Rats that were blocked with pentobarbital on the proestrus afternoon presented a similar critical period on the next day; if they were blocked at that time, the critical period was repeated on the third afternoon (152). Blockade for 3 days produced an anovulatory cycle, followed by a short diestrus and return of cyclic ovulation. Thus, circadian rhythmicity was clearly evident. This had been implied by the 24-hour advance of ovulation through treatment of rats with progesterone (161), by the fact that this advance could be blocked pharmacologically on the preceding afternoon (151), and by the temporal dependency of the LH surge on regularity of environmental lighting (162). As noted later, although the close relationship of the spontaneous ovulatory surge of LH to time of day has been amply demonstrated in rats and hamsters, this feature appears to be limited to the small polyestrous rodents.

Delayed Pseudopregnancy

A serendipitous finding occurred during experiments that were intended to search for reflex ovulation in rats blocked with pentobarbital (163,164). They were caged with fertile bucks overnight, and proof of copulation was shown next morning by presence of vaginal plugs and spermatozoa. A few had ovulated, but when similarly treated rats were blocked again on the second afternoon, the usual result was early follicular atresia, a short diestrus of 2 to 3 days, then a new proestrus and estrus with spontaneous ovulation, followed by pseudopregnancy of normal duration. The influence of the copulatory stimulus had been retained for over a week, supposedly through some functional change in the central nervous system. Such a long-delayed effect could not be obtained by stimulating the cervix, but shorter delay had been shown by Greep and Hisaw (165) when cervical stimulation during late diestrus caused pseudopregnancy beginning 2 days later, after completion of the cycle. Quinn and Everett (166) produced delayed pseudopregnancy by electrically stimulating the dorsomedial–ventromedial hypothalamus, a stimulation site not conducive to ovulation. As with genital stimulation, there was a quantitative effect, depending in this case on duration of the stimulus: 10-minute stimulation was adequate for short-term delay, but 30-minute stimulation was needed for long-term delay. This was confirmed by Beach et al. (167,168), who also recorded twice-daily surges of prolactin secretion during the long-delay interval. It is known, however, that these surges during the interval are not essential to the subsequent pseudopregnancy (169). The mnemonic influence of genital or hypothalamic stimulation suggests a protracted change in brain chemistry somehow expressed in the prolactin surges.

Brain Stimulation of Gonadotropin in Rats

In spite of the strong indirect evidence that the central nervous system controls spontaneous as well as reflex ovulation, until 1957 there were no reports of gonadotropin release induced in spontaneous ovulators by brain stimulation. In that year, Anand et al. (170) appear to have had some success from daily hypothalamic stimulation of monkeys early in their menstrual cycles. Bunn and Everett (171), using rats in constant estrus under continuous illumination, consistently induced ovulation and luteinization by electrical stimulation of the amygdala and, in one case, the lateral septum. Critchlow (172) discovered that rats blocked during the proestrus critical period with pentobarbital can be ovulated by stimulating the medial basal tuber (MBT) close to the median eminence. Such pharmacologically blocked rats have become favorite experimental subjects for artificial stimulation of gonadotropin release.

Attempts in the writer's laboratory to repeat Critchlow's work led to the demonstration that the medial preoptic area (MPOA) is an especially useful stimulation site. In contrast to the MBT, electrode placement in the MPOA is much less restricted. Everett and Radford (173) encountered the fact that passage of anodic direct current or anodic pulse trains through electrodes containing iron will produce an irritative focal lesion. Such an *electrochemical stimulus* in the MPOA or anterior hypothalamus can induce ovulation, its effectiveness being determined by the amount of iron deposited. A dose-response relationship between size of the electrochemical (EC) focus and the amount of LH released has been well documented, initially by its ovulatory effectiveness (174–177), and later by radioimmunoassay of the LH surge. Large bilateral EC foci involving most of the MPOA may produce enough gonadotropin within 30 minutes for full ovulation, as shown by hypophysectomy at that time (174). The amount of LH in the complete surge produced by the large bilateral stimulus reportedly resembles the normal proestrous surge in magnitude and duration (176).

Electrochemical stimulation is especially useful because of the very short time required for the passage of current (<60 seconds) and the long period of the stimulative action. Plasma LH typically rises slowly at first, then rapidly to reach peak concentration at 90 to 120 minutes, falling gradually thereafter. The stimulative processes near the EC lesion are poorly understood, but increased electrical activity in zones 0.4 to 0.8 mm from center of EC lesions has been reported (178). Spiking activity has also been observed 1 mm behind the MPOA, rising sharply after some 15 minutes and continuing for several hours (179). Although on the day after stimulation the EC lesion consists of a coagulated core of damaged tissue surrounded by an extensive halo devoid of nerve cells and fibers, on the preceding day soon after the passage of current, the region of the eventual halo shows no apparent neuronal damage. It is likely that this outlying region is the site of the increased neural activity. Although there has been some controversy over the possibility that EC lesions act by disinhibition rather than by direct neuronal excitation (180,181), the weight of evidence favors the latter alternative. After Hillarp et al. (182) placed electrolytic lesions with nichrome electrodes in the lateral preoptic areas of male rats there were pronounced behavioral manifestations for as long as 6 hours. Recent investigations by Willmore et al. (183,184) of the effects of ferrous ions introduced into the cerebral cortex indicate that the resulting epileptiform focus is produced by transient formation of free radical oxygen, hydroxyl radicals, and peroxides, causing neural lipid peroxidation.

A major limitation of electrochemical stimulation, aside from uncertainties about how it acts, is that a bulk of tissue is destroyed. That precludes repeated stimulation at the same site and may eliminate neurons important for effective stimulation elsewhere. When EC stimulation is applied near the median eminence, the damage itself may confound the experiment nonspecifically by emptying LHRH into the portal vessels. Knife cuts across the basal tuber (185) or radiofrequency lesions in the arcuate nucleus-median eminence (186) can cause release of an ovulatory quantity of LH. A further limitation of EC stimulation is that the duration of stimulation cannot be controlled.

Such problems are avoided by *electrical stimulation*, especially with platinum electrodes and nonlesioning current, such as matched biphasic pulse pairs. As with EC stimulation, electrical stimulation can induce ovulation in rats under blockade with pentobarbital (and several other drugs), except that it must continue much longer. The amount of LH released and the proportionate numbers of rats ovulating vary with microamperage of the pulses, overall duration of the stimulus, pulse frequency, and other features (187–192). The amount of LH released in a given time by MPOA electrical stimulation is less than expected after EC stimulation, perhaps because of involvement of fewer components of the preoptic-tuberal neuronal system, judged to be diffuse at that level (187–189). Where that system converges upon the arcuate nucleus-median eminence (ARC-ME), electrical stimulation consistently produces more LH than stimulation of the MPOA (174).

Curvilinear Pattern of Ovulatory Gonadotropin Surges

Comparison of the patterns of LH secretion induced by MPOA and ARC-ME stimulations (189) presents a distinct curvilinear parallel, with relatively slow increase during the first 30 to 60 minutes, followed by rapid increase thereafter. The greater amount of LH discharged during the second hour corresponds to the great excess commonly produced in the spontaneous proestrus surge and undoubtedly reflects the self-priming action of LHRH on the pars distalis (193–195). Aiyer et al. (193) disclosed that if two identical amounts of LHRH were injected intravenously 1 hour apart, plasma LH rose sixfold after the second injection. Comparable priming occurred in vitro (195). Fink et al. (194) noted a similar response pattern when the MPOA was electrically stimulated with two 15-minute pulse trains at a 45-minute interval. Grieg and Weisz (196) calculated that only approximately 15% of the normal surge is needed for ovulation. This conforms with the minimal ovulation quota released in the first 20 to 40 minutes as estimated from the results of hypophysectomy or atropine block during the critical period (156,157).

The curvilinear pattern of the ovulatory surge of LH (and FSH) is apparent in other species, both reflex ovulators and spontaneous ovulators, provided that frequent blood samples are assayed. Note, for example, the characteristic surges at midcycle in rhesus monkeys shown by Weick et al. (197). From beginning to end, the LH surge in monkeys lasts for 48 hours, but how much of this constitutes the minimal ovulation quota? That information is available only for the rat and, from the pioneer hypophysectomy experiments of Fee and Parkes (79,80) and others (198), for rabbits. While the quota is released within 60 minutes post coitum in rabbits, peak levels of plasma LH may be reached later and are said to continue for 1 or 2 hours longer (199). The function of the great excess is not understood, although one effect was noted in rats (156). After hypophysectomy or atropine blockade during the critical period, rats that had ovulated by the next morning characteristically lacked the depletion of cholesterol from the interstitial tissue normally shown at that time.

ACTIONS OF GONADAL SECRETIONS ON PITUITARY SECRETION

Functional mammalian corpora lutea suppress estrous cycles and ovulation. This view, expressed by Beard late in the 19th century (200), was supported experimentally by Loeb (201), who noted the early return of estrus and ovulation in guinea pigs after removal of corpora lutea during pregnancy or during the luteal phase of the cycle. Luteectomy in the pregnant goat (202) and cow (203) had the same effect. After confiming Loeb's finding in cycle guinea pigs (204), Papanicolaou (205) determined that administration of a lipid extract of corpora lutea had the same inhibitory action as the active luteal tissue. Similar findings were reported for the mouse (206) and the rat (207). Once progesterone became available, its daily administration gave the same result (208,209). According to Kennedy (210) and Mahnert (211) corpus luteum extracts prevented ovulation in rabbits if injected before coitus: that effect was later shown by Makepeace et al. (212) with pure progesterone. Mahnert, in fact, speculated on the possibility of female sterilization with luteal extracts, apparently the first expression of an idea that would emerge 30 years later as a primary means of population control, the Pincus Pill. Inhibition of gonadotropic potency of the AP by gonadal hormones was the basis for the Moore and Price (213) hypothesis of pituitary-gonadal reciprocity. The idea of negative feedback control was so firmly entrenched by the early 1930s that the equally important stimulative actions were slow to emerge.

The first indication of a positive stimulative influence of estrogen lay in its induction of early puberty (214). As mentioned earlier, Hohlweg and Chamorro (86) induced corpus luteum formation in prepubertal rats by treating them with estradiol benzoate, a result that could be prevented by hypophysectomy 2 days later. Confirming this finding, Westman and Jacobsohn (87–89) noted that the effect could also be prevented by pituitary stalk section. In adult rats, Hohlweg (215) observed that daily, month-long injections of estrogen resulted in great enlargement of the corpora lutea, an effect that can now be ascribed to the luteotropic action of prolactin. Although he states (216) that he observed estrogen-induced ovulation in adult rats, I am not aware of a published record.

Induction of ovulation by estrogen treatment of adult mammals was first recorded by Hammond et al. (217) for anestrous ewes and confirmed by Hammond (218) and Casida (219). Everett (220) reported that in rats having regular 5-day cycles ovulation was advanced 24 hours by administering estradiol benzoate or implanting an estradiol crystal on the second day of diestrus. In pregnant (221) or pseudopregnant rats (222), treatment with estradiol benzoate on day 4 or 5 of vaginal leukocytosis resulted in renewed ovulation and formation of new corpora lutea. These stimulative effects in the rat were subject to pharmacologic blockade with either atropine or Dibenamine (222,223).

The first indication that progesterone has a positive as well as a negative influence on gonadotropin secretion arose from Everett's observation that in rats persistently presenting spontaneous vaginal estrus and polycystic ovaries, ovulation could be induced by certain progesterone treatments. It was known that in normal rats the minimal daily subcutaneous dose for suppressing estrous cycles is 1.5 mg in oil. This was confirmed in the persistent-estrous rats, but daily injection of smaller amounts induced sequences of ovulatory cycles (224). Further study showed progesterone to be primarily important each time that the animal returned to proestrus-estrus, whereupon a single injection of 0.5 mg or more consistently induced ovulation (225). The positive action was next demonstrated in normal rats having 5-day cycles. Progesterone injection on diestrus day 3 induced 24-hour advancement of ovulation. As with advancement by estrogen, the progesterone effect was subject to pharmacologic blockade (226).

Biphasic Action of Progesterone: Interaction with Estrogen

There is an obvious interaction between progesterone and estrogen to produce the positive influence on gonadotropin secretion. In most such cases, progesterone acts acutely against a background of elevated estrogen. The action is biphasic (220): stimulation during the first several hours, followed by inhibition. Daily administration thereafter of a large amount of progesterone will suppress the next cyclic proestrus day-to-day until after treatment stops, thus reproducing the effect of functional corpora lutea. For example, injection for 2 days on diestrus days 1 and 2 of the rat cycle extends the normal diestrus exactly 2 days, proestrus occurring 3 days thereafter. However, omission of one daily injection allows the next injection of progesterone to exert its stimulative effect.

The timing of a progesterone injection on the day of proestrus is critical for determining whether it will stimulate or inhibit an ovulatory LH surge (227–229). Injection at 0200 hours will inhibit ovulation (227), whereas injection between 0900 and 1200 hours will produce an ovulatory surge of hormone in advance of the normal critical period (227–229). Comparable temporal relationships are seen in estrogen-primed ovariectomized rats (230). Progesterone's biphasic influence is well documented in the rabbit also (231), where for a few hours it enhances but later inhibits the coital ovulation reflex. Correlated biphasic effects were recorded (232) on behavior and on thresholds for the EEG afterreactions to electrical stimulation of the hypothalamus or the rhinencephalon.

In the special case of ovulation induced in pregnant or pseudopregnant rats by estrogen administration, progesterone serves as a background for the stimulative action of acutely rising estrogen. It is of some interest that the induced gonadotropin release occurs during an afternoon critical period like that in proestrus (222). The progesterone produced by the corpora lutea does not apparently advance the time of release under these circumstances.

Episodic Gonadotropin Release

Gonadectomized animals of either sex are valuable subjects in these inquiries. The finding of highly variable plasma levels of LH in ovariectomized rhesus monkeys led to the first demonstration that gonadotropins tend to be released in pulsatile fashion (233). Pulsatile (episodic, ultradian) release is now recognized as a general phenomenon of AP physiology in many species, including humans (234–237). The concentrations of circulating gonadotropins are known to depend on the frequency and amplitude of LHRH pulses discharged into the pituitary portal vessels; these, in turn, depend upon the endocrine status. The extensive studies of rhesus monkeys by Knobil and associates (236) show that while continuous infusion of LHRH fails to release LH, pulsatile infusion is effective, optimal results being obtained with a pulse frequency of 1/hr. In rats, while continuous infusion of LHRH will stimulate LH release (237), pulsatile infusion is nevertheless more effective (238). Castration of either sex increases the magnitude of episodic LHRH release, resulting in a rise of circulating gonadotropins (239–243). Replacement with gonadal steroids reduces the pulse frequency of LH at first, thus depressing plasma LH and FSH, but prolonged exposure to estrogen in gonadectomized females reverses the inhibition. There are correlated changes in electrical activity in the medial basal hypothalamus. Dufy et al. (244) recorded pronounced pulsatile multiunit events in the accurate nuclei of ovariectomized monkeys immediately preceding each LH pulse.

Since LHRH promotes secretion of both LH and FSH, and no specific FSH releaser has been identified, the means for differential regulation of these two secretions has been a mystery. A suggestion of an answer comes from observations that the LH/FSH ratios are influenced by changes in the LHRH pulse frequency (239,243), LH secretion being favored by higher frequencies and FSH by lower frequencies. Controls of the respective gonadotropins may thus be served entirely by the one neurohumor. (There is some evidence that certain regions of the hypothalamus contain a specific releaser for FSH, and the search for such an agent continues.)

Steroids and the Phasic Release of Gonadotropins

Whereas the episodic discharge of LH appears to be universally determined by hypothalamic signals mediated by LHRH, the relative involvement of the hypothalamus and the AP itself in the phasic (preovulatory) surge of gonadotropins varies greatly from species to species. In the rat, mouse, and hamster, whose reproductive processes are closely attuned to the photoperiod, the hypothalamus has the leading role. Under the influence of the gonadal steroids there is a rapid increase of LHRH content in the medial basal hypothalamus just before the critical afternoon period of proestrus (245). Multiple pulses of LHRH in the portal blood follow, accompanied by increased frequency of pulsatile LH release, at intervals of 16 to 25 seconds (246). The magnitude of these LH pulses is governed initially by the responsiveness of the AP after being primed by estrogen. As the plasma LH concentration rises, the responsiveness is enhanced, partly by the self-priming action of LHRH and probably also by the increased exposure of the gonadotrophs to progesterone. Whether progesterone or some other progestin is active at the very start of the surge has long been debated.

At the other extreme from the rat is the rhesus monkey, in which the role of the hypothalamus is judged to be more permissive and the timing of the LH surge is determined directly by the response of the AP to the rising tide of estrogen (236,247). Knobil et al. (248) have shown that in female monkeys bearing long-term lesions of the arcuate nuclei, the month-long infusion of the LHRH pulses once per hour sustained complete menstrual cycles. In long-term ovariectomized females having similar hypothalamic lesions and similarly treated with LHRH, administration of estrogen invoked LH surges. No relationship to the time of day was evident in such responses. Nevertheless, the amount of LHRH in the portal blood does vary in the rhesus monkey, being relatively high during the estrogen-induced LH surge (249). A modulating, though nonessential, influence of the hypothalamus has been suggested (247), and others have proposed a specific hypothalamic message (250).

The guinea pig is intermediate between the rat and monkey, such that in ovariectomized subjects the time of an LH surge induced by estrogen depends not only on the time of injection, but also on the time of day and dosage (251,252); the surge is larger in the dark than in the light phase of the daily rhythm. In intact animals, spontaneous surges are also more frequent in the dark phase. Furthermore, unlike the rat and hamster, the ovariectomized guinea pig receiving estrogen fails to present repeated daily surges of LH.

The ovariectomized rat, chronically supplied with estrogen, either by repeated injection of estradiol benzoate (253) or by implanted Silastic capsules containing estradiol (254,255), displays daily proestrus-like surges of LH secretion, confirming that there is an innate circadian periodicity in the control mechanism in this species. When progesterone is introduced into such a preparation early in the day, the amount of LH released is enhanced, but release on the days following is diminished or prevented (230,256). Both effects are dose dependent (257).

The female hamster presents interesting variations on the circadian manifestations displayed by the rat (258–260). During the anestrus induced by short photoperiods, when estrogen levels are low and progesterone is high, there are daily afternoon surges of LH (261,262). Estrogen treatment of long-day subjects intensifies the surges at first, but later suppresses them; that inhibition is hastened by progesterone.

Investigation of the positive and negative actions of the sex steroids on the hypothalamic-pituitary complex proceeds apace, resulting already in a voluminous literature that defies balanced analysis. A recent review (245) addressed selectively to the control of LH secretion in the laboratory rat cites over 400 articles, of which over 300 were published during the last decade. The broad perspective embraces similarities and differences among species, sex differences, developmental aspects, changes during pregnancy, and the influence of old age. There are concerns with anatomy of the LHRH nerve cells and fibers and their physical connections with steroid-concentrating neurons and other neural systems. Much interest focuses on the influence of the steroids on hypothalamic neurochemistry and on the participation of several neurotransmitters and hypothalamic enzymes that affect LHRH synthesis, transport, and release.

Nonsteroid Gonadal Feedback

The participation of nonsteroidal gonadal secretions in modulating the pars distalis responses to LHRH assumes increasing significance for the differential regulation of FSH and LH synthesis and discharge. Such material obtained from ovarian follicular fluid or testis extracts (inhibin, folliculostatin, gonadostatin) selectively inhibits FSH secretion in vivo and in vitro (263,264). On the other hand, inhibin is said to enhance the secretion in vitro of LH in response to LHRH, adding to the stimulative action of estradiol (265).

CONCLUSION

This chapter focuses primarily on background studies of the regulation of gonadotropin and prolactin secretion in adult female mammals. Little or no attention is given to a number of important subjects such as the pars intermedia and the classical neurosecretory system terminating in the neural lobe. Present knowledge of the mammalian hypothalamo-pars distalis apparatus is necessarily limited to a few readily available species. For the vast majority, distributed through more than 900 genera, the details of reproductive physiology are poorly known and for practical reasons will probably remain so. One can only assume that the range of specializations recognized among familiar species is representative. Each has contributed importantly in its own way. Thus, as principal representative of the reflex ovulators, the rabbit gave the first clues to the importance of the central nervous system and the hypothalamic-pars distalis connection for the ovulation process. The guinea pig gave the initial evidence that the corpora lutea suppress ovulation. The mouse, through the Allen-Doisy test, was influential in the purification and synthesis of estrogens. The rat, through the technique of hypophysectomy, greatly facilitated the isolation and purification of the several pars distalis hormones. From the rat also came the first proof of the generative capacity of the pituitary portal vessels and demonstration of the importance of the vascular supply to the gland from the median eminence, as well as the evidence for neurohumoral inhibition of prolactin secretion. The ovulation-blocking action of certain drugs in rabbits and rats gave the first clear evidence for an acute preovulatory surge of gonadotropins in spontaneous ovulators. The predictable time of this surge in rats, its dependence on the lighting rhythm, and failure of the surge in old rats led to disclosure of the biphasic action of progesterone and its interaction with estrogen in promoting the surge. Blockage of the surge pharmacologically or by exposure to continuous lighting fostered various studies of the central neural apparatus controlling spontaneous ovulation. Critical comparisons among rats, hamsters, guinea pigs, sheep, and monkeys, facilitated by radioimmunoassay and other modern techniques, have yielded interpretations that seem generally applicable to human subjects, a major goal of all research in reproductive biology.

ACKNOWLEDGMENTS

This research was supported in part by grants from the Research Council of Duke University and, since 1957, from the National Science Foundation. The author is also grateful to John Graves for typing the manuscript.

REFERENCES

1. Harris, G. W. (1972). Humours and hormones, the Sir Henry Dale lecture for 1971. *J. Endocrinol.* 53, ii–xxii.
2. Rathke, M. H. (1838). Uber die Entstehung der Glandula pituitaria. *Arch. Anat. Physiol. Wiss. Med.* 482–485. Cited by Medvei VC. *A history of endocrinology*. Boston: MTP Press, 1982.
3. Bayliss, W. M., *Principles of general physiology*. London: Longmans, Green, (1915).
4. Baylis, W. M., and Starling, E. H. (1904). The chemical regulation of the secretory process. *Proc. R. Soc. Lond (Biol).* 73, 310–322.
5. Ramón y Cajal, S. (1894). Algunas contribuciónes conociamento de los ganglios del encéfale. *Anal. Soc. Espan. Hist. Nat.* 23, 214–215.
6. Camus, J., and Roussy, G. (1920). Experimental researches on the pituitary body. Diabetes insipidus, glycosuria, and those dystrophies considered as hypophysial in origin. *Endocrinology* 4, 507–522.
7. Bailey, P., and Bremer, F. Experimental diabetes insipidus. *Arch. Intern. Med* (1921) 28, 773–803.
8. Evans, H. M., and Long, J. A. (1921). Effect of anterior lobe of hypophysis administered intraperitoneally upon growth, maturity, and oestrous cycles in the rat. *Anat. Rec.* 21, 61 (abstract).
9. Anderson, E. (1969). Earlier ideas of hypothalamic function, including irrelevant concepts. In: Haymaker, W., Anderson, E., and Nauta, W. J. H., eds. *The hypothalamus*. Springfield, IL: Charles C Thomas, 1–12.

10. Cushing, H. (1909). The hypophysis cerebri: chemical aspects of hyperpituitarism and hypopituitarism. *JAMA* 53, 249–255.
11. Crowe, S. J., Cushing, H., and Homans, J. (1910). Experimental hypophysectomy. *Bull. Johns. Hopkins. Hosp.* 21, 127–169.
12. Cushing, H. (1912). *The pituitary and its disorders.* Philadelphia: JB Lippincott.
13. Aschner, B. (1909). Demonstration von Hunden nach Extirpation der Hypophyse. *Wien. Klin. Wochenschr.* 22, 1730–1732.
14. Aschner, B. (1912). Ueber die Function der Hypophyse. *Pflugers. Arch. Ges. Physiol.* 146, 1–147.
15. Marshall, F. H. A. (1936). Sexual periodicity and the causes which determine it. The Croonian lecture. *Philos. Trans. R. Soc. Lond (Biol).* 226, 423–456.
16. Marshall, F. H. A. (1942). Exteroceptive factors in sexual periodicity. *Biol. Rev.* 17, 68–90.
17. Haighton, J. (1797). An experimental study concerning animal impregnation. *Philos. Trans. R. Soc.* 87, 157–196.
18. Cruikshank, W. (1797). Experiments in which, on the third day after impregnation, the ova of rabbits were found in the Fallopian tubes and on the fourth day after impregnation in the uterus itself with the first appearance of the foetus. *Philos. Trans. R. Soc.* 87, 197–214.
19. Barry, M. (1839). Researches in embryology. *Philos. Trans. R. Soc.* 129, 307–380.
20. Heape, W. (1905). Ovulation and degeneration of ova in rabbit. *Proc. R. Soc. Lond (Biol)* 76, 266–268.
21. Marshall, F. H. A. (1904). The oesstrous cycle of the common ferret. *Q. J. Microsc. Sci.* 48, 323–345.
22. Longley, W. H. (1911). Maturation of the egg and ovulation in the domestic cat. *Am. J. Anat.* 12, 139–172.
23. Fichera, G. (1905). Sur l'hypertrophie de la gland pituitaire consecutive à la castration. *Arch. Ital. Biol.* 43, 405–426.
24. Addison, W. H. F. (1917). The cell changes of the hypophysis of the albino rat after castration. *J. Comp. Neurol.* 28, 441–461.
25. Loeb, L. (1911). Über die Bedeutung des Corpus luteum für die Periodizität des sexuellen Zyklus beim weiblichen Säugetier-organismus. *Dtsch. Med. Wochenschr.* 37, 17–21.
26. Stockard, C. R., and Papanicolaou, G. N. (1917). The existence of a typical oestrous cycle in the guinea pig—with a study of its histological and physiological changes. *Am. J. Anat.* 22, 225–283.
27. Long, J. A., and Evans, H. M. (1922). The oestrous cycle in the rat and its associated phenomena. *Mem. Univ. Calif.* 6, 1–111.
28. Allen, E. (1922). The estrous cycle in the mouse. *Am. J. Anat.* 30, 297–371.
28a. Allen, E., and Doisy, E. A. (1923). An ovarian hormone a preliminary report on its localization, extraction and partial purification, and action in test animals. *JAMA* 81, 819–821.
29. Allen, E., Pratt, J. P., and Doisy, E. A. (1925). The ovarian follicular hormone. Its distribution in human genital tissues. *JAMA* 85, 399–405.
30. Allen, B. M. (1916). The results of extirpation of the anterior lobe of the hypophysis and of the thyroid of *Rana pipiens* larvae. *Science* 44, 755–757.
31. Allen, B. M. (1920). Experiments in the transplantation of the hypophysis of adult *Rana pipiens* to tadpoles. *Science.* 52, 274–276.
32. Smith, P. E. (1916). The effect of hypophysectomy in the early embryo upon growth and development of the frog. *Anat. Rec.* 11, 57–64.
33. Smith, P. E., and Smith, I. P. (1922). The effect of intraperitoneal injection of fresh anterior lobe substance in hypophysectomized tadpoles. *Ana. Rec.* 23, 38–39.
34. Smith, P. E., and Smith, I. P. (1922). The repair and activation of the thyroid in the hypophysectomized tadpole by the parenteral administration of fresh anterior lobe of the bovine hypophysis. *J. Med. Res.* 43, 267–283.
35. Smith, P. E. (1926). Ablation and transplantation of the hypophysis in the rat. *Anat. Rec.* 32, 221 (abstract).
36. Smith, P. E. (1927). The disabilities caused by hypophysectomy and their repair. *JAMA* 88, 158–161.
37. Smith, P. E. (1930). Hypophysectomy and replacement therapy in the rat. *Am. J. Anat.* 45, 205–274.
38. Smith, P. E., and Engle, E. T. (1927). Experimental evidence regarding the role of the anterior pituitary in the development and regulation of the genital system. *Am. J. Anat.* 40, 159–217.
39. Evans, H. M., Meyer, K., and Simpson, M. E. (1933). The growth and gonad-stimulating hormones of the anterior hypophysis. *Mem. Univ. Calif.* 11, 67–229.
40. Evans, H. M., Pencharz, R. I., Meyer, K., and Simpson, M. E., (1933) The growth and gonad-stimulating hormones of the anterior hypophysis. *Mem. Univ. Calif* 11, 315–334.
41. Greep, R. O. (1935). Separation of a thyrotropic from the gonadotropic substances of the pituitary. *Am. J. Physiol.* 110, 692–699.
42. Zondek, B. (1930). Über die Hormone des Hypophysenvorderlappens. I Wachstumshormon, Follikelreifungshormon (Prolan A), Luteinisierungshormon (Prolan B), Stoffwechselhormon. *Klin. Wochenschr.* 9, 245–248.
43. Zondek, B. (1930). Über die Hormone des Hypophysenvorderlappens. II. Follikelreifungshormon Prolan A-Klamakterium-Kastration. *Klin. Wochenschr.* 9, 393–396.
44. Reichert, F. L., Pencharz, R. I., Simpson, M. E., Meyer, K., and Evans, H. M. (1932). Relative ineffectiveness of Prolan in hypophysectomized animals. *Am. J. Physiol.* 100, 157–161.
45. Leonard, S. M., and Smith, P. E. (1934). Responses of the reproductive system of hypophysectomized rats in injections of pregnancy-urine extracts. II. The female. *Anat. Rec.* 58, 175–200.
46. Gey, G. O., Seegar, G. E., and Hellman, L. M. (1938). The production of gonadotropic substance (Prolan) by placental cells in tissue culture. *Science.* 88, 306–307.
47. Jones, G. E. S., Gey, G. O., and Cey, M. K. (1943). Hormone production by placental cells maintained in continuous culture. *Bull. Johns. Hopkins. Hosp.* 72, 26–38.
48. Fevold, H. L., Hisaw, F. L., and Leonard, S. L. (1931). The gonad-stimulating and the luteinzing hormones of the anterior lobe of the hypophysis. *Am. J. Physiol.* 97, 291–301.
49. Greep, R. O. (1974). History of research on anterior hypophysial hormones. In: *Handbook of Physiology, Section. 7, Endocrinology*, Vol. IV., part 2. Washington, DC: American Physiological Society, 1–27.
50. Astwood, E. B. (1941). The regulation of corpus leuteum function by by hypophysial luteotropin. *Endocrinology* 28, 309–320.
51. Evans, H. M., Simpson, M. E., Lyons, W. R., and Turpeinen, K. (1941). Anterior pituitary hormones which favor production of traumatic uterine placentoma. *Endocrinology* 28, 933–945.
52. Stricker, P., and Grueter, F. (1928). Action du lobe antérieur de l'hypophyse sur la montée laiteuse. *C. R. Soc. Biol (Paris)* 99, 1978–1980.
53. Stricker, P., and Grueter, F. (1929). Fonctions du lobe antérieur de l'hypophyse: influence des extraits du lobe antérieur sur l'appareil génitale de la lapine et sur la montée laiteuse. *Presse. Med.* 37, 1268–1271.
54. Corner, G. W. (1930). The hormonal control of lactation. I. Noneffect of the corpus luteum. II. Positive action of extracts of the hypophysis. *Am. J. Physiol.* 95, 43–55.
55. Evans, H. M., and Simpson, M. E. (1929). Hyperplasia of the mammary apparatus of adult virgin females induced by anterior hypophyseal hormones. *Proc. Soc. Exp. Biol. Med.* 26, 598.
56. Riddle, O. (1937). Physiological responses to prolactin. *Cold. Spring. Harbor. Symp. Quant. Biol.* 5, 218–228.
57. Riddle, O. (1938). Prolactin. *Assoc. Res. Nerv. Mental. Dis.* 17, 287–297.

58. Riddle, O., and Braucher, P. F. (1931). Studies on the physiology of reproduction in birds. XXX. Control of the special secretion of the crop gland in pigeons by an anterior pituitary hormone. *Am. J. Physiol.* 97, 617–625.
59. Riddle, O., Bates, R. W., and Dykshorn, S. W. (1933). The preparation, identification and assay of prolactin—a hormone of the anterior pituitary. *Am. J. Physiol.* 105, 191–216.
60. Dresl, L. (1935). The effect of prolactin on the estrus cycle of nonporous mice. *Science.* 82, 173.
61. Lahr, L., and Riddle, O. (1936). Temporary suppression of estrous cycles in the rat by prolactin. *Proc. Soc. Exp. Biol. Med.* 34, 880–893.
62. Bates, R. W., Miller, R. A., and Garrison, M. M. (1962). Evidence in the hypophysectomized pigeon of a synergism among prolactin, growth hormone, thyroxine and prednisone upon weight of the body, digestive tract, kidney and fat stores. *Endocrinology* 71, 345–360.
63. Schönemann, A. (1892). Hypophysis und Thyroidea. *Virchows. Arch (Pathol. Anat)* 129, 310–336.
64. Holmes, R. L., and Ball, J. N. (1974). *The pituitary gland—a comparative account*. Cambridge: Cambridge University Press.
65. Mallory, F. B. (1900). A contribution to staining methods. *J. Exp. Med.* 5, 15–20.
66. McManus, J. F. A. (1946). Histological demonstration of mucin after periodic acid. *Nature* 158, 202.
67. Herlant, M. (1960). Étude critique de deux techniques nouvelles destinées à mettre en evidence les différentes catégories cellulaires présente dans la glands pituitaire. *Bull. Microsc. Appl.* 10, 37–44.
68. Pearse, A. G. E. (1953). Cytological and cyto-chemical investigations on the foetal and adult hypophysis in various physiological and pathological states. *J. Pathol. Bacteriol.* 65, 355–370.
69. Coons, A. H. (1956). Histochemistry with labelled antibody. *Int. Rev. Cytol.* 5, 1–23.
70. Halmi, N. S., and Krieger, D. (1983). Immunocytochemistry of ACTH-related peptides in the hypophysis. In: Bhatnagar, A. S., ed. *The anterior pituitary gland*. New York: Raven Press, 1–15.
71. Hohlweg, W., and Junkmann, K. (1932). Die hormonal-nervöse Regulierung der Funktion des Hypophysenvorderlappens. *Klin. Wochenschr.* 11, 321–323.
72. Fröhlich, A. (1901). Ein Fall von Tumor der Hypophysis cerebri ohne Akromegalie. *Wien. Klin. Rundschau.* 15, 883–886; 906–908.
73. Hetherington, A. W., and Ranson, S. W. (1939). Experimental hypothalamico-hypophyseal obesity in the rat. *Proc. Soc. Exp. Biol. Med.* 41, 465–466.
74. Hetherington, A. W., and Ranson, S. W. (1940). Hypothalamic lesions and adiposity in the rat. *Anat. Rec.* 78, 149–172.
75. Hetherington, A. W., and Ranson, S. W. (1942). The relation of various hypothalamic lesions to adiposity in the rat. *J. Comp. Neurol.* 76, 475–499.
76. Bauer, H. G. (1954). Endocrine and other clinical manifestations of hypothalamic disease. *J. Clin. Endocrinol.* 14, 13–31.
77. Grafe, E., and Grünthal, E. (1929). Über isolierte Beinflussung des Gesamtstoffwechsels vom Zwischenhirn aus. *Klin. Wochenschr.* 8, 1013–1016.
78. Moore, C. R., and Price, D. (1932). Gonad hormone functions and the reciprocal influence between gonads and hypophysis with its bearing on the problem of sex-hormone antagonisms. *Am. J. Anat.* 50, 13–71.
79. Fee, A. R., and Parkes, A. S. (1929). The relation of the anterior pituitary body to ovulation in the rabbit. *J. Physiol (Lond).* 67, 383–388.
80. Fee, A. R., and Parkes, A. S. (1930). Effects of vaginal anesthesia on ovulation in the rabbit. *J. Physiol (Lond)* 70, 385–388.
81. Marshall, F. H. A., and Verney, E. B. (1936). The occurrence of ovulation and pseudopregnancy in the rabbit, as result of central nervous stimulation. *J. Physiol (Lond)* 86, 327–336.
82. Harris, G. W. (1937). The induction of ovulation in the rabbit by electrical stimulation of the hypothalamo-hypophysial mechanism. *Proc. R. Soc. Lond (Biol).* 122, 374–394.
83. Haterius, H. O., and Derbyshire, A. J. Jr. (1937). Ovulation in the rabbit upon stimulation of the hypothalamus. *Am. J. Physiol.* 119, 329–330.
84. Westman, A., and Jacobsohn, D. (1937). Experimentelle Untersuchungen über die Bedeutung des Hypophysen-Zwischenhirnsystems für die Produktion gonadotroper Hormone des Hypophysenvorderlappens. *Acta. Obstet. Gynecol. Scand.* 17, 235–265.
85. Rasmussen, A. T. (1938). Innervation of the hypophysis. *Endocrinology* 23, 263–278.
86. Hohlweg, W., and Chamorro, A. (1937). Über die luteinisierende Wirkung des Follikelhormons durch Beinflussung der endogenen Hypophysenvorderlappensekretion. *Klin. Wochenschr.* 16, 196–197.
87. Westman, A., and Jacobsohn, D. (1938). Endokrinologische Untersuchungen an Ratten mit durchtrenntem Hypophysenstiel. I. Hypophysenveränderungen nach Kastration und nach Oestrinbehandlungen. *Acta. Obstet. Gynecol. Scand.* 18, 99–108.
88. Westman, A., and Jacobsohn, D. (1938). Endokrinologische Untersuchungen an Ratten mit durchtrenntem Hypophysenstiel. III. Über die luteinisierende Wirkung des Follikelhormons. *Acta. Obstet. Gynecol. Saand.* 18, 115–123.
89. Westman, A., and Jacobsohn, D. (1938). Endokrinologische Untersuchungen an Ratten mit durchtrenntem Hypophysenstiel. VI. Produktion und Abgabe der gonadotropen Hormone. *Acta. Pathol. Microbiol. Scand.* 15, 445–453.
90. Pietsch, K. (1930). Aufbau und Entwicklung der Pars tuberalis des menschlichen Hirnanhangs in ihren Beziehung zu den übrigen Hypophysenteilen. *Z. Mikrosk. Anat. Forsch.* 22, 227–257.
91. Popa, G. T., and Fielding, U. (1930). A portal circulation from the pituitary to the hypothalamic region. *J. Anat (Lond).* 65, 88–91.
92. Popa, G. T., and Fielding, U. (1933). Hypophysio-portal vessels and their colloid accompaniment. *J. Anat (Lond).* 67, 227–232.
93. Houssay, B. A., Biosotti, A., and Sammartino, R. (1935). Modificationes fonctionelles de l'hypophyse après les lesions infundibulotubériennes chez le crapaud, *C. R. Soc. Biol. (Paris)* 120, 725–727.
94. Wislocki, G. B., and King, L. S. (1936). The permeability of the hypophysis and the hypothalamus to vital dyes, with a study of the hypophysial vascular supply. *Am. J. Anat.* 58, 421–472.
95. Wislocki, G. B. (1938). The vascular supply of the hypophysis cerebri of the rhesus monkey and man. *Res. Publ. Assoc. Nerv. Ment. Dis.* 17, 48–68.
96. Green, J. D. (1947). Vessels and nerves of the amphibian hypophysis: a study of the living circulation and of the histology of the hypophysial vessels and nerves. *Anat. Rec.* 99, 21–54.
97. Benoit, J., and Assenmacher, I. (1955). Le controle hypothalamique de l'activité préhypophysaire gonadotrope. *J. Physiol (Paris)* 47, 427–567.
98. Green, J. D., and Harris, G. W. (1947). The neurovascular link between the neurohypophysis and adenohypophysis. *J. Endocrinol.* 5, 136–146.
99. Green, J. D., and Harris, G. W. (1949). Observations of the hypophysial portal vessels of the living rat. *J. Physiol (Lond)* 108, 359–361.

100. Worthington, W. C. Jr. (1955). Some observations on the hypophyseal portal system in the living mouse. *Bull. Johns. Hopkins. Hosp.* 97, 343–357.
101. Worthington, W. C. Jr. (1960). Vascular responses in the pituitary stalk. *Endocrinology* 66, 19–31.
102. Török, B. (1954). Lebendbeobachtung des Hypophysenkreislaufes an Hunden. *Acta. Morphol. Hung.* 4, 83–89.
103. Török, B. (1962). Neue Angaben zum Blutkreislauf der Hypophyse. *Anat. Anz.* 109(Suppl):622–629.
104. Daniel, P. M. (1966). The anatomy of the hypothalamus and pituitary gland. In: Martini, L., and Ganong, W. F., eds. *Neuroendocrinology*, Vol. 1, New York: Academic Press, 15–80.
105. Page, R. B., and Bergland, R. M. (1977). The neurophysial capillary bed. I. Anatomy and arterial supply. *Am. J. Anat.* 148, 345–358.
106. Bergland, R. M., and Page, R. B. (1978). Can the pituitary secrete directly to the brain? (Affirmative anatomical evidence). *Endocrinology* 102, 1325–1338.
107. Oliver, C., Mical, R. S., and Porter, J. C. (1977). Hypothalamic-pituitary vasculature: evidence for retrograde blood flow in the pituitary stalk. *Endocrinology* 101, 598–604.
108. Daniel, P. M., and Prichard, M. M. L. (1975). Studies of the hypothalamus and the pituitary gland with special reference to the effects of transection of the pituitary stalk. *Acta. Endocrinol.* 80(Suppl. 201:1–216.
109. Pasteels, J. L. (1960). Étude expérimentale des differentes categories d'éléments chromophiles de l'hypophyse adulte de *Pleurodeles waltlii* et de leur controle par l'hypothalamus. *Arch. Biol. (Paris)* 71, 409–471.
110. Porter, J. C., Kamberi, I. A., and Grazia, Y. A. (1971). Pituitary blood flow and portal vessels. In: Martini, L., and Ganong, W. F., eds. *Frontiers in neuroendocrinology* New York: Oxford University Press. 145–175.
111. Hinsey, J. C., and Markee, J. E. (1933). Pregnancy following bilateral section of the cervical sympathetic trunks in the rabbits. *Proc. Soc. Exp. Biol. Med.* 31, 270–271.
112. Hinsey, J. C. (1937). The relationship of the nervous system to ovulation and other phenomena of the female reproductive tract. *Cold. Spring. Harbor. Symp. Quant. Biol.* 5, 269–279.
113. Wingstrand, K. G. (1966). Comparative anatomy and evolution of the hypophysis. In: Harris, G. W., and Donovan, B. T., eds. *The pituitary gland*, Vol. 1, London: Butterworths, 58–126.
114. Green, J. D. (1951). The comparative anatomy of the hypophysis, with special reference to its blood supply and innervation. *Am. J. Anat.* 88, 225–312.
115. Harris, G. W. (1950). Oestrous rhythm, pseudopregnancy and the pituitary stalk in the rat. *J. Physiol. (Lond)* 111, 347–360.
116. Harris, G. W., and Johnson, R. T. (1950). Regeneration of the hypophysial portal vessels after section of the hypophysial stalk, in the monkey *Macaca rhesus*. *Nature* 165, 819–820.
117. Jacobsohn, D. (1954). Regeneration of hypophysial portal vessels and grafts of anterior pituitary glands in rabbits. *Acta. Endocrinol. (Copenh)* 17, 187–197.
118. Greep, R. O. (1936). Functional pituitary grafts in rats. *Proc. Soc. Exp. Biol. Med.* 34, 754–755.
119. Harris, G. W., and Jacobsohn, D. (1952). Functional grafts of the anterior pituitary gland. *Proc. R. Soc. Lond (Biol).* 139, 263–276.
120. Nikitovitch-Winer, M., and Everett, J. W. (1958). Functional restitution of pituitary grafts re-transplanted from kidney to median eminence. *Endocrinology* 63, 916–930.
121. Nikitovitch-Winer, M., and Everett, J. W. (1959). Histocytologic changes in grafts of rat pituitary on the kidney and upon retransplantation under the diencephalon. *Endocrinology* 65, 357–368.
122. Smith, P. E. (1961). Postponed homotransplants of the hypophysis into the region of the median eminence in hypophysectomized male rats. *Endocrinology* 68, 130–143.
123. Smith, P. E. (1963). Postponed pituitary homotransplants into the region of the hypophysial portal circulation in hypophysectomized female rats. *Endocrinology* 73, 793–806.
124. Harris, G. W. (1955). *Neural control of the pituitary gland.* London: Arnold.
125. Saffran, M., Schally, A. V., and Benfry, B. G. (1955). Stimulation of the release of corticotropin from the adenohypophysis by a neurohypophysial factor. *Endocrinology* 57, 439–444.
126. Everett, J. W. (1954). Luteotrophic function of autografts of the rat hypophysis. *Endocrinology* 54, 685–690.
127. Everett, J. W. (1956). Functional corpora lutea maintained for months by autografts of rat hypophyses. *Endocrinology* 58, 786–796.
128. Desclin, L. (1950). A propos du méchanisme d'action des oestro-gènes sur le lobe antérieur de l'hypophyse chez le rat. *Ann. Endocrinol.* 11, 656–659.
129. Nikitovitch-Winer, M. B. (1965). Effect of hypophysial stalk transection on luteotropic hormone secretion in the rat. *Endocrinology* 77, 658–666.
130. Nicoll, C. S., and Meites, J. (1962). Prolactin secretion *in virtro*: comparative aspects. *Nature* 195, 606–607.
131. Nicoll, C. S. (1965). Neural regulation of adenohypophysial prolactin secretion in tetrapods: indications from *in vitro* studies. *J. Exp. Zool.* 158, 203–210.
132. Meites, J. (1967). Control of prolactin secretion. *Arch. Anat. Microsc. Morphol. Exp.* 56(Suppl):516–529.
133. Pasteels, J. L. (1961). Sécrétion de prolactine par l'hypophyse en culture de tissues. *C. R. Acad. Sci. (Paris)* 253, 2140–2142.
134. Pasteels, J. L. (1961). Premiers résultats de culture combinée *in vitro* d'hypophyis et d'hypothalamus dans le but d'en apprécier la sécrétion de prolactine. *C. R. Acad. Sci. (Paris)* 253, 3074–3075.
135. Wade, N. (1981). *The Nobel duel.* New York: Anchor Press/Doubleday.
136. MacLeod, R. M. (1976). Regulation of prolactin secretion. In: Martini, L., and Ganong, W. F., eds. *Frontiers in neuroendocrinology*. New York: Raven Press, 169–194.
137. Ben-Jonathan, N., Oliver, C., Weiner, H. J., Mical, R. S., and Porter, J. C. (1977). Dopamine in hypophysial portal plasma of the rat during the estrous cycle and throughout pregnancy. *Endocrinology* 100, 452–458.
138. Sawyer, C. H., Markee, J. E., and Hollinshead, W. H. (1947). Inhibition of ovulation in the rabbit by the adrenergic-blocking agent Dibenamine. *Endocrinology* 41, 395–402.
139. Markee, J. E., Sawyer, C. H., and Hollinshead, W. H. (1948). Adrenergic control of the release of luteinzing hormone from the hypophysis of the rabbit. *Recent. Prog. Horm. Res.* 2, 117–151.
140. Sawyer, C. H., Markee, J. E., and Townsend, B. F. (1949). Cholinergic and adrenergic components in the neurohumoral control of the release of LH in the rabbit. *Endocrinology* 44, 18–37.
141. Swayer, C. H. (1952). Stimulation of ovulation in the rabbit by the intraventricular injection of epinephrine or nerepinephrine. *Anat. Rec.* 112, 385 (abstract).
142. Sawyer, C. H. (1979). The Seventh Stevenson Lecture. Brain amines and pituitary gonadotrophin secretion. *Can. J. Physiol. Pharmacol.* 57, 667–680.
143. Tsou, R. C., Dailey, R. A., McLanahan, C. S., Parent, A. D., Tindall, G. T., and Neill, J. D. (1977). Luteinizing hormone releasing hormone (LHRH) levels in pituitary stalk plasma during the preovulatory gonadotropin surge of rabbits. *Endocrinology* 101, 534–539.
144. Dufy-Barbe, L., Franchimont, P., and Faure, J. M. A. (1973). Time courses of LH and FSH release in the female rabbit. *Endocrinology* 92, 1318–1321.

145. Kanematsu, S., Scaramuzzi, R. J., Hilliard, J., and Sawyer, C. H. (1974). Patterns of ovulation-inducing LH release following coitus, electrical stimulation and exogenous LH-RH in the rabbit. *Endocrinology* 95, 247–252.
146. Goodman, A. L., and Neill, J. D. (1976). Ovarian regulation of postcoital gonadotropin release in the rabbit: reexamination of a functional role for. 20α dihydro-progesterone. *Endocrinology* 99, 852–860.
147. Westman, A. (1942). Der Einfluss des Hypophysenzwishchenhirnsystems auf' die Sexualfunktionen. *Schweiz. Med. Wochenschr.* 72, 113–116.
148. Dempsey, E. W. (1937). Follicular growth rate and ovulation after various experimental procedures in the guinea pig. *Am. J. Physiol.* 120, 126–132.
149. Everett, J. W., Sawyer, C. H., and Markee, J. E. (1949). A neurogenic timing factor in control of the ovulating discharge of luteinizing hormone in the cyclic rat. *Endocrinology* 44, 234–250.
150. Sawyer, C. H., Everett, J. W., and Markee, J. E. (1949). A neural factor in the mechanism by which estrogen induces the release of luteinizing hormone in the rat. *Endocrinology* 44, 218–233.
151. Everett, J. W., and Sawyer, C. H. (1949). A neural timing factor in the mechanism by which progesterone advances ovulation in the cyclic rat. *Endocrinology* 45, 581–595.
152. Everett, J. W., and Sawyer, C. H. (1950). A 24-hour periodicity in the "LH-release apparatus" of female rats, disclosed by barbiturate sedation. *Endocrinology* 47, 198–218.
153. Barraclough, C. A., and Sawyer, C. H. (1955). Inhibition of the release of pituitary ovulatory hormone in the rat by morphine. *Endocrinology* 57, 329-337.
154. Barraclough, C. A., and Sawyer, C. H. (1957). Blockade of the release of pituitary ovulating hormone in the rat by chlorpromazine and reserpine: possible mechanisms of action. *Endocrinology* 61, 341–351.
155. Blake, C. A., and Sawyer, C. H. (1972). Ovulation blocking actions of urethane in the rat. *Endocrinology* 91, 87–94.
156. Everett, J. W., and Sawyer, C. H. (1953). Estimated duration of the spontenous activation which causes release of ovulating hormone from the rat hypophysis. *Endocrinology* 52, 83–92.
157. Everett, J. W. (1956). The time of release of ovulating hormone from the rat hypophysis. *Endocrinology* 59, 580–585.
158. Everett, J. W., and Tejasen, T. (1967). Time factor in ovulation blockade in rats under differing lighting conditions. *Endocrinology* 80, 790–792.
159. Blake, C. A. (1976). A detailed characterization of the proestrous luteinizing hormone surge. *Endocrinology* 98, 445–450.
160. Sarkar, D. K., Chiappa, S. A., Fink, G., and Sherwood, N. M. (1976). Gonadotropin-releasing hormone surge in pro-estrous rats. *Nature* 264, 461–463.
161. Everett, J. W. (1948). Progesterone and estrogen in the experimental control of ovulation time and other features of the estrous cycle in the rat. *Endocrinology* 43, 389–405.
162. Everett, J. W. (1970). Photoregulation of the ovarian cycle in the rat. In: Benoit, J., Assenmacher, I., eds. *La photoregulation de la reproduction chez les oiseaux et les mammiféres.* Paris: C.N.R.S., 387–403.
163. Everett, J. W. (1952). Presumptive hypothalamic control of spontaneous ovulation. *Ciba. Found. Coll. Endocrinol.* 4, 167–178.
164. Everett, J. W. (1967). Provoked ovulation of long-delayed pseudopregnancy from coital stimuli in barbiturate-blocked rats. *Endocrinology* 80, 145–154.
165. Greep, R. O., and Hisaw, F. L. (1938). Pseudopregnancies from electrical stimulation of the cervix in the diestrum. *Proc. Soc. Exp. Biol. Med.* 39, 359–360.
166. Quinn, D. L., and Everett, J. W. (1967). Delayed pseudopregnancy induced by selective hypothalamic stimulation. *Endocrinology* 80, 155–162.
167. Beach, J. E., Tyrey, L., and Everett, J. W. (1975). Serum prolactin and LH in early phases of delayed versus direct pseudopregnancy in the rat. *Endocrinology* 96, 1241–1246.
168. Beach, J. E., Tyrey, L., and Everett, J. W. (1978). Prolactin secretion preceding delayed pseudopregnancy in rats after electrical stimulation of the hypothalamus. *Endocrinology* 103, 2247–2251.
169. de Greef, W. J., and Zeilmaker, G. H. (1976). Prolactin and delayed pseudopregnancy in the rat. *Endocrinology* 98, 305–310.
170. Anand, B. K., Malkani, P. K., and Dua, S. (1957). Effect of electrical stimulation of the hypothalamus on menstrual cycle in monkey. *Indian. J. Med. Res.* 45, 499–502.
171. Bunn, J. P., and Everett, J. W. (1957). Ovulation in persistent-estrous rats after electrical stimulation of the brain. *Proc. Soc. Exp. Biol. Med.* 96, 369–371.
172. Critchlow, V. (1958). Ovulation induced by hypothalamic stimulation in the anesthetized rat. *Am. J. Physiol.* 195, 171–174.
173. Everett, J. W., and Radford, H. M. (1961). Irritative deposits from stainless steel electrodes in the preoptic rat brain causing release of pituitary gonadotropin. *Proc. Soc. Exp. Biol. Med.* 108–604–609.
174. Everett, J. W. (1964). Preoptic stimulative lesions and ovulation in the rat: "thresholds" and LH-release time in the late diestrus and proestrus. In: Bajusz, E., and Jasmin, G., eds. *Major problems in neuroendocrinology*. Basel: Karger, 346–366.
175. Everett, J. W., Krey, L. C., and Tyrey, L. (1973). The quantitative relationship between electrochemical preoptic stimulation and LH release in proestrous *versus* late diestrous rats. *Endocrinology* 93, 947–953.
176. Turgeon, J., and Barraclough, G. A. (1973). Temporal patterns of LH release following graded preoptic electrochemical stimulation in proestrous rats. *Endocrinology* 92, 755–761.
177. Velasco, M. E., and Rothchild, I. (1973). Factors influencing the secretion of luteinizing hormone and ovulation in response to electrochemical stimulation of the preoptic area in rats. *J. Endocrinol.* 58, 163–176.
178. Colombo, J. A., Whitmoyer, D. I., and Sawyer, C. H. (1974). Local changes in multiple unit activity induced by electrochemical means in preoptic and hypothalamic areas in the female rat. *Brain. Res.* 71, 1175—1183.
179. van der Schoot, P., Lincoln, D. W., and Clark, J. S. (1978). Activation of hypothalamic neuronal activity by electrolytic deposition of iron into the preoptic area. *J. Endocrinol.* 79, 107–120.
180. Dyer, R. G., and Burnet, F. (1976). Effects of ferrous ions on preoptic area neurons and luteinizing hormone secretion in the rat. *J. Endocrinol.* 69, 247–254.
181. Dyball, R. E., Dyer, R. G., MacLeod, N. K., Wright, R. J., and Yates, J. O. (1976). Effects of ferrous ions on secretion from incubated nerve terminals. *J. Endocrinol.* 72, 73P.
182. Hillarp, N. A., Olivecroxa, H., and Silferskiöld, W. (1954). Evidence for the participation of the preoptic area in male mating behaviour. *Experientia.* 10, 224–225.
183. Willmore, L. J., Hurd, R. W., and Sypert, G. W. (1978). Epileptiform activity initiated by pial iontophoresis of ferrous and ferric chloride on rat cerebral cortex. *Brain. Res.* 152, 406–410.
184. Willmore, L. J., Hiramatsu, M., Kochi, H., and Mori, A. (1983). Formation of superoxide redicals after $FeCl_3$ injection into rat isocortex. *Brain. Res.* 277, 393–396.

185. Tejasen, T., and Everett, J. W. (1967). Surgical analysis of the preoptico-tuberal pathway controlling ovulatory release of gonadotropins in the rat. *Endocrinology* 81, 1387–1396.
186. Everett, J. W., and Tyrev, L. (1977). Induction of LH release and ovulation in rats by radiofrequency lesions of the medial basal tuber cinereum. *Anat. Rec.* 187, 575 (abstract).
187. Everett, J. W., Quinn, D. L., and Tyrey, L. (1976). Comparative effectiveness of preoptic and tuberal stimulation for luteinizing hormone release and ovulation in two strains of rats. *Endocrinology* 98, 1302–1308.
188. Gosden, R. G., Everett, J. W., and Tyrey, L. (1976). Luteinizing hormone requirement for ovulation in the pentobarbital-treated proestrous rat. *Endocrinology* 99, 1046–1053.
189. Everett, J. W., and Tyrey, L. (1981). Comparative increments of circulating luteinizing hormone in rats with increasing duration of electrical stimulation in medial preoptic or medial basal tuberal sites. *Endocrinology* 109, 691–696.
190. Fink, G., and Jamieson, M. G. (1976). Immunoreactive luteinizing hormone releasing factor in rat pituitary stalk blood: effects of electrical stimulation of the medial preoptic area. *J. Endocrinol.* 68, 71–87.
191. Cramer, O. M., and Barraclough, C. A. (1971). Effect of electrical stimulation of preoptic area on plasma LH concentrations in proestrous rats. *Endocrinology* 88, 1175–1183.
192. Everett, J. W., and Tyrey, L. (1982). Similarity of luteinizing hormone surges induced by medial preoptic stimulation in female rats blocked with pentobarbital, morphine, chlorpromazine, or atropine, *Endocrinology* 111, 1979–1985.
193. Aiyer, M. S., Chiappa, S. A., and Fink, G. (1974). A priming effect of luteinzing hormone releasing factor on the anterior pituitary gland in the female rat. *J. Endocrinol.* 62, 573–588.
194. Fink, G., Chiappa, S. A., and Aiyer, M. S. (1976). Priming effect of luteinizing hormone releasing factor elicited by preoptic stimulation and by intravenous infusion and multiple injections of the synthetic peptide. *J. Endocrinol.* 69, 359–372.
195. Pickering, A. J. M. C., and Fink, G. (1979). Priming effect of luteinizing hormone releasing factor *in vitro*: role of protein synthesis contractile elements, Ca^{++} and cyclic AMP, *J. Endocrinol.* 81, 223-234.
196. Grieg, F., and Weisz, J. (1973). Preovulatory levels of luteinizing hormone, the critical period and ovulation in rats. *J. Endocrinol.* 57, 235–245.
197. Weick, R. F., Dierschke, D. J., Karsch, F. J., Butler, W. R., Hotchkiss, J., and Knobil, E. (1973). Periovulatory time courses of circulating gonadotropic and ovarian hormones in the rhesus monkey. *Endocrinology* 93, 1140–1147.
198. Westman, A., and Jacobsohn, D. (1936). Über Ovarialveränderungen beim Kaninchen nach Hypophysektomie. *Acta. Obstet. Gynecol. Scand.* 16, 483–508.
199. Hilliard, J., Haywood, J. N., and Sawyer, C. H. (1964). Postcoital patterns of secretion of pituitary gonadotropin and ovarian progestin in the rabbit. *Endocrinology* 75, 957–963.
200. Beard, J. (1898). The rhythm of reproduction in animals. *Anat. Anz.* 14, 97–102.
201. Loeb, L. (1911). Über die Bedeutung des Corpus luteum für die Periodizität des sexuellen Zyklus beim weiblichen Säugetieroganismus. *Dtsch. Med. Wochenschr.* 37, 17–21.
202. Drummond-Robinson, G., and Asdell, S. A. (1926). The relation between the corpus luteum and the mammary gland. *J. Physiol (Lond).* 61, 608–614.
203. Hammond, J. (1927). *The Physiology of reproduction in the cow*. Cambridge:Cambridge University Press.
204. Papnicolaou, G. N. (1920). Effect of removal of corpora lutea and ripe follicles on oestrous periodicity in guinea pigs. *Anat. Rec.* 19, 251 (abstract).
205. Papanicolaou, G. N. (1926). A specific inhibitory hormone of the corpus luteum. *JAMA* 86, 1422–1424.
206. Parkes, A. S., and Bellerby, C. W. (1928). Studies on the internal secretion of the ovary. V. The oestrus-inhibiting function of the corpus luteum. *J. Physiol (Lond).* 64, 233–245.
207. Gley, P. (1928). Sur l'inhibition de l'ovulation par le corps jaune. *C. R. Soc. Biol. (Paris)* 98, 504–505.
208. Selye, H., Browne, J. S. L., and Collip, J. B. (1936). Effects of large doses of progesterone in the female rat. *Proc. Soc. Exp. Biol. Med.* 34, 472–474.
209. Dempsey, E. W. (1937). Follicular growth rate and ovulation after various experimental procedures in the guinea pig. *Am. J. Physiol.* 120, 126–132.
210. Kennedy, W. P. (1925). Corpus luteum extracts and ovulation in the rabbit. *Q. J. Exp. Physiol.* 15, 103–112.
211. Mahnert, A. (1930). Weitere Untersuchungen über die Beziehungen zwischen Hypophysenvorderlappen und Ovarium. Zugleich ein Beitrag zur Frage der hormonalen Sterilisierung. *Zentralbl. Gynaekol.* 54, 2883–2887.
212. Makepeace, A. W., Weinstein, G. L., and Friedman, M. H. (1937). The effect of progestin and progesterone on ovulation in the rabbit. *Am. J. Physiol.* 119, 512–516.
213. Moore, C. R., and Price, D. (1932). Gonad hormone functions and the reciprocal influence between gonads and hypophysis. *Am. J. Anat.* 50, 13–72.
214. Engle, E. T. (1931). The pituitary gonadal relationship and the problem of precocious sexual maturity. *Endocrinology* 15, 405–420.
215. Hohlweg, W. (1934). Veränderungen des Hypophysenvorderlappens und des Ovariums nach Behandlungen mit grossen Dosen von Follikelhormonen. *Klin. Wochenschr.* 13, 92–95.
216. Hohlweg, W. (1975). The regulatory centers of endocrine glands in the hypothalamus. In: Meites, J., Donovan, B., and McCann, S. M., eds, *Pioneers in neuroendocrinology*, Vol. 1. New York: Plenum Press, 161–172.
217. Hammond, J. Jr, Hammond, J., and Parkes, A. S. (1942). Hormonal augmentation of fertility in sheep. I. Induction of ovulation, superovulation, and heat in sheep. *J. Agric. Sci.* 32, 308–323.
218. Hammond J. Jr. (1945). Induced ovulation and heat in anestrous sheep. *J. Endocrinol.* 4, 169–180.
219. Casida, L. E. (1946). Induction of ovulation and subsequent fertility in domestic animals. In: Engle, E. T., ed. *The problem of fertility*. Princeton: Princeton University Press, 49–59.
220. Everett, J. W. (1948). Progesterone and estrogen in the experimental control of ovulation time and other features of the estrous cycle in the rat. *Endocrinology* 43, 389–405.
221. Everett, J. W. (1947). Hormonal factors responsible for deposition of cholesterol in the corpus luteum of the rat. *Endocrinology* 41, 364–377.
222. Everett, J. W., and Nichols, D. C. (1968). The timing of ovulatory release of gonadotropin induced by estrogen in pseudopregnant and diestrous cyclic rats. *Anat. Rec.* 160, 346 (abstract).
223. Everett, J. W., Sawyer, C. H., and Markee, J. E. (1949). A neurogenic timing factor in control of the ovulatory discharge of luteinizing hormone in the cyclic rat. *Endocrinology* 44, 234–250.
224. Everett, J. W. (1940). The restoration of ovulatory cycles and corpus luteum formation in persistent-estrous rats by progesterone. *Endocrinology* 27, 681–686.
225. Everett, J. W. (1943). Further studies on the relationship of progesterone to ovulation and luteinization in the persistent-estrous rat. *Endocrinology* 32, 285–292.
226. Everett, J. W., and Sawyer, C. H. (1949). A neural timing factor in the mechanism by which progesterone advances ovulation in the cyclic rat. *Endocrinology* 45, 581–595.
227. Zeilmaker, G. H. (1966). The biphasic effect of progesterone on ovulation in the rat. *Acta. Endocrinol.* 51, 461–468.
228. Everett, J. W. (1951). Effects of estrogen-progesterone synergy on thresholds and timing of the "LH-release apparatus" of the female rat. *Anat. Rec.* 109, 291 (abstract).

229. Redmond, W. C. (1968). Ovulatory response to brain stimulation or exogenous luteinizing hormone in progesterone-treated rats. *Endocrinology* 83, 1013–1022.
230. Caligaris, L., Astrada, J. J., and Taleisnik, S. (1971). Biphasic effect of progesterone on the release of gonadotropin in rats. *Endocrinology* 89, 331–337.
231. Sawyer, C. H., and Everett, J. W. (1959). Stimulatory and inhibitory effects of progesterone on the release of pituitary ovulatory hormone in the rabbit. *Endocrinology* 65, 644–651.
232. Kawakami, M., and Sawyer, C. H. (1959). Neuroendocrine correlates of changes in brain activity thresholds by sex steroids and pituitary hormones. *Endocrinology* 65, 652–668.
233. Dierschke, D. J., Bhattacharya, A. N., Atkinson, L. E., and Knobi, E. (1970). Circhoral oscillations of plasma LH levels in the ovariectomized monkey. *Endocrinology* 87, 850–853.
234. Brinkley, H. J. (1981). Endocrine signaling and female reproduction. *Biol. Reprod.* 24, 22–43.
235. Knobil, E. (1981). Patterns of hypophysiotropic signals and gonadotropin secretions in the rhesus monkey. *Biol. Reprod.* 24, 44–49.
236. Knobil, E. (1980). The neuroendocrine control of the menstrual cycle. *Recent. Prog. Horm. Res.* 36, 53–88.
237. Blake, C. A. (1976). Simulation of the proestrous luteinizing hormone (LH) surge after infusion of LH-releasing hormone in Phenobarbital-blocked rats. *Endocrinology* 98 451–460.
238. Weick, R. F. (1981). The pulsatile nature of luteinizing hormone secretion. *Can. J. Physiol. Pharmacol.* 59, 779–785.
239. Wise, P. M., Rance, N., Barr, G. D., and Barraclough, C. A. (1979). Further evidence that luteinizing hormone-releasing hormone also is follicle-stimulating hormone-releasing hormone. *Endocrinology* 104, 940–947.
240. Carmel, P. W., Araki, S., and Ferin, M. (1976). Pituitary stalk portal blood collection in rhesus monkeys: evidence for pulsatile release of gonadotropin-releasing hormone. *Endocrinology* 99, 243–248.
241. Nett, T. M., Akbar, A. M., and Niswender, G. D. (1974). Serum levels of luteinizing hormone and gonadotropin-releasing hormone in cycling, castrated and anestrous ewes. *Endocrinology* 94, 713–718.
242. Savoy-Moore, R. T., and Schwartz, N. B. (1980). Differential control of FSH and LH secretion. In Greep, R. O., ed. *Reproductive physiology III, International review of physiology*, Vol. 22. Baltimore: University Park Press, 203–248.
243. Wildt, L., Häusler, A., and Marshall, G., et al. (1981). Frequency and amplitude of gonadotropin-releasing hormone stimulation and gonadotropin secretion in the rhesus monkey. *Endocrinology* 109, 376–385.
244. Dufy, B., Dufy-Barbe, L., Vincent, J. D., and Knobil, E. (1979). Étude électrophysiologique des neurons hypothalamiques et régulation gonadotrope chez de singe rhesus. *J. Physiol. (Paris)* 75, 105–108.
245. Kalra, S. P. (1986). Neural circuitry involved in the control of LHRH secretion: a model for preovulatory Lh release. In: Ganong, W. F., and Martini, L., eds. *Frontiers in neuroendocrinology*. New York: Raven Press, 203–246.
246. Gallo, R. V., and Pulsatile, L. H. (1981). release during the ovulatory LH surge on proestrus in the rat. *Biol. Reprod.* 24, 100–104.
247. Cogen, P. H., Antunes, J. L., Louis, K. M., Dyrenfurth, I., and Ferin, M. (1980). The effects of anterior hypothalamic disconnection on gonadotropin secretion in the female rhesus monkey. *Endocrinology* 107, 677–683.
248. Knobil, E., Plant, T. M., Wildt, L., Belchetz, P. E., and Marshall, G. (1980). Control of the rhesus monkey menstrual cycle: permissive role of hypothalamic gonadotropin-releasing hormone. *Science* 207, 1371–1373.
249. Neill, J. D., Patton, J. M., Dailey, R. A., Tsou, R. C., and Tindall, G. T. (1977). Luteinizing hormone releasing hormone (LHRH) in pituitary stalk blood of rhesus monkeys: relationship to level of LH release. *Endocrinology* 101, 430–434.
250. Norman, R. L., Gliessman, P., Lindstrom, S. A., Hill, J., and Spies, H. G. (1982). Reinitiation of ovulatory cycles in pituitary stalk-sectioned rhesus monkeys: evidence for a specific hypothalamic message for the preovulatory release of luteinizing hormone. *Endocrinology* 111, 1874–1882.
251. Terasawa, E., Rodriguez, J. S., Bridson, W. E., and Weigand, S. J. (1979). Factors influencing the positive feedback action of estrogen upon the luteinizing hormone surge in the ovariectomized guinea pig. *Endocrinology* 104, 680–686.
252. Terasawa, E., King, M. K., Wiegand, S. J., Bridson, W. E., and Goy, R. W. (1979). Barbiturate anesthesia blocks the positive feedback effect of progesterone, but not of estrogen, on luteinizing hormone release in ovariectomized guinea pigs. *Endocrinology* 104, 687–692.
253. Caligaris, L., Astrada, J. J., and Taleisnik, S. (1971). Release of luteinizing hormone induced by estrogen injection into ovariectomized rats. *Endocrinology* 88, 810–815.
254. Legan, S. J., Coon, G. A., and Karsch, F. J. (1975). Role of estrogen as initiator of daily LH surges in the ovariectomized rat. *Endocrinology* 96, 50–56.
255. Wise, P. M., Camp-Grossman, P., and Barraclough, C. A. (1981). Effects of estradiol and progesterone on plasma gonadotropins, prolactin, and LHRH in specific brain areas of ovariectomized rats. *Biol. Repord.* 24, 820–830.
256. Banks, J. A., and Feeman, M. E. (1978). The temporal requirement of progesterone on proestrus for extinction of the estrogen-induced daily signal controlling luteinizing hormone release in the rat. *Endocrinology* 102, 426–432.
257. DePaolo, L. V., and Barraclough, C. A. (1979). Dose-dependent effects of progesterone on the facilitation and inhibition of spontaneous gonadotropin surges in estrogen treated ovariectomized rats. *Biol. Reprod.* 21, 1015–1023.
258. Norman, R. L., Blake, C. A., and Sawyer, C. H. (1973). Estrogen-dependent twenty-four-hour periodicity in pituitary LH release in the female hamster. *Endocrinology* 93, 965–970.
259. Norman, R. L., and Spies, H. G. (1974). Neural control of the estrogen-dependent twenty-four-hour periodicity of LH release in the golden hamster. *Endocrinology* 95, 1367–1372.
260. Stetson, M. H., Watson-Whitmyre, M., and Matt, K. S. (1978). Cyclic gonadotropin release in the presence and absence of estrogenic feedback in ovariectomized golden hamsters. *Biol. Reprod.* 19, 40–50.
261. Seegal, R. F., and Goldman, B. D. (1975). Effects of photoperiod on cyclicity and serum gonadotropins in the Syrian hamster. *Biol. Reprod.* 12, 223–231.
262. Bridges, R. D., and Goldman, B. D. (1975). Diurnal rhythms in gonadotropins and progesterone in lactating and photoperiod induced acyclic hamsters. *Biol. Reprod.* 13, 617–622.
263. Schwartz, N. B. (1982). Role of ovarian inhibin (folliculostatin) in regulating FSH secretion in the female rat. In: Channing, C. P., and Segal, S. J., eds. *Intraovarian control mechanisms*. New York: Plenum Press, 15–36.
264. Thomas, C. L. Jr, and Nikitovitch-Winer, M. B. (1984). Complete suppression of plasma follicle-stimulating hormone in castrated male and female rats during continuous administration of porcine follicular fluid. *Biol. Reprod.* 30, 427–433.
265. Miller, W. L., and Huang, E. S. R. (1985). Secretion of ovine luteinizing hormone *in vitro*: differential positive control by 17β-estradiol and a preparation of porcine ovarian inhibin. *Endocrinology* 117, 907–911.

Knobil and Neill's Physiology of Reproduction, Third Edition
edited by Jimmy D. Neill,
Elsevier © 2006

CHAPTER 27

Anatomy of the Hypothalamo–Hypophysial Complex

Robert B. Page

Introduction, 1309
The Adenohypophysis, 1316
Epithelial Cells of the Adenohypophysis, 1316
Development of the Adenohypophysis, 1319
Organization of the Adenohypophysis, 1322
The Neurohypophysis, 1329
Nerve Terminals of the Neurohypophysis, 1329
Development of the Neurohypophysis, 1331
Organization of the Neurohypophysis, 1333
The Portal System, 1343
Capillaries of the Hypophysis, 1343
Development of the Portal System, 1343
Organization of the Portal System, 1344
The Hypothalamus, 1351
Cell Bodies of the Hypothalamus, 1351
Development of Neurosecretory Systems, 1355
Organization of Neurosecretory Systems, 1356
Input into the Periventricular Brain, 1373
References, 1380

INTRODUCTION

Consider the pituitary gland as did Harvey Cushing almost a century ago (1). The pituitary gland that Cushing (1) saw differed from that seen by others. It was described by a dictionary of the time as: "a small bilobed body of unknown function attached to the infundibulum at the base of the brain" (2,3). Common wisdom of the day held it to be a vestigial organ. He saw a structure composed of neural and epithelial elements lying in the sella turcica in the base of the skull. He held that the epithelial component developed from an oral pouch that was derived from the ectoderm of the primitive mouth and not from the endoderm of the primitive foregut as initially proposed by Rathke (4). He further realized that the epithelial portion, the anterior lobe, was organized in a glandular pattern (5,6) and that its cells could be differentiated into chromophobes and chromophiles on the basis of their staining affinity for hematoxylin and eosin dyes (1). He observed that the anterior lobe of the gland was very vascular and that its glandular cells assumed an intimate relationship with its sinusoids. His understanding of the blood supply to the pituitary was based on the studies of Dandy and Goetsh (7) who had described an arterial supply to the canine anterior lobe which arose from the vessels of the circle of Willis and coursed centripetally over the tuber cinereum to supply it at its junction with the anterior lobe. These authors (7) described the venous drainage of the anterior lobe to be directed toward the base of the brain. Based on the writings of Claude Bernard (1853) and of Brown-Sèquard (1856) who had proposed that one cell could secrete "on its own account certain products or special ferments which influence all other cells of the body by a mechanism other than the central nervous system"; and on his experience with patients suffering from acromegaly, Cushing (1) came to believe that the anterior lobe of the pituitary body was a gland of internal secretion and that a tumor of this apparently insignificant organ might cause profound changes in

Emeritus Professor of Neurosurgery, Department of Neurosurgery, H110, Milton S. Hershey Medical Center of the Pennsylvania State University, Hershey, Pennsylvania.

body habitus and metabolism because it released its secretions into the blood stream in excessive amounts.

Cushing recognized that acromegaly was not the uniform result of any tumor of the anterior lobe of the pituitary. Some patients with pituitary tumors developed adiposal-genital dystrophy and died (8) [cited by Cushing (1)]. To test the hypothesis that this syndrome might be due to destruction of the gland with a consequent reduction or cessation of its secretions, he began a series of ablative experiments on dogs in the newly established Hunterian Laboratory at Johns Hopkins Hospital (9). He was able to demonstrate that the anterior lobe was necessary to support the structure and function of the gonads, adrenals, and thyroid, to support the normal growth of the young animal and to support even life itself. The injection or ingestion of pituitary extracts did not reverse the effect of hypophysectomy. His attempt to reverse the deficits produced by hypophysectomy by transplanting the pituitary gland (to the rectus sheath, to bone marrow or to brain) met with only limited success (10). He was able to establish that in cases of partial ablation of the gland, the transplant seemed to support the dog (which otherwise would have died) until the remaining fragments could recover and hypertrophy. Although it was not until some 20 years later that P. E. Smith (11) was able to reverse all the deficits caused by hypophysectomy in rats by the injection of pituitary extract, Cushing (1) in 1912 clearly saw the anterior lobe of the pituitary as a ductless gland which synthesized hormones and secreted them into the blood stream which carried them toward the brain. These secretions supported the function of the other ductless glands and permitted normal body growth.

In 1912 Cushing (1) realized that most of the posterior lobe (the neural portion of the pituitary gland) was derived from the brain (5) but he included the pars intermedia with the neural lobe under the rubric of the "posterior lobe." In his morphological studies he saw "colloid" in the posterior lobe as had Herring (4) when he histologically examined the posterior lobe (1) (Fig. 1). He proposed, on the basis of the analogy with the thyroid, that colloid, now called *Herring bodies*, was a secretion of the posterior lobe; but he believed it was a secretion of the pars intermedia, not of the pars nervosa. The pars nervosa (along with the cervical sympathetics) served to innervate the pars intermedia (1,12). His visualization of colloid on microscopic examination within the pars intermedia, the neural lobe, beneath the ependyma of the diencephalic floor, and apparently bursting into the third ventricle on microscopic examination convinced him that the posterior lobe secreted directly to the brain (13). The paucity of blood vessels in the posterior lobe (7) reinforced his belief that the secretions from the pars intermedia of

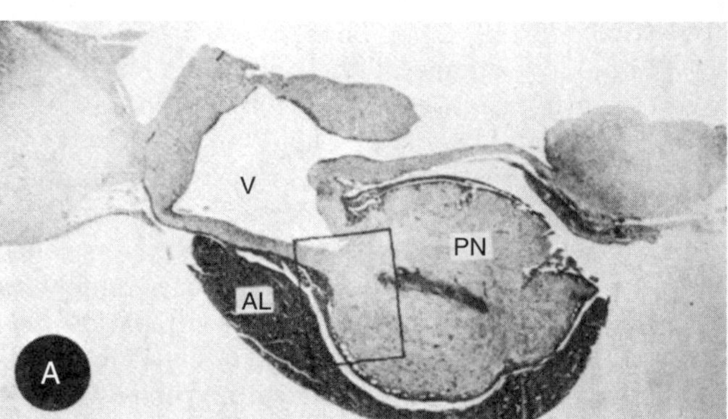

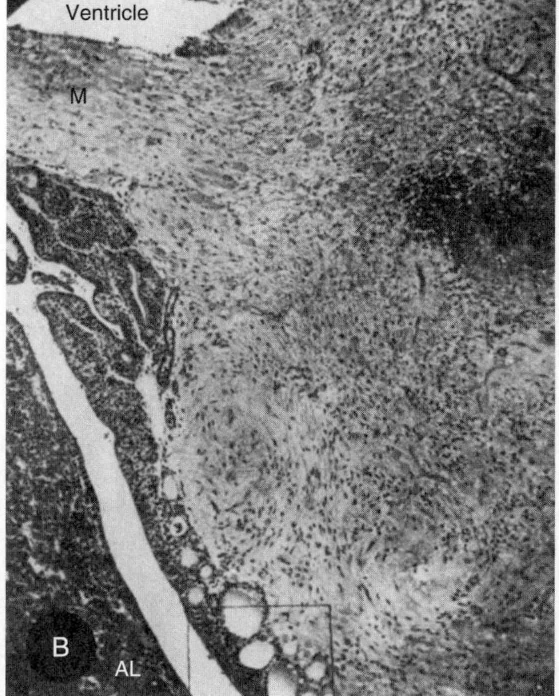

FIG. 1. A: Midsagittal section of canine pituitary gland. Note layer of colloid globules arising from the epithelial investment of the par nervosa (PN). Anterior lobe (AL) is separated from posterior lobe by cleft. V, third ventricle. **B:** Enlargement of squared-off area in A. Anterior lobe (AL) is separated by cleft from investment of pars nervosa containing colloid masses. [From (1).]

the posterior lobe entered the third ventricle, not the circulation.

To determine the function of the posterior lobe, he performed either hypophysectomies, selective ablations of the anterior or posterior lobes or selective stimulations of these regions of the pituitary gland in dogs. On the basis of these experiments, Cushing proposed that the posterior lobe secretions caused a fall in glucose tolerance (glucosuria) whereas their absence resulted in a rise in glucose tolerance (14). This surmise was strengthened by his finding that stimulation of the pars nervosa or of the superior cervical ganglion, both of which structures he envisioned as innervating the intermediate lobe (1,12), caused glucosuria (15). He was aware that a vasopressor effect of posterior lobe extract had been reported (16,17) as had an oxytocic effect (18). He accepted these observations as true but thought these roles were subordinate to the role of the posterior lobe in glucose metabolism. He did not accept the proposal that the posterior lobe secreted an antidiuretic substance. As late as 1930 he stated: "too much attention has been paid to the symptoms of thirst and polyuria, and too little to the symptoms of the opposite of these, oliguria having been observed not infrequently as a sequel of our early (1908–1910) canine hypophysectomies" (2). In 1912 Cushing (1) saw the posterior lobe as a gland (the pars intermedia) that was innervated principally by the pars nervosa. Hormones synthesized in the pars intermedia were carried through the pars nervosa to the third ventricle and functioned to regulate glucose metabolism, vascular tone, and uterine contractions.

In the intervening years from 1912 to 1930 when he delivered the Lister Memorial Lecture (2,3), Cushing clung to these two tenets: that the anterior lobe released its secretions into blood vessels draining toward the brain and the posterior lobe released its secretions into the third ventricle. The discovery of a portal system in the pituitary gland by Popa and Fielding (19,20) did not alter his convictions, which were based on his concept that the anterior and posterior lobes each received an independent blood supply and the posterior lobe drained to the cavernous sinus (7). A portal system supplied a route to Cushing by which anterior pituitary secretions could be delivered to the brain prior to being delivered to the remainder of the body (21). He continued to believe that the posterior lobe was a special brain gland whose secretory cells in the pars intermedia were innervated by the pars nervosa and that it released its secretions into the brain's ventricular system.

In those intervening years he had also become aware that in some unknown fashion the hypothalamus and the pituitary gland are inextricably joined. The accepted belief at that time held that the secretions of the posterior lobe were synthesized there in glial cells of the pars nervosa (22,23), in epithelial cells of the pars intermedia (21) or in glandular cells in the neural lobe (24) [for a contemporary review, see (25)]. As early as 1912, Cushing had been aware of the report of Cajal in 1894 describing a neural tract that originated in the hypothalamus and ended in the neural lobe. He subsequently cited reports of Greving (1926) and of Pines (1928) describing a fiber tract passing from the supraoptic nuclei (SON) and paraventricular nuclei (PVN) to the neural lobe but he persisted in thinking this tract innervated the pars intermedia (2,3). The true function of the supraoptico–hypophysial tract remained unsuspected (26,27). He acknowledged that diabetes insipidus could be caused by making tuberal lesions in rats that interrupted the supraoptico–hypophysial tract and could be cured by giving the rat's posterior lobe extract; but he did not accept the significance of this observation. Although he misunderstood the means by which the hypothalamus regulated posterior lobe function, he championed the concept that the hypothalamus regulated posterior pituitary function stating, "it is highly improbable that two corresponding effects should be produced, the one by a hypothalamic lesion, the other by removing the source of chemical messages, in the absence of any functional interaction" (2,3).

Cushing also noted that the hypothalamus apparently controlled certain aspects of anterior lobe function—such as ovulation in the rabbit because it regularly follows 10 hours after copulation in that species and hence resembles a reflex act. He became further aware of the role of the diencephalon in regulating sympathetic tone, alertness, body temperature, body habitus and sexual function from caring for patients with tumors which invaded the third ventricle and hypothalamus such as gliomas of the optic chiasm and craniopharyngiomas but which spared the pituitary gland. Their occurrence brought home to Cushing the realization that there is an "interdependence of the diencephalon and the pituitary body."

He explained this interdependence between the diencephalon and the gland beneath it in the following manner. Nerves in the diencephalon projecting to the pituitary gland through the supraoptico–hypophysial tract stimulate the release of substances from the pars intermedia in the posterior lobe which are transported directly into the third ventricle from which site they influence neural centers in its walls. Discharge from these diencephalic centers activates sympathetic mechanisms that result in the stimulation of the glandular cells in the anterior lobe by way of the cervical sympathetic plexus. In his picture, the peripheral sympathetic nervous system was the final common pathway to the epithelial cells of the anterior lobe of the pituitary and to the epithelial cells of the adrenal medulla, as well as to the smooth muscle cells of the intestines and the blood vessels (2,3). His seminal belief that the diencephalon coordinated the functions of the anterior pituitary gland and the sympathetic nervous system passed into the background as forces marshaled which would undo many of his concepts and change the way investigators see the pituitary gland.

To see the gland as most see it today, it is necessary to start with a consideration of the work of Wislocki and King (28). Published 6 years after Cushing's last public lecture on the pituitary gland, their work challenged several of his basic concepts. They stressed the observation that the pituitary gland and viscera stained after the intravascular injection of acid vital dyes into monkeys, cats, and rabbits whereas the brain (with the exception of the choroid plexus and several small regions surrounding the third ventricle and the area postrema of the IV ventricle) did not. The contents of the third ventricle were not stained following the intravascular injection with acid dyes even when the neural lobe was heavily stained. They thus demonstrated that substances could enter, and presumably leave, the neural portion of the hypophysis by vascular routes but that it was unlikely that substances in the neurohypophysis were discharged into the third ventricle. Wislocki and King (28) also studied the vascular anatomy of the pituitary. They made intravascular injections of monkeys, rabbits and rats with India ink and examined either serial sections of the injected gland or whole mounts after clearing the tissue. They concluded that they could not corroborate the observations of Popa and Fielding (19,20) that a large artery arising from each supraclinoid carotid artery supplies the pars distalis and that large venules connect the pars distalis to the hypothalamus. Their report erased a picture of the pituitary gland that had been held for almost 30 years. Neither the neural nor the glandular portion of the pituitary could discharge their contents directly to the brain. A new concept had to be erected.

Wislocki and King (28) provided the basis for it. First, they noted that the eminentia saccularis (the median eminence of the tuber cinereum or the infundibulum) was stained as were the anterior and posterior lobes, after the intravascular injection of acid dyes whereas the remainder of the tuber cinereum was not. They proposed that the median eminence should, as a consequence of this similarity with the pituitary gland, be classified as part of it. In this schema, based on the observation that the hypophysis lacks a blood–brain barrier, the neurohypophysis is composed of the infundibulum (median eminence, eminentia saccularis), infundibular stem and infundibular process (neural lobe). The adenohypophysis is composed of the pars tuberalis (applied to the infundibulum and infundibular stem), the pars intermedia (applied to the infundibular stem and infundibular process) and the pars distalis (anterior lobe) (28,29). This classification recognizes that the pituitary gland lies not only within the sella turcica (the lower infundibular stem and infundibular process with the pars intermedia and the pars distalis) but also in the subarachnoid space; and, it is applied to the base of the brain (the median eminence and upper infundibular stem with the pars tuberalis). Wislocki and King (28) observed, after the intravascular injection of vital dyes, that the boundary of the stained pituitary gland with the unstained hypothalamus lay at the level of the tubero–infundibular sulcus. This sulcus, which separates the tuber cinereum of the hypothalamus from the infundibulum, not the diaphragm sella, forms the rostral boundary of the pituitary gland (Fig. 2).

Second, they correctly surmised the direction of blood flow in the portal system and suggested its role in control of anterior pituitary function. They described superior hypophysial arteries in the monkey that arose from the supraclinoid internal carotid arteries and the

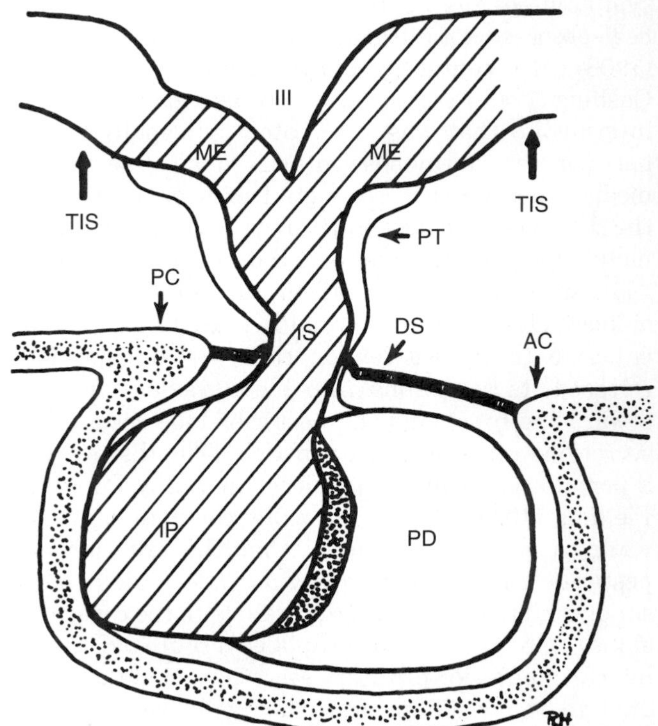

FIG. 2. Diagrammatic midsagittal section of the human pituitary gland. The neurohypophysis (hatching) consists of the median eminence (ME) (or infundibulum), infundibular stem (IS), and infundibular process (IP) (neural lobe). The adenohypophysis consists of the pars tuberalis (PT), which is applied to the ME and IS and lies in the subarachnoid space above the diaphragm sella (DS) and the pars distalis (PD), which lies within the sella turcica beneath the DS. The region corresponding to the pars intermedia of other forms is indicated by the *stippled* area between the PD and the IP. AC, anterior clinoid; PC, posterior clinoid; III, third ventricle; TIS, tubero–infundibular sulcus. [From (1170).]

vessels of the circle of Willis. These vessels approached the eminentia saccularis (median eminence) and pituitary stalk (pars tuberalis and infundibular stem) and the rostral pole of the pars distalis. The superior hypophysial arteries bifurcated sending one branch to the pars distalis and one branch to the infundibular stem and median eminence. The capillary bed in the median eminence and upper part of the infundibular stem was fed by the superior hypophysial arteries. Portal vessels that discharged their contents into a secondary capillary bed in the pars distalis drained it. The pars distalis in turn drained by lateral hypophysial veins into the adjacent cavernous sinuses. The pars distalis, according to the account of Wislocki and King (28) resembled the liver receiving both an arterial and a venous blood supply. They surmised that the direction of blood flow through the portal vessels had to be from the median eminence to the pars distalis as they could demonstrate no significant outflow routes from the median eminence to the vessels at the base of the brain. The only apparent outflow was from the median eminence to the pars distalis that could in turn drain to the adjacent cavernous sinuses through lateral hypophysial veins (Fig. 3).

Third, they saw the circulation of the neural lobe as isolated from that of the portal system. They described

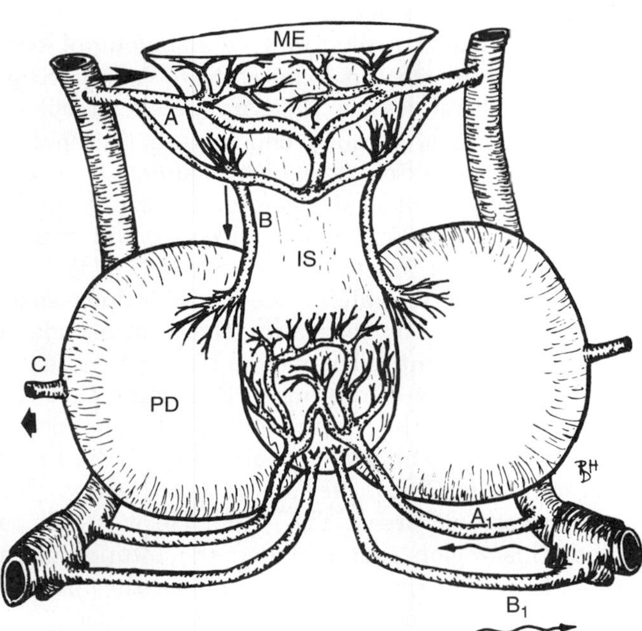

FIG. 3. Classical schema of pituitary blood flow proposed by Wislocki and King (28). Blood enters the median eminence (ME) at point A through superior hypophysial arteries (*large arrow*). It passes through the ME primary capillary plexus in the pars distalis (PD). From the PD, blood drains through lateral hypophysial veins (C) to the cavernous sinus (*arrowhead*). Blood enters the neural lobe through inferior hypophysial arteries (A_1). It drains from the neural lobe through inferior hypophysial veins (B_1). IS, infundibular stem. [From (673).]

the origin and course of the inferior hypophysial arteries from the intracavernous segment of the carotid arteries to the infundibular process and the paired venous structures (inferior hypophysial veins) that drained it. The function of the neural lobe had been clarified by 1936. In 1924 Starling and Verney (30) had reported that posterior lobe extract corrected diabetes insipidus, reduced urinary flow and raised urinary solute concentration. The posterior (neural) lobe was also recognized to regulate urine flow and blood pressure (16) and to stimulate uterine contractions (18). Its structure was also clear. The neural lobe contained axons and terminals of the supraoptico–hypophysial tract, specialized glial cells called pituicytes and blood vessels (26,27). With the demonstration by Ingram and his coworkers (31–33) that lesions of the supraoptico–hypophysial tract caused diabetes insipidus, atrophy of the neural lobe and degeneration of cell bodies in the supraoptic nucleus, it became apparent that the supraoptico–hypophysial tract regulated in some unknown manner the release of hormones from the neural lobe.

Ernest and Berte Scharrer provided the answer to this puzzle explaining how the brain regulates not only neural lobe function (34,35) but by extension of their ideas how the brain also regulates adenohypophysial function. The concept of neurosecretion, first promulgated by E. Scharrer, received support from the observations of Palay (36) who found the presence of a colloid material, stainable with silver or Masson's trichrome technique, that was unique to the preoptico–hypophysial tract of the goldfish (a tract analogous to the supraoptico–hypophysial tract of mammals). Bargmann and Hild (37) found that the chrome-alum-hematoxylin technique of Gomori selectively stained neurosecretory (colloid) material within the supraoptico–hypophysial tract, as it selectively stained secretory material in the islet cells of the pancreas. On the basis of the finding of stainable neurosecretory material in the cell bodies of large cells in the supraoptic nucleus and of finding this material in the axons of these cells and in their terminals in the neural lobe, Bargmann and Scharrer (38) proposed that hormones released from the posterior lobe of the pituitary were not synthesized there. They were synthesized in the supraoptic nucleus of the hypothalamus, transported down their axons to their terminals in the posterior lobe and were released from these terminals into the blood stream. Cushing had seen stainable (colloid) material in the neural lobe and proposed on the basis of analogy with the thyroid, that it was a secretion of the pars intermedia in the posterior lobe. Because he saw few vessels in the posterior lobe and because he saw colloid bursting into the third ventricle he came to believe that the neural lobe secreted to the brain.

Scharrer saw the same material within nerves lying in the hypothalamus and in their terminals in the neural lobe. Based on the analogy with the neurosecretory system of insects and the knowledge that the neural lobe lacked a blood–brain barrier but possessed direct venous connections to the cavernous sinuses, he proposed the (correct) mechanism by which the brain controls pituitary function. He saw that secretions of the neural lobe were made in the hypothalamus, transported to the neural lobe and were released into the circulation. Each had looked at the pituitary and each saw a different organ.

Scharrer broadened the scope of his histological observations (39–41) but it was the biochemical and physiological investigations of others that confirmed his seminal concept. The neural lobe hormone responsible for uterine contraction and for milk injection was identified as oxytocin (42,43). The neural lobe hormone responsible for antidiuresis was identified as vasopressin (44–46). These hormones were found not only in the neural lobe of the pituitary but also in the hypothalamic supraoptic (and paraventricular) nuclei (47). Subsequent anatomical and physiological studies have demonstrated that oxytocin and vasopressin are synthesized in separate cell bodies lying supraoptic and paraventricular nuclei, are transported down the axons of the supraoptico–hypophysial tract to their axon terminals in the neural lobe and are released into the systemic circulation in response to appropriate stimulation (48–63).

Harris (64) pursued the observation that, in the rabbit, ovulation reflexly follows copulation. He induced ovulation by stimulation of the hypothalamus but not the pituitary. The mechanism could not be a direct neural innervation of glandular cells as the pars distalis lacked nerves other than those terminating on vessels (65). In 1947 Green and Harris (66) proposed that there was a neurovascular link between axons terminating in the median eminence and glandular cells in the pars distalis. This link was provided by the portal vessels on the anterior surface of the pituitary stalk. Green and Harris supported their proposal with the following observations: (a) nerves terminating in the median eminence and upper infundibular stem are in close contact with capillaries that drain into portal vessels (66,67); (b) blood flowed through portal vessels of living animals (frogs and rats) from the median eminence to the pars distalis (68,69); (c) stimulation of the hypothalamus caused ovulation in the anesthetized (64) and unanesthetized rabbit (70); (d) stalk section disrupted trophic function of gland and disrupted estrus cycles but trophic function and estrus cycles returned if portal vessels were permitted to regenerate (71–73); (e) functional grafting of the excised pituitary gland only occurred if the excised portion was replaced beneath the median eminence and the portal vessels were permitted to regenerate (74,75). Harris (76) summarized the extant evidence that the anterior pituitary is regulated by the brain in 1962.

The findings of Green and Harris gained more significance when it was appreciated that a portal vascular system characteristically links the median eminence with the pars distalis in vertebrates (65,77–79) including humans (80–82). Although Harris (83) reported there was an independent arterial supply to the pars distalis of the rabbit (in addition to the portal [venous] supply), Green (65) could not identify a separate arterial supply to the pars distalis in 76 other species. Reexamination of the vascular supply of the rabbit pituitary employing scanning electron microscopy of vascular casts (84) failed to confirm Harris's findings. Contrary to Wislocki's belief (28,85–87), the pars distalis does not appear to have a dual blood supply as does the liver. It appears to be bathed entirely by blood that has passed through the neurohypophysis.

Stalk section experiments strengthened the case for a neurovascular link and further weakened the case for a separate arterial supply to the pars distalis. Such sections divided the supraoptico–hypophysial tract that linked cells of the supraoptic nucleus with their terminals in the neural lobe and the portal vessels that linked terminals in the median eminence with glandular cells in the pars distalis. Stalk section transiently disrupts neural lobe function until regeneration of divided axons, or establishment of new neurohemal contacts, can occur (88–91). With interruption of the portal vessels, the pars distalis atrophies but the degree of atrophy depends on the species studied and the extent to which channels from the neural lobe to the adjacent pars distalis are available (92–96). Estrus cycles are abolished (72) and trophic anterior pituitary function is destroyed (73,74).

Halasz sought to evaluate the neural influences on pituitary function by separating the pituitary gland with its attachment to the medial basilar hypothalamus from the rest of the brain. Reasoning that the only portion of the brain that could support a pituitary transplant lay in the region of the tuber cinereum (75,97,98), he set out to study this hypophysiotropic area (HTA) (99). He was able to isolate this region that included the infundibulum and entire pituitary gland as well as the arcuate nuclei, a portion of the ventromedial nucleus, the periventricular nuclei, the ventral premamillary, the median mammillary nuclei, and the retrochiasmatic area from the remainder of the brain. All neural input into the HTA was interrupted. However, neural output to the median eminence from the arcuate nucleus (the tubero–infundibular tract) (100) was spared (99). Isolation of the HTA and the pituitary gland from the brain produced (in the rat) a different picture than did isolation of the pars distalis from the median eminence by stalk section. Trophic pituitary function was not seriously disturbed (99). Basal thyroid function, adrenal output of corticosteroids, and testicular sperm production were maintained. The histologic picture of the thyroid, adrenals and testis was not markedly altered and compensatory hypertrophy of the remaining adrenal gland occurred after unilateral adrenalectomy (101–105). Modulating neural influences on pituitary function were disrupted, however, as the estrus cycle was abolished and an increased secretion of adrenal corticosteroids in response to stress did not occur (see [104] for summary). These observations focused attention on the HTA, now defined as comprising the periventricular nuclear groups in the hypothalamus, and particularly on the median eminence, as the focal point of converging neural systems for the humoral relay of information from the brain to the anterior pituitary gland.

The median eminence was found to contain substances that were capable of stimulating or inhibiting the release of anterior pituitary hormones (106,107). Saffran and Schally (108) and Guillemin (109) incubated pituitary tissue with hypothalamic tissue in vitro and were able to demonstrate support of adenocorticotropin (ACTH) production using bioassay techniques. Further studies employing such in vitro assays revealed hypothalamic factors capable of stimulating ACTH (110,111), thyroid-stimulating hormone (TSH) (112), growth hormone (GH) (113), and luteinizing hormone (LH) (114) release from incubated pituitary glands. Prolactin (PRL) production from anterior pituitary cells was inhibited by incubation with hypothalamic tissue (115–118). Anterior pituitary hormone release of ACTH (119,120), TSH (121), GH (122), follicle stimulating hormone (FSH) (123,124), and LH (125, 126) in vivo could also be stimulated by injection of hypothalamic extracts.

Porter et al. (127,128) developed a technique for perfusion of and sampling from the long portal vessels on the anterior surface of the rat's pituitary stalk. Perfusion of the long portal vessels with hypothalamic extract stimulated FSH and LH release from the anterior pituitary (129,130). TSH release was also stimulated by infusion of hypothalamic extracts into portal vessels (131). Gonadotropin-releasing activity (132) has been found in portal blood collected from the severed pituitary stalk; and GH releasing activity, TSH-releasing activity (133) and gonadotropin-releasing activity (130) have been found in blood sampled by cannulation of portal vessels.

The chemical composition of thyrotropin releasing hormone (TRH) was the first of the hypothalamic-releasing hormones to be revealed (134–136). TSH was released from the anterior pituitary following the infusion of TRH into a portal vessel on the surface of the rat pituitary stalk (137). Subsequently the structures of gonadotropic releasing hormone (GnRH) (138–140), corticotropin-releasing hormone (CRH) (141,142), somatotropin release inhibiting hormone (SRIF) or somatostatin (SS) (143), and growth hormone-releasing hormone (GRH) (144–146) have been elucidated. There is an increasing body of evidence supporting the role of dopamine as a prolactin inhibiting factor (147–153). The chemical characterization of hypothalamic hypophysiotropic hormones has permitted their localization within hypothalamic neurons by immunohistochemical techniques (154,155). The pattern revealed by their localization is one of convergence of axons containing peptide hormones (peptidergic neurons) on the median eminence from diverse medial preoptic (MPOA) and hypothalamic periventricular loci (156,157).

Immunohistochemistry, when combined with transmission electron microscopy, permitted the subcellular localization of hypothalamic hormones (158–160). GnRH-containing cells were believed by early investigators employing immunohistochemistry to originate in the hypothalamus and terminate in the median eminence of the rat. In the median eminence, GnRH was localized in granular vesicles in axon terminals (161) that lay in the perivascular space of median eminence capillaries. In cell bodies in the hypothalamus, GnRH was identified in granular vesicles associated with the Golgi apparatus. On the basis of morphologic evidence, it was proposed that GnRH is synthesized on polysomes in the cell body, packaged into dense core vesicles in the Golgi apparatus and transported in vesicles down the axon by axoplasmic flow to be stored within axon terminals and released upon appropriate stimulation (159,160,162, 163). The process of the synthesis of a hypothalamic-releasing (or -inhibiting) hormone in cell bodies lying in diverse hypothalamic nuclei and their delivery to and their release from neural terminals in the median eminence was generally held to be analogous to the synthesis of vasopressin or oxytocin in cell bodies lying in the supraoptic nuclei and their delivery to and release from axon terminals in the neural lobe.

Cushing (1) saw an organ composed of a neural and an epithelial portion lying in the sella turcica. His pars anterior released its secretions under regulation of the peripheral sympathetic system into veins that carried them to the brain and then to distant glands, which they supported. His posterior lobe released its secretions, which were synthesized in the pars intermedia, under the regulation of the diencephalon via the supraoptico–hypophysial tract. Secretions were carried through the neural lobe to the third ventricle, which they entered to influence periventricular hypothalamic centers. These in turn regulated the sympathetic system that controlled the tone of smooth muscle in the intestines and arteries, regulated adrenal medullary function and controlled the function of the anterior lobe of the pituitary. Some 100 years later we see a different organ: one that has a neural and an epithelial component. The rostral end of the neural region lies on the base of the brain and is the site where secretions manufactured outside the pituitary and behind the blood–brain barrier in the hypothalamus are released from nerve terminals and are carried by restricted portal routes to the epithelial portion of the pituitary gland (pars distalis within the sella turcica) to regulate its function. The caudal region of the neural portion lies within the sella turcica and is the site where secretions (synthesized in the hypothalamus) are released from terminals of the supraoptico–hypophysial tract and are carried by systemic routes to regulate the function of the kidneys, breast, and uterus. The means by which the brain controls pituitary function is now well understood.

Cushing's concept that the pituitary gland and the autonomic nervous system are inextricably linked through the diencephalon was seminal. Not only do the periventricular nuclear groups of the hypothalamus (HTA) and the medial preoptic area (MPOA) house cells that project to the (median eminence) to regulate pituitary function, they also house neurons that project to the intermedio-lateral cells column of the spinal cord and to brain stem visceral efferent nuclei to regulate autonomic function (164). The objective of this review is to explore in, further detail, (a) the structures employed by the brain to control pituitary function, and (b) the organization employed by the diencephalon to coordinate endocrine and neural responses to changes in the internal milieu and external environment.

THE ADENOHYPOPHYSIS

Epithelial Cells of the Adenohypophysis

Epithelial cells and smooth muscle cells are the motor elements of the neuroendocrine and autonomic nervous systems, which arise in the diencephalon. The epithelial cells of the neuroendocrine system are found in the adenohypophysis, the glandular portion of the pituitary body at the base of the brain. These epithelial cells carry out their role as effector elements of the neuroendocrine system by synthesizing peptide hormones and secreting them into nearby capillaries (Fig. 4A). They are round to polygonal in shape and are characterized by the presence of electron dense granules, lucent vesicles, rough endoplasmic reticulum and a Golgi apparatus in their cytoplasm. The electron dense granules are round to ovoid in shape and range from about 100 to 700 nm in diameter depending on the type of the cell, its age, and its functional state. They are the site of storage of pituitary hormones (165–167).

The Golgi apparatus in these cells is prominent and composed of a half-moon shaped system of stacked parallel cisternae. The cis face of the Golgi apparatus is the convex surface that faces the rough endoplasmic reticulum. The trans face is the concave surface that faces the plasmalemma. Plate like cisterna are closely associated with the trans elements of the Golgi (168) and are analogous to the GERL apparatus described in dorsal root ganglia (169,170). In this most trans region, often called the GERL, acid phosphatase has been demonstrated (170). Cytochemical staining techniques revealed specific enzymes localized to specific membranes of the Golgi stack [see (171) for review]. With fractionation of the Golgi membranes phosphorylation was demonstrated in the Cis region, glycosylation in the middle region and addition of silica acid or glycoside residue occurred in the trans region (168). These experiments support earlier Cytochemical observations. The existing evidence demonstrates that enzymatic functions in the Golgi apparatus are compartmentalized.

The electron lucent vesicles in the region of the cis face are about 40 nm in diameter. They lie between the rough endoplasmic reticulum and the Golgi apparatus and are believed to "shuttle" nascent proteins from the former to the latter. Farquhar and Wellings (172) first presented transmission electron microscopic (TEM) evidence which suggested that hormones were packaged into secretory granules within the Golgi apparatus. Electron micrographs revealed small vesicles surrounded by membranes budding from the end of the Golgi lamellae that progressively enlarged as they became increasingly displaced from the Golgi (172,173).

Pulse labeling experiments employing labeled amino acids revealed a path from rough endoplasmic reticulum via lucent vesicles to the Golgi apparatus and then to secretory granules (174–177). Further biochemical evidence suggested that at least two pituitary hormones (growth hormone and prolactin) are assembled on polyribosomes as pre-hormones with an excess 27 amino acids at the amino terminus. This segment serves as a signal sequence to permit passage of the hormone being assembled from the polysomes through the membranous lamellae of the rough endoplasmic reticulum into its channels (178,179). Membrane, pinched off from the channels of the endoplasmic reticulum, is believed to form the lucent vesicles that transport the protein to the Golgi apparatus for packaging into secretory granules [or alternatively to the plasmalemma for discharge under conditions of maximal stimulation (180). The Golgi apparatus is compartmentally organized and processing occurs in a cis-trans sequence (168). The proteins pass sequentially through channels in the Golgi from stack to stack by vesicular transport and during that transport undergo a series of modifications. In addition

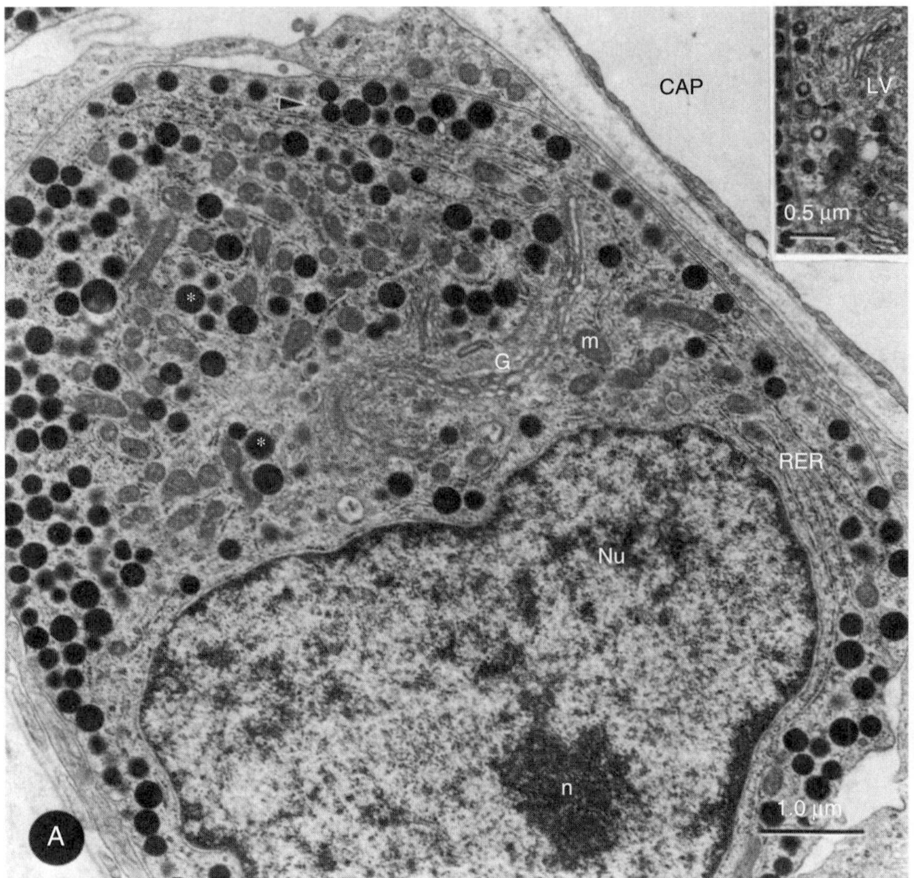

FIG. 4. A: Transmission electron micrograph of a rabbit pars distalis epithelial cell. The large nucleus (Nu) contains a single nucleolus (n). Nuclear chromatin in homogenously dispersed except at the nuclear membrane, where it is aggregated. The cytoplasm contains many electron-dense granules (*). Mitochondria (m) are abundant. The rough endoplasmic reticulum (RER) is plentiful. The Golgi apparatus (G) is found close to the nucleus. The cis border faces the nucleus. The trans (concave) border is associated with dense granules of varying size. A coated vesicle is indicated by the *small arrowhead*. CAP, capillary. Note fenestra in capillary endothelial tube. **Inset:** Extrusion of granule with formation of omega figure at *arrowhead*. Both dense granules and coated vesicles are present at cell plasmalemma. LV, lucent vesicles.

Continued

to the packaging of hormone and to their modification by phosphorylation, glycosylation or sialation (168,181), post-translational modification of pre-hormones by enzymatic cleavage of peptide bonds begins (182).

The large electron dense granules are usually aggregated near the concave trans face as well as dispersed throughout the cytoplasm. Membrane for the formation of secretory granules is provided by recycling the membrane of secretory granules after exocytosis (183). The site of formation of large secretory granules has been found to be the outermost trans elements of the Golgi (184). Cationic ferritin, endocytosed from the cell's surface by lucent coated vesicles is carried to the trans face of the Golgi. The marker membrane was observed to fuse with nascent secretory granules budding from the Golgi apparatus thus providing material for their enlargement. This region is characterized by the presence of plate-like cisternae (169) and stains with acid phosphatase. Novikoff et al. (170) attributed the function of lysosome formation to the GERL region because of the localization of acid phosphatase to it. It now appears that its function is complex and that the GERL is the site of both secretory granule and lysosome formation.

With stimulation of adenohypophysial cells (such as stimulation of lactotrophs by suckling or stimulation of somatotropes with GRH) the rough endoplasmic reticulum became more pronounced, vesicles increased in number and the granules discharged their contents by exocytosis (185,186). TEM examination of ultrathin sections and of freeze fracture material demonstrated migration of secretory granules to the cell periphery and fusion of the granular and plasma membranes (187). Microtubules have both been implicated in this migration of secretory granules to the periphery of the cell (188). Cortical filamentous actin has been

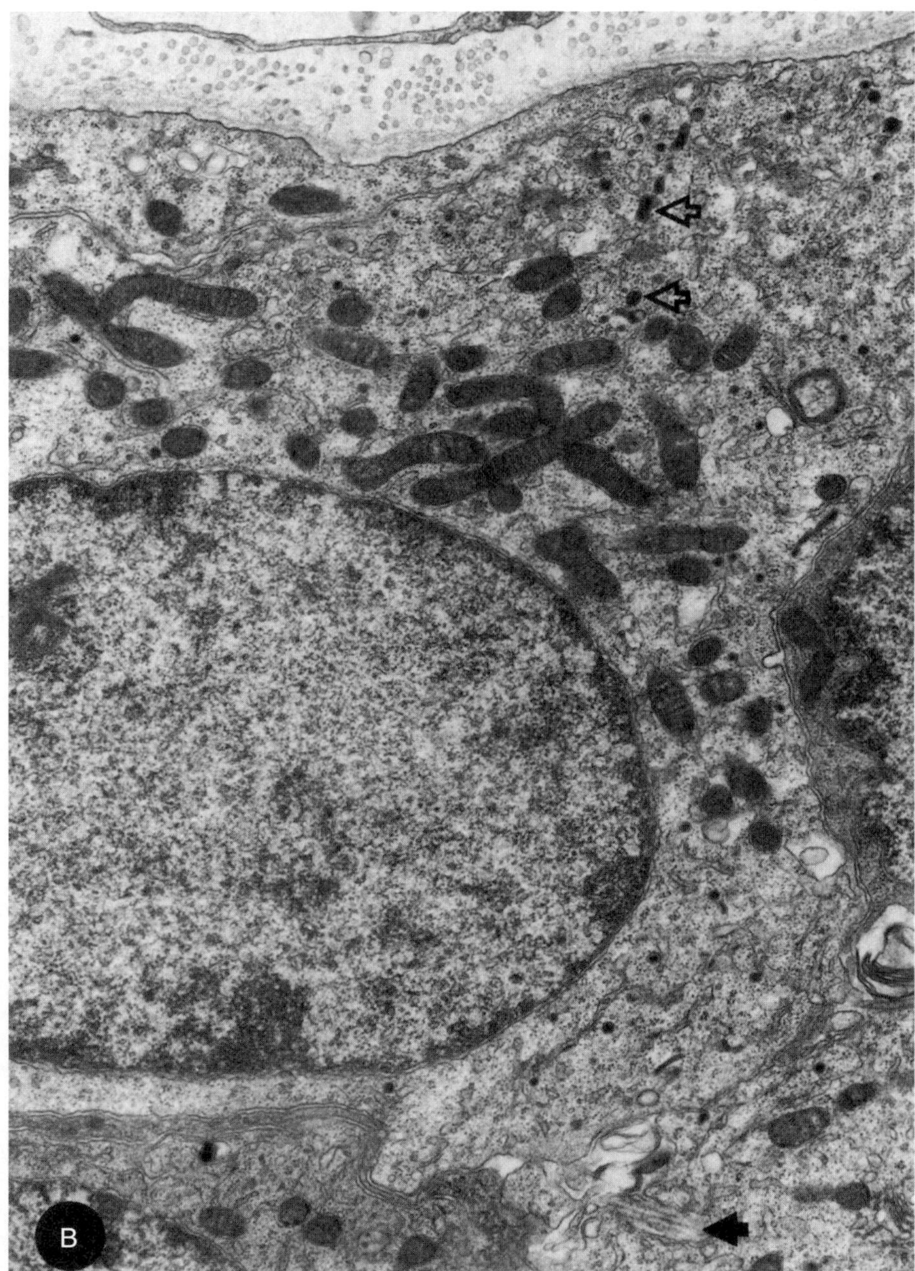

FIG. 4. cont'd B: Transmission electron micrograph of a rat par tuberalis cell. Many mitochondria and glycogen granules are present in the cytoplasm. Secretary granules are designated by *small open arrowheads* and are aggregated at the vascular pole of the cell. Lysosomes are not seen in this par tuberalis specific cell. A cilium is seen in an adjacent follicular cell (*black arrowhead*).

demonstrated by fluorescence microscopy to form a ring about the periphery of (cultured) lactotropes that diassembles in the presence of stimulatory factors such as TRH or neuropeptide Y (NPY) (189). Omega (Ω) figures form with discharge of hormone into the extracellular space (190). Membrane retrieval was demonstrated with internalization of membrane at the exocytotic site and its migration as coated vesicles to the trans Golgi region (183,184). Exocytotic events have been quantified by electron microscopy by freeze fracture techniques before and after stimulation of somatotropes with GRH. The number of exocytotic events paralleled the amount of growth hormone released (191). Although it appears that exocytosis is the primary mechanism for release of peptide hormones from pituitary cells, debate persists as to whether the only pathway employed in the synthesis and release of protein is from rough endoplasmic reticulum to vesicles to Golgi to granules to discharge at the plasma membrane (181), or whether under conditions of prolonged stimulation the formation of secretory granules may be short circuited and newly synthesized peptide hormone released from cytoplasmic sites such as the channels of the endoplasmic reticulum (Fig. 5) (180,192,193).

The release of protein hormones from adenohypophysial cells is in part regulated by secretions from distant sites that are carried to the glandular pituitary via the systemic circulation. Secretions of the target organs of the pituitary (the adrenal cortex, the gonads or the thyroid) regulate anterior pituitary function through feedback loops. In addition, ghrelin secreted from the stomach and leptin released from

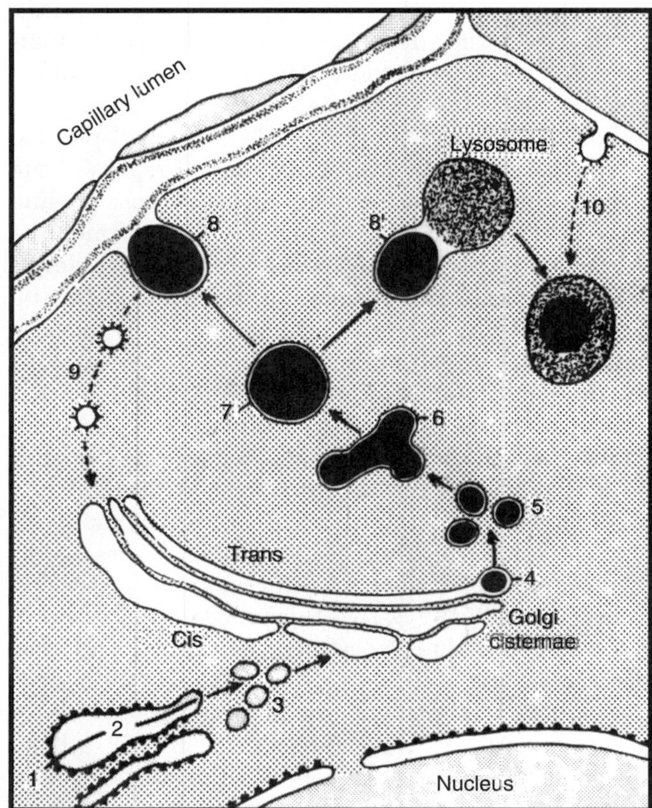

FIG. 5. Diagram illustrating the intracellular traffic that takes place in connection with the synthesis, packaging, and secretion of prolactin, as documented by work from many laboratories. Protein is synthesized on ribosomes (1), segregated into rough ER (2), and transported by small vesicles (3) to the Golgi complex; then it passes through the Golgi complex and is concentrated into small granules on the trans side of the stack (4). Several of these aggregate (5 and 6) to form the mature secretory granule (7). During active secretion, the latter fuses with the cell membrane by exocytosis (8), its content is discharged into the perivascular spaces, and the granule membrane recycles back to the Golgi cisternae (9). When secretory activity is suppressed and the cell must dispose of excess stored hormone, some granules fuse with lysosomes (8′) and their content is degraded. Besides these routes, an endocytic pathway from the cell surface to lysosomes has been demonstrated. (From Farquhar MG. Membrane traffic in prolactin and other secretory cells. In: MacLeod RM, Thorner MO, Seapagnini U, eds. *Prolactin, Basic and Clinical Correlates*. Fidia Research Services, Vol. 1. Padova: Liviana Press; 1985:3.)

adipose tissue can alter adenohypophysial function (194–196) as can catecholamines secreted from the adrenal medulla (197). Cytokines (198–200) released from multiple sites can also regulate the function of adenohypophysial cells.

The release of hormones is also regulated by peptide secretions of nerves that terminate in the neurohypophysis. These hypothalamic releasing and inhibiting hormones are carried to the adenohypophysial cells through a restricted portal circulation. There is ample evidence that peptide hypothalamic hypophysiotropic-releasing hormones cause degranulation of target adenohypophysial cells with consequent hormone release (201–205). Hypothalamic releasing hormones, catecholamines, and cytokines interact with receptors at the cell surface. The transduction of receptor binding at the cell's surface to hormone release at the cell surface is mediated by second messenger systems. G proteins, cyclic AMP, calcium, and phospatidylinositol have each been implicated as the second messengers mediating exocytosis and are the subjects of in depth discussion in later chapters.

Development of the Adenohypophysis

The adenohypophysis has classically been held to arise from the ectodermal tissue of the primitive mouth (5,206,207). The pouch of Rathke evaginates from the stomodeum and migrates dorsally. The floor of the diencephalon, the saccus infundibuli, migrates ventrally. This hollow diverticulum of brain is made up of a ventral (pituitary) and dorsal (saccular) wall. Caudally, these walls fuse to form the lower infundibular stem and infundibular process. As the pouch of Rathke migrates dorsally its aboral (posterior) wall becomes apposed to the caudal region of the saccus infundibuli (the presumptive lower infundibular stem and infundibular process). It will become the pars intermedia. In the account of Herring (5) the anterior, oral, wall of Rathke's pouch grows massively whereas the posterior, aboral, wall does not. Hence, the pars intermedia (aboral lobe) remains a thin band of glandular tissue separated from the expanded pars anterior (oral lobe) by Rathke's cleft except at the apex of the pouch where the oral and aboral walls are united. Wingstrand (78) agrees that the aboral wall becomes apposed to the lower infundibular stem and infundibular process to form the pars intermedia. The rostral region of the ventral diencephalic floor develops into the median eminence or infundibulum. The oral wall (anterior lobe) makes contact with the presumptive median eminence and becomes the site where portal connections between the median eminence and the pars distalis develop from interposed mesenchyme. From the anterior lobe two lateral buds develop and enlarge. These lateral lobes migrate cranially to approach the rostral portion of the ventral (or pituitary) surface of the evaginating saccus infundibuli. They become applied to it, fuse together and develop into the pars tuberalis that lies on the surface of the infundibulum (median eminence) and the posterior surface of the infundibular stem and into the zona tuberalis of the pars distalis that is continuous with the pars tuberalis and lies on the anterior surface of the infundibular stem (Fig. 6).

Takor and Pearse (208) proposed that the entire hypophysis (neural and glandular) is of neuroectodermal origin after observing the development of the white leghorn chick embryo. They found some adenohypophysial cells to possess amine precursor uptake and decarboxylation (APUD) activity and concluded they were of neural origin (209). They proposed that the ventral neural ridge is the origin of both pituitary components. Recent fate map studies support this idea by demonstrating that the site of the development of Rathke's pouch between the endoderm and ectoderm in the roof of the buccal cavity and immediately beneath the primitive notochord is derived from the most rostral region of the anterior neural plate (the anterior neural ridge [ANR]) that expresses *Six3*, a member of the *Six/sine oculis* homeobox family (210,211). The ANR will give rise to non-neural tissue such as the ectoderm of the nasal cavity and the anterior pituitary gland as well as to neural tissue such as the anterior hypothalamus, the olfactory placode, and optic vesicles (210). This scheme is generally accepted for pituitary development in fish, amphibians and birds but an alternative explanation of the events prior to the formation of Rathke's pouch in mammals is offered by Kawamura and

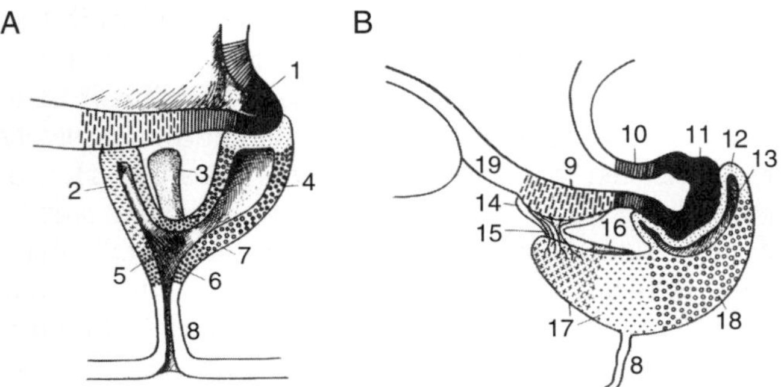

FIG. 6. Diagrams showing the structure of a generalized amniote pituitary **(B)** and its embryonic origin **(A)** as seen, particularly, in reptiles. (1) Saccus infundibuli; (2) anterior process; (3) lateral lobe; (4) aboral lobe; (5) opening of the lateral lobe cavity; (6) oral lobe; (7) constriction of Rathke's pouch; (8) epithelial stalk; (9) median eminence; (10) infundibular stem; (11) neural lobe; (12) pars intermedia; (13) hypophysial cleft; (14) juxtaneural pars tuberalis; (15) portotuberal tract; (16) pars tuberalis interna; (17) cephalic lobe of the pars distalis; (18) caudal lobe of the pars distalis; (19) pars oralis tuberis. [From (78).]

Kikuyma (212). On the basis of DiI tracer studies Kouki et al. (213) found that a small area just anterior to the anterior neural plate was the site of development of the future Rathke's pouch. These investigators state that there is "no discernible ANR (anterior neural ridge) in mammalian embryos" (214). With closure of the neural tube, elaboration of the forebrain and development of the cranial flexures, the tissue destined to become Rathke's pouch (the hypophysial placode) migrates from a position anterior to the neural plate to a position ventral to the developing diencephalon in the roof of the developing oral cavity where it is recognizable as a thickening in the in the ectoderm in the roof of the developing oral cavity (211). It is now well established that tissue destined to become the adenohypophysis originates in or adjacent to the anterior neural ridge and is contiguous with tissue destined to become the hypothalamus as early as the open neurula stage of development.

Further development of the hypophysial placode into Rathke's pouch and of the ventral region of the primitive diencephalon depends on reciprocal signaling between these contiguous regions. Commitment of the hypophysial placode the fate of pituitary development is induced by signaling molecules from the neuroepithelium of the diencephalon's ventral evagination, e.g., the infundibulum. *Rpx/Hesx1* (Rathke's pouch homeobox) expression is first found in the same distribution as *Six3* in the anterior neural plate (embryonic day 6 [E6] in the mouse) but becomes restricted solely to Rathke's pouch on E9 (215). As *Rpx* expression becomes confined to the hypophysial placode, sonic hedgehog (*Shh*), initially expressed in the ventral diencephalon and oral ectoderm is excluded from the developing Rathke's pouch creating a boundary between tissue expressing *Rpx/Hesx1* and that expressing *Shh* (216).

Further commitment by Rathke's pouch to a pituitary fate, requires expression of BMP4 (a bone morphogenic factor and a member of the transforming growth factor gamily), FGF8 a member of the fibroblast growth factor family and *Wnt5a* in the ventral diencephalon. BMP4 and FGF8 are extrinsic signaling molecules necessary for the further differentiation of the hypophysial placode. Blockade of the action of BMP4 in transgenic mice cause failure of development of the hypophysial placode into Rathke's pouch (217). A dorsal-ventral gradient of BMP4 and FGF8 is established in the invaginating pouch of Rathke. Concurrently BMP2 is produced in the ventral region of Rathke's pouch and a ventral-dorsal gradient of this signaling molecule is established opposing the dorsal ventral gradient of BMP4 and FGF8 (217,218). Positional cues are thus provided to each cell in the nascent pituitary as it is exposed to a unique spatiotemporal pattern of these signaling molecules.

The intrinsic expression of homeodomain transcription factors in the developing Rathke's pouch occurs in this unique environment as Rathke's pouch invaginates, enlarges and pinches off from the roof of the oral cavity. A member of the *Pitx* gene family produces PITX2, bicoid-related homeobox factor, that supports growth of Rathke's pouch through regulation of *Rpx* and *Prop1* (prophet of Pit1) expression (219). Members of the LIM homeobox family are also required to support expansion and development of Rathke's pouch.

The expression the LIM-homeobox genes *Lhx3*, *Lhx4* in the developing hypophysial placode (E9.5 in the mouse) is coincident with the expression of *FGF8* in the diencephalon; and the LIM-homeobox gene *Isl1* is expressed in the hypophysial placode concomitant with the diencephalic expression of *BMP4* (220). In the absence of these LIM-homeobox genes invagination of the hypophysial placode to form Rathke's pouch is arrested and no adenohypophysial cells develop (221). Supporting the conclusion that BMP4 and FGF8 serve as signaling molecules that determine the expression of LIM-homeodomain genes in the developing Rathke's pouch is the observation that cultured nascent pouch cells in the presence of FGF8 impregnated beads express *Lhx3* (222). Coculture of Rathke's pouch with BMP2 [which is first expressed at the boundary of *Shh* and *Rpx* domains in the roof of the oral cavity (216)] enhances the expression of *Isl1* (216,217). The expression of *Lhx4*, and of *Prop1*, are necessary for the expression of *Lhx3* which in turn is necessary to restrict the domain in which *Isl1* is expressed to the ventral region of the developing pouch (223). The establishment of unique dorsal and ventral fields in the developing Rathke's pouch is further reinforced by the transient expression of *Pax6,* a paired homeodomain transcription factor, in the dorsal region of Rathke's pouch. The antagonistic gradients of signaling molecules are necessary for the development of dorsal region in the anterior lobe characterized by the expression of *Lh3x* and a ventral region in the anterior lobe characterized by the expression of *Isl1* and provide a mechanism for the restricting "expression of transcription factors that specify cell fate through the activation of specific target genes" (218,224).

Prop1 appears to be essential for the exit of progenitor cells from the peri-luminary high proliferation zone of Rathke's pouch and the subsequent proliferation through the activation of Notch signaling pathways (224). Differentiation of progenitor cells also appears to be dependent on a diminishing expression of *Rpx* throughout Rathke's pouch beginning on E12.5 and extinction of expression of *Rpx* by E13.5 in the mouse coincident with the budding off of Rathke's pouch from the oral ectoderm to form the nascent adenohypophysis (215,225).

By this stage unique fields have been established in which cell specific differentiation is initiated by activation of intrinsic transcription factors. Melanotropes will develop in the intermediate lobe dorsal to the remaining Rathke's cleft. In the anterior lobe, beneath Rathke's cleft, a dorsal field contains presumptive somatotropes and lactotropes and a ventral field contains presumptive thyrotropes and gonadotropes. Corticotropes will develop at the caudal tip of the nascent pituitary gland (226).

Further differentiation of progenitor cells in the pars distalis depends on the ordered expression of linage specific transcription factors. Differentiation of POMC expressing cells is independent of Lhx3. Transgenic studies in the zebra fish identify the POMC promoter in two bilateral groups of cells in the anterior neural ridge in the open neurula stage (227). Differentiation of the remaining progenitor cells depends on the synergistic actions PITX1 and PITX2 to specify cell linage (219). Pit1, a POU-class homeodomain protein, is required for the terminal differentiation of somatotropes and lactotropes and of thyrotropes. Expression of *Pit1* requires PITX2 and PROP1. Thyrotropes require GATA as well as PIT1 to differentiate. GATA is required for the development of gonadotropes as is SF1 and ERG (218,226,228). *Gata* expression is supported by PITX2, as is the expression of *Sf1* and *Erg* (219). Differentiated cellular phenotypes of adenohypophysial cells sequentially appear in the anterior pituitary. POMC (Proopiomelanocortin) containing cells appear first followed by TSHβ (thyroid-stimulating hormone), MSH (melanocyte-stimulating hormone), GH (growth hormone), LHβ (luteinizing hormone), FSHβ (follicle-stimulating hormone), and PRL (prolactin)-containing cells (229).

The adenohypophysis is applied to the neurohypophysis in all chordates through the application of the pars tuberalis to the median eminence (the infundibulum) (77). There is considerable variation in the development of the neurohypophysis and in the relationship of the neurointermediate lobe to the pars distalis among classes of animals. For example, in fishes and in some birds, the saccus vasculosus is well developed. In birds, the intermediate lobe does not develop; and, the neural lobe is separated from the pars distalis by connective tissue, in some cases even by bone, of variable thickness. Neither neural nor vascular structures cross between the pars distalis and the infundibular process in birds. Furthermore, there is some dispute as to whether the pouch of Rathke is the analogue of the pituitary in all classes of vertebrates. Some feel that the pituitary may arise from Hatschek's pit in cyclostomes or from the nasal placode or dorsal lip in other lower species (78,79,230). These observations would not be unexpected in view of the origin of the hypophysial placode from the anterior neural ridge situated between the laterally placed olfactory placodes in the open neurula stage (211).

Even among mammals, species differences (for example, the degree of persistence of Rathke's cleft or amount of coaptation of the saccular and ventral surfaces of the saccus infundibuli and hence the depth of the infundibular recess) exist. Nevertheless, the apposition of glandular tissue to the rostral region of the evolving neurohypophysis with the consequent trapping of an interposed layer of vascular mesenchyme that will supply both the glandular and neural tissue at the base of the brain is a common characteristic of the (fetal) pituitary body in vertebrates (78). In most mammals the pouch of Rathke develops into the adenohypophysis that consists of a pars tuberalis applied to the median eminence at the base of the brain, a pars intermedia applied to the neural lobe (infundibular process) within the sella turcica and a pars distalis that lies in the sella turcica separated from the neurohypophysis.

Organization of the Adenohypophysis

The Pars Distalis

On the completion of its development, the pars distalis is comprised of epithelial cells, a connective tissue stroma, and many capillaries. Nerves do not terminate on or near pars distalis cells (77). There is no blood–brain barrier in the adenohypophysis (28,231). The capillaries in the adenohypophysis are not sinusoids (as believed by early investigators) because there are no phagocytic elements or large gaps in their walls. The capillaries of the adenohypophysis are fenestrated. Interposed between the epithelial cells and the fenestrated capillary tubes is a double basement membrane that on occasion is widely split with the space between the basal lamina containing connective tissue elements (232). The epithelial cells are arranged in cords (4,6). Both secretory cells (containing cytoplasmic secretory granules) and nonsecretory cells (lacking cytoplasmic granules) are present but secretory cells predominate. Adjacent granulated cells are united by desmosomes.

This anatomical organization of capillaries and cords of secretory epithelial cells has the obvious consequence that hormones released into the extracellular space of the adenohypophysis from any epithelial cell can reach nearby capillaries by diffusion or bulk flow and enter them through fenestrations in their endothelial tubes (Fig. 7). However, two other consequences can also be foreseen: (a) hormones can as easily reach adenohypophysial cells from the brain or from distant glands such as the thyroid, gonads, adrenal cortex or adrenal medulla; and (b) hormones released by one epithelial cell are free to interact with neighboring cells that contain appropriate receptors in their cell

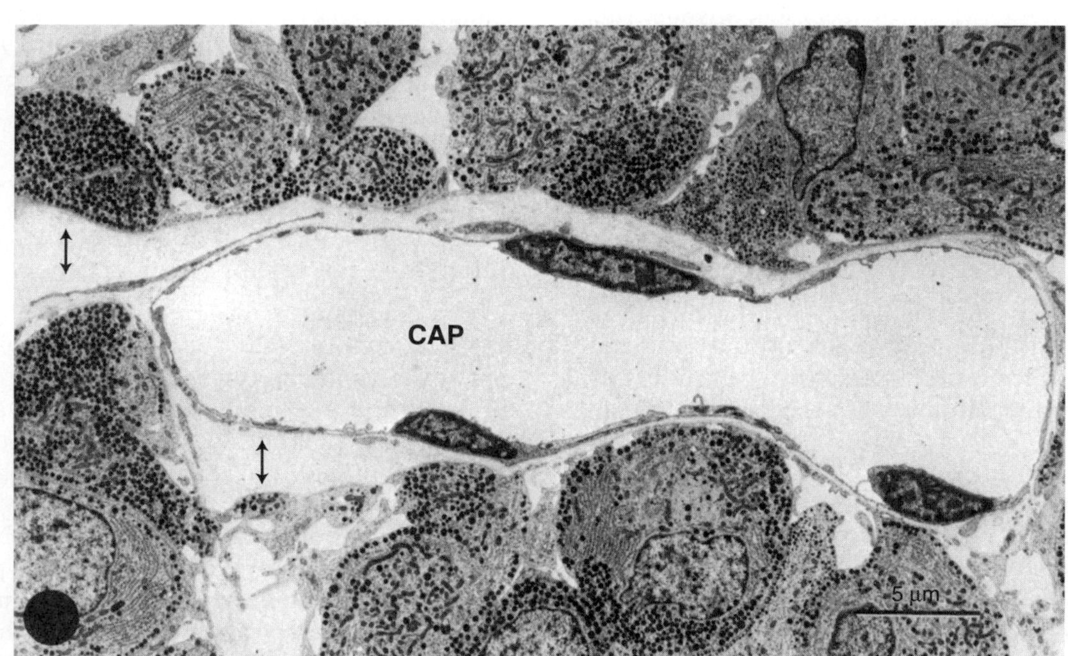

FIG. 7. Transmission electron micrograph of rabbit pars distalis. A fenestrated capillary (CAP) is surrounded by a double basement membrane (*arrows*) that defines the perivascular space upon which epithelial cell abut. Note occasional wide expansions of the perivascular space (↕) on which cell processes of connective tissue elements are seen. Glandular epithelial cells, identified by the presence of secretory granules, are closely related to the perivascular space and closely apposed to each other.

membranes (233). The structure of eight hormones secreted by the adenohypophysis has been identified. Although species differences occur frequently, TSH, FSH, LH (234); PRL (235–239), GH (240,241); ACTH (242–244), MSH (245–247), and beta endorphin (β-END) (248–252) have been isolated from several species and sequenced.

That each hormone secreted by the pituitary gland is synthesized in and released from a unique functionally specific group of cells (i.e., TSH only by thyrotropes, ACTH only by corticotropes, etc.) was initially postulated. This hypothesis was first explored by attempting to correlate clinical syndromes produced by pituitary tumors with pituitary histology. Subsequent attempts have employed histochemistry (253), immunohistochemistry (254), transmission electron microscopy (255), and localization of messenger RNA by light microscopic examination of histological sections after in situ hybridization (256); but, to date, not every pituitary hormone has been housed in its own unique cell.

The attempt to relate structure to function began with light microscopic examination of stained pituitary sections that revealed glandular cells with unstained cytoplasm (chromophobes) and glandular cells with stained cytoplasm (chromophiles). Those chromophiles that stained with acid dyes were called acidophiles (eosinophils) whereas those that stained with basic dyes were called basophiles. The human male gland was reported to be composed of 52% chromophobes, 37% acidophiles, and 11% basophils (257). In females the proportions differed in different stages of the menstrual cycle and with pregnancy (258). Acidophils lay in the lateral wings of the human pars distalis and basophils in the central (mucoid) wedge (257). It was recognized early that the staining of chromophiles was due to the staining of granules in their cytoplasm (259,260). The belief that these granules represented a stored source of hormone was based on the observed depletion of granules following castration, thyroidectomy or adrenalectomy of experimental animals; and, it has been subsequently substantiated (165–167). Cushing (261,262) had noted a preponderance of acidophiles in the pituitary tumors associated with acromegaly and he proposed that the substance responsible for excess growth in these individuals was released from these cells. His finding of a predominance of basophils in tumors associated with bilateral adrenal cortical hyperplasia (subsequently called Cushing's disease) convinced him that the hormone responsible for the support of the adrenal glands (ACTH) was made in and released from basophils (263).

This picture became muddied when it was found that the number of acidophiles increased in female animals at the time of parturition and that this increase was maintained during lactation (260). It was not known if acidophiles secrete two hormones (one to support lactation and one to support growth) or just one. In addition, changes in the morphology of both acidophiles and basophils occurred during the estrus cycle, during pregnancy, and following castration (260). Further, thyroidectomy produced a decrease in the number of acidophiles as well as a degranulation of basophils while adrenalectomy also produced degranulation of basophils. In order to clarify the situation, more histochemical techniques were employed based on modifications of the use of an acid dye such as eosin or orange G to stain acidophiles and of the periodic acid-Schiff (PAS) reaction to stain basophils. The result was a chaotic picture adding little to knowledge of the problem at hand—the localization of specific hormones within specific cell types as defined by histochemical staining. Confusion arose because of the vagaries of staining by different lots of aniline dyes, the lack of knowledge of the chemical reaction between the dye and the cellular or hormone elements, species differences in staining patterns and the use of the same terminology to describe cells stained by different dyes and so of different appearance. The subject is reviewed in detail by Romeis (264), Pearse (265), Herlant (266), Pearse and Noorden (253), and Purves (267). From the light microscopic examination of stained sections, the following firm conclusions were reached: (a) hormones are stored in granules in pituitary epithelial cells; (b) the hormone (or hormones) responsible for growth and lactation are found in eosinophils; (c) these eosinophils are localized predominantly in the lateral wings of the gland (257,258); (d) basophilic cells can be thyrotropic, gonadotrophic, or adrenocorticotrophic, and they lie predominantly in the middle of the gland (253,257,266–268); (e) the positive PAS reaction which characterizes basophils indicates glycosylation of the protein hormone molecule in these cells (253,265).

The nature of the chromophobes was resolved with electron microscopy. Observation of pars distalis cells with the transmission electron microscope revealed that all the epithelial cells, with the exception of folliculostellate cells, dendritic cells and macrophages contain electron dense granules in their cytoplasm (185,255,269). The granules seen in transmission electron microscopic study of the cytoplasm of chromophobes were too small, or too few, to be observed by light microscopy. Hence the large population of chromophobes, about 50%, observed by light microscopy in the human gland, could not represent a population of uncommitted cells (259,260).

The eosinophils were divided by light microscopists into two groups (alpha cells and eta [or epsilon cells]) (264): orangeophiles and carminophils (266). The color of the granules differed with the stain employed and the species studied; but, appropriate techniques revealed one type of acidophil that is found predominantly in the lateral wings of the pars distalis and does not vary with pregnancy and a second type which also lies in the lateral wings and becomes much more frequent at the time of pregnancy and parturition (267,269). Purves and Greisbach (270) described two classes of acidophiles in rats. Acidophiles found on the interior of cell cords became degranulated after castration or estrogen therapy and were believed to be lactotrophs. Columnar acidophiles arranged radially about capillaries became prominent following castration or estrogen therapy but became degranulated after thyroidectomy and were classified as somatotropes.

On electron microscopic examination somatotropes were described by Farquhar and Rinehart (271) as containing secretory granules of about 350 nm in their cytoplasm which lined up along the plasmalemma in a single row. Interestingly these cells became more prominent after castration (271) and less prominent (degranulated) after thyroidectomy (272). The somatotrope in the rat pituitary is described by Kurosumi (255) as a round, oval, or polygonal cell with distinctive short clubbed mitochondria and round dense secretory granules with a maximum diameter of 350 nm. Injection of a stalk median eminence extract rich in GRF activity results in rapid degranulation of these cells (201).

The use of immunoelectron microscopy (155) further strengthened the argument that growth hormone is produced by a functionally specific line of cells and is stored within cytoplasmic secretory granules (269,273–275). However, it cast doubt on the concept that all the somatotropes can be recognized solely by the size and shape of their secretory granules. Kurosumi and Tosaka (276) have described three morphologic appearances of somatotropes identified by immunogold electron microscopy. Type I cells contained granules of ~350 nm in diameter and corresponds to the cells described by Farquhar and Rinehart (271). Type II cells contained smaller granules but these were larger than those found in type III cells which were ~100 nm in diameter. In the late fetal stages type III cells predominated among somatotropes. As the animals matured the frequency distribution shifted and type I cells became the most frequent. They proposed that type I cells were mature somatotropes and that type III cells were immature. Type II cells were intermediate (276). The morphology of somatotropes is thus variable and depends, among other things, on their maturity.

The presumptive lactotrope was believed by Farquhar and Rinehart (271,272) to contain elliptical rather than round secretory vesicles that were as large as 600 to 900 nm in their longest diameter. Such cells became degranulated and developed large laminar areas of rough endoplasmic reticulum under the stimulus of suckling (185). Similar cells were identified by Kurosumi (255) as lactotrophs. Nakane (273,274) localized immunoreactive PRL within these granules. When immunohistochemistry was combined with electron microscopy the picture became more complex. The ultrastructural characteristics of cells that were immunostained for prolactin were not uniform. Granule size varied from ~150 nm to ~600 nm in diameter. The prominence of the Golgi apparatus, the degrees of dilatation of dilatation of the cisterns of the endoplasmic reticulum, the prominence of the nucleolus and the shape of the nucleus also varied among cells that were immunostained for PRL. Several classifications of lactotropes (as defined by immunohistochemistry) based principally on granule size and shape were made [see (277) for review]. Kurosumi et al. (278) classified lactotropes into three types. Type I is a crescent-shaped cell with an eccentric, oval nucleus. The Golgi and rough endoplasmic reticulum are well developed. Most of the mitochondria are perinuclear in location. Granules are large (~500 nm) and are mostly arranged along periphery. Type II is polygonal or elongated in shape. The nucleus is kidney shaped. Some lysosomes contain secretory granules that stain for PRL. These cells are usually more granulated than type I and the granules are smaller (~150–250 nm). Type III cells are small and polygonal. A few small secretory granules (~100 nm) are present in the cytoplasm. In adult males type I cells comprised 46% of the lactotropes, type II cells comprised 48%, and type III cells comprised 6%. In perinatal rats type III cells made up 83% of the total lactotrope population. In female rats, the type I cell comprised 91% of the lactotropes. Lactotrope morphology varied in females with the state of the estrus cycle (278) and in males with age (279). Growth hormone and prolactin are thus present in functionally distinct cell lines that are best differentiated by immunohistochemistry to identify the protein stored in their granules or by hybrid histochemistry to identify the nature of their messenger RNA (280,281). The ultrastructural morphology of the lactotropes varies with age, hormonal milieu and function and will be the subject of in depth discussion in a later chapter.

Basophiles may be thyrotropes, gonadotropes, and corticotropes. Thyrotropes and gonadotropes aggregate in the central region of the pars distalis: the mucoid wedge (so named because its cells contain glycoproteins). Corticotropes aggregate in the anterior portion of the mucoid wedge. The mucoid wedge occupies the zona tuberalis that, with the pars tuberalis, arises from the oral lobe of Rathke's pouch (78). On the basis of their reaction with such stains as Alcian Blue, Aldehyde Fuchsin, Resorcin Fuchsin and their staining characteristics with such dyes following oxidation with potassium permanganate, PAS positive cells have been subclassified into FSH, LH, TSH, and ACTH-containing cells (253,264–267,269). Such schemes are complex with patterns of staining that vary from species to species—they have never gained wide acceptance. Differentiation of cell types now relies principally on immunohistochemistry and hybrid histochemistry at the light and electron microscopic levels.

TSH is a glycoprotein which contains two chains designated α and β-TSH containing cells thus stain with PAS (267,282,283). The alpha chain is common to TSH, LH, and FSH; the beta chain is specific for TSH (234,284). Small angular cells in the rat's pituitary gland, which contained granules with a maximal diameter of 140 nm, were found on transmission electron microscopic examination by Farquhar and Rinehart (272) to exhibit a depletion of cytoplasmic dense granules and a marked increase in the number of lucent cytoplasmic vesicles following thyroidectomy and were thus identified as thyrotropes. Kawarai (285) published similar findings. Using antibodies to TSH, pre-absorbed with β-human chorionic gonadotropin and without cross reaction with FSH or LH, Baker et al. (284) were able to localize immunoreactive TSH to cells in the rat pituitary which corresponded to the thyroidectomy cells described by Farquhar and Rinehart (272). Moriarity and Tobin (286,287) employed antibodies to the β chain of TSH and obtained similar results. Immunohistochemically identified thyrotropes in human pituitaries resemble those described in the rat (288). The morphology of thyrotropes in the rat also varies with the state of development (289). In adult rats, angular cells resembling these described by Farquhar and Rinehart (272), and with cytoplasmic granules ranging between 120 to 180 nm, predominated. In neonatal rats, thyrotropes identified by immunoelectron microscopy contained few granules of 50 to 100 nm in diameter. By 20 days post partum, almost 70% of the thyrotropes were intermediate in type containing granules 80 to 120 nm in diameter. By 60 days, ~70% were mature in appearance.

FSH and LH are also glycoproteins which contain an alpha chain common to TSH, FSH, and LH and a beta chain specific for each gonadotropin (234). PAS-staining pituitary cells that become hyalinized following castration are termed "signet ring" or "castration cells" and are presumed to be gonadotropes. Farquhar and Rinehart (271) described two types of castration cell in the rat based on their transmission electron microscopic observations. The first appeared soon after castration and contained ovoid cytoplasmic vesicles. The second did not appear until later and contained irregularly shaped vesicles separated by cytoplasmic strands. It was described as resembling a filigree. The authors proposed that the former cells were "FSH producers" and the latter "LH producers." Kurosumi and Oota (290) proposed that FSH gonadotropes are large and round with two populations of granules in their cytoplasm: one of 200 nm diameter and the second as large as 700 nm in diameter. LH gonadotropes were described as smaller than FSH gonadotropes and frequently located along blood vessels (a site presumptively not occupied by FSH gonadotropes). LH gonadotropes were said to be polygonal with secretory vesicles measuring 250 nm situated at the periphery of the cell. In contrast to the FSH cell, the LH cell was said to have a small Golgi apparatus with little development of the rough endoplasmic reticulum. However, Moriarty (291) demonstrated the presence of immunoreactive LH-β in each of these cell types. Immunoreactive FSH and LH have been localized in the same cells in the human pituitary (1175). In male rats 80% of gonadotropes contained both FSH and LH, 10% contained only LH and 10% contained only FSH (292). Childs et al. (293) also have employed in situ hybridization techniques along with immunoelectron microscopy to study gonadotropes in normal and castrated male rats. In 80% of gonadotropes that contained FSH in secretory granules, LH mRNA was found in the cytoplasm.

The usual appearance of a gonadotrope was of large ovoid or polyhedral cell. The nucleus was spherical and the Golgi was prominent. The RER was abundant. Gonadotropes were divided into types depending granule size: type I harbored two granule sizes (200–250 nm, 450–700 nm); type II contained only the smaller granules. Lloyd and Childs (294) were able to separate gonadotropes into two populations (of large and small cells) on the basis of density centrifugation. Large cells tended to be multi-hormonal and small cells tended to be monohormonal. The ultrastructure of the gonadotrope is thus not predictive of the gonadotropic hormone(s) it contains (295,296). Gonadotropes may assume one of several appearances upon transmission electron microscopic examination. Their ultrastructural appearance varies as a function of the (estrus) cycle or cellular activity (synthesis, storage or release of hormone[s]) (278). This topic will be discussed in detail in a subsequent chapter.

Basophils (thyrotropes and gonadotropes) aggregate in the central portion of the pars distalis, i.e., the *mucoid wedge* (so called because its cells contain glycoproteins). The wedge occupies the zona tuberalis which, with the pars tuberalis, arises from the oral lobe of Rathke's pouch (78). ACTH is not a glycoprotein in humans (297). It is a single chain protein hormone 39 amino acids in length. However, in rats and mice, both glycosylated and nonglycosylated forms of ACTH 1-39 have been identified (298). ACTH containing cells have been described as basophilic as well as acidophilic, amphophilic and chromophobic [see (266, 267,269,275) for reviews]. In the human, ACTH cells are described by Baker (269) as intensely basophilic. His description of ACTH containing basophilic cells is consistent with the finding of basophiles in adenomas of the human pituitary that secrete ACTH and cause Cushing's disease (263). The source of the positive reaction obtained with PAS staining of human corticotropes has in the past been attributed to glycolipids (such as phosphatidylinositol), glycoproteins, or mucoproteins in the secretory granules or in their membranes (253,265,269). Proopiomelanocortin (POMC) (297,299), the 31K prohormone from which ACTH (and MSH) are derived, is glycosylated; and, POMC, not ACTH, may be the source of PAS positivity in corticotropes which contain the nonglycosylated form of ACTH in their granules (275).

The transmission electron microscopic description of corticotropes was at first based on the description of cells that showed signs of increased protein secretion following adrenalectomy. Adrenalectomy cells were described as large, irregular in outline and with a tendency "to insinuate cytoplasmic projections between neighboring cells in the direction of sinusoids" (300). Their cytoplasm contained many vesicles and their granules measured 200 nm on average. Following the administration of cortisol to normal rats dense granules accumulated in the cytoplasm in large angular cells believed to be corticotropes. Following the administration of cortisol to adrenalectomized rats an increase in the number of cytoplasmic granules was accompanied by a decrease in the number of cytoplasmic vesicles which are found in adrenalectomy cells of untreated animals (301,302). Employing an antibody to 17-39 or 25-39 ACTH, Moriarity and Halmi (303) were able to localize immunoreactive ACTH within granules in the cytoplasm of angular cells with long processes insinuated between neighboring and with granules of about 200 nm in diameter—cells which resembled those described by Siperstein and colleagues (300–302).

The adult human, unlike the rat, lacks a pars intermedia and thus corticotropes and melanotropes are present in the pars distalis. However, identification of each cell type by immunohistochemistry may be difficult. Alpha-MSH shares a common sequence of amino acids with ACTH (1-13) and β-MSH shares a common sequence with ACTH (4-10) (242–248). Furthermore, ACTH, α-MSH, β-lipotropin (β-LPH), and β-endorphin (β-END) are all derived from the same 31 kilodalton parent hormone by enzymatic cleavage of peptide bonds (297,303–306). In the rat, pars distalis corticotropes process this parent hormone principally to ACTH, β-LPH, and β-END whereas pars intermedia melanotropes further process (a) ACTH to α-MSH and corticotropin-like intermediate-peptide (CLIP) and (b) β-lipotropin to β-endorphin and α-lipotropin (which in turn can be processed to β-MSH) (297,306). Immunohistochemical studies to localize these hormones must be done with particular care because binding of the antibody to one sequence of amino acids may stain several protein hormones. For example an antibody to ACTH 4-9 may stain ACTH, α-MSH, β-MSH, and POMC; and antibodies to β-LPH may react with β-END as well as with the 31-kilodalton prohormone.

Immunohistochemical study of the human pars distalis with antibody against ACTH 17-39 demonstrates ACTH containing cells in the pars distalis which are basophilic (PAS positive) (307). These same cells also react with antibody against β-MSH (308) but this reaction may reflect LPH rather than β-MSH activity (309). β-LPH has been localized in the same secretory granules of corticotropes in the pars distalis as has ACTH by electron microscopic examination of human, monkey, ox, pig, and rat pituitaries (310). It also appears that ACTH and β-LPH (or β-END), are found in the same granules of corticotropes in the rat pars distalis (311).

Corticotropes in the rat's pars distalis rarely stain for melanotropin (312). Kurosumi et al. (313) employed immunogold labeling of POMC cells in the rat's pars distalis to identify putative corticotropes in fetal, neonatal, and adult rats. They found that in fetal pituitaries, POMC containing cells were polygonal and characterized by the presence of many mitochondria about the nucleus. The mitochondria were "thread-like" and sometimes bifurcated. Secretory granules were small. With advancing development the cell type became scarce and was replaced by angular cells with an average granular diameter of ~100 nm and then by cells with larger granules (average ~140 nm with a maximum diameter of 200 nm). The granule size of POMC containing cells was a function of the maturity of the animal as was the granule size of somatotropes, lactotropes, thyrotropes, and gonadotropes.

Colocalization of hormones from apparently different cell lines within a single cell type has been reported. A double label immunogold immunohistochemical of the rat pituitary has revealed a small number of cells that costore growth hormone and prolactin in the same secretory granules (314) and reflect the developmental stage when somatolactotropes have not progressed to terminal differentiation (226). These somatolactotropes do not resemble either lactotropes or somatotropes but contain small secretory granules (50–100 nm in diameter). Somatolactotropes have also been reported in the cow and human pituitary gland (288,314). The presence of somatotropes (identified by immunoelectron microscopic techniques) that contain TSH mRNA (identified by in situ hybridization) has been described in rats made hypothyroid by propylthiouracil treatment (315). This finding could not be confirmed in rat pituitaries following surgical ablation of the thyroid (289). Whether or not transdifferentiation of pituitary cells occurs in a functionally significant fashion is a matter of investigation.

The concept a pluripotent stem cell that can be recruited into a population of secreting adenohypophysial cells upon appropriate stimulation has received considerable support from the laboratory of Dr. Childs. She and her coworkers observed that the number of cells containing POMC mRNA in the dissociated anterior pituitary gland was increased after 15 minutes of cold stress (316). They also found an increased number of corticotropes (as identified by immunohistochemistry) and an increased number of cells binding biotinylated CRH (317). Dual immunohistochemical studies done under nonstimulated conditions have revealed a small number of adenohypophysial cells that costore ACTH and TSH. Upon appropriate stimulation, such as cold stress, these cells increased in number and add significantly to the population of actively secreting corticotropes and thyrotropes (317).

In addition to the granulated secretory cells described above, the pars distalis contains nongranulated cells: folliculostellate cells, dendritic cells, and macrophages (318,319). Folliculostellate cells are stellate-shaped cells with long cytoplasmic processes that extend between neighboring granulated cells. They are clustered to form follicles. These angular epithelial cells contain a few secretory granules in their cytoplasm. Microfilaments, lysosomes and lipid bodies are plentiful. They are frequently joined by gap junctions to each other to create a follicle with a central lumen (320). Microvilli and cilia protrude from the folliculostellate cell membrane lining the lumen which contains colloid (288). These cells are characterized by the presence of the immunoreactive S-100 neuronal protein and fibronectin in their cytoplasm (321,322). The role of these cells has not firmly been established but a role in intercellular communication has been postulated (323,324). On the basis of freeze fracture electron microscopic studies, Soji et al. (325) proposed that the folliculostellate cells form a scaffold that serves as a syncytium to permit rapid communication between cells that lack innervation. Immunohistochemical studies have demonstrated that folliculostellate cells can synthesize interleukin-6 (326,327)—a cytokine capable of stimulating ACTH release from corticotropes (328,329). It has even been proposed that folliculostellate cell can serve a pituitary stem cell (330).

The pars distalis contains somatotropes, lactotropes, thyrotropes, gonadotropes, cortico-melanotropes, folliculostellate cells, dendritic cells, and resident macrophages. In addition it appears to contain pluripotent cells that can contribute to population of secreting cells upon appropriate stimulation. Ultrastructural morphology of these cells alone is not a very good indicator of their function because the ultrastructure is not constant with time or with circumstance (331). The morphology of a given clone depends on its age, its state of function, and its hormonal milieu. Further, some clones contain more than one hormone, i.e., gonadotropes, corticomelanotropes and pluripotent cells. The epithelial cells of the pars distalis are not innervated. In some cases they are regulated by feedback mechanisms and/or by the interaction of catecholamines with receptors on their cell surfaces. In some cases they are regulated by cytokines or substances secreted from distant sites (e.g., leptin or ghrelin). In all cases they are regulated by hypothalamic releasing and inhibiting hormones that are released from nerve endings in the neurohypophysis and that, through restricted portal routes, reach receptors on the surface of the adenohypophysial cells.

The Pars Intermedia

This structure is prominent in many species, including the rat, but is lacking in others such as birds or humans with the exception of the pregnant human female and the human fetus. Although neither α-MSH nor β-MSH is a glycoprotein, pars intermedia cells have long been recognized as intensely basophilic and the source of a hormone which regulates skin color in amphibians (332). As in cortico-melanotropes in the pars distalis, the source of this PAS positivity may be the presence of proopiomelanocortin, a glycosylated protein, in the storage granules of pars intermedia cells.

The (rat's) pars intermedia is lobulated. Capillaries separate adjacent lobules but are sparse. The cells, with the exception of those that line the lumen of the hypophysial cleft, are polygonal. Light and dark cells have been distinguished by light microscopy (on the basis of their staining affinity for PAS) and by transmission electron microscopy. There are more light cells than dark ones. Light cells have large ovoid nuclei. Their cytoplasm contains many clear vesicles that are round and measure 200 to 300 nm. Round, dense granules measuring 200 to 300 nm in diameter lie in the Golgi region, which is well developed and lies near the nucleus. Dark cells are irregular in outline and contained lobulated nuclei. The cytoplasm contained closely packed particles measuring 20 nm in diameter that give the cells their dark appearance. The endoplasmic reticulum is well developed but vesicles are scant (333,334).

Alpha-MSH is localized in pars intermedia cells of the human fetus and pregnant human female (335). Both light and dark cells in the pars intermedia of the rat contain immunoreactive β-melanotropin (158,334). In cows, pigs, rabbits, and rats immunoreactive ACTH 17-39 was found in pars intermedia cells although the concentration of ACTH is believed to be much less than β-MSH in these cells (312). Moriarty and Halmi (336) localized immunoreactive ACTH 17-39 to granules of dark and light cells and subsequently Moriarty and Garner (309) found that the granules in cortico-melanotropes in the rat intermediate lobe immunostained for both ACTH (17-39) and for β-MSH. Similar findings have been reported in the mouse (158). β-MSH staining may also have indicated the presence of β-lipotropin as the entire sequence of β-MSH is found in β-LPH.

The pars intermedia is less well vascularized than the pars distalis (337). In the rat's pars intermedia, capillaries join with the capillaries of the pars distalis to form a portal system within the adenohypophysis (337). The extracellular space is well developed and communicates with the Rathke's cleft and the extracellular space of the adjacent infundibular process (neural lobe) and pars distalis (338). These vascular arrangement provide mechanisms to deliver the products of the pars intermedia to the neural lobe and pars distalis or alternatively to deliver secretions from the neural lobe and pars distalis to the pars intermedia. CRH, which releases melanotropin from pars intermedia cells as well as corticotropin from pars distalis corticotropes (339,340), can reach melanotropes from the extracellular spaces of the rat's pars distalis. The vascular architecture of the pars intermedia probably plays an important role in the regulation of its function.

Axons terminate in the pars intermedia and make synaptoid contact with its epithelial cells (333,341); and, hence, pars intermedia cells can be regulated by direct neural contact. β-adrenergic receptors have been demonstrated on melanotropes and β-adrenergic agonists stimulate the release of α-melanocyte stimulating hormone (342,343). Early evidence for the presence of a (nor)-adrenergic input to the neurointermediate lobe of the rat was obtained by microspectrofluorometric analysis of the neurointermediate lobe following formaldehyde fixation and exposure to hydrochloric acid vapors. The shifts in emission wavelength (and hence in color) were specific for dopamine and norepinephrine. Both dopaminergic and noradrenergic terminals were demonstrated in the pars intermedia (344). A direct innervation of melanocytes by noradrenergic or adrenergic systems arising in the brain stem was proposed on the basis of early ultrastructural studies (345).

Immunoreactive dopaminergic terminals have been demonstrated in the pars intermedia of the rat and cat (346,347). Dopamine has been shown to inhibit the release of POMC-derived products from dispersed pars intermedia cells and from the incubated neurointermediate lobe (347). GABAergic fibers also terminate in the pars intermedia and make synaptoid contact with pars intermedia cells (348–352). They (presumably) also inhibit MSH release from melanotropes. Immunoreactive GABA and ir-TH have been colocalized in the same terminals in the pars intermedia of the rat (353). Dopamine has also been reported to be costored with serotonin in terminals innervating melanotropes (354) as has colocalization of dopamine and GABA (355). Saland et al. (354) suggest that all three neurotransmitters may be costored together in one terminal type. GABA and dopamine are not costored in the pars intermedia of the rabbit or hare (355). Immunohistochemical studies have demonstrated ir-serotonergic fibers innervating pars intermedia cells in the rat and putatively stimulate MSH release (356).

Studies in amphibians, in species that adapt skin coloring to environmental background, have shed further light on the factors that regulate MSH release from melanotropes. CRH (357), TRH (358), and serotonin (359) stimulate MSH release from the perfused neurointermediate lobe of *Xenopus laevis*. Release of TRH from the skin may be important in the regulation of MSH secretion in amphibians (360). Dopamine, GABA (361), and NPY (362) inhibit MSH release. The control of MSH release from the amphibian pars intermedia is further reviewed by Tonon et al. (363) and by Tuinhof et al. (364).

The epithelial cells of the pars intermedia can thus be regulated by two different mechanisms—portal routes from the median eminence via the pars distalis, and synaptoid contact from nerve terminals resident in the hypothalamus and brain stem.

The Pars Tuberalis

The pars tuberalis as described is comprised of secretory glandular epithelial cells arranged in cords and containing dense cytoplasmic granules. The secretory cells are further subdivided into pars tuberalis specific cells and "invasive cells" from the pars distalis. Nonsecretory follicular cells also reside there (365). Pars tuberalis specific cells are the major cell type in the pars tuberalis and make up about 90% of the cell population (366). PT specific cells are spheroid in shape and are characterized by the presence of only a few secretory granules located at the vascular pole of the cell. These granules are generally smaller than those found in pars distalis cell. In the adult rat they range from 120 to 150 nm. A large number of lysosomes may be found in these cells. They typically consist of a dense core with a cup-like extension. Varying numbers of glycogen granules are present in the cytoplasm and are a distinguishing characteristic of pars tuberalis specific cells (Fig. 4B). PT specific cells express the α glycoprotein subunit common to TSH, FSH, and LH in the rat, mouse, and guinea pig (367). Evidence that PT specific cells also express the β-TSH chain in the species is summarized by Wittkowski (368). In the monkey, the ultrastructure of PT specific cells does not differ from other species and PT specific cells express both the common α subunit and the β-chain specific to TSH (369).

Invasive cells appear to be gonadotropes that have migrated into the pars tuberalis from the pars distalis. Gonadotropes have been localized within the pars tuberalis of the rat (370) and in the pars tuberalis of the monkey (369,371,372). With hypophysectomy of rats, there was an increase in the number and size of immunoreactive gonadotropes (373). Secretory cells in the pars tuberalis (invasive cells and PT specific cells) express the common α chain. This finding is consistent with the differential expression of *Pitx1* in a dorsal-ventral gradient from the pars tuberalis to the pars distalis and the maintenance of elevated levels of *Pitx1* in cells expressing the a subunit (374).

Follicular cells, in the human fetal pituitary, had irregular heterochromatic nuclei, no secretory granules and long processes that extended between secretory cells. At times they aggregated to form typical follicles. Follicles are lined with microvilli and cilia (365). In addition, connective tissue elements are present as are many (fenestrated) capillaries. Occasionally, an axon terminal is found adjacent to a glandular epithelial cell (375).

The pars tuberalis is rich in melatonin receptors but they are absent from the pars distalis (376,377). There is suggestive evidence that these receptors are found in pars tuberalis specific cells (366) whose morphology, in some species, varies with photoperiod and not with functional changes in endocrine organs that are targets of the pars distalis (378,379). Predictions by Wittkowski et al. (379) that the pars tuberalis modulates pars distalis function by secreting an as yet uncharacterized peptide from pars tuberalis specific cells that are regulated by melatonin secreted from the pineal gland have been largely vindicated. In rams FSH levels, testicular size, PRL levels, and pelage (wool growth) are influenced by photoperiod and by melatonin administration (380,381). Circadian fluctuations in the length of daylight provide seasonal cues to the pineal gland via retinohypothalamic pathways that regulate melatonin release (382,383). These responses are maintained after disconnection of the pituitary gland from the hypothalamus rostral to the median eminence. In this fashion all central neural inputs to the pituitary gland are severed but access to the adenohypophysial cells by humoral routes is maintained. Seasonal rhythms of melatonin levels (high during short days and low during long days) are maintained as are rhythms of plasma PRL (high during long days) and FSH (low during long days). Melatonin receptors are present in the pars tuberalis of hypothalamo–pituitary disconnected rams (384). The action of melatonin to suppress PRL secretion is independent of dopaminergic mechanisms (381). Evidence has been presented demonstrating the secretion of a prolactin stimulating factor (tuberalin) from cultured pars tuberalis cells (385–387). Melatonin appears to inhibit release of this factor from pars tuberalis cells by inhibiting the activation of cyclic AMP-dependent protein kinase and thus suppress PRL release from co-cultured pars distalis lactotropes (388,389). The pars tuberalis is believed to mediate seasonal rhythms of PRL secretion through the release of tuberalin which is delivered to pars distalis lactotropes via the local portal circulation. Tuberalin levels in turn are regulated by melatonin released from the pineal under the influence of photoperiod. The role of the pars tuberalis in humans is still unsettled as seasonal influences on PRL secretion are minimal and demonstration of melatonin receptors in the human pars tuberalis has been inconsistent (390,391).

THE NEUROHYPOPHYSIS

Nerve Terminals of the Neurohypophysis

Nerve terminals in the neurohypophysis release hormones into blood vessels to be carried to the adenohypophysis or to distant target organs to regulate their function. The neurohypophysis is composed of the median eminence (the infundibulum), the infundibular stem and the neural lobe (infundibular process) (29). The terminals there can be viewed as the terminals of the segmental motor neurons of the neuroendocrine system.

Three basic types of terminals can be recognized by transmission electron microscopy. The first is found predominantly in the neural lobe and in the rat measures 1.5 to 7.0 μm in diameter. These terminals of the magnocellular supraoptico–hypophysial tract contain mitochondria, large dense core vesicles that measure 180 nm on average and small lucent synaptic vesicles of 50 nm average diameter. The second type of terminal lies predominantly in the median eminence. These terminals of the parvicellular tubero–hypophysial tract are smaller, measuring 0.3 to 1.4 μm in the rat. They contain mitochondria, large granular vesicles measuring 90 nm on average and small lucent vesicles about 50 nm in diameter. The third type of terminal is the same size as the second but contains only lucent vesicles measuring about 50 nm in diameter (392). In the rabbit, terminals of the supraoptico–hypophysial tract in the neural lobe contain large dense core vesicles that are 250 to 300 nm in diameter as well as small lucent vesicles with an average diameter of 50 nm. In the rabbit's median eminence, some axon terminals contain large granular vesicles that average 120 nm in diameter and small lucent vesicles whereas others contain only small lucent vesicles with an average diameter of 50 nm (393,394). Similar observations have been made by others in the rat (395,396) and in the human (397,398), and even in submammalian species, for example, the toad (399,400). However, the number of groups into which the terminals are classified by the authors, based on the size of their vesicles, varies with the author and with the species (Fig. 8).

Magnocellular neurons that originate in the hypothalamic supraoptic and paraventricular nuclei course through the hypothalamus as the supraoptico–hypophysial tract and terminate in the neural lobe (32,33,401–403). This magnocellular neurosecretory system has been classified as peptidergic as it synthesizes, stores and secretes oxytocin and vasopressin with their associated neurophysins (I and II respectively) (56,61). These hormones are stored in the dense core vesicles in terminals of the supraoptico–hypophysial tract (62). Oxytocin and vasopressin have been isolated from bovine dense core vesicles (404). Vasopressin and neurophysin have been localized in terminals of the supraoptico–hypophysial tract of the guinea pig by combining transmission electron microscopic and immunohistochemical techniques (56). The large dense core vesicles containing peptide hormones are transported from nuclei in the hypothalamus by axoplasmic flow to terminals in the neural lobe (54,57).

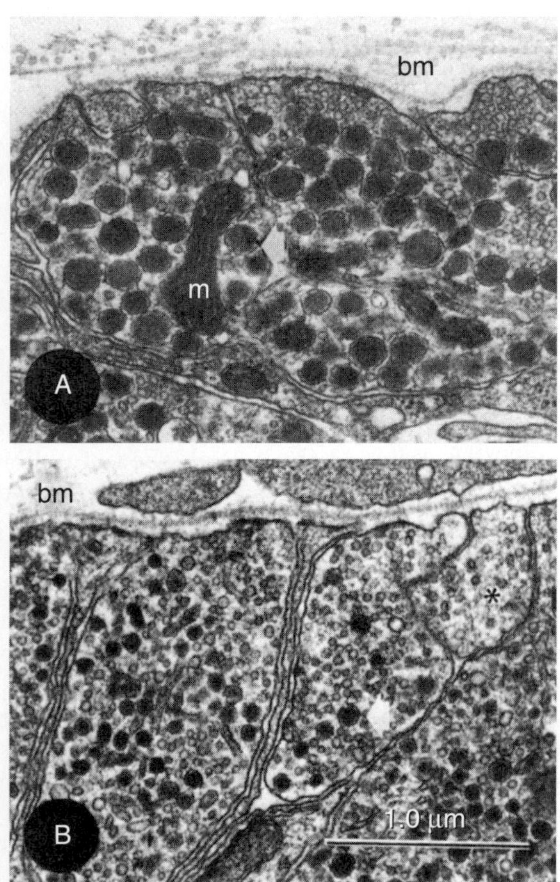

FIG. 8. Transmission electron micrograph of neural terminals in the rat neurohypophysis. **A:** Neural lobe. Large dense-core vesicles and small lucent vesicles fill neural terminals. bm, basement membrane; m, mitochondria. **B:** Median eminence. Smaller terminals contain large granular vesicles and lucent vesicles. One terminal (*) contains only lucent vesicles. Dense-core vesicles (A) are larger than large granular vesicles (B). Lucent vesicles in A resemble those in B.

The contents of the dense core vesicles are released from the nerve terminal by exocytosis (405). Transmission electron microscopy of thin sections (405,406) and of freeze fracture preparations (407) demonstrated fusion of the dense core vesicles with the terminal membrane and then the formation of omega figures (Ω) with opening of the membrane at the fusion site and release of the contents of the vesicles. Calculations based on the estimated hormone content of a single vesicle and the decrease in the number of dense core vesicles in terminals following stimulation suggested that all the hormone released can be accounted for by this mechanism (408).

Large, dense core vesicles in the terminals of the supraoptico–hypophysial tract of a single specimen are not always uniform in appearance when viewed by transmission electron microscopy (392,397,409–412). Frequently large vacuoles devoid of osmophilic material are found. Following the injection of histamine (409) or following ether treatment (410), the number of large dense core vesicles was reported to decrease and the number of large empty vacuoles to increase. Douglas et al. (413) stated that with optimal fixative techniques, empty vacuoles were few in number and believed them to be a fixation artifact. He reported that with stimulation the number of large empty vacuoles remained unchanged but the number of small "synaptic" vesicles increased.

The role of these small "synaptic" vesicles in large axon terminals of the supraoptico–hypophysial tract has been debated (408,413–416). Their nature has been clarified by the observation that their membranes have been found to contain synapsin, synaptophysin and protein III. These proteins were not found associated with the membrane of secretory granules (417). Meeker et al. (418,419) have presented evidence that glutamate is present as a neurotransmitter in microvesicles of the terminals in the neural lobe. Moriyama et al. (420) have demonstrated that microvesicles prepared from the neural lobe take up norepinephrine in vitro. The microvesicles in neurosecretory neuron terminals thus serve the same role that they do in axon terminals at synapses. They house and release neurotransmitters. The previous investigations cited demonstrated that following exocytosis, secretory granules appear empty and that membrane recycling, as indicated by labeling of the membranes with wheat-germ agglutinin occurs in the life cycles of the both secretory granules and microvesicles (416).

Axon terminals of parvicellular neurons contain granular vesicles measuring about 90 nm (in the rat) and small lucent vesicles. They have been until recently classified as aminergic or peptidergic (421–423). The colocalization of amines and peptides within the same neurosecretory neuron is now recognized as commonplace (424) and makes this method of classification less useful than it previously seemed to be. The localization of peptides and amines within separate organelles at the neurosecretory neuron's terminal has not been completely accomplished. TRH (425,426), SRIF (427,428), and GnRH (429) have been localized by immunohistochemical techniques to large granular vesicles in the terminals of parvicellular systems. Exocytosis characterized by fusion of the large granular vesicles with the terminal membrane and release of their granular material has been demonstrated (430–432).

The role of small lucent vesicles in parvicellular neurosecretory systems remains elusive although a preponderance of evidence suggests that they house neurotransmitters. Ajika (433) employed antibodies to LHRH and to tyrosine hydroxylase to identify LHRH and dopamine-containing cells in the same sections. LHRH was preferentially localized over large granular vesicles. Tyrosine hydroxylase was localized over small lucent vesicles of different cells. Small lucent vesicles in GnRH containing cells did not stain. A similar study localized TRH by immunohistochemistry and monoamines by autoradiography (after injection with tritiated monoamines) and demonstrated similar findings (434). Although uptake of the false neurotransmitter 5-hydroxydopamine (5-OH-DA) occurred in both the large granular and the small lucent vesicles (394,422,435–437), immunohistochemical techniques employing antibodies against specific enzymes in the biosynthetic pathway of amines localized them to lucent microvesicles in terminals (422,437–440). Synaptophysin has been demonstrated in the neurohemal contact zone of the gerbil median eminence where parvicellular terminals reside (441). As in the magnocellular neurosecretory neurons, microvesicles in parvicellular terminals appear to be synaptic vesicles.

Terminals containing only small lucent vesicles are also found on transmission electron microscopic examination of the neurohypophysis after aldehyde fixation (394). While some of these neurons may be aminergic, systems containing other neurotransmitters such as acetylcholine and GABA terminate in the neurohypophysis (348,349,355,442). Nerve terminals in the neurohypophysis are organized to release their contents into the perivascular space of fenestrated capillaries and not into a synaptic cleft between neuronal elements. Although terminals in the hypothalamus and other regions of the brain may contain peptides or amines, the distribution of vesicles within their terminals differs from that in the terminals in the neurohypophysis. In the cerebrum, presynaptic TRH neurons are characterized by the presence of small lucent vesicles aggregated near the thickening of the presynaptic terminals whereas the large granular vesicles are randomly distributed throughout the terminal (425). In the neurohypophysis this organizational pattern is seldom seen. The axon terminals are remarkably free of synaptic specializations and synapses are not frequently found on transmission electron microscopic examination. Both large and small vesicles are randomly distributed within the axon terminals. Thickening of an axon terminal, when present, occurs at its free surface facing the outer basal lamina and the perivascular space. The neural systems in the neurohypophysis are not organized for neurotransmission but are organized for neurosecretion and release the contents of their vesicles into the extracellular space of the neurohypophysis or the perivascular space of its fenestrated capillaries. The termination of a secretory neuron near a blood vessel without an interposed blood–brain barrier constitutes the fundamental organizational pattern of the neurohypophysis.

Development of the Neurohypophysis

The saccus infundibuli is discernible on day E15 in the fetal rat (443). At that time the floor of the third ventricle is made up only of 6 to 10 layers of round or oval cells. The presumptive infundibular process is an oval mass of cells with its long axis directed perpendicular (not parallel as in the adult rat) to the floor of the diencephalon (444,445). The aboral wall of Rathke's pouch has become apposed to the presumptive neural lobe and the oral wall has become apposed to the presumptive median eminence by this time. The capillaries, entrapped between the anterior (oral) wall of Rathke's pouch and the presumptive median eminence, display only a few fenestrations but pinocytotic vesicles are present in their endothelial cells. Ependymal terminals lie in the perivascular space of this developing tuberal (mantle) plexus.

On E16 neuronal processes can be clearly identified in the ventral region of the developing median eminence in the rat. These neurites contain only a few electron dense granules and neurotubules and hence their site of origin in the hypothalamus cannot be deduced from their morphology (443). Makarenko et al. (446) succeeded in placing DiI crystals in the developing median eminence of fetal rats. This technique permits post mortem retrograde diffusion of a fluorescent marker along neuronal processes. They were able to demonstrate connections between the developing median eminence and the paraventricular nucleus and preoptic area as early as day E15.

Neurites are oriented parallel to the ventral (oral) surface of the median eminence and lie between extended processes of ependymal cells (tanycytes) that stretch between the ventricular surface and the external surface of the median eminence. The nuclei of these stretched ependymal cells are indistinguishable from the nuclei of the primitive matrix cells found in abundance in the median eminence on day E15. During the next two days in the fetal life of the rat, the median eminence thickens as more axons enter it. Evidence is developing that axons from the hypothalamus are guided to the median eminence by soluble factors (447,448). The stretched ependymal cells lengthen in response to the thickening of the median eminence. The number of fenestra increase and the number of pinocytotic vesicles decrease in the endothelial tubes of the tuberal plexus at this time. Surface vascular connections are established by this plexus between the developing median eminence and the pars distalis.

Placement of DiI into the presumptive neural lobe at day E15 revealed fluorescence in the supraoptic nucleus suggesting that the magnocellular and parvicellular systems enter the neurohypophysis together. Fluorescence was not present in the paraventricular nucleus following placement of DiI crystals in the presumptive neural lobe until E17 (449). Neural lobe development requires the POU domain factor Brn-2 (450).

Neurohemal contact can be first demonstrated by transmission electron microscopy in the rat and in the mouse median eminence 2 to 4 days before birth. Abundant numbers of large granular and small clear vesicles and of mitochondria are found within the enlarged neural terminals which have displaced ependymal terminals and lie next to the outer basal lamina in the perivascular space of fenestrated capillaries near the surface of the median eminence (443, 445,451,452). Concurrently, dense core granules first appear in the cytoplasm of adenohypophysial cells (443). Immunoreactive GnRH, SRIF, and CRH containing terminals are first identified in the rat median eminence on the 19th fetal day by transmission electron microscopic examination of immunostained material (451–453). Setalo et al. (454) could not detect ir-GnRH in the median eminence of the rat with light microscopic techniques until the day of birth. However, transmission electron microscopic examination has demonstrated ir-GnRH containing fibers in the median eminence when light microscopic examination of immunostained sections failed (451). Similar findings are reported in mice where ir-GnRH was first detected in the median eminence three days prior to birth (455). Both neurohemal contact in the median eminence and the presence of immunoreactive LH in adenohypophysial cells could first be demonstrated on that day (444,456). The functional significance of these observations for the rodent fetus was questioned by Monroe and Paull (452) who demonstrated that neurohemal contact, although begun, was not "well established" until the day of birth. Freeze fracture studies first demonstrated configurations suggesting exocytosis from nerve terminals on the third postnatal day. However, the number of adequate preparations from earlier time periods was limited and exocytosis may have been initiated at an earlier stage (457). In the light of these limited and sometimes conflicting observations, it is not clear when the pituitary gland of the rat is first able to function.

The epithelium lining the saccus infundibuli also changes markedly with development. The infundibular recess in the rat brain is slit-like on the 15th and 16th days of gestation when viewed in coronal section. The ependymal cells of the saccus infundibuli are several layers thick and cannot be distinguished from the matrix cells. The floor of the ventricle is formed by round cells and by cells with apical processes, ovoid perikarya, and basal processes (458). Their apical processes of these tanycytes line the ventricular surface and are bathed by ventricular fluid. Their basal processes abut the outer basal lamina and lie in the perivascular space of capillaries of the tuberal plexus that are interposed between the rostral region of the saccus infundibuli and the oral lobe of Rathke's pouch. The tanycytes demonstrate polarity with the Golgi apparatus and mitochondria gathered in the apical process of each cell between the nucleus and the ventricular surface. The rough endoplasmic reticulum and ribosomes are poorly developed.

Between the days E16 and E18 the number of these cells (relative to the thickness of the median eminence) decreases with the in growth of axons from the rat's hypothalamus. The median eminence becomes thicker and the floor of the infundibular recess widens increasing its ventricular surface. The apical processes of tanycytes shorten, bringing the nuclei closer to the surface. A second type of ependymal cell, which is cuboidal and without apical or basal processes, appears in the expanding epithelial lining. The number and types of organelles in the cytoplasm of the tanycytes (mitochondria, polysomes, lipid bodies, lysosomes, and laminated dense bodies) increases until in cross section, the cells come to resemble those seen in adult animals (445). Gap junctions between ependymal cells are numerous on day 17 while the development of tight junctions is rudimentary. Tight junctions between ependymal cells of the ventricular floor in the infundibular recess are not mature until birth in the rat (459). The surface of the floor of the third ventricle remains flat without microvilli or apical blebs throughout gestation. The specializations at the ventricular surface that are characteristic of some tanycytes in the adult median eminence do not appear until after birth.

By the 18th day of gestation, 3 days before birth, the thickness of the median eminence has attained one half the thickness of the adult rat. Further differentiation of the median eminence will occur prior to and after birth and the median eminence will not assume its adult morphology for several weeks (443–445,460,461). However, in the rat by the 18th day of gestation, the basic organizational pattern of the adult neurohypophysis has been established. The neurohypophysis has differentiated into an infundibular process (neural lobe), an infundibular stem, and an infundibulum (median eminence). Neurohemal contact has been initiated. The median eminence has further differentiated into an ependymal zone, an internal layer defined by the presence of the supraoptico–hypophysial tract, and an external layer of small axons and terminals (462–464) (Fig. 9).

Organization of the Neurohypophysis

Infundibular Process

The principal elements of the neural lobe as ascertained by light microscopy are nerve terminals, glial cells called pituicytes and capillaries (26). Transmission electron microscopic studies of the neural lobe have been performed in the rat (392,409), mouse (465), rabbit (410), human (397), as well as in submammalian species (411). A basic picture of nerve terminals in the perivascular space of fenestrated capillaries is maintained and differences between mammalian species at the ultrastructural level are not marked. The axons of the supraoptico–hypophysial tract in the neural lobe are not uniform in diameter. Discrete swellings along the course of these axons give them a varicose appearance. Swellings contain dense core vesicles but not lucent vesicles, a feature that helps to differentiate them from terminals. Large swellings, filled with dense core vesicles, are called Herring bodies. Pulse labeling experiments suggest that newly synthesized hormone is first transported in dense core vesicles to terminals; but, subsequently the hormone is transported to the axon swellings if the hormone newly arrived at the axon terminal is not released (466).

Swellings have been viewed traditionally as a site for storage of hormones. More recently, Tweedle et al. (467) have made three-dimensional reconstructions of axons, axonal swellings, and axon terminals and neighboring terminals at the ultrastructural level. Individual axons made multiple contacts with a target capillary through axonal swellings before ending in a terminal. The authors proposed that with brief stimulation, hormone is only released from large dense core vesicles in the terminals; but with more prolonged stimulation, hormone release also occurs from the large dense core vesicles in the axonal swellings.

Most of the terminals in the neural lobe are large and contain dense core and lucent vesicles (411). Under resting conditions, oxytocin (OT) and vasopressin (AVP) are stored in large the dense core vesicles in separate terminals of the supraoptico–hypophysial tract (468). In the rat, vasopressin-containing fibers are aggregated centrally in the neural lobe whereas oxytocin containing fibers lie peripherally (469). Somewhat surprisingly in situ hybridization experiments have demonstrated OT mRNA in dense core granules in axons and terminals of the supraoptico–hypophysial tract (470). Stalk section abolished the signal confirming the conclusion the OT mRNA was present in the terminals and not in the pituicytes (471). The dogma that vasopressin and oxytocin are always to be found in separate terminals and hence synthesized in separate cell populations, has recently been challenged with the demonstration that both hormones can be found in the same in terminal (in the same dense core vesicle) under conditions of chronic stimulation such as lactation (472).

Opioid peptides have been demonstrated in the neural lobe. However the nature of the opioid peptides in the terminals in the neural lobe has been a matter of debate. Methionine (met) and leucine (leu)-enkephalin (ENK) were isolated from the rat neural lobe, and immunohistochemical light microscopic studies demonstrated ir-leu-enkephalin in varicose fibers in the neural lobe. Their cell bodies in the paraventricular and supraoptic nuclei were reported to have the characteristics of magnocellular neurons (473). Zamir (474) concluded on the basis of lesioning experiments that met- and leu-enkephalin have different cells of origin and that separate populations of magnocellular neurons contribute to the met- and leu-enkephalin content of the posterior pituitary. Martin and Voight (475) presented light microscopic studies which suggested (from examination of serial sections) that met-ENK is colocalized with OT and leu-ENK is colocalized with AVP. Martin et al. (476) subsequently reported colocalization of ir-OT and met-ENK in dense core vesicles within magnocellular terminals. Met-ENK, proenkephalin and OT have been localized by immunohistochemistry within the

FIG. 9. Diagram of the fine structure of the rat median eminence. (BC) Capillaries of the primary plexus; (EL) ependymal layer; (FL) fiber layer; (HL) hypendymal layer; (PL) palisade layer; (RL) reticular layer. (1) Cytoplasmic masses released from ependymal cells into the third ventricle; (2) marginal fold; (3) finger-like microvilli; (4) pinocytotic vesicles; (5) monoaminergic axon terminals protruding into the third ventricle; (6) ependymal cell; (7) hypendymal cell; (8) intercellular cavity; (9) commissure of monaminergic axons; (10) myelinated axon; (11) capillaries (note thin pericapillary space); (12) synaptoid contact; (13) large granule; (14) intermediate granule; (17 and 18) two types of unidentified processes; (19) fibroblast; (20) basement membranes; (21) collagen fiber; (22) endothelial cell; (23) red cell; (24) capillary lumen; (25) fenestration; (26) pars tuberalis cell; (27) glial cell; (28) terminal of ependymal, hypendymal, or glial process. [From (462).]

same magnocellular neurons in the bovine hypothalamus (477). Leu-ENK was colocalized in dense core vesicles with ir-AVP; but, because staining was enhanced by trypsin pretreatment, the leu-ENK was believed to be incorporated into a larger prohormone. Because α-neoendorphin and dynorphin have also been colocalized with AVP and because the sequence of leu-ENK is found at the N terminus of these peptides, the antigenic sequence of leu-ENK demonstrated in the AVP containing terminals was subsequently believed to be contained within these larger molecules (476). This conclusion is strengthened by the observation that the products of proenkephalin processing are not found in AVP containing neurons but dynorphin-A-(1-8) and the other products of preprodynorphin processing (including leu-ENK) are colocalized with vasopressin in terminals in the rat neural lobe (478). Colocalization of preprodynorphin with a subset of AVP-li neurons (but not with OT-li neurons) has been demonstrated in the guinea pig neural lobe where the kappa opioid receptor has been localized to the same terminals as dynorphin (479).

Whether met-enkephalin is colocalized in magnocellular neurons with oxytocin or is found in terminals of parvicellular neurons has been a matter of debate. Evidence for its colocalization with oxytocin within magnocellular terminals in the neural lobe has already been presented. VanLeeuwen et al. (480) have reported that enkephalins are localized in granular vesicles of about 120 nm in diameter and hence not in the terminals of magnocellular neurons which contain AVP or OT. These terminals made synaptoid contact with pituicytes. Merchanthaler et al. (481) could not demonstrate met-enkephalin in oxytocin containing magnocellular neurons in the SON or PVN or in large axons in the internal zone of the median eminence. Opiate systems terminating in the neural lobe are believed to play an inhibitory role in the regulation of AVP and OT secretion (482,483).

Other peptides have also been found to be colocalized with vasopressin or oxytocin. Neurons containing immunoreactive gastrin-like and cholecystokinin-like (CCK-li) peptide have been demonstrated to arise in the paraventricular and supraoptic nuclei and to terminate in the neural lobe of the rat (484). Based on light microscopic examination of immunostained sections of the rat hypothalamus, Kiss et al. (485) concluded that immunoreactive cholecystokinin is colocalized with oxytocin. Galanin (GAL) has been demonstrated by immunohistochemical techniques to be colocalized with oxytocin (486) and with vasopressin (487). In addition, immunoreactive (ir) substance P is present in terminals in the mouse neural lobe in the same distribution as vasopressin-containing terminals (488,489). Large terminals of the magnocellular system could thus contain: vasopressin with its associated neurophysin II (490), vasopressin, neurophysin II and/or dynorphin and/or substance P; or oxytocin with its associated neurophysin I (490); oxytocin, neurophysin I, and met-enkephalin or oxytocin, neurophysin I and gastrin-like, cholecystokinin-like peptide.

Parvicellular terminals in the neural lobe contain large granular and small lucent vesicles (411). TRH-containing fibers are abundant in the rat neural lobe (491). Somatostatin containing fibers are also present in the rat neural lobe but in fewer numbers (492). The axons of the dopaminergic tubero–hypophysial tract terminate in the neural lobe (345,493,494) in close proximity to AVP terminals (495). Colocalization of ir-serotonin within ir-TH terminals has been demonstrated (354). Immunoreactive GABA-containing terminals have been demonstrated in the neural lobe by light microscopy (348,349). Immunohistochemical studies at the ultrastructural level have demonstrated ir-GABA within clear microvesicles but also within large granular vesicles of parvicellular terminals (496,497). Colocalization with immunoreactive tyrosine hydroxylase (ir-TH) has been demonstrated by immuno-histochemistry (353,355). Thus dopamine, serotonin and GABA are colocalized within some parvicellular terminals in the neural lobe.

Norepinephrine has been isolated in the neurointermediate lobe of the rat pituitary. The concentration of norepinephrine decreased by only ~30% with bilateral superior cervical ganglionectomy (498). It decreased by ~70% following stalk section (495). The presence of ir-DBH in terminals in the neural lobe has been reported (499). A retrograde transport study employing injection of horseradish peroxidase into the neural lobe of the rat demonstrated HRP containing neurons in the dorsal vagal nucleus of the medulla (500). The presence of a central noradrenergic system that originates in the brain stem and terminates in the neural lobe is well accepted.

Both dopamine and GABA are inhibitory to the release of neurohormones in the neural lobe. Incubation of isolated neurointermediate lobes in medium containing dopamine reduced the release of milk ejecting factor into the media (501). The amount of oxytocin released from incubated neurointermediate lobes was increased when incubated with α-methyl-tyrosine (502). Dopamine agonists suppressed the release of AVP from incubated neurointermediate lobes that had been electrically stimulated (503).

Serotonin is also present in small terminals in the neural lobe and is colocalized with dopamine (354) in terminals of parvicellular neurons. Glutamate has been demonstrated in magnocellular terminals of the neural lobe as previously discussed. In addition, nitric oxide has recently been found in the neural lobe (504). Whereas the role of serotonin and glutamate is generally held to be stimulatory, the role of nitric oxide in the regulation of neurosecretion is yet to be clarified. The literature cited strongly suggests that peptides costored in magnocellular neurons and transmitters released from parvicellular terminals act locally in the neural lobe to regulate oxytocin and vasopressin secretion.

Dynorphin, co-released with AVP, is posited to suppress release of oxytocin from neighboring terminals in a paracrine fashion. This postulate is based on the observation that the administration of naloxone to dehydrated or stressed rats increased the plasma levels of oxytocin (482). In vitro study of the release of oxytocin and vasopressin from incubated neurointermediate lobes following electrical stimulation was carried out by Bondy et al. (483). Release of oxytocin and vasopressin was increased by about 20% during submaximal stimulation. Under these conditions the addition of a dynorphin κ-receptor agonist reduced the amount of oxytocin released into the medium. This reduction in oxytocin release was blocked by naloxone. These findings are consistent with a paracrine role for dynorphin that suppresses oxytocin release during stimulated release of vasopressin from magnocellular terminals.

Species variations in the distribution and content of the magnocellular and parvicellular systems terminating in the neural lobe, of course, exist. In fishes, arginine vasotocin is released from magnocellular terminals, whereas in amphibians isotocin and mesotocin are released from these terminals. Even among mammals, differences are present. GnRH, not present in the neural lobe of the rat (469), is present in the neural lobe of bats, ferrets, monkeys and humans (505). In the pig, CRH terminals are reported to lie in the perivascular space of the neural lobe in the same distribution as do vasopressin-containing fibers (506).

Until recently glial cells of the neural lobe have received little attention. In the 1930s they were considered to be the parenchymal elements of the neural lobe and the source of its secretions (23). When the source of neural lobe hormone was found to be the nerve terminals resident there, and not the glial cells, interest in pituicytes waned. Takei et al. (507) classified pituicytes into five major types. "Major pituicytes" resemble cerebral astrocytes and their processes are frequently found in the perivascular space of capillaries. "Dark pituicytes" resembled major pituicytes but are more electron dense. "Ependymal pituicytes" contain cilia. "Oncocytic pituicytes" contain abundant numbers of mitochondria in their cytoplasm and are only occasionally observed. "Granular pituicytes" contained numerous cytosegresomes (electron dense granules which are round or irregular in shape and are usually surrounded by a single membrane). These authors postulated that granular pituicytes are involved in the uptake and catabolism of extracellular material. The significance of ependymal and oncocytic pituicytes is not known (507).

The role of "major and dark" pituicytes has been evaluated by Hatton and Tweedle (508–514). They have shown that the relationship between terminals of the supraoptico–hypophysial tract and glial cells is not static (512,513). Under resting conditions, glial processes enclose terminals of the supraoptico–hypophysial tract and lie interposed between these terminals and the basal lamina of the perivascular space and capillaries perivascular space of capillaries. Synaptoid contact between nerve terminals and glial cells was occasionally found (508). With dehydration (deprivation of water for 24 hours), the number of enclosed axons decreases and more exposed terminals are present in the perivascular space (463). If the neural lobe was removed and incubated in hyperosmotic medium (as compared with incubation in a medium of low osmotic strength), the number of enclosed neurites was similarly decreased (514). In an analogous manner the number of enclosed neurites was decreased in postpartum and lactating rats when compared with control females (510,511). They went on to investigate the morphologic responses to chronic stimulation (prolonged lactation or hydration with 2% sodium chloride). With acute stimulation pituicytes withdrew leaving fewer enclosed and more exposed nerve terminals but the number of nerve terminals did not change. With chronic stimulation the pituicytes also withdrew leaving the exposed axon swellings and terminals in close proximity to the basal lamina. The number of terminals increased and they assumed a flattened shape (515). The authors postulated that withdrawal of the pituicyte processes exposed the axonal swellings to the basal lamina. The close relationship of the basal lamina to the axonal swellings induced the formation of small lucent vesicles in the axonal swellings. Axonal swellings were thus converted into axonal terminals and the number of terminals increased (467).

The change in shape of the pituicytes that both exposes and (indirectly) creates more (en passant) terminals could be initiated by β-adrenergic stimulation. β-adrenergic receptors have been demonstrated on pituicytes (516,517). Whether the source of β-agonists is the terminals of catecholaminergic neurons in the neural lobe (see above) or circulating catecholamines secreted by the adrenal is currently a matter of investigation. GABAergic and dopaminergic terminals make synaptoid contacts with pituicytes which express GABA and dopamine receptors. Double immunostaining for immunoreactive GABA (ir-GABA) and ir-TH revealed their colocalization in the immunostained cells. Incubation of the neural lobe with ir-GABA or dopamine receptor agonists or antagonists led to morphological changes in the pituicytes (518). Kappa (κ) opioid receptors are also present on the surface of pituicytes (519) and terminals containing an antigenic site common to enkephalins or dynorphins also make synaptoid contact with pituicytes (480). It seems likely that amines and/or opioids released from nerve terminals activate pituicytes to change their local spatial relationship with magnocellular neurites and capillaries (517,518,520).

Pituicytes can also influence release of neurohormones from the terminals of magnocellular neurons by chemical means. Kjeldsen et al. (521) present in vitro studies to demonstrate the presence of nitric oxide synthase in pituicytes and its release. Nitric oxide inhibits vasopressin release (522). The release of vasopressin in response to hypo-osmotic stimuli is blunted by taurine, a GABA-like amino acid that is concentrated in pituicytes (523). Hussy et al. (524) have demonstrated that the release of taurine from pituicytes is osmodependent and that taurine acts upon magnocellular terminals through glycine receptors to inhibit vasopressin release.

Whereas the interaction of opioid peptides and neurotransmitters serves to regulate selective secretion of oxytocin and vasopressin, the interactions between glial cells and nerve terminals (525), along with an increase in the release of hormone from nerve terminals (526), an increase in neural lobe blood flow (527) and metabolism (528) and an increase in the permeability of the neural lobe (529) serve to facilitate the movement of neurohypophysial hormones from the nerve terminal to the blood stream in response to specific physiologic stimuli.

Infundibular Stem

The principal elements of the infundibular stem are axons of the magnocellular systems and axons and terminals of the parvicellular systems. There is considerable variation in the length of the infundibular stem among mammals, for example, the length of the infundibular stem of the cat is very short whereas that of the ferret or human is very long (77). In addition to axon terminals containing TRH and SRIF, the rat's infundibular stem contains terminals that stain for GnRH (156), CRH (530), GRH (531), and neurotensin (532).

Infundibulum (Median Eminence)

The ependymal layer of the median eminence lines the infundibular recess. It is interposed between the neuropil of the median eminence and the ventricular fluid of the third ventricle. The ependymal cells comprising the epithelial lining of the infundibular recess differ in appearance from the extrachoroidal ependymal cells that comprise the epithelial lining of the remainder of the ventricular system. The ependymal layer of the median eminence is not ciliated. Scanning electron microscopic studies of the third ventricular surface and of the infundibular recess have been carried out in frogs (533), birds (534), rats (535,536), mice (537), cats (537,538), rabbits (539), mink (540), sheep (541), monkeys (542), humans (537, 542), and even in armadillos (543). The basic pattern is similar in each of these species. Ependyma overlying the dorsal third ventricle is ciliated. In the ventral regions of the third ventricle, the number of ciliated cells decreases at about the level of the ventral medial hypothalamic nuclei. As the infundibular recess is approached (at the level of the arcuate nuclei), ciliated cells become widely separated and sparse. Cells with abundant microvilli line the ventricular surface. This region between the ciliated ependyma of the dorsal third ventricle and the ependyma overlying the infundibular recess is called the transition zone. It overlies the ventromedial and arcuate hypothalamic nuclei. Beneath the level of the tubero–infundibular recess, the surface of the ventricular lining again changes. In the walls of the lateral recess, the cell surfaces become elevated and take on the appearance of cobblestones. Apical excrescences or blebs are seen at the cell surface (Fig. 10).

The floor of the infundibular recess shows regional variation in its appearance. Studies in the rabbit (544) demonstrate that the ependymal lining in the anterior third ventricular recess is a flattened, squamous epithelium with a smooth surface. In the middle third the surface of the cells is covered by microvilli. In the posterior third the surface of each cell is raised. Many apical blebs protrude from the elevated surface and give the cell a riveted appearance. A somewhat different regional variation is reported in the rat (536) but flat smooth, microvillated, and riveted cells with apical excrescences are also present. These blebs contain few organelles (535,545–547) and no secretory granules. Supraependymal cells (neurons and macrophages) lie on the ependymal surface in the floor of the infundibular recess (546–549).

Tanycytes contribute to the ependymal lining which overlies the hypothalamus in the ventral region of the third ventricle and the median eminence in the infundibular recess. Light microscopic studies of the distribution and morphology of tanycytes have been reported extensively in the rat (550–553) as well as the quail (554). Tanycytes are bipolar ependymal cells with a soma, a short apical process that extends to the ventricular surface and a long basal process. The basal process is subdivided into a neck, a tail and a terminal. Akmayev and Popov (555) designated tanycytes in the transition zone which overlies the hypothalamus of the rat as α-tanycytes. β-tanycytes are found in the ependyma overlying the lateral recess (β_1) or lining the floor of the infundibular recess (β_2).

α-tanycytes are present in the transition zone of the third ventricle and they stretch between the ventricular surface and the ventromedial and arcuate hypothalamic nuclei where they may terminate on neurons or capillaries in these hypothalamic regions (550,554) or they may terminate on the pial surface of the tuber cinereum (551). Their apical processes are insinuated between ciliated ependymal cells in the epithelium lining the transition zone. The ventricular surface of the apical processes contains many microvilli. Occasional apical protrusions are present. Mitochondria, ribosomes and numerous tubules are present in their apical cytoplasm. The soma of α-tanycytes contains the same complement of organelles and inclusions as epithelial cells in general. The ovoid nucleus contains a single large nucleolus.

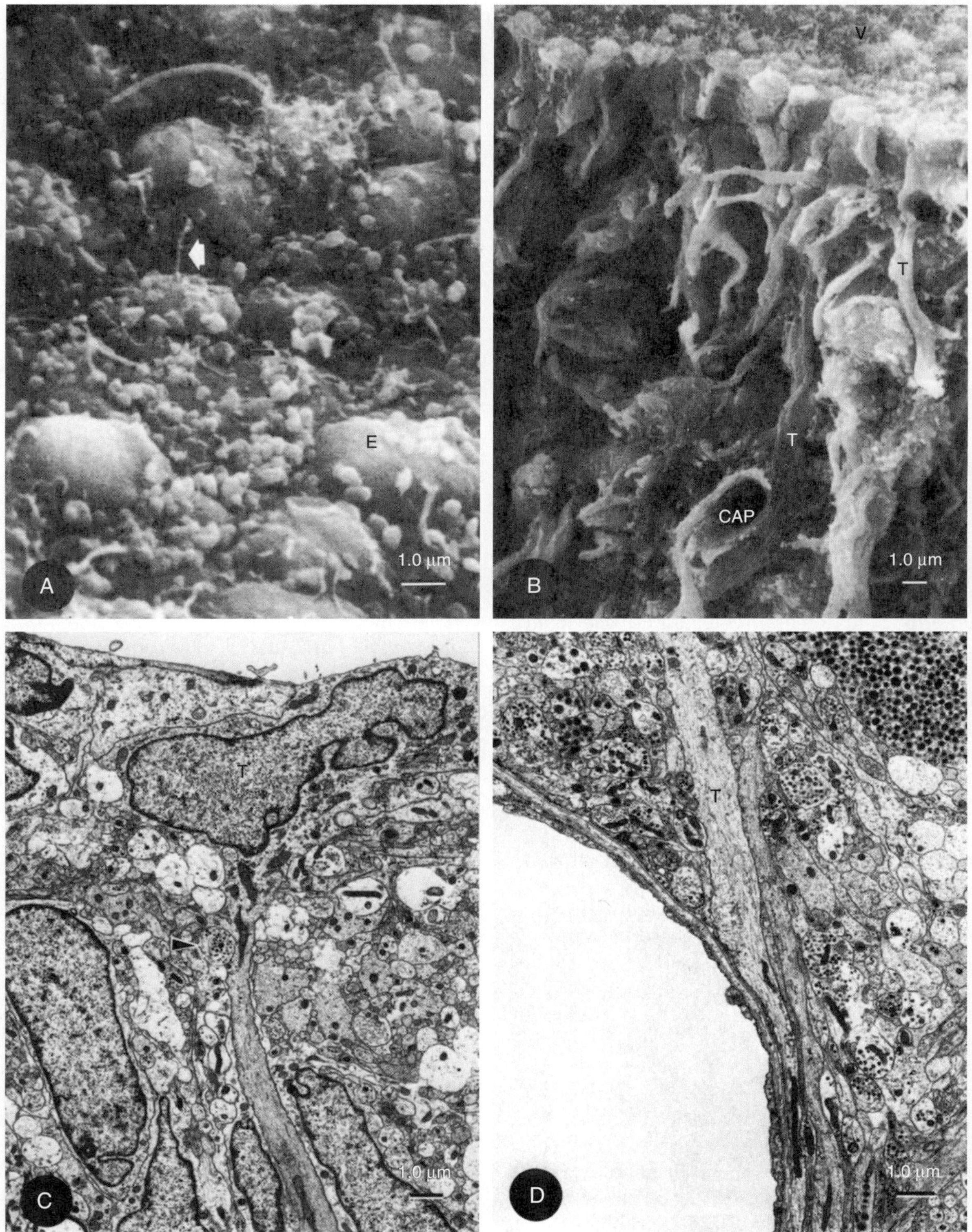

FIG. 10. A: Scanning electron micrograph of the floor of the rabbit third ventricle overlying the median eminence. *White arrowhead* points to microvilli. *Black arrow* points to an apical bleb. E designates the surface of one ependymal cell. Note the profusion of apical blebs in this (posterior) region of the infundibular recess. [From (1171).] **B:** Scanning electron micrograph of median eminence neuropil. Median eminence has been broken, thus exposing tissue between ventricular lumen (V) at top of picture and pial surface (not shown) at bottom. Tanycytes (T) lining the rabbit infundibular extend from the ventricular surface (V) to median eminence capillaries (CAP). Apical surface of tanycytes contains microvilli. [From (1172).] **C:** Transmission electron micrograph of rabbit median eminence. Tanycyte (T) sends basal process vertically through neuropil to the oral surface. In this region of the infundibular recess the ependymal cells are smooth. The organelles are concentrated in the cell body. A neurite (at *arrowhead*) contacts the basal process at its neck. Aggregated of microtublules give electron micrograph of rabbit median eminence. **D:** A tanycyte terminal (T) as well as neural terminals lies in the perivascular space of a fenestrated capillary. [From (1173).]

The Golgi apparatus, free ribosomes, and mitochondria are present in the soma. Vesicles associated with the Golgi and lysosomes are also found. The basal process contains a distinctive array of microtubules that are oriented parallel to the long axis of the process. In the neck, the endoplasmic reticulum elaborates "concentric shells of smooth cisternae" (556). The neck and tail of the basal processes contain a varied collection of inclusions and organelles including short tubules of endoplasmic reticulum and vesicles of differing shapes and sizes. Ependymal processes have been found to terminate on the basement membrane in the perivascular space of capillaries in the arcuate nucleus (556).

Adjacent apical processes are interdigitated and linked by gap junctions and desmosomes. Gotow and Hashimoto (557) divided the transition zone into two regions: (a) dorsal to the level of the arcuate nucleus and (b) ventral to the level of the arcuate nucleus. In the dorsal region, tight junctions were not found on transmission electron microscopic examination of freeze fracture replicas. Ventral to the level of the arcuate nucleus, tight junctions were found. They increased in number in the lateral recess and in the floor of the infundibular recess over the median eminence. Furthermore, the number of rows of particles that comprise the tight junctions in freeze fracture replicas and the density of particles per row increased as they examined specimens from the transition zone through the lateral recess onto the floor of the infundibular recess. These findings indicate an increasing tightness of the intercellular junctions in a centripetal pattern centered on the floor of the infundibular recess.

β-Tanycytes in the ependymal layer of the median eminence are stretched between the ventricular surface and the perivascular space of fenestrated capillaries on the surface of the median eminence (550). Akmayev and Popov (555) cited several ultrastructural differences between tanycytes in the epithelium lining the third ventricle (α-tanycytes) and tanycytes in the ependymal layer of the median eminence (β-tanycytes). At the surface of the β-tanycytes are apical blebs that are granular on transmission electron microscopic appearance and contain no secretory granules or organelles. The β-tanycytes lining the infundibular recess are more fibrillar in appearance than α-tanycytes lining the third ventricular walls due to the striking number of microtubules in their cytoplasm. They contain a well-developed system of membranous cavities and cisterns and a system of mitochondrion-borne tubules (558). Spine-like protrusions project from the basal process. Their inner structure differs from that of the α-tanycyte processes as they contain polysomes, lipid droplets, and a unique collection of vesicles. Smooth, dense core vesicles 40 to 50 nm in diameter lie in the Golgi area. Coated vesicles in the soma and in terminals form omega figures (Ω) with the plasmalemma. Dense core vesicles about 100 nm in diameter also lie in the soma, neck, tail, or terminal. Frequently rosettes or arrays of vesicles about 100 nm in diameter are present in terminals where glycogen bodies and lipid inclusions are prominent (559). Terminals of the β-tanycytes lie in the perivascular space of fenestrated capillaries on the surface of the median eminence. Their apical processes along with squamous and cuboidal ependymal cells form the ependymal layer of the median eminence.

Adjacent epithelial cells in the ependymal layer of the median eminence are joined by zonulae occludens as well as by gap junctions and desmosomes (560–562). Brightman et al. (561) found that junctions become increasingly tight (exhibited more particles per row and more rows per junction) as the infundibular floor is approached from the transition zone. They proposed that the lining of the ventricle formed a graded sieve, more porous at the periphery than at the center. Their findings were substantiated by Gotow and Hashimoto (557) who found that horseradish peroxidase could pass between ependymal cells in the transition zone but not between ependymal cells lining the lateral recess of the third ventricle. In this region, microperoxidase could pass between the ependymal cells. In the floor of the infundibular recess, microperoxidase could not pass between ependymal cells.

Kobayashi et al. (563,564) reported that horseradish peroxidase passed between the third ventricle and the neuropil of the median eminence. This passage was presumably accomplished by "active" pinocytotic transport (565). The observation of synaptoid contact between neurosecretory terminals and tanycyte processes (566,567) encouraged the concept that such transport could be controlled by hypothalamic (neural) projections. The observation that the surface of the ependymal lining changed concurrent with the reproductive cycles of the monkey (568) had previously rekindled the idea of a humoral link between the third ventricle and the pituitary gland (1).

While the transport role of tanycytes in the ependymal layer of the median eminence remains speculative, one role of this ependymal layer seems quite clear. This ependymal layer effectively separates the extracellular fluid of the median eminence from the ventricular fluid of the third ventricle. Although molecules as small as (5-hydroxy) dopamine could pass from the perivascular space of the internal zone of the median eminence into the third ventricle (557), this epithelial layer prohibits either the unregulated loss of hypothalamic hormones into the ventricular system or their dilution within the extracellular spaces of the median eminence (569). Rodriguez et al. (558) demonstrated that the basal processes of β-tanycytes in the rostral and post-infundibular palisade regions as well as in the lateral palisade region of the pre-infundibular median eminence form a continuous cuff at the ventral (oral) surface of the rat's median eminence which abuts the basal lamina about fenestrated capillaries. This cuff is composed of ependymal terminals linked by gap junctions and desmosomes that are interposed between neuronal terminals near the surface of the median eminence in its lateral thirds and the basal lamina. Only in the medial palisade zone of the rat median eminence do neurosecretory terminals directly contact fenestrated capillaries. The significance of this finding is not yet fully understood. Given the active role of pituicyte processes in neurosecretion in the neural lobe, an analogous role may be played by tanycytes in the median eminence (570,571).

The internal zone lies beneath the ependymal layer of the vertebrate median eminence (78). In many species, for example, the rat (462) or the rabbit (394), it can be subdivided into a hypendymal layer and a fiber layer (463). The hypendymal layer lies directly beneath the ependymal layer and contains subependymal cells and pituicytes. In the rat, parvicellular axons projecting from the hypothalamus (423,572) are present in the hypendymal layer of the internal zone. Catecholaminergic terminals of the reticulo-infundibular tract projecting from the brain stem to the median eminence and of the reticular–hypophysial tract projecting from the brain stem to the neuro-intermediate lobe were reported to be present in this zone on the basis of fluorescent microscopy studies (493). The fiber layer is defined by axons of the supraoptico–hypophysial tract (SOHT) (78,573). Nitric oxide synthase (NOS) had been demonstrated in the internal zone of the rat's median eminence perhaps in vasopressinergic neurons of the SOHT (574). Although axon terminals of the magnocellular supraoptico–hypophysial tract are not found in the fiber layer, evidence is accumulating on the basis of anatomic in vivo and in vitro studies that vasopressin can be released from axonal swellings in the fiber layer (406,432,575,576). Terminals of parvicellular neuron terminals have been demonstrated in the perivascular space of capillary loops penetrating into the internal zone of the rabbit median eminence (394). Peptidergic neurons that contain the processed products of pro-opiomelanocortin including ACTH, α-MSH, β-LPH, and β-END terminate in the internal zone of the rat median eminence (577,578).

The external zone is comprised of nerve terminals, pituicytes, ependymal (tanycyte) terminals and capillaries (462–464). Transmission electron microscopic studies of the external zone have been performed in the rat (392,396), rabbit (393), human (398), as well as in submammalian species (579). The external zone of the rat median eminence has been divided into a reticular layer and a palisade layer (462). The reticular layer lies between the fiber layer of the internal zone and the palisade layer of the external zone. It contains ependymal and glial processes and axons of parvicellular systems. The palisade layer lies between the reticular layer and the ventral surface of the median eminence. It is recognized by the presence of glial and tanycytic processes oriented at right angles to the ventricular surface of the infundibular recess on light microscopic examination (78,573) and by fascicles of axons and terminals (separated by tanycytic processes) in the perivascular space of median eminence capillaries on transmission electron microscopic examination (462,463,580). In the rat, the palisade layer may be further subdivided into a medial palisade layer (medial third) and a lateral palisade layers (lateral thirds). The medial palisade layer lies beneath the floor of the infundibular recess. Each lateral palisade layer lies lateral to it and extends to the tubero–infundibular sulcus (400).

Catecholamine containing fibers in the external zone of the median eminence have been identified by light microscopy using formaldehyde fluorescence (581–584) or glyoxylic acid induced fluorescence (493, 585–587). Fluorescence light microscopic studies demonstrated catecholamines in the palisade layer of the external zone (as well as in the hypendymal layer of the internal zone). Combining the fluorescence light microscopic technique with the use of selective enzyme inhibitors to halt synthesis of norepinephrine or dopamine, Loftstrom et al. (588) concluded that catecholamine containing fibers in the lateral palisade layer were predominantly dopaminergic and fibers in the medial palisade layer were both dopaminergic and noradrenergic. Ajika and Hokfelt (589) estimated the number of monaminergic terminals to be one third the total number of terminals in the median eminence. Over twice as many dopaminergic terminals are present in the lateral palisade layer as in the medial palisade layer of the rat's median eminence. A similar distribution of fluorescence has been found in the rabbit (590,591) and cat (592). As in the rat, fluorescence was maximal in the palisade layer of the external zone and the hypendymal layer of the internal zone in these animals. Immunohistochemical localization of tyrosine hydroxylase has been employed to identify dopaminergic terminals. However, not all ir-TH terminals are dopaminergic. Where as some may be noradrenergic or adrenergic because they contain DBH and PNMT others apparently lack dopa-decarboxylase and cannot convert dopa to dopamine (593).

Other neurotransmitters have been demonstrated in the external zone. Acetylcholine has been localized by immunohistochemical staining of acetylcholine esterase, choline acetyl transferase or the vesicular acetylcholine transporter (442,594,595). GABA has been localized by demonstrating glutamic acid decarboxylase (ir-GAD) in nerve terminals in the external zone (348,349). Glutamate has been demonstrated in the monkey median eminence (596). Nitric oxide has been localized at the ultrastructural levels in terminals in the rat's median eminence (597). Histamine has also been found in terminals in the median eminence of the cat (598).

Neurotransmitters are frequently colocalized. Some terminals that contain ir-TH, and which have been assumed to be dopaminergic, have been found to also contain GABA (identified by the presence of ir-GABA or ir-GAD) (355). Neuropeptides have been identified within classical aminergic terminals. Neurotensin (NT) has been identified in the palisade layer of the rat's median eminence (532); and, it has been colocalized with dopaminergic hypothalamic neurons (599). Galanin has been demonstrated in TH-li cells projecting to the median eminence (600).

Hypophysiotropic parvicellular systems terminate in the palisade layer of the external zone. Their distribution has been reviewed extensively (156). GnRH has been localized within large granular vesicles within terminals that, at infundibular levels of the median eminence, lie in the lateral palisade layer. In the guinea pig median eminence, delta sleep inducing peptide has been localized in the same terminals as GnRH. At the ultrastructural level the two immunoreactive hormones were in the same large granular vesicles (601). In the pre- and post-infundibular median eminence, GnRH fibers were more uniformly distributed between the medial and lateral palisade layers (156,429). If pre-embedding immunohistochemical techniques were employed on vibratome sections, more GnRH was found in the medial palisade zone than was found in paraffin-embedded thick sections and the differences between the medial and lateral palisade layers was less striking (602). Simultaneous localization of GnRH and dopamine containing terminals demonstrated GnRH fibers preferentially in the lateral palisade zone of the rat's median eminence in close relationship to dopamine containing fibers (603,604). On examination of their light microscopic preparations which were fluorescence stained for dopamine terminals and immunohistochemically stained for GnRH, McNeill and Sladek (603) found one population of aminergic terminals closely associated with portal vessels and a second population of aminergic terminals closely related to GnRH containing terminals. They proposed that dopaminergic terminals inhibit GnRH released from peptidergic terminals through axo-axonic synapses. Dopamine-containing terminals have been found closely apposed to GnRH-containing terminals on transmission electron microscopic studies of the rat median eminence (433) and synapses between TH immunoreactive terminals and GnRH immunoreactive terminals have been reported in the ewe (605). Although synapses were not reported in the rat, an interaction between dopamine and GnRH-containing terminals seems certain. In agreement with the findings of Rodriguez et al. (558), Ohtsuka et al. (428) found that GnRH terminals did not reach the basal lamina at the surface of the lateral palisade zone in the rat median eminence. Most lay at a distance greater than 0.5 μm from the perivascular space and were separated from it by a "glioependymal cuff." This arrangement was not found in the ewe's median eminence where the GnRH terminals contacted the basal lamina of the perivascular space (606). Anthony et al. (505) have stressed species differences in the distribution of GnRH terminals. In primates, GnRH staining fibers were found in the internal zone of the median eminence and could be followed into the infundibular stem and process. The authors pointed out that GnRH terminals in the neural lobe lay juxtaposed to regions in the pars distalis that are rich in gonadotropes and that "short portal" vascular connections between the neural lobe and pars distalis are ample in these animals. It is not clear whether a relationship of GnRH terminals to dopamine terminals is maintained in the neural lobe of these animals.

TRH terminals are aggregated in the medial palisade layer (156,491,607). These terminals lie in the perivascular space of capillaries in the median eminence (426). Somatostatin containing fibers are found throughout the external and internal zones of the median eminence in the rat (156,607). These terminals, like those that contain TRH, lie in the perivascular space of capillaries (428). GRH immunostained fibers are distributed in a pattern similar to somatostatin fibers in the rat median eminence (531, 608). CRH terminals also lie principally in the medial palisade layer of the external zone (609–611) in the rat and in an analogous site in the monkey (612). Vasopressin, angiotensin II, enkephalin, neurotensin, GABA, and CCK have been colocalized in CRH neurons by immunohistochemistry (613–617).

Vasopressin containing neurons also terminate in the palisade layer of the median eminence, principally in its medial region (58,618–620). In the guinea pig these terminals contain large granular vesicles (90–110 nm in diameter), not the large dense core vesicles (>150 nm in diameter) in which hormones are localized in terminals in the neural lobe (56,621). In the pre- and post-infundibular region of the rat's median eminence, there was considerable overlap between the distribution of immunoreactive CRH and AVP. In the infundibular region only a few AVP fibers were found in the palisade zone whereas CRH-containing fibers were plentiful there (622).

CRH can be costored with vasopressin in parvicellular neurons (623). Vasopressin and CRH are stored in the same secretory vesicles (624). Only about 50% of the CRH containing terminals in the rat's median eminence costored AVP (625). Whether all AVP containing terminals costore CRH has not as yet been determined. High concentrations of vasopressin (626) and of CRH (627) have been found in portal blood. The presence of high concentrations of AVP (626) in portal blood has, until recently, been used as evidence that vasopressinergic terminals are plentiful and have ample access to capillaries in the median eminence. However, the concentration of AVP in portal blood is higher than expected when compared to the concentration of CRH if it is assumed that AVP can only be released with CRH and then only from 50% of the CRH terminals (625). The demonstration that vasopressin is released from axons of passage in the internal zone (576) raises the possibility that AVP in the portal vessels draining the median eminence has several sources—one from parvicellular terminals in the external zone and another from magnocellular axons in the internal zone. Still another possibility is a retrograde flow of blood from the neural lobe to the median eminence (628).

Parvicellular AVP containing neurons were described as coursing ventrally, at right angles to the fiber layer, to gain access to capillaries in the external zone in this region. This pattern was accentuated after adrenalectomy (622). Immunogold electron microscopy has been employed to assess the changes in CRH/AVP terminals after adrenalectomy (629). Whereas immunogold staining of CRH remained constant, the density of particles over neurosecretory vesicles was increased when the tagged antibody was directed against AVP. Thus the amount of vasopressin in the terminals increased with respect to the amount of CRH following hypophysectomy. AVP potentiates CRH stimulated ACTH release from corticotropes (630,631). Increased vasopressin release from parvicellular terminals (and magnocellular axons of passage) provides a means to increase the output of ACTH in response to stress.

Other neuropeptides beside the classical neurosecretions and the hypophysiotropic peptides have been demonstrated in the median eminence. Terminals containing immunoreactive angiotensin II (ir-ANG II) have been identified by immunohistochemistry in the palisade zone of the rat's median eminence. Some ANG II cells co-stored CRH (see above) and their staining was selectively enhanced by adrenalectomy (632). Immunoreactive galanin is present in terminals in the external zone (633,634) and has been colocalized with neurotransmitters (ir-GAD and ir-TH (424,487) and a releasing hormone (GRH) (635). Galanin terminals contact terminals that contain large granular vesicles and those that contain only microvesicles as well as pituicytes and tanycyte processes (636). Substance P has been localized to the external zone of the rat (156), the mouse (488), the monkey and the human (637) median eminence. Immunoreactive enkephalin (ir-ENK) (156), vasoactive intestinal peptide (ir-VIP) (156), neurotensin (ir-NT) (532), neuropeptide y (ir-NPY) (638), ir-CCK (485), and gastrin (156) have been identified in the palisade layer of the rat's median eminence. CCK is frequently co-localized with CRH (617). Calcitonin gene-related peptide (CGRP) has been identified in the median eminence of the frog (639).

POMC derivatives are primarily localized by immunohistochemical techniques and light microscopy to the external zone of the median eminence in several species other than the rat. In the sheep and ox, β-LPH "projects to portal capillaries" (640). Bloch et al. (641) found that in humans, fibers containing immunoreactive β-END terminate "close to vessels in the median eminence." Such neurons are also reactive to antisera against ACTH, MSH, and β-LPH (642). Although species differences in the distribution of terminals containing the derivatives of POMC may be present, there seems to be general agreement that terminals in the median eminence are closely related to capillaries of the portal system because ACTH and β-END are present in portal blood (643,644).

The pattern of regional segregation of neurosecretory systems within the neurohypophysis has been most clarified in the rat. The temptation to regard this pattern as a prototype for the mammalian hypophysis should be resisted at this time because species differences may be marked. The distinction between the internal zone and the external zone is blurred in some mammalian species, including man, by invasion of the palisade layer into the internal zone (573). The site of termination of a particular neurosecretory system can also vary with the species studied (505). However, the basic organizational pattern of the mammalian neurohypophysis stands out clearly from the blurred background of differences in species detail. It is the relationship of individual neurosecretory hypophysiotropic terminals to capillaries and the regional segregation of systems of neurosecretory terminals within the neurohypophysis. Neurotransmitters such as dopamine, acetylcholine, GABA, serotonin, nitric oxide, and glutamate are co-released with hypophysiotropic neurosecretions to act locally in an autocrine or paracrine fashion to coordinate the release of a particular peptide with the function of neighboring terminals. These locally acting transmitters and peptides may alter the secretion of peptides from neighboring terminals, alter the relationship of terminals the ensheathing pituicytes, and potentially alter the permeability of the region and the blood flow into it. It would seem no accident that the compounds involved in the initiation and maintenance of a local inflammatory response (histamine, serotonin, CGRP, substance P, and VIP) are released from terminals in the median eminence. It is tempting to speculate that with release of hypophysiotropic secretions local blood flow and permeability will increase as a result of the release of peptides and transmitters from nerve terminals in the perivascular space because blood vessels, not other nerves, will carry the messages released from axon terminals in the median eminence and neural lobe.

THE PORTAL SYSTEM

Capillaries of the Hypophysis

A single capillary bed extends throughout the entire neurohypophysis (Fig. 11) (28,645). It communicates by capillary or portal routes with similar appearing capillaries in the pars tuberalis and pars distalis of the adenohypophysis (65,646); and, its perivascular space communicates extensively with the extracellular space of the pars intermedia (338). Hypophysial capillaries do not exclude acid vital dyes from the pituitary which lacks a blood–brain barrier (28). Neurohypophysial (392,393,409) and adenohypophysial (231) capillaries are fenestrated as are the portal vessels that unite them (393,646). These fenestrated capillaries are surrounded by a double basement membrane. The inner basal lamina lies next to each fenestrated endothelial tube whereas the outer basal lamina is separated from it and lies next to the parenchymal elements (Figs. 7 and 12). In the adenohypophysis, these parenchymal elements are epithelial cells arranged in cords. Because their capillaries are linearly arrayed (646), the configuration of the perivascular space is not complex (232). In the neurohypophysis these parenchymal elements are axons and their terminals, ependymal terminals, and glial cells. Pericytes, histiocytes, fibroblasts, microglia, and mast cells have been identified between the basal lamina in the perivascular space of human neural lobes (507). The inner and outer basal lamina define the perivascular space, which in the neural lobe of the rat is 0.3 to 0.5 μm in width and contains reticular fibers (647). In the median eminence of the rabbit it surrounds entire vascular formations (394). The perivascular space forms an extensive system of channels which are revealed with tannic acid staining (647) or silver impregnation (465). In the neural lobe these channels form a widespread system of connections between neighboring capillaries. In both the median eminence and neural lobe extensions of the perivascular space increase the surface area available for neurohemal contact with a minimal increase in neurohypophysial volume.

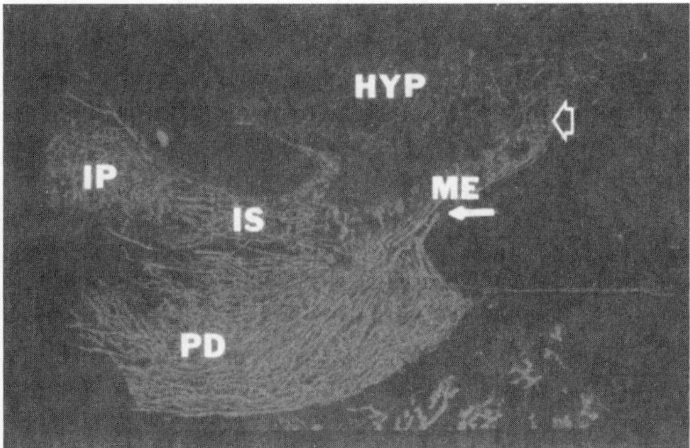

FIG. 11. Rabbit median eminence in sagittal section. A confluent capillary bed unites the median eminence (ME), infundibular stem (IS), and infundibular process (IP). Within the median eminence an external and an internal plexus may be discerned. Connections between median eminence and hypothalamic capillaries are seen far rostrally at *hollow arrowhead*. Portal connections between median eminence and par distalis (PD) are indicated by *white arrow*. ×25. HYP, hypothalamus. [From (646).]

Secretory elements in the hypophysis, neural or glandular, are disposed along the outer basal lamina of the perivascular space. Fenestrated capillaries are disposed along the inner basal lamina. Large protein molecules in the blood are relatively free to pass through fenestrae in the endothelial tube and interact with receptors at the cell's surface. Protein molecules sequestered in vesicles in the cell's cytoplasm are relatively free (after release into the perivascular space) to pass through the fenestrations in the capillaries into the blood stream. This freedom of passage is not absolute, however, as (for example) basic dyes do not enter the hypophysis easily. Simionescu et al. (648,649) demonstrated the presence of microdomains in fenestrated endothelial cells of mouse pancreatic capillaries and "expected that these structures will select permeant molecules according to charge, in addition to size." Similar findings can be anticipated in the fenestrated endothelial cells of the hypophysis.

Development of the Portal System

Vessels of the hypophysis develop from the mesenchyme that is entrapped between the advancing saccus infundibuli and Rathke's pouch and invades both structures according to Wislocki (650). Enemar (651) proposed that a (primary) plexus of capillaries lines the surface of the saccus infundibuli and a (secondary) plexus of capillaries lies on the surface of the Rathke's pouch in the mouse by the 14th fetal day. They are united by "linking" capillaries at the junction of the

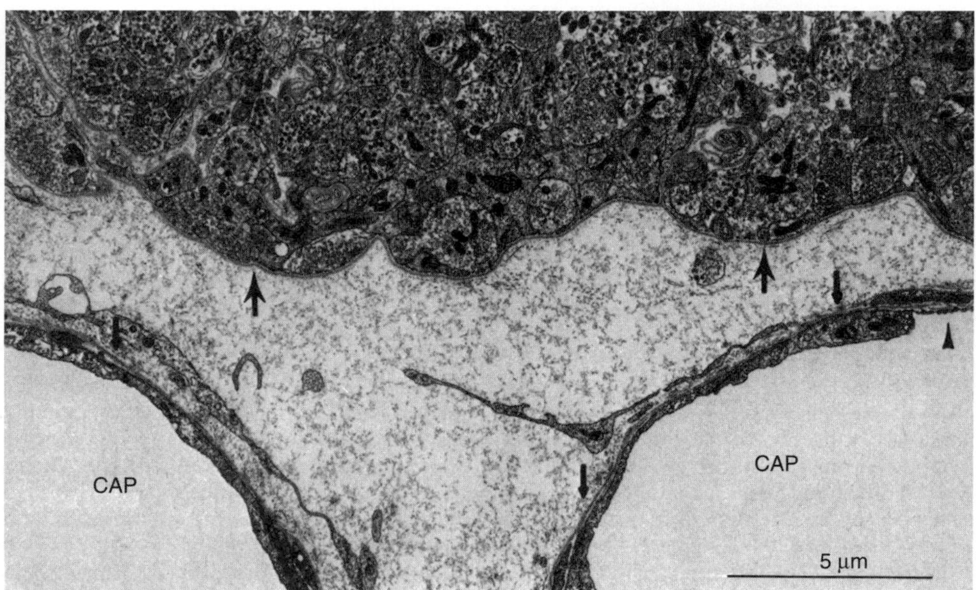

FIG. 12. Transmission electron micrograph of rabbit median-eminence-internal-zone-horizontal section. Two limbs of a capillary loop in the median eminence floor (CAP) are cut in horizontal section. One limb is fenestrated (*arrowhead*). The perivascular space is common to both. The outer basement membrane material is condensed at the margin of the contact zone common to both capillaries (*vertical large arrows*). The inner basement membrane is condensed about each individual capillary component (*vertical small arrows*). [From (394).]

rostral portion of the saccus infundibuli and the oral wall of Rathke's pouch. These linking capillaries become more numerous by the 15th fetal day, and by the 16th fetal day in the life of the mouse the primary plexus of capillaries overlying the saccus infundibuli has become very dense. As the saccus infundibuli enlarges and differentiates over the next 2 days, the continuous capillary network over the presumptive median eminence, infundibular stem, and infundibular process increases in thickness and begins to invade the developing neurohypophysis. Concurrently, the presumptive pars distalis becomes more densely vascularized by invasion of capillaries from its surface. By the 19th fetal day, portal vessels on the ventral surface of the median eminence (which unite the primary and secondary capillary plexuses) have become distinct. This appearance is unchanged on the 20th (birth) day of the mouse. The capillary loops of the internal (deep) plexus in the median eminence do not appear until 2 to 3 days after birth and do not complete their development for 2 weeks. The deep system of long capillary loops and subependymal vessels is completed by 7 days but the system of short capillary loops not completed until 14 days of age.

The development of the pituitary vessels is similar in the rat (652). On the 15th fetal day capillaries of the primary plexus on the surface of the saccus infundibuli and on the surface of the remainder of the neural tube are nonfenestrated endothelial tubes that lie free in the mesenchyme. On that day, fenestrae have been found only in capillaries entrapped between the presumptive median eminence and pars distalis (443). Fenestrae did not attain the size, density, or distribution of those found in the adult rat neurohypophysial and adenohypophysial capillary beds until the 20th fetal day (451,457). Ugrumov et al. (653) identified some short capillary loops in the rat's median eminence on the 18th fetal day. Few mitotic figures were found at this time. He explained the formation of capillary loops by the growth and attenuation of preexisting cells with buckling of the capillaries inward from the surface and invasion of the neuropil. A similar mechanism was reported by Page and Dovey-Hartman (394). In the rat, development of the internal plexus in the median eminence delayed until after birth as it is in the mouse (652); but this pattern is not representative of all mammals as in the rabbit it is completed before birth (654).

Organization of the Portal System

The neurohypophysial capillary bed extends from the median eminence to the neural lobe. It is supplied rostrally by superior hypophysial arteries and caudally by inferior hypophysial arteries (645). A third source of arterial supply has been identified and variously named the "anterior hypophysial artery" (83), "peduncular artery" (655), "loral artery" (656), or "trabecular artery" (80,96). The name "middle hypophysial artery" seems appropriate because this vessel arises from the internal carotid artery between the origins of the inferior and superior hypophysial arteries. The inferior hypophysial arteries arise as a pair of vessels from the intracavernous segment of the internal carotid arteries and supply the neural lobe (infundibular process). Frequently they unite to form an anastomosis between the left and right internal carotid arteries prior to supplying the neural lobe. The middle hypophysial arteries also arise from the intracavernous segment of the internal carotid arteries. They frequently unite (as in the rabbit, rat, and cat) to form a single artery that then supplies the infundibular stem (645). The superior hypophysial arteries arise from the intracranial carotid arteries and from the circle of Willis. These (multiple) arteries approach the median eminence centripetally to form an anastomotic ring about it and to supply the rostral region of the neurohypophysis (the median eminence) (645, 651,657). The arterial supply to the pituitary gland thus represents three ascending levels of anastomotic channels between the paired carotid arteries whose branches supply the neurohypophysial capillary bed.

Superior and inferior hypophysial arteries are innervated by postganglionic sympathetic nerves. Innervation of the middle hypophysial artery by these nerves is expected but has not as yet been reported. Arterioles that supply the neural lobe of the pig and rat (658) are also innervated by sympathetic fibers but a similar innervation of arterioles supplying the rat (581,582) or rabbit (659) median eminence has not been found. Some arterioles on the surface of the median eminence receive an innervation by VIP/PHI-li containing fibers that do not appear to be central in origin (660). Additional peptides and amines released from terminals in the external zone can reach precapillary arterioles that lie on the surface of the median eminence (659).

The venous drainage of the caudal region of the neurohypophysial capillary bed mirrors its arterial supply. Paired inferior hypophysial veins course from the neural lobe to the posterior intercavernous sinus and to the posterior regions of the paired cavernous sinuses which lie lateral to the midline (657). A less prominent route of drainage is from the neural lobe to the adjacent pars distalis by capillary and "short portal" routes. Short portal veins uniting the neural lobe with the pars distalis are abundant in some species (i.e., rat, mouse, and rabbit) but few in others (pig, dog, monkey) (84,646). The hypophysial cleft limits the number and disposition of vascular connections between the pars distalis and the neurointermediate lobe. The neural lobe has ample venous drainage routes to the systemic circulation and limited venous drainage routes to the pars distalis (657).

There are no direct venous drainage routes of the territory supplied by the middle hypophysial artery except for the short portal vessels that connect the infundibular stem with the pars distalis.

Venous drainage of the rostral region of the neurohypophysial capillary bed (the median eminence) does not mirror its arterial supply (Fig. 13) (28,65, 80,81). The primary plexus in the median eminence is drained by fenestrated long portal vessels that course to the pars distalis and then arborize into a secondary capillary plexus. Direct capillary connections may also be plentiful depending on the length of the infundibular stem and the species being examined (65,392,393,646). Drainage routes from the primary capillary plexus in the median eminence to veins at the surface of the brain were denied by Wislocki (28,85–87). However, Duvernoy et al. (661) demonstrated systemic drainage routes from the post-infundibular median eminence in man, and Ambach et al. (662) have demonstrated venous channels from the preinfundibular median eminence to the tuberal and chiasmatic veins in the rat. In both cases drainage routes from the median eminence to these systemic sites were poorly developed; and, they probably represent a regression of the ample venous connections between the saccus infundibuli and the remainder of the neural tube that were present prior to the apposition of the saccus infundibuli and Rathke's pouch and the development of portal vessels. Venous drainage routes from the rostral region of the neurohypophysial capillary bed, the median eminence, to the secondary capillary bed in the pars distalis are ample whereas venous drainage routes to the systemic circulation are not.

The adenohypophysis, contrary to the belief of Wislocki (28,85–87) does not appear to receive a direct arterial supply (65,77,80,81,84,645). Blood destined for the adenohypophysis passes through the neurohypophysis. Capillaries and veins bring blood to the pars tuberalis and pars distalis.

Veins connecting the pars distalis to the paired cavernous and posterior intercavernous sinuses provide routes of egress for blood in the pars distalis. Named "lateral hypophysial" veins by Wislocki (28,85–87), on the basis of his study of the venous drainage of the monkey pars distalis, these veins were found by Green (663) to be a constant feature of the mammalian pars distalis. Bergland and Page (657) found that these veins, which drained the pars distalis, united with the veins that drained the neural lobe and drained though a common stem to the cavernous sinus. These Y-shaped veins drained both the pars distalis and the neural lobe (Fig. 13).

The available patterns of blood flow in the pituitary gland are determined its the vascular anatomy but the patterns of blood flow employed in living animals are not easily deduced from anatomical studies of blood vessels. Observations of blood flow in the portal

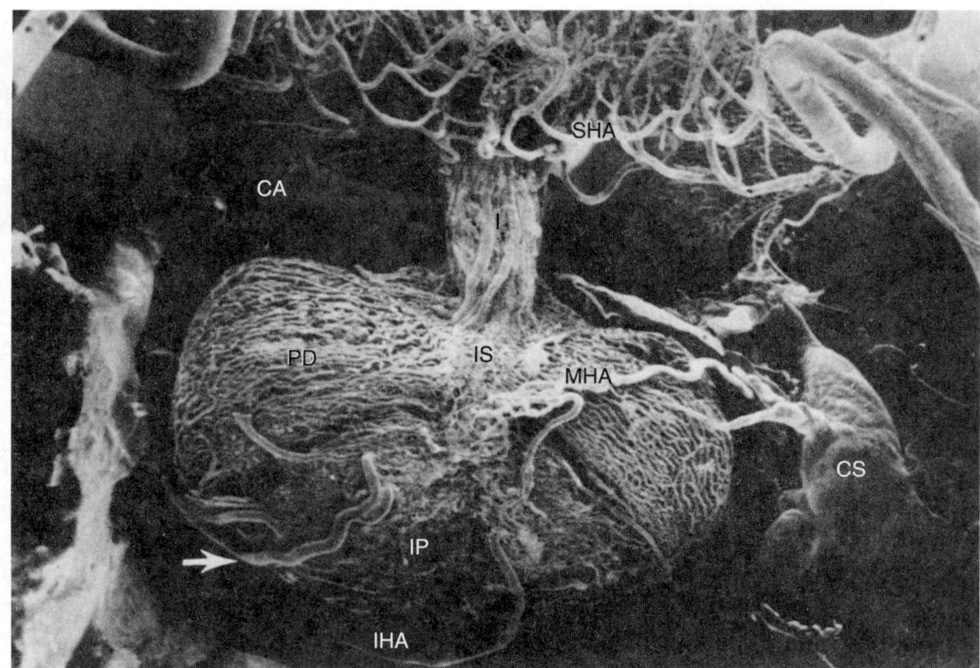

FIG. 13. Scanning electron micrograph of hypothalamic–pituitary vascular cast of the monkey as seen from behind. A common capillary bed extends throughout the infundibulum (I), infundibular stem (IS), and infundibular process (IP). It is supplied rostrally by superior hypophysial arteries (SHA) and caudally by inferior hypophysial arteries (IHA). Long portal vessels are present on the surface of the infundibulum. Lateral hypophysial veins from the pars distalis to the cavernous sinus (CS) are few in number. Y-shaped inferior hypophysial veins (*arrow*) drain both the pars distalis (PD) and the infundibular process. MHA, middle hypophysial artery; CA, carotid artery. [From (657).]

vessels of living animals leaves no doubt that blood flows from the median eminence to the pars distalis (68,69,664–670). The hypothalamic hypophysiotropic releasing factors in the terminals of neurons in the median eminence are carried to the pars distalis over restricted vascular routes.

Observation of blood flow in the porcine neurohypophysis revealed blood flow between adjacent neurohypophysial regions (between the neural lobe and lower infundibular stem and between the median eminence and the upper infundibular stem). It also revealed a major drainage route from the rostral neurohypophysis (the median eminence) to the pars distalis and the major drainage route of the caudal neurohypophysis (neural lobe) to be to the cavernous sinus. However, some blood did flow from the neural lobe to the adjacent pars distalis and neural hormones released in the neural lobe have access to glandular epithelium in this region (670). In addition, neural lobe hormones have access to the pars intermedia and adjacent regions of the pars distalis by vascular channels and through extensive connections between the perivascular space of the neural lobe and the extracellular space of the pars intermedia (338).

Rostrally, in the median eminence, the neurohypophysial capillary bed is specialized (Fig. 11). The primary plexus is subdivided into an external and an internal plexus (84,646). The external plexus corresponds to the superficial network of Duvernoy (671) and the mantle plexus of Romeis (264). The external plexus lies on the surface of the median eminence and is continuous with the capillaries of the infundibular stem and process. It receives the arterial supply to the median eminence. Its morphology differs little from species to species [compare the external plexus of the toad (672) to that of the rabbit (84)]. The external plexus forms a reticular network of fenestrated capillaries that lie partially buried within the oral (ventral) surface of the median eminence (Fig. 14) (84,394,646,673). These capillaries form a mosaic made up of multiple hexagonal units that is oriented parallel to the surface of the median eminence. Transmission electron microscopic studies (646,673) demonstrate that the capillaries are partially embedded within the external zone such that their oral surface is flush with the oral surface of the palisade layer. The central region of each hexagonal capillary unit is filled with "posts" of median eminence tissue. This capillary pattern resembles that found in two other organs where the exchange of materials between the blood and tissue is rapid—the gastric mucosa and the lung (674,675). Sobin et al. (676) described the pulmonary vascular space as resembling "an underground garage consisting

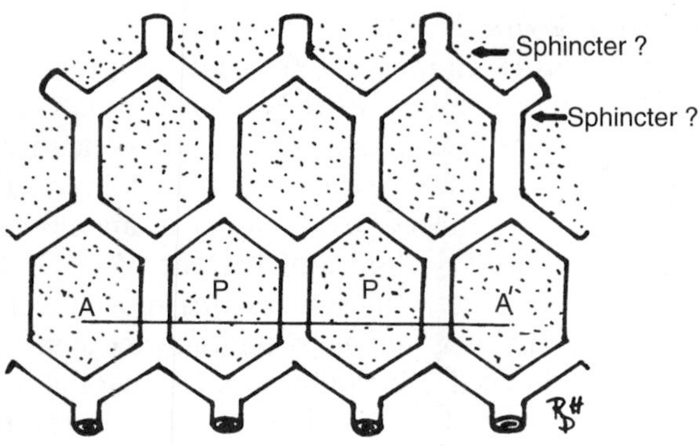

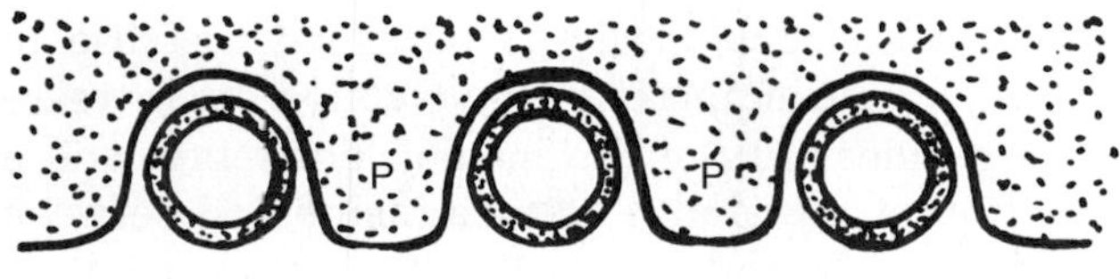

FIG. 14. Top: Schematic view of external plexus from below. Hexagonal arrays of capillaries are separated by posts (P) of median eminence tissue. **Bottom:** Schematic cross-sectional view of median eminence external plexus along plane A-A′ on top panel. Capillaries lie embedded in neuropil separated by posts (P) of median eminence tissue. [From (673).]

of a floor, ceiling and supportive pillars all covered with endothelium"; and, they argued that blood flows as a sheet through an essentially continuous space lined by endothelium. In the median eminence a similar pattern of flow may be present because the angioarchitecture of the external plexus resembles that in the lung.

The organization of the palisade layer and of the external plexus has functional consequences of importance to the regulation of the adenohypophysis. Each hexagonal capillary unit (microvascular module) encloses a post (microdomain) of neurosecretory axon terminals and glial tissue. With increased functional demand, glial processes can be expected to retract from positions interposed between axon terminals and capillaries to make more surface area available for the release of hypothalamic peptides and amines into the portal system to be carried to the pars distalis (570,571). Within each micro-domain, amines and peptides released from one terminal are free to act on neighboring terminals if appropriate receptors are present—a paracrine action of secretions released by terminals in the palisade layer on neighboring terminals is possible as well as an endocrine action on the epithelial cells of the pars distalis. Should sheet flow occur across the surface of the median eminence, the secretions of one microdomain would be free to interact with the terminals in other microdomains. The microcirculation in the palisade zone could permit interactions between microdomains separated from one another on the surface of the median eminence. Should future transmission electron microscopic studies reveal smooth muscle sphincters arrayed at the entrance to each hexagonal microvascular module, the opportunities to permit very regional adjustments in the distribution of blood flow at the surface of the palisade layer would be present; and, the capacity for neurons terminating in one microdomain to influence neurons terminating in another could be limited or regulated.

Whereas the external plexus receives the arterial blood supply from the superior hypophysial arteries, it distributes blood into portal vessels coursing to the pars distalis or into the internal plexus that invades the median eminence. The internal plexus arises from the external plexus (651–653). There is considerable variability in the anatomy of the internal plexus between species (compare the rat, the rabbit, and the monkey) (84,646). Viewed from within the ventricle, the median eminence forms a bowl or a funnel and hence the term infundibulum is employed to describe it. The bowl has a floor and walls. The basic pattern of the internal plexus is one of capillary loops in the floor of the infundibulum; and, in some animals, capillary coils in its walls. The capillary loops in the floor of the median eminence (of the rat) have been subdivided into long capillary loops which stretch into the hypendymal layer and short capillary loops which extend from the surface of the palisade layer only as far as the reticular layer.

The capillary loops of the internal plexus are well seen when the median eminence has been sectioned in the midsagittal plane following the intravascular injection of India ink (28,65,671,677–679), neoprene latex (96), or microfil (646) and viewed with a light microscope. They can be better appreciated by examining corrosion casts of the median eminence with the scanning electron microscope (SEM) (84,646). Such techniques revealed the loops of the internal plexus to be made up of an ascending limb, an apex, and a descending limb (Fig. 15). The ascending limb arises from the external plexus and passes through the palisade layer and reticular layer into the internal zone where it may pass as deep as the hypendymal layer. There may be extensive arborization beneath the ependyma at the apex of the long loops. The descending limb passes back through the internal zone, the reticular layer and the palisade layer to the surface of the median eminence where it joins either an external plexus capillary or a portal vessel (671).

The capillary loops of the internal plexus also form a series of microvascular modules, which is best appreciated on sagittal section (Fig. 11). They arise

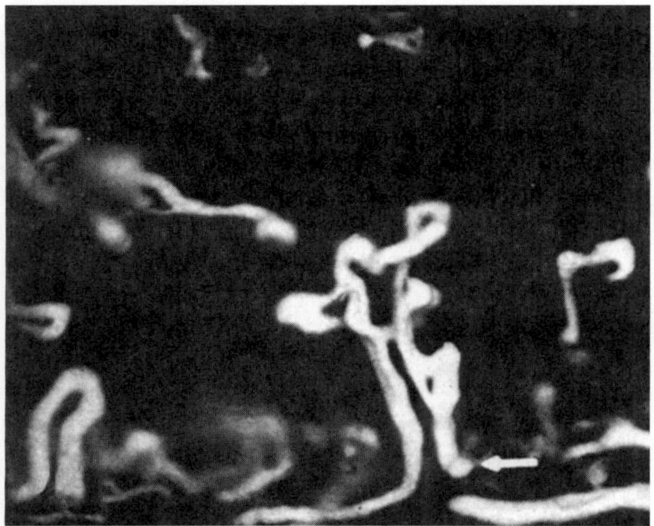

FIG. 15. Light-microscopic photograph of rabbit median eminence in sagittal section, Microfil injection. An internal-plexus capillary loop arises from the external plexus (*arrow*) to arborize beneath the ventricular surface and then drain to a portal vessel. Ependymal processes terminate about the apex of the loop, whereas small neurons make neurohemal contact along the limbs of these formations. If sphincters are present at the origin of these modules, a mechanism for regional control of blood flow within small (functional) regions of the median eminence will be present. ×100. [From (646).]

from the external plexus phylogenetically (they are not present in amphibians but are present in mammals) (672) and embryologically (651,652,654). As a loop invaginates into the median eminence, it carries with it an expanded perivascular space which surrounds the entire formation (394,653). Hence on serial sections elements of a single vascular module can be identified in light or TEM studies (394). The perivascular space between the ascending and descending limbs of each loop is continuous with the perivascular space on the surface of the palisade layer. It lies between the widely split basal lamina whose inner layer is applied to the convoluted endothelial tube and whose outer layer invests the entire formation in a fashion analogous to the disposition of the visceral and parietal peritoneum. Neurohemal contact has been documented along the course of these loops in the external and in the internal zone of the rabbit median eminence (394) and in the median eminence of the gerbil (441). Columns of capillary loops surrounded by parvicellular axon terminals thus stand on a base comprised of hexagonal arrays of capillaries each of which encloses a microdomain of neurosecretory terminals. Because each loop invaginated from the surface into the depths of the median eminence, it can be expected to have carried with it axon terminals from the surface of the palisade layer. Each vertical column should then be expected to contain terminals functionally similar to those at its base. The area for neurohemal contact will be increased but the endocrinotopic organization of the median eminence (394) will be preserved by such an arrangement. In this regard it is of interest that a regional difference in the concentration dopamine in portal vessels has been reported with the dopamine concentration higher in medial than in lateral long portal vessels of the rat (680). It is not known at present whether smooth muscle cells are arrayed at the origin of each capillary loop; and, thus it is not known if the flow into these capillary loops can be regulated by sphincters. The recruitment of vascular columns in functionally specific regions of the median eminence would markedly enhance the neuroendocrine response to a functional demand. Coupled with such mechanisms as an increase in neurosecretion, an increase in blood flow, an increase in the permeability of the median eminence, and a retraction of the glial processes insinuated between axon terminals and capillaries, the recruitment of microvascular modules would significantly add to the ability of the median eminence to respond to demands placed on it (Fig. 16).

A subependymal system of capillaries is present in the median eminence of some species; and, in addition to the external and internal plexuses forms a third capillary network in the median eminence that unites it with the hypothalamus in mammals. In birds (554) these capillaries appear to be hypothalamic vessels which unite the left and right sides of the hypothalamus and to be spatially separated from the capillaries of the primary plexus. In mammals, their relationship to capillaries in the median eminence is quite different. They unite the apices of adjacent long capillary loops in the median eminence and pass to the arcuate nucleus (646,662). Ambach et al. (662) employed light microscopy to examine serial sections of the rat's hypothalamus and pituitary gland after the intravascular injection of India ink. To differentiate arteries from veins and capillaries, they first injected the animals with a blue India ink mixed in a gelatin solution of low viscosity and then injected the animal with a red India ink in a gelatin solution of high viscosity. Vessels containing the red (high viscosity) mixture were identified as arteries for the high viscosity mixture were not expected to pass into smaller capillaries and thence to veins. Vessels containing the blue dye were identified as capillaries or veins. They employed this technique to identify subependymal vessels in the median eminence of the rat which united the apices of internal plexus capillary loops and stated that they had an independent arterial supply in the median eminence and drained to the hypothalamus. A similar conclusion has been reached by Murakami et al. (681) on the basis of scanning electron microscopy of vascular casts. Page et al. (646) also employed scanning electron microscopy to study the

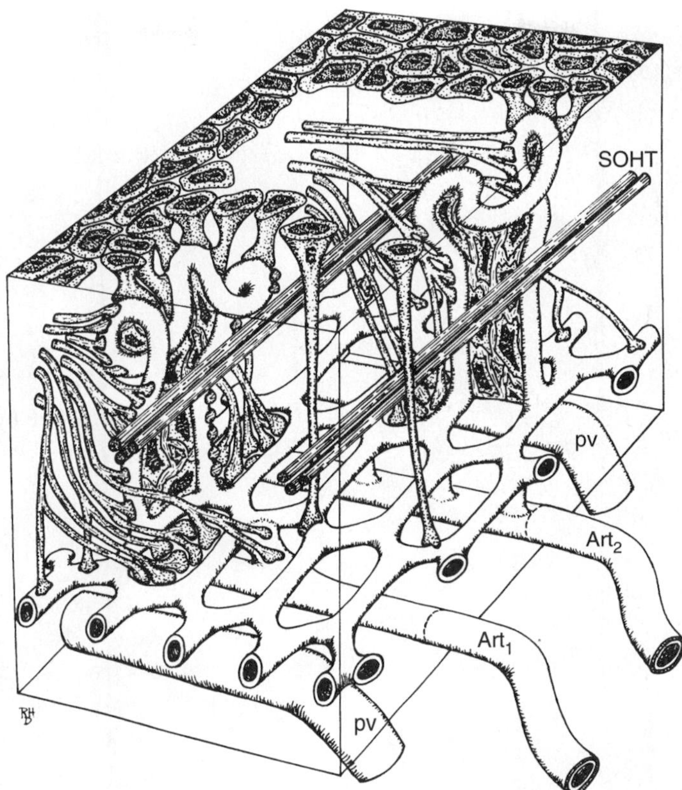

FIG. 16. Diagram of neurovascular organization in the medial third of the rabbit median eminence. The external plexus forms a hexagonal array oriented parallel to the median eminence external surface and receives arterial blood from arteriolar branches of the superior hypophysial arteries (Art_1, Art_2). Internal plexus capillary loops arise from the external plexus and arborize in the subependymal zone. Connective tissue fills the spaces between the ascending and descending limbs of the internal plexus loops and the ventral aspect of the median eminence between the surface of the external zone and the pars tuberalis. Nonciliated ependymal cells line the infundibular recess (E). The basal process of some ependymal cells end in the perivascular space about the apex of capillary loops and form an ependymal cuff about them in the hypendymal layer. Some ependymal processes extend to the external plexus. Parvicellular axons terminate on external plexus capillaries and about internal plexus capillary loops to from axonal cuff about them deep in the fiber layer and in the palisade zone. The axons are peptidergic and aminergic. Magnocellular systems (SOHT) pass through the fiber layer and do not send collaterals to the capillary loops. Arterioles that originate from superior hypophysial arteries lie on the median eminence surface and supply the external plexus by tapering down from terminal arterioles (Art_2). Arterioles lie in close proximity to the external plexus and to posts (microdomains) of neurosecretory and aminergic terminals. Portal vessels (pv) are fenestrated and do not contain a continuous layer of smooth muscles in their walls. [Modified from (394).]

angioarchitecture of the rat, dog, cat, sheep, and monkey median eminence. The subependymal plexus was found in the rat, dog, and sheep median eminence and was continuous with the subependymal capillaries in the hypothalamus but a direct arterial supply was not found. In the rabbit, cat, and monkey, a subependymal plexus was not present but connections between the internal plexus of the median eminence and subependymal capillaries about the third ventricle in the region of the hypothalamic arcuate nucleus were present (Fig. 17). Akmayev (682,683) had previously stressed the presence of these capillary connections between the median eminence and the arcuate nucleus. Ambach et al. (662) argued that the capillary beds of the arcuate nucleus and the neurohypophysis should be considered as a single unit; and, this concept has received support from physiological studies of neutral amino acid transport and transit time in the median eminence and adjacent tuberal hypothalamus in the rat. The ventral region of the arcuate nucleus adjacent to the median eminence displayed an increased permeability-surface product (PxS) and increased transport of the neutral amino acid α-aminoisobutyric acid when compared with the dorsal region of the arcuate nucleus (684). The hypophysial vascular system is not as isolated from the hypothalamus as originally

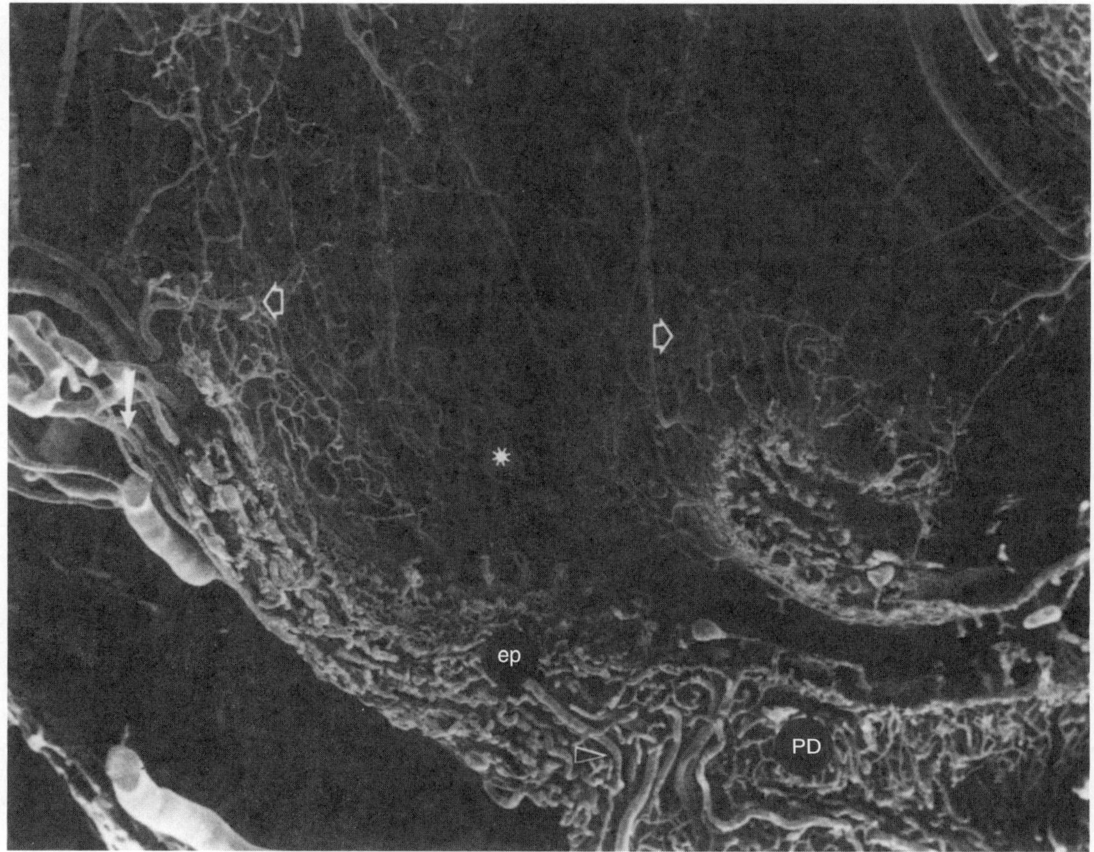

FIG. 17. Scanning electron micrograph of cat hypothalamus and median eminence, sagittal view. The rostral position is to the reader's left. The median eminence external plexus (ep) lies interposed between pars distalis (PD) capillaries below and internal plexus capillary loops above. Vertical arrow designates superior hypophysial arteries, which vascularize external plexus. External plexus capillaries unite to form portal vessel coursing to the pars distalis (*black arrowhead*). Capillary connections between the median eminence and hypothalamus are present anteriorly and posteriorly at hollow arrowheads and in the periventricular zone at the asterisk. Note that periventricular capillary connections uniting the lateral wall of the median eminence with the hypothalamus cannot be appreciated on Microfil section (compare with Fig. 11). [From (646).]

proposed and subsequently defended by Wislocki (28,85–87).

Torok (664,665) reported that blood flowed from the external plexus into capillary loops of the internal plexus and then into subependymal vessels and then to the hypothalamus. He further reported that on dissection of the pituitary stalk with separation of the infundibular stem from the pars distalis, he found some deep portal vessels in which blood appeared to be flowing from the pars distalis to the median eminence. This observation has not been independently verified. Neither the observation or its obvious consequences have been widely accepted. However, they cannot be easily dismissed. A labeled analog of ACTH [(^{3}H)-ACTH 4-9)] and (^{3}H) β-LPH have been injected into the pars distalis of the rat with subsequent recovery of label in the hypothalamus. If the pituitary stalk was severed just prior to injection, the label was not recovered in the hypothalamus. If 8 days were permitted to pass following stalk section (sufficient time for vascular but not neural regeneration), label was again recovered from the hypothalamus after injection into the pituitary (685,686). Injection of neurotensin into the rat pituitary produced a profound decrease in core body temperature. This effect is only seen if neurotensin reaches the hypothalamus in adequate amounts. The effect was abolished by prior stalk section but was regained if the vessels were permitted to regenerate (687). While it is possible that the act of injection of substances into the pars distalis could have reversed the normal direction of pituitary blood flow, the possibility of "two way transport in the hypothalamo–hypophysial system" (686) remains open. Such transport provides an anatomical route for the participation of hypothalamic neurons in the "autofeedback" regulation of GH proposed by Minami et al. (688).

The arterial supply to the median eminence is to its external plexus. Arterioles, which supply the median eminence, lie close to axon terminals and fenestrated capillaries of the external plexus containing vasoactive catecholamines and peptides (659). Hence, they will be exposed to high levels of dopamine and neurohormones released at the ventral surface of the median eminence. In addition, the descending limbs of capillary loops carry blood to the external plexus capillaries from deep in the internal zone. Terminals containing amines and peptides contact these loops in the internal zone (394). The noradrenergic reticulo–infundibular tract terminates in that zone. Blood returning to the external plexus through the internal plexus may contain vasopressin, norepinephrine and epinephrine (689) as well as dopamine. The resistance vessels lying at the surface of the median eminence regulate blood flow into its external plexus; but, they are in turn exposed to vasoactive amines and peptides secreted by the median eminence (659). The neurosecretory systems terminating in the median eminence may regulate blood flow into it (Fig. 16).

THE HYPOTHALAMUS

Cell Bodies of the Hypothalamus

Neurosecretory neurons are the motor neurons of the neuroendocrine system. They synthesize peptide hormones in their cell bodies in the hypothalamus behind the blood–brain barrier and transport them to nerve terminals in the neurohypophysis outside the blood–brain barrier. Magnocellular peptidergic neurons synthesize and secrete vasopressin or oxytocin. They project through the supraoptico–hypophysial tract to the neural lobe where they release these hormones into the systemic circulation. They have large cell bodies measuring 12 to 30 μm in diameter in the goldfish (50), 25 to 50 μm in diameter in the rabbit (690), and about 25 μm in diameter in the rat (691). These cells are multipolar or bipolar. Their cell bodies are oval or pear shaped and they contain an eccentrically placed nucleus that contains a single nucleolus. Light microscopic examination of cresyl violet preparations revealed a cytoplasm with a granular marginal zone rich in Nissl substance and a clear central zone. Golgi-Cox preparations revealed the soma of multipolar (but not bipolar) neurons to be spiny (690). LuQui and Fox (692) described many of them as having a "rough shaggy surface." A single axon and 3–5 (primary) dendrites emerge from the cell body. The dendrites bifurcate into secondary dendrites shortly after emerging. Spines are also prominent on both primary and secondary dendrites, which are bulbous and varicose. Tertiary branching is not frequent. Bipolar neurons are fusiform in shape with a single spiny dendrite arising from each pole of a smooth cell body. Axons typically emerge from a primary or a secondary dendrite (690) or from the cell body (693). Axons are thinner than dendrites and frequently have a beaded appearance (692–694). Collateral axons have been identified (692,693). They are presumed to be inhibitory collaterals on the basis of physiological studies (695). Immunohistochemical preparations revealed that oxytocin and vasopressin, with their associated neurophysins (I and II) are present in the cell bodies of separate magnocellular neurons (58,61, 696–699) as well as in their axons and dendrites (693).

The ultrastructural appearance of these magnocellular neurons was first described by Palay (50) who studied the preoptic nucleus of the goldfish. Subsequent descriptions of their ultrastructure in other species are not significantly different although perhaps more complete (700–702). These cells are found in the supraoptic (SON) and paraventricular (PVN) nuclei of mammals but their ultrastructural appearance is similar in the two locations (Fig. 18) (703,704). The eccentrically placed nucleus contains a prominent nucleolus. The nucleus is folded and it presents a smooth convex surface to the marginal zone and indented folded surface to the central zone. The marginal zone is rich in rough endoplasmic reticulum (RER) with flattened cisternae that may be arrayed, in places, parallel to the surface of the neuron. The central zone contains a random assortment of organelles including mitochondria, (shuttle) vesicles, Golgi, dense core vesicles, lysosomes, smooth endoplasmic reticulum (SER), and ribosomes. Occasional Golgi cisternae can sometimes be seen to pinch off a (new) dense core vesicle (neurosecretory granule).

Immunoelectron microscopy has been performed with pre-embedding techniques that expose vibratome sections to antiserum before embedding and sectioning; and, subsequently, with post-embedding techniques in which the antiserum has been applied to thin sections after embedding in plastic medium. The peroxidase-antiperoxidase (PAP) technique is frequently employed in pre-embedding techniques. The flocculent precipitate localizes the location of the antigen at the ultrastructural level but frequently obscures ultrastructural detail (705). Employing this technique Piekut (706) was able to localize oxytocin or vasopressin to large dense core vesicles in the somatic and dendritic cytoplasm of magnocellular neurons. Reaction product was also found over the rough endoplasmic reticulum (RER). Post-embedding techniques have employed colloidal gold as an electron dense marker to identify the antigen–antibody complex. The pattern of deposition is not flocculent and does not obscure the underlying ultrastructure. Castel et al. (707) employed dual immunolabeling to label oxytocin and vasopressin with different sized gold particles. They localized the hormone within large dense core vesicles and were able to demonstrate differences between the morphology of these vesicles in oxytocinergic and vasopressinergic neurons. The dense core vesicles in

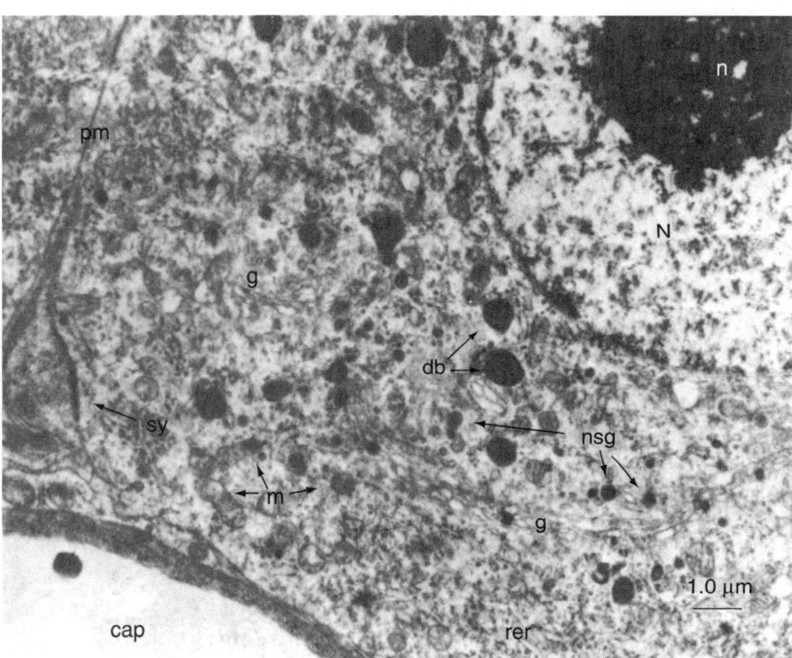

FIG. 18. A low-power electron micrograph of a typical large PVN neuron. The nucleus (N) is large, with a very prominent nucleolus (n). Around the nucleus there are extensive Golgi complexes (g). Rough-surfaced endoplasmic reticulum (rer) is situated mainly at the periphery of the neuron. Many mitochondria (m), neurosecretory granules (nsg), and lysosomal dense bodies (db) are seen in the cytoplasm. The neuron is separated from the lumen of a capillary (cap) by a thin rim of neuropil. On the left the neuron makes a close contact with another neuron, and below this there is a synapse (sy). pm, plasma membrane. [From (703).]

AVP-li magnocellular neurons were larger and denser than their counterparts in OT-li magnocellular neurons. In addition, the vesicles in ir-AVP magnocellular neurons were uniquely rendered pale (ghost like) by oxidation with sodium metaperiodate.

Synthesis of peptide hormones is carried out in the neuronal perikaryia in a fashion analogous to the synthesis of peptide hormones in adenohypophysial cells. In situ hybridization experiments have been carried out at the ultrastructural level to identify the presence of vasopressin and oxytocin mRNAs in the cytoplasm of different magnocellular neurons (708–710). The reverse transcriptase polymerase chain reaction has been applied to single magnocellular neurons by Xi et al. (711). Although AVP_{mRNA} and OT_{mRNA} were found in all neurons tested, quantitative analysis revealed three cell populations: (a) cells with a ratio of $AVP/OT_{mRNA} = 182$; (b) cells with a ratio of $OT/AVP_{mRNA} = 256$; (c) one of 11 cells tested with a ratio of $OT/AVP_{mRNA} = 2$. Thus by far the greatest preponderance of magnocellular neurons can be classified as OT or AVP neurons on the basis of their preponderant mRNA. Jirikowski et al. (470) have additionally localized oxytocin mRNA to dense core vesicles in axons and terminals of lactating female (but not in male) rats using transmission electron microscopy and in situ hybridization. The significance of this finding has yet to be determined.

Pulse labeling experiments showed that labeled amino acids are incorporated first into the rough endoplasmic reticulum, then into the Golgi, and then into dense core vesicles (712). Broadwell et al. (713) employed transmission electron microscopy and immunohistochemistry to localize neurophysin to sites of protein synthesis and storage: e.g., the nuclear envelope, RER, saccules of the Golgi apparatus, and dense core vesicles. Kozlowski et al. (714) and Piekut et al. (706) could localize neurophysin only to dense core vesicles. Unlike the adenohypophysial glandular cell that releases the hormone locally, the magnocellular neurosecretory neuron transports the hormone to a distant terminal for release. Biochemical and immunohistochemical findings demonstrated that newly synthesized hormone in magnocellular neurons is incorporated within a larger prohormone molecule which undergoes posttranslational modification and cleavage as it is carried within the dense core vesicle by axoplasmic flow to an axon terminal in the neural lobe (57,60,62,63,715,716). Events in this process has been witnessed in transgenic mice with confocal microscopic study of magnocellular neurons in which fluorescent green protein was fused to the C-terminus of OT neurophysin (717).

With stimulation, the changes in the ultrastructural morphology of vasopressin and oxytocin-containing neurons are similar. Kalimo (718) reported that in rats deprived of water for 4 to 6 days or suckled for 13 to 24 days, the magnocellular neurons in the paraventricular and supraoptic nuclei increase in size. The nucleus and nucleolus also enlarge. The RER increases as does the number of flattened cisternae. Many cisternae become dilated. The Golgi apparatus becomes hypertrophied. Increased numbers of small vacuoles were seen at both the cis and trans faces of the Golgi and dense core vesicles could be seen budding off from the Golgi in both the cis and trans regions. Broadwell and Oliver (719) made similar observations in mice that were dehydrated for 5 to 8 days. Whereas in control rats, reaction product localizing ir-AVP was only present over dense core vesicles in magnocellular neurons, additional reaction product was found over the RER in the marginal zone as well as over the Golgi apparatus and nearby small dense core vesicles after 9 days of dehydration (720). In addition, reaction product was found in the intercellular space separating neighboring somata. In the internal zone of the median eminence, axons of the supraoptico–hypophysial tract are dilated and contained amorphous collections of reaction product that were not associated with dense core vesicles which were absent from the axons in the median eminence or the terminals in the neural lobe. Krisch (721) proposed that vasopressin could be released from the cell soma into the neuropil where it could enter the ventricular fluid by passing between ependymal cells and then be transported to the floor of the infundibular recess to be carried by the ependyma of the median eminence to blood vessels. This proposal has received support from the observation exocytosis of immunoreactive OT and AVP neurophysins from neurosecretory granules in the perikaryia and dendrites of magnocellular neurons (722). The finding of autoreceptors for OT and AVP on the plasmalemma of magnocellular neurons suggests that the role of local release is a regulatory one and not a mechanism for ependymosecretion (723–725).

Krisch (721) also proposed that AVP could be transported down the axon through channels of the endoplasmic reticulum, incorporated into small vesicles which budded off these channels in axon terminals, and released. However, Kozlowski (726) cautioned that reaction product not associated with dense core vesicles may be artifactual. A specialized system of channels in the smooth endoplasmic reticulum was identified in the soma, axons and terminals of magnocellular neurons by Alonso and Assenmacher (727). They proposed that the electron-dense material found in the vacuoles budding off from its channels in the axon terminals was neurosecretory material (727). Broadwell and Brightman (728) also identified a system of channels in the magnocellular neurons of mice that was associated with a smooth endoplasmic reticulum. It was involved in anterograde transport of lysosomes from the cell body to the axon terminals as demonstrated by its movement of horseradish peroxidase (HRP) from the soma to the terminal and its association with acid phosphatase positive staining small vesicles. These vesicles contained a dense core that could not be distinguished from neurosecretory material. As the dense material in the vacuoles identified by Alonso and Assenmacher (727) was not submitted to immunostaining to confirm the presence of hormones and as the vesicles were not histologically examined for the presence of acid phosphatase to rule out lysosome formation, the case for transport of hormone from the soma to the terminal via the endoplasmic reticulum is not strong.

The percentage of magnocellular neurons in the rat's paraventricular nucleus that demonstrated dilatation of the cisternae of the RER rose from 12% to 24% after 12 hours of dehydration and remained elevated at 21% after 24 hours of dehydration (729). The number of cytoplasmic small dense core vesicles (<160 nm) per cell was decreased 4 and 12 hours after the start of dehydration but returned to normal levels at 24 hours. The number of large dense core vesicles (>160 nm) was decreased at four hours but then returned to normal values (730) suggesting that after an initial decrease the addition of new dense core vesicles to the cytoplasm is balanced by their removal into the axon and transport to and release from the terminal. Electrophysiological studies revealed that all putative vasopressinergic cells did not increase their (electrical) activity equally or in synchrony throughout a period of dehydration as they do in lactation. Increasing numbers of cells were progressively recruited over the duration of the stimulus (731). In situ hybridization experiments also showed recruitment of cells with increased AVP_{mRNA} content (732). Although radioimmunue assay studies have shown that the hypothalamic content of AVP decreases by 60% and the pituitary content by 88% in the mouse after 3 days of dehydration (733), these studies also demonstrated larger numbers of cells which showed the morphological characteristics of increased protein synthesis. It was concluded that new neurons were added to the population of actively secreting cells thus maintaining increased levels of secretion and release of AVP in the face of increased demand even though the hypothalamic and pituitary content of AVP was decreased.

Neurotransmitters and additional hormones are colocalized in AVP-li magnocellular neurons. Tyrosine hydroxylase has been identified in AVP-li cells (734–736) and in magnocellular neurons containing AVP_{mRNA} (737). Immunoreactive angiotensin II has also been demonstrated in magnocellular neurons (613,632) and was colocalized with AVP (632). Dynorphin has been colocalized with AVP in cell bodies (738) as well as terminals (739) of magnocellular neurons. Bilateral destruction of the paraventricular nucleus reduced the levels of immunoreactive dynorphin as well as immunoreactive OT and AVP in the neurointermediate lobe of the rat (740). Meister et al. (734) have demonstrated immunoreactive PHI and TRH in magnocellular vasopressinergic neurons. A small number of AVP-li cells contain immunoreactive VIP (741). Galanin is also co-stored with AVP in magnocellular neurons. Studies employing light microscopy, confocal microscopy, and transmission electron microscopy with immunohistochemistry and in situ hybridization have demonstrated compartmentalization GAL_{mRNA} and AVP_{mRNA} and differential routing of these peptides to preferred destinations (742,743). Whether this mechanism will serve as a paradigm for the processing of co-localized hormones remains to be seen.

Magnocellular oxytocinergic neurons also costore other peptides. Enkephalin has also been identified by immunohistochemistry in magnocellular neurons. Vanderhaeghen et al. (477) reported colocalization of Met-Enk and Pro-Enk in the same cells as OT in the magnocellular nuclei of the bovine hypothalamus. Gastrin-like, cholecystokinin-like peptide may also be colocalized with oxytocin (484,485). Substance P is present in some magnocellular neurons (488). CRH-li staining has also been demonstrated in OT-li magnocellular neurons (734,744,745). In addition, nitric oxide synthase has been demonstrated in magnocellular neurons in the SON and PVN of the hypothalamus and in magnocellular terminals in the neural lobe (746–749). Nitric oxide synthase was found in cells bodies that are distributed in the SON and PVN in the same fashion as OT-li magnocellular neurons (750). Staining intensity for nitric oxide synthase increased following salt loading suggesting that the synthesis of nitric oxide is related to activity (750) and raising the possibility that in OT-li magnocellular neurons, nitric oxide serves a transmitter function.

Parvicellular neurosecretory neurons synthesize and secrete hypophysiotropic releasing or inhibiting hormones. They project through the tubero–infundibular tract or the pre-optico–infundibular tract to the median eminence where they release their peptide hormones into a restricted circulation. They are found in the periventricular region in the arcuate, periventricular and paraventricular hypothalamic nuclei and in septal-preoptic area in mammals (100). Gross et al. (751) described neurons of the arcuate nucleus as having a lobulated nucleus with an evenly dispersed chromatin. The RER is well developed and free ribosomes and polysomes are abundant. The Golgi apparatus is frequently associated with a few dense core vesicles with a diameter of 80 to 120 nm. In a more detailed study, Brawer (752) characterized the neurons of the arcuate nucleus as being ellipsoid-shaped cells with an average diameter of 15 μm. Each nucleus contains a prominent nucleolus and its profile varies from smooth and round to highly folded. The Nissl substance is prominent in the cytoplasm. The rough endoplasmic reticulum is not confined to the Nissl substance where the cisternae were stacked in parallel rows but was found in other regions as a loosely arranged reticulum. Granular (neurosecretory) vesicles ranging from 100 to 150 nm are few in number. The Golgi apparatus is not distinctive. Mitochondria, lysosomes, multivesicular bodies, and microtubules are also present in the cytoplasm. Golgi impregnation demonstrated unipolar or bipolar neurons with fusiform-shaped perikaryia. The dendrites are varicose but not well endowed with regular spines. Except for two features, neurons in the arcuate nucleus conformed to textbook descriptions of cerebral neurons: (a) the nucleolus is frequently associated with the nuclear envelope at a location where the heterochromatin had formed into a large dense tuft, (b) a round "clump of densely intermeshed filaments" containing a central clear spot was frequently found in the cytoplasm. Bugnon et al. (753) studied the ultrastructure of Opioid-li cells in the arcuate nucleus. The peptide-secreting cells resembled the putative GnRH neurons described by Brawer et al. (752) as they contained a large nucleolus, tufts of heterochromatin along the wall of the nuclear envelop, and whorls of filaments in the cytoplasm as well as granular vesicles associated with the Golgi apparatus. Pre-embedding colloidal gold immunostaining of similar neurons in the arcuate nucleus with antibodies against β-END localized the reaction product to the granular vesicles (754). Transmission electron microscopic studies demonstrate that the morphology of neurosecretory neurons in the arcuate nucleus is modified by the level of circulating opiates (opioids) and/or gonadal steroids (755–757).

The CRH-li neuron in the rat's hypothalamic paraventricular nucleus is bipolar in shape with a fusiform cell body (758). Both dendrites and somata had spines. The long axis of the cell body was 20 to 28 μ; the short axis was 10 to 12 μ. The nucleus was infolded with a prominent nucleolus. Neurosecretory granules measured between 80 and 120 nm in diameter. The Golgi apparatus, rough endoplasmic reticulum, and polysomes were prominent. Immunoprecipitate identifying ir-CRH–stained free ribosomes in dendrites (which are an extension of the somatic cytoplasm) as well as neurosecretory granules in the cell body and axons. Following adrenalectomy CRH-li neurons showed morphologic commensurate with increased protein synthesis. The width of the cisternae of the endoplasmic reticulum was increased and progressed to the formation of vacuoles after several weeks. Neurosecretory granules were observed budding off the trans face of the Golgi apparatus (759). A similar description applies to the GRH neuron (760,761).

In the preoptic area GnRH-li cells are fusiform in shape with one or two major dendrites and a single axon (762). The surface of the perikaryia and dendrites may be smooth or spinney (763). Transmission electron microscopy has been employed to study ir-GnRH containing cells in the medial septum, nucleus of the diagonal band of Broca and medial preoptic area (764). The ultrastructure of their perikaryia does not differ markedly from that of POMC-li or putative GnRH-li–containing cells in the arcuate nucleus.

Development of Neurosecretory Systems

The basic neuroanatomic organization of the hypothalamus has been presented by Crosby and Showers (765), Nauta and Haymaker (766), and Knigge and Silverman (464). The development of the hypothalamus has been presented by Papez (767) and by Christ (768). A detailed description of diencephalic development in the rat is provided by Coggeshall (769). This discussion will focus on the development of neurosecretory systems in the hypothalamus.

Hyyppa (770) reported that on postcoital day 15, the rat diencephalon is composed of a germinal layer (lining the third ventricle), a mantle layer, and a marginal layer. On the basis of light microscopic examination of histologic sections, he proposed that nuclear groups, which are now known to contain neurosecretory neurons, such as the supraoptic nucleus, paraventricular nucleus, periventricular nucleus, and arcuate nucleus, originate from the germinal layer. A more detailed study was reported by Altman and Bayer (771). Employing pulse labeling of the developing rat's hypothalamus with tritiated thymidine, the authors were able to define the time at which cells in a given nuclear group ceased dividing (i.e., the birthday of these cells). They reasoned that all cells (and hence all nuclear groups) originated in the ventricular layer. By labeling separate animals with thymidine pulses on successive postcoital days, they could trace the migration of cells from the germinal layer to their final nuclear location. They determined that (in general) there was a lateral to medial pattern of migration. Cells destined for lateral nuclear sites migrated before cells destined for medial sites.

The cells contributing to the magnocellular components of the supraoptic and paraventricular nuclei as well as the cells contributing to the preoptic nuclei migrate concurrently but not from the same site. Magnocellular elements in the preoptic area originated from the germinal epithelium of the inferior horns of the lateral ventricles. These cells are not progenitor of neurosecretory cells. Those destined to become the magnocellular neurosecretory cells of the supraoptic nucleus and the magnocellular portion of the paraventricular nucleus were documented to arise from a single specialized locus of germinal epithelium located adjacent to the third ventricle at the level of the "adult" paraventricular nucleus. Internuclear magnocellular neurons (between the paraventricular and supraoptic nuclei) were found along the course taken by migrating cells from this germinal region to the site of the adult supraoptic nucleus over the optic tract. The neurons forming the supraoptic nucleus were "born" on the 13th through the 15th fetal days. They were recognized by light microscopy over the optic tract on postcoital day 16 and found in moderate numbers on day 17. Paraventricular large neurons were "born" on the same days and first seen on fetal day 16 in their adult locus. More cells appeared on E17 and by E18 growth into magnocellular neurons was discernible (772). Neurophysin was first localized in developing neurons of the paraventricular and supraoptic nuclei on the 18th fetal day. Vasopressin was identified in these nuclear groups on the E19 but oxytocin was not detectable until the 4th postnatal day. Neurophysin was also detected in the median eminence and the neural lobe on the 19th fetal day (773). It is of interest that Fink and Smith (443) first identified neural processes in the developing median eminence on the E16 at a time when magnocellular neurons first arrived in the region of the supraoptic and paraventricular nuclei.

More recently Markakis and Swanson (774) have proposed that the development and migration of neurosecretory cells are an exception to the model of the "outside-in" pattern described by Altman and Bayer (772). They studied the development and migration of hypothalamic neurosecretory cells in rats by labeling the developing cells with bromodeoxyuridine, identifying their phenotype with immunohistochemistry and documenting their neurosecretory state by their uptake of intravenously administered fast blue. They found that both magno- and parvicellular neurosecretory neurons were generated on E12-14. The earliest generated parvicellular neurosecretory neurons were located centrally along a rostro–caudal axis and are mostly GRH neurons in the arcuate nucleus. Cells generated on E12 fill the rostrodorsal region while cells generated on E14 fill the caudal region of the mantle layer beneath the third ventricular ependyma (100,774).

In situ hybridization studies POU III homeobox expression in the brain of the developing rat have characterized the birthdates and patterns of Brn1, Brn2, Brn4, and Tst1 expression (775). By examining patterns of expression of these regulatory transcription factors the authors were able to identify the presumptive supraoptic and paraventricular nuclei on day E12. They were also able to identify a unique region of the germinal third ventricular layer that gives rise to presumptive paraventricular nucleus and hence to CRH and TRH containing neurons and to a unique region of the ventricular lining that gives rise to the arcuate nucleus and hence to GRH. In addition, they noted that the early expression of Tst1 identified the future region of tanycytic ependyma in the ventral third ventricle. Brn2 appears to be necessary for the differentiation of the paraventricular and supraoptic nuclei (450,776) as does the Opt homeobox gene (777). Gonadotropin-stimulating neurons, in contrast, have an extracerebral origin (778). GnRH was first detected by immunohistochemistry in the rat on E15 in the nervus terminalis. On E17, the day when immunopositive cells were first present in the rhinencephalon, limbic lobe and preoptic area, gonadotropins were first detected in the pars distalis. About 60% of the total GnRH-containing cell population was localized in the ganglion cells of the nervous terminalis on that day. By E19 immunoreactive nerve terminals had appeared in the organum vasculosum of the lamina terminalis (OVLT) and the median eminence. The immunopositive cells in the nervus terminalis had decreased to about 30% of the total and the number of cells in the septum, olfactory tubercle, and preoptic area had doubled when compared with E17. The findings suggest that GnRH-containing cells originate outside the CNS and migrate inward. Subsequent studies have confirmed these observations that GnRH-li cells originate in the olfactory region (nervus terminalis or olfactory pit) outside the central nervous system (779–782).

Birth does not mark the end of development of hypothalamic nuclei of the rat (783,784) or hamster (461). Maturation of structure and a synaptic organization continues into the second postnatal week.

Organization of Neurosecretory Systems

Cell bodies of neurosecretory neurons are arranged in nuclear groups in the periventricular region of the preoptic area and the hypothalamus. These neurons form neurosecretory systems not only by virtue of the location of their cell bodies and the site of their terminals, but also by virtue of the hormone(s) they synthesize. This apparently simple organization is complicated by the following: (a) there is not a strict correspondence between a given nuclear group and specific cell type containing a single hormone; (b) a single nuclear group is seldom the unique location of cells synthesizing a given hormone; (c) a hypothalamic nuclear group may project to several different sites and any given peptide may play one role at one site and a different role at another; (d) a particular cell in a nuclear group may contain more than one peptide; (e) even Dale's hypothesis (785) that the same substance(s) secreted from one terminal of a neuron are released from all terminals of that neuron is unsure.

Supraoptic Nucleus

The SON "surmounts the lateral border of the optic tract" in mammals (765). It extends from the lamina terminalis at the level of the OVLT to midtuberal levels. In the mouse, the SON is divided into an anterior and retrochiasmatic portion. The anterior nucleus lies dorsolateral to the optic tract whereas the retrochiasmatic portion lies ventromedial to it (786). A similar parcellation was made in the rat by Peterson (787) and Rhodes et al. (788) save that the latter called the anterior nucleus "the principal nucleus" and the former considered the retrochiasmatic nucleus to be one of the accessory supraoptic nuclei. Three subdivisions of the SON were reported in the monkey. The dorsolateral, dorsomedial, and ventromedial subdivisions are all described by their relationship to the optic tract (61). Oxytocin and vasopressin with their associated neurophysins and colocalized hormones are synthesized in different cells in the supraoptic nuclei (58,61,490,697–699). Within the principal portion of the rat's supraoptic nucleus, oxytocin producing cells are aggregated in the dorsal and rostral portions of the nucleus (788,789). In the monkey, OT producing cells are aggregated in the dorsal and medial regions of the SON (61).

The SON is composed of magnocellular neurons (690,790,791) whose organization as well as morphology changes with increased functional demand. The cell bodies are closely related to the soma of other magnocellular neurosecretory cells. Adjacent cells are separated by glial processes, which in many instances are very attenuated (50,704). Actual soma-somatic contact was found in 4% of cells examined in the rat's supraoptic nucleus under normal conditions; and, the percentage of cells which directly contacted their neighbors increased with dehydration (704). Under conditions of chronic dehydration, 2% sodium chloride substituted for drinking water for 2 weeks, both AVP-li and OT-li neurons enlarge. The percentage somatic or dendritic membrane apposed to the soma or dendrites of other neurons also increases but the percentage of neuronal membrane apposed to glial processes does not (792). The number of synaptic profiles involving a single terminal that contacted two or more apposed (homotropic) neurons (number of multiple synapses) also increases. During lactation, the organization of OT-li magnocellular neurons, but not AVP-li neurons, is altered. The numbers of somato-somatic and dendritic contacts between OT-li magnocellular neurons in the rat's SON increases following lactation for 10 to 18 days as does the number of multiple or "shared" synapses (793). About 70% of the terminals in these double synapses are GABAergic (794). A significant increase in the synaptic density occurs upon enlarged oxytocinergic neurons (795). Both the number of apposed cells and the proportion of shared membrane on each OT-li neuron increases as a result of the retraction of astrocytic processes (796). Almost all the OT-li cells are juxtaposed following lactation or parturition (796). It is suggested that these changes facilitate the synchronous discharge of OT-li neurons during milk ejection and parturition (793). The finding that dye-coupling occurs between apposed cells and is indicative of electrical synapse formation between them supports this hypothesis (797). The vascularity of the supraoptic nucleus is dense (682,683) and occasionally neurosecretory cells were found to lie in direct contact with the basement membrane about nonfenestrated capillaries (720). The distance between magnocellular cell bodies and capillaries is diminished during lactation (798). This arrangement is a fitting one for a nuclear group whose cells must transport amino acids across the blood–brain barrier and incorporate them in order to synthesize proteins for delivery to the neural lobe for secretion into fenestrated neurohypophysial capillaries.

Synaptic relationships in the SON are now being investigated. AVP-li recurrent collaterals have been observed to originate from magnocellular neurons and to terminate on unidentified soma and dendrites in the SON (799,800). OT-li terminals have been demonstrated synapsing on OT-li soma or dendrites in the SON (801). Autoreceptors for AVP and OT have been demonstrated on magnocellular neurons (723, 725). GABA-li terminals make symmetrical synapses with the magnocellular neurons (794,802) and glutamate-li terminals make asymmetrical synapses with them (803). The source of these fibers has not been definitely established but presumably arises from within the hypothalamus. CRH-li terminals synapse with magnocellular neurons (804) and both vasopressinergic and oxytocinergic cells in the supraoptic nucleus express the mRNA for CRH receptors (805). Opioid containing (ACTH-li) terminals were reported to synapse with AVP-li soma and dendrites (806) in the SON of young monkeys. The former presumably originate in the parvicellular subdivisions of the PVN and the latter in the arcuate nucleus (Table 1). In rats β-endorphin-li neurons from the arcuate nucleus project to the supraoptic nucleus (807). The supraoptic nucleus also receives a dopaminergic input from the periventricular hypothalamic nucleus (808) and a noradrenergic innervation from the brain stem (809, 810). In the rat a direct synaptic input is also received median preoptic nucleus (811).

The supraoptico–hypophysial tract (SOHT) is the major projection of the SON. Up to 90% of its neurons degenerate after (high) section of the SOHT (402,812). Injection of horseradish-peroxidase (HRP) into the neural lobe of the rat resulted in a dense accumulation of HRP in cell bodies lying in the SON (403,813). Examination of autoradiographs of the rats hypothalamus and hypophysis after injection

TABLE 1. *Synapses on hypothalamic hypophysiotropic neurosecretory neurons*

	AVP	OT	GRH	POMC	DA	SRIF	TRH	CRH	GnRH
AVP	+								
OT		+							
GRH			+						
POMC	+			+	+		+@		+
SRIF			+			+		+	+
TRH			+						
CRH	+	+				+		+	+
GnRH									+
DA				+	+		+	+	+@
NE	+	+						+	
EPI	+				+	+	+		
5-HT			+	+	+	+	+	+	+
GLUT	+	+							+
GABA	+	+			+	+		+	+
NPY	+					+	+	+@	+
SP									+
NT					+				
GAL					+				
ENK					+				

The abbreviations are the same as those used in the text. The top row lists the neurosecretory neurons upon which synapses occur. The first column lists the transmitter or peptide that has been identified at the electron microscopic level in the synapsing terminal. @ indicates a confocal microscopic study in which "close appositions" have been observed. See text for references.

of tritiated leucine into the SON revealed the principal projection to be through the supraoptico–hypophysial tract to the neural lobe where it tended to be concentrated in the internal region (727). Nerve fibers in the supraoptico–hypophysial tract displayed irregular swellings and many had a beaded appearance in Golgi preparations. Collateral axons were found. Fibrous astrocytes and their processes were frequently found and gave the tract a "striking appearance" (692).

Projections to neural as well as to neurohypophysial sites have been identified by several authors employing immunohistochemistry. Labeled axons of magnocellular neurons course dorsally toward the stria medullaris (814). Zimmerman (696) described a projection from the rostral region of the SON to the OVLT, a circumventricular organ implicated in the regulation of salt and water homeostasis and of drinking behavior (815). Although it has been estimated that as many as 30% of the magnocellular neurons in the (hamster) SON do not terminate in the neural lobe (816), the major accepted pathway emerging from the SON is the supraoptico–hypophysial tract to the neural lobe of the pituitary gland.

Other magnocellular neurosecretory centers are present in the hypothalamus. Crosby and Showers (765) named these scattered small cell groups of neurosecretory cells that lie, for the most part, between the paraventricular and the supraoptic nuclei. These nuclear groups appear to, with two exceptions, lie very close to the SON or PVN or lie along the course taken by cells migrating from the germinal matrix in the periventricular region to their final location over the optic tract. The anterior commissural nucleus lies immediately posterior to the anterior commissure near the midline just rostral to the PVN. The nucleus circularis lies halfway between the SON and the PVN. The retrochiasmatic nucleus lies just ventral to the SON. The anterior and posterior fornical nuclei and the nucleus of the medial forebrain bundle are disposed more laterally than the presumed path of migration of the magnocellular neurons from the periventricular germinal epithelium to their site over the optic chiasm and tract (787). The retrochiasmatic nucleus was classified by Broadwell and Bleier (786) and by Rhodes et al. (788) as part of the SON. The anterior commissural nucleus was included by Swanson and Kuypers (817) as part of the paraventricular nucleus. Oxytocin or vasopressin has been localized in all of the accessory nuclear groups and their projection to the neural lobe has been demonstrated (403,691,693,788,813,818).

Paraventricular Nucleus

The paraventricular nucleus (PVN) contains both magnocellular neurons parvicellular neurons (791, 817,819). The territory of the PVN is described differently by different authors. Crosby and Showers (765) described a triangular-shaped nuclear group lying in the dorsal hypothalamic zone on either side of the third ventricle with its base parallel and adjacent to the third ventricle. Its anterior border lay in the plane of the posterior border of the anterior commissure and the posterior border of the optic chiasm. Its posterior extent was described as differing with different species. Swanson and Kuypers (817) described a nuclear group in the dorsal zone of the hypothalamus adjacent to the third ventricle that on coronal sections is polygonal in shape anteriorly and which becomes triangular posteriorly. Their periventricular cluster of parvicellular neurons is the periventricular nucleus of other authors (29). Their anterior and medial magnocellular divisions are in the anterior commissural nucleus of Peterson (Fig. 19) (787). Defendini and Zimmerman (791) described the PVN as fusiform in shape and oriented almost at right angles to the SON. It lies next to the wall of the third ventricle and extends from the inferior boundary of the nucleus reuniens of the thalamus to the superior border of the arcuate nucleus of the hypothalamus. Although it is not clear from their description how this ventral extension relates to the periventricular nucleus, the reader must assume that its lies next to the third ventricle surrounded anteriorly and posteriorly by the periventricular nucleus [compare Fig. 4 in Defendini and Zimmerman (791) with Figs. 7, 9, 10, and 17 in Rioch et al. (29)]. (This periventricular territory seems to be the source of considerable dispute as it is claimed by various authors as part of the periventricular, paraventricular, and arcuate nuclei.) Midway along the dorsoventral course of this nucleus

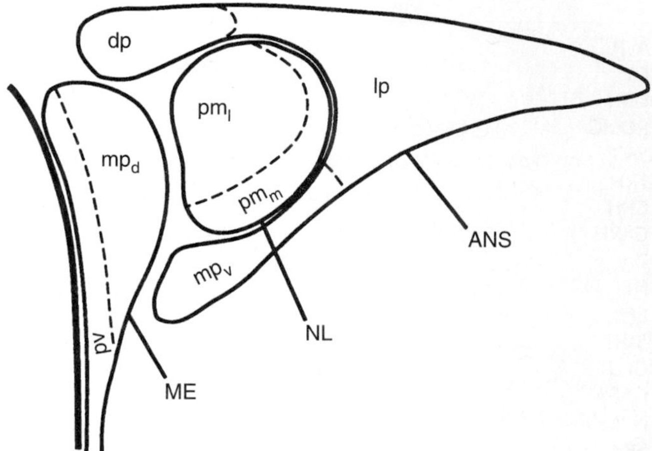

FIG. 19. Coronal section through the major portion of the paraventricular nucleus in the rat with identification of subdivisions. pm_1, posterior magnocellular (lateral) subdivision; pm_m, posterior magnocellular (medial) subdivision; dp, dorsal parvicellular; lp, lateral parvicellular; mp_d, medial parvicellular (dorsal); mp, medial parvicellular (ventral) subdivisions; pv, periventricular (parvicellular) subdivision. [From (836).]

is a swelling directed laterally and posteriorly [the PVN of Crosby and Showers (765)] extending from the level of the optic chiasm and the anterior commissure to the anterior (787) or mid-tuberal (29) regions.

The manner in which the PVN is parcellated by different investigators varies. Swanson and Kuypers (817) described three magnocellular clusters in the PVN (Fig. 19). Their anterior magnocellular and medial magnocellular clusters along with the anterior parvicellular cluster correspond to the anterior commissural nucleus of Peterson (787). Their posterior magnocellular division is partially surrounded by the periventricular, medial, lateral and dorsal parvicellular divisions of the PVN and this entire group corresponds to the PVN of other authors (786,788). Van den Pol (820) found that magnocellular neurons tended to be aggregated in the lateral region of the "PVN" on coronal section. His Figure 3a corresponds to Figure 1b in Swanson and Kuypers (817) and suggests that he, as others, considered the posterior magnocellular division of Swanson and Kuypers (817) to represent the magnocellular region of the PVN. Armstrong et al. (821) arranged the magnocellular clusters of neurons into the anterior commissural nucleus, the medial and lateral paraventricular nuclei and the posterior subnucleus of the PVN. Their medial and lateral paraventricular nuclei seem to correspond to the posterior magnocellular division of Swanson and Kuypers (817). The anterior commissural nucleus of Armstrong et al. (821) is the anterior magnocellular division of Swanson and Kuypers (817). In the following discussion the terminology of Swanson and Kuypers (817) will be employed in the descriptions of the paraventricular nucleus of the rat.

The magnocellular subnuclei in the PVN in the rat have been designated by Swanson and Kuypers (817) as the anterior, medial, and posterior magnocellular groups (Fig. 19). The anterior and medial subnuclei lie rostral to the main body of the PVN as described by van den Pol (820). The posterior magnocellular subnucleus lies in the main body of the PVN as described by van den Pol (820) and is surrounded by parvicellular clusters. As in the SON, OT, and AVP with their associated neurophysins have been found in separate neurons in the PVN (58,61,490, 697–699). The rat the anterior magnocellular nucleus contains primarily OT-li neurons, as does the medial magnocellular subnucleus. Hatton et al. (822) described two populations of magnocellular neurons in the posterior magnocellular subnucleus of the rat's PVN which could be distinguished on the basis of their size. Smaller cells were found in the anteromedial region of the PVN whereas in the dorsal lateral region the cells were larger and rounder. Their prediction that cells in the anterior-ventromedial region would primarily contain OT whereas those in the dorsolateral region would contain primarily AVP was subsequently confirmed with immunohistochemical techniques (788,823). Defendini and Zimmerman (791) on the basis of their own experience and review of the extant literature did not confirm the presence of a regional segregation of OT-li and AVP-li neurons in the PVN in humans, although segregation of these neurons was acknowledged in other animals.

As in the SON, in the PVN magnocellular neurons are closely packed, being separated from each other only by slender glial processes. As in the SON, under control conditions, cells in the PVN are occasionally seen to lie adjacent to each other without an intervening glial process (718,729). During lactation, the number of juxtaposed OT-li neurons increases as does the number of shared synapses (824). The changes in morphology of the PVN's magnocellular neurons and their associated astrocytes in response to dehydration or lactation mirror those seen in the supraoptic nucleus (797,825,826).

Terminals containing norepinephrine synapse with AVP-li magnocellular neurons in the PVN of the rat. Silverman et al. (827) labeled terminals with tritiated norepinephrine and AVP containing cells with an antibody to neurophysin II. Ochiai and Nakai (828) employed dual immunohistochemistry with antibodies against DBH and neurophysin II. The DBH was identified by pre-embedding techniques with the peroxidase-antiperoxidase technique and the neurophysin II was identified by immunogold post-embedding techniques. Both groups demonstrated synapses between noradrenergic terminals and AVP-containing neurites. Whereas Silverman et al. (827) were able to demonstrate such synapses only upon dendrites (lying in the medial parvicellular subdivision) of the magnocellular neurons, Ochiai and Nakai (828) reported noradrenergic synapses upon both the cell bodies in the posterior magnocellular subnucleus in the PVN and upon their dendrites. This same group has also demonstrated NPY-li terminals synapsing with magnocellular AVP-containing neurons (829). Triple-labeled experiments with the identification of monaminergic terminals by their uptake of the false neurotransmitter 5-hydroxydopamine, identification of NPY immunoreactivity by pre-embedding immunohistochemistry with the peroxidase-antiperoxidase technique, and the identification of AVP-containing elements by post-embedding immunohistochemistry has revealed that NPY is colocalized with a monoamine (presumably norepinephrine) in terminals that synapse upon AVP containing neurons (830). PNMT-li terminals have also been found to make asymmetric synapses with magnocellular neurons (831). Opioid (ACTH-li) terminals have been reported to make both symmetrical and asymmetrical synapses upon AVP-li magnocellular neurons and neurites in the PVN (806) as have NT-li terminals (832). GABA-li terminals have been demonstrated by dual labeling immunohistochemical techniques to synapse on magnocellular neurons in the PVN (833) as have glutamate-li terminals (803). As in the SON, most of the terminals involved in shared synapses with OT-li neurons during lactation were GABAergic (794).

Magnocellular neurons of the (posterior magnocellular subdivision of the) PVN project to the neurohypophysis and to the spinal cord. Axons destined for the neurohypophysis leave the PVN at its apex (820) and then course laterally and then ventrally to the level of the SON (691). There the fiber tract is joined by projections from the SON to form the hypothalamo–hypophysial tract. This fasciculus of magnocellular neurons approaches the median eminence through the lateral retrochiasmatic area and passes from lateral to medial as it moves caudally to enter the median eminence from its anterior aspect. In the median eminence its fibers lie in the internal zone. They pass on to the infundibular stem and process where they terminate (in the rat) near the ventral surface adjacent to the pars tuberalis and pars intermedia (727,821).

Although stalk section, even at high levels, resulted in the degeneration of only about 20% of large neurons in the PVN (402), Sherlock et al. (403) found that cells in "the magnocellular (i.e., lateral) part of the paraventricular nuclei were loaded with HRP reaction product" following HRP injection into the neural lobe of the rat. Subsequently injection of HRP into the rat's neural lobe for retrograde study of magnocellular neurons that project to the there, coupled with immunohistochemical staining for AVP or OT, demonstrated that the majority of immunohistochemically labeled neurons projected to the neural lobe. However some AVP-li and OT-li neurons in the caudal regions of the PVN were not retrogradely labeled (834). Nylen et al. (749) recently demonstrated a stout projection from the paraventricular nucleus to the spinal cord. Magnocellular OT-li neurons from the posterior magnocellular subdivisions contributed to this projection. Over 50% of the projections from the posterior magnocellular subdivision co-stored OT and NOS.

Parvicellular neurons in the PVN of the rat have been divided into five nuclear subdivisions (817). The anterior parvicellular subdivision lies at about the level of the anterior commissure and the descending limb of the fornix between the anterior and posterior magnocellular subdivisions. The medial parvicellular subdivision lies behind the anterior parvicellular subdivision and medial to the posterior magnocellular subdivision. In turn the periventricular subdivision lies medial to the medial parvicellular subdivision adjacent to the third ventricle. The lateral parvicellular subnucleus lies posterolateral to the posterior magnocellular subnucleus and the dorsal parvicellular subnucleus lies dorsal to the posterior magnocellular subnucleus. In coronal section at anterior tuberal levels where the PVN is triangular in shape, the posterior magnocellular subnucleus of the PVN is surrounded by the dorsal, lateral, and medial parvicellular subdivisions with the periventricular parvicellular subdivsion lying between the medial parvicellular subdivision and the third ventricle (Fig. 19).

Kiss et al. (835) have subdivided the PVN at this level on the basis of the average size of the predominant neuronal type. Magnocellular neurons (13–19 μm in diameter) were aggregated in a single magnocellular subdivision (analogous to the posterior magnocellular subdivision of Swanson and Kuypers (817). Mediocellular neurons (10–13 μm in diameter) predominated in two nuclear groups, a dorsal and a posterior subdivision. Parvicellular neurons (6–10 μm in diameter) predominated in the periventricular and medial subdivisions. These subdivisions assume importance because their cytoarchitectonic groupings have specific afferent and efferent pathways and because they subserve different functions that integrate autonomic and neuroendocrine responses to stress (836).

Neurons in the PVN project to the median eminence. This projection was first demonstrated in the monkey where destruction of the paraventricular nuclei resulted in loss of immunoreactive AVP, OT, and their associated neurophysins from the external zone of the median eminence (837,838). The projections to the external zone from the PVN are ipsilateral and arranged topographically. Following adrenalectomy in the rat the intensity of immunohistochemical staining for NP or AVP increases. This increase was blocked by pretreatment with dexamethasone (839). Following unilateral ablation of the PVN, bilateral adrenalectomy induces collateral sprouting of fibers of the hypothalamo–hypophysial tract from the intact side of the median eminence to the denervated side (840). Coupled with the evidence that AVP and CRH act synergistically to stimulate corticotropes (841), these anatomical observations leave little doubt that the vasopressinergic projections to the median eminence from the PVN play a role in the regulation of the pituitary–adrenal axis and in the response mounted by the animal to stress.

The source of OT and AVP terminals in the median eminence is the PVN. The site of these neuronal cell bodies within the PVN is being further clarified. The parvicellular divisions of the rat's PVN contain 31% of OT-stained cells, and 20% of the AVP-stained cells (823,842). The OT and AVP terminals and their neurosecretory granules are smaller in the external zone of the (guinea pig) median eminence than they are in the neural lobe (56,621) suggesting that terminals in the external zone arise from parvicellular neurons. Retrograde tracer studies employing injection of HRP or WGA injection into the median eminence have revealed heavier labeling in the parvicellular than in the magnocellular divisions (818, 843,844). Labeling of magnocellular neurons could have resulted from migration of tracer into the internal zone from the site of injection in the external zone. Because only 20% of the magnocellular neurons in the PVN project to the neural lobe, it was presumed by some authors, prior to the discovery of a projection to the spinal cord (749), that the other 80% project to the median eminence (791,845). Others presumed that the parvicellular neurons containing AVP or OT in the medial parvicellular nuclear group, along with other parvicellular neurons containing hypothalamic releasing and inhibiting hormones, project to the median eminence (836,846).

The largest aggregation of parvicellular neurons in the rats PVN lies medial to the (posterior) magnocellular division in the medial and periventricular subdivisions (817,820); and, some of these cells are hypophysiotropic neurosecretory neurons that project to the median eminence. Immunohistochemical studies have also demonstrated ir-AVP containing cells to be concentrated in the medial parvicellular division (817, 820). In contrast, immunoreactive parvicellular oxytocin (OT-li) neurons have been demonstrated "scattered throughout the parvicellular division" in the anterior, medial, periventricular, dorsal and lateral subdivisions (823). Catecholaminergic terminals have been demonstrated to synapse on their soma and dendrites (847).

The dorsal region of the medial parvicellular nucleus is characterized by the presence of CRH-li neurons in the Fischer rat strain (848). CRH has been found by immunohistochemical techniques in parvicellular neurons of the PVN of the cow (849) and sheep (850) as well as the rat (530,609,851,852). However, in the Long Evans and Sprague-Dawley strains, immunohistochemical staining of the PVN has revealed few CRH-li neurons in the absence of colchicine treatment. In colchicine treated Long-Evans or Sprague-Dawley rats pretreated with intraventricular injections (ivt) of colchicine CRH-li neurons were found to be concentrated in this region (853). Glucocorticoid receptors have been demonstrated in the nuclei of such cells by dual immunohistochemisty (854). Retrograde transport of materials from the surface of the rat's median eminence to the cell bodies of hypophysiotropic neurons has been combined with immunohistochemical identification of the neurons. Such studies have demonstrated that the majority of CRH-li cell bodies that project to the median eminence lie in the dorsal aspect of the medial parvicellular subnucleus of PVN and its periventricular subnucleus (855,856).

Other peptides and neurotransmitters are colocalized with CRH in parvicellular neurosecretory neurons. Kiss et al. (857) and Sawchenko et al. (858,859) reported that parvicellular ir-CRH containing cells in the PVN became vasopressin-positive after adrenalectomy. In rats treated with colchicine and adrenalectomized, CRH, AVP, and Ang II immunoreactivity were localized in the same parvicellular neurons by sequential immunostaining and elution procedures (613). Following adrenalectomy CRH, AVP, and CCK have been demonstrated in same parvicellular neurons in the medial parvicellular subnucleus of the PVN by serial immunostaining and elutions (860); and, glucocorticoid receptor was expressed in the cytoplasm (as opposed to nuclear expression) of cells immunostained for CRH (854).

Not all CRH neurons costore AVP. Whitnall et al. (625) identified two subsets of CRH-li parvicellular neurons. CRH+/AVP+ neurons were concentrated in the dorsal region of the medial parvicellular subnucleus whereas CRH–/AVP– neurons were localized more ventrally in the medial subnucleus. These authors concluded that CRH containing neurons are a source of AVP-staining parvicellular terminals in the external zone of the rat's median eminence.

Hokfelt et al. (861) described cell bodies in the parvicellular regions of the paraventricular nuclei which contained immunoreactive CRH, enkephalin, and PHI-27 (a peptide that with VIP shares a common precursor protein and which like VIP stimulates prolactin release); and, they envisioned release of these peptides from the same terminals in the median eminence could provide a mechanism for a coordinated response to stress. In their proposal, CRH stimulates ACTH release from corticotropes in the pars distalis, PHI-27 stimulates PRL release from lactotrophs in the pars distalis, enkephalin inhibits dopamine release from neighboring neurons by a paracrine action to further enhance PRL stimulation in lactotrophs and it also inhibits SRIF release from neighboring terminals to promote GH release from somatotropes in the pars distalis. Although subsequent studies have demonstrated that antibodies directed against the N-terminal of PHI cross-react with CRH, repeated analysis with different antibodies has confirmed the presence of a small population of CRH-li neurons in the medial parvicellular subnucleus that co-stores VIP/PHI and met-ENK. Hence the role of a neurosecretory neuron in the regulation of a particular pituitary response depends not only upon the release of a specific hypophysiotropic hormone to act in an endocrine function but also on the release of other neuropeptides (and amines) to act in paracrine and local transmitter fashion.

CRH-li neurons costore other peptides and monoamines. About 28 % costored neurotensin (NT) whereas less than 5% costored galanin in a study by Ceccatelli et al. (616). GABA was identified in a population of CRH-li neurons that did not costore AVP.

These cells demonstrated uptake of Fast Blue applied to the median eminence and hence projected there (614).

Neurotensin-li neurons have been demonstrated in the parvicellular subdivisions (532) of the PVN and about 40% of them co-stored CRH. However, Niimi et al. (862) could not demonstrate a significant projection of paraventricular NT-li neurons to the median eminence. ENK-li parvicellular neurons were prominent in the medial parvicellular subdivision and about 40% of them co-stored CRH (616). Immunoreactive CCK has also been found in the parvicellular neurons of the PVN (485) as has VIP/PHI but less than 10% of these cells costored CRH (616). Parvicellular neurons containing immunoreactive tyrosine hydroxylase have been demonstrated (809), but it is not yet known whether these cells are DOPAergic or dopaminergic. ANG II has also been identified in PVN cells (both large and small) by immunohistochemical techniques. As discussed above ANG II is colocalized with OT in magnocellular neurons and with AVP in parvicellular neurons. A projection to the internal zone and to the external zone of the median eminence has been identified and related to fluid balance and pituitary-adrenal function respectively (632,863).

CRH-li terminals have been found to synapse upon CRH cell bodies and dendrites forming local circuits (848,864). Terminals that take up 5-hydroxydopamine or tritiated norepinephrine were found to synapse with CRH-li neurons in the parvicellular regions of the PVN (865). Liposits et al. (866) employed simultaneous double labeling immunohistochemical technique to label CRH and tyrosine hydroxylase. They demonstrated TH-li terminals making asymmetrical synapses upon CRH neurons (866). Pre-embedding immunostaining for CRH coupled with post-embedding immunostaining for tyrosine hydroxylase yielded the same findings (758). These studies strongly suggest that catecholaminergic terminals make excitatory synapses with CRH neurons but do not differentiate whether these terminals originate from dopaminergic neurons within the hypothalamus or from noradrenergic neurons within the brain stem. This same group demonstrated asymmetrical synapse between serotonergic-li terminal and CRH-li soma and dendrites (867). GABAergic neurons make symmetric (inhibitory) synapses on CRH-li neurons (868). The source is not certain but likely to be from GABAergic interneurons within the PVN (596). NPY-li neurons from the arcuate nucleus project to the PVN where they make close appositions with CRH-li neurons as determined by confocal light microscopy (869). Cells in the medial parvicellular subdivision that contain CRH_{mRNA} also demonstrate angiotensin receptor mRNA (870). In addition to these connections, a reciprocal synaptic relationship has been found between SRIF-li and CRH-li neurons (871).

TRH-containing cells are also present in the medial parvicellular division (491). Further, cells which contain the presumptive precursor of TRH (872) and its mRNA (873) were found in the medial parvicellular division of the PVN in the same distribution as the TRH containing cells. This observation is compatible with the postulate that TRH is synthesized in neurons in this location. Fliers et al. (874) have reported the localization of TRH in the dorsocaudal region of the human paraventricular nucleus. A similar distribution of TRH_{mRNA} has been reported in the human PVN (875). Where as CRH-li neurons have a distribution that is restricted to the PVN, TRH-li neurons have a wide distribution throughout the diencephalon. The large number of TRH-li terminations on neurons emphasized the physiologic importance of TRH as a neuromodulator or neurotransmitter in addition to its role as a neurohormone (874). TRH neurons in the paraventricular nucleus frequently co-express cocaine- and amphetamine-regulated transcript (CART) (876,877).

Dual immunohistochemical studies at the light microscopic level with antibodies directed against either NPY or DBH and against TRH have demonstrated DBH-li terminals (as well as NPY-li terminals) in close proximity to TRH-li neurons (878). Ultrastructural studies have demonstrated TH-li terminals making synaptic contact with TRH-li dendrites and soma. After unilateral lesion placement in the median forebrain bundle, degeneration of TH-li neurons occurs but it is limited to synapses on distal dendrites. TH-li synapses on proximal dendrites and on the somata of TRH-li neurons were not altered. The authors propose that TRH-li neurons in the PVN receive a catecholaminergic innervation from the brain stem and a dopaminergic innervation from the arcuate nucleus. TRH-li neurons in the paraventricular nucleus of the rat are also contacted by NPY-li terminals (879). The pattern of TH synapses on TRH-li neurons resembled that of NPY-li synapses on TRH-li neurons (880). The arcuate nucleus is the major site of NPY-li projections to TRH-li neurons in the paraventricular nucleus (881) but medullary NPY-li neurons that co-localize PNMT also synapse on TRH-li neurons (882). TRH-li neurons are also contacted by serotonergic terminals in the rat paraventricular nucleus (883) and by GAD-ir terminals (884). Alpha-MSH-li neurons come into close apposition with TRH-li neurons in the PVN (Table 1) (885,886).

Axons of hormone-containing cells in the medial and lateral parvicellular divisions leave the PVN laterally, at its apex, to follow the course of magnocellular axons in the hypothalmo–hypophysial tract and to pass by the supraoptic nucleus where fibers from the supraoptic nucleus are added. This fasciculus of axons containing peptidergic projections from the hypothalamus enters the median eminence anterolaterally through the lateral retrochiasmatic zone (Fig. 20) (887). This projection brings together two systems that stimulate CRH release (the vasopressinergic and oxytocinergic systems and the CRH systems) with the system that regulates thyroid function. SRIF and TRH neurons in the periventricular division course ventrally in the periventricular zone to enter the median eminence at its lateral margin. These two pathways to the median eminence along with the pathway from the arcuate nucleus to the median eminence are included in the term "tubero–infundibular tract."

Parvicellular neurons in the paraventricular nucleus also project to the brain stem and spinal cord regions. These neurons are not the same neurons that project to the median eminence (842). Most such neurons contain ir-OT or ir-AVP and were reported to originate from parvicellular neurons in the caudal region of the PVN (821,888). Hallbeck and Blomqvist (889) reported that 40% of spinally projecting neurons, identified by retrograde transport of cholera toxin, contained AVP_{mRNA} ; and, that the majority of these cells arise in the lateral and in the ventral portion of the medial parvicellular subdivisions. Cells expressing mRNA for oxytocin, dynorphin and enkephalin also arise in the parvicellular subdivisions and project to the spinal cord (890). The lateral parvicellular subdivision contains the most cells projecting to the spinal cord but medial and dorsal parvicellular subdivisions also participate in this projection and each has a unique profile of mRNA neuropeptide expression. These finding contrast with those of Nylen et al. (749) who reported a robust magnocellular projection to the spinal cord that was characterized by the presence of oxytocin and NOS. Additional

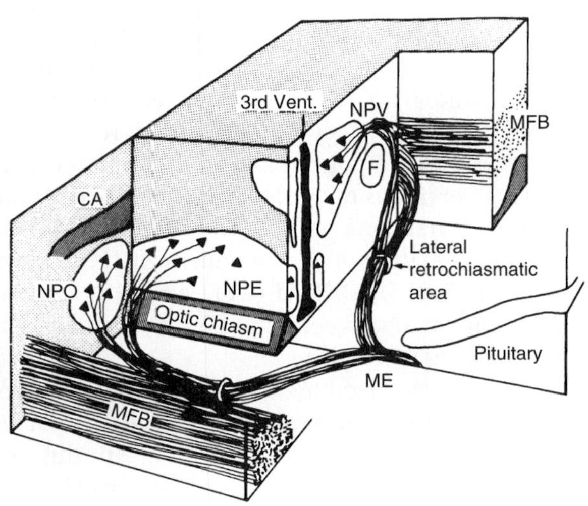

FIG. 20. Schematic drawing of the topography of the lateral retrochiasmatic area and cell groups that project axons through it to the median eminence. NPO, medial preoptic nucleus; NPE, periventricular nucleus; MFB, medial forebrain bundle; ME, median eminence; CA, anterior commissure. [From (887).]

descending projections of OT and AVP containing neurons in the PVN to the substantia nigra, the parabrachial nucleus, the locus ceruleus, the lateral reticular nucleus, and the commissural nucleus have been described (891).

Cells that project to the spinal cord lie in the dorsal parvicellular subnucleus and in the ventral region of the medial parvicellular division whereas cells that project to the dorsal vagal complex lie in the ventral region of the medial parvicellular division (823). Descending brain stem projections have been identified to the dorsal vagal complex (the nucleus of the solitary tract and the dorsal motor nucleus), and the intermediolateral cell column of the spinal cord by immunohistochemical techniques (892). Oxytocin containing fibers predominate (888). Injection of different markers at medullary and spinal levels demonstrated that about 15% of the labeled cells in the PVN were double-labeled and hence that these cells sent collaterals to both the dorsal vagal complex and the spinal cord (823). Collateral projections into the ventral lateral medulla from PVN projections to the spinal cord have also been demonstrated (893).

Oxytocinergic terminals have been found to synapse on norepinephrine-li cell bodies in the A_1 cell group in the ventrolateral medulla (894). Descending spinal projections have been identified in the intermediolateral cell column closely apposed to sympathetic preganglionic neurons and to preganglionic adrenal medullary neurons. The latter projection was shown to be vasopressinergic (895,896). An ir-OT projection from the PVN to a parasympathetic nucleus in the lumbar spinal cord that in turn projects to the corpus cavernosum has been identified employing the pseudorabies virus (897).

Paraventricular parvicellular neurons also project rostrally. OT containing cells project over the fornix to the hippocampus and also project to the septal area (898–900). Additional projections to the medial preoptic area (MPOA), bed nucleus of the stria terminalis and the periventricular nucleus of the thalamus have been revealed by autoradiographic studies following injection of labeled amino acids into the PVN (Table 2) (901).

Axons of parvicellular neurons also project to sites within the paraventricular nucleus. Whereas intrinsic axons were found not to arise from magnocellular neurons, they were found to arise from parvicellular neurons. Local axons terminated both in the medial parvicellular and lateral magnocellular regions of the PVN (820). Both GABAergic and glutaminergic neurons have been identified (596,902). The dendritic arbors of parvicellular neurons remain within the confines of the nucleus. They are oriented horizontally, parallel to the floor of the ventricle and to intrinsic axons in the medial portion of the nucleus. In the periventricular regions, the dendritic arbors are

TABLE 2. *Efferents from MPOA-HTA*

	SON	PVN(m)	PVN(p)	PvN	AN	MPOA
NL	+	+				
ME			+	+	+	+
OVLT						+
SON	+			+	+	
PVN(m)		+			+	
PVN(p)			+	+	+	
PvN				+	+	+
AN				+	+	
MPOA				+	+	+
VMN				+	+	
ScN				+		
Habenula				+		+
Hipp			+	+		
Septum			+	+	+	+
BNST			+		+	+
Amy			+		+	+
Ol. Tub.				+		+
IPN				+		+
SN			+			
Teg				+		+
PbN			+			+
LC			+			+
LRN			+			+
DVC			+			+
SC		+	+			

SON, supraoptic nucleus; PVN(m), magnocellular paraventricular nucleus; PVN(p), parvicellular paraventricular nucleus; PvN, periventricular nucleus; AN, arcuate nucleus; MPOA, medial preoptic area; NL, neural lobe; ME, median eminence; OVLT, organum vasculosum of the lamina terminalis; VMN, ventromedial nucleus; ScN, suprachiasmatic nucleus; Hipp, hippocampus; BNST, bed nucleus of the stria terminalis; Amy, amygdala; Ol. Tub., olfactory tubercle; IPN, interpeduncular nucleus; SN, substantia nigra; Teg, midbrain tegmentum; PbN, parabrachial nucleus; LC, locus ceruleus; LRN, lateral reticular nucleus; DVC, dorsal vagal complex; SC, spinal cord.

oriented vertically parallel to the ascending input from the suprachiasmatic and arcuate nuclei. The dendrites of parvicellular neurons surrounding the (posterior) magnocellular division peripherally are oriented parallel to the surface of the nucleus forming a network of dendrites about it. The parvicellular neurons are thus arrayed to serve as receptor sites for incoming afferent axons, as interneurons relaying afferent input into large and small peptidergic neurons, as neurosecretory cells releasing hormones in the median eminence to control anterior pituitary function, and as neurons projecting to autonomic centers in the brain stem and in the spinal cord to coordinate pituitary and endocrine function (Table 2) (820).

Periventricular Nucleus

The anterior periventricular nucleus is included by Rioch et al. (29) as a part of the periventricular region which includes the arcuate nucleus (AN) and the periventricular preoptic nucleus. Ingram (790) stated that it is characterized by "the presence of small cells arranged in vertical rows as seen in transverse sections stained with Nissl stain." It extends through the rostrocaudal extent of the hypothalamus and encircles the third ventricle. Anteriorly it is continuous with the preoptic periventricular system (29). At the level of the paraventricular nucleus it is sometimes ceded to the paraventricular nucleus (817). In the region between the AN and the PVN several peptides have been localized within cells in the periventricular nucleus. In this discussion the periventricular region will be subdivided (from anteriorly to posteriorly) into a preoptic periventricular nucleus, and anterior hypothalamic periventricular nucleus and the periventricular subnucleus of the paraventricular nucleus. A substantial proportion of somatostatin (SRIF-containing) cells lie in this region (156,903–908). Injection of horseradish peroxidase biotinylated wheat germ agglutinin into the external layer of the rat's median eminence with subsequent immunohistochemical staining for somatostatin demonstrated doubled labeled cells in the anterior periventricular nucleus and in the medial parvicellular subdivision of the paraventricular nucleus (909,910). Merchenthaler et al. (911) employed a similar technique and found SRIF-li cells that projected to the external zone of the median eminence to lie in the preoptic periventricular nucleus, the anterior periventricular nucleus and in the periventricular subnucleus of the paraventricular nucleus.

SRIF-li terminals have been found to synapse with SRIF somata and dendrites in the anterior hypothalamus. Because the somatostatin has been demonstrated in vitro experiments to inhibit its own release, it is presumed that this local circuitry is inhibitory in nature (912). The function of other inputs onto SRIF-li neurons has not been established. Immunoelectronmicroscopic studies of the periventricular nucleus in rats have demonstrated adrenergic (913) and NPY-li (883) synapses with SRIF-li neurons. It has not been determined if the NPY and epinephrine are colocalized in a single terminal. Serotonin-li terminals have also been shown to synapse with SRIF-li neurons (914). In addition GABA-li terminals make symmetrical synapses SRIF dendrites and soma (915,916). CRH-li (871) and GRH-li (917) terminals have also been shown to synapse with SRIF-li neurons (Table 1).

SRIF-li cell have been shown in lesion experiments to project to the median eminence by periventricular routes (918). Although this course is quite different from that taken by CRH-li axons on their way through the lateral retrochiasmatic zone to the median eminence and although the origin of this pathway is not the tuberal (arcuate) nuclei, this pathway is considered to be part of the tubero–infundibular tract.

In addition to projecting to the median eminence, SRIF-li neurons project to the preoptic and arcuate nuclei in the periventricular zone, to VMN, ScN, and premammillary nuclei in the hypothalamus and to the habenular nuclei of the epithalamus via the stria medullaris. They also project to the midbrain tegmentum through the stria medullaris via the fasciculus retroflexus to the interpeduncular nucleus and via the periventricular gray into the midbrain tegmentum (904). Ascending projections ir-SRIF to the olfactory tubercle, septum and hippocampus have also been demonstrated (887,904).

Immunohistochemical techniques have demonstrated immunoreactive tyrosine hydroxylase (919) in cell bodies of the anterior periventricular nucleus. This cluster of dopaminergic cell bodies is the A_{14} cell group (920). It has been shown project to the neurointermediate lobe in the cat and rat (347,921) and to the supraoptic nucleus in the rat (808). Neurotensin (922) and CRH (530) containing cells have been found in this region of the rat's hypothalamus and their projections to the median eminence have been identified.

Arcuate Nucleus

The arcuate nucleus (AN) (nucleus infundibularis) is a paired lateral expansion of the periventricular gray that lies juxtaposed to the most ventral region of the third ventricle at its entrance into the infundibular recess (768,786). It is bounded inferiorly by the tubero–infundibular recess and the median eminence and superiorly by the dorsal medial nucleus of the hypothalamus. Crosby and Woodburne (923), Crosby and Showers (765), Rioch et al. (29), Ingram (790), Papez (767), Christ (768), and Nauta and Haymaker (766) all considered the arcuate nucleus to be part of the periventricular system of neurons which, except at the level of the median eminence, encircles the third ventricle. The superior boundary of the arcuate nucleus in the periventricular zone is somewhat arbitrary because it is defined by the region where the condensation of neurons in the arcuate nucleus meets the less dense population of neurons in the periventricular hypothalamic nucleus; and, it is regarded differently by different authors in different species. van den Pol and Cassidy (924) described a triangular shaped nucleus (in the rat) with the side parallel to the third ventricle, the base parallel to the dorsal border of the median eminence and the tubero–infundibular sulcus, and the hypotenuse (dorsal lateral border) defined by the cell poor zone which separates the arcuate nucleus from the ventromedial nucleus of the hypothalamus. Others see a more pear-shaped nucleus with its concave lateral border defined by the cell poor zone separating the arcuate nucleus from the ventromedial nucleus. The nucleus extends as a column of cells from the anterior margin of the infundibular recess to a position slightly caudal to the posterior pole of the ventromedial nucleus and can thus be divided into a tuberal and a mammillary division (765,768). The nucleus is arcuate in shape anterior to the infundibulum in the rat (924) but not in the human (766); but, it changes from a paired to a single arcuate nucleus behind the infundibulum in the human (766). Retrograde transport of HRP and WGA has demonstrated the arcuate nucleus to be the source of many afferent fibers to the rostral region of the neurohypophysis—the median eminence (818,843,844).

Cells of the arcuate nucleus were described by Szentagothai et al. (100) as small, fusiform or triangular cells with several dendrites. The axons originated from the cell body or from the proximal portion of a dendrite. The ultrastructure of the neurons in the arcuate nucleus (462,751,752,925) does not differ from the description given for parvicellular neurons in the section on cell bodies of the hypothalamus. As discussed there, ultrastructural features of some neurons vary with the reproductive state of the animal and with age. Light microscopic examination of Golgi perpetrations has revealed three different kinds of neurons in the arcuate nucleus (924,926). Fusiform neurons with one apical dendrite lie medially in a juxtaventricular position. Fusiform neurons with two sparsely arborizing dendrites lie mainly in the medial and dorsal parts of the nucleus. Polygonal neurons with 4–5 repeatedly branching stem dendrites lie in the ventral and lateral regions of the nucleus. Dendritic morphology exhibited considerable variability with some dendrites containing many spines, some containing none, and some dendrites having a beaded appearance. Dendritic morphology of neurosecretory neurons is altered by castration of the male rat (927) and the density of axo-somatic synapses decreases with ovariectomy (928). The pattern of dendritic arborization in the arcuate nucleus differs from that in the adjacent hypothalamus (100,924,926). Cell bodies in the dorsal (and hence of necessity in the medial) region of the nucleus possessed vertically oriented dendritic trees while the cell bodies localized ventrally have horizontally oriented dendritic trees. In horizontal section, the course of dendrites in the periventricular zone is parallel to the third ventricle and orthagonal to the course of tanycytic processes. Immediately above the infundibular recess, the dendrites (in the medial portion of the nucleus) are oriented rostrocaudally, whereas more dorsally they are oriented dorsoventrally. In the periventricular zone of the AN, the neuropil is not separated from the ventricular system by a subependymal glial layer (550,558,924). Dendrites and axons course between laterally extending tanycytic processes.

Axons may terminate locally within the AN or leave the nucleus to terminate at distant sites. Recurrent collaterals may arise from these latter axons (100,924). Axons of the tubero–infundibular tract leave the tuberal division of the arcuate nucleus and project to the internal and external zones of the median eminence where they may collateralize (100). Arcuate neurons that project to the median eminence appear to be clustered in the dorsomedial and basolateral regions of the nucleus (818,844). Axons also project beneath the anterior third ventricle (rostral to the infundibulum of the rat) to terminate in the contralateral arcuate nucleus. Other axons appear to project dorsolaterally through the cell poor zone toward the ventromedial nucleus (924,926). However, a knife cut in the cell poor zone between the arcuate nucleus and the ventromedial nucleus of the rat produced axonal degeneration in the ventromedial nucleus and not in the arcuate nucleus (926). HRP injection into the medial and central nuclei of the amygdala demonstrated retrograde labeling in the arcuate nucleus of the rat and cat (929). These observations suggest that the arcuate nucleus sends projections to the ventromedial nucleus of the hypothalamus and to the amygdala. In addition, axons project dorsomedially to terminate in the periventricular division of the PVN (924).

Dopaminergic neurons in the AN project through the tubero–infundibular tract to the median eminence. Although not a peptide, dopamine is considered an important hypothalamic hormone because it acts as a physiologic prolactin inhibitor. The presence of these neurons was first revealed in the arcuate nucleus by the formaldehyde fluorescence technique (156,581, 582,584,930–932). Fluorescence of dopaminergic cell bodies in the AN and of dopaminergic terminals in the median eminence was preserved if the arcuate nucleus was surgically isolated from the remainder of the brain but not if it was lesioned (344,493,592, 933,934). Dopamine, as well as other catecholamines, has been identified in the arcuate nucleus by enzymatic assay (935). Tyrosine hydroxylase containing (presumably dopaminergic) cells have also been localized in the arcuate nucleus (919). Retrograde tracing experiments in the young monkey with application of tracer to the median eminence have demonstrated their projection to the median eminence (936). Neither dopamine as identified by enzymatic assay (935) nor dopamine-containing cells as identified by formaldehyde fluorescence (937) are homogeneously distributed in the rodent's arcuate nucleus (Fig. 21). Dopamine-containing cells were scarce in the medial aspect of the nucleus rostrally and in the ventral region of the tuberal division of the nucleus in the

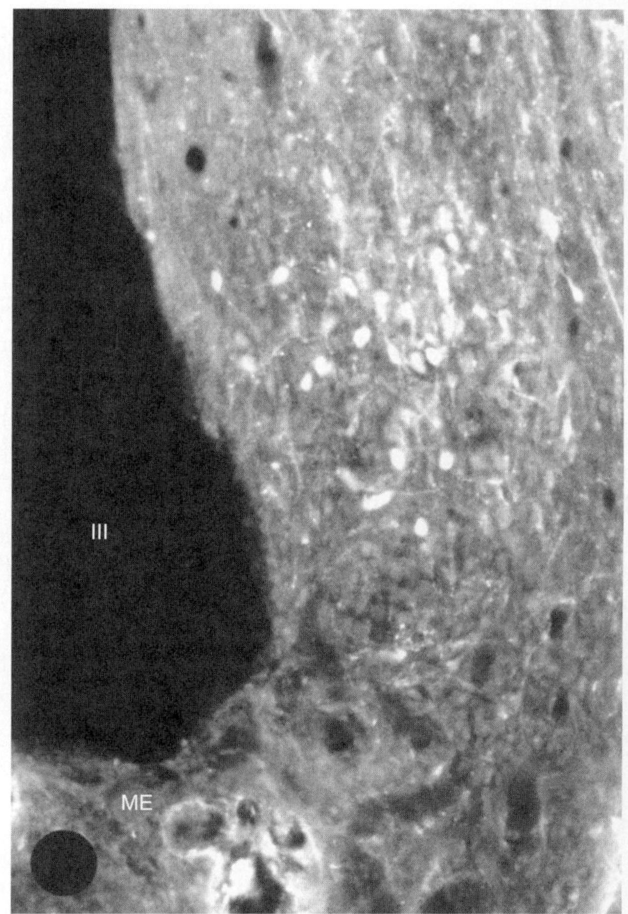

FIG. 21. Fluorescent dopaminergic neurons in the dorsal region of the rabbit arcuate nucleus. Glyoxylic acid treatment. III, third ventricle; ME, median eminence.

mouse (937). Bugnon et al. (938) localized dopamine (by fluorescence techniques) and ACTH (by immunohistochemistry) in different neurons. Dopaminergic neurons were aggregated medially and dorsally (in the distribution of the fusiform neurons) and ACTH-containing fibers were concentrated ventrally and laterally (in the distribution of the polygonal neurons). In contrast, Chan-Palay et al. (919) reported that TH-li neurons (considered to be the equivalent of dopamine containing cells) were present in the dorsomedial and ventrolateral regions of the AN. Tyrosine hydroxylase-containing cells in the AN were fusiform in shape and that cells near the ventricle were oriented vertically whereas more ventral (and lateral) cells were oriented horizontally. A similar distribution of TH-li neurons has been reported in the rat's AN by Everitt et al. (600). However, perhaps not all TH-ir neurons in the arcuate nucleus synthesize dopamine. L-aromatic amino acid decarboxylase (AADC) is necessary to complete the synthesis of dopamine from L-dopa. Meister et al. (939) have found that not all TH-ir neurons contained AADC. Cells containing both AADC and tyrosine hydroxylase and which fluoresced with formaldehyde treatment were confined to the dorsomedial region of the arcuate nucleus (940). The possibility must be considered that TH-li neurons in the ventrolateral region of the AN are DOPAergic and not dopaminergic.

TH-li neurons in the arcuate nucleus receive synaptic input from adrenergic terminals (941) and from Gal-li (942) and enkephalinergic terminals (943).

TH-li neurons in the arcuate nucleus costore other transmitters and peptides. Immunoreactive GAD and TH have been demonstrated in the same neurons of the rat's AN (355,600,944) mostly in the dorsomedial region of the nucleus. These observations at both the light and electron microscopic level suggest that dopamine and GABA are co-stored in the same neuron. Their costorage in the median eminence (940) suggests that both can be released into portal blood to inhibit PRL release from lactotropes. TH-li neurons in the arcuate nucleus have also been found that costore ir-NT (424,922,944), immunoreactive choline acetyl transferase (ir-CHAT) (940) and ir -GAL (487) as well as GRH (600,945).

Growth hormone releasing factor (GRF) has been localized by immunohistochemical techniques within arcuate neurons of the monkey (946,947), cat (948), rat, ox and monkey (949,950). Merchenthaler et al. (608) also found immunoreactive GRF in cells in the rat's arcuate nuclei and demonstrated that these cells project to the median eminence. In addition, these authors found immunoreactive cell bodies that projected to the median eminence scattered in the perifornical region of the lateral hypothalamus. Bloch et al. (951) reported that immunostaining of the arcuate nucleus and median eminence was abolished after treatment of neonatal rats with monosodium glutamate (952, 953), thus suggesting that most of the GRH innervation to the median eminence comes from the arcuate nucleus in the rat. Sawchenko et al. (531) also noted the presence of cells in the arcuate nucleus that contained immunoreactive GRH. Hypophysiotropic GRH neurons in the rat are predominantly located in the arcuate nucleus. They have been identified by applying True Blue to the surface of the median eminence and subsequent immunostaining for GRH. Double-labeled cells were found primarily in the arcuate nucleus (954). Although they were located throughout the nucleus they formed a particularly dense cluster in the ventrolateral portion of the nucleus (600). Balthasar et al. (955) confirmed these observations in a study of transgenic mice in which they targeted enhanced green fluorescent protein to secretory vesicles in growth hormone-releasing hormone neurons.

GRH projections to the periventricular nucleus and to the anterior and ventromedial regions of the medial parvicellular division of the PVN were described by Sawchenko et al. (531). A second projection of GRH fibers from the arcuate nucleus ascended and descended through the ventromedial hypothalamus just dorsal to the base of the brain. It projected into the anterior hypothalamic area, the preoptic region and also to the bed nucleus of the stria terminalis, the medial nucleus of the amygdala and the lateral septum. The descending tract projected to the posterior hypothalamus. A contribution to the tubero–infundibular tract was also identified and is the pathway by which GRH fibers reach the median eminence. In 20% to 40% of GRH-containing cells in the anterior arcuate nucleus, neurotensin could be co-localized.

Dual immunohistochemical staining has shown costorage of ir-NT, ir-GAL, or ir-TH (487,635,936). Although the major costorage is of ir-GRH with ir-TH (956), elution restaining techniques showed that some GRH-li cells co-store ir-NT and/or ir-GAL (635).

Pro-opiomelanocortin derived peptides have been isolated from the hypothalamus and identified by radioimmunoassay (577,957–959). In vitro synthesis of immunoreactive POMC and β-END by hypothalamic tissue has been reported (960). The highest hypothalamic concentrations of α-MSH, β-END, and ACTH were found in the arcuate nucleus (577). The arcuate nucleus of the sheep and ox contained cells staining for β-LPH (642). Immunoreactive β-LPH has been found in the arcuate neurons of the rat (961). ACTH and β-lipotropin and ACTH and β-END were localized in the same arcuate neurons by immunohistochemical techniques (961,962). Immunoreactive ACTH and β-END were found in the same cells in the rat's AN (963). Immunoreactive β-LPH, ACTH, β-END, and MSH have been localized in the same cells in the arcuate nucleus of the rat (964) and human (642,965,966). Cells containing POMC derived peptides were localized ventrally in the rat's AN whereas dopaminergic cells were localized dorsally (938). Cells containing $POMC_{mRNA}$ contain somatostatin receptors; and, β-endorphin-li and α-MSH-li neurites come into close contact with SRIF-li perikaryia (967). Cells containing POMC-derived peptides project to the median eminence (642,962,964). There they presumably have an endocrine function with respect to the adenohypophysis or a paracrine function with respect to neighboring terminals.

Cells containing POMC-derived peptides make synaptic contact with other cells in the arcuate nucleus (642,963) and innervate other hypothalamic regions. Eskay et al. (959) reported that lesions that destroyed the arcuate nucleus abolished immunostaining of hypothalamic fibers for α-MSH and resulted in a significant reduction in radioimmunoassayable intra and extra-hypothalamic α-MSH. Sawchenko et al. (968) demonstrated ACTH immunoreactive fibers leaving the rat's arcuate nucleus dorsomedially to course vertically in the periventricular zone and to enter the dorsal parvicellular division and the ventromedial aspect of the medial parvicellular division of the PVN. These regions contain OT-li and AVP-li parvicellular containing cells that project to the spinal cord and dorsal vagal complex (889,890). The ir-ACTH projection also terminated in the anterior magnocellular division of the PVN which contains only OT neurons and the anteroventral region of the posterior magnocellular division—an area with a high concentration of OT neurons. Similar results were reported by Mezey et al. (577). With retrograde labeling of the arcuate nucleus by injection of true blue into the ACTH innervated regions of the PVN, double-labeled (ir-ACTH and True Blue-containing) cells were found in the ventrolateral regions of the AN. Pro-opiomelanocortin was contained in about 8% arcuate neurons retrogradely labeled from the PVN (969). A β-endorphin-li projection to the supraoptic nucleus has also been reported (807). Retrograde tracers injected into the preoptic area of the rat were sequestered into ACTH-li cell bodies and β-endorphin-li neurons in the arcuate nucleus (970). In the monkey, ACTH-li terminals made axo-somatic and axo-axonic synapses with GnRH-li neurons near the infundibulum (971) and with AVP-li soma and neurites in the SON and PVN (806). Although the differences in the terminus of opioid-li arcuate projections to the magnocellular neurons in the rat and monkey remain to be worked out, it is clear that POMC-containing cells in the arcuate nucleus project to the median eminence and to preoptic and hypothalamic regions that are concerned with reproduction and with the response to stress.

NPY-li neurons are present in the rat's arcuate nucleus (972). They appear to be aggregated in the ventromedial region of the nucleus (969). About a third of these neurons co-localize GABA (973). They project to the paraventricular nucleus (973) and terminate ipsilaterally (972). NPY-li terminals from the arcuate nucleus are the major source of NPY terminals that synapse with TRH-li neurons and they come into close contact with CRH-li neurons in the PVN (869,881). NPY-li neurons, along with ACTH-li and β-endorphin-li neurons synapse with neurons in the median preoptic nucleus that in turn project to the PVN (970). Close associations between arcuate NPY-li fibers and preoptic GnRH-li neurons (and also GnRH-li terminals in the median eminence) were demonstrated by Li et al. (974).

The arcuate nucleus also contains GABAergic (349,350) neurons that project through the tubero–hypophysial tract to the median eminence. They have been identified by immunohistochemical techniques employing antibodies against glutamic acid decarboxylase (GAD). Ir-GAD was found colocalized with ir-TH, ir-GAL, ir-NT, and ir-GRH in varying complex combinations in the AN and median eminence of the rat by Meister and Hokfelt (940). For example, in the ventrolateral region of the arcuate nucleus ir-GAD/TH/NT/GAL neurons were demonstrated by combinations of double or triple labeling and elution techniques. Other neurons in the ventrolateral region showed a pattern of TH/NT/GAL/GRH immunoreactivity. In the dorsomedial region of the nucleus TH-li neurons co-stored ir-NT but not immunoreactive GRH, CHAT, or GAL. It is clear that the several transmitters and peptides released from a single terminal may interact with receptors on neighboring terminals and on distant pituitary cells.

GnRH has been found in the arcuate nucleus of the rat by microdissection and radioimmune assay (975). Immunohistochemistry has been difficult to employ to localize GnRH containing cells within the arcuate nucleus of the rat because the same antibodies that revealed GnRH in preoptic sites often failed to do so in hypothalamic sites (156,976). Immunoreactive GnRH has been identified in arcuate neurons in the mouse and rat (156,159,160,976–978), monkey (976, 979), rabbit (980), guinea pig (981), bat (982), and human (980,983,984).

Other peptides have been found in AN cells using immunohistochemistry. Substance P (488,489,985), Enk, DYN, NPY (600,607,986–988), Ghrelin (989) [a hormone with potent growth hormone stimulating activity (990,991)] and scattered cells containing SRIF (903) have been identified by immunohistochemistry and demonstrated to project to the external zone of the rat's median eminence. Neurotensin projections to the median eminence were demonstrated by a combination of retrograde tracing and immunohistochemistry. Double-labeled cells were localized within the median eminence (862).

Synaptic relations within the arcuate nucleus are now being studied with immunoelectron microscopy. Leranth et al. (992) have demonstrated homotropic TH-li synapses. TH-li neurons are contacted by GAD-li terminals (993,994). In view of the frequent colocalization of ir-TH and ir-GAD the possibility of homotropic synapses has to be kept in mind. Serotoninergic terminals (demonstrated by uptake of tritiated serotonin) synapsed with TH-li neurons (demonstrated by immunohistochemistry) (995). ACTH-li terminals made symmetrical synapses with TH-li neurons in the AN of the young monkey (996). GRH-li terminals have been found to make synaptic contact with GRH-li somata and dendrites (997). GRH-li neurons have also been shown to be contacted by SRIF-li terminals in the rat AN but of interest the synapses were asymmetric and not symmetric as might have been expected (424). Terminals that took up tritiated norepinephrine or 5-hydroxydopamine were also shown to synapse with GRH-li neurons (998). TRH-li terminals were also found to synapse with GRH neurons (999). POMC neurons have been identified by immunohistochemistry employing a number of antigens. ACTH-li neurons made homotropic synapses in the arcuate nucleus of the rat (1000). Both catecholaminergic and serotonergic synapses upon ACTH-li neurons have been demonstrated in the rat's arcuate nucleus (1001,1002). ENK-li terminals have also been reported to synapse with ACTH-li neurons (1003). Electron microscopic techniques coupled with immunohistochemistry (and autoradiography) have revealed a number of synaptic relationships. Arcuate neurons frequently made homotopic synapses to form local networks. Such network included TH-li, GRH-li, ACTH-li, and serotonin-li neurons (992,997,1000, 1004). In addition synaptic contact with presumptively inhibitory (GABAergic) elements from within the hypothalamus and with presumptive excitatory (catecholaminergic and serotonergic) element from outside the hypothalamus was common (Table 1).

Perhaps it would be advantageous to consider the parvicellular neurosecretory system by expanding on the original concepts of Szentagothai et al. (100). Consider the periventricular gray as composed of the periventricular (hypothalamic) nucleus along the border of the third ventricle. From this vertically oriented column of cells arise two lateral projections. The ventral lateral projection is the arcuate nucleus. The dorsal lateral projection is the parvicellular cluster of paraventricular neurons. In the arcuate nucleus are cells containing dopamine, GABA, POMC derivatives, GRH, neurotensin, substance P, NPY, and enkephalin. In the periventricular nucleus are cells that contain dopamine, SRIF, enkephalin, neurotensin, TRH, and CRH. In the parvicellular divisions of the paraventricular nucleus are cells that contain dopamine, TRH, CRH, SRIF, OT, and AVP. These periventricular cell groups are unique because they are retrogradely labeled by the application of HRP or wheat germ agglutinin–horseradish peroxidase (WGA-HRP) on the median eminence (818,843,844), i.e., these nuclear groups send projections to the median eminence. Cells located medially project to the median eminence by descending periventricular paths. Cells located more laterally in the PVN take a lateral and then a descending pathway with projections from the posterior magnocellular division of the PVN and from the region of the SON where they are joined by fibers of the SOHT. This fiber bundle comprised of axons from the SON and PVN passes through the lateral retrochiasmatic area to enter the median eminence anterolaterally (Fig. 20).

Cells in the periventricular gray also project to other (distant) neural targets. Further, cells in one region of the periventricular gray can project to another region of this system or contact cells within the same region. Parvicellular neurons can also project to the magnocellular neurons in the PVN. Although neurosecretory cells in the hypothalamic periventricular gray constitute a "hypophysiotropic area" within the hypothalamus, their cells project not only to the median eminence to regulate neuroendocrine function in response to changes in the environment or internal milieu but also to autonomic centers in the medulla and spinal cord and to reticular regions in the mesencephalon and to limbic regions of the forebrain.

Preoptic Area

The preoptic area (POA) also contains cells that are labeled following the topical application of wheat germ agglutinin (WGA) to the surface of the median eminence (843) and hence is a site where hypophysiotropic neurons reside. It is considered with the hypothalamus (and hence diencephalic) by Crosby and Woodburne (923), Rioch et al. (29), Crosby and Showers (765), but as part of the telencephalon by Nauta and Haymaker (766). Whether a telencephalic or a diencephalic derivative, it is intimately related to the hypothalamus (765). Papez (767) viewed the preoptic area as the gateway through which rostral innervation to the hypothalamus and to the epithalamus must pass. It has been implicated in temperature regulation (1005,1006), thirst and the maintenance of fluid balance (815,1007,1008), integration of adrenocortical response with autonomic responses (1009), cardiovascular regulation (1008,1010), sexual differentiation (1011), masculine sexual behavior (1012), scent marking behavior (1013), and cyclicity of gonadotropin secretions in female rodents and primates (1014,1015).

The anterior border to the POA is defined by the lamina terminalis and by the diagonal band of Broca (DBB). It is bordered inferiorly by the optic chiasm and the suprachiasmatic nucleus and superiorly by the anterior commissure. Posteriorly the boundary between the POA and the (remainder of) the hypothalamus in the adult mammal is difficult to define (765–768,923). A line drawn from the posterior border of the anterior commissure to the posterior border of the optic chiasm or the rostral extremity of the paraventricular nucleus would appear to be a good operational definition. The POA is continuous dorsomedially with the bed nucleus of the stria terminalis, which actually enters the POA in its most anterior extent. It is continuous anteriorly with the olfactory tubercle and anteriorly and superiorly with the septal region (766).

The POA is divided, cytoarchitectonic grounds, into a periventricular zone, a medial zone, a lateral zone, and a magnocellular preoptic nucleus (1016). The periventricular zone (periventricular preoptic nucleus, PVPOA) is a thin band of cells lying beneath the ependyma of the third ventricle that is continuous caudally with the anterior periventricular nucleus of the hypothalamus (765). Its cells are described by Swanson (1016) as being small and are often oriented parallel to the third ventricle.

The medial preoptic zone (medial preoptic area, MPOA) extends the length of the POA bordered dorsally by the anterior commissure and ventrally by the optic chiasm. Its rostral tip merges with the bed nucleus of the stria terminalis. The cells are of medium size and are densely packed (1016). Early transmission electron microscopic examination revealed light and dark neurons in the MPOA (1017). In light neurons the cytoplasm was the same density as the surrounding neuropil. The rough endoplasmic reticulum was dispersed throughout the cytoplasm. Free ribosomes and polysomes were numerous. The Golgi apparatus was well developed and associated with clear vesicles 40 to 80 nm in diameter and dense core vesicles about 100 nm in diameter. "Coated vesicles" were often seen. The rough endoplasmic reticulum extended into the dendrites. Occasional structures resembling the "whorls" or "ribbon rolls" in the cells of the arcuate nucleus were found. In dark neurons, some of the increased electron density was due to an increase in the number of free ribosomes. Free ribosomes lay between stacks of dilated endoplasmic reticulum. Many vesicles were associated with the Golgi apparatus. Both myelinated and unmyelinated axons were present in the neuropil of the MPOA with myelinated axons typically occurring in small clusters. A few axons contained dense core vesicles with diameters ranging from 125 to 165 nm. Dendrites were frequently beaded in appearance with large varicosities joined by narrow necks. Both symmetrical and asymmetrical synapses on dendritic shafts, dendritic spines, axons, and cell bodies were found. Greenough et al. (1018) reported that sex differences were present in the dendritic field patterns of the MPOA in hamsters. Dendrites in the MPOA of males tended to be clustered centrally whereas the dendrites in the MPOA of females were distributed more irregularly. These early ultrastructural observations suggested that (a) peptidergic cells were present in the MPOA, and (b) a differential input occurred in the MPOA of male and female animals.

The MPOA of the rat contains a sexually dimorphic nucleus (SDN-MPOA) that can be identified by visual inspection of Nissl-stained sections (1019). In males, the central region of the MPOA is more darkly stained and has sharper borders than in females. Its volume is larger in males than females. The different staining pattern is secondary to an increase in cell density in the SDN-POA. Differences can be found between males and females as early as the first day of (extra-uterine) life (1020). The cells of the SDN-MPOA migrate there from more ventral regions of the MPOA. A male pattern can be induced in females by exposure to androgens on the day of birth. Gonadectomy of males on postnatal day 1 reduced the size of the SDN-POA by more than 50% (1021). A similar sexual dimorphism has been found in the MPOA of the guinea pig (1022) and of the gerbil (1013), the toad (1023), and the human (1024). The role of this developmental difference between males and females is unknown. It is of interest that the sexually dimorphic region of the MPOA in the male toad is considered to be the center for mate calling and that its nuclear volume varies with the breeding season (1023).

Stimulation of the preoptic area of proestrus female rats caused an increase in plasma LH concentration and ovulation (1025). The integrity of the posterior region of the preoptic area is necessary for the demonstration of male sexual behavior in the rat (1012). It is to be expected then that this region (in the rodent) is related to ovulation and hence to the feedback relationships of estrogen on GnRH-containing neurons and to sexual behavior. Testosterone concentrating neurons have been localized in the arcuate and preoptic regions of adult rats (1026). Estrogen-concentrating neurons in the preoptic area have also been identified by autoradiographic means (1027–1029).

The bed nucleus of the stria terminalis enters the dorsal region of the MPOA. A sexually dimorphic pattern of nonstriatal synapses in the bed nucleus of the stria terminalis was first reported by Raisman and Field (1030). Female rats had more nonstriatal synapses on dendritic spines than did males. This pattern, along with the cyclic pattern of gonadotropin release and behavioral estrus was abolished if the females were treated on postnatal day 4 with testosterone. Males castrated within 12 hours of birth developed a female pattern of cyclic gonadotropin release and a female pattern of nonstriatal synapses in the bed nucleus of the stria terminalis. Cells in the lateral subdivision of the bed nucleus of the stria terminalis contained two or three dendrites that branch sparingly but might extend 300 to 400 microns across the field of incoming striatal neurons and were oriented perpendicular to them to form a reticular grid (1031). This arrangement should maximize contact between neurons of the bed nucleus and axons entering through the stria terminalis. Cells in the medial subdivision of the bed nucleus of the stria terminalis resembled those in the adjacent MPOA. Their dendrites were oriented parallel to incoming axons.

The magnocellular preoptic nucleus lies in the lateral zone, lateral to the diagonal band of Broca. Saper (1032) believed this nuclear group to be part of a system of large neurons 20 to 30 μm in diameter that are grouped into clusters in the medial septal nucleus, the magnocellular preoptic nucleus, substantia innominata, and globus pallidus. These neurons stain immunohistochemically for acetylcholine esterase.

The lateral zone (lateral preoptic area, LPOA) is characterized by the presence of medium-size neurons scattered among the fibers of the medial forebrain bundle (MFB). The bed nucleus of the stria terminalis is separated from the LPOA by a cell-free zone. The preoptic continuation of the bed nucleus of the stria terminalis occupies a position between the anterior hypothalamic area, the MPOA, and the LPOA.

Peptide hormones have been localized by immunohistochemistry within cells making up the nuclear groups of the POA, the medial septum, and in the bed nucleus of the stria terminalis of the rat. Somatostatin has been found within cells of the PVPOA (905–907, 1033). Cells containing immunoreactive CRH (530) and met-ENK (986) have been identified in the PVPOA and MPOA. Cells containing immunoreactive TRH have been localized within the MPOA and the diagonal band of Broca (491). Cells in the bed nucleus of the stria terminalis have been identified which contain immunoreactive VIP (1034), AVP (900), and CRH (530,611). Cells containing substance P and met-ENK have been identified in the MPOA and LPOA (988,1034).

GnRH is also found in the MPOA of the rat (975) but the cells containing this peptide are not confined within its borders. Immunoreactive cells are also present in the olfactory bulb, olfactory tubercle, nucleus of the diagonal band of Broca, medial septum, the periventricular preoptic nucleus and lateral hypothalamus (1035,1036). Unlike earlier investigators (159,160,976–978), subsequent investigators failed to confirm the presence of GnRH in the arcuate nucleus of the rat (1035,1036). King et al. (1035) suggest that previous studies employed antibodies that reacted against ACTH. The presence of GnRH in cells of the arcuate nucleus of the rabbit (1037), guinea pig (981), monkey (979,1038), and human (984) is generally accepted. Some GnRH-li neurons in the rabbit have been shown to costore delta sleep inducing peptide (1039).

Retrograde tracer application to the median eminence of the rat followed by immunohistochemical staining for GnRH has demonstrated GnRH-li neurons that project to the median eminence in the septal region, medial preoptic area and preoptic periventricular region (911,1040). In the monkey, GnRH-li cells were also found in the region of the arcuate nucleus (936). Not all the GnRH-li neurons in those regions project to the median eminence. There is no difference in morphology between those that projected to the median eminence and those that did not (1040).

GnRH-li terminals make symmetrical homotropic synapses with GnRH neurons (764,1041,1042). This circuit may mediate ultrashort loop feedback because in vitro studies have shown that incubation of hypothalamic fragments with LHRH-inhibited endogenous LHRH release. Terminals that take up tritiated norepinephrine have been shown to synapse with GnRH neurons in the medial preoptic area of the rat (1043). In addition TH-ir fibers come into close apposition with GnRH-li neurons in the preoptic area and infundibulum of humans although whether these monaminergic fibers arise from cell bodies in the brain stem or in the hypothalamus is uncertain (1044). NPY-li terminals have also been shown to synapse with GnRH neurons (1045). NPY-li projections, originating in the arcuate nucleus, come into close apposition with GnRH-li neurons in the preoptic area of the rat. These GnRH neurons demonstrate NPY receptors (974). In addition presumptive GABA terminals, shown by immunohistochemistry employing an antibody to GAD have been shown to synapse with GnRH neurons (1046). Immunoreactive glutamate terminals synapse with GnRH-li neurites in the monkey preoptic area (1047). CRH-li terminals make synapses with GnRH dendrites in the preoptic area (1048). Contacts between CRH-li fibers and GnRH-li neurites are numerous in humans (1049). Substance P-containing neurons from the arcuate nucleus have been found to synapse with GnRH-li neurons in the septo-preoptic area of the rat (1050). The distribution of these contacts in humans is presented by Dudas and Merchenthaler (1051). Neurons containing ir-POMC, or its processed opioid peptide products, also make synaptic contact with GnRH-li neurites (1052,1053). In the monkey, such terminals synapsed with GnRH neurons in the region of the arcuate nucleus (971).

GnRH pathways from the preoptic and septal regions of the rat brain have been established and the basic schema has been agreed on by several investigators. GnRH-containing perikaryia are arranged across the region of the rat's basilar forebrain in the shape of an inverted V. The apex is directed at midline nuclear groups such as the organum vasculosum of the lamina terminalis and the PVPOA and the diverging wings are bisected by the third ventricle. Two systems project caudally and ventrally to the median eminence, a periventricular system originating medially and a lateral system originating laterally and descending in the MFB. Although some differences exist in details, the systems following this basic pattern are described by King et al. (1035), Hoffman and Gibbs (1054), and Merchenthaler et al. (1036). Analogous systems are described in the guinea pig (981,1038), monkey (1055), baboon (1056), and human (984) except that GnRH-containing cells in the arcuate nucleus as well as in the MPOA project to the median eminence.

GnRH cells in the preoptic area of the baboon projected median eminence but also to the stria medullaris and the OVLT. Cells in the PVPOA projected to the median eminence. GnRH cells were also found in the medial septum, bed nucleus of the stria terminalis, lateral hypothalamus, and lateral preoptic area, and these cells did not project to the median eminence. They were believed to be involved with reproductive behavior other than endocrine function (1056). In the hamster, projections from the septum are reported to gain access to the hippocampus via the septum and to reach the amygdala via the stria terminalis. Projections from the medial septum also reach the OVLT and SFO. Projections from the olfactory tubercle reached the amygdala and piriform cortex via the ventroamygdalofugal pathway. Supracallosal projections reached the induseum griseum. Descending projections to the midbrain coursed along the stria medullaris and fasciculus retroflexus and some continued caudally into the pons (1057). A similar pattern of projections to the rhinencephalon and limbic system and brain stem was found in the rat (1058), bat (982), and guinea pig (981). The GnRH cells project cerebral and brainstem regions that can promote sexual behavior as well as to diencephalic regions that can promote ovulation.

The projections of the POA have been studied by autoradiography following the injection of labeled amino acids into the preoptic area of the rat brain (1016,1059) and following the injection of WGA-HRP (1060) and *Phaseolus vulgaris* leukoagglutinin (PHA-L) (1061). The PVPOA projections were principally descending ones in the periventricular zone to the "hypophysiotrophic area" of the medial basilar hypothalamus and the median eminence (1059). Swanson (1016) stated that dorsal, intermediate, and ventral parts of the PVPOA project to "characteristically different terminal fields." A lesser projection courses laterally to join the medial region of the medial forebrain bundle to terminate in the hypophysiotrophic area of the hypothalamus and the internal zone of the median eminence.

Neurons in the PVPOA and in the MPOA projected to the OVLT (1062). A projection of GnRH-containing neurons from the MPOA to the OVLT has been identified in the rat (1035). TRH (491), somatostatin (906), containing fibers in the OVLT also originated in the MPOA (1062). AVP-containing fibers which terminate in the OVLT probably arise from the suprachiasmatic nucleus and not from the supraoptic nucleus.

Ascending projections of the MPOA are dorsal to the stria medullaris (and hence via the habenular nuclei and fasciculus retroflexus to the midbrain tegmentum), to the diagonal band of Broca (and hence to the medial septum) and to the medial forebrain bundle (1059). More extensive ascending projections were reported by Chiba and Murata (1060) to the bed nucleus of the stria terminalis, nucleus of the diagonal band of Broca, medial amygdaloid nucleus, and septal regions. [The interested reader is referred to (901,1016,1059–1061,1064).]

The descending projections of the MPOA resemble those of the PVPOA except that the contribution to the lateral pathway through the MFB is more robust. Neither Conrad and Pfaff (901) or Swanson (1016) could trace descending pathways further caudally than the mesencephalon. Chiba and Murata (1060) were able to trace descending fibers from the MPOA to the lateral parabrachial nucleus of the pons as well as to the locus caeruleus and to the nucleus and tractus solitarius and the vagal complex of the medulla in the rat as well as using by WGA-HRP as a tracer. They propose that these projections are the basis of the functional relationship between the MPOA and autonomic control of blood pressure, of thermoregulation, and of angiotensin II-induced drinking behavior (Table 2).

In summary, concepts of the hypophysiotrophic area first proposed by Harris (72–75,1065) on the basis of his transplantation experiments and developed by Halaz and Pupp (99) on the basis of their hypothalamic islands have been expanded over the years. Cells which contain hypophysiotrophic peptides and neurotransmitters and which terminate in the median eminence are found (principally) in nuclear groups that extend along the periventricular area from the lamina terminalis to the mammillary body. Smaller populations of cells are dispersed laterally (in the septum, diagonal band of Broca, and the bed nucleus of the stria terminalis). Many peptidergic cells project to the median eminence, but others project to nuclear groups in the brainstem and spinal cord, as well as to terminal fields in rostral (limbic) regions.

The suprachiasmatic hypothalamic nucleus (ScN) is responsible for circadian rhythms of endocrine function because its cells display an intrinsic circadian periodicity and project to the MPOA and to the periventricular hypothalamus (383,1066). Situated in the anterior hypothalamus it caps the optic chiasm on either side of the third ventricle. It is comprised of a core composed primarily of VIP-ir neurons and a cap composed primarily of AVP-li neurons in the rat (1067), mouse (1068), mole (1069), and human (1070). In the rat, calretinin-li neurons, metENK-li neurons and SRIF-li neurons are also present in the shell along with AVP-li neurons but in the mouse SRIF-li neurons are absent. Their shell was comprised of AVP-li neurons, metENK-li neurons, and ANGII-li neurons. In both rat and mouse gastrin releasing peptide (GRP) was immunohistochemically identified in core neurons along with VIP-li neurons. NT-li neurons were identified in the core of the mouse but not the rat SCN. GABA-ir neurons were identified throughout the mouse and rat ScN (1067,1068).

The ScN receives input from diverse sources that provide visual, olfactory, auditory and humoral data. The ScN receives humoral input from the pineal gland. Melatonin receptors have been demonstrated in the SCN in radioligand binding studies employing 2-[125I]iodo-melatonin (376). Neurons in the SCN of the Siberian hamster that contain the mRNA for the MEL_{1a} receptor also contained AVP_{mRNA} (1071). CNS input to the rat's SCN is topographically organized (1072). The ScN receives photic input from the retina directly via the retinohypothalamic hypothalamic tract (RHT) (1073) to its core in the ventral region (1068). Glutamate, substance P and pituitary adenyl cyclase-activating polypeptide (PACAP) have been implicated as neurotransmitters in the RHT (1074). Photic input arises indirectly from the intergeniculate leaflet and also terminates in the ventral region of the ScN. The pattern of termination resembles the pattern of NPY endings in the ScN (1072). Pretectal regions including the olivary pretectal nucleus, perhaps a source of auditory input, also project to the ventral region of the ScN. The raphe nuclei in the brain stem send (serotoninergic) projections the same region (1075). Limbic input (a potential source of olfactory input) from the infralimbic cortex, lateral septal nucleus and ventral subiculum, terminates preferentially in the dorsal region as do projections several hypothalamic nuclei including the arcuate nucleus and the subparaventricular nucleus area. The thalamic projection from the thalamic paraventricular nucleus terminates more diffusely throughout the ScN (1072). Direct input from the subfornical organ, most preoptic regions, the sub-paraventricular zone and the dorsal medial nucleus of the hypothalamus has demonstrated by retrograde labeling with cholera toxin. Far more extensive input through secondary connections was documented by retrograde studies employing pseudorabies virus (1076).

Within the ScN there are frequent dendrodendritic and somato-somatic appostions (1077). VIP-li neurons and AVP-li neurons often make synaptic contact in the rat's ScN (1078). Astroglial processes were found by Tamada et al. (1079) were found to partially engulf VIP-li and AVP-li somata, and neuronal processes.

In an extensive study employing several retrograde tracers, Leak and Moore (1080) concluded that efferent projections from the shell of the rat's ScN are to the medial preoptic area, medial subparaventricular zone, the paraventricular and dorsal medial hypothalamic nuclei, and the paraventricular thalamic nucleus. The core projects to the lateral sub-paraventricular zone, ventral tuberal area, and peri-suprachiasmatic region. In the human the major sites of projection are the anteroventral hypothalamus, the ventral region of the paraventricular nucleus and the sub-PVN, and the dorsal motor nucleus (1081). The existence of projections from the ScN to the arcuate nucleus is supported by the physiologic studies of Saeb-Parsy et al. (1082). A direct connection between GnRH-li neurons and projections from the ScN has been demonstrated (1083). A projection of VIP neurons from the ScN has been demonstrated by confocal microscopy to come into close contact with retrogradely labeled spinal projecting neurons in the PVN (1084). It has yet to be established whether this projection forms a link in the pathway from the ScN to the pineal gland via the sympathetic system (1085).

INPUT INTO THE PERIVENTRICULAR BRAIN

Afferents enter the periventricular hypothalamus by one of four routes: the medial forebrain bundle, the periventricular system, the fornix, and the stria terminalis. The medial forebrain (MFB) lies in the lateral zone of the hypothalamus and extends from preoptic to mammillary levels. The cells within the lateral zone were called "path neurons" by Millhouse (1086). These cells are oriented perpendicular to the plane of fibers running in a rostrocaudal direction in the medial forebrain bundle and each dendritic tree radiates like a fan in the path of fibers in the MFB. Their axons project medially into the medial and periventricular zones and innervate the nuclear groups in the MPOA-HTA (1087). Collaterals from MFB neurons branch at right angles to the parent axon and course parallel to the dendrites of path neurons. Hence, each axon in the MFB passes through the dendritic arbor of several path neurons in series with the opportunity to send collaterals to each (1086).

Ascending and descending input enters into and passes through the MFB. Millhouse (1086) proposed that descending input tended to project collaterals into rostral areas (the MPOA and anterior hypothalamic area), whereas ascending input tended to send collaterals preferentially to caudal regions (the tubero-mammillary region). He also proposed that the fascicles of the MFB maintained constant and characteristic positions in the MFB that were related to their origin. Both these concepts have received support from subsequent autoradiographic studies on the MFB in the rat (1088,1089). Ascending components are confined to the dorsal half of the bundle and descending components to the ventral half. Major fiber groups that descend in the MFB arise (a) in the olfactory tubercle, amygdala, basal ganglia, and MPOA and lie in the ventral and ventrolateral regions, (b) in the septum and are confined to the ventromedial region of the bundle. The components arising from various regions in the POA are specifically arranged in the ventral region of the MFB. Ascending systems are probably similarly arranged with the ventral tegmental component near the center of the bundle and the ascending fibers of the parabrachial plexus in the dorsal and dorsolateral regions.

The dorsal tegmental noradreneregic bundle is held to arise from the locus caeruleus (A_6) cell group and to ascend in the central tegmental tract. There is disagreement regarding the pathway of catecholaminergic fibers that ascend from the brainstem in the MFB. According to the account of Lindvall and Bjorklund (1090), on the basis of fluorescence following glyoxylic acid treatment, fibers of the dorsal tegmental bundle enter the MFB. On the basis of immunohistochemistry of DBH, Swanson and Hartman (1091) concluded that noradrenergic fibers from the locus caeruleus travel with other ascending noradrenergic systems in the "principal noradrenergic system" "in the central tegmental tract" and entered the hypothalamus in a discrete bundle from a lateral position in the zona inserta. These fibers innervated the PVN by passing from lateral to medial and then passed down the periventricular area to the AN. The MPOA was innervated both from the periventricular area and from fibers coursing medially from the zona inserta.

The ventral tegmental noradrenergic bundle in the pons and mesencephalon receives input from the A1 (lateral reticular nucleus), A2 (dorsal medullary complex), A5 and A7 (parabrachial nucleus), and from the subcaeruleus region (920,1090). This projection ascends in the central tegmental tract; and, according to the account of Lindvall and Bjorklund (1090), contributes to the MFB at the level of the tegmental radiations. Swanson and Hartman (1091) disputed this interpretation on the basis of their DBH studies and stated that the dorsal and ventral tegmental bundles form the principal ascending noradrenergic bundle that enters the hypothalamus as a discrete system in the region of the zona inserta. They believed the fluorescent fibers seen in the MFB by Lindvall and Bjorklund (1090) to be dopaminergic and to arise from the ventral tegmental region, ascend in the MFB and terminate in hypothalamus and in the limbic forebrain (1092,1093). Palkovits et al. (1094) appears to accept the account of Lindvall and Bjorklund (1090) because lesions separating the MFB from the MPOA-HTA reduced the norepinephrine content of that region. At this time one can conclude that the MFB contains dopaminergic fibers, but the matter of noradrenergic neurons has yet to be settled.

The periventricular system (819) extends around the third ventricle. Axon terminals are oriented in a dorsal-ventral plane as are the axons and dendritic fields of neurons in this region (100). This region also carries fibers which project rostrally and caudally. Major descending input arises in the MPOA (1016, 1059,1095). Within the MPOA-HTA, the PVN contributes to this system (901). From outside the periventricular region, the ventral median nucleus of the hypothalamus sends ascending and descending fibers into the periventricular system (1096–1098). Ascending input into the periventricular system is from the brain stem through the dorsal longitudinal fasciculus, which receives input from the A_2 cell group in the dorsal medullary complex and to a lesser extent from the locus caeruleus and the subcaeruleus area (1090,1094,1099,1100).

The fornix also projects to the MPOA-HTA but the hippocampal fibers destined for hypothalamic terminals originate in the pro-subiculum, not in Ammon's horn (1101). Fibers originating in field CA1 of Ammon's horn project through the fornix to terminate (with or without relay in the mammillary nuclei) in the anterior nucleus of the thalamus. Field CA3 of Ammon's horn sends projections through the fimbria to the precommissural fornix where they terminate principally in the septal nuclei. The septum projects to the medial preoptic nucleus which in turn projects to the peri-ventricular hypothalamus (1061,1064, 1102). In the rat, fibers destined for the arcuate nucleus arise in the subiculum and pass in the dorsal region of the fornix to the postcommissural fornix where they depart from it and form the medial cortical hypothalamic tract that terminates in the arcuate nucleus. The dentate gyrus and Ammon's horn both receive a massive input from the rhinencephalon and sensory regions of the forebrain. The dentate gyrus projects to Ammon's horn, which in turn projects to the prosubiculum and thence to the arcuate nucleus (1103,1104). The medial cortico–hypothalamic tract, while prominent in rodents, is not found in primates (766).

The stria terminalis carries projections from the amygdala to the bed nucleus of the stria terminalis and the preoptic area as well as to the ventromedial nucleus of the hypothalamus. In the rat "the ventromedial nucleus in turn establishes massive intrahypothalamic projections to the medial preoptic, anterior ... and posterior hypothalamic nuclei but has much sparser inputs into the periventricular and lateral zones" (1105). For example projections from the ventromedial hypothalamic nucleus project the GnRH-li neurons in the medial preoptic region of the ewe (1106).

The SON receives input from several sources within pre-optic area and the hypothalamus including the arcuate nucleus, the MPOA and the OVLT (organum vasculosum of the lamina terminalis) and SFO (subfornical organ) (Table 3). The arcuate nucleus sends

TABLE 3. *Afferents to the MPOA-HTA*

	SON	PVN(m)	PVN(p)	PvN	AN	MPOA
OVLT	+	+				+
SFO	+	+				+
SON						
PVN(m)						
PVN(p)	+					+
PvN	+				+	
AN	+	+	+	+		+
MPOA	+	+	+	+	+	
VMN			+			+
ScN			+		+	+
Hipp	+	+			+	
Septum	+	+				+
BNST			+			+
Amy			+			+
Ol. Tub.	+					
IPN						
SN						
Teg	+					
PbN			+			+
LC			+		+	
Raphe	+		+	+		+
LRN	+	+	+	+	+	+
DVC			+	+	+	+

SON, supraoptic nucleus; PVN(m), magnocellular paraventricular nucleus; PVN(p), parvicellular paraventricular nucleus; PvN, periventricular nucleus; AN, arcuate nucleus; MPOA, medial preoptic area; SFO, subfornical organ; NL, neural lobe; ME, median eminence; OVLT, organum vasculosum of the lamina terminalis; VMN, ventromedial nucleus; ScN, suprachiasmatic nucleus; Hipp, hippocampus; BNST, bed nucleus of the stria terminalis; Amy, amygdala; Ol. Tub., olfactory tubercle; IPN, interpeduncular nucleus; SN, substantia nigra; Teg, midbrain tegmentum; PbN, parabrachial nucleus; LC, locus ceruleus; LRN, lateral reticular nucleus; DVC, dorsal vagal complex.

a projection of ACTH-containing axons to the SON which is distributed to regions where OT cells predominate (968). Input from the median preoptic nucleus to the SON has been demonstrated by both anterograde and retrograde tracing (1107). The OVLT is a circumventricular organ that lies above the optic chiasm. The SFO, also a circumventricular organ, lies above and behind the OVLT (just beneath the foramen of Monro). These circumventricular organs lie in the anteroventral region of the third ventricle, the AV3V region. The input from the SFO and the OVLT into the SON is of particular interest in view of the accepted role of the AV3V region in fluid and electrolyte balance (815,1007,1008). The OVLT is made up of specialized ependymal cells that are united by tight junctions, fenestrated capillaries, neurons, and axon terminals. In vitro recording from neurons in the region of the OVLT demonstrated that the majority of neurons increased their firing rates with an increase in the sodium concentration of the medium (1007). A projection from the OVLT to the SON has been demonstrated by retrograde tracing of HRP following injection into the SON (1108). Neurons in the SFO have been found to respond to circulating levels of ANG II rather than sodium (1109). The SFO has been shown to project to the Median Preoptic Nucleus and to the OVLT by precommissural fiber pathways and to the medial septum, diagonal band of Broca, and the magnocellular neurosecretory nuclei by postcommissural pathways (1110–1112). It should be remembered that the targets of the precommissural projection from the SFO also project to the SON. ANG II receptors have been localized in the SFO and some of its projections express ANG II (863). ANG II-li cell bodies have been found in and around the SFO and they project to nuclear groups in the periventricular zone of the hypothalamus (863). An ANG II-li projection to the SON has been reported (1113) and magnocellular neurons in the rat's SON and PVN immunostain for ANG II receptors (1114). These observations support the supposition that neurons in the SFO respond to circulating levels of ANG II and employ ANG II to convey that information directly to magnocellular neurons in the SON.

Descending input to the SON also comes from several sources in the rhinencephalon and in the limbic forebrain. Retrograde and anterograde tracing studies have demonstrated afferent input from the main and accessory olfactory bulbs to the ventral glial lamina—a region adjacent to the SON where dendrites of neurophysin-li neurons are plentiful (1115,1116). Electrophysiological studies have shown both monosynaptic and polysynaptic connections between the lateral olfactory tract and the SON (1116). Hatton et al. (1117,1118) have shown that stimulation of the lateral olfactory tract increased dye coupling between magnocellular neurons in lactating or maternally acting rats but not in virgin females. Hence, olfactory stimulation not only activates magnocellular neurons, it coordinates their firing in phase and thus, presumably, facilitates such functions as lactation.

The region surrounding the SON constitutes a dendritic shell and contains neurites that project into it (1108). Injections of HRP which extended slightly beyond the borders of the SON revealed an input that was more extensive than that revealed by injections that were confined within the boundaries of the SON (1108). Regions outside the MPOA-HTA project to regions about the SON. Fibers, descending from subiculum, septum, and nucleus of the diagonal band of Broca were labeled. Although Swanson and Cowan (1064) did not report labeling of the SON following injection of labeled amino acids into the septal region, Tribbolet et al. (1108) noted labeling in the region immediately surrounding the SON following the injection of tritiated amino acids into the septum and subiculum. The pattern of labeling probably signifies a functional connection between these limbic regions and the SON because (a) the dendritic field of the SON is not limited to the confines of the nucleus but extends beyond it particularly in the ventral glial lamina (1119); (b) TEM studies of the SON region following placement of septal lesions demonstrated degeneration of terminals (1120); (c) TEM studies following anterograde transport of HRP from the septum demonstrate labeled terminals synapsing on neurites at the border of the SON which stain immunohistochemically for either OT or AVP (1121); (d) electrophysiological studies demonstrated antidromic activation of septal neurons after stimulation of the SON (1122). Hence, descending afferent systems to the SON arise in the rhinencephalon and in the limbic forebrain. Although some projections (from the olfactory bulb and subiculum) are direct, others probably relay in the septum.

The ascending input from the brainstem enters the hypothalamus through the periventricular dorsal longitudinal fasciculus or the medial forebrain bundle (766). Both norepinephrine and dopamine are present in the SON (935,1099,1123). Catecholamine-containing terminals have been demonstrated in the SON of the rat (583), cat (1124), and monkey (1125), and several sources have been suggested. Small injections of HRP confined within the locus of the SON of the rat labeled the A_1 (lateral reticular nucleus) and the A_2 (nucleus of the solitary tract and the dorsal vagal complex) noradrenergic cell groups (1108). Similar observations were made in the rabbit where retrograde HRP labeling of catecholaminergic cells identified by fluoroscopic methods was demonstrated in the A_1 and A_2 cell groups (1126). Electrophysiological studies showed that stimulation of the carotid sinus nerves or baroreceptors activated neurons in the location of the A_2 cell group and stimulation of these noradrenergic neurons selectively facilitated activity of phasic vasopressin-secreting (but not oxytocin-secreting) neurons in the SON (1127,1128).

Whereas the evidence that the A_1 cell group projects to the SON is strong, evidence that the A_2 cell group directly innervates the SON is less compelling. Anterograde labeling experiments in the rat (1100, 1129) and the rabbit (1126) failed to confirm the presence of a direct innervation of the SON by the A_2 cell group. Stimulation of the nucleus of the solitary tract excited some SON neurons by "slow," presumably polysynaptic pathways. The authors concluded that stimulation of the A_2 cell group activated magnocellular neurons in the SON only after relay in the A_1 cell group (1130).

Projections from the lateral reticular nucleus in the ventrolateral medulla, as demonstrated by anterograde tracing techniques, terminate in the ventral and caudal regions of the SON where AVP-li magnocellular neurons are aggregated (1131). Noradrenergic terminals have been found within the ventral regions of the SON where AVP-li neurons are clustered (813,1132,1133). McNeill and Sladek (1132) noted that in the rat, CA terminals were most frequently found surrounding the SON and were particularly abundant in the ventral glial lamina (1119) where they contacted dendrites of SON-li neurons. These anatomic observations nicely complement the physiologic findings of Day and Renaud (1128) that stimulation of the A_1 cell groups selectively activated phasic AVP-secreting neurons.

Dopamine terminals are also present in the SON. They are homogeneously distributed throughout the nucleus and terminate on the soma of OT and AVP-containing cells (1134). In view of the close relationship of AVP terminals in the neural lobe of the rat to terminals of the dopaminergic tubero–hypophysial tract (494), it is of interest to determine the site of origin of dopaminergic terminals about the magnocellular neurons in the SON. Buijs et al. (1134) have suggested that they arise in the arcuate nucleus. Simon et al. (1092) identified a projection from the A_{10} dopaminergic cell group in the ventral tegmental area of Tsai to the SON using anterograde and retrograde tracing techniques. van Vulpen et al. (808) have demonstrated a dopaminergic projection from the PvN to the SON. Further work is necessary before the origin of the dopaminergic terminals in the SON can be considered established.

Scrotonin-li terminals in the SON originated in the raphe nuclei of the rat and entered the hypothalamus through the medial forebrain bundle (1135). The distribution of serotonin-li fibers within the SON and their mode of termination (axo-somatic, axo-dendritic, or axo-axonic) is not as yet known. In addition to these projections, a SP containing project has been reported by Bittencourt et al. (1136) to terminate in the SON of the rat. A projection from the dorsal raphe nucleus to the SON has been documented in the Syrian hamster (1075).

The PVN receives input from several structures within the MPOA-HTA region of the rat's hypothalamus. On the basis of studies employing retrograde transport of HRP following its injection into the PVN input from the MPOA, from the OVLT and SFO, and from within the HTA has been demonstrated (1137,1138). Anterograde autoradiographic tracer studies following the injection of ^{3}H-proline into the MPOA of the guinea pig revealed a projection into the PVN (1095). Chiba and Murata (1060) injected WGA-HRP into the MPOA of the rat and found labeling in multiple hypothalamic nuclei including the PVN. This finding has been confirmed in the rat where a projection from the medial preoptic nucleus to the parvicellular regions of the PVN has been shown by anterograde and retrograde tracing (1061). Input to the PVN from neurons in the median preoptic nucleus that in turn receive synaptic input from ascending catecholaminergic projections form the ventrolateral medulla has been demonstrated in rats (1139).

Projections to the magnocellular regions of the PVN from the OVLT have been identified by retrograde tracing but await confirmation by anterograde techniques (1138). A projection from the SFO to the PVN has been identified by retrograde and anterograde tracing (1110). Electrophysiological studies suggest that ANG II projections to the PVN activate magnocellular neurons that release AVP and parvicellular neurons that release CRH (1109,1140,1141). Silverman et al. (1138) proposed that input to the PVN from the SON arises from collaterals of the SOHT. Other projections from within the periventricular zone are difficult to demonstrate by conventional tracing techniques because the distances between the site of injection and uptake and the sites of labeled terminals is small. Sawchenko et al. (968) employed immunohistochemistry to demonstrate an ACTH projection from the AN of the rat to the magnocellular regions of the PVN where OT fibers predominated. Arcuate ACTH fibers also terminated about parvicellular neurons in the ventromedial region of the medial parvicellular division of the PVN. These regions also contain OT synthesizing neurons and project to the dorsovagal complex and to the spinal cord. GRH-li cells in the AN have also been observed to project to the anterior, periventricular and ventromedial parvicellular regions of the PVN (531). NPY and SRIF projections from the arcuate nucleus also terminate in the PVN (969).

Hypothalamic regions outside the MPOA-HTA also project to the PVN. Projections from the dorsal medial nucleus (DMN), ventral medial nucleus and suprachiasmatic nucleus have been revealed by studies of retrograde transport of HRP following its injection into the PVN (1137,1138). Anterograde tracer studies with labeled amino acids have been carried out to determine the efferent pathways of the VMN but the authors do not mention a projection from VMN to PVN in the rat (1059,1096,1097). Saper et al. (1097) stated that the anterograde labeling studies revealed no projection from the VMN to the PVN in the monkey. In the rat such projections were scant (1105). However, following placement of an electrolytic lesion in the VMN of the cat, Kaelber and Lesson (1142) found degenerating terminals in the PVN.

The PVN, like the SON, receives descending input from hippocampal formation. This input may be direct from the subiculum or from Ammon's horn with a relay in the septum and in the medial preoptic nucleus (1101,1137,1138). Oldfield et al. (1121) employed TEM techniques to examine the PVN after injection of HRP into the subiculum or septum and immunohistochemical staining of the PVN for ir-OT or ir-AVP. A projection from these limbic regions onto AVP and OT containing neurites just lateral or ventrolateral to the PVN was demonstrated. Unlike the SON, the PVN receives a projection from the medial and central nuclei of the amygdala (1137, 1138) and from the bed nucleus of the stria terminalis (1064,1112,1138). TEM studies demonstrated that the input from the amygdala is to the AVP containing neurites just lateral and ventrolateral to the nucleus—a site where dendrites from the PVN were found. Projections from the amygdala, the bed nucleus of the stria terminalis and from the medial preoptic nucleus and ventromedial nucleus of the hypothalamus, which in turn receive input from the amygdala (1102,1143–1145) are unique to the PVN when compared with the SON. It is tempting to relate these projections from limbic regions that contain glucocorticoid receptors to the parvicellular regions of the PVN that (a) contain AVP, CRH, and ANG II, (b) project to the median eminence (or to the medulla and spinal cord), and (c) are implicated in the regulation of the pituitary adrenal and autonomic neural axes (1138,1146–1148).

The PVN receives a much more elaborate ascending innervation than does the SON. Catecholamines have been identified in the PVN (935,1099,1123) and CA terminals have been demonstrated in the PVN of the rat (583), cat (1124), and monkey (1125) by fluorescence techniques. In the magnocellular divisions of the rat (1132) and monkey (1125), CA fibers are more abundant in regions rich in AVP cells than in regions where cells containing OT predominate. Swanson et al. (809) employed immunohistochemistry to demonstrate the terminals containing dopamine-beta-hydroxylase (DBH) but not phenylethanolamine-N-methyltransferase (PNMT) (and therefore noradrenergic terminals) were present predominantly in regions where AVP containing magnocellular neurons were found. However, these observations must be interpreted cautiously because Silverman et al. (1149) demonstrated on TEM study of the PVN that CA terminals primarily innervated the dendritic processes of nonvasopressinergic neurons. When noradrenergic terminals contacted AVP-containing neurons, contact was axo-dendritic, not axo-somatic. As the dendrites of peptidergic cells within the PVN have a large medial to lateral extension (820), the specific location of noradrenergic terminals within subdivisions of the PVN may not be predictive of the type of cell innervated.

Retrograde tracing experiments showed that the lateral reticular nucleus (A_1 cell group) projected to the PVN in rats (1100,1137) and rabbits (1126) (Fig. 22). Anterograde labeling experiments have confirmed this pattern in the rabbit (1126) and rat (1131) and further demonstrated that the A_1 projection is preferential to the magnocellular regions that contain

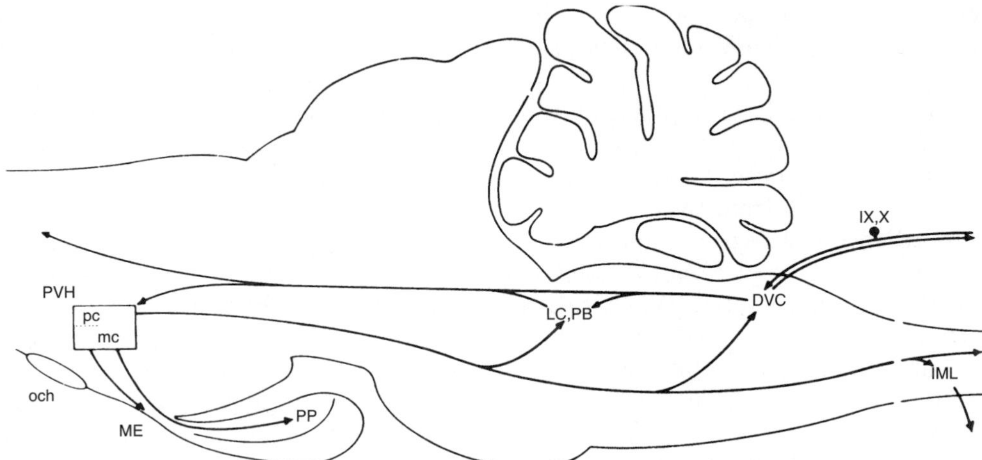

FIG. 22. A summary of the major longer connection of PVH in the rat. In this diagram, relationships between the PVH and cell groups associated with the autonomic nervous system are emphasized. DVC, dorsal vagal complex; IML, intermediolateral column; LC, locus ceruleus; mc, magnocellular division of PVH; ME, median eminence; och, optic chiasm; PB, parabrachial nucleus; pc, parvicellular division of PVH; PP, posterior pituitary; PVH, paraventricular nucleus; IX, glossopharyngeal nerve; X, vagus nerve. [From (1174).]

AVP-li neurons. In the rat, ir-NPY was frequently costored in the DBH-li terminals in the magnocellular regions of the PVN where AVP-li neurons were clustered. After injection of True Blue into the PVN, labeled cells in the A_1 cell group that immunostained for DBH and NPY were found as were neurons that immunostained for DBH alone (1150). The principle that A_1 neurons with or without costored NPY do preferentially innervate magnocellular neurons in the PVN which contain AVP is supported by the observation that stimulation in the region of the A_1 cell group activated phasic (AVP) neurons in the PVN (1151–1153) as it activated phasic (AVP) neurons in the SON (1128,1130). It is difficult not to place functional significance on the observation that large cells in the PVN and SON arise from the same germinal cell group, migrate to their final destinations at the same time, receive input from the same limbic, circumventricular, preoptic, and brain stem structures, and project to the same terminal field—the neural lobe.

Catecholaminergic terminals are also found in the parvicellular divisions of the PVN. Anterograde tracing studies with small selected injections in A_2 demonstrated preferential labeling in the dorsomedial parvicellular subdivision where CRH-li neurons were clustered. These terminals also immunostained for DBH leading the authors to the conclusion that the noradrenergic projection of the A_2 cell group in the dorsal motor nucleus innervated this region of the PVN (1131). Only about 10% to 15% of these noradrenergic neurons have been found to costore NPY (1150). The adrenergic cell groups in the lateral reticular nucleus and in the dorsal motor nucleus lie rostral to the noradrenergic cell groups and are termed the C_1 and C_2 cell groups, respectively. Anterograde transport studies revealed that the C_1 and C_2 cell groups, unlike the A_1 and A_2 cell groups, project to the same sites in the PVN—to all the parvicellular subgroups. The dorsal and dorsomedial subdivisions receive the strongest innervation. Terminals that contain the tracer frequently immunostained for PNMT (1154). Adrenergic projections from the C_1 and C_2 cell groups also contain NPY with the C_1 cell group having the higher proportion of doubly labeled cells (1150).

In addition to catecholaminergic input from A_2, C_1, and C_2, the parvicellular subdivisions received innervation from the A_6 cell group in the locus ceruleus. This noradrenergic input preferentially terminated in the periventricular subdivision of the PVN from where hypophysiotropic TRH-li neurons are clustered (1131). Only about 15% of the DBH-li neurons in the locus ceruleus costored NPY (1150). These medullary and pontine catecholaminergic sites projected via different pathways (the ventral and dorsal tegmental bundles, respectively), in the central tegmental area and then to the PVN via the medial forebrain bundle or via periventricular pathways (836).

Neurophysiologic studies demonstrated that stimulation of the PVN antidromically activated units in the A_1 region to which the PVN projects as confirmed by HRP tracing (1127). Non-noradrenergic pathways have been identified from the nucleus of the solitary tract to the A_1 cell group as have noradrenergic pathways from A_1 to A_2 and A_6 (1100).

These same PVN neurons were activated by polysynaptic pathways (presumably passing through the nucleus of the solitary tract to the lateral reticular nucleus) by stimulation of carotid sinus and aortodepressor nerves (1127).

Other studies reported a projection from the "pontine taste area" (the parabrachial nucleus-A_7-noradrenergic cell group) to the PVN (1155,1156). Anterograde tracing studies demonstrated that the A_7 cell group projects to the parvicellular divisions of the PVN, particularly to the dorsal and medial parvicellular regions (which project back to the dorsovagal region in the medulla) and to the intermediolateral cell column in the spinal cord (1157,1158). In addition, the A_7 cell group projected heavily to the central nucleus of the amygdala (1155) which in turn projected to the PVN (1137,1138) and the A_1 region, which, in turn, projected to the PVN (1157).

Other systems reportedly project to the PVN. Buijis et al. (1134) reported that dopaminergic fibers innervate the nucleus homogeneously. Liposits and Paull (1159) lesioned the ascending catecholaminergic bundle and subsequently found TH-li but not DBH-li terminals in the ipsilateral PVN. These terminals were closely associated with CRH-li neurons (1159) and may have arisen from the AN or the PvN. However, Liposits et al. (831) also reported the presence of TH-li cell bodies and fibers in the PVN of the rat. SP-li terminals have been found in all divisions of the PVN (1136). In the rat a projection from the medial raphe nucleus has been demonstrated where as in the Syrian hamster the projection arises in both the dorsal and median raphe nuclei (1075).

The periventricular nucleus (PvN) also receives input from within the MPOA-HTA. GRH-containing fibers from the arcuate nucleus synapse with SRIF-li neurons in the preoptic and hypothalamic periventricular sites (531,917). Descending projections in the rat and guinea pig from the MPOA to the periventricular nucleus by periventricular routes as well as by the MFB have been reported (1016,1059, 1095); as have projections from the bed nucleus of the stria terminalis (1064).

Other hypothalamic regions outside the MPOA-HTA project to the periventricular nucleus (901,1096–1098). Catecholamine terminals have been demonstrated by fluorescence techniques in the periventricular region (583,1124, 1125). Terminals containing ir-PNMT and those containing NPY have been demonstrated to synapse on SRIF-li neurons in the periventricular preoptic and hypothalamic regions (913). In addition GABA-containing terminals synapse with SRIF neurons (915). Although NPY and epinephrine may be stored in the same terminals and their source may be in the adrenergic cell groups in the medulla, the ascending fiber pathways to the narrow periventricular region of the preoptic area and the hypothalamus have not been worked out in detail.

The arcuate nucleus (AN) has been found to receive an input from SRIF-containing neurons in the periventricular nucleus of the hypothalamus (904,917). Hence, within the HTA reciprocal connections between GRH-li neurons in the AN and SRIF-li neurons in the PvN of the rat are present and presumably play a key role in the pulsatile release of growth hormone from the anterior pituitary. Both retrograde and anterograde tracing techniques have demonstrated projections from the MPOA to the AN in rat (1059,1061).

Input from medial temporal lobe structures into the AN in the rat is by direct and indirect routes. In the rat, a direct input from the subiculum to the arcuate nucleus by the corticohypothalamic tract has been proposed by Raisman (1103,1104) as discussed above. It is of interest that this fiber pathway was not described in the brains of cats, monkeys, and humans (766). Other input into the AN from the hippocampal formation via the fornix has been found to be relayed through the septum. Projections from Ammon's horn have been traced to the septum which in turn projected to the medial preoptic nucleus (1102). The medial preoptic nucleus has in turn been demonstrated to project to the periventricular hypothalamus including the arcuate nucleus by both the periventricular route and by the medial forebrain bundle (1061).

The VMN, which lies outside the MPOA-HTA, has been believed to serve as a relay through which the amygdala projects to the arcuate nucleus. Szentagothai et al. reported that "even very small foci in this region produced abundant signs of degeneration in all parts of the ipsi-lateral hypothalamus" (100). Golgi studies by the same group reveal axon projections to the periventricular gray. HRP injection of the median eminence labeled the VMN if the injection spread into the arcuate nucleus (818). Saper et al. (1098) described a projection from the VMN to the AN in the monkey and the cat. A significant projection to the AN in the rat was not found by Canteras et al. (1105) and Sutin and Eager (1160) did not describe any degeneration in the arcuate nucleus of the cat following placement of discrete lesions in the VMN. The axonal projections from the VMN were described by both Szentagothai et al. (100) and Millhouse (1161) as being primarily directed dorsally and posteriorly on the basis of studying Golgi preparations. However, both authors noted abundant collaterals.

Ascending projections to the arcuate nucleus have been identified and are believed to be principally noradrenergic (1162,1163). The norepinephrine content of the arcuate nucleus was reduced 75% by hypothalamic deafferentation (1099). Terminals containing immunoreactive dopamine beta hydroxylase were found in the arcuate nucleus of the rat (1091) and PNMT-li terminals synapse on ir-TH cells bodies there (941). Lesion experiments (1094) and anterograde tracing experiments (1129) suggest that the AN is, like the SON, the PVN, and the periventricular nucleus, predominantly innervated by the A_1 and A_2 cell groups in the medulla. Saper and Loewy (1157) described labeling of the AN following injection of ^{3}H-amino acids into the A_7 cell group in the parabrachial nucleus of the rat. However, such injections could have spread into the ventral tegmental system and further confirmatory studies are indicated. Palkovits et al. pointed out that "the median eminence and any of the hypothalamic nuclei may receive NA fibers from any of the NA cell groups, as monosynaptic communication between them have already been proved" (1094). Noradrenergic terminals have been shown to synapse with GRH-li neurons in the AN (998). Fibers from the raphe nuclei, believed to be indolaminergic, reach the AN via the medial forebrain bundle (1164) and serotonergic terminals have been shown to synapse with dopaminergic neurons and ACTH-li neurons in the AN (995,1000).

As reviewed above, the medial preoptic area is one of the sites in which hypophysiotropic GnRH cells reside in the rat. The MPOA is a complex region with considerable regional cytoarchitectonic specialization (1165). Input into the MPOA arises from several hypothalamic nuclei in the MPOA-HTA periventricular complex. Retrograde HRP studies demonstrated afferents from the AN and PVN (1060); and, synapses have been demonstrated between SP-li terminals and terminals containing opioid peptides (1052,1053) (presumably from the AN) and CRH-li terminals (presumably from the PVN) and GnRH li neurons (1048,1050). Afferents from hypothalamic nuclear groups outside the HTA have also been demonstrated and include projections from the VMN (1096,1105), DMN and mammillary complex (1060). Descending input from the lateral septal nucleus via the medial forebrain bundle and from the medial septal nucleus via the periventricular system have been demonstrated by ^{3}HHH-amino acid autoradiographic studies (1064). Afferents from the bed nucleus of the stria terminalis (1060) and from the medial and central nuclei of the amygdala have also been demonstrated (1166). Ascending input from the A_1 and A_2 CA cell groups in the medulla was shown by double-labeling with HRP injection into the MPOA and fluorescent staining for catecholamines of brainstem nuclear groups (1151). As reviewed above both NPY-li and presumptive noradrenergic terminals have been shown to synapse with GnRH neurons (1043,1045). The argument that this ascending noradrenergic pathway that costores NPY plays a role in stimulating GnRH neurons has been reviewed by Kalra (1167,1168). Input from the A_7 area of the parabrachial nucleus [the "pontine taste area" of Nogren (1155)] has been demonstrated by retrograde transport of HRP (1060); but this input is peptidergic, not noradrenergic, as some fibers contain IR-CRH and others contain IR-leu-enk (1169). The B_7, B_8, and B_9 raphe cell groups also projected to the MPOA (1164). The projection was serotonergic and sexually dimorphic (1165). Serotoninergic terminals have been shown to synapse with GnRH-li neurons (995).

The neuroendocrine motor units that regulate anterior pituitary function have turned out to be more complex than previously believed. They co-store other peptides and transmitters that (presumably) facilitate their hypophysiotropic function at the surface of anterior pituitary cells but which may also serve to regulate the function of neighboring terminals in the neurohemal contact zone by paracrine interactions and thus provide a coordinated discharge of neurohormones into the portal system. In addition, collateral axons of these neurons project to sites where other (neurosecretory) neurons reside and synapse with them to alter their function. The extrasegmental input on each of these neurosecretory cell groups arises from sites within the medial hypothalamus where cholinergic, GABAergic and glutaminergic cell bodies are thought to reside, from limbic and rhinencephalic sources, and from the brain stem. Further investigations as to the nature of this input to the GnRH motor neuron will be the subject of further discussions in this volume and of further research in the near and distant future.

REFERENCES

1. Cushing, H. (1912). *The Pituitary Body and Its Disorders.* J. B. Lippincott, Philadelphia.
2. Cushing, H. (1930). Neurohypophysial mechanisms from a clinical standpoint. Part 2. *Lancet.* 2, 175–184.
3. Cushing, H. (1930). Neurohypophysial mechanisms from a clinical standpoint. Part 1. *Lancet* 2, 119–127.
4. Herring, P. T. (1908). The histological appearances of the mammalian pituitary body. *Q. J. Exp. Physiol.* 1, 121–159.
5. Herring, P. T. (1908). The development of the mammalian pituitary and its morphological significance. *Q. J. Exp. Physiol.* 1, 161–185.
6. Tilney, F. (1911). Contribution to the study of the hypophysis cerebri with especial reference to its comparative histology. *Mem. Wistar Inst. Anat. Biol.* 2.
7. Dandy, W. E., and Goestsh, E. L. L. (1910). The blood supply of the pituitary body. *Am. J. Anat.* 11, 137–150.
8. Frolich, A. (1901). Ein fall von tumor der hypophysis cerebri ohne akromegalie. *Wein. Klin. Rundschau.* XV.
9. Crowe, S. J., Cushing, H., and Homans, J. (1910). Experimental hypophysectomy. *Bull. Johns Hopkins Hosp.* 21.
10. Crowe, S. J., Cushing, H., and Homans, J. (1909). Effects of hypophyseal transplantation following hypophysectomy in the canine. *Q. J. Exp. Physiol.* 2.

11. Smith, P. E. (1930). Hypophysectomy and a replacement therapy in the rat. *Am. J. Anat.* 45.
12. Dandy, W. E. (1913). The nerve supply to the pituitary body. *Am. J. Anat.* 15, 333–343.
13. Cushing, H., and Goetsch, E. (1910). Concerning the secretion of the infundibular lobe of the pituitary body and its presence in the cerebrospinal fluid. *Am. J. Physiol.* 27, 60–86.
14. Goetsch, E., Cushing, H., and Jacobson, C. (1911). Carbohydrate tolerance and the posterior lobe of the hypophysis cerebri. An experimental and clinical study. *Bull. Johns Hopkins Hosp.* 22, 165–190.
15. Weed, L. H., Cushing, H., and Jacobson, C. (1913). Further studies on the role of the hypophysis in the metabolism of carbohydrates. The autonomic control of the pituitary gland. *Johns Hopkins Hosp. Bull.* 24, 40–52.
16. Oliver, G., Schafer, E. A. (1895). On the physiological action of extracts of pituitary body and certain other glandular organs. *J. Physiol.* 18, 277–279.
17. Howell, W. H. (1898). The physiologic effects of extracts of the hypophysis cerebri and infundibular body. *J. Exp. Med.* 3, 254–258.
18. Dale, H. H. (1909). The action of extracts of the pituitary body. *Biochem. J.* 4, 427–447.
19. Popa, G., and Fielding, U. (1931). A portal circulation from the pituitary to the hypothalamic region. *J. Anat.* 65, 88–91.
20. Popa, G., and Fielding, U. (1930). The vascular link between the pituitary and the hypothalamus. *Lancet* 2, 238–240.
21. Cushing, H. (1933). Posterior pituitary activity from an anatomical standpoint. *Am. J. Path.* 9, 139–179.0
22. Gersh, I. (1938). Relation of histological structure to the active substances extracted from the posterior lobe of the hypophysis. *Res. Publ. Assoc. Res. Nerv. Ment. Dis.* 17, 433–436.
23. Gersh, I. (1939). The structure and function of the parenchymatous glandular cells in the neurohypophysis of the rat. *Am. J. Anat.* 64, 407–443.
24. Lewis, D., and Lee, F. C. (1927). On the glandular elements in the posterior lobe of the human hypophysis. *Bull. Johns Hopkins Hosp.* 41, 241–277.
25. Rioch, D. (1938). Paths of secretion from the hypophysis. *Res. Publ. Assoc. Res. Nerv. Ment. Dis.* XVII, 151–171.
26. Bucy, P. C. (1930). The pars nervosa of the bovine hypophysis. *J. Comp. Neurol.* 50, 505–511.
27. Bucy, P. C. (1932). The hypophysis cerebri. In *Cytology and Cellular Pathology of the Central Nervous System* (W. Penfield, Ed.), Vol. 2. Hoeber, New York.
28. Wislocki, G. B., and King, L. S. (1936). The permeability of the hypophysis and hypothalamus to vital dyes, with a study of the hypophyseal vascular supply. *Am. J. Anat.* 58, 421–472.
29. Rioch, D., Wislocki, G., and O'Leary, J. (1940). A precis of preoptic, hypothalamic and hypophyseal terminology with atlas. *Res. Publ. Assoc. Res. Nerv. Ment. Dis.* 20, 3–30.
30. Starling, E. H., and Verney, E. B. (1925). The secretion of urine as studied on the isolated kidney. *Proc. R. Soc. Lond.* 97, 321–363.
31. Fisher, C., Ingram, W. R., Hare, W. K., and Ranson, S. W. (1935). The degeneration of the supraoptico-hypophyseal system in diabetes insipidus. *Anat. Rec.* 63, 29–52.
32. Fisher, C., Ingram, W. R., and Ranson, S. W. (1935). The relation of the hypothalamo-hypophyseal system to diabetes insipidus. *Arch. Neurol. Psychiatr.* 34, 124–163.
33. Ingram, W. R., Fisher, C., and Ranson, S. W. (1936). Experimental diabetes insipidus in the monkey. *Arch. Int. Med.* 57, 1067–1080.
34. Scharrer, E., and Scharrer, B. (1939). Secretory cells within the hypothalamus. *Res. Publ. Assoc. Res. Nerve Ment. Dis.* 20, 170–194.
35. Scharrer, B., and Scharrer, E. (1944). Neurosecretion VI. A comparison between the intercerebralis-cardiacum-allatum of the insects and the hypothalamo-hypophyseal system of the vertebrates. *Biol. Bull.* 87, 242–251.
36. Palay, S. L. (1945). Neurosecretion. VII. The preoptico-hypophysial pathway in fishes. *J. Comp. Neurol.* 82, 129–143.
37. Bargmann, W., and Hild, W. (1949). Über die morphologie der neurosekretorischen verknupfung von hypothalamus und neurohypophyse. *Acta Anat.* 8, 264–280.
38. Bargmann, W., and Scharrer, E. (1951). The site of origin of the hormones of the posterior pituitary. *Amer. Sci.* 39, 255–259.
39. Scharrer, E., and Scharrer, B. (1954). Hormones produced by neurosecretory cells. *Rec. Prog. Horm. Res.* 10, 183–240.
40. Scharrer, E. (1965). The final common path in neuroendocrine integration. *Arch. d'Anatomie Microsc.* 54, 359–370.
41. Scharrer, B. (1967). The neurosecretory neuron in neuroendocrine regulatory mechanisms. *Am. Zoologist.* 7, 161–169.
42. Pierce, J. G., and Vignaud, Vd. (1950). Studies on high potency oxytotic materials from beef posterior pituitary lobes. *J. Biol. Chem.* 186, 77–84.
43. Vignaud, Vd., Ressler, C., Swan, J. M., Roberts, W., Katsoyannis, P. G., and Gordon, S. (1953). The synthesis of an octapeptide amide with the hormonal activity of oxytocin. *J. Am. Chem. Soc.* 75, 4879–4880.
44. Turner, R. A., Pierce, J. G., and Vignaud, Vd. (1951). The purification and the amino acid content of vasopressin preparations. *J. Biol. Chem.* 191, 21–28.
45. Vignaud, Vd., Lawler, H. C., and Papenea, E. A. (1953). Enzymatic cleavage of glycinamide from vasopressin and a proposed structure for this pressor-antidiuretic hormone of the posterior pituitary. *J. Am. Chem. Soc.* 75, 4880–4881.
46. Vignaud, Vd., Gish, D., and Katsayannis, P. G. (1954). A synthetic preparation possessing biological properties associated with arginine vasopressin. *J. Am. Chem. Soc.* 76, 4751–4752.
47. VanDyke, H. B., Adamsons, K., and Engel, S. L. (1955). Aspects of the biochemistry and physiology of the neurohypophyseal hormones. *Rec. Prog. Horm. Res.* XI, 1–42.
48. Verney, E. B. (1947). The antidiuretic hormone and the factors which determine its release. *Proc. R. Soc. Lond.* 135, 25–106.
49. Palay, S. L. (1955). An electron microscope study of the neurohypophysis in normal, hydrated, and dehydrated rats. *Anat. Rec.* 121, 348.
50. Palay, S. L. (1960). The fine structure of secretory neurons in the preoptic nucleus of the goldfish (Carassius auratus). *Anat. Rec.* 138, 417–444.
51. Sloper, J. C., Arnott, D., and King, B. C. (1960). Sulphur metabolism in the pituitary and hypothalamus of the rat: a study of radioisotope-uptake after the injection of 35S dl-cysteine, methionine, and sodium sulphate. *J. Endocrinol.* 20, 9–23.
52. Sloper, J. C. (1972). The validity of current concepts of hypothalamo-neurohypophyseal neurosecretion. *Prog. Brain Res.* 38, 123–143.
53. Bargmann, W. (1966). Neurosecretion. *Int. Rev. Cytol.* 19, 183–201.
54. Norstrom, A., Hansson, H.-A., and Sjostrand, J. (1971). Effects of colchicine on axonal transport and ultrastructure of the hypothalamo-neurohypophyseal system of the rat. *Z. Zellforsch.* 113, 271–293.
55. Lederis, K. (1974). Neurosecretion and the functional structure of the neurohypophysis. In *The Pituitary Gland and Its Neuroendocrine Control.* Part 1. Vol. IV. *Handbook of Physiology* (R. O. Greep and E. B. Astwood, Eds.), pp. 81–102. Am. Physiol. Soc., Washington, DC.
56. Silverman, A. J., and Zimmerman, E. A. (1975). Ultrastructural immunocytochemical localization of neurophysin and vasopressin in the median eminence and posterior pituitary of the guinea pig. *Cell. Tissue Res.* 159, 291–301.

57. Flament-Durand, J., Couck, A. M., and Dustin, P. (1975). Studies on the transport of secretory granules in the magnocellular hypothalamic neurons of the rat. II. Action of vincrestine on axonal flow and neurotubules in the paraventricular and supraoptic nuclei. *Cell. Tissue Res*. 164, 1–9.
58. Vandesande, F., Dierickx, K., and DeMay, J. (1975). Identification of separate vasopressin-neurophysin II and oxytocin-neurophysin I containing nerve fibers in the external region of the bovine median eminence. *Cell. Tissue Res*. 158, 509–516.
59. Weitzman, R. E., and Fisher, D. A. (1977). Log linear relationship between plasma arginine vasopressin and plasma osmolality. *Am. J. Physiol*. 233, E37–E40.
60. Gainer, H., Sarne, Y., and Brownstein, M. J. (1977). Biosynthesis and axonal transport of rat neurohypophysial proteins and peptides. *J. Cell. Biol*. 73, 366–381.
61. Antunes, J. L., and Zimmerman, E. A. (1978). The hypothalamic magnocellular system of the rhesus monkey: an immunocytochemical study. *J. Comp. Neurol*. 181, 539–565.
62. Brownstein, M. J., Russell, J. T., and Gainer, H. (1980). Synthesis, transport, and release of posterior pituitary hormones. *Science* 207, 373–378.
63. Russell, J. T., Brownstein, M. J., and Gainer, H. (1980). Biosynthesis of vasopressin, oxytocin, and neurophysins: isolation and characterization of two common precursors (propressophysin and prooxyphysin). *Endocrinology* 107, 1880–1891.
64. Harris, G. W. (1937). The induction of ovulation in the rabbit by electrical stimulation of the hypothalamo-hypophyseal mechanism. *Proc. R. Soc. Lond*. 122, 374–394.
65. Green, J. D. (1951). The comparative anatomy of the hypophysis, with special reference to its blood supply and innervation. *Am. J. Anat*. 88, 225–311.
66. Green, J. D., and Harris, G. W. (1947). The neurovascular link between the neurohypophysis and adenohypophysis. *J. Endocrinol*. 4, 136–146.
67. Green, J. D. (1948). The histology of the hypophyseal stalk and median eminence in man with special reference to blood vessels, nerve fibers and a peculiar neurovascular zone in this region. *Anat. Rec*. 100, 273–295.
68. Green, J. D. (1947). Vessels and nerves of amphibian hypophyses. *Anat. Rec*. 99, 21–53.
69. Green, J. D, and Harris, G. W. (1949). Observation of the hypophysio-portal vessels of the living rat. *J. Physiol*. 108, 359–361.
70. Harris, G. W. (1948). Electrical stimulation of the hypothalamus and the mechanism of neural control of the adenohypophysis. *J. Physiol*. 107, 418–429.
71. Harris, G. W. (1949). The relationship of the nervous system to (a) the neurohypophysis and (b) the adenohypophysis. *J. Endocrinol*. 6, xvii–xix.
72. Harris, G. W. (1950). Oestrous rhythm pseudopregnancy and the pituitary stalk in the rat. *J. Physiol. (Lond)*. 111, 347–360.
73. Harris, G. W., and Johnson, R. T. (1950). Regeneration of the hypophyseal portal vessels, after section of the hypophyseal stalk in the monkey (Macacus rhesus). *Nature (Lond)*. 165, 819–820.
74. Harris, G. W. (1949). Regeneration of the hypophysial portal vessels. *Nature (Lond)*. 163, 70.
75. Harris, G. W., and Jacobson, D. (1952). Functional grafts of the anterior pituitary gland. *Proc. R. Soc. Lond*. 139, 263–276.
76. Harris, G. W. (1962). Neuroendocrine relations. *Res. Publ. Assoc. Res. Nerv. Ment. Dis*. 40, 380–405.
77. Green, J. D. (1966). The comparative anatomy of the portal vascular system and of the innervation of the hypophysis. In *The Pituitary Gland* (G. W. Harris and B. T. Donovan, Eds.), Vol. 1. University of California Press, Berkley.
78. Wingstrand, K. G. (1966). Comparative anatomy and evaluation of the hypophysis. In *The Pituitary Gland* (G. W. Harris and B. T. Donovan, Eds.), Vol. 1. University of California Press, Berkley.
79. Jorgensen, C. B., and Larsen, L. O. (1967). Neuroendocrine mechanisms in lower vertebrates. In *Neuroendocrinology* (L. Martini and W. F. Ganong, Eds.), Vol. II. Academic Press, New York.
80. Xuereb, G. P., Prichard, M. M., and Daniel, P. M. (1954). The arterial supply and venous drainage of the human hypophysis cerebri. *Am. J. Exper. Physiol*. 39, 199–217.
81. Xuereb, G. P., Prichard, M. L., and Daniel, P. M. (1954). The hypophysial portal system of vessels in man. *Q. J. Exper. Physiol*. 39, 219–299.
82. Daniel, P. M. (1966). The blood supply of the hypothalamus and pituitary gland. *Br. Med. Bull*. 22, 202–208.
83. Harris, G. W. (1947). The blood vessels of the rabbit's pituitary gland, and the significance of the pars and zona tuberalis. *J. Anat*. 81, 343–351.
84. Page, R. B., Munger, B. L., and Bergland, R. M. (1976). Scanning microscopy of pituitary vascular casts: the rabbit pituitary portal system revisited. *Am. J. Anat*. 146, 273–301.
85. Wislocki, G. B. (1937). The vascular supply of the hypophysis cerebri of the cat. *Anat. Rec*. 69, 361–387.
86. Wislocki, G. (1938). Further observations on the blood supply of the hypophysis cerebri of the rhesus monkey. *Anat. Rec*. 72, 137–150.
87. Wislocki, G. (1938). The vascular supply of the hypophysis cerebri of the rhesus monkey and man. *Res. Publ. Assoc. Res. Nerv. Ment. Dis*. 17, 48–68.
88. Moll, J. (1957). Regeneration of the supraoptico-hypophyseal and paraventriculo-hypophyseal tracts in the hypophysectomized rat. *Z. Zellforsch*. 46, 686–709.
89. Moll, J., and DeWied, D. (1962). Observations on the hypothalamo-posthypophyseal system of the posterior lobectomized rat. *Gen. Comp. Endocrin*. 2, 215–228.
90. Adams, J. H., Daniel, P. M., and Prichard, M. M. (1969). Degeneration and regeneration of hypothalamic nerve fibers in the neurohypophysis after pituitary stalk section in the ferret. *J. Comp. Neurol*. 135, 121–144.
91. Daniel, P. M., and Prichard, M. M. (1970). Regeneration of hypothalamic nerve fibers after hypophysectomy in the goat. *Acta Endocrinol*. 64, 696–704.
92. Daniel, P. M., Duchen, L. W., and Prichard, M. M. (1964). The effect of transection of the pituitary stalk on the cytology of the pituitary gland of the rat. *Q. J. Exp. Physiol*. 49, 235–242.
93. Daniel, P. M., and Prichard, M. M. (1958). The effects of pituitary stalk section in the goat. *Am. J. Path*. 34, 433–469.
94. Adams, J. H., Daniel, P. M., and Prichard, M. M. (1969). The blood supply of the pituitary gland of the ferret with special reference to infarction after stalk section. *J. Anat*. 104, 209–225.
95. Daniel, P. M., and Prichard, M. M. (1972). The human hypothalamus and pituitary stalk after hypophysectomy or pituitary stalk section. *Brain (Oxf)*. 95, 813–824.
96. Daniel, P. M., and Prichard, M. M. (1975). Studies of the hypothalamus and the pituitary gland. With special reference to the effects of transection of the pituitary stalk. *Acta Endocrinol*. 80, 1–216.
97. Nikitovitch-Winer, M., and Everett, J. W. (1958). Functional restitution of pituitary grafts retransplanted from kidney to median eminence. *Endocrinology* 63, 916–930.
98. Nikitovitch-Winer, M., and Everett, J. W. (1959). Histologic changes in grafts of rat pituitary on the kidney and upon re-transplantation under the diencephalon. *Endocrinology* 65, 357–368.
99. Halasz, B., and Pupp, L. (1965). Hormone secretion of the anterior pituitary gland after physical interruption of all

nervous pathways to the hypophysiotropic area. *Endocrinology* 77, 553–562.
100. Szentagothai, J., Flerko, B., Mess, B., and Halasz, B. (1968). *Hypothalamic Control of Anterior Pituitary Function.* Akademiai Kiado, Budapest.
101. Halasz, B., and Gorski, R. A. (1967). Gonadotropic hormone secretion in female rats after partial or total interruption of neural afferents to the medial basal hypothalamus. *Endocrinology* 80, 608–622.
102. Halasz, B., Slusher, M., and Gorski, R. A. (1967). Adrenocorticotropic hormone secretion in rats after partial or total deafferentation of the medial basal hypothalamus. *Neuroendocrinology* 2, 43–55.
103. Halasz, B., Florsheim, W. H., Corcorran, N. L., and Gorski, R. A. (1967). Thyrotrophic hormone secretion after partial or total interruption of neural afferents to the medial basal hypothalamus. *Endocrinology* 80.
104. Halasz, B. (1969). The endocrine effects of isolation of the hypothalamus from the rest of the brain. *Frontiers in Neuroendocrinology*. Oxford University Press, New York.
105. Halasz, B. (1973). Neural control of pituitary ACTH secretion under resting conditions. *Acta Medica*. 29, 71–76.
106. Guillemin, R. (1964). Hypothalamic factors releasing pituitary hormones. *Rec. Prog. Horm. Res.* 20, 89–130.
107. Schally, A. V., Arimura, A., Bowers, C. Y., Kastin, A. J., Sawano, S., and Redding, T. W. (1968). Hypothalamic neurohormones regulating anterior pituitary function. *Rec. Prog. Horm. Res.* 24, 497–588.
108. Saffran, M., and Schally, A. V. (1955). Release of corticotropin by anterior pituitary tissue in vitro. *Can. J. Biochem. Physiol.* 33, 408–415.
109. Guillemin, R. (1955). Hypothalamic control of anterior pituitary study with tissue cultures techniques. *Fed. Proc.* 14, 65.
110. Guillemin, R., and Rosenberg, B. (1955). Humoral hypothalamic control of anterior pituitary: a study with combined tissue cultures. *Endocrinology* 57, 599–607.
111. Guillemin, R., Hearn, W. R., Cheek, W. R., and Housholder, D. E. (1957). Control of corticotropin release: further studies with in vitro methods. *Endocrinology* 60, 488–506.
112. Schreiber, V., Rybak, M., Eckertova, A., Jirgl, V., Koci, J., Franc, Z., and Kmentova, V. (1962). Isolation of a hypothalamic peptide with TRF (thyrotropin releasing factor) activity in vitro. *Experentia* 18, 338–340.
113. Deuben, R. R., and Meites, J. (1964). Stimulation of pituitary growth hormone release by a hypothalamic extract "in vitro." *Endocrinology* 74, 408–414.
114. Schally, A. V., and Bowers, C. Y. (1964). In vitro and in vivo stimulation of the release of luteinizing hormone. *Endocrinology* 75, 312–320.
115. Gala, R. R., and Reece, R. P. (1965). In vitro lactogen production by anterior pituitaries from various species. *Proc. Soc. Exper. Biol. Med.* 120, 263–264.
116. Meites, J., Kahn, R. H., and Nicole, C. S. (1961). Prolactin production by rat pituitary in vitro. *Pro. Soc. Exper. Biol. Med.* 108, 440–443.
117. Pasteels, J. L. (1962). Administration d'extraits hypothalamiques a' l'hypophyse de rat in vitro, dans le but d'en controler la secretion du prolactaire. *Comp. Rendu.* 254, 2664–2666.
118. Talwaker, P. K., Ratner, A., and Meites, J. (1963). In vitro inhibition of pituitary prolactin synthesis and release by hypothalamic extract. *Am. J. Physiol.* 205, 213–218.
119. Porter, J. C., and Jones, J. C. (1956). Effect of plasma from hypophyseal-portal vessel blood on adrenal-ascorbic acid. *Endocrinology* 58, 62–67.
120. Porter, J. C., Dhariwal, A. P. S., and McCann, S. M. (1967). Response of the anterior pituitary-adrenocortical axis to purified CRF. *Endocrinology* 80, 679–688.
121. Guillemin, R., Yamazaki, E., Jutisz, M., and Sakig, E. (1962). Presence dans un extrait de tissus hypothalamiques d'une substance stimulant le secretion de l'hormone hypophysaire threotrope. *Comp. Rendu.* 255, 1018–1020.
122. Garcia, J. F., and Geschwind, I. (1966). Increase in plasma growth hormone levels in the monkey following the administration of sheep hypothalamic extracts. *Nature (Lond).* 211, 372–374.
123. Dhariwal, A. P. S., Nallar, R., Batt, M., and McCann, S. M. (1965). Separation of follicle stimulating hormone-releasing factor from luteinizing hormone releasing factors. *Endocrinology* 76, 290–296.
124. Igarashi, M., and McCann, S. M. (1964). A hypothalamic follicle stimulating hormone-releasing factor. *Endocrinology* 74, 446–452.
125. McCann, S. M., Taleisnik, S., and Friedman, H. M. (1960). LH-releasing activity in hypothalamic extracts. *Proc. Soc. Exper. Biol. Med.* 104, 432–434.
126. McCann, S. M. (1962). A hypothalamic leutinizing-hormone-releasing factor. *Am. J. Physiol.* 202.
127. Porter, J. C., Mical, R. S., Kamberi, I. A., and Grazia, Y. R. (1970). A procedure for the cannulation of a pituitary stalk portal vessel and perfusion of the pars distalis in the rat. *Endocrinology* 87, 197–201.
128. Porter, J. C., Mical, R. S., Ondo, J. G., and Kamberi, I. A. (1972). Perfusion of the rat anterior pituitary via a cannulated portal vessel. *Acta Endocrinol. Suppl.* 158, 249–266.
129. Kamberi, I. A., Mical, R. S., and Porter, J. C. (1971). Pituitary portal infusion of hypothalamic extract and release of LH, FSH and prolactin. *Endocrinology* 88, 1294–1299.
130. Kamberi, I. A., Mical, R. S., and Porter, J. C. (1971). Hypophysial portal vessel infusion: in vivo demonstration of LRF, FRF, and PIF in pituitary stalk plasma. *Endocrinology* 89, 1042–1046.
131. Averill, R. L., and Kennedy, T. H. (1967). Elevation of thyrotropin release by intrapituitary infusion of crude hypothalamic extracts. *Endocrinology* 81, 113–120.
132. Fink, G., Nallar, R., and Worthington, W. C. (1967). The demonstration of luteinizing hormone releasing factor in hypophyseal portal blood of pro-estrus and hypophysectomized rats. *J. Physiol. (Lond).* 191, 401–426.
133. Wilbur, J., and Porter, J. C. (1970). Thyrotropin and growth hormone releasing activity in hypophyseal portal blood. *Endocrinology* 87, 807–811.
134. Boler, J., Enzmann, F., and Folkers, K. (1969). The identity of chemical and hormonal properties of the thyrotropin releasing hormone and pyroglutamyl-histidylproline amide. *Biochem. Biophys. Res. Comm.* 37, 705–710.
135. Burgus, R., Dunn, T. F., Desiderio, D. M., Ward, D. N., Vale, W., and Guillemin, R. (1970). Characterization of ovine hypothalamic hypophysiotropic TSH releasing factor. *Nature (Lond).* 226, 321–325.
136. Burgus, R., Dunn, T. F., Desiderio, D. M., Ward, D. N., Vale, W., Guillemin, R., Felix, A. M., Gillessen, D., and Stader, R. O. (1970). Biological activity of synthetic polypeptide derivatives related to the structure of hypothalamic TRF. *Endocrinology* 86, 573–582.
137. Porter, J. C., Vale, W., Burgus, R., Mical, R. S., and Guillemin, R. (1971). Release of TSH by TRF infused directly into a pituitary stalk portal vessel. *Endocrinology* 89, 1054–1056.
138. Geiger, R., Konig, W., Wissmann, H., Geisen, K., and Enzmann, F. (1971). Synthesis and characterization of a decapeptide having LH-RH/FSH-RH activity. *Biochem. Biophys. Res. Comm.* 45, 767–773.
139. Arimura, A., Matsuo, H., Baba, Y., Debeljuk, L., Sandow, J., and Schally, A. V. (1972). Stimulation of release of LH by

synthetic LH-RH in vivo. I. A comparative study of natural and synthetic hormones. *Endocrinology* 90, 163–168.
140. Schally, A. V., Arimura, A., Baba, Y., Nair, R. M. G., Matsuo, H., Redding, T. W., Debeljuk, L., and White, W. F. (1971). Isolation and properties of FSH and LH releasing hormone. *Biochem. Biophys. Res. Comm.* 43, 393–394.
141. Spiess, J., Rivier, J., Rivier, C., and Vale, W. (1981). Primary structure of corticotropin-releasing factor from ovine hypothalamus. *Proc. Natl. Acad. Sci. U S A* 78, 6517–6521.
142. Vale, W., Spiess, J., Rivier, C., and Rivier, J. (1981). Characterization of a 41-residue ovine hypothalamic peptide that stimulates secretion of corticotropin and B-endorphin. *Science* 213, 1394–1397.
143. Brazeau, P., Vale, W., Burgus, R., Ling, N., Butcher, M., Rivier, J., and Guillemin, R. (1973). Hypothalamic polypeptide that inhibits the secretion of immunoreactive pituitary growth hormone. *Science* 179, 77–79.
144. Rivier, J., Spiess, J., Thorner, M., and Vale, W. (1982). Characterization of a growth hormone-releasing factor from a human pancreatic islet tumor. *Nature (Lond).* 300, 276–278.
145. Thorner, M. O., Spiess, J., Vance, M. L., Rogol, A. D., Kaiser, D. L., Webster, J. D., Rivier, J., Borges, J. L., Bloom, S. R., Cronin, M. J., Evans, W. S., MacLeod, R. M., and Vale, W. (1983). Human pancreatic growth-hormone–releasing factor selectively stimulates growth-hormone secretion in man. *Lancet* 1.
146. Spiess, J., Rivier, J., and Vale, W. (1983). Characterization of rat hypothalamic growth hormone-releasing factor. *Nature (Lond).* 300, 276–278.
147. Shaar, C. J., and Clemens, J. H. (1974). The role of catecholamines in the release of anterior pituitary prolactin in vitro. *Endocrinology* 95, 1202–1212.
148. Gibbs, D. M., and Neill, J. D. (1978). Dopamine levels in hypophyseal stalk blood are sufficient to inhibit prolactin secretion in vivo. *Endocrinology* 102, 1895–1900.
149. Pilotte, N. S., Gudelsky, G. A., and Porter, J. C. (1980). Relationship of prolactin secretion to dopamine release into hypophyseal portal blood and dopamine turnover in the median eminence. *Brain Res.* 193, 284–288.
150. Gudelsky, G. A., and Porter, J. C. (1980). Release of dopamine from tubero-infundibular neurons into pituitary stalk blood after prolactin or haloperidol administration. *Endocrinology* 106, 526–529.
151. Selmanoff, M. (1981). The lateral and medial median eminence: distribution of dopamine, norepinephrine, and luteinizing hormone-releasing hormone and the effect of prolactin on catecholamine turnover. *Endocrinology* 108, 1716–1722.
152. Foord, S. M., Peters, J. R., Dieguez, C., Scanlon, M. F., and Hall, R. (1983). Dopamine receptors on intact anterior pituitary cells in culture: functional association with the inhibition of prolactin and thyrotropin. *Endocrinology* 112, 1567–1577.
153. Frawley, L. S., and Neill, J. D. (1984). Brief decreases in dopamine result in surges of prolactin secretion in monkeys. *Am. J. Physiol.* 247, E778–E780.
154. Sternberger, L. (1979). *Immunocytochemistry*. John Wiley & Sons, New York.
155. Nakane, P. K., and Pierce, G. B. (1967). Enzyme-labeled antibodies for the light and electron microscopic localization of tissue antigens. *J. Cell. Biol.* 33, 307–318.
156. Hokfelt, T., Elde, R., and Fuxe, K., et al. (1978). Aminergic and peptidergic pathways in the nervous system with special reference to the hypothalamus. In *The Hypothalamus* (S. Reichlin, R. J. Baldessarini, and J. B. Martin, Eds.). Raven Press, New York.
157. Fink, G., and Geffen, L. B. (1978). The hypothalamo-hypophysial system: model for central peptidergic and monoaminergic transmission. *Int. Rev. Physiol.* 17, 1–48.
158. Naik, D. V. (1973). Electron microscopic-immunocytochemical localization of adrenocorticotropin and melanocyte stimulating hormone in the pars intermedia cells of rats and mice. *Z. Zellforsch.* 142, 305–328.
159. Naik, D. V. (1975). Immunoreactive LH-RH neurons in the hypothalamus identified by light and fluorescent microscopy. *Cell. Tissue Res.* 157, 423–436.
160. Naik, D. V. Immunoelectron microscopic localization of luteinizing hormone-releasing hormone in the arcuate nuclei and median eminence of the rat. *Cell. Tissue Res.* 157, 437–455.
161. Styne, D. M., Goldsmith, P. C., Brustein, S. R., Kaplan, S. L., and Grumbach, M. M. (1977). Immunoreactive somatostatin and luteinizing hormone releasing hormone in median eminence synaptosomes of the rat: detection by immunohistochemistry and quantification by radioimmunoassay. *Endocrinology* 101, 1099–1103.
162. Barry, J., Dubois, M. P., and Poulain, P. (1973). LRF producing cells of the mammalian hypothalamus. A fluorescent antibody study. *Z. Zellforsch.* 146, 351–366.
163. Barry, J., Dubois, M. P., Poulain, P., and Leonardelli, J. (1973). Neuroendocrinologie-characterisation et topographic des neurones hypothalamiques immunoreactifs avec des anticorps anti-LRH de synthese. *C. R. Acad. Sci. Ser. D.* 276, 3191–3193.
164. Swanson, L. W. (1987). The Hypothalamus. In *Handbook of Chemical Neuroanatomy* (A. Bjorklund, T. Hokfelt, and L. W. Swanson, Eds.), Vol. 5, pp. 1–124. Elsevier, Amsterdam.
165. Hymer, W. C., and McShan, W. H. (1962). Isolation of cytoplasmic pituitary granules by column chromatography. *J. Cell. Biol.* 13, 350–354.
166. Hymer, W. C., and McShan, W. H. (1963). Isolation of rat pituitary granules and the study on their biological properties and hormonal activities. *J. Cell. Biol.* 17, 67–86.
167. Hymer, W. C. (1975). Separation of organelles and cells from the mammalian adenohypophysis. In *The Anterior Pituitary. Ultrastructure in Biological Systems* (A. Tixier-Vidal and M. G. Farquhar, Eds.), Vol. 7. Academic Press, New York.
168. Dunphy, W. G., and Rothman James, E. (1985). Compartmental organization of the Golgi Stack. *Cell* 42, 13–21.
169. Inoue, K., and Kurosumi, K. (1989). Ultrastructural observation of the trans-golgi associated plate-like cisterna in the secretory cells of the rat anterior pituitary gland with special reference to the intracisternal skeleton. *Anat. Rec.* 225, 272–278.
170. Novikoff, P. M., Novikoff, A. B., Quintana, N., and Hauw, J.-J. (1971). Golgi apparatus, gerl, and lysosomes of neurons in the rat dorsal ganglia. Studies by thick section and thin section cytochemistry. *J. Cell. Biol.* 50, 859–886.
171. Farquhar, M. G., and Palade, G. E. (1981). The Golgi apparatus (complex)—(1954–1981)—from artifact to center stage. *J. Cell. Biol.* 91, 77s–103s.
172. Farquhar, M. G., and Wellings, S. R. (1957). Electron microscopic evidence suggesting secretory granule formation within the Golgi apparatus. *J. Biophys. Biochem. Cytol.* 3, 319–321.
173. Farquhar, M. G. (1961). Origin and fate of secretory granules in cells of the anterior pituitary gland. *Trans. N Y Acad. Sci.* 23, 346.
174. Racadot, J., Olivier, L., Porcile, E., and Droz, B. (1965). Appareil de golgi et origine des graines de secretion dans les cellules adenohypophysaires chez le rat. Etude radioautographique en microscopie electronique apres injection de leucine tritiee. *Comp. Rendu.* 261, 2972–2974.
175. Tixier-Vidal, A., and Picart, R. (1967). Etude quantitative par radioautographie au microscope electronique de l'utilisation de la DL-leucine-3H par les cellules de l'hypophyse du canard en culture organotypiqe. *J. Cell. Biol.* 35, 501–519.
176. Howell, S. L., and Whitfield, M. (1973). Synthesis and secretion of growth hormone in the rat anterior pituitary.

I. The intracellular pathway, its time course and energy requirements. *J. Cell. Sci.* 12, 1–21.

177. Farquhar, M. G., Reid, J. J., and Daniell, L. W. (1978). Intracellular transport and packaging of prolactin: a quantitative electron microscope autoradiographic study of mammotrophs dissociated from rat pituitaries. *Endocrinology* 102, 296–311.
178. Blobel, G., and Dobberstein, B. (1975). Transfer of proteins across membranes. *J. Cell. Biol.* 67, 835–851.
179. Lingappa, V. R., Devillers-Thiery, A., and Blobel, G. (1977). Nascent prehormones are intermediates in the biosynthesis of authentic bovine pituitary growth hormone and prolactin. *Proc. Natl. Acad. Sci. U S A* 74, 2432–2436.
180. Gavier, M. F., Aoki, A., and Orgnero de Gaisan, E. (1999). Prolactin secretory bypath exposed in cultured lactotrophs. *Histochem. J.* 31, 661–670.
181. Rosenzweig, L. J., and Farquhar, M. G. (1980). Sites of sulfate incorporation into mammotrophs and somatotrophs of the rat pituitary as determined by quantitative electron microscopic autoradiography. *Endocrinology* 107, 422–431.
182. Schnabel, E., Mains, R. E., and Farquhar, M. G. (1989). Proteolytic processing of pro-ACTH/endorphin begins in the Golgi complex of pituitary corticotropes and AT-20 cells. *Mol. Endocrinol.* 3, 1223–1235.
183. Farquhar, M. G. (1978). Recovery of surface membrane in anterior pituitary cells. *J. Cell. Biol.* 77, R35–R42.
184. Komuro, M., Kiuchi, Y., and Shioda, T. (1987). Membrane modification during secretory granule formation in rat somatotrophs. *Eur. J. Cell. Biol.* 43, 98–103.
185. Farquhar, M. G. (1971). Processing of secretory products by cells of the anterior pituitary cell. *Mem. Soc. Endocrinol.* 19, 79–124.
186. Shimada, O., Tosaka-Shimada, H., and Ishikawa, H. (1990). Morphological effects of somatostatin on rat somatotrophs previously activated by growth hormone-releasing factor. *Cell. Tissue Res.* 261, 219–229.
187. Ishimura, K., Egawa, K., and Fujita, H. (1980). Freeze-fracture images of exocytosis and endocytosis in anterior pituitary cells of rabbits and mice. *Cell. Tissue Res.* 206, 233–241.
188. Senda, T., Fujita, H., Ban, T., Zhong, C., Ishimura, K., Kanda, K., and Sobue, K. (1989). Ultrastructural and immunocytochemical studies on the cytoskeleton in the anterior pituitary of rats, with special regard to the relationship between actin filaments and secretory granules. *Cell. Tissue Res.* 258, 25–30.
189. Carbajal, M. E., and Vitale, M. L. (1997). The cortical actin cytoskeleton of lactotropes as an intracellular target for the control of prolactin secretion. *Endocrinology* 138, 5374–5384.
190. Shimada, O., and Tosaka-Shimada, H. (1989). Morphological analysis of growth hormone release from rat somatotrophs into blood vessels by immunogold microscopy. *Endocrinology* 125, 2677–2682.
191. Draznin, B., Dahl, R., Sherman, N., Sussman, K. E., and Staehelin, L. A. (1988). Exocytosis in normal anterior pituitary cells. Quantitative correlation between growth hormone release and the morphological features of exocytosis. *J. Clin. Invest.* 81, 1042–1050.
192. Nikitovitch-Winer, M., Yu, S. M., and Papka, R. E. (1985). Soluble prolactin may be directly released from cellular compartments other than secretory granules. In *Prolactin, Basic and Clinical Correlates* (M. O. Thorner and U. Scapagnini, Eds.), Vol. 1 (Fidia Research Series). Liviana Press, Padova.
193. Torres, A. I., and Aoki, A. (1987). Release of big and small molecular forms of prolactin: Dependence upon dynamic state of the lactotroph. *J. Endocrinol.* 114, 213–220.
194. Kojima, M., Hosoda, H., Date, Y., Nakazato, M., Matsuo, H., and Kangawa, K. (1999). Ghrelin is a growth-hormone-releasing acylated peptide from stomach. *Nature* 402, 656–660.
195. Seoane, L. M., Tovar, S., Baldelli, R., Arvat, E., Ghigo, E., Casanueva, F. F., and Dieguez, C. (2000). Ghrelin elicits a marked stimulatory effect on GH secretion in freely-moving rats. *Eur. J. Endocrinol.* 143, R7–R9.
196. Pombo, M., Pombo, C. M., Garcia, A., Caminos, E., Gualillo, O., Alvarez, C. V., Casanueva, F. F., and Dieguez, C. (2001). Hormonal control of growth hormone secretion. *Horm. Res.* 55(Suppl 1), 11–16.
197. Mezey, E., Reisine, T. D., Brownstein, M. J., Palkovits, M., and Axelrod, J. (1984). B-Adrenergic mechanism of insulin-induced adrenocorticotropin release from the anterior pituitary. *Science* 226, 1085–1087.
198. Karanth, S., and McCann, S. M. (1991). Anterior pituitary hormone control by interleukin 2. *Proc. Natl. Acad. Sci. U S A* 88, 2961–2965.
199. Brown, S. L., Smith, L. R., and Blalock, J. E. (1987). Interleukin 1 and Interleukin 2 enhance proopiomelanocortin gene expression in pituitary cells. *J. Immunol.* 139, 3181–3183.
200. Watanabe, H., Sasaki, S., and Takebe, K. (1991). Evidence that intravenous administration of Interleukin-1 stimulates corticotropin releasing hormone secretion in the median eminence of freely moving rats: estimation by push-pull perfusion. *Neurosci. Lett.* 133, 7–10.
201. Couch, E. F., Arimura, A., Schally, A. V., Saito, M., and Sawano, S. (1969). Electron microscope studies of somatotrophs of rat pituitary after injection of purified growth hormone releasing hormone factor (GRF). *Endocrinology* 85, 1084–1091.
202. Coates, P. W., Ashby, E. A., Krulich, L., Dhariwal, A. P. S., and McCann, S. M. (1970). Morphologic alterations in somatotrophs of the rat adenohypophysis following administration of hypothalamic extracts. *Am. J. Anat.* 128, 389–412.
203. Shiino, M., Arimura, A., Schally, A. V., and Rennels, E. G. (1972). Ultrastructural observations of granule extrusion from rat anterior pituitary cells after injection of LH-releasing hormone. *Z. Zellforsch.* 128, 152–161.
204. Stratmann, I. E., Ezrin, C., Kovacs, K., and Sellers, E. A. (1973). Effect of TRH on the fine structure and replication of TSH and prolactin cells in the rat. *Z. Zellforsch.* 145, 23–37.
205. Westlund, K. N., Aguilera, G., and Childs, G. V. (1985). Quantification of morphological changes in pituitary corticotropes produced by in vivo corticotropin-releasing factor stimulation and adrenalectomy. *Endocrinology* 116, 439–445.
206. Atwell, W. J. (1926). The development of the hypophysis cerebri in man, with special reference to the pars tuberalis. *Am. J. Anat.* 37, 159–193.
207. Tilney, F. (1938). The glands of the brain with especial reference to the pituitary gland. *Res. Publ. Assoc. Res. Nerv. Ment. Dis.* 17.
208. Takor, T. T., and Pearse, A. G. (1975). Neuroectodermal origin of avian hypothalamo-hypophyseal complex: the role of the ventral neural ridge. *J. Embryol. Exp. Morphol.* 34, 311–325.
209. Pearse, A. G. E., and Takor, T. T. (1976). Neuroendocrine embryology and the APUD concept. *Clin. Endocrinol.* 5, 229s–244s.
210. Oliver, G., Mailhos, A., Wehr, R., Copeland, N. G., Jenkins, N. A., and Gruss, P. (1995). Six3, a murine homologue of the sine oculis gene, demarcates the most anterior border of the developing neural plate and is expressed during eye development. *Development* 121, 4045–4055.
211. Baker, C. V., and Bronner-Fraser, M. (2001). Vertebrate cranial placodes I. Embryonic induction. *Dev. Biol.* 232, 1–61.
212. Kawamura, K., and Kikuyama, S. (1998). Morphogenesis of the hypothalamus and hypophysis: their association, dissociation and reassociation before and after "Rathke." *Arch. Histol. Cytol.* 61, 189–198.
213. Kouki, T., Imai, H., Aoto, K., Eto, K., Shioda, S., Kawamura, K., and Kikuyama, S. (2001). Developmental origin of the rat

adenohypophysis prior to the formation of Rathke's pouch. *Development* 128, 959–963.
214. Kawamura, K., Kouki, T., Kawahara, G., and Kikuyama, S. (2002). Hypophyseal development in vertebrates from amphibians to mammals. *Gen. Comp. Endocrinol.* 126, 130–135.
215. Hermesz, E., Mackem, S., and Mahon, K. A. (1996). Rpx: a novel anterior-restricted homeobox gene progressively activated in the prechordal plate, anterior neural plate and Rathke's pouch of the mouse embryo. *Development* 122, 41–52.
216. Treier, M., O'Connell, S., Gleiberman, A., Price, J., Szeto, D. P., Burgess, R., Chuang, P. T., McMahon, A. P., and Rosenfeld, M. G. (2001). Hedgehog signaling is required for pituitary gland development. *Development* 128, 377–386.
217. Treier, M., Gleiberman, A. S., O'Connell, S. M., Szeto, D. P., McMahon, J. A., McMahon, A. P., and Rosenfeld, M. G. (1998). Multistep signaling requirements for pituitary organogenesis in vivo. *Genes Dev.* 12, 1691–1704.
218. Dasen, J. S., and Rosenfeld, M. G. (2001). Signaling and transcriptional mechanisms in pituitary development. *Annu. Rev. Neurosci.* 24, 327–355.
219. Suh, H., Gage, P. J., Drouin, J., and Camper, S. A. (2002). Pitx2 is required at multiple stages of pituitary organogenesis: pituitary primordium formation and cell specification. *Development* 129,329–337.
220. Sheng, H. Z., and Westphal, H. (1999). Early steps in pituitary organogenesis. *Trends Genet.* 15, 236–240.
221. Sheng, H. Z., Zhadanov, A. B., Mosinger, B. Jr., Fujii, T., Bertuzzi, S., Grinberg, A., Lee, E. J., Huang, S. P., Mahon, K. A., and Westphal, H. (1996). Specification of pituitary cell lineages by the LIM homeobox gene Lhx3. *Science* 272, 1004–1007.
222. Ericson, J., Norlin, S., Jessell, T. M., and Edlund, T. (1998). Integrated FGF and BMP signaling controls the progression of progenitor cell differentiation and the emergence of pattern in the embryonic anterior pituitary. *Development* 125, 1005–1015.
223. Raetzman, L. T., Ward, R., and Camper, S. A. (2002). Lhx4 and Prop1 are required for cell survival and expansion of the pituitary primordia. *Development* 129, 4229–4239.
224. Raetzman, L. T., Ross, S. A., Cook, S., Dunwoodie, S. L., Camper, S. A., and Thomas, P. Q. (2004). Developmental regulation of Notch signaling genes in the embryonic pituitary: Prop1 deficiency affects Notch2 expression. *Dev. Biol.* 265, 329–340.
225. Gage, P. J., Brinkmeier, M. L., Scarlett, L. M., Knapp, L. T., Camper, S. A., and Mahon, K. A. (1996). The Ames dwarf gene, df, is required early in pituitary ontogeny for the extinction of Rpx transcription and initiation of lineage-specific cell proliferation. *Mol. Endocrinol.* 10, 1570–1581.
226. Scully, K. M., and Rosenfeld, M. G. (2002). Pituitary development: regulatory codes in mammalian organogenesis. *Science* 295, 2231–2235.
227. Liu, N. A., Huang, H., Yang, Z., Herzog, W., Hammerschmidt, M., Lin, S., and Melmed, S. (2003). Pituitary corticotroph ontogeny and regulation in transgenic zebrafish. *Mol. Endocrinol.* 17, 959–966.
228. Savage, J. J., Yaden, B. C., Kiratipranon, P., and Rhodes, S. J. (2003). Transcriptional control during mammalian anterior pituitary development. *Gene* 319, 1–19.
229. Kioussi, C., Carriere, C., and Rosenfeld, M. G. (1999). A model for the development of the hypothalamic-pituitary axis: transcribing the hypophysis. *Mech. Dev.* 81, 23–35.
230. Farner, D. S., Wilson, F. E., and Oksche, A. (1967). Neuroendocrine mechanisms in birds. In Neuroendocrinology (L. Martini and W. F. Ganong, Eds.), Vol. 2. Academic Press, New York.
231. Dempsey, E. W., and Wislocki, G. B. (1955). An electron microscopic study of the blood-brain barrier in the rat, employing silver nitrate as a vital stain. *J. Biophys. Biochem. Cytol.* 1, 245–256.
232. Farquhar, M. G. (1961). Fine structure and function in capillaries of the anterior pituitary gland. *Angiology* 12, 270–292.
233. Denef, C., and Andries, M. (1983). Evidence for paracrine interaction between gonadotrophs and lactotrophs in pituitary cell aggregates. *Endocrinology* 112, 813–822.
234. Pierce, J. G., and Parsons, T. F. (1981). Glycoprotein hormones: structure and function. *Ann. Rev. Biochem.* 50, 465–495.
235. Li, C. H., Dixon, J. S., Lo, T.-B., Pankov, Y. A., and Schmidt, K. D. (1969). Amino-acid sequence of ovine lactogenic hormone. *Nature (Lond).* 224, 695–696.
236. Li, C. H., Dixon, J. S., Lo, T.-B., Schmidt, K. D., and Pankov, Y. A. (1970). Studies on pituitary lactogenic hormone. The primary structure of the sheep hormone. *Arch. Biochem. Biophys.* 141, 705–737.
237. Li, C. H. (1976). Studies on pituitary lactogenic hormone. The primary structure of the porcine hormone. *Int. J. Peptide Protein Res.* 8, 205–224.
238. Shome, B., and Parlow, A. F. (1977). Human pituitary prolactin (hPRL): the entire linear amino acid sequence. *J. Clin. Endocrinol. Metab.* 45, 1112–1115.
239. Wallis, M. (1964). The primary structure of bovine prolactin. *FEBS Lett.* 44, 205–208.
240. Li, C. H., Hayashida, T., Doneen, B. A., and Rao, A. J. (1976). Human somatotropin: biological characterization of the recombinant molecule. *Proc. Natl. Acad. Sci. U S A* 73, 3463–3465.
241. Lewis, U. J., Singh, R. N. P., Tutwiler, G. F., Sigel, M. B., VanderLaan, E. F., and VanderLaan, W. P. (1980). Human growth hormone: a complex of proteins. *Rec. Prog. Horm. Res.* 36, 377–508.
242. Howard, K. S., Shepherd, R. G., Eigner, E. A., Davies, D. S., and Bell, P. H. (1955). Structure of β-corticotropin: final sequence. *J. Am. Chem. Soc.* 77, 3419–3420.
243. Li, C. H., Geshwind, I. I., Cole, R. D., Raacke, I. D., Harris, J. I., and Dixon, J. S. (1955). Amino-acid sequence of alpha-corticotropin. *Nature (Lond).* 176, 687–689.
244. Li, C. H. (1959). Proposed system of terminology for preparations of adrenocorticotropic hormone. *Science* 129, 969–970.
245. Geshwind, I. I., Li, C. H., and Barnafi, L. (1956). Isolation and structure of melanocyte-stimulating hormone from porcine pituitary glands. *J. Am. Chem. Soc.* 79, 4494–4495.
246. Geschwind, I., and Li, C. H. (1957). The isolation and characterization of a melanocyte stimulating hormone (β-MSH) from hog pituitary glands. *J. Am. Chem. Soc.* 79, 615–620.
247. Geshwind, I. I., Li, C. H., and Barnafi, L. (1957). The structure of the β-melanocyte stimulating hormone. *J. Am. Chem. Soc.* 79, 620–625.
248. Li, C. H., and Chung, D. (1976). Isolation and structure of an unitriakontapeptide with opiate activity from camel pituitary glands. *Proc. Natl. Acad. Sci. U S A* 73, 1145–1148.
249. Li, C. H., and Chung, D. (1976). Primary structure of human β-lipotropin. *Nature (Lond).* 260, 622–624.
250. Li, C. H., Tan, L., and Chung, D. (1977). Isolation and primary structure of beta-endorphin and beta-lipotropin from bovine pituitary gland. *Biochem. Biophys. Res. Comm.* 77, 1088–1093.
251. Teschemacher, H., Opheim, K. E., Cox, B. M., and Goldstein, A. (1975). A peptide-like substance that acts like morphine. I. Isolation. *Life Sci.* 16, 1771–1776.
252. Teschemacher, H., Opheim, K. E., Cox, B. M., and Goldstein, A. (1975). A peptide-like substance that acts like morphine 2. Purification and properties. *Life Sci.* 16, 1777–1782.

253. Pearse, A. G. E., and Noorden, S. V. (1963). The functional cytology of the human adenohypophysis. *Can. Med. Assoc. J.* 88, 462–471.
254. Nakane, P. K. (1968). Simultaneous localization of multiple tissue antigens using the peroxidase-labeled antibody method: a study on pituitary glands of the rat. *J. Histochem. Cytochem.* 16, 557–559.
255. Kurosumi, K. (1968). Functional classification of cell types of the anterior pituitary gland accomplished by electron microscopy. *Arch. Histol. Jpn.* 29, 329–362.
256. Pochet, R., Brocas, H., Vassart, G., Toubeau, G., Seo, H., Refetoff, S., Dumont, J. E., and Pasteels, J. L. (1981). Radioautographic localization of prolactin messenger RNA on histological sections by in situ hybridization. *Brain Res.* 211, 433–438.
257. Rasmussen, A. T. (1929). The percentage of the different types of cells in the male human hypophysis. *Am. J. Path.* 5, 263.
258. Rasmussen, A. T. (1933). The percentage of different types of cells in the anterior lobe of the hypophysis of the adult female. *Am. J. Path.* 9, 459.
259. Severinghaus, A. E. (1938). The cytology of the pituitary gland. *Res. Publ. Assoc. Res. Nerv. Ment Dis.* 17, 96–117.
260. Severinghaus, A. E. (1939). Anterior hypophyseal cytology in relation to the reproductive hormones. In *Sex and Internal Secretions* (M. B. Allen, Ed.) p. 1045. Wood & Co., New York.
261. Cushing, H. (1927). Acromegaly from a surgical standpoint. *Br. Med. J.* 2, 1–9.
262. Cushing, H., and Davidoff, L. (1932). The pathologic findings in four cases of acromegaly with a discussion of their significance. Monograph Rockefeller Institute for Medical Research.
263. Cushing, H. (1932). The basophile adenomas of the pituitary body and their clinical manifestations. *Bull. Johns Hopkins Hosp.* 50, 137–195.
264. Romeis, B. (1940). Hypophyse. In *Handbuch der Mikroskopischen Anatomie des Menschen* (M. W. Van, Ed.). Springer, Berlin.
265. Pearse, A. G. (1952). Observations on the localization, nature and chemical constitution of some components of the anterior hypophysis. *J. Pathol. Bacteriol.* 64, 791–809.
266. Herlant, M. (1960). Etude critique de deux techniques nouvelles destinees a mettre en evidence les differentes categories cellulaires presented dans la glande pituitaire. *Bull. Micro. Appl.* 10, 37–44.
267. Purves, H. D. (1966). Cytology of the adenohypophysis. In *The Pituitary Gland* (G. W. Harris and B. T. Donovan, Eds.), Vol. 1. University of California Press.
268. Smith, P. E., and Smith, I. P. (1923). The topographical separation in the bovine anterior hypophysis of the principle reacting with the endocrine system from that controlling general body growth, with suggestions as to the cell types elaborating these encrations. *Anat. Rec.* 25, 150–151.
269. Baker, B. L. (1974). Functional cytology of the hypophyseal pars distalis and pars intermedia. In *The Pituitary Gland and Its Neuroendocrine Control.* Vol. 7. *Handbook of Physiology* (R. O. Greep and E. B. Astwood, Eds.), pp. 45–80. American Physiological Society, Washington, DC.
270. Purves, H. D., and Greisbach, W. E. (1952). Functional deafferentation in the acidophil cells and the gonadotropic basophil cells of the rat pituitary. *Proc. Univ. Otago Med. Sch.* 30, 27.
271. Farquhar, M. G., and Rinehart, J. F. (1954). Electron microscope studies of the anterior pituitary gland of castrate rats. *Endocrinology* 54, 516–541.
272. Farquhar, M. G., and Rinehart, J. F. (1954). Cytologic alterations in the anterior pituitary gland following thyroidectomy: an electron microscope study. *Endocrinology* 55, 857–876.
273. Nakane, P. K. (1970). Classifications of anterior pituitary cell types with immunoenzyme histochemistry. *J. Histochem. Cytochem.* 18, 9–20.
274. Nakane, P. K. (1971). Application of peroxidase-labelled antibodies to the intracellular localization of hormones. *Acta Endocrinol. Suppl.* 153, 190–204.
275. Moriarty, G. C. (1973). Adenohypophysis: ultrastructural cytochemistry. *J. Histochem. Cytochem.* 12, 855–894.
276. Kurosumi, K., and Tosaka, H. (1988). Prenatal development of growth hormone-producing cells in the rat anterior pituitary as studied by immunogold electron microscopy l. *Arch. Histol. Cytol.* 51, 183–204.
277. Kurosumi, K. (1991). Ultrastructural immunocytochemistry of the adenohypophysis in the rat: A review. *J. Electron Microsc. Tech.* 19, 42–56.
278. Kurosumi, K., Tanaka, S., and Tosaka, H. (1987). Changing ultrastructures in the estrous cycle and postnatal development of prolactin cells in the rat anterior pituitary as studied by immunogold electron microscopy. *Arch Histol. Jpn.* 50, 455–478.
279. Van Putten, J., and Kiliaan, A. (1988). Immuno-electron-microscopic study of the prolactin cells in the pituitary gland of male Wistar rats during aging. *Cell. Tissue Res.* 251, 353–358.
280. Hudson, P., Penschow, J., Shine, J., Ryan, G., Niall, H., and Coghlan, J. (1981). Hybridization histochemistry: use of recombination DNA as a "homing probe" for tissue localization of specific m-RNA populations. *Endocrinology* 108, 353–356.
281. Bauman, J. G. J., Wiegnant, J., and Van Duijn, P. (1981). Cytochemical hybridization with flourochrome-labelled RNA. *J. Histochem. Cytochem.* 29, 238–246.
282. Phifer, R. F., and Spicer, S. S. (1973). Immunohistochemical and histologic demonstration of thyrotropic cells of the human adenohypophysis. *J. Clin. Endocrinol. Metab.* 36, 1210–1221.
283. Girod, C., and Trouillas, J. (1980). Individualisation immunohistochimique des cellules thyreotropes antehypophysaires chez le Singe Macacus irus, a l'aide d'un anticorps anti-β-TSH humaine. *C. R. Acad. Sci. Ser. D.* 291.
284. Baker, B. L., Pierce, J. G., and Cornell, J. S. (1972). The utility of antiserums to subunits of TSH and LH for immunochemical staining of the rat hypophysis. *Am. J. Anat.* 135, 251–268.
285. Kawarai, Y. (1980). Identification of ACTH cells and TSH cells in rat anterior pituitary with the unlabeled antibody enzyme method on adjacent thin and thick sections. *Acta Histochem. Cytochem.* 13, 627–647.
286. Moriarty, G. C., and Tobin, R. B. (1976). Ultrastructural immunocytochemical characterization of the thyrotroph in rat and human pituitaries. *J. Histochem. Cytochem.* 24, 1131–1139.
287. Moriarty, G. C., and Tobin, R. B. (1976). An immunocytochemical study of TSHB storage in rat thyroidectomy cells with and without D or L thyroxine treatment. *J. Histochem. Cytochem.* 24, 1140–1149.
288. Horvath, E., and Kovacs, K. (1988). Fine structural cytology of the adenohypophysis in rat and man. *J. Electron Micros. Tech.* 8, 401–432.
289. Ozawa, H. (1991). Changing structure of thyrotrophs in the rat anterior pituitary after thyroidectomy as studied by immuno-electronmicroscopy and enzyme cytochemistry. *Cell. Tissue Res.* 263, 405–412.
290. Kurosumi, K., and Oota, Y. (1968). Electron microscopy of two types of gonadotrophs in the anterior pituitary gland of persistent estrous and diestrous rats. *Z. Zellforsch.* 85, 34–46.

291. Moriarty, G. C. (1975). Electron microscopic-immunocytochemical studies of rat pituitary gonadotrophs: a sex difference in morphology and cytochemistry of LH cells. *Endocrinology* 97, 1215–1225.
292. Childs, G. V., Ellison, D. G., and Gardner, L. (1980). An immunocytochemist's view of gonadotropin storage in the adult male rat. Cytochemical and morphological heterogeneity in serially stained gonadotropes. *Am. J. Anat.* 158, 397–409.
293. Childs, G. V., Lloyd, J. M., Unabia, G., Gharib, S. G., Wierman, M. E., and Chin, W. W. (1987). Detection of luteinizing hormone beta messenger ribonucleic acid (RNA) in individual gonadotropes after castration: Use of a new in situ hybridization method with a photobiotinylated complementary RNA probe. *Mol. Endocrinol.* 1, 926–932.
294. Lloyd, J. M., and Childs, G. V. (1988). Differential storage and release of luteinizing hormone and follicle-releasing hormone from individual gonadotropes separated by centrifugal elutriation. *Endocrinology* 122, 1282–1290.
295. Moriarty, G. C. (1976). Ultrastructural-immunocytochemical studies of rat pituitary gonadotrophs in cycling female rats. *Gunma Symposia Endocrinol.* 13, 207–209.
296. Childs, G. V., Ellison, D. G., Lorenzen, J. R., Collins, T. J., and Schwartz, N. B. (1982). Immunocytochemical studies of gonadotrophin storage in developing castration cells. *Endocrinology* 111, 1318–1225.
297. Eipper, B. A., and Mains, R. E. (1980). Structure and biosynthesis of pro-ACTH/ endorphin and related peptides. *Endocrin. Rev.* 1, 1–27.
298. Eipper, B. A., and Mains, R. E. (1977). Peptide analysis of glycoprotein form of adrenocorticotropic hormone. *J. Biol. Chem.* 252, 8821–8832.
299. Hope, J., and Lowry, P. J. (1981). Pro-opiocortin: the ACTH/LPH common precursor protein. *Front. Horm. Res.* 8, 44–61.
300. Siperstein, E. R., and Allison, V. F. (1965). Fine structure of the cells responsible for the secretion of adrenocorticotropin in the adrenalectomized rat. *Endocrinology* 76, 70–79.
301. Siperstein, E. R., and Miller, K. J. (1970). Further cytophysiologic evidence for the identity of the cells that produce adrenocorticotrophic hormone. *Endocrinology* 86, 451–486.
302. Siperstein, E. R., and Miller, K. J. (1973). Hypertrophy of the ACTH-producing cell following adrenalectomy: a quantitative electron microscopic study. *Endocrinology* 93, 1257–1268.
303. Moriarty, G. C., and Halmi, N. S. (1972). Electron microscopic study of the adreno-corticotropin producing cell with the use of unlabeled antibody and the soluble peroxidase-antiperoxidase complex. *J. Histochem. Cytochem.* 20, 590–603.
304. Mains, R. E., Eipper, B. A., and Ling, N. (1977). Common precursor to corticotropins and endorphins. *Proc. Natl. Acad. Sci. U S A* 74, 3014–3018.
305. Roberts, J. L., Seeburg, P. H., and Shine, J., et al. (1979). Corticotropin and β-endorphin: construction and analysis of recombinant DNA complementary to mRNA for the common precursor. *Proc. Natl. Acad. Sci. U S A* 76, 2153–2157.
306. Smith, A. I., and Funder, J. W. (1988). Proopiomelanocortin processing in the pituitary, central nervous system, and peripheral tissues. *Endocr. Rev.* 9, 159–179.
307. Phifer, R. F., Spicer, S. S., and Orth, D. N. (1970). Specific demonstration of the human hypophyseal cells which produce adrenocorticotropic hormone. *J. Clin. Endocrinol. Metab.* 31, 347–361.
308. Phifer, R. F., Orth, D. N., and Spicer, S. S, eds. (1972). Immunohistologic evidence that β-melanocyte stimulating hormone (β-MSH) and adrenocorticotropin (ACTH) are produced in the same human hypophyseal cells. Excerpta Medica, Amsterdam.
309. Moriarty, G. C., and Garner, L. L. (1977). Immunoelectronmicroscopical localization of ACTH/MSH peptides in rat and human pituitaries. *Front. Horm Res.* 4, 26–41.
310. Pelletier, G., Leclerc, R., Labrie, F., Cote, J., Chretien, M., and Lis, M. (1977). Immunohistochemical localization of B-lipotropic hormone in the pituitary gland. *Endocrinology* 100, 770–776.
311. Weber, E., Voigt, K. H., and Martin, R. (1978). Concomitant storage of ACTH and endorphin-like immunoreactivity in the secretory granules of anterior pituitary corticotrophs. *Brain Res.* 157, 386–390.
312. Baker, B. L., and Drummond, T. (1972). The cellular origins of corticotropin and melanotropin as revealed by immunochemical staining. *Am. J. Anat.* 134, 395–400.
313. Kurosumi, K., Tosaka, H., and Ijima, K. (1989). The immature type of pro-opiomelanocortin cell of the rat anterior pituitary as observed by immunogold electron microscopy. *Arch. Histol. Cytol.* 52, 135–150.
314. Hashimoto, S., Fumagalli, G., Zanini, A., and Meldolesi, J. (1987). Sorting of three secretory proteins to distinct secretory granules in acidophilic cells of the cow anterior pituitary. *J. Cell. Biol.* 105, 1579–1586.
315. Horvath, E., Lloyd, R. B., and Kovacs, K. (1990). Propylthiouracyl-induced hypothyroidism results in reversible transdifferentiation of somatotrophs into thyroidectomy cells. A morphologic study of the rat pituitary including immunoelectron microscopy. *Lab. Invest.* 63, 511–520.
316. Wu, P., and Childs, G. V. (1991). Changes in rat pituitary POMC mRNA after exposure to cold or a novel environment, detected by in situ hybridization. *J. Histochem. Cytochem.* 39, 843–852.
317. Sasaki, F., Wu, P., Rougeau, D., Unabia, G., and Childs, G. V. (1990). Cytochemical studies of responses of corticotropes and thyrotropes to cold and novel environment stress. *Endocrinology* 127, 285–297.
318. Allaerts, W., Fluitsma, D. M., Hoefsmit, E. C., Jeucken, P. H., Morreau, H., Bosman, F. T., and Drexhage, H. A. (1996). Immunohistochemical, morphological and ultrastructural resemblance between dendritic cells and folliculo-stellate cells in normal human and rat anterior pituitaries. *J. Neuroendocrinol.* 8, 17–29.
319. Sato, T., and Inoue, K. (2000). Dendritic cells in the rat pituitary gland evaluated by the use of monoclonal antibodies and electron microscopy. *Arch. Histol. Cytol.* 63, 291–303.
320. Garcia-Navarro, F., Porter, D., Garcia-Navarro, S., and Licht, P. (1989). Immunocytochemical and ultrastructural study of the frog (Rana Pipiens) pars distalis with special reference to folliculo-stellate cell function in vitro superfusion. *Cell. Tissue Res.* 256, 623–630.
321. Nakajima, T., Yamaguchi, H., and Takahashi, K. (1980). S100 protein in folliculostellate cells of the rat pituitary anterior lobe. *Brain Res.* 191, 523–531.
322. Liu, Y. C., Tanaka, S., Inoue, K., and Kurosumi, K. (1989). Localization of fibronectin in the folliculo-stellate cells of the rat anterior pituitary by the double bridge-antiperoxidase method. *Histochemistry* 92, 34–45.
323. Soji, T., and Herbert, D. C. (1989). Intercellular communication between rats anterior pituitary cells. *Anat. Rec.* 224, 523–533.
324. Soji, T., Yashiro, T., and Herbert, D. C. (1990). Intercellular communication in the rat anterior pituitary gland. I. Postnatal development and changes after injection of luteinizing hormone-releasing hormone (LH-RH) or testosterone. *Anat Rec.* 226, 337–341.
325. Soji, T., Mabuchi, Y., Kurono, C., and Herbert, D. C. (1997). Folliculo-stellate cells and intercellular communication within the rat anterior pituitary gland. *Microsc. Res. Tech.* 39, 138–149.
326. Spangelo, B. L ., Macleod, R. M., and Isakson, P. C. (1991). Production of Interleukin-6 by anterior pituitary cells in vitro. *Endocrinology* 126, 582–586.

327. Vankelecom, H., Carmeliet, P., Vandamme, J., Billiau, A., and Denef, C. (1989). Production of Interleukin-6 by folliculo-stellate cells of the anterior pituitary gland in a histiotypic cell aggregate culture system. *Neuroendocrinology* 49, 102–106.
328. Spangelo, B. L., Judd, A. M., Isakson, P. C., and MacLeod, R. M. (1989). Interleukin-6 stimulates anterior pituitary hormone release in vitro. *Endocrinology* 125, 575–577.
329. Naitoh, Y., Fukata, J., Tominaga, T., Nakai, Y., Tamai, S., Mori, K., and Imura, H. (1988). Interleukin-6 stimulates the secretion of adrenocorticotropic hormone in conscious, freely moving rats. *Biochem. Biophys. Res. Comm.* 155, 1459–1463.
330. Horvath, E., and Kovacs, K. (2002). Folliculo-stellate cells of the human pituitary: a type of adult stem cell? *Ultrastruct. Pathol.* 26, 219–228.
331. Malagon, M. M., Garrido-Gracia, J. C., Torronteras, R., Dobado-Berrios, P. M., Ruiz-Navarro, A., and Gracia-Navarro, F. (1998). Cell heterogeneity as a reflection of the secretory cell cycle. *Ann. N Y Acad. Sci.* 839, 244–248.
332. Smith, P. E., and Smith, I. P. (1923). The response of the hypophysectomized tadpole to the intraperitoneal injection of the various lobes and colloid of the bovine hypophysis. *Anat. Rec.* 25, 150.
333. Girod, C. (1984). Fine structure of the pituitary pars distalis. In *Ultrastructure of Endocrine Cells and Tissues* (P. M. Motta, Ed.). Nijhoff, Boston.
334. Kurosumi, K., Matsuzawa, T., and Shibasaki, S. (1961). Electron microscope studies on the fine structures of the pars nervosa and pars intermedia, and their morphological interrelation in the normal rat hypophysis. *Gen. Comp. Endocrinol.* 1, 433–452.
335. Visser, M., and Swaab, D. F. (1977). aMSH in the human pituitary. *Front. Horm. Res.* 4, 42–45.
336. Moriarty, G. C., and Halmi, N. S. (1972). Adrenocorticotropin production by the intermediate lobe of the rat pituitary. An electron microscopic-immunocytochemical study. *Z. Zellforsch.* 132, 1–14.
337. Murakami, T., Ohtsuka, A., Taguchi, T., and Ohtani, O. (1985). Blood vascular bed of the rat pituitary intermediate lobe, with special reference to its development and portal drainage into the anterior lobe. A scanning electron microscopic study of vascular casts. *Arch. Histol. Jpn.* 48, 69–87.
338. Saland, L. C. (1980). Extracellular spaces of the rat pars intermedia as outlined by lanthanum tracer. *Anat. Rec.* 196, 355–361.
339. Sakly, M., Schmitt, G., and Koch, B. (1982). CRF enhances release of both aMSH and ACTH from anterior and intermediate pituitary. *Neuroendocrinol. Lett.* 4, 289–293.
340. Proulx-Ferland, L., Labrie, F., Dumont, D., Cote, J., Coy, D. H., and Sveiraf, J. (1982). Corticotropin-releasing factor stimulates secretion of melanocyte-stimulating hormone from the rat pituitary. *Science* 217, 62–63.
341. Kobayashi, Y. (1965). Functional morphology of the pars intermedia of the rat hypophysis as revealed with the electron microscope. II. Correlation of the pars intermedia with the hypophyseo-adrenal axis. *Z. Zellforsch.* 68, 155–171.
342. Cote, T., Munemura, M., Eskay, R. L., and Kebabian, J. W. (1980). Biochemical identification of the β-adrenoceptor and evidence for the involvement of an adenosine 3′, 5′-monophosphate system in the β′-adrenergically induced release of a-melanocyte-stimulating hormone in the intermediate lobe of the rat pituitary gland. *Endocrinology* 107, 108–116.
343. Tilders, F. J. H., Post, M., Jackson, S., Lowry, P. J., and Smelik, P. G. (1981). Beta-adrenergic stimulation of the release of ACTH and LPH-related peptides from the pars intermedia of the rat pituitary gland. *Acta Endocrinol.* 97, 343–351.
344. Bjorklund, A., Falck, B., Hromek, F., Owman, C., and West, K. A. (1970). Identification and terminal distribution of the tubero-hypophyseal monoamine fibre systems in the rat by means of stereotaxic and microspectrofluorimetric techniques. *Brain Res.* 17, 1–23.
345. Baumgarten, H. G., Bjorklund, A., Holstein, A. F., and Nobin, A. (1972). Organization and ultrastructural identification of the catecholamine nerve terminals in the neural lobe and pars intermedia of the rat pituitary. *Z. Zellforsch.* 126, 483–517.
346. Tilders, F. J. H., and Smelik, P. G. (1977). Direct neural control of MSH secretion in mammals: the involvement of dopaminergic tubero-hypophysial neurones. *Front. Horm. Res.* 4, 80–93.
347. Luppi, P. H., Sakai, K., Salvert, D., Berod, A., and Jouvet, M. (1986). Periventricular dopaminergic neurons terminating in the neurointermediate lobe of the cat hypophysis. *J. Comp. Neurol.* 244, 204–212.
348. Vincent, S. R., Hokfelt, T., and Wu, J.-Y. (1982). GABA neuron systems in hypothalamus and the pituitary gland. *Neuroendocrinology* 34, 117–125.
349. Tappaz, M. L., Wassef, M., Oertel, W. H., Paut, L., and Pujol, J. F. (1983). Light- and electron-microscopic immunocytochemistry of glutamic acid decarboxylase (GAD) in the basal hypothalamus: morphological evidence for neuroendocrinegamma-aminobutyrate (GABA). *Neuroscience* 9, 271–287.
350. Verburg-Van Kemenade, B. M., Tappaz, M. L., Paut, L., and Jenks, B. G. (1986). GABAergic regulation of melanocyte-stimulating hormone secretion from the pars intermedia of Xenopus Laevis: immunocytochemical and physiological evidence. *Endocrinology* 118, 260–267.
351. Rabhi, M., Onteniente, B., Kah, O., Geffard, M., and Calas, A. (1987). Immunocytochemical study of the mouse pituitary by use of antibodies against gamma-aminobutyric acid (GABA). *Cell. Tissue Res.* 247, 33–48.
352. Oertel, W. H., Mugnaini, E., and Tappaz, M. L., et al. (1982). Central gabaergic innervation of the neurointermediate pituitary lobe: biochemical and immunohistochemical study in the rat. *Proc Natl Acad Sci U S A* 79, 675–679.
353. Vuillez, P., Carbajo Perez, S., and Stoeckel, M. E. (1987). Colocalization of gaba and tyrosine hydroxylase immunoreactivities in the axons innervating the neurointermediate lobe of the rat pituitary: An ultrastructural immunogold study. *Neurosci. Lett.* 79, 53–58.
354. Saland, L., Wallace, J., Samora, A., and Guitierrez, L. (1988). Colocalization of tyrosine hydroxylase (TH)– and serotonin (5-HT) immunoreactive innervation in the rat pituitary gland. *Neurosci. Lett.* 94, 39–45.
355. Schimochowitsch, S., Vuillez, P., Tappaz, M. L., Klein, M. J., and Stoeckel, M. E. (1991). Systematic presence of GABA-immunoreactivity in the tubero-infundibular and tubero-hypophyseal dopaminergic systems: an ultrastructural immunogold study on several mammals. *Exp. Brain Res.* 83, 575–586.
356. Westlund, K. N., and Childs, G. V. (1982). Localization of serotonin fibers in the rat adenohypophysis. *Endocrinology* 111, 1761–1763.
357. Verburg-Van Kemenade, B. M., Jenks, B. G., Cruijsen, P. M., Dings, A., Tonon, M. C., and Vaudry, H. (1987). Regulation of MSH release from the neurointermediate lobe of Xenopus laevis by CRF-like peptides. *Peptides* 8, 1093–1100.
358. Verburg-van Kemenade, B. M., Jenks, B. G., Visser, T. J., Tonon, M. C., and Vaudry, H. (1987). Assessment of TRH as a potential MSH release stimulating factor in Xenopus laevis. *Peptides* 8, 69–76.
359. Ubink, R., Buzzi, M., Cruijsen, P. M., Tuinhof, R., Verhofstad, A. A., Jenks, B. G., and Roubos, E. W. (1999). Serotonergic innervation of the pituitary pars intermedia of xenopus laevis. *J. Neuroendocrinol.* 11, 211–219.

360. Vaudry, H., Chartrel, N., Desrues, L., Galas, L., Kikuyama, S., Mor, A., Nicolas, P., and Tonon, M. C. (1999). The pituitary-skin connection in amphibians. Reciprocal regulation of melanotrope cells and dermal melanocytes. *Ann. N Y Acad. Sci.* 885, 41–56.
361. Verburg-Van Kemenade, B. M., Jenks, B. G., and Driessen, A. G. (1986). GABA and dopamine act directly on melanotropes of Xenopus to inhibit MSH secretion. *Brain Res. Bull.* 17, 697–704.
362. Verburg-van Kemenade, B. M., Jenks, B. G., Danger, J. M., Vaudry, H., Pelletier, G., and Saint-Pierre, S. (1987). An NPY-like peptide may function as MSH-release inhibiting factor in Xenopus laevis. *Peptides* 8, 61–67.
363. Tonon, M. C., Desrues, L., Lamacz, M., Chartrel, N., Jenks, B., and Vaudry, H. (1993). Multihormonal regulation of pituitary melanotrophs. *Ann. N Y Acad. Sci.* 680, 175–187.
364. Tuinhof, R., Gonzalez, A., Smeets, W. J., Scheenen, W. J., and Roubos, E. W. (1994). Central control of melanotrope cells of Xenopus laevis. *Eur. J. Morphol.* 32, 307–310.
365. Dellmann, H. D., Stoeckel, M. E., Hindelang-Gertner, C., Porte, A., and Stutinsky, F. (1974). A comparative ultrastructural study of the pars tuberalis of various mammals, the chicken and the newt. *Cell. Tissue Res.* 148, 313–329.
366. Morgan, P. J., King, T. P., Lawson, W., Slater, D., and Davidson, G. (1991). Ultrastructure of melatonin-responsive cells in the ovine pars tuberalis. *Cell. Tissue Res.* 263, 529–534.
367. Stoeckel, M. E., Hindelang, C., Klein, M. J., Poissonnier, M., and Felix, J. M. (1994). Expression of the alpha-subunit of glycoprotein hormones in the pars tuberalis-specific glandular cells in rat, mouse and guinea-pig. *Cell. Tissue Res.* 278, 617–624.
368. Wittkowski, W., Bockmann, J., Kreutz, M. R., and Bockers, T. M. (1999). Cell and molecular biology of the pars tuberalis of the pituitary. *Int. Rev. Cytol.* 185, 157–194.
369. Bock, N., Bockers, T. M., Bockmann, J., Nowak, P., Buse, E., and Wittkowski, W. (2001). The Pars tuberalis of the monkey (Macaca fascicularis) hypophysis: cell types and hormone expression. *Cells Tissues Organs* 169, 55–63.
370. Baker, B. L., and Yu, Y. Y. (1975). Immunocytochemical analysis of cells in the pars tuberalis of the rat hypophysis with antisera to hormones of the pars distalis. *Cell. Tissue Res.* 156, 443–449.
371. Baker, B. L., Karsch, F. J., Hoffman, D. L., and Beckman, W. C. (1977). The presence of gonadotropic and thyrotropic cells in the pituitary pars tuberalis of the monkey (Macaca mulatta). *Biol. Reprod.* 17, 232–240.
372. Girod, C., Dubois, M. P., and Trouillas, J. (1980). Immunohistochemical study of the pars tuberalis of the adenohypophysis in the monkey, Macaca irus. *Cell. Tissue Res.* 210, 191–203.
373. Gross, D. S. (1983). Hormone production in the hypophysial pars tuberalis of intact and hypophysectomized rats. *Endocrinology* 112, 733–744.
374. Lanctot, C., Gauthier, Y., and Drouin, J. (1999). Pituitary homeobox 1 (Ptx1) is differentially expressed during pituitary development. *Endocrinology* 140, 1416–1422.
375. Cameron, E., and Foster, C. L. (1972). Some light- and electron-microscopical observations on the pars tuberalis of the pituitary gland of the rabbit. *J. Endocrinol.* 54, 505–511.
376. Morgan, P. J., Barrett, P., Howell, H. E., and Helliwell, R. (1994). Melatonin receptors: localization, molecular pharmacology and physiological significance. *Neurochem. Int.* 24, 101–146.
377. Stankov, B., Capsoni, S., Lucini, V., et al. (1993). Autoradiographic localization of putative melatonin receptors in the brains of two Old World primates: Cercopithecus aethiops and Papio ursinus. *Neuroscience* 52, 459–468.
378. Wittkowski, W., Hewing, M., Hoffmann, K., Bergmann, M., and Fechner, J. (1984). Influence of photoperiod on the ultrastructure of the hypophysial pars tuberalis of the Djungarian hamster, Phodopus sungorus. *Cell. Tissue Res.* 238, 213–216.
379. Wittkowski, W. H., Schulze-Bonhage, A. H., and Bockers, T. M. (1992). The pars tuberalis of the hypophysis: a modulator of the pars distalis? *Acta Endocrinol. (Copenh).* 126, 285–290.
380. Lincoln, G. A., and Clarke, I. J. (1994). Photoperiodically-induced cycles in the secretion of prolactin in hypothalamo-pituitary disconnected rams: evidence for translation of the melatonin signal in the pituitary gland. *J. Neuroendocrinol.* 6, 251–260.
381. Lincoln, G. A., and Clarke, I. J. (1995). Evidence that melatonin acts in the pituitary gland through a dopamine-independent mechanism to mediate effects of daylength on the secretion of prolactin in the ram. *J. Neuroendocrinol.* 7, 637–643.
382. Morgan, P. J. (2000). The pars tuberalis: the missing link in the photoperiodic regulation of prolactin secretion? *J. Neuroendocrinol.* 12, 287–295.
383. Stehle, J. H., von Gall, C., and Korf, H. W. (2003). Melatonin: a clock-output, a clock-input. *J. Neuroendocrinol.* 15, 383–389.
384. Williams, L. M., Lincoln, G. A., Mercer, J. G., Barrett, P., Morgan, P. J., and Clarke, I. J. (1997). Melatonin receptors in the brain and pituitary gland of hypothalamo-pituitary disconnected Soay rams. *J. Neuroendocrinol.* 9, 639–643.
385. Morgan, P. J., Webster, C. A., Mercer, J. G., Ross, A. W., Hazlerigg, D. G., MacLean, A., and Barrett, P. (1996). The ovine pars tuberalis secretes a factor(s) that regulates gene expression in both lactotropic and nonlactotropic pituitary cells. *Endocrinology* 137, 4018–4026.
386. Hazlerigg, D. G., Hastings, M. H., and Morgan, P. J. (1996). Production of a prolactin releasing factor by the ovine pars tuberalis. *J. Neuroendocrinol.* 8, 489–492.
387. Graham, E. S., Webster, C. A., Hazlerigg, D. G., and Morgan, P. J. (2002). Evidence for the biosynthesis of a prolactin-releasing factor from the ovine pars tuberalis, which is distinct from thyrotropin-releasing hormone. *J. Neuroendocrinol.* 14, 945–954.
388. Morgan, P. J., Lawson, W., Davidson, G., and Howell, H. E. (1989). Melatonin inhibits cyclic AMP production in cultured ovine pars tuberalis cells. *J. Mol. Endocrinol.* 3, R5–R8.
389. Hazlerigg, D., Gonzalez-Brito, A., Lawson, W., and Hastings, M. H. (1991). Melatonin inhibits the activation of cyclic AMP-dependent protein kinase in cultures pars tuberalis cells from ovine pituitary. *J. Neuroendocrinol.* 3, 597–603.
390. Weaver, D. R., Stehle, J. H., Stopa, E. G., and Reppert, S. M. (1993). Melatonin receptors in human hypothalamus and pituitary: implications for circadian and reproductive responses to melatonin. *J. Clin. Endocrinol. Metab.* 76, 295–301.
391. Thomas, L., Purvis, C. C., Drew, J. E., Abramovich, D. R., and Williams, L. M. (2002). Melatonin receptors in human fetal brain: 2-[(125)I]iodomelatonin binding and MT1 gene expression. *J. Pineal Res.* 33, 218–224.
392. Monroe, B. G. (1967). A comparative study of the ultrastructure of the median eminence, infundibular stem and neural lobe of the hypophysis of the rat. *Z. Zellforsch.* 76, 405–432.
393. Duffy, P. E., and Menefee, M. (1965). Electron microscopic observations of neurosecretory granules, nerve and glial fibers and blood vessels in the median eminence of the rabbit. *Am. J. Anat.* 117, 251–286.
394. Page, R. B., and Dovey-Hartman, B. J. Neurohemal contact in the internal zone of the rabbit median eminence. *J. Comp. Neurol.* 226, 274–288.

395. Kobayashi, H., Oota, Y., Uemura, H., and Hirano, T. (1966). Electron microscopic and pharmacological studies on the rat median eminence. *Z. Zellforsch.* 71, 387–404.
396. Rinne, U. K. (1966). Ultrastructure of the median eminence of the rat. *Z. Zellforsch.* 74, 98–122.
397. Lederis, K. (1965). An electron microscopical study of the human neurohypophysis. *Z. Zellforsch.* 65, 847–868.
398. Bergland, R. M., and Torack, R. M. (1969). An electron microscopic study of the human infundibulum. *Z. Zellforsch.* 99, 1–12.
399. Rodriguez, E. M. (1969). Ependymal specializations. I. Fine structure of the neural (internal) region of the toad median eminence, with particular reference to the connections between ependymal cells and the subependymal capillary loops. *Z. Zellforsch.* 102, 153–171.
400. Rodriguez, E. M. (1969). Ultrastructure of the neurohemal region of the toad median eminence. *Z. Zellforsch.* 93, 182–212.
401. Magoun, H. W, and Ranson, S. W. (1939). Retrograde degeneration of the supraoptic nuclei after section of the infundibular stalk in the monkey. *Anat. Rec.* 75, 107–123.
402. Rasmussen, A. T. (1940). Effects of hypophysectomy and hypophysial stalk resection of the hypothalamic nuclei of animals and man. *Res. Publ. Assoc. Res. Nerv. Ment. Dis.* 20, 245–269.
403. Sherlock, D. A., Field, P. M., and Raisman, G. (1975). Retrograde transport of horse–radish peroxidase in the magnocellular neurosecretory system of the rat. *Brain Res.* 88, 403–414.
404. Barer, R., Heller, H., and Lederis, K. (1963). The isolation, identification and properties of the hormonal granules of the neurohypophysis. *Proc. R. Soc. Lond.* 158, 388–416.
405. Douglas, W. W., and Nagasawa, J. (1971). Membrane vesiculation at sites of exocytosis in the neurohypophysis, adenohypophysis and adrenal medulla: a device for membrane conservation. *J. Physiol.* 218, 94P–95P.
406. Buma, P., and Nieuwenhuys, R. (1987). Ultrastructural demonstration of oxytocin and vasopressin release sites in the neural lobe and median eminence of the rat by tannic acid and immunogold methods. *Neurosci. Lett.* 74, 151–157.
407. Theodosis, D. T., Dreifuss, J. J., and Orci, L. (1978). A freeze-fracture study of membrane events during neurohypophysial secretion. *J. Cell. Biol.* 78, 542–553.
408. Morris, J. F., and Nordmann, J. J. (1980). Membrane recapture after hormone release from nerve endings in the neural lobe of the rat pituitary gland. *Neuroscience* 5, 639–649.
409. Hartmann, J. F. (1958). Electron microscopy of the neurohypophysis in normal and histamine-treated rats. *Z. Zellforsch.* 48, 291–301.
410. Barer, R., and Lederis, K. (1966). Ultrastructure of the rabbit neurohypophysis with special reference to the release of hormones. *Z. Zellforsch.* 75, 201–239.
411. Rodriguez, E. M. (1971). The comparative morphology of neural lobes of species with different neurohypophysial hormones. *Mem. Soc. Endocrinol.* 19, 263–292.
412. Seyama, S., Pearl, G. S., and Takei, Y. (1980). Ultrastructural study of the human neurohypophysis. I. Neurosecretory axons and their dilatations in the pars nervosa. *Cell. Tissue Res.* 205, 253–271.
413. Douglas, W. W., Nagasawa, J., and Schulz, R. (1971). Electron microscope studies on the mechanism of secretion of posterior pituitary hormones and significance of microvesicles (synaptic vesicles): evidence of secretion by exocytosis and formation of microvesicles as a by-product of this process. *Mem. Soc. Endocrinol.* 19, 353–377.
414. Nordmann, J. J., and Chevallier, J. (1980). The role of microvesicles in buffering (Ca2+) in the neurohypophysis. *Nature (Lond).* 287, 54–56.
415. Shaw, F. D., and Morris, J. F. (1980). Calcium localization in the rat neurohypophysis. *Nature (Lond).* 287, 56–58.
416. Broadwell, R. D., Cataldo, A. M., and Balin, B. J. (1984). Further studies of the secretory process in hypothalamo-neurohypophysial neurons: an analysis using immunocytochemistry, wheat germ agglutinin-peroxidase, and native peroxidase. *J. Comp. Neurol.* 228, 155–167.
417. Navone, R., Di Gioia, G., Jahn, R., Browning, M., Greengard, P., and De Camilli, P. (1989). Microvesicles of the neurohypophysis are biochemically related to small synaptic vesicles of presynaptic terminals. *J. Cell. Biol.* 109, 3425–3433.
418. Meeker, R. B., Swanson, D. J., and Hayward, J. N. (1991). Light and electron microscopic localization of glutamate immunoreactivity in the supraoptic nucleus of the rat hypothalamus. *Neuroscience* 333, 157–167.
419. Meeker, R. B., Swanson, D. J., Greenwood, R. S., and Hayward, J. N. (1991). Ultrastructural distribution of glutamate immunoreactivity within neurosecretory endings and pituicytes of the rat neurohypophysis. *Brain Res.* 564, 181–193.
420. Moriyama, Y., Yamamoto, A., and Yamada, H., et al. (1995). Microvesicles isolated from bovine posterior pituitary accumulate norepinephrine. *J. Biol. Chem.* 270, 11424–11429.
421. Bloom, F. E., and Aghajanian, G. K. (1968). An electron microscopic analysis of large granular synaptic vesicles of the brain in relation to monoamine content. *J. Pharmacol. Exp. Ther.* 159, 261–273.
422. Bloom, F. E. (1970). The fine structural localization of biogenic monoamines in nervous tissue. *Int. Rev. Neurobiol.* 13, 27–66.
423. Ajika, K. (1980). Relationship between catecholaminergic neurons and hypothalamic hormone-containing neurons in the hypothalamus. In *Frontiers of Neuroendocrinology* (L. Martini and W. F. Ganong, Eds.), Vol. 6, 1–32. Raven Press, New York.
424. Hokfelt, T., Meister, B., Melander, T., and Everitt, B. (1987). Coexistence of classical transmitters and peptides with special reference to the arcuate nucleus. *Adv. Biochem. Pharmacol.* 93, 21–34.
425. Johansson, O., Hokfelt, T., Jeffcoate, S. L., White, N., and Sternberger, L. (1980). Ultrastructural localization of TRH-like immunoreactivity. *Exp. Brain Res.* 38, 1–10.
426. Shioda, T., and Nakai, Y. (1983). Immunocytochemical localization of TRH and autoradiographic determination of 3H-TRH–binding sites in the arcuate nucleus-median eminence of the rat. *Cell. Tissue Res.* 228, 475–487.
427. Pelletier, G., Labrie, F., Arimura, A., and Schally, A. V. (1974). Electron microscopic immunohistochemical localization of growth hormone-release inhibiting hormones (somatostatin) in the rat median eminence. *Am. J. Anat.* 140, 583–588.
428. Ohtsuka, M., Yamamoto, Y., and Daikoku, S. (1983). Topography and ultrastructure of LHRH- and somatostatin-containing axonal terminals in the median eminence of rats. *Arch. Histol. Jpn.* 46, 203–211.
429. Silverman, A. J., and Desnoyers, P. (1976). Ultrastructural immunocytochemical localization of luteinizing hormone-releasing hormone (LH-RH) in the median eminence of the guinea pig. *Cell. Tissue Res.* 169, 157–166.
430. Stoeckart, R., Jansen, H. G., and Kreike, A. J. (1972). Ultrastructural evidence for exocytosis in the median eminence of the rat. *Z. Zellforsch.* 131, 99–107.
431. Daikoku, S., Takahashi, T., Kojimoto, H., and Watanabe, Y. G. (1973). Secretory surface phenomena in freeze-etched preparations of the adenohypophysial cells and neurosecretory fibers. *Z. Zellforsch.* 136, 207–214.
432. Buma, P., and Nieuwenhuys, R. (1988). Ultrastructural characterization of exocitotic release sites in different layers

of the median eminence of the rat. *Cell. Tissue Res.* 252, 107–114.
433. Ajika, K. (1979). Simultaneous localization of LHRH and catecholamines in rat hypothalamus. *J. Anat.* 128, 331–347.
434. Nakai, Y., Shioda, S., Ochiai, H., Kudo, J., and Hashimoto, A. (1983). Ultrastructural relationship between monoamine- and TRH-containing axons in the rat median eminence as revealed by combined autoradiography and immunocytochemistry in the same tissue section. *Cell. Tissue Res.* 230, 1–14.
435. Tranzer, J. P., and Thoenen, H. (1967). Electronmicroscopic localization of 5-hydroxydopamine (3,4,5-trihydroxyphenyl-ethylamine), a new "false" sympathetic transmitter. *Experentia* 23, 743–745.
436. Richards, J. G., and Tranzer, J. P. (1970). The ultra-structural localization of amine storage sites in the central nervous system with the aid of a specific marker, 5-hydroxydopamine. *Brain Res.* 17, 463–469.
437. Tranzer, J. P., and Richards, J. G. (1976). Ultrastructural cytochemistry of biogenic amines in nervous tissue: methodologic improvements. *J. Histochem. Cytochem.* 24, 1178–1193.
438. Pickel, V. M., Joh, T. H., and Reis, D. J. (1975). Ultrastructural localization of tyrosine hydroxylase in noradrenergic neurons of the brain. *Proc. Natl. Acad. Sci. U S A* 72, 659–663.
439. Pickel, V. M., Joh, T. H., and Reis, D. J. (1976). Monoamine-synthesizing enzymes in central dopaminergic, noradrenergic and serotonergic neurons. Immunocytochemical localization by light and electron microscopy. *J. Histochem. Cytochem.* 24, 792–306.
440. Pickel, V. M., Joh, T. H., Field, P. M., Becker, C. G., and Reis, D. J. (1975). Cellular localization of tyrosine hydroxylase by immunohistochemistry. *J. Histochem. Cytochem.* 23, 1–12.
441. Redecker, P. (1991). Ultrastructural demonstration of neurohemal contacts in the internal zone of the median eminence of the mongolian gerbil (meriones unguiculatus): Correlation with synaptophysin immunohistochemistry. *Histochemistry* 95, 503–511.
442. Carson, K. A., Nemcroff, C. B., Rone, M. S., Youngblood, W. W., Prange, A. J., Hanber, J. S., and Kizer, J. S. (1977). Biochemical and histochemical evidence for the existence of a tubero-infundibular cholinergic pathway in the rat. *Brain Res.* 129, 169–173.
443. Fink, G., and Smith, G. C. (1971). Ultrastructural features of the developing hypothalamo-hypophysial axis in the rat. *Z. Zellforsch.* 119, 208–226.
444. Paull, W. K. (1973). A light and electron microscopic study of the development of the neurohypophysis of the fetal rat. *Anat. Rec.* 175, 407–408.
445. Eurenius, L., and Jarskar, R. (1971). Electron microscope studies on the development of the external zone of the mouse median eminence. *Z. Zellforsch.* 122, 488–502.
446. Makarenko, I. G., Ugrumov, M. V., and Calas, A. (2001). Axonal projections from the hypothalamus to the median eminence in rats during ontogenesis: DiI tracing study. *Anat. Embryol. (Berl).* 204, 239–252.
447. Rogers, M. C., Silverman, A. J., and Gibson, M. J. (1997). Gonadotropin-releasing hormone axons target the median eminence: in vitro evidence for diffusible chemoattractive signals from the mediobasal hypothalamus. *Endocrinology* 138, 3956–3966.
448. Gibson, M. J., Ingraham, L., and Dobrjansky, A. (2000). Soluble factors guide gonadotropin-releasing hormone axonal targeting to the median eminence. *Endocrinology* 141, 3065–3071.
449. Makarenko, I. G., Ugrumov, M. V., Derer, P., and Calas, A. (2000). Projections from the hypothalamus to the posterior lobe in rats during ontogenesis: 1,1'-dioctadecyl-3,3,3', 3'-tetramethylindocarbocyanine perchlorate tracing study. *J. Comp. Neurol.* 422, 327–337.
450. Schonemann, M. D., Ryan, A. K., McEvilly, R. J., O'Connell, S. M., Arias, C. A., Kalla, K. A., Li, P., Sawchenko, P. E., and Rosenfeld, M. G. (1995). Development and survival of the endocrine hypothalamus and posterior pituitary gland requires the neuronal POU domain factor Brn-2. *Genes Dev.* 9, 3122–3135.
451. Monroe, B. G., Newman, B. L., and Schapiro, S. (1972). Ultrastructure of the median eminence of neonatal and adult rats. In *Brain–Endocrine Interaction. Median Eminence Structure and Function* (K. M. Knigge, D. E. Scott, and A. Weindl, Eds.). Karger, New York.
452. Monroe, B. G., and Paull, W. K. (1974). Ultrastructural changes in the hypothalamus during development and hypothalamic activity: the median eminence. *Prog. Brain Res.* 41, 185–208.
453. Kawano, H., Watanabe, Y. G., and Daikoku, S. (1980). Light and electron microscopic observation on the appearance of immunoreactive LHRH in perinatal rat hypothalamus. *Cell. Tissue Res.* 213, 465–474.
454. Setalo, G., Antalicz, M., Saarossy, K., Arimura, A., Schally, A. V., and Flerko, B. (1978). Ontogenesis of LH–RH containing neuronal elements in the hypothalamus of the rat. *Acta Biol. Acad. Sci. Hung.* 29, 285–290.
455. Gross, D. S., and Baker, B. L. (1977). Immunohistochemical localization of gonadotropin-releasing hormone (GnRH) in the fetal and early postnatal mouse brain. *Am. J. Anat.* 148, 195–216.
456. Gross, D. S., and Baker, B. L. (1979). Developmental correlation between hypothalamic gonadotropin-releasing hormone and hypophysial luteinizing hormone. *Am. J. Anat.* 154, 1–10.
457. Monroe, B. G., and Holmes, E. M. (1983). The freeze-fractured median eminence. II. Developmental changes in the neurohemal contact zone of the median eminence of the rat. *Cell. Tissue Res.* 233, 81–97.
458. Ugrumov, M. V., Chandrasekhar, K., Borisova, N. A., and Mitsekevich, M. S. (1979). Light and electron microscopical investigations on the tanycyte differentiation during the perinatal period in the rat. *Cell. Tissue Res.* 201, 295–303.
459. Monroe, B. G., and Holmes, E. M. (1982). The freeze-fractured median eminence. 1. Development of intercellular junctions in the ependyma of the 3rd ventricle of the rat. *Cell. Tissue Res.* 222, 389–408.
460. Silverman, A. J., and Desnoyers, P. (1975). Post-natal development of the median eminence of the guinea pig. *Anat. Rec.* 183, 459–475.
461. Lamperti, A., and Mastovich, J. (1983). Morphological changes in the hypothalamic arcuate nucleus and median eminence in the golden hamster during the neonatal period. *Am. J. Anat.* 166, 173–185.
462. Kobayashi, H., and Matsui, T. (1969). Fine structure of the median eminence and its functional significance. In *Frontiers in Neuroendocrinology* (W. F. Ganong and L. Martini, Eds.), Vol. 1. Oxford University Press, Oxford.
463. Kobayashi, H., Matsui, T., and Ishii, S. (1970). Functional electron microscopy of the hypothalamic median eminence. *Int. Rev. Cytol.* 29, 281–381.
464. Knigge, K. M., and Silverman, A.-J. (1974). Anatomy of the endocrine hypothalamus. In Section 7: *Endocrinology. The Pituitary Gland and Its Neuroendocrine Control.* Vol. IV. *Handbook of Physiology* (R. O. Greep and E. B. Astwood, Eds.), pp. 1–32. Am Phys Soc, Washington, DC.
465. Enemar, A., and Eurenius, L. (1979). Organization and development of the perivascular space system in the neuro-hypophysis of the laboratory mouse. *Cell. Tissue Res.* 199, 66–116.
466. Heap, P. F., Jones, C. W., Morris, J. F., and Pickering, B. T. (1975). Movement of neurosecretory product through the

anatomical compartments of the neural lobe of the pituitary gland. (An electron microscopic autoradiographic study). *Cell. Tissue Res.* 156, 483–497.

467. Tweedle, C. D., Smithson, K. G., and Hatton, G. I. (1989). Neurosecretory endings in the rat neurohypophysis are en passant. *Exp. Neurol.* 106, 20–26.
468. Aspeslagh, M.-R., Vandesande, F., and Dierickx, K. (1976). Electron microscopic immunocytochemical demonstration of separate neurophysin-vasopressinergic and neurophysin-oxytocinergic nerve fibres in the neural lobe of the rat hypophysis. *Cell. Tissue Res.* 171, 31–37.
469. VanLeeuwen, F. W., de Raay, C., Swaab, D. F., and Fisser, B. (1979). The localization of oxytocin, vasopressin, somatostatin and luteinizing hormone releasing hormone in the rat neurohypophysis. *Cell. Tissue Res.* 202, 189–201.
470. Jirikowski, G. F., Sanna, P. P., and Bloom, F. E. (1990). mRNA coding for oxytocin is present in axons of the hypothalamo-neurohypophysial tract. *Proc. Natl. Acad. Sci. U S A* 87, 7400–7404.
471. Mohr, E., Zhou, A., Thorn, N. A., and Richter, D. (1990). Rats with physically disconnected hypothalamo-pituitary tracts no longer contain vasopressin-oxytocin gene transcripts in the posterior pituitary lobe. *FEBS Lett.* 263, 332–336.
472. Mezey, E., and Kiss, J. Z. (1990). Coexpression of vasopressin and oxytocin in hypothalamic supraoptic neurons of lactating rats. *Endocrinology* 129, 1814–1820.
473. Rossier, J., Pittman, Q., Bloom, F. E., and Guillemin, R. Distribution of opioid peptides in the pituitary: a new hypothalamic–pars nervosa enkephalinergic pathway. *Fed. Proc.* 39, 2555–2560.
474. Zamir, N. (1985). On the origin of leu-enkephalin and met-enkephalin in the rat neurohypophysis. *Endocrinology* 117, 1687–1692.
475. Martin, R., and Voigt, K. H. (1981). Enkephalins co-exist with oxytocin and vasopressin in nerve terminals of rat neurohypophysis. *Nature (Lond).* 289, 502–504.
476. Martin, R., Geis, R., Holl, R., Schafer, M., and Voigt, K. H. (1983). Co-existence of unrelated peptides in oxytocin and vasopressin terminals of rat neurohypophyses: immunoreactive methionine5-enkephalin-, leucine5- enkephalin- and cholecystokinin-like substances. *Neuroscience* 8, 213–227.
477. Vanderhaeghen, J. J., Lotstra, F., Liston, D. R., and Rossier, J. (1983). Proenkephalin, (Met)enkephalin, and oxytocin immunoreactivities are colocalized in bovine hypothalamic magnocellular neurons. *Proc. Natl. Acad. Sci. U S A* 80, 5139–5143.
478. Weber, E., Roth, K. A., Evans, C. J., Chang, J. K., and Barchas, J. D. (1982). Immunohistochemical localization of dynorphin (1-8) in hypothalamic magnocellular neurons: evidence for absence of proenkephalin. *Life Sci.* 31, 1761–1764.
479. Shuster, S. J., Riedl, M., Li, X., Vulchanova, L., and Elde, R. (2000). The kappa opioid receptor and dynorphin co-localize in vasopressin magnocellular neurosecretory neurons in guinea-pig hypothalamus. *Neuroscience* 96, 373–383.
480. VanLeeuwen, F. W., Pool, C. W., and Sluiter, A. A. (1983). Enkephalin immunoreactivity in synaptoid elements on glial cells in the rat neural lobe. *Neuroscience* 8, 229–241.
481. Merchenthaler, I., Maderdrut, J. L., Altshculer, R. A., and Petrusz, P. (1986). Immunocytochemical localization of proenkephalin-derived peptides in the central nervous system of the rat. *Neuroscience* 17, 325–348.
482. Summy-Long, J. Y., Miller, D. S., and Rosella-Dampman, L. M., et al. (1984). A functional role for opioid peptides in the differential secretion of vasopressin and oxytocin. *Brain Res.* 309, 362–366.
483. Bondy, C. A., Gainer, H., and Russell, J. T. (1988). Dynorphin A inhibits and naloxone increases the electrically stimulated release of oxytocin but not vasopressin from terminals of the neural lobe. *Endocrinology* 122, 1321–1327.
484. Vanderhaeghen, J. J., Lotstra, F., DeMey, J., and Gilles, C. (1980). Immunohistochemical localization of cholecystokinin and gastrin–like peptides in the brain and hypophysis of the rat. *Proc. Natl. Acad. Sci. U S A* 77, 1190–1194.
485. Kiss, J. Z., Williams, T. H., and Palkovits, M. (1984). Distribution and projections of cholecystokinin-immunoreactive neurons in the hypothalamic paraventricular nucleus of rat. *J. Comp. Neurol.* 227, 173–181.
486. Gaymann, W., Martin, R. (1989). Immunoreactive galanin-like material in magnocellular hypothalamo-neurohypophysial neurones of the rat. *Cell. Tissue Res.* 255, 139–147.
487. Melander, T., Hokfelt, T., Rokaeus, A., Cuello, A. C., Oertel, W. H., Verhofstad, A., and Goldstein, M. (1986). Coexistence of galanin-like immunoreactivity with catecholamines, 5-hydroxytryptamine, GABA and neuropeptides in the rat CNS. *J. Neurosci.* 6, 3640–3654.
488. Stoeckel, M. E., Porte, A., Klein, M. J., and Cuello, A. C. (1982). Immunocytochemical localization of substance P in the neurohypophysis and hypothalamus of the mouse compared with the distribution of other neuropeptides. *Cell. Tissue Res.* 223, 533–544.
489. Mikkelsen, J. D., Larsen, P. J., Vilhardt, H., and Soermark, T. (1989). Substance P in the median eminence and pituitary of the rat: demonstration of immunoreactive fibers and specific binding sites. *Neuroendocrinology* 50, 100–108.
490. Zimmerman, E. A., Robinson, A. G., Husain, M. K., Acosta, M., Frantz, A. G., and Sawyer, W. H. (1974). Neurohypophysial peptides in the bovine hypothalamus: the relationship of neurophysin I to oxytocin, and neurophysin II to vasopressin in supraoptic and paraventricular regions. *Endocrinology* 95, 931–936.
491. Lechan, R. M., and Jackson, I. (1982). Immunohistochemical localization of thyrotropin-releasing hormone in the rat hypothalamus and pituitary. *Endocrinology* 111, 55–65.
492. Mikkelsen, J. D., Bersani, M., Holst, J. J., and Larsen, P. J. (1991). Nerve fibers in the rat posterior pituitary lobe contain prosomatostatin (1-64). *Neuroendocrinology* 54, 469–476.
493. Bjorklund, A., Moore, R. Y., Nobin, A., and Stenevi, U. (1973). The organization of tubero-hypophyseal and reticulo-infundibular catecholamine neuron systems in the rat brain. *Brain Res.* 51, 171–191.
494. Pelletier, G. (1983). Identification of endings containing dopamine and vasopressin in the rat posterior pituitary by a combination of radioautography and immunocytochemistry at the ultrastructural level. *J. Histochem. Cytochem.* 31, 562–564.
495. Saavedra, J. M. (1985). Central and peripheral catecholamine innervation of the rat intermediate and posterior pituitary lobes. *Neuroendocrinology* 40, 281–284.
496. Brustle, O., Pilgrim, C. H., Gaymann, W., and Reisert, I. (1988). Abundant gabaergic innervation of rat posterior pituitary revealed by inhibition of gaba-transaminase. *Cell. Tissue Res.* 251, 59–64.
497. Buijs, R. M., Van Vulpen, E. H. S., and Geffard, M. (1987). Ultrastructural localization of gaba in the supraoptic nucleus and neural lobe. *Neuroscience* 20, 347–355.
498. Alper, R. H., Demarest, K. T., and Moore, K. E. (1980). Effects of surgical sympathectomy on catecholamine concentrations in the posterior pituitary of the rat. *Experentia* 35, 134–135.
499. Back, N., Soinila, S., Joh, T. H., and Rechardt, L. (1987). Catecholamine-synthesizing enzymes in the rat pituitary. An immunohistochemical study. *Histochemistry* 86, 459–464.
500. Bicknell, R. J., Dyball, R. E., Garten, L. L., Heavens, R. P., Sirinathsinghji, D. J. S., and Zahao, B.-G. (1988). Evidence

for a direct noradrenergic projection from the brainstem to the neural lobe in the rat. *J. Physiol.* 396, 127P.
501. Barnes, P. R. J., and Dyball, R. E. (1982). Inhibition of neurohypophysial hormone release by dopamine in the rat. *J. Physiol.* 327, 85P–86P.
502. Vizi, E. T., and Volbekas, V. (1980). Inhibition by dopamine of oxytocin release from isolated posterior lobe of the hypophysis of the rat: disinhibitory effect of beta-endorphin/enkephalin. *Neuroendocrinology* 31, 46–52.
503. Racke, K., Ritzel, H., Trapp, B., and Muscholl, E. (1982). Dopaminergic modulation of evoked vasopressin release from the isolated neurohypophysis of the rat. Possible involvement of the endogenous opioids. *Naunyn-Schmiedeberg's Arch. Pharmac.* 319, 56–68.
504. Bredt, D. S., Glatt, C. E., Hwang, P. M., Fotuhi, M., Dawson, T. M., and Snyder, S. H. (1991). Nitric oxide synthase protein and mRNA are discretely localized in neuronal populations of the mammalian CNS together with NADPH diaphorase. *Neuron.* 7, 615–624.
505. Anthony, E. L. P., King, J. C., and Stopa, E. G. (1984). Immunocytochemical localization of LHRH in the median eminence, infundibular stalk, and neurohypophysis. (Evidence for multiple sites of releasing hormone secretion in humans and other mammals). *Cell. Tissue Res.* 236, 5–14.
506. Kawata, M., Hashimoto, K., Takahara, J., and Sano, Y. (1983). Immunohistochemical identification of the corticotropin releasing factor (CRF)-containing nerve fibers in the pig hypophysis, with special reference to the relationship between CRF and posterior lobe hormones. *Arch. Histol. Jpn.* 46, 183–190.
507. Takei, Y., Seyama, S., Pearl, G. S., and Tindall, G. T. (1980). Ultrastructural study of the human neurohypophysis. II. Cellular elements of neural parenchyma, the pituicytes. *Cell. Tissue Res.* 205, 273–287.
508. Tweedle, C. D., and Hatton, G. I. (1980). Glial cell enclosure of neurosecretory endings in the neurohypophysis of the rat. *Brain Res.* 192, 555–559.
509. Tweedle, C. D., and Hatton, G. I. (1980). Evidence for dynamic interactions between pituicytes and neurosecretory axons in the rat. *Neuroscience.* 5, 661–667.
510. Hatton, G. I., and Tweedle, C. D. (1982). Magnocellular neuropeptidergic neurons in hypothalamus: increases in membrane apposition and number of specialized synapses from pregnancy to lactation. *Brain Res. Bull.* 8, 197–204.
511. Tweedle, C. D., and Hatton, G. I. (1982). Magnocellular neuropeptidergic terminals in the neurohypophysis: rapid glial release of enclosed axons during parturition. *Brain Res. Bull.* 8, 205–209.
512. Tweedle, C. D. (1983). Ultrastructural manifestations of increased hormone release in the neurohypophysis. *Prog. Brain Res.* 60, 259–272.
513. Hatton, G. I., Perlmutter, L. S., Salm, A. K., and Tweedle, C. D. (1984). Dynamic neuronal-glial interactions in hypothalamus and pituitary: implications for control of hormone synthesis and release. *Peptides* 5, 121–138.
514. Perlmutter, L. S., Hatton, G. I., and Tweedle, C. D. (1984). Plasticity in the in vitro neurohypophysis: effects of osmotic changes on pituicytes. *Neuroscience* 12, 503–511.
515. Tweedle, C. D., and Hatton, G. I. (1987). Morphological adaptability at neurosecretory axonal endings on the neurovascular contact zone of the rat neurohypophysis. *Neuroscience* 20, 241–246.
516. Luckman, S. M., and Bicknell, R. J. (1990). Morphological plasticity that occurs in the neurohypophysis following activation of the magnocellular neurosecretory system can be mimicked in vitro by beta-adrenergic stimulation. *Neuroscience* 39, 701–709.
517. Bicknell, R. J., Luckman, S. M., Inenga, K., Mason, W. T., and Hatton, G. I. (1989). Beta-adrenergic and opioid receptors on pituicytes cultured from adult rat neurohypophysis: regulation of cell morphology. *Brain Res. Bull.* 22, 379–385.
518. Alonso, G., Runquist, M., and Hussy, N., et al. (2003). Age-related modifications of the morphological organization of pituicytes are associated with alteration of the GABAergic and dopaminergic innervation afferent to the neurohypophysial lobe. *Eur. J. Neurosci.* 18, 1889–1903.
519. Bunn, S. J., Hanley, M. R., and Wilkin, G. P. (1985). Evidence for kappa opioid receptor on pituitary astrocytes: an autoradiographic study. *Neurosci. Lett.* 55, 317–323.
520. Bicknell, R. J. (1985). Endogenous opioid peptides and hypothalamic neuroendocrine neurones. *J. Endocr.* 107, 437–446.
521. Kjeldsen, T. H., Rivier, C., Lee, S., Hansen, E. W., Christensen, J. D., and Moesby, L. (2003). Inducible nitric oxide synthase is responsible for nitric oxide release from murine pituicytes. *J. Neuroendocrinol.* 15, 250–255.
522. Lutz-Bucher, B., and Koch, B. (1994). Evidence for an inhibitory effect of nitric oxides on neuropeptide secretion from isolated neural lobe of the rat pituitary gland. *Neurosci. Lett.* 165, 48–50.
523. Miyata, S., Matsushima, O., and Hatton, G. I. (1997). Taurine in rat posterior pituitary: localization in astrocytes and selective release by hypoosmotic stimulation. *J. Comp. Neurol.* 381, 513–523.
524. Hussy, N., Bres, V., Rochette, M., Duvoid, A., Alonso, G., Dayanithi, G., and Moos, F. C. (2001). Osmoregulation of vasopressin secretion via activation of neurohypophysial nerve terminals glycine receptors by glial taurine. *J. Neurosci.* 21, 7110–7116.
525. Hatton, G. I. (1990). Emerging concepts of structure-function dynamics in adult brain: the hypothalamo-neurohypophysial system. *Prog. Neurobiol.* 34, 437–504.
526. Negro-Vilar, A., and Samson, W. K. (1979). Dehydration-induced changes in immunoreactive vasopressin levels in specific hypothalamic structures. *Brain Res.* 169, 585–589.
527. Ziedonis, D. M., Severs, W. B., Brennan, R. W., and Page, R. B. (1986). Blood flow and functional responses correlate in the ovine neural lobe. *Brain Res.* 373, 27–34.
528. Kadekaro, M., Gross, P. M., and Sokoloff, L. (1986). Local cerebral glucose utilization in Long-Evans and Brattleboro rats during acute dehydration. *Neuroendocrinology* 42, 203–210.
529. Gross, P. M., Blasberg, R. G., Fenstermacher, J. D., and Patlak, C. S. (1985). Rapid amino acid uptake in rat pituitary neural lobe during functional stimulation by chronic dehydration. *J. Cereb. Blood Flow Metab.* 5, 151–155.
530. Merchenthaler, I., Vigh, S., Petrusz, P., and Schally, A. V. (1982). Immunocytochemical localization of corticotropin-releasing factor (CRF) in the rat brain. *Am. J. Anat.* 165, 385–396.
531. Sawchenko, P. E., Swanson, L. W., Rivier, J., and Vale, W. (1985). The distribution of growth-hormone-releasing factor (GRF) immunoreactivity in the central nervous system of the rat: an immunohistochemical study using antisera directed against rat hypothalamic GRF. *J. Comp. Neurol.* 237, 100–115.
532. Kahn, D., Abrams, G. M., Zimmerman, E. A., Carraway, R., and Leeman, S. E. (1980). Neurotensin neurons in the rat hypothalamus: an immunohistochemical study. *Endocrinology* 107, 47–54.
533. Dierickx, K., and De Waele, G. (1975). Scanning electron microscopy of the wall of the third ventricle of the brain of Rana temporia. II. Electron microscopy of the ventricular surface of the pars ventralis of the tuber cinereum. *Cell. Tissue Res.* 159, 81–90.
534. Mikami, S.-I. (1975). A correlative ultrastructural analysis of the ependymal cells of the third ventricle of Japanese quail,

Coturnix japonica. In *Brain-Endocrine Interaction* (K. M. Knigge, D. E. Scott, H. Kobayashi, and S. Ishii, Eds.), Vol. II. Karger, Basel.

535. Martinez, P. M., and de Weerd, H. (1977). The fine structure of the ependymal surface of the recessus infundibularis in the rat. *Anat. Embryol. (Berl).* 151, 241–265.
536. Paull, W. K., Martin, H., and Scott, D. E. (1977). Scanning electron microscopy of the third ventricular floor of the rat. *J. Comp. Neurol.* 175, 301–310.
537. Bruni, J. E., Montemurro, D. G., Clattenburg, R. E., and Singh, R. P. (1972). A scanning electron microscopic study of the ependymal surface of the third ventricle of the rabbit, rat, mouse and human brain. *Anat. Rec.* 174, 407–420.
538. Climenti, F., and Marini, D. (1972). The surface fine structure of the walls of cerebral ventricles and of choroid plexus in cat. *Z. Zellforsch.* 123, 82–95.
539. Bruni, J. E., Clattenburg, R. E., and Montemurro, D. G. (1974). Ependymal tanycytes of the rabbit third ventricle: a scanning electron microscopic study. *Brain Res.* 73, 145–150.
540. Scott, D. E. (1973). A comparative ultrastructural analysis of the third cerebral ventricle of the North American Mink. *Anat. Rec.* 175, 155–168.
541. Kozlowski, G. P., Scott, D. E., and Dudley, G. K. (1973). Scanning electron microscopy of the third ventricle of sheep. *Z. Zellforsch. Mikrosk. Anat.* 136, 169–176.
542. Scott, D. E., Krobisch-Dudley, G., Paull, W. K., Kozlowski, G. P., and Ribas, J. (1975). The primate median eminence. I. Correlative scanning-transmission electron microscopy. *Cell. Tissue Res.* 162, 61–73.
543. Jacobs, J. J., and Monroe, K. D. (1977). A scanning electron microscopic survey of the brain ventricular system of the female armadillo. *Cell. Tissue Res.* 183, 531–539.
544. Bruni, J. E., Montemurro, D. G., and Clattenburg, R. E. (1977). Morphology of the ependymal lining of the rabbit third ventricle following intraventricular administration of synthetic luteinizing hormone–releasing hormone (LH-RH): a scanning electron microscopic investigation. *Am. J. Anat.* 150, 411–425.
545. Matsui, T., and Kobayashi, H. (1968). Surface protrusions from the ependymal cells of the median eminence. *Arch. Anat. Histol. Embryol.* 51, 429–436.
546. Dierickx, K., and DeWaele, G. (1975). Scanning electron microscopy of the wall of the third ventricle on Rana temporaria. III. Electron microscopy of the ventricular surface of the median eminence. *Cell. Tissue Res.* 161, 343–349.
547. Scott, D. E., Krobisch-Dudley, G., Paull, W. K., and Kozlowski, G. P. (1977). The ventricular system in neuroendocrine mechanisms. III. Supraependymal neuronal networks in the primate brain. *Cell. Tissue Res.* 179, 235–254.
548. Bleier, R., Albrecht, R., and Cruce, J. A. (1975). Supraependymal cells of hypothalamic third ventricle: identification as resident phagocytes of the brain. *Science* 189, 299–301.
549. Bleier, R. (1977). Ultrastructure of supraependymal cells and ependyma of hypothalamic third ventricle of mouse. *J. Comp. Neurol.* 174, 359–376.
550. Bleier, R. (1971). The relations of ependyma to neurons and capillaries in the hypothalamus: a Golgi-Cox study. *J. Comp. Neurol.* 142, 439–463.
551. Millhouse, O. E. (1971). A Golgi study of third ventricle tanycytes in the adult rodent brain. *Z. Zellforsch. Mikrosk. Anat.* 121, 1–13.
552. Millhouse, O. E. (1972). Light and electron microscopic studies of the ventricular wall. *Z. Zellforsch. Mikrosk. Anat.* 127, 149–174.
553. Card, J. P., Rafols, J. A. (1978). Tanycytes of the third ventricle of the neonatal rat: a Golgi study. *Am. J. Anat.* 151, 173–190.
554. Sharp, P. J. (1972). Tanycyte and vascular patterns in the basal hypothalamus of Coturnix quail with reference to their possible neuroendocrine significance. *Z. Zellforsch. Mikrosk. Anat.* 127, 552–569.
555. Akmayev, I. G., and Popov, A. P. (1977). Morphological aspects of the hypothalamic- hypophyseal system. VII. The tanycytes: their relation to the hypophyseal adrenocorticotrophic function. An ultrastructural study. *Cell. Tissue Res.* 180, 263–282.
556. Brawer, J. R. (1972). The fine structure of the ependymal tanycytes at the level of the arcuate nucleus. *J. Comp. Neurol.* 145, 25–42.
557. Gotow, T., and Hashimoto, P. H. (1981). Graded differences in tightness of ependymal intercellular junctions within and in the vicinity of the rat median eminence. *J. Ultrastruct. Res.* 76, 293–311.
558. Rodriguez, E. M., Gonzalez, C. B., and Delannoy, L. (1979). Cellular organization of the lateral and postinfundibular regions of the median eminence in the rat. *Cell. Tissue Res.* 201, 377–408.
559. Brawer, J. R., and Walsh, R. J. (1982). Response of tanycytes to aging in the median eminence of the rat. *Am. J. Anat.* 163, 247–256.
560. Brightman, M. W., and Reese, T. S. (1969). Junctions between intimately apposed cell membranes in the vertebrate brain. *J. Cell. Biol.* 40, 648–677.
561. Brightman, M. W., Prescott, L., and Reese, T. S. (1975). Intercellular junctions of special ependyma. In *Brain-Endocrine Interaction* (K. M. Knigge, D. E. Scott, H. Kobayashi, and S. Ishii, Eds.) Vol. II, pp. 146–165. Karger, Basel.
562. Nakai, Y., Ochiai, H., and Uchida, M. (1977). Fine structure of ependymal cells in the median eminence of the frog and mouse revealed by freeze-etching. *Cell. Tissue Res.* 181, 311–318.
563. Kobayashi, H., Wada, M., and Uemura, H. (1972). Uptake of peroxidase from the 3rd ventricle by ependymal cells of the median eminence. *Z. Zellforsch.* 127, 545–551.
564. Kobayashi, H. (1975). Absorption of cerebrospinal fluid by ependymal cells of the median eminence. In *Brain–Endocrine Interaction* (K. M. Knigge, D. E. Scott, H. Kobayashi, and S. Ishii, Eds.) Vol. II, pp. 109–122. Karger, Basel.
565. Nakai, Y., and Naito, N. (1975). Uptake and bidirectional transport of peroxidase injected into the blood and cerebrospinal fluid by ependymal cells of the median eminence. In *Brain–Endocrine Interaction* (K. M. Knigge, D. E. Scott, H. Kobayashi, and S. Ishii, Eds.) Vol. II, pp. 94–108. Karger, Basel.
566. Guldner, F.-H., and Wolff, J. R. (1973). Neurono-glial synaptoid contacts in the median eminence of the rat: ultrastructure, staining properties and distribution on tanycytes. *Brain Res.* 61, 217–234.
567. Scott, D. E., and Paull, W. K. (1979). The tanycyte of the rat median eminence. I. Synaptoid contacts. *Cell. Tissue Res.* 200, 329–334.
568. Knowles, F., and Kumar, T. C. A. (1969). Structural changes, related to reproduction, in the hypothalamus and in the pars tuberalis of the rhesus monkey. *Phil. Trans. Roy. Soc. Lond.* 256, 357–375.
569. Krisch, B., Leonhardt, H., and Buchheim, W. (1978). The functional and structural border of the neurohemal region of the median eminence. *Cell. Tissue Res.* 192, 327–339.
570. Wittkowski, W., and Scheuer, A. (1974). Functional changes of the neuronal and glial elements at the surface of the external layer of the median eminence. *Z. Anat. Entwicklungsgesch.* 143, 255–262.
571. Prevot, V. (2002). Glial-neuronal-endothelial interactions are involved in the control of GnRH secretion. *J. Neuroendocrinol.* 14, 247–255.

572. Bugnon, C., Fellmann, D., Gouget, A., and Cardot, J. (1982). Corticoliberin in rat brain: immunocytochemical identification and localization of a novel neuroglandular system. *Neurosci. Lett.* 30, 25–30.
573. Hanstrom, B. (1953). The neurohypophysis in the series of mammals. *Z. Zellforsch.* 39, 241–259.
574. Alm, P., Skagerberg, G., Nylen, A., Larsson, B., and Andersson, K. E. (1997). Nitric oxide synthase and vasopressin in rat circumventricular organs. An immunohistochemical study. *Exp. Brain Res.* 117, 59–66.
575. Holmes, M. C., Antoni, F. A., Aguilera, G., and Catt, K. J. (1986). Magnocellular axons in passage through the median eminence release vasopressin. *Nature (Lond).* 319, 326–329.
576. Wotjak, C. T., Kubota, M., Kohl, G., and Landgraf, R. (1996). Release of vasopressin from supraoptic neurons within the median eminence in vivo. A combined microdialysis and push-pull perfusion study in the rat. *Brain Res.* 726, 237–241.
577. Mezey, E., Kiss, J. Z., Mueller, G. P., Eskay, R. L., O'Donohue, T. L., and Palkovits, M. (1985). Distribution of the pro-opiomelanocortin derived peptides, adrenocorticotrope hormone, α-melanocyte-stimulating hormone and β-endorphin (ACTH, α-MSH, β-End) in the rat hypothalamus. *Brain Res.* 328, 341–347.
578. Kiss, J. Z., Mezey, E., Cassell, M. D., Williams, T. H., Mueller, G. P., O'Donohue, T. L., and Palkovits, M. (1985). Topographical distribution of pro-opiomelanocortin-derived peptides (ACTH/β-End/a-MSH) in the rat median eminence. *Brain Res.* 329, 169–176.
579. Rodriguez, E. M. (1972). Comparative and functional morphology of the median eminence. In *Brain–Endocrine Interaction* (K. M. Knigge, D. E. Scott, and A. Weindl, Eds.), Vol. I, pp. 319–334. Karger, Basel.
580. Knigge, K. M., and Scott, D. E. (1970). Structure and function of the median eminence. *Am. J. Anat.* 129, 223–244.
581. Fuxe, K. (1963). Cellular localization of monoamines in the median eminence and infundibular stem of some mammals. *Acta Physiol. Scand.* 58.
582. Fuxe, K. (1964). Cellular localization of monoamines in the median eminence and infundibular stem of some mammals. *Z. Zellforsch.* 61, 710–724.
583. Fuxe, K. (1965). Evidence for the existence of monoamine neurons in the central nervous system. III. The monoamine nerve terminal. *Z. Zellforsch.* 65, 573–596.
584. Fuxe, K. (1965). Evidence for the existence of monoamine neurons in the central nervous system. IV. The distribution of monoamine nerve terminals in the central nervous system. *Acta Physiol. Scand. Suppl.* 247, 39–85.
585. Bjorklund, A., Lindvall, O., and Svenson, L. A. (1972). Mechanisms of fluorophore formation in the histochemical glyoxylic acid method for monoamines. *Histochemie* 32, 113–131.
586. Lindvall, O., and Bjorklund, A. (1974). The glyoxylic acid fluorescence histochemical method. A detailed account of the methodology for the visualization of central catecholamine neurons. *Histochemistry* 39, 97–127.
587. Chiba, T., Hwang, B. H., and Williams, T. H. (1976). A method for studying glyoxylic acid induced fluorescence and ultrastructure of monoamine neurons. *Histochemistry* 49, 95–106.
588. Lofstrom, A., Jonsson, G., and Fuxe, K. (1976). Microfluorimetric quantitation of catecholamine fluorescence in rat median eminence. I. Aspects on the distribution of dopamine and noradrenaline nerve terminals. *J. Histochem. Cytochem.* 24, 415–429.
589. Ajika, K., and Hokfelt, T. (1973). Ultrastructural identification of catecholamine neurones in the hypothalamic periventricular-arcuate nucleus-median eminence complex with special reference to quantitative aspects. *Brain Res.* 57, 97–117.
590. Bensch, C., Lescure, H., Robert, J., and Faure, J. M. (1975). Catecholamine histofluorescence in the median eminence of female rabbits activated by mating. *J. Neural. Transm.* 36, 1–18.
591. Bensch, C., Lescure, H., Dufy, B., and Gross, C. (1979). Histofluorometrie des catecholamines de la couche externe de l'eminence mediane hypothalamique de la lapine. *Annals d'Endocrin.* 39, 281–302.
592. Nojyo, Y., Ibata, Y., and Sano, Y. (1976). Demonstration of tuberinfundibular tract of the cat. Fluorescence histochemistry and electron microscopy. *Cell. Tissue Res.* 168, 289–301.
593. Okamura, H., Kitahama, K., Mons, N., Ibata, Y., Jouvet, M., and Geffard, M. (1988). L-dopa-immunoreactive neurons in the rat hypothalamic tuberal region. *Neurosci. Lett.* 95, 42–46.
594. Tago, H., McGreer, P. L., Gruce, G., and Hersh, L. B. (1987). Distribution of choline acetyltransferase-containing neurons of the hypothalamus. *Brain Res.* 415, 49–62.
595. Schafer, M. K., Eiden, L. E., and Weihe, E. (1998). Cholinergic neurons and terminal fields revealed by immunohistochemistry for the vesicular acetylcholine transporter. I. Central nervous system. *Neuroscience* 84, 331–359.
596. Thind, K. K., and Goldsmith, P. C. (1995). Glutamate and GABAergic neurointeractions in the monkey hypothalamus: a quantitative immunomorphological study. *Neuroendocrinology* 61, 471–485.
597. Kawakami, S., Ichikawa, M., Yokosuka, M., Tsukamura, H., and Maeda, K. (1998). Glial and neuronal localization of neuronal nitric oxide synthase immunoreactivity in the median eminence of female rats. *Brain Res.* 789, 322–326.
598. Yoshimoto, Y., Sakai, K., Panula, P., Salvert, D., Stuart, M., and Jouvet, M. (1989). Cells of origin of histaminergic afferents to the cat median eminence. *Brain Res.* 504, 149–153.
599. Ibata, Y., Fukui, K., Okamura, H., Kawakami, T., Tanaka, M., Obata, H. L., Tsuto, T., Terubayashi, H., Yanaihara, C., and Yanaihara, N. (1983). Coexistence of dopamine and neurotensin in hypothalamic arcuate and periventricular neurons. *Brain Res.* 269, 177–179.
600. Everitt, B. J., Meister, B., and Hokfelt, T., et al. (1986). The hypothalamic arcuate nucleus-median eminence complex: immunohistochemistry of transmitters, peptides and DARPP–32 with special reference to coexistence in dopamine neurons. *Brain Res.* 396, 97–155.
601. Pu, L.-P., Charmay, Y., Leduque, P., Morel, G., and Dubois, P. M. (1991). Light and electron microscopic immunocytochemical evidence that delta sleep-inducing peptide and gonadotropin-releasing hormone are coexpressed in the same nerve structures in the guinea pig median eminence. *Neuroendocrinology* 53, 332–338.
602. Joseph, S. A., Piekut, D. T., and Knigge, K. M. (1981). Immunocytochemical localization of luteinizing hormone-releasing hormone (LHRH) in vibratome-sectioned brain. *J. Histochem. Cytochem.* 29, 247–254.
603. McNeill, T. H., and Sladek, J. R. (1978). Fluorescence-immunocytochemistry. Simultaneous localization of catecholamines and gonadotropin-releasing hormone. *Science* 200, 72–74.
604. Ibata, Y., Watanabe, K., Kinoshita, H., Kubo, S., Sano, Y., Sin, S., Hashimura, E., and Imagewa, K. (1979). Detection of catecholamine and luteinizing hormone-releasing hormone (LH-RH) containing nerve endings in the median eminence and the organum vasculosum laminae terminalis by fluorescence histochemistry and immunohistochemistry on the same microscopic sections. *Neurosci. Lett.* 11, 181–186.
605. Kuljis, R. O., and Advis, J. P. (1989). Immunocytochemical and physiological evidence of a synapse between dopamine

and luteinizing hormone releasing hormone releasing hormone-containing neurons in the ewe median eminence. *Endocrinology* 124, 1579–1581.
606. Lehman, M. N., Karsch, F. J., Robinson, J. E., and Silverman, A.-J. (1988). Ultrastructural and synaptic organization of luteinizing hormone-releasing hormone in the anestrous ewe. *J. Comp. Neurol.* 273, 447–548.
607. Johansson, O., and Hokfelt, T. (1980). Thyrotropin releasing hormone, somatostatin, and enkephalin: distribution studies using immunohistochemical techniques. *J. Histochem. Cytochem.* 28, 364–366.
608. Merchenthaler, I., Vigh, S., Schally, A. V., and Petrusz, P. (1984). Immunocytochemical localization of growth hormone-releasing factor in the rat hypothalamus. *Endocrinology* 114, 1082–1085.
609. Fellmann, D., Bugnon, C., Gouget, A., and Cardot, J. (1982). Les neurones a corticoliberine (CRF) du cerveau de rat. *Comp. Rendu.* 176, 511–516.
610. Bugnon, C., Fellmann, D., Gouget, A., and Cardot, J. (1982). Immunocytochemical detection of the CRF-containing neurons in the rat brain. *C. R. Acad. Sci. Paris* 294, 279–284.
611. Swanson, L. W., Sawchenko, P. E., Rivier, J., and Vale, W. (1983). Organization of ovine corticotropin-releasing factor immunoreactive cells and fibers in the rat brain: an immunohistochemical study. *Neuroendocrinology* 36, 165–186.
612. Kawata, H., Hashimoto, K., Takahara, J., and Sano, Y. (1982). Immunohistochemical demonstration of corticotropin releasing factor containing nerve fibers in the median eminence of the rat and monkey. *Histochemistry* 76, 15–19.
613. Lind, R. W., Swanson, L. W., Chin, D. A., Bruhn, T. O., and Ganten, D. (1984). Angiotensin II: an immunohistochemical study of its distribution in the paraventriculo-hypophysial system and its co-localization with vasopressin and CRF in parvocellular neurons. *Neuroscience Abstr.* 10, 88.
614. Meister, B., Hokfelt, T., Geffard, M., and Oertel, W. H. (1988). Glutamic acid decarboxylase-and gamma-aminobutyric acid-like immunoreactivities in corticotropin-releasing factor-containing parvocellular neurons of the hypothalamic paraventricular nucleus. *Neuroendocrinology* 48, 516–526.
615. Hisano, S., Tsuruo, Y., Katoh, S., Daikoku, S., Yanaihara, S., and Shibasaki, N. T. (1987). Intragranular colocalization of arginine vasopressin and methionine-enkephalin-octapeptide in CRF-axons in the rat median eminence. *Cell. Tissue Res.* 249, 497–507.
616. Ceccatelli, S., Eriksson, M., and Hokfelt, T. (1989). Distribution and coexistence of CRF-, neurotensin-enkephalin-, cholecystokinin-, galanin-, and VIP/PHI-like peptides in the parvocellular part of the paraventricular nucleus. *Neuroendocrinology* 49, 309–323.
617. Juaneda, C., Dubourg, P., Ciofi, P., Corio, M., and Tramu, G. (1999). Ultrastructural colocalization of vesicular cholecystokinin and corticoliberin in the periportal nerve terminals of the rat median eminence. *J. Neuroendocrinol.* 11, 203–209.
618. Dierickx, K., Vandesande, F., and DeMay, J. (1976). Identification, in the external region of the rat median eminence, of separate neurophysin-vasopressin and neurophysin-oxytocin containing nerve fibers. *Cell. Tissue Res.* 168, 141–151.
619. Dierickx, K., and Vandesande, F. (1977). Immunocytochemical demonstration, in the external region of the amphibian median eminence, of separate vasotocinergic and mesotocinergic nerve fibres. *Cell. Tissue Res.* 177, 47–56.
620. Zimmerman, E. A., and Antunes, J. L. (1976). Organization of the hypothalamic- pituitary system: current concepts from immunohistochemical studies. *J. Histochem. Cytochem.* 24, 807–815.
621. Silverman, A.-J. (1976). Ultrastructural studies on the localization of neurohypophyseal hormones and their carrier proteins. *J. Histochem. Cytochem.* 24, 816–827.
622. Kawata, M., Hashimoto, K., Takahara, J., and Sano, Y. (1983). Differences in the distributional pattern of CRF-, oxytocin-, and vasopressin-immunoreactive nerve fibers in the median eminence of the rat. *Cell. Tissue Res.* 230, 247–258.
623. Whitnall, M. H. (1990). Subpopulations of corticotropin-releasing hormone neurosecretory cells distinguished by presence or absence of vasopressin: confirmation with multiple corticotropin-releasing hormone antisera. *Neuroscience* 36, 201–205.
624. Whitnall, M. H., Mezey, E., and Gainer, H. (1985). Co-localization of corticotropin-releasing factor and vasopressin in median eminence neurosecretory vesicles. *Nature (Lond).* 317, 248–250.
625. Whitnall, M. H., Smyth, D., and Gainer, H. (1987). Vasopressin coexists in half of the corticotropin-releasing factor axons in the external zone of the median eminence. *Neuroendocrinology* 45, 420–424.
626. Zimmerman, E. A., Carmel, P. W., Husain, M. K., Ferin, M., Tannenbaum, M., Frantz, A. G., and Robinson, A. G. (1973). Vasopressin and neurophysin: high concentrations in monkey hypophyseal portal blood. *Science* 182, 925–927.
627. Gibbs, D. M., and Vale, W. (1982). Presence of corticotropin releasing factor-like immunoreactivity in hypophysial portal blood. *Endocrinology* 111, 1418–1420.
628. Oliver, C., Mical, R. S., and Porter, J. C. (1977). Hypothalamic pituitary vasculature: evidence for retrograde blood flow in the pituitary stalk. *Endocrinology* 101, 598–604.
629. Bertini, L. T., and Kiss, J. Z. (1991). Hypophysiotrophic neurons are capable of altering the ratio of co-package neurohormones. *Neuroscience* 42, 237–244.
630. Gillies, G. E., Linton, E. A., and Lowry, P. J. (1982). Corticotropin releasing activity of the new CRF is potentiated several times by vasopressin. *Nature (Lond).* 299, 355–367.
631. Rivier, C., Rivier, J., Mormede, P., and Vale, W. (1984). Studies of the nature of the interaction between vasopressin and corticotropin-releasing factor on adrenocorticotropin release in the rat. *Endocrinology* 115, 882–886.
632. Lind, R. W., Swanson, L. W., Bruhn, T. O., and Ganten, D. (1985). The distribution of angiotensin II-immunoreactive cells and fibers in the paraventriculo-hypophysial system of the rat. *Brain Res.* 338, 81–89.
633. Melander, T., Hokfelt, T., and Rokaeus, A. (1986). Distribution of galanin-like immunoreactivity in the rat central nervous system. *J. Comp. Neurol.* 248, 475–515.
634. Merchenthaler, I. (1991). The hypophysiotropic galanin system of the rat brain. *Neuroscience* 44, 643–654.
635. Niimi, M., Takahara, J., Sato, M., and Kawanishi, K. (1990). Immunohistochemical identification of galanin and growth hormone-releasing factor-containing neurons projecting to the median eminence of the rat. *Neuroendocrinology* 51.
636. Arai, R., and Calas, A. (1991). Ultrastructural localization of galanin immunoreactivity in the rat median eminence. *Brain Res.* 562.
637. Hokfelt, T., Pernow, B., Nilsson, G., Wetterberg, L., Goldstein, M., and Jeffcoate, S. L. (1978). Dense plexus of substance P immunoreactive nerve terminals in eminentia medialis of the primate hypothalamus. *Proc. Natl. Acad. Sci. U S A* 75, 339–343.
638. Gray, T. S., and Morley, J. E. (1986). Neuropeptide Y: anatomical distribution and possible function in mammalian nervous system. *Life Sci.* 38, 389–401.

639. Mulatero, B., and Fasolo, A. (1991). Calcitonin-gene related peptide immunoreactivity in the hypothalamo-hypophysial system of the green frog, rana esculenta. *Gen. Comp. Endocrinol.* 81.
640. Zimmerman, E. A., Liotta, A., and Krieger, D. T. (1978). β-lipotropin in brain: localization in hypothalamic neurons by immunoperoxidase technique. *Cell. Tissue Res.* 186, 393–398.
641. Bloch, B., Bugnon, C., Lenys, D., and Fellmann, D. (1978). Description des neurones immunoreactifs a un immunoserum anti B-endorphin presents dans le noyau infundibulaire chez l'homme. *C. R. Acad. Sci. Paris* 287, 309–312.
642. Bugnon, C., Bloch, B., Lenys, D., and Fellmann, D. (1979). Infundibular neurons of the human hypothalamus simultaneously reactive with antisera against endorphins, ACTH, MSH, and B-LPH. *Cell. Tissue Res.* 199, 177–196.
643. Newman, C. B., Wardlaw, S. L., and Van Vugt, D. A., et al. (1984). Adrenocorticotropin immunoactivity in monkey hypophyseal portal blood. *J. Clin. Endocrinol. Metab.* 59, 108–112.
644. Wardlaw, S. L., Wehrenberg, W. B., Ferin, M., Carmel, P. W., and Frantz, A. G. (1980). High levels of beta-endorphin in hypophyseal portal blood. *Endocrinology* 106, 1323–1326.
645. Page, R. B., and Bergland, R. M. (1977). The neurohypophyseal capillary bed. Part I. Anatomy and arterial supply. *Am. J. Anat.* 148, 345–358.
646. Page, R. B., Leure-duPree, A. E., and Bergland, R. M. (1978). The neurohypophyseal capillary bed. Part II. Specializations within median eminence. *Am. J. Anat.* 153, 33–66.
647. Livingston, A., and Wilks, P. N. (1976). Perivascular regions of the rat neural lobe. *Cell. Tissue Res.* 174, 273–280.
648. Simionescu, M., Simionescu, N., Silbert, J. E., and Palade, G. E. (1981). Differentiated microdomains on the luminal surface of the capillary endothelium. II. Partial characterization of their anionic sites. *J. Cell. Biol.* 90, 614–621.
649. Simionescu, N., Simionescu, M., and Palade, G. E. (1981). Differentiated microdomains on the luminal surface of the capillary endothelium. I. Preferential distribution of anionic sites. *J. Cell. Biol.* 90, 505–613.
650. Wislocki, G. (1937). The meningeal relations of the hypophysis cerebri. II. An embryological study of the meninges and blood vessels of the human hypophysis. *Am. J. Anat.* 61, 95–129.
651. Enemar, A. (1961). The structure and development of the hypophysial portal system in the laboratory mouse, with particular regard to the primary plexus. *Arkiv Zool.* 13, 203–252.
652. Glydon, R. (1957). The development of the blood supply of the pituitary in the albino rat, with special reference to the portal vessels. *J. Anat.* 91, 237–244.
653. Ugrumov, M. V., Ivanova, I. P., and Mitskevich, M. S. (1983). Light- and electron-microscopic study on the maturation of the primary portal plexus during the perinatal period in rats. *Cell. Tissue Res.* 234, 179–191.
654. Campbell, H. J. (1966). The development of the primary portal plexus in the median eminence of the rabbit. *J. Anat.* 100, 381–387.
655. Landsmeer, J. M. E. (1951). Vessels of the rat's hypophysis. *Acta Anat.* 12, 82–109.
656. McConnell, E. M. (1953). The arterial blood supply of the human hypophysis cerebri. *Anat. Rec.* 115, 175–203.
657. Bergland, R. M., and Page, R. B. (1978). Can the pituitary secrete directly to the brain? (Affirmative anatomical evidence). *Endocrinology* 102, 1325–1338.
658. Bjorklund, A. (1968). Monamine-containing fibers in the neuro-intermediate lobe of the pig and rat. *Z. Zellforsch.* 89, 573–589.
659. Page, R. B., and Dovey-Hartman, B. J. (1984). Resistance vessels in the tuber cinereum of the rabbit, rat and cat. *Anat. Rec.* 210, 647–655.
660. Ceccatelli, S., Fahrenkrug, J., Villar, M., and Hokfelt, T. (1991). Vasoactive intestinal polypeptide/peptide histidine isoleucine immunoreactive neuron systems in the basal hypothalamus of the rat with special reference to the portal vasculature: an immunohistochemical an in situ hybridization study. *Neuroscience* 43, 483–502.
661. Duvernoy, H., Koritke, J. G., and Monnier, G. (1971). Sur la vascularisation du tuber posterieur chez l'homme et sur les relations vasculaires tubero-hypophysaires. *J. Neuro-Visc. Relations* 32:112.
662. Ambach, G., Palkovits, M., and Szentagothai, J. (1976). Blood supply of the rat hypothalamus. IV. Retrochiasmatic area, median eminence, arcuate nucleus. *Acta Morph. Acad. Sci. Hung.* 24, 93–119.
663. Green, H. T. (1957). The venous drainage of the human hypophysis cerebri. *Am. J. Anat.* 100, 435–469.
664. Torok, B. (1954). Lebendbeobachtung des hypophysenkreislaufes an hunden. *Acta Morph. Acad. Sci. Hung.* 4, 83–89.
665. Torok, B. (1964). Structure of the vascular connections of the hypothalamo-hypophyseal region. *Acta Anat.* 59, 84–99.
666. Houssay, B. A., Biasotti, A., and Sammartino, R. (1935). Modifications fonctionnelles du l'hypophyse apres les lesions infundibulo-tuberiennes chez le crapaud. *C. R. Soc. Biol.* 120, 725–727.
667. Barnett, R. J., and Greep, R. O. (1951). The direction of blood flow in the blood vessels of the infundibular stalk. *Science* 113, 185.
668. Worthington, W. C. (1955). Some observations on the hypophyseal portal system in the living mouse. *Bull. Johns Hopkins Hosp.* 97, 343–357.
669. Worthington, W. C. (1960). Vascular responses in the pituitary stalk. *Endocrinology* 66, 19–31.
670. Page, R. B. (1983). Directional pituitary blood flow: a microcinephotographic study. *Endocrinology* 112, 157–165.
671. Duvernoy, H. (1972). The vascular architecture of the median eminence. In *Brain–Endocrine Interaction* (K. M. Knigge, D. E. Scott, and A. Weindl, Eds.), pp. 79–108. Karger, Basel.
672. Lametschwandtner, A., and Simonsberger, P. (1975). Light and scanning electron microscopical studies of the hypothalamo-adenohypophysial portal vessels of the toad bufo bufo. *Cell. Tissue Res.* 162, 131–139.
673. Page, R. B. (1982). Pituitary blood flow. *Am. J. Physiol.* 243, E427–E442.
674. Baez, S. (1977). Skeletal muscle and gastrointestinal microvascular morphology. In *Microcirculation* (G. Kalsz and B. Altura B, Eds.), Vol. 1. University Park Press, Baltimore.
675. Sobin, S. S., and Tremer, H. M. (1977). Three-dimensional organization of microvascular beds as related to function. In *Microcirculation* (G. Kalsz and B. Altura, Eds.), Vol. 1, pp. 43–57. University Park Press, Baltimore.
676. Sobin, S. S., Tremer, H. M., and Fung, Y. C. (1970). The morphometric basis of the sheet-flow concept of the pulmonary alveolar microcirculation in the cat. *Circ. Res.* 26, 397–414.
677. Duvernoy, H., and Koritke, J. G. (1964). Contribution de l'etude de l'angioarchitectonie des organes circumventriculaires. *Arch. Biol. Suppl.* 75, 849–904.
678. Duvernoy, H., and Koritke, J. G. (1968). Les vaisseaux sous-ependymaires due recessus hypophysaire. *J. Hirnforsch.* 10, 227–245.
679. Duvernoy, H. (1969). Considerations sur la vascularisation de l'hypophyse. *Acta Neurol. Belg.* 69, 469–481.
680. Reymond, M. J., Speciale, S. G., and Porter, J. C. (1983). Dopamine in plasma of lateral and medial hypophysial portal

vessels: evidence for regional variation in the release of hypothalamic dopamine into hypophysial portal blood. *Endocrinology* 112, 1958–1963.

681. Murakami, T., Kikuta, A., Taguchi, T., Ohtsuka, A., and Ohtani, O. (1987). Blood vascular architecture of the rat cerebral hypophysis and hypothalamus. A dissection/ scanning electron microscopy of vascular casts. *Arch Histol. Jpn.* 50, 133–176.
682. Akmayev, I. G. (1971). Morphological aspects of the hypothalamic-hypophyseal system II. Functional morphology of pituitary microcirculation. *Z. Zellforsch.* 116, 178–194.
683. Akmayev, I. G. (1971). Morphological aspects of the hypothalamic-hypophyseal system. III. Vascularity of the hypothalamus, with special reference to its quantitative aspects. *Z. Zellforsch.* 116, 195–204.
684. Shaver, S. W., Pang, J. J., Wainman, D. S., Wall, K. W., and Gross, P. M. (1992). Morphology and function of capillary networks in subregions of the rat tuber cinereum. *Cell. Tissue Res.* 267, 437–448.
685. Mezey, E., Palkovits, M., Dekloet, E. R., Verhoef, J., and DeWied, D. (1978). Evidence for pituitary-brain transport of a behaviorally potent ACTH analog. *Life Sci.* 22, 831–838.
686. Mezey, E., and Palkovits, M. (1982). Two way transport in the hypothalamo-hypophysial system. In *Frontiers in Neuroendocrinology* (W. F. Ganong and L. Martini, Eds.), Vol. 7, p. 1. Raven Press, New York.
687. Dorsa, D. M., Dekloet, E. R., Mezey, E., and DeWied, D. (1979). Pituitary-brain transport of neurotensin: functional significance of retrograde transport. *Endocrinology* 104, 1663–1666.
688. Minami, S., Kamegai, J., Sugihara, H., Suzuki, N., and Wakabayashi, I. (1998). Growth hormone inhibits its own secretion by acting on the hypothalamus through its receptors on neuropeptide Y neurons in the arcuate nucleus and somatostatin neurons in the periventricular nucleus. *Endocr. J.* 45 (Suppl), S19–S26.
689. Gibbs, D. M. (1985). Hypothalamic epinephrine is released into hypophysial portal blood during stress. *Brain Res.* 335, 360–364.
690. Felten, D. L., and Cashner, K. A. (1979). Cytoarchitecture of the supraoptic nucleus. A Golgi study. *Neuroendocrinology* 29, 312–230.
691. Fisher, A. W. F., Price, P. G., Burford, G. D., and Lederis, K. (1979). A 3-dimensional reconstruction of the hypothalamo-neurohypophysial system of the rat. The neurons projecting to the neuro-intermediate lobe and those containing vasopressin and somatostatin. *Cell. Tissue Res.* 204, 343–354.
692. LuQui, I. J., and Fox, C. A. (1976). The supraoptic nucleus and the supraoptico-hypophysial tract in the monkey (Macaca mulatta). *J. Comp. Neurol.* 168, 7–40.
693. Sofroniew, M. V., and Glasmann, W. (1981). Golgi-like immunoperoxidase staining of hypothalamic magnocellular neurons that contain vasopressin, oxytocin or neurophysin in the rat. *Neuroscience* 6, 619–643.
694. Dyball, R. E. J., Howard, M., and Kemplay, S. K. (1979). A Golgi study of the neurosecretory neurons in the supraoptic nucleus of the rat. *J. Anat.* 128, 417.
695. Renaud, L. P. (1978). Neurophysiological organization of the endocrine hypothalamus. In *The Hypothalamus* (S. Reichlin, R. J. Baldessarini, and J. B. Martin, Eds.), pp. 269–300. Raven Press, New York.
696. Zimmerman, E. A. (1981). The organization of oxytocin and vasopressin pathways. In *Neurosecretion and Brain Peptides* (J. B. Martin, S. Reichlin, and K. L. Bick, Eds.), pp. 63–75. Raven Press, New York.
697. DeMay, J., Vandesande, F., and Dierickx, K. (1974). Identification of neurophysin producing cells. II. Identification of the neurophysin I and the neurophysin II producing neurons in the bovine hypothalamus. *Cell. Tissue Res.* 153, 531–543.
698. Dierickx, K., and Vandesande, F. (1979). Immunocytochemical demonstration of separate vasopressin-neurophysin and oxytocin-neurophysin neurons in the human hypothalamus. *Cell. Tissue Res.* 196, 203–212.
699. Kawata, M., and Sano, Y. (1982). Immunohistochemical identification of the oxytocin and vasopressin neurons in the hypothalamus of the monkey (Macaca fuscata). *Anat. Embryol. (Berl).* 165, 151–167.
700. Sloper, J. C., and Bateson, R. G. (1965). Ultrastructure of neurosecretory cells in the supraoptic nucleus of the dog and rat. *J. Endocr.* 31, 139–150.
701. Flament-Durand, J. (1971). Ultrastructural aspects of the paraventricular nuclei in the rat. *Z. Zellforsch.* 116, 61–69.
702. Morris, J. F., and Dyball, R. E. (1974). A quantitative study of the ultrastructural changes in the hypothalamo-neurohypophysial system during and after experimentally induced hypersecretion. *Cell. Tissue Res.* 149, 525–535.
703. Kalimo, H. (1971). Ultrastructural studies on the hypothalamic neurosecretory neurones of the rat. I. The paraventricular neurones of the non-treated rat. *Z. Zellforsch.* 122, 283–300.
704. Tweedle, C. D., and Hatton, G. I. (1976). Ultrastructure comparisons of neurons of supraoptic and circularis nuclei in normal and dehydrated rats. *Brain Res. Bull.* 1, 103–121.
705. Piekut, D. T. (1986). Ultrastructural characteristics of peptidergic neurons using pre-embedding immunocytochemical methods. *Am. J. Anat.* 175, 197–216.
706. Piekut, D. T. (1983). Ultrastructural characteristics of vasopressin-containing neurons in the paraventricular nucleus of the hypothalamus. *Cell. Tissue Res.* 234, 125–134.
707. Castel, M., Morris, J. F., Whitnall, M. H., and Sivan, N. (1986). Improved visualization of the immunoreactive hypothalamo-hypophysial system by use of immuno-gold techniques. *Cell. Tissue Res.* 243, 193–204.
708. Guitteny, A.-F., and Bloch, B. (1989). Ultrastructural detection of the vasopressin messenger RNA in normal and Brattleboro rat. *Histochemistry* 92, 277–281.
709. Trembleau, A., Calas, A., and Fevre-Montange, M. (1990). Ultrastructural localization of oxytocin mRNA in the rat hypothalamus by in situ hybridization using a synthetic oligonucleotide. *Brain Res. Mol. Brain Res.* 8, 37–45.
710. Kiyama, H., and Emson, P. C. (1990). Evidence for the co-expression of oxytocin and vasopressin messenger ribonucleic acids in magnocellular neurosecretory cells: simultaneous demonstration of two neurophysin messenger ribonucleic acids by hybridization histochemistry. *J. Neuroendocrinol.* 2, 257–259.
711. Xi, D., Kusano, K., and Gainer, H. (1999). Quantitative analysis of oxytocin and vasopressin messenger ribonucleic acids in single magnocellular neurons isolated from supraoptic nucleus of rat hypothalamus. *Endocrinology* 140, 4677–4682.
712. Nishioka, R. S., Zambrano, D., and Bern, H. A. (1970). Electron microscope radioautography of amino acid incorporation by supraoptic neurons of the rat. *Gen. Comp. Endocrinol.* 15, 477–483.
713. Broadwell, R. D., Oliver, C., and Brightman, M. W. (1979). Localization of neurophysin within organelles associated with protein synthesis and packaging in the hypothalamo-neurohypophysial system: an immunocytochemical study. *Proc. Natl. Acad. Sci. U S A* 76, 5999–60003.
714. Kozlowski, G. P., Frenk, S., and Brownfield, M. S. (1977). Localization of neurophysin in the rat supraoptic nucleus. I. Ultrastructural immunocytochemistry using the post-embedding technique. *Cell. Tissue Res.* 179, 467–473.
715. Lu, C.-L., Cantin, M., Seidah, N. G., and Chretien, M. (1982). Distribution pattern in the human pituitary and hypothalamus

of a new neuropeptide: the C-terminal glycoprotein-fragment of human pro-pressophysin (CPP). *Histochemistry* 75, 319–326.

716. Pickering, B. T., Swann, R. W., Birkett, S. D., O'Shaughnessy, P., Wathes, D. C., and Porter, D. G. (1984). Precursors and products in the formation of neurohypophysial hormones. In *Endocrinology* (F. Labrie and L. Proulx-Ferland, Eds.), pp. 653–657. Elsevier, New York.
717. Zhang, B. J., Kusano, K., Zerfas, P., Iacangelo, A., Young, W. S. III, and Gainer, H. (2002). Targeting of green fluorescent protein to secretory granules in oxytocin magnocellular neurons and its secretion from neurohypophysial nerve terminals in transgenic mice. *Endocrinology* 143, 1036–1046.
718. Kalimo, H. (1975). Ultrastructural studies on the hypothalamic neurons of the rat. III. Paraventricular and supraoptic neurons during lactation and dehydration. *Cell. Tissue Res.* 163, 151–168.
719. Broadwell, R. D., and Oliver, C. (1981). Golgi apparatus, GERL, and secretory granule formation within neurons of the hypothalamo-neurohypophysial system of control and hyperosmotically stressed mice. *J. Cell. Biol.* 90, 474–484.
720. Krisch, B. (1977). Electronmicroscopic immunocytochemical study on the vasopressin-containing neurons of the thirsting rat. *Cell. Tissue Res.* 184, 237–247.
721. Krisch, B. (1979). Indication for a granule-free form of vasopressin in immobilization-stressed rats. *Cell. Tissue Res.* 197, 95–104.
722. Pow, D. V., and Morris, J. F. (1989). Dendrites of hypothalamic magnocellular neurons release neurohypophysial peptides by exocytosis. *Neuroscience* 32, 435–439.
723. Freund-Mercier, M. J., Stoeckel, M. E., and Klein, M. J. (1994). Oxytocin receptors on oxytocin neurones: histoautoradiographic detection in the lactating rat. *J. Physiol.* 480 (Pt 1), 155–161.
724. Ludwig, M. (1998). Dendritic release of vasopressin and oxytocin. *J. Neuroendocrinol.* 10, 881–895.
725. Hurbin, A., Orcel, H., Alonso, G., Moos, F., and Rabie, A. (2002). The vasopressin receptors colocalize with vasopressin in the magnocellular neurons of the rat supraoptic nucleus and are modulated by water balance. *Endocrinology* 143, 456–466.
726. Kozlowski, G. P. (1983). Comparative ultrastructure of neuropeptide-containing cells of the parvo- and magnocellular neurosecretory system. In *Structure and Function of Aminergic Neurons* (Y. Sano, Y. Ibata, and E. A. Zimmerman, Eds.), pp. 73–84. Japan Sci Soc Press, Tokyo.
727. Alonso, G., and Assenmacher, I. (1979). The smooth endoplasmic reticulum in neurohypophysial axons of the rat: possible involvement in transport, storage and release of neurosecretory material. *Cell. Tissue Res.* 199, 415–419.
728. Broadwell, R. D., and Brightman, M. W. (1979). Cytochemistry of undamaged neurons transporting exogenous protein in vivo. *J. Comp. Neurol.* 185, 31–74.
729. Gregory, W. A., Tweedle, C. D., and Hatton, G. I. (1979). Ultrastructure of neurons in the paraventricular nucleus of normal, dehydrated and rehydrated rats. *Brain Res. Bull.* 5, 301–306.
730. Tweedle, C. D., and Hatton, G. I. (1977). Ultrastructural changes in rat hypothalamic neurosecretory cells and their associated glia during minimal dehydration and rehydration. *Cell. Tissue Res.* 181, 59–72.
731. Wakerley, J. B., Poulain, D. A., and Brown, D. (1978). Comparison of firing patterns in oxytocin- and vasopressin-releasing neurones during progressive dehydration. *Brain Res.* 148, 425–440.
732. Meeker, R. B., Greenwood, R. S., and Hayward, J. N. (1991). Vasopressin mRNA expression in individual magnocellular neuroendocine cells of the supraoptic and paraventricular nucleus in response to water deprivation. *Neuroendocrinology* 54, 236–247.
733. Epstein, Y. M. C., Glick, S. M., Sivan, N., and Ravid, R. (1983). Changes in hypothalamic and extra-hypothalamic vasopressin content of water-deprived rats. *Cell. Tissue Res.* 233, 99–111.
734. Meister, B., Villar, M., Ceccatelli, S., and Hokfelt, T. (1990). Localization of chemical messengers in magnocellular neurons of the hypothalamic supraoptic and paraventricular nuclei: An immunohistochemical study using experimental manipulations. *Neuroscience* 37, 603–633.
735. Abramova, M., Calas, A., Thibault, J., and Ugrumov, M. (2000). Tyrosine hydroxylase in vasopressinergic axons of the pituitary posterior lobe of rats under salt-loading as a manifestation of neurochemical plasticity. *Neural. Plast.* 7, 179–191.
736. Abramova, M. A., Calas, A., Mailly, P., Thibault, J., and Ugryumov, M. V. (2000). The responses of vasopressin- and tyrosine hydroxylase-expressing neurons of the supraoptic nucleus in rats to chronic osmotic stimulation. *Neurosci. Behav. Physiol.* 30, 617–624.
737. Panayotacopoulou, M. T., Goudsmit, E., Van Heerikhuize, J. J., and Swaab, D. F. (2000). Simultaneous detection of tyrosine hydroxylase-immunoreactivity and vasopressin mRNA in neurons of the human paraventricular and supraoptic nucleus. *Brain Res.* 855, 181–185.
738. Watson, S. J., Akil, H., Fischli, W., Goldstein, A., Zimmerman, E. A., Nilaver, G., and Greidanus, T. B. V. W. (1982). Dynorphin and vasopressin: common localization in magnocellular neurons. *Science* 216, 85–87.
739. Whitnall, M. H., Gainer, H., Cox, B. M., and Molineaux, C. I. (1983). Dynorphin-A-(1-8) is contained within vasopressin neurosecretory vesicles in the rat pituitary. *Science* 222, 1137–1139.
740. Millan, M. H., Millan, M. J., and Herz, A. (1984). The hypothalamic paraventricular nucleus: relationship to brain and pituitary pools of vasopressin and oxytocin as compared to dynorphin, B-endorphin and related opioid peptides in the rat. *Neuroendocrinology* 38, 108–116.
741. Romijn, H. J., van Uum, J. F., Emmering, J., Goncharuk, V., and Buijs, R. M. (1999). Colocalization of VIP with AVP in neurons of the human paraventricular, supraoptic and suprachiasmatic nucleus. *Brain Res.* 832, 47–53.
742. Landry, M., and Hokfelt, T. (1998). Subcellular localization of preprogalanin messenger RNA in perikarya and axons of hypothalamo-posthypophyseal magnocellular neurons: an in situ hybridization study. *Neuroscience* 84, 897–912.
743. Landry, M., Vila-Porcile, E., Hokfelt, T., and Calas, A. (2003). Differential routing of coexisting neuropeptides in vasopressin neurons. *Eur. J. Neurosci.* 17, 579–589.
744. Burlet, A., Tonon, M.-C., Tankosic, P., Coy, D. H., and Vaudry, H. (1983). Comparative immunocytochemical localization of corticotropin releasing factor (CRF-41) and neurohypophysial peptides in the brain of Brattleboro and Long-Evans rats. *Neuroendocrinology* 37, 64–72.
745. Brownstein, M. J., and Mezey, E. (1986). Multiple chemical messengers in hypothalamic magnocellular neurons. *Prog. Brain Res.* 68, 161–168.
746. Bredt, D. S., Hwang, P. M., and Snyder, S. H. (1990). Localization of nitric oxide synthase indicating a neural role for nitric oxide. *Nature (Lond).* 347, 768–770.
747. Arevalo, R., Sanchez, F., Alonso, J. R., Carretero, J., Vazquez, R., and Aijon, J. (1992). NADPH-Diaphorase activity in the hypothalamic magnocellular neurosecretory nuclei of the rat. *Brain Res. Bull.* 28, 599–603.

748. Vanhatalo, S., and Soinila, S. (1995). Nitric oxide synthase in the hypothalamo-pituitary pathways. *J. Chem. Neuroanat.* 8, 165–173.
749. Nylen, A., Skagerberg, G., Alm, P., Larsson, B., Holmqvist, B., and Andersson, K. E. (2001). Nitric oxide synthase in the hypothalamic paraventricular nucleus of the female rat; organization of spinal projections and coexistence with oxytocin or vasopressin. *Brain Res.* 908, 10–24.
750. Pow, D. (1992). NADPH-Diaphorase (nitric oxide synthase) staining in the rat supraoptic nucleus is activity dependent: possible functional indications. *J. Neuroendocrinol.* 4, 377–380.
751. Gross, J. H., Knigge, K. M., and Sheridan, M. N. (1976). Fine structure of neurons of the arcuate nucleus and median eminence of the hypothalamus of the golden hamster following immobilization. *Cell. Tissue Res.* 168.
752. Brawer, J. R. (1971). The role of the arcuate nucleus in the brain-pituitary-gonad axis. *J. Comp. Neurol.* 143, 411–446.
753. Bugnon, C., Bloch, B., and Lenys, D. (1981). Ultrastructural study of presumptive pro-opiocortin producing neurons in the rat hypothalamus. *Neuroscience* 6, 1299–1313.
754. Lamberts, R., and Goldsmith, P. C. (1985). Preembedding colloidal gold immunostaining of hypothalamic neurons: light and electron microscopic localization of beta-endorphin-immunoreactive perikarya. *J. Histochem. Cytochem.* 33, 499–507.
755. King, J. C., Williams, T. H., and Gerall, A. A. (1974). Transformations of hypothalamic arcuate neurons. I. Changes associated with stages of the estrous cycle. *Cell. Tissue Res.* 153, 497–515.
756. Price, M. T., Olney, J. W., and Cicero, T. J. (1976). Proliferation of lamellar whorls in arcuate neurons of the hypothalamus of castrated and morphine-treated male rats. *Cell. Tissue Res.* 171, 277–284.
757. Garcia-Segura, L. M., Hernandez, P., Olmos, G., Tranque, P. A., and Naftolin, F. (1988). Neuronal membrane remodelling during the oestrus cycle: a freeze-fracture study in the arcuate nucleus of the rat hypothalamus. *J. Neurocytol.* 17, 377–383.
758. Liposits, Z. (1990). Ultrastructural immunocytochemistry of the hypothalamic corticotropin releasing hormone synthesizing system. Anatomical basis of neuronal and humoral regulatory mechanisms. *Prog. Histochem. Cytochem.* 21, 1–98.
759. Liposits, Z., and Paull, W. K. (1985). Ultrastructural alterations of the paraventriculo-infundibular corticotropin releasing factor (GRF)-immunoreactive neuronal system in long term adrenalectomized rats. *Peptides* 6, 1021–1036.
760. Beauvillain, J. C., Tramu, G., and Mazzuca, M. (1987). Fine structural studies of growth-hormone-releasing-factor (GRF)-immunoreactive neurons and their synaptic connections in the guinea pig arcuate nucleus. *J. Comp. Neurol.* 255, 110–123.
761. Ibata, Y., Okamura, H., Makino, S., Kawakami, F., Morimote, N., and Chihara, K. (1986). Light and electron microscopic immunocytochemistry of GRF-like immunoreactive neurons and terminals in the rat hypothalamic arcuate nucleus and median eminence. *Brain Res.* 370, 136–143.
762. Lehman, M. N., Silverman, A.-J., Witkin, J. W., and Millar, R. P. (1990). Ultrastructure of luteinizing hormone-releasing hormone (LHRH) neurons and their projections in the golden hamster. *Brain Res. Bull.* 20, 211–221.
763. Silverman, A.-J., Witkin, J. W., and Millar, R. P. (1990). Light and electron microscopic immunocytochemical analysis of antibodies directed against GNRH and its precursor in hypothalamic neurons. *J. Histochem. Cytochem.* 38, 803–813.
764. Jennes, L., Stumpf, W. E., and Sheedy, M. E. (1985). Ultrastructural characterization of gonadotropin-releasing hormone (GnRH)-producing neurons. *J. Comp. Neurol.* 232, 534–547.
765. Crosby, E. C., and Showers, M. J. C. (1969). Comparative anatomy of the preoptic and hypothalamic areas. In *The Hypothalamus* (W. Haymaker, E. Anderson, and W. J. H. Nauta, Eds.), p. 61. Charles C Thomas, Springfield, IL.
766. Nauta, W. J. H.,and Haymaker, W. (1969). Hypothalamic nuclei and fiber connections. In *The Hypothalamus* (W. Haymaker, E. Anderson, and W. J. H. Nauta, Eds.), p. 136. Charles C Thomas, Springfield, IL.
767. Papez, J. W. (1940). The embryologic development of the hypothalamic area in mammals. *Res. Publ. Assoc. Res. Nerv. Ment. Dis.* 20, 31–51.
768. Christ, J. F. (1969). Derivation and boundaries of the hypothalamus with atlas of hypothalamic grisea. In *The Hypothalamus* (W. Haymaker, E. Anderson, and W. J. H. Nauta, Eds.), pp. 13–60. Charles C Thomas, Springfield, IL.
769. Coggeshall, R. E. (1964). A study of diencephalic development in the albino rat. *J. Comp. Neurol.* 122, 241–254.
770. Hyyppa, M. (1969). Differentiation of the hypothalamic nuclei during ontogenetic development in the rat. *Z. Anat. Entwickl-Gesch.* 129, 41–52.
771. Altman, J., and Bayer, S. A. (1978). Development of the diencephalon in the rat. I. Autoradiographic study of the time of origin and settling patterns of neurons of the hypothalamus. *J. Comp. Neurol.* 182, 945–972.
772. Altman, J., and Bayer, S. A. (1978). Development of the diencephalon in the rat. II. Correlation of the embryonic development of the hypothalamus with the time of origin of its neurons. *J. Comp. Neurol.* 182, 973–994.
773. Choy, V. J., and Watkins, W. B. (1979). Maturation of the hypothalamo-neurohypophysial system. I. Localization of neurophysin, oxytocin and vasopressin in the hypothalamus and neural lobe of the developing rat brain. *Cell. Tissue Res.* 197, 325–336.
774. Markakis, E. A., and Swanson, L. W. (1997). Spatiotemporal patterns of secretomotor neuron generation in the parvicellular neuroendocrine system. *Brain Res. Brain Res. Rev.* 24, 255–291.
775. Alvarez-Bolado, G., Rosenfeld, M. G., and Swanson, L. W. (1995). Model of forebrain regionalization based on spatiotemporal patterns of POU-III homeobox gene expression, birthdates, and morphological features. *J. Comp. Neurol.* 355, 237–295.
776. Nakai, S., Kawano, H., and Yudate, T., et al. (1995). The POU domain transcription factor Brn-2 is required for the determination of specific neuronal lineages in the hypothalamus of the mouse. *Genes Dev.* 9, 3109–3121.
777. Wang, W., and Lufkin, T. (2000). The murine Otp homeobox gene plays an essential role in the specification of neuronal cell lineages in the developing hypothalamus. *Dev Biol.* 227, 432–449.
778. Schwanzel-Fukuda, M., Morrell, J. I., and Pfaff, D. W. (1985). Ontogenesis of neurons producing luteinizing hormone-releasing hormone (LHRH) in the nervus terminalis of the rat. *J. Comp. Neurol.* 238, 348–364.
779. Setalo, G., Hagino, N., and Dittrich, E. (1992). Ontogenesis of the GnRh neuron system in the rat. A quantitative immunohistochemical study with special reference to the extra cerebral GnRH-positive cells and the occupation of intracerebral termination fields. *Neuropeptides* 21, 93–102.
780. Schwanzel-Fukuda, M., and Pfaff, D. W. (1989). Origins of luteinizing hormone-releasing hormone neurons. *Nature (Lond).* 338, 161–164.
781. Wray, S., Grant, P. H., and Gainer, H. (1989). Evidence that cells expressing luteinizing hormone-releasing hormone

mRNA are derived from progenitor cells in the olefactory placode. *Proc. Natl. Acad. Sci. U S A* 86, 8132–8136.
782. Wu, T. J., Gibson, M. J., Rogers, M. C., and Silverman, A. J. (1997). New observations on the development of the gonadotropin-releasing hormone system in the mouse. *J. Neurobiol.* 33, 983–998.
783. Krisch, B. (1980). Electron microscopic immunocytochemical investigation on the postnatal development of the vasopressin system in the rat. *Cell. Tissue Res.* 205, 453–471.
784. Koritsanszky, S. (1979). Cyto- and synaptogenesis in the arcuate nucleus of the rat hypothalamus during fetal and early postnatal life. *Cell. Tissue Res.* 200, 135–146.
785. Dale, H. H. (1935). Pharmacology and nerve-endings. *Proc. R. Soc. Lond.* 28, 319–332.
786. Broadwell, R. D., and Bleier, R. (1976). A cytoarchitectonic atlas of the mouse hypothalamus. *J. Comp. Neurol.* 167, 315–340.
787. Peterson, R. P. (1966). Magnocellular neurosecretory centers in the rat hypothalamus. *J. Comp. Neurol.* 167, 181–190.
788. Rhodes, C. H., Morrell, J. I., and Pfaff, D. W. (1981). Immunohistochemical analysis of magnocellular elements in rat hypothalamus: distribution and numbers of cells containing neurophysin, oxytocin and vasopressin. *J. Comp. Neurol.* 198, 45–64.
789. Swaab, D. F., Nijveldt, F., and Pool, C. W. (1975). Distribution of oxytocin and vasopressin the rat supraoptic and paraventricular nucleus. *J. Endocr.* 67, 461–462.
790. Ingram, W. R. (1940). Nuclear organization and chief connections of the primate hypothalamus. *Res. Publ. Assoc. Res. Nerv. Ment. Dis.* 20, 195–244.
791. Defendini, R., and Zimmerman, E. A. (1978). The magnocellular neurosecretory system of the mammalian hypothalamus. In *The Hypothalamus* (S. Reichlin, R. J. Baldessarini, J. B. Martin, Eds.). Raven Press, New York.
792. Marzbanm, F., Tweedle, C. D., and Hatton, G. I. (1992). Reevaluation of the plasticity in the rat supraoptic nucleus after chronic dehydration using immunogold for oxytocin and vasopressin at the ultrastructural level. *Brain Res. Bull.* 28, 756–766.
793. Theodosis, D. T., Chapman, D. B., Montagnese, C., Poulain, D. A., and Morris, J. F. (1986). Structural plasticity in the hypothalamic supraoptic nucleus at lactation affects oxytocin- but not vasopressin-secreting neurons. *Neuroscience* 17, 661–678.
794. Theodosis, D. T., Paut, L., and Tappaz, M. L. (1986). Immunocytochemical analysis of the gabaergic innervation of oxytocin- and vasopressin-secreting neurons in the rat supraoptic nucleus. *Neuroscience* 19, 207–222.
795. El Majdoubi, M., Poulain, D. A., and Theodosis, D. T. (1997). Lactation-induced plasticity in the supraoptic nucleus augments axodendritic and axosomatic GABAergic and glutamatergic synapses: an ultrastructural analysis using the disector method. *Neuroscience* 80, 1137–1147.
796. Montagnese, C., Poulain, D. A., Vincent, J.-D., and Theodosis, D. T. (1988). Synaptic and neuronal-glial plasticity in the adult oxytocinergic system in response to physiological stimuli. *Brain Res. Bull.* 20, 681–692.
797. Miyata, S., and Hatton, G. I. (2002). Activity-related, dynamic neuron-glial interactions in the hypothalamo-neurohypophysial system. *Microsc. Res. Tech.* 56, 143–157.
798. Blanco, E., Pilgrim, C., Vazquez, R., and Jirikowski, G. F. (1991). Plasticity of the interface between oxytocin neurons and the vasculature in late pregnant rats: an ultrastructural morphometric study. *Acta Histochem.* 91, 165–172.
799. Ray, P. K., and Choudhury, S. R. (1990). Vasopressinergic axon collaterals and axon terminals in the magnocellular neurosecretory nuclei of the rat hypothalamus. *Acta Anat. (Basel)* 137, 37–44.
800. Choudhury, S. R., and Ray, P. K. (1990). Ultrastructural features of presumptive vasopressinergic synapses in the hypothalamic magnocellular secretory nuclei of the rat. *Acta Anat. (Basel)* 137, 252–256.
801. Theodosis, D. T. (1985). Oxytocin-immunoreactive terminals synapse on oxytocin neurones in the supraoptic nucleus. *Nature* 313, 682–684.
802. van den Pol, A. N. (1985). Dual ultrastructural localization of two neurotransmitter-related antigens: colloidal gold-labeled neurophysin-immunoreactive supraoptic neurons receive peroxidase-labeled glutamate decarboxylase- or gold-labeled GABA-immunoreactive synapses. *J. Neurosci.* 5, 2940–2954.
803. van den Pol, A. N. (1991). Glutamate and aspartate immunoreactivity in hypothalamic presynaptic axons. *J. Neurosci.* 11, 2087–2101.
804. Shioda, S., Nakai, Y., Kitazawa, S., and Sunayama, H. (1985). Immunocytochemical observations of corticotropin-releasing-factor-containing neurons in the rat hypothalamus with special reference to neuronal communication. *Acta Anat. (Basel)* 124, 58–64.
805. Arima, H., and Aguilera, G. (2000). Vasopressin and oxytocin neurones of hypothalamic supraoptic and paraventricular nuclei co-express mRNA for Type-1 and Type-2 corticotropin-releasing hormone receptors. *J. Neuroendocrinol.* 12, 833–842.
806. Goldsmith, P. C., Boggan, J. E., and Thind, K. K. (1991). Opioid synapses on vasopressin neurons in the paraventricular and supraoptic nuclei of juvenile monkeys. *Neuroscience* 45, 709–719.
807. Douglas, A. J., Bicknell, R. J., Leng, G., Russell, J. A., and Meddle, S. L. (2002). Beta-endorphin cells in the arcuate nucleus: projections to the supraoptic nucleus and changes in expression during pregnancy and parturition. *J. Neuroendocrinol.* 14, 768–777.
808. van Vulpen, E. H., Yang, C. R., Nissen, R., and Renaud, L. P. (1999). Hypothalamic A14 and A15 catecholamine cells provide the dopaminergic innervation to the supraoptic nucleus in rat: a combined retrograde tracer and immunohistochemical study. *Neuroscience* 93, 675–680.
809. Swanson, L. W., Sawchenko, P. E., Berod, A., Hartman, B. K., Helle, K. B., and Vanorden, D. E. (1981). An immunohistochemical study of the organization of catecholamine cells and terminal fields in the paraventricular and supraoptic nuclei of the hypothalamus. *J. Comp. Neurol.* 196, 271–285.
810. Michaloudi, H. C., el Majdoubi, M., Poulain, D. A., Papadopoulos, G. C., and Theodosis, D. T. (1997). The noradrenergic innervation of identified hypothalamic magnocellular somata and its contribution to lactation-induced synaptic plasticity. *J. Neuroendocrinol.* 9, 17–23.
811. Armstrong, W. E., Tian, M., and Wong, H. (1996). Electron microscopic analysis of synaptic inputs from the median preoptic nucleus and adjacent regions to the supraoptic nucleus in the rat. *J. Comp. Neurol.* 373, 228–239.
812. Raisman, G. (1973). An ultrastructural study of the effects of hypophysectomy on the supraoptic nucleus of the rat. *J. Comp. Neurol.* 147, 181–208.
813. Price, P., and Fisher, A. W. (1978). Electron microscopical study of retrograde axonal transport of horseradish peroxidase in the supraoptico-hypophyseal tract in rat. *J. Anat.* 125, 137–147.
814. DeVries, G. J., Buijs, R. M., Van Leeuwen, F. W., Caffe, A. R., and Swaab, D. F. (1985). The vasopressinergic innervation of the brain in normal and castrated rats. *J. Comp. Neurol.* 233, 236–254.
815. Simpson, J. B. (1981). The circumventricular organs and the central actions of angiotensin. *Neuroendocrinology* 32, 248–256.
816. Mahoney, P. D., Koh, E. T., Irvin, R. W., and Ferris, C. F. (1990). Computer-aided mapping of vasopressin neurons in

the hypothalamus of the male golden hamster: evidence of magnocellular neurons that do not project to the hypothalamus. *J. Neuroendocrinol.* 2, 113–122.
817. Swanson, L. W., and Kuypers, H. G. (1980). The paraventricular nucleus of the hypothalamus: cytoarchitectonic subdivisions and organization of projections to the pituitary, dorsal vagal complex, and spinal cord as demonstrated by retrograde fluorescence double-labeling methods. *J. Comp. Neurol.* 194, 555–570.
818. Wiegand, S. L., and Price, J. L. (1980). Cells of origin of the afferent fibers to the median eminence in the rat. *J. Comp. Neurol.* 192, 1–19.
819. Krieg, W. J. S. (1932). The hypothalamus of the albino rat. *J. Comp. Neurol.* 55, 19–89.
820. van den Pol, A. N. (1982). The magnocellular and parvocellular paraventricular nucleus of rat: intrinsic organization. *J. Comp. Neurol.* 206, 317–345.
821. Armstrong, W. E., Warach, S., Hatton, G. I., and McNeill, T. H. (1980). Subnuclei in the rat hypothalamic paraventricular nucleus: a cytoarchitectural, horseradish peroxidase and immunocytochemical analysis. *Neuroscience* 5, 1931–1958.
822. Hatton, G. I., Hutton, U. E., Hoblitzell, E. R., and Armstrong, W. E. (1976). Morphological evidence for two populations of magnocellular elements in the rat paraventricular nucleus. *Brain Res.* 108, 187–193.
823. Sawchenko, P. E., and Swanson, L. W. (1982). Immunohistochemical identification of neurons in the paraventricular nucleus of the hypothalamus that project to the medulla or to the spinal cord in rat. *J. Comp. Neurol.* 205, 260–272.
824. Theodosis, D. T., and Poulain, D. A. (1989). Neuronal-glial and synaptic plasticity in the adult rat paraventricular nucleus. *Brain Res.* 484, 361–366.
825. Hatton, G. I., and Li, Z. H. (1998). Neurophysiology of magnocellular neuroendocrine cells: recent advances. *Prog. Brain Res.* 119, 77–99.
826. Theodosis, D. T. (2002). Oxytocin-secreting neurons: a physiological model of morphological neuronal and glial plasticity in the adult hypothalamus. *Front. Neuroendocrinol.* 23, 101–135.
827. Silverman, A.-J., Hou-Yu, A., and Oldfield, B. J. (1983). Ultrastructural identification of noradrenergic nerve terminals and vasopressin-containing neurons of the paraventricular nucleus in the same thin section. *J. Histochem. Cytochem.* 31, 1151–1156.
828. Ochiai, H., and Nakai, Y. (1990). Ultrastructural demonstration of dopamine-beta-hydroxylase immunoreactive nerve terminals on vasopressin neurons in the paraventricular nucleus of the rat by double-labeling immunocytochemistry. *Neurosci. Lett.* 120, 87–90.
829. Iwai, C., Ochiai, H., and Nakai, Y. (1989). Electron-microscopic immunocytochemistry of neuropeptide Y immunoreactive innervation of vasopressin neurons in the paraventricular nucleus of the rat hypothalamus. *Acta Anat.* 136, 279–284.
830. Kagotani, Y., Tsuruo, Y., Hisano, S., Daikoku, S., and Chihara, K. (1989). Synaptic regulation of paraventricular arginine vasopressin-containing neurons by neuropeptide Y-containing monoaminergic neurons in rats, electron microscopic triple labeling. *Cell. Tissue Res.* 257, 269–278.
831. Liposits, Z., Phelix, C., and Paull, W. K. (1986). Electron microscopic analysis of tyrosine hydroxylase dopamine-B-hydroxylase and phenylethanolamine-N-methyl-transferase immunoreactive innervation of the hypothalamic paraventricular nucleus in the rat. *Histochemistry* 84, 105–120.
832. Wang, Q. P., Guan, J. L., Ochiai, H., and Nakai, Y. (1999). The neurotensinergic synaptic innervation of vasopressin containing neurons in the rat hypothalamic paraventricular nucleus. *Brain Res.* 834, 25–31.
833. Decavel, C., Dubourg, P., Leon-Henri, B., Geffard, M., and Calas, A. (1989). Simultaneous immunogold labeling of gabaergic terminals and vasopressin-containing neurons in the rat peraventricular nucleus. *Cell. Tissue Res.* 255, 77–80.
834. Taniguchi, Y., Yoshida, M., Ishikawa, K., Suzuki, M., and Kurosumi, K. (1988). The distribution of vasopressin- or oxytocin-neurons projecting to the posterior pituitary as revealed by a combination of retrograde transport of horseradish peroxidase and immunohistochemistry. *Arch. Histol. Cytol.* 51, 83–89.
835. Kiss, J. Z., Martos, J., and Palkovits, M. (1991). Hypothalamic paraventricular nucleus: a quantitative analysis of cytoarchitectonic subdivisions in the rat. *J. Comp. Neurol.* 313, 563–573.
836. Swanson, L. W., and Sawchenko, P. E. (1983). Hypothalamic integration: organization of the paraventricular and supraoptic nuclei. *Annu. Rev. Neurosci.* 6, 269–324.
837. Antunes, J. L., Carmel, P. W., and Zimmerman, E. A. (1977). Projections from the paraventricular nucleus to the zona externa of the median eminence of the rhesus monkey: an immunohistochemical study. *Brain Res.* 137, 1–10.
838. Vandesande, F., Dierickx, K., and DeMay, J. (1977). The origin of the vasopressinergic and oxytocinergic fibres of the external region of the median eminence of the rat hypophysis. *Cell. Tissue Res.* 180, 443–452.
839. Silverman, A.-J., Hoffman, D. L., Gadde, C. A., Krey, L. C., and Zimmerman, E. A. (1981). Adrenal steroid inhibition of the vasopressin-neurophysin neurosecretory system to the median eminence of the rat. *Neuroendocrinology* 32, 129–133.
840. Silverman, A.-J., and Zimmerman, E. A. (1982). Adrenalectomy increases sprouting in a peptidergic neurosecretory system. *Neuroscience* 7, 2705–2714.
841. Rivier, C., and Vale, W. (1983). Interaction of corticotropin-releasing factor and arginine vasopressin on adrenocorticotropin secretion in vivo. *Endocrinology* 113, 939–942.
842. Swanson, L. W., Sawchenko, P. E., Wiegand, S. L., and Price, J. L. (1980). Separate neurons in the paraventricular nucleus project to the median eminence and to the medulla or spinal cord. *Brain Res.* 198, 190–195.
843. Lechan, R. M., Nestler, J. L., Jacobson, S., and Reichlin, S. (1980). The hypothalamic 'tuberoinfundibular' system of the rat as demonstrated by horseradish peroxidase (HRP) microiontophoresis. *Brain Res.* 195, 13–27.
844. Lechan, R. M., Nestler, J. L., and Jacobson, S. (1982). The tuberoinfundibular system of the rat as demonstrated by immunohistochemical localization of retrogradely transported wheat germ agglutinin (WGA) from the median eminence. *Brain Res.* 245, 1–15.
845. Zimmerman, E. A., Hou-Yu, A., and Nilaver, G., et al. (1983). Organization of the oxytocin and vasopressin systems of the hypothalamus: intra- and extra-hypothalamic projections. In *Structure and Function of Peptidergic and Aminergic Neurons* (Y. Sano, Y. Ibata, and E. A. Zimmerman, Eds.). Japan Sci Soc Press, Tokyo.
846. Armstrong, W. E., and Hatton, G. I. (1980). The localization of projection neurons in the rat hypothalamic paraventricular nucleus following vascular and neurohypophysial injections of HRP. *Brain Res. Bull.* 5, 473–477.
847. Yamano, M., Bai, F.-L., Tohyama, M., and Shiotani, Y. (1985). Ultrastructural evidence of direct synaptic contact of catecholamine terminals with oxytocin-containing neurons in the parvocellular portion of the rat hypothalamic paraventricular nucleus. *Brain Res.* 336, 176–179.
848. Silverman, A.-J., Hou-Yu, A., and Chen, W.-P. (1989). Corticotropin-releasing factor synapses within the paraventricular nucleus of the hypothalamus. *Neuroendocrinology* 49, 291–299.

849. Paull, W. K., Scholer, J., and Arimura, A., et al. (1982). Immunocytochemical localization of CRF in the ovine hypothalamus. *Peptides* 1, 183–191.
850. Kolodziejczyk, E., Baertschi, A. J., and Tramu, G. (1983). Corticoliberin-immunoreactive cell bodies localised in two distinct areas of the sheep hypothalamus. *Neuroscience* 9, 261–270.
851. Joseph, S. A., and Knigge, K. M. (1983). Corticotropin releasing factor: immunocytochemical localization in rat brain. *Neurosci. Lett.* 35, 135–141.
852. Paull, W. K., and Gibbs, F. P. (1983). The corticotropin releasing factor (CRF) neurosecretory system in intact, adrenalectomized, and adrenalectomized-dexamethasone treated rats. An immunocytochemical analysis. *Histochemistry* 78, 303–316.
853. Daikoku, S., Okamura, Y., Kawano, H., Tsuruo, Y., Maegawa, M., and Shibasaki, T. (1985). CRF-containing neurons of the rat hypothalamus. *Cell. Tissue Res.* 240, 575–584.
854. Liposits, Z., Uht, R. M., Harrison, R. W., Gibbs, F. P., Paull, W. K., and Bohn, M. C. (1987). Ultrastructural localization of glucocorticoid receptor (GR) in hypothalamic paraventricular neurons synthesizing corticotropin releasing factor (CRF). *Histochemistry* 87, 407–412.
855. Kawano, H., Daikoku, S., and Shibasaki, T. (1988). CRF-containing neuron systems in the rat hypothalamus: retrograde tracing and immunohistochemical studies. *J. Comp. Neurol.* 272, 260–268.
856. Kawano, H., Tsuruo, Y., Gando, H., and Daikoku, S. (1991). Hypophysiotropic TRH-producing neurons identified by combining immunohistochemistry for pro-TRH and retrograde tracing. *J. Comp. Neurol.* 307, 531–538.
857. Kiss, J. Z., Mezey, E., and Skirboll, L. (1984). Corticotropin-releasing factor-immunoreactive neurons of the paraventricular nucleus become vasopressin positive after adrenalectomy. *Proc. Natl. Acad. Sci. U S A* 81, 1854–1858.
858. Sawchenko, P. E., Swanson, L. W., and Vale, W. (1984). Corticotropin-releasing factor: coexpression within distinct subsets of oxytocin-, vasopressin-, and neurotensin-immunoreactive neurons in the hypothalamus of the male rat. *J. Neurosci.* 4, 1118–1129.
859. Sawchenko, P. E., Swanson, L. W., and Vale, W. (1984). Coexpression of corticotropin releasing factor and vasopressin immunoreactivity in parvocellular neurosecretory neurons of the adrenalectomized rat. *Proc. Natl. Acad. Sci. U S A* 81, 1883–1887.
860. Whitnall, M. H., and Gainer, H. (1988). Major pro-vasopressin–expressing and pro-vasopressin–deficient subpopulations of corticotropin-releasing hormone neurons in normal rats. Differential distribution in the paraventricular nucleus. *Neuroendocrinology* 47, 176–180.
861. Hokfelt, T., Fahrenkrug, J., Tatemoto, K., Mutt, V., Werner, S., Hulting, A.-L., Terenius, L., and Chang, K. J. (1983). The PHI (PHI-27)/corticotropin-releasing factor/enkephalin immunoreactive hypothalamic neuron: possible morphological basis for integrated control of prolactin, corticotropin, and growth hormone secretion. *Proc. Natl. Acad. Sci. U S A* 80, 895–898.
862. Niimi, M., Takahara, J., Sato, M., and Kawanishi, K. (1991). Neurotensin and growth hormone-releasing factor-containing neurons projecting to the median eminence of the rat: a combined retrograde tracing and immunohistochemical study. *Neurosci. Lett.* 144, 183–186.
863. Lind, R. W., Swanson, L., and Ganten, D. (1985). Organization of angiotensin II immunoreactive cells and fibers in the rat central nervous system: an immunohistochemical study. *Neuroendocrinology* 40, 1–24.
864. Liposits, Z., Paull, W. K., Setalo, G., and Bigh, S. (1985). Evidence for local corticotropin releasing factor (CRF)-immunoreactive neuronal circuits in the paraventricular nucleus of the rat hypothalamus. An electron microscopic immunohistochemical analysis. *Histochemistry* 83, 5–16.
865. Kitazawa, S., Shioda, S., and Nakai, Y. (1987). Catecholaminergic innervation of neurons containing corticotropin-releasing factor in the paraventricular nucleus of the rat hypothalamus. *Acta Anat.* 129, 337–343.
866. Liposits, Z., Sherman, D., Phelix, C., and Paull, W. K. (1986). A combined light and electron microscopic immunocytochemical method for the simultaneous localization of multiple tissue antigens. Tyrosine hydroxylase immunoreactive innervation of corticotropin releasing factor synthesizing neurons in the paraventricular nucleus. *Histochemistry* 85, 95–106.
867. Liposits, Z., Phelix, C., and Paull, W. K. (1987). Synaptic interaction of serotonergic axons and corticotropin releasing factor (CRF) synthesizing neurons in the hypothalamic paraventricular nucleus of the rat. A light and electronmicroscopic study. *Histochemistry* 86, 541–549.
868. Miklos, I. H., and Kovacs, K. J. (2002). GABAergic innervation of corticotropin-releasing hormone (CRH)-secreting parvocellular neurons and its plasticity as demonstrated by quantitative immunoelectron microscopy. *Neuroscience* 113, 581–592.
869. Li, C., Chen, P., and Smith, M. S. (2000). Corticotropin releasing hormone neurons in the paraventricular nucleus are direct targets for neuropeptide Y neurons in the arcuate nucleus: an anterograde tracing study. *Brain Res.* 854, 122–129.
870. Aguilera, G., Young, W. S., Kiss, A., and Bathia, A. (1995). Direct regulation of hypothalamic corticotropin-releasing-hormone neurons by angiotensin II. *Neuroendocrinology* 61, 437–444.
871. Hisano, S., and Kaikoku, S. (1991). Existence of mutual synaptic relations between corticotropin-releasing-factor-containing and somatostatin-containing neurons in the rat hypothalamus. *Brain Res.* 545, 265–275.
872. Jackson, I., Wu, P., and Lechan, R. M. (1985). Immunohistochemical localization in the rat brain of the precursor for thyrotropin-releasing hormone. *Science* 229, 1097–1099.
873. Lechan, R. M., Wu, P., Jackson, I., Wolf, H., Cooperman, S., Mandel, G., and Goodman, R. H. (1986). Thyrotropin-releasing hormone precursor: characterization in rat brain. *Science* 231, 159–161.
874. Fliers, E., Noppen, N. W., Wiersinga, W. M., Visser, T. J., and Swaab, D. F. (1994). Distribution of thyrotropin-releasing hormone (TRH)-containing cells and fibers in the human hypothalamus. *J. Comp. Neurol.* 350, 311–323.
875. Guldenaar, S. E., Veldkamp, B., Bakker, O., Wiersinga, W. M., Swaab, D. F., and Fliers, E. (1996). Thyrotropin-releasing hormone gene expression in the human hypothalamus. *Brain Res.* 743, 93–101.
876. Elias, C. F., Lee, C. E., Kelly, J. F., Ahima, R. S., Kuhar, M., Saper, C. B., and Elmquist, J. K. (2001). Characterization of CART neurons in the rat and human hypothalamus. *J. Comp. Neurol.* 432, 1–19.
877. Larsen, P. J., Seier, V., Fink-Jensen, A., Holst, J. J., Warberg, J., and Vrang, N. (2003). Cocaine- and amphetamine-regulated transcript is present in hypothalamic neuroendocrine neurones and is released to the hypothalamic-pituitary portal circuit. *J. Neuroendocrinol.* 15, 219–226.
878. Liao, N., Bulant, M., Nicolas, P., Vaudry, H., and Pelletier, G. (1991). Anatomical interactions of proopiomelanocortin (POMC)-related peptides, neuropeptide Y (NPY) and dopamine beta-hydroxylase (D beta H) fibers and thyrotropin-releasing hormone (TRH) neurons in the paraventricular nucleus of rat hypothalamus. *Neuropeptides* 18, 63–67.

879. Toni, R., Jackson, I. M., and Lechan, R. M. (1990). Neuropeptide-Y–immunoreactive innervation of thyrotropin-releasing hormone-synthesizing neurons in the rat hypothalamic paraventricular nucleus. *Endocrinology* 126, 2444–2453.
880. Diano, S., Naftolin, F., Goglia, F., and Horvath, T. L. (1998). Segregation of the intra- and extrahypothalamic neuropeptide Y and catecholaminergic inputs on paraventricular neurons, including those producing thyrotropin-releasing hormone. *Regul. Pept.* 75–76, 117–126.
881. Legradi, G., and Lechan, R. M. (1998). The arcuate nucleus is the major source for neuropeptide Y-innervation of thyrotropin-releasing hormone neurons in the hypothalamic paraventricular nucleus. *Endocrinology* 139, 3262–3270.
882. Wittmann, G., Liposits, Z., Lechan, R. M., and Fekete, C. (2002). Medullary adrenergic neurons contribute to the neuropeptide Y-ergic innervation of hypophysiotropic thyrotropin-releasing hormone-synthesizing neurons in the rat. *Neurosci. Lett.* 324, 69–73.
883. Kiss, J. Z., and Halasz, B. (1990). Ultrastructural analysis of the innervation of TRH-immunoreactive neuronal elements located in the periventricular subdivision of the paraventricular nucleus of the rat hypothalamus. *Brain Res.* 532, 107–114.
884. Fekete, C., Wittmann, G., Liposits, Z., and Lechan, R. M. (2002). GABA-ergic innervation of thyrotropin-releasing hormone-synthesizing neurons in the hypothalamic paraventricular nucleus of the rat. *Brain Res.* 957, 251–258.
885. Mihaly, E., Fekete, C., Tatro, J. B., Liposits, Z., Stopa, E. G., and Lechan, R. M. (2000). Hypophysiotropic thyrotropin-releasing hormone-synthesizing neurons in the human hypothalamus are innervated by neuropeptide Y, agouti-related protein, and alpha-melanocyte-stimulating hormone. *J. Clin. Endocrinol. Metab.* 85, 2596–2603.
886. Fekete, C., Legradi, G., Mihaly, E., Huang, Q. H., Tatro, J. B., Rand, W. M., Emerson, C. H., and Lechan, R. M. (2000). α-Melanocyte-stimulating hormone is contained in nerve terminals innervating thyrotropin-releasing hormone-synthesizing neurons in the hypothalamic paraventricular nucleus and prevents fasting-induced suppression of prothyrotropin-releasing hormone gene expression. *J. Neurosci.* 20, 1550–1558.
887. Palkovits, M. (1982). Neuropeptides in the median eminence: their sources and destinations. *Peptides* 3, 299–303.
888. Sofroniew, M. V., and Schrell, U. (1981). Evidence for a direct projection from oxytocin and vasopressin neurons in the hypothalamic paraventricular nucleus to the medulla oblongata: immunohistochemical visualization of both the horseradish peroxidase. *Neurosci. Lett.* 22, 211–217.
889. Hallbeck, M., and Blomqvist, A. (1999). Spinal cord-projecting vasopressinergic neurons in the rat paraventricular hypothalamus. *J. Comp. Neurol.* 411, 201–211.
890. Hallbeck, M., Larhammar, D., and Blomqvist, A. (2001). Neuropeptide expression in rat paraventricular hypothalamic neurons that project to the spinal cord. *J. Comp. Neurol.* 433, 222–238.
891. Sofroniew, M. V. (1980). Projections from vasopressin, oxytocin and neurophysin neurons to neural targets in the rat and human. *J. Histochem. Cytochem.* 28, 475–478.
892. Swanson, L. W. (1977). Immunohistochemical evidence for a neurophysin-containing autonomic pathway arising in the paraventricular nucleus of the hypothalamus. *Brain Res.* 128, 346–353.
893. Shafton, A. D., Ryan, A., and Badoer, E. (1998). Neurons in the hypothalamic paraventricular nucleus send collaterals to the spinal cord and to the rostral ventrolateral medulla in the rat. *Brain Res.* 801, 239–243.
894. Buijis, R. M., Van Der Beck, E. M., Renaud, L. P., Day, T. A., and Jhamandas, J. (1990). Oxytocin localization and function of the A1 noradrenergic cell group: ultrastructural and electrophysiological studies. *Neuroscience* 39, 717–725.
895. Ranson, R. N., Motawei, K., Pyner, S., and Coote, J. H. (1998). The paraventricular nucleus of the hypothalamus sends efferents to the spinal cord of the rat that closely appose sympathetic preganglionic neurones projecting to the stellate ganglion. *Exp. Brain Res.* 120, 164–172.
896. Motawei, K., Pyner, S., Ranson, R. N., Kamel, M., and Coote, J. H. (1999). Terminals of paraventricular spinal neurones are closely associated with adrenal medullary sympathetic preganglionic neurones: immunocytochemical evidence for vasopressin as a possible neurotransmitter in this pathway. *Exp. Brain Res.* 126, 68–76.
897. Veronneau-Longueville, F., Rampin, O., Freund-Mercier, M. J., Tang, Y., Calas, A., Marson, L., McKenna, K. E., Stoeckel, M. E., Benoit, G., and Giuliano, F. (1999). Oxytocinergic innervation of autonomic nuclei controlling penile erection in the rat. *Neuroscience* 93, 1437–1447.
898. Buijis, R. M. (1978). Intra- and extrahypothalamic vasopressin and oxytocin pathways in the rat. *Cell. Tissue Res.* 192, 423–435.
899. Buijis, R. M., and Swaab, D. F. (1979). Immuno-electron microscopical demonstration of vasopressin and oxytocin synapses in the limbic system of the rat. *Cell. Tissue Res.* 204, 355–365.
900. DeVries, G. J., and Buijis, R. M. (1983). The origin of the vasopressinergic and oxytocinergic innervation of the rat brain with special reference to the lateral septum. *Brain Res.* 273, 307–371.
901. Conrad, L. C., and Pfaff, D. W. (1976). Efferents from medial basal forebrain and hypothalamus in the rat. II. An autoradiographic study of the anterior hypothalamus. *J. Comp. Neurol.* 169, 221–261.
902. Csaki, A., Kocsis, K., Halasz, B., and Kiss, J. (2000). Localization of glutamatergic/aspartatergic neurons projecting to the hypothalamic paraventricular nucleus studied by retrograde transport of [3H]D-aspartate autoradiography. *Neuroscience* 101, 637–655.
903. Dierickx, K., and Vandesande, F. (1979). Immunocytochemical localization of somatostatin-containing neurons in the rat hypothalamus. *Cell. Tissue Res.* 201, 349–359.
904. Krisch, B. (1978). Hypothalamic and extrahypothalamic distribution of somatostatin immunoreactive elements in the rat brain. *Cell. Tissue Res.* 195, 499–513.
905. Elde, R. P., and Parsons, J. A. (1975). Immunocytochemical localization of somatostatin in cell bodies of the rat hypothalamus. *Am. J. Anat.* 144, 541–548.
906. Bennett-Clarke, C., Romagnano, M. A., and Joseph, S. A. (1980). Distribution of somatostatin in the rat brain: telencephalon and diencephalon. *Brain Res.* 188, 473–486.
907. Alpert, L. C., Brawer, J. R., Patel, Y. C., and Reichlin, S. (1976). Somatostatinergic neurons in anterior hypothalamus: immunohistochemical localization. *Endocrinology* 98, 255–258.
908. Crowley, W. R., and Terry, L. C. (1980). Biochemical mapping of somatostatinergic systems in rat brain: effects of periventricular hypothalamic and medial basal amygdaloid lesions on somatostatin-like immunoreactivity in discrete brain nuclei. *Brain Res.* 200, 283–291.
909. Ishikawa, K., Taniguchi, Y., Kurosumi, K., Suzuki, M., and Shinoda, M. (1987). Immunohistochemical identification of somatostatin-containing neurons projecting to the median eminence of the rat. *Endocrinology* 121, 94–97.
910. Kawano, H., and Daikoku, S. (1988). Somatostatin-containing neuron systems in the rat hypothalamus: retrograde tracing and immunohistochemical studies. *J. Comp. Neurol.* 271, 293–299.

911. Merchenthaler, I., Setalo, G., Csontos, C., Petrusz, P., Flerko, B., and Negro-Vilar, A. (1989). Combined retrograde tracing and immunocytochemical identification of luteinizing hormone-releasing hormone- and somatostatin-containing neurons projecting to the median eminence of the rat. *Endocrinology* 125, 2812–2821.
912. Epelbaum, J., Tapia-Arancibia, L., Alonso, G., Astier, H., and Kordon, C. (1986). The anterior periventricular hypothalamus is the site of somatostatin inhibition of its own release: an in vitro and immunocytochemical study. *Neuroendocrinology* 44, 255–259.
913. Liposits, Z., Kallo, I., Barkovics-Kallo, M., Bohn, M. C., and Paull, W. K. (1990). Innervation of somatostatin synthesizing neurons by adrenergic, phenylethanolamine-N-methyltransferase (PNMT)-immunoreactive axons in the anterior periventricular nucleus of the rat hypothalamus. *Histochemistry* 94, 13–20.
914. Kiss, J. Z., Csaky, A., and Halasz, B. (1988). Demonstration of serotonergic axon terminals on somatostatin-immunoreactive neurons of the anterior periventricular nucleus of the rat hypothalamus. *Brain Res.* 442, 23–32.
915. Willoughby, J. O., Beroukas, D., and Blessing, W. W. (1987). Ultrastructural evidence for gamma aminobutyric acid-immunoreactive synapses on somatostatin-immunoreactive perikarya in the periventricular anterior hypothalamus. *Neuroendocrinology* 46, 268–272.
916. Kakucska, I., Tappaz, M. L., Gaal, G. Y., Stoickel, M. E., and Makara, G. B. (1988). Gabaergic innervation of somatostatin-containing neurosecretory cells of the anterior periventricular hypothalamic area: a light and electron microscopy double immunolabelling study. *Neuroscience* 25, 585–593.
917. Horvath, E., Palkovits, M., Gorcs, T., and Arimura, A. (1989). Electron microscopic immunocytochemical evidence for the existence of bidirectional synaptic connections between growth hormone-releasing hormone-and somatostatin-containing neurons in the hypothalamus of the rat. *Brain Res.* 481, 585–593.
918. Jew, J. Y., Leranth, C., Arimura, A., and Palkovits, M. (1984). Preoptic LH-RH and somatostatin in the rat median eminence. An experimental light and electron microscopic immunocytochemical study. *Neuroendocrinology* 38, 169–175.
919. Chan-Palay, V., Zaborszky, L., Kohler, C., Goldstein, M., and Palay, S. L. (1984). Distribution of tyrosine-hydroxylase-immunoreactive neurons in the hypothalamus of rats. *J. Comp. Neurol.* 227, 467–496.
920. Dahlstrom, A., and Fuxe, K. (1964). Evidence for the existence of monoamine-containing neurons in the central nervous system. I. Demonstration of monoamines in the cell bodies of brain stem neurons. *Acta Physiol. Scand.* 62, 1–55.
921. Goudreau, J. L., Falls, W. M., Lookingland, K. J., and Moore, K. E. (1995). Periventricular-hypophysial dopaminergic neurons innervate the intermediate but not the neural lobe of the rat pituitary gland. *Neuroendocrinology* 62, 147–154.
922. Ibata, Y., Kawakami, F., Fukui, K., Obata-Tsuto, H. L., Tanaka, M., Kubo, T., Okamura, H., Morimoto, N., Yanaihara, C., and Yanaihara, N. (1984). Light and electron microscopic immunocytochemistry of neurotensin-like immunoreactive neurons in the rat hypothalamus. *Brain Res.* 302, 221–230.
923. Crosby, E. C., and Woodburne, R. T. (1966). The comparative anatomy of the preoptic area and the hypothalamus. *Res. Publ. Assoc. Res. Nerv. Ment. Dis.* 20, 52–169.
924. van den Pol, A. N., and Cassidy, J. R. (1982). The comparative anatomy of the preoptic area and the hypothalamus. *J. Comp. Neurol.* 204, 65–98.
925. Walsh, R. J., and Brawer, J. R. (1979). Cytology of the arcuate nucleus in newborn male and female rats. *J. Anat.* 128, 121–133.
926. Bodoky, M., and Rethelyi, M. (1977). Dendritic arborization and axon trajectory of neurons in the hypothalamic arcuate nucleus of the rat. *Exp. Brain Res.* 28, 543–555.
927. Danzer, S. C., McMullen, N. T., and Rance, N. E. (1998). Dendritic growth of arcuate neuroendocrine neurons following orchidectomy in adult rats. *J. Comp. Neurol.* 390, 234–246.
928. Parducz, A., Zsarnovszky, A., Naftolin, F., and Horvath, T. L. (2003). Estradiol affects axo-somatic contacts of neuroendocrine cells in the arcuate nucleus of adult rats. *Neuroscience* 117, 791–794.
929. Ottersen, O. P. (1980). Afferent connections to the amygdaloid complex of the rat and cat. II. Afferents from the hypothalamus and the basal telencephalon. *J. Comp. Neurol.* 194, 267–289.
930. Fuxe, K., and Hokfelt, T. (1966). Further evidence for the existence of tubero-infundibular dopamine neurons. *Acta Physiol. Scand.* 66, 245–246.
931. Fuxe, K., and Hokfelt, T. (1969). Catecholamines in the hypothalamus and the pituitary gland. In *Frontiers in Neuroendocrinology* (W. F. Ganong and L. Martini, Eds.), Vol. 1, pp. 47–96. Oxford University Press, New York.
932. Bjorklund, A., and Nobin, A. (1973). Fluorescence histochemical and microspectrofluorometric mapping of dopamine and noradrenaline cell groups in the rat diencephalon. *Brain Res.* 51, 193–205.
933. Jonsson, G., Fuxe, K., and Hokfelt, T. (1972). On the catecholamine innervation of the hypothalamus, with special reference to the median eminence. *Brain Res.* 40, 271–281.
934. Smith, G. C., and Fink, G. (1972). Experimental studies on the origin of monoamine-containing fibres in the hypothalamo-hypophysial complex of the rat. *Brain Res.* 43, 37–51.
935. Palkovits, M., Brownstein, M. J., Saavedra, J. M., and Axelrod, J. (1974). Norepinephrine and dopamine content of hypothalamic nuclei of the rat. *Brain Res.* 77,137–149.
936. Goldsmith, P. C., Thind, K. K., and Boggan, J. E. (1988). Neuroendocrine GNRH and dopamine neurons in the monkey hypothalamus identified by retrograde staining and immunostaining. *Neurosci. Abstr.* 14, 439.
937. Nishizuka, M. (1979). Topography of the dopamine neurons in the arcuate nucleus of the mouse hypothalamus. *Acta Anat.* 103, 34–38.
938. Bugnon, C., Bloch, B., Lenys, D., Gouget, A., and Fellmann, D. (1979). Comparative study of the neuronal populations containing B-endorphin, corticotropin and dopamine in the arcuate nucleus of the rat hypothalamus. *Neurosci. Lett.* 14, 43–48.
939. Meister, B., Hokfelt, T., Steinbusch, H. W., Skagerberg, G., Lindvall, O., Geffard, M., Joh, T. H., Cuello, A. C., and Goldstein, M. (1988). Do tyrosine hydroxylase-immunoreactive neurons in the ventrolateral arcuate nucleus produce dopamine or only L-dopa? *J. Chem. Neuroanat.* 1, 59–64.
940. Meister, B., and Hokfelt, T. (1988). Peptide- and transmitter-containing neurons in the mediobasal hypothalamus and their relation to GABAergic systems: possible roles in control of prolactin and growth hormone secretion. *Synapse* 2, 585–605.
941. Hrabovszky, E., and Liposits, Z. (1994). Adrenergic innervation of dopamine neurons in the hypothalamic arcuate nucleus of the rat. *Neurosci. Lett.* 182, 143–146.
942. Hrabovszky, E., and Liposits, Z. (1994). Galanin-containing axons synapse on tyrosine hydroxylase-immunoreactive neurons in the hypothalamic arcuate nucleus of the rat. *Brain Res.* 652, 49–55.
943. Magoul, R., Dubourg, P., Kah, O., and Tramu, G. (1994). Ultrastructural evidence for synaptic inputs of enkephalinergic nerve terminals to target neurons in the rat arcuate nucleus. *Peptides* 15, 883–892.
944. Everitt, B., Hokfelt, T., Wu, J.-Y., and Goldstein, M. (1984). Coexistence of tyrosine hydrolase-like and

gamma-aminobutyric acid-like immunoreactivities in neurons of the arcuate nucleus. *Neuroendocrinology* 39, 189–191.
945. Okamura, H., Murakami, S., Chihara, M., and Jouvet, M. (1985). Coexistence of growth hormone releasing factor-like and tyrosine hydroxylase-like immunoreactivities in neurons of the rat arcuate nucleus. *Neuroendocrinology* 41, 177–179.
946. Bloch, B., Brazeau, P., Bloom, F. E., and Ling, N. (1983). Topographical study of the neurons containing hyGRF immunoreactivity in monkey hypothalamus. *Neurosci. Lett.* 37, 23–28.
947. Lechan, R. M., Lin, H. D., Ling, N., Jackson, I., Jacobson, S., and Reichlin, S. (1984). Distribution of immunoreactive growth hormone releasing factor (1-44)NH2 in the tuberoinfundibular system of the rhesus monkey. *Brain Res.* 309, 55–61.
948. Bugnon, C., Gouget, A., Fellmann, D., and Clavequin, M. C. (1983). Immunocytochemical demonstration of a novel peptidergic neurone system in the cat brain with an anti-growth hormone-releasing factor serum. *Neurosci. Lett.* 38, 131–137.
949. Bloch, B., Brazeau, P., Ling, N., Bohlen, P., Esch, F., Wehrenberg, W. B., Benoit, R., Bloom, F. E., and Guillemin, R. (1983). Immunohistochemical detection of growth hormone-releasing factor in brain. *Nature (Lond).* 301, 607–608.
950. Fellmann, D., Bugnon, C., and Lavry, G. N. (1985). Immunohistochemical demonstration of a new neurone system in rat brain using antibodies against human growth hormone-releasing factor (1-37). *Neurosci. Lett.* 58, 91–96.
951. Bloch, B., Ling, N., Benoit, R., Wehrenberg, W. B., and Guillemin, R. (1984). Specific depletion of immunoreactive growth hormone-releasing factor by monosodium glutamate in rat median eminence. *Nature (Lond).* 307, 272–273.
952. Holzwarth-McBride, M. A., Hurst, E. M., and Knigge, K. M. (1976). Monosodium glutamate induced lesions of the arcuate nucleus. I. Endocrine deficiency and ultrastructure of the median eminence. *Anat. Rec.* 186, 185–196.
953. Holzwarth-McBride, M. A., Sladek, J. R., and Knigge, K. M. (1976). Monosodium glutamate induced lesions of the arcuate nucleus. II. Fluorescence histochemistry of catecholamines. *Anat. Rec.* 186, 197–206.
954. Niimi, M., Takahara, J., Sato, A., and Kawanishi, K. (1989). Sites of origin of growth hormone-releasing factor-containing neurons projecting to the stalk-median eminence of the rat. *Peptides* 10, 197–206.
955. Balthasar, N., Mery, P. F., Magoulas, C. B., Mathers, K. E., Martin, A., Mollard, P., and Robinson, I. C. (2003). Growth hormone-releasing hormone (GHRH) neurons in GHRH-enhanced green fluorescent protein transgenic mice: a ventral hypothalamic network. *Endocrinology* 144, 2728–2740.
956. Meister, B., Hokfelt, T., Vale, W. W., Sawchenko, P. E., Swanson, L., and Goldstein, M. (1986). Coexistence of tyrosine hydroxylase and growth hormone-releasing factor in a subpopulation of tubero-infundibular neurons of the rat. *Neuroendocrinology* 42, 237–247.
957. Rossier, J., Vargo, T. M., Minick, S., Ling, N., Bloom, F. E., and Guillemin, R. (1977). Regional dissociation of B-endorphin and enkephalin contents in rat brain and pituitary. *Proc. Natl. Acad. Sci. U S A* 74, 5162–5165.
958. Oliver, C., and Porter, J. C. (1978). Distribution and characterization of a-melanocyte-stimulating hormone in the rat brain. *Endocrinology* 102, 697–705.
959. Eskay, R. L., Giraud, P., Oliver, C., and Brownstein, M. J. (1979). Distribution of a-melanocyte-stimulating hormone in the rat brain: evidence that a-MSH-containing cells in the arcuate region send projections to extrahypothalamic areas. *Brain Res.* 178, 55–67.
960. Liotta, A., Gildersleeve, D., Brownstein, M. J., and Krieger, D. T. (1979). Biosynthesis in vitro of immunoreactive 31,000-dalton corticotropin/β-endorphin–like material by bovine hypothalamus. *Proc. Natl. Acad. Sci. U S A* 76, 1448–1452.
961. Watson, S. J., Barchas, J. D., and Li, C. H. (1977). β-lipotropin: localization of cells and axons in rat brain by immunocytochemistry. *Proc. Natl. Acad. Sci. U S A* 74, 5155–5158.
962. Nilaver, G., Zimmerman, E. A., Defendini, R., Liotta, A., Krieger, D. T., and Brownstein, M. J. (1979). Adrenocorticotropin and β-lipotropin in the hypothalamus. Localization in the same arcuate neurons by sequential immunocytochemical procedures. *J. Cell. Biol.* 81, 50–58.
963. Hisano, S., Kawano, H., Nishiyama, T., and Daikoku, S. (1982). Immunoreactive ACTH/β-endorphin neurons in the tubero-infundibular hypothalamus of rats. *Cell. Tissue Res.* 224, 303–314.
964. Bloch, B., Bugnon, C., Fellmann, D., Lenys, D., and Gouget, A. (1979). Neurons of the rat hypothalamus reactive with antisera against endorphins, ACTH, MSH, and β-LPH. *Cell. Tissue Res.* 204, 1–15.
965. Bloch, B., Bugnon, C., Fellmann, D., and Lenys, D. (1978). Immunocytochemical evidence that the same neurons in the human infundibular nucleus are stained with anti-endorphins and antisera of other related peptides. *Neurosci. Lett.* 10, 147–152.
966. Bloch, B., Bugnon, C., Lenys, D., and Fellmann, D. (1978). Presence de determinants antigeniques de la B-LPH, de la β-MSH, de l'a-endorphine, de l'ACTH et de l'a-MSH dans les neurones reveles par l'anti-β-endorphine au niveau du noyau infundibulaire de l'Homme. *C. R. Acad. Sci. Paris* 287, 1019–1022.
967. Fodor, M., Csaba, Z., Epelbaum, J., Vaudry, H., and Jegou, S. (1998). Interrelations between hypothalamic somatostatin and proopiomelanocortin neurons. *J. Neuroendocrinol.* 10, 75–78.
968. Sawchenko, P. E., Swanson, L. W., and Joseph, S. A. (1982). The distribution and cells of origin of ACTH(1-39)-stained varicosities in the paraventricular and supraoptic nuclei. *Brain Res.* 232, 365–374.
969. Baker, R. A., and Herkenham, M. (1995). Arcuate nucleus neurons that project to the hypothalamic paraventricular nucleus: neuropeptidergic identity and consequences of adrenalectomy on mRNA levels in the rat. *J. Comp. Neurol.* 358, 518–530.
970. Kawano, H., and Masuko, S. Beta-endorphin-, adrenocorticotrophic hormone- and neuropeptide y-containing projection fibers from the arcuate hypothalamic nucleus make synaptic contacts on to nucleus preopticus medianus neurons projecting to the paraventricular hypothalamic nucleus in the rat. *Neuroscience* 98, 555–565.
971. Thind, K. K., and Goldsmith, P. C. (1988). Infundibular gonadotropin-releasing hormone neurons are inhibited by direct opioid and autoregulatory synapses in juvenile monkeys. *Neuroendocrinology* 47, 203–216.
972. Bai, F.-L., Yamano, M., Shiotani, Y., Emson, P. C., Smith, A. D., Powell, J. F., and Tohyama, M. (1985). An arcuato-paraventricular and dorsomedial hypothalamic neuropeptide Y-containing system which lacks noradrenaline in the rat. *Brain Res.* 331, 172–175.
973. Horvath, T. L., Bechmann, I., Naftolin, F., Kalra, S. P., and Leranth, C. (1997). Heterogeneity in the neuropeptide Y-containing neurons of the rat arcuate nucleus: GABAergic and non-GABAergic subpopulations. *Brain Res.* 756, 283–286.
974. Li, C., Chen, P., and Smith, M. S. (1999). Morphological evidence for direct interaction between arcuate nucleus neuropeptide Y (NPY) neurons and gonadotropin-releasing hormone neurons and the possible involvement of NPY Y1 receptors. *Endocrinology* 140, 5382–5390.
975. Selmanoff, M., Wise, P. M., and Barraclough, C. A. (1980). Regional distribution of luteinizing hormone-releasing

hormone (LH-RH) in rat brain determined by microdissection and radioimmunoassay. *Brain Res.* 192, 421–432.
976. Silverman, A.-J., and Zimmerman, E. A. (1978). Pathways containing luteinizing hormone-releasing hormone (LHRH) in the mammalian brain. In *Brain–Endocrine Interaction* (D. E. Scott, G. P. Kozlowski, and A. Weindl, Eds.), Vol. 3, pp. 83–96. Karger, Basel.
977. Kozlowski, G. P., Nett, T. M., and Zimmerman, E. A. (1975). Immunocytochemical localization of gonadotropin-releasing hormone (Gn-RH) and neurophysin in the brain. In *Anatomical Neuroendocrinology* (W. E. Stumpf and L. D. Grant, Eds.), pp. 185–191. Karger, Basel.
978. Kawano, H., and Daikoku, S. (1981). Immunohistochemical demonstration of LHRH neurons and their pathways in the rat hypothalamus. *Neuroendocrinology* 32, 179–186.
979. Silverman, A.-J., Antunes, J., Ferin, M., and Zimmerman, E. A. (1977). The distribution of luteinizing hormone-releasing hormone (LHRH) in the hypothalamus of the rhesus monkey. light microscopic studies using immunoperoxidase technique. *Endocrinology* 101, 134–142.
980. Barry, J. (1976). Characterization and topography of LH-RH neurons in the human brain. *Neurosci. Lett.* 3, 287–291.
981. Silverman, A.-J., and Krey, L. C. (1978). The luteinizing hormone-releasing hormone (LH-RH) neuronal networks of the guinea pig brain. I. Intra- and extra-hypothalamic projections. *Brain Res.* 157, 233–246.
982. King, J. C., Anthony, E. L. P., Gustafson, A. W., and Damassa, D. A. (1984). Luteinizing hormone-releasing hormone (LH-RH) cells and their projections in the forebrain of the bat Myotis lucifugus. *Brain Res.* 298, 289–301.
983. Barry, J. (1977). Immunofluorescence study of LRF neurons in man. *Cell. Tissue Res.* 181, 1–14.
984. King, J. C., Anthony, E. L. P., Fitzgerald, D. M., and Stopa, E. G. (1985). Luteinizing hormone-releasing hormone neurons in human preoptic/hypothalamus: differential intraneuronal localization of immunoreactive forms. *J. Clin. Endocrin. Metab.* 60, 88–97.
985. Tsuruo, Y., Kawano, H., Nishiyama, T., Hisano, S., and Daikoku, S. (1983). Substance P-like immunoreactive neurons in the tuberinfundibular area of rat hypothalamus, light and electron microscopy. *Brain Res.* 289, 1–9.
986. Hokfelt, T., Elde, R., Johansson, O., Terenius, L., and Stein, L. (1977). Distribution of enkephalin-like immunoreactivity in the rat central nervous system. *Neurosci. Lett.* 5, 25–31.
987. Finley, J. C. W., Maderdrut, J. L., and Petrusz, P. (1981). The immunocytochemical localization of enkephalin in the central nervous system of the rat. *J. Comp. Neurol.* 198, 541–565.
988. Khachaturian, H., Lewis, M. E., and Watson, S. J. (1983). Enkephalin systems in diencephalon and brainstem of the rat. *J. Comp. Neurol.* 220, 310–320.
989. Lu, S., Guan, J. L., Wang, Q. P., Uehara, K., Yamada, S., Goto, N., Date, Y., Nakazato, M., Kojima, M., Kangawa, K., and Shioda, S. (2002). Immunocytochemical observation of ghrelin-containing neurons in the rat arcuate nucleus. *Neurosci. Lett.* 321, 157–160.
990. Broglio, F., Gottero, C., Arvat, E., and Ghigo, E. (2003). Endocrine and non-endocrine actions of ghrelin. *Horm. Res.* 59, 109–117.
991. Tannenbaum, G. S., Epelbaum, J., and Bowers, C. Y. (2003). Interrelationship between the novel peptide ghrelin and somatostatin/growth hormone-releasing hormone in regulation of pulsatile growth hormone secretion. *Endocrinology* 144, 967–974.
992. Leranth, C., Sakamoto, H., Maclusky, N. J., Shanabrough, M., and Naftolin, F. (1985). Intrinsic tyrosine hydroxylase (TH) immunoreactive axons synapse with TH immunopositive neurons in the rat arcuate nucleus. *Brain Res.* 331, 371–375.
993. van den Pol, A. N. (1986). Tyrosine hydroxylase immunoreactive synapses: a double pre-embedding immunocytochemical study with particulate silver and HRP. *J. Neurosci.* 6, 877–891.
994. Tappaz, M. L., Bosler, O., Paut, L., and Berod, A. (1985). Glutamate decarboxylase-immunoreactive boutons in synaptic contacts with hypothalamic dopaminergic cells: a light and electron microscopic study combining immunohistochemistry and radioautography. *Neuroscience* 16, 112–122.
995. Kiss, J. Z., and Halasz, B. (1986). Synaptic connections between serotoninergic axon terminals and tyrosine hydroxylase-immunoreactive neurons in the arcuate nucleus of the rat hypothalamus. A combination of electron microscopic autoradiography and immunocytochemistry. *Brain Res.* 364, 284–294.
996. Goldsmith, P. C., Boggan, J. E., and Thind, K. K. (1989). Opioid neurons synapse on tuberoinfundibular dopamine neurons in the arcuate nucleus of juvenile monkeys. *Neurosci. Abstr.* 15, 722.
997. Horvath, E., and Palkovits, M. (1988). Synaptic interconnections among growth hormone-releasing hormone (GHRH)-containing neurons in the arcuate nucleus of the rat hypothalamus. *Neuroendocrinology* 48, 471–476.
998. Sato, A., Shioda, S., and Nakai, Y. (1989). Catecholaminergic innervation of GRF-containing neurons in the rat hypothalamus revealed by electron-microscopic cytochemistry. *Cell. Tissue Res.* 258, 31–34.
999. Shioda, S., Kohara, H., and Nakai, Y. (1987). TRH axon terminals in synapsis with GRF neurons in the arcuate nucleus of the rat hypothalamus as revealed by double labeling immunocytochemistry. *Brain Res.* 302, 355–358.
1000. Kiss, J. Z., and Williams, T. H. (1983). ACTH-immunoreactive boutons form synaptic contacts in the hypothalamic arcuate nucleus of rat: evidence for local opiocortin connections. *Brain Res.* 263, 142–146.
1001. Kiss, J. Z., Leranth, C., and Halasz, B. (1984). Serotonergic endings on VIP-neurons in the suprachiasmatic nucleus and on ACTH-neurons in the arcuate nucleus of the rat hypothalamus. A combination of high resolution autoradiography and electron microscopic immunohistochemistry. *Neurosci. Lett.* 44, 119–124.
1002. Kozasa, K., and Nakai, Y. (1987). Electron-microscopic cytochemistry of the catecholaminergic innervation of ACTH-containing neurons in the rat hypothalamic arcuate nucleus. *Acta Anat.* 128, 243–249.
1003. Zhang, R., Hisano, S., Chikamori-Aoyama, M., and Kaikoku, S. (1987). Synaptic association between enkephalin-containing axon terminals and proopiomelanocortin-containing neurons in the arcuate nucleus of rat hypothalamus. *Neurosci. Lett.* 82, 151–156.
1004. Tsuruo, Y., Hisano, S., and Daikoku, S. (1984). Morphological evidence for synaptic junctions between substance P-containing neurons in the arcuate nucleus of the rat. *Neurosci. Lett.* 46, 65–69.
1005. Hammel, H. T. (1986). Regulation of internal body temperature. *Ann. Rev. Physiol.* 30, 641–710.
1006. Squires, R. D., and Jacobson, F. H. (1968). Chronic deficits of temperature regulation produced in cats by preoptic lesions. *Am. J. Physiol.* 214, 549–560.
1007. Ramsay, D. J., Thrasher, T. N., and Keil, L. C. (1983). The organum vasculosum laminae terminalis: a critical area for osmoreception. *Prog. Brain Res.* 60, 91–98.
1008. Brody, M. J., and Johnson, A. K. (1980). Role of the anteroventral third ventricle region in fluid and electrolyte balance, arterial pressure regulation, and hypertension. In *Frontiers in Neuroendocrinology* (L. Martini and W. F. Ganong, Eds.), Vol. 6, pp. 249–292. Raven Press, New York.

1009. Saphier, D., and Feldman, S. (1986). Effects of stimulation of the preoptic area on hypothalamic paraventricular nucleus unit activity and corticosterone secretion in freely moving rats. *Neuroendocrinology* 42, 167–173.
1010. Boudier, H. S., Smeets, G., Brouwer, G., and Van Rossum, J. M. (1975). Central nervous system alpha-adrenergic mechanisms and cardiovascular regulation in rats. *Arch. Int. Pharmacodyn. Ther.* 213, 285–293.
1011. Gorski, R. A. (1984). Critical role for the medial preoptic area in the sexual differentiation of the brain. *Prog. Brain Res.* 61, 129–146.
1012. Arendash, G. W., and Gorski, R. A. (1983). Effects of discrete lesions of the sexually dimorphic nucleus of the preoptic area or other medial preoptic regions on the sexual behavior of male rats. *Brain Res. Bull.* 10, 147–1254.
1013. Commins, D., and Yahr, P. (1984). Adult testosterone levels influence the morphology of a sexually dimorphic area in the mongolial gerbil brain. *J. Comp. Neurol.* 224, 132–140.
1014. Masken, J. F., Kragt, C. L., Gallo, R. V., and Ganong, W. F. (1974). Release of luteinizing hormone by electrical stimulation of the medial preoptic area and arcuate nucleus in the male rat. *Neuroendocrinology* 15, 249–254.
1015. Knobil, E., Plant, T. M., Wildt, L., Belchetz, P. E., and Marshall, G. (1980). Control of the rhesus monkey menstrual cycle: permissive role of hypothalamic gonadotropin-releasing hormone. *Science* 207, 1371–1373.
1016. Swanson, L. (1976). An autoradiographic study of the efferent connections of the preoptic region in the rat. *J. Comp. Neurol.* 167, 227–256.
1017. Prince, F. P., and Jones-Witters, P. H. (1974). The ultrastructure of the medial preoptic area of the rat. *Cell. Tissue Res.* 153, 517–530.
1018. Greenough, W. T., Carter, C. S., Steerman, C., and DeVoogd, T. J. (1977). Sex differences in dendritic patterns in hamster preoptic area. *Brain Res.* 126, 63–72.
1019. Gorski, R. A., Harlan, R. E., Jacobson, C. D., Shryne, J. E., and Southam, A. M. (1980). Evidence for the existence of a sexually dimorphic nucleus in the preoptic area of the rat. *J. Comp. Neurol.* 193, 529–539.
1020. Jacobson, C., Shryne, J. E., Shapiro, F., and Gorski, R. A. (1980). Ontogeny of the sexually dimorphic nucleus of the preoptic area. *J. Comp. Neurol.* 193, 541–548.
1021. Dohler, K.-D., Coquelin, A., Davis, F., Hines, M., Shryne, J. E., and Gorski, R. A. (1982). Differentiation of the sexually dimorphic nucleus in the preoptic area of the rat brain is determined by the perinatal hormone environment. *Neurosci. Lett.* 33, 295–298.
1022. Hines, M., Davis, F. C., Goy, R. W., and Gorski, R. A. (1985). Sexually dimorphic regions in the medial preoptic area and the bed nucleus of the stria terminalis of the guinea pig brain: a description and an investigation of their relationship to gonadal steroids in adulthood. *J. Neurosci.* 5, 40–47.
1023. Takami, S., and Urano, A. (1984). The volume of the toad medial amygdala-anterior preoptic complex is sexually dimorphic and seasonally variable. *Neurosci. Lett.* 44, 253–258.
1024. Swaab, D. F., and Fliers, E. (1984). A sexually dimorphic nucleus in the human brain. *Science* 228, 1112–1115.
1025. Cramer, O. M., and Barraclough, C. A. (1971). Effect of electrical stimulation of the preoptic area on plasma LH concentrations in proestrous rats. *Endocrinology* 88, 1175–1183.
1026. Sar, M., and Stumpf, W. E. (1973). Autoradiographic localization of radioactivity in the rat brain after the injection of 1,2-3H-testosterone. *Endocrinology* 92, 251–256.
1027. Stumpf, W. E. (1970). Estrogen-neurons and estrogen-neuron systems in the periventricular brain. *Am. J. Anat.* 129, 207–218.
1028. Pfaff, D. W., and Keiner, M. (1973). Atlas of estradiol-concentrating cells in the central nervous system of the female rat. *J. Comp. Neurol.* 151, 121–158.
1029. Pfaff, D. W., Gerlach, J. L., McEwen, B. S., Ferin, M., Carmel, P. W., and Zimmerman, E. A. (1976). Autoradiographic localization of hormone-concentrating cells in the brain of the female rhesus monkey. *J. Comp. Neurol.* 170, 279–293.
1030. Raisman, G., and Field, P. M. (1973). Sexual dimorphism in the neuropil of the preoptic area of the rat and its dependence on neonatal androgen. *Brain Res.* 54, 1–29.
1031. McDonald AJ. Neurons of the bed nucleus of the stria terminalis: a golgi study in the rat. *Brain Res Bull.* 1983;10:111–120.
1032. Saper, C. B. (1984). Organization of the cerebral cortical afferent systems in the rat. II. Magnocellular basal nucleus. *J. Comp. Neurol.* 222, 313–342.
1033. Epelbaum, J., Arancibia, L. T., Herman, J. P., Kordon, C., and Palkovits, M. (1981). Topography of median eminence somatostatinergic innervation. *Brain Res.* 230, 412–416.
1034. Woodhams, P. L., Roberts, G. W., Polak, J. M., and Crow, T. J. (1983). Distribution of neuropeptides in the limbic system of the rat: the bed nucleus of the stria terminalis, septum and preoptic area. *Neuroscience* 8, 677–703.
1035. King, J. C., Tobet, S. A., Snavely, F. L., and Arimura, A. (1982). LHRH immunopositive cells and their projections to the median eminence and organum vasculosum of the lamina terminalis. *J. Comp. Neurol.* 209, 287–300.
1036. Merchenthaler, I., Gorcs, T., Setalo, G., Petrusz, P., and Flerko, B. (1984). Gonadotropin-releasing hormone (GnRH) neurons and pathways in the rat brain. *Cell. Tissue Res.* 237, 15–29.
1037. Barry, J. (1976). Characterization and topography of LH-RH neurons in the rabbit. *Neurosci. Lett.* 2, 201–205.
1038. Krey, L. C., and Silverman, A.-J. (1978). The luteinizing hormone-releasing hormone (LH-RH) neuronal networks of the guinea pig brain. II. The regulation of gonadotropin secretion and the origin of terminals in the median eminence. *Brain Res.* 157, 247–255.
1039. Charney, Y., Bouras, C., Vallet, P. G., Golaz, J., Guntern, R., and Constandinidis, J. (1989). Immunohistochemical distribution of delta sleep-inducing peptide in the rabbit brain and hypophysis. *Neuroendocrinology* 49, 169–175.
1040. Silverman, A.-J., Jhamandas, J., and Renaud, L. P. (1987). Localization of luteinizing hormone-releasing hormone (LHRH) neurons that project to the median eminence. *J. Neurosci.* 7, 2312–2319.
1041. Pelletier, G. (1987). Demonstration of contacts between neurons staining for LHRH in the preoptic area of the rat brain. *Neuroendocrinology* 46, 457–459.
1042. Leranth, C., Segura, L. M. G., Palkovits, M., Maclusky, N. J., Shanabrough, M., and Naftolin, F. (1985). The LHRH-containing neuronal network in the preoptic area of the rat: demonstration of LH-RH containing nerve terminals in synaptic contact with LH-RH neurons. *Brain Res.* 345, 332–336.
1043. Watanabe, T., and Nakai, Y. (1987). Electron microscopic cytochemistry of catecholaminergic innervation of LHRH neurons in the medial preoptic area of the rat. *Arch. Histol. Jpn.* 50, 103–112.
1044. Dudas, B., and Merchenthaler, I. (2001). Catecholaminergic axons innervate LH-releasing hormone immunoreactive neurons of the human diencephalon. *J. Clin. Endocrinol. Metab.* 86, 5620–5626.
1045. Tsuruo, Y., Kawano, H., Kagotani, Y., Hisano, S., Kaikoku, S., Chihara, K., Shang, T., and Yanaihara, C. (1990). Morphological evidence for neuronal regulation of luteinizing

hormone-releasing hormone-containing neurons by neuropeptide Y in the rat septopreoptic area. *Neurosci. Lett.* 110, 261–266.
1046. Leranth, C., Maclusky, N. J., Sakamoto, H., Shanobrough, M., and Naftolin, F. (1985). Glutamic acid decarboxylase-containing axons synapse on LHRH neurons in the rat medial preoptic area. *Neuroendocrinology* 40, 536–539.
1047. Goldsmith, P. C., Thind, K. K., Perera, A. D., and Plant, T. M. (1994). Glutamate-immunoreactive neurons and their gonadotropin-releasing hormone-neuronal interactions in the monkey hypothalamus. *Endocrinology* 134, 858–868.
1048. Maclusky, N. J., Naftolin, F., and Leranth, C. (1988). Immunocytochemical evidence for direct synaptic connections between corticotropin-releasing factor (CRF) and gonadotropin-releasing hormone (GNRH)-containing neurons in the preoptic area of the rat. *Brain Res.* 439, 391–395.
1049. Dudas, B., and Merchenthaler, I. (2002). Close juxtapositions between luteinizing hormone-releasing hormone-immunoreactive neurons and corticotropin-releasing factor-immunoreactive axons in the human diencephalon. *J. Clin. Endocrinol. Metab.* 87, 5778–5784.
1050. Tsuruo, Y., Kawano, H., Hisano, S., Kagotani, Y., Daikoku, S., Zhang, T., and Yanaihara, N. (1991). Substance P-containing neurons innervating LHRH-containing neurons in the septo-preoptic area of rats. *Neurosci. Lett.* 110, 261–266.
1051. Dudas, B., and Merchenthaler, I. (2002). Close juxtapositions between LHRH immunoreactive neurons and substance P immunoreactive axons in the human diencephalon. *J. Clin. Endocrinol. Metab.* 87, 2946–2953.
1052. Chen, W.-P., Witkin, J. W., and Silverman, A.-J. (1989). Beta-endorphin and gonadotropin-releasing hormone synaptic input to gonadotropin-releasing hormone neurosceretory cells in the male rat. *J. Comp. Neurol.* 286, 85–95.
1053. Leranth, C., Maclusky, N. J., Shanabrough, M., and Naftolin, F. (1988). Immunohistochemical evidence for synaptic connections between proopiomelanocortin-immunoreactive axons and LH-RH neurons in the preoptic area of the rat. *Brain Res.* 449, 167–176.
1054. Hoffman, G. E., and Gibbs, D. M. (1982). LHRH pathways in rat brain: deafferentation spares a sub-chiasmatic LHRH projection to the median eminence. *Neuroscience* 7, 1979–1993.
1055. Silverman, A.-J., Antunes, J., Abrams, G. M., Nilaver, G., Thau, R., Robinson, J. A., Ferin, M., and Krey, L. C. (1982). The luteinizing hormone-releasing hormone pathways in rhesus (Macaca mulatta) and pigtailed (Macaca nemestrina) monkeys: new observations on thick, unembedded sections. *J. Comp. Neurol.* 211, 309–317.
1056. Marshall, P. E., and Goldsmith, P. C. (1980). Neuroregulatory and neuroendocrine GnRH pathways in the hypothalamus and forebrain of the baboon. *Brain Res.* 193, 353–372.
1057. Jennes, L., and Stumpf, W. E. (1980). LHRH-systems in the brain of the golden hamster. *Cell. Tissue Res.* 209, 239–256.
1058. Witkin, J. W., Paden, C. M., and Silverman, A.-J. (1982). The luteinizing hormone- releasing hormone (LHRH) systems in the rat brain. *Neuroendocrinology* 35, 429–438.
1059. Conrad, L. C., and Pfaff, D. W. (1976). Efferents from medial basal forebrain and hypothalamus in the rat. I. An autoradiographic study of the medial preoptic area. *J. Comp. Neurol.* 169, 185–220.
1060. Chiba, T., and Murata, Y. (1985). Afferent and efferent connections of the medial preoptic area in the rat: a WGA-HRP study. *Brain Res. Bull.* 14, 261–272.
1061. Simerly, R. B., and Swanson, L. (1988). Projections of the medial preoptic nucleus: a phaseolus vulgaris leucoagglutinin anterograde tract-tracing study in the rat. *J. Comp. Neurol.* 270, 209–242.
1062. Palkovits, M., Mezey, E., Ambach, G., and Kivovics, P. (1978). Neural and vascular connections between the organum vasculosum laminae terminalis and preoptic nuclei. In *Brain–Endocrine Interaction* (D. E. Scott, G. P. Kozlowski, and A. Weindl, Eds.), Vol. 3, pp. 302–312. Karger, Basel.
1063. Hoorneman, E. M. D., and Buijis, R. M. (1982). Vasopressin fiber pathways in the rat brain following suprachiasmatic nucleus lesioning. *Brain Res.* 243, 235–241.
1064. Swanson, L., and Cowan, W. M. (1979). The connections of the septal region in the rat. *J. Comp. Neurol.* 186, 621–656.
1065. Harris, G. W. (1955). Neural control of the pituitary gland. *Physiol. Rev.* 28, 139–179.
1066. Lincoln, G., Messager, S., Andersson, H., and Hazlerigg, D. (2002). Temporal expression of seven clock genes in the suprachiasmatic nucleus and the pars tuberalis of the sheep: evidence for an internal coincidence timer. *Proc. Natl. Acad. Sci. U S A* 99, 13890–13895.
1067. Moore, R. Y., Speh, J. C., and Leak, R. K. (2002). Suprachiasmatic nucleus organization. *Cell. Tissue Res.* 309, 89–98.
1068. Abrahamson, E. E., and Moore, R. Y. (2001). Suprachiasmatic nucleus in the mouse: retinal innervation, intrinsic organization and efferent projections. *Brain Res.* 916, 172–191.
1069. Negroni, J., Nevo, E., and Cooper, H. M. (1997). Neuropeptidergic organization of the suprachiasmatic nucleus in the blind mole rat (Spalax ehrenbergi). *Brain Res. Bull.* 44, 633–639.
1070. Hofman, M. A., Zhou, J. N., and Swaab, D. F. (1996). Suprachiasmatic nucleus of the human brain: an immunocytochemical and morphometric analysis. *Anat. Rec.* 244, 552–562.
1071. Song, C. K., Bartness, T. J., Petersen, S. L., and Bittman, E. L. (2000). Co-expression of melatonin (MEL1a) receptor and arginine vasopressin mRNAs in the Siberian hamster suprachiasmatic nucleus. *J. Neuroendocrinol.* 12, 627–634.
1072. Moga, M. M., and Moore, R. Y. (1997). Organization of neural inputs to the suprachiasmatic nucleus in the rat. *J. Comp. Neurol.* 389, 508–534.
1073. Reuss, S., and Decker, K. (1997). Anterograde tracing of retinohypothalamic afferents with Fluoro-Gold. *Brain Res.* 745, 197–204.
1074. Hannibal, J. (2002). Neurotransmitters of the retino-hypothalamic tract. *Cell. Tissue Res.* 309, 73–88.
1075. Morin, L. P., and Meyer-Bernstein, E. L. (1999). The ascending serotonergic system in the hamster: comparison with projections of the dorsal and median raphe nuclei. *Neuroscience* 91, 81–105.
1076. Krout, K. E., Kawano, J., Mettenleiter, T. C., and Loewy, A. D. (2002). CNS inputs to the suprachiasmatic nucleus of the rat. *Neuroscience* 110, 73–92.
1077. Guldner, F. H., and Wolff, J. R. (1996). Complex synaptic arrangements in the rat suprachiasmatic nucleus: a possible basis for the "Zeitgeber" and non-synaptic synchronization of neuronal activity. *Cell. Tissue Res.* 284, 203–214.
1078. Jacomy, H., Burlet, A., and Bosler, O. (1999). Vasoactive intestinal peptide neurons as synaptic targets for vasopressin neurons in the suprachiasmatic nucleus. Double-label immunocytochemical demonstration in the rat. *Neuroscience* 88, 859–870.
1079. Tamada, Y., Tanaka, M., and Munekawa, K., et al. (1998). Neuron-glia interaction in the suprachiasmatic nucleus: a double labeling light and electron microscopic immunocytochemical study in the rat. *Brain Res. Bull.* 45, 281–287.
1080. Leak, R. K., and Moore, R. Y. (2001). Topographic organization of suprachiasmatic nucleus projection neurons. *J. Comp. Neurol.* 433, 312–334.
1081. Dai, J., Swaab, D. F., and Buijs, R. M. (1997). Distribution of vasopressin and vasoactive intestinal polypeptide (VIP)

fibers in the human hypothalamus with special emphasis on suprachiasmatic nucleus efferent projections. *J. Comp. Neurol.* 383, 397–414.

1082. Saeb-Parsy, K., Lombardelli, S., Khan, F. Z., McDowall, K., Au-Yong, I. T., and Dyball, R. E. (2000). Neural connections of hypothalamic neuroendocrine nuclei in the rat. *J. Neuroendocrinol.* 12, 635–648.
1083. Van der Beek, E. M., Horvath, T. L., Wiegant, V. M., Van den Hurk, R., and Buijs, R. M. (1997). Evidence for a direct neuronal pathway from the suprachiasmatic nucleus to the gonadotropin-releasing hormone system: combined tracing and light and electron microscopic immunocytochemical studies. *J. Comp. Neurol.* 384, 569–579.
1084. Teclemariam-Mesbah, R., Kalsbeek, A., Pevet, P., and Buijs, R. M. (1997). Direct vasoactive intestinal polypeptide-containing projection from the suprachiasmatic nucleus to spinal projecting hypothalamic paraventricular neurons. *Brain Res.* 748, 71–76.
1085. Csernus, V., and Mess, B. (2003). Biorhythms and pineal gland. *Neuroendocrinol. Lett.* 24, 404–411.
1086. Millhouse, O. E. (1969). A golgi study of the descending medial forebrain bundle. *Brain Res.* 15, 341–363.
1087. Van Cuc, H., Leranth, C., and Palkovits, M. (1980). Light and electron microscopic studies on the medial forebrain bundle in rat: III. Degenerated nerve elements in the medial hypothalamic nuclei following surgical transections of the medial forebrain bundle. *Brain Res. Bull.* 5, 13–22.
1088. Nieuwenhuys, R., Geeraedts, L. M. G., and Veening, J. G. (1982). The medial forebrain bundle of the rat. I. General introduction. *J. Comp. Neurol.* 206, 49–81.
1089. Veening, J. G., Swanson, L., Cowan, W. M., Nieuwenhuys, R., and Geeraedts, L. M. G. (1982). The medial forebrain bundle of the rat. II. An autoradiographic study of the topography of the major descending and ascending components. *J. Comp. Neurol.* 206, 82–102.
1090. Lindvall, O., and Bjorklund, A. (1974). The organization of the ascending catecholamine neuron systems in the rat brain as revealed by the glyoxylic acid fluorescence method. *Acta Physiol. Scand. Suppl.* 412, 1–48.
1091. Swanson, L., and Hartman, B. K. (1975). The central adrenergic system. An immunofluorescence study of the location of cell bodies and their efferent connections in the rat utilizing dopamine-B-hydroxylase as a marker. *J. Comp. Neurol.* 163, 467–506.
1092. Simon, H., LeMoal, M., and Calas, A. (1979). Efferents and afferents of the ventral tegmental-A10 region studied after local injection of (3H)Leucine and horseradish peroxidase. *Brain Res.* 178, 17–40.
1093. Moore, K. E., and Demarest, K. T. (1982). Tuberoinfundibular and tuberohypophyseal dopaminergic neurons. In *Frontiers in Neuroendocrinology* (W. F. Ganong and L. Martini, Eds.), Vol. 7, pp. 616–690. Raven Press, New York.
1094. Palkovits, M., Zaborszky, L., Feminger, A., Mezey, E., Fekete, M. I. K., Herman, J. P., Kanyicska, B., and Szabo, D. (1980). Noradrenergic innervation of the rat hypothalamus: experimental biochemical and electron microscopic studies. *Brain Res.* 191, 161–171.
1095. Anderson, C. H., and Shen, C. L. (1980). Efferents of the medial preoptic area in the guinea pig: an autoradiographic study. *Brain Res. Bull.* 5, 257–265.
1096. Krieger, M. S., Conrad, L. C., and Pfaff, D. W. (1979). An autoradiographic study of the efferent connections of the ventromedial nucleus of the hypothalamus. *J. Comp. Neurol.* 183, 785–816.
1097. Saper, C. B., Swanson, L., and Cowan, W. M. (1976). The efferent connections of the ventromedial nucleus of the hypothalamus of the rat. *J. Comp. Neurol.* 169, 409–442.
1098. Saper, C. B., Swanson, L., and Cowan, W. M. (1979). Some efferent connections of the rostral hypothalamus in the squirrel monkey (Saimiri sciureus) and cat. *J. Comp. Neurol.* 184, 205–242.
1099. Palkovits, M., Fekete, M. I. K., Makara, G. B., and Herman, J. P. (1977). Total and partial hypothalamic deafferentations for topographical identification of catecholaminergic innervations of certain preoptic and hypothalamic nuclei. *Brain Res.* 127, 127–136.
1100. Sawchenko, P. E., and Swanson, L. (1981). Central noradrenergic pathways for the integration of hypothalamic neuroendocrine and autonomic responses. *Science* 214, 685–687.
1101. Swanson, L., and Cowan, W. M. (1975). Hippocampo-hypothalamic connections: origin in subicular cortex, not Ammon's horn. *Science* 189, 303–304.
1102. Simerly, R. B., and Swanson, L. (1986). The organization of neural inputs to the medial preoptic nucleus of the rat. *J. Comp. Neurol.* 246, 312–342.
1103. Raisman, G., Cowan, W. M., and Powell, T. P. S. (1966). An experimental analysis of the efferent projection of the hippocampus. *Brain (Oxf).* 89, 83–108.
1104. Raisman, G. (1970). An evaluation of the basic pattern of connections between the limbic system and the hypothalamus. *Am. J. Anat.* 129, 197–202.
1105. Canteras, N. S., Simerly, R. B., and Swanson, L. W. (1994). Organization of projections from the ventromedial nucleus of the hypothalamus: a Phaseolus vulgaris-leucoagglutinin study in the rat. *J. Comp. Neurol.* 348, 41–79.
1106. Goubillon, M. L., Caraty, A., and Herbison, A. E. (2002). Evidence in favour of a direct input from the ventromedial nucleus to gonadotropin-releasing hormone neurones in the ewe: an anterograde tracing study. *J. Neuroendocrinol.* 14, 95–100.
1107. Sawchenko, P. E., and Swanson, L. (1983). The organization of forebrain afferents to the paraventricular and supraoptic nuclei of the rat. *J. Comp. Neurol.* 218, 121–144.
1108. Tribollet, E., Armstrong, W. E., Dubois-Dauphin, M., and Dreifuss, J. J. (1985). Extra-hypothalamic afferent inputs to the supraoptic nucleus area of the rat as determined by retrograde and anterograde tracing techniques. *Neuroscience* 15, 135–148.
1109. Tanaka, J., Saito, H., and Seto, K. (1988). Involvement of the septum in the regulation of paraventricular vasopressin neurons by the subfornical organ in the rat. *Neurosci. Lett.* 92, 187–191.
1110. Lind, R. W., Hoesen, G. W. V., and Johnson, A. K. (1982). An HRP Study of the Connections of the Subfornical Organ of the Rat. *J. Comp. Neurol.* 210, 265–277.
1111. Miselis, R. R. (1982). The subfornical organ's neural connections and their role in water balance. *Peptides* 3, 501–502.
1112. Sawchenko, P. E., and Swanson, L. (1983). The organization and biochemical specificity of afferent projections to the paraventricular and supraoptic nuclei. *Prog. Brain Res.* 60, 19–29.
1113. Jhamandas, J., Lind, R. W., and Renaud, L. P. (1989). Angiotensin II may mediate excitatory neurotransmission from the subfornical organ to the hypothalamic supraoptic nucleus: an anatomical and electrophysiological study in the rat. *Brain Res.* 487, 52–61.
1114. Pfister, J., Spengler, C., Grouzmann, E., Raizada, M. K., Felix, D., and Imboden, H. (1997). Intracellular staining of angiotensin receptors in the PVN and SON of the rat. *Brain Res.* 754, 307–310.
1115. Smithson, K. G., Weiss, M. L., and Hatton, G. I. (1989). Supraoptic nucleus afferents from the main olfactory bulb-I. Anatomical evidence from anterograde and retrograde tracers in rat. *Neuroscience* 31, 277–287.

1116. Hatton, G. I., and Yang, Q. Z. (1989). Supraoptic nucleus afferents from the manin olefactory bulb–II. Intracellularly recorded responses to lateral olefactory tract stimulation in rat brain slices. *Neuroscience* 31, 277–287.

1117. Hatton, G. I., and Yang, Q. Z. (1990). Activation of excitatory amino acid inputs to supraoptic neurons. I. Induced increases in dye-coupling in lactating, but not virgin or male rats. *Brain Res.* 513, 264–269.

1118. Modney, B. K., Yang, Q. Z., and Hatton, G. I. (1990). Activation of excitatory amino acid inputs to supraoptic neurons. II. Increased dye-coupling in maternally behaving virgin rats. *Brain Res.* 513, 270–273.

1119. Armstrong, W. E., Scholer, J., and McNeill, T. H. (1982). Immunocytochemical, golgi and electron microscopic characterization of putative dendrites in the ventral glial lamina of the rat supraoptic nucleus. *Neuroscience* 7, 679–694.

1120. Zaborszky, L., Leranth, C., Makara, G. B., and Palkovits, M. (1975). Quantitative studies on the supraoptic nucleus in the rat. II. Afferent fiber connections. *Exp. Brain Res.* 42, 260–268.

1121. Oldfield, B. J., Hou-Yu, A., and Silverman, A.-J. (1985). A combined electron microscopic HRP and immunocytochemical study of the limbic projections to rat hypothalamic nuclei containing vasopressin and oxytocin neurons. *J. Comp. Neurol.* 231, 221–231.

1122. Poulain, D. A., Lebrun, C. J., and Vincent, J. D. (1981). Electrophysiological evidence for connections between septal neurones and the supraoptic nucleus of the hypothalamus of the rat. *Exp. Brain Res.* 42, 260–268.

1123. Versteeg, D. H. G., Van der Gugten, J., De Jong, W., and Palkovits, M. (1976). Regional concentrations of noradrenaline and dopamine in rat brain. *Brain Res.* 113, 563–564.

1124. Cheung, Y., and Sladek, J. R. (1975). Catecholamine distribution in feline hypothalamus. *J. Comp. Neurol.* 164, 339–360.

1125. Hoffman, G. E., Felten, D. L., and Sladek, J. R. (1976). Monoamine distribution in primate brain. III. Catecholamine-containing varicosities in the hypothalamus of macaca mulatta. *Am. J. Anat.* 147, 501–514.

1126. Blessing, W. W., Jaeger, C. B., Ruggeiero, D. A., and Reis, D. J. (1982). Hypothalamic projections of medullary catecholamine neurons in the rabbit: a combined catecholamine fluorescence and HRP transport study. *Brain Res. Bull.* 9, 279–286.

1127. Ciriello, J., and Caverson, M. M. (1984). Direct pathway from neurons in the ventrolateral medulla relaying cardiovascular afferent information to the supraoptic nucleus in the cat. *Brain Res.* 292, 221–228.

1128. Day, T. A., and Renaud, L. P. (1984). Electrophysiological evidence that noradrenergic afferents selectively facilitate the activity of supraoptic vasopressin neurons. *Brain Res.* 303, 233–240.

1129. Ricardo, J. A., and Koh, E. T. (1978). Anatomical evidence of direct projections from the nucleus of the solitary tract to the hypothalamus, amygdala, and other forebrain structures in the rat. *Brain Res.* 153, 1–26.

1130. Day, T. A., and Sibbald, J. R. (1989). A1 cell group mediates solitary nucleus excitation of supraoptic vasopressin cells. *Am. J. Physiol.* 257, R1020–1026.

1131. Cunningham, E. T., and Sawchenko, P. E. (1988). Anatomical specificity of noradrenergic inputs to the paraventricular and supraoptic nuclei of the rat hypothalamus. *J. Comp. Neurol.* 193, 1023–1033.

1132. McNeill, T. H., and Sladek, J. R. (1980). Simultaneous monoamine histofluorescence and neuropeptide immunocytochemistry: II. Correlative distribution of catecholamine varicosities and magnocellular neurosecretory neurons in the rat supraoptic and paraventricular nuclei. *J. Comp. Neurol.* 193, 1023–1033.

1133. Sladek, J. R., and Zimmerman, E. A. (1982). Simultaneous monoamine histofluorescence and neuropeptide immunocytochemistry: VI. Catecholamine innervation of vasopressin and oxytocin neurons in the rhesus monkey hypothalamus. *Brain Res. Bull.* 9, 431–440.

1134. Buijis, R. M., Geffard, M., Pool, C. W., and Hoorneman, E. M. D. (1984). The dopaminergic innervation of the supraoptic and paraventricular nucleus: a light and electron microscopical study. *Brain Res.* 323, 65–72.

1135. Moore, R. Y., Halaris, A. E., and Jones, B. E. (1978). Serotonin neurons of the midbrain raphe: ascending projections. *J. Comp. Neurol.* 180, 417–438.

1136. Bittencourt, J. C., Benoit, R., and Sawchenko, P. E. (1991). Distribution and origins of substance P-immunoreactive projections to the paraventricular and supraoptic nuclei: partial overlap with ascending catecholaminergic projections. *J. Chem. Neuroanat.* 4, 63–78.

1137. Tribollet, E., and Dreifuss, J. J. (1981). Localization of neurones projecting to the hypothalamic paraventricular nucleus area of the rat: a horseradish peroxidase study. *Neuroscience* 6, 1315–1328.

1138. Silverman, A.-J., Hoffman, D. L., and Zimmerman, E. A. (1981). The descending afferent connections of the paraventricular nucleus of the hypothalamus (PVN). *Brain Res. Bull.* 6, 47–61.

1139. Kawano, H., and Masuko, S. (1999). Synaptic contacts between nerve terminals originating from the ventrolateral medullary catecholaminergic area and median preoptic neurons projecting to the paraventricular hypothalamic nucleus. *Brain Res.* 817, 110–116.

1140. Ferguson, A. V., Day, T. A., and Renaud, L. P. (1984). Subfornical organ efferents influence the excitability of neurohypophyseal and tuberoinfundibular nucleus neurons in the rat. *Neuroendocrinology* 39, 423–428.

1141. Ferguson, A. V. (1988). Systemic angiotensin acts at the subfornical organ to control the activity of paraventricular nucleus neurons with identified projection to the median eminence. *Neuroendocrinology* 47, 489–497.

1142. Kaelber, W. W., and Leeson, C. R. (1967). A degeneration and electron microscopic study of the nucleus hypothalamicus ventromedialis of the cat. *J. Anat.* 101, 209–221.

1143. Heimer, L., and Nauta, W. J. H. (1969). The hypothalamic distribution of the stria terminalis in the rat. *Brain Res.* 13, 284–297.

1144. McBride, R. L., and Sutin, J. (1977). Amygdaloid and pontine projections to the ventromedial nucleus of the hypothalamus. *J. Comp. Neurol.* 174, 377–396.

1145. Krettek, J. E., and Price, J. L. (1978). Amygdaloid projections to subcortical structures within the basal forebrain and brainstem in the rat and cat. *J. Comp. Neurol.* 178, 225–244.

1146. Herman, J. P., Schafer, M. K. H., Young, E. A., Thompson, R., Douglass, J., Akil, H., and Watson, S. J. (1989). Evidence for hippocampal regulation of neuroendocrine neurons of the hypothalamo-pituitary-adrenocortical axis. *J. Neurosci.* 9, 3072–3082.

1147. Saphier, D., and Feldman, S. (1989). Catecholaminergic projections to tuberinfundibular neurons of the paraventricular nucleus: II. Effects of stimulation of the ventral noradrenergic ascending bundle: evidence for cotransmission. *Brain Res. Bull.* 23, 397–404.

1148. Sapolsky, R. M., Armanini, M. P., Sutton, S. W., and Plotsky, P. M. (1989). Elevation of hypophysial portal concentrations of adrenocorticotropin secretagogues after fornix transection. *Endocrinology* 125, 2881–2887.

1149. Silverman, A.-J., Oldfield, B. J., Hou-Yu, A., and Zimmerman, E. A. (1985). The noradrenergic innervation of vasopressin neurons in the paraventricular nucleus of the hypothalamus: an ultrastructural study using

radioautography and immunocytochemistry. *Brain Res.* 325, 215–229.
1150. Sawchenko, P. E., Swanson, L., Grzanna, R., Howe, P. R. C., Bloom, S. R., and Olak, J. M. (1985). Colocalization of neuropeptide Y immunoreactivity in brainstem catecholaminergic neurons that project to the paraventricular nucleus. *J. Comp. Neurol.* 241, 138–153.
1151. Day, T. A., Blessing, W. W., and Willoughby, J. O. (1980). Noradrenergic and dopaminergic projections to the medial preoptic area of the rat. A combined horseradish peroxidase/catecholamine fluorescence study. *Brain Res.* 193, 543–548.
1152. Kannan, H., Yamashita, H., and Osaka, T. (1984). Paraventricular neurosecretory neurons: synaptic inputs from the ventrolateral medulla in rats. *Neurosci. Lett.* 51, 183–188.
1153. Tanaka, J., Kaba, H., Saito, H., and Seko, K. (1985). Inputs from the A1 noradrenergic region to hypothalamic paraventricular neurons in the rat. *Brain Res.* 335, 368–371.
1154. Cunningham, E. T., Bohn, M. C., and Sawchenko, P. E. (1990). Organization of adrendergic inputs to the paraventricular and supraoptic nuclei of the hypothalamus in the rat. *J. Comp. Neurol.* 292, 651–667.
1155. Norgren, R. (1976). Taste pathways to hypothalamus and amygdala. *J. Comp. Neurol.* 166, 17–30
1156. Takeuchi, Y., and Hopkins, D. A. (1984). Light and electron microscopic demonstration of hypothalamic projections to the parabrachial nuclei in the cat. *Neurosci. Lett.* 46, 53–58.
1157. Saper, C. B., and Loewy, A. D. (1980). Efferent connections of the parabrachial nucleus in the rat. *Brain Res.* 197, 291–317.
1158. McKellar, S., and Loewy, A. D. (1981). Organization of some brain stem afferents to the paraventricular nucleus of the hypothalamus in the rat. *Brain Res.* 217, 351–357.
1159. Liposits, Z., and Paull, W. K. (1989). Association of dopaminergic fibers with corticotropin releasing hormone (CRH)-synthesizing neurons in the paraventricular nucleus of the rat hypothalamus. *Histochemistry* 93, 119–127.
1160. Sutin, J., and Eager, R. P. (1969). Fiber degeneration following lesions in the hypothalamic ventromedial nucleus. *Ann. N Y Acad. Sci.* 157, 610–628.
1161. Millhouse, O. E. (1973). The organization of the ventromedial hypothalamic nucleus. *Brain Res.* 55, 71–87.
1162. Lindvall, O., and Bjorklund, A. (1983). Dopamine- and norepinephrine-containing neuron systems: their anatomy in the rat brain. In *Chemical Neuroanatomy* (P. C. Emson, Ed.), pp. 229–255. Raven Press, New York.
1163. Ajika, K., and Hokfelt, T. (1975). Projections to the median eminence and the arcuate nucleus with special reference to monoamine systems: effects of lesions. *Cell. Tissue Res.* 158, 15–35.
1164. Steinbusch, H. W., and Nieuwenhuys, R. (1983). The raphe nuclei of the rat brain stem: a cytoarchitectonic and immunohistochemical study. In *Chemical Neuroanatomy* (P. C. Emson, Ed.), pp. 131–207. Raven Press, New York.
1165. Simerly, R. B., Swanson, D. J., and Gorski, R. A. (1984). The cells of origin of a sexually dimorphic serotonergic input to the medial preoptic nucleus of the rat. *Brain Res.* 324, 185–193.
1166. Berk, M. L., and Finkelstein, J. A. (1981). Afferent projections to the preoptic area and hypothalamic regions in the rat brain. *Neuroscience* 6, 1601–1624.
1167. Kalra, S. P. (1986). Neural circuitry involved in control of LHRH secretion: a model for the preovulatory LH release. In *Frontiers in Neuroendocrinology* (W. F. Ganong and L. Martini, Eds.), Vol. 9, pp. 31–75. Raven Press, New York.
1168. Kalra, S. P., Allen, L. G., Sahu, A., Kalra, P. S., and Crowley, W. R. (1988). Gonadal steroids and neuropeptide Y-opioid-LHRH axis: interactions and diversities. *J. Steroid Biochem.* 30, 185–193.
1169. Lind, R. W., and Swanson, L. (1984). Evidence for corticotropin releasing factor and leu-enkephalin in the neural projection from the lateral parabrachial nucleus to the median preoptic nucleus: a retrograde transport, immunohistochemical double labeling study in the rat. *Brain Res.* 321, 217–224.
1170. Page, R. B. (1985). Hypothalamic control of anterior pituitary function: surgical implications. In *Neurosurgery* (R. Wilkins and S. Rengachary, Eds.), p. 791. McGraw-Hill, New York.
1171. Page, R. B. (1975). Scanning electron microscopy of ventricular system in normal and hydrocephalic rabbits. *J. Neurosurg.* 42, 646.
1172. Page, R. B., and Leure-duPree, A. E, eds. (1983). Ependymal alterations in hydrocephalus. In *Neurobiology of Cerebrospinal Fluid* (J. Wood, Ed.), pp. 789–820. Plenum Press, New York.
1173. Page, R. B. (1986). The pituitary portal system. In *Current Topics in Neuroendocrinology* (D. W. Pfaff and D. Ganten, Eds.), Vol. 7, pp. 1–47. Springer-Verlag, New York.
1174. Swanson, L., and Sawchenko, P. E. (1980). Paraventricular nucleus: a site for the integration of neuroendocrine and autonomic mechanisms. *Neuroendocrinology* 31, 410–447.
1175. Phifer, R. F., Midgley, A. R., and Spicer, S. S (1973). Immunohistologic and histologic evidence that follicle-stimulating hormone and luteinizing hormone are present in the same cell type in the human pars distalis. *J. Clin. Endocrinol. Metab.* 36, 125–141.

Knobil and Neill's Physiology of Reproduction, Third Edition
edited by Jimmy D. Neill,
Elsevier © 2006

CHAPTER 28

Physiology of the Gonadotropin-Releasing Hormone Neuronal Network

Allan E. Herbison

Introduction, 1415
Multiple GnRH Molecules for Neuroendocrine Control of Reproduction?, 1416
GnRH-II, 1416
Other Molecules?, 1417
GnRH Neuron, 1417
Neuroanatomy, 1417
Electrical Properties, 1421
GnRH Biosynthesis, 1422
Coexpressed Neuropeptides, 1426
GnRH as a Neuropeptide Transmitter, 1428
GnRH Neuronal Network, 1429
Components, 1429
Synchronization and Pulsatility, 1440
Estrous Cycle Plasticity, 1446
Heterogeneity within the GnRH Neuronal Phenotype, 1455
Acknowledgments, 1456
References, 1456

INTRODUCTION

This chapter attempts to present the current state of knowledge regarding the properties and operating features of the gonadotropin-releasing hormone (GnRH) neuronal network as it exists in the adult mammal. The GnRH neuronal network is defined as the GnRH neurons and associated assembly of brain cells responsible for controlling GnRH release into the pituitary portal circulation. The GnRH neuronal network integrates multiple internal homeostatic and external factors to achieve levels of fertility appropriate to the organism. However, this review focuses only on the functioning of the GnRH neuronal network under normal physiological conditions in reproductively mature adults. The fascinating developmental neurobiology of the GnRH neurons has been covered in detail (1,2), and the effects of aging are the subject of an independent chapter. Other chapters in this series address the effects of physiological and pathophysiological states such as lactation, stress, seasonality, and malnutrition on reproduction.

It is now more than 10 years since the publication of the second edition in 1994, and before going forward, it is interesting to touch on the perspectives in GnRH research at that time. Kordon et al. (3) and Silverman et al. (4) were writing when microdialysis, in situ hybridization, and dual-labeling histochemical technologies had been embraced by workers in the field and the enormously influential immortalized GT1 cell lines had just appeared. All shared the sentiment that these new tools would provide a clearer picture of both the GnRH neuron and the roles of the individual elements within the network. It would not have been possible to predict then that by 2004 we would be able make direct electrical recordings from native GnRH neurons in both the mouse (5) and rat (6), determine hundreds of genes expressed in a GnRH neuron simultaneously (7), and undertake genetic manipulations that alter the expression of a single molecule specifically within GnRH neurons in vivo (8). These technical developments have a major impact on our understanding of the GnRH network and, consequently, change substantially the nature

Center for Neuroendocrinology and Department of Physiology, Otago School of Medical Sciences, Dunedin, New Zealand.

of any review of this topic. Whereas much discussion of the properties of the GnRH neuronal network was by necessity inferred from in vivo whole animal studies or in vitro cell lines, we are now in the fortunate situation of being able to consider data obtained directly from the GnRH neuron. Accordingly, this review does not attempt to catalogue all past work within the field. Instead, this review, wherever possible, concentrates on primary or direct information concerning the GnRH neuronal network and considers these data in the light of past literature to provide a physiological perspective.

MULTIPLE GnRH MOLECULES FOR NEUROENDOCRINE CONTROL OF REPRODUCTION?

Multiple forms of the GnRH decapeptide have now been identified in vertebrates (9,10). However, only two or possibly three (11,12) distinct GnRH decapeptides are thought to be expressed in mammalian species. Mammalian GnRH (or GnRH-I) has an unquestionable role in the neural control of fertility and is the focus of this chapter. The only mammal presently known to not use GnRH-I is the guinea pig, which appears to have its own unique hypophysiotropic GnRH decapeptide (13,14).

GnRH-II

The other principal GnRH decapeptide found in mammals, differing in 3 amino acids to GnRH-I (pGlu-His-Trp-Ser-**His**-Gly-**Trp**-**Tyr**-Pro-Gly), was initially isolated from the chicken brain and termed chicken GnRH (15). Now renamed GnRH-II, this decapeptide represents the most ancient and conserved form of GnRH (9,10) and is expressed in multiple body organs in mammalian species (16–20). As a consequence of an early duplication event, the genomic and mRNA structures of GnRH-II parallel those of GnRH-I, although significant disparity exists in the GnRH-associated peptide (GAP) sequences of the two genes (21). The function of GAP is unknown in any species (see section on Coexpressed Neuropeptides). Alongside the cloning of the GnRH-II gene has come the recent discovery of a second seven transmembrane G-protein–coupled GnRH receptor in humans and monkeys with high selectivity for GnRH-II (22,23). Although present in primates, the GnRH-II receptor has not yet been identified in rodents and may not be present in the mouse genome (24).

The potential role of GnRH-II in the neural control of fertility is presently under examination. Initial efforts to determine the distribution of GnRH-II within the brain of lower vertebrates delineated only a small population of expressing cell bodies within the midbrain (19). However, analyses using different GnRH-II antibodies and mRNA in situ hybridization have revealed a more widespread expression of GnRH-II-expressing cell bodies within the midbrain, hippocampus, thalamus, and hypothalamus, with GnRH-II–immunoreactive nerve fibers even more widely distributed (16–18,20,25). Of special relevance to reproductive control has been the observation of GnRH-II–expressing cell bodies located in the mediobasal hypothalamus (MBH) and magnocellular nuclei and GnRH-II immunoreactive fibers in the median eminence of the monkey (25). Immunoreactivity for GnRH-II is also reported in the median eminence of the mouse (17) and shrew (18). Because GnRH-II is not expressed by GnRH-I neurons (16,26), these observations suggest that GnRH-II may be released into the portal circulation or the posterior pituitary by a separate population of hypophysiotropic neurons.

The proposal that a GnRH-II–expressing neuronal population is involved in reproductive control gains some support from reports that exogenous GnRH-II can stimulate the secretion of luteinizing hormone (LH) and follicle-stimulating hormone (FSH) in monkeys, sheep, and rodents (20,22,27,28). Whether GnRH-II is as potent as GnRH-I in stimulating gonadotropin secretion remains unclear; GnRH-II is reported to be as effective as GnRH-I in female monkeys (27) but significantly less potent in male monkeys, rats, sheep, and shrews (18,22,28). In keeping with the near ubiquitous expression of the GnRH-II receptor throughout the body, this receptor is synthesized by gonadotrophs (22,29). However, recent studies in the monkey and sheep have clearly shown that bolus injections of GnRH-II (and GnRH-I) act via the GnRH-I receptor and not the GnRH-II receptor to stimulate gonadotropin secretion (27,28,30).

Although it is early days in the GnRH-II field, it is plausible that GnRH-II released by hypophysiotropic neurons may reach the gonadotrophs either directly through the portal system or after diffusion from the posterior pituitary and activate the GnRH-I receptor to stimulate LH and FSH release. However, many aspects of this scenario, not least the presence of GnRH-II in portal blood, remain to be established (31). Furthermore, the case against GnRH-II being a critical physiological regulator of LH secretion also seems persuasive. On one hand, there may be no room for another significant releasing hormone driving pituitary LH secretion. There is a near perfect correlation between pulsatile GnRH-I and LH secretion over the estrous cycle (32), and one does not need to invoke a new releasing factor to explain LH release under normal conditions. Further, if GnRH-II only regulates gonadotropin release through the GnRH-I receptor, it is hard to envisage how GnRH-II, with only ~20% efficacy at this receptor compared with GnRH-I (33), could make a significant impact on gonadotroph function. Finally, the inability of GnRH-II to compensate for the selective loss of GnRH-I in the hypogonadal *hpg* mouse (17,34) indicates, at best, a minor role for GnRH-II in the control of gonadotropin secretion in this species.

One interesting further possibility, fueled by evidence for a possible separate FSH-releasing factor (11,35), is that GnRH-II may preferentially stimulate the release of FSH from gonadotrophs (36). Approximately one-third of FSH pulses are not preceded by a GnRH-I pulse, and recent work has shown that these pulses do not involve GnRH-I or the GnRH-I receptor (36). However, present data on whether GnRH-II can preferentially stimulate FSH secretion are equivocal. A study undertaken in the ram has shown that GnRH-II releases more FSH than LH compared with GnRH-I (22), whereas others have found no FSH selectivity of GnRH-II in the male and female monkey (27,28) or rat (11). Thus, at present, the balance of data indicates that GnRH-II is not likely to be a selective FSH-releasing factor and that, likewise, it may not have a critical direct role in the control of gonadotropin secretion. Nevertheless, it will be intriguing to examine the phenotype of a GnRH-II knockout animal, and, no doubt, further investigations will establish if it has any physiological role in the neuroendocrine regulation of LH and FSH.

The expression of GnRH-II and its receptor in most if not all organs of the body is a strong clue that its name is somewhat misleading, and as with most respectable signaling peptides, we might expect GnRH-II to subserve many different functions completely unrelated to gonadotropin-releasing activity. Unlike GnRH-I, it appears as though neurons using GnRH-II are very widespread within the brain, and GnRH-II is therefore likely to be involved in a large number of different neuronal networks. Electrophysiological studies have demonstrated it to be a potent modulator of potassium channel activity in sympathetic ganglion neurons (37,38), and exogenous GnRH-II administration has already been reported to modulate food intake and reproductive behavior in the musk shrew (39,40).

Other Molecules?

The only other GnRH decapeptide for which there is any suggestion for a role in gonadotropin regulation in the mammal is lamprey GnRH-III. This molecule was identified in human hypothalami in 1988 (41) and was shown to have a distribution in the brain very similar to that of GnRH-I in both humans (41) and rats (42). More recent work has shown that most cells expressing lamprey GnRH-III are in fact the GnRH-I neurons (43), and it is possible that this variant is secreted from GnRH-I nerve terminals in the median eminence. Again, much of the focus on the physiological role of this decapeptide has been related to its potential selective FSH-releasing ability. Although McCann and colleagues (11) reported a selective FSH-releasing ability of lamprey GnRH-III in the rat, this has not been independently confirmed (44), and there is also no FSH-selective action of this decapeptide in the sheep (36). Furthermore, because lamprey GnRH-III is only a very weak agonist at the GnRH-I receptor (44), it is again questionable whether it might have any significant role in the control of gonadotropin release. Indeed, its very nature remains to be established because, although originally identified in human tissue (41), lamprey GnRH-III does not appear as a unique gene in the human genome (29).

Because the rest of the review focuses on GnRH-I neurons alone, the molecule is referred to "GnRH" for simplicity.

GnRH NEURON

Neuroanatomy

Distribution of GnRH Neuron Cell Bodies

As a consequence of the unique developmental origins of GnRH neurons and their subsequent migration into brain (1), the GnRH cell bodies come to reside as a scattered "continuum" along their migratory pathway within the mammalian forebrain (Fig. 1). The GnRH cell bodies can be found to reside anywhere along an axis extending from the olfactory bulbs to the medial septal nuclei, diagonal band of Broca, and medial preoptic area through to the MBH. This distribution is sometimes referred to as the "inverted Y" pattern, whereby the bottom of the "Y" is the rostral-most midline GnRH soma in the medial septum (Fig. 1A) with the two top poles of the "Y" representing the more caudally placed GnRH cells lying on either side of the third ventricle (Fig. 1B). Although this continuum appears to exist in all mammalian species, there are species differences in the locations along this pathway at which most GnRH neurons come to reside (4,45–47). In species such as the sheep, rat, and mouse, most GnRH soma are located at the rostral end of the continuum and are most numerous at the junction of the inverted "Y" existing nearby the organum vasculosum of the lamina terminalis (OVLT) (Fig. 1B) (48–54). Whereas small numbers of GnRH neurons do exist further

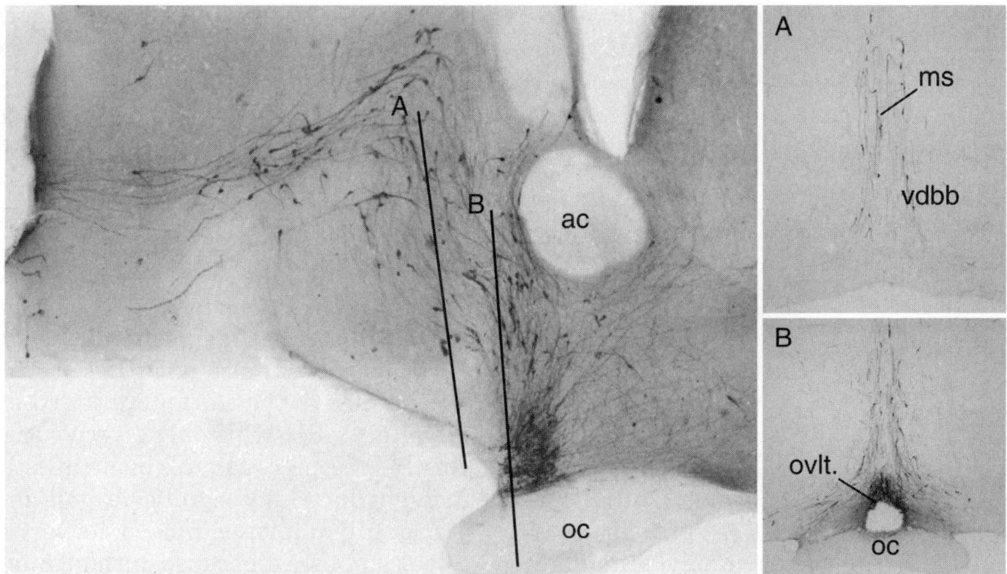

FIG. 1. Distribution of GnRH neurons in the mouse brain. Low-power photomicrograph of a parasagittal section of adult mouse brain immunostained for GnRH showing the scattered distribution of the "GnRH continuum" extending from the olfactory bulbs (*left*) into the hypothalamus (*lower right*). Coronal sections taken at the two levels indicated are shown in **A** and **B**. Note the "inverted Y" distribution with the midline GnRH neurons in the medial septum (ms) and vertical limb of the diagonal band of Broca (vdbb) forming the rostral pole (**A**) and the more caudal split in the GnRH continuum to both sides of the third ventricle around the organum vasculosum of the lamina terminals (ovlt) (**B**). Note the dense fiber innervation of the ovlt. ac, anterior commissure; oc, optic chiasm.

caudal in the sheep, it is debatable whether any GnRH soma exists in the arcuate nucleus (ARN) of the rat and mouse. In contrast, a more even distribution of GnRH soma throughout this rostrocaudal continuum is evident in guinea pigs, monkeys, and humans (47,55–58).

Although the olfactory to mediobasal hypothalamic continuum accounts for nearly all GnRH neurons in mammals, individual cell bodies can occasionally be found outside this distribution in regions such as the bed nucleus of the stria terminalis, subfornical organ, and even the corpus callosum and cingulate cortex (48,59). The locations of these "aberrant" cells may result from disrupted migratory mechanisms on an individual basis. Disordered netrin deleted in colorectal cancer signaling at the time of GnRH neuron entry into the forebrain results in their failure to turn ventrally, resulting in their final location within the cortex (60).

It is also worth noting that multiple extrahypothalamic cell populations express GnRH-1 during development in mice and primates (61,62). The expression of GnRH mRNA and GnRH peptide within these other cell types declines soon after birth in the mouse (61). However, this does not appear to be the case in the primate where widespread GnRH-1 mRNA-expressing cells remain evident in the adult human (63). Whether these cells use the GnRH-1 decapeptide for signaling is doubtful, however, because studies have suggested that these extrahypothalamic neurons express only fragments of the decapeptide such as $GnRH_{1-5}$ (64). The function of these GnRH-1 fragments is unclear.

For an individual GnRH neuron, the functional consequence of having its cell body located in one region of the GnRH continuum as opposed to another has been difficult to resolve experimentally. Lesion studies in the monkey (65) and guinea pig (55) have indicated that GnRH neurons located in the MBH may be of particular relevance to gonadotropin secretion in those species. This subpopulation of GnRH neurons may also have a different pattern of innervation compared with GnRH neurons in the preoptic area in monkeys (66). In contrast, the cluster of GnRH neurons around the OVLT in the preoptic area of rats has been demonstrated to be of particular importance in the generation of the GnRH/LH surge (67). This problem of understanding the impact of GnRH cell body location upon function impinges directly on the important unresolved issue of heterogeneity within the GnRH neuronal population.

Projections of GnRH Neurons

The principal projection site of the GnRH neurons is the external zone of the median eminence.

Retrograde tracing studies have revealed that between 50% and 70% of all GnRH neurons project to the median eminence (58,68,69). In the rat and sheep, the soma of the GnRH neurons innervating the median eminence are scattered throughout the GnRH continuum (68–70). A similar situation exists in the monkey, although the percentage of all GnRH neurons projecting to the median eminence is highest within the MBH, and none of the rostral-most GnRH neurons in the diagonal band of Broca is hypophysiotropic (58). The axons of GnRH neurons targeting the median eminence take either a midline subchiasmatic and periventricular route along the floor and wall of the third ventricle or pass out laterally to run near the median forebrain bundle before turning ventrally and medially to reach the median eminence (45,49,53,54,56,71,72). The GnRH axons taking a midline periventricular route often run within specialized canaliculi formed by the ependymal cells lining the third ventricle (73).

The next most substantial projection of GnRH neurons is to the OVLT (Fig. 1), a brain region outside the blood–brain barrier. Although individual fibers in the OVLT have been traced to nearby GnRH soma (49), the locations of GnRH neurons innervating the OVLT have not been determined. Similarly, the function of GnRH fibers in the OVLT is unknown. One role may be to enable the release of GnRH directly into the cerebrospinal fluid, possibly to facilitate estrous behavior (74,75). Strikingly, GnRH fibers are also found in the subfornical organ, subcommissural organ, and area postrema that, alongside the OVLT, represent all the circumventricular organs with significant non-blood–brain barrier elements. The large population of GnRH neurons shown to take up retrograde tracers, such as Fluorogold and horseradish peroxidase after intravenous or intraperitoneal administration, reflects this unusual projection pattern (59,76). These molecules are taken up through micropinocytosis by axon terminals lying outside the blood–brain barrier and are retrogradely transported back to the cell bodies. Studies in rats have shown that over 90% of GnRH neurons have a projection outside the blood–brain barrier (76). Only around 65% of GnRH neurons in the mouse were shown to be labeled following intravenous horseradish peroxidase (59), but nearly all GnRH neurons are labeled in this species after intraperitoneal administration of Fluorogold (Pape and Herbison, unpublished data). Thus, it would appear that most GnRH neurons project to brain regions outside the blood–brain barrier. The purpose of this is not known. Whereas plasma levels of GnRH are reported to be minimal (77), cerebrospinal fluid levels of GnRH are high (74) and may result from GnRH released at circumventricular sites.

Low densities of GnRH fibers are distributed in a heterogeneous manner in the brain. In the rat, brain regions reported to contain GnRH fibers include the medial and cortical amygdaloid complex, stria terminalis, habenula, hippocampus, interpeduncular nucleus, neocortex, and periaqueductal gray (48,53). These areas are similar to the brain regions reported to express GnRH-I receptor in the rat (78). Whether the intracerebral projections of GnRH neurons arise from specific subpopulations of GnRH neurons or represent collateral innervation from hypophysiotropic GnRH neurons is an important unanswered question. The observation that approximately 30% of GnRH neurons do not target the median eminence (68) clearly indicates that a subpopulation of these neurons is involved in networks unrelated to the direct control of gonadotropin secretion. However, the identification of nonhypophysiotropic GnRH neurons is problematic, because they do not appear to exist in any specific region of the continuum. Retrograde labeling from the amygdala and interpeduncular nuclei in the rat has shown that the GnRH cell bodies projecting to these two areas are spread throughout the GnRH continuum (79). Further, approximately half of the GnRH neurons projecting to the interpeduncular nucleus are also known to target non-blood–brain barrier sites (80). Although this suggests that individual GnRH neurons can release the decapeptide into the vascular space as well as intracerebrally, it does not resolve the issue of whether the GnRH neurons targeting the median eminence also release the decapeptide at other sites within the brain. Clearly, there is much that we do not yet know about GnRH neuron innervation patterns. Perhaps the only thing that can be stated with certainty is that subpopulations of GnRH neurons with different projection patterns must exist alongside the hypophysiotropic GnRH neurons.

GnRH Neuron Cytology

In most species, the cell body of the GnRH neuron exhibits a predominantly bipolar morphology (Fig. 2A,B) with an oblique orientation that may reflect the direction of its embryonic migration through the brain (Fig. 1). In some species, such as the sheep, the orientation and scattered distribution is maintained, but GnRH neurons often display complex multipolar morphologies (Fig. 2C) (52). At the level of the light microscope, the bipolar GnRH neurons in the rat, mouse, and monkey have been classified as having either a "smooth" or "spiny" profile (Fig. 2A,B) (48,49, 81–83). Ultrastructurally, the spiny GnRH neurons are characterized by multiple spine-like cytoplasmic protrusions (84,85). Despite an initial suggestion that

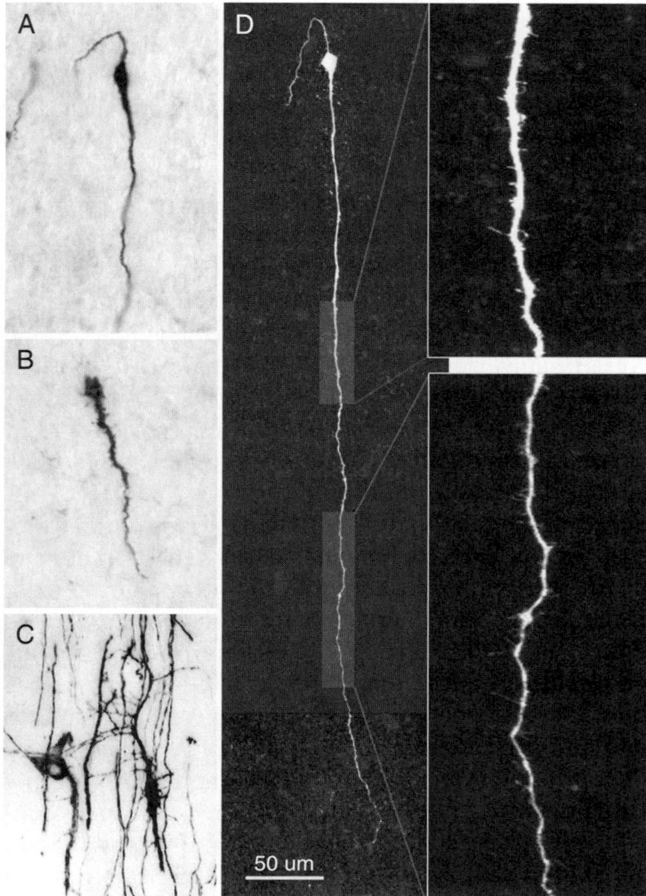

FIG. 2. GnRH neuron morphology in the mouse and sheep. The typical bipolar nature of GnRH neurons is shown by the cells in **A** and **B,** with B exhibiting a clear "spiny" morphology after GnRH immunocytochemistry. The two cells in **C** exhibit the more complex and variable morphology of GnRH cell bodies in the sheep. **D:** A montage showing the full cell body and dendritic structure of an adult mouse GnRH neuron filled with biocytin in situ. Note the length of the primary dendrite (over 500 μm) and the high density of spines throughout the length of the dendrite (two *highlighted insets*).

spiny cells might receive more synaptic input than smooth GnRH neurons (84), a subsequent quantitative study reported no differences in the numbers of synapses upon the cell bodies of smooth and spiny GnRH neurons (86). The functional relevance of smooth versus spiny morphologies remains unknown. The two cell types are distributed in an intermingled manner throughout the GnRH neuron continuum, and the only hint as to their respective roles has been through the observation that the ratio of spiny to smooth GnRH neurons increases up until the time of puberty (81) and may be modified by gonadal steroids (83,87).

The GnRH neuron cell bodies appear remarkable in that they have been reported to receive relatively few synaptic inputs. Whereas an early electron microscopic analysis reported no axosomatic synapses on rat GnRH cell bodies (85), other investigators did find these elements but reported that they were far fewer in number compared with unidentified neighboring neurons (66,84,88,89). An ultrastructural serial reconstruction of GnRH neuron cell bodies in the rat and monkey found a total of between 2 and 12 synapses on each cell body (90). However, recent studies in which mouse GnRH neurons have been examined after filling with small molecular weight dyes (Fig. 2D) suggest that previous electron microscopic studies, limited to the analysis of only those elements of the cell containing GnRH peptide, may have underestimated the connectivity of GnRH neurons (91). In addition to discovering that GnRH neurons have extensive dendrites, often over 1,000 μm in length, these studies revealed large numbers of spines throughout the length of the dendrites (Fig. 2D). Thus, although it is likely true that the soma of the GnRH neuron receives relatively few inputs compared with surrounding preoptic area neurons, this may be compensated for by extensive dendritic profiles that each receive many hundreds of synapses.

A second notable feature of GnRH neurons at the ultrastructural level has been that of substantial glial cell contact or "wrapping" (66,84,89,92). In the monkey and sheep in particular, a wide range (0%–90%) of glial ensheathment of individual GnRH neurons may exist (66,89). Importantly, glial cell invaginations can be seen to separate presynaptic terminals from postsynaptic densities on the GnRH neuron (84,89). Such observations suggest an important role for glia in modulating the afferent inputs to GnRH neurons. Ultrastructural evidence for the pubertal (93) and gonadal steroid (66) modulation of GnRH neuron glial cell wrapping in the monkey further suggests a potential physiological role for glial–GnRH neuron interactions.

Although the GnRH neuron cell bodies are typically scattered along the "GnRH continuum," in several instances GnRH soma can be observed in close association with each other (48,49,72,84). Using serial reconstructions, Witkin and colleagues (90) showed that a small number of GnRH neurons in the rat (~5%) formed a cytoplasmic bridge with another GnRH neuron and suggested that these structures may function to coordinate activity. In the mouse, however, the dye filling of GnRH neurons has not, as yet, provided any supportive evidence for gap junctional or other forms of direct communication between GnRH neurons or GnRH neurons and neighboring cells (91,94).

Upon reaching the median eminence, the GnRH fibers form a close association with tanycytes, the specialized ependymal cells of the third ventricle, such that the GnRH neurovascular terminals are often encapsulated by tanycytic end-feet (73,95–97). In addition, the GnRH terminals appear to be unusual compared with the terminals of other hypophysiotropic neurons in that they do not sit directly adjacent to the endothelial basal lamina, ready to release GnRH into the pericapillary space of the portal system (89,98–100). Instead, the tanycyte-encapsulated GnRH terminals are located several microns from the basal lamina (98), and astrocytic processes may exist as a further barrier between the terminal and portal vasculature (101). This arrangement suggests that access of the decapeptide to the portal system upon release from the GnRH terminal is not straightforward and that ultrastructural variations may represent an additional mechanism for regulating portal GnRH profiles (73,95,99).

The other interesting ultrastructural characteristic of the GnRH terminals in the median eminence is that they appear to receive very few, if any, direct neural inputs. Whereas GnRH terminals are often observed adjacent to other axons in the median eminence, evidence for synaptic specializations has only been observed rarely (89,101,102). The one exception has been with dopamine, where evidence for dopaminergic inputs synapsing on GnRH terminals has been found in both the rat and sheep (97,103,104). However, in the face of substantial in vitro evidence for the regulation of GnRH secretion at the level of the median eminence by many different neurotransmitters (see section on Neuronal Elements of the GnRH Network), the absence of synaptic input to GnRH terminals suggests that this regulation occurs in a nonclassical manner. Volume transmission, whereby neurotransmitters are released into the surrounding extracellular fluid to then diffuse short distances to the relevant neural element (105), is one possibility. Molecules capable of influencing GnRH secretion in this way may also arise from nonneuronal sources such as glial and endothelial cells in the vicinity of the GnRH nerve terminals (106). However, obtaining GnRH terminal specificity for any such signal would need to rely on GnRH terminals expressing receptors that other nearby hypophysiotropic nerve terminals do not.

Electrical Properties

Direct electrical recordings of native GnRH neurons from adult mammals have now been achieved in brain slices obtained from guinea pigs (107), mice (5,94,108), and, recently, rats (6). These studies show that with the exception of relatively high input resistances in the order of 1 GΩ, GnRH neurons exhibit conductances and general membrane properties similar to that of most other neurons in the brain (5,94,107,109,110). No clear electrophysiological "fingerprint" of the GnRH neuron has emerged as yet.

To date, channels identified to be operational in native GnRH neurons include voltage-dependent sodium channels critical for action potential generation, a variety of potassium channels, and both high (L,N,P/Q,R-types) and low (T-type) voltage-activated calcium channels. Among the potassium channels, the rapidly inactivating voltage-dependent I_A current is critical alongside the delayed rectifier I_K in determining membrane repolarization and interspike interval in GnRH neurons (109,110). Small conductance calcium-activated potassium (SK) channels, responsible for after-hyperpolarization, have also been identified in GnRH neurons and contribute to the control of cell firing rate (111). The presence of I_H, a cation channel activated by hyperpolarization, is of particular interest because this has been suggested to be an important pacemaker current playing a role in the ability of neurons to fire phasically (107,109,112). Equally, the low-voltage activated T-type calcium channel (6), also activated by hyperpolarization and responsible for after-depolarization potentials, may act to aid phasic firing (113,114). However, at this stage our understanding of the role of any specific current in contributing to the electrical characteristics of the GnRH neuron is rudimentary and awaits further investigation. Discussion of the potential role of GnRH membrane properties in generating pulsatility can be found in the section on Generation of a GnRH Pulse.

One characteristic of GnRH neurons that is readily apparent from the recordings undertaken so far is that a wide variety of different firing patterns are displayed by these cells (5,107,109,115–118). Under various patch-clamp or sharp electrode recording conditions, mean firing rates have been found to extend from 0 up to ~5 Hz. The pattern of activity ranges from "silent" neurons, which do not fire spontaneous action potentials (Fig. 3A), to neurons firing in a regular manner (Fig. 3B) through to those exhibiting phasic patterns of firing (Fig. 3C). During brief periods of phasic activity, the firing frequency of GnRH neurons can reach 10 Hz (119), compatible with significant neuropeptidergic release from neuroendocrine nerve terminals (120). The firing rate of GnRH neurons appears to be reduced after intracellular dialysis with patch-clamp electrode solution, indicating that the intracellular milieu of the GnRH neurons is critical in determining the firing pattern of the cell (116). In interpreting current findings, it needs to be borne in mind that all such electrophysiological data on native GnRH neurons come from studies in the acute brain slice preparation where the axons and dendrites of GnRH neurons and their afferent inputs are almost certainly damaged. Thus, the degree of firing rate variability may reflect different degrees of damage to the GnRH neuronal network in the slice. However, it is interesting to note that recordings from acutely isolated GnRH

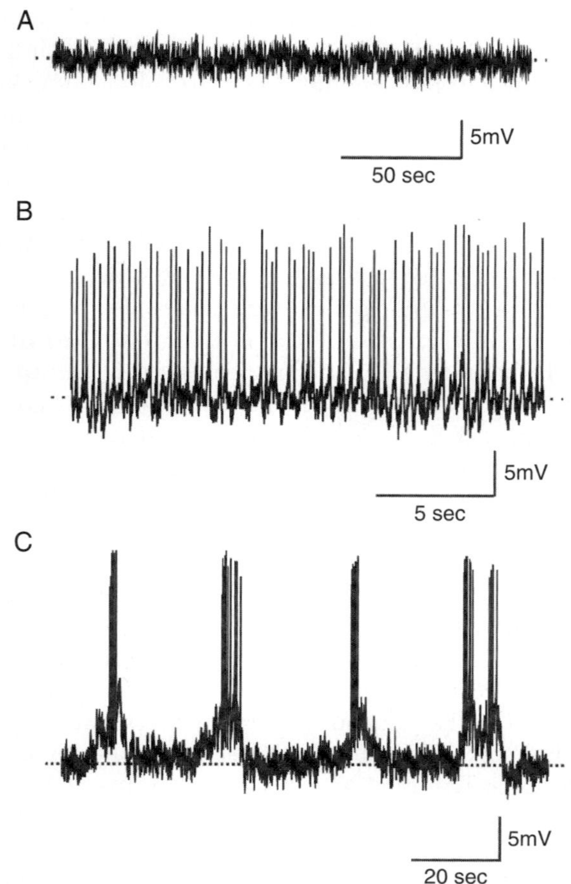

FIG. 3. Heterogeneity in GnRH firing patterns. Gramicidin-perforated-patch recordings of GnRH neurons from adult GnRH-GFP transgenic mice have revealed these cells to exhibit silent (**A**), regular (**B**), and phasic (**C**) patterns of spontaneous activity.

neurons uniformly stripped of their dendrites continue to display marked heterogeneity in firing rates and patterns (119). Hence, it seems likely that heterogeneity is a dominant feature of the GnRH neuronal phenotype and that this is reflected in both their channel expression (109) and firing patterns (Fig. 3) (121). Understanding the functional significance of this electrical heterogeneity is a major challenge for the future.

GnRH Biosynthesis

GnRH Gene and Transcription

The GnRH coding sequence is found on chromosome 8 in the human and its syntenic region on chromosome 14 in the mouse (122). The gene is comprised of four short exons and three long introns spanning approximately 4.5 kb (Fig. 4). The GnRH decapeptide is ultimately derived from exon 2, whereas GAP is transcribed from exons 2, 3, and 4 (Fig. 4) (123,124). One intriguing feature of the GnRH locus is that the DNA strand "antisense" to that from which GnRH is transcribed is responsible for the transcription of multiple so-called SH RNAs in the heart and brain (125,126). The functional relevance of these complementary transcripts has not been determined. As with other genes, transcription of the GnRH gene has been found to begin at multiple sites. The organization of the human gene has been well documented in this respect with transcription start sites identified at +1 for GnRH neurons and upstream at –579 for reproductive tissues such as the placenta (Fig. 1) (127,128).

Much work has been undertaken examining the cell-specific and enhancer elements of 5′ sequence controlling transcription of the GnRH coding region (129–133). Evaluation of the rat gene has received most attention, and critical regions conferring neuron-specific GnRH expression have been identified in vitro to include the first 175 bp of the promoter and a 300-bp enhancer region located ~1.8 kb upstream (–1,571 to –1,863) (Fig. 4) (134,135). Transgenic mice in which reporter constructs composed of just these two 175- and 300-bp elements fused to *lacZ* also target GnRH neurons, suggesting that these elements are sufficient to direct expression in vivo (136). Substantial progress has been made in defining the transcription factors such as POU and GATA family members active at these two sites in the rat promoter, and reviews of this subject are available (129,131,133).

Less is known about the cell-specific and critical enhancer regions of the human GnRH gene. Initial studies using fragments of human GnRH 5′ sequence in murine GT1 cells indicated that the human GnRH neuron-specific elements were located between –535 and –479 bp (137). However, subsequent studies using fragments of human GnRH sequence as transgenes in mice revealed that elements between –795 and –992 bp were critical for hypothalamic expression in vivo (138). This is a similar region to that shown to be important for enhancing transcription at the upstream/placental transcription start site of the human GnRH gene (127). Although caution is required when interpreting results obtained using human sequence in a mouse transcriptional environment, these data suggested that quite different regions of the rat and human GnRH promoter are used to confer GnRH neuron specificity (Fig. 4). This is a little surprising considering the reasonably well-defined role of the first 175 bp of rat promoter sequence, at least in vitro, and its close homology with the same 175-bp regions of the mouse and human promoter (134). However, species-specific differences in the structural organization of the GnRH promoter have now been documented on several occasions (137,139).

The functional organization of the mouse GnRH gene has been examined principally using a transgenic

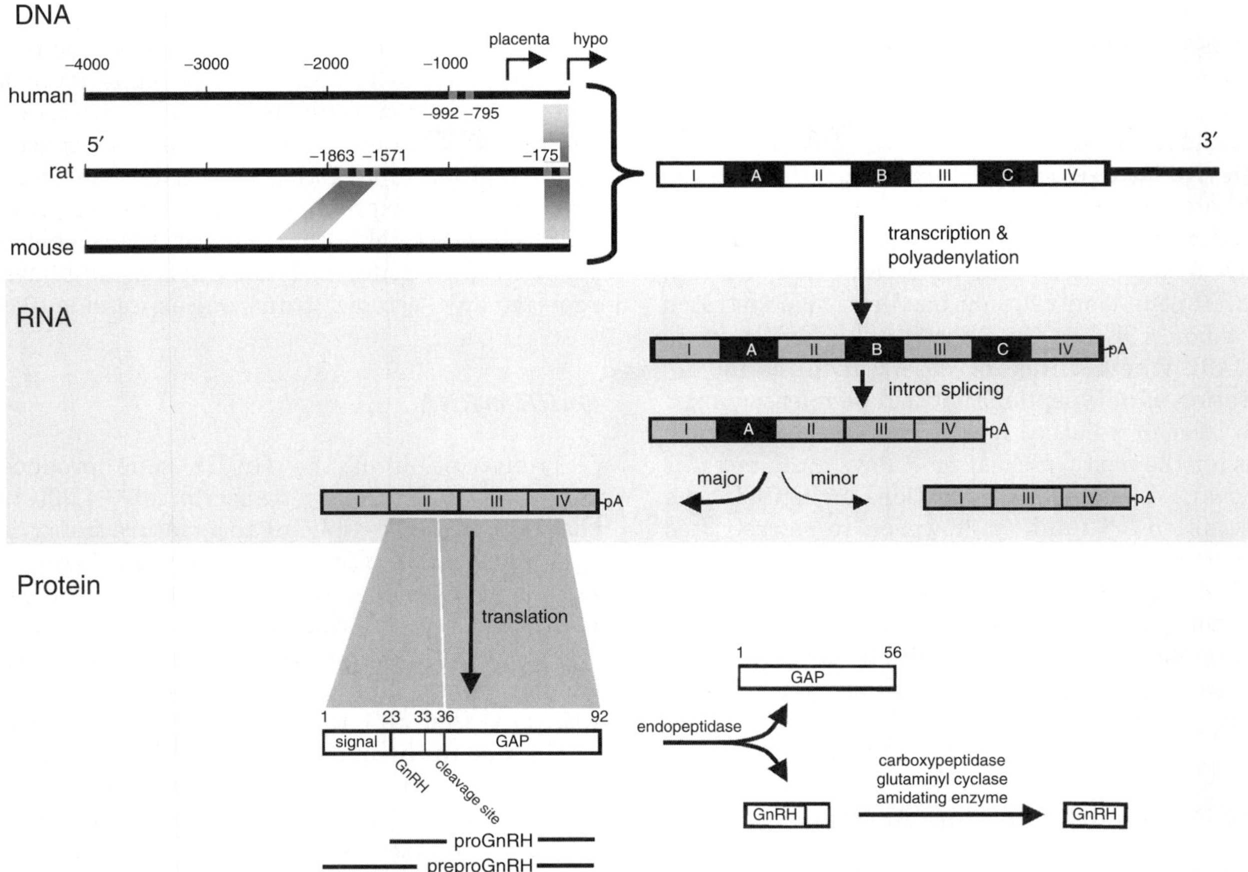

FIG. 4. Biosynthesis of GnRH and GAP. *Top panel* depicts structure of the GnRH gene with the 5′ regulatory sequence of human, rat, and mouse overlayed and specific regions of known function highlighted (see text). *Middle panel* depicts GnRH RNA and various splicing and processing intermediates. *Bottom panel* shows translation and post-translational modifications that result in the final GnRH decapeptide.

approach. Studies using transgenes comprised of mouse GnRH sequence coupled to the *lacZ* reporter have demonstrated that nearly all GnRH neurons are targeted using 5.2 kb of the 5′ flanking sequence but that expression levels decrease with only 2.1 kb of the 5′ sequence and then decline to only low levels with 1.7 kb of the flanking sequence (140,141). A study examining 3.5, 2.1, and 1.0 kb of the 5′ murine GnRH flanking sequence coupled to luciferase similarly showed that the levels of reporter expression in the hypothalamus of transgenic mice decreased with smaller amounts of flanking sequence (142). These studies have not as yet defined any critical cell-specific regions and leave open the question as to whether the conserved first 175 bp of promoter are as critical in the mouse (143) as they are in the rat (Fig. 4). Equally, it would seem that important enhancer elements reside at least –3.5 kb, and possibly up to –5.2 kb, away in the 5′ flank of the murine gene. The defined rat 300-kb enhancer exists as a homologous but nonidentical region between –2.4 and –2.0 kb in the mouse 5′ flank (131), and a role for this region as one of several critical enhancers would be compatible with the transgenic studies undertaken so far.

The role of the 3′ flanking sequence has received little attention in terms of the functional organization or regulation of GnRH gene expression. One study has shown that although the first 3.5 kb of the 3′ flanking sequence does not contribute substantially to the targeting of expression to the GnRH neurons, it does appear to contain sequences that suppress GnRH expression in non-GnRH neurons (140). This is apparent as increased amounts of "ectopic" transgene expression observed in transgenic mice bearing constructs without any 3′ flanking sequence (140).

In terms of transcriptional regulation, the relatively short half-lives of the GnRH primary (~18 minutes) and mature (~10 minutes) transcripts (144,145) suggest that a high basal level of gene transcription exists to maintain the very high GnRH mRNA copy number of GnRH neurons (146). There is also evidence that GnRH gene transcription does not occur at a constant rate but fluctuates on an approximately 2-hour basis in GT1 cells (147). Whereas studies in cell lines have indicated that a number of hormones and peptides can alter GnRH gene transcription in vitro (129,131,133), alterations in GnRH gene transcription in vivo have seldom been examined. Perhaps the best characterized example of regulated GnRH gene transcription in vivo is that associated with estrogen positive feedback in the rat where an ~40% increase in GnRH gene transcription has been demonstrated 2 to 4 hours before the onset of the GnRH surge (148,149). Whether this increase is required for the generation of new peptide that will be released a few hours later or whether it may be used to replenish stores for the next GnRH surge 4 days away remains unknown. Physiological alterations in GnRH gene transcription are unlikely, however, to be restricted solely to positive feedback in the rat, and future technical improvements will doubtless enhance our understanding of GnRH gene transcription in vivo. For example, estrogen negative feedback was recently found to decrease GnRH gene transcription in the male mouse (141).

GnRH Promoter Transgenic Animals

Aside from deriving a comprehensive knowledge of GnRH gene organization, an understanding of critical regulatory elements within the GnRH gene locus is important for the generation of appropriate GnRH promoter transgenic rodent models. The usefulness of these transgenic mice and rats relies on (a) the random insertion of the transgene in the genome finding a "favorable" region where it can be expressed but not influenced by flanking DNA sequence and (b) the GnRH gene construct having sufficient elements to enable its appropriate activation by the transcriptional environment of the GnRH neurons but not other cells. Just how much 5′ and 3′ flanking sequence is required to achieve the latter is not yet established. One additional unforeseen complication has been that multiple populations of cells outside that of the "classical" GnRH neurons have been found to express GnRH (and transgene reporter) during embryonic development (61,64). Whereas expression of the endogenous GnRH gene is suppressed in these other cell populations shortly after birth in the mouse, even large GnRH transgenes continue to be expressed in the various limbic structures in adults (61). It is thought that repressor elements outside the first 5 kb of the 5′ flank and 3.5 kb of the 3′ flank are used by limbic cells to switch off GnRH gene expression postnatally (150). Hence, although not always appreciated initially, it is now apparent that all the murine GnRH promoter transgenic mice produced to date exhibit reporter expression outside the classical GnRH neurons. The same problem of ectopic expression would appear to be true for transgenic mice bearing 5′ rat GnRH promoter constructs as, in addition to targeting the GnRH neurons, low-level transgene expression is widespread in the brain (136). Interestingly, the two rat transgenic lines reported so far using rat 5′ GnRH constructs have not reported any "ectopic" transgene expression (6,8).

GnRH mRNA

Transcription of the GnRH gene produces a polyadenylated primary transcript of ~4,200 bases that is processed by splicing to a mature transcript of 560 bases (123) (Fig. 4). Maurer and Wray (146) estimated that single GnRH neurons contain >1,000 GnRH mRNAs, making them a high copy number transcript with levels comparable with that of housekeeping genes. One unusual feature of the GnRH neuron is that 30% to 40% of all GnRH transcripts are located in the nucleus at any one time (144,151), and it has been suggested that this acts as a large steady-state pool of mRNA available for translocation into the cytoplasm when required (152).

The processing of the primary transcript appears to be rapid, with a half-life in the order of 18 minutes (144). Although initial studies in the mouse suggested no specific order of intron splicing for GnRH (144), other studies in the rat and mouse indicate that intron B and then intron C are rapidly excised from the primary transcript, with intron A being spliced out more slowly (151,153). Indeed, the work of Kim and colleagues (154) has demonstrated that the removal of intron A is a key regulatory event ultimately enabling efficient translation of the mature GnRH mRNA. Whereas introns B and C contain appropriate consensus splice sites, intron A exhibits a suboptimal 3′ splice site that does not allow efficient splicing by the spliceosome complex (153). Intron A can only be spliced out after the formation of a splice enhancer complex comprised of the serine/arginine-rich RNA-binding protein Tra2α and, as yet unresolved, neuron-specific interacting proteins. This complex must interact with exonic splicing enhancers located in exons 3 and 4 to achieve the splicing of intron A (153,155,156). Thus, it is postulated that introns B and C are removed in the normal manner, thereby bringing exons 3 and 4 physically closer to intron A to allow the recruitment of the splice enhancer complex that facilitates splicing of the suboptimal 3′ splice site in intron A (154). The relative difficulty in excision of intron A explains why intron A–containing transcripts are found alongside the mature mRNA in the brain (151) and also provides an explanation for the presence of low levels of a further GnRH RNA splice variant missing exon 2 (Fig. 4) (153,157). In this case, it is likely that splicing occurs between the normal 5′ splice site of intron A and the 3′ site of intron B, thus excising the whole of exon 2 (154).

The functional relevance of intron A splicing was revealed by demonstrating that the translation of intron A–containing transcripts is very poor compared with the mature RNA (158). These observations regarding intron A removal may also provide a nice explanation for the failure of the *hpg* mouse to produce GnRH decapeptide (159). The *hpg* mouse is infertile as a result of a spontaneous ~30-kb deletion in chromosome 14 that involves exons 3 and 4 of the GnRH gene (159). Despite the presence of RNA transcribed from exon 2, it was unclear why *hpg* mice did not produce the GnRH peptide. It now seems likely that the absence of exons 3 and 4 in these mice results in truncated RNAs unable to recruit the splice enhancer complex required to splice out intron A and thus results in the presence of an intron A–containing RNA that cannot be translated efficiently (160). Together, these studies indicate that pre-RNA processing and the excision of intron A, in particular, is critical for the biosynthesis of GnRH decapeptide. Whether this is an actively regulated facet of GnRH biosynthesis is less clear, although a gradual increase in intron A excision rate has been shown to occur with postnatal development in mice (160).

The mechanisms underlying the degradation of cytoplasmic GnRH mRNA are not established, but the half-life of transcript decay in primary cultured neurons has an initial rapid phase of 5 to 13 minutes followed by a much slower phase lasting around 300 hours (145). A primary role of poly(A+) tail shortening in transcript decay is suspected (152), and the 3′ untranslated region of the mature mRNA is known to exhibit two highly conserved motifs that may bind factors regulating degradation (161). As GnRH mRNA content has been shown to change in the presence of unaltered primary transcript during puberty (162) and in response to neurotransmitters like *N*-methyl-D-aspartate (NMDA) (163), the regulation of GnRH mRNA degradation or stability appears to be a major form of biosynthetic regulation in vivo. However, it is worth noting the considerable species and model differences that exist in this regard. Whereas NMDA receptor activation uses posttranscriptional mechanisms to induce a rapid increase in GnRH mRNA levels in vivo in male rats (163,164), NMDA uses posttranscriptional mechanisms to rapidly inhibit expression in male mice (165). In contrast, in GT1 cells, NMDA uses transcriptional mechanisms to either stimulate (166) or inhibit (167) GnRH mRNA content.

GnRH Translation and Peptide Processing

Translation of the mature GnRH transcript results in the production of a pre-prohormone consisting of 92 amino acids (Fig. 4). The first 23 N-terminal amino acids comprise a signal sequence with the remainder representing the prohormone in which GnRH is located at the N-terminal end and separated from GAP by a 3 amino acid cleavage site (Fig. 4). Because protein kinase C activation results in a significant decrease in the numbers of ribosomes associated with GnRH mRNA (168), it is possible that the translation of prepro-GnRH is itself regulated. At least four different enzymatic steps are involved in the processing of the prohormone, and these have been reviewed in detail by Wetsel and Srinivasan (169). The initial step involves cleavage of the prohormone to create a GnRH intermediate and mature GAP (Fig. 4). This is achieved by one or more endopeptidases, and both prohormone convertase 1/3 and 2 are implicated in this process (170–172). The GnRH intermediate then undergoes cleavage of its C-terminal basic residues by carboxypeptidase E (169,170). The importance of carboxypeptidase E in pro-GnRH processing in vivo was demonstrated recently in the infertile Cpe^{fat} mouse where a point mutation rendered carboxypeptidase E inactive and a large increase in unprocessed C-terminal GnRH intermediates was found (173). The final steps in generating the biologically active decapeptide require the conversion of the N-terminal glutamine (Gln^1-GnRH) to pyroglutamate ($pGlu^1$-GnRH) by glutaminyl cyclase and the amidation of the C-terminal glycine by peptidylglycine α-amidating monooxygenase (174). Immunocytochemical studies suggest that the processing of the GnRH prohormone occurs within vesicles during its transport down the axon and while resident within the nerve terminals (175–177). It remains unclear whether the enzymatic processing of GnRH represents an opportunity for regulating GnRH biosynthesis in a physiological context.

One further posttranslational modification of GnRH is that of hydroxylation of the proline residue to create hydroxyproline (Hyp^9) GnRH. Gautron and colleagues (178) showed that as much as 10% of GnRH-like peptide in the hypothalamus exists as (Hyp^9) GnRH in adult rats. Although not as potent as normal GnRH, the hydroxyproline moiety is able to stimulate the release of LH and FSH (179). The physiological significance and roles of (Hyp^9) GnRH are not known. It may be of importance that it represents upward of

40% of all GnRH molecules in the neonatal rat (178) and is reported to be located preferentially in GnRH nerve terminals found outside the hypothalamus (180).

GnRH Peptide Degradation

Multiple enzymes are involved in the degradation of the GnRH decapeptide with the most clearly implicated being zinc metalloendopeptidase EC 3.4.24.15 (EP24.15) and propyl endopeptidase. Whereas EP24.15 cleaves GnRH at the Tyr^5-Gly^6 bond to create the $GnRH^{1\text{-}5}$ and $GnRH^{6\text{-}10}$ fragments (181,182), propyl endopeptidase is responsible for removal of the amidated C-terminal glycine that may be a prerequisite for fast cleavage by EP24.15 (183).

Studies examining the distribution of EP24.15 have shown that it is located in the perivascular space around the GnRH nerve terminals in the median eminence and that it can be secreted into the portal circulation (184). Investigations have further shown that a selective antagonist of EP24.15 reduces GnRH degradation and enhances LH secretion in vivo (181, 184,185). Thus, it seems likely that endopeptidases represent a further factor capable of determining the profile or amounts of GnRH reaching the gonadotrophs. The degradation of GnRH may also be regulated physiologically because a small reduction in propyl endopeptidase levels was observed at the time of the LH surge in the sheep, suggesting that a decrease in peptidase activity may contribute to enhanced GnRH release at that time (186). In contrast, levels of EP24.15 were not found to fluctuate over the estrous cycle of rats (184) or in sheep (186). It is intriguing to note that EP24.15 may have a more extensive role than that of simply degrading GnRH. The $GnRH^{1\text{-}5}$ fragment has been shown to function as an NMDA receptor antagonist, raising the possibility that the products of GnRH enzymatic cleavage may act in a feedback autoinhibitory control mechanism at the level of the GnRH nerve terminals (184,187).

Regulation

It is clear from the description above that in addition to the electrical activity of a GnRH neuron, multiple biosynthetic and catabolic processes are responsible for determining the final levels of GnRH entering the portal circulation (Fig. 4). Whereas the various biosynthetic steps must clearly be operational to enable GnRH secretion from the nerve terminal, the physiological impact of regulating their efficiency at any one point is much less clear. For example, a variety of different stimuli has been shown to alter GnRH mRNA levels by 20% to 30% (188,189), indicating that transcription and mRNA processing can be regulated. However, the critical issue that remains to be resolved is whether these relatively modest changes in mRNA levels have any functional impact upon the long-term functioning of a cell type that maintains "housekeeping gene" levels of GnRH mRNA content.

Because GnRH neurons maintain a very high level of GnRH decapeptide (~1 pg or 10^9 molecules/cell) (146), it might be argued that a high constitutive level of GnRH biosynthesis is sufficient without the need for complex regulation. If biosynthesis were to be regulated, however, it would seem sensible to have biosynthesis coupled to secretion in a neuroendocrine neuron. Whereas evidence for this does exists before the GnRH surge in the rat (149,188,190), it is not always the case, with large increases in GnRH secretion sometimes occurring in the face of decreases in GnRH mRNA expression (165,188,191). Thus, it would appear that a tight relationship does not necessarily exist between secretion and biosynthesis in GnRH neurons, with the level of coupling likely depending on the stimulus presented to the cell. The development of experimental models in which the electrical or secretory activity of GnRH neurons can be monitored alongside their biosynthetic capacity should help considerably in examining this important issue.

Coexpressed Neuropeptides

Electron microscopic investigations of the GnRH neurons have identified both dense core and clear synaptic vesicles within these cells (84,89). Dense core vesicles are typically associated with neuropeptidergic transmission, and many such vesicles in GnRH neurons are labeled by both GnRH and GAP antisera, indicating that these peptides are stored together in dense core vesicles in these cells. The presence of clear vesicles, not labeled by GnRH (84), is intriguing and suggests that other neurotransmitters may be synthesized and released by GnRH neurons.

The first neuropeptides suggested to be coexpressed by GnRH neurons were pro-opiomelanocortin-related peptides (192,193), delta sleep-inducing peptide (DSIP) (194), and galanin (195,196). More recently, it has been suggested that over half of GnRH neurons located around the OVLT express cholecystokinin and that some of these further express neurotensin immunoreactivity (197). Similarly, Foradori and colleagues (198) provided preliminary data suggesting that nearly all GnRH neurons in the sheep express the opioid peptide orphanin FQ. The validation and functional significance of these latter findings await further investigations. The role of galanin in GnRH neuron physiology has received considerable attention, whereas the case for GAP and DSIP remains open.

Galanin

Galanin is likely to be used as a signaling molecule by multiple elements within the GnRH neuronal network, and it is important to distinguish between roles of galanin secreted by GnRH neurons and galanin released by other neuronal elements within the network. However, it is usually impossible to distinguish between these two origins of galanin release in physiological experiments. The neuropeptide galanin was first identified to be coexpressed by GnRH neurons in the rat (195,196). Subsequent studies showed that galanin was synthesized in a highly sexually dimorphic manner, being present in ~70% of GnRH neuron soma and terminals in the adult female but only around 20% in the male (199,200). Whereas a similar pattern of sexual dimorphic expression is seen in the mouse (201), this does not exist in the sheep where apparently all GnRH neurons express galanin regardless of sex (202). In contrast, no galanin was reported in GnRH neurons of the macaque monkey (203). In the rat, GnRH and galanin can be colocalized to the same secretory vesicles in median eminence terminals (200), and galanin is secreted into the portal vasculature in a pulsatile manner similar to that of GnRH (204). Although the GnRH neurons expressing galanin are found throughout the GnRH continuum, there is a slight over-representation of colocalization in the subpopulation that is believed to be intensely activated at the time of the GnRH/LH surge (205).

Thus, it would seem that galanin co-released from GnRH terminals is poised to exert a significant physiological role in the control of gonadotroph activity in several species. This concept has been reinforced further by evidence for the gonadal steroid and activity-dependent regulation of galanin gene expression in GnRH neurons, with peak synthesis rates being achieved around the time of the GnRH/LH surge (206–208). With evidence that galanin facilitates GnRH-stimulated LH secretion directly from gonadotrophs (195,204,209) as well as GnRH release from the median eminence (210,211), it is proposed that co-released galanin aids the generation of the LH surge by acting at both brain and pituitary sites (212). Because some GnRH neurons express galanin receptor mRNA (7,213), it is conceivable that galanin acts directly back on its own cosecreting nerve terminal in the median eminence to facilitate GnRH secretion.

When reviewing galanin biology, Shine (214) remarked on the similarities between galanin and a politician in stating, "It [galanin] has not been found to be primarily responsible for any important activity, but rather to modulate many and thus take credit for a wide range of outcomes." So it may be for the GnRH neuron. Whereas a plausible modulatory role can be suggested, evidence for a critical role of coexpressed galanin in GnRH neurobiology is lacking. The galanin knockout mouse is fertile and exhibits no reproductive phenotype attributable to abnormal GnRH function (215,216). Equally, mice with a functional deletion of the *GalR1* gene are reported to breed normally (217). Against this argument, however, an in vivo study in the rat reported complete abolition of the LH surge after intracerebroventricular administration of a galanin receptor antagonist, galantide (211). Whether galantide is acting to block effects of galanin secreted by GnRH neurons or galanin released by other cells is unknown in in vivo studies of this nature. It will soon be possible to distinguish between these two sources of galanin by engineering mice with a GnRH neuron-selective knockout of galanin. However, given the normal phenotype of the global galanin knockout, investigators may not be rushing to do this work, and the definitive experiment may well require an inducible GnRH neuron-specific galanin knockout mouse. At present, it seems reasonable to suggest that galanin co-released from GnRH neurons plays an important but dispensable role in facilitating GnRH and LH secretion, particularly at the time of the surge. How well this translates to primates and humans remains to be determined.

Delta Sleep-Inducing Peptide

The nonapeptide DSIP was discovered in 1977 for its ability to stimulate slow or delta wave sleep (218). Immunocytochemical studies over the following 5 years showed that DSIP immunoreactivity was restricted mostly to the hypothalamus of multiple species, including humans, and that a high degree of overlap existed between DSIP and GnRH immunostaining (194,219–221). Although varying degrees of coexpression between DSIP and GnRH immunoreactivities were identified, ranging from complete to partial, all studies found large numbers of coexpressing fibers in the median eminence (194,219,222). In one case DSIP and GnRH immunoreactivities were found to exist in the same dense core vesicles (220).

Studies in which DSIP was injected intracerebroventricularly revealed a stimulatory effect upon LH secretion that was not mediated by the pituitary (223,224). The effect was not selective for LH, however, because the same treatment also evoked growth hormone release (225). Because DSIP was found to stimulate GnRH release from a median eminence preparation (224), it is possible that DSIP is released from GnRH nerve terminals and acts in a manner similar to galanin to enhance GnRH release. However, the physiological role of DSIP in the GnRH neuronal network remains unknown, and it is now over 10 years since the last DSIP-GnRH publication.

GnRH-Associated Peptide

The 56 amino acid GAP is generated from the prohormone after endopeptidase processing (Fig. 4). It is present in GnRH nerve terminals and co-released with GnRH into the portal system in vivo (226,227). Soon after the cloning of the GnRH/GAP gene in 1984, Nikolics and colleagues (228) reported that human recombinant GAP stimulated LH and FSH release while inhibiting prolactin secretion from rat pituitary cultures. This was soon followed by a report that the first 13 amino acids of GAP stimulated LH and FSH release from primate pituitary cultures in a manner independent of the GnRH receptor (229). Over the latter half of the 1980s, several independent investigators examined the role of GAP in the regulation of gonadotropin and prolactin release. However, no clear consensus on its function emerged, and activity in the field appears to have died.

Whereas GAP was reported to inhibit prolactin secretion from human, monkey, and rat pituitary cells in vitro (228,230,231), others found no effects of GAP on prolactin in rats (232) or sheep (233) in vivo. Likewise, short GAP-derived peptides did not alter prolactin under any conditions (229,234). Studies of prolactin-secreting pituitary cell lines have reported that GAP reduces intracellular calcium levels through a cAMP-dependent mechanism compatible with a decrease in prolactin secretion (235,236). However, from a physiological perspective, it is difficult to envisage how GAP, released at the same time as GnRH, could be exerting a substantial suppressive effect upon prolactin secretion. Prolactin and LH pulses occur coincidentally in humans (237,238), and even though the mid-cycle prolactin surge begins earlier than the GnRH/LH surge, both surges peak at approximately the same time (239).

A better, albeit imperfect, consensus is apparent in terms of GAP's actions on gonadotropin secretion. Both the full GAP peptide and shorter fragments such as GAP^{1-13} have been found to stimulate LH and FSH secretion weakly in vivo (232,240) and in vitro (228,229,231) in rats and primates. No effects of GAP on LH secretion were found in sheep (229), and one in vivo study was unable to repeat the GAP stimulation of LH (240), suggesting instead that GAP^{1-13} may have selective FSH-releasing ability (241). The mechanism of GAP action on gonadotrophs remains unknown apart from the initial observation that it may not involve the GnRH-1 receptor. Considering the general agreement that GAP stimulates gonadotropin secretion, it is perhaps surprising that the physiological significance of GAP in the control of LH and FSH remains unclear.

GnRH as a Neuropeptide Transmitter

Extensive work has documented clear neuromodulatory effects of GnRH upon sympathetic ganglia in amphibians. In this system, activation of the GnRH receptor leads to closure of a non-inactivating potassium M current that results in a slow depolarizing effect on postganglionic neurons (242,243). Recent work further indicates that GnRH receptor activation in these cells has longer term consequences for maintaining calcium channel activity (244). The role of GnRH as a neuropeptide in the mammalian brain is less well defined. In vivo and in vitro studies have reported a range of excitatory, inhibitory, and modulatory effects of GnRH on the firing of neurons in several brain regions, including the amygdala (245), hippocampus (246,247), hypothalamus (248–252), midbrain central gray (253–256), and cerebellum (257). These studies clearly document that GnRH is capable of modulating neuronal activity in multiple different neuronal networks in the mammalian brain. However, the mechanism(s) of action is unknown (does it involve M channels?) and the physiological relevance of GnRH neuropeptidergic signaling is unclear.

The one brain function where evidence supports a potential physiological role for GnRH as a neuropeptide transmitter is in the control of female reproductive behavior (258). Early studies in rats (259,260), and later in monkeys (261), indicated that GnRH could modulate elements of reproductive behavior. Subsequent work indicated that the midbrain central gray was a key site in this response as, in addition to evidence for GnRH fibers and receptors in this area, direct infusions of GnRH and GnRH antisera into midbrain central gray were found to alter lordosis behavior in the rat (253). However, the question as to whether GnRH actions in the midbrain are obligatory for lordosis has been challenged by the apparently normal sexual behavior of the *hpg* mouse lacking GnRH (262) and studies in the sheep (75) that suggest a more likely facilitatory role for GnRH in sexual behavior. Interestingly, a role for GnRH in sexual behavior implies that a mechanism may exist to coordinate release of GnRH at the median eminence and the midbrain central gray on late proestrus. One possibility would be for hypophysiotropic GnRH neurons to send collateral inputs to the central gray.

Another would be for GnRH overflow into the cerebrospinal fluid from the median eminence at the time of the surge to modulate the activity of reproductive neural networks in a global manner (74,75).

GnRH NEURONAL NETWORK

Components

The GnRH neuronal network is defined here as the GnRH neurons and associated assembly of brain cells responsible for controlling GnRH release into the pituitary portal circulation. Broadly, these "brain cells" can be defined as either neurons or glial cells and each is dealt with separately.

Neuronal Elements of the GnRH Network

Different subclasses of interneurons in the hippocampus receive anywhere between 2,000 and 16,000 synaptic inputs per cell (263,264). If one were to make a conservative assumption that an individual GnRH neuron received inputs from 1,000 cells (91) and that each of these interneurons, in turn, received ~5,000 inputs, the activity of any one GnRH neuron could be influenced directly or indirectly by approximately 5 million neurons. If third-order neuronal connections were considered, then the number of potential neurons involved becomes substantial. Because the 5 million primary and secondary afferent inputs to a GnRH neuron are likely to represent many, if not all, of the neuronal phenotypes existing within the brain, it is perhaps not surprising to find that essentially any neurochemical administered into the ventricular system of an animal alters LH secretion (Tables 1 and 2). As such, it would seem that the challenge is not to catalogue every possible neurotransmitter involved in the network but rather to define the functional hierarchy of these inputs.

Clearly, the definition of those neural inputs that represent the primary afferents to the GnRH neurons are of most interest, because these should have the greatest influence on the behavior of the GnRH neuron. Ideally, the definition of a neurotransmitter used by a functionally significant primary afferent neuron within this network would require (a) evidence that it alters LH and/or FSH secretion when administered into brain in vivo; (b) the demonstration that it can alter GnRH secretion either in vivo or in vitro, thereby ruling out any non–GnRH-mediated actions upon gonadotrophs; (c) anatomical evidence for direct neuronal inputs on GnRH neurons, particularly if this were electron microscopic in nature and corroborated with evidence for the expression of an appropriate receptor by the GnRH neurons; and (d) electrophysiological evidence for a direct effect of the particular neurotransmitter on the electrical excitability of a GnRH neuron. A review of the existing literature using these criteria (Tables 1 and 2) shows several neurotransmitter candidates that have fulfilled many of the criteria listed above.

Having defined a neurotransmitter used by primary afferents to the GnRH neuron, it is then critical to establish its physiological role and relative contribution to the final output of the GnRH neuron; some primary inputs to GnRH neurons might be expected to have only subtle modulatory or permissive effects, whereas others would represent powerful tonically active drivers of their excitability. The development of transgenic animal lines with GnRH neuron-specific deletions of appropriate receptors should be of great use in this regard. Equally, the definition and characterization of the neuronal cell bodies synthesizing the neurotransmitter under question should be insightful. Unfortunately, present tracing methodologies do not enable the cell body characteristics of primary afferents to specific neuronal phenotypes to be determined. Nevertheless, there is good evidence to suggest that neurons with cell bodies located in a variety of locations, including the suprachiasmatic nucleus (SCN), medial preoptic area, MBH, and brainstem, project directly to the GnRH neurons in the rat (265–269) and sheep (270–274).

As is evident from Tables 1 and 2, there are a large number of neurotransmitters likely to be used by neuronal elements within the GnRH network. A good case can be made for some of these representing primary afferents to the GnRH neurons. Extensive evidence indicates that neurons using the amino acids glutamate and γ-aminobutyric acid (GABA) are important primary afferent inputs to the GnRH neurons (Table 1). Although less complete, evidence accumulated over several years also suggests that neurons using norepinephrine (NE), GnRH, and galanin are primary afferents, and a case may also be made for β-endorphin, neurotensin, and neuropeptide Y (NPY) (Table 2). Some neurochemicals such as vasoactive intestinal peptide (VIP), orexin, and endothelin have achieved recent prominence as potential primary inputs, mostly as a result of dual-labeling studies showing the expression of VIP-2, orexin-1, and endothelin-B receptors by GnRH neurons (Table 2). The ability to make direct electrical recordings from GnRH neurons in situ has further enhanced the speed with which suspected and novel primary afferents can be established. This is exemplified by bombesin, where all three bombesin receptor mRNA family members were identified recently in adult GnRH

TABLE 1. *Catalogue of known classical and non-neuropeptidergic neurotransmitter actions on LH secretion and GnRH neurons in adult mammals*

	ICV or IC→ LH/FSH release	In vitro → GnRH release	ICC evidence of fiber apposition	EM evidence for synapses	Receptors expressed by GnRH neurons	Direct effects on GnRH neuron firing
GABA	r, p see (3,406) s see (728)	r (286,312 313,729)	r (730) s (324,325)	r (297,298)	GABA$_A$ r (293,294,296) m (108,295) GABA$_B$ gp (107)	m (5,108, 116,118,292) gp (107)
Glutamate	r, p see (406,731)	r (285,286) gp (732)	s (733)	p (282) r (281)	AMPA r (275) NMDA r (102,276,277) m (280) kainate r (102,279)	m (5,119)
Norepinephrine	r, s, p see (3,328,406, 585,734)	r (334,335)	r (585) s (325) m (376,735)	m (331)	α_{1B} r (329) α_{2A} r (330)	
Dopamine	r see (3) s see (736,737)	r (334,738–740) s (103)	r (730) s (741,742)	r (269,743, 744) s (103)		
Serotonin	r see (3) s (745)	r (746)		r (747)		
Histamine	r (748–750)	r (751)	r, p (748)			
Epinephrine	r see (3,585)		r (585)			
Acetylcholine	r (752–757)	r (758,759)				
Cannabinoids	r (760,761)	r (762)				

The definition of a neurotransmitter used by a primary afferent input to the GnRH neurons requires six criteria to be filled. This includes evidence that the neurotransmitter (1) alters gonadotropin secretion when injected intracerebrally or intraventricularly (ICV), (2) can alter GnRH secretion, (3) is present in fibers observed by immunocytochemistry (ICC) to be in the vicinity of the GnRH cell bodies or terminals, (4) is present in terminals observed by electron microscopy (EM) to synapse upon GnRH neurons, (5) has cognate receptor protein or mRNA expressed by GnRH neurons, and (6) evokes direct actions on GnRH neurons to alter their electrical excitability. The prefix indicates the species in which the evidence exists: r, rat; m, mouse; s, sheep; p, primate; gp, guinea pig. References are given in parentheses.

neurons by single-cell microarray and the presence of functional receptors shown by electrophysiology (7). Thus, despite the almost complete absence of prior data on bombesin (Table 2), it seems likely to represent a primary afferent input to GnRH neurons in the mouse.

Glutamate, GABA, NE, GnRH, opioid peptides, and NPY have been consistent front-runners as physiologically important primary afferents to the GnRH neurons. A review of the evidence in favor of this and their potential role within the network is provided below.

Glutamate. Substantial evidence implicates glutamatergic neurons as being an important class of primary afferents to the GnRH neurons in multiple mammalian species. Direct effects of glutamate, AMPA, NMDA, and kainate have been found on the electrical excitability of GnRH neuron cell bodies in the mouse (5,119), and the relevant ionotropic receptors are expressed by rodent GnRH neurons (275–280). Furthermore, electron microscopic evidence for glutamatergic synapses on GnRH neuron cell bodies has been found in both the rat and monkey (281,282). Whereas all GnRH neurons in the mouse express functional glutamate receptors (5,119), it seems that the individual AMPA, NMDA, and kainite subunits may be expressed differentially by subpopulations of GnRH neurons. Early dual-labeling studies indicated that very few if any GnRH neurons express NMDA receptors (275,278,283), but improvements in reagents and techniques have now enabled investigators to identify NMDA (NR1, NR2A, NR2 subunits) and kainate (KA2 and GluR5 subunits) receptors in >50% of GnRH neurons (277,279,284). In one electrophysiological study, approximately 50% of adult GnRH neurons were found to respond to NMDA (5), whereas in another study 100% of juvenile GnRH neurons responded to NMDA (119). Interestingly, it has been reported that GnRH neurons located medially within the GnRH neuronal continuum are more likely to express NMDA receptors compared with those located more laterally (276). Although metabotropic glutamate

TABLE 2. *Catalogue of known neuropeptide actions on LH secretion and GnRH neurons in adult mammals*

	ICV or IC → LH/FSH release	→ GnRH release	ICC evidence of fiber apposition	EM evidence for synapses	Receptors expressed by GnRH neurons	Direct effects on GnRH neuron firing
GnRH	r (364,369) s (362)	r (363,365,367) s (362)	r (48,72,84) s (52)	r (356,358,763) p (416)	m (7,359)	m (7)
Galanin	see (212,394)	r (211,764)	r (210) m (722) p (765)		r (213) m (7)	m (7)
Somatostatin	r (766,767) s (768)	r (769)			m (7)	m (7)
Bombesin	r (770)				m (7)	m (7)
β-Endorphin	r (420,423)	r (424)	r (771) s (324) p (772)	r (414,763) s (415) p (416)		gp (107)
Neuropeptide Y/AGRP	r see (3,394,773) p see (406)	r (773–775) p (404)	r (372,373,381) m (376) s (324,325,742) p (776)	r (375) m (376)	Y1 r (373) Y5 r (372)	
Neurotensin	see (394,777)		m (735)		NT1 r (633)	
VIP	r (778–780)	r (781,782)	r (562,563,783)		VIP2 r (563)	
Orexin	r (784–788)		r (784,789) s (790)		OXR1 r (789)	
Endothelin		r (791)			ET_B r (792)	
Dynorphin	r (421,423,432)		r (771) s (415)	s (415)		
Tachykinins (substance P, NKB)	see (394)	r (793)	r (794) m (735) s (795) p (796)	r (794)		
CRF	r (797–800) s (801)	r (798,800,802)	p (803)	r (804)		
Vasopressin	r (805–807) p (808)	r (809)		p (810)		
CART		r (811,812)	r (813) hmst (814)			
GALP	r (815) m (816,817) p (818)	r (819)	r (820)			
Enkephalin	r (423,821)	r (428)	r (771) p (822)			
MSH	r (419)	r (419)				
Natriuretic peptides	r (823–825)	r (823)				
Oxytocin	r (826)	r (827,828)				
Angiotensin	see (394)	r (829)				
Bradykinin	r (830)	r (830)				
Orphanin FQ		r (831)				
Ghrelin	r (832)					
Cholecystokinin	r (833)					
CGRP	r (834)					
Kisspeptin	r (835,836) m (837)					

The definition of a functionally relevant neuropeptide used by a primary afferent input to the GnRH neurons requires six criteria to be filled. This includes evidence that the neuropeptide (1) alters gonadotropin secretion when injected intracerebrally (IC) or intraventricularly (ICV), (2) can alter GnRH secretion, (3) is present in fibers observed by immunocytochemistry (ICC) to be in the vicinity of the GnRH cell bodies or terminals, (4) is present in terminals observed by electron microscopy (EM) to synapse upon GnRH neurons, (5) has cognate receptor protein or mRNA expressed by GnRH neurons, and (6) evokes direct actions on GnRH neurons to alter their electrical excitability. The prefix indicates the species in which the evidence exists: r, rat; m, mouse; s, sheep; p, primate; gp, guinea pig; hmst, hamster. References are given in parentheses.

receptor subunit mRNAs have been detected in single GnRH neurons (7), there is no evidence as yet that GnRH neurons express functional metabotropic glutamate receptors. The locations of the glutamatergic cell bodies projecting to the GnRH cell bodies remain unknown.

Evidence also exists suggesting a potential role for glutamate actions at the level of the GnRH nerve terminal to regulate secretion. In vitro preparations of GnRH nerve terminals that are devoid of GnRH cell bodies have been found to release GnRH in response to glutamatergic agonists (285,286). Whether this represents a direct effect of glutamate on the GnRH nerve terminal is not known, but immunocytochemical evidence exists for the expression of both NMDA and kainate receptors by GnRH nerve terminals (102). There are abundant glutamatergic fibers within the median eminence, and some lie directly adjacent to the GnRH nerve terminals, but, as is the case generally, there have been no demonstrations of conventional glutamatergic synapses upon GnRH nerve terminals (102,282).

The relative contribution of primary glutamatergic inputs to the electrical behavior of the GnRH neuron is, as yet, unclear. Electrophysiological studies have demonstrated the presence of a tonic glutamatergic input at the level of the GnRH neuron cell bodies, indicating that glutamate is likely to be involved in determining the excitability of a GnRH neuron. However, compared with direct GABAergic inputs (see GABA, below), the level of tonic glutamatergic receptor activity is relatively low (108,109). Further, the glutamate- and NMDA-evoked currents in GnRH neurons are small in magnitude (119,287). It is important to recognize that these recordings have been made from GnRH neurons in in vitro preparations in which there is a likelihood that afferent glutamatergic inputs to the GnRH neurons have been damaged. Thus, the impact of glutamate on GnRH neurons may be more potent than has been suggested to date.

What function the primary afferent glutamatergic inputs may subserve is not certain. Antagonist studies in vivo have documented the necessity of AMPA/kainate and NMDA signaling within the brain in enabling both pulsatile and surge modes of LH secretion (288–290), and a role in the initiation of puberty is suspected (see Chapters 38 and 40). However, given the abundance and importance of glutamatergic signaling for all neurons in the central nervous system (291), whole brain manipulations of glutamatergic signaling are not especially informative about the role of the primary afferent glutamatergic inputs to the GnRH neurons.

GABA. The amino acid GABA is the principal inhibitory neurotransmitter in the adult nervous system and acts on ionotropic $GABA_A$ and metabotropic $GABA_B$ receptors. All GnRH neurons have been found to express functional $GABA_A$ receptors in mice (5,108,116,292). Although the $GABA_B$ receptor does not appear to play a significant role in mediating rapid GABA actions upon the GnRH soma in the mouse (5,108), guinea pig GnRH neurons have been shown to express functional $GABA_B$ receptors (107). Dual-labeling immunocytochemical and single-cell reverse transcriptase polymerase chain reaction analyses have demonstrated that a range of different $GABA_A$ receptor subunits are expressed by rodent GnRH neurons and that this exists in a sexually differentiated and developmentally regulated manner (108,293–296). Taken together with electron microscopic evidence for GABA synapses on GnRH neurons (297,298) and extensive in vivo data (Table 1), there can be little doubt that GnRH neurons receive direct inputs from GABAergic neurons.

Electrophysiological studies have demonstrated that GnRH neurons are subjected to a substantial tonic "barrage" of GABA (108,109). The $GABA_A$ receptor is a GABA-activated chloride channel, and under artificial whole cell recording conditions that magnify chloride ion channel activity, it can be seen that the $GABA_A$ receptors on a GnRH neurons are activated continuously by GABA (Fig. 5A). In fact, the removal of $GABA_A$ receptor-mediated postsynaptic potentials with the antagonist bicuculline reveals that GABAergic inputs to a GnRH neuron are substantially more numerous than any other input, including that of glutamate (Fig. 5A). Thus, GABAergic inputs to GnRH neurons are direct and substantial.

One interesting and potentially important feature of GABA release upon GnRH neurons is that a large component of $GABA_A$ receptor activation in these cells results from action potential-independent GABA release in the slice preparation. After the application of tetrodotoxin, which blocks all action potential related activity, substantial $GABA_A$ receptor activation continues in GnRH neurons (Fig. 5A). This phenomenon of action potential-independent or "spontaneous" GABA release is observed in a cell-specific manner in several brain regions, including other hypothalamic neurons (299,300). The mechanisms underlying action potential-independent GABA release are not clear, and although it is responsible for much of the $GABA_A$ receptor current in a cell such as the GnRH neuron, its physiological relevance is unknown (301,302). One possible role for spontaneous GABA release may be to provide a "tonic" background of extrasynaptic $GABA_A$ receptor activation outside the synapse itself that is distinct from the temporally coded "phasic" action potential-dependent GABA release (302). Tonic GABA actions in hippocampal neurons are thought to occur through extrasynaptic-located $\alpha 5$ subunit-containing

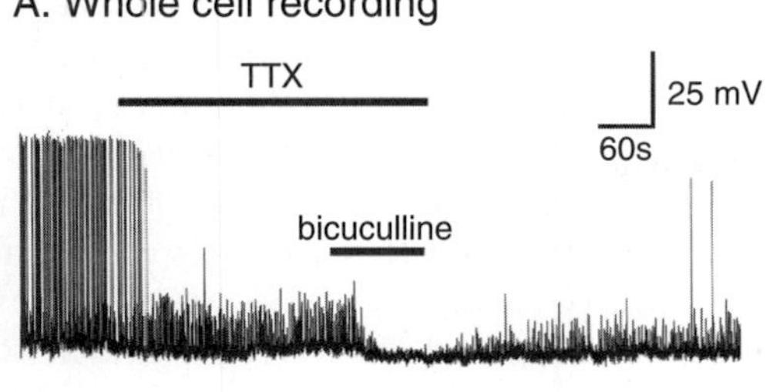

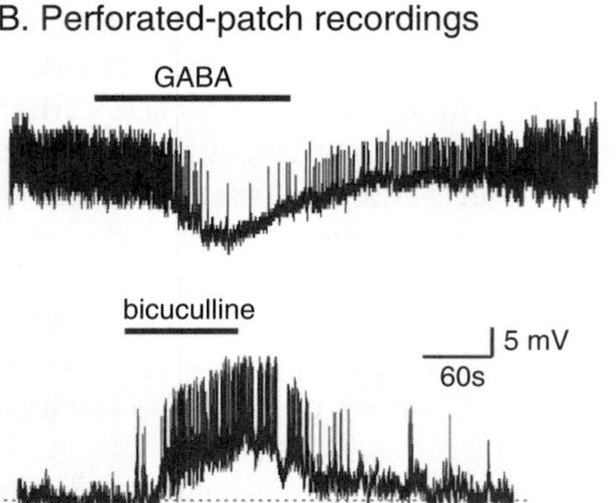

FIG. 5. Potent inhibitory influence of GABA upon GnRH neuronal activity. **A:** Recording of an adult GnRH neuron under artificial whole-cell recording conditions in which $GABA_A$ receptor activation induces chloride ion outflow (i.e., depolarization). Note that a large amount of postsynaptic activity exists in the presence of the sodium channel blocker tetrodotoxin (TTX), indicating substantial action potential-independent neurotransmitter release upon GnRH neurons. Subsequent application of the $GABA_A$ receptor blocker bicuculline reveals that most of this is GABA release. **B:** Gramicidin-perforated-patch recordings (that maintain normal chloride ion homeostasis) from two adult GnRH neurons showing that GABA inhibits and the $GABA_A$ receptor antagonist bicuculline excites GnRH neuron activity. [Adapted with permission from (108,116,118).]

$GABA_A$ receptors (303), and the GnRH neurons are unusual in that they are one of the few other cell types in the brain to express the α5 subunit (108). The importance of this particular mode of GABA signaling to GnRH neurons is not yet fully realized, but evidence to date shows that it has a role in regulating their electrical excitability (118) and, intriguingly, is increased after perinatal androgen treatment (304).

The amino acid GABA exerts a stimulatory action upon neurons during embryogenesis, but for most neurons, this then switches in the perinatal period to its established role as a potent inhibitory neurotransmitter (305,306). Whereas there is agreement that GABA exerts a direct excitatory effect upon developing GnRH neurons right up until the time of puberty, electrophysiological studies have reported disparate effects of direct $GABA_A$ receptor activation on GnRH neurons in the adult mouse (116,292). One laboratory found exogenous GABA to inhibit the electrical activity of adult GnRH neurons (116), whereas the other reported excitatory actions (292). The reasons for this are unclear, but potential critical differences between the two studies involve the use of different GnRH transgenic animal models, concentrations of exogenous GABA and the modulation of $GABA_A$ receptor activity in the presence or absence of glutamatergic receptor blockers. Recent work showing that *endogenous* GABA release in the absence of glutamatergic blockade inhibits the activity of 80% of adult male and female GnRH neurons in two different transgenic models (Fig. 5B) indicates that the overall effect of GABA on GnRH neurons is indeed inhibitory (118). However, excitatory GABA actions at specific synapses on GnRH neurons or under certain conditions cannot be ruled out at present. The membrane chloride ion gradient is the principal determinant of whether opening of the $GABA_A$ receptor depolarizes or hyperpolarizes a cell. As such, an understanding of the regulation of transporters and channels responsible for determining chloride ion equilibrium within GnRH neurons is critical. At present, the potassium-chloride cotransporter 2 (KCC2) is thought to be the key protein maintaining a low intracellular chloride concentration compatible with hyperpolarizing effects of GABA in adult neurons (307). Although KCC2 mRNA was initially not detected in adult GnRH neurons (292), a recent study has demonstrated KCC2 immunoreactivity in approximately one-third of adult mouse GnRH neurons (308).

Alongside potent effects of GABA acting through the $GABA_A$ receptor on the GnRH cell bodies, it is also possible that GABAergic inputs influence the activity of GnRH terminals in the median eminence. Once again, the relative absence of synapses upon GnRH terminals suggests that any direct GABA action at this level occurs through volume transmission. It is clear that GABA acts upon $GABA_B$ receptors located within the MBH of the sheep to stimulate LH and GnRH secretion in vivo (309,310). However, in vitro studies using MBH explants from other species have produced a much less consistent picture with $GABA_A$ and/or $GABA_B$ receptor activation having both stimulatory and inhibitory effects upon terminal GnRH release (311–313). As indicated by some of these studies, the effects of GABA modulation upon GnRH release might not be direct, and considering the importance of GABAergic transmission in regulating the activity of all neurons, it is perhaps not surprising that subtly different MBH preparations give different results. It remains to be determined whether GnRH nerve terminals express $GABA_A$ or $GABA_B$ receptors.

The physiological roles of GABAergic primary afferents to GnRH neurons are not yet established with certainty. The high sustained level of tonic GABA release upon GnRH neurons suggests that GABA transmission is critical for multiple aspects of GnRH neuron behavior. Indeed, the acute removal of $GABA_A$ receptor-mediated inputs to GnRH neurons has a marked effect on their electrical excitability in vitro (118). Further, in vivo investigations with specific antagonists have shown that the tonic activation of $GABA_A$ receptors in the immediate vicinity of the GnRH cell bodies is necessary for both pulsatile and surge modes of LH secretion to occur normally (314–317). Although the presence or not of GABAergic inputs on GnRH nerve terminals is unknown, ovulation is disrupted if GABA release within the MBH is perturbed (318). Thus, GABAergic afferents to the cell body, as well as indirect GABA interactions influencing GnRH terminal release, are likely to be relevant. The direct inputs to the GnRH cell bodies seem certain to be core elements of the GnRH neuronal network involved in maintaining ongoing GnRH neuron pulsatility.

In addition to roles in maintaining GnRH pulsatility, it has been suggested that physiological alterations in GABAergic afferents to GnRH neurons are involved in regulating GnRH neuron activation at puberty (108,319–322), relaying information regarding the gonadal steroid status of the animal (see section on Estrogen-Negative Feedback), the seasonal regulation of GnRH neuron activity (323–325), and relaying metabolic cues to the GnRH neurons (326,327). Together, these observations suggest that GABAergic primary afferents to GnRH neurons represent the most important transsynaptic modulator identified to date.

Norepinephrine. Although controversial in the past (328), it is now evident that NE terminals do synapse directly upon the GnRH cell soma. Investigations in the mouse, rat, and sheep have all found dopamine-β-hydroxylase immunoreactive fibers in close apposition to the GnRH neuron (Table 1), and the presence of α_{1B} as well as α_{2A} receptor immunoreactivity has been reported in rat GnRH neurons (329,330). In the mouse, an electron microscopic study has shown dopamine-β-hydroxylase immunoreactive terminals synapsing upon GnRH neurons (331). Further, a recent single-cell microarray analysis revealed the presence of α_{1A}, α_{1B}, and β_1 receptor mRNAs in GnRH neurons obtained from the adult female mouse (7), and there is recent evidence for direct effects of NE on GnRH neuron excitability (Han and Herbison, unpublished data). Thus, NE neurons represent a primary afferent input to the GnRH neurons.

The locations of the NE cell bodies giving rise to these primary afferents are reasonably well established. Retrograde labeling studies that allow the definition of neurons that have axons terminating in the vicinity of the GnRH cell bodies suggest that the NE neurons located principally within the brainstem A1 and A2 cell groups project to the GnRH neurons in the rat (266,267) and sheep (271,274,332). However, lesions of the locus coeruleus, the other source of hypothalamic NE, have been found to reduce hypothalamic NE levels and block the LH surge, suggesting that these A6 cells might also be involved (333).

Early studies showed that NE modulated GnRH release from the isolated median eminence of the rat (334). Subsequent work in both the rat and monkey suggested that this occurred through an α-adrenergic receptor mechanism that used both nitric oxide and prostaglandin E_2 as intermediaries (335–337), indicating no NE primary afferent to the GnRH nerve terminals. The precise locations of NE neuron cell bodies that project to the median eminence to influence GnRH terminal release in an indirect manner are unknown. However, given the present lack of evidence for effects of MBH NE upon GnRH release in the rat in vivo (338,339), it is unclear what physiological role the indirect presynaptic modulation of GnRH release by NE may have.

Studies undertaken in ovariectomized rats show that NE inputs within the vicinity of the GnRH cell bodies are important in enabling pulsatile GnRH release to exist. Infusions of NE or adrenergic receptor agonists into the third ventricle of OVX rats suppress LH pulsatility (340,341), and microinfusion data suggest that this results from actions of NE within the vicinity of the preoptic area GnRH cell bodies (342). Interestingly, infusions of α-adrenergic receptor antagonists into the vicinity of the GnRH cell bodies in rats and monkeys have the *same* effect of reducing pulsatile GnRH and LH secretion (339, 343–345). Thus, the acute pharmacological "clamping" of adrenergic receptor activity in the vicinity of the GnRH cell bodies at either a constant high (agonist) or constant low (antagonist) level has the same effect of reducing GnRH pulsatility in the rat.

One possible explanation for these somewhat paradoxical results in the rat, whereby both increases and decreases in adrenergic activation have the same effect, may be that NE release itself must be episodic to entrain GnRH neuron pulsatility. Although a clear correlation between fluctuating NE release and GnRH pulses has been observed in the vicinity of the GnRH cell bodies in the monkey (344), no similar temporal relationship has been identified between preoptic area NE levels and pulsatile LH release in the rat (339,346). However, these microdialysis and push–pull experiments are technically challenging and may lack the appropriate spatial resolution to resolve this issue satisfactorily in the rat.

Another explanation for the apparent dependence of GnRH pulsatility upon ongoing NE release is that this neurotransmitter may represent a permissive or "enabling" influence within the GnRH neuronal network. In other words, when present in appropriate amounts, NE may simply enable the GnRH neurons to operate efficiently in a pulsatile mode without exerting any specific inhibitory or stimulatory influence on their excitability. The effects of any permissive regulator within a network are constrained by the properties of that network but, nevertheless, may still be able to set the efficiency of the network within these constraints (Fig. 6). Thus, small changes in the patterning or amount of NE neurotransmission may alter GnRH firing output within limits, whereas any large changes, in either direction, would collapse the network (Fig. 6). The idea of NE as a permissive, rather than a critical, component within the GnRH network is compatible with evidence for the reemergence of pulsatile LH release several weeks after complete destruction of the ascending NE pathways (347,348). Clearly, NE is not absolutely necessary for the network to function and, as a permissive element, can be eliminated.

As a permissive component within the GnRH network, NE would be expected to have an important role in enabling GnRH neuron pulsatility. However, as outlined above, small shifts in adrenergic input should be able to facilitate or slow pulsatility within certain limits. Studies in the rabbit, an induced ovulator, clearly suggest the involvement of brainstem NE neurons in the activation of the GnRH neurons after mating (349–351). Similarly, in the rat, there is abundant evidence that a gradual increase in NE transmission is involved in the generation of the estrogen-induced GnRH surge (see section on Estrogen-Positive Feedback). Thus, an increase in NE release may promote "high output" states of the GnRH neurons and possibly promote high-frequency firing that would be compatible with the GnRH surge. However, again in support of a putative permissive role for NE, an increase in preoptic area NE release begins several hours before the abrupt onset of the GnRH surge on proestrous (352), making it unlikely that the increment in NE activity represents the actual trigger for GnRH release at the time of the surge.

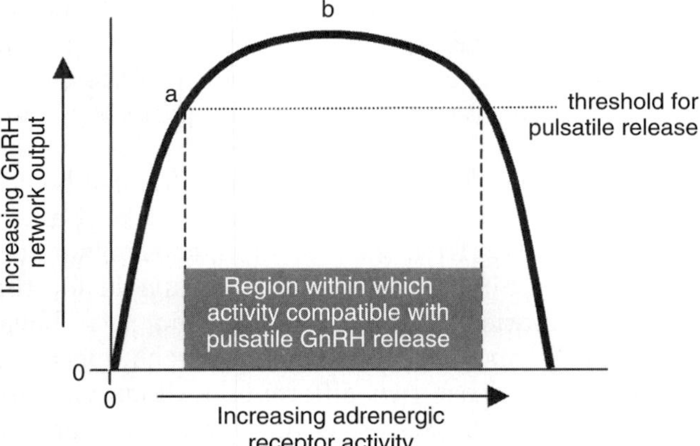

FIG. 6. Model proposing the dynamics of permissive regulation of GnRH neuron output by NE. For pulsatile GnRH release to occur it is proposed that NE must activate network adrenergic receptors within a specific range (*gray box*). Activation outside of this range, either too low or too high, is incompatible with episodic GnRH release. Although an increment in adrenergic receptor activation from a to b may enhance GnRH output, NE acts principally as a permissive influence, as long as adrenergic receptor activation remains within the gray boundary it enables pulsatile release without dictating its amplitude or frequency.

Together these studies indicate an important role for NE primary afferents in modulating GnRH neurons activity. Subtle, but important, differences may exist between its role in primates and rodents. In the former, available evidence suggests an important "driver" role of NE for GnRH pulsatility. In the latter, it seems more appropriate to consider NE as a permissive influence within the GnRH neuronal network. Whether this permissive influence of NE occurs through primary afferents to GnRH soma or terminals or involves parallel indirect actions of brainstem NE neurons within the GnRH network (353,354) is unknown. Equally, the influence of co-released molecules such as NPY (see below) and, possibly, ATP (355) on NE actions remains to be established.

GnRH. There is electron microscopic evidence for GnRH-immunoreactive terminals comprising up to 10% of all synapses observed on GnRH cell bodies in the rat (356–358). Recent studies in the mouse have further identified GnRH-1 receptor mRNA in subpopulations of GnRH neurons (7,359), and electrophysiological investigations have shown GnRH to exert direct depolarizing actions on approximately 50% of GnRH neurons (7). An investigation using cultured adult rat hypothalamic cells has further provided evidence for GnRH receptor immunoreactivity in GnRH neurons (360). Together, these data indicate that GnRH neurons receive direct inputs from GnRH neurons. Whether this represents the recurrent collateral innervation or connections between different GnRH neurons is a key unanswered question, as is the functional significance of only a subpopulation of GnRH neurons being innervated by GnRH fibers.

The function of GnRH inputs upon GnRH neurons is currently under investigation. Studies in vivo have shown that intravenous or i.c.v. GnRH inhibits portal GnRH or plasma LH secretion (361–364), suggesting a potential autocrine short-loop negative feedback role for these contacts. This has been supported by studies in vitro where GnRH has been found to decrease and GnRH antagonists to increase the frequency of GnRH pulses in fetal and adult hypothalamic cell cultures (360,365). Interestingly experiments using fetal hypothalamic cultures and GT1 cells have indicated that the concentration of GnRH may be critical in determining the response of GnRH neurons. The type of G protein activated by the GnRH receptor in these cells depends on agonist concentration, with low (nM) levels of GnRH ultimately resulting in a decrease in cAMP levels and high (μM) levels stimulating cAMP production (366).

Although there is a reasonable consensus among in vitro and in vivo studies for an inhibitory autocrine effect of GnRH, recent electrophysiological studies have provided a different picture. One study undertaken in gonadectomized male mice reported that low (20 nM) concentrations of GnRH added to the bathing medium inhibited action currents in GnRH neurons, whereas higher concentrations excited these cells (359). Unfortunately, it was not possible in that study to determine whether the effects of GnRH were direct upon the recorded GnRH neurons. However, a further study, undertaken in intact adult female mice, showed that GnRH evoked consistent depolarizing actions on GnRH neurons at 20 nM to μM concentrations and that these effects resulted from the direct activation of GnRH-1 receptors located on the GnRH cell body (7). Thus, the predominant suppressive influence of GnRH administered into the brain upon GnRH/LH secretion is not mirrored by its stimulatory effects upon the GnRH neurons themselves.

One explanation for the difference between the effects of GnRH on GnRH neurons and LH secretion may lie in the observation that GnRH inhibits GnRH release from tissue fragments containing only GnRH axons and terminals (363,367,368), whereas it stimulates LH secretion when applied directly to the region of the GnRH cell bodies in the preoptic area (369). Thus, it is possible that GnRH acts at more than one site to regulate GnRH secretion with possible inhibitory actions at the terminals and stimulatory effects at the cell bodies. At present, the mechanisms of GnRH action on the GnRH terminals are completely unknown.

What might be the function of a stimulatory GnRH ultrashort-loop feedback at the level of the GnRH soma? One possibility would be in the coordination of the anatomically dispersed GnRH cell bodies to enable synchronized GnRH secretion. For example, the LH surge is abolished after GnRH-R antisense treatment (370). As outlined in the section on Synchronization and Pulsatility, it is possible that a depolarizing influence of one GnRH neuron on the next could build quickly to a generalized activation of the 40% or so GnRH neurons thought to be stimulated at the time of the GnRH surge (371). However, whether the ~50% of GnRH neurons with GnRH inputs use them for feedback autoregulation or communication between one another, or both, remains to be resolved. These important issues will need to be resolved before the physiological relevance of GnRH inputs onto GnRH neurons is established.

Neuropeptide Y. Investigations in mouse, rat, sheep, and primates have all found NPY fibers to exist in close apposition to the GnRH cell bodies and terminals (Table 2). Studies in the rat have further demonstrated that GnRH nerve terminals exhibit NPY Y1 receptor immunoreactivity and that GnRH cell soma are immunoreactive for the Y5 receptor subtype (372,373). Together, these data provide evidence that GnRH neurons receive direct inputs from NPY neurons. However, it is worth noting that published electron microscopic investigations that have examined this issue have not found abundant evidence for NPY terminals synapsing upon GnRH neurons. A study in the monkey was unable to find NPY synapses on any GnRH neurons (374), and an investigation in rats only detected NPY terminals synapsing upon "one or two" GnRH neurons in total (375). In contrast, terminals immunoreactive for agouti-related peptide (AGRP), a neuropeptide coexpressed exclusively by NPY neurons, were "frequently" detected to form synapses with mouse GnRH neurons (376). Definitive evidence of direct functional NPY inputs to the GnRH cell bodies awaits electrophysiological investigation.

The NPY neurons that impact upon GnRH neurons are thought to have their cell bodies located primarily in the ARN and the brainstem. Nearly all the NPY neurons in the ARN coexpress AGRP (377), and a third of these cells further express GABA (378). In the brainstem, most NPY neurons coexpress NE and epinephrine (379). Tracing studies in the mouse and rat indicate that NPY neurons located in both the brainstem ventrolateral medulla and ARN project to the GnRH cell bodies as well as the vicinity of the GnRH nerve terminals (266,373,376,380). In particular, Turi and coworkers (376) reported recently that around 50% of putative NPY inputs to the GnRH cell bodies in the mouse arise from the ARN with at least a further 25% coming from the brainstem catecholaminergic neurons. Similar data are not available for the NPY innervation of the median eminence. Thus, direct NPY inputs to the GnRH neurons appear to come from at least two different locations and are likely to cosecrete AGRP, GABA, and NE upon the GnRH cell body. It is important to note that NPY is also likely to be used by other nonprimary afferent elements of the GnRH network (381,382).

Studies in the rat indicate that NPY acts at multiple levels to regulate gonadotropin secretion. In addition to direct Y1 receptor-mediated facilitatory effects on the gonadotroph itself (383–387), NPY modulates GnRH release through multiple NPY receptor subtypes at the level of the GnRH nerve terminals (388,389) and other locations (390–393). As is often the case with the third ventricular administration of neurochemicals (394), NPY inhibits LH secretion when given to OVX rats but stimulates LH secretion when given to intact or OVX animals with estrogen replacement (395,396). Whereas subsequent studies in rats have provided little evidence for a physiologically significant role of NPY in the generation or maintenance of pulsatile LH secretion, a clear facilitatory role has been elucidated in terms of the steroid-induced GnRH/LH surge (see section on Generation of the GnRH Surge). This concept has been supported, in part, by work in the global NPY knockout mouse where LH levels in OVX and OVX estrogen-replaced mice are normal but the magnitude of the LH surge is halved compared with wild-type mice (385,397). These observations would be compatible with a permissive influence of NPY within the GnRH network.

The information obtained in the rat for NPY has not translated well to other species. Although the central infusion of NPY was similarly found to inhibit LH secretion in the OVX sheep (398,399), no stimulatory effects of NPY have been found in the presence of gonadal steroids, and it is doubtful whether NPY plays any physiological role in the ovine GnRH surge (400). In the monkey, the infusion of NPY into the median eminence stimulates GnRH secretion irrespective of the gonadal status of the animal (401–403). Importantly, NPY levels fluctuate in synchrony with GnRH release within the monkey median eminence, and the reduction of NPY activity through immunoneutralization or antisense was found to suppress GnRH pulsatility (404,405). It is interesting that NE release fluctuates in the same manner as NPY within the median eminence of the monkey (344), suggesting that it may be the brainstem NE neurons coexpressing NPY that provide this input.

Together, these observations indicate that NPY has an important physiological role in enabling pulsatile GnRH release in the monkey and that NPY is most likely to act at the level of the GnRH nerve terminals to achieve this regulation (406). There is further evidence suggesting that, as in the rat, NPY exerts a positive steroid-dependent influence on GnRH terminals to facilitate the primate GnRH surge (405). Interestingly, the infusion of NPY into the ventricular system rather than the median eminence has a completely different inhibitory action on LH secretion in the monkey, and this is mediated in part by the Y1 receptor (403,407). This clearly suggests that there are multiple sites of action for NPY within the GnRH neuronal network of the monkey.

It seems reasonable from the above description to suggest that NPY primary afferents to the GnRH neurons are likely to represent an important permissive regulatory element in the rodent (408) and a potential integral "driver" of pulsatility in the monkey. Additional roles for NPY in the activation of the GnRH neurons at puberty (see Chapter 40) and in the integration of metabolic cues with reproduction (409) have been suggested. In all species, the effects of NPY on the GnRH nerve terminals and gonadotrophs appear to dominate compared with effects of NPY on the GnRH cell bodies. This is a little surprising considering evidence suggestive of direct NPY inputs onto GnRH cell bodies. However, it is likely that the physiological effects of primary afferent NPY inputs to the GnRH network will be difficult to determine. Aside from the likely different physiological roles of ARN versus brainstem NPY inputs to the GnRH neurons and the unique effects of the different NPY receptors (410), there remains the issue of establishing the significance of co-released AGRP/GABA and NE. Looking at NPY actions independently of AGPR, GABA, and NE may be misleading when investigating a neuropeptide that is an established neuromodulator of other neurotransmitter actions (411–413).

Opioid Peptides. The endogenous opioid peptides, β-endorphin, enkephalin, and dynorphin, act through μ, κ, and δ opiate receptors. At present, the evidence that neurons using opioid peptides are primary afferents to the GnRH neurons is incomplete. Fibers immunoreactive for all three opioid peptides have been detected surrounding GnRH neurons (Table 2), and there is electron microscopic evidence that β-endorphin– and dynorphin-immunoreactive terminals synapse upon the GnRH neuron soma in the rat (357,414), sheep (415), and monkey (416). However, studies that have examined whether receptor mRNAs are expressed by GnRH neurons have, curiously, been unable to detect opiate receptor expression by these cells (417,418). As yet the only electrophysiological analysis of opiate action has been undertaken in the guinea pig, where a μ opioid agonist was found to inhibit the excitability of GnRH neurons (107). Thus, present data make it highly likely that neurons expressing β-endorphin, and possibly dynorphin, represent primary afferents in the GnRH network. The apparent absence of conventional μ, κ, and δ opioid receptors is intriguing and needs to be addressed directly by electrophysiology. One possible explanation for the apparent mismatch is that the "β-endorphin" terminals synapsing on GnRH neurons are using other pro-opiomelanocortin–derived peptides such melanocortin-stimulating hormone (419) as their active transmitter.

Studies in vivo in the OVX rat have clearly established that exogenous opioid peptides suppress pulsatile LH secretion (420–424) and, more importantly, that endogenous opioids are involved in the maintenance and/or generation of pulsatile GnRH/LH secretion (425–427). The specific opioids and receptors mediating these effects are less clear because β-endorphin (420,423), enkephalin (428), dynorphin (421,423), as well as μ, κ, and δ agonists (422–424) have all been shown to suppress LH or GnRH secretion in the OVX rat. The anatomical location of opioid actions upon the GnRH neurons has been elucidated in part. Mu receptor agonists, but not δ or κ, act within the vicinity of the GnRH neuron cell bodies to suppress LH secretion (429), whereas the suppressive effects of β-endorphin occur within the preoptic area or MBH (430). Thus, studies in the OVX rat indicate that opioid peptides act at multiple locations within the GnRH network to suppress GnRH release and that an endogenous opioid tone exerts an ongoing restraint upon pulsatility. Although unproven, the data suggest a scenario in which a direct β-endorphin input to the GnRH cell bodies activates μ opioid receptors to restrain GnRH neuron excitability.

A similar situation exists in the intact or OVX estrogen-replaced rat where the central administration of naloxone, an opiate receptor antagonist, reveals the presence of an ongoing restraining influence of opiates upon GnRH secretion (425). Although it is doubtful whether opioid peptides are involved in mediating the negative feedback actions of gonadal steroids upon the GnRH neurons in the rat (425,426), there is considerable evidence that a reduction in opioid restraint occurs to enable the generation of the GnRH surge on proestrus (427,431–433). Again, there are results in support of all three opioid receptors being involved in this mechanism (427,431–433), and the specific role of any direct β-endorphin or dynorphin inputs to the GnRH neurons are unknown.

The sheep is the other experimental model in which significant research has been undertaken on the potential physiological role of opioid peptides in regulating GnRH neurons. As found in the rat, endogenous opioid peptides exert an ongoing suppressive influence on pulsatile GnRH/LH secretion (434–439). Using the portal bleeding approach, investigators have been able to show that tonic opioid actions are involved in defining the sharp profile of a GnRH pulse and also in restraining interpulse secretion (434,438). Such effects may only exist during the breeding season (437), although it is unclear whether they may be more prominent in the follicular or luteal phases of the cycle (436,437,440). Unfortunately, no studies appear to have examined whether endogenous opioid peptides are involved in the generation of the GnRH surge in the sheep, although μ opioid agonists given centrally can clearly abolish its appearance (441).

Less data exist in the sheep on the specific opioid peptides and receptors involved in GnRH neuron modulation and their sites of action. Nevertheless, it is reasonably clear that μ receptors within both the preoptic area and MBH are involved in the suppression of the GnRH surge (442), whereas κ receptors located in the same regions are the predominant receptors responsible for the tonic suppressive effects of opioids upon LH pulsatility in the luteal phase (415). In addition, it has been shown that β-endorphin infused into the median eminence of the sheep suppresses GnRH release and naloxone stimulates GnRH secretion (440). Thus, as for the rat, there appear to be multiple sites within the GnRH neuronal network at which opioid peptides can act to regulate GnRH secretion. Again, it is plausible that a direct β-endorphin projection to the GnRH cell bodies exists and is critically involved in determining the pulsatile activity of the GnRH neurons. Whereas, like the rat, it seems unlikely that opioid peptides are involved in the estrogen negative feedback pathway to GnRH neurons (434), a case has recently been made for dynorphin inputs to the more caudally placed GnRH neurons to mediate progesterone negative feedback in the ewe (415,443).

In summary, there are a large number of potential primary afferents to the GnRH neurons (Tables 1 and 2) but sufficient data indicating a physiological role is restricted to only a few. The above review suggests that glutamate, GABA, and, possibly, β-endorphin represent critical or core components of the network responsible for shaping and maintaining pulsatile GnRH neuronal activity (Fig. 7). These inputs are in

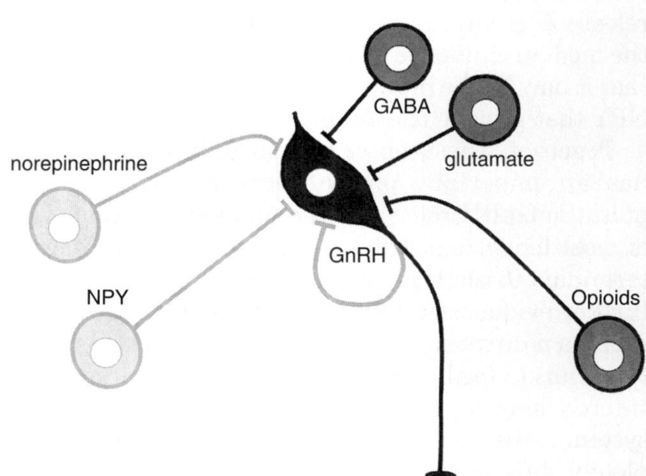

FIG. 7. Schematic diagram of established primary afferent neuronal inputs to GnRH neurons. Those inputs in black are considered to represent core elements of the network responsible for "driving" the activity of the GnRH neurons, whereas those in gray are more likely to represent modulatory or permissive regulators within the network. Not shown on the schematic is the likelihood that each of these inputs may independently regulate GnRH release at the level of the GnRH nerve terminals.

an excellent position to regulate the pulsatility of the GnRH neurons and might therefore also mediate specific hormonal and other cues to the GnRH neurons. If GnRH itself were found to be involved in the synchronization of GnRH neurons or in the feedback regulation of their activity, it would also likely represent a core component of the pulse machinery. In contrast, it is suggested that NE acts as permissive regulator within the network facilitating, but not determining, the episodic release of GnRH. At present it appears that NPY has a distinct permissive role in facilitating GnRH release principally at the level of the GnRH nerve terminal (Fig. 7). It is increasingly apparent that any one neurotransmitter can exert multiple effects on GnRH secretion through direct actions on the cell body and often different effects at the GnRH nerve terminals.

Glial Elements of the GnRH Network

As is evident from the ultrastructural work undertaken on GnRH neurons (see section on GnRH Neuron Cytology), there is a close physical relationship between glial cells and GnRH neurons at the level of their cell bodies and terminals. Of the macroglia family, this appears to involve principally astrocytes and tanycytes. It is now recognized that these cell types can have an impact on the activity of neurons in a multitude of ways, ranging from direct physical interactions at the synapse to the modulation of the extracellular environment and release of specific neuromodulators (444). In relation to the regulation of the GnRH neuron, most attention has focused on the ability of astrocytes and tanycytes to form physical barriers that regulate synaptic contacts or GnRH release and to control GnRH neurons through growth factor–mediated glial-neuronal signaling.

Direct Glial–Neuronal Contacts. A feature of the GnRH neuron under the electron microscope has been that of extensive but variable glial cell wrapping (see section on GnRH Neuron Cytology). For example, individual GnRH neurons in the monkey can be observed to have anything from 0% to 90% of their cell soma in direct apposition to glial cell elements (66). Specifically, fine protrusions from glial cells can be seen to interdigitate between presynaptic terminals and the postsynaptic membrane of the GnRH neuron (84,89), indicating that glial cells may control directly the connectivity of GnRH neurons. These ultrastructural arrangements have suggested that even subtle modulations of the physical relationship between glial cells and GnRH neurons may exert a substantial impact on GnRH neuron connectivity. Unfortunately, it is very difficult to test this hypothesis, and studies are restricted to correlative investigations. Nevertheless, evidence has been found for seasonal- (324), pubertal- (93), and gonadal hormone–dependent (66) correlations between glial cell apposition and GnRH neuron activity. It is interesting to note that the changes in glial cell ensheathment have not always occurred in the manner predicted because more active states of GnRH neuron activity have not necessarily correlated with decreased glial ensheathment (92,324).

The other site at which an intimate physical relationship exists between glial cells and the GnRH neurons is at the level of their terminals in the median eminence (see section on GnRH Neuron Cytology). Electron and light microscopic studies have shown that the GnRH terminals course down beside the tanycytes streaming out from the third ventricle and that the terminals themselves are often enveloped by tanycytic end-feet in the median eminence (73,95–97). As a consequence, the GnRH terminals do not sit adjacent to the endothelial basal lamina of the portal vessels. Again, this has provided the impetus for suggesting that glial cells may represent a physical barrier influencing GnRH secretion by allowing more or less GnRH to reach the portal vasculature depending on the degree of glial cell interposition (95,106). Good evidence to support this idea has come from correlative studies showing clear decreases in both the degree of tanycytic envelopment of GnRH nerve terminals and their distance from the endothelial basal lamina at times of GnRH hypersecretion (96,98,99). Whether this mechanism is relevant only for the GnRH surge or also involved more generally in fluctuating GnRH secretion is unknown.

Growth Factor–Dependent Glial Signaling. Substantial interest has been generated in the possibility that a range of growth factors secreted by glial cells may play an active role in determining GnRH secretion, and this has been reviewed in detail (106,445,446). The signaling pathways that have received the most attention are those related to the actions of epidermal growth factor family members, transforming growth factor (TGF)-α and the neuregulins, and TGF-β family members.

Ojeda and colleagues (447–450) provided evidence that TGF-α and neuregulin secreted by glial cells act on erbB1 and erbB4 receptors located on astrocytes to induce the release of prostaglandin E_2 that, in turn, regulates GnRH neurons directly, principally at the level of the GnRH nerve terminal. This glial–glial–GnRH neuron pathway is though to play a key role in the activation of GnRH neurons at puberty in the rat (see Chapter 38). How important it may be to the regulation of GnRH neurons in the adult is unclear, although a study has shown that disruption of hypothalamic erbB4/2 signaling abolishes estrous cyclicity in the female rat (451).

The other growth factor pathway that has been elucidated most clearly involves the TGF-β family (106,446). In this case it is thought that TGF-β secreted by astrocytes acts directly on GnRH neurons to regulate gene expression and excitability at the level of the cell body (452–456). It has also become apparent that significant cross-talk exists between the TGF-α-erbB receptor and TGF-β pathways because tanycytes were shown recently to secrete TGF-β in response to TGF-α-erbB1 activation (457), extending further a glial–tanycyte–GnRH terminal pathway.

In general, a case can be made for a physiological role of structural and paracrine glial–GnRH neuron interactions within the GnRH network. However, technical difficulties have, as yet, prohibited any definitive role for glia being elucidated in the mature animal in vivo.

Synchronization and Pulsatility

Generation of a GnRH Pulse

It is clear that GnRH is released into the portal vasculature in a predominantly pulsatile fashion. Portal bleeding and microdialysis approaches in a variety of species have documented the pulsatile nature of GnRH release and, under most circumstances, its tight correlation with pulsatile LH secretion (344,458–466). Detailed analyses of GnRH secretion in the sheep (Fig. 8) have shown that each GnRH pulse begins abruptly, plateaus for 5 to 6 minutes, and then falls precipitously (434,462). Studies investigating the electrical stimulation of the GnRH nerve terminals in the median eminence of the rat have indicated that >5 spikes/s activation of the terminals is required to evoke GnRH release and that 10 Hz is the most efficient stimulus frequency for generating GnRH release per stimulus pulse (467). The frequency of GnRH pulses in vivo has been found to range between 40 and 60 minutes in the gonadectomized monkey, sheep, and rat, with longer intervals usually reported in intact animals (344,434, 463–465,468–470). Together, these observations suggest that a 5- to 10-Hz activation of the GnRH nerve terminals lasting for ~5 minutes occurs every 40 to 60 minutes to generate a pulsatile pattern of gonadotropin secretion in the ovariectomized mammal.

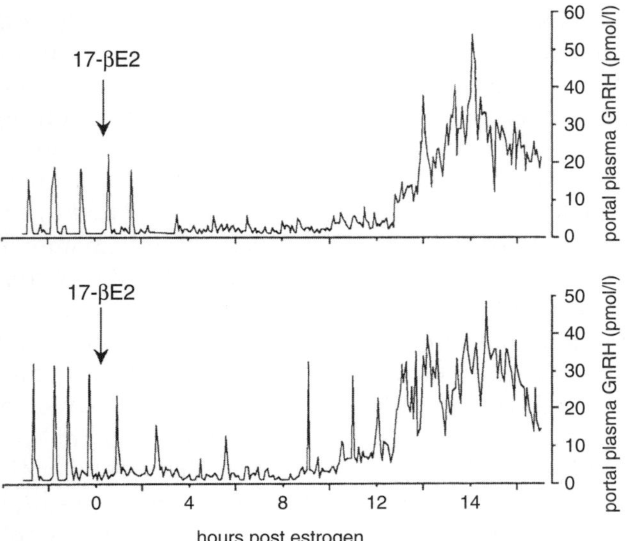

FIG. 8. Portal plasma GnRH concentrations in two OVX ewes showing their response to parenteral 17β-estradiol administration. Note that GnRH secretion is clearly pulsatile with sharp episodes of GnRH secretion. After estrogen there is a relatively rapid phase of negative feedback with reduced amplitude and frequency of GnRH pulses, and this is then followed approximately 12 hours later by an increase in pulsatile and baseline GnRH resulting in the GnRH surge. (Modified with permission.)

A fundamental question within GnRH neurobiology is that of how the dispersed GnRH neurons generate a sharp 5- to 6-minute pulse of GnRH release into the portal vasculature on an intermittent basis. This issue has been reviewed (406,471), and there is a growing consensus that pulsatile GnRH release originates from inherent pulsatility within the GnRH neurons that is then entrained and regulated by neural afferents to these cells.

Inherent Pulsatility within the GnRH Neuron. The various GnRH-secreting GT1 cell lines release GnRH in a pulsatile manner with an interpulse interval of 20 to 40 minutes (472–474). Because these cultures contain only GnRH-secreting neurons, it is clear that GT1 cells can generate a pulsatile release profile by themselves. Indeed, isolated GT1 cells can be found to exhibit spontaneous oscillations in their electrical excitability on an ~30-minute basis (475). The slightly faster pulse frequency observed in GT1 cells compared with in vivo observations (~50 minutes) may be revealing or simply reflect the in vivo pulse generator frequency of the mouse, which is unknown. Thus, although GT1 cells are immortalized and have little choice other than to communicate with one another, these observations raise the possibility that GnRH neurons in vivo may also possess mechanisms that engender inherent pulsatility.

Arguably better models of native GnRH neurons are those derived from the culture of embryonic nasal placodes. After 2–3 weeks of culture, these preparations exhibit pulsatile GnRH release with a frequency of ~50 minutes in the monkey (476) and sheep (477) or ~30 minutes in the rat (478). Although enriched in GnRH neurons, these preparations do contain other neuronal and nonneuronal cell types, and it is likely that the GnRH neurons interact with these other cells to generate a pulsatile pattern of release (479).

Electrophysiological recordings of native GnRH-GFP neurons have not as yet shed significant light on the issue of inherent pulsatility within these cells. When GnRH neurons are recorded in situ, they exhibit a range of different firing patterns ranging from complete silence to regular and phasic patterns of firing (Fig. 3). These patterns may originate in part from the damage each neuron undergoes as part of the brain slice preparation and patching procedure or may reflect the situation in vivo. Those GnRH neurons that demonstrate phasic patterns of firing have been found to exhibit almost random interburst intervals ranging from seconds to minutes and have bursts comprised of 2 to >100 spikes (119,121). Nothing resembling the uniform pattern of episodic electrical activity observed in magnocellular oxytocin and vasopressin neurons (480) has been revealed as yet in GnRH neurons. Juvenile GnRH-GFP neurons disassociated and cultured for 24 hours were found to have similar patterns of silent to highly variable episodic activity (119), suggesting that an individual GnRH neuron has the ability to fire in an episodic manner. However, the absence of any regular pattern of phasic activity in isolated GnRH neurons suggests that communication among the GnRH neurons is critical to pattern electrical firing appropriately.

Intracellular Mechanisms Underlying Inherent Pulsatility. If GnRH neurons are, indeed, able to generate episodic firing patterns, what might be the underlying intracellular "machinery"? Terasawa and colleagues (481,482) showed that GnRH neurons exhibit calcium oscillations with intervals of ~8 minutes and that these become synchronized every 50 minutes in monkey placode cultures. A similar scenario, with approximately 20-minute intervals between calcium synchronizations, was reported in GnRH neurons of mouse olfactory placode cultures (483). How the individual calcium events become synchronized every 50 minutes is not yet understood (482). Whether these calcium oscillations are the generators of pulsatile activity in GnRH neurons or a consequence of episodic electrical activation is unclear, although studies in GT1 cells support the latter (484). Thus, increased episodic firing of GnRH neurons leading to increased episodic intracellular calcium may then be related to pulsatile secretion (485). If this is the case, the calcium fluctuations themselves are not likely to be involved in generating the episodic release rhythm.

Another avenue of investigation has focused on the role cAMP may play in generating episodic activity within the GnRH neurons. Weiner and colleagues (486,487) showed that cAMP activates cyclic nucleotide-gated channels to permit cation entry and facilitate the excitability of GT1 cells. This group went on to show that cyclic nucleotide-gated channels are expressed in rat GnRH neurons in situ (488) and demonstrated, using a GnRH promoter transgenic strategy, that the likely depletion of cAMP levels within GnRH neurons reduces pulsatile LH secretion in vivo in the rat (8). These observations suggest that either the baseline electrical activity and/or episode generator within native GnRH neurons depends on cAMP levels being unperturbed.

Another line of investigation has centered on the role of circadian proteins in the generation of pulsatile GnRH secretion. Many cells throughout the body express circadian genes of the Period, Cryptochrome, and Clock families on an approximately 24-hour basis (489), and investigations in the GT1 cell lines have shown the same phenomenon (490–492). Interestingly, disruption of the *Clock* gene in GT1 cells results in a modest decrease in GnRH pulse frequency from these cells (490), suggesting that one or several of the many genes controlled by clock proteins (493) are involved in the electrical activity and/or episodic pattern generator of GT1 cells. The relationship between circadian genes that cycle on an ~24-hour basis and episodic GnRH secretion that occurs on a 30-minute basis has not been resolved. However, evidence that circadian genes may have a role in GnRH pulsatility in vivo is supported by the observation of increased LH pulse frequency in the *tau* mutant hamster in which the phosphorylation of Per2 is abnormal (494). Also, mice with a *Clock* gene mutation show reduced fertility (490), but it is not known if this relates to abnormal pulsatile GnRH release or defects in the circadian-driven GnRH surge.

Fluctuations in ion channel activation may also be responsible for episodic activity in GnRH neurons (see section on Electrical Properties). Kelly and Wagner (114) suggested that phasic activity in GnRH neurons may originate from the presence of specific conductances such as the T-type Ca^{2+} current (I_T) and the hyperpolarization-activated cation current (I_h). Both of these hyperpolarization-activated conductances are implicated in the phasic firing of neurons (112,113,495) and could conceivably play a role in enabling burst firing in the GnRH neurons. Studies in teleosts have shown that GnRH-secreting neurons located in the terminal nerve exhibit a spontaneous regular discharge that is dependent on a tetrodotoxin-insensitive sodium current (496,497). How relevant this may be to the GnRH neurons controlling fertility and mammalian GnRH neurons is unknown. However, overall, it seems unlikely that a purely electrical interplay could account for the >60-minute pulse interval that exists in GnRH neurons in vivo.

GnRH Neuron Synchronization. How GnRH neurons may communicate with one another to generate synchronized activity is still unknown. There are at least three fundamentally different ways in which this could occur (Fig. 9). The first two

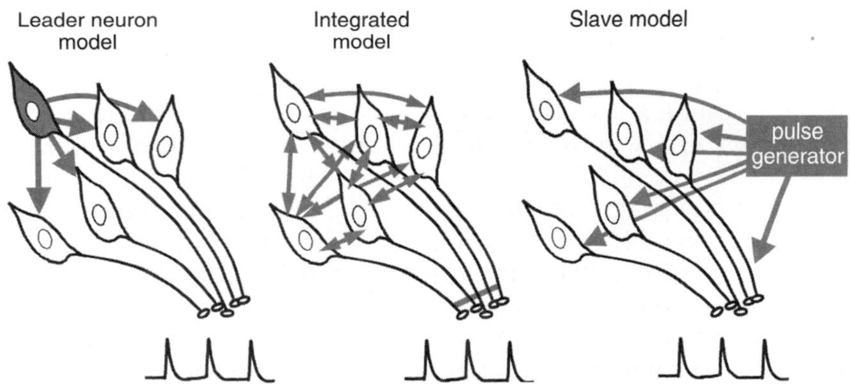

FIG. 9. Schematic diagram showing three potential models of GnRH neuron synchronization. In the Leader Model, a small number of GnRH neurons (*gray*) have the ability to synchronize other GnRH neurons. In the Integrated Model, all GnRH neurons are interconnected in a reciprocal manner so that excitability in any one cell is transferred to the others enabling the slow build-up of network excitability. In the Slave Model, the synchronization of the GnRH neurons is generated and controlled by an entirely different and remote set of neurons with pulse generating properties (see text).

models rely on intrinsic episodic activity within the GnRH neurons themselves, whereas the latter requires the pulse generator to exist principally within other elements of the GnRH network.

Leader Neuron Model. In this model, a small subset of GnRH neurons would have pulse generator capabilities and be able to drive the activity of all the others (Fig. 9). Recent studies of the magnocellular oxytocin neurons have provided evidence that before the beginning of each oxytocin burst, "leader" oxytocin neurons become activated and then drive or recruit the activity of the rest of the oxytocin neurons to generate a period of synchronous high-frequency firing (498). Although GnRH neurons may not share the same dendrodendritic mechanisms of coordination exhibited by oxytocin neurons (499), they may operate in a similar manner in having a subpopulation of leader neurons that drive the rest of the neurons. Electrophysiological studies of native GnRH neurons have found substantial heterogeneity in the electrical properties of GnRH neurons and the firing patterns they exhibit (109,121). This includes a rather small number of GnRH neurons that display clearly phasic firing patterns (Fig. 3C).

In many respects, this model begs the question of just how many GnRH neurons need to be activated to generate sufficient GnRH release at the median eminence to evoke a gonadotropin pulse. If it was just a small number, then it might be sufficient for those cells with inherent pulsatility to function in a coordinated manner alone. Studies involving the transplantation of GnRH neurons into the ventricular system of the *hpg* mouse indicate that less than 20 GnRH neurons are sufficient to generate pulsatile LH secretion (500). The true number of GnRH neurons in these grafts may be underestimated, however, and GnRH itself is likely acting on a hypersensitive pituitary gland. A mouse line with only 60 to 70 GnRH neurons can support normal reproductive activity in males, implying their ability to release LH in a pulsatile manner (501). Studies using cFos as an indicator of neuronal activation to examine pulsatile GnRH release have reported that a small number of ~100 GnRH neurons located in the MBH of the sheep are activated before an LH pulse (502). Similarly, the most caudal GnRH neurons in the rat express Fos after naloxone treatment (503). Whether the Fos expression reflects the total numbers of GnRH neurons activated or those activated in a particular manner is not clear, but it remains possible that the caudal-most GnRH neurons have some special role in episodic GnRH secretion. In the context of this model, they may represent "driver" neurons for the rest of the population. However, key questions such as the numbers of GnRH neurons involved in each pulse and whether it is the same GnRH neurons triggering each pulse remain to be answered.

Integrated Model. In this model, synchronization would result from extensive interconnectedness among GnRH neurons that enables reciprocal activation (Fig. 9). Although there may be individual neurons within the network that are more likely than others to initiate activation, they would not, themselves, determine its final output. Studies in the GT1-1 cell line have shown that calcium waves propagate through a cell network in complex spatial and temporal dimensions providing high levels of synchronization (504). The same may be true in vivo where a depolarizing stimulus would gradually gain momentum both spatially and temporally within the

GnRH neurons and reach a level upon which coordinated bursting activity of large numbers of GnRH neurons would occur.

The key question for such a model is that of how individual GnRH neurons may be extensively interconnected. Studies to date indicate that gap junctional, humoral, and synaptic forms of communication may all be relevant. Functional gap junctions have been identified in GT1 cell lines and have been proposed to be of critical importance to the synchronization of these cells (504–508). Immunocytochemical studies in the rat have reported the presence of connexin-32 in GnRH neurons (509). However, there is only very limited evidence for the presence of dye transfer between mouse GnRH neurons in situ (91,115). Whereas it is not difficult to understand how the closely apposed GT1 cells can use gap junctions for synchronization, it is more challenging to perceive how gap junctions between individual GnRH neurons could be of physiological importance given the scattered distribution of their cell bodies in vivo.

One possibility, however, may be that gap junctions exist at the level of the GnRH axons and nerve terminals as they bundle together within the median eminence. Interestingly, connexin-43 immunoreactivity has been reported in GnRH nerve terminals in the rat (510). Further, theoretical and experimental work indicates that gap junctions between axons may play an important role in the synchronized output of pyramidal cells (511,512). The presence of axo-axonic gap junctions between GnRH fibers would provide one plausible explanation for synchronized GnRH release at the median eminence.

Humoral forms of intercommunication between GnRH neurons have also been proposed. Intriguingly, acutely disassociated adult hypothalami secrete GnRH in a pulsatile manner that is not different from that of intact hypothalami maintained in vitro (365,513). It is of note, however, that this phenomenon is not seen with embryonic tissue. Olfactory placode preparations and late embryonic dispersed hypothalami all require 2 to 3 weeks in vitro before steady episodic GnRH release is observed (360,474, 476,477). Nevertheless, separate GT1 culture dishes perifused together in the same chamber give a pulsatile GnRH release profile indistinguishable from that of a single culture dish (472). Such data suggest that diffusible molecules are involved in the synchronization of GnRH release in vitro. The molecules that may be involved in this communication are not known, but roles for nitric oxide at the GnRH terminals (514,515) and ATP at the cell bodies (482) have been suggested. In both cases, the source of the diffusible molecule would not appear to be the GnRH neurons themselves, indicating an external element in pulse generation.

The third possible mechanism for GnRH neuron intercommunication would be that of synaptic transmission. The most likely mediator of this would be GnRH itself. As outlined above (see section on GnRH), there is good evidence for GnRH terminals synapsing upon GnRH neurons in vivo, and it is likely that the GnRH neurons are interconnected directly with one another. The finding that GnRH evokes a direct depolarizing influence on GnRH neuronal excitability (7) would be compatible with the integrated model proposed here of gradual and progressive mutual excitation. In addition, it is important to recognize that GnRH neurons are likely to co-release other neurotransmitters such as galanin (see section on Coexpressed Neuropeptides), and it is equally possible that one or several of these may be involved in GnRH neuron intercommunication. For example, galanin also evokes a modest depolarizing influence on GnRH neuron excitability (7), and the two neuropeptides, GnRH and galanin, may work synergistically to move networks of GnRH neurons progressively toward a bursting pattern of firing.

Slave Model. This model is based on the possibility that GnRH neurons have no powerful direct interactions with one another and that the synchronizing influence is generated by an entirely separate group of neurons within the GnRH neuronal network (Fig. 9). There are, in fact, several blends of this model that extend from GnRH neurons having a modest degree of episodic activity that is enabled or further entrained by network interactions, through to the complete slave model in which GnRH neurons are simply driven by an external pulse generator. In terms of the former, the best analogy might be that of multiple free-running "idling engines" upon which network elements can engage to coordinate and generate different patterns of functional pulsatility relevant to the physiological status of the animal. In terms of the network elements involved, it is important to distinguish between those that may only be permissive to GnRH pulsatility (e.g., NE) and those that may form a core integrated component directly responsible for pulse dynamics within the network (e.g., glutamate, GABA, and opioids; Fig. 7). For example, GABA neurons have been found to represent critical episode generators and modulators in other neuronal networks within the forebrain (512,516, 517). However, the identity of network elements that might be critical for episodic GnRH release remains unclear (see section on Components).

Some studies in the rat have suggested that neurons located in the MBH comprise a critical element of the pulse generating network. Deafferentation experiments in which the preoptic area is disconnected from the MBH, or a complete "hypothalamic island" created, were found to be compatible with normal pulsatile LH secretion (518–520). However, because there is an almost complete absence of GnRH neuron cell bodies in the MBH of the rat, it is difficult to understand how the terminals alone could maintain GnRH secretion for anything more than a few hours. Importantly, subchiasmatic projections from GnRH neurons to the median eminence were discovered, and there is a very high probability that these were left intact in these lesion studies (71). Thus, those investigations may have revealed, instead, that the preoptic GnRH neurons using a subchiasmatic axonal pathway are sufficient to enable pulsatile GnRH secretion (71). Nevertheless, Soper and Weick (519) found that when combined with anterior hypothalamic deafferentation, specific lesions of the rostral-most pole of the ARN completely abolished pulsatile LH release. A similar effect was found when deafferentation procedures involved the rostral arcuate (521), suggesting that a group of cells in this area, or fibers passing through it, was critical for pulsatile LH release. Electrolytic or chemical lesions of the ARN itself had no effect on pulsatile LH secretion with or without anterior hypothalamic deafferentation (519,522), suggesting further that it may be fibers passing through this area that are critical. All these data would be compatible with the notion that a population of GnRH neurons with projections passing beneath the optic chiasm and through the rostral ARN on their way to the median eminence is critical and sufficient to drive pulsatile LH secretion.

Because monkeys and sheep have GnRH neuron cell bodies spread throughout the preoptic area and MBH, similar types of lesion studies have not been especially revealing in terms of defining the location of neurons involved in the pulse generator. In both species, anterior hypothalamic deafferentation studies indicate that GnRH neurons within the MBH are sufficient to maintain pulsatile GnRH secretion (65,523,524), although they do not exclude a similar role for more rostrally positioned GnRH neurons. Electrical multiunit recording studies have shown an excellent correlation between the occurrence of action potential volleys in the MBH and pulsatile LH secretion in several species (525–528). The identity of the cell types generating these multiunit recordings has not been determined (529), and they may represent direct recordings of action potentials passing down the GnRH axons (528) or the responses of MBH neurons to GnRH collateral input within the ARN (252,530). At present, these recordings provide no direct evidence for the location of a pulse generator in the MBH.

One curious observation in terms of the origins of GnRH pulsatility has been that of episodic GnRH release continuing for several hours in vitro in the absence of the GnRH cell bodies. Perhaps most surprising has been the report that episodic GnRH release can be obtained from the isolated median eminence of adult male rats (531). In that study, erratic but nonetheless pulsatile release of GnRH was observed that was partially dependent on extracellular calcium levels. As suggested by Rasmussen (531), regardless of whether this episodic release results from damage-induced action potential generation or represents a fundamental element of GnRH pulsatility, some degree of terminal synchronization must be going on. However, a subsequent study was unable to detect episodic GnRH release from the adult male rat median eminence (532). Nevertheless, the group of Bourguignon (187,286,532) demonstrated that rat hypothalamic explants that do not contain GnRH cell bodies exhibit clearly pulsatile GnRH secretion. The pulse interval of this secretion depends on the sampling frequency in their static incubation model and can range from 12 to 37 minutes (532). These observations are perhaps the most convincing that a pulse generator independent of the GnRH cell bodies exists within the MBH. The nature of these cells is unknown. However, it is interesting to come back to the speculation that axo-axonic gap junctions may be relevant for GnRH neurons and could, conceivably, be involved in the apparent synchronization of GnRH release observed in the absence of GnRH cell bodies. Equally, studies have shown that nitric oxide release is pulsatile within the median eminence and that it stimulates GnRH secretion from the terminals (514,515). Thus, nitric oxide may represent a further mechanism for synchronizing GnRH release at the level of their nerve terminals (106).

To summarize, there is a good, but not yet definitive, case for inherent intrinsic pulsatility within adult GnRH neurons. The studies in GT1 cells and olfactory placode explants indicate that in vitro preparations composed of mostly GnRH neurons can exhibit episodic release of GnRH with a pulse frequency approximating that of pulsatile GnRH release in vivo. Whether this episodic behavior also exists in adult GnRH neurons in situ is unknown. The intracellular mechanics and basis of episodic release are not yet established, and the definition of pathways responsible for maintaining GnRH neuron excitability, as distinct from those that may underlie episodic activity, has not been addressed. The fact that the individual profiles of episodic electrical and calcium dynamics are often found to vary widely among the GnRH neurons suggests that they must communicate with one another to achieve coordinated release. How individual GnRH neurons may function together to achieve this remains unclear. Intercommunication between individual GnRH neurons may involve combinations of gap junctional, humoral, and/or synaptic forms of interaction. The proposed pattern of GnRH neuronal activity underlying each pulse (>5 Hz for 5 minutes every 60 minutes) is quite different from that displayed by other neuroendocrine neurons displaying pulsatile release, such as the oxytocin neurons (>50 Hz for 3 seconds every few minutes). This suggests that insights regarding the synchronization mechanisms of oxytocin neurons may not translate well to the GnRH neurons. Evidence for a pulse generator remote from the GnRH neurons themselves in the MBH is only partial at best. Nevertheless, it seems likely that synchronization mechanisms may be in place both at the level of the GnRH cell body and terminals. Although proposed models are highly speculative, the most plausible concept is that of GnRH neurons exhibiting episodic activity (the idling engine) that is then entrained to a defined episodic pulse output by other network elements to suit the physiological status of the organism.

Generation of the GnRH Surge

It is now established that the preovulatory LH surge in females is generated by an increase in GnRH release within the portal vasculature (462,463,465,468,533–540). This GnRH surge in spontaneous ovulators typically results from critical interactions between estrogen and circadian inputs (see section on Estrous Cycle Plasticity) that evoke a marked change in the mode of GnRH secretion from the GnRH neurons. Detailed analyses of portal GnRH secretion in the ewe have shown that GnRH release changes from a strictly episodic pattern to one of significant nonepisodic GnRH release intermingled with high frequency pulsatile GnRH secretion before the surge (Fig. 8) (539,541,542). Ultimately, the GnRH surge itself appears to be composed of high-frequency pulsatile events superimposed on a high constant level of GnRH outpouring (Fig. 8). This massive increase in GnRH secretion continues for a period of 24 hours, considerably beyond the duration of the LH surge it induces, before returning to a strictly episodic pattern of release (533,538). These data from the sheep suggest that GnRH neurons exhibit extreme plasticity in their behavior.

The results of GnRH measurement studies in other species are similar, but because of technical constraints, less detailed information is available compared with ewes. In the monkey, both median eminence and third ventricular monitoring of GnRH concentrations have shown that a marked increase in GnRH release occurs at the time of the LH surge (536,537). The precise profile of GnRH release over this time has not been studied on a minute by minute basis but seems likely to comprise pulsatile events superimposed on high basal outpourings of GnRH (536). Like the sheep, the GnRH surge in the monkey long outlasts the LH surge (536). Investigations of this kind in the rat have used push–pull or microdialysis approaches to sample GnRH in the median eminence. Again, hampered by a lack of good temporal resolution, these studies have, nevertheless, reported a clear increase in GnRH secretion correlated with the LH surge (461,534,535). The dynamics of the GnRH surge in the rat appears to take the form of a marked increase in GnRH pulse amplitude with more minor variations in GnRH pulse frequency and interpulse secretion (408,461,469,543). Although remote from the GnRH signal itself, deconvolution analysis of the LH surge in rats also suggests that an increase in LH pulse frequency, amplitude, and duration exists at this time (544). Whether the proestrous increase in GnRH secretion persists for several hours after the end of the LH surge, as it does in the sheep and monkey, is not clear.

It seems likely that the GnRH neurons exhibit a fundamentally different pattern of activity to evoke the LH surge. Although it is clear that estrogenic and circadian influences are responsible for initiating these surges, the nature of their effects is unknown (see section on Estrous Cycle Plasticity). Previous hypotheses have included the possibility that the GnRH neurons responsible for the surge are a subset of the cells independent from those involved in pulsatile release (545). Certainly, it is clear in the rat that only a subpopulation of GnRH neurons clustered mostly around the OVLT express c-Fos at the time of the surge (67), and it seems reasonable to assume that these cells are activated to release GnRH at this time (371). Furthermore, biosynthetic changes in GnRH neurons also appear to exist preferentially in this region in rats (546). Whereas similar studies in the sheep and mouse have also found that only a subpopulation of GnRH neurons expresses c-Fos at the time of the surge, no preferential location of the c-Fos–expressing GnRH neurons was detected (547). There is no increase in the numbers of GnRH neurons expressing this immediate early gene at the time of the surge in the monkey (548). Although it remains possible that a subgroup of GnRH neurons not involved in pulsatile GnRH release become intensely activated to generate the surge, definitive evidence is lacking.

Another possibility is that the GnRH neurons projecting to the median eminence represent a single functional unit that changes their pattern of activity from purely episodic to one of continuous or high frequency release. This might result from increased coordination and coupling between the GnRH neurons so that under the integrated model, mutual stimulation continues unabated for many hours.

Equally, however, the uncoupling of the GnRH neurons so that they release GnRH asynchronously as independent units could also be envisaged to raise net GnRH secretion at the median eminence (541). Clearly, a key area for future investigation is that of the nature of GnRH network plasticity that enables the GnRH surge to occur.

Estrous Cycle Plasticity

Whereas pulsatile GnRH secretion occurs in a regular and continuous manner in males (464,549), spontaneously ovulating females exhibit cycles of episodic GnRH secretion over the course of the ovarian cycle (see Chapters 43 and 44). Although the rising follicular phase concentrations of estradiol appear to be sufficient to evoke the GnRH surge in sheep and primates (32,536), there is a clear requirement for a circadian clock signal to interact with feedback actions of estrogen to generate a GnRH surge in rodents (550–552). It is most convenient, therefore, to consider circadian and gonadal steroid inputs separately. In passing, it is worth noting that the GnRH surge can be generated in multiple different ways and that, in induced ovulators such as rabbits and ferrets, circadian and gonadal steroids inputs play little, if any, role in the activation of the GnRH neurons before ovulation (553).

Circadian Inputs within the Rodent Network

Lesions involving the SCN result in a complete abolition of the preovulatory LH surge and clearly document the critical role of this structure in generating a circadian signal within the GnRH network of rodents (554–558). How the SCN relays this information to the GnRH neurons is under investigation, with both direct and indirect pathways likely. The involvement of diffusible molecules originating from the SCN, as has been found for wheel running rhythms (559), seems unlikely to be relevant for GnRH neuron activation (555,560).

Anterograde tracing studies have shown that SCN neurons project directly to the GnRH neurons in rodents (268,561) and that this involves VIP as a neurotransmitter in the rat (268,562). There is evidence implicating VIP neurons as primary afferents within the GnRH network (Table 2), including the expression of the VIP-2 receptor by GnRH neurons in the rat (563). Interestingly, neonatal cocultures of the SCN and preoptic area enable a circadian pattern of GnRH release to occur, although this pattern was found to correlate best with vasopressin secretion rather than that of VIP (478). In addition to direct projections, SCN inputs have also been observed on estrogen receptor (ER)-expressing cells within the preoptic area (561,564), including the anteroventral periventricular area (AVPV), a brain area of critical importance in transmitting estrogen signals to the GnRH neurons (565,566). Furthermore, a circadian rhythm in cAMP levels exists within cells of the AVPV of the rat (567). Thus, it seems likely that the SCN may project both directly and indirectly to the GnRH neurons and that these pathways are highly lateralized (560). In the case of indirect pathways in the rodent, ER-expressing cells in the AVPV would represent a strong candidate for integrating steroid and circadian input within the GnRH network.

Estrogen Interactions within the Network

In spontaneously ovulating species, the rising plasma estrogen levels of the mid to late follicular phase of the cycle evoke a positive feedback action on the GnRH network to induce the gonadotropin surge. At other times of the cycle, estrogen is responsible for exerting a suppressive effect on gonadotropin secretion. It now appears that at least four fundamentally different modes of estrogen action work together to influence the output of the GnRH neuronal network (Fig. 10) (566,568).

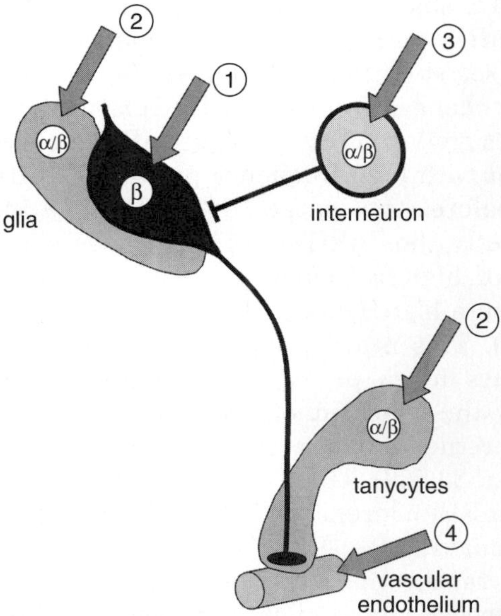

FIG. 10. Schematic diagram showing four possible modes of estrogen action within the GnRH neuron network. 1, Direct estrogen actions through classical transcriptional regulation utilizing ER-β or direct nongenomic influences; 2, indirect actions through the estrogenic modulation of glial cells interacting with the GnRH soma or GnRH terminals in the median eminence; 3, indirect actions through the estrogenic modulation of ER-α and ER-β neurons that are primary afferents to the GnRH cells; 4, indirect actions through the estrogenic modulation of vascular endothelial cells that act in a paracrine manner to regulate GnRH terminal function.

Modes of Estrogen Action. *Direct Estrogen Actions on GnRH Neurons.* With one exception (569), immunocytochemical and in situ hybridization studies have failed to detect ER-α protein or mRNA in the GnRH neurons of multiple species [see (566)]. In contrast, ER-β protein and mRNA have now been detected in the GnRH neurons of embryonic, prepubertal, and adult mice (570–572) and in adult rats (573–576). This suggests that GnRH neurons express functional ER-β receptors, although it should be borne in mind that the levels of ER-β may be quite low. Whereas an initial study failed to provide evidence for estrogen binding to GnRH neurons in the rat (577), a reevaluation of this issue by Hrabovszky and colleagues (575) demonstrated estradiol accumulation in approximately 10% of GnRH neurons. In addition to these classical ERs, an orphan receptor from the nuclear steroid hormone receptor family, estrogen-related receptor α, was found inadvertently in mouse GnRH neurons (572).

It appears that rapid nongenomic effects of estrogen may exist alongside classical genomic mechanisms of estrogen action within GnRH neurons. Electrophysiological studies in the guinea pig have reported a rapid hyperpolarizing influence of estrogen on GnRH neurons (107). Estrogen has also been shown to phosphorylate cAMP-response element binding protein in mouse GnRH neurons in a rapid manner in vivo (578). In that study, the effect of estrogen was direct on the GnRH neurons and abolished in the ERβKO mouse, indicating that ER-β within the GnRH neurons was involved in rapid nongenomic estrogen effects on intracellular signaling (578). Because the level of cAMP-response element binding protein phosphorylation induced by estrogen in GnRH neurons was similar to that observed in intact females, it is possible that these rapid actions of estrogen are persistent.

Transsynaptic Transmission of Estrogen Actions. The present technical difficulties of defining the cellular characteristics of neurons that innervate other neurons have prevented any definitive description of the identity and somal locations of ER-expressing neurons that project to the GnRH neurons. For example, it has been possible to define the locations of ER-expressing neurons that project to the general vicinity of the GnRH cell bodies (266,271) and terminals (273), but it has not been possible to confirm that these cells make direct synaptic contacts with GnRH neurons. Nevertheless, on the basis of those studies, it can be suggested that ER-α–expressing neurons located within the AVPV and other medial preoptic area nuclei, as well as within the MBH and caudal brainstem, project to the GnRH cell bodies.

In considering the likely primary afferents to the GnRH neurons (Fig. 7), it is possible to speculate on the neurochemical identity of estrogen-receptive primary afferents to GnRH neurons. Neurons synthesizing GABA are located throughout the hypothalamus, and those in the vicinity of the GnRH soma and preoptic area express ER-α (579–581). Thus, although not yet proven, it is possible that some of the GABAergic primary afferents to the GnRH neurons also express ERs. The same argument can be made for glutamatergic neurons (582). Interestingly, Petersen and colleagues (568) suggested recently that some of the ER-expressing AVPV inputs may release both GABA and glutamate. Although the precise nature of NPY's influence on GnRH secretion is unclear, a small subpopulation of NPY neurons in the ARN expresses ER-α (266,583,584), and, again, it is possible that estrogen influences NPY neuronal activity directly to modulate GnRH release. Because a subpopulation of pro-opiomelanocortin cells that project to the preoptic area express ER-α (266), they are further likely contenders in the transsynaptic transmission of estrogen status. Finally, it has been shown that some of the ER-expressing brainstem neurons that project to the preoptic area of the rat are NE neurons (266), and there is growing evidence for NE neurons representing direct inputs to GnRH neurons. Interestingly, these NE cells are located principally in the nucleus tractus solitarius and do not coexpress NPY (266). It is important to emphasize that this list certainly does not include all the potential ER-expressing primary afferents to GnRH neurons (585). The simple fact that many neurons in the brain exhibit a rapid nongenomic response to estrogen (586) increases substantially the potential numbers of estrogen "receptive" primary afferents within the GnRH network.

Glial and Endothelial Cell Transmission of Estrogen Actions. There is growing evidence that glial cells participate in gonadal steroid-dependent plasticity within the hypothalamus [for a recent review, see (587)]. Astrocytes and ependymal cells express both classical and membrane ERs (588–590) and may serve as important targets for estrogen in altering the activity of the GnRH network (Fig. 10). Estrogen could act on glia to alter their intimate ultrastructural relationships with GnRH neurons and/or change their pattern of growth factor secretion (106). A good example of the former situation is that of the estrogen-induced plasticity that occurs within the median eminence at the time of the GnRH surge. In this case there is evidence for a rearrangement of the relationship between tanycytic end-feet and GnRH terminals that results in increased numbers of GnRH terminals within the pericapillary space (98,99). This is envisaged to promote access of released GnRH to the portal vasculature (106,591). Gonadal steroid-dependent changes in glial cell ultrastructural appositions with the GnRH cell bodies have also been reported (66). It should be noted that the estrogen-dependent alterations in glial cell morphology are not necessarily slow and that significant changes may be observed within 5 hours in vitro (592).

There is also evidence suggesting that estrogen alters the secretion of glial cell–derived growth factors to modulate GnRH neuron activity. Studies in vitro have shown that estrogen can increase TGF-β synthesis and release from astrocytes and that this may then act to stimulate GnRH secretion from GT1 cells (446,453,593). As noted in the section on Glial Elements of the GnRH Network, there is now evidence for the expression of TGF-β receptors by GnRH neurons in vivo (456). Important interactions between estrogen and insulin-like growth factors are likely to exist in terms of the development and pubertal activation of the GnRH neurons (see Chapter 38), and this may extend into adulthood. Indeed, the estrogen-induced LH surge was shown to be blocked after insulin-like growth factor-1 receptor antagonism (594). However, recent data showing that the GnRH neurons themselves may synthesize or accumulate insulin-like growth factors (595) question whether the estrogen–insulin-like growth factor interactions necessarily involve glial cells. A further glial-neuronal pathway implicated in mediating estrogen actions upon GnRH neurons is that involving the secretion of unknown glial cell factors to up-regulate prostaglandin receptor expression by GnRH neurons (447). Although not related directly to adult gonadal steroid actions within the GnRH network, it is interesting to note the importance of microglia in the development of the GnRH neuron response to estrogen in the adult mouse (596).

Studies by Prevot (106) indicated a further non-neuronal pathway of estrogen action within the GnRH network. Nitric oxide originating from epithelial cells in the median eminence is thought to a play a role in modulating GnRH secretion (336,514, 597,598), and estrogen has been shown to act in both nongenomic and classical modes to stimulate nitric oxide from these cells (515,597,599). A further study has shown that estrogen acts in a nongenomic manner to stimulate GnRH release from the MBH (600). Thus, it is possible that estrogen stimulates or even synchronizes GnRH secretion by regulating nitric oxide release from the endothelial cells adjacent to the GnRH nerve terminals in the median eminence.

Estrogen-positive Feedback. It is first important to reiterate the positive feedback effects of estrogen on the GnRH neurons. As noted above (section on Generation of the GnRH Surge), for the rat this may represent a relatively "simple" increase in GnRH pulse frequency and amplitude. However, the most detailed analyses of GnRH secretion over the surge come from portal bleeding studies in the ewe. A series of investigations by Karsch and colleagues (32,470,541) clearly demonstrated that the first change from the strictly episodic pattern of GnRH secretion observed under estrogen's negative feedback influence begins 15–20 hours after estradiol administration when significant interpulse GnRH secretion occurs. This is followed by an increase in GnRH pulse frequency and amplitude and further enhanced by elevated interpulse secretion in the 2- to 4-hour period immediately before the onset of the GnRH surge. There also evidence for increased irregularity of pulsatile GnRH secretion before the surge in the rat (408). The surge itself in ewes comprises a massive continuous release of GnRH with likely superimposed pulses. A sudden increment in GnRH secretion is important for the proper initiation of the LH surge (32). Hence, in the ewe, estrogen's positive feedback effects are to evoke a gradual increase in episodic GnRH secretion alongside the initiation of a continuous pattern of release that ultimately "explodes" into the prolonged GnRH surge (Fig. 8).

How does estrogen bring about these changes in GnRH neuron secretion? One clue lies in the dynamics of the estrogen signal that is interpreted by the GnRH network to evoke the surge. In primates, sheep, and rodents, the LH/GnRH surge is only induced by an estrogen signal of sufficiently high or increasing levels that lasts for several hours (463,536,538,550, 601–603). Attempts to refine the exact period of estrogen exposure required in the ewe have suggested that a 7- to 14-hour period is the absolute minimum (604). Critically, as found earlier for the LH surge in the rat (550), these studies also reported that estrogen did not need to be present at the actual time of surge initiation for a normal GnRH surge to occur (604). This indicates that the GnRH network "reads" a prolonged high concentration estrogen state as the appropriate signal to initiate positive feedback. The length of the estrogen exposure and the fact that estrogen does not need to be present at the time of surge onset itself would be compatible with the idea that estrogen acts in a classical genomic manner to alter gene expression profiles within the network to induce the surge.

Which cell types within the network might be regulated by estrogen in this manner to induce the changing patterns of GnRH release at the time of the surge? Two lines of evidence strongly suggest that estrogen uses a transsynaptic mode of action for driving positive feedback within the network.

Lesion, Implant, and Neurochemical Studies in the Rat. Most lesioning and estradiol implant studies addressing the sites of action of estrogen in regulating LH secretion have been undertaken in the rat. In terms of the LH surge, there is good agreement that the AVPV, lying medial to the GnRH neuron continuum, is the critical brain region within which estrogen acts to evoke the GnRH/LH surge (565,566). Lesions of the AVPV, but not neighboring nuclei, were found to result in persistent estrous and the abolition of the estrogen-induced LH surge or estrogen plus progesterone–induced LH surge (554,605,606). Although not administered specifically into the AVPV, estradiol placed in the medial preoptic area is able to generate an LH surge in the rat (607,608), whereas similarly positioned implants of antiestrogens inhibit the estrogen-induced LH surge (609,610). Implants of estradiol into the MBH have not been found to elicit a surge-like alteration in LH secretion (607,608). There is evidence for abundant ER-α and ER-β expression within neurons of the AVPV (611–613) and a very high likelihood that ER-expressing AVPV neurons project directly to the GnRH neurons (265,266). This, coupled with evidence for the activation of AVPV neurons at the time of the LH surge (614), strongly suggests that estrogen regulates that activity of AVPV neurons to activate the GnRH neurons. Because ER-α–containing neurons of the AVPV receive direct inputs from the SCN in the rat (564), they may also represent a site of integration of circadian and estrogen inputs in the regulation of GnRH neurons in this species.

The neurochemical identity of the ER-expressing neurons of the AVPV regulating the GnRH neurons in the rat is unknown. A variety of classical neurotransmitters and neuropeptides is synthesized by AVPV neurons, and several of these are known to express ERs, including glutamate (582), GABA (580), dynorphin (615), enkephalin (615,616), galanin (617), substance P (618), calcitonin gene-related peptide (619), and neurotensin (613). At present, a strong case for their involvement in the generation of the GnRH surge can only be made for glutamate, GABA, and neurotensin:

1. *Glutamate.* Antagonists of glutamatergic receptors administered into the ventricular system of rats block the occurrence of the LH surge (289,290, 620), and glutamate levels within the vicinity of the GnRH neuron cell bodies, but not their terminals, increase at the time of the surge (338,621). Although not addressing the AVPV glutamatergic cells themselves, present data are consistent with the idea that estrogen could activate these cells to stimulate the GnRH neurons.
2. *GABA.* There is a consistent set of data that show GABA levels within the vicinity of the GnRH cell bodies rise in the morning and then fall precipitously in the afternoon just before the onset of the LH surge (338,622–625). Pharmacological studies with $GABA_A$ receptor antagonists have further indicated that this fall in GABA release is necessary for the GnRH surge to proceed (316,626,627). These data are consistent with the idea that a reduction in inhibitory GABAergic input to the GnRH neurons (disinhibition) must occur to enable GnRH neuron activation to generate the surge. However, it is unclear how this afternoon decline in GABA release occurs, particularly as estrogen has been shown to exert a predominant stimulatory effect on preoptic GABAergic neurons (628). If indeed the GABAergic neurons of the AVPV are responsible for this effect, then it may well be that suppressive circadian inputs to the AVPV ER-expressing cells (564) act to counteract the stimulatory effects of estrogen on GABAergic neurons projecting to the GnRH neurons (566,624).
3. *Neurotensin.* The infusion of neurotensin into the preoptic area of the rat increases the magnitude of the LH surge (629,630), whereas the administration of neurotensin antisera reduces the size of the LH surge (631). Neither treatment influences the timing of the LH surge, indicating that neurotensin neurons are not involved in the circadian component of surge generation in the rat. Neurotensin neurons in the AVPV express ER-α (613), and estrogen increases their biosynthesis of neurotensin (631,632). Because GnRH neurons were shown to express neurotensin receptors (633), it is possible that AVPV neurotensin neurons are stimulated by estrogen to help activate the GnRH neurons before the surge.

Implant Studies in the Sheep. Investigations in the ewe demonstrated that estradiol-filled cannulae implanted within the MBH adjacent to the ER-α–expressing cells of the ventromedial nucleus are sufficient to generate a normal GnRH surge (634). Earlier studies examining the LH surge in OVX ewes found the same result (635). Estradiol implants positioned within the medial preoptic area, including a region similar to the rat AVPV, were unable to elicit a GnRH surge (634). Estrogen-receptive neurons in the VMN of the ewe project to the vicinity of the GnRH cell bodies (271,272) and VMN cells are activated by estrogen in the sheep (636), suggesting that estrogen regulates the activity of VMN cells projecting to the GnRH neurons to help evoke the GnRH surge.

Although it is surprising that such major species differences should exist in the locations of estrogen-receptive neurons used to evoke the GnRH surge, all the data are consistent in terms of their demonstration that transsynaptic pathways are critical for positive feedback. Of course, these data do not exclude effects of estrogen at other locations within the network using different modes of action, but they do show that estrogen actions on specific AVPV and VMN transsynaptic inputs to the GnRH neurons are sufficient to induce a normal LH/GnRH surge.

Put another way, actions of estrogen on tanycytes and endothelial cells within the median eminence, directly on the GnRH neurons themselves, or on putative ER-expressing primary afferents located in the brainstem or ARN are unlikely to be sufficient for the generation the GnRH surge.

ER Knockout Studies in Mice. Experiments undertaken in ER knockout mice are entirely consistent with the proposal that estrogen uses transsynaptic pathways to induce the GnRH surge. The global ERαKO female mice is infertile (637–639), and recent studies have shown that these mice are unable to exhibit an estrogen-induced LH surge and that this correlates with failed activation of neurons in the AVPV and the GnRH cells themselves (640). In contrast, the ERβKO mice exhibit a normal LH surge and the correct pattern of both AVPV and GnRH neuron activation in response to estrogen (640). This indicates that the deletion of ER-β from the GnRH neurons and all other cells does not prevent mice exhibiting a positive feedback response to estrogen. Thus, it would appear that ER-β expressed by GnRH neurons is redundant, or at least replaceable, in terms of estrogen-positive feedback. Instead, transsynaptic inputs using ER-α appear to be critical, at least in the mouse. It is of note, however, the infusion of an ER-β antisense oligonucleotide into the third ventricle of intact rats was found to prolong their estrous cycle duration (612).

Together, these studies on the mechanism of positive feedback support the hypothesis that estrogen generates the preovulatory GnRH surge by regulating the gene expression of ER-α–expressing neurons located in the AVPV (rodents) or MBH (ewe) that project to the GnRH neurons (Fig. 11). The estrogen-induced changes in these neurons result in the transsynaptic relay of a signal to the GnRH neurons that both increases their pulsatile activity and initiates a pattern of continuous release that, together, culminates in the GnRH surge. Precisely how the transsynaptic inputs modulate the firing of the GnRH neurons is not known, although there is preliminary evidence for an up-regulation of the pacemaker I_H channel in mouse GnRH neurons on proestrus (641).

Estrogen-negative Feedback. The mechanisms of estrogen-negative feedback are much less clear compared with current knowledge for the positive feedback. For example, it had long been debated whether or not a clear suppressive influence of estrogen on GnRH secretion existed in primates. However, data in the human and monkey now indicate that estrogen does have a central action in suppressing LH secretion (642–644). Furthermore, investigations in the OVX rat and sheep demonstrated a clear inhibitory effect of bolus estrogen administration on pulsatile GnRH release (Fig. 8) (463,602,645). In general, the suppressive effects of estrogen on GnRH secretion occur much faster (1–2 hours) (463,602, 644) than the positive feedback actions and involve, principally, a decrease in GnRH pulse amplitude, although changes in frequency have also been observed (Fig. 8) (434,644,645).

What are the cellular mechanisms through which estrogen induces a relatively rapid suppression of GnRH pulse amplitude? At present it would seem that all three modes of direct GnRH, transsynaptic,

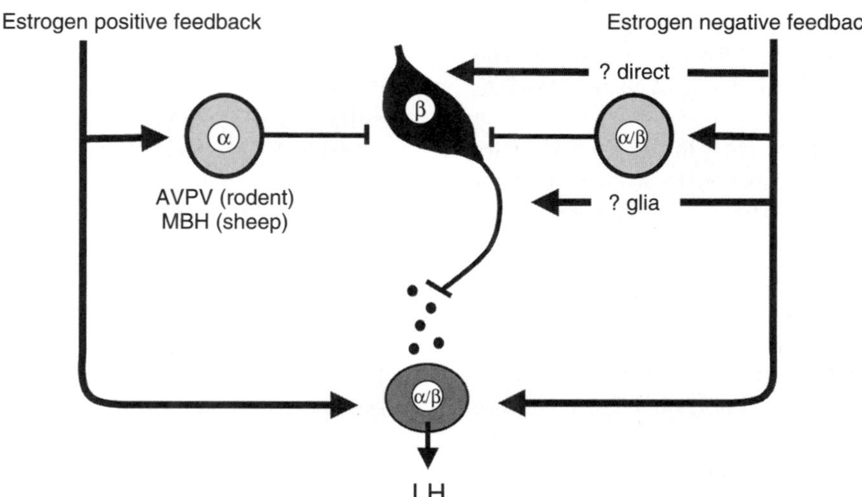

FIG. 11. Schematic diagram showing proposed mechanisms of positive and negative estrogen feedback within the GnRH neuronal network. Work in the rat and mouse suggests that estrogen acts through ER-α–expressing primary afferents with cell bodies located in the AVPV to induce positive feedback. Studies in the sheep suggest that cells within the MBH have the same role. No clearly defined mode of estrogen action is known to be responsible for negative feedback in any species and it may involve multiple modes of estrogen action.

and glial cell–dependent actions might be involved (Fig. 11). In contrast to definition of the AVPV/MBH for positive feedback, lesion and implant studies have indicated that multiple brain regions may be involved in negative feedback. Estrogen implants within the vicinity of the GnRH cell bodies (634,646) as well as several nuclei within the MBH (634,647–651) have all been shown to suppress LH or GnRH secretion in rats, sheep, and monkeys with no clear species differences. Whereas some of the MBH results may have been confounded by diffusion of estrogen to the pituitary, the suppressive effects of estrogen within the preoptic area do not agree well with the observations that lesions of this area do not prevent negative feedback (554,606). Thus, it remains very unclear where within the brain estrogen acts to suppress GnRH release.

Studies in ERKO mice have only been partially successful in furthering our understanding of the mechanisms of negative feedback. The observation of direct nongenomic estrogen actions on GnRH neurons in situ (107,578) suggested that ER-β expressed by these cells might be involved in negative feedback actions. However, it is the global ERαKO mouse, rather than the ERβKO line, that exhibits clearly defective negative feedback mechanisms. Basal LH levels in ERαKO mice are similar to those of OVX wild-type mice and do not suppress after estrogen treatment (637,652). This clearly suggests a major role for ER-α somewhere within the pituitary and/or GnRH network. Indeed, the observation that the modest suppressive influence of gonadal steroids on GnRH mRNA expression is absent in ERαKO mice (653) indicates that it is ER-α–dependent elements within the GnRH network that are also involved in the suppression of GnRH biosynthesis.

In contrast, the ERβKO mouse exhibits normal (637) or modestly impaired (653) LH levels and normal GnRH mRNA negative feedback (653). Interestingly, it was recently reported that the infusion of genistein, an ER-β–selective agonist, into the ventricular system of the OVX ewe suppresses pulsatile LH secretion (654). However, an earlier study comparing the effects of estrogen and genistein infusion into the hypothalamic retrochiasmatic region of the ewe found that genistein was unable to replicate the suppressive actions of estrogen in this area (655). Thus, at present, the role of ER-β, and ER-β expressed by GnRH neurons in estrogen-negative feedback is unclear.

The identity of ER-α–expressing neurons within the network involved in the transsynaptic suppression of GnRH neuron secretion and biosynthesis is unknown. As noted above, estradiol implant studies have not been especially helpful in dissecting these pathways and suggested only that several different ER-α–expressing primary afferents may have the capacity to suppress GnRH neuron secretion, possibly under different physiological circumstances (566). However, some headway has been made with pharmacological and neurochemical profiling investigations where several studies have suggested that GABAergic signaling, in particular, may be important for estrogen-negative feedback. In contrast, investigations have now excluded the β-endorphin neurons from any major role in estrogen-negative feedback in the rat and sheep (425,426,434).

The GABAergic network of the preoptic area is highly sensitive to fluctuations in circulating estrogen levels. These GABAergic neurons express ER-α (579, 581,628,656), and estrogen has been shown to increase GABA release (657–659), GABA reuptake mechanisms (660), and $GABA_A$ receptor subunit expression (661) in the preoptic area. Thus, in the presence of estrogen, GABAergic neurotransmission within the preoptic area is increased, and, accordingly, it has been hypothesized that estrogen acts to enhance the activity of GABAergic primary afferents to the GnRH neurons to suppress their electrical firing. This concept is supported by evidence that (a) an inverse correlation exists between endogenous GABA levels within the vicinity of the GnRH cell bodies and circulating LH levels in rats and primates (319,657,659), (b) increases in $GABA_A$ receptor activation within the preoptic area suppress pulsatile LH secretion in OVX animals (314,315,317), (c) the pharmacological blockade of $GABA_A$ receptors elevate LH levels in OVX estrogen-treated rats and sheep (323,662), and (d) GABAergic primary afferents suppress GnRH neuron activity in the mouse (118).

Interestingly, electrophysiological recordings from GnRH neurons in situ have not as yet clarified the mechanisms of estrogen-negative feedback. Nunemaker and colleagues (117) recorded extracellular action currents from GnRH neurons in the acute brain slice preparation and found that the mean firing rate of GnRH neurons obtained from OVX mice was the same as that from estrogen-treated OVX mice. However, the pattern of firing was found to be different with GnRH neurons in OVX mice firing in a more continuous manner than those obtained from OVX-estrogen animals. Although it is very likely that the dynamics of action potential patterning is critical in determining GnRH secretion, these data do not provide an easy correlate with the in vivo evidence for decreased GnRH pulse amplitude in response to estrogen-negative feedback. Furthermore, other studies in GnRH-GFP mice revealed that long-term estrogen treatment that suppressed LH secretion resulted in GnRH neurons displaying altered potassium channel dynamics that would have made them *more* excitable compared with GnRH neurons in OVX animals (110). Perhaps the only electrophysiological finding that "makes sense" in this context is the observation that estrogen rapidly hyperpolarizes GnRH neurons in the OVX guinea pig (107). These types of data suggest that either the acute brain slice preparation from the mouse is not appropriate for examining estrogen-negative feedback actions on GnRH neurons or that our perception of estrogen-negative feedback in this network needs to be completely revised.

Together, these results indicate that ER-α–expressing components of the GnRH network are critical for normal negative feedback to occur. Because there are multiple brain sites at which estrogen may act to induce negative feedback, it remains possible that ER-α–dependent transsynaptic and/or glial cell modes of estrogen transmission may be involved. Indeed, evidence that negative feedback correlates with increased numbers of synapses upon GnRH neurons (66,663,664) suggests that transsynaptic influences may be critical. It is not clear whether direct estrogen actions upon the GnRH neurons through ER-β are important for the suppression of GnRH release by estrogen. The role of ER-β in estrogen feedback upon the GnRH neurons remains something of a mystery, although it is of note that the ERβKO female is quite clearly subfertile (639,665).

Progesterone Interactions within the Network

Progesterone is the dominant gonadal steroid regulating GnRH secretion during the luteal phase of the ovarian cycle. Although there has been some debate as to whether progesterone levels increase coincident with or slightly before the LH surge (666–668), it is clear that the elevated levels of circulating progesterone through the luteal phase exhibit a potent suppressive influence on GnRH and LH secretion in spontaneously ovulating species, including humans (669–671). The principal role of this luteal-phase increase in progesterone is thought to be in the prevention of GnRH surges on the days subsequent to ovulation (672). In addition to this inhibitory action, it is also evident that progesterone can enhance the time of onset and amplitude of the GnRH/LH surge in rodent and primate species (673–675).

Modes of Progesterone Action. The mechanisms through which progesterone regulates the GnRH neurons are currently under intense investigation. The GnRH neurons themselves seem unlikely to express significant progesterone receptor (PR) immunoreactivity in the sheep (676) and monkey (677), although a small population of medially located GnRH neurons was observed to express PR immunoreactivity in the guinea pig (678). Thus, a transsynaptic mode of communicating progesterone levels to the GnRH neurons appears to be likely (Fig. 12). PRs have been visualized in hypothalamic glutamatergic (656), GABAergic (656,679,680), dopaminergic (681,682), and neuropeptidergic (443,683,684) cells as well as in brainstem NE neurons (685). However, as for ER, the PR-expressing neurons that project directly to the GnRH neurons are not known.

In addition, to classical PR-dependent transcriptional actions, membrane effects of progesterone have been reported upon GnRH release in the rat (686–688). The nature of these membrane actions is unknown, although a G-protein–coupled membrane PR was recently identified in multiple species (689). However, so far, whole-cell recordings from native GnRH neurons in situ have not shown progesterone to have any rapid effects on their electrical excitability (690),

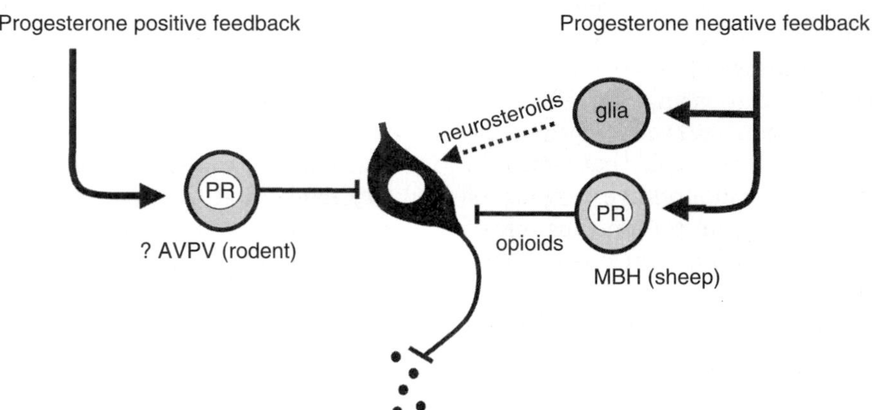

FIG. 12. Schematic diagram showing proposed mechanisms of positive and negative progesterone feedback within the GnRH neuronal network. Positive feedback may occur through estrogen-inducible PR-expressing cells in the AVPV of rats. The suppressive effects of progesterone observed in the luteal phase may result from a combination of transsynaptic regulation and glial cell–dependent synthesis of progesterone-derived neurosteroids such as allopregnanolone.

suggesting that any membrane actions of progesterone within the network are not occurring at the level of the GnRH cell bodies. Because progesterone is metabolized to a variety of neuroactive steroids by glial cells (691,692), a further mode of indirect progesterone action on GnRH neurons is possible. Indeed, electrophysiological studies of GnRH neurons (690, 693) have shown that the 5α-pregnan-3α-ol-20-one (allopregnanolone) metabolite of progesterone has direct allosteric modulatory actions on $GABA_A$ receptors expressed by adult GnRH neurons. Together, these data suggest that both transsynaptic and indirect glial cell modes of progesterone transmission may exist within the GnRH network (Fig. 12).

Progesterone-negative Feedback. How does progesterone suppress GnRH secretion? Studies in the sheep have demonstrated that progesterone exerts a potent suppressive influence on GnRH pulse frequency during the luteal phase (694). Investigations by Skinner and colleagues (695) in the ewe have shown that progesterone can act relatively quickly (~50 minutes) to suppress GnRH pulse frequency and that this requires the presence of estrogen-induced PRs. The precise neural pathway through which progesterone suppresses GnRH pulse frequency in the ewe is not yet certain. Investigators have shown a clear role for endogenous opioid peptides in progesterone-negative feedback (434,694,696,697), and recent work has highlighted the potential importance of PR-expressing dynorphin neurons, in particular within the GnRH network in this role (415,443). There is also evidence to suggest that opioid peptides are involved in the progesterone-induced suppression of LH secretion in humans (698).

The direct and rapid modulation of GnRH neurons by progesterone metabolites derived from nearby glial cells may play a role in suppressing pulsatile LH secretion during the luteal phase of the cycle. Allopregnanolone has been shown to enhance directly the $GABA_A$ receptor channel activity of GnRH neurons (690,693) and to suppress GnRH secretion (699), LH release (700), and ovulation (701) in the mature rat. Because endogenous $GABA_A$ receptor activation inhibits the firing of adult female GnRH neurons (118), it is possible that the high circulating progesterone concentrations of the luteal phase are metabolized to allopregnanolone by glial cells in the vicinity of the GnRH cell bodies and that this then acts to enhance inhibitory GABAergic drive to these neurons (Fig. 12). This scenario may play a role alongside membrane and/or nuclear PR-dependent transsynaptic mechanisms in suppressing GnRH neuron firing. It is important to note, however, that there is little direct evidence for allopregnanolone to suppress GnRH release in the sheep (695), suggesting that major species differences may exist in the nature of progesterone's suppressive actions.

Progesterone-positive Feedback. How does progesterone facilitate the LH surge? Because progesterone does not exert typical positive feedback actions in the sheep (702), there are no detailed analyses of its direct effects on GnRH portal blood secretion. Nevertheless, the examination of LH levels in the rodent and primate indicate that progesterone can advance the time of onset of the GnRH/LH surge and also greatly magnify its amplitude (673–675, 703). Both of these effects have a critical requirement for estrogen-induced PRs (704–706), suggesting that classical mechanisms of estrogen-dependent gene transcription are involved. Further, PR antagonists given before the onset of the proestrous or even the estrogen only–induced LH surge result in its marked attenuation or abolition (706–708). However, a major difficulty in understanding the physiological significance of these stimulatory effects of progesterone has revolved around the lack of evidence for significant circulating progesterone secretion before the onset of the GnRH/LH surge (394,666–668).

Two solutions to this issue have been proposed. The first involves the possibility that estrogen may induce progesterone secretion de novo within the brain and thereby provide sufficient presurge progesterone exposure within the GnRH network in the absence of significant circulating levels of progesterone (709). The second proposal suggests that ligand-independent PR activation may underlie the stimulatory roles of PRs in the surge generation (408). In that model, the rise in follicular phase estrogen would serve to initiate or enhance PR expression in AVPV cells projecting to the GnRH neurons. If these PRs were then transactivated by stimuli other than progesterone, they could initiate surge-appropriate AVPV gene regulation. This attractive model would explain why PR blockade in the absence of circulating progesterone inhibits the LH surge. It also provides a mechanism for how ovarian progesterone secreted concurrent with the LH surge could feed back and now activate these PRs in a ligand-dependent manner to further enhance AVPV gene transcription and bolster the GnRH surge. The signal-inducing ligand-independent activation of the PRs is unknown but is suggested to be a circadian input activating cAMP within the AVPV (567).

The neurochemical identity of AVPV or other GnRH network neurons involved directly in the transsynaptic activation of the GnRH neurons by progesterone is not clear. Again, however, the endogenous opioid peptides have been favored as having a key role. There is abundant evidence that progesterone can facilitate the estrogen/circadian-induced decline in hypothalamic β-endorphin levels and opioid receptor expression before the onset of the LH surge in the rat, thereby enhancing the disinhibition of the GnRH neurons (433,710–716). It has also been suggested that the role of progesterone to suppress the occurrence of daily LH surges after ovulation results from the activation of opioid peptide signaling within the GnRH network (717).

An Integrated Model of Gonadal Steroid-Induced GnRH Plasticity

Estrogen is the principal gonadal steroid determining the cyclical pattern of GnRH secretion throughout the estrous cycle and forms the basis for the present model. The simplest model that could explain the complex feedback effects of estrogen on GnRH neuron output is one in which two independent pathways are responsible for the negative and positive feedback actions of estrogen (Fig. 13) (566). As reviewed in the section on Estrogen Interactions within the Network, it is readily apparent that the negative and positive feedback effects have different mechanisms of operation and, consequently, different dynamics. The positive feedback actions of estrogen require a prolonged period of estrogen exposure

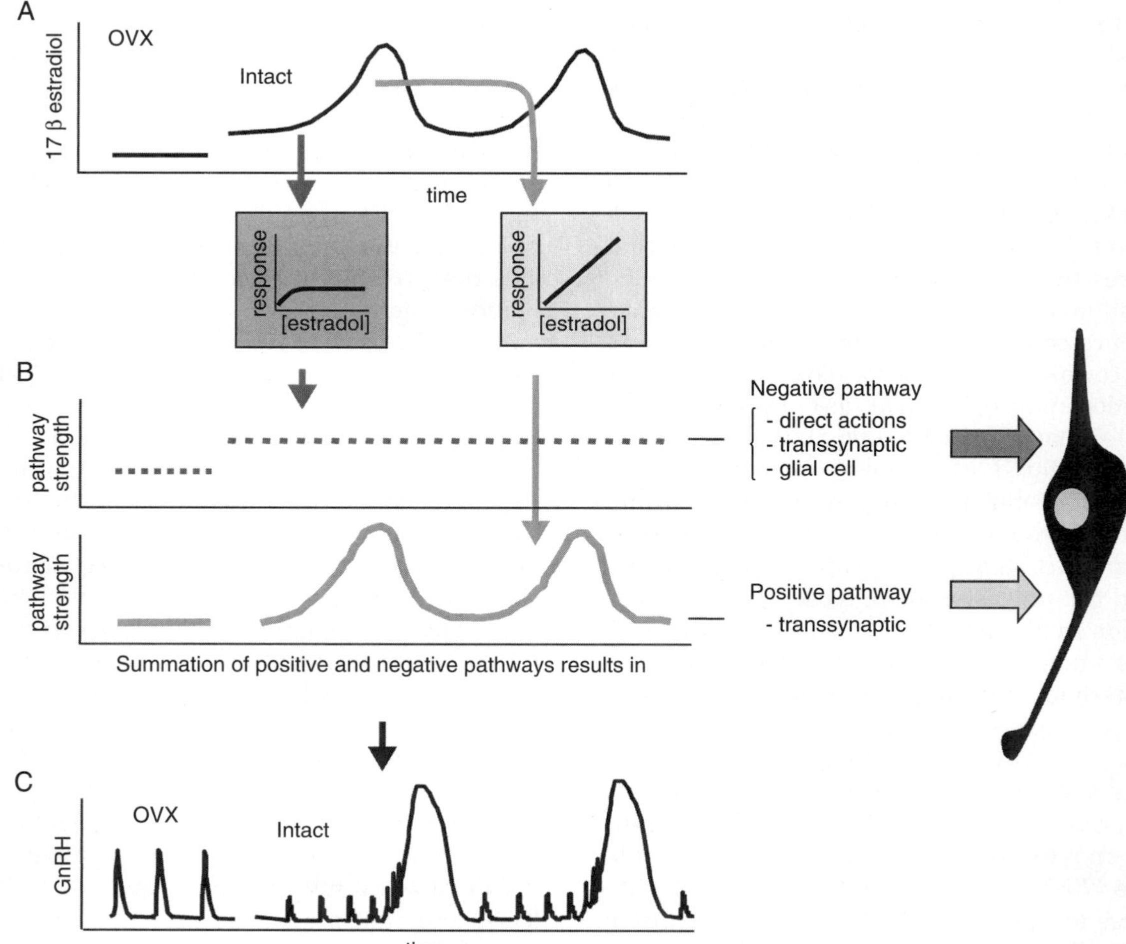

FIG. 13. Dual-pathway model of estrogen regulation of the GnRH neuronal network. The relationship between circulating estrogen levels (**A**) and GnRH output (**C**) is explained by two independent estrogen-regulated pathways (**B**) that are responsible for negative (*red*) and positive (*green*) influences on GnRH neurons. Estrogen modulates the activity of the multimodal (direct, glial, endothelial, and transsynaptic) negative pathway in such a way that a maximum inhibitory response is achieved at low levels of estrogen (*dark gray box*). This results in a constant suppressive influence on the output of the GnRH neuronal network. The cyclic fluctuations in GnRH output, including the GnRH surge, are proposed to arise from actions through the positive ER-α–dependent transsynaptic pathway (*green*). In this case, there is a linear relationship between estrogen concentrations and the stimulatory input (*light gray box*). Thus, the rising estrogen levels in diestrus activate neurons within the positive pathway progressively so that their influence "breaks through" the inhibitory pathway to generate the GnRH surge. At any one time the profile of GnRH output reflects the summation of the negative and positive pathway influences. [Modified with permission from (566).] (See color insert.)

compatible with a classical transcriptional mode of action to bring about changes in gene expression. Evidence in rodents for a key role of ER-α–expressing neurons in the AVPV as transducers of this positive feedback signal is compelling. This region may also represent the locus at which circadian inputs, critical in rodents, are integrated with estrogen actions. The negative feedback actions of estrogen have been much harder to elucidate. Intriguingly, lesion and implant studies have not provided any clear view of which brain regions may be involved in mediating negative feedback. This may be indicative of the multimodal nature of estrogen action for negative feedback, as opposed to the unimodal pathway for positive feedback. Again in contrast to positive feedback, suppressive actions of estrogen occur relatively quickly (Fig. 8), suggesting the possibility of rapid estrogen mechanisms in the negative feedback pathway. Thus, it could be speculated that estrogen treatment of an OVX animal may suppress GnRH secretion initially by rapid direct actions that are then reinforced by transsynaptic events and consolidated by glial cell rearrangements.

Evidence that the negative and positive feedback pathways may exist as independent entities comes from their differing dynamics as well as from studies that have been able to evoke positive feedback in the absence of negative feedback (634,718,719). For this model to work, it is important that the two pathways have different input–output relationships. The positive pathway is modeled as having a linear relationship (Fig. 13, green box) whereby changing estrogen levels result in a proportionally linear effect on pathway output. For example, a linear stimulatory relationship exists between circulating estrogen concentrations and neurotensin (631,632) and the PR (706,720) in the preoptic area. The negative pathway is modeled as having a "capped" maximal suppressive effect on GnRH output (Fig. 13, red box). This key, but unproven, relationship results in estrogen-suppressing GnRH neuron output at a similar level regardless of the level of plasma estrogen concentration. Thus, under these conditions, circulating estrogen activates the negative pathway maximally and the positive pathway in a concentration-dependent manner. The prediction from this model is that both positive and negative feedback pathways are active within the network at most times of the cycle (Fig. 13). Interestingly, estrogen has been shown to alter GnRH neuron potassium channel activity in a manner compatible with positive feedback despite the animals exhibiting negative feedback (110). Thus, the net effect of estrogen on GnRH neurons at any one time reflects the balance of positive and negative pathway influences. For example, the increasing estrogen levels of the late follicular phase drives the positive pathway to overcome the influence of the relatively constant negative pathway and generate the GnRH surge (Fig. 13).

Although this model is centered on estrogen actions, the effects of progesterone are easily integrated into it. Current knowledge of how the positive feedback actions of progesterone occur in the rodent (see section on Progesterone-Positive Feedback) (408) overlaps entirely onto the role of ER-α–expressing AVPV neurons and the positive pathway proposed. The suppressive effects of progesterone during the luteal phase can be envisaged as a further layer of the multimodal negative pathway involving transsynaptic and glial neurosteroid effects (Fig. 12). Because pulsatile GnRH secretion is at it lowest during the luteal phase, it is possible that the progesterone-negative feedback actions are simply an additional luteal phase–specific component of the negative feedback pathway.

Heterogeneity within the GnRH Neuronal Phenotype

It seems that the only properties common to all GnRH neurons are that of GnRH biosynthesis and the presence of functional GABA and glutamatergic inputs. Virtually everything else that has been described so far in GnRH neurons has been detected only in subgroups of these cells. As highlighted throughout the chapter, this heterogeneity exists across multiple parameters, including GnRH neuron projection patterns, light and electron microscopic appearance, coexpression of neuropeptides, firing patterns, rates of GnRH biosynthesis, ion channel, and neurotransmitter receptor expression. Of these parameters, only two have functional correlates. The first is their classification on whether they project to the median eminence or not; those that do are hypophysiotropic GnRH neurons and those that do not are involved in other ill-defined actions of GnRH within the brain. The second is their classification on whether or not they express immediate early genes at the time of the GnRH surge, where it is presumed that those that do are involved in GnRH surge generation.

The relationship of these individual heterogeneous features to each other has been difficult to ascertain. The one exception has been with regard to the GnRH neurons that express c-Fos at the time of the surge. This subpopulation of GnRH neurons is more likely to project outside the blood–brain barrier (721); to receive more galanin (722), kainate (279), and VIP (723) inputs than other GnRH neurons; and to be more transcriptionally active in terms of GnRH mRNA (724) and more likely to coexpress galanin (205). In the rat, these c-Fos–expressing GnRH neurons are found predominantly, but not exclusively, in the vicinity of the OVLT (67) where gonadal steroids exert their most profound effects on GnRH biosynthesis (725–727). Together, these observations suggest that a functionally distinct subpopulation within the GnRH population is located preferentially around the OVLT and is involved in the generation of the GnRH surge. These cells appear to receive greater numbers of specific neurochemical inputs and exhibit the greatest rates of steroid-dependent GnRH biosynthesis.

The functional significance of other properties of the GnRH neuron expressed in a heterogenous manner is not known. On an anatomical basis, exceptions have been in the observations that the rostral-most GnRH neurons in the DBB/medial septum of the monkey do not project to the median eminence (58) and that only GnRH neurons in the medial septum of the rodent express the GalR1 receptor (213) and specific $GABA_A$ receptor subunit mRNAs (295). Viewed from a different neuroanatomical orientation, it has been reported that the GnRH neurons located closest to the third ventricle express NMDA receptors, whereas those placed more laterally do not (276). However, as noted above, many of the heterogeneous characteristics of GnRH neurons occur in a topographically independent manner. This is perhaps best highlighted by the observation that the hypophysiotropic GnRH neurons are scattered throughout the GnRH continuum of rodents and sheep (68–70) and the preoptic area and MBH of monkeys (58).

What is clear is that it is not appropriate to consider this population as a homogeneous group of cells with a unitary function. A major challenge for future research is to establish the functional significance of this extensive level of heterogeneity within the GnRH neuron population. As such, with new techniques allowing better cellular and molecular definition of the GnRH neurons, it will likely become increasingly important to refine investigations to particular subsets of GnRH neurons. Just how these subpopulations will be defined remains to be determined.

ACKNOWLEDGMENTS

Many thanks to Drs. Christine Jasoni, Rebecca Campbell, and Dave Grattan, who provided many valuable comments. Dr. Seong-Kyu Han, Dr. Rebecca Campbell, and Ms. Lizzy Cottrell kindly allowed unpublished data to be included in the figures. Supported by the UK Wellcome Trust.

REFERENCES

1. Wray, S. (2002). Development of gonadotropin-releasing hormone-1 neurons. *Front. Neuroendocrinol.* 23, 292–316.
2. Tobet, S. A., Bless, E. P., and Schwarting, G. A. (2001). Developmental aspect of the gonadotropin-releasing hormone system. *Mol. Cell. Endocrinol.* 185, 173–84.
3. Kordon, C., Drouva, S. V., Martinez de la Escalera, G., and Weiner, R. I. (1994). Role of classic and peptide neuromodulators in the neuroendocrine regulation of luteinizing hormone and prolactin. In *The Physiology of Reproduction* (E. Knobil and J. D. Neill, Eds.), pp. 1621–1681. Raven Press, New York.
4. Silverman, A., Livne, I., and Witkin, J. W. (1994). The gonadotrophin-releasing hormone (GnRH), neuronal systems: immunocytochemistry and in situ hybridization. In *The Physiology of Reproduction* (E. Knobil and J. D. Neill, Eds.), pp. 1683–1706. Raven Press, New York.
5. Spergel, D. J., Kruth, U., Hanley, D. F., Sprengel, R., and Seeburg, P. H. (1999). GABA- and glutamate-activated channels in green fluorescent protein-tagged gonadotropin-releasing hormone neurone in transgenic mice. *J. Neurosci.* 19, 2037–2050.
6. Kato, M., Ui-Tei, K., Watanabe, M., and Sakuma, Y. (2003). Characterization of voltage-gated calcium currents in gonadotropin-releasing hormone neurons tagged with green fluorescent protein in rats. *Endocrinology* 144, 5118–5125.
7. Todman, M. G., Han, S. K., Abraham, I. M., Allen, J. P., and Herbison, A. E. (2004). Profiling neurotransmitter receptor expression in mouse GnRH neurons using GFP-promoter transgenics and microarrays. *Neuroscience* 132, 703–712.
8. Paruthiyil, S., M., M., Conti, M., and Weiner, R. I. (2002). Phosphodiesterase expression targeted to gonadotropin-releasing hormone neurons inhibits luteinizing hormone pulses in transgenic rats. *Proc. Natl. Acad. Sci. U S A* 99, 17191–17196.
9. Sherwood, N. M., Lovejoy, D. A., and Coe, I. R. (1993). Origin of mammalian gonadotropin-releasing hormones. *Endocr. Rev.* 14, 241–254.
10. Fernald, R. D., and White, R. B. (1999). Gonadotropin-releasing hormone genes: phylogeny, structure, and functions. *Front. Neuroendocrinol.* 20, 224–240.
11. Yu, W. H., Karanth, S., Walczewska, A., Sower, S. A., and McCann, S. M. (1997). A hypothalamic follicle-stimulating hormone-releasing decapeptide in the rat. *Proc. Natl. Acad. Sci. U S A* 94, 9499–9503.
12. Yahalom, D., Chen, A., Ben-Aroya, N., Rahimipour, S., Kaganovsky, E., Okon, E., Fridkin, M., and Koch, Y. (1999). The gonadotropin-releasing hormone family of neuropeptides in the brain of human, bovine and rat: identification of a third isoform. *FEBS Lett.* 463, 289–294.
13. Grove-Strawser, D., Sower, S. A., Ronsheim, P. M., Connolly, J. B., Bourn, C. G., and Rubin, B. S. (2002). Guinea pig GnRH: localization and physiological activity reveal that it, not mammalian GnRH, is the major neuroendocrine form in guinea pigs. *Endocrinology* 143, 1602–1612.
14. Jimenez-Linan, M., Rubin, B. S., and King, J. C. (1997). Examination of guinea pig luteinizing hormone-releasing hormone gene reveals a unique decapeptide and existence of two transcripts in the brain. *Endocrinology* 138, 4123–4130.
15. Miyamoto, K., Hasegawa, Y., Nomura, M., Igarashi, M., Kangawa, K., and Matsuo, H. (1984). Identification of the second gonadotropin-releasing hormone in chicken hypothalamus: evidence that gonadotropin secretion is probably controlled by two distinct gonadotropin-releasing hormones in avian species. *Proc. Natl. Acad. Sci. U S A* 81, 3874–3878.

16. Gestrin, E. D., White, R. B., and Fernald, R. D. (1999). Second form of gonadotropin-releasing hormone in mouse: immunocytochemistry reveals hippocampal and periventricular distribution. *FEBS Lett.* 448, 289–291.
17. Chen, A., Yahalom, D., Ben-Aroya, N., Kaganovsky, E., Okon, E., and Koch, Y. (1998). A second isoform of gonadotropin-releasing hormone is present in the brain of human and rodents. *FEBS Lett.* 435, 199–203.
18. Rissman, E. F., Alones, V. E., Craig-Veit, C. B., and Millam, J. R. (1995). Distribution of chicken-II gonadotropin-releasing hormone in mammalian brain. *J. Comp. Neurol.* 357, 524–531.
19. Kasten, T. L., White, S. A., Norton, T. T., Bond, C. T., Adelman, J. P., and Fernald, R. D. (1996). Characterization of two new preproGnRH mRNAs in the tree shrew: first direct evidence for mesencephalic GnRH gene expression in a placental mammal. *Gen. Comp. Endocrinol.* 104, 7–19.
20. Lescheid, D. W., Terasawa, E., Abler, L. A., Urbanski, H. F., Warby, C. M., Millar, R. P., and Sherwood, N. M. (1997). A second form of gonadotropin-releasing hormone (GnRH) with characteristics of chicken GnRH-II is present in the primate brain. *Endocrinology* 138, 5618–5629.
21. White, R. B., Eisen, J. A., Kasten, T. L., and Fernald, R. D. (1998). Second gene for gonadotropin-releasing hormone in humans. *Proc. Natl. Acad. Sci. U S A* 95, 305–9.
22. Millar, R., Lowe, S., Conklin, D., Pawson, A., Maudsley, S., Troskie, B., Ott, T., Millar, M., Lincoln, G., Sellar, R., Faurholm, B., Scobie, G., Kuestner, R., Terasawa, E., and Katz, A. (2001). A novel mammalian receptor for the evolutionarily conserved type II GnRH. *Proc. Natl. Acad. Sci. U S A* 98, 9636–9641.
23. Neill, J. D., Duck, L. W., Sellers, J. C., and Musgrove, L. C. (2001). A gonadotropin-releasing hormone (GnRH) receptor specific for GnRH II in primates. *Biochem. Biophys. Res. Commun.* 282, 1012–1018.
24. Neill, J. D. (2002). GnRH II receptor is encoded in genomes of human, monkey and pig but not the mouse. *Proc. 84th Ann. Meeting Endo. Society, San Fransisco,* Abstract P1–P97.
25. Urbanski, H. F., White, R. B., Fernald, R. D., Kohama, S. G., Garyfallou, V. T., and Densmore, V. S. (1999). Regional expression of mRNA encoding a second form of gonadotropin-releasing hormone in the macaque brain. *Endocrinology* 140, 1945–1948.
26. Latimer, V. S., Rodrigues, S. M., Garyfallou, V. T., Kohama, S. G., White, R. B., Fernald, R. D., and Urbanski, H. F. (2000). Two molecular forms of gonadotropin-releasing hormone (GnRH-I and GnRH-II) are expressed by two separate populations of cells in the rhesus macaque hypothalamus. *Brain Res. Mol. Brain Res.* 75, 287–292.
27. Densmore, V. S. and Urbanski, H. F. (2003). Relative effect of gonadotropin-releasing hormone (GnRH)-I and GnRH-II on gonadotropin release. *J. Clin. Endocrinol. Metab.* 88, 2126–2134.
28. Okada, Y., Murota-Kawano, A., Kakar, S. S., and Winters, S. J. (2003). Evidence that gonadotropin-releasing hormone (GnRH) II stimulates luteinizing hormone and follicle-stimulating hormone secretion from monkey pituitary cultures by activating the GnRH I receptor. *Biol. Reprod.* 69, 1356–1361.
29. Neill, J. D. (2002). GnRH and GnRH receptor genes in the human genome. *Endocrinology* 143, 737–743.
30. Gault, P. M., Maudsley, S., and Lincoln, G. A. (2003). Evidence that gonadotropin-releasing hormone II is not a physiological regulator of gonadotropin secretion in mammals. *J. Neuroendocrinol.* 15, 831–839.
31. Terasawa, E. (2003). Gonadotropin-releasing hormone II: is this neuropeptide important for mammalian reproduction? *Endocrinology* 144, 3–4.
32. Karsch, F. J., Bowen, J. M., Caraty, A., Evans, N. P., and Moenter, S. M. (1997). Gonadotropin-releasing hormone requirements for ovulation. *Biol. Reprod.* 56, 303–309.
33. Sealfon, S. C., Weinstein, H., and Millar, R. P. (1997). Molecular mechanisms of ligand interaction with the gonadotropin-releasing hormone receptor. *Endocr. Rev.* 18, 180–205.
34. Cattanach, B. M., Iddon, C. A., Charlton, H. M., Chiappa, S. A., and Fink, G. (1977). Gonadotrophin-releasing hormone deficiency in a mutant mouse with hypogonadism. *Nature* 269, 338–340.
35. Padmanabhan, V., and McNeilly, A. S. (2001). Is there an FSH-releasing factor? *Reproduction* 121, 21–30.
36. Padmanabhan, V., Brown, M. B., Dahl, G. E., Evans, N. P., Karsch, F. J., Mauger, D. T., Neill, J. D., and Van Cleeff, J. (2003). Neuroendocrine control of follicle-stimulating hormone (FSH) secretion: III. Is there a gonadotropin-releasing hormone-independent component of episodic FSH secretion in ovariectomized and luteal phase ewes? *Endocrinology* 144, 1380–1392.
37. Jones, S. W. (1987). Chicken II luteinizing hormone-releasing hormone inhibits the M-current of bullfrog sympathetic neurons. *Neurosci. Lett.* 80, 180–184.
38. Bosma, M. M., Bernheim, L., Leibowitz, M. D., Pfaffinger, P. J., and Hille, B. (1990). Modulation of M current in frog sympathetic ganglion cells. In *G Protein and Signal Transduction* (N. M. Nathanson and T. K. Harden, Eds.), pp. 43–59. The Rockefeller University Press, New York.
39. Temple, J. L., Millar, R. P., and Rissman, E. F. (2003). An evolutionarily conserved form of gonadotropin-releasing hormone coordinates energy and reproductive behavior. *Endocrinology* 144, 13–19.
40. Kauffman, A. S. and Rissman, E. F. (2004). The evolutionarily conserved gonadotropin-releasing hormone II modifies food intake. *Endocrinology* 145, 686–691.
41. Stopa, E. G., Sower, S. A., Svendsen, C. N., and King, J. C. (1988). Polygenic expression of gonadotropin-releasing hormone (GnRH) in human? *Peptides* 9, 419–423.
42. Dees, W. L., Hiney, J. K., Sower, S. A., Yu, W. H., and McCann, S. M. (1999). Localization of immunoreactive lamprey gonadotropin-releasing hormone in the rat brain. *Peptides* 20, 1503–1511.
43. Hiney, J. K., Sower, S. A., Yu, W. H., McCann, S. M., and Dees, W. L. (2002). Gonadotropin-releasing hormone neurons in the preoptic-hypothalamic region of the rat contain lamprey gonadotropin-releasing hormone III, mammalian luteinizing hormone-releasing hormone, or both peptides. *Proc. Natl. Acad. Sci. U S A* 99, 2386–2391.
44. Kovacs, M., Seprodi, J., Koppan, M., Horvath, J. E., Vincze, B., Teplan, I., and Flerko, B. (2002). Lamprey gonadotropin hormone-releasing hormone-III has no selective follicle-stimulating hormone-releasing effect in rats. *J. Neuroendocrinol.* 14, 647–655.
45. Barry, J. (1979). Immunohistochemistry of luteinizing hormone-releasing hormone-producing neurons of the vertebrates. *Int. Rev. Cytol.* 60, 179–221.
46. Silverman, A. J., Krey, L. C., and Zimmerman, E. A. (1979). A comparative study of the luteinizing hormone releasing hormone (LHRH) neuronal networks in mammals. *Biol. Reprod.* 20, 98–110.
47. King, J. C., and Anthony, E. L. (1984). LHRH neurons and their projections in humans and other mammals: species comparisons. *Peptides* 5 Suppl 1, 195–207.
48. Witkin, J. W., Paden, C. M., and Silverman, A. J. (1982). The luteinizing hormone-releasing hormone (LHRH) systems in the rat brain. *Neuroendocrinology* 35, 429–438.

49. King, J. C., Tobet, S. A., Snavely, F. L., and Arimura, A. A. (1982). LHRH immunopositive cells and their projections to the median eminence and organum vasculosum of the lamina terminalis. *J. Comp. Neurol.* 209, 287–300.
50. Hoffman, G. E., Knigge, K. M., Moynihan, J. A., Melnyk, V., and Arimura, A. (1978). Neuronal fields containing luteinizing hormone releasing hormone (LHRH) in mouse brain. *Neuroscience* 3, 219–231.
51. Schwanzel-Fukuda, M., Garcia, M. S., Morrell, J. I., and Pfaff, D. W. (1987). Distribution of luteinizing hormone-releasing hormone in the nervus terminalis and brain of the mouse detected by immunocytochemistry. *J. Comp. Neurol.* 255, 231–44.
52. Lehman, M. N., Robinson, J. E., Karsch, F. J., and Silverman, A. J. (1986). Immunocytochemical localization of luteinizing hormone-releasing hormone (LHRH) pathways in the sheep brain during anestrus and the mid-luteal phase of the estrous cycle. *J. Comp. Neurol.* 244, 19–35.
53. Merchenthaler, I., Gorcs, T., Setalo, G., Petrusz, P., and Flerko, B. (1984). Gonadotropin-releasing hormone (GnRH) neurons and pathways in the rat brain. *Cell. Tissue Res.* 237, 15–29.
54. Kawano, H., and Daikoku, S. (1981). Immunohistochemical demonstration of LHRH neurons and their pathways in the rat hypothalamus. *Neuroendocrinology* 32, 179–86.
55. Krey, L. C., and Silverman, A. J. (1978). The luteinizing hormone-releasing hormone (LH-RH) neuronal networks of the guinea pig brain. II. The regulation on gonadotropin secretion and the origin of terminals in the median eminence. *Brain Res.* 157, 247–255.
56. Silverman, A. J., Antunes, J. L., Abrams, G. M., Nilaver, G., Thau, R., Robinson, J. A., Ferin, M., and Krey, L. C. (1982). The luteinizing hormone-releasing hormone pathways in rhesus (*Macaca mulatta*) and pigtailed (*Macaca nemestrina*) monkeys: new observations on thick, unembedded sections. *J. Comp. Neurol.* 211, 309–317.
57. Barry, J. (1979). Immunofluorescence study of the preoptico-terminal LRH tract in the female squirrel monkey during the estrous cycle. *Cell. Tissue Res.* 198, 1–13.
58. Goldsmith, P. C., Thind, K. K., Song, T., Kim, E. J., and Boggan, J. E. (1990). Location of the neuroendocrine gonadotropin-releasing hormone neurons in the monkey hypothalamus by retrograde tracing and immunostaining. *J. Neuroendocrinol.* 2, 157–168.
59. Jennes, L., and Stumpf, W. E. (1986). Gonadotropin-releasing hormone immunoreactive neurons with access to fenestrated capillaries in mouse brain. *Neuroscience* 18, 403–416.
60. Schwarting, G. A., Raitcheva, D., Bless, E. P., Ackerman, S. L., and Tobet, S. (2004). Netrin 1-mediated chemoattraction regulates the migratory pathway of LHRH neurons. *Eur. J. Neurosci.* 19, 11–20.
61. Skynner, M. J., Slater, R., Sim, J. A., Allen, N. D., and Herbison, A. E. (1999). Promoter transgenics reveal multiple gonadotropin-releasing hormone-1–expressing cell populations of different embryological origin in mouse brain. *J. Neurosci.* 19, 5955–5966.
62. Quanbeck, C., Sherwood, N. M., Millar, R. P., and Terasawa, E. (1997). Two populations of luteinizing hormone-releasing hormone neurons in the forebrain of the rhesus macaque during embryonic development. *J. Comp. Neurol.* 380, 293–309.
63. Rance, N. E., Young, W. S. III, and McMullen, N. T. (1994). Topography of neurons expressing luteinizing hormone-releasing hormone gene transcripts in the human hypothalamus and basal forebrain. *J. Comp. Neurol.* 339, 573–586.
64. Terasawa, E., Busser, B. W., Luchansky, L. L., Sherwood, N. M., Jennes, L., Millar, R. P., Glucksman, M. J., and Roberts, J. L. (2001). Presence of luteinizing hormone-releasing hormone fragments in the rhesus monkey forebrain. *J. Comp. Neurol.* 439, 491–504.
65. Krey, L. C., Butler, W. R., and Knobil, E. (1975). Surgical disconnection of the medial basal hypothalamus and pituitary function in the rhesus monkey. I. Gonadotropin secretion. *Endocrinology* 96, 1073–1087.
66. Witkin, J. W., Ferin, M., Popilskis, S. J., and Silverman, A. (1991). Effects of gonadal steroids on the ultrastructure of GnRH neurons in the rhesus monkey: synaptic input and glial apposition. *Endocrinology* 129, 1083–1092.
67. Lee, W. S., Smith, M. S., and Hoffman, G. E. (1990). Luteinizing hormone-releasing hormone neurons express Fos protein during the proestrous surge of luteinizing hormone. *Proc. Natl. Acad. Sci. U S A* 87, 5163–5167.
68. Silverman, A. J., Jhamandas, J., and Renaud, L. P. (1987). Localization of luteinizing hormone-releasing hormone (LHRH) neurons that project to the median eminence. *J. Neurosci.* 7, 2312–2319.
69. Merchenthaler, I., Setalo, G., Csontos, C., Petrusz, P., Flerko, B., and Negro-Vilar, A. (1989). Combined retrograde tracing and immunocytochemical identification of luteinizing hormone-releasing hormone- and somatostatin-containing neurons projecting to the median eminence of the rat. *Endocrinology* 125, 2812–2821.
70. Jansen, H. T., Hileman, S. M., Lubbers, L. S., Kuehl, D. E., Jackson, G. L., and Lehman, M. N. (1997). Identification and distribution of neuroendocrine gonadotropin-releasing hormone neurons in the ewe. *Biol. Reprod.* 56, 655–662.
71. Hoffman, G. E., and Gibbs, F. P. (1982). LHRH pathways in rat brain: "deafferentation" spares a sub-chiasmatic LHRH projection to the median eminence. *Neuroscience* 7, 1979–1993.
72. Liposits, Z., Setalo, G., and Flerko, B. (1984). Application of the silver-gold intensified 3,3′-diaminobenzidine chromogen to the light and electron microscopic detection of the luteinizing hormone-releasing hormone system of the rat brain. *Neuroscience* 13, 513–525.
73. Kozlowski, G. P., and Coates, P. W. (1985). Ependymoneuronal specializations between LHRH fibers and cells of the cerebroventricular system. *Cell. Tissue Res.* 242, 301–311.
74. Skinner, D. C., Malpaux, B., Delaleu, B., and Caraty, A. (1995). Luteinizing hormone (LH)-releasing hormone in third ventricular cerebrospinal fluid of the ewe: correlation with LH pulses and the LH surge. *Endocrinology* 136, 3230–3237.
75. Caraty, A., Delaleu, B., Chesneau, D., and Fabre-Nys, C. (2002). Sequential role of e2 and GnRH for the expression of estrous behavior in ewes. *Endocrinology* 143, 139–145.
76. Witkin, J. W. (1990). Access of luteinizing hormone-releasing hormone neurons to the vasculature in the rat. *Neuroscience* 37, 501–506.
77. Caraty, A., de Reviers, M. M., Pelletier, J., and Dubois, M. P. (1980). Reassessment of LRF radioimmunoassay in the plasma and hypothalamic extracts of rats and rams. *Reprod. Nutr. Dev.* 20, 1489–1501.
78. Jennes, L., and Conn, P. M. (1994). Gonadotropin-releasing hormone and its receptors in rat brain. *Front. Neuroendocrinol.* 15, 51–77.
79. Jennes, L. (1987). Sites of origin of gonadotropin releasing hormone containing projections to the amygdala and the interpeduncular nucleus. *Brain Res.* 404, 339–344.
80. Jennes, L. (1991). Dual projections of gonadotropin releasing hormone containing neurons to the interpeduncular nucleus and to the vasculature in the female rat. *Brain Res.* 545, 329–333.
81. Wray, S., and Hoffman, G. (1986). Postnatal morphological changes in rat LHRH neurons correlated with sexual maturation. *Neuroendocrinology* 43, 93–97.

82. Krisch, B. (1980). Two types of luliberin-immunoreactive perikarya in the preoptic area of the rat. *Cell. Tissue Res.* 212, 443–455.
83. Witkin, J. W. (1996). Effects of ovariectomy on GnRH neuronal morphology in rhesus monkey (macaca mulatta). *J. Neuroendocrinol.* 8, 601–604.
84. Jennes, L., Stumpf, W. E., and Sheedy, M. E. (1985). Ultrastructural characterization of gonadotropin-releasing hormone (GnRH)-producing neurons. *J. Comp. Neurol.* 232, 534–547.
85. Kozlowski, G. P., Chu, L., Hostetter, G., and Kerdelhue, B. (1980). Cellular characteristics of immunolabeled luteinizing hormone releasing hormone (LHRH) neurons. *Peptides* 1, 37–46.
86. Witkin, J. W., and Demasio, K. (1990). Ultrastructural differences between smooth and thorny gonadotropin-releasing hormone neurons. *Neuroscience* 34, 777–783.
87. Witkin, J. W. (1989). Morphology of luteinizing hormone-releasing hormone neurons as a function of age and hormonal condition in the male rat. *Neuroendocrinology* 49, 344–348.
88. Witkin, J. W., and Silverman, A. J. (1985). Synaptology of luteinizing hormone-releasing hormone neurons in rat preoptic area. *Peptides* 6, 263–271.
89. Lehman, M. N., Karsch, F. J., and Silverman, A. J. (1988). Ultrastructure and synaptic organization of luteinizing hormone-releasing hormone (LHRH) neurons in the anestrous ewe. *J. Compar. Neurol.* 273, 447–458.
90. Witkin, J. W., O'Sullivan, H., and Silverman, A. J. (1995). Novel associations among gonadotropin-releasing hormone neurons. *Endocrinology* 136, 4323–4330.
91. Campbell, R. E., Han, S. K., and Herbison, A. E. (2005). Biocytin filling of adult gonadotropin-releasing hormone neurons in situ reveals extensive, spiny, dentritic processes. *Endocrinology* 146, 1163–1169.
92. Perera, A. D., and Plant, T. M. (1997). Ultrastructural studies of neuronal correlates of the pubertal reaugmentation of hypothalamic gonadotropin-releasing hormone (GnRH) release in the rhesus monkey (*Macaca mulatta*). *J. Comp. Neurol.* 385, 71–82.
93. Witkin, J. W., O'Sullivan, H., Miller, R., and Ferin, M. (1997). GnRH perikarya in medial basal hypothalamus of pubertal female rhesus macaque are ensheathed with glia. *J. Neuroendocrinol.* 9, 881–885.
94. Suter, K. J., Song, W. J., Sampson, T. L., Wuarin, J. P., Saunders, J. T., Dudek, F. E., and Moenter, S. M. (2000). Genetic targeting of green fluorescent protein to gonadotropin-releasing hormone neurons: characterization of whole-cell electrophysiological properties and morphology. *Endocrinology* 141, 412–419.
95. King, J. C., and Rubin, B. S. (1995). Dynamic alterations in luteinizing hormone-releasing hormone (LHRH) neuronal cell bodies and terminals of adult rats. *Cell. Mol. Neurobiol.* 15, 89–106.
96. Prevot, V., Dutoit, S., Croix, D., Tramu, G., and Beauvillain, J. C. (1998). Semi-quantitative ultrastructural analysis of the localization and neuropeptide content of gonadotropin releasing hormone nerve terminals in the median eminence throughout the estrous cycle of the rat. *Neuroscience* 84, 177–191.
97. Ugrumov, M., Hisano, S., and Daikoku, S. (1989). Topographic relations between tyrosine hydroxylase- and luteinizing hormone-releasing hormone-immunoreactive fibers in the median eminence of adult rats. *Neurosci. Lett.* 102, 159–164.
98. King, J. C., and Letourneau, R. J. (1994). Luteinizing hormone-releasing hormone terminals in the median eminence of rats undergo dramatic changes after gonadectomy, as revealed by electron microscopic image analysis. *Endocrinology* 134, 1340–1351.
99. Prevot, V., Croix, D., Bouret, S., Dutoit, S., Tramu, G., Stefano, G. B., and Beauvillain, J. C. (1999). Definitive evidence for the existence of morphological plasticity in the external zone of the median eminence during the rat estrous cycle: implication of neuro-glio-endothelial interactions in gonadotropin-releasing hormone release. *Neuroscience* 94, 809–819.
100. Goldsmith, P. C., and Ganong, W. F. (1975). Ultrastructural localization of luteinizing hormone-releasing hormone in the median eminence of the rat. *Brain Res.* 97, 181–193.
101. Durrant, A. R. and Plant, T. M. (1999). A study of the gonadotropin releasing hormone neuronal network in the median eminence of the rhesus monkey (*Macaca mulatta*) using a post-embedding immunolabelling procedure. *J. Neuroendocrinol.* 11, 813–821.
102. Kawakami, S. I., Hirunagi, K., Ichikawa, M., Tsukamura, H., and Maeda, K. I. (1998). Evidence for terminal regulation of GnRH release by excitatory amino acids in the median eminence in female rats: a dual immunoelectron microscopic study. *Endocrinology* 139, 1458–1461.
103. Kuljis, R. O. and Advis, J. P. (1989). Immunocytochemical and physiological evidence of a synapse between dopamine- and luteinizing hormone-releasing hormone-containing neurons in the ewe median eminence. *Endocrinology* 124, 1579–1581.
104. Ajika, K. (1979). Simultaneous localization of LHRH and catecholamines in rat hypothalamus. *J. Anat.* 128, 331–347.
105. Agnati, L. F., Zoli, M., Stromberg, I., and Fuxe, K. (1995). Intercellular communication in the brain: wiring versus volume transmission. *Neuroscience* 69, 711–726.
106. Prevot, V. (2002). Glial-neuronal-endothelial interactions are involved in the control of GnRH secretion. *J. Neuroendocrinol.* 14, 247–255.
107. Lagrange, A. H., Rønnekleiv, O. K., and Kelly, M. J. (1995). Estradiol-17b and m-opioid peptides rapidly hyperpolarize GnRH neurons: a cellular mechanism of negative feedback. *Endocrinology* 136, 2341–2344.
108. Sim, J. A., Skynner, M. J., Pape, J.-R., and Herbison, A. E. (2000). Late postnatal reorganization of $GABA_A$ receptor signalling in native GnRH neurons. *Eur. J. Neurosci.* 12, 3497–3504.
109. Sim, J. A., Skynner, M. J., and Herbison, A. E. (2001). Heterogeneity in the basic membrane properties of postnatal gonadotropin-releasing hormone neurons in the mouse. *J. Neurosci.* 21, 1067–1075.
110. DeFazio, R. A., and Moenter, S. M. (2002). Estradiol feedback alters potassium currents and firing properties of gonadotropin-releasing hormone neurons. *Mol. Endocrinol.* 16, 2255–2265.
111. Bosch, M. A., Kelly, M. J., and Ronnekleiv, O. K. (2002). Distribution, neuronal colocalization, and 17beta-E2 modulation of small conductance calcium-activated K(+) channel (SK3) mRNA in the guinea pig brain. *Endocrinology* 143, 1097–1107.
112. Pape, H. C. (1996). Queer current and pacemaker: the hyperpolarization-activated cation current in neurons. *Annu. Rev. Physiol.* 58, 299–327.
113. Erickson, K. R., Ronnekleiv, O. K., and Kelly, M. J. (1993). Role of a T-type calcium current in supporting a depolarizing potential, damped oscillations, and phasic firing in vasopressinergic guinea pig supraoptic neurons. *Neuroendocrinology* 57, 789–800.
114. Kelly, M. J., and Wagner, E. J. (2002). GnRH neurons and episodic bursting activity. *Trends Endocrinol. Metab.* 13, 409–410.
115. Suter, K. J., Wuarin, J. P., Smith, B. N., Dudek, F. E., and Moenter, S. M. (2000). Whole-cell recordings from

preoptic/hypothalamic slices reveal burst firing in gonadotropin-releasing hormone neurons identified with green fluorescent protein in transgenic mice. *Endocrinology* 141, 3731–3736.

116. Han, S. K., Abraham, I. M., and Herbison, A. E. (2002). Effect of GABA on GnRH neurons switches from depolarization to hyperpolarization at puberty in the female mouse. *Endocrinology* 143, 1459–1466.
117. Nunemaker, C. S., DeFazio, R. A., and Moenter, S. M. (2002). Estradiol-sensitive afferents modulate long-term episodic firing patterns of GnRH neurons. *Endocrinology* 143, 2284–2292.
118. Han, S. K., Todman, M. G., and Herbison, A. E. (2004). Endogenous GABA release inhibits the firing of adult gonadotropin-releasing hormone neurons. *Endocrinology* 145, 495–499.
119. Kuehl-Kovarik, M. C., Pouliot, W. A., Halterman, G. L., Handa, R. J., Dudek, F. E., and Partin, K. M. (2002). Episodic bursting activity and response to excitatory amino acids in acutely dissociated gonadotropin-releasing hormone neurons genetically targeted with green fluorescent protein. *J. Neurosci.* 22, 2313–2322.
120. Bicknell, R. J., Brown, D., Chapman, C., Hancock, P. D., and Leng, G. (1984). Reversible fatigue of stimulus-secretion coupling in the rat neurohypophysis. *J. Physiol.* 348, 601–613.
121. Nunemaker, C. S., Straume, M., DeFazio, R. A., and Moenter, S. M. (2003). Gonadotropin-releasing hormone neurons generate interacting rhythms in multiple time domains. *Endocrinology* 144, 823–831.
122. Williamson, P., Lang, J., and Boyd, Y. (1991). The gonadotropin-releasing hormone (Gnrh) gene maps to mouse chromosome 14 and identifies a homologous region on human chromosome 8. *Somat. Cell. Mol. Genet.* 17, 609–615.
123. Adelman, J. P., Mason, A. J., Hayflick, J. S., and Seeburg, P. H. (1986). Isolation of the gene and hypothalamic cDNA for the common precursor of gonadotropin-releasing hormone and prolactin release-inhibiting factor in human and rat. *Proc. Natl. Acad. Sci. U S A* 83, 179–183.
124. Seeburg, P. H. and Adelman, J. P. (1984). Characterization of cDNA for precursor of human luteinizing hormone releasing hormone. *Nature* 311, 666–668.
125. Bond, C. T., Hayflick, J. S., Seeburg, P. H., and Adelman, J. P. (1989). The rat gonadotropin-releasing hormone: SH locus: structure and hypothalamic expression. *Mol. Endocrinol.* 3, 1257–1262.
126. Adelman, J. P., Bond, C. T., Douglass, J., and Herbert, E. (1987). Two mammalian genes transcribed from opposite strands of the same DNA locus. *Science* 235, 1514–1517.
127. Dong, K. W., Yu, K. L., Chen, Z. G., Chen, Y. D., and Roberts, J. L. (1997). Characterization of multiple promoters directing tissue-specific expression of the human gonadotropin-releasing hormone gene. *Endocrinology* 138, 2754–2762.
128. Dong, K. W., Yu, K. L., and Roberts, J. L. (1993). Identification of a major up-stream transcription start site for the human progonadotropin-releasing hormone gene used in reproductive tissues and cell lines. *Mol. Endocrinol.* 7, 1654–1666.
129. Nelson, S. B., Eraly, S. A., and Mellon, P. L. (1998). The GnRH promoter: target of transcription factors, hormones, and signaling pathways. *Mol. Cell. Endocrinol.* 140, 151–155.
130. Wierman, M. E., Fang, Z., and Kepa, J. K. (1996). GnRH gene expression in neuronal cell lines. *TEM* 7, 60–65.
131. Chandran, U. R., and DeFranco, D. B. (1999). Regulation of gonadotropin-releasing hormone gene transcription. *Behav. Brain Res.* 105, 29–36.
132. Wolfe, A., Kim, H. H., and Radovick, S. (2002). The GnRH neuron: molecular aspects of migration, gene expression and regulation. *Prog. Brain Res.* 141, 243–257.
133. Lawson, M. A. and Mellon, P. L. (2002). GnRH gene expression: lessons learned from immortalized cells. In *Neuroplasticity, Development, and Steroid Hormone Action* (R. J. Handa, S. Hayashi, E. Terasawa, and M. Kawata, Eds.), pp. 141–152. CRC Press, Boca Raton, FL.
134. Eraly, S. A., and Mellon, P. L. (1995). Regulation of gonadotropin-releasing hormone transcription by protein kinase C is mediated by evolutionarily conserved promoter-proximal elements. *Mol. Endocrinol.* 9, 848–859.
135. Whyte, D. B., Lawson, M. A., Belsham, D. D., Eraly, S. A., Bond, C. T., Adelman, J. P., and Mellon, P. L. (1995). A neuron-specific enhancer targets expression of the gonadotropin-releasing hormone gene to hypothalamic neurosecretory neurons. *Mol. Endocrinol.* 9, 467–477.
136. Lawson, M. A., Macconell, L. A., Kim, J., Powl, B. T., Nelson, S. B., and Mellon, P. L. (2002). Neuron-specific expression in vivo by defined transcription regulatory elements of the GnRH gene. *Endocrinology* 143, 1404–1412.
137. Kepa, J. K., Spaulding, A. J., Jacobsen, B. M., Fang, Z., Xiong, X., Radovick, S., and Wierman, M. E. (1996). Structure of the distal human gonadotropin releasing hormone (hGnrh) gene promoter and functional analysis in Gt1-7 neuronal cells. *Nucleic Acids Res.* 24, 3614–3620.
138. Wolfe, A. M., Wray, S., Westphal, H., and Radovick, S. (1996). Cell-specific expression of the human gonadotropin-releasing hormone gene in transgenic animals. *J. Biol. Chem.* 271, 20018–20023.
139. Zakaria, M., Dunn, I. C., Zhen, S., Su, E., Smith, E., Patriquin, E., and Radovick, S. (1996). Phorbol ester regulation of the gonadotropin-releasing hormone (GnRH) gene in GnRH-secreting cell lines: a molecular basis for species differences. *Mol. Endocrinol.* 10, 1282–1291.
140. Pape, J.-R., Skynner, M. J., Allen, N. D., and Herbison, A. E. (1999). Transgenics identify distal 5′- and 3′ sequences specifying gonadotropin-releasing hormone expression in adult mice. *Mol. Endocrinol.* 13, 2203–2211.
141. Thanky, N. R., Slater, R., and Herbison, A. E. (2003). Sex differences in estrogen-dependent transcription of gonadotropin-releasing hormone (GnRH) gene revealed in GnRH transgenic mice. *Endocrinology* 144, 3351–3358.
142. Kim, H. H., Wolfe, A., Smith, G. R., Tobet, S. A., and Radovick, S. (2002). Promoter sequences targeting tissue-specific gene expression of hypothalamic and ovarian gonadotropin-releasing hormone in vivo. *J. Biol. Chem.* 277, 5194–5202.
143. Sun, W., Choe, Y. S., Lee, Y. J., and Kim, K. (1997). Suppression of GnRH gene expression in GT1-1 hypothalamic neuronal cells: action of protein kinase C. *Neuroreport* 8, 3541–3546.
144. Yeo, T. T., Gore, A. C., Jakubowski, M., Dong, K. W., Blum, M., and Roberts, J. L. (1996). Characterization of gonadotropin-releasing hormone gene transcripts in a mouse hypothalamic neuronal GT1 cell line. *Brain Res. Mol. Brain Res.* 42, 255–262.
145. Maurer, J. A., and Wray, S. (1997). Luteinizing hormone-releasing hormone (LHRH) neurons maintained in hypothalamic slice explant cultures exhibit a rapid LHRH mRNA turnover rate. *J. Neurosci.* 17, 9481–9491.
146. Maurer, J. A., and Wray, S. (1999). Luteinizing hormone-releasing hormone quantified in tissues and slice explant cultures of postnatal rat hypothalami. *Endocrinology* 140, 791–799.
147. Nunez, L., Faught, W. J., and Frawley, L. S. (1998). Episodic gonadotropin-releasing hormone gene expression revealed

by dynamic monitoring of luciferase reporter activity in single, living neurons. *Proc. Natl. Acad. Sci. U S A* 95, 9648–9653.
148. Gore, A. C., and Roberts, J. L. (1995). Regulation of gonadotropin-releasing hormone gene expression in the rat during the luteinizing hormone surge. *Endocrinology* 136, 889–896.
149. Petersen, S. L., Gardner, E., Adelman, J., and McCrone, S. (1996). Examination of steroid-induced changes in LHRH gene transcription using ^{33}P- and ^{33}S-labeled probes specific for Intron 2. *Endocrinology* 137, 234–29.
150. Herbison, A. E., Pape, J. R., Simonian, S. X., Skynner, M. J., and Sim, J. A. (2001). Molecular and cellular properties of GnRH neurons revealed through transgenics in the mouse. *Mol. Cell. Endocrinol.* 185, 185–194.
151. Jakubowski, M., and Roberts, J. L. (1994). Processing of gonadotropin-releasing hormone gene transcripts in the rat brain. *J. Biol. Chem.* 269, 4078–4083.
152. Gore, A. C., and Roberts, J. L. (1997). Regulation of gonadotropin-releasing hormone gene expression in vivo and in vitro. *Front. Neuroendocrinol.* 18, 209–245.
153. Seong, J. Y., Park, S., and Kim, K. (1999). Enhanced splicing of the first intron from the gonadotropin-releasing hormone (GnRH) primary transcript is a prerequisite for mature GnRH messenger RNA: presence of GnRH neuron-specific splicing factors. *Mol. Endocrinol.* 13, 1882–1895.
154. Han, J., Son, G. H., Seong, J. Y., and Kim, K. (2002). GnRH pre-mRNA splicing: role of exonic splicing enhancer. *Prog. Brain Res.* 141, 209–219.
155. Han, J., Seong, J. Y., Kim, K., Wuttke, W., and Jarry, H. (2001). Analysis of exonic splicing enhancers in the mouse gonadotropin-releasing hormone (GnRH) gene. *Mol. Cell. Endocrinol.* 173, 157–166.
156. Seong, J. Y., Han, J., Park, S., Wuttke, W., Jarry, H., and Kim, K. (2002). Exonic splicing enhancer-dependent splicing of the gonadotropin-releasing hormone premessenger ribonucleic acid is mediated by tra2alpha, a 40-kilodalton serine/arginine-rich protein. *Mol. Endocrinol.* 16, 2426–2438.
157. Zhen, S., Dunn, I. C., Wray, S., Liu, Y., Chappell, P. E., Levine, J. E., and Radovick, S. (1997). An alternative gonadotropin-releasing hormone (GnRH) RNA splicing product found in cultured GnRH neurons and mouse hypothalamus. *J. Biol. Chem.* 272, 12620–12625.
158. Son, G. H., Jung, H., Seong, J. Y., Choe, Y., Geum, D., and Kim, K. (2003). Excision of the first intron from the gonadotropin-releasing hormone (GnRH) transcript serves as a key regulatory step for GnRH biosynthesis. *J. Biol. Chem.* 278, 18037–18044.
159. Mason, A. J., Hayflick, J. S., Zoeller, R. T., Young, W. S., 3rd, Phillips, H. S., Nikolics, K., and Seeburg, P. H. (1986). A deletion truncating the gonadotropin-releasing hormone gene is responsible for hypogonadism in the hpg mouse. *Science* 234, 1366–1371.
160. Seong, J. Y., Kim, B. W., Park, S., Son, G. H., and Kim, K. (2001). First intron excision of GnRH pre-mRNA during postnatal development of normal mice and adult hypogonadal mice. *Endocrinology* 142, 4454–4461.
161. Maurer, J. A., and Wray, S. (1997). Luteinizing hormone-releasing hormone (LHRH) neurons maintained in hypothalamic slice explant cultures exhibit a rapid LHRH mRNA turnover rate. *J. Neurosci.* 17, 9481–9491.
162. Gore, A. C., Roberts, J. L., and Gibson, M. J. (1999). Mechanisms for the regulation of gonadotropin-releasing hormone gene expression in the developing mouse. *Endocrinology* 140, 2280–2287.
163. Gore, A. C., and Roberts, J. L. (1994). Regulation of gonadotropin-releasing hormone gene expression by the excitatory amino acids kainic acid and N-methyl-D,L-aspartate in the male rat. *Endocrinology* 134, 2026–2031.
164. Petersen, S. L., McCrone, S., Keller, M., and Gardner, E. (1991). Rapid increase in LHRH mRNA levels following NMDA. *Endocrinology* 129, 1679–1681.
165. Wu, T. J., Gibson, M. J., and Roberts, J. L. (2000). Effect of N-methyl-D,L-aspartate (NMA) on gonadotropin-releasing hormone (GnRH) gene expression in male mice. *Brain Res.* 862, 238–241.
166. Jung, N., Sun, W., Lee, H., Cho, S., Shim, C., and Kim, K. (1998). Gonadotropin-releasing hormone (GnRH) gene regulation by N-methyl-D-aspartic acid in GT1-1 neuronal cells: differential involvement of c-fos and c-jun protooncogenes. *Brain Res. Mol. Brain Res.* 61, 162–169.
167. Belsham, D. D., Wetsel, W. C., and Mellon, P. L. (1996). NMDA and nitric oxide act through the cGMP signal transduction pathway to repress hypothalamic gonadotropin-releasing hormone gene expression. *EMBO J.* 15, 538–547.
168. Gore, A. C., Ho, A., and Roberts, J. L. (1995). Translational efficiency of gonadotropin-releasing hormone messenger ribonucleic acid is negatively regulated by phorbol ester in GT1-7 cells. *Endocrinology* 136, 1620–1625.
169. Wetsel, W. C., and Srinivasan, S. (2002). Pro-GnRH processing. *Prog. Brain Res.* 141, 221–241.
170. Wetzel, C. H., Hermann, B., Behl, C., Pestel, E., Rammes, G., Zieglgansberger, W., Holsboer, F., and Rupprecht, R. (1998). Functional antagonism of gonadal steroids at the 5-hydroxytryptamine type 3 receptor. *Mol. Endocrinol.* 12, 1441–1451.
171. Jackson, R. S., Creemers, J. W., Ohagi, S., Raffin-Sanson, M. L., Sanders, L., Montague, C. T., Hutton, J. C., and O'Rahilly, S. (1997). Obesity and impaired prohormone processing associated with mutations in the human prohormone convertase 1 gene. *Nat. Genet.* 16, 303–306.
172. Rangaraju, N. S. and Harris, R. B. (1993). GAP-releasing enzyme is a member of the pro-hormone convertase family of precursor protein processing enzymes. *Life Sci.* 52, 147–153.
173. Srinivasan, S., Bunch, D. O., Feng, Y., Rodriguiz, R. M., Li, M., Ravenell, R. L., Luo, G. X., Arimura, A., Fricker, L. D., Eddy, E. M., and Wetsel, W. C. (2004). Deficits in reproduction and pro-gonadotropin-releasing hormone processing in male Cpefat mice. *Endocrinology* 145, 2023–2034.
174. Wetsel, W. C., Liposits, Z., Seidah, N. G., and Collins, S. (1995). Expression of candidate pro-GnRH processing enzymes in rat hypothalamus and an immortalized hypothalamic neuronal cell line. *Neuroendocrinology* 62, 166–177.
175. Rubin, B. S., King, J. C., Millar, R. P., Seeburg, P. H., and Arimura, A. (1987). Processing of luteinizing hormone-releasing hormone precursor in rat neurons. *Endocrinology* 121, 305–309.
176. King, J. C., and Anthony, E. L. (1983). Biosynthesis of LHRH: inferences from immunocytochemical studies. *Peptides* 4, 963–970.
177. Rangaraju, N. S., Xu, J. F., and Harris, R. B. (1991). Pro-gonadotropin-releasing hormone protein is processed within hypothalamic neurosecretory granules. *Neuroendocrinology* 53, 20–28.
178. Rochdi, L., Theraulaz, L., Enjalbert, A., and Gautron, J. P. (2000). Differential in vitro secretion of gonadotropin-releasing hormone (GnRH) and (hydroxyproline) GnRH from the rat hypothalamus during postnatal development. *J. Neuroendocrinol.* 12, 919–926.
179. Gautron, J. P., Leblanc, P., Bluet-Pajot, M. T., Pattou, E., L'Heritier, A., Mounier, F., Ponce, G., Audinot, V., Rasolonjanahary, R., and Kordon, C. (1992). A second endogenous molecular form of mammalian hypothalamic

luteinizing hormone-releasing hormone (LHRH), (hydroxyproline9) LHRH, releases luteinizing hormone and follicle-stimulating hormone in vitro and in vivo. *Mol. Cell. Endocrinol.* 85, 99–107.

180. Gautron, J. P., Pattou, E., Leblanc, P., L'Heritier, A., and Kordon, C. (1993). Preferential distribution of C-terminal fragments of (hydroxyproline9) LHRH in the rat hippocampus and olfactory bulb. *Neuroendocrinology* 58, 240–250.
181. Molineaux, C. J., Lasdun, A., Michaud, C., and Orlowski, M. (1988). Endopeptidase-24.15 is the primary enzyme that degrades luteinizing hormone releasing hormone both in vitro and in vivo. *J. Neurochem.* 51, 624–633.
182. Krause, J. E., Advis, J. P., and McKelvy, J. F. (1982). In vivo biosynthesis of hypothalamic luteinizing hormone releasing hormone in individual free-running female rats. *Endocrinology* 111, 344–346.
183. Lew, R. A., Tetaz, T. J., Glucksman, M. J., Roberts, J. L., and Smith, A. I. (1994). Evidence for a two-step mechanism of gonadotropin-releasing hormone metabolism by prolyl endopeptidase and metalloendopeptidase EC 3.4.24.15 in ovine hypothalamic extracts. *J. Biol. Chem.* 269, 12626–12632.
184. Wu, T. J., Pierotti, A. R., Jakubowski, M., Sheward, W. J., Glucksman, M. J., Smith, A. I., King, J. C., Fink, G., and Roberts, J. L. (1997). Endopeptidase EC 3.4.24.15 presence in the rat median eminence and hypophysial portal blood and its modulation of the luteinizing hormone surge. *J. Neuroendocrinol.* 9, 813–822.
185. Advis, J. P., Krause, J. E., and McKelvy, J. F. (1983). Evidence that endopeptidase-catalyzed luteinizing hormone releasing hormone cleavage contributes to the regulation of median eminence LHRH levels during positive steroid feedback. *Endocrinology* 112, 1147–1149.
186. Lew, R. A., Cowley, M., Clarke, I. J., and Smith, A. I. (1997). Peptidases that degrade gonadotropin-releasing hormone: influence on LH secretion in the ewe. *J. Neuroendocrinol.* 9, 707–712.
187. Bourguignon, J. P., Alvarez Gonzalez, M. L., Gerard, A., and Franchimont, P. (1994). Gonadotropin releasing hormone inhibitory autofeedback by subproducts antagonist at N-methyl-D-aspartate receptors: a model of autocrine regulation of peptide secretion. *Endocrinology* 134, 1589–1592.
188. Gore, A. C., and Roberts, J. L. (1997). Regulation of gonadotropin-releasing hormone gene expression in vivo and in vitro. *Front. Neuroendocrinol.* 18, 209–245.
189. Sagrillo, C. A., Grattan, D. R., McCarthy, M. M., and Selmanoff, M. (1996). Hormonal and neurotransmitter regulation of GnRH gene expression and related reproductive behaviours. *Behavior Genetics* 26, 241–277.
190. Jimenez-Linan, M., and Rubin, B. S. (2001). Dynamic changes in luteinizing hormone releasing hormone transcriptional activity are associated with the steroid-induced LH surge. *Brain Res.* 922, 71–79.
191. Harris, T. G., Robinson, J. E., Evans, N. P., Skinner, D. C., and Herbison, A. E. (1998). GnRH mRNA expression changes prior to the onset of the estradiol-induced LH surge in the ewe. *Endocrinology* 139, 57–64.
192. Beauvillain, J. C., Tramu, G., and Dubois, M. P. (1981). Ultrastructural immunocytochemical evidence of the presence of a peptide related to ACTH in granules of LHRH nerve terminals in the median eminence of the guinea pig. *Cell. Tissue Res.* 218, 1–6.
193. Leonardelli, J., and Tramu, G. (1979). Immunoreactivity for beta-endorphin in LH-RH neurons of the fetal human hypothalamus. *Cell. Tissue Res.* 203, 201–207.
194. Charnay, Y., Bouras, C., Vallet, P. G., Golaz, J., Guntern, R., and Constantinidis, J. (1989). Immunohistochemical colocalization of delta sleep-inducing peptide and luteinizing hormone-releasing hormone in rabbit brain neurons. *Neuroscience* 31, 495–505.
195. Coen, C. W., Montagnese, C., and Opacka-Juffry, J. (1990). Coexistence of gonadotrophin-releasing hormone and galanin: immunohistochemical and functional studies. *J. Neuroendocrinol.* 2, 107–111.
196. Merchenthaler, I., Lopez, F. J., and Negro-Vilar, A. (1990). Colocalization of galanin and luteinizing hormone-releasing hormone in a subset of preoptic hypothalamic neurons: anatomical and functional correlates. *Proc. Natl. Acad. Sci. U S A* 87, 6326–6330.
197. Ciofi, P. (2000). Phenotypical segregation among female rat hypothalamic gonadotropin-releasing hormone neurons as revealed by the sexually dimorphic coexpression of cholecystokinin and neurotensin. *Neuroscience* 99, 133–147.
198. Foradori, C. D., Goodman, R. L., and Lehman, M. N. (2003). Colocalization of orphanin FQ in GnRH neurons of the ovine preoptic area and hypothlamus. *Soc. Neurosci. Abstr.* 237.1.
199. Merchenthaler, I., Lopez, F. J., Lennard, D. E., and Negro-Vilar, A. (1991). Sexual differences in the distribution of neurons coexpressing galanin and luteinizing hormone-releasing hormone in the rat brain. *Endocrinology* 129, 1977–1986.
200. Liposits, Z., Reid, J. J., Negro-Vilar, A., and Merchenthaler, I. (1995). Sexual dimorphism in copackaging of luteinizing hormone-releasing hormone and galanin into neurosecretory vesicles of hypophysiotrophic neurons: estrogen dependency. *Endocrinology* 136, 1987–1992.
201. Rajendren, G., and Gibson, M. J. (1999). Expression of galanin immunoreactivity in gonadotropin-releasing hormone neurons in mice: a confocal microscopic study. *Brain Res.* 821, 270–276.
202. Dufourny, L., Schofield, N., and Skinner, D. C. (2003). Immunoreactive galanin expression in ovine gonadotropin-releasing hormone neurones: no effects of gender or reproductive status. *J. Neuroendocrinol.* 15, 1062–1069.
203. Finn, P. D., Pau, K. Y., Spies, H. G., Cunningham, M. J., Clifton, D. K., and Steiner, R. A. (2000). Galanin's functional significance in the regulation of the neuroendocrine reproductive axis of the monkey. *Neuroendocrinology* 71, 16–26.
204. Lopez, F. J., Merchenthaler, I., Ching, M., Wisniewski, M. G., and Negro-Vilar, A. (1991). Galanin: a hypothalamic-hypophysiotropic hormone modulating reproductive functions. *Proc. Natl. Acad. Sci. U S A* 88, 4508–4512.
205. Hrabovszky, E., Vrontakis, M. E., and Petersen, S. L. (1995). Triple-labeling method combining immunocytochemistry and in situ hybridization histochemistry: demonstration of overlap between Fos-immunoreactive and galanin mRNA-expressing subpopulations of luteinizing hormone-releasing hormone neurons in female rats. *J. Histochem. Cytochem.* 43, 363–370.
206. Marks, D. L., Lent, K. L., Rossmanith, W. G., Clifton, D. K., and Steiner, R. A. (1994). Activation-dependent regulation of galanin gene expression in gonadotropin-releasing hormone neurons in the female rat. *Endocrinology* 134, 1991–1998.
207. Marks, D. L., Smith, M. S., Vrontakis, M., Clifton, D. K., and Steiner, R. A. (1993). Regulation of galanin gene expression in gonadotropin-releasing hormone neurons during the estrous cycle of the rat. *Endocrinology* 132, 1836–1844.
208. Rossmanith, W. G., Marks, D. L., Clifton, D. K., and Steiner, R. A. (1996). Induction of galanin mRNA in GnRH neurons by estradiol and its facilitation by progesterone. *J. Neuroendocrinol.* 8, 185–191.
209. Splett, C. L., Scheffen, J. R., Desotelle, J. A., Plamann, V., and Bauer-Dantoin, A. C. (2003). Galanin enhancement of gonadotropin-releasing hormone-stimulated luteinizing

hormone secretion in female rats is estrogen dependent. *Endocrinology* 144, 484–490.

210. Merchenthaler, I., Lopez, F. J., and Negro-Vilar, A. (1990). Colocalization of galanin and luteinizing hormone-releasing hormone in a subset of preoptic hypothalamic neurons: anatomical and functional correlates. *Proc. Natl. Acad. Sci. U S A* 87, 6326–6330.
211. Sahu, A., Xu, B., and Kalra, S. P. (1994). Role of galanin in stimulation of pituitary luteinizing hormone secretion as revealed by a specific receptor antagonist, galantide. *Endocrinology* 134, 529–536.
212. Cheung, C. C., Clifton, D. K., and Steiner, R. A. (1996). Galanin, an unassuming neuropeptide moves to center stage in reproduction. *Trends Endo. Metab.* 7, 301–306.
213. Mitchell, V., Bouret, S., Prevot, V., Jennes, L., and Beauvillain, J. C. (1999). Evidence for expression of galanin receptor Gal-R1 mRNA in certain gonadotropin releasing hormone neurones of the rostral preoptic area. *J Neuroendocrinol* 11, 805–812.
214. Shine, J. (1994). Structural conservation and functional diversity—profile of a political peptide. *Endocrinology* 134, 1989–1990.
215. Wynick, D., Small, C. J., Bacon, A., Holmes, F. E., Norman, M., Ormandy, C. J., Kilic, E., Kerr, N. C., Ghatei, M., Talamantes, F., Bloom, S. R., and Pachnis, V. (1998). Galanin regulates prolactin release and lactotroph proliferation. *Proc. Natl. Acad. Sci. U S A* 95, 12671–12676.
216. Hohmann, J. G., Krasnow, S. M., Teklemichael, D. N., Clifton, D. K., Wynick, D., and Steiner, R. A. (2003). Neuroendocrine profiles in galanin-overexpressing and knockout mice. *Neuroendocrinology* 77, 354–366.
217. Jacoby, A. S., Hort, Y. J., Constantinescu, G., Shine, J., and Iismaa, T. P. (2002). Critical role for GALR1 galanin receptor in galanin regulation of neuroendocrine function and seizure activity. *Brain Res. Mol. Brain Res.* 107, 195–200.
218. Schoenenberger, G. A. and Monnier, M. (1977). Characterization of a delta-electroencephalogram (-sleep)-inducing peptide. *Proc. Natl. Acad. Sci. U S A* 74, 1282–1286.
219. Charnay, Y., Golaz, J., Vallet, P. G., and Bouras, C. (1992). Production and immunohistochemical application of monoclonal antibodies against delta sleep-inducing peptide. *J. Chem. Neuroanat.* 5, 503–509.
220. Vallet, P. G., Charnay, Y., Boura, C., and Kiss, J. Z. (1991). Colocalization of delta sleep inducing peptide and luteinizing hormone releasing hormone in neurosecretory vesicles in rat median eminence. *Neuroendocrinology* 53, 103–106.
221. Vallet, P. G., Charnay, Y., and Bouras, C. (1990). Distribution and colocalization of delta sleep-inducing peptide and luteinizing hormone-releasing hormone in the aged human brain: an immunohistochemical study. *J. Chem. Neuroanat.* 3, 207–214.
222. Pu, L. P., Charnay, Y., Leduque, P., Morel, G., and Dubois, P. M. (1991). Light and electron microscopic immunocytochemical evidence that delta sleep-inducing peptide and gonadotropin-releasing hormone are coexpressed in the same nerve structures in the guinea pig median eminence. *Neuroendocrinology* 53, 332–338.
223. Sahu, A., and Kalra, S. P. (1987). Delta sleep-inducing peptide (DSIP) stimulates LH release in steroid-primed ovariectomized rats. *Life Sci.* 40, 1201–1206.
224. Iyer, K. S., and McCann, S. M. (1987). Delta sleep inducing peptide (DSIP) stimulates the release of LH but not FSH via a hypothalamic site of action in the rat. *Brain Res. Bull.* 19, 535–538.
225. Iyer, K. S., and McCann, S. M. (1987). Delta sleep-inducing peptide (DSIP) stimulates growth hormone (GH) release in the rat by hypothalamic and pituitary actions. *Peptides* 8, 45–48.
226. Clarke, I. J., Cummins, J. T., Karsch, F. J., Seeburg, P. H., and Nikolics, K. (1987). GnRH-associated peptide (GAP) is cosecreted with GnRH into the hypophyseal portal blood of ovariectomized sheep. *Biochem. Biophys. Res. Commun.* 143, 665–671.
227. Sarkar, D. K., and Mitsugi, N. (1990). Correlative changes of the gonadotropin-releasing hormone and gonadotropin-releasing-hormone-associated peptide immunoreactivities in the pituitary portal plasma in female rats. *Neuroendocrinology* 52, 15–21.
228. Nikolics, K., Mason, A. J., Szonyi, E., Ramachandran, J., and Seeburg, P. H. (1985). A prolactin-inhibiting factor within the precursor for human gonadotropin-releasing hormone. *Nature* 316, 511–517.
229. Millar, R. P., Wormald, P. J., and Milton, R. C. (1986). Stimulation of gonadotropin release by a non-GnRH peptide sequence of the GnRH precursor. *Science* 232, 68–70.
230. Wormald, P. J., Abrahamson, M. J., Seeburg, P. H., Nikolics, K., and Millar, R. P. (1989). Prolactin-inhibiting activity of GnRH associated peptide in cultured human pituitary cells. *Clin. Endocrinol.* 30, 149–155.
231. Milton, S. C., Brandt, W. F., Schnolzer, M., and Milton, R. C. (1992). Total solid-phase synthesis and prolactin-inhibiting activity of the gonadotropin-releasing hormone precursor protein and the gonadotropin-releasing hormone associated peptide. *Biochemistry* 31, 8799–8809.
232. Chandrashekar, V., Bartke, A., and Browning, R. A. (1988). Assessment of the effects of a synthetic gonadotropin-releasing hormone associated peptide on hormone release from the in situ and ectopic pituitaries in adult male rats. *Brain Res. Bull.* 21, 95–99.
233. Thomas, G. B., Cummins, J. T., Doughton, B. W., Griffin, N., Millar, R. P., Milton, R. C., and Clarke, I. J. (1988). Gonadotropin-releasing hormone associated peptide (GAP) and putative processed GAP peptides do not release luteinizing hormone or follicle-stimulating hormone or inhibit prolactin secretion in the sheep. *Neuroendocrinology* 48, 342–350.
234. Yu, W. H., Arisawa, M., Millar, R. P., and McCann, S. M. (1989). Effects of the gonadotropin-releasing hormone associated peptides (GAP) on the release of luteinizing hormone (LH), follicle stimulating hormone (FSH) and prolactin (PRL) in vivo. *Peptides* 10, 1133–1138.
235. Vacher, P., Mariot, P., Dufy-Barbe, L., Nikolics, K., Seeburg, P. H., Kerdelhue, B., and Dufy, B. (1991). The gonadotropin-releasing hormone associated peptide reduces calcium entry in prolactin-secreting cells. *Endocrinology* 128, 285–294.
236. Van Chuoi, M. T., Vacher, P., and Dufy, B. (1993). GnRH-associated peptide decreases cyclic AMP accumulation in the GH3 pituitary cell line. *Neuroendocrinology* 58, 251–257.
237. Veldhuis, J. D., Johnson, M. L., and Seneta, E. (1991). Analysis of the copulsatility of anterior pituitary hormones. *J. Clin. Endocrinol. Metab.* 73, 569–576.
238. Dunger, D. B., Matthews, D. R., Edge, J. A., Jones, J., and Preece, M. A. (1991). Evidence for temporal coupling of growth hormone, prolactin, LH and FSH pulsatility overnight during normal puberty. *J. Endocrinol.* 130, 141–149.
239. Freeman, M. E. (1994). The neuroendocrine control of the ovarian cycle in the rat. In *The Physiology of Reproduction* (E. Knobil and J. D. Neill, Eds.), pp. 613–638. Raven Press, New York.
240. Yu, W. H., Seeburg, P. H., Nikolics, K., and McCann, S. M. (1988). Gonadotropin-releasing hormone-associated peptide exerts a prolactin-inhibiting and weak gonadotropin-releasing activity in vivo. *Endocrinology* 123, 390–395.
241. Yu, W. H., Millar, R. P., Milton, S. C., Milton, R. C., and McCann, S. M. (1990). Selective FSH-releasing activity of (D-Trp9)GAP1–13: comparison with gonadotropin-releasing

abilities of analogs of GAP and natural LHRHs. *Brain Res. Bull.* 25, 867–873.
242. Jan, Y. N., and Jan, L. Y. (1983). A LHRH-like peptidergic neurotransmitter capable of "action at a distance" in autonomic ganglia. *Trends Neurosci.* 6, 320–325.
243. Brown, D. (1988). M-currents: an update. *Trends Neurosci.* 11, 294–299.
244. Ford, C. P., Dryden, W. F., and Smith, P. A. (2003). Neurotrophic regulation of calcium channels by the peptide neurotransmitter luteinizing hormone releasing hormone. *J. Neurosci.* 23, 7169–7175.
245. Dudley, C. A., Lee, Y., and Moss, R. L. (1990). Electrophysiological identification of a pathway from the septal area to the medial amygdala: sensitivity to estrogen and luteinizing hormone-releasing hormone. *Synapse* 6, 161–168.
246. Palovcik, R. A., and Phillips, M. I. (1986). A biphasic excitatory response of hippocampal neurons to gonadotropin-releasing hormone. *Neuroendocrinology* 44, 137–141.
247. Wong, M., Eaton, M. J., and Moss, R. L. (1990). Electrophysiological actions of luteinizing hormone-releasing hormone: intracellular studies in the rat hippocampal slice preparation. *Synapse* 5, 65–70.
248. Pan, J. T., Kow, L. M., and Pfaff, D. W. (1988). Modulatory actions of luteinizing hormone-releasing hormone on electrical activity of preoptic neurons in brain slices. *Neuroscience* 27, 623–628.
249. Moss, R. L., and Dudley, C. A. (1978). Changes in responsiveness of medial preoptic neurons to the microelectrophoresis of releasing hormones as a function of ovarian hormones. *Brain Res.* 149, 511–515.
250. Dyer, R. G., and Dyball, R. E. (1974). Evidence for a direct effect of LRF and TRF on single unit activity in the rostral hypothalamus. *Nature* 252, 486–488.
251. Kawakami, M., and Sakuma, Y. (1974). Responses of hypothalamic neurons to the microiontophoresis of LH-RH, LH and FSH under various levels of circulating ovarian hormones. *Neuroendocrinology* 15, 290–307.
252. Herbison, A. E., Hubbard, J. I., and Sirett, N. E. (1984). LH-RH in picomole concentrations evokes excitation and inhibition of rat arcuate neurones in vitro. *Neurosci. Lett.* 46, 311–315.
253. Sakuma, Y., and Pfaff, D. W. (1980). LH-RH in the mesencephalic central gray can potentiate lordosis reflex of female rats. *Nature* 283, 566–567.
254. Ogawa, S., Kow, L. M., and Pfaff, D. W. (1992). Effects of lordosis-relevant neuropeptides on midbrain periaqueductal gray neuronal activity in vitro. *Peptides* 13, 965–975.
255. Schiess, M. C., Dudley, C. A., and Moss, R. L. (1987). Estrogen priming affects the sensitivity of midbrain central gray neurons to microiontophoretically applied LHRH but not beta-endorphin. *Neuroendocrinology* 46, 24–31.
256. Chan, A., Dudley, C. A., and Moss, R. L. (1985). Hormonal modulation of the responsiveness of midbrain central gray neurons to LH-RH. *Neuroendocrinology* 41, 163–168.
257. Renaud, L. P., Martin, J. B., and Brazeau, P. (1975). Depressant action of TRH, LH-RH and somatostatin on activity of central neurones. *Nature* 255, 233–235.
258. Sakuma, Y. (2002). GnRH in the regulation of female rat sexual behavior. *Prog. Brain Res.* 141, 293–301.
259. Pfaff, D. W. (1973). Luteinizing hormone-releasing factor potentiates lordosis behavior in hypophysectomized ovariectomized female rats. *Science* 182, 1148–1149.
260. Moss, R. L., and McCann, S. M. (1973). Induction of mating behavior in rats by luteinizing hormone-releasing factor. *Science* 181, 177–179.
261. Kendrick, K. M., and Dixson, A. F. (1985). Luteinizing hormone releasing hormone enhances proceptivity in a primate. *Neuroendocrinology* 41, 449–453.
262. Ward, B. J., and Charlton, H. M. (1981). Female sexual behaviour in the GnRH deficient, hypogonadal (hpg) mouse. *Physiol. Behav.* 27, 1107–1109.
263. Gulyas, A. I., Megias, M., Emri, Z., and Freund, T. F. (1999). Total number and ratio of excitatory and inhibitory synapses converging onto single interneurons of different types in the CA1 area of the rat hippocampus. *J. Neurosci.* 19, 10082–10097.
264. Matyas, F., Freund, T. F., and Gulyas, A. I. (2004). Convergence of excitatory and inhibitory inputs onto CCK-containing basket cells in the CA1 area of the rat hippocampus. *Eur. J. Neurosci.* 19, 1243–1256.
265. Gu, G. B., and Simerly, R. B. (1997). Projections of the sexually dimorphic anteroventral periventricular nucleus in the female rat. *J. Compar. Neurol.* 384, 142–164.
266. Simonian, S. X., Spratt, D. P., and Herbison, A. E. (1999). Identification and characterization of estrogen receptor a-containing neurons projecting to the vicinity of the gonadotropin-releasing hormone perikarya in the rostral preoptic area of the rat. *J. Compar. Neurol.* 411, 346–358.
267. Wright, D. E., and Jennes, L. (1993). Origin of noradrenergic projections to GnRH perikarya-containing areas in the medial septum-diagonal band and preoptic area. *Brain Res.* 621, 272–278.
268. Van der Beek, E. M., Horvath, T. L., Wiegant, V. M., Van den Hurk, R., and Buijs, R. M. (1997). Evidence for a direct neuronal pathway from the suprachiasmatic nucleus to the gonadotropin-releasing hormone system: combined tracing and light and electron microscopic immunocytochemical studies. *J. Comp. Neurol.* 384, 569–579.
269. Horvath, T. L., Naftolin, F., and Leranth, C. (1993). Luteinizing hormone-releasing hormone and gamma-aminobutyric acid neurons in the medial preoptic area are synaptic targets of dopamine axons originating in anterior periventricular areas. *J. Neuroendocrinol.* 5, 71–79.
270. Tillet, Y., Batailler, M., and Thibault, J. (1993). Neuronal projections to the medial preoptic area of the sheep, with special reference to monoaminergic afferents: immunohistochemical and retrograde tract tracing studies. *J. Compar. Neurol.* 330, 195–220.
271. Goubillon, M., Delaleu, B., Tillet, Y., Caraty, A., and Herbison, A. E. (1999). Localization of estrogen-receptive neurons projecting to the GnRH neuron-containing rostral preoptic area of the ewe. *Neuroendocrinology* 70, 228–236.
272. Goubillon, M. L., Caraty, A., and Herbison, A. E. (2002). Evidence in favour of a direct input from the ventromedial nucleus to gonadotropin-releasing hormone neurones in the ewe: an anterograde tracing study. *J. Neuroendocrinol.* 14, 95–100.
273. Jansen, H. T., Hileman, S. M., Lubbers, L. S., Jackson, G. L., and Lehman, M. N. (1996). A subset of estrogen receptor-containing neurons project to the median eminence in the ewe. *J. Neuroendocrinol.* 8, 21–927.
274. Rawson, J. A., Scott, C. J., Pereira, A., Jakubowsak, A., and Clarke, I. J. (2001). Noradrenergic projections from the A1 field of preoptic area in the brain of the ewe and Fos reponses to oestrogen in the A1 cells. *J. Neuroendocrinol.* 13, 129–138.
275. Eyigor, O., and Jennes, L. (1996). Identification of glutamate receptor subtype mRNAs in gonadotropin-releasing hormone neurons in rat brain. *Endocrine* 4, 133–139.
276. Ottem, E. N., Godwin, J. G., and Petersen, S. L. (2002). Glutamatergic signaling through the N-methyl-D-aspartate

receptor directly activates medial subpopulations of luteinizing hormone-releasing hormone (LHRH) neurons, but does not appear to mediate the effects of estradiol on LHRH gene expression. *Endocrinology* 143, 4837–4845.

277. Miller, B. H., and Gore, A. C. (2002). N-Methyl-D-aspartate receptor subunit expression in GnRH neurons changes during reproductive senescence in the female rat. *Endocrinology* 143, 3568–3574.
278. Gore, A. C., Wu, T. J., Rosenberg, J. J., and Roberts, J. L. (1996). Gonadotropin-releasing hormone and NMDA receptor gene expression and colocalization change during puberty in female rats. *J. Neurosci.* 16, 5281–5289.
279. Eyigor, O., and Jennes, L. (2000). Kainate receptor subunit-positive gonadotropin-releasing hormone neurons express c-Fos during the steroid-induced luteinizing hormone surge in the female rat. *Endocrinology* 141, 779–786.
280. Simonian, S. X., and Herbison, A. E. (2001). Differing, spatially restricted roles of ionotropic glutamate receptors in regulating the migration of GnRH neurons during embryogenesis. *J. Neurosci.* 21, 934–943.
281. Kiss, J., Kocsis, K., Csaki, A., and Halasz, B. (2003). Evidence for vesicular glutamate transporter synapses onto gonadotropin-releasing hormone and other neurons in the rat medial preoptic area. *Eur. J. Neurosci.* 18, 3267–3278.
282. Goldsmith, P. C., Thind, K. K., Perera, A. D., and Plant, T. M. (1994). Glutamate-immunoreactive neurons and their gonadotropin-releasing hormone-neuronal interactions in the monkey hypothalamus. *Endocrinology* 134, 858–868.
283. Abbud, R., and Smith, M. S. (1995). Do GnRH neurons express the gene for the NMDA receptor? *Brain Res.* 690, 117–120.
284. Jennes, L., Lin, W., and Lakhlani, S. (2002). Glutamatergic regulation of gonadotropin-releasing hormone neurons. *Prog. Brain Res.* 141, 183–192.
285. Lopez, F. J., Donoso, A. O., and Negro-Vilar, A. (1992). Endogenous excitatory amino acids and glutamate receptor subtypes involved in the control of hypothalamic luteinizing hormone-releasing hormone secretion. *Endocrinology* 130, 1986–1992.
286. Bourguignon, J. P., Gerard, A., Purnelle, G., Czajkowski, V., Yamanaka, C., Lemaitre, M., Rigo, J. M., Moonen, G., and Franchimont, P. (1997). Duality of glutamatergic and GABAergic control of pulsatile GnRH secretion by rat hypothalamic explants: I. Effects of antisense oligodeoxynucleotides using explants including or excluding the preoptic area. *J. Neuroendocrinol.* 9, 183–191.
287. Spergel, D. J., Kruth, U., Shimshek, D. R., Sprengel, R., and Seeburg, P. H. (2001). Using reporter genes to label selected neuronal populations in transgenic mice for gene promoter, anatomical, and physiological studies. *Prog. Neurobiol.* 63, 673–686.
288. Ping, L., Mahesh, V. B., and Brann, D. W. (1994). A physiological role for N-methyl-D-aspartic acid and non-N-methyl-D-aspartic acid receptors in pulsatile gonadotropin secretion in the adult female rat. *Endocrinology* 135, 113–118.
289. Lopez, F., Donoso, A., and Negro-Vilar, A. (1990). Endogenous excitatory amino regulates the estradiol-induced LH surge in ovariectomized rats. *Endocrinology* 126, 1771–1773.
290. Brann, D. W., and Mahesh, V. B. (1991). Endogenous excitatory amino acid involvement in the preovulatory and steroid-induced surge of gonadotropins in the female rat. *Endocrinology* 128, 541–1547.
291. Monaghan, D. T., Bridges, R. J., and Cotman, C. W. (1989). The excitatory amino acid receptors: their classes, pharmacology, and distinct properties in the function of the central nervous system. *Annu. Rev. Pharmacol. Toxicol.* 29, 365–402.
292. DeFazio, R. A., Heger, S., Ojeda, S. R., and Moenter, S. M. (2002). Activation of A-Type gamma-aminobutyric acid receptors excites gonadotropin-releasing hormone neurons. *Mol. Endocrinol.* 16, 2872–2891.
293. Jung, H., Shannon, E. M., Fritschy, J.-M., and Ojeda, S. R. (1998). Several $GABA_A$ receptor subunits are expressed in LHRH neurons of juvenile female rats. *Brain Res.* 780, 218–229.
294. Petersen, S. L., McCrone, S., Coy, D., Adelman, J. P., and Mahan, L. C. (1993). $GABA_A$ receptor subunit mRNAs in cells of the preoptic area: colocalization with LHRH mRNA using dual-label in situ hybridization histochemistry. *Endocrine J* 1, 29–34.
295. Pape, J. R., Skynner, M. J., Sim, J. A., and Herbison, A. E. (2001). Profiling gamma-aminobutyric acid (GABA(A)) receptor subunit mRNA expression in postnatal gonadotropin-releasing hormone (GnRH) neurons of the male mouse with single cell RT-PCR. *Neuroendocrinology* 74, 300–308.
296. Moragues, N., Ciofi, P., Lafon, P., Tramu, G., and Garret, M. (2003). GABAA receptor epsilon subunit expression in identified peptidergic neurons of the rat hypothalamus. *Brain Res.* 967, 285–289.
297. Leranth, C., MacLusky, N. J., Sakamoto, H., Shanabrough, M., and Naftolin, F. (1985). Glutamic acid decarboxylase-containing axons synapse on LHRH neurons in the rat medial preoptic area. *Neuroendocrinology* 40, 536–539.
298. Witkin, J. W. (1992). Increased synaptic input to gonadotropin-releasing hormone neurons in aged, virgin, male Sprague-Dawley rats. *Neurobiol. Aging* 13, 681–686.
299. Otis, T. S., Staley, I. J., and Mody, I. (1991). Perpetual inhibitory activity in mammalian brain slices generated by spontaneous GABA release. *Brain Res.* 545, 142–150.
300. Pinto, S., Roseberry, A. G., Liu, H., Diano, S., Shanabrough, M., Cai, X., Friedman, J. M., and Horvath, T. L. (2004). Rapid rewiring of arcuate nucleus feeding circuits by leptin. *Science* 304, 110–115.
301. Staley, K. J. (1999). Quantal GABA release: noise or not? *Nat. Neurosci.* 2, 494–495.
302. Semyanov, A., Walker, M. C., Kullmann, D. M., and Silver, R. A. (2004). Tonically active GABA A receptors: modulating gain and maintaining the tone. *Trends Neurosci.* 27, 262–269.
303. Caraiscos, V. B., Elliott, E. M., You-Ten, K. E., Cheng, V. Y., Belelli, D., Newell, J. G., Jackson, M. F., Lambert, J. J., Rosahl, T. W., Wafford, K. A., MacDonald, J. F., and Orser, B. A. (2004). Tonic inhibition in mouse hippocampal CA1 pyramidal neurons is mediated by alpha5 subunit-containing gamma-aminobutyric acid type A receptors. *Proc. Natl. Acad. Sci. U S A* 101, 3662–3667.
304. Sullivan, S. D., and Moenter, S. M. (2004). Prenatal androgens alter GABAergic drive to gonadotropin-releasing hormone neurons: implications for a common fertility disorder. *Proc. Natl. Acad. Sci. U S A* 101, 7129–7134.
305. Ben-Ari, Y., Khazipov, R., Leinekugel, X., Caillard, O., and Gaiarsa, J.-L. (1997). $GABA_A$, NMDA and AMPA receptors: a developmentally regulated "menage a trios." *Trends Neurosci.* 20, 523–529.
306. McCarthy, M. M., Auger, A. P., and Perrot-Sinal, T. S. (2002). Getting excited about GABA and sex differences in the brain. *Trends Neurosci.* 25, 307–312.
307. Rivera, C., Voipio, J., Payne, J. A., Ruusuvuori, E., Lahtinen, H., Lamsa, K., Pirvola, U., Saarma, M., and Kaila, K. (1999). The K^+/Cl^- co-transporter KCC2 renders GABA hyperpolarizing during neuronal maturation. *Nature* 397, 251–255.
308. Leupen, S. M., Tobet, S. A., Crowley, W. F. Jr., and Kaila, K. (2003). Heterogeneous expression of the potassium-chloride

cotransporter KCC2 in gonadotropin-releasing hormone neurons of the adult mouse. *Endocrinology* 144, 3031–3036.
309. Jackson, G. L., Wood, S. G., and Kuehl, D. E. (2000). A gamma-aminobutyric acidB agonist reverses the negative feedback effect of testosterone on gonadotropin-releasing hormone and luteinizing hormone secretion in the male sheep. *Endocrinology* 141, 3940–3945.
310. Jackson, G. L., and Kuehl, D. (2004). Effects of applying gamma-aminobutyric acid(B) drugs into the medial basal hypothalamus on basal luteinizing hormone concentrations and on luteinizing hormone surges in the female sheep. *Biol. Reprod.* 70, 334–339.
311. Bourguignon, J. P., Gerard, A., Purnelle, G., Czajkowski, V., Yamanaka, C., Lemaitre, M., Rigo, J. M., Moonen, G., and Franchimont, P. (1997). Duality of glutamatergic and GABAergic control of pulsatile GnRH secretion by rat hypothalamic explants: II. Reduced NR2C- and GABAA-receptor-mediated inhibition at initiation of sexual maturation. *J. Neuroendocrinol.* 9, 193–199.
312. Nikolarakis, K. E., Loeffler, J. P., Almeida, O. F., and Herz, A. (1988). Pre- and postsynaptic actions of GABA on the release of hypothalamic gonadotropin-releasing hormone (GnRH). *Brain Res. Bull.* 21, 677–683.
313. Masotto, C., Wisniewski, G., and Negro-Vilar, A. (1989). Different g-aminobutyric acid receptor subtypes are involved in the regulation of opiate-dependent and independent luteinizing hormone-releasing hormone secretion. *Endocrinology* 125, 548–553.
314. Jarry, H., Leonhardt, S., and Wuttke, W. (1991). Gamma-aminobutyric acid neurons in the preoptic/anterior hypothalamic area synchronize the phasic activity of the gonadotropin-releasing hormone pulse generator in ovariectomized rats. *Neuroendocrinology* 53, 261–267.
315. Herbison, A. E., Chapman, C., and Dyer, R. G. (1991). Role of medial preoptic GABA neurones in regulating luteinizing secretion in the ovariectomised rat. *Exp. Brain Res.* 87, 345–352.
316. Herbison, A. E., and Dyer, R. G. (1991). Effect on luteinizing hormone secretion of GABA receptor modulation in the medial preoptic area at the time of proestrous luteinizing hormone surge. *Neuroendocrinology* 53, 317–320.
317. Scott, C. J., and Clarke, I. J. (1993). Inhibition of luteinizing hormone secretion in ovariectomized ewes during the breeding season by g-aminobutyric acid (GABA) is mediated by GABA-A receptors, but not GABA-B receptors. *Endocrinology* 132, 1789–1796.
318. Bilger, M., Heger, S., Brann, D. W., Paredes, A., and Ojeda, S. R. (2001). A conditional tetracycline-regulated increase in gamma amino butyric acid production near luteinizing hormone-releasing hormone nerve terminals disrupts estrous cyclicity in the rat. *Endocrinology* 142, 2102–2114.
319. Mitsushima, D., Hei, D. L., and Terasawa, E. (1994). Gamma-aminobutyric acid is an inhibitory neurotransmitter-restricting the release of luteinizing hormone-releasing hormone before the onset of puberty. *Proc. Natl. Acad. Sci. U S A* 91, 395–399.
320. Mitsushima, D., Marzban, F., Luchansky, L. L., Burich, A. J., Keen, K. L., Durning, M., Golos, T. G., and Terasawa, E. (1996). Role of glutamic acid decarboxylase in the prepubertal inhibition of the luteinizing hormone releasing hormone release in female rhesus monkeys. *J. Neurosci.* 16, 2563–2573.
321. Feleder, C., Jarry, H., Leonhardt, S., Wuttke, W., and Moguilevsky, J. A. (1996). The GABAergic control of gonadotropin-releasing hormone secretion in male rats during sexual maturation involves effects on hypothalamic excitatory and inhibitory amino acid systems. *Neuroendocrinology* 64, 305–312.
322. Moguilevsky, J. A., Carbone, S., Szwarcfarb, B., and Rondina, D. (1991). Sexual maturation modifies the GABAergic control of gonadotrophin secretion in female rats. *Brain Res.* 563, 12–16.
323. Scott, C. J., and Clarke, I. J. (1993). Evidence that changes in the function of the subtypes of the receptors for gamma-aminobutyric acid may be involved in the seasonal changes in the negative-feedback effects of oestrogen on gonadotropin-releasing hormone secretion and plasma luteinizing hormone levels in the ewe. *Endocrinology* 133, 2904–2912.
324. Jansen, H. T., Cutter, C., Hardy, S., Lehman, M. N., and Goodman, R. L. (2003). Seasonal plasticity within the gonadotropin-releasing hormone (GnRH) system of the ewe: changes in identified GnRH inputs and glial association. *Endocrinology* 144, 3663–3676.
325. Pompolo, S., Pereira, A., Kaneko, T., and Clarke, I. J. (2003). Seasonal changes in the inputs to gonadotropin-releasing hormone neurones in the ewe brain: an assessment by conventional fluorescence and confocal microscopy. *J. Neuroendocrinol.* 15, 538–545.
326. Sullivan, S. D., DeFazio, R. A., and Moenter, S. M. (2003). Metabolic regulation of fertility through presynaptic and postsynaptic signaling to gonadotropin-releasing hormone neurons. *J. Neurosci.* 23, 8578–8585.
327. Sullivan, S. D., and Moenter, S. M. (2004). Gamma-aminobutyric acid neurons integrate and rapidly transmit permissive and inhibitory metabolic cues to gonadotropin-releasing hormone neurons. *Endocrinology* 145, 1194–1202.
328. Herbison, A. E. (1997). Noradrenergic regulation of cyclic GnRH secretion. *Rev. Reprod.* 2, 1–6.
329. Hosny, S., and Jennes, L. (1998). Identification of alpha1B adrenergic receptor protein in gonadotropin releasing hormone neurones of the female rat. *J. Neuroendocrinol.* 10, 687–92.
330. Lee, A., Talley, E., Rosin, D. L., and Lynch, K. R. (1995). Characterization of a_{2A}-adrenergic receptors in GT1 neurosecretory cells. *Neuroendocrinology* 62, 215–225.
331. Miller, M. M., and Zhu, L. (1995). Ovariectomy and age alter gonadotropin hormone releasing hormone-noradrenergic interactions. *Neurobiol. Aging* 16, 613–625.
332. Scott, C. J., Rawson, J. A., Pereira, A. M., and Clarke, I. J. (1999). Oestrogen receptors in the brainstem of the female sheep: relationship to noradrenergic cells and cells projecting to the medial preoptic area. *J. Neuroendocrinol.* 11, 745–755.
333. Anselmo-Franci, J. A., Franci, C. R., Krulich, L., Antunes-Rodrigues, J., and McCann, S. M. (1997). Locus coeruleus lesions decrease norepinephrine input into the medial preoptic area and medial basal hypothalamus and block the LH, FSH and prolactin preovulatory surge. *Brain Res,* 767, 289–296.
334. Negro-Vilar, A., Ojeda, S. R., and McCann, S. M. (1979). Catecholaminergic modulation of luteinizing hormone-releasing hormone release by median eminence terminals in vitro. *Endocrinology* 104, 1749–1757.
335. Ojeda, S. R., Negro-Vilar, A., and McCann, S. M. (1982). Evidence for involvement of a-adrenergic receptors in norepinephrine-induced prostaglandin E2 and luteinizing hormone-releasing hormone release from the median eminence. *Endocrinology* 110, 409–411.
336. Rettori, V., Gimeno, M., Lyson, K., and McCann, S. M. (1992). Nitric oxide mediates norepinephrine-induced prostaglandin E2 release from the hypothalamus. *Proc. Natl. Acad. Sci. U S A* 89, 11543–11546.
337. Gearing, M., and Terasawa, E. (1991). Prostaglandin E2 mediates the stimulatory effect of methoxamine on in vivo luteinizing hormone-releasing hormone (LH-RH) release in the ovariectomized female rhesus monkey. *Brain Res.* 560, 276–281.

338. Jarry, H., Leonhardt, S., Schwarze, T., and Wuttke, W. (1995). Preoptic rather than mediobasal hypothalamic amino acid neurotransmitter release regulates GnRH secretion during the estrogen-induced LH surge in the ovariectomized rat. *Neuroendocrinology* 62, 479–486.
339. Jarry, H., Leonhardt, S., and Wuttke, W. (1990). A norepinephrine-dependent mechanism in the preoptic/anterior hypothalamic area but not in the mediobasal hypothalamus is involved in the regulation of the gonadotropin-releasing hormone pulse generator in ovariectomized rats. *Neuroendocrinology* 51, 337–344.
340. Gallo, R. V., and Drouva, S. V. (1979). Effect of intraventricular infusion of catecholamines on luteinizing hormone release in ovariectomized and ovariectomized steroid-primed rats. *Neuroendocrinology* 29, 149–162.
341. Leung, P. C. K., Whitmoyer, D. I., Arendash, G. W., and Sawyer, C. H. (1982). Differential effects of central adrenoceptor agonists on luteinizing hormone release. *Neuroendocrinology* 34, 207–214.
342. Leipheimer, R. E., and Gallo, R. V. (1985). Medial preoptic area involvement in norepinephrine-induced suppression of pulsatile luteinizing hormone release in ovariectomized rats. *Neuroendocrinology* 40, 345–351.
343. Gearing, M., and Terasawa, E. (1991). The alpha-1-adrenergic neuronal system is involved in the pulsatile release of luteinizing hormone-releasing hormone in the ovariectomized female rhesus monkey. *Neuroendocrinology* 53, 373–381.
344. Terasawa, E., Krook, C., Hei, D. L., Gearing, M., Schultz, N. J., and Davis, G. A. (1988). Norepinephrine is a possible neurotransmitter stimulating pulsatile release of luteinizing hormone-releasing hormone in the rhesus monkey. *Endocrinology* 123, 1808–1816.
345. Pau, K. Y. F., Hess, D. L., Kaynard, A. H., Ji, W., Gliessman, P. M., and Spies, H. G. (1989). Suppression of mediobasal hypothalamic gonadotropin-releasing hormone and plasma luteinizing hormone pulsatile patterns by phentolamine in ovariectomized rhesus macaques. *Endocrinology* 124, 891–898.
346. Legan, S. J., and Callahan, W. H. (1999). Suppression of tonic luteinizing hormone secretion and norepinephrine release near the GnRH neurons by estradiol in ovariectomized rats. *Neuroendocrinology* 70, 237–245.
347. Clifton, D. K., and Sawyer, C. H. (1979). LH release and ovulation in the rat following depletion of hypothalamic norepinephrine, chronic vs acite effects/. *Neuroendocrinology* 28, 442–449.
348. Leonhardt, S., Jarry, H., Falkenstein, G., Palmer, J., and Wuttke, W. (1991). LH release in ovariectomized rats is maintained without noradrenergic neurotransmission in the preoptic/anterior hypothalamic area: extreme functional plasticity of the GnRH pulse generator. *Brain Res.* 562, 105–110.
349. Caba, M., Bao, J., Pau, K. Y., and Spies, H. G. (2000). Molecular activation of noradrenergic neurons in the rabbit brainstem after coitus. *Brain Res. Mol. Brain Res.* 77, 222–231.
350. Yang, S. P., Pau, K. Y., Airhart, N., and Spies, H. G. (1998). Attenuation of gonadotropin-releasing hormone reflex to coitus by alpha1-adrenergic receptor blockade in the rabbit. *Proc. Soc. Exp. Biol. Med.* 218, 204–9.
351. Spies, H. G., Pau, K. Y., and Yang, S. P. (1997). Coital and estrogen signals: a contrast in the preovulatory neuroendocrine networks of rabbits and rhesus monkeys. *Biol. Reprod.* 56, 310–319.
352. Mohankumar, P. S., Thyagarajan, S., and Quadri, S. K. (1994). Correlations of catecholamine release in the medial preoptic area with proestrous surges of luteinizing hormone and prolactin: effects of aging. *Endocrinology* 135, 119–126.
353. Conde, G. L., Herbison, A. E., Fernandez-Galaz, C., and Bicknell, R. J. (1996). Estrogen uncouples noradrenergic activation of Fos expression in the female rat preoptic area. *Brain Res.* 735, 197–207.
354. Herbison, A. E., Heavens, R. P., and Dyer, R. G. (1990). Oestrogen modulation of excitatory A1 noradrenergic input to rat medial preoptic gamma aminobutyric acid neurones demonstrated by microdialysis. *Neuroendocrinology* 52, 161–168.
355. Sperlagh, B., Sershen, H., Lajtha, A., and Vizi, E. S. (1998). Co-release of endogenous ATP and (3H) noradrenaline from rat hypothalamic slices: origin and modulation by alpha2-adrenoceptors. *Neuroscience* 82, 511–520.
356. Pelletier, G. (1987). Demonstration of contacts between neurons staining for LHRH in the preoptic area of the rat brain. *Neuroendocrinology* 46, 457–459.
357. Chen, W. P., Witkin, J. W., and Silverman, A. J. (1989). Beta-endorphin and gonadotropin-releasing hormone synaptic input to gonadotropin-releasing hormone neurosecretory cells in the male rat. *J. Comp. Neurol.* 286, 85–95.
358. Leranth, C., Segura, L. M., Palkovits, M., MacLusky, N. J., Shanabrough, M., and Naftolin, F. (1985). The LH-RH-containing neuronal network in the preoptic area of the rat: demonstration of LH-RH-containing nerve terminals in synaptic contact with LH-RH neurons. *Brain Res.* 345, 332–336.
359. Xu, C., Xu, X. Z., Nunemaker, C. S., and Moenter, S. M. (2004). Dose-dependent switch in response of gonadotropin-releasing hormone (GnRH) neurons to GnRH mediated through the type I GnRH receptor. *Endocrinology* 145, 728–735.
360. Krsmanovic, L. Z., Martinez-Fuentes, A. J., Arora, K. K., Mores, N., Navarro, C. E., Chen, H. C., Stojilkovic, S. S., and Catt, K. J. (1999). Autocrine regulation of gonadotropin-releasing hormone secretion in cultured hypothalamic neurons. *Endocrinology* 140, 1423–1431.
361. Sarkar, D. K. (1987). In vivo secretion of LHRH in ovariectomized rats is regulated by a possible autofeedback mechanism. *Neuroendocrinology* 45, 510–513.
362. Padmanabhan, V., Evans, N. P., Dahl, G. E., McFadden, K. L., Mauger, D. T., and Karsch, F. J. (1995). Evidence for short or ultrashort loop negative feedback of gonadotropin-releasing hormone secretion. *Neuroendocrinology* 62, 248–258.
363. DePaolo, L. V., King, R. A., and Carrillo, A. J. (1987). In vivo and in vitro examination of an autoregulatory mechanism for luteinizing hormone-releasing hormone. *Endocrinology* 120, 272–279.
364. Bedran de Castro, J. C., Khorram, O., and McCann, S. M. (1985). Possible negative ultra-short loop feedback of luteinizing hormone releasing hormone (LHRH) in the ovariectomized rat. *Proc. Soc. Exp. Biol. Med.* 179, 132–135.
365. Woller, M. J., Meyer, S., Ada-Nguema, A., and Waechter-Brulla, D. (2004). Dissecting autocrine effects on pulsatile release of gonadotropin-releasing hormone in cultured rat hypothalamic tissue. *Exp. Biol. Med.* 229, 56–64.
366. Krsmanovic, L. Z., Mores, N., Navarro, C. E., Arora, K. K., and Catt, K. J. (2003). An agonist-induced switch in G protein coupling of the gonadotropin-releasing hormone receptor regulates pulsatile neuropeptide secretion. *Proc. Natl. Acad. Sci. U S A* 100, 2969–2974.
367. Valenca, M. M., Johnston, C. A., Ching, M., and Negro-Vilar, A. (1987). Evidence for a negative ultrashort loop feedback mechanism operating on the luteinizing hormone-releasing hormone neuronal system. *Endocrinology* 121, 2256–2259.
368. Zanisi, M., Messi, E., Motta, M., and Martini, L. (1987). Ultrashort feedback control of luteinizing hormone-releasing hormone secretion in vitro. *Endocrinology* 121, 2199–2204.

369. Hiruma, H., Funabashi, T., and Kimura, F. (1989). LHRH injected into the medial preoptic area potentiates LH secretion in ovariectomized estrogen-primed and proestrous rats. *Neuroendocrinology* 50, 421–426.
370. Seong, J. Y., Kang, S. S., Kam, K., Han, Y. G., Kwon, H. B., Ryu, K., and Kim, K. (1998). Differential regulation of gonadotropin-releasing hormone (GnRH) receptor expression in the posterior mediobasal hypothalamus by steroid hormones: implication of GnRH neuronal activity. *Brain Res. Mol. Brain Res.* 53, 226–235.
371. Hoffman, G. E., Smith, M. S., and Verbalis, J. G. (1993). c-Fos and related immediate early gene products as markers of activity in neuroendocrine systems. *Front. Neuroendocrinol.* 14, 173–213.
372. Campbell, R. E., ffrench-Mullen, J. M., Cowley, M. A., Smith, M. S., and Grove, K. L. (2001). Hypothalamic circuitry of neuropeptide Y regulation of neuroendocrine function and food intake via the Y5 receptor subtype. *Neuroendocrinology* 74, 106–119.
373. Li, C., Chen, P., and Smith, M. S. (1999). Morphological evidence for direct interaction between arcuate nucleus neuropeptide Y (NPY) neurons and gonadotropin-releasing hormone neurons and the possible involvement of NPY Y1 receptors. *Endocrinology* 140, 5382–5390.
374. Thind, K. K., Boggan, J. E., and Goldsmith, P. C. (1993). Neuropeptide Y system of the female monkey hypothalamus: retrograde tracing and immunostaining. *Neuroendocrinology* 57, 289–298.
375. Tsuruo, Y., Kawano, H., Kagotani, Y., Hisano, S., Daikoku, S., Chihara, K., Zhang, T., and Yanaihara, N. (1990). Morphological evidence for neuronal regulation of luteinizing hormone-releasing hormone-containing neurons by neuropeptide Y in the rat septo-preoptic area. *Neurosci. Lett.* 110, 261–266.
376. Turi, G. F., Liposits, Z., Moenter, S. M., Fekete, C., and Hrabovszky, E. (2003). Origin of neuropeptide Y-containing afferents to gonadotropin-releasing hormone neurons in male mice. *Endocrinology* 144, 4967–4974.
377. Broberger, C., Johansen, J., Johansson, C., Schalling, M., and Hokfelt, T. (1998). The neuropeptide Y/agouti gene-related protein (AGRP) brain circuitry in normal, anorectic, and monosodium glutamate-treated mice. *Proc. Natl. Acad. Sci. U S A* 95, 15043–15048.
378. Horvath, T. L., Bechmann, I., Naftolin, F., Kalra, S. P., and Leranth, C. (1997). Heterogeneity in the neuropeptide Y-containing neurons of the rat arcuate nucleus: GABAergic and non-GABAergic subpopulations. *Brain Res.* 756, 283–286.
379. Everitt, B. J., Hokfelt, T., Terenius, L., Tatemoto, K., Mutt, V., and Goldstein, M. (1984). Differential co-existence of neuropeptide Y (NPY)-like immunoreactivity with catecholamines in the central nervous system of the rat. *Neuroscience* 11, 443–462.
380. Sahu, A., Kalra, S. P., Crowley, W. R., and Kalra, P. S. (1988). Evidence that NPY-containing neurons in the brainstem project into selected hypothalamic nuclei: implication in feeding behavior. *Brain Res.* 457, 376–378.
381. Guy, J., Li, S., and Pelletier, G. (1988). Studies on the physiological role and mechanism of action of neuropeptide Y in the regulation of luteinizing hormone secretion in the rat. *Regul. Pept.* 23, 209–216.
382. Horvath, T. L., Naftolin, F., Leranth, C., Sahu, A., and Kalra, S. P. (1996). Morphological and pharmacological evidence for neuropeptide Y-galanin interaction in the rat hypothalamus. *Endocrinology* 137, 3069–3078.
383. Crowley, W. R., Hassid, A., and Kalra, S. P. (1987). Neuropeptide Y enhances the release of luteinizing hormone (LH) induced by LH-releasing hormone. *Endocrinology* 120, 941–945.
384. Hill, J. W., Urban, J. H., Xu, M., and Levine, J. E. (2004). Estrogen induces neuropeptide Y (NPY) Y1 receptor gene expression and responsiveness to NPY in gonadotrope-enriched pituitary cell cultures. *Endocrinology* 145, 2283–2290.
385. Xu, M., Hill, J. W., and Levine, J. E. (2000). Attenuation of luteinizing hormone surges in neuropeptide Y knockout mice. *Neuroendocrinology* 72, 263–271.
386. Woller, M. J., Campbell, G. T., Liu, L., Steigerwalt, R. W., and Blake, C. A. (1993). Estrogen alters the effects of neuropeptide-Y on luteinizing hormone and follicle-stimulating hormone release in female rats at the level of the anterior pituitary gland. *Endocrinology* 133, 2675–2681.
387. Leupen, S. M., Besecke, L. M., and Levine, J. E. (1997). Neuropeptide Y Y1-receptor stimulation is required for physiological amplification of preovulatory luteinizing hormone surges. *Endocrinology* 138, 2735–2739.
388. Crowley, W. R., and Kalra, S. P. (1987). Neuropeptide Y stimulates the release of luteinizing hormone-releasing hormone from medial basal hypothalamus in vitro: modulation by ovarian hormones. *Neuroendocrinology* 46, 97–103.
389. Besecke, L. M., and Levine, J. E. (1994). Acute increase in responsiveness of luteinizing hormone (LH)-releasing hormone nerve terminals to neuropeptide-Y stimulation before the preovulatory LH surge. *Endocrinology* 135, 63–66.
390. Jain, M. R., Pu, S., Kalra, P. S., and Kalra, S. P. (1999). Evidence that stimulation of two modalities of pituitary luteinizing hormone release in ovarian steroid-primed ovariectomized rats may involve neuropeptide Y Y1 and Y4 receptors. *Endocrinology* 140, 5171–5177.
391. Kalra, S. P., Fuentes, M., Fournier, A., Parker, S. L., and Crowley, W. R. (1992). Involvement of the Y-1 receptor subtype in the regulation of luteinizing hormone secretion by neuropeptide Y in rats. *Endocrinology* 130, 3323–3330.
392. Raposinho, P. D., Broqua, P., Pierroz, D. D., Hayward, A., Dumont, Y., Quirion, R., Junien, J. L., and Aubert, M. L. (1999). Evidence that the inhibition of luteinizing hormone secretion exerted by central administration of neuropeptide Y (NPY) in the rat is predominantly mediated by the NPY-Y5 receptor subtype. *Endocrinology* 140, 4046–4055.
393. Raposinho, P. D., Broqua, P., Hayward, A., Akinsanya, K., Galyean, R., Schteingart, C., Junien, J., and Aubert, M. L. (2000). Stimulation of the gonadotropic axis by the neuropeptide Y receptor Y1 antagonist/Y4 agonist 1229U91 in the male rat. *Neuroendocrinology* 71, 2–7.
394. Kalra, S. P. (1993). Mandatory neuropeptide-steroid signaling for the preovulatory luteinizing hormone-releasing hormone discharge. *Endocr. Rev.* 14, 507–538.
395. Kalra, S. P., and Crowley, W. R. (1984). Norepinephrine-like effects of neuropeptide Y on LH release in the rat. *Life Sci.* 35, 1173–1176.
396. Sahu, A., Crowley, W. R., Tatemoto, K., Balasubramaniam, A., and Kalra, S. P. (1987). Effects of neuropeptide Y, NPY analog (norleucine4-NPY), galanin and neuropeptide K on LH release in ovariectomized (ovx) and ovx estrogen, progesterone-treated rats. *Peptides* 8, 921–926.
397. Hill, J. W., and Levine, J. E. (2003). Abnormal response of the neuropeptide Y-deficient mouse reproductive axis to food deprivation but not lactation. *Endocrinology* 144, 1780–1786.
398. Malven, P. V., Haglof, S. A., and Degroot, H. (1992). Effects of intracerebral administration of neuropeptide-Y on secretion of luteinizing hormone in ovariectomized sheep. *Brain Res. Bull.* 28, 871–875.
399. Barker-Gibb, M. L., Scott, C. J., Boublik, J. H., and Clarke, I. J. (1995). The role of neuropeptide Y (NPY) in the control of LH secretion in the ewe with respect to season, NPY receptor

subtype and the site of action in the hypothalamus. *J. Endocrinol.* 147, 565–579.
400. Estrada, K. M., Pompolo, S., Morris, M. J., Tilbrook, A. J., and Clarke, I. J. (2003). Neuropeptide Y (NPY) delays the oestrogen-induced luteinizing hormone (LH) surge in the ovariectomized ewe: further evidence that NPY has a predominant negative effect on LH secretion in the ewe. *J. Neuroendocrinol.* 15, 1011–1020.
401. Woller, M. J., and Terasawa, E. (1991). Infusion of neuropeptide Y into the stalk-median eminence stimulates in vivo release of luteinizing hormone-release hormone in gonadectomized rhesus monkeys. *Endocrinology* 128, 1144–1150.
402. Woller, M. J., and Terasawa, E. (1992). Estradiol enhances the action of neuropeptide Y on in vivo luteinizing hormone-releasing hormone release in the ovariectomized rhesus monkey. *Neuroendocrinology* 56, 921–925.
403. Pau, K. Y., Berria, M., Hess, D. L., and Spies, H. G. (1995). Hypothalamic site-dependent effects of neuropeptide Y on gonadotropin-releasing hormone secretion in rhesus macaques. *J. Neuroendocrinol.* 7, 63–67.
404. Woller, M. J., McDonald, J. K., Reboussin, D. M., and Terasawa, E. (1992). Neuropeptide Y is a neuromodulator of pulsatile luteinizing hormone-releasing hormone release in the gonadectomized rhesus monkey. *Endocrinology* 130, 2333–2342.
405. Mizuno, M., Gearing, M., and Terasawa, E. (2000). The role of neuropeptide Y in the progesterone-induced luteinizing hormone-releasing hormone surge in vivo in ovariectomized female rhesus monkeys. *Endocrinology* 141, 1772–1779.
406. Terasawa, E. (2001). Luteinizing hormone-releasing hormone (LHRH) neurons: mechanism of pulsatile LHRH release. *Vitam. Horm.* 63, 91–129.
407. Shahab, M., Balasubramaniam, A., Sahu, A., and Plant, T. M. (2003). Central nervous system receptors involved in mediating the inhibitory action of neuropeptide Y on luteinizing hormone secretion in the male rhesus monkey (*Macaca mulatta*). *J. Neuroendocrinol.* 15, 965–970.
408. Levine, J. E. (1997). New concepts of the neuroendocrine regulation of gonadotropin surges in rats. *Biol. Reprod.* 56, 293–302.
409. Smith, M. S., and Grove, K. L. (2002). Integration of the regulation of reproductive function and energy balance: lactation as a model. *Front. Neuroendocrinol.* 23, 225–256.
410. Thorsell, A., and Heilig, M. (2002). Diverse functions of neuropeptide Y revealed using genetically modified animals. *Neuropeptides* 36, 182–193.
411. Illes, P., and Regenold, J. T. (1990). Interaction between neuropeptide Y and noradrenaline on central catecholamine neurons. *Nature* 344, 62–63.
412. Parker, S. L., and Crowley, W. R. (1993). Central stimulation of oxytocin release in the lactating rat: interaction of neuropeptide Y with alpha-1-adrenergic mechanisms. *Endocrinology* 132, 658–666.
413. Widdowson, P. S., Masten, T., and Halaris, A. E. (1991). Interactions between neuropeptide Y and alpha 2-adrenoceptors in selective rat brain regions. *Peptides* 12, 71–75.
414. Leranth, C., MacLusky, N. J., Shanabrough, M., and Naftolin, F. (1988). Immunohistochemical evidence for synaptic connections between pro-opiomelanocortin-immunoreactive axons and LHRH neurones in the preoptic area of the rat. *Brain Res.* 449, 167–176.
415. Goodman, R. L., Coolen, L. M., Anderson, G. M., Hardy, S. L., Valent, M., Connors, J. M., Fitzgerald, M. E., and Lehman, M. N. (2004). Evidence that dynorphin plays a major role in mediating progesterone negative feedback on gonadotropin-releasing hormone neurons in sheep. *Endocrinology* 145, 2959–2967.
416. Thind, K. K., and Goldsmith, P. C. (1988). Infundibular gonadotropin-releasing hormone neurons are inhibited by direct opioid and autoregulatory synapses in juvenile monkeys. *Neuroendocrinology* 47, 203–216.
417. Sannella, M. I., and Petersen, S. L. (1997). Dual label in situ hybridization studies provide evidence that luteinizing hormone-releasing hormone neurons do not synthesize messenger ribonucleic acid for mu, kappa, or delta opiate receptors. *Endocrinology* 138, 1667–1672.
418. Mitchell, V., Prevot, V., Jennes, L., Aubert, J. P., Croix, D., and Beauvillain, J. C. (1997). Presence of mu and kappa opioid receptor mRNAs in galanin but not in GnRH neurons in the female rat. *Neuroreport* 8, 3167–3172.
419. Stanley, S. A., Davies, S., Small, C. J., Gardiner, J. V., Ghatei, M. A., Smith, D. M., and Bloom, S. R. (2003). Gamma-MSH increases intracellular cAMP accumulation and GnRH release in vitro and LH release in vivo. *FEBS Lett.* 543, 66–70.
420. Wiesner, J. B., Koenig, J. I., Krulich, L., and Moss, R. L. (1985). Possible delta receptor mediation of the effect of beta-endorphin on luteinizing hormone (LH) release, but not on prolactin (PRL) release, in the ovariectomized rat. *Endocrinology* 116, 475–477.
421. Kinoshita, F., Nakai, Y., Katakami, H., and Imura, H. (1982). Suppressive effect of dynorphin-(1-13) on luteinizing hormone release in conscious castrated rats. *Life Sci.* 30, 1915–1919.
422. Leadem, C. A., and Yagenova, S. V. (1987). Effects of specific activation of m-, d - and k-opioid receptors on the secretion of luteinizing hormone and prolactin in ovariectomized rats. *Neuroendocrinology* 145, 109–117.
423. Leadem, C. A., and Kalra, S. P. (1985). Effects of endogenous opioid peptides and opiates on luteinizing hormone and prolactin secretion in ovariectomized rats. *Neuroendocrinology* 41, 342–352.
424. Pfeiffer, D. G., Pfeiffer, A., Almeida, O. F., and Herz, A. (1987). Opiate suppression of LH secretion involves central receptors different from those mediating opiate effects on prolactin secretion. *J. Endocrinol.* 114, 469–476.
425. Babu, G. N., Marco, J., Bona-Gallo, A., and Gallo, R. V. (1987). Steroid independent endogenous opioid peptide suppression of pulsatile luteinizing hormone release between estrus and diestrus I in the rat estrous cycle. *Brain Res.* 416, 235–242.
426. Karahalios, D. G., and Levine, J. E. (1988). Naloxone stimulation of in vivo LHRH release is not diminished following ovariectomy. *Neuroendocrinology* 47, 504–510.
427. Lustig, R. H., Pfaff, D. W., and Fishman, J. (1988). Opioidergic modulation of the oestradiol-induced LH surge in the rat: roles of ovarian steroids. *J. Endocrinol.* 116, 55–69.
428. Drouva, S. V., Epelbaum, J., Tapia-Arancibia, L., Laplante, E., and Kordon, C. (1980). Met-enkephalin inhibition of K+-induced LHRH and ARIF release from rat mediobasal hypothalamic slices. *Eur. J. Pharmacol.* 61, 411–412.
429. Mallory, D. S., and Gallo, R. V. (1990). Medial preoptic-anterior hypothalamic area involvement in the suppression of pulsatile LH release by a μ-opioid agonist in the ovariectomized rat. *Brain Res. Bull.* 25, 251–257.
430. Wiesner, J. B., Koenig, J. I., Krulich, L., and Moss, R. L. (1984). Site of action for beta-endorphin-induced changes in plasma luteinizing hormone and prolactin in the ovariectomized rat. *Life Sci.* 34, 1463–1473.
431. Hashimoto, R., and Kimura, F. (1991). Effects of intraventricular injection of delta receptor antagonist ICI 154, 129 on the secretion of luteinizing hormone and prolactin in the proestrous rat. *Endocrinol. Jpn.* 38, 213–218.
432. Zhang, Q., and Gallo, R. V. (2002). Effect of prodynorphin-derived opioid peptides on the ovulatory luteinizing hormone surge in the proestrous rat. *Endocrine* 18, 27–32.

433. Masotto, C., Sahu, A., Dube, M. G., and Kalra, S. P. (1990). A decrease in opioid tone amplifies the luteinizing hormone surge in estrogen-treated ovariectomised rats: comparisons with progesterone effects. *Endocrinology* 126, 18–25.
434. Goodman, R. L., Parfitt, D. B., Evans, N. P., Dahl, G. E., and Karsch, F. J. (1995). Endogenous opioid peptides control the amplitude and shape of gonadotropin-releasing hormone pulses in the ewe. *Endocrinology* 136, 2412–2420.
435. Weesner, G. D., and Malven, P. V. (1990). Intracerebral immunoneutralization of beta-endorphin and met-enkephalin disinhibits release of pituitary luteinizing hormone in sheep. *Neuroendocrinology* 52, 382–388.
436. Whisnant, S. C., Havern, R. L., and Goodman, R. L. (1991). Endogenous opioid suppression of luteinizing hormone pulse frequency and amplitude in the ewe: hypothalamic sites of action. *Neuroendocrinology* 54, 587–593.
437. Horton, R. J., Francis, H., and Clarke, I. J. (1989). Seasonal and steroid-dependent effects on the modulation of LH secretion in the ewe by intracerebroventricularly administered beta-endorphin or naloxone. *J. Endocrinol.* 122, 509–517.
438. Horton, R. J., Cummins, J. T., and Clarke, I. J. (1987). Naloxone evokes large-amplitude GnRH pulses in luteal-phase ewes. *J. Reprod. Fertil.* 81, 277–286.
439. Brooks, A. N., Haynes, N. B., Yang, K., and Lamming, G. E. (1986). Ovarian steroid involvement in endogenous opioid modulation of LH secretion in seasonally anoestrous mature ewes. *J. Reprod. Fertil.* 76, 709–715.
440. Conover, C. D., Kuljis, R. O., Rabii, J., and Advis, J. P. (1993). Beta-endorphin regulation of luteinizing hormone-releasing hormone release at the median eminence in ewes: immunocytochemical and physiological evidence. *Neuroendocrinology* 57, 1182–1195.
441. Walsh, J. P., and Clarke, I. J. (1996). Effects of central administration of highly selective opioid mu-, delta- and kappa-receptor agonists on plasma luteinizing hormone (LH), prolactin, and the estrogen-induced LH surge in ovariectomized ewes. *Endocrinology* 137, 3640–3648.
442. Walsh, J. P., and Clarke, I. J. (1998). Blockade of the oestrogen-induced luteinizing hormone surge in ovariectomized ewes by a highly selective opioid mu-receptor agonist: evidence for site of action. *Neuroendocrinology* 67, 164–170.
443. Foradori, C. D., Coolen, L. M., Fitzgerald, M. E., Skinner, D. C., Goodman, R. L., and Lehman, M. N. (2002). Colocalization of progesterone receptors in parvicellular dynorphin neurons of the ovine preoptic area and hypothalamus. *Endocrinology* 143, 4366–4374.
444. Araque, A., Carmignoto, G., and Haydon, P. G. (2001). Dynamic signaling between astrocytes and neurons. *Annu. Rev. Physiol.* 63, 795–813.
445. Ojeda, S. R., Prevot, V., Heger, S., Lomniczi, A., Dziedzic, B., and Mungenast, A. (2003). Glia-to-neuron signaling and the neuroendocrine control of female puberty. *Ann. Med.* 35, 244–255.
446. Melcangi, R. C., Martini, L., and Galbiati, M. (2002). Growth factors and steroid hormones: a complex interplay in the hypothalamic control of reproductive functions. *Prog. Neurobiol.* 67, 421–449.
447. Rage, F., Lee, B. J., Ma, Y. J., and Ojeda, S. R. (1997). Estradiol enhances prostaglandin E2 receptor gene expression in luteinizing hormone-releasing hormone (LHRH) neurons and facilitates the LHRH response to PGE2 by activating a glia-to-neuron signaling pathway. *J. Neurosci.* 17, 9145–9156.
448. Ma, Y. J., Berg-Von der Emde, K., Rage, F., Wetsel, W. C., and Ojeda, S. R. (1997). Hypothalamic astrocytes respond to transforming growth factor-a with the secretion of neuroactive substances that stimulate the release of luteinizing hormone-releasing hormone. *Endocrinology* 138, 19–25.
449. Ma, Y. J., Hill, D. F., Creswick, K. E., Costa, M. E., Cornea, A., Lioubin, M. N., Plowman, G. D., and Ojeda, S. R. (1999). Neuregulins signaling via a glial erbB-2-erbB-4 receptor complex contribute to the neuroendocrine control of mammalian sexual development. *J. Neurosci.* 19, 9913–9927.
450. Ma, Y. J., Junier, M. P., Costa, M. E., and Ojeda, S. R. (1992). Transforming growth factor-alpha gene expression in the hypothalamus is developmentally regulated and linked to sexual maturation. *Neuron* 9, 657–670.
451. Hou, J., Li, B., Yang, Z., Fager, N., and Ma, M. Y. (2002). Functional integrity of ErbB-4/-2 tyrosine kinase receptor complex in the hypothalamus is required for maintaining normal reproduction in young adult female rats. *Endocrinology* 143, 1901–1912.
452. Melcangi, R. C., Galbiati, M., Messi, E., Piva, F., Martini, L., and Motta, M. (1995). Type 1 astrocytes influence luteinizing hormone-releasing hormone release from the hypothalamic cell line GT1-1: is transforming growth factor-beta the principle involved? *Endocrinology* 136, 679–686.
453. Buchanan, C. D., Mahesh, V. B., and Brann, D. W. (2000). Estrogen-astrocyte-luteinizing hormone-releasing hormone signalling: a role for transforming growth factor-β(1). *Biol. Reprod.* 62, 1710–1721.
454. Galbiati, M., Zanisi, M., Messi, E., Cavarretta, I., Martini, L., and Melcangi, R. C. (1996). Transforming growth factor-beta and astrocytic conditioned medium influence luteinizing hormone-releasing hormone gene expression in the hypothalamic cell line GT1. *Endocrinology* 137, 5605–5609.
455. Prevot, V., Bouret, S., Croix, D., Takumi, T., Jennes, L., Mitchell, V., and Beauvillain, J. C. (2000). Evidence that members of the TGFb superfamily play a role in regulation of the GnRH neuroendocrine axis: expression of a type I serine-threonine kinase receptor for TGRb and activin in GnRH neurones and hypothalamic areas of the female rat. *J. Neuroendocrinol.* 12, 677–680.
456. Bouret, S., De Seranno, S., Beauvillain, J. C., and Prevot, V. (2004). Transforming growth factor beta1 may directly influence gonadotropin-releasing hormone gene expression in the rat hypothalamus. *Endocrinology* 145, 1794–1801.
457. Prevot, V., Cornea, A., Mungenast, A., Smiley, G., and Ojeda, S. R. (2003). Activation of erbB-1 signaling in tanycytes of the median eminence stimulates transforming growth factor beta1 release via prostaglandin E2 production and induces cell plasticity. *J. Neurosci.* 23, 10622–10632.
458. Knobil, E. (1980). The neuroendocrine control of the menstrual cycle. *Rec. Prog. Horm. Res.* 36, 53–88.
459. Carmel, P. W., Araki, S., and Ferin, M. (1976). Pituitary stalk portal blood collection in rhesus monkeys: evidence for pulsatile release of gonadotropin-releasing hormone (GnRH). *Endocrinology* 99, 243–248.
460. Neill, J. D., Patton, J. M., Dailey, R. A., Tsou, R. C., and Tindall, G. T. (1977). Luteinizing hormone releasing hormone (LHRH) in pituitary stalk blood of rhesus monkeys: relationship to level of LH release. *Endocrinology* 101, 430–434.
461. Levine, J. E. and Ramirez, V. D. (1982). Luteinizing hormone-releasing hormone release during the rat estrous cycle and after ovariectomy, as estimated with push-pull cannulae. *Endocrinology* 111, 1439–1446.
462. Moenter, S. M., Brand, R. M., Midgley, A. R., and Karsch, F. J. (1992). Dynamics of gonadotropin-releasing hormone release during a pulse. *Endocrinology* 130, 503–510.
463. Caraty, A., Locatelli, A., and Martin, G. B. (1989). Biphasic response in the secretion of gonadotrophin-releasing hormone in ovariectomized ewes injected with oestradiol. *J. Endocrinol.* 123, 375–382.
464. Levine, J. E., and Duffy, M. T. (1988). Simultaneous measurement of luteinizing hormone (LH)-releasing hormone,

LH, and follicle-stimulating hormone release in intact and short-term castrate rats. *Endocrinology* 122, 2211–2221.

465. Levine, J. E., Pau, K. Y., Ramirez, V. D., and Jackson, G. L. (1982). Simultaneous measurement of luteinizing hormone-releasing hormone and luteinizing hormone release in unanesthetized, ovariectomized sheep. *Endocrinology* 111, 1449–1455.
466. Clarke, I. J., and Cummins, J. T. (1982). The temporal relationship between gonadotropin releasing hormone (GnRH) and luteinizing hormone (LH) secretion in ovariectomized ewes. *Endocrinology* 111, 1737–179.
467. Dyer, R. G., Mansfield, S., and Yates, J. O. (1980). Discharge of gonadotrophin-releasing hormone from the mediobasal part of the hypothalamus: effect of stimulation frequency and gonadal steroids. *Exp. Brain Res.* 39, 453–460.
468. Clarke, I. J., and Cummins, J. T. (1985). Increased gonadotropin-releasing hormone pulse frequency associated with estrogen-induced luteinizing hormone surges in ovariectomized ewes. *Endocrinology* 116, 2376–2383.
469. Park, O. K., and Ramirez, V. D. (1989). Spontaneous changes in LHRH release during the rat estrous cycle, as measured with repetitive push-pull perfusions of the pituitary gland in the same female rats. *Neuroendocrinology* 50, 66–72.
470. Evans, N. P., Dahl, G. E., Mauger, D., and Karsch, F. J. (1995). Estradiol induces both qualitative and quantitative changes in the pattern of gonadotropin-releasing hormone secretion during the presurge period in the ewe. *Endocrinology* 136, 1603–1609.
471. Moenter, S. M., Anthony DeFazio, R., Pitts, G. R., and Nunemaker, C. S. (2003). Mechanisms underlying episodic gonadotropin-releasing hormone secretion. *Front. Neuroendocrinol.* 24, 79–93.
472. Martinez de la Escalera, G., Choi, A. L., and Weiner, R. I. (1992). Generation and synchronization of gonadotropin-releasing hormone (GnRH) pulses: intrinsic properties of the GT1-1 GnRH neuronal cell line. *Proc. Natl. Acad. Sci. U S A* 89, 1852–1855.
473. Wetsel, W. C., Valenca, M. M., Merchenthaler, I., Liposits, Z., Lopez, F. J., Weiner, R. I., Mellon, P. L., and Negro-Vilar, A. (1992). Intrinsic pulsatile secretory activity of immortalized luteinizing hormone-releasing hormone-secreting neurons. *Proc. Natl. Acad. Sci. U S A* 89, 4149–4153.
474. Krsmanovic, L. Z., Stojilkovic, S. S., Merelli, F., Dufour, S. M., Virmani, M. A., and Catt, K. J. (1992). Calcium signaling and episodic secretion of gonadotropin-releasing hormone in hypothalamic neurons. *Proc. Natl. Acad. Sci. U S A* 89, 8462–8466.
475. Bosma, M. M. (1993). Ion channel properties and episodic activity in isolated immortalized gonadotropin-releasing hormone (GnRH) neurons. *J. Membr. Biol.* 136, 85–96.
476. Terasawa, E., Keen, K. L., Mogi, K., and Claude, P. (1999). Pulsatile release of luteinizing hormone-releasing hormone (LHRH) in cultured LHRH neurons derived from the embryonic olfactory placode of the rhesus monkey. *Endocrinology* 140, 1432–1441.
477. Duittoz, A. H., and Batailler, M. (2000). Pulsatile GnRH secretion from primary cultures of sheep olfactory placode explants. *J. Reprod. Fertil.* 120, 391–396.
478. Funabashi, T., Daikoku, S., Shinohara, K., and Kimura, F. (2000). Pulsatile gonadotropin-releasing hormone (GnRH) secretion is an inherent function of GnRH neurons, as revealed by the culture of medial olfactory placode obtained from embryonic rats. *Neuroendocrinology* 71, 138–144.
479. Richter, T. A., Keen, K. L., and Terasawa, E. (2002). Synchronization of Ca(2+) oscillations among primate LHRH neurons and nonneuronal cells in vitro. *J. Neurophysiol.* 88, 1559–1567.
480. Leng, G., Brown, C. H., and Russell, J. A. (1999). Physiological pathways regulating the activity of magnocellular neurosecretory cells. *Prog. Neurobiol.* 57, 625–655.
481. Terasawa, E., Schanhofer, W. K., Keen, K. L., and Luchansky, L. (1999). Intracellular Ca(2+) oscillations in luteinizing hormone-releasing hormone neurons derived from the embryonic olfactory placode of the rhesus monkey. *J. Neurosci.* 19, 5898–5909.
482. Terasawa, E., Richter, T. A., and Keen, K. L. (2002). A role for non-neuronal cells in synchronization of intracellular calcium oscillations in primate LHRH neurons. *Prog. Brain Res.* 141, 283–291.
483. Moore, J. P. Jr., Shang, E., and Wray, S. (2002). In situ GABAergic modulation of synchronous gonadotropin releasing hormone-1 neuronal activity. *J. Neurosci.* 22, 8932–8941.
484. Costantin, J. L., and Charles, A. C. (1999). Spontaneous action potentials initiate rhythmic intercellular calcium waves in immortalized hypothalamic (GT1-1) neurons. *J. Neurophysiol.* 82, 429–435.
485. Nunez, L., Villalobos, C., Boockfor, F. R., and Frawley, L. S. (2000). The relationship between pulsatile secretion and calcium dynamics in single, living gonadotropin-releasing hormone neurons. *Endocrinology* 141, 2012–2017.
486. Charles, A., Weiner, R., and Costantin, J. (2001). cAMP modulates the excitability of immortalized H=hypothalamic (GT1) neurons via a cyclic nucleotide gated channel. *Mol. Endocrinol.* 15, 997–1009.
487. Vitalis, E. A., Costantin, J. L., Tsai, P. S., Sakakibara, H., Paruthiyil, S., Iiri, T., Martini, J. F., Taga, M., Choi, A. L., Charles, A. C., and Weiner, R. I. (2000). Role of the cAMP signaling pathway in the regulation of gonadotropin-releasing hormone secretion in GT1 cells. *Proc. Natl. Acad. Sci. U S A* 97, 1861–1866.
488. El-Majdoubi, M., and Weiner, R. I. (2002). Localization of olfactory cyclic nucleotide-gated channels in rat gonadotropin-releasing hormone neurons. *Endocrinology* 143, 2441–2444.
489. Balsalobre, A. (2002). Clock genes in mammalian peripheral tissues. *Cell. Tissue Res.* 309, 193–199.
490. Chappell, P. E., White, R. S., and Mellon, P. L. (2003). Circadian gene expression regulates pulsatile gonadotropin-releasing hormone (GnRH) secretory patterns in the hypothalamic GnRH-secreting GT1–7 cell line. *J. Neurosci.* 23, 11202–11213.
491. Gillespie, J. M., Chan, B. P., Roy, D., Cai, F., and Belsham, D. D. (2003). Expression of circadian rhythm genes in gonadotropin-releasing hormone-secreting GT1-7 neurons. *Endocrinology* 144, 5285–5292.
492. Olcese, J., Domagalski, R., Bednorz, A., Weaver, D. R., Urbanski, H. F., Reuss, S., and Middendorff, R. (2003). Expression and regulation of mPer1 in immortalized GnRH neurons. *Neuroreport* 14, 613–618.
493. Panda, S., Antoch, M. P., Miller, B. H., Su, A. I., Schook, A. B., Straume, M., Schultz, P. G., Kay, S. A., Takahashi, J. S., and Hogenesch, J. B. (2002). Coordinated transcription of key pathways in the mouse by the circadian clock. *Cell* 109, 307–320.
494. Loudon, A. S., Wayne, N. L., Krieg, R., Iranmanesh, A., Veldhuis, J. D., and Menaker, M. (1994). Ultradian endocrine rhythms are altered by a circadian mutation in the Syrian hamster. *Endocrinology* 135, 712–718.
495. Womack, M. D., and Khodakhah, K. (2004). Dendritic control of spontaneous bursting in cerebellar Purkinje cells. *J. Neurosci.* 24, 3511–3521.
496. Oka, Y. (1995). Tetrodotoxin-resistant persistent Na+ current underlying pacemaker potentials of fish gonadotrophin-releasing hormone neurones. *J. Physiol.* 482 (Pt 1), 1–6.
497. Oka, Y. and Matsushima, T. (1993). Gonadotropin-releasing hormone (GnRH)-immunoreactive terminal nerve cells have

intrinsic rhythmicity and project widely in the brain. *J. Neurosci.* 13, 2161–2176.

498. Moos, F., Fontanaud, P., Mekaouche, M., and Brown, D. (2004). Oxytocin neurones are recruited into co-ordinated fluctuations of firing before bursting in the rat. *Neuroscience* 125, 391–410.
499. Ludwig, M. (1998). Dendritic release of vasopressin and oxytocin. *J. Neuroendocrinol.* 10, 881–895.
500. Kokoris, G. J., Lam, N. Y., Ferin, M., Silverman, A. J., and Gibson, M. J. (1988). Transplanted gonadotropin-releasing hormone neurons promote pulsatile luteinizing hormone secretion in congenitally hypogonadal (hpg) male mice. *Neuroendocrinology* 48, 45–52.
501. Herbison, A. E., Gamble, J. R., Pape, J. R., Karunadasa, D. K., Skynner, M. J., and Allen, J. P. (2002). Embryonic migration, heterogeneity and redundancy in the GnRH neuronal population: Insights from a unique transgenic mouse. *Proc. 5th Int. Cong. Neuroendocrinol., Bristol.*
502. Boukhliq, R., Goodman, R. L., Berriman, S. J., Adrian, B., and Lehman, M. N. (1999). A subset of gonadotropin-releasing hormone neurons in the ovine medial basal hypothalamus is activated during increased pulsatile luteinizing hormone secretion. *Endocrinology* 140, 5929–5936.
503. Funabashi, T., Jinnai, K., and Kimura, F. (1997). Fos expression by naloxone in LHRH neurons of the mediobasal hypothalamus and effects of pentobarbital sodium in the proestrous rat. *J. Neuroendocrinol.* 9, 87–92.
504. Charles, A. C., Kodali, S. K., and Tyndale, R. F. (1996). Intercellular calcium waves in neurons. *Mol. Cell. Neurosci.* 7, 337–353.
505. Matesic, D. F., Hayashi, T., Trosko, J. E., and Germak, J. A. (1996). Upregulation of gap junctional intercellular communication in immortalized gonadotropin-releasing hormone neurons by stimulation of the cyclic AMP pathway. *Neuroendocrinology* 64, 286–297.
506. Hu, L., Olson, A. J., Weiner, R. I., and Goldsmith, P. C. (1999). Connexin 26 expression and extensive gap junctional coupling in cultures of GT1-7 cells secreting gonadotropin-releasing hormone. *Neuroendocrinology* 70, 221–227.
507. Funabashi, T., Suyama, K., Uemura, T., Hirose, M., Hirahara, F., and Kimura, F. (2001). Immortalized gonadotropin-releasing hormone neurons (GT1-7 cells) exhibit synchronous bursts of action potentials. *Neuroendocrinology* 73, 157–165.
508. Vazquez-Martinez, R., Shorte, S. L., Boockfor, F. R., and Frawley, L. S. (2001). Synchronized exocytotic bursts from gonadotropin-releasing hormone-expressing cells: dual control by intrinsic cellular pulsatility and gap junctional communication. *Endocrinology* 142, 2095–2101.
509. Hosny, S. and Jennes, L. (1998). Identification of gap junctional connexin-32 mRNA and protein in gonadotropin-releasing hormone neurons of the female rat. *Neuroendocrinology* 67, 101–108.
510. Tsukahara, S., Maekawa, F., Tsukamura, H., Hirunagi, K., and Maeda, K. (1999). Morphological characterization of relationship between gap junctions and gonadotropin releasing hormone nerve terminals in the rat median eminence. *Neurosci. Lett.* 261, 105–108.
511. Traub, R. D., Pais, I., Bibbig, A., LeBeau, F. E., Buhl, E. H., Hormuzdi, S. G., Monyer, H., and Whittington, M. A. (2003). Contrasting roles of axonal (pyramidal cell) and dendritic (interneuron) electrical coupling in the generation of neuronal network oscillations. *Proc. Natl. Acad. Sci. U S A* 100, 1370–1374.
512. Whittington, M. A., and Traub, R. D. (2003). Interneuron diversity series: inhibitory interneurons and network oscillations in vitro. *Trends Neurosci.* 26, 676–682.
513. Melrose, P., Gross, L., Cruse, I., and Rush, M. (1987). Isolated gonadotropin-releasing hormone neurons harvested from adult male rats secrete biologically active neuropeptide in a regular repetitive manner. *Endocrinology* 121, 182–189.
514. Rettori, V., Belova, N., Dees, W. L., Nyberg, C. L., Gimeno, M., and McCann, S. M. (1993). Role of nitric oxide in the control of luteinizing hormone-releasing hormone release in vivo and in vitro. *Proc. Natl. Acad. Sci. U S A* 90, 10130–10134.
515. Knauf, C., Prevot, V., Stefano, G. B., Mortreux, G., Beauvillain, J. C., and Croix, D. (2001). Evidence for a spontaneous nitric oxide release from the rat median eminence: influence on gonadotropin-releasing hormone release. *Endocrinology* 142, 2343–23450.
516. Sohal, V. S., Keist, R., Rudolph, U., and Huguenard, J. R. (2003). Dynamic GABA(A) receptor subtype-specific modulation of the synchrony and duration of thalamic oscillations. *J. Neurosci.* 23, 3649–3657.
517. Serafin, M., Williams, S., Khateb, A., Fort, P., and Muhlethaler, M. (1996). Rhythmic firing of medial septum non-cholinergic neurons. *Neuroscience* 75, 671–675.
518. Blake, C. A., and Sawyer, C. H. (1974). Effects of hypothalamic deafferentation on the pulsatile rhythm in plasma concentrations of luteinizing hormone in ovariectomized rats. *Endocrinology* 94, 730–736.
519. Soper, B. D., and Weick, R. F. (1980). Hypothalamic and extrahypothalamic mediation of pulsatile discharges of luteinizing hormone in the ovariectomized rat. *Endocrinology* 106, 348–355.
520. Halasz, B., and Gorski, R. A. (1967). Gonadotrophic hormone secretion in female rats after partial or total interruption of neural afferents to the medial basal hypothalamus. *Endocrinology* 80, 608–622.
521. Ohkura, S., Tsukamura, H., and Maeda, K. (1992). Effects of transplants of fetal mediobasal hypothalamus on luteinizing hormone pulses impaired by hypothalamic deafferentation in adult ovariectomized rats. *Neuroendocrinology* 55, 422–426.
522. Rose, P. A., and Weick, R. F. (1986). Effects of anterior hypothalamic deafferentation and neonatal monosodium-L-glutamate treatment on pulsatile LH secretion in the castrated rat. *Neuroendocrinology* 43, 12–17.
523. Pau, K. F., Kuehl, D. E., and Jackson, G. L. (1982). Effect of frontal hypothalamic deafferentation on luteinizing hormone secretion and seasonal breeding in the ewe. *Biol. Reprod.* 27, 999–1009.
524. Whisnant, C. S., and Goodman, R. L. (1994). Effect of anterior hypothalamic deafferentation on the negative feedback of gonadal steroids on luteinizing hormone pulse frequency in the ewe. *Domest. Anim. Endocrinol.* 11, 151–159.
525. Mori, Y., Nishihara, M., Tanaka, T., Shimizu, T., Yamaguchi, M., Takeuchi, Y., and Hoshino, K. (1991). Chronic recording of electrophysiological manifestation of the hypothalamic gonadotropin-releasing hormone pulse generator activity in the goat. *Neuroendocrinology* 53, 392–395.
526. Wilson, R. C., Kesner, J. S., Kaufman, J. M., Uemura, T., Akema, T., and Knobil, E. (1984). Central electrophysiologic correlates of pulsatile luteinizing hormone secretion in the rhesus monkey. *Neuroendocrinology* 39, 256–260.
527. Thiery, J. C., and Pelletier, J. (1981). Multiunit activity in the anterior median eminence and adjacent areas of the hypothalamus of the ewe in relation to LH secretion. *Neuroendocrinology* 32, 217–224.
528. Kimura, F., Nishihara, M., Hiruma, H., and Funabashi, T. (1991). Naloxone increases the frequency of the electrical activity of luteinizing hormone-releasing hormone pulse generator in long-term ovariectomized rats. *Neuroendocrinology* 53, 97–102.

529. Silverman, A. J., Wilson, R., Kesner, J. S., and Knobil, E. (1986). Hypothalamic localization of multiunit electrical activity associated with pulsatile LH release in the rhesus monkey. *Neuroendocrinology* 44, 168–171.
530. Hiruma, H., and Kimura, F. (1995). Luteinizing hormone-releasing hormone is a putative factor that causes LHRH neurons to fire synchronously in ovariectomized rats. *Neuroendocrinology* 61, 509–516.
531. Rasmussen, D. D. (1993). Episodic gonadotropin-releasing hormone release from the rat isolated median eminence in vitro. *Neuroendocrinology* 58, 511–518.
532. Purnelle, G., Gerard, A., Czajkowski, V., and Bourguignon, J. P. (1997). Pulsatile secretion of gonadotropin-releasing hormone by rat hypothalamic explants without cell bodies of GnRH neurons (corrected). *Neuroendocrinology* 66, 305–312.
533. Caraty, A., Antoine, C., Delaleu, B., Locatelli, A., Bouchard, P., Gautron, J. P., Evans, N. P., Karsch, F. J., and Padmanabhan, V. (1995). Nature and bioactivity of gonadotropin-releasing hormone (GnRH) secreted during the GnRH surge. *Endocrinology* 136, 3452–3460.
534. Sarkar, D. K., Chiappa, S. A., Fink, G., and Sherwood, N. M. (1976). Gonadotropin-releasing hormone surge in pro-oestrous rats. *Nature* 264, 461–463.
535. Ching, M. (1982). Correlative surges of LHRH, LH and FSH in pituitary stalk plasma and systemic plasma of rat during proestrus. Effect of anaesthetics. *Neuroendocrinology* 34, 279–285.
536. Xia, L., Van Vugt, D., Alston, E. J., Luckhaus, J., and Ferin, M. (1992). A surge of gonadotropin-releasing hormone accompanies the estradiol-induced gonadotropin surge in the rhesus monkey. *Endocrinology* 131, 2812–2820.
537. Pau, K. F., Berria, M., Hess, D. L., and Spies, H. G. (1993). Preovulatory gonadotropin-releasing hormone surge in ovarian-intact rhesus macaques. *Endocrinology* 133, 1650–1656.
538. Moenter, S. M., Caraty, A., and Karsch, F. J. (1990). The estradiol-induced surge of gonadotropin-releasing hormone in the ewe. *Endocrinology* 127, 1375–1384.
539. Moenter, S. M., Brand, R. C., and Karsch, F. J. (1992). Dynamics of gonadotropin-releasing hormone (GnRH) secretion during the GnRH surge: insights into the mechanism of GnRH surge induction. *Endocrinology* 130, 2978–2984.
540. Levine, J. E., Norman, R. L., Gliessman, P. M., Oyama, T. T., Bangsberg, D. R., and Spies, H. G. (1985). In vivo gonadotropin-releasing hormone release and serum luteinizing hormone measurements in ovariectomized, estrogen-treated rhesus macaques. *Endocrinology* 117, 711–721.
541. Evans, N. P., Dahl, G. E., Mauger, D. T., Padmanabhan, V., Thrun, L. A., and Karsch, F. J. (1995). Does estradiol induce the preovulatory gonadotropin-releasing hormone (GnRH) surge in the ewe by inducing a progressive change in the mode of operation of the GnRH neurosecretory system. *Endocrinology* 136, 5511–559.
542. Clarke, I. J. (1993). Variable patterns of gonadotropin-releasing hormone secretion during the estrogen-induced luteinizing hormone surge in ovariectomized ewes. *Endocrinology* 133, 1624–1632.
543. Sisk, C. L., Richardson, H. N., Chappell, P. E., and Levine, J. E. (2001). In vivo gonadotropin-releasing hormone secretion in female rats during peripubertal development and on proestrus. *Endocrinology* 142, 2929–2936.
544. Veldhuis, J. D., Johnson, M. L., and Gallo, R. V. (1993). Reanalysis of the rat proestrous LH surge by deconvolution analysis. *Am. J. Physiol.* 265, R240–R245.
545. Kimura, F., and Funabashi, T. (1998). Two subgroups of gonadotropin releasing hormone neurons control gonadotropin secretion in rats. *News Physiol. Sci.* 13, 225–231.
546. Rubin, B. S., and King, J. C. (1994). The number and distribution of detectable luteinizing hormone (LH)-releasing hormone cell bodies changes in association with the preovulatory LH surge in the brains of young but not middle-aged female rats. *Endocrinology* 134, 467–474.
547. Moenter, S. M., Karsch, F. J., and Lehman, M. N. (1993). Fos expression during the estradiol-induced gonadotropin-releasing hormone (GnRH) surge of the ewe: induction in GnRH and other neurons. *Endocrinology* 133, 896–903.
548. Witkin, J. W., Xiao, E., Popilskis, S., Ferin, M., and Silverman, A. (1994). Fos expression in the gonadotropin-releasing hormone (GnRH) neuron does not increase during the ovarian steroid-induced GnRH surge in the rhesus monkey. *Endocrinology* 135, 956–961.
549. Caraty, A., and Locatelli, A. (1988). Effect of time after castration on secretion of LHRH and LH in the ram. *J. Reprod. Fertil.* 82, 263–269.
550. Legan, S. J., Coon, G. A., and Karsch, F. J. (1975). Role of estrogen as initiator of daily LH surges in the ovariectomized rat. *Endocrinology* 96, 50–56.
551. Everett, J., and Sawyer, C. H. (1950). A 24h periodicity in the "LH release apparatus" of female rats, disclosed by barbiturate sedation. *Endocrinology* 46, 198–216.
552. Legan, S. J., and Karsch, F. J. (1975). A daily signal for the LH surge in the rat. *Endocrinology* 96, 57–62.
553. Bakker, J., and Baum, M. J. (2000). Neuroendocrine regulation of GnRH release in induced ovulators. *Front. Neuroendocrinol.* 21, 220–262.
554. Wiegand, S. J., and Terasawa, E. (1982). Discrete lesions reveal functional heterogeneity of suprachiasmatic structures in regulation of gonadotropin secretion in the female rat. *Neuroendocrinology* 34, 395–404.
555. Meyer-Bernstein, E. L., Jetton, A. E., Matsumoto, S. I., Markuns, J. F., Lehman, M. N., and Bittman, E. L. (1999). Effects of suprachiasmatic transplants on circadian rhythms of neuroendocrine function in golden hamsters. *Endocrinology* 140, 207–218.
556. Kawakami, M., Arita, J., and Yoshioka, E. (1980). Loss of estrogen-induced daily surges of prolactin and gonadotropins by suprachiasmatic nucleus lesions in ovariectomized rats. *Endocrinology* 106, 1087–1092.
557. Gray, G. D., Soderstein, P., Tallentire, D., and Davidson, J. M. (1978). Effects of lesions in various structures of the suprachiasmatic-preoptic region on LH regulation and sexual behavior in female rats. *Neuroendocrinology* 25, 174–191.
558. Watts, A. G., Sheward, W. J., Whale, D., and Fink, G. (1989). The effects of knife cuts in the sub-paraventricular zone of the female rat hypothalamus on oestrogen-induced diurnal surges of plasma prolactin and LH, and circadian wheel-running activity. *J. Endocrinol.* 122, 593–604.
559. Silver, R., LeSauter, J., Tresco, P. A., and Lehman, M. N. (1996). A diffusible coupling signal from the transplanted suprachiasmatic nucleus controlling circadian locomotor rhythms. *Nature* 382, 810–813.
560. de la Iglesia, H. O., Meyer, J., and Schwartz, W. J. (2003). Lateralization of circadian pacemaker output: Activation of left- and right-sided luteinizing hormone-releasing hormone neurons involves a neural rather than a humoral pathway. *J. Neurosci.* 23, 7412–7414.
561. de la Iglesia, H. O., Blaustein, J. D., and Bittman, E. L. (1995). The suprachiasmatic area in the female hamster projects to neurons containing estrogen receptors and GnRH. *Neuroreport.* 6, 1715–1722.
562. van der Beek, E. M., Wiegant, V. M., van der Donk, H. A., van den Hurk, R., and Buijs, R. M. (1993). Lesions of the suprachiasmatic nucleus indicate the presence of a direct vasoactive intestinal polypeptide-containing projection to

gonadotrophin-releasing hormone neurons in the female rat. *J. Neuroendocrinol.* 5, 137–144.
563. Smith, M. J., Jennes, L., and Wise, P. M. (2000). Localization of the VIP2 receptor protein on GnRH neurons in the female rat. *Endocrinology* 141, 4317–4320.
564. Watson, R. E., Langub, M. C., Engle, M. G., and Maley, B. E. (1995). Estrogen-receptive neurons in the anteroventral periventricular nucleus are synaptic targets of the suprachiasmatic nucleus and peri-suprachiasmatic region. *Brain Res.* 689, 254–264.
565. Simerly, R. B. (2002). Wired for reproduction: organization and development of sexually dimorphic circuits in the mammalian forebrain. *Annu. Rev. Neurosci.* 25, 507–536.
566. Herbison, A. E. (1998). Multimodal influence of estrogen upon gonadotropin-releasing hormone neurons. *Endocr. Rev.* 19, 302–330.
567. Chappel, P. E., Lee, J., and Levine, J. E. (2000). Stimulation of gonadotropin-releasing hormone surges by estrogen. II. Role of cyclic andrenosine 3´,5´-monophosphate. *Endocrinology* 141, 1486–1492.
568. Petersen, S. L., Ottem, E. N., and Carpenter, C. D. (2003). Direct and indirect regulation of gonadotropin-releasing hormone neurons by estradiol. *Biol. Reprod.* 69, 1771–1778.
569. Butler, J. A., Sjoberg, M., and Coen, C. W. (1999). Evidence for oestrogen receptor a-immunoreactivity in gonadotrophin-releasing hormone-expressing neurones. *J. Neuroendocrinol.* 11, 331–335.
570. Skynner, M. J., Sim, J. S., and Herbison, A. E. (1999). Detection of estrogen receptor a and b messenger ribonucleic acids in adult gonadotropin-releasing hormone neurons. *Endocrinology* 140, 5195–5201.
571. Sharifi, N., Reuss, A. E., and Wray, S. (2002). Prenatal LHRH neurons in nasal explant cultures express estrogen receptor beta transcript. *Endocrinology* 143, 2503–2507.
572. Herbison, A. E., and Pape, J. R. (2001). New evidence for estrogen receptors in gonadotropin-releasing hormone neurons. *Front. Neuroendocrinol.* 22, 292–308.
573. Kallo, I., Butler, J. A., Barkovics-Kallo, M., Goubillon, M. L., and Coen, C. W. (2001). Oestrogen receptor beta-immunoreactivity in gonadotropin releasing hormone-expressing neurones: regulation by oestrogen. *J. Neuroendocrinol.* 13, 741–748.
574. Hrabovszky, E., Shughrue, P. J., Merchenthaler, I., Hajszan, T., Carpenter, C. D., Liposits, Z., and Petersen, S. L. (2000). Detection of estrogen receptor-b messenger ribonucleic acid and ^{125}I-estrogen binding sites in luteinizing hormone-releasing hormone neurons of the rat brain. *Endocrinology* 141, 3506–3509.
575. Hrabovszky, E., Steinhauser, A., Barabas, K., Shughrue, P. J., Petersen, S. L., Merchenthaler, I., and Liposits, Z. (2001). Estrogen receptor-b immunoreactivity in luteinizing hormone-releasing hormone neurons of the rat brain. *Endocrinology* 142, 3261–3264.
576. Legan, S. J., and Tsai, H. W. (2003). Oestrogen receptor-alpha and -beta immunoreactivity in gonadotropin-releasing hormone neurones after ovariectomy and chronic exposure to oestradiol. *J. Neuroendocrinol.* 15, 1164–1170.
577. Shivers, B. D., Harlan, R. E., Morrell, J. I., and Pfaff, D. W. (1983). Absence of estradiol concentration in cell nuclei of LHRH-immunoreactive neurons. *Nature* 304, 345–347.
578. Abraham, I. M., Han, K., Todman, M. G., Korach, K. S., and Herbison, A. E. (2003). Estrogen receptor b mediates rapid estrogen actions on gonadotropin-releasing hormone neurons *in vivo*. *J. Neurosci.* 23, 5771–5777.
579. Flugge, G., Oertel, W. H., and Wuttke, W. (1986). Evidence for estrogen-receptive GABAergic neurons in the preoptic/anterior hypothalamic area of the rat brain. *Neuroendocrinology* 43, 1–5.
580. Herbison, A. E., Fenelon, V. S., and Brussaard, A. B. (1998). Gonadal steroid regulation of GABAergic transmission. *Eur. J. Neurosci.* 10, 2.
581. Herbison, A. E., Robinson, J. E., and Skinner, D. C. (1993). Distribution of estrogen receptor-immunoreactive cells in the preoptic area of the ewe: co-localization with glutamic acid decarboxylase but not luteinizing hormone-releasing hormone. *Neuroendocrinology* 57, 751–759.
582. Eyigor, O., Lin, W., and Jennes, L. (2004). Identification of neurones in the female rat hypothalamus that express oestrogen receptor-alpha and vesicular glutamate transporter-2. *J. Neuroendocrinol.* 16, 26–31.
583. Skinner, D. C., and Herbison, A. E. (1997). Effects of photoperiod on estrogen receptor, tyrosine hydroxylase, neuropeptide Y, and β-endorphin immunoreactivity in the ewe hypothalamus. *Endocrinology* 138, 2585–2595.
584. Sar, M., Sahu, A., Crowley, W. R., and Kalra, S. P. (1990). Localization of neuropeptide-Y immunoreactivity in estradiol-concentrating cells in the hypothalamus. *Endocrinology* 127, 2752–2756.
585. Smith, M. J., and Jennes, L. (2001). Neural signals that regulate GnRH neurones directly during the oestrous cycle. *Reproduction* 122, 1–10.
586. Kelly, M. J., and Levin, E. R. (2001). Rapid actions of plasma membrane estrogen receptors. *Trends Endo. Metab.* 12, 152–156.
587. Garcia-Segura, L. M. and McCarthy, M. M. (2004). Minireview: role of glia in neuroendocrine function. *Endocrinology* 145, 1082–1086.
588. Langub, M. C., and Watson, R. E. (1992). Estrogen receptor-immunoreactive glia, endothelia, and ependyma in guinea pig preoptic area and median eminence: electron microscopy. *Endocrinology* 130, 364–372.
589. Garcia-Ovejero, D., Veiga, S., Garcia-Segura, L. M., and Doncarlos, L. L. (2002). Glial expression of estrogen and androgen receptors after rat brain injury. *J. Comp. Neurol.* 450, 256–271.
590. Chaban, V. V., Lakhter, A. J., and Micevych, P. (2004). A membrane estrogen receptor mediates intracellular calcium release in astrocytes. *Endocrinology* 145, 3788–3795.
591. King, J. C., Ronshei, P. M., and Rubin, B. S. (1995). Dynamic relationships between LHRH neuronal terminals and the end-feet of tanycytes in cycling rats revealed by confocal microscopy. *Soc. Neurosci. Abst.* 21, 112.10.
592. Garcia-Segura, L. M., Luquin, S., Parducz, A., and Naftolin, F. (1994). Gonadal hormone regulation of glial fibrillary acidic protein immunoreactivity and glial ultrastructure in the rat neuroendocrine hypothalamus. *Glia* 10, 59–69.
593. Zwain, I. H., Arroyo, A., Amato, P., and Yen, S. S. (2002). A role for hypothalamic astrocytes in dehydroepiandrosterone and estradiol regulation of gonadotropin-releasing hormone (GnRH) release by GnRH neurons. *Neuroendocrinology* 75, 375–383.
594. Quesada, A. and Etgen, A. M. (2002). Functional interactions between estrogen and insulin-like growth factor-I in the regulation of alpha 1B-adrenoceptors and female reproductive function. *J. Neurosci.* 22, 2401–2408.
595. Daftary, S. S., and Gore, A. C. (2003). Developmental changes in hypothalamic insulin-like growth factor–1: relationship to gonadotropin-releasing hormone neurons. *Endocrinology* 144, 2034–2045.
596. Cohen, P. E., Zhu, L., Nishimura, K., and Pollard, J. W. (2002). Colony-stimulating factor 1 regulation of neuroendocrine pathways that control gonadal function in mice. *Endocrinology* 143, 1413–1422.
597. Prevot, V., Croix, D., Rialas, C. M., Poulain, P., Fricchione, G. L., Stegano, G. B., and Beauvillain, J.-C. (1999). Estradiol coupling to endothelial nitric oxide stimulates

gonadotropin-releasing hormone release from rat median eminence via a membrane receptor. *Endocrinology* 140, 652–659.

598. Aguan, K., Mahesh, V. B., Ping, L., Bhat, G., and Brann, D. W. (1996). Evidence for a physiological role for nitric oxide in the regulation of the LH surge: effect of central administration of antisense oligonucleotides to nitric oxide synthase. *Neuroendocrinology* 64, 449–455.
599. Knauf, C., Ferreira, S., Hamdane, M., Mailliot, C., Prevot, V., Beauvillain, J. C., and Croix, D. (2001). Variation of endothelial nitric oxide synthase synthesis in the median eminence during the rat estrous cycle: an additional argument for the implication of vascular blood vessel in the control of GnRH release. *Endocrinology* 142, 4288–4294.
600. Drouva, S. V., Laplante, E., Gautron, J.-P., and Kordon, C. (1984). Effects of 17b-estradiol on LH-RH release from rat mediobasal hypothalamic slices. *Neuroendocrinology* 38, 152–157.
601. Bronson, F. H. (1981). The regulation of luteinizing hormone secretion by estrogen: relationships among negative feedback, surge potential, and male stimulation in juvenile, peripubertal, and adult female mice. *Endocrinology* 108, 506–516.
602. Sarkar, D. K., and Fink, G. (1980). Luteinizing hormone releasing factor in pituitary stalk plasma from long-term ovariectomized rats: effects of steroids. *J. Endocrinol.* 86, 511–524.
603. Yamaji, T., Dierschke, D. J., Hotchkiss, J., Bhattacharya, A. N., Surve, A. H., and Knobil, E. (1971). Estrogen induction of LH release in the rhesus monkey. *Endocrinology* 89, 1034–1041.
604. Evans, N. P., Dahl, G. E., Padmanabhan, V., Thrun, L. A., and Karsch, F. J. (1997). Estradiol requirements for induction and maintenance of the gonadotropin-releasing hormone surge: implications for neuroendocrine processing of the estradiol signal. *Endocrinology* 138, 5408–5414.
605. Wiegand, S. J., Terasawa, E., Bridson, W. E., and Goy, R. W. (1980). Effects of discrete lesions of preoptic and suprachiasmatic structures in the female rat. *Neuroendocrinology* 31, 147–157.
606. Ronnekleiv, O. K., and Kelly, M. J. (1988). Plasma prolactin and luteinizing hormone profiles during the estrous cycle of the female rat: effects of surgically induced persistent estrus. *Neuroendocrinology* 47, 133–141.
607. Kalra, P. S., and McCann, S. M. (1975). The stimulatory effect on gonadotropin release of implants of estradiol or progesterone in certain sites in the central nervous system. *Neuroendocrinology* 19, 289–302.
608. Goodman, R. L. (1978). The site of the positive feedback action of estradiol in the rat. *Endocrinology* 102, 151–159.
609. Petersen, S. L., and Barraclough, C. A. (1989). Suppression of spontaneous LH surges in estrogen-treated ovariectomized rats by microimplants of antiestrogens into the preoptic brain. *Brain Res.* 484, 279–289.
610. Petersen, S. L., Cheuk, C., Hartman, R. D., and Barraclough, C. A. (1989). Medial preoptic microimplants of the antiestrogen, keoxifene, affect luteinizing hormone-releasing hormone mRNA levels, median eminence luteinizing hormone-releasing hormone concentrations and luteinizing hormone release in ovariectomized, estrogen-treated rats. *J. Neuroendocrinol.* 1, 279–283.
611. Shughrue, P. J., Lane, M. V., and Merchenthaler, I. (1997). Comparative distribution of estrogen receptor-a and -b mRNA in the rat central nervous system. *J. Compar. Neurol.* 388, 507–525.
612. Orikasa, C., Kondo, Y., Hayashi, S., McEwen, B. S., and Sakuma, Y. (2002). Sexually dimorphic expression of estrogen receptor beta in the anteroventral periventricular nucleus of the rat preoptic area: implication in luteinizing hormone surge. *Proc. Natl. Acad. Sci. U S A* 99, 3306–3311.
613. Herbison, A. E., and Theodosis, D. T. (1992). Localisation of oestrogen receptors in preoptic neurons containing neurotensin but not tyrosine hydroxylase, cholecystokinin or luteinizing hormone-releasing hormone in the male and female rat. *Neuroscience* 50, 283–298.
614. Le, W. W., Berghorn, K. A., Rassnick, S., and Hoffman, G. E. (1999). Periventricular preoptic area neurons coactivated with luteinizing hormone (LH)-releasing hormone (LHRH) neurons at the time of the LH surge are LHRH afferents. *Endocrinology* 140, 510–519.
615. Simerly, R. B. (1991). Prodynorphin and proenkephalin gene expression in the anteroventral periventricular nucleus of the rat: sexual differentiation and hormonal regulation. *Mol. Cell. Neurosci.* 2, 473–484.
616. Yuri, K., and Kawata, M. (1994). Estrogen receptor-immunoreactive neurons contain calcitonin gene-related peptide, methionine-enkephalin or tyrosine hydroxylase in the female rat preoptic area. *Neurosci. Res.* 21, 135–141.
617. Bloch, G. J., Kurth, S. M., Akesson, T. R., and Micevych, P. E. (1992). Estrogen-concentrating cells within cell groups of the medial preoptic area: sex differences and co-localization with galanin-immunoreactive cells. *Brain Res.* 595, 301–308.
618. Okamura, H., Yokosuka, M., and Hayashi, S. (1994). Induction of substance P-immunoreactivity by estrogen in neurons containing estrogen receptors in the anteroventral periventricular nucleus of female but not male rats. *J. Neuroendocrinol.* 6, 609–615.
619. Herbison, A. E., and Theodosis, D. T. (1992). Immunocytochemical identification of oestrogen receptors in preoptic neurones containing calcitonin gene-related peptide in the male and female rat. *Neuroendocrinology* 56, 761–764.
620. Brann, D. W., Ping, L., and Mahesh, V. B. (1993). Role of non-NMDA receptor neurotransmission in steroid and preovulatory gonadotropin surge expression in the female rat. *Mol. Cell. Neurosci.* 4, 92–297.
621. Ping, L., Mahesh, V. B., Wiedmeier, V. T., and Brann, D. W. (1994). Release of glutamate and aspartate from the preoptic area during the progesterone-induced LH surge: in vivo microdialysis studies. *Neuroendocrinology* 59, 318–324.
622. Jarry, H., Perschl, A., and Wuttke, W. (1988). Further evidence that preoptic anterior hypothalamic GABAergic neurons are part of the GnRH pulse and surge generator. *Acta Endocrinol.* 118, 573–579.
623. Robinson, J. E., Kendrick, K. M., and Lambart, C. E. (1991). Changes in the release of gamma-aminobutyric acid and catecholamines in the preoptic/septal area prior to and during the preovulatory surge of luteinizing hormone in the ewe. *J. Neuroendocrinol.* 3, 393–399.
624. Mitsushima, D., Shwe, T. T., Funabashi, T., Shinohara, K., and Kimura, F. (2002). GABA release in the medial preoptic area of cyclic female rats. *Neuroscience* 113, 109–114.
625. Tin Tin Win, S., Mitsushima, D., Shinohara, K., and Kimura, F. (2004). Sexual dimorphism of GABA release in the medial preoptic area and luteinizing hormone release in gonadectomized estrogen-primed rats. *Neuroscience* 127, 243–250.
626. Seltzer, A. M., and Donoso, A. O. (1992). Restraining action of GABA on estradiol-induced LH surge in the rat: GABA activity in brain nuclei and effects of GABA mimetics in the medial preoptic nucleus. *Neuroendocrinology* 55, 28–34.
627. Kimura, F., and Jinnai, K. (1994). Bicuculline infusions advance the timing of luteinizing hormone surge in proestrous rats: comparisons with naloxone effects. *Horm. Behav.* 28, 424–430.
628. Herbison, A. E. (1997). Estrogen regulation of GABA transmission in rat preoptic area. *Brain Res. Bull.* 44, 321–326.
629. Ferris, C. F., Pan, J. X., Singer, E. A., Boyd, N. D., Carraway, R. E., and Leeman, S. E. (1984). Stimulation of luteinizing hormone release after stereotaxic microinjection of neurotensin

into the medial preoptic area of rats. *Neuroendocrinology* 38, 144–151.

630. Akema, T., Praputpittaya, C., and Kimura, F. (1987). Effects of preoptic microinjection of neurotensin on luteinizing hormone secretion in unanesthetized ovariectomized rats with or without estrogen priming. *Neuroendocrinology* 46, 345–349.
631. Alexander, M. J., Mahoney, P. D., Ferris, C. G., Carraway, R. E., and Leeman, S. E. (1989). Evidence that neurotensin participates in the central regulation of the preovulatory surge of luteinizing hormone in the rat. *Endocrinology* 124, 783–788.
632. Alexander, M. J., and Leeman, S. E. (1994). Estrogen-inducible neurotensin immunoreactivity in the preoptic area of the female rat. *J. Compar. Neurol.* 345, 496–509.
633. Smith, M. J., and Wise, P. M. (2001). Neurotensin gene expression increases during proestrus in the rostral medial preoptic nucleus: potential for direct communication with gonadotropin-releasing hormone neurons. *Endocrinology* 142, 3006–3013.
634. Caraty, A., Fabre-Nys, C., Delaleu, B., Locatelli, A., Bruneau, G., Karsch, F. J., and Herbison, A. (1997). Evidence that the mediobasal hypothalamus is the primary site of action of estradiol in inducing the preovulatory GnRH surge in the ewe. *Endocrinology* 139, 1752–1760.
635. Blache, D., Fabre-Nys, C. J., and Venier, G. (1991). Ventromedial hypothalamus as a target for oestradiol action on proceptivity, receptivity and luteinizing hormone surge of the ewe. *Brain Res.* 546, 241–249.
636. Clarke, I. J., Pompolo, S., Scott, C. J., Rawson, J. A., Caddy, D., Jakubowska, A. E., and Pereira, A. M. (2001). Cells of the arcuate nucleus and ventromedial nucleus of the ovariectomized ewe that respond to oestrogen: a study using Fos immunohistochemistry. *J. Neuroendocrinol.* 13, 934–941.
637. Couse, J. F., Yates, M. M., Walker, V. R., and Korach, K. S. (2003). Characterization of the hypothalamic-pituitary-gonadal (HPG) axis in estrogen receptor null mice reveals hypergonadism and endocrine sex-reversal in females lacking ERβ but not ERβ. *Mol. Endocrinol.* 17, 1039–1053.
638. Lubahn, D. B., Moyer, J. S., Golding, T. S., Couse, J. F., Korach, K. S., and Smithies, O. (1993). Alteration of reproductive function but not prenatal sexual development after insertional disruption of the mouse estrogen receptor gene. *Proc. Natl. Acad. Sci. U S A* 90, 11162–11166.
639. Dupont, S., Krust, A., Gansmuller, A., Dierich, A., Chambon, P., and Mark, M. (2000). Effect of single and compound knockouts of estrogen receptors alpha (ERalpha) and beta (ERbeta) on mouse reproductive phenotypes. *Development* 127, 4277–4291.
640. Herbison, A. E., Abraham, I. M., Korach, K. S., Dorling, A., and Todman, M. G. (2004). Definition of new and old mechanisms of estrogen action on GnRH neurons. *Endocr. Soc. Abstr.* S18–S21.
641. Herbison, A. E. (2002). Electrical properties of postnatal GnRH neurons in the mouse and their regulation by gonadal steroids. In *Neuroplasticity, Development and Steroid Hormone Action*.
642. Cemeroglu, A. P., Kletter, G. B., Guo, W., Brown, M. B., Kelch, R. P., Marshall, J. C., Padmanabhan, V., and Foster, C. M. (1998). In pubertal girls, naloxone fails to reverse the suppression of luteinizing hormone secretion by estradiol. *J. Clin. Endocrinol. Metab.* 83, 3501–3506.
643. Hayes, F. J., Seminara, S. B., Decruz, S., Boepple, P. A., and Crowley, W. F. J. (2000). Aromatase inhibition in the human male reveals a hypothalamic site of estrogen feedback. *J. Clin. Endocrinol. Metab.* 85, 3027–3035.
644. Chongthammakun, S., and Terasawa, E. (1993). Negative feedback effects of estrogen on luteinizing hormone-releasing hormone release occur in pubertal, but not prepubertal, ovariectomized female rhesus monkeys. *Endocrinology* 132, 735–743.
645. Evans, N. P., Dahl, G. E., Glover, B. H., and Karsch, F. J. (1994). Central regulation of pulsatile gonadotropin-releasing hormone (GnRH) secretion by estradiol during the period leading up to the preovulatory GnRH surge in the ewe. *Endocrinology* 134, 1806–1811.
646. Akema, T., Takokoro, Y., and Kimura, F. (1984). Regional specificity in the effect of estrogen implantation within the forebrain on the frequency of pulsatile luteinizing hormone secretion in the ovariectomized rat. *Neuroendocrinology* 39, 517–523.
647. Nagatani, S., Tsukamura, H., and Maeda, K. (1994). Estrogen feedback needed at the paraventricular nucleus or A2 to suppress pulsatile luteinizing hormone release in fasting female rats. *Endocrinology* 135, 870–875.
648. Ferin, M., Carmel, P. W., Zimmerman, E. A., Warren, M., Perez, R., and Vande Wiele, R. L. (1974). Location of intrahypothalamic estrogen-responsive sites influencing LH secretion in the female Rhesus monkey. *Endocrinology* 95, 1059–1068.
649. Blake, C. A. (1977). A medial basal hypothalamic site of synergistic action of estrogen and progesterone on the inhibition of pituitary luteinizing hormone release. *Endocrinology* 101, 1130–1134.
650. Smith, E. R., and Davidson, J. M. (1974). Location of feedback receptors: effects of intracranially implanted steroids on plasma LH and LRF response. *Endocrinology* 95, 1566–1573.
651. Gallegos-Sanchez, J., Delaleu, B., Caraty, A., Malpaux, B., and Thiery, J. C. (1997). Estradiol acts locally within the retrochiasmatic area to inhibit pulsatile luteinizing-hormone release in the female sheep during anestrus. *Biol. Reprod.* 56, 1544–9.
652. Wersinger, S. R., Haisenleder, D. J., Lubahn, D. B., and Rissman, E. F. (1999). Steroid feedback on gonadotropin release and pituitary gonadotropin subunit mRNA in mice lacking a functional estrogen receptor alpha. *Endocrine* 11, 137–143.
653. Dorling, A. A., Todman, M. G., Korach, K. S., and Herbison, A. E. (2003). Critical role for estrogen receptor alpha in negative feedback regulation of gonadotropin-releasing hormone mRNA expression in the female mouse. *Neuroendocrinology* 78, 204–209.
654. Romanowicz, K., Misztal, T., and Barcikowski, B. (2004). Genistein, a phytoestrogen, effectively modulates luteinizing hormone and prolactin secretion in ovariectomized ewes during seasonal anestrus. *Neuroendocrinology* 79, 73–81.
655. Hardy, S. L., Anderson, G. M., Valent, M., Connors, J. M., and Goodman, R. L. (2003). Evidence that estrogen receptor alpha, but not beta, mediates seasonal changes in the response of the ovine retrochiasmatic area to estradiol. *Biol. Reprod.* 68, 846–852.
656. Thind, K. K., and Goldsmith, P. C. (1997). Expression of estrogen and progesterone receptors in glutamate and GABA neurons of the pubertal female monkey hypothalamus. *Neuroendocrinology* 65, 314–324.
657. Mansky, T., Mestres-Ventura, P., and Wuttke, W. (1982). Involvement of GABA in the feedback action of estradiol on gonadotropin and prolactin release: hypothalamic GABA and catecholamine turnover rates. *Brain Res.* 231, 353–364.
658. Ondo, J., Mansky, T., and Wuttke, W. (1982). In vivo GABA release from the medial preoptic area of diestrous and ovariectomized rats. *Exp. Brain Res.* 46, 69–72.
659. Herbison, A. E., Heavens, R. P., Dye, S., and Dyer, R. G. (1991). Acute action of oestrogen on medial preoptic gamma-aminobutyric acid neurons: correlation with oestrogen

negative feedback on luteinizing hormone secretion. *J. Neuroendocrinol.* 3, 101–106.

660. Herbison, A. E., Augood, S. J., Simonian, S. X., and Chapman, C. (1995). Regulation of GABA transporter activity and mRNA expression by estrogen in rat preoptic area. *J. Neurosci.* 15, 8302–8309.
661. Herbison, A. E., and Fenelon, V. S. (1995). Estrogen regulation of $GABA_A$ receptor subunit mRNA expression in preoptic area and bed nucleus of the stria terminalis of female rat brain. *J. Neurosci.* 15, 2328–2337.
662. Akema, T., Chiba, A., and Kimura, F. (1990). On the relationship between noradrenergic stimulatory and GABAergic inhibitor systems in the control of luteinizing hormone secretion in female rats. *Neuroendocrinology* 52, 566–572.
663. Rajendren, G., and Gibson, M. J. (2001). A confocal microscopic study of synaptic inputs to gonadotropin-releasing hormone cells in mouse brain: regional differences and enhancement by estrogen. *Neuroendocrinology* 73, 84–90.
664. Zsarnovszky, A., Horvath, T. L., Garcia-Segura, L. M., Horvath, B., and Naftolin, F. (2001). Oestrogen-induced changes in the synaptology of the monkey (Cercopithecus aethiops) arcuate nucleus during gonadotropin feedback. *J. Neuroendocrinol.* 13, 22–28.
665. Krege, J. H., Hodgin, J. B., Couse, J. F., Enmark, E., Warner, M., Magler, J. F., Sar, M., Korach, K. S., Gustafsson, J.-A., and Smithies, O. (1998). Generation and reproductive phenotypes of mice lacking estrogen receptor b. *Proc. Natl. Acad. Sci. U S A* 95, 15677–15682.
666. Feder, H. H., Brown-Grant, K., and Corker, C. S. (1971). Pre-ovulatory progesterone, the adrenal cortex and the "critical period" for luteinizing hormone release in rats. *J. Endocrinol.* 50, 29–39.
667. Smith, M. S., Freeman, M. E., and Neill, J. D. (1975). The control of progesterone secretion during the estrous cycle and early pseudopregnancy in the rat: prolactin, gonadotropin and steroid levels associated with rescue of the corpus luteum of pseudopregnancy. *Endocrinology* 96, 219–226.
668. Kalra, S. P., and Kalra, P. S. (1974). Temporal interrelationships among circulating levels of estradiol, progesterone and LH during the rat estrous cycle: effects of exogenous progesterone. *Endocrinology* 95, 1711–1718.
669. Karsch, F. J., Cummins, J. T., Thomas, G. B., and Clarke, I. J. (1987). Steroid feedback inhibition of pulsatile secretion of gonadotropin-releasing hormone in the ewe. *Biol. Reprod.* 36, 1207–1218.
670. Goodman, R. L., Bittman, E. L., Foster, D. L., and Karsch, F. J. (1981). The endocrine basis of the synergistic suppression of luteinizing hormone by estradiol and progesterone. *Endocrinology* 109, 1414–1417.
671. Gibson, M., Nakajima, S. T., and McAuliffe, T. L. (1991). Short-term modulation of gonadotropin secretion by progesterone during the luteal phase. *Fertil. Steril.* 55, 522–528.
672. Freeman, M. C., Dupke, K. C., and Croteau, C. M. (1976). Extinction of the estrogen-induced daily signal for LH release in the rat: a role for the proestrous surge of progesterone. *Endocrinology* 99, 223–229.
673. Clifton, D. K., Steiner, R. A., Resko, J. A., and Spies, H. G. (1975). Estrogen-induced gonadotropin release in ovariectomized rhesus monkeys and its advancement by progesterone. *Biol. Reprod.* 13, 190–194.
674. Levine, J. E., and Ramirez, V. D. (1980). In vivo release of luteinizing hormone-releasing hormone estimated with push-pull cannulae from the mediobasal hypothalami of ovariectomized, steroid-primed rats. *Endocrinology* 107, 1782–1790.
675. Krey, L. C., Tyrey, L., and Everett, J. W. (1973). The estrogen-induced advance in the cyclic LH surge in the rat: dependency on ovarian progesterone secretion. *Endocrinology* 93, 385–390.
676. Skinner, D. C., Caraty, A., and Allingham, R. (2001). Unmasking the progesterone receptor in the preoptic area and hypothalamus of the ewe: no colocalization with gonadotropin-releasing neurons. *Endocrinology* 142, 573–579.
677. Leranth, C., MacLusky, N. J., Brown, T. J., Chen, E. C., Redmond, D. E., and Naftolin, F. (1992). Transmitter content and afferent connections of estrogen-sensitive progestin receptor-containing neurons in the primate hypothalamus. *Neuroendocrinology* 55, 667–682.
678. King, J. C., Tai, D. W., Hanna, I. K., Pfeiffer, A., Haas, P., Ronsheim, P. M., Mitchell, S. C., Turcotte, J. C., and Blaustein, J. D. (1995). A subgroup of LHRH neurons in guinea pigs with progestin receptors is centrally positioned within the total population of LHRH neurons. *Neuroendocrinology* 61, 265–275.
679. Leranth, C., MacLusky, N. J., Brown, T. J., Chen, E. C., Redmond, D. E. Jr., and Naftolin, F. (1992). Transmitter content and afferent connections of estrogen-sensitive progestin receptor-containing neurons in the primate hypothalamus. *Neuroendocrinology* 55, 667–682.
680. Olster, D. H., and Blaustein, J. D. (1990). Immunocytochemical colocalization of progestin receptors and beta-endorphin or enkephalin in the hypothalamus of female guinea pigs. *J. Neurobiol.* 21, 768–780.
681. Sar, M. (1988). Distribution of progestin-concentrating cells in rat brain:colocalization of (^{3}H)ORG.2058, a synthetic progestin, and antibodies to tyrosine hydroxylase in hypothalamus by combined autoradiography and immunocytochemistry. *Endocrinology* 123, 1110–1118.
682. Kohama, S. G., Freesh, F., and Bethea, C. L. (1992). Immunocytochemical colocalization of hypothalamic progestin receptors and tyrosine hydroxylase in steroid-treated monkeys. *Endocrinology* 131, 509–517.
683. Dufourny, L., Warembourg, M., and Jolivet, A. (1999). Quantitative studies of progesterone receptor and nitric oxide synthase colocalization with somatostatin, or neurotensin, or substance P in neurons of the guinea pig ventrolateral hypothalamic nucleus: an immunocytochemical triple-label analysis. *J. Chem. Neuroanat.* 17, 33–43.
684. Bethea, C. L., and Widmann, A. A. (1996). Immunohistochemical detection of progestin receptors in hypothalamic beta-endorphin and substance P neurons of steroid-treated monkeys. *Neuroendocrinology* 63, 132–141.
685. Haywood, S. A., Simonian, S. X., Beek van der, E. M., Bicknell, R. J., and Herbison, A. E. (1999). Fluctuating estrogen and progesterone receptor expression in brainstem norepinephrine neurons through the rat estrous cycle. *Endocrinology* 140, 3255–3263.
686. Kim, K., and Ramirez, V. D. (1982). *In vitro* progesterone stimulates the release of luteinizing hormone-releasing hormone from superfused hypothalamic tissue from ovariectomized estradiol-primed prepuberal rats. *Endocrinology* 111, 750–757.
687. Drouva, S. V., Laplante, E., and Kordon, C. (1985). Progesterone-induced LHRH release *in vitro* is an estrogen – as well as Ca^{++} – and calmodulin-dependent secretory process. *Neuroendocrinology* 40, 325–331.
688. Ke, F. C., and Ramirez, V. D. (1987). Membrane mechanism mediates progesterone stimulatory effect on LHRH release from superfused rat hypothalami in vitro. *Neuroendocrinology* 45, 514–517.
689. Zhu, Y., Bond, J., and Thomas, P. (2003). Identification, classification, and partial characterization of genes in humans and other vertebrates homologous to a fish membrane progestin receptor. *Proc. Natl. Acad. Sci. U S A* 100, 2237–2242.
690. Sim, J. A., Skynner, M. J., and Herbison, A. E. (2001). Direct regulation of postnatal GnRH neurons by the progesterone

derivative allopregnanolone in the mouse. *Endocrinology* 142, 4448–4453.
691. Baulieu, E. E. (1998). Neurosteroids: a novel function of the brain. *Psychoneuroendocrinology* 23, 963–987.
692. Mensah-Nyagan, A. G., Do-Rego, J.-L., Beaujean, D., Luu-The, V., Pelletier, G., and Vaudry, H. (1999). Neurosteroids: expression of steroidogenic enzymes and regulation of steroid biosynthesis in the central nervous system. *Pharmacol. Rev.* 51, 63–81.
693. Sullivan, S. D., and Moenter, S. M. (2003). Neurosteroids alter gamma-aminobutyric acid postsynaptic currents in gonadotropin-releasing hormone neurons: a possible mechanism for direct steroidal control. *Endocrinology* 144, 4366–4375.
694. Goodman, R. L., Gibson, M., Skinner, D. C., and Lehman, M. N. (2002). Neuroendocrine control of pulsatile GnRH secretion during the ovarian cycle: evidence from the ewe. *Reprod. Suppl.* 59, 41–56.
695. Skinner, D. C., Evans, N. P., Delaleu, B., Goodman, R. L., Bouchard, P., and Caraty, A. (1998). The negative feedback actions of progesterone on gonadotropin-releasing hormone secretion are transduced by the classical progesterone receptor. *Proc. Natl. Acad. Sci. U S A* 95, 10978–10983.
696. Whisnant, C. S., and Goodman, R. L. (1988). Effects of an opioid antagonist on pulsatile luteinizing hormone secretion in the ewe vary with changes in steroid negative feedback. *Biol. Reprod.* 39, 1032–1038.
697. Yang, K., Haynes, N. B., Lamming, G. E., and Brooks, A. N. (1988). Ovarian steroid hormone involvement in endogenous opioid modulation of LH secretion in mature ewes during the breeding and non-breeding seasons. *J. Reprod. Fertil.* 83, 129–139.
698. Ferin, M., Van Vugt, D., and Wardlaw, S. (1984). The hypothalamic control of the menstrual cycle and the role of endogenous opioid peptides. *Rec. Prog. Horm. Res.* 40, 441–485.
699. Calogero, A. E., Palumbo, M. A., Bosboom, A. M. J., Burrello, N., Ferrara, E., Palumbo, G., Petraglia, F., and D'Agata, R. (1998). The neuroactive steroid allopregnanolone suppresses hypothalamic gonadotropin-releasing hormone release through a mechanism mediated by gamma-aminobutyric $acid_A$ receptor. *J. Endocrinol.* 158, 121–125.
700. Nuti, K. M., and Karavolas, H. J. (1977). Effect of progesterone and its 5a-reduced metabolites on gonadotropin levels in estrogen-primed ovariectomized rats. *Endocrinology* 100, 777–781.
701. Genazzani, A. R., Palumbo, M. A., de Micherouz, A. A., Artini, P. G., Criscuolo, M., Ficarra, G., Guo, A.-L., Benelli, A., Bertolini, A., Petraglia, F., and Purdy, R. H. (1995). Evidence for a role for the neurosteroid allopregnanolone in the modulation of reproductive function in female rats. *Eur. J. Endocrinol.* 133, 375–380.
702. Caraty, A., and Skinner, D. C. (1999). Progesterone priming is essential for the full expression of the positive feedback effect of estradiol in inducing the preovulatory gonadotropin-releasing hormone surge in the ewe. *Endocrinology* 140, 165–170.
703. Lustig, R. H., Pfaff, D. W., and Fishman, J. (1988). Induction of LH hypersecretion in cyclic rats during the afternoon of oestrus by oestrogen in conjunction with progesterone antagonism or opioidergic blockade. *J. Endocrinol.* 117, 229–235.
704. Mahesh, V. B., and Brann, D. W. (1998). Regulation of the preovulatory gonadotropin surge by endogenous steroids. *Steroids* 63, 616–629.
705. Chappel, P. E., Lydon, J. P., Conneely, O. M., O'Malley, B. T., and Levine, J. E. (1997). Endocrine defects in mice carrying a null mutation for the progesterton receptor gene. *Endocrinology* 138, 4147–4152.
706. Chappel, P. E., and Levine, J. E. (2000). Stimulation of gonadotropin-releasing hormone surges by estrogen. I. Role of hypothalamic progesterone receptors. *Endocrinology* 141, 1477–1485.
707. Rao, I. M., and Mahesh, V. B. (1986). Role of progesterone in the modulation of the preovulatory surge of gonadotropins and ovulation in the pregnant mare's serum gonadotropin-primed immature rat and the adult rat. *Biol. Reprod.* 35, 1154–1161.
708. Bauer-Dantoin, A. C., Tabesh, B., Norgle, J. R., and Levine, J. E. (1993). RU486 administration blocks neuropeptide Y potentiation of luteinizing hormone (LH)-releasing hormone-induced LH surges in proestrous rats. *Endocrinology* 133, 2418–2423.
709. Micevych, P., Sinchak, K., Mills, R. H., Tao, L., LaPolt, P., and Lu, J. K. (2003). The luteinizing hormone surge is preceded by an estrogen-induced increase of hypothalamic progesterone in ovariectomized and adrenalectomized rats. *Neuroendocrinology* 78, 29–35.
710. Ieiri, T., Chen, H. T., Campbell, G. A., and Meites, J. (1980). Effect of naloxone and morphine on the proestrous surge of prolactin and gonadotropins in the rat. *Endocrinology* 106, 1568–1570.
711. Gabriel, S. M., Berglund, L. A., and Simpkins, J. W. (1986). A decline in endogenous opioid influence during the steroid-induced hypersecretion of luteinizing hormone in the rat. *Endocrinology* 118, 558–561.
712. Allen, L. G., and Kalra, S. P. (1986). Evidence that a decrease in opioid tone may evoke preovulatory luteinizing hormone release in the rat. *Endocrinology* 118, 2375–2381.
713. Sarkar, D. K., and Yen, S. S. C. (1985). Changes in β-endorphin-like immunoreactivity in pituitary portal blood during the estrous cycle and after ovariectomy in rats. *Endocrinology* 116, 2075–2079.
714. Jacobson, W., and Kalra, S. P. (1989). Decreases in mediobasal hypothalamic and preoptic area opioid ([^{3}H)]naloxone) binding are associated with the progesterone-induced luteinizing hormone surge. *Endocrinology* 124, 199–206.
715. Weiland, N. G., and Wise, P. M. (1990). Estrogen and progesterone regulate opiate receptor densities in multiple brain regions. *Endocrinology* 126, 804–808.
716. Lustig, R. H., Pfaff, D. W., and Fishman, J. (1988). Opioidergic modulation of the oestradiol-induced LH surge in the rat: roles of ovarian steroids. *J. Endocrinol.* 116, 55–69.
717. Lustig, R. H., Pfaff, D. W., and Fishman, J. (1988). Induction of LH hypersecretion in cyclic rats during the afternoon of oestrus by oestrogen in conjunction with progesterone antagonism or opioidergic blockade. *J. Endocrinol.* 117, 229–235.
718. Gibson, M. J., and Silverman, A. J. (1989). Effects of gonadectomy and treatment with gonadal steroids on luteinizing hormone secretion in hypogonadal male and female mice with preoptic area implants. *Endocrinology* 125, 1525–1532.
719. Gibson, M. J., Wu, T. J., Miller, G. M., and Silverman, A. (1997). What nature's knockout teaches us about GnRH activity: hypogonadal mice and neuronal grafts. *Horm. Behav.* 31, 212–220.
720. Shughrue, P. J., Lane, M. V., and Merchenthaler, I. (1997). Regulation of progesterone receptor messenger ribonucleic acid in the rat medial preoptic nucleus by estrogenic and antiestrogenic compounds: an in situ hybridization study. *Endocrinology* 138, 5476–5484.
721. Rajendren, G. (2001). Subsets of gonadotropin-releasing hormone (GnRH) neurons are activated during a steroid-induced luteinizing hormone surge and mating in mice: a combined retrograde tracing double immunohistochemical study. *Brain Res.* 918, 74–79.

722. Rajendren, G. (2002). Increased galanin synapses onto activated gonadotropin-releasing hormone neuronal cell bodies in normal female mice and in functional preoptic area grafts in hypogonadal mice. *J Neuroendocrinol* 14, 435–441.

723. van der Beek, E. M., van Oudheusden, H. J., Buijs, R. M., van der Donk, H. A., van den Hurk, R., and Wiegant, V. M. (1994). Preferential induction of c-fos immunoreactivity in vasoactive intestinal polypeptide-innervated gonadotropin-releasing hormone neurons during a steroid-induced luteinizing hormone surge in the female rat. *Endocrinology* 134, 2636–2644.

724. Wang, H., Hoffman, G. E., and Smith, M. S. (1995). Increased GnRH mRNA in the GnRH neurons expressing cFos during the proestrous LH surge. *Endocrinology* 136, 3673–3676.

725. Petersen, S. L., McCrone, S., Keller, M., and Shores, S. (1995). Effects of estrogen and progesterone on luteinizing hormone-releasing hormone messenger ribonucleic acid levels: consideration of temporal and neuroanatomical variables. *Endocrinology* 136, 3604–3610.

726. Porkka-Heiskanen, T., Urban, J. H., Turek, F. W., and Levine, J. E. (1994). Gene expression in a subpopulation of luteinizing hormone-releasing hormone (LHRH) neurons prior to the preovulatory gonadotropin surge. *J. Neurosci.* 14, 5548–5558.

727. Hiatt, E. S., Brunetta, P. G., Seiler, G. R., and King, J. C. (1992). Subgroups of luteinizing hormone-releasing hormone perikarya defined by computer analyses in the basal forebrain of intact female rats. *Endocrinology* 130, 1030–1043.

728. Jackson, G. L., and Kuehl, D. (2002). Gamma-aminobutyric acid (GABA) regulation of GnRH secretion in sheep. *Reprod. Suppl.* 59, 15–24.

729. McRee, R. C., and Meyer, D. C. (1993). GABA control of LHRH release is dependent on the steroid milieu. *Neurosci. Lett.* 157, 227–230.

730. Jennes, L., Stumpf, W. E., and Tappaz, M. L. (1983). Anatomical relationships of dopaminergic and GABAergic systems with the GnRH-systems in the septo-hypothalamic area. Immunohistochemical studies. *Exp. Brain Res.* 50, 91–99.

731. Brann, D. W., and Mahesh, V. B. (1997). Excitatory amino acids: evidence for a role in the control of reproduction and anterior pituitary hormone secretion. *Endocr. Rev.* 18, 678–700.

732. Giri, M., and Kaufman, J. M. (1995). Involvement of neuroexcitatory amino acids in the control of gonadotropin-releasing hormone release from the hypothalamus of the adult male guinea pig: predominantly inhibitory action of N-methyl-D-aspartate-mediated neurotransmission and its reversal after orchidectomy. *Endocrinology* 136, 2404–2407.

733. Pompolo, S., Pereira, A., Scott, C. J., Fujiyma, F., and Clarke, I. J. (2003). Evidence for estrogenic regulation of gonadotropin-releasing hormone neurons by glutamatergic neurons in the ewe brain: An immunohistochemical study using an antibody against vesicular glutamate transporter-2. *J. Comp. Neurol.* 465, 136–144.

734. Clarke, I. J., and Scott, C. J. (1993). Studies on the neuronal systems involved in the oestrogen-negative feedback effect on gonadotrophin releasing hormone neurons in the ewe. *Hum. Reprod.* 8, 2–6.

735. Hoffman, G. E. (1985). Organization of LHRH cells: differential apposition of neurotensin, substance P and catecholamine axons. *Peptides* 6, 439–461.

736. Thiery, J. C., Chemineau, P., Hernandez, X., Migaud, M., and Malpaux, B. (2002). Neuroendocrine interactions and seasonality. *Domest. Anim. Endocrinol.* 23, 87–100.

737. Hileman, S. M. and Jackson, G. L. (1999). Regulation of gonadotrophin-releasing hormone secretion by testosterone in male sheep. *J. Reprod. Fertil. Suppl.* 54, 231–242.

738. Lacau-Mengido, I. M., Becu-Villalobos, D., Thyssen, S. M., Rey, E. B., Lux-Lantos, V. A., and Libertun, C. (1993). Antidopaminergic-induced hypothalamic LHRH release and pituitary gonadotrophin secretion in 12 day-old female and male rats. *J. Neuroendocrinol.* 5, 705–709.

739. Jarjour, L. T., Handelsman, D. J., Raum, W. J., and Swerdloff, R. S. (1986). Mechanism of action of dopamine on the in vitro release of gonadotropin-releasing hormone. *Endocrinology* 119, 1726–1732.

740. Rotsztejn, W. H., Charli, J. L., Pattou, E., and Kordon, C. (1977). Stimulation by dopamine of luteinizing hormone-releasing hormone (LHRH) release from the mediobasal hypothalamus in male rats. *Endocrinology* 101, 1475–1483.

741. Lehman, M. N., Karsch, F. J., and Silverman, A. J. (1988). Potential sites of interaction between catecholamines and LHRH in the sheep brain. *Brain Res. Bull.* 20, 49–58.

742. Tillet, Y., Caldani, M., and Batailler, M. (1989). Anatomical relationships of monoaminergic and neuropeptide Y-containing fibers with luteinizing hormone-releasing hormone systems in the preoptic area of the sheep brain: immunohistochemical studies. *J. Chem. Neuroanat.* 2, 319–326.

743. Chen, W., Witkin, J. W., and Silverman, A. (1989). Gonadotropin-releasing hormone (GnRH) neurons are directly innervated by catecholamine terminals. *Synapse* 3, 288–290.

744. Leranth, C., MacLusky, N. J., Shanabrough, M., and Naftolin, F. (1988). Catecholaminergic innervation of luteinizing hormone-releasing hormone and glutamic acid decarboxylase immunopositive neurons in the rat medial preoptic area. An electron-microscopic double immunostaining and degeneration study. *Neuroendocrinology* 48, 591–602.

745. Riggs, B. L., and Malven, P. V. (1974). Effects of intraventricular infusion of serotonin, norepinephrine, and dopamine on spontaneous LH release in castrate male sheep. *Biol. Reprod.* 11, 587–92.

746. Meyer, D. C. (1989). Serotonin stimulation of the period of in vitro LHRH release is estradiol dependent. *Brain Res. Bull.* 22, 525–530.

747. Kiss, J., and Halasz, B. (1985). Demonstration of serotoninergic axons terminating on luteinizing hormone-releasing hormone neurons in the preoptic area of the rat using a combination of immunocytochemistry and high resolution autoradiography. *Neuroscience* 14, 69–78.

748. Fekete, C. S., Strutton, P. H., Cagampang, F. R., Hrabovszky, E., Kallo, I., Shughrue, P. J., Dobo, E., Mihaly, E., Baranyi, L., Okada, H., Panula, P., Merchenthaler, I., Coen, C. W., and Liposits, Z. S. (1999). Estrogen receptor immunoreactivity is present in the majority of central histaminergic neurons: evidence for a new neuroendocrine pathway associated with luteinizing hormone-releasing hormone-synthesizing neurons in rats and humans. *Endocrinology* 140, 4335–4341.

749. Donoso, A. O., Banzan, A. M., and Borzino, M. I. (1976). Prolactin and luteinizing hormone release after intraventricular injection of histamine in rats. *J. Endocrinol.* 68, 171–172.

750. Libertun, C., and McCann, S. M. (1976). The possible role of histamine in the control of prolactin and gonadotropin release. *Neuroendocrinology* 20, 110–120.

751. Miyake, A., Ohtsuka, S., Nishizaki, T., Tasaka, K., Aono, T., Tanizawa, O., Yamatodani, A., Watanabe, T., and Wada, H. (1987). Involvement of H1 histamine receptor in basal and estrogen-stimulated luteinizing hormone-releasing hormone secretion in rats in vitro. *Neuroendocrinology* 45, 191–196.

752. Piva, F., Borrell, J., Limonta, P., Gavazzi, G., and Martini, L. (1980). Cholinergic inputs to the amygdala and the control of gonadotrophin release. *Acta Endocrinol.* 93, 1–6.

753. Vijayan, E., and McCann, S. M. (1980). Effect of blockade of dopaminergic receptors on acetylcholine (Ach)-induced

alterations of plasma gonadotropin and prolactin (Prl) in conscious ovariectomized rats. *Brain Res. Bull.* 5, 23–29.
754. Kalash, J., Romita, V., and Billiar, R. B. (1989). Third ventricular injection of alpha-bungarotoxin decreases pulsatile luteinizing hormone secretion in the ovariectomized rat. *Neuroendocrinology* 49, 462–470.
755. Billiar, R. B., Kalash, J., Romita, V., Tsuji, K., and Kosuge, T. (1988). Neosurugatoxin: CNS acetylcholine receptors and luteinizing hormone secretion in ovariectomized rats. *Brain Res. Bull.* 20, 315–322.
756. Libertun, C., and McCann, S. M. (1973). Blockade of the release of gonadotropins and prolactin by subcutaneous or intraventricular injection of atropine in male and female rats. *Endocrinology* 92, 1714–1724.
757. Libertun, C., and McCann, S. M. (1976). Blockade of the postorchidectomy increase in gonadotropins by implants of atropine into the hypothalamus. *Proc. Soc. Exp. Biol. Med.* 152, 143–146.
758. Richardson, S. B., Prasad, J. A., and Hollander, C. S. (1982). Acetylcholine, melatonin, and potassium depolarization stimulate release of luteinizing hormone-releasing hormone from rat hypothalamus in vitro. *Proc. Natl. Acad. Sci. U S A* 79, 2686–2689.
759. Koren, D., Egozi, Y., and Sokolovsky, M. (1992). Muscarinic involvement in the regulation of gonadotropin-releasing hormone in the cyclic rat. *Mol. Cell. Endocrinol.* 90, 87–93.
760. Ayalon, D., Nir, I., Cordova, T., Bauminger, S., Puder, M., Naor, Z., Kashi, R., Zor, U., Harell, A., and Lindner, H. R. (1977). Acute effect of delta1-tetrahydrocannabinol on the hypothalamo-pituitary-ovarian axis in the rat. *Neuroendocrinology* 23, 31–42.
761. Wenger, T., Rettori, V., Snyder, G. D., Dalterio, S., and McCann, S. M. (1987). Effects of delta-9-tetrahydrocannabinol on the hypothalamic-pituitary control of luteinizing hormone and follicle-stimulating hormone secretion in adult male rats. *Neuroendocrinology* 46, 488–493.
762. Rettori, V., Aguila, M. C., Gimeno, M. F., Franchi, A. M., and McCann, S. M. (1990). In vitro effect of delta 9-tetrahydrocannabinol to stimulate somatostatin release and block that of luteinizing hormone-releasing hormone by suppression of the release of prostaglandin E2. *Proc. Natl. Acad. Sci. U S A* 87, 10063–10066.
763. Chen, W. P., Witkin, J. W., and Silverman, A. J. (1990). Sexual dimorphism in the synaptic input to gonadotropin releasing hormone neurons. *Endocrinology* 126, 695–702.
764. Lopez, F. J., and Negro-Vilar, A. (1990). Galanin stimulates luteinizing hormone-releasing hormone secretion from arcuate nucleus-median eminence fragments in vitro: involvement of an alpha-adrenergic mechanism. *Endocrinology* 127, 2431–2436.
765. Dudas, B., and Merchenthaler, I. (2004). Bi-directional associations between galanin and luteinizing hormone-releasing hormone neuronal systems in the human diencephalon. *Neuroscience* 127, 695–707.
766. Starcevic, V., Milosevic, V., Brkic, B., and Severs, W. B. (2002). Somatostatin affects morphology and secretion of pituitary luteinizing hormone (LH) cells in male rats. *Life Sci.* 70, 3019–3027.
767. Van Vugt, H. H., Swarts, H. J., Van De Heijning, B. J., and Van Der Beek, E. M. (2004). Centrally applied somatostatin inhibits the estrogen-induced luteinizing hormone surge via hypothalamic gonadotropin-releasing hormone cell activation in female rats. *Biol. Reprod.* 71, 813–819.
768. Pillon, D., Caraty, A., Fabre-Nys, C., Lomet, D., Cateau, M., and Bruneau, G. (2004). Regulation by estradiol of hypothalamic somatostatin gene expression: possible involvement of somatostatin in the control of luteinizing hormone secretion in the ewe. *Biol. Reprod.*
769. Rotsztejn, W. H., Drouva, S. V., Epelbaum, J., and Kordon, C. (1982). Somatostatin inhibits in vitro release of luteinizing hormone releasing hormone from rat mediobasal hypothalamic slices. *Experientia* 38, 974–975.
770. Pinski, J., Yano, T., and Schally, A. V. (1992). Inhibitory effects of the new bombesin receptor antagonist RC–3095 on the luteinizing hormone release in rats. *Neuroendocrinology* 56, 831–837.
771. Hoffman, G. E., Fitzsimmons, M. D., and Watson, R. E. (1989). Relationship of endogenous opioid peptide axons to GnRH neurones in the rat. In *Brain Opioid Systems in Reproduction* (R. G. Dyer and R. J. Bicknell, Eds.), pp. 125–134. Oxford University Press, Oxford.
772. Dudas, B., and Merchenthaler, I. (2004). Close anatomical associations between beta-endorphin and luteinizing hormone-releasing hormone neuronal systems in the human diencephalon. *Neuroscience* 124, 221–229.
773. Stanley, S. A., Small, C. J., Kim, M. S., Heath, M. M., Seal, L. J., Russell, S. H., Ghatei, M. A., and Bloom, S. R. (1999). Agouti related peptide (Agrp) stimulates the hypothalamo pituitary gonadal axis in vivo & in vitro in male rats. *Endocrinology* 140, 5459–5462.
774. Sabatino, F. D., Collins, P., and McDonald, J. K. (1989). Neuropeptide-Y stimulation of luteinizing hormone-releasing hormone secretion from the median eminence in vitro by estrogen-dependent and extracellular Ca2+-independent mechanisms. *Endocrinology* 124, 2089–2098.
775. Urban, J. H., Das, I., and Levine, J. E. (1996). Steroid modulation of neuropeptide Y-induced luteinizing hormone releasing hormone release from median eminence fragments from male rats. *Neuroendocrinology* 63, 112–119.
776. Dudas, B., Mihaly, A., and Merchenthaler, I. (2000). Topography and associations of luteinizing hormone-releasing hormone and neuropeptide Y-immunoreactive neuronal systems in the human diencephalon. *J. Comp. Neurol.* 427, 593–603.
777. Rostene, W. H., and Alexander, M. J. (1997). Neurotensin and neuroendocrine regulation. *Front. Neuroendocrinol.* 18, 115–173.
778. Alexander, M. J., Clifton, D. K., and Steiner, R. A. (1985). Vasoactive intestinal polypeptide effects a central inhibition of pulsatile luteinizing hormone secretion in ovariectomized rats. *Endocrinology* 117, 2134–2139.
779. Stobie, K. M., and Weick, R. F. (1990). Effects of lesions of the suprachiasmatic and paraventricular nuclei on the inhibition of pulsatile luteinizing hormone release by exogenous vasoactive intestinal peptide in the ovariectomized rat. *Neuroendocrinology* 51, 649–657.
780. Vijayan, E., Samson, W. K., Said, S. I., and McCann, S. M. (1979). Vasoactive intestinal peptide: evidence for a hypothalamic site of action to release growth hormone, luteinizing hormone, and prolactin in conscious ovariectomized rats. *Endocrinology* 104, 53–57.
781. Samson, W. K., Burton, K. P., Reeves, J. P., and McCann, S. M. (1981). Vasoactive intestinal peptide stimulates luteinizing hormone-releasing hormone release from median eminence synaptosomes. *Regul. Pept.* 2, 253–264.
782. Ohtsuka, S., Miyake, A., Nishizaki, T., Tasaka, K., and Tanizawa, O. (1988). Vasoactive intestinal peptide stimulates gonadotropin-releasing hormone release from rat hypothalamus in vitro. *Acta Endocrinol* 117, 399–402.
783. Kriegsfeld, L. J., Silver, R., Gore, A. C., and Crews, D. (2002). Vasoactive intestinal polypeptide contacts on gonadotropin-releasing hormone neurones increase following puberty in female rats. *J. Neuroendocrinol.* 14, 685–690.
784. Small, C. J., Goubillon, M. L., Murray, J. F., Siddiqui, A., Grimshaw, S. E., Young, H., Sivanesan, V., Kalamatianos, T., Kennedy, A. R., Coen, C. W., Bloom, S. R., and Wilson, C. A.

(2003). Central orexin A has site-specific effects on luteinizing hormone release in female rats. *Endocrinology* 144, 3225–3236.

785. Pu, S., Jain, M. R., Kalra, P. S., and Kalra, S. P. (1998). Orexins, a novel family of hypothalamic neuropeptides, modulate pituitary luteinizing hormone secretion in an ovarian steroid-dependent manner. *Regul. Pept.* 78, 133–136.
786. Furuta, M., Funabashi, T., and Kimura, F. (2002). Suppressive action of orexin A on pulsatile luteinizing hormone secretion is potentiated by a low dose of estrogen in ovariectomized rats. *Neuroendocrinology* 75, 151–157.
787. Tamura, T., Irahara, M., Tezuka, M., Kiyokawa, M., and Aono, T. (1999). Orexins, orexigenic hypothalamic neuropeptides, suppress the pulsatile secretion of luteinizing hormone in ovariectomized female rats. *Biochem. Biophys. Res. Commun.* 264, 759–762.
788. Irahara, M., Tamura, T., Matuzaki, T., Saito, S., Yasui, T., Yamano, S., Kamada, M., and Aono, T. (2001). Orexin-A suppresses the pulsatile secretion of luteinizing hormone via beta-endorphin. *Biochem. Biophys. Res. Commun.* 281, 232–236.
789. Campbell, R. E., Grove, K. L., and Smith, M. S. (2003). Gonadotropin-releasing hormone neurons coexpress orexin 1 receptor immunoreactivity and receive direct contacts by orexin fibers. *Endocrinology* 144, 1542–1548.
790. Iqbal, J., Pompolo, S., Sakurai, T., and Clarke, I. J. (2001). Evidence that orexin-containing neurones provide direct input to gonadotropin-releasing hormone neurones in the ovine hypothalamus. *J. Neuroendocrinol.* 13, 1033–1041.
791. Moretto, M., Lopez, F. J., and Negro-Vilar, A. (1993). Endothelin-3 stimulates luteinizing hormone-releasing hormone (LHRH) secretion from LHRH neurons by a prostaglandin-dependent mechanism. *Endocrinology* 132, 789–794.
792. Yamamoto, T., Suzuki, H., and Uemura, H. (1997). Endothelin B receptor-like immunoreactivity is associated with LHRH-immunoreactive fibers in the rat hypothalamus. *Neurosci. Lett.* 223, 117–120.
793. Ohtsuka, S., Miyake, A., Nishizaki, T., Tasaka, K., Aono, T., and Tanizawa, O. (1987). Substance P stimulates gonadotropin-releasing hormone release from rat hypothalamus in vitro with involvement of oestrogen. *Acta Endocrinol.* 115, 247–252.
794. Tsuruo, Y., Kawano, H., Hisano, S., Kagotani, Y., Daikoku, S., Zhang, T., and Yanaihara, N. (1991). Substance P-containing neurons innervating LHRH-containing neurons in the septo-preoptic area of rats. *Neuroendocrinology* 53, 236–245.
795. Goubillon, M.-L., Forsdike, R. A., Robinson, J. E., Ciofi, P., Caraty, A., and Herbison, A. E. (2000). Identification of neurokinin B-expression neurons as an highly estrogen-receptive, sexually dimorphic cell group in the ovine arcuate nucleus. *Endocinology* 141, 4218–4225.
796. Dudas, B., and Merchenthaler, I. (2002). Close juxtapositions between LHRH immunoreactive neurons and substance P immunoreactive axons in the human diencephalon. *J. Clin. Endocrinol. Metab.* 87, 2946–2953.
797. Rivier, C., and Vale, W. (1984). Influence of corticotropin-releasing factor on reproductive functions in the rat. *Endocrinology* 114, 914–921.
798. Rivest, S., Plotsky, P. M., and Rivier, C. (1993). CRF alters the infundibular LHRH secretory system from the medial preoptic area of female rats: possible involvement of opioid receptors. *Neuroendocrinology* 57, 236–246.
799. Ono, N., Lumpkin, M. D., Samson, W. K., McDonald, J. K., and McCann, S. M. (1984). Intrahypothalamic action of corticotrophin-releasing factor (CRF) to inhibit growth hormone and LH release in the rat. *Life Sci.* 35, 1117–1123.
800. Petraglia, F., Sutton, S., Vale, W., and Plotsky, P. (1987). Corticotropin-releasing factor decreases plasma luteinizing hormone levels in female rats by inhibiting gonadotropin-releasing hormone release into hypophysial-portal circulation. *Endocrinology* 120, 1083–1088.
801. Tilbrook, A. J., Canny, B. J., Stewart, B. J., Serapiglia, M. D., and Clarke, I. J. (1999). Central administration of corticotrophin releasing hormone but not arginine vasopressin stimulates the secretion of luteinizing hormone in rams in the presence and absence of testosterone. *J. Endocrinol.* 162, 301–311.
802. Nikolarakis, K. E., Almeida, O. F., Sirinathsinghji, D. J., and Herz, A. (1988). Concomitant changes in the in vitro and in vivo release of opioid peptides and luteinizing hormone-releasing hormone from the hypothalamus following blockade of receptors for corticotropin-releasing factor. *Neuroendocrinology* 47, 545–550.
803. Dudas, B., and Merchenthaler, I. (2002). Close juxtapositions between luteinizing hormone-releasing hormone-immunoreactive neurons and corticotropin-releasing factor-immunoreactive axons in the human diencephalon. *J. Clin. Endocrinol. Metab.* 87, 5778–5784.
804. MacLusky, N. J., Naftolin, F., and Leranth, C. (1988). Immunocytochemical evidence for direct synaptic connections between corticotrophin-releasing factor (CRF) and gonadotrophin-releasing hormone (GnRH)-containing neurons in the preoptic area of the rat. *Brain Res.* 439, 391–395.
805. Funabashi, T., Aiba, S., Sano, A., Shinohara, K., and Kimura, F. (1999). Intracerebroventricular injection of arginine-vasopressin V1 receptor antagonist attenuates the surge of luteinizing hormone and prolactin secretion in proestrous rats. *Neurosci. Lett.* 260, 37–40.
806. Cates, P. S., Forsling, M. L., and O'Byrne K, T. (1999). Stress-induced suppression of pulsatile Luteinising hormone release in the female rat: role of vasopressin. *J. Neuroendocrinol.* 11, 677–683.
807. Palm, I. F., van der Beek, E. M., Wiegant, V. M., Buijs, R. M., and Kalsbeek, A. (2001). The stimulatory effect of vasopressin on the luteinizing hormone surge in ovariectomized, estradiol-treated rats is time-dependent. *Brain Res.* 901, 109–116.
808. Chen, M. D., Ordog, T., O'Byrne, K. T., Goldsmith, J. R., Connaughton, M. A., and Knobil, E. (1996). The insulin hypoglycemia-induced inhibition of gonadotropin-releasing hormone pulse generator activity in the rhesus monkey: roles of vasopressin and corticotropin-releasing factor. *Endocrinology* 137, 2012–2021.
809. Funabashi, T., Shinohara, K., Mitsushima, D., and Kimura, F. (2000). Gonadotropin-releasing hormone exhibits circadian rhythm in phase with arginine-vasopressin in co-cultures of the female rat preoptic area and suprachiasmatic nucleus. *J. Neuroendocrinol.* 12, 521–528.
810. Thind, K. K., Boggan, J. E., and Goldsmith, P. C. (1991). Interactions between vasopressin- and gonadotropin-releasing-hormone-containing neuroendocrine neurons in the monkey supraoptic nucleus. *Neuroendocrinology* 53, 287–297.
811. Parent, A. S., Lebrethon, M. C., Gerard, A., Vandersmissen, E., and Bourguignon, J. P. (2000). Leptin effects on pulsatile gonadotropin releasing hormone secretion from the adult rat hypothalamus and interaction with cocaine and amphetamine regulated transcript peptide and neuropeptide Y. *Regul. Pept.* 92, 17–24.
812. Lebrethon, M. C., Vandersmissen, E., Gerard, A., Parent, A. S., and Bourguignon, J. P. (2000). Cocaine and amphetamine-regulated-transcript peptide mediation of leptin stimulatory effect on the rat gonadotropin-releasing hormone pulse generator in vitro. *J. Neuroendocrinol.* 12, 383–385.
813. Rondini, T. A., Baddini, S. P., Sousa, L. F., Bittencourt, J. C., and Elias, C. F. (2004). Hypothalamic cocaine- and amphetamine-regulated transcript neurons project to areas

expressing gonadotropin releasing hormone immunoreactivity and to the anteroventral periventricular nucleus in male and female rats. *Neuroscience* 125, 735–748.

814. Leslie, R. A., Sanders, S. J., Anderson, S. I., Schuhler, S., Horan, T. L., and Ebling, F. J. (2001). Appositions between cocaine and amphetamine-related transcript- and gonadotropin releasing hormone-immunoreactive neurons in the hypothalamus of the Siberian hamster. *Neurosci. Lett.* 314, 111–114.
815. Matsumoto, H., Noguchi, J., Takatsu, Y., Horikoshi, Y., Kumano, S., Ohtaki, T., Kitada, C., Itoh, T., Onda, H., Nishimura, O., and Fujino, M. (2001). Stimulation effect of galanin-like peptide (GALP) on luteinizing hormone-releasing hormone-mediated luteinizing hormone (LH) secretion in male rats. *Endocrinology* 142, 3693–3696.
816. Krasnow, S. M., Fraley, G. S., Schuh, S. M., Baumgartner, J. W., Clifton, D. K., and Steiner, R. A. (2003). A role for galanin-like peptide in the integration of feeding, body weight regulation, and reproduction in the mouse. *Endocrinology* 144, 813–822.
817. Krasnow, S. M., Hohmann, J. G., Gragerov, A., Clifton, D. K., and Steiner, R. A. (2004). Analysis of the contribution of galanin receptors 1 and 2 to the central actions of galanin-like Peptide. *Neuroendocrinology* 79, 268–277.
818. Cunningham, M. J., Shahab, M., Grove, K. L., Scarlett, J. M., Plant, T. M., Cameron, J. L., Smith, M. S., Clifton, D. K., and Steiner, R. A. (2004). Galanin-like peptide as a possible link between metabolism and reproduction in the macaque. *J. Clin. Endocrinol. Metab.* 89, 1760–1766.
819. Seth, A., Stanley, S., Jethwa, P., Gardiner, J., Ghatei, M., and Bloom, S. (2004). Galanin-like peptide stimulates the release of gonadotropin-releasing hormone in vitro and may mediate the effects of leptin on the hypothalamo-pituitary-gonadal axis. *Endocrinology* 145, 743–750.
820. Takatsu, Y., Matsumoto, H., Ohtaki, T., Kumano, S., Kitada, C., Onda, H., Nishimura, O., and Fujino, M. (2001). Distribution of galanin-like peptide in the rat brain. *Endocrinology* 142, 1626–1634.
821. Motta, M., and Martini, L. (1982). Effect of opioid peptides on gonadotrophin secretion. *Acta Endocrinol.* 99, 321–325.
822. Dudas, B., and Merchenthaler, I. (2003). Topography and associations of leu-enkephalin and luteinizing hormone-releasing hormone neuronal systems in the human diencephalon. *J. Clin. Endocrinol. Metab.* 88, 1842–1848.
823. Samson, W. K., Aguila, M. C., and Bianchi, R. (1988). Atrial natriuretic factor inhibits luteinizing hormone secretion in the rat: evidence for a hypothalamic site of action. *Endocrinology* 122, 1573–1582.
824. Zhang, J., Yuen, B. H., Currie, W. D., and Leung, P. C. (1991). Suppression of luteinizing hormone secretion by atrial and brain natriuretic peptides in ovariectomized rats. *Endocrinology* 129, 801–806.
825. Huang, F. S., Skala, K. D., and Samson, W. K. (1992). Hypothalamic effects of C-type natriuretic peptide on luteinizing hormone secretion. *J. Neuroendocrinol.* 4, 325–330.
826. Johnston, C. A., Waes, J. G., and Templin, M. V. (1992). Physiological significance and interactions between oxytocin and central neuropeptide and monoamine neurotransmitters in the regulation of the preovulatory secretion of luteinizing hormone. *Ann. N Y Acad. Sci.* 652, 440–442.
827. Selvage, D., and Johnston, C. A. (2001). Central stimulatory influence of oxytocin on preovulatory gonadotropin-releasing hormone requires more than the median eminence. *Neuroendocrinology* 74, 129–134.
828. Rettori, V., Canteros, G., Renoso, R., Gimeno, M., and McCann, S. M. (1997). Oxytocin stimulates the release of luteinizing hormone-releasing hormone from medial basal hypothalamic explants by releasing nitric oxide. *Proc. Natl. Acad. Sci. U S A* 94, 2741–2744.
829. Steele, M. K., Stephenson, K. N., Meredith, J. M., and Levine, J. E. (1992). Effects of angiotensin II on LHRH release, as measured by in vivo microdialysis of the anterior pituitary gland of conscious female rats. *Neuroendocrinology* 55, 276–281.
830. Shi, B., Mahesh, V. B., Bhat, G. K., Ping, L., and Brann, D. W. (1998). Evidence for a role of bradykinin neurons in the control of gonadotropin-releasing hormone secretion. *Neuroendocrinology* 67, 209–218.
831. Dhandapani, K. M., and Brann, D. W. (2002). Orphanin FQ inhibits GnRH secretion from rat hypothalamic fragments but not GT1–7 neurons. *Neuroreport* 13, 1247–1249.
832. Fernandez-Fernandez, R., Tena-Sempere, M., Aguilar, E., and Pinilla, L. (2004). Ghrelin effects on gonadotropin secretion in male and female rats. *Neurosci. Lett.* 362, 103–107.
833. Hashimoto, R., and Kimura, F. (1986). Inhibition of gonadotropin secretion induced by cholecystokinin implants in the medial preoptic area by the dopamine receptor blocker, pimozide, in the rat. *Neuroendocrinology* 42, 32–37.
834. Li, X. F., Bowe, J. E., Mitchell, J. C., Brain, S. D., Lightman, S. L., and O'Byrne, K. T. (2004). Stress-induced suppression of the gonadotropin-releasing hormone pulse generator in the female rat: a novel neural action for calcitonin gene-related peptide. *Endocrinology* 145, 1556–1563.
835. Navarro, V. M., Castellano, J. M., Fernandez-Fernandez, R., Barreiro, M. L., Roa, J., Sanchez-Criado, J. E., Aguilar, E., Dieguez, C., Pinilla, L., and Tena-Sempere, M. (2004). Developmental and hormonally regulated messenger ribonucleic acid expression of KiSS-1 and its putative receptor GPR54 in rat hypothalamus and potent LH releasing activity of KiSS-1 peptide*. *Endocrinology*.
836. Matsui, H., Takatsu, Y., Kumano, S., Matsumoto, H., and Ohtaki, T. (2004). Peripheral administration of metastin induces marked gonadotropin release and ovulation in the rat. *Biochem. Biophys. Res. Commun.* 320, 383–388.
837. Gottsch, M. L., Cunningham, M. J., Smith, J. T., Popa, S. M., Acohido, B. V., Crowley, W. F., Seminara, S., Clifton, D. K., and Steiner, R. A. (2004). A role for kisspeptins in the regulation of gonadotropin secretion in the mouse. *Endocrinology* 145, 4073–4037.

Knobil and Neill's Physiology of Reproduction, Third Edition
edited by Jimmy D. Neill,
Elsevier © 2006

CHAPTER 29

Gonadotropes and Lactotropes

Gwen V. Childs

Introduction, 1483
Discovery and Historical Perspective, 1484
Early Identification of Gonadotropes and Lactotropes, 1484
How Immunocytochemists Bridged the Early Studies to the Present, 1485
Embryological Origin of Gonadotropes and Lactotropes, 1486
Origin and Formation of the Anterior Pituitary, 1486
How the Individual Pituitary Cell Types Develop, 1488
Spatial Organization of Transcription Factors, 1490
Monohormonal Cells Become More Complex, Multihormonal Cells, 1492
Gonadotropes, 1492
Early Studies, 1492
Gonadotrope Responses to Different Physiological States, 1497
Understanding Gonadotrope Diversity, 1513
Expression of GnRH Receptors, 1530
Autocrine and Paracrine Influences on Gonadotropes, 1539
Gonadotrope and Lactotrope Turnover, 1540
Proliferation of Differentiated Gonadotropes and Lactotropes, 1540
Effect of Age and Reproductive Hormones, 1541
Effects of Growth Factors, 1542
Regulation of Apoptosis in Gonadotropes and Lactotropes, 1543
Lactotropes, 1546
Early Studies Defined the Lactotrope Population, 1546
Significance of Functional Heterogeneity Expressed by Lactotropes, 1548
Prolactin Cells as a Model for Studies of the Cell Secretory Pathway, 1551
Thyroid-Releasing Hormone Regulation of Lactotropes, 1553
Regulation of Lactotropes by Dopamine, 1554
Neurointermediate Lobe Prolactin Regulatory Factors, 1555
Integrative Studies of Gonadotropes and Lactotropes, 1556
Regulating the Transition from Birth to the Adult State, 1556
Detecting Receptor Populations in Gonadotropes and Lactotropes, 1556
Communication between Lactotropes and Gonadotropes, 1559
How Past Lessons May Shape Future Experiments, 1560
Acknowledgments, 1562
References, 1562

INTRODUCTION

Gonadotropes are cells in the anterior pituitary (pars distalis of the adenohypophysis) that produce gonadotropins to regulate the ovary and testes (1). Gonadotropins include luteinizing hormone (LH), which is important in regulating ovulation and luteinization of the ovarian follicles in the female. It regulates Leydig cells in the testes of the male. Another gonadotropin is follicle-stimulating hormone (FSH), which regulates the development of follicles in the female and sperm production in the male. The gonadotropins are glycoproteins that are made of two subunits, an α and a β subunit. LH, FSH, thyroid-stimulating hormone (TSH), and human chorionic gonadotropin (hCG) have nearly identical α subunits. The β subunit is unique to each of these glycoprotein hormones and confers biological specificity (2–6).

Department of Neurobiology and Developmental Sciences, College of Medicine, University of Arkansas for Medical Sciences, Little Rock, Arkansas.

Lactotropes are also specialized cells in the anterior pituitary. They may have a number of functions in the body, including the development of the mammary gland and the regulation of milk production during lactation (7). Prolactin is a protein that shares amino acid sequences in common with growth hormone (GH) (8–10). Other chapters in this text discuss the chemistry and functions of gonadotropins and prolactin in great detail. This chapter focuses on the organization and structure of the cells themselves.

The chapter begins with a brief review of the discovery of gonadotropes and lactotropes. This historical perspective shows how our modern concept of these cell types evolved, including the original beliefs and discoveries that drove the design of experiments in the field. The chapter then summarizes the state-of-the-art research on the embryological development of the lactotropes and gonadotropes. This background is important because it improves our understanding of early regulators of development and differentiation of these cell types. The remaining sections focus on how the gonadotropes and lactotropes are structured and organized to perform their functions, including further remodeling of populations after puberty as adult lactotropes and gonadotropes continue to differentiate in a complex postpubertal environment. Along the way, this chapter also highlights the many ways that gonadotropic and lactotropic functions are assessed at the cellular level, as we seek a better understanding of how these cells function and the regulatory pathways involved in their functions.

Discovery and Historical Perspective

The concept that anterior pituitary hormones may be required for the regulation of lactation and gonadal function developed in the 1920s. Evans and Long (11) showed that daily treatment of rats with fresh anterior pituitary lobe extracts produced significant growth and also luteinization of the ovaries. Evans and Long believed at that point, however, that the "growth hormone" in the extracts regulated the ovary. The actual origin of the stimulatory substance was confirmed in 1926 by P. E. Smith, who showed that ovarian atrophy, after removal of the pituitary, was reversed by pituitary transplants (12,13). However, 1 year later, Zondek and Ascheim reported that urine extracts from pregnant women stimulated the formation of corpora lutea (thereby showing luteinizing activity) (14). Zondek (15) later reported that extracts of urine from postmenopausal women influenced follicular development. This introduced the concept that two different hormones might regulate the events of the reproductive cycle.

This two-hormone concept was reinforced in a publication from a student dissertation in the laboratory of Dr. F. Hisaw (16) in collaboration with a chemist, Dr. Fevold. In Dr. Leonard's dissertation project, they were able to isolate two protein hormones from the anterior pituitary that had distinct actions on the rat ovary. The FSH extract caused precocious sexual maturity in immature rats and stimulated ovarian follicle development. The LH extract had no effect on its own; however, it caused luteinization of the follicles that had been stimulated by FSH. Thus, the modern concept of the two hormones was defined.

For several years, however, this finding fueled a controversy as the Evans group continued to maintain that GH was responsible for luteinization, even after they themselves had produced gonadotropin extracts (17,18). In fact, today we know that Evans may have been partially correct. There are GH receptors in the ovary (19,20). GH potentiates the actions of LH and FSH and has significant effects on early follicular development on its own (21–23). These concepts will be developed further as the modern characterization of the gonadotrope is presented.

Prolactin was discovered as the growth-promoting and gonadotropic properties of the anterior pituitary were being explored (24). Dr. Oscar Riddle described lactation-inducing effects of the anterior pituitary and developed the pioneering pigeon crop-sac assay to detect the hormone (24,25). This assay allowed further purification and isolation of fractions rich in prolactin. He was able to distinguish the prolactin activity from that of growth-promoting or gonad-stimulating activities. This crop-sac assay proved to be the main assay for prolactin until the radioimmunoassay was developed in the 1970s (26). This has allowed more detailed studies of prolactin and the discovery of over 300 separate functions for this hormone in a number of species (27).

Early Identification of Gonadotropes and Lactotropes

The earliest light microscopic studies distinguished pituitary cell types by their reactions with different-colored dyes (28). The classic Mallory trichrome stain differentiated three types of cells: acidophils (red), basophils (blue), and chromophobes (colorless or pale) (28). One of the earliest descriptions of an acidophilic tumor was in 1900, by Benda (29), and Cushing helped to establish that this overactive tumor caused acromegaly (30). Based on his patient's symptoms, he speculated correctly that the tumor may be secreting a growth-promoting substance that could promote an abnormal increase in size. During this same period, Evans and Long experimentally proved this hypothesis in animals (31) and isolated the growth-promoting substance.

The appearance of different-colored cells fueled a controversy about whether the cells of the anterior lobe secreted more than one substance. Alternatively, they could secrete one substance with multiple functions. Cushing believed in the latter hypothesis for over 20 years (32). He postulated that the abundant acidophilic cells were "functionally mature," whereas the basophilic cells were "immature" (32). His later studies were to describe tumors made up of basophils, however (33). Because these produced an entirely different set of symptoms, Cushing realized that the basophils might have unique functions of their own. He speculated that basophils might regulate the adrenal gland, a hypothesis that was proved in 1932 (33).

Pituitary historians now recognize the unfortunate fact that the term "basophil" was actually a misnomer because the early trichrome stains included *only acid dyes* (28,34). True basophilia depends on abundant rough endoplasmic reticulum and other base-loving cytoplasmic domains as well as the presence of *basic dyes* in the staining sets. Thus, these blue cells were misnamed from the beginning. Later electron microscopic evidence showed that "true basophils" characterized by abundant rough endoplasmic reticulum and stained by "true basic dyes," would likely include the acidophils (35), and prolactin cells in particular. Nevertheless, the term "basophil" has been maintained, even in the nomenclature for tumors.

In spite of these and other later problems with nomenclature, the anterior pituitary cells were sorted and identified further, in situ, with special dyes developed to reveal subsets of basophils and acidophils (28,36). The dyes were applied to pituitaries from animals in different physiological states to study changes, if any, in subsets of the various cell types. Pioneering studies by Romeis in the 1940s (36) used Kresazan dyes to distinguish subsets of pituitary cells. Later studies published during the 1950s used combinations of aldehyde fuchsin or aldehyde thionin, Alcian blue, orange G, carmosine or erythrosine, and periodic acid-Schiff (PAS) to differentiate the cell types (28,34–45). Responses by cells with particular tinctorial staining properties were correlated with changes in their physiological state. In this way, the investigators formulated hypotheses for the identity of each cell type. For example, prolactin cells stained bright red with erythrosine or carmosine. They were identified because these bright red cells enlarged greatly during lactation and pregnancy, which pointed to the involvement of this cell type in lactation. They were difficult to detect in normal animals (36).

Gonadotropes stained blue or violet with aldehyde fuchsin, Alcian blue, or aldehyde thionin and they were PAS positive, which could later be linked to their glycoprotein hormone content (37–44). They were infrequent during childhood and enlarged, becoming filled with one or more vacuoles and taking on a "signet ring" or "Swiss cheese" appearance, after castration (37–44).

Thus, by the 1930s the idea that all pituitary cells produced a factor with multiple functions changed because of the recognition that there were a number of different hormones produced from this region. As researchers linked changes in each cell type with responses to endocrine events, the hypothesis emerged that there was a separate cell type for each hormone. This "one-cell/one-hormone" hypothesis became a driving force for the design and interpretation of experiments for over 70 years.

The next major breakthrough in the identification of gonadotropes and prolactin cells came with the pioneering work by Farquhar and Rinehart in the mid-1950s (45). Their work combined electron microscopy with the classic cytophysiological approaches of their predecessors. They were able to distinguish specific cell types at the electron microscopic level, based on the morphology of the secretory granules (size, shape, and distribution) as well as other morphological features. A number of electron microscopic studies then emerged, building an information base from different physiological tests, and expanding information about different species. This information is summarized in the later, more detailed sections on gonadotropes and lactotropes. However, first we focus on how cytochemical studies added to the field.

How Immunocytochemists Bridged the Early Studies to the Present

Immunolabeling was a technological breakthrough that ultimately linked the tinctorial staining and ultrastructure of the cells with their identity based on hormone content (46–49). In our review that appeared during this time (49), the linkages for each cell type were described, including the fact that the pioneers initially retraced the steps of the earlier workers, taking care to link immunolabeling evidence back to the tinctorial labeling characteristics. For the most part, the early immunolabeling confirmed the tinctorial staining in terms of the general categories of hormones in the different cell types. When they added the dimension of changing physiological state to study the immunolabeled cells, this provided the first potential for semiquantitative analyses of numbers and sizes of identified gonadotropes and prolactin cells.

The immunocytochemists continued to advance the field by developing double labeling techniques that detected more than one hormone per field (47,50,51). This added information about the distribution and possible association of gonadotropes and prolactin cells (47). It also presented the earliest evidence for coexpression of more than one hormone by a cell, which challenged the one-cell/one-hormone hypothesis (50,51).

During the mid-1970s to the mid-1980s, immunolabeling was extended to the electron microscopic levels, which confirmed and expanded most of the early work of Farquhar and Rinehart [reviewed in Childs and Ellison (49)]. However, the electron microscopic view also added information about subsets of cells, based on hormone content and different morphologies, which are described in detail in later sections. Thus, by the early 1980s the gonadotropes and lactotropes had been characterized with respect to their size and the shape of their secretory granules, as well as any unique responses to endocrine states that involve these cells. However, as the labeling techniques became more sensitive and precise, we found that we had much to learn about these cells. The individual cell types began to be divided into subsets based on morphology and responses to the physiological state. As with any breakthrough in methodology (or sensitivity), unexpected information was obtained. For example, we now know that a significant percentage of pituitary cells are multipotential, coexpressing more than one pituitary hormone (23,52–54). Whereas this new concept led workers to question the exclusivity of the one-cell/one-hormone hypothesis, it also has stimulated studies designed to test factors that would signal the terminal differentiation of a given cell type, or, possibly the transdifferentiation of one cell type to another.

The remaining sections in this chapter focus on the modern multifaceted view of the gonadotropes and lactotropes. They show how immunolabeling and other cytochemical protocols are used to define different cell types and their subsets. However, before each of the cell types is discussed, their embryologic origin is reviewed. This body of research contains information critical to our understanding of factors that drive development and differentiation of gonadotropic or lactotropic functions. We will learn about the specific factors that drive the birth of each cell type in the embryo, which will provide the background for ongoing studies of how the postnatal pituitary cells mature and become fully differentiated.

EMBRYOLOGICAL ORIGIN OF GONADOTROPES AND LACTOTROPES

Origin and Formation of the Anterior Pituitary

Rathke first described the origin of the pituitary gland in 1839 (55). It originates at approximately the third week of life in the human (or at approximately day 8 in the mouse or rat). Long before it appears, the cells in the anterior part of the neural plate ("anterior neural ridge") are precommitted to become the anterior pituitary, the nasal cavity ectoderm, and the olfactory placode. Regions adjacent to the anterior neural ridge in the diencephalon will become the hypothalamus, infundibulum, and posterior pituitary (56–62). These regions form a partnership in that the hypothalamus will produce important factors that drive the organogenesis of the pituitary.

This discussion focuses on factors leading to the development of gonadotropes and lactotropes. The key factors and their times of origin are diagrammed in Figs. 1 and 2, beginning with those important for organogenesis of the pituitary itself. The regions in the diencephalon adjacent to the developing pituitary stimulate the initial formation of Rathke's pouch by a series of signaling molecules. One of these is from a homeobox gene expressed in the ventral diencephalon next to the pituitary anlagen, called thyroid transcription factor or *T/ebp* (also called *Nkx1.1* or *TitfI*). Takuma et al. have studied downstream events resulting from a null mutation of this gene (56). They found that Rathke's pouch forms, but it remains single layered and the cells fail to become columnar. Eventually the pouch cells are eliminated by apoptosis. Only two of the normal gene products that appear in the pouch are found, *Isl1* and *Ptx1*. Downstream gene products are never detected because of aborted development of the pouch cells.

Takuma et al. (56) studied these null mutants for expression of two important genes in the diencephalon that are known to be needed for full Rathke's pouch development. The first is a member of the transforming growth factor-β (TGF-β) family, bone morphogenetic protein-4 (*Bmp4*). This is normally expressed in the ventral diencephalon adjacent to the Rathke's pouch at embryonic day (E) 8.5 and is extinguished by E11.5. Another vital signaling molecule is fibroblast growth factor-8 (*Fgf8*), normally produced by the ventral diencephalon and developing infundibulum on E9.25. They found normal expression of *Bmp4*, which drives the expression of *Isl1* by Rathke's pouch cells. However, the null mutants failed to express *Fgf8*, thereby preventing definitive formation of the pouch. Thus, they concluded that *T/ebp* must drive the expression of *Fgf8*. The function of *Fgf8* is described later.

At the same time, their study brought out the fact that *Bmp4* promotes the development of the pouch and expression of one of its first early gene products, *Isl1*. The latter product is important for the formation of the pouch itself and its ventral expression is vital for the appearance of the first cell type, which produces α subunit (common to gonadotropes and thyrotropes). *Isl1* null mutants have only a primitive pouch with flattened epithelium. Thus, this reveals the first set of factors needed to produce gonadotropes. However, before they can emerge, the anterior pituitary itself must form, stimulated by *Fgf8*.

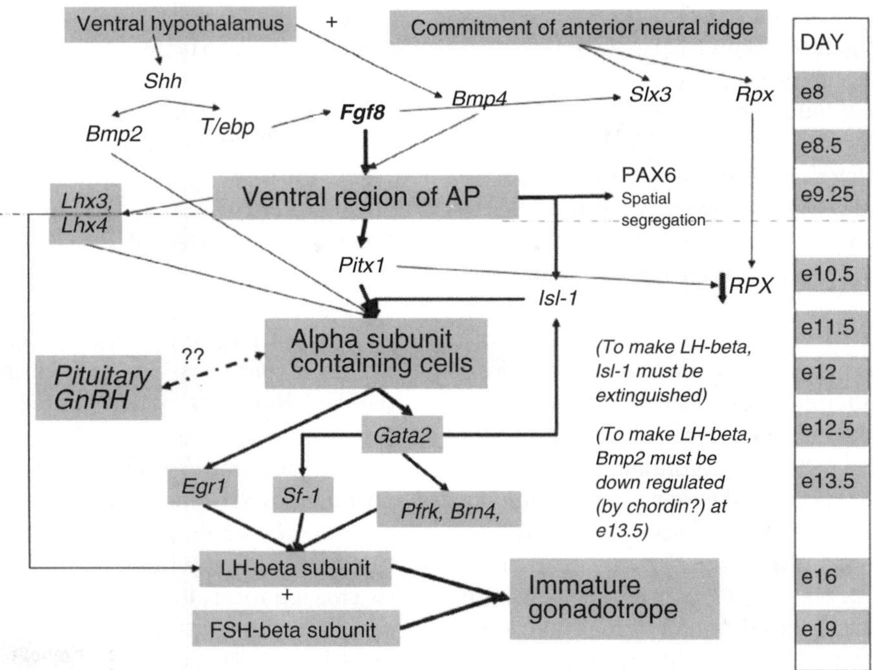

FIG. 1. Flow diagram showing the critical regions, transcription factors, and cells that are activated in the processes leading to the birth of a gonadotrope. The time line to the right indicates the approximate time of appearance of individual cells or factors.

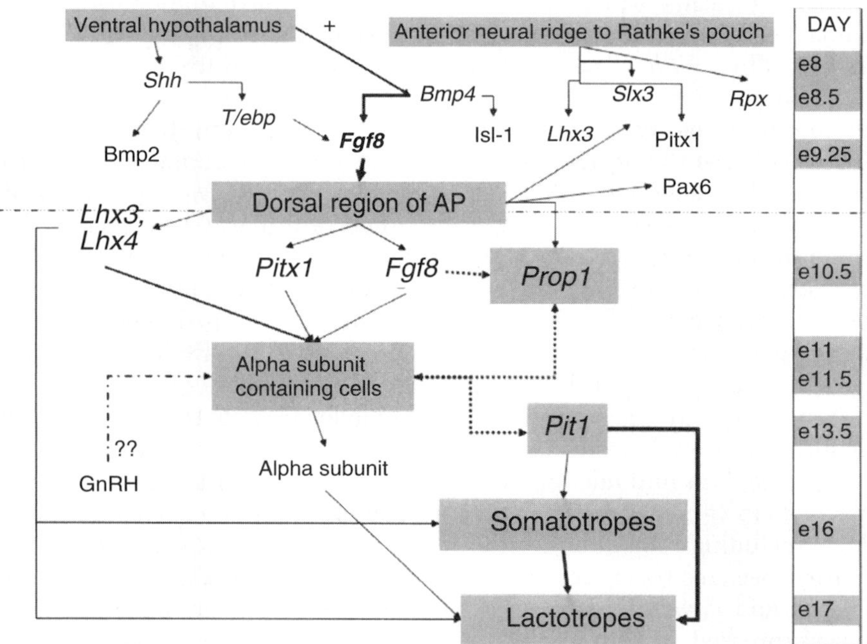

FIG. 2. Flow diagram showing the critical regions, transcription factors, and cells that are activated in the processes leading to the birth of a lactotrope. The time line to the right indicates the approximate time of appearance of individual cells or factors.

At E9.25, *Fgf8* stimulates the thickened pituitary placode and the Rathke's pouch growing up from the pharynx to meet the ventral hypothalamus. *Fgf8* also activates downstream genes in the pouch that produce *Lhx3* and *Lhx4,* which are vital for further proliferation and development of specific anterior pituitary cells (61,62). Other workers have shown that *Bmp4* is also important for expression of *Lhx3* and *Lhx4* (63), but their expression is *dependent* on *Fgf8* (64).

If the expression of these genes proceeds in the right sequence and timing, the definitive Rathke's pouch grows up and its anterior wall is formed by the rapid proliferation of cells, forming the anterior pituitary by E12.5 (56–62). Another gene in the ventral diencephalon, *Wnt5a*, is believed to serve as a promoter of this growth in a way that dictates the species-specific shape of the pituitary (65). Indeed, in *Wnt5a*-deficient mice, the Rathke's pouch appears to branch abnormally and form a dorsal pouch, which thickens with development, along with a normal-appearing anterior lobe. The defect in anterior lobe–derived tissues is not obvious and each of the cell types appears normally. Thus, this ventral diencephalon product is primarily responsible for the shape changes in early development of the pouch (65).

Another important gene in the oral epithelium and ventral diencephalon is sonic hedgehog (*Shh*), which influences early development of the pouch as well as later regional patterns of development of specific pituitary cell types (66). This distinctive gene is expressed by ventral hypothalamus and oral ectoderm *around* the forming Rathke's pouch, but is strikingly absent from the ectodermal cells destined to become Rathke's pouch. Thus, *Shh* is an external driving force.

Studies of *Shh* null mice by Treier et al. show that they do not express *T/ebp* because the entire ventral diencephalon does not develop normally. This, as described previously (56), prevents expression of *Fgf8* and genes critical to the formation of the definitive Rathke's pouch. However, *Shh* also works during pituitary organogenesis, as described by Treier et al. (66). They produced transgenes that knocked down *Shh* expression in a way that does not interfere with expression of *T/epb*. Thus, in the absence of *Shh*, Rathke's pouch formed normally, although it was hypoplastic. Furthermore, *Shh* null mice showed a reduction in gene products in ventral regions of the developing anterior lobe, including *Bmp2* and *Gata2*. This is critical to our story, because this is the region from which gonadotropes and thyrotropes emerge. When Treier et al. overexpressed *Shh* in the pituitary itself, the proportion of gonadotropes and thyrotropes increased, at the expense of somatotropes (66). The workers concluded that *Shh* is needed to promote the development of *Bmp2* expression in ventral regions of the anterior pituitary, which in turn leads to the production of a cascade of factors that produce thyrotropes and gonadotropes. These are discussed in more detail in the next section.

How the Individual Pituitary Cell Types Develop

In the foregoing sections, we discussed four gene products that were expressed during the early development of Rathke's pouch that are actually stimulated by products from the adjacent diencephalon (*Isl1*, *Lhx3*, *Lhx4*, and *Bmp2*). We also discussed the notable absence of *Shh* expression from cells destined to become Rathke's pouch. Treier et al. (66) speculate that this too may be driven by *Bmp4*. Of importance to this chapter is the fact that *Isl1* and *Bmp2* are early products of the ventral cells destined to be gonadotropes (Fig. 1). *Isl1* expression is rapidly downregulated in dorsal regions (possibly by *Bmp4*), becoming restricted to ventral regions (64,66,67), and *Bmp*2 expression is induced in ventral regions by *Shh* (67). This promotes the development of the ventral gonadotropes. In the meantime, patterns of expression of other factors develop to isolate dorsal regions and their products so that prolactin and GH cells can develop. These are discussed later.

First, it is important to note that there are additional products that appear in the Rathke's pouch cells, as the oral epithelium becomes committed, and these have unique functions in the early appearance of the cell types. The genes that are expressed include *Six3*, *R*athke's *p*ouch homeobo*x* gene (*Rpx* or *Hesx1*), and *Ptx*.

The *Six3* gene belongs to a family of homeobox genes that are homologs of the *Drosophila* sine oculis genes (*so*) (68). Most members of the gene family are important in eye development, but some products are also produced by other developing cells, such as the pituitary. The anterior neural plate/anterior neural ridge cells destined to become Rathke's pouch express *Six3* as early as E6.5 (68). *Rpx* (69), which is identical to the gene *Hesx1* (70), appears in the anterior visceral endoderm with the beginning of gastrulation. *Rpx* expression migrates to the oral ectoderm and then concentrates in the pituitary placode. *Rpx* and *Six3* are found in all cells committed to become Rathke's pouch by E8.5. *Rpx* is then downregulated gradually from E11.5 to E15.5 (69). The specific target genes for *Rpx* or *Six3* are not known, although studies show that *Rpx* represses the transactivation of α subunit and LHβ genes, leading to the birth of gonadotropes (71). Rathke's pouch is absent or deformed in *Rpx* null mice. However, it is clear that *Rpx* expression later in development *must be attenuated* to allow expression of gonadotropin genes (71).

Another set of transcription factors, *Pitx1* and *Pitx2*, are expressed in the developing pituitary, initially in cells with the α-glycoprotein subunits (αGSU) and corticotropes (60,72,73). Like *Isl1*, *Pitx1* production is ventrally focused in these early αGSU cells. However *Isl1* becomes extinguished after the αGSU cells appear, whereas *Pitx1* is expressed even in the adult state. *Pitx1* messenger RNA (mRNA) first appears in epithelial layers moving forward to lie along a line in the mandibular component by E9, and then appears in Rathke's pouch ectoderm by E9.5 (60). Its expression seems to follow the wave of differentiation of cell types that moves from ventral to dorsal regions of the anterior pituitary. *Pitx1* expression is highest in cells that produce the αGSU, which are the first to appear at E11.5 in the mouse. These cells form the precursor population for all other pituitary cell types. *Pitx1* directly activates the α subunit promoter and is important for its continued function even in the adult (60,72,73). It also modulates the activity of the orphan nuclear receptor transcription factor SF-1 (74,75) and synergizes with epidermal growth factor (EGF)-1 on the LHβ promoter (76). *Pitx1* is also highly expressed in the pars tuberalis, early in development; the significance of this may relate to the concentration of basophils found in this region. The repression of gonadotropin expression by *Rpx* has been mapped to the *Pitx1* transactivation site (71). Thus, the appearance of *Pitx1* products heralds the downregulation of *Rpx*. To summarize, to develop the early gonadotropes, *Rpx* must be repressed and *Isl1* and *Pitx1* enhanced, which transactivates the genes producing the α subunit (Fig. 1).

In dorsal regions, *Pitx1* transactivates the GH promoter (77), although it is not needed for normal somatotrope development (78). It does synergize with Pit-1 to activate the prolactin promoter (77). The main impact of its absence is seen in vivo in a reduced population of gonadotropes and thyrotropes in dorsal regions (60).

Pitx2 appears to be expressed more widely than *Pitx1* in both the embryo and the adult. It is found in nascent Rathke's pouch as well as in adult thyrotropes, lactotropes, and gonadotropes and somatotropes (60,77,79). It may be vital for the development of one or more pituitary cell types; at least three isoforms have been described. It is thought that *Pitx2* works synergistically with other gene products to activate target genes (79).

Lhx3 and Lhx4 are two Lim-type homeodomain proteins, discussed early in this presentation as Rathke's pouch genes stimulated by the ventral diencephalon and *Fgf8*. Their expression is also important during differentiation and development of both gonadotropes and lactotropes (60). These Lim proteins are characterized by two cysteine/histidine Lim domains located between the N-terminus and the DNA-binding domain (80–82). They are involved in transcriptional regulation, although the Lim domains do not bind to DNA. The Lim protein most important for gonadotrope and prolactin expression is Lhx3. One isoform (Lhx3a) is first seen at E8.5 and the second isoform appears after E9.5. At that time, Lhx3 is found in the Rathke's pouch, and expression extends to the anterior and intermediate lobes of the pituitary, with levels of expression highest just before the initial detection of αGSU (the α subunit) at E11.5 (82).

Lhx3 is vital for the normal appearance of several pituitary cell types. In *Lhx3* knockout mice, Rathke's pouch can develop and *Rpx* appears on schedule, as described earlier. However, continued expression of *Rpx* is not seen after E12.5, indicating that its maintenance for the next few days must depend on the appearance of *Lhx3* (60). Other genes that do not appear in the knockout include those for the α subunit, TSHβ, and GH. The appearance of the LHβ gene is also delayed. Another important gene that never is expressed if *Lhx3* is knocked out is *Pit1* (61). Thus, workers have concluded that *Lhx3* is vital for the production of gonadotropes and the *Pit1*-dependent cells: somatotropes, lactotropes and thyrotropes. Indeed, other studies have shown that *Lhx3* and *Pit1* synergize to promote activation of the TSHβ and prolactin promoters (82). Humans deficient in *Lhx3* show losses in GH, prolactin, TSH, and gonadotropins (83).

With regard to *Lhx4*, Rathke's pouch can form normally in mice that do not have this gene. After it appears at E9.5 in the pouch, its expression is then focused to the anterior lobe region by E12.5. It then diminishes by day E15.5, whereas *Lhx3* expression is maintained throughout adulthood (62,69). *Lhx4* null mice pituitaries can make α subunit, Pit-1, GH, and TSHβ. However, the numbers of gonadotropes are greatly reduced (62). After development, all anterior pituitary cell types show reduced numbers (62). The exact role of this gene is unclear; there is little obvious phenotype in the heterozygous state.

Prop1 is a gene expressed during development of the anterior pituitary, at E10.5. It is particularly prominent in the dorsal portion of the anterior pituitary and expression is maximal at E12, especially in caudal and medial regions, where another transcription factor, Pit-1, is expressed (84). *Prop1* was in fact named as the "prophet of Pit-1" because its appearance precedes that of *Pit1*. The Ames mouse is an important example of a mutation in *Prop1* that shows how vital this gene is to the development of *Pit1*-dependent cells, as well as gonadotropes (85,86). The adult Ames mouse pituitary gland is reduced by 85%, and there are major reductions in gonadotropes, somatotropes, and lactotropes (86). Thus, it is believed that *Prop1* may direct the proliferation of precursors for these cell types and eventually enhance the production of Pit-1, which leads to the differentiation of somatotropes, thyrotropes, and lactotropes. These hormones are clearly deficient in humans with *Prop1* mutations; however, there may be some gonadotrope function at low levels, suggesting that *Prop1* may not be absolutely required for gonadotrope differentiation (87).

Pit1 (*POU1F1*) is a member of the POU family of genes that also includes *Oct1*, *Oct2*, and *Unc86*. *Pit1* proteins bind to AT-rich elements in the rat prolactin and GH promoters. Its N-terminal transactivational domain contains abundant hydroxylated amino acids which bind and activate the genes (60). *Pit1* appears on E13.5 in the mouse, after the anterior lobe has developed, in the central and medial regions, and spreads to the entire lobe the following day (88). *Pit1* proteins are found in somatotropes and lactotropes 1 day before their expression of GH and prolactin on E16 and E17, respectively (89). *Pit1* knockout mice have hypoplastic pituitaries. The Snell dwarf mouse, a mutant strain, shows the normal appearance of *Pit1* transcripts, but the protein product is defective (90). Not only do the target cells (somatotropes, thyrotropes, and prolactin cells) fail to develop in normal numbers, *Pit1* expression itself is significantly reduced, indicating that it must be maintained by some autoregulatory loop.

Finally, another member of the Wnt family, *Wnt4*, appears during the development of anterior pituitary cells. Studies of *Wnt4* null mice show that the pituitary is small and hypoplastic, although all cell types are represented (67). Treier et al. have proposed that *Wnt4* may be a signal for proliferation and expansion of the pituitary cell type precursors.

Spatial Organization of Transcription Factors

It is clear that, in the differentiated anterior pituitary, many cells synthesize, store, and secrete only one pituitary hormone. As reviewed by Rosenfeld et al., they start from a precursor set of cells that produces transcription factors that promote the development of more than one cell type (91). In addition, all Rathke's pouch cells are also initially marked by expression of αGSU by E11. αGSU is a glycoprotein subunit that is expressed by differentiating and mature gonadotropes and thyrotropes. Thus, the subsequent development of differentiated cell types begins in precursor cells that are all producing αGSU.

The process by which different monohormonal cells are induced from this precursor group of cells involves a fascinating series of events that spatially segregates the different cell types (60,63,67,89,91,92). This separation and segregation is driven by different patterns and timing of expression of key growth or transcription factors, which then dictate the timing of the appearance of each pituitary cell type as well as its original location in the pituitary (89,92).

First, as described in the foregoing section on sonic hedgehog proteins, early in development of the anterior pituitary there is a ventral-to-dorsal graded pattern of *Bmp2* expression, stimulated by *Shh* from adjacent non–Rathke's pouch oral epithelial cells (66,91). This opposes a dorsal-to-ventral graded pattern of *Fgf8* expression in Rathke's pouch itself (64,66,91). The presence of high FGF-8 in dorsal regions and high BMP-2 in ventral regions promotes the production of unique sets of transcription factors. *Fgf8* in the dorsal regions promotes the expression of *Nkx-3.1* (93), *Pax-6* (94), *Six3* (68), and *Prop1* (84–87). *Prop1* then promotes the expression of *Pit1*, as described previously, which is responsible for the development of early somatotropes, prolactin cells, and thyrotropes (84–91). A completely different set of factors gives rise to corticotropes (which are mentioned briefly later) (95–99).

Isl1's signal is attenuated in dorsal regions so that it appears in αGSU cells only in ventral regions (60,63,67,91). Coculture experiments by Ericson et al. have shown that *Fgf8* produced by the dorsal infundibulum blocks the production of Isl1 and thus focuses its expression to ventral gonadotropes (64). These Isl1/αGSU–expressing cells also develop a ventral-most layer of differentiated thyrotropes that are not dependent on *Pit1* (64).

As stated previously, *Bmp2* also leads to the development of the αGSU expression throughout the pituitary, although the aforementioned segregation events focus this expression to ventral regions (67). Continued expression of *Bmp2* prevents differentiation of lactotropes, somatotropes, and thyrotropes, so the dorsal factors that downregulate *Bmp2* allow differentiation of these Pit-1–dependent cell types. *Bmp2* must also then be downregulated in ventral regions to elicit full differentiation of gonadotropes. Studies have shown that the signal from this gene becomes attenuated by E13.5 (67). One mechanism behind this may be the expression of a BMP antagonist, chordin, in the developing mesenchyme in regions caudal to the ventral zone of Rathke's pouch (67). If *Bmp2* is overexpressed, some downstream genes important for gonadotrope expression are seen, although the β subunit is not expressed.

Gonadotropes are then differentiated after the attenuation of *Bmp2* and upregulation of *Pfrk* (67), *Brn4*, and *Gata2* (100) in ventral regions. *Pfrk* is a member of the winged-helix (forkhead) transcription factor family (*p*ituitary *fork*head factor). It is expressed at the most ventral part of Rathke's pouch, coincident with the appearance of the ventral Shh/BMP-2 boundary, and remains restricted to the ventral-rostral part of the gland (67). It thus subdivides the *Brn4* expression domain, which is more broadly distributed. The *Brn4* gene belongs to a family of POU domain transcription factors that are needed for development of neuroendocrine neurons in the hypothalamus (101). Its appearance in the pituitary is focused in the ventral regions of the gland, and it is also needed for development of ventral-most cells.

Gata2 expression drives the appearance of the early gonadotropes. Its expression is inhibited in dorsal regions (100). This is significant because *Gata2* actually inhibits the expression of *Pit1*. *Gata2* ushers in production of the other transcription factors needed for gonadotrope expression, including SF1, PFrk, and Isl1 (67). Thus, the gradient, originally established by FGF-8 and BMP-2 proteins, isolates the optimal environment for the selective development and differentiation of the early *Pit1*-derived lineage (somatotrope/lactotrope and thyrotrope) and the *Gata2*-derived lineage, the gonadotrope. It permits the development of the ventral Pit-1–independent thyrotropes (60,63,67,89,91,92).

Another gene product should be added to the list of those that may regulate differentiation of gonadotropes. Van Bael et al. (102) detected gonadotropin-releasing hormone (GnRH) mRNA in crude and pure explants of Rathke's pouch taken on E12. GnRH is a primary regulator for gonadotropes and also has been shown to regulate proliferation of prolactin cells by paracrine mechanisms (requiring gonadotropes) (103–108). In 1990, Jennes had reported GnRH receptors in pituitaries from fetal rats on E13. He also assayed GnRH in amniotic fluid of such fetuses (109).

Van Bael et al. showed that a GnRH antagonist would decrease prolactin antigen expression (detected by image analysis) in the 9-day culture period of the crude 12-day explants, but not in the pure explants devoid of diencephalon tissue (102). GnRH may be produced by Rathke's pouch cells early in development to drive the development of the αGSU-bearing cells, perhaps by autocrine mechanisms, or it may stimulate the β subunits. It may also work on the prolactin cells, perhaps after critical diencephalon factors are produced to begin the spatial segregation needed for *Pit1* expression (60,63,67,89,91,92). We have no information about spatial organization of the GnRH-producing cells themselves. Clearly, they appear as some of these organizational patterns are still being set. Because of its importance to neuroendocrine neurons, *Brn4* may play a role in development of the expression of the GnRH gene by early Rathke's pouch cells (101), which may then facilitate development of the gonadotropes.

What is particularly elegant about this gradient is how the boundaries between the two regions are maintained, to allow sequestration and normal development of the separate cell lines. This is facilitated by FGF-8 induction of *Pax6* in the dorsal regions (60,63,67,89, 91,92). The *Pax6* product then reinforces the boundary between the two regions, insuring that *Gata2* expression is suppressed in dorsal domains (94). This allows the somatotropes/lactotropes to develop normally. *Gata2* in thyrotropes is also sufficiently suppressed in the dorsal regions to allow the development of the *Pit1*-dependent form of thyrotropes. Researchers have shown that if the *Pax6* gene is missing, the ventral cells proliferate, producing abundant *Gata2*, which inhibits Pit-1 expression (94). The ventral TSH cells take over the dorsal regions, and *Pax6* null mice become deficient in somatotropes and lactotropes (94).

These regions of specification are prominent in the embryological pituitary; however, they do not necessarily persist in the adult pituitary of all species. In the human pituitary, for example, basophils (including gonadotropes, thyrotropes, and corticotropes) are found in the anteromedial regions (equivalent to ventral-medial), whereas acidophils are concentrated in the posterolateral regions (dorsal-lateral wings) (28). In other mammals, like rodents, the regions are not as obvious in either the neonate or the adult and, although there are some concentrations in key areas, they do not seem to be related to their region of origin. The cells have mostly scattered throughout the anterior lobe.

Finally, a brief mention of corticotrope development is important because later studies showed some coexpression of adrenocorticotropic hormone (ACTH) with gonadotropins. Corticotropes develop from a lineage completely separate from that of other cell types. They appear at the ventral base of Rathke's pouch at E13. A fraction of these cells produce Pitx1, which works synergistically with a T box factor, Tpit, which is present only in pro-opiomelanocortin (POMC)–expressing cells. This action initiates POMC cell differentiation and POMC transcription (95). Other factors are involved in the enhancement of POMC cells, including leukemia inhibitory factor (LIF) (96) and the NEUROD1/PanI heterodimer (97–99). Corticotropes also appear to need Lhx3b isoform to proliferate (but not for differentiation) (61).

The findings describing these early development events have strongly reinforced the one-cell/one-hormone hypothesis, while allowing for some crossover as families of hormone-bearing cells are identified by their dependence on common transcription factors (e.g., somatotropes, thyrotropes, and lactotropes). The findings also have emphasized the importance of cells that have differentiated to a monohormonal state, and the need for such cells. However, they have also discovered heterogeneity among cell types, such as the ventral- and dorsal-derived thyrotropes.

Monohormonal Cells Become More Complex, Multihormonal Cells

Sensitive assay technology applied since the early 1990s has demonstrated that multihormonal cells are more abundant than expected in the adult (23, 52–54). Reconciling the elegant findings in the embryo with studies of multihormonal expression in pituitaries from the adult has been a challenge. Some of this reconciliation comes with the recognition that all pituitary cells developed from a background of cells that expressed αGSU. The conditions that segregated the populations and repressed gene expression to allow monohormonal cells to appear have been attenuated in the adult. This alone could allow for coexpression of different combinations of hormones. However, as later sections of this chapter will show, there are significant remodeling processes in the neonatal and adult pituitary that involve maturation, enhanced storage, proliferation, apoptosis, and expression of multihormonal phenotypes. All of these steps are needed to support differentiated functions in the postpubertal pituitary. In later sections, we show how the pituitary is remodeled through apoptosis and proliferation to support different reproductive needs and functions, such as pregnancy, lactation, weaning, and the aging process itself.

More discussion will be presented in the individual sections describing the lactotropes and gonadotropes. Before we leave this section, findings by Seuntjens et al. are presented because they have provided important clues about the changes in anterior pituitary cells. Seuntjens et al. used reverse transcriptase-polymerase chain reaction (RT-PCR) and single-cell analysis of gene expression in pituitary cells at key times during development (110,111). These highly sensitive techniques showed that on E16, 44% of the hormone-expressing cells were monohormonal and 14% were multihormonal. The analysis showed a steady increase in the proportion of multihormonal cells with age. When the values were expressed as a percentage of total pituitary cells, 6% were multihormonal at E16, 11% at postnatal day (P) 1, and 25% at P38. A number of different combinations were seen. For example, cells coexpressing POMC and GH or prolactin mRNAs increased to 8% of the population between E16 and P1. Cells expressing GH/prolactin and αGSU remained at approximately 4% to 6% of the population during this developmental period. This last combination of course involves a mixture of hormones normally produced by cells originating in different regions. This may reflect geographical overlap in the transcription factor gradients, allowing for some coexpression of genes in regions of overlap. Seuntjens et al. (110,111) also showed that these multihormonal cells were difficult to detect by antigen content, requiring the highest levels of sensitivity (e.g., RT-PCR). Thus, they would be missed in studies that relied on immunolabeling. Later sections of this chapter present studies that show the functional significance of some of these multihormonal cells.

In conclusion, the original specification early in development serves an important purpose in providing significant numbers of active monohormonal pituitary cells that support a single function during early development. We have proposed that multihormonal phenotypes can add a new dimension, providing coordinated support for a set of functions or backup support during periods of enhanced remodeling (23,52). Some of these multihormonal cells may have common regulators, providing an efficient regulatory circuit.

Although the aforementioned rationale may provide an explanation for their appearance, this does not explain how subsets of monohormonal lactotropes and gonadotropes become capable of multihormonal expression. Does this reflect the reawakening of former stem cells in their lineage? Or does it represent further differentiation or maturation? These questions are revisited later in the chapter.

GONADOTROPES

Early Studies

Electron Microscopy Detects Subsets of Gonadotropes

By the 1950s, gonadotropes had been described as relatively large, PAS-positive, blue-staining oval cells in anterior and lateral regions of the pars distalis (36–44). In addition, a separate group of smaller gonadotropes was found in the pars tuberalis, near the intermediate lobe and posterior regions of the pars distalis (112,113). Thus, in the rodent at least, the adult pattern of distribution does not match the region of origin in the embryo, indicating that further migration and differentiation must occur.

As stated in the foregoing sections, the combined use of cytophysiology and electron microscopy, coupled with immunolabeling since the early 1970s, allowed researchers to identify subsets of gonadotropes as well as link functional significance to different morphologies. In the beginning, most experiments were designed and interpreted with assumptions based on the one-cell/one-hormone hypothesis. The first such assumption was that there were two gonadotropes, one producing LH and one producing FSH. A number of workers looked for changes in pituitary morphology that fit the sole production of one of these hormones. In the 1950s, Purves and Griesbach described a cell type that showed increased granulation after estrogen treatment (37). This cell type also enlarged rapidly after castration and became filled with vacuoles or one large vacuole, giving it a "Swiss cheese" or a "signet ring" appearance. Because the rapid changes were associated with a rapid rise in FSH, Purves and Griesbach suggested that this might be the "FSH cell." Farquhar and Rinehart (45) then characterized this cell type at the electron microscopic level under the conditions that caused this postcastration rise in FSH. The enlarged gonadotrope was filled with granules and dilated vesicles of rough endoplasmic reticulum by 6 days after surgery and, by 20 days, the Golgi complex was prominent and the vacuoles of rough endoplasmic reticulum had coalesced into the larger vacuoles that gave the cells their characteristic appearance. Fewer granules were apparent in these castration cells. Farquhar and Rinehart suggested that the cells with the more vacuolated endoplasmic reticulum might be the FSH cells. Figure 3 shows an example of three "castration cells" taken from female rats, 30 days after surgery.

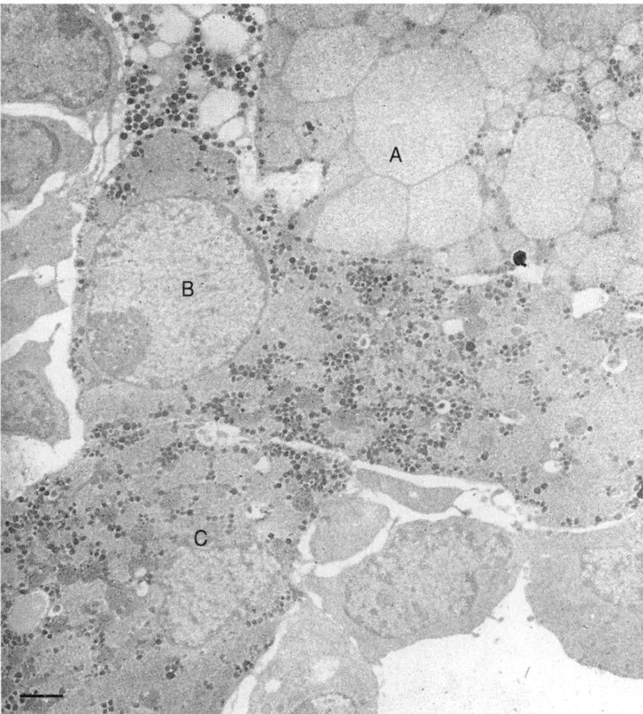

FIG. 3. Example of "classic FSH cell" first described by early electron microscopic studies (A) (45). However, they are immunolabeled with anti-bovine LHβ serum and the avidin–biotin–peroxidase technique. Two other castration cells are also shown (B, C). They are hypertrophied, but the expansion of the rough endoplasmic reticulum is not as striking. Light labeling is seen in the rer. Bar = 1 μm.

The cells were dispersed and enriched by size and then immunolabeled for LHβ subunits. The top cell (A) in the figure resembles the heavily vacuolated gonadotropes first described by Farquhar and Rinehart (classic "FSH cell").

LH cells were designated by Purves and Griesbach at the light microscopic level as being the most prominent cells before puberty (37). They also responded during the first ovulation by rapid degranulation. The LH cells involuted after testosterone treatment and also appeared later after castration, when pituitary LH was highest. Taking these criteria into consideration, Farquhar and Rinehart described gonadotropes at the electron microscopic level that were heavily granulated and contained endoplasmic reticulum that was filigreed (45). They agreed that castration cells with filigreed endoplasmic reticulum appeared later after castration and suggested that these cells were "LH cells." The two lower cells (B and C) in Fig. 3 are more heavily granulated and also label for LHβ subunits. These cells more nearly resemble this Purves-Griesbach LH cell. Note that cell B has more of an angular shape, typical of the "classic" LH cell.

Subsequent studies followed the same theme (114–116). Barnes's pioneering work on the mouse reported changes in an FSH gonadotrope after castration, describing the cells as having two types of granules (115). LH cells that were more angular had only one type of granule and predominated when LH levels were rising. Kurosumi and Oota applied these same classification criteria to gonadotropes in the rat and presented more information about these cell types (116). FSH cells were distinguished by two sizes of granules scattered among the vesicular rough endoplasmic reticulum. One type was large, 700 to 1,000 nm in diameter, and the other was small and ovoid, 200 nm in diameter. LH cells, however, were more angular and contained a uniform population of granules approximately 250 nm in diameter. Examples are shown in Figs. 4 and 5.

Researchers also reported that intermediate forms of these morphologically distinct "gonadotropes" could be found and that LH cells often were difficult to distinguish from GH cells, which had some of the same features (35,49,116). Clearly, cytochemical confirmation was needed to detect the true hormone content of these cell types.

Immunolabeling Challenges the One-Cell/One-Hormone Hypothesis

Pioneering immunolabeling studies by Nakane used dual labeling for LH and FSH and showed that many of the gonadotropes contained both hormones (47,114). Nakane's findings were confirmed

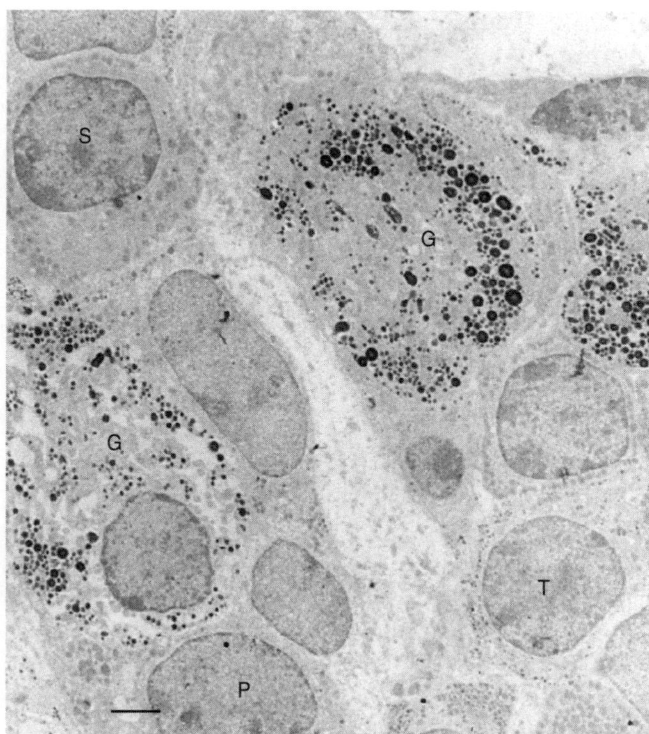

FIG. 4. Type I gonadotrope (G) from male rat, immunolabeled for LHβ with peroxidase–antiperoxidase technique. Note the distinctive oval-round shape and two sizes of secretory granules. Unlabeled prolactin cells (P) are also shown. Bar = 1 μm.

by Phifer et al. (50) in the human. Careful controls were done to prove that the antisera were not reacting with identical sequences in the α subunits of both gonadotropins. This early evidence thus became the first to contradict the one-cell/one-hormone hypothesis. However, in fact the cell population is mixed: monohormonal gonadotropes have been found in a number of species (35,49).

Several groups used electron microscopic immunolabeling to identify gonadotropes (48,117–127). For the most part they reported labeling in the same subtypes of gonadotropes described by electron microscopists (45,115,116) (Figs. 3 to 5). The workers used antisera to the specific β subunits to avoid detecting cross-reactive α subunit sequences. Because these subtypes could contain FSH and LH, they were renamed type A (118–121), the former FSH cell, or types I (1) or II (2), the former FSH or LH cell types, respectively (122–127). In addition, the laboratories discovered a third type that was angular or stellate, containing peripheral granules. Nakane called it the type B gonadotrope (118–121), and we named it the type III gonadotrope (124–127). Initially this cell was described based on its FSH content (Fig. 6). However, because it resembled the cell that produces ACTH, we immunolabeled serial sections for FSHβ and ACTH and discovered that half of these cells contained FSHβ and immunoreactive C-terminal 17–25 ACTH sequences (127). The significance of this discovery is discussed in greater detail later in this chapter. Suffice it to say, this was one of the more surprising examples of a cell type that did not follow the one-cell/one-hormone hypothesis, even assuming

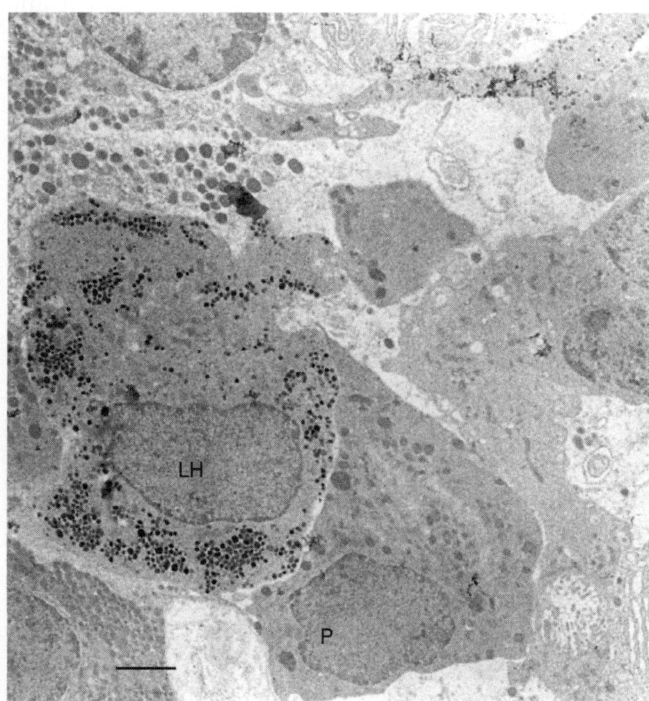

FIG. 5. Type II gonadotrope (LH), immunolabeled for LHβ as in Fig. 4. Taken from a proestrous female rat. Note the angular shape, process, and homogeneous population of secretory granules. Unlabeled prolactin cell (P) is seen. Bar = 1 μm.

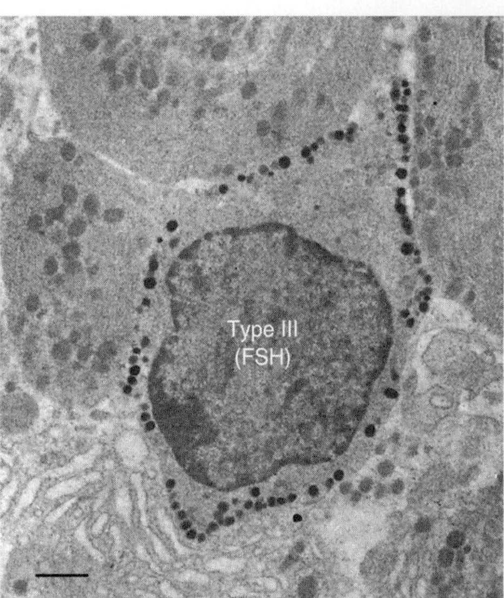

FIG. 6. Type III gonadotrope, immunolabeled for FSHβ as in Fig. 4. Taken from a proestrous rat. Note the angular shape and peripherally organized secretory granules, features shared by adrenocorticotropic hormone cells. Bar = 1 μm.

the modifications that allowed for the coexpression of "families" of hormones like gonadotropins.

Identified gonadotropes could be studied in different physiological states, and this provided the first functional correlates for the different morphologies. For example, the morphology of the gonadotropes changed as the cells approached the active secretory stages of proestrus (124–126). Many extended cellular processes and thus changed from ovoid type I cells or angular type II to more stellate cells as they approached stages of higher secretory activity (proestrus or estrus). Examples of such a change are seen in the LH-bearing cell in Fig. 5 and the FSH-bearing cell in Fig. 7, both of which are from a proestrous female rat. Furthermore, counts of the varying cell types in the serial sections showed that the changes in LH- or FSH-bearing cells were not in parallel, which indicated that subsets of these cells may be responding differently. This and the serial section evidence (described later) confirmed early studies by Nakane (47,114,118–121) that there were both monohormonal and bihormonal gonadotropes in the population.

Early workers detected gonadotropes in a number of mammalian species, including cow (128,129), pig (130–132), sheep (133–136), monkeys (135–138), and human (137–142), and found that they shared many similar structural features. Some of these workers also tested coexpression of gonadotropins by the same cells. The reported percentages of bihormonal gonadotropes varied from 0% in bovine pituitaries (129) to 100% in rat (123) or ovine pituitaries (132–134). Humans had only a few monohormonal FSH gonadotropes (139–140).

Studies in our laboratory have shown that the monohormonal state varies significantly with the physiological state of the animal. This has been reviewed (125) and is described in more detail later in this chapter. By the early 1980s, there was a sense that the research challenges had moved from a need to justify the presence of cells that were bihormonal to

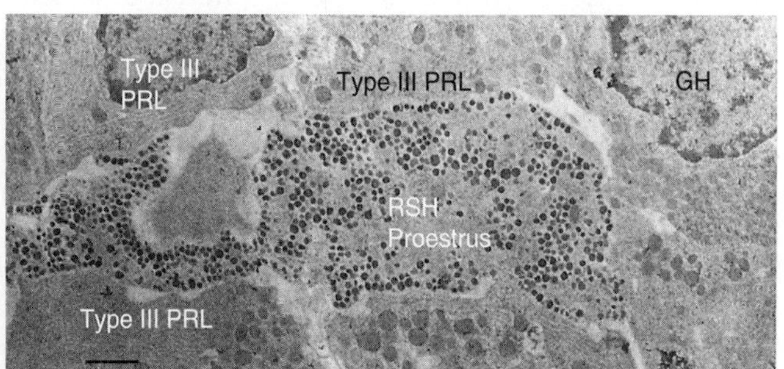

FIG. 7. Type II cell from proestrous female rat immunolabeled for FSHβ as in Fig. 4, showing a process stretching to a sinusoid. Granules are filling the process. Unlabeled prolactin (PRL) cells are also seen. Bar = 1 μm.

a need to explain the significance of monohormonal subtypes. Both needs were particularly obvious as information about GnRH regulation of gonadotropes was growing (143). GnRH was reported to stimulate both LH and FSH in early experiments, which made a bihormonal gonadotrope a convenient target cell. The search for a separate FSH-releasing factor (FSH-RF) has been ongoing since 1964 (144–149), but the results have been the subject of controversy (150–152). Candidate hormones are still being characterized, and a discussion of the state of this field is presented in the ongoing research topics presented at the end of this chapter. In any event, without a definitive FSH-RF, investigators were challenged to explain how one cell and one releasing hormone could regulate nonparallel release of LH and FSH from bihormonal cells during the estrous cycle. This challenge has been the theme of experiments since the mid-1980s (117,125), and it will continue as a thread in many of the studies described in subsequent sections.

Gonadotrope Secretory Responses to GnRH

As immunolabeling studies confirmed the identity of the gonadotrope at the electron microscopic level, there was clear agreement about the cell that contained two types of secretory granules, the so-called "type I gonadotrope" (Fig. 4). This subtype was easily identified by morphology alone and thus provided an excellent model for early studies of the gonadotrope responses to a secretagogue. Note the distinctive small and large granules scattered in the cytoplasm.

Gonadotropes secrete by a classic pattern of exocytosis, which involves fusion of the granule membrane to the plasma membrane and extrusion of granule contents. Exocytosis could be seen after stimulation by hypothalamic stalk median eminence extract (153) or within 1 to 15 minutes after an injection of purified GnRH, or synthetic GnRH (154). In 1973, Luborsky-Moore et al. described the complete process of exocytosis by gonadotropes in the male rat seen within 1 to 2 minutes after an injection of GnRH (154). They discovered that gonadotropes formed compound exocytosis profiles from the extrusion of secretory granules at a single site on the cell membrane. The compound exocytosis profiles provided a morphological correlate for a burst of LH or FSH that, when multiplied by a number of cells, could support surge secretion. Figure 8 illustrates an example of a compound exocytosis profile in a gonadotrope secreting after a stimulus, in vitro, with a biotinylated analog of GnRH. The field shows a cluster of granules that are freshly secreted (exocytosis), and deeper in the cell is a region that shows a larger, forming compound exocytosis profile 10 minutes after a stimulus with

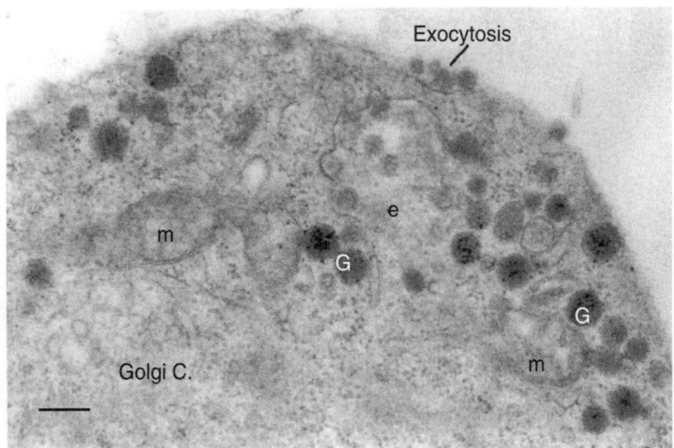

FIG. 8. Compound exocytosis profile in gonadotrope that was stimulated for 10 minutes with biotinylated gonadotropin-releasing hormone (GnRH). Multiple granules are seen at the periphery, and under this region is a developing compound exocytosis profile (e) that is collecting the contents of additional granules. It is likely continuous with the outside in another plane of the section. In this region are granules labeled for biotinylated GnRH (G), which may be a site for recycling, or may represent an early endosome. Bar = 0. 5 μm.

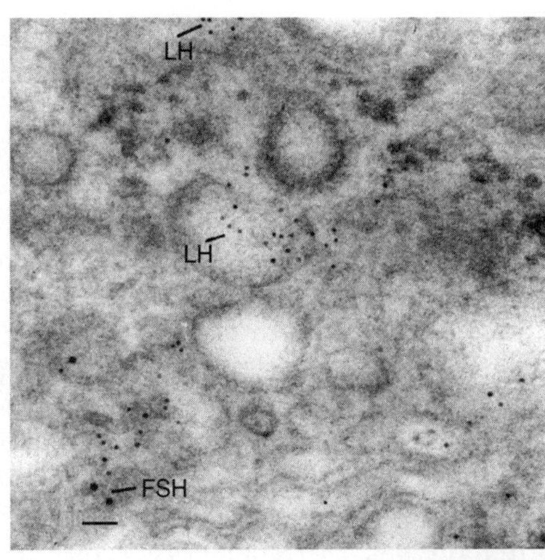

FIG. 9. Golgi region in a cell stimulated for 10 minutes with biotinylated gonadotropin-releasing hormone (GnRH). Dual labeled for LHβ (10 nm gold, L) and FSHβ (50 nm gold, F). Patches of gold label show small monohormonal vesicles in the *trans* Golgi region. Bar = 0. 25 μm.

biotinylated GnRH. The label for biotinylated GnRH is found on a subset of secretory granules. Its significance is discussed in later sections.

Luborsky-Moore et al. (154) also reported changes in synthesis and packaging domains in the secreting gonadotropes. Initially, the cells showed enhancement of their rough endoplasmic reticulum, suggesting expansion of synthetic processes as secretion was being stimulated. During the first 15 minutes, the Golgi complex began to fill with newly packaged granules. At the 30-minute time point, the rough endoplasmic reticulum was not as dilated, but the granules in the cytoplasm were very abundant. Clearly, the cell had been primed by GnRH to support another burst of gonadotropin secretion. Figure 9 illustrates stimulated gonadotropes labeled for LH or FSH in the Golgi complex region. Vesicles labeled for LH (10-nm gold) and FSH (50-nm gold) are evident. This cell had been stimulated with GnRH 10 minutes previously.

Several studies correlated these events with serum LH (155–157). Römmler et al. (155) injected female rats with GnRH and noted an initial depletion of secretion granules after 3 to 15 minutes coinciding with a rapid rise in serum LH and a loss in pituitary LH content. The compound exocytosis profiles formed elaborate tubules and pouches, the space of which was continuous with the extracellular space but invaginated deeply into the cytoplasm (Fig. 8). Changes in the rough endoplasmic reticulum (dilation) and the Golgi complex (active packaging) were similar to those described by Luborsky-Moore et al. (154) for the male. However, the regranulation process in the female appeared to be slower and evident only after 60 minutes. After a second injection of GnRH 2 hours later, serum LH rose to a peak that was twice the level of the previous rise, showing the priming effect of the first bolus of GnRH. Obviously, the expansion in the secretory and storage apparatus provided the support for that higher secretion. However, 3 hours later, the cells expressed normal levels of granules (155). The investigators did find more lysosomes in these cells, suggesting that some granules may have been degraded by autosomal processes (termed *crinophagy* by Farquhar [156]).

Garner and Blake (157) studied the effects of a *continuous* exposure to GnRH in vivo in rats that also had a phenobarbital block of endogenous GnRH. After exposure to 15 minutes of continuous infusion, the pituitary LH content was maintained and the gonadotropes did not lose secretory granules. This suggests that the brief continuous exposure stimulated synthesis at a rate that replenished stores. After 2 hours of continuous exposure, however, there was a loss in secretory granules coinciding with a reduction in pituitary LH content. After 3 to 5 hours of continuous infusion, the cells became refractory to further stimulation. Serum LH and the number of secretory granules fell. This early report thus showed the consequences of continuous exposure to GnRH for long periods and the fact that these conditions are not favorable for a continued rise in LH secretion.

The hypogonadal mouse is an animal model with responsive gonadotropes, but no functional GnRH neurons (158–160). Thus, one can study early responses to GnRH by unprimed gonadotropes. McDowell et al. (158) reported an increase in large vacuoles filled with flocculent material 5 minutes after a GnRH injection. Higher LH and FSH content and gonadotrope hypertrophy were evident after 20 days of GnRH treatment. Lewis et al. showed that GnRH treatment of hypogonadal mice caused the movement of granules to the cell periphery so that 60% of the granules were found in this outer marginal zone after a GnRH stimulus (1 hour) (159). They also noted that the mean diameter of the granules in that zone was smaller than that of granules in the central zone. They reported depletion of pituitary LH and granules by GnRH in intact mice. One explanation for this movement was seen in a study of microfilaments by the same group (160). Morphometric analyses showed that microfilaments lengthened and became more oriented to the plasma membrane as a result of GnRH stimulus. Cytochalasin (which prevents the formation of microfilaments) blocked this change. It did not block the early release of LH, during the first hour after GnRH stimulation. However, it did block the marked increase in LH seen after a second stimulation with GnRH (termed the "priming effect"). These cytoskeletal filaments may be used in the sorting and trafficking process needed to move sufficient LH granules to the periphery after the initial stores are released during the first pulse of GnRH. Granules already at the periphery may be the ones that are released after the first GnRH stimulus, and their release may not be dependent on microfilament transport.

Collectively, these studies provided concrete evidence that correlated serum and pituitary gonadotropin levels with changes in each domain belonging to the gonadotrope secretory cycle. They provided the first morphological correlates for surge secretion, enhanced responses to a priming dose of GnRH, and the fact that gonadotrope responses can become refractory to continuous exposure to GnRH. The morphological data provide a strong basis for our understanding of how gonadotropes organize their cytoplasmic domains, including the cytoskeletal system. This theme of cytoplasmic organization is reintroduced in later sections that focus on sorting strategies in bihormonal gonadotropes.

Mystery of the Diverse Morphologies

Most of the foregoing studies of secreting gonadotropes relied on morphology, and the focus was mainly on the type I gonadotropes, which are easiest to identify (Fig. 4). In fact, the cytochemical detection of gonadotropins or GnRH binding sites revealed that the population of gonadotropes included cells with diverse morphologies, which would be difficult to distinguish from GH cells (type II gonadotropes) (124–126) or ACTH cells (type III gonadotropes) (127). Figure 7 is a good example of a gonadotrope that would be difficult to distinguish by morphology as an FSH cell. Thus, a number of laboratories used cytochemical labeling and a variety of approaches to learn more about the significance of the different morphologies.

Several hypotheses were proposed to explain these morphological differences. First, the type III gonadotropes could belong to a class of multihormonal cell that is in an earlier stage of a cell lineage. A study of gonadotrope maturation was therefore conducted to test this hypothesis. Second, either type II or III gonadotropes could be in different stages of the secretory cycle, becoming more angular or changing distribution of granules as they prepare to secrete. Several figures are included to show examples of such changes. Figures 5 and 7 illustrate more stellate "type II cells" preparing for proestrous secretion. In contrast, LH cells from estrous rats can be so poorly granulated that they are impossible to detect because the content of LHβ has declined significantly and the secretory granules are sparse and small (49,125).

Further clarification of the significance of changing subtypes came from studies of gonadotrope morphology during fetal and postnatal maturation. Animals were also studied throughout the estrous cycle or after castration, both experimental models associated with changing steroid levels. To learn if the different morphologies correlated with different responses, workers also separated the different subsets of gonadotropes by size to study possible differences in secretory patterns. The following sections discuss the studies since the early 1980s that have approached the mystery of the diverse morphologies and contributed to our understanding of the significance of the diversity of gonadotropes. In these sections, we establish the importance of the bihormonal gonadotrope to the population, exploring the challenge of having to release only one gonadotropin in cases where nonparallel release is occurring. However, at the end, we reintroduce the monohormonal gonadotrope and show how it could contribute to the overall function of the population as a "stand-alone" subset.

Gonadotrope Responses to Different Physiological States

Maturation of the Gonadotrope

In the previous section on embryology, we discussed how the sequence, timing, and location of specific transactivation factors are important for organogenesis as well as the birth of each of the pituitary cell types (Fig. 1). The first hormonal product to appear is one expressed by adult gonadotropes and thyrotropes, the α subunit. Later in fetal development, the two β subunits are expressed in cells in the ventral regions of the gland. Sétáló and Nakane (161) reported that LHβ is seen first on E17 in the ventral regions of the rat developing anterior pituitary. At the electron microscopic level, however, immunoreactive LHβ may first be seen on E16 in the rat (162). FSHβ appears later, only 2 days before birth (E19 to E20).

Pituitary embryologists have designated a gonadotrope as fully differentiated once it expresses the β subunit. However, workers agreed that these newly developed gonadotropes do not resemble any of the "mature gonadotropes" in the adult pituitary. Tougard et al. (162) reported that the fetal gonadotropes were round or angular and, even at birth, did not look like those in the adult. LHα and β subunits were detected on endoplasmic reticulum vesicles and small, sparsely scattered secretory granules. Yoshimura et al. (163) agreed that immature gonadotropes contained small, scarce secretory granules, no larger than 200 nm in diameter. Further postnatal differentiation is needed for full maturity.

We initiated studies of postnatal maturation of gonadotropes in the male rat and reported remarkable changes in hormone expression and morphology of the gonadotropes during the first week of life (164,165). In agreement with Tougard et al. (162), gonadotropes 1 to 2 days after birth are small, mostly monohormonal, and sparsely granulated (Figs. 10 and 11). Counts of LH cells identified by immunolabeling showed that they represented only 6% of pituitary cells (approximately half the number found in the adult), whereas FSH cells were even more scarce at 3% of pituitary cells (or a third of adult FSH antigen-bearing cells). Serial sections were labeled for LHβ and FSHβ and counts showed that 66% of the cells were bihormonal, compared with over 79% in the older age groups (Fig. 12). Comparing the morphology of these newborn gonadotropes with those of the adult, it was clear that some resembled the type III cells; however, most did not fit any of the adult classification criteria, although they looked more like type I cells based on the mixed granules.

During the first 2 weeks after birth, the gonadotropes enlarged and became better granulated (Figs. 13 and 14). By the end of the week, the percentages of LH or FSH cells were comparable with those in the adult, and nearly 80% stored both gonadotropins (Fig. 13). The gonadotropes continued to fill with granules and the relative percentages of bihormonal cells persisted to weaning (Fig. 12). The only difference between the adult and weaning was a slightly lower percentage of bihormonal gonadotropes (70%) in the adult. The cells clearly resembled the adult population in the male rat by 2 weeks of age (Fig. 14). However, as is evident from the data in

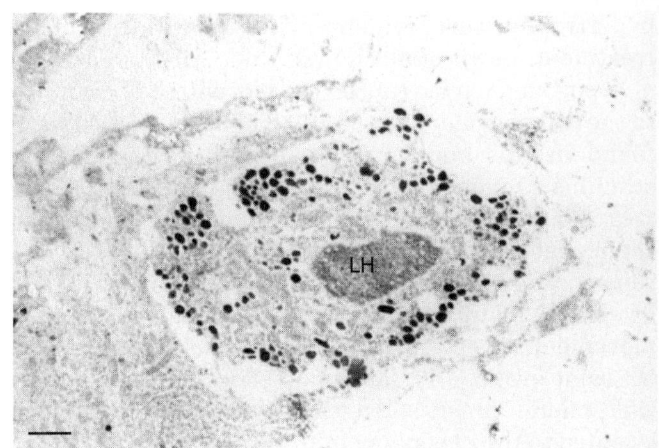

FIG. 10. Small LH cell immunolabeled as with anti-bovine LHβ absorbed with FSHβ and the peroxidase–antiperoxidase complex. Taken from 2-day-old male rat. Granules are sparse and small. Bar = 1 μm.

Fig. 12, their storage patterns are unusual because two peaks of ACTH-like immunoreactivity were detected in gonadotropes during the first 2 weeks of life. This did not depend on any morphological subtype. The significance of this finding is discussed later in the chapter.

Rats are relatively immature at birth and their stage in the neonate is more like that of human fetuses. In fetal human pituitaries, similar evidence is seen. Currie et al. (166) reported that LH and FSH labeling could be detected as early as 9.5 to 10 weeks of gestation in humans in small cells with only a few small granules. The numbers of granules and their labeling intensity increased progressively with age.

As stated previously, the dual labeling in the rat pituitary revealed that gonadotropes are a mixed population at birth with respect to costorage of LHβ and FSHβ. Whereas many are monohormonal, as might

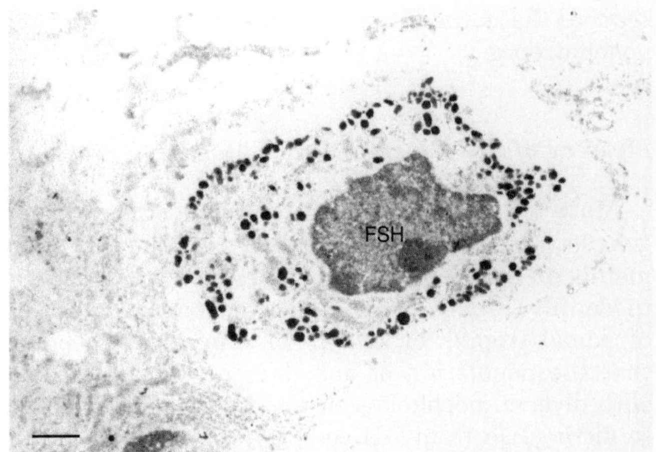

FIG. 11. Section serial to that in Fig. 10, except the cell was labeled with anti-human FSHβ, and absorbed with LHβ and the peroxidase–antiperoxidase technique. Bar = 1 μm.

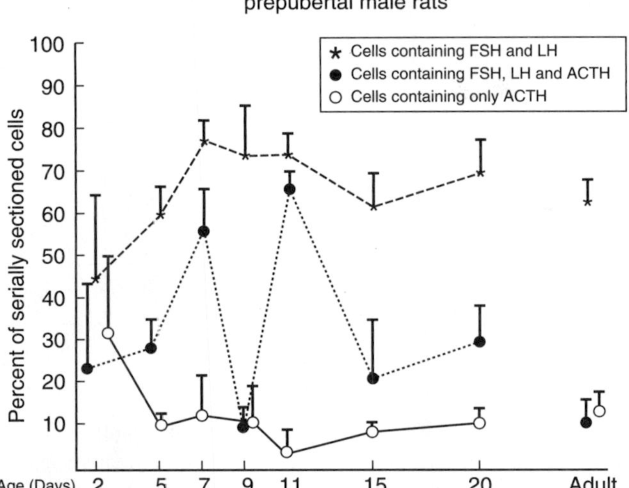

FIG. 12. Counts of percentages of bihormonal gonadotropes during the first 3 weeks of postnatal development. There is a steady increase in percentages of bihormonal gonadotropes as FSHβ cells expand, to reach a peak that exceeds that in the adult. The graph also shows that subsets of gonadotropes express adrenocorticotropic hormone (ACTH)–like peptides in two peaks, one at 7 days of age and a second at 11 days. This expression may reflect the production of pro-opiomelanocortin (POMC) to make a 16-kDa polypeptide that stimulates expansion of the prolactin cell population. Note that numbers of cells containing only ACTH decrease during this time and do not return to adult levels for 3 weeks. The nadir point is during a period known as the "stress-nonresponsive period."

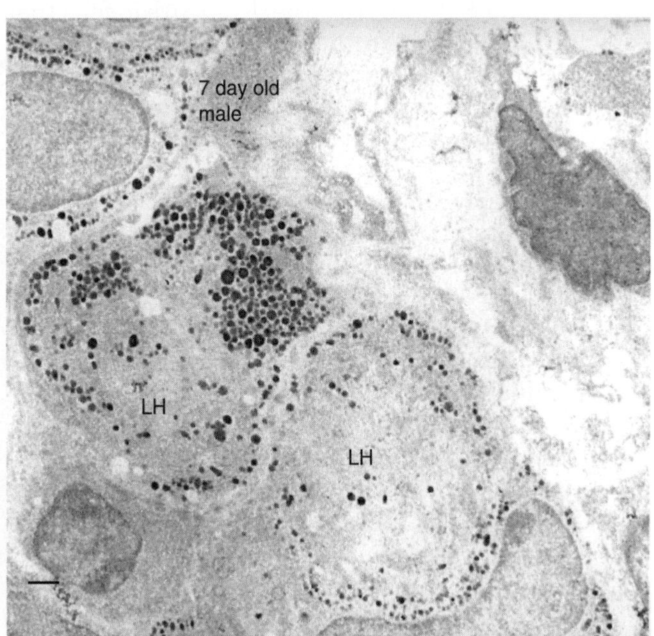

FIG. 13. Two gonadotropes immunolabeled for LHβ (see Fig. 10) from a 1-week-old male rat. The cells are better granulated and larger. Bar = 1 μm.

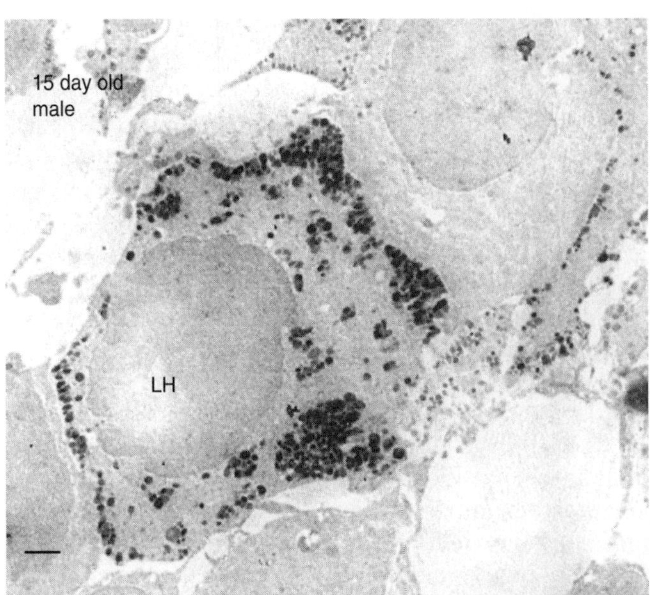

FIG. 14. Gonadotrope immunolabeled for LHβ from a 2-week-old male rat. These cells more nearly resemble the adult gonadotrope. They label intensely for LHβ (see Fig. 10 for protocol). Bar = 1 μm.

be predicted by the early embryology studies, those that are bihormonal increased proportionately only after birth, as the population increased overall (164). Also, at this time the gonadotropes become more scattered than they were in the early embryo. Clearly, more studies are needed to learn what factors regulate the full maturation of the prepubertal gonadotrope to allow it fully to support reproductive functions after puberty.

Gonadotrope Responses to Changing Levels of Gonadal Steroids

A time-honored way of evaluating structure/function correlations has been to study gonadotropes in animals after removal of the gonads. As described previously, this was how the gonadotropes were first discovered, and Farquhar and Rinehart (45) provided the first electron microscopic view of their ultrastructure. The gonadotropes hypertrophied and became filled with dilated rough endoplasmic reticulum (Fig. 3). Farquhar and Rinehart noted that the rough endoplasmic reticulum contained dense regions and suggested that it might condense and package the hormone, taking on Golgi complex functions. This was confirmed by Pelletier (167), who compared gonadotropes from normal and castrated rats and reported a shift in the uptake of tritiated fucose from the Golgi complex to the rough endoplasmic reticulum after castration. Fucose is a terminal sugar normally taken up by the Golgi complex, and this shift suggested that the rough endoplasmic reticulum had taken over some of the functions of the Golgi in the castration cells. Figures 3, 15, and 16 show examples of labeling in the rough endoplasmic reticulum.

The advent of immunolabeling allowed workers to characterize the changes after castration in immunocytochemically identified gonadotropes. Garner and Blake (168) also correlated an increase in secretion granules per cell profile with a 22-fold increase in serum LH and a fivefold increase in pituitary LH. Initial studies in our laboratory (169) reported a morphometric view of the development of a castration cell within the first 24 hours after surgery. Figure 15 shows a developing castration cell labeled for LHβ. By 24 hours, it is already filled with rough endoplasmic reticulum that is labeled intensely for LH. Scattered secretory granules are few. During this brief period after surgery, we reported a fourfold increase in serum LH and nearly a twofold increase in serum FSH. There was a 1.5-fold increase in percentages of LH or FSH cells by 24 hours, along with increases in area and volume fraction of LH cells (Table 1). After castration, LH cell areas were significantly increased from 133 μm^2 to 198 μm^2. FSH cells identified by antigen content tended to be larger in the normal animal (179 μm^2) and their areas overlapped with those of developing castration cells (186 μm^2) during the first 24 hours. However, because of the increase in percentages, the volume fraction of FSH cells was increased 1.5-fold after castration (Table 1). When serially sectioned cells labeled for LHβ or FSHβ proteins were analyzed, castration produced an increase in the percentages of cells that expressed both gonadotropins

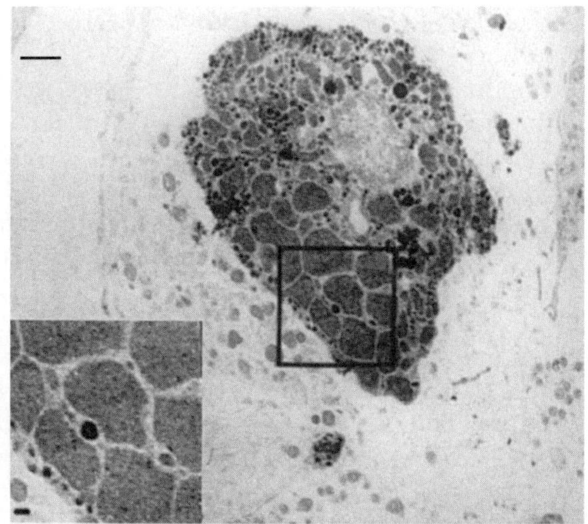

FIG. 15. Castration cell, 24 hours after surgery, showing expansion of rough endoplasmic reticulum. It is immunolabeled for LHβ as described in Fig. 10. The label for LHβ is in the rough endoplasmic reticulum (seen as large dense vacuoles in the field and the *inset*) and secretory granules. Bar = 1 μm. *Inset* shows a higher magnification of the rough endoplasmic reticulum and granules. Bar = 0. 4 μm.

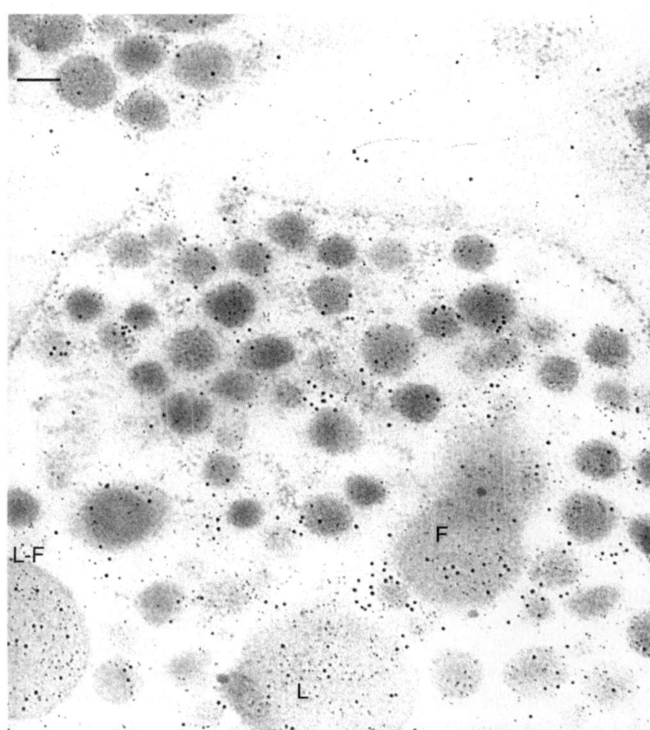

FIG. 16. An ovariectomy cell, 30 days postsurgery, dual labeled for LHβ (10 nm) and FSHβ (50 nm) with immunogold techniques. Some larger profiles contain both LH and FSH (L-F). Others contain mostly one of the hormones. Bar = 0. 5 μm.

from 70% in the normal animals to 91.6% in the 12- to 14-hour castrates.

All of the foregoing studies showed that castration cells were more numerous with time; however, questions about their origin arose. Did they come from dividing cells, or were there precursor cells that expanded and added to their numbers after this differentiation? This section briefly reviews the evidence for mitosis in castrated animals. A later section of this chapter focuses on proliferation and gonadotrope turnover.

By the early 1980s, most workers agreed that expanding castration cells included mitotic figures, which suggested that not all mature-appearing gonadotropes were postmitotic. Inoue and Kurosumi used autoradiography and immunolabeling and reported that as soon as 1 day after castration, there was a tenfold increase both in the number of cells labeled with tritiated thymidine and the number of mitotic LH cells (170). Smith and Keefer (171) reported increases in mitotic ovariectomy cells; however, they also described mitotic figures in small cells that did not label well for LH. This suggests the presence of small precursor cells that can divide and then differentiate to the mature castration cell. Sakuma et al. (172) also found mitotic cells in castrated rat pituitaries, although there were 1.4-fold more mitotic figures in cells that looked immature.

TABLE 1. *Changes in gonadotropes after castration (male)*

	Time after surgery							
Parameter tested	**0**	**12 h**	**24 h**	**1 wk**	**2 wk**	**4 wk**	**12 wk**	**24 wk**
LH ag*	10 ± 1	11 ± 2	14.4 ± 1.6		19 ± 2	25 ± 6	30 ± 2	22 ± 2
LH mRNA*	10.4 ± 1				19 ± 2	21 ± 1		
FSH ag*	8.8 ± 0.4	12.8 ± 1	15 ± 1		21 ± 2	26 ± 3	30 ± 3	21 ± 1
FSHβ mRNA*	10.5 ± 2			12.8 ± 3	21 ± 2	14 ± 2	20 ± 3	
Size changes								
Vv LH ag	0.05 ± 0.004	0.056 ± 0.01	0.067 ± 0.005					
Vv FSH ag	0.05 ± 0.003	0.067 ± 0.004	0.07 ± 0.004					
Area LH ag	133 ± 12	164 ± 4	198 ± 10			200 ± 2	280 ± 3	250 ± 2
Area LH mRNA	141 ± 4.5				319 ± 17	306 ± 16		
Area FSH ag	179 ± 14	162 ± 6	186 ± 7			225 ± 4	295 ± 2	250 ± 1
Area FSH mRNA	120 ± 10			480 ± 20	240 ± 6	300 ± 2	240 ± 2	

ag, antigen; FSH, follicle-stimulating hormone; LH, luteinizing hormone; mRNA, messenger RNA; Vv, volume fraction of cells with LH or FSH antigen; *, percentage of total pituitary cells shown in the top 4 rows.
Blank cells indicate that the parameter was not tested.

The mitotic index of LH cells rose five- to sixfold in both immature and mature animals, indicating that there is much potential for expansion of the population by mitosis even in the adult state.

Most of these studies of castration effects were focused on a period 24 hours to 3 months after the surgery. Two studies reported that the gonadotrope population continued to evolve later after castration. Kihara (173) showed that the large, expansive castration cell began to disappear after 3 months. Kihara reported that cells with only a few small secretory granules comprised 2% of the population at 3 months, but increased to 52.5% of the population by 18 months after surgery. This expansion was at the expense of the large castration cells, normally seen for up to 3 months after castration. Because the serum levels of LH remained high in spite of the major change in morphology, the author assumed that the morphology reflected abundant LH cells that were secreting LH too rapidly for a buildup of stores. The interpretation was that the cells were "immature." However, although they may resemble immature cells seen at birth, their secretory responses clearly do not support lack of maturity.

A study in our laboratory by Ibrahim et al. correlated morphometric measurements of immunolabeled LH or FSH cells at 1, 3, and 6 months after surgery (174) with serum levels. If LH and FSH were detected by immunolabeling, there was a two- to threefold increase in the percentages of LH and FSH cells 1 month after surgery in males, reaching as high as 30% of the population by 3 months after surgery. The average area of LH or FSH cells increased steadily to a peak at 3 months after surgery. At that point, it began to decline and was significantly decreased in the group taken at 6 months. This decline correlates well with the findings by Kihara (173), who reported a steady increase in the small LH cells along with a decrease in the larger castration cells, thus causing the overall decline in average area. Serum levels of LH followed the changes in area, although serum FSH levels continued to rise during the entire 6 months, in spite of the changes in the cells' size. Obviously, the high serum levels of FSH and LH can be maintained by morphologically distinct cell types; the typical "castration cell" morphology is not required for high secretory activity.

We continued these studies of castration cells, collaborating with Dr. William Chin and his laboratory to add the dimension of nonradioactive in situ hybridization for the detection of LH or FSH mRNAs. We correlated changes in expression of these transcripts with assays of pituitary transcripts. New biotinylated complementary RNA probes for LHβ or FSHβ mRNA were developed and the hybrid was detected by avidin–biotin peroxidase complexes (175,176). The morphometric data and cell counts (Table 1) were correlated with measurements of LH or FSH mRNA content. When cells with LHβ mRNA were counted, the percentages of labeled cells rose from 10% of anterior pituitary cells to nearly 25% 2 to 4 weeks after castration (175). There was also a twofold increase in cell area and label for mRNA per LH cell. This pattern correlated well with the steady increase in LHβ mRNA in the pituitary.

A biphasic pattern was seen when FSHβ mRNA was assayed (176). After a two- to fourfold rise 7 to 14 days after castration, there was a decline in FSHβ mRNA to near-normal levels at 1 month. This was followed by another rise to fourfold normal by 96 days. Analysis of the cells with FSHβ mRNA in intact rats showed that most (nearly 79%) were small, bringing the average area lower than that of cells with LHβ mRNA, LHβ antigens, or FSHβ antigens (Table 1). Seven days after castration, the average area of FSHβ mRNA–bearing cells increased fourfold, but there was no significant increase in percentage of FSH cells. Densitometric measurements showed a sixfold increase in levels of mRNA per cell. The percentages of cells with FSH mRNA began to rise later than those of cells bearing LHβ mRNA; they were increased twofold by 14 days after surgery (Table 1).

These morphometric data show that the early rise in mRNA is mainly due to the increased mRNA/cell and expansion in cell area. The decline in FSHβ mRNA assayed in the pituitary between 14 and 28 days could be explained by a drop in both the mRNA/cell and the percentages of FSH cells with mRNA. By the nadir point, most of the remaining cells with FSHβ mRNA were small. However, after this 1-month dip, there were twofold increases both in the number of FSH mRNA–bearing cells and the density of the label in the individual cells. These morphological correlates explained the increased FSH mRNA assayed at the latest time points (176).

Dual labeling for antigens for one gonadotropin and the mRNA for the other allowed quantification of multihormonal expression based on expression of transcripts for one of the gene products. This was important because the nonparallel changes in LH and FSH cell populations suggested that there were subpopulations of gonadotropes responding independently to the castration. In the earlier studies, when LH and FSH antigens were detected in dual immunolabeling, castration produced an increase to 91% bihormonal cells in 24 hours, with proportional decreases in the percentages of monohormonal cells (169). However, when the mRNA was used for one of the gonadotropins in the dual labeling, monohormonal cells producing only the mRNA appeared. The most important changes were seen when FSHβ mRNA was detected with LHβ antigens. The analysis revealed a selective expansion in monohormonal FSH cells. The proportion of monohormonal FSH cells increased from 23% to 37% of detectable gonadotropes 1 to 2 weeks after castration. The data thus correlate well with the morphometric data showing cryptic FSHβ-bearing cells that could be detected only by their content of mRNA. It also emphasizes the importance of independent subsets of monohormonal gonadotropes to future expansion.

In contrast, when FSHβ antigens were detected in dual labeling protocols with LHβ mRNA, the values were more comparable with those from the dual immunolabeling experiments (169), and over 90% of gonadotropes were bihormonal. At later times after castration (after 2 weeks), the proportion of monohormonal FSHβ mRNA–bearing gonadotropes returned to normal values (23%).

The dual labeling also showed variability in expression of each of the gene products. Figure 17 illustrates several castration cells from a male rat, 7 days postsurgery. Some are bihormonal and others are labeled only for FSHβ mRNA (black) or LHβ antigens (orange). The inset shows a large LHβ castration cell that is mostly monohormonal, based on the low level of FSHβ

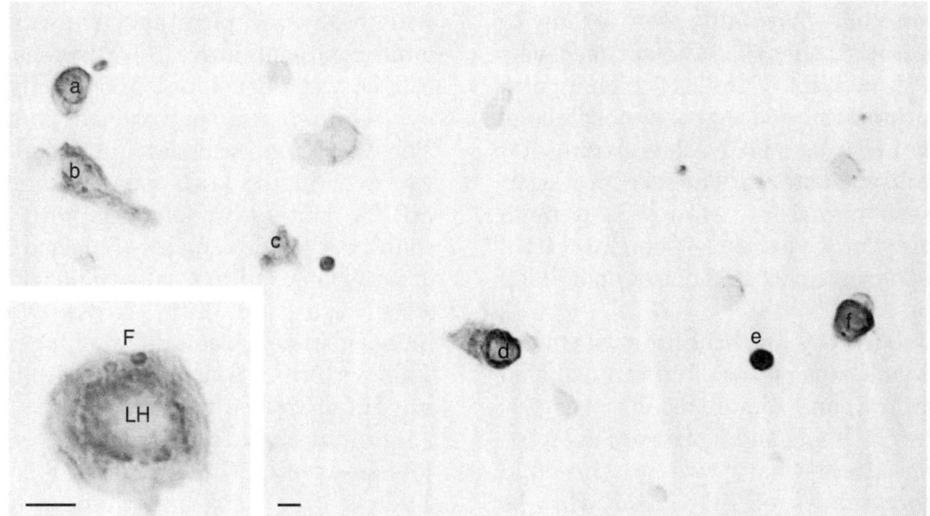

FIG. 17. Dual in situ hybridization for FSHβ mRNA (*black*) and LHβ antigens (orange amber or lighter gray in the black-and-white photograph) in a male rat, 7 days postcastration. The technique is described elsewhere (175,176); the probes were made in Dr. Chin's laboratory and are biotinylated cRNA probes. They were applied to freshly dissociated pituitary cells in monolayer culture and detected by the avidin–biotin complex (ABC) technique. Cells **a, d,** and **f** are bihormonal. Cells **b** and **c** contain only LHβ antigens, and cell **e** is a small, monohormonal FSH cell. *Inset* shows a castration cell that is filled with labeling for LHβ antigens, except for the region of the nucleus. There is one vacuole that has some label for FSHβ mRNA just below **f**; otherwise, it is mostly a monohormonal LH cell. In the black-and-white version of this color photograph, the light gray is the amber label (LH) and the very dense regions (note cells **a, d,** or **f**) are the black label for FSH mRNA. Bar = 10 μm. (See color insert.)

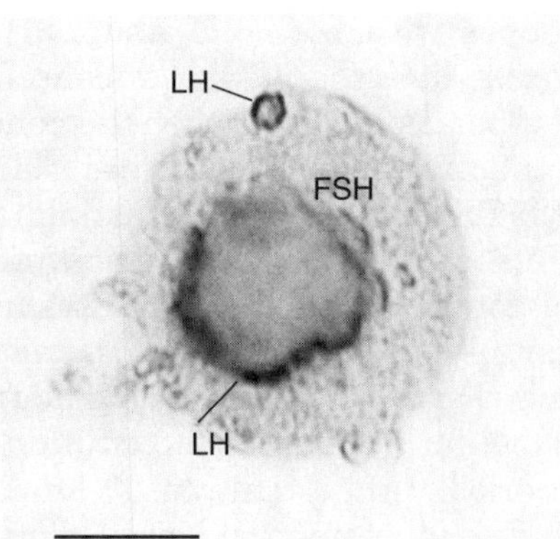

FIG. 18. Dual labeling for LHβ mRNA and FSHβ antigens in castration cell, 7 days postsurgery. The technique is described elsewhere (175,176) and in Fig. 17. The cell has prominent black labeling for LHβ mRNA on the nuclear envelope and in one vacuole, which is likely a profile of rough endoplasmic reticulum. Figures in Childs et al. (176) illustrate labeling at the electron microscopic level. Label for FSHβ mRNA is amber or gray in the black-and-white photograph scattered in granules or diffusely throughout the cytoplasm. Bar = 10 μm. (See color insert.)

mRNA expressed. One small vacuole appears to have some dark label for FSH mRNA. Figure 18 illustrates a dual-labeled bihormonal castration cell from a male rat, 35 days after surgery, with LHβ mRNA label focused around the nucleus and in one rough endoplasmic reticulum vacuole, whereas the FSHβ antigen labeling is scattered throughout the cytoplasm or over the nuclear envelope (rough endoplasmic reticulum). Recall that the dual gold labeling illustrated in Fig. 16 shows that the rough endoplasmic reticulum vacuoles may contain both LHβ and FSHβ, although some profiles are enriched in one of the gonadotropins. This variability is seen in the whole cells labeled for mRNA and antigens as well. Electron micrographs in Childs et al. (176) show labeling for the mRNA on profiles of rough endoplasmic reticulum.

Clearly, the bihormonal state is promoted by castration, although dual immunolabeling does not account for all of the gonadotropes present. Significant numbers of monohormonal cells might be missed in a given dual-labeling protocol that depends on immunolabeling for both antigens. Add to this the fact that the FSH cell population defined by antigen content does not change in parallel with that defined by mRNA content. This becomes strikingly obvious when one views the differences between the areas of FSH mRNA–bearing cells (which remain small) and those bearing FSH antigens, which grow larger (Table 1). The application of in situ hybridization revealed new populations of small gonadotropes that predominated when mRNA was detected. This may be a precursor group that would form the mitotic group contributing to the expansion in the population. As we will learn in studies of changes during the estrous cycle in the next section, such cells may also have functions in the normal intact animal.

Performance of Gonadotropes during the Estrous Cycle

Changes in Cell Shape and Gonadotropin Stores with Secretory Activity. The estrous cycle provides an excellent model to use for a study correlating gonadotrope structure and function. Early studies had shown changes in gonadotrope morphology suggesting that the gonadotropes became more irregularly shaped as they began their active proestrous secretory activity (124–126). This is illustrated in Figs. 5 and 7. Workers have suggested that the different cell types might be in different phases of secretory activity. This correlated well with the evidence from studies of exocytotic profiles that showed expansion of invaginating regions formed from the secreted granules, described in earlier sections of this chapter (153–157). Such regions could expand to become cellular processes, which would change both the shape and the juxtaposition to blood vessels. Figures 5 and 7 illustrate examples of this change in LH and FSH cells just before the proestrous surge. This is reviewed later in this chapter in sections that discuss sorting and trafficking.

More comprehensive studies correlated changes in LH cell shape and morphology with the stage and time of the cycle, as well as serum and pituitary LH. Blake (177) reported increased LH content on proestrus, with peak levels seen at 1400 hours. The LH cells were more irregularly shaped during the LH surge. After the surge (at 2200 hours), the number of granules were decreased. Early in estrus, degranulated LH cells were evident, storing only a few small granules. Our studies concur and an example is illustrated in Fig. 19.

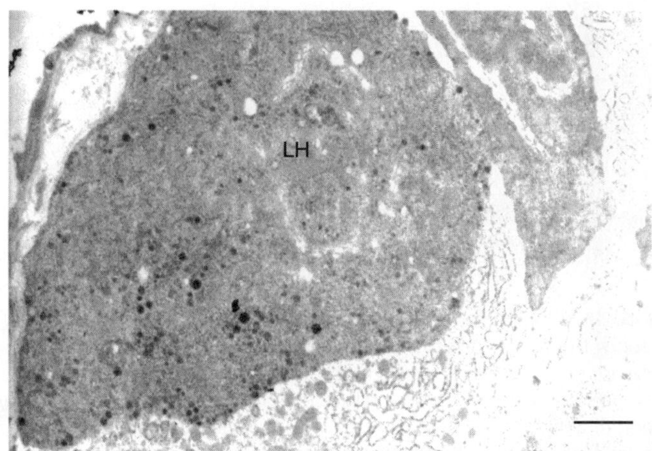

FIG. 19. Cell labeled for LHβ from an estrous rat, showing loss of granules. The remaining granules are small and not labeled intensely. Not all LH cells are this poorly granulated, however, this is a good example of one that has actively secreted its stores. Labeling is with anti-bovine LHβ and the peroxidase-antiperoxidase technique. Bar = 1 μm.

Yoshimura et al. (163) also correlated changes in optical density of label for LH with the cycle. They reported that percentages of intensely labeled LH cells increased at 1200 hours proestrus and decreased by 1200 hours estrus. They also showed the accumulation of small secretory granules near blood vessels by 1200 hours proestrus and the degranulation of these cells at estrus.

Studies since the late 1980s continued to correlate serum levels of LH and FSH with changes in storage granule content, cell size and shape, and expression of mRNA in rats (178–181) or sheep (133,134). Colloidal gold techniques detected changes in the sorting process (133,134,178). In situ hybridization detected changes in expression of transcripts for the β subunits (179–181).

A view of the changes during the estrous cycle in rats revealed that the preovulatory state was associated with an increased percentage of gonadotropes, overall, as well as an increase in the percentage of gonadotropes with both LH and FSH. The full bars in Fig. 20 show the results of counts of LH or FSH antigen-bearing cells during the estrous cycle. The rise in the LH and FSH cell types is nonparallel, with a rise in FSHβ-bearing cells lasting throughout the proestrous PM to support the estrous rise in FSH. Both sets of gonadotropes showed losses in numbers by estrus, largely because of their loss in stores.

Densitometric analyses revealed a gradual increase in area of label for LHβ, with a continuing increase during the period of the LH surge. The density of the LH label decreased during the surge, suggesting loss of stores as a result of secretion. In contrast, the area and density of label for FSH increased during preovulatory periods, but decreased with the proestrous and estrous surges in FSH (178).

After the surge, there were fewer gonadotropes of both types seen by immunolabeling (in estrus and metestrus). However, the gonadotropes did not simply disappear. Figure 21 shows the changes in percentages of gonadotropes bearing LHβ or FSHβ mRNA transcripts (overall percentages are shown by the

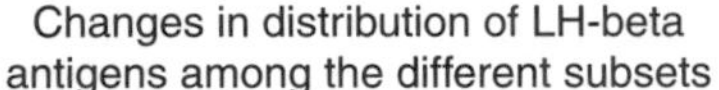

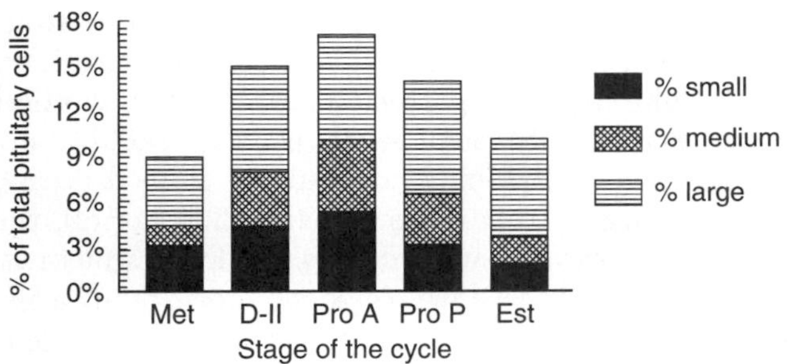

Changes in distribution of FSH-beta antigens among the different subsets

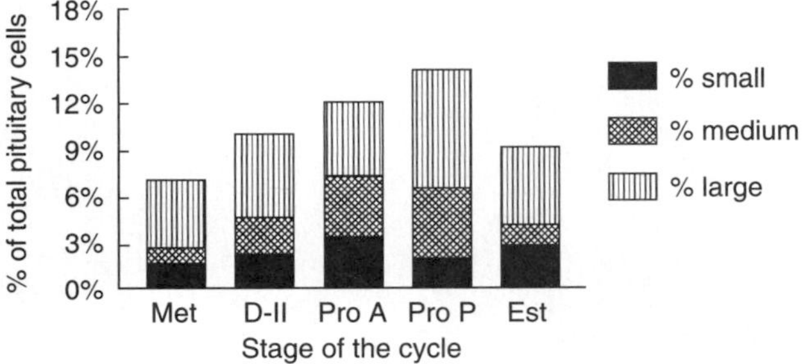

FIG. 20. Changes in percentages of cells with LHβ or FSHβ antigens during the cycle. The rats were taken at 10 A.M. on each day of the cycle except proestrus, when two groups were taken at 10 A.M. and 2 P.M. Dispersed cells were plated and fixed within 1 hour of plating for detection of antigens or mRNA. The full bars graph the changes in percentages. Metestrous and estrous values are the lowest (p <0.05). There is a rise in percentage of each type of gonadotrope, with a peak during diestrus and proestrus (LH) or proestrous P.M. (FSH). Note that the curves for each hormone are not parallel. The sections in each bar are the data from the elutriation fractions showing the distribution of total gonadotropes into small, medium, or large subsets of gonadotropes.

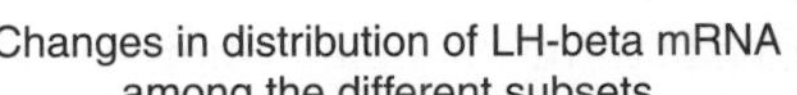

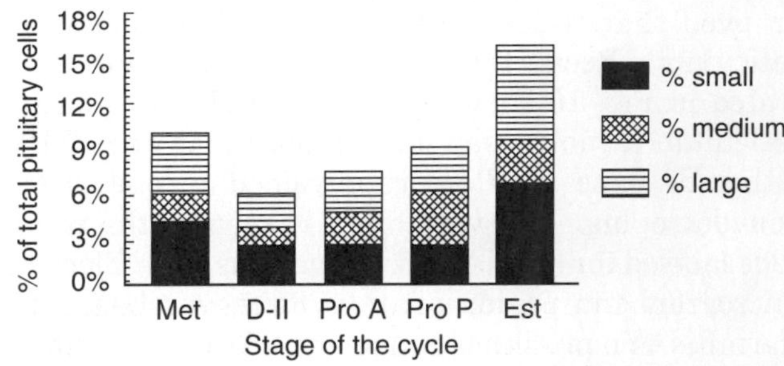

Changes in distribution of FSH-beta mRNA among the different subsets

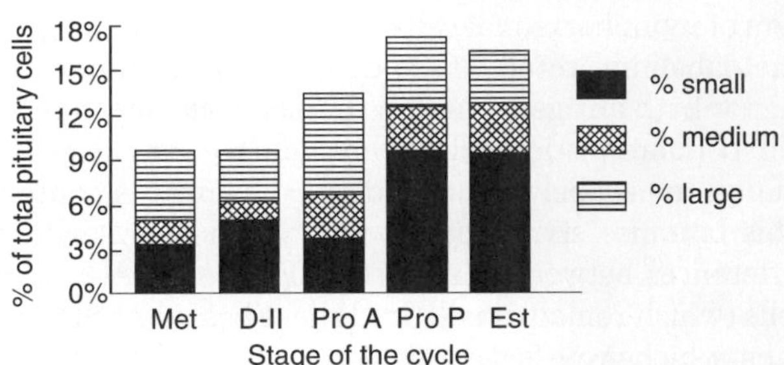

FIG. 21. Counts of cells from the experiment in Fig. 20, except they were labeled for LHβ or FSHβ mRNA. Note that peak expression (*full bars*) for LH is in estrus and that for FSH is in proestrus and estrus. The sections in the bars represent the percentages of gonadotropes that were small, medium, or large. The greatest increases in mRNA in both populations appear to be in the small subsets.

full bars). Because peak transcription is detected during the LH or FSH surges, the analysis of gonadotropes bearing mRNA showed that they could still be detected by their content of LHβ or FSHβ mRNA (179,181) late in proestrus and early in estrus, extending into metestrus. These are times when they are depleted of gonadotropin stores.

Figure 20 shows that the protein appears again in the cells by diestrus as the cells reduce transcription. This coincides with a temporary reduction in cells with mRNAs at this point in the cycle. Transcription and translation are out of phase, with peak levels of transcripts seen during secretion and the protein products replenished 1 to 2 days later (during diestrus).

More recent studies by Currie and McNeilly (133) reported striking changes in LH cells during the cycle in sheep. They enlarged and became more polarized, with granules concentrated next to a sinusoid. As the sheep approached the LH surge, this polarization increased from 20% of LH cells during the luteal phase to 90% of the LH cells during the middle of the LH surge. LHβ antigens could be detected in rough endoplasmic reticulum after the surge, as the cells became degranulated. Currie and McNeilly also reported the preferential exocytosis of small (130- to 150-nm) granules from LH cells. Unlike in the rat, LHβ mRNA levels were not increased during the preovulatory LH surge. The cells appear more synchronized in releasing their hormone, although the evidence for labeling in the rough endoplasmic reticulum after the surge suggests that synthesis is occurring.

Follow-up studies in this same laboratory focused on LH gonadotropes after a GnRH-induced LH surge (182). They reported depletion of granules in irregularly shaped gonadotropes 24 hours after the surge, with colloidal gold labeling for LH restricted to vesicles and presumptive endoplasmic reticulum elements. Whereas the levels of mRNA remained constant, the numbers of secretory granules increased. Initially, as the granules were repopulating, they appeared in the polarized state (mostly localized along the plasma membrane); however, with time the cell again filled with secretory granules. These morphological data were correlated with a return to pulsatile secretion as the granule population was stored. Crawford et al. (183) suggested that the polarized granule locus near blood vessels provided an optimal site from which to support the release of pulses of LH. This may be analogous to the sites of compound exocytosis profiles that form in the rat after a pulse of GnRH (153–157), and the marginalization of granules seen in the normal or hypogonadal mouse (158–160) (Fig. 8).

Differential Sorting and Trafficking of LH and FSH, and the Role of Granins. The general consensus that arose from investigations of gonadotrope morphology during the cycle was that cells changed shape and size and degree of granulation as they were stimulated to store or secrete. The presence of granules of mixed sizes in the type I gonadotrope (Fig. 4) has remained enigmatic, however. Our early studies had reported that this morphology predominated during times of active translation of gonadotropins, a day before the surge in secretory activity, suggesting that it was related to a state of synthesis and storage. In contrast, actively secreting gonadotropes tended to have more uniform granules that were distributed in the periphery of processes (124–126,134,159,160,182) (Figs. 5 and 7). Yoshimura et al. (163) found that smaller granules were more abundant in LH cells as they entered proestrus; larger granules appeared in estrous animals. They suggested that the smaller granules might be a conduit for LH during the surge. Other studies concur with this finding (134,159,182).

In agreement with our studies, Yoshimura et al. reported that, in diestrous rats, the range in size was mixed and somewhat intermediate (120 to 220 nm) (163). They also correlated granule volume and number with pituitary and serum LH and FSH. Not surprisingly, granule volume was highest before the proestrous surge (on the proestrous PM) and lowest in estrus. After the surge, there may be more small granules, (124–126,177), which could signify depletion of hormone as well as selective storage.

During the cycle, an analysis of storage patterns in our laboratory revealed a significant increase in bihormonal gonadotropes as the population approached surge secretion period, from 45% of all gonadotropes (detected by immunolabeling) to nearly 70% by early proestrus (125). These early findings, based on dual immunolabeling alone, led to the recognition that such cells must initiate sorting mechanisms to effect nonparallel secretion of FSH during the estrous surge. Dual colloidal gold techniques, correlated with serum levels of LH and FSH, proved valuable in the quantification of stores of LH and FSH in granules of gonadotropes (178). When fields from rats in diestrus were compared with those in proestrus, there was a significant three- to fourfold rise in the percentage of monohormonal LH or FSH granules in bihormonal gonadotropes (Fig. 22). The first drop in monohormonal LH granules then occurred during the LH surge (1600 hours) and a later drop in monohormonal FSH granules was seen during its rise at 1900 hours, followed by another decline evident at 0400 hours during the second FSH rise. These data suggest that sorting factors may selectively drive packaging of LH into separate granules to support the LH surge, which is steeper than the proestrous rise in FSH. This is followed by a selective loss,

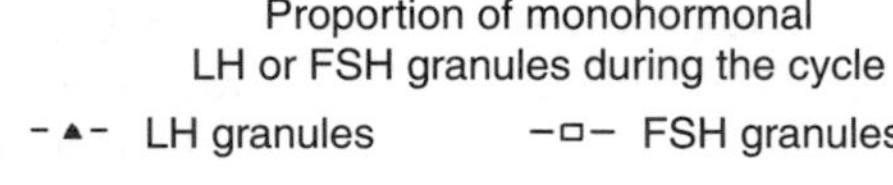

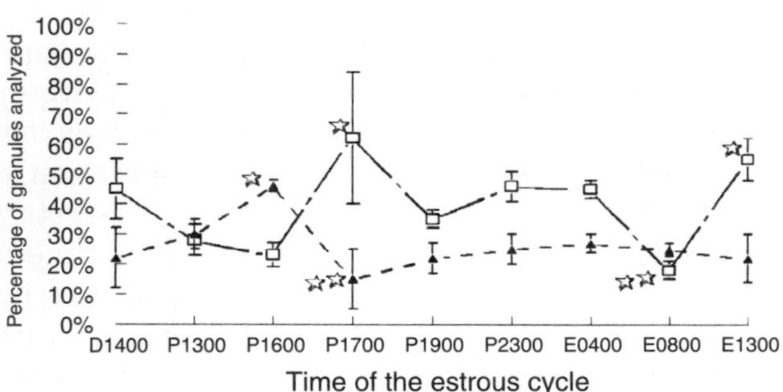

FIG. 22. Dual immunogold applied to ultrathin sections from rats at different times of the estrous cycle. Techniques are described in Childs et al. (178). The counts of monohormonal granules showed that percentages increased just before active secretory activity (LH surge, FSH rise). They fell after the surge secretion was completed.

indicating further sorting possibly at the level of the plasma membrane. Similarly, the striking increase in large monohormonal FSH granules seen during estrus suggested a special sorting mechanism for this hormone to support its surge during early estrus. In particular, the increase in proportion of large granules agrees with the earlier studies by Yoshimura et al. (163), who reported more large granules early in estrus. Figures 9 and 23 illustrate monohormonal packaging in the Golgi complex (Fig. 9) or in the secretory granules (Fig. 23).

Thomas and Clarke used ovariectomized ewes treated with estrogen as a model to study differential sorting of LH and FSH granules detected by immunolabeling and confocal microscopy (134). Their model is characterized by a biphasic effect of estrogen on the gonadectomy cells. It first produces a decrease in secretion and then it produces an increase, like that of a preovulatory surge. At the same time, it suppresses FSH secretion. This provided a way to look at trafficking of one set of granules, correlating it with release of LH or FSH. In vehicle controls, nearly 84% of the gonadotropes had LH granules mostly distributed throughout the cytoplasm, and only 16% showed movement to the periphery. FSHβ immunoreactivity appeared diffusely throughout the cell. In the estradiol-injected animals, only the LHβ-containing granules moved to the periphery, and after 16 hours, over 84% of the gonadotropes showed this peripheral distribution of LHβ granules, without movement of the FSH stores. These results provide excellent morphological correlates for nonparallel release from bihormonal gonadotropes.

This is an area for state-of-the-art cell biological studies, especially in light of what we know about sorting and targeting of vesicles in cells. Figures 9 and 23 illustrate gonadotropes in the process of sorting LH and FSH stores in Golgi regions. What factors regulate this sorting?

Granins are acidic proteins that have been investigated for possible roles in sorting and trafficking

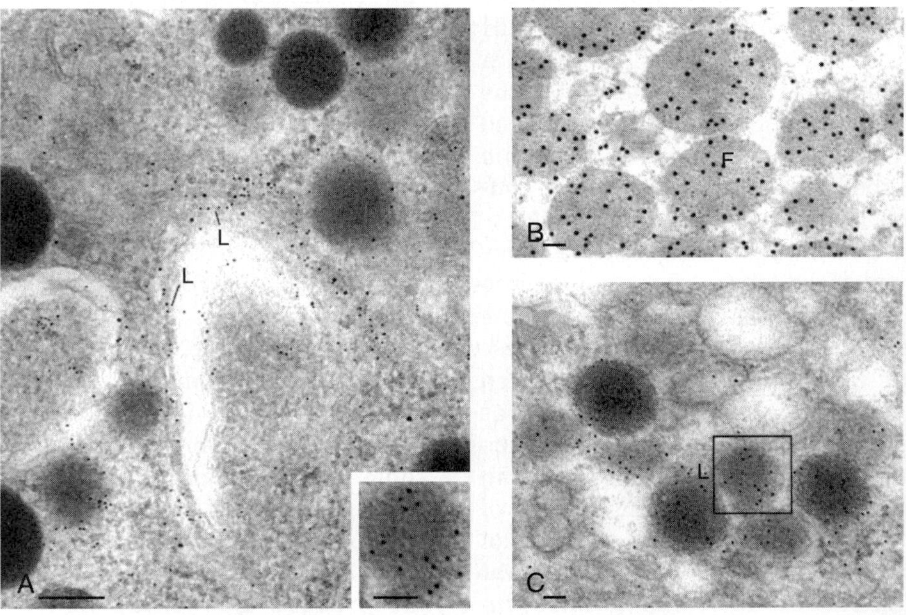

FIG. 23. A: Production of monohormonal granules in regions of the Golgi complex labeled for LHβ (10 nm gold) and FSHβ (50 nm gold). Label is aligned in linear arrays in the cisternae. **B:** Group of monohormonal FSHβ-bearing granules that are prominent early in estrous, just before the FSH rise. They are larger than their counterparts in other stages of the cycle. **C:** Cluster of small, monohormonal granules labeled for LHβ. They are difficult to visualize, so the box delineates an area that is magnified in an *inset* in **A. A,** *inset*, and **C:** bar = 0. 1 μm; **B:** bar = 0. 35 μm.

(182–195), especially in gonadotropes. Secretogranin II (SgII) expression also varies with the cycle, estrogen treatment, GnRH treatment, and age (183–186). SgIIp is a unique form that was isolated from female rat pituitaries and also found to be released by GnRH (187).

Watanabe et al. (188–190) reported that, in male rats, SgII was coexpressed in small secretory granules with LH, whereas chromogranin A (CgA) was found in larger granules with FSH. The same distribution was not seen in the female; dual colloidal gold studies in our laboratory (178) would concur that, just before the LH surge, small monohormonal granules are abundant and, just before the rise in FSH, large monohormonal FSH granules are frequent (Fig. 23). In the studies by Watanabe et al., however, some granules contained only one of the granins, and others contained both CgA and SgII, which could correlate with the bihormonal granules seen in our studies. The size of the CgA-bearing granules did vary with the stage of the cycle, but the significance of this was not apparent. One interesting finding was that an intermediate type of granule appeared just before ovulation or after stimulation in male rats. This granule type contained LHβ and SgII labeling in the center and CgA labeling in the outer regions. The significance of the CgA could relate to interactions with second messenger products, and this is discussed in more detail later in this section.

Wei et al. (191) measured SgII release in a reverse hemolytic plaque assay (RHPA), correlating it with LH labeling. RHPA is a complement-mediated reaction using protein-A–coated sheep red blood cells that allows the detection of a hormone-secreting cell by adding antibodies to the hormone in question. These antibodies bind the hormone around the secreting cell, the complex activates complement, and the blood cells around the secreting cell are lysed. Thus, a patch of lysis forms around an individual LH cell that is proportional to the amount of hormone secreted.

SgII release was found in all plaques that contained LH cells and was highest at proestrus. All plaques with LH cells from proestrous rats secreted SgII. In contrast, only some of the LH cells secreted SgII in other stages of the cycle. GnRH also strongly stimulated more SgII release in these plaque assays.

Crawford et al. (192) reported enhanced coaggregation of LHβ with SgII in mice that had been inhibited by an injection of antiserum to GnRH at 12-hour intervals for 2 days. When they replaced the GnRH with exogenous buserelin (a potent GnRH agonist), the numbers of these dual-labeled granules were reduced as the LH was secreted (assayed in serum). This suggested that the SgII was sorted with LHβ in a regulated secretory pathway and available for secretion by the buserelin stimulation. Those LH granules that did not contain SgII did not change with the stimulation, suggesting they may be a source for constitutive (basal) secretion.

Kakar et al. studied the regulated expression of SgII mRNA throughout the estrous cycle of rats, comparing it with that of LHβ mRNA (193). SgII mRNA decreases as the rat approaches mid-cycle and as LHβ mRNA reaches a peak at mid-cycle. Thus, the expression of the two mRNAs appears out of phase. This likely relates to the fact that there is a 1- to 2-day lag period from LHβ mRNA synthesis to peak LHβ translation and storage. Thus, SgII mRNA is likely not transcribed and translated until it is needed during diestrus, a time when LHβ mRNA is at a nadir, but LHβ stores are at a peak. These same studies showed parallel decreases in LHβ mRNA and SgII mRNA after ovariectomy or treatment with a GnRH antagonist. These data suggest that SgII mRNA may be regulated by estrogens, which rise during diestrus.

Crawford and McNeilly (194) presented a comprehensive study of changing expression of LHβ and FSHβ mRNAs, serum LH and FSH, and coexpression of LHβ and FSHβ proteins with granins during the estrous cycle of sheep. They reported losses in monohormonal granules as well as bihormonal granules after the proestrous surge. Granules with LHβ tended to label for SgII and those with FSHβ tended to label for CgA, although there was some crossover. In contrast to the results in the rodent (193), neither CgA nor SgII mRNAs changed with the stage of the cycle. Also, changes in gonadotropin mRNAs appeared mostly to reflect improved stability of the mRNAs.

Thus, whereas the granins appear to be synthesized and sorted with LH and FSH granules and are under regulatory controls by estrogens or GnRH, their exact roles are uncertain. Studies by Hosaka et al. (195) have identified a sorting sequence for CgA, which may include binding to another granin found in large secretory granules, secretogranin III (SgIII). When they mutated the binding domain of CgA, the protein was mis-sorted to the constitutive pathway. They cited evidence by Yoo et al. (196,197) showing that CgA binds to the inositol 1,4,5-trisphosphate (IP_3) receptor as well as vesicle matrix proteins. The IP_3 receptor may be needed for secretion of a particular granule or set of granules and, hence, CgA may be part of a mechanism that facilitates, or prevents the secretion of FSH granules. Its detection by Watanabe et al. on the periphery of LH-containing granules may signify a site available to interact with this receptor and other proteins (188,189).

Other sorting and trafficking proteins have been identified and investigated in pituitary cells to learn if their distribution changes in parallel with specific sets of pituitary hormone–bearing granules. Rab3a is a guanosine triphosphate (GTP)–binding protein important in targeting and fusion of vesicles and granules. Davidson et al. (198) tested the effect of a synthetic peptide related to the rab effector domain on secretion of LH in ovine pituitary cells. The synthetic 18–amino acid peptide inhibited LH exocytosis that had been stimulated by both calcium-dependent and calcium-independent pathways. It also inhibited secretion stimulated by phorbol myristate acetate plus cyclic adenosine monophosphate (cAMP). Removing only three amino acids from this peptide abolished its activity. These data suggest that LH granules are sorted and targeted by rab3 proteins (198). Thomas et al. used the ovariectomized rat model, treated with estradiol (199), to learn if there were parallel changes in distribution of sorting and trafficking proteins associated with a regulated secretory pathway. This comprehensive study found that both gonadotropes and somatotropes expressed synaptosomal-associated protein (SNAP)-25, vesicle-associated membrane protein (VAMP)-2, rab3A, Munc 18-1, α/β-SNAP, cysteine string protein (csp), and SgII. Estrogen caused the expected migration of granules to the periphery [described earlier (134)], but there were no well-defined changes in any of the exocytotic proteins. Only SgII appeared to migrate with the LH-bearing secretory granules. Rab3A may have migrated with LH granules in some cells, although the pattern was not as obvious as that of SgII. Csps are lipidated synaptic vesicle proteins that may be involved in the uncoating of clathrin-coated endocytotic vesicles. There was a tendency for csp to be coexpressed with LH-bearing granules. These studies show that the pituitary contains many of the proteins known to regulate sorting, targeting and trafficking. However, the experimental model did not show major changes associated with estrogen stimulation. It would be interesting to look for potential changes in normal gonadotropes after the stimulation of different signaling pathways, including those activated by GnRH.

Purified fractions of gonadotropes could provide a valuable cellular model for a study of differential trafficking of gonadotropins. The challenge, of course, is to learn what pathways are active that might drive selective sorting of LH and FSH in bihormonal gonadotropes. Today, we know that rapid pulses of GnRH (15 minutes every hour) favor LH synthesis and secretion, whereas slow, lower-amplitude pulses of GnRH (15 minutes every 3 hours) favor FSH synthesis and secretion (200,201). Therefore, this question could be addressed with cells exposed to different pulse amplitudes and frequencies. A study from our laboratory provides support for this hypothesis. Figure 24 graphs the changes in the proportion of monohormonal LH and FSH granules after a pulse of GnRH (10 and 60 minutes). After 10 minute, there is a loss in LH granules and a rise in FSH-bearing

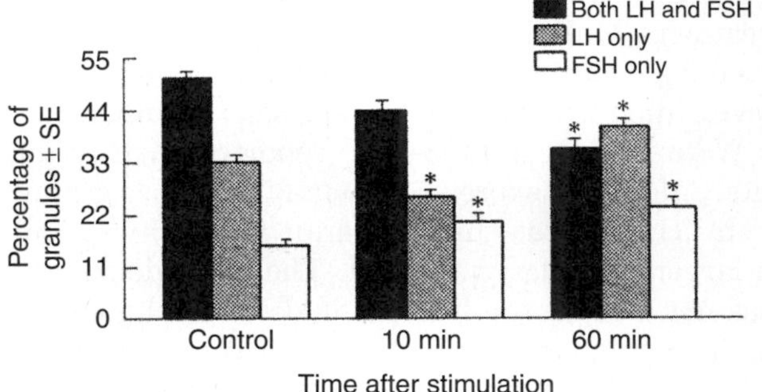

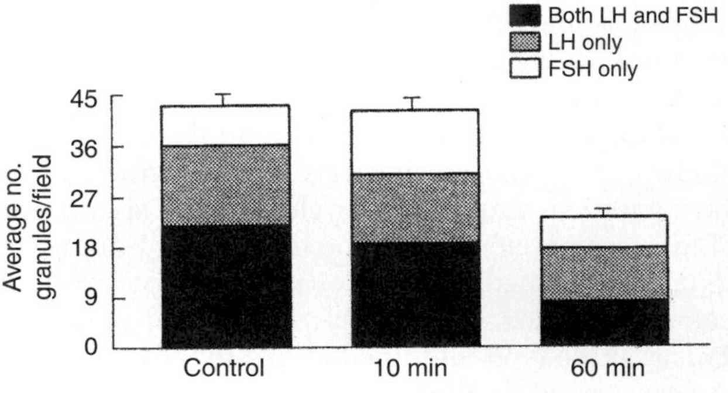

FIG. 24. Pituitary cells were dispersed and stimulated for 10 or 60 min in 1 nM GnRH. They were then fixed and embedded for electron microscopy and labeled for LH and FSH beta subunits with immunogold techniques, as described in Childs et al. (178). Counts of granules containing only LH and FSH (monohormonal) or both gonadotropins (bihormonal) were done on the first 20 gonadotropes/section × 5 sections per group; the smaller (10 nm) gold identified LH and the larger (15 nm gold) identified FSH. There is a reduction in monohormonal LH granules after 10 min as the proportion of monohormonal FSH granules increases, suggesting preferential secretion of LH from this group. After 60 min, the overall number of granules is reduced along with the bihormonal subtype and monohormonal FSH. More of the remaining granules are classified as monohormonal LH.

granules, with no changes in the bihormonal subtype. After 60 minutes, there is a significant overall loss in granules, but a rise in the proportion of LH granules and a loss in the bihormonal subtype, as if the treatment favored the sorting of LH.

In Vitro Studies: Differential Responses to GnRH by Subsets of Gonadotropes

In vitro studies provide another way to investigate the significance of the different morphologies and storage patterns. Tougard and Tixier-Vidal traced the fate of cultured gonadotropes, showing that cells storing LHβ persisted long after cell culture was initiated compared with those storing FSHβ (202,203). The cells that persisted had few small secretory granules, sparse endoplasmic reticulum, and a small

Golgi complex. Denef et al. (204) added a new dimension by separating gonadotropes from 14-day-old females into fractions containing small, medium, or large cells. They then correlated cell size with different secretory patterns in response to GnRH. The largest cells secreted more FSH than LH after exposure to GnRH. More LH than FSH was secreted from the fractions with the smallest cells. When hormone stores were detected, over 70% of the hormones were recovered from large cells, yet significant amounts were still found in medium-sized gonadotropes. Each fraction appeared to contain subsets of gonadotropes that responded to GnRH and steroids in a different pattern in terms of the amounts and proportions of LH and FSH. This suggested that not all gonadotropes had equal secretory potential.

This was reinforced by pioneering studies by Smith et al. (205) using RHPA and antisera to LH. As described previously, the plaque area of lysis around the secreting cell that forms in this complement-mediated reaction is proportional to the amount of hormone secreted. The assays allow the correlation of synthesis, storage, secretion, and receptivity in the same cells.

This assay was applied to learn more about secretory patterns of LH gonadotropes throughout the estrous cycle or in response to GnRH. Smith et al. (205) reported that, under basal conditions, only a few small LH plaques were detected. GnRH (100 nM) added to the assay stimulated a 10-fold increase in the size and number of plaques. When they compared GnRH treated cells from diestrous rats that stored LH with those that formed plaques, they discovered that only half of the identified LH cells were secretory. However, the same comparison in proestrous rats showed that nearly all LH cells were secretory. They then showed that overnight estrogen treatment of cultures from diestrous rats increased the secretory LH cells to levels nearly as high as those from proestrous animals. Collectively, these studies were the first to show heterogeneous secretory responses among individual gonadotropes.

Our laboratory collaborated with a group from Dr. Catt's laboratory to learn more about gonadotrope subsets from females rats (mixed cycles) separated by counterflow centrifugation in an elutriator (206). Fractions containing the largest cells also had the most abundant gonadotropes (a fivefold enrichment over the starting populations). As expected, they responded well to GnRH. However, fractions containing fewer gonadotropes (small and medium-sized cells) responded surprisingly well to GnRH, suggesting that there were different subsets that could be discriminated on the basis of secretory responsiveness. Figure 25 illustrates the changes in percentages of gonadotropes in each of these fractions and the relative changes in the monohormonal and bihormonal

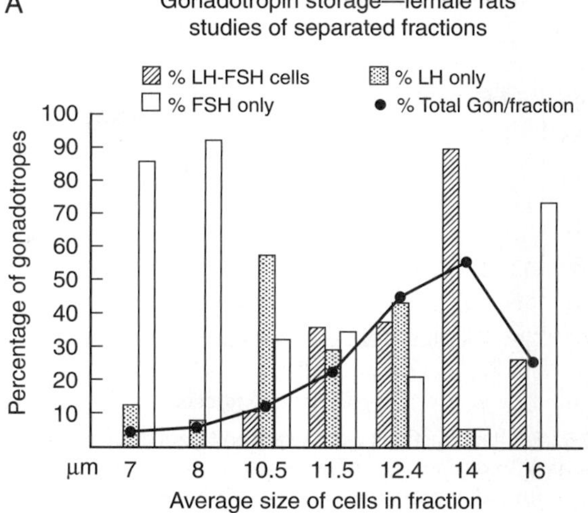

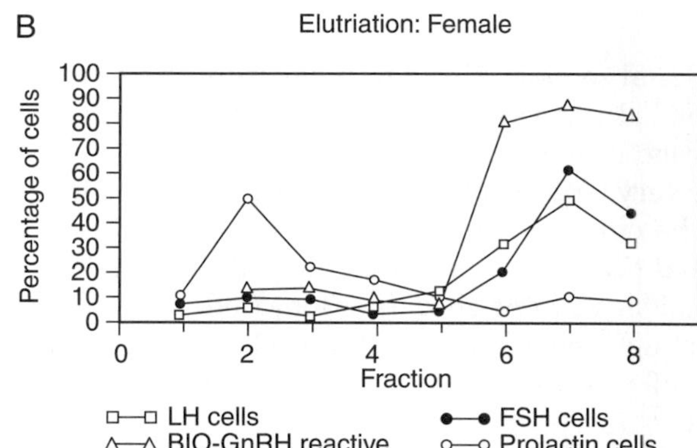

FIG. 25. Pituitaries from 50 female rats (mixed cycles) were dispersed and separated by size and density in the standard elutriator chamber. Cells were plated and fixed for dual immunolabeling for LHβ and FSHβ with avidin–biotin complex immunoperoxidase techniques. **A:** Counts of bihormonal cells showed that gonadotrope-enriched fractions (average diameter, 12–14 μm) contained the most bihormonal cells. **B:** Comparison of the overall percentages of gonadotropins, prolactin, and target cells for biotinylated gonadotropin-releasing hormone (GnRH) in the same fractions. The small cells are enriched in lactotropes. The largest cells are enriched in gonadotropes.

subtypes. Figure 25b shows the relative distribution of prolactin cells and gonadotropes defined by their content of LH, FSH antigens, or GnRH binding sites. Figure 26 graphs differential secretory patterns exhibited by these same fractions, demonstrating a correlation between the amount of secreted LH and the degree of enrichment in gonadotropes.

A morphological analysis showed that the largest fractions contained the well-developed type I and II gonadotropes (more of the former) (207). Medium-sized fractions also contained both types, although there were more of the type II subset. These were the fractions that were more responsive to GnRH. The smallest fractions looked most like type III gonadotropes, or even the immature forms of the cells seen in neonatal rats (164). These three types are illustrated in Figs. 27 to 29.

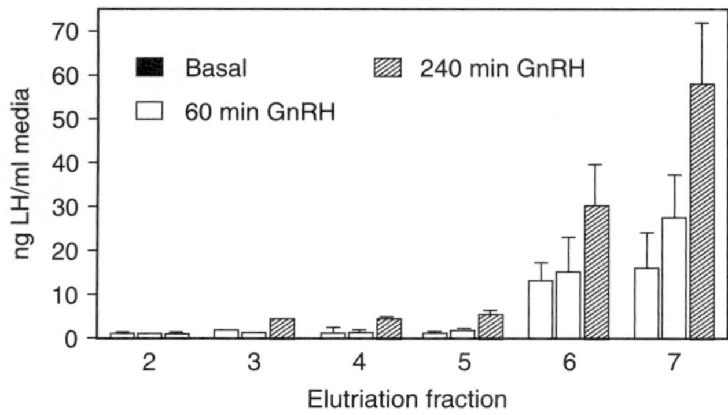

FIG. 26. Pituitary fractions were exposed to gonadotropin-releasing hormone (GnRH) for 60 minutes or 4 hours and media were collected for radioimmunoassay for LH. The largest cells secrete the most LH, basally and after GnRH stimulation.

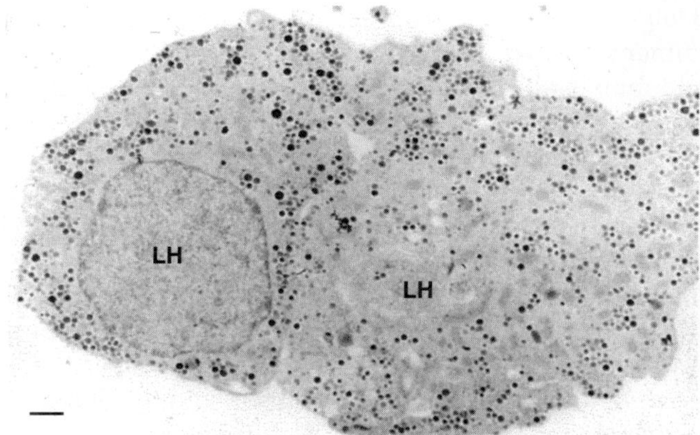

FIG. 28. Two gonadotropes from a medium-sized fraction, immunolabeled for LH as in Fig. 27. These cells are well granulated and may be classified as type I or II. Bar = 0. 4 μm.

Serial sections through fields from these fractions provided important information about the hormone storage properties of these gonadotropes (Fig. 25a). The very largest cells, found in the fraction remaining in the chamber, tended to be enriched in monohormonal FSH cells. The moderately large gonadotropes, found in the enriched fraction, had 78% cells with both LH and FSH and 10% to 12% cells with only one of the hormones. As the gonadotrope size was reduced, their bihormonal storage patterns changed. Only 15.2% of medium-sized cells contained both LH and FSH, leaving over 30% to 40% storing only LH and 40% to 50% storing only FSH. The fractions with the smallest cells were mostly monohormonal (207).

Thus, there was a clear correlation between size and bihormonal state. However, these cells were from a mixed population of cycling rats, and therefore it

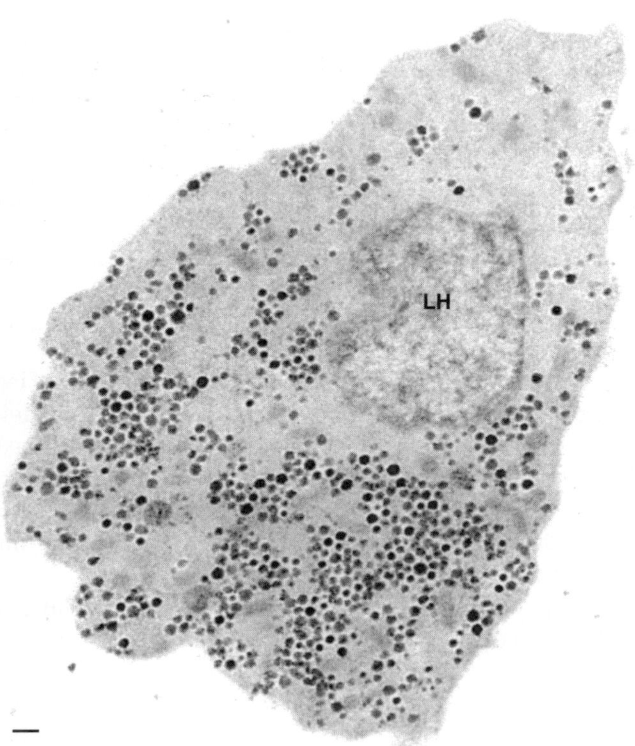

FIG. 27. Gonadotrope from largest cell fractions, immunolabeled for LHβ with anti-bovine LHβ and the avidin–biotin peroxidase technique. The cell has some large granules, but most of the granules are approximately 200 to 250 nm diameter. Bar = 0. 4 μm.

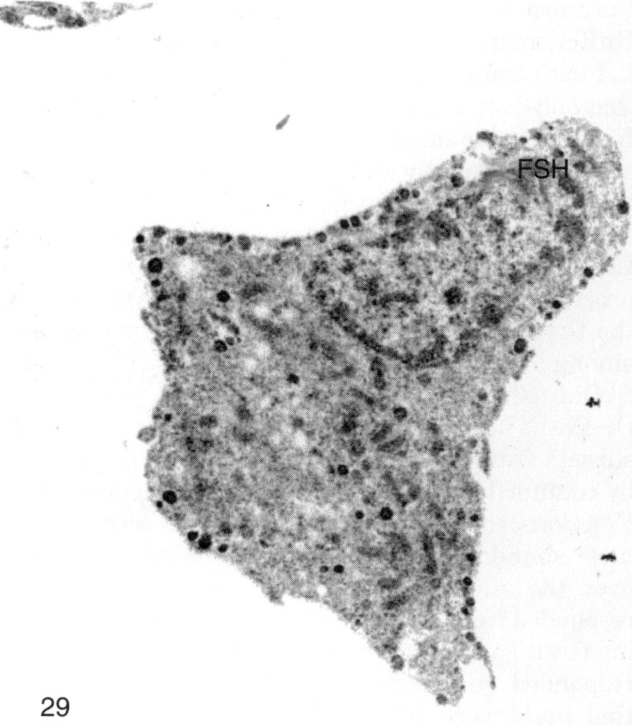

FIG. 29. Gonadotrope from the fraction containing the smallest cells, immunolabeled for FSHβ. It is sparsely granulated and resembles a type III gonadotrope. Bar = 0. 4 μm.

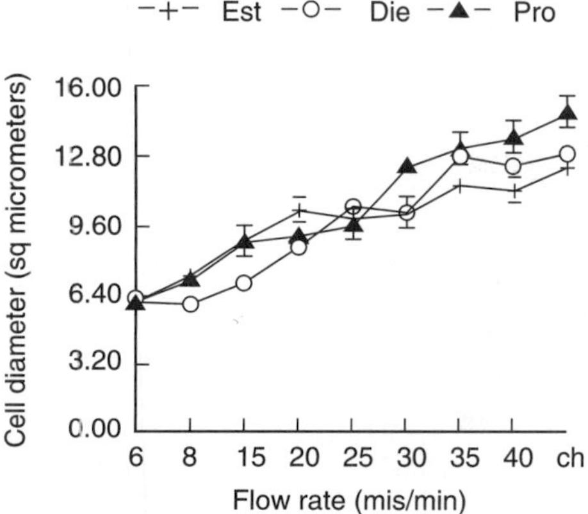

FIG. 30. Pituitaries from four to six rats in different stages of the estrous cycle were dispersed and separated by size or density by centrifugal elutriation with the Sanderson chamber. Cell areas were measured by image analysis, and plots of the average area in each fraction are shown. Note that the largest gonadotrope-enriched fractions from proestrous rats have, on average, significantly larger cells.

was possible that we were seeing differences based on the stages of the estrous cycle represented in the rat population. One might postulate that the largest cells contained well-developed gonadotropes from proestrous animals, for example. This could be proved only by a careful study of gonadotrope storage properties and morphology during the cycle, as well as the application of these cell separation techniques to homogeneous populations of pituitary cells from rats in one stage of the cycle. Figure 30 shows the correlation between cell size and fraction in populations of cells from rats in one stage of the cycle. Note that the proestrous female rat population had larger cells, possibly reflecting the increasing size or density of LH and FSH cells.

Lloyd et al. (208) from our laboratory designed experiments to learn if estrogen or GnRH effected conversion of monohormonal to bihormonal gonadotropes. Small cells from estrous rats (from the merged fractions 1 through 3—flow rates 6, 8, and 15 mls/min seen in Fig. 30) increased their proportion of bihormonal gonadotropes in response to GnRH alone, or with estrogen (100 pM; Fig. 31). This illustrates their potential for further maturation or differentiation. However, their responses depended on the estrous cycle stage of origin, indicating the need for a state of readiness for such stimulation.

Bihormonal large gonadotropes (taken from fractions 6 through 8—flow rates 30–40 mls/min, Fig. 30) from estrous and diestrous rats were clearly responsive to GnRH, and the diestrous group also showed increased bihormonal cells after estrogen alone (Fig. 32). Only estrogen and GnRH together increased bihormonal cells in the proestrous group. It had no effect on large cells from diestrous rats.

To summarize, by the early 1990s, gonadotrope heterogeneity in terms of size had several physiological correlates. First, small gonadotropes appeared during states of immaturity and they are largely monohormonal in the intact rat. They may be associated with high secretory activity months after castration; however, in the intact adult, they are not highly responsive when GnRH is given (Fig. 26). They respond to GnRH and change storage patterns from a monohormonal to a bihormonal state, which shows their potential. Large and medium-sized gonadotropes appear to be a secretory subset, as seen both from morphology (Fig. 27 and 28) and responses in separated fractions (Fig. 26). These subsets may also contain monohormonal gonadotropes, which suggests that the monohormonal state is not necessarily immature.

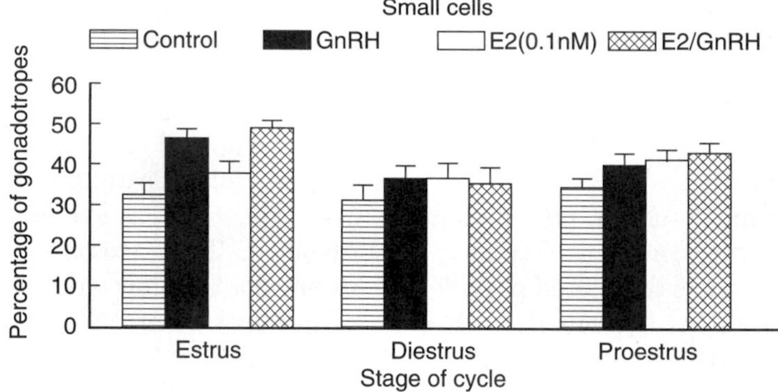

FIG. 31. Pituitaries from groups of rats described in Fig. 30 were separated, and the three smallest fractions were collected and plated. They were then treated for 3 days in 1 nM gonadotropin-releasing hormone (GnRH) with and without 0.1 nM estradiol. Most cells in this fraction were monohormonal. GnRH and estrogen plus GnRH increased the proportion of small gonadotropes that were bihormonal, but only in the populations from estrous animals.

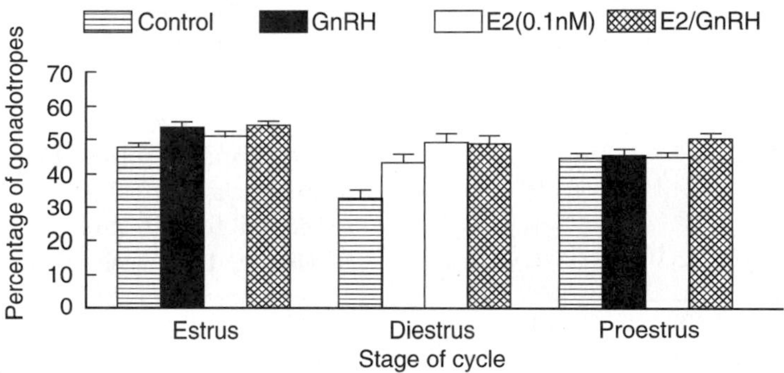

FIG. 32. Large cell fractions collected and treated as in Fig. 31. Only populations from diestrous animals showed an increase in the proportion of bihormonal gonadotropes with the different treatments.

Role of Matrix Proteins and Gap Junction Proteins

Gonadotropes can be stimulated to produce components of their matrix and secrete it in the pathway used by the Golgi complex and secretory granules. Vila-Porcile et al. (209–211) reported that laminin was one important matrix protein produced by gonadotropes and was found in secretory granules with LH (209). Other important matrix components included heparin sulfate proteoglycan, entactin, and type IV collagen. Expression of integrins that react with laminin may stimulate local differentiation into one or more gonadotrope subsets.

Gonadotropes also produce regulatory proteins that function in the matrix, matrix metalloproteinases (MMPs) 2 and 9 (212). These MMPs (called gelatinases) cause the release of EGF-like ligands bound in transmembrane precursor proteins. Such ligands activate EGF receptors (EGFRs) and, as discussed later in this section, EGF potentiates stimulatory actions of GnRH. Roelle et al. reported that there was cross-talk between GnRH and EGFRs involving the MMP-dependent transactivation of EGFRs (212). Specific inhibitors of MMP-2 and MMP-9 were used to show that the EGFR was activated only after the MMPs had released the EGF-like ligand. Other products of the EGFR second messenger pathways also depended on the MMPs.

Another route to modulating gonadotrope function would be through the use of gap junctions. Gap junctions are transmembrane channels that could provide a route for stimulatory factors throughout the pituitary. They were reported in pituitary cells as early as 1975 (213–215). These junctions allow cell-to-cell exchange of cytoplasmic molecules smaller than 1,000 Da, which could include second messengers (calcium, cAMP) and nutrients. They allow the cells to form a functional syncytium providing communication pathways that could pass signaling molecules or other solutes from cell to cell. Gap junction formation was seen both in intact pituitaries and cultured rat pituitary cells.

Initially, gap junctions were discovered between hormone-producing cells (213–215). However, a series of papers by Soji et al. reported gap junctions only in folliculostellate cells, in vivo. Their studies of castrated, GnRH-treated, cycling, and pregnant rats reported that the number of gap junctions depended on sex steroids early in development and that their overall numbers in the adult varied with the reproductive state (217–224). In general, the numbers increased with age (217,219) and later pregnancy (220), and decreased after castration (221–223). Estrogen or testosterone replacement therapy restored gap junction numbers (221–224).

Whereas the foregoing studies focused on folliculostellate cells as the major site for the formation of gap junction syncytium (217–224), other studies have also shown gap junction formation by hormone-producing cells. Yamamoto et al. reported that gonadotropes could make gap junction proteins. Connexin43 (Cx43) was actually detected in the largest secretory granules of LH cells (225). Meda et al. reported the presence of both Cx43 and Cx26 in subsets of rat anterior pituitary cells (226). Morand et al. reported evidence for gap junction coupling between folliculostellate cells and hormone-producing cell types (227). Frequently, prolactin cells were involved. They exposed a cut pituitary section to fluorescent yellow and traced the route and extent of passage from cell to cell in the pituitary itself. They reported that not all cells were coupled to their neighbors and that the coupling formed specific routes of communication. They postulated that folliculostellate cells could form a network with these endocrine cells that would link sets of different cell types and regulate their functions by supplying growth factors, nutrients, or other cellular activators along the route, passing the small activators through gap junctions (227). These syncytia could form functional units that might activate the cells in the syncytia to deliver cocktails of hormones to the blood stream, which are needed for particular functions. This cellular pathway could activate diverse populations of pituitary cells to provide more than one hormone. This topic is revisited in later sections of this chapter that highlight paracrine or juxtacrine interactions.

Understanding Gonadotrope Diversity

Through the 1980s, it became clear that the pituitary gonadotropes were made up of subsets with different morphologies and functions (125). The importance of the bihormonal gonadotropes was assumed by their relative abundance in highly stimulated states such as after castration (169,175,176) or at mid-cycle (178). The importance of the small gonadotropes as an immature or reserve subset was assumed from studies of their appearance in fetal or prepubertal periods and when they responded well to estrogen and GnRH priming (125,208) (Figs. 31 and 32). As stated previously, their reserve characteristics extend to the fact that they are mostly monohormonal (Fig. 25), leading one to suspect that they are subject to further differentiation to the larger subsets that are mainly bihormonal (125,208). It seemed as if a simple model could be drawn showing how the smaller cells added to the medium-sized and larger subsets, thereby populating the gonadotropes, as needed, during proestrus and estrus (125).

With the advent of dual-labeling protocols that combined the detection of mRNA and antigens came the recognition that no one dual-labeling protocol may detect all of the gonadotropes (175,176). Therefore, whereas most cells may appear bihormonal by antigen content after castration, the labeling may not detect the relative abundance of one of the monohormonal subtypes if it contains only mRNA (176). Also, the more sophisticated techniques that detected antigen content and secretion began to discover cells that looked like gonadotropes, but did not secrete like gonadotropes (205). In short, studies began to discover new levels of diversity that rendered the previous questions about morphological diversity almost moot.

The interpretation of these studies was also modulated by the recognition that some of the functional diversity may be due to paracrine interactions or even interactions between contiguous cell types (228–230). Some of the diversity may appear regionally as gap junction syncytia regulate clusters of cells or MMP functions are activated. In this context, the foregoing discussion about matrix proteins and gap junctions assumes even greater importance.

Questions about how gonadotropes are regulated are admittedly difficult to address in a pituitary population that already is functionally diverse based on different cell types. Add to that the fact that the gonadotropes themselves are diverse. Therefore, workers seeking mechanisms behind signaling pathways and gene expression began to turn to approaches that separated or enriched the populations. The early 1990s also saw the development of immortalized gonadotrope cell lines.

We recognized that the enriched fractions or cell lines represented only a subset of gonadotropes. Also, any enrichment did not allow for the potential cross-talk among the different cell types that may occur in the normal pituitary. Nevertheless, with these caveats in mind, these simpler systems, especially those with the immortalized gonadotropes, allowed the systematic dissection of steps in regulatory pathways and have led to many discoveries about mechanisms behind gonadotrope regulatory hormones. The following sections focus on how these populations were developed.

Targeted Tumorigenesis Produces Valuable Gonadotrope Cell Lines

A report in 1990 by Windle et al. (231), from Dr. P. Mellon's laboratory, described several immortalized cell lines produced from tumors driven by an SV40 T-antigen oncogene (*Tag*) linked to the human glycoprotein α subunit promoter. These *Tag*- driven tumors were specific to the anterior pituitary, and the most rapidly growing tumor, the αT3-1 line, proved to contain immortalized gonadotropes caught in an early stage of differentiation. The tumor cells were purified and cloned and tested for their hormone content and responses to GnRH and thyrotropin-releasing hormone (TRH). No other pituitary hormone or LH, FSH, or TSH β-subunit activity was detected. In addition, they responded to GnRH and not TRH, which indicated that they likely belonged to the ventral population of α-subunit–responsive cells destined to become gonadotropes. Recall that during embryonic development, virtually all of the cells produce α subunits by E11.5, but only the ventral α-subunit–producing cells (αGSU) become gonadotropes (60,72). Also, GnRH mRNA is detected (102), along with GnRH receptors, around E12 (109). Hence, the αT3 cells could have been derived from the *Pitx1*-sensitive population developing during E11 to E12 that also expressed the GnRH receptors. There is no evidence that αT3 cells express GnRH mRNA.

Studies designed to immortalize cells with β subunits progressed in this same laboratory and reports of cell lines that expressed both α subunits and LHβ subunits appeared in 1996 (232). Called the LβT2 cells, they were initially found to produce only LHβ and α subunit, but not FSHβ. However, later studies with more sensitive RT-PCR techniques detected FSHβ mRNA expression and also responses to activin (233). In fact, the entire activin–inhibin–follistatin circuit is in the LβT2 cells, which makes them an ideal model in which to study factors that regulate secretion in bihormonal gonadotropes. These tumor cell lines contain all of the signaling pathways used by adult gonadotropes and have been extremely valuable for studies of regulation of gene expression, signaling, and cellular responses that would be impossible to interpret in a mixed population. This will be seen throughout the remaining sections of this chapter, as studies that use these cells are described along with studies of normal populations of pituitary cells.

Diversity and a Division of Labor

Those workers studying primary cultures of gonadotropes in the early 1990s actually focused on questions about their diversity, looking for explanations and evidence for a "division of labor" in the gonadotrope population. Some of these studies sought to enrich subpopulations, comparing their responses with each other and with their counterparts in a mixed culture.

In previous sections of this chapter, different secretory patterns of LH and FSH were described, when fractions containing different-sized gonadotropes were compared (204,206–208). Smaller cells tended to be monohormonal and larger gonadotropes tended to be bihormonal (206-208) (Fig. 26). The fact that the gonadotropes shifted to express more bihormonal cells just before proestrous surge secretion (178), after castration, or after GnRH and estrogen stimulation (125,208) suggested that the bihormonal state might be a more mature state (in terms of differentiation). This initially led us to focus on mechanisms in those cells that might subserve nonparallel secretion. However, it did not explain the significance of the remaining, monohormonal cells, which could represent as much as 40% of the antigen-bearing population depending on the physiological state and types of products assayed (234).

There were several possible explanations for these monohormonal cells, none of them mutually exclusive. The first, as stated earlier, might be the fact that they are immature or poorly differentiated. Indeed, the ones in the smallest cell fractions resembled the neonatal gonadotropes. They could be gonadotrope precursors, used to replenish the population as needed. Another interpretation is that they have secreted their stores of the other gonadotropin and that they are in a resting state, ready to replenish stores during the diestrous translational activity. This would be logical, especially for those gonadotropes in the medium-sized fractions that appear mature in terms of numbers and sizes of granules. Figure 33 shows an example of a monohormonal gonadotrope detected through analysis of serial sections. It is storing LH and appears to have all of the attributes of an active, functioning gonadotrope. Yet it stores little, if any FSH in the field imaged by electron microscopy. (Of course, one could challenge these data by stating that the FSH stores could have been in another region of the cell, outside the plane of the section. For this reason, we correlate our studies of sectioned gonadotropes with those of dispersed, whole cells. We see the same proportion of bihormonal and monohormonal cells in dispersed, whole gonadotropes, and previous studies have shown no changes even after different fixation conditions.)

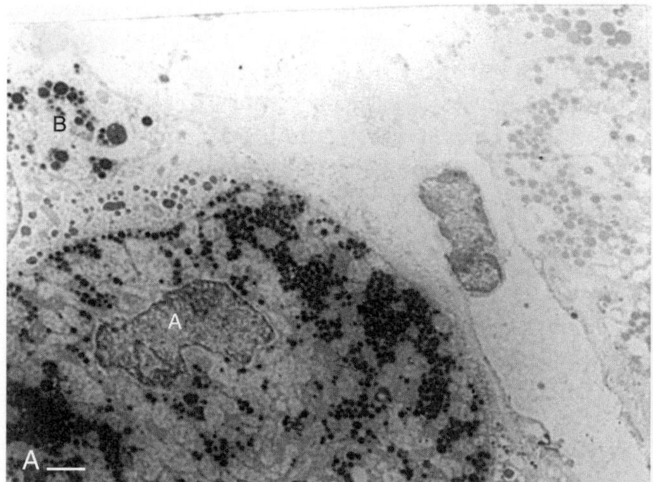

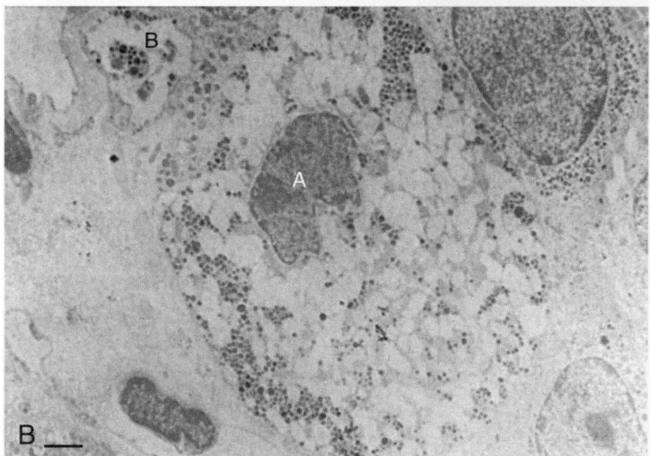

FIG. 33. Monohormonal gonadotrope seen in ultrathin serial sections labeled for LHβ and FSHβ. **A:** The cell contains only LHβ, and not FSHβ. **B:** A nearby process is bihormonal, however, which proves that the labeling for FSHβ was working. Bar = 0. 5 μm.

A third possible explanation for monohormonal cells is that they subserve a separate function, to support the surge activity as a group that might be linked by unique signaling pathways. This hypothesis is supported by the disproportionately high responses to GnRH from medium-sized cells, in light of their lower numbers in the fraction (206).

In the late 1980s and early 1990s, Lloyd from our laboratory began to look at differential storage and secretory activities of these different fractions of gonadotropes (208), using the RHPA technique for LH developed by Smith et al. (205). We applied both an LH and an FSH RHPA to cultures of gonadotropes from each of the fractions collected by Hyde and her colleagues (206). These cells were from rats in mixed cycles, which provided ample cell numbers for a complete study. The cultures were then stimulated with GnRH and the RHPA was run to detect secreting gonadotropes. Our findings agreed with the pioneering work reported by Smith et al. (205) in that one could not always predict secretory activity on the basis of detectable stores of hormone. Furthermore, we introduced a new meaning to the concept "monohormonal" because we detected bihormonal gonadotropes that secreted only one of the gonadotropins.

Lloyd began his studies by testing for hormone expression after GnRH stimulation for 3 hours. Only the fractions with the smallest cells showed an increase in percentages of total gonadotropes (storing LH or FSH), from 14% to 25% of the population. This increase pointed to the significance of the smallest gonadotropes in supporting the population, assuming the appropriate GnRH stimulation was given. Correlate this with data shown in Fig. 31, in which the smallest cells, which were largely monohormonal, changed their storage patterns to levels of bihormonal cells nearly like those of the total population (46% of gonadotropes). Again, this showed the potential for these gonadotropes.

However, the cells did not show high levels of secreted product when tested by the RHPA. There were modest or no increases in the number or size of LH or FSH plaques produced around these cells, suggesting that these immature gonadotropes must build storage levels and perhaps develop more mature signaling pathways to become highly secretory. This fits with the data from the LH radioimmunoassays (Fig. 26). These data supported the hypothesis that the small gonadotropes had potential to be stimulated to produce gonadotropins. It is possible that purification of these reserve cells would provide an important cellular model for the study of factors that regulate gonadotrope differentiation, to learn what factors might be needed for full secretory activity.

The largest gonadotropes remained bihormonal even after GnRH stimulation (Fig. 32). They did not become monohormonal as a result of secretion of either gonadotropin, suggesting that the monohormonal state was not always the result of secretion of stores of one of the hormones. In addition, as shown in Fig. 32, GnRH stimulation did increase the proportion of bihormonal gonadotropes, especially in large cell fractions from diestrous and proestrous animals (125).

In large gonadotropes from mixed cycling female rats, the overall numbers of LH or FSH cells did not change. However, the largest of the cells (from fraction 7) responded well to GnRH by increasing numbers and sizes of LH or FSH plaques. Therefore, the largest gonadotropes presented response patterns consistent with what one might expect of "mature" gonadotropes.

The enigma was presented, however, in the medium-sized cells in fraction 6 that contained 25% to 35% gonadotropes. Over half of these gonadotropes were bihormonal and all expressed intense labeling for LH and FSH stores. However, when the cells were challenged to secrete in the basal state, an expected 37% secreted LH, but only 2% of the same group of cells secreted FSH. Essentially, the FSH in these bihormonal cells was not being secreted basally, despite the fact that the fraction contained nearly 30% FSH cells. GnRH stimulation promoted an increase in secretion of gonadotropins in both types of assays, but the LH plaques increased only slightly (to 43%). There was a fivefold increase in FSH plaques (to 10.5%), but the final number of FSH plaques was still only 25% of what might be expected from the bihormonal storage patterns seen in this population (208). Thus, unknown factors were needed to promote the secretion of FSH from the majority of these medium-sized FSH cells. One could not blame the result on the plaque assays because FSH secretion was readily detected from the initial cell suspension as well as from other fractions (large or small). This introduced the concept of a bihormonal or monohormonal cell that is either not mature enough to secrete FSH, or is inhibited by the environment in some way. In other words, a bihormonal gonadotrope can selectively sort its stores and secrete as a monohormonal cell. It provided another, somewhat unexpected mechanism for nonparallel release of LH, but not FSH from the same cells.

The ongoing concern in these studies, however, was that the cells from the elutriation fractions came from mixed cycling female rats. A more homogeneous group of gonadotropes was needed to correlate a size with storage or secretory patterns. Therefore, fractions of cells from homogeneous groups of female rats in the same stage of the cycle were separated, this time with the Sanderson elutriation chamber, which accommodated less than 10 million cells (179–181). This facilitated the use of small populations of rats (two to four pituitaries) at a time, which was necessary for studies of homogeneous populations. The cells were cultured and stimulated with GnRH. Media were collected for LH or FSH radioimmunoassay and the cells fixed and labeled for LH or FSH antigens and mRNA. We were thus able to compare the levels of LH or FSH secreted by a fraction with the numbers of gonadotropes present in that fraction, detected by their content of mRNA. We reasoned that if a set of gonadotropes was actively secreting, then the response to GnRH (LH or FSH secreted) should increase proportionately with their enrichment in the population.

Figure 34 compares the responses of large LH gonadotropes throughout the estrous cycle with their counterparts in the initial cell suspension (179,180). As in previous studies, the large cell fractions are enriched in LH or FSH cells. This figure shows that this enrichment persists throughout the estrous cycle. In other words, large gonadotropes are not selectively lost at any time in the cycle, in spite of the rapid loss in granules. It was somewhat surprising to see that, in the estrous rats, there were proportionately more large gonadotropes compared with the initial cell suspension. This may reflect expansion of their surface area as a result of membranes added to the surface during exocytosis, and these data correlate well with the early evidence reporting compound exocytosis profiles developing after GnRH stimulation (153–158).

However, if one compares the secretory responses of large gonadotropes with their counterparts in a mixed population, their responses did not always correlate with the degree of enrichment, which was three- to fivefold. In other words, basal or GnRH-stimulated LH levels are not threefold those secreted by the initial cell suspension except in rats taken on the morning of diestrus or proestrus; the last two groups of rats have been primed to optimize secretory activity during the period leading up to the proestrous surge.

Figure 35 shows the same large cell fractions assayed for FSH activity (181). As in previous studies, the largest cell fractions are highly enriched in FSH cells, particularly those from estrous or metestrous rats. Such fractions and stages of the cycle would provide an excellent source of FSH cells for future studies. However, our tests of secretion show that basal FSH and GnRH-stimulated FSH secretion do not match the levels expected from the numbers of FSH cells found, except on the morning of proestrus. They may not be "mature" enough to respond to GnRH. Alternatively, there may be factors missing in the large cell fraction, or the fractions themselves may contain an unusual level of inhibitory factors. These studies show once more that a gonadotrope's secretory capacity cannot be predicted simply by detecting its hormone content.

Figures 36 and 37 graph the findings for the medium-sized LH and FSH gonadotropes, respectively (179–181). Note that in populations from all stages of the cycle, there were never more than 2.5-fold more LH gonadotropes compared with the initial cell suspension. However, GnRH stimulated the secretion of levels of gonadotropins that were expected on the basis of this enrichment. What is interesting, however, are the surprisingly high responses seen in medium-sized LH cells from estrous rats. Such cells secreted over threefold more LH than the initial cell suspension from the same group of rats, indicating that these medium-sized LH cells may be particularly primed to produce LH, even in the period after the LH surge.

Similarly, medium-sized FSH cells showed two- to 2.5-fold enrichment over the initial cell suspension (181) (Fig. 37). However, their secretory responses were mixed. Basal and GnRH-stimulated secretion from estrous rat populations was higher than expected, as was basal FSH secretion from proestrous PM rats. However, medium-sized FSH cells from diestrous and proestrous rats were poorly responsive to GnRH. These could be the same medium-sized cells seen in the populations from mixed cycling female rats in Lloyd's RHPA study (208). Similarly, as in the case of the larger FSH cells, the factors required to elicit FSH secretion from these diestrous or proestrous medium-sized cells may have been missing, or the conditions in the cultures themselves may have inhibited FSH.

Finally, as shown in Figs. 38 and 39, the small, presumptive immature gonadotropes were tested (179–181). Note that in most stages, this subset does not enrich the small cell population more than 1.4-fold. This adds to evidence from Lloyd's studies showing that small gonadotropes from mixed cycling female rats form small plaques (208). Figure 38 confirms our previous studies in that most small LH cells did not secrete well in response to GnRH in populations from most stages of the cycle (see also Fig. 26, which graphs responses from mixed populations). However there was a surprisingly good response to this neuropeptide by the small LH cells from estrous animals. As in the medium-sized subset, this group of cells may have been primed during the LH surge to respond well the night before.

Probably the most surprising result in this series of tests came when we looked for FSH activity in the same small fractions (Fig. 39). Again, there was not much enrichment over initial cell suspension in FSH cells. However, those small FSH cells from rats in the proestrous PM secreted surprisingly well both basally and in response to GnRH, especially in view of the fact that they represented only half of the numbers

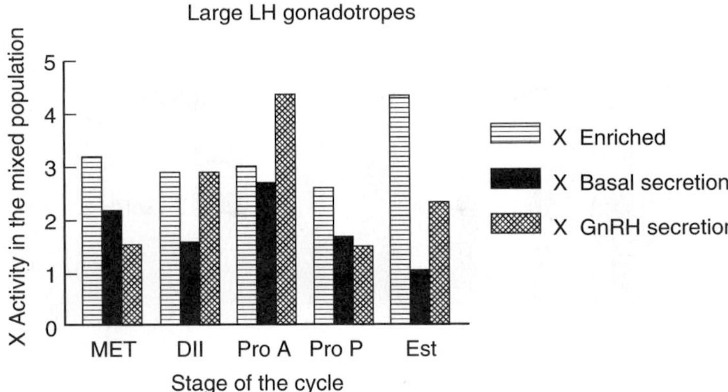

FIG. 34

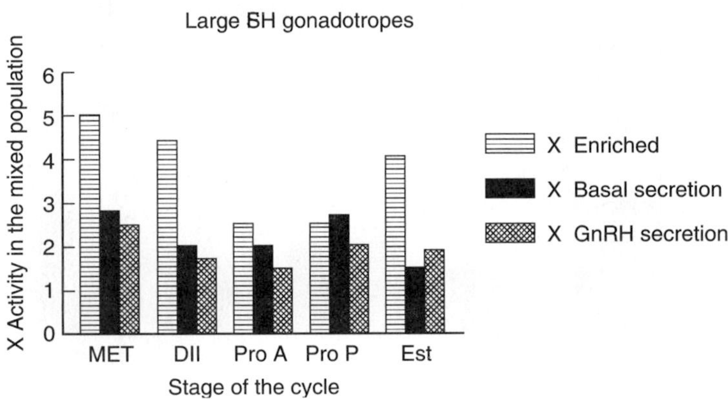

FIG. 35

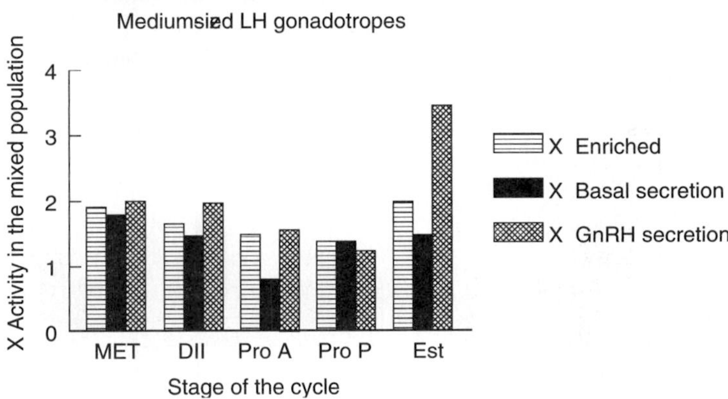

FIG. 36

FIGS. 34–39. Pituitaries were collected from rats in different stages of the cycle and separated into fractions by size. The percentage of gonadotropes in each of the pooled fractions was compared with that in the initial cell suspension. The cells were challenged for 3 hours with 1 nM gonadotropin-releasing hormone (GnRH) and their secretory responses compared with those from the initial cell suspension. The rationale was that secretory responses should correlate with degree of enrichment in gonadotropes. Figures 34 and 35 compare the enrichment and responses to control media or GnRH in the largest fractions (multiplied by that of initial cell suspension); Figs. 36 and 37 compare medium-sized cells; and Figs. 38 and 39 compare responses in the largest cell fractions.

Continued on next page

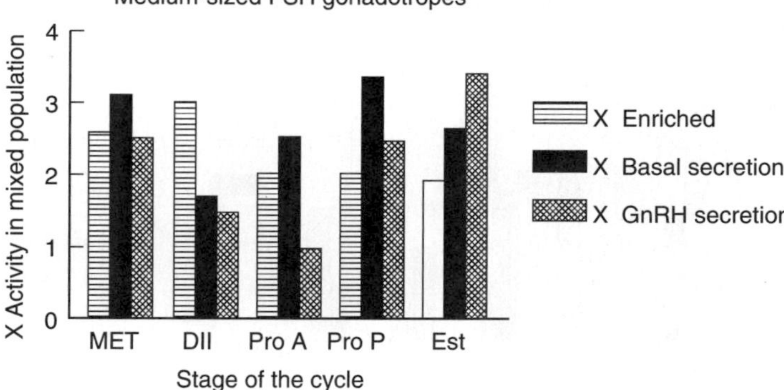

FIG. 37

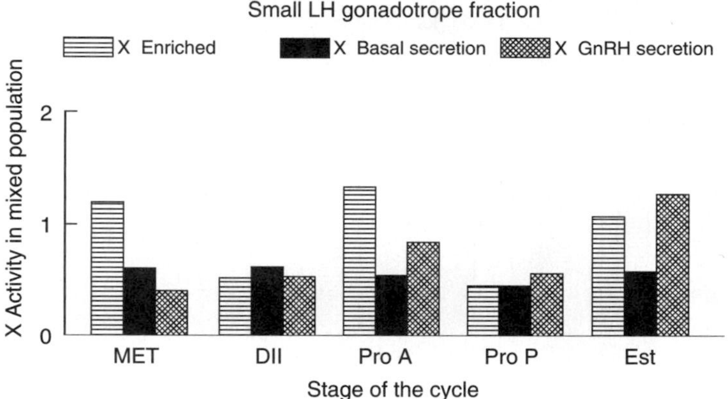

FIG. 38

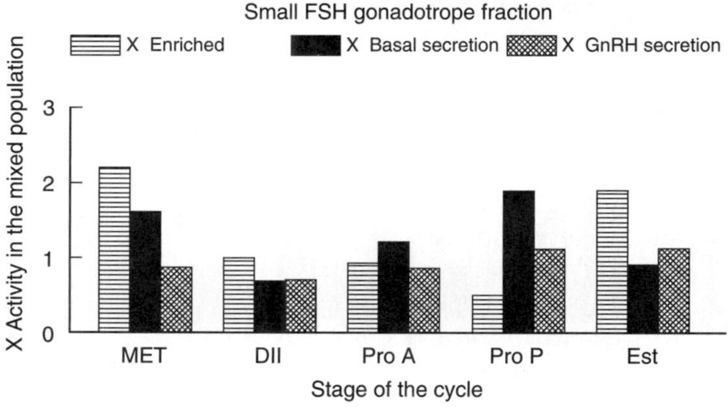

FIG. 39

found in the initial cell suspension (181). Recall that most of these cells are monohormonal (207) (Fig. 25), although their bihormonal potential can be brought out by GnRH and estrogen (125) (Fig. 31). If this evidence for enhanced secretory activity indicates their potential, in vivo, there may be a role for the small gonadotropes beyond that of a resting or reserve cell population. This is reinforced by the earlier studies in which we showed the most striking GnRH-mediated changes in storage patterns (monohormonal to bihormonal) are in this subset (taken from mixed cycling females) (208). Finally, recall that after castration there is a striking increase in FSHβ mRNA by small gonadotropes (176).

We recognized that the process of separating and culturing individual fractions might create an environment that either eliminated or introduced regulatory factors, or perhaps produced an abnormal concentration of such factors (180,181). Thus, we also may be showing the effects of the presence or absence of paracrine regulators, and the next phase of the study clearly needs to use immunoneutralization or antagonist experiments to learn if follistatins, activins, or inhibins are acting in these systems (181). Later sections of this chapter focus on these factors and how paracrine influences may affect the population. Although it is intriguing to know that not all small gonadotropes are incapable of responding to GnRH and not all large, well-granulated gonadotropes respond as expected, we must interpret the data with caution and recognize that some responses may be the result of the cells' environment, and not their potential.

We also recognized from our previous work (206,207) that changing sizes and density of the gonadotropes during their secretory cycle may affect their position in a particular fraction. Thus, a small, potentially active LH cell may enlarge and become denser as it translates LH and FSH and enter a medium-sized or larger fraction, thus increasing the population of larger cells (179–181). Figure 30 shows that proestrous rats have larger gonadotropes, and this may reflect enlargement of the cells in that pool and movement of similarly enlarged medium-sized cells into the larger cell pool. Similarly, after secretion, a large gonadotrope may become lighter from secreted stores, and, after the membranes are endocytosed, may belong to a smaller cell fraction.

Therefore, we used these data to trace the potential flow of gonadotropes from small to large fractions (179,181). Counts of numbers of LH and FSH gonadotropes in each fraction were derived from the percentages. These were translated to real numbers based on cell yield in that fraction, which could be added and expressed as a percentage of the total gonadotropes recovered. This was compared with the recovery in the initial cell suspension to validate the recovery of starting numbers of gonadotropes. Once this validation was completed, the numbers were used to do a distribution analysis of gonadotropes expressing LH or FSH antigens or mRNA in the three fractions, comparing different stages of the cycle. The bar graphs in Fig. 20a plot the total percentages of LH-antigen bearing gonadotropes recovered in all three fractions. The subsections of each bar show how they were distributed. Figure 20a shows that, during peak expression of LHβ antigens in diestrus and proestrus, most of the LH antigen-bearing gonadotropes that were recovered separated in the large cell fraction. Very few LH cells were found in small and medium-sized fractions during estrus and metestrus.

However, we know that there is an increase in overall percentages of LH cells during diestrus (178,179). The distribution analysis shows that this increase is focused in the subset of small and medium-sized LH cells (179). As predicted, these cells may have matured, or recovered their stores after secretion. To test if this maturation or recovery comes from cells with new mRNA, the same distribution analysis was run on all gonadotropes identified by the presence of LHβ mRNA (Fig. 21). In this group, again, the biggest increase in cells with LHβ mRNA was seen in the medium-sized fractions, especially on the proestrous PM (179). In estrus, another increase in small LH cells bearing mRNA could come from recruitment of new cells, or it could reflect degranulated LH cells that entered the smaller pool after secretion. To summarize, the new mRNA that appeared in proestrous gonadotropes was distributed in smaller as well as the largest subsets. This contrasts to the findings for FSHβ, as discussed in the following.

When the same distribution analysis was applied to the FSH cell population, during peak expression of FSH antigens (diestrus and proestrus), most of the cells are in medium and large cell fractions (181) (Fig. 20b). In contrast, during peak expression of the FSH mRNA (estrus), most of the FSH cells are small (Fig. 21b). What follows is a rather clear-cut series of events suggesting that, after the FSH rise in estrus, small cells are stimulated to produce the mRNA. As they translate it, they grow or become denser and thus enter larger cell pools. This sequence is identical to that first described by us for small castration cells identified by their content of FSHβ mRNA (176). Collectively, the evidence for the castration and estrous cycle responses by FSH cells shows that subsets of FSHβ-bearing cells mature to an actively secreting cell through a pathway distinct from that for LHβ. It is clearly subserved by abundant monohormonal FSH cells.

These studies continue to highlight the importance of the small gonadotropes, which has remained an undeveloped area of research, mainly because they cannot be identified easily by morphology, or they are thrown out because they do not populate the gonadotrope-enriched fractions. Collectively, all of the studies agree that they exhibit the most plasticity in response to stimulation by estrogen or GnRH (179–181,208) (Fig. 32). The signaling pathways that direct their development and maturation to a monohormonal FSH gonadotrope may be different from those that direct the pathways to maturation of a bihormonal gonadotrope.

The Gonadotrope as a Multihormonal Cell

The point about bihormonal gonadotropes changing with stimulation has been made repeatedly in this chapter. They increase proportionately with the increase in overall percentages of LH and FSH. One logical explanation for this change is that monohormonal cells begin to produce the other gonadotropin, thereby adding to the population of bihormonal cells. Working under this assumption, we used dual immunolabeling/in situ hybridization to address this question (235). The top line in the graph in Fig. 40 shows the overall percentages of LHβ mRNA–bearing cells and the lower lines show the cells that contributed to the population. Not surprisingly, up to 90% of LH antigen–bearing cells contributed to the population of cells with LHβ mRNA (Fig. 40). However, the data showing relatively low contributions by FSH cells were surprising. During peak expression of the LHβ mRNA, less than 40% of cells with FSHβ antigens expressed LHβ transcripts. This has stimulated the hypothesis that significant percentages of cells with LHβ mRNA may belong to a monohormonal subset that is activated before the LH surge.

To learn if any other cells contributed to the rise in LHβ mRNA, dual labeling was used to detect it with other pituitary hormones. Figure 40 shows that, during peak expression of LHβ mRNA, most of the cells also contained GH antigens (Figs. 40 and 41). Thus, the principal contributors to the rise in LHβ mRNA during diestrus and proestrus are cells bearing LH or GH antigens. Figure 40 also shows that, during estrus, there is a significant decline in expression of LHβ mRNA by cells with GH antigens, along with a significant rise in percentages of cells with LHβ mRNA and prolactin antigens. Other contributors

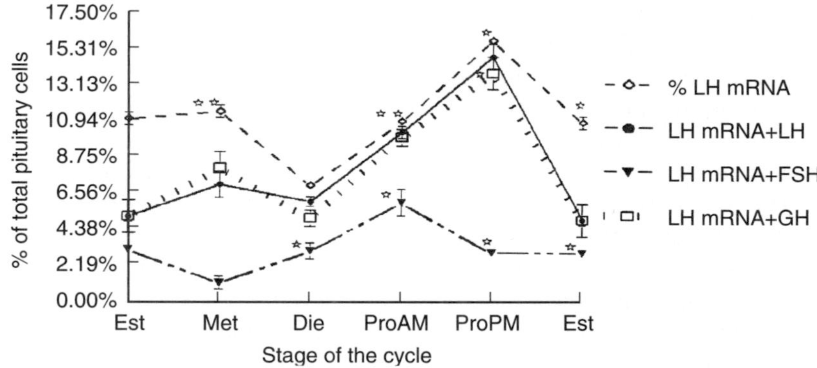

FIG. 40

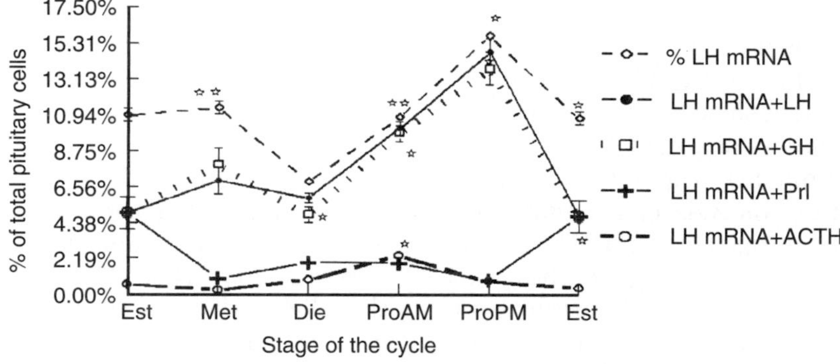

FIGS. 40 and 41. Pituitary cells from female rats at 10 A.M. on estrus (est), metestrus (met), diestrus (die), or proestrus (proAM), or 2 P.M. on proestrus (proPM) were dispersed, plated on coverslips, and fixed the same day. They were dual labeled with in situ hybridization for LHβ mRNA and immunolabeling for all pituitary hormones. The techniques are described in Childs et al. (234). Figure 40 compares the overall percentages of anterior pituitary (AP) cells with LHβ mRNA (*top line*) with the percentages of AP cells that also contain LH, FSH, or growth hormone (GH). Figure 41 adds data from the dual labeling for prolactin and adrenocorticotropic hormone (ACTH). *Significantly different from estrus; **significantly different from nadir point (diestrus).

are shown in Fig. 41. Note that approximately half of the LHβ mRNA–bearing cells in estrus are small prolactin cells. Of course, the loss in LH antigens is mainly due to degranulation. With respect to the other hormones, there is no significant expression of TSH antigens by cells with LHβ mRNA at any point in the cycle (data not shown); however, there is a significant rise in the expression of ACTH antigens by cells with LHβ mRNA on the morning of proestrus, to 20% to 25% of the LH-bearing gonadotropes (or 3% of the pituitary cells; Fig. 41).

When cells with FSHβ mRNA were detected, most also contained FSHβ antigens (Fig. 42). In addition, during peak expression of the FSHβ mRNA, 75% to 80% of the mRNA-bearing cells coexpress LHβ or FSHβ antigens. This left approximately 25% of FSHβ mRNA–bearing cells as monohormonal during peak expression of the transcript and antigens.

Contributions by GH cells to FSHβ mRNA were also seen (Fig. 42). On the afternoon of proestrus, nearly all of the cells with FSHβ mRNA contained GH antigens. With respect to the other cell types, 5% of anterior pituitary cells expressed prolactin and FSHβ during diestrus and proestrus, which dropped to 3% of pituitary cells during the other times of the cycle (Fig. 43). Expression of FSHβ and ACTH or TSHβ remained just above the significance level ($p = .05$) at 3% of the population (data not shown). The significance of the expression of ACTH by either LH- or FSH-bearing cells is discussed in greater detail in later sections.

Figures 44 and 45 show the percentages of each cell type that express LHβ mRNA or FSHβ mRNA. During peak expression of LHβ mRNA, over 90% of LH cells express LHβ mRNA. Cells with FSH, GH, and prolactin antigens contribute at different times in the cycle. During the morning of proestrus, 30% of cells with FSH antigens are transcribing LHβ mRNA. During peak expression in proestrous PM, 40% of GH cells are represented, and 50% of prolactin cells contribute LHβ mRNA during estrus. Figure 45

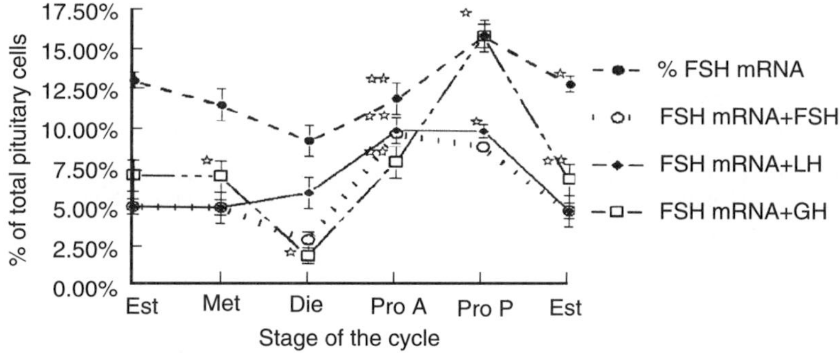

FIG. 42

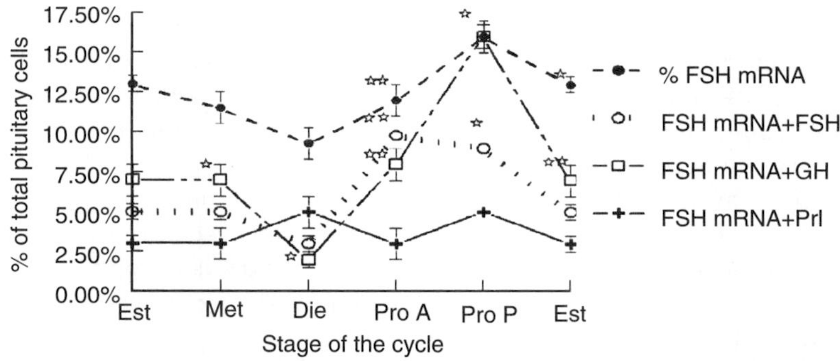

FIGS. 42 and 43. The same populations as in Figs. 40 and 41, except the labeling was for FSH mRNA (*top line*) and dual labeling for FSH, LH, growth hormone (GH), and prolactin. The technique is described in Childs et al. (234). *Significantly different from estrus; **significantly different from nadir point (diestrus).

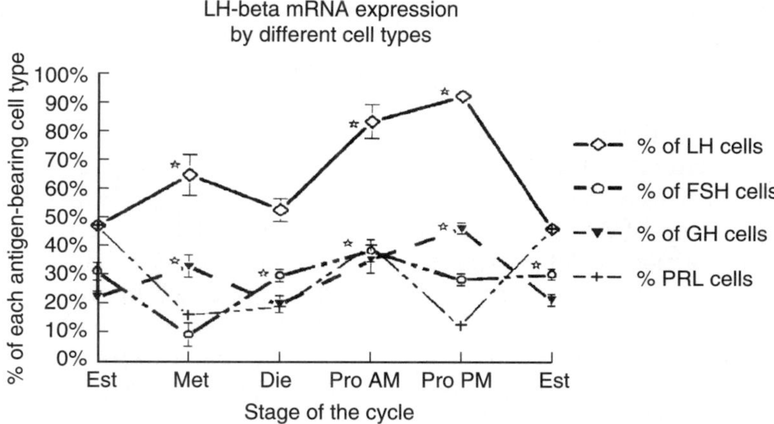

FIG. 44

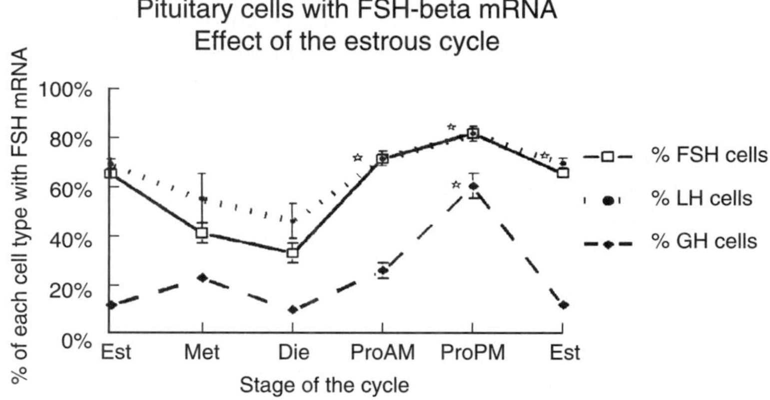

FIGS. 44 and 45. To show the impact of multihormonal expression on each population, these figures graph the percentages of each hormone-bearing cell type that contained LHβ mRNA (Fig. 44) or FSHβ mRNA (Fig. 45) in dual in situ immunolabeling protocols (234).

focuses on cells with FSHβ mRNA. Transcription of this gene product involved over 80% of cells with LH or FSH antigens. Over 60% of GH cells are represented in the FSHβ mRNA–bearing population during proestrous PM. Not shown is the fact that estrous contributors involve 30% of prolactin cells. These data provided the first evidence that other cell types (defined by their antigen content) can contribute to the rapid expansion of cells bearing gonadotropin transcripts during the cycle (179–181). This expression could reflect contributions from a multihormonal cell capable of supporting more than one system. The parallel rise in contributions by prolactin and GH cells suggests that this may be from the mammosomatotrope lineage, which is described in more detail in the section on prolactin cells.

There have been many questions about the significance of the coexpression of GH antigens in gonadotropes. GH is known to regulate LH secretion (236,237), and GH-binding proteins have been detected in gonadotropes (238,239). Thus, the GH antigens could have been bound to receptors or binding proteins in gonadotropes. To address these questions, a series of subsequent studies focused on expression of other genes and regulators of this expression.

These studies showed that cells with GH antigens bind GnRH receptors (240) and that this expression can be regulated by inhibin (241), activin (242), and estrogens (Fig. 46). GH cells increase their expression of GnRH receptors to a peak of 40% of somatotropes by proestrus (240). Inhibin has no reductive effects on GH cells overall, but it does cause a slight reduction in expression of GnRH receptors by GH cells (241). Activin stimulates a significant increase in GH cells with GnRH receptors both in diestrous or proestrous rat pituitary cell populations (242). The increase in the proestrous rat cell population is to nearly 70% of GH cells, which is much higher than expected from the levels normally seen during proestrus (40% of GH cells) (240).

Dual labeling showed that gonadotropes with LHβ or FSHβ antigens bound biotinylated analogs of

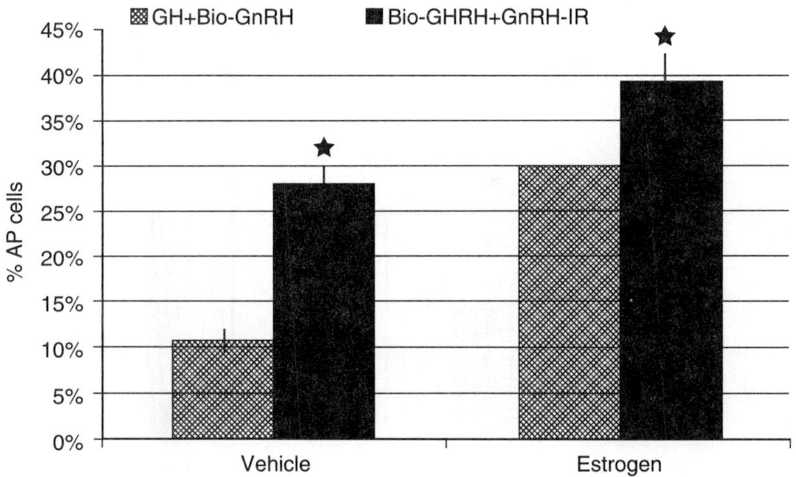

FIG. 46. Growth hormone (GH) cells expressed receptors for biotinylated gonadotropin-releasing hormone (GnRH) during the peak periods of somatogonadotrope expression (239). Activin upregulated expression of GnRH binding sites (241). This graph shows data from diestrous rat pituitary cells plated for 15 hours with and without 100 pM estradiol (unpublished data). They were then exposed to biotinylated GnRH or biotinylated GH-releasing hormone (GHRH) for 10 minutes and fixed for dual labeling for biotinylated GnRH and GH, or biotinylated GHRH and GnRH receptor proteins. Vehicle controls show that over 25% of diestrous rat pituitary cells bear receptors for GnRH and GHRH. Estrogen stimulates coexpression of GnRH receptors by GH cells or coexpression of GnRH and GHRH receptor proteins. Childs et al. (244) describe the technology and estrogen effect on GH cells themselves.

GH-releasing hormone (GHRH), and binding was highest during the times when expression of somatogonadotropes was highest (242). Even more definitive evidence for the involvement of somatotropes came from in situ hybridization experiments (243). The dual labeling for GH mRNA and LH or FSH antigens showed coexpression by over 60% of gonadotropes late in proestrus. Some of these data are shown in Fig. 47.

The presence of GH mRNA and GHRH receptors signified that the GH proteins were not in gonadotropes simply as a result of binding to receptors or binding proteins (238,239). In addition, the somatotropes themselves were undergoing cyclic changes, with an increase in GH mRNA during diestrus (243). All of the changes coincided with the rise in estrogen levels early in the cycle, and subsequent studies showed that low concentrations of estrogen stimulated an increase in GH mRNA and GHRH receptor expression (244). Estrogen administered to diestrous rat populations also stimulated a significant increase in the percentage of GH cells with GnRH receptors (Fig. 46). In this latest group of studies, GnRH receptor proteins were detected with antiserum to the external binding domain of the GnRH receptor protein, which was applied, in dual-labeling studies, to target cells labeled with biotinylated GHRH.

Figure 46 graphs the data from the counts of cells with the two receptor subtypes showing a surprisingly high concentration of multipotential somatogonadotropes as defined by their expression of GHRH receptors and GnRH receptor proteins. These cells represent nearly 25% of a diestrous rat pituitary cell population. This indicates that the potential for conversion is higher than one could predict on the basis of dual labeling for the antigens or the antigens and mRNA. Figure 46 also shows the significant increase in somatogonadotropes defined by GnRH receptor expression after 24 hours in estradiol.

These findings confirm those published by La Rosa et al. (245), who reported labeling for GnRH receptor proteins in over 70% of somatotropes in human pituitaries. GnRH receptor mRNA is also found in all gonadotropes and thyrotropes. No thyrotropes or corticotropes showed expression of GnRH receptor proteins. In pituitary adenomas, GnRH receptor labeling was found in 7 of 12 gonadotrope, 6 of 10 α-subunit, and 8 of 10 GH-secreting tumors. No immunolabeling was found in prolactin, TSH, or ACTH tumors. Their findings for the detection of GnRH receptor in somatotrope and gonadotrope, but not prolactin tumors confirm findings reported by other investigators for coexpression in human tumors [reviewed in Seuntjens et al. (246)].

Somatogonadotropes have also been detected by other sensitive molecular assays. Roudbaraki et al. (53) reported that GH mRNA could be detected in over

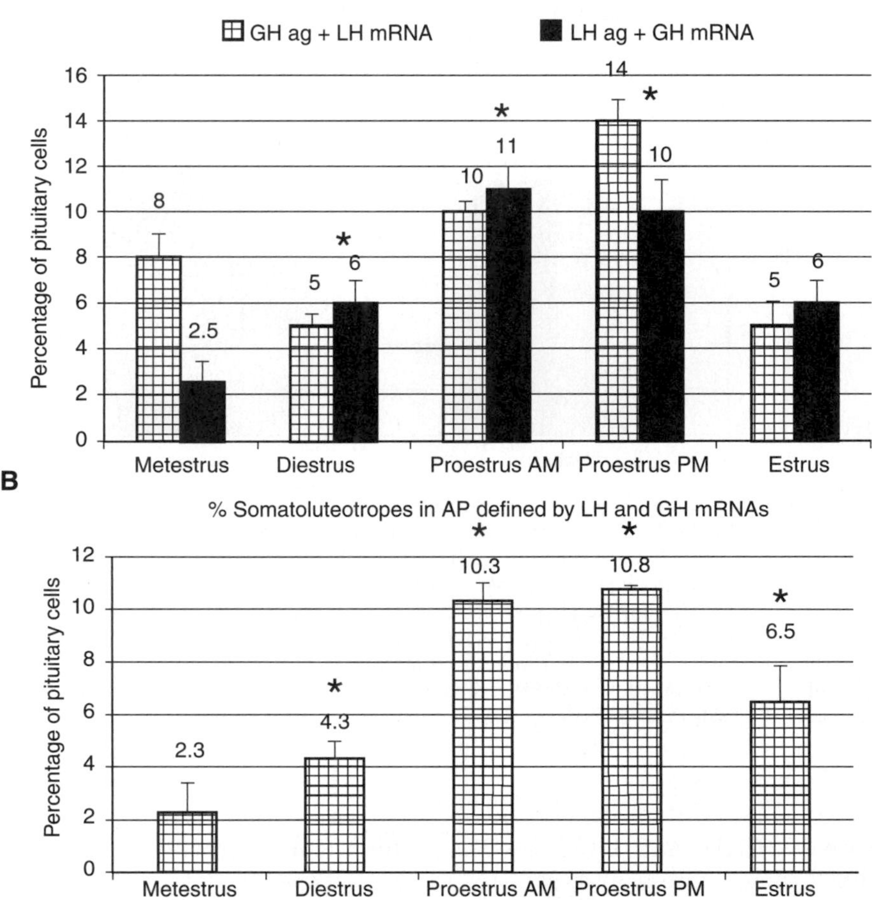

FIG. 47. **A** compares the changes in percentages of somatoluteotropes detected by different combinations of growth hormone (GH) and LH antigens and mRNAs. Losses in one gene product can affect overall detection of the population of somatoluteotropes. **B** shows the percentages of somatoluteotropes detected by LHβ and GH mRNAs with new biotinylated and digoxigenin oligonucleotide probes and simultaneous in situ hybridization (252). These fields were labeled after the simultaneous application of digoxigenin probes for GH mRNA and biotinylated probes for LHβ mRNA; probes were developed by GeneDetect.com. Comparing the two graphs shows that because of the decline in GH mRNA from proestrus to metestrus, dual in situ hybridization underestimates the somatoluteotropes population during the PM of proestrus.

half of the pituitary population; 32% contained only GH mRNA and the remaining cells coexpressed GH mRNA with prolactin, LH, POMC, or TSH. Subsequent studies (246) showed that these multihormonal cells were particularly responsive to fragments of POMC (1–26) as a regulator of calcium influx, indicating the presence of unique signaling pathways in these multipotential pituitary cells.

Okada et al. used RT-PCR to test coexpression of androgen receptor mRNA with LHβ mRNA in gonadotropes from primate pituitaries (247). Their tests included assays for GH mRNA, and they reported that of the 12 cells positive for LHβ mRNA, 7 also expressed GH mRNA, confirming the existence of the somatogonadotropes in primate pituitaries.

More recently, Wu et al. used a clever and innovative method to isolate pure FSH-secreting gonadotropes from ovariectomized or male mice (248). Transgenic mice were produced with a gene construct that put a cell surface tag on FSH cells. These cells were then purified by antibodies to the surface tag and magnetic beads, which resulted in a fraction containing 90% FSH cells. As expected, this also greatly enriched LH cells, and the fraction will be invaluable for future studies of primary cultures of gonadotropes. In addition, when they labeled for GH, they reported that 10% of the gonadotrope population expressed GH antigens. The source of their starting material was ovariectomized or male mice (by necessity, because other groups did not yield high numbers of FSH cells); they might have seen more coexpression of GH and FSH or LH had they been able to isolate gonadotropes from an estrogen-rich environment (Figs. 40 to 46).

This multihormonal cell type can also be detected by its responses to multiple neuropeptides. In 1994, Kasahara et al. were the first to report that individual anterior pituitary cells could respond to a sequence of stimuli with different releasing hormones, by increasing intracellular calcium (249). Some pituitary cells responded to combinations of releasing hormones that included both GnRH and GHRH (249). Villalobos et al. (250,251) also tested calcium influx in single, identified anterior pituitary cells in response to a series of releasing hormones. They reported that nearly 30% of GH cells responded to GnRH and 28% of FSH cells responded to combinations that included GHRH. In later studies (251), they reported that the multipotential gonadotropes were found only in female mice. Such gonadotropes responded to GHRH, but not GnRH, and they also contained ACTH.

In previous sections, the importance of detecting transcripts in dual cytochemical assays was emphasized, mainly because this approach brings out expression that is not evident by dual immunolabeling. Another important consideration is the use of technology that has no components that could interact during the protocols. Digoxigenin and biotin complementary oligonucleotide probes are excellent tools for the simultaneous detection of two mRNAs (gonadotropin β subunits and GH) (252). Highly sensitive probes that have a high concentration of signaling molecules per probe molecule have been developed commercially by GeneDetect.com (Bradenton, FL, and Auckland, NZ). They were successfully applied, simultaneously, to the dual detection of LHβ and GH mRNAs (252). Noninteractive, nonradioactive detection systems can then be applied, including anti-biotin and streptavidin peroxidase for the biotinylated probe and anti-digoxigenin labeled with alkaline phosphatase for the digoxigenin-labeled probes.

Figure 47 compares the coexpression of GH and LHβ with the use of dual in situ hybridization–immunolabeling or dual in situ hybridization with digoxigenin- and biotin-labeled probes. As is evident, the blue label for the digoxigenin probe for GH mRNA is easy to distinguish from the orange label for the biotinylated probe for LHβ mRNA. The data illustrate the importance of using multiple combinations of labeling methods to get the full picture of the population.

In the dual-labeling sequences featured in Fig. 47a, the detection of LH proteins in somatoluteotropes is limited in metestrus and proestrous PM by the fact that LH cells have exhausted their stores. In the dual labeling sequences that rely on GH mRNA in Fig. 47a and b, the detection is limited by the fact that GH mRNA has dropped to a nadir by the AM of metestrus (243). However, in Fig. 47a, somatoluteotropes can still be detected by their content of GH antigens. In fact, these graphs show that only the combination that includes GH antigens and LH mRNA will detect the full expression of the somatoluteotropes during proestrous PM. This is an excellent example of why one must use multiple cytochemical methods to get the full and accurate view of expression of the somatoluteotrope population. However, the data in Fig. 46, in which somatogonadotropes were detected by dual receptor expression, show that the full potential is often seen only by the detection of GHRH and GnRH receptors. Somatoluteotropes detected by dual in situ hybridization and oligonucleotide probes for LHβ and GH mRNAs are depicted in Fig. 48.

Finally, the use of probes for LHβ heteronuclear RNA (hnRNA) allows one to detect nascent gonadotropin RNAs and thus time the "birth" of the somatoluteotrope during the cycle (252). Earlier transcription assays that detect nascent LHβ mRNA show that the transcription rates are highest during the LH surge, on the afternoon of proestrus (253). The analysis of nascent LHβ mRNA with heteronuclear probes supports these findings, showing that significant nascent LHβ mRNA is detected early in estrus and continues to be evident until a significant decline on the morning of proestrus (Fig. 49). Nascent LHβ mRNA is seen in the nuclei of somatotropes (Fig. 50), and Fig. 49 shows that the appearance of the LH hnRNA in GH cells parallels that in the overall LH cell population. These latest findings thus show that the somatotrope is an important site for this new transcription. As we learn more about the timing, we will be in a better position to study regulators for this transitional coexpression.

The findings for this somatogonadotropic partnership have roots that date back as early as 1921. Evans and Long (11,17,18) maintained that it was the GH in their preparation that caused luteinization of the follicles. Now we know that GH can indeed increase LH receptors, and its presence may in fact facilitate luteinization (21–23). Early electron microscopists argued that the so-called "LH gonadotrope" looked too much like a GH cell to allow confident discrimination (reviewed in references 35, 49, 117, and 125). Perhaps type II LH cells are GH cells. As is discussed in a later section, in 1977, Hopkins and Gregory reported that ferritin-labeled GnRH analog bound GH cells (254), a finding that was not confirmed until 1994 (239), nearly 17 years later.

Although these data add to others that show the surprisingly high concentration of multihormonal cells in the anterior pituitary, critics raise questions based on studies of embryonic development (53,54, 246–249). Recall that whereas there is a common

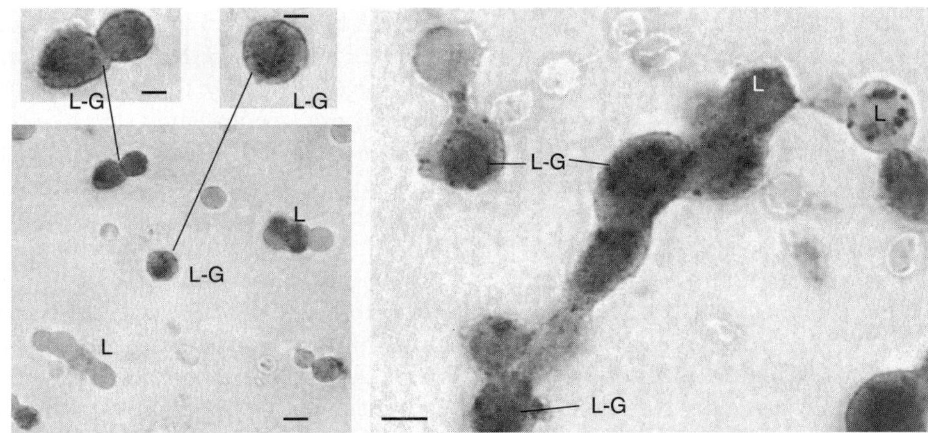

FIG. 48. Simultaneous dual in situ hybridization with a biotinylated oligonucleotide probe for LHβ mRNA and a digoxigenin-labeled oligonucleotide probe for growth hormone (GH; produced by GeneDetect.com). The biotinylated probe was detected by mouse anti-biotin and streptavidin peroxidase (*orange*) and the digoxigenin probe was detected by sheep anti-digoxigenin and alkaline phosphatase (*blue*). Monohormonal LH and bihormonal cells, LH-GH (L-G), are evident. The *left* panels are from rats in proestrous AM, and the *right* photograph is from a rat in proestrous PM. Bar = 10 μm; upper photographs on *left*, bar = 5 μm. (See color insert.)

lineage for somatotropes and gonadotropes early in development (the common α-subunit–producing cell and the *Prop1*-responsive cell; Figs. 1 and 2), the downstream regulators—*pit1* for somatotropes, thyrotropes, and lactotropes, and *Gata2* for gonadotropes—are expressed in separate dorsal and ventral regions of the pituitary, to ensure the production of monohormonal cells (Figs. 1 and 2). Indeed, the spatial separation of these cell types during embryonic development allows gonadotropes and somatotropes to develop in relative seclusion so that *Gata2* expression does not inhibit proper formation of *pit1*-dependent cells. This provides a significant base of GH cells and gonadotropes to support the population and any unique functions during growth and development after birth.

However, a review of data on prepubertal development points to striking changes in gonadotropes that occur in receptivity, response patterns, and hormone storage patterns that mature the cells for function in a postpubertal environment (117,125). In other words, the embryonic developmental events facilitate

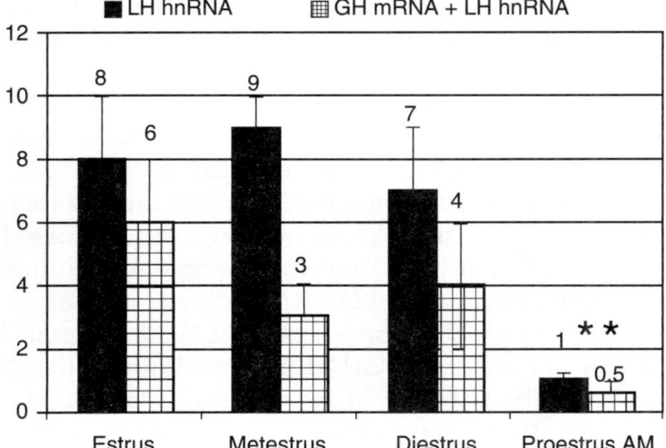

FIG. 49. Dual in situ hybridization with biotinylated oligonucleotide probes for heteronuclear RNA (intronic sequences) for LHβ (hnRNA) and digoxigenin-labeled probes for growth hormone (GH) mRNA (GeneDetect.com). Detected as described in Fig. 48 (252). Very little LHβ hnRNA is detected in the AM of proestrus, and its rise is first seen in estrus (which is a peak period of expression of LHβ mRNA). Proestrous PM was not tested in these studies. When the two RNAs were detected, the rise in coexpression paralleled that seen for hnRNA, suggesting that the somatoluteotrope is born as nascent LHβ RNA is transcribed.

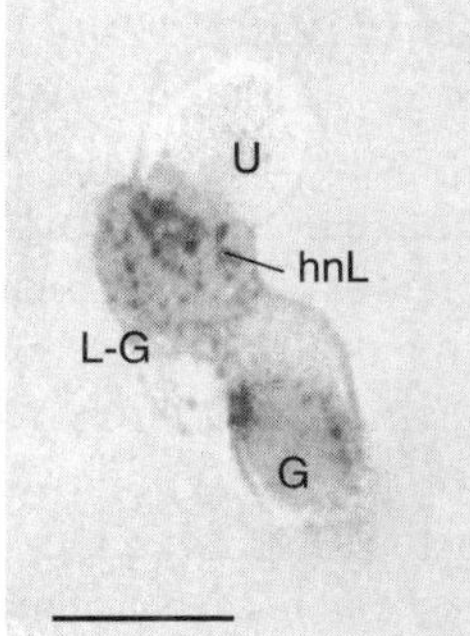

FIG. 50. Dual label for LHβ heteronuclear RNA (hnRNA) with biotinylated probe (hnL) and anti-biotin/streptavidin peroxidase, and growth hormone (GH) mRNA with digoxigenin probe detected by anti-digoxigenin and alkaline phosphatase (*blue*). Field shows monohormonal GH cell and bihormonal somatoluteotrope with GH mRNA in the cytoplasm and LH hnRNA in the nucleus. In the black-and-white photograph, the hnLH RNA is dark gray (see hnL) and the GH mRNA is light gray. It is impossible to distinguish dark blue in the cell labeled G from the label for hnL, however. Unlabeled cell (U) is also seen. Bar = 10 μm. (See color insert.)

and permit the birth of the cells, but not their final maturation to the adult state. Cyclic expression of estrogens may be needed to bring somatotrope functions online in a reproductive context so that both GH and gonadotropins can be secreted in support of activities associated with reproduction (Fig. 46).

However, this raises another set of questions about the significance of GH in reproduction. Most scientists would agree about the importance of GH in the adult because of its effects on body composition and general metabolic support, which, in themselves, contribute to reproductive health. However, are there unique contributions to the reproductive system that would warrant functions from a multifunctional cell? As it turns out, there is evidence that GH promotes reproduction, and a number of excellent reviews of the subject are available (21,22). They cite ample evidence that animal and human models with low GH reproduce poorly, if at all. A mid-cycle surge or rise in GH is now recognized for many species. GH has been shown to have receptors in the ovary (19,20), although the presence of GH receptors in the testis has been shown by some (255–257), but not all (20,258) workers. GH may also work by stimulating local production of insulin-like growth factor (IGF)-1, which has clear reproductive functions in the ovary and testis (21,22). GH has stimulatory effects on preantral follicles and Leydig cell development. Some workers have called GH a "cogonadotropin" because it elevates LH or FSH receptors in the ovary and allows lower concentrations of gonadotropins to work more effectively [reviewed in Giustina and Veldhuis (21), Hull and Harvey (22), and Childs (23)].

It is possible that a background level of GH facilitates reproduction because it provides significant support for the overall metabolic needs of the body. The significance of the somatogonadotrope therefore would be to support this effort.

A paradigm shift may be needed to promote recognition of the fact that the birth of the cells during embryonic development is not a final maturation event. The adult pituitary functions are further differentiated, including their coordinated organization into subsets of multihormonal cells, subject to multiple stimuli and affected by multiple receptor types. The research focus needs to shift to learn how these cell types are regulated and inter-related, including any transitions between the subtypes. There are no tumor cell models for the somatogonadotropes, which has made their study very challenging.

Roles for Monohormonal Gonadotropes

As the sensitive assays bring out more evidence for multihormonal cells, questions arise about the significance of monohormonal gonadotropes. Previous sections of this chapter have emphasized the importance of the bihormonal gonadotropes because most of the approaches used techniques that showed significant increases in their proportions or numbers at times of heightened secretion. However, significant increases in percentages of gonadotropes that bear only FSHβ mRNA (to 37% of gonadotropes) and not LHβ antigens are seen in a subset of small cells 14 days after castration (176). Also, when LHβ mRNA is detected at mid-cycle, only 40% of the cells contain FSHβ antigens, suggesting that nearly 60% of the LH cells appearing during the surge are monohormonal LH cells (234). A similar case may arise when FSHβ mRNA is detected during the estrous rise in serum FSH, although this has yet to be tested. The clear difference in expression of FSHβ mRNA (in small gonadotropes) and FSHβ antigens (in large and medium-sized gonadotropes) during estrus (179–181) suggests that a new subset of FSH cells may be recruited to synthesize the transcripts at that time. Even the expression of GH by gonadotropes is not the same. Late in proestrus, there are significantly more somatotropes with FSHβ mRNA than with LHβ mRNA.

Another experimental model that brings out more monohormonal gonadotropes involves the dual surgeries of adrenalectomy and castration. This resulted in a reduction in bihormonal gonadotropes to only 30% of the gonadotrope population, with a corresponding increase in monohormonal subtypes (259). Do factors from the adrenal (corticosterone, dehydroepiandrosterone) regulate expression of one or both gonadotropins in the bihormonal subset? Why would the dual surgeries prevent expression of the bihormonal cells?

Collectively, this information may also show the limitations of cytochemical techniques when only antigens are detected. There are times when we may be underestimating or possibly overestimating the monohormonal subsets. The surge secretion causes an apparent "loss" in percentages of antigen-bearing gonadotropes so that levels are less than half of those found during peak storage periods (178–181). Indeed, the only way some gonadotropes can be identified during estrus or metestrus is by in situ hybridization. This has led to studies with the new protocols that detect two mRNAs, in situ, with the use of biotinylated and digoxigenin probes for LHβ and FSHβ mRNAs (252). The two probes are applied simultaneously and detected by antisera to biotin or digoxigenin, with peroxidase used for the detection of biotin and alkaline phosphatase used for the detection of digoxigenin. The following is a brief comparison of the expression of bihormonal and monohormonal gonadotropes detected by antigens or dual mRNAs.

In most dual-labeling studies that involve immunolabeling as one of the detection systems, the overall percentages of gonadotropes are 15% to 17% which is the sum of bihormonal cells and monohormonal cells bearing one of the two products detected. It is limited by the fact that the product that was not detected is not counted in the totals.

Detection of the two gonadotropin antigens in the rat has yielded approximately 50% to 70% bihormonal cells and 15% to 25% monohormonal cells, depending on the sex and physiological state (178–181). Because new mRNA-bearing cells may add to the population during the surge secretion, we hypothesized that one might be able to determine the full impact of monohormonal gonadotropes on the population only by applying dual in situ hybridization. When this was applied to male rat pituitary cells, we discovered more total gonadotropes than were found with either of the former dual-labeling techniques. The counts of labeled cells showed that 15%±4% and 16%±1% of the anterior pituitary cells contained LHβ or FSHβ mRNA, respectively, which is comparable with previous counts. However, when dual labeled, the subsets of gonadotropes detected included so many monohormonal cells that there was an overall increase to 27%±1% total gonadotropes, compared with 15% ±1% in methods that detected two antigens or an antigen and mRNA. Of the total gonadotropes detected by mRNAs, half were monohormonal LH or FSH, and the other half were bihormonal. Nearly half of either LH or FSH cell populations were monohormonal (49%±5% of LH cells and 46%±6% of FSH cells). Thus, the previous dual labeling, which has relied on antigens for one or both detected products, has underestimated the proportion of gonadotropes that are monohormonal, based on their transcripts. Such cells may serve as reserve subsets. They may be derived from undifferentiated stem cells or preexisting pituitary cell types. The big question for the future is, do they persist as a separate subset, regulated by unique factors and capable of supporting nonparallel secretion events? Or can all of them be converted to the bihormonal state?

The Gonadotrope as a Regulatory Cell

The strongest evidence for regulatory functions by gonadotropes comes from studies that have produced conditional knockouts of one or both gonadotropins, or the gonadotropes themselves. Kendall et al. (260) used the bovine α-subunit promoter to direct expression of a diphtheria toxin gene and founded a line of mice that do not make gonadotropes. These mice are hypogonadal with no gonadal development or differentiation; LH is not detected. All other functions are normal except prolactin, which is significantly reduced. More recently, this experiment was repeated, only the focus was on changes in prolactin gene expression, mainly because it was clear that over 70% of developing prolactin cells were in fact multihormonal (54). However, prolactin cells and somatotropes were significantly reduced (106,261), and none of the other cell types was affected by the loss of gonadotropes. Their studies showed that gonadotropes regulate prolactin at the level of expression of the prolactin gene. The multihormonal cell type of origin continued to produce normal levels of ACTH because their cell numbers are not reduced (54).

A number of workers have focused on learning what factors might be missing in the α-subunit knockout that would regulate prolactin cells. The α subunit was considered a likely candidate during fetal development (262). More recently, Stahl et al. have shown that thyroid hormone is needed in these knockouts to bring out stem somatotropes that can give rise to the prolactin cells (263).

A systematic study by Tilemans et al. (264) has shown that gonadotropes produce substances that regulate proliferation of lactotropes, corticotropes, and somatotropes. When GnRH- or neuropeptide Y (NPY)–stimulated gonadotrope aggregates were cocultured with the remaining fractions, [^{3}H]-thymidine uptake was enhanced in lactotropes or corticotropes (265–267). They began to look for factors in the gonadotrope-conditioned media and found a fraction that stimulated mitoses in these cells and also two factors that inhibited mitoses in somatotropes. These factors are identified and discussed in the following sections.

Expression of Pro-opiomelanocortin Peptides. The earliest studies of immunolabeled gonadotropes reported that the adult gonadotrope population contained a subset that was small and angular with peripherally organized granules (127,164,165). These cells resembled immature gonadotropes, or ACTH cells. When antisera to LHβ or FSHβ and the C-terminal fragment of ACTH containing amino acids 17-39 were used to detect coexpression, gonadotropes with ACTH-like activity were found, particularly early in prepubertal development (165). Over 3,000 photographs of serial ultrathin sections (some with 6 fields) were collected. Analysis of these fields showed that expression of ACTH in gonadotropes increased during development to reach a peak by approximately 2 to 3 weeks of age. Then, in the adult, such cells were reduced to less than 1% to 3% of the pituitary population (Fig. 12). Figure 51 illustrates two of these cells in the 11-day-old rat. Cell A is a type I gonadotrope and cell B is a type III gonadotrope. Both contain all three hormones, however.

The significance and regulation of the production of ACTH-like peptides by gonadotropes were unclear

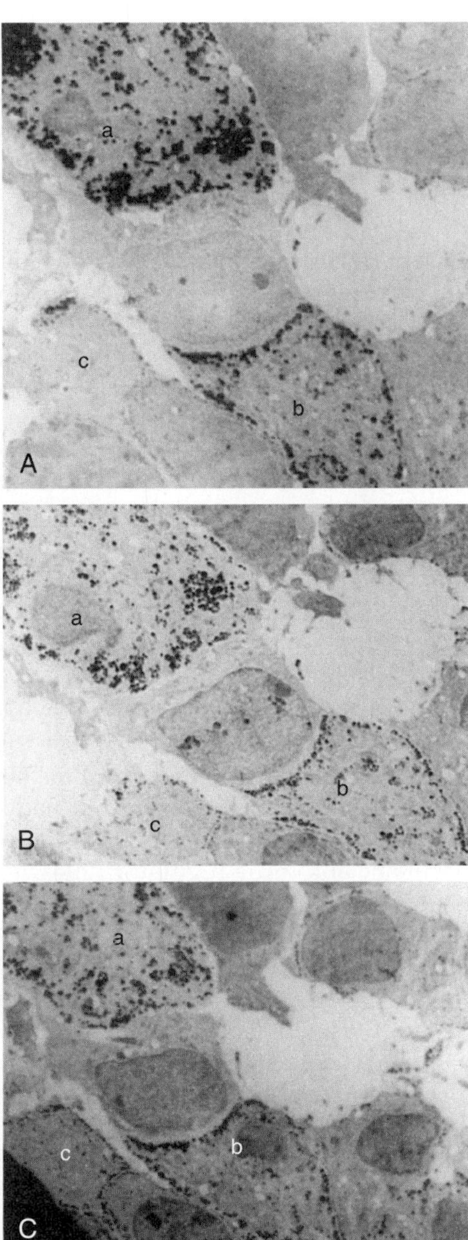

FIG. 51. Serial sections through the same field labeled for LHβ (**A**), FSHβ (**B**), and the 17-39 adrenocorticotropic hormone (ACTH) C-terminal fragment (**C**) with the peroxidase–antiperoxidase technique. Sections taken from 1-week-old male rat. Cells **a** and **b** are multihormonal, showing labeling for LH, FSH, and ACTH. Cell **c** is labeled only for ACTH. Bar = 1 μm. Consult Fig. 12 for counts of these different subsets of cells.

and have been the subject of a number of studies since the late 1970s. They would not be predicted from the studies of pituitary embryology that described a distinct region of corticotrope development (59,60, 92,95–99). In the early 1980s, studies of rats that had combined castration and adrenalectomy were begun to learn if corticogonadotrope cells appeared to support both systems. The results showed the opposite effect; the small percentages of corticogonadotropes found in the adult actually were reduced by the combined surgeries (259).

Clues to this mystery finally emerged when Tilemans et al. discovered that one of the factors from enriched, GnRH-stimulated gonadotropes was capable of stimulating the release of prolactin from reaggregate cultures of purified gonadotropes and lactotropes (267). A series of experiments, in vitro, identified this as a 16-kDa fragment from the POMC precursor product. They used aggregate cultures of lactotropes and gonadotropes to show the dependence of this activity on gonadotropes.

Lactotropes are the last group of cells to appear during embryonic development, and they do not present in significant numbers until 1 to 2 weeks after birth. Seuntjens et al. reported that over 70% of the cells with prolactin mRNA are multihormonal, containing GH and ACTH (54,106,107,111,112). Thus, the rise in coexpression of the 17-39 ACTH fragment by neonatal gonadotropes, depicted in Figs. 12 and 51, likely reflected production of the precursor molecule, POMC. These multihormonal gonadotropes are producing the POMC to release the 16-kDa prolactin stimulatory factor. Collectively, these findings point to an example of an important postnatal maturation step needed to drive a paracrine communication pathway leading to full prolactin function. They reconcile the information from the gonadotrope knockout mice showing significantly reduced expression of prolactin mRNA (260,261,263) with the findings that gonadotropes produce POMC sequences (164,165).

Expression of the Renin–Angiotensin System. Since the mid-1980s, components of the renin–angiotensin system have been detected in all pituitary cells (268), particularly in gonadotropes (269,270). Angiotensin II immunoreactivity has been colocalized with LH in secretion granules (269,270). Cultured pituitary cells continue to produce angiotensin II after 2 weeks, indicating an endogenous source along with angiotensin-converting enzyme. Angiotensinogen may be produced by another cell type, although the exact identity is uncertain. Thus, the entire system is available and angiotensin II is known to have effects on corticotropes, thyrotropes, and somatotropes as well as gonadotropes themselves [reviewed in Robberecht et al. (270)]. Its role as an autocrine or paracrine factor deserves more investigation because tests of blockers, when gonadotropes are stimulated by GnRH, show little effect on prolactin secretion (270). Thus, the current belief is that angiotensin II is not the main paracrine factor that stimulates prolactin cell development. It may play a role in mediating other functions, however.

Expression of Inhibin, Activin, and Follistatin. Functions for these polypeptides are discussed in a later section on regulation of nonparallel gonadotropin release. The hormones are discussed in this section because mRNAs for the subunits of inhibin and activin (271–276), as well as follistatin (277,278), have been found in gonadotropes. Furthermore, dual immunolabeling for activin or inhibin subunits has detected coexpression in granules that also contain LH, FSH, or α subunits (273). The nature and regulation of the paracrine or autocrine functions for these hormones are poorly understood (275). Multiple roles for pituitary activin include an inhibition of GH and prolactin release (279,280) as well as the synthesis and secretion of ACTH from AtT20 tumor cells (281). Thus, these gonadotrope-derived molecules could produce activin to regulate somatotropes, prolactin cells, or corticotropes. Alternatively, they could be autoregulatory.

Follistatin expression changes in a cyclic manner with dramatic increases just before the proestrus secretory activity (278). The source may be from somatogonadotropes to bind activin and limit expression. Alternatively, Halvorson et al. suggest that follistatin could also concentrate activin in a way that makes it more available to FSH cells early in the cycle (275). The exact cellular source of this surge of follistatin expression has not yet been determined.

Expression of GnRH, GHRH, and TRH by Gonadotropes. One of the earliest products to be found in gonadotropes, by our laboratory, was GnRH (282). This was thought to reflect peptide taken up after binding to the target cells. However, May et al. (283) reported that GnRH-like immunolabeling persisted for up to 3 weeks in culture. These data correlate well with more recent studies reporting GnRH mRNA expression in fetal pituitaries (102). The significance of a local site of GnRH production might relate to the gap junction linkages found in the pituitary, discussed in a previous section. It is possible that autocrine mechanisms may be used to spread the effects of a GnRH pulse either directly by GnRH, or indirectly by the passage of second messengers through the functional syncytium created by these junctions. This coordinated effort could be one mechanism that facilitates a surge in gonadotropin secretion, for example.

TRH has also been found by workers in our laboratory in gonadotropes (284–286) and somatotropes (287,288); TRH mRNA and proteins have also persisted in long-term culture (284,287,288). The significance of this neuropeptide could be related to regulatory functions of gonadotropes or somatotropes. Brar et al. (289) reported that transgenic mice expressing the human GHRH gene expressed it in somatotropes and mammotropes. However, what was unexpected was the expression in gonadotropes and thyrotropes as well. Again, assuming these neuropeptides are expressed as secretory products, they could be paracrine effectors or autocrine regulators of somatogonadotropes.

Expression of Other Peptides. The neuroendocrine chaperone polypeptide, 7B2, has been detected, by dual labeling, in LH gonadotropes (290). This polypeptide is processed to a C-terminal 18- to 21-kDa secretory product and it interacts with prohormone convertases in the *trans* region of the Golgi apparatus to regulate cleavage of proteins (291,292); for example, it is believed to regulate the cleavage of POMC by this mechanism (292). The detection of this chaperone protein may reflect POMC production by gonadotropes and the cleavage may be designed to release the 16-kDa N-terminal fragment that regulates prolactin cells (268).

Neurotensin (293) has also been detected by dual labeling in gonadotropes. Castration reduces expression of neurotensin, and there is an age-related increase in expression coinciding with a rise in serum estradiol (293), suggesting that neurotensin may be regulated by gonadal steroids. Neurotensin is believed to stimulate polyphosphoinositide breakdown, which may provide second messengers for pituitary hormone release. Also, expression of neurotensin by female rat pituitary cells is cyclic. The presence of neurotensin receptors in the pituitary suggests that neurotensin is a paracrine or autocrine regulator (294).

Endothelin is another peptide, normally produced by vascular endothelial cells and found in gonadotropes (295,296). Because endothelin-3 is stimulatory to gonadotropes, its presence may reflect an autoregulatory role. Perhaps it activates the endothelin receptors, which can mobilize calcium and activate phospholipase C–, D–, and mitogen-activated protein (MAP) kinase–dependent pathways, as reviewed by Stojilkovic and Catt (297). It may also regulate prolactin during the estrous cycle, as discussed by Kanyicska et al. (298).

Expression of GnRH Receptors

Detection of Ligand–Receptor Binding by a Variety of Approaches

Gonadotropes may be identified cytochemically and thereby characterized by their capacity to bind the decapeptide, GnRH. Virtually every major laboratory in the field has developed methods to detect GnRH binding to gonadotropes, with a variety of signaling molecules, including biotin (299,300), ferritin (251,301), gold (302,303), rhodamine (304,305), photoreactive amino acids (306–309), and radioactive amino acids (310–314).

Before the findings are described, a word about the technology is important to the interpretation of the results. Each approach developed since the late 1970s has had advantages and disadvantages (299,300), and some have allowed a multidimensional view of functioning gonadotropes. The most sensitive techniques involved the use of biotin or radioactive or photoreactive amino acids; these techniques allowed the use of physiological concentrations of GnRH because the ligand remained biologically potent (300). Both rhodamine (304,305) and biotin (315) also allowed the visualization of GnRH uptake by *living cells* (Fig. 52). In many of the studies, the ligands could also be used with dual detection methods to detect the hormone content of GnRH target cells, mapping hormone stores (301,313,316). The labeled ligands could also be used with RHPAs to detect GnRH bound to secreting gonadotropes (317).

As our understanding of the process of receptor-mediated endocytosis and the structure and function of the GnRH receptor became more sophisticated, the most sensitive assays for GnRH receptors developed since the late 1990s have consisted of those that detect the receptor itself, often with constructs that allow cells to produce a GnRH receptor attached to fluorescent molecules that can be traced through the endocytotic pathways (318,319).

Whatever the signaling molecule, the following considerations are important in interpreting the cytochemical data. First, if exogenous ligands were used, the use of physiological concentrations similar to those used in receptor assays or secretion experiments added credibility to the protocol. The media also must include the use of protease inhibitors and, in most cases, protease-resistant analogs of GnRH (such as D-Lys^6 GnRH analogs) as hardy ligands (320). If a ligand was being detected at the receptor binding site, saturation of binding sites (detected by image analysis) was seen with concentrations in the physiological range (0.01 to 1 nM). A morphological study of time after exposure in living cells showed classic signs of receptor-mediated endocytosis (described later). Maximal numbers of binding sites are seen within the first 20 minutes after exposure with potent ligands. Finally, the ligand binding and labeling could be eliminated with increasing concentrations of unlabeled GnRH or other peptides that bind to the GnRH receptors. Other neuropeptides that do not bind the GnRH receptor should have no effects on labeling. Examples of peptides that have been used include TRH, corticotropin-releasing hormone, GHRH, and lamprey GnRH III.

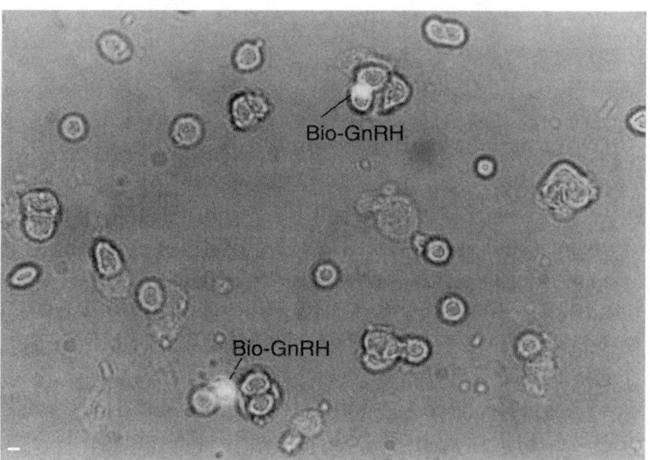

FIG. 52. Cells from proestrous female rats were exposed to biotinylated gonadotropin-releasing hormone, while living, for 3 minutes, followed by detection with cell sorter–grade avidin fluorescein. Labeling covers cells and cellular processes that form rapidly after stimulation. Bar = 10 μm.

Once binding was established, the labeled ligand, or label for the receptor, is traced into the cell to learn if the trafficking patterns resembled those described for classic receptor-mediated endocytosis. The steps and fate of the GnRH and its receptor are described in the following sections.

Membrane Binding and Microaggregation

Electron microscopic studies agreed that one could visualize GnRH binding to membranes of pituitary gonadotropes very rapidly (1 to 3 minutes) (299–314). GnRH ligands will bind at 0°C (304,305,320), and imaging of this early event shows diffuse labeling over the cell, particularly if the cells are kept at cool temperatures. This initial binding event often reflects a relatively widespread distribution of the receptors; however, most studies in our laboratory of cells exposed to GnRH at 37°C show that receptors are already clustered in macroaggregates (299,315,316) (Fig. 53). One can force a diffuse labeling pattern by first cooling the cells to 4°C. Studies by Naor et al. (304) and Hazum et al. (305) used rhodamine-labeled (D-Lys^6) GnRH and video-intensified fluorescent microscopy to demonstrate binding in a diffuse pattern to 10% to 20% of cooled pituitary cells. When the cells were warmed, the pattern changed to punctate patches (macroaggregates) within minutes. Our studies show that it may also concentrate on cellular processes that begin to project from the stimulated gonadotrope (240–242,299,300,315,316) (Figs. 52 and 53). This can be visualized by video micrography, and the response to stimulation is often an identifying marker in itself for a gonadotrope.

Studies with different ligands show that two critical early events are important for the stimulation of secretion. First, LH secretion is assayed within 1 minute after binding, which indicates that secretion depends on GnRH binding, but not GnRH receptor internalization (306,321). This is reinforced by the

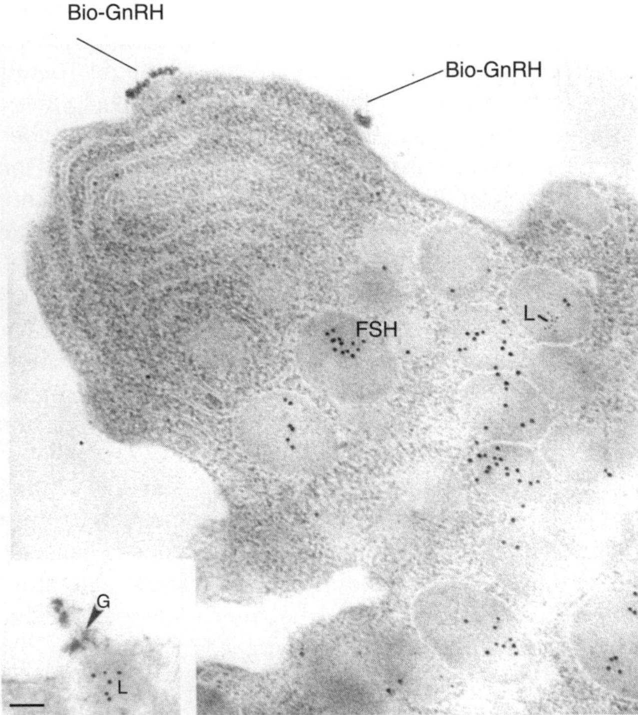

FIG. 53. Gonadotrope stimulated with biotinylated gonadotropin-releasing hormone (GnRH) for 3 minutes, followed by fixation and labeling with avidin–biotin peroxidase complex for biotinylated GnRH and immunolabeling with dual colloidal gold techniques for FSHβ (F, 50 nm gold) or LHβ (L, 10 nm gold). Labeling for biotinylated GnRH is on patches at the cell surface, and that for FSHβ and LHβ is in secretory granules or rough endoplasmic reticulum, which fill the process. Bar = 0. 4 μm. *Inset* shows a region from another cell, magnified to show the small gold labeling for LHβ better. The region shows an exocytosis profile near a site of labeling for biotinylated GnRH. The granule contains LHβ. Bar = 0. 1 μm.

finding that GnRH bound to small agarose fragments can stimulate LH secretion in spite of the cells' inability to internalize the bead (321).

The aggregation process is also important for secretion; however, it is likely driven by microaggregation of receptors. Conn et al. (322) reported that an antagonist can become an agonist if forced to aggregate by the introduction of bivalent antibodies that would bring the receptors together. More recently, Janovick and Conn (323) reported that GnRH agonists caused receptors to move closely enough to allow the transfer of a radioiodine molecule from the agonist (bound covalently to its receptor) to an adjacent receptor. Cornea and Conn (324) used fluorescence resonance energy transfer (FRET) to show GnRH-mediated receptor microaggregation. GH3 cells were transfected with GnRH receptor constructs containing green fluorescent protein and red fluorescent protein signaling molecules. They showed that agonist-induced microaggregation promoted the transfer of energy from the green fluorescent protein to the red fluorescent protein, with subsequent loss in green signal and gain in the red signal. The rapid time course correlated well with the timing of the GnRH-mediated LH secretion. Drugs that inhibit the microfilament-mediated patching and capping of the receptors (macroaggregation), such as vinblastine, had no effect on the FRET signal for microaggregation. Similarly, other drugs that inhibit internalization, such as cytochalasin (which destabilizes microtubules) or EGTA (which lowers calcium in the culture medium), blocked macroaggregation and internalization, but not the FRET signal (324). They concluded that receptor microaggregation was important for activation of the receptor, but that it was not dependent on cytoskeletal elements. They estimated that the distance between the agonist binding sites was less than 120 Å. They also observed that microaggregation is restricted to particular areas of the membrane.

Our ultrastructural studies of biotinylated GnRH binding concur and show that these areas may form processes and also appear to attract a high concentration of product (LH or FSH) destined for secretion. The ultrastructural view of these processes shows them to be filled with rough endoplasmic reticulum as well as secretory granules (Fig. 53). In some cells, the area around the nucleus empties as the gonadotropin stores move to the area of processes containing sites of binding for biotinylated GnRH (240,241,299,300, 316,317,325,326).

A number of workers have also shown that GnRH antagonists do not aggregate and are not internalized rapidly, if at all (327–329). Thus, this aggregation and eventual internalization are promoted only by biologically active receptors. Wynn et al. (327) reported that a radioactive D-Lys^6 antagonist of GnRH was internalized very slowly, if at all, and the labeling remained on the membrane for hours. This contrasted sharply to the pattern exhibited by the radioactive agonist. These data suggested that, whereas internalization may not be needed for LH secretion, it did follow from the rapid receptor activation by the agonist. This group also brought out the importance of glutaraldehyde fixation and the use of a D-Lys^6 analog of the GnRH to preserve the ligand at the membrane binding site, issues that also had been reported by Duello et al. (314).

Early Internalization Events

Most G-protein–coupled receptors are brought into an early endosome compartment after being sequestered in clathrin-coated pits, which bind and trap the ligand–receptor complex. A set of amino acid sequences on the C-terminal tail of the receptor binds to adaptin in the clathrin coat. β-Arrestins assist in the guiding of these receptors to the clathrin-coated pits and help potentiate the binding to adaptin (330–333). This pit then invaginates and remains connected to the plasma membrane by a narrow stem. Dynamin, a GTPase, becomes associated with the stemlike connection to the plasma membrane (333). Hydrolysis of GTP provides the energy needed for constriction and loss of this connection and ultimate formation of the clathrin-coated vesicle carrying the receptor and ligand as cargo. The clathrin-coated vesicle then loses its coat and fuses, by a specific sorting signal, with other vesicles to form the early endosome. This domain may either serve as a site for further sorting to late endosomes or it may recycle the receptors to the plasma membrane.

All of the early studies that detected labeled GnRH ligands reported internalization by small vesicles within minutes of exposure (300–314,325,326). However, the mammalian GnRH receptor expresses two characteristics that do not fit the classic pattern for G-protein–coupled receptors. First, there are detectable ligands and receptors that remain on the surface even at 20 minutes. Second, in many of the electron microscopic studies, investigators were unable to determine if GnRH–receptor complexes were internalized specifically by clathrin-coated pits and vesicles. A review of these studies shows that GnRH receptors appear to be associated with both uncoated and coated pits (302,303,310,314,325,326).

In the 1990s, the structure of the GnRH receptor was elucidated and workers reported that the mammalian GnRH receptor is unusual in that it lacks the C-terminal adaptin binding site needed for sequestration in clathrin-coated pits (334,335). Although it may be internalized in this pit, mechanisms other than adaptin may be used for the initial sorting and sequestration. This explained the reports suggesting entry by noncoated pits (302,303,310,314,325,326).

After the structure of the GnRH receptor was revealed, a number of studies focused on internalization events, comparing mammalian GnRH receptors with nonmammalian species that had GnRH receptors with C-terminal tails (318,333,336–338). Workers created constructs that included mammalian GnRH receptors with artificial C-terminal peptides, or a hybrid of GnRH receptors with the TRH receptor C-terminal peptide. They also looked at its dependency on β-arrestins and dynamin. Several facts became apparent. First, the presence of a C-terminal binding domain promoted the rapid internalization of the nonmammalian GnRH receptor and its desensitization. Furthermore, hybrid GnRH with a TRH C-terminal peptide tail also internalized rapidly.

However in spite of the absence of a C-terminal tail, Vrecl et al. (318) showed parallel internalization of transferrin receptors and GnRH receptors, suggesting that the same domains were used. Because the transferrin receptor enters through clathrin-coated pits after binding to adaptin, it was evident that the GnRH receptors also entered by this pathway. Heding et al. (336) further reported that internalization was dependent on dynamin, but not on β-arrestins.

Thus far, workers believe that the early internalization events bring the receptor into the early endosome domain in the first 3 to 10 minutes. However, the GnRH–receptor complex remains on the membrane longer and there is evidence for recycling of the receptor as the pH of the early endosome domain drops. It is unclear if this is the only route for recycling, however. Alternate routes are discussed in the following sections.

Trafficking to the Golgi Complex, Granule, and Lysosomal Domains

The earliest studies with labeled ligands showed that the GnRH–receptor complex moves to the Golgi complex and late endosome–lysosomal domains (251,300,301,310–312,314,325–327), including multivesicular bodies (Fig. 54). This clearly could degrade some of the receptors, and quantitative evidence showed losses in numbers of detectable target cells (310,314,325,326) in 30 minutes. Studies by Schvartz and Hazum (339) used lysosomotropic agents to raise the pH of the lysosomal compartment. These agents blocked degradation of the GnRH–receptor complex, but not its internalization, suggesting it can move to other domains.

This fits with reports from most of the earliest electron microscopic studies that showed that, 30 minutes after exposure, the labeled ligands were also found in secretory granules (325,326,310–312,314). These granules might provide another route for recycling. Evidence for this comes from the fact that, after LH release by low concentrations of GnRH, there is an increase in GnRH receptors. Hence, internalization into a nonacidic domain, such as secretory granules, may be one more key intracellular pathway for GnRH or its receptor. It might provide a source of cryptic receptors that are made available through secretion (325,326,339,340). One can connect this evidence with the earliest ultrastructural evidence showing the rapid formation of compound exocytosis profiles after GnRH stimulation, in vivo or in vitro (155–160) (Fig. 8). Figure 8 also shows labeling for biotinylated GnRH on some of the granules. The expansion in membrane surface area provided by the compound exocytosis profiles would provide an abundant source of new receptors or recycled receptors for the cell.

There is evidence that GnRH receptors are turned over regularly. Braden et al. showed that 6 hours is required for the synthesis and transit of newly formed

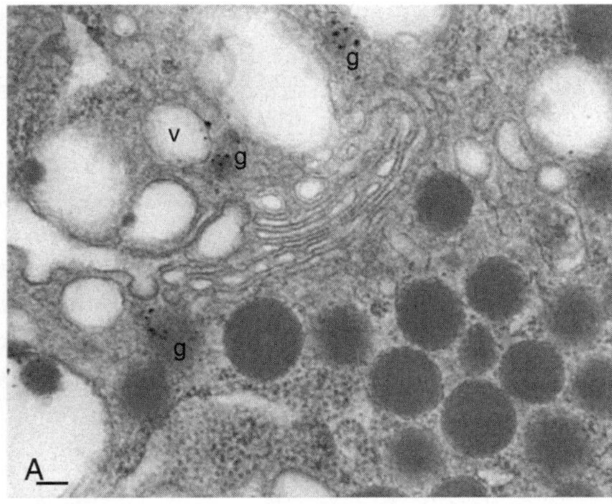

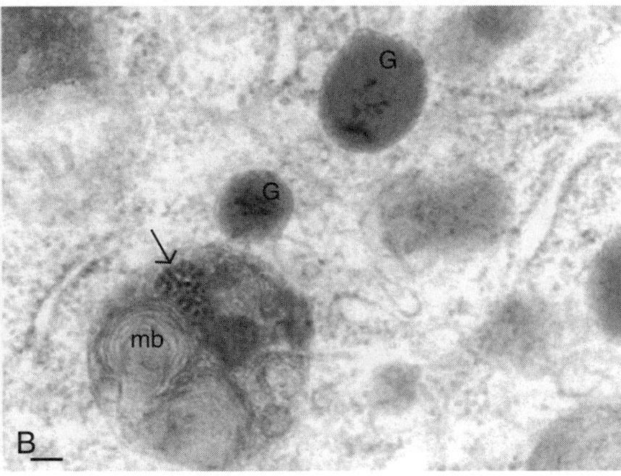

FIG. 54. Biotinylated gonadotropin-releasing hormone (GnRH) is seen in the Golgi complex (**A**) and in multivesicular bodies (**B**) with time after exposure. Some granule labeling (G) is also evident in **b.** Bar = 0. 1 μm.

GnRH receptors and that 28 hours is required to replace half of the GnRH receptor population (341). Starling et al. (342) estimated a 100% turnover of unstimulated receptors every 150 minutes; however, they indicated that turnover is much faster for the stimulated GnRH receptor. Finally, it is also clear that stimulation with protein kinase C (329) or addition of EGF (340) brings out cryptic receptors with independent properties in terms of their links to second messenger pathways. It is unclear if this population of cryptic receptors appears in new target cells, or in alternative domains in existing target cells.

Pituitary GnRH versus the Trafficking of the GnRH–Receptor Complex

The earliest reports of GnRH in gonadotropes found immunolabeling with anti-GnRH in secretory granules (283,284) (Fig. 8) and in the *trans* Golgi region (Fig. 54). This was presumed to have detected GnRH at receptor sites, and these early results were confirmed by those who reported uptake of the labeled ligand to a secretory granule domain (310–312,314, 325,326). During this time, reports of GnRH binding to nuclei emerged (310–312). At this point, there is no supporting evidence for a GnRH receptor nuclear binding site, but it is worth further exploration.

With regard to the binding to granules, studies in our laboratory reported the detection of GnRH immunoreactivity in cultured cells, even after 3 weeks in culture (284). This would presumably have been beyond the time of any exposure to the exogenous neuropeptide GnRH. Furthermore, as described previously in the section on studies of fetal development, Van Bael et al. (102) showed that GnRH transcripts are produced by Rathke's pouch cells early in development. They may promote differentiation and function of pituitary gonadotropes and any gonadotrope-dependent cell types, like lactotropes. Collectively, this evidence may herald the discovery of a local GnRH circuit that works in a paracrine fashion to increase gonadotrope functions. It will become more relevant as future workers learn more about the gap junction–formed syncytia that may carry the information (second messengers?) through this circuit (213–227).

Identification of Hormone Stores in GnRH Target Cells

The dogma about GnRH having only one target cell population has been strong since the mid-1970s. Only limited tests to prove or disprove these assumptions were done until the early 1990s (23,125,240–241). In the first studies of GnRH binding by Hopkins and Gregory, in 1977, labeling was reported on cells with the morphological characteristics of large gonadotropes, thyrotropes, and somatotropes (254). The findings were largely ignored because they did not fit assumptions at the time about how the pituitary was organized. During the next 20 years, most workers, including our own laboratory, simply used dual labeling for gonadotropic markers to identify and characterize the target cells. The earliest dual-labeling studies by Hopkins et al. showed that GnRH was taken up in cells that contained granules immunolabeled for LH and FSH (301). Light microscopic studies of radioactive GnRH binding by Duello and Nett (313) combined cytochemical labeling with immunolabeling to detect LH in the target cells. Early studies in our laboratory detected anti-GnRH antibodies and showed coexpression of GnRH and LHβ in serial sections through the same gonadotropes (283,284). Workers also correlated gonadotrope size or morphology as identifying markers, especially after castration or ovariectomy, which enlarged the target gonadotropes selectively (304,305,307–312,314).

Our studies in 1983 were the first to dual-label for biotinylated GnRH and LH or FSH β subunits, thus showing that cells with FSH were indeed targets for this ligand (316). Dual labeling for biotinylated GnRH and ACTH showed no significant coexpression in these early studies of male rat pituitary cells. The dual labeling has been validated over the years by a variety of methods. Most recently, the avidin detection system for biotinylated GnRH was followed by a new ImmPRESS (Vector Laboratories, Burlingame, CA) system that uses peroxidase nanopolymers attached to second antibody. This avoids the use of biotin-avidin reagents in the second labeling sequence (244).

In cycling female rats, the dual labeling for biotinylated GnRH and LH or FSH supported the concept that there were monohormonal LH or FSH cells. When biotinylated GnRH was detected with LH, 88% of GnRH target cells coexpressed receptors and LH, and 12% expressed only the GnRH binding sites. Similarly, when the ligand was detected with FSH, 65% of the GnRH-labeled cells contained FSH, whereas 35% expressed only GnRH. The cells expressing only GnRH binding may well belong to a monohormonal subset that was not detected by the immunolabeling protocol (the other gonadotropin) (316). Overall, there were approximately 15% to 21% cells with GnRH receptors, depending on the physiological state of the animal.

After a more thorough investigation of the hormone content of cells that bound biotinylated GnRH, our quantitative analysis showed the cells that contributed to the increase in target cells during diestrus (240,241). The top line of the graph in Fig. 55 demonstrates a significant increase in percentages of biotinylated GnRH-bound cells, and the lower two lines graph the dual labeling for gonadotropins. Comparing the three top lines shows that most of these target cells contain either LH or FSH. However, when other hormones were tested, significant percentages of GnRH-bound cells (40%) contained GH when the populations were taken from diestrous PM or proestrous AM samples. Figure 56 illustrates this binding with the use of avidin to detect biotinylated GnRH and the ImmPRESS nanopolymers to detect GH.

Counts showed that the percentages of cells with GnRH receptors and TSH, prolactin, or ACTH were less than 3% during the entire cycle. These findings confirm the first report by Hopkins and Gregory (254) which identified GnRH–receptor binding, and described target cells that resembled somatotropes or thyrotropes.

Regulation of Expression of GnRH Receptors

Dihydrotestosterone and Corticosterone. Cytochemical tools that detect GnRH binding to target cells have been combined with cytophysiological tests to learn more about the morphological correlates of

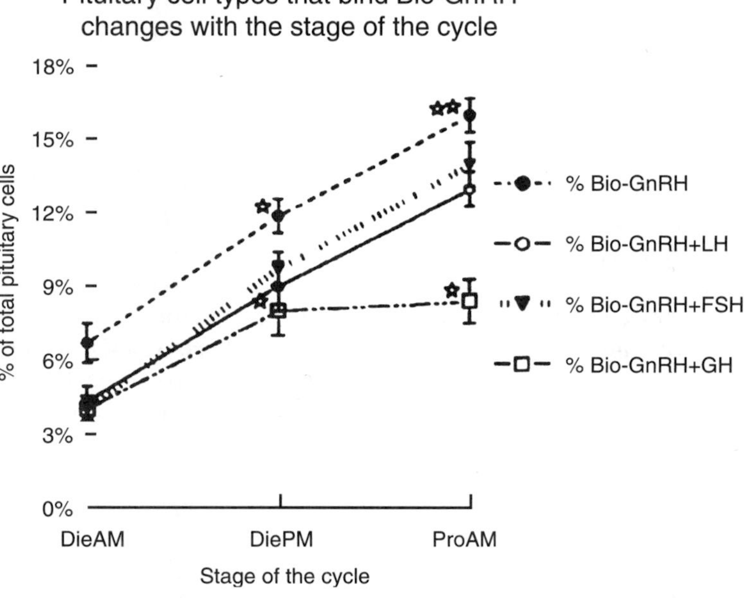

FIG. 55. Pituitary cells from cycling female rats were dispersed and plated after which they were stimulated with 0.5 nm biotinylated gonadotropin-releasing hormone (GnRH). The technique is described in Childs et al. (239). They were then fixed and dual labeled for biotinylated GnRH and LHβ, FSHβ, and growth hormone (GH) antigens. The *top line* shows the overall changes in the expression of GnRH with the stage of the cycle. The *middle two lines* show that most of the target gonadotropes contain LHβ and FSHβ because these are the percentages of anterior pituitary cells that coexpress GnRH receptors and gonadotropins. The *bottom line* shows that nearly 40% of cells with GH antigens express GnRH receptors near the time of the LH surge.

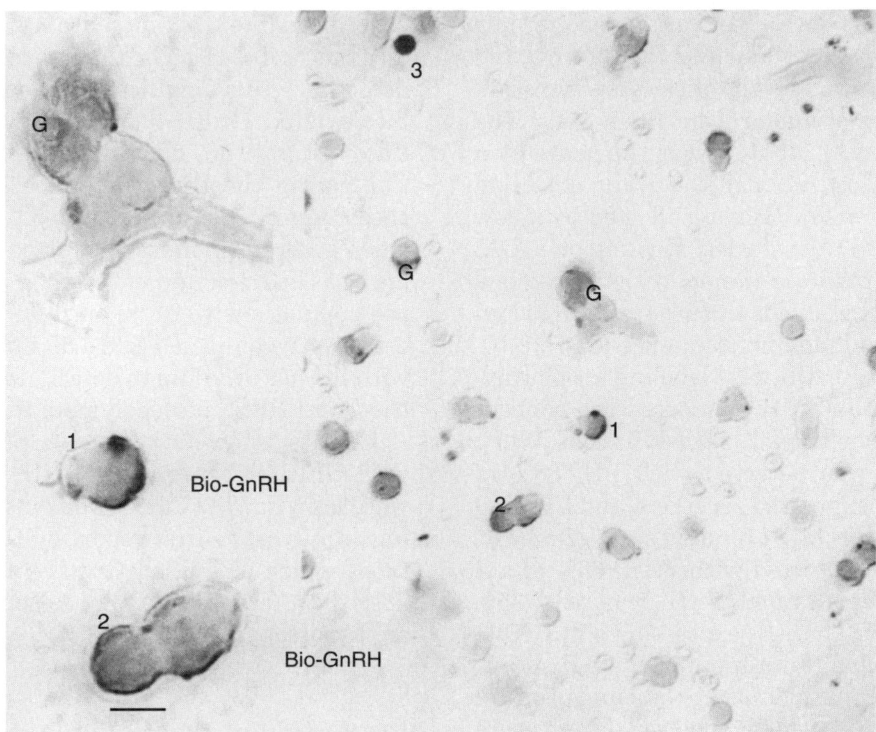

FIG. 56. Cells from diestrous female rats were plated overnight and stimulated for 10 minutes with 1 nM biotinylated gonadotropin-releasing hormone (GnRH). They were then fixed and labeled with the avidin–biotin–peroxidase complex (ABC) and nickel-intensified diaminobenzidine for the biotinylated GnRH (*black*, biotinylated GnRH). They were then immunolabeled with 1:125,000 anti-growth hormone (GH) followed by goat anti-rabbit immunoglobulin G linked to a peroxidase nanopolymer (ImmPRESS reagent). This sequence was detected with amber DAB (3,3′-diaminobenzidine tetrahydrochloride). The orange GH cells (seen as shades of gray here) are seen scattered throughout the population, and approximately 30% of them show binding to biotinylated GnRH. The higher magnifications show both dual-labeled and single-labeled GH cells. Bars = 10 μm (low magnification), 5 μm (high magnification). (See color insert.)

changes in GnRH binding or expression with the physiological state. In 1985, biotinylated GnRH was used in studies of male rats to learn about responses of GnRH target cells to dihydrotestosterone (DHT). DHT was known to cause a decrease in GnRH-mediated LH or FSH secretion. At the same time, we tested responses to corticosterone, which also had a potent inhibitory effect on LH and a less potent effect on FSH secretion (343). When cells with GnRH binding sites were counted, only DHT treatment resulted in a significant decline (from 16% to 9% of pituitary cells). Corticosterone did not appear to affect gonadotrope expression of GnRH receptors. Dual labeling showed that DHT reduced the percentage of LH-bearing cells with GnRH receptors by 50%, but it had no effect on percentages of FSH-bearing cells with GnRH receptors. This suggested a selective effect of DHT on a subset of gonadotropes in the male rat, perhaps on monohormonal LH cells. When the percentages of cells labeled *only with biotinylated GnRH* were counted in these dual-labeled fields, they were much higher than those in the fields from female rats. In vehicle-treated rats, in a given dual stain, 45% to 50% of GnRH-labeled cells did not contain either LH or FSH (343).

Estrogen Regulation and the Estrous Cycle. During the early to mid-1980s, workers reported significant increases in expression of GnRH receptors with the estrous cycle [reviewed in Lloyd and Childs (315)]. There was a rise in expression from metestrus to late diestrus or early proestrus that paralleled the rise in serum estrogen. We studied the expression of biotinylated GnRH binding in pituitary cells to learn if this rise reflected changes in numbers of GnRH receptor–expressing cells, or changes in expression of biotinylated GnRH binding per cell (315). Counts of GnRH-receptive cells showed the lowest values (5% of pituitary cells) at metestrus, followed by a gradual rise to peak levels (20% of pituitary cells) by the morning of proestrus (Fig. 57).

Pituitary cultures from metestrous, diestrous, and proestrous rats were treated for 3 days with 0.001 to 1 nM estradiol benzoate. The metestrous and diestrous cells showed enhanced expression (to >10% of pituitary

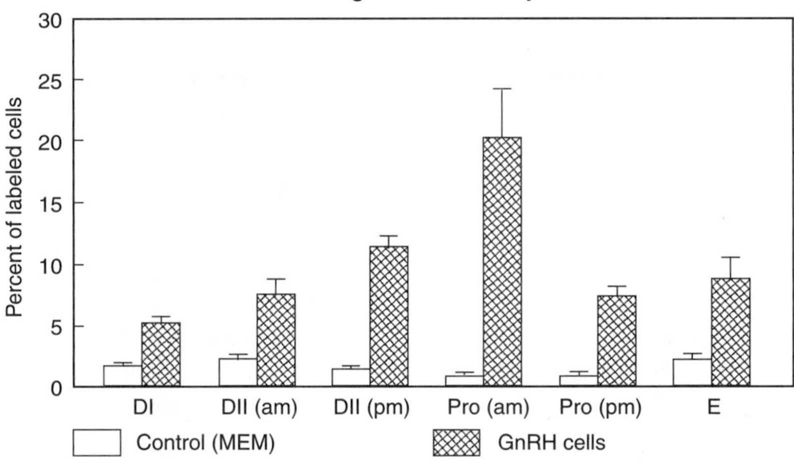

FIG. 57. Changes in expression of gonadotropin-releasing hormone (GnRH) receptivity overall during the estrous cycle. Cells were collected from cycling female rats at 10 A.M. on each day, plated on coverslips, stimulated with biotinylated GnRH, and then labeled with avidin–biotin–peroxidase complex and compared with fields treated with minimum essential medium only (control MEM). There was a steady rise in expression of GnRH-receptive cells, reaching a peak by the morning of proestrus followed by a decline to reach a nadir in diestrous I (metestrus).

cells) with the lowest concentrations of estrogen. In contrast, the proestrous cells, which started with 20% GnRH target cells, showed a dose-dependent decline in expression to only 5% of pituitary cells after estradiol treatment (Fig. 58). This bipotential action of estrogen reflects both the positive and negative feedback effects of this steroid, and the data showed that the changes in receptor numbers reported by other workers [reviewed in Lloyd and Childs (315)] correlated with an actual change in number of receptive cells (315). Another example of the enhancing effect of estrogen on GnRH receptors is graphed in Fig. 46.

Lloyd also focused on cell fractions separated by size and mapped responses to estradiol in the largest gonadotropes, comparing them with the smallest group. Figure 25b shows that small cell fractions from mixed cycling female rats do not contain abundant GnRH target cells. However, when biotinylated

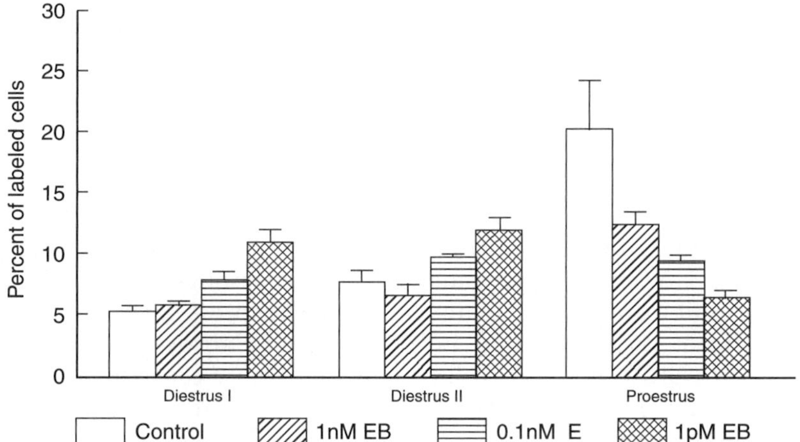

FIG. 58. Cells from diestrous or proestrous rats were treated for 3 days with 1,100, and 1,000 pM estradiol benzoate (EB) and then exposed to biotinylated gonadotropin-releasing hormone (GnRH), followed by labeling with avidin–biotin–peroxidase complex. Significant increases in GnRH-receptive cells were seen after the lowest concentrations of EB in populations from diestrous rats, indicating a bipotential effect of EB similar to that described for growth hormone cells. In contrast, EB in all concentrations was inhibitory to expression of GnRH receptors in proestrous rats.

GnRH binding sites were detected, a surprisingly high percentage of small cells from estrous animals bound the neuropeptide (315). These results support the suggestion that the small cell fractions from estrous rats include gonadotropes that are lighter because they have secreted their stores, although they may still retain the GnRH receptor capacity. They do not respond to estrogen, however. In contrast, small cells from diestrous rats are more responsive to estrogen. The steroid caused nearly a twofold increase in the percentage of biotinylated GnRH-bound target cells in this population.

When large gonadotropes from estrous rats were tested, treatment with 0.1 nM estrogen caused a significant increase in percentages of biotinylated GnRH target cells. In contrast, estrogen had no effects on the largest gonadotropes from the diestrous rats. These data suggest that estrogen promotes GnRH receptor synthesis selectively in the smaller gonadotropes and, as described previously, is most effective during diestrus.

Inhibin and Activin Regulation of GnRH Receptors. Inhibin was first discovered in 1932 as a factor that suppressed hypertrophy of gonadotropes after castration. It belongs to the TGF-β superfamily and is a heterodimer of an α subunit and one of two β subunits. It may be produced in the ovary or testis, as well as by cells in the pituitary (271–275). Inhibin blocks basal and GnRH-mediated FSH synthesis and release, and although it blocks LH secretion, it does not block its synthesis. This action might be consistent with an effect on secretion from bihormonal gonadotropes with specific effect on FSH synthesis in all cells containing FSH. Farnworth et al. (344) and Wang et al. (345) reported that inhibin reduced binding to GnRH and also blocked upregulation of GnRH binding sites by the calcium ionophore, A23187. Braden et al. (346) showed that inhibin did not block the rate of GnRH receptor synthesis; however, it did block the GnRH stimulation of synthesis of its own receptors (347,348).

Cytochemical studies in our laboratory were conducted in collaboration with Dr. W. Miller to learn if inhibin's effects were focused on FSH gonadotropes and if they affected binding by biotinylated analogs of GnRH (241). Counts of FSH from diestrous rats (after a 24-hour exposure to inhibin, or porcine follicular fluid) showed a significant 50% reduction in the area of label for biotinylated GnRH on target cells as well as a significant reduction in the density of that label. When gonadotropes were further identified, there was a 30% reduction in percentages of FSH-bearing cells, a 10% decrease in average FSH cell area, and significant decreases in the density of FSH stores in those cells. However, there were no changes in overall percentages of LH cells and only small decreases in area of LH cells or density of LH stores. In addition, there was a reduction in the percentages of FSH cells that bound GnRH from 83% of all FSH-bearing cells to 32%. Thus, only one third of the remaining FSH cells showed evidence of binding sites for GnRH. Inhibin also reduced the percentages of LH cells with GnRH receptors from 84% to 40%, indicating that one of its actions indeed focused on the bihormonal gonadotropes, with likely additional actions on monohormonal FSH cells (241).

Activin is another member of the TGF-β superfamily and is a dimer of the two inhibin β subunits. It is produced by cells in the gonads and the pituitary and thus, like inhibin, is both an endocrine and paracrine or autocrine regulator [reviewed in Childs and Unabia (241)]. Activin stimulates the synthesis and release of FSH and also the synthesis of GnRH receptor mRNA and proteins (347–349). Activin stimulated significant (40% to 50%) increases in GnRH binding to target cells after 24 hours of treatment (241). In addition, activin increased the density of labeling for biotinylated GnRH on the target cells. In all dual-labeling experiments on diestrous rats, activin increased the percentages of LH or FSH cells that bound GnRH. However, there were no activin-mediated increases in binding when proestrous rat populations were studied. These cells already show peak GnRH binding activity by diestrus.

There were also increases in the overall percentages of cells with FSH or LH antigens in diestrous rats (241). In proestrous populations, activin stimulated an increase only in percentages of FSH cells. Activin stimulation also resulted in a decrease in average areas of FSH cells, suggesting that activin may be regulating the appearance of a population of small FSH gonadotropes (241). Collectively, the data for the proestrous rats suggest that activin stimulated an expansion in monohormonal FSH gonadotropes. The decrease in average area suggests that the expansion is from a smaller subset of FSH cells. This correlates well with our previous studies of FSH cells during the cycle (181) and after ovariectomy (176). Activin may be the mediator that selectively regulates the appearance of a small subset of FSH cells to support the population. This provides more evidence for selective regulation of subsets of gonadotropes to differentiate and express gonadotropins and GnRH receptors.

Other Factors that Increase GnRH Receptors. A number of conditions and factors that upregulate GnRH receptors are beyond the purview of this chapter. Some of these arise from the pituitary, and are briefly mentioned. Sutton et al. (350) showed that immunoneutralization of NPY blocked the steroid-induced LH surge in ovariectomized rats and suggested that NPY may potentiate a priming effect of GnRH on the gonadotrope. Activators of protein kinase C (and other kinases) can bring out more GnRH receptors in the population (329). Finally, EGF, produced by populations of pituitary cells, was also shown to bring out GnRH binding sites (340). Conditioned media and immunoneutralization approaches were used to show that EGF and TGF-α were the likely growth factors that increased GnRH binding activity from the conditioned media of pituitary cells. Further work is needed to learn if secreted pituitary GnRH also plays a role in self-priming of its receptors.

Autocrine and Paracrine Influences on Gonadotropes

Activin, Inhibin, and Follistatin and Nonparallel Release

In the foregoing sections, we discussed a number of factors that could affect nonparallel release of gonadotropins, including monohormonal cells or monohormonal granules in bihormonal cells. However, this event could also be modulated by endocrine or paracrine factors other than GnRH. It is well established that pituitary cells can synthesize and secrete α-subunit proteins or FSH β subunits in a manner that is independent of GnRH. An excellent review of this topic by Winters and Moore (351) is available. Culture experiments have shown that factors regulating this activity are present in pituitary cells [reviewed in Winters and Moore (351)]. Activin-β antisera added to pituitary cultures decreases secretion of FSHβ, FSH, and also activin/inhibin-βB. It also decreases follistatin mRNA production. Hence, it is believed that FSH cells are regulated heavily by local factors produced by pituitary cells. Activin genes are expressed in pituitaries of several species (271–275). Dual labeling in the human suggests that activin-βB is found in gonadotropes and thyrotropes, whereas activin-βA is found in gonadotropes, somatotropes and lactotropes (352). Inhibin subunits are found variably in pituitaries of different species. Anti-porcine inhibin-α labeling has been detected in a number of pars distalis cells in the primate (271–275).

Follistatin is the third molecule in the circuit, and it acts to limit the actions of activin (276–279,351). It may be a paracrine or autocrine regulator because it is produced, to a varying extent, by subsets of all pituitary cells (278,279). Thus, regulation of FSH by steroids or neuropeptides could be indirect and reflect the regulation of expression of activin (to increase FSH production and release) or inhibin or follistatin (to limit FSH production and release).

Some of the nonparallel changes in LH and FSH secretion may well be based on the combination of the frequency of the GnRH pulses (slow pulses favor FSH secretion and rapid pulses favor LH secretion) and the regulation of activin, inhibin, or follistatin at the level of the pituitary or gonads. Examples of how this regulation may occur, taken from a review (351), are given in the next paragraphs.

During prepubertal development in the male rat, there is a rapid rise in FSH β subunits and GnRH receptors with no changes in LH secretion. Whereas the factors driving this nonparallel change are not known, workers have presented evidence that the rise in FSH might be permitted by a decline in inhibin or the significant reduction in the mRNA encoding a potent small form of the follistatin molecule (353). The reduction in follistatin would facilitate actions by activin on both FSH and GnRH receptor genes (351,353).

Changes in LH and FSH after castration occur in a nonparallel fashion, and different species also may exhibit different response patterns. For example, species differences in the expression of follistatin after castration may cause the different rates of rise in FSH after surgery (354). The slow rise in follistatin expression after surgery in the rat contributes to a slower postcastration rise in FSH mRNA and may also contribute to the lower FSH mRNA seen by 28 days after surgery. In contrast, there is no postcastration rise in follistatin in the primate, which permits a more rapid rise in FSH. These differences may reflect the source or the regulatory circuitry for follistatin in the different species (351).

One mechanism behind the relationship between GnRH pulse frequency and LH or FSH secretion may also relate to follistatin or activin. In rats, rapid GnRH pulses have been shown to favor LH secretion because they also elevate follistatin, which would limit the supply of activin needed to stimulate FSH (355). This does not explain the effect of GnRH pulses on FSH secretion in primates, however (351).

Gonadal steroids may also regulate LH and FSH differentially through this circuitry. Gharib et al. (356) showed that testosterone negative feedback inhibits GnRH, GnRH receptors, and LH secretion, but increases FSHβ and FSHβ mRNA. This rise may be due to androgen inhibition of follistatin, which permits activin-mediated increases in FSH (356). In contrast, androgens stimulate follistatin in primates, which would have an inhibitory effect on FSH (357).

More recent studies by Leal et al. (358) highlighted the different and complicated effects of glucocorticoids and testosterone on FSH secretion and expression of FSHβ mRNA in pituitary cultures. Whereas both glucocorticoids and testosterone stimulated FSH production, the actual mechanisms behind the enhanced FSH appear to be different because the adrenal steroids stimulated and testosterone inhibited the production of follistatin mRNA transcripts.

These are only a few examples illustrating key concepts used to interpret studies of nonparallel release of LH and FSH by paracrine or autocrine mechanisms. A number of considerations go into interpreting these data. First, nonparallel release mechanisms may include follistatin, activin, or inhibin as mediators. Second, the regulation of follistatin may vary from species to species and, although it may be a mediator, there may be different consequences because of species-specific changes in expression. Third, it is evident that the same final response may actually be elicited through different pathways in this circuitry.

Epidermal Growth Factor as a Paracrine or Autocrine Regulator

Studies in our laboratory first discovered a potential role for EGF and its receptor in gonadotrope function when EGFR was upregulated, after cold stress, in cells with FSH antigens (359). We therefore began follow-up studies of EGFR expression in gonadotropes to learn if it was regulated by events in the estrous cycle (360). EGFRs were most abundant (45% of pituitary cells) during metestrus; 2% to 3% of pituitary cells expressed LH or FSH antigens and 6% expressed the mRNAs. Expression of EGFRs slowly declined to reach a nadir by proestrus (20% of pituitary cells). However, nearly half of these EGF target cells expressed LH or FSH antigens by proestrus. Thus, as EGFRs overall are decreased, expression of EGFRs by gonadotropes is increased to 80% of the gonadotropes by proestrus.

This led to a search for potential regulators for EGFR expression by gonadotropes. Studies showed that increases in expression were seen after the use of serum in the media as well as EGF itself. Immunoneutralization of the media with anti-EGF reduced EGFR expression by gonadotropes to basal levels. This suggested that the pituitary cells were the source of the EGF that upregulated EGFRs on gonadotropes (360).

One function for these receptors was described in previous sections of this chapter. Locally produced EGF appears to increase the GnRH binding capacity of pituitary cells, unmasking cryptic GnRH receptors (340). Other workers have shown an EGF-mediated increase in LH release either directly (361) or in tandem with hypothalamic tissue, or by increasing responsiveness to estradiol (362). Thus, EGF may potentiate the GnRH-mediated actions on LH. Our studies have also reported that 10 ng/mL EGF causes nearly a fourfold increase in expression of c-fos in FSH cells, but not LH cells (363). These findings correlate well with the fact that the activator protein-1 (AP-1) signaling pathway is used for the activation of the FSH gene, but not the LH gene. Finally, EGF has potent effects on gonadotrope proliferation, as described in the next section. In summary, EGF may be an important paracrine or autocrine regulator for gonadotropes. Findings for the production of MMPs by gonadotropes suggest that this mechanism could cleave the EGF-like ligands bound to membranes (212). Thus, EGF may also work as a juxtacrine regulator.

GONADOTROPE AND LACTOTROPE TURNOVER

Proliferation of Differentiated Gonadotropes and Lactotropes

This section is interposed between sections on gonadotropes and lactotropes because of the evidence that gonadotropes influence lactotrope proliferation. Transgenic mice without gonadotropes do not show prolactin cell expansion (257–259), and the factor that regulates this proliferation postnatally may be a 16-kDa protein fragment from the POMC molecule, produced by gonadotropes (268).

Historically, mitoses by pituitary cells were studied as early as 1917 (364). These workers used morphology and "mitotic figures" as their major evidence and reported mitotic figures in only 1% to 2% of the population (364–366). Thus, early work suggested that mitotic activity in the pituitary gland is relatively low. However, as more sensitive methods were developed, information about the extent of mitotic activity grew.

Early cytochemical studies detected [^{3}H]-thymidine uptake by nuclei during the S phase of the cell cycle and agreed that mitotic activity increased during estrus in mature female rats (367–369). Oishi et al. (370) further reported that female rats renew their pituitary cell populations twice as fast as male rats.

Subsets of all pituitary cell types have been identified cytochemically (by dual labeling for [^{3}H]-thymidine, bromodeoxyuridine [BrDU], or proliferating cell nuclear antigen) (367,368,369–371). BrDU is incorporated into new DNA during the S phase (372–374), and proliferating cell nuclear antigen appears just before S phase, moves to the nucleus, and facilitates DNA polymerase (374–376). Both of these products can be detected cytochemically. Dual labeling detects pituitary hormones and hence the identity of the dividing cell.

Tilmans et al. (264,265) reported [^{3}H]-thymidine uptake by prepubertal female rat pituitary cells in aggregate culture. [^{3}H]-Thymidine uptake was seen in a number of cell types and was highest on day 6 of culture. In the pituitary aggregates, 15.5% ± 1% were lactotropes and 5.3% ± 1.1% of the cells contained gonadotropins. After redispersion and culture in monolayers, 19.2% ± 0.7% were lactotropes and 4.5% ± 1.7% were gonadotropes. Because GnRH has been shown to stimulate the gonadotrope-derived factor that enhances lactotropes, 1 nM of GnRH was tested for 24 hours and then during a 16-hour labeling period with [^{3}H]-thymidine. The analysis showed an increase in [^{3}H]-thymidine uptake by lactotropes from 17% in the control to over 22%. No changes were seen in gonadotropes. There were significant GnRH-mediated decreases in mitotic somatotropes.

Lactotropes are small and they separate in a fraction that contains less than half of the gonadotropes found in the initial cell suspension (264,265). Figure 25b shows this relationship and peak percentages of lactotropes in small cell fractions. Gonadotropes enrich the large cell fractions (206). Tilmans et al. separated fractions enriched in gonadotropes or prolactin cells and then combined the enriched fractions, creating aggregate cultures. When lactotropes were cultured with gonadotropes, GnRH stimulated significant increases in [^{3}H]-thymidine labeling, from 30% of identified lactotropes to over 42%. GnRH had no effect on lactotropes in aggregates made from gonadotrope-deficient fractions. NPY also had a dose-dependent stimulatory effect on mitotic lactotropes in these cultures, although the magnitude was not as great as that in gonadotropes (264,265).

This evidence adds to that showing a postnatal expansion in prolactin cells, mediated by gonadotropes and regulated by GnRH. Perhaps the increase in lactotropes is at the expense of transformed somatotropes, which may explain the GnRH-mediated reduction in GH cells. This relationship deserves further exploration in studies focused on factors that regulate the somatogonadotropes.

Regulators of mitotic activity point to additional mediators that could change mitoses in gonadotropes and lactotropes. Szijan et al. (377) reported [^{3}H]-thymidine uptake stimulated by estrogen along with the increase in coexpression of the two proto-oncogenes, c-*myc* and c-*fos*, as well as an increase in transcription of the prolactin gene. Suganuma et al. (378) described an increase in DNA synthesis and poly adenosine ribosylation (ADP-ribose) of chromatin proteins in prolactin cells from diestrus to proestrus. Their results showed that these events were decreased by estrus. Also, the increase in prolactin mRNA from diestrus to proestrus was abolished by nicotinamide, an inhibitor of poly (ADP-ribose) synthesis, indicating that both cell proliferation and prolactin gene expression are regulated by events that stimulate poly (ADP-ribose) synthesis.

Chaidarun et al. reported that estradiol reduced the expression of gonadotropin β-subunit mRNAs in sheep gonadotropes, in vitro, as it increased the uptake of [^{3}H]-thymidine in pituitary cells overall (379). Many of the new mitotic cells were not identified by any pituitary hormones, and it is possible that some of these cells have the potential to become gonadotropes. Katayama et al. (380) and Chaidarun et al. (379) have both shown that activin stimulates mitotic activity of FSH cells in vitro. Finally, Jones et al. (381) concur with the detection of BrDU uptake in gonadotropes, only they report uptake by only 1% to 2% of the gonadotropes.

The physiological state that has shown the most abundant mitotic gonadotropes is that after gonadectomy. Inoue and Kurosumi (170), Sakuma et al. (172), Smith and Keefer (171), and Romano et al. (382) reported an increased mitotic index in identified gonadotropes after castration, indicating that the new gonadotropes were derived from mitoses. Castration reduces mitotic prolactin cells (172). Some of these workers also reported an increase in small mitotic cells, which were not identifiable by their hormone content (170,171). Romano et al. (382) reported that only 20% of mitotic cells could be identified as gonadotropes. We have shown up to a threefold increase in numbers of gonadotropes after gonadectomy as well as conversion to a small subtype at longer time points after the surgery (174). Finally, in studies of female (383) and male (384) animals, the castration-mediated increase in mitotic cells was blocked by DHT and not affected by injection of a single dose of GnRH.

Effect of Age and Reproductive Hormones

As the techniques for detection of proliferative activity became more sensitive, information about stimulatory factors has emerged. Several stimulatory factors have been shown to regulate gonadotrope proliferation, suggesting that this may be a mechanism for replenishing the population. Williams et al. (385) reported an estradiol-induced increase in uptake of [^{3}H]-thymidine incorporation in the αT3-1 gonadotrope cell line. They also reported that estradiol increased the proportion of cells in the S and G_2/M phases of the cycle and reduced the doubling time from 73 to 39 hours. Chaidarun et al. (379) reported an estrogen-induced, 33% increase in [^{3}H]-thymidine labeling in sheep pituitary cells as it reduced gonadotropin mRNA. Collectively, these studies suggested that estrogen drives gonadotropic cells to a less differentiated state, permitting mitoses.

Some of the earliest studies on mitotic pituitary cells focused on prolactin cells (386,387). Kurosumi noted that mitotic cells included differentiated cells and illustrated this by showing a photograph of a mitotic and obviously identifiable prolactin cell (387). Shirasawa and Yoshimura reported that there was a progressive increase in mitotic growth rates in GH and prolactin cells with postnatal age (388), with little increase in other cell types except for ambiguous cells. Again, this suggests the presence of a subset of pituitary cells that divide and differentiate to one of the other cell types. Takahashi et al. (389) reported changes in mitotic activity of lactotropes with age and also with the estrous cycle. The mitotic index of prolactin cells was higher at estrus than at all other stages of the cycle and also higher than the index in male rats. The mitotic index of prolactin cells peaked at 60 days of age and then decreased as the females aged. In contrast, in males, the peak index was at 20 days, with a progressive decrease with age.

Estrogen is a well-known stimulator for prolactin cells (discussed in more detail in later sections). Estradiol stimulated proliferation as it increased prolactin release (390), and bromocriptine, a well-known prolactin inhibitor, abolished the stimulatory effects of estradiol. In all of these treated rats, 80% of the mitotic cells were lactotropes, indicating that this cell type has great potential for expansion. Yonezawa et al. (391) also reported the inhibitory effect of bromocriptine on cell proliferation in estrogen-stimulated pituitaries.

Effects of Growth Factors

The earliest studies by Chaidarun et al. (379) reported that a number of growth factors stimulated proliferative activity of undifferentiated sheep pituitary cells, labeled only by keratin. This group tested bovine FGF and EGF as stimulatory agents. Thus, a primary role for growth factors may be to stimulate a population of cells that could serve as stem cells. TGF-β increased the growth of gonadotropes, especially increasing FSHβ gene expression. This is consistent with its similarity to activin.

Proesmans et al. (392) have reported mitogenic effects of nerve growth factor (NGF) on the reaggregate cell cultures of 14-day female rat pituitaries. NGF increased the number of mitotic lactotropes with [^{3}H]-thymidine in a dose-dependent manner and increased the number of cells expressing prolactin mRNA, but not those with prolactin antigens. NGF also stimulated more mitotic corticotropes, but it had no effects on somatotropes, thyrotropes, or gonadotropes. The overall mitotic index stimulated by NGF was much larger than that from the corticotropes or lactotropes alone, suggesting that NGF has a potent effect on progenitor or stem cell mitoses. Many of these cells were also labeled for Pit-1, suggesting that they belong to the line producing somatotropes, lactotropes, and thyrotropes. This fits with a role for Pit-1 in expansion of this cell line.

Their study may reveal a function for pituitary-based NGF, characterized by Patterson et al. from our laboratory (393,394). Our studies showed that biologically active NGF was expressed by a number of pituitary cells, stimulated by interleukin-1 and inhibited by GHRH. Perhaps it is part of a circuit that regulates the numbers of stem cells for somatotropes, prolactin cells, or thyrotropes in support of the body's responses to disease or inflammation.

More recent studies in our laboratory have evaluated the efficacy of GnRH, EGF, and activin as mitogens for gonadotropes (395,396). We first developed techniques to enrich gonadotropes to greater than 90% (397) and then applied MTT cell growth/cell death assays to detect changes in cell number. All three factors were mitogenic when added alone, and no additive effects of two of the factors together were seen (in any combination). Because the MTT assays cannot differentiate changes in cell number due to mitoses from those due to survival, mitotic gonadotropes were also detected by BrDU uptake during the S phase of the cell cycle. BrDU labeling after treatment with GnRH or EGF showed that only 2% of pituitary cells were mitotic. However, both stimulatory factors increased percentages of mitotic pituitary cells to 12% to 15% of the population (395,396) (Fig. 59). When dual labeling was applied, only 10% of LH or FSH cells were mitotic (or <1% of pituitary cells). However, both stimulatory factors showed equal efficacy in stimulating mitotic activity in LH or FSH cells to 25% to 30% of gonadotropes, or 3% to 4% of anterior pituitary cells (Fig. 60). Thus, gonadotropes identified by their content of antigens are only a subset of cells stimulated by EGF or GnRH.

Studies by Shah et al. (398) identified signaling pathways leading from GnRH or EGFR to second messengers that mediate proliferation. They transfected HEK293 cells, which have endogenous EGFRs, with GnRH receptors. Their results showed that GnRH causes sustained extracellular-signal–regulated kinase-1 and 2 (ERK1/2) phosphorylation (in the MAP kinase pathway) that is independent of Src and activation of the EGFR. This correlates well with our findings that there are no additive effects of GnRH and EGF signals on mitotic gonadotropes (395,396). They also traced the mitogenic signal to the nucleus, detecting ribosomal S6 kinase-1 (RSK1), which is transported to the nucleus after phosphorylation by ERK1/2. RSKs are cytoplasmic serine/threonine kinases that phosphorylate transcription factors, including c-fos and c-jun. In pituitary LβT2 gonadotropes, Liu et al. reported nuclear accumulation of ERK1/2 along with the activation of c-fos and LH genes (399). It is possible that GnRH's and EGF's mitotic effects are mediated through these kinases.

One important site of EGF ligands includes those bound to transmembrane precursors. Such EGF ligands could stimulate neighboring pituitary cells by "juxtacrine stimulation" only after they are released

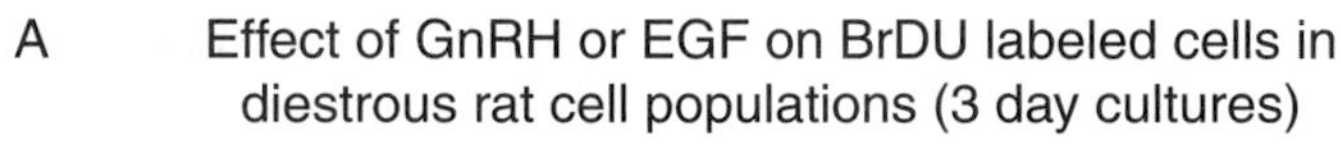

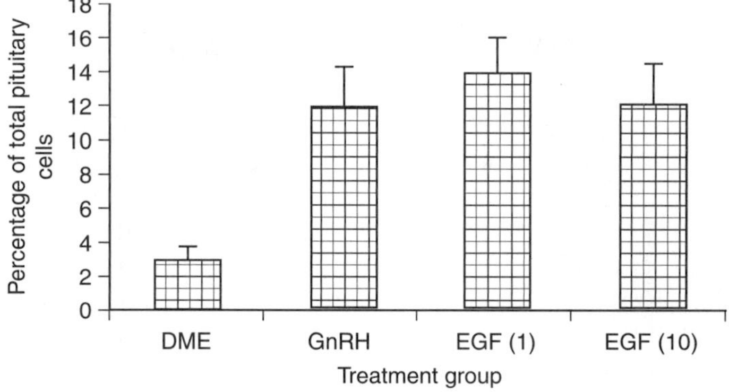

B

Effect of GnRH or EGF on numbers of dividing gonadotropes in the pituitary cell population: Duel labeling

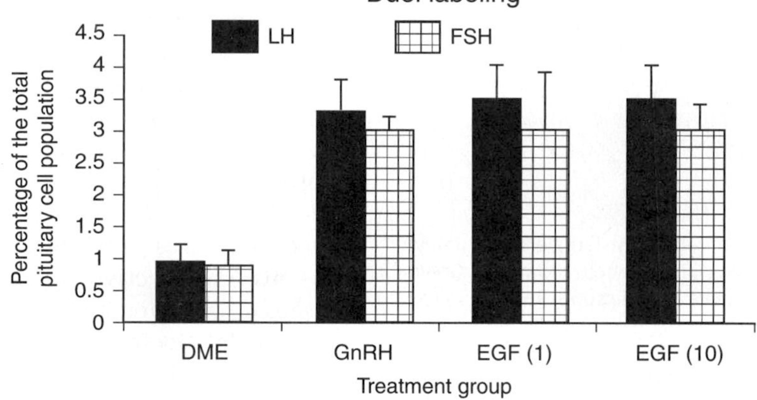

FIG. 59. Pituitary cells from diestrous rats were dispersed and grown for 3 days in media containing defined media (DME), epidermal growth factor (EGF), or gonadotropin-releasing hormone (GnRH). Cells were then treated with bromodeoxyuridine (BrDU) for 1 hour before fixation and dual-labeled for BrDU and LH or FSH. Each of the factors increased mitotic cells in the anterior pituitary by over fourfold (**A**). After dual labeling for gonadotropins, all factors increased mitotic gonadotropes from 1% of the population to 3.5% (**B**). EGF and GnRH did not have additive effects. See Childs (396) and Childs and Unabia (397) for technique and illustrations of labeled cells.

from their transmembrane precursor by MMPs. In a previous section, studies by Roelle et al. were described in which the EGFR was activated by GnRH through ERK1/2, which caused the shedding of MMP-2 and MMP-9, and thereby the release of EGF-like ligands (212). For further information, the role of these MMPs has been reviewed by Shah and Catt (400).

Studies in our laboratory have also shown that EGF stimulates c-fos expression in FSH cells, whereas GnRH stimulated c-fos expression only in GH or ACTH cells (401). Thus, these factors may have more heterogeneous effects in a normal population of pituitary cells than in transfected cells or tumor cell lines. Finally, IGF-1 is now being studied widely as a stimulatory factor for pituitary cell populations; however, it is largely being studied for its antiapoptotic effects, and therefore is discussed in this context in the next section.

Regulation of Apoptosis in Gonadotropes and Lactotropes

The realization that apoptosis can play a role in pituitary cell function stems from basic evidence that lipopolysaccharide (LPS), an endotoxin of gram-negative bacteria, is able to reduce prolactin release from both male and female rat pituitaries (402) in a dopamine-inducing manner (403). As will be discussed in the section on prolactin, dopamine is a potent inhibitor of prolactin cells. Specific tests show that that some of this reduction results from apoptosis of key pituitary cells. Studies of apoptosis in the pituitary have proliferated rapidly since the mid-1990s, along with information about regulators. One of the earliest reports, by Drewett et al., focused on apoptotic events regulating populations of prolactin cells made

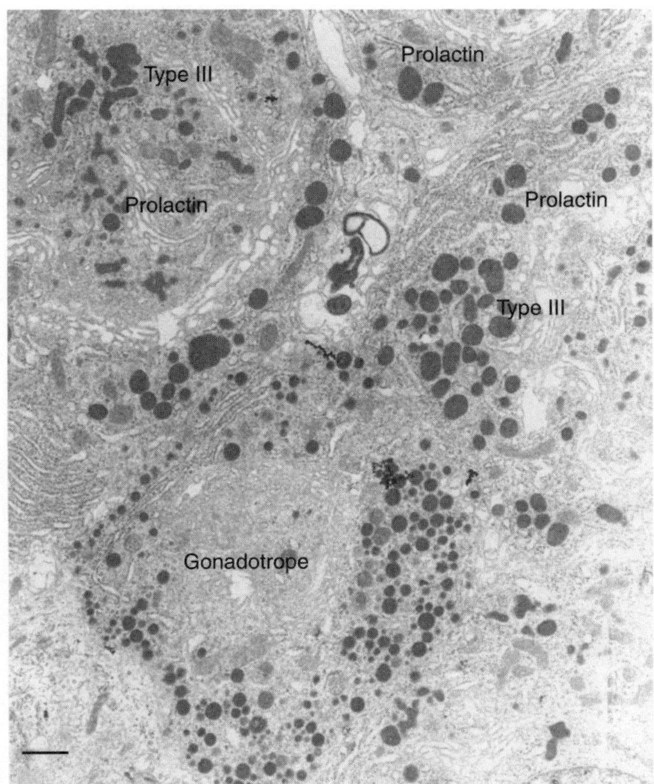

FIG. 60. Type III, mature prolactin cells from female rat pituitary characterized by large pleomorphic granules and a prominent Golgi complex. They surround a presumptive gonadotrope. Bar = 0. 5 μm.

hyperplastic by estrogen implants. The prolactin cell population was reduced after estrogen withdrawal with and without bromocriptine treatment. The reduction was mediated by apoptosis (404).

Several proapoptotic factors have been found in more recent studies. Cadmium is a critical apoptosis-inducing factor in a number of cells, including anterior pituitary cells (405). It is found in tobacco smoke and other pollutants (406). Glutamate, found in dietary supplements, is also proapoptotic (407). Peroxisome proliferators activated receptor (PPAR γ) was reported to be antiproliferative and also increases apoptosis in GH adenomas. (408).

A number of growth factors and second messengers are protective. A second messenger that prevents cadmium-induced apoptosis in the pituitary is nitric oxide (406). Nitric oxide is synthesized by nitric oxide synthase (NOS) from L-arginine. The NOS enzymes have two calcium/calmodulin-dependent forms, neural NOS (nNOS) and endothelial NOS (eNOS) and a calcium-independent, inducible form, called inducible NOS (iNOS). All three forms of the enzyme are found in the pituitary (148,409–414). Nitric oxide is a messenger that protects against apoptosis, although it has been shown to be cytotoxic (415), depending on the experimental conditions. nNOS in particular has been found in gonadotropes (411–413). It varies with the estrous cycle, reaching a peak in the proestrous evening (416). Somatotropes express iNOS and nNOS and lactotropes express all three forms (412,414).

IGF-1 prevents apoptosis in pituitary cells, and prolactin cells in particular (417,418). Genes sensitive to apoptotic or proliferative stimuli are being sought. Spengler et al. (419) reported the discovery of a new gene that produces a zinc finger protein, called Zac1. It uses different pathways to regulate apoptosis and cell cycle arrest and is found in highest levels in GH and prolactin cells.

Some studies have focused on apoptosis as a way to "remodel" the pituitary cell population during or after critical endocrine events (404,420–423). In the field of reproduction, this would include weaning (404,420), inhibition of egg laying in chickens (421), or supporting gonadotrope or lactotrope functions during the estrous cycle. As discussed in the preceding section, proliferative activity is increased during estrus, particularly among prolactin cells (389,390). However, apoptotic processes may be used to maintain homeostasis and cell numbers during the cycle, or regression from a hyperplastic state. For example, Ahlbom et al. have shown that termination of lactation brings about apoptosis and increases the proapoptotic genes of the *Bcl-2* family (420). Thus, the most recent investigative efforts have promoted the discovery of hormones that might be active in this process, including the possibility that apoptosis plays a role in "remodeling" the gonadotrope or lactotrope population during the estrous cycle. The remainder of this section focuses on factors that regulate apoptosis in gonadotropes and lactotropes, including estrogens, progesterone, GnRH, EGF, and tumor necrosis factor-α (TNF-α).

First, it should be recognized that a basic connection with the reproductive axis was initially seen with the discovery that the LPS-induced apoptotic events in the pituitary are more pronounced in female rats, indicating modulation by sex steroids (402,403). Watanabe and Yoneda (424) reported that this sex-based difference can be abolished by gonadectomy and restored by estrogen replacement. These data pointed to estrogen as an important mediator or inducer of apoptotic events, remodeling estrogen-receptive target cells during the cycle. However, estrogen does not work alone. It appears to work in partnership with TNF-α (425,426).

TNF-α is a critical proapoptotic factor in rat pituitaries (426). It is produced by and secreted from pituitary cells and its expression is induced by LPS (426). TNF-α is therefore a paracrine factor that reduces proliferation of pituitary cells, inhibits prolactin secretion, and promotes apoptosis of lactotropes and somatotropes, without increasing apoptosis of gonadotropes or corticotropes (426). Pisera et al. (425) studied its expression during the estrous cycle and found the highest expression of TNF-α in female rats taken at proestrus. Moreover, TNF-α actions appeared to depend on estrogens because it was proapoptotic for lactotropes in ovariectomized animals only in the presence of estrogens.

More recently, Candolfi et al. (427) described an antagonistic role for progesterone in this process, suggesting that this steroid is protective. They used the same model that focused on apoptosis induced by estradiol and TNF-α and added identified somatotropes to the cell types being studied. When they added progesterone, they found that it reversed the permissive effect of estradiol on TNF-α–induced apoptosis of somatotropes. This was blocked by an antiprogestin, ZK 98 299. Their studies showed that the progesterone effects were not mediated by dexamethasone receptors. However, in correlating their data with the well-known sites of progesterone receptors (solely on gonadotropes), they concluded that these somatotropes might have been multihormonal somatogonadotropes (239–243). The fact that progesterone expression is limited in the pituitary cells also led to the suggestion that it might be acting indirectly through paracrine factors.

Candolfi et al. (428) also studied the effect of nitric oxide on TNF-α induction of apoptosis, correlating the results with changes in serum estrogen during the cycle and after ovariectomy. There was an overall increase in NOS activity, with peak levels in proestrus or after stimulation by estradiol or TNF-α. TNF-α and estrogen both stimulated NOS in pituitaries from ovariectomized rats, and the effects of both factors were additive. When they focused on mRNA expression, iNOS expression was highest in proestrus and was also stimulated by TNF-α. This is the form found in somatotropes and lactotropes. nNOS, found in gonadotropes and folliculostellate cells, did not vary with the stage of the cycle, and TNF-α did not modify its expression. eNOS, found in prolactin cells, was higher in proestrus, and TNF-α decreased its expression. TNF-α also upregulated iNOS gene expression and downregulated an estrogen-induced eNOS expression. They suggested that nitric oxide may limit the apoptotic effects of TNF-α, especially on prolactin cells. However, it was difficult to fit a pattern of NOS expression into regulation of gonadotropes at this point, apart from the indication that nNOS may not be subject to TNF-α regulation (425).

The most direct evidence for regulators of apoptotic events in gonadotropes comes from studies using GnRH. However, most in vitro studies have shown that GnRH may induce cell cycle arrest, increase apoptotic cells, or limit the antiapoptotic effects of IGF-1 (418). Rose et al. reviewed the subject, providing evidence for signaling pathways and genes mediating the effects of GnRH (418). Miles et al. have also seen proapoptotic effects in gonadotrope tumor cell lines or cell lines transfected with the GnRH receptor (429). They reported a reduction in cell numbers, although they concluded that not all of the reduction was due to apoptosis. GnRH increased the number of cells in G_0/G_1 and decreased the numbers of cells in S phase of the cell cycle (429). These findings may support the concept that GnRH mediates differentiation of gonadotropes by stopping mitotic activity. Whereas this does not agree with GnRH affects seen in primary cultures (395,396), it is possible that subsets of gonadotropes and tumor cell lines respond differently. The response patterns may also be specific to the physiological state of the cell population (e.g., stage of the cycle).

GnRH has interactive effects on IGF-1 protective actions (418). IGF-1 promoted proliferation and rescued αT3 gonadotrope cell lines from apoptosis caused by serum starvation. IGF-1 induced the protein kinase Cα–mediated phosphorylation of Akt, which in turn prevented the phosphorylation of Bad (normally leading to apoptosis). GnRH cotreatment reduced the rescue efforts of IGF-1 on αT3 cells by inhibiting the protein kinase Cα signaling pathway, resulting in lower Akt (418).

An in vivo study also shows GnRH's protective effects, however. Yin and Arita (2002) reported that GnRH is able to rescue apoptotic gonadotropes in vivo (430). They found that higher numbers of TUNEL (terminal deoxynucleotidyl transferase–mediated dUTP nick end-labeling)–positive and Bax-positive cells appeared on the morning of proestrus compared with those for estrous rats. When Bax is present in a cell, this increases its susceptibility to apoptotic stimuli, and if Bax predominates over Bcl-2, cell death is accelerated (420).

Yin and Arita used dual labeling for Bax and pituitary hormones and found that only subsets of cells with LHβ immunoreactivity showed significant expression (430). The same proportions of gonadotropes were dual-labeled for LH and Bax when proestrous and estrous populations were compared. This difference is accentuated with the recognition that there are three- to fourfold more LH antigen–bearing cells in proestrus than estrus (180,235). Most of the LH cells in estrous animals can be detected only by their content of LHβ mRNA (180). Dual labeling for Bax and LH mRNA would be necessary to estimate total numbers of apoptotic LH gonadotropes in estrus. The overall increases in apoptotic cells followed the mid-cycle increase in serum estradiol. The reduction in apoptotic cells suggested that pituitary cells were being "rescued" late in proestrus, perhaps by the higher GnRH stimulation or the lower estrogen. Treating the animals with estrogen late in proestrus did not affect the number of apoptotic cells. When Yin and Arita blocked the GnRH surge with pentobarbital on the afternoon of proestrus, the numbers of TUNEL- and Bax-positive cells remained high on the morning of estrus. This was further tested by an injection of GnRH to mimic the surge that promoted the reduction in apoptotic cells and presumably the rescue of gonadotropes. Also, tests of progesterone administered on the afternoon of proestrus to pentobarbital-blocked animals showed that it could not substitute for GnRH, suggesting that GnRH worked directly as a mediator in the rescue effort, helping directly or indirectly to maintain the population of prolactin cells and gonadotropes needed for the next cycle.

There are several interesting facts from this study. First, the relatively high expression of Bax, a proapoptotic protein marker, in LH gonadotropes compared with other cell types may reflect an unusual susceptibility in the population of gonadotropes to remodeling and other changes that might be needed to support various endocrine systems. Second, the exclusive expression of nNOS by gonadotropes (411–413) may also provide a route that could be regulated to produce nitric oxide to rescue these susceptible cells. The expression of nNOS in somatotropes may signify the subset of cells that become somatogonadotropes.

LACTOTROPES

Early Studies Defined the Lactotrope Population

Lactotropes Defined by Immunolabeling and Electron Microscopy

Lactotropes were first identified on the basis of their pink erythrosine staining in the Herlant tetrachrome reaction (40,35). Pioneers in the field then confirmed their identity in a number of species with immunolabeling (35,46,47,49,51,431–442). Prolactin cells are among the smallest of the pituitary cells, showing an angular, polyhedral, or "cup" shape (35, 431–442). Their percentages vary with the species and physiological state. There are clear-cut gender differences, and many workers report that there are twofold more prolactin cells in the female than the male (35). An interesting feature of prolactin cells is their tendency to wrap themselves in a cuplike conformation around gonadotropes, which could facilitate some of the paracrine interactions discussed in the foregoing sections (35,46,47).

The earliest identification of prolactin cells came from pregnant animals, which have an abundant population of lactotropes (40). The subtypes of lactotropes that are easiest to identify have a well-developed secretory apparatus, including stacks of parallel cisternae of rough endoplasmic reticulum, a prominent Golgi region, and pleomorphic secretory granules that may include both large and small subtypes (Figs. 60 and 61). In the human, the secretory granules tend to be smaller (443).

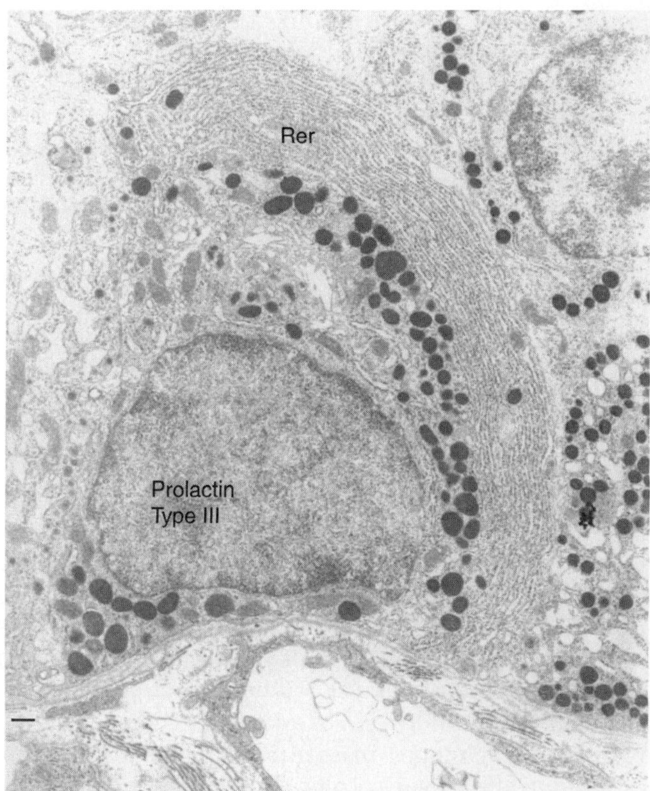

FIG. 61. Type III, mature prolactin cell showing the stacks of rough endoplasmic reticulum (Rer) often seen at one pole. The cell is characterized by its population of large, pleomorphic granules. Bar = 0. 5 μm.

In the earliest period of their study at the electron microscopic level, pituitary cytologists confirmed the identity of prolactin cells by the striking group of pleomorphic granules depicted in Figs. 60 and 61 (431,433,437–443). However, later studies discovered subsets of cells with more homogeneous granules that were round and smaller. They resembled a GH cell, except for the overall size of the granules (437,444,445). As the studies progressed, workers identified three types of prolactin cells. Type I cells had a population of small, spherical granules (Fig. 62). Type II cells contained the medium-sized polymorphic granules (Figs. 63; see Fig. 65). Type III cells were distinguished by larger, pleomorphic granules (Figs. 60, 61, and 64). This same diversity was seen in other species, leading workers to seek specific functions for the different morphological phenotypes (438–440).

Early Work Differentiated Functions of Small and Large Prolactin Cells

As with gonadotropes, cell separation techniques were able to isolate subsets of lactotropes based on

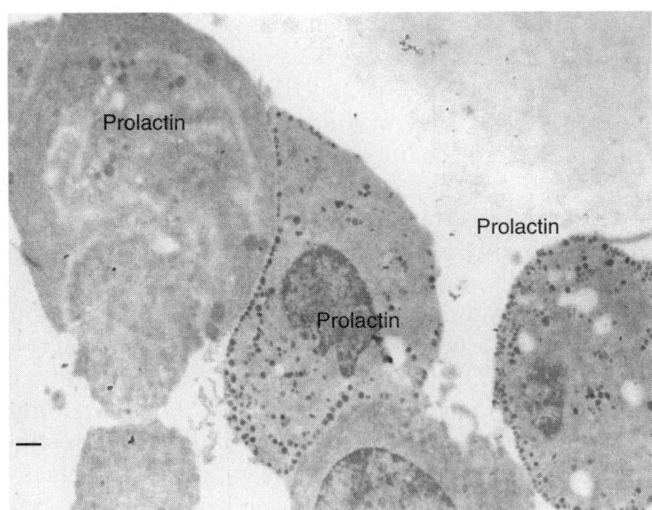

FIG. 62. Type I, immature prolactin cell, in one of the smallest fractions separated by centrifugal elutriation. These cells contain very small, oval granules and are impossible to distinguish by morphology alone. These cells were immunolabeled for prolactin with 1:10,000 anti-rat prolactin and the avidin–biotin peroxidase technique. One of the cells in the field has the morphological characteristics of a lactotrope, but is very poorly labeled. Bar = 0.5 μm.

content of secretory granules as well as size. Working with 14-day rats, Snyder and colleagues' pioneering studies correlated levels of prolactin released into the media with size or the fraction of origin (446). The cells retained the memory of that particular subset for 14 days in culture. The lighter cells (smaller or low-density) contained more prolactin mRNA and fewer prolactin-bearing granules. Figure 62 shows an example of prolactin cells from small cell elutriation fractions. They were thought to be actively synthesizing the hormone. However, when tested for their

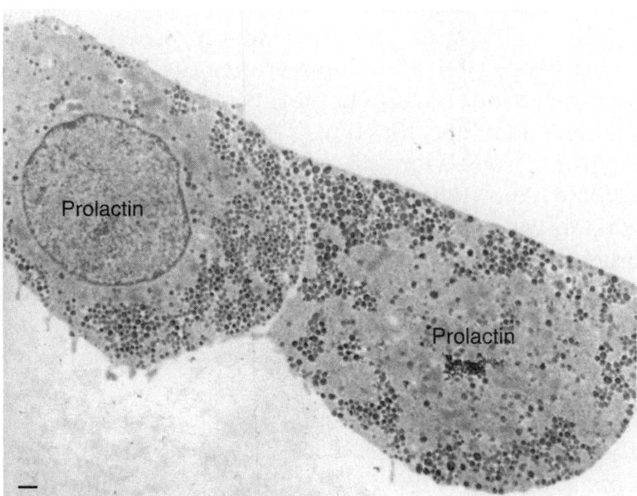

FIG. 63. Prolactin cells from medium-sized cell fractions separated by centrifugal elutriation. The cells are immunolabeled for prolactin. They contain numerous small granules, some of which may be pleomorphic, but most are round. The cells have the characteristics of type II or maturing prolactin cells. Bar = 0.5 μm.

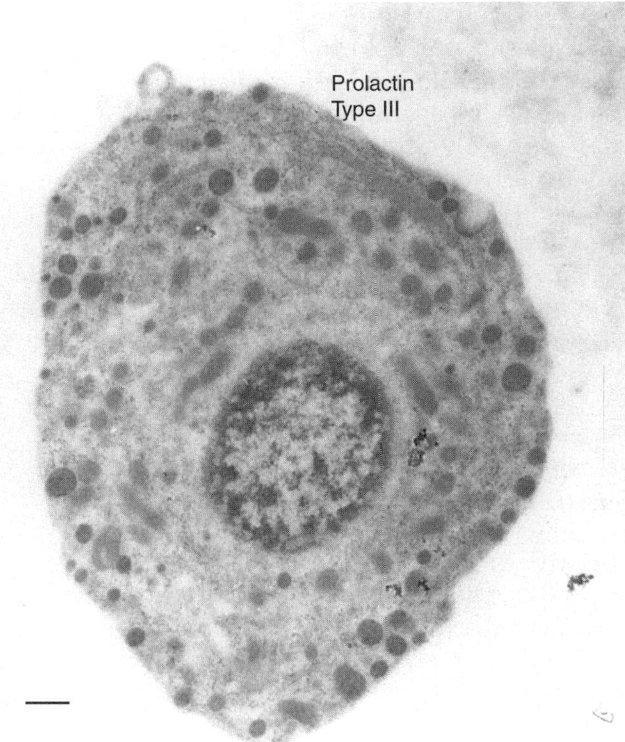

FIG. 64

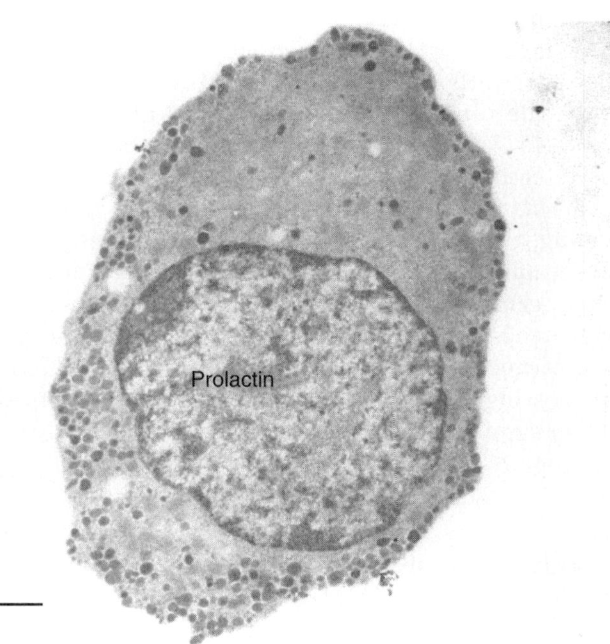

FIGS. 64 and 65. Lactotropes from the largest cell fractions separated by centrifugal elutriation. Figure 64 shows a type I lactotrope characterized by large pleomorphic granules that label lightly for anti-rat prolactin. Figure 65 shows a cell with the characteristics of a type I or II lactotrope. Prolactin cells are not numerous in the largest cell fractions (Fig. 25). Bar = 0.5 μm.

secretory activity, they were also highly responsive in vitro. This indicated that these cells could secrete prolactin as rapidly as they made it. In contrast, the prolactin cells separating in the larger or more dense fractions had more prolactin (more granules), but they contained lower levels of prolactin mRNA and were not as responsive to secretagogues (Figs. 64 and 65). This might indicate they were building stores and perhaps in a reserve state. This functional diversity is revisited in later sections. The earliest studies showed that it reflected diversity in hormone storage patterns as well, as is shown in the next section.

Discovery of Bihormonal Somatomammotropes

Early studies of prolactin cells also challenged the one-cell/one-hormone theory. Shortly after prolactin cells were distinguished by immunolabeling, Halmi et al. reported that there were cells in human pituitary adenomas that coexpressed both prolactin and GH (51). This had been suspected because patients with acromegaly frequently also suffered from hyperprolactinemia. Shortly thereafter, the development of clonal cell lines in rats derived from the MtT/W5 tumor showed some cells secreting only GH or prolactin and others coexpressing both hormones (447,448). Because the tumor cells store little hormone, it has been difficult to detect their coexpression by dual immunolabeling.

The pioneering use of sequential RHPAs provided definitive proof of their dual secretory capacity, along with the dual secretory capacity of human tumors (448–451). Indeed, this was the most reliable means of detecting evidence for the coexpression of GH and prolactin. In spite of the difficulties with immunolabeling, the presence of the mammosomatotrope was eventually confirmed in a number of species (452–456). Furthermore, the pioneering studies by Frawley and colleagues showed that fluctuation of mammosomatotropes depended on the steroid environment, suggesting that they served as a set of reserve cells. Their significance is discussed further in later sections (457–459).

Significance of Functional Heterogeneity Expressed by Lactotropes

Lactotrope Dependence on Somatotropes and Gonadotropes

To understand fully the heterogeneity among prolactin cells, their origin must be reviewed. As described previously in the section detailing embryonic development, prolactin cells are the last to appear in the cell line that was derived from *Prop1* and *pit-1* gene expression (57,60,62,63,67,72,82,84–91) (Fig. 1). In the embryo, they originate in the dorsal regions, separated from the lines that give rise to the GATA2-responsive gonadotropes (73–76,91,94,99,100). This sequestration is important to allow full development of the Pit-1–responsive cell lines because GATA2 inhibits expression of Pit-1 (100).

Pit-1 mRNA is detected at 15.5 days of gestation in the rat, approximately 1 day before detection of the prolactin and GH mRNAs (88,89). All of the pituitary cells can produce Pit-1 transcripts; however, only prolactin, GH, or thyrotropes actually translate the protein. As stated earlier, strains of dwarf mice lacking *Pit1* gene expression have no somatotropes, lactotropes, or thyrotropes (84–86).

The importance of stem cells in the lineage of prolactin cell development was shown in elegant studies by Borrelli et al., who developed transitional transgenic mice that expressed the rat GH or prolactin promoters fused to herpes simplex virus-1 thymidine kinase (460). This construct allowed the *postnatal*, cell-specific destruction of mitotic cells producing GH or prolactin. When mitosis in prolactin cells was prevented after birth, there was no effect on GH or prolactin cell populations or on pituitary size. However, prevention of postnatal mitoses in GH cells resulted in dwarfism and a decrease in both GH and prolactin cells. These data showed the importance of the GH cell line to the development of prolactin cells and clearly showed that subsets of somatotropes serve as stem cells for the more mature lactotropes (460). The data also showed that postnatal expansion of the population by mitosis is critical to the production of a normal population of prolactin cells.

The lactotropes also depend on products from gonadotropes for their appearance, as described in earlier sections of this chapter. Studies have reported that gonadotrope α-subunit knockout mice showed a great reduction in numbers of lactotropes (260,261,263). This led to a number of studies seeking gonadotrope-derived factors that stimulate prolactin cell mitoses. Although a number of factors themselves will stimulate lactotropes, including EGF-like peptides (262), angiotensin II [reviewed in Schwartz and Cherny (230), Steele et al. (269), and Robberecht et al. (270)], or gonadotropin α subunit itself (262), a more recent study has shown that the lactotrope-inducing factor is a peptide derived from the POMC glycoprotein (267). This study also links our studies showing that postnatal gonadotropes express unusually abundant POMC peptides (164,165). The timing coincides with the postnatal increase and maturation of prolactin cells.

Prolactin proteins or mRNA are detected on fetal days 16 to 19 or even as late as the newborn, depending on the authors and the products detected (161,461–467) (Fig. 2). As stated previously, the cells emerge from a stem cell that also produces GH, although this is difficult to prove by immunolabeling because the antigens are not always detected (54). Indeed, the disagreement about the timing of their appearance depends both on the method used and the gene products detected. Some workers have shown that if mRNA is used to detect the cells, the timing of appearance of the prolactin transcript may be as early as 17 days, along with that of GH mRNA (89). There are more cells with prolactin mRNA than with antigens in the population before birth (54). Workers have suggested that this may be because many prolactin cells secrete their stores as rapidly as they are synthesized.

Like gonadotropes, prolactin cells mature during neonatal development. One maturation step may be a movement by some prolactin cells to areas where they become cupped around gonadotropes, maintaining a close relationship during adult life. Figure 66 illustrates a prolactin cell (type III) "cupped" around a presumptive small gonadotrope. Of course, the morphology of the gonadotrope resembles that of other cells with small granules, including type I prolactin cells, and one cannot be certain of its identity.

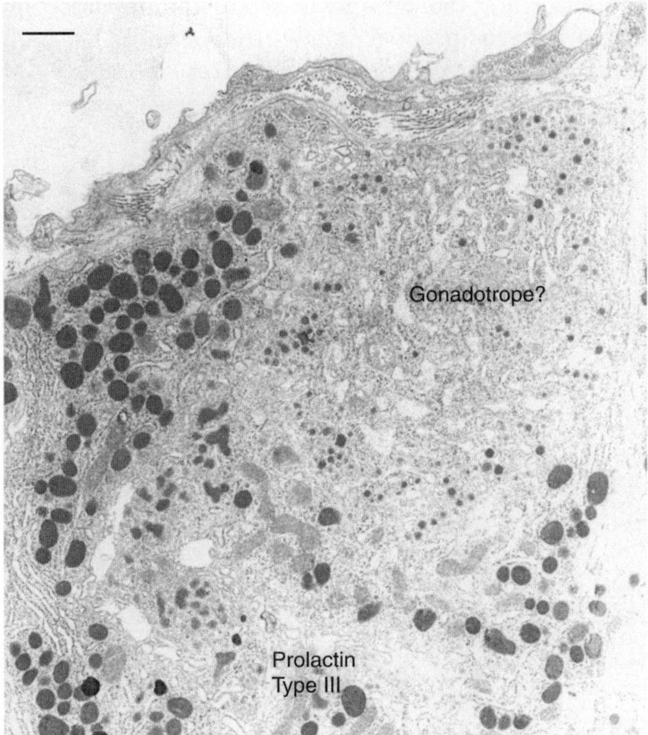

FIG. 66. Since they were first discovered by immunolabeling, lactotropes were often seen in juxtaposition to gonadotropes, cupped around the cell. Such a configuration around a presumptive gonadotrope is illustrated. The relationship between the two cells may be synergistic and important to the function of both. Bar = 0.5 μm.

Immunolabeling has established this relationship, however.

A number of additional factors may be involved in further prolactin cell maturation (461–467). Dopamine chronically inhibits prolactin and estrogen stimulates the population in females (467). Prolactin cell secretory capacity increases during the first 7 postnatal days; this is a better method to detect the cells than immunolabeling (467). The numbers of prolactin cells are not at the adult level in the rat for at least 1 month (461,463). This contrasts with developing gonadotropes, which populate the anterior pituitary rapidly after birth, attaining normal numbers in 1 week (164,165).

Frawley and colleagues have explored prolactin cell secretion after birth with the RHPA (467). They could not detect prolactin secretion before 4 days. However, after that time, nearly 36% of the cells secreted both prolactin and GH (467,468).

Studies of human fetuses also show few prolactin cells from 14 to 23 weeks of gestation (436,468). In anencephalic fetuses, prolactin cells appear more numerous; the chronic dopaminergic regulation of prolactin cells likely is absent in fetuses without a brain (468). Studies of human fetal pituitary cells in culture show that they secrete prolactin as early as 18 weeks, although most of them secrete both prolactin and GH (469). Because of the sparse storage capacity of these actively secreting prolactin cells, it was difficult to detect cells with both prolactin and GH stores in these groups.

To summarize, several elements appear to be needed for full expansion of prolactin cells, including (a) the set of Prop1- and Pit-1–responsive stem cells that give rise to the lactotropes (57,60,62,63,67,72,82,84–91); (b) functioning stem somatotropes to expand the population after birth (460); (c) functioning gonadotropes before birth (260–262) [producing α subunit (262)] and under the influence of pituitary GnRH (264); and (d) functioning gonadotropes after birth to produce the 16-kDa fragment of POMC peptides (267).

Thyroid hormone was identified as another critical factor for full expansion of prolactin cells. Stahl et al. (263) studied α-subunit knockout mice and noted that, in addition to being hypogonadal, these animals were also hypothyroid. This is not surprising because α subunit is also used to make TSH. Also, as expected, in response to the lack of thyroxine feedback, the thyrotropes undergo hyperplasia and hypertrophy and TRH levels are high. Stahl et al. (263) wondered why prolactin cells were hypoplastic, especially because TRH is stimulatory for lactotrope proliferation and function. They tested the hypothesis that thyroxine might be needed to expand somatotropes and lactotropes after birth by treating young male knockout mice with thyroxine for 40 days.

Thyroxine reversed TSH cell hypertrophy and mediated a recovery to normal numbers of prolactin and GH cells, and the dwarfism typical of the knockout was cured.

An interesting feature of the α-subunit knockout is the fact that gonadotropes do not hypertrophy, which is a normal response to lack of gonadal feedback. In the thyroxine-treated model, typical castration cells are seen (263), suggesting that thyroid hormone plays a role in the gonadotrope response to castration.

Stahl et al. reasoned that the recovery had several explanations (263). First, thyroxine may have stimulated expansion of the stem cell somatotropes needed to produce the prolactin cells. Alternatively, by limiting thyrotrope expansion, thyroxine may have allowed the existing stem cells to develop into somatotropes. They also hypothesized that α subunit was not an absolute requirement for prolactin cell expansion. This does not rule out the importance of α subunit during embryonic development. However, normal thyroid function is required for complete expansion of both somatotropes and lactotropes, after birth. In addition, it may be required to develop normal gonadotrope responses to gonadal steroid feedback. Figure 67 illustrates the rapid expansion in gonadotropes and thyrotropes after birth, to show how the pituitary provides the cells needed to expand the prolactin cell population during the first month of life. Gonadotropes are producing the 16-kDa fragment of POMC and thyrotropes are needed to stimulate thyroid hormone, which in turn stimulates the stem cell somatotropes for expansion of the lactotrope populations.

Significance of Prolactin Cells with a Rapidly Releasable Pool of Hormone

The aforementioned studies of prolactin cell maturation and heterogeneity pointed to the fact that the prolactin cell population included a subset that secreted more rapidly than it stored (446). This made it difficult to detect the cells by immunolabeling and caused some controversy about the timing of appearance of prolactin, and also whether there were cells that coexpressed prolactin and GH [reviewed in Tougard and Tixier-Vidal (470)]. These findings echoed the earliest studies by Swearingen in 1971, which had shown evidence for a heterogeneous turnover of prolactin both in vivo and in vitro (471).

Snyder et al. had reported that cells from adult animals included those that stored abundant prolactin, but did not respond well to secretagogues (446). Thus, as in the gonadotrope population, prolactin cells are not homogeneous and one cannot predict a cell's ability to secrete its stores solely on the basis of morphology (abundant secretory granules) or prolactin content. In a sense, these findings were similar to those of Lloyd et al. (208) in which cells with FSH were evident by immunolabeling, but the cells did not respond well in an RHPA.

Walker and Farquhar (472) looked more carefully at the events surrounding prolactin release with high-resolution autoradiography linked to immunodetection of the secreted product. Their studies confirmed that some prolactin cells exhibited a rapid turnover of prolactin, which allowed them to release the newly

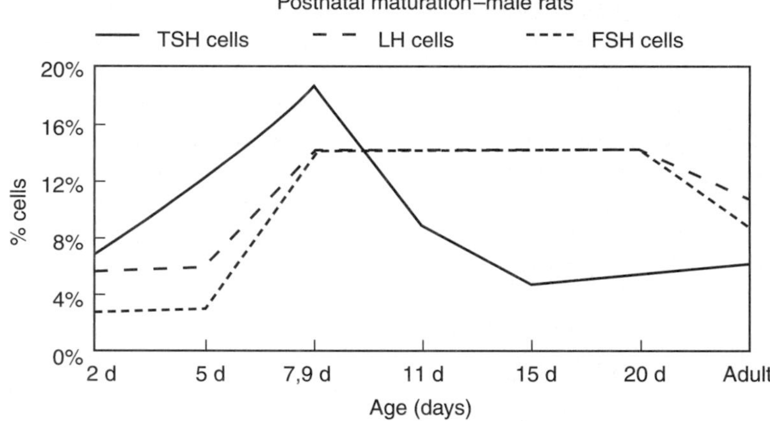

FIG. 67. Studies of the postnatal development of gonadotropes and thyrotropes show a dramatic increase in percentages of both cell types during the first week of postnatal life. Both cell types produce products that ultimately are vital for the postnatal expansion of lactotropes (see text). Graph shows the increase in gonadotropes and thyrotropes identified by immunolabeling.

synthesized prolactin after only 15 to 30 minutes. Other cells expressed a slower turnover. These investigators did not correlate the functionally distinct subsets with a particular morphological subtype, however.

The GH3 tumor cell line actually consists of mammosomatotropes, but the cells resemble the subset of prolactin cells that store few secretory granules (469,473–476). In this group of cells, there are two subsets of cells based on turnover time. Also, because GH3 cells have a uniform morphology, the heterogeneous responses do not depend on a particular morphology. Morin et al. suggest that the responses may be linked to the size of the intracellular prolactin stores in individual cells (473,474). The turnover time of the prolactin pools in GH3 cells, however, is eightfold greater than that in normal cells.

Single-Cell Secretion Assays Also Detect Heterogeneity among Prolactin Cells

The RHPA has been used to quantify prolactin release from individual pituitary cells (448). Plaque areas are directly proportional to the amount of radioimmunoassayable prolactin. Studies have shown at least two functionally distinct subpopulations of prolactin cells, in agreement with previous reports (477). The cell immunoblot assay also detected secretion from individual cells (478–480). Interestingly, prolactin cells were taken from different regions of the anterior pituitary, and the RHPA showed that their releasing capacity depended on their original location in the pituitary. Again, there were heterogeneous responses; not all prolactin cells secreted well.

Further RHPA tests of responses after 21 hours in protein synthesis inhibitors showed that half of the population actually depends on newly synthesized prolactin for its basal secretion (480). Attempts to correlate these heterogeneous responses with mRNA levels or morphology have failed to find specific landmarks that predict a particular response beyond that originally defined in the studies of separated prolactin cells. Collectively, the foregoing studies predict that some prolactin cells may be detected only by their content of prolactin mRNA or secretion of prolactin, and not by their content of prolactin stores (446). The fully functioning prolactin cell may be the most morphologically distinctive of the subsets, identified by its content of pleomorphic granules and stacks of rough endoplasmic reticulum. Clearly, however, there are populations that may be reserve or resting cells, some of which might belong to the subset that coexpresses prolactin and GH.

Prolactin Cells as a Model for Studies of the Cell Secretory Pathway

The rapid turnover of prolactin and the striking appearance of the actively secreting prolactin cell have added value to studies of its secretory cycle. Over the years, such studies were facilitated by the remarkable changes in the cells during pregnancy, lactation, and cessation of lactation. Prolactin cells are distinguished by stacks of rough endoplasmic reticulum, much like those found in other protein-synthesizing cells, such as pancreatic acinar cells (Fig. 61). These stacks are often in one pole of the cell and studies show that they communicate by vesicles with the *cis* face of the Golgi complex (158). Figure 61 illustrates a good example of a polarized prolactin cell, with abundant rough endoplasmic reticulum at the pole opposite to that of the vascular pole. In the Golgi complex, the granules mature and attain their pleomorphic shape in the *trans* Golgi zone. The secretory granules are then stored in the cytoplasm or in prolactin cell processes before being released by exocytosis.

Prolactin Cells Respond Rapidly to the Stimulus of Pregnancy

A number of studies have shown dramatic changes in prolactin cell organelles during pregnancy and lactation (434,481,482). There is a general increase in number and size of acidophils that correlates well with enhanced serum levels of prolactin. The identity of the cells has been confirmed by immunolabeling in the human (435) and rat (482). In the pregnant rat, the cells that enlarged and became prominent appeared more like the type II prolactin cells (483) (containing more uniform and spherical small granules; Figs. 62 and 63). However, later in gestation (after day 14), the cells were distinguished by an expansion in the rough endoplasmic reticulum and the Golgi zone. Secretory granules were few and sparse (484).

Goluboff and Ezrin also reported that GH cells were reduced during pregnancy in humans (484). It has therefore been tempting to speculate that the expansion in prolactin cells may have been at the expense of the somatotropes. Therefore, Porter et al. used the RHPA to learn if the number of GH- and prolactin-secreting plaques changed during pregnancy and lactation (485). They found that the number of cells secreting prolactin increased during gestation as the number of GH secreting cells declined. A subset of the cells secreted both hormones. There was no overall change in numbers of total plaques, which provided strong evidence that the same cell population was interconverting to support the prolactin secretion needed for pregnancy and lactation. After lactation and weaning, the population converted to one that was predominantly GH secreting (485). These data are supported by the evidence for expansion of fetal and neonatal prolactin cells from multihormonal somatotropes (59,89–91,103–106). They further suggest that the mammosomatotrope population maintains this plasticity and ability to support the reproductive system in the adult state as well (459,485).

Responses to Lactation and Weaning

In previous sections, we discussed the sequelae of proliferation or apoptosis including cell remodeling needed during lactation and weaning. We reviewed the evidence for the increase in mitotic rates of prolactin cells with postnatal age (388), the estrous cycle (389) and estrogen treatment (390). Most of the pioneering studies found that prolactin cells were among the few cell types that typically showed mitotic activity, especially in females during estrus. The expansion in lactotropes during lactation may also be due to mitoses, although the work with RHPAs suggests that interconversion from multipotential somatotropes is also a source for the new cells. Lactating rat pituitaries have prolactin cells with dilated Golgi complexes, numerous exocytotic profiles (158,486–489), and a striking expansion in the rough endoplasmic reticulum, which is usually seen in stacks of flattened cisternae at one cellular pole. The morphology is similar to that seen in actively secreting prolactin cells, where secretion is occurring too rapidly for the buildup of stores.

Weaning brings about the need to remodel the pituitary and reduce the number of actively secreting prolactin cells. This likely eventually involves apoptosis (402–404). However, the changes in response to weaning occur in several stages. First, removing litters from their mothers resulted in a dramatic increase in numbers of secretory granules as well as the numbers of immunolabeled cisternae in the Golgi region. In other words, without the neural stimulus associated with suckling, the prolactin cells do not secrete and there is a buildup of granules (158,290). In a subset of studies, pups were returned to their dams and allowed to suckle for as little as 5 minutes (488–490). This produced a rapid increase in granule exocytosis, suggesting rapid neural links to stimuli for prolactin cells similar to those leading to oxytocin cell bodies.

Regulation of storage was evident in prolactin cells from lactating rats during the weaning process. The storage granules accumulated during the first 10 hours and then became the repository for lysosomal enzymes, which may digest the excess prolactin once acidified. This process, called *crinophagy*, reduces stores of a product once secretion is inhibited or stimulation has ceased (158). It is likely that the sorting mechanisms that normally regulate trafficking of lysosomal enzymes are regulated to prevent buildup of granular stores. Even though secretion was not being stimulated as rapidly, the Golgi complex still showed evidence of stimulation during the first 6 to 50 hours after weaning. The cisternae were increased and there was immunostainable material in the Golgi complex. Forming granules were still found, and exocytotic granules were seen (490).

Responses during the Estrous Cycle and Estrogen Treatment

Sexual dimorphism in numbers of prolactin cells have been described in a number of species, although the differences are often seen only when certain parameters are measured (436,445,456). Prolactin cells tend to be more numerous in the female than the male. Sasaki and Iwama reported that there were over 45% prolactin cells in female mice compared with 23.8% in the males (491). Ibrahim et al., from our laboratory (174), reported 35% prolactin cells in female and 26% prolactin cells in male rats by immunolabeling. This may reflect the changes in prolactin cells seen during the cycle, largely driven by estrogens. Studies have also shown that estrogen may drive the conversion of the mammosomatotropes to mammotropes, thus supporting the reserve nature of the bihormonal cells (459,485). Clearly, these mechanisms can operate as the population is remodeled to prepare for pregnancy and lactation.

It has long been recognized that estrogen is a primary stimulator for prolactin cell hyperplasia and hypertrophy in a number of species. Studies of short-term treatment have demonstrated a transition in morphology between the three subtypes of prolactin cells that suggests each represents a different stage of development or differentiation (445). Prolactin cells with the large pleomorphic secretory granules (type III; Figs. 60, 61, 64, and 66) became more numerous with estrogen treatment. This increase in type III cells was correlated with an increase in prolactin content as well as serum prolactin, and a decrease in the types I and II prolactin cells (those with smaller, homogeneous granules) (445). Figures 62 and 63 show examples. Because the type III cells appear during states of active secretion, they are considered more "mature" secretory cells.

The opposite effect is seen after removal of the ovaries and estrogen. This causes a decrease in prolactin cells and the remaining cells contain small secretory granules, thus reverting to a type II configuration (492). They were also reduced in number (178). The type III cell then appeared in these castrated rats only 4 days after they were treated with replacement estrogen, as if maturation is associated with recovery.

Studies have compared expression of prolactin mRNA and secretory capacity in cells taken from ovariectomized rats that also received estrogen for 4 days before death. Estrogen clearly increased the number of prolactin cells that formed plaques in an RHPA (493,494). Scarbrough et al. correlated prolactin secretory activity (by RHPA) with expression of mRNA (by in situ hybridization) and reported dissociated expression of mRNA and secretory activity (495). The nonsecreting prolactin cells expressed the most abundant mRNA. These may be cells caught in an early stage of differentiation to an actively secreting prolactin cell. Estrogen did stimulate the percentages of plaque-forming cells, but the steroid did not stimulate the percentages of mRNA-bearing lactotropes, which suggests that its effects may be downstream of regulation of transcription (495).

Antakly et al. have provided information about the effects of long-term treatment with 17β-estradiol on prolactin cells (496). After 4 to 8 days of treatment in primary cultures, the lactotropes hypertrophied, but there was no change in their percentages in the anterior pituitary. The increased prolactin secretion in response to 10^{-8} M estrogen was correlated with an enlarged Golgi apparatus and an increased number of vesicles and immature secretory granules. The cisternae themselves were dilated. There was also an expansion in the area containing rough endoplasmic reticulum. After 4 days of treatment, the large granules typical of the type III prolactin cells were evident. The cell surface also showed increased microvilli projections.

Additional evidence for estrogen-mediated stimulation of prolactin cell subtypes comes from studies of prolactin cells during postnatal development (455). One week after birth, most (82% to 83%) of the prolactin cells are of the immature, type I variety, filled with small granules. At puberty in the female, there is a great increase in the number of type III prolactin cells, comprising 91% of prolactin cells. In contrast, the pubertal male has an equal number of type II and type III prolactin subtypes. This fits well with the known luteotropic functions of prolactin cells in the female rat. It also suggests conversion of prolactin cells into more mature subtypes is driven by estrogen.

Cessation of reproductive cycles during aging produces changes that vary with sex. Chuknyiska et al. compared responses to estrogen by short term cultures of dissociated pituitary cells taken from estrous or diestrous female rats, or 22- to 24-month-old (aged) rats (497). The most striking difference was the 80% to 120% increase in secretory responses to a 4-day treatment with estradiol seen in the aged rat populations (497). The responses were out of proportion to the numbers of immunolabeled prolactin cells. However, it is possible that many prolactin cells are secreting more rapidly than they can build up stores needed for detection (497). Console et al. showed changes in cell density, volume density, or surface density that varied with sex (498). Cell and surface density of prolactin cells appear to increase in females and decrease in males. However, in females, prolactin production per cell was reduced (498). The human pituitary shows a decrease in prolactin cell number with advanced age in both sexes (436).

Thyroid-Releasing Hormone Regulation of Lactotropes

Shortly after the tri-amino acid peptide was discovered as the stimulating hormone for TSH cells (499), pioneers reported that TRH also stimulated release of prolactin in vitro (500) and in vivo (501). TRH also stimulates proliferation of GH3 tumor cells as well as prolactin release from these cells (502). We have already discussed the unexpected reduction in prolactin cell number in the α-subunit knockouts, which have hypertrophied and hyperplastic thyrotropes due to the absence of thyroid hormone feedback and stimulation by TRH (460). In spite of its importance as a secretagogue, these studies show that high TRH is not sufficient for prolactin release or even the maintenance of the population. Stahl et al. showed convincingly that thyroid hormone was needed for normal prolactin cell expansion and function (263).

This information helps with the interpretation of early studies of prolactin cells after thyroidectomy by Ozawa and Kurosumi (503). Not surprisingly, prolactin cells atrophied after thyroidectomy, showing reductions in the cytoplasm and Golgi complex as well as a reduction in the size and number of secretion granules. Most of the prolactin cells resembled the type I (immature) subtype (filled with small, spherical secretory granules). Chronic treatment with thyroxine prevented atrophy of the prolactin cells and increased the percentage area fraction of rough endoplasmic reticulum to levels beyond control levels. There was also the expected increase in the number and size of the secretory granules. Treatment of thyroidectomized rats with both thyroxine and TRH caused further hypertrophy of prolactin cells, including an expansion in the rough endoplasmic reticulum and the Golgi apparatus. The stimulated prolactin cells began to take on the appearance of the mature, type III subtype and the changes were correlated with increases in serum and pituitary prolactin. Thus, in agreement with Stahl et al. (263), thyroxine is needed to maintain prolactin cell numbers and normal morphology. However, TRH appears to have additive effects to thyroxine on further differentiation to a mature secretory subtype.

Studies by Boockfor and Frawley (478) showed that TRH stimulates the formation of large plaques in a subset of prolactin cells. In the largest plaque-forming cells, the rate of plaque formation is accelerated by TRH. There is a gradual TRH-mediated increase in secreting prolactin cells. However, TRH does not appear to recruit more cells than are detected by immunolabeling. This group also reported that cells from peripheral regions of the pituitary gland secrete more prolactin in response to TRH than those from the central regions (478). This difference did not vary with the physiological state of lactation, however. Arita et al. (477) also demonstrated a heterogeneous response to TRH with the use of sequential cell immunoblot assays. They reported that cells secreting the lowest basal levels of prolactin showed the highest responses to TRH.

Further work on the turnover of prolactin was done in pulse chase experiments, and two pools of prolactin were found (473,474,504). TRH appears preferentially to release the stored prolactin (rather than the radiolabeled form), suggesting that its subcellular target is the secretory granules. However, it appears that the newly synthesized prolactin is preferentially released under basal conditions, even in the presence of dopamine. Also, prolactin cells can respond to TRH even if basal secretion is blocked by inhibitors of protein synthesis (505).

The pathway for secretion of these two sources of prolactin may stem from different cells, or from different routes from the same cell. Walker and Farquhar suggested that the two prolactin pools may belong to different subsets of cells (472). Rapid responses to TRH would thereby depend on the subset with the most TRH receptors. However, one could also argue for differential secretion through different subcellular routes based on work with monensin and GH3 cells (506). Monensin induces a buildup of prolactin in Golgi vesicles, upstream from the *trans* Golgi network that condenses the secretory granules. Monensin also decreases basal prolactin release, and pulse chase experiments have shown that newly synthesized prolactin release is reduced. In contrast, monensin does not affect the release stimulated by TRH. Collectively, these data suggest that basal release may be affected through the Golgi complex, but that the pathway may involve elements upstream from the site blocked by monensin. The granule route would be in the TRH-mediated pathway (507). These two pathways may be found in the same cells, or different cells. It is uncertain if they explain the heterogeneous responses completely, however.

Regulation of Lactotropes by Dopamine

Inhibition

Dopamine receptor agonists are potent suppressors of the release of prolactin, both in vivo and in vitro (508–511). Dopamine is produced by neurons located in the medial basal hypothalamus (511). Dopamine inhibits prolactin release tonically through the D_2 subclass of dopamine receptors, which inhibit adenylyl cyclase (512). Lactotropes respond initially to dopamine by increasing stores, as shown by enhanced immunolabeling (513). This is likely due to an initial dopamine block of secretion. After several days, stores are decreased. There is an associated increase in lysosomal enzyme activities, as expected from the crinophagic destruction of the secretory granules after their buildup (514). Indeed, if cells are treated with [^{3}H]-leucine for 3 hours, in the presence of 1 μM dopamine, there is an initial increase in specific activity of the released prolactin, suggesting that dopamine may have a preferential effect on the release of stored prolactin, rather than the newly synthesized proteins (515). Alternatively, there may be cells that are not responsive to dopamine, or are responding to its stimulatory effects (see later). To summarize, the cytophysiological information on inhibitory effects suggests that dopamine acts on the secretory granules, possibly preventing exocytosis.

However, not all prolactin cells respond to dopamine with an identical pattern, as seen with the RHPA (478). Dopamine causes a shift in the proportion of plaques from a mixture of small and large plaques, to plaques that are mostly small. This points to dopamine-mediated selective action on the lactotropes that are secreting the most prolactin (and forming large plaques). Frawley and Clark also showed that dopamine transiently slowed the rate of plaque formation, and TRH reversed its actions (514).

One of the most interesting set of findings emerged when the responses of prolactin cells were correlated with their original location in the anterior pituitary. This could well be related to functional syncytia set up by gap junctions and folliculostellate cells (213–227). Such networks could begin at sites contiguous to regions in the pars tuberalis/stalk, which would contain nerve endings and capillaries bearing neurotransmitters and neuropeptides. Developing cellular routes or networks from these capillaries could effect site-specific regulation that depends on types of nerve endings at that site. Prolactin cells from central regions of the gland are most sensitive to inhibitory effects of dopamine (479). Recall that the opposite effect was seen when the cells of central origin were tested for their responses to TRH. However, there is plasticity in their responses. If the same zone is sampled in lactating rats, after suckling, the central prolactin cells are resistant to the inhibitory effects of dopamine (515). Data on heterogeneous responses were also reported by Arita et al. (477), who used a sequential immunoblot assay to detect secretion from individual prolactin cells. They reported that there were subsets of cells that do not respond to dopamine.

The concept that there are some prolactin cells that do not respond to dopamine was studied further by Kazemzadeh et al., who compared responses by different subsets of prolactin cells separated by size on a Percoll gradient (516). After 3 days in culture with or without 18 hours in dopamine (1 μM), in situ hybridization studies showed that prolactin mRNA expression was lower, although the percentages of mRNA-expressing cells were not changed. However, when different subsets were compared, dopamine inhibited prolactin mRNA expression in those cells with the highest rate of secretion and lowest prolactin stores (the smallest prolactin cells). Those cells with the highest stores (and lowest basal release), the larger prolactin cells, were actually stimulated by dopamine. This provided early evidence that dopamine might have diverse effects on prolactin cells.

Stimulation

As early as 1977, several groups reported evidence that dopamine may stimulate prolactin secretion, depending on the physiological state of the animal (517–522). The mechanism behind this action may act through a different class of receptor that is known to stimulate adenylyl cyclase, and it also may be dose dependent, occurring with lower concentrations [reviewed in Porter et al. (522)]. Porter et al. studied the types of receptors that might be involved in the stimulation of prolactin release in subclones of a GH_4 C_1 mammotropic cell line that were stably transfected with these stimulatory dopamine receptors (D_1 and D_5), or the inhibitory form D_2 (522). Dopamine increased prolactin release from subclones with the stimulatory receptors, but not from those transfected with the inhibitory class. When they investigated dopamine receptors in normal anterior pituitary tissue, they found evidence for expression of the D_5 receptor. This evidence is further strengthened in studies reported by Schoors et al. (523), who detected D_1-like dopamine receptors with a dopamine antagonist selective for D_1 and D_5 receptors (SCH 23390). They used dual immunolabeling for prolactin and detected the binding for this antagonist in a subpopulation of prolactin cells.

Thus, it appears that the stimulatory form of dopamine receptor may be present on subsets of prolactin cells. Their unique functions are unclear at this point; however, to complicate matters further, it is important to note that the inhibitory dopamine receptor can also stimulate. Burris et al. (520) reported that eticlopride, a D_2 antagonist, blocked the stimulatory effect of dopamine. This could have occurred only through the D_2 receptor class, which normally mediates the inhibitory functions. Denef et al. (518) reported that the actions of stimulatory dopamine receptors are not detected until the inhibitory receptors are antagonized. Clearly, the mechanisms behind the actions of these two receptor classes are complex.

Neurointermediate Lobe Prolactin Regulatory Factors

The rapid response of prolactin cells to neural stimuli, such as suckling, has led workers to continue the search for the elusive prolactin regulatory factor (PRF) since the early 1980s. These stimuli clearly decrease the inhibitory effects of dopamine. Therefore, they may work both on prolactin cells and on dopaminergic neurons in the hypothalamus.

In 1987, Murai and Ben-Jonathan discovered that removal of the posterior lobe prevented a suckling-induced rise in prolactin (524). Therefore, researchers turned to cells in this lobe for a potential source of PRF (525–531). Perchloric acid extracts of the neurointermediate lobe stimulated the release of prolactin in vitro (528,529) and in vivo (530,531). Cocultures of neurointermediate lobes and anterior pituitary cells stimulated prolactin mRNA levels (532), and Steinmetz et al. identified a subpopulation of intermediate lobe cells as the source of PRFs (533). The population of neurointermediate lobe cells is relatively small. Therefore, to scale up the material needed to learn about the PRF, Allen et al. used targeted tumorigenesis with the POMC gene promoter ligated to the simian virus-40 large T antigens (POMC-TAG) to produce tumor cell lines in nude mice. They were able to isolate fractions containing PRF activity and identified both a large and a small class of PRF (534). They also generated cell lines that produced the PRFs (535).

Although the exact identity of the PRF peptides has proven elusive, one group has isolated significant prolactin-releasing activity in a neurointermediate lobe fraction containing salsolinol (SAL) (536,537). This compound belongs to the tetrahydroisoquinoline family and is derived from dopamine by (R)-salsolinol synthase. They reported that SAL releases prolactin in a selective and dose-dependent manner. They also found high concentrations of SAL in neurointermediate extracts with suckling, and the fact that it acts quickly on the release of prolactin. Radnai et al. have further reported that an analog of SAL can block its effects on stress- or suckling-induced prolactin release (537). Thus, SAL may be one of the putative PRFs in the neurointermediate lobe. Tóth et al. reported a bipotential dose–response curve when SAL was applied in RHPAs (536). There was a rise with low concentrations of SAL ($<10^{-10}$) and a decline in the presence of higher concentrations (10^{-6}), a response similar to that seen with dopamine (536). Using competitive inhibitors, they also showed evidence that D_2 receptors are not the site of action for SAL. Therefore, it is unlikely that SAL acts through blocking dopamine actions.

Thus, altthough the search for peptide PRFs continues, it is clear that a derivative of dopamine has potent, rapid stimulatory effects that can be driven by neural inputs, such as stress or suckling. The findings for the origin of the PRFs add significance to the studies that have shown different prolactin responses based on the region of the pituitary (478,515). Collectively, these data suggest a direct stimulatory route from the pars intermedia to neighboring prolactin cells, a route that could be enhanced by a network of cells with communicating gap junctions.

INTEGRATIVE STUDIES OF GONADOTROPES AND LACTOTROPES

Regulating the Transition from Birth to the Adult State

Many of the studies described in this chapter have led to more questions than answers. The remaining sections focus on some of the cutting-edge questions that are currently being addressed. Gonadotropes and lactotropes are born during embryonic development in families of cells that emerge from a precursor cell identified by the presence of α subunits. Whereas somatotropes are needed for the expansion of lactotropes, especially after birth, it is not known if lactotropes can be born in the embryo as a monohormonal cell type. Furthermore, although the appearance of LHβ heralds the appearance of the gonadotrope, it is not known if all FSH cells are derived from this subset.

Information about the regulation of the postnatal development of lactotropes is better developed than that for gonadotropes. Lactotrope expansion after birth depends on a number of cellular and hormonal factors, including some from gonadotropes and functioning thyrotropes. Gonadotropes may expand through the synergistic actions of EGF and GnRH, although activin may play a role in expansion of monohormonal FSH cells. More information is needed about how gonadotropes transition through puberty to the functional, heterogeneous population seen in the adult.

Multihormonal cells tend to become more abundant with age, which raises a number of questions about their significance and regulation. For many years, workers have suggested they were simply residual cells from embryonic development. However, studies today that show more rapid turnover and factors regulating both apoptosis and proliferation suggest that the pituitary cell population may be constantly remodeled and in need of a population of reserve or stem cells. Perhaps these multihormonal cells are sensitive to specific combinations of growth factors that drive them to support a system, as needed. Alternatively, they could function to serve a "cocktail" of hormones from a group of cells that can be regulated synchronously. Such cells may be responsive only to a group of regulatory factors that activate multiple signaling pathways. As the studies of sorting and targeting mechanisms become more sophisticated, we will learn more about how a multihormonal cell sends one set of granules to the sinusoidal pole, leaving another set in the cytoplasm. Alternatively, the cell may be called upon, by activating multiple pathways, to send out multihormonal granules, or at least a multihormonal granule population.

Detecting Receptor Populations in Gonadotropes and Lactotropes

As stated in the foregoing sections, studies of individual binding and responses to releasing and inhibitory hormones have shown that gonadotropes and lactotropes are heterogeneous in their binding and responses. Often pituitary functions may appear to be subserved by a subpopulation of cells; perhaps the other cells are in reserve for enhanced secretion as needed. The heterogeneity may also be based on the receptor populations that are expressed by individual subsets. Several cutting-edge questions related to these populations are briefly reviewed in the following sections.

The Search for the Prolactin-Releasing Factor

Most of this work has been reviewed in the section on lactotropes (524–537). It is evident that candidate factors have been identified, along with an interesting source for such stimulation. The pars intermedia cells as a source for PRF (or PRFs) helps explain the rapid responses to suckling and other environmental clues that stimulate prolactin release because these cells are directly innervated. Thus, as workers continue to isolate candidate factors (such as SAL), an important task will be to trace the route (neural and vascular pathways) leading to the prolactin cells. This also raises the importance of communicating vessels between the neurointermediate and anterior lobes and possibly juxtacrine or gap junction communication.

The Search for the FSH-Releasing Factor

As stated earlier, GnRH releases both LH and FSH, and there are a number of factors or conditions that might drive "nonparallel release" of gonadotropins. These include monohormonal cells; sorting mechanisms for monohormonal granules in bihormonal cells; rapid pulses of GnRH favoring LH over FSH; and inhibin, activin, or follistatin. Nevertheless, the presence of a separate releasing factor for FSH has been postulated, and the search for the elusive FSH-releasing factor has been ongoing since 1964 (144–152,538–544). Because of problems with assays and separation activities, the exact identity of an FSH-releasing factor has remained controversial (144–152). Beginning in the mid-1990s, optimal assay conditions were developed and fractions free of LH-releasing hormone activity were identified from ovine hypothalami (145–151,539–542). These fractions stimulated FSH release and appeared to be similar to lamprey GnRH III (150,543,544).

Not all workers have been able to detect selective FSH release from cells taken from rats or other species with lamprey GnRH III peptides (150–152), and the GnRH III receptor mRNA has not been found in rats (545). We have explored binding sites for a biotinylated analog of lamprey III GnRH to learn conditions needed for binding, as well as the target cells that bind the peptide (546). The biotin was linked to lysine 14 by a 14-atom spacer (546). Biotinylated lamprey GnRH III bound to cells containing either LH or FSH. Over 80% of FSH cells expressed binding, in contrast to 48% of LH cells. Binding by this lamprey GnRH III ligand was not affected by 100-fold GnRH I, added to compete for binding sites, nor was it affected by 100-fold corticotropin-releasing hormone or GHRH. Binding is reduced with increasing concentrations of competing, unlabeled lamprey GnRH III. Thus, the studies show that this peptide does bind specifically to pituitary cells, some of which are monohormonal FSH cells. However an important finding may explain some of the discrepant results about the GnRH III. In our studies, the cells quickly lose their ability to bind biotinylated lamprey GnRH III in culture. No binding was detected after 24 hours of culture. Thus, to detect biotinylated lamprey GnRH III binding, one must always use freshly dispersed cells. The latter finding may explain some of the controversy. Some workers report no FSH-releasing actions by GnRH III in cultured cells. Dual labeling for GnRH III and GnRH I receptors is ongoing. The findings thus far show that some cells express both receptors; others express only one. Whether lamprey GnRH III is binding to a GnRH III receptor or a splice variant of another type remains to be determined. It clearly shows a preference for cells with FSH antigens.

Detecting Gonadal Steroid Receptor Expression

Estrogen Receptors. Detection of estrogen receptors (ER) in the anterior pituitary has been reported since the mid-1960s, with different methods. As antisera to ER proteins became available, the data on isoform expression showed considerable variation (547). These differences may reflect species differences, physiological state, the specificity of the antisera or probe used (selectivity for the isoform), or the tissue preparation method (547,548). Estrogen is essential for reproductive functions in the female, stimulating LH, FSH, and prolactin and sensitizing the gonadotropes to GnRH stimuli. Estrogen stimulates the synthesis of GnRH receptors and induces expression of pituitary progesterone receptors in gonadotropes, thereby eliciting GnRH self-priming. These events are vital for the preovulatory LH surge (549).

Early studies used antisera to ER and found labeling that varied with the cycle; peak expression was seen in gonadotropes and GH cells during diestrus (550). Later studies developed antibodies to the α and β isoforms of the ER and reported findings in a number of species. (547). The ERα isoform was found to be most abundant in lactotropes and gonadotropes, although there was some activity in other cell types (551–555). Pelletier reported that over 90% of pituitary secretory cells express ERα (555). Studies of actions of a specific agonist for ERα, propylpyrazole triol (PPT), showed that it mimicked some restorative actions of estrogen on gonadotropes or prolactin cells in ovariectomized rats—namely, the hypertrophied gonadectomy cells regressed and prolactin cell secretion recovered. In addition, PPT increased progesterone receptor mRNA and immunolabeling in gonadotropes (556). Studies of fetal development showed changes in expression of ERα that are parallel to those of LHβ in sheep (557) and in rats (558). In the latter of the two studies, ERα did not become as prominent until after birth and its increase paralleled the increase in gonadotropes seen during the first week of postnatal life. Thus, the consensus has been that ERα may play a major role in modulating estrogen functions in gonadotropes and lactotropes.

ER and ERα expression has been found in pituitary tumors, particularly abundantly in those that produced prolactin (547,552,559). Friend et al. (553) reported that 70% of the ER-positive tumors produced FSH, 83% produced LH, and 50% produced prolactin. Very few of these tumors produced the other pituitary hormones (<4%). Other workers reported that ER mRNA was expressed in a number of adenoma types and all pituitary cell types were represented, including some multihormonal subtypes like mammosomatotropes (559–562). More recently, Pereira-Lima et al. reported that ERα proteins were expressed at a higher level in larger or more invasive adenomas; however, ERα was found in only a subset of the tumors in this study (563).

ERα knockout mice have been used to investigate the importance of this receptor isoform to reproduction (551,552). Both males and females are infertile, again emphasizing the importance of this isoform. Pelletier et al. (564) reported that the percentage of prolactin cells in the ERα knockout mice was reduced from 40% in the wild type to 5% in the female and 9% in the male. They also reported that the knockout mice had atrophied type III prolactin cells (Figs. 61 and 64; although note that their terminology is type 1) and very poorly granulated type II cells (Figs. 62, 63, and 65; note that their terminology is type 2).

The other isoform, ERβ, is somewhat of an enigma in the pituitary. ERβ knockouts are fertile, although the females have reduced litter sizes (547,551). Pelletier et al. (564) reported that ERβ knockout animals had normal prolactin cells. ERβ appears prominently during day 17 of fetal development (558) and was colocalized with LHβ, which also appears at that time (Fig. 1). It may appear in parallel with LHβ during fetal life, after which it becomes less prominent. Thus, it may be involved in early gonadotrope development events. ERβ was never found in prolactin cells (558). In other studies of pituitaries from adult animals, ERβ labeling has either been absent (555) or relatively sparse (554). ERβ is more frequently found in human GH or ACTH cells, suggesting it may modulate estrogen actions on these cell types. Using RT-PCR, Chaidarun et al. (565) reported that 89% of somatotrope adenomas produced ERβ. Three of these also produced ERα; however, these tumors were distinguished by scattered coexpression of prolactin. Gittoes et al. (566) and Shupnik et al. (559) used RT-PCR to show that GH-secreting tumors expressed ERβ.

Because ERβ knockouts are fertile (38), one can question the importance of this isoform to reproduction, although it clearly has been detected in gonadotropes, especially during development (559). ERβ has also been reported to be a dominant negative regulator for ERα (567–569).

Progesterone Receptor. As stated in the previous section, estrogen upregulates expression of progesterone receptors (PR) during the cycle. PR is like a transcription factor that can be activated by ligand or ligand-independent mechanisms. It stimulates gene expression in gonadotropes and mediates estrogen's positive feedback effects. Levine suggested that activation of the PR may enhance or amplify the estrogen signal by permitting cross-talk between signaling pathways (549). A proestrous rise in progesterone, just before the LH surge, amplifies the surge. This may involve the combined recruitment of more secreting LH cells and an enhancement of signaling pathways in existing cells. PR were found only in gonadotropes in early studies of rats and primates (570,571). PR exists as two isoforms, PR_A and PR_B; the smaller A form is a truncated version of the B form. Cells express different ratios of types A and B and because the isoforms have different promoters, it is believed that they subserve different functions. PR knockout mice have several reproductive deficiencies (572,573): there are no gonadotropin surges, and vaginal smears indicate no normal estrous cycles. Both LH and prolactin levels are elevated and ovariectomy resulted in a higher than normal rise in LH secretion after surgery (573). Turgeon et al. investigated these knockouts to learn if PR was needed for the normal development and expansion of pituitary cell types (574). They reported that the percentages of cells bearing LH, GH, or prolactin antigens were comparable in the PR knockout, indicating that PR had no influence on development or expansion of these cells. However, they also found PR expression in both gonadotropes and prolactin cells of wild-type mice. There was also a less abundant subset of GH cells that expressed PR, and they suggested that these might be from either mammosomatotrope or somatogonadotrope populations. Recall that this same reasoning was presented by Candolfi et al. (427), who reported that, in rats, progesterone antagonized the apoptotic effects of estrogen and TNF-α on somatotropes. Because PR was reported to be found exclusively in gonadotropes (570,571), they suggested that this PR protective effect on somatotropes was actually affecting somatogonadotropes. Most workers have suggested that any PR effects on cells other than gonadotropes are indirect, mediated through neural or paracrine interactions (575). Taking lessons learned from the past into consideration, one might also postulate that lactotropes with PR may be difficult to identify in some species, largely because of the reports that some subsets of lactotropes are difficult to identify except through RHPA (452–456). Studies that combine PR labeling (or PR mRNA detection) and RHPA might be needed to learn if there are lactotropes with PR in other species.

Androgen Receptors. Androgens feed back on gonadotropes and limit gonadotropin secretion by suppressing expression of the α subunit. They also reduce GnRH binding sites, as described in a previous section [reviewed in Tibolt and Childs (343)]. Studies by Thieulant and Duval (576) reported that androgen (DHT) binding was highest in pituitary cell fractions that were enriched in gonadotropes and lowest in fractions containing the small prolactin cells. The ratio of androgen receptors (AR) to ER was 0.86 in gonadotrope-enriched fractions and 0.35 in fractions enriched in lactotropes. Another important regulatory effect of AR is to suppress expression of the gonadotropin α-subunit gene (577,578). Recall that this is the subunit shared by gonadotropins, TSH, and human chorionic gonadotropin.

Studies in primates have shown that androgens may act either indirectly, or only after conversion to estrogens [reviewed in Okada et al. (247)]. The nonconvertible androgen, DHT, appears to have little direct effect. In rats, however, androgen effects are more direct, and can be seen on gonadotropes in culture.

Studies by Okada et al. detected AR in gonadotropes of primate or rat pituitaries to learn if the species differences reflect different expression of nuclear AR by gonadotropes (247). Dual labeling for LHβ and FSHβ antigens showed that 80% of primate gonadotropes were bihormonal, 17% contained only LHβ, and 3% were monohormonal FSHβ. They detected AR in 22% of monkey FSHβ cells and 15% of rat FSHβ cells. After treatment with testosterone (which moves the activated receptor to the nucleus), the percentages of gonadotropes bearing AR rose to 79% in the monkey and 81% in the rat. Thus, they concluded that the failure of primate cells to respond to androgens directly is not due to the absence of the receptor in gonadotropes. Furthermore, it can be activated in the presence of AR.

Neither GnRH nor testosterone affected levels of AR mRNA in rat or monkey cultures. RT-PCR on single cells also showed coexpression of AR mRNA and LHβ mRNA in most gonadotropes tested. AR were also found on somatotropes in their studies, 7 of 12 of which also expressed LHβ mRNA (247).

Heckert et al. detected key elements in the promoter sequence of the α-subunit gene, α-basal element, and the cAMP-regulatory element, which are important in AR-mediated suppression of the α subunit (578). Jorgensen and Nilson (579) reported that factors that activated these gene regulatory sites included cJun and activation transcription factor 2.

Communication between Lactotropes and Gonadotropes

Since they were first identified by immunolabeling, the juxtaposition of lactotropes around gonadotropes was noted (47) and studied further (580,581). With the growing understanding of the factors produced by gonadotropes that influence lactotropes, this cellular organization has developed greater significance. However, lactotropes also influence gonadotropes in a variety of species (580,582). An increase in prolactin secretion is inhibitory to gonadotropes and is often associated with infertility or, during lactation, with amenorrhea. This appears to be mediated directly at the level of the pituitary because an increase in prolactin, in vitro, reduces the percentages of gonadotropes secreting LH in an RHPA (582). During lactation, there is a reduced LH response to GnRH (583,584).

Evidence that prolactin acts directly on gonadotropes comes from studies that detected prolactin binding to gonadotropes (585), or prolactin receptor mRNA and proteins in the ovine pituitary (586). The latter study detected prolactin receptor proteins by dual labeling in all LHβ gonadotropes and most FSHβ gonadotropes, indicating that the receptor is expressed by bihormonal gonadotropes and some monohormonal LH gonadotropes (585). [Note that previous studies had not found monohormonal gonadotropes in the sheep (133,134), but this study brings out a small subpopulation.] During the course of their investigation of prolactin receptors, they also noted the remarkable clusters of prolactin cells around individual gonadotropes in ovine pituitaries. A later study by the same group reported similar clusters of lactotropes around gonadotropes in the mare pituitary. However, gonadotropes in the horse did not express prolactin receptors (587). More recently, Townsend et al. (588) focused on changes in subsets of gonadotropes in the male horse after different physiological states. Overall, they found that during the breeding season there were greater numbers of monohormonal LH and FSH cells and bihormonal gonadotropes than in the nonbreeding season. Castration abolished any seasonal differences in expression of these populations. Monohormonal LH gonadotropes were nearly 50% of the population, with bihormonal and monohormonal FSH gonadotropes splitting the remaining 50%.

When lactotropes were detected, there was a significantly higher percentage of them in castrated animals during the breeding season compared with castrated animals in the nonbreeding season. In addition, there were greater percentages of lactotropes in fields from intact horses during the breeding season compared with those in the nonbreeding season. When lactotrope/gonadotrope associations were counted, there was a twofold increase in horses that were castrated and in the breeding season over intact horses or castrated horses in the nonbreeding season. The lack of prolactin receptors suggested that an intermediate cell type might be involved in mediating prolactin's effect on gonadotropes.

The significance of the associations may indeed relate to the cross-talk between the two populations that may begin as early as during fetal development (260–263) and is expressed prominently during postnatal life (263,265–268). The delay in expansion of the lactotropes after birth (relative to that of other cell types; Fig. 67) may be significant in that it allows the expansion of the gonadotropes. It would be interesting to learn if the limiting effects of prolactin on gonadotropes begin before puberty. It is also tempting to speculate that gap junctions play a role in this interaction. As was discussed earlier, Morand et al. (227) reported the presence of functioning gap junctions between endocrine cells, including prolactin cells and gonadotropes. Whereas it is doubtful that the protein hormones could get through this passage, activation of signaling pathways in one of the cells could send small signaling molecules (like cAMP or calcium) across a gap junction to the other cells. Collectively, these data point to the importance of future studies that attend to the contiguous association between pituitary cells either by testing them in vivo, or by studying cell aggregates that are allowed to form the communicating junctions.

How Past Lessons May Shape Future Experiments

In designing experiments for future cell biology studies of gonadotropes and lactotropes, it will be important to benefit from lessons learned from the past. Our experimental designs should not be limited or dictated by dogma or assumptions about pituitary cell functions. The most obvious of these is the "one-cell/one-hormone" assumption, which drives workers to think about the pituitary cell population as simple sets of monohormonal or (at most) bihormonal cells that possess only one set of releasing hormones and are never interconvertible. The lessons learned from the past will help us design more complete experiments and better interpret our results. Some caveats are presented here.

First, we have learned that not all cytochemical tools will detect cells that are multihormonal. Immunolabeling has been a wonderful standard, a significant step up from the tinctorial stains; however, it may or may not be able to detect the full potential of the cells. This is especially true when the cells are actively secreting the stored hormones. Unless they can be detected in the rough endoplasmic reticulum, they may virtually "disappear" from the population until stores are replenished. A concept cannot be proved by immunolabeling alone, and the following paragraphs show why.

Immunolabeling may also detect hormones at their site of binding. This may produce a "false-positive" result in terms of number of hormone-producing cells if one pituitary cell has bound and endocytosed the hormone in question. A classic example is seen when one recognizes that GH-binding proteins are abundant in the pituitary and the presence of GH may reflect its binding, rather than its production at that site. Finding prolactin receptors raises the same caveat when interpreting dual immunolabeling for LH, FSH, and prolactin.

Detection of the mRNA is therefore an important confirmatory step to identify a site of production. However, in situ hybridization may not be sensitive enough to detect the true potential, and RT-PCR applied to single cells will likely provide more information about a subpopulation of cells. Advances in the sensitivity of in situ hybridization may improve its results, especially with the production of probes that have high concentrations of signaling molecules (244), as are available from GeneDetect.com. Also, this technology provides a way to detect the earliest transcriptional activity by detecting hnRNA. However, one must recognize that hnRNA may be quickly degraded and take into consideration the probability that timing is critical to its detection.

One lesson that has been learned is the fact that the production of the transcripts and proteins may be out of phase so that some subsets of cells are detectable only by one of the products. This is particularly true in an actively secreting cell with rate-limiting gene products that are produced in a cyclic manner (such as gonadotropin β subunits). There may be cyclic variations in mRNA, hnRNA, or proteins that would render subsets of cells undetectable by one cytochemical protocol. Most important, we should not expect the different gene products always to be in phase, especially in an actively secreting cell that expresses high mRNA turnover. Studies of cyclic expression should be done, especially when dealing with gene products that are regulated by steroid hormones, such as estrogens. All of this variability may cause workers to use the immortalized gonadotrope cell lines to avoid detecting the net effect of cells in different stages of the secretory cycle or the estrous cycle.

A big challenge in interpreting and analyzing data is to know how to quantify the cytochemical fields, as well as what each type of quantification really means to the population. For example, changes in percentages of a particular cell type are a measure of the concentration of those cells in the population. The concentration could go up or down, but unless some absolute numbers are used along with the percentages, one does not know the actual mechanism. For example, a concentration of one cell type could appear to increase only because of the death of other cell types. Similarly, its concentration could decrease only because of the rapid proliferation of another cell type. Thus, the challenge in mixed pituitary cell populations is to learn if the change in concentration of a given cell type is due to mitotic activity (see previous sections for those tests), to apoptosis of this cell type, or to the previously mentioned factors. The change could also be due to differentiation, and there may be no changes in absolute cell numbers associated with this activity. However, an evaluation of other cell types may reveal more multihormonal cells or a loss of the contributing cell type. Thus, there are a number of interpretations of data on cell percentages that are important in the overall story.

Another approach to quantification is the use of image analysis and densitometry. The equipment will likely allow the detection of changes in area or density of label in a given cell type, or there may be subroutine or formula that allows the integration of label area and density in a field. This would take into account changes in intensity and area of label as well as the concentration of labeled cells. Thus, image analysis and densitometry can be used to correlate changes in a gene product in test tube assays with changes in expression of cellular product in situ. One may find changes in the amount of label per cell (expressed as area or density) or overall changes in the labeled cells (integrated optical density). If the fields sampled are uniform and the lighting conditions for the camera optimized, this can provide a rapid readout of the effect of a change in physiological state on expression of a gene product. A given state can bring out more cells with gene product or more gene product per cell. In the latter case, the existing cells are being stimulated to produce the gene product. In the former state, additional cells are differentiating to produce the product. Both sets of data are important to the overall interpretation of cellular correlates behind a change in a test tube assay. Finally, a correlate that we demonstrated in our studies of small versus large FSH gonadotropes after castration is that cell size can contribute to the overall decline in expression of a gene product (176). There may even be an increase in the number of such cells, but their smaller area yields lower levels of FSHβ mRNA.

Tests for secretory activity by RHPA or immunoblot protocols have proven invaluable in the detection of cellular function and potential. The RHPA can be combined with tests of any gene product or ligand-binding capacity. One lesson learned is that the RHPA may be the only way accurately to identify the bihormonal secretory capacity of many mammosomatotropes.

Tests for ligand-binding activity help to identify a target cell population and predict its responsiveness. Correlations between numbers of cells bound or molecules bound per cell (image analysis) and the cellular response can be made, assuming the ligand is biologically active and has not downregulated the cells in question. It is important to pick ligands with reporter molecules that allow them to be used in physiological concentrations so the experiments will mimic those normally done to detect responses to this regulator. Correlations between binding and secretory activity should be obtained to prove potency (if the ligand is an agonist).

Also, it is important to recognize facts about any cyclic changes in ligand binding. For example, because GnRH receptor expression changes with the stage of the cycle, the use of a population of pituitary cells from mixed cycling female rats combines gonadotropes that have different levels of receptivity. The results are then the net effect of heterogeneous responses. One may not receive a clear answer about how GnRH functions at a given time in the cycle unless cells from that period in the cycle are used. Again, some questions may be better answered with the immortalized cell lines, comparing data with primary cultures from rats in one stage of the cycle.

An important lesson is that the presence of the hormone or its transcript in a cell does not necessarily mean the cell will be secretory. Populations of lactotropes and gonadotropes have subsets that do not respond in any cell secretion assay, in spite of their detectable hormone content.

Another caveat relates to species differences. Throughout the chapter these have been noted, and although many general features are evident across species, there are some striking differences. For example, one that has not been mentioned is the fact that inhibin induces GnRH receptor expression in the sheep, but inhibits it in other species (589). Species differences may also play a role as we try to apply information learned from transgenic mice to rats or primates.

To summarize, we have learned that whereas the pituitary cell population started as a simple monohormonal (or bihormonal) set of cells in the embryo, it expresses more heterogeneity in the adult, even in one stage of the cycle. Subsets of cells do not respond alike to a number of secretagogues or inhibiting factors. A general rule of thumb is that more cells tend to come "online" when secretion approaches a peak, although the factors that cause these heterogeneous responses in a normal population are unknown. Thus, the complexity seen in the adult pituitary suggests further differentiation or maturation. Some of the regulation may be driven by gonadal steroids during and after puberty. Such actions may drive the development of subsets of cells that can respond in the adult environment. This fact makes the study of cell function more challenging in the mixed pituitary cell population.

As stated earlier in this chapter, major discoveries about receptors and signaling pathways, gene regulation, and other cellular functions have been made thanks to the availability of valuable immortalized tumor lines. However, additional new techniques that allow us to look at identified gonadotropes, in situ, are emerging. An example is the study published by Wu et al. (248), which tags FSH gonadotropes so they can be followed and studied in their natural environment. Thus, we will be in a position to balance reductive approaches with those that acknowledge the fact that some regulatory circuits may be fully understood only in a population of gonadotropes in situ, surrounded by lactotropes and communicating in a functional syncytium.

ACKNOWLEDGMENTS

The author acknowledges the significant contributions to research on gonadotropes described in this chapter by students: Laura Garner, Ph.D., Thomas Bauer, M.D., Ph.D., Dayle Ellison Cole, Ph.D., Victor May, Ph.D., Robert Tibolt, M.D., Ph.D., Samir Ibrahim, Ph.D., Jon Lloyd, Ph.D., Ping Wu, M.D., Ph.D., James Patterson, Ph.D., Xuemo Fan, M.D., Ph.D., Jennifer Armstrong, Ph.D., and Mary Iruthayanathan, Ph.D.; and fellows and visiting faculty: Byung Lan Lee, Ph.D., Noor Ahkter, Ph.D., and Akihiko Kudo, Ph.D. She is also grateful for the excellent technical assistance by Linda Foster, Geda Unabia, Diana Rougeau, and Brandy Whitehead Johnson. Finally, she thanks her husband, Gary D. Jones, for much support, patience, and understanding.

The author is grateful to the Hormone Distribution Program and Dr. A. Parlow for the ongoing supply of antisera to LHβ, FSHβ, and rat GH as well as radioimmunoassay kits for these hormones. She thanks Dr. J. G. Pierce for the antiserum to bovine LHβ used throughout these studies. Finally, recent studies with highly sensitive biotin and digoxigenin oligonucleotide mRNA and hnRNA probes have been designed by GeneDetect.com. She thanks Paul Hughes, Ph.D., for many helpful suggestions during the developmental phases of these studies.

The studies of gonadotropes were originally funded by a National Institutes of Health (NIH) grant that began at the University of Nebraska in 1974 as R01 HD08842. It was moved to Northwestern University as R01 HD10930 from 1978–1980, and then to the University of Texas Medical Branch as NIH R01 HD15472 from 1981–1994. Support from an RCDA NIH HD00395 for these studies also was made possible from 1979–1984. Funding for somatogonadotropes was supported by HD 33915 from 1994–2003. Current studies of gonadotropes are supported by R03 HD 44875 and R21 HD 047467. Additional sources of support came from National Science Foundation IBN 9724066.

REFERENCES

1. Greep, R. O. (1973). The gonadotrophins and their releasing factors [review]. *J. Reprod. Fertil. Suppl.* 20, 1–9.
2. Li, C. H. (1973). Background of the development of the work in isolating interstitial cell-stimulating hormone (luteinizing hormone) and follicle-stimulating hormone. *Am. J. Obstet. Gynecol.* 115, 1150–1152.
3. Closset, J., Maghuin-Rogister, G., Combarnous, Y., and Hennen, G. (1973). Primary structure of the subunits of luteinizing hormone (LH): a comparison of amino acid sequence of the alpha and beta chains of LH of ovine, bovine, porcine and human origin. *Arch. Int. Physiol. Biochim.* 81, 174–176.
4. Reichert, L. E. Jr., and Ward, D. N. (1974). On the isolation and characterization of the alpha and beta subunits of human pituitary follicle-stimulating hormone. *Endocrinology* 94, 655–664.
5. Reichert, L. E. Jr., Lawson, G. M. Jr., and Leidenberger, F. L. (1973). Influence of alpha- and beta-subunits on the kinetics of formation and activity of native and hybrid molecules of LH and human chorionic gonadotropin. *Endocrinology* 93, 938–946.
6. Bahl, O. P., Carlsen, R. B., Bellisario, R., and Swaminathan, N. (1972). Human chorionic gonadotropin: amino acid sequence of the alpha and beta subunits. *Biochem. Biophys. Res. Commun.* 48, 416–422.
7. Meites, J., Lu, K. H., Wuttke, W., Welsch, C. W., Nagasawa, H., and Quadri, S. K. (1972). Recent studies on functions and control of prolactin secretion in rats. *Recent Prog. Horm. Res.* 28, 471–526.
8. Dixon, J. D., and Li, C. H. (1964). Chemistry of prolactin. *Metabolism* 13 (Suppl.), 1093–1101.
9. Catt, K., and Moffat, B. (1965). Fractionation of rat pituitary extract by starch gel electrophoresis and identification of growth hormone and prolactin. *Endocrinology* 76, 678–685.
10. Apostolakis, M. (1965). The extraction of prolactin from human pituitary glands *Acta Endocrinol. (Copenh).* 49, 1–16.
11. Evans, H. M., and Long, J. A. (1922). Characteristic effects upon growth, oestrus and ovulation induced by the intraperitoneal administration of fresh anterior hypophyseal substance. *Proc. Natl. Acad. Sci. U S A* 8, 38–39.
12. Smith, P. E. (1926). Ablation and transplantation of the hypophysis of the rat. *Anat. Rec.* 32, 221.
13. Smith, P. E. (1926). Hastening development of female genital system by daily hypoplastic pituitary transplants. *Proc. Soc. Exp. Biol. Med.* 24, 131–132.
14. Ascheim, S., and Zondheim, B. (1927). Hypophysenvorderlappenhormon und Ovarial-hormone im Harn von Schwangeren. *Klin. Wochenschr.* 6, 1322–1327.
15. Zondek, B. (1930). Über die Hormone des Hypophysenvorderlappens: II. Follikelreifungshormon Prolan A-Klimakterium-Kastration. *Klin. Wochenschr.* 9, 393–396.
16. Fevold, H. L., Hisaw, F. L. and Leonard, S. L. (1931). The gonad stimulating and the luteinizing hormones of the anterior lobe of the hypophesis. *Am. J. Physiol.* 97, 291–301.
17. Evans, H. M., Korpi, K., Simpson, M. E., Pencharz, R. I., and Wonder, D. H. (1936). On the separation of the interstitial cell stimulating, luteinizing and follicle stimulating fractions in the anterior pituitary gonadotropic complex. *Univ. Calif. Berkeley Publ. Anat.* 1, 255–273.
18. Evans, H. M., and Simpson, M. E. (1930). Hormones of the anterior hypophysis. *Am. J. Physiol.* 98, 511–546.

19. Lobie, P. E., Breipohl, W., Aragon, J. G., and Waters, M. J. (1990). Cellular localization of the growth hormone receptors/binding protein in the male and female reproductive systems. *Endocrinology* 126, 2214–2221.
20. Tiong, T. S., and Herington, A. C. (1991). Tissue distribution, characterization and regulation of messenger ribonucleic acid for growth hormone receptor and serum binding protein in the rat. *Endocrinology* 129, 1628–1634.
21. Giustina, A., and Veldhuis, J. D. (2000) Pathophysiology of the neuroregulation of growth hormone secretion in experimental animals and the human. *Endocr. Rev.* 19, 717–797.
22. Hull, K. L., and Harvey, S. (2002). GH as a co-gonadotropin: the relevance of correlative changes in GH secretion and the reproductive state [review]. *J. Endocrinol.* 172, 1–19.
23. Childs, G. V. (2000). Growth hormone cells as co-gonadotropes: partners in the regulation of the reproductive system. *Trends Endocrinol. Metab.* 11, 168–174.
24. Riddle, O., and Braucher, P. F. (1931). Control of the special secretion of the crop gland in pigeons by an anterior pituitary hormone. *Am. J. Physiol.* 97, 617–625.
25. Riddle, O., Bates, R. W., and Dykshorn, S. W. (1933). The preparation, identification and assay of prolactin: a hormone of the anterior pituitary. *Am. J. Physiol.* 105, 191–216.
26. Neill, J. D., and Reichert, L. E. (1971). Development of a radioimmunoassay for rat prolactin and evaluation of the NIAMD rat prolactin radioimmunoassay. *Endocrinology* 88, 548–555.
27. Bern, H. A., and Nicoll, C. A. (1968). The comparative endocrinology of prolactin. *Recent Prog. Horm. Res.* 24, 681–720.
28. Halmi, N. S., and Moriarty, G. C. (1977). The hypophysis (pituitary gland). In *Histology* (R. Greep and L. Weiss, Eds.), pp. 1093–1967. McGraw-Hill, New York.
29. Benda, C. (1900). Beitrage zur normalen und pathologischen Histologie der menschlichen Hypophysis Cerebri. *Berl. Klin. Wochenshr.* 37, 1205–1210.
30. Cushing, H. (1909). Partial hypophysectomy for acromegaly with remarks on the function of the hypophysis. *Ann. Surg.* 50, 1002–1017.
31. Evans, H. M., and Long, J. A. (1921). The effect of the anterior lobe administered intraperitoneally upon growth, maturity, and oestrous cycles of the rat. *Anat. Rec.* 21, 62–63.
32. Cushing, H. (1909). The hypophysis cerebri: clinical aspects of hypopituitarism and hyperpituitarism. *JAMA* 53, 249–255.
33. Cushing, H. (1932). The basophil adenomas of the pituitary body and their clinical manifestations (pituitary basophilism). *Johns Hopkins Hosp. Bull.* 50, 137–195.
34. Halmi, N. S. (1974). The current status of human pituitary cytophysiology. *N. Z. Med. J.* 80, 551–556.
35. Moriarty, G. (1973). Adenohypophysis: ultrastructural cytochemistry [review]. *J. Histochem. Cytochem.* 21, 855–894.
36. Romeis, B. (1940). Hypophyse. In *Handbuch der Mikroskopischen Anatomie des Menschen* (W. Mollendorff von, Ed.), Vol. 6, Part 3. J. Springer, Berlin.
37. Purves, H. D., and Griesbach, W. E. (1954). The site of follicle-stimulating and luteinizing hormone production in the rat pituitary. *Endocrinology* 55, 785–793.
38. Ezrin, C., Swanson, H. D., Humphrey, J. G., Dawson, J. W., and Wilson, W. D. (1958). The delta cell of the human adenohypophysis: its response to acute and chronic illness. *J. Clin. Endocrinol. Metab.* 18, 917–936.
39. Wilson, W. D., and Ezrin, C. (1954). Three types of chromophil cells of the adenohypophysis demonstrated by a modification of the periodic acid-Schiff test technique. *Am. J. Pathol.* 30, 891–899.
40. Herlant, M. (1964). The cells of the adenohypophysis and their functional significance. *Int. Rev. Cytol.* 17, 299–382.
41. Halmi, N. S. (1950). Two types of basophils in the anterior pituitary of the rat and their respective cytophysiological significance. *Endocrinology* 47, 289–299.
42. Halmi, N. S. (1952). Two types of basophils in the rat pituitary: thyrotrophs and gonadotrophs vs. beta and delta cells. *Endocrinology* 50, 140–142.
43. Halmi, N. S. (1952). Differentiation of two types of basophils in the adenohypophysis of the rat and the mouse. *Stain Technol.* 27, 61–64.
44. Halmi, N. S., and McCormick, W. R. (1969). The delta cell of the human hypophysis in childhood. *J. Clin. Endocrinol. Metab.* 29, 1036–1041.
45. Farquhar, M. G., and Rinehart, J. F. (1954). Electron microscopic studies of the anterior pituitary gland of castrate rats. *Endocrinology* 54, 516–541.
46. Baker, B. L., Midgley, A. R., Gersten, B. E., and Yu, Y. Y. (1969). Differentiation of growth hormone- and prolactin-containing acidophils with peroxidase-labeled antibody. *Anat. Rec.* 164, 163–172.
47. Nakane, P. K. (1970). Classifications of anterior pituitary cell types with immunoenzyme histochemistry. *J. Histochem. Cytochem.* 18, 9–20.
48. Tixier-Vidal, A., Tougard, C., Kerdelhué, B., and Jutisz, M. M. (1975). Light and electron microscopic studies on immunocytochemical localization of gonadotropic hormones in the rat pituitary gland with antisera against ovine FSH, LH, LHa, and LHβ. *Ann. N Y Acad. Sci.* 254, 433–460.
49. Childs, G. V., and Ellison, D. G. (1980). An immunocytochemist's view of gonadotropin storage in the adult male rat. Cytochemical and morphological heterogeneity in serially sectioned gonadotropes. *Am. J. Anat.* 158, 397–410.
50. Phifer, R. F., Midgley, A. R., and Spicer, S. S. (1973). Immunohistologic and histologic evidence that follicle-stimulating hormone and luteinizing hormone are present in the same cell types in the human pars distalis. *J. Clin. Endocrinol.* 17, 125–141.
51. Halmi, N. S. (1982). Occurrence of both growth hormone- and prolactin-immunoreactive material in the cells of human somatotropic pituitary adenomas containing mammotropic elements. *Virchows Arch. A Pathol. Anat. Histopathol.* 398, 19–31.
52. Childs, G. V. (1991). Multipotential pituitary cells that contain ACTH and other pituitary hormones. *Trends Endocrinol. Metab.* 2, 112–117.
53. Roudbaraki, M., Lorsignol, A., Langouche, L., Callewaert, G., Vankelecom, H., and Denef, C. (1999). Target cells of γ3-melanocyte-stimulating hormone detected through intracellular Ca^{2+} responses in immature rat pituitary constitute a fraction of all main pituitary cell types, but mostly express multiple hormone phenotypes at the messenger ribonucleic acid level: refractoriness to melanocortin-3 receptor blockade in the lacto-somatotroph lineage. *Endocrinology* 140, 4874–4885.
54. Seuntjens, E., Hauspie, A., Vankelecom, H., and Denef, C. (2002). Ontogeny of plurihormonal cells in the anterior pituitary of the mouse, as studied by means of hormone mRNA detection in single cells. *J. Neuroendocrinol.* 14, 611–609.
55. Rathke, H. (1839). *Entwicklungsgeschichte der Natter.* Konigsberg, Germany.
56. Takuma, N., Sheng, H. Z., Furuta, Y., Ward, J. M., Sharma, K., Hogan, B. L. M., Pfaff, S. L., Westphal, H., Kimura, S., and Mahon, K. A. (1998). Formation of Rathke's pouch requires dual induction from the diencephalons. *Development* 125, 4835–4840.
57. Westphal, H. (2002). Genes that fashion the pituitary gland [review]. *Mol. Cell. Endocrinol.* 197, 45–46.
58. Dattani, M. T., and Robinson, I. C. (2000). The molecular basis for developmental disorders of the pituitary gland in man. *Clin. Genet.* 57, 337–346.
59. Watkins-Chow, D. E., and Camper, S. A. (1998). How many homeobox genes does it take to make a pituitary gland? *Trends Genet.* 14, 284–289.

60. Cohen, L. E., and Radovick, S. (2002). Molecular basis of combined pituitary hormone deficiencies *Endocr. Rev.* 23, 431–442.
61. Sheng, H. Z., Zhadanov, A. B., Mosinger, A. B., Fuji, T., Bertuzzi, S., Grinberg, A., Lee, E. J., Huang, S. P., Mahon, K. A., and Westphal, H. (1996). Specification of pituitary cell lineages by the LIM homeobox gene *Lhx3*. *Science* 278, 1004–1007.
62. Sheng, H. Z., Moriyama, K., Yamashita, T., Li, H., Potter, S. S., Mahon, K. A., and Westphal, H. (1997). Multistep control of pituitary organogenesis. *Science* 278, 1809–1812.
63. Treier, M., and Rosenfeld, M. G. (1996). The hypothalamic-pituitary axis: co-development of two organs. *Curr. Opin. Cell Biol.* 8, 833–842.
64. Ericson, J., Norlin, S., Jessell, T. M., and Edlund, T. (1998). Integrated FGF and BMP signaling controls the progression of progenitor cell differentiation and the emergence of pattern in the embryonic anterior pituitary. *Development* 125, 1005–1015.
65. Cha, K. B., Douglas, K. R., Potok, M. S., Liang, H., Jones, S. N., and Camper, S. A. (2004). WNT5A signaling affects pituitary gland shape. *Mech. Dev.* 121, 183–194.
66. Treier, M., O'Connell, S., Gleiberman, A., Price, J., Szeto, D. P., Burgess, R., Chuang, P-T., McMahon, A. P., and Rosenfeld, M. G. (2001). Hedgehog signaling is required for pituitary gland development. *Development* 128, 377–366.
67. Treier, M., Gleiberman, A. S., O'Connell, S. M., Szeto, D. P., McMahon, J. A., McMahon, A. P., and Rosenfeld, M. G. (1998). Multistep signaling requirements for pituitary organogenesis in vivo. *Genes Dev.* 12, 1691–1704.
68. Jean, D., Berneir, G., and Gruss, P. (1999). Six6 (Optx2) is a novel murine Six3-related homeobox gene that demarcates the presumptive pituitary/hypothalamic axis and the ventral optic stalk. *Mech. Dev.* 84, 31–40.
69. Hermesz, E., Mackem, S., and Mahon, K. A. (1995). Rpx: a novel anterior-restricted homeobox gene progressively activated in the prechordal plate, anterior neural plate and Rathke's pouch of the mouse embryo. *Development* 122, 41–52.
70. Thomas, P. Q., Johnson, B. V., Rathjen, J., and Rathjen, P. (1995). Sequence, genomic organization, and expression of the novel homeobox gene Hesx1. *J. Biol. Chem.* 270, 3869–3875.
71. Quirk, J., and Brown, P. (2002). Hesx1 homeodomain protein represses transcription as a monomer and antagonizes trans-activation of specific sites as a homodimer. *J. Mol. Endocrinol.* 28, 193–205.
72. Szeto, D. P., Ryan, A. K., O'Connell, S. M., and Rosenfeld, M. G. (1996). P-OTX: a PIT-1 interacting homeodomain factor expressed during anterior pituitary gland development. *Proc. Natl. Acad. Sci. U S A* 93, 7706–7710.
73. Lanctot, C., Gauthier, Y., and Drouin, J. (1999). Pituitary homeobox 1 (Ptx1) is differentially expressed during pituitary development. *Endocrinology* 140, 1416–1422.
74. Tremblay, J. J., Lanctot, C., and Drouin, J. (1988). The pan-pituitary activator of transcription, Ptx1 (pituitary homeobox 1), acts in synergy with SF-1 and Pit1 and is an upstream regulator of the Lim-homeodomain gene Lim3/Lhx3. *Mol. Endocrinol.* 12, 428–441.
75. Tremblay, J. J., Marcil, A., Gauthier, Y., and Drouin, J. (1999). Ptx1 regulates SF-1 activity by an interaction that mimics the role of the ligand-binding domain. *EMBO J.* 18, 3431–3441.
76. Tremblay, J. J., and Drouin, J. (1999). Egr-1 is a downstream effector of GnRH and synergizes by direct interaction with Ptx and SF-1 to enhance luteinizing hormone β gene transcription. *Mol. Cell. Biol.* 19, 2567–2576.
77. Gage, P. J., and Camper, S. A. (1977). Pituitary homeobox 2, a novel member of the bicoid-related family of homeobox genes, is a potential regulator of anterior structure formation. *Hum. Mol. Genet.* 6, 457–464.
78. Szeto, D. P., Rodriguiez-Estaban, C., Ryan, A. K., O'Connell, S. M., Liu, F., Kioussi, C., Gleiberman, A. S., Izpisua-Belmonte, J. C., and Rosenfeld, M. G. (1999). Role of the Bicoid-related homeodomain factor Pitx1 in specifying hindlimb morphogenesis and pituitary development. *Genes Dev.* 13, 484–494.
79. Gage, P. J., Suh, H., and Camper, S. A. (1999). Dosage requirement of Pitx2 for development of multiple organs. *Development* 126, 4643–4651.
80. Zhadanov, A. B., Bertuzzi, S., Taira, M., Dawid, I. B., and Westphal, H. (1995). Expression pattern of the murine LIM class homeobox gene Lhx3 in subsets of neural and neuro-endocrine tissues. *Dev. Dyn.* 202, 354–364.
81. Mbikay, M., Tadros, H., Seidah, N. G., and Simpson, E. M. (1995). Linkage mapping of the gene for the LIM-homeoprotein LIM3 (locus lhx3) to mouse chromosome 2. *Mamm. Genome* 6, 818–819.
82. Bach, I., Rhodes, S. J., Pearse, R. V., Heinzel, T., Gloss, B., Scully, K. M., Sawchenko, P. E., and Rosenfeld, M. G. (1995). P-Lim, a LIM homeodomain factor, is expressed during pituitary organ and cell commitment and synergizes with Pit-1. *Proc. Natl. Acad. Sci. U S A* 92, 2720–2724.
83. Netchine, I. Sobrier, M.-L. Krude, H., Schnabel, D., Maghnie, M., Marcos, E., Duriez, B., Cacheux, V., Moers, A., Goossens, M., Gruters, A., and Amselem, S. (2000). Mutations in Lhx3 result in a new syndrome revealed by combined pituitary hormone deficiency. *Nat. Genet.* 25, 182–186.
84. Fluck, C., Deladoey, J., Rutishauser, K., Eble, A., Marti, U., Wu, W., and Mullis, P. E. (1998). Phenotypic variability in familial combined pituitary hormone deficiency caused by a PROP1 gene mutation resulting in the substitution of Arg→Cys at codon 120 (R120C). *J. Clin. Endocrinol. Metab.* 83, 3727–3734.
85. Andersen, B., Pearse, R. V., Jenne, K., Sornson, M., Lin, S. C., Bartke, A., and Rosenfeld, M. G. (1995). The Ames dwarf gene is required for Pit-1 gene activation. *Dev. Biol.* 172, 495–503.
86. Sornson, M. W., Wu, W., Dasen, J. S., Flynn, S. E., Norman, D. J., O'Connell, S. M., Gukovsky, I., Carriere, C., Ryan, A. K., Miller, A. P., Wuo, L., Gleiberman, A. S., Andersen, B., Beamer, W. G., and Rosenfeld, M. G. (1996). Pituitary lineage determination by the Prophet of Pit-1 homeodomain factor defective in Ames dwarfism. *Nature* 384, 327–333.
87. Wu, W., Cogan, J. D., Pfaffle, R. W., Dasen, J. S., Frisch, H. O'Connell, S. M., Flynn, S. E., Brown, M. R., Mullis, P. E., Parks, J. S., Phillips, J. A., and Rosenfeld, M. G. (1998). Mutations in PROP1 cause familial combined pituitary hormone deficiency. *Nat. Genet.* 18, 147–149.
88. Dolle, P., Castrillo, J. L., Theill, L. E., Deerinck, T., Ellisman, M., and Karin, M. (1990). Expression of GHR-1 protein in mouse pituitaries correlates both temporally and spatially with the onset of growth hormone gene activity. *Cell* 60, 809–820.
89. Simmons, D. M., Voss, J. W., Ingraham, H. A., Holloway, J. M., Broide, R. S., Rosenfeld, M. G., and Swanson, L. W. (1990). Pituitary cell phenotypes involved cell-specific Pit-1 mRNA translation and synergistic interactions with other classes of transcription factors. *Genes Dev.* 4, 695–711.
90. Rhodes, S. J., Chen, R., DiMattia, G. E., Scully , K. M., Kalla, K. A., Lin, S. C., Yu, V. C., and Rosenfeld, M. G. (1993). A tissue-specific enhancer confers Pit-1-dependent morphogen inducibility and autoregulation on the pit-1 gene. *Genes Dev.* 7, 913–932.
91. Rosenfeld, M. G., Briata, P., Dasen, J., Gleiberman, A. S., Kiussi, C., Lin, C., O'Connell, S. M., Ryan, A., Szeto, D. P., and Treier, M. (2000). Multistep signaling and transcriptional requirements for pituitary organogenesis in vivo. *Recent Prog. Horm. Res.* 55, 1–13.
92. Japon, M. A., Rubinstein, M., and Low, M. J. (1994). In situ hybridization analysis of anterior pituitary hormone gene

expression during fetal mouse development. *J. Histochem. Cytochem.* 42, 1117–1125.

93. Bieberich, C., Fujita, K., He, W., and Jay, G. (1996). Prostate-specific and androgen-dependent expression of a novel homeobox gene. *J. Biol. Chem.* 271, 31779–31782.
94. Kioussi, C., O'Connell, S., St-Onge, L., Treier, M., Gleiberman, A. S., Gruss, P., and Rosenfeld, M. G. (1999). *Pax6* is essential for establishing ventral-dorsal cell boundaries in pituitary gland development. *Proc. Natl. Acad. Sci. U S A* 96, 14378–14382.
95. Lamolet, B., Pulichino, A.-M., Lamonerie, T., Gauthier, Y., Brue, T., Enjalbert, A., and Drouin, J., (2001). A pituitary cell-restricted T box factor, Tpit, activates POMC transcription in cooperation with the pitx homeoproteins. *Cell* 104, 849–859.
96. Yano, H., Readhead, C., Nakashima, M., Ren, S. G., and Melmed, S. (1998). Pituitary-directed leukemia inhibitory factor transgene causes Cushing's syndrome: neuro-immune-endocrine modulation of the pituitary. *Mol. Endocrinol.* 12, 1708–1720.
97. Lamonerie, T., Tremblay, J. J., Lanctot, C., Therrien, M., Gauthier, Y., and Drouin, J. (1996). Ptx1, a bicoid-related homeo box transcription factor involved in transcription of the pro-opiomelanocortin gene. *Genes Dev.* 10, 1284–1295.
98. Poulin, G., Turgeon, B., and Drouin, J. (1997). NeuroD1/BETA2 contributes to cell-specific transcription of the POMC gene. *Mol. Cell. Biol.* 17, 6673–6682.
99. Poulin, G. Lebel, M., Chamberland, M. Paradis, F. W., and Drouin, J. (2000). Specific protein: protein interaction between basic helix-loop-helix transcription factors and homeoproteins of the Pitx family. *Mol. Cell Biol.* 20, 4826–4837.
100. Dasen, J. S., O-Connell, S. M., Flynn, S. E., Treier, M., Gleiberman, A. S., Szeto, D. P., Hooshmand, F., Aggarwal, A. K., and Rosenfeld, M. G. (1999). Reciprocal interactions of Pit1 and GATA2 mediate signaling of gradient induced determination of pituitary cell types. *Cell* 97, 587–598.
101. Mathis, J. M., Simmons, D. M., He, X., Swanson, L. W., and Rosenfeld, M. G. (1992). Brain 4: a novel mammalian POU domain transcription factor exhibiting restricted brain-specific expression. *EMBO J.* 11, 2551–2561.
102. Van Bael, A. Seuntjens, E., Proesmans, M., and Denef, C. (1998). Presence of gonadotropin-releasing hormone (GnRH) mRNA in Rathke's pouch and effect of the GnRH-antagonist ORG 30276 on lactotroph development *in vitro*. *J. Neuroendocrinol.* 10, 437–445.
103. Tilemans, D., Andries, M., and Denef, C. (1992). Luteinizing hormone-releasing hormone and neuropeptide Y influence deoxyribonucleic acid replication in three anterior pituitary cell types: evidence for mediation by growth factors released from gonadotrophs. *Endocrinology* 130, 882–894.
104. Van Bael, A., Huygen, R., Himpens, B., and Denef, C. (1994). In vitro evidence that LHRH stimulates the recruitment of prolactin mRNA-expressing cells during the postnatal period in the rat. *J. Mol. Endocrinol.* 12, 107–118.
105. Andries, M., and Denef, C. (1995). Gonadotropin-releasing hormone influences the release of prolactin and growth hormone from intact rat pituitary in vitro during a limited period in neonatal life. *Peptides* 16, 527–532.
106. Seuntjens, E., Vankelecom, H., Quaegebeur, A., Vande Vijver, V., and Denef, C. (1999). Targeted ablation of gonadotrophs in transgenic mice affects embryonic development of lactotrophs. *Mol. Cell. Endocrinol.* 150, 129–139.
107. Hauspie, A., Seuntjens, E., Vankelecom, H., and Denef, C. (2003). Stimulation of combinatorial expression of prolactin and glycoprotein hormone alpha-subunit genes by gonadotropin-releasing hormone and estradiol-17β in single rat pituitary cells during aggregate cell culture. *Endocrinology* 144, 388–399.
108. Denef, C. (2003). Paracrine control of lactotrope proliferation and differentiation [review]. *Trends Endocrinol. Metab.* 14, 188–195.
109. Jennes, L. (1990). Prenatal development of gonadotropin-releasing hormone receptors in the rat anterior pituitary. *Endocrinology* 126, 942–947.
110. Seuntjens, E., Hauspie, A., Roudbaraki, M., Vankelecom, H., and Denef, C. (2002). Combined expression of different hormone genes in single cells of normal rat and mouse pituitary [review]. *Arch. Physiol. Biochem.* 110, 12–15.
111. Seuntjens, E., Hauspie, A., Vankelecom, H., and Denef, C. (2002). Ontogeny of plurihormonal cells in the anterior pituitary of the mouse, as studied by means of hormone mRNA detection in single cells. *J. Neuroendocrinol.* 14, 611–619.
112. Baker, B. L. (1977). Cellular composition of the human pituitary pars tuberalis as revealed by immunocytochemistry. *Cell. Tissue Res.* 182, 151–163.
113. Baker, B. L., Karsch, F. J., Hoffman, D. L., and Beckman, W. C. Jr. (1977). The presence of gonadotropic and thyrotropic cells in the pituitary pars tuberalis of the monkey (*Macaca mulatta*). *Biol. Reprod.* 17, 232–240.
114. Nakane, P. K. (1968). Simultaneous localization of multiple tissue antigens using the peroxidase-labeled antibody method: a study on pituitary glands of the rat. *J. Histochem. Cytochem.* 16, 557–560.
115. Barnes, B. G. (1962). Electron microscope studies on the secretory cytology of the mouse anterior pituitary. *Endocrinology* 71, 618–628.
116. Kurosumi, K., and Oota, Y. (1980). Electron microscopy of two types of gonadotrophs in the anterior pituitary glands of persistent estrous and diestrous rats. *Z. Zellforsch.* 85, 34–46.
117. Childs, G. V. (1986). Functional ultrastructure of gonadotropes: a review. *Curr. Top. Neuroendocrinol.* 7, 49–99.
118. Kawarai, Y., and Nakane, P. K. (1970). Localization of tissue antigens on the ultrathin sections with peroxidase-labeled antibody method. *J. Histochem. Cytochem.* 18, 161–166.
119. Nakane, P. K. (1971). Application of peroxidase-labelled antibodies to the intracellular localization of hormones. *Acta Endocrinol. Suppl. (Copenh).* 153, 190–204.
120. Mazurkiewicz, J. E., and Nakane, P. K. (1972). Light and electron microscopic localization of antigens in tissues embedded in polyethylene glycol with a peroxidase labeled antibody method. *J. Histochem. Cytochem.* 20, 969–974.
121. McLean, I. W., and Nakane, P. K. (1974). Periodate-lysine-paraformaldehyde fixative: a new fixation for immunoelectron microscopy. *J. Histochem. Cytochem.* 22, 1077–1083.
122. Tougard, C., Kerdelhué, B., Tixier-Vidal, A., and Jutisz, M. (1973). Light and electron microscope localization of binding sites of antibodies against ovine luteinizing hormone and its two subunits in rat adenohypophysis using peroxidase-labeled antibody technique. *J. Cell Biol.* 58, 503–521.
123. Tougard, C., Picart, R., and Tixier-Vidal, A. (1980). Immunocytochemical localization of glycoprotein hormones in the rat anterior pituitary. *J. Histochem. Cytochem.* 28, 101–114.
124. Moriarty, G. C. (1975). Electron microscopic-immunocytochemical studies of rat pituitary gonadotrophs: a sex difference in morphology and cytochemistry of LH cells. *Endocrinology* 97, 1215–1225.
125. Childs, G. V. (1995). Division of labor among gonadotropes. *Vitam. Horm.* 50, 215–286.
126. Moriarty, G. C. (1976b). Immunocytochemistry of the pituitary glycoprotein hormones. *J. Histochem. Cytochem.* 24, 846–863.
127. Moriarty, G. C., and Garner, L. L. (1977). Immunocytochemical studies of cells in the rat adenohypophysis containing both ACTH and FSH. *Nature* 265, 356–358.
128. Dacheux, F., and Dubois, M. P. (1976). Ultrastructural localization of prolactin, growth hormone, and luteinizing

hormone by immunocytochemical techniques in the bovine pituitary. *Cell. Tissue Res.* 174, 245–260.

129. Bastings, E., Beckers, A., Reznik, M., and Beckers, J. F. (1991). Immunocytochemical evidence for production of luteinizing hormone and follicle-stimulating hormone in separate cells in the bovine. *Biol. Reprod.* 45, 788–796.
130. Dacheux, F. (1978). Ultrastructural localization of gonadotrophic hormones in the porcine pituitary using the immunoperoxidase technique. *Cell. Tissue Res.* 191, 219–232.
131. Dacheux, F., (1984). Subcellular localization of gonadotropic hormones in pituitary cells of the castrated pig with the use of pre- and post-embedding immunocytochemical methods. *Cell. Tissue Res.* 236, 153–160.
132. Batten, T. F. C., and Hopkins, C. R. (1978). Discrimination of LH, FSH, TSH, and ACTH in dissociated porcine anterior pituitary cells by light and electron microscope immunocytochemistry. *Cell. Tissue Res.* 192, 107–120.
133. Currie, R. J., and McNeilly, A. S. (1995). Mobilization of LH secretory granules in gonadotropes in relation to gene expression, synthesis and secretion of LH during the preovulatory phase of the sheep oestrous cycle. *J. Endocrinol.* 147, 259–270.
134. Thomas, S. G., and Clarke, I. J. (1997). The positive feedback action of estrogen mobilizes LH-containing, but not FSH-containing secretory granules in ovine gonadotropes. *Endocrinology* 138, 1347–1350.
135. Dacheux, F., and Dubois, M. P. (1978). LH-producing cells in the ovine pituitary: an electron microscopic immunocytochemical study. *Cell. Tissue Res.* 188, 449–463.
136. Shirasawa, N., Kihara, H., and Yoshimura, F. (1985). Fine structural and immunohistochemical studies of goat adenohypophyseal cells. *Cell. Tissue Res.* 240, 315–321.
137. Herbert, D. C. (1976). Immunocytochemical evidence that luteinizing hormone (LH) and follicle-stimulating hormone (FSH) are present in the same cell type in the rhesus monkey pituitary gland. *Endocrinology* 98, 1554–1557.
138. Herbert, D. C. (1978). Identification of the LH- and TSH-secreting cells in the pituitary gland of the Rhesus monkey. *Cell. Tissue Res.* 190, 151–161.
139. Girod, C., Dubois, M. P., and Trouillas, J. (1980). Immunohistochemical study of the pars tuberalis of the adenohypophysis in the monkey, *Macaca irus*. *Cell. Tissue Res.* 210, 191–203.
140. Girod, C., Dubois, M. P., and Trouillas, J. (1981). Immunohistochemical localization of FSH and LH in the pars distalis of vervet (*Cercopithecus aethiops*) and baboon (*Papio hamadryas*) pituitaries. *Cell. Tissue Res.* 217, 245–257.
141. Pelletier, G., Leclerc, R., and Labrie, F. (1976). Identification of gonadotropic cells in the human pituitary by immunoperoxidase technique. *Mol. Cell. Endocrinol.* 6, 123–128.
142. Pelletier, G., Robert, F., and Hardy, J. (1978). Identification of human anterior pituitary cells by immunoelectron microscopy. *J. Clin. Endocrinol. Metab.* 46, 534–542.
143. Schally, A. V., Arimura, A., Kastin, A. J., Matsuo, H., Baba, Y., Redding, T. W., Nair, R. M., Debeljuk, G., and White, W. F. (1971). The gonadotropin-releasing hormone: a single hypothalamic polypeptide regulates the secretion of both LH and FSH. *Science* 173, 1036–1037.
144. Igarashi, M., and McCann, S. M. (1964). A hypothalamic follicle stimulating hormone releasing factor. *Endocrinology* 74, 446–452.
145. Lumpkin, M. D., Moltz, J. H., Yu, W. H., Samson, W. K., and McCann, S. M. (1987). Purification of FSH-releasing factor: its dissimilarity from LHRH of mammalian, avian, and piscian origin. *Brain Res. Bull.* 18, 175–178.
146. McCann, S. M., Marubayashi, U., Sun, H. Q., and Yu, W. H. (1993). Control of follicle-stimulating hormone and luteinizing hormone release by hypothalamic peptides [review]. *Ann. N Y Acad. Sci.* 687, 55–59.
147. Koppan, M., Kovacs, M., Mezo, I., and Flerko, B. (1998). Electrochemical stimulation of the median eminence evokes FSH but not LH release after LHRH antagonist treatment in vivo and in vitro. *J. Neuroendocrinol.* 10, 231–236.
148. McCann, S. M., Kimura, M., Walczewska, A., Karanth, S., Rettori, V., and Yu, W. H. (1998). Hypothalamic control of FSH and LH by FSH-RF, LHRH, cytokines, leptin and nitric oxide [review]. *Neuroimmunomodulation* 5, 193–202.
149. Hiney, J. K., Sower, S. A., Yu, W. H., McCann, S. M., and Dees, W. L. (2002). Gonadotropin-releasing hormone neurons in the preoptic-hypothalamic region of the rat contain lamprey gonadotropin-releasing hormone III, mammalian luteinizing hormone-releasing hormone, or both peptides. *Proc. Natl. Acad. Sci. U S A* 99, 2386–2391.
150. Padmanabhan, V., and Sharma, T. P. (2001). Neuroendocrine vs. paracrine control of follicle-stimulating hormone [review]. *Arch. Med. Res.* 32, 533–543.
151. Kovacs, M., Seprodi, J., Koppan, M., Horvath, J. E., Vincze, B., Teplan, I., and Flerko, B. (2002). Lamprey gonadotropin hormone-releasing hormone-III has no selective follicle-stimulating hormone-releasing effect in rats. *J. Neuroendocrinol.* 14, 647–655.
152. Amstalden, M., Zieba, D. A., Garcia, M. R., Stanko, R. L., Welsh, T. H. Jr., Hansel, W. H., and Williams, G. L. (2004). Evidence that lamprey GnRH-III does not release FSH selectively in cattle. *Reproduction* 127, 35–43.
153. Pelletier, G., Peillon, F., and Vila-Porcile, E. (1971). An ultrastructural study of sites of granule extrusion in the anterior pituitary of the rat. *Z. Zellforsch.* 115, 501–507.
154. Luborsky-Moore, J. L., Poliakoff, S. J., and Worthington, W. C. (1975). Ultrastructural observation of anterior pituitary gonadotrophs following hypophysial portal vessel infusion of luteinizing hormone-releasing hormone. *Am. J. Anat.* 144, 549–555.
155. Römmler, A., Seinsch, W., Hasan, A. S., and Haase, F. (1978). Ultrastructure of rat pituitary LH gonadotrophs in relation to serum and pituitary LH levels following repeated LH-RH stimulation. *Cell. Tissue Res.* 190, 135–149.
156. Farquhar, M. G., (1971). Processing of secretory products by cells of the anterior pituitary gland. *Mem. Soc. Endocrinol.* 19, 79–124.
157. Garner, L. L., and Blake, C. A. (1981). Ultrastructural, immunocytochemical study of the LH-secreting cell of the rat anterior pituitary gland: changes occurring after ovariectomy. *Biol. Rep.* 24, 461–474.
158. McDowell, I. F. W., Morris, J. F., Charlton, H. M., and Fink, G. (1982). Effects of luteinizing hormone-releasing hormone on the gonadotrophs of hypogonadal (hpg). mice. *J. Endocrinol.* 95, 331–340.
159. Lewis, C. E., Morris, J. F., Fink, G., and Johnson, M. (1986). Changes in the granule population of gonadotrophs of hypogonadal (hpg) and normal female mice associated with the priming effect of LH-releasing hormone in vitro. *J. Endocrinol.* 109, 35–44.
160. Lewis, C. E., Morris, J. F., and Fink, G. (1985). The role of microfilaments in the priming effect of LH-releasing hormone: an ultrastructural study using cytochalasin B. *J. Endocrinol.* 106, 1–8.
161. Sétáló, G., and Nakane, P. K. (1976). Functional differentiation of the fetal anterior pituitary cells in the rat. *Endocrinol. Exp. (Bratis).* 10, 155–166.
162. Tougard, C., Picart, R., and Tixier-Vidal, A. (1977). Cytogenesis of immunoreactive gonadotropic cells in the fetal rat pituitary at light and electron microscope levels. *Dev. Biol.* 58, 148–163.

163. Yoshimura, F., Nakamura, F., Nogami, H., and Suzuke, K. (1984). In *Endocrine Correlates of Reproduction* (K. Ochiai, Y. Arai, T. Shioda, and M. Takahashi, Eds.), pp. 41–58. Japanese Science Society Press, Tokyo.
164. Childs, G. V., Ellison, D. G., Foster, L., and Ramaley, J. A. (1981). Postnatal maturation of gonadotropes in the male rat pituitary. *Endocrinology* 109, 1683–1693.
165. Childs, G. V., Ellison, D. G., and Ramaley, J. A. (1982). Adrenocorticotropin storage in corticotropes and a subpopulation of gonadotropes during the stress non-responsive period in the neonatal male rat. *Endocrinology* 110, 1676–1692.
166. Currie, R. W., Faiman, C., and Thliveris, J. A., (1981). An immunocytochemical and routine electron microscopic study of LH and FSH cells in the human fetal pituitary. *Am. J. Anat.* 161, 281–297.
167. Pelletier, G. (1974). Autoradiographic studies of synthesis and intracellular migration of glycoproteins in the rat anterior pituitary gland. *J. Cell. Biol.* 62, 185–192.
168. Garner, L. L., and Blake, C. A. (1981). Ultrastructural, immunocytochemical study of the LH-secreting cell of the rat anterior pituitary gland: changes occurring after ovariectomy. *Biol. Rep.* 24, 461–474.
169. Childs, G. V., Ellison, D. G., Lorenzen, J. R., Collins, T. J., and Schwartz, N. B. (1982b). Immunocytochemical studies of gonadotropin storage in developing castration cells. *Endocrinology* 111, 1318–1328.
170. Inoue, K., and Kurosumi, K. (1981). Mode of proliferation of gonadotrophic cells of the anterior pituitary after castration-immunocytochemical and autoradiographic studies. *Arch. Histol. Jpn.* 44, 71–85.
171. Smith, P. F., and Keefer, D. A. (1982). Immunocytochemical and ultrastructural identification of mitotic cells in the pituitary gland of ovariectomized rats. *J. Reprod. Fertil.* 66, 383–388.
172. Sakuma, S., Shirasawa, N., and Yoshimura, F. (1983). A histometrical study of immunohistochemically identified mitotic adenohypophysial cells in immature and mature castrated rats. *J. Endocrinol.* 100, 322–328.
173. Kihara, H. (1984). Compensatory development of immature LH cells after long-term gonadectomy: an immunocytochemical electron microscopic and cell count study. *Endocrinol. Jpn.* 31, 395–406.
174. Ibrahim, S. N., Moussa, S. M., and Childs, G. V. (1986). Morphometric studies of rat anterior pituitary cells after gonadectomy: correlation of changes in gonadotropes with serum levels of gonadotropins. *Endocrinology* 119, 629–637.
175. Childs, G. V., Lloyd, J., Unabia, G., Gharib, S. D., Wierman, M. E., and Chin, W. W. (1987a). Detection of LHβ mRNA in individual gonadotropes after castration: use of a new *in situ* hybridization method with a photobiotinylated cRNA probe. *Mol. Endocrinol.* 1, 926–932.
176. Childs, G. V., Unabia, G., Weirman, M. E., Gharib, S. D., and Chin, W. W. (1990). Castration induces time-dependent changes in the FSHβ mRNA-containing gonadotrope cell population. *Endocrinology* 126, 2205–2213.
177. Blake, C. A. (1980). Correlative study of changes in the morphology of the LH gonadotroph and anterior pituitary gland LH secretion during the 4-day rat estrous cycle. *Biol. Reprod.* 23, 1097–1108.
178. Childs, G. V., Unabia, G., Tibolt, R., and Lloyd, J. M. (1987b). Cytological factors that support nonparallel secretion of LH and FSH during the estrous cycle. *Endocrinology* 121, 1801–1813.
179. Childs, G. V., Unabia, G., and Lloyd, J. (1992). Recruitment and maturation of small subset of luteinizing hormone (LH) gonadotropes during the estrous cycle. *Endocrinology* 130, 335–345.
180. Childs, G. V., Unabia, G., Lee, B. L., and Rougeau, D. (1992). Heightened secretion by small and medium-sized luteinizing hormone (LH) gonadotropes late in the cycle suggests contributions to the LH surge or possible paracrine interactions. *Endocrinology* 130, 345–352.
181. Childs, G. V., Unabia, G., and Lloyd, J. M. (1992). Maturation of FSH gonadotropes during the rat estrous cycle. *Endocrinology* 131, 29–36.
182. Rundle, S., Somogyi, P., Fischer-Colbrie, R., Hagn, C., Winkler, H., Chubb, I. W. (1986). Chromogranin A, B, and C: immunohistochemical localization in ovine pituitary and the relationship with hormone-containing cells. *Regul. Pept.* 16, 217–233.
183. Cozzi, M. G., and Zanini, A. (1986). Sulfated LH subunits and a tyrosine-sulfated secretory protein (secretogranin II) in female rat adenohypophyses: changes with age and stimulation of release by LHRH. *Mol. Cell. Endocrinol.* 44, 47–54.
184. Sion, B., Chanat, E., Duval, J., and Thieulant, M. L. (1988). Peptides co-released with luteinizing hormone by perfused pituitary cell aggregates. *Mol. Cell. Endocrinol.* 60, 151–161.
185. Anouar, Y., Benie, T., deMonti, M., Counis, R., and Duval, J. (1991). Estradiol negatively regulates secretogranin II and chromogranin A messenger ribonucleic acid levels in the female rat pituitary, but not in the adrenal. *Endocrinology* 129, 2393–2399.
186. Anouar, Y., and Dubal, J. (1991). Differential expression of secretogranin II and chromogranin A genes in the female rat pituitary through sexual maturation and estrous cycle. *Endocrinology* 128, 1374–1380.
187. Conn, P. M., Janovick, J. A., Braden, T. D., Maurer, R. A., and Jennes, L. (1992). SIIp: a unique secretogranin/chromogranin of the pituitary released in response to gonadotropin-releasing hormone. *Endocrinology* 130, 3033–3040.
188. Watanabe, T., Jeziorowski, T., Wuttke, W., and Grube, D. (1993). Secretory granules and granins in hyperstimulated male rat gonadotropes. *J. Histochem. Cytochem.* 41, 1801–1812.
189. Watanabe, T., Azuma, T., Banno, T., Jeziorowski, T., Ohwasa, Y., Waguri, S., and Grube, D. (1998). Immunocytochemical localization of chromogranin A and secretogranin II in female rat gonadotropes. *Arch. Histol. Cytol.* 61, 99–113.
190. Watanabe, T., Banno, T., Jeziorowski, T., Ohsawa, Y., Waguri, S., Grube, D., and Uchiyama, Y. (1998). Effects of sex steroids on secretory granule formation in gonadotropes of castrated male rats with respect to granin expression. *Endocrinology* 139, 2765–2773.
191. Wei, N., Kakar, S. S., and Neill, J. D. (1995). Measurement of secretogranin II release from individual adenohypophysial gonadotropes. *Am. J. Physiol.* 268, E145–E152.
192. Crawford, J. L., McNeilly, J. R., Nicol, L., and McNeilly, A. S. (2002). Promotion of intragranular co-aggregation with LH by enhancement of secretogranin-II storage resulted in increased intracellular granule storage in gonadotrophs of GnRH-deprived male mice. *Reproduction* 124, 267–277.
193. Kakar, S. S., Wei, N., Mulchahey, J. J., LeBoeuf, R. D., and Neill, J. D. (1993). Regulation of expression of secretogranin II mRNA in female rat pituitary and hypothalamus. *Neuroendocrinology* 57, 422–431.
194. Crawford, J. L., and McNeilly, A. S. (2002). Co-localisation of gonadotrophins and granins in gonadotrophs at different stages of the oestrous cycle in sheep. *J. Endocrinol.* 174, 179–194.
195. Hosaka, M., Watanabe, T., Sakai, Y., Uchiyama, Y., and Takeuchi, T. (2002). Identification of a chromogranin A domain that mediates binding to secretogranin III and targeting to secretory granules in pituitary cells and pancreatic β cells. *Mol. Biol. Cell* 13, 3388–3399.
196. Yoo, S. H. (2000). Coupling of the IP_3 receptor/Ca^{2+} channel with Ca^{2+} storage proteins chromogranins A and B in secretory granules. *Trends Neurosci.* 23, 424–428.

197. Yoo, S. H., and Lewis, M. S. (2000). Interaction of chromogranin B and the near N-terminal region of chromogranin B with an intraluminal loop peptide of the inositol 1,4,5-triphosphate receptor. *J. Biol. Chem.* 275, 30293–30300.
198. Davidson, J. S., Eales, A., Roeske, R. W., and Millar, R. P. (1993). Inhibition of pituitary hormone exocytosis by a synthetic peptide related to the rab effector domain. *FEBS Lett.* 326, 219–221.
199. Thomas, S. G., Takahashi, Neill, J. D., and Clarke, I. J. (1998). Components of the neuronal exocytotic machinery in the anterior pituitary of the ovariectomised ewe and the effects of oestrogen in gonadotropes as studied with confocal microscopy. *Neuroendocrinology* 67, 244–259.
200. Marshall, J. C. (1993). The role of changing pulse frequency in the regulation of ovulation. *Hum. Reprod.* 8 (Suppl. 2), 57–61.
201. Ishizaka, K., Kitahara, S., Oshima, H., Troen, P., Attardi, B., and Winters, S. J. (1992). Effect of gonadotropin-releasing hormone pulse frequency on gonadotropin secretion and subunit messenger ribonucleic acids in perifused pituitary cells. *Endocrinology* 130, 1467–1474.
202. Tougard, C., Tixier-Vidal, A., Kerdelhué, B., and Jutisz, M. (1977). Étude immunocytochimique de l'evolution des cellules gonadotropes dans des cultures primaries de cellules antéhypophysaires de rat: aspects quantitatifs et ultrastructuraux. *Biol. Cell.* 28, 251–260.
203. Tixier-Vidal, A., Gourdji, D., and Tougard, C. (1975). A cell culture approach to the study of the anterior pituitary. *Int. Rev. Cytol.* 41, 173–239.
204. Denef, C., Hautekeete, E., DeWolf, A., and Vanderschuren, B. (1978). Pituitary basophils from immature male and female rats: distribution of gonadotrophs and thyrotrophs as studied by unit gravity sedimentation. *Endocrinology* 103, 724–738.
205. Smith, P. F., Frawley, L. S., and Neill, J. D. (1984). Detection of LH release from individual pituitary cells by the reverse hemolytic plaque assay: estrogen increases the fraction of gonadotropes responding to GnRH. *Endocrinology* 115, 2484–2496.
206. Hyde, C. L., Childs, G. V., Wahl, L. M., Naor, Z., and Catt, K. (1982). Preparation of gonadotropin-enriched cell populations from adult rat pituitary by centrifugal elutriation. *Endocrinology* 111, 1421–1423.
207. Childs, G. V., Hyde, C., Naor, Z., and Catt, K. (1983). Heterogeneous LH and FSH storage patterns in subtypes of gonadotropes separated by centrifugal elutriation. *Endocrinology* 113, 2120–2128.
208. Lloyd, J. M., and Childs, G. V. (1988). Differential storage and release of LH and FSH from individual gonadotropes separated by centrifugal elutriation. *Endocrinology* 122, 1282–1290.
209. Vila-Porcile, E., Picart, R., Tixier-Vidal, A., and Tougard, C. (1987). Cellular and subcellular distribution of laminin in adult rat anterior pituitary. *J. Histochem. Cytochem.* 35, 287–299.
210. Vila-Porcile, E., Picart, R., Vigny, M., Tixier-Vidal, A., and Tougard, C. (1992). Immunolocalization of laminin, heparin-sulfate proteoglycan, entactin, and type IV collagen in the rat anterior pituitary: I. An in vivo study. *Anat. Rec.* 232, 432–492.
211. Vila-Porcile, E., Picart, R., Vigny, M., Tixier-Vidal, A., and Tougard, C. (1992). Immunolocalization of laminin, heparin-sulfate proteoglycan, entactin, and type IV collagen in the rat anterior pituitary: II. An in vitro study on primary cultures. *Anat. Rec.* 233, 1–12.
212. Roelle, S., Grosse, R., Aigner, A., Krell, H. W., Czubayko, F., and Gudermann, T. (2003). Matrix metalloproteinases 2 and 9 mediate epidermal growth factor receptor transactivation by gonadotropin-releasing hormone. *J. Biol. Chem.* 47, 47307–47318.
213. Fletcher, W. H., Anderson, N. C., and Everett, J. W. (1975). Intercellular communication in the rat anterior pituitary gland: an in vivo and in vitro study. *J. Cell. Biol.* 67, 469–476.
214. Wilfinger, W. W., Larsen, W. J., Downs, T. R., and Wilbur, D. L. (1984). An in vitro model for studies of intercellular communication in cultured rat anterior pituitary cells. *Tissue Cell* 16, 483–497.
215. Herbert, D. C. (1979). Intercellular junctions in the rhesus monkey pars distalis. *Anat. Rec.* 195, 1–6.
216. Soji, T., and Herbert, D. C., (1989). Intercellular communication between rat anterior pituitary cells. *Anat. Rec.* 224, 523–533.
217. Soji, T., Yashiro, T., and Herbert, D. C. (1990). Intercellular communication within the rat anterior pituitary gland: I. Postnatal development and changes after injection of luteinizing hormone-releasing hormone (LH-RH) or testosterone. *Anat. Rec.* 226, 337–341.
218. Soji, T., and Herbert, D. C. (1990). Intercellular communication within the rat anterior pituitary gland: II. Castration effects and changes after injection of luteinizing hormone-releasing hormone (LH-RH) or testosterone. *Anat. Rec.* 22, 342–346.
219. Soji, T., Nishizono, H., Yashiro, T., and Herbert, D. C. (1991). Intercellular communication within the rat anterior pituitary gland: III. Postnatal development and periodic changes of cell-to-cell communications in female rats. *Anat. Rec.* 231, 351–357.
220. Soji, T., Nishizono, H., Yashiro, T., and Herbert, D. C. (1992). Intercellular communication within the rat anterior pituitary gland: IV. Changes of cell-to-cell communications during pregnancy. *Anat. Rec.* 233, 97–102.
221. Nishizono, H., Soji, T., and Herbert, D. C. (1993). Intercellular communication within the rat anterior pituitary gland: V. Changes in cell-to-cell communications as a function of the timing of castration in male rats. *Anat. Rec.* 235, 577–582.
222. Sakuma, E., Herbert, D. C., and Soji, T. (2003). The effects of sex steroids on the formation of gap junctions between folliculo-stellate cells: a study in castrated male rats and ovariectomized female rats. *Arch. Histol. Cytol.* 66, 229–238.
223. Kurono, C. (1996). Intercellular communication within the rat anterior pituitary gland: VI. Development of gap junctions between folliculo-stellate cells under the influence of ovariectomy and sex steroids in the female rat. *Anat. Rec.* 244, 366–373.
224. Soji, T., Mabuchi, Y., Kurono, C., and Herbert, D. C. (1997). Folliculo-stellate cells and intercellular communication within the rat anterior pituitary gland. *Microsc. Res. Tech.* 39, 138–149.
225. Yamamoto, T., Hossain, M. Z., Hertsberg, E. L., Uemura, H., Murphy, L. J., and Nagy, J. I. (1993). Connexin43 in rat pituitary: localization at pituicyte and stellate cell gap junctions and within gonadotrophs. *Histochemistry* 100, 53–64.
226. Meda, P., Petter, M. S., Traub, O., Willecke, K., Gros, D., Beyer, E., Nicholson, B., Paul, D., and Orci, L. (1993). Differential expression of gap junction connexins in endocrine and exocrine glands. *Endocrinology* 133, 2171–2178.
227. Morand, I., Fonlupt, P., Guerrier, A., Trouillas, J., Calle, A., Remy, C., Rousset, B., and Munari-Silem, Y. (1996). Cell-to-cell communication in the anterior pituitary: evidence for gap junction-mediated exchanges between endocrine cells and folliculostellate cells. *Endocrinology* 137, 3356–3367.
228. Denef, C. (1986). Paracrine interactions in the anterior pituitary. *J. Clin. Endocrinol. Metab.* 15, 1–31.
229. Denef, C., and Andries, M. (1983). Evidence for paracrine interaction between gonadotrophs and lactotrophs in pituitary cell aggregates. *Endocrinology* 112, 813–822.

230. Schwartz, J., and Cherny, R. (1992). Intercellular communication within the anterior pituitary influencing the secretion of hypophysial hormones. *Endocr. Rev.* 13, 453–475.
231. Windle, J. J., Weiner, R. I., and Mellon, P. L. (1990). Cell lines of the pituitary gonadotrope lineage derived by targeted oncogenesis in transgenic mice. *Mol. Endocrinol.* 4, 597–603.
232. Alarid, E. T., Windle, J. J., Whyte, D. B., and Mellon, P. L. (1996). Immortalization of pituitary cells at discrete stages of development by directed oncogenesis in transgenic mice. *Development* 122, 3319–3329.
233. Pernasetti, F., Vasilyev, V., Rosenberg, S. B., Bailey, J. S., Huang, H-J., Miller, W. L., and Mellon, P. L. (2001). Cell-specific transcriptional regulation of follicle-stimulating hormone-β by activin and gonadotropin-releasing hormone in the LβT2 pituitary gonadotrope cell model. *Endocrinology* 142, 2284–2295.
234. Childs, G. V., Unabia, G., and Rougeau, D. (1994). Cells that express luteinizing hormone (LH) and follicle stimulating hormone (FSH) beta (β) subunit mRNAs during the estrous cycle: the major contributors contain LHβ, FSHβ and/or growth hormone. *Endocrinology* 134, 990–997.
235. Chandrashekar, V., and Bartke, A. (1993). Effects of age and endogenously secreted human GH on the regulation of gonadotropin secretion in female and male transgenic mice expressing the human growth hormone gene. *Endocrinology* 132, 1482–1488.
236. Chandrashekar, V., and Bartke, A. (1993). Induction of endogenous insulin-like growth factor I secretion alters the hypothalamic-pituitary-testicular function in GH-deficient adult dwarf mice. *Biol. Reprod.* 48, 544–551.
237. Fraser, R. A., and Harvey, S. (1992). Ubiquitous distribution of growth hormone receptors and/or binding proteins in adenohypophyseal tissue. *Endocrinology* 130, 3593–3600.
238. Harvey, S., Baumbach, W. R., Sadeghi, H., and Sanders, E. J. (1993). Ultrastructural colocalization of growth hormone binding protein and pituitary hormones in adenohypophyseal cells of the rat. *Endocrinology* 133, 1125–1130.
239. Childs, G. V., Unabia, G., and Miller, B. T. (1994). Cytochemical detection of GnRH binding sites on rat pituitary cells with LH, FSH and GH antigens during diestrous upregulation. *Endocrinology* 134, 1943–1951.
240. Childs, G. V., Miller, B., and Miller, W. (1997). Differential effects of inhibin on gonadotropin stores and gonadotropin releasing hormone binding to pituitary cells from cycling female rats. *Endocrinology* 138, 1577–1584.
241. Childs, G. V., and Unabia, G. (1997). Cytochemical studies of the effects of activin on gonadotropin releasing hormone (GnRH) binding by pituitary gonadotropes and growth hormone cells. *J. Histochem. Cytochem.* 45, 1603–1610.
242. Childs, G. V., Unabia, G., Miller, B., and Collins, T. J. (1999). Differential expression of gonadotropin and prolactin antigens by GHRH target cells from male and female rats. *J. Endocrinol.* 162, 177–187.
243. Childs, G. V., Unabia, G., and Wu, P. (2000). Differential expression of growth hormone messenger ribonucleic acid by somatotropes and gonadotropes in male and cycling female rats. *Endocrinology* 141, 1560–1570.
244. Childs, G. V., Iruthayanathan, M., Akhter, N., Unabia, G., and Whitehead-Johnson, B. (2005). Bipotential effects of estrogen on growth hormone synthesis and storage, in vitro. *Endocrinology* 146, 1780–1788.
245. La Rosa, S., Celato, N., Uccella, S., and Capella, C. (2000). Detection of gonadotropin-releasing hormone receptor in normal human pituitary cells and pituitary adenomas using immunohistochemistry. *Virchows Arch.* 437, 264–269.
246. Seuntjens, E., Hauspie, A., Roudbaraki, M., Vankelecom, H., and Denef, C. (2002). Combined expression of different hormone genes in single cells of normal rat and mouse pituitary. *Arch. Physiol. Biochem.* 110, 12–15.
247. Okada, Y., Fujii, Y., Moore, J. P., and Winters, S. J. (2003). Androgen receptors in gonadotrophs in pituitary cultures from adult male monkeys and rats. *Endocrinology* 144, 387–373.
248. Wu, J. C., Su, P., Safwat, S. W., Sebastian, J., and Miller, W. L. (2004). Rapid efficient isolation of murine gonadotropes and their use in revealing control of follicle-stimulating hormone by paracrine pituitary factors. *Endocrinology* 145, 5832–5839.
249. Kasahara, K., Tasaka, K., Masumoto, N., Mizuki, J., Tahara, M., Miyake, A., and Tanizawa, O. (1994). Characterization of rat pituitary cells by their responses to hypothalamic releasing hormones. *Biochem. Biophys. Res. Commun.* 199, 1436–1441.
250. Villalobos, C., Nunez, L., Frawley, L. S., Garcia-Sancho, J., and Sanchez, A. (1997). Multi-responsiveness of single anterior pituitary cells to hypothalamic-releasing hormones: a cellular basis for paradoxical secretion. *Proc. Natl. Acad. Sci. U S A* 94, 14132–14137.
251. Villalobos, C., Nunez, L., and Garcia-Sancho, J. (2004). Phenotypic characterization of multi-functional somatotropes, mammotropes and gonadotropes of the mouse anterior pituitary. *Pflugers Arch.* 449, 257–264.
252. Childs, G. V. (2004). Dual *in situ* hybridization detects anterior pituitary (AP) cells that express growth hormone (GH) and gonadotropin mRNA or hnRNA. *J. Histochem. Cytochem.* Proceedings of the 12th International Congress of the Society for Histochemistry and Cytochemistry. P12-3; p. 52, 2004.
253. Shupnik, M. A., Gharib, S. D., and Chin, W. W. (1989). Divergent effects of estradiol on gonadotropin gene transcription in pituitary fragments. *Mol. Endocrinol.* 3, 474–480.
254. Hopkins, C. R., and Gregory, H. (1977). Topographical localization of the receptors for luteinizing hormone-releasing hormone on the surface of dissociated pituitary cells. *J. Cell. Biol.* 75, 528–540.
255. Kanzaki, M., and Morris, P. (1999). Growth hormone regulates steroidogenic acute regulatory protein expression and steroidogenesis in Leydig cell progenitors. *Endocrinology* 140, 1681–1686.
256. Shoba, L., An, M. R., Frank, L. R., and Lowe, W. L. Jr. (1999). Developmental regulation of insulin-like growth factor-I and growth hormone receptor gene expression. *Mol. Cell. Endocrinol.* 152, 125–136.
257. N'Diaye, M., Sun, S., Fanua, S., Loseth, K., Shaw, W. E., and Crabo, B. (2002). Growth hormone receptors in the porcine testis during prepuberty. *Reprod. Domest. Anim.* 37, 305–309.
258. Menon, R. K., Shaufl, A., Yu, J. H., Stephan, D. A., and Friday, R. P. (2001). Identification and characterization of a novel transcript of the murine growth hormone receptor gene exhibiting development and tissue specific expression. *Mol. Cell. Endocrinol.* 172, 135–146.
259. Childs, G. V., Ellison, D. G., Collins, T. J., Lorenzen, J. R., and Schwartz, N. B. (1983). Retardation of the development of castration cells by adrenalectomy or sham adrenalectomy. *Endocrinology* 113, 166–177.
260. Kendall, S. K., Saunders, T. L., Jin, L., Lloyd, R. V., Glade, L. M., Nett, T. M., Keri, R. A., Nilson, J. H., and Camper, S. A. (1991). Targeted ablation of pituitary gonadotropes in transgenic mice. *Mol. Endocrinol.* 5, 2025–2036.
261. Vankelcom, H., Seuntjens, E., Hauspie, A., and Denef, C. (2003). Targeted ablation of gonadotrophs in transgenic mice depresses prolactin but not growth hormone gene expression at birth as measured by quantitative mRNA detection. *J. Biomed. Sci.* 10, 805–812.
262. Begeot, M., Hemming, F., DuBois, P., Combarnous, Y., DuBois, M., and Aubert, M. (1984). Induction of pituitary

lactotrope differentiation by luteinizing hormone a subunit. *Science* 226, 566–568.
263. Stahl, J. H., Kendall, S. K., Brinkmeier, M. L., Greco, T. L., Watkins-Chow, D. E., Campos-Barros, A., Lloyd, R. V., and Camper, S. A. (1999). Thyroid hormone is essential for pituitary somatotropes and lactotropes. *Endocrinology* 140, 1884–1892.
264. Tilemans, D., Andries, M., and Denef, C. (1992). Luteinizing hormone-releasing hormone and neuropeptide Y influence deoxyribonucleic acid replication in three anterior pituitary cell types: evidence for mediation by growth factors released from gonadotrophs. *Endocrinology* 130, 882–894.
265. Tilemans, D., Andries, M., and Denef, C., (1993). Possible involvement of an EGF-like mitogen in the postnatal development of lactotrophs in rat pituitary. *Endocr. J.* 1, 329–335.
266. Van Bael, A., Huygen, H., Himpens, B., and Denef, C. (1994). *In vitro* evidence that LHRH stimulates the recruitment of prolactin mRNA-expressing cells during the postnatal period in the rat. *J. Mol. Endocrinol.* 12, 107–118.
267. Tilemans, D., Andries, M., Proost, P., Devreese, B., Van Beeumen, J., and Denef, C. (1994). In vitro evidence that an 11-kilodalton N-terminal fragment of proopiomelanocortin is a growth factor specifically stimulating the development of lactotrophs in rat pituitary during postnatal life. *Endocrinology* 135, 168–174.
268. Vila-Porcile, E., and Corvol, P. (1998). Angiotensinogen, prorenin, and renin are co-localized in the secretory granules of all glandular cells of the rat anterior pituitary: an immunoultrastructural study. *J. Histochem. Cytochem.* 46, 301–311.
269. Steele, M. K., Brownfield, M. S., and Ganong, W. F. (1992). Immunocytochemical localization of angiotensin immunoreactivity in gonadotropes and lactotropes of the rat anterior pituitary gland. *Neuroendocrinology* 35, 155–158.
270. Robberecht, W., Andries, M., and Denef, C. (1992). Stimulation of prolactin secretion from rat pituitary by luteinizing hormone releasing hormone: evidence against mediation by angiotensin II acting through a Sar^1-Ala^8 angiotensin II sensitive receptor. *Neuroendocrinology* 56, 185–194.
271. Meunier, H., Rivier, C., Evans, R. M., and Vale, W. (1988). Gonadal and extragonadal expression of inhibin α, βA and βB subunits in various tissues predicts diverse functions. *Proc. Natl. Acad. Sci. U S A* 85, 247–251.
272. Roberts, V., Carroll, R. S., Corrigan, A. Z., Gharib, S. D., Vale, W., and Chin, W. W. (1989). Inhibin, activin and follistatin: regulation of follicle stimulating hormone messenger ribonucleic acid levels. *Mol. Endocrinol.* 3, 1969–1976.
273. Roberts, V. J., Peto, C. A., Vale, W., and Sawchenko, P. I. (1992). Inhibin/activin subunits are costored with FSH and LH in secretory granules of the rat anterior pituitary gland. *Neuroendocrinology* 56, 214–224.
274. Schlatt, S., Weinbauer, G. F., and Nieschlag, E. (1991). Inhibin-like and gonadotropin-like immunoreactivity in pituitary cells of male monkeys (*Macaca fascicularis, Macaca mulatta*). *Cell. Tissue Res.* 265, 203–209.
275. Halvorson, L. M., Weiss, J., Bauer-Dantoin, A. C., and Jameson, J. L. (1994). Dynamic regulation of pituitary follistatin messenger ribonucleic acids during the rat estrous cycle. *Endocrinology* 134, 1247–1253.
276. Kogawa, K., Nakamura, T., Sugino, T., Takio, K., Titani, K., and Sugino, H. (1991). Activin-binding protein is present in pituitary. *Endocrinology* 128, 1434–1440.
277. Kaiser, U., Lee, B. L., Unabia, G., Chin, W. W., and Childs, G. V. (1992). Follistatin gene expression in gonadotropes and folliculostellate cells of diestrous rats. *Endocrinology* 130, 3048–3056.
278. Lee, B. L., Unabia, G., and Childs, G. (1993). Expression of follistatin mRNA in somatotropes and mammotropes early in the estrous cycle. *J. Histochem. Cytochem.* 41, 955–960.
279. Bilezikjian, L. M., Corrigan, A. Z., and Vale, W. (1990a). Activin-A modulates growth hormone secretion from cultures of rat anterior pituitary cells. *Endocrinology* 126, 2369–2376.
280. Bilezikjian, L. M., Gonzales-Manchon, C., Potter, E., and Vale, W. (1990b). Inhibition of somatotroph growth and growth hormone biosynthesis by activin, in vitro. *Mol. Endocrinol.* 4, 356–362.
281. Bilezikjian. L. M., Blount, A. L., Campen, C. A., Gonzalez-Mancon, C., and Vale, W. (1991). Activin-A inhibits proopiomelanocortin messenger RNA accumulation and adrenocorticotropin secretion of AtT20 cells. *Mol. Endocrinol.* 5, 1389–1395.
282. Bauer, T. W., and Childs, G. V. (1981). Studies of immunoreactive gonadotropin releasing hormone (GnRH). in the rat anterior pituitary. *J. Histochem. Cytochem.* 29, 1171–1178.
283. May, V., Wilber, J. F., U'Prichard, D. C., and Childs, G. V. (1987). Persistence of immunoreactive TRH and GnRH in long-term primary anterior pituitary culture. *Peptides* 8, 543–558.
284. Childs, G. V., Cole, D., Kubek, M., Tobin, R. B., and Wilber, J. F. (1978). Endogenous thyrotropin releasing hormone in the anterior pituitary: sites of activity as identified by immunocytochemical staining. *J. Histochem. Cytochem.* 26, 901–908.
285. Badiu, M. D., Ham, C. Scanlon, J., Moller, M., and Coculescu, M. (1999). Expression of thyrotropin-releasing hormone messenger RNA in human pituitary adenomas with follicle-stimulating hormone immunoreactivity. *Endocrinology* 5, 10–16.
286. Bruhn, T. O., Mclean, D. B., Bolduc, T. G., and Jackson, I. M. D. (1990). Pro-TRH peptides are synthesized and secreted by anterior pituitary cells in long term culture. *Endocrinology* 129, 556–558.
287. Bruhn, T. O., Rondeel, J. M. M., Bolduc, T. G., and Jackson, I. M. D. (1994). TRH gene expression in the anterior pituitary: I. Presence of pro-TRH messenger RNA and pro-TRH-derived peptide in a subpopulation of somatotrophs. *Endocrinology* 134, 815–820.
288. Luo, L., and Jackson, I. M. (1995). Thyrotropin-releasing hormone and c-fos/c-jun genes are colocalized in rat anterior pituitary cells: stimulation of transcription by glucocorticoids. *Endocrinology* 136, 2705–2710.
289. Brar, A. K., Brinster, R. L., and Frohman, L. A. (1989). Immunohistochemical analysis of human growth hormone-releasing hormone gene expression in transgenic mice. *Endocrinology* 126, 801–809.
290. Marcinkiewicz, M., Touraine, P., Mbikay, M., and Chretien, M. (1993). Expression of neuroendocrine secretory protein 7B2 mRNA in the mouse and rat pituitary gland. *Neuroendocrinology* 58, 86–93.
291. Braks, J. A., and Martens, G. J. (1994). 7B2 is a neuroendocrine chaperone that transiently interacts with prohormone convertase PC2 in the secretory pathway. *Cell* 78, 263–273.
292. Braks, J. A. M., and Martens, G. J. M. (1995). The neuroendocrine chaperone 7B2 can enhance in vitro POMC cleavage by prohormone convertase PC2. *FEBS Lett.* 371, 154–158.
293. Bello, A. R., Dubourg, P., Kah, O., and Tramu, G. (1992). Identification of neurotensin-immunoreactive cells in the anterior pituitary of normal and castrated rats: a double immunocytochemical investigation at the light and electron-microscopic levels. *Neuroendocrinology* 55, 714–723.
294. Bello, A. R., Reyes, R., Hernandez, G., Negrin, I., Gonzalez, M., Tramu, G., and, Alonso, R. (2004). Developmental expression of neurotensin in thyrotropes and gonadotropes of male and female rats. *Neuroendocrinology* 79, 90–99.

295. Stojilkovic, S. S., Merelli, F., Lida, T., Krsmanovic, L. Z., and Catt, K. J. (1990). Endothelin stimulation of cytosolic calcium and gonadotropin secretion in anterior pituitary cells. *Science* 248, 1663–1666.
296. Naruse, K., Naruse, M., Obana, K., Demura, R., Demura, H., Inagami, T., and Shizume, K. (1992). Endothelin-3 immunoreactivity in gonadotrophs of the human anterior pituitary. *Endocrinology* 118, 2470–2476.
297. Stojilkovic, S. S., and Catt, K. J. (1996). Expression and signal transduction pathways of endothelin receptors in neuroendocrine cells. *Front. Neuroendocrinol.* 17 327–369.
298. Kanyicska, B., Sellix, M. T., and Freeman, M. E. (2003). Autocrine regulation of prolactin secretion by endothelins throughout the estrous cycle. *Endocrine* 20, 53–58.
299. Childs, G. V., Naor, Z., Hazum, E., Tibolt, R., Westlund, K. M., and Hancock, M. B. (1983). Localization of biotinylated gonadotropin releasing hormone on pituitary monolayer cells with avidin-biotin peroxidase complexes. *J. Histochem. Cytochem.* 31, 1422–1425.
300. Childs, GV, Westlund, K. N., Tibolt, R. E., and Lloyd, J. M. (1991). Hypothalamic regulatory peptides and their receptors: cytochemical studies of their role in regulation at the adenohypophyseal level. *J. Electron Microsc. Tech.* 19, 21–41.
301. Hopkins, C. R., Semoff, S., and Gregory, H. (1981). Regulation of gonadotropin secretion in the anterior pituitary. *Philos. Trans. R. Soc. Lond. B Biol. Sci.* 296, 73–81.
302. Jennes, L. Stumpf, W. E., and Conn, P. M. (1984). Receptor-mediated binding and uptake of GnRH agonist and antagonist by pituitary cells. *Peptides* 5 (Suppl. 1), 215–220.
303. Suarez-Quian, C. A., Wynn, P. C., and Catt, K. J. (1986). Receptor-mediated endocytosis of GnRH analogs: differential processing of gold-labeled agonist and antagonist derivatives. *J. Steroid Biochem.* 24, 183–192.
304. Naor, Z., Atlas, D., Clayton, R. N., Foman, D. S., Amsterdam, A., and Catt, K. J. (1981). Interaction of fluorescent gonadotropin releasing hormone with receptors in cultured pituitary cells. *J. Biol. Chem.* 257, 689–696.
305. Hazum, E., Cuatrecasas, P., Marian, J., and Conn, P. M. (1980). Receptor mediated internalization of fluorescent gonadotropin-releasing hormone by pituitary gonadotropes. *Proc. Natl. Acad. Sci. U S A* 77, 6692–6695.
306. Conn, P. M., and Hazum, E. (1981). Luteinizing hormone release and gonadotropin-releasing hormone (GnRH) receptor internalization: independent actions of GnRH. *Endocrinology* 109, 2040–2045.
307. Hazum, E., Meidan, R., Keinam, E., Okon, E., Koch, Y., Lindner, H. R., and Amsterdam, A. (1982). A novel method for localization of gonadotropin releasing hormone receptors. *Endocrinology* 111, 2135–2137.
308. Hazum, E., Medan, R., Liscovitch, M., Keinan, D., Lindner, H. R., and Koch, Y. (1983). Receptor-mediated internalization of LHRH antagonists by pituitary cells. *Mol. Cell. Endocrinol.* 30, 291–301.
309. Hazum, E., Koch, Y., Liscovitch, M., and Amsterdam, A. (1985). Intracellular pathways of receptor-bound GnRH agonist in pituitary gonadotropes. *Cell. Tissue Res.* 239, 3–8.
310. Pelletier, G., Dubé, D., Guy, J., Séguin, C., and Lefebvre, F. A. (1982). Binding and internalization of a luteinizing hormone-releasing hormone agonist by rat gonadotropic cells: a radioautographic study. *Endocrinology* 111, 1068–1076.
311. Morel, G., Dihl, F., Aubert, M. L., and Dubois, P. M. (1987). Binding and internalization of native gonadoliberin (GnRH) by anterior pituitary gonadotrophs of the rat: a quantitative autoradiographic study after cryoultramicrotomy. *Cell. Tissue Res.* 248, 541–550.
312. Morel, G., Aubert, M. L., and Dubois, P. M. (1994). Intracellular gonadotropin-releasing hormone immunoreactivity I gonadotrophs of intact or castrated male rats: semiquantitative estimation of testosterone and GnRH antagonist treatment. *Acta Anat. (Basel)* 149, 46–54.
313. Duello, T. M., and Nett, T. M. (1980). Uptake, localization and retention of gonadotropin-releasing hormone and gonadotropin-releasing hormone analogs in rat gonadotrophs. *Mol. Cell. Endocrinol.* 19, 101–112.
314. Duello, T. M., Nett, T. M., and Farquhar, M. G. (1983). Fate of a gonadotropin-releasing hormone agonist internalized by rat pituitary gonadotrophs. *Endocrinology* 112, 1–10.
315. Lloyd, J. M., and Childs, G. V. (1988). Changes in the number of GnRH-receptive cells during the rat estrous cycle: biphasic effects of estradiol. *Neuroendocrinology* 48, 138–146.
316. Childs, G. V., Naor, Z., Hazum, E., Tibolt, R., Westlund, K. N., and Hancock, M. B. (1983). Cytochemical characterization of pituitary target cells for biotinylated gonadotropin releasing hormone. *Peptides* 4, 549–555.
317. Smith, P. F., and Neill, J. D. (1984). Simultaneous measurement of hormone release and secretagogue binding by individual pituitary cells. *Proc. Natl. Acad. Sci. U S A* 84, 5501–5505.
318. Vrecl, M., Anderson, L., Hanyaloglu, A., McGregor, A. M., Groarke, A. D., Milligan, G., Taylor, P. L., and Eidne, K. A. (1996). Agonist-induced endocytosis and recycling of the gonadotropin-releasing hormone receptor: effect of β-arrestin on internalization kinetics. *Mol. Endocrinol.* 12, 1818–1829.
319. Cornea, A. Janovick, J. A., Maya-Núnez, G., and Conn, P. M. (2001). Gonadotropin-releasing hormone receptor microaggregation: rate monitored by fluorescence resonance energy transfer. *J. Biol. Chem.* 276, 2153–2158.
320. Catt, K. J., Loumaye, E., Katikineni, M., Hyde, C. L., Childs, G., Amsterdam, A., and Naor, Z. (1983). Receptors and actions of gonadotropin releasing hormone (GnRH) on pituitary gonadotropes. In *Role of Peptides and Proteins in Control of Reproduction* (S. M. McCann and D. S. Dhindsa, Eds.), pp. 33–61. Elsevier Science, New York.
321. Conn, P. M., Smith, R. G., and Rogers, D. C. (1981). Stimulation of pituitary gonadotropin release does not require internalization of gonadotropin-releasing hormone. *J. Biol. Chem.* 256, 1098–1100.
322. Conn, P. M., Rogers, D. C., Stewart, J. M., Niedel, J., and Sheffield, T. (1992). Conversion of a gonadotropin-releasing hormone antagonist to an agonist: implication for a receptor microaggregate as the functional unit for signal transduction. *Nature* 296, 653–655.
323. Janovick, J. A., and Conn, P. M. (1996). Gonadotropin releasing hormone agonist provokes homologous receptor microaggregation: an early event in seven-transmembrane receptor mediated signaling. *Endocrinology* 137, 3602–3605.
324. Cornea, A., and Conn, P. M. (2002). Measurement of changes in fluorescence resonance energy transfer between gonadotropin-releasing hormone in response to agonists. *Methods* 27, 333–339.
325. Childs, G. V. (1998). Identification of biotinylated ligands on specific target cells in the pituitary: studies of regulation of binding. In *Methods in Neuroendocrinology* (L. D. Van de Kar, Ed.), pp. 31–48. CRC Press, Boca Raton, FL.
326. Childs, GV, Hazum, E., Amsterdam, A., Limor, R., and Naor, Z. (1986). Cytochemical evidence for different routes of gonadotropin-releasing hormone processing by large gonadotropes and granulosa cells. *Endocrinology* 119, 1329–1338.
327. Wynn, P. C., Suarez-Quian, C. A., Childs, G. V., and Catt, K. J. (1986). Pituitary binding and internalization of radioiodinated

gonadotropin-releasing hormone agonist and antagonist ligands, in vitro and in vivo. *Endocrinology* 119, 1852–1863.
328. Loumaye, E., Wynn, P. C., Coy, D., and Catt, K. J., (1984). Receptor-binding properties of gonadotropin-releasing hormone derivatives. *J. Biol. Chem.* 259, 12663–12671.
329. Huckle, W. R., McArdle, C. A., and Conn, P. M. (1987). Differential sensitivity of agonist-and antagonist-occupied gonadotropin-releasing hormone receptors to protein kinase C activators: a marker for receptor activation. *J. Biol. Chem.* 268, 3296–3302.
330. Ferguson, S. S. G., Barak, L. S., Zhang, J., and Caron, M. G. (1996). G-protein coupled receptor regulation: role of G-protein-coupled receptor kinases and arrestins. *Can. J. Physiol. Pharmacol.* 74, 1095–1110.
331. Urrutia, R., Henley, J. R., Cook, T., and McNiven, M. A. (1997). The dynamins: redundant or distinct functions for an expanding family of related GTPases? *Proc. Natl. Acad. Sci. U S A* 94, 377–384.
332. Goodman, O. B., Krupnick, J. G., Santini, F., Gurevich, V. V., Penn, R. B., Gagnon, A. W., Keen, J. H., and Benovic, J. L. (1996). β-Arrestin acts as a clathrin adapter in endocytosis of the β_2-adrenergic receptor. *Nature* 383, 447–450.
333. Neill, J. D., Duck, L. W., Musgrove, L. C., and Sellers, J. C. (1998). Potential regulatory roles for G protein-coupled receptor kinases and β-arrestins in gonadotropin-releasing hormone receptor signaling. *Endocrinology* 139, 1781–1788.
334. Tsutsumi, M., Zhou, W., Millar, R. P., Mellon, P. L., Roberts, J. L., Flanagan, C. A., Dong, K., Gillo, B., and Sealfon, S. C. (1992). Cloning and functional expression of a mouse gonadotropin-releasing hormone receptor. *Mol. Endocrinol.* 6, 1163–1169.
335. Eidne, K. A., Sellar, R. E., Couper, G., Anderson I, and Taylor, P. I. (1992). Molecular-cloning and characterization of the rat pituitary gonadotropin-releasing hormone (GnRH) receptor. *Mol. Cell. Endocrinol.* 90, 85–89.
336. Heding, A., Vrecl, M., Hanyaloglu, A. C., Sellar, R., Taylor, P. L., and Eidne, K. A. (2000). The rat gonadotropin-releasing hormone receptor internalizes via a β-arrestin-independent, but dynamin-dependent, pathway, addition of a carboxy-terminal tail confers β-arrestin dependency. *Endocrinology* 141, 299–306.
337. Hislop, J. N., Madziva, M. T., Everest, H. M., Harding, T., Uney, J. B., Willars, G. B., Millar, R. P., Troskie, B. E., Davidson, J. S., and McArdle, C. A. (2000). Desensitization and internalization of human and *Xenopus* gonadotropin-releasing hormone receptors expressed in a T4 pituitary cells using recombinant adenovirus. *Endocrinology* 141, 4564–4575.
338. King, J. A., Fidler, A., Lawrence, S., Adam, T., Millar, R. P., and Katz, A. (2000). Cloning and expression, pharmacological characterization and internalization kinetics of the pituitary GnRH receptor in a metatherian species of mammal. *Gen. Comp. Endocrinol.* 117, 439–448.
339. Schvartz, I., and Hazum, E. (1987). Internalization and recycling of receptor-bound gonadotropin-releasing hormone agonist in pituitary gonadotropes. *J. Biol. Chem.* 262, 17046–17050.
340. Leblanc, P., L'Héritier, A., and Kordon, C. (1997). Cryptic gonadotropin-releasing hormone receptors of rat pituitary cells in culture are unmasked by epidermal growth factor. *Endocrinology* 138, 575–579.
341. Braden, T. C., Farnworth, P. G., Burger, H. G., and Conn, P. M. (1990). Regulation of the synthetic rate of gonadotropin-releasing hormone receptors in rat pituitary cell cultures by inhibin. *Endocrinology* 127, 2387–2392.
342. Starling, L., McIntosh, J. E., and McIntosh, R. P. (1988). Estimating the rate of externalization of gonadotrophin-releasing hormone receptors in ovine anterior pituitary cells, in vitro. *J. Endocrinol.* 117, 97–107.
343. Tibolt, R. E., and Childs, G. V. (1985). Cytochemical and cytophysiological studies of GnRH target cells in the male rat pituitary: differential effects of androgens and corticosterone on GnRH binding and gonadotropin release. *Endocrinology* 117, 396–404.
344. Farnworth, P. G., Findlay, J. K., and Buger, H. G. (1988). Effect of purified 31K bovine inhibin on the specific binding of gonadotropin-releasing hormone to rat anterior pituitary cells in culture. *Endocrinology* 123, 2161–2166.
345. Wang, Q. F., Farnworth, P. G., Findlay, J. K., and Burger, H. G. (1989). Inhibitory effect of pure 31-kilodalton bovine inhibin on gonadotropin-releasing hormone (GnRH)-induced up-regulation of GnRH binding sites in cultured rat anterior pituitary cells. *Endocrinology* 127, 363–368.
346. Braden, T. D., Farnworth, P. G., Burger, H. G., and Conn, P. M. (1990). Regulation of the synthetic rate of gonadotropin-releasing hormone receptors in rat pituitary cell cultures by inhibin. *Endocrinology* 127, 2387–2392.
347. Braden, T. D., and Conn, P. M. (1990). Activin A stimulates the synthesis of gonadotropin-releasing hormone receptors. *Endocrinology* 130, 2102–2105.
348. Braden, T. D., and Conn, P. M. (1991). The 1990 James A. F. Stevenson Memorial Lecture. Gonadotropin-releasing hormone and its actions. *Can. J. Physiol. Pharmacol.* 69, 445–458.
349. Fernandez-Vazquez, G., Kaiser, U. B., Albarracin, C. T., and Chin, W. W. (1996). Transcriptional activation of the gonadotropin-releasing hormone receptor gene by activin A. *Mol. Endocrinol.* 10, 356–366.
350. Sutton, S. W., Toyama, T. T., Otto, S., and Plotsky, P. M. (1988). Evidence that neuropeptide Y (NPY) released into the hypophysial-portal circulation participates in priming gonadotropes to the effects of gonadotropin releasing hormone (GnRH). *Endocrinology* 123, 1208–1210.
351. Winters, S. J., and Moore, J. P. (2004). Intra-pituitary regulation of gonadotrophs in male rodents and primates. *Reproduction* 128, 13–23.
352. Uccella, S., La Rosa, S., Genasetti, A., and Capella, C. (2000). Localization of inhibin/activin subunits in normal pituitary and in pituitary adenomas. *Pituitary* 3, 131–139.
353. Moore, J. P. Jr., Wilson, L., Dalkin, A. C., and Winters, S. J. (2003). Differential expression of the pituitary gonadotropin subunit genes during male rat sexual maturation: reciprocal relationship between hypothalamic pituitary adenylate cyclase activating polypeptide and follicle stimulating hormone beta expression. *Biol. Reprod.* 69, 234–241.
354. Winters, S. J., Kawakami, S., Sahu, A., and Plant, T. M. (2001). Pituitary follistatin and activin gene expression and the testicular regulation of FSH in the adult rhesus monkey (*Macaca mulatta*). *Endocrinology* 142, 2874–2878.
355. Besecke, L. M., Guendner, M. J., Schneyer, A. L., Bauer, D. A., Jameson, J. L., and Weiss, J. (1996). Gonadotropin-releasing hormone regulates follicle stimulating hormone-beta gene expression through an activin/follistatin autocrine or paracrine loop. *Endocrinology* 137, 3667–3673.
356. Gharib, S. D., Leung, P. C., Carroll, R. S., and Chin, W. W. (1990). Androgens positively regulate follicle-stimulating hormone beta-subunit mRNA levels in rat pituitary cells. *Mol. Endocrinol.* 4, 1620–1626.
357. Kawakami, S., and Winters, S. J. (1999). Regulation of luteinizing hormone secretion and subunit messenger ribonucleic acid expression by gonadal steroids in perifused pituitary cells form male monkeys and rats. *Endocrinology* 140, 3587–3593.
358. Leal, A. M., Blount, A. L., Donaldson, C. J., Bilezikjian, L. M., and Vale, W. W. (2003). Regulation of follicle-stimulating hormone secretion by the interactions of activin-A, dexamethasone and testosterone in anterior pituitary cell cultures of male rats. *Neuroendocrinology* 77, 298–304.

359. Fan, X., and Childs, G. V. (1995). EGF and TGFβ mRNA and their receptors in the rat anterior pituitary: localization and regulation. *Endocrinology* 136, 2284–2324.
360. Armstrong, J. L., and Childs, G. (1997). Regulation of expression of epidermal growth factor receptors in gonadotropes by epidermal growth factor and estradiol: studies in cycling female rats. *Endocrinology* 138, 5434–5441.
361. Przylipiak, A., Kiesel, L., Rabe, T., Helm, K., Przylipiak, M., and Runnebaum, B., (1988). Epidermal growth factor stimulates luteinizing hormone and arachidonic acid release in rat pituitary cells. *Mol. Cell. Endocrinol.* 57, 157–162.
362. Miyake, A., Tasaka, K., Otsuka, S., Kohmura, H., Wakimoto, H., and Anno, T. (1985). Epidermal growth factor stimulates secretion of rat pituitary luteinizing hormone, in vitro. *Acta Endocrinol. (Copenh).* 108, 175–178.
363. Armstrong, J., and Childs, G. V. (1997). Differential expression of c-fos *in vitro* by all anterior pituitary cell types during the estrous cycle: enhanced expression by luteinizing hormone but not follicle stimulating hormone cells. *J. Histochem. Cytochem.* 45, 785–794.
364. Addison, W. H. F., (1917). The cell-changes in the hypophysis of the albino rat after castration. *J. Comp. Neurol.* 28, 441–463.
365. Pomerat, G. R., (1941). Mitotic activity in the pituitary of the white rat following castration. *Am. J. Anat.* 69, 89–121.
366. Hunt, T. E., (1943). Mitotic activity in the anterior hypophysis of female rats of different age groups and at different periods of the day. *Endocrinology* 32, 334–339.
367. Crane, W. A. J., and Loomes, R. S. (1967). Effect of age, sex and hormonal state on tritiated thymidine uptake by rat pituitary. *Br. J. Cancer* 21, 787–792.
368. Mastro, A., Shelton, E., and Hymer, W. C. (1969). DNA synthesis in the rat anterior pituitary: an electron microscope radioautographic study. *J. Cell. Biol.* 43, 626–629.
369. Hunt, T., and Hunt, E. (1966). A radioautographic study of the proliferative activity of adrenocortical and hypophyseal cells of the rat at different periods of the estrous cycle. *Anat. Rec.* 156, 361–368.
370. Oishi, Y., Okuda, M., Takahashi, H., Fuji, T., and Morii, S. (1993). Cellular proliferation in the anterior pituitary gland of normal adult rats: influences of sex, estrous cycle and circadian change. *Anat. Rec.* 235, 111–120.
371. Carbajo-Perez, E., and Watanabe, Y. G. (1990). Cellular proliferation in the anterior pituitary of the rat during the postnatal period. *Cell. Tissue Res.* 261, 333–338.
372. Gratzner, H. G., (1982). Monoclonal antibody to 5-bromo and 5-iododeoxyuridine: a new reagent for detection of DNA replication. *Science* 218, 474–475.
373. Hardonk, N., and Harms, G. (1990). The use of 5′-bromodeoxyuridine in the study of cell proliferation. *Acta Histochem.* 5, 99–109.
374. Yu, C., Woods, A., and Levison, D. (1992). The assessment of cellular proliferation by immunohistochemistry: a review of currently available methods and their applications. *Histochem. J.* 24, 121–131.
375. Bravo, R. (1986). Synthesis of the nuclear protein cyclin (PCNA) and its relationship with DNA replication. *Exp. Cell. Res.* 163, 287–293.
376. Kurki, P., Vanderlaan, M., Dolbeare, F., Gray, J., and Tan, E. (1986). Expression of proliferating cell nuclear antigen (PCNA)/cyclin during the cell cycle. *Exp. Cell. Res.* 166, 209–219.
377. Szijan, I., Parma, D. L., and Engel. N. I. (1992). Expression of c-myc and c-fos protooncogenes in the anterior pituitary gland of the rat: effect of estrogen. *Horm. Metab. Res.* 24, 154–157.
378. Suganuma, N., Kikkawa, F., Seo, H., Matusi, N., and Tomoda, Y., (1993). Poly (adenosine diphosphate-ribose) synthesis in the anterior pituitary of the female rat throughout the estrous cycle: study of possible relation to cell proliferation and prolactin gene expression. *J. Endocrinol. Invest.* 16, 475–480.
379. Chaidarun, S. S., Eggo, M. C., Stewart, P. M., Barber, P. C., and Sheppard, M. C. (1994). Role of growth factors and estrogen as modulators of growth, differentiation, and expression of gonadotropin subunit genes in primary cultured sheep pituitary cells. *Endocrinology* 134, 935–944.
380. Katayama, T., Siota, K., and Takahashi, M. (1990). Activin A increases the number of follicle stimulating hormone cells in anterior pituitary cultures. *Mol. Cell. Endocrinol.* 69, 179–187.
381. Jones, H. B., Harbottle, S. J., and Bowdler, A. L. (1994). Assessment of the labeling index of cohorts of the anterior pituitary cell population in phenobarbital-treated male rats by a double immunohistochemical technique for bromodeoxyuridine and pituitary hormones. *J. Histochem. Cytochem.* 42, 543–549.
382. Romano, M., Machiavelli, G., Perez, R., Carricarte, V., and Burdman, J. (1984). Correlation between LH secretion in castrated rats with cellular proliferation and synthesis of DNA in the anterior pituitary gland. *J. Endocrinol.* 102, 13–18.
383. Machiavelli, G. A., Romano, M. I., and Burdman, J. A. (1985). Relationship between release of LH and incorporation of tritiated thymidine in the anterior pituitary gland of the castrated female rat. *Horm. Metab. Res.* 17, 298–300.
384. Romano, M. I., Machiavelli, G. A., Alonso, G. E., and Burdman, J. A. (1986). Relationship between release of LH and incorporation of tritiated thymidine in the anterior pituitary gland of the castrated male rat: effect of LHRH and its highly active analogue buserelin. *Horm. Metab. Res.* 18, 31–33.
385. Williams, B., Brooks, A. N., Aldridge, T. C., Pennie, W. D., Stephenson, R., and McArdle, C. A. (2000). Oestradiol is a potent mitogen and modulator of GnRH signalling in alphaT3-1 cells: are these effects causally related? *J. Endocrinol.* 164, 31–43.
386. MacLeod, R. M., Lehmeyer, J. E., and Bruni, C. (1973). Effect of anti-mitotic drugs on the in vitro secretory activity of mammotrophs and somatotrophs and on their microtubules. *Proc. Soc. Exp. Biol. Med.* 144, 259–267.
387. Kurosumi, K. (1979). Formation and release of secretory granules during mitosis in the anterior pituitary gland. *Arch. Histol. Jpn.* 42, 481–486.
388. Shirasawa, N., and Yoshimura, F. (1982). Immunohistochemical and electron microscopical studies of mitotic adenohypophysial cells in different ages of rats. *Anat. Embryol.* 165, 51–61.
389. Takahashi, S., Okazaki, K., and Kawashima, S. (1984). Mitotic activity of prolactin cells in the pituitary glands of male and female rats of different ages. *Cell. Tissue Res.* 235, 497–502.
390. Perez, R. L., Machiavelli, G. A., Romano, M. I., and Burdman, J. A. (1986). Prolactin release, oestrogens and proliferation of prolactin-secreting cells in the anterior pituitary gland of adult male rats. *J. Endocrinol.* 108, 399–403.
391. Yonezawa, K., Tamaki, N., and Kokunai, T. (1997). Effects of bromocriptine and terguride on cell proliferation and apoptosis in the estrogen-stimulated anterior pituitary gland of the rat. *Neurol. Med. Chir.* 37, 901–906.
392. Proesmans, M., Van Bael, A., Andries, M., and Denef, C. (1997). Mitogenic effects of nerve growth factor on different cell types in reaggregate cell cultures of immature rat pituitary. *Mol. Cell. Endocrinol.* 143, 119–127.
393. Patterson, J. C., and Childs, G. V. (1994). Nerve growth factor and its receptor in the anterior pituitary. *Endocrinology* 135, 1689–1697.

394. Patterson, J. C., and Childs, G. V. (1994). Nerve growth factor in the anterior pituitary: regulation of secretion. *Endocrinology* 135, 1697–1704.
395. Childs, G. V., and Unabia, G., (2001). Epidermal growth factor and gonadotropin releasing hormone stimulate proliferation of enriched populations of pituitary gonadotropes. *Endocrinology* 142, 847–854.
396. Childs, G. V. (2001). Sites of epidermal growth factor synthesis and action in the anterior pituitary: paracrine and autocrine interactions. *Clin. Exp. Pharmacol. Physiol.* 28, 249–252.
397. Childs, G. V., and Unabia, G. (2001). The use of counterflow centrifugation to enrich gonadotropes and somatotropes. *J. Histochem. Cytochem.* 49, 663–664.
398. Shah, B. H., Farshori, M. P., Jambusaria, A., and Catt, K. J. (2003). Roles of Src and epidermal growth factor receptor transactivation in transient and sustained ERK1/2 responses to gonadotropin-releasing hormone receptor activation. *J. Biol. Chem.* 278, 19118–19126.
399. Liu, F., Austin, D. A., Mellon, P. L., Olefsky, J. M., and Webster, N. J. (2002). GnRH activates ERK1/2 leading to the induction of c-fos and LHβ protein expression in LβT2 cells. *Mol. Endocrinol.* 16, 419–434.
400. Shah, B. H., and Catt, K. J. (2004). Matrix metalloproteinases in reproductive endocrinology. *Trends Endocrinol. Metab.* 15, 47–49.
401. Armstrong, J., and Childs, G. V. (1998). Regulation of C-fos mRNA and protein by EGF and GnRH within proestrous female rat anterior pituitary glands. *J. Histochem. Cytochem.* 46, 935- 943.
402. Theas, M. W., De Laurentiis, A., Lasaga, M., Pisera, D., Duvillanski, B., and Seilcovich, A. (1998). Effect of lipopolysaccharide on tumor necrosis factor and prolactin secretion from rat anterior pituitary cells. *Endocrine* 8, 241–245.
403. De Laurentiis, A., Piscera, D., Caruso, C., Candolifi, M., Mohn, C., Rettori, V., and Seilicovich, A. (2002). Lipopolysaccharide- and tumor necrosis factor-alpha induced changes in prolactin secretion and dopaminergic activity in the hypothalamic-pituitary axis. *Neuroimmunomodulation* 10, 30–39.
404. Drewett, N., Jacobi, H. M., Willgoss, D. A., and Lloyd, H. M. (1993). Apoptosis in the anterior pituitary gland of the rat: studies with estrogen and bromocriptine. *Neuroendocrinology* 57, 89–95.
405. Poliandri, A. H. B., Cabilla, J. P., Velardez, M. O., Bodo, C. C. A., and Duvilanski, B. H. (2003). Cadmium induces apoptosis in anterior pituitary cells that can be reversed by treatment with antioxidants. *Toxicol. Appl. Pharmacol.* 190, 17–24.
406. Poliandri, A. H. B., Velardez, M. O., Cabilla, J. P., Bodo, C. C. A., Machiavelli, L. I., Quinteros, A. F., and Duvilanski, B. H. (2004). Nitric oxide protects anterior pituitary cells from cadmium-induced apoptosis. *Free Radic. Biol. Med.* 37, 1463–1471.
407. Caruso, C., Bottino, M. C., Pampillo, M., Pisera, D., Jaita, G., Duvilanski, B., Seilicovich, A., and Lasaga, M. (2004). Glutamate induces apoptosis in anterior pituitary cells through group II metabotropic glutamate receptor activation. *Endocrinology* 145, 4677–4684.
408. Bogazzi, F., Ultimieri, F., Raggi, F., Russo, D., Vanacore, R., Guida, C., Viacava, P., Cecchetti, D., Acerbi, G., Brogioni, S., Cosci, C., Gasperi, M., Bartalene, L., and Martino, E. (2004). PPARγ inhibits GH synthesis and secretion and increases apoptosis of pituitary GH-secreting adenomas. *Eur. J. Endocrinol.* 150, 863–875.
409. Vankelecom, H., Matthys, P., and Denef, C. (1997). Inducible nitric oxide synthase in the anterior pituitary gland: induction by interferon-gamma in a subpopulation of folliculostellate cells and in an unidentifiable population of non-hormone-secreting cells. *J. Histochem. Cytochem.* 45, 847–857.
410. Duvilansi, B. H., Zambruno, C., Seilicovich, A., Pisera, D., Lasagaa, M., Diaz, M. C., Belova, N., Rettori, V., and McCann, S. M. (1995). Role of nitric oxide in control of prolactin release by the adenohypophysis. *Proc. Natl. Acad. Sci. U S A* 92, 170–174.
411. Ceccatelli, S., Hulting, A. L., Zhang, X., Gustafsson, L., Villar, M., and Hökfelt, T. (1993). Nitric oxide synthase in rat anterior pituitary gland and the role of nitric oxide in regulation of luteinizing hormone secretion. *Proc. Natl. Acad. Sci. U S A* 90, 11292_11296.
412. Lloyd, R. V., Jin, L., Zian, X., Zhang, S. Y., and Scheithauer, B. W. (1995). Nitric oxide synthase in the human pituitary gland. *Am. J. Pathol.* 146, 86–94.
413. Qian, X., Jin, L., and Lloyd, R. V. (1999). Estrogen downregulates neuronal nitric oxide synthase in rat anterior pituitary cells and GH3 tumors. *Endocrine* 1, 123–130.
414. Kostic, T. S., Andric, S. A., and Stojilkovic, S. S. (2001). Spontaneous and receptor-controlled soluble guanylyl cyclase activity in anterior pituitary cells. *Mol. Endocrinol.* 15, 1010–1022.
415. Velardez, M. W., Poliandri, A. H., Cabilla, J. P., Bodo, C. C., and Duvilanski, B. H. (2003). Long term treatment of anterior pituitary cells with nitric oxide induces programmed cell death. *Endocrinology* 145, 2064–2070.
416. Lozach, A., Garrel, G., Lerrant, Y., Berault, A., and Counis, R. (1998). GnRH-dependent upregulation of nitric oxide synthase I level in pituitary gonadotrophs mediates cGMP elevation during rat proestrus. *Mol. Cell. Endocrinol.* 143, 43–51.
417. Fernández, M., Sánchez-Franco, F., Palacios, N., Sánchez, I., Fernández, C., and Cacicedo, L. (2004). IGF-1 inhibits apoptosis through the activation of the phosphatidylinositol 3-kinase/Akt pathway in pituitary cells. *J. Mol. Endocrinol.* 33, 155–163.
418. Rose, A., Froment, P., Perrot, V., Quon, M. J., LeRoith, D., and Dupont, J. (2004). The luteinizing hormone-releasing hormone inhibits the anti-apoptotic activity of insulin-like growth factor-1 in pituitary αT3 cells by protein kinase Cα-mediated negative regulation of Akt. *J. Biol. Chem.* 279, 52500–52516.
419. Spengler, D., Villalba, M., Hoffmann, A., Pantaloni, C., Houssami, S., Bockaert, J., and Journot, L. (1997). Regulation of apoptosis and cell cycle arrest by Zac1, a novel zinc finger protein expressed in the pituitary gland and the brain. *EMBO J.* 16, 2814–2825.
420. Ahlbom, E., Gandison, L., Zhivotovsky, B., and Ceccatelli, S. (1998). Termination of lactation induces apoptosis and alters the expression of the Bcl-2 family members in the rat anterior pituitary. *Endocrinology* 139, 2465–2471.
421. Chowdhury, V. S., and Yoshimura, Y. (2002). Cell Proliferation and apoptosis in the anterior pituitary of chicken during inhibition and resumption of laying. *Gen. Comp. Endocrinol.* 125, 134–141.
422. Aoki, M. P., Aoki, A., and Maldonado, C. A. (2001). Sexual dimorphism of apoptosis in lactotrophs induced by bromocryptine. *Histochem. Cell. Biol.* 116, 215–222.
423. Kulig, E., Camper, S. A., Kuecker, S., Jin, L., and Lloyd, R. V. (1998). Remodeling of hyperplastic pituitaries in hypothyroid alpha-subunit knockout mice after thyroxine and 17β estradiol treatment: role of apoptosis. *Endocr. Pathol.* 9, 261–274.
424. Watanabe, H., and Yoneda, M. (2003). A mechanism underlying the sexually dimorphic ACTH response to lipopolysaccharide in rats: sex steroid modulation of

cytokine binding sites in the hypothalamus. *J. Physiol. (Lond).* 547, 221–232.

425. Pisera, D., Candolfi, M., Navarra, S., Ferraris, J., Zaldivar, V., Jaita, G., Castro, M. G., and Seilicovich, A. (2004). Estrogens sensitize anterior pituitary gland to apoptosis. *Am. J. Physiol. Endocrinol. Metab.* 287, E767–E771.
426. Candolfi, M., Zaldivar, V., De Laurentiis, A., Jaita, G., Pisera, D., and Seilicovich, A. (2002). TNF-α induces apoptosis of lactotropes from female rats. *Endocrinology* 143, 3611–3617.
427. Candolfi, M., Jaita, G., Zaldivar, V., Zárate, S., Ferrari, L., Pisera, D., Castro, M. G., and Seilicovich, A. (2005). Progesterone antagonizes the permissive action of estradiol on TNF-α induced apoptosis of anterior pituitary cells. *Endocrinology* 146, 736–743.
428. Candolfi, M., Jaita, G., Zaldivar, V., Zárate, S., Pisera, D., and Seilicovich, A. (2004). Tumor necrosis factor-alpha-induced nitric oxide restrains the apoptotic response of anterior pituitary cells. *Neuroendocrinology* 80, 83–91.
429. Miles, L. E. C., Hanyaloglu, A. C., Dromey, J. R., Pfleger, K. D. G., and Eidne, K. A. (2004). Gonadotropin-releasing hormone receptor-mediated growth suppression of immortalized LβT2 gonadotrope and stable HEK293 cell lines. *Endocrinology* 145, 194–204.
430. Yin, P., and Arita, J. (2002). Proestrous surge of gonadotropin-releasing hormone secretion inhibits apoptosis of anterior pituitary cells in cycling female rats. *Neuroendocrinology* 76, 272–282.
431. Nakane, P. K. (1975). Identification of anterior pituitary cells by immunoelectron microscopy. In *The Anterior Pituitary* (M. G. Farquhar and A. Tixier-Vidal, Eds.), pp. 45–61. Academic Press, New York.
432. Baker, B. L., and Gross, D. S. (1978). Cytology and distribution of secretory cell types in mouse hypophysis as demonstrated with immuno-cytochemistry. *Am. J. Anat.* 153, 193–215.
433. Farquhar, M. G., Skutelsky, E. H., and Hopkins, C. R. (1975). Structure and function of the anterior pituitary and dispersed pituitary cells: in vitro studies. In *The Anterior Pituitary* (M. G. Farquhar and A. Tixier-Vidal, Eds.), pp. 83–135. Academic Press, New York.
434. Pasteels, J. L., Gausset, P., Danguy, A., Ectors, F., Nicoll,. C. S., and Varavudhi, P. (1972). Morphology of the lactotropes and somatotropes of man and rhesus monkeys. *J. Clin. Endocrinol. Metab.* 34, 959–967.
435. Halmi, N. S., Parsons, J. A., Erlandsen, S. L., and Duello, T. (1975). Prolactin and growth hormone cells in the human hypophysis: a study with immunoenzyme histochemistry and differential staining. *Cell. Tissue Res.* 158, 497–507.
436. Baker, B. L., and Yu, Y. Y. (1977). An immunocytochemical study of human pituitary mammotropes from fetal life to old age. *Am. J. Anat.* 148, 217–240.
437. Nogami, H., and Yoshimura, F. (1982). Fine structural criteria of prolactin cells identified immunohistochemically in the male rat. *Anat. Rec.* 202, 261–274.
438. Beauvillain, J. C., Mazucca, M., and Dubois, M. P. (1977). The prolactin and growth hormone producing cells of the guinea pig pituitary: electron microscopic study using immunocytochemical means. *Cell. Tissue Res.* 184, 343–338.
439. Dacheux, F., and Dubois, M. P. (1976). Ultrastructural localization of prolactin, growth hormone, and luteinizing hormone by immunocytochemical techniques in the bovine pituitary. *Cell. Tissue Res.* 174, 245–260.
440. Dacheux, F. (1980). Ultrastructural immunocytochemical localization of prolactin and growth hormone in the porcine pituitary. *Cell. Tissue Res.* 207, 277–286.
441. Martin-Comin, J., and Robyn, C. (1976). Comparative immunoenzymatic localization of prolactin and growth hormone in human and rat pituitaries. *J. Histochem. Cytochem.* 24, 1012–1016.
442. Pasteels, J. L., Ectors, F., Danguy, A., Robyn, C., L'Hermite, M., and Dujardin, M. (1973). Histological immunofluorescent and electron microscopic identification of prolactin producing cells in the human pituitary. *Exerpta Med.* 373, 616–621.
443. Duello, T. M., and Halmi, N. S. (1979). Ultrastructural-immunocytochemical localization of growth hormone and prolactin in human pituitaries. *J. Clin. Endocrinol. Metab.* 49, 189–196.
444. Nogami, H., and Yoshimura, F. (1980). Prolactin immunoreactivity of acidophils of the small granule type. *Cell. Tissue Res.* 2, 1–4.
445. Nogami, H. (1984). Fine structural heterogeneity and morphological changes in rat pituitary prolactin cells after estrogen and testosterone treatment. *Cell. Tissue Res.* 237, 195–202.
446. Snyder, J. M., Wilfinger, W., and Hymer, W. C. (1976). Maintenance of separated rat pituitary mammotrophs in cell culture. *Endocrinology* 98, 25–32.
447. Inoué, K., Hattori, M. S., Sakai, T., Inukai,, S., Fujimoto, N., and Ito, A. (1990). Establishment of a series of pituitary clonal cell lines differing in morphology, hormone secretion and response to estrogen. *Endocrinology* 126, 2313–2320.
448. Neill, J. D., and Frawley, L. S. (1983). Detection of hormone release from individual cells in mixed populations using a reverse hemolytic plaque assay. *Endocrinology* 112, 1135–1137.
449. Bookfor, F. R., and Schwartz, L. (1987). Cultures of GH3 cells contain both single and dual hormone secretors. *Endocrinology* 122, 762–767.
450. Frawley, L. S., and Bookfor, F. R. (1991). Mammosomatotropes: presence and functions in normal and neoplastic pituitary tissue. *Endocr. Rev.* 12, 337–335.
451. Lloyd, R. V., Anajnostou, D., Cano, M., Barkan, A. L., and Chandler, W. F. (1988). analysis of mammosomatotropic cells in normal and neoplastic human pituitary tissues by the reverse hemolytic plaque assay and immunocytochemistry. *J. Clin. Endocrinol. Metab.* 66, 1103–1110.
452. Fumagalli, G., and Zanini, A., (1985). In cow anterior pituitary, growth hormone and prolactin can be packed in separate granules of the same cell. *J. Cell. Biol.* 100, 2019–2024.
453. Nikitovitch-Winer, M. B., Atkin, J., and Maley, B. E. (1987). Colocalization of prolactin and growth hormone within specific adenohypophyseal cells in male, female, a lactating female rats. *Endocrinology* 121, 625–635.
454. Smets, G., Welkeniers, B., Finne, E., Baldys, A., Gepts, W., and Vanhaelst, L. (1987). Postnatal development of growth hormone and prolactin cells in male and female pituitary: an immunocytochemical light and electron microscopic study. *J. Histochem. Cytochem.* 35, 335–342.
455. Kurosumi, K. (1988). Ultrastructural modifications in prolactin-producing cells in the adenohypophysis during postnatal development and functional variations. In *Prolactin Gene Family and Its Receptors: Molecular Biology to Clinical Probems* (K. Hoshino, Ed.), pp. 289–297. Elsevier, Amsterdam.
456. Leong, D. A., Lau, S. K., Sinha, Y. N., Kaiser, D. L., and Thorner, M. O. (1985). Enumeration of lactotropes and somatotropes among male and female pituitary cells in culture: evidence in favor of a mammosomatotrope subpopulation in the rat. *Endocrinology* 116, 1371–1378.
457. Frawley, L. S., Boockfor, F. R., and Hoeffler, J. P. (1991). Identification by plaque assays of a pituitary cell type that secretes both growth hormone and prolactin. *Endocrinology* 116, 734–737.

458. Kineman, R. A. D., Henricks, D. M., Faught, W. J., and Frawley, L. S. (1991). Fluctuations in the proportions of growth hormone and prolactin secreting cells during the bovine estrous cycle. *Endocrinology* 129, 1221–1225.
459. Porter, T. E., Wiles, C. D., and Frawley, L. S. (1991). Evidence for bidirectional interconversion of mammotropes and somatotropes: rapid reversion of acidophilic cell types to pregestational proportions after weaning. *Endocrinology* 129, 1215–1220.
460. Borrelli, E., Heyman, R. A., Arias, C., Sawchenko, P. E., and Evans, R. M. (1989). Transgenic mice with inducible dwarfism. *Nature* 339, 543–541.
461. Nemeskeri, A., Grouselle, D., Tixier-Vidal, A., and Halasz, B. (1976). Ontogeny of pro1actin synthesizing cells in fetal and early postnatal rat pituitary: in vivo and in vitro studies. In *Prolactin Gene Family and Its Receptors: Molecular Biology to Clinical Probems* (K. Hoshino, Ed.), pp. 319–325. Elsevier, Amsterdam.
462. Tong, Y. A., Zhao, H. F., Labrie, F., and Pelletier, G. (1989). Ontogeny of prolactin mRNA in the rat pituitary gland as evaluated by in situ hybridization. *Mol. Cell. Endocrinol.* 67, 11–16.
463. Smets, G., Velkeniers, B., Herregodts, P., Vanhaelst, L., Gepts, W., and Hooghe-Peters, E. L. (1989). Ontogeny of hormone-secreting cells of the rat pituitary gland: an immunocytochemical study on dissociated cells. *Histochem. J.* 21, 337–342.
464. Chatelain, A., Dupouy, J. P., and Dubois, M. P. (1979). Ontogenesis of cell producing polypeptide hormones (ACTH, MSH, LPH, GH, prolactin) in the fetal hypophysis of the rat: influence of the hypothalamus. *Cell. Tissue Res.* 196, 409–427.
465. Watanabe, Y. G., and Daikoku, S. (1979). An immunohistochemical study on the cytogenesis of adenohypophysial cells in fetal rats. *Dev. Biol.* 68, 557–567.
466. Hooghe-Peters, E. L., Belayew, A., Herregodts, P., Velkeniers, B., Smets, G., Martial, J. A., and Vanhaelst, L. (1988). Discrepancy between prolactin (PRL) messenger ribonucleic acid and PRL content in rat fetal pituitary cells: possible role of dopamine. *Mol. Endocrinol.* 2, 1163–1168.
467. Frawley, L. S., and Miller III, H. A. (1989). Ontogeny of prolactin secretion in the neonatal rats is regulated postranscriptionally. *Endocrinology* 124, 3–6.
468. Begeot, M., Dubois, M. P., and Dubois, P. M. (1984). Evolution of lactotropes in normal and anencephalic human fetuses. *J. Clin. Endocrinol. Metab.* 8, 726–730.
469. Mulchahey, J. J., and Jaffe, R. D. (1987). Detection of a potential progenitor cell in the human fetal pituitary that secretes both growth hormone and prolactin. *J. Clin. Endocrinol. Metab.* 62, 24–31.
470. Tougard, C., and Tixier-Vidal, A. (1994). Lactotropes and gonadotropes. In *The Physiology of Reproduction* (E. Knobil and J. D. Neill, Eds.), pp. 1711–1747. Raven Press, New York.
471. Swearingen, K. C. (1971). Heterogeneous turnover of adenohypophysial prolactin. *Endocrinology* 89, 1380–1388.
472. Walker, A. M., and Farquhar, M. G. (1980). Preferential release of newly synthesized prolactin granules is the result of functional heterogeneity among mammotrophs. *Endocrinology* 107, 1095–1104.
473. Morin, A., Rosenbaum, E., and Tixier-Vidal, A. (1984). Effects of thyrotropin releasing hormone on prolactin compartments in normal rat pituitary cells in primary cultures. *Endocrinology* 115, 2278–2284.
474. Morin, A., Rosenbaum, E., and Tixier-Vidal, A. (1984). Effects of thyrotropin releasing hormone on prolactin compartments in clonal rat pituitary tumor cells. *Endocrinology* 115, 2271–2277.
475. Hoyt, R. F., and Tashjian, A. H. (1980). Immunocytochemical analysis of prolactin production by monolayer cultures of GH3 rat anterior pituitary tumor cells: I. Long term effects of stimulation with thyrotropin-releasing hormone (TRH). *Anat. Rec.* 197, 153–162.
476. Luque, E. H., Munoz de Toro, M., Smith, P. F., and Neill, J. D. (1986). Subpopulations of lactotropes detected with the reverse hemolytic plaque assay show differential responsiveness to dopamine. *Endocrinology* 118, 2120–2124.
477. Arita, J., Kojima, Y., and Kimura, F. (1991). Identification by the sequential cell immunoblot assay of a subpopulation of rat dopamine unresponsive lactotrophs. *Endocrinology* 128, 1887–1894.
478. Boockfor, F. R., and Frawley, L. S. (1987). Functional variations among prolactin cells from different pituitary regions. *Endocrinology* 120, 874–879.
479. Mukherjee, P., Salada, T., and Hymer, W. C. (1991). Function of prolactin cells in the individual rat pituitary gland is location dependent. *Mol. Cell. Endocrinol.* 76, 35–44.
480. Chen, T. T., Kineman, R. D., Betts, J. G., Hill, J. B., and Frawley, L. S. (1989). Relative importance of newly synthesized and stored hormone to basal secretion by growth hormone and prolactin cells. *Endocrinology* 125, 1904–1909.
481. Everett, N. B., and Baker, B. L. (1945). The distribution of cell types in the anterior hypophysis during late pregnancy and lactation. *Endocrinology* 37, 83–88.
482. Merchant, F. W. (1974). Prolactin and luteinizing hormone cells of pregnant rat anterior pituitary as studied by immunogold electron microscopy. *Am. J. Anat.* 139, 245–268.
483. Ozawa, H., and Kurosumi, K. Ultrastructure of prolactin cells in the pregnant rat anterior pituitary as studied by immunogold electron microscopy. In *Proceedings of the Kyoto Prolactin Conferences* (K. Hoshino, Ed.), Vol. 4, pp. 58–67. Shinko, Kyoto.
484. Goluboff, L. G., and Ezrin, C. (1969). Effect of pregnancy on the somatotroph and the prolactin cell of the human adenohypophysis. *J. Clin. Endocrinol. Metab.* 29, 1533–1539.
485. Porter, T. E., Hill, J. B., Wiles, C. D., and Frawley, L. S. (1990). Is the mammosomatotrope a transitional cell for the functional interconversion of growth hormone- and prolactin-secreting cells? Suggestive evidence from virgin, gestating and lactating rats. *Endocrinology* 127, 2789–2794.
486. Grosvenor, C. E., and Turner, C. W. (1958). Pituitary lactogenic hormone and milk secretion in lactating rats. *Endocrinology* 63, 535–539.
487. Shiino, M., Williams, G., and Rennels, E. G. (1972). Ultrastructural observation of pituitary release of prolactin in the rat by suckling stimulus. *Endocrinology* 90, 176–181.
488. Pasteels, J. L. (1963). Recherches morphologiques et expérimentales sur la secretion de prolactine. *Arch. Biol. Liege* 74, 439–453.
489. Vila-Porcile, E., and Olivier, L. (1980). Exocytosis and related membrane events. In *Synthesis and Release of Adenohypophyseal Hormones* (M. Jutisz and W. McKerns, Eds.), pp. 67–104. Plenum, New York.
490. Vila-Porcile, E., Picart, R., Olivier, L., Tixier-Vidal, A., and Tougard, C. (1988). Subcellular distribution of laminin and prolactin in stimulated and blocked prolactin cells in the pituitary of lactating rats. *Cell. Tissue Res.* 254, 617–627.
491. Sasaki, F., and Iwama, Y. (1988). Sex difference in prolactin and growth hormone cells in mouse adenohypophysis: stereological, morphometric, and immunohistochemical studies by light and electron microscopy. *Endocrinology* 123, 905–912.
492. Osamura, R. Y., Komatsu, N., Izumi, S., Yoshimura, S., and Watanabe, K. (1982). Ultrastructural localization of prolactin in the rat anterior pituitary glands by preembedding peroxidase-labeled antibody method: observations in normal, castrated, or estrogen stimulated specimen. *J. Histochem. Cytochem.* 30, 919–925.

493. Ellerkmann, E., Nagy, G. M., and Frawley, L. S. (1991). Rapid augmentation of prolactin cell number and secretory capacity by an estrogen-induced factor released from the neurointermediate lobe. *Endocrinology* 129, 629–637.
494. Porter, T. E., Ellerkmann, E., and Frawley, L. S. (1992). Acute recruitment of prolactin-secreting cells is regulated posttranscriptionally. *Mol. Cell. Endocrinol.* 84, 23–31.
495. Scarbrough, K., Weiland, N. G., Larson, G. H., Sortino, M. A., Shiu, S. F., Hirshfield, A. N., and Wise, P. M. (1991). Measurement of peptide secretion and gene expression in the same cell. *Mol. Endocrinol.* 5, 134–142.
496. Antakly, T., Pelletier, G., Zeytinoglu, F., and Labrie, F. (1980). Changes of cell morphology and prolactin secretion induced by 2-BR α-ergocryptine, estradiol, and thyrotropin-releasing hormone in rat anterior pituitary cells in culture. *J. Cell Biol.* 86, 377–387.
497. Chuknyiska, S. R., Blackman, M. R., Hymer, W. C., and Roth, S. R. (1986). Age related alterations in the number and function of pituitary lactotropic cells from intact and ovariectomized rat. *Endocrinology* 118, 1856–1862.
498. Console, G. M., Gomez-Dumm, C. L., Brown, O. A., Ferese, C., and Goya, R. G. (1997). Sexual dimorphism in the age changes of the pituitary lactotrophs in rats. *Mech. Ageing Dev.* 95, 157–166.
499. Burgus, R., and Guillemin, R. (1970). Hypothalamic releasing factors. *Annu. Rev. Biochem.* 39, 499–526.
500. Vale, W., Blackwell, R., Grant, G., and Guillemin, R. (1973). TRF and thyroid hormones on prolactin secretion by rat anterior pituitary cells in vitro. *Endocrinology* 93, 26–33.
501. Takahara, J., Arimura, A., and Schally, A. V. (1974). Stimulation of prolactin and growth hormone release by TRH infused into a hypophysial portal vessel. *Proc. Soc. Exp. Biol. Med.* 146, 831–835.
502. Brunet, N., Rizzino, A., Gourdji, D., and Tixier-Vidal, A. (1981). Effects of thyroliberin (TRH) on cell proliferation and prolactin secretion by GH3/B6 rat pituitary cells: a comparison between serum-free and serum-supplemented media. *J. Cell. Physiol.* 109, 363–372.
503. Ozawa, H., and Kurosumi, K. (1993). Morphofunctional study of prolactin-producing cells of the anterior pituitaries in adult male rats following thyroidectomy, thyroxine treatment and/or thyrotropin releasing hormone treatment *Cell. Tissue Res.* 272, 41–47.
504. Stirling, R. G., and Shin, S. H. (1990). A high concentration of dopamine preferentially permitted release of newly synthesized prolactin. *Mol. Cell. Endocrinol.* 70, 65–72.
505. Chen, T. T., Kineman, R. D., Betts, J. G., Hill, J. B., and Frawley, L. S. (1989). Relative importance of newly synthesized and stored hormone to basal secretion by growth hormone and prolactin cells. *Endocrinology* 125, 1904–1909.
506. Tougard, C., Picart, R., Morin, A., and Tixier-Vidal, A. (1983). Effect of monensin on secretory pathway in GH3 prolactin cells: a cytochemical study. *J. Histochem. Cytochem.* 31, 745–754.
507. Morin, A., Rosenbaum, E., and Tixier-Vidal, A. (1984). Effects of thyrotropin releasing hormone on prolactin compartments in clonal rat pituitary tumor cells. *Endocrinology* 115, 2271–2277.
508. MacLeod, R. M., and Lehmeyer, J. E. (1974). Studies on the mechanism of dopamine-mediated inhibition of prolactin secretion. *Endocrinology* 94, 1077–1085.
509. Neill, J. D., and Nagy, G. M. (1994). Prolactin secretion and its control. In *The Physiology of Reproduction* (E. Knobil and J. D. Neill, Eds.), pp. 1833–1860. Raven Press, New York.
510. Freeman, M. R., Kanyiecska, B., Lérant, A., and Nagy, G. M. (2000). Prolactin: structure, function, and regulation of secretion. *Physiol. Rev.* 80, 1523–1631.
511. Ben-Jonathan, N., and Hnasko, R. (2001). Dopamine as a prolactin (PRL) inhibitor. *Endocr. Rev.* 22, 744–763.
512. Albert, P. R., Neve, K. A., Bunzow, J. R., and Civelli, O. (1990). Coupling of a cloned rat dopamine-D2 receptor to inhibition of adenylyl cyclase and prolactin secretion. *J. Biol. Chem.* 265, 2098–2104.
513. Nansel, D. D., Gudelsky, G. A., Reymond, M. J., Neaves, W. B., and Poster, J. C. (1981). A possible role for lysosomes in the inhibitory action of dopamine on PRL release. *Endocrinology* 108, 896–902.
514. Frawley, L. S., and Clark, C. L. (1986). Ovine prolactin and dopamine preferentially inhibit PRL release from the same subpopulation of rat mammotropes. *Endocrinology* 119, 1462–1466.
515. Nagy, G. M., Boockfor, F. R., and Frawley, L. S. (1991). The suckling stimulus increases the responsiveness of mammotropes located exclusively within the central region of the adenohypophysis. *Endocrinology* 128, 761–764.
516. Kazemzadeh, M., Velkeniers, B., Herregodts, P., Collumbien, R., Finne, E., Derde, M. P., Vanhaelst, L., and Hooghe-Peters, E. L. (1992). Differential dopamine-induced prolactin mRNA levels in various prolactin secreting cell (sub)populations. *J. Endocrinol.* 132, 401–409.
517. Shin, S. H. (1977). Dopamine-induced inhibition of prolactin release from cultured adenohypophysial cells: spare receptEors for dopamine. *Life Sci.* 22, 67–74.
518. Denef, C., Manet, D., and Dewals, R. (1980). Dopaminergic stimulation of prolactin release. *Nature* 285, 243–246.
519. Kramer, I. M., and Hopkins, C. R. (1982). Studies on the kinetics of dopamine-regulated prolactin secretion. *Mol. Cell. Endocrinol.* 28, 191–198.
520. Burris, T. F., Stringer, L. C., and Freeman, M. E. (1991). Pharmacologic evidence that a D2 receptor subtype mediates dopaminergic stimulation of prolactin secretion from the anterior pituitary gland. *Neuroendocrinology* 54, 175–183.
521. Hill, J. B., Nagy, G. M., and Frawley, L. S. (1991). Suckling unmasks the stimulatory effect of dopamine on prolactin release: possible role for a-melanocyte stimulating hormone as a mammotrope responsiveness factor. *Endocrinology* 129, 843–847.
522. Porter, T. E., Grandy, D., Bunzow, J., Wiles, C. D., Civelli, O., and Frawley, S. (1994). Evidence that stimulatory dopamine receptors may be involved in the regulation of prolactin secretion. *Endocrinology* 134, 1263–1268.
523. Schoors, D. F., Vauquelin, G. P., De Vos, H., Smets, G., Veleniers, B., Vanhaelst, L., and Dupont, A. G. (1991). Identification of a D1 dopamine receptor, not linked to adenylate cyclase on lactotroph cells. *Br. J. Pharmacol.* 102, 1928–1934.
524. Murai, I., and Ben-Jonathan, N. (1987). Posterior pituitary lobectomy abolishes the suckling-induced rise in prolactin (PRL): evidence for a prolactin-releasing factor in the posterior pituitary. *Endocrinology* 121, 205–211.
525. Averill, R. L., Grattan, D. R., and Norris, S. K. (1991). Posterior pituitary lobectomy chronically attenuates the nocturnal surge of prolactin in early pregnancy. *Endocrinology* 128, 705–709.
526. Vecsemyés, M., Krempels, K., Tóth, B. E., Julesz, J., Makara, G. B., and Nagy, G. M. (1998). Effect of a posterior pituitary denervation (PPD) on prolactin (PRL) and α-melanocyte-stimulating hormone (aMSH) secretion of lactating rats. *Brain Res. Bull.* 43, 313–319.
527. Thomas, G. B., Cummins, J. T., Canny, B. J., Rundle, S. E., Griffin, N., Katasahambas, S., and Clarke, I. J. (1989). The posterior pituitary regulates prolactin, but not adrenocorticotropin or gonadotropin secretion in the sheep. *Endocrinology* 125, 2204–2211.

528. Hyde, J. F., Murai, I., and Ben-Jonathan, N. (1987). The rat posterior pituitary contains a potent prolactin-releasing factor: studies with perifused anterior pituitary cells. *Endocrinology* 121, 1531–1539.
529. Hyde, J. F., and Ben-Jonathan, N. (1988). Characterization of prolactin-releasing factor in the rat posterior pituitary. *Endocrinology* 122, 2433–2439.
530. Hyde, J. F., and Ben-Jonathan, N. (1989). The posterior pituitary contains a potent prolactin-releasing factor: in vivo studies. *Endocrinology* 125, 736–741.
531. Samson, W. K., Martin, L., Mogg, R. J., and Fulton, R. J. (1990). A nonoxytocinergic prolactin releasing factor and a nondopaminergic prolactin inhibiting factor in bovine neurointermediate lobe extracts: in vitro and in vivo studies. *Endocrinology* 126, 1616–1617.
532. Dymshitz, J., and Ben-Jonathan, N. (1991). Co-culture of anterior and posterior pituitary cells: selective stimulation of lactotrophs. *Endocrinology* 128, 2469–2475.
533. Steinmetz, R., Liu, J., and Ben-Jonathan, N. (1993). Posterior pituitary cells: isolation of subpopulations producing prolactin releasing factor. *Endocr. J.* 1, 373–379.
534. Allen, D. L., Low, M. J., Allen, R. G., and Ben-Jonathan, N. (1995). Identification of two classes of prolactin-releasing factors in intermediate lobe tumors form transgenic mice. *Endocrinology* 137, 3093–3099.
535. Hnasko, R., Khurana, S., Shackleford, N., Steinmetz, R., Low, M. J., and Ben-Jonathan, N. (1997). Two distinct pituitary cell lines from mouse intermediate lobe tumors: a cell that produces prolactin regulating factor and a melanotroph. *Endocrinology* 138, 5589–5596.
536. Tóth, B. E., Homicskó, K., Radnai, B., Maruyama, W., DeMaria, J. E., Vecsemyés, M., Fekete, M. I. K., Fülüp, F., Naoi, M., Freeman, M. E., and Nagy, G. M. (2001). Salsolinol is a putative endogenous neuro-intermediate lobe prolactin releasing factor. *J. Neuroendocrinol.* 13, 1042–1050.
537. Radnai, B., Mravec, B., Bodnár, I., Kubovcakova, L., Fülüp, F., Fekete, M. I. K., Nagy, G. M., and Kvetnansky, R. (2004). Pivotal role of an endogenous tetrahydroisoquinoline, salsolinol, in stress- and suckling-induced release of prolactin. *Ann. N Y Acad. Sci.* 1018, 184–191.
538. Yu, W. H., Karanth, S., Mastronardi, C. A., Sealfon, S., Dean, C., Dees, W. L., and McCann, S. M. (2002). Lamprey GnRH-III acts on its putative receptor via nitric oxide to release follicle-stimulating hormone specifically. *Exp. Biol. Med.* 227, 786–793.
539. Dhariwal, A. P. S., Nallar, R., Batt, M., and McCann, S. M. (1965). Separation of FSH-releasing factor from LH-releasing factor. *Endocrinology* 76, 290–294.
540. Dhariwal, A. P. S., Watanabe, S., Antunes-Rodrigues, J., and McCann, S. M. (1967). Chromatographic behavior of follicle stimulating hormone releasing factor on Sephadex and carboxymethyl cellulose. *Neuroendocrinology* 2, 294–303.
541. Mizunuma, H., Samson, W. K., Lumpkin, M. D., Moltz, J. H., Fawcett, C. P., and McCann, S. M. (1983). Purification of a bioactive FSH-releasing factor (FSHRF). *Brain Res. Bull.* 10, 623–639.
542. McCann, S. M., Karanth, S., Mastronardi, C. A., Dees, W. L., Childs, G., Miller, B., and Sower, S., and Yu, W. H. (2001). Control of gonadotropin secretion by follicle-stimulating hormone-releasing factor, luteinizing hormone-releasing hormone, and leptin [review]. *Arch. Med. Res.* 32, 476–485.
543. Yu, W. H., Karanth, S., Walczewska, A., Sower, S. A., and McCann, S. M. (1997). A hypothalamic follicle stimulating hormone releasing decapeptide in the rat. *Proc. Natl. Acad. Sci. U S A* 84, 9499–9503.
544. Yu, W. H., Karanth, S., Sower, S. A., Parlow, A. F., and McCann, S. M. (2000). The similarity of FSH-releasing factor to lamprey gonadotropin-releasing hormone III (L-GnRH-III). *Proc. Soc. Exp. Biol. Med.* 224, 87.
545. Neill, J. D. (2002). Mammalian gonadotropin-releasing hormone (GnRH) receptor subtypes *Arch. Physiol. Biochem.* 110, 129–136.
546. Childs, G. V., Miller, B. T., Chico, D. E., Unabia, G. C., Yu, W. H., and McCann, S. M. (2001). Preferential expression of receptors for lamprey gonadotropin releasing hormone-III (GnRH-III) by follicle-stimulating hormone (FSH) cells: support for its function as an FSH-RF. In *Proceedings of the 83rd Annual Meeting of the Endocrine Society, Denver, June 22, 2001*. P3-268.
547. Shupnik, M. A. (2002). Oestrogen receptors, receptor variants and oestrogen actions in the hypothalamic-pituitary axis. *J. Neuroendocrinol.* 14, 85–94.
548. Watson, C. S., Campbell, C. H., and Gametchu, B. (2002). The dynamic and elusive membrane estrogen receptor-alpha. *Steroids* 67, 429–437.
549. Levine, J. E. (1997). New concepts of the neuroendocrine regulation of gonadotropin surges in rats. *Biol. Reprod.* 56, 293–302.
550. Kikuta, T., Yamamoto, K., Namiki, H., and Hayashi, S. (1993). Immunocytochemical localization of estrogen receptor in various anterior pituitary hormone cells of adult male and female rats. *Acta Histochem. Cytochem.* 26, 609–614.
551. Couse, J. F., and Korach, K. S. (1999). Estrogen receptor null mice: what have we learned and where will they lead us? *Endocr. Rev.* 20, 358–417.
552. Scully, K. M., Bleiberman, A. S., Lindsey, J., Lubahn, D. B., Korach, K. S., and Rosenfeld, M. G. (1997). Role of estrogen receptor-a in the anterior pituitary gland. *Mol. Endocrinol.* 11, 674–681.
553. Friend, K. E., Chiou, Y. K., Lopes, M. B., Laws, E. R. Jr., Hughes, K. M., and Shupnik, M. A. (1994). Estrogen receptor expression in human pituitary: correlation with immunohistochemistry in normal tissue, and immunohistochemistry and morphology in macroadenomas. *J. Clin. Endocrinol. Metab.* 78, 1497–1504.
554. Mitchner, N. A., Garlick, C., and Ben-Jonathan, N. (1998). Cellular distribution and gene regulation of estrogen receptors alpha and beta in the rat pituitary gland. *Endocrinology* 139, 3976–3983.
555. Pelletier, G. (2000). Localization of androgen and estrogen receptors in rat and primate tissues. *Histol. Histopathol.* 15, 1261–1270.
556. Sánchez-Criado, J. E., De las Mulas, J. M., Bellido, C., Tena-Sempere, M., Aguilar, R., and Bianco, A. (2004). Biological role of pituitary estrogen receptors ERα and ERβ on progesterone receptor expression and action and on gonadotropin and prolactin secretion in the rat. *Neuroendocrinology* 79, 247–258.
557. Sheng, C., McNeilly, A. S., and Brooks, A. N. (1998). Immunohistochemical distribution of oestrogen receptor and luteinizing hormone B subunit in the ovine pituitary gland during foetal development. *J. Neuroendocrinol.* 10, 713–718.
558. Nishihara, E., Nagayama, Y., Inoue, S., Hiroi, H., Muramatsu, M., Yamashita, S., and Koji, T. (2000). Ontogenetic changes in the expression of estrogen receptor α and β in rat pituitary gland detected by immunohistochemistry. *Endocrinology* 141, 615–620.
559. Shupnik, M. A., Pitt, L. K., Soh, A. Y., Anderson, A., Lopes, M. B., and Laws, E. R. (1998). Selective expression of estrogen receptor α and β isoforms in human pituitary tumors. *J. Clin. Endocrinol. Metab.* 83, 3965–3972.
560. Stefaneau, L., Kovacs, K., Horvath, E., Lloyd, R. V., Buchfelder, M., Fahlbusch, R., and Smyth, H. (1994). *In situ*

hybridization study of estrogen receptor messenger ribonucleic acid in human adenohypophysial cells and pituitary adenomas. *J. Clin. Endocrinol. Metab.* 78, 83–88.

561. Zafar, M., Ezzat, S., Ramyar, L., Pan, N., Smyth, H. S., and Asa, S. L. (1995). Cell-specific expression of estrogen receptor in the human pituitary and its adenomas. *J. Clin. Endocrinol. Metab.* 80, 3621–3627.
562. Chaidarun, S. S., Klibanski, A., and Alexander, J. M. (1997). Tumor-specific expression of alternatively spliced estrogen receptor messenger ribonucleic acid variants in human pituitary adenomas. *J. Clin. Endocrinol. Metab.* 82, 1058–1065.
563. Pereira-Lima, J. F., Marroni, C. P., Pizzaro, C. B., Barbosa-Coutinho, L. M., Ferriera, N. P., and Oliveiria, M. C. (2004). Immunohistochemical detection of estrogen receptor alpha in pituitary adenomas and its correlation with cellular replication. *Neuroendocrinology* 79, 119–124.
564. Pelletier, G., Li, S., Phaneuf, D., Martel, C., and Labrie, F. (2003). Morphological studies of prolactin-secreting cells in estrogen receptor a and estrogen receptor β knockout mice. *Neuroendocrinology* 77, 324–333.
565. Chaidarun, S. S., Swearingen, B., and Alexander, J. M. (1998). Differential expression of estrogen receptor-β (ERβ) in human pituitary tumors: functional interactions with ERα and a tumor-specific splice variant. *J. Clin. Endocrinol. Metab.* 83, 3308–3315.
566. Gittoes, N. J., McCabe, C. J., Sheppard, M. C., and Franklyn, J. A. (1999). Estrogen receptor beta mRNA expression in normal and adenomatous pituitaries. *Pituitary* 1, 99–104.
567. Hall, J. M., and McDonnell, D. P. 1999 The estrogen receptor β-isoform (ERβ). of the human estrogen receptor modulates ERα transcriptional activity and is a key regulator of the cellular response to estrogens and antiestrogens. *Endocrinology* 140, 5566–5578.
568. Weihua, Z., Saji, S., Makinen, S., Cheng, G., Jensen, E. V., Warner, M., and Gustafsson, J. A. (2000). Estrogen receptor (ER) beta, a modulator of ER alpha in the uterus. *Proc. Natl. Acad. Sci. U S A* 97, 5936–5941.
569. Pettersson, K., Delaunay, F., and Gustafsson, J. A. (2000). Estrogen receptor beta acts as a dominant regulator of estrogen signaling. *Oncogene* 19, 4970–4978.
570. Fox, S. R., Harlan, R. E., Shivers, B. D., and Pfaff, D. W. (1990). Chemical characterization of neuroendocrine targets for progesterone in the female rat brain and pituitary. *Neuroendocrinology* 51, 276–283.
571. Bethea, C. L., and Widmann, A. A. (1998). Differential expression of progestin receptor isoforms in the hypothalamus, pituitary, and endometrium of rhesus macaques. *Endocrinology* 139, 677–687.
572. Levine, J. E., Chappell, P. E., Schneider, J. S., Sleiter, N. C., and Szabo, M. (2001). Progesterone receptors as neuroendocrine integrators. *Front. Neuroendocrinol.* 22, 69–106.
573. Chappell, P. E., Lydon, J. P., Conneely, O. M., O'Malley, B. W., and Levine, J. E. (1997). Endocrine defects in mice carrying a null mutation for the progesterone receptor gene. *Endocrinology* 138, 4147–4152.
574. Turgeon, J. L., Shyamala, G., and Waring, D. W. (2001). PR localization and anterior pituitary cell populations in vitro in ovariectomized wild-type and PR-knockout mice. *Endocrinology* 142, 4479–4485.
575. Bethea, C. L. (2001). Large lessons from little lactotropes [editorial]. *Endocrinology* 142, 4170–4172.
576. Thieulant, M. L., and Duval, J. (1985). Differential distribution of androgen and estrogen receptors in rat pituitary cell populations separated by centrifugal elutriation. *Endocrinology* 116, 1299–1303.
577. Kawakami, S., and Winters, S. J. (1999). Regulation of luteinizing hormone secretion a subunit messenger ribonucleic acid expression by gonadal steroids in perifused pituitary cells from male monkeys and rats. *Endocrinology* 140, 3587–3593.
578. Heckert, L. L., Wilson, E. M., and Nilson, J. H. (1997). Transcriptional repression of the a-subunit gene by androgen receptor occurs independently of DNA binding by requires the DNA-binding and ligand-binding domains of the receptor. *Endocrinology* 11, 1497–1506.
579. Jorgensen, J. S., and Nilson, J. H. (2001). AR suppresses transcription of the a glycoprotein subunit gene through protein-protein interactions with cJun and activation transcription factor 2. *Mol. Endocrinol.* 15, 1496–1504.
580. McNeilly, A. S., (1987). Prolactin and the control of gonadotrophin secretion. *J. Endocrinol.* 115, 1–5.
581. Horvath, E., Kovacs, E., and Ezrin, C. (1977). Functional contact between lactotrophs and gonadotrophs in rat pituitary. *IBCS Med. Sci.* 5, 511.
582. Sortino, M. A., and Wise, P. M. (1989). Effect of hyperprolactinemia on luteinizing hormone and prolactin secretion assessed using the reverse hemolytic plaque assay. *Biol. Reprod.* 41, 618–625.
583. Smith, M. S. (1978). A comparison of pituitary responsiveness to luteinizing hormone releasing hormone during lactation and the estrous cycle of the rat. *Endocrinology* 102, 114–120.
584. Smith, M. S. (1982). Effect of pulsatile gonadotropin-releasing hormone on the release of luteinizing hormone and follicle stimulating hormone in vitro by anterior pituitaries from lactating and cycling rats. *Endocrinology* 110, 882–889.
585. Franz, W. L., Payne, P., and Dombrooke, O. (1975). Binding of [^{3}H] PRL to cultured anterior pituitary tumor cells and normal cells. *Nature* 255, 636–638.
586. Tortonese, D. J., Brooks, J., Ingleton, P. M., and McNeilly, A. S. (1998). Detection of prolactin receptor gene expression in the sheep pituitary gland and visualization of the specific translation of the signal in gonadotrophs. *Endocrinology* 138, 5215–5223.
587. Gregory, S. J., Brooks, J., McNeilly, A. S., Ingleton, P. M., and Tortonese, D. J. (2000). Gonadotroph-lactotroph associations and expression of prolactin receptors in the equine pituitary gland throughout the seasonal reproductive cycle. *J. Reprod. Fertil.* 119, 223–231.
588. Townsend, J., Sneddon, C. L., and Tortonese, D. J. (2004). Gonadotroph heterogeneity, density and distribution, and gonadotroph-lactotroph associations in the pars distalis of the male equine pituitary gland. *J. Neuroendocrinol.* 16, 432–440.
589. Ghosh, B. R., Wu., J. C., Strahl, B. D., Childs, G. V., and Miller, W. L. (1996). Inhibin and estradiol alter gonadotropes differentially in ovine pituitary cultures: changing gonadotrope numbers and calcium responses to gonadotropin-releasing hormone. *Endocrinology* 137, 5144–5154.

Knobil and Neill's Physiology of Reproduction, Third Edition
edited by Jimmy D. Neill,
Elsevier © 2006

CHAPTER 30

Gonadotropins: Chemistry and Biosynthesis

George R. Bousfield, Li Jia, and Darrell N. Ward[1]

Introduction, 1581
Structural Features of the Gonadotropins, 1582
Cystine Knot Proteins, 1582
Three-Dimensional Structure, 1583
The α or Common Subunit, 1583
The β or Hormone-Specific Subunit, 1595
Allocation of Function to Structural Areas of the Glycoprotein Hormones, 1599
Aberrant Gonadotropin Molecules, 1601
Genetic Relationships of the Subunits, 1602
The α-Subunit Gene, 1602
The Chorionic Gonadotropin and Luteinizing Hormone β-Subunit Gene Families, 1603
The Follicle-Stimulating Hormone β-Subunit Gene, 1603
The Thyroid-Stimulating Hormone β-Subunit Gene, 1603
Orphan Subunit Genes, 1603
Carbohydrate Structure and Function, 1603
Gonadotropin Glycosylation Sites, 1605
Carbohydrate Structure, 1607
Carbohydrate Function, 1609
Deglycosylated Forms, 1611
Hormone Receptors, 1613
Luteinizing Hormone/Chorionic Gonadotropin Receptors, 1613
Follicle-Stimulating Hormone Receptors, 1619
Thyroid-Stimulating Hormone and Orphan Leucine-Rich Repeat/G-Protein–Coupled Receptors, 1619
Biosynthesis of the Gonadotropins, 1619
References, 1625

INTRODUCTION

In 1975, Dr. Frank Zeller began the gonadotropin section of a reproductive biology course at Indiana University with the statement, "Knowledge of these [hormones] advanced along with advances in protein chemistry. The first breakthrough was the demonstration that gonadotropins were composed of two subunits." In the present chapter, we summarize increased understanding of these important hormones thanks to advances in several scientific fields, including continued progress in protein chemistry as well as in carbohydrate chemistry, molecular biology, genomics, and bioinformatics. Emphasis is given the human gonadotropins, but information from other mammalian and lower vertebrate gonadotropins is freely used to develop our comparative understanding of these important hormones and their genetic control. Several extensive reviews have been used for this study, as are acknowledged in the text. We particularly note our own earlier review (1) and previous chapters on this subject in the first and second editions of the present volume (2,3).

In 1994, when the second edition of this work appeared, two landmark studies reported the crystal structures of chemically deglycosylated urinary human chorionic gonadotropin (hCG) and deglycosylated recombinant selenomethionyl-hCG (4,5). One consequence of these reports was to place the glycoprotein hormones in a broader biological context. The cystine knot motif found in both subunits identified the glycoprotein hormones as the fourth family in the cystine knot growth factor (CKGF) superfamily. Other families, including, the nerve growth factor

Department of Biological Sciences, Wichita State University, Wichita, Kansas.
[1]Retired.

(NGF) and platelet-derived growth factor (PDGF) families, as well as the transforming growth factor (TGF)-β superfamily, comprise this superfamily. The latter includes the reproductively relevant hormones, inhibin, activin, several bone morphogenic proteins, and müllerian inhibiting hormone. Although distinguishing features of the various families have been reviewed (6), we begin this chapter by pointing out some of the features that make the glycoprotein hormones unusual members of this superfamily.

STRUCTURAL FEATURES OF THE GONADOTROPINS

Cystine Knot Proteins

The glycoprotein hormone family includes the pituitary hormones, follicle-stimulating hormone (FSH; follitropin) and luteinizing hormone (LH; lutropin), of direct interest to reproductive physiology, and thyroid-stimulating hormone (TSH; thyrotropin), as well as the chorionic gonadotropins (also of interest for reproductive physiology, but whose presence has been demonstrated only in primates and equids). Figure 1 conveys some of the features of the glycoprotein hormone subunit protein organization compared with other members of the CKGF superfamily. Subunits found in the other three families, NGF, PDGFA, and TGF-β, generally possess an N-terminal proprotein domain, as well as the functional C-terminal Cys knot domain. In this regard, they resemble the mucins, which also possess a C-terminal Cys knot domain, but have a very much larger N-terminal domain that remains attached to the Cys knot domain (Fig. 1, *inset*). The pattern is not absolute because PDGF-B possesses a second prodomain located at the C-terminal end of the Cys knot domain (7). The CKGF prodomain is known to keep TGF-β Cys knot domains in an inactive state until proteases denature the prodomain, releasing the functional Cys knot domain dimer (8). In contrast, because glycoprotein hormone subunits consist of just the signal peptide and Cys knot domain, the functional form of the hormone is secreted by gonadotropes. Although all CKGFs function as dimers, the glycoprotein hormones are exclusively heterodimeric glycoproteins (Fig. 2). The other CKGF families consist of both homodimers and heterodimers. Although glycosylation of the Cys knot domain is relatively rare in the CKGF superfamily as a whole, and typically is restricted to the N-terminal prodomain, all glycoprotein hormone subunits are glycosylated. The mechanism for gonadotropin subunit association involves another unique feature of

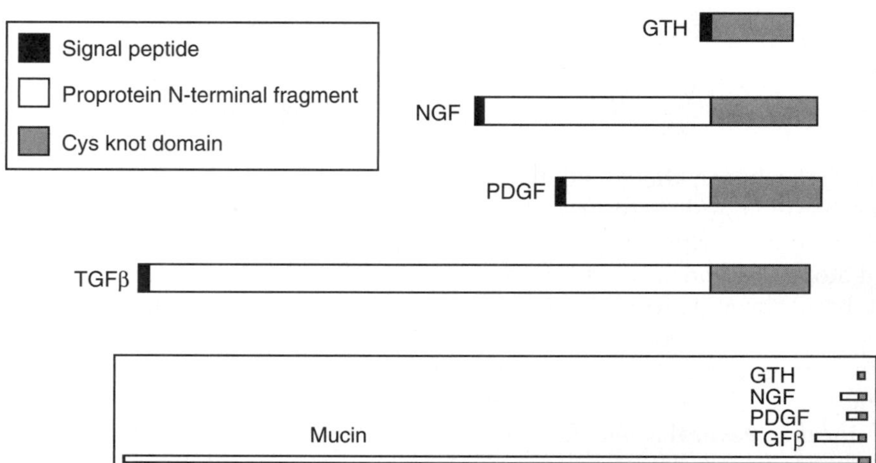

FIG. 1. Cys knot protein families: protein subunit organization for each of the four Cys knot growth factor families. Each precursor growth factor subunit is drawn to scale. The N-terminal prodomain found in most other Cys knot growth factor superfamily members is absent from gonadotropin subunits. Variations on this pattern exist, such as platelet-derived growth factor (PDGF) β subunit, which possesses a C-terminal prodomain that is removed along with the N-terminal prodomain. *Inset:* Cys knot growth factor subunits drawn on the same scale as the much larger mucin preprotein, which possesses a C-terminal Cys knot domain. The nerve growth factor (NGF), PDGFA, and transforming growth factor (TGF)-β proprotein structure is a greatly reduced version of the mucin organization, in which the Cys knot domain is located at the C-terminal end of a very large molecule.

these hormones. Instead of one or two intermolecular disulfide bonds, as in the case of TGF-β and PDGF family members, respectively, an additional disulfide-stabilized loop, designated the "seatbelt loop" (4) or "cystine noose" (9), embraces the α-subunit long loop, αL2 (Fig. 3). The hCGβ, hFSHβ, and hLHβ seatbelt loops carry essential elements of the LH receptor and FSH receptor recognition domains (10). NGF subunits are also associated noncovalently; however, their orientation is head-to-head, rather than head-to-tail (Fig. 2B), and the mechanism appears to be a back-to-back interaction rather than front-to-back observed with glycoprotein hormones. In fact, subunit interacting regions of the four families do not significantly overlap (Fig. 2C). Receptors provide an additional defining factor for each family: the NGF family bind Cys-TK class V and tumor necrosis factor receptor class p75 and p55, the PDGF family bind Ig-TK receptors, and the TGF-β superfamily bind type I, II, and III serine/threonine kinase receptors. Glycoprotein hormones bind G-protein–coupled receptors that possess a large extracellular domain, which functions as the high-affinity binding site (11–14).

Three-Dimensional Structure

In addition to revealing that both subunits shared a Cys knot motif, the crystal structures for hCG provided correct disulfide bond placements and clarified the mechanism for heterodimer formation (4,5). Absent from both structures were distal portions of the N-linked oligosaccharides, which were removed by the deglycosylation procedure, as well as the C-terminal extension of hCG along with its four O-linked oligosaccharides (Fig. 3). A subsequent report of the structure for intact hCG cocrystallized with monoclonal antibody fragments directed against the α and β subunits also lacked these features (15), indicating heterogeneity and flexibility rather than acidic cleavage (16) were responsible for the lack of electron density for these features. It was suggested that the other hormones would have essentially the same fold, and this was confirmed by the crystal structure for recombinant insect cell–expressed hFSH (17). Significant conformational differences between hCG and hFSH occurred at one end of the molecule involving two β-subunit loops, βL1 and βL3, the seatbelt loop, and one α-subunit loop, αL2 (Fig. 3). Distal monosaccharides in the intact recombinant hFSH oligosaccharides were also missing, providing additional supporting evidence for oligosaccharide flexibility.

Jirgensons and Ward (18) proposed that the β subunit provided a more rigid framework to organize the conformation of the more flexible α subunit. This was partly borne out in the comparison of the hCG and hFSH structures, because the root mean squared deviations (r.m.s.d.) between hCGβ and hFSHβ were greater (1.5 Å and 1.6 Å for FSH1 and FSH2, respectively) than those between hCGα and hFSHα (1.1 Å and 0.9 Å). Two hFSH forms were observed in the crystal structure, which defined some of the conformational flexibility that can occur in these molecules (17). The r.m.s.d. between hFSH1 and hFSH2 were overall quite similar (0.9 Å), although the r.m.s.d. increased to 1.6 Å when the α subunits were superimposed, suggesting a difference in the angle between the α and β subunits in the two forms. Other differences were conformational changes at the ends of the loops. The FSH2 αL2 loop was less helical than in FSH1, resulting in a 2.4-Å displacement of residue 48. Other differences between the two forms occurred at βL2 and αL3. The former indicated flexibility of βL2, whereas the latter moved as a consequence of being disulfide bonded to βL1. Conformational differences between the two hFSH forms occurred at the opposite end of the molecule from where the differences between hFSH and hCG were noted.

Most differences between hCG and hFSH occurred at βL1, βL2, βL3, and the seatbelt loop (Fig. 3). Although βL2 and the seatbelt loop are known to be involved in LH receptor recognition, biochemical data implicating βL1 and βL3 in receptor recognition have not been reported. The various hCG structures report some variations in αL2 helical content, as if there is some flexibility in this part of the molecule. Indeed, we attempted to use changes in helical content detected by circular dichroism as a probe for the influence of carbohydrate structure on eLH and eFSH conformation (19). Helical or β structure content varied among the hybrids, but no clear-cut association could be made between conformational changes and hormone activity.

The α or Common Subunit

The glycoprotein hormones are now known to comprise two dissimilar subunits, although the first indication of noncovalently linked subunits was not obtained until the studies of Li and Starman (20), who showed that acid dissociation of ovine LH reduced the molecular weight by approximately one half. Our laboratory first proposed the dissimilar nature of the glycoprotein hormone subunits (21) as a result of our first structural studies. The abstract of this report concluded, "From these observations we propose a model of LH composed of two different chains of similar molecular dimensions." This conclusion was largely borne out by the crystal structures reported for hCG because the portion of that molecule that could be discerned by x-ray crystallography approximated

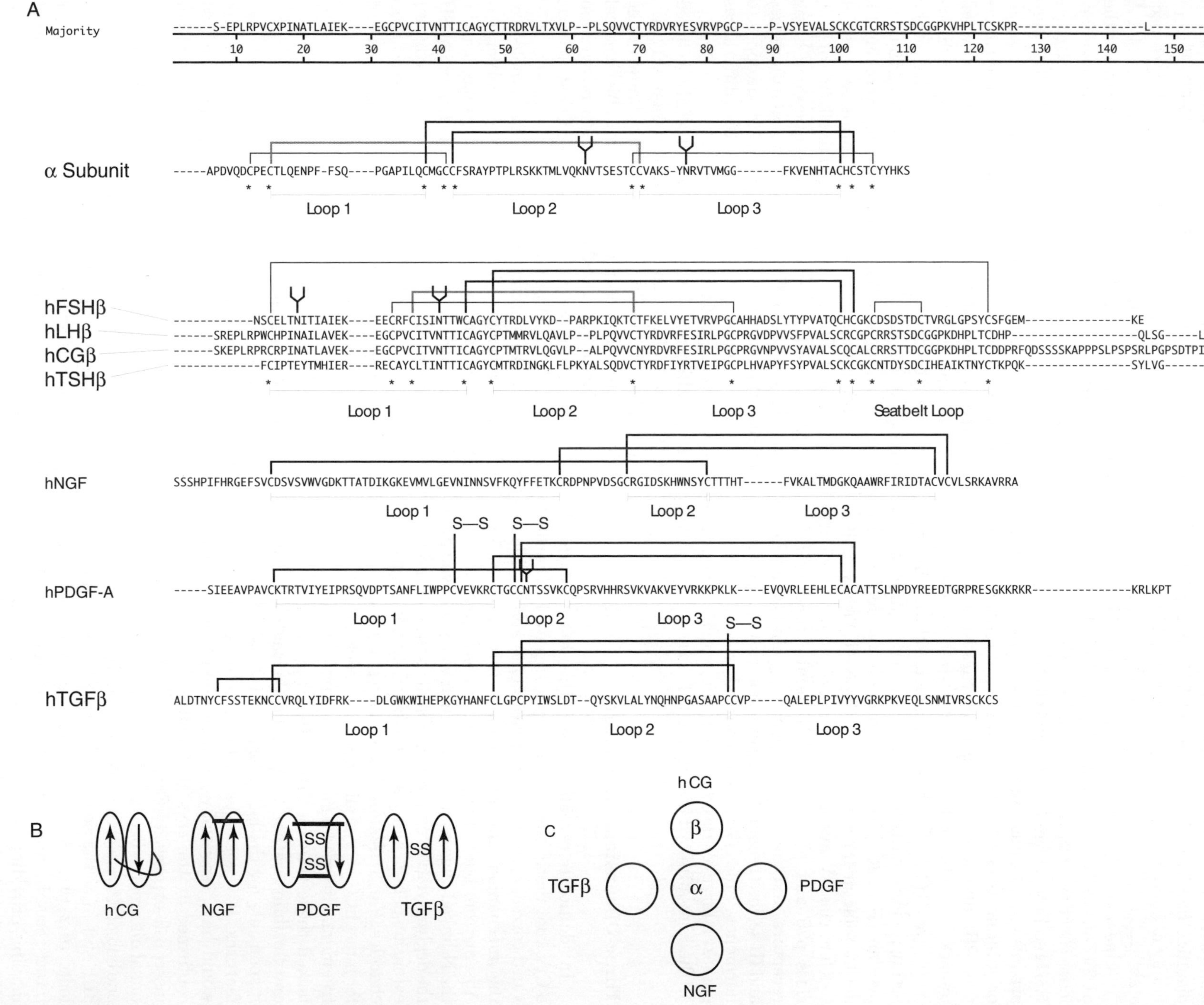

A
Majority
------S-EPLRPVCXPINATLAIEK----EGCPVCITVNTTICAGYCTTRDRVLTXVLP--PLSQVVCTYRDVRYESVRVPGCP----P-VSYEVALSCKCGTCRRSTSDCGGPKVHPLTCSKPR-----------------L----------
10 20 30 40 50 60 70 80 90 100 110 120 130 140 150
α Subunit
-----APDVQDCPECTLQENPF-FSQ----PGAPILQCMGCCFSRAYPTPLRSKKTMLVQKNVTSESTCCVAKS-YNRVTVMGG-------FKVENHTACHCSTCYYHKS
Loop 1
Loop 2
Loop 3
hFSHβ
hLHβ
hCGβ
hTSHβ
------------NSCELTNITIAIEK----EECRFCISINTTWCAGYCYTRDLVYKD--PARPKIQKTCTFKELVYETVRVPGCAHHADSLYTYPVATQCHCGKCDSDSTDCTVRGLGPSYCSFGEM----------------KE
------SREPLRPWCHPINAILAVEK----EGCPVCITVNTTICAGYCPTMMRVLQAVLP--PLPQVVCTYRDVRFESIRLPGCPRGVDPVVSFPVALSCRCGPCRRSTSDCGGPKDHPLTCDHP------------------QLSG-----LLFL
------SKEPLRPRCRPINATLAVEK----EGCPVCITVNTTICAGYCPTMTRVLQGVLP--ALPQVVCNYRDVRFESIRLPGCPRGVNPVVSYAVALSCQCALCRRSTTDCGGPKDHPLTCDDPRFQDSSSSKAPPPSLPSPSRLPGPSDTPILPQ
------------FCIPTEYTMHIER----RECAYCLTINTTICAGYCMTRDINGKLFLPKYALSQDVCTYRDFIYRTVEIPGCPLHVAPYFSYPVALSCKCGKCNTDYSDCIHEAIKTNYCTKPQK----------------SYLVG------FSV
Loop 1
Loop 2
Loop 3
Seatbelt Loop
hNGF
SSSHPIFHRGEFSVCDSVSVWVGDKTTATDIKGKEVMVLGEVNINNSVFKQYFFETKCRDPNPVDSGCRGIDSKHWNSYCTTTHT------FVKALTMDGKQAAWRFIRIDTACVCVLSRKAVRRA
Loop 1
Loop 2
Loop 3
S—S
S—S
hPDGF-A
-----SIEEAVPAVCKTRTVIYEIPRSQVDPTSANFLIWPPCVEVKRCTGCCNTSSVKCQPSRVHHRSVKVAKVEYVRKKPKLK----EVQVRLEEHLECACATTSLNPDYREEDTGRPRESGKKRKR--------------KRLKPT
Loop 1
Loop 2
Loop 3
S—S
hTGFβ
ALDTNYCFSSTEKNCCVRQLYIDFRK----DLGWKWIHEPKGYHANFCLGPCPYIWSLDT--QYSKVLALYNQHNPGASAAPCCVP-----QALEPLPIVYYVGRKPKVEQLSNMIVRSCKCS
Loop 1
Loop 2
Loop 3
B
SS
SS
SS
hCG
NGF
PDGF
TGFβ
C
hCG
β
α
TGFβ
PDGF
NGF

FIG. 2. Comparison of Cys knot growth factor disulfide bond arrangements and subunit relationships. **A:** Disulfide bonds in glycoprotein hormone subunits and representative Cys knot growth factors. The human glycoprotein hormone α subunit and human chorionic gonadotropin (hCG) β, luteinizing hormone (hLH) β, follicle-stimulating hormone (hFSH) β, and thyroid-stimulating hormone (hTSH) β sequences were aligned along with the Cys knot domains for human nerve growth factor (hNGF), platelet-derived growth factor A (hPDGFA), and transforming growth factor (hTGF)-β sequences using the Clustal W algorithm as implemented in the MegAlign v5.53 module of the DNA Star software suite (DNASTAR, Inc., Madison, WI). The disulfide bonds are shown as found in the crystal structures for these proteins. The three loops created by the Cys knot are indicated by thin lines below the sequences. Note that whereas the α subunit has five disulfide bonds, and the β subunit six, the NGF and PDGF families have only the three Cys knot disulfide bonds. Although PDGF possesses two additional Cys residues, these are involved in intermolecular disulfide bonds that stabilize the dimeric form of the growth factor. TGF-β has four intramolecular disulfide bonds and a single Cys residue that forms a single intermolecular disulfide in the dimer. Another difference is the relatively short loop 2 in the NGF and PDGF families compared with TGF-β and the glycoprotein hormone subunits. **B:** The orientation of the subunits with respect to each other. Parallel and antiparallel subunit orientation indicated by *arrows*. Subunit association through disulfides indicated by "SS." Interconnecting peptides and loops are indicated by *lines*. **C:** Subunits aligned to hCGα Cys knot motif show the absence of overlap in the complementary subunit contact regions [adapted from Wu et al. (5)]. The amino acid single-letter code has been used in this compilation: A, alanine; B, either asparagine or aspartic acid, but not determined which; C, cysteine or half-cystine; D, aspartic acid; E, glutamic acid; F, phenylalanine; G, glycine; H, histidine; I, isoleucine; K, lysine; M, methionine; N, asparagine; P, proline; Q, glutamine; R, arginine; S, serine; T, threonine; V, valine; W, tryptophan; Y, tyrosine; Z, either glutamine or glutamic acid, but not determined which.

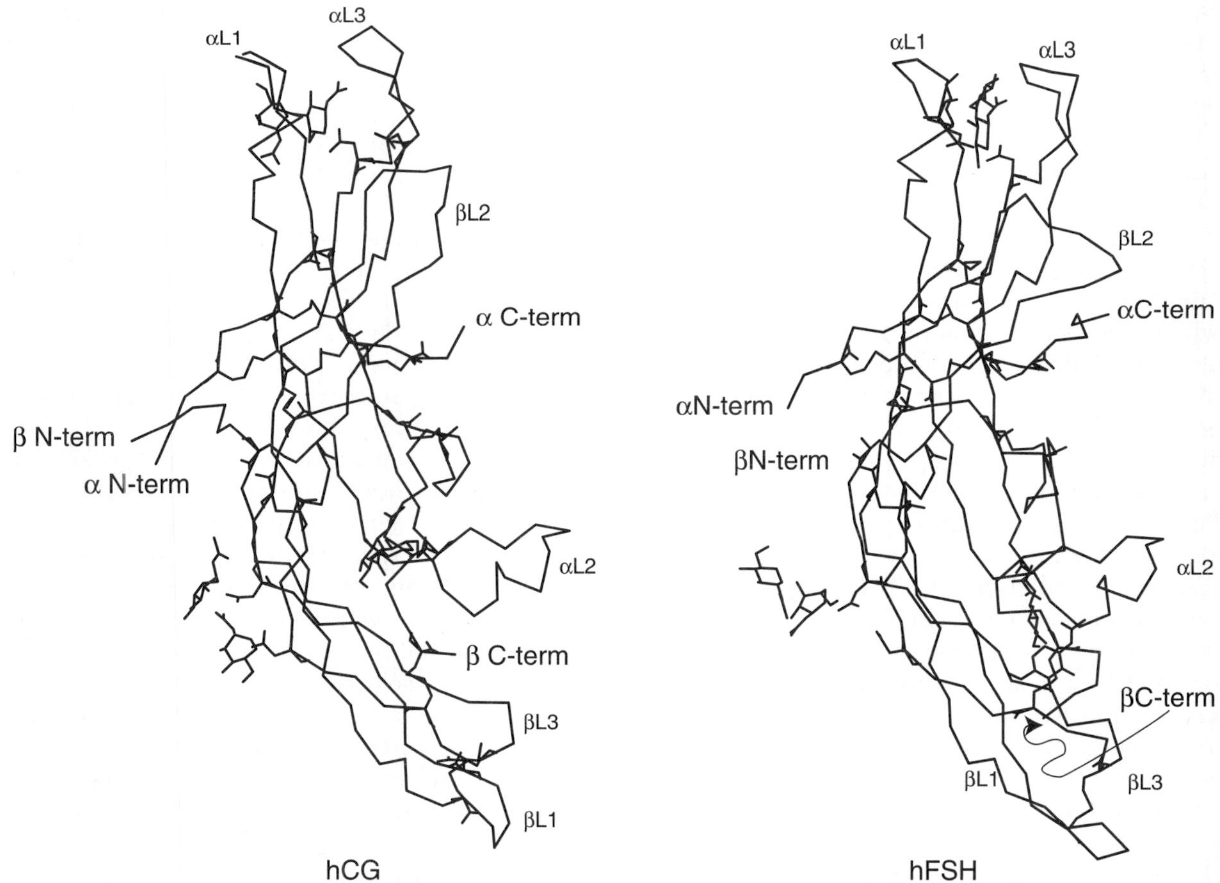

FIG. 3. Crystal structures for human chorionic gonadotropin (hCG) and follicle-stimulating hormone (hFSH). The backbones of each subunit are represented by *solid lines*. Cys residues are represented as *wire diagrams* to illustrate disulfide bonds. Oligosaccharide fragments are also shown as *wire diagrams*. The subunits are oriented in an antiparallel manner and the seatbelt loop of the β subunit is wrapped around α-subunit loop 2 to stabilize the heterodimer. Coordinate files 1HRP, and 1FL7, respectively, were obtained from the Protein Data Bank (298) and visualized with the software package iMol (www.pirx.com/iMol/) using a Macintosh PowerBook G4 computer.

the size of oLH. Thus, hCGα measured 60 × 25 × 15 Å, very similar to the 65 × 25 × 20 Å dimensions reported for the β subunit (5).

The next important information on the subunit story came from Papkoff and Samy (22), who provided a two-phase system for separation of the dissimilar subunits by countercurrent distribution. They obtained subunits with activity on the order of only 7% to 8% of the original oLH, whereas recombination of the separated subunits recovered approximately 20% of the original activity. These figures demonstrated important trends, although as this and other systems of separation were subsequently improved, the intrinsic potency of the separated subunits has approached zero activity and the activity of the recombined subunits has approximated 100% of the original hormone potency. Other important observations from that study were the differences in the amino acid composition for the two subunits. About that same time, DeLaLlosa and Jutisz (23) were studying urea or guanidine hydrochloride dissociation of oLH and found that the subunits could be dissociated by these chaotropic agents with almost complete loss of activity. The activity was largely recovered on reassociation of the subunit dimer (e.g., over 98% in one experiment).

Our laboratory followed with an additional subunit separation procedure (24), which was useful in our early structure studies. This allowed a comparison of the subunits obtained by this procedure or the Papkoff-Samy procedure (22), as well as C- and N-terminal amino acid studies on the subunits. Meanwhile, others were finding that other glycoprotein hormones also were composed of dissimilar subunits. Reports on hCG were provided by two laboratories (25,26). The study of bTSH by countercurrent distribution after propionic acid dissociation also demonstrated the subunit nature of this hormone (27). This report by Liao and Pierce was a milestone in our understanding of the glycoprotein hormones for it also demonstrated that the TSHα subunit of Liao and Pierce was nearly identical with the S-subunit of Ward (24) or the CI-subunit of Papkoff and Samy (22) for oLH. For this conclusion, the amino acid sequence around the carbohydrate moieties of oLH (28) and those of TSHα found by Liao and Pierce were indicative. Most important, however, was the demonstration that the CI or S-subunit of LH could recombine with the TSHβ to produce an active form of TSH just as well as could the TSHα. This discovery established the "common subunit" model for glycoprotein hormones that has been so useful in our understanding of the chemistry and physiology of these hormones. In fact, much of the balance of this chapter is devoted to enlarging this point. With this information available, Pierce and coworkers, with concurrence of others in the field, proposed the designation of α and β subunits as the common subunit and hormone-specific subunits, respectively (29). This nomenclature for the glycoprotein hormones has been maintained since that time.

The initial sequence studies on the glycoprotein hormones prompted the examination of several other species. However, the early methodology for protein sequence study necessitated production of considerable quantities of the hormone, and thus the species of pituitary readily available was an early limitation to these studies. The early pituitary fractionation studies have been reviewed (30). Advances in DNA sequencing have produced an abundance of protein sequence information deduced initially from complementary DNA (cDNA) sequences, but increasingly from genomic sequencing. Although the previous edition of this chapter provided a list of references, the increasing use of databases to access sequences prompted us to tabulate gonadotropin accession numbers (Table 1). This highlights the inconsistent coverage of vertebrate glycoprotein hormones because mammals and fish are the most highly represented by 35 and 49 species, respectively, whereas birds, reptiles, and amphibians are represented by a handful of species. Moreover, of the 80 species represented by α-subunit sequences, only 14 species have all 3 pituitary glycoprotein hormone β-subunit sequences characterized.

Representative α-subunit sequences are shown in Fig. 4A. Because the 80 sequences are simply too many to illustrate conveniently, the sequences shown are drawn from model organisms representing the various vertebrate orders. Thus, humans represent the mammalian gonadotropins because these hormones are clinically important and because humans possess all four glycoprotein hormones. Likewise, the zebrafish *Danio rerio* and the frog *Xenopus laevis* are important model organisms representing fish and amphibians. From the limited reptilian and avian sequences available, chicken and turtle were used. The sequences were aligned using the Clustal W algorithm (31). Because of the relatively uniform size of the leader sequences and mature sequences, this alignment was as satisfactory as the laborious manual alignments used in previous reviews (1,3,32). Only the mature sequences are shown because attention is now directed at the tertiary structures of these molecules. The highly conserved protein sequences from mammals down to teleosts are apparent from the homologous placement of the half-cystine residues in these molecules. This uniformity of the half-cystine locations implied a uniformity of the secondary structure generated by the formation of the disulfide bonds between these residues that was confirmed by x-ray crystallography, whereas the conflicting

TABLE 1. *Glycoprotein hormone subunit sequences accession numbers*

Species	α Subunit	LHβ	FSHβ	TSHβ	CGβ
Fish					
Acanthopagrus latus (yellowfin sea bream)	gi|399546| sp|P30970|	gi|11133051| sp|Q90225|			
Acipenser baerii (Siberian sturgeon)	gi|14589313| emb|CAC43060.1|	gi|8250128| emb|CAB93502.1|	gi8250132 embCAB93504.1	gi|8250134| emb|CAB93505.1|	
Acipenser gueldenstaedtii (Russian *sturgeon)*		gi|33151281| gb|AAP97490.1|			
Acipenser schrenckii (Amur sturgeon)		gi|46399189| gb|AAS92239.1|	gi46399187 gbAAS92238.1		
Anguilla japonica (Japanese eel)	gi|46093514| dbj|BAD14301.1|	gi|20384912| gb|AAL93618.1|	gi|11133227| sp|Q9YGK3|	gi|30313578| gb|AAO17791.1|	
Anguilla anguilla (European eel)	gi|121309| sp|P27794|		gi|25528803| gb|AAN73407.1|	gi|431913| emb|CAA51908.1|	
Aristichthys nobilis (bighead carp)				gi|11359153| gb|AAD51753.2|	
Carassius auratus (goldfish)	gi|1469836| dbj|BAA13111.1| gi|1469838| dbj|BAA13112.1|	gi|1644243| dbj|BAA13531.1|	FSH1gi|1644241| dbj|BAA13530.1| FSH2gi|4140033| dbj|BAA36975.1|	gi|2114098| dbj|BAA20081.1|	
Channa maculata (snakehead mullet)	gi|32745117| gb|AAP87114.1|	gi|41387512| gb|AAS01609.1|	gi|41387514| gb|AAS01610.1|		
Clarias gariepinus (African catfish)	gi|1707947| sp|P53542|	gi|1321782| emb|CAA66359.1|	gi|28625515| gb|AAO49013.1|		
Clupea harengus (Atlantic herring)		gi|4200297| emb|CAA63038.1|			
Conger conger (conger eel)			gi|8250136| emb|CAB93518.1|		
Conger myriaster (Japanese conger eel)		gi|21322748| dbj|BAB97391.1|	gi|21322746| dbj|BAB97390.1|		
Coregonus autumnalis (Arctic cisco)		gi|1346202| sp|P48251|	gi|858749| gb|AAA68208.1|		
Ctenopharyngodon idella (grass carp)	gi|399547| sp|P30983|				
Cyprinus carpio (common carp)	gi|121303| sp|P01221| gi|121306| sp|P18857| GLH2_CYPCA2	gi|62622| emb|CAA42543.1|	gi|2114096| dbj|BAA20080.1|	gi|2114100| dbj|BAA20082.1|	
Danio rerio (zebrafish)	gi|40317182| gb|AAR84285.1|	gi|49170064| ref|NP_991186.1|	gi|45387671| ref|NP_991187.1|	gi|28848612| gb|AAN08914.1|	
Dicentrarchus labrax (sea bass)	gi|13925744| gb|AAK49431.1| AF269157_1	gi|23955477| gb|AAN40507.1|	gi|23955475| gb|AAN40506.1|		
Epinephelus coioides (orange-spotted grouper)	gi|23380193| gb|AAN18038.1|	gi|20805920| gb|AAM28896.1| AF507939_1|			
Epinephelus akaara (Hong Kong grouper)	gi|28566196| gb|AAO43056.1|				
Epinephelus septemfasciatus (seven-band grouper)	gi|32400651| dbj|BAC78811.1|	gi|32400655| dbj|BAC78813.1|			
Fundulus heteroclitus (killifish)	gi|1346138| sp|P47744|	gi|382579| prf||1819439A|			
Gasterosteus aculeatus (three-spined *stickleback)*		gi|26986055| emb|CAD59185.1|			
Hippoglossus hippoglossus (Atlantic halibut)	gi|16604714| emb|CAD10503.1|	gi|16604712| emb|CAD10502.1|	gi|16604710| emb|CAD10501.1|		
Hypophthalmichthys molitrix (silver carp)	gi|544390| sp|P37037|	gi|544444| sp|P37038|			
		gi|11177152|	gi|11177150|gb|		

(Continued)

TABLE 1. *Glycoprotein hormone subunit sequences accession numbers—cont'd*

Species	α Subunit	LHβ	FSHβ	TSHβ	CGβ
Ictalurus punctatus (channel catfish)	gi|11132583| sp|Q9YGP3|	gb|AAG32156.1| AF112192_1|	AAG32155.1| AF112191_1|		
Katsuwonus plelamis (bonito)	gi|298264| gb|AAB25414.1|				
Monopterus albus (swamp eel)	gi|25992130| gb|AAN77069.1| gi|3913744| sp|Q91119|				
Morone saxatilis (striped sea bass)	gi|2147428| pir||I50992|	gi|598255| gb|AAC38019.1|	gi|2267222| gb|AAC38035.1|		
Muraenesox cinereus (daggertooth pike conger)	gi|121315| sp|P12836|				
Mylopharyngodon piceus (black carp)		gi|12746254| gb|AAK07414.1|	gi|12746256| gb|AAK07415.1|		
Neoceratodus forsteri (Australian lungfish)	gi|10934062| dbj|BAB16881.1|	gi|40644819| emb|CAE17335.1|	gi|40644823| emb|CAE17337.1|	gi|40644821| emb|CAE17336.1|	
Odontesthes bonariensis			gi|32493150| gb|AAP85606.1|		
Oncorhynchus keta (chum salmon)	gi|121307| sp|P13153| gi|121304| sp|P13152|	gi|121734| sp|P10256|	gi|121731| sp|P10257|		
Oncorhynchus kisutch (Coho salmon)	gi|29289969| gb|AAO72301.1|1 gi|29289971| gb|AAO72302.1|2				
Oncorhynchus masou (cherry salmon)	gi|2133952| pir||I512291 gi|2133953| pir||I512302	gi|546264| gb|AAB30424.1|	gi|546262| gb|AAB30423.1|		
Oncorhynchus mykiss (rainbow trout/summer salmon)	gi|11127607| dbj|BAB17685.1|	gi|11127611| dbj|BAB17687.1|	gi|11127609| dbj|BAB17686.1|	gi|586132| sp|P37240|	
Ophisternon bengalense	gi|25992128| gb|AAN77068.1|				
Oreochromis mossambicus (Mozambique tilapia)	gi|10946373| gb|AAG24881.1|				
Oreochromis niloticus (Nile tilapia)		gi|31559053| gb|AAP49576.1|			
Pagrus major (red sea bream)	gi|11414885| dbj|BAB18562.1|	gi|11414889| dbj|BAB18564.1|	gi|11414887| dbj|BAB18563.1|		
Paralichthys olivaceus (Japanese flounder)	gi|14279439| gb|AAK58600.1| AF268692_1	gi|14279443| gb|AAK58602.1|			
Plecoglossus altivelis (ayu)		gi|22203360| gb|AAM92270.1|	gi|22203358| gb|AAM92269.1|		
Salmo salar (Atlantic salmon)				gi|3091282| gb|AAC77908.1|	
Scyliorhinus canicula (smallspotted catshark)	gi|14589315| emb|CAC43234.1	gi|14589319| emb|CAC43236.1|			
Thunnus obesus (bigeye tuna)	gi|585199| sp|P37204|				
Trichogaster trichopterus (blue *gourami)*		gi|5805301| gb|AAD51935.1|	gi|5805299| gb|AAD51934.1|		
Amphibians					
Bufo japonicus (Japanese toad)		gi|21202821| *dbj|BAB93556.1|*	gi|21202825| *dbj|BAB93558.1|*	gi|21202831| *dbj|BAB93561.1|*	
Cynops pyrrhogaster (Japanese firebelly newt)		gi|20975249| dbj|BAB92959.1|	gi|20975247| dbj|BAB92958.1|		
Rana catesbeiana (bullfrog)	gi|121319| sp|P80051|				

TABLE 1. *Glycoprotein hormone subunit sequences accession numbers*—cont'd

Species	α Subunit	LHβ	FSHβ	TSHβ	CGβ
Rana japonica (Japanese brown frog)	gi|46575869| dbj|BAD16758.1|	gi|46575865| dbj|BAD16756.1|	gi|46575867| dbj|BAD16757.1|		
Rana ridbunda (lake frog)		gi|14250924| emb|CAC39252.1|			
Xenopus laevis (African clawed frog)	gi|47937801| gb|AAH72372.1| MGC84501	gi|13936900| gb|AAK49986.1|	gi|46093518| dbj|BAD14295.1|		
Reptiles					
Chinemys reevesii (Reeve's turtle)	gi|20975237| dbj|BAB92946.1|	gi|20975243| dbj|BAB92949.1|	gi|20975241| dbj|BAB92948.1|		
Pelodiscus sinensis (Chinese softshell turtle)				gi|50058984| gb|AAT69236.1|	
Sphenedon punctatus	gi|20975235| dbj|BAB92945.1|				
Takydromus tachydromoides (lizard)	gi|20975239| dbj|BAB92947.1|				
Birds					
Coturnix japonica (Japanese quail)	gi|546920| gb|AAB30866.1|	gi|546922| gb|AAB30867.1|	gi|21624380| dbj|BAC01164.1|		
Gallus gallus (chicken)		gi|482732| pir||A61091|	gi|20384652| gb|AAK31580.1|	gi|45384224| ref|NP_990394.1|	
Meleagris gallopavo (turkey)	gi|544391| sp|P37035|	gi|2134444| pir||I51373|			
Nipponia nippon (crested ibis)	gi|22164058| dbj|BAC07315.1|		gi|22164056| dbj|BAC07314.1|	gi|22164054| dbj|BAC07313.1|	
Struthio camelus (ostrich)	gi|1707948| sp|P80665|				
Mammals					
Ailuropoda melanoleuca (giant panda)	gi|17646742| gb|AAL41020.1| AF448453_1	gi|17646746| gb|AAL41022.1| AF448455_1	gi|17646744| gb|AAL41021.1| AF448454_1		
Ailurus fulgens (red panda)	gi|23451976| gb|AAN32897.1| AF488737_1	gi|23451979| gb|AAN32898.1| AF488738_1			
Balaenoptera acutorostrata (Minke whale)	gi|544389| sp|P37036|				
Bos taurus (cow)	gi|27806913| ref|NP_776326.1|	gi|126475| sp|P04651|	gi|120550| sp|P04837|	gi|136442| sp|P01223|	
Bubalus bubalis (water buffalo)	gi|11182274| emb|CAC16185.1|		gi|38193970| gb|AAR13163.1|		
Callithrix jacchus (white-tufted-ear marmoset)	gi|1707946| sp|P51499|				gi|606607| gb|AAC00029.1|
Camelus dromedarius (camel)					
Canis familiaris (dog)	gi|50950217| ref|NP_001002988.1|			gi|50979102| ref|NP_001003290.1|	
Capra hircus (domestic goat)	gi|18390095| gb|AAL68841.1| AF464001_1		gi|26891799| gb|AAN84783.1|		
Cavia porcellus (domestic guinea pig)	gi|14194760| sp|Q9JK68|	gi|34576985| gb|AAQ75732.1|	gi|7739708| gb|AAF68975.1| AF257212_1		
Ceratotherium simum (white rhinoceros)		gi|3617825| gb|AAC36049.1|			
Cervus nippon (sika deer)	gi|18086541| gb|AAL57755.1|		gi|26891797| gb|AAN84782.1|		
Equus asinus (ass/donkey)	gi|2494807| sp|Q28365|	gi|1129135| emb|CAA56422.1|			
Equus burchellii (Burchell's zebra)	gi|9910681| sp|O46642|	gi|2808686| emb|CAA76146.1|			
Equus caballus (horse)	gi|2851612| sp|P01220|	gi|252741| gb|AAB22775.1|	gi|12644092| sp|P01226|	gi|1262916| gb|AAA96826.1|	

(Continued)

TABLE 1. *Glycoprotein hormone subunit sequences accession numbers*—cont'd

Species	α Subunit	LHβ	FSHβ	TSHβ	CGβ
Felis catus (cat)		gi|3747119| *gb|AAC64196.1|*			
Homo sapiens (human)	gi|4502787| ref|NP_000726.1| gi|15012103| gb|AAH10957.1|	gi|4504989| ref|NP_000885.1|	gi|4503791| ref|NP_000501.1|	gi|7690113| gb|AAB30828.2|	
Lama glama (llama)				gi|1872550| gb|AAB49315.1|	
Macaca fascicularis (long-tailed macaque)	gi|13027712| gb|AAK08642.1|	gi|50079994| emb|CAH03730.1|	gi|50079992| emb|CAH03729.1|		gi|13027716| gb|AAK08644.1| gi|13027714| gb|AAK08643.1|
Macaca mulatta (rhesus monkey)	gi|121313| sp|P22762|				
Macropus rufus (red kangaroo)	gi|3924633| sp|O46687|				
Mastomys coucha (southern multimammate mouse)	gi|14194756| sp|Q9ERG4|	gi|33868591| gb|AAQ55237.1|	gi|38490000| gb|AAR21602.1|		
Meriones unguiculatus (Mongolian gerbil)	gi|14194758| sp|Q9ERJ6|	gi|34555681| gb|AAQ74976.1|	gi|34978847| gb|AAQ83633.1|		
Mesocricetus auratus (golden hamster)	gi|14194757| sp|Q9ERG5|	gi|33868593| gb|AAQ55238.1|			
Microtus montebelli (Japanese grass vole)	gi|14194755| sp|Q9ERG3|				
Monodelphis domestica (gray short-tailed opossum)	gi|15777853| gb|AAL05939.1|	gi|16186292| gb|AAL13337.1|	gi|15217164| gb|AAK92541.1| AF406610_1|	gi|15777851| gb|AAL05938.1|	
Mus musculus (house mouse)	gi|6753414| ref|NP_034019.1|	gi|930345| gb|AAA92841.1|	gi|6679865| ref|NP_032071.1|	gi|6678443| ref|NP_033458.1|	
Oryctolagus cuniculus (rabbit)	gi|12744485| gb|AAK06653.1| AF318299_1	gi|47680441| gb|AAT37165.1|	gi|47680443| gb|AAT37166.1|		
Ovis aries (sheep)	gi|229457| prf||731693A| gi|121320| sp|P01218|	gi|1170836| sp|P01231|	gi|120555| sp|P01227|		
Panthera tigris altaica (Amur tiger)	gi|13561974| gb|AAK30590.1| AF354939_1	gi|23344109| gb|AAN28376.1|	gi|23344120| gb|AAN28381.1|		
Papio anubis (olive baboon)					gi|69183| pir|KTBAB|
Phodopus sungorus (Russian dwarf hamster)	gi|6539650| gb|AAF15967.1| AF106916_1	gi|6539648| gb|AAF15966.1|	gi|11132432| sp|Q9QYB0|		
Physeter catodon (sperm whale)	gi|121316| sp|P25329|				
Rattus norvegicus (rat)	gi|16758786| ref|NP_446370.1| gi|206111| gb|AAA97425.1|	gi|6981158| ref|NP_036990.1|	gi|34856638| ref|XP_342486.1|	gi|37231732| gb|AAH58488.1|	
Sus scrofa (pig)	gi|47523944| ref|NP_999611.1 gi|229562| prf||760571A|	gi|126480| sp|P01232|	gi|47523108| ref|NP_999040.1|	gi|136445| sp|P01224|	
Trichosurus vulpecula (silver-gray brushtail possum)	gi|3719213| gb|AAC63900.1|	gi|2738807| gb|AAC96019.1|	gi|6016054| sp|O46430|		

CG, chorionic gonadotropin; FSH, follicle-stimulating hormone; LH, luteinizing hormone; TSH, thyroid-stimulating hormone.

```
                           10         20         30         40         50         60         70         80         90
Homo sapiens      AP--D--VQDCPECTLQENPFFSQ--PGAPILQCMGCCFSRAYPTPLRSKKTMLVQKNVTSESTCCVAKSYNRVTVMGGFKVENHTACHCSTCYYHKS
Gallus gallus     FPDGEFLMQGCPECKLGENRFFSK--PGAPIYQCTGCCFSRAYPTPMRSKKTMLVPKNITSEATCCVAKAFTKITLKDNVKIENHTDCHCSTCYYHKS
Chinemys reevesii FPDGEFLTQGCPECKLGENRFFSK--PGAPIYQCTGCCFSRAYPTPMRSKKTMLVPKNITSEATCCVAKAFTKITLKDNVKIENHTDCHCSTCYYHKS
Xenopus laevis    FPEGDLMVQGCPECKLKENTYFTKRLGKALIFQCTGCCFSRAYPTPMRSKKTMLVPKNITSEATCCVAKASTRVTVIDNLKIENHTDCHCSTCYYHKS
Danio rerio       YSRNDVSNYGCEECKLKMNERFSK--PGAPVYQCVGCCFSRAYPTPLRSKKTMLVPKNITSEATCCVAKESK--MVATNIPLYNHTDCHCSTCYYHKS
                            *  *                  *   **                            **                        *  *  *
                      NTP              Loop 1                     Loop 2                     Loop 3               CTP
```

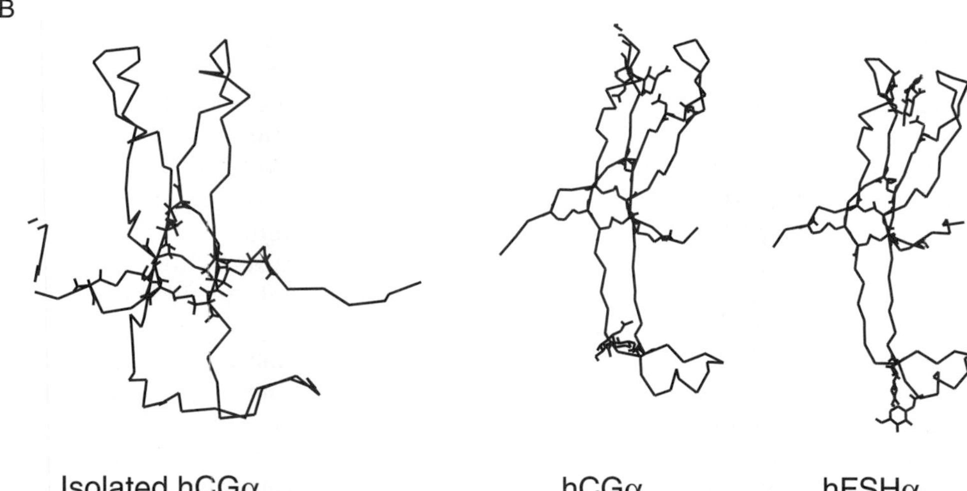

FIG. 4. Primary and tertiary α-subunit structures. **A:** Representative vertebrate α-subunit sequences showing mammalian, avian, reptilian, amphibian, and piscine. The species were chosen on the basis of clinical significance (*Homo sapiens*), model species (*Xenopus laevis* and *Danio rerio*), availability (*Chinemys reevsii*), or author bias (*Gallus gallus*). **B:** Three-dimensional structures showing the solution structure of free human chorionic gonadotropin (hCG) α as determined by nuclear magnetic resonance spectroscopy (1DZ7), the crystal structure for hCGα in its heterodimeric conformation (hCGβ not shown), and the crystal structure for human follicle-stimulating hormone (hFSH) α in its heterodimeric conformation (hFSHβ not shown). The backbone is represented by a *solid ribbon*. Cys residues are represented as *wire diagrams* to illustrate disulfide bonds. Oligosaccharide fragments are also shown as *wire diagrams*. Note that Asn^{52} oligosaccharide is directed outward from hCGα and downward, along the long axis in hFSHα.

disulfide placements in the literature were all found to include some incorrect pairings, usually those residues involved in forming the Cys knot. Figure 2A presents the disulfide linkages, which from the best available evidence appear to be the same in all glycoprotein hormones (6). Three disulfides constitute the Cys knot and create a pair of hairpin loops on one side of the knot, designated αL1 and αL3 (Fig. 4A). The intervening loop, αL2, is located on the opposite side of the knot. The other two disulfides constrain the juxta N-terminal and juxta C-terminal peptides to the molecule. Partial reduction and site-directed mutagenesis studies have indicated that these two disulfides are not absolutely essential for biological activity (33,34). There are notable departures from the exactly homologous location of these half-cystine residues. The first observed results from the gap of four residues at positions 4 to 7 in the human α sequence due to a deletion in the human gene, possibly resulting from a mutation of the intron–exon splice junction (35). This results in a foreshortened N-terminal tail on the human α subunit, before the first half-cystine, and does not alter the size of any of the disulfide loops (Fig. 4A), but it does shift the numbering for the human subunit. This is the only gap in the mammalian series. Thus, all of the disulfide loops in the mammalian α subunits are identical in size as determined by the amino acid residues in each loop. In the sequences from lower vertebrates, there are two common gaps in the sequence from fish represented by *D. rerio*. These are in locations that necessarily shorten disulfide loops αL1 and αL3, whereas in amphibians a one- or two-residue insertion lengthens αL1. The shortened loops may be important to the observation that the carp α subunit failed to produce an active dimeric hormone when associated with oLHβ (36). The first reported amphibian sequence, the bullfrog, had an arginine residue (R) inserted between position 28 and 29 compared with the other species of α subunits. This means that the bullfrog has one extra residue in αL1. This insertion was common to two of the three additional amphibian α-subunit sequences. However, the African clawed frog, *X. laevis*, possesses a leucine residue (L) as well as the extra R in this loop. These insertions occur at the end of the loop, where the conformational flexibility noted previously can more readily accommodate insertions and deletions. Likewise, the deletion in αL3 occurs at the end of the loop. These gaps and insertions alter the actual numbering of the α subunit, but for reasons of simplicity we have ignored this numbering difference in the figure presented. The reader should be aware that in the original reports the appropriate number shifts are used. Indeed, some of the early studies of the α subunit were confounded by an N-terminal heterogeneity involving the first seven residues that led to variances in the numbering and a slower realization of the very high homology of the α subunits. The work of Liu et al. (37) first clarified this for the mature protein forms of the α subunit. Confusion in numbering continues to this day because virtually all known mammalian α subunits possess 96 residues, whereas the clinically important 92-residue human α subunit lacks 4 N-terminal residues, as mentioned. Recombinant DNA results have made it clear that N-terminal heterogeneity is a consequence of the isolation procedures used because a single cleavage site is predicted in virtually all cases for signal peptidase (38), which is in agreement with the experimentally determined amino terminus. Thus, α-subunit N-terminal heterogeneity is likely a proteolysis artefact. Interestingly, this also tells us that for most of the common measures of biopotency (see discussion of assay systems, later), the first seven amino acid residues are not essential. Indeed, single-chain glycoprotein hormones created by expressing cDNAs encoding the β subunit tandemly linked in phase with α-subunit cDNA create functional heterodimers, whereas the reverse orientation produced nonfunctional molecules (39). However, for the renotropic activity of LH studied by Nomura et al. (40), the complete α subunit was required for the active LH isoform. On the other hand, quantitative Edman degradation revealed very low (<10%) initial yields for conventionally isolated oLHα subunit preparations. Modification of the pituitary extraction procedure, eliminating acetate buffers, had the result of changing the initial yield of phenylthiohydantoin (PTH)-Phe^{1} from negligible levels to 50%, suggesting acetylation of the N-terminal Phe during the classic Koenig and King (41) extraction procedure obscured the α-subunit amino terminus (G. R. Bousfield, unpublished data).

The hCG and hFSH crystal structures define five regions of the α-subunit primary structure: an N-terminal end, three Cys knot loops (αL1, αL2, αL13), and a C-terminal region. At each end, the N-terminal and C-terminal peptide regions are delimited by Cys^{11} and Cys^{93}, which anchor these regions to opposite sides of the molecule by disulfide bonds to Cys^{40} and Cys^{57}, respectively (Fig. 4B). Flexibility at each end was indicated by the absence of residues 1 to 4 and 90 to 92 in hCGα, whereas in both hFSHα structures residues 1 to 4 were missing and in one hFSH structure residues 91 to 92 were also missing. Peptide walking and deletion experiments have indicated the α-subunit C-terminal residues are important for receptor activation, receptor binding, and TSH heterodimer stability (42–44). The NGF structure also showed flexible terminal regions, which adopted α-helical secondary structure upon receptor binding. A similar conformational change in the hFSHα C-terminus may also accompany receptor binding because Schmidt et al. (45) noted a small increase in helical content of the hFSH–FSH receptor extracellular domain by circular dichrographic (CD) spectroscopy. The rest of the α structure consists of the three loops formed by the Cys knot motif. Loops αL1 and αL3 are on the same side of the molecule, and side chains near the tips of both loops form hydrogen bonds with oligosaccharide attached to Asn^{78}. Both loops are twisted hairpins. Although αL1 is fairly straight, there is a definite curve in αL3, occurring at Lys^{63} in the outgoing strand and at Thr^{80} in the incoming strand. On the other side of the Cys knot, αL2 consists of two extended β strands up to residue Pro^{38}, where a pronounced "kink" occurs, bending the loop almost 90 degrees. From this point, the backbone loops around, forms the only consensus helical region in either hormone, and bends back, beginning a β strand antiparallel to the residue 33 to 39 β strand at Asn^{52}. A potentially significant difference in the Asn^{52} side chain orientation can be noted between hCGα and hFSHα. This has the result of orienting this critical oligosaccharide such that the hCGα oligosaccharide extends out along the narrow axis of the molecule, effectively increasing its diameter, whereas the hFSHα oligosaccharide extends along the long axis. The different orientations of the oligosaccharide in the two molecules may have consequences for oligosaccharide processing and the relative size of oligosaccharide that can be accommodated in each hormone.

The solution structures for free α differed significantly from those of FSHα and CGα associated with their complementary β subunits (46–48). The most striking changes occurred in αL2, which lost most of its secondary structure, including the only consensus helical region in either hCG or hFSH (Fig. 4B). Perhaps this accounts for the dramatic changes in the CD spectrum that prompted Jirgensons and Ward (18) to propose greater flexibility for this subunit over that of the β subunit. The ensemble of solution structures shows a high degree of motion in αL2 compared with the two other loops (49). Two solution structures showed no oligosaccharide at Asn^{52}, and the kink in the loop moved toward the Cys knot from residues Pro^{38} and Asn^{52} in the heterodimer to residues Ser^{34} or Ala^{36} and Ser^{55} or Thr^{54}. This flexibility might facilitate the threading mechanism for subunit association (50) if it can be maintained in the presence of intact oligosaccharide. Because there is no interloop disulfide bond connecting the αL1 and αL3 loops, the two loops show more conformational flexibility than their counterparts in the β subunit. The αL3 bend occurs at residues Lys^{63} and Thr^{80} in both the free subunit and in the heterodimer. However, in the latter, βL2 appears to limit the extent of the αL3 bend, which is more pronounced in the free α-subunit structure.

There are two N-glycosylation sites at positions 52 to 58 and 77 to 84, depending on the species. These sites are maintained throughout the series. Although it appears from site-directed mutagenesis studies (51) that the Asn^{52} glycosylation site plays the primary role for full expression of biological activity, the same study revealed a secondary role for βAsn^{13} oligosaccharide that was apparent only after eliminating glycosylation at Asn^{52}. The Asn^{78} site is involved in α-subunit folding because the proximal GlcNAc residues interact with side chains of αL1 and αL3 loops (52), and bacterially expressed bovine α subunit failed to fold under conditions that obtained efficient native bLHα folding (53). The pike eel sequence was first reported with an aspartic acid residue at Asn^{58} rather than an asparagine required for the N-linked glycosylation. However, this sequence was obtained by conventional sequencing in which placement of amide forms may be technically difficult. Subsequent listings for this sequence (gi121315, spP12836) show it as the conventional Asn residue, presumably glycosylated. Similar proposals were initially made for the goldfish α2 and bonito LHα subunit. These also eventually changed to conform to the classic Asn-Ile/Val-Thr sequon associated with this position (54). The Thr residue at position 43 is an O-glycosylation site for the free α subunit isolated from the pituitary, but not placenta, and this threonine residue is found in all known mature α-subunit proteins.

The structural significance of the half-cystine placements has been alluded to earlier as sites that define the disulfide positions, which provide rigidity to the three-dimensional structure. In fact, because these hormones lack a significant hydrophobic core, the disulfide bonds determine the three-dimensional structure for each subunit. We previously reviewed the several studies of disulfide linkages in both the α and β subunits of the glycoprotein hormones (1). Because of the technical difficulties associated with disulfide placement identification, these reports did not all agree, and all eventually proved partially erroneous. Before their correct elucidation, two assumptions were made regarding their importance: "(1) The uniformity of half-cystine placements indicates a uniformity of disulfide linkages between these residues characteristic of the subunit (i.e., whether α or β) and regardless of the hormone (i.e., whether LH, FSH, CG, or TSH). (2) The 3-dimensional structure of all the glycoprotein hormones will be analogous for all the glycoprotein hormones" (3). The second assumption lacked direct experimental support at that time, but has now been substantiated by comparison of the hCG and hFSH structures. Moreover, the first assumption is now thought also likely to be true because disulfide bonds in hFSH and hCG were the same. The α-subunit disulfide placements are based on the crystal structures for hCG and hFSH (4,5,17). There are 10 half-cystines in the α subunit, which provide the 5 disulfides indicated in Fig. 2A. The α subunits have sufficient structural information in their linear sequence to specify proper recombination of the indicated half-cystines after reduction (to destroy the disulfide bonds) and reoxidation (to reform the disulfide bonds). Pierce and colleagues first called attention to this property of the reduced α subunits (55). This process has been studied in detail for mammalian α subunits (56), the equine special case (57), as well as oLHα subunit fragments (58). Further discussion of the β-subunit disulfide bonds is provided in the section on The β or Hormone-Specific Subunit.

The half-cystine residues in the leader sequence are present as the reduced form, cysteine, and are never oxidized to a disulfide form before removal from the mature protein, and therefore these signal peptide half-cystine residues are not involved in the disulfide bond formation under consideration. The term *signal peptide* is derived from the hypothesis originating from the seminal work of Blobel and colleagues (59,60). From this work, it has been established that the signal peptide serves as a targeting means for a signal recognition particle, which directs the newly formed precursor protein to the receptor on the membrane, which the protein is destined to cross through a protein channel. In the course of this membrane transport, the signal peptide is removed by a signal peptidase. For some glycoprotein hormones, notably FSHβ, signal peptidase appears to contribute to N-terminal heterogeneity (61).

Apart from the structural importance of the subunit, the possibility that the free form of the subunit may play other, separate physiological roles has received little study, largely because the requisite assays have not been devised. It has been reported that the free α subunit stimulates lactotrope differentiation in the immature pituitary (62). The lactotropes were characterized by their ability to secrete prolactin after differentiation. Closely related to this observation is the report (63) that the α subunit (free α or hormone derived) stimulated the secretion of prolactin in human decidual cell cultures derived from the placenta. Both observations implicate the free α subunit in the physiology of prolactin. These observations merit further study with respect to the physiological role of the α subunit.

Since the advent of suitable radioimmunoassays for the glycoprotein hormones and their subunits, it has been known that the α subunit and, to a lesser extent, the free β subunits may be found in the serum or the secretory products of pituitary cell or chorionic cell cultures [see, for example, the review by Cole (64) or the study by Keel et al. (65)]. Kourides and coworkers (66) observed that the majority of the "free" α subunit had a molecular weight greater than "hormone-derived" α subunit (i.e., subunit obtained from dissociation of TSH in their study—or from hCG, LH, or FSH in similar studies by others subsequently). It was shown (66) that this increased molecular weight was the result of increased glycosylation. Parsons and Pierce (67) isolated free α subunit from bovine pituitaries and showed that the increased glycosylation was attributable to O-linked glycosylation of the Thr^{43} residue, and that this glycosylation prevented recombination with the bLHβ subunit. Enzymatic removal of the O-linked carbohydrate permitted formation of the α-β dimer. The O-glycosylation is a late-occurring processing step that has no influence on the regular α-β dimer formation because it occurs in the Golgi, a post–endoplasmic reticulum (ER) cellular compartment (68). Placental-derived free hCGα possesses less than 10% O-linked carbohydrate. Increased glycosylation takes the form of increased branching, particularly at hCGα-Asn^{52}, with triantennary and tetra-antennary oligosaccharides constituting the majority released by peptide N-glycanase (PNGase) digestion (69,70).

The secretion of free α subunit results from a higher rate of biosynthesis of this subunit compared with the β subunit. This difference is particularly evident in some systems (71). Moreover, the α subunit is much more resistant to biodegradation (72), whereas the β subunit is subjected to an efficient removal process (72,73). This difference appears to result from the α subunit rapidly folding into its native conformation and thereby eluding the ER quality control mechanisms. In contrast, the β subunit folds much more slowly, permitting trapping of partially folded intermediates and rendering it susceptible to elimination by the ER quality control machinery.

It was reported (74) that there was a tight correlation of free α-subunit levels with LH levels, but not with FSH, in gonadotropin-releasing hormone (GnRH)–deficient patients. However, in a perfused pituitary cell system, the same laboratory (75) showed that free α subunit, LH, and FSH responded in a similar manner to pulsatile administration of GnRH. In studies quantifying free α subunit and its Thr^{43} O-linked form as well as bLH in a bovine pituitary slice preparation (76), it was shown that GnRH increased LH 40-fold but free α-subunit levels only doubled. Moreover, they showed that over 75% of the free α subunit was of the Thr^{43} O-linked form. Thus, we may conclude that the in vivo relationship of the free α-subunit form to the intact hormone is a complex interrelationship of biosynthesis, degradation, secretion, and stimulation. The picture is further complicated by the fact that the various isoforms (isohormones) present an extensive heterogeneity (see further discussion in the section on Biosynthesis of the Gonadotropins). For studies in which the carbohydrate complexity was measured directly for α subunits of porcine origin, see Maghuin-Rogister et al. (77), and for hCG, hLH, hFSH, and hTSH, see Nilsson et al. (78) and Renwick and colleagues (79–81). From these and other studies it is clear that the α subunit shows an identity for the protein portion within a given species, but the different types of hormone synthesized by separate cell types provide great variation in the carbohydrate moieties, as is described further in the section on Carbohydrate Structure and Function.

As noted previously, Pierce and colleagues produced hybrid molecules to demonstrate the common subunit. These hybrid hormones, in which the α subunit of one species is combined with the β subunit from another species, have been very useful in the study of hormone activity. By this means it has been established that the α subunits are generally interchangeable. However, there are certain exceptions. The inability of carp α subunit to combine with oLHβ to produce an active hormone was noted previously (36). The putative carp α2, which differs by only four residues from the carp α1, did not produce an active hormone when expressed in a baculovirus system with the carp gonadotropic hormone β subunit (82), whereas the carp α1 produced the expected active hormone.

In the case of equine gonadotropins, there seems to be an absolute dependence on an equine α subunit to obtain an active hormone with any of the equine β subunits, but the equine α subunit combines readily with other species, often to produce more potent hormones than the native hormone from the species in question (83). All of these examples of inactive hybrids tell us that there are some very critical three-dimensional positions in the hormone complex that potentially alter hormone–receptor sensitivity. Defining these subtle differences should prove a very fruitful area for future research.

The β or Hormone-Specific Subunit

There are three types of β subunit in all mammalian species (the pituitary LH, FSH, and TSH β subunits), and in the primates there is an additional β subunit type, the CGβ, which is structurally very closely related to the LHβ. Indeed, they both bind to the same receptor to initiate hormone action, but with different kinetics. In the equids, the only other mammalians that have a CG, there is a single gene that produces the LH and CG proteins in the pituitary and chorion, respectively (84), which differ only in their glycosylation and tissue of origin. The search for CGs in other species is also complicated because separate LHβ and CGβ genes may not exist. For example, a putative CG was isolated from guinea pig placenta (85), whereas a single LH/CGβ gene was reported (86). In the lower vertebrates there are probably also three different types of β subunit, although it has long been held that there is only a single primordial gonadotropin. This has been designated *gonadotropic hormone* (GTH). However, Ng and Idler (87,88) found two types of gonadotropin in plaice, flounder, salmon, and carp pituitaries. Kawauchi and colleagues (89) isolated two different forms of salmon GTH with different steroidogenic activities. In subsequent studies of the amino acid sequences of these GTH I and GTH II forms (90), the closer similarity of GTH I to bFSH and of GTH II to bLH became apparent. Sekine et al. (91), by means of recombinant DNA methodology for the two types of GTH in *Oncorhynchus keta*, the chum salmon, extended these findings and established primitive LH and FSH types of gonadotropin in species as low as the teleost fish (Table 1). The sGTH-1β has the greatest homology to the FSH types, and would be the putative salmon FSH. The sGTH-2β is almost identical to the sGTH sequence described by Trinh et al. (92) for Chinook salmon, and this bears the closest homology to LH. We will use LH and FSH to identify the fish gonadotropins for the remainder of the chapter.

Alignment of the β subunits by the Clustal W method was less satisfactory than for the α subunits because of greater differences between the different hormones as well as species-specific differences between the same hormone. For example, avian LHβ subunits possess an unusually long leader sequence, and mature ostrich LHβ may also possess an extended amino terminus. Moreover, the C-terminal extension in the primate chorionic gonadotropin β subunits and the equid LH/CGβ subunit results in gaps inserted in the nonconserved C-terminal sequences beyond the second seatbelt Cys. Alignment of the mature sequences was not much better because signal peptidase cleavage predictions produce extended amino termini for the avian β subunits and disagree with known TSHβ sequences, reducing confidence in the predictions. Because the C-terminal extension is retained in the mature protein, the algorithm introduces gaps in the C-terminal regions of the shorter pituitary hormone β subunits. Previously, to make the sequence comparisons, we took advantage of the alignment of the half-cystine residues, which maintain an identical spacing throughout the series from eels to humans. Alignment by the Clustal W method also maintained the regular spacing of the Cys residues in the core regions of the β subunits. There are several irregularities to this conformity of half-cystine placements. The first was identified in the FSHβ from chum salmon, *O. keta*. This β subunit had an "irregular" half-cystine at position 5, which seemed to be translocated from the normal half-cystine at position 26, where a serine was placed. The Cys^{26} is normally disulfide linked to position Cys^{110} (Fig. 2) and latches the seatbelt loop that embraces the α subunit (Fig. 3). Does this mean that the salmon FSHβ is instead linked Cys^{5} to Cys^{110}? Additional sequences have revealed this variant Cys arrangement in FSHβ subunits of two other salmon species, nine other fish species, and in a turtle FSHβ sequence. The variant is not present in the other 37 fish sequences, including *D. rerio*, shown here.

The second irregularity to the half-cystine placement rule is found on the C-terminus of the donkey CGβ (93), where the cDNA analysis of a partial clone (lacking the first 84 residues and the signal peptide) indicated an "extra" half-cystine at position 118 where all other CGβ molecules have a serine. This implies either that the donkey CG either has a reduced cysteine at this position (which would in fact be a sulfur analog of serine) or the –SH is involved in some other chemical combination. The balance of the disulfide placements are not out of the ordinary for donkey CG (94). It seems unlikely this residue, donkey CGβ-118, is involved in an unusual disulfide bond, although we do not have sufficient information to state this with certainty. We do know that Aggarwal et al. (95) isolated donkey CG and found no unusual properties for the molecular volume compared with other CG preparations. (If the Cys^{118} were disulfide linked to a second Cys^{118}, for example, the resulting dimeric form would be much larger.) The complete zebra LH/CGβ sequence has a Ser residue at this position, as do two other partial zebra sequences. As shown in Fig. 10, LH/CGβ glycosylation is much more highly conserved in equids than is CGβ glycosylation in primates (96).

The third irregularity results from a two-residue insertion into βL2, in the TSHβ series. As in the case of similar differences in size between the α-subunit loops αL1 and αL3, the insertions occur near the end of the loop. Interestingly, this places all the insertions and deletions in Cys knot loops on the same end of the glycoprotein hormone molecule. The fourth insertion is a one- or two-residue insertion in the seatbelt loop of TSHβ.

The Cys knot organizes the β subunit into three loop regions, as in the α subunit. The N-terminal region is found only in the LH and CG β subunits. It is largely missing in full-length FSHβ and TSHβ and completely missing in FSHβ isoforms in which Cys^3 is the amino terminus. A fourth loop that embraces the αL2 loop in the heterodimer has been dubbed the seatbelt loop. Within this region is the octapeptide determinant loop, which provided the first testable hypothesis for hormone specificity.

Luteinizing Hormone (Lutropin) and Chorionic Gonadotropin (Choriogonadotropin) β Subunits

Inasmuch as the β subunits define the activity of the α-β complex, it is not surprising that the amino acid sequences among the LHβ series (Fig. 5A) show greater homology within the series than the FSHβ series or the TSHβ series. Moreover, because both LH and CG interact with the same hormone receptor, it is to be expected that the LHβ and CGβ series show extensive homology with each other. There are 12 half-cystine residues in the β subunits (Fig. 5A, locations indicated by asterisks), and in the mature protein subunit these are all oxidized to form 6 disulfide bonds (Fig. 2A). Three disulfides form the Cys knot motif, creating the pair of loops βL1 and βL3 on one side, with βL2 on the opposite side. One disulfide connects the tips of loops βL1 and βL3, providing a constraint missing in the corresponding α-subunit loops. The other two disulfides play an important role in heterodimer formation. The disulfide between Cys^{93} and Cys^{100} creates an octapeptide loop, the so-called determinant loop, that plays an important role in LH receptor binding. The Cys^{26}-Cys^{110} disulfide latches the seatbelt loop that was initially proposed to stabilize glycoprotein hormone heterodimers. Deletion experiments involving hCGβ supported this role (97); however, recombinant eLH/CGβ mutants lacking the seatbelt latch disulfide remained stable (98). An intact eLHβ determinant loop is necessary because Lys^{96}-Thr^{97}–nicked eLHβ does not reassociate with eLHα (99).

Early CD spectra of oLH before and after dissociation of the subunits revealed only a moderate change in the CD spectra of oLHβ, but a transformation to more random spectra for oLHα. This led to the proposal that the β subunit was much more rigid than the α subunit and that the β subunit therefore served as the greater determinant for the three-dimensional structure of the α-β complex (18). The subunit association process was subsequently studied in greater detail using CD spectra (100,101). We lack the luxury of the structural information for the β subunit that exists for the α subunit. No published results exist for the hCGβ subunit. Although the Spectra Stable Isotopes website (www.spectrastableisotopes.com/Technical_Overview.asp) shows a picture representing a solution structure for this subunit that appears similar to the hCGβ structure in hCG, no systematic comparison with the crystal structure has been reported. The disulfide bond between Cys^{23} in βL1 and Cys^{72} in βL3 is consistent with greater β-subunit rigidity. However, the lack of secondary structure in βL2 probably also contributed to the relatively smaller change in CD spectrum associated with β-subunit dissociation.

The sequences for LH and CG (Fig. 5A) have much in common up to the point of the last half-cystine (Cys^{110}). Beyond that point the homologies diminish markedly, as indicated by the gaps inserted in the LHβ C-terminal sequences. We were able to remove chemically this part of the molecule from eLHβ and show that the biological activity (after recombination with the α subunit) was not altered (102). By site-directed mutagenesis, Matzuk et al. (103) showed that loss of the C-terminus of hCGβ beyond residue 115 did not interfere with receptor binding or signal transduction in vivo for the reconstituted hormone. On the other hand, for eCG and to a lesser extent eLH, the FSH, but not LH biological activity of these molecules with

→

FIG. 5. Primary and tertiary β-subunit structures. **A:** Representative vertebrate chorionic gonadotropin (CG), luteinizing hormone (LH), follicle-stimulating hormone (FSH), and thyroid-stimulating hormone (TSH) β-subunit sequences using the same species examples as in Fig. 4A. **B:** Crystal structures for recombinant, deglycosylated hCGβ and recombinant, insect cell–expressed hFSHβ shown in their heterodimeric conformations (α subunit not shown). The backbone is represented by a *solid line*. Cys residues are represented as *wire diagrams* to illustrate disulfide bonds. Oligosaccharide fragments are also shown as *wire diagrams*. Note that only one hFSHβ oligosaccharide fragment is present because the Asn^{24} glycosylation site was silenced by a T54A mutation (17).

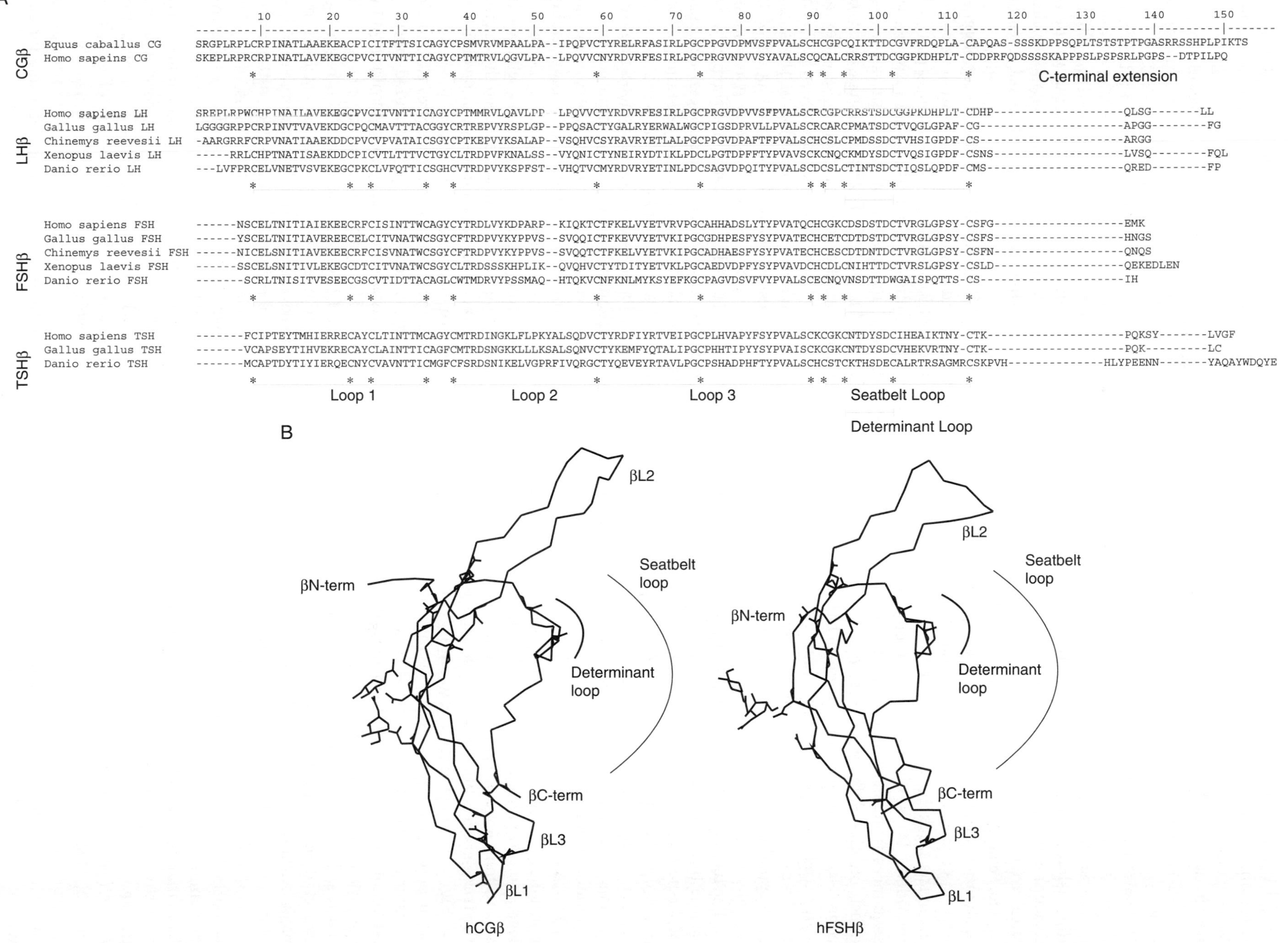
A
10 20 30 40 50 60 70 80 90 100 110 120 130 140 150
CGβ
Equus caballus CG SRGPLRPLCRPINATLAAEKEACPICITFTTSICAGYCPSMVRVMPAALPA--IPQPVCTYRELRFASIRLPGCPPGVDPMVSFPVALSCHCGPCQIKTTDCGVFRDQPLA-CAPQAS-SSSKDPPSQPLTSTSTPTPGASRRSSHPLPIKTS
Homo sapeins CG SKEPLRPRCRPINATLAVEKEGCPVCITVNTTICAGYCPTMTRVLQGVLPA--LPQVVCNYRDVRFESIRLPGCPRGVNPVVSYAVALSCQCALCRRSTTDCGGPKDHPLT-CDDPRFQDSSSSKAPPPSLPSPSRLPGPS--DTPILPQ
C-terminal extension
LHβ
Homo sapiens LH
Gallus gallus LH
Chinemys reevesii LH
Xenopus laevis LH
Danio rerio LH
FSHβ
Homo sapiens FSH
Gallus gallus FSH
Chinemys reevesii FSH
Xenopus laevis FSH
Danio rerio FSH
TSHβ
Homo sapiens TSH
Gallus gallus TSH
Danio rerio TSH
Loop 1
Loop 2
Loop 3
Seatbelt Loop
Determinant Loop
B
βL2
βN-term
Seatbelt loop
Determinant loop
βC-term
βL3
βL1
hCGβ
βL2
βN-term
Seatbelt loop
Determinant loop
βC-term
βL3
βL1
hFSHβ

dual LH/FSH function seems to involve the equine C-terminal extension (104,105). The CGs are known to have much longer half-lives in vivo, which is attributable at least in part to the C-terminal extension. Taking advantage of this fact, Fares et al. (106) have designed a long-acting follitropin agonist by fusing the C-terminal sequence of hCGβ to the follitropin β subunit. This demonstrates that all of a given protein serves some purpose, from specification of three-dimensional architecture to particular physiological functions, but the relative importance of each structural area may vary.

The comparative sequences for the consensus sequences of the CG and LH in the mammalian series are much more homologous than those obtained when the chicken, turkey, and bullfrog sequences are included. It is obvious that inclusion of these sequences from the lower vertebrates has a marked effect on the detectable differences (compare the consensus LH versus the consensus LH, mammalian, sequences, Fig. 4). The 39-residue signal sequences for the avian species are significantly larger than the corresponding 20-residue mammalian LH and CG signal sequences.

The high degree of homology for the β subunits from eels to humans cannot be interpreted as a reflection of the potency of the parent hormones across species. Several laboratories have provided comparative studies of preparations of glycoprotein hormones in various assay systems; notable among these are the laboratories of Licht (107), Papkoff (108), and Fontaine (109). As a general statement, the application of hormones across species in selected assay systems may lead to no response, to instances in which the potency of the homologous species hormone is exceeded by the trans-species hormone. The sequence homologies in Fig. 5 provide only rudimentary indications of what potencies may be expected upon interchanging hormones. The absence of basic residues in positions 94 to 96 of the chicken LHβ determinant loop is consistent with reduced potency in mammalian LH bioassays (110). The same reduced potency would be expected for turtle and fish gonadotropins on this basis. Although *Xenopus* LHβ possesses a Lys residue at position 94, the Asp at position 96 is not consistent with high biological activity. On the other hand, hCG is routinely used to stimulate ovulation in this species. Fontaine (109,111) found that an activity obtained from mammalian pituitaries that was effective in fish as a thyrotropin could be traced to the mammalian gonadotropic activity. This gave rise to the designation *heterothyrotropin*. In studying this thyrotropic activity in fish, MacKenzie (112) showed all the mammalian pituitary glycoprotein hormones possessed significant thyrotropic activity in fish, but their subunits were inactive. Although chicken LH exhibited scant LH activity in mammalian LH bioassays, it displayed significant FSH activity in rodent-based assay systems.

Follicle-Stimulating Hormone (Follitropin) β *Subunits*

The relative abundance of the full-length isoforms in pituitary FSH preparations ranges from 20% in hFSHβ to 50% in eFSHβ and corresponds to the relative strengths of the predicted signal peptidase cleavage sites. The same algorithm predicts a C-terminal Cys for TSHβ.

Follitropin (FSH) stimulates follicular development in the ovary, and in particular the granulosa cells of the follicle. Follitropin in concert with lutropin (LH) prepares the follicle for ovulation and luteinization. For extensive discussion of these processes, see Chapters 10, 11, and 12. FSH stimulates granulosa cell proliferation and then promotes the differentiated function of these steroidogenic cells (113). FSH specifically stimulates the production of the steroid hormone, progesterone, and the production of the enzyme system (aromatase) for converting testosterone to estrogen (114). FSH also stimulates the production of the enzymes involved in plasminogen activation (115). Both the aromatase and the plasminogen activator systems have been adapted for sensitive assays of FSH. Study on the induction of follicular development by exogenous gonadotropin administration in women and domesticated animals has been accelerated by its application in in vitro fertilization and embryo transfer (116). In the male, FSH acts on the Sertoli cells of the seminiferous tubules in the testis, whereas LH acts principally on the interstitial cells. See the chapters on the testis and spermatogenesis in this volume for more extensive discussion.

One of the problems for the study of FSH action has been the scarcity of highly purified hormone preparations, because most species of pituitary contain approximately 1/20th the amount of FSH (or TSH) compared with LH on a molar basis. This has obliged investigators to use crude preparations of FSH or to substitute. In several species of laboratory animal, the equine CG has a potent FSH-like effect, although it is greatly reduced in the equine (117). Thus, particularly in the older literature, one finds eCG of varying degrees of purity as obtained from pregnant mare serum (thus designated PMSG, for "pregnant mare serum gonadotropin") used as an FSH substitute. As is now known, the various isoforms of FSH have circulating half-lives that depend on the degree of sialylation (118). Moreover, these half-lives are usually much shorter than those of PMSG, and thus uncertainties are introduced into the dose considerations for such studies.

As shown in Fig. 5, the FSH sequences for the β subunit are shorter, only 111 residues in length for the mammalian series, and only 95 residues for *Odontesthes bonariensis* FSH. With the half-cystines in register this numbering, to be consistent with the other series' numbering, brings a position 7 to 122 span for the FSH series. Although the mammalian FSHβ subunit sequences exhibit a very high sequence identity of 86% to 95%, this drops when other species are added. The vertebrate-wide comparisons show 47.9% sequence identity for FSH, 75.7% for CG (with 13 full comparisons available: 10 primate, 3 equid), and 55.4% for LH. These calculations are for the mature protein only (i.e., excluding the signal sequences).

The FSHβ protein is encoded by a single gene. However, Guzman et al. (119) found a gene in the sheep that is 87% homologous to the FSHβ gene, indicating there is an FSHβ-like gene or a pseudogene in the sheep. The physiological significance of this finding remains to be determined.

Allocation of Function to Structural Areas of the Glycoprotein Hormones

The allocation of functional roles to specific areas of the protein structure requires innovative approaches and usually considerable effort. Investigators have taken the following broad approaches to this task:

1. Structural comparisons
2. Functional group substitutions
3. Immunological approaches
4. Peptide-walk analysis
5. Site-directed mutagenesis
6. Genetic engineering
7. Three-dimensional analysis of hormone and receptor

The use of *structural comparisons* implies the availability of several structures to compare. For the glycoprotein hormones the sequence information is now extensive enough to use this approach. We applied this to the eCG molecule as the sequence information became available and it led us to the determinant loop area as the one potentially responsible for the LH-FSH–like role of this molecule in various biological systems. The determinant loop hypothesis was thus proposed (120,121), which posits a correlation of net charge and structural features in the 90 to 105 region of the molecule as determinants of the LH, FSH, or TSH activity [see Moore et al. (121) for details]. Although other areas of structure have also been shown to be involved in specification of hormonal activity, the general importance of this area, which includes the 93 to 100 disulfide loop, for gonadotropic activity has been substantiated by several more recent lines of evidence, as noted later. Recent studies have implicated the eCG C-terminal extension as contributing to its FSH activity. The first evidence along this line was a report by the late Francesca Stewart that as the carbohydrate content of eCG increased, the FSH/LH ratio also increased (122). Min et al. (104) reported that mutating recombinant eLH/CG to eliminate the αAsn^{56} glycosylation site impaired LH activity, but not FSH activity when the C-terminal extension was present. We reported that elimination of eLHα Asn^{56} carbohydrate and eLHβ residues 121 to 149 was necessary to eliminate FSH biological activity. The FSH antagonist activity of this eLH derivative was reduced tenfold by the presence of the C-terminus in eLHβ and was eliminated by using eCGβ (105).

Combarnous has also relied heavily on comparative sequence information to develop his "negative specificity" model for glycoprotein hormone action. This proposal has been reviewed in detail (123). The major elements of the negative specificity model of hormone action are the assumptions that the α subunit is responsible for the majority of the high-affinity binding of the hormones to their receptors, that the β subunits induce the functional conformation of the α subunit, and that the β subunits specifically inhibit (hence, "negative specificity") the binding of that particular α-β complex (LH/CG, FSH, TSH) to receptors for the other hormones. These areas of negative specificity have not yet been defined, but the hypothesis should provide the basis for much fruitful research because experiments can be designed to explore these possibilities. (In particular, see comments on peptide-walk analysis and site-directed mutagenesis, later.) The studies of Moyle and colleagues to produce β-subunit chimeras are good examples of the application of this approach to define areas essential for hormone specificity for the β subunit (124).

Functional group substitutions have long been used to modify a particular group in a molecule to determine its role in a chemical reaction. As applied to the glycoprotein hormones, this implies the availability of a reagent that will selectively attack certain functional groups in the sequence to produce a derivative that alters the chemical nature of that particular group. The biopotency of the derivative is then tested to deduce the nature of the effect. In practice, several functional groups (e.g., amino groups, carboxyl groups, aromatic hydroxyls) of the type involved by the given reagent may be present. Thus, reaction rates and accessibility to the reagent must be established to determine selectivity of the reaction and groups actually involved. This, in turn, requires further structure studies of the derivatized products. Finally, it is never certain how much of the effect(s) on biopotency to attribute to the functional group derivatized, to the steric requirements of the derivatizing group, or to long-range steric effects introduced by the group added. In spite of these limitations, useful information has been obtained. For a review of this approach and the reagents used with the glycoprotein hormones, see Ward (125). In a modification of this approach, the carbodiimide reaction has been used to establish the close proximity of carboxyl and amino groups by their reactivity to produce amide bond cross-linking of the α-β subunits through the α-Lys^{49} and β-Asp^{111} of bovine LH (126) or porcine LH (127).

Immunological methods to define active areas in a glycoprotein hormone are numerous and reflect the ingenuity of the immunologists and endocrinologists involved in the individual studies. Both polyclonal antibodies (generally of broader specificity) and monoclonal antibodies (highly specific and selective) have been used. We limit our consideration to a few examples of the approaches available. To introduce greater specificity into the polyclonal antibodies, Atassi and colleagues (128,129) used synthetic peptides as antigens to produce highly specific polyclonal antibodies to selected epitopes (defined by the peptide synthesized as antigen). These antibodies were the basis for highly selective antibodies that would not cross-react with other glycoprotein hormones, and could be used to identify surface areas in the α-β complex or in the receptor-bound hormones. Moyle and colleagues (130,131) have used a well-characterized series of monoclonal antibodies to epitopes of both the intact hormone and isolated subunits to measure conformational changes upon interaction with the receptor. Birken et al. (132) showed that removal of sialic acid from the carbohydrate moieties of the C-terminal peptide of hCGβ produced a more antigenic peptide and thus were able to improve the sensitivity of an hCG-specific antibody. Bidart and colleagues (133) have used monoclonal antibodies to both human and equine CG to elucidate fine differences between the two and refine the specificity of a series of monoclonal antibodies directed at discontinuous epitopes on the surface of hCG. This group has also used monoclonal antibodies to three distinct antigenic domains that allow them to distinguish between intact and proteolytically nicked forms of hCG (134).

The approach to the analysis of functional areas using synthetic peptides we have termed *peptide-walk analysis* was introduced to glycoprotein hormone studies by Ryan and associates (43). This analysis involves the synthesis of peptides approximately 15 residues in length. By starting at the N-terminus of the linear amino acid sequence and designing the peptides to overlap the sequence of the previous peptide by approximately five residues, a series of peptides is generated for testing in selected assay systems. The peptides are used to compete with the intact hormone or subunits to determine areas that appear to be associated with, for example, the ability to combine with the counterpart subunit, or the binding of the native hormone to its receptor. In a system such as the latter, the native hormone is effective at approximately 10^{-10} M, whereas the peptides are effective, if at all, only at approximately 10^{-5} M. Thus the native hormone may be 100,000-fold more effective on a molar basis. The early studies using this approach have been reviewed by Ryan and colleagues (135). Although the low sensitivity of the assay systems to the peptides has been a criticism of this analysis, nonetheless it has been demonstrated that peptides implicated in the receptor binding area of the hormone were also capable of stimulating testosterone production in a Leydig cell system (136). Moreover, the four putative receptor binding regions identified in this manner formed a contiguous patch on the hCG three-dimensional structure. Reichert and colleagues have effectively applied this analysis to FSH (137,138), as well as demonstrating that peptides based on the hFSH sequence that showed homology to calmodulin stimulated calcium transport and would bind calcium (139,140). These findings suggest a role for specific areas of the FSH sequence involved in the requirement for calcium for maximum binding to the FSH receptor (141). The peptide walk analyses have provided us some of our most specific information about the functional areas of the glycoprotein hormones.

Site-directed mutagenesis is another methodology that provides very specific information about functional areas of the glycoprotein hormones. This method requires an available clone of the hormone subunit to be mutated. By insertion, deletion, or more often replacement of specific codons, a single amino acid may be changed as desired in a given subunit. Then, by expressing the mutated subunit, combining it with the required counterpart subunit, and testing for biological activity (or lack thereof) in selected assay systems, one learns about the essentiality of that particular residue in hormone function. The extension of this approach to larger areas of structure brings us to *genetic engineering*, which is essentially the same as site-directed mutagenesis but on a grander scale. In the glycoprotein hormones, the production of a series of chimeric molecules, mixing TSH, FSH, or LH/CG structural features to determine effects of these structural features on the final hormone activity of the chimeric product, has been extensively applied by Moyle and colleagues (124,142). Examples of the use of site-specific mutagenesis can be found in the demonstration of the role of the N-linked carbohydrate moieties on the α subunit by Matzuk and Boime (51,143), whereas Ji and colleagues (144) used the method to confirm the essential nature of the penultimate lysine residue for biological activity. Puett and colleagues used site-directed mutagenesis to study the effects of removal of residues 122 to 145 from the β subunit of hCG (145), or to study the role of the invariant Asp^{99} (Fig. 4) present in all β subunits (146). The latter study showed that this Asp group could be replaced without complete loss of activity in the resulting heterodimer unless this negatively charged functional group was substituted with a positive group (e.g., Arg), in which case the product was totally inactive. Many such examples could be cited, but the foregoing suffice to illustrate the great versatility and selectivity of genetic engineering. The only criticism or reservation that might be leveled at this approach comes from the lack of specificity in the carbohydrate moieties applied in most gene expression systems used to generate the final product, but the evidence so far suggests that this is usually not a serious problem to the interpretation of the results. A major problem with this research has been the requirement to obtain a stable heterodimer in the face of destabilizing mutations, particularly those probing disulfide bond function. Single-chain gonadotropin molecules created by fusing β-subunit and α-subunit cDNAs in frame with a linker, usually the hCGβ C-terminus, pioneered by Irv Boime's laboratory (39,147), have provided a way around this limitation.

Ultimately, the availability of an exact three-dimensional model of the heterodimeric glycoprotein hormones in juxtaposition with their respective receptors should provide us with powerful tools for the study of hormone action. Significant progress has been made in the form of crystal structures for hCG (4,5,15) and hFSH (17), along with the solution structures for various forms of free hCGα (46–48). Reports of soluble bacteria-expressed FSH receptor extracellular domain expression (148) and studies of insect cell–expressed hFSH receptor extracellular domain in association with hFSH (45) suggest that hormone binding to its receptor will be defined in the near future.

Aberrant Gonadotropin Molecules

The production of aberrant molecules has been demonstrated in a few instances for the glycoprotein hormones. Nishimura et al. (149) studied an α subunit secreted by an ectopic tumor in a patient. This subunit was unable to combine with the β subunit of hCG to produce a normal α-β complex. By peptide analysis, they established that the only sequence difference from normal α subunit was the conversion of Glu^{56} (Glu^{60} in Fig. 3) to an Ala. This single change produced a serious defect in the folding ability, tertiary structure, or accessibility for proper glycosylation that rendered the product unsuitable for normal function as an α subunit.

Cox (150), studying the α subunits secreted by several cell lines of human tumors, found that glycoprotein subunits secreted by trophoblastic tumors (JAR, JEG) or nontrophoblastic tumors (HeLa, ChaGo) were phosphorylated on the α subunit. The JAR line, which also secretes hCGβ, phosphorylated only the hCGα subunit. Phosphorylation of the hCGβ C-terminus was reported (151). In the cytoplasm, phosphorylation competes with O-glycosylation and may play a role in regulating the activity of RNA polymerase II. It is possible that phosphorylation is a prelude to O-glycosylation.

The N-terminal heterogeneity of the oLHα subunit was first described by Liu et al. (37), but this same type of heterogeneity has also been described for several species of α subunit (see previous discussion). It has not been clearly established whether this is simply an isolation artefact or normal physiological processing, but proteolysis during isolation would seem to be the most likely source of this type of aberrant molecule.

In the first reports (152) of the amino acid sequence of hLHβ, some proteolytic "nicking" was observed around the 45 to 50 area of the sequence. It was later shown by two laboratories (153–155) that this proteolytic cleavage was commonly observed in hLH preparations (which are almost always derived from pooled pituitary extracts). These reports allowed estimates of the percentage cleavage at positions between residues 44, 45, 46, 47, and 48. The potency of the nicked hormones was considerably diminished.

The heterogeneity of hCG collected from single individuals was studied by Kardana et al. (156). They found substantial variation in the types of heterogeneity. They observed nicks in the β subunits, particularly at the 47 to 48 position (11 of 13), but also at the 44 to 45 and 46 to 47 positions in some samples. Other evidence of aberrant molecules observed in their study included N-terminal heterogeneity of the α subunit and loss of the C-terminal portion of the β subunit (2 of 13). One preparation had an α-subunit nick at position 70 to 71 (or 74 to 75 in Fig. 3 because of the four-residue genomic deletion in the human α subunit). Peptide heterogeneity therefore was found to be extensive among individual preparations, a variance that cannot be appreciated in the pooled sample preparations.

A fragment that is frequently generated by proteolytic action on the β subunit of hCG during pregnancy, and in some patients with cancer secreting CGβ subunit, is termed the "β-subunit core fragment." This fragment is secreted in the urine and may be detected by many anti-hCGβ antibodies. This fragment is a disulfide-linked, two-polypeptide residuum of the intact hCGβ, representing the 6 to 40 and 55 to 92 polypeptide portions of the β subunit (157). In pregnancy urine, levels of the β core fragment may exceed 90 nmol/L of urine in the 12th to 15th week of pregnancy, which represents a significant portion of the immunoreactivity detectable in pregnancy urine. Because serum hCGβ is intact, the nicked form appears to be a product of processing in the kidney (158).

GENETIC RELATIONSHIPS OF THE SUBUNITS

The α-Subunit Gene

The human α-subunit gene is located on chromosome 6 at 6q12–q21. The common α subunit is encoded by a single gene, which has been isolated and characterized in numerous vertebrate species (Table 1). All of the α-subunit genes isolated thus far have a very similar genomic organization consisting of four exons separated by three introns of varying size (Fig. 6). The first intron is the largest, ranging from 6.4 to 14.7 kilobases in size, and separates exon 1, containing 5′ untranslated sequences, from exon 2, which encodes the leader sequence and the first 9 amino acids of the mature polypeptide. Exon 3 encodes amino acid residues 10 to 71, and exon 4 corresponds to the

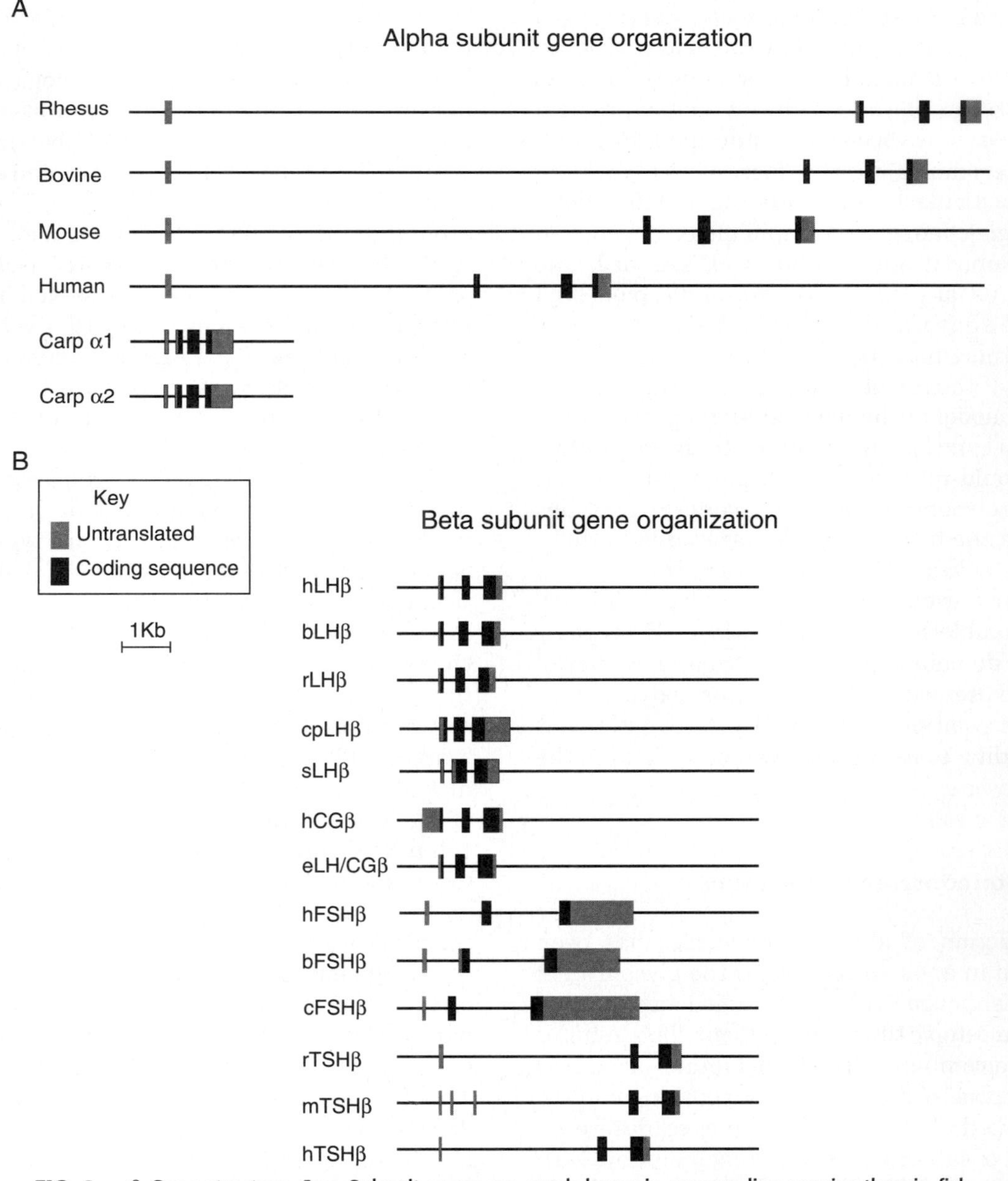

FIG. 6. α/β Gene structure. **A:** α-Subunit genes are much larger in mammalian species than in fish species. The former vary largely in the size of intron 1. Carp α-subunit genes are both very compact. **B:** β-Subunit genes are consistent in structure for each hormone. All the luteinizing hormone (LH) and chorionic gonadotropin (CG) β-subunit genes are compact. The follicle-stimulating hormone (FSH) β-subunit genes show variation in the size of intron 1 and the 3′ untranslated region, whereas the thyroid-stimulating hormone (TSH) β-subunit genes show variation in the size of exon 1 and in the number of exons (mTSHβ is composed of two additional exons).

C-terminal residues 72 to 96 and the 3′ untranslated region. Fish α-subunit genes possess small introns, like those associated with the β-subunit genes. Mature α-subunit transcripts range in size from 730 to 800 nucleotides, according to species.

The Chorionic Gonadotropin and Luteinizing Hormone β-Subunit Gene Families

The human CG and LH β-subunit genes are located on chromosome 19 at 19q13.32. The LHβ subunit is synthesized in the gonadotropes of the pituitary gland, whereas the related polypeptide, the CGβ subunit, is synthesized in the syncytiotrophoblast of the placenta of primates and in the chorion-derived endometrial cups of horses. Unlike the α subunit and the β subunits of the other members of the gonadotropin family, in the human the CGβ and LHα subunits are encoded by genes in a multigene cluster that contains seven sequences with extensive homology (159–161). Subsequent studies (162) demonstrated that CGβ6 is an allele of gene 7 with differences in the 5′ nontranslated sequence. Thus, the CGβ subunit is encoded by six genes or pseudogenes, whereas the LHβ subunit is encoded by a single gene. Transcripts from at least five of the CGβ genes are present over a wide range of concentrations in choriocarcinoma cells, with three of the CGβ genes being expressed preferentially (162). The human CGβ subunit genes appear to have evolved from an ancestral LHβ gene by multiple gene duplications and rearrangements. DNA sequence comparisons suggest that the divergence of the LHβ/CGβ genes is a relatively recent evolutionary event (163). The equid CGβ gene appears to have evolved independently because a single gene encodes the equine LHβ and CGβ subunits (84). Although the human CGβ and LHβ genes have highly conserved structures, the CGβ gene transcription initiation site occurs at a position more than 350 base pairs upstream of the analogous site in the LHβ promoter, suggesting that very different regulatory elements are responsible for the activity of these promoters. LHβ gene transcripts are further distinguished by an unusually short 5′ untranslated region. For example, the LHβ gene transcript contains a 6- to 11-nucleotide long 5′ untranslated region (164). In contrast to the situation in the human, transcription of the equine LHβ/CGβ gene in the placenta and anterior pituitary is controlled by a single promoter that corresponds structurally to the pituitary-specific LHβ gene promoter in humans (84). These results indicate that different regulatory mechanisms are involved in placenta-specific expression of the CGβ genes in humans and equids (84).

The Follicle-Stimulating Hormone β-Subunit Gene

The human FSHβ-subunit gene is located on chromosome 11 at 11p13. The FSHβ subunit is encoded by a single gene as isolated from the human, bovine, pig, and rat. In sheep there appear to be two distinct FSHβ genes; however, it is not known if both genes are expressed (119). Like the other β-subunit genes, the genomic FSHβ genes in each species are evolutionarily conserved and contain three exons and two introns. Among the glycoprotein hormone β-subunit genes, the FSHβ gene is unique because of its rather long 3′ untranslated region (Fig. 7).

The Thyroid-Stimulating Hormone β-Subunit Gene

The human TSHβ-subunit gene is located on chromosome 1 at 1p13. This is the largest of the β-subunit genes, spanning up to 5 kilobase pairs (kbp). As with the α-subunit genes, increased intron 1 size accounted for most of the increased gene size.

Orphan Subunit Genes

A fifth glycoprotein hormone has been proposed (165). It is composed of a second α subunit capable of combining with several classic β subunits. A similar second α subunit has been identified in mouse and rat genomes. Unlike the second α subunits identified in various fish species, which conserve the conventional α-subunit Cys organization and N-glycosylation patterns, this putative molecule lacks one of the Cys residues of the disulfide that passes through the Cys knot. Two additional Cys residues would be located in extended loops αL1 and αL3 and might, therefore, form a disulfide bond (Fig. 8). It possesses the Asn^{52} glycosylation site homolog, but lacks the Asn^{78} site. In a yeast two-hybrid assay, it complemented hCGβ, hFSHβ to the same extent, LHβ to a lesser extent, and TSHβ to a limited extent. A TSH-like β subunit was also identified that lacked the seatbelt loop. Because this determinant is critical for gonadotropin receptor binding, it is bound only to TSH receptors.

CARBOHYDRATE STRUCTURE AND FUNCTION

The carbohydrate possessed by the glycoprotein hormones distinguishes them from the other pituitary

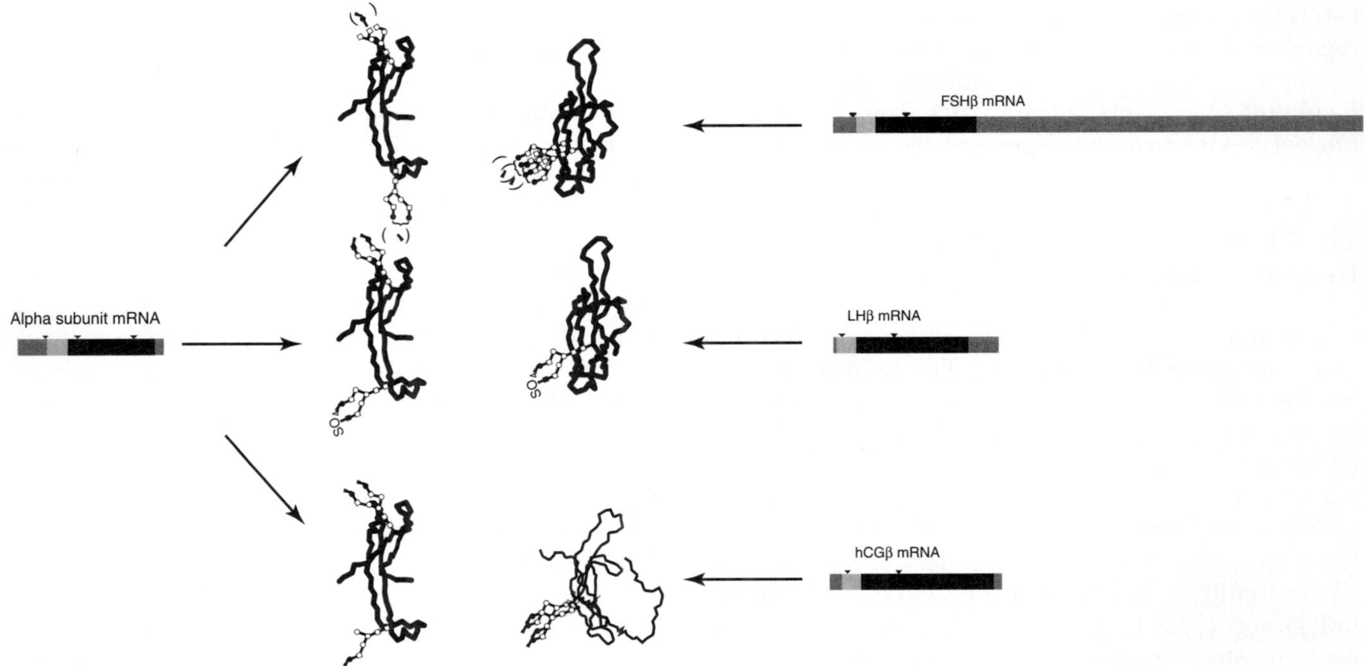

FIG. 7. Messenger RNA and protein structure. A common α subunit provides the same polypeptide component of each glycoprotein hormone. The hormone-specific β subunit is provided by expression of a second gene in the pituitary and in primate species by a placentally expressed gene or gene family. The α and β subunits combine to form unique heterodimers, and a hormone-specific pattern of glycosylation distinguishes one α subunit from another.

hormones, although glycosylated forms of such hormones as prolactin and growth hormone have been reported (166,167). Variations in gonadotropin molecular weight, which were correlated with the reproductive cycle, have been attributed to changes in carbohydrate structure (168). These changes in physicochemical properties and the observation that the ratios of biological to immunological levels of gonadotropins in serum also varied suggested that changes in glycosylation could affect their biological activity. Oligosaccharide structures have been found to be highly variable (169,170), and there is a growing body of evidence for carbohydrate involvement in signal transduction (51,171–174).

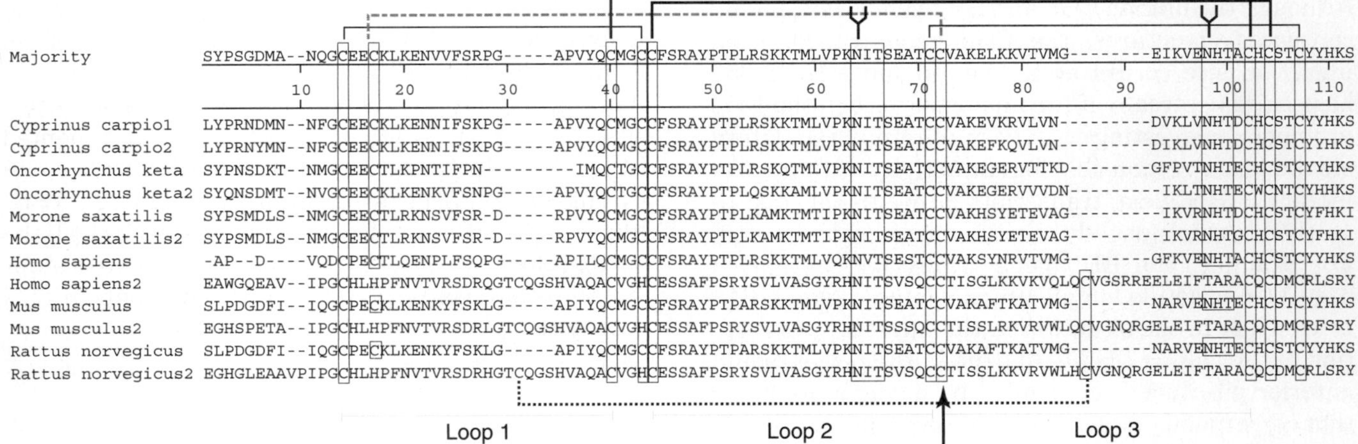

FIG. 8. Comparison of second α-subunit proteins in fish and mammals. The fish α subunits possess the same Cys pattern of all vertebrate α subunits. Note the missing Cys knot Cys, the odd Cys, and the putative disulfide between two Cys residues in the αL1-αL3 inserts in the putative thyrostimulin α subunits.

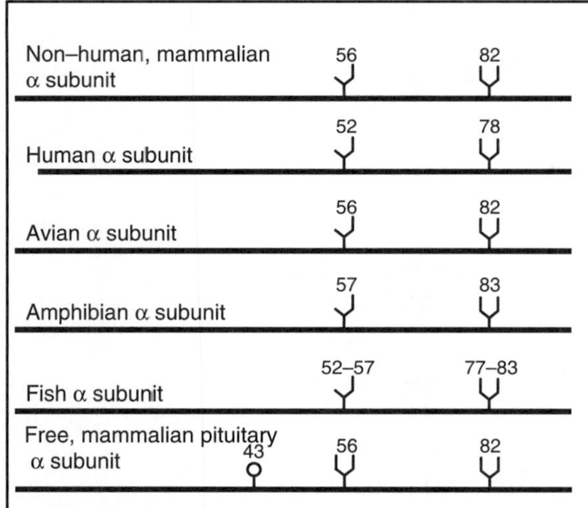

FIG. 9. Glycosylation patterns: α subunit. The *bars* represent the mature protein sequences for various α subunits. The *tuning fork* symbols represent N-glycosylation sites, whereas the *lollipop* symbols represent O-glycosylation sites. Two invariant N-glycosylation sites have been identified in all α-subunit sequences to date. A potential O-glycosylation site exists at Thr^{43} for free α subunit that does not associate with a β subunit. This site apparently is glycosylated in the pituitary, but not in the placenta.

Gonadotropin Glycosylation Sites

Unlike other heterodimeric glycoprotein hormones, such as inhibin, in which a single protein chain is glycosylated, both the α and β subunits of the glycoprotein hormones are glycosylated. Figures 9 and 10 show the patterns of glycosylation for the α and β subunits of the glycoprotein hormones. For the α subunits, N-linked oligosaccharides are attached to asparagines 56 and 82. The α-subunit Asn^{56} oligosaccharide is essential for biological activity (51). The human α-subunit glycosylation pattern is slightly different from other species because of a 12-base deletion in the second exon, which results in a 4–amino acid deletion near the N-terminus, as noted previously. The asparagines are thus at positions 52 and 78 in the human sequence. However, for the sake of clarity, all the numbering of the glycoprotein hormone sequences will be based on that of ovine LHα. All the other sequences will be aligned to this sequence by their highly conserved half-cystines to facilitate comparisons of homologous parts of the hormones. Thus, in Fig. 8, with the half-cystines in register, the human glycosylation sites line up with those in all the other α subunits. A similar convention, using the oLHβ numbering, is followed with the glycosylation sites in the β subunits (Fig. 10).

Free α subunit is found in all glycoprotein hormone–producing tissues. It differs from α subunit associated with a β subunit in that it may be O-glycosylated at Thr^{43} (175) and its N-linked oligosaccharides may be more completely processed (176) and more highly branched (70,177). Secretion of free α subunit results from its higher rate of biosynthesis, especially in term placenta (71), as well as its resistance to intracellular degradation (72). Even in in vitro systems in which α subunit is synthesized at nearly the same rate as β subunit (72), the uncombined β subunit is improperly folded (178) and therefore is degraded, whereas most of the α subunit is completely folded and is secreted either as part of a hormone heterodimer or in the uncombined state (72,73). The functional significance of free α subunit is not known. There are reports that the α subunit stimulates differentiation of lactotropes in vitro (62) and stimulates prolactin secretion by human decidual cell cultures derived from placenta (63).

Glycosylation of the β subunit occurs either at one or two N-linked glycosylation sites located at residues 13 or 30. Most LHβ subunits have only a single glycosylation site at Asn^{13}. Two exceptions are hLH and eLH. The most common human LHβ allele lacks the glycosylation signal sequence at Asn^{13}, but this is compensated by the presence of the glycosylation site at Asn^{30}. The sequence at Asn^{13} in nonhuman LHβ subunits is –Asn–Ala–Thr–, which meets the requirement for N-glycosylation signal sequence, –Asn–Xaa–Thr/Ser, where Xaa means any amino acid except proline can be substituted and Thr/Ser means that the third position can be either Thr or Ser. The human LHβ sequence at Asn^{13} is –Asn–Ala–Ile– (Fig. 5). The substitution of Ile for Thr prevents glycosylation. Glycosylation is possible at position 30 as a result of the substitution of Asn for Thr^{30}, resulting in the sequence –Asn–Thr–Thr–. An alternative hLHβ allele has been found that includes, among two other mutations, a Thr residue at position 15, which restores the glycosylation sequon. Accordingly, this form of hLHβ carries two N-linked oligosaccharides like hCGβ (179).

Equine LHβ glycosylation differs from that of other LHβ subunits because in addition to the N-linked oligosaccharide attached to Asn^{13}, the former also possesses a dozen O-linked oligosaccharides attached to a C-terminal extension, a structure characteristic of CGβ subunits (96). This "characteristic structure," however, is based on only one example, the human, where both hLH and hCG had been sequenced. The amino acid sequences for eLHβ and eCGβ are identical (180,181) because there is a single gene in the horse that is expressed in both the pituitary and the placenta (84), whereas in the human there are separate genes for hLHβ and hCGβ that are differentially expressed

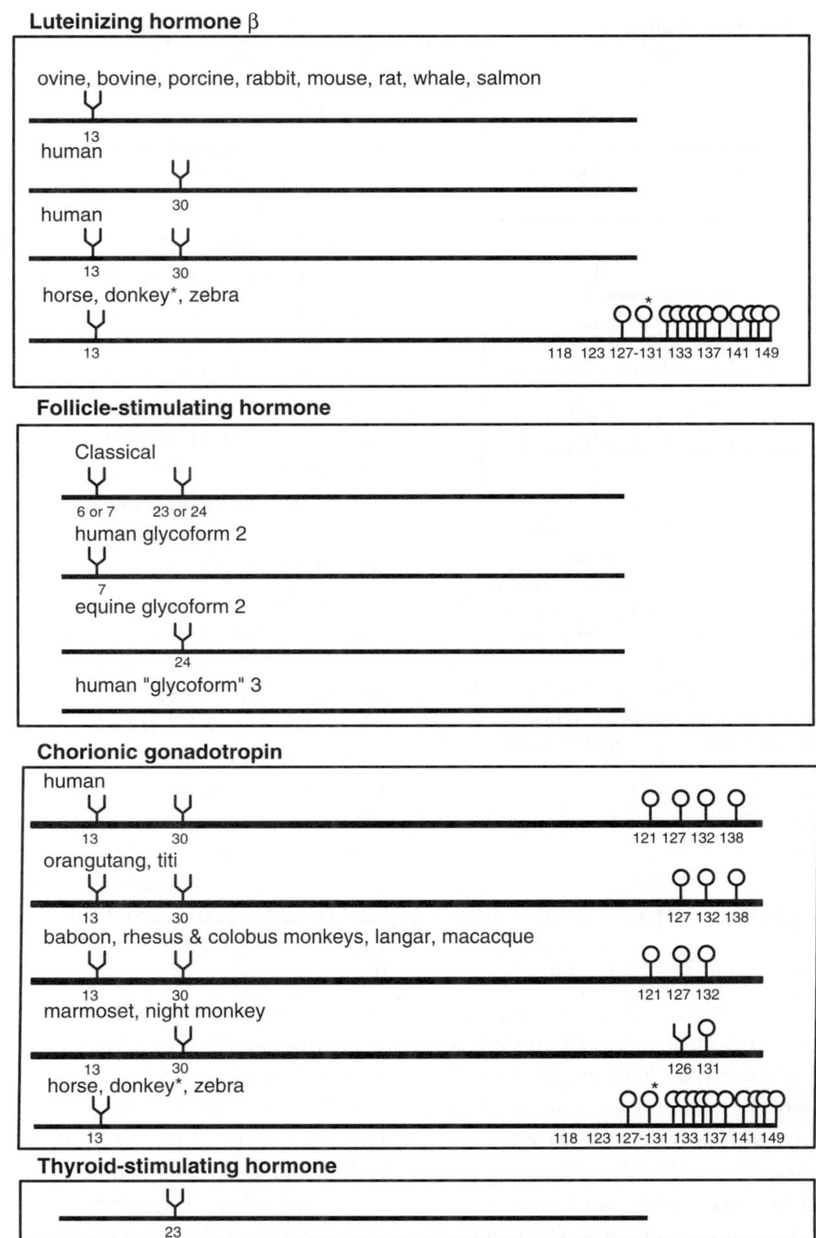

FIG. 10. Glycosylation patterns: β subunits. The *bars* represent the primary structure of each β subunit. The *tuning fork* symbols represent N-glycosylation sites, whereas the lollipop symbols represent O-glycosylation sites. Changes in the classic glycosylation patterns illustrated in the previous edition include a human luteinizing hormone (hLH) β allele possessing the Asn^{13} glycosylation site. Partial glycosylation of follicle-stimulating hormone (FSH) β is shown, including one form lacking N-linked carbohydrate altogether. Additional chorionic gonadotropin (CG) β-subunit sequences revealed heterogeneity in the primate glycosylation patterns as opposed to glycosylation site conservation observed in the equids. Because carbohydrate is the only distinction between equid LH and equid CG, these glycosylation sites may be maintained by selection.

in the pituitary and placenta, respectively (160). Twelve O-glycosylation sites were identified in eLHβ and the same 12 were found in eCGβ, although a higher occupancy rate and larger oligosaccharide size were associated with the latter (96). With the exception of the possible Cys^{118} in donkey LHβ, both sequences predicted for donkey and zebra possess the same potential N- and O-glycosylation sites as the horse (Fig. 10).

FSHβ subunits possess both N-linked glycosylation sites, Asn^7 and Asn^{24} (these positions are homologous to LH/CGβ Asn^{13} and Asn^{30}, respectively), whereas TSHβ is glycosylated only at Asn^{23}. Several glycoform variants have emerged for FSHβ. Recombinant hFSH, expressed in insect cells, possessed an underglycosylated glycoform in which Asn[24] was not glycosylated. Mutation to eliminate this partially occupied site facilitated crystallization of this hormone (17). Horse FSH preparations comprise two eFSHβ glycoforms. One possesses both N-linked oligosaccharides, the other lacks oligosaccharide attached to Asn[7]. Although the abundance of both eFSHβ glycoforms is roughly equivalent in purified eFSHβ preparations, Western blot analysis of intact eFSHβ preparations indicated the partially glycosylated eFSHβ glycoform is the more abundant form (W. J. Walton and G. R. Bousfield, unpublished data). A third hFSH glycoform is hFSHβ, which lacks both N-linked oligosaccharides (61). A similar pattern of all-or-none N-glycosylation has been observed for two other primate species (182). There are two mechanisms accounting for the absence of carbohydrate at a potential N-glycosylation site. The first is that carbohydrate is not attached during biosynthesis, whereas the second involves carbohydrate transfer in the ER, but subsequent removal by PNGase. Because this enzyme converts the Asn residue to Asp, Edman degradation can readily distinguish between the two mechanisms. In the case of FSHβ glycoforms, the presence of phenylthiohydantoin-Asn indicated carbohydrate was never attached to underglycosylated eFSHβ or nonglycosylated primate FSHβ preparations (182). The CGs show two different patterns of glycosylation for the two known examples, hCG and eCG. hCGβ is N-glycosylated at both Asn[13] and Asn[30] and O-glycosylated at Ser residues 121, 127, 132, and 138 (183–185). The amino acid sequences deduced from nucleotide sequences derived from nine other primate species predict a variable pattern of glycosylation. All but two species possess Asn[13] and all possess Asn[30] glycosylation sites. Most nonhuman primate CGβ subunits possess only three of the four human CGβ O-glycosylation sites. The Ser[138] site is most commonly absent owing to an Ala substitution. The Ser[121] site occurs in the fourth position of a series of four Ser residues. In some species, only three residues are present, which might not present the proper glycosylation signal. In contrast, glycosylation patterns for the equids may be more highly conserved. The horse subunit, eCGβ, possesses a single N-linked glycosylation site at Asn[13]. The major N-linked oligosaccharides of eCGβ are predominantly biantennary complex sialylated structures, similar to structure 7 in Fig. 11 (186,187). However, eCG is more heavily glycosylated (45% carbohydrate) than hCG (31% carbohydrate). Most of the difference in carbohydrate content results from more extensive glycosylation of the eCGβ C-terminus. Equine CGβ possesses the same 12 O-glycosylation sites in the C-terminal extension as eLHβ (Fig. 10). The more extensive glycosylation is due to a higher percentage of glycosylation sites decorated with larger oligosaccharides (96).

Carbohydrate Structure

Proposals for glycoprotein glycosylation reflect the available technologies. The gel filtration method of Kobata et al. (188) was restricted to neutral oligosaccharides, requiring sialic acid removal before oligosaccharide characterization. Anion exchange chromatographic methods could deal with charged oligosaccharides, but not neutral oligosaccharides (189,190). Mass spectrometric methods initially were restricted to neutral oligosaccharides, but can now deal with charged oligosaccharides (191), although the bias toward neutral oligosaccharides is readily apparent when mixtures of both oligosaccharides are analyzed. For structure determination, nuclear magnetic resonance (NMR) spectroscopy is the most informative, although the large amounts of highly purified oligosaccharide limit the usefulness of the method (192).

It is well established that the glycoprotein hormones possess highly diverse oligosaccharide structures (169,170). A single N-linked oligosaccharide structure was initially proposed for hCG (structure 7 in Fig. 11). All the carbohydrate heterogeneity was assumed to be due to incomplete processing (193). At the same time, however, Endo et al. (194) reported five charged and three neutral oligosaccharide structures isolated from hCG. One structure was the same as structure 7 in Fig. 11, and others were similar to it in that they were missing only one residue such as the fucose or one or both of the terminal sialic acid residues. Some variations were consistent between subunits. For example, there was no fucose on any of the oligosaccharides isolated from the α subunit (193). A site-specific analysis of all four hCG N-glycosylation sites was conducted by Renwick and colleagues (80). Only eight oligosaccharide structures were identified by NMR spectroscopy. Three were hybrid structures, including one like structure 8 in Fig. 11, and the other two possessing one and two more Man residues, respectively. The fourth was a biantennary oligosaccharide terminated with sialic acid at the nonreducing end of each branch (structure 7).

More recent investigations suggest additional oligosaccharide structures associated with the gonadotropins (176,195). Parsons and Pierce (196) found that bLH was sulfated on its terminal sugars, which prevented exoglycosidases from hydrolyzing terminal sugars, thus hindering structural characterization. Baenziger's laboratory subsequently determined the structure of the disulfated oligosaccharides of bLH (Fig. 11, structure 12) (197). They next examined the

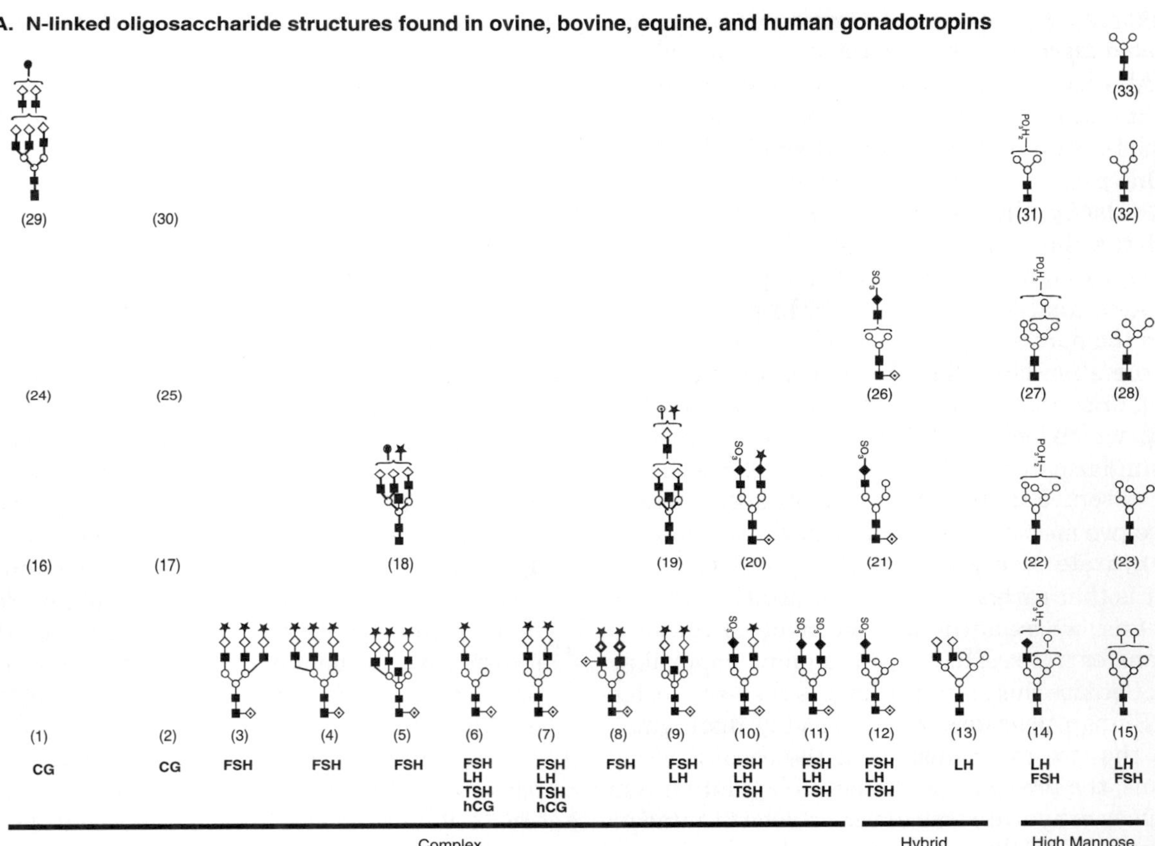

B. O-linked oligosaccharide structures found in equine and human gonadotropins

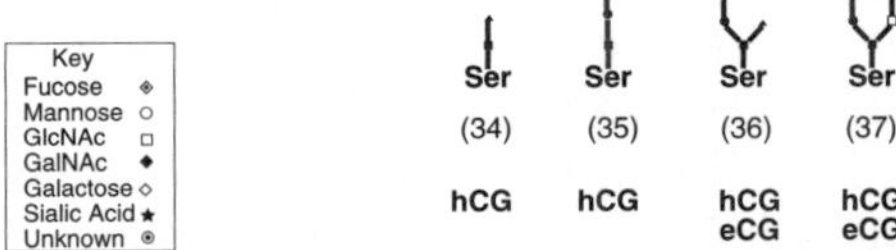

FIG. 11. Oligosaccharide structures found in gonadotropins. N-linked oligosaccharides fall into three major classes: high-mannose, hybrid, and complex. All three are found in gonadotropins, although their relative abundance varies with hormone, tissue source, and species.

oligosaccharides on the ovine, bovine, and human pituitary hormones and found that each hormone possessed several oligosaccharide structures (189,190). The complete structures are shown in Fig. 11. Some hormones, such as oFSH, had 16 charged structures, accounting for 68% of the oligosaccharides released by PNGase digestion. Numerous neutral oligosaccharides were also released, but were too sparse to characterize. The other characterized structures were variations on the structures shown lacking one or more sialic acid residues, or lacking sulfate or fucose. Based on these and other studies, the gonadotropin oligosaccharides fell into three categories: high-mannose, hybrid, and complex. The terminal groups were sialic acid and sulfate. Figure 11 shows the qualitative distribution of the oligosaccharide structures. This may vary from species to species, from hormone to hormone, and from glycosylation site to glycosylation site. For example, hCG and hFSH possessed only sialylated oligosaccharides, and most species of LH possessed predominantly sulfated oligosaccharides or hybrids (Fig. 12). Ovine FSH, on the other hand, had a little bit of everything. For more detailed description of the distribution of oligosaccharide structures, see the review by Baenziger and Green (169). Figure 13, based on this review, summarizes the biosynthetic buildup of the gonadotropin oligosaccharides. (See also the section on Biosynthesis of the Gonadotropins for further comment.)

Oligosaccharide structures have been characterized for eCG and eLH, which differ from each other only in their carbohydrate moieties (180,181,198). Ten N-linked structures were initially characterized from eCG (199), demonstrating that the variability of oligosaccharides obtained from tissue-derived hormone is also characteristic of circulating gonadotropins. Oligosaccharide heterogeneity in eCG was confirmed in a report (186) in which 24 N-linked oligosaccharides were isolated from eCGβ, and even more from eCGα. Only one set of four oligosaccharides was completely characterized by NMR spectroscopy. These oligosaccharides were similar to structure 7 in Fig. 11, except that either one or both of the sialic acids were O-acetylated. The N-linked oligosaccharides of eCGβ have been compared with those of eLHβ (198). Only sialylated oligosaccharides were obtained from eCGβ, whereas eLHβ possessed both sialylated and sulfated oligosaccharides, with the latter being the most abundant. The development of suitable mass spectrometric methods (191) permitted examination of αAsn^{56} oligosaccharides for the equine gonadotropins (19). At least 105 oligosaccharide structures were identified from the oligosaccharide pool released from eLHα at Asn^{56}. The high-mannose oligosaccharides included many small structures that appeared to be the result of continued mannosidase digestion beyond the typical $GlcNAc_2Man_5$ substrate for GlcNAc transferase I (Fig. 11, structures 15, 23, 28, 32, and 33). These species were particularly abundant in the eLHα oligosaccharide population, less abundant in the eFSHα population, and absent in the eCGα population. Not all the high-mannose oligosaccharides were neutral, however; phosphorylated forms were also detected in the eLHα oligosaccharides and were particularly abundant in the eFSHα oligosaccharide population (Fig. 11, structures 12, 22, 27, and 31).

Renwick and colleagues have examined the charged oligosaccharide structures for each glycosylation site of oLH and hLH (79,200). In oLH, 61% of the β-subunit oligosaccharides were of the disulfated type (Fig. 11, structure 12), whereas only 16% of the Asn^{56} and 7% of the Asn^{82} α-subunit oligosaccharides were represented by this structure. The major (35%) oligosaccharide structure attached to Asn^{82} was structure 14, whereas the major structures attached to Asn^{56} were structures 13 and 14 (23% and 27%, respectively). The β-subunit oligosaccharides were 85% to 95% fucosylated, whereas the α-subunit oligosaccharides were partially fucosylated: 10% to 15% at Asn^{56} and 45% to 55% at Asn^{82}. The hLH glycosylation was more heterogeneous than that of oLH, but the oligosaccharides fell into three basic categories based on their Manα1-3 branches, with variations in the structures of the Manα1-6 branches. The major species (18% of αAsn^{56} oligosaccharides and 15% of βAsn^{30} oligosaccharides) was the hybrid sulfated/sialylated oligosaccharide (Fig. 11, structure 10), and various sulfated forms comprised half of the oligosaccharides attached to these asparagine residues. A minor, new oligosaccharide (structure 10, Fig. 11) was detected predominantly at Asn^{82}. This apparently resulted from the presence of an α2-6GalNAcβ1-4GlcNAc sialyltransferase in the human pituitary. This enzyme has been reported to be present in bovine colostrum (201). The remaining charged oligosaccharides were variations on structure 9. Most of the hLHβ-subunit oligosaccharides were fucosylated, whereas only 15% to 25% of αAsn^{56} and 0% to 10% of the αAsn^{82} oligosaccharides were fucosylated, similar to the situation for oLHα and hCGα described previously. These results suggest that the β subunit reduces the accessibility of the α-subunit oligosaccharides to fucosyltransferase once the subunits are associated.

LH and CG produced by recombinant DNA methodology do not possess the same oligosaccharides as the native hormones (202,203). Recombinant LH lacks the sulfated oligosaccharides because CHO cells used for the in vitro synthesis lack GalNAc transferase and sulfotransferase (202). Recombinant hCGβ, produced in a baculovirus expression system, possesses high-mannose N-linked oligosaccharides and O-linked disaccharides, based on composition analysis of the recombinant subunit (203). Recombinant hFSH appears to possess similar types of oligosaccharides compared with the native hormone, but some variations in extent of fucosylation and other structural details have been reported (204).

Carbohydrate Function

The best-understood role of carbohydrate is in determining the circulatory half-lives of the gonadotropins because this fits into the current model for oligosaccharide function involving binding to specific lectins. The classic asialo-glycoprotein receptor (205) has been joined by a more recently identified receptor that recognizes sulfated oligosaccharides (206). Together, these two receptor systems are involved in the rapid clearance of these forms of the hormones from the circulation. Baenziger and colleagues (206) have proposed that a component of the pulsatile levels of LH in serum is the rapid clearance by the latter receptor, which recognizes newly released hormone, whereas the asialo-glycoprotein receptor removes glycoproteins only after they lose terminal sialic acid residues. This proposal is supported by the observation that native LH bearing sulfated oligosaccharides is removed from circulation four to five times more quickly than recombinant bLH having only complex sialylated oligosaccharides (207). Additional supporting evidence was provided when sulfated oligosaccharides were demonstrated in lower vertebrate gonadotropins (208). The sulfated oligosaccharide receptor was identified as the Cys-rich N-terminal domain of the macrophage mannose

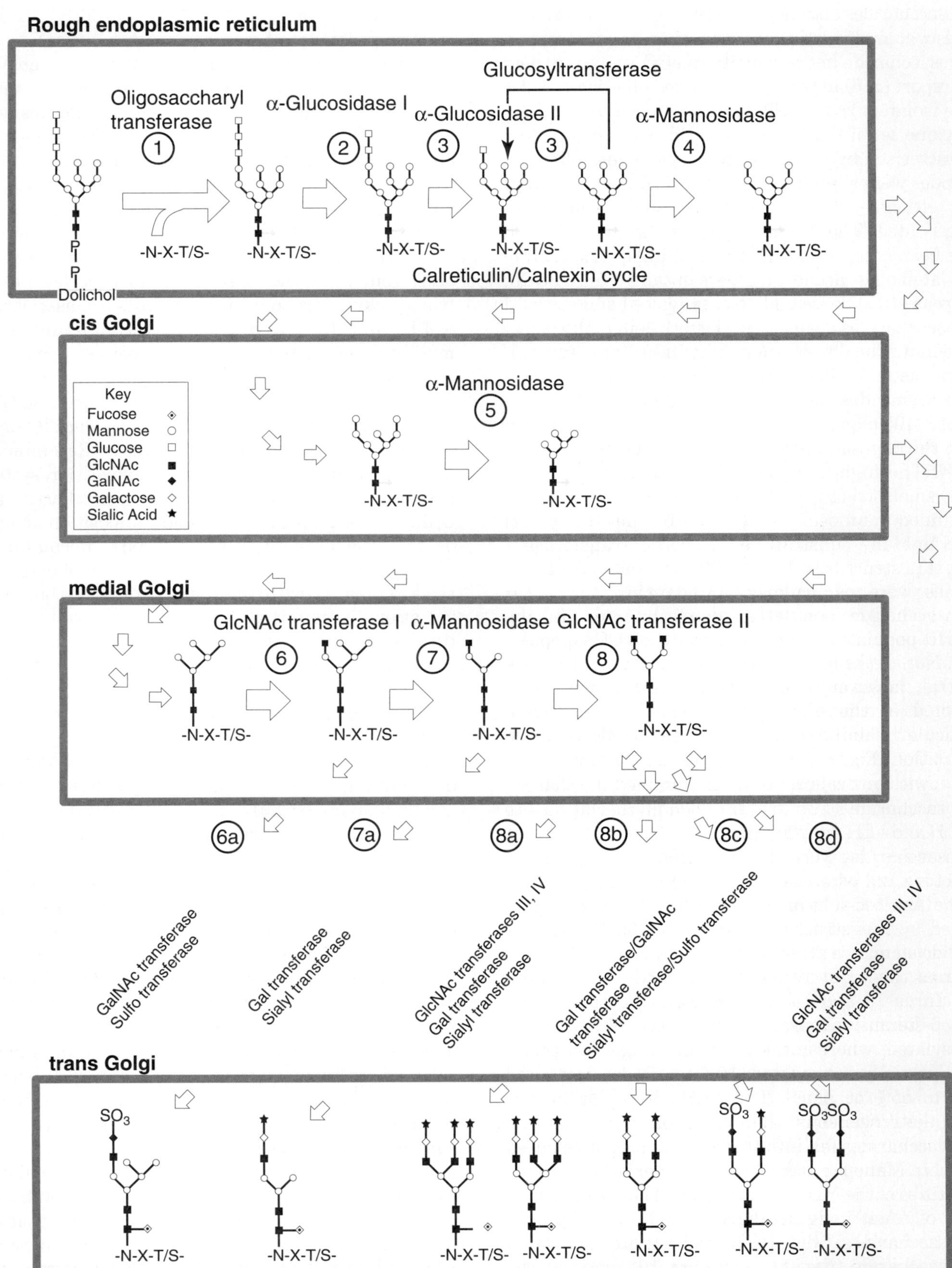

Rough endoplasmic reticulum
Oligosaccharyl transferase
1
α-Glucosidase I
2
α-Glucosidase II
3
Glucosyltransferase
3
α-Mannosidase
4
-N-X-T/S-
P
P
Dolichol
Calreticulin/Calnexin cycle
cis Golgi
Key
Fucose
Mannose
Glucose
GlcNAc
GalNAc
Galactose
Sialic Acid
α-Mannosidase
5
medial Golgi
GlcNAc transferase I
6
α-Mannosidase
7
GlcNAc transferase II
8
6a
7a
8a
8b
8c
8d
GalNAc transferase
Sulfo transferase
Gal transferase
Sialyl transferase
GlcNAc transferases III, IV
Gal transferase
Sialyl transferase
Gal transferase/GalNAc transferase
Sialyl transferase/Sulfo transferase
GlcNAc transferases III, IV
Gal transferase
Sialyl transferase
trans Golgi
SO_3
SO_3SO_3

FIG. 12. Biosynthetic pathway for N-linked oligosaccharides leading to completed N-linked oligosaccharide structures shown in Fig. 11. This is based on the pathway for N-linked glycosylation (299) and pathways suggested by structures on glycoprotein hormones (169).

1. Oligosaccharyl transferase catalyzed transfer of dolichol-linked $Glc_3Man_9GlcNAc_2$ oligosaccharide to asparagine at glycosylation site –Asn–Xaa–Ser/Thr–.
2. α-Glucosidase I removes terminal glucose.
3. α-Glucosidase II removes remaining two glucose residues.
4. Endoplasmic reticulum α-1,2-mannosidase removes one mannose residue.
5. Golgi α-mannosidase I removes another three mannose residues, producing the precursor for N-acetylglucosamine transferase I.
6. N-Acetylglucosamine transferase I adds N-acetylglucosamine, creating precursor for S-1 type having only one completed branch or leading to continued processing.
   a. Addition of N-acetylgalactosamine diverts processing to complex biantennary or triantennary oligosaccharide, resulting in this single-branch sulfated structure. Signal for GalNAc transferase recognition appears to be –Pro–Xaa–Arg/Lys– (291).
7. Golgi α-mannosidase II removes two mannose residues.
   a. Addition of N-acetylgalactosamine could lead to this structure.
8. N-Acetylglucosamine transferase II adds second N-acetylglucosamine, leading to further processing that produces the following oligosaccharide structures:
   a. Triantennary sialylated
   b. Biantennary sialylated
   c. Hybrid sialylated-sulfated
   d. Disialylated, due to α2-6GalNAcβ1-4GlcNAc sialyltransferase in human pituitaries (79)
   e. Disulfated

receptor (209). In hepatocytes it reportedly existed as a homodimer, which primarily bound GalNAc-sulfate–terminated oligosaccharides. Knockout experiments in different laboratories provided conflicting results. Lee et al. (210) first reported that knocking out the macrophage mannose receptor had no effect on pLH clearance. In contrast, Mi et al. (211) reported mannose/GalNAc-4-SO_4 receptor knockout was embryonic lethal, but were able to study the heterozygotes, which displayed reduced LH clearance and a defect in implantation.

In the human, the placenta lacks GalNAc transferase and sulfotransferase; therefore, hCG possesses only sialylated oligosaccharides and in part for this reason is cleared more slowly from the circulation. The other reason for the slow clearance of hCG is the presence of the glycosylated C-terminal extension on hCGβ. This glycopeptide is not required for biological activity because it can be removed by mild acid cleavage (212) and by mutagenesis of the hCGβ cDNA without affecting receptor binding activity (103,145). The C-terminal extension has been attached to hormones that lack this moiety with the effect of prolonging their circulatory half-lives (106). A genetic model for chronically elevated LH has been devised by knocking in the C-terminal peptide on mouse LHβ (213).

In addition to its effects on circulating glycoprotein hormone half-lives, carbohydrate also modulates their specific activities (118,206,214). In this regard, in vitro studies have suggested that deglycosylated gonadotropins can act as antagonists. However, in vivo experiments do not support this idea.

Deglycosylated Forms

Chemically or enzymatically deglycosylated CG, LH, FSH, and TSH bind their receptors with a higher apparent affinity than the native hormones (173, 215–217). However, the activities of deglycosylated glycoprotein hormones are much lower in assays that measure a target cell response such as steroidogenesis. By deglycosylating α or β subunit and then recombining the deglycosylated subunit with the intact complementary subunit, it was demonstrated that only the α-subunit carbohydrate was required for these responses (172–174). In the course of these studies, some investigators reported that the deglycosylated gonadotropins acted as antagonists to the native hormones (218). Deglycosylated hormone occupied receptors and formed inactive hormone–receptor complexes that prevented native hormone from binding to receptors and activating the target cells. Liu et al. (174) showed deglycosylated hormone had a faster on-rate than native hormone, but the off-rates of both were comparable. This probably contributes to the poor in vivo antagonism of the deglycosylated hormone.

The use of site-directed mutagenesis of hCG subunit cDNAs to eliminate single glycosylation sites has further localized the roles of individual glycosylation sites. Of the two α-subunit glycosylation sites, the Asn^{52} site is essential for biological activity (51). As long as this oligosaccharide is present, elimination of the other glycosylation sites has no effect on biological activity. However, elimination of the Asn^{52}

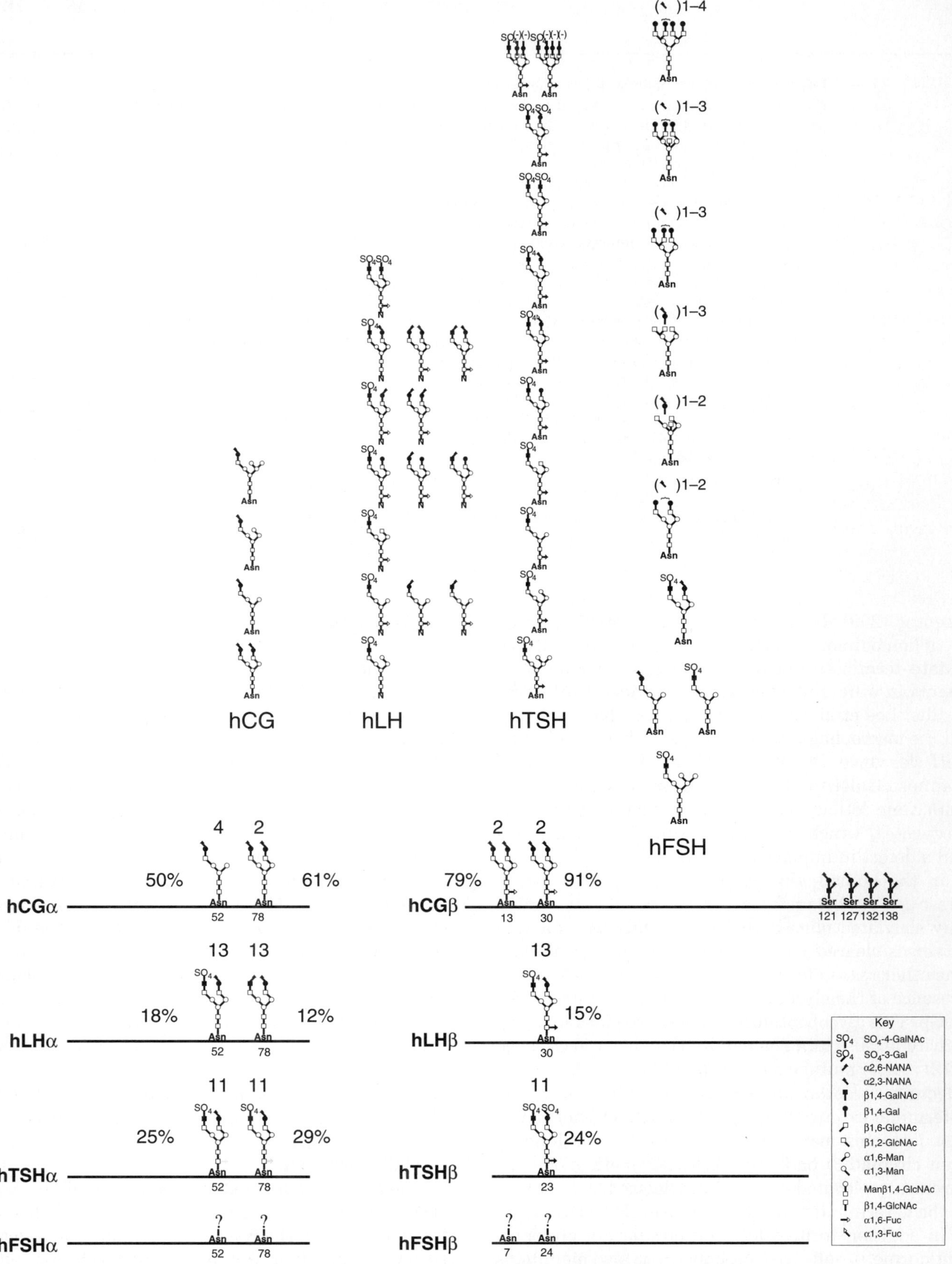

FIG. 13. Human gonadotropin oligosaccharide heterogeneity. **A:** Populations of oligosaccharide structures found at N-glycosylation sites of human glycoprotein hormone preparations: chorionic gonadotropin (hCG), luteinizing hormone (hLH), thyroid-stimulating hormone (hTSH), and follicle-stimulating hormone (hFSH). Most structures are found at all glycosylation sites, although their relative abundance varies from site to site. **B:** Glycosylation patterns for human glycoprotein hormones, as indicated, showing the most abundant structure found at each site for hCG, hLH, and hTSH (79–81). No site-specific information exists for hFSH (190,300).

glycosylation site only partially impairs hCG biological activity. In the absence of α-subunit Asn^{52} oligosaccharide, the β-subunit Asn^{13} oligosaccharide contributes to the biological activity of hCG, because elimination of both oligosaccharides reduces biological activity to that of totally deglycosylated hCG (51). Similar experiments have been carried out with FSHβ; however, deletion of either or both glycosylation sites did not affect in vitro biological activity in one study, whereas deletion of both sites affected biological activity in the second (219,220). Expression of the recombinant hFSH derivatives in different cell lines might have decorated the hormones with different populations of oligosaccharides. A pituitary hFSH glycoform that possessed only α-subunit carbohydrate exhibited very high biological activity in a rat granulosa cell bioassay (61).

In an attempt to interfere with LH maintenance of postovulatory corpora lutea, deglycosylated hCG was administered to normal cycling women (221). However, there was no effect on the luteal phase except to elevate progesterone levels. This prompted the investigators to suggest that agonist activity in the preparation interfered with its antagonist activity. The activities of native, desialylated, and deglycosylated hCG were compared in male cynomolgus monkeys (222). hCG was deglycosylated, and partially deglycosylated hCG removed by concanavalin A–Sepharose chromatography. In vitro antagonism of hCG stimulation of cyclic adenosine monophosphate (cAMP) production by this deglycosylated hCG preparation was demonstrated in rat testis Leydig cells. The serum half-life of deglycosylated hCG was 23 minutes, 23 times that of asialo-hCG, which is rapidly cleared by the galactose-binding, asialo-glycoprotein receptor in the liver (205), but less than 10% that of native hCG. However, the early (48-hour) responses to stimulation by hCG, asialo-hCG, and deglycosylated hCG were identical. No in vivo antagonistic activity by deglycosylated hCG was observed. After 48 hours, testosterone levels of hCG-treated monkeys remained elevated whereas those of asialo-hCG– and deglycosylated hCG–treated animals had returned to normal levels, suggesting carbohydrate may affect long-acting hormone stimulation in vivo.

Circulating forms of hFSH with in vitro antagonistic effects have been reported (223), although no in vivo studies have been performed to determine the physiological relevance of this observation. Chappel and colleagues (118) reviewed evidence by a number of investigators that physicochemical changes occur in FSH in response to changing physiological conditions. These changes in apparent molecular size and distribution of isoelectric forms were attributed principally to variations in FSH sialic acid content. Desialylation of FSH shortened its circulatory half-life, but elevated its potency in receptor binding assays in vitro. A model was proposed in which FSH was modified by neuraminidase action, which converted relatively more acidic, long-acting, low-potency forms into more basic, more potent, short-lived forms. However, subsequent studies using a different in vitro FSH assay reported a different FSH activity profile. Hsueh and colleagues (224) have developed an in vitro steroidogenesis assay sensitive enough to detect FSH in serum. The more acidic forms of FSH (Chappel's less potent, long-acting forms) were determined to be the active circulating forms. Moreover, the more basic forms, which were released in response to a GnRH antagonist, were found to be FSH antagonists in the steroidogenesis bioassay (223). In a rare study involving purified hFSH glycoforms, Stanton et al. (225) reported that the most active preparations possessed the highest sialic acid content, 13.5 mol sialic acid/mol hFSH. This suggested three of four N-glycosylation sites were triantennary, whereas at least one was tetra-antennary. The nonglycosylated hFSHβ isoform complicated this simple oligosaccharide branch accounting because even the most acidic hFSH fractions possessed the nonglycosylated hFSHβ, suggesting maximal branching of hFSHα oligosaccharides (61).

HORMONE RECEPTORS

The glycoprotein hormone receptors have been cloned (226–232) and found to share a similar structure consisting of a large extracellular domain, seven transmembrane regions, and a cytoplasmic tail (Fig. 14). Except for the large extracellular domains, these receptors are similar to other G-protein–coupled receptors that bind much smaller ligands. Comparison of the glycoprotein hormone receptor gene organization of 10 to 11 exons with the single exon of the β-adrenergic receptor gene suggests that the former arose by insertion of a leucine-rich repeat domain into the single exon of a β-adrenergic–like receptor (233–235).

Luteinizing Hormone/Chorionic Gonadotropin Receptors

The human LH receptor gene is located on chromosome 2 at 2p21. It is the smallest of the three receptor genes, spanning 68.9 kbp (Fig. 14). The receptor for LH also recognizes CG, as to be expected by their similar or identical amino acid sequences. Receptors for LH/hCG have been demonstrated on a variety of tissues in the reproductive system, including Leydig cells, granulosa cells, and luteal cells (236,237). LH receptors have also been detected in nonovarian

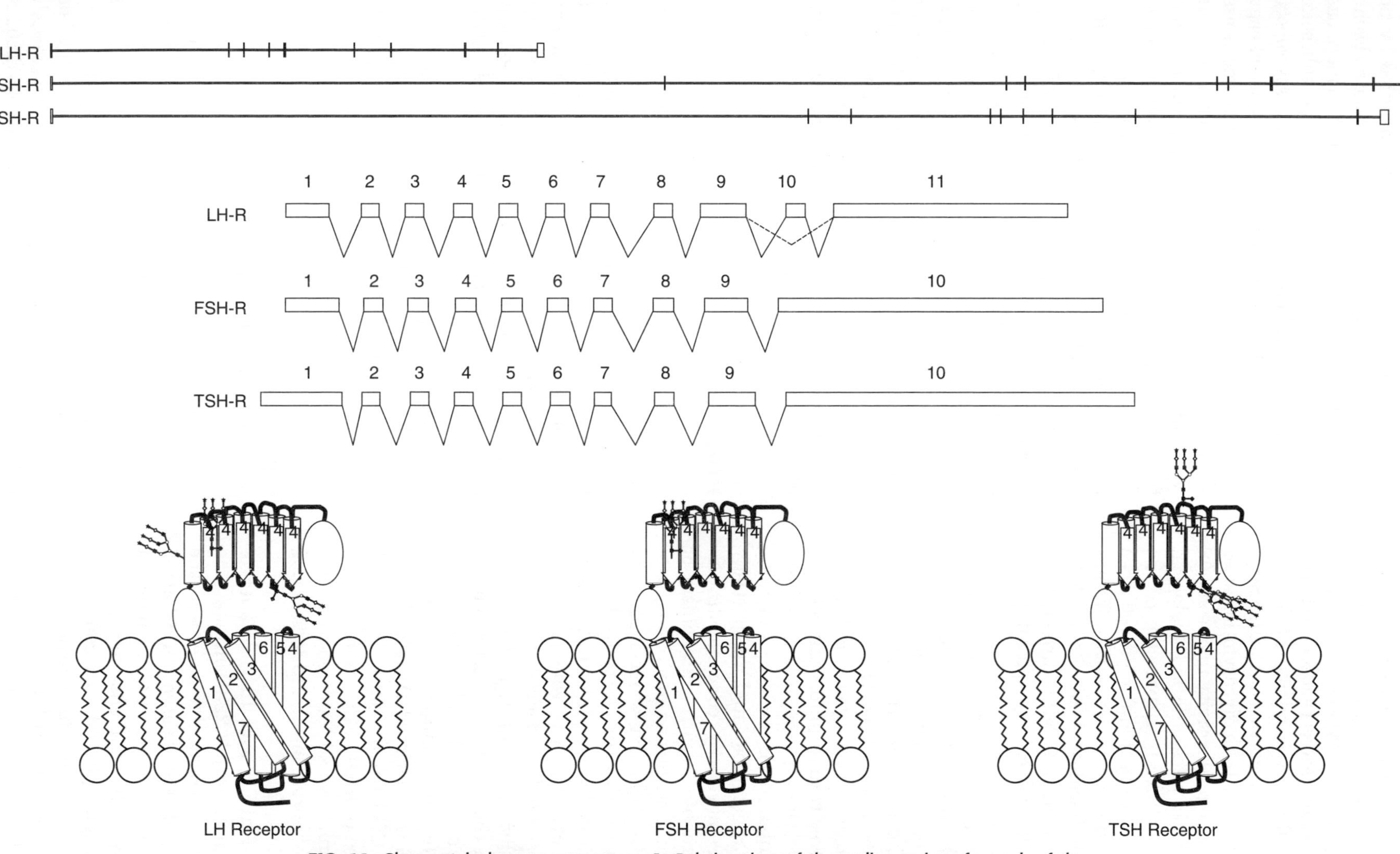

FIG. 14. Glycoprotein hormone receptors. **A:** Relative sizes of the coding regions for each of the glycoprotein hormone receptor genes. Both luteinizing hormone receptor (LHR) and follicle-stimulating hormone receptor (FSHR) are located in the same region of human chromosome 2 at 2p21 and 2p21-p16, respectively. The thyroid-stimulating hormone receptor (TSHR) lies on human chromosome 14 at 14q31. **B:** Enlarged exon structures for each gene, showing alternative splicing for LHR. Note that exons 2 to 9 encode individual leucine-rich repeats. **C:** Diagrammatic depiction of the receptors, showing putative leucine-rich repeat structure in the extracellular domain and seven transmembrane helices in the transmembrane domains. The connections between the extracellular and transmembrane domains are unknown, as are the relative orientations of the two domains.

tissues [see Lincoln et al. (238) for references]. LH receptors have been cloned from a variety of species using recombinant DNA methodology, including pigs (228), rats (227), humans (232), mice (239,240), a frog (241), and a variety of other species (Table 2). The rat LH receptor gene has been cloned (234,242) and consists of 11 exons. The first 10 encode the extracellular domain, whereas the 11th encodes the transmembrane domain (Fig. 14). The LH receptor consists of a single polypeptide chain ranging in size from 669 residues in the pig (227) to 742 residues in the rat (228). All of these receptors have a long extracellular sequence, the hormone-binding site (11), that contains six potential N-linked glycosylation sites (227,228, 232,243). All species' receptors also possess seven predicted transmembrane segments, which are similar to other receptors known to be coupled to G proteins, the adenylyl cyclase second messenger system, and

TABLE 2. *Glycoprotein hormone receptor sequence accession numbers*

Species	LHR	FSHR	TSHR
Fish			
Acanthopagrus latus (yellowfin sea bream) *Acipenser baerii* (Siberian sturgeon) *Acipenser gueldenstaedtii* (Russian sturgeon) *Acipenser schrenckii* (Amur sturgeon) *Anguilla japonica* (Japanese eel) *Anguilla anguilla* (European eel) *Aristichthys nobilis* (bighead carp) *Carassius auratus* (goldfish) *Channa maculata* (snakehead mullet)			
Clarias gariepinus (African catfish)	gi|25987159| gb|AAN75752.1| AF324540_1	gi|27262930| emb|CAB51907.2|	gi|30061315| gb|AAN01360.1|
Clupea harengus (Atlantic herring) *Conger conger* (conger eel) *Conger myriaster* (Japanese conger eel) *Coregonus autumnalis* (Arctic cisco) *Ctenopharyngodon idella* (grass carp) *Cyprinus carpio* (common carp)			
Danio rerio (zebrafish)	gi|45387709| ref|NP_991188.1|	gi|30692148| gb|AAP33512.1|	
Dicentrarchus labrax (sea bass) *Epinephelus coioides* (orange-spotted grouper) *Epinephelus akaara* (Hong Kong grouper) *Epinephelus septemfasciatus* (seven-band grouper) *Fundulus heteroclitus* (killifish) *Gasterosteus aculeatus* (three-spined stickleback) *Hippoglossus hippoglossus* (Atlantic halibut) *Hypophthalmichthys molitrix* (Silver carp)			
Ictalurus punctatus (channel catfish)	gi|13236165| gb|AAK16066.1| AF285181_1	gi|13236167| gb|AAK16067.1| AF285182_1	gi|42767647| gb|AAS45557.1|
Katsuwonus plelamis (bonito) *Monopterus albus* (swamp eel)			
Morone saxatilis (striped sea bass)			gi|8886877| gb|*AAF80596.1*| AF239761_1|
Muraenesox cinereus (daggertooth pike conger) *Mylopharyngodon piceus* (black carp) *Neoceratodus forsteri* (Australian lungfish) *Odontesthes bonariensis* *Oncorhynchus keta* (chum salmon) *Oncorhynchus kisutch* (Coho salmon) *Oncorhynchus masou* (cherry salmon)			
Oncorhynchus mykiss (rainbow trout/summer salmon)	gi|33315749| gb|AAQ04550.1| AF439404_1		

(Continued)

TABLE 2. *Glycoprotein hormone receptor sequence accession numbers—cont'd*

Species	LHR	FSHR	TSHR
Oncorhynchus rhodurus	gi|5926645| dbj|BAA84638.1|	gi|33315752| gb|AAQ04551.1| AF439405_1	gi|9711260| dbj|BAB07800.1|
Ophisternon bengalense			
Oreochromis mossambicus (Mozambique tilapia)			
Oreochromis niloticus (Nile tilapia)	gi|10567246| dbj|BAB16107.1|	gi|10567244| dbj|BAB16106.1|	gi|13365517| dbj|BAB39132.1|
Oryctolagus cuniculus (Nile tilapia)			
Pagrus major (red sea bream)			
Paralichthys olivaceus (Japanese flounder)			
Plecoglossus altivelis (ayu)			
Salmo salar (Atlantic salmon)	gi|33438476| emb|CAE30288.1|	gi|32127688| emb|CAD98923.1|	
Scyliorhinus canicula (smallspotted catshark)			
Sparus aurata (gilthead seabream)	gi|46577733| gb|AAT01412.1|	gi|46577735| gb|AAT01413.1|	
Thunnus obesus (bigeye tuna)			
Trichogaster trichopterus (blue gourami)			
Amphibians			
Bufo japonicus (Japanese toad)			
Cynops pyrrhogaster (Japanese firebelly newt)		gi|25295803| pir||JC7361|	
Rana catesbeiana (bullfrog)			
Rana japonica (Japanese brown frog)			
Rana ridbunda (lake frog)			
Xenopus laevis (African clawed frog)			
Reptiles			
Bothrops jararaca		gi|37778925| gb|AAO72730.1|	
Chinemys reevesii (Reeve's turtle)			
Pelodiscus sinensis (Chinese softshell turtle)			
Podarcis sicula		gi|16580628| emb|CAC82173.1|	
Sphenedon punctatus			
Takydromus tachydromoides (lizard)			
Birds			
Cairina moschata		gi|30313394| gb|AAM34797.1|	
Coturnix japonica (Japanese quail)			
Gallus gallus (chicken)	gi|45384388| ref|NP_990267.1|	gi|45384184| ref|NP_990410.1|	gi|50749032| ref|XP_426455.1|
Meleagris gallopavo (turkey)			
Nipponia nippon (crested ibis)			
Struthio camelus (ostrich)			
Mammals			
Ailuropoda melanoleuca (giant panda)			
Ailurus fulgens (red panda)			
Balaenoptera acutorostrata (Minke whale)			
Bos taurus (cow)	gi|20067177| gb|AAM09535.1|	gi|41386766| ref|NP_776486.1|	gi|27806373| ref|NP_776631.1|
Bubalus bubalis (water buffalo)			
Callithrix jacchus (white-tufted-ear marmoset)	gi|3122375| sp|O02721|		
Camelus dromedarius (camel)			
Canis familiaris (dog)	gi|854| emb|CAA35027.1|		gi|50979096| ref|NP_001003285.1|
Capra hircus (domestic goat)			
Cavia porcellus (domestic guinea pig)		gi|19849629| gb|AAL92577.1|	
Ceratotherium simum (white rhinoceros)			

TABLE 2. *Glycoprotein hormone receptor sequence accession numbers—cont'd*

Species	LHR	FSHR	TSHR
Cercopithecus aethiops (African green monkey)			gi|27413473| gb|AAO11782.1|
Cervus nippon (sika deer)			
Equus asinus (ass/donkey)		gi|1658014| gb|AAB18245.1|	
Equus burchellii (Burchell's zebra)			
Equus caballus (horse)		gi|1346042| sp|P47799|	
Felis catus (cat)			gi|12642092| gb|AAK00133.1| AF218264_1|
Homo sapiens (human)	gi|1225984| emb|CAA59234.1|	gi|31657138| ref|NP_000136.2|	gi|38016895| gb|AAR07906.1|
Lama glama (llama)			
Macaca fascicularis (long-tailed macaque)		gi|539528| pir||JN0898|	
Macaca mulatta (rhesus monkey)			gi|27413475| gb|AAO11783.1|
Macropus rufus (red kangaroo)			
Macropus eugenii		gi|37543266| gb|AAL99292.1|	
Mastomys coucha (southern multimammate mouse)			
Meriones unguiculatus (Mongolian gerbil)			
Mesocricetus auratus (golden hamster)		gi|41019342| gb|AAR98576.1|	
Microtus montebelli (Japanese grass vole)			
Monodelphis domestica (gray short-tailed opossum)			
Mus musculus (house mouse)	gi|7305233| ref|NP_038610.1|	gi|31980789| ref|NP_038551.2|	gi|575922| gb|AAA53209.1|
Oryctolagus cuniculus (rabbit)			
Ovis aries (sheep)		gi|544352| sp|P35379|	gi|3024766| sp|P56495|
Panthera tigris altaica (Amur tiger)			
Papio anubis (olive baboon)			
Phodopus sungorus (Russian dwarf hamster)			
Physeter catodon (sperm whale)			
Rattus norvegicus (rat)	gi|6981160| ref|NP_037110.1|	gi|40385885| ref|NP_954707.1|	gi|6981680| ref|NP_037020.1|
Sus scrofa (pig)	gi|47523950| ref|NP_999614.1|	gi|1066833| gb|AAA86933.1|	gi|47523664| ref|NP_999462.1|
Trichosurus vulpecula (silver-gray brushtail possum)			

FSHR, follicle-stimulating hormone receptor; LHR, luteinizing hormone receptor; TSHR, thyroid-stimulating hormone receptor.

cAMP. The transmembrane segments are predicted by hydropathy analysis that identifies hydrophobic regions of the protein sequence, and by comparison with rhodopsin (228). In addition, there is a short cytoplasmic tail consisting of 68 to 72 residues that may possess a plasma membrane–targeting signal sequence (244).

Antibodies raised against synthetic peptides corresponding to the N-terminal and C-terminal portions of the LH receptor have been used to confirm that the N-terminal portion is exposed extracellularly whereas the C-terminus is intracellular, as predicted by homology to other G-protein–coupled receptors (245). Expression of a truncated form of the rat ovarian LH receptor in COS1 cells resulted in secretion of soluble receptor (242). The amount of soluble receptor expressed was approximately 16% that of full-length receptor transfected into the same cell line. Expression of human kidney 293 cells of a cDNA encoding an experimentally truncated rat LH receptor gene consisting of only the extracellular domain demonstrated that the high-affinity binding to hCG by the receptor was attributable to this portion of the molecule alone (11). The experiment also indicated that the truncated form of the receptor was not secreted by the cell, so that receptor-binding studies were possible only by using detergent extracts of the cells. Perhaps this is why LH receptor-binding inhibitor was detected only after freeze-thawing of rat ovaries (246). Entrapment of modified receptors occurs frequently. When chimeric molecules consisting of the LH receptor extracellular domain and either the β_2-adrenergic receptor transmembrane domain or the vesicular stomatitis virus–G-protein transmembrane domain were transfected into COS-7 cells, the chimeric receptors were not transported to the cell surface, but could be detected in detergent extracts of the transfected cells (142). Native LH receptor and a chimera containing the C-terminal domain of the β_2-adrenergic receptor were transported to the surface of these cells. Only chimeric receptors containing the extracellular domain of the LH receptor were capable of binding ^{125}I-hCG. A solubilized chimera consisting of the LH receptor extracellular domain and β_2-adrenergic receptor transmembrane domain could bind both hCG and β-adrenergic ligands. The LH receptor transmembrane domain could not bind β-adrenergic ligands.

Construction of chimeric receptor molecules containing portions of the LH, FSH, and TSH receptor extracellular domains is currently in progress to characterize the determinants responsible for recognition of the different ligands. Substitution of TSH receptor residues 82 to 170 and 260 to 360 for the corresponding regions of the LH receptor eliminated binding to hCG (247). Substitution of residues 170 to 260 resulted in a chimera that could bind both hCG and TSH, although the affinity of the chimeric receptor for each hormone was a 1/10th that of each native receptor for its respective ligand. Expression of truncated LH receptors indicated that leucine-rich repeats 1 to 8 could bind hCG with an affinity similar to that of the native receptor (248). Substitution of the N-terminal six leucine-rich repeats of the LH receptor for the corresponding regions of the FSH receptor resulted in a chimeric receptor that could bind hCG and stimulate cAMP accumulation when expressed in human embryonic kidney 293 cells. Conversely, substitution of this region on the LH receptor by the corresponding region of the FSH receptor abolished hCG binding but did not confer FSH binding. Leucine-rich repeats 1 to 11 had to be substituted to enable a chimeric LH/FSH receptor to bind FSH and stimulate cAMP production.

A role for carbohydrate in the functioning of the LH receptor is not known. When site-directed mutagenesis was used to eliminate N-glycosylation sites of the TSH receptor, elimination of two of the six resulted in abolishment or reduction of TSH binding, whereas loss of the remainder had no effect on TSH binding (249). Neither of the two glycosylation sites that affect TSH binding are present in LH or FSH receptors. Removal of carbohydrate from the rat LH receptor by PNGase digestion had no effect on the binding of hCG (250). Two products of enzymatic deglycosylation of the receptor were observed by sodium dodecyl sulfate (SDS)–polyacrylamide gel electrophoresis followed by transfer to nitrocellulose and ligand blotting. The native receptor migrated with an M_r of 90,000 and the deglycosylated receptor forms migrated at M_r 67,000 and M_r 62,000. Digestion of hormone–receptor complexes removed oligosaccharide from the receptor and from both subunits of hCG, suggesting that all the oligosaccharides faced outward from the hormone–receptor interface.

The transmembrane domain contains loops exposed to the cytoplasm, which may be important in receptor function. One of the receptors that is also linked to G protein is the β_2-adrenergic receptor. The sites of interaction of this receptor with G protein were localized to the third cytoplasmic loop and a portion of the cytoplasmic tail close to the cell membrane (251). The amino acid sequences of these regions are not conserved between the β_2-adrenergic receptor and the LH receptor, and conversion of lysines 541, 544, and 547 (Lys-653, -656, and -659, respectively, in Fig. 14) had no effect on G-protein coupling (252). Membrane trafficking was affected because most of the mutant receptor remained inside the cell.

All LH receptors contain a C-terminal domain that extends into the cell cytoplasm and may be a further site for the regulation of hormone–receptor function because it contains potential phosphorylation sites. Two potential protein kinase-C phosphorylation sites have been identified, along with a third in the third cytoplasmic loop of the transmembrane domain (227). In other G-protein–linked receptors, phosphorylation of the receptor leads to uncoupling of the receptor from G protein and subsequent systems (253). It is interesting to speculate that one mechanism for the regulation of LH at the target cell could be phosphorylation of the receptor, induced by the phosphorylating enzymes stimulated by the hormone–receptor complex. The phosphorylated receptor would be ineffective in stimulating further cAMP production, shutting off the initiating signal for biological activity.

Mutations in the cytoplasmic C-terminal tail of the rat LH receptor suggested that residues 616 to 631 (residues 728 to 742 in Fig. 12) were important for membrane trafficking because mutants lacking the last C-terminal 58 residues were not expressed on the plasma membrane, whereas those lacking the last 43 and 21 amino acid residues were expressed on the plasma membrane when transfected in human kidney 293 cells (244). The amount of cAMP stimulated by hCG binding to mutant LH receptors was double that stimulated by hCG binding to full-length LH receptor, even though the same amount of hCG was bound by each cell type. Furthermore, the mutant receptors were internalized more rapidly than the full-length receptor. Subsequently, clonal 293 cells derived from transiently transfected cells were compared. Cells transfected with wild-type LH receptor expressed three to four times more receptor than those cells expressing a mutant LH receptor lacking the C-terminal 48 amino acid residues (254). However, the stimulation of cAMP accumulation in response to hCG binding in cells with the mutant receptor was three times greater than that in cells expressing the wild-type receptor. In addition, hCG-induced uncoupling of the cAMP response was reduced, indicating a role for the C-terminus in this phenomenon.

In general, throughout their sequences the rat LH receptor and porcine LH receptor exhibit 84% homology. The major difference in the length of the two receptors is accounted for by a 24–amino acid insert in the extracellular portion of the rat LH receptor compared with the porcine LH receptor. The most homologous region is the transmembrane portion, whereas the least similar region is the intracellular domain. The human LH receptor is overall 85% identical to the rat LH receptor and 87% identical to the porcine LH receptor (243). Despite the high degree of homology between the human and rat LH receptors, the human LH receptor has a high degree of species specificity, binding hLH and hCG, but not eLH, eCG, rLH, or oLH (232). The rat and porcine LH receptors bind all of these hormones.

Shortly after the crystal structure for hCG appeared, the crystal structure for ribonuclease inhibitor was determined. This protein consists of 12 leucine-rich repeats, the predominant structural element in the glycoprotein hormone extracellular domains. These are organized in a horseshoe-shaped arrangement with the α helices on the outside and the β strands lining the inner face (255). Several groups have built models of the LH extracellular domain assuming anywhere from 7 to 11 leucine-rich repeat motifs. Moyle and colleagues (256) suggested that hCG bound to the loops connecting the inner β strands to the outer helices and pushed the ends of the molecule apart. They speculated that the αAsn^{52} oligosaccharide had to be larger than a minimal size to push the tips far enough apart to cause an activating conformational change in the transmembrane domain. The model was supported by monoclonal antibody binding evidence that indicated most of the hCG molecule was exposed when bound to the LH/CG receptor. Jiang et al. (257) built models for the extracellular domains for the LH, FSH, and TSH receptors that incorporated seven leucine-rich repeats. The model of the LHR included hCG facing the β strands. Wu et al. (5) had noted a positive surface charge in the region responsible for LH receptor-binding activity and suggested its complement might be found in the receptor. This model was tested by mutating charged residues in hCG and finding receptor binding could be restored by compensatory mutations of charged residues in the LH/CG receptor (258).

Follicle-Stimulating Hormone Receptors

The receptor for FSH has been studied in both males and females. The 192-kbp human FSH receptor gene is located on human chromosome 2 at 2p21–p16. The major location for FSH receptor in males is the seminiferous tubule, specifically on Sertoli cells (259). In females, the granulosa cell FSH receptor has been extensively studied (114). In both sexes, the action of FSH through its receptor is mediated through the cAMP second messenger system, induced by adenylyl cyclase and G protein.

The FSH receptor has been cloned from rat (231), human (260), mouse (239), and frog (241) cDNA libraries. It possesses the same large extracellular domain, seven transmembrane helices, and cytoplasmic tail structure as the LH receptor (Fig. 14). The extracellular domain consists of 14 leucine-rich repeats, similar to those described for the LH receptor (228). Leucine-rich repeats 1 to 11 from the FSH receptor were required to create a chimeric LH/FSH receptor that could respond to FSH stimulation (248). There are three N-glycosylation sites on the extracellular domain of the rat FSH receptor and four in the human FSH receptor. An alternative form of the FSH receptor has been reported that consisted of the extracellular domain attached to a single-pass transmembrane domain (261). Protein expression has been detected by immunological means (262). This result suggests that the FSH receptor gene has not yet been completely defined. The FSH receptor has not been as extensively investigated as the LH receptor, but this in part reflects the absence of readily available FSH preparations for use as ligand in binding studies.

Thyroid-Stimulating Hormone and Orphan Leucine-Rich Repeat/G-Protein–Coupled Receptors

The TSH receptor is located on human chromosome 14 at 14q31. Orphan G-protein–coupled receptors possessing an extracellular leucine-rich repeat domain have been reported in vertebrates and invertebrates.

BIOSYNTHESIS OF THE GONADOTROPINS

Once the genetic controls of the pituitary hormones initiate transcription of the required mRNA, protein translation occurs on the polysomes bound to the rough ER (RER). The completed linear subunit from this synthetic step carries a "signal peptide," and has no formed disulfide bonds. The signal peptide (also called the *leader sequence*) at the N-terminus of the subunit precursor is characteristic of the particular subunit. For example, the LH/CGβ subunit prehormone has a 20–amino acid residue leader sequence that is highly homologous throughout the known mammalian species. [The leader sequence on the salmon gonadotropic hormone β subunit has 23 residues but still maintains a considerable homology. The other known leader sequences from the lower vertebrates include the chicken (263) and turkey, which have exceptionally long leader sequences (39 residues).] Most currently known α subunits have a 24-residue leader sequence with a high homology among species. The TSHβ signal peptide for those known consists of 20 residues. The known FSHβ leader sequences have 18 residues for the known cases, but these have sometimes been reported as 19 or 20 residues owing to uncertainties about the exact cleavage point of the signal protease, a confusion that arises from some N-terminal heterogeneity of the isolated gene products, in the authors' opinion. The correct cleavage site in all probability will be shown to be between cysteine and asparagine, arginine, or histidine. This requires an 18-residue leader sequence. Prediction of potential signal peptidase cleavage sites for hFSHβ and eFSHβ suggested the most likely cleavage of the former occurred between residues 2 and 3. This was consistent with 80% possessing N-terminal Cys^3 instead of Asn^1 (61). Cleavage probabilities were roughly equivalent for eFSHβ at both sites, and this corresponded to 50% of the preparation possessing the N-terminal Asn^1 residue and 50% Cys^3.

The signal peptide, which may range from 15 to 40 amino acid residues, characteristically has a very hydrophobic cluster of amino acids 5 to 20 residues toward the N-terminus from the site of cleavage (processing) by the signal peptidase. This step of the protein processing probably occurs *cotranslationally* at about the time the C-terminal portion of the peptide chain coded by the mRNA is being completed. Thus, the signal peptide removal marks the removal of the completed protein chain from the ribosomes and release of the protein into the cisternal space of RER.

In the comments that follow, the reader is alerted to the fact that the processes related to glycosylation have been studied only to a limited degree for the glycoprotein hormones, and thus much of what we propose is based on analogy to other systems. The reasons for this are twofold. First, the gonadotropes or thyrotropes of the pituitary are not readily obtained for study. Second, there are few suitable cell lines that provide reasonable systems to study this glycosylation. In this respect, those that have received the most extensive study are trophoblast-derived cell lines, particularly the JAR or BeWo human choriocarcinoma cell lines, originally isolated by Dr. Roland Patillo of the Medical College of Wisconsin. The JAR cell line, for example, has been through several hundred passages and maintained an essentially constant production of hCG throughout. This property is important for their utility in studies of biosynthesis. However, hCG is somewhat unusual among the glycoprotein hormones in that, besides an extra N-linked glycosylation site on the β subunit, it also has O-linked glycosylation sites on the C-terminus (see earlier). This O-linked glycosylation may be characteristic among all the CGs. However, we have shown it is not a unique feature because equine pituitary LH also has this C-terminal extension on the β subunit, and, as in the CGs, it is O-glycosylated. (See the structural summaries, the β or Hormone-Specific Subunit section, and Fig. 5, earlier.)

For the glycoprotein hormones, there is another important cotranslational event (N-glycosylation) that also occurs in the RER. N-glycosylation takes place at two sites on the α subunit and one or two sites on the β subunit (see the diagrammatic summary in Figs. 9 and 10). O-glycosylation, if it occurs, happens later as a post-translational event in the Golgi apparatus.

N-glycosylation (Fig. 12) is mediated through a "high-mannose" dolichol phosphate intermediate, enzymatically targeted to specific N-glycosylation sites. Not all sites are glycosylated in those proteins that contain this structural feature, but in the glycoprotein hormones all potential N-glycosylation sites are usually N-glycosylated. Exceptions include partial glycosylation of some FSHβ subunits. In the case of the FSHβ subunit, partial glycosylation of recombinant bFSHβ has been reported (264). We have observed partial glycosylation of a preparation of oFSH in which only one of the two potential glycosylation sites was occupied by carbohydrate, giving the preparation an appearance on SDS gels identical to that of oLH, which has only a single N-linked glycosylation site (unpublished data). In eFSHβ, evidence during amino acid sequence determination revealed that the first N-glycosylation site was only partially glycosylated (265). In several primate species, including humans, FSHβ glycosylation appears to be all-or-none (182).

The carbohydrate moiety transferred from the dolichol phosphate intermediate has the general composition, $Glc_3Man_9GlcNAc_2$. The two N-acetylglucosamine residues form the link to the asparagine amide nitrogen in the N-glycosylation site (steps 1 to 3, Fig. 12). On the nonreducing end of this characteristic disaccharide attachment to the asparagine nitrogen amide is a mannose residue linked through positions 1 to 4 of the last N-acetylglucosamine residue. This mannose residue, in turn, is linked (through positions

3 and 6) to two additional mannose residues. This generates a two-armed branch point in the carbohydrate, which creates a "biantennary" structure (see diagram Fig. 11, above). In some of the more complex carbohydrate structures associated with the glycoprotein hormones, a "triantennary" branching is required (see diagrammatic examples, Fig. 11). Whether a biantennary or triantennary carbohydrate will be produced in the "mature" glycoprotein is determined in the Golgi apparatus, but a structure that provides for this possibility is contained in the high-mannose form or precursor form generated in the RER. Characteristically, the three glucose residues in a linear array are attached to the longest mannose chain on one arm of the triantennary high-mannose precursor from dolichol phosphate. While the glycoprotein is still in the RER the three glucose residues are removed; α-glucosidase I removes the terminal glucose residue, then α-glucosidase II sequentially removes the remaining two glucose residues. After the first of these two glucoses is removed, calnexin or calreticulin can bind the remaining glucose residue and recruit chaperones, such as ERp57, to assist in folding (266). Binding is terminated when α-glucosidase II removes the last glucose. An RER glucosyltransferase with a denatured protein sensor can restore the glucose to incompletely folded glycoproteins, enabling them to interact with calnexin or calreticulin for another round of folding. This enzyme seems to act as a folding sensor. Although calreticulin or calnexin complexes with the α subunit or various β subunits have not yet been reported, the reduced yields of glycosylation site-silencing mutants suggest these proteins may be involved in glycoprotein hormone subunit folding (143,267–269). No mature glycoproteins from vertebrates contain glucose in their carbohydrate moiety, with the exception of inhibin (270). A slow-acting α-1,2-exomannosidase I cleaves the middle terminal Man residue, creating a marker for binding to a membrane-bound protein called ER degradation-enhancing α-mannosidase-like (EDEM) protein, which directs glycoproteins to the retrotranslocation and cellular degradation pathway. An endomannosidase can cleave the glucosylated mannose branch, permitting glycoproteins missed by the glucosidases to exit the ER. Lectins, such as ERGIC-53, are responsible for transporting glycoproteins to the Golgi.

A third major development that occurs in the RER is the formation of disulfide bonds in the α and β subunits. Ruddon and colleagues (178) have demonstrated that formation of all the α-subunit disulfide bonds and formation of most of the β-subunit disulfide bonds precedes subunit association. No intermediate, partially folded forms of the α subunit have been detected (178). This is consistent with rapid refolding of denatured α subunit in vitro. Intermediate forms of hCGβ have permitted disulfide bond formation to be followed. Disulfides detected in these studies differed from those in the hCG crystal structure and might indicate disulfide exchange during hCGβ folding (271).

The combination of the α and β subunits is initiated in the RER with the high-mannose forms of the subunits (272). Ruddon and coworkers (178) have detected discrete intermediates in this process. The first of these, designated pβ1, has two disulfide bonds, 34–88 and 38–57, formed. The next detectable form, designated pβ2-free, results from a rate-limiting folding process that results in formation of disulfides 23–72 and 9–90. After formation of disulfide bond 93–100, the intermediate, pβ2-combined, is associated with the α subunit. Formation of the last disulfide bond, 26–110, apparently takes place after association with the α subunit (273). This observation is supported by experiments involving site-directed mutagenesis of hCGβ in which residues 101 to 145 were deleted, yet the mutant β subunit still associated with α subunit (97). On the other hand, substitution of cysteines 26 and 100 with alanine resulted in a β-subunit mutant with reduced ability to associate with α subunit (274). From these folding data, Ruddon and colleagues (273) proposed that residues 1 to 90 comprise one folding domain and that the remaining residues of the C-terminus, where the last two disulfide bonds are formed, comprise a second folding domain. In vitro dimerization of gonadotropin subunits involves a threading mechanism in which the glycosylated αL2 loop is threaded through the seatbelt loop (50). This requires high concentrations of subunits for dimer formation to occur and places limitations on the size of the oligosaccharide attached to αL2 (69,275). Not surprisingly, most subunit dimerization studies have involved hCG and oLH, which possess small Asn^{52} or Asn^{56} oligosaccharides (80,200). Blithe and colleagues reported low dimerization efficiency for free hCGα preparations, which they attributed to the more extensively branched oligosaccharides attached to Asn^{52} (69). Oligosaccharide mapping of Asn^{52} oligosaccharides released from similar preparations showed the relative abundance of triantennary and tetra-antennary oligosaccharides was consistent with the amount of nonassociable free hCGα (70). Results with equine α-subunit reassociation experiments showed a similar inhibition of dimer formation by increasing Asn^{56} oligosaccharide size (276). Ruddon and colleagues (277) reported that formation of the seatbelt disulfide blocked hCG subunit dimerization, presumably because the triantennary oligosaccharide could not be threaded through the latched seatbelt loop. Indeed, use of redox conditions resembling those of the ER, inclusion of protein disulfide isomerase, or addition of low concentrations of β-mercaptoethanol to reassociation mixtures facilitates subunit dimerization (48,278,279). Several more recent reports involving an extensive series of Cys mutants, however, have suggested that in vivo dimer formation also proceeds through the threading mechanism because only limited amounts of alternative disulfide bonds were detected that should have formed if a latching mechanism played a role (280–283).

It is at this point in the secretory pathway that pituitary β subunits diverge from CGβ. hCG, hCGα, and hCGβ are secreted by the placenta (284,285), by cultured choriocarcinoma cells (284), and by cells into which the hCGβ gene has been transfected (73). For the remainder of the β subunits, it has been demonstrated that without cotransfection with α subunit the products of transfected genes for these β subunits do not appear in the medium and are eventually degraded (73,143,286). Several mechanisms have been proposed to explain retention of β subunit by the RER. The first is that not all β subunits fold properly, and these subunits are retained by the calnexin/calreticulin cycle (266). Partially folded β subunits, which do not associate with α subunit, have been detected (178). Evidence for a second mechanism has been reported in pituitaries of thyroidectomized mice, which produce more β subunit and less α subunit than pituitaries of intact animals (287). The excess β subunit ends up in the intracisternal granules that develop in the RER and appear to be transformed into lysosomes in which the β subunits are degraded (288). A third mechanism suggests the β-subunit C-terminal residues are responsible for retention of free β subunit. The basis for this possibility results from the finding that the sequences of hLH and hTSH β subunits as predicted from their respective cDNAs are as much as six to seven residues longer than those found on the mature proteins by conventional protein sequencing methods. (Figure 5 presents the complete cDNA-predicted sequences.) Possibly these residues are responsible for β-subunit retention by the ER. Support for this alternative was provided when the C-terminal extension of hCGβ was attached to hFSHβ and the resulting chimeric protein was found to be secreted as efficiently as hCGβ when transfected into CHO cells alone (106).

However the α-β dimeric precursor of the glycoprotein hormones may get from the RER to the Golgi, it is in the Golgi that a variety of post-translational events occur that have an important bearing on the biochemical and physiological properties of the mature protein hormone produced.

The Golgi complex is a cellular organelle that is best examined at the electron microscope. It is a structure in close juxtaposition with the nucleus. This organelle or complex appears to be composed of parallel, flattened saccules, vesicles, and vacuoles. The portion nearest to the nucleus is designated the *cis* Golgi network, and it is here the protein from the RER enters the Golgi, as followed by pulse-chase labeling (289). Proteins next enter the Golgi stacks, where processing of oligosaccharides continues. This portion of the Golgi consists of at least three compartments, *cis*, medial, and *trans*. Finally proteins enter the *trans*-Golgi network where they are distributed to granules that will carry them to their ultimate destinations.

Thus, the high-mannose form (nine, eight, or six mannose residues at this point) enters the Golgi for post-translational processing. In the *cis* Golgi, α-1, 2-mannosidase may remove an additional three mannose residues, resulting in the structure that is the substrate for N-acetylglucosamine transferase I, which adds N-acetylglucosamine. This structure can be further processed to yield a single branched sulfated structure or go on to be processed into biantennary or triantennary oligosaccharide. The deciding factor is whether the pituitary enzyme N-acetylgalactosamine transferase adds N-acetylgalactosamine. This enzyme recognizes the peptide sequence, –Pro–Xaa–Arg/Lys–, which is present in the α-subunit (residues 44 to 46, –Pro–Ala/Leu–Arg–), in LHβ and CGβ (residues 4 to 6, –Pro–Ala–Arg–) on the N-terminal side of N-glycosylation sites, and in FSHβ (residues 48 to 50, –Pro–Ala–Arg–) on the C-terminal side of the N-glycosylation sites, and is absent in TSHβ (Figs. 4 and 5). Glycosylation of the α subunit is affected by the β subunit with which it is associated (Fig. 13). Therefore, it has been suggested that the β subunit must mask the recognition site on the α subunit. However, only hFSH fits neatly into this model. Although only 7% of the oligosaccharides isolated from hFSH have been found to be sulfated (189,190), this represents the low end of the range of sulfated oligosaccharides attached to FSH (13% of bFSH oligosaccharides and 40% of those of oFSH are sulfated). Furthermore, TSHβ, which also lacks the recognition sequence for N-acetylgalactosamine transferase, possesses 64% to 80% sulfated oligosaccharides. Although the N-acetylgalactosamine transferase recognition sequence has proved useful in predicting the presence of sulfated oligosaccharides in glycoproteins (290), the relative importance of the α and β subunit for determining the activity of N-acetylgalactosamine transferase in glycoprotein hormone biosynthesis remains to be established. Perhaps a more important determinant is the observation that the αAsn^{52} oligosaccharides in hCG and hFSH are oriented differently (Fig. 4B), which might alter accessibility for a protein-bound glycosyl transferase.

Because LHβ and CGβ possess the N-acetylgalactosamine transferase recognition sequence, both LH and CG oligosaccharides are potentially capable of being modified by N-acetylgalactosamine, if it is present.

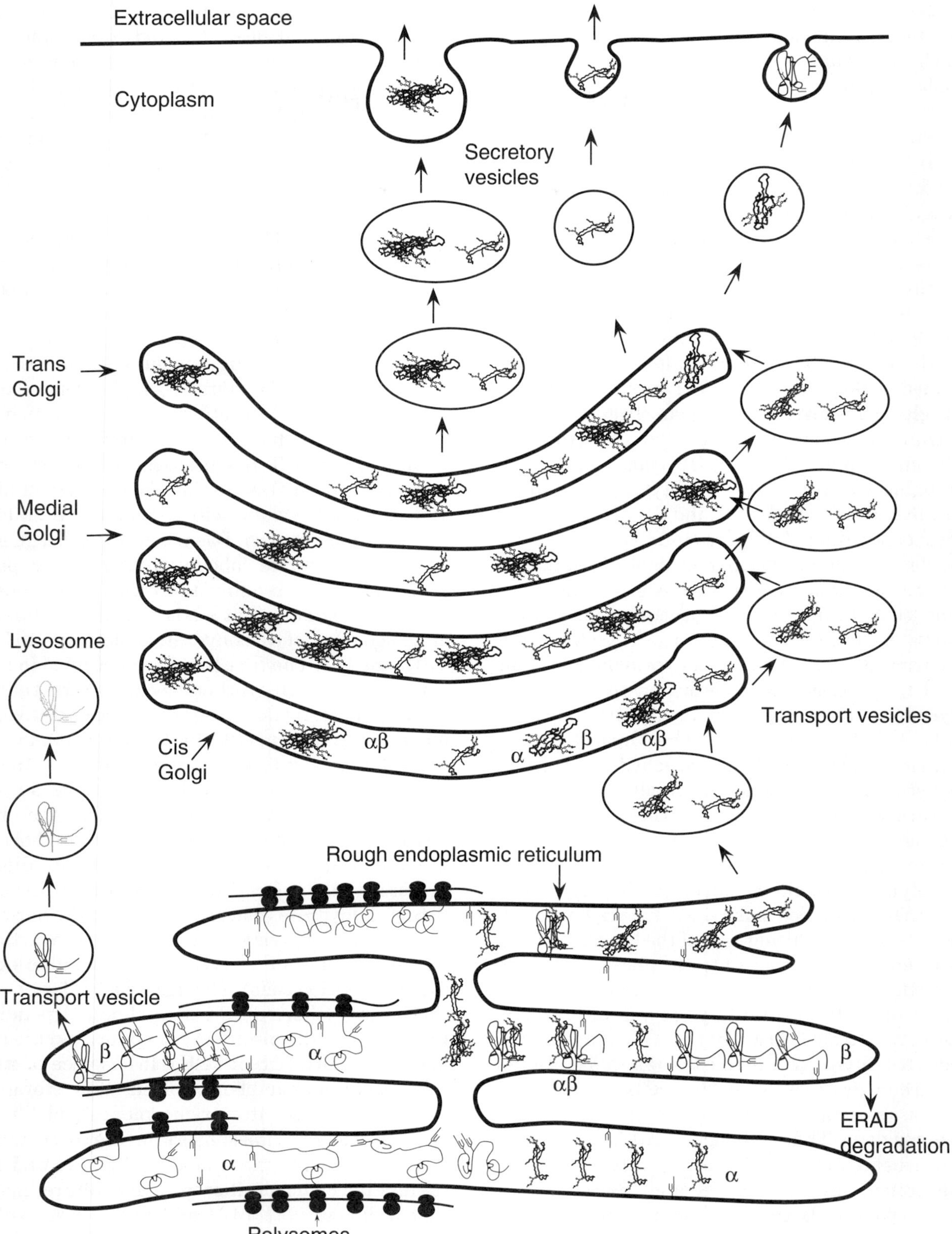

FIG. 15. Biosynthetic pathway for glycoprotein hormones. This illustrates the constitutive pathway because most of the available data have been obtained from cell lines that make human chorionic gonadotropin (hCG), in which only the constitutive secretory pathway is operative. The subunits are synthesized in the rough endoplasmic reticulum (ER), where N-glycosylation occurs cotranslationally. Folding of the α subunit probably occurs cotranslationally as well. Folding of the β subunit occurs post-translationally and appears to be completed after association with the α subunit. Improperly or partially folded β subunit is retained by the rough ER and eventually shunted to lysosomes, where it is degraded. Hormone dimer, properly folded free β subunit, and free α subunit are transferred to the Golgi, where their oligosaccharides are processed before secretion.

The enzyme has been found in the pituitary gland of various species, but not in human placenta (291). If N-acetylgalactosamine is present and the glycoprotein substrate possesses the recognition sequence, –Pro–Xaa–Arg/Lys–, the enzyme may add N-acetylgalactosamine, and the monosulfated structure will result. If not, then two more mannose residues are removed. The resulting oligosaccharide may be galactosylated and processed as monosialylated oligosaccharide or processed further by addition of a second N-acetylglucosamine. The latter structure is the precursor of triantennary sialylated oligosaccharides, as well as biantennary sialylated, or hybrid sulfated-sialylated, or bisulfated oligosaccharides.

The glycoprotein hormone subunits are then moved to the medial Golgi for addition of specific sugars by specific glycosyl transferases. The substrates for these sugar transferases are synthesized in the cytoplasm (e.g., uridine diphosphate [UDP]-galactose, UDP-N-acetylglucosamine, UDP-N-acetylgalactosamine, guanosine diphosphate-fucose, and the sulfate donor, 3′-phosphoadenosine-5′-phosphosulfate [PAPS]) or in the nucleus (cytidine 5′-monophospho-N-acetyl-β-D [CMP]-neuraminic acid). There are selective transport proteins that facilitate each of these substrates' movement into the Golgi, where the glycosyl transferases bring about the buildup of hybrid intermediate forms of the glycosylated moieties on the subunits. (For review of the topography of glycosylation, see Hirschberg and Snider [292].) The "hybrid" intermediates in the formation of the carbohydrate moieties of the glycoprotein hormones are finally processed to the mature secreted form of the carbohydrate in the *trans* Golgi. This is designated the "complex" carbohydrate moiety after addition of N-acetylglucosamine or N-acetylgalactosamine, galactose, N-acetyl neuraminic acid (also called sialic acid), fucose, or sulfate groups. The hCG-producing cells of the human chorion lack the enzyme system, sulfotransferase, in their Golgi apparatus for the addition of sulfate to the carbohydrate (293). In the pituitary there is an N-acetylgalactosamine transferase in the Golgi apparatus that adds N-acetylgalactosamine to the peripheral ends of the carbohydrate moieties (291). The sulfation in the pituitary glycoprotein hormones is on the 4-position of these residues. Thus, the lack of sulfation in hCG is a consequence of the absence of these two relatively unique transferases for PAPS and N-acetylgalactosamine. This leads to a heavier concentration of neuraminic acid termini on the hCG carbohydrate compared with the carbohydrate on pituitary glycoprotein hormones. This, in turn, is a factor in the longer in vivo half-life of hCG compared with the pituitary LH, TSH, or FSH. The sulfated oligosaccharides determine the short half-lives of pituitary hormones possessing these carbohydrates because there is a liver receptor, similar to the asialoglycoprotein receptor, that clears sulfated glycoproteins (206). This rapid clearing of the gonadotropins may contribute to the pulsatile pattern of their circulating levels. The longer half-life of hCG leads to a much longer presentation of it to the LH/CG receptor than is the case for LH. Moreover, the structural differences in LH and CG lead to a much longer binding of CG to the receptor (e.g., a much slower off-rate) than the binding time of LH on the receptor (294,295). Thus, although CG and LH interact with the same receptor, their net physiological (or pharmacological) effect may be considerably different.

In the conversion of the carbohydrate moieties on the glycoprotein hormone subunits from the high-mannose type to the mature complex type, there are several degrees of structural buildup possible. Thus it is possible to have some biantennary structures with two, one, or zero sulfate groups; where there is no sulfate group, it is possible that a neuraminic acid and galactose residue will complete the buildup of the antennary branch. Baenziger and colleagues have studied the sulfate labeling patterns and possible combinations to produce the complex carbohydrate structural buildup that leads to a complex and heterogeneous set of carbohydrate moieties on LH, FSH, TSH, and CG. The first three were studied in bovine, ovine, and human synthetic systems, and the CG in human cell systems. These studies culminated in two outstanding reports (189,190). Their extensive series of studies was reviewed comprehensively (169), and this review is recommended for a detailed statement of the structures and diversity (heterogeneity) encountered in the carbohydrate portion of these hormones.

After the carbohydrate of the α-β dimer is processed in the Golgi to the several forms of complex carbohydrate (mature hormone), the hormone is transported to storage vacuoles in the cytoplasm (pituitary cells) or directly to the cellular exterior (human chorion cells). The degree of sulfation among the pituitary hormones varies widely depending on the species of the pituitary cell donor, and the type of hormone (FSH, LH, or TSH). It has been postulated that sulfation may be involved with storage and secretion of these pituitary hormones (169). There is as yet no experimental support for this concept. In fact, a study involving pulse-labeling of bLH and free α subunit in bovine pituitary primary cultures indicated that although both bLH and free α were sulfated, free α was secreted exclusively by the constitutive pathway, whereas bLH was secreted both constitutively and by the regulatory pathway (296).

In the case of hCG compared with the human pituitary glycoprotein hormones, only the latter are sulfated (to widely varying degrees). This does not hold for equine CG (eCG), for which our laboratory has been able to detect sulfation both in eCG and eFSH or eLH. We have not examined eTSH. Thus, between species marked differences in the degree and form of carbohydrate post-translational processing are found. The question of the heterogeneity of these carbohydrate forms as well as protein heterogeneity has been extensively examined in a monograph edited by Keel and Grotjan (297).

There are multiple forms of the glycoprotein hormones, which are for the most part attributable to the degree of processing of the carbohydrate during biosynthesis. These forms have been designated in a variety of ways, but usually as *isoforms* of the particular hormone. The designation *isohormones* has also been used. The relative concentration of these isoforms in the serum has been shown to correlate with physiological status of the individual in the case of all the pituitary hormones [see, for example, the review of the FSH isoforms (118)]. It is beyond the scope of this chapter to consider the physiological and structural correlation of these isoforms, although some reference to the physiological parameters is noted in other chapters of the text.

As summarized in this section, the biosynthesis of the glycoprotein hormones involves all the complexities of the protein biosynthetic mechanisms provided in nature, from genomic transcription, mRNA translation, cotranslational processing, to the final post-translational processing. The numerous intermediates in the post-translational processing in the Golgi apparatus, several of which may become final forms of the complex carbohydrate moieties, produce a significant heterogeneity in the secreted hormone. This heterogeneity is reflected in several isoforms of that particular glycoprotein hormone. These isoforms, to a degree, also correlate with the physiological state of the individual organism. Thus, the protein biosynthesis of glycoprotein hormones is a very complex process subject to many physiological controls, few of which are fully understood.

REFERENCES

1. Ward, D. N., Bousfield, G. R., and Moore, K. H. (1991). Gonadotropins. In *Reproduction in Domestic Animals* (P. T. Cupps, Ed.), pp. 25–80. Academic Press, San Diego.
2. Pierce, J. G. (1988). Gonadotropins: chemistry and biosynthesis. In *The Physiology of Reproduction* (E. Knobil and J. Neill, Eds.), Vol. 1, pp. 1335–1348. Raven Press, New York.
3. Bousfield, G. R., Perry, W. M., and Ward, D. N. (1994). Gonadotropins: chemistry and biosynthesis. In *The Physiology of Reproduction* (E. Knobil and J. D. Neill, Eds.), Vol. 1, pp. 1749–1792. Raven Press, New York.
4. Lapthorn, A. J., Harris, D. C., Littlejohn, A., Lustbader, J. W., Canfield, R. E., Machin, K. J., Morgan, F. J., and Isaacs, N. W. (1994). Crystal structure of human chorionic gonadotropin. *Nature* 369, 455–461.
5. Wu, H., Lustbader, J. W., Liu, Y., Canfield, R. E., and Hendrickson, W. A. (1994). Structure of human chorionic gonadotropin at 2.6 Å resolution from MAD analysis of the selenomethionyl protein. *Structure* 2, 545–558.
6. Hearn, M. T. W., and Gomme, P. T. (2000). Molecular architecture and biorecognition processes of the cystine knot protein superfamily: I. The glycoprotein hormones. *J. Mol. Recognit.* 13, 223–278.
7. Chiu, I. M., Reddy, E. P., Givol, D., Robbins, K. C., Tronick, S. R., and Aaronson, S. A. (1984). Nucleotide sequence analysis identifies the human c-sis proto-oncogene as a structural gene for platelet-derived growth factor. *Cell* 37, 123–129.
8. Sporn, M. B., Roberts, A. B., Wakefield, L. M., and de Crombrugghe, B. (1987). Some recent advances in the chemistry and biology of transforming growth factor-beta. *J. Cell. Biol.* 105, 1039–1045.
9. Isaacs, N. (1995). Cystine knots. *Curr. Opin. Struct. Biol.* 5, 391–395.
10. Moyle, W. R., Campbell, R. K., Myers, R. V., Bernard, M. P., Han, Y., and Wang, X. (1994). Co-evolution of ligand-receptor pairs. *Nature* 368, 251–255.
11. Xie, Y.-B., Wang, H., and Segaloff, D. L. (1990). Extracellular domain of lutropin/choriogonadotropin receptor expressed in transfected cells binds choriogonadotropin with high affinity. *J. Biol. Chem.* 35, 21411–21414.
12. Ji, I., and Ji, T. H. (1991). Exons 1-10 of the rate LH receptor encode a high affinity hormone binding site and exon 11 encodes G-protein modulation and a potential second hormone binding site. *Endocrinology* 128, 2648–2650.
13. Segaloff, D. L., and Ascoli, M. (1993). The lutropin/choriogonadotropin receptor…4 years later. *Endocr. Rev.* 14, 324–347.
14. Ji, T., Ryu, K., Gilchrist, R., and Ji, I. (1997). Interaction, signal generation, signal divergence, and signal transduction of LH/CG and the receptor. *Recent Prog. Horm. Res.* 52, 431–454.
15. Tegoni, M., Spinelli, S., Verhoeyen, M., Davis, P., and Camillau, C. (1999). Crystal structure of a ternary complex between human chorionic gonadotropin (hCG) and two Fv fragments specific for the α and β-subunits. *J. Mol. Biol.* 289, 1375–1385.
16. van Zuylen, C. W. E. M., De Beer, T., Rademaker, G. J., Haverkamp, J., Thomas-Oates, J. E., Hård, K., Kamerling, J. P., and Vliegenthart, J. F. G. (1995). Site-specific and complete enzymatic deglycosylation of the native human chorionic gonadotropin α-subunit. *Eur. J. Biochem.* 231, 754–760.
17. Fox, K. M., Dias, J. A., and Van Roey, P. (2001). Three-dimensional structure of human follicle-stimulating hormone. *Mol. Endocrinol.* 15, 378–389.
18. Jirgensons, B., and Ward, D. N. (1970). Circular dichroism of ovine luteinizing hormone and its subunits. *Tex. Rep. Biol. Med.* 28, 553–559.
19. Bousfield, G. R., Butnev, V. Y., Butnev, V. Y., Nguyen, V. T., Gray, C. M., Dias, J. A., MacColl, R., Eisele, L., and Harvey, D. J. (2004). Differential effects of α-subunit asparagine^{56} oligosaccharide structure on equine lutropin and follitropin hybrid conformation and receptor-binding activity. *Biochemistry* 43, 10817–10833.
20. Li, C. H., and Starman, B. (1964). Molecular weight of sheep pituitary interstitial cell-stimulating hormone. *Nature* 202, 291–292.
21. Ward, D. N., Fujino, M., and Lamkin, W. M. (1966). Evidence for two carbohydrate moieties in ovine luteinizing hormone (LH). *Fed. Proc.* 25, 348.
22. Papkoff, H., and Samy, T. S. (1967). Isolation and partial characterization of the polypeptide chains of ovine interstitial cell-stimulating hormone. *Biochim. Biophys. Acta* 147, 175–177.

23. DeLaLlosa, P., and Jutisz, M. (1969). Reversible dissociation into subunits and biological activity of ovine luteinizing hormone. *Biochim. Biophys. Acta* 181, 426–436.
24. Lamkin, W. M., Fujino, M., Mayfield, J. D., Holcomb, G. N., and Ward, D. N. (1970). Separation of the subunits of ovine luteinizing hormone by a chromatographic procedure and comparison with a countercurrent distribution procedure. *Biochim. Biophys. Acta* 214, 290–298.
25. Swaminathan, N., and Bahl, O. P. (1970). Dissociation and recombination of the subunits of human chorionic gonadotropin. *Biochem. Biophys. Res. Commun.* 40, 422–427.
26. Morgan, F. J., and Canfield, R. E. (1971). Nature of the subunits of human chorionic gonadotropin. *Endocrinology* 88, 1045–1053.
27. Liao, T.-H., and Pierce, J. G. (1970). The presence of a common type of subunit in bovine thyroid-stimulating and luteinizing hormones. *J. Biol. Chem.* 245, 3275–3281.
28. Ward, D. N., Sweeney, C. M., Holcomb, G. N., Lamkin, W. M., and Fujino, M. (1969). Recent studies on the structure of ovine luteinizing hormone. In *Progress in Endocrinology: Proceedings of the 3rd International Congress of Endocrinology* (C. Gual, Ed.), pp. 385–393. Excerpta Medica, Amsterdam.
29. Pierce, J. G., Liao, T.-H., Howard, S. M., Shome, B., and Cornell, J. S. (1971). Recent studies on the chemistry of TSH. *Recent Prog. Horm. Res.* 27, 165–212.
30. Liu, W. K., and Ward, D. N. (1975). The purification and chemistry of pituitary glycoprotein hormones. *Pharmacol. Ther. B.* 1, 545–570.
31. Thompson, J. D., Higgins, D. G., and Gibson, T. J. (1994). CLUSTAL W: improving the sensitivity of progressive multiple sequence alignment through sequence weighting, positions-specific gap penalties and weight matrix choice. *Nucleic Acids Res.* 22, 4673–4680.
32. Ward, D. N., Bousfield, G. R., Gordon, W. L., and Sugino, H. (1989). Chemistry of the peptide components of glycoprotein hormones. In *Microheterogeneity of Glycoprotein Hormones* (B. A. Keel and H. E. Grotjan Jr., Eds.), pp. 1–21. CRC Press, Boca Raton, FL.
33. Giudice, L. C., and Pierce, J. G. (1979). Studies on the disulfide bonds of glycoprotein hormones: formation and properties of 11,35-bis(S-alkyl) derivatives of the α subunit. *J. Biol. Chem.* 254, 1164–1169.
34. Darling, R. J., Ruddon, R. W., Perini, F., and Bedows, E. (2000). Cystine knot mutations affect the folding of the glycoprotein hormone α-subunit. *J. Biol. Chem.* 275, 15413–15421.
35. Gharib, S. D., Wierman, M. E., Shupnik, M. A., and Chin, W. W. (1990). Molecular biology of the pituitary gonadotropins. *Endocr. Rev.* 11, 177–199.
36. Burzawa-Gerard, É., and Fontaine, Y.-A. (1976). Formation d'une molécule hybride douée d'une activité gonadotrope sur la grenouille, à partir de la sous-unité a de l'hormone lutéinisante bovine et d'une sous-unité de l'hormone gonadotrope d'un poisson téléostéen. *C. R. Acad. Sci.* 282, 97–100.
37. Liu, W. K., Nahm, H. S., Sweeney, C. M., Lamkin, W. M., Baker, H. N., and Ward, D. N. (1972). The primary structure of ovine luteinizing hormone: I. The amino acid sequence of the reduced and S-aminoethylated S-subunit (LH-α). *J. Biol. Chem.* 247, 4351–4364.
38. von Heijne, G. (1986). A new method for predicting signal sequence cleavage sites. *Nucleic Acids Res.* 14, 4683–4690.
39. Sugahara, T., Pixley, M. R., Minami, S., Perlas, E., Ben-Menahem, D., Hsueh, A. J., and Boime, I. (1995). Biosynthesis of a biologically active single peptide chain containing the human common α and chorionic gonadotropin β subunits in tandem. *Proc. Natl. Acad. Sci. U S A* 92, 2041–2045.
40. Nomura, K., Tsunasawa, T., Ohmura, K., Sakiyama, F., and Shizume, K. (1988). Renotropic activity in ovine luteinizing hormone isoform(s). *Endocrinology* 123, 700–712.
41. Koenig, V. L., and King, E. (1950). Extraction studies of sheep pituitary gonadotropic and lactogenic hormones in alcoholic acetate buffers. *Arch. Biochem.* 26, 219–229.
42. Cheng, K. W., Glazer, A. N., and Pierce, J. G. (1973). The effects of modification of the COOH-terminal regions of bovine thyrotropin and its subunits. *J. Biol. Chem.* 248, 7930–7937.
43. Ryan, R. J., Charlesworth, M. C., Erickson, L. D., McCormick, D. J., Milius, R. P., and Morris, J. C. III. (1990). Structure-function relationships of the α-subunit of the glycoprotein hormones. In *Gonadotropins: Structure, Synthesis, and Biologic Function* (W. W. Chin and I. Boime, Eds.), pp. 71–80. Serono Symposia, USA, Norwell, MA.
44. Yoo, J., Zeng, H., Ji, I., Murdoch, W. J., and Ji, T. H. (1993). COOH-terminal amino acids of the α subunit play common and different roles in human choriogonadotropin and follitropin. *J. Biol. Chem.* 268, 13034–13042.
45. Schmidt, A., MacColl, R., Lindau-Shepard, B., Buckler, D. R., and Dias, J. A. (2001). Hormone-induced conformational change of the purified soluble hormone binding domain of follitropin receptor complexed with single chain follitropin. *J. Biol. Chem.* 276, 23373–23381.
46. Erbel, P. J., Karimi-Nejad, Y., De Beer, T., Boelens, R., Kamerling, J. P., and Vliegenthart, J. F. (1999). Solution structure of the alpha-subunit of human chorionic gonadotropin. *Eur. J. Biochem.* 260, 490–498.
47. Erbel, P. J. A., Karimi-Nejad, Y., van Kuik, J. A., Boelens, R., Kamerling, J. P., and Vliegenthart, J. F. G. (2000). Effects of the N-linked glycans on the 3D structure of the free α-subunit of human chorionic gonadotropin. *Biochemistry* 39, 6012–6021.
48. Erbel, P. J. A., Haseley, S. R., Kamerling, J. P., and Vliegenthart, J. F. G. (2002). Studies on the relevance of the glycan at Asn-52 of the α-subunit of human chorionic gonadotropin in the αβ dimer. *Biochem. J.* 364, 485–495.
49. Leeflang, B. R., Faber, E. J., Erbel, P., and Vliegenthart, J. F. G. (2000). Structure elucidation of glycoprotein glycans and of polysaccharides by NMR spectroscopy. *J. Biotechnol.* 77, 115–122.
50. Xing, Y., Williams, C., Campbell, R. K., Cook, S., Knoppers, M., Addona, T., Altarocca, V., and Moyle, W. R. (2001). Threading of a glycosylated protein loop through a protein hole: implications for combination of human chorionic gonadotropin subunits. *Protein Sci.* 10, 226–235.
51. Matzuk, M. M., Keene, J. L., and Boime, I. (1989). Site specificity of the chorionic gonadotropin N-linked oligosaccharides in signal transduction. *J. Biol. Chem.* 264, 2409–2414.
52. van Zuylen, C. W. E. M., de Beer, T., Leeflang, B. R., Boelens, R., Kaptein, R., Kamerling, J. P., and Vliegenthart, J. F. (1998). Mobilities of the inner three core residues and the Man(α1-6) branch of the glycan at Asn^{78} of the α-subunit of human chorionic gonadotropin are restricted by the protein. *Biochemistry* 37, 1933–1940.
53. Strickland, T. W., Thomason, A. R., Nilson, J., and Pierce, J. G. (1985). The common alpha-subunit of bovine glycoprotein hormones: limited formation of native structure by the totally nonglycosylated polypeptide chain. *J. Cell. Biochem.* 29, 225–237.
54. Bousfield, G. R. (1999). LH (luteinizing hormone). In *Encyclopedia of Reproduction* (E. Knobil and J. D. Neill, Eds.), Vol. 2, pp. 1034–1054. Academic Press, San Diego.
55. Giudice, L. C., and Pierce, J. G. (1976). Studies on the disulfide bonds of glycoprotein hormones: complete reduction and reoxidation of the disulfide bonds of the a subunit of bovine luteinizing hormone. *J. Biol. Chem.* 251, 6392–6399.
56. Giudice, L. C., and Pierce, J. G. (1978). Glycoprotein hormones: some aspects of secondary and tertiary structure. In *Structure and Function of the Gonadotropins* (K. W. McKerns, Ed.), pp. 81–110. Plenum, New York.

57. Bousfield, G. R., and Ward, D. N. (1992). Reduction and reoxidation of equine gonadotropin a subunits. *Endocrinology* 131, 2986–2998.
58. Bousfield, G. R., and Ward, D. N. (1994). Evidence for two folding domains in glycoprotein hormone a subunits. *Endocrinology* 135, 624–635.
59. Blobel, G., and Dobberstein, B. (1975). Transfer of proteins across membranes: II. Reconstruction of functional rough microsomes from heterologous components. *J. Cell. Biol.* 67, 852–863.
60. Jackson, R. C., and Blobel, G. (1980). Posttranslational processing of full-length presecretory proteins with canine pancreatic signal peptidase. *Ann. N Y Acad. Sci.* 343, 391–404.
61. Walton, W. J., Nguyen, V. T., Butnev, V. Y., Singh, V., Moore, W. T., and Bousfield, G. R. (2001). Characterization of human follicle-stimulating hormone isoforms reveals a non-glycosylated β-subunit in addition to the conventional glycosylated β-subunit. *J. Clin. Endocrinol. Metab.* 86, 3675–3685.
62. Begeot, M., Hemming, F. J., Dubois, P. M., Combarnous, Y., Dubois, M. P., and Aubert, M. L. (1984). Induction of pituitary lactotrope differentiation by luteinizing hormone alpha subunit. *Science* 226, 566–568.
63. Blithe, D. L., Richards, R. G., and Skarulis, M. C. (1991). Free alpha molecules from pregnancy stimulate secretion of prolactin from human decidual cells: a novel function for free alpha in pregnancy. *Endocrinology* 129, 2257–2259.
64. Cole, L. A. (1989). Occurrence and properties of glycoprotein hormone free subunits. In *Microheterogeneity of Glycoprotein Hormones* (B. A. Keel and H. E. Grotjan Jr., Eds.), pp. 59–79. CRC Press, Boca Raton, FL.
65. Keel, B. A., Schanbacher, B. D., and Grotjan, H. E. Jr. (1987). Ovine luteinizing hormone: II. Effects of castration and steroid administration on the levels of uncombined subunits within the pituitary. *Biol. Reprod.* 36, 1114–1124.
66. Kourides, I. A., Hoffman, B. J., and Landon, M. B. (1980). Difference in glycosylation between secreted and pituitary free alpha subunit of the glycoprotein hormones. *J. Clin. Endocrinol. Metab.* 51, 1372–1377.
67. Parsons, T. F., and Pierce, J. G. (1984). Free α-like material from bovine pituitaries: removal of its O-linked oligosaccharide permits combination with lutropin-β. *J. Biol. Chem.* 259, 2662–2666.
68. Peters, B. P., Krzesicki, R. F., Perini, F., and Ruddon, R. W. (1989). O-glycosylation of the α-subunit does not limit the assembly of chorionic gonadotropin α-β dimer in human malignant and non-malignant trophoblast cells. *Endocrinology* 124, 1602–1612.
69. Blithe, D. L. (1990). N-linked oligosaccharides on free α interfere with its ability to combine with human chorionic gonadotropin-β subunit. *J. Biol. Chem.* 265, 21951–21956.
70. Bousfield, G. R., Baker, V. L., Gotschall, R. R., Butnev, V. Y., and Butnev, V. Y. (2000). Carbohydrate analysis of glycoprotein hormones. *Methods* 21, 15–39.
71. Boime, I., Boothby, M., Hoshina, M., Daniels-McQueen, S., and Darnell, R. (1982). Expression and structure of human placental hormone genes as a function of placental development. *Biol. Reprod.* 26, 73–91.
72. Peters, B. P., Krzesicki, R. F., Hartle, R. J., Perini, F., and Ruddon, R. W. (1984). A kinetic comparison of the processing and secretion of the αβ dimer and the uncombined α and β subunits of chorionic gonadotropin synthesized by human choriocarcinoma cells. *J. Biol. Chem.* 259, 15123–15130.
73. Corless, C. L., Matzuk, M. M., Ramabhadran, T. V., Krichevsky, A., and Boime, I. (1987). Gonadotropin beta subunits determine the rate of assembly and the oligosaccharide processing of hormone dimer in transfected cells. *J. Cell. Biol.* 104, 1173–1181.
74. Spratt, D. I., Chin, W. W., Ridgway, E. C., and Crowley, W. F. Jr. (1986). Administration of low dose pulsatile gonadotropin-releasing hormone (GnRH) to GnRH-deficient men regulates free alpha subunit secretion. *J. Clin. Endocrinol. Metab.* 62, 102–108.
75. Weiss, J., Duca, K. A., and Crowley, W. F. Jr. (1990). Gonadotropin-releasing hormone-induced stimulation and desensitization of free alpha subunit secretion mirrors luteinizing hormone and follicle-stimulating hormone in perifused rat pituitary cells. *Endocrinology* 127, 2364–2371.
76. Corless, C. L., and Boime, I. (1985). Differential secretion of O-glycosylated gonadotropin alpha subunit and luteinizing hormone (LH) in the presence of LH-releasing hormone. *Endocrinology* 117, 1699–1706.
77. Maghuin-Rogister, G., Closset, J., and Hennen, G. (1975). Differences in the carbohydrate portion of the α subunit of porcine lutropin (LH), follitropin (FSH) and thyrotropin (TSH). *FEBS Lett.* 60, 263–266.
78. Nilsson, B., Rosen, S. W., Weintraub, B. D., and Zopf, D. A. (1986). Differences in the carbohydrate moieties of the common α-subunits of human chorionic gonadotropin, luteinizing hormone, follicle-stimulating hormone, and thyrotropin: preliminary structural inferences from direct methylation analysis. *Endocrinology* 119, 2737–2743.
79. Weisshaar, G., Hiyama, J., Renwick, A. G. C., and Nimtz, M. (1991). NMR investigations of the N-linked oligosaccharides at individual glycosylation sites of human lutropin. *Eur. J. Biochem.* 195, 257–268.
80. Weisshaar, G., Hiyama, J., and Renwick, A. G. C. (1991). Site-specific N-glycosylation of human chorionic gonadotropin: structural analysis of glycopeptides by one- and two-dimensional 1H NMR spectroscopy. *Glycobiology* 1, 393–404.
81. Hiyama, J., Weisshaar, G., and Renwick, A. G. C. (1992). The asparagine-linked oligosaccharides at individual glycosylation sites in human thyrotropin. *Glycobiology* 2, 401–409.
82. Huang, C.-J., Huang, F.-L., Chang, G.-D., Chang, Y.-S., Lo, C.-F., Fraser, M. J., and Lo, T.-B. (1991). Expression of two forms of carp gonadotropin α subunit in insect cells by recombinant baculovirus. *Proc. Natl. Acad. Sci. U S A* 88, 7486–7490.
83. Bousfield, G. R., Liu, W.-K., and Ward, D. N. (1985). Hybrids from equine LH: alpha enhances, beta diminishes activity. *Mol. Cell. Endocrinol.* 40, 69–77.
84. Sherman, G. B., Wolfe, M. W., Farmerie, T. A., Clay, C. M., Threadgill, D. S., Sharp, D. C., and Nilson, J. H. (1992). A single gene encodes the β-subunits of equine luteinizing hormone and chorionic gonadotropin. *Mol. Endocrinol.* 6, 951–959.
85. Bambra, C. S., Lynch, S. S., Foxcroft, G. R., Robinson, G., and Amoroso, E. C. (1984). Purification and characterization of guinea-pig chorionic gonadotrophin. *J. Reprod. Fertil.* 71, 227–233.
86. Sherman, G., Heilman, D., Hoss, A., Bunick, D., and Lund, L. (2001). Messenger RNAs encoding the beta subunits of guinea pig (*Cavia porcellus*) luteinizing hormone (gpLH) and putative chorionic gonadotropin (gpCG) are transcribed from a single-copy gpLH/CGbeta gene. *J. Mol. Endocrinol.* 26, 267–280.
87. Ng, T. B., and Idler, D. R. (1979). Studies on two types of gonadotropins from both American plaice and winter flounder pituitaries. *Gen. Comp. Endocrinol.* 38, 410–420.
88. Idler, D. R., and Ng, T. B. (1979). Studies on two types of gonadotropins from both salmon and carp pituitaries. *Gen. Comp. Endocrinol.* 38, 421–440.
89. Suzuki, K., Nagahama, Y., and Kawauchi, H. (1988). Steroidogenic activities of two distinct salmon gonadotropins. *Gen. Comp. Endocrinol.* 71, 452–458.
90. Itoh, H., Suzuki, K., and Kawauchi, H. (1988). The complete amino acid sequences of β-subunits of two distinct chum salmon GTHs. *Gen. Comp. Endocrinol.* 71, 438–451.

91. Sekine, S., Saito, A., Itoh, H., Kawauchi, H., and Itoh, S. (1989). Molecular cloning and sequence analysis of chum salmon gonadotropin cDNAs. *Proc. Natl. Acad. Sci. U S A* 86, 8645–8649.
92. Trinh, K. Y., Wang, N. C., Hew, C. L., and Crim, L. W. (1986). Molecular cloning and sequencing of salmon gonadotropin β subunit. *Eur. J. Biochem.* 159, 619–624.
93. Leigh, S. E. A., and Stewart, F. (1990). Partial cDNA sequence for the donkey chorionic gonadotrophin-β subunit suggests evolution from an ancestral LH-β gene. *J. Mol. Endocrinol.* 4, 143–150.
94. Chopineau, M., Martinat, N., Troispoux, C., Marichatou, H., Combarnous, Y., Stewart, F., and Guillou, F. (1997). Expression of horse and donkey LH in COS-7 cells: evidence for low FSH activity in donkey LH compared with horse LH. *J. Endocrinol.* 152, 371–377.
95. Aggarwal, B. B., Farmer, S. W., Papkoff, H., Stewart, F., and Allen, W. R. (1980). Purification and characterization of donkey chorionic gonadotrophin. *J. Endocrinol.* 85, 449–455.
96. Bousfield, G. R., Butnev, V. Y., and Butnev, V. Y. (2001). Identification of twelve-O-glycosylation sites in eCGβ and eLHβ by solid-phase Edman degradation. *Biol. Reprod.* 64, 136–147.
97. Chen, F., and Puett, D. (1991). Delineation via site-directed mutagenesis of the carboxyl-terminal region of human choriogonadotropin β required for subunit assembly and biological activity. *J. Biol. Chem.* 266, 6904–6908.
98. Galet, C., Chopineau, M., Martinat, N., Combarnous, Y., and Guillou, F. (2000). Expression of an in vitro biologically active equine LH/CG without C-terminal peptide (CTP) and/or β26-110 disulphide bridge. *J. Endocrinol.* 167, 117–124.
99. Bousfield, G. R., and Ward, D. N. (1989). Determinant loop involvement in α-β association of equine LH. In International Symposium on Glycoprotein Hormones (W. W. Chin and I. Boime, Eds.). Serono Symposia USA, Newport Beach, CA.
100. Garnier, J. (1978). Molecular aspects of the subunit assembly of glycoprotein hormones. In *Structure and Function of the Gonadotropins* (K. W. McKerns, Ed.), pp. 381–414. Plenum, New York.
101. Bewley, T. A. (1979). Circular dichroism of pituitary hormones. *Recent Prog. Horm. Res.* 35, 155–210.
102. Bousfield, G. R., Liu, W.-K., and Ward, D. N. (1989). Effects of removal of carboxy-terminal extension from equine luteinizing hormone (LH) β-subunit on LH and follicle-stimulating hormone receptor binding activities and LH steroidogenic activity in rat testicular Leydig cells. *Endocrinology* 124, 379–387.
103. Matzuk, M. M., Hsueh, A. J. W., Lapolt, P., Tsafriri, A., Keene, J. L., and Boime, I. (1990). The biological role of the carboxyl-terminal extension of human chorionic gonadotropin β-subunit. *Endocrinology* 126, 376–383.
104. Min, K.-S., Hattori, N., Aikawa, J.-I., Shiota, K., and Ogawa, T. (1996). Site-directed mutagenesis of recombinant equine chorionic gonadotropin/luteinizing hormone: differential role of oligosaccharides in luteinizing hormone- and follicle-stimulating hormone-like activities. *Endocr. J.* 43, 585–593.
105. Butnev, V. Y., Singh, V., Nguyen, V. T., and Bousfield, G. R. (2002). Truncated eLHβ and asparagine^{56}-deglycosylated eLHα combine to produce a potent follicle-stimulating hormone antagonist. *J. Endocrinol.* 172, 545–555.
106. Fares, F. A., Suganuma, N., Nishimori, K., LaPolt, P. S., Hsueh, A. J. W., and Boime, I. (1992). Design of a long-acting follitropin agonist by fusing the C-terminal sequence of the chorionic gonadotropin β subunit to the follitropin β subunit. *Proc. Natl. Acad. Sci. U S A* 89, 4304–4308.
107. Bona Gallo, A., and Licht, P. (1979). Differences in the properties of FSH and LH binding sites in the avian gonad revealed by homologous radioligands. *Gen. Comp. Endocrinol.* 37, 521–532.
108. Licht, P., and Papkoff, H. (1976). Species specificity in the response of an in vitro amphibian (*Xenopus laevis*) ovulation assay to mammalian luteinizing hormones. *Gen. Comp. Endocrinol.* 29, 552–555.
109. Fontaine, Y. A., and Burzawa-Gerard, E. (1977). Esquisse de l'evolution des hormones gonadotropes et thyreotropes des vertebres. *Gen. Comp. Endocrinol.* 32, 341–347.
110. Bousfield, G. R. (1981). Purification and in vitro biological characterization of equine and chicken gonadotropins: demonstration of intrinsic FSH activity of equine and chicken LH in the rat. p. 217. Indiana University, Bloomington, IN.
111. Fontaine, Y.-A. (1969). La specificite zoologique des proteines hypophysaires capables de stimuler la thyroide. *Acta Endocrinol. Suppl.* 136, 1–154.
112. MacKenzie, D. S. (1982). Stimulation of the thyroid gland of a teleost fish, *Gillichthys mirabilis*, by tetrapod pituitary glycoprotein hormones. *Comp. Biochem. Physiol.* 72A, 477–482.
113. Robker, R. L., and Richards, J. S. (1998). Hormonal control of the cell cycle in ovarian cells: proliferation versus differentiation. *Biol. Reprod.* 59, 476–482.
114. Hsueh, A. J. W., Adashi, E. Y., Jones, P. B. C., and Welsh, T. H. (1984). Hormonal regulation of the differentiation of cultured ovarian granulosa cells. *Endocr. Rev.* 5, 76–127.
115. Beers, W. H., and Strickland, S. (1978). A cell culture assay for follicle-stimulating hormone. *J. Biol. Chem.* 253, 3877–3881.
116. Barbieri, R. L., and Hornstein, M. D. (1999). Assisted reproduction-in vitro fertilization success is improved by ovarian stimulation with exogenous gonadotropins and pituitary suppression with gonadotropin-releasing hormone analogues. *Endocr. Rev.* 20, 249–252.
117. Squires, E. L., and Ginther, O. J. (1975). Follicular and luteal development in pregnant mares. *J. Reprod. Fertil. Suppl.* 23, 429–433.
118. Chappel, S. C., Ulloa-Aguirre, A., and Coutifaris, C. (1983). Biosynthesis and secretion of follicle-stimulating hormone. *Endocr. Rev.* 4, 179–211.
119. Guzman, K., Miller, C. D., Phillips, C. L., and Miller, W. L. (1991). The gene encoding ovine follicle-stimulating hormone beta: isolation, characterization, and comparison to a related ovine genomic sequence. *DNA Cell. Biol.* 10, 593–601.
120. Ward, D. N., and Moore, W. T. Jr. (1979). Comparative study of mammalian glycoprotein hormones. In *Animal Models for Research in Fertility and Contraception* (N. J. Alexander, Ed.), pp. 151–164. Harper and Row, Baltimore.
121. Moore, W. T. Jr., Burleigh, B. D., and Ward, D. N. (1980). Chorionic gonadotropins: comparative studies and comments on relationships to other glycoprotein hormones. In *Chorionic Gonadotropin* (S. J. Segal, Ed.), pp. 89–126. Plenum, New York Press.
122. Stewart, F., Allenm W. R., and Moore, R. M. (1976). Pregnant mare serum gonadotrophin: ratio of follicle-stimulating hormone and luteinizing hormone activities measured by radioreceptor assay. *J. Endocrinol.* 71, 371–382.
123. Combarnous, Y. (1992). Molecular basis of the specificity of binding of glycoprotein hormones to their receptors. *Endocr. Rev.* 13, 670–691.
124. Campbell, R. K., Matzuk, M. M., Dean-Emig, D. M., Cogliani, E., Myers-Shamy, R. V., Krichevsky, A., Boime, I., Barnett, R., and Moyle, W. R. (1990). Use of β-subunit chimeras to study the structures of glycoprotein hormones and to develop a model of the β-subunit. In *Gonadotropins: Structure, Synthesis, and Biologic Function* (W. W. Chin and I. Boime, Eds.), pp. 37–43. Serono Symposia USA, Norwell, MA.

125. Ward, D. N. (1978). Chemical approaches to the structure-function relationships of luteinizing hormone (lutropin). In *Structure and Function of the Gonadotropins* (K. W. McKerns, Ed.), pp. 31–45. Plenum, New York.
126. Weare, J. A., and Reichert, L. E. Jr. (1979). Studies with carbodiimide-cross-linked derivatives of bovine lutropin: II. Location of the cross-link and implication for interaction with receptors in testes. *J. Biol. Chem.* 254, 6972–6979.
127. van Dijk, S., and Ward, D. N. (1993). Chemical cross-linking of porcine luteinizing hormone: location of the cross-link and consequences for stability and biological activity. *Endocrinology* 132, 534–538.
128. Torres, J. V., Yoshioka, N., and Atassi, M. Z. (1987). Antigenic regions on the β chain of human chorionic gonadotropin and development of hormone specific antibodies. *Immunol. Invest.* 16, 697–618.
129. Atassi, M. Z., Manshouri, T., and Sakata, S. (1991). Localization and synthesis of the hormone-binding regions of the human thyrotropin receptor. *Proc. Natl. Acad. Sci. U S A* 88, 3613–3617.
130. Moyle, W. R., Ehrlich, P. H., and Canfield, R. E. (1982). Use of monoclonal antibodies to subunits of human chorionic gonadotropin to examine the orientation of the hormone in its complex with receptor. *Proc. Natl. Acad. Sci. U S A* 79, 2245–2249.
131. Moyle, W. R., Pressey, A., Dean-Emig, D., Anderson, D. M., Demeter, M., Lustbader, J., and Ehrlich, P. (1987). Detection of conformational changes in human chorionic gonadotropin upon binding to rat gonadal receptors. *J. Biol. Chem.* 262, 16920–16926.
132. Birken, S., Canfield, R., Lauer, R., Agosto, G., and Gabel, M. (1980). Immunochemical determinants unique to human chorionic gonadotropin: importance of sialic acid for antisera generated to the human chorionic gonadotropin β-subunit COOH-terminal peptide. *Endocrinology* 106, 1659–1664.
133. Bidart, J.-M., Troalen, F., Bousfield, G. R., Bohuon, C., and Bellet, D. (1989). Monoclonal antibodies directed to human and equine chorionic gonadotropins as probes for the topographic analysis of epitopes on the human α-subunit. *Endocrinology* 124, 923–929.
134. Bidart, J.-M., Troalen, F., Lazar, V., Berger, P., Marcillac, I., Lhomme, C., Droz, J. P., and Bellet, D. (1992). Monoclonal antibodies to the free β-subunit of human chorionic gonadotropin define three distinct antigenic domains and distinguish between intact and nicked molecules. *Endocrinology* 131, 1832–1840.
135. Ryan, R. J., Keutmann, H. T., Charlesworth, M. C., McCormick, D. J., Milius, R. P., Calvo, F. O., and Vutyavanich, T. (1987). Structure-function relationships of gonadotropins. *Recent Prog. Horm. Res.* 43, 383–429.
136. Erickson, L. D., Rizza, S. A., Bergert, E. R., Charlesworth, M. C., McCormick, D. J., and Ryan, R. J. (1990). Synthetic α-subunit peptides stimulate testosterone production in vitro by rat Leydig cells. *Endocrinology* 126, 2555–2560.
137. Schneyer, A. L., Sluss, P. M., Huston, J. S., Ridge, R. J., and Reichert, L. E. Jr. (1988). Identification of a receptor binding region on the beta subunit of human follicle-stimulating hormone. *Biochemistry* 27, 666–671.
138. Santa-Coloma, T. A., and Reichert, L. E. Jr. (1991). Determination of α-subunit contact regions of human follicle-stimulating hormone β-subunit using synthetic peptides. *J. Biol. Chem.* 266, 2759–2762.
139. Grasso, P., Santa-Coloma, T. A., and Reichert, L. E. Jr. (1991). Synthetic peptides corresponding to human follicle-stimulating hormone (hFSH)-β-(1-15) and hFSH-β-(51-65) induce uptake of 45Ca^{++} by liposomes: evidence for calcium-conducting transmembrane channel formation. *Endocrinology* 128, 2745–2751.
140. Santa-Coloma, T. A., Grasso, P., and Reichert, L. E. Jr. (1992). Synthetic human follicle-stimulating hormone-β-(1-15) peptide-amide binds Ca^{++} and possesses sequence similarity to calcium binding sites of calmodulin. *Endocrinology* 130, 1103–1107.
141. Andersen, T. T. (1982). Follitropin binding to receptors in testis. *J. Biol. Chem.* 257, 11551–11557.
142. Moyle, W. R., Bernard, M. P., Myers, R. V., Marko, O. M., and Strader, C. D. (1991). Leutropin/beta-adrenergic receptor chimeras bind choriogonadotropin and adrenergic ligands but are not expressed at the cell surface. *J. Biol. Chem.* 266, 10807–10812.
143. Matzuk, M. M., and Boime, I. (1988). Site-specific mutagenesis defines the intracellular role of the asparagine-linked oligosaccharides of chorionic gonadotropin β subunit. *J. Biol. Chem.* 263, 17106–17111.
144. Yoo, J., Ji, I., and Ji, T. H. (1991). Conversion of lysine 91 to methionine or glutamic acid in human choriogonadotropin a results in the loss of cAMP inducibility. *J. Biol. Chem.* 266, 17747–17743.
145. El-Deiry, S., Kaetzel, D., Kennedy, G., Nilson, J., and Puett, D. (1989). Site-directed mutagenesis of the human chorionic gonadotropin β-subunit: bioactivity of a heterologous hormone, bovine α-human des-(122-145)β. *Mol. Endocrinol.* 3, 1523–1528.
146. Chen, F., Wang, Y., and Puett, D. (1991). Role of the invariant aspartic acid 99 of human choriogonadotropin β in receptor binding and biological activity. *J. Biol. Chem.* 266, 19357–19361.
147. Garcia-Campayo, V., Sato, A., Hirsch, B., Sugahara, T., Muyan, M., Hsueh, A. J., and Boime, I. (1997). Design of stable biologically active recombinant lutropin analogs. *Nat. Biotechnol.* 15, 663–667.
148. Lobel, L. I., Pollak, S., Lustbader, B., Klein, J., and Lustbader, J. W. (2002). Bacterial expression of a natively folded extracellular domain fusion protein of the hFSH receptor in the cytoplasm of *Escherichia coli*. *Protein Expr. Purif.* 25, 124–133.
149. Nishimura, R., Shin, J., Ji, I., Middaugh, C. R., Kruggel, W., Lewis, R. V., and Ji, T. H. (1986). A single amino acid substitution in an ectopic α subunit of a human carcinoma choriogonadotropin. *J. Biol. Chem.* 261, 10475–10477.
150. Cox, G. S. (1986). Phosphorylation of the glycoprotein hormone α-subunit by human tumor cell lines. *Biochem. Biophys. Res. Commun.* 140, 143–150.
151. Lentz, S. R., Birken, S., Lustbader, J., and Boime, I. (1984). Posttranslational modification of the carboxy-terminal region of the β subunit of human chorionic gonadotropin. *Biochemistry* 23, 5330–5337.
152. Shome, B., and Parlow, A. F. (1973). The primary structure of the hormone-specific, beta subunit of human pituitary luteinizing hormone (hLH). *J. Clin. Endocrinol. Metab.* 36, 618–621.
153. Hartree, A. S., Lester, J. B., and Shownkeen, R. C. (1985). Studies of the heterogeneity of human pituitary LH by fast protein liquid chromatography. *J. Endocrinol.* 105, 405–414.
154. Ward, D. N., Glenn, S. D., Nahm, H. S., and Wen, T. (1986). Characterization of cleavage products in selected human lutropin preparations. *Int. J. Pept. Protein Res.* 27, 70–78.
155. Hartree, A. S., and Shownkeen, R. C. (1991). Studies of human pituitary lutropin containing internally cleaved beta subunit. *J. Mol. Endocrinol.* 6, 101–109.
156. Kardana, A., Elliott, M. M., Gawinowicz, M. A., Birken, S., and Cole, L. A. (1991). The heterogeneity of human chorionic gonadotropin (hCG): I. Characterization of peptide heterogeneity in 13 individual preparations of hCG. *Endocrinology* 129, 1541–1550.

157. Cole, L. A., and Birken, S. (1988). Origin and occurrence of human chorionic gonadotropin β-subunit core fragment. *Mol. Endocrinol.* 2, 825–830.
158. Birken, S., Gawinowicz, M. A., Kardana, A., and Cole, L. A. (1991). The heterogeneity of human chorionic gonadotropin (hCG): II. Characteristics and origins of nicks in hCG reference standards. *Endocrinology* 129, 1551–1558.
159. Fiddes, J. C., and Talmadge, K. (1984). Structure, expression and evolution of the genes for the human glycoprotein hormones. *Recent Prog. Horm. Res.* 40, 43–74.
160. Talmadge, K., Boorstein, W. R., and Fiddes, J. C. (1983). The human genome contains seven genes for the β-subunit of chorionic gonadotropin but only one gene for the β-subunit of luteinizing hormone. *DNA* 2, 281–289.
161. Talmadge, K., Boorstein, W. R., Vamvakopolous, N. C., Gething, M.-J., and Fiddes, J. C. (1984). Only three of the seven human chorionic gonadotropin beta subunit genes can be expressed in the placenta. *Nucleic Acids Res.* 12, 8414–8436.
162. Bo, M., and Boime, I. (1992). Identification of the transcriptionally active genes of the chorionic gonadotropin β gene cluster in vivo. *J. Biol. Chem.* 267, 3179–3184.
163. Talmadge, K., Vamvakopolous, N. C., and Fiddes, J. C. (1984). Evolution of the genes for the β subunits of human chorionic gonadotropin and luteinizing hormone. *Nature* 307, 37–40.
164. Virgin, J. B., Silver, B. J., Thomason, A. R., and Nilson, J. H. (1985). The gene for the β subunit of bovine luteinizing hormone encodes a gonadotropin mRNA with an unusually short 5′-untranslated region. *J. Biol. Chem.* 260, 7072–7077.
165. Nakabayashi, K., Matsumi, H., Bhalla, A., Bae, J., Mosselman, S., Hsu, S. Y., and Hsueh, A. J. (2002). Thyrostimulin, a heterodimer of two new human glycoprotein hormone subunits, activates the thyroid-stimulating hormone receptor. *J. Clin. Invest.* 109, 1445–1452.
166. Markoff, E., Sinha, Y. N., Lewis, U. J. (1989). Prolactin microheterogeneity. In *Microheterogeneity of Glycoprotein Hormones* (B. A. Keel and H. E. Grotjan Jr., Eds.), pp. 99–106. CRC Press, Boca Raton, FL.
167. Sinha, Y. N., and Lewis, U. J. (1986). A lectin-binding immunoassay indicates a possible glycosylated growth hormone in the human pituitary gland. *Biochem. Biophys. Res. Commun.* 140, 491–497.
168. Bogdanov, E. M., and Nansel, D. D. (1978). Biological and immunological distinctions between pituitary and serum LH in the rat. In *Structure and Function of the Gonadotropins* (K. W. McKerns, Ed.), pp. 415–430. Plenum, New York.
169. Baenziger, J. U., and Green, E. D. (1988). Pituitary glycoprotein hormone oligosaccharides: structure, synthesis and function of the asparagine-linked oligosaccharides on lutropin, follitropin and thyrotropin. *Biochim. Biophys. Acta* 947, 287–306.
170. Grotjan, H. E. Jr. (1989). Oligosaccharide structures in pituitary and placental glycoprotein hormones. In *Microheterogeneity of Glycoprotein Hormones* (B. A. Keel and H. E. Grotjan Jr., Eds.), pp. 23–52. CRC Press, Boca Raton, FL.
171. Moyle, W. R., Bahl, O. P., and Marz, L. (1975). Role of the carbohydrate of human chorionic gonadotropin in the mechanism of hormone action. *J. Biol. Chem.* 250, 9163–9169.
172. Sairam, M. R. (1980). Deglycosylation of ovine pituitary lutropin subunits: effects on subunit interaction and hormone activity. *Arch. Biochem. Biophys.* 204, 199–206.
173. Governman, J. M., Parsons, T. F., and Pierce, J. G. (1982). Enzymatic deglycosylation of the subunits of chorionic gonadotropin: effects on formation of tertiary structure and biological activity. *J. Biol. Chem.* 257, 15059–15064.
174. Liu, W. K., Young, J. D., and Ward, D. N. (1984). Deglycosylated ovine lutropin: preparation and characterization by in vitro binding and steroidogenesis. *Mol. Cell. Endocrinol.* 37, 29–39.
175. Parsons, T. F., Bloomfield, G. A., and Pierce, J. G. (1983). Purification of an alternate form of the α subunit of the glycoprotein hormones from bovine pituitaries and identification of its O-linked oligosaccharide. *J. Biol. Chem.* 258, 240–244.
176. Blithe, D. L., and Nisula, B. C. (1985). Variations in the oligosaccharides on free and combined alpha-subunits of human choriogonadotropin in pregnancy. *Endocrinology* 117, 2218–2228.
177. Gotschall, R. R., and Bousfield, G. R. (1996). Oligosaccharide mapping reveals hormone-specific glycosylation patterns on equine gonadotropin α-subunit Asn^{56}. *Endocrinology* 137, 2543–2557.
178. Ruddon, R. W., Krzesicki, R. F., Norton, S. E., Beebe, J. S., Peters, B. P., and Perini, F. (1987). Detection of a glycosylated, incompletely folded form of chorionic gonadotropin β subunit that is a precursor of hormone assembly in trophoblastic cells. *J. Biol. Chem.* 262, 12533–12540.
179. Manna, P., Joshi, L., Reinhold, V., Aubert, M. L., Suganuma, N., Pettersson, K., and Huhtaniemi, I. T. (2002). Synthesis, purification and structural and functional characterization of recombinant form of a common genetic variant of human luteinizing hormone. *Hum. Mol. Genet.* 11, 301–315.
180. Sugino, H., Bousfield, G. R., Moore, W. T. Jr., and Ward, D. N. (1987). Structural studies on equine gonadotropins: amino acid sequence of equine chorionic gonadotropin β-subunit. *J. Biol. Chem.* 262, 8603–8609.
181. Bousfield, G. R., Liu. W.-K., Sugino, H., and Ward, D. N. (1987). Structural studies on equine glycoprotein hormones: amino acid sequence of equine lutropin β subunit. *J. Biol. Chem.* 262, 8610–8620.
182. Bousfield, G. R., Butnev, V. Y., and Rance, N. (2003). Primate FSH β-subunit glycosylation occurs in an all-or-none manner. In *36th Annual Meeting of the Society for the Study of Reproduction*. Cincinnati, OH. Abstract 617.
183. Birken, S., and Canfield, R. E. (1977). Isolation and amino acid sequence of COOH-terminal fragments from the β subunit of human choriogonadotropin. *J. Biol. Chem.* 252, 5386–5392.
184. Keutmann, H. T., and Williams, R. M. (1977). Human chorionic gonadotropin: amino acid sequence of the hormone-specific COOH-terminal region. *J. Biol. Chem.* 252, 5393–5397.
185. Kessler, M. J., Mise, T., Ghai, R. D., and Bahl, O. P. (1979). Structure and location of the O-glycosidic carbohydrate units of human chorionic gonadotropin. *J. Biol. Chem.* 254, 7909–7914.
186. Damm, J. B. L., Hard, K., Kammerling, J. P., van Dedem, G. W. K., and Vliegenthart, F. G. (1990). Structure determination of the major N- and O-linked carbohydrate chains of the β subunit from equine chorionic gonadotropin. *Eur. J. Biochem.* 189, 175–183.
187. Matsui, T., Sugino, H., Miura, M., Bousfield, G. R., Ward, D. N., Titani, K., and Mizuochi, T. (1991). β-Subunits of equine chorionic gonadotropin and luteinizing hormone with an identical amino acid sequence have different asparagine-linked oligosaccharide chains. *Biochem. Biophys. Res. Commun.* 174, 940–945.
188. Yamashita, K., Mizuochi, T., and Kobata, A. (1982). Analysis of oligosaccharides by gel filtration. *Methods Enzymol.* 83, 105–126.
189. Green, E. D., and Baenziger, J. U. (1988). Asparagine-linked oligosaccharides on lutropin, follitropin, and thyrotropin: I. Structural elucidation of the sulfated and sialylated oligosaccharides on bovine, ovine and human pituitary glycoprotein hormones. *J. Biol. Chem.* 263, 25–35.
190. Green, E. D., and Baenziger, J. U. (1988). Asparagine-linked oligosaccharides on lutropin, follitropin, and thyrotropin: II. Distributions of sulfated and sialylated oligosaccharides on

bovine, ovine, and human pituitary glycoprotein hormones. *J. Biol. Chem.* 263, 36–44.

191. Harvey, D. J. (1999). Matrix-assisted laser desorption/ionization mass spectrometry of carbohydrates. *Mass Spectrom. Rev.* 18, 349–451.
192. Vliegenthart, J. F. G., Dorland, L., and van Halbeek, H. (1983). High-resolution, ^{1}H-nuclear magnetic resonance spectroscopy as a tool in the structural analysis of carbohydrates related to glycoproteins. *Adv. Carbohydr. Chem. Biochem.* 41, 209–374.
193. Kessler, M. J., Reddy, M. S., Shah, R. H., and Bahl, O. P. (1979). Structures of the N-glycosidic carbohydrate units of human chorionic gonadotropin. *J. Biol. Chem.* 254, 7901–7908.
194. Endo, Y., Yamashita, K., Tachibana, Y., Tojo, S., and Kobata, A. (1979). Structures of the asparagine-linked sugar chains of human chorionic gonadotropin. *J. Biochem.* 85, 669–679.
195. Grotjan, H. E. Jr. (1989). Oligosaccharide structures of the anterior pituitary and placental glycoprotein hormones. In *Microheterogeneity of Glycoprotein Hormones* (B. A. Keel and H. E. Grotjan Jr., Eds.), pp. 23–52. CRC Press, Boca Raton, FL.
196. Parsons, T. F., and Pierce, J. G. (1980). Oligosaccharide moieties of glycoprotein hormones: bovine lutropin resists enzymatic deglycosylation because of terminal O-sulfated N-acetylhexosamines. *Proc. Natl. Acad. Sci. U S A* 77, 7089–7093.
197. Green, E. D., van Halbeek, H., Boime, I., and Baenziger, J. U. (1985). Structural elucidation of the disulfated oligosaccharide from bovine lutropin. *J. Biol. Chem.* 260, 15623–15630.
198. Matsui, T., Mizouchi, T., Titani, K., Okinaga, T., Hoshi, M., Bousfield, G. R., Sugino, H., and Ward, D. N. (1994). Structural analysis of N-linked oligosaccharides from equine chorionic gonadotropin and lutropin β-subunits. *Biochemistry* 33, 14039–14048.
199. Bahl, O. P., and Wagh, P. V. (1986). Characterization of glycoproteins: carbohydrate structures of glycoprotein hormones. In *Molecular and Cellular Aspects of Reproduction* (D. S. Dhindsa and O. P. Bahl, Eds.), Vol. 205, pp. 1–51. Plenum, New York.
200. Weisshaar, G., Hiyama, J., and Renwick, A. G. C. (1990). Site-specific N-glycosylation of ovine lutropin: structural analysis by one- and two-dimensional 1H-NMR spectroscopy. *Eur. J. Biochem.* 192, 741–751.
201. Nemansky, M., and Van Den Eijnden, D. H. (1992). Bovine colostrum CMP-NeuAc:Galβ(1-4)GlcNAc-R α(2-6)-sialyltransferase is involved in the synthesis of the terminal NeuAcα(2-6)GalNAcβ(1-4)GlcNAc sequence occurring on N-linked glycans of bovine milk glycoproteins. *Biochem. J.* 287, 311–316.
202. Smith, P. L., Kaetzel, D., Nilson, J., and Baenziger, J. U. (1990). The sialylated oligosaccharides of recombinant bovine lutropin modulate hormone bioactivity. *J. Biol. Chem.* 265, 874–881.
203. Chen, W., Shen, Q.-X., and Bahl, O. P. (1991). Carbohydrate variant of the recombinant β-subunit of human choriogonadotropin expressed in baculovirus expression system. *J. Biol. Chem.* 266, 4081–4087.
204. Hard, K., Mekking, A., Damm, J. B., Kamerling, J. P., de Boer, W., Wijnands, R. A., and Vliegenthart, J. F. (1990). Isolation and structure determination of the intact sialylated N-linked carbohydrate chains of recombinant human follitropin expressed in Chinese hamster ovary cells. *Eur. J. Biochem.* 193, 263–271.
205. Morell, A. G., Gregoriadis, G., Scheinberg, I. H., Hickman, J., and Ashwell, G. (1971). The role of sialic acid in determining the survival of glycoproteins in the circulation. *J. Biol. Chem.* 246, 1461–1467.
206. Fiete, D., Srivastava, V., Hindsgaul, O., and Baenziger, J. U. (1991). A hepatic reticuloendothelial cell receptor specific for SO_4–4GalNAcβ1,4GlcNAcβ1,2Manα that mediates rapid clearance of lutropin. *Cell* 67, 1103–1110.
207. Baenziger, J. U., Kumar, S., Brodbeck, R. M., Smith, P. L., and Beranek, M. C. (1992). Circulatory half-life but not interaction with the lutropin/chorionic gonadotropin receptor is modulated by sulfation of bovine lutropin oligosaccharides. *Proc. Natl. Acad. Sci. U S A* 89, 334–338.
208. Manzella, S. M., Dharmesh, S. M., Beranek, M. C., Swanson, P., and Baenziger, J. U. (1995). Evolutionary conservation of the sulfated oligosaccharides on vertebrate glycoprotein hormones that control circulatory half-life. *J. Biol. Chem.* 270, 21665–21671.
209. Fiete, D. J., Beranek, M. C., and Baenziger, J. U. (1998). A cysteine-rich domain of the "mannose" receptor mediates GalNAc-4-SO_4 binding. *Proc. Natl. Acad. Sci. U S A* 95, 2089–2093.
210. Lee, S. J., Evers, S., Roeder, D., Parlow, A. F., Risteli, J., Risteli, L., Lee, Y. C., Feizi, T., Langen, H., and Nussenzweig, M. C. (2002). Mannose receptor-mediated regulation of serum glycoprotein homeostasis. *Science* 295, 1898–1901.
211. Mi, Y., Shapiro, S. D., and Baenziger, J. U. (2002). Regulation of lutropin circulatory half-life by the mannose/N-acetylgalactosamine-4–SO_4 receptor is critical for implantation in vivo. *J. Clin. Invest.* 108, 269–276.
212. Bousfield, G. R., and Ward, D. N. (1986). Biologic activity of eLH and hCG following removal of the C-terminal glycopeptide. In *68th Annual Meeting of the Endocrine Society*. Anaheim, CA. Abstract 530.
213. Risma, K., Clay, C. M., Nett, T. M., Wagner, T., Yun, J., and Nilson, J. H. (1995). Targeted overexpression of luteinizing hormone in transgenic mice leads to infertility, polycystic ovaries, and ovarian tumors. *Proc. Natl. Acad. Sci. U S A* 92, 1322–1326.
214. Thotakura, N. R., Desai, R. K., Bates, L. G., Cole, E. S., Pratt, B. M., and Weintraub, B. D. (1991). Biological activity and metabolic clearance of a recombinant human thyrotropin produced in Chinese hamster ovary cells. *Endocrinology* 128, 341–348.
215. Sairam, M. R., and Schiller, P. W. (1979). Receptor binding, biological, and immunological properties of chemically deglycosylated pituitary lutropin. *Arch. Biochem. Biophys.* 197, 294–301.
216. Berman, M. I., Thomas, C. G. Jr., Manjunath, P., Sairam, M. R., and Nayfeh, S. N. (1985). The role of the carbohydrate moiety in thyrotropin action. *Biochem. Biophys. Res. Commun.* 133, 680–687.
217. Calvo, F. O., Keutmann, H. T., Bergert, E. R., and Ryan, R. J. (1986). Deglycosylated human follitropin: characterization and effects on adenosine cyclic 3′,5′-phosphate production in porcine granulosa cells. *Biochemistry* 25, 3938–3943.
218. Sairam, M. R. (1983). Hormonal antagonistic properties of chemically deglycosylated human choriogonadotropin. *J. Biol. Chem.* 258, 445–449.
219. Flack, M. R., Froehlich, J., Bennet, A. P., Anasti, J., and Nisula, B. C. (1994). Site-directed mutagenesis defines the individual roles of the glycosylation sites on follicle-stimulating hormone. *J. Biol. Chem.* 269, 14015–14020.
220. Bishop, L. A., Nguyen, T. V., and Schofield, P. R. (1995). Both of the β-subunit carbohydrate residues of follicle-stimulating hormone determine the metabolic clearance rate and in vivo potency. *Endocrinology* 136, 2635–2640.
221. Patton, P. E., Calvo, F. O., Fujimoto, V. Y., Bergert, E. R., Kempers, R. D., and Ryan, R. J. (1988). The effect of deglycosylated human chorionic gonadotropin on corpora luteal function in healthy women. *Fertil. Steril.* 49, 620–625.

222. Liu, L., Southers, J. L., Banks, S. M., Blithe, D. L., Wehmann, R. E., Brown, J. H., Chen, H. C., and Nisula, B. C. (1989). Stimulation of testosterone production in the cynomolgus monkey in vivo by deglycosylated and desialylated human choriogonadotropin. *Endocrinology* 124, 175–180.
223. Dahl, K. D., Bicsak, T. A., and Hsueh, A. J. W. (1988). Naturally occurring antihormones: secretion of FSH antagonists by women treated with a GnRH analog. *Science* 239, 72–74.
224. Jia, X.-C., and Hsueh, A. J. W. (1986). Granulosa cell aromatase bioassay for follicle-stimulating hormone: validation and application of the method. *Endocrinology* 119, 1570–1577.
225. Stanton, P. G., Burgon, P. G., Hearn, M. T. W., and Robertson, D. M. (1996). Structural and functional characterization of hFSH and hLH isoforms. *Mol. Cell. Endocrinol.* 125, 133–141.
226. Libert, F., Lefort, A., Gerard, C., Parmentier, M., Perret, J., Ludgate, M., Dumont, J. E., and Vassart, G. (1989). Cloning, sequencing and expression of the human thyrotropin (TSH) receptor: evidence for binding of autoantibodies. *Biochem. Biophys. Res. Commun.* 165, 1250–1255.
227. Loosfelt, H., Misrahi, M., Atger, M., Salesse, R., Vu Hai-Thi, M. T., Joliver, A., Guiochon-Mantel, A., Sar, S., Jallal, B., Garnier, J., and Milgrom, E. (1989). Cloning and sequencing of porcine LH-hCG receptor cDNA: variants lacking transmembrane domain. *Science* 245, 525–528.
228. McFarland, K. C., Sprengel, R., Phillips, H. S., Kohler, M., Rosemblit, N., Nikolics, K., Segaloff, D. L., and Seeburg, P. H. (1989). Lutropin-choriogonadotropin receptor: an unusual member of the G protein-coupled receptor family. *Science* 245, 494–499.
229. Nagayama, Y., Kaufman, K. D., Seto, P., and Rappaport, B. (1989). Molecular cloning, sequence and functional expression of the cDNA for the human thyrotropin receptor. *Biochem. Biophys. Res. Commun.* 165, 1184–1190.
230. Parmentier, M., Libert, F., Maenhout, C., Lefort, A., Gérard, C., Perret, J., Van Sande, J., Dumont, J. E., and Vassart, G. (1989). Molecular cloning of the thyrotropin receptor. *Science* 246, 1620–1622.
231. Sprengel, R., Braun, T., Nikolics, K., Segaloff, D. L., and Seeburg, P. H. (1990). The testicular receptor for follicle stimulating hormone: structure and functional expression of cloned cDNA. *Mol. Endocrinol.* 4, 525–530.
232. Jia, X.-C., Oikawa, M., Bo, M., Tanaka, T., Ny, T., Boime, I., and Hsueh, A. J. (1991). Expression of human luteinizing hormone (LH) receptor: interaction with LH and chorionic gonadotropin from human but not equine, rat, and ovine species. *Mol. Endocrinol.* 5, 759–768.
233. Gross, B., Misrahi, M., Sar, S., and Milgrom, E. (1991). Composite structure of human thyrotropin receptor gene. *Biochem. Biophys. Res. Commun.* 177, 679–687.
234. Koo, Y. B., Ji, I., Slaughter, R. G., and Ji, T. H. (1991). Structure of the luteinizing hormone receptor gene and multiple exons of the coding sequence. *Endocrinology* 128, 2297–2308.
235. Heckert, L. L., Daley, I. J., and Griswold, M. D. (1992). Structural organization of the follicle-stimulating hormone receptor gene. *Mol. Endocrinol.* 6, 70–80.
236. Ascoli, M., and Segaloff, D. L. (1989). On the structure of the luteinizing hormone/chorionic gonadotropin receptor. *Endocr. Rev.* 10, 27–44.
237. Akamizu, T., Ikuyama, S., Saji, M., Kosugi, S., Kozak, C., McBride, O. W., and Kohn, L. D. (1990). Cloning, chromosomal assignment, and regulation of the rat thyrotropin receptor: expression of the gene is regulated by thyrotropin, agents that increase cAMP levels, and thyroid antibodies. *Proc. Natl. Acad. Sci. U S A* 87, 5677–5681.
238. Lincoln, S. R., Lei, Z. M., Rao, C. V., and Yussman, M. A. (1992). The expression of human chorionic gonadotropin/human luteinizing hormone receptors in ectopic endometrial implants. *J. Clin. Endocrinol. Metab.* 75, 1140–1144.
239. Huhtaniemi, I. T., Eskola, V., Pakarinen, P., Matikainen, T., and Sprengel, R. (1992). Molecular cloning of the murine FSH and LH receptor (R) gene promoter sequences, identification of transcription initiation sites and demonstration of promoter activity. In *74th Annual Meeting of the Endocrine Society*, p. 1288. San Antonio, TX.
240. Gudermann, T., Birnbaumer, M., and Birnbaumer, L. (1992). Evidence for dual coupling of the murine luteinizing hormone receptor to adenylyl cyclase and phosphoinositide breakdown and Ca^{2+} mobilization: studies with the cloned murine luteinizing hormone receptor expressed in L cells. *J. Biol. Chem.* 267, 4479–4488.
241. Oates, E., Jin. S. X., McKenzie, J. M., and Zakarija, M. (1992). Unique conserved regions of the glycoprotein hormone receptors allow *Xenopus* FSH and LH/CG receptor gene fragment cloning by PCR. In *74th Annual Meeting of the Endocrine Society*, p. 1265. San Antonio, TX.
242. Tsai-Morris, C. H., Buczko, E., Wang, W., and Dufau, M. L. (1990). Intronic nature of the rat luteinizing hormone receptor gene defines a soluble receptor subspecies with hormone binding activity. *J. Biol. Chem.* 265, 19385–19388.
243. Minegishi, T., Nakamura, K., Takakura, Y., Miyamoto, K., Hasegawa, Y., Ibuki, Y., and Igarashi, M. (1990). Cloning and sequencing of human LH/hCG receptor cDNA. *Biochem. Biophys. Res. Commun.* 172, 1049–1054.
244. Rodriguez, M. C., Xie, Y. B., Wang, H., Collison, K., and Segaloff, D. L. (1992). Effects of truncations of the cytoplasmic tail of the luteinizing hormone/chorionic gonadotropin receptor on receptor-mediated hormone internalization. *Mol. Endocrinol.* 6, 327–336.
245. Rodriguez, M. C., and Segaloff, D. L. (1990). The orientation of the lutropin/choriogonadotropin receptor in rat luteal cells as revealed by site-specific antibodies. *Endocrinology* 127, 674–681.
246. Yang, K. P., Samaan, N. A., and Ward, D. N. (1976). Characterization of an inhibitor for luteinizing hormone receptor site binding. *Endocrinology* 98, 233–241.
247. Nagayama, Y., Russo, D., Chazenbalk, G. D., Wadsworth, H. L., and Rapoport, B. (1990). Extracellular domain chimeras of the TSH and LH/CG receptors reveal the mid-region (amino acids 171-260) to play a vital role in high affinity TSH binding. *Biochem. Biophys. Res. Commun.* 173, 1150–1156.
248. Braun, T., Schofield, P. R., and Sprengel, R. (1991). Amino-terminal leucine-rich repeats in gonadotropin receptors determine hormone selectivity. *EMBO J.* 10, 1885–1890.
249. Russo, D., Chazenbalk, G. D., Nagayama, Y., Wadsworth, H. L., and Rapoport, B. (1991). Site-directed mutagenesis of the human thyrotropin receptor: role of asparagine-linked oligosaccharides in the expression of a functional receptor. *Mol. Endocrinol.* 5, 29–33.
250. Petäjä-Repo, U. E., Merz, W. E., and Rajaniemi, H. J. (1991). Significance of the glycan moiety of the rat ovarian luteinizing hormone/chorionic gonadotropin (CG) receptor and human CG for receptor-hormone interaction. *Endocrinology* 128, 1209–1217.
251. O'Dowd, B. F., Hnatowich, M., Regan, J. W., Leader, W. M., Caron, M. G., and Lefkowitz, R. J. (1988). Site-directed mutagenesis of the cytoplasmic domains of the human β2-adrenergic receptor: localization of regions involved in G protein-receptor coupling. *J. Biol. Chem.* 263, 15985–15992.
252. Wang, H., Jaquette, J., Collison, K., and Segaloff, D. (1993). Positive charges in a putative amphiphilic helix in the carboxyl-terminal region of the third intracellular loop of the

luteinizing hormone/chorionic gonadotropin receptor are not required for hormone-stimulated cAMP production but are necessary for expression of the receptor at the plasma membrane. *Mol. Endocrinol.* 7, 1437–1444.

253. Jha, P. K., Pal, R., Nakhai, B., Sridar, P., and Hasnain, S. E. (1992). Simultaneous synthesis of enzymatically active luciferase and biologically active β subunit of human chorionic gonadotropin in caterpillars infected with a recombinant baculovirus. *FEBS Lett.* 310, 148–152.

254. Sánchez-Yagüe, J., Rodríguez, M. C., Segaloff, D. L., and Ascoli, M. (1992). Truncation of the cytoplasmic tail of the lutropin/choriogonadotropin receptor prevents agonist-induced uncoupling. *J. Biol. Chem.* 267, 7217–7220.

255. Kobe, B., and Diesenhofer, J. (1993). Crystal structure of porcine ribonuclease inhibitor, a protein with leucine-rich repeats. *Nature* 366, 751–756.

256. Moyle, W. R., Campbell, R. K., Rao, S. N. V., Ayad, N. G., Bernard, M. P., Han, Y., and Wang, Y. (1995). Model of human chorionic gonadotropin and lutropin receptor interaction that explains signal transduction of the glycoprotein hormones. *J. Biol. Chem.* 270, 20020–20031.

257. Jiang, X., Dreano, M., Buckler, D. R., Cheng, S., Ythier, A., Wu, H., Hendrickson, W. A., and el Tayar, N. (1995). Structural predictions for the ligand-binding region of glycoprotein hormone receptors and the nature of hormone-receptor interactions. *Structure* 3, 1341–1353.

258. Bhowmick, N., Huang, J., Puett, D., Isaacs, N. W., and Lapthorn, A. J. (1996). Determination of residues important in hormone binding to the extracellular domain of the luteinizing hormone/chorionic gonadotropin receptor by site-directed mutagenesis and modeling. *Mol. Endocrinol.* 10, 1147–1159.

259. Reichert, L. E. Jr., Grasso, P., Boniface, J. J., Santa Coloma, T. A., Zhang, S. B., and Dattatreyamurty, B. (1989). Studies on the mechanism of follitropin-receptor interaction and signal transduction in testis. In *Structure–Function Relationship of Gonadotropins* (D. Bellet and J.-M. Bidart, Eds.), Vol. 65, pp. 237–246. Raven Press, New York.

260. Minegishi, T., Nakamura, K., Takakura, Y., Ibuki, Y., and Igarashi, M. (1991). Cloning and sequencing of human FSH receptor cDNA. *Biochem. Biophys. Res. Commun.* 175, 1125–1130.

261. Babu, P., Jiang, L., Sairam, A., Touyz, R., and Sairam, M. (1999). Structural features and expression of an alternatively spliced growth factor type I receptor for follitropin signaling in the developing ovary. *Mol. Cell. Biol. Res. Commun.* 2, 21–27.

262. Babu, P., Danilovich, N., and Sairam, M. (2001). Hormone-induced receptor gene splicing: enhanced expression of the growth factor type I follicle-stimulating hormone receptor motif in the developing mouse ovary as a new paradigm in growth regulation. *Endocrinology* 142, 381–389.

263. Noce, T., Ando, H., Ueda, T., Kubokawa, K., Higashinakagawa, T., and Ishii, S. (1989). Molecular cloning and nucleotide sequence analysis of the putative cDNA for the precursor molecule of the chicken LH-β subunit. *J. Mol. Endocrinol.* 3, 129–137.

264. Chappel, S., Beck, A., Nugent, N., Hyman, L., Zabrecky, J., and Maurer, R. (1987). Production of bovine follicle stimulating hormone (FSH) by recombinant DNA technology. In *69th Annual Meeting of the Endocrine Society*. Indianapolis, IN. Abstract 130.

265. Bousfield, G. R., Butnev, V. Y., Gotschall, R. R., Baker, V. L., and Moore, W. T. (1996). Structural features of mammalian gonadotropins. *Mol. Cell. Endocrinol.* 125, 3–19.

266. Helenius, A., and Aebi, M. (2004). Roles of N-linked glycans in the endoplasmic reticulum. *Annu. Rev. Biochem.* 73, 1019–1049.

267. Matzuk, M. M., and Boime, I. (1988). The role of the asparagine-linked oligosaccharides of the α subunit in the secretion and assembly of human chorionic gonadotropin. *J. Cell Biol.* 106, 1049–1059.

268. Flack, M. R., Bennet, A. P., Froehlich, J., and Anasti, J. N. (1992). Selective mutagenesis of the glycosylation sites of FSH; effects on secretion, conformation, and biologic potency. In *74th Annual Meeting of the Endocrine Society*, p. 1296. San Antonio, TX.

269. Grossmann, M., Szkudlinski, M. W., Tropea, J. E., Bishop, L. A., Thotakura, N. R., Schofield, P. R., and Weintraub, B. D. (1995). Expression of human thyrotropin in cell lines with different glycosylation patterns combined with mutagenesis of specific glycosylation sites: characterization of a novel role for the oligosaccharides in the in vitro and in vivo bioactivity. *J. Biol. Chem.* 270, 29378–29385.

270. Ward, D. N., Hines, K. K., Gordon, W. L., and Bousfield, G. R. (1988). The purification of native inhibin and chemical characterization. In *Nonsteroidal Gonadal Factors: Physiological Roles and Possibilities in Contraceptive Development* (G. D. Hodgen, Z. Rosenwaks, and J. M. Spieler, Eds.), pp. 1–16. The Jones Institute Press, Norfolk, VA.

271. Ruddon, R. W., Sherman, S. A., and Bedows, E. (1996). Protein folding in the endoplasmic reticulum: lessons from the human chorionic gonadotropin β subunit. *Protein Sci.* 5, 1443–1452.

272. Hoshina, H., and Boime, I. (1982). Combination of rat lutropin subunits occurs early in the secretory pathway. *Proc. Natl. Acad. Sci. U S A* 79, 7649–7653.

273. Huth, J. R., Mountjoy, K., Perini, F., and Ruddon, R. W. (1992). Intracellular folding pathway of human chorionic gonadotropin β subunit. *J. Biol. Chem.* 267, 8870–8879.

274. Suganama, N., Matzuk, M. M., and Boime, I. (1989). Elimination of disulfide bonds affects assembly and secretion of the human chorionic gonadotropin β subunit. *J. Biol. Chem.* 264, 19302–19307.

275. Parsons, T. F., Strickland, T. W., and Pierce, J. G. (1985). Disassembly and assembly of glycoprotein hormones. *Methods Enzymol.* 109, 736–749.

276. Butnev, V. Y., Gotschall, R. R., Butnev, V. Y., Baker, V. L., Moore, W. T., and Bousfield, G. R. (1998). Hormone-specific inhibitory influence of α-subunit Asn^{56} oligosaccharide on in vitro subunit association and FSH receptor binding of equine gonadotropins. *Biol. Reprod.* 58, 458–469.

277. Feng, W., Huth, J. R., Norton, S. E., and Ruddon, R. W. (1995). Asparagine-linked oligosaccharides facilitate human chorionic gonadotropin β-subunit folding but not assembly of prefolded β with α. *Endocrinology* 136, 52–61.

278. Huth, J. R., Feng, W., and Ruddon, R. W. (1994). Redox conditions for stimulation of in vitro folding and assembly of the glycoprotein hormone chorionic gonadotropin. *Biotechnol. Bioeng.* 44, 66–72.

279. Cosowsky, L., Rao, S. N. V., Macdonald, G. J., Papkoff, H., Campbell, R. K., and Moyle, W. R. (1995). The groove between the α- and β-subunits of hormones with lutropin (LH) activity appears to contact the LH receptor, and its conformation is changed during hormone binding. *J. Biol. Chem.* 270, 20011–20019.

280. Xing, Y., Myers, R., Cao, D., Lin, W., Jiang, M., Bernard, M. P., and Moyle, W. R. (2004). Glycoprotein hormone assembly in the endoplasmic reticulum: I. The glycosylated end of human alpha-subunit loop 2 is threaded through a beta-subunit hole. *J. Biol. Chem.* 279, 35426–35436.

281. Xing, Y., Myers, R., Cao, D., Lin, W., Jiang, M., Bernard, M. P., and Moyle, W. R. (2004). Glycoprotein hormone assembly in the endoplasmic reticulum: II. Multiple roles of a redox sensitive beta-subunit disulfide switch. *J. Biol. Chem.* 279, 35437–35448.

282. Xing, Y., Myers, R., Cao, D., Lin, W., Jiang, M., Bernard, M. P., and Moyle, W. R. (2004). Glycoprotein hormone assembly in the endoplasmic reticulum: III. The seatbelt and its latch site determine the assembly pathway. *J. Biol. Chem.* 279, 35449–35457.
283. Xing, Y., Myers, R., Cao, D., Lin, W., Jiang, M., Bernard, M. P., and Moyle, W. R. (2004). Glycoprotein hormone assembly in the endoplasmic reticulum: IV. Probable mechanism of subunit docking and completion of assembly. *J. Biol. Chem.* 279, 35458–35468.
284. Cole, L. A., Hartle, R. J., Laferla, J. J., and Ruddon, R. W. (1983). Detection of the free beta subunit of human chorionic gonadotropin (hCG) in cultures of normal and malignant trophoblast cells, pregnancy sera, and sera of patients with choriocarcinoma. *Endocrinology* 113, 1176–1178.
285. Ozturk, M., Bellet, D., Manl, L., Hennen, G., Frydman, R., and Wands, J. (1987). Physiological studies of human chorionic gonadotropin (hCG), αhCG, and βhCG as measured by specific monoclonal immunoradiometric assays. *Endocrinology* 120, 549–558.
286. Keene, J. L., Matzuk, M. M., Otani, T., Fauser, B. C., Galway, A. B., Hsueh, A. J., Boime, I. (1989). Expression of biologically active human follitropin in Chinese hamster ovary cells. *J. Biol. Chem.* 264, 4769–4775.
287. Ross, D. S., Downing, M. F., Chin, W. W., Kieffer, J. D., and Ridgway, E. C. (1983). Divergent changes in murine pituitary concentration of free α- and thyrotropin-β subunits in hypothyroidism and after thyroxine administration. *Endocrinology* 112, 187–193.
288. Noda, T., and Farquhar, M. G. (1992). A non-autophagic pathway for diversion of ER secretory proteins to lysosomes. *J. Cell. Biol.* 119, 85–97.
289. Rothman, J. E., and Orci, L. (1992). Molecular dissection of the secretory pathway. *Nature* 355, 409–415.
290. Smith, P. L., and Baenziger, J. U. (1992). Molecular basis of recognition by the glycoprotein hormone-specific N-acetylgalactosamine-transferase. *Proc. Natl. Acad. Sci. U S A* 89, 329–333.
291. Smith, P. L., and Baenziger, J. U. (1988). A pituitary N-acetylgalactosamine transferase that specifically recognizes glycoprotein hormones. *Science* 242, 930–933.
292. Hirschberg, C. B., and Snider, M. D. (1987). Topography of glycosylation in the rough endoplasmic reticulum and Golgi apparatus. *Annu. Rev. Biochem.* 56, 63–87.
293. Green, E. D., Gruenebaum, J., Bielinska, M., Baenziger, J. U., and Boime, I. (1984). Sulfation of lutropin oligosaccharides with a cell-free system. *Proc. Natl. Acad. Sci. U S A* 81, 5320–5324.
294. Mock, E. J., Papkoff, H., and Niswender, G. D. (1983). Internalization of ovine luteinizing hormone/human chorionic gonadotropin recombinants: differential effects of the α- and β-subunits. *Endocrinology* 113, 265–269.
295. Niswender, G. D., Roess, D. A., Sawyer, H. R., Silvia, W. J., and Barisas, B. G. (1985). Differences in the lateral mobility of receptors for luteinizing hormone (LH) in the luteal cell plasma membrane when occupied by ovine LH *versus* human chorionic gonadotropin. *Endocrinology* 116, 164–169.
296. Blomquist, J. F., and Baenziger, J. U. (1992). Differential sorting of lutropin and the free α-subunit in cultured bovine pituitary cells. *J. Biol. Chem.* 267, 20798–20803.
297. Keel, B. A., and Grotjan, H. E. Jr., Eds. (1989). *Microheterogeneity of Glycoprotein Hormones*. CRC Press, Boca Raton, FL.
298. Berman, H. M., Westbrook, J., Feng, Z., Gilliland, G., Bhat, T. N., Weissig, H., Shindyalov, I. N., and Bourne, P. E. (2000). The protein data bank. *Nucleic Acids Res.* 28, 235–242.
299. Kornfeld, R., and Kornfeld, S. (1985). Assembly of asparagine-linked oligosaccharides. *Annu. Rev. Biochem.* 54, 631–634.
300. Renwick, A. G. C., Mizuochi, T., Kochibe, N., and Kobata, A. (1987). The asparagine-linked sugar chains of human follicle-stimulating hormone. *J. Biochem.* 101, 1209–1221.

Knobil and Neill's Physiology of Reproduction, Third Edition
edited by Jimmy D. Neill,

CHAPTER 31

Gonadotropin-Releasing Hormone Regulation of Gonadotropin Biosynthesis and Secretion

Kyeong-Hoon Jeong and Ursula B. Kaiser

Introduction, 1635
Structure and Function of GnRH Receptor, 1636
Physical Characteristics, 1636
Ligand Affinity and Expression in the Cell Membrane, 1637
Internalization, 1638
Desensitization, 1638
Effects of Mutations, 1640
Type II GnRH Receptor, 1641
GnRH Signal Transduction and Gonadotropin Secretion, 1643
Guanosine Triphosphate–Binding Proteins, 1643
Phospholipase C and Phosphoinositides, 1643
Intracellular Calcium, 1646
Diacylglycerol, 1647
Protein Kinase C Pathways, 1648
Mitogen-Activated Protein Kinase Cascade, 1649
Immediate–Early Gene Expression, 1650
Protein Kinase A Pathways, 1651
Gonadotropin Secretion, 1651
Physiological Regulators of Gonadotropin Biosynthesis and Secretion, 1653
GnRH Pulsatility, 1653
Estrous Cycle, 1655
Inhibin, 1657
Activin, 1658
Follistatin, 1659
Gonadectomy, 1659
Estrogen and Progesterone, 1660
Testosterone, 1661
GnRH Regulation of Gonadotropin Gene Expression, 1661
Glycoprotein α-Subunit Gene Expression, 1661
Luteinizing Hormone β-Subunit Gene Expression, 1666
Follicle-Stimulating Hormone β-Subunit Gene Expression, 1669
GnRH Receptor Gene Expression, 1672
References, 1677

INTRODUCTION

Synthesis and secretion of the pituitary gonadotropic hormones, luteinizing hormone (LH) and follicle-stimulating hormone (FSH), are under tight control by gonadotropin-releasing hormone (GnRH). This neuronal decapeptide is synthesized and stored in GnRH neurons located in the medial basal hypothalamus and the preoptic area. Upon neuronal stimulation, GnRH is released in a pulsatile manner through the nerve terminals located in the median eminence, and enters the hypophysial portal circulation to reach the gonadotropes in the anterior pituitary gland. In the gonadotropes, this pulsatile GnRH stimulates the synthesis and secretion of LH and FSH (1). These gonadotropins then enter the systemic circulation to reach their target organs, the gonads, to stimulate gonadal steroidogenesis and gametogenesis. LH stimulates testosterone secretion in males and ovulation and corpus luteum formation in females, whereas FSH stimulates spermatogenesis in males and ovarian follicle maturation and estrogen secretion in females. The gonadal steroids, testosterone in males and estrogen in females, in turn act on to the hypothalamus and pituitary to modulate GnRH and gonadotropin synthesis, completing a feedback loop. The gonadotropic

Brigham and Women's Hospital, Harvard Medical School, Boston, Massachusetts.

hormones are composed of two different subunits. The glycoprotein α subunit (αGSU) is common to both LH and FSH, whereas the β subunits for each of these hormones, LHβ and FSHβ, are distinct from each other and determine hormonal specificity for each glycoprotein hormone. GnRH is the most potent stimulator of expression of these gonadotropin subunit genes (2).

Research into the neuroendocrine regulation of reproductive function by GnRH has undergone an explosion since the early 1970s, sparked by the isolation and biochemical characterization of GnRH (3–6). This led to the development of both agonist and antagonist analogs, resulting in rapid advances in the basic understanding of GnRH physiology as well as clinical applications to the treatment of disorders such as prostate cancer, endometriosis, precocious puberty, and infertility (7,8). These advances were later followed by the molecular cloning of complementary DNAs (cDNAs) encoding the GnRH receptor (GnRHR). The availability of the GnRHR cDNA has sparked a second explosion of investigation, leading to a better understanding of the mechanisms underlying GnRH action (9,10).

Historically, studies of the mechanisms of neuroendocrine control of reproduction at the hypothalamic–pituitary level have been performed in vivo in animal models and in vitro in dispersed primary pituitary cell cultures. These studies are limited by the heterogeneity of anterior pituitary secretory cells in adult animals (11). In addition, anterior pituitary secretory cells cannot be propagated in culture systems, limiting the feasibility of many studies.

The development of immortalized murine gonadotrope-derived cell lines, notably αT3-1 (12) and LβT2 cells (13), has enabled the acceleration of studies of the mechanisms of GnRH and GnRHR biology and action. These cell lines were derived from pituitary tumor cells of transgenic mice, expressing a fusion gene containing 5′-flanking sequences of either the human αGSU (αT3-1) or the rat LHβ (LβT2) gene linked to the protein-coding sequences of the simian virus-40 (SV-40) T antigen oncogene to target pituitary gonadotropes. Expression of these transgenes resulted in development of tumors in the anterior pituitary. The αT3-1 cell line does not support LHβ and FSHβ gene expression; nonetheless, these two cell lines provide extremely valuable model systems in the study of GnRH and GnRHR action. More recently, a better and more efficient method for isolation of gonadotrope cells from mouse pituitaries has been developed (14). Using a transgenic mouse model in which the gonadotropes are specifically tagged with a cell surface antigen (H-2K^k), it is now possible to isolate gonadotropes, with up to 95% purity, that secrete gonadotropins robustly, whereas the gonadotrope population in the pituitary comprises only 3% to 15% of the total cell population (15,16).

STRUCTURE AND FUNCTION OF GnRH RECEPTOR

Physical Characteristics

GnRH signaling begins with recognition by its receptor. Initial studies identified the GnRHR in the pituitaries of several species (17,18) as well as in the αT3-1 cell line (19). Additional minor sites/cell lines of GnRHR expression include hypothalamic GnRH neurons (20), the mouse GnRH neuronal cell line GT1-7 (21), testicular Leydig cells (22), ovarian granulosa and luteal cells (23,24), placental cells (25), endometrial cancer cell lines (26), as well as prostate and breast cells and a breast tumor cell line (27).

The mouse GnRHR cDNA was first isolated from the αT3-1 cell line (28–30). These isolates became templates for the cloning of the GnRHR from other species, based on homology with the mouse cDNA sequences. The GnRHR was thus cloned from mouse (28,29), followed by human (27,31,32), rat (30,33,34), cow (35), sheep (36,37), pig (38), dog (39), and marsupials (40,41) (Fig. 1). In amino acid sequence, the mouse GnRHR is 97% homologous to the rat GnRHR, 89% to the human GnRHR, and 87% to the ovine GnRHR. The human GnRHR is 89% homologous to the ovine GnRHR.

The GnRHR belongs to the family of guanosine triphosphate (GTP)-binding (G) protein–coupled receptors (GPCRs). General features of GPCRs include an extracellular N-terminus and an intracellular C-terminus connected by seven transmembrane helices (TMHs), which are linked by three extracellular loops (ECLs) and three intracellular loops (ICLs) (42,43). The GnRHR gene is a single-copy gene, comprising three exons and two introns, approximately 20 to 25 kilobases (kb) in size (32,44–46). It is located in chromosome 5 in the mouse (47), chromosome 6 in the sheep (48), and chromosome 4q13.2-21.1 in the human (47). The GnRHR gene in rodents, including mouse and rat, encodes a 327–amino acid protein, whereas human, bovine, and ovine GnRHR genes encode a 328–amino acid protein, having 1 additional amino acid, Lys^{191}, in the ECL2 (31) (Fig. 1). This extra residue has been implicated as an important determinant for GnRHR function, by reducing receptor expression and ligand-binding affinity, and by increasing the internalization rate (49). The GnRHR does not have a complete C-terminal intracellular tail, which has implications for a lack of rapid internalization and desensitization of ligand-occupied receptor (50–52), a mechanism dependent on a receptor-interacting clathrin adaptor protein, β-arrestin (53). Rather, this receptor has been suggested to interact with other receptor-interacting proteins, dynamin and c-Src, which have slow internalization kinetics (54,55). However, a dynamin-insensitive internalization has

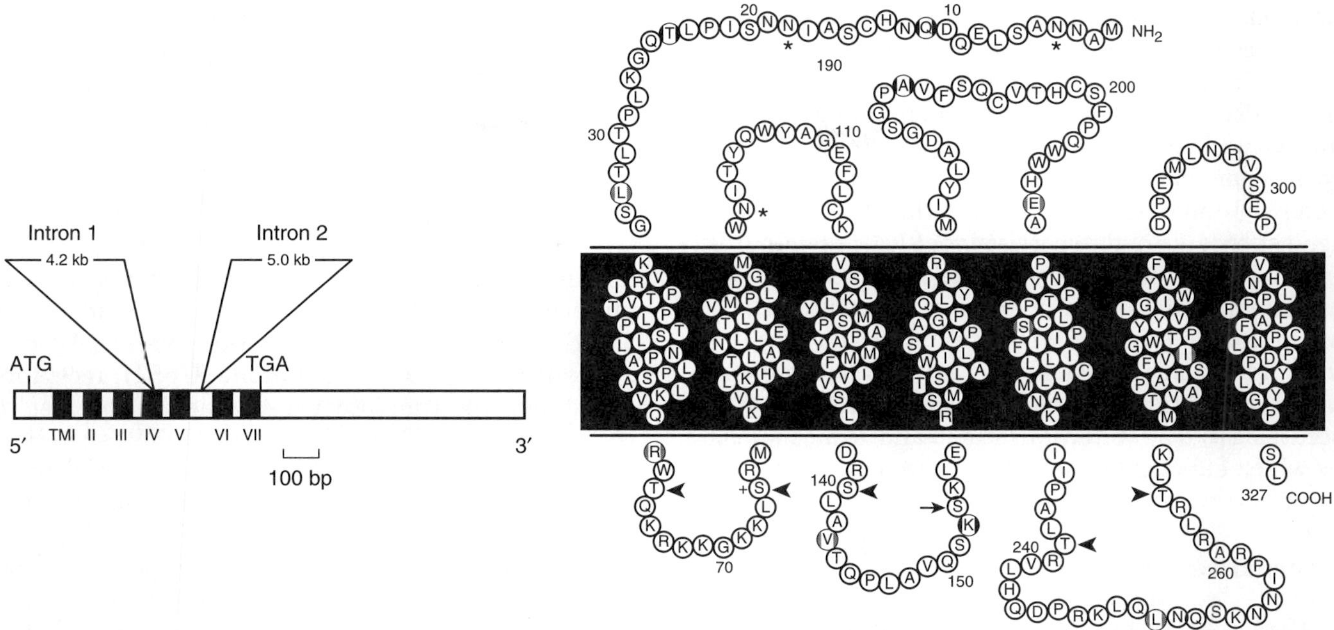

FIG. 1. *Left:* The structure of the human gonadotropin-releasing hormone receptor (GnRHR) cDNA. *Open boxes* indicate the protein coding region and *solid boxes* the putative transmembrane domains. The location and sizes of the introns are shown. (Reproduced with permission from Kakar, S. S. [1997]. Molecular structure of the human gonadotropin-releasing hormone receptor gene. *Eur. J. Endocrinol.* 137, 183–192 © Society of the European Journal of Endocrinology [1997].) *Right:* Model of the rat GnRHR. Amino acid residues in *black* represent nonconserved amino acids between the rat and mouse GnRHRs, and *shaded* amino acid residues are nonidentical but conserved between the two species. *Asterisks* denote potential glycosylation sites. Potential phosphorylation sites are indicated for protein kinase C (PKC; *arrowheads*), casein kinase II (*arrow*), and protein kinase A (PKA; *cross*). (Reproduced with permission from Kaiser, U. B., Zhao, D., Cardona, G. R., and Chin, W. W. [1992]. Isolation and characterization of cDNAs encoding the rat pituitary gonadotropin-releasing hormone receptor. *Biochem. Biophys. Res. Commun.* 189, 1645–1652. Copyright 1992, with permission from Elsevier.)

also been suggested (56,57). Thus, GnRHRs are unique in that G-protein coupling as well as ligand-induced desensitization and internalization are related to other regions than the C-terminal tail. The lack of a C-terminal intracellular tail has also been suggested to decrease the GnRHR expression level, receptor downregulation, and G-protein coupling (58,59). There are several potential phosphorylation and G-protein coupling sites in the ICLs (60).

Ligand Affinity and Expression in the Cell Membrane

Specific and high-affinity binding sites for GnRH are present in the rodent pituitary and αT3-1 cell membranes (19). The dissociation constant (K_d) of the GnRHR has been determined in a number of species to be between 2.8 to 4.9 nM for GnRH, and between 0.2 to 0.9 nM for a GnRH agonist, des-Gly^{10}-(D-Ala^{6})N-GnRH-N-ethylamide (18,19,27–31,37). The number of binding sites for GnRH in the membranes is 0.19 to 0.67 pmol/mg in the mouse pituitary, 0.24 to 0.41 pmol/mg in the rat pituitary, and 1.3 to 1.9 pmol/mg in the αT3-1 cell line (19). Given the facts that the pituitary gonadotropes comprise only 3% to 15% of the total pituitary cell population (16) and that the αT3-1 cell line represents a homogenous cell population, the estimated number of GnRH binding sites on αT3-1 cells is approximately 50% of the number on primary gonadotropes (9).

The number of GnRHR present in the cell membranes of pituitary gonadotropes varies during ontogenesis, the estrous cycle, pregnancy, and lactation, and as a consequence of hypothalamic lesions, gonadectomy, administration of GnRH agonists or antagonists, or gonadal steroid hormones (61–65). GnRH-mediated responses of pituitary gonadotropes correlate directly with the number of GnRHR present in the cell membrane, which in turn is regulated primarily by GnRH itself (64,66,67). Homologous regulation of the GnRHR by its ligand occurs in vivo in rats (68) as well as in vitro in rat anterior pituitary cell cultures (66). At physiological concentrations, GnRH has been shown to stimulate GnRHR gene expression and GnRHR number in primary pituitary cell cultures, as well as in the αT3-1 and LβT2 cell lines (69–71). The GnRH-induced increase in GnRHR number in the αT3-1 cell model occurs in part at a post-transcriptional level, where GnRH alters GnRHR messenger RNA (mRNA) translational efficiency without changing mRNA levels (70,72–74).

GnRH agonist, but not antagonist, occupancy of the GnRHR has been shown to promote physical intimacy (microaggregation) between receptors in a dose-dependent manner (75,76). This microaggregation has been observed to start within 1 minute after agonist binding and to persist for at least 80 minutes. In addition, the microaggregation process is not inhibited by disruption of actin or microfilament, or by depletion of Ca^{2+}, which is different from the internalization process. These studies suggest that microaggregation is an early event of GnRH binding, occurring before internalization, and a mechanism by which effector regulation may occur.

Internalization

Many GPCRs are downregulated by their ligands (77). Receptors are sequestered from the plasma membrane and internalized, followed by proteolytic degradation. This leads to a reduction in receptor number (homologous downregulation) over a period of hours. High concentrations of GnRH, as occur during initial proestrus phase in the rodent estrous cycle, can cause classic receptor-mediated internalization of the GnRH–receptor complexes, resulting in downregulation of the GnRHR (66,78). The internalization of GnRH–receptor complexes has been shown to take place over 2 hours after GnRH binding, beginning within 5 to 10 minutes and with kinetics depending on ligand potency and incubation temperature (51,79,80). In addition, GnRHR internalization has been shown to require G-protein coupling (81). Internalization occurs as an endocytosis pathway in the form of clathrin-coated small vesicles (pits) containing GnRH–receptor complexes, by forming patches and caps (82), separated from the plasma membrane by a GTPase dynamin (83).

The internalized GnRH–receptor complexes are subjected to one of two different fates: either GnRHR is released from the ligand in the Golgi complexes and recycled to the cell surface through secretory granules and multivesicular bodies (84,85), or the complexes are dissociated, processed, and ultimately degraded in lysosomes (78). These secretory granules and multivesicular bodies may provide a means not only for GnRHR recycling but also for the upregulation of GnRHR number upon GnRH stimulation. Although GnRH agonists have been shown to induce rapid internalization of GnRHR (66), some potent GnRH antagonists remain at the cell surface for a longer time (86). GnRH antagonists are also internalized at a slower rate and appear to follow a different intracellular route compared with agonists (87). These observations are consistent with the possibility that internalized antagonists are simply riding along with plasma membrane undergoing routine cycling. The internalization of GnRH-bound receptor is not required for stimulation of gonadotropin secretion in the pituitary (88).

Desensitization

GnRH is secreted from the hypothalamus in a pulsatile fashion, and pulsatile GnRH stimulates LH and FSH biosynthesis and secretion (89–91). Agonist occupancy of plasma membrane GnRH receptors is required to stimulate LH release. In contrast to the stimulatory effects of pulsatile GnRH, sustained exposure to high concentrations of GnRH reduces the response of gonadotropes to subsequent stimulation with GnRH (homologous desensitization), leading to suppression of gonadotropin secretion (7,8,89). Homologous desensitization of gonadotropes can be defined as a reduction in the ability of GnRH to elicit LH release after a prior exposure to GnRH.

The mechanism of this desensitization is not known, and both receptor (92) and postreceptor (93–95) mechanisms have been proposed. For a number of other GPCRs, early desensitization events are thought to involve the uncoupling of the receptor from its regulatory G protein, with loss of downstream signaling events (50). Rapid desensitization appears to involve phosphorylation by specific intracellular GPCR kinases of the ICL3 or the C-terminal tail (77,96). However, this is unlikely the case for GnRHR because the GnRHR lacks the C-terminal cytoplasmic tail as well as the ICL3 sequences implicated in the desensitization of other receptors (42). Homologous desensitization is likely to result, initially, from loss of cell surface receptors and is maintained by loss of a functional Ca^{2+} channel (97). Other mechanisms of desensitization are evident, however, because measurable desensitization occurs even when receptor internalization is blocked (98), and a desensitized state is manifest even after GnRH receptor numbers have returned to control levels (97).

Desensitization of gonadotrope cells to GnRH cannot be explained solely by loss of surface receptors. This early loss of responsiveness appears to be regulated initially by receptor loss and then maintained by loss of functional receptor–effector coupling or receptor-coupled Ca^{2+} channels, owing to a GnRHR-specific profile rather than to cell-specific regulatory features (99). Nonetheless, desensitization of the GnRHR in αT3-1 cells occurs slowly compared with other GPCRs, such as the muscarinic receptor. Similarly, sensitization or self-priming of gonadotropes occurs within a time period insufficient for substantive receptor upregulation (100). It has been suggested that postreceptor factors, such as the efficiency of receptor–effector coupling, rates of second messenger turnover, or responsiveness to second messengers, may govern aspects of the self-priming action of GnRH and may be regulated by protein kinase C (PKC).

A subset of G_α protein levels ($G_{q\alpha}$ and $G_{11\alpha}$) has been shown to decrease during prolonged GnRH exposure in αT3-1 cells (101) without changes in their mRNA levels (102), suggesting that enhanced proteolysis of the activated G_α proteins may contribute to gonadotrope desensitization. This GnRH-mediated downregulation of G_α proteins is not mimicked by sustained activation of PKC with a PKC activator, the phorbol ester 12-O-tetradecanoyl-phorbol-13-acetate (TPA) or phorbol 12-myristate 13-acetate (PMA), nor is it blocked by a PKC inhibitor, indicating that downregulation of the G_α-protein levels is a direct result of the activation of the G proteins themselves, not a result of downstream kinase activation. The rate of decay of these G_α proteins during GnRH stimulation is biphasic, having an initial rapid decay rate and a slower secondary decay phase. It is likely that the initial decay rate on receptor occupancy is reduced to a lower rate with desensitization of the receptor response, or that the fast decay phase may depend on the fraction of the cellular G protein that becomes activated on receptor occupancy, whereas the slow decay phase depends on the residual G-protein pool (9).

Regulators of G-protein signaling (RGS) proteins, with at least 16 homologs in mammals, interact directly with active G_α proteins to shorten their half-life by promoting GTPase activity of G_α proteins and GTP hydrolysis, and thus accelerating inactivation of G_α proteins and reassociation with βγ-subunit dimers (103,104). One of RGS family members, RGS3, is expressed in αT3-1 cells (104), and has been shown to be involved in the stimulation of GnRHR desensitization through GnRH occupancy–dependent palmitoylation and attenuation of $G_{q\alpha}$-protein activation of phospholipase C (PLC) (104,105) (Fig. 2). In addition, RGS10 stimulates GnRH-induced GnRHR desensitization by constitutive palmitoylation (105), and the lack of a C-terminal cytoplasmic tail has been implicated in preventing RGS10 interaction (106), thus resulting in a slow desensitization.

Treatment of pituitary cell cultures with physiological concentrations of GnRH results in a biphasic response by the cells with respect to GnRH receptor number (92). Initially, a downregulation of receptors is observed (0.5 to 4 hours post-treatment), followed by an increase in the number of GnRH receptors (9 hours post-treatment). The initial downregulation of receptors for GnRH is temporally associated and coincident with desensitization of gonadotropes to GnRH (107). Homologous downregulation of receptors occurs after continuous treatment with high concentrations of GnRH (97) and involves physical internalization

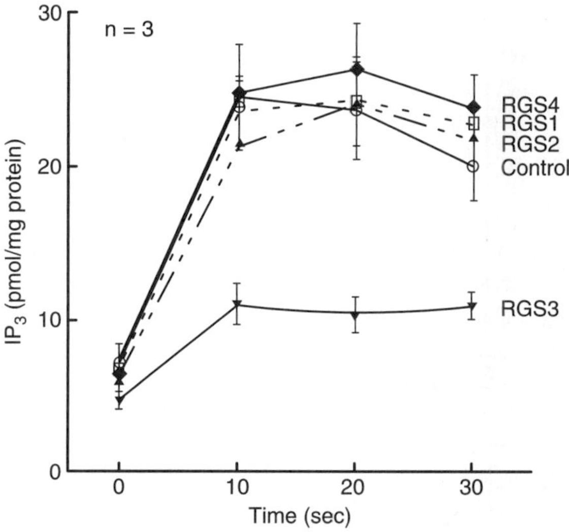

FIG. 2. Inhibition of gonadotropin-releasing hormone (GnRH)–induced cellular inositol 1,4,5-trisphosphate (IP_3) concentration by regulator of G-protein signaling protein-3 (RGS3). (Reproduced with permission from Neill, J. D., Duck, L. W., Sellers, J. C., Musgrove, L. C., Scheschonka, A., Druey, K. M., and Kehrl, J. H. [1997]. Potential role for a regulator of G protein signaling [RGS3] in gonadotropin-releasing hormone [GnRH] stimulated desensitization. *Endocrinology* 138, 843–846. Copyright 1997, The Endocrine Society.)

of agonist-occupied receptors (66,78). Homologous downregulation of GnRH receptors appears to be independent of extracellular Ca^{2+} concentration ($[Ca^{2+}]_o$), whereas upregulation of GnRH receptors depends on $[Ca^{2+}]_o$ (108). Neither homologous receptor downregulation nor desensitization appears to require PKC because phorbol esters do not mimic the acute effects of GnRH on receptor number or responsiveness, and cells depleted of measurable PKC exhibit normal downregulation and desensitization in response to GnRH (109).

Measurement of second messengers may lead to insights into early or short-term desensitization events. GnRH treatment leads to a linear increase in total inositol phosphate (IP) production in αT3-1 cells over 0 to 15 minutes, and GnRH pretreatment for 5 minutes does not alter subsequent stimulation of inositol 1,4,5-trisphosphate (IP_3) production by GnRH 15 minutes later (110,111). These data indicate a lack of desensitization of the rapid GnRH-induced IP_3 response in αT3-1 cells. Pretreatment with GnRH for 1 hour reduces subsequent cellular total IP (including IP, phosphatidylinositol 4-phosphate [PIP] and phosphatidylinositol 4,5-bisphosphate [PIP_2], and IP_3) accumulation in response to GnRH (Fig. 3), but this may be attributable to a reduction in GnRHR numbers. GnRH pretreatment of αT3-1 cells for short times (5 to 15 minutes) had no effect on GnRHR number, and treatment for 1 hour with GnRH reduced GnRHR number by 48% without altering the affinity for GnRH. Desensitization of both the extracellular Ca^{2+}-dependent and -independent phases of the Ca^{2+}

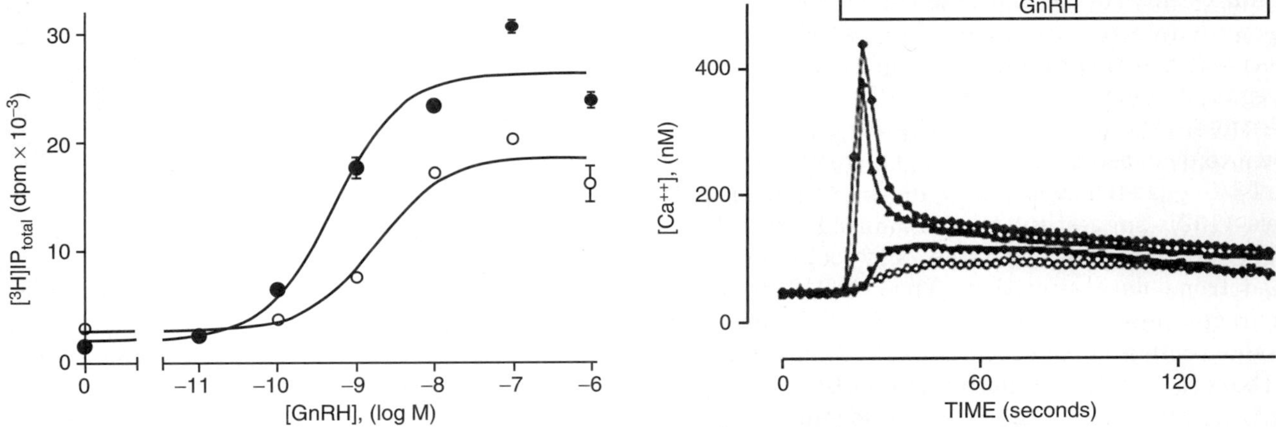

FIG. 3. Influence of gonadotropin-releasing hormone (GnRH) pretreatment on GnRH-stimulated total inositol phosphate (IP) accumulation (*left*) and elevation of cytosolic Ca^{2+} (*right*). *Left:* αT3-1 cells were pretreated with 0 (*closed circle*) or 10^{-7} (*open circle*) M GnRH. *Right:* αT3-1 cells were pretreated with 0 (*closed circle*), 10^{-10} (*triangle*), 10^{-8} (*inverted triangle*), or 10^{-6} (*open circle*) M GnRH. (Reproduced with permission from McArdle, C. A., Forrest-Owen, W., Willars, G., Davidson, J., Poch, A., and Kratzmeier, M. [1995]. Desensitization of gonadotropin-releasing hormone action in the gonadotrope-derived αT3-1 cell line. *Endocrinology* 136, 4864–4871. Copyright 1995, The Endocrine Society.)

response to GnRH has been observed after pretreatment with GnRH for 1 hour (111). Thus, one mechanism of intermediate desensitization to GnRH may be receptor loss. However, this does not account for rapid or early desensitization or the degree of desensitization of the Ca^{2+} response. An additional uncoupling event may occur during the pretreatment, which reduces the ability of the agonist-occupied GnRHR to elevate intracellular Ca^{2+}. Treatment of αT3-1 cells with 5-minutes pulses of GnRH every 15 minutes resulted in desensitization of the Ca^{2+} response after the first pulse in a dose-dependent manner (110). The mechanisms underlying this desensitization are likely to include loss of IP_3 receptors, depletion of intracellular Ca^{2+} stores, and inactivation of Ca^{2+} channels, as has been suggested in studies of primary pituitary cells (93) and αT3-1 cells (112,113). The dissociation and uncoupling of total IP production and Ca^{2+} stimulation suggests that desensitization of GnRH-induced Ca^{2+} mobilization is a postreceptor phenomenon occurring distal to PLC activation. The lack of the C-terminal cytoplasmic tail, implicated in the desensitization of other GPCRs, in the GnRHR therefore appears to correlate with a lack of receptor desensitization and, rather, desensitization to GnRH appears to be primarily a postreceptor phenomenon (112). Alternatively, αT3-1 cells may be lacking a factor(s) necessary for mediating rapid receptor desensitization in primary gonadotropes.

A study using LβT2 cells has observed that chronic GnRH or phorbol ester treatment induces desensitization throughout the signaling pathways, including GnRHR number, $G_{q/11\alpha}$ levels, cyclic adenosine 5′-monophosphate (cAMP) and Ca^{2+} mobilization, PKC, extracellular signal–regulated kinase (ERK), and p38 activation, c-*fos* induction, and PKC, LHβ, and GnRHR synthesis (114).

Effects of Mutations

Numerous in vitro mutational analysis studies using heterologous cell systems have identified many amino acid residues critical for functional activity of the GnRHR. The rodent GnRHR has a disulfide bridge that stabilizes receptor structure, between Cys^{114} (ECL1) and Cys^{195} (ECL2) residues, and Cys^{14} (N-terminal extracellular chain) and Cys^{199} (ECL2) are important for ligand binding or signal transduction, although they are not likely to form a disulfide bond (115). Cys^{14} is also a part of the residues that form a ligand-recognition pocket of the receptor, along with Asp^{98}, Trp^{101}, Asn^{102} (TMH2), Lys^{121} (TMH3), Asn^{211} (TMH5), Tyr^{289} (TMH6), and Glu^{301} (Asp^{302} in human, ECL3), with Trp^{279} (TMH6) at the bottom of the pocket (60,116–123). Also, residues flanking Glu^{301} (Asp^{302} in human), Ser^{300}, and Pro^{302} have been suggested to be important for differential ligand selectivity by stabilizing the conformation of ECL3 (124,125), and Trp^{279} (TMH6) and Phe^{310} (TMH7) are critical for agonist-induced IP production (126). In addition, several residues in the ICL2 (Asp^{138} and Leu^{147}) and ICL3 (Leu^{237} and Ala^{260}), as well as at the end of the TMH7 (Asp^{318}, Tyr^{322}, and Ser^{326}), have been shown to be involved in $G_{q/11\alpha}$-protein coupling and agonist-stimulated IP production (60,81,127–132). Also, several residues in the TMH1 (Leu^{58}) and ICL1 (Leu^{73}, Ser^{74}, Arg^{75}, and Leu^{80}) have been implicated for $G_{s\alpha}$-protein coupling, while not affecting $G_{q/11\alpha}$-protein coupling (60,133).

The mouse GnRHR has two N-linked glycosylation sites in the N-terminal extracellular domains (Asn^{4} and Asn^{18}) (134). These N-linked glycosylation sites have been shown to be important for maintaining GnRHR expression levels, presumably by improving receptor trafficking or stability (134,135). Furthermore, residues Asp^{138} (ICL2), Leu^{237}, Ser^{253}, Thr^{264} (ICL3), and Phe^{271} (TMH6) also have been identified to be important for GnRHR expression (130,131,136,137), residues Arg^{139}, Ser^{140} (ICL2), Thr^{238} (ICL3), Val^{299} (Leu^{300} in human, ECL3), and Phe^{325} (TMH7) for agonist binding affinity (119,127,132,136,138), and residues Asp^{138}, Arg^{139}, Ser^{140}, Leu^{147} (ICL2), Leu^{237}, Ala^{260} (ICL3), Asp^{318}, and Tyr^{322} (TMH7) for agonist-induced internalization (81,127–131). These studies indicate that many residues contribute to the structural integrity of the GnRHR (10,60).

In most GPCRs, there are highly conserved Asp and Asn residues located in the TMH2 and TMH7, respectively, and the Asp residue in the TMH2 is essential for normal ligand binding and G-protein coupling (42). In the GnRHR, they are substituted for each other, Asp being located in the TMH7 at position 318 and Asn in the TMH2 at position 87 of the rodent GnRHR, another aspect making the GnRHR unique among the known GPCRs. Interestingly, whereas the Asn^{87}Asp mutation results in impaired ligand binding and signal transduction (139), the Asp^{318}Asn mutation or a reciprocal double mutation (Asn^{87}Asp/Asp^{318}Asn) restores them moderately (140). It is therefore possible that TMHs 2 and 7 are in close spatial proximity and act in a complementary manner so that the functional conformation necessary for maintaining the ligand binding pocket and G-protein coupling is preserved (141).

A number of loss-of-function GnRHR mutations have been identified in a subset of human patients with idiopathic hypogonadotropic hypogonadism (IHH). Seventeen mis-sense mutations in the coding sequence have been reported to date. IHH is characterized as hypogonadism associated with low or undetectable gonadotropin secretion levels and a disruption of the usual pulsatile pattern, associated with complete or partial failure of sexual maturation (142–148). IHH in these patients is transmitted in an autosomal recessive pattern of inheritance, and either compound heterozygous mutations (different mutations in each allele) or homozygous mutations have been identified. The cellular and molecular phenotypes of each mutation have been extensively studied in vitro, using heterologous cell types.

Arg^{262}Gln (ICL3) and Tyr^{284}Cys (TMH6) mutations do not modify the binding of GnRH, but decrease PKC activation and total IP production (149,150); Asn^{10}Lys, Asn^{10}Lys+Gln^{11}Lys, Thr^{32}Ile (N-terminal extracellular chain), Gln^{106}Arg (ECL1), Cys^{200}Tyr (ECL2), and Ser^{217}Arg (TMH5) mutations dramatically decrease, but do not completely abolish, GnRH binding and IP production (149,151–155); Ala^{129}Asp (TMH3) and Ser^{168}Arg (TMH4) mutations cause a complete loss of GnRH-mediated IP production (156,157); and Glu^{90}Lys (TMH2), Arg^{139}His (ICL2), Ala^{171}Thr (TMH4), Leu^{266}Arg (ICL3), Cys^{279}Tyr (TMH6), Leu^{314}Stop, and Pro^{320}Leu (TMH7) mutations result in a complete loss of receptor function (59,152–155, 158,159). A splice junction mutation (G-to-A transversion) at the intron 1–exon 2 boundary has also been described, resulting in a lack of exon 2 transcription, and causing a frame shift that adds three more amino acids followed by a stop codon (160). This mutation results in a nonfunctional truncated GnRHR. Interestingly, despite greatly attenuated IP production, Asn^{10}Lys+Gln^{11}Lys, Thr^{32}Ile, Gln^{106}Arg, and Arg^{262} Gln mutations are able to stimulate gonadotropin and GnRHR gene promoter activities as well as ERK activation, albeit with markedly reduced and attenuated levels (154,155,161). On the other hand, the mutants with complete loss of function, Leu^{266}Arg, Cys^{279}Tyr, and Pro^{320}Leu, fail to activate the gonadotropin gene promoters (154,155).

The loss of function of some mis-sense mutant GnRHRs can be rescued in vitro. For example, 11 of the identified mis-sense mutations have been shown to recover partial function in response to a nonspecific antagonist or pharmacological chaperone, which functions as a folding template. This finding suggests that some of the GnRHR mutants have not lost intrinsic function but rather that the loss of function results from receptor misfolding and misrouting, likely to endoplasmic reticulum (162,163). Not all mutations result in misrouting (132). Furthermore, the lack of GnRH-induced IP response by the Glu^{90}Lys mutant GnRHR can be rescued by deletion of Lys^{191}, the residue absent in the rodent GnRHR, suggesting that deletion of Lys^{191} restores the loss of spatial arrangement between receptor and G protein provoked by the Glu^{90}Lys mutation (59). It also demonstrates that the loss of function can be rescued by stabilizing receptor folding and thus improving membrane expression. These studies not only confirm previous in vitro mutation analyses, but provide insights into the genetic abnormalities underlying such conditions and define distinct regions in the GnRHR structure critical for normal GnRH signal transduction and gonadotropin secretion as well as gene expression.

Type II GnRH Receptor

Although the hypothalamic decapeptide GnRH (GnRH I, pyroGlu-His-Trp-Ser-Tyr-Gly-Leu-Arg-Pro-Gly·NH_2) is considered to be the principal ligand for the classic pituitary gonadotrope GnRHR (type I GnRHR) in mammals (6,164), it has become apparent that most vertebrates have one or more additional forms of GnRH. One of these is GnRH II (pyroGlu-His-Trp-Ser-His-Gly-Trp-Tyr-Pro-Gly·NH_2), a product of a second GnRH gene (165). GnRH II is conserved among species, including human (except mouse), is widely expressed throughout the brain (60), functions as a neuromodulator (166), coordinates reproductive behavior as appropriate to the energy condition (167), and modifies food intake (168).

The existence of two GnRH subtypes in mammals has suggested the existence of additional GnRHR subtypes. Type I GnRHRs from a number of nonmammalian vertebrates are different from the mammalian type I GnRHR, having a C-terminal intracellular tail, and having different GnRH subtypes as their cognate ligands (169–175). Subsequently, a second GnRHR, referred to as type II GnRHR, has been cloned from limited number of nonhuman primates, including marmoset (176), rhesus monkey (macaque), and African green monkey (177) (Fig. 4). A rodent type II GnRHR has not yet been identified.

The type II GnRHR gene encodes a 380–amino acid protein that includes a cytoplasmic C-terminal tail, which is phosphorylated, rapidly internalized, and desensitized on ligand binding (178). Interestingly, the Asp^{86} residue in TMH2, replaced with Asn^{87} in type I GnRHR, is conserved in type II GnRHR (176,177). It has been shown that type II GnRHR can internalize through either a dynamin- or β-arrestin–dependent mechanism, and that $Ser^{338,339}$ residues are critical for the internalization as phosphorylation sites by GPCR kinases (52,54,57). The primate type II gene is expressed almost ubiquitously throughout the body, including the brain, pituitary gonadotropes, and reproductive system, with mRNA 1.8 kb in size (176,179). This receptor has a higher affinity for GnRH II than for GnRH I (over 40-fold) (176,177), whereas type I receptor has a higher affinity for GnRH I (tenfold) (118). Interestingly, type I mammalian GnRHR full antagonists behave as agonists at the type II GnRHR and at chicken or *Xenopus* type I GnRHR, although with reduced potency compared even with GnRH I (173,176).

Type II GnRHR is also coupled to the $G_{q/11\alpha}$ subfamily of G proteins, and mediates the GnRH II–induced intracellular IP_3 synthesis and activation of

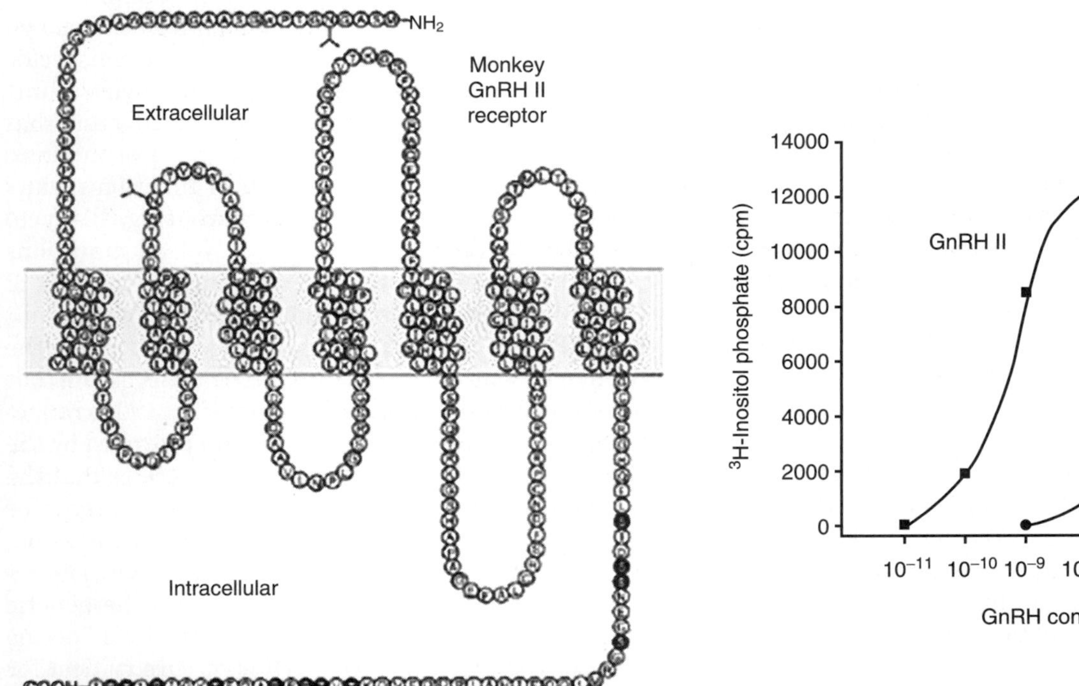

FIG. 4. *Left:* Seven transmembrane G-protein–coupled receptor (GPCR) structure of African green monkey type II gonadotropin-releasing hormone receptor (GnRHR). The *darkened* Ser (S) and Thr (T) residues in the C-terminal tail of the type II GnRHR are predicted sites of phosphorylation. (Reproduced with permission from Neill, J. D. [2002]. Minireview: GnRH and GnRH receptor genes in the human genome. *Endocrinology* 143, 737–743. Copyright 2002, The Endocrine Society.) *Right:* Functionality of the type II GnRHR and its selectivity for GnRH II over GnRH I. COS-1 cells were transfected with African green monkey type II GnRHR cDNA, and total inositol phosphate (IP) accumulation in response to GnRH II and GnRH I was measured. (Reproduced with permission from Neill, J. D., Duck, L. W., Sellers, J. C., and Musgrove, L. C. [2001]. A gonadotropin-releasing hormone [GnRH] receptor specific for GnRH II in primates. *Biochem. Biophys. Res. Commun.* 282, 1012–1018. Copyright 2001, with permission from Elsevier.)

the PKC pathway, ERK, and p38. These activations are more sustained than for type I receptor with GnRH I (180). Studies have demonstrated that PKC activation can stabilize the conformation of type II GnRHR but not type I GnRHR, causing an increase in GnRH II binding (57,181). This conformational stabilization is selective for GnRH II and suggests that the C-terminal intracellular tail of the type II receptor may mediate this phenomenon.

A putative human type II receptor gene is located on chromosome 1q.12, and shares 40% sequence homology with the type I receptor, composed of three exons approximately 7.5 kb in size (177,179,180). This gene is transcribed throughout human tissues with mRNAs of 2.4, 3.0, 6.0, or 7.5 kb in size (176,177,179,180,182). However, this human gene contains a frame shift and a C-to-T substitution in the open reading frame, which changes the codon for Arg^{179} in the primate sequences to an in-frame premature stop codon (179). Moreover, this type of disruption or silencing, of the type II GnRHR gene has also been reported in the chimpanzee, mouse, rat, cow, and sheep (179, 183,184). In addition, this gene overlaps a retro-transposed pseudogene, and a truncated form of the type II GnRHR gene lacking the first exon and part of the first intron is also present on chromosome 14q.22, which also overlaps another similar retro-transposed pseudogene (179,182,185,186). Thus, despite apparent active gene transcription and immunolocalization of type II GnRHR in the human pituitary (176), whether the transcripts of the human type II GnRHR encode a functional receptor protein remains largely unresolved (187). Therefore, it is likely that GnRH II in humans also mediates its action through type I GnRHR, perhaps a more recently evolved form of receptor than the type II GnRHR and nonmammalian type I GnRHR.

GnRH SIGNAL TRANSDUCTION AND GONADOTROPIN SECRETION

Guanosine Triphosphate–Binding Proteins

G proteins are heterotrimeric proteins, composed of α, β, and γ subunits, located on the cytoplasmic side of cell membranes (188). During unstimulated conditions, the α subunit is coupled with guanosine diphosphate (GDP), forming the intact heterotrimeric G protein. When extracellular ligands bind to their seven transmembrane receptors, the conformational change in the receptor leads the G proteins coupled with those receptors to propagate intracellular signal transduction pathways by replacing GDP with GTP, which dissociates the βγ subunits from the α subunit. The free G_α subunit has an intrinsic GTPase activity, and uses GTP transformation to GDP to activate sequentially the downstream signal cascade. There are several distinct G_α subunits: G_s, G_{olf}, G_i, G_q, $G_{11\text{-}16}$, G_o, G_z, G_t, G_k, G_e, and G_g. Several G_α subtypes, G_i, G_o, and G_t, are sensitive to pertussis toxin (PTX), which adenosine diphosphate (ADP)-ribosylates and dissociates the G_α subunit. Other G_α subunits, notably G_s, G_q, and G_{11}, are PTX insensitive. The G_α subtype determines the particular downstream signaling pathways activated. For example, $G_{s\alpha}$ and $G_{i\alpha}$ mediate stimulation and inhibition, respectively, of adenylate cyclase and protein kinase A (PKA), $G_{q\alpha}$ and $G_{11\alpha}$ mediate stimulation of PLCβ1 and β2, $G_{o\alpha}$ and $G_{z\alpha}$ Ca^{2+} and potassium channels, $G_{t\alpha}$ transducin light-activated photoreceptors and cyclic guanosine monophosphate-phosphodiesterase, $G_{k\alpha}$ ion channel conductance, and $G_{e\alpha}$ exocytosis (78,188–190).

Initial studies using GTP analogs have indicated that GnRH activity is mediated by a GPCR and that they enhance total IP formation and LH release from the pituitary in a time- and dose-dependent manner (191,192). GnRH signal transduction is mediated primarily by a subset of G proteins, notably PTX- and cholera toxin–insensitive $G_{q\alpha}$ and $G_{11\alpha}$ (193,194). The ICL2 and ICL3 of the GnRHR are considered critical in the $G_{q/11\alpha}$-protein coupling (33,60,195). Through coupling to $G_{q/11\alpha}$, GnRH binding to GnRHR activates PLCβ, which results in the rapid phosphodiesteric hydrolysis, breakdown, and turnover of PIP_2 to generate IP_3 and the backbone moiety of the phosphoinositides, 1,2-diacylglycerol (DAG), within 10 seconds (196–199). These products, in turn, act as second messengers (Fig. 5).

Several studies demonstrate, in addition, that at least a subset of GnRHR signaling is mediated by $G_{s\alpha}$ and $G_{i\alpha}$ through modulation of the PKA pathway (94, 200–203), suggesting GnRHR can couple to multiple G proteins. In addition, dissociated β and γ subunits of the G proteins have also been implicated in mediating GnRH-stimulated signal transduction and cAMP and IP production (204).

Phospholipase C and Phosphoinositides

There are at least 16 PLC isoenzymes falling into three major types, PLCβ, PLCγ, and PLCδ, and all of these three types of PLC catalyze hydrolysis of PIP_2 to IP_3 and DAG. Ligand-induced conformational changes in GPCRs of the $G_{q/11\alpha}$ family activate the β-forms of PLC (189). It has been shown that $G_{q\alpha}$ and $G_{11\alpha}$ activate PLCβ1, but not PLCγ1 or PLCδ1 (205,206). Substrates for PLC range from IP to PIP to PIP_2. The formation of IP_3 is a rapid and transient process because of its subsequent conversion to a more stable but inactive form, inositol 1,3,4-trisphosphate, or to

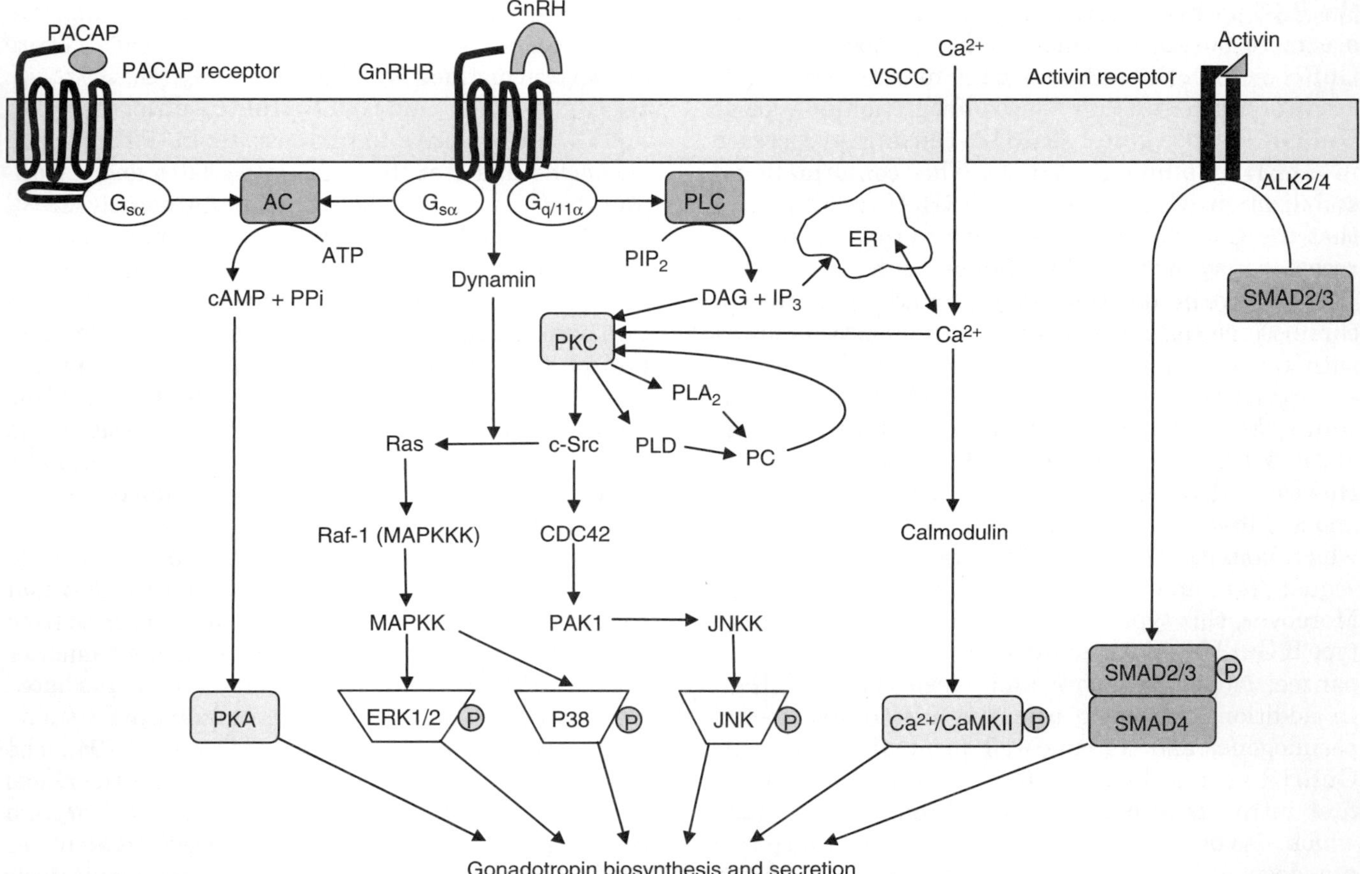

FIG. 5. Schematic summary of the gonadotropin-releasing hormone (GnRH)–mediated signal transduction pathways in pituitary gonadotropes. GnRH binds to the GnRH receptor (GnRHR), which is coupled primarily to $G_{q/11\alpha}$. Activation of $G_{q/11\alpha}$ activates phospholipase C (PLC), which stimulates the hydrolysis of phosphatidylinositol 4,5-bisphosphate (PIP_2) to diacylglycerol (DAG), which activates protein kinase C (PKC), and to inositol 1,4,5-trisphosphate (IP_3), which triggers Ca^{2+} release from the intracellular stores, such as endoplasmic reticulum (ER). GnRH binding to GnRHR also activates T-type and L-type voltage-sensitive Ca^{2+} channels (VSCCs). Phospholipases D and A_2 (PLD and PLA_2) are activated by PKC and metabolize other membrane phospholipids, such as phosphatidylcholine (PC). Three major mitogen-activated protein kinase (MAPK) cascades are activated by GnRHR signaling, leading to activation of extracellular signal–regulated kinase (ERK), c-Jun-N-terminal kinase (JNK), and p38. Coupling to $G_{s\alpha}$ stimulates adenylate cyclase (AC), which produces cyclic adenosine monophosphate (cAMP) and activates the protein kinase A (PKA) pathway. Ca^{2+}-mediated activation of Ca^{2+}/calmodulin–dependent protein kinase type II (Ca^{2+}/CaMKII) is also involved in GnRHR signaling. SMAD (Sma/mothers against decapentaplegic homolog) activation and translocation by activin and PACAP (pituitary adenylate cyclase–activating polypeptide-38) signaling are also illustrated, which are important pathways for potentiation of the GnRH response.

PIP_2, PIP, and IP from inositols and DAGs (78). The PLCβ1 isoenzyme has the greatest selectivity for PIP_2 as substrate, whereas PLCγ1 has the least selectivity (189). The action of PLC on PIP_2 is a key step in the accumulation of IP_3, and IP_3 is the major second messenger for Ca^{2+} mobilization.

The maintenance of IP_3 and its signaling activity is regulated by its rates of synthesis and metabolism, as determined by PLC activity, dephosphorylation by IP_3 5-phosphomonoesterase (IP_3-phosphatase), and phosphorylation by inositol 1,4,5-triphosphate 3-kinase (IP_3-kinase) (190). IP_3-phosphatase degrades IP_3 to inositol 1,4-bisphosphate, serving as the major pathway of inactivation. This IP_3-phosphatase activity is regulated by GPCR activation as well as by several inositol phosphate intermediate metabolites, including inositol 1,3,4,5-tetraphosphate (IP_4). IP_4 may compete with IP_3 as a phosphatase substrate and thereby sustain IP_3 levels and hence intracellular Ca^{2+} mobilization (207). IP_3-kinase converts IP_3 to IP_4. Agonist-induced intracellular Ca^{2+} mobilization further stimulates IP_3-kinase activity (208), and PKC inhibits this kinase activity (209). Protein tyrosine kinases are also thought to regulate IP_3-kinase activity (210). IP_4 can also be converted to IP_3 by the action of IP_4 3-phosphomonoesterase (IP_4-phosphatase) instead of inositol 1,3,4-trisphosphate (211). Biologically inactive inositol 1,3,4-trisphosphate is also generated from IP_4 by the action of IP_3-phosphatase, and inositol 1,3,4-trisphosphate can undergo further dephosphorylation or recruit the phosphorylation process.

Many earlier studies have noted that phosphoinositide metabolism (turnover, hydrolysis, and phosphorylation) in the pituitary is induced by GnRH (196,197), and that PIP_2 is the major substrate of PLC activated by GnRH in pituitary gonadotropes and is converted to IP_3 (212). The changes in phosphoinositide turnover in pituitary gonadotropes after GnRH stimulation indicate that IP_3 production is biphasic, with an early transient rise, followed by a nadir between 1 to 5 minutes, and a slower and higher subsequent elevation between 5 to 20 minutes, suggesting that IP_3 production is the result of both an initial acute and a subsequent sustained response to GnRH stimulation (198) (Fig. 6).

In αT3-1 cells, GnRH causes an increase in phosphoinositide turnover and in translocation of PKC to membranes, and enhances activation of voltage-sensitive Ca^{2+} channels (VSCCs) (19). The increase in IP_3 is rapid, occurring within 30 seconds, and IP_3 continues to accumulate, reaching a maximum after 20 minutes. IP_3 catabolism, including dephosphorylation and phosphorylation pathways, is also functional after GnRH stimulation in this cell line (19).

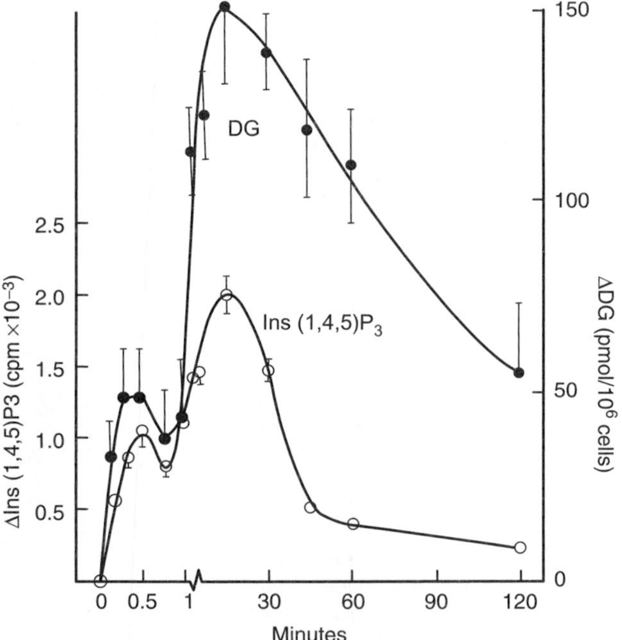

FIG. 6. Effects of gonadotropin-releasing hormone (GnRH) on inositol 1,4,5-trisphosphate (IP_3) and diacylglycerol (DG) production in αT3-1 cells. Cells were prelabeled with *myo*-[^{3}H]inositol for 48 hours and then incubated with 100 nM GnRH for the indicated time periods. Phosphoinositides were extracted and quantified. (Reproduced with permission from Zheng, L., Stojilkovic, S. S., Hunyady, L., Krsmanovic, L. Z., and Catt, K. J. [1994]. Sequential activation of phospholipase-C and -D in agonist-stimulated gonadotrophs. *Endocrinology* 134, 1446–1454. Copyright 1994, The Endocrine Society.)

IP_3 acts through specific IP_3 receptors to induce a rapid biphasic increase in the intracellular Ca^{2+} concentration ($[Ca^{2+}]_i$) by mobilizing Ca^{2+} from intracellular stores and by enhancing Ca^{2+} entry from the extracellular pool, and DAG and elevated $[Ca^{2+}]_i$ translocate PKC subspecies from the cytosol to the membrane to activate them (7,213–216) (Fig. 5). Intracellular Ca^{2+} and PKC serve in turn as messengers for propagation of signal transduction. Specific, saturable, and reversible binding of IP_3 to its receptors has been characterized in membrane preparations from the anterior pituitary (214) and other cell systems (217). IP_3 binding reaches a maximum by 15 minutes after addition of IP_3 and is sustained for up to 30 minutes. Scatchard analysis of IP_3 binding to its receptors indicates that there is a single set of high-affinity binding sites with a K_d of approximately 1 nM. Addition of IP_3 to a pituitary membrane preparation causes a dose-dependent release of Ca^{2+} from the vesicular component of membrane fractions (214), such as endoplasmic reticulum, and injection of IP_3 into single gonadotropes in a patch-clamp experiment has induced Ca^{2+} mobilization in an extracellular Ca^{2+}-independent manner that mimics the effect of GnRH (218). These studies demonstrate the functional significance of IP_3 receptors in triggering GnRH-induced Ca^{2+} mobilization in pituitary gonadotropes.

IP_3 receptors have been identified to be ligand-gated Ca^{2+}-permeable channels (219). The IP_3 receptor is a homotetrameric complex with four identical glycosylated subunits, each of which contains a binding site for one IP_3 molecule in the N-terminal region. The C-terminal region contains a cluster of hydrophobic sequences, which may comprise eight TMHs to form the Ca^{2+} channel intrinsic to the IP_3 receptor. Each subunit of the IP_3 receptor is intermolecularly associated through the TMHs and the successive C-terminus. The interaction between receptor subunit and IP_3 may cause a conformational change in the tetrameric complex, resulting in the opening of the Ca^{2+} channel. It has been suggested that binding of one IP_3 molecule to the receptor is sufficient fully to activate the Ca^{2+} channel (220). The IP_3 receptors have four conductance states because of the four subunits, with a mean open time of approximately 10 msec (221). The N-terminal IP_3 binding domain and C-terminal TMHs are separated by a stretch of 1,800–amino acid residues that contain regulation sites for cAMP kinase, PKC, and $[Ca^{2+}]_i$ (222,223). It is therefore possible that IP_3 receptors may also participate in signal transduction. The GnRH-induced release of Ca^{2+} through these receptors into the cytosol is electrically compensated by allowing exchange with a fast inward K^+ uptake through Ca^{2+}-sensitive K^+ channels (224,225).

Intracellular Calcium

As predicted by the presence of a site for regulation on IP_3 receptors, cytoplasmic Ca^{2+} is also a regulator for Ca^{2+} mobilization (Fig. 7). Intracellular Ca^{2+} directly modulates IP_3 receptor activity (226), and may also act through modulation of IP_3 binding (227) and activation of Ca^{2+}-sensitive proteins (228). The effect of $[Ca^{2+}]_i$ on IP_3 receptor activity is biphasic, having both positive and negative feedback effects. The initial GnRH-induced increase in $[Ca^{2+}]_i$ facilitates the further elevation of $[Ca^{2+}]_i$, whereas the higher $[Ca^{2+}]_i$ attained in the second phase of the intracellular Ca^{2+} increase reduces IP_3 receptor activity (220,226, 229–231). Thus, the initiation step for the GnRH-induced increase of $[Ca^{2+}]_i$ by release from intracellular stores occurs through IP_3, and this is potentiated by moderate increases in $[Ca^{2+}]_i$. Once initiated, Ca^{2+} release continues until the IP_3-sensitive Ca^{2+} pool is depleted, and the regeneration step begins to replenish Ca^{2+} stores from the extracellular Ca^{2+} pool in the absence of increases in IP_3 production (232,233). Elevated $[Ca^{2+}]_i$ facilitates IP_3-dependent activation of the cytoplasmic oscillator, along with the coagonist actions of luminal Ca^{2+} (Ca^{2+} present in the lumen of the endoplasmic reticulum) and cytosolic IP_3 (233). IP_3 receptors are sensitive to luminal Ca^{2+} (234). Coagonist actions of luminal Ca^{2+} and cytosolic IP_3 are suggested, where these two molecules act as coagonists to increase the sensitivity of the intracellular Ca^{2+} pool to IP_3, so that IP_3 receptors can readily open when the intracellular pool contains less Ca^{2+} (235). Decreases in luminal Ca^{2+} also reduce IP_3-induced Ca^{2+} release by a direct negative action of discharged luminal $[Ca^{2+}]$ through IP_3 receptors (236). The inhibitory effect of high levels of luminal $[Ca^{2+}]$ on Ca^{2+} release has been suggested to have a stabilizing effect on resting gonadotropes by reducing their

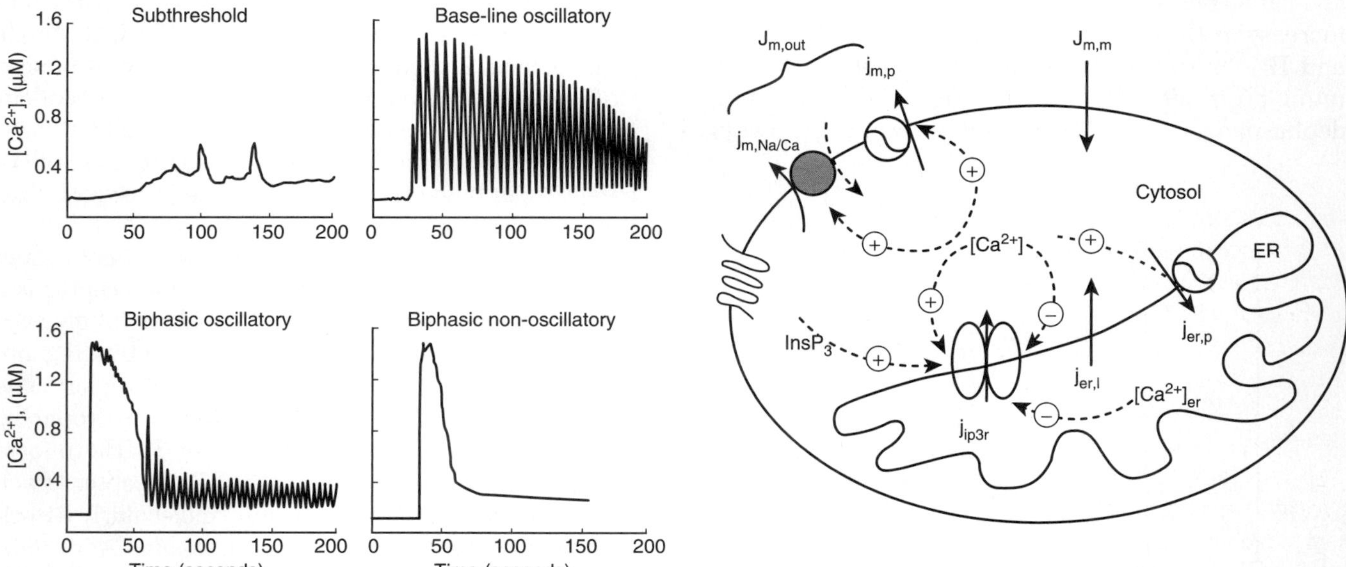

FIG. 7. *Left:* Typical patterns of Ca^{2+} signaling in gonadotropes from ovariectomized animals stimulated with gonadotropin-releasing hormone (GnRH). Agonist was added in 1-mL volumes and was present until the end of the recording. (Reproduced with permission from Tomic, M., Cesnajaj, M., Catt, K. J., and Stojilkovic, S. S. [1994]. Developmental and physiological aspects of Ca^{2+} signaling in agonist-stimulated pituitary gonadotrophs. *Endocrinology* 135, 1762–1771. Copyright 1994, The Endocrine Society.) (237). *Right:* Schematic representation of the model for agonist-induced Ca^{2+} dynamics. The essential elements for this model are (a) the endoplasmic reticulum (ER) is the primary intracellular Ca^{2+} pool; (b) the inositol 1,4,5-trisphosphate (IP_3) receptor is a major physiological pathway of Ca^{2+} release from ER into the cytosol; (c) Ca^{2+} uptake into the ER is modulated by a Ca^{2+}-ATPase; (d) $[Ca^{2+}]_i$ is responsible for both activation and inactivation of the IP_3 receptor; and (e) Ca^{2+} flux through the IP_3 receptor is enhanced at low $[Ca^{2+}]_{er}$. The IP_3 receptor and the ER Ca^{2+}-ATPase are the key elements of the oscillatory machinery, with $[Ca^{2+}]_i$, $[IP_3]$, and $[Ca^{2+}]_{er}$ being the major gating regulators of the channel. Phosphorylation of the ER Ca^{2+}-ATPase by ATP provides the primary driving force for the oscillations by maintaining the ER/cytosol $[Ca^{2+}]$ gradient. $J_{m,out}$, Ca^{2+} fluxes out of the plasma membrane; $J_{m,in}$, Ca^{2+} fluxes into the plasma membrane; $J_{m,p}$, Ca^{2+} pump on the plasma membrane; $J_{m,Na/Ca}$, Na^+/Ca^{2+} exchanger on the plasma membrane; $J_{er,p}$, Ca^{2+} pump on the ER; J_{IP3r}, IP_3 receptor; $[Ca^{2+}]_{er}$, ER luminal Ca^{2+} concentration. (Reproduced with permission from Li, Y.-X., Rinzel, J., Keizer, J., and Stojilkovic, S. S. [1994]. Calcium oscillations in pituitary gonadotrophs: comparison of experiment and theory. *Proc. Natl. Acad. Sci. U S A* 91, 58–62. Copyright 1994, National Academy of Sciences, USA.)

sensitivity to the background concentration of IP_3 (229). Once a threshold level of Ca^{2+} is released by increased levels of IP_3, lower levels of luminal $[Ca^{2+}]$ now may facilitate further Ca^{2+} release even when the IP_3 levels decrease. Thus, there are three phases in GnRH-induced Ca^{2+} oscillation in the pituitary gonadotropes: fast activation of IP_3 receptors at low, subthreshold cytoplasmic $[Ca^{2+}]_i$, followed by slow inactivation at high cytoplasmic $[Ca^{2+}]_i$ (baseline oscillation), and enhanced Ca^{2+} channel opening by depleted luminal $[Ca^{2+}]$ (biphasic response) (232,237) (Fig. 7).

The IP_3-sensitive intracellular Ca^{2+} pool also contains luminal Ca^{2+}/calmodulin–dependent adenosine triphosphatases (ATPases) (238), suggesting that Ca^{2+} channels and Ca^{2+}/calmodulin–dependent ATPases can function as interacting pairs of Ca^{2+} regulators in the control of cytoplasmic and intraluminal Ca^{2+} (190). Ca^{2+}/calmodulin–dependent ATPase is an integral element of the IP_3-sensitive Ca^{2+} store in coagonist actions of IP_3 and $[Ca^{2+}]_i$ during regeneration of the Ca^{2+} response (239). This enzyme compensates for the inhibitory effect of $[Ca^{2+}]_i$ on its own release by replenishing Ca^{2+} in the lumen of the endoplasmic reticulum in favor of reinitiation of a Ca^{2+} response. Along with a $[Ca^{2+}]_o$ influx system through VSCCs, this system may be a part of cytoplasmic Ca^{2+} oscillator in the pituitary gonadotropes, through which the intracellular Ca^{2+} pool discharges its contents into the cytoplasm and ultimately reaccumulates Ca^{2+} from the cytoplasm (240,241) (Fig. 7).

Earlier studies using cloned GnRHRs showed them to mediate GnRH-induced $[Ca^{2+}]_i$ elevations in various heterologous systems (27,28), and these findings have been confirmed in pituitary cells (215), in αT3-1 cells (242), and in LβT2 cells (243), where GnRH induces a dose-dependent biphasic change of $[Ca^{2+}]_i$, a rapid increase followed by a slower but more sustained elevation, that matches the time scale of the GnRH-induced IP_3 production (244). An initial GnRH-induced, IP_3-mediated rapid intracellular Ca^{2+} peak occurs within 10 seconds, followed by a more sustained increase in $[Ca^{2+}]_i$ and by a gradual decline over a period of 30 minutes (215). The initial phase of the $[Ca^{2+}]_i$ response is independent of $[Ca^{2+}]_o$, whereas the second phase depends on Ca^{2+} influx through VSCCs (241,242). Activation of an L-type VSCC has been suggested for GnRH-induced influx of extracellular Ca^{2+} in gonadotropes, which may be necessary to refill intracellular Ca^{2+} stores (19,215,244,245). Detailed characterization of the Ca^{2+} currents has uncovered the existence of two types of Ca^{2+} channels in rat pituitary gonadotropes, low and high VSCCs (246). The transient, low-voltage–activated, and rapidly inactivated current is attributed to a T-type channel, and the delayed, high-voltage–activated, and noninactivated current corresponds to the L-type channel in gonadotropes as well as in αT3-1 cells (19, 246,247). Pituitary gonadotropes also have voltage- and tetrodotoxin-sensitive Na^+ channels that allow opening of VSCCs by the action potentials (248). The $[Ca^{2+}]_o$-independence of the initial elevation of $[Ca^{2+}]_i$ reflects the release of Ca^{2+} from intracellular stores, activated by IP_3 and IP_3–receptor interaction (Fig. 7). The temporal correlation between the second phase of IP_3 production (212) and the $[Ca^{2+}]_o$-dependent second-phase elevation of $[Ca^{2+}]_i$ (245) suggests an IP_3-induced Ca^{2+} entry into gonadotropes. Ca^{2+} influx has also been shown to be induced by the DAG-activated PKC pathway in rat gonadotropes (240,245, 249) and αT3-1 cells (247).

Diacylglycerol

GnRH-induced hydrolysis of PIP_2 by PLC produces DAG as well as IP_3. The released DAG is rapidly metabolized to phosphatidic acid by DAG kinase, and eventually to IP. It has been shown that GnRH induces protein–protein interaction between DAG kinase-ζ and c-Src to facilitate the activation of PKC pathway in LβT2 cells (250). As a result of the two phases of increased IP_3 production, the initial peak in GnRH-induced DAG production within 45 seconds is followed by a more sustained peak between 15 and 30 minutes after GnRH stimulation of pituitary gonadotropes (251) and αT3-1 cells (252) (Fig. 6). The amplitude of the sustained peak in DAG is larger than the corresponding IP_3 peak, suggesting involvement of additional pathways for DAG production in gonadotropes, in addition to PLC (252). The second phase of DAG production also results from several other phospholipase pathways, such as phospholipase D (PLD) and phospholipase A_2 (PLA_2), which metabolize other membrane phospholipids, such as phosphatidylcholine, to enhance and maintain prolonged activation of PKC (253–256), which appears to occur in αT3-1 cells on GnRH stimulation (252).

These studies suggest that GnRH activates dual phospholipase pathways in a sequential and synchronized way in gonadotropes, in which PLC initiates the biphasic induction of IP_3 and DAG, and PKC mediates the interaction of PLD with the signaling pathways during the second phase of the DAG response (252). It also has been suggested that membrane receptor–coupled G proteins directly activate both PLD and PLC, that PKC switches its regulation from initial activation of PLC to sustained activation of PLD by differential phosphorylation of G proteins, and that $[Ca^{2+}]_i$ is required for translocation of PLA_2 to the plasma membrane and activation by PKC or by a G protein (255). The products of phosphatidylcholine hydrolysis by PLA_2 have also been suggested to enhance PKC activation by interacting with the PKC pathway and by increasing the sensitivity of PKC to Ca^{2+} during the sustained phase of agonist stimulation (257,258).

Protein Kinase C Pathways

PKC, a protein kinase sensitive to the tumor-promoting reagents, phorbol esters, is one of the key regulators involved in a variety of cellular signaling, which includes the PLC-mediated cellular cascades (259). At least 10 different subspecies of PKC have been identified in mammalian cells (256,260). All are single-polypeptide kinases containing an N-terminal regulatory domain, a C-terminal catalytic domain (also referred to as protein kinase M [PKM]), and several highly variable (V) and conserved (C) domains. There are ATP and substrate binding motifs in the catalytic domain, as well as an autoinhibitory pseudosubstrate motif. This motif binds to the catalytic domain to keep the enzyme inactive. The regulatory domain has binding sites for DAG; divalent cations, including Ca^{2+}; membrane phospholipids, including phosphatidylserine; and phorbol ester, and the binding of these factors liberates the catalytic domain from autoinhibition, leading to activation of the enzyme (256). IP_3-induced Ca^{2+} mobilization changes the conformation of PKC, exposing the hydrophobic regulatory domain to interact with DAGs, which serve as anchors and enable the translocation of PKC from the cytosol to the membrane phosphatidylserine, while the hydrophilic catalytic domain faces the cytosol (261). This translocation step is considered to be the primary mechanism of PKC activation, and GnRH has been shown to induce translocation of PKC from the cytosol to the plasma membrane in pituitary primary culture, αT3-1 cells, and in vivo (19,213,262). Activated PKC phosphorylates substrate proteins that participate in gonadotropin gene expression.

The type I group of PKC consists of four isoenzymes, PKCα, βI, βII, and γ, all of which contain five variable (V_1 to V_5) and four conserved (C_1 to C_4) domains. Their enzyme activities are regulated by DAG, $[Ca^{2+}]_i$, phosphatidylserine, and phorbol ester, and the C_2 domain functionally binds to Ca^{2+} (263). The type II group also consists of four isoenzymes, PKCδ, ε, η, and θ, all of which lack the C_2 domain and therefore are independent of Ca^{2+} for their activation. PKCδ and ε can be present in a phosphorylated state in native tissues, and can be phosphorylated by other protein kinases. PKCη is activated by DAG, phosphatidylserine, and phorbol ester (190). The type III group consists of two atypical isozymes, PKCλ and ζ, both of which are sensitive to phosphatidylserine, but not to DAG, Ca^{2+}, or phorbol ester (259,263). PKCα, βI, and βII are expressed in most tissues, whereas PKCγ is expressed only in central nervous system hypothalamic tissues. The rat pituitary contains PKCα and βII (264). Most of the rat pituitary gonadotropes containing PKCβII also contain FSH (265). Human pituitary adenomas have been shown to express all of the type I isozymes, PKCα, βI, βII, and γ, and normal and neoplastic human pituitaries also express mRNAs for the type II and III isozymes, PKCε and ζ (266). The αT3-1 cells contain PKCα, βI, βII, δ, ε, η, and ζ isozymes (267–269), and LβT2 cells contain PKCα, β, δ, ε, λ, and θ isozymes (270). A pituitary cell line, GH_4C_1, has also been shown to express PKCα, βI, βII, δ, ε, and ζ, in association with thyrotropin-releasing hormone action in and prolactin secretion from this cell line (271,272). The diversity of these PKC subfamily members in expression, responsiveness, and fate suggests specificity in activity for each member in different tissues.

PKC activity is increased by DAG, which potentiates affinity for Ca^{2+} and phosphatidylserine (263). The products of phospholipases (PLC, PLD, and PLA_2) and a ligand-induced elevation of $[Ca^{2+}]_i$ also regulate PKC activity, and PKC in turn regulates the enzyme activity of these phospholipases by negative and positive feedback control mechanisms, inhibiting PLC activity and stimulating PLD and PLA_2 activities (273). PKC also regulates Ca^{2+} release and influx by controlling IP_3 receptor activity and voltage-sensitive Ca^{2+} and Na^+ channels (274). In gonadotropes, phorbol ester pretreatment cause a dose-dependent reduction in the amplitude of GnRH-stimulated Ca^{2+} response, suggesting a negative feedback of PKC on PLC signaling (240), and PKCα has been found to mediate feedback inhibition of PLC signaling (275). Involvement of PKC in Ca^{2+} influx in gonadotropes is suggested by the actions of DAG and phorbol ester in elevating VSCC-dependent increases in $[Ca^{2+}]_i$ (249, 276), and by the similar effect of PMA and GnRH on the amplitude of VSCC responses (247). A GnRH agonist or PMA stimulation results in a dose-dependent increase in PKCβ, but not PKCα, mRNA levels in αT3-1 cells. A Ca^{2+} ionophore, ionomycin, stimulates the expression of both PKCα and β mRNA levels, and removal of Ca^{2+} or inhibition of PKC abolishes the effect of GnRH, indicating that GnRH-induced PKCβ gene expression is Ca^{2+} sensitive and autoregulated by PKC (277). PKC has diverse phosphorylation targets in the GnRH-mediated signal transduction pathways (278,279), in particular the mitogen-activated protein kinase (MAPK) pathway, which affect diverse cell functions, including target gene expression, as well as cell proliferation and differentiation.

Mitogen-Activated Protein Kinase Cascade

MAPK, also known as ERK, is a family of serine/threonine protein kinases that are rapidly activated in response to a wide variety of stimuli (280,281). Several members of the MAPK family have been identified, notably Ras ($p21^{ras}$) GTPase, Raf-1 kinase, ERK1 ($p44^{mapk}$), ERK2 ($p42^{mapk}$), c-Jun-N-terminal kinase (JNK), also known as stress-activated protein kinase (SAPK), and p38 (282). Stimuli for their activation include growth factors, many of which have receptors with intrinsic tyrosine kinase activity. MAPKs are involved in transmitting extracellular growth and differentiation signals into the cell nucleus, resulting in an array of transcriptional and mitogenic effects. Previous evidence indicates that some GPCRs can activate the MAPK family of enzymes and that MAPKs may also be involved in nonproliferative signaling cascades (283,284). GPCRs appear to activate MAPK through both monomeric GTPase Ras-dependent and -independent pathways, and both G_{α} and $G_{\beta\gamma}$ subunits appear to be variably involved.

These findings have led to several studies demonstrating the ability of the GnRHR to activate MAPK and the role of MAPK in mediating cellular effects of GnRH (268,285–287). Stimulation of αT3-1 or LβT2 cells with GnRH results in phosphorylation of tyrosine residues on both ERK1 and ERK2 through activated Ras, and rapid and sustained activation of both proteins, which leads to activation of the MAPK effector Elk1 (268,285,287,288) (Fig. 8). Activation of ERK1 and ERK2 by GnRH in αT3-1 cells was blocked by treatment with a GnRHR antagonist, confirming that activation of the MAPK signal transduction cascade by GnRH is receptor mediated (285). Activation of MAPK by GnRH was comparable with that observed in response to PMA and, furthermore, PMA pretreatment to deplete phorbol ester–sensitive forms of PKC (PKCα and ε) blocked the activation of ERK1 by GnRH. These data suggest that the activation of MAPK by GnRH involves activation of PKC (268). MAPK activity was also stimulated by GnRH in rat pituitary cells, although to a lesser extent, likely reflecting the heterogeneity of the pituitary cell population. Thus, it appears that the MAPK signal transduction pathway is activated by GnRH in both αT3-1 cells and primary pituitary gonadotropes.

The entire MAPK cascade includes several steps involving the receptor-interacting protein dynamin, protein tyrosine kinases, Ras, Raf-1 (also known as MAPK kinase kinase [MAPKKK] or mitogen-activated ERK-activating kinase [MEK or MKK] kinase [MEKK]), and MAPK kinase (MAPKK, also known as MEK), and phosphorylated MAPKKs become activated to phosphorylate their substrate kinases, MAPKs (ERK1/2) (282). Phosphorylated ERK1/2 are then translocated into the nucleus to interact with and activate MAPK effectors, such as Elk1 (289), ultimately to regulate target gene expression, usually immediate–early genes such as *jun*, *fos*, and *egr-1* (55,290–292) (Fig. 9).

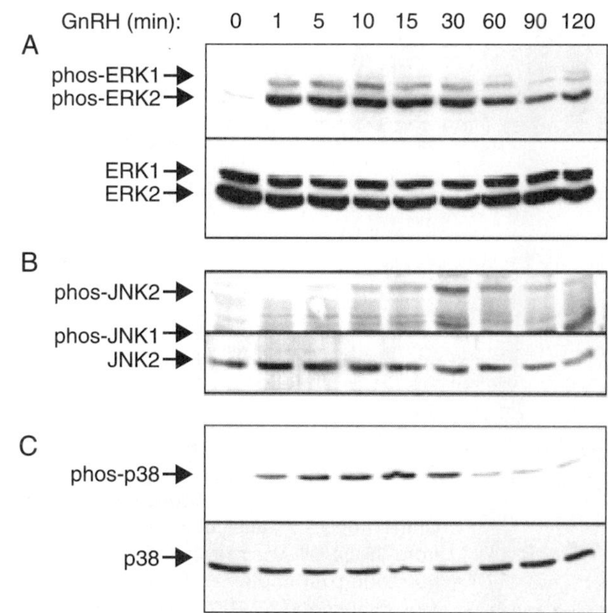

FIG. 8. Gonadotropin-releasing hormone (GnRH) activates mitogen-activated protein kinase (MAPK) subfamilies in LβT2 cells. **A:** Time course of GnRH-stimulated extracellular signal–regulated kinase (ERK) activation in LβT2 cells. LβT2 cells were treated with 100 nM GnRH for 0 to 120 minutes. Whole-cell lysates were separated by sodium dodecyl sulfate–polyacrylamide gel electrophoresis (SDS-PAGE) and immunoblotted with an antibody to phospho-ERK1/2 (*top panel*) and an antibody to ERK1/2 (*bottom panel*). **B:** Time course of GnRH-stimulated c-Jun-N-terminal kinase (JNK) activation in LβT2 cells. Whole-cell lysates were separated by SDS-PAGE and immunoblotted with an antibody to phospho-JNK1/2 (*top panel*) and an antibody to JNK2 (*bottom panel*). **C:** Time course of GnRH-stimulated p38 kinase activity in LβT2 cells. Whole-cell lysates were separated by SDS-PAGE and immunoblotted with an antibody to phospho-p38 MAPK (*top panel*) and an antibody to p38 MAPK (*bottom panel*). (Reproduced with permission from Liu, F., Austin, D. A., Mellon, P. L., Olefsky, J. M., and Webster, N. J. G. [2002]. GnRH activates ERK1/2 leading to the induction of c-*fos* and LHb protein expression in LβT2 cells. *Mol. Endocrinol.* 16, 419–434. Copyright 2002, The Endocrine Society.)

Although GnRH-induced ERK activation requires PKC activation and Ca^{2+} influx through L-type VSCCs (287,293), JNK activation by GnRH seems to require a differential PKC-mediated cascade. GnRH rapidly phosphorylates and activates JNK in a time- and dose-dependent manner in rat pituitary cells, αT3-1 cells, and LβT2 cells (287,294) (Fig. 8). However, the time course of JNK phosphorylation is different from that of ERK. ERK phosphorylation by a GnRH agonist peaks within 5 minutes and falls thereafter, whereas JNK phosphorylation by a GnRH agonist is detectable at 5 minutes, reaches a plateau at 30 minutes,

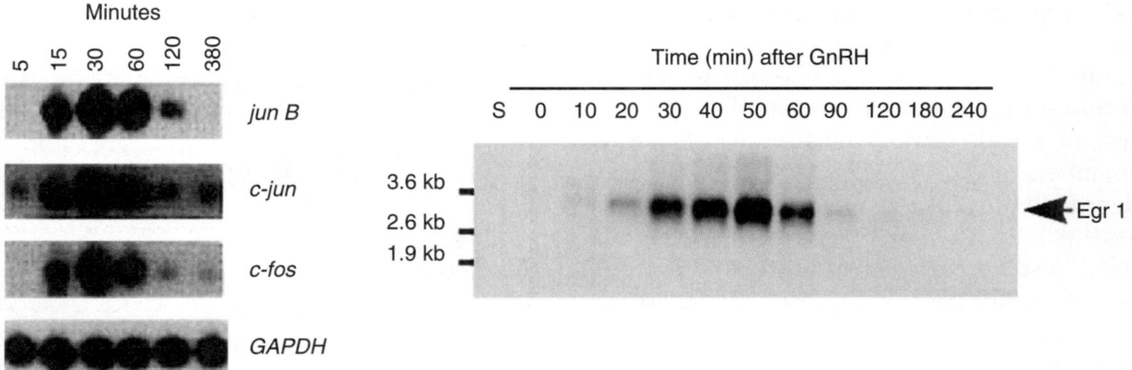

FIG. 9. *Left:* Time-dependent effects of gonadotropin-releasing hormone (GnRH) on *jun/fos* expression in αT3-1 cells. (Reproduced with permission from Cesnjaj, M., Catt, K. J., and Stojilkovic, S. S. [1994]. Coordinate actions of calcium and protein kinase-C in the expression of primary response genes in pituitary gonadotrophs. *Endocrinology* 135, 692–701. Copyright 1994, The Endocrine Society.) *Right:* GnRH induces a transient increase in expression of *egr-1* in αT3-1 cells. (Reproduced with permission from Wolfe, M. W., and Call, G. B. [1999]. Early growth response protein 1 binds to the luteinizing hormone-β promoter and mediates gonadotropin-releasing hormone-stimulated gene expression. *Mol. Endocrinol.* 13, 752–763. Copyright 1999, The Endocrine Society.)

and declines thereafter (287,293). JNK activation is mediated by receptor-interacting protein kinase c-Src, a GTPase CDC42, $p21^{ras}$-activated kinase (PAK) 1, and JNK kinase (JNKK), rather than Ras/Raf and MEK (55,282), and is not dependent on extracellular Ca^{2+}, but is sensitive to intracellular Ca^{2+} mobilization (294), contrary to ERK activation (293). GnRH also stimulates rapid phosphorylation and activation of p38 with a peak in 5 minutes in a time- and dose-dependent manner in αT3-1 and LβT2 cells, and this activation is blocked by depletion of PKC (286,287) (Fig. 8), suggesting an involvement of PKC-dependent activation of the p38 cascade in the GnRH signal transduction pathway.

GnRH stimulation has been shown to induce the expression of a MAPK phosphatase (MKP), MKP-2, which selectively inactivates MAPKs, in rat pituitary cells and αT3-1 cells (295). GnRH-stimulated MKP-2 induction is PKC and ERK/JNK dependent, but not p38 dependent, and both Ca^{2+} influx through L-type VSCCs and elevation of $[Ca^{2+}]_i$ are required for the activation of MKP-2. These studies suggest that MAPKs activated by GnRH in the gonadotropes can be quickly inactivated by MKP-2 in physiological conditions.

Immediate–Early Gene Expression

Perhaps the best-studied gene targets of the PKC pathway are several proto-oncogenes, representing mitogenic immediate–early genes, such as c-*jun* and c-*fos* (296). The major function of these gene products is to initiate the transcription of late gene responses (297). These gene products and others form a family of transcription factors, activating protein-1 (AP-1), and have been shown to be important for GnRH-induced GnRHR and FSHβ gene expression in gonadotropes (298–302). The AP-1 family includes the Jun subfamily (c-Jun, JunB, and JunD) and the Fos subfamily (c-Fos, FosB, Fos-related antigen-1 [Fra-1], and Fra-2), as well as the activating transcription factor (ATF) subfamily (ATF-1 and 2) (299,303). These factors become active by forming heterodimers (between Jun and Fos or ATF subfamily members) or homodimers (among Jun subfamily members). Transcriptional activation of an AP-1 target gene is mediated by site-specific phosphorylation of the AP-1 factors (c-Jun and ATF-2) bound to their cognate binding *cis*-element, the TPA response element, by JNK (304).

Basal expression levels of these genes are almost undetectable, but they are rapidly transcribed upon extracellular stimulation through the PKC pathway, often reaching peak mRNA levels within 30 minutes (305). PLC-induced PKC activation of c-*fos* expression is mediated by the serum response factor, which binds to a serum response element located in the c-*fos* promoter (306). GnRH or PMA stimulation induces rapid and transient expression of c-*jun*, c-*fos*, and another member of the *jun* subfamily, *junB*, in a dose-dependent manner in αT3-1 cells, reaching a peak 30 minutes after GnRH treatment (307,308), followed by an increase in the protein levels (269) (Fig. 9). In PKC-depleted cells, the PMA response is blunted and the GnRH response is diminished, supporting the involvement of PKC in GnRH-induced *jun/fos* expression. A biphasic dose-dependent modulation of expression of these genes during GnRH or PMA stimulation in response to changes in $[Ca^{2+}]_i$ has been observed in rat pituitary gonadotropes, in which initial additive stimulation of gene expression becomes suppressed by further increases in $[Ca^{2+}]_i$ (307).

These studies suggest that GnRH-mediated induction of immediate–early gene expression is mediated by PKC and that this regulation is bidirectionally modulated by changes in $[Ca^{2+}]_i$.

Although c-*fos* is the best known target for ERK, JNK seems to target and activate c-Jun by rapid phosphorylation as well as c-*jun* expression (294). GnRH-induced PKC pathway activation and PKC-mediated Ca^{2+} influx through L-type VSCCs are required for ERK activation and c-*fos* expression, but not for JNK activation and c-Jun activation/c-*jun* expression in both αT3-1 and LβT2 cells, as well as in rat pituitary cells (293,308). Also, overexpression of p38 results in activation of c-*jun* and c-*fos* promoter activities in αT3-1 cells, and inhibition of p38 activity attenuates GnRH-induced c-*fos* promoter activity (286), suggesting an involvement of p38 in GnRH-mediated c-*fos* expression. These studies indicate that at least three MAPK family members are activated by GnRH stimulation and involved in the divergent signal transduction pathways with differential sensitivities to PKC and Ca^{2+} mobilization through differential cascades.

GnRH also induces *egr-1* expression in a PKC/ERK-dependent manner, through four serum response elements and an oncogene E2b transformation-specific (Ets) factor binding site of the 5′-flanking region (309) (Fig. 9). Similar to c-*fos* induction (289), it has been suggested that GnRH-induced, ERK-mediated Elk-1 phosphorylation enables Elk-1 to form a ternary complex with serum response factors and Ets factor and rapidly activate *egr-1* expression (309).

Protein Kinase A Pathways

The PKA pathway is initiated by GPCRs coupled with $G_{s\alpha}$ or $G_{i\alpha}$ upon ligand binding, which activate or inhibit membrane-associated adenylate cyclase, respectively. Adenylate cyclase produces cAMP from ATP, and cAMP binding to the regulatory subunits of the PKA causes conformational changes of the catalytic subunits. Activated PKA catalytic subunits then activate effector transcription factors by phosphorylation. A large number of studies have demonstrated that at least a subset of GnRH signaling is mediated by $G_{s\alpha}$ and $G_{i\alpha}$ through the PKA pathway (94,200–203,310). During prolonged stimulation, GnRH has been shown to increase cAMP production in pituitary gonadotropes (311), suggesting that the GnRHR is coupled to $G_{s\alpha}$ in pituitary gonadotropes (Fig. 5).

In αT3-1 cells, however, no significant change in cAMP levels by GnRH has been detected even in the presence of a phosphodiesterase inhibitor to prevent the degradation of cAMP (19), suggesting a lack of supporting mechanisms for GnRH-induced activation of the PKA pathway in this cell line. Some studies have shown that activation of PKC can alter PKA subunit levels in the absence of cAMP elevation in αT3-1 cells, suggesting cross-talk between the PKA and PKC pathways (312). On the other hand, the LβT2 cells respond to GnRH by increasing cAMP accumulation, and forskolin treatment activates ERK and c-*fos* independently of and additively to PKC activation (202), suggesting that this cell line can support GnRH activation of the $G_{s\alpha}$-mediated PKA pathway in addition to the PKC pathway.

Expression of c-*fos* is not only stimulated by the PKC pathway, but is induced by the PKA pathway, in a mechanism including cAMP/Ca^{2+} kinase–dependent phosphorylation of cAMP response element (CRE)–binding protein (CREB) (313,314). The phosphorylated CREB in turn binds to a CRE located on the c-*fos* promoter and activates its gene expression. In addition to PKC activation, the activation of adenylate cyclase by forskolin also induces *egr-1* expression in LβT2 cells, although to a lesser extent (310), possibly through a CRE on the 5′-flanking region (309). The p38 MAPK also has been shown to stimulate *egr-1* expression through mediation by a CRE (315). Thus, all the GnRH-induced immediate–early genes involved in gonadotropin subunit gene expression are stimulated by the PKA pathway.

Gonadotropin Secretion

GnRH regulates gonadotropin release by means of the rapid increase in $[Ca^{2+}]_i$ after its binding to the GnRHR. Through G protein–coupled signal transduction, GnRH binding rapidly activates PLC, and PLC produces IP_3 and DAG from PIP_2, which in turn rapidly mobilizes transient intracellular Ca^{2+} to trigger a burst initiation of exocytosis, causing rapid LH secretion within 10 seconds and lasting for 100 seconds (196,230, 243,316,317). Each Ca^{2+} spike (approximately 500 nM) caused by release of Ca^{2+} from the intracellular luminal Ca^{2+} stores is above the threshold for activation of exocytosis, and induces a small burst of exocytosis. The temporal changes in Ca^{2+} signaling affect the rate of exocytosis. The initial rapid LH secretion is mediated by transiently mobilized intracellular Ca^{2+}, rather than by replenished Ca^{2+} influx from the extracellular Ca^{2+} pool, based on its independence of $[Ca^{2+}]_o$ (215,318–320) (Fig. 10). The second phase of $[Ca^{2+}]_i$ increase resulting from extracellular Ca^{2+} influx through VSCCs has a role in the sustained phase of GnRH-induced LH release rather than in the rapid initial LH secretion (215,244,245,316,318,321).

One potential intracellular mediator of the Ca^{2+} signal in gonadotropes is calmodulin, which, after binding to Ca^{2+}, alters the activity of several enzymes

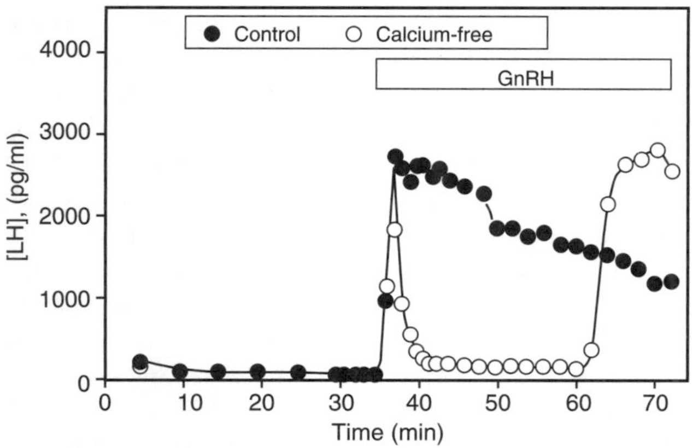

FIG. 10. Ca^{2+} and time dependence of gonadotropin-releasing hormone (GnRH)–stimulated luteinizing hormone (LH) release from superfused gonadotropes. Rat pituitary cells were either superfused continuously with control medium (*closed circle*) or received Ca^{2+}-free medium (*open circle*), as indicated. (Reproduced with permission from McArdle, C. A., and Poch, A. [1992]. Dependence of gonadotropin-releasing hormone-stimulated luteinizing hormone release upon intracellular Ca^{2+} pools is revealed by desensitization and thapsigargin blockade. *Endocrinology* 130, 3567–3574. Copyright 1992, The Endocrine Society.)

and cytoskeletal proteins implicated in the secretory process (Fig. 5). Calmodulin is localized in association with the GnRHR patch after GnRH treatment (322), and inhibition of the activity of Ca^{2+}/calmodulin blocks LH release in response to GnRH or Ca^{2+} ionophores (323). Several gonadotrope proteins, spectrin, caldesmon, and calcineurin, which bind to and are regulated by Ca^{2+}/calmodulin, have been identified (324). Neutralization of caldesmon potentiates LH release in response to GnRH (325), whereas a calmodulin antagonist decreases GnRH-stimulated LH release (326). The granin proteins, secretogranin II and chromogranin A, are commonly found associated with LH or FSH within specialized secretory granules in gonadotropes, and may play an important role in the differential secretion of the gonadotropins (327,328).

In gonadotropes, the extent of the changes in $[Ca^{2+}]_i$ and exocytosis in response to GnRH depend on the steroid hormone background. The stimulation of LH secretion in response to GnRH is enhanced by steroids, 17β-estradiol (E_2) or glucocorticoids, in LβT2 cells as well as in primary rat pituitary cells (71,327,329). E_2 and glucocorticoids cause an increase in the peak $[Ca^{2+}]_i$ stimulated by GnRH as well as an increase in sensitivity of exocytosis to Ca^{2+} in LβT2 cells (243). The increased $[Ca^{2+}]_i$ response may be due to an increase in GnRHR numbers or to other second messenger pathways activated by GnRH or steroids, independent of changes in the GnRHR.

PKC is also involved in GnRH-stimulated gonadotropin release (213,216,330). Initial studies showed that DAG and phorbol ester can stimulate LH secretion, implying a role of GnRH, intracellular Ca^{2+}, and PKC in this process (196). LH release by PKC activators is largely insensitive not only to $[Ca^{2+}]_o$, but to blockade of Ca^{2+} channels or inhibition of calmodulin, and is synergistically enhanced in the presence of a Ca^{2+} ionophore, A23187 (331). PMA pretreatment produces a leftward shift in GnRH dose–response curves for LH release, whereas curves for receptor occupancy and total IP production remain unchanged (332). The ionophore dose–response curves for intracellular Ca^{2+} are also left-shifted, suggesting that cellular responsiveness to mobilized Ca^{2+} is enhanced by prior activation of PKC. It is suggested that an increase in both intracellular Ca^{2+} and PKC activity is required for eliciting the full GnRH-induced LH secretion (249,333,334), and that the PKCα and β isoenzymes, present in the pituitary cells, are used in this process (78,275,335). The activated PKC isoenzymes are also likely to phosphorylate substrate proteins to activate the secretory process. Altogether, it is likely that initial LH release depends on changes in $[Ca^{2+}]_i$, but enhancement of LH release after periods of elevated GnRH concentrations may depend on PKC. PKC has also been shown to induce FSH secretion in rat pituitary gonadotropes (330).

The role of the PKA pathway in the secretion of gonadotropins is unclear. Activation of this pathway may be involved strictly in the biosynthesis of the gonadotropin subunits (200). However, previous studies have shown that the activation of adenylate cyclase by forskolin results in an increase in LH secretion as well as LHβ gene expression in rat pituitary cells (201). Moreover, pulsatile administration of pituitary adenylate cyclase–activating polypeptide-38 (PACAP), a member of the vasoactive intestinal polypeptide, secretin, and glucagon family of peptides, stimulates LH secretion by increasing LH pulse amplitude, and this effect is potentiated by GnRH in rat pituitary cells (336), suggesting a possible involvement of the PKA pathway in gonadotropin secretion.

Gonadotropin secretion is also regulated by gonadal steroids and peptides in vivo. Both androgen and estrogen maintain an inhibitory effect on LH and FSH secretion (337–339). This action of gonadal steroids is primarily through the inhibition of GnRH pulsatile release, although direct effects on gonadotropes have also been observed (340,341). The gonadal peptides, inhibins, activins, and follistatin, predominantly regulate FSH secretion, producing both stimulatory (activins) and inhibitory (inhibins, follistatin) effects (339,342–345), and may be important in the differential regulation of LH and FSH (346). All three gonadal peptides are also produced in the pituitary, suggesting not only an endocrine action but a paracrine/autocrine role in gonadotropin regulation (347–350).

PHYSIOLOGICAL REGULATORS OF GONADOTROPIN BIOSYNTHESIS AND SECRETION

GnRH Pulsatility

Besides stimulating the acute release of gonadotropins, GnRH regulates long-term maintenance of pituitary responsiveness to the hormone. The pulse pattern of GnRH administration to the pituitary is a critical determinant of gonadotropin release over periods of time (351). For example, delivery of GnRH in a pulsatile fashion approximating normal hypothalamic GnRH release produces pulses of LH release of consistent magnitude (89) (Fig. 11). Thus, pulsatile exposure to GnRH is viewed as essential for maintenance of normal gonadotrope function (91,352). Constant exposure to GnRH, in contrast, results in a pituitary that is refractory to subsequent administration of GnRH with respect to LH release (89). These relationships between exposure pattern and pituitary responsiveness have permitted GnRH and GnRH agonists to be used clinically for either the restoration (353) or biochemical ablation of gonadotropin release.

The effects of GnRH pulse frequency and amplitude on gonadotropin subunit gene expression and gonadotropin secretion have been well characterized (354–360). LHβ and FSHβ mRNA levels are markedly induced by pulsatile GnRH in the rat pituitary and in LβT2 cells, and αGSU mRNA levels are also increased by continuous GnRH stimulation (71, 361–363). Interestingly, different frequencies of pulsatile GnRH have different effects on gonadotropin secretion and subunit gene expression. Induction of αGSU and LHβ subunit gene expression by pulsatile GnRH has different sensitivity and different kinetics (364,365). A high pulse frequency of GnRH (30-minute interval) stimulates preferentially αGSU and LH biosynthesis and secretion, whereas a lower pulse frequency of GnRH (2-hour interval) stimulates preferentially FSH biosynthesis and secretion, both in vivo and in vitro (355,361,362,365–368) (Fig. 12). These data suggest that the mechanisms by which GnRH regulates αGSU and LHβ gene expression are

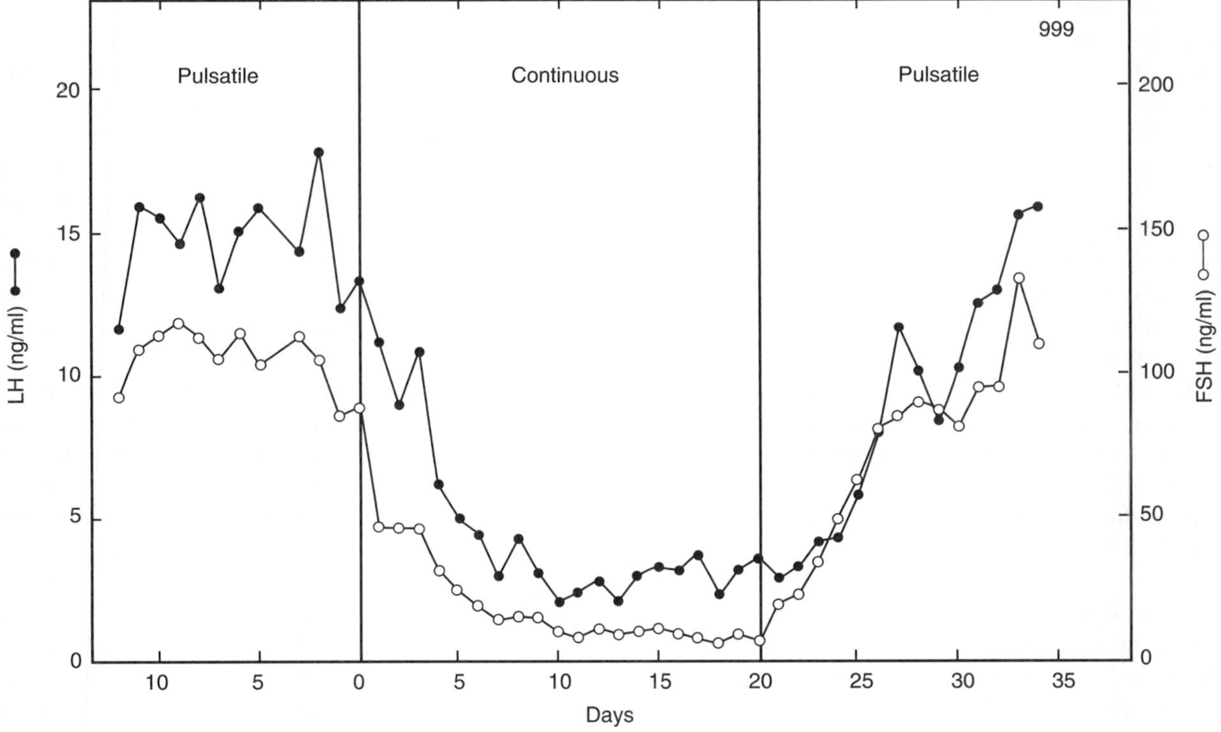

FIG. 11. Suppression of plasma luteinizing hormone (LH) and follicle-stimulating hormone (FSH) concentrations after initiation, on day 0, of a continuous gonadotropin-releasing hormone (GnRH) infusion (1 μg/min) in an ovariectomized rhesus monkey with a radiofrequency lesion in the hypothalamus. Gonadotropin secretion had been reestablished by the intermittent (pulsatile) administration of the decapeptide (1 μg/min for 6 minutes once per hour). The inhibition of gonadotropin secretion was reversed after reinstitution of the intermittent mode of GnRH stimulation on day 20. (Reproduced with permission from Belchetz, P. E., Plant, T. M., Nakai, Y., Keogh, E. J., and Knobil, E. [1978]. Hypophysial responses to continuous and intermittent delivery of hypothalamic gonadotropin-releasing hormone. *Science* 202, 631–633. Copyright 1978, AAAS.)

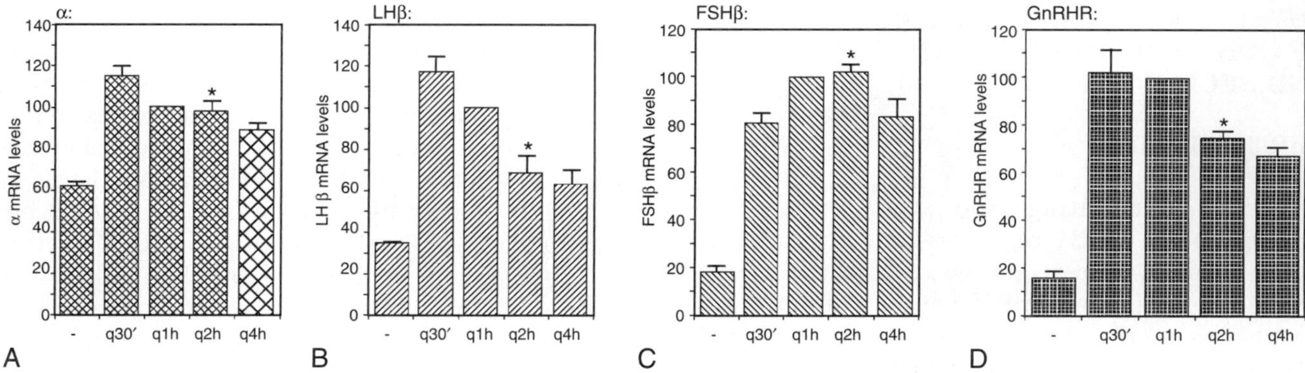

FIG. 12. Effects of gonadotropin-releasing hormone (GnRH) pulse frequency on gonadotropin subunit and GnRHR mRNA levels in primary rat pituitary cell cultures. Glycoprotein α-subunit (αGSU) (**A**), luteinizing hormone β-subunit (LHβ) (**B**), follicle-stimulating hormone β-subunit (FSHβ) (**C**), and GnRH receptor (GnRHR) (**D**) mRNA levels are shown for each frequency of pulsatile GnRH tested. (Reproduced with permission from Kaiser, U. B., Jakubowiak, A., Steinberger, A., and Chin, W. W. [1997]. Differential effects of gonadotropin-releasing hormone [GnRH] pulse frequency on gonadotropin subunit and GnRH receptor messenger ribonucleic acid levels *in vitro*. *Endocrinology* 138, 1224–1231. Copyright 1997, The Endocrine Society.)

distinct from those by which FSHβ gene expression is regulated.

Pulsatile GnRH also increases GnRHR gene expression and GnRHR number on the cell surface. Rat pituitary primary cells treated with hourly pulses of GnRH increase GnRHR mRNA levels by over tenfold, whereas continuous GnRH stimulation is not effective (69). This pattern is also observed in vivo using gonadectomized rats (64,69). LβT2 cells respond to pulsatile GnRH administration with an increase in GnRHR mRNA levels (71). However, this increase is only approximately twofold, compared with the much greater increase observed in primary pituitary cells (69). This difference may be due to differences in the experimental paradigm of pulsatile GnRH delivery. Pulsatile GnRH with low concentrations induces upregulation, whereas continuous GnRH with high concentrations causes downregulation and desensitization of GnRHR in cultured pituitary cells (66).

Pulsatile GnRH administered at differential pulse frequencies has differential effects on the expression of GnRHR (67,362,367). Specifically, an interval of 30 minutes between GnRH pulses is associated with the highest concentration of GnRHR and GnRH-binding activity, whereas an interval of 2 hours between GnRH pulses is associated with a lower concentration of GnRHR (67,367–369) (Fig. 12). This view is further supported by the increase in GnRH sensitivity of the gonadotropes during the preovulatory surge of LH, which is mediated by many factors, including an increase in GnRHR number in the late follicular phase (370). These observations support the hypothesis that varying GnRH pulse frequencies differentially regulate LH and FSH biosynthesis and secretion in vivo by regulating pituitary GnRHR numbers (367–369) (Fig. 13). Furthermore, the signal transduction pathways activated by GnRH may be different at low versus high GnRH receptor numbers. The details of the different intracellular signaling pathways activated by GnRH at low versus high GnRHR numbers remain to be fully elucidated.

Pulsatile influx of Ca^{2+}, pulsatile PKC activation, or pulsatile cAMP can mimic the effect of pulsatile GnRH on gonadotropin gene expression and secretion (357,358). An interval of 1 hour or less between Ca^{2+} channel activator pulses preferentially increases αGSU and LHβ mRNA levels, and an interval of 3 hours between pulses maximally increases FSHβ and GnRHR mRNA levels in rat pituitary cells, suggesting that an intermittent increase in intracellular Ca^{2+} is an important step in the transmission of GnRH pulse signals from the plasma membrane to the nucleus (359). Furthermore, pulsatile GnRH activates ERK phosphorylation more effectively than constant GnRH in a dose-dependent manner in vivo, maintained for over 8 hours, and an interval of 2 hours between GnRH pulses results in maximal increases in ERK activation (360) (Fig. 14). These results suggest that a pulsatile GnRH signal is required to maintain MAPK activation, and that divergent MAPK responses to alterations in the GnRH signal pattern may be one mechanism involved in differential regulation of gonadotrope gene expression.

The amplitude of the pulsatile GnRH signal can exert differential effects on subunit mRNA concentrations. Earlier studies demonstrated that GnRH increased the number of GnRH receptors, with maximum responses seen after 25 ng/pulse, and both higher and lower doses were less effective (364). In male rats, αGSU and FSHβ mRNA concentrations are increased by all pulse amplitudes. In contrast, LHβ mRNA responses parallel those of the GnRHRs.

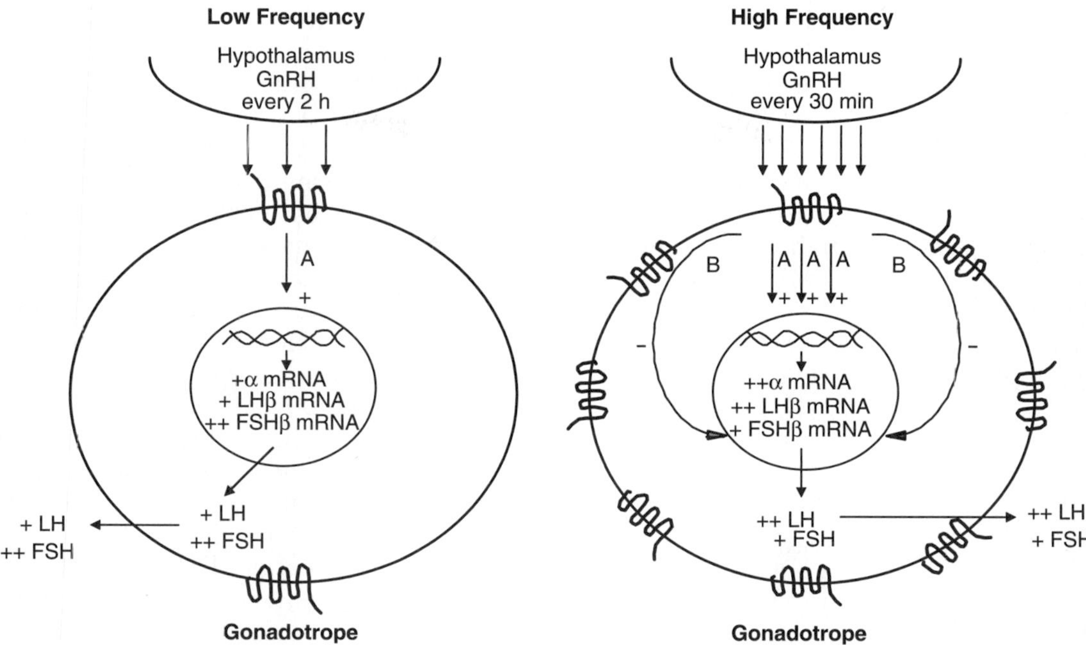

FIG. 13. Hypothesis for the mechanism of differential regulation of the gonadotropin subunit genes by gonadotropin-releasing hormone (GnRH) at low (*left*) and high (*right*) frequencies of pulsatile GnRH. *Left:* At low GnRH pulse frequencies, cell surface GnRH receptor (GnRHR) concentrations are low, and when GnRH binds to its receptors, a signal transduction pathway, pathway A, is activated, resulting in the stimulation of the expression of all three of the gonadotropin subunit genes, glycoprotein α subunit (αGSU), luteinizing hormone β subunit (LHβ), and follicle-stimulating hormone β subunit (FSHβ). *Right:* At higher GnRH pulse frequencies, cell surface GnRHR concentrations are higher, and when GnRH binds to the now greater receptor numbers, signal transduction pathway A is activated to an even greater extent, resulting in the greater stimulation of αGSU and LHβ genes. In addition, a second signal transduction pathway, pathway B, is now activated. Activation of pathway B results in the specific inhibition of expression of the FSHβ gene, with no effects on the αGSU and LHβ genes. The net effect is that αGSU and LHβ gene expression is maximally stimulated at relatively high GnRH pulse frequencies and GnRHR concentrations, whereas FSHβ gene expression is optimally stimulated at lower GnRH pulse frequencies and cell surface GnRHR concentrations.

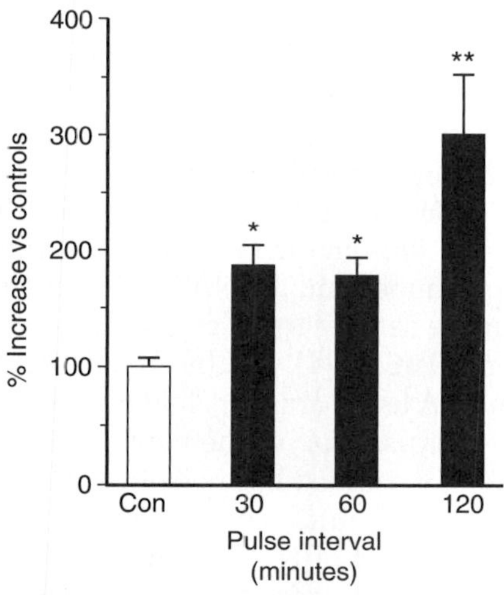

FIG. 14. The effect of gonadotropin-releasing hormone (GnRH) pulse frequency on pituitary mitogen-activated protein kinase (MAPK) activity. Groups of male rats received GnRH pulses (50 ng) at 30-, 60-, or 120-minute intervals for 8 hours. Controls (Con) received bovine serum albumin–saline pulses. (Reproduced with permission from Haisenleder, D. J., Cox, M. E., Parsons, S. J., and Marshall, J. C. [1998]. Gonadotropin-releasing hormone pulses are required to maintain activation of mitogen-activated protein kinase: role in stimulation of gonadotrope gene expression. *Endocrinology* 139, 3104–3111. Copyright 1998, The Endocrine Society.)

LHβ mRNA is elevated only after doses of 10 to 75 ng/pulse, with maximal responses occurring after 25 ng/pulse (365,371). In female rats, LHβ and FSHβ mRNAs rise after 0.5 to 25 ng/pulse of GnRH, whereas changes in αGSU mRNA are small and seen only at higher pulse doses (25 to 250 ng), demonstrating that the dependence of subunit mRNA responses on GnRH pulse dose is less evident in the female model (356). In pituitary cells in vitro, LHβ mRNA is increased by a narrower range of GnRH pulse amplitudes than are αGSU and FSHβ mRNAs (372), supporting the view that LHβ mRNA expression is more sensitive to GnRH amplitude than either αGSU or FSHβ, which both rise in response to a wider range of GnRH pulse doses.

Estrous Cycle

In rats, preovulatory surges of pituitary gonadotropins occur on the evening of the proestrus phase of the 4-day cycle (373). The LH surge lasts approximately 6 to 8 hours and is primarily under the control of GnRH (374). The FSH surge also begins on the evening of proestrus but continues into estrus on the following morning. The biphasic pattern of the FSH surge may be separated into an initial GnRH-regulated phase (coincident with the LH surge) and a later inhibin-regulated phase (374–376) (Fig. 15).

During the proestrus phase, both LHβ and FSHβ mRNA concentrations rise during the preovulatory gonadotropin surge, and although the increases in mRNA concentrations are generally parallel, there are differences in onset and duration. The rise in LHβ mRNA begins at 1400 hours (2 hours before the onset of LH surge), peaks at 1700 hours (2.5-fold increase),

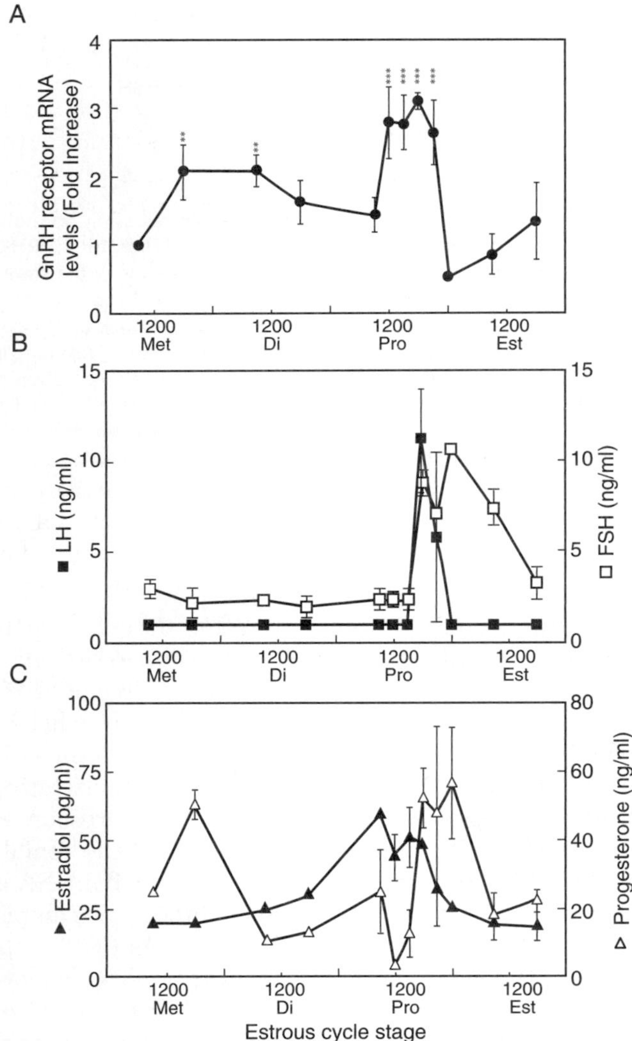

FIG. 15. Changes in gonadotropin-releasing hormone receptor (GnRHR) mRNA expression, serum gonadotropins, and gonadal steroid levels during the rat estrous cycle. GnRHR mRNA levels (**A**), serum luteinizing hormone (LH) and follicle-stimulating hormone (FSH) values (**B**), and serum estradiol and progesterone levels (**C**). Met, metestrus; Di, diestrus; Pro, proestrus; Est, estrus. (Reproduced with permission from Bauer-Dantoin, A. C., Hollenberg, A. N., and Jameson, J. L. [1993]. Dynamic regulation of gonadotropin-releasing hormone receptor mRNA levels in the anterior pituitary gland during the rat estrous cycle. *Endocrinology* 133, 1911–1914. Copyright 1993, The Endocrine Society.)

and then returns to basal levels by 2200 hours. In contrast, increases in FSHβ mRNA are not seen until 2000 hours, 2 hours after the initial rise in circulating FSH. Maximal (fourfold) increases in FSHβ mRNA are seen during the early morning of estrus (0200), with levels returning to baseline 6 hours later (0800). αGSU mRNA does not increase significantly during the proestrus gonadotropin surge. A selective rise in FSHβ mRNA levels is seen during the metestrus phase, when measurable increases in FSH secretory activity are not observed. On the morning of diestrus, a parallel rise (twofold) in αGSU and LHβ mRNAs is present, which is not associated with a measurable rise in LH secretion. Thus, during the rat estrous cycle, both coordinate and differential increases in gonadotropin subunit mRNA expression are seen. LHβ and FSHβ mRNAs both increase during the proestrus gonadotropin surges, FSHβ selectively increases during metestrus, and αGSU and LHβ increase in parallel during the diestrus phase, when LH and FSH secretion remain stable (377,378).

The rise in LHβ and FSHβ mRNA levels seen in proestrus appears to be the result of altered synthesis because LHβ and FSHβ gene transcriptional rates increase at the time of the surge (379) in response to the increase in GnRH release on the afternoon of proestrus (380,381). The selective increase in FSHβ mRNA in metestrus may reflect input from ovarian peptides (inhibins, activins, and follistatin) that selectively influence FSHβ gene expression (343). In support, the decline in FSHβ mRNA during the late afternoon of metestrus is coincident with a rise in circulating inhibin levels (382). The increase in αGSU and LHβ in diestrus may be related to the rise in E_2 or to increased pulsatile GnRH secretion, both of which occur during diestrus (373,374,381,383) (Fig. 15).

The number as well as mRNA level of GnRHR in female rodent pituitary also changes during the estrous cycle. This change in GnRHR is tightly regulated, their number and mRNA levels correlating with changes in the sensitivity of gonadotrope responsiveness to GnRH (61,384). In the female rat pituitary, GnRHR mRNA levels are threefold higher on the afternoon of proestrus than on the morning of metestrus. This peak begins only 2 hours before the onset of the LH surge and simultaneously drops to its lowest levels with the end of the LH surge (374) (Fig. 15). There are several factors involved in these GnRHR changes, including gonadal steroids and GnRH.

Lactation has also been shown to be a factor influencing GnRHR expression levels (385). During lactation postpartum in rats, pituitary GnRHR mRNA levels decrease to less than half of the diestrus levels, and removal of the sucking stimulus by removing pups restores GnRHR mRNA back up to the diestrus level.

Inhibin

The pleiotropic cytokine transforming growth factor-β (TGF-β) superfamily of growth factors is composed of related compounds with diverse biological actions, including the regulation of cell growth and differentiation as well as cell–cell signaling. These actions appear to be exerted through a family of specific membrane receptors with serine/threonine kinase activity (386,387). Upon ligand binding to the type II receptor, the type I receptor is *trans*-phosphorylated by the type II receptor, and in turn activates Sma/mothers against decapentaplegic homolog (SMAD) signal transduction by phosphorylation (388,389). There are a number of proteins in this family sharing a similar dimeric structure, including TGF-β, müllerian inhibiting substance (MIS), bone morphogenetic proteins (BMPs), and the inhibins/activins (388,390). The biological action of these products may be altered by noncovalent binding to other proteins; for example, follistatin binds to activins.

TGF-β and activin receptors phosphorylate SMADs 2 and 3 (391,392), whereas BMP receptors phosphorylate SMADs 1, 5, and 8 (393,394), suggesting ligand-specific activation of the SMADs. There are two common-mediator SMADs, SMAD4 and 10, which associate with activated receptor-regulated SMADs (SMAD1, 2, 3, 5, and 8) and translocate into the nucleus (395,396) (Fig. 5). Two vertebrate inhibitory SMADs also have been identified: SMAD6 represses the BMP pathway, and SMAD7 inhibits the activin/TGF-β pathway (397–399).

The inhibins and activins consist of different combinations of α and β subunits. The dimerization of an inhibin α and either an inhibin β_A or β_B subunit results in the formation of inhibin A or inhibin B, respectively. If two heterologous or homologous β subunits are combined, activin A ($\beta_A\beta_A$), AB ($\beta_A\beta_B$), or B ($\beta_B\beta_B$) is produced (400,401). Inhibins A and B and activins A and AB have been isolated in gonadal extracts (402–405). Inhibin subunit α and β_B, but not β_A, mRNAs are present in the pituitary gonadotropes and, interestingly, mRNA concentrations of these two inhibin subunits increase after removal of the gonads (349).

The idea of the existence of a nonsteroidal gonadal factor that exerts an endocrine effect was first conceived over 80 years ago (406). The inhibins were initially shown to be secreted from the gonads (407), and subsequently have also been shown to be secreted from the pituitary, adrenal gland, and brain (408). Inhibin reduces FSH synthesis and cell content, as well as basal and GnRH-stimulated FSH release in dispersed pituitary cells (388,389,401,409,410). In vivo studies in gonadectomized male and female rats have shown a dose-related decrease in FSH secretion in response to inhibin, and it usually begins 4 to 6 hours after inhibin administration and persists for 8 to 12 hours (339,344,411–413). Furthermore, inhibin α-subunit–deficient mice show increased levels of FSH secretion as well as pituitary FSH content (414). Inhibin can also further decrease FSH secretion over E_2- or GnRH antagonist–induced decreases in FSH secretion (339), suggesting an independent mechanism of inhibin action. Inhibin exerts a profound inhibitory effect on FSHβ mRNA concentrations both in vivo and in vitro. Inhibin reduces levels of FSHβ mRNA within 2 to 4 hours, followed by a less marked decline in αGSU mRNA (343,415), further supporting direct actions of inhibin at the pituitary. Inhibin does not alter LHβ mRNA levels directly and, overall, the primary action of inhibin is likely to regulate FSH secretion and synthesis.

The effects of inhibin on LH synthesis and secretion are less clear than those on FSH. Some studies have reported a decline in basal LH release and proestrus surge (409), whereas others have reported an effect only on GnRH-stimulated release (416), and still others have shown no effect of inhibin on LH secretion (339,411).

The mechanism by which inhibin acts remains unclear. In addition to the direct effects of inhibin on FSH, it may exert effects on the gonadotropins through indirect mechanisms. One such mechanism is through regulation of GnRHR expression and number, which can determine gonadotrope responsiveness to GnRH. Inhibin does not appear to alter hypothalamic GnRH release in rats (417), but may change the number of GnRHRs, thereby resulting in a reduced GnRH-binding capacity (410,418). Inhibin has been shown to decrease number and mRNA levels of GnRHR in the rat pituitary cells (418,419) and to block GnRH-stimulated upregulation of GnRHR (420).

Inhibins can interact with activin type II receptors, which may block the recruitment of type I receptors, but the binding affinity is approximately 10 times lower than for activins (421,422), thus making it difficult to antagonize activin action through the same receptors. Rather, inhibins bind to the TGF-β type III receptor (coreceptor for type II receptor, also known as betaglycan) with high affinity, and form cross-linked complexes with betaglycan and activin type II receptor (423), thus preventing association of type I receptors. Rat pituitary gonadotropes and LβT2 cells express betaglycan (423-425). In contrast, an increase in both GnRHR number and GnRH binding as well as mRNA levels has been demonstrated using ovine pituitary cell cultures incubated with inhibin (401,426). These data suggest that the mechanisms by which inhibin, activin, or both regulate GnRHR number and expression may differ between species.

Activin

The activins were initially isolated from gonadal tissues (427,428), are produced in multiple tissues throughout the body, including the pituitary, and have diverse biological actions in multiple tissues both during development and in the adult (347,348,408). Both αT3-1 and LβT2 cells bind activin A and express the activin receptor types IA, IB, IIA, and IIB, as well as the inhibin β_B subunit (421,424,429–431). Unlike αT3-1 cells, LβT2 cells also express the inhibin α subunit. Activin binds to its specific type II receptors, and type II receptor then recruits and forms a heterodimeric complex with the type IA or IB receptors (also known as activin receptor–like kinases, ALK2 and 4, respectively) (388,389) (Fig. 5).

Numerous studies have shown that activin A and B increase FSHβ primary transcript synthesis and mRNA levels, followed by stimulation of FSH secretion (343,345,388,389,411,432). A constitutively active form of ALK4 stimulates FSHβ mRNA levels in LβT2 cells (392), and activin receptor type II–deficient mice exhibit suppression of FSH levels both in the circulation and in the pituitary, decreased FSHβ mRNA levels, defects in gonadal function, and infertility (414,433). Like inhibin, activin acts independently of GnRH and can induce FSH secretion in GnRH-desensitized pituitary cells (434). Activin also enhances the GnRH response in gonadotrope cells and synergizes in the induction of FSHβ gene transcriptional activity with GnRH (435,436). These enhancements have been shown in vivo as well, where concomitant treatment with activin A and GnRH enhances both FSH secretion and FSHβ mRNA levels in rats (437). Activin is also able to alter the gonadotrope cell population and increases the number of FSH-secreting cells and perhaps the amount of FSH secreted by certain subpopulations of gonadotropes (438).

Activin also has a functional role in modulating gonadotrope responsiveness to GnRH by increasing the expression of GnRHR. Activin A increases GnRHR mRNA levels and GnRHR gene promoter activity in αT3-1 cells in a time- and dose-dependent manner (429,439,440), and increases GnRHR synthesis in rat pituitary cells (441). Activin has also been shown to synergize with GnRH signaling pathways to augment the GnRH-mediated transcriptional activation of the GnRHR gene (429,439,440). Therefore, it is possible that activin not only exerts a direct effect on FSH but stimulates GnRHR gene expression, which can enhance the biosynthesis of LH. By this mechanism, activin may play a role in the regulation of both gonadotropin hormones.

Pretreatment with activin A enhances GnRH-induced activation of αGSU promoter in αT3-1 cells. On the other hand, activin A has been shown to block the stimulatory effect of GnRH on αGSU promoter activity in this cell model (442). Nonetheless, changes in LH secretion and αGSU or LHβ subunit mRNA concentration have not been observed on a consistent basis. For example, activin receptor type II–deficient mice have shown to have normal pituitary LH content as well as circulating LH levels (414).

It has been shown that the β_B subunit is produced uniquely in pituitary gonadotropes (347), suggesting that activin B may act as an autocrine stimulator of FSH synthesis (348). Supporting this, neutralization of activin B, but not activin A, in vitro results in a decline in FSHβ mRNA levels, implying that basal FSHβ gene expression depends on locally produced activin B (348). Also, in perifused pituitary cells, FSHβ mRNA levels decline over time when the cells are perifused with media alone. However, when activin is replaced in the perifused media, FSHβ mRNA levels return to original values (50-fold induction). This initial decline in FSHβ gene expression does not occur in static culture, suggesting that perifusion removes locally produced activin, which is necessary for basal FSHβ mRNA concentrations (421,443). However, in β_B-subunit–deficient mice, FSH secretion does not decrease but shows a modest elevation, possibly because of upregulation of β_A subunit in the ovaries, and hence increased activin A (444), lack of inhibin B ($\alpha\beta_B$), or both. Thus, the exact roles of gonadal and pituitary activins remain to be established.

Another group of TGF-β superfamily proteins, BMP-6, 7, and 15, stimulate ovine FSHβ gene promoter activity in pituitary cells from transgenic mice carrying an ovine FSHβ reporter gene and in LβT2 cells, and also increase FSH secretion from LβT2 cells, although less potently than activin (445,446). The mRNAs for these proteins are present in mouse pituitary and LβT2 cells, and inhibition of BMP-6 and 7 decreases basal ovine FSHβ promoter activity and FSH secretion in transgenic pituitary cells, whereas inhibition of activins has no such effects, suggesting a specific paracrine/autocrine role for BMP-6 and 7 in basal FSHβ gene expression (445). Similarly, BMP-15 stimulates FSHβ gene expression selectively without affecting LHβ or GnRHR gene expression, and, like activin, all these BMPs stimulate FSH but not LH secretion (446). BMP-7 has also been shown to interact with type II activin receptor and transduce an activin-like signal (447), and therefore BMPs are sensitive to the effects of inhibin and its coreceptor betaglycan, which can also block BMP signaling through activin and BMP type II receptors (448).

Another TGF-β family member, MIS, has also been shown to stimulate promoter activities of both LHβ and FSHβ genes in LβT2 cells (449). Type II MIS receptors are present in gonadotropes, and MIS can also enhance or synergize GnRH stimulation of rat FSHβ or LHβ gene promoter activities, respectively. In addition, TGF-β has been shown to compete with inhibin for binding to betaglycan, although it does not have any direct effect on FSHβ or GnRHR gene promoter activation (424). By interacting with betaglycan, TGF-β reduces the inhibin activity that antagonizes the activin stimulation of the ovine FSHβ gene promoter activity. Together, BMPs, MIS, and TGF-β all may play important roles in the regulation of gonadotropin gene expression.

Follistatin

Follistatin is a glycosylated, monomeric protein with FSH-suppressing activity (450), and the amino acid sequence of follistatin is highly conserved (98%) between species (451–453). Follistatin reduces FSHβ mRNA levels and FSH release in vitro (366,454).

Similar to inhibin, follistatin reduces FSHβ mRNA levels within 2 hours and FSH secretion within 8 hours in rat pituitary cells (343), and the effects of follistatin and inhibin on FSH secretion appear to be additive (454). Also, follistatin lowers ovine FSHβ gene promoter activity in pituitary cells from transgenic mice carrying an ovine FSHβ reporter gene and in LβT2 cells (430,435). Follistatin is more effective than inhibin in blocking the effects of activin in increasing FSHβ mRNA concentrations and FSH secretion (413). In vivo, follistatin decreases FSH secretion within 4 hours, without altering circulating LH or LH responses to exogenous GnRH (412). Follistatin also inhibits GnRH-induced FSHβ gene promoter activity in LβT2 cells (430).

The addition of follistatin to rat pituitary cells or LβT2 cells in static culture can reduce FSHβ gene expression and FSH secretion in the absence of exogenous activin, and pretreatment with follistatin allows further enhancement of FSHβ gene promoter activity by activin (412,430), suggesting that either follistatin can act independently of activin/inhibin or that activin is produced locally by the pituitary. These data raise the possibility that follistatin may regulate gonadotrope function through locally produced activin (350,455). Supporting this, follistatin expression is present in pituitary gonadotropes and folliculostellate cells in the pituitary, αT3-1 cells, and LβT2 cells (349,350,430), and activin, GnRH, and PACAP all stimulate the synthesis of follistatin (366,456–460), whereas inhibin and testosterone decrease follistatin gene transcription (455,457,461). Follistatin itself also has an inhibitory effect on follistatin expression (457,460), perhaps through its binding of activin.

Interestingly, follistatin gene expression in rat pituitary cells is increased by high frequencies of pulsatile GnRH. This GnRH pulse frequency dependency of follistatin is the inverse of that of FSHβ expression and may play a role in differential effects of GnRH pulse frequency on LHβ and FSHβ gene expression (366). Follistatin can also bind BMP-7 and 15 and antagonize BMP activities (447,462), suggesting that follistatin may decrease FSH synthesis by incapacitating a number of TGF-β family members other than activin.

Follistatin may also exert effects on the gonadotropins through indirect mechanisms by changing GnRHR expression and number, thus determining gonadotrope responsiveness to GnRH. Follistatin blocks the effect of activin on the GnRHR gene by binding to and inactivating activin in pituitary gonadotropes and in αT3-1 cells (441). In addition, follistatin decreases basal activity of the mouse GnRHR gene promoter (429,463), suggesting that endogenously released activin can act in a paracrine/autocrine fashion to stimulate basal GnRHR gene activity.

A follistatin-related protein (FSRP), also referred as follistatin-like protein or follistatin-like 3, a product of follistatin-related gene, with two, rather than three, follistatin domains, has been identified (464–466). FSRP binds to activins and BMP-2, but not BMP-4, 6, and 7, with high affinity, but less affinity than follistatin (464,467). Interestingly, FSRP is less potent than follistatin in interfering with endogenous activin activity, whereas its effect on exogenous activin is comparable with follistatin's, and it is unable to suppress FSH secretion in rat pituitary cells (467). These studies suggest a differential role of FSRP from follistatin, but its physiological function is yet to be determined.

Gonadectomy

Removal of the gonads results in a rise in LH and FSH secretion (468,469). Data obtained from rat, mouse, and sheep have shown that αGSU, LHβ, and FSHβ mRNA concentrations also increase after gonadectomy, likely because of lack of steroid negative feedback on GnRH (461,470–475), but the magnitude and time course of changes for each subunit vary. The removal of feedback inhibition of GnRH secretion by gonadal steroids is predominantly responsible for the postgonadectomy rise in gonadotropin gene expression. Supporting this, estrogen receptor (ER) α–deficient mice have been shown to exhibit elevated pituitary αGSU, LHβ, and FSHβ mRNA levels, similar to the effect of gonadectomy, indicative of the reduced feedback inhibition by E_2 (476). This positive effect of gonadectomy on αGSU and LHβ mRNA synthesis is reversed by steroid replacement (472). In rats, this appears to involve both an elevation of subunit gene transcriptional rates (477,478) as well as the recruitment of active gonadotrope cells in the pituitary (479).

In male rats, circulating gonadotropin levels increase rapidly, and αGSU and LHβ subunit mRNAs are significantly increased within 24 hours of gonadectomy. αGSU mRNA attains maximal levels (three- to fourfold increase) within 10 days, whereas LHβ plateaus (fivefold increase) after 14 days. FSHβ mRNA shows smaller-magnitude changes and increases only twofold after 7 days, thereafter becoming stable or even declining. In females, the smaller increase in circulating LH is accompanied by similar changes in mRNAs. αGSU and LHβ subunit mRNAs rise slowly, with significant increases not being seen until 4 to 7 days postgonadectomy. αGSU mRNA (fivefold increase) plateaus after 14 days, whereas LHβ mRNA continues to rise through 30 days (15-fold increase). In contrast, circulating FSH and FSHβ mRNA increased within 12 hours of gonadectomy, with maximal concentrations stable after 4 days (fourfold increase) (472,475).

Replacement of physiological concentrations of E_2 in ovariectomized female rats inhibits GnRH secretion (480) and results in suppression of all three gonadotropin subunit mRNAs, although by different degrees (471,473–475). The rise in LHβ mRNA after gonadectomy can be completely blocked by administration of E_2 at the time of gonadectomy, but αGSU mRNA concentrations still increase (approximately 50%) above levels seen in intact animals. This action of E_2 on increasing αGSU mRNA requires the presence of GnRH because it does not occur if the animals also received a GnRH antagonist (475), and E_2 does not alter αGSU mRNA in vitro (481). In contrast to LHβ, the rapid increase in FSHβ mRNA is only partially suppressed by E_2 at the time of gonadectomy, suggesting that the loss of nonsteroidal factor(s) from the gonad (inhibins and follistatin) is likely to be causally related to the increase in FSHβ mRNA expression (475).

Gonadectomy also causes a rise in GnRHR mRNA levels, both in female and male rats (34,69). This is presumably due to gonadal steroids because when E_2 or testosterone is replaced after gonadectomy in female or male rats, respectively, pituitary GnRHR mRNA levels are reduced to baseline (69). However, removal of the gonads can increase the number of GnRHRs in vivo when hypothalamic–pituitary connections are severed. In the absence of hypothalamic input, E_2 has been shown to increase the number of GnRHRs (482,483), suggesting a feedback inhibition by steroid hormones on GnRH synthesis and a direct effect of E_2 on GnRHR expression. This steroid action is slow, requiring several days or even weeks to see a clear induction of GnRHR mRNA by gonadectomy or a reduction by steroids. This regulation of GnRHR by gonadectomy or gonadal steroids has also been observed in the ovine pituitary (37). E_2 alone has also been shown to stimulate and potentiate pulsatile GnRH stimulation of the rat GnRHR gene expression in vivo in gonadectomized rats (64).

Estrogen and Progesterone

Gonadal steroids can influence the rate of gonadotropin subunit mRNA synthesis directly at the level of the pituitary. For example, E_2 selectively stimulates LHβ gene transcription (379) and LH synthesis and accumulation (329) in rat pituitary cells. Further, a segment of the 5′-flanking region of the rat LHβ gene plays a role in mediating responsiveness to E_2 (484). E_2 also has been shown to increase basal as well as GnRH-induced LHβ mRNA levels in LβT2 cells (327). In ovine pituitary cells, both E_2 and progesterone can suppress the rate of FSHβ gene transcription and gene promoter activity (485,486). The bovine FSHβ gene contains a sequence 80% homologous to the consensus estrogen response element (ERE) in the distal 5′-flanking region, and expression of the gene can be suppressed by E_2 (487). Human αGSU mRNA synthesis can also be inhibited by both E_2 and glucocorticoids (488,489), although high-affinity E_2 and glucocorticoid receptor binding sites have not been characterized in the human αGSU gene.

In ovine pituitary gonadotrope cell cultures, E_2 increases and progesterone decreases the number of GnRHRs (490–492). E_2 has also been shown to hyperglycosylate ovine GnRHR (492), and the ovine GnRHR promoter is activated directly by E_2 in vivo in a transgenic mouse model where the pituitaries are targeted by the ovine GnRHR gene promoter, but not in vitro in αT3-1 or LβT2 cells (493). αT3-1 cells express ERα and ERβ (65). E_2 has a time- and dose-dependent inhibitory effect on GnRHR number in αT3-1 cells (494), and both acute inhibitory and chronic stimulatory biphasic effects on rat pituitary gonadotropes (495) or chronic stimulatory effects only (329) have been reported. GnRHR mRNA levels in LβT2 cells are increased by E_2 and glucocorticoids (71,327). These discrepancies may be attributable to differences in cellular responses between αT3-1, LβT2, and primary gonadotropes, to indirect influences from other cell types in the anterior pituitary, and to species-specific regulation. In αT3-1 cells, GnRH-induced IP_3 accumulation is inhibited by E_2 pretreatment, indicating that gonadal steroids modulate the coupling of GnRHR to second messengers (494). The maximum level of IP_3 is decreased, and the cells require a higher GnRH concentration to reach the same levels of IP_3 as the control group. These data suggest that E_2 reduces GnRHR number and the efficiency of PLC activation, and that downregulation of $G_{q\alpha}$ and $G_{11\alpha}$ by E_2 may contribute to this effect (494).

The effects of progesterone depend on the prevailing gonadal steroid milieu. Progesterone replacement alone does not prevent increases in gonadotropin subunit gene expression. In contrast, the addition of progesterone to E_2 (for 2 to 7 days) is more effective than E_2 alone in reducing the rise in αGSU, LHβ, and FSHβ mRNAs after gonadectomy in females (475). Progesterone and E_2 replacement at the time of gonadectomy maintains LHβ subunit mRNA concentrations at or below intact values. The effects are identical to those of a GnRH antagonist, suggesting that progesterone and E_2 act by suppressing GnRH secretion.

Higher concentrations of progesterone can inhibit FSHβ gene transcription in ovine pituitary cells (485), and in the presence of E_2, progesterone reduces gonadotropin subunit gene expression by inhibiting GnRH secretion (496), suggesting that progesterone acts to reduce gonadotropin gene expression both by reducing the GnRH stimulus and by a direct action on the gonadotrope. However, an in vivo study suggests that short-term (<24 hours) progesterone in immature female rats can selectively increase FSHβ mRNAs (497), and an in vitro study shows that progesterone increases ovine FSHβ gene promoter activity (498). Thus, in the rat, progesterone may act directly at the pituitary selectively to increase FSHβ mRNA concentrations, although the mechanism by which this occurs is not well understood. In addition, progesterone decreases human GnRHR gene promoter activity in αT3-1 cells (499).

Testosterone

Testosterone replacement in gonadectomized male rats reduces gonadotropin subunit mRNAs to intact levels. However, the effect of testosterone on FSHβ mRNA appears to be complex. Testosterone replacement at physiological levels at the time of orchidectomy completely prevents the postcastration rise in FSHβ mRNA levels (475,478). The inhibitory effect of testosterone on FSHβ mRNA levels has also been shown in vitro, suggesting that testosterone inhibits the FSHβ subunit gene directly at the pituitary gland in addition to indirect mechanisms of GnRH suppression (500). However, other studies have found that pharmacological doses of testosterone do not fully suppress FSHβ mRNA levels in gonadectomized male rats (473,474). Furthermore, in gonadectomized male rats treated with a GnRH antagonist, the addition of testosterone increased FSHβ mRNA and circulating FSH levels (478,501–503), and testosterone can increase FSHβ mRNA levels in a dose-dependent manner by a direct action in cultured rat pituitary cells (481). Thus, in the rat, lower doses of testosterone probably inhibit FSHβ mRNA expression by the suppression of GnRH release, whereas higher doses may stimulate FSHβ mRNA through a direct action at the pituitary. As testosterone levels rise, GnRH secretion is inhibited with a consequent reduction in gonadotropin, especially LH, secretion. Thus, the direct action of testosterone on FSHβ mRNA and the close relationship between FSHβ mRNA concentration and basal FSH secretion (478) would favor maintaining FSH secretion during periods of reduced GnRH stimuli. Regardless, testosterone can increase FSHβ mRNA concentrations in both male and female rats (504), suggesting that posttranscriptional regulation of gonadotropin subunit mRNAs can occur in both sexes.

GnRH REGULATION OF GONADOTROPIN GENE EXPRESSION

Glycoprotein α-Subunit Gene Expression

αGSU is the common subunit for all glycoprotein hormones, LH and FSH in gonadotropes, thyroid-stimulating hormone (TSH) in thyrotropes, and chorionic gonadotropin (CG) in the placental syncytiotrophoblasts of primates and horses (505). Unique β subunits confer the physiological specificity of each glycoprotein hormone. Distinct combinations of a large number of regulatory elements located on the αGSU gene promoter enable cell- and tissue-specific regulation of gene expression.

αGSU genes have been cloned from several species, including the mouse, rat, cow, horse, monkey, and human (2,506–509). In all species studied to date, the αGSU is coded by a single gene that contains four exons and three introns. In primate and equine species, the αGSU gene is expressed in placental trophoblasts as well as in pituitary cells, specifically gonadotropes and thyrotropes (2,509). The size of the αGSU gene varies from 8 to 16.5 kb, largely because of variation in the size of the first intron, and the coding region is highly conserved. Mature αGSU mRNA is 0.7 to 0.8 kb in length and codes for a polypeptide that contains a 24–amino acid signal peptide followed by a 96–amino acid protein in mouse, rat, and cow. In the human, the αGSU is 92 amino acids in length because of the deletion of 4 residues near the N-terminus.

αT3-1 cells express αGSU mRNA as well as GnRHR, and produce and secrete αGSU, but do not express LHβ or FSHβ subunit. These cells respond to GnRH stimulation with an increase in αGSU mRNA levels in a time- and dose-dependent manner (12).

A similar increase of αGSU mRNA levels has been observed in response to PMA or ionomycin, and this increase is not additive with GnRH, suggesting a role of the PKC pathway (19,510). Induction of αGSU gene promoter activity has also been observed in LβT2 cells, with lower basal activity (511). The GnRH-induced increase in αGSU mRNA levels is maximal at 12 to 24 hours and maintained for a further 24 hours, and this increase in mRNA levels is mediated by both an increase in αGSU gene transcription and mRNA stability, in which the αGSU gene transcription rate peaks within 1 hour of GnRH treatment and returns back to baseline by 12 hours, whereas the half-life of the mRNA increases from 1.2 to 8 hours (512). Stimulation of αGSU promoter activity peaks 4 to 6 hours after exposure to GnRH but thereafter declines, returning to the levels in unstimulated control cells by 12 to 24 hours (510). This promoter activity is further stimulated by a Ca^{2+} channel agonist, and an activator of the PKA pathway, 8-bromo-cAMP, but not by TPA. These data suggest that the transcriptional stimulation of the αGSU gene by GnRH is mediated primarily by the PKC pathway (512). Thus, GnRH appears to stimulate a burst of αGSU gene transcription lasting less than 4 to 6 hours, and the persistent elevation of αGSU mRNA levels for at least 48 hours suggests that the mRNA has a long half-life or that GnRH stabilizes the mRNA in addition to its transcriptional effects. Whether this mechanism also occurs in primary gonadotropes is unclear because the half-life of αGSU mRNA in primary pituitary cultures is 6.5 hours (513); however, in this case both gonadotropes and thyrotropes contribute to αGSU mRNA levels.

To date, the promoter regions of mouse and human αGSU genes have been studied the most among species, and the *cis*-elements and *trans*-factors necessary for gonadotrope-specific gene expression are quite similar in these two species. The mouse αGSU gene contains at least three types of sequences for transcription factor interaction in the proximal 5′-flanking region: cell type–specific (gonadotropes versus thyrotropes or trophoblasts), basal, and GnRH-responsive *cis*-elements. These include a GnRH-response element (GnRH-RE) at position –406 to –399, a pituitary homeobox-1 (Pitx-1) binding site at –398 to –385, a pituitary glycoprotein hormone basal element (PGBE) at –344 to –300, and a gonadotrope-specific element (GSE) at –215 to –207, relative to the major transcriptional start site (514). Several studies have also identified a distal enhancer activity located between –4,600 and –3,700 (515,516) (Fig. 16).

The element in the mouse αGSU promoter that has been best characterized as a basal, tissue-specific enhancer is the GSE. The GSE sequence, TGACCTTGT, occurs upstream of the placenta-specific elements, and is highly conserved among species, including mouse, human, rat, cow, and horse (517). The GSE is bound by a 54-kDa protein, steroidogenic factor-1 (SF-1, also known as Ad4BP) (518). An orphan nuclear receptor, SF-1 was first identified by its ability to bind to and coordinately regulate the expression

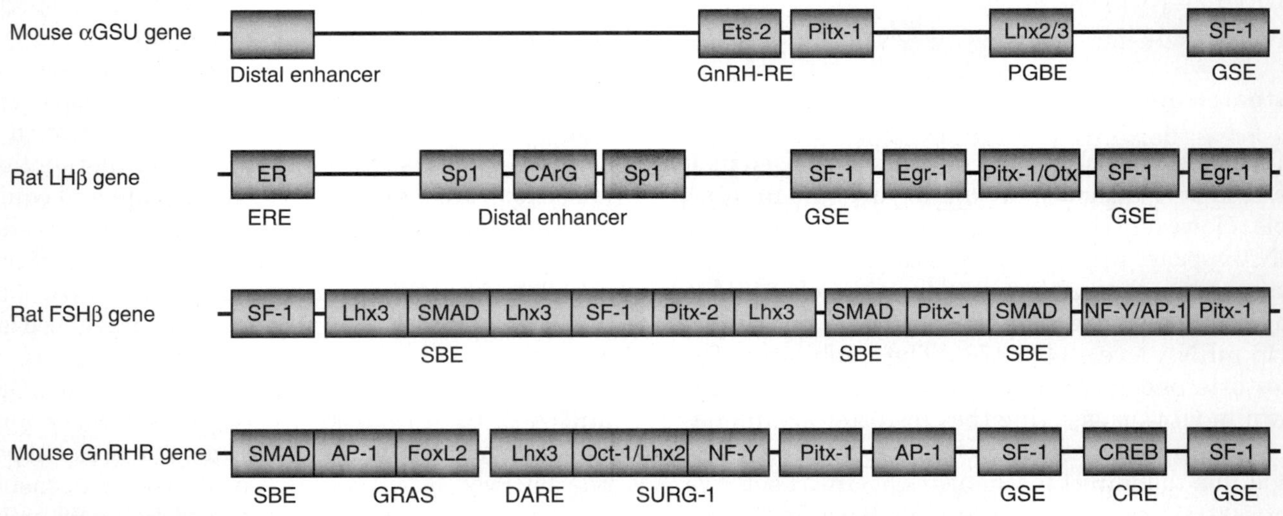

FIG. 16. Schematic diagram of 5′-flanking regions of the rodent glycoprotein α-subunit (αGSU), luteinizing hormone β-subunit (LHβ), follicle-stimulating hormone β-subunit (FSHβ), and gonadotropin-releasing hormone receptor (GnRHR) genes that contain *cis*-acting elements and transcription factors important for cell-specific and GnRH-regulated expression of each gene. Note that this illustration is simplified for clarity. AP-1, activating protein-1; CArG, $CC(A/T)_6GG$ factor; CRE, cyclic adenosine monophosphate (cAMP) response element; CREB, CRE binding protein; DARE, downstream activin regulatory element; Egr-1, early growth response protein-1; ER, estrogen receptor; ERE, estrogen response element; FoxL2, forkhead/winged-helix family protein; GnRH-RE, GnRH response element; GRAS, GnRHR activating sequence; GSE, gonadotrope-specific element; Lhx, LIM-homeodomain protein; NF-Y, nuclear transcription factor-Y; Oct-1, octamer-1; Otx, orthodenticle; PGBE, pituitary glycoprotein hormone basal element; Pitx, pituitary homeobox; SBE, SMAD binding element; SF-1, steroidogenic factor-1; SMAD, Sma/mothers against decapentaplegic homolog; Sp1, specificity protein 1; SURG-1, sequence underlying responsiveness to GnRH-1.

of genes encoding enzymes in the corticosteroid biosynthetic pathway (519). Subsequently, it has also been shown to bind to and regulate the aromatase and MIS genes in gonadal tissues (520,521). It is expressed in the ventromedial hypothalamus, pituitary gonadotropes, adrenal cortex, and gonadal tissues (522–524), and is considered as an authentic defining factor for pituitary-specific expression of αGSU gene (292). Disruption of the *fushi tarazu* factor 1 homolog gene (*Ftz-F1* or *Nr5a1*) encoding SF-1 in mice precludes adrenal and gonadal development and also results in the selective loss of expression of gonadotrope-specific markers, including LHβ, FSHβ, and GnRHR, and a reduction in αGSU levels (523, 525–529), and a mutation in human *Ftz-F1* has been shown to cause similar defects (530). Treatment of SF-1–deficient mice with exogenous GnRH partially restores expression of LHβ and FSHβ, suggesting that SF-1 is not critical for GnRH stimulation of gonadotropin gene expression (529,531). Thus, SF-1 appears to be important for function of the reproductive axis at multiple levels (532).

An additional putative basal enhancer of the mouse αGSU gene is the PGBE (533). The PGBE supports basal transcription, and is able to direct expression of the αGSU gene promoter to cells of both gonadotrope and thyrotrope lineages, but not to placental trophoblasts (292,534,535). A member of the LIM (lin-11, isl-1, mec-3, and Lmx-1)-homeodomain (HD) family of transcription factors, Lhx2 (LH2), binds to a 14-base pair (bp) imperfect palindrome in the PGBE domain (536–538). LIM-HD proteins contain both a zinc finger (the LIM domain) and an HD (539). The HD of these factors is sufficient for specific DNA binding, and the LIM domains do not appear to be DNA-binding domains, but rather may function as protein–protein interaction domains to facilitate homodimer or heterodimer formation. Lhx2 has a restricted tissue distribution, being most abundant in αT3-1 and αTSH cells, cell lines of gonadotropic and thyrotropic origin, respectively, and in mouse brain; less abundant in whole rat pituitaries, corticotropic AtT-20 cells, and somatolactotropic GH_3 cells; and undetectable in placental JEG-3 cells and in mouse liver. Lhx2 is able to activate specifically the αGSU gene promoter in a heterologous cell system, suggesting that the LIM-HD protein Lhx2 is an activator of αGSU gene expression in gonadotropes and thyrotropes (9).

Another candidate factor for a role in mediating αGSU gene expression by binding to PGBE is Lhx3, a related member of the family of LIM-HD proteins. Lhx3, also known as P-Lim or mLim-3, is expressed in the mouse pituitary throughout development and in the adult, as well as transiently in the spinal cord, pons, and medulla oblongata, but with no detectable expression elsewhere (540). Lhx3 expression was detected in cell lines of pituitary origin, including cells representative of somatolactotropes (GH_3, GH_4C1, GC), thyrotropes (αTSH), gonadotropes (αT3-1 and LβT2), and corticotropes (AtT-20), but not in cell lines derived from peripheral, other endocrine, or neural tissues (540–542). Lhx3 is able to bind to the PGBE sequence (core motif ATTA) and is a strong activator of transcriptional activity of the αGSU gene promoter, as well as the prolactin, TSHβ, and pituitary-specific Pit-Oct-Unc (POU) domain factor (Pit-1) (543) gene promoters (536,540,542,544). In addition, a point mutation in the LIM domain of Lhx3 results in a reduction of αGSU promoter activity (545). Interestingly, targeted disruption of the Lhx3 gene in mice leads to failure of growth and differentiation of the anterior and intermediate lobes of the pituitary, and the development of all pituitary cell lineages, except the corticotropes, was affected (546). Similar phenotypes have been reported in human patients with combined pituitary hormone deficiency due to Lhx3 gene mutations (547–549). Similarly, another LIM-HD protein, Lhx4 (also known as LIM4 or GSH4), has been shown to act with Lhx3 for pituitary development, with Lhx4-deficient mice exhibiting a reduced number of all the anterior pituitary cell types (550,551). This suggests that Lhx3 plays an important role not only in αGSU gene expression, but in differentiation and proliferation of nearly all the pituitary cell lineages. Activin has been shown to suppress the mouse αGSU gene promoter activity through the PGBE in αT3-1 cells (442).

A pan-pituitary *bicoid*-related HD protein, Pitx-1 (also known as Ptx-1, Pix-1, or P-OTX), has been shown to interact with a Pitx-1 binding site (552), and has been implicated in cooperative function specifically with Lhx3 for mouse αGSU gene expression through protein–protein interaction (553). Inhibition of endogenous Pitx-1 expression in αT3-1 cells results in a loss of both Lhx3 and αGSU gene transcription, although Lhx2 gene expression remains intact (554). Pitx-1 was initially isolated from AtT-20 cells (555), but is expressed in all pituitary cell types, including gonadotropes (556). Pitx-1–deficient mice exhibit diminished numbers of gonadotropes and reduced expression of gonadotrope-specific genes, including αGSU (557). These studies suggest an importance of Pitx-1 in αGSU gene expression. In fact, Pitx-1 is capable of potentiating the expression of almost all the pituitary-specific genes, including αGSU, LHβ, FSHβ, and GnRHR (552,554,558,559).

A second Pitx family member, Pitx-2, has also been shown to stimulate the expression of these genes by mechanisms similar to those of Pitx-1, perhaps interacting with Pitx-1 binding sites and with other transcription factors (560). Pitx-2–deficient mice exhibit an arrest of pituitary growth and differentiation, and are embryonic lethal (561). Homozygous Pitx-2 hypomorphic (reduced function) mice, generated by insertion of a neomycin resistance gene cassette into one of the introns, thereby interfering with Pitx-2 gene transcriptional efficiency, can live until postnatal day 1, and exhibit strikingly similar pituitary gonadotrope phenotypes to Pitx-1-deficiency (562). The pituitary gonadotropes are nearly absent, judged by LHβ and FSHβ gene expression, and expression of *egr-1*, GATA, SF-1, and GnRHR are all abolished in this mouse model, whereas Pitx-1 expression is normal, suggesting an independent requirement for both Pitx-1 and Pitx-2 in maintaining normal gonadotrope function. On the other hand, another HD protein, a homeobox repressor Hesx1 (also known as Rpx), has been shown to suppress mouse αGSU gene promoter activity through the Pitx-1 binding site, possibly by recruiting a nuclear corepressor (N-CoR) (563). Hesx1 is likely to interact with the Pitx-1 binding site as a monomer, and its repressor activity can be superseded by Pitx-1 overexpression.

Like the mouse αGSU gene, the human αGSU gene also contains a complex array of *cis*-elements in the proximal 5′-flanking region (564–569). A PGBE is located at position –329 to –320, α-basal element 1 and 2 (αBE1 and 2) at –316 to –302 and –296 to –285, respectively, a GSE at –219 to –211, an α-activating element (αACT) at –161 to –141, two CREs at –146 to –111, a Pitx-1–responsive element at –80 to –65, and two E boxes at –51 to –45 and –21 to –16. These sequences serve as gonadotrope-specific *cis*-elements or as common *cis*-elements for expression in gonadotropes, thyrotropes, or trophoblasts (514).

Two CREs have been identified in the 5′-flanking region of the human αGSU gene (534,564,565), and are the most important contributors to human promoter activity. The region, –146/–111, contains two identical copies of an 18-bp nucleotide sequence, with an 8-bp core palindromic sequence identical to the consensus CRE (TGACGTCA). Of interest, these tandem CREs have been shown to regulate αGSU gene transcription in a synergistic manner (568). These two CREs play a role not only in regulation by cAMP but in basal expression and tissue specificity. Because CREBs are known to form homodimers through a basic leucine zipper (bZIP) domain in the active state, a specific spatial orientation between the two CREs is required for maximal enhancer activity (568,570). Other bZIP CREB proteins, such as CRE modulator (CREM), c-Jun, and ATF-1 and 2, also have been shown to bind to these elements (538). Human αGSU promoter activity is suppressed by c-Jun in placenta-derived cells through the CREs (571), but is slightly activated in pituitary-derived cells. Also, gonadotrope transcriptional activity of the αGSU gene is suppressed by overexpression of CREB (572), suggesting that c-Jun/ATF-2 heterodimers are the preferred binding pair for CRE sequences in gonadotropes to determine cell specificity.

Unlike the human, the αGSU genes of lower primates (rhesus monkey, baboon) contain a single CRE, and the αGSU genes of other mammalian species, including the mouse, cow, and sheep, also have a single CRE but with a 1-bp substitution from the primate CRE (TGATGTCA) (509,569). This 1-bp substitution in these species has been shown to decrease the binding affinity for CREB, whereas heterodimers of c-Jun and ATF-2 bind to this nonprimate variant CRE with much higher affinity than CREB (538). A pair of the primate CREs targets αGSU gene expression to both gonadotrope- and trophoblast-derived cell lines, whereas a nonprimate CRE targets expression only to gonadotrope-derived cells (538,566). Also, the cAMP-induced stimulation of αGSU gene transcription seen in nonprimate species may be the result of interactions between CREB and other nuclear proteins at the CRE site (570). Thus, αGSU gene responsiveness to cAMP has diverged between primate and nonprimate species. In primates, CREB appears to play a central role, whereas in nonprimates, other nuclear proteins may mediate both tissue specificity and responsiveness to cAMP.

Other elements identified as playing a role in expression of the human αGSU gene include the αACT element, which binds to the zinc finger proteins GATA2 and 3 (573), and two E boxes, which bind members of the family of basic-helix–loop-helix zipper proteins, E box proteins (αEB) 1 and 2 or upstream stimulatory factor (574). The αBE1 and 2 elements have been suggested to be cognate binding sites for αBE-binding protein 1 and 2 (534). The upstream tissue-specific response element (URE) or trophoblast-specific element, located between –182 and –141, appears to induce placental cell specificity to the αGSU gene (566,568,569,575,576). Basal expression appears to be primarily mediated through the URE and CRE regions (567,569,577,578). Also, a CCAAT box (579) and a junctional regulatory element (576, 580) have been suggested to be trophoblast-specific enhancer elements.

The optimum level of αGSU gene expression in gonadotropes is probably determined by the combined actions of widely expressed, pituitary-restricted, and gonadotrope-specific transcriptional activators that act in combination and synergistically. Human αGSU gene expression in gonadotropes is controlled by an array of weaker elements through a complicated combinatorial interplay among these elements and cognate binding proteins, rather than by one dominant element (514). It has been shown that transcription factors binding to PGBE, αBE1 and 2, and the CREs influence each other's activities, whereas GSE-binding

SF-1 seems to have an autonomous effect independent of these factors (534), suggesting that the transcription factors for PGBE, αBEs, and CREs all converge and work together synergistically to stimulate transcriptional activity, possibly by sharing a common coactivator or adapter complex. The mouse αGSU gene is also regulated similarly for gonadotrope-specific expression by the PGBE and GSE, along with distal enhancer activity (515,516,535).

The mouse αGSU gene promoter confers a robust GnRH-induced activity compared with that of other species. An upstream element of the mouse αGSU gene, the GnRH-RE, confers responsiveness to GnRH, as well as to PMA and to cAMP, and GnRH responsiveness requires the cooperative interaction of GnRH-RE and PGBE (533). The rat αGSU gene also contains this GnRH-RE, and its promoter activity is also stimulated by PMA in both rat primary pituitary culture and αT3-1 cells (581). The need for a complex response unit for the mediation of GnRH stimulation may provide a mechanism for the maintenance of appropriate, tissue-specific expression and regulation of the αGSU gene, and this mechanism may require different sets of signal transduction pathways that distinguish gonadotropes from thyrotropes (582). The involvement of a tissue-specific basal element may restrict αGSU gene expression to the appropriate cell type (559), and the involvement of two elements in mediating GnRH responses may prevent the αGSU gene from responding to activation of the PKC signaling pathway in nongonadotrope cells and tissues (9). A core Ets factor has been identified as a GnRH-RE–binding factor candidate, which appears to be important in mediating GnRH stimulation of αGSU gene expression (285). The DNA-binding domain of Ets-2 is able to bind specifically to the GnRH-RE, and a dominant negative form of Ets-2 reduces the ability of GnRH to stimulate αGSU gene promoter activity. These findings suggest that the Ets factor binding site in the GnRH-RE may contribute to transcriptional stimulation of mouse αGSU gene by GnRH through activation of the MAPK pathway. It is possible that an Ets factor, binding to the GnRH-RE, may interact with Lhx2 bound to the PGBE to mediate GnRH-induced expression of the αGSU gene.

The PGBE and CREs, along with the αBE1 and 2, appear to confer the responsiveness of human αGSU gene promoter to GnRH synergistically (292,534,578), whereas the two tandem CREs do not mediate GnRH responsiveness by themselves. A potential Ets factor binding sequence similar to the mouse GnRH-RE has been localized between the PGBE and αBE (534). In addition, the GSE and URE have been suggested to play minor roles in GnRH responsiveness (292,511). Thus, as for gonadotrope-specific gene expression, it is likely that a complex combinatorial interplay among these elements and cognate binding factors confers GnRH responsiveness, possibly by sharing common coactivators, such as CREB-binding protein (CBP) (514).

The nuclear androgen receptor (AR) has an important role in modulation of gonadotropin subunit gene expression, mediating feedback inhibition by testosterone (475,478). Previous studies have shown that AR represses human αGSU gene promoter activity by either directly or indirectly binding to the CRE and αBE region in a ligand-dependent manner, suggesting that activated AR may interfere with the binding of cognate factors, thereby leading to an attenuation in transcription of human αGSU gene (583,584). In contrast, estrogen and ER do not have this effect. Subsequently, c-Jun and ATF-2 were found to interact with AR and functionally block the AR-mediated suppression of human αGSU gene promoter activity, suggesting that AR-mediated suppression of human αGSU gene expression occurs through protein–protein interaction with c-Jun and ATF-2 (572). Thus, as a common feature for both the mouse and human αGSU genes, it has been suggested that, in the presence of testosterone, αGSU gene expression is under repression because cognate factors are prevented from binding to CREs, the PGBE, or αBEs by activated AR, thereby interfering with functional and combinatorial interactions through a common coactivator or adapter complex. Moreover, GnRH stimulation, or the absence of testosterone, releases the AR bound to the c-Jun/ATF-2 heterodimer to allow interactions between these protein complexes and result in synergistic activation of αGSU gene transcription (572).

Dominant negative mutant forms of Ras, ERK1, and ERK2 have been shown to reduce basal expression of the human αGSU gene but have no effect on GnRH-stimulated expression (268), suggesting an involvement of the MAPK cascade in the regulation of basal expression but not in GnRH-induced expression. However, in another study, inhibition of MAPK activity or overexpression of MKP-2 led to the attenuation of GnRH-induced activation of the mouse αGSU gene promoter (285). In addition, a constitutively active mutant of Raf-1 has been shown to increase MAPKK and MAPK activities as well as αGSU gene promoter activity in αT3-1 cells, acting through the GnRH-RE and PGBE (285). ERK has been shown to activate the human αGSU gene promoter through the PGBE and αBEs, and c-Src increases αGSU gene expression through the GSE (292). ERK activation of human αGSU may also be mediated by GATA, the αACT-binding factor, in LβT2 cells (511). Moreover, activation of PKC stimulates human αGSU gene promoter activity, and in contrast to the rat αGSU (581), Ca^{2+} channel blockade inhibits GnRH-induced αGSU gene promoter activity in GGH_3-1′ cells (585), a GH_3 cell line stably transfected with the rat GnRHR (586). The GnRH-RE is also responsive to PMA, further supporting the role of the PKC pathway in mediating the effects of GnRH on expression of the αGSU gene. Thus, these studies strongly support an involvement of the MAPK cascade through a Ca^{2+}-sensitive PKC pathway in GnRH-mediated stimulation of αGSU gene expression, through modulation of activities of combinatorial enhancer elements and their cognate factors (Fig. 5).

PACAP has been implicated in the stimulation of αGSU and LHβ subunit gene expression. PACAP is mainly produced from the hypothalamus, and interacts with its seven transmembrane GPCR and signals as a potent stimulator of αGSU gene transcription through both PKC and PKA pathways in the rat pituitary and in αT3-1 cells (336,587–589). The effect of PACAP on human αGSU gene promoter activity is potent and sustained, with a peak within 8 hours and a slow reduction by 24 hours post-treatment, and is mediated through the GSE and CRE, possibly by activation of both MEK/MAPK and cAMP/CREB cascades (589,590).

Luteinizing Hormone β-Subunit Gene Expression

The LHβ and CGβ peptides are similar in structure and amino acid sequence and are believed to be derived from a single ancestral gene (505,591). The LHβ gene is expressed in pituitary gonadotropes and has been characterized in multiple species, including mouse, rat, cow, and human (591–595). The LHβ gene contains three exons and two introns, is considerably smaller than the αGSU gene (approximately 1.5 kb), and encodes an mRNA that is approximately 0.7 kb in length. In the human, the mRNA encodes a polypeptide that contains a 22–amino acid signal peptide and a 121–amino acid protein.

CGβ is selectively produced in primates and equine species in the syncytiotrophoblast cells of the placenta, and its mRNA is approximately 1 kb in length (593). Unlike the other gonadotropin subunits, CGβ is not encoded by a single gene. In the human, a cluster of seven CGβ-like genes or pseudogenes is located on chromosome 19 (596,597), and the LHβ gene is also located in this region. Evidence suggests that only one or two of the seven human CG (hCG) β–like genes are expressed (593,596). The hCGβ peptide contains a 24–amino acid C-terminus extension not seen in LHβ. Nucleotide sequence homology upstream from the transcriptional start site is greater than 90%, but the 5′-untranslated region of the CGβ mRNA is considerably longer (366 bases) than that of LHβ (9 bases). This difference appears to be due to the use of alternate transcriptional start sites, with CGβ using a promoter approximately 350 bp upstream from the TATA box (598). Despite the homology between the two genes, LHβ is not expressed in placenta-derived cell lines (2), suggesting that the LHβ gene either contains a repressor element(s) that requires gonadotrope-specific transcription factors to override basal suppression or lacks a placental tissue–specific enhancer element(s).

Almost all the regulatory *cis*-elements identified from each species to date as important for LHβ gene expression are located in the 5′-flanking region, within 500 bp of the transcriptional start site. In this region, the proximal 140 bp are conserved across species (599), implying their functional significance. The rat LHβ gene contains two specificity protein-1 (Sp1) binding sites located at positions –450 to –434 and –366 to –354, a $CC(A/T)_6GG$ factor (CArG) binding site at –443 to –434, two GSEs at –127 to –119 and –58 to –50, two early growth response protein-1 (Egr-1) binding sites at –112 to –104 and –49 to –41, a Pitx-1 binding site at –99 to –94, and a TATA box at –30 to –26 (514). Among them, five elements, namely the Pitx-1 binding site, two GSEs, and two Egr-1 binding sites, where transcription factors Pitx-1 (600), SF-1 (601,602), and Egr-1 (602,603) bind, respectively, form a 140-bp proximal promoter core cassette. Two Sp1 and the CArG binding sites, where the GC box–binding protein Sp1, a three–zinc finger transcription factor (604,605), and a serum response factor–related protein CArG (606), bind, respectively, form a distal enhancer (Fig. 16).

The bovine LHβ gene contains a remarkably and strikingly similar set of proximal *cis*-elements to the rat LHβ gene, including two GSEs at –128 to –120 and –59 to –51, two Egr-1 binding sites at –111 to –104 and –50 to –42, a Pitx-1 binding site at position –100 to –95, and a TATA box at –31 to –27 (607). In addition, instead of the distal elements present in the rat LHβ gene, the bovine LHβ gene contains a bipartite nuclear transcription factor-Y (NF-Y) binding site located at positions –400 to –391 and –337 to –328, in the region corresponding to the Sp1 binding sites in the rat LHβ gene. NF-Y (also known as CBF/CP1), a ubiquitous heterotrimeric CCAAT box (Y box)–binding factor, binds to these elements (608). All three NF-Y subunits (NF-YA, B, and C) are required for DNA binding, and the NF-YA subunit is responsible for the sequence-specific contacts (609). The distal elements seem to be necessary for both gonadotrope-specific and GnRH-induced expression of the LHβ gene in both rat and cow, and both the distal and proximal sets of the elements are required for the full GnRH response (514,610). Proximal 5′-flanking regions of mouse (611), equine (612,613), and human (611) LHβ genes also contain a strikingly similar set of proximal elements to the rat and bovine LHβ genes, namely, two GSEs and two Egr-1 binding sites with an intervening Pitx-1 binding site (611).

The zinc finger protein Egr-1, also known as nuclear growth factor I-A (NGFI-A), Krox-24, or Zif268, is an immediate–early serum response gene *egr-1* product expressed in various tissues in response to a range of physiological states. Egr-1 mRNA and protein levels are stimulated by GnRH in αT3-1 and LβT2 as well as GGH_3-1′ cells, and Egr-1 has been shown to be critical in mediating GnRH-induced LHβ gene expression by binding to its cognate binding sites (291,309,599,605,612,614,615). Disruption of *egr-1* in mice causes defects in reproductive function as well as growth (603,616). In the pituitary, gonadotropes are normal in numbers but specifically fail to express the LHβ subunit, whereas FSHβ production is evident. These findings suggest that Egr-1 influences reproductive capacity through its regulation of LHβ gene transcription. Disruption of gonadotropin expression in SF-1–deficient mice can be rescued by exogenous GnRH treatment (529,531), suggesting that SF-1 is critical for gonadotrope-specific or basal, but not for GnRH-induced, expression of the LHβ gene. SF-1 has also been shown to augment basal but not GnRH induced rat LHβ gene promoter activity in GGH_3-1′ cells (610). Pitx-1–deficient mice also exhibit reduced expression of gonadotrope-specific genes, including the LHβ gene (557).

Mutation of GSE (601,617), Egr-1 (614), Pitx-1 (607), or NF-Y (608) binding sites reduces LHβ gene promoter activity as well as the binding activities of the cognate factors. The zinc finger domains of Egr-1 are necessary for protein–protein interaction with SF-1 in the regulation of LHβ gene expression, and Egr-1 binding sites and GSEs in the LHβ gene are required for synergistic activation of promoter activity, basally and on GnRH stimulation (599,602, 605,612–614,616). Pitx-1 also synergizes with SF-1 in the stimulation of LHβ gene promoter activity by causing a conformational change of SF-1 through protein–protein interaction. Both Egr-1 and SF-1 interact with Pitx-1, forming a tripartite protein complex to activate the LHβ gene promoter synergistically (599,600). This functional synergism suggests the importance of interactions among these factors, directly and cooperatively, for both gonadotrope-specific and GnRH-induced expression of the LHβ gene. The synergistic effect of Pitx-1 does not require the Pitx-1 binding site because mutation of the Pitx-1 binding site does not prevent Pitx-1 from synergistically increasing LHβ gene promoter activity with SF-1 (600). However, it remains possible that an additional low-affinity Pitx-1 binding site is present (563), much as low-affinity Pitx-1 binding occurs in the regulation of mouse GnRHR gene promoter activity (618).

In addition to Pitx-1, an LβT2 cell–specific HD protein has been suggested to interact with the Pitx-1 binding site of the rat LHβ gene promoter (619). This transcription factor appears to be a *bicoid*-related, orthodenticle (Otx)-like protein that belongs to the same HD subfamily as Pitx-1, and is likely not present in αT3-1 cells. This factor may mediate rat LHβ gene promoter activity in LβT2 cells that is not observed in αT3-1 cells. The identity of this protein remains largely unknown. In contrast, the homeobox repressor Hesx1 inhibits LHβ gene promoter activity by a mechanism different from its effects on αGSU regulation (563). It antagonizes Pitx-1 activation of LHβ gene promoter in a dose-dependent manner, which requires both the Pitx-1 binding site (core sequence TAAT) and a potential proximal Pitx-1 binding site (core sequence TAAC) juxtaposed to the Pitx-1 binding site. Rather than interacting directly with the Pitx-1 binding site, Hesx1 binds to the proximal site as a homodimer and interferes with the interaction of Pitx-1 with its binding site. The sequences of these two sites are well conserved across species, suggesting their importance.

The upstream Sp1 binding sites are also required for GnRH responsiveness. Mutations of these elements alone, preventing Sp1 binding, can reduce GnRH-induced or SF-1/Egr-1–mediated rat LHβ gene promoter activity, and combined mutations of the Sp-1, GSE, and Egr-1 binding sites further eliminate GnRH responsiveness (604,605). Thus, it is likely that both gonadotrope-specific and GnRH-induced expression of the LHβ gene are mediated by the same set of distal and proximal elements and their cognate factors (605,610).

An Egr-1 (NGFI-A)–binding protein 1 (NAB1) is stimulated by GnRH in αT3-1 cells (612). Interestingly, this protein has a species-specific function in regulation of LHβ gene promoter activity. NAB1 suppresses Egr-1 activation of the nonrodent LHβ gene expression by interacting with Egr-1 through a conserved domain of Egr-1 (620), whereas it enhances Egr-1 stimulation of rodent LHβ gene expression (621). It has been suggested that in the nonrodent LHβ gene promoter, including cow, horse, and human, the distal Egr-1 binding site binds specifically to Egr-1 only, whereas in the rodent LHβ gene promoter, the same element is open for competition among several zinc finger proteins, including Egr-1, Sp1, and Sp3, and the proximal element is capable of binding to all of these zinc finger proteins in all the species tested (611). Thus, specific Egr-1 binding to the distal element in the nonrodent LHβ gene promoter may recruit NAB1 to suppress promoter activity (611).

DAX-1, a product of a dosage-sensitive, sex-reversal, adrenal hypoplasia congenita critical region on the chromosome X gene 1 (*DAX-1* or *Nr0b1*), is an orphan nuclear receptor, expressed in the same distribution as SF-1, including pituitary gonadotropes, adrenals, and gonads (622,623). DAX-1 has been implicated in the pathogenesis of adrenal hypoplasia congenita and hypogonadotropic hypogonadism (624–626). DAX-1 reduces synergy between Egr-1 and SF-1, and diminishes GnRH stimulation of the LHβ gene promoter as an SF-1 repressor (614,627,628).

A *c-myc* promoter–binding protein-1 (MBP-1)–interacting protein-2A (MIP-2A) is a small cellular protein, initially suggested to associate with the transcription factor MBP-1 to relieve and antagonize its activity (629). MIP-2A is expressed in the pituitary and in GH_3 and LβT2 cells, and inhibits basal human LHβ gene promoter activity in a dose-dependent manner. SF-1–or Pitx-1–mediated transactivation of LHβ is disrupted through physical interaction of MIP-2A with SF-1 and Pitx-1 (630). Thus, MIP-2A may also play a role as a corepressor in the regulation of LHβ gene expression.

The rat LHβ gene promoter region harbors an ERE between –1,173 and –1,159 that binds to ER and confers a direct stimulatory response to E_2 (484,631). This ERE contains a 15-base imperfect palindromic sequence, and ER binds to this region with high affinity. The rat LHβ gene promoter also confers an inhibitory response to testosterone between sequences –617 and +44 (632). This inhibitory response to testosterone suppresses GnRH stimulation of LHβ gene promoter activity, and this AR-mediated suppression requires the distal Sp1 binding sites, suggesting a direct interaction between AR and Sp1 in suppression of rat LHβ gene transcription. In addition, AR is also able to interact with Egr-1, perhaps indicative of a contribution by the proximal elements in AR-mediated suppression. Similarly, the bovine LHβ gene promoter confers inhibitory E_2 and testosterone responsiveness between –779 and +10, but it lacks an ERE or AR element (ARE) with high-affinity binding (633), and ligand-bound AR has been shown to suppress bovine LHβ gene promoter activity by interacting with SF-1 alone (634). However, in the presence of both Egr-1 and Pitx-1, testosterone-mediated suppression of the bovine LHβ gene promoter activity disappears.

Like the regulation of αGSU gene transcription, a general model has been proposed for the regulation of LHβ gene transcription by GnRH and testosterone. In the presence of testosterone, LHβ gene expression is under repression by the binding of activated AR to SF-1 or Sp1, preventing the cognate binding factors of the distal and proximal elements from functional and combinatorial interaction through a common coactivator or adapter complex. Upon stimulation by GnRH, increased concentrations of Egr-1 form a ternary protein complex with SF-1 and Pitx-1, causing the release of AR. Distal elements and their cognate factors (Sp1 and CArG in the rat and NF-Y in the cow) are now able to integrate with the tripartite proximal elements and their cognate factors (SF-1, Egr-1, and Pitx-1) to allow interactions between the protein complexes and synergistically activate LHβ gene transcription by sharing a common coactivator or adapter complex (605,607,634). Supporting this hypothesis, a coactivator, small nuclear RING finger protein (SNURF), has been suggested as a specific coactivator for rat LHβ gene expression in LβT2 cells (635). SNURF increases both basal and GnRH-induced LHβ gene promoter activity and interacts with Sp1 and SF-1, but not with Egr-1, and these interactions are required for stimulation of LHβ gene promoter activity. SNURF also interacts with AR, and overexpression of SNURF releases androgen suppression of GnRH-induced LHβ, but not αGSU, gene promoter activity.

A PKC-mediated stimulatory effect of GnRH on LHβ gene transcription has been well characterized. ERK1 and 2, JNK, and p38 all have shown to be involved in mediating GnRH-induced LHβ gene expression, downstream of PKC, suggesting an involvement of a multitude of diverse and complex signaling pathways that may eventually participate in a complicated cross-talk network to maximize signaling amplification. Stimulation of L-type VSCCs enhances basal LHβ gene expression, whereas blockade of these Ca^{2+} channels inhibits GnRH-induced gene expression in primary rat pituitary cells and αT3-1 cells (581), suggesting a role for Ca^{2+} influx in mediating GnRH-induced LHβ gene transcription. GnRH treatment or PKC activation can stimulate a rise in LHβ mRNA levels and gene promoter activity. Furthermore, cellular depletion of PKC by prolonged exposure to phorbol esters or inhibition of PKC blocks the effect of GnRH in rat pituitary cells and in GGH_3-1′ cells (201,333,585). GnRH and PMA also synergize with Egr-1 or SF-1 in the induction of LHβ gene expression (599,615), further suggesting an involvement of the PKC pathway in signaling GnRH stimulation of LHβ gene expression (Fig. 5).

On the other hand, inhibition of MEK does not have any effect on LHβ gene promoter activity, demonstrating at least partial independence of LHβ gene expression from the ERK cascade (293,581). Blockade of Ca^{2+} channels or extracellular Ca^{2+} chelation does not completely prevent GnRH stimulation of rat LHβ gene promoter activity (290,293), possibly because of the involvement of alternative signaling pathways, such as JNK. A dominant negative form of JNK or c-Jun significantly inhibits the induction of rat LHβ gene promoter activity by GnRH in LβT2 cells. This suggests that GnRH differentially activates ERK and JNK, and that a JNK cascade is also necessary to elicit rat LHβ gene expression in response to GnRH in a c-Jun–dependent mechanism (293). In addition, GnRH also activates p38 activity, and p38 induces LHβ gene expression through a Ca^{2+}- and MEK-dependent mechanism (286,287). Finally, it has been suggested that GnRH may use PKC-stimulated pathways for acute induction, whereas Ca^{2+} signaling is responsible for long-term repression of LHβ gene expression by prolonged, continuous GnRH. Such acute induction and long-term repression may use different signaling systems, possibly different PKC isoenzymes, as well as different target sequences for regulating the LHβ gene (636). Also, activation of differential signal transduction pathways for αGSU and LHβ gene expression may in part be a mechanism for preferential stimulation of transcription for each gonadotropin subunit gene (585).

Ca^{2+}/calmodulin–dependent protein kinase type II (Ca^{2+}/CaMKII) has been suggested to mediate LHβ gene induction by GnRH in LβT2 cells and in rat pituitary cells (637,638). After GnRH-induced phosphorylation and nuclear translocation, Ca^{2+}/CaMKII may mediate Ca^{2+} signaling through phosphorylation of transcription factors involved in LHβ gene expression, and perhaps may influence the expression of other gonadotropin subunits and gonadotropin secretion as well (Fig. 5).

A cAMP-responsive PKA pathway is also involved in GnRH-mediated stimulation of rat LHβ gene expression (639). Activation of adenylate cyclase by forskolin results in the elevation of LHβ mRNA synthesis, but not mRNA stability, to the same level as GnRH- or PKC-induced synthesis in rat pituitary cells (201), as well as stimulation of rat LHβ gene promoter activity in LβT2 cells (310). These findings strongly suggest an involvement of the PKA pathway in GnRH-mediated LHβ gene transcription. Forskolin stimulates *egr-1* expression and Egr-1 levels, as well as Sp1, which stimulate LHβ gene expression together with SF-1. PACAP has been shown to stimulate LHβ gene promoter activity to a similar extent as forskolin in LβT2 cells (310).

Follicle-Stimulating Hormone β-Subunit Gene Expression

The FSHβ subunit is encoded by a single gene in mouse, rat, cow, and human that contains three exons and two introns (487,640–642). FSHβ mRNA is approximately 1.7 kb in length, and is much larger than the other gonadotropin subunit mRNAs because of a large 3′-untranslated region of more than 1 kb (642). The significance of this 3′-region is not clear, but elements located in this region may play a role in determining FSHβ mRNA stability. The FSHβ mRNA nucleotide and polypeptide amino acid sequences are highly conserved between species (approximately 80%). In rat and cow, only one mRNA species has been demonstrated, but the human FSHβ gene produces four mRNA size variations. The different mRNA sizes appear to be due to the use of two different transcriptional start sites and two different polyadenylation sites, but it is unknown if all four mRNA transcripts are translated or hormonally regulated (641).

GnRH stimulates FSHβ expression in gonadotropes (2,443). Earlier in vivo studies showed that GnRH antagonist treatment in gonadectomized rats results in a dramatic decrease in FSHβ mRNA and FSH secretion levels (643), and pulsatile GnRH increases the transcriptional rate of the FSHβ gene (362), suggesting that the effects of GnRH on FSHβ synthesis are, at least in part, at the transcriptional level. Along with GnRH, activin from paracrine/autocrine sources is considered to be a major regulator of FSHβ gene expression. Activin acts independently of GnRH and can induce FSH secretion in GnRH-desensitized pituitary cells (434). Actinomycin-D treatment prevents an activin-stimulated increase in FSHβ primary transcript as well as mRNA levels, suggesting that the action of activin occurs at the transcriptional level (392,644). Activin also appears to regulate FSHβ gene expression through nontranscriptional mechanisms by stabilizing the mRNA (645,646). Activin also enhances GnRH responsiveness in gonadotrope cells and synergizes with GnRH in the stimulation of FSHβ gene transcriptional activity (435,436).

Initial studies using transgenic mouse models have shown that 10 kb of the human FSHβ gene containing 4 kb of the 5′-flanking region (647) or 5.5 kb of the 5′-flanking region of the ovine FSHβ gene (648) is sufficient to direct gonadotrope-specific expression. Nevertheless, relatively little is known about regulation of the FSHβ gene compared with αGSU and LHβ, primarily because of the lack of an FSHβ-expressing gonadotrope cell model until the late 1990s. Unlike αT3-1 cells, LβT2 cells have been found to express the FSHβ gene in response to treatment with physiological levels of activin A (649). These findings have suggested that LβT2 cells may serve as an appropriate in vitro cell model and have encouraged initiation of a large number of studies for regulatory mechanisms of FSHβ gene expression.

The 5′-flanking region of the ovine FSHβ gene contains several *cis*-elements identified by sequence homology, including four SMAD-binding elements (SBEs) located at positions –973 to –959 (repeated half-sites, AGAC, separated by a 7-bp gap), –453 to –439 (inverted palindromic consensus, GTCT and AGAC, each half-site motif separated by a 7-bp gap), –163 to –160 (half-site, AGAC), and –136 to –125 (imperfect inverted palindrome, GTCTATCC) (650).

A potential tripartite Lhx3 binding site at –297 to –287, –262 to –256, and –222 to –214 (542), two putative AREs at –245 to –231 and –153 to –139 (651), and two potential AP-1 binding sites at –111 to –105 and –74 to –68 (652) have also been identified. The most proximal SBE overlaps with two imperfect Pbx1/Prep1 consensus (CTGTCA) elements at position –136 to –125, where a three-amino acid loop extension (TALE) HD protein Pbx1 and its cofactor Prep1 as well as SMAD4 have been shown to bind (650). A tripartite progesterone response element (PRE) has been initially identified at positions –245 to –231, –212 to –198, and –153 to –139 (498), two of which are also AREs. These three putative PREs have been shown to mediate the stimulatory effect of progesterone on ovine FSHβ gene promoter activity by interacting with ligand-bound progesterone receptor (PR). In fact, all three of these elements contain a half-site with homology to the known consensus core motif (TGTTCT) for a nuclear hormone receptor subfamily, including ARE, PRE, and glucocorticoid response element (GRE) (653). The C/G bases known to be critical for the function of the consensus ARE/PRE/ GRE are perfectly conserved in all of these potential elements among several mammalian species. Thus, it is possible that all ligand-bound ARs, PRs, and glucocorticoid receptors (GRs) can interact with some or all of these elements. Initial studies, however, have shown that both E_2 and progesterone can suppress the rate of the ovine FSHβ gene transcription and promoter activity (485,486). Sequences between –105 and –84 have been shown to mediate negative E_2 responsiveness, which suppresses FSHβ gene promoter activity, but ER fails to interact with this region, suggesting an indirect action of ER, perhaps through protein–protein interaction with other factors that interact with this region (486). Two far-upstream enhancer elements are also present, between –4,152 and –2,878 and between –2,550 and –1,089, each of which partially mediates GnRH responsiveness (270).

A major difference between the 5′-flanking regions of the ovine and rat FSHβ genes is that there is no apparent AP-1 binding site in rat FSHβ gene promoter. The rat FSHβ gene promoter *cis*-elements identified to date include a potential tripartite Lhx3 binding site located at positions –287 to –277, –244 to –238, and –205 to –197 (542), three SBEs at –266 to –259 (inverted palindromic consensus, GTCTAGAC) (436,650,654), –150 to –147 (half-site, AGAC), and –123 to –112 (repeated half-sites, GTCTGTCT, also two possible Pbx1/Prep1 binding sites) (650), a Pitx-2 binding site at –230 to –210 (654), two potential AREs at –228 to –214 and –140 to –126 (651), and two Pitx-1 binding sites at –134 to –129 and –54 to –48 (655) (Fig. 16).

The *cis*-elements in the mouse FSHβ gene promoter region are similar to the rat elements. These include bipartite GSEs located at positions –341 to –333 and –239 to –231 (656), three SBEs at –270 to –263 (inverted palindromic consensus, GTCTAGAC), –151 to –148 (half-site, AGAC), and –124 to –113 (imperfect repeated half-sites, GTCCGTCT, also two possible Pbx1/Prep1 binding sites) (650), a potential bipartite ARE at –230 to –216 and –141 to –127 (651), an imperfect NF-Y binding site at –76 to –70 (656), an AP-1 half-site at –72 to –69 (302), and a Pitx-1 binding site at –54 (656). The proximal GSE is conserved in sheep, cow, mouse, rat, and human; the distal GSE is conserved in mouse and rat; the distal SBE is conserved in mouse and rat; two proximal SBEs and bipartite AREs are conserved in sheep, cow, pig, mouse, rat, and human; and the NF-Y binding site/ AP-1 half-site is conserved in mouse, rat, and human.

A report (2004) has identified six potential Lhx3 binding sites in the porcine FSHβ gene promoter located at positions –5,057 to –5,030, –1,442 to –1,423, –838 to –809, –295 to –284, –259 to –253, and –219 to –209, and the three proximal elements are well conserved in sheep, pig, rat, and human (542). All six elements bind to Lhx3, apparently through the HD, and several Lhx3 isoforms or Lhx4 functionally activate basal porcine and human FSHβ gene promoter activity in both heterologous and homologous (LβT2) cell systems. Given that Pitx-1 can interact with Lhx3 (553) and that the rat FSHβ gene promoter contains both Pitx-1 and Lhx3 binding sites, a logical analogy is that these factors may interact with each other to synergize in the tissue-specific expression of the FSHβ gene. Whether functional Lhx3 elements are also present in the mouse FSHβ gene promoter has yet to be determined. Interestingly, Egr-1 overexpression represses the porcine FSHβ gene promoter activity in these studies (542), opposite to its effects on the LHβ gene.

SF-1 and NF-Y are important basal regulators in many systems. Although SF-1 is able to bind to both GSE sequences, a stronger affinity of SF-1 for the proximal GSE is observed in both LβT2 and αT3-1 cells, consistent with this sequence being a more highly conserved GSE among the species. Despite the observation that SF-1 alone has only a minimal effect on mouse FSHβ gene promoter activity, NF-YA and SF-1 functionally interact to regulate the mouse FSHβ gene expression in a cell type–specific manner, which requires intact GSE and NF-Y binding sites (656). SF-1 has been shown to interact physically with transcription factor IIB (TFIIB), c-Jun (657), Pitx-1 (599), AR (634), and a number of coactivators, including p300/CBP/cointegrator–associated protein (p/CIP), transcription intermediary factor-2 (658), and general control nonderepressed 5 (GCN5), which has histone acetyltransferase activity (659). NF-Y is also able to interact with other transcription factors and coactivators, including c-Jun (302), ERα (660), p300/CBP (661), p300/CBP-associated factor (P/CAF), GCN5 (662), and transcription factor IID (TFIID) complex, which contains TATA-binding protein (TBP) and TBP-associated factors (609). Thus, it is possible that SF-1 and NF-Y recruit these factors to maintain and regulate basal and gonadotrope-specific expression of the mouse FSHβ gene.

Pitx-1 has been shown to activate the rat FSHβ gene promoter both basally and in synergy with GnRH in the GGH_3-1′ cell line (655). Although Pitx-1 binds to both distal and proximal Pitx-1 binding sites, the binding affinity with the distal site is relatively weak and the function of this distal site is unclear. A mutation of the proximal Pitx-1 binding site markedly reduces Pitx-1 responsiveness in GGH_3-1′ cells. However, there is residual activation of the promoter by Pitx-1 independent of the Pitx-1 binding sites, suggesting possible Pitx-1 activity independent of direct Pitx-1 binding. Mutation of the Pitx-1 binding sites does not inhibit GnRH responsiveness in LβT2 cells.

The AP-1 half-site (GTCA) present in the mouse FSHβ gene promoter is located in juxtaposition to an NF-Y binding site (CCAAT box), and both overlapping sites are required for AP-1 binding (302). Several members of the AP-1 family (c-Jun, JunB, c-Fos, and FosB) can interact with this AP-1 half-site. Mutation of this AP-1 half-site reduces GnRH responsiveness and eliminates induction of transcriptional activity by c-Jun/c-Fos in LβT2 cells, consistent with an important role for this site in the GnRH induction of mouse FSHβ gene promoter activity.

In contrast to the αGSU and LHβ genes, androgen has a stimulatory effect on ovine FSHβ gene promoter activity in a dose-dependent manner in LβT2 cells (651). Only the distal ARE contains a perfect consensus half-site (TGTTCT) and thus is able to interact with ligand-bound AR. Nonetheless, mutation of either of two AREs eliminates the androgen-mediated promoter activity. The mechanism of androgen induction of ovine FSHβ gene expression through the proximal ARE remains unknown. Androgen stimulation of ovine FSHβ gene expression is synergistic with activin, and mutation of the distal ARE abolishes this synergism, whereas mutation of the proximal SBE abolishes androgen's effect completely, suggesting a dependence of androgen induction on the proximal SBE. Androgen also potentiates GnRH stimulation of ovine FSHβ gene promoter activity in a dose-dependent manner that does not require the distal ARE, suggesting that the enhancement of GnRH stimulation is independent of DNA binding of ligand-bound AR.

GnRH stimulation of FSHβ gene expression involves activation of the MAPK cascade through the PKC pathway when tested in rat pituitary cells (360). AP-1 has been shown to be involved in mediating GnRH stimulation of ovine FSHβ gene expression through the PKC/MAPK signaling cascade in a heterologous cell system (663). Activation of MAPK has also been shown to be necessary for induction of the mouse FSHβ gene promoter as well as of *junB* in LβT2 cells (302). Although both GnRH and TPA are able to induce ovine FSHβ gene promoter activity, inhibition of MAPK blocks only induction by GnRH but not by TPA in LβT2 cells (270). Moreover, GnRH and TPA activate overlapping but distinct pools of PKC isoforms, suggesting that GnRH and TPA stimulation of ovine FSHβ gene expression may be mediated through different cascades. Ca^{2+} influx is also necessary for both GnRH and TPA induction of ovine FSHβ gene promoter activity. Moreover, Ca^{2+}/CaMKII has been suggested to mediate induction of the FSHβ gene by GnRH in rat pituitary cells (638), similar to LHβ gene regulation (Fig. 5).

However, in LβT2 cells, mutation of this bipartite AP-1 site reduces ovine FSHβ gene promoter activity only slightly (270). In addition, in pituitary cells from transgenic mice harboring a reporter gene regulated by the ovine FSHβ gene promoter (648), activin has been shown to induce transgene activity regardless of whether the AP-1 sites are intact. GnRH alone has no effect, but a concomitant treatment with GnRH and activin further increases promoter activity, largely attenuated by mutation of the AP-1 sites (435). These findings suggest that activin is a major regulator of ovine FSHβ gene promoter activity, and that AP-1 is likely to mediate the GnRH response only when activin is also present in homologous cell systems. Consistent with these observations, inhibition of MAPK activity does not reduce activin- or concomitant activin/GnRH–mediated stimulation of rat FSHβ gene transcriptional activity, yet it eliminates the GnRH response, and interference with activin signaling decreases synergy between activin and GnRH in LβT2 cells (436).

Intracellular signaling molecules SMAD2, 3, and 4 have all been shown to mediate activin-stimulated activity of the rat FSHβ promoter as well as FSHβ mRNA synthesis in LβT2 cells (392,431,436, 650, 654), although the role of SMAD2 is controversial. LβT2 cells express SMAD2, 3, 4, and the inhibitory SMAD7, which prevents SMAD2 and 3 phosphorylation (431). Both SMAD3 and 4 bind to a palindromic SMAD binding sequence (GTCTAGAC), but there is no evidence for direct binding of SMAD2 to this sequence (664). A unique domain in SMAD2 interferes with DNA binding (665); however, SMAD2 is functionally active in many systems and recruits transcriptional cofactors to stimulate transcriptional activity and to stabilize protein–DNA interactions, without direct

DNA binding (666). Upon activin stimulation, activin-specific SMAD2 or 3 is rapidly phosphorylated within 10 minutes by the serine/threonine kinase activin receptors, which allows their association with a common-mediator SMAD4, and subsequent translocation of the SMAD complex into the nucleus (392,431) (Fig. 5). Interestingly, nuclear translocation of SMADs can also be detected subsequent to GnRH treatment in αT3-1 and LβT2 cells (440), suggesting cross-talk between GnRH and activin signaling pathways and the ability of GnRH to stimulate activin signaling pathways in gonadotropes. Once in the nucleus, the SMAD complex interacts with the SBE to transactivate the FSHβ gene. SMAD3 and 4 bind to the distal SBE (–266/–259) of the rat FSHβ gene promoter, and deletion of this SBE abolishes activin- or concomitant activin/GnRH–mediated FSHβ gene transcriptional activity in LβT2 cells (436,654). Depletion of endogenous SMAD2 or 3 attenuates activin-mediated FSHβ gene transcription, and overexpression of the inhibitory SMAD7 blocks activin-stimulated FSHβ gene expression in LβT2 cells (392). Overexpression of SMAD3 and 4 but not SMAD2 further enhances rat FSHβ gene transcriptional activity (436).

There are four SBEs in the ovine FSHβ gene promoter required for full activin stimulation, the two proximal sites being well conserved across species (650). Mutation of the most distal SBE reduces activin responsiveness of ovine FSHβ gene expression only marginally, whereas mutation of either of two proximal SBEs in the ovine or mouse FSHβ gene promoter completely abolishes activin stimulation of gene promoter activity in LβT2 cells. Thus, the two proximal SBEs are likely to confer activin responsiveness of FSHβ gene expression universally, as two critical elements across species. Two distal ovine SBEs and the most proximal SBE bind to SMAD4, but the identity of protein complex shown to bind to a third SBE (–163/–160) is not known (650). Pitx-2 has been shown to upregulate basal and activin-mediated rat FSHβ gene promoter activity in a dose-dependent manner, and elimination of the Pitx-2 binding site results in a loss of activin-regulated FSHβ gene promoter activity (654).

Activin-regulated FSHβ gene expression is highly specific and selective compared with other gonadotropin genes. This signal specificity and selectivity is likely to be due to the interaction of SMADs with other tissue- and cell-specific partners or coregulators (667). For example, the activin-specific SMADs can interact with DNA-binding adaptors such as the forkhead/winged-helix family members, forkhead activin signal transducer (FAST) molecules (also known as FoxH), to bind to DNA (668,669). This complex can also recruit coactivators such as p300/CBP (670,671) or corepressors such as TGF-β–induced factor (TGIF) (672), E2F4/5, p107 (673), and a splice variant of CAATT-binding factor subunit C (CBF-C), CBF-Cb (674), thereby selectively regulating target genes in a cell type–specific manner. It has been shown that expression of an immediate–early gene, *junB*, is regulated by SMAD signaling through an SBE in the *junB* promoter (675). SMAD3 and 4 have also been shown to interact or associate with a variety of Jun/Fos proteins, and promoter activities for genes containing AP-1 sites, such as c-*jun*, are strongly enhanced by activin (676–678), suggesting synergistic associations between SMADs and AP-1 proteins, especially on concomitant activin and GnRH stimulation. In addition, the most proximal ovine SBE binds SMAD4 in association with Pbx1 and Prep1 through protein–protein interactions between them and SMAD2/3, through direct DNA binding, or both (650). Pbx1/ Prep1 increase the DNA-binding affinity of the homeotic selector protein, Hox, and modulate its target site selection (679), thus raising the possibility that constitutive binding of Pbx1/Prep1 in the absence of activin stimulation may enable recruitment of SMADs or stabilize the weak binding of SMADs upon activin stimulation.

Alternatively, activin can act through SMAD-independent pathways, or SMADs can be activated by pathways other than through activin receptors. Activin has been shown to activate JNK (680) and p38 (681), and activated MEK/ERK and p38 cascades have been shown to induce SMAD3 phosphorylation through different phosphorylation sites, which can also cause nuclear translocation of SMAD3/4 complex (682,683). Given the fact that GnRH also activates MAPK pathways in gonadotropes, it is possible that FSHβ gene expression is activated through this cross-talk mechanism between activin and GnRH signaling pathways. Taken together, interactions between basal transcription factors, SF-1, NF-Y, Pitx-1, Pbx1/Prep1, and AR may recruit a corepressor or coactivator, depending on the ligand occupancy of AR, to limit and maintain basal expression of the FSHβ gene when there is no GnRH or activin stimulation. Upon GnRH or activin stimulation, however, activated AP-1 and SMADs are integrated into this network of protein complexes to stimulate FSHβ gene expression synergistically, possibly by sharing a common coactivator or adaptor complex (436).

GnRH Receptor Gene Expression

In the mouse pituitary, there are two major type I GnRHR transcripts, with the predominant transcript approximately 4.0 to 4.5 kb in size and a minor transcript of about 1.6 to 1.8 kb; in the rat pituitary, there is one additional transcript of approximately 5.0 to 5.5 kb (28,29,34,72,298). In the human pituitary, there is a predominant transcript of approximately 4.7 to 5.0 kb in size and two minor transcripts of 2.5 kb and 1.5 kb (31,140). In the ovine pituitary, there are at least four transcripts for GnRHR, 5.0 to 6.0 kb, 3.6 to 4.0 kb, 2.3 kb, and 1.3 kb in size, although the 2.3-kb transcript does not include the entire coding sequence (36,37,45). Alternative splicing products encoding C-terminal truncated proteins have been reported in the mouse (684). The implications and abundance of these transcripts, however, are not clear.

GnRHR gene expression is stimulated by several hormonal inputs, most notably by its own ligand, GnRH, in gonadotropes (69,298,685–687). In addition, activin A increases GnRHR gene expression or synthesis in the rat pituitary (441), in αT3-1 cells (429,439), and in LβT2 cells (430). Steroid hormones, estrogen (64,71,327,482,483), progesterone (499), androgen (651), and glucocorticoids (688), have also been shown to regulate GnRHR gene expression.

Since the 5′-flanking region of the mouse GnRHR gene promoter was cloned (689,690), significant work has been performed to elucidate the molecular mechanisms underlying the regulation of GnRHR gene transcription. Initial studies of the mouse gene have demonstrated that the major transcriptional start site is located 62 nucleotides upstream of the translational start site, without clear involvement of a TATA box, and that 1.2 kb of the 5′-flanking sequences of the GnRHR gene are sufficient for both basal and GnRH-stimulated gene expression (689,690). This 1.2-kb region possesses promoter activity for gonadotrope-specific as well as GnRH-responsive GnRHR gene expression, directing high reporter gene expression in αT3-1 but not in heterologous cell lines (298,689). This view has been further supported by transgenic mouse studies, in which use of this region (691) or a larger 1.9-kb fragment that encompasses this region (692) has resulted in tissue-specific targeting of transgenes to the pituitary and, more specifically, to the gonadotrope. More detailed 5′-deletion analyses indicate the presence of sequences between –440 and –340 relative to the transcriptional start site that activate GnRHR gene expression in αT3-1 cells (690).

There are several *cis*-elements in the mouse GnRHR gene promoter that have been identified and characterized, including two CCAAT boxes located at positions –379 to –375 and –288 to –284 (46,693), a GnRHR activating sequence (GRAS) at –329 to –318 (694,695), a downstream activin regulatory element (DARE) at –303 to –286 (696), a sequence underlying responsiveness to GnRH-1 (SURG-1) at –292 to –285 (298), putative repressor elements at –281 to –273 and –264 to –220 (298,697), a consensus AP-1 binding site (TGAGTCA) at –274 to –268 (298,689,695), a bipartite GSE at –181 to –173 and +48 to +56 (689, 694,698), a CRE at –47 to –40 (697), and a TATA box at –36 to –31 (689) (Fig. 16).

The 1.2-kb section of the 5′-flanking region of the rat GnRHR gene promoter has also been demonstrated to be sufficient for tissue-specific gene expression; sequences are more than 82% homologous to the corresponding mouse sequences (46,699,700), including several putative *cis*-elements. Based on sequence homologies, there are two putative Myc-responsive elements located at positions –1,137 to –1,131 and –209 to –203 (699), GnRHR-specific enhancer (GnSE) elements at –983 to –962 and –871 to –862 (701), a putative Pit-1 binding site at –657 to –651, two CCAAT boxes at –458 to –454 and –366 to –362 (46), a GRAS-like element at –412 to –395 (701), a SURG-1–like element at –370 to –363 (298), a consensus AP-1 binding site at –352 to –346 (46,699,701), a putative polyomavirus enhancer activator 3 (PEA3) binding site at –314 to –309 (699), two putative CREs at –295 to –290 (46) and –110 to –103 (702), two PACAP response elements (PARE I and II) at –260 to –237 and –136 to –101, respectively (702), a GSE at –245 to –237 (46,699,701), and a TATA box at –126 to –121 (46). Also, a junctional regulatory element–like element has been located at position –734 to –718 (699), which may be a trophoblast-specific enhancer element.

The 5′-flanking region of the human GnRHR gene has also been cloned and sequenced and appears to be more complex than that of other species, with multiple transcriptional start sites (32,45). This complexity may provide means for the regulatory multiplicity of human GnRHR gene expression. It contains a large number of putative *cis*-elements, including a putative AP-2 binding site located at position –3,993 to –3,987 (45), three putative thyroid hormone response elements at –3,745 to –3,739, –2,011 to –2,005, and –1,030 to –1,024, three putative CREs at –1,614 to –1,607 (32,45), +11 to +18, and +239 to +246 (703), a putative bipartite PEA3 binding site at –1,384 to –1,379 and –1,353 to –1,348 (32,45), four CCAAT boxes at –1,125 to –1,121, –511 to –507, –388 to –384, and –297 to –293, three pyrimidine-rich initiator (Inr) elements at –1,104 to –1,098 (704), –24 to –18, and –11 to –5 (705), a GnRHR gene PACAP-responsive element at –1,097 to –1,069 (706), a silencer element at –1,056 to –1,036, a SURG-1–like element at –982 to –975 (298), three putative AP-1 binding sites at –940 to –933 (32,45), –420 to –414 (269), and +450 to +456 (707), an octamer-1 (Oct-1) binding site at –438 to –430 (708), a putative Pit-1 binding site at –364 to –357 (32,45), a putative Engrailed binding site at –281 to –276, a putative PRE/GRE at –215 to –201 (32,45) and +44 to +58 (499), a bipartite putative CAAT/enhancer–binding protein (C/EBP) site

(CAAT box) at –197 to –194 and –152 to –149, a GSE at +436 to +444 (709), and seven TATA boxes at –3,858 to –3,854, –2,351 to –2,346, –855 to –850, –842 to –837, –273 to –268, –234 to –229, and –137 to –132 (32,45), relative to the most proximal transcriptional start site (45).

The PEA3 element is an oncogene-, serum growth factor–, and phorbol ester–responsive element, first recognized in the polyomavirus enhancer (710). The PEA3 protein is an Ets oncogene family member implicated to play a regulatory role during mouse embryogenesis. In the mouse, however, PEA3 expression is highly restricted to the epididymis and brain only (710). Thus, PEA3 may play a role in brain-specific, but not pituitary, human GnRHR gene expression. Pit-1 (also identified as growth hormone factor-1, GHF-1) is an important pituitary cell type–specific transcription factor involved in regulation of growth hormone (711), prolactin (712), and TSHβ (713) gene expression whose expression is restricted to somatotrope, lactotrope, and some thyrotrope cell types. It is unlikely that Pit-1 can have any functional role in gonadotrope-specific GnRHR gene expression because Pit-1 is not known to be expressed in gonadotropes.

Unlike the mouse, at least 7 consensus TATA boxes were identified in the human GnRHR gene promoter, and multiple (up to 18 in human pituitary) transcriptional start sites were detected associated with these TATA sequences (45), in contrast to the more limited number of transcriptional start sites (up to 5) found in the human brain (32). These transcriptional start sites reside further upstream than the major transcriptional start site identified in the mouse, in which there is one putative TATA box (689). These findings suggest the possibility of the use of species- or tissue-specific promoters and transcriptional start sites. The distal Inr element (–1,104/–1,098) and its immediate upstream CCAAT box (–1,125/–1,121) have been shown to be critical elements for the upstream transcriptional start site activity in the human GnRHR gene promoter (704), and the two proximal Inr elements (–24/–19 and –11/–5) are important for the downstream transcriptional start site activity in αT3-1 cells (705). These Inr elements interact with TFIID complex, implicated as a regulatory mechanism for dual (distal and proximal) transcriptional start sites.

The 3′-end of the human GnRHR gene contains five classic polyadenylation signals (32). The large 3′-untranslated sequence likely accounts for the greatest portion of the major mRNA species observed by Northern blot analysis. In addition, there are several ATTTA elements in the 3′-untranslated region that may serve as signals for mRNA instability, suggesting a possible mechanism for rapid degradation of the human GnRHR mRNA (32).

There is only approximately 42% to 63% homology among the 5′-flanking regions of the mouse, rat, human, and ovine GnRHR genes (44,699). Compared with those of the mouse and human GnRHR gene promoters, however, the mechanisms underlying regulation of expression of the ovine GnRHR gene promoter are less well defined. Indeed, although the 5′-flanking sequence from the mouse GnRHR gene is responsive to both GnRH and activin in vitro, the same region of the ovine GnRHR gene promoter does not display transcriptional activity in the αT3-1 or LβT2 cell lines (493). In a transgenic mouse model in which a reporter gene is targeted to pituitary gonadotropes by the ovine GnRHR promoter, 2.7 kb of the 5′-flanking region of the ovine GnRHR gene has been shown to be sufficient to target gonadotropes specifically and to confer GnRH responsiveness. However, the *cis*-elements mediating the GnRH response have not been determined in the sheep; for example, the ovine GnRHR gene promoter does not appear to contain an AP-1 binding site (44). These studies suggest that the mechanisms by which GnRH regulates GnRHR number and expression may differ between species.

Involvement of tripartite *cis*-elements, the GRAS, the AP-1 binding site, and the GSE, has been suggested for cell-specific expression of the mouse GnRHR gene in αT3-1 cells (695). Mutation of any of these elements alone results in loss of 60% of promoter activity, combined mutation of any two decreases activity by 80%, and mutation of all three elements completely eliminates promoter activity (694,695). Subsequently, the GRAS has been shown to mediate activin responsiveness (463), and closer inspection of the GRAS has revealed that it consists of an imperfect SBE (GTCTAGTC) at –331 to –324 (439,440,714), a nonconsensus AP-1 binding site (AGTCAC) at –327 to –322 (440,714), and a forkhead/winged-helix family protein FoxL2 (initially known as pituitary forkhead factor, PFrk) binding core motif (ACAACA) at –323 to –318 (714). FoxL2 is expressed in the pituitary gonadotrope and thyrotrope lineages (715,716) and in αT3-1 cells, and physically interacts with SMAD3 and FosB in vivo (714).

The AP-1 binding site and the GSE have been shown to be important for gonadotrope-specific activity of the rat GnRHR gene promoter as well (701). Despite the presence of a GRAS-like element at –412 to –395 in the rat GnRHR gene promoter, however, it is not responsive to activin (696,701). The rat GRAS differs by a single nucleotide from the mouse GRAS, and a change of the mouse sequence to reflect the rat sequence eliminates the activin response (696), suggesting that this single nucleotide difference may be the primary reason for the lack of activin responsiveness of the rat promoter. However, mutation of the rat sequence to reflect the mouse GRAS sequence fails to confer activin responsiveness to the rat promoter, suggesting that GRAS is necessary but not sufficient for activin responsiveness.

GnSE elements located at positions –983 to –962 and –871 to –862 in the rat GnRHR gene promoter have been identified to be essential for activation of the rat GnRHR in αT3-1 and LβT2 cells (701). These activities require an intact GSE at –245 to –237 for full activation. A mutation of the GSE abolishes the activity of the GnSE elements, suggesting a potential functional interaction between the GnSE and GSE elements. The GnSE contains several putative *cis*-elements, including two GATA binding sites at –970 to –966 and –966 to –961 and two palindromic consensus (CTAATTAG) Lhx2 half-sites at –869 to –863 and –868 to –862 (699,701). In addition, PACAP has also been shown to be involved in the regulation of rat GnRHR gene promoter activity through the GSE (PARE I) and the proximal CRE (–110/–103, PARE II), which can be bound by SF-1 and by CREB/CREM/ATF, respectively (702). These studies suggest that, despite the high sequence homology between the mouse and rat GnRHR gene promoters, their activities are regulated by quite different mechanisms.

Although the GSEs are important for basal expression, they are not likely to be critical elements for GnRH responsiveness of the mouse GnRHR gene (300). SF-1–deficient mice exhibit a complete loss of gonadotrope markers, including GnRHR (523,529). However, disruption of gonadotropin expression in SF-1–deficient mice can be partially rescued by exogenous GnRH treatment, and GnRHR mRNAs are detectable (529,531), indicating that a low level of GnRHR is likely still to be present on the cell surface in this mouse model. The GSE in the human GnRHR gene promoter has also been shown to be important for basal human GnRHR gene expression by interacting with SF-1, and mutation of this element results in an 80% reduction in promoter activity in αT3-1 cells (709). A nuclear receptor protein, Nur77 (also known as NGFI-B) (717), has been shown to compete with SF-1 for binding to the two GSEs in the mouse GnRHR gene promoter (698). Nur77 is expressed in αT3-1 cells and acts as a negative regulator of SF-1 action on PKA-mediated mouse GnRHR gene promoter activity.

Pitx-1 has been shown to be important for cell type–specific expression of the mouse GnRHR gene (554,618). This protein confers GnRHR gene transcriptional activation by interacting both with the proximal promoter region near the AP-1 binding site and with c-Jun protein through its HD (618). A previous study has suggested six putative Pitx-1 binding sites on the mouse GnRHR gene promoter (554). Unlike other gonadotropin subunit genes, however, Pitx-1 does not seem to have a definite binding site on the mouse GnRHR gene promoter. Rather, both low-affinity DNA-binding activity of the Pitx-1 HD and full Pitx-1 transactivation of the mouse GnRHR gene require an intact proximal promoter region that includes the SURG-1 and AP-1 binding sites. Moreover, Pitx-1 synergistically activates the mouse GnRHR gene promoter together with c-Jun, suggesting a specific cooperative role between Pitx-1 and c-Jun in GnRH-induced GnRHR gene expression (618).

GnRH responsiveness of the mouse GnRHR gene promoter is primarily localized to two *cis*-elements, the AP-1 binding site and SURG-1 (298,299). Mutation of the AP-1 binding site results in complete abrogation of the GnRH-mediated GnRHR gene promoter activity in αT3-1 cells, whereas mutation of SURG-1 significantly diminishes, but does not abrogate, the GnRH response of the promoter. Also, SURG-1 potentiates the AP-1–mediated GnRH response. The mouse, rat, and human GnRHR gene promoters all contain a perfect consensus AP-1 binding site, and there is a high degree of homology (76% to 88%) in the SURG-1 sequences, located in close proximity distal to the AP-1 binding site. However, whether these SURG-1–like elements in the rat and human GnRHR gene promoters operate in a similar way as in the mouse GnRHR gene has not yet been determined. GnRH-stimulated GnRHR gene expression is induced primarily through the activation of Jun and Fos that bind to the AP-1 site (298–301,688). Upon GnRH stimulation, *jun/fos* expression is quickly induced (307,308), and increasing amounts of dimeric AP-1 complexes bind to the AP-1 binding site, resulting in GnRH-induced activation of GnRHR gene expression (298). Among several AP-1 family members, JunD and FosB have been shown to be the major factors, with a minor contribution by c-Jun and c-Fos, binding selectively to the AP-1 binding site upon GnRH stimulation in αT3-1 cells (301).

SURG-1 consists of binding sites for transcription factors Oct-1 and NF-Y at positions –290 to –287 (TAAT) and –288 to –284 (inverted CCAAT), respectively, and mediates basal and GnRH-induced GnRHR gene transcriptional activity in a cooperative manner with AP-1 (693). Mutation of the Oct-1, NF-Y, or AP-1 binding site decreases both basal and GnRH-induced mouse GnRHR gene promoter activity, and the combined mutation of the Oct-1 and AP-1 binding sites results in further reduction of basal and complete elimination of GnRH-induced promoter activity (693). The ubiquitously expressed Oct-1 is a member of the POU domain transcription factor family, and has been shown to be an important regulator in neuronal-specific GnRH gene expression and maintaining the pulsatile nature of GnRH gene promoter activity in GnRH neuronal cell lines, including GT1-7 cells (718–720). A study has shown that the DARE element, which consists of paired TAAT core motifs (–298/–295 and –290/–287), with the proximal one being the same TAAT motif as the Oct-1 binding site, serves as an auxiliary element to the GRAS for full activin responsiveness (696). The Lhx2 HD is able to bind to the proximal TAAT motif (696), although another study has shown that Lhx2 does not bind to this motif (693). Another study has shown that Lhx3 binds to the distal TAAT motif, and that this binding is functionally important for basal mouse GnRHR gene promoter activity (721). Interestingly, Oct-1 has been shown to be involved in transcriptional repression of the human GnRHR gene through an upstream binding site in αT3-1 cells (708), suggesting a diverse and species-specific action of Oct-1. In addition, Oct-1 has been shown to interact physically with steroid hormone receptors, such as AR, PR, and GR, in synergistic stimulation of target genes to facilitate DNA binding (722,723) and recruitment of coactivators such as steroid receptor coactivator-1 (SRC-1) (724).

The components of GRAS, an SBE that binds SMAD3 and 4 proteins and mediates the activin response, and a nonconsensus AP-1 site that binds both Jun and Fos proteins, have also been shown to be important for mediating full GnRH stimulation of mouse GnRHR gene expression (439,440,463,714). Activin A–mediated SMAD binding has been shown to augment GnRH-induced GnRHR gene promoter activity in αT3-1 cells, whereas activin alone has no effect (439,440). Mutation of the SBE has been shown to block activin responsiveness of mouse GnRHR gene promoter activity (463,714). In addition, it prevents activin from augmenting GnRH-stimulated transcriptional activation of the GnRHR gene, but does not affect the responsiveness to GnRH alone (439). Furthermore, the nonconsensus AP-1 binding site in GRAS can confer GnRH responsiveness in the absence of the proximal consensus AP-1 binding site. This GnRH response is potentiated by activin treatment or SMAD3/4 overexpression, which is completely abolished by a mutation in this element (440,714). These data suggest that both activin and GnRH alone can stimulate transcriptional activity of the GnRHR gene. However, the magnitude of transcriptional activation is enhanced in the presence of activin and GnRH together. Moreover, FoxL2 binding to GRAS requires the presence of SMAD3, and the presence of both DNA-bound FoxL2 and SMADs is necessary for FoxL2 activation of GRAS (714). These observations suggest that SMAD binding to the SBE may recruit the FoxL2, which in turn may interact with AP-1 and SMADs to maximize activin or GnRH responsiveness. Taken together, these studies suggest that GnRH is the major stimulator of mouse GnRHR gene expression and that activin potentiates this GnRH responsiveness (435). Thus, transcriptional activation of the mouse GnRHR gene by activin may serve as a mechanism to modulate gonadotrope responsiveness to GnRH.

The mouse GnRHR gene promoter has no consensus ERE or PRE/GRE (689), whereas two putative PREs/GREs have been identified in the human GnRHR gene promoter (32,45,499). Previous studies have shown a direct involvement of E_2 in stimulation of GnRHR gene expression in LβT2 cells (71,327), although the mechanisms of these effects of E_2 on GnRHR gene transcriptional activity are not known. It has been shown that AP-1 and ER can share a coactivator (coactivator of AP-1 and ER, CAPER) (725), suggesting possible cooperation between these two factors. Interestingly, E_2-bound ERα has been suggested to repress the human GnRHR gene promoter without direct DNA binding through the most proximal AP-1 binding site (+450/+456) in an ovarian cell line. This effect is antagonized by PMA treatment, and overexpression of the coactivator CBP attenuates this E_2-mediated repression (707). Moreover, ERα has been shown to interact physically with promoter-bound c-Jun (726), and phosphorylated c-Jun interacts with CBP (727). Therefore, these studies suggest that the interaction between ERα and AP-1 may interfere with human GnRHR gene promoter activity, and that PKC-mediated AP-1 activation may enable the release of ERα to allow recruitment of p300/CBP, resulting in transcriptional activation. However, it is not yet clear if similar regulation occurs in pituitary gonadotropes. It has been shown that progesterone decreases human GnRHR gene promoter activity through binding of PR to the proximal PRE (+44/+58) in αT3-1 cells (499). Androgen has a stimulatory effect on mouse GnRHR gene promoter activity in a dose-dependent manner, with peak stimulation at 10^{-8} M dihydrotestosterone after 48 hours in LβT2 cells (651). In addition, glucocorticoids have been suggested to induce mouse GnRHR gene promoter activity in the GGH_3-1′ cells, an effect that appears to be mediated by the proximal AP-1 binding site (688). This action suggests a possible indirect mechanism of action of ligand-mediated GR through GR/AP-1 interaction, through stimulation of *jun/fos* expression, or through a nongenomic action of GR. In this regard, GR has been shown to be able to interact physically with c-Jun/c-Fos (728), TFIID complex (729), and Oct-1 (730). These interactions are, however, known to repress their target gene expression. Nevertheless, as for FSHβ gene regulation, cooperative interactions between and among the basal transcription factors FoxL2, Pitx-1, Oct-1, NF-Y, and SF-1 in association with their *cis*-elements and steroid receptors may contribute to basal and gonadotrope-specific expression of the mouse GnRHR gene. In response to GnRH or activin stimulation, activated AP-1 and SMADs may confer the response by recruiting and sharing common coactivators or adaptor complexes, such as p300/CBP, P/CAF, p/CIP, and SRC-1.

GnRH-induced activation of the PKC pathway is a mediator of the regulation of GnRHR gene expression (298,300,301,731). The GnRH-induced upregulation of GnRHR gene expression can be mimicked by activation of the PKC pathway as well as by cellular depolarization. PMA treatment, but not forskolin, potentiates GnRH responsiveness of the mouse GnRHR gene promoter in αT3-1 cells (298,300), which is likely to be mediated by the activated AP-1 through JNK and ERK cascades without involvement of L-type VGCCs (298,300,301). Other studies, however, have suggested an involvement of Ca^{2+} influx in GnRH-stimulated mouse GnRHR gene promoter activity in GGH_3-1′ cells (732). Thus, the PKC pathway activates both extracellular Ca^{2+}-independent JNK and extracellular Ca^{2+}-dependent ERK cascades for the activation of GnRHR gene expression (Fig. 5).

Studies have shown that activation of the PKA pathway by forskolin treatment not only reduces GnRHR mRNA levels in a time-dependent manner in αT3-1 cells (73), but blocks GnRH- or PMA-stimulated mouse GnRHR gene promoter activity. This effect may be due to inhibition of JNK activation and thus AP-1 activation, while not affecting MAPK activation (300,301). However, other studies have shown that the same treatment enhances mouse or rat GnRHR gene promoter activity as well as mRNA levels in αT3-1 cells and GT1-7 cells (46,698,702), and the rodent GnRHR gene promoters contain a putative CRE that appears to be functional (46,697,702,733). Furthermore, PKA stimulation of mouse GnRHR gene promoter activity is potentiated by SF-1 overexpression, and Nur77 overexpression blocks this potentiation in αT3-1 cells (698). This PKA stimulation is modulated positively and negatively by the distal (–181/–173) and proximal (+48/+55) GSEs, respectively, possibly through Nur77 antagonism, and forskolin treatment increases SF-1 mRNA levels in a dose-dependent manner. The mouse GnRHR has been shown to couple with $G_{s\alpha}$ as well as $G_{q/11\alpha}$ in GGH_3-1′ cells, deletion of the CRE reduces basal and GnRH-induced mouse GnRHR gene promoter activity, and basal and GnRH-induced mouse or rat GnRHR gene promoter activity is enhanced by forskolin or a cAMP analog, dibutyryl cAMP and reduced by an adenylate cyclase inhibitor (697,732,733). PACAP has been shown to activate the rat GnRHR gene promoter through the PKA pathway in αT3-1 cells (698,702). The PKA pathway can also stimulate human GnRHR gene expression through two proximal CREs (+11/+18 and +239 to +246) in αT3-1 cells (703). These CREs are partially homologous to the consensus CRE (TGACGTCA) as well as to the consensus AP-1 binding site (TGAGTCA), and although CREB seems to be a primary partner for the middle CRE (+11/+18), an AP-1–like molecule appears to bind to the most proximal CRE (+239/+246), being replaced by CREB when stimulated by forskolin. Together, these studies suggest an involvement of the PKA pathway, at least in part, in regulation of GnRHR gene expression through both GSE and CRE elements (Fig. 5). Nonetheless, the PKA response is slow, and whether GnRH stimulation is mediated through the PKA pathway is not clear.

REFERENCES

1. Vale, W., Rivier, C., and Brown, M. (1977). Regulatory peptides of the hypothalamus. *Annu. Rev. Physiol.* 39, 473–527.
2. Gharib, S. D., Wierman, M. E., Shupnik, M. A., and Chin, W. W. (1990). Molecular biology of the pituitary gonadotropins. *Endocr. Rev.* 11, 177–199.
3. Baba, Y., Matsuo, H., and Schally, A. V. (1971). Structure of the porcine LH- and FSH-releasing hormone: II. Confirmation of the proposed structure by conventional sequential analyses. *Biochem. Biophys. Res. Commun.* 44, 459–463.
4. Matsuo, H., Baba, Y., Nair, R. M., Arimura, A., and Schally, A. V. (1971). Structure of the porcine LH- and FSH-releasing hormone: I. The proposed amino acid sequence. *Biochem. Biophys. Res. Commun.* 43, 1334–1339.
5. Schally, A. V., Arimura, A., Kastin, A. J., Matsuo, H., Baba, Y., Redding, T. W., Nair, R. M., Debeljuk, L., and White, W. F. (1971). Gonadotropin-releasing hormone: one polypeptide regulates secretion of luteinizing and follicle-stimulating hormones. *Science* 173, 1036–1038.
6. Burgus, R., Butcher, M., Amoss, M., Ling, N., Monahan, M., Rivier, J., Fellows, R., Blackwell, R., Vale, W., and Guillemin, R. (1972). Primary structure of the ovine hypothalamic luteinizing hormone-releasing factor (LRF). *Proc. Natl. Acad. Sci. U S A* 69, 278–282.
7. Conn, P. M. (1986). The molecular basis of gonadotropin-releasing hormone action. *Endocr. Rev.* 7, 3–10.
8. Conn, P. M., and Crowley, W. F., Jr. (1991). Gonadotropin-releasing hormone and its analogues. *N. Engl. J. Med.* 324, 93–103.
9. Kaiser, U. B., Conn, P. M., and Chin, W. W. (1997). Studies of gonadotropin-releasing hormone (GnRH) action using GnRH receptor-expressing pituitary cell lines. *Endocr. Rev.* 18, 46–70.
10. Sealfon, S. C., Weinstein, H., and Millar, R. P. (1997). Molecular mechanisms of ligand interaction with the gonadotropin-releasing hormone receptor. *Endocr. Rev.* 18, 180–205.
11. Ibrahim, S. N., Moussa, S. M., and Childs, G. V. (1986). Morphometric studies of rat anterior pituitary cells after gonadectomy: correlation of changes in gonadotropes with the serum levels of gonadotropins. *Endocrinology* 119, 629–637.
12. Windle, J. J., Weiner, R. I., and Mellon, P. L. (1990). Cell lines of the pituitary gonadotrope lineage derived by targeted oncogenesis in transgenic mice. *Mol. Endocrinol.* 4, 597–603.
13. Alarid, E. T., Windle, J. J., Whyte, D. B., and Mellon, P. L. (1996). Immortalization of pituitary cells at discrete stages of development by directed oncogenesis in transgenic mice. *Development* 122, 3319–3329.
14. Wu, J. C., Su, P., Safwat, N. W., Sebastian, J., and Miller, W. L. (2004). Rapid, efficient isolation of murine gonadotropes and

their use in revealing control of follicle-stimulating hormone by paracrine pituitary factors. *Endocrinology* 145, 5832–5839.
15. Pelletier, G., Leclerc, R., and Labrie, F. (1976). Identification of gonadotropic cells in the human pituitary by immunoperoxidase technique. *Mol. Cell. Endocrinol.* 6, 123–128.
16. Childs, G. V., Unabia, G., Tibolt, R., and Lloyd, J. M. (1987). Cytological factors that support nonparallel secretion of luteinizing hormone and follicle-stimulating hormone during the estrous cycle. *Endocrinology* 121, 1801–1813.
17. Naor, Z., Clayton, R. N., and Catt, K. J. (1980). Characterization of gonadotropin-releasing hormone receptors in cultured rat pituitary cells. *Endocrinology* 107, 1144–1152.
18. Wormald, P. J., Eidne, K. A., and Millar, R. P. (1985). Gonadotropin-releasing hormone receptors in human pituitary: ligand structural requirements, molecular size, and cationic effects. *J. Clin. Endocrinol. Metab.* 61, 1190–1194.
19. Horn, F., Bilezikjian, L. M., Perrin, M. H., Bosma, M. M., Windle, J. J., Huber, K. S., Blount, A. L., Hille, B., Vale, W., and Mellon, P. L. (1991). Intracellular responses to gonadotropin-releasing hormone in a clonal cell line of the gonadotrope lineage. *Mol. Endocrinol.* 5, 347–355.
20. Krsmanovic, L. Z., Martinez-Fuentes, A. J., Arora, K. K., Mores, N., Navarro, C. E., Chen, H.-C., Stojilkovic, S. S., and Catt, K. J. (1999). Autocrine regulation of gonadotropin-releasing hormone secretion in cultured hypothalamic neurons. *Endocrinology* 140, 1423–1431.
21. Krsmanovic, L. Z., Stojilkovic, S. S., Mertz, L. M., Tomic, M., and Catt, K. J. (1993). Expression of gonadotropin-releasing hormone receptors and autocrine regulation of neuropeptide release in immortalized hypothalamic neurons. *Proc. Natl. Acad. Sci. U S A* 90, 3908–3912.
22. Bourne, G. A., Regiani, S., Payne, A. H., and Marshall, J. C. (1980). Testicular GnRH receptors: characterization and localization on interstitial tissue. *J. Clin. Endocrinol. Metab.* 51, 407–409.
23. Clayton, R. N., Harwood, J. P., and Catt, K. J. (1979). Gonadotropin-releasing hormone analogue binds to luteal cells and inhibits progesterone production. *Nature* 282, 90–92.
24. Hazum, E., and Nimrod, A. (1982). Photoaffinity-labeling and fluorescence-distribution studies of gonadotropin-releasing hormone receptors in ovarian granulosa cells. *Proc. Natl. Acad. Sci. U S A* 79, 1747–1750.
25. Cheng, K. W., Nathwani, P. S., and Leung, P. C. K. (2000). Regulation of human gonadotropin-releasing hormone receptor gene expression in placental cells. *Endocrinology* 141, 2340–2349.
26. Chatzaki, E., Bax, C. M. R., Eidne, K. A., Anderson, L., Grudzinskas, J. G., and Gallagher, C. J. (1996). The expression of gonadotropin-releasing hormone and its receptor in endometrial cancer, and its relevance as an autocrine growth factor. *Cancer Res.* 56, 2059–2065.
27. Kakar, S. S., Musgrove, L. C., Devor, D. C., Sellers, J. C., and Neill, J. D. (1992). Cloning, sequencing, and expression of human gonadotropin-releasing hormone (GnRH) receptor. *Biochem. Biophys. Res. Commun.* 189, 289–295.
28. Reinhart, J., Mertz, L. M., and Catt, K. J. (1992). Molecular cloning and expression of cDNA encoding the murine gonadotropin-releasing hormone receptor. *J. Biol. Chem.* 267, 21281–21284.
29. Tsutsumi, M., Zhou, W., Millar, R. P., Mellon, P. L., Roberts, J. L., Flanagan, C. A., Dong, K., Gillo, B., and Sealfon, S. C. Cloning and functional expression of a mouse gonadotropin-releasing hormone receptor. *Mol. Endocrinol.* 6, 1163–1169.
30. Perrin, M. H., Bilezikjian, L. M., Hoeger, C., Donaldson, C. J., Rivier, J., Haas, Y., and Vale, W. W. (1993). Molecular and functional characterization of GnRH receptors cloned from rat pituitary and a mouse pituitary tumor cell line. *Biochem. Biophys. Res. Commun.* 191, 1139–1144.
31. Chi, L., Zhou, W., Prikhozhan, A., Flanagan, C., Davidson, J. S., Golembo, M., Illing, N., Millar, R. P., and Sealfon, S. C. (1993). Cloning and characterization of the human GnRH receptor. *Mol. Cell. Endocrinol.* 91, R1–R6.
32. Fan, N. C., Peng, C., Krisinger, J., and Leung, P. C. (1995). The human gonadotropin-releasing hormone receptor gene: complete structure including multiple promoters, transcription initiation sites, and polyadenylation signals. *Mol. Cell. Endocrinol.* 107, R1–R8.
33. Eidne, K. A., Sellar, R. E., Couper, G., Anderson, L., and Taylor, P. L. (1992). Molecular cloning and characterisation of the rat pituitary gonadotropin-releasing hormone (GnRH) receptor. *Mol. Cell. Endocrinol.* 90, R5–R9.
34. Kaiser, U. B., Zhao, D., Cardona, G. R., and Chin, W. W. (1992). Isolation and characterization of cDNAs encoding the rat pituitary gonadotropin-releasing hormone receptor. *Biochem. Biophys. Res. Commun.* 189, 1645–1652.
35. Kakar, S. S., Rahe, C. H., and Neill, J. D. (1993). Molecular cloning, sequencing, and characterizing the bovine receptor for gonadotropin releasing hormone (GnRH). *Domest. Anim. Endocrinol.* 10, 335–342.
36. Brooks, J., Taylor, P. L., Saunders, P. T., Eidne, K. A., Struthers, W. J., and McNeilly A S. (1993). Cloning and sequencing of the sheep pituitary gonadotropin-releasing hormone receptor and changes in expression of its mRNA during the estrous cycle. *Mol. Cell. Endocrinol.* 94, R23–R27.
37. Illing, N., Jacobs, G. F., Becker, I. I., Flanagan, C. A., Davidson, J. S., Eales, A., Zhou, W., Sealfon, S. C., and Millar, R. P. (1993). Comparative sequence analysis and functional characterization of the cloned sheep gonadotropin-releasing hormone receptor reveal differences in primary structure and ligand specificity among mammalian receptors. *Biochem. Biophys. Res. Commun.* 196, 745–751.
38. Weesner, G. D., and Matteri, R. L. (1994). Rapid communication: nucleotide sequence of luteinizing hormone-releasing hormone (LHRH) receptor cDNA in the pig pituitary. *J. Anim. Sci.* 72, 1911.
39. Cui, J., Smith, R. G., Mount, G. R., Lo, J.-L., Yu, J., Walsh, T. F., Singh, S. B., DeVita, R. J., Goulet, M. T., Schaeffer, J. M., and Cheng, K. (2000). Identification of Phe^{313} of the gonadotropin-releasing hormone (GnRH) receptor as a site critical for the binding of nonpeptide GnRH antagonists. *Mol. Endocrinol.* 14, 671–681.
40. King, J. A., Fidler, A., Lawrence, S., Adam, T., Millar, R. P., and Katz, A. (2000). Cloning and expression, pharmacological characterization, and internalization kinetics of the pituitary GnRH receptor in a metatherian species of mammal. *Gen. Comp. Endocrinol.* 117, 439–448.
41. Cheung, T. C., and Hearn, J. P. (2002). Molecular cloning and tissue expression of the gonadotrophin-releasing hormone receptor in the tammar wallaby (*Macropus eugenii*). *Reprod. Fertil. Dev.* 14, 157–164.
42. Probst, W. C., Snyder, L. A., Schuster, D. I., Brosius, J., and Sealfon, S. C. (1992). Sequence alignment of the G-protein coupled receptor superfamily. *DNA Cell. Biol.* 11, 1–20.
43. Ji, T. H., Grossmann, M., and Ji, I. (1998). G protein-coupled receptors: I. Diversity of receptor-ligand interactions. *J. Biol. Chem.* 273, 17299–17302.
44. Campion, C. E., Turzillo, A. M., and Clay, C. M. (1996). The gene encoding the ovine gonadotropin-releasing hormone (GnRH) receptor: cloning and initial characterization. *Gene* 170, 277–280.
45. Kakar, S. S. (1997). Molecular structure of the human gonadotropin-releasing hormone receptor gene. *Eur. J. Endocrinol.* 137, 183–192.
46. Reinhart, J., Xiao, S., Arora, K. K., and Catt, K. J. (1997). Structural organization and characterization of the promoter region of the rat gonadotropin-releasing hormone receptor gene. *Mol. Cell. Endocrinol.* 130, 1–12.

47. Kaiser, U. B., Dushkin, H., Altherr, M. R., Beier, D. R., and Chin, W. W. (1994). Chromosomal localization of the gonadotropin-releasing hormone receptor gene to human chromosome 4q13.1-q21.1 and mouse chromosome 5. *Genomics* 20, 506–508.
48. Montgomery, G. W., Penty, J. M., Lord, E. A., Brooks, J., and McNeilly A S. (1995). The gonadotrophin-releasing hormone receptor maps to sheep chromosome 6 outside of the region of the FecB locus. *Mamm. Genome.* 6, 436–438.
49. Arora, K. K., Chung, H.-O., and Catt, K. J. (1999). Influence of a species-specific extracellular amino acid on expression and function of the human gonadotropin-releasing hormone receptor. *Mol. Endocrinol.* 13, 890–896.
50. Dohlman, H. G., Thomer, J., Caron, M. G., and Lefkowitz, R. J. (1991). Model systems for the study of seven transmembrane segment receptors. *Annu. Rev. Biochem.* 60, 653–688.
51. Vrecl, M., Anderson, L., Hanyaloglu, A., McGregor, A. M., Groarke, A. D., Milligan, G., Taylor, P. L., and Eidne, K. A. (1998). Agonist-induced endocytosis and recycling of the gonadotropin-releasing hormone receptor: effect of b-arrestin on internalization kinetics. *Mol. Endocrinol.* 12, 1818–1829.
52. Ronacher, K., Matsiliza, N., Nkwanyana, N., Pawson, A. J., Adam, T., Flanagan, C. A., Millar, R. P., and Katz, A. A. (2004). Serine residues 338 and 339 in the carboxyl-terminal tail of the type II gonadotropin-releasing hormone receptor are critical for b-arrestin-independent internalization. *Endocrinology* 145, 4480–4488.
53. Goodman, O. B., Jr., Krupnick, J. G., Santini, F., Gurevich, V. V., Penn, R. B., Gagnon, A. W., Keen, J. H., and Benovic, J. L. (1996). β-Arrestin acts as a clathrin adaptor in endocytosis of the b_2-adrenergic receptor. *Nature* 383, 447–450.
54. Heding, A., Vrecl, M., Hanyaloglu, A. C., Sellar, R., Taylor, P. L., Eidne, K. A. The rat gonadotropin-releasing hormone receptor internalizes via a b-arrestin-independent, but dynamin-dependent, pathway: addition of a carboxyl-terminal tail confers b-arrestin dependency. *Endocrinology* 2000;141, 299–306.
55. Benard, O., Naor, Z., and Seger, R. (2001). Role of dynamin, Src, and Ras in the protein kinase C-mediated activation of ERK by gonadotropin-releasing hormone. *J. Biol. Chem.* 276, 4554–4563.
56. Hislop, J. N., Everest, H. M., Flynn, A., Harding, T., Uney, J. B., Troskie, B. E., Millar, R. P., and McArdle, C. A. (2001). Differential internalization of mammalian and non-mammalian gonadotropin-releasing hormone receptors: uncoupling of dynamin-dependent internalization from mitogen-activated protein kinase signaling. *J. Biol. Chem.* 276, 39685–39694.
57. Caunt, C. J., Hislop, J. N., Kelly, E., Matharu, A.-L., Green, L. D., Sedgley, K. R., Finch, A. R., and McArdle, C. A. (2004). Regulation of gonadotropin-releasing hormone receptors by protein kinase C: inside out signalling and evidence for multiple active conformations. *Endocrinology* 145, 3594–3602.
58. Blomenrohr, M., Heding, A., Sellar, R., Leurs, R., Bogerd, J., Eidne, K. A., and Willars, G. B. (1999). Pivotal role for the cytoplasmic carboxyl-terminal tail of a nonmammalian gonadotropin-releasing hormone receptor in cell surface expression, ligand binding, and receptor phosphorylation and internalization. *Mol. Pharmacol.* 56, 1229–1237.
59. Maya-Nunez, G., Janovick, J. A., Ulloa-Aguirre, A., Soderlund, D., Conn, P. M., and Mendez, J. P. (2002). Molecular basis of hypogonadotropic hypogonadism: restoration of mutant ($E^{90}K$) GnRH receptor function by a deletion at a distant site. *J. Clin. Endocrinol. Metab.* 87, 2144–2149.
60. Millar, R. P., Lu, Z.-L., Pawson, A. J., Flanagan, C. A., Morgan, K., and Maudsley, S. R. (2004). Gonadotropin-releasing hormone receptors. *Endocr. Rev.* 25, 235–275.
61. Savoy-Moore, R. T., Schwartz, N. B., Duncan, J. A., and Marshall, J. C. (1980). Pituitary gonadotropin-releasing hormone receptors during the rat estrous cycle. *Science* 209, 942–944.
62. Chan, V., Clayton, R. N., Knox, G., and Catt, K. J. (1981). Ontogeny of pituitary GnRH receptors in the rat. *Endocrinology* 108, 2086–2092.
63. Clayton, R. N., Channabasavaiah, K., Stewart, J. M., and Catt, K. J. (982). Hypothalamic regulation of pituitary gonadotropin-releasing hormone receptors: effects of hypothalamic lesions and a gonadotropin-releasing hormone antagonist. *Endocrinology* 110, 1108–1115.
64. Yasin, M., Dalkin, A. C., Haisenleder, D. J., Kerrigan, J. R., and Marshall, J. C. (1995). Gonadotropin-releasing hormone (GnRH) pulse pattern regulates GnRH receptor gene expression: augmentation by estradiol. *Endocrinology* 136, 1559–1564.
65. Williams, B., Brooks, A. N., Aldridge, T. C., Pennie, W. D., Stephenson, R., and McArdle, C. A. (2000). Oestradiol is a potent mitogen and modulator of GnRH signalling in αT3-1 cells: are these effects causally related? *J. Endocrinol.* 164, 31–43.
66. Loumaye, E., and Catt, K. J. (1982). Homologous regulation of gonadotropin-releasing hormone receptors in cultured pituitary cells. *Science* 215, 983–985.
67. Katt, J. A., Duncan, J. A., Herbon, L., Barkan, A., and Marshall, J. C. (1985). The frequency of gonadotropin-releasing hormone stimulation determines the number of pituitary gonadotropin-releasing hormone receptors. *Endocrinology* 116, 2113–2115.
68. Frager, M. S., Pieper, D. R., Tonetta, S., Duncan, J. A., and Marshall, J. C. (1981). Pituitary gonadotropin-releasing hormone (GnRH) receptors: effects of castration, steroid replacement, and the role of GnRH in modulating receptors in the rat. *J. Clin. Invest.* 67, 615–623.
69. Kaiser, U. B., Jakubowiak, A., Steinberger, A., and Chin, W. W. (1993). Regulation of rat pituitary gonadotropin-releasing hormone receptor mRNA levels *in vivo* and *in vitro*. *Endocrinology* 133, 931–934.
70. Tsutsumi, M., Laws, S. C., Rodic, V., and Sealfon, S. C. (1995). Translational regulation of the gonadotropin-releasing hormone receptor in αT_3-1 cells. *Endocrinology* 136, 1128–1136.
71. Turgeon, J. L., Kimura, Y., Waring, D. W., and Mellon, P. L. (1996). Steroid and pulsatile gonadotropin-releasing hormone (GnRH) regulation of luteinizing hormone and GnRH receptor in a novel gonadotrope cell line. *Mol. Endocrinol.* 10, 439–450.
72. Tsutsumi, M., Laws, S. C., and Sealfon, S. C. (1993). Homologous up-regulation of the gonadotropin-releasing hormone receptor in αT_3-1 cells is associated with unchanged receptor messenger RNA (mRNA) levels and altered mRNA activity. *Mol. Endocrinol.* 7, 1625–1633.
73. Alarid, E. T., and Mellon, P. L. (1995). Down-regulation of the gonadotropin-releasing hormone receptor messenger ribonucleic acid by activation of adenylyl cyclase in αT3-1 pituitary gonadotrope cells. *Endocrinology* 136, 1361–1366.
74. Nguyen, K. A., Santos, S. J., Kreidel, M. K., Diaz, A. L., Rey, R., and Lawson, M. A. (2004). Acute regulation of translation initiation by gonadotropin-releasing hormone in the gonadotrope cell line LbT2. *Mol. Endocrinol.* 18, 1301–1312.
75. Cornea, A., Janovick, J. A., Maya-Nunez, G., and Conn, P. M. (2001). Gonadotropin-releasing hormone receptor microaggregation: rate monitored by fluorescence resonance energy transfer. *J. Biol. Chem.* 276, 2153–2158.
76. Horvat, R. D., Roess, D. A., Nelson, S. E., Barisas, B. G., and Clay, C. M. (2001). Binding of agonist but not antagonist leads to fluorescence resonance energy transfer between intrinsically fluorescent gonadotropin-releasing hormone receptors. *Mol. Endocrinol.* 15, 695–703.
77. Lefkowitz, R. J., Hausdorff, W. P., and Caron, M. G. (1990). Role of phosphorylation in desensitization of the b-adrenoreceptor. *Trends Pharmacol. Sci.* 11, 190–194.

78. Naor, Z. (1990). Signal transduction mechanisms of Ca^{2+} mobilizing hormones: the case of gonadotropin-releasing hormone. *Endocr. Rev.* 11, 326–353.
79. Hazum, E., Cuatrecasas, P., Marian, J., and Conn, P. (1980). Receptor-mediated internalization of fluorescent gonadotropin-releasing hormone by pituitary gonadotropes. *Proc. Natl. Acad. Sci. U S A* 77, 6692–6695.
80. Naor, Z., Atlas, D., Clayton, R. N., Forman, D. S., Amsterdam, A., and Catt, K. J. (1981). Interaction of fluorescent gonadotropin-releasing hormone with receptors in cultured pituitary cells. *J. Biol. Chem.* 256, 3049–3052.
81. Myburgh, D. B., Millar, R. P., and Hapgood, J. P. (1998). Alanine-261 in intracellular loop III of the human gonadotropin-releasing hormone receptor is crucial for G-protein coupling and receptor internalization. *Biochem. J.* 331, 893–896.
82. Suarez-Quian, C. A., Wynn, P. C., and Catt, K. J. (1986). Receptor-mediated endocytosis of GnRH analogs: differential processing of gold-labeled agonist and antagonist derivatives. *J. Steroid Biochem.* 24, 183–192.
83. Urrutia, R., Henley, J. R., Cook, T., and McNiven, M. A. (1997). The dynamins: redundant or distinct functions for an expanding family of related GTPases? *Proc. Natl. Acad. Sci. U S A* 94, 377–384.
84. Schvartz, I., and Hazum, E. (1987). Internalization and recycling of receptor-bound gonadotropin-releasing hormone agonist in pituitary gonadotropes. *J. Biol. Chem.* 262, 17046–17050.
85. Cornea, A., Janovick, J. A., Lin, X., and Conn, P. M. (1999). Simultaneous and independent visualization of the gonadotropin-releasing hormone receptor and its ligand: evidence for independent processing and recycling in living cells. *Endocrinology* 140, 4272–4280.
86. Loumaye, E., Wynn, P. C., Coy, D., and Catt, K. J. (1984). Receptor-binding properties of gonadotropin-releasing hormone derivatives: prolonged receptor occupancy and cell-surface localization of a potent antagonist analog. *J. Biol. Chem.* 259, 12663–12671.
87. Jennes, L., Coy, D., and Conn, P. M. (1986). Receptor-mediated uptake of GnRH agonist and antagonists by cultured gonadotropes: evidence for differential intracellular routing. *Peptides* 7, 459–463.
88. Conn, P. M., Smith, R. G., and Rogers, D. C. (1981). Stimulation of pituitary gonadotropin release does not require internalization of gonadotropin-releasing hormone. *J. Biol. Chem.* 256, 1098–1100.
89. Belchetz, P. E., Plant, T. M., Nakai, Y., Keogh, E. J., and Knobil, E. (1978). Hypophysial responses to continuous and intermittent delivery of hypothalamic gonadotropin-releasing hormone. *Science* 202, 631–633.
90. Pohl, C. R., Richardson, D. W., Hutchison, J. S., Germak, J. A., and Knobil, E. (1983). Hypophysiotropic signal frequency and the functioning of the pituitary-ovarian system in the rhesus monkey. *Endocrinology* 112, 2076–2080.
91. Marshall, J. C., and Kelch, R. P. (1986). Gonadotropin-releasing hormone: role of pulsatile secretion in the regulation of reproduction. *N. Engl. J. Med.* 315, 1459–1468.
92. Conn, P. M., Rogers, D. C., and Seay, S. (1984). Biphasic regulation of the gonadotropin-releasing hormone receptor by receptor microaggregation and intracellular Ca^{+2} levels. *Mol. Pharmacol.* 25, 51–55.
93. Conn, P. M., Staley, D. D., Yasumoto, T., Huckle, W. R., and Janovick, J. A. (1987). Homologous desensitization with gonadotropin-releasing hormone (GnRH) also diminishes gonadotrope responsiveness to maitotoxin: a role for the GnRH receptor-regulated calcium ion channel in mediation of cellular desensitization. *Mol. Endocrinol.* 1, 154–159.
94. Janovick, J. A., and Conn, P. M. (1993). A cholera toxin-sensitive guanyl nucleotide binding protein mediates the movement of pituitary luteinizing hormone into a releasable pool: loss of this event is associated with the onset of homologous desensitization to gonadotropin-releasing hormone. *Endocrinology* 132, 2131–2135.
95. Davidson, J. S., Wakefield, I. K., and Millar, R. P. (1994). Absence of rapid desensitization of the mouse gonadotropin-releasing hormone receptor. *Biochem. J.* 300, 299–302.
96. Lefkowitz, R. J. (1993). G protein-coupled receptor kinases. *Cell* 74, 409–412.
97. Conn, P. M., Rogers, D. C., and Seay, S. G. (1983). Structure-function relationship of calcium ion channel antagonists at the pituitary gonadotrope. *Endocrinology* 113, 1592–1595.
98. Gorospe, W. C., and Conn, P. M. (1987). Agents that decrease gonadotropin-releasing hormone (GnRH) receptor internalization do not inhibit GnRH-mediated gonadotrope desensitization. *Endocrinology* 120, 222–229.
99. Willars, G. B., McArdle, C. A., and Nahorski, S. R. (1998). Acute desensitization of phospholipase C-coupled muscarinic M3 receptors but not gonadotropin-releasing hormone receptors co-expressed in αT3-1 cells: implications for mechanisms of rapid desensitization. *Biochem. J.* 333, 301–308.
100. Waring, D. W., and Turgeon, J. L. (1980). Luteinizing hormone-releasing hormone-induced luteinizing hormone secretion *in vitro*: cyclic changes in responsiveness and self-priming. *Endocrinology* 106, 1430–1436.
101. Shah, B. H., and Milligan, G. (1994). The gonadotrophin-releasing hormone receptor of αT3-1 pituitary cells regulates cellular levels of both of the phosphoinositidase C-linked G proteins, G_qa and $G_{11}a$, equally. *Mol. Pharmacol.* 46, 1–7.
102. Shah, B. H., MacEwan, D. J., and Milligan, G. (1995). Gonadotrophin-releasing hormone receptor agonist-mediated down-regulation of Gqa/G11a (pertussis toxin-insensitive) G proteins in αT3-1 gonadotroph cells reflects increased G protein turnover but not alterations in mRNA levels. *Proc. Natl. Acad. Sci. U S A* 92, 1886–1890.
103. Koelle, M. R. (1997). A new family of G-protein regulators: the RGS proteins. *Curr. Opin. Cell. Biol.* 9, 143–147.
104. Neill, J. D., Duck, L. W., Sellers, J. C., Musgrove, L. C., Scheschonka, A., Druey, K. M., and Kehrl, J. H. (1997). Potential role for a regulator of G protein signaling (RGS3) in gonadotropin-releasing hormone (GnRH) stimulated desensitization. *Endocrinology* 138, 843–846.
105. Castro-Fernandez, C., Janovick, J. A., Brothers, S. P., Fisher, R. A., Ji, T. H., and Conn, P. M. (2002). Regulation of RGS3 and RGS10 palmitoylation by GnRH. *Endocrinology* 143, 1310–1317.
106. Castro-Fernandez, C., and Conn, P. M. (2002). Regulation of the gonadotropin-releasing hormone receptor (GnRHR) by RGS proteins: role of the GnRHR carboxyl-terminus. *Mol. Cell. Endocrinol.* 191, 149–156.
107. Zilberstein, M., Zakut, H., and Naor, Z. (1983). Coincidence of down-regulation and desensitization in pituitary gonadotrophs stimulated by gonadotropin releasing hormone. *Life Sci.* 32, 663–669.
108. Young, L. S., Naik, S. I., and Clayton, R. N. (1984). Adenosine 3′,5′-monophosphate derivatives increase gonadotropin-releasing hormone receptors in cultured pituitary cells. *Endocrinology* 114, 2114–2122.
109. McArdle, C. A., Gorospe, W. E., Huckle, W. R., and Conn, P. M. (1987). Homologous down-regulation of gonadotropin-releasing hormone receptors and desensitization of gonadotropes: lack of dependence on protein kinase C. *Mol. Endocrinol.* 1, 420–429.
110. Anderson, L., McGregor, A., Cook, J. V., Chilvers, E., and Eidne, K. A. (1995). Rapid desensitization of GnRH-stimulated intracellular signalling events in αT3-1 and HEK-293 cells expressing the GnRH receptor. *Endocrinology* 136, 5228–5231.

111. McArdle, C. A., Forrest-Owen, W., Willars, G., Davidson, J., Poch, A., and Kratzmeier, M. (1995). Desensitization of gonadotropin-releasing hormone action in the gonadotrope-derived αT3-1 cell line. *Endocrinology* 136, 4864–4871.
112. McArdle, C. A., Davidson, J. S., and Willars, G. B. (1999). The tail of the gonadotrophin-releasing hormone receptor: desensitization at, and distal to, G protein-coupled receptors. *Mol. Cell. Endocrinol.* 151, 129–136.
113. Willars, G. B., Royall, J. E., Nahorski, S. R., El-Gehani, F., Everest, H., and McArdle, C. A. (2001). Rapid down-regulation of the type I inositol 1,4,5-trisphosphate receptor and desensitization of gonadotropin-releasing hormone-mediated Ca^{2+} responses in αT3-1 gonadotropes. *J. Biol. Chem.* 276, 3123–3129.
114. Liu, F., Austin, D. A., and Webster, N. J. G. (2003). Gonadotropin-releasing hormone-desensitized LβT2 gonadotrope cells are refractory to acute protein kinase C, cyclic AMP, and calcium-dependent signaling. *Endocrinology* 144, 4354–4365.
115. Cook, J. V. F., and Eidne, K. A. (1997). An intramolecular disulfide bond between conserved extracellular cysteines in the gonadotropin-releasing hormone receptor is essential for binding and activation. *Endocrinology* 138, 2800–2806.
116. Flanagan, C. A., Becker, I. I., Davidson, J. S., Wakefield, I. K., Zhou, W., Sealfon, S. C., and Millar, R. P. (1994). Glutamate 301 of the mouse gonadotropin-releasing receptor confers specificity for arginine 8 of mammalian gonadotropin-releasing hormone. *J. Biol. Chem.* 269, 22636–22641.
117. Zhou, W., Rodic, V., Kitanovic, S., Flanagan, C. A., Chi, L., Weinstein, H., Maayani, S., Millar, R. P., and Sealfon, S. C. (1995). A locus of the gonadotropin-releasing hormone receptor that differentiates agonist and antagonist binding sites. *J. Biol. Chem.* 270, 18853–18857.
118. Davidson, J. S., McArdle, C. A., Davies, P., Elario, R., Flanagan, C. A., and Millar, R. P. (1996). Asn^{102} of the gonadotropin-releasing hormone receptor is a critical determinant of potency for agonists containing C-terminal glycinamide. *J. Biol. Chem.* 271, 15510–15514.
119. Chauvin, S., Berault, A., Lerrant, Y., Hibert, M., and Counis, R. (2000). Functional importance of transmembrane helix 6 Trp^{279} and exoloop 3 Val^{299} of rat gonadotropin-releasing hormone receptor. *Mol. Pharmacol.* 57, 625–633.
120. Flanagan, C. A., Rodic, V., Konvicka, K., Yuen, T., Chi, L., Rivier, J. E., Millar, R. P., Weinstein, H., and Sealfon, S. C. (2000). Multiple interactions of the $Asp^{2.61(98)}$ side chain of the gonadotropin-releasing hormone receptor contribute differentially to ligand interaction. *Biochemistry* 39, 8133–8141.
121. Hoffmann, S. H., ter Laak, T., Kuhne, R., Reilander, H., and Beckers, T. (2000). Residues within transmembrane helices 2 and 5 of the human gonadotropin-releasing hormone receptor contribute to agonist and antagonist binding. *Mol. Endocrinol.* 14, 1099–1115.
122. Fromme, B. J., Katz, A. A., Roeske, R. W., Millar, R. P., and Flanagan, C. A. (2001). Role of aspartate7.32(302) of the human gonadotropin-releasing hormone receptor in stabilizing a high-affinity ligand conformation. *Mol. Pharmacol.* 60, 1280–1287.
123. Hovelmann, S., Hoffmann, S. H., Kuhne, R., ter Laak, T., Reilander, H., and Beckers, T. (2002). Impact of aromatic residues within transmembrane helix 6 of the human gonadotropin-releasing hormone receptor upon agonist and antagonist binding. *Biochemistry* 41, 1129–1136.
124. Fromme, B. J., Katz, A. A., Millar, R. P., and Flanagan, C. A. (2004). $Pro^{7.33(303)}$ of the human GnRH receptor regulates selective binding of mammalian GnRH. *Mol. Cell. Endocrinol.* 219, 47–59.
125. Wang, C., Yun, O., Maiti, K., Oh, D. Y., Kim, K. K., Chae, C. H., Lee, C. J., Seong, J. Y., and Kwon, H. B. (2004). Position of Pro and Ser near $Glu^{7.32}$ in the extracellular loop 3 of mammalian and nonmammalian gonadotropin-releasing hormone (GnRH) receptors is a critical determinant for differential ligand selectivity for mammalian GnRH and chicken GnRH-II. *Mol. Endocrinol.* 18, 105–116.
126. Chauvin, S., Hibert, M., Berault, A., and Counis, R. (2001). Critical implication of transmembrane Phe310, possibly in conjunction with Trp279, in the rat gonadotropin-releasing hormone receptor activation. *Biochem. Pharmacol.* 62, 329–334.
127. Arora, K. K., Sakai, A., and Catt, K. J. (1995). Effects of second intracellular loop mutations on signal transduction and internalization of the gonadotropin-releasing hormone receptor. *J. Biol. Chem.* 270, 22820–22826.
128. Arora, K. K., Cheng, Z., and Catt, K. J. (1996). Dependence of agonist activation on an aromatic moiety in the DPLIY motif of the gonadotropin-releasing hormone receptor. *Mol. Endocrinol.* 10, 979–986.
129. Awara, W. M., Guo, C.-H., and Conn, P. M. (1996). Effects of Asn^{318} and $Asp^{87}Asn^{318}$ mutations on signal transduction by the gonadotropin-releasing hormone receptor and receptor regulation. *Endocrinology* 137, 655–662.
130. Arora, K. K., Cheng, Z., and Catt, K. J. (1997). Mutations of the conserved DRS motif in the second intracellular loop of the gonadotropin-releasing hormone receptor affect expression, activation, and internalization. *Mol. Endocrinol.* 11, 1203–1212.
131. Chung, H.-O., Yang, Q., Catt, K. J., and Arora, K. K. (1999). Expression and function of the gonadotropin-releasing hormone receptor are dependent on a conserved apolar amino acid in the third intracellular loop. *J. Biol. Chem.* 274, 35756–35762.
132. Brothers, S. P., Janovick, J. A., Maya-Nunez, G., Cornea, A., Han, X.-B., and Conn, P. M. (2002). Conserved mammalian gonadotropin-releasing hormone receptor carboxyl terminal amino acids regulate ligand binding, effector coupling and internalization. *Mol. Cell. Endocrinol.* 190, 19–27.
133. Arora, K. K., Krsmanovic, L. Z., Mores, N., O'Farrell, H., and Catt, K. J. (1998). Mediation of cyclic AMP signaling by the first intracellular loop of the gonadotropin-releasing hormone receptor. *J. Biol. Chem.* 273, 25581–25586.
134. Davidson, J. S., Flanagan, C. A., Zhou, W., Becker, I. I., Elario, R., Emeran, W., Sealfon, S. C., and Millar, R. P. (1995). Identification of N-glycosylation sites in the gonadotropin-releasing hormone receptor: role in receptor expression but not ligand binding. *Mol. Cell. Endocrinol.* 107, 241–245.
135. Davidson, J. S., Flanagan, C. A., Davies, P. D., Hapgood, J., Myburgh, D., Elario, R., Millar, R. P., Forrest-Owen, W., and McArdle, C. A. (1996). Incorporation of an additional glycosylation site enhances expression of functional human gonadotropin-releasing hormone receptor. *Endocrine* 4, 207–212.
136. Lin, X., Janovick, J. A., and Conn, P. M. (1998). Mutations at the consensus phosphorylation sites in the third intracellular loop of the rat gonadotropin-releasing hormone receptor: effects on receptor ligand binding and signal transduction. *Biol. Reprod.* 59, 1470–1476.
137. Myburgh, D. B., Pawson, A. J., Davidson, J. S., Flanagan, C. A., Millar, R. P., and Hapgood, J. P. (1998). A single amino acid substitution in transmembrane helix VI results in overexpression of the human GnRH receptor. *Eur. J. Endocrinol.* 139, 438–447.
138. Ballesteros, J., Kitanovic, S., Guarnieri, F., Davies, P., Fromme, B. J., Konvicka, K., Chi, L., Millar, R. P., Davidson, J. S., Weinstein, H., and Sealfon, S. C. (1998). Functional microdomains in G-protein-coupled receptors: the conserved arginine-cage motif in the gonadotropin-releasing hormone receptor. *J. Biol. Chem.* 273, 10445–10453.

139. Cook, J. V., Faccenda, E., Anderson, L., Couper, G. G., Eidne, K. A., and Taylor, P. L. (1993). Effects of Asn87 and Asp318 mutations on ligand binding and signal transduction in the rat GnRH receptor. *J. Endocrinol.* 139, R1–R4.
140. Zhou, W., Flanagan, C., Ballesteros, J. A., Konvicka, K., Davidson, J. S., Weinstein, H., Millar, R. P., and Sealfon, S. C. (1994). A reciprocal mutation supports helix 2 and helix 7 proximity in the gonadotropin-releasing hormone receptor. *Mol. Pharmacol.* 45, 165–170.
141. Flanagan, C. A., Zhou, W., Chi, L., Yuen, T., Rodic, V., Robertson, D., Johnson, M., Holland, P., Millar, R. P., Weinstein, H., Mitchell, R., and Sealfon, S. C. (1999). The functional microdomain in transmembrane helices 2 and 7 regulates expression, activation, and coupling pathways of the gonadotropin-releasing hormone receptor. *J. Biol. Chem.* 274, 28880–28886.
142. Bertherat, J. (1998). Gonadotropin-releasing hormone receptor gene mutation: a new cause of hereditary hypogonadism and another mutated G-protein-coupled receptor. *Eur. J. Endocrinol.* 138, 621–622.
143. Seminara, S. B., Hayes, F. J., and Crowley, W. F., Jr. (1998). Gonadotropin-releasing hormone deficiency in the human (idiopathic hypogonadotropic hypogonadism and Kallmann's syndrome): pathophysiological and genetic considerations. *Endocr. Rev.* 19, 521–539.
144. Layman, L. C., McDonough, P. G., Cohen, D. P., Maddox, M., Tho, S. P. T., and Reindollar, R. H. (2001). Familial gonadotropin-releasing hormone resistance and hypogonadotropic hypogonadism in a family with multiple affected individuals. *Fertil. Steril.* 75, 1148–1155.
145. Pitteloud, N., Boepple, P. A., DeCruz, S., Valkenburgh, S. B., Crowley, W. F., Jr., and Hayes, F. J. (2001). The fertile eunuch variant of idiopathic hypogonadotropic hypogonadism: spontaneous reversal associated with a homozygous mutation in the gonadotropin-releasing hormone receptor. *J. Clin. Endocrinol. Metab.* 86, 2470–2475.
146. Soderlund, D., Canto, P., de la Chesnaye, E., Ulloa-Aguirre, A., and Mendez, J. P. (2001). A novel homozygous mutation in the second transmembrane domain of the gonadotrophin releasing hormone receptor gene. *Clin. Endocrinol.* 54, 493–498.
147. Dewailly, D., Boucher, A., Decanter, C., Lagarde, J. P., Counis, R., and Kottler, M.-L. (2002). Spontaneous pregnancy in a patient who was homozygous for the Q106R mutation in the gonadotropin-releasing hormone receptor gene. *Fertil. Steril.* 77, 1288–1291.
148. Layman, L. C., Cohen, D. P., Xie, J., and Smith, G. D. (2002). Clinical phenotype and infertility treatment in a male with hypogonadotropic hypogonadism due to mutations Ala129Asp/Arg262Gln of the gonadotropin-releasing hormone receptor. *Fertil. Steril.* 78, 1317–1320.
149. de Roux, N., Young, J., Misrahi, M., Genet, R., Chanson, P., Schaison, G., Milgrom, E. (1997). A family with hypogonadotropic hypogonadism and mutations in the gonadotropin-releasing hormone receptor. *N. Engl. J. Med.* 337, 1597–1602.
150. Layman, L. C., Cohen, D. P., Jin, M., Xie, J., Li, Z., Reindollar, R. H., Bolbolan, S., Bick, D. P., Sherins, R. R., Duck, L. W., Musgrove, L. C., Sellers, J. C., and Neill, J. D. (1998). Mutations in gonadotropin-releasing hormone receptor gene cause hypogonadotropic hypogonadism. *Nat. Genet.* 18, 14–15.
151. de Roux, N., Young, J., Brailly-Tabard, S., Misrahi, M., Milgrom, E., and Schaison, G. (1999). The same molecular defects of the gonadotropin-releasing hormone receptor determine a variable degree of hypogonadism in affected kindred. *J. Clin. Endocrinol. Metab.* 84, 567–572.
152. Beranova, M., Oliveira, L. M., Bedecarrats, G. Y., Schipani, E., Vallejo, M., Ammini, A. C., Quintos, J. B., Hall, J. E., Martin, K. A., Hayes, F. J., Pitteloud, N., Kaiser, U. B., Crowley, W. F., Jr., and Seminara, S. B. (2001). Prevalence, phenotypic spectrum, and modes of inheritance of gonadotropin-releasing hormone receptor mutations in idiopathic hypogonadotropic hypogonadism. *J. Clin. Endocrinol. Metab.* 86, 1580–1588.
153. Costa, E. M. F., Bedecarrats, G. Y., Mendonca, B. B., Arnhold, I. J. P., Kaiser, U. B., and Latronico, A. C. (2001). Two novel mutations in the gonadotropin-releasing hormone receptor gene in Brazilian patients with hypogonadotropic hypogonadism and normal olfaction. *J. Clin. Endocrinol. Metab.* 86, 2680–2686.
154. Bedecarrats, G. Y., Linher, K. D., Janovick, J. A., Beranova, M., Kada, F., Seminara, S. B., Conn, P. M., and Kaiser, U. B. (2003). Four naturally occurring mutations in the human GnRH receptor affect ligand binding and receptor function. *Mol. Cell. Endocrinol.* 205, 51–64.
155. Meysing, A. U., Kanasaki, H., Bedecarrats, G. Y., Acierno, J. S., Jr., Conn, P. M., Martin, K. A., Seminara, S. B., Hall, J. E., Crowley, W. F., Jr., and Kaiser, U. B. (2004). *GnRHR* mutations in a woman with idiopathic hypogonadotropic hypogonadism highlight the differential sensitivity of luteinizing hormone and follicle-stimulating hormone to gonadotropin-releasing hormone. *J. Clin. Endocrinol. Metab.* 89, 3189–3198.
156. Caron, P., Chauvin, S., Christin-Maitre, S., Bennet, A., Lahlou, N., Counis, R., Bouchard, P., and Kottler, M.-L. (1999). Resistance of hypogonadic patients with mutated GnRH receptor genes to pulsatile GnRH administration. *J. Clin. Endocrinol. Metab.* 84, 990–996.
157. Pralong, F. P., Gomez, F., Castillo, E., Cotecchia, S., Abuin, L., Aubert, M. L., Portmann, L., and Gaillard, R. C. (1999). Complete hypogonadotropic hypogonadism associated with a novel inactivating mutation of the gonadotropin-releasing hormone receptor. *J. Clin. Endocrinol. Metab.* 84, 3811–3816.
158. Kottler, M.-L., Chauvin, S., Lahlou, N., Harris, C. E., Johnston, C. J., Lagarde, J.-P., Bouchard, P., Farid, N. R., and Counis, R. (2000). A new compound heterozygous mutation of the gonadotropin-releasing hormone receptor (L314X, Q106R) in a woman with complete hypogonadotropic hypogonadism: chronic estrogen administration amplifies the gonadotropin defect. *J. Clin. Endocrinol. Metab.* 85, 3002–3008.
159. Karges, B., Karges, W., Mine, M., Ludwig, L., Kuhne, R., Milgrom, E., and de Roux, N. (2003). Mutation Ala^{171}Thr stabilizes the gonadotropin-releasing hormone receptor in its inactive conformation, causing familial hypogonadotropic hypogonadism. *J. Clin. Endocrinol. Metab.* 88, 1873–1879.
160. Silveira, L. F. G., Stewart, P. M., Thomas, M., Clark, D. A., Bouloux, P. M. G., and MacColl, G. S. (2002). Novel homozygous splice acceptor site GnRH receptor (GnRHR) mutation: human GnRHR "knockout." *J. Clin. Endocrinol. Metab.* 87, 2973–2977.
161. Bedecarrats, G. Y., Linher, K. D., and Kaiser, U. B. (2003). Two common naturally occurring mutations in the human gonadotropin-releasing hormone (GnRH) receptor have differential effects on gonadotropin gene expression and on GnRH-mediated signal transduction. *J. Clin. Endocrinol. Metab.* 88, 834–843.
162. Janovick, J. A., Maya-Nunez, G., and Conn, P. M. (2002). Rescue of hypogonadotropic hypogonadism-causing and manufactured GnRH receptor mutants by a specific protein-folding template: misrouted proteins as a novel disease etiology and therapeutic target. *J. Clin. Endocrinol. Metab.* 87, 3255–3262.
163. Leanos-Miranda, A., Janovick, J. A., and Conn, P. M. (2002). Receptor-misrouting: an unexpectedly prevalent and rescuable

etiology in gonadotropin-releasing hormone receptor-mediated hypogonadotropic hypogonadism. *J. Clin. Endocrinol. Metab.* 87, 4825–4828.

164. Conn, P. M., and Crowley, W. F., Jr. (1994). Gonadotropin-releasing hormone and its analogs. *Annu. Rev. Med.* 45, 391–405.
165. White, R. B., Eisen, J. A., Kasten, T. L., and Fernald, R. D. (1998). Second gene for gonadotropin-releasing hormone in humans. *Proc. Natl. Acad. Sci. U S A* 95, 305–309.
166. Sherwood, N. M., Lovejoy, D. A., and Coe, I. R. (1993). Origin of mammalian gonadotropin-releasing hormones. *Endocr. Rev.* 14, 241–254.
167. Temple, J. L., Millar, R. P., and Rissman, E. F. (2003). An evolutionarily conserved form of gonadotropin-releasing hormone coordinates energy and reproductive behavior. *Endocrinology* 144, 13–19.
168. Kauffman, A. S., and Rissman, E. F. (2004). The evolutionarily conserved gonadotropin-releasing hormone II modifies food intake. *Endocrinology* 145, 686–691.
169. Troskie, B., Illing, N., Rumbak, E., Sun, Y.-M., Hapgood, J., Sealfon, S., Conklin, D., and Millar, R. (1998). Identification of three putative GnRH receptor subtypes in vertebrates. *Gen. Comp. Endocrinol.* 112, 296–302.
170. Illing, N., Troskie, B. E., Nahorniak, C. S., Hapgood, J. P., Peter, R. E., and Millar, R. P. (1999). Two gonadotropin-releasing hormone receptor subtypes with distinct ligand selectivity and differential distribution in brain and pituitary in the goldfish (*Carassius auratus*). *Proc. Natl. Acad. Sci. U S A* 96, 2526–2531.
171. Troskie, B. E., Hapgood, J. P., Millar, R. P., and Illing, N. (2000). Complementary deoxyribonucleic acid cloning, gene expression, and ligand selectivity of a novel gonadotropin-releasing hormone receptor expressed in the pituitary and midbrain of *Xenopus laevis*. *Endocrinology* 141, 1764–1771.
172. Okubo, K., Nagata, S., Ko, R., Kataoka, H., Yoshiura, Y., Mitani, H., Kondo, M., Naruse, K., Shima, A., and Aida, K. (2001). Identification and characterization of two distinct GnRH receptor subtypes in a teleost, the medaka *Oryzias latipes*. *Endocrinology* 142, 4729–4739.
173. Sun, Y.-M., Flanagan, C. A., Illing, N., Ott, T. R., Sellar, R., Fromme, B. J., Hapgood, J., Sharp, P., Sealfon, S. C., and Millar, R. P. (2001). A chicken gonadotropin-releasing hormone receptor that confers agonist activity to mammalian antagonists: identification of D-Lys^6 in the ligand and extracellular loop two of the receptor as determinants. *J. Biol. Chem.* 276, 7754–7761.
174. Wang, L., Bogerd, J., Choi, H. S., Seong, J. Y., Soh, J. M., Chun, S. Y., Blomenrohr, M., Troskie, B. E., Millar, R. P., Yu, W. H., McCann, S. M., and Kwon, H. B. (2001). Three distinct types of GnRH receptor characterized in the bullfrog. *Proc. Natl. Acad. Sci. U S A* 98, 361–366.
175. Seong, J. Y., Wang, L., Oh, D. Y., Yun, O., Maiti, K., Li, J. H., Soh, J. M., Choi, H. S., Kim, K., Vaudry, H., and Kwon, H. B. (2003). Ala/Thr^{201} in extracellular loop 2 and Leu/Phe^{290} in transmembrane domain 6 of type 1 frog gonadotropin-releasing hormone receptor confer differential ligand sensitivity and signal transduction. *Endocrinology* 144, 454–466.
176. Millar, R., Lowe, S., Conklin, D., Pawson, A., Maudsley, S., Troskie, B., Ott, T., Millar, M., Lincoln, G., Sellar, R., Faurholm, B., Scobie, G., Kuestner, R., Terasawa, E., and Katz, A. (2001). A novel mammalian receptor for the evolutionarily conserved type II GnRH. *Proc. Natl. Acad. Sci. U S A* 98, 9636–9641.
177. Neill, J. D., Duck, L. W., Sellers, J. C., and Musgrove, L. C. (2001). A gonadotropin-releasing hormone (GnRH) receptor specific for GnRH II in primates. *Biochem. Biophys. Res. Commun.* 282, 1012–1018.
178. Pawson, A. J., Katz, A., Sun, Y.-M., Lopes, J., Illing, N., Millar, R. P., and Davidson, J. S. (1998). Contrasting internalization kinetics of human and chicken gonadotropin-releasing hormone receptors mediated by C-terminal tail. *J. Endocrinol.* 156, R9–R12.
179. Morgan, K., Conklin, D., Pawson, A. J., Sellar, R., Ott, T. R., and Millar, R. P. (2003). A transcriptionally active human type II gonadotropin-releasing hormone receptor gene homolog overlaps two genes in the antisense orientation on chromosome 1q.12. *Endocrinology* 144, 423–436.
180. Neill, J. D. (2002). Minireview: GnRH and GnRH receptor genes in the human genome. *Endocrinology* 143, 737–743.
181. Millar, R. P., and Pawson, A. J. (2004). Outside-in and inside-out signaling: the new concept that selectivity of ligand binding at the gonadotropin-releasing hormone receptor is modulated by the intracellular environment. *Endocrinology* 145, 3590–3593.
182. Millar, R., Conklin, D., Lofton-Day, C., Hutchinson, E., Troskie, B., Illing, N., Sealfon, S. C., and Hapgood, J. (1999). A novel human GnRH receptor homolog gene: abundant and wide tissue distribution of the antisense transcript. *J. Endocrinol.* 162, 117–126.
183. Pawson, A. J., Morgan, K., Maudsley, S. R., and Millar, R. P. (2003). Type II gonadotrophin-releasing hormone (GnRH-II) in reproductive biology. *Reproduction* 126, 271–278.
184. Gault, P. M., Morgan, K., Pawson, A. J., Millar, R. P., and Lincoln, G. A. (2004). Sheep exhibit novel variations in the organization of the mammalian type II gonadotropin-releasing hormone receptor gene. *Endocrinology* 145, 2362–2374.
185. Salicioni, A. M., Xi, M., Vanderveer, L. A., Balsara, B., Testa, J. R., Dunbrack, R. L., Jr., and Godwin, A. K. (2000). Identification and structural analysis of human RBM8A and RBM8B: two highly conserved RNA-binding motif proteins that interact with OVCA1, a candidate tumor suppressor. *Genomics* 69, 54–62.
186. Faurholm, B., Millar, R. P., and Katz, A. A. (2001). The genes encoding the type II gonadotropin-releasing hormone receptor and the ribonucleoprotein RBM8A in humans overlap in two genomic loci. *Genomics* 78, 15–18.
187. Cheng, C. K., and Leung, P. C. K. (2005). Molecular biology of gonadotropin-releasing hormone (GnRH)-I, GnRH-II and their receptors in humans. *Endocr. Rev.* 26, 283–306.
188. Cabrera-Vera, T. M., Vanhauwe, J., Thomas, T. O., Medkova, M., Preininger, A., Mazzoni, M. R., and Hamm, H. E. (2003). Insights into G protein structure, function, and regulation. *Endocr. Rev.* 24, 765–781.
189. Spiegel, A. M., Shenker, A., and Weinstein, L. S. (1992). Receptor-effector coupling by G proteins: implications for normal and abnormal signal transduction. *Endocr. Rev.* 13, 536–565.
190. Stojilkovic, S. S., Reinhart, J., and Catt, K. J. (1994). Gonadotropin-releasing hormone receptors: structure and signal transduction pathways. *Endocr. Rev.* 15, 462–499.
191. Perrin, M. H., Haas, Y., Porter, J., Rivier, J., and Vale, W. (1989). The gonadotropin-releasing hormone pituitary receptor interacts with a guanosine triphosphate-binding protein: differential effects of guanyl nucleotides on agonist and antagonist binding. *Endocrinology* 124, 798–804.
192. Hawes, B. E., Marzen, J. E., Waters, S. B., and Conn, P. M. (1992). Sodium fluoride provokes gonadotrope desensitization to gonadotropin-releasing hormone (GnRH) and gonadotrope sensitization to A23187, evidence for multiple G proteins in GnRH action. *Endocrinology* 130, 2465–2475.
193. Stanislaus, D., Janovick, J. A., Brothers, S., and Conn, P. M. (1997). Regulation of $G_{q/11}a$ by the gonadotropin-releasing hormone receptor. *Mol. Endocrinol.* 11, 738–746.

194. Grosse, R., Schmid, A., Schoneberg, T., Herrlich, A., Muhn, P., Schultz, G., and Gudermann, T. (2000). Gonadotropin-releasing hormone receptor initiates multiple signaling pathways by exclusively coupling to $G_{q/11}$ proteins. *J. Biol. Chem.* 275, 9193–9200.
195. Ulloa-Aguirre, A., Stanislaus, D., Arora, V., Vaananen, J., Brothers, S., Janovick, J. A., and Conn, P. M. (1998). The third intracellular loop of the rat gonadotropin-releasing hormone receptor couples the receptor to G_s- and $G_{q/11}$-mediated signal transduction pathways: evidence from loop fragment transfection in GGH_3 cells. *Endocrinology* 139, 2472–2478.
196. Naor, Z., and Catt, K. J. (1981). Mechanism of action of gonadotropin-releasing hormone. Involvement of phospholipid turnover in luteinizing hormone release. *J. Biol. Chem.* 256, 2226–2229.
197. Andrews, W. V., and Conn, P. M. (1986). Gonadotropin-releasing hormone stimulates mass changes in phosphoinositides and diacylglycerol accumulation in purified gonadotrope cell cultures. *Endocrinology* 118, 1148–1158.
198. Naor, Z., Azrad, A., Limor, R., Zakut, H., and Lotan, M. (1986). Gonadotropin-releasing hormone activates a rapid Ca^{2+}-independent phosphodiester hydrolysis of polyphosphoinositides in pituitary gonadotrophs. *J. Biol. Chem.* 261, 12506–12512.
199. Naor, Z., Harris, D., and Shacham, S. (1998). Mechanism of GnRH receptor signaling: combinatorial cross-talk of Ca^{2+} and protein kinase C. *Front. Neuroendocrinol.* 19, 1–19.
200. Starzec, A., Jutisz, M., and Counis, R. (1989). Cyclic adenosine monophosphate and phorbol ester, like gonadotropin-releasing hormone, stimulate the biosynthesis of luteinizing hormone polypeptide chains in a nonadditive manner. *Mol. Endocrinol.* 3, 618–624.
201. Park, D., Kim, C., Cheon, M., Kim, K., and Ryu, K. (1997). cAMP and protein kinase C elevate LH beta mRNA levels by activating transcription rather than stabilizing mRNA in rat pituitary cells. *Mol. Cells* 7, 98–103.
202. Liu, F., Usui, I., Evans, L. G., Austin, D. A., Mellon, P. L., Olefsky, J. M., and Webster, N. J. G. (2002). Involvement of both $G_{q/11}$ and G_s proteins in gonadotropin-releasing hormone receptor-mediated signaling in LbT2 cells. *J. Biol. Chem.* 277, 32099–32108.
203. Castro-Fernandez, C., Brothers, S. P., and Conn, P. M. (2003). A $G\alpha_s$ mutation ($D^{229}S$) differentially effects gonadotropin-releasing hormone receptor regulation by RGS10, RGS3 and RGS3T. *Mol. Cell. Endocrinol.* 200, 119–126.
204. Guo, C.-H., Janovick, J. A., Kuphal, D., and Conn, P. M. (1995). Transient transfection of GGH_3-1′ cells (GH3 cells stably transfected with the gonadotropin-releasing hormone (GnRH) receptor complementary deoxyribonucleic acid) with the carboxyl-terminal of β-adrenergic receptor kinase 1 blocks prolactin release: evidence for a role of the G protein βγ-subunit complex in GnRH signal transduction. *Endocrinology* 136, 3031–3036.
205. Martin, T. F. J., Lewis, J. E., and Kowalchyk, J. A. (1991). Phospholipase C-β_1 is regulated by a pertussis toxin-insensitive G-protein. *Biochem. J.* 280, 753–760.
206. Rhee, S. G., and Choi, K. D. (1992). Regulation of inositol phospholipid-specific phospholipase C isozymes. *J. Biol. Chem.* 267, 12393–12396.
207. Hoer, A., and Oberdisse, E. (1991). Inositol 1,3,4,5,6-pentakisphosphate and inositol hexakisphosphate are inhibitors of the soluble inositol 1,3,4,5-tetrakisphosphate 3-phosphatase and the inositol 1,4,5-trisphosphate/1,3,4,5-tetrakisphosphate 5-phosphatase from pig brain. *Biochem. J.* 278, 219–224.
208. Biden, T. J., Comte, M., Cox, J. A., and Wollheim, C. B. (1987). Calcium-calmodulin stimulates inositol 1,4,5-trisphosphate kinase activity from insulin-secreting RINm5F cells. *J. Biol. Chem.* 262, 9437–9440.
209. Sim, S. S., Kim, J. W., and Rhee, S. G. (1990). Regulation of D-*myo*-inositol 1,4,5-trisphosphate 3-kinase by cAMP-dependent protein kinase and protein kinase C. *J. Biol. Chem.* 265, 10367–10372.
210. Johnson, R. M., Wasilenko, W. J., Mattingly, R. R., Weber, M. J., and Garrison, J. C. (1989). Fibroblasts transformed with v-*src* show enhanced formation of an inositol tetrakisphosphate. *Science* 246, 121–124.
211. Oberdisse, E., Nolan, R. D., and Lapetina, E. G. (1990). Thrombin and phorbol ester stimulate inositol 1,3,4,5-tetrakisphosphate 3-phosphomonoesterase in human platelets. *J. Biol. Chem.* 265, 726–730.
212. Morgan, R. O., Chang, J. P., and Catt, K. J. (1987). Novel aspects of gonadotropin-releasing hormone action on inositol polyphosphate metabolism in cultured pituitary gonadotrophs. *J. Biol. Chem.* 262, 1166–1171.
213. Naor, Z., Zer, J., Zakut, H., and Hermon, J. (1985). Characterization of pituitary calcium-activated phospholipid-dependent protein kinase: redistribution by gonadotropin-releasing hormone. *Proc. Natl. Acad. Sci. U S A* 82, 8203–8207.
214. Guillemette, G., Balla, T., Baukal, A. J., and Catt, K. J. (1987). Inositol 1,4,5-trisphosphate binds to a specific receptor and releases microsomal calcium in the anterior pituitary gland. *Proc. Natl. Acad. Sci. U S A* 84, 8195–8199.
215. Naor, Z., Capponi, A. M., Rossier, M. F., Ayalon, D., and Limor, R. (1988). Gonadotropin-releasing hormone-induced rise in cytosolic free Ca^{2+} levels: mobilization of cellular and extracellular Ca^{2+} pools and relationship to gonadotropin secretion. *Mol. Endocrinol.* 2, 512–520.
216. Stojilkovic, S. S., Chang, J. P., Ngo, D., Tasaka, K., Izumi, S.-I., and Catt, K. J. (1989). Mechanism of action of GnRH: the participation of calcium mobilization and activation of protein kinase C in gonadotropin secretion. *J. Steroid Biochem.* 33, 693–703.
217. Spat, A., Bradford, P. G., McKinney, J. S., Rubin, R. P., and Putney, J. W., Jr. (1986). A saturable receptor for ^{32}P-inositol-1,4,5-triphosphate in hepatocytes and neutrophils. *Nature* 319, 514–516.
218. Tse, A., and Hille, B. (1992). GnRH-induced Ca^{2+} oscillations and rhythmic hyperpolarizations of pituitary gonadotropes. *Science* 255, 462–464.
219. Ferris, C. D., Huganir, R. L., and Snyder, S. H. (1990). Calcium flux mediated by purified inositol 1,4,5-trisphosphate receptor in reconstituted lipid vesicles is allosterically regulated by adenine nucleotides. *Proc. Natl. Acad. Sci. U S A* 87, 2147–2151.
220. Finch, E. A., Turner, T., and Goldin, S. M. (1991). Calcium as a coagonist of inositol 1,4,5-trisphosphate-induced calcium release. *Science* 252, 443–446.
221. Watras, J., Bezprozvanny, I., and Ehrlich, B. E. (1991). Inositol 1,4,5-trisphosphate-gated channels in cerebellum: presence of multiple conductance states. *J. Neurosci.* 11, 3239–3245.
222. Furuichi, T., Yoshikawa, S., Miyawaki, A., Wada, K., Maeda, N., and Mikoshiba, K. (1989). Primary structure and functional expression of the inositol 1,4,5-trisphosphate-binding protein P_{400}. *Nature* 342, 32–38.
223. Mignery, G. A., Newton, C. L., Archer, B. T., III, and Sudhof, T. C. (1990). Structure and expression of the rat inositol 1,4,5-trisphosphate receptor. *J. Biol. Chem.* 265, 12679–12685.
224. Muallem, S., Schoeffield, M., Pandol, S., and Sachs, G. (1985). Inositol trisphosphate modification of ion transport

in rough endoplasmic reticulum. *Proc. Natl. Acad. Sci. U S A* 82, 4433–4437.
225. Kukuljan, M., Stojilkovic, S. S., Rojas, E., and Catt, K. J. (1992). Apamin-sensitive potassium channels mediate agonist-induced oscillations of membrane potential in pituitary gonadotrophs. *FEBS Lett.* 301, 19–22.
226. Bezprozvanny, I., Watras, J., and Ehrlich, B. E. (1991). Bell-shaped calcium-response curves of Ins(1,4,5)P_3- and calcium-gated channels from endoplasmic reticulum of cerebellum. *Nature* 351, 751–754.
227. Pietri, F., Hilly, M., and Mauger, J. P. (1990). Calcium mediates the interconversion between two states of the liver inositol 1,4,5-trisphosphate receptor. *J. Biol. Chem.* 265, 17478–17485.
228. Dupont, G., and Goldbeter, A. (1993). One-pool model for Ca^{2+} oscillations involving Ca^{2+} and inositol 1,4,5-trisphosphate as co-agonists for Ca^{2+} release. *Cell. Calcium* 14, 311–322.
229. Iida, T., Stojilkovic, S. S., Izumi, S.-I., and Catt, K. J. (1991). Spontaneous and agonist-induced calcium oscillations in pituitary gonadotrophs. *Mol. Endocrinol.* 5, 949–958.
230. Leong, D. A., and Thorner, M. O. (1991). A potential code of luteinizing hormone-releasing hormone-induced calcium ion responses in the regulation of luteinizing hormone secretion among individual gonadotropes. *J. Biol. Chem.* 266, 9016–9022.
231. Iino, M., and Endo, M. (1992). Calcium-dependent immediate feedback control of inositol 1,4,5-triphosphate-induced Ca^{2+} release. *Nature* 360, 76–78.
232. Li, Y.-X., Rinzel, J., Keizer, J., and Stojilkovic, S. S. (1994). Calcium oscillations in pituitary gonadotrophs: comparison of experiment and theory. *Proc. Natl. Acad. Sci. U S A* 91, 58–62.
233. Stojilkovic, S. S., Tomic, M., Kukuljan, M., and Catt, K. J. (1994). Control of calcium spiking frequency in pituitary gonadotrophs by a single-pool cytoplasmic oscillator. *Mol. Pharmacol.* 45, 1013–1021.
234. Taylor, C. W., and Marshall, I. C. (1992). Calcium and inositol 1,4,5-trisphosphate receptors: a complex relationship. *Trends Biochem. Sci.* 17, 403–407.
235. Missiaen, L., De Smedt, H., Droogmans, G., and Casteels, R. (1992). Ca^{2+} release induced by inositol 1,4,5-trisphosphate is a steady-state phenomenon controlled by luminal Ca^{2+} in permeabilized cells. *Nature* 357, 599–602.
236. Swillens, S., and Mercan, D. (1990). Computer simulation of a cytosolic calcium oscillator. *Biochem. J.* 271, 835–838.
237. Tomic, M., Cesnajaj, M., Catt, K. J., and Stojilkovic, S. S. (1994). Developmental and physiological aspects of Ca^{2+} signaling in agonist-stimulated pituitary gonadotrophs. *Endocrinology* 135, 1762–1771.
238. Sagara, Y., Fernandez-Belda, F., de Meis, L., and Inesi, G. (1992). Characterization of the inhibition of intracellular Ca^{2+} transport ATPases by thapsigargin. *J. Biol. Chem.* 267, 12606–12613.
239. De Young, G. W., and Keizer, J. (1992). A single-pool inositol 1,4,5-trisphosphate-receptor-based model for agonist-stimulated oscillations in Ca^{2+} concentration. *Proc. Natl. Acad. Sci. U S A* 89, 9895–9899.
240. Stojilkovic, S., Kukuljan, M., Iida, T., Rojas, E., and Catt, K. (1992). Integration of cytoplasmic calcium and membrane potential oscillations maintains calcium signaling in pituitary gonadotrophs. *Proc. Natl. Acad. Sci. U S A* 89, 4081–4085.
241. Kukuljan, M., Rojas, E., Catt, K. J., and Stojilkovic, S. S. (1994). Membrane potential regulates inositol 1,4,5-trisphosphate-controlled cytoplasmic Ca^{2+} oscillations in pituitary gonadotrophs. *J. Biol. Chem.* 269, 4860–4865.
242. Merelli, F., Stojilkovic, S. S., Iida, T., Krsmanovic, L. Z., Zheng, L., Mellon, P. L., and Catt, K. J. (1992). Gonadotropin-releasing hormone-induced calcium signaling in clonal pituitary gonadotrophs. *Endocrinology* 131, 925–932.
243. Thomas, P., Mellon, P. L., Turgeon, J. L., and Waring, D. W. (1996). The LβT2 clonal gonadotrope: a model for single cell studies of endocrine cell secretion. *Endocrinology* 137, 2979–2989.
244. Tasaka, K., Stojilkovic, S. S., Izumi, S.-I., and Catt, K. J. (1988). Biphasic activation of cytosolic free calcium and LH responses by gonadotropin-releasing hormone. *Biochem. Biophys. Res. Commun.* 154, 398–403.
245. Stojilkovic, S. S., Stutzin, A., Izumi, S.-I., Dufour, S., Torsello, A., Virmani, M. A., Rojas, E., and Catt, K. J. (1990). Generation and amplification of the cytosolic calcium signal during secretory responses to gonadotropin-releasing hormone. *New Biol.* 2, 272–283.
246. Stutzin, A., Stojilkovic, S. S., Catt, K. J., and Rojas, E. (1989). Characteristics of two types of calcium channels in rat pituitary gonadotrophs. *Am. J. Physiol. Cell. Physiol.* 257, C865–C874.
247. Bosma, M. M., and Hille, B. (1992). Electrophysiological properties of a cell line of the gonadotrope lineage. *Endocrinology* 130, 3411–3420.
248. Tse, A., and Hille, B. (1993). Role of voltage-gated Na^+ and Ca^{2+} channels in gonadotropin-releasing hormone-induced membrane potential changes in identified rat gonadotropes. *Endocrinology* 132, 1475–1481.
249. Stojilkovic, S. S., Iida, T., Merelli, F., Torsello, A., Krsmanovic, L. Z., and Catt, K. J. (1991). Interactions between calcium and protein kinase C in the control of signaling and secretion in pituitary gonadotrophs. *J. Biol. Chem.* 266, 10377–10384.
250. Davidson, L., Pawson, A. J., de Maturana, R. L., Freestone, S. H., Barran, P., Millar, R. P., and Maudsley, S. (2004). Gonadotropin-releasing hormone-induced activation of diacylglycerol kinase-z and its association with active c-Src. *J. Biol. Chem.* 279, 11906–11916.
251. Chang, J. P., Morgan, R. O., and Catt, K. J. (1988). Dependence of secretory responses to gonadotropin-releasing hormone on diacylglycerol metabolism: studies with a diacylglycerol lipase inhibitor, RHC 80267. *J. Biol. Chem.* 263, 18614–18620.
252. Zheng, L., Stojilkovic, S. S., Hunyady, L., Krsmanovic, L. Z., and Catt, K. J. (1994). Sequential activation of phospholipase-C and -D in agonist-stimulated gonadotrophs. *Endocrinology* 134, 1446–1454.
253. Exton, J. H. (1990). Signaling through phosphatidylcholine breakdown. *J. Biol. Chem.* 265, 1–4.
254. Billah, M. M., Anthes, J. C., and Mullmann, T. J. (1991). Receptor-coupled phospholipase D: regulation and functional significance. *Biochem. Soc. Trans.* 19, 324–329.
255. Liscovitch, M. (1992). Crosstalk among multiple signal-activated phospholipases. *Trends Biochem. Sci.* 17, 393–399.
256. Nishizuka, Y. (1992). Intracellular signaling by hydrolysis of phospholipids and activation of protein kinase C. *Science* 258, 607–614.
257. Shinomura, T., Asaoka, Y., Oka, M., Yoshida, K., and Nishizuka, Y. (1991). Synergistic action of diacylglycerol and unsaturated fatty acid for protein kinase C activation: its possible implications. *Proc. Natl. Acad. Sci. U S A* 88, 5149–5153.
258. Asaoka, Y., Nakamura, S., Yoshida, K., and Nishizuka, Y. (1992). Protein kinase C, calcium and phospholipid degradation. *Trends Biochem. Sci.* 17, 414–417.
259. Nishizuka, Y. (1984). Turnover of inositol phospholipids and signal transduction. *Science* 225, 1365–1370.

260. Huang, K. P. (1989). The mechanism of protein kinase C activation. *Trends Neurosci.* 12, 425–432.
261. Kraft, A. S., and Anderson, W. B. (1983). Phorbol esters increase the amount of Ca^{2+}, phospholipid-dependent protein kinase associated with plasma membrane. *Nature* 301, 621–623.
262. McArdle, C. A., and Conn, P. M. (1986). Hormone-stimulated redistribution of gonadotrope protein kinase C *in vivo*: dependence on Ca^{2+} influx. *Mol. Pharmacol.* 29, 570–576.
263. Grunicke, H. H., and Uberall, F. (1992). Protein kinase C modulation. *Semin. Cancer Biol.* 3, 351–360.
264. Naor, Z. (1990). Further characterization of protein kinase C subspecies in the hypothalamo-pituitary axis: differential activation by phorbol esters. *Endocrinology* 126, 1521–1526.
265. Ohmichi, M., Hirota, K., Koike, K., Miyake, A., Tanizawa, O., Sato, M., and Tohyama, M. (1992). Immunohistochemical evidence that rat FSH cells contain βII-subspecies of protein kinase C. *Endocrinol. Jpn.* 39, 609–613.
266. Jin, L., Maeda, T., Chandler, W. F., and Lloyd, R. V. (1993). Protein kinase C (PKC) activity and PKC messenger RNAs in human pituitary adenomas. *Am. J. Pathol.* 142, 569–578.
267. Johnson, M. S., MacEwan, D. J., Simpson, J., and Mitchell, R. (1992). Characterisation of protein kinase C isoforms and enzymic activity from the αT3-1 gonadotroph-derived cell line. *FEBS Lett.* 333, 67–72.
268. Sundaresan, S., Colin, I. M., Pestell, R. G., and Jameson, J. L. (1996). Stimulation of mitogen-activated protein kinase by gonadotropin-releasing hormone: evidence for the involvement of protein kinase C. *Endocrinology* 137, 304–311.
269. Cheng, K. W., Ngan, E. S. W., Kang, S. K., Chow, B. K. C., and Leung, P. C. K. (2000). Transcriptional down-regulation of human gonadotropin-releasing hormone (GnRH) receptor gene by GnRH: role of protein kinase C and activating protein 1. *Endocrinology* 141, 3611–3622.
270. Vasilyev, V. V., Pernasetti, F., Rosenberg, S. B., Barsoum, M. J., Austin, D. A., Webster, N. J. G., and Mellon, P. L. (2002). Transcriptional activation of the ovine follicle-stimulating hormone-b gene by gonadotropin-releasing hormone involves multiple signal transduction pathways. *Endocrinology* 143, 1651–1659.
271. Kiley, S. C., Parker, P. J., Fabbro, D., and Jaken, S. (1991). Differential regulation of protein kinase C isozymes by thyrotropin-releasing hormone in GH_4C_1 cells. *J. Biol. Chem.* 266, 23761–23768.
272. Kiley, S. C., Parker, P. J., Fabbro, D., and Jaken, S. (1992). Hormone- and phorbol ester-activated protein kinase C isozymes mediate a reorganization of the actin cytoskeleton associated with prolactin secretion in GH_4C_1 cells. *Mol. Endocrinol.* 6, 120–131.
273. Nishizuka, Y. (1986). Studies and perspectives of protein kinase C. *Science* 233, 305–312.
274. Shearman, M. S., Sekiguchi, K., and Nishizuka, Y. (1989). Modulation of ion channel activity: a key function of the protein kinase C enzyme family. *Pharmacol. Rev.* 41, 211–237.
275. Ozawa, K., Yamada, K., Kazanietz, M. G., Blumberg, P. M., and Beaven, M. A. (1993). Different isozymes of protein kinase C mediate feedback inhibition of phospholipase C and stimulatory signals for exocytosis in rat RBL-2H3 cells. *J. Biol. Chem.* 268, 2280–2283.
276. Stojilkovic, S. S., Chang, J. P., Izumi, S.-I., Tasaka, K., and Catt, K. J. (1988). Mechanisms of secretory responses to gonadotropin-releasing hormone and phorbol esters in cultured pituitary cells: participation of protein kinase C and extracellular calcium mobilization. *J. Biol. Chem.* 263, 17301–17306.
277. Shraga-Levine, Z., Ben-Menahem, D., and Naor, Z. (1994). Activation of protein kinase C β gene expression by gonadotropin-releasing hormone in αT3-1 cell line. *J. Biol. Chem.* 269, 31028–31033.
278. Turgeon, J. L., and Cooper, R. H. (1986). Protein kinase C and an endogenous substrate associated with adenohypophyseal secretory granules. *Biochem. J.* 237, 53–61.
279. Strulovici, B., Tahilramani, R., and Nestor, J. J., Jr. (1987). Phosphorylation substrates for protein kinase C in intact pituitary cells: characterization of a receptor-mediated event using novel gonadotropin-releasing hormone analogues. *Biochemistry* 26, 6005–6011.
280. Hunter, T. (1995). Protein kinases and phosphatases: the Yin and Yang of protein phosphorylation and signaling. *Cell* 80, 225–236.
281. Seger, R., and Krebs, E. G. (1995). The MAPK signaling cascade. *FASEB J.* 9, 726–735.
282. Pearson, G., Robinson, F., Beers Gibson, T., Xu, B.-E., Karandikar, M., Berman, K., and Cobb, M. H. (2002). Mitogen-activated protein (MAP) kinase pathways: regulation and physiological functions. *Endocr. Rev.* 22, 153–183.
283. Alblas, J., van Corven, E. J., Hordijk, P. L., Milligan, G., and Moolenaar, W. H. (1993). G_i-mediated activation of the $p21^{ras}$-mitogen-activated protein kinase pathway by α_2-adrenergic receptors expressed in fibroblasts. *J. Biol. Chem.* 268, 22235–22238.
284. Crespo, P., Xu, N., Simonds, W. F., and Gutkind, J. S. (1994). Ras-dependent activation of MAP kinase pathway mediated by G-protein βγ subunits. *Nature* 369, 418–420.
285. Roberson, M. S., Misra-Press, A., Laurance, M. E., Stork, P. J., and Maurer, R. A. (1995). A role for mitogen-activated protein kinase in mediating activation of the glycoprotein hormone alpha-subunit promoter by gonadotropin-releasing hormone. *Mol. Cell. Biol.* 15, 3531–3539.
286. Roberson, M. S., Zhang, T., Li, H. L., and Mulvaney, J. M. (1999). Activation of the p38 mitogen-activated protein kinase pathway by gonadotropin-releasing hormone. *Endocrinology* 140, 1310–1318.
287. Liu, F., Austin, D. A., Mellon, P. L., Olefsky, J. M., and Webster, N. J. G. (2002). GnRH activates ERK1/2 leading to the induction of c-*fos* and LHb protein expression in LβT2 cells. *Mol. Endocrinol.* 16, 419–434.
288. Grosse, R., Roelle, S., Herrlich, A., Hohn, J., and Gudermann, T. (2000). Epidermal growth factor receptor tyrosine kinase mediates Ras activation by gonadotropin-releasing hormone. *J. Biol. Chem.* 275, 12251–12260.
289. Gille, H., Kortenjann, M., Thomae, O., Moomaw, C., Slaughter, C., Cobb, M. H., and Shaw, P. E. (1995). ERK phosphorylation potentiates Elk-1-mediated ternary complex formation and transactivation. *EMBO J.* 14, 951–962.
290. Harris, D., Bonfil, D., Chuderland, D., Kraus, S., Seger, R., and Naor, Z. (2002). Activation of MAPK cascades by GnRH: ERK and Jun N-terminal kinase are involved in basal and GnRH-stimulated activity of the glycoprotein hormone LHβ-subunit promoter. *Endocrinology* 143, 1018–1025.
291. Yuen, T., Wurmbach, E., Ebersole, B. J., Ruf, F., Pfeffer, R. L., and Sealfon, S. C. (2002). Coupling of GnRH concentration and the GnRH receptor-activated gene program. *Mol. Endocrinol.* 16, 1145–1153.
292. Harris, D., Chuderland, D., Bonfil, D., Kraus, S., Seger, R., and Naor, Z. (2003). Extracellular signal-regulated kinase and c-Src, but not Jun N-terminal kinase, are involved in basal and gonadotropin-releasing hormone-stimulated activity of the glycoprotein hormone α-subunit promoter. *Endocrinology* 144, 612–622.
293. Yokoi, T., Ohmichi, M., Tasaka, K., Kimura, A., Kanda, Y., Hayakawa, J., Tahara, M., Hisamoto, K., Kurachi, H., and Murata, Y. (2000). Activation of the luteinizing hormone β promoter by gonadotropin-releasing hormone requires c-Jun

NH_2-terminal protein kinase. *J. Biol. Chem.* 275, 21639–21647.

294. Mulvaney, J. M., and Roberson, M. S. (2000). Divergent signaling pathways requiring discrete calcium signals mediate concurrent activation of two mitogen-activated protein kinases by gonadotropin-releasing hormone. *J. Biol. Chem.* 275, 14182–14189.
295. Zhang, T., Mulvaney, J. M., and Roberson, M. S. (2001). Activation of mitogen-activated protein kinase phosphatase 2 by gonadotropin-releasing hormone. *Mol. Cell. Endocrinol.* 172, 79–89.
296. Sheng, M., and Greenberg, M. E. (1990). The regulation and function of c-fos and other immediate early genes in the nervous system. *Neuron* 4, 477–485.
297. McMahon, S. B., and Monroe, J. G. (1992). Role of primary response genes in generating cellular responses to growth factors. *FASEB J.* 6, 2707–2715.
298. Norwitz, E. R., Cardona, G. R., Jeong, K.-H., and Chin, W. W. (1999). Identification and characterization of the gonadotropin-releasing hormone response elements in the mouse gonadotropin-releasing hormone receptor gene. *J. Biol. Chem.* 274, 867–880.
299. Norwitz, E. R., Jeong, K.-H., and Chin, W. W. (1999). Molecular mechanisms of gonadotropin-releasing hormone receptor gene regulation. *J. Soc. Gynecol. Investig.* 6, 169–178.
300. White, B. R., Duval, D. L., Mulvaney, J. M., Roberson, M. S., and Clay, C. M. (1999). Homologous regulation of the gonadotropin-releasing hormone receptor gene is partially mediated by protein kinase C activation of an activator protein-1 element. *Mol. Endocrinol.* 13, 566–577.
301. Ellsworth, B. S., White, B. R., Burns, A. T., Cherrington, B. D., Otis, A. M., and Clay, C. M. (2003). c-Jun N-terminal kinase activation of activator protein-1 underlies homologous regulation of the gonadotropin-releasing hormone receptor gene in αT3-1 cells. *Endocrinology* 144, 839–849.
302. Coss, D., Jacobs, S. B. R., Bender, C. E., and Mellon, P. L. (2004). A novel AP-1 site is critical for maximal induction of the follicle-stimulating hormone β gene by gonadotropin-releasing hormone. *J. Biol. Chem.* 279, 152–162.
303. Lee, W., Haslinger, A., Karin, M., and Tjian, R. (1987). Activation of transcription by two factors that bind promoter and enhancer sequences of the human metallothionein gene and SV40. *Nature* 325, 368–372.
304. Livingstone, C., Patel, G., and Jones, N. (1995). ATF-2 contains a phosphorylation-dependent transcriptional activation domain. *EMBO J.* 14, 1785–1797.
305. Morgan, J. I., and Curran, T. (1991). Stimulus-transcription coupling in the nervous system: involvement of the inducible proto-oncogenes *fos* and *jun*. *Annu. Rev. Neurosci.* 14, 421–451.
306. Treisman, R. (1992). The serum response element. *Trends Biochem. Sci.* 17, 423–426.
307. Cesnjaj, M., Catt, K. J., and Stojilkovic, S. S. (1994). Coordinate actions of calcium and protein kinase-C in the expression of primary response genes in pituitary gonadotrophs. *Endocrinology* 135, 692–701.
308. Mulvaney, J. M., Zhang, T., Fewtrell, C., and Roberson, M. S. (1999). Calcium influx through L-type channels is required for selective activation of extracellular signal-regulated kinase by gonadotropin-releasing hormone. *J. Biol. Chem.* 274, 29796–29804.
309. Duan, W. R., Ito, M., Park, Y., Maizels, E. T., Hunzicker-Dunn, M., and Jameson, J. L. (2002). GnRH regulates early growth response protein 1 transcription through multiple promoter elements. *Mol. Endocrinol.* 16, 221–233.
310. Horton, C. D., and Halvorson, L. M. (2004). The cAMP signaling system regulates LHb gene expression: roles of early growth response protein-1, SP1 and steroidogenic factor-1. *J. Mol. Endocrinol.* 32, 291–306.
311. Borgeat, P., Chavancy, G., Dupont, A., Labrie, F., Arimura, A., and Schally, A. V. (1972). Stimulation of adenosine 3′:5′-cyclic monophosphate accumulation in anterior pituitary gland *in vitro* by synthetic luteinizing hormone-releasing hormone. *Proc. Natl. Acad. Sci. U S A* 69, 2677–2681.
312. Garrel, G., McArdle, C. A., Hemmings, B. A., and Counis, R. (1997). Gonadotropin-releasing hormone and pituitary adenylate cyclase-activating polypeptide affect levels of cyclic adenosine 3′,5′-monophosphate-dependent protein kinase A (PKA) subunits in the clonal gonadotrope αT3-1 cells: evidence for cross-talk between PKA and protein kinase C pathways. *Endocrinology* 138, 2259–2266.
313. Ransone, L. J., and Verma, I. M. (1990). Nuclear proto-oncogenes *fos* and *jun*. *Annu. Rev. Cell. Biol.* 6, 539–557.
314. Meyer, T. E., and Habener, J. F. (1993). Cyclic adenosine 3′,5′-monophosphate response element binding protein (CREB) and related transcription-activating deoxyribonucleic acid-binding proteins. *Endocr. Rev.* 14, 269–290.
315. Rolli, M., Kotlyarov, A., Sakamoto, K. M., Gaestel, M., and Neininger, A. (1999). Stress-induced stimulation of early growth response gene-1 by p38/stress-activated protein kinase 2 is mediated by a cAMP-responsive promoter element in a MAPKAP kinase 2-independent manner. *J. Biol. Chem.* 274, 19559–19564.
316. Smith, C. E., Wakefield, I., King, J. A., Naor, Z., Millar, R. P., and Davidson, J. S. (1987). The initial phase of GnRH-stimulated LH release from pituitary cells is independent of calcium entry through voltage-gated channels. *FEBS Lett.* 225, 247–250.
317. Tse, A., Tse, F. W., Almers, W., and Hille, B. (1993). Rhythmic exocytosis stimulated by GnRH-induced calcium oscillations in rat gonadotropes. *Science* 260, 82–84.
318. Hansen, J. R., McArdle, C. A., and Conn, P. M. (1987). Relative roles of calcium derived from intra- and extracellular sources in dynamic luteinizing hormone release from perifused pituitary cells. *Mol. Endocrinol.* 1, 808–815.
319. Chang, J. P., Stojilkovic, S. S., Graeter, J. S., and Catt, K. J. (1988). Gonadotropin-releasing hormone stimulates luteinizing hormone secretion by extracellular calcium-dependent and -independent mechanisms. *Endocrinology* 123, 87–97.
320. McArdle, C. A., and Poch, A. (1992). Dependence of gonadotropin-releasing hormone-stimulated luteinizing hormone release upon intracellular Ca^{2+} pools is revealed by desensitization and thapsigargin blockade. *Endocrinology* 130, 3567–3574.
321. Bates, M. D., and Conn, P. M. (1984). Calcium mobilization in the pituitary gonadotrope: relative roles of intra- and extracellular sources. *Endocrinology* 115, 1380–1385.
322. Jennes, L., Bronson, D., Stumpf, W. E., and Conn, P. M. (1985). Evidence for an association between calmodulin and membrane patches containing gonadotropin-releasing hormone–receptor complexes in cultured gonadotropes. *Cell. Tissue Res.* 239, 311–315.
323. Conn, P. M., Chafouleas, J. G., Rogers, D., and Means, A. R. (1981). Gonadotropin releasing hormone stimulates calmodulin redistribution in rat pituitary. *Nature* 292, 264–265.
324. Natarajan, K., Ness, J., Wooge, C. H., Janovick, J. A., and Conn, P. M. (1991). Specific identification and subcellular localization of three calmodulin-binding proteins in the rat gonadotrope: spectrin, caldesmon, and calcineurin. *Biol. Reprod.* 44, 43–52.
325. Janovick, J. A., Natarajan, K., Longo, F., and Conn, P. M. (1991). Caldesmon: a bifunctional (calmodulin and actin) binding protein which regulates stimulated gonadotropin release. *Endocrinology* 129, 68–74.

326. Huckle, W. R., and Conn, P. M. (1987). The relationship between gonadotropin-releasing hormone-stimulated luteinizing hormone release and inositol phosphate production: studies with calcium antagonists and protein kinase C activators. *Endocrinology* 120, 160–169.
327. Nicol, L., McNeilly, J., R., Stridsberg, M., Crawford, J. L., and McNeilly, A. S. (2002). Influence of steroids and GnRH on biosynthesis and secretion of secretogranin II and chromogranin A in relation to LH release in LβT2 gonadotroph cells. *J. Endocrinol.* 174, 473–483.
328. Nicol, L., McNeilly, J., R., Stridsberg, M., and McNeilly, A. S. (2004). Differential secretion of gonadotrophins: investigation of the role of secretogranin II and chromogranin A in the release of LH and FSH in LβT2 cells. J. *Mol. Endocrinol.* 32, 467–480.
329. Tang, L. K., Martellock, A. C., and Horiuchi, J. K. (1982). Estradiol stimulation of LH response to LHRH and LHRH binding in pituitary cultures. *Am. J. Physiol. Endocrinol. Metab.* 242, E392–E397.
330. Johnson, M. S., Mitchell, R., and Fink, G. (1988). The role of protein kinase C in LHRH-induced LH and FSH release and LHRH self-priming in rat anterior pituitary glands *in vitro*. *J. Endocrinol.* 116, 231–239.
331. Harris, C. E., Staley, D., and Conn, P. M. (1985). Diacylglycerols and protein kinase C: potential amplifying mechanism for Ca^{2+}-mediated gonadotropin-releasing hormone-stimulated luteinizing hormone release. *Mol. Pharmacol.* 27, 532–536.
332. McArdle, C. A., Huckle, W. R., Johnson, L. A., and Conn, P. M. (1988). Enhanced responsiveness of gonadotropes after protein kinase-C activation: postreceptor regulation of gonadotropin releasing hormone action. *Endocrinology* 122, 1905–1914.
333. Stojilkovic, S. S., Chang, J. P., Ngo, D., and Catt, K. J. (1988). Evidence for a role of protein kinase C in luteinizing hormone synthesis and secretion: impaired responses to gonadotropin-releasing hormone in protein kinase C-depleted pituitary cells. *J. Biol. Chem.* 263, 17307–17311.
334. Stojilkovic, S. S., and Catt, K. J. (1992). Calcium oscillations in anterior pituitary cells. *Endocr. Rev.* 13, 256–280.
335. Naor, Z., Dan-Cohen, H., Hermon, J., and Limor, R. (1989). Induction of exocytosis in permeabilized pituitary cells by α- and β-type protein kinase C. *Proc. Natl. Acad. Sci. U S A* 86, 4501–4504.
336. Tsujii, T., Ishizaka, K., and Winters, S. J. (1994). Effects of pituitary adenylate cyclase-activating polypeptide on gonadotropin secretion and subunit messenger ribonucleic acids in perifused rat pituitary cells. *Endocrinology* 135, 826–833.
337. Gay, V. L., and Bogdanove, E. M. (1969). Plasma and pituitary LH and FSH in the castrated rat following short-term steroid treatment. *Endocrinology* 84, 1132–1142.
338. Schanbacher, B. D., and Ford, J. J. (1977). Gonadotropin secretion in cryptorchid and castrate rams and the acute effects of exogenous steroid treatment. *Endocrinology* 100, 387–393.
339. Rivier, C., and Vale, W. (1991). Effect of recombinant inhibin on follicle-stimulating hormone secretion by the female rat: interaction with a gonadotropin-releasing hormone antagonist and estrogen. *Endocrinology* 129, 2160–2165.
340. Kennedy, J., and Chappel, S. (1985). Direct pituitary effects of testosterone and luteinizing hormone-releasing hormone upon follicle-stimulating hormone: analysis by radioimmuno- and radioreceptor assay. *Endocrinology* 116, 741–748.
341. Strobl, F. J., and Levine, J. E. (1988). Estrogen inhibits luteinizing hormone (LH), but not follicle-stimulating hormone secretion in hypophysectomized pituitary-grafted rats receiving pulsatile LH-releasing hormone infusions. *Endocrinology* 123, 622–630.
342. Rivier, C., Rivier, J., and Vale, W. (1986). Inhibin-mediated feedback control of follicle-stimulating hormone secretion in the female rat. *Science* 234, 205–208.
343. Carroll, R. S., Corrigan, A. Z., Gharib, S. D., Vale, W., and Chin, W. W. (1989). Inhibin, activin, and follistatin: regulation of follicle-stimulating hormone messenger ribonucleic acid levels. *Mol. Endocrinol.* 3, 1969–1976.
344. Rivier, C., Schwall, R., Mason, A., Burton, L., Vaughan, J., and Vale, W. (1991). Effect of recombinant inhibin on luteinizing hormone and follicle-stimulating hormone secretion in the rat. *Endocrinology* 128, 1548–1554.
345. Rivier, C., and Vale, W. (1991). Effect of recombinant activin-A on gonadotropin secretion in the female rat. *Endocrinology* 129, 2463–2465.
346. Campen, C. A., and Vale, W. (1988). Interaction between purified ovine inhibin and steroids on the release of gonadotropins from cultured rat pituitary cells. *Endocrinology* 123, 1320–1328.
347. Roberts, V., Meunier, H., Vaughan, J., Rivier, J., Rivier, C., Vale, W., and Sawchenko, P. (1989). Production and regulation of inhibin subunits in pituitary gonadotropes. *Endocrinology* 124, 552–554.
348. Corrigan, A. Z., Bilezikjian, L. M., Carroll, R. S., Bald, L. N., Schmelzer, C. H., Fendly, B. M., Mason, A. J., Chin, W. W., Schwall, R. H., and Vale, W. (1991). Evidence for an autocrine role of activin B within rat anterior pituitary cultures. *Endocrinology* 128, 1682–1684.
349. Kogawa, K., Nakamura, T., Sugino, K., Takio, K., Titani, K., and Sugino, H. (1991). Activin-binding protein is present in pituitary. *Endocrinology* 128, 1434–1440.
350. Kaiser, U. B., Lee, B. L., Carroll, R. S., Unabia, G., Chin, W. W., and Childs, G. V. (1992). Follistatin gene expression in the pituitary: localization in gonadotropes and folliculostellate cells in diestrous rats. *Endocrinology* 130, 3048–3056.
351. Blum, J. J., Reed, M. C., Janovick, J. A., and Conn, P. M. (2000). A mathematical model quantifying GnRH-induced LH secretion from gonadotropes. *Am. J. Physiol. Endocrinol. Metab.* 278, E263–E272.
352. Marshall, J. C., Eagleson, C. A., and McCartney, C. R. (2001). Hypothalamic dysfunction. *Mol. Cell. Endocrinol.* 183, 29–32.
353. Hoffman, A. R., and Crowley, W. F., Jr. (1982). Induction of puberty in men by long-term pulsatile administration of low-dose gonadotropin-releasing hormone. *N. Engl. J. Med.* 307, 1237–1241.
354. Haisenleder, D. J., Khoury, S., Zmeili, S. M., Papavasiliou, S., Ortolano, G. A., Dee, C., Duncan, J. A., and Marshall, J. C. (1987). The frequency of gonadotropin-releasing hormone secretion regulates expression of a and luteinizing hormone β-subunit messenger ribonucleic acids in male rats. *Mol. Endocrinol.* 1, 834–838.
355. Dalkin, A. C., Haisenleder, D. J., Ortolano, G. A., Ellis, T. R., and Marshall, J. C. (1989). The frequency of gonadotropin-releasing-hormone stimulation differentially regulates gonadotropin subunit messenger ribonucleic acid expression. *Endocrinology* 125, 917–924.
356. Haisenleder, D. J., Ortolano, G. A., Dalkin, A. C., Ellis, T. R., Paul, S. J., and Marshall, J. C. (1990). Differential regulation of gonadotropin subunit gene expression by gonadotropin-releasing hormone pulse amplitude in female rats. *Endocrinology* 127, 2869–2875.
357. Haisenleder, D. J., Yasin, M., Yasin, A., and Marshall, J. C. (1993). Regulation of prolactin, thyrotropin subunit, and gonadotropin subunit gene expression by pulsatile or continuous calcium signals. *Endocrinology* 133, 2055–2061.

358. Haisenleder, D. J., Yasin, M., and Marshall, J. C. (1995). Regulation of gonadotropin, thyrotropin subunit, and prolactin messenger ribonucleic acid expression by pulsatile or continuous protein kinase-C stimulation. *Endocrinology* 136, 13–19.
359. Haisenleder, D. J., Yasin, M., and Marshall, J. C. (1997). Gonadotropin subunit and gonadotropin-releasing hormone receptor gene expression are regulated by alterations in the frequency of calcium pulsatile signals. *Endocrinology* 138, 5227–5230.
360. Haisenleder, D. J., Cox, M. E., Parsons, S. J., and Marshall, J. C. (1998). Gonadotropin-releasing hormone pulses are required to maintain activation of mitogen-activated protein kinase: role in stimulation of gonadotrope gene expression. *Endocrinology* 139, 3104–3111.
361. Shupnik, M. A. (1990). Effects of gonadotropin-releasing hormone on rat gonadotropin gene transcription *in vitro*: requirement for pulsatile administration for luteinizing hormone-β gene stimulation. *Mol. Endocrinol.* 4, 1444–1450.
362. Haisenleder, D. J., Dalkin, A. C., Ortolano, G. A., Marshall, J. C., and Shupnik, M. A. (1991). A pulsatile gonadotropin-releasing hormone stimulus is required to increase transcription of the gonadotropin subunit genes: evidence for differential regulation of transcription by pulse frequency *in vivo*. *Endocrinology* 128, 509–517.
363. Jakubowiak, A., Tong, D., Janecki, A., Sanborn, B. M., and Steinberger, A. (1991). Pulsatile GnRH stimulation increases steady-state mRNA levels for FSHβ, LHβ, and α subunits in superfused pituitary cell cultures. *Mol. Cell. Neurosci.* 2, 277–283.
364. Papavasiliou, S. S., Zmeili, S., Khoury, S., Landefeld, T. D., Chin, W. W., and Marshall, J. C. (1986). Gonadotropin-releasing hormone differentially regulates expression of the genes for luteinizing hormone α and β subunits in male rats. *Proc. Natl. Acad. Sci. U S A* 83, 4026–4029.
365. Haisenleder, D. J., Katt, J. A., Ortolano, G. A., el-Gewely, M. R., Duncan, J. A., Dee, C., and Marshall, J. C. (1988). Influence of gonadotropin-releasing hormone pulse amplitude, frequency, and treatment duration on the regulation of luteinizing hormone (LH) subunit messenger ribonucleic acids and LH secretion. *Mol. Endocrinol.* 2, 338–343.
366. Besecke, L. M., Guendner, M. J., Schneyer, A. L., Bauer-Dantoin, A. C., Jameson, J. L., and Weiss, J. (1996). Gonadotropin-releasing hormone regulates follicle-stimulating hormone-β gene expression through an activin/follistatin autocrine or paracrine loop. *Endocrinology* 137, 3667–3673.
367. Kaiser, U. B., Jakubowiak, A., Steinberger, A., and Chin, W. W. (1997). Differential effects of gonadotropin-releasing hormone (GnRH) pulse frequency on gonadotropin subunit and GnRH receptor messenger ribonucleic acid levels *in vitro*. *Endocrinology* 138, 1224–1231.
368. Bedecarrats, G. Y., and Kaiser, U. B. (2003). Differential regulation of gonadotropin subunit gene promoter activity by pulsatile gonadotropin-releasing hormone (GnRH) in perifused LβT2 cells: role of GnRH receptor concentration. *Endocrinology* 144, 1802–1811.
369. Kaiser, U. B., Sabbagh, E., Katzenellenbogen, R. A., Conn, P. M., and Chin, W. W. (1995). A mechanism for the differential regulation of gonadotropin subunit gene expression by gonadotropin-releasing hormone. *Proc. Natl. Acad. Sci. U S A* 92, 12280–12284.
370. Bauer-Dantoin, A. C., Weiss, J., and Jameson, J. L. (1995). Roles of estrogen, progesterone, and gonadotropin-releasing hormone (GnRH) in the control of pituitary GnRH receptor gene expression at the time of the preovulatory gonadotropin surges. *Endocrinology* 136, 1014–1019.
371. Iliff-Sizemore, S. A., Ortolano, G. A., Haisenleder, D. J., Dalkin, A. C., Krueger, K. A., and Marshall, J. C. (1990). Testosterone differentially modulates gonadotropin subunit messenger ribonucleic acid responses to gonadotropin-releasing hormone pulse amplitude. *Endocrinology* 127, 2876–2883.
372. Haisenleder, D. J., Ortolano, G. A., Yasin, M., Dalkin, A. C., and Marshall, J. C. (1993). Regulation of gonadotropin subunit messenger ribonucleic acid expression by gonadotropin-releasing hormone pulse amplitude *in vitro*. *Endocrinology* 132, 1292–1296.
373. Butcher, R. L., Collins, W. E., and Fugo, N. W. (1974). Plasma concentration of LH, FSH, prolactin, progesterone and estradiol-17β throughout the 4-day estrous cycle of the rat. *Endocrinology* 94, 1704–1708.
374. Bauer-Dantoin, A. C., Hollenberg, A. N., and Jameson, J. L. (1993). Dynamic regulation of gonadotropin-releasing hormone receptor mRNA levels in the anterior pituitary gland during the rat estrous cycle. *Endocrinology* 133, 1911–1914.
375. Rivier, C., Roberts, V., and Vale, W. (1989). Possible role of luteinizing hormone and follicle-stimulating hormone in modulating inhibin secretion and expression during the estrous cycle of the rat. *Endocrinology* 125, 876–882.
376. Woodruff, T. K., D'Agostino, J., Schwartz, N. B., and Mayo, K. E. (1989). Decreased inhibin gene expression in preovulatory follicles requires primary gonadotropin surges. *Endocrinology* 124, 2193–2199.
377. Zmeili, S. M., Papavasiliou, S. S., Thorner, M. O., Evans, W. S., Marshall, J. C., and Landefeld, T. D. (1986). Alpha and luteinizing hormone beta subunit messenger ribonucleic acids during the rat estrous cycle. *Endocrinology* 119, 1867–1869.
378. Ortolano, G. A., Haisenleder, D. J., Dalkin, A. C., Iliff-Sizemore, S. A., Landefeld, T. D., Maurer, R. A., and Marshall, J. C. (1988). Follicle-stimulating hormone beta subunit messenger ribonucleic acid concentrations during the rat estrous cycle. *Endocrinology* 123, 2149–2151.
379. Shupnik, M. A., Gharib, S. D., and Chin, W. W. (1989). Divergent effects of estradiol on gonadotropin gene transcription in pituitary fragments. *Mol. Endocrinol.* 3, 474–480.
380. Levine, J. E., and Ramirez, V. D. (1982). Luteinizing hormone-releasing hormone release during the rat estrous cycle and after ovariectomy, as estimated with push-pull cannulae. *Endocrinology* 111, 1439–1448.
381. Fox, S. R., and Smith, M. S. (1985). Changes in the pulsatile pattern of luteinizing hormone secretion during the rat estrous cycle. *Endocrinology* 116, 1485–1492.
382. Haisenleder, D. J., Ortolano, G. A., Jolly, D., Dalkin, A. C., Landefeld, T. D., Vale, W. W., and Marshall, J. C. (1990). Inhibin secretion during the rat estrous cycle: relationships to FSH secretion and FSH beta subunit mRNA concentrations. *Life Sci.* 47, 1769–1773.
383. Smith, M. S., Freeman, M. E., and Neill, J. D. (1975). The control of progesterone secretion during the estrous cycle and early pseudopregnancy in the rat: prolactin, gonadotropin and steroid levels associated with rescue of the corpus luteum of pseudopregnancy. *Endocrinology* 96, 219–226.
384. Childs, G. V., Unabia, G., and Miller, B. T. (1994). Cytochemical detection of gonadotropin-releasing hormone-binding sites on rat pituitary cells with luteinizing hormone, follicle-stimulating hormone, and growth hormone antigens during diestrous up-regulation. *Endocrinology* 134, 1943–1951.
385. Smith, M. S., and Reinhart, J. (1993). Changes in pituitary gonadotropin-releasing hormone receptor messenger ribonucleic acid content during lactation and after pup removal. *Endocrinology* 133, 2080–2084.
386. Mathews, L. S., and Vale, W. W. (1991). Expression cloning of an activin receptor, a predicted transmembrane serine kinase. *Cell* 65, 973–982.
387. Attisano, L., Wrana, J. L., Cheifetz, S., and Massague, J. (1992). Novel activin receptors: distinct genes and alternative

mRNA splicing generate a repertoire of serine/threonine kinase receptors. *Cell* 1992;68, 97–108.
388. Welt, C., Sidis, Y., Keutmann, H., and Schneyer, A. (2002). Activins, inhibins, and follistatins: from endocrinology to signaling. A paradigm for the new millennium. *Exp. Biol. Med.* 227, 724–752.
389. Gregory, S. J., and Kaiser, U. B. (2004). Regulation of gonadotropins by inhibin and activin. *Semin. Reprod. Med.* 22, 253–267.
390. Massague, J. (1998). TGF-β signal transduction. *Annu. Rev. Biochem.* 67, 753–791.
391. Nakao, A., Imamura, T., Souchelnytskyi, S., Kawabata, M., Ishisaki, A., Oeda, E., Tamaki, K., Hanai, J.-I., Heldin, C.-H., Miyazono, K., and ten Dijke, P. (1997). TGF-β receptor-mediated signalling through Smad2, Smad3 and Smad4. *EMBO J.* 16, 5353–5362.
392. Bernard, D. J. (2004). Both SMAD2 and SMAD3 mediate activin-stimulated expression of the follicle-stimulating hormone β subunit in mouse gonadotrope cells. *Mol. Endocrinol.* 18, 606–623.
393. Hoodless, P. A., Haerry, T., Abdollah, S., Stapleton, M., O'Connor, M. B., Attisano, L., and Wrana, J. L. (1996). MADR1, a MAD-related protein that functions in BMP2 signaling pathways. *Cell* 85, 489–500.
394. Kretzschmar, M., Liu, F., Hata, A., Doody, J., and Massague, J. (1997). The TGF-β family mediator Smad1 is phosphorylated directly and activated functionally by the BMP receptor kinase. *Genes Dev.* 11, 984–995.
395. Lagna, G., Hata, A., Hemmati-Brivanlou, A., and Massague, J. (1996). Partnership between DPC4 and SMAD proteins in TGF-β signalling pathways. *Nature* 383, 832–836.
396. Howell, M., Itoh, F., Pierreux, C. E., Valgeirsdottir, S., Itoh, S., ten Dijke, P., and Hill, C. S. (1999). *Xenopus* Smad4b is the co-Smad component of developmentally regulated transcription factor complexes responsible for induction of early mesodermal genes. *Dev. Biol.* 214, 354–369.
397. Heldin, C.-H., Miyazono, K., and ten Dijke, P. (1997). TGF-β signalling from cell membrane to nucleus through SMAD proteins. *Nature* 390, 465–471.
398. Bai, S., Shi, X., Yang, X., and Cao, X. (2000). Smad6 as a transcriptional corepressor. *J. Biol. Chem.* 275, 8267–8270.
399. Bilezikjian, L. M., Corrigan, A. Z., Blount, A. L., Chen, Y., and Vale, W. W. (2001). Regulation and actions of Smad7 in the modulation of activin, inhibin, and transforming growth factor-β signaling in anterior pituitary cells. *Endocrinology* 142, 1065–1072.
400. Mason, A. J., Berkemeier, L. M., Schmelzer, C. H., and Schwall, R. H. (1989). Activin B: precursor sequences, genomic structure and *in vitro* activities. *Mol. Endocrinol.* 3, 1352–1358.
401. Gregg, D. W., Schwall, R. H., and Nett, T. M. (1991). Regulation of gonadotropin secretion and number of gonadotropin-releasing hormone receptors by inhibin, activin-A, and estradiol. *Biol. Reprod.* 44, 725–732.
402. Ling, N., Ying, S.-Y., Ueno, N., Esch, F., Denoroy, L., and Guillemin, R. (1985). Isolation and characterization of a M_r 32,000 protein with inhibin activity from porcine follicular fluid. *Proc. Natl. Acad. Sci. U S A* 82, 7217–7221.
403. Rivier, J., Spiess, J., McClintock, R., Vaughan, J., and Vale, W. (1985). Purification and partial characterization of inhibin from porcine follicular fluid. *Biochem. Biophys. Res. Commun.* 133, 120–127.
404. Robertson, D. M., Foulds, L. M., Leversha, L., Morgan, F. J., Hearn, M. T. W., Burger, H. G., Wettenhall, R. E. H., and de Kretser, D. M. (1985). Isolation of inhibin from bovine follicular fluid. *Biochem. Biophys. Res. Commun.* 126, 220–226.
405. Sugino, K., Nakamura, T., Takio, K., Miyamoto, K., Hasegawa, Y., Igarashi, M., Titani, K., and Sugino, H. (1992). Purification and characterization of high molecular weight forms of inhibin from bovine follicular fluid. *Endocrinology* 130, 789–796.
406. Mottram, J. C., and Cramer, W. (1923). On the general effects of exposure to radium on metabolism and tumour growth in the rat and the special effects on the testis and pituitary. *Q. J. Exp. Physiol. Cogn. Med. Sci.* 13, 209–229.
407. McCullagh, D. R. (1932). Dual endocrine activity of the testes. *Science* 76, 19–20.
408. Meunier, H., Rivier, C., Evans, R. M., and Vale, W. (1988). Gonadal and extragonadal expression of inhibin α, βA, and βB subunits in various tissues predicts diverse functions. *Proc. Natl. Acad. Sci. U S A* 85, 247–251.
409. Farnworth, P. G., Robertson, D. M., de Kretser, D. M., and Burger, H. G. (1988). Effects of 31 kilodalton bovine inhibin on follicle-stimulating hormone and luteinizing hormone in rat pituitary cells *in vitro*: actions under basal conditions. *Endocrinology* 122, 207–213.
410. Childs, G. V., Miller, B. T., and Miller, W. L. (1997). Differential effects of inhibin on gonadotropin stores and gonadotropin-releasing hormone binding to pituitary cells from cycling female rats. *Endocrinology* 138, 1577–1584.
411. Carroll, R. S., Kowash, P. M., Lofgren, J. A., Schwall, R. H., and Chin, W. W. (1991). *In vivo* regulation of FSH synthesis by inhibin and activin. *Endocrinology* 129, 3299–3304.
412. DePaolo, L. V., Shimonaka, M., Schwall, R. H., and Ling, N. (1991). *In vivo* comparison of the follicle-stimulating hormone-suppressing activity of follistatin and inhibin in ovariectomized rats. *Endocrinology* 128, 668–674.
413. Robertson, D. M., Prisk, M., McMaster, J. W., Irby, D. C., Findlay, J. K., and de Kretser, D. M. (1991). Serum FSH-suppressing activity of human recombinant inhibin A in male and female rats. *J. Reprod. Fertil.* 91, 321–328.
414. Kumar, T. R., Agno, J., Janovick, J. A., Conn, P. M., and Matzuk, M. M. (2003). Regulation of FSHβ and GnRH receptor gene expression in activin receptor II knockout male mice. *Mol. Cell. Endocrinol.* 212, 19–27.
415. Attardi, B., Keeping, H. S., Winters, S. J., Kotsuji, F., Maurer, R. A., and Troen, P. (1989). Rapid and profound suppression of messenger ribonucleic acid encoding follicle-stimulating hormone β by inhibin from primate Sertoli cells. *Mol. Endocrinol.* 3, 280–287.
416. Kotsuji, F., Winters, S. J., Keeping, H. S., Attardi, B., Oshima, H., and Troen, P. (1988). Effects of inhibin from primate Sertoli cells on follicle-stimulating hormone and luteinizing hormone release by perifused rat pituitary cells. *Endocrinology* 122, 2796–2802.
417. de Greef, W. J., Eilers, G. A., de Koning, J., Karels, B., and de Jong, F. H. (1987). Effects of ovarian inhibin on pulsatile release of gonadotrophins and secretion of LHRH in ovariectomized rats: evidence against a central action of inhibin. *J. Endocrinol.* 113, 449–455.
418. Wang, Q. F., Farnworth, P. G., Findlay, J. K., and Burger, H. G. (1988). Effect of purified 31K bovine inhibin on the specific binding of gonadotropin-releasing hormone to rat anterior pituitary cells in culture. *Endocrinology* 123, 2161–2166.
419. Winters, S. J., Pohl, C. R., Adedoyin, A., and Marshall, G. R. (1996). Effects of continuous inhibin administration on gonadotropin secretion and subunit gene expression in immature and adult male rats. *Biol. Reprod.* 55, 1377–1382.
420. Wang, Q. F., Farnworth, P. G., Findlay, J. K., and Burger, H. G. (1989). Inhibitory effect of pure 31-kilodalton bovine inhibin on gonadotropin-releasing hormone (GnRH)-induced up-regulation of GnRH binding sites in cultured rat anterior pituitary cells. *Endocrinology* 124, 363–368.

421. Weiss, J., Crowley, W. F., Jr., Halvorson, L. M., and Jameson, J. L. (1993). Perifusion of rat pituitary cells with gonadotropin-releasing hormone, activin, and inhibin reveals distinct effects on gonadotropin gene expression and secretion. *Endocrinology* 132, 2307–2311.
422. Martens, J. W. M., de Winter, J. P., Timmerman, M. A., McLuskey, A., van Schaik, R. H. N., Themmen, A. P. N., and de Jong, F. H. (1997). Inhibin interferes with activin signaling at the level of the activin receptor complex in Chinese hamster ovary cells. *Endocrinology* 138, 2928–2936.
423. Lewis, K. A., Gray, P. C., Blount, A. L., MacConell, L. A., Wiater, E., Bilezikjian, L. M., and Vale, W. (2000). Betaglycan binds inhibin and can mediate functional antagonism of activin signalling. *Nature* 404, 411–414.
424. Ethier, J.-F., Farnworth, P. G., Findlay, J. K., and Ooi, G. T. (2002). Transforming growth factor-b modulates inhibin A bioactivity in the LβT2 gonadotrope cell line by competing for binding to betaglycan. *Mol. Endocrinol.* 16, 2754–2763.
425. MacConell, L. A., Leal, A. M. O., and Vale, W. W. (2002). The distribution of betaglycan protein and mRNA in rat brain, pituitary, and gonads: implications for a role for betaglycan in inhibin-mediated reproductive functions. *Endocrinology* 143, 1066–1075.
426. Ghosh, B. R., Wu, J. C., Strahl, B. D., Childs, G. V., and Miller, W. L. (1996). Inhibin and estradiol alter gonadotropes differentially in ovine pituitary cultures: changing gonadotrope numbers and calcium responses to gonadotropin-releasing hormone. *Endocrinology* 137, 5144–5154.
427. Ling, N., Ying, S. Y., Ueno, N., Shimasaki, S., Esch, F., Hotta, M., and Guillemin, R. (1986). Pituitary FSH is released by a heterodimer of the β-subunits from the two forms of inhibin. *Nature* 321, 779–782.
428. Vale, W., Rivier, J., Vaughan, J., McClintock, R., Corrigan, A., Woo, W., Karr, D., and Spiess, J. (1986). Purification and characterization of an FSH releasing protein from porcine ovarian follicular fluid. *Nature* 321, 776–779.
429. Fernandez-Vazquez, G., Kaiser, U. B., Albarracin, C. T., and Chin, W. W. (1996). Transcriptional activation of the gonadotropin-releasing hormone receptor gene by activin A. *Mol. Endocrinol.* 10, 356–366.
430. Pernasetti, F., Vasilyev, V. V., Rosenberg, S. B., Bailey, J. S., Huang, H.-J., Miller, W. L., and Mellon, P. L. (2001). Cell-specific transcriptional regulation of follicle-stimulating hormone-β by activin and gonadotropin-releasing hormone in the LβT2 pituitary gonadotrope cell model. *Endocrinology* 142, 2284–2295.
431. Dupont, J., McNeilly, J., Vaiman, A., Canepa, S., Combarnous, Y., and Taragnat, C. (2003). Activin signaling pathways in ovine pituitary and LβT2 gonadotrope cells. *Biol. Reprod.* 68, 1877–1887.
432. Woodruff, T. K., Krummen, L. A., Lyon, R. J., Stocks, D. L., and Mather, J. P. (1993). Recombinant human inhibin A and recombinant human activin A regulate pituitary and ovarian function in the adult female rat. *Endocrinology* 132, 2332–2341.
433. Matzuk, M. M., Kumar, T. R., and Bradley, A. (1995). Different phenotypes for mice deficient in either activins or activin receptor type II. *Nature* 374, 356–360.
434. Schwall, R. H., Szonyi, E., Mason, A. J., and Nikolics, K. (1988). Activin stimulates secretion of follicle-stimulating hormone from pituitary cells desensitized to gonadotropin-releasing hormone. *Biochem. Biophys. Res. Commun.* 151, 1099–1104.
435. Huang, H.-J., Sebastian, J., Strahl, B. D., Wu, J. C., and Miller, W. L. (2001). Transcriptional regulation of the ovine follicle-stimulating hormone-β gene by activin and gonadotropin-releasing hormone (GnRH): involvement of two proximal activator protein-1 sites for GnRH stimulation. *Endocrinology* 142, 2267–2274.
436. Gregory, S. J., Lacza, C. T., Detz, A. A., Xu, S., Petrillo, L. A., and Kaiser, U. B. (2004). Synergy between activin A and gonadotropin-releasing hormone in transcriptional activation of the rat follicle-stimulating hormone-b gene. *Mol. Endocrinol.* 19, 237–254.
437. Gajewska, A., Siawrys, G., Bogacka, I., Przala, J., Lerrant, Y., Counis, R., and Kochman, K. (2002). *In vivo* modulation of follicle-stimulating hormone release and β subunit gene expression by activin A and the GnRH agonist buserelin in female rats. *Brain Res. Bull.* 58, 475–480.
438. Katayama, T., Shiota, K., and Takahashi, M. (1990). Activin A increases the number of follicle-stimulating hormone cells in anterior pituitary cultures. *Mol. Cell. Endocrinol.* 69, 179–185.
439. Norwitz, E. R., Xu, S., Jeong, K.-H., Bedecarrats, G. Y., Winebrenner, L. D., Chin, W. W., and Kaiser, U. B. (2002). Activin A augments GnRH-mediated transcriptional activation of the mouse GnRH receptor gene. *Endocrinology* 143, 985–997.
440. Norwitz, E. R., Xu, S., Xu, J., Spiryda, L. B., Park, J. S., Jeong, K.-H., McGee, E. A., and Kaiser, U. B. (2002). Direct binding of AP-1 (Fos/Jun) proteins to a SMAD binding element facilitates both gonadotropin-releasing hormone (GnRH)- and activin-mediated transcriptional activation of the mouse GnRH receptor gene. *J. Biol. Chem.* 277, 37469–37478.
441. Braden, T. D., and Conn, P. M. (1992). Activin-A stimulates the synthesis of gonadotropin-releasing hormone receptors. *Endocrinology* 130, 2101–2105.
442. Attardi, B., Klatt, B., and Little, G. (1995). Repression of glycoprotein hormone α-subunit gene expression and secretion by activin in αT3-1 cells. *Mol. Endocrinol.* 9, 1737–1749.
443. Weiss, J., Harris, P. E., Halvorson, L. M., Crowley, W. F., Jr., and Jameson, J. L. (1992). Dynamic regulation of follicle-stimulating hormone-β messenger ribonucleic acid levels by activin and gonadotropin-releasing hormone in perifused rat pituitary cells. *Endocrinology* 131, 1403–1408.
444. Vassalli, A., Matzuk, M. M., Gardner, H. A., Lee, K. F., and Jaenisch, R. (1994). Activin/inhibin β_B subunit gene disruption leads to defects in eyelid development and female reproduction. *Genes Dev.* 8, 414–427.
445. Huang, H.-J., Wu, J. C., Su, P., Zhirnov, O., and Miller, W. L. (2001). A novel role for bone morphogenetic proteins in the synthesis of follicle-stimulating hormone. *Endocrinology* 142, 2275–2283.
446. Otsuka, F., and Shimasaki, S. (2002). A novel function of bone morphogenetic protein-15 in the pituitary: selective synthesis and secretion of FSH by gonadotropes. *Endocrinology* 143, 4938–4941.
447. Yamashita, H., ten Dijke, P., Huylebroeck, D., Sampath, T. K., Andries, M., Smith, J. C., Heldin, C.-H., and Miyazono, K. (1995). Osteogenic protein-1 binds to activin type II receptors and induces certain activin-like effects. *J. Cell. Biol.* 130, 217–226.
448. Wiater, E., and Vale, W. (2003). Inhibin is an antagonist of bone morphogenetic protein signaling. *J. Biol. Chem.* 278, 7934–7941.
449. Bedecarrats, G. Y., O'Neill, F. H., Norwitz, E. R., Kaiser, U. B., and Teixeira, J. (2003). Regulation of gonadotropin gene expression by mullerian inhibiting substance. *Proc. Natl. Acad. Sci. U S A* 100, 9348–9353.
450. Ueno, N., Ling, N., Ying, S.-Y., Esch, F., Shimasaki, S., and Guillemin, R. (1987). Isolation and partial characterization of follistatin: a single-chain M_r 35,000 monomeric protein that inhibits the release of follicle-stimulating hormone. *Proc. Natl. Acad. Sci. U S A* 84, 8282–8286.

451. Esch, F. S., Shimasaki, S., Mercado, M., Cooksey, K., Ling, N., Ying, S.-Y., Ueno, N., and Guillemin, R. (1987). Structural characterization of follistatin: a novel follicle-stimulating hormone release-inhibiting polypeptide from the gonad. *Mol. Endocrinol.* 1, 849–855.
452. Robertson, D. M., Klein, R., de Vos, F. L., McLachlan, R. I., Wettenhall, R. E., Hearn, M. T., Burger, H. G., and de Kretser, D. M. (1987). The isolation of polypeptides with FSH suppressing activity from bovine follicular fluid which are structurally different to inhibin. *Biochem. Biophys. Res. Commun.* 149, 744–749.
453. Shimasaki, S., Koga, M., Esch, F., Cooksey, K., Mercado, M., Koba, A., Ueno, N., Ying, S.-Y., Ling, N., and Guillemin, R. (1988). Primary structure of the human follistatin precursor and its genomic organization. *Proc. Natl. Acad. Sci. U S A* 85, 4218–4222.
454. Ying, S.-Y., Becker, A., Swanson, G., Tan, P., Ling, N., Esch, F., Ueno, N., Shimasaki, S., and Guillemin, R. (1987). Follistatin specifically inhibits pituitary follicle stimulating hormone release *in vitro*. *Biochem. Biophys. Res. Commun.* 149, 133–139.
455. Kaiser, U. B., and Chin, W. W. (1993). Regulation of follistatin messenger ribonucleic acid levels in the rat pituitary. *J. Clin. Invest.* 91, 2523–2531.
456. Kirk, S. E., Dalkin, A. C., Yasin, M., Haisenleder, D. J., and Marshall, J. C. (1994). Gonadotropin-releasing hormone pulse frequency regulates expression of pituitary follistatin messenger ribonucleic acid: a mechanism for differential gonadotrope function. *Endocrinology* 135, 876–880.
457. Bilezikjian, L. M., Corrigan, A. Z., Blount, A. L., and Vale, W. W. (1996). Pituitary follistatin and inhibin subunit messenger ribonucleic acid levels are differentially regulated by local and hormonal factors. *Endocrinology* 137, 4277–4284.
458. Dalkin, A. C., Haisenleder, D. J., Yasin, M., Gilrain, J. T., and Marshall, J. C. (1996). Pituitary activin receptor subtypes and follistatin gene expression in female rats: differential regulation by activin and follistatin. *Endocrinology* 137, 548–554.
459. Winters, S. J., Dalkin, A. C., and Tsujii, T. (1997). Evidence that pituitary adenylate cyclase activating polypeptide suppresses follicle-stimulating hormone-β messenger ribonucleic acid levels by stimulating follistatin gene transcription. *Endocrinology* 138, 4324–4329.
460. Dalkin, A. C., Haisenleder, D. J., Gilrain, J. T., Aylor, K., Yasin, M., and Marshall, J. C. (1999). Gonadotropin-releasing hormone regulation of gonadotropin subunit gene expression in female rats: actions on follicle-stimulating hormone β messenger ribonucleic acid (mRNA) involve differential expression of pituitary activin (β-B) and follistatin mRNAs. *Endocrinology* 140, 903–908.
461. Dalkin, A. C., Haisenleder, D. J., Gilrain, J. T., Aylor, K., Yasin, M., and Marshall, J. C. (1998). Regulation of pituitary follistatin and inhibin/activin subunit messenger ribonucleic acids (mRNAs) in male and female rats: evidence for inhibin regulation of follistatin mRNA in females. *Endocrinology* 139, 2818–2823.
462. Iemura, S.-I., Yamamoto, T. S., Takagi, C., Uchiyama, H., Natsume, T., Shimasaki, S., Sugino, H., and Ueno, N. (1998). Direct binding of follistatin to a complex of bone-morphogenetic protein and its receptor inhibits ventral and epidermal cell fates in early *Xenopus* embryo. *Proc. Natl. Acad. Sci. U S A* 95, 9337–9342.
463. Duval, D. L., Ellsworth, B. S., and Clay, C. M. (1999). Is gonadotrope expression of the gonadotropin releasing hormone receptor gene mediated by autocrine/paracrine stimulation of an activin response element? *Endocrinology* 140, 1949–1952.
464. Tsuchida, K., Arai, K. Y., Kuramoto, Y., Yamakawa, N., Hasegawa, Y., and Sugino, H. (2000). Identification and characterization of a novel follistatin-like protein as a binding protein for the TGF-β family. *J. Biol. Chem.* 275, 40788–40796.
465. Schneyer, A., Tortoriello, D., Sidis, Y., Keutmann, H., Matsuzaki, T., and Holmes, W. (2001). Follistatin-related protein (FSRP): a new member of the follistatin gene family. *Mol. Cell. Endocrinol.* 180, 33–38.
466. Schneyer, A., Sidis, Y., Xia, Y., Saito, S., del Re, E., Lin, H. Y., and Keutmann, H. (2004). Differential actions of follistatin and follistatin-like 3. *Mol. Cell. Endocrinol.* 225, 25–28.
467. Sidis, Y., Tortoriello, D. V., Holmes, W. E., Pan, Y., Keutmann, H. T., and Schneyer, A. L. (2002). Follistatin-related protein and follistatin differentially neutralize endogenous *vs.* exogenous activin. *Endocrinology* 143, 1613–1624.
468. Gay, V. L., and Midgley, A. R., Jr. (1969). Response of the adult rat to orchidectomy and ovariectomy as determined by LH radioimmunoassay. *Endocrinology* 84, 1359–1364.
469. Wise, P. M., and Ratner, A. (1980). Effect of ovariectomy on plasma LH, FSH, estradiol, and progesterone and medial basal hypothalamic LHRH concentrations old and young rats. *Neuroendocrinology* 30, 15–19.
470. Abbot, S. D., Docherty, K., Roberts, J. L., Tepper, M. A., Chin, W. W., and Clayton, R. N. (1985). Castration increases luteinizing hormone subunit messenger RNA levels in male rat pituitaries. *J. Endocrinol.* 107, R1–R4.
471. Gharib, S. D., Bowers, S. M., Need, L. R., and Chin, W. W. (1986). Regulation of rat luteinizing hormone subunit messenger ribonucleic acids by gonadal steroid hormones. *J. Clin. Invest.* 77, 582–589.
472. Papavasiliou, S. S., Zmeili, S., Herbon, L., Duncan-Weldon, J., Marshall, J. C., and Landefeld, T. D. (1986). α And luteinizing hormone β messenger ribonucleic acid (RNA) of male and female rats after castration: quantitation using an optimized RNA dot blot hybridization assay. *Endocrinology* 119, 691–698.
473. Gharib, S. D., Wierman, M. E., Badger, T. M., and Chin, W. W. (1987). Sex steroid hormone regulation of follicle-stimulating hormone subunit messenger ribonucleic acid (mRNA) levels in the rat. *J. Clin. Invest.* 80, 294–299.
474. Wierman, M. E., Gharib, S. D., LaRovere, J. M., Badger, T. M., and Chin, W. W. (1988). Selective failure of androgens to regulate follicle stimulating hormone β messenger ribonucleic acid levels in the male rat. *Mol. Endocrinol.* 2, 492–498.
475. Dalkin, A. C., Haisenleder, D. J., Ortolano, G. A., Suhr, A., and Marshall, J. C. (1990). Gonadal regulation of gonadotropin subunit gene expression: evidence for regulation of follicle-stimulating hormone-β messenger ribonucleic acid by nonsteroidal hormones in female rats. *Endocrinology* 127, 798–806.
476. Scully, K. M., Gleiberman, A. S., Lindzey, J., Lubahn, D. B., Korach, K. S., and Rosenfeld, M. G. (1997). Role of estrogen receptor-α in the anterior pituitary gland. *Mol. Endocrinol.* 11, 674–681.
477. Shupnik, M. A., Gharib, S. D., and Chin, W. W. (1988). Estrogen suppresses rat gonadotropin gene transcription *in vivo*. *Endocrinology* 122, 1842–1846.
478. Paul, S. J., Ortolano, G. A., Haisenleder, D. J., Stewart, J. M., Shupnik, M. A., and Marshall, J. C. (1990). Gonadotropin subunit messenger RNA concentrations after blockade of gonadotropin-releasing hormone action: testosterone selectively increases follicle-stimulating hormone β-subunit messenger RNA by posttranscriptional mechanisms. *Mol. Endocrinol.* 4, 1943–1955.
479. Childs, G. V., Lloyd, J. M., Unabia, G., Gharib, S. D., Wierman, M. E., and Chin, W. W. (1987). Detection of luteinizing hormone β messenger ribonucleic acid (RNA) in

individual gonadotropes after castration: use of a new *in situ* hybridization method with a photobiotinylated complementary RNA probe. *Mol. Endocrinol.* 1, 926–932.

480. Sarkar, D. K., and Fink, G. (1980). Luteinizing hormone releasing factor in pituitary stalk plasma from long-term ovariectomized rats: effects of steroids. *J. Endocrinol.* 86, 511–524.
481. Gharib, S. D., Leung, P. C. K., Carroll, R. S., and Chin, W. W. (1990). Androgens positively regulate follicle-stimulating hormone β-subunit mRNA levels in rat pituitary cells. *Mol. Endocrinol.* 4, 1620–1626.
482. Clarke, I. J., Cummins, J. T., Crowder, M. E., and Nett, T. M. (1988). Pituitary receptors for gonadotropin-releasing hormone in relation to changes in pituitary and plasma gonadotropins in ovariectomized hypothalamo/pituitary-disconnected ewes: II. A marked rise in receptor number during the acute feedback effects of estradiol. *Biol. Reprod.* 39, 349–354.
483. Gregg, D. W., and Nett, T. M. (1989). Direct effects of estradiol-17 β on the number of gonadotropin-releasing hormone receptors in the ovine pituitary. *Biol. Reprod.* 40, 288–293.
484. Shupnik, M. A., Weinmann, C. M., Notides, A. C., and Chin, W. W. (1989). An upstream region of the rat luteinizing hormone β gene binds estrogen receptor and confers estrogen responsiveness. *J. Biol. Chem.* 264, 80–86.
485. Phillips, C. L., Lin, L. W., Wu, J. C., Guzman, K., Milsted, A., and Miller, W. L. (1988). 17β-Estradiol and progesterone inhibit transcription of the genes encoding the subunits of ovine follicle-stimulating hormone. *Mol. Endocrinol.* 2, 641–649.
486. Miller, C. D., and Miller, W. L. (1996). Transcriptional repression of the ovine follicle-stimulating hormone-β gene by 17β-estradiol. *Endocrinology* 137, 3437–3446.
487. Kim, K. E., Gordon, D. F., and Maurer, R. A. (1988). Nucleotide sequence of the bovine gene for follicle-stimulating hormone b-subunit. DNA. 7, 227–233.
488. Chatterjee, V. K., Madison, L. D., Mayo, S., and Jameson, J. L. (1991). Repression of the human glycoprotein hormone α-subunit gene by glucocorticoids: evidence for receptor interactions with limiting transcriptional activators. *Mol. Endocrinol.* 5, 100–110.
489. Keri, R. A., Andersen, B., Kennedy, G. C., Hamernik, D. L., Clay, C. M., Brace, A. D., Nett, T. M., Notides, A. C., and Nilson, J. H. (1991). Estradiol inhibits transcription of the human glycoprotein hormone α-subunit gene despite the absence of a high affinity binding site for estrogen receptor. *Mol. Endocrinol.* 5, 725–733.
490. Laws, S. C., Beggs, M. J., Webster, J. C., and Miller, W. L. (1990). Inhibin increases and progesterone decreases receptors for gonadotropin-releasing hormone in ovine pituitary culture. *Endocrinology* 127, 373–380.
491. Laws, S. C., Webster, J. C., and Miller, W. L. (1990). Estradiol alters the effectiveness of gonadotropin-releasing hormone (GnRH) in ovine pituitary cultures: GnRH receptors versus responsiveness to GnRH. *Endocrinology* 127, 381–386.
492. Gardner, D. B., Sebastian, J., and Miller, W. L. (2000). Estradiol induces and hyperglycosylates the receptor for ovine gonadotropin-releasing hormone. *Endocrinology* 141, 91–99.
493. Duval, D. L., Farris, A. R., Quirk, C. C., Nett, T. M., Hamernik, D. L., and Clay, C. M. (2000). Responsiveness of the ovine gonadotropin-releasing hormone receptor gene to estradiol and gonadotropin-releasing hormone is not detectable in vitro but is revealed in transgenic mice. *Endocrinology* 141, 1001–1010.
494. McArdle, C. A., Schomerus, E., Groner, I., and Poch, A. (1992). Estradiol regulates gonadotropin-releasing hormone receptor number, growth and inositol phosphate production in αT3-1 cells. *Mol. Cell. Endocrinol.* 87, 95–103.
495. Emons, G., Hoffman, H. G., Brack, C., Ortmann, O., Sturm, R., Ball, P., and Knuppen, R. (1988). Modulation of gonadotropin releasing hormone receptor concentration in cultured female rat pituitary cells by estradiol treatment. *J. Steroid Biochem.* 31, 751–756.
496. Karsch, F. J., Cummins, J. T., Thomas, G. B., and Clarke, I. J. (1987). Steroid feedback inhibition of pulsatile secretion of gonadotropin-releasing hormone in the ewe. *Biol. Reprod.* 36, 1207–1218.
497. Attardi, B., Vaughan, J., and Vale, W. (1992). Regulation of FSHβ messenger ribonucleic acid levels in the rat by endogenous inhibin. *Endocrinology* 130, 557–559.
498. Webster, J. C., Pedersen, N. R., Edwards, D. P., Beck, C. A., and Miller, W. L. (1995). The 5′-flanking region of the ovine follicle-stimulating hormone-β gene contains six progesterone response elements: three proximal elements are sufficient to increase transcription in the presence of progesterone. *Endocrinology* 136, 1049–1058.
499. Cheng, K. W., Cheng, C.-K., and Leung, P. C. K. (2001). Differential role of PR-A and -B isoforms in transcription regulation of human GnRH receptor gene. *Mol. Endocrinol.* 15, 2078–2092.
500. Kumar, T. R., and Low, M. J. (1995). Hormonal regulation of human follicle-stimulating hormone-β subunit gene expression: GnRH stimulation and GnRH-independent androgen inhibition. *Neuroendocrinology* 61, 628–637.
501. Bhasin, S., Fielder, T. J., and Swerdloff, R. S. (1987). Testosterone selectively increases serum follicle-stimulating hormone (FSH) but not luteinizing hormone (LH) in gonadotropin-releasing hormone antagonist-treated male rats: evidence for differential regulation of LH and FSH secretion. *Biol. Reprod.* 37, 55–59.
502. Perheentupa, A., and Huhtaniemi, I. (1990). Gonadotropin gene expression and secretion in gonadotropin-releasing hormone antagonist-treated male rats: effect of sex steroid replacement. *Endocrinology* 126, 3204–3209.
503. Wierman, M. E., and Wang, C. (1990). Androgen selectively stimulates follicle-stimulating hormone-β mRNA levels after gonadotropin-releasing hormone antagonist administration. *Biol. Reprod.* 42, 563–571.
504. Dalkin, A. C., Paul, S. J., Haisenleder, D. J., Ortolano, G. A., Yasin, M., and Marshall, J. C. (1992). Gonadal steroids effect similar regulation of gonadotrophin subunit mRNA expression in both male and female rats. *J. Endocrinol.* 132, 39–45.
505. Pierce, J. G., and Parsons, T. F. (1981). Glycoprotein hormones: structure and function. *Annu. Rev. Biochem.* 50, 465–495.
506. Chin, W. W., Kronenberg, H. M., Dee, P. C., Maloof, F., and Habener, J. F. (1981). Nucleotide sequence of the mRNA encoding the pre-α-subunit of mouse thyrotropin. *Proc. Natl. Acad. Sci. U S A* 78, 5329–5333.
507. Fiddes, J. C., and Goodman, H. M. (1981). The gene encoding the common alpha subunit of the four human glycoprotein hormones. *J. Mol. Appl. Genet.* 1, 3–18.
508. Goodwin, R. G., Moncman, C. L., Rottman, F. M., and Nilson, J. H. (1983). Characterization and nucleotide sequence of the gene for the common α subunit of the bovine pituitary glycoprotein hormones. *Nucleic Acids Res.* 11, 6873–6882.
509. Fenstermaker, R. A., Farmerie, T. A., Clay, C. M., Hamernik, D. L., and Nilson, J. H. (1990). Different combinations of regulatory elements may account for expression of the glycoprotein hormone α-subunit gene in primate and horse placenta. *Mol. Endocrinol.* 4, 1480–1487.
510. Kay, T. W. H., Chedrese, P. J., and Jameson, J. L. (1994). Gonadotropin-releasing hormone causes transcriptional

stimulation followed by desensitization of the glycoprotein hormone a promoter in transfected αT3 gonadotrope cells. *Endocrinology* 134, 568–573.

511. Fowkes, R. C., King, P., and Burrin, J. M. (2002). Regulation of human glycoprotein hormone α-subunit gene transcription in LβT2 gonadotropes by protein kinase C and extracellular signal-regulated kinase 1/2. *Biol. Reprod.* 67, 725–734.
512. Chedrese, P. J., Kay, T. W. H., and Jameson, J. L. Gonadotropin-releasing hormone stimulates glycoprotein hormone α-subunit messenger ribonucleic acid (mRNA) levels in αT3 cells by increasing transcription and mRNA stability. *Endocrinology* 134, 2475–2481.
513. Bouamoud, N., Lerrant, Y., Ribot, G., and Counis, R. (1992). Differential stability of mRNAs coding for alpha and gonadotropin beta subunits in cultured rat pituitary cells. *Mol. Cell. Endocrinol.* 88, 143–151.
514. Jorgensen, J. S., Quirk, C. C., and Nilson, J. H. (2004). Multiple and overlapping combinatorial codes orchestrate hormonal responsiveness and dictate cell-specific expression of the genes encoding luteinizing hormone. *Endocr. Rev.* 25, 521–542.
515. Kendall, S. K., Gordon, D. F., Birkmeier, T. S., Petrey, D., Sarapura, V. D., O'Shea, K. S., Wood, W. M., Lloyd, R. V., Ridgway, E. C., and Camper, S. A. (1994). Enhancer-mediated high level expression of mouse pituitary glycoprotein hormone α-subunit transgene in thyrotropes, gonadotropes, and developing pituitary gland. *Mol. Endocrinol.* 8, 1420–1433.
516. Brinkmeier, M. L., Gordon, D. F., Dowding, J. M., Saunders, T. L., Kendall, S. K., Sarapura, V. D., Wood, W. M., Ridgway, E. C., and Camper, S. A. (1998). Cell-specific expression of the mouse glycoprotein hormone α-subunit gene requires multiple interacting DNA elements in transgenic mice and cultured cells. *Mol. Endocrinol.* 12, 622–633.
517. Horn, F., Windle, J. J., Barnhart, K. M., and Mellon, P. L. (1992). Tissue-specific gene expression in the pituitary: the glycoprotein hormone α-subunit gene is regulated by a gonadotrope-specific protein. *Mol. Cell. Biol.* 12, 2143–2153.
518. Barnhart, K. M., and Mellon, P. L. (1994). The orphan nuclear receptor, steroidogenic factor-1, regulates the glycoprotein hormone α-subunit gene in pituitary gonadotropes. *Mol. Endocrinol.* 8, 878–885.
519. Lala, D. S., Rice, D. A., and Parker, K. L. (1992). Steroidogenic factor I, a key regulator of steroidogenic enzyme expression, is the mouse homolog of *fushi tarazu*-factor I. *Mol. Endocrinol.* 6, 1249–1258.
520. Lynch, J. P., Lala, D. S., Peluso, J. J., Luo, W., Parker, K. L., and White, B. A. (1993). Steroidogenic factor 1, an orphan nuclear receptor, regulates the expression of the rat aromatase gene in gonadal tissues. *Mol. Endocrinol.* 7, 776–786.
521. Shen, W.-H., Moore, C. C. D., Ikeda, Y., Parker, K. L., and Ingraham, H. A. (1994). Nuclear receptor steroidogenic factor 1 regulates the Mullerian inhibiting substance gene: a link to the sex determination cascade. *Cell* 1994;77, 651–661.
522. Ikeda, Y., Lala, D. S., Luo, X., Kim, E., Moisan, M.-P., and Parker, K. L. (1993). Characterization of the mouse *FTZ-F1* gene, which encodes a key regulator of steroid hydroxylase gene expression. *Mol. Endocrinol.* 7, 852–860.
523. Ingraham, H. A., Lala, D. S., Ikeda, Y., Luo, X., Shen, W.-H., Nachtigal, M. W., Abbud, R., Nilson, J. H., and Parker, K. L. (1994). The nuclear receptor steroidogenic factor 1 acts at multiple levels of the reproductive axis. *Genes Dev.* 8, 2302–2312.
524. Stallings, N. R., Hanley, N. A., Majdic, G., Zhao, L., Bakke, M., and Parker, K. L. (2002). Development of a transgenic green fluorescent protein lineage marker for steroidogenic factor 1. *Mol. Endocrinol.* 16, 2360–2370.
525. Luo, X., Ikeda, Y., and Parker, K. L. (1994). A cell-specific nuclear receptor is essential for adrenal and gonadal development and sexual differentiation. *Cell* 77, 481–490.
526. Sadovsky, Y., Crawford, P. A., Woodson, K. G., Polish, J. A., Clements, M. A., Tourtellotte, L. M., Simburger, K., and Milbrandt, J. (1995). Mice deficient in the orphan receptor steroidogenic factor 1 lack adrenal glands and gonads but express P450 side-chain-cleavage enzyme in the placenta and have normal embryonic serum levels of corticosteroids. *Proc. Natl. Acad. Sci. U S A* 92, 10939–10943.
527. Shinoda, K., Lei, H., Yoshii, H., Nomura, M., Nagano, M., Shiba, H., Sasaki, H., Osawa, Y., Ninomiya, Y., Niwa, O., Morohashi, K.-I., and Li, E. (1995). Developmental defects of the ventromedial hypothalamic nucleus and pituitary gonadotroph in the *Ftz-F1* disrupted mice. *Dev. Dyn.* 204, 22–29.
528. Zhao, L., Bakke, M., and Parker, K. L. (2001). Pituitary-specific knockout of steroidogenic factor 1. *Mol. Cell. Endocrinol.* 185, 27–32.
529. Zhao, L., Bakke, M., Krimkevich, Y., Cushman, L. J., Parlow, A. F., Camper, S. A., and Parker, K. L. (2001). Steroidogenic factor 1 (SF1) is essential for pituitary gonadotrope function. *Development* 128, 147–154.
530. Achermann, J. C., Ito, M., Hindmarsh, P. C., and Jameson, J. L. (1999). A mutation in the gene encoding steroidogenic factor-1 causes XY sex reversal and adrenal failure in humans. *Nat. Genet.* 22, 125–126.
531. Ikeda, Y., Luo, X., Abbud, R., Nilson, J. H., and Parker, K. L. (1995). The nuclear receptor steroidogenic factor 1 is essential for the formation of the ventromedial hypothalamic nucleus. *Mol. Endocrinol.* 9, 478–486.
532. Parker, K. L., and Schimmer, B. P. (1997). Steroidogenic factor 1: a key determinant of endocrine development and function. *Endocr. Rev.* 18, 361–377.
533. Schoderbek, W. E., Roberson, M. S., and Maurer, R. A. (1993). Two different DNA elements mediate gonadotropin releasing hormone effects on expression of the glycoprotein hormone α-subunit gene. *J. Biol. Chem.* 268, 3903–3910.
534. Heckert, L. L., Schultz, K., and Nilson, J. H. (1995). Different composite regulatory elements direct expression of the human α subunit gene to pituitary and placenta. *J. Biol. Chem.* 270, 26497–26504.
535. Wood, W. M., Dowding, J. M., Sarapura, V. D., McDermott, M. T., Gordon, D. F., and Ridgway, E. C. (1998). Functional interactions of an upstream enhancer of the mouse glycoprotein hormone α-subunit gene with proximal promoter sequences. *Mol. Cell. Endocrinol.* 142, 141–152.
536. Schoderbek, W. E., Kim, K. E., Ridgway, E. C., Mellon, P. L., and Maurer, R. A. (1992). Analysis of DNA sequences required for pituitary-specific expression of the glycoprotein hormone α-subunit gene. *Mol. Endocrinol.* 6, 893–903.
537. Roberson, M. S., Schoderbek, W. E., Tremml, G., and Maurer, R. A. (1994). Activation of the glycoprotein hormone α-subunit promoter by a LIM-homeodomain transcription factor. *Mol. Cell. Biol.* 14, 2985–2993.
538. Heckert, L. L., Schultz, K., and Nilson, J. H. (1996). The cAMP response elements of the α subunit gene bind similar proteins in trophoblasts and gonadotropes but have distinct functional sequence requirements. *J. Biol. Chem.* 271, 31650–31656.
539. Sanchez-Garcia, I., and Rabbitts, T. H. (1994). The LIM domain: a new structural motif found in zinc-finger-like proteins. *Trends Genet.* 10, 315–320.
540. Bach, I., Rhodes, S. J., Pearse, R. V., II, Heinzel, T., Gloss, B., Scully, K. M., Sawchenko, P. E., and Rosenfeld, M. G. (1995). P-Lim, a LIM homeodomain factor, is expressed during pituitary organ and cell commitment and synergizes with Pit-1. *Proc. Natl. Acad. Sci. U S A* 92, 2720–2724.

541. Seidah, N. G., Barale, J. C., Marcinkiewicz, M., Mattei, M. G., Day, R., and Chretien, M. (1994). The mouse homeoprotein mLIM-3 is expressed early in cells derived from the neuroepithelium and persists in adult pituitary. *DNA Cell. Biol.* 13, 1163–1180.
542. West, B. E., Parker, G. E., Savage, J. J., Kiratipranon, P., Toomey, K. S., Beach, L. R., Colvin, S. C., Sloop, K. W., and Rhodes, S. J. (2004). Regulation of the follicle-stimulating hormone β gene by the LHX3 LIM-homeodomain transcription factor. *Endocrinology* 145, 4866–4879.
543. Latchman, D. S. (1999). POU family transcription factors in the nervous system. *J. Cell. Physiol.* 179, 126–133.
544. Sloop, K. W., Meier, B. C., Bridwell, J. L., Parker, G. E., McCutchan Schiller, A., and Rhodes, S. J. (1999). Differential activation of pituitary hormone genes by human Lhx3 isoforms with distinct DNA binding properties. *Mol. Endocrinol.* 13, 2212–2225.
545. Howard, P. W., and Maurer, R. A. (2001). A point mutation in the LIM domain of Lhx3 reduces activation of the glycoprotein hormone α-subunit promoter. *J. Biol. Chem.* 276, 19020–19026.
546. Sheng, H. Z., Zhadanov, A. B., Mosinger, B., Jr., Fujii, T., Bertuzzi, S., Grinberg, A., Lee, E. J., Huang, S.-P., Mahon, K. A., and Westphal, H. (1996). Specification of pituitary cell lineages by the LIM homeobox gene *Lhx3*. *Science* 272, 1004–1007.
547. Netchine, I., Sobrier, M.-L., Krude, H., Schnabel, D., Maghnie, M., Marcos, E., Duriez, B., Cacheux, V., Moers, A., Goossens, M., Gruters, A., and Amselem, S. (2000). Mutations in *LHX3* result in a new syndrome revealed by combined pituitary hormone deficiency. *Nat. Genet.* 25, 182–186.
548. Sloop, K. W., Walvoord, E. C., Showalter, A. D., Pescovitz, O. H., and Rhodes, S. J. (2000). Molecular analysis of *LHX3* and *PROP-1* in pituitary hormone deficiency patients with posterior pituitary ectopia. *J. Clin. Endocrinol. Metab.* 85, 2701–2708.
549. Sloop, K. W., Parker, G. E., Hanna, K. R., Wright, H. A., and Rhodes, S. J. (2001). LHX3 transcription factor mutations associated with combined pituitary hormone deficiency impair the activation of pituitary target genes. *Gene* 265, 61–69.
550. Sheng, H. Z., Moriyama, K., Yamashita, T., Li, H., Potter, S. S., Mahon, K. A., and Westphal, H. (1997). Multistep control of pituitary organogenesis. *Science* 278, 1809–1812.
551. Raetzman, L. T., Ward, R., and Camper, S. A. (2002). *Lhx4* and *Prop1* are required for cell survival and expansion of the pituitary primordia. *Development* 129, 4229–4239.
552. Szeto, D. P., Ryan, A. K., O'Connell, S. M., and Rosenfeld, M. G. (1996). P-OTX: a PIT-1-interacting homeodomain factor expressed during anterior pituitary gland development. *Proc. Natl. Acad. Sci. U S A* 93, 7706–7710.
553. Bach, I., Carriere, C., Ostendorff, H. P., Andersen, B., and Rosenfeld, M. G. (1997). A family of LIM domain-associated cofactors confer transcriptional synergism between LIM and Otx homeodomain proteins. *Genes Dev.* 11, 1370–1380.
554. Tremblay, J. J., Lanctot, C., and Drouin, J. (1998). The pan-pituitary activator of transcription, Ptx1 (pituitary homeobox 1), acts in synergy with SF-1 and Pit1 and is an upstream regulator of the Lim-homeodomain gene Lim3/Lhx3. *Mol. Endocrinol.* 12, 428–441.
555. Lamonerie, T., Tremblay, J. J., Lanctot, C., Therrien, M., Gauthier, Y., and Drouin, J. (1996). Ptx1, a *bicoid*-related homeo box transcription factor involved in transcription of the pro-opiomelanocortin gene. *Genes Dev.* 10, 1284–1295.
556. Lanctot, C., Gauthier, Y., and Drouin, J. (1999). Pituitary homeobox 1 (Ptx1) is differentially expressed during pituitary development. *Endocrinology* 140, 1416–1422.
557. Szeto, D. P., Rodriguez-Esteban, C., Ryan, A. K., O'Connell, S. M., Liu, F., Kioussi, C., Gleiberman, A. S., Izpisua-Belmonte, J. C., and Rosenfeld, M. G. (1999). Role of the Bicoid-related homeodomain factor Pitx1 in specifying hindlimb morphogenesis and pituitary development. *Genes Dev.* 13, 484–494.
558. Poulin, G., Turgeon, B., and Drouin, J. (1997). NeuroD1/β2 contributes to cell-specific transcription of the proopiomelanocortin gene. *Mol. Cell. Biol.* 17, 6673–6682.
559. Sarapura, V. D., Strouth, H. L., Wood, W. M., Gordon, D. F., and Ridgway, E. C. (1998). Activation of the glycoprotein hormone α-subunit gene promoter in thyrotropes. *Mol. Cell. Endocrinol.* 146, 77–86.
560. Tremblay, J. J., Goodyer, C. G., and Drouin, J. (2000). Transcriptional properties of Ptx1 and Ptx2 isoforms. *Neuroendocrinology* 71, 277–286.
561. Gage, P. J., Suh, H., and Camper, S. A. (1999). Dosage requirement of *Pitx2* for development of multiple organs. *Development* 126, 4643–4651.
562. Suh, H., Gage, P. J., Drouin, J., and Camper, S. A. (2002). *Pitx2* is required at multiple stages of pituitary organogenesis: pituitary primordium formation and cell specification. *Development* 129, 329–337.
563. Quirk, J., and Brown, P. (2002). Hesx1 homeodomain protein represses transcription as a monomer and antagonises transactivation of specific sites as a homodimer. *J. Mol. Endocrinol.* 28, 193–205.
564. Delegeane, A. M., Ferland, L. H., and Mellon, P. L. (1987). Tissue-specific enhancer of the human glycoprotein hormone α-subunit gene: dependence on cyclic AMP-inducible elements. *Mol. Cell. Biol.* 7, 3994–4002.
565. Silver, B. J., Bokar, J. A., Virgin, J. B., Vallen, E. A., Milsted, A., and Nilson, J. H. (1987). Cyclic AMP regulation of the human glycoprotein hormone α-subunit gene is mediated by an 18-base-pair element. *Proc. Natl. Acad. Sci. U S A* 84, 2198–2202.
566. Bokar, J. A., Keri, R. A., Farmerie, T. A., Fenstermaker, R. A., Andersen, B., Hamernik, D. L., Yun, J., Wagner, T., and Nilson, J. H. (1989). Expression of the glycoprotein hormone α-subunit gene in the placenta requires a functional cyclic AMP response element, whereas a different *cis*-acting element mediates pituitary-specific expression. *Mol. Cell. Biol.* 9, 5113–5122.
567. Fuh, V. L., Burrin, J. M., and Jameson, J. L. (1989). Cyclic AMP (cAMP) effects on chorionic gonadotropin gene transcription and mRNA stability: labile proteins mediate basal expression whereas stable proteins mediate cAMP stimulation. *Mol. Endocrinol.* 3, 1148–1156.
568. Andersen, B., Kennedy, G. C., Hamernik, D. L., Bokar, J. A., Bohinski, R., and Nilson, J. H. (1990). Amplification of the transcriptional signal mediated by the tandem cAMP response elements of the glycoprotein hormone α-subunit gene occurs through several distinct mechanisms. *Mol. Endocrinol.* 4, 573–582.
569. Steger, D. J., Altschmied, J., Buscher, M., and Mellon, P. L. (1991). Evolution of placenta-specific gene expression: comparison of the equine and human gonadotropin α-subunit genes. *Mol. Endocrinol.* 5, 243–255.
570. Drust, D. S., Troccoli, N. M., and Jameson, J. L. (1991). Binding specificity of cyclic adenosine 3′,5′-monophosphate-responsive element (CRE)-binding proteins and activating transcription factors to naturally occurring CRE sequence variants. *Mol. Endocrinol.* 5, 1541–1551.
571. Pestell, R. G., Hollenberg, A. N., Albanese, C., and Jameson, J. L. (1994). c-Jun represses transcription of the human chorionic gonadotropin α and β genes through distinct types of CREs. *J. Biol. Chem.* 269, 31090–31096.

572. Jorgensen, J. S., and Nilson, J. H. (2001). AR suppresses transcription of the a glycoprotein hormone subunit gene through protein-protein interactions with cJun and activation transcription factor 2. *Mol. Endocrinol.* 15, 1496–1504.
573. Steger, D. J., Hecht, J. H., and Mellon, P. L. (1994). GATA-binding proteins regulate the human gonadotropin α-subunit gene in the placenta and pituitary gland. *Mol. Cell. Biol.* 14, 5592–5602.
574. Jackson, S. M., Gutierrez-Hartmann, A., and Hoeffler, J. P. (1995). Upstream stimulatory factor, a basic-helix-loop-helix-zipper protein, regulates the activity of the α-glycoprotein hormone subunit gene in pituitary cells. *Mol. Endocrinol.* 9, 278–291.
575. Pittman, R. H., Clay, C. M., Farmerie, T. A., and Nilson, J. H. (1994). Functional analysis of the placenta-specific enhancer of the human glycoprotein hormone α subunit gene: emergence of a new element. *J. Biol. Chem.* 269, 19360–19368.
576. Budworth, P. R., Quinn, P. G., and Nilson, J. H. (1997). Multiple characteristics of a pentameric regulatory array endow the human α-subunit glycoprotein hormone promoter with trophoblast specificity and maximal activity. *Mol. Endocrinol.* 11, 1669–1680.
577. Jameson, J. L., Jaffe, R. C., Deutsch, P. J., Albanese, C., and Habener, J. F. (1988). The gonadotropin α-gene contains multiple protein binding domains that interact to modulate basal and cAMP-responsive transcription. *J. Biol. Chem.* 263, 9879–9886.
578. Kay, T. W. H., and Jameson, J. L. (1992). Identification of a gonadotropin-releasing hormone-responsive region in the glycoprotein hormone α-subunit promoter. *Mol. Endocrinol.* 6, 1767–1773.
579. Kennedy, G. C., Andersen, B., and Nilson, J. H. (1990). The human α subunit glycoprotein hormone gene utilizes a unique CCAAT binding factor. *J. Biol. Chem.* 265, 6279–6285.
580. Andersen, B., Kennedy, G. C., and Nilson, J. H. (1990). A *cis*-acting element located between the cAMP response elements and CCAAT box augments cell-specific expression of the glycoprotein hormone α subunit gene. *J. Biol. Chem.* 265, 21874–21880.
581. Weck, J., Fallest, P. C., Pitt, L. K., and Shupnik, M. A. (1998). Differential gonadotropin-releasing hormone stimulation of rat luteinizing hormone subunit gene transcription by calcium influx and mitogen-activated protein kinase-signaling pathways. *Mol. Endocrinol.* 12, 451–457.
582. Kaiser, U. B., Katzenellenbogen, R. A., Conn, P. M., and Chin, W. W. (1994). Evidence that signalling pathways by which thyrotropin-releasing hormone and gonadotropin-releasing hormone act are both common and distinct. *Mol. Endocrinol.* 8, 1038–1048.
583. Clay, C. M., Keri, R. A., Finicle, A. B., Heckert, L. L., Hamernik, D. L., Marschke, K. M., Wilson, E. M., French, F. S., and Nilson, J. H. (1993). Transcriptional repression of the glycoprotein hormone α subunit gene by androgen may involve direct binding of androgen receptor to the proximal promoter. *J. Biol. Chem.* 268, 13556–13564.
584. Heckert, L. L., Wilson, E. M., and Nilson, J. H. (1997). Transcriptional repression of the α-subunit gene by androgen receptor occurs independently of DNA binding but requires the DNA-binding and ligand-binding domains of the receptor. *Mol. Endocrinol.* 11, 1497–1506.
585. Saunders, B. D., Sabbagh, E., Chin, W. W., and Kaiser, U. B. (1998). Differential use of signal transduction pathways in the gonadotropin-releasing hormone-mediated regulation of gonadotropin subunit gene expression. *Endocrinology* 1998;139, 1835–1843.
586. Stanislaus, D., Janovick, J. A., Jennes, L., Kaiser, U. B., Chin, W. W., and Conn, P. M. (1994). Functional and morphological characterization of four cell lines derived from GH_3 cells stably transfected with gonadotropin-releasing hormone receptor complementary deoxyribonucleic acid. *Endocrinology* 135, 2220–2227.
587. Miyata, A., Arimura, A., Dahl, R. R., Minamino, N., Uehara, A., Jiang, L., Culler, M. D., and Coy, D. H. (1989). Isolation of a novel 38 residue-hypothalamic polypeptide which stimulates adenylate cyclase in pituitary cells. *Biochem. Biophys. Res. Commun.* 164, 567–574.
588. Rawlings, S. R., and Hezareh, M. (1996). Pituitary adenylate cyclase-activating polypeptide (PACAP) and PACAP/vasoactive intestinal polypeptide receptors: actions on the anterior pituitary gland. *Endocr. Rev.* 17, 4–29.
589. Burrin, J. M., Aylwin, S. J. B., Holdstock, J. G., and Sahye, U. (1998). Mechanism of action of pituitary adenylate cyclase-activating polypeptide on human glycoprotein hormone α-subunit transcription in αT3-1 gonadotropes. *Endocrinology* 139, 1731–1737.
590. Fowkes, R. C., Desclozeaux, M., Patel, M. V., Aylwin, S. J. B., King, P., Ingraham, H. A., and Burrin, J. M. (2003). Steroidogenic factor-1 and the gonadotrope-specific element enhance basal and pituitary adenylate cyclase-activating polypeptide-stimulated transcription of the human glycoprotein hormone α-subunit gene in gonadotropes. *Mol. Endocrinol.* 17, 2177–2188.
591. Talmadge, K., Vamvakopoulos, N. C., and Fiddes, J. C. (1984). Evolution of the genes for the β subunits of human chorionic gonadotropin and luteinizing hormone. *Nature* 307, 37–40.
592. Chin, W. W., Godine, J. E., Klein, D. R., Chang, A. S., Tan, L. K., and Habener, J. F. (1983). Nucleotide sequence of the cDNA encoding the precursor of the β subunit of rat lutropin. *Proc. Natl. Acad. Sci. U S A* 80, 4649–4653.
593. Fiddes, J. C., and Talmadge, K. (1984). Structure, expression, and evolution of the genes for the human glycoprotein hormones. *Recent Prog. Horm. Res.* 40, 43–78.
594. Jameson, L., Chin, W. W., Hollenberg, A. N., Chang, A. S., and Habener, J. F. (1984). The gene encoding the β-subunit of rat luteinizing hormone: analysis of gene structure and evolution of nucleotide sequence. *J. Biol. Chem.* 259, 15474–15480.
595. Virgin, J. B., Silver, B. J., Thomason, A. R., and Nilson, J. H. (1985). The gene for the β subunit of bovine luteinizing hormone encodes a gonadotropin mRNA with an unusually short 5′-untranslated region. *J. Biol. Chem.* 260, 7072–7077.
596. Boorstein, W. R., Vamvakopoulos, N. C., and Fiddes, J. C. (1982). Human chorionic gonadotropin β-subunit is encoded by at least eight genes arranged in tandem and inverted pairs. *Nature* 300, 419–422.
597. Policastro, P., Ovitt, C. E., Hoshina, M., Fukuoka, H., Boothby, M. R., and Boime, I. (1983). The β subunit of human chorionic gonadotropin is encoded by multiple genes. *J. Biol. Chem.* 258, 11492–11499.
598. Jameson, J. L., Lindell, C. M., and Habener, J. F. (1986). Evolution of different transcriptional start sites in the human luteinizing hormone and chorionic gonadotropin β-subunit genes. *DNA* 5, 227–234.
599. Tremblay, J. J., and Drouin, J. (1999). Egr-1 is a downstream effector of GnRH and synergizes by direct interaction with Ptx1 and SF-1 to enhance luteinizing hormone β gene transcription. *Mol. Cell. Biol.* 19, 2567–2576.
600. Tremblay, J. J., Marcil, A., Gauthier, Y., and Drouin, J. (1999). Ptx1 regulates SF-1 activity by an interaction that mimics the role of the ligand-binding domain. *EMBO J.* 18, 3431–3441.
601. Halvorson, L. M., Kaiser, U. B., and Chin, W. W. (1996). Stimulation of luteinizing hormone β gene promoter activity

by the orphan nuclear receptor, steroidogenic factor-1. *J. Biol. Chem.* 271, 6645–6650.

602. Halvorson, L. M., Ito, M., Jameson, J. L., and Chin, W. W. (1998). Steroidogenic factor-1 and early growth response protein 1 act through two composite DNA binding sites to regulate luteinizing hormone β-subunit gene expression. *J. Biol. Chem.* 273, 14712–14720.
603. Topilko, P., Schneider-Maunoury, S., Levi, G., Trembleau, A., Gourdji, D., Driancourt, M.-A., Rao, C. V., and Charnay, P. (1998). Multiple pituitary and ovarian defects in *Krox-24* (*NGFI-A*, *Egr-1*)-targeted mice. *Mol. Endocrinol.* 12, 107–122.
604. Kaiser, U. B., Sabbagh, E., Chen, M. T., Chin, W. W., and Saunders, B. D. (1998). Sp1 binds to the rat luteinizing hormone β (LHβ) gene promoter and mediates gonadotropin-releasing hormone-stimulated expression of the LHβ subunit gene. *J. Biol. Chem.* 273, 12943–12951.
605. Kaiser, U. B., Halvorson, L. M., and Chen, M. T. (2000). Sp1, steroidogenic factor 1 (SF-1), and early growth response protein 1 (Egr-1) binding sites form a tripartite gonadotropin-releasing hormone response element in the rat luteinizing hormone-β gene promoter: an integral role for SF-1. *Mol. Endocrinol.* 14, 1235–1245.
606. Weck, J., Anderson, A. C., Jenkins, S., Fallest, P. C., and Shupnik, M. A. (2000). Divergent and composite gonadotropin-releasing hormone-responsive elements in the rat luteinizing hormone subunit genes. *Mol. Endocrinol.* 14, 472–485.
607. Quirk, C. C., Lozada, K. L., Keri, R. A., and Nilson, J. H. (2001). A single Pitx1 binding site is essential for activity of the LHβ promoter in transgenic mice. *Mol. Endocrinol.* 15, 734–746.
608. Keri, R. A., Bachmann, D. J., Behrooz, A., Herr, B. D., Ameduri, R. K., Quirk, C. C., and Nilson, J. H. (2000). An NF-Y binding site is important for basal, but not gonadotropin-releasing hormone-stimulated, expression of the luteinizing hormone β subunit gene. *J. Biol. Chem.* 275, 13082–13088.
609. Mantovani, R. (1999). The molecular biology of the CCAAT-binding factor NF-Y. *Gene* 239, 15–27.
610. Kaiser, U. B., Sabbagh, E., Saunders, B. D., and Chin, W. W. (1998). Identification of *cis*-acting deoxyribonucleic acid elements that mediate gonadotropin-releasing hormone stimulation of the rat luteinizing hormone β-subunit gene. *Endocrinology* 139, 2443–2451.
611. Call, G. B., and Wolfe, M. W. (2002). Species differences in GnRH activation of the LHβ promoter: role of Egr1 and Sp1. *Mol. Cell. Endocrinol.* 189, 85–96.
612. Wolfe, M. W., and Call, G. B. (1999). Early growth response protein 1 binds to the luteinizing hormone-β promoter and mediates gonadotropin-releasing hormone-stimulated gene expression. *Mol. Endocrinol.* 13, 752–763.
613. Wolfe, M. W. (1999). The equine luteinizing hormone b-subunit promoter contains two functional steroidogenic factor-1 response elements. *Mol. Endocrinol.* 13, 1497–1510.
614. Dorn, C., Ou, Q., Svaren, J., Crawford, P. A., and Sadovsky, Y. (1999). Activation of luteinizing hormone β gene by gonadotropin-releasing hormone requires the synergy of early growth response-1 and steroidogenic factor-1. *J. Biol. Chem.* 274, 13870–13876.
615. Halvorson, L. M., Kaiser, U. B., and Chin, W. W. (1999). The protein kinase C system acts through the early growth response protein 1 to increase LHβ gene expression in synergy with steroidogenic factor-1. *Mol. Endocrinol.* 13, 106–116.
616. Lee, S. L., Sadovsky, Y., Swirnoff, A. H., Polish, J. A., Goda, P., Gavrilina, G., and Milbrandt, J. (1996). Luteinizing hormone deficiency and female infertility in mice lacking the transcription factor NGFI-A (Egr-1). *Science* 273, 1219–1221.
617. Keri, R. A., and Nilson, J. H. (1996). A steroidogenic factor-1 binding site is required for activity of the luteinizing hormone β subunit promoter in gonadotropes of transgenic mice. *J. Biol. Chem.* 271, 10782–10785.
618. Jeong, K.-H., Chin, W. W., and Kaiser, U. B. (2004). Essential role of the homeodomain for pituitary homeobox 1 activation of mouse gonadotropin-releasing hormone receptor gene expression through interactions with c-Jun and DNA. *Mol. Cell. Biol.* 24, 6127–6139.
619. Rosenberg, S. B., and Mellon, P. L. (2002). An Otx-related homeodomain protein binds an LHβ promoter element important for activation during gonadotrope maturation. *Mol. Endocrinol.* 16, 1280–1298.
620. Russo, M. W., Sevetson, B. R., and Milbrandt, J. (1995). Identification of NAB1, a repressor of NGFI-A- and Krox20-mediated transcription. *Proc. Natl. Acad. Sci. U S A* 92, 6873–6877.
621. Sevetson, B. R., Svaren, J., and Milbrandt, J. (2000). A novel activation function for NAB proteins in EGR-dependent transcription of the luteinizing hormone β gene. *J. Biol. Chem.* 275, 9749–9757.
622. Ikeda, Y., Swain, A., Weber, T. J., Hentges, K. E., Zanaria, E., Lalli, E., Tamai, K. T., Sassone-Corsi, P., Lovell-Badge, R., Camerino, G., and Parker, K. L. (1996). Steroidogenic factor 1 and Dax-1 colocalize in multiple cell lineages: potential links in endocrine development. *Mol. Endocrinol.* 10, 1261–1272.
623. Swain, A., Zanaria, E., Hacker, A., Lovell-Badge, R., and Camerino, G. (1996). Mouse Dax1 expression is consistent with a role in sex determination as well as in adrenal and hypothalamus function. *Nat. Genet.* 12, 404–409.
624. Zanaria, E., Muscatelli, F., Bardoni, B., Strom, T. M., Guioli, S., Guo, W., Lalli, E., Moser, C., Walker, A. P., McCabe, E. R. B., Meitinger, T., Monaco, A. P., Sassone-Corsi, P., and Camerino, G. (1994). An unusual member of the nuclear hormone receptor superfamily responsible for X-linked adrenal hypoplasia congenita. *Nature* 372, 635–641.
625. Muscatelli, F., Strom, T. M., Walker, A. P., Zanaria, E., Recan, D., Meindl, A., Bardoni, B., Guioli, S., Zehetner, G., Rabl, W., Schwartz, H. P., Kaplan, J. C., Camerino, G., Meitinger, T., and Monaco, A. P. (1994). Mutations in the DAX-1 gene give rise to both X-linked adrenal hypoplasia congenita and hypogonadotropic hypogonadism. *Nature* 372, 672–676.
626. Lalli, E., and Sassone-Corsi, P. (2003). DAX-1, an unusual orphan receptor at the crossroads of steroidogenic function and sexual differentiation. *Mol. Endocrinol.* 17, 1445–1453.
627. Ito, M., Yu, R., and Jameson, J. L. (1997). DAX-1 inhibits SF-1-mediated transactivation via a carboxy-terminal domain that is deleted in adrenal hypoplasia congenita. *Mol. Cell. Biol.* 17, 1476–1483.
628. Crawford, P. A., Dorn, C., Sadovsky, Y., and Milbrandt, J. (1998). Nuclear receptor DAX-1 recruits nuclear receptor corepressor N-CoR to steroidogenic factor 1. *Mol. Cell. Biol.* 18, 2949–2956.
629. Ghosh, A. K., Majumder, M., Steele, R., White, R. A., and Ray, R. B. (2001). A novel 16-kilodalton cellular protein physically interacts with and antagonizes the functional activity of c-*myc* promoter-binding protein 1. *Mol. Cell. Biol.* 21, 655–662.
630. Ghosh, A. K., Steele, R., and Ray, R. B. (2003). Modulation of human luteinizing hormone β gene transcription by MIP-2A. *J. Biol. Chem.* 278, 24033–24038.
631. Shupnik, M. A., and Rosenzweig, B. A. (1991). Identification of an estrogen-responsive element in the rat LHβ gene. *J. Biol. Chem.* 266, 17084–17091.
632. Curtin, D., Jenkins, S., Farmer, N., Anderson, A. C., Haisenleder, D. J., Rissman, E., Wilson, E. M., and Shupnik,

M. A. (2001). Androgen suppression of GnRH-stimulated rat LHβ gene transcription occurs through Sp1 sites in the distal GnRH-responsive promoter region. *Mol. Endocrinol.* 15, 1906–1917.

633. Keri, R. A., Wolfe, M. W., Saunders, T. L., Anderson, I., Kendall, S. K., Wagner, T., Yeung, J., Gorski, J., Nett, T. M., Camper, S. A., and Nilson, J. H. (1994). The proximal promoter of the bovine luteinizing hormone β-subunit gene confers gonadotrope-specific expression and regulation by gonadotropin-releasing hormone, testosterone, and 17β-estradiol in transgenic mice. *Mol. Endocrinol.* 8, 1807–1816.
634. Jorgensen, J. S., and Nilson, J. H. (2001). AR suppresses transcription of the LHβ subunit by interacting with steroidogenic factor-1. *Mol. Endocrinol.* 15, 1505–1516.
635. Curtin, D., Ferris, H. A., Hakli, M., Gibson, M., Janne, O. A., Palvimo, J. J., and Shupnik, M. A. (2004). Small nuclear RING finger protein stimulates the rat luteinizing hormone-β promoter by interacting with Sp1 and steroidogenic factor-1 and protects from androgen suppression. *Mol. Endocrinol.* 18, 1263–1276.
636. Vasilyev, V. V., Lawson, M. A., Dipaolo, D., Webster, N. J. G., and Mellon, P. L. (2002). Different signaling pathways control acute induction versus long-term repression of LHβ transcription by GnRH. *Endocrinology* 143, 3414–3426.
637. Haisenleder, D. J., Ferris, H. A., and Shupnik, M. A. (2003). The calcium component of gonadotropin-releasing hormone-stimulated luteinizing hormone subunit gene transcription is mediated by calcium/calmodulin-dependent protein kinase type II. *Endocrinology* 144, 2409–2416.
638. Haisenleder, D. J., Burger, L. L., Aylor, K. W., Dalkin, A. C., and Marshall, J. C. (2003). Gonadotropin-releasing hormone stimulation of gonadotropin subunit transcription: evidence for the involvement of calcium/calmodulin-dependent kinase II (Ca/CAMK II) activation in rat pituitaries. *Endocrinology* 144, 2768–2774.
639. Clayton, R. N., Lalloz, M. R. A., Salton, S. R. J., and Roberts, J. L. (1991). Expression of luteinising hormone-β subunit chloramphenicol acetyltransferase (LH-β-CAT) fusion gene in rat pituitary cells: induction by cyclic 3′-adenosine monophosphate (cAMP). *Mol. Cell. Endocrinol.* 80, 193–202.
640. Maurer, R. A. (1987). Molecular cloning and nucleotide sequence analysis of complementary deoxyribonucleic acid for the β-subunit of rat follicle stimulating hormone. *Mol. Endocrinol.* 1, 717–723.
641. Jameson, J. L., Becker, C. B., Lindell, C. M., and Habener, J. F. (1988). Human follicle-stimulating hormone β-subunit gene encodes multiple messenger ribonucleic acids. *Mol. Endocrinol.* 2, 806–815.
642. Gharib, S. D., Roy, A., Wierman, M. E., and Chin, W. W. (1989). Isolation and characterization of the gene encoding the β-subunit of rat follicle-stimulating hormone. *DNA* 8, 339–349.
643. Wierman, M. E., Rivier, J. E., and Wang, C. (1989). Gonadotropin-releasing hormone-dependent regulation of gonadotropin subunit messenger ribonucleic acid levels in the rat. *Endocrinology* 124, 272–278.
644. Weiss, J., Guendner, M. J., Halvorson, L. M., and Jameson, J. L. (1995). Transcriptional activation of the follicle-stimulating hormone β-subunit gene by activin. *Endocrinology* 136, 1885–1891.
645. Carroll, R. S., Corrigan, A. Z., Vale, W., and Chin, W. W. (1991). Activin stabilizes follicle-stimulating hormone-beta messenger ribonucleic acid levels. *Endocrinology* 129, 1721–1726.
646. Attardi, B., and Winters, S. J. (1993). Decay of follicle-stimulating hormone-b messenger RNA in the presence of transcriptional inhibitors and/or inhibin, activin, or follistatin. *Mol. Endocrinol.* 7, 668–680.
647. Kumar, T. R., Fairchild-Huntress, V., and Low, M. J. (1992). Gonadotrope-specific expression of the human follicle-stimulating hormone β-subunit gene in pituitaries of transgenic mice. *Mol. Endocrinol.* 6, 81–90.
648. Huang, H.-J., Sebastian, J., Strahl, B. D., Wu, J. C., and Miller, W. L. (2001). The promoter for the ovine follicle-stimulating hormone-β gene (FSHβ) confers FSHβ-like expression on luciferase in transgenic mice: regulatory studies *in vivo* and *in vitro*. *Endocrinology* 142, 2260–2266.
649. Graham, K. E., Nusser, K. D., and Low, M. J. (1999). LβT2 gonadotroph cells secrete follicle stimulating hormone (FSH) in response to activin A. *J. Endocrinol.* 162, R1–R5.
650. Bailey, J. S., Rave-Harel, N., McGillivray, S. M., Coss, D., and Mellon, P. L. (2004). Activin regulation of the follicle-stimulating hormone β-subunit gene involves Smads and the TALE homeodomain proteins Pbx1 and Prep1. *Mol. Endocrinol.* 18, 1158–1170.
651. Spady, T. J., Shayya, R., Thackray, V. G., Ehrensberger, L., Bailey, J. S., and Mellon, P. L. (2004). Androgen regulates follicle-stimulating hormone β gene expression in an activin-dependent manner in immortalized gonadotropes. *Mol. Endocrinol.* 18, 925–940.
652. Strahl, B. D., Huang, H.-J., Pedersen, N. R., Wu, J. C., Ghosh, B. R., and Miller, W. L. (1997). Two proximal activating protein-1-binding sites are sufficient to stimulate transcription of the ovine follicle-stimulating hormone-β gene. *Endocrinology* 138, 2621–2631.
653. Nelson, C. C., Hendy, S. C., Shukin, R. J., Cheng, H., Bruchovsky, N., Koop, B. F., and Rennie, P. S. (1999). Determinants of DNA sequence specificity of the androgen, progesterone, and glucocorticoid receptors: evidence for differential steroid receptor response elements. *Mol. Endocrinol.* 13, 2090–2107.
654. Suszko, M. I., Lo, D. J., Suh, H., Camper, S. A., and Woodruff, T. K. (2003). Regulation of the rat follicle-stimulating hormone β-subunit promoter by activin. *Mol. Endocrinol.* 17, 318–332.
655. Zakaria, M. M., Jeong, K.-H., Lacza, C., and Kaiser, U. B. (2002). Pituitary homeobox 1 activates the rat FSHβ (rFSHβ) gene through both direct and indirect interactions with the rFSHβ gene promoter. *Mol. Endocrinol.* 16, 1840–1852.
656. Jacobs, S. B. R., Coss, D., McGillivray, S. M., and Mellon, P. L. (2003). Nuclear factor Y and steroidogenic factor 1 physically and functionally interact to contribute to cell-specific expression of the mouse follicle-stimulating hormone-β gene. *Mol. Endocrinol.* 17, 1470–1483.
657. Li, L.-A., Chiang, E. F.-L., Chen, J.-C., Hsu, N.-C., Chen, Y.-J., and Chung, B. (1999). Function of steroidogenic factor 1 domains in nuclear localization, transactivation, and interaction with transcription factor TFIIB and c-Jun. *Mol. Endocrinol.* 13, 1588–1598.
658. Borud, B., Hoang, T., Bakke, M., Jacob, A. L., Lund, J., and Mellgren, G. (2002). The nuclear receptor coactivators p300/CBP/cointegrator-associated protein (p/CIP) and transcription intermediary factor 2 (TIF2) differentially regulate PKA-stimulated transcriptional activity of steroidogenic factor 1. *Mol. Endocrinol.* 16, 757–773.
659. Jacob, A. L., Lund, J., Martinez, P., and Hedin, L. (2001). Acetylation of steroidogenic factor 1 protein regulates its transcriptional activity and recruits the coactivator GCN5. *J. Biol. Chem.* 276, 37659–37664.
660. Farsetti, A., Narducci, M., Moretti, F., Nanni, S., Mantovani, R., Sacchi, A., and Pontecorvi, A. (2001). Inhibition of ERα-mediated *trans*-activation of human coagulation factor XII gene by heteromeric transcription factor NF-Y. *Endocrinology* 142, 3380–3388.

661. Faniello, M. C., Bevilacqua, M. A., Condorelli, G., de Crombrugghe, B., Maity, S. N., Avvedimento, V. E., Cimino, F., and Costanzo, F. (1999). The β subunit of the CAAT-binding factor NFY binds the central segment of the co-activator p300. *J. Biol. Chem.* 274, 7623–7626.
662. Currie, R. A. (1998). NF-Y is associated with the histone acetyltransferases GCN5 and P/CAF. *J. Biol. Chem.* 273, 1430–1434.
663. Strahl, B. D., Huang, H.-J., Sebastian, J., Ghosh, B. R., and Miller, W. L. (1998). Transcriptional activation of the ovine follicle-stimulating hormone β-subunit gene by gonadotropin-releasing hormone: involvement of two activating protein-1-binding sites and protein kinase C. *Endocrinology* 139, 4455–4465.
664. Zawel, L., Le Dai, J., Buckhaults, P., Zhou, S., Kinzler, K. W., Vogelstein, B., and Kern, S. E. (1998). Human Smad3 and Smad4 are sequence-specific transcription activators. *Mol. Cells* 1, 611–617.
665. Shi, Y., Wang, Y.-F., Jayaraman, L., Yang, H., Massague, J., and Pavletich, N. P. (1998). Crystal structure of a Smad MH1 domain bound to DNA: insights on DNA binding in TGF-β signaling. *Cell* 94, 585–594.
666. Massague, J., and Wotton, D. (2000). Transcriptional control by the TGF-β/Smad signaling system. *EMBO J.* 19, 1745–1754.
667. Chen, X., Rubock, M. J., and Whitman, M. (1996). A transcriptional partner for MAD proteins in TGF-β signalling. *Nature* 383, 691–696.
668. Chen, X., Weisberg, E., Fridmacher, V., Watanabe, M., Naco, G., and Whitman, M. (1997). Smad4 and FAST-1 in the assembly of activin-responsive factor. *Nature* 389, 85–89.
669. Labbe, E., Silvestri, C., Hoodless, P. A., Wrana, J. L., and Attisano, L. (1998). Smad2 and Smad3 positively and negatively regulate TGFβ-dependent transcription through the forkhead DNA-binding protein FAST2. *Mol. Cells* 2, 109–120.
670. Pouponnot, C., Jayaraman, L., and Massague, J. (1998). Physical and functional interaction of SMADs and p300/CBP. *J. Biol. Chem.* 273, 22865–22868.
671. Shen, X., Hu, P. P., Liberati, N. T., Datto, M. B., Frederick, J. P., and Wang, X.-F. (1998). TGF-β-induced phosphorylation of Smad3 regulates its interaction with coactivator p300/CREB-binding protein. *Mol. Biol. Cell.* 9, 3309–3319.
672. Wotton, D., Lo, R. S., Lee, S., and Massague, J. (1999). A Smad transcriptional corepressor. *Cell* 97, 29–39.
673. Chen, C.-R., Kang, Y., Siegel, P. M., and Massague, J. (2002). E2F4/5 and p107 as Smad cofactors linking the TGFβ receptor to c-*myc* repression. *Cell* 110, 19–32.
674. Chen, F., Ogawa, K., Liu, X., Stringfield, T. M., and Chen, Y. (2002). Repression of Smad2 and Smad3 transactivating activity by association with a novel splice variant of CCAAT-binding factor C subunit. *Biochem. J.* 364, 571–577.
675. Jonk, L. J. C., Itoh, S., Heldin, C.-H., ten Dijke, P., and Kruijer, W. (1998). Identification and functional characterization of a Smad binding element (SBE) in the *JunB* promoter that acts as a transforming growth factor-β, activin, and bone morphogenetic protein-inducible enhancer. *J. Biol. Chem.* 273, 21145–21152.
676. Zhang, Y., Feng, X.-H., and Derynck, R. (1998). Smad3 and Smad4 cooperate with c-Jun/c-Fos to mediate TGF-β-induced transcription. *Nature* 394, 909–913.
677. Liberati, N. T., Datto, M. B., Frederick, J. P., Shen, X., Wong, C., Rougier-Chapman, E. M., and Wang, X.-F. (1999). Smads bind directly to the Jun family of AP-1 transcription factors. *Proc. Natl. Acad. Sci. U S A* 96, 4844–4849.
678. Wong, C., Rougier-Chapman, E. M., Frederick, J. P., Datto, M. B., Liberati, N. T., Li, J.-M., and Wang, X.-F. (1999). Smad3-Smad4 and AP-1 complexes synergize in transcriptional activation of the c-Jun promoter by transforming growth factor β. *Mol. Cell. Biol.* 19, 1821–1830.
679. Mann, R. S., and Chan, S.-K. (1996). Extra specificity from *extradenticle*: the partnership between HOX and PBX/EXD homeodomain proteins. *Trends Genet.* 12, 258–262.
680. Zhang, L., Wang, W., Hayashi, Y., Jester, J. V., Birk, D. E., Gao, M., Liu, C.-Y., Kao, W. W.-Y., Karin, M., and Xia, Y. (2003). A role for MEK kinase 1 in TGF-β/activin-induced epithelium movement and embryonic eyelid closure. *EMBO J.* 22, 4443–4454.
681. Cocolakis, E., Lemay, S., Ali, S., and Lebrun, J.-J. (2001). The p38 MAPK pathway is required for cell growth inhibition of human breast cancer cells in response to activin. *J. Biol. Chem.* 276, 18430–18436.
682. Furukawa, F., Matsuzaki, K., Mori, S., Tahashi, Y., Yoshida, K., Sugano, Y., Yamagata, H., Matsushita, M., Seki, T., Inagaki, Y., Nishizawa, M., Fujisawa, J., and Inoue, K. (2003). p38 MAPK mediates fibrogenic signal through Smad3 phosphorylation in rat myofibroblasts. *Hepatology* 38, 879–889.
683. Hayashida, T., Decaestecker, M., and Schnaper, H. W. (2003). Cross-talk between ERK MAP kinase and Smad signaling pathways enhances TGF-β-dependent responses in human mesangial cells. *FASEB J.* 17, 1576–1578.
684. Zhou, W., and Sealfon, S. C. (1994). Structure of the mouse gonadotropin-releasing hormone receptor gene: variant transcripts generated by alternative processing. *DNA Cell. Biol.* 13, 605–614.
685. Braden, T. D., and Conn, P. M. (1990). Altered rate of synthesis of gonadotropin-releasing hormone receptors: effects of homologous hormone appear independent of extracellular calcium. *Endocrinology* 126, 2577–2582.
686. Mercer, J. E., and Chin, W. W. (1995). Regulation of pituitary gonadotrophin gene expression. *Hum. Reprod. Update* 1, 363–384.
687. Cheon, M., Park, D., Kim, K., Park, S. D., and Ryu, K. (1999). Homologous upregulation of GnRH receptor mRNA by continuous GnRH in cultured rat pituitary cells. *Endocrine* 11, 49–55.
688. Maya-Nunez, G., and Conn, P. M. (2003). Transcriptional regulation of the GnRH receptor gene by glucocorticoids. *Mol. Cell. Endocrinol.* 200, 89–98.
689. Albarracin, C. T., Kaiser, U. B., and Chin, W. W. (1994). Isolation and characterization of the 5′-flanking region of the mouse gonadotropin-releasing hormone receptor gene. *Endocrinology* 135, 2300–2306.
690. Clay, C. M., Nelson, S. E., DiGregorio, G. B., Campion, C. E., Wiedemann, A. L., and Nett, R. J. (1995). Cell-specific expression of the mouse gonadotropin-releasing hormone (GnRH) receptor gene is conferred by elements residing within 500 bp of proximal 5′ flanking region. *Endocrine* 3, 615–622.
691. Albarracin, C. T., Frosch, M. P., and Chin, W. W. (1999). The gonadotropin-releasing hormone receptor gene promoter directs pituitary-specific oncogene expression in transgenic mice. *Endocrinology* 140, 2415–2421.
692. McCue, J. M., Quirk, C. C., Nelson, S. E., Bowen, R. A., and Clay, C. M. (1997). Expression of a murine gonadotropin-releasing hormone receptor-luciferase fusion gene in transgenic mice is diminished by immunoneutralization of gonadotropin-releasing hormone. *Endocrinology* 138, 3154–3160.
693. Kam, K.-Y., Jeong, K.-H., Norwitz, E. R., Jorgensen, E. M., and Kaiser, U. B. (2004). Oct-1 and nuclear factor Y bind to the SURG-1 element to direct basal and gonadotropin-releasing hormone (GnRH)-stimulated mouse GnRH receptor gene transcription. *Mol. Endocrinol.* 19, 148–162.
694. Duval, D. L., Nelson, S. E., and Clay, C. M. (1997). A binding site for steroidogenic factor-1 is part of a complex enhancer

that mediates expression of the murine gonadotropin-releasing hormone receptor gene. *Biol. Reprod.* 56, 160–168.
695. Duval, D. L., Nelson, S. E., and Clay, C. M. (1997). The tripartite basal enhancer of the gonadotropin-releasing hormone (GnRH) receptor gene promoter regulates cell-specific expression through a novel GnRH receptor activating sequence. *Mol. Endocrinol.* 11, 1814–1821.
696. Cherrington, B. D., Farmerie, T. A., Lents, C. A., Cantlon, J. D., Roberson, M. S., and Clay, C. M. (2005). Activin responsiveness of the murine gonadotropin releasing hormone receptor gene is mediated by a composite enhancer containing spatially distinct regulatory elements. *Mol. Endocrinol.* 19, 898–912.
697. Maya-Nunez, G., and Conn, P. M. (1999). Transcriptional regulation of the gonadotropin-releasing hormone receptor gene is mediated in part by a putative repressor element and by the cyclic adenosine 3′,5′-monophosphate response element. *Endocrinology* 140, 3452–3458.
698. Sadie, H., Styger, G., and Hapgood, J. (2003). Expression of the mouse gonadotropin-releasing hormone receptor gene in αT3-1 gonadotrope cells is stimulated by cyclic 3′,5′-adenosine monophosphate and protein kinase A, and is modulated by steroidogenic factor-1 and Nur77. *Endocrinology* 144, 1958–1971.
699. Pincas, H., Forrai, Z., Chauvin, S., Laverriere, J.-N., and Counis, R. (1998). Multiple elements in the distal part of the 1.2 kb 5′-flanking region of the rat GnRH receptor gene regulate gonadotrope-specific expression conferred by proximal domain. *Mol. Cell. Endocrinol.* 144, 95–108.
700. Granger, A., Ngo-Muller, V., Bleux, C., Guigon, C., Pincas, H., Magre, S., Daegelen, D., Tixier-Vidal, A., Counis, R., and Laverriere, J.-N. (2004). The promoter of the rat gonadotropin-releasing hormone receptor gene directs the expression of the human placental alkaline phosphatase reporter gene in gonadotrope cells in the anterior pituitary gland as well as in multiple extrapituitary tissues. *Endocrinology* 145, 983–993.
701. Pincas, H., Amoyel, K., Counis, R., and Laverriere, J.-N. (2001). Proximal *cis*-acting elements, including steroidogenic factor 1, mediate the efficiency of a distal enhancer in the promoter of the rat gonadotropin-releasing hormone receptor gene. *Mol. Endocrinol.* 15, 319–337.
702. Pincas, H., Laverriere, J.-N., and Counis, R. (2001). Pituitary adenylate cyclase-activating polypeptide and cyclic adenosine 3′,5′-monophosphate stimulate the promoter activity of the rat gonadotropin-releasing hormone receptor gene via a bipartite response element in gonadotrope-derived cells. *J. Biol. Chem.* 276, 23562–23571.
703. Cheng, K. W., and Leung, P. C. K. (2001). Human gonadotropin-releasing hormone receptor gene transcription: up-regulation by 3′,5′-cyclic adenosine monophosphate/protein kinase A pathway. *Mol. Cell. Endocrinol.* 181, 15–26.
704. Ngan, E. S. W., Leung, P. C. K., and Chow, B. K. C. (2000). Identification of an upstream promoter in the human gonadotropin-releasing hormone receptor gene. *Biochem. Biophys. Res. Commun.* 270, 766–772.
705. Hoo, R. L. C., Ngan, E. S. W., Leung, P. C. K., and Chow, B. K. C. (2003). Two *Inr* elements are important for mediating the activity of the proximal promoter of the human gonadotropin-releasing hormone receptor gene. *Endocrinology* 144, 518–527.
706. Ngan, E. S. W., Leung, P. C. K., and Chow, B. K. C. (2001). Interplay of pituitary adenylate cyclase-activating polypeptide with a silencer element to regulate the upstream promoter of the human gonadotropin-releasing hormone receptor gene. *Mol. Cell. Endocrinol.* 176, 135–144.
707. Cheng, C. K., Chow, B. K. C., and Leung, P. C. K. (2003). An activator protein 1-like motif mediates 17β-estradiol repression of gonadotropin-releasing hormone receptor promoter via an estrogen receptor α-dependent mechanism in ovarian and breast cancer cells. *Mol. Endocrinol.* 17, 2613–2629.
708. Cheng, C. K., Yeung, C. M., Hoo RL C., Chow, B. K. C., and Leung, P. C. K. (2002). Oct-1 is involved in the transcriptional repression of the gonadotropin-releasing hormone receptor gene. *Endocrinology* 143, 4693–4701.
709. Ngan, E. S. W., Cheng PK W., Leung, P. C. K., and Chow, B K. C. (1999). Steroidogenic factor-1 interacts with a gonadotrope-specific element within the first exon of the human gonadotropin-releasing hormone receptor gene to mediate gonadotrope-specific expression. *Endocrinology* 140, 2452–2462.
710. Xin, J. H., Cowie, A., Lachance, P., and Hassell, J. A. (1992). Molecular cloning and characterization of PEA3, a new member of the Ets oncogene family that is differentially expressed in mouse embryonic cells. *Genes Dev.* 6, 481–496.
711. Bodner, M., Castriilo, J.-L., Theill, L. E., Deerinck, T., Ellisman, M., and Karin, M. (1988). The pituitary-specific transcription factor GHF-1 is a homeobox-containing protein. *Cell* 55, 505–518.
712. Ingraham, H. A., Flynn, S. E., Voss, J. W., Albert, V. R., Kapiloff, M. S., Wilson, L., and Rosenfeld, M. G. (1990). The POU-specific domain of Pit-1 is essential for sequence-specific, high affinity DNA binding and DNA-dependent Pit-1-Pit-1 interactions. *Cell* 61, 1021–1033.
713. Steinfelder, H. J., Radovick, S., Mroczynski, M. A., Hauser, P., McClaskey, J. H., Weintraub, B. D., and Wondisford, F. E. (1992). Role of a pituitary-specific transcription factor (Pit-1/GHF-1) or a closely related protein in cAMP regulation of human thyrotropin-β subunit gene expression. *J. Clin. Invest.* 89, 409–419.
714. Ellsworth, B. S., Burns, A. T., Escudero, K. W., Duval, D. L., Nelson, S. E., and Clay, C. M. (2003). The gonadotropin releasing hormone (GnRH) receptor activating sequence (GRAS) is a composite regulatory element that interacts with multiple classes of transcription factors including Smads, AP-1 and a forkhead DNA binding protein. *Mol. Cell. Endocrinol.* 206, 93–111.
715. Treier, M., Gleiberman, A. S., O'Connell, S. M., Szeto, D. P., McMahon, J. A., McMahon, A. P., and Rosenfeld, M. G. (1998). Multistep signaling requirements for pituitary organogenesis in vivo. *Genes Dev.* 12, 1691–1704.
716. Kioussi, C., O'Connell, S., St-Onge, L., Treier, M., Gleiberman, A. S., Gruss, P., and Rosenfeld, M. G. (1999). *Pax6* is essential for establishing ventral-dorsal cell boundaries in pituitary gland development. *Proc. Natl. Acad. Sci. U S A* 96, 14378–14382.
717. Wilson, T. E., Fahrner, T. J., and Milbrandt, J. (1993). The orphan receptors NGFI-B and steroidogenic factor 1 establish monomer binding as a third paradigm of nuclear receptor-DNA interaction. *Mol. Cell. Biol.* 13, 5794–5804.
718. Clark, M. E., and Mellon, P. L. (1995). The POU homeodomain transcription factor Oct-1 is essential for activity of the gonadotropin-releasing hormone neuron-specific enhancer. *Mol. Cell. Biol.* 15, 6169–6177.
719. Eraly, S. A., Nelson, S. B., Huang, K. M., and Mellon, P. L. (1998). Oct-1 binds promoter elements required for transcription of the GnRH gene. *Mol. Endocrinol.* 12, 469–481.
720. Vazquez-Martinez, R., Leclerc, G. M., Wierman, M. E., and Boockfor, F. R. (2002). Episodic activation of the rat GnRH promoter: role of the homeoprotein Oct-1. *Mol. Endocrinol.* 16, 2093–2100.
721. McGillivray, S. M., Bailey, J. S., Ramezani, R., Kirkwood, B. J., and Mellon, P. L. (2005). Mouse GnRH receptor gene expression is mediated by the LHX3 homeodomain protein. *Endocrinology* 146, 2180–2185.

722. Bruggemeier, U., Kalff, M., Franke, S., Scheidereit, C., and Beato, M. (1991). Ubiquitous transcription factor OTF-1 mediates induction of the MMTV promoter through synergistic interaction with hormone receptors. *Cell* 64, 565–572.
723. Prefontaine, G. G., Walther, R., Giffin, W., Lemieux, M. E., Pope, L., and Hache, R. J. G. (1999). Selective binding of steroid hormone receptors to octamer transcription factors determines transcriptional synergism at the mouse mammary tumor virus promoter. *J. Biol. Chem.* 274, 26713–26719.
724. Lee, S.-K., Kim, H.-J., Na, S.-Y., Kim, T. S., Choi, H.-S., Im, S.-Y., and Lee, J. W. (1998). Steroid receptor coactivator-1 coactivates activating protein-1-mediated transactivations through interaction with the c-Jun and c-Fos subunits. *J. Biol. Chem.* 273, 16651–16654.
725. Jung, D.-J., Na, S.-Y., Na, D. S., and Lee, J. W. (2002). Molecular cloning and characterization of CAPER, a novel coactivator of activating protein-1 and estrogen receptors. *J. Biol. Chem.* 277, 1229–1234.
726. Jakacka, M., Ito, M., Weiss, J., Chien, P.-Y., Gehm, B. D., and Jameson, J. L. (2001). Estrogen receptor binding to DNA is not required for its activity through the nonclassical AP1 pathway. *J. Biol. Chem.* 276, 13615–13621.
727. Bannister, A. J., Oehler, T., Wilhelm, D., Angel, P., and Kouzarides, T. (1995). Stimulation of c-Jun activity by CBP: c-Jun residues Ser63/73 are required for CBP induced stimulation *in vivo* and CBP binding *in vitro*. *Oncogene* 11, 2509–2514.
728. Yang-Yen, H.-F., Chambard, J.-C., Sun, Y.-L., Smeal, T., Schmidt, T. J., Drouin, J., and Karin, M. (1990). Transcriptional interference between c-Jun and the glucocorticoid receptor: mutual inhibition of DNA binding due to direct protein-protein interaction. *Cell* 62, 1205–1215.
729. Ford, J., McEwan, I. J., Wright, A. P. H., and Gustafsson, J.-A. (1997). Involvement of the transcription factor IID protein complex in gene activation by the N-terminal transactivation domain of the glucocorticoid receptor *in vitro*. *Mol. Endocrinol.* 11, 1467–1475.
730. Chandran, U. R., Warren, B. S., Baumann, C. T., Hager, G. L., and DeFranco, D. B. (1999). The glucocorticoid receptor is tethered to DNA-bound Oct-1 at the mouse gonadotropin-releasing hormone distal negative glucocorticoid response element. *J. Biol. Chem.* 274, 2372–2378.
731. Conn, P. M. (1989). Does protein kinase C mediate pituitary actions of gonadotropin-releasing hormone? *Mol. Endocrinol.* 3, 755–757.
732. Lin, X., and Conn, P. M. (1999). Transcriptional activation of gonadotropin-releasing hormone (GnRH) receptor gene by GnRH: involvement of multiple signal transduction pathways. *Endocrinology* 140, 358–364.
733. Maya-Nunez, G., and Conn, P. M. (2001). Cyclic adenosine 3′,5′-monophosphate (cAMP) and cAMP responsive element-binding protein are involved in the transcriptional regulation of gonadotropin-releasing hormone (GnRH) receptor by GnRH and mitogen-activated protein kinase signal transduction pathway in GGH_3 cells. *Biol. Reprod.* 65, 561–567.

Knobil and Neill's Physiology of Reproduction, Third Edition
edited by Jimmy D. Neill,
Elsevier © 2006

CHAPTER 32

Prolactin: Structure, Function, and Regulation of Secretion

Karen A. Gregerson

Introduction, 1703
Evolution of PRL, 1703
Pituitary PRL, 1703
Placental Lactogens, 1704
Nonlactogenic PRL-like Peptides, 1704
Extrapituitary PRL, 1705
Biochemistry of PRL, 1705
Prolactin, 1705
Variants of PRL, 1706
PRL Receptors and PRL Signaling, 1707
Receptors, 1707
Signal Transduction, 1708
Regulators of PRL Secretion, 1709
Dopamine, 1710
Other Inhibitors of PRL, 1713
Estrogen, 1713
PRL-Releasing Factors, 1714
Episodic PRL Release Associated with Female Reproduction, 1714
Estrous Cycle, 1715
Pregnancy, 1715
Lactation, 1715
Reproductive Physiology and Pathophysiology of PRL in Mammals, 1716
Physiological Levels of PRL, 1716
PRL Actions, 1716
Pathologies of PRL Secretion, 1717
Summary, 1718
References, 1719

INTRODUCTION

Prolactin (PRL) is a protein hormone secreted from the anterior pituitary gland that has been identified in all vertebrate classes. PRL has numerous and diverse biological functions, including water and electrolyte balance, growth and development, metabolism and endocrine regulation, behavior, immune function, and reproduction. In most species, and especially in mammals, PRL has developed highly specialized roles in female reproduction. It is in this category that PRL has been demonstrated to be irreplaceable. PRL-related knockout models exhibit female infertility and lactational failure. These models do not, however, exhibit major disruptions in growth, metabolism, immune function, and behavior, suggesting that PRL serves a modulatory rather than an essential role in these nonreproductive functions. This chapter discusses PRL within the context of reproductive physiology and focuses attention particularly on PRL actions in mammals.

EVOLUTION OF PRL

Pituitary PRL

PRL was the first of the pituitary hormones to be purified and identified nearly 80 years ago (1,2) and was so named ("pro-lactin") for its stimulatory actions on mammary gland development and lactation. The amino acid sequence and tertiary structure of PRL is closely related to that of growth hormone (GH) (3), also secreted from the anterior pituitary gland. It has

Division of Pharmaceutical Sciences, University of Cincinnati Medical Center, Cincinnati, Ohio.

been inferred that the PRL and GH genes arose from a common ancestral gene at least 400 million years ago, with the origin of vertebrates (4). PRL and GH are distinct proteins present in all vertebrate groups except the cyclostomes, in which neither protein has been described (5). Proteins related to PRL and GH presumably had a prevertebrate origin, but this has not been demonstrated. Neither peptide nor indeed other members of the cytokine family have been convincingly described in any invertebrate. This remains the case, despite the recent complete genome sequencing of two invertebrates, *Caenorhabditis elegans* and *Drosophila melanogaster*. In fish, a third pituitary hormone, somatolactin, arose that shows about equal similarity to PRL and GH (6).

In some mammalian groups a variety of derivative genes has appeared through duplication of the PRL and GH genes. These duplications appear to have occurred independently, yet in both cases the expression of these PRL- and GH-like genes has been largely confined to the placenta. The discovery of these related genes has involved numerous laboratories, and therefore a less than unified nomenclature has developed (7). This nomenclature is based on biological activity, structural similarity to PRL, or a combination of structure and proliferative actions.

Placental Lactogens

The best known of the genes arising from PRL or GH duplication are the placental lactogens (PL) (8,9), so named for their lactogenic activity. The PLs are synthesized in placental trophoblast cells during pregnancy in most, but not all, eutherian mammals. Those mammals that do not produce PLs appear in a variety of families and include common species such as dogs, horses, and pigs.

In primates, including humans, PLs arose from gene duplication within the GH locus (10,11). In humans, this locus contains a cluster of five GH-related genes, including two PL genes, known as the human chorionic somatomammotropin (CS) genes, *hCS-A* and *hCS-B*. The *hCS-A* and *hCS-B* genes differ only slightly at the nucleotide sequence level and encode identical proteins. Both genes are coexpressed in the placenta, and the levels of their products are highest in the maternal circulation during the last half of gestation (12).

In ruminants, PL genes arose from duplication of the PRL gene. There exist at least eight PRL-like genes expressed in the bovine placenta, and they are arranged as a cluster on chromosome 23 (13–15). Bovine PL is one product of these genes and is secreted as a glycosylated protein that, as in primates, circulates at high concentrations in the second half of gestation (12). Goats and sheep also produce PLs that are clearly orthologous with the bovine PL (16,17), but these ruminants do not appear to express the other placental PRL-like genes found in the cow.

Rodents (rats and mice) also express a large family of placental genes, which phylogenetic analysis and gene clustering indicate are products of PRL gene duplication (13,18–20). Two of these code for PLs and are clearly orthologous. Placental lactogen-I (PL-I) is a glycosylated protein, synthesized early during gestation, appearing immediately after implantation. PL-I expression reaches maximum levels at about mid-gestation, after which it is quickly extinguished and replaced by predominant secretion of PL-II, a nonglycosylated protein. PL-II secretion falls near term and declines rapidly as soon as the placenta is delivered (12).

Nonlactogenic PRL-like Peptides

In nonprimate mammals, many of the proteins produced from the placental PRL-like genes do not possess lactogenic activity. The nature and activities of these nonlactogenic gene products have not been established. Yet, the expression of these proteins appears to be tightly regulated during pregnancy (9,21,22), suggesting that they are physiologically important. In cattle, seven nonlactogenic PRL-like genes have been described (13,14), although most of them have been characterized only as cDNA sequences. In rodents, both mice and rats, the placenta produces at least five nonlactogenic PRL-like proteins, known as PLP-A through PLP-E. Two additional placental proteins have been described in mice. These two proteins, proliferin and proliferin-related protein, have been implicated as regulators of angiogenesis in the placenta (23).

A feature of the PRL-like gene family is the apparent variable evolutionary rates in its history, a feature similar in nature to that seen for GH (5). PRL sequences across species provide evidence for a slow basal rate of evolution. The marsupial PRL sequence differs from that deduced for the ancestral placental mammal by only 7% of all residues (24). Also, PRLs from birds and amphibia are a strongly conserved divergence of members of this family (25). Interestingly, this slow rate of change bears amazingly little correlation with the enormous range of biological actions across these species. In contrast, within mammals the major biological actions of PRL and the PLs are remarkably conserved, whereas the structure shows periods of rapid evolution (25). The PL proteins generally share less than 25% sequence identity with PRL. However, they all share two pairs of cysteine residues that are conserved throughout the PRL and GH superfamily (9). The PLs bind to and activate the PRL receptor (PRL-R) (20,26) and appear not to have their own unique receptor. Little is known about receptors for the nonlactogenic PRL-related peptides, although it has been shown that proliferin binds to the mannose-6-phosphate-insulin-like growth factor-2 II receptor (27).

Extrapituitary PRL

Although most circulating PRL is of anterior pituitary origin, the PRL gene is expressed in several extrapituitary sites in mammals, including mammary gland (28,29), decidua (30), prostate (31), immune cells (32,33), adipose tissue (34), and umbilical endothelial cells (35). PRL is synthesized by both the uterine decidua and myometrium in humans (36). High concentrations of PRL are present in amniotic fluid, which can be traced to both decidually synthesized hormone and plasma PRL that is transported across the placenta. The mammary gland is another important site of PRL synthesis and secretion (28,29,34). As in amniotic fluid, significant concentrations of PRL are present in milk, which can be traced to both locally produced PRL and circulating pituitary PRL transported across the mammary epithelium (36). PRL in milk is absorbed by the neonatal gut and exerts important changes in the maturation of the hypothalamic neuroendocrine system of the offspring (37).

The human decidua and lymphocyte PRL are controlled by a promoter different from that of pituitary PRL. The decidua PRL mRNA has a distinct 5′ untranslated sequence corresponding to an additional exon (exon 1A) (38) (Fig. 1). Exon 1A and its associated promoter elements are located approximately 8,000 bp distal to the initiation site for pituitary PRL transcription. In rodents, there is not yet evidence for a distinct extrapituitary PRL promoter. It is clear, however, that the control of extrapituitary PRL is different from that of pituitary PRL. Foremost, extrapituitary PRL is not inhibited by dopamine (DA), and progesterone is an important regulator of decidua PRL secretion (36). It may be that growth factors that can regulate the conventional pituitary PRL promoter also provide control of extrapituitary PRL in rodents.

Locally synthesized PRL may act as an autocrine or paracrine factor in extrapituitary tissues (36). It has been suggested that locally synthesized PRL in the mammary gland might act as a growth factor for both normal breast epithelium and breast cancer cells (39–41). Similarly, locally produced PRL has been proposed to contribute to morphological and functional changes that occur in the uterine decidua during pregnancy (42,43). Determining the functional roles of locally produced PRL is difficult, either under physiological or pathological conditions, because there is a lack of clear models of isolated disruption of extrapituitary PRL.

BIOCHEMISTRY OF PRL

Prolactin

In mammals, there appear to be single unique genes for PRL and GH, located on different chromosomes. Both native genes contain five exons split by four introns (Fig. 1). However, the PRL gene is much larger (~10 kb) than the GH gene (~2 kb), almost entirely due to much longer introns. The PRL gene expressed in decidua also contains an alternative first exon (1a). The PRL gene is transcribed as a single unit to produce a precursor mRNA that is then spliced to form the mature PRL mRNA with an open reading frame of 684 bases. Human PRL is synthesized from this mRNA as a prehormone ("pre-PRL") that is 227 amino acids in length, with a deduced molecular weight of nearly 26,000 Da. Cleaving the signal peptide (28 amino acids) from the N-terminus of

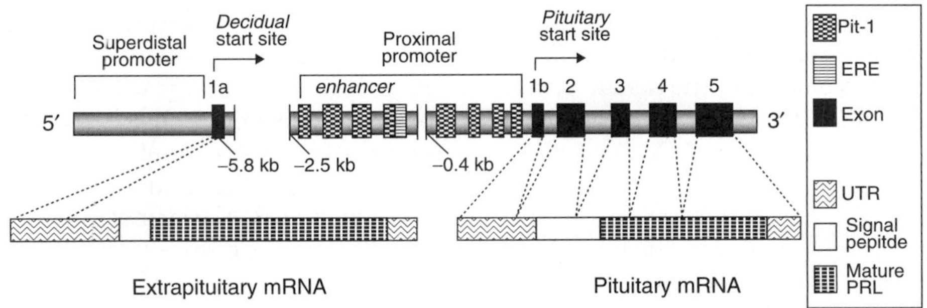

FIG. 1. Schematic diagrams of the human genes (*top*) and transcripts (*bottom*) for extrapituitary and pituitary PRL. Exon1a is used as the transcription start site in decidual tissue and produces a transcript with a longer 5′ untranslated region than the pituitary transcript. ERE, estrogen response element. (Adapted from Zinger, M., McFarland, M., and Ben-Jonathan, N. [2003]. Prolactin expression and secretion by human breast glandular and adipose tissue explants. *J. Clin. Endocrinol. Metab.* 88, 689–696.)

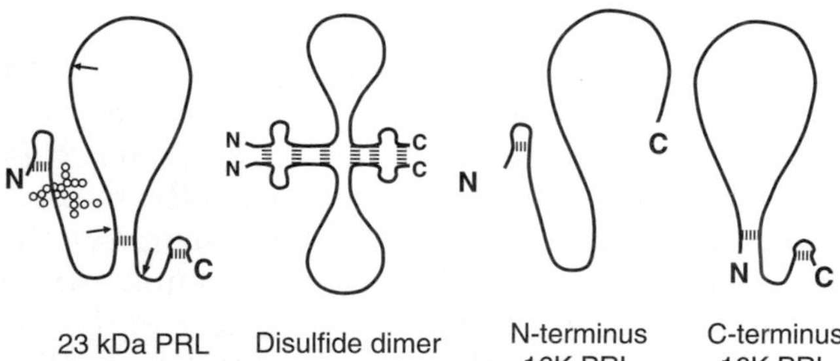

FIG. 2. Models of the mature 23-kDa PRL protein (*far left*) and the dimerized and cleaved variants. The striped bands represent the disulfide bonds. The circles on the 23-kDa PRL represent the glycosylation at amino acids 31–33. The arrows identify the major sites of phosphorylation.

pre-PRL results in the mature single polypeptide chain of 199 residues with a molecular weight of nearly 23,000 Da (23k PRL).

The PRL polypeptide chain is folded by three disulfide bridges. These occur between C58 and C174, linking distant parts of the polypeptide chain to form a large loop, and one at each of the chain's ends, forming small loops near the amino (C4 to C11) and carboxy (C191 to C199) termini (Fig. 2). The bridges forming the large loop and the C-terminus loop are conserved in all members of the PRL–GH family. The third bridge, linking residues 4 and 11 in the N-terminus, is unique to PRL and its closest relatives (3). Both specific reduction experiments and mutational analyses have demonstrated that neither of the terminal end bridges is necessary for full biological activity of PRL (44,45). In contrast, the central disulfide bond forming the large loop is absolutely necessary for biological activity (44,45).

Physical–chemical studies have suggested that the PRL and GH molecules contain a high proportion of α-helix bundles, and this has been confirmed by x-ray crystallography of pig GH (46) and human GH complexed to the extracellular domain of its receptor (47). The three-dimensional structure has not yet been reported for PRL, but homology models (48,49) have predicted four α-helical domains arranged in an up–up–down–down topology, as was described for GH. This molecular architecture identifies PRL- and GH-like proteins as members of the hematopoietic cytokine superfamily (48,50).

Variants of PRL

Several biochemical variants of 23k PRL have been identified (51). Many of these are due to post-translational modification of the peptide, including glycosylation, phosphorylation, and cleavage. As discussed above, although most circulating PRL is of pituitary origin, PRL is also synthesized in a variety of other tissues. The sequence of mature PRL is the same whether synthesized in the pituitary or in peripheral tissues (36), and it is not known to what extent glycosylation or phosphorylation of the peptide occurs in the extrapituitary sites. Cleavage of PRL, however, appears to occur primarily in peripheral tissues, although it is unknown whether the PRL so processed is of pituitary or local origin (52).

Glycosylation

Glycosylated PRL was first reported for ovine PRL (53) and was shown to be an asparagine N-linked oligosaccharide at a single consensus sequence (NXS) in positions 31–33 on the polypeptide chain (Fig. 2). This site of glycosylation has been subsequently confirmed in numerous species, including human, monkey, baboon, pig, sheep, goat, and horse (51). Glycosylated PRL has also been demonstrated in species, such as rodents, that do not contain the NXS consensus sequence (54). However, N-glycosylation may occur at another site on the hormone, possibly the NXC sequence at positions 56–58 (54–56 in rodents) that is present in all PRLs except hamster, whale, and fish (52).

Glycosylation of PRL differs widely among species. For example, only a small fraction of human PRL is glycosylated, whereas in pigs a large proportion of both circulating and stored PRL (pituitary content) is glycosylated (51). Glycosylation alters the relative potency of PRL either by modifying its receptor binding characteristics or by changing its plasma half-life. Generally, glycosylated PRL has been shown to have diminished activity in a variety of biological assays, including pigeon crop sac, casein synthesis, and Nb2 lymphocyte proliferation assays (51). Little is understood about the physiological regulation and function of glycosylated PRL.

Phosphorylation

PRL may also be phosphorylated on its serine and/or threonine residues (Fig. 2), although this represents a very small fraction of the hormone pool in both humans and rodents. The primary site of phosphorylation is on serine 179 (177 in rodents), which is found in all PRLs, and is surrounded by the most conserved amino acid motif among all PRLs. Other minor sites of phosphorylation on the PRL molecule include threonine 58/60 or 63/65 and serine 90 (52).

As with glycosylation, phosphorylation of PRL generally has a negative effect on its activity. Phosphorylated PRL has reduced potency in several bioassays (55) and acts as an antagonist to Nb2 cell proliferation stimulated by unmodified PRL. Moreover, dephosphorylating standard PRL produces a more biologically active molecule in the Nb2 cell proliferation assay (56). As predicted by these in vitro assays, phosphorylated PRL has been shown to antagonize endogenous PRL actions on rat maternal behavior (57) and pup development (58) in vivo. Phosphorylation at threonine 58/60 or 63/65 and at serine 90 may interfere with receptor binding. On the other hand, the effect of phosphorylation at serine 179 is more complex. S179D PRL acts as an antagonist of endogenous PRL in the in vitro and in vivo assays mentioned above, but it acts as a superagonist in promoting casein gene expression in pregnant rats (59). These different biological activities may be due to different signaling pathways involved (60).

Although the "PRL kinase" has not been definitively identified, both protein kinase A (PKA) and p21-activated protein kinase gamma-PAK**2** have been demonstrated to phosphorylate S177/179 on the PRL molecule (61,62). S177/179 kinase activity appears to be regulated by estrogen, a potent physiological modulator of lactotroph function. Estrogen exposure reduces kinase activity, resulting in a decreased proportion of phosphorylated PRL both in vitro and in vivo (63,64). Thus, during periods of elevated estrogen, there is both increased synthesis of PRL and an increase in the growth/proliferative potency of the PRL.

Multimeric Forms of PRL

High-molecular-weight variants of PRL, sometimes referred to as "big" PRLs, have also been described. PRL is stored in secretory vesicles as oligomeric forms (65) that depolymerize during exocytosis. Dimers and higher multimers of PRL may be released into the circulation during periods of very high secretory activity. PRL will also aggregate and form intermolecular disulfide bridges spontaneously when in solution at high concentrations (Fig. 2). Aside from this multimerization, high-molecular-weight PRL may arise from cross-linking with other molecules such as IgG (66) or heparin (67). There have been several reports of patients with asymptomatic hyperprolactinemia and prolactinomas. These patients exhibit high circulating levels of multimeric PRL (68,69) or PRL complexed with IgG (70) and are asymptomatic, presumably due to the low bioactivity of these "big" PRL variants.

Cleaved PRL

Fragments of the PRL molecule can be produced by proteolysis, and a variety of cleaved PRLs has been described in various tissues and in the circulation (51). An N-terminal 16-kDa PRL fragment, first described 25 years ago (71), has received a great deal of attention in recent years because it has been proposed to have antiangiogenic bioactivity (72). To produce such a fragment, the PRL molecule must be cleaved in the large disulfide loop (Fig. 2) to produce an open-loop variant. This cleavage has been demonstrated in various PRL target tissues, including the liver and mammary gland, where it occurs under acidic conditions, suggesting it takes place intracellularly (73). Cathepsin D mimics the cleavage of rat PRL that occurs in mammary tissue (74), but this enzyme does not cleave human PRL (75). To produce a free N-terminal 16-kDa fragment, the disulfide bridge between C58 and C174 must also be reduced (Fig. 2). How this is accomplished is not known, although, again, it seems more likely to be an intracellular process.

The 16-kDa PRL (16K PRL) exhibits antiangiogenic activity as demonstrated in vitro by the inhibition of angiogenesis in chick chorioallantoic membranes (72) and in basic fibroblast growth factor–stimulated corneal membranes (76). 16K PRL also inhibits basic fibroblast growth factor–stimulated endothelial cell proliferation (72), and a specific binding site for 16K PRL has been identified in endothelial cells (77), although a receptor protein has not been cloned.

Thrombin has been demonstrated to cleave both human and rat PRL into a 16-kDA fragment (75). This is a C-terminal fragment (Fig. 2) that does not bind to the PRL-R but also has no angiostatic activity. Nevertheless, this thrombin cleavage is likely to occur in the circulation, and the formation of the C-terminal 16K fragment does not require the further reduction of a disulfide bond (Fig. 2). Thus, studies examining circulating 16K PRL fragments and potential physiological regulation necessarily need to distinguish between the N- and C-terminus fragments.

The potential physiological functions of the various modifications of PRL are far from clear, particularly with regard to activation and signaling through the PRL-R. Generally, modifications are detrimental to PRL bioactivity (51). Glycosylation reduces PRL bioactivity; phosphorylation, in most cases, reduces bioactivity to such a degree that it serves as a PRL antagonist; and the 16k PRL cleavage products do not bind to the PRL-R at all. Based on these observations and the fact that the bacterially synthesized recombinant 23k PRL monomer binds to the PRL-R and transduces functional signals (78), it is clear that no additional modifications are required for the core functions of PRL.

PRL RECEPTORS AND PRL SIGNALING

Receptors

The PRL-R was first identified in 1974 as a specific, high-affinity, saturable, membrane-bound protein (79). The sequence of the PRL-R was elucidated in 1988 (80) and is very closely related to that of the GH receptor (81). Both the PRL-R and the GH receptor are members of the type 1 cytokine receptor family, which also includes the receptors for erythropoietin and most of the interleukins (82,83). All the members of this family share three common features. Two of these signature motifs are found in the extracellular domain and include two disulfide bonds and a tandem repeat of tryptophan-serine interrupted by a single amino acid (the WSXWS motif). The third conserved feature resides in the intracellular domain and is an 8-amino acid proline-rich motif, referred to as "box 1," which interacts directly with the tyrosine kinases that are activated upon ligand binding (84).

Various isoforms of the PRL-R exist and are referred to as long, short, or intermediate, depending on their overall length. The isoforms differ only in the cytoplasmic tail, with the extracellular (ligand-binding) domain being identical within a species (84) (Fig. 3). The long isoform (~350 intracellular residues) has been identified in all vertebrate species to date. The short isoforms (<100 intracellular residues) have been identified in several mammalian species, including humans and rodents. The short forms of the PRL-R include box 1, but they lack the tyrosine residues of the distal portion that are required for normal signal transduction. The functional significance of these short isoforms of the PRL-R is not known (Fig. 3), but it has been proposed that they may act as decoy receptors and/or transport molecules. The latter idea is supported by the observation that the choroid plexus has a preponderance of short-form receptors and the levels of PRL circulating in the vascular system are mirrored in the levels of the hormone in cerebrospinal fluid in both rats and humans (85). The proposal that short PRL-R isoforms may act as decoy receptors arises from the fact that the PRL-R is expressed in virtually every organ or tissue, albeit at widely varying levels. It has been suggested that some tissues and cells may need to be protected from exaggerated PRL signaling during periods of physiological hyperprolactinemia, such as pregnancy and lactation. The fact that expression of the various isoforms in many tissues changes during these physiological states supports the idea that these isoforms play a dynamic role in the regulation of PRL signaling.

The PRL-R binds three types of ligands: PRL, PLs, and primate GHs (20). Receptor activation begins with ligand binding and involves two regions, binding sites 1 and 2, on a single molecule of ligand, each interacting with one molecule of PRL-R and leading to their dimerization (Fig. 3). Biochemical and crystallographic mapping of hormone–receptor complexes for human GH and PRL receptors have verified this ligand and dimeric receptor interaction (26,47,86). Transcriptional activation requires homodimerization of the long-form PRL-R. Heterodimers of short and long receptors, or short homodimers, do not mediate normal signal transduction (87).

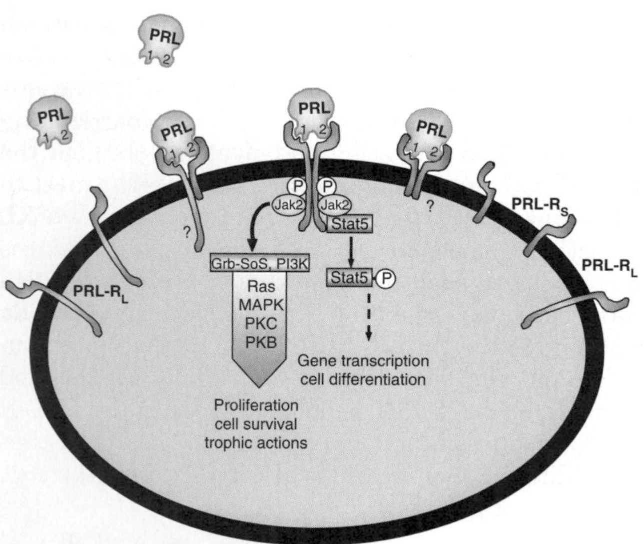

FIG. 3. Schematic representation of the PRL-R signaling pathways. Both the long and short isoforms of the PRL receptor are illustrated. PRL binding and dimerization of the long PRL-R isoforms lead to activation of the Jak2-Stat5 pathways, potentially regulating a variety of cell functions. It is currently unknown whether the short PRL-R isoform (either in homodimers or in heterodimers with the long isoform) can activate these phosphorylation pathways.

Signal Transduction

Ligand binding and PRL-R dimerization triggers a cascade of intracellular signaling events. Like all cytokine receptors, the PRL-R lacks intrinsic enzymatic activity and transduces its signal through numerous kinases that in turn activate other downstream effectors.

Janus Kinase-2

Janus kinase-2 (JAK2) activation is the first intracellular event in PRL-R signaling. JAK2 is a protein kinase that binds to the box 1 motif of the PRL-R's intracellular domain. Both biochemical and genetic experiments have demonstrated that JAK2 is essential for signaling by PRL (88,89) and signaling by other cytokines (90). It is presumed that JAK2 binds directly to the PRL-R (Fig. 3), although it is possible that this association is mediated by another protein. This possibility has been raised because it has been found that JAK2 is associated with unliganded unactivated PRL-Rs. This is quite different from the closely related GH receptor, where ligand binding is necessary before JAK2 can bind to the GH receptor. Ligand-induced dimerization of PRL-Rs initiates phosphorylation by JAK2 of specific tyrosine residues on the receptor intracellular domain. Tyrosine residues within the kinase are also autophosphorylated. These phosphorylated tyrosines then serve as docking sites for additional signal transduction proteins (Fig. 3). Multiple tyrosine phosphatases rapidly dephosphorylate these proteins to end signaling and maintain basal tyrosine phosphorylation at very low levels in the absence of hormonal stimulation. Indeed, the importance of the dephosphorylation step in this cycle is shown by the observation that dominant-negative mutants of the SHP-23 tyrosine phosphatase inhibit PRL activation of JAK2 (91).

STAT5 Transcription Factors

The principal and most studied cascade from JAK2 is the STAT pathway. STATs are a family of *cis*-acting transcription factors, and the recognition for a role of these factors in PRL signaling came from the study of PRL-induced genes (92,93). These genes were found to contain conserved motifs in their promoter regions that bound STAT proteins (94). From lactating sheep mammary glands, a novel STAT protein, STAT5 was cloned (95), and it was subsequently shown that mammals synthesize two STAT5 proteins (a and b) that are encoded by two closely related genes. Targeted disruption of these genes in mice results in phenotypes remarkably similar to those of mice in which either PRL or the PRL-R have been knocked out (96–99). These findings establish the STAT5 proteins as primary mediators of the physiologic actions of PRL. Both STAT5a and STAT5b contribute to PRL signaling in the ovaries and mammary glands. STAT5a plays a predominant role in mediating PRL effect in the mammary gland (96), whereas STAT5b plays a more important role in the ovaries (97,100).

STAT5 is phosphorylated on an essential tyrosine residue (Tyr694) in its C-terminus by JAK2 (101). Once phosphorylated, STAT5 dimerizes through interactions between this phosphotyrosine and src homology 2 (SH2) domains. STAT dimers translocate into the nucleus where they bind to specific sites in the promoters of PRL-regulated genes and activate transcription of those genes. The exact mechanism by which STAT5 is transported into the nucleus and how it interacts with the transcription machinery remain to be elucidated.

In addition to dephosphorylation by the numerous intracellular tyrosine phosphatases, STAT activation by PRL is regulated by an intracellular negative feedback mechanism. CIS (cytokine-inducible SH2 protein), and SOCS (suppressor of cytokine signaling) are members of a class of proteins that are transcriptionally regulated by activated STAT proteins. CIS and SOCS feed back on the PRL-R complex to inhibit the coupling of JAK to either the receptor or to the STAT proteins (102).

Alternative PRL-R Signaling Pathways

In addition to the STAT-dependent events triggered by JAK2 activation, there exist other PRL-R signaling pathways activated by ligand binding that do not involve STAT proteins (Fig. 3). Phosphotidylinositol-3-kinase, mitogen-activated protein kinases, and protein kinase C are among those STAT-independent factors demonstrated to be activated by PRL (84). The physiological relevance of the STAT5-independent signaling pathways is not yet clear, although they have been proposed as mechanisms in the mitogenic actions of PRL (103). That these pathways do play some role in PRL signaling in mammalian cells is supported by the existence of subtle differences between the PRL or PRL-R knockout mice and mice with STAT5 null mutations.

Another signaling pathway for PRL recently proposed involves the direct transport of the PRL–PRL-R complex to the nucleus. Although internalization of receptor-bound PRL has been clearly demonstrated, transport of this complex to the nucleus remains controversial. One group reported anti-PRL immunoreactivity in the nuclei of target cells (104), but others failed to detect either the ligand or the receptor within the nucleus (105). The PRL-R lacks any consensus nuclear localization sequence, but it has been suggested that cyclophilin B (which does possess a nuclear localization sequence) may act as a chaperone protein. Cyclophilin B does complex with PRL, as demonstrated in yeast two-hybrid assays, and treatment of cells with exogenous cyclophilin B was shown to increase PRL immunoreactivity in the nuclei. Cyclophilin B also enhanced the mitogenic activity of PRL on these cells (106).

REGULATORS OF PRL SECRETION

PRL secretion is regulated in a complex manner by a variety of hormones and neurotransmitters (107,108) and is influenced by photoperiod, sleep patterns, and stress. In nonmammalian vertebrates, the principal form of regulatory input is stimulatory with specific releasing factors (PRL-releasing factors [PRFs]) required for PRL secretion. Vasoactive intestinal peptide (VIP) appears to be the primary PRF in birds and other nonmammals (109,110). In mammals, however, quite the opposite situation is in effect. In mammals, lactotrophs are "spontaneous" secretors, requiring no acute stimulatory input to maintain secretion. This is evidenced by the synthesis and release of copious amounts of PRL from anterior pituitary tissue that has been removed from hypothalamic control, such as in transplants to ectopic sites and removal to cell culture. Thus, the major regulatory input to mammalian lactotrophs is inhibitory. This inhibition of PRL secretion is mediated almost entirely by DA produced by the tuberoinfundibular neurons in the arcuate nucleus of the hypothalamus (Fig. 4). This DA is released from nerve terminals in the median eminence and delivered to the anterior pituitary gland via the hypophyseal portal vessels (111). PRL, in turn, feeds back on the hypothalamic neurons to regulate the release of DA and, ultimately, its own release (112,113) (Fig. 4). Superimposed onto this system is the influence of estrogen, discussed below. In addition, there exists an extensive list of factors that demonstrate a stimulatory action on acute PRL release (108), and the episodic release of pituitary PRL, including robust surges of PRL, argues for the input of such PRFs. However, none of these releasing factors has emerged as a required input for normal PRL secretion. What is clear is that normally PRL secretion is under tonic inhibition by hypothalamic DA, and a reduction in this dopaminergic inhibition is associated with the periodic physiological surges of PRL in females. Regulation of these episodic surges of PRL is addressed in more detail below. Because of its major role in the regulation of PRL in mammals, the mechanisms of DA is considered in detail.

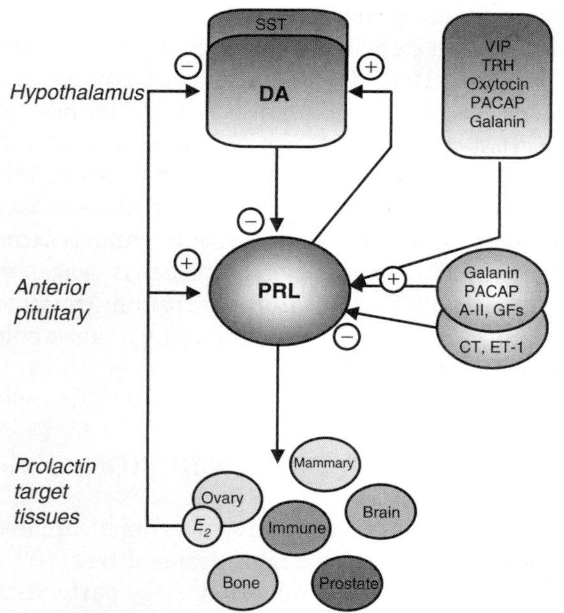

FIG. 4. Schematic illustration of the interplay of hypothalamic, pituitary, and target tissue factors in the regulation of pituitary PRL secretion in mammals. See text for discussion.

Dopamine

DA acts on the lactotroph at various levels of cell function. In addition to inhibiting acute release of PRL, DA also inhibits transcription of the hormone (114). DA is also a potent antimitotic factor in pituitary lactotrophs (115,116). Thus, in addition to the broad diversity of potential PRL regulators, the actions of this one modulator of lactotroph function are impressively diverse.

DA Receptors

Dopaminergic actions on the lactotroph are mediated via seven transmembrane spanning D_2-like DA receptors (117,118). Modern molecular cloning techniques have identified five structurally distinct DA receptors, D_1 to D_5 (119,120). Still, they are grouped into two families (D_1-like and D_2-like) based on pharmacological and functional, as well as molecular, similarities. D_1-like receptors (D_1 and D_5) are defined as being linked to the stimulation of adenylate cyclase, whereas D_2-like receptors (D_2, D_3, and D_4) either inhibit or have no effect on this enzyme. Neither of the adenylate cyclase stimulating DA receptors are present in the anterior pituitary. Both the long and short forms of the D_2 receptor (produced by alternative splicing) and the D_4 receptor transcript are present in the anterior pituitary (121,122), but the D_3 receptor is not (121,123). These D_2-like DA receptors are both physically and functionally coupled to a pertussis toxin–sensitive G protein (Gi/o) (124–127) (Fig. 5), and this receptor–G protein coupling plays a necessary role in the inhibitory actions of DA on the lactotroph (127–130). The D_2 receptor–G protein coupling, in turn, can regulate a number of effectors in the lactotroph, most notably adenylate cyclase and potassium channels.

Adenylate Cyclase/Cyclic AMP

D_2 receptor activation leads to inhibition of adenylate cyclase and reduction of cAMP in normal lactotrophs (Fig. 5). This pathway is understood to be an important mediator of DA's inhibition of PRL synthesis, and certainly elevations in cAMP will increase PRL synthesis and release. But the inhibition of

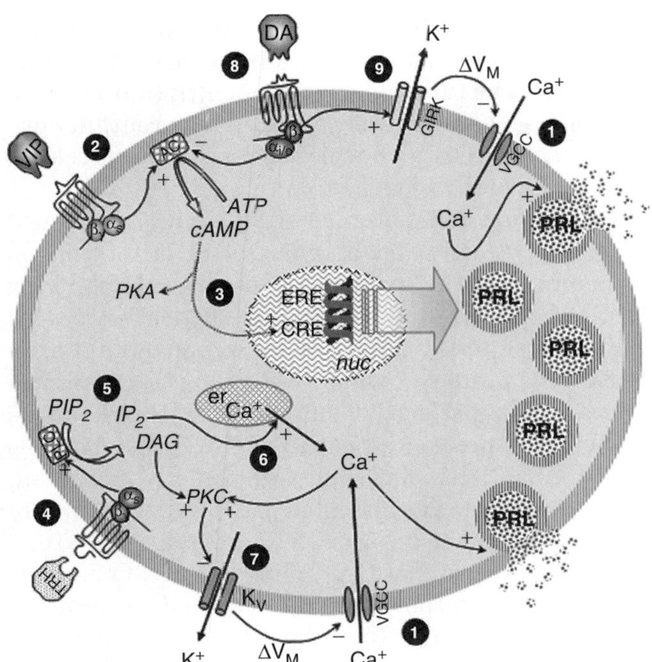

FIG. 5. Schematic representation of regulatory signaling pathways in the lactotroph. In the absence of regulatory input, lactotrophs spontaneously release large amounts of PRL and most of this release is driven by influx of extracellular Ca^{2+} through VGCCs (1). VIP, a PRL-releasing peptide, binds to VIP receptors (2) and leads to activation of adenylate cyclase (AC) via the α_s G protein subunit. Increased cAMP production (3) can increase PKA activity and affect PRL synthesis via the cAMP response element (CRE). TRH, also a PRL-releasing peptide, binds to TRH receptors (4) and activates phospholipase C via the α_q G protein subunit, leading to the generation of IP_3 and diacyglycerol (DAG, 5). IP_3 causes release of Ca^{2+} from the endoplasmic reticulum (er, 6) and DAG activates PKC, which can phosphorylate and inactivate some voltage-gated K^+ channels (7). DA, the major inhibitory input to the lactotroph, binds to D_2 DA receptors (8) that are negatively coupled to AC via the α_i G protein subunit. The reduction in cAMP leads to reduced PKA and CRE activation. The $\beta\gamma$ subunit complex directly interacts with and activates a GIRK channel (9). The resulting membrane hyperpolarization closes VGCC and blocks Ca^{2+} influx. Estrogen stimulates PRL synthesis through the estrogen response element (ERE) on the PRL gene.

cAMP as a mediator of DA inhibition of acute PRL release is controversial at best. Correlative studies suggest a relationship between the two events. However, investigators have also been able to dissociate these two actions of DA. Thorner and colleagues (131) first demonstrated that elevations in cytosolic [Ca^{2+}] using selective ionophores could maintain PRL secretion in the presence of DA despite effective suppression of cAMP. Moreover, DA could inhibit PRL release even when cAMP levels were elevated (131–134). It has also been demonstrated that elevated extracellular K^+ can block the inhibitory effect of DA on PRL release without diminishing the effect of DA on cAMP (135). Further evidence for the dissociation of DA inhibition of cAMP and PRL release was found in rat pituitary cells treated with antisense oligonucleotides recognizing both the long and short forms of the D_2 receptor but not the D_4 receptor (121). After this treatment, the D_2 agonist, bromocriptine, lost its ability to suppress cAMP accumulation, whereas inhibition of PRL release was unaffected. Taken together, these data indicate that DA-induced suppression of cAMP is neither sufficient nor necessary for dopaminergic inhibition of PRL release.

Phospholipase/Phosphoinositols

Evidence that DA's inhibitory action on PRL release involved decreases in cytosolic [Ca^{2+}] (131,132) led to the proposal that the D_2 receptor may also be negatively coupled to phospholipase C in lactotrophs (136,137). The idea was that inhibition of phospholipase C would reduce phosphoinositol hydrolysis and its product, IP_3, leading to reduced mobilization of intracellular stores of Ca^{2+}. However, it is now well accepted that acute administration of DA does not alter the basal levels of inositol phosphates in pituitary cells (135,138). DA does reduce inositol phosphate production induced by thyrotropin-releasing hormone (TRH), but only during the late phase (>1 min) of this stimulation (139), which is associated with enhanced influx of extracellular Ca^{2+}. The transient generation of IP_3 in response to TRH activation of phospholipase C (5–60 seconds) is unaffected by DA. In addition, cell-free pituitary membranes exposed to TRH showed no inhibition of inositol phosphate generation by DA (139). These data support the conclusion that D_2-like receptors in lactotrophs are not direct inhibitors of phosphoinositol hydrolysis.

Ion Channels

Lactotrophs can display spontaneous electrical activity, including spontaneous membrane potential fluctuations and calcium-dependent action potentials (140). The ionic conductances underlying this electrical activity have been characterized in lactotrophs (141,142). These electrical events play a major role in both stimulated and basal PRL release (143,144). In particular, the high rate of spontaneous PRL release from mammalian lactotrophs, normally suppressed by DA, is dependent on the influx of extracellular Ca^{2+}. Block of Ca^{2+} influx through voltage-gated Ca^{2+} channels (VGCC) is as potent an inhibitor of this release as a maximal dose of DA.

When examined on the single-cell level, unchallenged lactotrophs fall into two functionally distinct groups (145,146). One population exhibits a stable membrane potential and intracellular [Ca^{2+}] that is not acutely sensitive to extracellular Ca^{2+} ("quiescent" cells). The other exhibits spontaneous membrane depolarizations supporting oscillations in $[Ca^{2+}]_i$ that are dependent on influx of extracellular Ca^{2+} ("active" cells). This functional difference is also evident in secretory activity of single cells as measured by reverse hemolytic plaque assay (147). Quiescent lactotrophs secrete PRL at a low basal rate, whereas the spontaneous membrane depolarizations in active cells result in influxes of Ca^{2+}, producing elevations of cytosolic $[Ca^{2+}]$ and supporting a high rate of PRL release (148). DA causes a rapid inhibition of these depolarizations followed by a rapid and dramatic fall in cytosolic $[Ca^{2+}]$ in spontaneously active lactotrophs. Direct block of Ca^{2+} channels with verapamil mimics this effect (145,146). Neither DA nor verapamil, however, cause further decreases in the low and stable cytosolic $[Ca^{2+}]$ in quiescent cells. DA also preferentially inhibits PRL release from spontaneously active cells while having little or no effect on release from quiescent cells (148).

One mechanism by which DA could block spontaneous action potential activity and cytosolic (Ca^{2+}) in lactotrophs is by directly inhibiting VGCCs. There is some experimental support for this in cultured lactotrophs derived from lactating rats and in GH_4C_1 cells expressing human D_2 receptors ("GH_4C_1/D_2-DAR") (143,149). In these studies, DA treatment of the PRL-secreting cells produced a decrease in calcium current (I_{Ca}) measured under voltage-clamped conditions. A 27% decrease in I_{Ca} was observed after 1 or more minutes of DA exposure, whereas prolonged treatment (≥24 hours) with DA was required for maximal (57%–61%) reductions in I_{Ca} (143,149). Voltage-independent inhibition of I_{Ca} is not seen in response to acute applications of DA (150), although this would be expected if the D_2 receptor were directly coupled with VGCCs. The reduction in I_{Ca} observed after longer exposures to DA cannot account for the immediate inhibition of $[Ca^{2+}]_i$ oscillations observed in response to DA application (145,146). Reduced I_{Ca} after DA treatment most likely reflects changes that occur downstream of other actions of DA, such as changes in cAMP-dependent protein kinase (PKA)-mediated phosphorylation (151,152) that regulates VGCC activation (153).

Experimental evidence has made it clear that K^+ channels in primary lactotrophs play an integral role in mediating dopaminergic inhibition of both spontaneous and stimulated PRL release. DA-induced increases in K^+ permeability result in membrane hyperpolarization that rapidly shuts down VGCCs (Fig. 5). This indirect block of Ca^{2+} influx and the resulting decline in intracellular $[Ca^{2+}]$ inhibits PRL secretion. This model is consistent with the experimental data obtained from single-cell studies. The activation of K^+ current is immediate, causing a rapid hyperpolarization and suppression of Ca^{2+} influx (145,154). The functional consequences (inhibition of PRL release) are greater, or more apparent, in electrically active cells from which the spontaneous release of PRL is much higher (148).

Biophysical and molecular studies have determined that the DA-activated K^+ channel in lactotrophs is a member of the superfamily of inward rectifying K^+ (IRK) channels (154,155). The activation of this channel requires mediation by a pertussin toxin–sensitive G protein (154,156), which further identifies it as a member of the 3.0 subfamily of IRK channels, known as G-protein–activated IRK (GIRK) channels (157,158). GIRK channels are directly activated, through a membrane-delimited pathway, by the $\beta\gamma$ subunit of pertussis toxin–sensitive G proteins (159–161). Current theory is that G-protein–gated IRK channels function as heteromultimers of the GIRK proteins (162,163). Although the identities of the subunits comprising the DA-activated channel in lactotrophs have not been unequivocally determined, transcript and protein analyses of anterior pituitary membranes, along with functional analyses of D_2R–GIRK coupling, support the idea that the channel is a GIRK1–GIRK4 oligomer (155). Expression of a GIRK1 dominant-negative mutant in primary cultured pituitary cells reverses dopaminergic inhibition of PRL release without altering basal release (148), indicating that the GIRK channel in lactotropes plays a critical role in the inhibition of PRL release induced by application of DA.

Anomalous Stimulation of PRL Secretion by DA

Under certain conditions, changes in DA input to the lactotroph have been demonstrated to produce elevations in PRL secretion. One condition is an acute response, in which subnanomolar concentrations of DA ($\leq 10^{-10}$ M) directly stimulate PRL release (164,165). These low concentrations of DA have been shown to induce elevations in cytosolic $[Ca^{2+}]$ (166) that are believed to underlie the stimulatory effect on PRL release. What is puzzling is that the same concentrations shown to stimulate PRL release have also shown either no effect or small inhibitory effects on PRL release in other studies, even by the same investigators (164,167). Although the experimental conditions appear identical in these conflicting reports, details of the preparation of the DA solutions are not presented. DA oxidizes rapidly at physiological pH (168) and is subject to enzymatic degradation in the presence of cells (169). In concentrations of 10^{-8} M or greater, enough unmetabolized DA should remain, even over periods of hours, to inhibit PRL release through cellular mechanisms discussed above. However, oxidation of lower concentrations of DA would decrease the levels of intact DA below those required for activation of the classic DA receptors. A survey of the literature on stimulatory low concentrations of DA indicates that this phenomenon is studied either under conditions when pituitary cells are exposed to DA for long periods of time (reverse hemolytic plaque assay or static culture) or under conditions when it is unknown how quickly the DA solutions are used after preparation in physiological buffer. Thus, the possibility exists that a metabolite, and not DA per se, is acting to stimulate $[Ca^{2+}]_i$ and PRL release. Toth and colleagues (170) reported that an enzymatic product of DA, salsolinol, could stimulate PRL release. In addition, low concentrations of DA left in physiological buffer for 60 minutes or more before exposure to cells has been found to stimulate rises in $[Ca^{2+}]_i$, whereas freshly prepared DA either has no effect or reduces cytosolic $[Ca^{2+}]$ (148). Whether the active component is produced by enzymatic or nonenzymatic processes, the possibility that a metabolic product stimulates the same cell functions that are normally inhibited by the parent compound is an exciting avenue for future research.

Disruption of dopaminergic tone in vivo or withdrawal of DA from lactotrophs in vitro also elicits transient "surges" of PRL release, as has been demonstrated by numerous groups (167,171–173). Investigations into the mechanisms by which this phenomenon occurs have demonstrated that the secretory rebound is dependent on the influx of Ca^{2+} (135,146) and is distinct from the stimulatory actions of low concentrations of DA (174).

Based on the mechanisms known for DA's inhibition of PRL release, it was suggested that the DA-induced changes in membrane potential might play a role in the secretory rebound after its withdrawal. Single-cell studies demonstrate that recovery from DA-induced membrane hyperpolarization leads to activation of Ca^{2+}-spiking activity in lactotrophs that were electrically quiescent before application of DA (174). The amplitude of Ca^{2+} action potentials in spontaneously active cells was also increased by ~20% after recovery from DA. Both responses can be explained by an increased number of VGCCs available for activation after DA withdrawal, leading to the hypothesis that the DA-induced hyperpolarization "recruits" previously inactivated VGCCs by allowing recovery from inactivation. Upon DA washout and recovery of the resting membrane potential, the enhanced influx of Ca^{2+} through these channels supports the rebound release of PRL. This hypothesis is supported by the demonstration that direct hyperpolarization (bypassing D_2 receptor activation) of the lactotroph membrane mimics both the inhibitory and rebound phases of PRL secretion (174).

Other Inhibitors of PRL

Additional factors have also been shown to inhibit PRL secretion. Somatostatin, secreted from the hypothalamus, inhibits PRL secretion and acts through both cAMP-dependent and -independent mechanisms (175). Calcitonin has also been shown to inhibit PRL secretion and may also be of hypothalamic origin (176). Endothelin-1 is produced by lactotrophs themselves and inhibits PRL secretion through an autocrine mechanism (177). Another locally synthesized factor, transforming growth factor-β1, can act as a paracrine inhibitor of PRL (178). However, hypothalamic DA is still sufficient and required for the normal pituitary development and control of lactotrophs, whereas the physiological significance of these other factors is not yet established.

Estrogen

Estrogen is the other major regulator of PRL in mammals and is stimulatory to lactotrophs. It is the cause for the greater serum levels and pituitary content of PRL in females than in males. Estrogen exerts its effects on lactotrophs through its classic nuclear receptor. Although some responses appear to be quite rapid, definitive involvement of a membrane estrogen receptor in these cells has not been demonstrated. Pituitary lactotrophs express the α form of the nuclear estrogen receptor (179). Mice that have targeted disruption of estrogen receptor β are reported to have normal mammary gland histology and lactation (180), suggesting that estrogen receptor β plays little or no role in PRL regulation.

Estrogen acts directly on the lactotroph to increase PRL gene expression (181,182). Although it appears that estrogen does not directly stimulate exocytotic release of PRL, the dramatic increase in PRL synthesis that is supported by estrogen leads to both a higher rate of spontaneous PRL release and a larger storage pool of PRL that can be released upon stimulation by releasing factors or removal of DA.

Estrogen also regulates lactotrophs by modifying their responsiveness to other regulators (183–185). For example, estrogen increases the density of TRH receptors in the membrane of lactotrophs (186) and, at high concentrations (late proestrus), *decreases* the density of D_2 receptors (187). Estrogen also dramatically alters the density of VGCCs and electrical excitability of lactotrophs (188,189), another mechanism by which this steroid increases both basal activity and responsiveness of these cells.

Direct mitotic actions of estrogen on PRL cells also have been demonstrated in both in vivo and in vitro studies (190–192). Again, these actions are the basis for the gender difference in the size of the lactotroph population, that being greater in females than in males. In the female, transient increases in the number of pituitary lactotrophs also occur during periods of physiological elevations in estrogen, such as late follicular phase of the reproductive cycle and pregnancy. Prolonged periods of exposure to high concentrations of estrogen can lead to hyperplasia of lactotrophs and, eventually, tumorigenesis of prolactinomas. Estrogen-induced galanin secretion from lactotrophs may mediate these mitotic actions on lactotrophs (193,194) and involves signaling through the classic estrogen receptor isoform α (195).

PRL-Releasing Factors

A wide variety of PRFs has been identified in mammals over the years (108), and, no doubt, additional PRFs will continue to be identified. The known stimulators of acute PRL secretion include, but are not limited to, hypothalamic peptides [VIP, TRH, oxytocin (196,197), pituitary adenylate cyclase activating peptide (198), and galanin (199)] and local pituitary factors [growth factors such as epidermal growth factor (200) and fibroblast growth factor-2 (201), angiotensin II (202), and, again, pituitary adenylate cyclase activating peptide (203) and galanin (193)]. Each is discussed briefly. However, it should be noted that despite decades of work, demonstration of a clear physiological role for any of these PRL secretagogues has remained elusive.

VIP is produced in the hypothalamus and acts directly on lactotrophs to stimulate PRL synthesis and release. VIP stimulates cAMP production in lactotrophs via a G_s coupled receptor (Fig. 5) and acts on an intermediate to long-term basis. Two lines of evidence support VIP as an important PRF. Passive immunization against VIP reduces the secretion of PRL (204). Also, VIP appears to be the primary PRF in nonmammalian vertebrates (205,206), animals in which PRL secretion requires stimulation. It has been suggested that this stimulatory mechanism may have preceded the evolution of the dopaminergic inhibitory system in mammals. TRH is a potent and rapid stimulator of PRL release in vitro. Lactotrophs express TRH receptors, which are coupled to the G_q heterotrimeric G protein. Activation of these TRH receptors leads to a rapid and sustained release of PRL via calcium-mediated pathways, including rapid mobilization of intracellular calcium followed by a prolonged increase in the influx of extracellular calcium through VGCCs (Fig. 5) (207). Oxytocin can also stimulate PRL release in vitro and exhibits similar release patterns during lactation because both hormones are secreted in response to nipple stimulation. Although not released at the median eminence, oxytocin can reach the anterior pituitary through both the short portal vessels and the general circulation. Antagonism of oxytocin does partially suppress PRL secretion (204), so it appears to provide some portion of the physiologic stimulus for PRL release in response to suckling. Pituitary adenylate cyclase activating protein, produced both in the hypothalamus and the anterior pituitary, can stimulate PRL synthesis and release. Galanin is another peptide that is synthesized in both the hypothalamus and the pituitary and acts by autocrine and paracrine mechanisms to stimulate lactotroph function (193). It has been proposed that galanin mediates much of the estrogen-induced stimulation of PRL (195,208).

A putative PRL-releasing peptide from the hypothalamus was identified by searching for ligands that bound to an orphan G-protein–coupled receptor expressed in the anterior pituitary gland. The original report identifying this 20-amino acid peptide in bovine hypothalamus demonstrated that PRL-releasing peptide causes a rapid secretion of PRL from isolated pituitary cells (209). However, subsequent studies have failed to confirm these findings and suggest, rather, that PRL-releasing peptide may act within the hypothalamus to indirectly elevate PRL by inhibiting DA release (210).

EPISODIC PRL RELEASE ASSOCIATED WITH FEMALE REPRODUCTION

In the mammalian female, PRL release is dynamic, with episodic release, or surges, during different reproductive states (108). These states include estrous and menstrual cycles, lactation, and mating and pregnancy. The stimuli for each of these physiological PRL surges is different: a rise in circulating estrogen producing the mid-cycle PRL release, the suckling stimulus eliciting PRL release in lactation, and uterine cervical stimulation initiating the PRL surges of early pregnancy or pseudopregnancy in the rodent. Whereas the initiating stimuli are known, the ultimate hypothalamic control of these PRL surges is far from clear. Many plausible PRFs have been implicated in each state, although none has emerged as an undisputed primary regulator. In all three states, there occurs a withdrawal of dopaminergic tone. But this occurs to varying degrees in each of the different states. In addition, regulated changes in PRL transcription and lactotroph proliferation occur during these physiological phases of increased demand for PRL. Again, the integrated actions of estrogen and DA appear to be the major factors regulating these lactotroph functions. However, the mechanisms coordinating this integration are not understood.

Estrous Cycle

The secretion in rodents of PRL throughout most of the estrous cycle is low and stable from the afternoon of estrus through the morning of proestrus. During the afternoon of proestrus, secretion of PRL rises to produce a preovulatory surge that is coincident with the luteinizing hormone surge (Fig. 6) (211,212). It is clear that the rising serum levels of estradiol, beginning on diestrus 2, signal the hypothalamus–hypophyseal axis to release this surge of PRL. Passive immunization of estradiol on diestrus 2 blocks the proestrous surge of PRL (213). Estrogen replacement in ovariectomized rats induces daily surges of PRL with afternoon timing similar to that of the normal proestrous surge (214). As described above, the DA-induced changes in membrane potential play critical roles in initiating rebound secretion of PRL upon withdrawal of DA. Furthermore, the ability of DA to induce membrane hyperpolarization changes dramatically during the estrous cycle (215). Thus, in the cycling female rodent, the functional expression of this DA mechanism in the lactotroph coincides with the time of functional withdrawal of hypothalamic DA input to the pituitary. This implicates the DA-activated GIRK channel as a critical component of the regulatory machinery responsible for the unique secretory profile of PRL on proestrus.

Pregnancy

Mating, or artificial, mechanical, or electrical stimulation of the uterine cervix, switches the proestrous pattern of PRL secretion to a pattern of twice-daily surges of PRL. These surges occur in the latter half of the dark and light phases of the light cycle and are called the nocturnal and diurnal surges, respectively (216). In the pregnant female rodent, these daily surges occur for 9 to 10 days after mating and are required for luteal progesterone secretion during this period. Subsequent to the termination of these PRL surges, PLs support luteal function through the end of pregnancy (8). If the mating is infertile or the female receives artificial cervical stimulation, then the PRL surges persist for 12 to 13 days (216), defining the period of "pseudopregnancy." There is evidence that DA plays a role in governing the expression of both the nocturnal and diurnal surges of PRL. Tyrosine hydroxylase activity (the rate-limiting enzyme in the synthesis of DA) is lower in the median eminence during both surges (217), and DA concentrations in the hypophyseal portal blood are lower during both surges than during the intersurge period (218). However, the endogenous surges are considerably smaller than PRL secretory responses to maximal D_2 receptor antagonism in ovariectomized rats at similar times of the day (216), and the necessity of disruption of DA tone is not clear. Also, PRL responses to acute D_2 receptor antagonism in pseudopregnant rats are dramatically different during each of the two surges (216), indicating they may be differentially regulated.

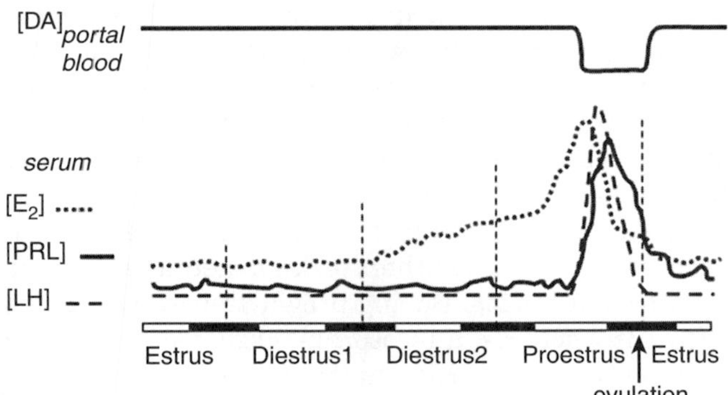

FIG. 6. Profiles of serum estrogen, PRL, luteinizing hormone, and hypophysial portal blood DA concentrations throughout the estrous cycle of the rodent. Note the gradual rise in circulating estrogen beginning on diestrus 2. This rise is required for the occurrence of both the luteinizing hormone and PRL surges on the afternoon of proestrus.

Lactation

Suckling-induced PRL secretion during lactation is the best known episodic release of PRL and is considered a classic neuroendocrine reflex. In rodents, the concentration of PRL in serum begins to rise within a few minutes of the onset of suckling, peaks by 10 minutes, and remains elevated until the cessation of suckling (219). The amount of PRL released correlates with the intensity of the stimulus (i.e., number of pups) (220). The suckling stimulus causes a reduction in tuberoinfundibular DA activity and a transient diminution of the concentration of DA in the portal blood (221). Thus, suckling induces a withdrawal of dopaminergic inhibitory tone on the lactotroph. However, the amount of PRL released in response to suckling lasts much longer than the decreased DA input to the pituitary (221) and is far greater than that elicited by pharmacological or surgical disruption of DA input to the pituitary. Although the decrease in hypothalamic DA secretion during suckling is transient, it is profound and is believed to play an important role in sensitizing the lactotrophs to subsequent release of PRFs from the brain.

REPRODUCTIVE PHYSIOLOGY AND PATHOPHYSIOLOGY OF PRL IN MAMMALS

Physiological Levels of PRL

As described above, PRL release in the mammalian female is dynamic, fluctuating rapidly during normal physiological conditions. Circulating PRL concentrations in normal adult women who are neither pregnant nor lactating range from 4 to less than 20 ng/mL. Late pregnancy and lactational levels are normally in the range of 100 to 200 ng/mL, with the highest levels occurring after active bouts of nursing. Similar increases by orders of magnitude are seen in the proestrous-, pregnancy-, and suckling-induced surges in rodents. In normal adult men the average values are several units lower than those seen in women (1–4 ng/mL), and such robust surges do not occur. Both sexes experience smaller bouts of PRL release during rapid-eye-movement sleep (222), after orgasm (223), and in response to acute stress (224), but, again, the amplitudes of these PRL release episodes are generally lower in males than in females. These values are based on measurement by radioimmunoassay. Although posttranslational modification of PRL can affect its immunoreactivity, both physiological and pathological changes in PRL concentrations are readily detected by radioimmunoassay. PRL was originally measured by bioassay of the growth of the pigeon crop sac mucosal epithelium (pigeon crop sac assay) (225), and it is this assay that is used for the international standardization of PRL bioactivity.

It has become clear that the episodic release of PRL, separated by periods of low circulating levels of the hormone, is necessary for proper mammalian female reproductive function (Fig. 7). Hypersecretion of PRL is the major neuroendocrine-related pathology associated with female infertility in humans. Similarly, chronically elevated PRL also disrupts cyclicity and ovulation in most rodents. Yet, the episodes, or surges, of PRL are required for successful pregnancy and lactation. Isolated PRL deficiency, although rare, results in lactational failure and subfertility in women (226–229). This phenotype is consistent with that in mice in which either the PRL gene or the PRL-R gene has been disrupted by targeted mutation. In these genetic models, females are infertile and mammary gland development is defective (98,99).

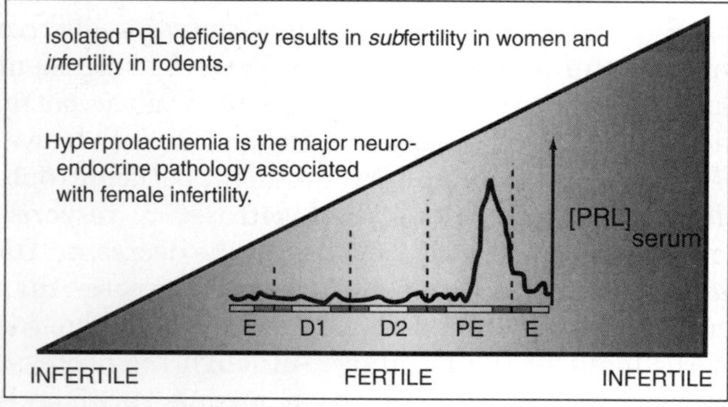

FIG. 7. Episodic release of PRL, separated by periods of low circulating levels, is required for proper mammalian female reproductive function and female fertility. Excess circulating PRL leads to disruption of estrous and menstrual cyclicity and infertility. Isolated PRL deficiency results in subfertility in women and infertility in female rodents.

PRL Actions

Female Reproductive Tissues

PRL is essential for lactation in all mammals. Mammary gland organogenesis begins prenatally with the mammary ductal rudiment. This epithelial rudiment grows with the fat pad until puberty, at which time the ductal system expands rapidly under the influence of estrogen, GH, and insulin-like growth factor 1 (230). It is during the latter stages of puberty that PRL begins to exert its influence. In synergy with progesterone, PRL stimulates the branching of lobular buds from the ductal system. It appears that progesterone induces ductal arborization, whereas PRL induces the differentiation and growth of alveolar progenitor cells from the ductal epithelium (231). With repeated estrous or menstrual cycles, the complexity of the mammary ductal branching increases progressively and the epithelial cells undergo cyclic changes. In mice bearing a null mutation of either PRL or the PRL-R, mammary organogenesis is arrested at an immature pubertal state, with a basic ductal system and terminal end buds but no lobuloalveolar development (98,99).

During pregnancy the lobuloalveolar epithelium undergoes extensive proliferation under the influence of PRL, PLs, progesterone, and local growth factors such as RANK-ligand and insulin-like growth factor II (232,233). At parturition, progesterone, estrogen, and PLs fall precipitously and PRL rises. It is this combination of hormone changes that leads to functional lactogenesis and lactation. PRL stimulates the synthesis of several secreted milk proteins, including the caseins, lactalbumin, and whey acidic protein. PRL also induces enzymes essential for milk synthesis, such as lactose synthetase, lipoprotein lipase, and fatty acid synthase. During pregnancy and lactation, PRL also acts as an essential survival factor for lobuloalveolar cells (234). Involution of the lobuloalveolar system occurs in response to falling systemic lactogens. Thus, PRL and PLs, each of which bind to the PRL-R, act during three stages of mammary gland development: lobule budding during organogenesis, lobuloalveolar expansion during pregnancy, and lactational differentiation after parturition.

High levels of PRL inhibit female reproductive function in mammals by actions at multiple levels. Elevated PRL decreases the hypothalamic drive for pulsatile luteinizing hormone secretion (235), thereby inhibiting ovarian folliculogenesis (236), and estradiol synthesis (237,238). PRL also acts directly on the ovary and contributes to the breakdown of the corpus luteum in many mammalian species, including humans. These antigonadal effects are manifest during lactation in humans and in clinical hyperprolactinemia. Persistent amenorrheas are one consequence of high PRL secretion in women who are breast-feeding. This is referred to as lactational infertility and, although breast-feeding has been promoted as a natural means of contraception, it is very unreliable for most women.

In rodents, PRL is luteotrophic and essential to corpus luteum maintenance in early pregnancy. The luteal phase of the estrous cycle in rodents is transient, and implantation cannot occur unless the corpus luteum is maintained by high levels of PRL. Cervical stimulation during mating establishes a stereotypic pattern of twice-daily surges of PRL. A well-characterized mechanism of the luteotrophic action of PRL is inhibition of 20α-hydroxysteroid dehydrogenase activity (239). This action prevents the conversion of progesterone to 20α-hydroxyprogesterone and increases progesterone secretion from the corpus luteum.

Male Reproductive Tissues

High levels of PRL are also inhibitory to male reproductive function, with the most common presenting symptoms being loss of libido and impotence. These symptoms may or may not be associated with galactorrhea. PRL inhibits GnRH and luteinizing hormone secretion in males, as well as in females (240). PRL increases the number of androgen receptors in the prostate, and high levels of PRL cause prostate hyperplasia in mice (241). This has led to the hypothesis that PRL may be a contributing factor in human prostate disease.

Brain and Behavior

PRL has numerous actions in the vertebrate brain, many of which are related directly to the parental care of offspring. In birds, systemic or intracranial PRL infusion stimulates behaviors associated with brooding and migration (206). In rats, central PRL infusions decrease the latency to onset and increase the intensity of parental behaviors (242). Mice that lack the PRL-R are profoundly deficient in maternal behaviors (243). However, mice with a null mutation for the ligand do not exhibit such a strong deficit. These PRL knockout mice have functional PRL-Rs, and it has been suggested that prenatal exposure to maternal PLs may account for this difference in phenotype. The neuroanatomical and neurochemical substrates that mediate the PRL-regulated parental behaviors in mammals are not yet known. However, sensory stimuli are clearly important cues for these behaviors, and elevated PRL, such as that seen during pregnancy, stimulates neurogenesis in the olfactory lobe of mice (244).

Maternal behavior patterns in animals such as birds, mice, and rats are highly stereotypic and, therefore, have been quantified and studied objectively. Such behaviors in humans are not amenable to rigorous characterization. Thus, it has not been determined whether PRL has a similar role in human parenting.

Pathologies of PRL Secretion

Hypersecretion of PRL

Hypersecretion of PRL is among the most common of pituitary disorders, and hyperprolactinemia that manifests clinical symptoms is most commonly a consequence of a lactotroph adenoma. These tumors may secrete high levels of PRL alone or both PRL and GH. Pituitary adenomas are quite common, with more than 20% of individuals harboring tumors of at least 3 mm at autopsy (245). The precise defects underlying these prolactinomas are not clear, although reduced sensitivity to DA is a common feature. Hyperprolactinemia can also be caused by any intracranial mass or trauma that causes compression or disruption of the pituitary stalk and results in the loss of dopaminergic tone on the anterior pituitary. The results of disrupting dopaminergic tone to the lactotroph are dramatically demonstrated in three murine models with genetic deletions of PRL (98), the PRL-R (99), or the D_2 receptor (246,247). All three models have reduced (or absent) dopaminergic input to the anterior pituitary, and all three models develop lactotroph hyperplasia (247–249) (Fig. 8). The hyperplasia in the PRL knockouts and PRL-R knockouts can be blocked by treatment with the D_2 agonist, bromocriptine (250). Correlatively, excessive DA during development leads to lactotroph hypoplasia, as seen in mice with the DA transporter null mutation (251) (Fig. 8).

The role of estrogen in the lactotroph hyperplasia in these models is evident in the much greater rate of pituitary growth in the females as compared with the males. In fact, the increased cell proliferation leads, ultimately, to the development of adenomas in the

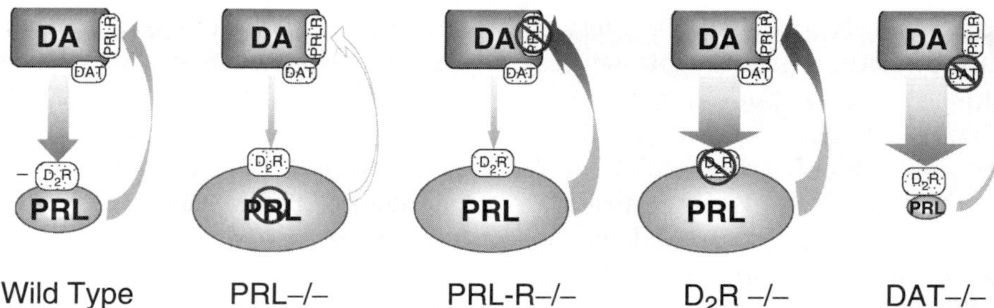

FIG. 8. Schematic illustrations of the tuberoinfundibular DA (TIDA) and pituitary PRL axis in normal wild-type mice and in various genetically altered mice. In wild-type mice, DA binds to D_2 receptors and inhibits PRL secretion. Elevations in circulating PRL feed back to the TIDA neurons, binding to PRL receptors to stimulate the synthesis and release of DA. In mice with null mutations of either PRL or its receptor, the feedback of PRL is lost and the TIDA neurons dramatically reduce their production of DA. The resulting drop in DA acting on the lactotroph leads to hyperproliferation of the lactotrophs and, eventually, adenoma formation. Although the PRL–/– mice do not make intact PRL, the PRL-R–/– mice have hyperprolactinemia. In D_2R knockout mice, DA cannot exert its inhibitory effects on lactotrophs. These mice will develop hyperprolactinemia, lactotroph hyperplasia, and, eventually, prolactinomas. Mice that bear a null mutation of the DA transporter (DAT) have an increased dopaminergic tone on the pituitary during development. These mice have hypoprolactinemia and hypoplasia of lactotrophs.

female mice (249,252), and this can be blocked by ovariectomy (250,252). Yet, these females have only about 50% of normal circulating estrogen levels as compared with wild-type or the asymptomatic heterozygous controls (249,253). Thus, even half the normal levels of estrogen can lead to uncontrolled lactotroph proliferation if the inhibitory tone of DA is compromised. Clearly, there is an interactive balance between DA and estrogen that underlies normal lactotroph growth and function.

D_2 agonists are commonly used to treat hypersecreting prolactinomas. In most cases, these agonists effectively correct the hyperprolactinemia. They also stop the growth and reduce the size of prolactinomas (254). This reduction in tumor size is primarily through loss of cytoplasmic and nuclear volume, with involution of rough endoplasmic reticulum and Golgi apparatus, rather than reduction of cell number (254,255). Although DA does not appear to induce cell death, the antimitotic effect of the DA agonists has been known for some time. However, the mechanisms through which DA exerts this effect have not been delineated.

Isolated PRL Deficiency

When PRL deficiency occurs, it is normally one component of a combined pituitary hormone deficiency. Such conditions make it impossible to know the specific role of PRL in the resulting symptoms. However, several cases of PRL deficiency without evidence of other pituitary defects have been reported in women. Isolated PRL deficiency results in lactational failure and reproductive difficulty, but no other obvious health problems (226–229). Isolated PRL deficiency has not been reported in men except as members of an inherited PRL disruption brought to light by the reproductive difficulties in the female members of the family (228). The male with PRL deficiency had no reproductive deficits. These results in a few humans are consistent with the phenotype of mice in which the PRL gene, or its receptor, has been disrupted by a targeted mutation. Mammary gland development in these mice is defective and the females fail to reproduce, but the males do not have any overt symptoms (98,99). The concordance of these phenotypes in humans and mice is remarkable, considering the well-known differences between PRL actions in primates and rodents. Most notably, the rodent corpus luteum requires PRL but the human does not. This luteotrophic action may explain why PRL deficient mice are infertile whereas PRL-deficient women are only subfertile.

SUMMARY

PRL may not perform an indispensable function for survival of the *individual*, but gestation and lactation lie at the core of the mammalian life cycle, placing PRL at the center of survival of the *species*. These reproductive states also place extreme demands on the physiology of the individual. Although not discussed in the present chapter, other actions of PRL on immunological, metabolic, and behavioral functions are therefore integral to the biology of all mammals. Much remains to be revealed about this complex and adaptive hormone, regarding both its functions and its regulation. With regard to the physiology of reproduction, a deeper understanding of the mechanisms controlling the various phases of

PRL secretion, particularly throughout female reproductive life, should lead to more effective and precise treatments of PRL disease while keeping intact the physiology of PRL.

REFERENCES

1. Riddle, O., Bates, R. W., and Dykshorn, S. W. (1933). The preparation, identification and assay of prolactin—a hormone of the anterior pituitary. *Am. J. Physiol.* 105, 191–216.
2. Stricker, P., and Grueter, R. (1928). Action du lobe antérieur de l'hypophyse sur la montée laiteuse. *C. R. Soc. Biol.* 99, 1978–1980.
3. Li, C. H. (1980). The chemistry of prolactin. In *Hormonal Proteins and Peptides* (C. H. Li, Ed.), Vol. 8, p. 2. Academic Press, New York.
4. Cooke, N. E., and Liebhaber, S. A. (1995). Molecular biology of the growth hormone-prolactin gene system. *Vitam. Horm.* 50, 385–459.
5. Forsyth, I. A., and Wallis, M. (2002). Growth hormone and prolactin—molecular and functional evolution. *J. Mammary Gland Biol. Neopl.* 7, 291–312.
6. May, D., Alrubaian, J., Patel, S., Dores, R. M., and Rand-Weaver, M. (1999). Studies on the GH/SL gene family: cloning of African lungfish (*Protopterus annectens*) growth hormone and somatolactin and toad (*Bufo marinus*) growth hormone. *Gen. Comp. Endocrinol.* 113, 121–135.
7. Soares, M. J., and Linzer, D. I. H. (2001). Rodent prolactin family and pregnancy. In *Prolactin* (N. D. Horseman, Ed.), pp. 139–167. Kluwer Academic Publishers, Boston.
8. Soares, M. J., Faria, T. N., Roby, K. F., and Deb, S. (1991). Pregnancy and the prolactin family of hormones: coordination of anterior pituitary, uterine, and placental expression. *Endocr. Rev.* 12, 402–423.
9. Soares, M. J., Muller, H., Orwig, K. E., Peters, T. J., and Dai, G. (1998). The uteroplacental prolactin family and pregnancy. *Biol. Reprod.* 58, 273–284.
10. Chen, E. Y., Liao, Y.-C., Smith, D. H., Barrera-Saldana, H. A., Gelinas, R. E., and Seeburg, P. H. (1989). The human growth hormone locus: Nucleotide sequence, biology, and evolution. *Genomics* 4, 479–497.
11. Wallis, O. C., Zhang, Y.-P., and Wallis, M. (2001). Molecular evolution of growth hormone (GH) in primates. Characterization of the GH genes from slow loris and marmoset defines an episode of rapid evolutionary change. *J. Mol. Endocrinol.* 26, 249–258.
12. Forsyth, I. A. (1994). Comparative aspects of placental lactogens—structure and function. *Exp. Clin. Endocrinol.* 102, 244–251.
13. Wallis, M. (1992). The expanding growth hormone/prolactin family. *J. Mol. Endocrinol.* 9, 185–188.
14. Schuler, L. A., and Kessler, M. A. (1992). Bovine placental prolactin-related hormones. *Trends Endocrinol. Metab.* 3, 334–338.
15. Dietz, A. B., Georges, M., Threadgill, D. W., Womack, J. E., and Schuler, L. A. (1992). Somatic-cell mapping, polymorphism, and linkage analysis of bovine prolactin-related proteins and placental-lactogen. *Genomics* 14, 137–143.
16. Colosi, P., Thordarson, G., Hellmiss, R., Singh, K., Forsyth, I. A., Gluckman, P., and Wood, W. I. (1989). Cloning and expression of ovine placental lactogen. *Mol. Endocrinol.* 3, 1462–1469.
17. Sakal, E., Bignon, C., Chapnik-Cohen, N., Daniel, N., Paly, J., Belair, L., Djiane, J., and Gertler, A. (1998). Cloning, preparaton and characterization of biologically active recombinant caprine placental lactogen. *J. Endocrinol.* 159, 509–518.
18. Shah, P., Sun, Y. X., Szpirer, C., and Duckworth, M. L. (1998). Rat placental lactogen II gene: characterization of gene structure and placental-specific expresssion. *Endocrinology* 139, 967–973.
19. Orwig, K. E., Ishimura, R., Muller, H., Liu, B., and Soares, M. J. (1997). Identification and characterization of mouse homolog for decidual/trophoblast prolactin-related protein. *Endocrinology* 138, 5511–5517.
20. Goffin, V., Shiverick, K. T., Kelly, P. A., and Martial, J. A. (1996). Sequence-function relationships within the expanding family of prolactin, growth hormone, placental lactogen, and related proteins in mammals. *Endocr. Rev.* 17, 385–410.
21. Talamantes, F., Ogren, L., Markoff, E., Woodard, S., and Madrid, J. (1980). Phylogenetic distribution, regulation of secretion, and prolactin-like effects of placental lactogens. *Fed. Proc.* 39, 2582–2587.
22. Handwerger, S. (1991). Clinical counterpoint: the physiology of placental lactogen in human pregnancy. *Endocr. Rev.* 12, 329–336.
23. Jackson, D., Volpert, O. V., Bouck, N., and Linzer, D. I. (1994). Stimulation and inhibition of angiogenesis by placental proliferin and proliferin-related protein. *Science* 266, 1581–1584.
24. Curlewis, J. D., Saunders, M. C., Kuang, J., Harrison, G. A., and Cooper, D. W. (1998). Cloning and sequence analysis of a pituitary prolactin cDNA from the brushtail possum (*Trichosurus vulpecula*). *Gen. Comp. Endocrinol.* 111, 61–67.
25. Wallis, M. (2000). Episodic evolution of protein hormones: molecular evolution of pituitary prolactin. *J. Mol. Evol.* 50, 465–473.
26. Gertler, A., Grosclaude, J., and Djiane, J. (1998). Interaction of lactogenic hormones with prolactin receptors. *Ann. N Y Acad. Sci.* 839, 177–181.
27. Lee, S. J., and Nathans, D. (1988). Proliferin secreted by cultured cells binds to mannose 6-phosphate receptors. *J. Biol. Chem.* 263, 3521–3527.
28. Kurtz, A., Bristol, L. A., Toth, B. E., Lazar-Wesley, E., Takacs, L., and Kacsoh, B. (1993). Mammary epithelial cells of lactating rats express prolactin messenger ribonucleic acid. *Biol. Reprod.* 48, 1095–1103.
29. Shaw-Bruha, C. M., Pirrucello, S. J., and Shull, J. D. (1997). Expression of the prolactin gene in normal and neoplastic human breast tissues and human mammary cell lines: promoter usage and alternative mRNA splicing. *Breast Cancer Res. Treat.* 44, 243–253.
30. Prigent-Tessier, A., Tessier, C., Hirosawa-Takamori, M., Boyer, C., Ferguson-Gottschall, S., and Gibori, G. (1999). Rat decidual prolactin. Identification, molecular cloning, and characterization. *J. Biol. Chem.* 274, 37982–37989.
31. Nevalainen, M. T., Valve, E. M., Ingleton, P. M., Nurmi, M., Martikainen, P. M., and Harkonen, P. L. (1997). Prolactin and prolactin receptors are expressed and functioning in human prostate. *J. Clin. Invest.* 99, 618–27.
32. Shah, G. N., Laird, H. E. Jr., and Russell, D. H. (1991). Identification and characterization of a prolactin-like polypeptide synthesized by mitogen-stimulated murine lymphocytes. *Int. Immunol.* 3, 297–304.
33. Kooijman, R., Gerlo, S., Coppens, A., and Hooghe-Peters, E. L. (2000). Growth hormone and prolactin expression in the immune system. *Ann. N Y Acad. Sci.* 917, 534–540.
34. Zinger, M., McFarland, M., and Ben-Jonathan, N. (2003). Prolactin expression and secretion by human breast glandular and adipose tissue explants. *J. Clin. Endocrinol. Metab.* 88, 689–696.
35. Corbacho, A. M., Macotela, Y., Nava, G., Torner, L., Duenas, Z., Noris, G., Morales, M. A., Martinez de la Escalera, G., and Clapp, C. (2000). Human umbilical vein endothelial cells express multiple prolactin isoforms. *J. Endocrinol.* 166, 53–62.

36. Ben-Jonathan, N., Mershon, J. L., Allen, D. L., and Steinmetz, R. W. (1996). Extrapituitary prolactin: distribution, regulation, functions, and clinical aspects. *Endocr. Rev.* 17, 639–669.
37. Kacsóh, B., Veress, Z., Tóth, B. E., Avery, L. M., and Grosvenor, C. E. (1993). Bioactive and immunoreactive variants of prolactin in milk and serum of lactating rats and their pups. *J. Endocrinol.* 138, 243–257.
38. Gellerson, B., Dimattia, G. E., Friesen, H. G., and Bohnet. H. G. (1989). Prolactin (PRL) mRNA from human decidua differs from pituitary PP mRNA but resembles the IM-9-P3 lymphoblast PRL transcript. *Mol. Cell. Endocrinol.* 64, 127–130.
39. Clevenger, C. V., Furth, P. A., Hankinson, S. E., and Schuler, L. A. (2003). The role of prolactin in mammary carcinoma. *Endocr. Rev.* 24, 1–27.
40. Ginsburg, E., and Vonderhaar, B. K. (1995). Prolactin synthesis and secretion by human breast cancer cells. *Cancer Res.* 55, 2591–2595.
41. Clevenger, C. V., and Plank, T. L. (1997). Prolactin, an autocrine/paracrine factor in breast cancer. *J. Mammary Gland Biol. Neopl.* 2, 59–68.
42. Jabbour, H. N., and Critchley, H. O. (2001). Potential roles of decidual prolactin in early pregnancy. *Reproduction* 121, 197–205.
43. Prigent-Tessier, A., Barkai, U., Tessier, C., Cohen, H., and Gibori, G. (2001). Characterization of a rat uterine cell line, U(III) cells: prolactin (PRL) expression and endogenous regulation of PRL-dependent genes; estrogen receptor beta, alpha(2)-macroglobulin, and decidual PRL involving the Jak2 and Stat5 pathway. *Endocrinology* 142, 1242–1250.
44. Doneen, B. A., Bewley, T. A., and Li, C. H. (1979). Studies on prolactin. Selective reduction of the disulfide bonds of the ovine hormone. *Biochemistry* 18, 4851–4860.
45. Luck, D. N., Gout, P. W., Sutherland, E. R., Fox, K., Huyer, M., and Smith, M. (1992). Analysis of disulphide bridge function in recombinant bovine prolactin using site-specific mutagenesis and renaturation under mild alkaline conditions: a crucial role for the central disulphide bridge in the mitogenic activity of the hormone. *Prot. Eng.* 5, 559–567.
46. Abdel-Meguid, S. S., Shieh, H. S., Smith, W. W., Dayringer, H. E., Violand, B. N., and Bentle, L. A. (1987). Three-dimensional structure of a genetically engineered variant of porcine growth hormone. *Proc. Natl. Acad. Sci. U S A* 84, 6434–6437.
47. de Vos, A. M., Ultsch, M., and Kossiakoff, A. A. (1992). Human growth hormone and extracellular domain of its receptor: crystal structure of the complex. *Science* 255, 306–312.
48. Goffin, V., Martial, J. A., and Summers, N. L. (1995). Use of a model to understand prolactin and growth hormone specificities. *Protein Eng.* 8, 1215–1231.
49. Halaby, D., Thoreau, E., Djiane, J., and Mornon, J. P. (1997). Homology modeling of rabbit prolactin hormone complexed with its receptor. *Proteins* 27, 459–468.
50. Kossiakoff, A. A., and de Vos, A. M. (1998). Structural basis for cytokine hormone-receptor recognition and receptor activation. *Adv. Protein Chem.* 52, 67–108.
51. Sinha, Y. N. (1995). Structural variants of prolactin: occurrence and physiological significance. *Endocr. Rev.* 16, 354–369.
52. Lorenson, M. Y., and Walker, A. M. (2001). Structure–function relationships in prolactin. In *Prolactin* (N. D. Horseman, Ed.), pp. 189–217. Kluwer Academic Publishers, Boston.
53. Lewis, U. J., Singh, R. N., Lewis, L. J., Seavey, B. K., and Sinha, Y. N. (1984). Gycosylated ovine prolactin. *Proc. Natl. Acad. Sci. U S A* 81, 385–389.
54. Bollengier, F., Velkeniers, Houghe-Peters, E., Mahler, A., and Vanhaelst, L. (1989). Multiple forms of rat prolactin and growth hormone in pituitary cell subpopulations separated using a Percoll gradient system: disulphide-bridged dimers and glycosylated variants. *J. Endocrinol.* 120, 201–206.
55. Chen, T. J., Kuo, C. B., Tsai, K. F., Liu, J. W., Chen, D. Y., and Walker, A. M. (1998). Development of recombinant human prolactin receptor antagonists by molecular mimicry of the phosphorylated hormone. *Endocrinology* 139, 609–616.
56. Wang, Y.-F., and Walker, A. M. (1993). Dephosphorylation of standard prolactin produces a more biologically active molecule: Evidence for antagonism between nonphosphorylated and phosphorylated prolactin in the stimulation of Nb2 cell proliferation. *Endocrinology* 133, 2156–2160.
57. Bridges, R. S., Rigero, B. A., Byrnes, E. M., Yang, L., and Walker, A. M. (2001). Central infusions of the recombinant human prolactin receptor antagonist, S179D-PRL, delay the onset of maternal behavior in steroid-primed, nulliparous female rats. *Endocrinology* 142, 730–739.
58. Yang, L., Kuo, C. B., Liu, Y., Coss, D., Xu, X., Chen, C., Oster-Granite, M. L., and Walker, A. M. (2001). Administration of unmodified prolactin (U-PRL) and a molecular mimic of phosphorylated prolactin (PP-PRL) during rat pregnancy provides evidence that the U-PRL:PP-PRL ratio is crucial to the normal development of pup tissues. *J. Endocrinol.* 168, 227–238.
59. Kuo, C. B., Wu, W., Xu, X., Yang, L., Chen, C., Coss, D., Birdsall, B., Nasseri, D., and Walker, A. M. (2002). Pseudophosphorylated prolactin (S179D PRL) inhibits growth and promotes beta-casein gene expression in the rat mammary gland. *Cell. Tissue Res.* 309, 429–437.
60. Coss, D., Kuo, C. B., Yang, L., Ingleton, P., Luben, R., and Walker, A. M. (1999). Dissociation of Janus kinase 2 and signal transducer and activator of transcription 5 activation after treatment of Nb2 cells with a molecular mimic of phosphorylated prolactin. *Endocrinology* 140, 5087–5094.
61. Oetting, W. S., Tuazon, P. T., Traugh, J. A., and Walker, A. M. (1986). Phosphorylation of prolactin. *J. Biol. Chem.* 261, 1649–1652.
62. Tuazon, P. T., Lorenson, M. Y., Walker, A. M., and Traugh, J. A. (2002). p21-activated protein kinase gamma-PAK in pituitary secretory granules phosphorylates prolactin. *FEBS Lett.* 515, 84–88.
63. Liu, J. W., and Walker, A. M. (1994). Long term effect of estradiol and thyrotropin releasing hormone of the release on non-phosphorylated and phosphorylated monomer prolactin in vitro. Abstract 1363, 76th Annual Meeting of the Endocrine Society, Anaheim, CA.
64. Ho, T. W., Leong, F. S., Olaso, C. H., and Walker, A. M. (1993). Secretion of specific nonphosphorylated and phosphorylated rat prolactin isoforms at different stages of the estrous cycle. *Neuroendocrinology* 58, 160–165.
65. Lorenson, M. Y., Miska, S. P., and Jacobs, L. S. (1984). Molecular mechanisms of prolactin release from pituitary secretory granules. In *Prolactin Secretion: A Multidisciplinary Approach* (F. Mena and C. M. Valverde, Eds.), pp. 141–160. Academic Press, New York.
66. Cohen, H., Coeh, O., and Gagnon, J. (1994). Serum prolactin-binding protein (PRL-BP) of human and rat are identified as IgG. *C. R. Acad. Sci. Paris* 317, 293–298.
67. Khurana, S., Kuns, R., and Ben-Jonathan, N. (1999). Heparin-binding property of human prolactin: a novel aspect of prolactin biology. *Endocrinology* 140, 1026–1029.
68. Miyai, K., Ichihara, K., Kondo, K., and Mori, S. (1986). Asymptomatic hyperprolactinemia and prolactinoma in the general population: mass screening by paired assays of serum prolactin. *Clin. Endocrinol.* 35, 549–554.
69. Tritos, N. A., Guay, A. T., and Malarkey, W. B. (1998). Asymptomatic "big" hyperprolactinemia in two men with pituitary adenomas. *Eur. J. Endocrinol.* 138, 82–85.
70. Cavaco, B., Leite, V., Santos, M. A., Arranhado, E., and Sobrinho, L. G. (1995). Some forms of (big, big) prolactin

behave as a complex of monomeric prolactin with an immunoglobulin G in patients with macroprolactinemia or prolactinoma. *J. Clin. Endocrinol. Metab.* 80, 2342–2349.

71. Mittra, I. (1980). A novel "cleaved prolactin" in the rat pituitary. Part 1. Biosynthesis, characterization and regulatory control. *Biochem. Biophys. Res. Commun.* 95, 1750–1759.
72. Clapp, C. Martial, J. A., Guzman, R. C., Rentier-Delrue, R., and Weiner, R. I. (1993). The 16-kilodalton N-terminal fragment of human prolactin is a potent inhibitor of angiogenesis. *Endocrinology* 133, 1292–1299.
73. Baldocchi, R. A., Tan, L., and Nicoll, C. S. (1992). Processing of rat prolactin by rat tissue explants and serum in vitro. *Endocrinology* 130, 1653–1659.
74. Baldocchi, R. A., Tan, L., King, D. S., and Nicoll, C. S. (1993). Mass spectrometric analysis of the fragments produced by cleavage and reduction of rat prolactin: evidence that the cleaving enzyme is cathepsin D. *Endocrinology* 133, 935–938.
75. Khurana, S., Liby, K., Buckley, A. R., and Ben-Jonathan, N. (1999). Proteolysis of human prolactin: resistance to cathepsin D and formation of a nonangiostatic, C-terminal 16K fragment by thrombin, *Endocrinology* 140, 4127–4132.
76. Dueñas, Z., Torner, K., Corbacho, A. M., Ochoa, A., Gutierrez-Ospina, G., López-Barrera, F., Barrios, F. A., Berger, P., Martínez de la Escalera, G., and Clapp, C. (1999). Inhibition of rat corneal angiogenesis by 16kDa prolactin and by endogenous prolactin-like molecules. *Invest. Ophthalmol. Vis. Sci.* 40, 2498–2503.
77. Clapp, C., and Weiner, R. I. (1992). A specific, high-affinity, saturable binding site for the 16-kilodalton fragment of prolactin on capillary endothelial cells. *Endocrinology* 130, 1380–1386.
78. Paris, N., Rentier-Delrue, F., Defontaine, A., Goffin, V., Lebrun, J. J., Mercier, L., and Martial, J. A. (1990). Bacterial production and purification of recombinant human prolactin. *Biotechnol. Appl. Biochem.* 12, 436–449.
79. Posner, B. I., Kelly, P. A., Shiu, R. P., and Friesen, H. G. (1974). Studies of insulin, growth hormone and prolactin binding: tissue distribution, species variation and characterization. *Endocrinology* 95, 521–531.
80. Boutin, J. M., Jolicoeur, C., Okamura, H., Gagnon, J., Edery, M., Shirota, M., Banville, D., Dusanter-Fourt, I., Djiane, J., and Kelly, P. A. (1988). Cloning and expression of the rat prolactin receptor, a member of the growth hormone/prolactin receptor gene family. *Cell* 53, 69–77.
81. Leung, D. W., Spencer, S. A., Cachianes, G., Hammonds, R. G., Collins, C., Henzel, W. J., Barnard, R., Waters, M. J., and Wood, W. I. (1987). Growth hormone receptor and serum binding protein: purification, cloning and expression. *Nature* 330, 537–543.
82. Bazan, F. (1989). A novel family of growth factor receptors: a common binding domain in the growth hormone, prolactin, erythropoietin and IL-6 receptors, and p75 IL-2 receptor β-chain. *Biochem. Biophys. Res. Commun.* 164, 788–795.
83. Cosman, D., Lyman, S. D., Idzerda, R. L., Beckmann, M. P., Park, L. S., Goodwin, R. G., and March, C. J. (1990). A new cytokine receptor superfamily. *Trends Biochem. Sci.* 15, 265–270.
84. Bole-Feysot, C., Goffin, V., Edery, M., Binart, N., and Kelly, P. A. (1998). Prolactin and its receptor: actions, signal transduction pathways and phenotypes observed in prolactin receptor knockout mice. *Endocr. Rev.* 19, 225–268.
85. Login, I. S., and MacLeod, R. M. (1977). Prolactin in human and rat serum and cerebrospinal fluid. *Brain Res.* 132, 477–483.
86. Somers, W., Ultsch, M., De Vos, A. M., and Kossiakoff, A. A. (1994). The x-ray structure of a growth hormone-prolactin receptor complex. *Nature* 372, 478–481.
87. Chang, W.-P., and Clevenger, C. V. (1996). Modulation of growth factor receptor function by isoform heterodimerization. *Proc. Natl. Acad. Sci. U S A* 93, 5947–5952.
88. Campbell, G. S., Argetsinger, L. S., Ihle, J. N., Kelly, P. A., Rillema, J. A., and Carter-Su, C. (1994). Activation of JAK2 tyrosine kinase by prolactin receptors in Nb2 cells and mouse mammary gland explants. *Proc. Natl. Acad. Sci. U S A* 91, 5232–5236.
89. Gao, J., Hughes, J. P., Auperin, B., Buteau, H., Edery, M., Zhuang, H., Wojochowski, D. M., and Horseman, N. D. (1995). Interaction among JANUS kinases and the prolactin (PRL) receptor in the regulation of a PRL response element. *Mol. Endocrinol.* 10, 847–856.
90. Parganas, E., Wang, D., Stravopodis, D., Topham, D. J., Marine, J. C., Teglund, S., Vanin, E. F., Boddner, S., Colamonici, O. R., van Deursen, J. M., Grosveld, G., and Ihle, J. N. (1998). Jak2 is essential for signaling through a variety of cytokine receptors. *Cell* 93, 385–395.
91. Berchtold, S., Volarevic, S., Moriggl, R., Mercep, M., and Groner, B. (1998). Dominant negative variants of the SHP-2 tyrosine phosphatase inhibit prolactin activation of Jak2 (janus kinase 2) and induction of Stat5 (signal transducer and activator of transcription 5)-dependent transcription. *Mol. Endocrinol.* 12, 556–567.
92. Horseman, N. D., and Yu-Lee, L.-Y. (1994). Transcriptional regulation by the helix bundle peptide hormones: GH, PRL, and hematopoietic cytokines. *Endocr. Rev.* 15, 627–649.
93. Darnell, J. E. Jr., Kerr, I. M., and Stark, G. R. (1994). Jak-Stat pathways and transcriptional activation in response to IFNs and other extracellular signaling proteins. *Science* 264, 1415–1421.
94. Sidis, Y., and Horseman, N. D. (1994). Prolactin induces rapid p95/p70 tyrosine phosphorylation, and protein binding to GAS-like sites in the *anx* I_{cp35} and *c-fos* genes. *Endocrinology* 134, 1979–1985.
95. Wakao, H., Gouilleux, F., and Groner, B. (1994). Mammary gland factor (MGF) is a novel member of the cytokine regulated transcription factor gene family and confers the prolactin response. *EMBO J.* 13, 2182–2191. [Erratum in (1995). *EMBO J.* 14, 854–855.]
96. Liu, X., Robinson, G. W., Wagner, K.-U., Garrett, L., Wynshaw-Boris, A., and Hennighausen, L. (1997). Stat5a is mandatory for adult mammary gland development and lactogenesis. *Genes Dev.* 11, 179–186.
97. Udy, G. B., Towers, R. P., Snell, R. G., Wilkins, R. J., Park, S. H., Ram, P. A., Waxman, D. J., and Davey, H. W. (1997). Requirement of STAT5b for sexual dimorphism of body growth rates and liver gene expression. *Proc. Natl. Acad. Sci. U S A* 94, 7239–7244.
98. Horseman, N. D., Zhao, W., Montecino-Rodriguez, E., Tanaka, M., Nakashima, K., Engle, S. J., Smith, F., Markoff, E., and Dorshkind, K. (1997). Defective mammopoiesis, but normal hematopoiesis, in mice with a targeted disruption of the prolactin gene. *EMBO J.* 16, 6926–6935.
99. Ormandy, C. J., Camus, A., Barra, J., Damotte, D., Lucas, B., Buteau, H., Edery, M., Brousse, N., Babinet, C., Binart, N., and Kelly, P. A. (1997). Null mutation of the prolactin receptor gene produces multiple reproductive defects in the mouse. *Genes Dev.* 11, 167–178.
100. Teglund, S., McKay, C., Schuetz, E., van Deursen, J. M., Stravopodis, D., Wang, D., Brown, M., Bodner, S., Grosveld, G., and Ihle, J. N. (1998). Stat5a and Stat5b proteins have essential and nonessential, or redundant, roles in cytokine responses. *Cell* 93, 841–850.
101. Gouilleux, F., Wakao, H., Mundt, M., and Groner, B. (1994). Prolactin induces phosphorylation of tyr694 of Stat5 (MGF),

a prerequisite for DNA binding and induction of transcription. *EMBO J.* 13, 4361–4369.

102. Helman, D., Sandowski, Y., Cohen, Y., Matsumoto, A., Yoshimura, A., Merchav, S., and Gertler, A. (1998). Cytokine-inducible SH2 protein (CIS3) and Jak2 binding protein (JAB) abolish prolactin receptor-mediated Stat5 signaling. *FEBS Lett.* 441, 287–291.
103. Das, R., and Vonderhaar, B. K. (1997). Prolactin as a mitogen in mammary cells. *J. Mammary Gland Biol. Neoplasia.* 2, 29–39.
104. Clevenger, C. V., and Rycyzyn, M. A. (2000). Translocation and action of polypeptide hormones within the nucleus. Relevance to lactogenic transduction. *Adv. Exp. Med. Biol.* 480, 77–84.
105. Perrot-Applanat, M., Gualillo, O., Buteau, H., Edery, M., and Kelly, P. A. (1997). Internalization of prolactin receptor and prolactin transfected cells does not involve nuclear translocation. *J. Cell. Sci.* 110, 1123–1132.
106. Rycyzyn, M. A., Reilly, S. C., O'Malley, K., and Clevenger, C. V. (2000). Role of cyclophilin B in prolactin signal transduction and nuclear retrotranslocation. *Mol. Endocrinol.* 14, 1175–1186.
107. Lamberts, S. W. J., and MacLeod, R. M. (1990). Regulation of prolactin secretion at the level of the lactotroph. *Physiol. Rev.* 70, 279–318.
108. Freeman, M. E., Kanyicska, B., Lerant, A., and Nagy, G. (2000). Prolactin: structure, function and regulation of secretion. *Physiol. Rev.* 80, 1523–1631.
109. El Halawani, M. E., Burke, W. H., Millam, J. R., Fehrer, S. C., and Hargis, B. M. (1984). Regulation of prolactin and its role in gallinaceous bird reproduction. *J. Exp. Zool.* 232, 521–529.
110. Horseman, N. D., and Buntin, J. D. (1995). Regulation of pigeon crop milk secretions and parental behaviors by prolactin. *Annu. Rev. Nutr.* 15, 213–238.
111. Ben-Jonathan, N., and Hnasko, R. (2001). Dopamine as a prolactin (PRL) inhibitor. *Endocr. Rev.* 22, 724–763.
112. Hökfelt, T., and Fuxe, K. (1972). Effects of prolactin and ergot alkaloids on the tuberoinfundibular dopamine (DA) neurons. *Neuroendocrinology* 9, 100–122.
113. Gudelsky, G. A., and Porter, J. C. (1980). Release of dopamine from tuberoinfundibular neurons into pituitary stalk blood after prolactin or haloperidol administration. *Endocrinology* 106, 526–529.
114. Maurer, R. A. (1980). Dopaminergic inhibition of prolactin synthesis and prolactin messenger RNA accumulation in cultured pituitary cells. *J. Biol. Chem.* 255, 8092–8097.
115. Lloyd, H. M., Meares, J. D., and Jacobi, J. (1975). Effects of oestrogen and bromocryptine on in vivo secretion and mitosis in prolactin cells. *Nature* 255, 497.
116. Melmed, S. (1981). Bromocriptine inhibits colony formation by rat pituitary tumor cells in a double-layered agar clonogenic assay. *Endocrinology* 109, 2258–2260.
117. Caron, M. G., Beaulieu, M., Raymond, V., Gagne, B., Drouin, J., Lefkowitz, R., and Labrie, F. Dopaminergic receptors in the anterior pituitary gland. *J. Biol. Chem.* 253, 2244–2253.
118. Enjalbert, A., and Bockaert, J. (1983). Pharmacological characterization of the D_2 dopamine receptor negatively coupled with adenylate cyclase in rat anterior pituitary. *Mol. Pharmacol.* 23, 576–584.
119. Andersen, P. H., Gingrich, J. A., Bates, M. D., Dearry, A., Falardeau, P., Senogles, S. E., and Caron, M. (1990). Dopamine receptor subtypes: beyond the D1/D2 classification. *Trends Pharmacol. Sci.* 11, 231–236.
120. Civelli, O., Bunzow, J. R., Grandy, D. K., Zhou, Q.-Y., and van Tol, H. H. M. (1991). Molecular biology of the dopamine receptors. *Eur. J. Pharmacol. Mol. Pharmacol. Sect.* 207, 277–286.
121. Valerio, A., Alberici, A., Tinti, C., Spano, P. F., and Memo, M. (1994). Antisense strategy unravels a dopamine receptor distinct from the D2 subtype, uncoupled with adenylyl cyclase, inhibiting prolactin release from rat pituitary cells. *J. Neurochem.* 62, 1260–1266.
122. Matsumoto, M., Hidaka, K., Tada, S., Tasaki, Y., and Yamaguchi, T. (1995). Full-length cDNA cloning and distribution of human dopamine D4 receptor. *Mol. Brain Res.* 29, 157–162.
123. Sokoloff, P., Giros, B., Martres, M.-P., Bouthenet, M.-L., and Schwartz, J.-C. (1990). Molecular cloning and characterization of a novel dopamine receptor (D_3) as a target for neuroleptics. *Nature* 347, 146–151.
124. DeLean, A., Kilpatrick, B., and Caron M. (1982). Dopamine receptor of the porcine anterior pituitary gland. *Mol. Pharmacol.* 22, 290–297.
125. Kilpatrick, B. F., and Caron, M. G. (1983). Agonist binding promotes a guanine nucleotide reversible increase in the apparent size of the bovine anterior pituitary dopamine receptors. *J. Biol. Chem.* 258, 13528–13534.
126. George, S. R., Watanabe, M., Di Paolo, T., Falardeau, P., Labrie, F., and Seeman, P. (1985). The functional state of the dopamine receptor in the anterior pituitary is in the high affinity form. *Endocrinology* 117, 690–697.
127. Senogles, S. E., Benovic, J. L., Amlaiky, N., Unson, C., Milligan, G., Vinitsky, R., Spiegel, A. M., and Caron, M. G. (1987). The D_2-dopamine receptor of anterior pituitary is functionally associated with a pertussis toxin-sensitive guanine nucleotide binding protein. *J. Biol. Chem.* 262, 4860–4867.
128. Cronin, M. J., Myers, G. A., MacLeod, R. M., and Hewlett, E. L. (1983). Pertussis toxin uncouples dopamine agonist inhibition of prolactin release. *Am. J. Physiol.* 244, E499–E504.
129. Enjalbert, A., Mussett, F., Chenard, C., Priam, M., Kordon, C., and Heisler, S. (1988). Dopamine inhibits prolactin secretion stimulated by the calcium channel agonist Bay-K-8644 through a pertussis toxin-sensitive G protein in anterior pituitary cells. *Endocrinology* 123, 406–412.
130. Schofield, J. G., Khan, A. I., and Wood, A. (1988). Modification by pertussis toxin of the responses of bovine anterior pituitary cells to acetylcholine and dopamine: effects on hormone secretion and ^{86}Rb efflux. *J. Endocrinol.* 116, 393–401.
131. Thorner, M. O., Hackett, J. T., Murad, F., and MacLeod, R. M. (1980). Calcium rather than cyclic AMP as the physiological intracellular regulator of prolactin release. *Neuroendocrinology* 31, 390–402.
132. Delbeke, D., Scammell, J. G., Martinez-Campos, A., and Dannies, P. S. (1986). Dopamine inhibits prolactin release when cyclic adenosine 3',5'-monophosphate levels are elevated. *Endocrinology* 118, 1271–1277.
133. Lafond, J., Ducharme, J. R., and Collu, R. (1986). Inhibition of prolactin release and blockade of adenohypophyseal cell cyclic AMP accumulation are two dissociable effects of dopaminergic and non-dopaminergic drugs. *Mol. Cell. Endocrinol.* 44, 219–225.
134. Ray, K. P., Gomm, J. J., Law, G. J., Sigournay, C., and Wallis, M. (1986). Dopamine and somatostatin inhibit forskolin-stimulated prolactin and growth hormone secretion but not stimulated cyclic AMP levels in sheep anterior pituitary cell cultures. *Mol. Cell. Endocrinol.* 45, 175–182.
135. Gregerson, K. A., Chuknyiska, R., and Golesorkhi, N. (1994). Stimulation of prolactin release by dopamine withdrawal: role of calcium influx. *Am. J. Physiol.* 267, E789–E794.
136. Canonico, P. L., Valdenegro, C. A., and MacLeod, R. M. (1982). Dopamine inhibits ^{32}P incorporation into phosphatidylinositol in the anterior pituitary gland of the rat. *Endocrinology* 111, 347–349.
137. Enjalbert, A., Sladeczek, F., Guillon, G., Bertrand, P., Shu, C., Epelbaum, J., Garcia-Sainz, A., Jard, S., Lombard, C., and

Kordon, C. (1986). Angiotensin II and dopamine modulate both cAMP and inositol phosphate productions in anterior pituitary cells. Involvement in prolactin secretion. *J. Biol. Chem.* 261, 4071–4075.

138. Martínez de la Escalera, G., and Weiner, R. I. (1988). Effect of dopamine withdrawal on activation of adenylate cyclase and phospholipase C in enriched lactotrophs. *Endocrinology* 123, 1682–1687.
139. Vallar, L., Vicentini, L. M., and Meldolesi, J. (1988). Inhibition of inositol phosphate production is a late, Ca^{2+}-dependent effect of D2 dopaminergic receptor activation in rat lactotroph cells. *J. Biol. Chem.* 263, 10127–10134.
140. Ozawa, S., and Sand, O. (1986). Electrophysiology of excitable endocrine cells. *Physiol. Rev.* 66, 887–952.
141. Lingle, C. J., Sombati, S., and Freeman, M. E. (1986). Membrane currents in identified lactotrophs of rat anterior pituitary. *J. Neurosci.* 6, 2995–3005.
142. Lledo, P. M., Legendre, P., Israel, J. M., and Vincent, J. D. (1990). Dopamine inhibits two characterized voltage-dependent calcium currents in identified rat lactotrophs cells. *Endocrinology* 127, 990–1001.
143. Law, G. J., Pachter, J. A., and Dannies, P. S. (1989). Ability of repetitive Ca^{2+} spikes to stimulate prolactin release is frequency dependent. *Biochem. Biophys. Res. Commun.* 158, 811–816.
144. Ray, K. P., and Wallis, M. (1982). Involvement of calcium ions in dopamine inhibition of prolactin secretion from sheep pituitary cells. *Mol. Cell. Endocrinol.* 28, 691–703.
145. Malgaroli, A., Vallar, L., Elahi, F. R., Pozzan, T., Spada, A., and Meldolesi, J. (1986). Dopamine inhibits cytosolic Ca^{2+} increases in rat lactotroph cells. Evidence of a dual mechanism of action. *J. Biol. Chem.* 262, 13920–13927.
146. Ho, M.-Y., Kao, J. P. Y., and Gregerson, K. A. (1996). Dopamine withdrawal elicits prolonged calcium rise to support prolactin rebound release. *Endocrinology* 137, 3513–3521.
147. Luque, E. H., de Toro, M. M., Smith, P. F., and Neill, J. D. (1986), Subpopulations of lactotrophs detected with the reverse hemolytic plaque assay show differential responsiveness to dopamine. *Endocrinology* 118, 2120–2124.
148. Gregerson, K. A. (2001). Mechanisms of dopamine action on the lactotroph. In *Prolactin* (N. D. Horseman, Ed.), pp. 45–61. Kluwer Academic Publishers, Boston.
149. Lledo, P.-M., Israel, J. M., and Vincent, J.-D. (1991). Chronic stimulation of D_2 dopamine receptors specifically inhibits calcium but not potassium currents in rat lactotrophs. *Brain Res.* 558, 231–238.
150. Rendt, J., and Oxford, G. S. (1994). Absence of coupling between D_2 dopamine receptors and calcium channels in lactotrophs from cycling female rats. *Endocrinology* 135, 501–508.
151. Curtis, B. M., and Catterall, W. A. (1985). Phosphorylation of the calcium antagonist receptor of the voltage-sensitive calcium channel by cAMP-dependent protein kinase. *Proc. Natl. Acad. Sci. U S A* 82, 2528–2532.
152. Nastainczyk, W., Röhrkasten, A., Sieber, M., Rudolph, C., Schächtele, C., Marmé, D., and Hofmann, F. (1987). Phosphorylation of the purified receptor for calcium channel blockers by cAMP kinase and protein kinase C. *Eur. J. Biochem.* 169, 137–142.
153. Armstrong, D., and Eckert, R. (1987). Voltage-activated calcium channels that must be phosphorylated to respond to membrane depolarization. *Proc. Natl. Acad. Sci. U S A* 84, 2519–2522.
154. Gregerson, K. A., Einhorn, L., Smith, M. M., and Oxford, G. S. (1989). Modulation of potassium channels by dopamine in rat pituitary lactotrophs: a role in the regulation of prolactin secretion? In *Secretion and Its Control* (G. S. Oxford and C. M. Armstrong, Eds.), pp. 123–141. Rockefeller University Press, New York.
155. Gregerson, K., Flagg, T., O'Neill, T. J., Anderson, M., Lauring, O., Horel, J. S., and Welling, P. A. (2001). Identification of the G-protein-coupled, inward rectifying potassium channel gene products in rat anterior pituitary gland. *Endocrinology* 142, 2820–2832.
156. Einhorn, L. C., and Oxford, G. S. (1993). Guanine nucleotide binding proteins mediate D_2 dopamine receptor activation of a potassium channel in rat lactotrophs. *J. Physiol.* 462, 563–578.
157. Aldrich, R. (1993). Advent of a new family. *Nature* 362, 107–108.
158. Doupnik, C. A., Davidson, N., and Lester, H. A. (1995). The inward rectifier potassium channel family. *Curr. Opin. Neurobiol.* 5, 268–277.
159. Krapivinsky, G., Krapivinsky, L., Wickman, K., and Clapham, D. E. (1995). G beta gamma binds directly to the G protein-gated K^+ channel, I_{KAch}. *J. Biol. Chem.* 270, 29059–29062.
160. Reuveny, E., Slesinger, P. A., Inglese, J., Morales, J. M., Inlguez-Lluhl, Lefkowitz, R. J., Bourne, H. R., Jan, Y. N., and Jan, L. Y. (1994). Activation of the cloned muscarinic potassium channel by G protein $\beta\gamma$ subunits. *Nature* 370, 143–146.
161. Takao, K., Yoshii, M., Kanda, A., Kokubun, S., and Nukada, T. (1994). A region of the muscarinic-gated atrial K^+ channel critical for activation by G protein beta gamma subunits. *Neuron* 13, 747–755.
162. Kofuji, P., Davidson, N., and Lester, H. A. (1995). Evidence that neuronal G-protein-gated inwardly rectifying K^+ channels are activated by $G\beta\gamma$ subunits and function as heteromultimers. *Proc. Natl. Acad. Sci. U S A* 92, 6542–6546.
163. Krapivinsky, G., Gordon, E. A., Wickman, K., Velimirovic, B., Krapivinsky, L., and Clapham, D. E. (1995). The G-protein-gated atrial K^+ channel I_{KAch} is a heteromultimer of two inwardly rectifying K^+ channel proteins. *Nature* 374, 135–141.
164. Denef, C., Manet, D., and Dewals, R. (1980). Dopaminergic stimulation of prolactin release. *Nature* 285, 243–246.
165. Kramer, I. M., and Hopkins, C. R. (1982). Studies on the kinetics of dopamine-regulated prolactin secretion. *Mol. Cell. Endocrinol.* 28, 191–198.
166. Burris, T. P., and Freeman, M. E. (1993). Low concentrations of dopamine increase cytosolic calcium in lactotrophs. *Endocrinology* 133, 63–68.
167. Denef, C., Baes, M., and Schramme, C. (1984). Stimulation of prolactin secretion after short term or pulsatile exposure to dopamine in superfused anterior pituitary cell aggregates. *Endocrinology* 114, 1371–1378.
168. Senoh, S., and Witkop, B. (1959). Non-enzymatic conversions of dopamine to norepinephrine and trihydroxyphenethylamines. *J. Am. Chem. Soc.* 81, 6222–6231.
169. Senoh, S., Creveling, C. R., Udenfriend, S., and Witkop, B. (1959). Chemical, enzymatic and metabolic studies on the mechanisms of oxidation of dopamine. *J. Am. Chem. Soc.* 81, 6236–6245.
170. Toth, B. E., Homicsko, K., Radnai, B., Maruyama, W., DeMaria, J. E., Vecsemyes, M., Fekete, M. I., Fulop, F., Naoi, M., Freeman, M. E., and Nagy, G. M. (2001). Salsolinol is a putative endogenous neuro-intermediate lobe prolactin-releasing factor. *J. Neuroendocrinol.* 13, 1042–1050.
171. Frawley, L. S., and Neill, J. D. (1984). Brief decreases in dopamine result in surges of prolactin secretion in monkeys. *Am. J. Physiol.* 247, E778–E780.
172. Haisenleder, D. J., Moy, J. A., Gala, R. R., and Lawson, D. M. (1986). The effect of transient dopamine antagonism on thyrotropin-releasing hormone-induced prolactin release in

ovariectomized rats treated with estradiol and/or progesterone. *Endocrinology* 119, 1996–2003.

173. Martinez de la Escalera, G., and Weiner, R. I. (1992). Dissociation of dopamine from its receptor as a signal in the pleiotropic hypothalamic regulation of prolactin secretion. *Endocr. Rev.* 13, 241–255.
174. Gregerson, K. A., Golesorkhi, N., and Chuknyiska, R. (1994). Stimulation of prolactin release by dopamine withdrawal: role of membrane hyperpolarization. *Am. J. Physiol.* 267, E781–E788.
175. Koch, B. D., Blalock, J. B., and Schonbrunn, A. (1988). Characterization of the cyclic AMP-independent actions of somatostatin in GH cells. I. An increase in potassium conductance is responsible for both the hyperpolarization and the decrease in intracellular free calcium produced by somatostatin. *J. Biol. Chem.* 263, 216–225.
176. Shah, G. V., Pedchenko, V., Stanley, S., Li, Z., and Samson, W. K. (1996). Calcitonin is a physiological inhibitor of prolactin secretion in ovariectomized female rats. *Endocrinology* 137, 1814–1822.
177. Kanyicska, B., Lerant, A., and Freeman, M. E. (1998). Endothelin is an autocrine regulator of prolactin secretion. *Endocrinology* 139, 5164–5173.
178. Sarkar, D. K., Kim, K. H., and Minami, S. (1992). Transforming growth factor ß1 messenger RNA and protein expression in the pituitary gland: Its action on prolactin secretion and lactotropic growth. *Mol. Endocrinol.* 6, 1825–1833.
179. Couse, J. F., Lindzey, J., Grandien, K., Gustafsson, J.-Å., and Korach, K. S. (1997). Tissue distribution and quantitative analysis of estrogen receptor-α (ERα) and estrogen receptor-β (ERβ) messenger ribonucleic acid in the wild-type and ERα-knockout mouse. *Endocrinology* 138, 4613–4621.
180. Krege, J. H., Hodgin, J. B., Couse, J. F., Enmark, E., Warner, M., Mahler, J. F., Sar, M., Korach, K. S., Gustafsson, J.-Å., and Smithies, O. (1998). Generation and reproductive phenotypes of mice lacking estrogen receptor ß. *Proc. Natl. Acad. Sci. U S A* 95, 15677–15682.
181. Lieberman, M. E., Maurer, R. A., Claude, P., Wiklund, J., Wertz, N., and Gorski, J. (1981). Regulation of pituitary growth and prolactin gene expression by estrogen. *Adv. Exp. Med. Biol.* 138, 151–163.
182. Maurer, R. A. (1982). Estradiol regulates the transcription of the prolactin gene. *J. Biol. Chem.* 257, 2133–2136.
183. Raymond, V., Beaulieu, M., Labrie, F., and Boissier, J. (1978). Potent antidopaminergic activity of estradiol at the pituitary level on prolactin release. *Science* 200, 1173–1175.
184. Dufy, B., Vincent, J. D., Fleury, H., Du Pasquier, P., Gourdji, D., and Tixior-Vidal, A. (1979). Dopamine inhibition of action potentials in a prolactin secreting cell line is modulated by oestrogen. *Nature* 282, 855–857.
185. West, B., and Dannies, P. S. (1980). Effects of estradiol on prolactin production and dihydroergocryptine-induced inhibition of prolactin production in primary cultures of rat pituitary cells. *Endocrinology* 106, 1108–1113.
186. Lean, A. D., Ferland, L., Drouin, J., Kelly, P. A., and Labrie, F. (1977). Modulation of pituitary thyrotropin releasing hormone receptor levels by estrogens and thyroid hormones. *Endocrinology* 100, 1496–1504.
187. Pilotte, N. S., Burt, D. R., and Barraclough, C. A. (1984). Ovarian steroids modulate the release of dopamine into hypophysial portal blood and the density of anterior pituitary (^{3}H)spiperone-binding sites in ovariectomized rats. *Endocrinology* 114, 2306–2311.
188. Ritchie, A. K. (1993). Estrogen increases low voltage-activated calcium current density in GH3 anterior pituitary cells. *Endocrinology* 132, 1621–1629.
189. Dufy, B., Vincent, J. D., Fleury, H., Du Pasquier, P., Gourdji, D., and Tixier-Vidal, A. (1979). Membrane effects of thyrotropin-releasing hormone and estrogen shown by intracellular recording from pituitary cells. *Science* 204, 509–511.
190. Perez, R. L., Machiavelli, G. A., Romano, M. I., and Burdman, J.A. (1986). Prolactin release, oestrogens and proliferation of prolactin-secreting cells in the anterior pituitary gland of adult male rats. *J. Endocrinol.* 108, 399–403.
191. Chun, T. Y., Gregg, D., Sarkar, D. K., and Gorski, J. (1998). Differential regulation by estrogens of growth and prolactin synthesis in pituitary cells suggest that only a small pool of estrogen receptors is required for growth. *Proc. Natl. Acad. Sci. U S A* 95, 2325–2330.
192. Spady, T. J., McComb, R. D., and Shull, J. D. (1999). Estrogen action in the regulation of cell proliferation, cell survival, and tumorigenesis in the rat anterior pituitary gland. *Endocrine* 11, 217–233.
193. Cai, A., Bowers, R. C., Moore, J. P., and Hyde, J. F. (1998). Function of galanin in the anterior pituitary of estrogen-treated Fischer 344 rats: autocrine and paracrine regulation of prolactin secretion. *Endocrinology* 139, 2452–2458.
194. Wynick, D., Small, C. J., Bacon, A., Holmes, F. E., Norman, M., Ormandy, C. J., Kilic, E., Kerr, N. C. H., Ghatei, M., Talamantes, F., Bloom, S. R., and Pachnis, V. (1998). Galanin regulates prolactin release and lactotroph proliferation. *Proc. Natl. Acad. Sci. U S A* 95, 12671–12676.
195. Shen, E. S., Hardenburg, J. L., Meade, E. H., Arey, B. J., Merchenthaler, I., and López, F. J. (1999). Estradiol induces galanin gene expression in the pituitary of the mouse in an estrogen receptor alpha-dependent manner. *Endocrinology* 140, 2628–2631.
196. Yan, G.-Z., Pan, W. T., and Bancroft, C. (1991). Thyrotropin-releasing hormone action on the prolactin promoter is mediated by the POU protein Pit-1. *Mol. Endocrinol.* 5, 535–541.
197. Bredow, S., Kacsóh, B., Obál, F. Jr., Fang, J., and Krueger, J. M. (1994). Increase of prolactin mRNA in the rat hypothalamus after intracerebroventricular injection of VIP or PACAP. *Brain Res.* 660, 301–308.
198. Arbogast, L. A., and Voogt, J. L. (1994). Progesterone suppresses tyrosine hydroxylase messenger ribonucleic acid levels in the arcate nucleus on proestrus. *Endocrinology* 135, 343–350.
199. López, F. J., Merchenthaler, I., Ching, M., Wisniewski, M. G., and Negor-Vilar, A. (1991). Galanin: a hypothalamic-hypophysiotropic hormone modulating reproductive functions. *Proc. Natl. Acad. Sci. U S A* 88, 4508–4512.
200. Pickett, C. A., and Gutierrez-Hartmann, A. (1994). Ras mediates Src but not epidermal growth factor-receptor tyrosine kinase signaling pathways in GH_4 neuroendocrine cells. *Proc. Natl. Acad. Sci. U S A* 91, 8612–8616.
201. Porter, T. E., Wiles, C. D., and Frawley, L. S. (1994). Stimulation of lactotrope differentiation in vitro by fibroblast growth factor. *Endocrinology* 134, 164–168.
202. Aguilera, G., Hyde, C. L., and Catt, K. J. (1982). Angiotensin II receptors and prolactin release in pituitary lactotrophs. *Endocrinology* 111, 1045–1050.
203. Koves, K., Molnar, J., Kantor, O., Gorcs, T. J., Lakatos, A., and Arimua, A. (1996). New aspects of the neuroendocrine role of PACAP. *Ann. N Y Acad. Sci.* 805, 648–654.
204. Ben-Jonathan, N. (1994). Regulation of prolactin secretion. In *The Pituitary Gland* (H. Imura, Ed.), p. 261. Raven Press, New York.
205. Lea, R. W., Talbot, R. T., and Sharp, P. J. (1991). Passive immunization against chicken vasoactive intestinal polypeptide suppresses plasma prolactin and crop sac development in incubating ring doves. *Horm. Behav.* 25, 283–294.

206. Horseman, N. D., and Buntin, J. D. (1995). Regulation of pigeon crop milk secretions and parental behaviors by prolactin. *Annu. Rev. Nutr.* 15, 213–238.
207. Bjoro, T., Sand, O., Ostberg, B. C., Gordeladze, J. O., Torjesen, P., Gautvik, K. M., and Haug, E. (1990). The mechanisms by which vasoactive intestinal peptide (VIP) and thyrotropin releasing hormone (TRH) stimulate prolactin release from pituitary cells. *Biosci. Rep.* 10, 189–199.
208. Wynick, D., Hammond, P. J., Akinsanya, K. O., and Bloom, S. R. (1993). Galanin regulates basal and oestrogen-stimulated lactotroph function. *Nature* 364, 529–532.
209. Hinuma, S., Habata, Y., Fujii, R., Kawamata, Y., Hosoya, M., Fukusumi, S., Kitada, C., Masuo, Y., Asano, T., Matsumoto, H., Sekiguchi, M., Kurokawa, T., Nishimura, O., Onda, H., and Fujino, M. (1998). Prolactin-releasing peptide in the brain. *Nature* 393, 272–276. [Erratum in (1998). *Nature* 394, 302.]
210. Samson, W. K., Keown, C., Samson, C. K., Samson, H. W., Lane, B., Baker, J. R., and Taylor, M. M. (2003). Prolactin-releasing peptide and its homolog RFRP-1 act in hypothalamus but not in anterior pituitary gland to stimulate stress hormone secretion. *Endocrine* 20, 59–66.
211. Butcher, R. L., Collins, W. E., and Fugo, N. W. (1974). Plasma concentrations of LH, FSH, prolactin, progesterone and estradiol-17β throughout the 4-day estrous cycle of the rat. *Endocrinology* 94, 1704–1708.
212. Smith, M. S., Freeman, M. E., and Neill, J. D. (1975). The control of progesterone secretion during the estrous cycle and early pseudopregnancy in the rat: prolactin, gonadotropin and steroid levels associated with rescue of the corpus luteum of pseudopregnancy. *Endocrinology* 96, 219–226.
213. Neill, J. D., Freeman, M. E., and Tillson, S. A. (1971). Control of the proestrous surge of prolactin and luteinizing hormone secretion by estrogens in the rat. *Endocrinology* 89, 1448–1453.
214. Legan, S. J., Coon, G. A., and Karsch, F. J. (1975). Role of estrogen as initiator of daily LH surges in the ovariectomized rat. *Endocrinology* 96, 50–56.
215. Gregerson, K. A. (2003). Dopamine activation of a G-protein-coupled, inward rectifying potassium channel in pituitary lactotropes: regulation during the estrous cycle. *Endocrine* 20, 67–74.
216. Erskine, M. S. (1995). Prolactin release after mating and genitosensory stimulation in females. *Endocr. Rev.* 16, 508–528.
217. Voogt, J. L., and Carr, L. A. (1981). Tyrosine hydroxylase activity in the median eminence area of the pseudopregnant rat during surges of prolactin secretion. *Proc. Soc. Exp. Biol. Med.* 166, 277–281.
218. de Greef, W. J., and Neill, J. D. (1979). Dopamine levels in hypophyseal stalk plasma of the rat during surges of prolactin secretion induced by cervical stimulation. *Endocrinology* 105, 1093–1099.
219. Grosvenor, C. E., Mena, F., and Whitworth. (1977). The secretion rate of prolactin in the rat during suckling and it metabolic clearance rate after increasing intervals on non-suckling. *Endocrinology* 104, 372–376.
220. Mena, F., and Grosvenor, C. E. (1968). Effect of number of pups upon suckling-induced fall in pituitary prolactin concentration and milk ejection in the rat. *Endocrinology* 82, 623–626.
221. de Greef, W. J., Plotsky, P. M., and Neill, J. D. (1981). Dopamine levels in hypophysial stalk plasma and prolactin levels in peripheral plasma of the lactating rat: effects of a simulated suckling stimulus. *Neuroendocrinology* 32, 229–233.
222. Roky, R., Obal, F., Valatx, J.-L., Bredow, S., Fang, J., Pagano, L. P., and Krueger, J. M. (1995). Prolactin and rapid eye movement sleep regulation. *Sleep* 18, 536–542.
223. Exton, M. S., Kruger, T. H., Koch, M., Paulson, E., Knapp, W., Hartmann, U., and Schedlowski, M. (2001). Coitus-induced orgasm stimulates prolactin secretion in healthy subjects. *Psychoneuroendocrinology* 26, 287–294.
224. Nicoll, C. S., Talwalker, P. K., and Meites, J. (1960). Initiation of lactation in rats by nonspecific stresses. *Am. J. Physiol.* 198, 1103–1106.
225. Nicoll, C. S. (1967). Bioassay of prolactin. Analysis of the pigeon crop-sac response to local protein injection by objective and quantitative methods. *Endocrinology* 80, 641.
226. Kauppila, A., Chatelain, P., Kirkinen, P., Kivinen, S., and Ruokonen, A. (1987). Isolated prolactin deficiency in a woman with puerperal alactogenesis. *J. Clin. Endocrinol. Metab.* 64, 309–312.
227. Falk, R. J. (1992). Isolated prolactin deficiency: a case report. *Fertil. Steril.* 58, 1060–1062.
228. Zargar, A. H., Masoodi, S. R., Laway, B. A., Shah, N. A., and Salahudin, M. (1997). Familial puerperal alactogenesis: possibility of a genetically transmitted isolated prolactin deficiency. *Br. J. Obstet. Gynaecol.* 104, 629–631.
229. Douchi, T., Nakae, M., Yamamoto, S., Iwamoto, I., Oki, T., and Nagata, Y. (2001). A woman with isolated prolactin deficiency. *Acta Obstet. Gynaecol. Scand.* 80, 368–370.
230. Topper, Y. J., and Freeman, C. S. (1980). Multiple hormone interactions in the developmental biology of the mammary gland. *Physiol. Rev.* 60, 1049–1106.
231. Smith, G. H. (1996). Experimental mammary epithelial morphogenesis in an in vivo model: evidence for distinct cellular progenitors of the ductal and lobular phenotype. *Breast Cancer Res. Treat.* 39, 21–31.
232. Srivastava, S., Matsuda, M., Hou, Z., Bailey, J. P., Kitazawa, R., Herbst, M. P., and Horseman, N. D. (2003). Receptor activator of NF-kappaB ligand induction via Jak2 and Stat5a in mammary epithelial cells. *J. Biol. Chem.* 278, 46171–46178.
233. Hovey, R. C., Harris, J., Hadsell, D. L., Lee, A. V., Ormandy, C. J., and Vonderhaar, B. K. (2003). Local insulin-like growth factor-II mediates prolactin-induced mammary gland development. *Mol. Endocrinol.* 17, 460–471.
234. Travers, M. T., Barber, M. C., Tonner, E., Quarrie, L., Wilde, C. J., and Flint, D. J. (1996). The role of prolactin and growth hormone in the regulation of casein gene expression and mammary cell survival: relationships to milk synthesis and secretion. *Endocrinology* 137, 1530–1539.
235. Cohen-Becker, I., Selmanoff, M., and Wise, P. (1986). Hyperprolactinemia alters the frequency and amplitude of pulsatile luteinizing hormone secretion in the ovariectomized rat. *Neuroendocrinology* 42, 328–333.
236. Larsen, J., Bhanu, A., and Odell, W. (1990). Prolactin inhibition of pregnant mare's serum stimulated follicle development in the rat ovary. *Endocr. Res.* 16, 449–459.
237. Tsai-Morris, C., Ghosh, M., Hirshfield, A. N., Wise, P. M., and Brodie, A. M. (1983). Inhibition of ovarian aromatase by prolactin in vivo. *Biol. Reprod.* 29, 342–346.
238. Krasnow, J., Hickey, G., and Richards, J. (1990). Regulation of aromatase mRNA and estradiol biosynthesis in rat ovarian granulosa and luteal cells by prolactin. *Mol. Endocrinol.* 4, 13–22.
239. Albarracin, C. T., Parmer, T. G., Duan, W. R., Nelson, S. E., and Gibori, G. (1994). Identification of a major prolactin-regulated protein as 20α-hydroxysteroid dehydrogenase: coordinate regulation of its activity, protein content, and messenger ribonucleic acid expression. *Endocrinology* 134, 2453–2460.
240. Voogt, J. L., de Greef, W. J., Visser, T. J., de Koning, J., Vreeburg, J. T., and Weber, R. F. (1987). In vivo release of dopamine, luteinizing hormone-releasing hormone and

thyrotropin-releasing hormone in male rats bearing a prolactin-secreting tumor. *Neuroendocrinology* 46, 110–116.
241. Wennbo, H., Kindblom, J., Isaksson, O. G., and Tornell, J. (1997). Transgenic mice overexpressing the prolactin gene develop dramatic enlargement of the prostate gland. *Endocrinology* 138, 4410–4415.
242. Bridges, R. S. (1994). The role of lactogenic hormones in maternal behavior in female rats. *Acta Paediatr. Suppl.* 397, 33–39.
243. Lucas, B. K., Ormandy, C. J., Binart, N., Bridges, R. S., and Kelly, P. A. (1998). Null mutation of the prolactin receptor gene produces a defect in maternal behavior. *Endocrinology* 139, 4102–4107.
244. Shingo, T., Gregg, C., Enwere, E., Fujikawa, H., Hassam, R., Geary, C., Cross, J. C., and Weiss, S. (2003). Pregnancy-stimulated neurogenesis in the adult female forebrain mediated by prolactin. *Science* 299, 117–120.
245. Asa, S. L., and Ezzat, S. (1998). The cytogenesis and pathogenesis of pituitary adenomas. *Endocr. Rev.* 19, 798–827.
246. Baik, J. H., Picetti, R., Saiardi, A., Thiriet, G., Dierich, A., Depaulis, A., Le Meur, M., and Borrelli, E. (1995). Parkinsonian-like locomotor impairment in mice lacking dopamine D2 receptors. *Nature* 377, 424–428.
247. Kelly, M. A., Rubinstein, M., Asa, S. L., Zhang, G., Saez, C., Bunzow, J. R., Allen, R. G., Hnasko, R., Ben-Jonathan, N., Grandy, D. K., and Low, M. J. (1997). Pituitary lactotroph hyperplasia and chronic hyperprolactinemia in dopamine D2 receptor-deficient mice. *Neuron* 19, 103–113.
248. Saiardi, A., Bozzi, Y., Baik, J.-H., and Borrelli, E. (1997). Antiproliferative role of dopamine: loss of D_2 receptors causes hormonal dysfunction and pituitary hyperplasia. *Neuron* 19, 115–126.
249. Cruz-Soto, M. E., Scheiber, M. D., Gregerson, K. A., Boivin, G. P., and Horseman, N. D. (2002). Pituitary tumorigenesis in prolactin gene-disrupted mice. *Endocrinology* 143, 4429–4436.
250. Cruz-Soto, M. E., Herbst, M. P., Nieport, K. M., and Horseman, N. D. (2003). Pituitary tumor transforming gene (PTTG) responds to bromocriptine treatment in the prolactin (PRL)-deficient mouse. Endocrine Society Annual Meeting, P2-101.
251. Bosse, R., Fumagalli, F., Jaber, M., Giros, B., Gainetdinov, R. R., Wetsel, W. C., Missale, C., and Caron, M. G. (1997). Anterior pituitary hypoplasia and dwarfism in mice lacking the dopamine transporter. *Neuron* 19, 127–138.
252. Hentges, S. T., and Low, M. J. (2002). Ovarian dependence for pituitary tumorigenesis in D2 dopamine receptor-deficient mice. *Endocrinology* 143, 4536–4543.
253. Clement-Lacroix, P., Ormandy, C., Lepescheux, L., Ammann, P., Damotte, D., Goffin, V., Bouchard, B., Amling, M., Gaillard-Kelly, M., Binart, N., Baron, R., and Kelly, P. A. (1999). Osteoblasts are a new target for prolactin: analysis of bone formation in prolactin receptor knockout mice. *Endocrinology* 140, 96–105.
254. Bevan, J. S., Webster, J., Burke, C. W., and Scanlon, M. F. (1992). Dopamine agonists and pituitary tumor shrinkage. *Endocr. Rev.* 13, 220–240.
255. Tindall, G. T., Kovacs, K., Horvath, E., and Thorner, M. O. (1982). Human prolactin-producing adenomas and bromocriptine: a histological, immunocytochemical, ultrastructural, and morphometric study. *J. Clin. Endocrinol. Metab.* 55, 1178–1183.